CATALOG OF HUMAN CANCER GENES

CATALOG OF HUMAN CANCER GENES

McKusick's *Mendelian Inheritance in Man* for Clinical and Research Oncologists

(Onco-MIM)

John J. Mulvihill, M.D.

Kimberly V. Talley Professor, Section of Genetics, Department of Pediatrics

University of Oklahoma Health Sciences Center

Oklahoma City, Oklahoma

with a foreword by Victor A. McKusick, M.D.

THE JOHNS HOPKINS UNIVERSITY PRESS
Baltimore and London • 1999

Printed in the United States of America on acid-free paper
9 8 7 6 5 4 3 2 1

The Johns Hopkins University Press
2715 North Charles Street
Baltimore, Maryland 21218-4363
www.press.jhu.edu

Library of Congress Cataloging-in-Publication Data

Mulvihill, John J. 1943–
Catalog of Human cancer genes: McKusick's Mendeliam inheritance in man for clinical and research oncologists (onco-MIM) / John J. Mulvihill: with a foreword by Victor McKusick.
p. cm.
Includes bibliographical references and index.
ISBN 0-8018-4799-0 (alk. paper)
1. Oncogenes—Catalogs and collections. 2. Human gene libraries. I OMIM. II. Title. III Title: Onco-MIM.
[DNLM: 1. Oncogenes bibliography. 2. Neoplasms—genetics encyclopedias. ZQZ 202 M961c 1999]
RC268.42.M84 1999
616.99″4042—dc21
DNLM/DLC 98-54978
for Library of Congress CIP

A catalog record for this book is available from the British Library.

To my wife, Dr. Charlotte K. Mulvihill

Contents

Foreword

John Mulvihill's *Onco-MIM* is an annotated selection of neoplasia-related entries in OMIM (on-line *Mendelian Inheritance in Man,* otherwise known as *MIM*). *MIM* is a computerized knowledgebase of human genes and genetic disorders. The on-line version of *MIM* is updated almost daily; the entries selected for presentation here as *Onco-MIM* were down-loaded on December 7, 1998. *Mendelian Inheritance in Man* has been ''on the computer'' since 1964. The first print edition, a pioneer in computer-based publication, appeared in 1966; the twelfth print edition was published July 1, 1998. *MIM* became available on-line locally at Johns Hopkins as an aid in the authoring and editing process in late 1985; it was made generally accessible on-line under the designation *OMIM* in 1987. *OMIM* is distributed on the World Wide Web by the National Center for Biotechnology Information of the National Library of Medicine. The URL for *OMIM* is as follows: http://www.ncbi.nlm.nih.gov/omim.

When *MIM* was initiated in the 1960s, it was a catalog of phenotypes or, more accurately stated, three catalogs of phenotypes arranged according to the mode of inheritance: autosomal dominant, autosomal recessive, and X-linked. Always, the purpose was to characterize specific genes. Virtually the only way to do so in the 1960s was by Mendelizing phenotype. A focus on the gene from the very first is indicated by the fact that the first catalog to be assembled, the catalog of X-linked phenotypes, was in an effort to answer the question, What genetic information is carried by the X chromosome? As pointed out then, the list of disorders showing X-linked inheritance was like a photographic negative from which a positive picture of the genetic constitution of the X chromosome could be created.

In the 1970s and thereafter, methods of somatic cell hybridization permitted the chromosomal mapping of specific human genes, even though for some (e.g., that encoding thymidine kinase) no Mendelizing phenotype was known (then or since). All such genes were given an entry in *MIM*. About 1980, the advent of molecular genetics permitted the further cloning, mapping, and sequencing, and sometimes functional characterization, of genes with which no phenotypic variation, abnormal or pathologic, had been associated. All such genes were likewise incorporated into the *MIM* knowledgebase.

In general, consistent with the focus on the gene, an attempt has been made to create only one asterisked *OMIM* entry per gene locus. Thus, in the case of the *RET* gene, there is only one entry (under number 164761). The various distinct clinical phenotypes due to mutations in that gene have their main discussion in entry *164761, with listings of specific mutations, or ''allelic variants,'' in subentries under the main entry. Four of the disorders due to mutations in the *RET* gene are described in separate ''phenotype entries'' (identified as such by a number sign (#) preceding the entry number): multiple endocrine neoplasia IIA (#171400), multiple endocrine neoplasia IIB (#162300), medullary thyroid carcinoma (#155240), and Hirschsprung disease (#142623).

In 1988 two additional chromosome-specific catalogs joined the X chromosome catalog: the catalog for the Y chromosome and that for the mitochondrial chromosome. Since May 1994, autosomal entries have not been broken down into autosomal dominant and autosomal recessive, since these designations have become relatively meaningless. All new autosomal entries have been lumped together in a catalog with 6----- ID numbers.

Following is the system for numbering entries:

Autosomal	1. . . . . (100000) — autosomal dominant before 1994
	2. . . . . (200000) — autosomal recessive before 1994
	6. . . . . (600000) — all since 1994
X chromosome	3. . . . . (300000)
Y chromosome	4. . . . . (400000)
Mitochondrial chromosome	5. . . . . (500000)

OMIM is based almost exclusively on the periodical biomedical literature, which, for the most part, is peer reviewed. Some 75–80% of the material comes from 20 journals; the remainder comes from more than 200 other journals. An attempt is made to incorporate material from the 20 high-impact journals within a few days or two or three weeks of publication.

Beginning with the third print edition in 1971, *MIM* has attempted a comprehensive cataloging of information on the human gene map. This is a valuable feature of the on-line version as well. When one is reading the entry concerning a particular disorder or gene, one can easily ''go into'' the *OMIM* version of the gene map for an answer to the question, Has this gene been mapped, and, if so, where?

The second area in which *OMIM* attempts to be not only absolutely up to date but also comprehensive is that of the molecular basis of clinical disorders. One can ask, Has the molecular defect been identified in this disorder? Or, Has a disorder been related to mutation in this gene? Specific mutations causing specific genetic disorders are listed as so-called allelic variants and described in subentries with a unique 10-digit ID number, after the text of the entry describing the gene. More than 850 gene entries have at least one specific mutation listed as an allelic variant; the beta globin locus (HGB; 141900) lists more than 475 allelic variants. (Sickle hemoglobin has the allelic variant number 141900.0243.) The total number of cataloged mutations approaches 7,000.

Relevant to this catalog of human cancer genes, it should be noted that representative somatic mutations underlying cancers are cataloged as well as germline mutations. For a specific gene in which many mutations, germline and somatic, have been described, only representative or particularly instructive ones have been cataloged in *OMIM*. For example, many thousands of mutations in the p53 gene have been described, but the specific mutations that are listed as allelic variants (totaling only 33) are those causing the inherited Li-Fraumeni syndrome and selected ones causing sporadic cancers with features of particular note [e.g., the Arg249Ser mutation of the p53 gene (*191170.0249), which is characteristic of aflatoxin-induced hepatocellular carcinoma].

For reasons of space, Dr. Mulvihill has chosen not to include the allelic variant subentries. It is hoped, however, that *Onco–MIM* will pique an interest in *OMIM* on the part of cancer clinicians and researchers, and will prompt them to survey the information on cancer genes that has been added to *OMIM* since December 7, 1998. Naturally, *OMIM* has advantages over any print version of *Mendelian Inheritance in Man*, namely, its powerful searchability and its timeliness (it is updated daily). On the other hand, the print version has archival value; the successive 12 editions between 1966 and 1998 represent cross sections of the progress of human and medical genetics during that period. Furthermore, with its frequent editions, the print version has reasonably up-to-date information and, significantly, is available to users in a non-electronic setting.

Cancer in *OMIM*

All cancer is genetic disease; at the same time, much of cancer has defined environmental cause(s). As in much of medical genetics, the distinction between nature and nurture has only relative significance.

As a gene-based knowledgebase, *OMIM* easily accommodates the information on the genes that have been identified in neoplasms of various types, through their involvement either in the germline or as somatic mutations. In many instances, gene discovery in cancer has started with clinical observations. For example, first, the finding of recurrent reciprocal translocations in hematologic malignancies and some forms of solid tumors has led to the isolation of genes from each chromosome involved in the translocation, the genes that caused malignancy when fused. The prototypes are the *ABL* gene (*189980) on chromosome 9 and the *BCR1* (*151410) gene on chromosome 22, which are joined in the 9;22 translocation found in chronic myeloid leukemia.

Second, the Knudson two-hit model has been of great heuristic value in defining the change in both familial forms of cancer and their sporadic counterparts. This approach led to the identification of tumor-suppressor genes. Referred to as ''gatekeeper'' genes by Vogelstein and his colleagues, these include the retinoblastoma (*RB*; *180200) and adenomatous polyposis coli (*APC*; *175100) genes, two of the most widely involved cancer genes. The definition of a gene that is mutant in the germline in familial cancer has often led to the discovery of its important role also in sporadic cancers of the same type. For example, the *APC* gene, which was found to be mutant in adenomatous polyposis coli (what we used to call familial polyposis coli), is mutated as the initiating event in the great majority of sporadic colorectal cancers.

The Vogelstein multistep, multigene, preferred order, clonal expansion and evolution concept of carcinogenesis has been worked out particularly well for colorectal cancer. The individual genes involved therein are cataloged in *OMIM*. Defects in mismatch repair genes (''caretaker'' genes, in the designation of Vogelstein and colleagues) drive the process of multigenic change. Several different mismatch repair genes, first identified in yeast, have been defined in humans on the basis of family cancer syndromes, such as Lynch syndrome I of site-specific colorectal cancer (114500), also known as hereditary nonpolyposis colorectal cancer (HNPCC).

Mulvihill and *Onco-MIM*

Dr. Mulvihill has chosen to organize the neoplasia-related entries in *OMIM* according to systems. This at times leads to seemingly strange bedfellows: breast cancer is discussed under the integument system with malignancies of the skin.

Many cystic disorders are included by Mulvihill. One cystic disorder that is not included is polycystic kidney disease (PKD). In most cases, the kidney changes of PKD have not yet developed at birth. Later, multiple cysts develop and increase progressively in number and size with age. A justification for including PKD in this compendium might be the demonstration that the development of cysts follows the Knudson two-hit model: the patient with autosomal dominant PKD is born heterozygous for a mutation in a gene on the chromosome 16 from one parent; for the development of cysts, mutation in the chromosome 16 from the other parent must occur. The same mechanism is operative in the cysts of liver and other organs that may be affected in PKD.

Mulvihill comments on examples of the crossover (or overlap) between a congenital defect and neoplasia. Gorlin syndrome, otherwise known as the basal cell nevus syndrome (#109400), is an example. Some of the features, such as bifid ribs, are congenital malformation; others, such as basal cell carcinomas, are neoplasms. Here, again, is an example of a gene that as a germline mutation produces a characteristic family cancer syndrome: as a somatic mutation, it causes sporadic basal cell carcinoma. The gene that is the site of the mutations, ''patched'' (PTCH; *601309), was identified by study of the familial disorder by the process of map-based gene discovery (positional cloning).

We are indebted to Dr. Mulvihill for assembling this compendium of neoplasia-related information from *OMIM*, organizing it, and giving the value-added comments that make it all the more worthwhile for an important group of users.

Victor A. McKusick, M.D.
University Professor of Medical Genetics
The Johns Hopkins University School of Medicine

Acknowledgments

This volume is an experiment in synthesis and publication. Many people helped from beginning to end.

Victor A. McKusick wrote most of the words on these pages over the course of more than three decades. If a cliché ever needed an example, here it is. I stand on the shoulders of a giant in producing this *Catalog* as a possible increment in our understanding of cancer genes and genetics. The more I worked with him, with his hard copy volume, *Mendelian Inheritance in Man (MIM),* and its on-line version, *OMIM,* the more I was in awe of his incredible achievement in keeping up with and ahead of the literature on human and medical genetics. He has accomplished in *MIM* and *OMIM* is a Herculean encyclopedia that has well served medical geneticists, genetic researchers, and patients and families touched by genetic disease—and, aren't we all?—for the last three decades and will for decades to come. In a real sense, before the international genome project existed, he was it! Besides writing the original text, he made helpful comments on my ''JJM Notes.''

I may be so bold as to count Victor McKusick among the wonderful blend of mentors I have enjoyed in my career.

In making this *Catalog,* I recognize lessons from other mentors. Kurt Benirschke, now president of the San Diego Zoological Society and professor of pathology at the University of California at San Diego, catalyzed my interest in medical genetics as my professor of pathology at Dartmouth Medical School in the 1960s. His *Pathology of the Human Placenta* (1995) set an example of the merits of a total literature review and the value of cataloging observations that otherwise might be lost. He also steered me to Seattle to finish medical school at the University of Washington, when my classmates in the two-year Dartmouth Medical School were transferring to Boston. In Seattle, I met Arno Motulsky, whose book review of the eighth edition of *MIM* set the ball rolling for this prototype of what may be other specialty catalogs of *OMIM.* David W. Smith, the founder of dysmorphology, guided my M.D. thesis in Seattle (''The Genesis of Dermatoglyphics''). His enormous but prematurely truncated corpus of writing illustrates well what one person with a broad mind and attention to detail can achieve without a large laboratory or staff or computer. His gifts of simplifying and cataloging are major elements of this volume.

I have been privileged to write at length (Mulvihill 1998) about my most sustained mentor, Robert W. Miller, who defined the clinical and biologic overlap between carcinogenesis, teratogenesis, and mutagenesis. He honed the laterospective (looking sideward) strategy of clinical epidemiology; that is, he taught me to spot the occasional association of birth defects and cancer in the same patient. The overlap is clinically important to recognize and can shed light on the biology of all three pathologic processes. Finally, his prolific protégé, Joseph F. Fraumeni, Jr., was another major influence on me and is unequaled for his recall and incorporation of literature into creative synthesis.

Directly involved with various aspects of this effort were two former students. Kenneth R. Fromkin, now a cardiology fellow, volunteered for a genetics tutorial with me when he was a senior undergraduate at the Johns Hopkins University. Using the precursor of these pages, he helped prepare a rough tabulation of these entities and showed me that a full compendium was indeed possible. Sean Davis, now a pediatric resident, transferred the rough table to a computer database, and also illustrated the uncommon focus I needed to bring this task to conclusion. With my brutal administrative burden at the University of Pittsburgh, his empathy augmented the luxurious scholastic pleasure of pursuing the preparation of this *Catalog*.

Many secretaries helped all along. The real hero was Ms. Anita Socci, in Pittsburgh, who kept track of numerous versions of tables, notes, and modifications to this volume. I appreciate my new colleagues at a warm institution, the University of Oklahoma Health Sciences Center, and the availability of time during my transition that allowed putting this project to bed. Which thought brings me to thank my wife, Dr. Charlotte K. Mulvihill, a lifelong partner and supporter in this and all my endeavors.

A special note of gratitude needs to be made to Ms. Wendy Harris, medical editor of the Johns Hopkins University Press. She catalyzed the conception of this volume. Over the years, she worked patiently in extending deadlines and in crystallizing the final formatting of the book. I very much appreciate her professional support over the years. Ms. Debbie Tourtlotte of Columbia, Md., produced the index—a major entry point to the world of cancer genetics.

Benirschke, K., and Kaufmann, P. 1995. *Pathology of the Human Placenta.* New York: Springer-Verlag.

Mulvihill, J. J. 1998. Encomium: Robert Warwick Miller: Mentor, synthesizer, and international interdisciplinary initiator. *Am. J. Med. Genet.* 76:1–8.

Introduction

The Burden and Causes of Human Cancer

In the United States, one in three to four persons gets cancer in a lifetime, apart from skin cancer (basal cell and squamous cell carcinomas), which affects one in two persons. In 1999 an estimated 1,382,000 Americans will develop cancer (plus 900,000 with skin cancer), and 560,000 will die of cancer (Landis et al. 1998). Cancer is the second most common cause of childhood death, after accidents. In addition to frank malignant neoplasia (cancer), nearly everyone has some neoplastic or preneoplastic lesion. Sometimes, such lesions become cancer for reasons not well understood.

Benign and malignant neoplasia is a group of more than 100 diseases, all marked by disordered cellular growth. At the level of the cell, cancer is a genetic disease, for no one would die of metastatic cancer if a cancer cell did not pass on to offspring cells the mutation(s) that set new rules for uncontrolled proliferation and spread. Usually, one cannot say why a person gets a specific cancer at a specific age. In human populations, the causes or contributing determinants are known in detail for only a few cancers. Twenty-seven substances, groups of substances, medical treatments, and occupational exposures are clearly carcinogenic in humans, and another 130 may reasonably be anticipated to be carcinogenic, according to the National Toxicology Program's *Report on Carcinogens* (U.S. DHHS 1998) (Table 1). The environment is a well-known cause of cancer, and efforts should be maintained to identify and control exogenous carcinogens, including the use of tobacco and imprudent lifestyles that allow excessive exposure to the sun.

Yet, few black Africans in North America and elsewhere (i.e., persons of very dark skin) get skin cancers, namely, squamous cell and basal cell carcinomas. Their melanocytes and melanin granules absorb ultraviolet energy before it can reach the nuclei of skin cells and initiate the process of carcinogenesis. The genetic trait of dark skin no doubt protects against skin cancer; white skin is a genetic predisposition to skin cancer. At the other extreme, individuals with the genetic trait of oculocutaneous albinism, whether of the Negroid or Caucasian race, lack protective melanin pigments and have an enormous risk for skin cancer. The list of genetic traits with marked susceptibility to cancer is short (Table 2). But, for most categories of human carcinogens, whether physical, biologic, or chemical agents, there is an illustrative example of ecogenetics—that is, variation in susceptibility to cancer. It is inaccurate to claim that 90% of cancer is due to the environment or that 70% is due to genetic susceptibility. Rather, a more credible hypothesis is that every person's cancer arises from an inherent susceptibility and resistance to a cumulative lifelong history of personal and environmental exposures.

Table 1. Human Carcinogens

Aflatoxins
4-Aminobiphenyl
Analgesic mixtures containing phenacetin
Arsenic and certain arsenic compounds
Asbestos
Azathioprine
Benzene
Benzidine
Bis(chloromethyl) ether
Chlorambucil
1-(2-Chloroethyl)-3-(4-methylcyclohexyl)-1-nitrosourea (meCCNU)
Chloromethyl methyl ether
Chromium and certain chromium compounds
Coke oven emissions
Conjugated estrogens
Cyclophosphamide
Cyclosporin
Diethylstilbestrol
Erionite
Melphalan
Mineral oils
Mustard gas
Radon
Soots, tars, and mineral oils
Thiotepa
Thorium dioxide
Vinyl chloride

Origins and Development of This *Catalog*

This volume enumerates the hereditary traits and genes that seem to alter the risk for neoplasia in humans. This *Catalog* is, in a sense, the full repertory of human variation in predisposition or resistance to tumor growth, both benign and malignant.

The work derives from Victor A. McKusick's *Mendelian Inheritance in Man*, widely called *MIM*, at least among geneticists. In 1999 the title is truly a misnomer: not all the entries are truly inherited; many are not Mendelian or single-gene traits; and most affect women as well as men. But, the first edition of *MIM* appeared in hard copy a third of a century ago (McKusick 1966). Its origin was a tabulation of the 60 X-linked traits in human beings, as they were known in 1962 (McKusick 1962). In 1966 the first print edition of *MIM* was published. In 1987 the on-line version, designated *OMIM*, became available.

Geneticists can skip this paragraph and the next, but oncologists need to know how McKusick developed and maintains *MIM*, now in its twelfth edition (McKusick 1998) and available on-line at http://www.ncbi.nlm.nih.gov/omim. The present print edition is 4,518 pages in three large-format volumes printed on bible-weight paper; indeed, it is the holy book of medical and human geneticists. McKusick faithfully reads 20 genetics journals known to have a high impact or yield from the counts made of citations in the database beginning in 1975. He dictates updated information for established entries or makes a new entry. Other journals are scanned in *Current Contents*, and likely articles are retrieved and summarized if relevant. Geneticists around the world, some trained by McKusick, have offered suggestions over the years. Recently, human genome program funds have permitted the establishment of a board of contributing authors and editors; I was a charter member. *MIM* and *OMIM* set a standard for nomenclature, such that most medical genetics journals use the *MIM* entry number to indicate beyond doubt the exact genetic disorder or trait that is being discussed. Print translations of *MIM* have appeared in Russian, Spanish, and Mandarin Chinese, and *OMIM* is accessed by nearly 4,000 unique users per day. The most common traits of human beings—handedness, baldness, and eye color—are explained, as are the common conditions determined largely by a single-gene mutation, such as the hemoglobinopathies, hemophilias, chondrodysplasia, and Marfan disease. Syndromes of multiple congenital malformations, even those represented by a single case report, are included. Also included are genes, some without a clear phenotype: protooncogenes, receptors, antigens, tumor-suppressor genes, and proteins differentially expressed by cancer cell lines and not by normal human cells.

A typical *OMIM* entry has information arranged into nine areas:

1. *number:* the unique six-digit *MIM* number, with or without a preceding marker * or #;
2. *title:* and, often, ''alternate titles'' or ''included titles'';
3. *symbol:* the gene symbol found in the *OMIM* gene map, if any;
4. *locus:* the cytogenetic location (e.g., 1p31.6) found in the OMIM gene map, if any;
5. *text:* the main entry or discussion covering clinical features, genetic aspects, gene mapping and structure, pathogenesis, diagnosis, management, annual models, and sometimes history;
6. *subentries for allelic variants*: (mutations) with a 10-digit ID number (*MIM* number with four other digits beginning with .0001) and with text annotating the mutation and the associated clinical phenotype;
7. *references;*
8. *clinical synopsis:* key words summarizing clinical features;
9. *contributors:* authors or editors of the entry.

This *Catalog* extracts from *OMIM* those entries that are associated with neoplasia. It omits alternative titles, the description of allelic variants, and contributors. Before the text written by McKusick and coworkers, I have inserted a ''JJM Note,'' based on my personal experience or other literature, sometimes not cited in the McKusick text. The title of some entries has been changed to speed comprehension by clinical and research oncologists.

An asterisk (*) before a *MIM* number indicates that an entry describes a distinct gene or phenotype. The absence of an asterisk preceding the number indicates that McKusick feels that the trait or gene is not well characterized. A number sign (#) before a *MIM* number indicates that the phenotype is caused by a mutation in a gene explained in one or several other entries. Whenever the number sign is used, the reason is stated at the outset of the entry. The number sign is also used for phenotypes that result from specific chromosomal aberrations and for contiguous gene syndromes.

The Focus on Cancer Genes

In 1974 the National Cancer Institute and the American Cancer Society sponsored a meeting on risk factors for human cancer. Although the emphasis at the time was strongly on the environmental determinants of cancer, the scientific organizer of the meeting, Joseph F. Fraumeni, Jr., asked me, his junior colleague, to do the opening lecture and book chapter, a necessarily short note on ''Congenital and Genetic Diseases [that predispose to cancer]'' (Mulvihill 1975). In an effort to offer some original synthesis to the conference and its proceedings, I took the opportunity to read, from cover to cover, the third edition of *Mendelian Inheritance in Man* (McKusick 1971), spotting, as well as could be done by eye, all traits that had benign or malignant neoplasia as a sole feature, a frequent concomitant, or just a rare complication, sometimes only a single case report. Of the 866 proved single-gene traits and 1,010 others with possible mendelian behavior in that edition, 161 conditions were associated with neoplasia—some 9%.

Two years after the survey meeting on cancer risk factors, a follow-up meeting under the same sponsorship was held to focus exclusively on cancer genetics. In 1974, when I founded the Clinical Genetics Section of the Epidemiology Branch of the National Cancer Institute, an early task was to convene a national conference on cancer genetics. The topic was out of popular favor at the time, when the focus was on the environmental determinants of cancer. The conference, held in 1975, was the first substantial meeting among cancer researchers and medical geneticists, and, in a sense, revalidated the merit of genetic studies of cancer in humans. As I and my colleagues edited the proceedings volume (Mulvihill, Miller, and Fraumeni 1977), it became clear that many mendelian traits predisposing to cancer were mentioned throughout the volume. Again, the 1975 tabulation enumerated 161 mendelian traits that were associated with neoplasia, representing 9% of monogenic disorders known in 1971 (Mulvihill 1975). Subsequent lists brought the number to 200 in 1977 (Mulvihill 1977; McKusick 1974), 339 in 1989 (Mulvihill 1989; McKusick 1988), and 467 in 1995 (Mulvihill, Fromkin, and Davis 1997; McKusick 1994).

For several years, the ongoing tally has been increasingly difficult to maintain, as the international human genome project to map and sequence all human genes was launched and results soon began arriving.

The final catalyst for producing this *Catalog* was a book review of the eighth edition of *MIM* (Motulsky 1989). The suggestion was made that the rapidly growing full book was unwiedly. Furthermore, *MIM* was not sufficiently well known to clinical specialists other than geneticists who should be aware of the material in their own areas but who could afford to pass over the details in other special areas. At the following annual meeting of the American Society of Human Genetics, Wendy Harris, medical editor for the Johns Hopkins University Press, raised the issues with me. We agreed to proceed and have—at long last—produced this *Catalog*.

Preparing the *Catalog*

The *Catalog* was prepared by cover-to-cover reading of the eighth edition of *Mendelian Inheritance in Man* (McKusick 1988) and an ongoing monitoring of the additions and deletions in subsequent editions, supplemented by MEDLINE searches and reading of original journal articles. The search capability of *OMIM* was quite helpful in the final screening for eligible entries. As of November 18, 1998, the count in *OMIM* was 9,964 entries, with 6,856 as gene loci and 548 as clinical phenotypes. A simple word search of *OMIM* titles, and clinical synopses texts (not references) produced 667 entries with the word *tumor,* 430 with *cancer,* 346 with *carcinoma,* and 223 with *neoplas** (to pick up *neoplasia, neoplasm,* and *neoplastic*), but much overlap occurs.

Again, an *OMIM* entry is included in this *Catalog* when benign or malignant neoplasia or tumor occurs as a sole feature, a frequent concomitant, or only a rare complication of the trait. Because of their rarity, some disorders are represented by only single case reports. Certain entities may be contestable and others may be missing (e.g., not all known reports of familial aggregation of neoplasms of the same or diverse cell types are included). A few conditions are included that present as masses but are not true neoplasms, such as thyroid goiter and cysts of various organs.

The mode of presentation used here—admittedly somewhat arbitrary—is by organ system, not strictly of the major features of the trait, but rather by the organ system of the associated neoplasia. Hence, cystic fibrosis is not in the chapter on the cardiorespiratory system, its major site of morbidity and mortality, but rather is in the chapter on the alimentary tract, because the major associated neoplasm is adenocarcinoma of the small intestine.

Table 2. Ecogenetics of Human Cancer

Environmental Agent	Genetic Trait
Ionizing radiation	Retinoblastoma
	Ataxia-telangiectasia with lymphoma
	Nevoid basal cell carcinoma
Ultraviolet radiation	Oculocutaneous albinism
	Xeroderma pigmentosum
	Hereditary dysplastic nevus syndrome
Stilbestrol	Turner syndrome (45,X)
Androgen	Fanconi pancytopenia
Iron	Hemochromatosis
Tyrosine	Tyrosinemia
Monosaccharides	Glycogen storage disease 1
Epstein-Barr virus	Purtilo X-linked lymphoproliferative syndrome
Papillomavirus 5	Epidermodysplasia verruciformis

Source: Modified from Mulvihill (1994).

Lessons from the *Catalog*

In summary, 635 (6.2%) entries in McKusick's *Mendelian Inheritance in Man* represent genes and/or disorders that predispose to or are associated with neoplasia. They include 424 traits that are clinical phenotypes, disorders, or conditions and 213 other genes or proteins.

Biology of Cancer

Several points need to be made about the *Catalog*. First, it is lengthy. A substantial portion of all known single-gene traits in human beings is associated with neoplasia. In many, neoplasia is only an occasional feature that may arise at various ages; therefore, other factors in the environment or other genes must be interacting to produce a tumor.

Second, the *Catalog* represents the mutant genetic repertory of neoplasia in humans. It emphasizes the large number of genes that might be involved in susceptibility to cancer and, by inference, the number of normal genes that contribute to resistance to neoplasia.

Third, nearly all histologic types of tumor are represented, including the most common malignancies—those of the skin, breast, colon, and lung. These, and other tumors occurring in monogenic traits, are often morphologically indistinguishable from sporadically occurring tumors. Lung cancers of various types, for example, have been reported in antecedent pulmonary disorders that are sometimes familial (and probably genetic), in familial cancer syndromes, and in association with genetic biochemical markers. In these three categories, examples include, respectively, fibrocystic pulmonary dysplasia, idiopathic pulmonary fibrosis, cystic disease of the lung, sarcoidosis, and Kartagener syndrome; site-specific familial lung cancer mesothelioma and the Li-Fraumeni cancer family syndrome; and aryl hydrocarbon hydroxylase inducibility, a cytochrome P450, and gastrin-releasing polypeptide. In short, histologic type is not a reliable flag for a neoplasm of single-gene origin. Rather, Mendelian traits associated with neoplasia may be identified by recognizing other features of the trait or syndrome in the patient or in family members or, ultimately, by genetic testing.

Of the preneoplastic phenotypic traits in the *Catalog,* most have been mapped to a chromosomal band or at least assigned to a specific chromosome. Genes on every chromosome have been associated with a possibly increased risk of neoplasia. As the international genome program produces increasingly dense genetic-marker maps, all will be mapped and cloned. And, the list of genes, proteins, and breakpoints will grow rapidly. The remaining challenges for clinical research will be to associate the newly discovered genes with a clinical condition.

Applications

The cancer clinician who diagnoses a person's cancer but not the syndrome of which it is a manifestation makes an incomplete diagnosis. Every surgeon or radiotherapist who has a patient with breast cancer should look for small papillomas of the lips and gums, and palpebral fissures or hyperkeratotic papules of the hands, which would suggest the Cowden multiple hamartoma syndrome. Failure to recognize a rare syndrome is a disservice to the patient and perhaps his or her relatives, who may be denied the possible benefits of counseling for the prevention and early detection of cancer. Also, an opportunity for further research into pathogenesis is lost. Some of these traits were discovered because an unusual aggregation of cancer was seen and investigated in a family. Genes in the *Catalog* could provide an explanation for a family's apparent excess of cancer, even genes that, to date, have been found because they are mutated only in cancer tissue itself. In others, genes now known because of somatic cell mutations may eventually be seen as germ cell mutations and hence be heritable.

The epidemiologist investigating cancer etiology, perhaps with a case-control study design, must be aware that histopathologically identical tumors may have different etiologies. Geneticists have long recognized the phenomenon of genetic heterogeneity, namely, the fact that one phenotype (such as tall stature and long digits) can be attributed to a number of different causes and, specifically, to various gene mutations. Pheochromocytoma can be a feature of neurofibromatosis 1, von Hippel-Lindau syndrome, multiple endocrine neoplasia types 2a and 2b (the Sipple syndrome and the multiple mucosal neuroma syndrome, respectively), and familial paraganglioma. Sarcomas of various types have been described in association with neurofibromatosis 1, retinoblastoma, congenital adrenal hyperplasia, meningioma, lymphedema with distichiasis, Paget disease of bone, multiple exostosis, OSLAM syndrome, osteochondromatosis, the Li-Fraumeni cancer family syndrome, familial cancer with in vitro radioresistance, Rothmund-Thomson syndrome, and Werner syndrome.

The epidemiologist concerned with cancer control might guess that the conditions in the *Catalog* are rare and probably account for a small fraction of human cancer. However, a few, such as the intestinal polyposes, dysplastic nevus syndrome, and the breast cancer genes (*BRCA1* and *BRCA2*), clearly have public health implications. When these dominant traits are discovered in a patient, an opportunity arises to identify individuals at enormous risk for cancer. Such high-risk persons are likely to be especially motivated to understand and comply with a personalized prescription for cancer surveillance. As set forth by a National Institutes of Health workshop (Parry et al. 1987a, 1987b), it is reasonable to suggest that the clinical geneticist assume responsibility for alerting families to the hereditary and familial nature of the disorders that predispose to cancer, and to offer surveillance. Research in cancer prevention is best piloted in such individuals at highest risk by protocols involving intensive counseling and informed consent, especially when laboratory testing for major predisposing genes is planned. The broader significance of even the rarer forms of genetic cancer disorders is the possibility that intense study of these conditions may give insight into the etiology and pathogenesis of cancers in general.

Landis, S. H. et al. 1998. Cancer statistics, 1998. *CA: Cancer J. Clin.* 48:6–29.

McKusick, V. A. 1962. On the X chromosome of man. *Quart. Rev. Biol.* 37:69–175.

—. 1966. *Mendelian Inheritance in Man: Catalogs of Autosomal Dominant, Autosomal Recessive, and X-Linked Phenotypes,* 1st ed. Baltimore: Johns Hopkins Press.

—. 1971. *Mendelian Inheritance in Man: Catalogs of Autosomal Dominant, Autosomal Recessive, and X-Linked Phenotypes,* 3rd ed. Baltimore: Johns Hopkins Press, ix.

—. 1974. *Mendelian Inheritance in Man: Catalogs of Autosomal Dominant, Autosomal Recessive, and X-Linked Phenotypes*, 4th ed. Baltimore: Johns Hopkins University Press, xii.

—. 1988. *Mendelian Inheritance in Man: Catalogs of Autosomal Dominant, Autosomal Recessive, and X-Linked Phenotypes,* 8th ed. Baltimore: Johns Hopkins University Press, xi.

—. 1994. *Mendelian Inheritance in Man: A Catalog of Human Genes and Genetic Disorders,* 11th ed. Baltimore: Johns Hopkins University Press, vii.

—. 1998. *Mendelian Inheritance in Man: A Catalog of Human Genes and Genetic Disorders,* 12th ed. Baltimore: Johns Hopkins University Press.

Motulsky, A. 1989. Review of McKusick, V. A., *Mendelian Inheritance in Man: Catalogs of Autosomal Dominant, Autosomal Recessive, and X-Linked Phenotypes,* 8th ed. *JAMA* 261:3177–78.

Mulvihill, J. J. 1975. Congenital and genetic diseases. In Fraumeni, J. F., Jr., Ed., *Persons at High Risk of Cancer: An Approach to Cancer Etiology and Control.* New York: Academic 3.

—. 1977. Genetic repertory of human neoplasia. In Mulvihill, J. J. et al., *Genetics of Human Cancer.* New York: Raven 137.

—. 1989. Prospects for cancer control and prevention through genetics. *Clin. Genet.* 36:313–19.

—. 1994. Clinical ecogenetics of human cancer. *Hem./Onc. Ann.* 2:157.

Mulvihill, J. J. et al. 1997. Genetic repertory of human neoplasia. In Weber, W. et al., *Familial Cancer Management.* Boca Raton FL: CRC Press, 203–37.

Mulvihill, J. J. et al. 1977. *Genetics of Human Cancer.* New York: Raven.

Parry, D. M., Berg, K., Mulvihill, J. J., Carter, C. L., and Miller, R. W. 1987a. Strategies for controlling cancer through genetics: report of a workshop. *Am. J. Hum. Genet.* 41:63–69.

Parry, D. M., Mulvihill, J. J., Miller, R. W., Berg, K., and Carter, C. L. 1987b. Strageties for controlling cancer through genetics. *Cancer Res.* 47:6814–17.

U.S. Department of Health and Human Services (U.S.D. HHS), National Toxicology Program. 1998. *Report on Carcinogens,* 8th ed. Washington, DC: U.S. Government Printing Office, 89–239.

1. PHACOMATOSES

The Neurofibromatoses

NEUROFIBROMATOSIS 1 (VON RECKLINGHAUSEN DISEASE) (*162200)

JJM NOTE

The estimates of sarcomatous degeneration and other malignancies in neurofibromotosis 1 cannot come from hospital series (e.g., the highest rate of sarcomas comes from Memorial Hospital, New York, famous for its management of sarcomas). When you read that 24% of patients with neurofibromatosis 1 get cancer, remember that one in three persons gets a cancer in his or her lifetime anyway. For counseling (e.g., of a newly diagnosed child with only café-au-lait macules and axillary freckling), I express the risk of cancer this way: "Our study in Denmark (Sorensen et al., 1985) showed rates of cancer in neurofibromatosis 1 persons that were 1 times higher than average. It was higher in women than in men and in those who were in the study because they had to be hospitalized at one time. For boys with a mild course, the risk was no different from that of the general population; for girls, it was less than double the risk." The confusion arises from the concept of *relative risk,* which for some rare cancers is enormous, such as for neurofibrosarcoma. But, the background rate is minute, so the *absolute risk* is also very low.

Description

Neurofibromatosis is an autosomal dominant disorder characterized particularly by cafe-au-lait spots and fibromatous tumors of the skin.

Clinical Features

The only consistent features of this disorder are cafe-au-lait spots and fibromatous skin tumors. Crowe et al. (1956) suggested that the presence of 6 spots, each more than 1.5 cm in diameter, is necessary for the diagnosis of neurofibromatosis. Crowe (1964) considered axillary freckling to be an especially useful diagnostic clue. Occasional features include scoliosis, pseudoarthrosis of the tibia, pheochromocytoma, meningioma, glioma, acoustic neuroma, optic neuroma, mental retardation, hypertension, and hypoglycemia. Central neurofibromatosis (101000), characterized by bilateral acoustic neuroma and meningioma but few skin lesions or neurofibromas, is a distinct nosologic entity which has come to be known as neurofibromatosis type II.

The patients described by Hashemian (1953) apparently had von Recklinghausen neurofibromatosis, although the skin changes were not as striking as in some patients. Hayes et al. (1961) reported hypoglycemia associated with massive intraperitoneal tumor of mesodermal origin in a patient with typical cutaneous lesions. Neurofibromata of the bowel leading to gastrointestinal bleeding were described by Manley and Skyring (1961) in a patient (G.R., P368525) with striking skin changes. (An 'intestinal' form of neurofibromatosis (162220) in which many neurofibromas involve the entire length of the gastrointestinal tract is probably distinct from ordinary von Recklinghausen disease.) Fibromas may occur in the iris and glaucoma occurs in rare instances (Grant and Walton, 1968). Lisch nodules in the iris, a frequent finding in adults, are true tumors, not merely hyperpigmented patches. Zehavi et al. (1986) found Lisch nodules in 73% of 30 cases. They concluded that their presence correlated directly with the severity of skin manifestations.

Unusual clinical manifestations were described by Diekmann et al. (1967): hypertension due to renal artery stenosis, and hypertrophy of the clitoris. Sutphen et al. (1995) described clitoromegaly in 4 patients with NF1 and reviewed the literature documenting 26 NF1 patients with clitoral involvement. Involvement of the heart in neurofibromatosis was described and reviewed by Rosenquist et al. (1970), who also reviewed involvement of the abdominal aorta and renal, carotid and other arteries.

Nicolls (1969) described 2 cases of sectorial (or segmental) neurofibromatosis which he plausibly interpreted as representing somatic mutation. One had a mediastinal neurofibroma and, in the skin area corresponding segmentally to the site of the internal lesion, five small neurofibromas. Miller and Sparkes (1977) also reported on this phenomenon. Riccardi and Eichner (1986) referred to the segmental form as neurofibromatosis type V. Combemale et al. (1994) presented 2 new cases of segmental NF1 and reviewed reports concerning 88 cases. One of their patients was a 71-year-old woman with multiple cutaneous tumors limited to the left side of the trunk and present since the age of 41 years. Wolkenstein et al. (1995) concluded that segmental NF is about 30 times less frequent than NF type 1.

Crowe et al. (1956) found 6 secondary malignant lesions in 168 patients with neurofibromatosis. D'Agostino et al. (1963) discovered 21 cases of secondary neoplasms in his study of 678 cases of neurofibromatosis.

Neurofibromatosis is associated with a tendency to malignant degeneration of the neurofibromas in an estimated 3 to 15% of cases. Knight et al. (1973) reviewed 69 patients with single and 45 patients with multiple neurofibromas. Five patients in the group were found to have a total of 11 secondary malignant lesions including 3 fibrosarcomas, 3 squamous cell carcinomas, and 1 neurofibrosarcoma, among other forms. Some earlier studies have reported mainly sarcomas associated with neurofibromatosis. Hunerbein et al. (1996) described a 56-year-old man with NF1 who had had a 6-month history of recurrent epigastric pain and was found to have a multifocal malignant schwannoma of the duodenum causing biliary obstruction.

Adornato and Berg (1977) observed the diencephalic syndrome in 2 infants who had neurofibromatosis and hypothalamic tumors.

Although neurofibromatosis type 1 had been called peripheral neurofibromatosis, it has been associated with tumors of the central nervous system, which include astrocytomas of the visual pathways, ependymomas, meningiomas, and some primitive neuroectodermal tumors. The most common neuroimaging abnormality in neurofibromatosis type 1 has been a high signal intensity lesion in the basal ganglia, thalamus, brain stem, cerebellum, or subcortical white matter referred to as an 'unidentified bright object' (UBO). These UBOs are thought to represent sites of vacuolar change. Parazzini et al. (1995) documented spontaneous regression of optic pathway lesions in 4 patients with neurofibromatosis type 1. They cautioned against diagnosis of optic nerve glioma without evidence of progression. Molloy et al. (1995) studied 17 NF1 patients with brain stem tumors, which also presented increased T2 signal abnormality on MRI scanning. Fifteen of the these 17 patients had neurologic signs and symptoms indicative of brain stem dysfunction and 35% of them had evidence of radiographic tumor progression. In the 2 patients that had partial surgical resection, pathology demonstrated either a fibrillary or anaplastic astrocytoma. As 15 of these 17 patients remained alive after a 52-month follow-up, this suggested that these are much less aggressive than typical pontine tumors which should be distinguished from the UBOs seen elsewhere in the brains of neurofibromatosis patients.

Robertson (1979) reported a patient with neurofibromatosis and grotesque, massive overgrowth of one leg. Benedict et al. (1968) studied the pigmentary anomaly of neurofibromatosis in relation to that of Albright polyostotic fibrous dysplasia. Gross appearance of the pigmented areas was not always reliable. However, special microscopic studies showed giant pigment granules in malpighian cells or melanocytes of normal skin and of neurofibromatosis spots but rarely in Albright syndrome.

Erickson et al. (1980) described 2 sisters with neurofibromatosis and intracranial arterial occlusive disease leading to the moyamoya pattern of collateral circulation (252350). Four other members of their sibship of 8, and members of 2 previous generations, including the mother, had neurofibromatosis.

Zonana and Weleber (1984) illustrated a patient who had multiple cafe-au-lait spots of von Recklinghausen type only on the right side of the body. Iris hamartomata (Lisch nodules) were present in the right eye only.

Clark and Hutter (1982) reported an apparent association between the rare entity juvenile chronic myelogenous leukemia and neurofibromatosis. They suggested that other types of nonlymphocytic leukemia have an increased frequency, but Riccardi (1982) raised the question as to whether these are families with only cafe-au-lait spots. Voutsinas and Wynne-Davies (1983) suggested that the risk of malignant change in NF has been exaggerated and that the true value is 2.0% (or 4.2% of those over 21 years).

Crawford (1986) reported on a study of 116 patients under 12 years of age and reviewed the literature. Among the unusual presentations was rhabdomyosarcoma projecting from the urethra in a girl who also had congenital pseudoarthrosis of the tibia. Crawford (1986) stated that 'most of the rhabdomyosarcomas associated with neurofibromatosis involve the genitourinary tract.'

Sorensen et al. (1986) conducted a highly valuable follow-up study of natural history in a nationwide cohort of 212 cases (and their families) identified in Denmark by Borberg (1951). Follow-up information was obtained in 99%. All 76 probands had been ascertained through hospitals and were more severely affected than their incidentally identified relatives. Relatives had poorer survival rates than persons in the general population. The worst prognosis was shown by female probands. Malignant neoplasms or benign CNS tumors occurred in 45% of the probands, giving a relative risk of 4.0 compared with expected numbers. Pheochromocytoma occurred in 3 of 212 patients.

Senveli et al. (1989) reported 6 patients with NF-1 who had aqueductal stenosis and hydrocephalus requiring surgical intervention. Ages varied from 14 to 24. Twenty-two similar cases were found in the literature. Westerhof et al. (1983) found hypertelorism in 24% of patients with neurofibromatosis.

Zochodne (1984) reported the case of a 16-year-old female with aneurysm of the superior mesenteric artery complicating renovascular hypertension associated with coarctation of the abdominal aorta from above the celiac trunk to above the origin of the inferior mesenteric artery. The coarctation was associated with stenosis of the renal, celiac and superior mesenteric arteries. The patient had typical skin signs of neurofibromatosis and had had a right below-knee amputation at age 5 for nonunion of a tibial fracture. The mother and 2 sibs were affected. A very similar patient with neurofibromatosis vasculopathy, or vascular neurofibromatosis, was reported by Lehrnbecher et al. (1994). The 4-year-old boy presented with congenital pseudarthrosis of the right tibia (suggesting the vascular origin of this well-known complication of NF1), multiple cafe-au-lait spots, short stature, and mild systemic arterial hypertension. The mother and grandmother had NF1. Subsequent complications of the vasculopathy were hypertension, septic infection of an aneurysm in the deltoid muscle, infarction of a segment of colon, sudden appearance of multiple arterial aneurysms, and venous thrombosis. Histologic examination of the bowel specimen confirmed the clinical diagnosis of vascular NF1. The vascular changes were not secondary to the initially mild arterial hypertension lasting less than 4 months. Reubi (1945) first described vascular NF1. The pathogenesis of the vascular lesions has been subject to controversy. In the case of Lehrnbecher et al. (1994), the proliferating cells seemed to have originated from myoblasts or myofibroblasts and not, as has been speculated, from Schwann cells. Benatar (1994) described a 27-year-old man with neurofibromatosis who presented with 3 intracranial fusiform aneurysms. He referred to 3 previous descriptions of large intracranial fusiform aneurysms in patients with neurofibromatosis type I, which he considered to be considerably less common than renal an d gastrointestinal vascular lesions in this disorder.

Nopajaroonsri and Lurie (1996) described venous aneurysm, arterial dysplasia, and near-fatal hemorrhages in a 62-year-old who was said to have familial neurofibromatosis (no family history was given). The patient presented with an aneurysm of the internal jugular vein which was associated with dysplasia of cervical arteries. Neurofibromatous tissue was found in the wall of the aneurysm as well as in small veins. During and after surgical excision of the aneurysm, the patient developed massive hemorrhages that required reexploration and evacuation of cervical hematomas. During surgery, bleeding was difficult to control because of excessive friability of blood vessels. Despite the vascular invasion by a tumor, there was no evidence of malignancy or malignant transformation in the patient after a 10-year follow-up.

Uren et al. (1988) found a congenital left atrial wall aneurysm in a patient with neurofibromatosis; the association may be coincidence. Fitzpatrick and Emanuel (1988) observed the association of typical autosomal dominant neurofibromatosis with hypertrophic cardiomyopathy in a brother and sister. Kousseff and Gilbert-Barness (1989) reported what they referred to as 'vascular neurofibromatosis' in 2 patients who as infants developed idiopathic gangrene with vascular changes resembling those of NF1. An additional review of 105 patients uncovered a 27-month-old boy with NF1 and extensive vascular changes with renal hypertension. They discussed the possible relationship to fibromuscular dysplasia. Stanley (1975) found that 5 of 25 children with fibromuscular dysplasia had NF1 as well. Massaro and Katz (1966) established the association of interstitial pulmonary fibrosis (fibrosing alveolitis) with von Recklinghausen neurofibromatosis on the basis of studies of 76 patients. Porterfield et al. (1986) described pulmonary hypertension secondary to interstitial pulmonary fibrosis.

Among 18 cases of neurofibromatosis with hypertension, Kalff et al. (1982) found pheochromocytoma in 10. Age at diagnosis ranged from 15 to 62 years. The clinical characteristics of the neurofibromatosis did not predict the presence of pheochromocytoma. Younger patients tended to have causes of hypertension other than pheochromocytoma. Several causes of hypertension may coexist. The pheochromocytomas secreted epinephrine as well as norepinephrine and resided in or next to the adrenal gland. Control of hypertension was less successful in the patients without surgically resected pheochromocytoma. One patient without pheochromocytoma had coarctation of the aorta and 1 had renal artery stenosis; this patient was described as having the Turner phenotype. At least 2 of the pheochromocytoma patients had renal artery stenosis. Three had small-bowel and/or stomach neurofibromata. One patient with pheochromocytoma also had hypernephroma with metastases and another had disseminated metastases from an undifferentiated leiomyosarcoma thought to originate from her upper gastrointestinal tract. Horwich et al. (1983) presented evidence that aqueductal stenosis occurs in neurofibromatosis. Sayed et al. (1987) described malignant schwannoma in 3 brothers who had inherited neurofibromatosis from their mother. Two of the brothers had been reported by Herrmann (1950). Sakaguchi et al. (1996) described a 48-year-old man with NF1 and paroxysmal hypertension in progressive respiratory insufficiency. Clinical investigation displayed calcified tumors in the anterior mediastinum and pararenal region. Histologic examination at autopsy revealed composite tumors consisting of pheochromocytoma and malignant peripheral nerve sheath tumor at 2 sites: the left adrenal gland and the region surrounding the inferior vena cava, probably corresponding to the right adrenal gland. In addition, the gastrointestinal tract was involved with mesenchymal tumors showing neurogenic differentiation.

In 9 cases of neurofibromatosis with a carcinoid tumor studied by Griffiths et al. (1987), all carcinoid tumors were in the duodenum, were distinctive histologically, and had widespread somatostatin immunoreactivity. Furthermore, the duodenum was the primary site in 18 of 20 published cases of carcinoid tumor and neurofibromatosis. Pheochromocytoma was also present in 6 of the 27 cases with neurofibromatosis and duodenal carcinoid tumor. In cases of von Hippel-Lindau syndrome (193300), with which pheochromocytoma also occurs, Griffiths et al. (1987) found no carcinoid tumors, but did find islet cell tumor in association with pheochromocytoma. Swinburn et al. (1988) reported 2 patients with neurofibromatosis and duodenal carcinoid tumor, bringing the total number of cases of this association to 18. Their 2 cases as well as 5 others were positively identified as somatostatinomas. The histologic finding of psammoma bodies is important in the diagnosis of duodenal somatostatinomas. One patient also had a parathyroid adenoma which was found at postmortem.

Konishi et al. (1991) described the case of a 40-year-old woman with NF1 and typical hypophosphatemic osteomalacia. Bone pain, multiple pseudofractures, marked increase in osteoid by bone biopsy, and hypophosphatemia with renal phosphate wasting were features. Treatment with oral phosphate and vitamin D was effective. They found reports of 34 similar cases and pointed out that of the 67 patients collected by Dent (1952), 2 had neurofibromatosis.

By linkage analysis, Pulst et al. (1991) demonstrated that an atypical family with spinal neurofibromatosis and cafe-au-lait spots almost certainly had a mutation at the NF1 locus; see 162210. Poyhonen et al. (1997) described a family in which 7 members in 3 generations had rare spinal neurofibromatosis. The affected adults showed, at the ages of 32, 37, 38, and 61, respectively, multiple spinal neurofibromas symmetrically affecting all spinal roots. Two patients were operated on for histopathologically proven cervical spinal neurofibromas. All patients had cafe-au-lait spots, 1 had several freckles in the axillary area, and 2 had possible dermal neurofibromas, but iris Lisch nodules were not present. Other signs of neurofibromatosis type I and type II were absent. Linkage study of the family suggested close linkage to the NF1 locus and excluded linkage with the NF2 locus. DNA analysis of histopathologically verified spinal neurofibromas in 2 patients showed no evidence of loss of heterozygosity (LOH) at 17q11.2. The findings in this family, together with those in the the family reported by Pulst et al. (1991), suggested to Poyhonen et al. (1997) the existence of a clinically distinct form of neurofibromatosis with extensive spinal involvement and cafe-au-lait macules, which may be allelic to classic NF1. Ars et al. (1998) described a 3-generation family that included 5 members, all female, with spinal neurofibromatosis. All presented with multiple spinal neurofibromas and cafe-au-lait spots. Linkage analysis showed segregation with markers within and flanking the NF1 gene. Mutation analysis identified a frameshift mutation (162200.0018) that was predicted to result in a truncated NF1 protein.

In a father and 3 children by 2 different women, Schotland et al. (1992) described cosegregation of NF1 and osseous fibrous dysplasia. In the 4 individuals with NF1, cafe-au-lait spots and neurofibromata were present in all 4, Lisch nodules and macrocrania in 3, and scoliosis and curvature of the long bones in 2. Schotland et al. (1992) found at least 8 reports of NF1 and osseous fibrous dysplasia associated in individuals but no previous description of a familial association. The osseous dysplasia consisted of multiple lesions at the distal ends of the shafts of the femurs and in the tibias and fibulas, with bowing of the fibulas.

Friedman et al. (1993) described a central database designed to collect information on NF1 from 16 centers around the world. The aspects of the disorder for which information was being collected included renal artery stenosis and cerebral artery stenosis.

Ragge et al. (1993) provided a comprehensive discussion of Lisch nodules accompanied by colored photographs in irides of different colors. They pointed out that iris nodules were reported by several workers in the decade before the paper by Lisch (1937). In particular, Sakurai (1935) published a beautifully illustrated paper linking characteristic iris nodules with von Recklinghausen neurofibromatosis. They suggested that the lesions be renamed Sakurai-Lisch nodules in her honor.

Kurotaki et al. (1993) described the case of a 13-year-old Japanese boy who was found to have small nodules in the lung on chest radiography. He was asymptomatic. Although there was no family history of NF1, he had multiple cafe-au-lait spots over the whole body since birth, and soft subcutaneous tumors of the forehead and back were noticed from the age of 7 years. On biopsy the lung lesions were found to be papillary adenomas of type II pneumocytes. The patient had remained asymptomatic for 6 years thereafter.

Easton et al. (1993) studied variation in expression of 3 quantitative traits (number of cafe-au-lait patches, number of cutaneous neurofibromas, and head circumference) and 5 binary traits (presence or absence of plexiform neurofibromas, optic gliomas, scoliosis, epilepsy, and referral for remedial education). For cafe-au-lait patches and neurofibromas, correlation was highest between MZ twins, less high between first-degree relatives, and lower still between more distant relatives. The higher correlation between MZ twins suggested a strong genetic component in variation of expression, but the low correlation between distant relatives suggested that the type of mutation at the NF1 locus itself plays only a minor role. All 5 binary traits, with the exception of plexiform neurofibromas, also showed significant familial clustering. The familial effects for these traits were consistent with polygenic effects, but there were insufficient data to rule out other models, including a significant effect of NF1 mutations. There was no evidence of any association between different traits in affected individuals. Easton et al. (1993) concluded that the phenotypic expression of NF1 is to a large extent determined by the genotype at other 'modifying' loci and that these modifying genes are trait specific.

All the lesions of NF1, the benign and malignant tumors, the cafe-au-lait spots, the Lisch iris nodules, etc., are presumably the result of 2 mutations, the inherited mutation and a second mutation on the normal homolog. Collins (1993) suggested that the wide variability of clinical manifestations in members of the same family is related to the element of chance in determining what cells are involved by the second mutation and at what stage of development. The progressive nature of the disorder is also indicated.

Hofman et al. (1994) conducted a study to determine whether the presence of the NF1 gene results in a global cognitive deficit, as measured by lowering of IQ, or in a more specific cognitive deficit or learning disability. In addition, they sought to establish whether learning disabilities could be correlated with brain MRI scan findings. Families were informed concerning the study by NF centers and organizations. Of those expressing interest, 12 families with the appropriate structure were chosen. Each comprised 1 child with NF1, an unaffected sib, and both natural parents. NF children with known intracranial problems were excluded, but family members with known learning disabilities or hyperactivity disorders were not, making some of the results difficult to interpret. Full scale IQs ranged from 70 to 130 among children with NF1 and from 99 to 139 among unaffected sibs. Scores of parents with NF1 ranged from 85 to 114 compared to 80 to 134 in unaffected parents. Children with NF1 showed significant deficits in language and reading abilities compared to sibs, but not in mathematics. They also had impaired visuospatial and neuromotor skills. In 11 of 12 NF1 children but in none of the unaffected sibs, foci of high signal intensity on T2-weighted MRI scan images were observed. A statistically significant correlation was found between lowering of IQ and visuospatial deficits and the number of foci seen on scan.

Legius et al. (1994) studied the neuropsychologic profiles of 46 children with NF1. They found a reduction in total IQ, but a significantly better verbal rating than performance rating in all age groups. Concentration problems were especially significant in children with a higher IQ. Legius et al. (1994) suggested that these children may benefit from the use of Ritalin.

T2 'unidentified bright objects' are seen in 50 to 75% of children with neurofibromatosis type 1, most frequently in the basal ganglia, corpus cerebellum, and brain stem. Legius et al. (1995) found no difference in the mean intelligence of 18 children with such lesions and 10 neurofibromatosis children who did not show such lesions.

Winter (1991) described dural ectasia in neurofibromatosis causing bony erosion that was sufficiently severe to destroy spinal stability. Eichhorn et al. (1995) described dural ectasia in a 20-year-old woman with NF1 who presented with back and leg pain. Increasingly severe back pain led to investigations which showed multiple fractures of the pedicles of L1 to L4 with dural ectasia penetrating the body of L2. The transverse diameter of the dura was twice that of the vertebral body at that level, reaching and lifting the psoas.

Dugoff and Sujansky (1996) reported outcome data of 247 pregnancies in 105 women with NF1. The 247 pregnancies resulted in 44 first trimester spontaneous abortions. The cesarean section rate (36%) was greater than in the general population (9.1 to 23.5%). In 7 of the patients, cesarean section was required because of maternal complications of NF1 including pelvic neurofibromas, pelvic bony abnormality with or without kyphoscoliosis, pheochromocytoma, and spinal cord neurofibroma. Dugoff and Sujansky (1996) reported that 80% of the women in their study experienced either the appearance of new neurofibromas, growth of existing neurofibromas, or both. Thirty-three percent of these women noted a decrease in the size of their neurofibromas in the postpartum period. Eighteen percent of the women reported no changes in neurofibromas and no appearance of new neurofibromas during pregnancy.

Wu et al. (1997) described a father and son with NF1 due to deletion of the entire NF1 gene detected by fluorescence in situ hybridization (FISH). Both had severe NF1, including a large number of cutaneous neurofibromas, facial anomalies, large hands, feet, and head, and developmental impairment. Only the 15-year-old son had seizures and plexiform neurofibromas.

Precocious puberty in NF1 occurs, as a generalization, in children with tumors of the optic chiasm. A longitudinal study of 219 patients with NF1 reported that clinical precocious puberty developed in 7 children, all of whom had optic chiasmal tumors (124,125:Listernick et al., 1994, 1995). On the other hand, Zacharin (1997) described precocious puberty in a 5-year-old girl and 8-year-old boy with NF1 in whom magnetic resonance imaging on 2 occasions failed to demonstrate any abnormality of the optic tracts or optic chiasm. Previous studies have indicated that optic tract lesions develop at a mean age of 3.6 years, and longitudinal studies have failed to demonstrate symptomatic optic tract tumors occurring after age 6 years. The 2 patients of Zacharin (1997) were aged 11 and 14.7 years at the time of the report. Thus, the possibility of new lesions developing in these patients is unlikely.

Biochemical Features

To study the NF1 gene product, Gutmann et al. (1991) raised antibodies against both fusion proteins and synthetic peptides. A specific protein of about 250 kD was identified by both immunoprecipitation and immunoblotting. The protein was found in all tissues and cell lines examined and was detected in human, rat, and mouse tissues. Based on the homology between the NF1 gene product and members of the GTPase-activating protein (GAP) superfamily, the name NF1-GAP-related protein (NF1GRP) was suggested. DeClue et al. (1991) raised rabbit antisera to a bacterially synthesized peptide corresponding to the GAP-related domain of NF1 (NF1-GRD). The sera specifically detected a 280-kD protein in lysates of HeLa cells. This protein corresponded to the NF1 gene product, as shown by several criteria. NF1 was present in a large molecular mass complex in fibroblast and Schwannoma cell lines and appeared to associate with a very large (400-500 kD) protein in both cell lines.

Basu et al. (1992) presented evidence supporting the hypothesis that NF1 is a tumor-suppressor gene whose product acts upstream of RAS. They showed that the RAS proteins in malignant tumor cell lines from patients with NF1 were in a constitutively activated state as measured by the ratio of the guanine nucleotides bound to them, i.e., the ratio of GTP (active) to GDP (inactive). Transforming mutants of p21(ras) bind large amounts of GTP, whereas wildtype p21(ras) is almost entirely GDP-bound. Daston et al. (1992) raised antibodies against peptides coded by portions of the NF1 cDNA. These antibodies specifically recognized a 220-kD protein, called neurofibromin, in both human and rat spinal cord. Neurofibromin was most abundant in the nervous system. Immunostaining of tissue sections indicated that neurons, oligodendrocytes, and nonmyelinating Schwann cells contained neurofibromin, whereas astrocytes and myelinating Schwann cells did not. In schwannoma cell lines from patients with neurofibromatosis, loss of neurofibromin is associated with impaired regulation of GTP/RAS. Analysis of other neural crest-derived tumor cell lines showed that some melanoma and neuroblastoma cell lines established from tumors occurring in patients without neurofibromatosis also contained reduced or undetectable levels of neurofibromin, with concomitant genetic abnormalities of the NF1 locus. In contrast to the schwannoma cell lines, however, GTP/RAS was appropriately regulated in the melanoma and neuroblastoma lines that were deficient in neurofibromin, even when HRAS was overexpressed (Johnson et al., 1993). These results demonstrated that some neural crest tumors not associated with neurofibromatosis have acquired somatically inactivated NF1 genes and suggested a tumor-suppressor function for neurofibromin that is independent of RAS GTPase activation.

Nakafuku et al. (1993) took advantage of the yeast RAS system to isolate mutants in the RAS GTPase activating protein-related domain of the NF1 gene product (NF1-GRD) that can act as antioncogenes specific for oncogenic RAS. They demonstrated that these mutant NF1-GRDs, when expressed in mammalian cells, were able to induce morphologic reversion of RAS-transformed NIH 3T3 cells.

Other Features

Schenkein et al. (1974) reported increased nerve growth stimulating activity in the serum of patients with von Recklinghausen disease. Kanter et al. (1980) showed an increase only in antigenic activity of nerve growth factor in central neurofibromatosis and only in functional activity in peripheral neurofibromatosis. Thus, these disorders may involve different defects in NGF synthesis and/or regulation.

Fialkow et al. (1971) concluded from analysis of neurofibromas from G6PD A-B heterozygotes with von Recklinghausen disease that each tumor must originate in many cells, perhaps at least 150. Although the benign tumors of neurofibromatosis are multiclonal in nature, the malignant lesion (neurofibrosarcoma) is monoclonal (Friedman et al., 1982).

In 8 of 30 unrelated females with NF1, Skuse et al. (1989) found heterozygosity for a PGK (311800) RFLP which could be used to test for clonality. In all 8 cases the neurofibromas appeared to be monoclonal in origin. These results supported the suggestion that benign neurofibromas in NF1 arise by a mechanism that is different from that of the malignant tumors. In a neural fibrosarcoma from a patient with NF1, Legius et al. (1993) found a somatic deletion of the NF1 gene on one chromosome and loss of heterozygosity for all chromosome 17 polymorphisms. Thus, homozygous inactivation of NF1 was demonstrated at the molecular level, providing strong support for the view that NF1 is a tumor suppressor gene.

Colman et al. (1995) examined the '2-hit' hypothesis in relation to benign neurofibromas in NF1. Using both NF1 intragenic polymorphisms as well as markers from flanking and more distal regions of chromosome 17, they investigated loss of heterozygosity (LOH) in 22 neurofibromas from 5 unrelated NF1 patients. Eight of these tumors revealed somatic deletions involving NF1, indicating that inactivation of NF1 is associated with at least some neurofibromas. On the other hand, Stark et al. (1995) found single-cell PCR on neurofibroma Schwann cells and found that both alleles of the NF1 gene were present; i.e., there was no evidence of loss of heterozygosity by a nondisjunction, large deletions, or somatic recombination. They granted that small mutations inactivating the wildtype allele could not be excluded.

Based on the International Database maintained by the National NF Foundation (NNFF), which contained information on 1,479 probands and 249 of their affected relatives with NF1 at the time of analysis, Friedman and Birch (1997) summarized clinical information about the population. The age of diagnosis of NF1 was 8 years younger in the probands than in the affected relatives. Many of the manifestations of NF1 were more frequent in the probands than in their affected relatives. The age-specific prevalence of most manifestations of NF1 increased with age. Despite biases inherent in a convenience sample from specialist clinics, the frequency of manifestations of NF1 in many of the series was similar to those in 2 smaller population-based studies. Lisch nodules were said to be present in 57% of probands and 69.9% of affected relatives.

Inheritance

Miller and Hall (1978) found that patients born of affected mothers had more severe disease than those born of affected fathers. (A similar maternal effect was known to occur in myotonic dystrophy (160900) and subsequently a maternal effect on severity was noted in neurofibromatosis, type II (NF2; see 101000).) In their series of 62 patients from 54 families, only 16 were new mutations, as contrasted with the figure of 50% arrived at by Crowe et al. (1956). Crowe et al. (1956) estimated the relative fertility of affected males and females to be 0.41 and 0.75, respectively.

Samuelsson and Akesson (1988) estimated that the relative fertility of neurofibromatosis cases is 78% and the mutation rate somewhere between 2.4 and 4.3 x 10(-5). Ritter and Riccardi (1985) studied 111 3-generation families with NF and found no instance of skipped generation. They suggested that penetrance of NF is complete and that previous impressions to the contrary have failed to recognize heterogeneity, minimal NF expression, and nonpaternity.

Clementi et al. (1990) used the methods of classic segregation analysis to test whether there was a deviation from the expected mendelian segregation rate in a sample of 129 Italian sibships. With this approach, they obtained a maximum likelihood estimate of the proportion of sporadic cases, and hence they estimated the mutation rate to be 6.5 x 10(-5) gametes per generation.

Jadayel et al. (1990) used molecular methods to identify the parental origin of new mutations in NF1. They found that in 12 out of 14 families analyzed, the new mutation was of paternal origin. The estimated mutation rate, 1 in 10,000 gametes, is one of the highest for a human disorder (Huson et al., 1989) and suggests that the NF1 gene is large or has some other structural peculiarity. The same bias toward paternal origin of new mutations has been demonstrated for retinoblastoma (180200). In both of these disorders, however, there is little or no evidence of paternal age effect in the incidence of mutations. (Riccardi et al. (1984) found increased paternal age.)

The high mutation rate of NF1 may reflect the fact that the gene is large like that of dystrophin (310200) and/or that it has an unusual internal structure predisposing to deletions and other mutations. Predominant paternal derivation suggests that mutation may occur in the mitotic divisions that take place in male gametogenesis but not in female gametogenesis. Since there is little or no evidence of accumulation of mutations reflected by paternal age effect, mutation may be occurring in cells not involved in the process of replenishment of the germ cell 'bank.'

In 10 families with an NF1 mutation, Stephens et al. (1992) found that the mutation had occurred in the paternally derived chromosome 17. The probability of observing this result by chance was estimated as less than 0.001, assuming an equal frequency of mutation of paternal and maternal NF1 genes. They suggested a role for genomic imprinting that may either enhance mutation of the paternal NF1 gene or confer protection from mutation to the maternal NF1 gene.

Lazaro et al. (1994) observed a family in which completely normal parents had a son and daughter with a clinically severe form of NF1. The sibs showed no inheritance of paternal alleles for a marker in intron 38 of the NF1 gene, whereas they received alleles from both parents for other NF1 markers. Analysis with probes from this region of the NF1 gene showed a 12-kb deletion involving exons 32 to 39, in the affected offspring. In the father's spermatozoa, 10% were found to carry the same NF1 deletion, but the abnormality was not detected in DNA from his lymphocytes. Thus, this appeared to be an example of gonadal mosaicism. Colman et al. (1996) identified a new mutation in an NF1 patient who was somatically mosaic for a large maternally derived deletion in the NF1 gene region. The deletion extended at least from exon 4 near the 5-prime end of the gene to intron 39 near the 3-prime end. Thus, 100 kb or more of the gene was lost. Colman et al. (1996) suggested that the deletion occurred at a relatively early developmental timepoint, since signs of NF1 in this patient were not segmental and both mesodermally and ectodermally derived cells were affected.

Shannon et al. (1992) reviewed the occurrence of leukemia in NF1. In 16 of 21 cases of juvenile chronic myelogenous leukemia in children with familial NF1, the genetic disorder was inherited from the mother. Of the 21 children, 17 were boys. Myeloid leukemia developed in 12 boys and 4 girls who inherited NF1 from their mothers, and in 5 boys who inherited the disease from their fathers. Father-to-daughter transmission was not observed. Shannon et al. (1992) found that among 5 children with bone marrow monosomy 7 (Mo7), 3 had NF1 and 2 others had suggestive evidence of NF1. Studying DNA extracted from the bone marrow of 11 children with NF1 in whom malignant myeloid disorders had developed, Shannon et al. (1994) found that in samples from 5 there was loss of heterozygosity. In each case, the NF1 allele was inherited from a parent with NF1 and the normal allele was deleted. Loss of constitutional heterozygosity had not been reported in the benign tumors associated with NF1 and had been detected only in a few malignant neural crest tumors and in some tumor-derived cell lines. The data from the study of children with myeloid disorders provided evidence that NF1 may function as a tumor-suppressor allele in malignant myeloid diseases and that neurofibromin is a regulator of RAS in early myelopoiesis.

Cytogenetics

Krone and Hogemann (1986) found monosomy 22 as a predominant numerical anomaly in cultured cells grown from peripheral neurofibromas in patients described simply as suffering 'from sporadic peripheral NF.' Duncan et al. (1987) observed a ring chromosome 22 in a man with an atypical form of neurofibromatosis. He lacked a family history of NF, cafe-au-lait spots, and axillary freckling. He had multiple neurofibromas and a plexiform neuroma. By in situ hybridization, Duncan et al. (1987) showed that both the normal chromosome 22 and the ring chromosome 22 carried this gene.

Kaneko et al. (1989) found no microscopically detectable chromosome changes in the juvenile chronic myelogenous leukemia associated with neurofibromatosis. However, deletion of the whole or part of certain chromosomes, such as chromosomes 6 or 7, may be an important step towards the evolution of the accelerated blast phase or the development of de novo acute leukemia in a patient. The increased risk of leukemia in NF was thought by the authors to be 'quite low.'

Upadhyaya et al. (1996) claimed to have provided the first physical cytogenetic deletion involving the NF1 gene in a patient with sporadic neurofibromatosis, dysmorphic features, and marked developmental delay. Combined evidence of molecular and cytogenetic techniques predicted that the deletion was approximately 7 Mb.

Mapping

Using 2 RFLPs related to the beta-nerve growth factor gene (162030), Darby et al. (1985) excluded the gene for nerve growth factor as the site of the mutation in 4 families with neurofibromatosis of the classic type. About half of cases are sporadic.

Family studies by Dunn et al. (1985) excluded close linkage of NF1 (lod score less than -2.0) with 8 markers (ABO, Rh, MNSs, GC, PGP, ACP, GPT, and HP). Negative lod scores at all values of theta were obtained with both GC (on 4) and Se (on 19), which others had proposed were linked to NF. Dietz et al. (1985) excluded linkage of NF with GC. Findings of DiLiberti et al. (1982) brought the total lod score over 3.0 for linkage of NF with myotonic dystrophy. Huson et al. (1986) excluded linkage with chromosome 19 markers linked to myotonic dystrophy. Thus, the reports of coinheritance of DM and NF cannot be explained by close linkage of the 2 loci.

Ledbetter et al. (1989) described a patient in whom a balanced translocation between chromosomes 17 and 22 was found in association with von Recklinghausen neurofibromatosis. The breakpoint on chromosome 17 in this patient was at 17q11.2. Creation of a human-mouse somatic cell hybrid containing the derivative chromosome 22 but not the derivative 17 or normal 17 from this patient allowed rapid localization of ERBA1, ERBB1, and granulocyte colony-stimulating factor (CSF3) distal to the breakpoint, and HHH202 (D17S33) and beta crystallin (CRYB1) proximal to the breakpoint. By linkage analysis of 15 kindreds, Barker et al. (1987) demonstrated that a gene responsible for NF is located near the centromere on chromosome 17. No evidence for heterogeneity was found.

Because of the high mutation rate in NF, Barker et al. (1987) suggested that the NF1 gene may be unusually large, a situation comparable to that with Duchenne muscular dystrophy (310200). The results of Barker et al. (1987) suggested a genetic distance of approximately 4 cM between NF1 and the centromere. Since recombination is reduced near the centromere, a longer sequence of DNA than one would predict from the usual equation of 1 million bases per cM may separate the 2 landmarks in this instance. The results leave a region of at least 10 megabases on either side of the centromere for the physical location of NF1. Using an alphoid DNA probe that maps to the centromeric region of chromosome 17, Barker et al. (1987) found close linkage of NF1 (theta = 0.04; lod = 4.21). Seizinger et al. (1987) presented evidence that the NF gene is linked to the locus for nerve growth factor receptor (NGFR; 162010) in the region 17q12-q22. A peak lod score of 4.41 was obtained at a theta of 0.14. However, crossovers between the 2 loci suggested that a mutation in NGFR is not the fundamental defect (Seizinger et al., 1987). No loss of alleles at chromosome 17 loci has thus far been found in NF1 tumors (Gusella, 1987).

On the basis of the occurrence of neurofibromatosis and galactokinase deficiency in a family reported by Fanconi (1933), Stambolian and Zackai (1988) suggested that the NF1 locus may be closely linked to that of galactokinase (230200). One of the affected sibs in this family was the first enzymatically identified case of galactokinase deficiency (Gitzelmann, 1965). The parents of this sibship were first cousins and the mother had NF.

Vance et al. (1989) reported linkage studies in 6 multigenerational families with NF1 using 9 markers known to map in the pericentromeric region of chromosome 17. The closest marker was HHH202, with a lod score of 3.86 at theta = 0. Two-point lod scores for NF1 versus all the markers studied were presented, and the most likely gene order determined. Similar studies were reported by Seizinger et al. (1989), who performed a multipoint linkage analysis using 6 closely linked markers on chromosome 17. In this study, as in the one reported by Vance et al. (1989), the only probe showing no recombination at a theta of 0 was HHH202, with a lod score of 3.83. The authors concluded, on the basis of the linkage data, that the NF1 gene maps to the long arm rather than the short arm of chromosome 17.

Further linkage studies involving the NF1 locus and pericentromeric markers on chromosome 17 were reported by Diehl et al. (1989), Mathew et al. (1989), Upadhyaya et al. (1989), Kittur et al. (1989), Goldgar et al. (1989), and Stephens et al. (1989). Goldgar et al. (1989) summarized the results of the international consortium for NF1 linkage. The 8 teams of researchers studied 142 families with more than 700 affected persons, using 31 markers in the pericentric region of chromosome 17. The best gene order derived from these studies was pter—pA10-41—EW301—cen—pHHH202—NF1—EW206—EW207—EW203—CRI-L581—CRI-L946—HOX2—NGFR—qter.

Physical mapping data concerning the NF1 region on chromosome 17 were reported by O'Connell et al. (1989), Fountain et al. (1989), and Fain et al. (1989). Menon et al. (1989) studied further the translocation t(1;17) described by Schmidt et al. (1987). In a somatic cell hybrid line containing only the derivative chromosome 1, they showed the breakpoint occurred between SRC2 (164940) and D1S57, which are separated by 14 cM. The translocation breakpoint was located on chromosome 17 between D17S33 and D17S58, markers which also flank NF1 within a region of 4 cM. The findings were considered consistent with the possibility that the translocation event was the cause of NF1 in this family.

Korenberg et al. (1989) and 156,157:Pulst et al. (1990, 1991) studied markers flanking the NF1 locus in multiplex families with achondroplasia. By linkage analysis, they excluded the achondroplasia locus from the region between the 2 groups of markers flanking NF1. Thus, the concurrence of achondroplasia and NF1 is a single patient was a matter of chance.

Fain et al. (1989) and Upadhyaya et al. (1989) described linkage studies of markers around the neurofibromatosis type 1 locus.

Molecular Genetics

Skuse et al. (1989) observed loss of DNA markers from the NF1 region of 17q in DNA from malignant tumors from patients with NF1, compared with DNA from nontumor tissue from the same patients. Further, in hereditary cases, they found that the NF1 allele remaining in the tumor was derived from the affected parent. The findings suggested that malignant tumors in NF1 arise as a result of homozygous deficiency of a tumor-suppressor gene. These studies, however, did not detect loss of heterozygosity for DNA markers in neurofibromas, the benign tumors of NF1. This finding suggested that neurofibromas are either polyclonal or monoclonal in origin, but arise by a mechanism different from that of NF1 malignancies.

In a search for deletions in the proximal region of 17q in NF1, Menon et al. (1990) found no deletions in NF1-derived tumor specimens. However, both neurofibrosarcomas from patients with 'atypical' NF and 5 of 6 neurofibrosarcomas from NF1 patients displayed loss of alleles for polymorphic DNA markers on 17p outside the area of mapping of NF1. Since the common region of deletion included the site of the p53 gene (191170), they searched for p53 alterations in neurofibrosarcomas by direct sequencing of PCR-amplified DNA. In 2 of 7 neurofibrosarcomas they found point mutations in exon 4 of the p53 gene.

Wallace et al. (1989) described a NotI fragment from human chromosome 17q11.2 which detected breakpoints in 2 patients with NF1. Fountain et al. (1989) mapped a series of chromosome 17 NotI-linking clones to proximal 17q and studied them by pulsed field gel electrophoresis in order to define the region of the breakpoint involved in a 17q11.2 balanced translocation present in 2 NF1 patients. One clone, D17S133, identified the breakpoint in 1 of the 2 patients. A pulsed field map indicated that the breakpoint was within 10 to 240 kb of the cloned segment. Similarly, O'Connell et al. (1989) isolated human cosmids and mapped them to the immediate vicinity of NF1. One cosmid probe demonstrated that the breakpoint in both patients (and presumably the NF1 gene) was contained within a 600-kb NruI fragment.

Yagle et al. (1989) and Xu et al. (1989) described the isolation of cosmid clones flanking the acute promyelocytic leukemia and von Recklinghausen neurofibromatosis loci on chromosome 17.

Yagle et al. (1990) isolated 5 cosmids that map directly proximal to 2 NF1 translocations and 11 cosmids that map directly distal to them. Of these, 2 cosmids in each region were found to be linked to the disease locus and 3 of these 4 cosmids showed no recombination. One distal cosmid detected the 2 NF1 translocations by pulsed field gel analysis and was used by Yagle et al. (1990) to produce a long-range restriction map that covered the translocations. Wallace et al. (1990) identified an Alu insert in the putative NF1 gene. Edwards (1990) commented that such a mechanism of mutation might be a factor in a disorder with a high rate of mutation such as neurofibromatosis. Another disorder in which an Alu-insertion has been identified as causative is ADA deficiency (102700). Most of the ADA mutations have been single nucleotide substitutions, however. Using pulsed field gel electrophoresis, Upadhyaya et al. (1990) identified a 90-kb deletion in the proximal portion of 17q in 1 of 90 unrelated patients with NF1. Wallace et al. (1990) identified a large transcript from 17q11.2 disrupted in 3 patients with NF1. Viskochil et al. (1990) detected deletions of 190, 40, and 11 kb in the gene located at the translocation breakpoint in 3 patients with NF1.

Xu et al. (1990) extended the known open reading frame of the NF1 gene by cDNA walking and sequencing. The new sequence predicted 2,485 amino acids of the NF1 peptide. A 360-residue region showed significant similarity to the catalytic domains of both human and bovine GAP (GTPase activating protein; 139150). Xu et al. (1990) also pointed out the similarity to the yeast IRA1 product. The findings suggest that NF1 encodes a cytoplasmic GAP-like protein that may be involved in the control of cell growth by interacting with proteins such as the RAS gene product. Mapping of the cDNA clones confirmed that NF1 spans the breakpoint in a t(1;17) translocation. Furthermore, 3 active genes, called OMGP, EVI2B and EVI2A, lie within an intron of NF1 but in opposite orientation. Xu et al. (1992) found a pseudogene of the adenylate kinase-3 gene (103030) in an intron of the NF1 gene. It appeared to be a processed pseudogene since it lacked introns and contained a polyadenylate tract; it nevertheless retained coding potential because the open reading frame was not impaired by any observed base substitutions.

Marchuk et al. (1991) reported an extensive cDNA walk resulting in the cloning of the complete coding region of the NF1 transcript. Analysis of the sequences revealed an open reading frame of 2,818 amino acids, although alternatively spliced products may code for different protein isoforms. The gene occupies approximately 300 kb on chromosome 17, with its promoter in a CpG-rich island. Heim et al. (1994) cited evidence that the NF1 gene spans approximately 350 kb of genomic DNA, encodes an mRNA of 11 to 13 kb, and contains at least 56 exons.

Legius et al. (1992) characterized an NF1-related locus on chromosome 15. The nonprocessed NF1 pseudogene (NF1P1) can produce additional fragments in Southern blotting, pulsed field gel, and PCR experiments with some NF1 cDNA probes or oligonucleotides. In addition, certain regions of the NF1 gene cross-hybridize with a locus on chromosome 14. These loci can cause confusion in the mutation analysis of patients with NF1.

Using Southern blotting and PCR amplification of exons in the 3-prime region of the NF1 gene, followed by single-strand conformation polymorphism (SSCP) analysis, Weiming et al. (1992) found only 2 unequivocal mutations: a 571-bp deletion that removed exon 6 and resulted in a frameshift in exon 7, and a 2-bp deletion in exon 1. They detected mutations in at most 3% of subjects (there was a third mutation of questionable pathologic significance) in an analysis that covered 17% of the coding sequence by SSCP and a larger region by Southern blotting. This experience accorded with that of others. The results suggested that most NF1 mutations lie elsewhere in the coding sequence or outside it.

DeClue et al. (1992) presented evidence implicating the NF1 protein as a tumor suppressor gene product that negatively regulates p21(ras) and defined a 'positive' growth role for RAS activity in NF1 malignancies. Li et al. (1992) referred to the NF1 gene product as neurofibromin. Neurofibromin contains a GTPase-activating protein (GAP)-related domain, otherwise known as NF1 GRD, that can downregulate p21(ras) by stimulating its intrinsic GTPase. Li et al. (1992) reasoned that since p21(ras)-GTP is a major regulator of growth and differentiation, mutant neurofibromins resulting from somatic mutations in the NF1 gene might interfere with RAS signaling pathways and contribute to the development of tumors. In support of this hypothesis, they found an amino acid substitution in the NF1 GRD, altering lys1423 (to either glu or gln) in a colon adenocarcinoma, in anaplastic astrocytoma, and in a myelodysplastic syndrome. A lys1423-to-glu mutation was found in a patient with NF1 and affected members of his family.

Collins (1993) developed FISH techniques to detect large deletions in the gene.

Kayes et al. (1994) investigated the contribution to variability in the clinical phenotype of NF1 by genes either contiguous to or contained within the NF1 gene, by screening 6 NF1 patients with mild facial dysmorphism, mental retardation, and/or learning disabilities for DNA rearrangement of the NF1 region. Five of the 6 patients carried a deletion of more than 700 kb on one chromosome 17. Minimally, each of the deletions involved the entire 350-kb NF1 gene, the 3 genes (EVI2A, EVI2B, and OMG) contained within an NF1 intron, and considerable flanking DNA. In 4 of the patients, the deletion mapped to the same interval; the deletion in the fifth patient was larger, extending farther in both directions. The remaining NF1 allele appeared to be producing functional neurofibromin. The data provided compelling evidence that NF1 results from haploid insufficiency of neurofibromin. Of the 3 documented de novo deletion cases, 2 involved the paternal NF1 allele and 1 the maternal allele. All 5 patients were remarkable for the large number of neurofibromas for their age, suggesting that deletion of an unknown gene in the NF1 region may affect tumor initiation or development. All had plexiform neurofibromas. Four had hypertelorism, 4 had ptosis, and all had micrognathia. Cnossen et al. (1997) also examined the question of a particular phenotype with NF1 deletion. They studied DNA from 84 unrelated patients with NF1, unselected for clinical features or severity, screening for deletion with intragenic polymorphic repeat markers and Southern analysis with cDNA probes. Deletion of the entire gene was detected in 5 patients from 4 unrelated families. Their phenotype resembled that of 5 previously reported patients with deletions, including intellectual impairment and dysmorphic features, but they could not confirm the existence of an excessive number of dermal neurofibromas. Postnatal overgrowth suggesting Weaver syndrome (277590) and manifestations somewhat like Noonan syndrome were commented on. Slight micrognathia and extreme overbite of the maxilla were noted in individual cases.

Using FISH with intragenic probes, Wu et al. (1995) looked for deletions in 13 unrelated individuals with NF1. Among 6 with severe manifestations, 4 were found to have deletions of the entire gene. All 4 had severe developmental delay, minor and major anomalies (including 1 with bilateral iris colobomas), and multiple cutaneous neurofibromas or plexiform neurofibromas which were present before age 5 years.

By denaturing gradient gel electrophoresis (DGGE), Valero et al. (1994) screened 70 unrelated NF1 patients for mutations in exons 29 and 31. Of the 4 mutations that were identified, 3 consisted of C-to-T transitions resulting in nonsense mutations, 2 in exon 29 (C5242T and C5260T) and 1 in exon 31 (C5839T). The fourth mutation consisted of a 2-bp deletion in exon 31, 5843delAA, resulting in a premature stop codon. The C5839T mutation had previously been reported in 3 independent studies, suggesting that this position is a mutational hotspot within the NF1 gene. It occurs in a CpG residue.

Heim et al. (1994) stated that although mutations had been sought in several hundred patients, by August 1994, only 70 mutations had been reported in a total of 78 individuals; only the R1947X mutation had been seen in as many as 6 unrelated patients. The NF1 mutations that had been identified included 14 large (more than 25 bp) deletions, 3 large insertions, 18 small (less than 25 bp) deletions, 8 small insertions, 6 nonsense mutations, 14 missense mutations, and 7 intronic mutations. At least 56 of the 70 (80%) potentially encode a truncated protein because of premature translation termination. Therefore, Heim et al. (1994) analyzed polypeptides synthesized in vitro from RT-PCR products, using an approach that had been applied successfully to mutation detection in dystrophin and APC genes.

Li et al. (1995) showed that the 5-prime end of the NF1 gene is embedded in a CpG island containing a NotI restriction site and that the remainder of the gene lies in the adjacent 35-kb NotI fragment. In their efforts to develop a comprehensive screen for NF1 mutations, they isolated genomic DNA clones that together contain the entire NF1 cDNA sequence. They identified all intron-exon boundaries of the coding region and established that it is composed of 59 exons. The 3-prime untranslated region of the NF1 gene was found to span approximately 3.5 kb and to be continuous with the stop codon.

Heim et al. (1995) used a protein truncation assay to identify abnormal polypeptides synthesized in vitro from 5 RT-PCR products that represented the entire NF1 coding region. Truncated polypeptides were observed in 14 of 20 patients with familial or sporadic NF1 diagnosed clinically and in 1 patient with only cafe-au-lait spots and no other diagnostic criterion. Mutations responsible for the generation of abnormal polypeptides were characterized by DNA sequencing; 13 previously unpublished mutations were characterized in the 14 individuals. The mutation 2027insC was observed in 2 unrelated individuals; the other 12 mutations were unique. Thus, mutations were identified in two-thirds of the patients studied using the protein truncation assay. Because the entire NF1 coding region was spanned in each individual, the distribution of NF1 truncating mutations was discerned for the first time. The mutations were relatively evenly distributed throughout the coding region.

Upadhyaya et al. (1995) stated that fewer than 90 mutations had been reported to the NF1 mutation analysis consortium and details of only 76 of these mutations had been published. They had previously reported the identification of 14 mutations and described 5 new mutations identified by SSCP analysis and heteroduplex analysis. Three intragenic deletions were also identified by analyzing their families with intron-specific microsatellite markers.

Vogel et al. (1995) used a targeted disruption of the NF1 gene in mice to examine the role of neurofibromin in the acquisition of neurotrophin dependence in embryonic neurons. They showed that both neural crest- and placode-derived sensory neurons isolated from NF1 -/- embryos develop, extend neurites, and survive in the absence of neurotrophins, whereas their wildtype counterparts die rapidly unless nerve growth factor (162030) or brain-derived neurotrophic factor (113505) is added to the culture median. Moreover, NF1 -/- sympathetic neurons survive for extended periods and acquire mature morphology in the presence of NGF-blocking antibodies. These results were considered by Vogel et al. (1995) as consistent with a model wherein neurofibromin acts as a negative regulator of neurotrophin-mediated signaling for survival of embryonic peripheral neurons.

Shen et al. (1996) summarized the molecular genetics of NF1 as follows: the gene spans over 350 kb of genomic DNA and encodes an mRNA of 11 to 13 kb containing at least 59 exons. It is widely expressed in a variety of human tissues. Four alternatively spliced NF1 transcripts have been identified. Three of these transcript isoforms (each with an extra exon: 9br, 23a, and 48a, respectively) show differential to some extent in various tissues, while the fourth isoform (2.9 kb in length) remains to be examined. The protein encoded by NF1, neurofibromin, has a domain homologous to the GTPase activating protein (GAP) family and downregulates RAS activity. Over 80% of germline mutations appear to predict severe truncation of neurofibromin. The actual mechanisms of tumorigenesis, in which NF1 is involved, remained unknown. No evidence of LOH had been observed in neurofibromas. Shen et al. (1996) speculated that it may be that a second mutation in another gene is required for genesis of neurofibromas, or that they may arise because of the loss of 1 allele. Another possibility is that a second mutation in the NF1 gene is required. In NF2 (101000), loss of the wildtype allele is demonstrable in tumors, consistent with the classic Knudson theory of tumor suppressor genes. On the other hand, the effects of the APC gene (175100) in the formation of colonic polyps is apparently not dependent on loss of the wildtype allele.

Although several observations support the contention that the NF1 gene product is a tumor suppressor involved in the RAS signal transduction pathway, mutations had not been identified in both NF1 alleles in dermal neurofibromas until the report by Sawada et al. (1996). Their patient was previously shown to have large submicroscopic deletions of the NF1 locus by both somatic cell hybrid analysis (Kayes et al., 1994) and FISH of lymphoblastoid cells (Leppig et al., 1996). The deletion extended at least 125 kb centromeric and 135 kb telomeric to NF1. As part of her medical care, the patient electively had a scalp neurofibroma removed surgically. Sawada et al. (1996) showed that the tumor DNA harbored a 4-bp deletion in NF1 exon 4b in the other allele. The authors stated that this was the first reported definitive identification of a somatic mutation limited to the NF1 locus in a benign neurofibroma from an NF1 individual in whom the constitutional NF1 mutation was known.

The NF1 gene product, neurofibromin, displays partial homology to GAP (139150), or GTPase-activating protein. The GAP-related domain (GRD), encoded by exons 20-27a, is the only region of neurofibromin to which a biologic function has been ascribed. Upadhyaya et al. (1997) screened 320 unrelated NF1 patients for mutations in the GRD-encoding region of the NF1 gene. Sixteen different lesions in the NF1 GRD region were identified in a total of 20 patients. Of these lesions, 14 were novel and together comprised 3 missense, 2 nonsense, and 3 splice site mutations plus 6 deletions of between 1 and 4 bp.

For the most part the NF1 tumor suppressor acts through the interaction of its GRD with the product of the RAS protooncogene. Skuse et al. (1996) discovered an mRNA editing site within the NF1 mRNA. Editing at this site changes a cytidine at nucleotide 2914 to a uridine, creating an in-frame translation stop codon. The edited transcript, if translated, would produce a protein truncated in the N-terminal region of the GRD, thereby inactivating the NF1 tumor-suppressor function. Analysis of RNA from a variety of cell lines, tumors, and peripheral blood cells revealed that the NF1 mRNA undergoes editing, to different extents, in every cell type studied. Three tumors analyzed as part of their study, an astrocytoma, a neurofibroma, and a neurofibrosarcoma, each had levels of NF1 mRNA editing substantially higher than did peripheral blood leukocytes. To investigate the role played by editing in NF1 tumorigenesis, Cappione et al. (1997) analyzed RNA from 19 NF1 and 4 non-NF1 tumors. (The authors referred to the editing site as nucleotide 3916.) They observed varying levels in NF1 mRNA editing in different tumors, with a higher range of editing in more malignant tumors (e.g., neurofibrosarcomas) compared to benign tumors (cutaneous neurofibromas). Plexiform neurofibromas had an intermediate range of levels of NF1 mRNA editing. The constitutional levels of NF1 mRNA editing varied slightly in NF1 individuals but were consistent with the levels observed in non-NF1 individuals. In every case, there was a greater level of NF1 mRNA editing in the tumor than in the nontumor tissue from the same patient. These results suggested to Cappione et al. (1997) that inappropriately high levels of NF1 mRNA editing indeed plays a role in NF1 tumorigenesis and that editing may result in the functional equivalent of biallelic inactivation of the NF1 tumor suppressor.

The NF1 tumor suppressor protein is thought to restrict cell proliferation by functioning as a Ras-specific guanosine triphosphatase-activating protein. However, The et al. (1997) found that Drosophila homozygous for null mutations of an NF1 homolog show no obvious signs of perturbed RAS1-mediated signaling. Loss of NF1 resulted in a reduction in size of larvae, pupae, and adults. This size defect was not modified by manipulating RAS1 signaling but was restored by expression of activated adenosine 3-prime, 5-prime-monophosphate-dependent protein kinase (176892; 188830). Thus, NF1 and PKA appear to interact in a pathway that controls the overall growth of Drosophila. Guo et al. (1997) showed, from a study of Drosophila NF1 mutants, that NF1 is essential for the cellular response to the neuropeptide PACAP38 (pituitary adenylyl cyclase-adenosine activating polypeptide) at the neuromuscular junction. The peptide induced a 3-prime, 5-prime-monophosphate (cAMP) pathway. This response was eliminated in NF1 mutants. NF1 appeared to regulate the rutabaga-encoded adenylyl cyclase rather than the RAS-RAF pathway. Moreover, the NF1 defect was rescued by the exposure of cells to pharmacologic treatment that increased concentrations of cAMP.

The risk of malignant myeloid disorders in young children with NF1 is 200 to 500 times the normal risk. Neurofibromin, the protein encoded by the NF1 gene, negatively regulates signals transduced by Ras proteins. Genetic and biochemical data support the hypothesis that NF1 functions as a tumor-suppressor gene in immature myeloid cells. This hypothesis was further supported by the demonstration by Side et al. (1997) that both NF1 alleles were inactivated in bone marrow cells from children with NF1 complicated by malignant myeloid disorders. Using an in vitro transcription and translation system, they screened bone marrow samples from 18 such children for NF1 mutations that cause a truncated protein. Mutations were confirmed by direct sequencing of genomic DNA from the patients, and from the affected parents in cases of familial NF1. Side et al. (1997) found that the normal NF1 allele was absent in bone marrow samples from 5 of 8 children who had truncating mutations of the NF1 gene.

Abernathy et al. (1997) stated that about half of NF1 cases represent new mutations and fewer than 100 constitutional mutations have been reported up to that time, and analyzed the NF1 gene for mutation using a combined heteroduplex/SSCP approach. In a set of 67 unrelated NF1 patients, they identified 26 mutations and/or variants in 45 of the 59 exons tested. Disease-causing mutations were found in 19% (13 of 67) of cases studied. The mutations included splice mutations, insertions, deletions, and point changes.

Maynard et al. (1997) screened exon 16 of the NF1 gene in 465 unrelated NF1 patients. Nine novel mutations were identified: 3 nonsense, 2 single-base deletions, 1 7-bp duplication, 2 missense, and 1 recurrent splice site mutation. No mutations had been reported previously in exon 16, which is the largest exon (441 bp) of NF1. The previous absence of mutation identification in exon 16 suggested to the authors that codons in this region may have a lower propensity to mutate.

As reviewed earlier, submalignant tumors occurring in NF1 patients have been found to show loss of heterozygosity (LOH) consistent with the 2-hit hypothesis of Knudson, with 1 allele constitutionally inactivated and the other somatically mutated. Somatic NF1 deletions in benign neurofibromas were described by Colman et al. (1995) and mutations in both copies of the NF1 gene in a dermal neurofibroma were reported by Sawada et al. (1996). Serra et al. (1997) performed LOH analysis on 60 neurofibromas derived from 17 patients, of whom 9 had a family history of the disease and 8 represented sporadic cases. LOH was found in 25% of the neurofibromas (corresponding to 53% of the patients). In addition, they found that in the neurofibromas of patients from familial cases, the deletions occurred in the allele that was not transmitted with the disease, indicating that both copies of the NF1 gene were inactivated in these tumors. Thus, the authors concluded that there appears to be double inactivation of the NF1 gene in benign neurofibromas.

Riva et al. (1996) characterized a 12-year-old male patient with sporadic NF1, dysmorphism, mental retardation, and skeletal anomalies (162200.0017). Karyotyping of the patient revealed a cytogenetically visible deletion at 17q11.2. Analysis of microsatellite markers demonstrated that the patient was hemizygous, due to loss of the paternal allele, at several sites within the NF1 gene and at an extragenic marker distal to the 3-prime end of NF1. The 9-cM deletion in the interval between D17S841 and D17S250 was in agreement with that originally detected cytogenetically. The karyotypes of the parents were normal. The patient had no neurofibromas; the authors attributed this fact to his genetic background, i. e., to the influence of modifying genes.

Skuse and Cappione (1997) reviewed the possible molecular basis of the wide clinical variability in NF1 observed even among affected members of the same family (Huson et al., 1989). The complexities of alternative splicing and RNA editing may be involved. Skuse and Cappione (1997) suggested that the classical 2-hit model for tumor suppressor inactivation used to explain NF1 tumorigenesis can be expanded to include post-transcriptional mechanisms that regulate NF1 gene expression. Aberrations in these mechanisms may play a role in the observed clinical variability.

Stop, or nonsense, mutations can have a number of effects. In the case of several genes, they affect mRNA metabolism and reduce the amount of detectable mRNA. Also, in the NF1 gene, a correlation between a high proportion of stop mutations and unequal expression of the 2 alleles is demonstrable. A second, less common outcome is that mRNA containing a nonsense mutation is translated and results in a truncated protein. A third possible outcome is an abnormally spliced mRNA induced by a premature-termination codon (PTC) in the skipped exon. This was demonstrated in several disease genes, including the CFTR gene (Hull et al., 1994) and the fibrillin gene (Dietz et al., 1993). Hoffmeyer et al. (1998) characterized several stop mutations localized within a few basepairs in exons 7 and 37 of the NF1 gene and noticed complete skipping of either exon in some cases. Because skipping of exons 7 and 37 does not lead to a frameshift, premature termination codons (PTCs) are avoided. Hoffmeyer et al. (1998) found that some other stop mutations in the same general region did not lead to a skip. Calculations of minimum-free-energy structures of the respective regions suggested that both changes in the secondary structure of mRNA and creation or disruption of exonic sequences relevant for the splicing process may in fact cause these different splice phenomena observed in the NF1 gene.

Using a novel multitrack screening strategy, Upadhyaya et al. (1998) studied 67 NF1 families (54 2-generation, 13 3-generation) with a de novo mutation in the germline of the first generation; 2 extragenic and

11 intragenic markers were employed. The pathologic lesion was identified in 31 cases. Loss of heterozygosity (LOH) in the affected individual revealed a gross gene deletion in 15 of the 2-generation families; in 12 (80%) of them, the deletion was maternally derived. Eleven patients with a gross deletion exhibited developmental delay, 10 had dysmorphic features, and 6 manifested a learning disability. No gross deletion was apparent in any of the 13 3-generation families, suggesting that such lesions are subject to more intense selection. In these 13 families, the new mutation was of paternal origin in 11 and the underlying mutational event could be characterized in 3 of them.

Rasmussen et al. (1998) studied 67 patients with NF1 and their parents. Five patients showed loss of heterozygosity, suggesting NF1 gene deletion. These patients did not have severe NF1 manifestations, mental retardation, or dysmorphic features. All 5 deletions were de novo and occurred on the maternal chromosome. Two patients showed partial loss of heterozygosity, consistent with somatic mosaicism for NF1 deletion.

Juvenile myelomonocytic leukemia (JMML) is a pediatric myelodysplastic syndrome that is associated with neurofibromatosis type I. The NF1 gene regulates the growth of immature myeloid cells by accelerating guanosine triphosphate hydrolysis on RAS proteins. Side et al. (1998) undertook a study to determine if the NF1 gene is involved in the pathogenesis of JMML in children without a clinical diagnosis of NF1. An in vitro transcription and translation system was used to screen JMML marrows from 20 children for NF1 mutations that resulted in a truncated protein. SSCP analysis was used to detect RAS point mutations in these samples. Side et al. (1998) confirmed mutations of NF1 in 3 cases of JMML, 1 of which also showed loss of the normal NF1 allele. An NF1 mutation was detected in normal tissue from the only patient tested, suggesting that JMML may be the presenting feature of NF1 in some children. Activating RAS mutations were found in 4 patients; as expected, none of these samples harbored NF1 mutations. Because 10 to 14% of children with JMML had a clinical diagnosis of NF1, these data were consistent with the existence of NF1 mutations in approximately 30% of JMML cases.

Most mutations underlying NF1 result in the loss of 1 allele at the DNA, mRNA, or protein level and thus in loss of any function of the gene product neurofibromin. Simultaneous loss of several different neurofibromin functions was postulated to explain the pleiotropic effects of its loss. However, Klose et al. (1998) identified a novel missense mutation in a family with a classic multisymptomatic NF1 phenotype, including a malignant schwannoma. The mutation specifically abolished the Ras-GTPase-activating function of neurofibromin. The arg1276-to-pro mutation (R127P; 162200.0022) in this family was shown to involve the arginine finger of the neurofibromin GAP-related domain (GRD), which is the most essential catalytic element for Ras-GAP activity. Klose et al. (1998) presented data demonstrating that the R1276P mutation, unlike previously reported missense mutations of the GRD region, did not impair the secondary and tertiary protein structure. It neither reduced the level of cellular neurofibromin nor influenced its binding to Ras substantially, but it did completely disable GAP activity. The findings provided direct evidence that failure of neurofibromin GAP activity is a critical element in NF1 pathogenesis. Thus, therapeutic approaches aimed at the reduction of the Ras-GTP levels in neural crest-derived cells can be expected to relieve most of the NF1 symptoms.

Diagnosis

Johnson and Charneco (1970) suggested that the cafe-au-lait spot of neurofibromatosis can be distinguished from the innocent spot that occurs in normal persons and from the pigmented areas of Albright disease by the presence of a large number of DOPA-positive melanocytes that have giant pigment granules in the cytoplasm. The plexiform neuroma is specific to von Recklinghausen disease. Only on this can the histopathologist make a definitive diagnosis.

Ward et al. (1990) estimated that tightly linked, flanking DNA markers available to them permitted prediction of NF1 in a child with greater than 98% accuracy. They predicted that even after the NF1 gene is cloned, linkage testing will probably remain important. This has been the experience with factor VIII deficiency and 21-hydroxylase deficiency, in which linked markers complement the use of direct genetic probes. Linked markers may remain more cost-effective than screening for 1 genetic event among a large number of possible mutations that could be responsible for NF1 in a particular family.

Gutmann et al. (1997) provided guidelines for the diagnostic evaluation and multidisciplinary management of both NF1 and NF2.

Population Genetics

Littler and Morton (1990) reviewed data from 4 studies, with the following results: the carrier incidence at birth is 0.0004; the gene frequency is 0.0002; and the proportion of cases due to fresh mutation is 0.56. Lazaro et al. (1994) gave the incidence of NF1 as approximately 1 in 3,500 and stated that about half of cases are the result of new mutations.

Garty et al. (1994) found an unusually high frequency of NF1 in young Israeli adults. They surveyed 374,440 17-year-old Jewish recruits for military service and concluded that 390 of them had NF1. The prevalence was 1.04/1,000 (0.94/1,000 for males and 1.19/1,000 for females). This value is 2 to 5 times greater than the previously reported prevalence. NF1 was more common in young adults whose parents were of North African and Asian origin (1.81/1,000 and 0.95/1,000, respectively) and less common in those of European and North American origin (0.64/1,000). All these differences were statistically significant; Garty et al. (1994) suggested that they may be partially explained by the more advanced parental age of the NF group (as suggested by the larger number of children in the North African and Asian families) or by founder effect or both.

In a survey of 56,183 young men, aged 17 and 18 years, Ingordo et al. (1995) found 11 cases of NF1 and 1 case of segmental NF. In this group, the relative frequency was 0.02% for NF and 0.0018% for segmental NF. Wolkenstein et al. (1995), in a period from November, 1988 through August, 1995, saw 308 patients with NF type I according to the criteria of the National Institutes of Health Consensus Development Conference (1988) and 9 patients with segmental NF (so-called NF type V), according to the classification of Riccardi (1982). These findings and those of Ingordo et al. (1995) suggest that segmental NF is about 30 times more infrequent than in NF type I.

Animal Model

Hinrichs et al. (1987) showed that the TAT gene of human T-lymphotropic virus type 1 (HTLV-1) under control of its own long terminal repeat is capable of inducing tumors in transgenic mice. The morphologic and biologic properties of these tumors indicated a close resemblance to NF1. Multiple tumors developed simultaneously in the transgenic tat mice at approximately 3 months of age, and the phenotype was successfully passed through 3 generations. The tumors arose from the nerve sheaths of peripheral nerves and were composed of perineural cells and fibroblasts. Evidence of HTLV-1 infection in patients with neural and other soft tissue tumors is needed in order to establish a link between infection by this human retrovirus and von Recklinghausen disease.

Buchberg et al. (1990) sequenced a portion of the murine NF-1 gene and showed that the predicted amino acid sequence is nearly the same as the corresponding region of the human NF1 gene product. Computer searches identified homology between the mouse NF-1 gene and the IRA-1 and IRA-2 genes identified in Saccharomyces cerevisiae, which negatively regulate the RAS-cyclic AMP pathway. RAS proteins are involved in the control of proliferation and differentiation in mammalian cells. Their activity is modulated by their ability to bind and hydrolyze guanine nucleotides. GTP-binding activates RAS, whereas GTP hydrolysis inactivates RAS. Mutant forms of RAS found in human tumors have greatly decreased GTPase activity, resulting in accumulation of RAS in the GTP-bound active form.

Silva et al. (1997) stated that learning disabilities are said to occur in 30 to 45% of patients with NF1, even in the absence of any apparent neuropathology. The learning disabilities may include a depression in mean IQ scores, visuoperceptual problems, and impairment in spatial cognitive abilities. They cited several studies that suggested a role of neurofibromin in brain function. The expression of the NF1 gene is largely restricted to neuronal tissues in the adult. This GTPase-activating protein may act as a negative regulator of neurotrophin-mediated signaling. They also noted immunohistochemical studies that suggested that activation of astrocytes may be common in the brain of NF1 patients. Silva et al. (1997) described a mouse model (Nf1+/-) for the learning and memory deficits associated with NF1. Importantly related to these observations is the demonstration that mice heterozygous for an Nf1 knockout show a deficit of learning and memory. As in humans, the learning and memory deficits of Nf1+/- mice are restricted to specific types of learning, they are fully penetrant, they can be compensated for with extended training, and they do not involve deficits in simple associative learning.

History

Although the Elephant Man (Howell and Ford, 1980) has often been thought to have had von Recklinghausen disease, it has been suggested (Pyeritz, 1987) that Proteus syndrome (176920) is a more likely diagnosis. After considering several diagnostic possibilities, Cohen (1988) also concluded that the skeletal findings in Joseph Merrick are most consistent with Proteus syndrome. He pointed out that the 'moccasin' lesions of the feet are particularly characteristic of that disorder. See the study of the case of Joseph Merrick by Graham and Oehlschlaeger (1992).

See also Abeliovich et al. (1995); Allen and Davies (1970); Barker et al. (1987); Bidot-Lopez and Frankel (1983); Boudin et al. (1970); Buntin and Fitzgerald (1970); Cawthon et al. (1990); Charron and Gariepy (1970); Clark et al. (1977); Cotlier (1977); Dunn et al. (1976); Fabricant and Todaro (1981); Fain et al. (1989); Fienman and Yakovac (1970); Fountain et al. (1989); Hochberg et al. (1974); Holt (1978); Izumi et al. (1971); Kaplan et al. (1982); Kohn (1979); Lund and Skovby (1991); Miles et al. (1969); Miller and Hall (1978); Muller-Wiefel (1978); Nager (1964); Newman and So (1971); O'Connell et al. (1989); Obringer et al. (1989); Pellock et al. (1980); Perry and Font (1982); Philippart (1961); Riccardi (1981); Riccardi and Mulvihill (1981); Rockower et al. (1982); Sands et al. (1975); Satran et al. (1980); Seizinger et al. (1987); Siggers et al. (1975); Skuse et al. (1991); Smith et al. (1970); Taylor (1962); Upadhyaya et al. (1989); von Recklinghausen (1882); Wallace et al. (1990); Wallis et al. (1970)

Abeliovich, D.; Gelman-Kohan, Z.; Silverstein, S.; Lerer, I.; Chemke, J.; Merlin, S.; Zlotogora, J.: Familial cafe au lait spots: a variant of neurofibromatosis type 1. J. Med. Genet. 32: 985-986, 1995.

Abernathy, C. R.; Rasmussen, S. A.; Stalker, H. J.; Zori, R.; Driscoll, D. J.; Williams, C. A.; Kousseff, B. G.; Wallace, M. R.: NF1 mutation analysis using a combined heteroduplex/SSCP approach. Hum. Mutat. 9: 548-554, 1997.

Adornato, B. T.; Berg, B.: Diencephalic syndrome and von Recklinghausen's disease. Ann. Neurol. 2: 159-160, 1977.

Ainsworth, P. J.; Rodenhiser, D. I.; Costa, M. T.: Identification and characterization of sporadic and inherited mutations in exon 31 of the neurofibromatosis (NF1) gene. Hum. Genet. 91: 151-156, 1993.

Allen, T. N. K.; Davies, E. R.: Neurofibromatosis of the renal artery. Brit. J. Radiol. 43: 906-908, 1970.

Ars, E.; Kruyer, H.; Gaona, A.; Casquero, P.; Rosell, J.; Volpini, V.; Serra, E.; Lazaro, C.; Estivill, X.: A clinical variant of neurofibromatosis type 1: familial spinal neurofibromatosis with a frameshift mutation in the NF1 gene. Am. J. Hum. Genet. 62: 834-841, 1998.

Barker, D.; Wright, E.; Nguyen, K.; Cannon, L.; Fain, P.; Goldgar, D.; Bishop, D. T.; Carey, J.; Baty, B.; Kivlin, J.; Willard, H.; Waye, J. S.; Greig, G.; Leinwand, L.; Nakamura, Y.; O'Connell, P.; Leppert, M.; Lalouel, J.-M.; White, R.; Skolnick, M.: Gene for von Recklinghausen neurofibromatosis is in the pericentromeric region of chromosome 17. Science 236: 1100-1102, 1987.

Barker, D. F.; Wright, E.; Nguyen, K.; Cannon, L.; Fain, P.; Goldgar, D.; Bishop, D. T.; Carey, J.; Baty, B.; Kivlin, J.; Willard, H.; Waye, J. S.; Greig, G.; Leinwand, L.; Nakamura, Y.; O'Connell, P.; Leppert, M.; Lalouel, J.-M.; White, R.; Skolnick, M.: The gene for NF1 (von Recklinghausen neurofibromatosis) is on chromosome 17 near the centromere. (Abstract) Cytogenet. Cell Genet. 46: 576, 1987.

Basu, T. N.; Gutmann, D. H.; Fletcher, J. A.; Glover, T. W.; Collins, F. S.; Downward, J.: Aberrant regulation of ras proteins in malignant tumour cells from type 1 neurofibromatosis patients. Nature 356: 713-715, 1992.

Benatar, M. G.: Intracranial fusiform aneurysms in von Recklinghausen's disease: case report and literature review.(Letter) J. Neurol. Neurosurg. Psychiat. 63: 1279-1280, 1994.

Benedict, P. H.; Szabo, G.; Fitzpatrick, T. B.; Sinesi, S. J.: Melanotic macules in Albright's syndrome and in neurofibromatosis. J.A.M.A. 205: 618-626, 1968.

Bidot-Lopez, P.; Frankel, J. W.: Enhanced viral transformation of skin fibroblasts from neurofibromatosis patients. Ann. Clin. Lab. Sci. 13: 27-32, 1983.

Borberg, A.: Clinical and genetic investigations into tuberous sclerosis and Recklinghausen's neurofibromatosis: contribution to elucidation of interrelationship and eugenics of the syndromes. Acta Psychiat. Neurol. 71 (suppl.): 1-239, 1951.

Boudin, G.; Pepin, B.; Vernant, C.: Les tumeurs multiples du systeme nerveux au cours de la maladie de Recklinghausen: a propos d'une observation anatomo-clinique avec adenome chromophobe de l'hypophyse. Presse Med. 78: 1427-1430, 1970.

Buchberg, A. M.; Cleveland, L. S.; Jenkins, N. A.; Copeland, N. G.: Sequence homology shared by neurofibromatosis type-1 gene and IRA-1 and IRA-2 negative regulators of the RAS cyclic AMP pathway. Nature 347: 291-294, 1990.

Buntin, P. T.; Fitzgerald, J. F.: Gastrointestinal neurofibromatosis: a rare cause of chronic anemia. Am. J. Dis. Child. 119: 521-523, 1970.

Cappione, A. J.; French, B. L.; Skuse, G. R.: A potential role for NF1 mRNA editing in the pathogenesis of NF1 tumors. Am. J. Hum. Genet. 60: 305-312, 1997.

Cawthon, R. M.; Viskochil, D.; O'Connell, P.; Buchberg, A.; Andersen, L.; Wallace, M.; Marchuk, D.; Stevens, J.; Culver, M.; Xu, G.; Collins, F. S.; Jenkins, N.; Copeland, N.; White, R.: Identification and characterization of several genes between or nearby neurofibromatosis type I translocation breakpoint.(Abstract) Am. J. Hum. Genet. 47 (suppl.): A109, 1990.

Cawthon, R. M.; Weiss, R.; Xu, G.; Viskochil, D.; Culver, M.; Stevens, J.; Robertson, M.; Dunn, D.; Gesteland, R.; O'Connell, P.; White, R.: A major segment of the neurofibromatosis type 1 gene: cDNA sequence, genomic structure, and point mutations. Cell 62: 193-201, 1990.

Charron, J. W.; Gariepy, G.: Neurofibromatosis of bladder: case report and review of the literature. Canad. J. Surg. 13: 303-306, 1970.

Clark, R. D.; Hutter, J. J., Jr.: Familial neurofibromatosis and juvenile chronic myelogenous leukemia. Hum. Genet. 60: 230-232, 1982.

Clark, S. S.; Marlett, M.; Prudencio, R.; Dasgupta, T. K.: Neurofibromatosis of the bladder in children: case report and literature review. J. Urol. 118: 654-656, 1977.

Clementi, M.; Barbujani, G.; Turolla, L.; Tenconi, R.: Neurofibromatosis-1: a maximum likelihood estimation of mutation rate. Hum. Genet. 84: 116-118, 1990.

Cnossen, M. H.; van der Est, M. N.; Breuning, M. H.; van Asperen, C. J.; Breslau-Siderius, E. J.; van der Ploeg, A. T.; de Goede-Bolder, A.; van den Ouweland, A. M. W.; Halley, D. J. J.; Niermeijer, M. F.: Deletions spanning the neurofibromatosis type 1 gene: implications for genotype-phenotype correlations in neurofibromatosis type 1? Hum. Mutat. 9: 458-464, 1997.

Cohen, M. M., Jr.: Further diagnostic thoughts about the Elephant Man. Am. J. Med. Genet. 29: 777-782, 1988.

Collins, F. S.: Personal Communication. Bethesda, Md. 11/17/1993.

Colman, S. D.; Rasmussen, S. A.; Ho, V. T.; Abernathy, C. R.; Wallace, M. R.: Somatic mosaicism in a patient with neurofibromatosis type 1. Am. J. Hum. Genet. 58: 484-490, 1996.

Colman, S. D.; Williams, C. A.; Wallace, M. R.: Benign neurofibromas in type 1 neurofibromatosis (NF1) show somatic deletions of the NF1 gene. Nature Genet. 11: 90-92, 1995.

Combemale, P.; Abitan, R.; Kanitakis, J.: Segmental neurofibromatosis: report of two cases and critical review of the literature. Europ. J. Derm. 4: 194-201, 1994.

Cotlier, E.: Cafe-au-lait spots of the fundus in neurofibromatosis. Arch. Ophthal. 95: 1990-1992, 1977.

Crawford, A. H.: Neurofibromatosis in children. Acta Orthop. Scand. 57 (suppl. 218): 9-60, 1986.

Crowe, F. W.: Axillary freckling as a diagnostic aid in neurofibromatosis. Ann. Intern. Med. 61: 1142-1143, 1964.

Crowe, F. W.; Schull, W. J.; Neel, J. V.: A Clinical, Pathological and Genetic Study of Multiple Neurofibromatosis. Springfield, Ill.: Charles C Thomas (pub.) 1956.

D'Agostino, A. N.; Soule, E. H.; Miller, R. H.: Sarcomas of the peripheral nerves and somatic soft tissue associated with multiple neurofibromatosis. Cancer 16: 1015-1027, 1963.

Darby, J. K.; Feder, J.; Selby, M.; Riccardi, V.; Ferrell, R.; Siao, D.; Goslin, K.; Rutter, W.; Shooter, E. M.; Cavalli-Sforza, L. L.: A discordant sibship analysis between beta-NGF and neurofibromatosis. Am. J. Hum. Genet. 37: 52-59, 1985.

Daston, M. M.; Scrable, H.; Nordlund, M.; Sturbaum, A. K.; Nissen, L. M.; Ratner, N.: The protein product of the neurofibromatosis type 1 gene is expressed at highest abundance in neurons, Schwann cells, and oligodendrocytes. Neuron 8: 415-428, 1992.

DeClue, J. E.; Cohen, B. D.; Lowy, D. R.: Identification and characterization of the neurofibromatosis type 1 protein product. Proc. Nat. Acad. Sci. 88: 9914-9918, 1991.

DeClue, J. E.; Papageorge, A. G.; Fletcher, J. A.; Diehl, S. R.; Ratner, N.; Vass, W. C.; Lowy, D. R.: Abnormal regulation of mammalian p21(ras) contributes to malignant tumor growth in von Recklinghausen (type 1) neurofibromatosis. Cell 69: 265-273, 1992.

Dent, C. E.: Rickets and osteomalacia from renal tubular defects. J. Bone Joint Surg. 34B: 266-274, 1952.

Diehl, S. R.; Boehnke, M.; Erickson, R. P.; Ploughman, L. M.; Seiler, K. A.; Lieberman, J. L.; Clarke, H. B.; Bruce, M. A.; Schorry, E. K.; Pericak-Vance, M.; O'Connell, P.; Collins, F. S.: A refined genetic map of the region of chromosome 17 surrounding the von Recklinghausen neurofibromatosis (NF1) gene. Am. J. Hum. Genet. 44: 33-37, 1989.

Diekmann, L.; Huther, W.; Pfeiffer, R. A.: Ungewoehnliche Erscheinungsformen der Neurofibromatose (von Recklinghausensche Krankheit) im Kindesalter. Z. Kinderheilk. 101: 191-222, 1967.

Dietz, H. C.; Valle, D.; Francomano, C. A.; Kendzior, R. J., Jr.; Pyeritz, R. E.; Cutting, G. R.: The skipping of constitutive exons in vivo induced by nonsense mutations. Science 259: 680-683, 1993.

Dietz, J. N.; Robbins, T.; Schwartz, C.; Cannon, L.; McLellan, T.; Williamson, R.; Carey, J.; Johnson, J.; Kivlin, J.; Skolnick, M. H.: Linkage analysis of neurofibromatosis: chromosomes 4 and 19.(Abstract) Am. J. Hum. Genet. 37: A193, 1985.

DiLiberti, J. H.; Buist, N. R. M.; Rosenberg, N.; Andrews, T.: Myotonic dystrophy, hyperparathyroidism and multiple endocrine adenomatosis type 2A.(Abstract) Am. J. Hum. Genet. 34: 86A, 1982.

Dugoff, L.; Sujansky, E.: Neurofibromatosis type 1 and pregnancy. Am. J. Med. Genet. 66: 7-10, 1996.

Duncan, A. M. V.; Partington, M. W.; Soudek, D.: Neurofibromatosis in a man with a ring 22: in situ hybridization studies. Cancer Genet. Cytogenet. 25: 169-174, 1987.

Dunn, B. G.; Ferrell, R. E.; Riccardi, V. M.: A genetic linkage study in 15 families of individuals with von Recklinghausen neurofibromatosis. Am. J. Med. Genet. 22: 403-407, 1985.

Dunn, F. G.; DeCarvalho, J. G. R.; Kem, D. C.; Higgins, J. R.; Frohlich, E. D.: Pheochromocytoma crisis induced by saralasin: relation of angiotensin analogue to catecholamine release. New Eng. J. Med. 295: 605-607, 1976.

Easton, D. F.; Ponder, M. A.; Huson, S. M.; Ponder, B. A. J.: An analysis of variation in expression of neurofibromatosis (NF) type I (NF1): evidence for modifying genes. Am. J. Hum. Genet. 53: 305-313, 1993.

Edwards, J. H.: Personal Communication. Oxford, England 10/1990.

Eichhorn, C.; Wendt, G.; Staudte, H.-W.; Gilsbach, J.-M.: Dural ectasia in Von Recklinghausen's disease of the lumbar spine: a case report. J. Bone Joint Surg. 77: B834-B835, 1995.

Erickson, R. P.; Woolliscroft, J.; Allen, R. J.: Familial occurrence of intracranial arterial occlusive disease (moyamoya) in neurofibromatosis. Clin. Genet. 18: 191-196, 1980.

Estivill, X.; Lazaro, C.; Casals, T.; Ravella, A.: Recurrence of a nonsense mutation in the NF1 gene causing classical neurofibromatosis type 1. Hum. Genet. 88: 185-188, 1991.

Fabricant, R. N.; Todaro, G. J.: Increased serum levels of nerve growth factor in von Recklinghausen's disease. Arch. Neurol. 38: 401-405, 1981.

Fain, P. R.; Goldgar, D. E.; Wright, E. C.; Andersen, W. L.; Willard, H. F.; van Tuinen, P.; Ledbetter, D. H.; Barker, D. F.: Markers tightly linked to the NF1 locus on chromosome 17.(Abstract) Cytogenet. Cell Genet. 51: 996, 1989.

Fain, P. R.; Wright, E.; Willard, H. F.; Stephens, K.; Barker, D. F.: The order of loci in the pericentric region of chromosome 17, based on evidence from physical and genetic breakpoints. Am. J. Hum. Genet. 44: 68-72, 1989.

Fanconi, G.: Hochgradige Galaktose-intoleranz (Galaktose-Diabetes) bei einem Kinde mit Neurofibromatosis Recklinghausen. Jahrb. Kinderheilk. 138: 1-8, 1933.
Fialkow, P. J.; Sagebiel, R. W.; Gartler, S. M.; Rimoin, D. L.: Multiple cell origin of hereditary neurofibromas. New Eng. J. Med. 284: 298-300, 1971.
Fienman, N. L.; Yakovac, W. C.: Neurofibromatosis in childhood. J. Pediat. 76: 339-346, 1970.
Fitzpatrick, A. P.; Emanuel, R. W.: Familial neurofibromatosis and hypertrophic cardiomyopathy. Brit. Heart J. 60: 247-251, 1988.
Fountain, J. W.; Wallace, M. R.; Brereton, A. M.; O'Connell, P.; White, R. L.; Rich, D. C.; Ledbetter, D. H.; Leach, R. J.; Fournier, R. E. K.; Menon, A. G.; Gusella, J. F.; Barker, D.; Stephens, K.; Collins, F. S.: Physical mapping of the von Recklinghausen neurofibromatosis region on chromosome 17. Am. J. Hum. Genet. 44: 58-67, 1989.
Fountain, J. W.; Wallace, M. R.; Bruce, M. A.; Seizinger, B. R.; Menon, A. G.; Gusella, J. F.; Michels, V. V.: Schmidt, M. A.; Dewald, G. W. and Collins, F. S.: Physical mapping of a translocation breakpoint in neurofibromatosis. Science 244: 1085-1087, 1989.
Friedman, J. M.; Birch, P.; Greene, C.; NNFF International Database Participants: National Neurofibromatosis Foundation International Database. Am. J. Med. Genet. 45: 88-91, 1993.
Friedman, J. M.; Birch, P. H.: Type 1 neurofibromatosis: a descriptive analysis of the disorder in 1,728 patients. Am. J. Med. Genet. 70: 138-143, 1997.
Friedman, J. M.; Fialkow, P. J.; Greene, G. L.; Weinberg, M. N.: Probable clonal origin of neurofibrosarcoma in a patient with hereditary neurofibromatosis. J. Nat. Cancer Inst. 69: 1289-1292, 1982.
Garty, B. Z.; Laor, A.; Danon, Y. L.: Neurofibromatosis type 1 in Israel: survey of young adults. J. Med. Genet. 31: 853-857, 1994.
Gitzelmann, R.: Deficiency of erythrocyte galactokinase in a patient with galactose diabetes. Lancet II: 670-671, 1965.
Goldgar, D. E.; Green, P.; Parry, D. M.; Mulvihill, J. J.: Multipoint linkage analysis in neurofibromatosis type 1: an international collaboration. Am. J. Hum. Genet. 44: 6-12, 1989.
Graham, P. W.; Oehlschlaeger, F. H.: Articulating the Elephant Man: Joseph Merrick and His Interpreters. Baltimore: Johns Hopkins Univ. Press (pub.) 1992.
Grant, W. M.; Walton, D. S.: Distinctive gonioscopic findings in glaucoma due to neurofibromatosis. Arch. Ophthal. 79: 127-134, 1968.
Griffiths, D. F. R.; Williams, G. T.; Williams, E. D.: Duodenal carcinoid tumours, phaeochromocytoma and neurofibromatosis: islet cell tumour, phaeochromocytoma and the von Hippel-Lindau complex: two distinctive neuroendocrine syndromes. Quart. J. Med. 245: 769-782, 1987.
Guo, H.-F.; The, I.; Hannan, F.; Bernards, A.; Zhong, Y.: Requirement of Drosophila NF1 for activation of adenylyl cyclase by PACAP38-like neuropeptides. Science 276: 795-798, 1997.
Gusella, J. F.: Personal Communication. Boston, Mass. 9/12/1987.
Gutmann, D.; Aylsworth, A.; Carley, J.; Korf, B.; Marks, J.; Pyeritz, R.; Rubenstein, A.; et al.: The diagnostic evaluation and multidisciplinary management of neurofibromatosis 1 and neurofibromatosis 2. J.A.M.A. 278: 51-57, 1997.
Gutmann, D. H.; Wood, D. L.; Collins, F. S.: Identification of the neurofibromatosis type 1 gene product. Proc. Nat. Acad. Sci. 88: 9658-9662, 1991.
Hashemian, H.: Familial fibromatosis of small intestine. Brit. J. Surg. 40: 346-350, 1953.
Hayes, D. M.; Spurr, C. L.; Felts, J. H.; Miller, E. C., Jr.: Von Recklinghausen's disease with massive intra-abdominal tumor and spontaneous hypoglycemia: metabolic studies before and after perfusion of abdominal cavity with nitrogen mustard. Metabolism 10: 183-199, 1961.
Heim, R. A.; Kam-Morgan, L. N. W.; Binnie, C. G.; Corns, D. D.; Cayouette, M. C.; Farber, R. A.; Aylsworth, A. S.; Silverman, L. M.; Luce, M. C.: Distribution of 13 truncating mutations in the neurofibromatosis 1 gene. Hum. Molec. Genet. 4: 975-981, 1995.
Heim, R. A.; Silverman, L. M.; Farber, R. A.; Kam-Morgan, L. N. W.; Luce, M. C.: Screening for truncated NF1 proteins.(Letter) Nature Genet. 8: 218-219, 1994.
Herrmann, J.: Sarcomatous transformation in multiple neurofibromatosis (von Recklinghausen's disease). Ann. Surg. 131: 206-217, 1950.
Hinrichs, S. H.; Nerenberg, M.; Reynolds, R. K.; Khoury, G.; Jay, G.: A transgenic mouse model for human neurofibromatosis. Science 237: 1340-1343, 1987.
Hochberg, F. H.; DaSilva, A. B.; Galdabini, J.; Richardson, E. P., Jr.: Gastrointestinal involvement in von Recklinghausen's neurofibromatosis. Neurology 24: 1144-1151, 1974.
Hoffmeyer, S.; Nurnberg, P.; Ritter, H.; Fahsold, R.; Leistner, W.; Kaufmann, D.; Krone, W.: Nearby stop codons in exons of the neurofibromatosis type 1 gene are disparate splice effectors. Am. J. Hum. Genet. 62: 269-277, 1998.
Hofman, K. J.; Harris, E. L.; Bryan, R. N.; Denckla, M. B.: Neurofibromatosis type 1: the cognitive phenotype. J. Pediat. 124: S1-S8, 1994.
Holt, J. F.: Neurofibromatosis in children. Am. J. Roentgen. 130: 615-639, 1978.
Horiuchi, T.; Hatta, N.; Matsumoto, M.; Ohtsuka, H.; Collins, F. S.; Kobayashi, Y.; Fujita, S.: Nonsense mutations at arg-1947 in two cases of familial neurofibromatosis type 1 in Japanese. Hum. Genet. 93: 81-83, 1994.
Horwich, A.; Riccardi, V. M.; Francke, U.: Aqueductal stenosis leading to hydrocephalus—an unusual manifestation of neurofibromatosis. Am. J. Med. Genet. 14: 577-581, 1983.
Howell, M.; Ford, P.: The True History of the Elephant Man. New York: Penguin (pub.) 1980.
Hull, J.; Shackleton, S.; Harris, A.: The stop mutation R553X in the CFTR gene results in exon skipping. Genomics 19: 362-364, 1994.
Hunerbein, M.; Dohmoto, M.; Haensch, W.; Schlag, P. M.: Biliary obstruction caused by a multifocal duodenal neurosarcoma in a patient with von Recklinghausen's disease. Encoscopy 28: 405 only, 1996.
Huson, S. M.; Compston, D. A. S.; Clark, P.; Harper, P. S.: A genetic study of von Recklinghausen neurofibromatosis in south east Wales. I. Prevalence, fitness, mutation rate, and effect of parental transmission on severity. J. Med. Genet. 26: 704-711, 1989.
Huson, S. M.; Meredith, A. L.; Sarfarazi, M.; Shaw, D. J.; Compston, D. A. S.; Harper, P. S.: Linkage analysis of peripheral neurofibromatosis (Von Recklinghausen disease) and chromosome 19 markers linked to myotonic dystrophy. J. Med. Genet. 23: 55-57, 1986.
Ingordo, V.; D'Andria, G.; Mendecini, S.; Grecucci, M.; Baglivo, A.: Segmental neurofibromatosis: is it uncommon or underdiagnoses? Arch. Derm. 131: 959-960, 1995.
Izumi, A. K.; Rosato, F. E.; Wood, M. G.: Von Recklinghausen's disease associated with multiple neurolemmomas. Arch. Derm. 104: 172-176, 1971.
Jadayel, D.; Fain, P.; Upadhyaya, M.; Ponder, M. A.; Huson, S. M.; Carey, J.; Fryer, A.; Mathew, C. G. P.; Barker, D. F.; Ponder, B. A. J.: Paternal origin of new mutations in Von Recklinghausen neurofibromatosis. Nature 343: 558-559, 1990.
Johnson, B. L.; Charneco, D. R.: Cafe-au-lait spot in neurofibromatosis and in normal individuals. Arch. Derm. 102: 442-446, 1970.
Johnson, M. R.; Look, A. T.; DeClue, J. E.; Valentine, M. B.; Lowy, D. R.: Inactivation of the NF1 gene in human melanoma and neuroblastoma cell lines without impaired regulation of GTP-Ras. Proc. Nat. Acad. Sci. 90: 5539-5543, 1993.
Kalff, V.; Shapiro, B.; Lloyd, R.; Sisson, J. C.; Holland, K.; Nakajo, M.; Beierwaltes, W. H.: The spectrum of pheochromocytoma in hypertensive patients with neurofibromatosis. Arch. Intern. Med. 142: 2092-2098, 1982.
Kaneko, Y.; Maseki, N.; Sakurai, M.; Shibuya, A.; Shinohara, T.; Fujimoto, T.; Kanno, H.; Nishikawa, A.: Chromosome pattern in juvenile chronic myelogenous leukemia, myelodysplastic syndrome, and acute leukemia associated with neurofibromatosis. Leukemia 3: 36-41, 1989.
Kanter, W. R.; Eldridge, R.; Fabricant, R.; Allen, J. C.; Koerber, T.: Central neurofibromatosis with bilateral acoustic neuroma: genetic, clinical and biochemical distinctions from peripheral neurofibromatosis. Neurology 30: 851-859, 1980.
Kaplan, J.; Cushing, B.; Chang, C.-H.; Poland, R.; Roscamp, J.; Perrin, E.; Bhaya, N.: Familial T-cell lymphoblastic lymphoma: association with von Recklinghausen neurofibromatosis and Gardner syndrome. Am. J. Hemat. 12: 247-250, 1982.
Kayes, L. M.; Burke, W.; Riccardi, V. M.; Bennett, R.; Ehrlich, P.; Rubenstein, A.; Stephens, K.: Deletions spanning the neurofibromatosis I gene: identification and phenotype of five patients. Am. J. Hum. Genet. 54: 424-436, 1994.
Kittur, S. D.; Bagdon, M. M.; Lubs, M. L.; Phillips, J. A., III; Murray, J. C.; Slaugenhaupt, S. A.; Chakravarti, A.; Adler, W. H.: Linkage analysis of neurofibromatosis type 1, using chromosome 17 DNA markers. Am. J. Hum. Genet. 44: 48-50, 1989.
Klose, A.; Ahmadian, M. R.; Schuelke, M.; Scheffzek, K.; Hoffmeyer, S.; Gewies, A.; Schmitz, F.; Kaufmann, D.; Peters, H.; Wittinghofer, A.; Nurnberg, P.: Selective disactivation of neurofibromin GAP activity in neurofibromatosis type 1 (NF1). Hum. Molec. Genet. 7: 1261-1268, 1998.
Knight, W. A., III; Murphy, W. K.; Gottlieb, J. A.: Neurofibromatosis associated with malignant neurofibromas. Arch. Derm. 107: 747-750, 1973.
Kohn, S. R.: The elephant man.(Letter) New Eng. J. Med. 301: 947, 1979.
Konishi, K.; Nakamura, M.; Yamakawa, H.; Suzuki, H.; Saruta, T.; Hanaoka, H.; Davatchi, F.: Case report: hypophosphatemic osteomalacia in von Recklinghausen neurofibromatosis. Am. J. Med. Sci. 301: 322-328, 1991.
Korenberg, J. R.; Barker, D.; Fain, P.; Graham, J. M.; Pribyl, T.; Pulst, S.-M.: Achondroplasia is not tightly linked to the locus for neurofibromatosis 1.(Abstract) Cytogenet. Cell Genet. 51: 1025, 1989.
Kousseff, B. G.; Gilbert-Barness, E. F.: 'Vascular neurofibromatosis' and infantile gangrene. Am. J. Med. Genet. 34: 221-226, 1989.
Krone, W.; Hogemann, I.: Cell culture studies on neurofibromatosis (von Recklinghausen): V. Monosomy 22 and other chromosomal anomalies in cultures from peripheral neurofibromas. Hum. Genet. 74: 453-455, 1986.
Kurotaki, H.; Kamata, Y.; Kimura, M.; Nagai, K.: Multiple papillary adenomas of type II pneumocytes found in a 13-year-old boy with von Recklinghausen's disease. Virchows Arch. A 423: 319-322, 1993.
Lazaro, C.; Kruyer, H.; Gaona, A.; Estivill, X.: Two further cases of mutation R1947X in the NF1 gene: screening for a relatively common recurrent mutation. Hum. Genet. 96: 361-363, 1995.
Lazaro, C.; Ravella, A.; Gaona, A.; Volpini, V.; Estivill, X.: Neurofibromatosis type 1 due to germ-line mosaicism in a clinically normal father. New Eng. J. Med. 331: 1403-1407, 1994.
Ledbetter, D. H.; Rich, D. C.; O'Connell, P.; Leppert, M.; Carey, J. C.: Precise localization of NF1 to 17q11.2 by balanced translocation. Am. J. Hum. Genet. 44: 20-24, 1989.
Legius, E.; Descheemaeker, M.-J.; Spaepen, A.; Casaer, P.; Fryns, J.-P.: Neuropsychological profile in children with NF1. (Abstract) Genet. Counsel. 5: 213-214, 1994.
Legius, E.; Descheemaeker, M. J.; Steyaert, J.; Spaepen, A.; Vlietinck, R.; Casaer, P.; Demaerel, P.; Fryns, J. P.: Neurofibromatosis type 1 in childhood: correlation of MRI findings with intelligence. J. Neurol. Neurosurg. Psychiat. 59: 638-640, 1995.
Legius, E.; Marchuk, D. A.; Collins, F. S.; Glover, T. W.: Somatic deletion of the neurofibromatosis type 1 gene in a neurofibrosarcoma supports a tumour suppressor gene hypothesis. Nature Genet. 3: 122-125, 1993.
Legius, E.; Marchuk, D. A.; Hall, B. K.; Andersen, L. B.; Wallace, M. R.; Collins, F. S.; Glover, T. W.: NF1-related locus on chromosome 15. Genomics 13: 1316-1318, 1992.
Lehrnbecher, T.; Gassel, A. M.; Rauh, V.; Kirchner, T.; Huppertz, H.-I.: Neurofibromatosis presenting as a severe systemic vasculopathy. Europ. J. Pediat. 153: 107-109, 1994.
Leppig, K. A.; Viskochil, D.; Neil, S.; Rubenstein, A.; Johnson, V. P.; Zhu, X. L.; Brothman, A. R.; Stephens, K.: The detection of contiguous gene deletions at the neurofibromatosis 1 locus with fluorescence in situ hybridization. Cytogenet. Cell Genet. 72: 95-98, 1996.
Li, Y.; Bollag, G.; Clark, R.; Stevens, J.; Conroy, L.; Fults, D.; Ward, K.; Friedman, E.; Samowitz, W.; Robertson, M.; Bradley, P.; McCormick, F.;

White, R.; Cawthon, R.: Somatic mutations in the neurofibromatosis 1 gene in human tumors. Cell 69: 275-281, 1992.
Li, Y.; O'Connell, P.; Huntsman Breidenbach, H.; Cawthon, R.; Stevens, J.; Xu, G.; Neil, S.; Robertson, M.; White, R.; Viskochil, D.: Genomic organization of the neurofibromatosis 1 gene (NF1). Genomics 25: 9-18, 1995.
Lisch, K.: Ueber Beteiligung der Augen, insbesondere das Vorkommen von Irisknotchen bei der Neurofibromatose (Recklinghausen). Augenheilkunde 93: 137-143, 1937.
Listernick, R.; Charrow, J.; Greenwald, M.; Mets, M.: Natural history of optic pathway tumors in children with neurofibromatosis type 1: a longitudinal study. J. Pediat. 125: 63-66, 1994.
Listernick, R.; Darling, C.; Greenwald, M.; Strauss, L.; Charrow, J.: Optic pathway tumors in children: the effect of neurofibromatosis type 1 on clinical manifestations and natural history. J. Pediat. 127: 718-722, 1995.
Littler, M.; Morton, N. E.: Segregation analysis of peripheral neurofibromatosis (NF1). J. Med. Genet. 27: 307-310, 1990.
Lund, A. M.; Skovby, F.: Optic gliomas in children with neurofibromatosis type 1. Europ. J. Pediat. 150: 835-838, 1991.
Manley, K. A.; Skyring, A. P.: Some heritable causes of gastrointestinal disease: special reference to hemorrhage. Arch. Intern. Med. 107: 182-203, 1961.
Marchuk, D. A.; Saulino, A. M.; Tavakkol, R.; Swaroop, M.; Wallace, M. R.; Andersen, L. B.; Mitchell, A. L.; Gutmann, D. H.; Boguski, M.; Collins, F. S.: cDNA cloning of the type 1 neurofibromatosis gene: complete sequence of the NF1 gene product. Genomics 11: 931-940, 1991.
Massaro, D.; Katz, S.: Fibrosing alveolitis: its occurrence, roentgenographic, and pathologic features in von Recklinghausen's neurofibromatosis. Am. Rev. Resp. Dis. 93: 934-942, 1966.
Mathew, C. G. P.; Thorpe, K.; Easton, D. F.; Chin, K. S.; Jadayel, D.; Ponder, M.; Moore, G.; Wallis, C. E.; Slater, C. P.; De Jong, G.; O'Connell, P.; White, R.; Barker, D.; Ponder, B. A. J.: Linkage analysis of chromosome 17 markers in British and South African families with neurofibromatosis type 1. Am. J. Hum. Genet. 44: 38-40, 1989.
Maynard, J.; Krawczak, M.; Upadhyaya, M.: Characterization and significance of nine novel mutations in exon 16 of the neurofibromatosis type 1 (NF1) gene. Hum. Genet. 99: 674-676, 1997.
Menon, A. G.; Anderson, K. M.; Riccardi, V. M.; Chung, R. Y.; Whaley, J. M.; Yandell, D. W.; Farmer, G. E.; Freiman, R. N.; Lee, J. K.; Li, F. P.; Barker, D. F.; Ledbetter, D. H.; Kleider, A.; Martuza, R. L.; Gusella, J. F.; Seizinger, B. R.: Chromosome 17p deletions and p53 gene mutations associated with the formation of malignant neurofibrosarcomas in von Recklinghausen neurofibromatosis. Proc. Nat. Acad. Sci. 87: 5435-5439, 1990.
Menon, A. G.; Ledbetter, D. H.; Rich, D. C.; Seizinger, B. R.; Rouleau, G. A.; Michels, V. F.; Schmidt, M. A.; Dewald, G.; DallaTorre, C. M.; Haines, J. L.; Gusella, J. F.: Characterization of a translocation within the von Recklinghausen neurofibromatosis region of chromosome 17. Genomics 5: 245-249, 1989.
Miles, J.; Pennybacker, J.; Sheldon, P.: Intrathoracic meningocele. Its development and association with neurofibromatosis. J. Neurol. Neurosurg. Psychiat. 32: 99-110, 1969.
Miller, M.; Hall, J. G.: Possible maternal effect of severity of neurofibromatosis. Lancet II: 1071-1073, 1978.
Miller, M. E.; Hall, J. G.: Maternal effect in neurofibromatosis.(Abstract) Am. J. Hum. Genet. 30: 60A, 1978.
Miller, R. M.; Sparkes, R. S.: Segmental neurofibromatosis. Arch. Derm. 113: 837-838, 1977.
Molloy, P. T.; Bilaniuk, L. T.; Vaughan, S. N.; Needle, M. N.; Liu, G. T.; Zackai, E. H.; Phillips, P. C.: Brainstem tumors in patients with neurofibromatosis type 1: a distinct clinical entity. Neurology 45: 1897-1902, 1995.
Muller-Wiefel, D. E.: Renovaskulare Hypertension bei Neurofibromatose von Recklinghausen. Mschr. Kinderheilk. 126: 113-118, 1978.
Nager, G. T.: Association of bilateral VIIIth nerve tumors with meningiomas in von Recklinghausen's disease. Laryngoscope 74: 1220-1261, 1964.
Nakafuku, M.; Nagamine, M.; Ohtoshi, A.; Tanaka, K.; Toh-e, A.; Kaziro, Y.: Suppression of oncogenic Ras by mutant neurofibromatosis type 1 genes with single amino acid substitutions. Proc. Nat. Acad. Sci. 90: 6706-6710, 1993.
National Institutes of Health Consensus Development Conference : Neurofibromatosis: conference statement. Arch. Neurol. 45: 575-578, 1988.
Newman, A.; So, S. K.: Bilateral neurofibroma of the intrathoracic vagus associated with von Recklinghausen's disease. Am. J. Roentgen. 112: 389-392, 1971.
Nicolls, E. M.: Somatic variation and multiple neurofibromatosis. Hum. Hered. 19: 473-479, 1969.
Nopajaroonsri, C.; Lurie, A. A.: Venous aneurysm, arterial dysplasia, and near-fatal hemorrhages in neurofibromatosis type 1. Hum. Path. 27: 982-985, 1996.
O'Connell, P.; Leach, R.; Cawthon, R. M.; Culver, M.; Stevens, J.; Viskochil, D.; Fournier, R. E. K.; Rich, D. C.; Ledbetter, D. H.; White, R.: Two NF1 translocations map within a 600-kilobase segment of 17q11.2. Science 244: 1087-1088, 1989.
O'Connell, P.; Leach, R. J.; Ledbetter, D. H.; Cawthon, R. M.; Culver, M.; Eldridge, J. R.; Frej, A.-K.; Holm, T. R.; Wolff, E.; Thayer, M. J.; Schafer, A. J.; Fountain, J. W.; Wallace, M. R.; Collins, F. S.; Skolnick, M. H.; Rich, D. C.; Fournier, R. E. K.; Baty, B. J.; Carey, J. C.; Leppert, M. F.; Lathrop, G. M.; Lalouel, J.-M.; White, R.: Fine structure DNA mapping studies of the chromosomal region harboring the genetic defect in neurofibromatosis type 1. Am. J. Hum. Genet. 44: 51-57, 1989.
Obringer, A. C.; Meadows, A. T.; Zackai, E. H.: The diagnosis of neurofibromatosis-1 in the child under the age of 6 years. Am. J. Dis. Child. 143: 717-719, 1989.
Parazzini, C.; Triulzi, F.; Bianchini, E.; Agnetti, V.; Conti, M.; Zanolini, C.; Maninetti, M. M.; Rossi, L. N.; Scotti, G.: Spontaneous involution of optic pathway lesions in neurofibromatosis type 1: serial contrast MR evaluation. Am. J. Neuroradiol. 16: 1711-1718, 1995.
Pellock, J. M.; Kleinman, P. K.; McDonald, B. M.; Wixson, D.: Childhood hypertensive stroke with neurofibromatosis. Neurology 30: 656-659, 1980.
Perry, H. D.; Font, R. L.: Iris nodules in von Recklinghausen's neurofibromatosis: electron microscopic confirmation of their melanocytic origin. Arch. Ophthal. 100: 1635-1640, 1982.
Philippart, M.: Neurofibromatose hereditaire a large spectre phenotypique (famille SN). J. Genet. Hum. 10: 338-346, 1961.
Porterfield, J. K.; Pyeritz, R. E.; Traill, T. A.: Pulmonary hypertension and interstitial fibrosis in von Recklinghausen neurofibromatosis. Am. J. Med. Genet. 25: 531-535, 1986.
Poyhonen, M.; Leisti, E.-L.; Kytola, S.; Leisti, J.: Hereditary spinal neurofibromatosis: a rare form of NF1? J. Med. Genet. 34: 184-187, 1997.
Pulst, S.-M.; Graham, J. M., Jr.; Fain, P.; Barker, D.; Pribyl, T.; Korenberg, J. R.: The achondroplasia gene is not linked to the locus for neurofibromatosis 1 on chromosome 17. Hum. Genet. 85: 12-14, 1990.
Pulst, S.-M.; Pribyl, T.; Barker, D. F.; Riccardi, V. M.; Ren, M.; Yaari, H.; Korenberg, J. R.: Molecular analysis of a patient with neurofibromatosis 1 and achondroplasia. Am. J. Med. Genet. 40: 84-87, 1991.
Pulst, S.-M.; Riccardi, V. M.; Fain, P.; Korenberg, J. R.: Familial spinal neurofibromatosis: clinical and DNA linkage analysis. Neurology 41: 1923-1927, 1991.
Purandare, S. M.; Lanyon, W. G.; Arngrimsson, R.; Connor, J. M.: Characterisation of a novel splice donor mutation affecting position +1 in intron 18 of the NF-1 gene. Hum. Molec. Genet. 4: 767-768, 1995.
Pyeritz, R. E.: Personal Communication. Baltimore, Md. 7/20/1987.
Ragge, N. K.; Falk, R. E.; Cohen, W. E.; Murphree, A. L.: Images of Lisch nodules across the spectrum. Eye 7: 95-101, 1993.
Rasmussen, S. A.; Colman, S. D.; Ho, V. T.; Abernathy, C. R.; Arn, P. H.; Weiss, L.; Schwartz, C.; Saul, R. A.; Wallace, M. R.: Constitutional and mosaic large NF1 gene deletions in neurofibromatosis type 1. J. Med. Genet. 35: 468-471, 1998.
Reubi, F.: Neurofibromatose et lesions vasculaires. Schweiz. Med. Wochenschr. 75: 463-465, 1945.
Riccardi, V. M.: Von Recklinghausen neurofibromatosis. New Eng. J. Med. 305: 1617-1626, 1981.
Riccardi, V. M.: Personal Communication. Houston, Tex. 1982.
Riccardi, V. M.: Neurofibromatosis: clinical heterogeneity. Curr. Probl. Cancer 7: 1-34, 1982.
Riccardi, V. M.; Dobson, C. E., II; Chakraborty, R.; Bontke, C.: The pathophysiology of neurofibromatosis: IX. Paternal age as a factor in the origin of new mutations. Am. J. Med. Genet. 18: 169-176, 1984.
Riccardi, V. M.; Eichner, J. E.: Neurofibromatosis: Phenotype, Natural History and Pathogenesis. Baltimore: Johns Hopkins Univ. Press (pub.) 1986.
Riccardi, V. M.; Mulvihill, J. J.: Neurofibromatosis (von Recklinghausen Disease): Genetics, Cell Biology, and Biochemistry. Advances in Neurology. New York: Raven Press (pub.) 29: 1981. Pp. 304 only.
Ritter, J. L.; Riccardi, V. M.: Von Recklinghausen neurofibromatosis (NF-I): an argument for very high penetrance and a comparison of sporadic and inherited cases.(Abstract) Am. J. Hum. Genet. 37: A135, 1985.
Riva, P.; Castorina, P.; Manoukian, S.; Dalpra, L.; Doneda, L.; Marini, G.; den Dunnen, J.; Larizza, L.: Characterization of a cytogenetic 17q11.2 deletion in an NF1 patient with a contiguous gene syndrome. Hum. Genet. 98: 646-650, 1996.
Robertson, D.: Personal Communication. Nashville, Tenn. 8/1979.
Robinson, P. N.; Buske, A.; Neumann, R.; Tinschert, S.; Nurnberg, P.: Recurrent 2-bp deletion in exon 10c of the NF1 gene in two cases of von Recklinghausen neurofibromatosis. Hum. Mutat. 7: 85-88, 1996.
Rockower, S.; McKay, D.; Nason, S.: Dislocation of the spine in neurofibromatosis: a report of two cases. J. Bone Joint Surg. 64: 1240-1242, 1982.
Rosenquist, G. C.; Krovetz, L. J.; Haller, J. A., Jr.; Simon, A. L.; Bannayan, G. A.: Acquired right ventricular outflow obstruction in a child with neurofibromatosis. Am. Heart J. 79: 103-108, 1970.
Sakaguchi, N.; Sano, K.; Ito, M.; Baba, T.; Fukuzawa, M.; Hotchi, M.: A case of von Recklinghausen's disease with bilateral pheochromocytoma-malignant peripheral nerve sheath tumors of the adrenal and gastrointestinal autonomic nerve tumors. Am. J. Surg. Path. 20: 889-897, 1996.
Sakurai, T.: Multiple neurofibroma patient showing multiple flecks on the anterior surface of the iris. Acta Soc. Ophthal. Jpn. 39: 87-93, 1935.
Samuelsson, B.; Akesson, H. O.: Relative fertility and mutation rate in neurofibromatosis. Hereditas 108: 169-171, 1988.
Sands, M. J.; McDonough, M. T.; Cohen, A. M.; Rutenberg, H. L.; Eisner, J. W.: Fatal malignant degeneration in multiple neurofibromatosis. J.A.M.A. 233: 1381-1382, 1975.
Satran, L.; Letson, R. D.; Seljeskog, E. L.: Neurofibromatosis with congenital glaucoma and buphthalmos in a newborn. Am. J. Dis. Child. 134: 182-183, 1980.
Sawada, S.; Florell, S.; Purandare, S. M.; Ota, M.; Stephens, K.; Viskochil, D.: Identification of NF1 mutations in both alleles of a dermal neurofibroma. Nature Genet. 14: 110-112, 1996.
Sayed, A. K.; Bernhardt, B.; Perez-Atayde, A. R.; Bannerman, R. M.: Malignant schwannoma in siblings with neurofibromatosis. Cancer 59: 829-835, 1987.
Schenkein, I.; Bueker, E. D.; Helson, L.; Axelrod, F.; Dancis, J.: Increased nerve-growth stimulating activity in disseminated neurofibromatosis. New Eng. J. Med. 290: 613-614, 1974.
Schmidt, M. A.; Michels, V. V.; Dewald, G. W.: Cases of neurofibromatosis with rearrangements of chromosome 17 involving band 17q11.2. Am. J. Med. Genet. 28: 771-777, 1987.
Schotland, H. M.; Eldridge, R.; Sommer, S. S.; Malawar, M.: Neurofibromatosis 1 and osseous fibrous dysplasia in a family. Am. J. Med. Genet. 43: 815-822, 1992.
Seizinger, B. R.; Farmer, G. E.; Haines, J. L.; Ozelius, L. J.; Anderson, K.; Korf, B. R.; Parry, D. M.; Pericak-Vance, M. A.; Mulvihill, J. J.; Menon, A.; Hobbs, W. J.; Martuza, R. L.; Gusella, J. F.: Flanking markers for the gene causing von Recklinghausen neurofibromatosis (NF1). Am. J. Hum. Genet. 44: 30-32, 1989.

Seizinger, B. R.; Rouleau, G. A.; Ozelius, L. J.; Lane, A. H.; Faryniarz, A. G.; Chao, M. V.; Huson, S.; Korf, B. R.; Parry, D. M.; Pericak-Vance, M. A.; and 21 others: Genetic linkage of von Recklinghausen neurofibromatosis to the nerve growth factor receptor gene. Cell 49: 589-594, 1987.
Seizinger, B. R.; Rouleau, G. A.; Ozelius, L. J.; Lane, A. H.; Faryniarz, A. G.; Chao, M. V.; Huson, S.; Korf, B. R.; Parry, D. M.; Pericak-Vance, M. A.; Collins, F. S.; Hobbs, W. J.; and 21 others : Genetic linkage of von Recklinghausen neurofibromatosis to the nerve growth factor receptor gene.(Abstract) Cytogenet. Cell Genet. 46: 690 only, 1987.
Senveli, E.; Altinors, N.; Kars, Z.; Arda, N.; Turker, A.; Cinar, N.; Yalniz, Z.: Association of von Recklinghausen's neurofibromatosis and aqueduct stenosis. Neurosurgery 24: 99-101, 1989.
Serra, E.; Puig, S.; Otero, D.; Gaona, A.; Kruyer, H.; Ars, E.; Estivill, X.; Lazaro, C.: Confirmation of a double-hit model for the NF1 gene in benign neurofibromas. Am. J. Hum. Genet. 61: 512-519, 1997.
Shannon, K. M.; O'Connell, P.; Martin, G. A.; Paderanga, D.; Olson, K.; Dinndorf, P.; McCormick, F.: Loss of the normal NF1 allele from the bone marrow of children with type 1 neurofibromatosis and malignant myeloid disorders. New Eng. J. Med. 330: 597-601, 1994.
Shannon, K. M.; Watterson, J.; Johnson, P.; O'Connell, P.; Lange, B.; Shah, N.; Steinherz, P.; Kan, Y. W.; Priest, J. R.: Monosomy 7 myeloproliferative disease in children with neurofibromatosis, type 1: epidemiology and molecular analysis. Blood 79: 1311-1318, 1992.
Shen, M. H.; Harper, P. S.; Upadhyaya, M.: Molecular genetics of neurofibromatosis type 1 (NF1). J. Med. Genet. 33: 2-17, 1996.
Side, L.; Taylor, B.; Cayouette, M.; Conner, E.; Thompson, P.; Luce, M.; Shannon, K.: Homozygous inactivation of the NF1 gene in bone marrow cells from children with neurofibromatosis type 1 and malignant myeloid disorders. New Eng. J. Med. 336: 1713-1720, 1997.
Side, L. E.; Emanuel, P. D.; Taylor, B.; Franklin, J.; Thompson, P.; Castleberry, R. P.; Shannon, K. M.: Mutations of the NF1 gene in children with juvenile myelomonocytic leukemia without clinical evidence of neurofibromatosis, type 1. Blood 92: 267-272, 1998.
Siggers, D. C.; Boyer, S. H.; Eldridge, R.: Nerve-growth factor in disseminated neurofibromatosis.(Letter) New Eng. J. Med. 292: 1134, 1975.
Silva, A. J.; Frankland, P. W.; Marowitz, Z.; Friedman, E.; Lazlo, G.; Cioffi, D.; Jacks, T.; Bourtchuladze, R.: A mouse model for the learning and memory deficits associated with neurofibromatosis type I. Nature Genet. 15: 281-284, 1997.
Skuse, G. R.; Cappione, A. J.: RNA processing and clinical variability in neurofibromatosis type 1 (NF1). Hum. Mol. Genet. 6: 1707-1712, 1997.
Skuse, G. R.; Cappione, A. J.; Sowden, M.; Metheny, L. J.; Smith, H. C.: The neurofibromatosis type I messenger RNA undergoes base-modification RNA editing. Nucleic Acids Res. 24: 478-485, 1996.
Skuse, G. R.; Kosciolek, B. A.; Rowley, P. T.: Molecular genetic analysis of tumors in von Recklinghausen neurofibromatosis: loss of heterozygosity for chromosome 17. Genes Chromosomes Cancer 1: 36-41, 1989.
Skuse, G. R.; Kosciolek, B. A.; Rowley, P. T.: The neurofibroma in von Recklinghausen neurofibromatosis has a unicellular origin. Am. J. Hum. Genet. 49: 600-607, 1991.
Smith, C. J.; Hatch, F. E.; Johnson, J. G.; Kelly, B. J.: Renal artery dysplasia as a cause of hypertension in neurofibromatosis. Arch. Intern. Med. 125: 1022-1026, 1970.
Sorensen, S. A.; Mulvihill, J. J.; Nielsen, A.: Long-term follow-up of von Recklinghausen neurofibromatosis: survival and malignant neoplasms. New Eng. J. Med. 314: 1010-1015, 1986.
Stambolian, D.; Zackai, E. H.: Gene location in neurofibromatosis.(Letter) Am. J. Med. Genet. 29: 963-965, 1988.
Stanley, J. C.: Arterial fibrodysplasia. Arch. Surg. 110: 561-566, 1975.
Stark, M.; Assum, G.; Kaufmann, D.; Kehrer, H.; Krone, W.: Analysis of segregation and expression of an identified mutation at the neurofibromatosis type 1 locus. Hum. Genet. 90: 356-359, 1992.
Stark, M.; Assum, G.; Krone, W.: Single-cell PCR performed with neurofibroma Schwann cells reveals the presence of both alleles of the neurofibromatosis type 1 (NF1) gene. Hum. Genet. 96: 619-623, 1995.
Stark, M.; Assum, G.; Krone, W.: A small deletion and an adjacent base exchange in a potential stem-loop region of the neurofibromatosis 1 gene. Hum. Genet. 87: 685-687, 1991.
Stephens, K.; Green, P.; Riccardi, V. M.; Ng, S.; Rising, M.; Barker, D.; Darby, J. K.; Falls, K. M.; Collins, F. S.; Willard, H. F.; Donis-Keller, H.: Genetic analysis of eight loci tightly linked to neurofibromatosis 1. Am. J. Hum. Genet. 44: 13-19, 1989.
Stephens, K.; Kayes, L.; Riccardi, V. M.; Rising, M.; Sybert, V. P.; Pagon, R. A.: Preferential mutation of the neurofibromatosis type 1 gene in paternally derived chromosomes. Hum. Genet. 88: 279-282, 1992.
Sutphen, R.; Galan-Gomez, E.; Kousseff, B. G.: Clitoromegaly in neurofibromatosis. Am. J. Med. Genet. 55: 325-330, 1995.
Swinburn, B. A.; Yeong, M. L.; Lane, M. R.; Nicholson, G. I.; Holdaway, I. M.: Neurofibromatosis associated with somatostatinoma: a report of two patients. Clin. Endocr. 28: 353-359, 1988.
Tassabehji, M.; Strachan, T.; Sharland, M.; Colley, A.; Donnai, D.; Harris, R.; Thakker, N.: Tandem duplication within a neurofibromatosis type I (NF1) gene exon in a family with features of Watson syndrome and Noonan syndrome. Am. J. Hum. Genet. 53: 90-95, 1993.
Taylor, P. E.: Encapsulated glioma of the sylvian fissure associated with neurofibromatosis. Report of a case with histopathological comparison of surgical lesion and autopsy specimen following recurrence. J. Neuropath. Exp. Neurol. 21: 566-578, 1962.
The, I.; Hannigan, G. E.; Cowley, G. S.; Reginald, S.; Zhong, Y.; Gusella, J. F.; Hariharan, I. K.; Bernards, A.: Rescue of a Drosophila NF1 mutant phenotype by protein kinase A. Science 276: 791-794, 1997.
Upadhyaya, M.; Cheryson, A.; Broadhead, W.; Fryer, A.; Shaw, D. J.; Huson, S.; Wallace, M. R.; Andersen, L. B.; Marchuk, D. A.; Viskochil, D.; Black, D.; O'Connell, P.; Collins, F. S.; Harper, P. S.: A 90 kb DNA deletion associated with neurofibromatosis type 1. J. Med. Genet. 27: 738-741, 1990.
Upadhyaya, M.; Maynard, J.; Osborn, M.; Huson, S. M.; Ponder, M.; Ponder, B. A. J.; Harper, P. S.: Characterisation of germline mutations in the neurofibromatosis type 1 (NF1) gene. J. Med. Genet. 32: 706-710, 1995.
Upadhyaya, M.; Osborn, M. J.; Maynard, J.; Kim, M. R.; Tamanoi, F.; Cooper, D. N.: Mutational and functional analysis of the neurofibromatosis type 1 (NF1) gene. Hum. Genet. 99: 88-92, 1997.
Upadhyaya, M.; Roberts, S. H.; Maynard, J.; Sorour, E.; Thompson, P. W.; Vaughan, M.; Wilkie, A. O. M.; Hughes, H. E.: A cytogenetic deletion, del(17)(q11.22q21.1), in a patient with sporadic neurofibromatosis type 1 (NF1) associated with dysmorphism and developmental delay. J. Med. Genet. 33: 148-152, 1996.
Upadhyaya, M.; Ruggieri, M.; Maynard, J.; Osborn, M.; Hartog, C.; Mudd, S.; Penttinen, M.; Cordeiro, I.; Ponder, M.; Ponder, B. A. J.; Krawczak, M.; Cooper, D. N.: Gross deletions of the neurofibromatosis type 1 (NF1) gene are predominantly of maternal origin and commonly associated with a learning disability, dysmorphic features and developmental delay. Hum. Genet. 102: 591-597, 1998.
Upadhyaya, M.; Sarfarazi, M.; Huson, S. M.; Broadhead, W.; Fryer, A.; Harper, P. S.: Close flanking markers for neurofibromatosis type 1 (NF1). Am. J. Hum. Genet. 44: 41-47, 1989.
Upadhyaya, M.; Sarfarazi, M.; Huson, S. M.; Broadhead, W.; Fryer, A. E.; Harper, P. S.: Genetic linkage mapping of chromosome 17 markers and neurofibromatosis type I (NF1).(Abstract) Cytogenet. Cell Genet. 51: 1094, 1989.
Upadhyaya, M.; Shen, M.; Cherryson, A.; Farnham, J.; Maynard, J.; Huson, S. M.; Harper, P. S.: Analysis of mutations at the neurofibromatosis 1 (NF1) locus. Hum. Molec. Genet. 1: 735-740, 1992.
Uren, N.; Been, M.; Guzman, F.: Congenital left atrial wall aneurysm in a patient with neurofibromatosis. Brit. Heart J. 59: 391-394, 1988.
Valero, M. C.; Velasco, E.; Moreno, F.; Hernandez-Chico, C.: Characterization of four mutations in the neurofibromatosis type 1 gene by denaturing gradient gel electrophoresis (DGGE). Hum. Molec. Genet. 3: 639-641, 1994.
Vance, J. M.; Pericak-Vance, M. A.; Yamaoka, L. H.; Speer, M. C.; Rosenwasser, G. O. D.; Small, K.; Gaskell, P. C., Jr.; Hung, W.-Y.; Alberts, M. J.; Haynes, C. S.; Gilbert, J. R.; Aylsworth, A. S.; Roses, A. D.: Genetic linkage mapping of chromosome 17 markers and neurofibromatosis type 1. Am. J. Med. Genet. 44: 25-29, 1989.
Viskochil, D.; Buchberg, A. M.; Xu, G.; Cawthon, R. M.; Stevens, J.; Wolff, R. K.; Culver, M.; Carey, J. C.; Copeland, N. G.; Jenkins, N. A.; White, R.; O'Connell, P.: Deletions and a translocation interrupt a cloned gene at the neurofibromatosis type 1 locus. Cell 62: 187-192, 1990.
Vogel, K. S.; Brannan, C. I.; Jenkins, N. A.; Copeland, N. G.; Parada, L. F.: Loss of neurofibromin results in neurotrophin-independent survival of embryonic sensory and sympathetic neurons. Cell 82: 733-742, 1995.
von Recklinghausen, F.: Ueber die multiplen Fibrome der Haut und ihre Beziehung zu den multiplen Neuromen. Berlin: August Hirschwald (pub.) 1882.
Voutsinas, S.; Wynne-Davies, R.: The infrequency of malignant disease in diaphyseal aclasis and neurofibromatosis. J. Med. Genet. 20: 345-349, 1983.
Wallace, M.; Fountain, J.; Ledbetter, D. H.; Leach, R.; Fournier, R. E. K.; O'Connell, P.; White, R.; Seizinger, B.; Menon, A.; Gusella, J. F.; Collins, F. S.: A NotI linking clone detects the 17q11.2 breakpoints in two patients with von Recklinghausen neurofibromatosis (NF1).(Abstract) Cytogenet. Cell Genet. 51: 1101, 1989.
Wallace, M. R.; Andersen, L. B.; Marchuk, D. A.; Mitchell, A. L.; Letcher, R.; Odeh, H.; Saulino, A. M.; Brereton, A.; Collins, F. S.: Candidate genes in the von Recklinghausen neurofibromatosis (NF1) region.(Abstract) Am. J. Hum. Genet. 47 (suppl.): A240, 1990.
Wallace, M. R.; Andersen, L. B.; Saulino, A. M.; Gregory, P. E.; Glover, T. W.; Collins, F. S.: A de novo Alu insertion results in neurofibromatosis type 1. Nature 353: 864-866, 1991.
Wallace, M. R.; Marchuk, D. A.; Andersen, L. B.; Letcher, R.; Odeh, H. M.; Saulino, A. M.; Fountain, J. W.; Brereton, A.; Nicholson, J.; Mitchell, A. L.; Brownstein, B. H.; Collins, F. S.: Type 1 neurofibromatosis gene: identification of a large transcript disrupted in three NF1 patients. Science 249: 181-186, 1990.
Wallis, K.; Deutsch, V.; Azizi, E.: Hypertension in a case of Von Recklinghausen's neurofibromatosis. Helv. Paediat. Acta 25: 147-153, 1970.
Ward, K.; O'Connell, P.; Carey, J. C.; Leppert, M.; Jolley, S.; Plaetke, R.; Ogden, B.; White, R.: Diagnosis of neurofibromatosis 1 by using tightly linked, flanking DNA markers. Am. J. Hum. Genet. 46: 943-949, 1990.
Weiming, X.; Yu, Q.; Lizhi, L.; Ponder, M.; Wallace, M.; Gangfeng, X.; Ponder, B.: Molecular analysis of neurofibromatosis type 1 mutations. Hum. Mutat. 1: 474-477, 1992.
Westerhof, W.; Delleman, J. W.; Wolters, E. C. M. J.; Dijkstra, P. F.: Neurofibromatosis and hypertelorism. Brit. J. Derm. 109: 475-477, 1983.
Winter, R. B.: Spontaneous dislocation of a vertebra in a patient who had neurofibromatosis: report of a case with dural ectasia. J. Bone Joint Surg. 73A: 1402-1404, 1991.
Wolkenstein, P.; Mahmoudi, A.; Zeller, J.; Revuz, J.: More on the frequency of segmental neurofibromatosis.(Letter) Arch. Derm. 131: 1465, 1995.
Wu, B.-L.; Austin, M. A.; Schneider, G. H.; Boles, R. G.; Korf, B. R.: Deletion of the entire NF1 gene detected by FISH: four deletion patients associated with severe manifestations. Am. J. Med. Genet. 59: 528-535, 1995.
Wu, B.-L.; Schneider, G. H.; Korf, B. R.: Deletion of the entire NF1 gene causing distinct manifestations in a family. Am. J. Med. Genet. 69: 98-101, 1997.
Wu, R.; Legius, E.; Robberecht, W.; Dumoulin, M.; Cassiman, J.-J.; Fryns, J.-P.: Neurofibromatosis type I gene mutation in a patient with features of LEOPARD syndrome. Hum. Mutat. 8: 51-56, 1996.
Xu, G.; O'Connell, P.; Stevens, J.; White, R.: Characterization of human adenylate kinase 3 (AK3) cDNA and mapping of the AK3 pseudogene to an intron of the NF1 gene. Genomics 13: 537-542, 1992.

Xu, G.; O'Connell, P.; Viskochil, D.; Cawthon, R.; Robertson, M.; Culver, M.; Dunn, D.; Stevens, J.; Gesteland, R.; White, R.; Weiss, R.: The neurofibromatosis type 1 gene encodes a protein related to GAP. Cell 62: 599-608, 1990.
Xu, W.; Lizhi, L.; Yuengde, B.; Yagle, M.; Solomon, E.: Enrichment of regional cosmid clones from a human chromosome 17 transfectant PLT-6B constructed by chromosome mediated gene transfer.(Abstract) Cytogenet. Cell Genet. 51: 1112, 1989.
Yagle, M.; Rider, S. H.; Parruti, G.; Nicolai, H.; Borrow, J.; Mathew, C.; Ponder, B.; Sheer, D.; Ledbetter, D.; Solomon, E.: Isolation of cosmids flanking the translocations for acute promyelocytic leukaemia (APL) and von Recklinghausen neurofibromatosis (NF1).(Abstract) Cytogenet. Cell Genet. 51: 1112, 1989.
Yagle, M. K.; Parruti, G.; Xu, W.; Ponder, B. A. J.; Solomon, E.: Genetic and physical map of the von Recklinghausen neurofibromatosis (NF1) region on chromosome 17. Proc. Nat. Acad. Sci. 87: 7255-7259, 1990.
Zacharin, M.: Precocious puberty in two children with neurofibromatosis type I in the absence of optic chiasmal glioma. J. Pediat. 130: 155-157, 1997.
Zehavi, C.; Romano, A.; Goodman, R. M.: Iris (Lisch) nodules in neurofibromatosis. Clin. Genet. 29: 51-55, 1986.
Zochodne, D.: Von Recklinghausen's vasculopathy. Am. J. Med. Sci. 287: 64-65, 1984.
Zonana, J.; Weleber, R. G.: Segmental neurofibromatosis and iris hamartomata (Lisch nodules).(Abstract) Proc. Greenwood Genet. Center 3: 140-141, 1984.

WATSON SYNDROME (#193520)

JJM NOTE

This clinically distinct but overlapping syndrome adds mental retardation and pulmonic stenosis to the usual phenotype of neurofibromatosis 1. It is included because of the occurrence of neurofibromas; I am not aware of a case of frank malignancy complicating it.

A number sign (#) is used with this entry because of evidence that the phenotype is caused by mutation in the NF1 gene (162200).

Watson (1967) described 15 persons from 2 generations of each of 3 families with pulmonic stenosis (8/15), cafe-au-lait spots (15/15) and low normal or dull intelligence (12/15). There were 8 males and 7 females; male-to-male transmission was noted. There were no signs of neurofibromata. Partington et al. (1985) described a father (aged 57 years), his daughter (aged 20) and his son (aged 18), all with pulmonary stenosis, cafe-au-lait spots, and dull intelligence. The daughter also had soft tissue limitation of movement of the knees and ankles and the father had ectasia of the coronary arteries. None had neurofibromas, Lisch nodules, lentigines, or deafness. Partington et al. (1985) insisted that the Watson syndrome is distinct from both neurofibromatosis (NF1) and the LEOPARD syndrome (151100). Although it was not obvious from the original description, short stature is a universal feature of the Watson syndrome.

Allanson et al. (1989) reviewed the 2 largest reported families including members of the extended family who had not previously been examined. They expanded the clinical phenotype to include relative macrocephaly and Lisch nodules in most affected individuals and neurofibromata in at least 4 family members. Because the additional clinical findings enhance the similarity between Watson syndrome and neurofibromatosis, linkage studies were performed using probes known to flank the NF1 gene on chromosome 17. Tight linkage with Watson syndrome was found (maximum lod = 3.29 at theta = 0.0). 6,7:Upadhyaya et al. (1989, 1990) did a linkage study of a 3-generation family with Watson syndrome. Close linkage with DNA marker D17S33 was found; maximum lod = 3.28 at theta = 0.00. This marker is also the closest marker to NF1. Thus, Watson syndrome and NF1 may be allelic, or it is possible that pulmonic stenosis is the result of a change in an adjacent gene. Allanson et al. (1991) extended their review to an additional family. Neurofibromas were found in about one-third of affected persons. Using probes flanking the NF1 gene on chromosome 17, Allanson et al. (1991) found tightest linkage with probe HHH202; maximum lod score = 3.59 at theta = 0.0. They interpreted this to indicate that either the Watson syndrome and NF1 are allelic or that there is a group of contiguous genes responsible for the several features of the Watson syndrome. Supporting the conclusion that Watson syndrome is allelic to NF1 is the finding by Upadhyaya et al. (1992) of an 80-kb deletion in the NF1 gene in a patient with Watson syndrome. Similarly, Tassabehji et al. (1993) demonstrated an almost perfect inframe tandem duplication of 42 bases in exon 28 of the NF1 gene in 3 members of a family with Watson syndrome (162200.0010). Relevant to the question of whether the Watson syndrome is a contiguous gene syndrome resulting from deletion of both the NF1 gene and a gene for Noonan syndrome, Sharland et al. (1992), in a linkage study in 11 families with Noonan syndrome in 2 or 3 generations, excluded the proximal region of 17q as the location of the gene.

Allanson, J. E.; Upadhyaya, M.; Watson, G. H.; Partington, M.; Harper, P.; Huson, S.: Molecular linkage analysis of Watson syndrome. (Abstract) Am. J. Hum. Genet. 45 (suppl.): A38 only, 1989.
Allanson, J. E.; Upadhyaya, M.; Watson, G. H.; Partington, M.; MacKenzie, A.; Lahey, D.; MacLeod, H.; Sarfarazi, M.; Broadhead, W.; Harper, P. S.; Huson, S. M.: Watson syndrome: is it a subtype of type 1 neurofibromatosis?. J. Med. Genet. 28: 752-756, 1991.
Partington, M. W.; Burggraf, G. W.; Fay, J. E.; Frontini, E.: Pulmonary stenosis, cafe au lait spots and dull intelligence: the Watson syndrome revisited. (Abstract) Proc. Greenwood Genet. Center 4: 105 only, 1985.
Sharland, M.; Taylor, R.; Patton, M. A.; Jeffery, S.: Absence of linkage of Noonan syndrome to the neurofibromatosis type 1 locus. J. Med. Genet. 29: 188-190, 1992.
Tassabehji, M.; Strachan, T.; Sharland, M.; Colley, A.; Donnai, D.; Harris, R.; Thakker, N.: Tandem duplication within a neurofibromatosis type I (NF1) gene exon in a family with features of Watson syndrome and Noonan syndrome. Am. J. Hum. Genet. 53: 90-95, 1993.
Upadhyaya, M.; Sarfarazi, M.; Broadhead, W.; Huson, S. M.; Allanson, J.; Fryer, A. E.; Harper, P. S.: Linkage of Watson's syndrome to chromosome 17 markers. (Abstract) Cytogenet. Cell Genet. 51: 1094 only, 1989.
Upadhyaya, M.; Sarfarazi, M.; Huson, S.; Broadhead, W.; Allanson, J.; Fryer, A.; Harper, P. S.: Linkage of Watson's syndrome to chromosome 17 markers. (Abstract) J. Med. Genet. 27: 209 only, 1990.
Upadhyaya, M.; Shen, M.; Cherryson, A.; Farnham, J.; Maynard, J.; Huson, S. M.; Harper, P. S.: Analysis of mutations at the neurofibromatosis 1 (NF1) locus. Hum. Molec. Genet. 1: 735-740, 1992.
Watson, G. H.: Pulmonary stenosis, cafe-au-lait spots, and dull intelligence. Arch. Dis. Child. 42: 303-307, 1967.

NEUROFIBROMATOSIS-NOONAN SYNDROME (601321)

JJM NOTE

Although the jury is out on whether this is a distinct locus, nonetheless, the occurrence of strange retroperitoneal and visceral ganglioneurofibromas merits the entry of this disorder.

Allanson et al. (1985) reported 4 unrelated patients with neurofibromatosis (NF1; 162200) who had manifestations of the Noonan syndrome (163950), including short stature, ptosis, midface hypoplasia, webbed neck, learning disabilities, and muscle weakness. Family history was negative in each case. Average paternal and maternal ages were 37 and 28 years, respectively, at the birth of the patients, suggesting new dominant mutation. The chromosomes, including prometaphase preparations in 3 of the 4, were normal. The authors suggested that this is a distinct entity. Opitz and Weaver (1985) likewise favored the distinctness of what they called the neurofibromatosis-Noonan syndrome. They suggested that males are more likely to have fusiform swelling of nerve strands, while females more often show the classic neurofibromata seen in von Recklinghausen disease. Lisch nodules of the iris are uncommon. The tendency to develop retroperitoneal or visceral (ganglio-) neurofibromatosis is strong. Abuelo and Meryash (1988) described an 18-year-old man with neurofibromatosis and classic manifestations of Noonan syndrome, including atrial septal defect of the secundum type and valvular and supravalvular pulmonic stenosis, for which cardiac surgery was performed. The father also had neurofibromatosis and was described by the authors as having 'some of the characteristics of Noonan syndrome.' These appear to have been only 'prominent nasolabial folds and apparently low-set ears.' There are several possibilities to explain the combination: (1) it is due to mutation at a locus different from that of NF1 on chromosome 17; (2) it is due to an allele at the NF1 locus; (3) it is the coincidence of 2 relatively frequent conditions; or (4) it represents associated disorders due to mutations at closely linked loci. In connection with the last possibility, it is of note that Abuelo and Meryash (1988) found no abnormality in the prophase chromosome analysis, performed with special attention to chromosome 1 7, in their propositus. It is uncertain whether neurofibromatosis with Noonan syndrome-like features is a separate entity (Meinecke, 1987; Quattrin et al., 1987). Colley et al. (1996) examined 94 persons with neurofibromatosis for features of the Noonan syndrome and found that 12, including some familial cases, had diagnostic criteria of the Noonan syndrome. One of the families showed independent segregation of NF1 and Noonan syndrome; in other families, half of those affected with NF1 had manifestations of the Noonan syndrome. Linkage of the Noonan syndrome gene to chromosome 12 (Jamieson et al., 1994) provides evidence that the genes for NF1 and Noonan syndrome are neither allelic nor contiguous. Nevertheless, the possibility that another gene or genes are involved in Noonan syndrome remains open. Colley et al. (1996) suggested that some alterations of the NF1 gene may predispose to the Noonan syndrome phenotype, but considered it unlikely that a neurofibromatosis-Noonan syndrome phenotype is a distinct disorder or occurs as part of classic Noonan syndrome.

Abuelo, D. N.; Meryash, D. L.: Neurofibromatosis with fully expressed Noonan syndrome. Am. J. Med. Genet. 29: 937-941, 1988.
Allanson, J. E.; Hall, J. G.; Van Allen, M. I. Noonan phenotype associated with neurofibromatosis. Am. J. Med. Genet. 21: 457-462, 1985.
Colley, A.; Donnai, D.; Evans, D. G. R.: Neurofibromatosis/Noonan phenotype: a variable feature of type 1 neurofibromatosis. Clin. Genet. 49: 59-64, 1996.
Jamieson, C. R.; van der Burgt, I.; Brady, A. F.; van Reen, M.; Elsawi, M. M.; Hol, F.; Jeffery, S.; Patton, M. A.; Mariman, E.: Mapping a gene for Noonan syndrome to the long arm of chromosome 12. Nature Genet. 8: 357-360, 1994.
Meinecke, P.: Evidence that the neurofibromatosis-Noonan syndrome is a variant of von Recklinghausen neurofibromatosis. Am. J. Med. Genet. 26: 741-745, 1987.
Opitz, J. M.; Weaver, D. D.: The neurofibromatosis-Noonan syndrome. Am. J. Med. Genet. 21: 477-490, 1985.
Quattrin, T.; McPherson, E.; Putnam, T.: Vertical transmission of the neurofibromatosis/Noonan syndrome. Am. J. Med. Genet. 26: 645-649, 1987.

NEUROFIBROMATOSIS 2 (BILATERAL VESTIBULAR SCHWANNOMA) (*101000)

JJM NOTE

A patient with a "mild" or "atypical" case of neurofibromatosis 1 should be suspected of having neurofibromatosis 2, because occasional features of neurofibromatosis 1, such as café-au-lait spots and cutaneous neurofibromas, are seen, albeit in fewer numbers. A 1991 National Institutes of Health (NIH) Consensus Development Conference on Acoustic Neuroma slightly revised the prior consensus definition of neurofibromatosis 2 in this way: Neurofibromatosis 2 may be diagnosed when one of the following is present: bilateral eighth cranial nerve masses seen by MRI with gadolinium *or* a parent, sibling, or child with neurofibromatosis 2 and *either* a unilateral eighth cranial nerve mass *or* any one of the following: neurofibroma, meningioma, glioma, schwannoma, or posterior capsular cataract or opacity at a young age. Later, an ad hoc expert group (Gutmann et al., 1997) suggested another definition of definite neurofibromatosis 2: Bilateral vestibular schwannomas *or* a first-degree relative with neurofibromatosis 2, plus either a unilateral vestibular schwannoma under age 30 years or any two of the following: meningioma, glioma, schwannoma, or juvenile posterior subcapsular lenticular opacities/juvenile cortical cataracts.

The central form of neurofibromatosis, characterized by tumors of the eighth cranial nerve (usually bilateral), meningiomas of the brain and schwannomas of the dorsal roots of the spinal cord, has few of the hallmarks of the peripheral form of neurofibromatosis (162200). Most patients with the central form have no cafe-au-lait spots or peripheral neurofibromata and no patients in one large series had 6 or more cafe-au-lait spots (Eldridge, 1981). The term von Recklinghausen disease should be reserved for the peripheral form of neurofibromatosis. Gardner and Frazier (1933) reported a family of 5 generations in which 38 members were deaf because of acoustic neuromas; of these, 15 later became blind. The average age of onset of deafness was 20 years. The average age at death of affected persons in the second generation was 72, in the third generation 63, in the fourth 42, and in the fifth 28. Follow-up of this family (Gardner and Turner, 1940; Young et al., 1970) revealed no evidence of the systemic manifestations of von Recklinghausen disease. Other families with no evidence of the latter disease were reported by Worster-Drought et al. (1937), Feiling and Ward (1920), and Moyes (1968). Worster-Drought et al. (1937) pointed out that Wishart reported the first case of bilateral acoustic neuroma in 1822. Wishart's patient, Michael Blair, was 21 years old when he consulted Mr. Wishart, president of the Royal College of Surgeons of Edinburgh, because of bilateral deafness. He had a peculiarly shaped head from infancy, and blindness in the right eye was discovered at about 4 months after birth. He became completely blind and deaf toward the end of his life. Autopsy revealed tumors of the dura mater and brain and also a 'tumour of the size of a small nut, and very hard, being attached to each of them (auditory nerves), just where they enter the meatus auditorius internus.'

Nager (1969) showed that in about 4% of cases acoustic neuroma is bilateral. In addition to their autosomal dominant inheritance and association with neurofibromatosis, bilateral tumors differ from unilateral ones in that they can reach a remarkably large size with extensive involvement of the temporal bone and the nerves therein. More than 30 kindreds with 'central neurofibromatosis' have been reported (Fabricant et al., 1979). Kanter et al. (1980), who reviewed 9 personally studied kindreds and 15 reported ones, with a total of 130 cases, showed an increase only in antigenic activity of nerve growth factor (NGF) in central neurofibromatosis and only in functional activity in peripheral neurofibromatosis. Thus, these disorders may involve different defects in NGF synthesis and/or regulation. In a review of NF2, Martuza and Eldridge (1988) defined criteria for the diagnosis of both NF1 and NF2. An NIH Consensus Development Conference (1988) concluded that the criteria for NF2 are met if a person is found to have '(1) bilateral eighth nerve masses seen with appropriate imaging techniques (e.g., CT or MRI); or (2) a first-degree relative with NF2 and either unilateral eighth nerve mass, or two of the following: neurofibroma, meningioma, glioma, schwannoma, or juvenile posterior subcapsular lenticular opacity.' Pearson-Webb et al. (1986) pointed out that Lisch nodules, which are iris hamartomas, are not found in NF2. They found, however, an apparently high frequency of presenile posterior subcapsular and nuclear cataracts which sometimes required surgery and/or predated the symptoms of bilateral acoustic neurofibromatosis. Kaiser-Kupfer et al. (1989) found posterior capsular lens opacities in 20 NF2 patients in 11 families. Parry et al. (1991) extended these observations. In 26 persons who were first-degree relatives of an affected individual, they found posterior capsular cataracts in 21. Of 14 at-risk individuals, i.e., persons with mild changes of NF but not NF1, persons under age 40 with unilateral acoustic neuroma, a child with meningioma and/or schwannoma, and a person with multiple meningioma, they found posterior capsular lens opacities in 13. These patients probably represented new mutations. The presence of posterior capsular opacities in a relative of persons with NF2 was suggestive of NF2. Furthermore, NF2 should be considered in young persons without NF1 but with mild skin findings of NF or CNS tumors with posterior capsular opacities. Bouzas et al. (1993) found posterior subcapsular/capsular cataracts in 36 (80%) of 45 affected individuals in 29 families. In addition, the association of peripheral cortical lens opacities with NF2 was found to be statistically significant: such cataracts were found in 17 of the patients (37.8%) but in none of the unaffected family members (p less than 0.0001). In 3 patients, peripheral cortical opacities were present despite the absence of posterior subcapsular/capsular cataracts. Bouzas et al. (1993), reporting further on the NIH experience, reviewed visual impairment in 54 NF2 patients, 51 of whom had bilateral vestibular schwannomas. Causes of decreased vision were cataracts, damage in the optic pathways, macular hamartomas, and corneal opacities. Although lens opacities are an important marker for NF2, they usually do not interfere with vision; some progress, requiring cataract extraction. In 6 patients, decreased visual acuity was due to corneal opacifications secondary to either seventh or fifth cranial nerve damage, or both. Damage to the seventh cranial nerve caused lagophthalmos and decreased lacrimal secretion; damage to the fifth cranial nerve caused corneal hypesthesia. The nerves were damaged by the growth of vestibular tumors in 1 patient, but in most patients they were damaged during neurosurgical procedures.

Pastores et al. (1991) demonstrated that small (less than 8 mm) acoustic neuromas can be detected in asymptomatic individuals by the use of gadolinium-enhanced MRI. They demonstrated such neuromas in 2 asymptomatic children, aged 7 and 11 years, one of whom had normal audiometric and brain stem-evoked response testing. Landau et al. (1990) described combined pigment epithelial and retinal hamartoma (CEPRH) in NF2. In a series reported by Mrazek et al. (1988), 1 of 41 acoustic neurinoma cases was bilateral. This was in a 10-year-old girl with von Recklinghausen neurofibromatosis, whose first tumor had been diagnosed at age 6. Mayfrank et al. (1990) studied 10 patients with NF2 and found that all were sporadic cases, each presumably the result of a new mutational event. From a survey of these patients and those in the literature, they concluded that sporadic cases are characterized by a high incidence of multiple meningiomas and spinal tumors in addition to bilateral acoustic neurinomas. Pulst et al. (1991) described a family with spinal neurofibromatosis without cafe-au-lait spots or other manifestations of either NF1 or NF2 such as cutaneous tumors, Lisch nodules, or acoustic tumors. Mutation at the NF1 locus was excluded with odds greater than 100,000:1. Markers with the NF2 locus were uninformative in this family.

Evans et al. (1992) studied 150 patients. The mean age at onset was 21.57 years (n = 110) and no patient presented after 55 years of age. Patients presented with symptoms attributable to vestibular schwannomas (acoustic neuroma), cranial meningiomas, and spinal tumors. In 100 patients studied personally by the authors, 44 presented with deafness, which was unilateral in 35. Deafness was accompanied by tinnitus in 10. Muscle weakness or wasting was the first symptom in 12%. In 3 of the 100 patients, there was a distal symmetrical sensorimotor neuropathy, confirmed by nerve conduction studies and electromyography. Although similar features may result from the multiple spinal and intracranial tumors that occur in this condition, a generalized and isolated neuropathy appears to be a relatively common feature of NF2. Cafe-au-lait spots occurred in 43 of the 100 patients but only 1 had as many as 6 spots. Cataract was detected in 34 of 90 patients. Cataracts were probably congenital in 4 patients in this study. Three types of skin tumors were recognized. The first and least common was similar to the intradermal papillary skin neurofibroma with violaceous coloring occurring in NF1. The second type comprised subcutaneous well-circumscribed, often spherical, tumors that appeared to be located on peripheral nerves; the thickened nerve could often be palpated at either end of the tumor, the skin being mobile and separate from the tumor. The third and most frequent type, first described by Martuza and Eldridge (1988), was represented by discrete well-circumscribed, slightly raised, roughened areas of skin often pigmented and accompanied by excess hair. Skin tumors of some kind were found in 68% of patients, type 1 being present in 20%, type 2 in 33%, and type 3 in 47%. Evans et al. (1992) divided their 120 cases of NF2 into 2 types: the Wishart (1822) type, with early onset, rapid course, and multiple other tumors in addition to bilateral vestibular schwannomas, and the Gardner type (19 30, 1933, 1940), with late onset, more benign course, and usually only bilateral vestibular schwannomas. This classification had been suggested by Eldridge et al. (1991). Evans et al. (1992) found no evidence for the existence of a third type of generalized meningiomatosis that might be designated the Lee-Abbott type (Lee and Abbott, 1969). They could find no evidence that either pregnancy or contraceptive pill has adverse effects on vestibular schwannomas or other manifestations. Evans et al. (1992) provided useful advice on the follow-up of persons identified as having NF2 and the management of persons at risk of developing NF2. The age at onset of deafness and the age at diagnosis were almost identical in the 2 sexes. Birth incidence of NF2 was estimated to be 1 in 33,000-40,562. Evans et al. (1992) considered 49% of the 150 cases to represent new mutations. The mutation rate was estimated to be 6.5 x 10(-6). A maternal effect on severity was noted in that age of onset was 18.17 years in 36 maternally inherited cases and 24.5 years in 20 paternally inherited cases (p = 0.027). A preponderance of maternally inherited cases was also significant (p = 0.03). (A maternal effect on severity had been noted also for neurofibromatosis, type I (NF1; 162200).)

Parry et al. (1994) assessed possible heterogeneity in NF2 by evaluating 63 affected members of 32 families. In addition to skin and neurologic examinations, workup included audiometry, complete ophthalmologic examination with slit-lamp biomicroscopy of the lens and fundus, and gadolinium-enhanced MRI of the brain and, in some, of the spine. Mean age-at-onset in 58 individuals was 20.3 years; initial symptoms were related to vestibular schwannomas (44.4%), other CNS tumors (22.2%), skin tumors (12.7%), and ocular manifestations including cataracts and retinal hamartomas (12.7%). Screening uncovered 5 affected but asymptomatic family members; vestibular schwannomas were demonstrated in 62 (98.4%). Other findings included cataracts (81.0%), skin tumors (67.7%), spinal tumors (67.4%), and meningiomas (49.2%). As a rule, clinical manifestations and clinical course were similar within families but differed among families. Parry et al. (1994) concluded that 2 subtypes but not 3 can be defined.

Ragge et al. (1995) concluded that the most common ocular abnormalities in NF2 are posterior subcapsular or capsular, cortical, or mixed lens opacities, found in 33 of 49 patients (67%), and retinal hamartomas found in 11 of 49 patients (22%). The types of cataract that were most suggestive of NF2 were plaque-like posterior subcapsular or capsular cataract and cortical cataract with onset under the age of 30 years.

Seizinger et al. (1986) found loss of genes on chromosome 22 in acoustic neuromas; i.e., whereas normal tissue was heterozygous, tumor tissue was hemizygous (or homozygous) for the polymorphic markers SIS (190040), IGLC (147220), and the anonymous DNA locus D22S1. They were prompted to undertake the study by analogy to retinoblastoma and Wilms tumor and by the facts that meningioma occurs in association with familial acoustic neuroma and that cytologic change in chromosome 22 is frequent in meningioma (see 156100). Seizinger et al. (1987) found specific loss of alleles from chromosome 22 in 2 acoustic neuromas, 2 neurofibromas, and 1 meningioma from patients with bilateral acoustic neurofibromatosis. In each case, a partial deletion occurred with a breakpoint distal to the D22S9 locus in band 22q11. Wertelecki et al. (1988) confirmed localization of the gene on chromosome 22 (22q11.21-q13.1) by demonstration of linkage in family studies to markers on chromosome 22. Wertelecki et al. (1988) also presented the clinical data on 15 affected male and 8 affected female members of the 1 large kindred they studied for linkage data. Rouleau et al. (1990) identified markers bracketing the NF2 gene which are therefore useful for accurate presymptomatic and prenatal diagnosis, as well as for isolating the defective gene. Narod et al. (1992) concluded that there is no evidence of genetic heterogeneity in NF2. They indicated that the presence of bilateral vestibular schwannomas, as they termed the acoustic neuromas, is sufficient for the diagnosis. Using 8 polymorphic loci on chromosome 22 to study tumor and constitutional DNAs isolated from 39 unrelated patients with sporadic or NF2-associated acoustic neuromas, meningiomas, schwannomas, and ependymomas, Wolff et al. (1992) found 2 tumors with loss of heterozygosity (LOH) patterns consistent with the presence of chromosome 22 terminal deletions. By use of additional polymorphic markers, the terminal deletion breakpoi nt in one of the tumors, an acoustic neuroma from an NF2 patient, was mapped within the previously defined NF2 region. In addition, they identified a sporadic acoustic neuroma with an LOH pattern consistent with mitotic recombination or deletion and reduplication. The findings lent further support to the recessive tumor-suppressor model for the NF2 gene. Arai et al. (1992) described a patient with bilateral acoustic neurinomas and other tumors in the central nervous system and a constitutional translocation t(4;22)(q12;q12.2). Thus, 22q12.2 is a refined localization for the NF2 gene. The same karyotype that was seen in cultured peripheral lymphocytes was found in a paraspinal neurinoma. The patient's father was also a carrier of the translocation but he had no clinical symptoms of NF2, nor did other relatives. As explanation for the failure of expression in the father, Arai et al. (1992) suggested various possibilities including nonpenetrance, mosaicism, or genetic imprinting. They quoted Kanter et al. (1980) as demonstrating earlier onset of symptoms when NF2 is transmitted by the mother. In a family with the mild or so-called Gardner type of neurofibromatosis type 2, Watson et al. (1993) defined a submicroscopic deletion which involved the neurofilament heavy chain locus (NEFH; 162230) but did not extend as far as the Ewing sarcoma region (EWS; 133450) proximally or the leukemia inhibitory factor locus (LIF; 159540) distally. They estimated that the deletion was about 700 kb long.

Claudio et al. (1994) demonstrated that the mouse homolog of the NF2 gene is located in the proximal region of chromosome 11. The localization was achieved by analysis of allele distribution in recombinant inbred strains using a simple sequence repeat polymorphism in the 3-prime untranslated region of the mouse NF2 cDNA. The region of chromosome 11 also contains genes for leukemia inhibitory factor (LIF; 159540) and neurofilament heavy chain polypeptide (NFH; 162230), both of which map to the same region of human chromosome 22 as does NF2.

Trofatter et al. (1993) identified a candidate gene for the NF2 tumor suppressor that had suffered nonoverlapping deletions in DNA from 2 independent NF2 families as well as alterations in the meningiomas from 2 unrelated NF2 patients. The candidate gene encoded a 587-amino acid protein with striking similarity to several members of a family of proteins proposed to link cytoskeletal components with proteins in the cell membrane; these included moesin (309845), ezrin (123900), and radixin (179410). Because of the resemblance to these 3 proteins (45-47% identity), Trofatter et al. (1993) called the NF2 gene product merlin. The NF2 gene may represent a novel class of tumor suppressor genes. Schwannomin (symbol = SCH) was the designation used by Rouleau et al. (1993), who likewise isolated a gene bearing homology to erythrocyte protein 4.1 and the ezrin/moesin/talin family of genes. They provided incontrovertible evidence that this gene is the site of the mutations causing NF2 by demonstrating germline and somatic SCH mutations in NF2 patients and in NF2-related tumors. To isolate the gene, they cloned the region between 2 flanking polymorphic markers in which they found several genes, only one of which carried mutations in NF2. Rouleau et al. (1993) found 16 mutations, 15 of which were predicted to result in truncated proteins. Consistent with the classic Knudson theory of tumor suppressor genes, loss of the wildtype allele at the NF2 locus was demonstrated in 6 of 8 tumors containing NF2 mutations (Trofatter et al., 1993; Rouleau et al., 1993). For example, in a meningioma in a patient without features of NF2, they found deletion of 2 nucleotides, TC, from codon 61 resulting in a frameshift; the normal allele on the other chromosome had been lost. In 2 instances of schwannoma occurring in patients without evidence of NF2, Rouleau et al. (1993) found nonsense mutations that were absent in the patient's blood DNA; in these instances also the normal allele ha d been lost.

Using polymorphic DNA markers in a study of 13 NF2 kindreds, Ruttledge et al. (1993) concluded that it is possible to determine, with a high degree of certainty, the carrier status of about 85% of persons at risk. Risk prediction was possible in every case in which DNA was available from both parents. In 76% of informative individuals, it was possible to assign a decreased risk of being carriers. Thus, the use of probes for construction of chromosome 22 haplotypes for risk assessment should result in a greatly reduced number of individuals who will require periodic screening.

Bianchi et al. (1994) described a novel isoform of the NF2 transcript that shows differential tissue expression and encodes a modified C terminus of the predicted protein. Mutations affecting both isoforms of the NF2 transcript were detected in multiple tumor types including melanoma and breast carcinoma. These findings provided evidence that alterations in the NF2 transcript occurred not only in the hereditary brain neoplasms typically associated with NF, but also as somatic mutations in their sporadic counterparts.

By November 1993, 24 mutations, including both germline and somatic mutations, had been detected in schwannomin (Thomas, 1993). Most of the mutations cause the synthesis of a truncated schwannomin protein. After examining 8 of the 16 known NF2 exons in 151 meningiomas, Ruttledge et al. (1994) characterized 24 inactivating mutations. Significantly, these aberrations were detected exclusively in tumors that lost the other chromosome 22 allele. These results provided strong evidence that the suppressor gene on chromosome 22, frequently inactivated in meningioma, is the NF2 gene. The same group had found loss of heterozygosity (LOH) for polymorphic DNA markers flanking NF2 on chromosome 22 in 102 (60%) of 170 primary sporadic meningiomas. Thus, another gene may be involved in the development of 40% of meningiomas. It is probably noteworthy that all 24 of the inactivating mutations found by Ruttledge et al. (1994) in sporadic meningiomas were nonsense, frameshift (due to small deletions), or splice site mutations; there were no missense mutations. Wellenreuther et al. (1995) likewise concluded that NF2 represents the meningioma locus on chromosome 22. There was a significant association of loss of heterozygosity on chromosome 22 with mutations in the NF2 gene. They analyzed the entire coding region of the NF2 gene in 70 sporadic meningiomas and identified 43 mutations in 41 patients. These resulted predominantly in immediate truncation, splicing abnormalities, or an altered reading frame of the predicted protein product. All mutations occurred in the first 13 exons, the region of homology with the filopodial proteins moesin, ezrin, and radixin.

Parry et al. (1996) used SSCP analysis to screen for mutations in DNA from 32 unrelated NF2 patients. Mutations were identified in 66% of patients and 20 different mutations were found in 21 patients. They reported that their results confirm the association between nonsense and frameshift mutations and clinical manifestations compatible with severe disease. Parry et al. (1996) stated that their data raise questions regarding the role of factors, other than the intrinsic properties of individual mutations, that might influence the phenotype. Ruttledge et al. (1996) reported that when individuals harboring protein-truncating mutations are compared with patients having single codon alterations, a significant correlation (p less than 0.001) with clinical outcome is observed. They noted that 24 of 28 patients with mutations that cause premature truncation of the NF2 protein present with severe phenotypes. In contrast, all 16 cases from 3 families with mutations that affect only a single amino acid have mild NF2. Evans et al. (1998) reported 42 cases of NF2 from 38 families with truncating mutations. The average age of onset of symptoms was 19 years and age at diagnosis 22.4 years. Fifty-one cases from 16 families (15 with splice site mutations, 18 with missense mutations, and 18 with large deletions) had an average age of onset of 27.8 years and age at diagnosis of 33.4 years. Subjects with truncating mutations were significantly more likely to develop symptoms before 20 years of age (p less than 0.001) and to develop at least 2 symptomatic CNS tumors in addition to vestibular schwannoma before 30 years (p less than 0.001). There were significantly fewer multigenerational families with truncating mutations.

Sainz et al. (1994) performed mutational analysis in 30 vestibular schwannomas and found 18 mutations, 7 of which contained loss or mutation of both NF2 alleles. Most mutations predicted a truncated protein. Mutational hotspots were not identified. Only 1 of the mutations was in a tumor from a patient with NF2. Immunocytochemical studies using antibodies to the NF2 protein showed complete absence of staining in tumor Schwann cells, whereas staining was observed in normal vestibular nerve. These data indicated that loss of NF2 protein function is a necessary step in schwannoma pathogenesis and that the NF2 gene functions as a recessive tumor suppressor gene. In studies of 34 vestibular schwannomas and 14 schwannomas at other locations, Bijlsma et al. (1994) found that the SCH gene is implicated in the development of these tumors in all locations of the nervous system. Using a screening method based on denaturing gradient gel electrophoresis, which allows the detection of mutations in 95% of the coding sequence, Merel et al. (1995) observed mutations in 17 of 57 meningiomas and in 30 of 89 schwannomas. All of the meningiomas and half of the schwannomas with identified NF2 mutations demonstrated chromosome 22 allelic losses. No mutations were observed in 17 ependymomas, 70 gliomas, 23 primary melanomas, 24 pheochromocytomas, 15 neuroblastomas, 6 medulloblastomas, 15 colon cancers, and 15 breast cancers. This led Merel et al. (1995) to conclude that the involvement of the NF2 gene is restricted to schwannomas and meningiomas, where it is frequently inactivated by a 2-hit process.

Neurilemmomatosis, first reported by Niimura (1973) as neurofibromatosis type 3, is characterized by multiple cutaneous neurilemmomas and spinal schwannomas, without acoustic tumors or other signs of NF1 or NF2. In neurilemmomas, the tumor consists of Schwann cells. Honda et al. (1995) analyzed the peripheral leukocytes and tissue from cutaneous neurilemmomas of 7 patients with neurilemmomatosis using DNA markers for different regions of chromosome 22. They detected allele losses in 3 of 7 tumors from 7 patients with a probe for the NF2 region and the germline mutations in 2 of 3 tumors from the same 3 patients. They described 2 mutations in the NF2 gene (101000.0017; 101000.0018). They concluded that neurilemmomatosis is a form of NF2.

Malignant mesotheliomas (MMs) are aggressive tumors that develop most frequently in the pleura of patients exposed to asbestos. In contrast to many other cancers, relatively few molecular alterations had been described in MMs. The most frequent numerical cytogenetic abnormality in MMs is loss of chromosome 22. This prompted Bianchi et al. (1995) to investigate the status of the NF2 gene in these tumors. In studies of cDNAs from 15 MM cell lines and genomic DNAs from 7 matched primary tumors, NF2 mutations predicting either interstitial in-frame deletions or truncation of the NF2-encoded protein (merlin) were detected in 8 cell lines (53%), 6 of which were confirmed in primary tumor DNAs. In 2 samples that showed NF2 gene transcript alterations, no genomic DNA mutations were detected, suggesting that aberrant splicing may constitute an additional mechanism for merlin inactivation. Unlike previously described NF2-related tumors, MM derived from the mesoderm; malignancies of this origin had not previously been associated with frequent alterations of the NF2 gene. In a commentary in the same journal issue, Knudson (1995) wrote: 'We are left wondering why mesothelioma is not a feature of the hereditary disease NF2.'

Kehrer-Sawatzki et al. (1997) reported a patient with NF2 and a ring chromosome 22 (46,XX,r(22)/45,XX,-22). Severe manifestations included multiple meningiomas, spinal and peripheral neurinomas, and bilateral vestibular schwannomas. The patient was also severely mentally retarded, a feature not usually associated with NF2. The authors hypothesized that a mutation in the NF2 gene of the normal chromosome 22, in addition to the loss of the ring 22 in many cells during mitosis, could explain the presence of multiple tumors. Using a meningioma cell line lacking the ring chromosome, Kehrer-Sawatzki et al. (1997) searched for deletions, rearrangements, or other mutations of the NF2 gene on the normal chromosome 22; no such alterations were found. The authors concluded that the loss of the entire chromosome 22 and its multiple tumor suppressor genes may have led to the severe phenotype in this patient.

Jacoby et al. (1997) investigated the molecular genetic basis of schwannomatosis (162091) in patients with multiple schwannomas without vestibular schwannomas, which has been postulated to represent a distinct subclass of neurofibromatosis. They found the unusual situation that in studies of 20 unrelated schwannomatosis patients and their affected relatives, tumors from the patients frequently harbored typical truncating mutations of the NF2 gene and loss of heterozygosity of the surrounding region of chromosome 22. Surprisingly, unlike patients with NF2, no heterozygous NF2 gene changes were seen in normal tissues. Furthermore, examination of multiple tumors from the same patients revealed that some schwannomatosis patients are somatic mosaics for NF2 gene changes. By contrast, other individuals, particularly those with a positive family history, appeared to have an inherited predisposition to formation of tumors that carry somatic alterations of the NF2 gene.

Gutmann et al. (1997) provided guidelines for the diagnostic evaluation and multidisciplinary management of both NF1 and NF2. The criteria for definite NF2 were bilateral vestibular schwannomas; or family history of NF2 in 1 or more first-degree relative(s) plus (a) unilateral vestibular schwannomas at age less than 30 years, or (b) any 2 of the following: meningioma, glioma, schwannoma, or juvenile posterior subcapsular lenticular opacities/juvenile cortical scataract. The criteria for presumptive or probable NF2 was unilateral vestibular schwannomas at age less than 30 years, plus at least 1 of the following: meningioma, glioma, schwannoma, or juvenile posterior subcapsular lenticular opacities/juvenile cortical cataract; or multiple meningiomas (2 or more) plus (a) unilateral vestibular schwannomas at age less than 30 years, or (b) 1 of the following: glioma, schwannoma, or juvenile posterior subcapsular lenticular opacities/juvenile cortical cataract.

Evans et al. (1998) sought mutations in the NF2 gene in 125 families with classic NF2 with bilateral vestibular schwannomas; causative mutations were identified in 52 families. In 5 families, the first affected individual in the family was a mosaic for a disease-causing mutation. Only 1 of the 9 children from the 3 mosaic cases with children were affected. Four of these 9 children inherited the allele associated with the disease-causing mutation yet did not inherit the mutation. NF2 mutations were identified in only 27 of 79 (34%) sporadic cases, compared with 25 of 46 (54%) familial cases (P less than 0.05). In 48 families in which a mutation had not been identified, the index cases had 125 children, of whom only 29 were affected with NF2 and of whom only a further 21 cases would be predicted to be affected by use of life curves. The 50 of 125 (40%) cases is significantly less than the 50% expected eventually to develop NF2 (P less than 0.05). Somatic mosaicism is likely to be a common cause of classic NF2 and may well account for a low detection rate for mutations in sporadic cases. Degrees of gonosomal mosaicism mean that recurrence risks may well be less than 50% in the offspring of the index case when a mutation was not identified in lymphocyte DNA.

Stereotactic radiosurgery is the principal alternative to microsurgical resection for acoustic neuromas. The goals of radiosurgery are the long-term prevention of tumor growth, maintenance of neurologic function, and prevention of new neurologic deficits. Kondziolka et al. (1998) evaluated 162 consecutive patients who underwent radiosurgery for acoustic neuromas between 1987 and 1992, surveying the results between 5 and 10 years after the procedure. Resection had been performed previously in 42 patients; in 13 patients, the tumor represented a recurrence of disease after a previous total resection. The rate of tumor control (with no resection required) was 98%. Radiosurgery was believed to have been successful by all 30 patients who had undergone surgery previously and by 81 (95%) of the 85 who had not. Pitts and Jackler (1998) pointed out that when radiotherapy is considered for a benign, surgically curable tumor in a young patient, the risk of inducing a secondary tumor must be seriously weighed. The risk of intracranial arterial occlusion from external-beam irradiation must also be considered, although there had been no reports of accelerated atherosclerosis after radiosurgery. The anterior inferior cerebellar artery, which is the primary source of blood supply to the lateral pons and upper medulla, lies right next to the surface of acoustic neuromas.

See also Evans et al. (1992); Evans et al. (1992); Gardner and Frazier (1930); Martuza and Ojemann (1982); Nager (1964); Perez Demoura et al. (1969); Rouleau et al. (1987); Rouleau et al. (1987); Siggers et al. (1975)

Arai, E.; Ikeuchi, T.; Karasawa, S.; Tamura, A.; Yamamoto, K.; Kida, M.; Ichimura, K.; Yuasa, Y.; Tonomura, A.: Constitutional translocation t(4;22)(q12;q12.2) associated with neurofibromatosis type 2. Am. J. Med. Genet. 44: 163-167, 1992.

Bianchi, A. B.; Hara, T.; Ramesh, V.; Gao, J.; Klein-Szanto, A. J. P.; Morin, F.; Menon, A. G.; Trofatter, J. A.; Gusella, J. F.; Seizinger, B. R.; Kley, N.: Mutations in transcript isoforms of the neurofibromatosis 2 gene in multiple human tumour types. Nature Genet. 6: 185-192, 1994.

Bianchi, A. B.; Mitsunaga, S.-I.; Cheng, J. Q.; Klein, W. M.; Jhanwar, S. C.; Seizinger, B.; Kley, N.; Klein-Szanto, A. J. P.; Testa, J. R.: High frequency of inactivating mutations in the neurofibromatosis type 2 gene (NF2) in primary malignant mesotheliomas. Proc. Nat. Acad. Sci. 92: 10854-10858, 1995.

Bijlsma, E. K.; Merel, P.; Bosch, D. A.; Westerveld, A.; Delattre, O.; Thomas, G.; Hulsebos, T. J. M.: Analysis of mutations in the SCH gene in schwannomas. Genes Chromosomes Cancer 11: 7-14, 1994.

Bourn, D.; Carter, S. A.; Mason, S.; Evans, D. G. R.; Strachan, T.: Germline mutations in the neurofibromatosis type 2 tumor suppressor gene. Hum. Molec. Genet. 3: 813-816, 1994.

Bouzas, E. A.; Freidlin, V.; Parry, D. M.; Eldridge, R.; Kaiser-Kupfer, M. I.: Lens opacities in neurofibromatosis 2: further significant correlations. Brit. J. Ophthal. 77: 354-357, 1993.

Bouzas, E. A.; Parry, D. M.; Eldridge, R.; Kaiser-Kupfer, M. I.: Visual impairment in patients with neurofibromatosis 2. Neurology 43: 622-623, 1993.

Claudio, J. O.; Malo, D.; Rouleau, G. A.: The mouse neurofibromatosis type 2 gene maps to chromosome 11. Genomics 21: 437-439, 1994.

Eldridge, R.: Central neurofibromatosis with bilateral acoustic neuroma. Adv. Neurol. 29: 57-65, 1981.

Eldridge, R.; Parry, D. M.; Kaiser-Kupfer, M. I.: Neurofibromatosis 2 (NF2): clinical heterogeneity and natural history based on 39 individuals in 9 families and 16 sporadic cases. (Abstract) Am. J. Hum. Genet. 49 (suppl.): 133, 1991.

Evans, D. G. R.; Bourn, D.; Wallace, A.; Ramsden, R. T.; Mitchell, J. D.; Strachan, T.: Diagnostic issues in a family with late onset type 2 neurofibromatosis. J. Med. Genet. 32: 470-474, 1995.

Evans, D. G. R.; Huson, S. M.; Donnai, D.; Neary, W.; Blair, V.; Newton, V.; Harris, R.: A clinical study of type 2 neurofibromatosis. Quart. J. Med. 84: 603-618, 1992.

Evans, D. G. R.; Huson, S. M.; Donnai, D.; Neary, W.; Blair, V.; Newton, V.; Strachan, T.; Harris, R.: A genetic study of type 2 neurofibromatosis in the United Kingdom. II. Guidelines for genetic counselling. J. Med. Genet. 29: 847-852, 1992.

Evans, D. G. R.; Huson, S. M.; Donnai, D.; Neary, W.; Blair, V.; Teare, D.; Newton, V.; Strachan, T.; Ramsden, R.; Harris, R.: A genetic study of type 2 neurofibromatosis in the United Kingdom. I. Prevalence, mutation rate, fitness, and confirmation of maternal transmission effect on severity. J. Med. Genet. 29: 841-846, 1992.

Evans, D. G. R.; Trueman, L.; Wallace, A.; Collins, S.; Strachan, T.: Genotype/phenotype correlations in type 2 neurofibromatosis (NF2): evidence for more severe disease associated with truncating mutations. J. Med. Genet. 35: 450-455, 1998.

Evans, D. G. R.; Wallace, A. J.; Wu, C. L.; Trueman, L.; Ramsden, R. T.; Strachan, T.: Somatic mosaicism: a common cause of classic disease in tumor-prone syndromes? Lessons from type 2 neurofibromatosis. Am. J. Hum. Genet. 63: 727-736, 1998.

Fabricant, R. N.; Todaro, G. J.; Eldridge, R.: Increased levels of a nerve-growth factor cross-reacting protein in 'central' neurofibromatosis. Lancet I: 4-7, 1979.

Feiling, A.; Ward, E.: A familial form of acoustic tumour. Brit. Med. J. 1: 496-497, 1920.

Gardner, W. J.; Frazier, C. H.: Hereditary bilateral acoustic tumors. J. Hered. 22: 7-8, 1933.

Gardner, W. J.; Frazier, C. H.: Bilateral acoustic neurofibromas. A clinical study and survey of a family of five generations with bilateral deafness in 38 members. Arch. Neurol. Psychiat. 23: 266-302, 1930.

Gardner, W. J.; Turner, O.: Bilateral acoustic neurofibromas: further clinical and pathologic data on hereditary deafness and Recklinghausen's disease. Arch. Neurol. Psychiat. 44: 76-99, 1940.

Gutmann, D.; Aylsworth, A.; Carley, J.; Korf, B.; Marks, J.; Pyeritz, R.; Rubenstein, A.; et al.: The diagnostic evaluation and multidisciplinary management of neurofibromatosis 1 and neurofibromatosis 2. J.A.M.A. 278: 51-57, 1997.

Honda, M.; Arai, E.; Sawada, S.; Ohta, A.; Niimura, M.: Neurofibromatosis 2 and neurilemmomatosis gene are identical. J. Invest. Derm. 104: 74-77, 1995.

Jacoby, L. B.; Jones, D.; Davis, K.; Kronn, D.; Short, M. P.; Gusella, J.; MacCollin, M.: Molecular analysis of the NF2 tumor-suppressor gene in schwannomatosis. Am. J. Hum. Genet. 61: 1293-1302, 1997.

Kaiser-Kupfer, M. I.; Freidlin, V.; Datiles, M. B.; Edwards, P. A.; Sherman, J. L.; Parry, D.; McCain, L. M.; Eldridge, R.: The association of posterior capsular lens opacities with bilateral acoustic neuromas in patients with neurofibromatosis type 2. Arch. Ophthal. 107: 541-544, 1989.

Kanter, W. R.; Eldridge, R.; Fabricant, R.; Allen, J. C.; Koerber, T.: Central neurofibromatosis with bilateral acoustic neuroma: genetic, clinical and biochemical distinctions from peripheral neurofibromatosis. Neurology 30: 851-859, 1980.

Kehrer-Sawatzki, H.; Udart, M.; Krone, W.; Baden, R.; Fahsold, R.; Thomas, G.; Schmucker, B.; Assum, G.: Mutational analysis and expression studies of the neurofibromatosis type 2 (NF2) gene in a patient with a ring chromosome 22 and NF2. Hum. Genet. 100: 67-74, 1997.

Kluwe, L.; Mautner, V.-F.: A missense mutation in the NF2 gene results in moderate and mild clinical phenotypes of neurofibromatosis type 2. Hum. Genet. 97: 224-227, 1996.

Knudson, A.: Asbestos and mesothelioma: genetic lessons from a tragedy. Proc. Nat. Acad. Sci. 92: 10819-10820, 1995.

Kondziolka, D.; Lunsford, L. D.; McLaughlin, M. R.; Flickinger, J. C.: Long-term outcomes after radiosurgery for acoustic neuromas. New Eng. J. Med. 339: 1426-1433, 1998.

Landau, K.; Dossetor, F. M.; Hoyt, W. F.; Muci-Mendoza, R.: Retinal hamartoma in neurofibromatosis 2. Arch. Ophthal. 108: 328-329, 1990.

Lee, D. K.; Abbott, M. L.: Familial central nervous system neoplasia: case report of a family with von Recklinghausen's neurofibromatosis. Arch. Neurol. 20: 154-160, 1969.

MacCollin, M.; Ramesh, V.; Jacoby, L. B.; Louis, D. N.; Rubio, M.-P.; Pulaski, K.; Trofatter, J. A.; Short, M. P.; Bove, C.; Eldridge, R.; Parry, D. M.; Gusella, J. F.: Mutational analysis of patients with neurofibromatosis 2. Am. J. Hum. Genet. 55: 314-320, 1994.

Martuza, R. L.; Eldridge, R.: Neurofibromatosis 2 (bilateral acoustic neurofibromatosis). New Eng. J. Med. 318: 684-688, 1988.

Martuza, R. L.; Ojemann, R. G.: Bilateral acoustic neuromas: clinical aspects, pathogenesis and treatment. Neurosurgery 10: 1-12, 1982.

Mayfrank, L.; Wullich, B.; Wolff, G.; Finke, J.; Gouzoulis, E.; Gilsbach, J. M.: Neurofibromatosis 2: a clinically and genetically heterogeneous disease? Report on 10 sporadic cases. Clin. Genet. 38: 362-370, 1990.

Merel, P.; Hoang-Xuan, K.; Sanson, M.; Moreau-Aubry, A.; Bijlsma, E. K.; Lazaro, C.; Moisan, J. P.; Resche, F.; Nishisho, I.; Estivill, X.; Delattre, J. Y.; Poisson, M.; Theillet, C.; Hulsebos, T.; Delattre, O.; Thomas, G.: Predominant occurrence of somatic mutations of the NF2 gene in meningiomas and schwannomas. Genes Chromosomes Cancer 13: 211-216, 1995.

Moyes, P. D.: Familial bilateral acoustic neuroma affecting 14 members from four generations. J. Neurosurg. 29: 78-82, 1968.

Mrazek, J.; Fiser, Z.; Mrazkova, D.: Diagnosis, size, and operation results in 41 acoustic neurinomas. Neoplasma 35: 467-474, 1988.

Myers, R. M.; Fischer, S. G.; Lerman, L. S.; Maniatis, T.: Nearly all single base substitutions in DNA fragments joined to a GC-clamp can be detected by denaturing gradient gel electrophoresis. Nucleic Acids Res. 13: 3131-3145, 1985.

Nager, G. T.: Association of bilateral VIIIth nerve tumors with meningioma in von Recklinghausen's disease. Laryngoscope 74: 1220-1261, 1964.

Nager, G. T.: Acoustic neuromas: pathology and differential diagnosis. Arch. Otolaryng. 89: 252-279, 1969.

Narod, S. A.; Parry, D. M.; Parboosingh, J.; Lenoir, G. M.; Ruttledge, M.; Fischer, G.; Eldridge, R.; Martuza, R. L.; Frontali, M.; Haines, J.; Gusella, J. F.; Rouleau, G. A.: Neurofibromatosis type 2 appears to be a genetically homogeneous disease. Am. J. Hum. Genet. 51: 486-496, 1992.

NIH Consensus Development Conference: Neurofibromatosis: conference statement. Arch. Neurol. 45: 475-578, 1988.

Niimura, M.: Neurofibromatosis (3). Rinsho Derma (Tokyo) 15: 653-663, 1973.

Papi, L.; De Vitis, L. R.; Vitelli, F.; Ammannati, F.; Mennonna, P.; Montali, E.; Bigozzi, U.: Somatic mutations in the neurofibromatosis type 2 gene in sporadic meningiomas. Hum. Genet. 95: 347-351, 1995.

Parry, D. M.; Eldridge, R.; Kaiser-Kupfer, M. I.; Bouzas, E. A.; Pikus, A.; Patronas, N.: Neurofibromatosis 2 (NF2): clinical characteristics of 63 affected individuals and clinical evidence for heterogeneity. Am. J. Med. Genet. 52: 450-461, 1994.

Parry, D. M.; Kaiser-Kupfer, M. I.; Eldridge, R.: Neurofibromatosis 2 (NF2): lens findings in 40 patients in 5 high risk groups. (Abstract) Am. J. Hum. Genet. 49 (suppl.): 155, 1991.

Parry, D. M.; MacCollin, M. M.; Kaiser-Kupfer, M. I.; Pulaski, K.; Nicholson, H. S.; Bolesta, M.; Eldridge, R.; Gusella, J. F.: Germ-line mutations in the neurofibromatosis 2 gene: correlations with disease severity and retinal abnormalities. Am. J. Hum. Genet. 59: 529-539, 1996.

Pastores, G. M.; Michels, V. V.; Jack, C. R., Jr.: Early childhood diagnosis of acoustic neuromas in presymptomatic individuals at risk for neurofibromatosis 2. Am. J. Med. Genet. 41: 325-329, 1991.

Pearson-Webb, M. A.; Kaiser-Kupfer, M. I.; Eldridge, R.: Eye findings in bilateral acoustic (central) neurofibromatosis: association with presenile lens opacities and cataracts but absence of Lisch nodules. (Letter) New Eng. J. Med. 315: 1553-1554, 1986.

Perez Demoura, L. F.; Hayden, R. C., Jr.; Conner, G. H.: Bilateral acoustic neurinoma and neurofibromatosis. Arch. Otolaryng. 90: 28-34, 1969.

Pitts, L. H.; Jackler, R. K.: Treatment of acoustic neuromas. (Editorial) New Eng. J. Med. 339: 1471-1473, 1998.
Pulst, S.-M.; Riccardi, V. M.; Fain, P.; Korenberg, J. R.: Familial spinal neurofibromatosis: clinical and DNA linkage analysis. Neurology 41: 1923-1927, 1991.
Ragge, N. K.; Baser, M. E.; Klein, J.; Nechiporuk, A.; Sainz, J.; Pulst, S.-M.; Riccardi, V. M.: Ocular abnormalities in neurofibromatosis 2. Am. J. Ophthal. 120: 634-641, 1995.
Rouleau, G. A.; Merel, P.; Lutchman, M.; Sanson, M.; Zucman, J.; Marineau, C.; Hoang-Xuan, K.; Demczuk, S.; Desmaze, C.; Plougastel, B.; Pulst, S. M.; Lenoir, G.; Bijlsma, E.; Fashold, R.; Dumanski, J.; de Jong, P.; Parry, D.; Eldrige, R.; Aurias, A.; Delattre, O.; Thomas, G.: Alteration in a new gene encoding a putative membrane-organizing protein causes neuro-fibromatosis type 2. Nature 363: 515-521, 1993.
Rouleau, G. A.; Seizinger, B. R.; Wertelecki, W.; Haines, J. L.; Superneau, D. W.; Martuza, R. L.; Gusella, J. F.: Flanking markers bracket the neurofibromatosis type 2 (NF2) gene on chromosome 22. Am. J. Hum. Genet. 46: 323-328, 1990.
Rouleau, G. A.; Wertelecki, W.; Haines, J. L.; Hobbs, W. J.; Trofatter, J. A.; Seizinger, B.; Martuza, R. L.; Superneau, D. W.; Conneally, P. M.; Gusella, J. F.: Genetic linkage of bilateral acoustic neurofibromatosis to DNA markers on chromosome 22. (Abstract) Cytogenet. Cell Genet. 46: 684-685, 1987.
Rouleau, G. A.; Wertelecki, W.; Haines, J. L.; Hobbs, W. J.; Trofatter, J. A.; Seizinger, B. R.; Martuza, R. L.; Superneau, D. W.; Conneally, P. M.; Gusella, J. F.: Genetic linkage of bilateral acoustic neurofibromatosis to a DNA marker on chromosome 22. Nature 329: 246-248, 1987.
Ruttledge, M. H.; Andermann, A. A.; Phelan, C. M.; Claudio, J. O.; Han, F.; Chretien, N.; Rangaratnam, S.; MacCollin, M.; Short, P.; Parry, D.; Michels, V.; Riccardi, V. M.; Weksberg, R.; Kitamura, K.; Bradburn, J. M.; Hall, B. D.; Propping, P.; Rouleau, G. A.: Type of mutation in the neurofibromatosis type 2 gene (NF2) frequently determines severity of disease. Am. J. Hum. Genet. 59: 331-342, 1996.
Ruttledge, M. H.; Narod, S. A.; Dumanski, J. P.; Parry, D. M.; Eldridge, R.; Wertelecki, W.; Parboosingh, J.; Faucher, M.-C.; Lenoir, G. M.; Collins, V. P.; Nordenskjold, M.; Rouleau, G. A.: Presymptomatic diagnosis for neurofibromatosis 2 with chromosome 22 markers. Neurology 43: 1753-1760, 1993.
Ruttledge, M. H.; Sarrazin, J.; Rangaratnam, S.; Phelan, C. M.; Twist, E.; Merel, P.; Delattre, O.; Thomas, G.; Nordenskjold, M.; Collins, V. P.; Dumanski, J. P.; Rouleau, G. A.: Evidence for the complete inactivation of the NF2 gene in the majority of sporadic meningiomas. Nature Genet. 6: 180-184, 1994.
Sainz, J.; Huynh, D. P.; Figueroa, K.; Ragge, N. K.; Baser, M. E.; Pulst, S.-M.: Mutations of the neurofibromatosis type 2 gene and lack of the gene product in vestibular schwannomas. Hum. Molec. Genet. 3: 885-891, 1994.
Scoles, D. R.; Baser, M. E.; Pulst, S.-M.: A missense mutation in the neurofibromatosis 2 gene occurs in patients with mild and severe phenotypes. Neurology 47: 544-546, 1996.
Seizinger, B. R.; Martuza, R. L.; Gusella, J. F.: Loss of genes on chromosome 22 in tumorigenesis of human acoustic neuroma. Nature 322: 644-647, 1986.
Seizinger, B. R.; Rouleau, G.; Ozelius, L. J.; Lane, A. H.; St. George-Hyslop, P.; Huson, S.; Gusella, J. F.; Martuza, R. L.: Common pathogenetic mechanism for three tumor types in bilateral acoustic neurofibromatosis. Science 236: 317-319, 1987.
Siggers, D. C.; Boyer, S. H.; Eldridge, R.: Nerve-growth factor in disseminated neurofibromatosis. (Letter) New Eng. J. Med. 292: 1134, 1975.
Thomas, G.: Personal Communication. Paris, France 11/17/1993.
Trofatter, J. A.; MacCollin, M. M.; Rutter, J. L.; Murrell, J. R.; Duyao, M. P.; Parry, D. M.; Eldridge, R.; Kley, N.; Menon, A. G.; Pulaski, K.; Haase, V. H.; Ambrose, C. M.; Munroe, D.; Bove, C.; Haines, J. L.; Martuza, R. L.; MacDonald, M. E.; Seizinger, B. R.; Short, M. P.; Buckler, A. J.; Gusella, J. F.: A novel moesin-, ezrin-, radixin-like gene is a candidate for the neurofibromatosis 2 tumor suppressor. Cell 72: 791-800, 1993.
Watson, C. J.; Gaunt, L.; Evans, G.; Patel, K.; Harris, R.; Strachan, T.: A disease-associated germline deletion maps the type 2 neurofibromatosis (NF2) gene between the Ewing sarcoma region and the leukaemia inhibitory factor locus. Hum. Molec. Genet. 2: 701-704, 1993.
Wellenreuther, R.; Kraus, J. A.; Lenartz, D.; Menon, A. G.; Schramm, J.; Louis, D. N.; Ramesh, V.; Gusella, J. F.; Wiestler, O. D.; von Deimling, A.: Analysis of the neurofibromatosis 2 gene reveals molecular variants of meningioma. Am. J. Path. 146: 827-832, 1995.
Wertelecki, W.; Rouleau, G. A.; Superneau, D. W.; Forehand, L. W.; Williams, J. P.; Haines, J. L.; Gusella, J. F.: Neurofibromatosis 2: clinical and DNA linkage studies of a large kindred. New Eng. J. Med. 319: 278-283, 1988.
Wishart, J. H.: Case of tumours in the skull, dura mater, and brain. Edinburgh Med. Surg. J. 18: 393-397, 1822.
Wolff, R. K.; Frazer, K. A.; Jackler, R. K.; Lanser, M. J.; Pitts, L. H.; Cox, D. R.: Analysis of chromosome 22 deletions in neurofibromatosis type 2-related tumors. Am. J. Hum. Genet. 51: 478-485, 1992.
Worster-Drought, C.; Dickson, W. E. C.; McMenemey, W. H.: Multiple meningeal and perineural tumors with analogous changes in the glia and ependyma (neurofibroblastomatosis). Brain 60: 85-117, 1937.
Young, D. F.; Eldridge, R.; Gardner, W. J.: Bilateral acoustic neuroma in a large kindred. J.A.M.A. 214: 347-353, 1970.

SCHWANNOMATOSIS (162091)

JJM NOTE

When multiple cutaneous neurilemmomas are seen, especially with a loss of auditory or vestibular function, neurofibromatosis 2 must be ruled out. If the patient is old enough and the MRI is negative, this rare disorder could be considered.

Neurilemmomatosis or schwannomatosis has been reported mainly from Japan. Sasaki and Nakajima (1992) described multiple cutaneous neurilemmomas in an 8-year-old Japanese girl and a 5-year-old Japanese boy. The boy's father had bilateral acoustic neuromas and also had multiple skin tumors that were diagnosed as neurilemmomas histopathologically. There were no pigmented macules. The multiple skin tumors had been present in both children since birth. The relationship of neurilemmomatosis to NF2 (101000) is not clear.

Neurilemmomatosis, first reported by Niimura (1973) as neurofibromatosis type 3, is characterized by multiple cutaneous neurilemmomas and spinal schwannomas, without acoustic tumors or other signs of NF1 or NF2. In neurilemmomas, the tumor consists of Schwann cells. Honda et al. (1995) analyzed the peripheral leukocytes and tissue from cutaneous neurilemmomas of 7 patients with neurilemmomatosis using DNA markers for different regions of chromosome 22. They detected allele losses in 3 of 7 tumors from 7 patients with a probe for the NF2 region and the germline mutations in 2 of 3 tumors from the same 3 patients. They described 2 mutations in the NF2 gene (101000.0017; 101000.0018). They concluded that neurilemmomatosis is a form of NF2.

Schwannomas are benign tumors of the peripheral nerve sheath that usually occur singly in otherwise normal individuals. Multiple schwannomas in the same individual suggest an underlying tumor-predisposition syndrome. The most common such syndrome is neurofibromatosis 2. The hallmark of NF2 is the development of bilateral vestibular nerve schwannomas; but two-thirds or more of all NF2-affected individuals develop schwannomas in other locations, and dermal schwannomas (or neurilemmonmas) may precede vestibular tumors in NF2-affected children (Evans et al., 1992; Mautner et al., 1993; Parry et al., 1994). MacCollin et al. (1996) reviewed reports of individuals with multiple schwannomas who do not show evidence of vestibular schwannomas, and suggested that schwannomatosis is a clinical entity distinct from other forms of neurofibromatosis. Jacoby et al. (1997) undertook a molecular genetic investigation of the relationship of schwannomatosis to NF2. They examined the NF2 locus in 20 unrelated schwannomatosis patients and their affected relatives. Tumors from these patients frequently harbored typical truncating mutations of the NF2 gene and loss of heterozygosity of the surrounding region of chromosome 22. Surprisingly, unlike patients with NF2, no heterozygous NF2 gene changes were seen in normal tissues. Examination of multiple tumors from the same patient revealed that some schwannomatosis patients were somatic mosaics for NF2 gene changes. In contrast, other individuals, particularly those with a positive family history, appeared to have an inherited predisposition to formation of tumors that carry somatic alterations of the NF2 gene.

The criteria used by Jacoby et al. (1997) for schwannomatosis were as follows: 2 or more pathologically proved schwannomas and lack of radiographic evidence of vestibular nerve tumor at age more than 18 years was taken as evidence of definite schwannomatosis. For presumptive or probable schwannomatosis the criteria were 2 or more pathologically proved schwannomas, without symptoms of eighth-nerve dysfunction at an age of more than 30 years; or 2 or more pathologically proved schwannomas in an anatomically limited distribution (single limb or segment of the spine), without symptoms of eighth-nerve dysfunction at any age. In the study of Jacoby et al. (1997), the findings suggested an inherited tendency to formation of tumors that carry somatic alteration of the NF2 gene. This tendency was biased toward the occurrence of different mutations in the same, coinherited allele within a given family, combined with loss of the trans allele in any given individual. Although the family data were consistent with linkage of this trait to the NF2 locus, the studies implied that the primary event in the tumors lay outside the coding region of the NF2 gene.

Evans, D. G. R.; Huson, S. M.; Donnai, D.; Neary, W.; Blair, V.; Newton, V.; Harris, R.: A clinical study of type 2 neurofibromatosis. Quart. J. Med. 84: 603-618, 1992.
Honda, M.; Arai, E.; Sawada, S.; Ohta, A.; Niimura, M.: Neurofibromatosis 2 and neurilemmomatosis gene are identical. J. Invest. Derm. 104: 74-77, 1995.
Jacoby, L. B.; Jones, D.; Davis, K.; Kronn, D.; Short, M. P.; Gusella, J.; MacCollin, M.: Molecular analysis of the NF2 tumor-suppressor gene in schwannomatosis. Am. J. Hum. Genet. 61: 1293-1302, 1997.
MacCollin, M.; Woodfin, W.; Kronn, D.; Short, M. P.: Schwannomatosis: a clinical and pathologic study. Neurology 46: 1072-1079, 1996.
Mautner, V. F.; Tatagiba, M.; Guthoff, R.; Samii, M.; Pulst, S. M.: Neurofibromatosis 2 in the pediatric age group. Neurosurgery 33: 92-96, 1993.
Niimura, M.: Neurofibromatosis. Rinsho Derma 15: 653-663, 1973.
Parry, D. M.; Eldridge, R.; Kaiser-Kupfer, M. I.; Bouzas, E. A.; Pikus, A.; Patronas, N.: Neurofibromatosis 2 (NF2): clinical characteristics of 63 affected individuals and clinical evidence for heterogeneity. Am. J. Med. Genet. 52: 450-461, 1994.

Sasaki, T.; Nakajima, H.: Congenital neurilemmomatosis. J. Am. Acad. Derm. 26: 786-787, 1992.

NEUROFIBROMATOSIS, FAMILIAL SPINAL (#162210)

JJM NOTE

The upper spine is the common battleground of neurofibromatosis 1 and neurofibromatosis 2; that is, similar spinal neurofibromas can appear in both diseases. The two types can be distinguished based on ophthalmic examination, other signs, or, as in this report, molecular testing. A family, linked to neurofibromatosis 1 with a lod score of 1.7, met NIH criteria for neurofibromatosis 1, but lacked Lisch nodules and, hence, was considered distinct from but allelic to neurofibromatosis 1 (Poyhonen M et al.: Hereditary spinal neurofibromatosis: a rare form of NF1? J. Med. Genet. 34:184–87, 1997).

Pulst et al. (1991) studied 2 families with spinal neurofibromatosis. The first family had cafe-au-lait spots (CLS); the second family had no CLS. Other signs of NF1 (162200) or NF2 (101000), such as cutaneous tumors, Lisch nodules or acoustic tumors, were absent in both families. The inheritance pattern was consistent with autosomal dominant inheritance, with at least 1 instance of male-to-male transmission in both. Using genetic linkage analysis with DNA markers tightly linked to the NF1 and NF2 loci, Pulst et al. (1991) determined that the likely location for the mutation in the first family was in the NF1 gene with odds of 97:1, whereas the mutation in the second family was excluded from the NF1 locus with odds of more than 100,000:1. It seems quite possible that the second family in fact was a variant of NF2; markers for the NF2 locus were uninformative in the family.

Pulst, S.-M.; Riccardi, V. M.; Fain, P.; Korenberg, J. R.: Familial spinal neurofibromatosis: clinical and DNA linkage analysis. Neurology 41: 1923-1927, 1991.

NEUROFIBROMATOSIS, TYPE 3 OF RICCARDI (162260)

JJM NOTE

My current view is that *any* neurofibromatosis-like syndrome with acoustic neuroma (vestibular schwannoma) must be assumed to be neurofibromatosis 2 until proven otherwise (e.g., by molecular assays).

Riccardi and Eichner (1986) designated the classic von Recklinghausen disease (162200) as NF-I and familial acoustic neuromas (101000) as NF-II. They suggested the existence of an entity they labeled NF-III that combines features of both with some additional distinctive features. Cafe-au-lait spots (CLS), freckling, and cutaneous neurofibromas occur as in NF-I, although typically the CLS are few in number and pale and may be relatively large, and cutaneous neurofibromas are few in number. The palm of the hands is a characteristic site for cutaneous neurofibromas in NF-III. Iris Lisch nodules do not occur. Bilateral acoustic neuromas, posterior fossa and upper cervical meningiomas, and spinal/paraspinal neurofibromas are the predominant features of NF-III, but optic gliomas have not been seen. The CNS tumors make their presence known usually by the second or early third decade and develop rapidly thereafter. The natural history and therefore the prognosis are different from that of classic von Recklinghausen disease. A relatively rapid and fatal course from multiple CNS tumors is usual.

Nomenclature: Cohen (1997) suggested that the intestinal type of neurofibromatosis, type 3 (162220), should be designated as NF3, intestinal type (or NF3B), whereas the type 3 neurofibromatosis discussed here should be referred to as neurofibromatosis, type 3, Riccardi type (or NF3A).

See also Martuza and Ojemann (1982); Rodriguez and Berthrong (1966)

Cohen, M. M., Jr.: Personal Communication. Halifax, Nova Scotia, Can. 12/18/1997.
Martuza, R. L.; Ojemann, R. G.: Bilateral acoustic neuromas: clinical aspects, pathogenesis, and treatment. Neurosurgery 10: 1-12, 1982.
Riccardi, V. M.; Eichner, J. E.: Neurofibromatosis: Phenotype, Natural History and Pathogenesis. Baltimore: Johns Hopkins Univ. Press (pub.) 1986.
Rodriguez, H. A.; Berthrong, M.: Multiple primary intracranial tumors in von Recklinghausen's neurofibromatosis. Arch. Neurol. 14: 467-475, 1966.

NEUROFIBROMATOSIS, TYPE 4 OF RICCARDI (162270)

JJM NOTE

The rationale for this subgroup is weakening with the molecular characterization. The burden is on the investigator to prove that individuals with this clinical variant lack mutations in the genes for neurofibromatosis 1 and neurofibromatosis 2.

Riccardi (1982) described cases of neurofibromatosis that are sufficiently variant that they seem to warrant separation from the classic von Recklinghausen NF-I (162200), the acoustic neuroma type, NF-II (101000), and the mixed type, NF-III (162260). The group still is undoubtedly heterogeneous. Iris Lisch nodules, one of the most specific features of NF-I, are usually absent in NF-IV. The importance of a separate category for these cases is related to the probable difference in prognosis and genetic counseling and the desirability of avoiding confusion of studies of the natural history and pathogenesis of NF-I.

See also Riccardi and Eichner (1986)

Riccardi, V. M.: Neurofibromatosis: clinical heterogeneity. Curr. Probl. Cancer 7(2): 1-34, 1982.
Riccardi, V. M.; Eichner, J. E.: Neurofibromatosis: Phenotype, Natural History and Pathogenesis. Baltimore: Johns Hopkins Univ. Press (pub.) 1986.

NEUROFIBROMATOSIS, FAMILIAL INTESTINAL (162220)

JJM NOTE

This variant must be considered tentative. Linkage analysis or mutation detection on the family should first rule out neurofibromatosis 1 and neurofibromatosis 2.

Neurofibromata of the intestine are a recognized though rare feature of von Recklinghausen neurofibromatosis (see photograph in Manley and Skyring, 1961). Sporadic cases of multiple intestinal neurofibromatosis without cutaneous features of von Recklinghausen disease have been observed. Hashemian (1952) described 'familial fibromatosis' of the small intestine, but at least 2 of the 3 cases had other stigmata of von Recklinghausen neurofibromatosis. Lipton and Zuckerbrod (1966) described familial intestinal neurofibromatosis (without other features of von Recklinghausen disease). Verhest et al. (1981) described a Belgian family in which affected persons appeared to have an interstitial inversion of 12q21-12q24.2. Full description of this family (Heimann et al., 1988) indicated that onset of symptoms was in adulthood and that some gene carriers remained asymptomatic into middle and late adulthood. Four females (3 sisters and a paternal aunt) were symptomatic; a male presumed gene carrier was asymptomatic. Restudy of the family (Verhest et al., 1988) led to a reinterpretation of the chromosomal aberration to be a reciprocal translocation t(12;14)(q13;q13). The translocation was present in all the persons thought to have intestinal neurofibromatosis or to carry the determinant for the disorder, except for 1 of the 3 affected sisters. Thus, the translocation may be a fortuitous finding in this family.

Nomenclature: Cohen (1997) suggested that the intestinal type of neurofibromatosis, which has been designated NF3 in the past, should be called NF, type 3, intestinal type to distinguish it from neurofibromatosis, type 3, Riccardi type (162260). These are additionally given the symbols NF3B for the intestinal type and NF3A for the Riccardi type.

Cohen, M. M., Jr.: Personal Communication. Halifax, Nova Scotia, Can. 12/18/1997.
Hashemian, H.: Familial fibromatosis of small intestine. Brit. J. Surg. 40: 346-350, 1952.
Heimann, R.; Verhest, A.; Verschraegen, J.; Grosjean, W.; Draps, J. P.; Hecht, F.: Hereditary intestinal neurofibromatosis. I. A distinctive genetic disease. Neurofibromatosis 1: 26-32, 1988.
Lipton, S.; Zuckerbrod, M.: Familial enteric neurofibromatosis. Med. Times 94: 544-548, 1966.
Manley, K. A.; Skyring, A. P.: Some heritable causes of gastrointestinal disease: special reference to hemorrhage. Arch. Intern. Med. 107: 182-203, 1961.
Verhest, A.; Verschraegen, J.; Grosjean, W.; Draps, J. P.; Vamos, E.; Heimann, R.; Hecht, F.: Hereditary intestinal neurofibromatosis. II. Translocation between chromosomes 12 and 14. Neurofibromatosis 1: 33-36, 1988.
Verhest, A.; Verschraegen, J.; Grosjean, W.; Heimann, R.: Transmissible chromosome abnormality in familial intestinal neurofibromatosis. (Abstract) Sixth Int. Cong. Hum. Genet., Jerusalem 176 only, 1981.

NEUROFIBROMATOSIS-PHEOCHROMOCYTOMA-DUODENAL CARCINOID SYNDROME (162240)

JJM NOTE

Given the emerging but rare association of intestinal carcinoid with neurofibromatosis 1, the hypothesis should be addressed that, in fact, this entity is simply a variant of neurofibromatosis 1.

From their own experience and from the literature, Griffiths et al. (1983) collected 3 patients with duodenal carcinoid tumor in association with neurofibromatosis and pheochromocytoma and 4 patients with duodenal carcinoid with either von Recklinghausen disease or pheochromocytoma. The duodenal carcinoids had an unusual morphology. They reacted only weakly to normal silver impregnation techniques and had an unusual glandular pattern, and 3 of them contained psammoma bodies. Three were strongly positive on immunolocalization with an antibody to somatostatin (which stains normal pancreatic D cells). Griffiths et al. (1983) suggested that neurofibromatosis, pheochromocytoma, and duodenal carcinoid constitute a distinct and specific multiple endocrine neoplasia syndrome. They suggested that it might be termed MEN IIIa, with MEN IIIb being assigned to the von Hippel-Lindau syndrome (193300) with pheochromocytoma and islet cell tumors. In only 1 of the cases was the neurofibromatosis familial. It appears that the cutaneous involvement was typical of von Recklinghausen disease (162200) in all the cases. Obstructive jaundice (due to the duodenal carcinoid) was the manner of clinical presentation in 2 of the cases. Symptoms of somatostatin secretion by the carcinoid were not observed.

Griffiths, D. F. R.; Williams, G. T.; Williams, E. D.: Multiple endocrine neoplasia associated with von Recklinghausen's disease. Brit. Med. J. 287: 1341-1343, 1983.

Other Classic Phacomatoses

MCCUNE-ALBRIGHT SYNDROME (#174800)

JJM NOTE

This entity is included because of its superficial overlap with neurofibromatosis 1 and the spectrum of associated tumor types: pituitary adenoma, thyroid neoplasm, multinodular adrenal disease, and sarcomatous degeneration.

Description

A number sign (#) is used with this entry because this phenotype is associated with mutations in the GNAS1 gene (139320).

The McCune-Albright syndrome, usually caused by mosaicism for a mutation in the GNAS1 gene, is characterized by polyostotic fibrous dysplasia, pigment patches of the skin and endocrinologic abnormalities, including precocious puberty, thyrotoxicosis, pituitary gigantism, and Cushing syndrome.

Clinical Features

This disorder is called McCune-Albright syndrome or simply Albright syndrome, but should not be confused with Albright hereditary osteodystrophy, or pseudohypoparathyroidism (103580). The predominant features of the syndrome occur in 3 areas: the bony skeleton, the skin, and the endocrine system. In all 3 systems, the extent of the abnormality and, in the case of the endocrine system, the nature of the abnormality, are highly variable from case to case, depending on the specific tissues involved in the mosaicism and the extent of involvement.

Skeletal

No bone is spared. There is a strong tendency to asymmetry. Involvement of the skull and facial bones can be striking, and in the case of these bones also, asymmetry is the rule. Pathologic fracture or bone deformity may be presenting manifestations and pseudoarthrosis occurs frequently. Deafness and blindness can result from impingement of the bony process on the cranial foramina. Shepherd's crook deformity of the proximal femur is particularly characteristic of the bony involvement. (The bone lesions of neurofibromatosis are usually less extensive than are those in polyostotic fibrous dysplasia, but may be difficult to distinguish on radiologic grounds alone.)

Hypophosphatemic osteomalacia ('rickets') has been observed in some cases of polyostotic fibrous dysplasia. Dent and Gertner (1976) suggested that this may represent a situation comparable to 'tumor rickets' which is associated with mesenchymal tumors and regresses when the tumor is removed. McArthur et al. (1979) described 4 patients with Albright syndrome, hypophosphatemia, and inappropriately low renal tubular reabsorption of phosphate. Three of the patients had radiologic evidence of rickets. They postulated that a substance elaborated by the dysplastic bone interfered with phosphate reabsorption in the renal tubule.

Kaplan et al. (1988) reported the case of a 31-year-old woman with this disorder in whom bone lesions progressed rapidly during pregnancy. A study of these lesions showed the presence of estrogen and progesterone receptors in osteogenic cells. Whether estrogen receptors are unique to the McCune-Albright syndrome or, alternatively, a property of any activated, dedifferentiated, or neoplastic bone cells is unclear.

Viljoen et al. (1988) described a severely affected man, aged about 30, with massive craniofacial hyperostosis. They raised the question of whether this represents the severe end of the scale for polyostotic fibrous dysplasia, or whether it is a distinct entity. McKusick is of the opinion that it is a severe expression of polyostotic fibrous dysplasia; a patient of his (M.G., JHH 276369) has severe craniofacial hyperostosis (see 139320.0009 for the molecular lesion in the GNAS1 gene in this patient). He has other, more characteristic features of polyostotic fibrous dysplasia, including cafe-au-lait pigmentation, which was absent in the case of Viljoen et al. (1988). There are other definite cases of patients with very severe craniofacial involvement (Nager et al., 1982; Nager and Holliday, 1984). Such is also evident from inspection of the descriptions in the classic monograph of Albright and Reifenstein (1948). For example, they produced a picture (Fig. 145) of a skeleton thought by von Recklinghausen (1891) to represent hyperparathyroidism (osteitis fibrosa generalisata of von Recklinghausen). That this was severe Albright polyostotic fibrous dysplasia was suggested by the shepherd's crook deformity of the femora and the asymmetric bulge of the occipital area of the skull. Taconis (1988) reported 2 cases of osteosarcomatous transformation, 1 in the skull of a 25-year-old man and 1 in the iliac bone of a 38-year-old man.

Taconis (1988) stated that the incidence of malignancy of the bone lesions may be lower than previously suggested. Malignant transformation occurs in both monostotic and polyostotic disease but is more common in the latter form. Although radiation therapy is often an inciting cause, this was not the case in either of these patients, who had received no radiation therapy.

Skin

The involvement of the skin consists predominantly of large cafe-au-lait spots with irregular margins, giving them a 'coast of Maine' configuration as opposed to the more regularly outlined 'coast of California' cafe-au-lait spots of neurofibromatosis (162200). Like the bony lesions, the pigmentary lesions of the skin may be limited predominantly to one side and stop sharply at the midline. The nape of the neck is a commonly involved site. (The cafe-au-lait spots of neurofibromatosis are for the most part smaller, more regular, and more uniformly distributed, and are accompanied by axillary freckling and usually by skin tumors.)

Endocrine

The leading endocrinologic feature is precocious puberty, which occurs in over half of female cases. Menstruation may occur in the first months of life. Precocious puberty has been reported also in males with this syndrome; testicular biopsy has revealed the full process of spermatogenesis with mature sperm in patients as young as 6 years (Benedict, 1966; Giovannelli et al., 1978). Probably the second most common endocrinopathy is hyperthyroidism (Lichtenstein and Jaffe, 1942; Hamilton and Maloof, 1973). Excessive secretion of growth hormone with gigantism, hyperadrenocorticism with Cushing syndrome, and gynecomastia have been observed. In specific instances, it has been demonstrated that the Cushing syndrome is due to multinodular change in the adrenals and the hyperthyroidism to multinodular toxic goiter (Hamilton and Maloof, 1973).

DiGeorge (1975) reviewed unusual features of the syndrome and the evidence that the endocrinopathy represents autonomous function of the endocrine glands. Two main hypotheses had been stated: (1) excessive secretion of the hypothalamic releasing hormones is involved in the endocrinopathy of this disorder (Hall and Warrick, 1972); and (2) this disorder represents multiple, circumscribed embryonic alterations in a variety of tissues resulting from clones of cells characterized by autonomous behavior, and perhaps aberrant behavior, toward otherwise normal stimuli. The latter hypothesis appeared to be the more compatible with the skeletal, cutaneous, and endocrinologic features and with the mosaic hypothesis of Happle (1986); its validity was proved by demonstration of mosaicism for mutations in the GNAS1 gene.

Majzoub and Scully (1993) described a 6-year-old boy who had apocrine sweat, facial acne, Tanner stage 2 pubic hair, and midpubertalsized testes and penis, all indicative of precocious puberty, in association with osseous changes of this disorder.

We have observed a patient (M.G., JHH 276369), aged 50 years, with a large syrinx (syringomyelia) causing neurologic manifestations in the arms and legs. Because of a linear midline pigmentation in the high cervical area posteriorly, in the same area of the syrinx, the possibility can be raised that the syrinx, which extended from C2 to T11, was a primary manifestation of the syndrome. The patient also had extensive involvement of the base of the skull with Arnold-Chiari syndrome (i.e., extension of cerebellar tissue through the foramen magnum). He had had hyperthyroidism, as reported by Hamilton and Maloof (1973). (The same patient had severe craniofacial hyperostosis (see earlier) and was demonstrably mosaic for a specific GNAS1 mutation.)

Abs et al. (1990) described a case of atypical McCune-Albright syndrome; a 36-year-old woman had acromegaly due to a pituitary adenoma, a toxic multinodular goiter that was associated with spontaneous normalization of thyroid function, and asymptomatic polyostotic fibrous dysplasia. There was no skin pigmentation and no sexual precocity. See 139320.0009 for discussion of the molecular basis of the association of pituitary tumor with McCune-Albright syndrome. Chanson et al. (1994) reported 5 patients with McCune-Albright syndrome and acromegaly. In all, acromegaly began before the age of 20 years and was recognized after the diagnosis of fibrous dysplasia, which was polyostotic in 3 and monostotic in 2. Bone fibrous dysplasia always involved the base of the skull and in 4 patients prevented surgical removal of the pituitary adenoma, which was visualized easily by MRI.

Yoshimoto et al. (1991) described the extraordinary case of a female infant who at birth had cutaneous pigmentation, hyperthyroidism, and Cushing syndrome.

Inheritance

Few convincing instances of familial occurrence of POFD have been reported. Hibbs and Rush (1952) reported the case of a 50-year-old woman with typical skin pigmentation and involvement of multiple bones. The daughter had no skin pigmentation (which is absent in some cases) but had a pathologic fracture of the left radius and radiologic and histologic changes interpreted as those of fibrous dysplasia. Firat and Stutzman (1968) described hyperthyroidism in 1 patient who also had pituitary gigantism and hyperparathyroidism in 2 others. The last 2 cases were mother and daughter. The fibrous dysplasia was limited to the jaw.

Reitzik and Lownie (1975) described a family in which many members had craniofacial PFD in an autosomal dominant pedigree pattern. This may have represented, however, an entity distinct from MAS. Alvarez-Arratia et al. (1983) presented a family that had several members in at least 3 generations with the bony and cutaneous lesions of polyostotic fibrous dysplasia.

Happle (1986) made the intriguing suggestion that this disorder is caused by an autosomal dominant lethal gene that is compatible with viability of the conceptus only when it occurs in the mosaic state, having arisen by somatic mutation. Endo et al. (1991) described monozygotic twin girls of whom one showed major signs of MAS: precocious puberty, cafe-au-lait nevi, and polyostotic fibrous dysplasia. The lack of fully convincing familial cases, except for the occurrence in monozygotic twins (Lemli, 1977), is consistent with the Happle hypothesis. (In the twins reported by Lemli (1977), one had classic signs of McCune-Albright syndrome and the other had only radiologic signs of bone disease and elevated serum alkaline phosphatase.) The frequency of the disorder is about equal in males and females.

Feuillan et al. (1991) reported the cases of 2 girls with precocious puberty initially diagnosed at the ages of 1 and 4 years. Both were unresponsive to treatment with the LHRH (luteinizing hormone-releasing hormone; 152760) agonist deslorelin for 5 years. In this respect they resembled patients with the McCune-Albright syndrome, but they had none of the other manifestations of MAS. Feuillan et al. (1991) suggested that although the underlying defect was absent in bone and skin, it was expressed in the ovaries of the 2 girls. This suggestion is consistent with the view that the McCune-Albright syndrome is a somatic mutation disorder.

Molecular Genetics

The mystery of the etiology and pathogenesis of polyostotic fibrous dysplasia appears to have been solved by the identification of activating mutations in the GNAS1 gene (139320) which render the gene functionally constitutive (Weinstein et al., 1991; Schwindinger et al., 1992). Similar mutations leading to constitutive activation of this gene had been identified in some human growth hormone secreting pituitary tumors and human thyroid tumors. Furthermore, the demonstration of the mutation in peripheral blood leukocytes but not in DNA from biopsies of clinically normal skin supports the proposal of Happle (1986) that this is a disorder of mosaicism resulting from postzygotic somatic cell mutation.

Bianco et al. (1998) isolated progenitor cells from the stromal system of the bone marrow involved in fibrous dysplasia in patients with McCune-Albright syndrome and analyzed these cells in culture. Analysis of the GNAS1 gene from individual colonies provided direct evidence for the presence of 2 different genotypes within single fibrous dysplastic lesions: marrow stromal cells containing 2 normal GNAS1 alleles, and those containing a normal allele and an allele with an activating mutation. Transplantation of clonal populations of normal cells into the subcutis of immunocompromised mice resulted in normal ossicle formation. In contrast, transplantation of clonal populations of mutant cells always led to the loss of transplanted cells from the transplantation site and no ossicle formation. However, transplantation of a mixture of normal and mutant cells reproduced an abnormal ectopic ossicle recapitulating human fibrous dysplasia and providing an in vivo cellular model of this disease. The results provided experimental evidence for the need of both normal and mutant cells in the development of McCune-Albright syndrome fibrous dysplastic lesions in bone. This study confirmed the hypothesis of Happle (1986). It also confirmed the editorial comment of DiGeorge (1975) concluding with the remark that MAS is 'a rare disorder, yes; an unimportant one, never' (Olsen, 1998).

History

This disorder was first described by McCune and Bruch (1937) and 2,4:Albright et al. (1937, 1938). Affected persons came to the attention of Fuller Albright because of the similarity of the skeletal changes to those of osteitis fibrosa cystica resulting from hyperparathyroidism (Axelrod, 1970). Lichtenstein (1938) introduced the designation 'polyostotic fibrous dysplasia' for the skeletal aspect of the syndrome.

See also Comite et al. (1984); Feuillan et al. (1986); Nakagawa et al. (1985); Shires et al. (1979); Wirth et al. (1971)

Abs, R.; Beckers, A.; Van de Vyver, F. L.; De Schepper, A.; Stevenaert, A.; Hennen, G.: Acromegaly, multinodular goiter and silent polyostotic fibrous dysplasia: a variant of the McCune-Albright syndrome. J. Endocr. Invest. 13: 671-675, 1990.

Albright, F.; Butler, A. M.; Hampton, A. O.; Smith, P.: Syndrome characterized by osteitis fibrosa disseminata, areas of pigmentation and endocrine dysfunction, with precocious puberty in females: report of five cases. New Eng. J. Med. 216: 727-746, 1937.

Albright, F.; Reifenstein, E. C., Jr.: The Parathyroid Glands and Metabolic Bone Disease: Selected Studies. Baltimore: Williams and Wilkins (pub.) 1948.

Albright, F.; Scoville, B.; Sulkowitch, H. W.: Syndrome characterized by osteitis fibrosa disseminata, areas of pigmentation, and a gonadal dysfunction: further observations including the report of two more cases. Endocrinology 22: 411-421, 1938.

Alvarez-Arratia, M. C.; Rivas, F.; Avila-Abundis, A.; Hernandez, A.; Nazara, Z.; Lopez, C.; Castillo, A.; Cantu, J. M.: A probable monogenic form of polyostotic fibrous dysplasia. Clin. Genet. 24: 132-139, 1983.

Axelrod, L.: Bones, stones and hormones: the contributions of Fuller Albright. New Eng. J. Med. 283: 964-970, 1970.

Benedict, P. H.: Sex precocity and polyostotic fibrous dysplasia: report of a case in a boy with testicular biopsy. Am. J. Dis. Child. 111: 426-429, 1966.

Bianco, P.; Kuznetsov, S. A.; Riminucci, M.; Fisher, L. W.; Spiegel, A. M.; Robey, P. G.: Reproduction of human fibrous dysplasia of bone in immunocompromised mice by transplanted mosaics of normal and Gs-alpha-mutated skeletal progenitor cells. J. Clin. Invest. 101: 1737-1744, 1998.

Chanson, P.; Dib, A.; Visot, A.; Derome, P. J.: McCune-Albright syndrome and acromegaly: clinical studies and responses to treatment in five cases. Europ. J. Endocr. 131: 229-234, 1994.

Comite, F.; Shawker, T. H.; Pescovitz, O. H.; Loriaux, D. L.; Cutler, G. B., Jr.: Cyclical ovarian function resistant to treatment with an analogue of luteinizing hormone releasing hormone in McCune-Albright syndrome. New Eng. J. Med. 311: 1032-1036, 1984.

Dent, C. E.; Gertner, J. M.: Hypophosphataemic osteomalacia in fibrous dysplasia. Quart. J. Med. 45: 411-420, 1976.

DiGeorge, A. M.: Albright syndrome: is it coming of age? (Editorial) J. Pediat. 87: 1018-1020, 1975.

Endo, M.; Yamada, Y.; Matsuura, N.; Niikawa, N.: Monozygotic twins discordant for the major signs of McCune-Albright syndrome. Am. J. Med. Genet. 41: 216-220, 1991.

Feuillan, P. P.; Foster, C. M.; Pescovitz, O. H.; Hench, K. D.; Shawker, T.; Dwyer, A.; Malley, J. D.; Barnes, K.; Loriaux, D. L.; Cutler, G. B., Jr.: Treatment of precocious puberty in the McCune-Albright syndrome with the aromatase inhibitor testolactone. New Eng. J. Med. 315: 1115-1119, 1986.

Feuillan, P. P.; Jones, J.; Oerter, K. E.; Manasco, P. K.; Cutler, G. B., Jr.: Luteinizing hormone-releasing hormone (LHRH)-independent precocious puberty unresponsive to LHRH agonist therapy in two girls lacking features of the McCune-Albright syndrome. J. Clin. Endocr. Metab. 73: 1370-1373, 1991.

Firat, D.; Stutzman, L.: Fibrous dysplasia of the bone: review of twenty-four cases. Am. J. Med. 44: 421-429, 1968.

Giovannelli, G.; Bernasconi, S.; Banchini, G.: McCune-Albright syndrome in a male child: a clinical and endocrinologic enigma. J. Pediat. 92: 220-226, 1978.

Hall, R.; Warrick, C.: Hypersecretion of hypothalamic releasing hormones: a possible explanation of the endocrine manifestations of polyostotic fibrous dysplasia (Albright's syndrome). Lancet I: 1313-1316, 1972.

Hamilton, C. R., Jr.; Maloof, F.: Unusual types of hyperthyroidism. Medicine 52: 195-213, 1973.

Happle, R.: The McCune-Albright syndrome: a lethal gene surviving by mosaicism. Clin. Genet. 29: 321-324, 1986.

Hibbs, R. E.; Rush, H. P.: Albright's syndrome. Ann. Intern. Med. 37: 587-593, 1952.
Kaplan, F. S.; Fallon, M. D.; Boden, S. D.; Schmidt, R.; Senior, M.; Haddad, J. G.: Estrogen receptors in bone in a patient with polyostotic fibrous dysplasia (McCune-Albright syndrome). New Eng. J. Med. 319: 421-425, 1988.
Lemli, L.: Fibrous dysplasia of bone: report of female monozygotic twins with and without the McCune-Albright syndrome. J. Pediat. 91: 947-949, 1977.
Lichtenstein, L.: Polyostotic fibrous dysplasia. Arch. Surg. 36: 874-898, 1938.
Lichtenstein, L.; Jaffe, H. L.: Fibrous dysplasia of the bone: a condition affecting one, several or many bones, the graver cases of which may present abnormal pigmentation of skin, premature sexual development, hyperthyroidism or still other extraskeletal abnormalities. Arch. Path. 33: 777-816, 1942.
Majzoub, J. A.; Scully, R. E.: A six-year-old boy with multiple bone lesions, repeated fractures, and sexual precocity. New Eng. J. Med. 328: 496-502, 1993.
McArthur, R. G.; Hayles, A. B.; Lambert, P. W.: Albright's syndrome with rickets. Mayo Clin. Proc. 54: 313-320, 1979.
McCune, D. J.; Bruch, H.: Progress in pediatrics: osteodystrophia fibrosa. Am. J. Dis. Child. 54: 806-848, 1937.
Nager, G. T.; Holliday, M. J.: Fibrous dysplasia of the temporal bone: update with case reports. Ann. Otol. Rhinol. Laryng. 93: 630-633, 1984.
Nager, G. T.; Kennedy, D. W.; Kopstein, E.: Fibrous dysplasia: a review of the disease and its manifestations in the temporal bone. Ann. Otol. Rhinol. Laryng. 91 (suppl. 92): 1-52, 1982.
Nakagawa, H.; Nagasaka, A.; Sugiura, T.; Nakagawa, K.; Yabe, Y.; Nihei, N.; Hirooka, M.; Itoh, M.; Nakai, A.; Ohyama, T.; Aono, T.; Gerich, J. E.: Gigantism associated with McCune-Albright's syndrome. Hormone Metab. Res. 17: 522-527, 1985.
Olsen, B. R.: A rare disorder, yes; an unimportant one, never. (Editorial) J. Clin. Invest. 101: 1545-1546, 1998.
Reitzik, M.; Lownie, J. F.: Familial polyostotic fibrous dysplasia. Oral Surg. 40: 769-774, 1975.
Schwindinger, W. F.; Francomano, C. A.; Levine, M. A.: Identification of a mutation in the gene encoding the alpha subunit of the stimulatory G-protein of adenylyl cyclase in McCune-Albright syndrome. Proc. Nat. Acad. Sci. 89: 5152-5156, 1992.
Shires, R.; Whyte, M. P.; Avioli, L. V.: Idiopathic hypothalamic hypogonadotropic hypogonadism with polyostotic fibrous dysplasia. Arch. Intern. Med. 139: 1187-1189, 1979.
Taconis, W. K.: Osteosarcoma in fibrous dysplasia. Skeletal Radiol. 17: 163-170, 1988.
Viljoen, D. L.; Versfeld, G. A.; Losken, W.; Beighton, P.: Polyostotic fibrous dysplasia with cranial hyperostosis: new entity or most severe form of polyostotic fibrous dysplasia? Am. J. Med. Genet. 29: 661-667, 1988.
von Recklinghausen, F. D.: Die Fibrose oder deformirende Ostitis, die Osteomalacie und die Osteoplastische Carcinose in ihren gegenseitigen Beziehungen. Festschrift f. Rudolf Virchow. Berlin (pub.) 1891.
Weinstein, L. S.; Shenker, A.; Gejman, P. V.; Merino, M. J.; Friedman, E.; Spiegel, A. M.: Activating mutations of the stimulatory G protein in the McCune-Albright syndrome. New Eng. J. Med. 325: 1688-1695, 1991.
Wirth, W. A.; Leavitt, D.; Enzinger, F. M.: Multiple intramuscular myxomas: another extraskeletal manifestation of fibrous dysplasia. Cancer 27: 1167-1173, 1971.
Yoshimoto, M.; Nakayama, M.; Baba, T.; Uehara, Y.; Niikawa, N.; Ito, M.; Tsuji, Y.: A case of neonatal McCune-Albright syndrome with Cushing syndrome and hyperthyroidism. Acta Paediat. Scand. 80: 984-987, 1991.

GUANINE NUCLEOTIDE-BINDING PROTEIN, ALPHA-STIMULATING ACTIVITY POLYPEPTIDE 1; GNAS1 (*139320)

JJM NOTE

Germline or at least major mosaic cell line mutations have been seen in patients with McCune-Albright syndrome. Somatic cell mutations have been seen in various secreting pituitary adenomas, ovarian tumors, and Leydig cell tumors of the testes, all of a sporadic (i.e., nonfamilial or syndromic) nature.

Description

Mutations in the GNAS1 gene are associated with pseudohypoparathyroidism, type IA (103580) and McCune-Albright syndrome (174800). The change in the gene is an inactivating mutation or an activating mutation, respectively. The inactivating mutation in the McCune-Albright syndrome is usually present in the mosaic state; the nonmosaic state for most activating mutations is presumably lethal to the embryo.

The activity of hormone-sensitive adenylate cyclase is regulated by at least 2 guanine nucleotide-binding proteins (G proteins), 1 stimulatory (Gs) and 1 inhibitory (Gi; see 139310). Each is a heterotrimer; each has a unique alpha chain, but the beta and gamma chains are apparently identical polypeptides of 35,000 and 8,000 daltons, respectively. The alpha subunit of the Gs protein exists in 2 predominant species of molecular weights 52,000 and 45,000, which are thought to be products of a single gene.

A third G protein, Go (139311), is abundant in brain. Its alpha subunit has a molecular weight of 39,000 and its beta and gamma subunits are indistinguishable from those of Gs and Gi. The G protein family includes transducin (189970).

Mapping

Using a cDNA probe in connection with a mouse/human somatic cell hybrid panel, Sparkes et al. (1987) mapped the alpha-stimulating polypeptide of G protein to chromosome 20. (See also Blatt et al. (1988).) Ashley et al. (1987) mapped the corresponding gene in the mouse to chromosome 2 which, by the argument of homology of synteny, supports the assignment of the human stimulatory G protein gene to chromosome 20. They mapped an inhibitory G protein to mouse chromosome 9.

Gejman et al. (1991) mapped the GNAS1 gene to the distal long arm of chromosome 20 by linkage studies using a polymorphism detected by denaturing gradient gel electrophoresis (DGGE). By in situ hybridization, Gopal Rao et al. (1991) assigned the GNAS1 gene to 20q12-q13.2. Using the same method, Levine et al. (1991) mapped GNAS1 to 20q13.2-q13.3.

The GTP-binding protein Gs couples hormone-receptor binding to adenylyl cylase activation. Mutations of the GNAS1 gene, encoding the Gs-alpha subunit, occur in the hormone resistance syndrome pseudohypoparathyroidism type Ia (e.g., 103580).

This syndrome is virtually always maternally inherited, although its skeletal features alone may occur after paternal transmission, giving the picture referred to as pseudopseudohypoparathyroidism. This suggests that GNAS1 may be a maternally expressed imprinted gene. However, Gs-alpha encoding mRNA is biallelically derived. By restriction landmark genomic scanning, Hayward et al. (1998) identified a differentially methylated locus containing a previously undescribed GNAS1 exon.

This exon was included within transcripts homologous to rat mRNAs encoding the large alternatively spliced Gs variant protein XLas. Two restriction sites flanking this exon were methylated on a maternal allele and unmethylated on a paternal allele. RT-PCR of human fetal tissues showed that in contrast to Gs-encoding transcripts, mRNAs encoding XLas are derived exclusively from the paternal allele.

This confirms that GNAS1 is indeed an imprinted gene, which is expressed simultaneously from both monoallelically and biallelically active promoters, each resulting in the synthesis of different protein species. Nevertheless, the pattern of GNAS1 imprinting is opposite that predicted from the inheritance of PHP-Ia.

Thus, the role of XLas in the pathogenesis of PHP requires further exploration.

Using linkage studies, Juppner et al. (1998) demonstrated that pseudohypoparathyroidism type Ib (PHP1B; 603233) is paternally imprinted and maps to 20q13.3, the same region as the GNAS1 gene.

Animal Model

Yu et al. (1998) generated mice with a null allele of the Gnas gene.

Homozygous Gs deficiency was embryonically lethal. Heterozygotes with maternal (m-/+) and paternal (+/p-) inheritance of the Gnas null allele had distinct phenotypes, suggesting that Gnas is an imprinted gene.

Parathyroid hormone (PTH) resistance is present in m-/+ but not +/p- mice.

Expression of the alpha subunit in the renal cortex (the site of PTH action) was markedly reduced in m-/+ but not in +/p- mice, demonstrating that the Gnas paternal allele is imprinted in this tissue. Gnas was also imprinted in brown and white adipose tissue. The maximal physiologic response to vasopression (urinary concentrating ability) was normal in both m-/+ and +/p- mice and Gnas was not imprinted in the renal inner medulla, the site of vasopression action. Tissue-specific imprinting of Gnas was likely the mechanism for variable and tissue-specific hormone resistance in the knockout mice and a similar mechanism might explain the variable phenotype in AHO.

A suspicion of the existence of one or more imprinted genes on distal mouse chromosome 2 had been raised by Cattanach and Kirk (1985) and Peters et al. (1994): paternal uniparental disomy (UPD)/maternal deletion and maternal UPD/paternal deletion of a region between breakpoints T2Wa and T28H on distal mouse chromosome 2 resulted in distinct phenotypes and early lethality.

Neuronatin (NNAT; 603106) is an imprinted gene on distal mouse chromosome 2 that maps outside the T2Wa-T28H imprinted region (Kikyo et al., 1997).

Given the large distance and the presence of multiple nonimprinted genes between Gnas and Nnat, it is likely that they lie within distinct imprinting domains.

The tissue-specific imprinting of Gnas observed by Yu et al. (1998) has been demonstrated for other imprinted genes (e.g., DeChiara et al. (1991) demonstrated tissue-specific parental imprinting in the case of the insulin-like growth factor II gene by study of targeted disruption of the gene in mice).

The observation that pseudohypoparathyrodisim type Ia is almost always inherited maternally has led to the suggestion that GNAS1 may be an imprinted gene.

Hayward et al. (1998) showed that, although G(s)-alpha expression (directed by the promoter upstream of exon 1) is biallelic, GNAS1 is indeed imprinted in a promoter-specific fashion. They used parthenogenetic lymphocyte DNA to screen by restriction landmark genomic scanning for loci showing differential methylation between paternal and maternal alleles.

This screen identified a region that was found to be methylated exclusively on a maternal allele and was located approximately 35 kb upstream of GNAS1 exon 1.

This region contains 3 novel exons that are spliced into alternative GNAS1 mRNA species, including 1 exon that encodes the human homolog of the large G protein XL-alpha(s) of the rat (Kehlenbach et al., 1994).

Transcription of these novel mRNAs was exclusively from the paternal allele in all tissues examined.

The differential imprinting of separate protein products of GNAS1 may therefore contribute to the anomalous inheritance of PHP Ia.

See also Carter et al. (1987); Harris et al. (1985); Kozasa et al. (1988); Lin et al. (1992); Mattera et al. (1989); Shenker et al. (1995); Shenker et al. (1993)

Ashley, P. L.; Ellison, J.; Sullivan, K. A.; Bourne, H. R.; Cox, D. R. : Chromosomal assignment of the murine Gi and Gs genes. (Abstract) Am. J. Hum. Genet. 41: A155 only, 1987.

Billestrup, N.; Swanson, L. W.; Vale, W. : Growth hormone-releasing factor stimulates proliferation of somatotrophs in vitro. Proc. Nat. Acad. Sci. 83: 6854-6857, 1986.

Blatt, C.; Eversole-Cire, P.; Cohn, V. H.; Zollman, S.; Fournier, R. E. K.; Mohandas, L. T.; Nesbitt, M.; Lugo, T.; Jones, D. T.; Reed, R. R.; Weiner, L. P.; Sparkes, R. S.; Simon, M. I. : Chromosomal localization of genes encoding guanine nucleotide-binding protein subunits in mouse and human. Proc. Nat. Acad. Sci. 85: 7642-7646, 1988.

Candeliere, G. A.; Glorieux, F. H.; Prud'Homme, J.; St.-Arnaud, R. : Increased expression of the c-fos proto-oncogene in bone from patients with fibrous dysplasia. New Eng. J. Med. 332: 1546-1551, 1995.

Carter, A.; Bardin, C.; Collins, R.; Simons, C.; Bray, P.; Spiegel, A. : Reduced expression of multiple forms of the alpha subunit of the stimulatory GTP-binding protein in pseudohypoparathyroidism type Ia. Proc. Nat. Acad. Sci. 84: 7266-7269, 1987.

Cattanach, B. M.; Kirk, M. : Differential activity of maternally and paternally derived chromosome regions in mice. Nature 315: 496-498, 1985.

Cremonini, N.; Graziano, E.; Chiarini, V.; Sforza, A.; Zampa, G. A. : Atypical McCune-Albright syndrome associated with growth hormone-prolactin pituitary adenoma: natural history, long-term follow-up, and SMS 201-995—bromocriptine combined treatment results. J. Clin. Endocr. Metab. 75: 1166-1169, 1992.

DeChiara, T.; Robertson, E. J.; Efstratiadis, A. : Parental imprinting of the mouse insulin-like growth factor II gene. Cell 64: 849-859, 1991.

Falconer, M. A.; Cope, C. L.; Robb-Smith, A. H. T. : Fibrous dysplasia of bone with endocrine disorders and cutaneous pigmentation (Albright's disease). Quart. J. Med. 11: 121-154, 1942.

Fischer, J. A.; Egert, F.; Werder, E.; Born, W. : An inherited mutation associated with functional deficiency of the alpha-subunit of the guanine nucleotide-binding protein G(s) in pseudo- and pseudopseudohypoparathyroidism. J. Clin. Endocr. Metab. 83: 935-938, 1998.

Fragoso, M. C. B. V.; Latronico, A. C.; Carvalho, F. M.; Zerbini, M. C. N.; Marcondes, J. A. M.; Araujo, L. M. B.; Lando, V. S.; Frazzatto, E. T.; Mendonca, B. B.; Villares, S. M. F. : Activating mutation of the stimulatory G protein (gsp) as a putative cause of ovarian and testicular human stromal Leydig cell tumors. J. Clin. Endocr. Metab. 83: 2074-2078, 1998.

Gejman, P. V.; Weinstein, L. S.; Martinez, M.; Spiegel, A. M.; Cao, Q.; Hsieh, W.-T.; Hoehe, M. R.; Gershon, E. S. : Genetic mapping of the Gs-alpha subunit gene (GNAS1) to the distal long arm of chromosome 20 using a polymorphism detected by denaturing gradient gel electrophoresis. Genomics 9: 782-783, 1991.

Gopal Rao, V. V. N.; Schnittger, S.; Hansmann, I. : G protein Gs-alpha (GNAS1), the probable candidate gene for Albright hereditary osteodystrophy, is assigned to human chromosome 20q12-q13.2. Genomics 10: 257-261, 1991.

Happle, R. : The McCune-Albright syndrome: a lethal gene surviving by mosaicism. Clin. Genet. 29: 321-324, 1986.

Harris, B. A.; Robishaw, J. D.; Mumby, S. M.; Gilman, A. G. : Molecular cloning of complementary DNA for the alpha subunit of the G protein that stimulates adenylate cyclase. Science 229: 1274-1277, 1985.

Hayward, B.; Kamiya, M.; Takada, S.; Moran, V.; Strain, L.; Hayashizaki, Y.; Bonthron, D. T. : XL alpha s is a paternally derived protein product of the human GNAS1 gene. (Abstract) Europ. J. Hum. Genet. 6 (suppl. 1): 36 only, 1998.

Hayward, B. E.; Kamiya, M.; Strain, L.; Moran, V.; Campbell, R.; Hayashizaki, Y.; Bonthron, D. T. : The human GNAS1 gene is imprinted and encodes distinct paternally and biallelically expressed G proteins. Proc. Nat. Acad. Sci. 95: 10038-10043, 1998.

Juppner, H.; Schipani, E.; Bastepe, M.; Cole, D. E. C.; Lawson, M. L.; Mannstadt, M.; Hendy, G. N.; Plotkin, H.; Koshiyama, H.; Koh, T.; Crawford, J. D.; Olsen, B. R.; Vikkula, M. : The gene responsible for pseudohypoparathyroidism type Ib is paternally imprinted and maps in four unrelated kindreds to chromosome 20q13.3. Proc. Nat. Acad. Sci. 95: 11798-11803, 1998.

Kehlenbach, R. H.; Matthey, J.; Huttner, W. B. : XL alpha S is a new type of G protein. Nature 372: 804-809, 1994.

Kikyo, N.; Williamson, C. M.; John, R. M.; Barton, S. C.; Beechey, C. V.; Ball, S. T.; Cattanach, B. M.; Surani, M. A.; Peters, J. : Genetic and functional analysis of neuronatin in mice with maternal or paternal duplication of distal chr. 2. Dev. Biol. 190: 66-77, 1997.

Kinard, R. E.; Walton, J. E.; Buckwalter, J. A. : Pseudohypoparathyroidism. Arch. Intern. Med. 139: 204-207, 1979.

Kozasa, T.; Itoh, H.; Tsukamoto, T.; Kaziro, Y. : Isolation and characterization of the human G(s)-alpha gene. Proc. Nat. Acad. Sci. 85: 2081-2085, 1988.

Landis, C. A.; Masters, S. B.; Spada, A.; Pace, A. M.; Bourne, H. R.; Vallar, L. : GTPase inhibiting mutations activate the alpha chain of G(s) and stimulate adenylyl cyclase in human pituitary tumours. Nature 340: 692-696, 1989.

Levine, M. A.; Deily, J. R. : Identification of multiple mutations in the gene encoding the alpha subunit of G(s) in patients with pseudohypoparathyroidism type IA. (Abstract) J. Bone Miner. Res. 5: S142 only, 1990.

Levine, M. A.; Modi, W. S.; O'Brien, S. J. : Mapping of the gene encoding the alpha subunit of the stimulatory G protein of adenylyl cyclase (GNAS1) to 20q13.2-q13.3 in human by in situ hybridization. Genomics 11: 478-479, 1991.

Levine, M. A.; Vechio, J. D. : Personal Communication. Baltimore, Md. 8/1/1990.

Lin, C. K.; Hakakha, M. J.; Nakamoto, J. M.; Englund, A. T.; Brickman, A. S.; Scott, M. L.; Van Dop, C. : Prevalence of three mutations in the Gs-alpha gene among 24 families with pseudohypoparathyroidism type Ia. Biochem. Biophys. Res. Commun. 189: 343-349, 1992.

Malchoff, C. D.; Reardon, G.; MacGillivray, D. C.; Yamase, H.; Rogol, A. D.; Malchoff, D. M. : An unusual presentation of McCune-Albright syndrome confirmed by an activating mutation of the G(s) alpha-subunit from a bone lesion. J. Clin. Endocr. Metab. 78: 803-806, 1994.

Mattera, R.; Graziano, M. P.; Yatani, A.; Zhou, Z.; Graf, R.; Codina, J.; Birnbaumer, L.; Gilman, A. G.; Brown, A. M. : Splice variants of the alpha subunit of the G protein G(8) activate both adenylyl cyclase and calcium channels. Science 243: 804-807, 1989.

Nerlich, A.; Peschel, O.; Lohrs, U.; Parsche, F.; Betz, P. : Juvenile gigantism plus polyostotic fibrous dysplasia in the Tegernsee giant. (Letter) Lancet 338: 886-887, 1991.

Patten, J. L.; Johns, D. R.; Valle, D.; Eil, C.; Gruppuso, P. A.; Steele, G.; Smallwood, P. M.; Levine, M. A. : Mutation in the gene encoding the stimulatory G protein of adenylate cyclase in Albright's hereditary osteodystrophy. New Eng. J. Med. 322: 1412-1419, 1990.

Patten, J. L.; Smallwood, P. M.; Eil, C.; Johns, D. R.; Valle, D.; Steel, G.; Levine, M. A. : An initiator codon mutation in the gene encoding the alpha subunit of Gs in pseudohypoparathyroidism type IA (PHP IA). (Abstract) Am. J. Hum. Genet. 45 (suppl.): A212 only, 1989.

Peters, J.; Beechey, C. V.; Ball, S. T.; Evans, E. P. : Mapping studies of the distal imprinting region of mouse chromosome 2. Genet. Res. 63: 169-174, 1994.

Premawardhana, L. D. K. E.; Vora, J. P.; Mills, R.; Scanlon, M. F. : Acromegaly and its treatment in the McCune-Albright syndrome. Clin. Endocr. 36: 605-608, 1992.

Schwindinger, W. F.; Francomano, C. A.; Levine, M. A. : Identification of a mutation in the gene encoding the alpha subunit of the stimulatory G-protein of adenylyl cyclase in McCune-Albright syndrome. Proc. Nat. Acad. Sci. 89: 5152-5156, 1992.

Schwindinger, W. F.; Francomano, C. A.; Levine, M. A.; McKusick, V. A. : DNA light on the Tegernsee giant. (Letter) Lancet 338: 1454-1455, 1991.

Shenker, A.; Chanson, P.; Weinstein, L. S.; Chi, P.; Spiegel, A. M.; Lomri, A.; Marie, P. J. : Osteoblastic cells derived from isolated lesions of fibrous dysplasia contain activating somatic mutations of the G(S)-alpha gene. Hum. Molec. Genet. 4: 1675-1676, 1995.

Shenker, A.; Weinstein, L. S.; Moran, A.; Pescovitz, O. H.; Charest, N. J.; Boney, C. M.; Van Wyk, J. J.; Merino, M. J.; Feuillan, P. P.; Spiegel, A. M. : Severe endocrine and nonendocrine manifestations of the McCune-Albright syndrome associated with activating mutations of stimulatory G protein Gs. J. Pediat. 123: 509-518, 1993.

Sparkes, R. S.; Cohn, V. H.; Mohandas, T.; Zollman, S.; Cire-Eversole, P.; Amatruda, T. T.; Reed, R. R.; Lochrie, M. A.; Simon, M. I. : Mapping of genes encoding the subunits of guanine nucleotide-binding protein (G-proteins) in humans. (Abstract) Cytogenet. Cell Genet. 46: 696 only, 1987.

Vallar, L.; Spada, A.; Giannattasio, G. : Altered Gs and adenylate cyclase activity in human GH-secreting pituitary adenomas. Nature 330: 566-568, 1987.

Warner, D. R.; Gejman, P. V.; Collins, R. M.; Weinstein, L. S.: A novel mutation adjacent to the switch III domain of Gs-alpha in a patient with pseudohypoparathyroidism. Molec. Endocr. 11: 1718-1727, 1997.

Warner, D. R.; Weng, G.; Yu, S.; Matalon, R.; Weinstein, L. S.: A novel mutation in the switch 3 region of Gs-alpha in a patient with Albright hereditary osteodystrophy impairs GDP binding and receptor activation. J. Biol. Chem. 273: 23976-23983, 1998.

Weinstein, L. S.; Gejman, P. V.; Collins, R. M.; Friedman, E.; Gershon, E. S.; Spiegel, A. M. : Mutations of the gene for the alpha-subunit of Gs in Albright's hereditary osteodystrophy: detection by denaturing gradient gel electrophoresis. (Abstract) J. Bone Miner. Res. 5: S142 only, 1990.

Weinstein, L. S.; Gejman, P. V.; de Mazancourt, P.; American, N.; Spiegel, A. M. : A heterozygous 4-bp deletion mutation in the G(s)-alpha gene (GNAS1) in a patient with Albright hereditary osteodystrophy. Genomics 13: 1319-1321, 1992.

Weinstein, L. S.; Gejman, P. V.; Friedman, E.; Kadowaki, T.; Collins, R. M.; Gershon, E. S.; Spiegel, A. M. : Mutations of the Gs alpha-subunit gene in Albright hereditary osteodystrophy detected by denaturing gradient gel electrophoresis. Proc. Nat. Acad. Sci. 87: 8287-8290, 1990.

Weinstein, L. S.; Shenker, A.; Gejman, P. V.; Merino, M. J.; Friedman, E.; Spiegel, A. M. : Activating mutations of the stimulatory G protein in the McCune-Albright syndrome. New Eng. J. Med. 325: 1688-1695, 1991.

Williamson, E. A.; Ince, P. G.; Harrison, D.; Kendall-Taylor, P.; Harris, P. E. : G-protein mutations in human pituitary adrenocorticotrophic hormone-secreting adenomas. Europ. J. Clin. Invest. 25: 128-131, 1995.
Wrong, O. : Tegernsee giant. (Letter) Lancet 339: 194 only, 1992.
Yang, I.; Park, S.; Ryu, M.; Woo, J.; Kim, S.; Kim, J.; Kim, Y.; Choi, Y. : Characteristics of gsp-positive growth hormone-secreting pituitary tumors in Korean acromegalic patients. Europ. J. Endocr. 134: 720-726, 1996.
Yu, S.; Yu, D.; Lee, E.; Eckhaus, M.; Lee, R.; Corria, Z.; Accili, D.; Westphal, H.; Weinstein, L. S. : Variable and tissue-specific hormone resistance in heterotrimeric Gs protein alpha-subunit (Gs-alpha) knockout mice is due to tissue-specific imprinting of the Gs-alpha gene. Proc. Nat. Acad. Sci. 95: 8715-8720, 1998.

VON HIPPEL-LINDAU SYNDROME (*193300)

JJM NOTE

When a clinician sees a patient with any one of the many features of this syndrome, he or she should conduct targeted examinations to rule out the full syndrome. Protocols for following individuals with the mutation (that could be modified for diagnostic evaluation of the person suspected of having von Hippel-Lindau disease) include annual physical examination; urine analysis; direct ophthalmoscopy; fluorescein angiography from age 10 to 60 years; 24-hour urine for VMA, renal ultrasound (age 20–65); CT scan of the kidneys every 3 years from age 20 to 65 (more often if there are multiple renal cysts); and MRI of the brain from age 15 to 40 years, then every fifth year until age 60 (Maher ER, Kealin WG Jr.: von Hippel-Lindau Disease. Medicine 76:381–91, 1997; Bender BU et al.: Functioning thoracic paraganglioma: association with Von Hippel-Lindau syndrome. J. Clin. Endocrinol. Metab. 82:3356–60, 1997).

The cardinal features are angiomata of the retina and hemangioblastoma of the cerebellum. Hemangioma of the spinal cord has also been observed. Pheochromocytoma occurs in some patients. The combination of hypertension with angioma may lead to subarachnoid hemorrhage. Hypernephroma-like renal tumors occur in some patients. Polycythemia may be due to either the hemangioblastoma of the cerebellum or the hypernephroma. Hemangiomas of the adrenals, lungs and liver and multiple cysts of the pancreas and kidneys have been observed in some instances. The condition of arteriovenous aneurysm of retina and midbrain with facial nevus, described by Bonnet et al. (1938) and by Wyburn-Mason (1943), is of uncertain relationship to this condition. Metastatic renal cancer occurs in some instances (Kranes and Balogh, 1966). Goldberg and Duke (1968) examined the eyes of an affected 51-year-old black male whose mother died of cerebellar tumor at age 26 years. The same case was described by McKusick et al. (1961). In addition to the association of tumors of the brain and adrenal medulla that occurs in neurofibromatosis and in von Hippel-Lindau disease, cerebellar tumors sometimes produce paroxysmal hypertension similar to that of pheochromocytoma. Urinary catecholamines are normal in such cases (Cameron and Doig, 1970). Shokeir (1970) described a pedigree he interpreted as indicative of autosomal dominant inheritance with incomplete penetrance. I observed a skipped generation in the large kindred reported by Silver (1954). In another kindred Shokeir (1970) suggested autosomal recessive inheritance, but incomplete penetrance of the dominant disorder seems equally plausible.

Melmon and Rosen (1964) reviewed the literature and reported studies of an extensively affected kindred. They pointed out that von Hippel's first publication on this subject concerned 2 patients, Otto Meyer and Otto Mobius by name, who figured in several later publications as well. Following this historic lead, Melmon and Rosen (1964) gave the full name of each of their patients beginning with the proband Bruno Bernardini. Cushing and Bailey (1928) pointed out that Lindau (1927) was mainly responsible for showing the connection between retinal angiomas and CNS lesions. They reported the rather long-term follow-up of a patient named Frank McA. with the full syndrome. The family history of involvement in several generations was obtained by John F. Fulton (1899-1960), Yale neurophysiologist who worked with Cushing for a year in the 1920s, and drawings of the fundus lesions were obtained by W. H. Wilmer of Johns Hopkins. As pointed out by Cushing and Bailey (1928), Treacher Collins (1894) was the first to recognize the angiomatous nature of the retinal lesions; he reported brother and sister. Then, von Hippel (1904) reported his 2 patients. In his biography of Cushing, Fulton (1946) gave an interesting account of the 'young Swedish pathologist named Arvid Lindau (who) had attracted Cushing's attention in 1927 [Lindau, 1927] through his description of an important new disease entity, which Cushing appropriately christened Lindau's disease' (p. 579).

Cysts and 'hypernephroid' tumors of the epididymis have been described (see review of Grossman and Melmon, 1972). Tsuda et al. (1976) observed as part of this syndrome the occurrence of bilateral papillary cystadenoma of the epididymis in 3 brothers. Male patients may have papillary cystadenoma of the epididymis, an unusual tumor that is bilateral when it occurs in von Hippel-Lindau disease and is not familial when unilateral (Price, 1971). (This is a generalization with some other mendelizing tumors, e.g., pheochromocytoma in this same disorder.) The experience of Lamiell (1987) differed, however; 7 of 21 affected males in 1 kindred had an epididymal mass and 5 of these were unilateral Bilateral papillary cystadenomas of the broad ligament, presumably of mesonephric origin, is the probable homologous tumor of the female (Erbe, 1978; see later for a discussion of broad ligament papillary cystadenoma in women with von Hippel-Lindau syndrome). See hypernephroma (144700). Keeler and Klauber (1992) described renal cell carcinoma (RCC) in a 16-year-old boy, probably the youngest reported example of hypernephroma in VHL disease. Atuk et al. (1979) described hypercalcemia with pheochromocytoma in von Hippel-Lindau disease. Hypercalcemia was corrected in all by removal of the tumor. In several patients, pheochromocytoma antedated development of retinal lesions. Fishman and Bartholomew (1979) described 3 related patients with striking pancreatic involvement. One had exocrine pancreatic insufficiency. In an extensively affected kindred, Fill et al. (1979) found renal cell carcinoma in 16 of 42 cases and carcinoma of the pancreas in 4 of 42. Griffiths et al. (1987) found reports of 6 patients with von Hippel-Lindau syndrome, pheochromocytoma, and islet cell tumor. A further 11 patients showed pheochromocytoma and islet cell tumor. No patient with von Hippel-Lindau syndrome had a carcinoid tumor (which is a feature of neurofibromatosis with pheochromocyt oma; see 162200). No cases of neurofibromatosis had islet cell tumor. In Cardiff, Wales, 20 patients with cerebellar hemangioblastoma were seen between 1972 and 1985. In 8 of these, Huson et al. (1986) subsequently established the diagnosis of von Hippel-Lindau disease. Although the diagnosis had not previously been considered, in retrospect, 7 of the 8 cases were known to be at risk for the syndrome. Lamiell et al. (1989) identified 43 affected members in a large kindred, which was exceptional for absence of pheochromocytoma and erythrocythemia, for more renal and pancreatic cysts and malignancies, and for somewhat fewer eye or CNS lesions. Bilateral renal adenocarcinoma was found presymptomatically in 5 young subjects who had bilateral nephrectomy and hemodialysis. Three survived long-term after renal transplants. Five members of the family had pancreatic malignancy. Jennings et al. (1988) demonstrated the usefulness of family studies in determining asymptomatic lesions requiring treatment, e.g., renal cell carcinoma. They also reported the occurrence of a spermatic cord mesenchymal hamartoma in this disorder. Horbach et al. (1989) suggested that the combination of adrenal pheochromocytoma and ipsilateral renal cell carcinoma may represent a forme fruste of von Hippel-Lindau disease. Lenz et al (1992) demonstrated that norepinephrine-producing adrenal pheochromocytoma in von Hippel-Lindau disease can produce the clinical syndrome of hypertension associated with severe hypokalemia and hyperreninemic hyperaldosteronism. The hyperreninemic hyperaldosteronism was rapidly improved by beta-blockade and was completely reversed by tumor removal. Kerr et al. (1995) described hemangioblastoma of the optic nerve in a 27-year-old woman with von Hippel-Lindau syndrome, the tenth such reported case.

James (1998) tabulated reports of 4 women with VHL and broad ligament papillary cystadenoma published between 1988 and 1994 (Gersell and King, 1988; Funk and Heiken, 1989; Korn et al., 1990; Gaffey et al., 1994; Karsdorp et al., 1994) and added a fifth case. These are mesosalpinx cysts, which are the equivalent of epididymal cysts in the male. The cysts were unilateral in at least 3 of the 5 cases. They occurred along the full course of the mesonephric duct, in the mesosalpinx close to the ovary, over the uterine tubes, and near the vaginal fornix in a remnant of the Gartner duct (the female counterpart to the duct of the epididymis). At least 3 of the patients had multiple renal cysts and bilateral renal cell carcinoma. In the patient reported by Korn et al. (1990), screening for VHL after the papillary cystadenomas were diagnosed revealed pancreatic cysts and lesions of the cerebellum and kidney; renal cell carcinoma was diagnosed during follow-up surgery. The unilateral cyst in the patient reported by Gaffey et al. (1994) was preceded by the finding of a middle ear papillary tumor. The combined presentation of mesonephric cystadenoma and ear tumor was noted in reports of epididymal cysts in men (Price, 1971). Gaffey et al. (1994) suggested that the ear tumor and the adnexal tumor may represent 'major visceral manifestations of VHL.' (Nomenclature: According to the VHL Family Alliance, a genetic support group, the approved terminology is 'adnexal papillary cystadenoma of probable mesonephric origin,' abbreviated APMO (Graff, 1998).)

Endolymphatic sac tumors (ELSTs) are in general very rare. Manski et al. (1997) undertook a study to determine if hearing loss and ELSTs are a component of VHL. Brain magnetic resonance images (MRIs) from 374 patients screened for VHL were reviewed for evidence of ELSTs. VHL patients with MRI evidence suggestive of ELSTs or with a history of hearing loss, tinnitus, or vertigo underwent additional radiologic and audiologic evaluations. MRI revealed evidence of 15 ELSTs in 13 (11%) of 121 patients with VHL, but in none of 253 patients without evidence of VHL (P less than 0.001). Clinical findings in these 13 patients included hearing loss in 13, tinnitus in 12, vertigo in 8, and facial paresis in 1. Mean age at onset of hearing loss was 22 years (range, 12 to 50 years).

Maher et al. (1990) compared age incidence curves for sporadic cerebellar hemangioblastoma and sporadic renal cell carcinoma to those for familial forms of these tumors that occur as part of von Hippel-Lindau disease. The curves for tumors in VHL disease were compatible with a single mutation model, whereas the age incidence curves for sporadic tumors suggested a 2-stage mutation process. On the whole, the findings suggested that the VHL gene functions as a recessive tumor suppressor gene. Crossey et al. (1994) presented convincing evidence that the VHL disease gene functions as a recessive tumor suppressor gene and that inactivation of both alleles of the VHL gene is a critical event in the pathogenesis of VHL neoplasms. Studies of loss of heterozygosity (LOH) showed that in 7 tumors from 7 familial cases in which the parental origin of the 3p26-p25 allele loss could be determined, the allele had been lost from the chromosome inherited from the unaffected parent. In 4 VHL tumors, LOH on other chromosomes (5q21, 13q, 17q) was found, indicating that homozygous VHL gene mutations may be required but not sufficient for tumorigenesis in von Hippel-Lindau disease.

Seizinger et al. (1991) pointed out that visceral cysts of the kidney, pancreas, and epididymis occur not only as features of VHL but also in the general population, and that the presence of such cysts, unaccompanied by other more typical lesions such as retinal and cerebellar hemangioblastoma, may represent a major diagnostic problem. The application of flanking markers for the VHL gene for presymptomatic diagnostic testing confirmed that epididymal cysts are indeed not suitable as a diagnostic criterion. The genetic studies suggested that VHL with or without pheochromocytomas is caused by defects within the same gene. Renal cell carcinoma occurs as part of VHL; a second more proximal region of chromosome 3, 3p14.2, is responsible for 'pure familial renal cell carcinoma' (144700).

Go et al. (1984) found no linkage of VHL with any of 31 marker loci. Forty-one affected persons were found among 220 descendants of an ancestral couple. Wells et al. (1987) found linkage with a segregating minisatellite band at a recombination fraction of 0.15 (lod = about 3.0). In studies of 9 families, Seizinger et al. (1988) found linkage with RAF1 (164760) on chromosome 3—combined maximum lod = 4.38 at theta = 0.11 (1 lod unit confidence interval, the approximate equivalent of 95% confidence interval, of theta = 0.04 to 0.23). The finding of recombination between VHL and RAF1 indicated that the site of the VHL mutation is not in the RAF1 gene. Kiechle-Schwarz et al. (1989) found rearrangements resulting in partial or total trisomy of 3p in 3 cell clones from pheochromocytomas derived from patients with von Hippel-Lindau syndrome. Seizinger et al. (1989) reported a multipoint linkage analysis of the VHL gene using a battery of polymorphic markers in the region 3p26-p25. Vance et al. (1990) confirmed the linkage of VHL to 3p. Hosoe et al. (1990) concluded that VHL is located between RAF1 (3p25) and D3S18 (3p26). Kovacs and Kung (1991) analyzed the DNA from 28 nonpapillary renal cell carcinomas arising in 2 patients with VHL disease. They used both karyotypic and RFLP for evidence of allelic recombination on chromosomes 3 and 5. Two distinct breakpoint clusters were identified, each associated with different karyotypic alterations. The first type involves a breakpoint at chromosome 3p13, with the common nonreciprocal translocations occurring between 3p and 5q or 1q, resulting in the net loss of 3p and gain of 5q or 1q segments. In the second form of translocation, which was less common, the breakpoint on 3p and on the partner chromosome is near the centromere at bands p11 or q11. This kind of rearrangement was observed in 10 tumors with a nonrandom involvement of chromosome 3. In each case the derivative chromosome carrying the translocated 3p segment was preferentially eliminated from the tumor cells. Maher et al. (1990) confirmed the assignment to 3p by linkage studies in 12 British families. They suggested that the VHL locus is telomeric to THRB. No evidence for genetic heterogeneity was found. Seizinger et al. (1991) reported the location of the VHL locus at 3p26-p25 on the basis of studies in 28 pedigrees comprising 164 affected persons. By multipoint linkage analysis, Maher et al. (1991) demonstrated flanking markers. They found no evidence of locus heterogeneity; families with and without pheochromocytoma showed linkage to D3S18 with which no recombination was observed (maximum lod score = 6.6 at theta = 0.0; confidence interval 0.00-0.06). By multipoint linkage analysis, Richards et al. (1993) narrowed the target region for isolation of the VHL disease gene by positional cloning techniques to a 4-cM interval between D3S1250 and D3S18.

Maher et al. (1991) estimated the point prevalence of heterozygotes in East Anglia to be 1 in 53,000, with an estimated birth incidence of 1 in 36,000 live births. Reproductive fitness was 0.83. Direct and indirect estimates of the mutation rate were 4.4 per million gametes per generation and 2.32 per million gametes per generation, respectively. No significant association was found between parental age or birth order and new mutations. In the Freiburg district of Germany, Neumann and Wiestler (1991) calculated the prevalence of this disorder to be 1 in 38,951. They found a striking tendency for familial clustering of particular VHL features. Both angiomatosis retinae and hemangioblastoma of the CNS occurred in most families, whereas the occurrence of renal lesions and/or pancreatic cysts was mutually exclusive with pheochromocytoma. Neumann and Wiestler (1991) interpreted these findings to indicate that the VHL locus is complex, with the existence of different mutations in different families or the occurrence of additional genetic lesions which cooperate with the VHL gene on chromosome 3p. They suggested a linear sequence of features as follows: pheochromocytoma, angiomatosis retinae, hemangioblastoma of the CNS, renal lesions, pancreatic cysts, and epididymal cystadenoma. In the course of an evaluation of 41 families with this disorder from the United States and Canada, Glenn et al. (1991) found 1 large family with a distinctive phenotype: the most common disease manifestation was pheochromocytoma occurring in 57% (27 of 47) of affected members; few (4 of 47) had symptomatic spinal or cerebellar hemangioblastomas; no affected family member had renal cell carcinoma or pancreatic cysts. Genetic analysis demonstrated, however, that the disorder in this family was linked to the same markers found to be linked to typical VHL. The observations are clearly relevant to the descriptions of families with 'pure' pheochromocytoma (171300); they may be instances of allelism at the VHL locus.

Eng et al. (1995) screened 7 exons of the RET oncogene (164761) and the VHL gene for mutations in 48 sporadic pheochromocytomas and found a RET mutation in exons 10, 11, and 16 in 5 of the 48. Of these, 1 was proven to be germline and 2 were proven to be somatic mutations. Mutations of the VHL gene were found in 4 of the 48 sporadic pheochromocytomas; these 4 included the 2 bilateral cases in the series, of which 1 was proven to be a germline mutation, and 2 others, of which 1 was proven to be somatic.

Using DNA polymorphic markers, Glenn et al. (1992) studied 16 families with VHL disease. Of 48 asymptomatic persons at risk of developing this illness because of an affected parent or sib, DNA polymorphism analysis predicted that 9 were carriers of the disease gene and 33 had the wildtype allele. The test was not informative in 6 persons. All 9 persons predicted to carry the VHL gene had evidence of occult disease on clinical examination. There was no clinical evidence of VHL disease in 32 of 33 persons predicted to carry the wildtype allele.

Davies et al. (1994) described a 65-year-old woman who was an obligate carrier of the gene for von Hippel-Lindau disease. Her father, 2 brothers, 2 sisters, and 3 sons had hemangioblastomas and renal carcinomas. Careful examination of the woman showed only a small benign renal cyst. Such cysts are very common in the general population. Therefore, obligate gene carriers may not exhibit any features of the disease beyond the age of 60 years.

By positional cloning, Latif et al. (1993) identified a gene on 3p as the VHL tumor suppressor gene. A restriction fragment encompassing the gene showed rearrangements in 28 of 221 VHL kindreds. Eighteen of these rearrangements were due to deletion in the candidate gene, including 3 large nonoverlapping deletions. Intragenic mutations were detected in all cell lines derived from VHL patients and from sporadic renal cell carcinomas. The VHL gene was found to be evolutionarily conserved and to encode 2 widely expressed transcripts of approximately 6 and 6.5 kb. The partial sequence of the inferred gene product showed no homology to other proteins except for an acidic repeat domain found in the procyclic surface membrane glycoprotein of Trypanosoma brucei.

Latif et al. (1993) performed positional cloning by first obtaining nearly complete genomic coverage of the area to which the VHL gene mapped in overlapping yeast artificial chromosomes (YACs) and cosmid-phage contigs. In parallel with the cloning efforts, a physical map of the region by pulsed field gel electrophoresis was established and gross rearrangements affecting the region sought. These efforts resulted in the discovery of nested constitutional deletions in 3 unrelated VHL patients. This finding and the availability of cloned DNA provided rapid access to the VHL gene.

Latif et al. (1993) reasoned that the smallest of the 3 deletions should either encompass or interrupt the gene, and they identified a cosmid mapping to the commonly deleted region for use in the study of other cases. In addition to looking for deletions, they looked for other alterations by SSCP and RT-PCR in sporadic RCC cell lines and VHL lymphoblastoid cell lines. RCC was studied because the Knudson model predicted that sporadic cancers should be associated with mutations in the same locus affected in the corresponding hereditary cancer. Aberrant patterns were found in 5 RCC cell lines. In 4 of them, the pattern was due to small, 1- to 10-nucleotide deletions that created frameshift mutations and, presumably, truncated proteins. In the fifth RCC line, the change was a nonsense mutation, resulting from a C-to-A transversion at nucleotide 761. One VHL patient was found to have an 8-nucleotide insertion at nucleotide 714. Another patient had an in-frame 3-nucleotide deletion at nucleotide 434, predicted to remove ile146 in the gene product. A third patient had an in-frame 9-nucleotide deletion at nucleotide 456, predicted to remove arg153/val154/val155. Richards et al. (1993) constructed a long range physical map around D3S601, which had been shown to be tightly linked to the VHL locus, and screened 91 VHL patients from 80 kindreds for germline rearrangements using pulsed field gel electrophoresis. Two patients were found to have germline deletions, approximately 120 kb and 50 kb, respectively, telomeric to D3S601. The results established the position of the VHL locus with respect to D3S601, refined its localization to a small region of approximately 50 kb, and excluded the plasma membrane Ca(2+)-transporting ATPase isoform 2 gene (108733) as the site of the VHL mutation.

93,92:Richards et al. (1993, 1994) found that large germline deletions could be detected by Southern analysis and pulsed field gel electrophoresis in 19% and 3% of VHL patients, respectively. Crossey et al. (1994) used single-strand conformation polymorphism and heteroduplex analysis to investigate 94 VHL patients without large deletions for intragenic mutations. In 55 unrelated kindreds, 40 different mutations were identified: 19 missense mutations, 6 nonsense mutations, 12 frameshift deletions or insertions, 2 in-frame deletions, and 1 splice donor site mutation. The 2 most frequent mutations were arg238 to gln and arg238 to trp, which were detected in 5 and 4 unrelated kindreds, respectively. Although pheochromocytoma occurs in only about 7% of VHL patients, marked interfamilial differences were observed. Examining the relationship between VHL gene mutations and phenotype in 65 kindreds, Crossey et al. (1994) found that large deletions or intragenic mutations predicted to cause a truncated protein were found in 36 of 53 families without pheochromocytoma but in only 2 of 12 families with pheochromocytoma (P less than 0.01). Of 12 families with pheochromocytoma, 10 had missense mutations compared with 13 of 53 kindreds without pheochromocytoma (P less than 0.001). In particular, the arg238-to-trp and arg238-to-gln mutations were associated with a high risk (62%) of pheochromocytoma.

Vogelstein (1995) referred to the VHL gene as a gatekeeper gene for cancers such as those of the kidney. Like the retinoblastoma gene (180200), the APC gene (175100) appears to serve a similar gatekeeper role in connection with a multistep process of tumorigenesis. Rubenstein and Yaari (1994) suggested that the VHL gene may serve that role in relation to astrocytoma. They presented the pedigree of a Puerto Rican family in which at least 9 members had von Hippel-Lindau syndrome and 2 of these had astrocytoma.

Zhuang et al. (1996) analyzed VHL gene alterations in sporadic human colon carcinomas and adenomas using techniques that allowed for procurement and analysis of selected sub-populations of cells from paraffin embedded and frozen human tumor specimens. Allelic loss of the VHL gene was detected in 7 of 11 (64%) of informative patients with sporadic colon carcinoma. No allelic loss was shown in colon adenomas from 8 informative patients. The authors suggested that VHL gene loss may represent a relatively late event in colonic neoplasia progression.

The VHL cDNA encodes 852 nucleotides distributed over 3 exons. The VHL gene is unrelated to any known gene family. Chen et al. (1995) identified germline mutations in 85 of 114 VHL families (75%). They found that the types of mutations responsible for VHL without pheochromocytoma (VHL type 1) differed from those responsible for VHL with pheochromocytoma (VHL type 2). Microdeletions/insertions, nonsense mutations, or deletions were found in 56% of families with VHL type 1; missense mutations accounted for 96% of those responsible for VHL type 2. Specific mutations in codon 238 accounted for 43% of the mutations responsible for VHL type 2 (see 193300.0003-193300.0005). The homologous rat gene encodes a 185-amino acid protein that is 88% identical in sequence to the aligned 213-amino acid human VHL gene product. Duan et al. (1995) found that, when expressed in COS-7 cells, both the human and the rat VHL proteins showed predominant nuclear, nuclear and cytosolic, or predominant cytosolic VHL staining by immunofluorescence. A complicated pattern of cellular proteins were seen that could be specifically coimmunoprecipitated with the introduced VHL protein. A complex containing VHL and proteins with apparent molecular masses of 16 and 9 kD was the most consistently observed. Certain naturally VHL missense mutations demonstrated either complete or partial loss of the p16-p9 complex. Thus, Duan et al. (1995) concluded that the VHL tumor suppressor gene product is a nuclear protein, perhaps capable of specifically translocating between the nucleus and the cytosol. It is likely that VHL executes its function via formation of specific multiprotein complexes.

Iliopoulos et al. (1995) demonstrated that the product of the VHL gene is an approximately 30-kD protein. They showed further that the renal cell carcinoma cell line 786-0, which is known to harbor a VHL mutation, fails to produce a wildtype VHL protein. Reintroduction of wildtype, but not mutant, VHL into these cells had no demonstrable effect on their growth in vitro but inhibited their ability to form tumors in nude mice.

Ohh et al. (1998) showed that fibronectin coimmunoprecipitated with normal VHL protein but not tumor-derived VHL mutants. Immunofluorescence and biochemical fractionation experiments showed that fibronectin colocalized with a fraction of VHL associated with the endoplasmic reticulum, and cold competition experiments suggested that complexes between fibronectin and VHL protein exist in intact cells. Assembly of an extracellular fibronectin matrix by VHL -/- renal carcinoma cells, as determined by immunofluorescence and ELISA assays, was grossly defective compared with VHL +/+ renal carcinoma cells. Reintroduction of wildtype, but not mutant, VHL protein into VHL -/- renal carcinoma cells partially corrected this defect. Extracellular fibronectin matrix assembly by VHL -/- mouse embryos and mouse embryo fibroblasts, unlike their VHL +/+ counterparts, was grossly impaired. Ohh et al. (1998) concluded that VHL protein is important in fibronectin matrix assembly.

To elucidate the biochemical mechanisms underlying tumor suppression by the VHL protein, Duan et al. (1995) and Kibel et al. (1995) searched for cellular proteins that bind to wildtype VHL protein but not to tumor-derived VHL protein mutants. They found that 2 transcriptional elongation factors, elongin B (600787) and elongin C (600788), bind in vitro and in vivo to a short, colinear region of the VHL protein that is frequently mutated in human tumors. Kibel et al. (1995) showed that a peptide replica of this region inhibited binding of VHL protein to elongin B and elongin C, whereas a point-mutant derivative, corresponding to a naturally occurring VHL missense mutation, had no effect. Duan et al. (1995) showed that recombinant VHL competes with elongin A (600786) for elongin B and C binding in vitro. The results were interpreted as indicating that the normal tumor suppression function of VHL protein involves the inhibition of transcription elongation by its binding to elongin B and elongin C.

Maddock et al. (1996) reported on a VHL register set up in the northwest of England in 1990. There was information on 83 affected persons. In addition, the effectiveness of the screening program used and the occurrence of CNS hemangioblastomas in the general populations were examined. The diagnostic point prevalence of heterozygotes in the region was 1 in 85,000 persons, with an estimated birth incidence of 1 in 45,500 live births. The mutation rate was estimated directly to be 1.4 x 10 (-6)/gene/generation (1 in 714,200). The mean age at onset of first symptoms was 26.25, with cerebellar hemangioblastoma being the most common presenting manifestation (34.9% of cases). The mean age at diagnosis of VHL was 30.87 years. Overall, 50 patients (60.2%) developed a cerebellar hemangioblastoma, 34 (41%) a retinal angioma, 21 (25.3%) a renal cell carcinoma, 12 (14.5%) a spinal hemangioblastoma, and 12 (14.5%) a pheochromocytoma. Mean age at death was 40.9 years with cerebellar hemangioblastoma being the most common cause (47.7% of deaths). In addition to the 83 clinically affected subjects, Maddock et al. (1996) identified 3 obligate carriers who were considered to be lesion free on extensive screening tests. In the regionally based cancer registry, 14% of all CNS hemangioblastomas were found to occur as part of VHL, but investigations for VHL in apparently sporadic cases appeared to have been limited.

Lee et al. (1996) demonstrated that there is a tightly regulated, cell-density-dependent transport of the VHL protein into and/or out of the nucleus. In densely grown cells, it is predominantly in the cytoplasm, whereas in sparse cultures, most of the protein can be detected in the nucleus. They identified a putative nuclear localization signal in the first 60 and first 28 amino acids of the human and rat VHL protein, respectively. Sequences in the C-terminal region of VHL protein may also be required for localization to the cytosol. The findings indicated a novel cell-density-dependent pathway responsible for the regulation of VHL cellular localization.

Richards et al. (1996) used in situ hybridization to investigate the principal sites of VHL expression during embryogenesis. They also analyzed a variety of fetal tissues for levels of the 2 VHL isoforms, isoform I, which contains all 3 exons, and isoform II, which contains only exons 1 and 3. Although VHL expression was found to be ubiquitous, particularly high levels of expression were detected in the urogenital system, brain, spinal cord, sensory ganglia, eyes, and bronchial epithelium. Richards et al. (1996) noted that this expression pattern correlated to some extent with the pattern of organ involvement but that there were significant differences. Both isoforms of VHL were detected in all tissues, and the ratio of isoforms was similar between tissues.

Oberstrass et al. (1996) found abnormalities of the VHL gene in 10 of 20 capillary hemangioblastomas of the CNS. Seven tumors had a frameshift mutation due either to deletion of 1 or more basepairs (6 cases) or to insertion of 1 basepair (1 case). The remaining 3 tumors had either point mutations with intron splice site sequences (2 cases) or a point mutation resulting in an amino acid substitution (1 case). Evidence for germline alterations of the VHL gene was found in 2 patients who showed identical mutations in both tumors and corresponding leukocyte DNA. Oberstrass et al. (1996) noted that it is significant that one of the 2 tumors with a germline mutation was in an 18-year-old male, and the other in a 40-year-old female. The non-germline mutations included tumors from individuals 70, 62, 60, 55, and 52 years old. Kenck et al. (1996) investigated 91 different parenchymal tumors of the kidney for mutation in the VHL gene by SSCP and/or heteroduplex techniques. Evidence of mutation of the VHL gene was associated exclusively with nonpapillary renal cell carcinoma.

The highly vascular tumors associated with von Hippel-Lindau disease overproduce angiogenic peptides such as vascular endothelial growth factor/vascular permeability factor (VEGF/VPF; 192240). Iliopoulos et al. (1996) found that renal carcinoma cells lacking wildtype VHL protein produce mRNAs encoding VEGF/VPF, the glucose transporter GLUT1 (SLC2A1; 138140), and the platelet-derived growth factor B chain (190040) under both normoxic and hypoxic conditions. Reintroduction of wildtype, but not mutant, VHL protein into these cells specifically inhibited production of these mRNAs under normoxic conditions, thus restoring their previously described hypoxia-inducible profile. Thus, Iliopoulos et al. (1996) concluded that the VHL protein appears to play a critical role in the transduction of signals generated by changes in ambient oxygen tension.

Zbar et al. (1996) performed germline mutation analysis in 469 VHL families from North America, Europe, and Japan. Germline mutations were identified in 300 (63%) of the families tested; a total of 137 distinct intragenic germline mutations were detected. Most (124 of 137) of the mutations occurred in 1 or 2 families; a few occurred in 4 or more families. In this large series, it was possible to compare the effects of identical germline mutations in different populations. Germline VHL mutations produce similar cancer phenotypes in Caucasian and Japanese VHL families. Germline VHL mutations were identified that produced 3 distinct cancer phenotypes: (1) renal carcinoma without pheochromocytoma, (2) renal carcinoma with pheochromocytoma (e.g., 193300.0010), and (3) pheochromocytoma alone (e.g., 193300.0012). The catalog of VHL germline mutations with phenotype information provided by Zbar et al. (1996) should be useful for diagnostic and prognostic studies of VHL and for studies of genotype-phenotype correlations in this disorder.

Interfamilial differences in predisposition to pheochromocytoma in VHL reflect allelic heterogeneity such that there is a strong association between missense mutations and risk of pheochromocytoma. Prowse et al. (1997) investigated the mechanism of tumorigenesis in VHL tumors to determine whether there were differences between tumor types or classes of germline mutations. They studied 53 tumors (30 renal cell carcinomas, 15 hemangioblastomas, 5 pheochromocytomas, and 3 pancreatic tumors) from 33 patients (27 kindreds) with VHL. Overall, 51% of 45 informative tumors showed LOH at the VHL locus. In 11 cases, it was possible to distinguish between loss of the wildtype and mutant alleles, and in each case the wildtype allele was lost. LOH was detected in all tumor types and occurred in the presence of both germline missense mutations and other types of germline mutation associated with a low risk of pheochromocytoma. Intragenic somatic mutations were detected in 3 tumors (all hemangioblastomas) and in 2 of these could be shown to occur in the wildtype allele. Their study provided the first example of homozygous inactivation of the VHL gene by small intragenic mutations in this type of tumor. Hypermethylation of the VHL gene was detected in 33% (6 of 18) of tumors without LOH, including 2 renal cell carcinomas and 4 hemangioblastomas. Prowse et al. (1997) stated that, although hypermethylation of the VHL gene had been reported previously in nonfamilial RCC and although methylation of tumor-suppressor genes had been implicated in the pathogenesis of other sporadic cancers, this was the first report of somatic methylation in a familial cancer syndrome. Herman et al. (1994) observed hypermethylation of the VHL gene in 19% of sporadic RCCs. Versteeg (1997) provided a general discussion of aberrant methylation in cancer.

To determine whether the pheochromocytoma-associated syndromes VHL and MEN2 (171400) play a role in the development of thoracic functioning paragangliomas, Bender et al. (1997) analyzed germline DNA from 5 unselected patients with this tumor for mutations in the genes that predispose to VHL and MEN2. Molecular and clinical data revealed that 3 (60%) had VHL, with 2 different germline mutations of the VHL gene, but no individual was affected by MEN2. Two of these 3 patients with VHL did not show any additional VHL-associated lesions. Bender et al. (1997) suggested that VHL should be considered in the differential diagnosis of thoracic pheochromocytoma, and that in VHL patients suspected of a catecholamine-secreting tumor, thoracic localization should be considered if an adrenal pheochromocytoma cannot be detected.

Vascular endothelial growth factor (VEGF; 192240) mRNA is upregulated in von Hippel-Lindau disease-associated tumors. Mukhopadhyay et al. (1997) assessed the effect of the VHL gene product on VEGF expression. Using a VEGF promoter-luciferase construct for cotransfection with a wildtype VHL vector in embryonic kidney and renal cell carcinoma cell lines, they showed that wildtype VHL protein inhibited VEGF promoter activity in a dose-dependent manner up to 5- to 10-fold. Deletion analysis defined a 144-bp region of the VEGF promoter necessary for VHL repression. This VHL-responsive element is GC rich and specifically bound the transcription factor Sp1 (189906) in crude nuclear extracts. They further demonstrated that VHL and Sp1 directly interact with an inhibitory effect on Sp1, suggesting that loss of Sp1 inhibition may be important in the pathogenesis of von Hippel-Lindau disease and renal cell carcinoma.

Fearon (1997) reviewed more than 20 different hereditary cancer syndromes that had been defined and attributed to specific germline mutations in various inherited cancer genes. In a useful diagram, he illustrated the roles of allelic variation ('1 gene - different syndromes') and genetic heterogeneity ('different genes - 1 syndrome') in inherited cancer syndromes. VHL mutations were used as an example of the former: inactivating mutations, such as nonsense mutations or deletions, predisposed to clear-cell renal carcinoma, retinal angioma, and cerebellar and spinal hemangioblastoma; missense mutations, e.g., in codon 167, predisposed to these tumors and pheochromocytoma in addition.

It is a physiologic characteristic of normal and immortalized cells in culture to exit the cell cycle upon contact inhibition or withdrawal of serum. Most cancer cells lose this ability and continue to proliferate at high density or in low serum. This suggests that the inability of cells to exit the cell cycle either by contact inhibition or serum deprivation may be a prerequisite for malignant transformation. VHL-negative 786-0 renal cell carcinoma (RCC) cells are tumorigenic in nude mice and tumorigenicity is suppressed by the reintroduction of VHL. This particular cell line, like many cancer cells, fails to exit the cell cycle upon serum withdrawal. Pause et al. (1998) showed that reintroduction of the wildtype VHL gene restores the ability of VHL-negative RCC cells to exit the cell cycle and enter G0/quiescence in low serum. The cyclin-dependent kinase inhibitor p27 (CDKN1B; 600778) accumulates upon serum withdrawal, only in the presence of VHL, as a result of an increase in protein stability. Pause et al. (1998) proposed that loss of the wildtype VHL gene results in a specific cellular defect in serum-dependent growth control, which may initiate tumor formation. Thus, VHL appears to be the first tumor suppressor involved in the regulation of cell cycle exit, which is consistent with its gatekeeper function in the kidney.

To understand the mechanisms through which VHL mediates tumor suppression, biochemical analyses of VHL gene products are necessary. However, the translation start site was not evident from the initial cloning of the human VHL gene (Latif et al., 1993). The first AUG in the human VHL open reading frame (ORF) was designated as the translation start site, i.e., amino acid 1, on the basis of homology between the mouse, rat, and human VHL sequences, and a product initiated at this start site was detected in cell lines. This site, however, contains a suboptimal Kozak consensus sequence. A second AUG is present in the VHL ORF at codon 54. This region contains a more conserved Kozak consensus sequence, and thus may serve as a second, internal, translation initiation site. The significance of a second translation start site is underscored by the lack of mutations found between the first and second methionine codons of the VHL gene in both sporadic and VHL-associated renal carcinomas. This observation suggested that mutation in this region may not lead to VHL inactivation if translation could be initiated at the second methionine codon, producing a functional VHL protein. Furthermore, both the rat and mouse contain only 19 of the 53 amino acids present in this region in the human VHL ORF. Thus, this 53-amino acid region makes little or no contribution to the biologic activity of VHL protein.

Schoenfeld et al. (1998) identified a second native VHL gene product. They showed that this protein is initiated from the second translation start site and contains the biologic activity of the VHL gene. Iliopoulos et al. (1998) demonstrated that in addition to the protein encoded by the VHL gene consisting of 213 amino acid residues with an apparent molecular mass of 30 kD (VHL30), cells also produce a VHL protein (VHL19) that appears to arise as a result of internal translation from the second methionine within the VHL ORF. VHL30 resides primarily in the cytosol, with lower amounts found in the nucleus or associated with cell membranes. In contrast, VHL19 is equally distributed between the nucleus and cytosol and is not found in association with membranes. VHL19, like VHL30, can bind to elongin B (600787), elongin C (600788), and Hs-Cul2 in coimmunoprecipitation assays and can inhibit the production of hypoxia-inducing proteins such as VEGF and GLUT1 when reintroduced into renal carcinoma cells that lack a wildtype VHL allele. Thus, cells contain 2 biologically active VHL gene products.

To discover genes involved in VHL-mediated carcinogenesis, Ivanov et al. (1998) used renal cell carcinoma cell lines stably transfected with wildtype VHL-expressing transgenes. Large-scale RNA differential display technology applied to these cell lines identified several differentially expressed genes, including an alpha carbonic anhydrase gene, termed CA12 (603263). The deduced protein sequence was classified as a one-pass transmembrane carbonic anhydrase possessing an apparently intact catalytic domain in the extracellular CA module. Reintroduced wildtype VHL strongly inhibited the overexpression of the CA12 gene in the parental renal cell carcinoma cell lines. Similar results were obtained with CA9 (603179) which encodes another transmembrane carbonic anhydrase with an intact catalytic domain. Although both domains of the VHL protein contributed to regulation of CA12 expression, the elongin binding domain alone could effectively regulate CA9 expression. By fluorescence in situ hybridization, Ivanov et al. (1998) mapped CA12 and CA9 to chromosome bands 15q22 and 17q21.2, respectively, regions prone to amplification in some human cancers. Ivanov et al. (1998) stated that additional experiments were necessary to define the role of CA IX and CA XII enzymes in the regulation of pH in the extracellular microenvironment and its potential impact on cancer cell growth.

See also Brown et al. (1973); Cahill et al. (1942); Chapman and Diaz-Perez (1962); Christoferson et al. (1961); Crossey et al. (1994); Decker et al. (1988); Duan et al. (1995); Green et al. (1986); Hagler et al. (1971); Hennessy et al. (1967); Horton et al. (1976); Kaplan et al. (1961); King et al. (1987); Maher et al. (1991); Maher et al. (1990); Maher et al. (1990); Nibbelink et al. (1969); Otenasek and Silver (1961); Probst et al. (1978); Rho (1969); Richards et al. (1993); Sander et al. (1970); Schechterman (1961); Sharp and Platt (1971); Thomas and Burnside (1961); Tory et al. (1989); Wesolowski et al. (1981); Wise and Gibson (1971)

Atuk, N. O.; McDonald, T.; Wood, T.; Carpenter, J. T.; Walzak, M. P.; Donaldson, M.; Gillenwater, J. Y.; Turner, S. M.; Westfall, V.: Familial pheochromocytoma, hypercalcemia, and von Hippel-Lindau disease: a ten-year study of a large family. Medicine 58: 209-218, 1979.

Bender, B. U.; Altehofer, C.; Januszewicz, A.; Gartner, R.; Schmidt, H.; Hoffmann, M. M.; Heidemann, P. H.; Neumann, H. P. H.: Functioning thoracic paraganglioma: association with Von Hippel-Lindau syndrome. J. Clin. Endocr. Metab. 82: 3356-3360, 1997.

Bonnet, P.; Dechaume, J.; Blanc, E.: L'anevrisme cirsoide de la retine l'anevrisme racemeux, ses relations avec l'anevrisme cirsoide de la face et l'anevrisme cirsoid du cerveau. Bull. Soc. Ophtal. Franc. 51: 521-524, 1938.

Brauch, H.; Kishida, T.; Glavac, D.; Chen, F.; Pausch, F.; Hofler, H.; Latif, F.; Lerman, M. I.; Zbar, B.; Neumann, H. P. H.: Von Hippel-Lindau (VHL) disease with pheochromocytoma in the Black Forest region of Germany: evidence for a founder effect. Hum. Genet. 95: 551-556, 1995.

Brown, D. G.; Hilal, S. K.; Tenner, M. S.: Wyburn-Mason syndrome. Arch. Neurol. 28: 67-68, 1973.

Cahill, G. F.; Melicow, M. M.; Guerry, D.: The renal lesions in von Hippel-Lindau's disease. Trans. Am. Assoc. Genitourinary Surg. 35: 271-281, 1942.

Cameron, S. J.; Doig, A.: Cerebellar tumors presented with clinical features of phaeochromocytoma. Lancet I: 492-494, 1970.

Chapman, R. C.; Diaz-Perez, R.: Pheochromocytoma associated with cerebellar hemangioblastoma. J.A.M.A. 182: 1014-1017, 1962.

Chen, F.; Kishida, T.; Yao, M.; Hustad, T.; Glavac, D.; Dean, M.; Gnarra, J. R.; Orcutt, M. L.; Duh, F. M.; Glenn, G.; Green, J.; Hsia, Y. E.; Lamiell, J.; Li, H.; Wei, M. H.; Schmidt, L.; Tory, K.; Kuzman, I.; Stackhouse, T.; Latif, F.; Linehan, W. M.; Lerman, M.; Zbar, B.: Germline mutations in the von Hippel-Lindau disease tumor suppressor gene: correlations with phenotype. Hum. Mutat. 5: 66-75, 1995.

Christoferson, L. A.; Gustafson, M. B.; Petersen, A. G.: Von Hippel-Lindau's disease. J.A.M.A. 178: 280-282, 1961.

Collins, E. T.: Intra-ocular growths (two cases, brother and sister, with peculiar vascular new growth, probably primarily retinal, affecting both eyes). Trans. Ophthal. Soc. U.K. 14: 141-149, 1894.

Crossey, P. A.; Foster, K.; Richards, F. M.; Phipps, M. E.; Latif, F.; Tory, K.; Jones, M. H.; Bentley, E.; Kumar, R.; Lerman, M. I.; Zbar, B.; Affara, N. A.; Ferguson-Smith, M. A.; Maher, E. R.: Molecular genetic investigations of the mechanism of tumourigenesis in von Hippel-Lindau disease: analysis of allele loss in VHL tumours. Hum. Genet. 93: 53-58, 1994.

Crossey, P. A.; Richards, F. M.; Foster, K.; Green, J. S.; Prowse, A.; Latif, F.; Lerman, M. I.; Zbar, B.; Affara, N. A.; Ferguson-Smith, M. A.; Maher, E. R.: Identification of intragenic mutations in the von Hippel-Lindau disease tumour suppressor gene and correlation with disease phenotype. Hum. Molec. Genet. 3: 1303-1308, 1994.

Cushing, H.; Bailey, P.: Hemangiomas of cerebellum and retina (Lindau's disease), with the report of a case. Arch. Ophthal. 57: 447-463, 1928.

Davies, D. R.; Norman, A. M.; Whitehouse, R. W.; Evans, D. G. R.: Non-expression of von Hippel-Lindau phenotype in an obligate gene carrier. Clin. Genet. 45: 104-106, 1994.

Decker, H.-J. H.; Neumann, H. P. H.; Walter, T. A.; Sandberg, A. A.: 3p involvement in a renal cell carcinoma in von Hippel-Lindau syndrome: region of tumor breakpoint clustering on 3p?. Cancer Genet. Cytogenet. 33: 59-65, 1988.

Duan, D. R.; Humphrey, J. S.; Chen, D. Y. T.; Weng, Y.; Sukegawa, J.; Lee, S.; Gnarra, J. R.; Linehan, W. M.; Klausner, R. D.: Characterization of the VHL tumor suppressor gene product: localization, complex formation, and the effect of natural inactivating mutations. Proc. Nat. Acad. Sci. 92: 6459-6463, 1995.

Duan, D. R.; Pause, A.; Burgess, W. H.; Aso, T.; Chen, D. Y. T.; Garrett, K. P.; Conaway, R. C.; Conaway, J. W.; Linehan, W. M.; Klausner, R. D.: Inhibition of transcription elongation by the VHL tumor suppressor protein. Science 269: 1402-1406, 1995.

Eng, C.; Crossey, P. A.; Mulligan, L. M.; Healey, C. S.; Houghton, C.; Prowse, A.; Chew, S. L.; Dahia, P. L. M.; O'Riordan, J. L. H.; Toledo, S. P. A.; Smith, D. P.; Maher, E. R.; Ponder, B. A. J.: Mutations in the RET proto-oncogene and the von Hippel-Lindau disease tumour suppressor gene in sporadic and syndromic phaeochromocytomas. J. Clin. Genet. 32: 934-937, 1995.

Erbe, R. W.: Cabot case. New Eng. J. Med. 298: 95-101, 1978.

Fearon, E. R.: Human cancer syndromes: clues to the origin and nature of cancer. Science 278: 1043-1050, 1997.

Fill, W. L.; Lamiell, J. M.; Polk, N. O.: The radiographic manifestations of von Hippel-Lindau disease. Radiology 133: 289-295, 1979.

Fishman, R. S.; Bartholomew, L. G.: Severe pancreatic involvement in three generations in von Hippel-Lindau disease. Mayo Clin. Proc. 54: 329-331, 1979.

Fulton, J. F.: Harvey Cushing: A Biography. Springfield, Illinois: Charles C Thomas (pub.) 1946.

Funk, K. C.; Heiken, J. P.: Papillary cystadenoma of the broad ligament in a patient with von Hippel-Lindau disease. Am. J. Radiol. 153: 527-528, 1989.

Gaffey, M. J.; Mills, S. E.; Boyd, J. C.: Aggressive papillary tumor of middle ear/temporal bone and adnexal papillary cystadenoma. Am. J. Surg. Path. 18: 1254-1260, 1994.

Garcia, A.; Matias-Guiu, X.; Cabezas, R.; Chico, A.; Prat, J.; Baiget, M.; De Leiva, A.: Molecular diagnosis of von Hippel-Lindau disease in a kindred with a predominance of familial phaeochromocytoma. Clin. Endocr. 46: 359-363, 1997.

Gersell, D. J.; King, T. C.: Papillary cystadenoma of the mesosalpinx in von Hippel-Lindau disease. Am. J. Surg. Path. 12: 145-149, 1988.

Gilcrease, M. Z.; Schmidt, L.; Zbar, B.; Truong, L.; Rutledge, M.; Wheeler, T. M.: Somatic von Hippel-Lindau mutation in clear cell papillary cystadenoma of the epididymis. Hum. Path. 26: 1341-1346, 1995.

Glenn, G. M.; Daniel, L. N.; Choyke, P.; Linehan, W. M.; Oldfield, E.; Gorin, M. B.; Hosoe, S.; Latif, F.; Weiss, G.; Walther, M.; Lerman, M. I.; Zbar, B.: Von Hippel-Lindau (VHL) disease: distinct phenotypes suggest more than one mutant allele at the VHL locus. Hum. Genet. 87: 207-210, 1991.

Glenn, G. M.; Linehan, W. M.; Hosoe, S.; Latif, F.; Yao, M.; Choyke, P.; Gorin, M. B.; Chew, E.; Oldfield, E.; Manolatos, C.; Orcutt, M. L.; Walther, M. M.; Weiss, G. H.; Tory, K.; Jensson, O.; Lerman, M. I.; Zbar, B.: Screening for von Hippel-Lindau disease by DNA polymorphism analysis. J.A.M.A. 267: 1226-1231, 1992.

Go, R. C. P.; Lamiell, J. M.; Hsia, Y. E.; Yuen, J. W.-M.; Paik, Y.: Segregation and linkage analyses of von Hippel Lindau disease among 220 descendants from one kindred. Am. J. Hum. Genet. 36: 131-142, 1984.

Goldberg, M. F.; Duke, J. R.: Von Hippel-Lindau disease: histopathologic findings in a treated and untreated eye. Am. J. Ophthal. 66: 693-705, 1968.

Graff, J. W.: Personal Communication. Brookline, Mass. 10/4/1998.

Green, J. S.; Bowmer, M. I.; Johnson, G. J.: Von Hippel-Lindau disease in a Newfoundland kindred. Canad. Med. Assoc. J. 134: 133-138 and 146, 1986.

Griffiths, D. F. R.; Williams, G. T.; Williams, E. D.: Duodenal carcinoid tumours, phaeochromocytoma and neurofibromatosis: islet cell tumour, phaeochromocytoma and the von Hippel-Lindau complex: two distinctive neuroendocrine syndromes. Quart. J. Med. 245: 769-782, 1987.

Gross, D. J.; Avishai, N.; Meiner, V.; Filon, D.; Zbar, B.; Abeliovich, D.: Familial pheochromocytoma associated with a novel mutation in the von Hippel-Lindau gene. J. Clin. Endocr. Metab. 81: 147-149, 1996.

Grossman, M.; Melmon, K. L.: Von Hippel-Lindau disease.In: Vinken, P. J.; Bruyn, G. W.: Handbook of Clinical Neurology. The Phakomatoses. Amsterdam: North Holland (pub.) 14: 1972. Pp. 241-259.

Hagler, W. S.; Hyman, B. N.; Waters, W. C., III: Von Hippel's angiomatosis retinae and pheochromocytoma. Trans. Am. Acad. Ophthal. Otolaryng. 75: 1022-1034, 1971.

Hennessy, T. G.; Stern, W. E.; Herrick, S. E.: Cerebellar hemangioblastoma: erythropoietic activity by radioion assay. J. Nucl. Med. 8: 601-606, 1967.

Herman, J. G.; Latif, F.; Weng, Y.; Lerman, M. I.; Zbar, B.; Liu, S.; Samid, D.; Duan, D.-S. R.; Gnarra, J. R.; Linehan, W. M.; Baylin, S. B.: Silencing of the VHL tumor-suppressor gene by DNA methylation in renal carcinomas. Proc. Nat. Acad. Sci. 91: 9700-9704, 1994.

Horbach, J. M. L. M.; Brenninkmeyer, S. J.; van de Velde, C. J. H.; Nieuwenhuyzen Kruseman, A. C.: A forme fruste of von Hippel-Lindau disease—a combination of adrenal pheochromocytoma and ipsilateral renal cell carcinoma: a case report. Surgery 105: 436-441, 1989.

Horton, W. A.; Wong, V.; Eldridge, R.: Von Hippel-Lindau disease—clinical and pathological manifestations in 9 families with 50 affected members. Arch. Intern. Med. 136: 769-777, 1976.

Hosoe, S.; Brauch, H.; Latif, F.; Glenn, G.; Daniel, L.; Bale, S.; Choyke, P.; Gorin, M.; Oldfield, E.; Berman, A.; Goodman, J.; Orcutt, M. L.; Hampsch, K.; Delisio, J.; Modi, W.; McBride, W.; Anglard, P.; Weiss, G.; Walther, M. M.; Linehan, W. M.; Lerman, M. I.; Zbar, B.: Localization of the von Hippel-Lindau disease gene to a small region of chromosome 3. Genomics 8: 634-640, 1990.

Huson, S. M; Harper, P. S.; Hourihan, M. D.; Cole, G.; Weeks, R. D.; Compston, D. A. S.: Cerebellar haemangioblastoma and von Hippel-Lindau disease. Brain 109: 1297-1310, 1986.

Iliopoulos, O.; Kibel, A.; Gray, S.; Kaelin, W. G., Jr.: Tumour suppression by the human von Hippel-Lindau gene product. Nature Med. 1: 822-826, 1995.

Iliopoulos, O.; Levy, A. P.; Jiang, C.; Kaelin, W. G., Jr.; Goldberg, M. A.: Negative regulation of hypoxia-inducible genes by the von Hippel-Lindau protein. Proc. Nat. Acad. Sci. 93: 10595-10599, 1996.

Iliopoulos, O.; Ohh, M.; Kaelin, W. G., Jr.: pVHL(19) is a biologically active product of the von Hippel-Lindau gene arising from internal translation initiation. Proc. Nat. Acad. Sci. 95: 11661-11666, 1998.
Ivanov, S. V.; Kuzmin, I.; Wei, M.-H.; Pack, S.; Geil, L.; Johnson, B. E.; Stanbridge, E. J.; Lerman, M. I.: Down-regulation of transmembrane carbonic anhydrases in renal cell carcinoma cell lines by wild-type von Hippel-Lindau transgenes. Proc. Nat. Acad. Sci. 95: 12596-12601, 1998.
James, G. P.: Personal Communication. Springfield, Ohio 8/15/1998.
Jennings, A. M.; Smith, C.; Cole, D. R.; Jennings, C.; Shortland, J. R.; Williams, J. L.; Brown, C. B.: Von Hippel-Lindau disease in a large British family: clinicopathological features and recommendations for screening and follow-up. Quart. J. Med. 66: 233-249, 1988.
Kanno, H.; Kondo, K.; Ito, S.; Yamamoto, I.; Fujii, S.; Torigoe, S.; Sakai, N.; Hosaka, M.; Shuin, T.; Yao, M.: Somatic mutations of the von Hippel-Lindau tumor suppressor gene in sporadic central nervous system hemangioblastomas. Cancer Res. 54: 4845-4847, 1994.
Kaplan, C.; Sayre, G. P.; Greene, L. F.: Bilateral nephrogenic carcinomas in Lindau-von Hippel disease. J. Urol. 86: 36-42, 1961.
Karsdorp, N.; Elderson, A.; Wittebol-Post, D.; Hene, R. J.; Vos, J.; Feldberg, M. A. M.; van Gils, A. P. G.; Jansen-Schillhorn van Veen, J. M.; Vroom, T. M.; Hoppener, J. W. M.; Lips, C. J. M.: Von Hippel-Lindau disease: new strategies in early detection and treatment. Am. J. Med. 97: 158-168, 1994.
Keeler, L. L., III; Klauber, G. T.: von Hippel-Lindau disease and renal cell carcinoma in a 16-year-old boy. J. Urol. 147: 1588-1591, 1992.
Kenck, C.; Wilhelm, M.; Bugert, P.; Staehler, G.; Kovacs, G.: Mutation of the VHL gene is associated exclusively with the development of non-papillary renal cell carcinomas. J. Path. 179: 157-161, 1996.
Kerr, D. J.; Scheithauer, B. W.; Miller, G. M.; Ebersold, M. J.; McPhee, T. J.: Hemangioblastoma of the optic nerve: case report. Neurosurgery 36: 573-581, 1995.
Kibel, A.; Iliopoulos, O.; DeCaprio, J. A.; Kaelin, W. G., Jr.: Binding of the von Hippel-Lindau tumor suppressor protein to elongin B and C. Science 269: 1444-1446, 1995.
Kiechle-Schwarz, M.; Neumann, H. P. H.; Decker, H.-J. H.; Dietrich, C.; Wullich, B.; Schempp, W.: Cytogenetic studies on three pheochromocytomas derived from patients with von Hippel-Lindau syndrome. Hum. Genet. 82: 127-130, 1989.
King, C. R.; Schimke, R. N.; Arthur, T.; Davoren, B.; Collins, D.: Proximal 3p deletion in renal cell carcinoma cells from a patient with von Hippel-Lindau disease. Cancer Genet. Cytogenet. 27: 345-348, 1987.
Korn, W. T.; Schatzki, S. C.; DiSciullo, A. J.; Scully, R. E.: Papillary cystadenoma of the broad ligament in von Hippel-Lindau disease. Am. J. Obstet. Gynec. 163: 596-598, 1990.
Kovacs, G.; Kung, H.: Nonhomologous chromatid exchange in hereditary and sporadic renal cell carcinomas. Proc. Nat. Acad. Sci. 88: 194-198, 1991.
Kranes, A.; Balogh, K., Jr.: Liver disease in a patient with von Hippel-Lindau disease. New Eng. J. Med. 275: 950-959, 1966.
Lamiell, J. M.: Personal Communication. San Francisco, Calif. 9/1987.
Lamiell, J. M.; Salazar, F. G.; Hsia, Y. E.: Von Hippel-Lindau disease affecting 43 members of a single kindred. Medicine 68: 1-29, 1989.
Latif, F.; Tory, K.; Gnarra, J.; Yao, M.; Duh, F.-M.; Orcutt, M. L.; Stackhouse, T.; Kuzmin, I.; Modi, W.; Geil, L.; Schmidt, L.; Zhou, F.; Li, H.; Wei, M. H.; Chen, F.; Glenn, G.; Choyke, P.; Walther, M. M.; Weng, Y.; Duan, D.-S. R.; Dean, M.; Glavac, D.; Richards, F. M.; Crossey, P. A.; Ferguson-Smith, M. A.; Le Paslier, D.; Chumakov, I.; Cohen, D.; Chinault, A. C.; Maher, E. R.; Linehan, W. M.; Zbar, B.; Lerman, M. I.: Identification of the von Hippel-Lindau disease tumor suppressor gene. Science 260: 1317-1320, 1993.
Lee, S.; Chen, D. Y. T.; Humphrey, J. S.; Gnarra, J. R.; Linehan, W. M.; Klausner, R. D.: Nuclear/cytoplasmic localization of the von Hippel-Lindau tumor suppressor gene product is determined by cell density. Proc. Nat. Acad. Sci. 93: 1770-1775, 1996.
Lenz, T.; Thiede, H. M.; Nussberger, J.; Atlas, S. A.; Distler, A.; Schulte, K. L.: Hyperreninemia and secondary hyperaldosteronism in a patient with pheochromocytoma and von Hippel-Lindau disease. Nephron 62: 345-350, 1992.
Lindau, A.: Zur Frage der Angiomatosis Retinae und Ihrer Hirncomplikation. Acta Ophthal. 4: 193-226, 1927.
Loeb, D. B.; Pericak-Vance, M. A.; Stajich, J. M.; Vance, J. M.: A novel mutation in the von Hippel-Lindau gene. Hum. Molec. Genet. 3: 1423-1424, 1994.
Maddock, I. R.; Moran, A.; Maher, E. R.; Teare, M. D.; Norman, A.; Payne, S. J.; Whitehouse, R.; Dodd, C.; Lavin, M.; Hartley, N.; Super, M.; Evans, D. G. R.: A genetic register for von Hippel-Lindau disease. J. Med. Genet. 33: 120-127, 1996.
Maher, E. R.; Bentley, E.; Yates, J. R. W.; Barton, D.; Jennings, A.; Fellows, I. W.; Ponder, M. A.; Ponder, B. A. J.; Benjamin, C.; Harris, R.; Ferguson-Smith, M. A.: Mapping of von Hippel-Lindau disease to chromosome 3p confirmed by genetic linkage analysis. J. Neurol. Sci. 100: 27-30, 1990.
Maher, E. R.; Bentley, E.; Yates, J. R. W.; Latif, F.; Lerman, M.; Zbar, B.; Affara, N. A.; Ferguson-Smith, M. A.: Mapping of the von Hippel-Lindau disease locus to a small region of chromosome 3p by genetic linkage analysis. Genomics 10: 957-960, 1991.
Maher, E. R.; Iselius, L.; Yates, J. R. W.; Littler, M.; Benjamin, C.; Harris, R.; Sampson, J.; Williams, A.; Ferguson-Smith, M. A.; Morton, N.: Von Hippel-Lindau disease: a genetic study. J. Med. Genet. 28: 443-447, 1991.
Maher, E. R.; Yates, J. R. W.; Ferguson-Smith, M. A.: Statistical analysis of the two stage mutation model in von Hippel-Lindau disease, and in sporadic cerebellar haemangioblastoma and renal cell carcinoma. J. Med. Genet. 27: 311-314, 1990.
Maher, E. R.; Yates, J. R. W.; Ferguson-Smith, M. A.: Statistical analysis of the two stage mutation model in von Hippel-Lindau disease, sporadic cerebellar haemangioblastoma, and renal cell carcinoma. (Abstract) J. Med. Genet. 27: 205, 1990.
Manski, T. J.; Heffner, D. K.; Glenn, G. M.; Patronas, N. J.; Pikus, A. T.; Katz, D.; Lebovics, R.; Sledjeski, K.; Choyke, P. L.; Zbar, B.; Linehan, W. M.; Oldfield, E. H.: Endolymphatic sac tumors: a source of morbid hearing loss in von Hippel-Lindau disease. J.A.M.A. 277: 1461-1466, 1997.
McKusick, V. A.; Abbey, H.; Bettersby, E. J.; Borhani, N. O.; Boyer, S. H., IV; Cohen, B. H.; Ferguson-Smith, M. A.; Franke, F. R.; Gordon, H.; Handmaker, S. D..; Harris, W. S.; Hawkins, M. R.; and 13 others: Medical genetics 1960. J. Chronic Dis. 14: 1-198, 1961. Fig. 71.
Melmon, K. L.; Rosen, S. W.: Lindau's disease: review of the literature and study of a large kindred. Am. J. Med. 36: 595-617, 1964.
Mukhopadhyay, D.; Knebelmann, B.; Cohen, H. T.; Ananth, S.; Sukhatme, V. P.: The von Hippel-Lindau tumor suppressor gene product interacts with Sp1 to repress vascular endothelial growth factor promoter activity. Molec. Cell. Biol. 17: 5629-5639, 1997.
Neumann, H. P. H.; Eng, C.; Mulligan, L. M.; Glavac, D.; Zauner, I.; Ponder, B. A. J.; Crossey, P. A.; Maher, E. R.; Brauch, H.: Consequences of direct genetic testing for germline mutations in the clinical management of families with multiple endocrine neoplasia, type II. J.A.M.A. 274: 1149-1151, 1995.
Neumann, H. P. H.; Wiestler, O. D.: Clustering of features of von Hippel-Lindau syndrome: evidence for a complex genetic locus. Lancet 337: 1052-1054, 1991.
Nibbelink, D. W.; Peters, B. H.; McCormick, W. F.: On the association of pheochromocytoma and cerebellar hemangioblastoma. Neurology 19: 455-460, 1969.
Oberstrass, J.; Reifenberger, G.; Reifenberger, J.; Wechsler, W.; Collins, V. P.: Mutation of the von Hippel-Lindau tumour suppressor gene in capillary haemangioblastomas of the central nervous system. J. Path. 179: 151-156, 1996.
Ohh, M.; Yauch, R. L.; Lonergan, K. M.; Whaley, J. M.; Stemmer-Rachamimov, A. O.; Louis, D. N.; Gavin, B. J.; Kley, N.; Kaelin, W. G., Jr.; Iliopoulos, O.: The von Hippel-Lindau tumor suppressor protein is required for proper assembly of an extracellular fibronectin matrix. Molec. Cell 1: 959-968, 1998.
Otenasek, F. J.; Silver, M. L.: Spinal hemangioma (hemangioblastoma) in Lindau's disease: report of six cases in a single family. J. Neurosurg. 18: 295-300, 1961.
Pause, A.; Lee, S.; Lonergan, K. M.; Klausner, R. D.: The von Hippel-Lindau tumor suppressor gene is required for cell cycle exit upon serum withdrawal. Proc. Nat. Acad. Sci. 95: 993-998, 1998.
Price, E. B., Jr.: Papillary cystadenoma of the epididymis: a clinicopathologic analysis of 20 cases. Arch. Path. 91: 456-470, 1971.
Probst, A.; Lotz, M.; Heitz, P.: Von Hippel-Lindau's disease, syringomyelia and multiple endocrine tumors: a complex neuroendocrinopathy. Virchows Arch. Path. Anat. Histol. 378: 265-272, 1978.
Prowse, A. H.; Webster, A. R.; Richards, F. M.; Richard, S.; Olschwang, S.; Resche, F.; Affara, N. A.; Maher, E. R.: Somatic inactivation of the VHL gene in Von Hippel-Lindau disease tumors. Am. J. Hum. Genet. 60: 765-771, 1997.
Rho, Y. M.: Von Hippel-Lindau's disease: a report of 5 cases. Canad. Med. Assoc. J. 101: 135-142, 1969.
Richards, F. M.; Crossey, P. A.; Phipps, M. E.; Foster, K.; Latif, F.; Evans, G.; Sampson, J.; Lerman, M. I.; Zbar, B.; Affara, N. A.; Ferguson-Smith, M. A.; Maher, E. R.: Detailed mapping of germline deletions of the von Hippel-Lindau disease tumour suppressor gene. Hum. Molec. Genet. 3: 595-598, 1994.
Richards, F. M.; Maher, E. R.; Latif, F.; Phipps, M. E.; Tory, K.; Lush, M.; Crossey, P. A.; Oostra, B.; Gustavson, K. H.; Green, J.; Turner, G.; Yates, J. R. W.; Linehan, W. M.; Affara, N. A.; Lerman, M.; Zbar, B.; Ferguson-Smith, M. A.: Detailed genetic mapping of the von Hippel-Lindau disease tumour suppressor gene. J. Med. Genet. 30: 104-107, 1993.
Richards, F. M.; Phipps, M. E.; Latif, F.; Yao, M.; Crossey, P. A.; Foster, K.; Linehan, W. M.; Affara, N. A.; Lerman, M. I.; Zbar, B.; Ferguson-Smith, M. A.; Maher, E. R.: Mapping the von Hippel-Lindau disease tumour suppressor gene: identification of germline deletions by pulsed field gel electrophoresis. Hum. Molec. Genet. 2: 879-882, 1993.
Richards, F. M.; Schofield, P. N.; Fleming, S.; Maher, E. R.: Expression of the von Hippel-Lindau disease tumour suppressor gene during human embryogenesis. Hum. Molec. Genet. 5: 639-644, 1996.
Rubenstein, J. L.; Yaari, H.: von Hippel-Lindau and the genetics of astrocytoma. J. Nat. Cancer Inst. 86: 142-143, 1994.
Sander, S.; Normann, T.; Mathisen, W.: Pheochromocytoma associated with von-Hippel-Lindau's disease in a family. Scand. J. Urol. Nephrol. 4: 259-263, 1970.
Schechterman, L.: Lindau's disease: report of an unusual case and two additional cases in a Negro family. Med. Ann. D.C. 30: 64-76, 1961.
Schoenfeld, A.; Davidowitz, E. J.; Burk, R. D.: A second major native von Hippel-Lindau gene product, initiated from an internal translation start site, functions as a tumor suppressor. Proc. Nat. Acad. Sci. 95: 8817-8822, 1998.
Seizinger, B. R.; Farmer, G.; Haines, J.; Anderson, K.; Whaley, J.; Hettlich, C.; Decker, J.; Rouleau, G.; Smith, D.; Drabkin, H.; Filling-Katz, M.; Neumann, H.; Collins, D.; Hsia, E.; Green, J.; Waziri, M.; Gusella, J.; Li, F.: Isolating the gene(s) for von Hippel-Lindau disease and renal cell carcinoma. (Abstract) Am. J. Hum. Genet. 45 (suppl.): A32, 1989.
Seizinger, B. R.; Rouleau, G. A.; Ozelius, L. J.; Lane, A. H.; Farmer, G. E.; Lamiell, J. M.; Haines, J.; Yuen, J. W.; Collins, D.; Majoor-Krakauer, D.; et al.: Von Hippel-Lindau disease maps to the region of chromosome 3 associated with renal cell carcinoma. Nature 332: 268-269, 1988.
Seizinger, B. R.; Smith, D. I.; Filling-Katz, M. R.; Neumann, H.; Green, J. S.; Choyke, P. L.; Anderson, K. M.; Freiman, R. N.; Klauck, S. M.; Whaley, J.; Decker, H.-J. H.; Hsia, Y. E.; Collins, D.; Halperin, J.; Lamiell, J. M.; Oostra, B.; Waziri, M. H.; Gorin, M. B.; Scherer, G.; Drabkin, H. A.; Aronin, N.; Schinzel, A.; Martuza, R. L.; Gusella, J. F.; Haines, J. L.: Genetic flanking markers refine diagnostic criteria and provide insights into the genetics of Von Hippel Lindau disease. Proc. Nat. Acad. Sci. 88: 2864-2868, 1991.
Sharp, W. V.; Platt, R. L.: Familial pheochromocytoma: association with von-Hippel-Lindau's disease. Angiology 22: 141-146, 1971.
Shokeir, M. H. K.: Von Hippel-Lindau syndrome: a report on three kindreds. J. Med. Genet. 7: 155-157, 1970.
Silver, M. L.: Hereditary vascular tumors of the nervous system. J.A.M.A. 156: 1053-1056, 1954.

Thomas, M.; Burnside, R. M.: Von Hippel-Lindau disease. Am. J. Ophthal. 51: 140-146, 1961.
Tisherman, S. E.; Gregg, F. J.; Danowski, T. S.: Familial pheochromocytoma. J.A.M.A. 182: 152-156, 1962.
Tisherman, S. E.; Tisherman, B. G.; Tisherman, S. A.; Dunmire, S.; Levey, G. S.; Mulvihill, J. J.: Three-decade investigation of familial pheochromocytoma: an allele of von Hippel-Lindau disease? Arch. Int. Med. 153: 2550-2556, 1993.
Tory, K.; Brauch, H.; Linehan, M.; Barba, D.; Oldfield, E.; Filling-Katz, M.; Seizinger, B.; Nakamura, Y.; White, R.; Marshall, F. F.; Lerman, M. I.; Zbar, B.: Specific genetic change in tumors associated with von Hippel-Lindau disease. J. Nat. Cancer Inst. 81: 1097-1101, 1989.
Tsuda, H.; Fukushima, S.; Takahashi, M.; Hikosaka, Y.; Hayashi, K.: Familial bilateral papillary cystadenoma of the epididymis: report of three cases in siblings. Cancer 37: 1831-1839, 1976.
Vance, J. M.; Small, K. W.; Jones, M. A.; Stajich, J. M.; Yamaoka, L. H.; Roses, A. D.; Hung, W.-Y.; Pericak-Vance, M. A.: Confirmation of linkage in von Hippel-Lindau disease. Genomics 6: 565-567, 1990.
Versteeg, R.: Aberrant methylation in cancer. (Editorial) Am. J. Hum. Genet. 60: 751-754, 1997.
Vogelstein, B.: Personal Communication. Baltimore, Md. 1/6/1995.
von Hippel, E.: Ueber eine sehr seltene Erkrankung der Netzhaut. Albrecht von Graefes Arch. Ophthal. 59: 83-106, 1904.
Wells, R. A.; Reeders, S. T.; Green, J.; Johnson, G. J.; Robson, K. J. H.: Linkage studies of von Hippel-Lindau syndrome using multiple-locus hypervariable DNA probes. (Abstract) Cytogenet. Cell Genet. 46: 714, 1987.
Wesolowski, D. P.; Ellwood, R. A.; Schwab, R. E.; Farah, J.: Hippel-Lindau syndrome in identical twins. Brit. J. Radiol. 54: 982-986, 1981.
Wise, K. S.; Gibson, J. A.: Von Hippel-Lindau's disease and phaeochromocytoma. Brit. Med. J. 1: 441, 1971.
Wyburn-Mason, R.: Arteriovenous aneurysm of mid-brain and retina, facial naevi and mental changes. Brain 66: 163-203, 1943.
Zbar, B.; Kishida, T.; Chen, F.; Schmidt, L.; Maher, E. R.; Richards, F. M.; Crossey, P. A.; Webster, A. R.; Affara, N. A.; Ferguson-Smith, M. A.; Brauch, H.; Glavac, D.; and 14 others: Germline mutations in the Von Hippel-Lindau disease (VHL) gene in families from North America, Europe, and Japan. Hum. Mutat. 8: 348-357, 1996.
Zhuang, Z.; Emmert-Buck, M. R.; Roth, M. J.; Gnarra, J.; Linehan, W. M.; Liotta, L. A.; Lubensky, I. A.: Von Hippel-Lindau disease gene deletion detected in microdissected sporadic human colon carcinoma specimens. Hum. Path. 27: 152-156, 1996.

TUBEROUS SCLEROSIS 1 (*191100)

JJM NOTE

Besides the defining benign tumors, many malignant tumor types have been associated with tuberous sclerosis: gliomas (specifically, subependymal giant-cell astrocytoma), Wilms tumor, renal cell carcinoma and breast carcinoma (Tsukada T, Pickren JW: Carcinoma of the breast and aysmptomatic tuberous sclerosis. NY State J. Med. 67:593–97, 1967). M. R. Gomez (*Tuberous Sclerosis,* 2nd ed., New York; Raven, 1988) emphasized that the classic triad of seizures, mental retardation, and facial angiofibroma is seen in only 29% of patients and is totally absent in 6%. A definitive diagnosis is made in the presence of any of the following: cortical tuber, subependymal glial nodules, retinal hamartoma, facial angiofibromas, ungual fibroma, forehead or scalp fibrous plaques or mutiple renal angiomyolipomas.

The preferred designation for this disorder refers to the changes observed in the brain. Adenoma sebaceum, a synonym which refers to the cutaneous features, is a misnomer; facial angiofibroma more accurately describes the lesions. Another synonym, epiloia, stands for 'epilepsy plus anoia'; anoia = mental retardation. Gorlin (1981) suggested that epiloia is a useful acronym and mnemonic for 'epilepsy, low intelligence, and adenoma sebaceum.' Rhabdomyoma of the myocardium and mixed tumor of the kidney also occur. Fitzpatrick et al. (1968) pointed out the diagnostic usefulness of white macules shaped like the leaf of a mountain ash in patients with TS. The white macules are probably present at birth in most cases, thus permitting early diagnosis. They may be evident only under Wood light. Kidney lesions are in the form of angiomyolipoma (Anderson and Tannen, 1969). Indeed, van Baal et al. (1989) found renal angiomyolipomas in 23 of 38 patients with proven tuberous sclerosis. Multiplicity and bilateral localization were important differences between the tuberous sclerosis cases and the isolated, usually solitary, angiomyolipomas. One of the parents of a patient with tuberous sclerosis had small renal angiomyolipomas without signs of tuberous sclerosis, suggesting that renal angiomyolipomas may be a forme fruste of tuberous sclerosis. Grether et al. (1987) described Wilms tumor (194070) in a 20-month-old child with tuberous sclerosis. Wilms tumor in this disorder appears to be rare, whereas angiomyolipomas and renal cysts are frequent.

Cook et al. (1996) reviewed 139 patients with TSC who had presented without renal symptoms but had been investigated by renal ultrasound. Renal lesions were found in 85 of the patients (61%). Forty patients had only angiomyolipomas and 17 had only cysts; 28 had both angiomyolipomas and cysts. Angiomyolipomas were multiple in 91% and bilateral in 84% of patients. The incidence of angiomyolipomas increased with age, but the incidence of renal cysts did not appear to be age related. In children under the age of 5 years, cysts were a more common lesion. Renal carcinomas were found in 3 patients.

In 3 affected infants, 2 of whom were newborns, de Leon et al. (1988) described olfactory hamartomas. They suggested that cardiac and olfactory hamartomas are particularly characteristic of tuberous sclerosis in infants and that olfactory involvement is not surprising because the hamartomas appear to arise from the subependymal germinal layer. Bender and Yunis (1981) reported 3 cases in neonates. Each had focal, frequently perivascular, collections of large cells in the spleen. These cells resembled those found in brain lesions of TS but did not stain for acidic protein. Norio (1981) showed me examples of enormous cystic kidneys in infants with tuberous sclerosis. Severe hypertension was present in some. Surgical marsupialization of large cysts appeared to be beneficial. The epithelium lining the cysts 'looked active' in a humoral or secretory way.

Marshall et al. (1959), described brain tumors in 2 families. In 1 family, ependymoma of the third ventricle was found, and in the second a mother and her 16-year-old son had astrocytoma of the third ventricle. Harvey Cushing had removed the mother's brain tumor in 1932. In 1959 the patient was still alive without evidence of recurrence. The incidence of childhood brain tumors in patients with tuberous sclerosis varies between 5 and 14%. More than 90% of these tumors are subependymal giant cell astrocytomas. The most common location of giant cell astrocytomas are the retina and the borders of the lateral ventricles, particularly the region of the foramen of Monro. Pascual-Castroviejo et al. (1995) reported a 23-year-old patient with giant cell astrocytomas in both the retina and the region of the foramen of Monro.

Teplick (1969) stated that adenoma sebaceum is absent in half the cases. He described a 53-year-old woman of normal intelligence with bone and pulmonary lesions misinterpreted as those of sarcoid. Bundey et al. (1970) described affected father and 3 children without adenoma sebaceum. Webb et al. (1996) found skin abnormalities in 126 of 131 English patients with tuberous sclerosis. Hypomelanotic macules were found in 80 patients; 32 of these had more than 5 such macules. Most macules were evident from birth, but some regressed in adulthood. Shagreen patches and facial angiofibromas appeared before 14 years of age, and their frequency remained constant in postpubertal patients (54 to 55% for shagreen patches and 81 to 88% for angiofibromas). The frequency of ungual fibromas increased with age: ungual fibromas were not found in children younger than 5 years, were found in 23% of children between 5 and 14 years, and were evident in 88% of patients older than 30 years.

Freycon et al. (1971) reported death due to rupture of an aortic aneurysm in a 3-year-old boy, and Larbre et al. (1971) described rupture of the ascending thoracic aorta in a 2.5-year-old boy.

Many pedigrees support autosomal dominant inheritance. Wilson and Carter (1978) described a family in which a son and daughter had full-blown TS, and the parents were nonconsanguineous and were normal by clinical examination and by computerized brain tomography. Thus, the conclusion that 86% of cases are fresh mutations (Bundey and Evans, 1969) must be tempered. Gonadal mosaicism of 1 parent might explain the findings of Wilson and Carter (1978). Rushton and Shaywitz (1979) reported a family in which 3 males, related as proband, maternal uncle and maternal great-uncle, had tuberous sclerosis but their mothers were unaffected. They postulated an independent dominant gene that modified expression of the gene for tuberous sclerosis. Sybert and Hall (1979) pointed out that expression is highly variable and involvement may be missed in the mildly affected individual. Cassidy et al. (1983) studied the 26 presumably unaffected parents of 13 patients with TS. Three fathers and 1 mother were found in fact to have signs of being affected; 3 had skin changes, 3 had intracranial calcification by computerized tomography, and 1 had renal cysts. Gomez (1979) reviewed the experiences of the Mayo Clinic from 1935 to 1979. Commenting on the monograph, Comings (1980) suggested that future studies may show that hamartomas in TS have a homozygous mutation of a cell surface protein, with heterozygosity in normal surrounding tissue.

Harding and Pagon (1990) concluded that between 51 and 86% of cardiac rhabdomyomas are associated with tuberous sclerosis. Cerebral and probable renal embolization from cardiac rhabdomyoma was reported by Kandt et al. (1985). Sugita et al. (1985) showed that intracranial calcification may be evident by computer-assisted cranial tomography within 1 week after birth. Journel et al. (1986) recorded the curious experience of making the diagnosis of familial tuberous sclerosis by finding cardiac tumors (rhabdomyomata) by routine ultrasonography. There was no evidence of TS in either of the parents, in the earlier-born sister, or in the maternal grandmother, but a maternal aunt was mentally deficient with seizures, numerous depigmented spots, and a shagreen patch on the back. The mother's grandmother died at age 67, following removal of a cerebral tumor of astrocytoma type. There were no signs of renal tumor in the fetus and the usual experience is that whereas rhabdomyomata are congenital, renal tumors develop only later in life and are never found on ultrasonography in the prenatal or neonatal period. Smith et al. (1989) concluded from an echocardiographic study that cardiac rhabdomyomata in tuberous sclerosis tend to regress in early infancy, remain the same size through childhood, and then again regress in adolescence. They stated that the 'prevalence of tumours in young infants with tuberous sclerosis is likely to be considerably above 50%.' As other signs of tuberous sclerosis are usually absent at this age, echocardiography may afford the most useful single diagnostic test in early infancy.

Fryer et al. (1990) concluded that the routine use of cranial CT scan, renal ultrasound, and skeletal survey is not indicated in the investigation of possibly affected parents and sibs of patients with tuberous sclerosis. They pointed to the fact that the diagnosis of tuberous sclerosis has been made in adults exclusively on the basis of a CT scan but felt that renal ultrasound or skeletal x-rays were not indicated in parents who had normal clinical examinations.

Lygidakis and Lindenbaum (1987) observed multiple enamel pits (pitted enamel hypoplasia) in 71% of persons with typical TS and in 1 of 10 'atypical' cases. One parent and one half-sib of persons with typical TS were found to have multiple enamel pits but no other sign of TS. Sampson et al. (1992) reported that of 23 patients with permanent teeth, 11 (48%) showed multiple enamel pits (mean 4.6 pits, range 3-9), but none was seen in 6 patients with deciduous teeth. Similar pitted enamel hypoplasia was found in 5 of 563 controls (0.88%). Webb et al. (1994) suggested that children with mental handicap due to causes other than tuberous sclerosis have a higher occurrence of dental pits than the normal controls reported by Sampson et al. (1992). Flanagan et al. (1997) studied 10 probands with tuberous sclerosis, 20 first-degree relatives, and 25 controls for evidence of pitted enamel hypoplasia. They found that 100% of TS patients, 65% of relatives, and 72% of controls had pits. While 70% of TS patients had more than 14 pits per person, only 5% of relatives and 4% of controls had a similar number. Eighty-five percent of relatives and 84% of controls had fewer than 6 pits per person. Flanagan et al. (1997) concluded that examination for the presence or absence of dental enamel pits is not a useful screening test for first-degree relatives to detect otherwise unsuspected tuberous sclerosis.

McGrae and Hashimoto (1996) described a patient with segmental expression of tuberous sclerosis in the form of unilateral facial angiofibromas and suggested that this resulted from a postzygotic mutation.

Scappaticci et al. (1988) described increased frequency of premature centromere disjunction (PCD) in cultured fibroblasts. Chromosome 3 appeared to be involved preferentially with both PCD and dicentric formation.

In a study of 5 families, Connor et al. (1987) found a maximum lod score of 1.46 at zero recombination with ABO. Fryer et al. (1987) found a maximum lod score of 3.85 for linkage of ABO and TS at zero recombination (confidence limits, 0-0.10). Smith et al. (1987) found no recombination between TS and ASSP3, a pseudogene of argininosuccinate synthetase (215700), which is located at 9q11-q22. The cumulative lod score was 2.2 at theta = 0.0. In the study by Smith et al. (1987), 7 out of 15 families were informative for ABO. The cumulative lod score for linkage to TS was negative at values of theta of 0.2 and lower. Included in this score were 3 families in which the lod score with ABO was positive. In these families the cumulative lod score between ABO and TS was highest at theta = 0.0 (lod = 1.22). Smith et al. (1987) pointed to the great variability of TS observed in a single family and to the fact that a substantial number of obligate heterozygotes have no clinical manifestations. The existence of 'secondary' genes that interact with the primary gene to modify the genotype seems likely. In 3 informative families, Northrup et al. (1987) found negative lod scores with ABO. They pointed out that the availability of multiple RFLPs at the ASS locus, which maps to 9q34, may be useful in the study of these families; furthermore, since the ASS locus is selectable and can be used for chromosome-mediated gene transfer, analysis of this portion of the genome may be possible. Renwick (1987) also advised caution in interpreting the data of Fryer et al. (1987). He presented data on the TSC/ABO linkage derived from the data published from Copenhagen on pedigrees of Borberg (1951). Combined data yielded a maximum lod score of 1.94 at theta = 0.16. Renwick (1987) pointed out that Fryer et al. (1987) found no recombinants out of 4 opportunities for the TSC/AK1 (103000) linkage. Thus, TSC might be nearer to AK1 than to ABO. Connor et al. (1987) found no recombination between TS and a RFLP detected by oncogene ABL (189980) in 13 informative meioses (4 phase known), giving a maximum lod score of 3.18 at theta = 0.0. It should be noted that rigorous diagnostic criteria were exercised with use of a protocol that included the latest methods of CNS and abdominal imaging. The likelihood of scoring persons as unaffected may have been reduced to a minimum.

Povey et al. (1988) reported recombinants between TSC and ABL in 2 of 3 families informative for the linkage; none of these was informative for ABO or AK1. All affected individuals fulfilled their criteria and all at-risk individuals classified as unaffected were rigorously investigated. Combined with the data of Connor et al. (1987), the new data make the maximum lod score 2.36 at theta = 0.1. Povey et al. (1988) suggested that it is impossible to distinguish between 2 possibilities: that TSC is loosely linked to ABL in all families, or that in some families the mutation for TSC is not in the 9q linkage group at all. They urged that ABL not be used as a marker for prenatal diagnosis of TSC because of these uncertainties. From study of 8 families, Kandt et al. (1988) concluded that they could exclude linkage of TS to ABO (lod equal to or less than

See also Bjornberg (1961); Connor et al. (1987); De Groot et al. (1981); De la Cruz and Laveck (1962); Dwyer et al. (1971); Fryer et al. (1987); Hudolin (1961); Lagos and Gomez (1967); Martin et al. (1976); McWilliam and Stephenson (1978); Milledge et al. (1966); Nickel and Reed (1962); O'Callaghan et al. (1975); Rattan et al. (1983); Sampson et al. (1992); Sampson et al. (1989); Scheig and Bornstein (1961); Schull and Crowe (1953); Schwartz et al. (1980); Smith et al. (1987); Stapleton et al. (1980); Stevenson and Fisher (1956)

Anderson, D. E.; Tannen, R. L.: Tuberous sclerosis and chronic renal failure: potential confusion with polycystic kidney disease. Am. J. Med. 47: 163-168, 1969.

Au, K. S.; Murrell, J.; Buckler, A.; Blanton, S. H.; Northrup, H.: Report of a critical recombination further narrowing the TSC1 region. J. Med. Genet. 33: 559-561, 1996.

Baraitser, M.; Patton, M. A.: Reduced penetrance in tuberous sclerosis. J. Med. Genet. 22: 29-31, 1985.

Bender, B. L.; Yunis, E. J.: Splenic involvement in tuberous sclerosis: report of three cases. Virchows Arch. Path. Anat. 391: 363-369, 1981.

Bjornberg, A.: Adenoma sebaceum: review, case reports and discussion of eugenic aspects. Acta Derm. Venerol. 41: 213-223, 1961.

Bjornsson, J.; Short, M. P.; Kwiatkowski, D. J.; Henske, E. P.: Tuberous sclerosis-associated renal cell carcinoma. Am. J. Path. 149: 1201-1208, 1996.

Borberg, A.: Clinical and Genetic Investigations into Tuberous Sclerosis and Recklinghausen's Neurofibromatosis. Copenhagen: Munksgaard (pub.) 1951.

Bosi, G.; Lintermans, J. P.; Pellegrino, P. A.; Svaluto-Moreolo, G.; Vliers, A.: The natural history of cardiac rhabdomyoma with and without tuberous sclerosis. Acta Pediat. 85: 928-931, 1996.

Bundey, S.; Dutton, G.; Wells, R. S.: Tuberose sclerosis without adenoma sebaceum. J. Ment. Defic. Res. 14: 243-249, 1970.

Bundey, S.; Evans, K.: Tuberous sclerosis—a genetic study. J. Neurol. Neurosurg. Psychiat. 32: 591-603, 1969.

Carbonara, C.; Longa, L.; Grosso, E.; Borrone, C.; Garre, M. G.; Brisigotti, M.; Migone, N.: 9q34 loss of heterozygosity in a tuberous sclerosis astrocytoma suggests a growth suppressor-like activity also for the TSC1 gene. Hum. Molec. Genet. 3: 1829-1832, 1994.

Carbonara, C.; Longa, L.; Grosso, E.; Mazzucco, G.; Borrone, C.; Garre, M. L.; Brisigotti, M.; Filippi, G.; Scabar, A.; Giannotti, A.; Falzoni, P.; Monga, G.; Garini, G.; Gabrielli, M.; Riegler, P.; Danesino, C.; Ruggieri, M.; Magro, G.; Migone, N.: Apparent preferential loss of heterozygosity at TSC2 over TSC1 chromosomal region in tuberous sclerosis hamartomas. Genes Chromosomes Cancer 15: 18-25, 1996.

Cassidy, S. B.; Pagon, R. A.; Pepin, M.; Blumhagen, J. D.: Family studies in tuberous sclerosis: evaluation of apparently unaffected parents. J.A.M.A. 249: 1302-1304, 1983.

Comings, D. E.: Review of Gomez's 'Tuberous Sclerosis.'. Am. J. Hum. Genet. 32: 285-286, 1980.

Connor, J. M.; Pirrit, L. A.; Yates, J. R. W.; Fryer, A. E.; Ferguson-Smith, M. A.: Linkage of the tuberous sclerosis locus to a DNA polymorphism detected by v-abl.(Letter) J. Med. Genet. 24: 544-546, 1987.

Connor, J. M.; Yates, J. R. W.; Mann, L.; Aitken, D. A.; Stephenson, J. B. P.: Tuberous sclerosis: analysis of linkage to red cell and plasma protein markers. Cytogenet. Cell Genet. 44: 63-64, 1987.

Cook, J. A.; Oliver, K.; Mueller, R. F.; Sampson, J.: A cross sectional study of renal involvement in tuberous sclerosis. J. Med. Genet. 33: 480-484, 1996.

De Groot, W. P.; Woerdeman, M. J.; Witkiewicz, I. M.; Fleury, P.; Delleman, J. W.; Verbeeten, B.: Tuberous sclerosis: an investigation into the ratio of sporadic versus familial cases. (Abstract) Brit. J. Derm. 104: 99-100, 1981.

De la Cruz, F. F.; Laveck, G. D.: Tuberous sclerosis: a review and report of eight cases. Am. J. Ment. Defic. 67: 369-380, 1962.

de Leon, G. A.; Zaeri, N.; Foley, C. M.: Olfactory hamartomas in tuberous sclerosis. J. Neurol. Sci. 87: 187-194, 1988.

Dwyer, J. M.; Hickie, J. B.; Garvan, J.: Pulmonary tuberous sclerosis. Report of three patients and a review of the literature. Quart. J. Med. 40: 115-126, 1971.

Fahsold, R.; Rott, H.-D.; Claussen, U.; Schmalenberger, B.: Tuberous sclerosis in a child with de novo translocation t(3;12)(p26.3;q23.3). Clin. Genet. 40: 326-328, 1991.

Fitzpatrick, T. B.; Szabo, G.; Hori, Y.; Simone, A. A.; Reed, W. B.; Greenberg, M. H.: White leaf-shaped macules. Arch. Derm. 98: 1-6, 1968.

Flanagan, N.; O'Connor, W. J.; McCartan, B.; Miller, S.; McMenamin, J.; Watson, R.: Developmental enamel defects in tuberous sclerosis: a clinical genetic marker? J. Med. Genet. 34: 637-639, 1997.

Freycon, F.; Mollard, P. C.; Hermier, M.; Guibaud, P.; Chazalette, J.-P.; Weill, B.; Flattot, M.; Jeune, M.: Aneurisme de l'aorte abdominale au cours d'une sclerose tubereuse de Bourneville. Pediatrie 26: 421-427, 1971.

Fryer, A. E.; Chalmers, A.; Connor, J. M.; Fraser, I.; Povey, S.; Yates, A. D.; Yates, J. R. W.; Osborne, J. P.: Evidence that the gene for tuberous sclerosis is on chromosome 9. Lancet I: 659-661, 1987.

Fryer, A. E.; Chalmers, A. H.; Osborne, J. P.: The value of investigation for genetic counselling in tuberous sclerosis. J. Med. Genet. 27: 217-223, 1990.

Fryer, A. E.; Osborne, J. P.; Schutt, W.: Forehead plaque: a presenting skin sign in tuberous sclerosis. Arch. Dis. Child. 62: 292-304, 1987.

Gomez, M. R.: Tuberous Sclerosis. New York: Raven Press (pub.) 1979.

Gorlin, R. J.: Personal Communication. Minneapolis, Minn. 7/1981.

Grasso, S.; Manusia, M.; Sciacca, F.: Unusual liver lesion in tuberous sclerosis. Arch. Path. Lab. Med. 106: 49 only, 1982.

Green, A. J.; Johnson, P. H.; Yates, J. R. W.: The tuberous sclerosis gene on chromosome 9q34 acts as a growth suppressor. Hum. Molec. Genet. 3: 1833-1834, 1994.

Green, A. J.; Sepp, T.; Yates, J. R. W.: Clonality of tuberous sclerosis hamartomas shown by non-random X-chromosome inactivation. Hum. Genet. 97: 240-243, 1996.

Grether, P.; Carnevale, A.; Pasquel, P.: Wilms' tumor in an infant with tuberous sclerosis. Ann. Genet. 30: 183-185, 1987

Gunther, M.; Penrose, L. S.: Genetics of epiloia. J. Genet. 31: 413-430, 1935.

Haines, J. L.: Personal Communication. Boston, Mass. 5/7/1990.

Haines, J. L.; Short, M. P.; Kwiatkowski, D. J.; Jewell, A.; Andermann, E.; Bejjani, B.; Yang, C.-H.; Gusella, J. F.; Amos, J. A.: Localization of one gene for tuberous sclerosis within 9q32-9q34, and further evidence for heterogeneity. Am. J. Hum. Genet. 49: 764-772, 1991.

Hall, J. G.; Byers, P. H.: Genetics of tuberous sclerosis. (Letter) Lancet I: 751, 1987.

Harding, C. O.; Pagon, R. A.: Incidence of tuberous sclerosis in patients with cardiac rhabdomyoma. Am. J. Med. Genet. 37: 443-446, 1990.

Henske, E. P.; Scheithauer, B. W.; Short, M. P.; Wollmann, R.; Nahmias, J.; Hornigold, N.; van Slegtenhorst, M.; Welsh, C. T.; Kwiatkowski, D. J.: Allelic loss is frequent in tuberous sclerosis kidney lesions but rare in brain lesions. Am. J. Hum. Genet. 59: 400-406, 1996.

Hudolin, V.: La sclerose tubereuse (compte rendu des cas Yougoslaves). J. Genet. Hum. 10: 128-155, 1961.

Hunt, A.; Lindenbaum, R. H.: Tuberous sclerosis: a new estimate of prevalence within the Oxford region. J. Med. Genet. 21: 272-277, 1984.

Janssen, L. A. J.; Sandkuijl, L. A.; Sampson, J. R.; Halley, D. J. J.: Computer simulation of linkage and heterogeneity in tuberous sclerosis: a critical evaluation of the collaborative family data. J. Med. Genet. 29: 867-874, 1992.

Janssen, L. A. J.; Sandkuyl, L. A.; Merkens, E. C.; Maat-Kievit, J. A.; Sampson, J. R.; Fleury, P.; Hennekam, R. C. M.; Grosveld, G. C.; Lindhout, D.; Halley, D. J. J.: Genetic heterogeneity in tuberous sclerosis. Genomics 8: 237-242, 1990.

Jones, A. C.; Daniells, C. E.; Snell, R. G.; Tachataki, M.; Idziaszczyk, S. A.; Krawczak, M.; Sampson, J. R.; Cheadle, J. P.: Molecular genetic and phenotypic analysis reveals differences between TSC1 and TSC2 associated familial and sporadic tuberous sclerosis. Hum. Molec. Genet. 6: 2155-2161, 1997.

Journel, H.; Roussey, M.; Plais, M. H.; Milon, J.; Almange, C.; Le Marec, B.: Prenatal diagnosis of familial tuberous sclerosis following detection of cardiac rhabdomyoma by ultrasound. Prenatal Diag. 6: 283-289, 1986.

Kandt, R. S.; Gebarski, S. S.; Goetting, M. G.: Tuberous sclerosis with cardiogenic cerebral embolism: magnetic resonance imaging. Neurology 35: 1223-1225, 1985.

Kandt, R. S.; Haines, J. L.; Smith, M.; Northrup, H.; Gardner, R. J. M.; Short, M. P.; Dumars, K.; Roach, E. S.; Steingold, S.; Wall, S.; Blanton, S. H.; Flodman, P.; Kwiatkowski, D. J.; Jewell, A.; Weber, J. L.; Roses, A. D.; Pericak-Vance, M. A.: Linkage of a major gene locus for tuberous sclerosis to a chromosome 16 marker for polycystic kidney disease. (Abstract) Am. J. Hum. Genet. 51 (suppl.): A4, 1992.

Kandt, R. S.; Pericak-Vance, M. A.; Hung, W.-Y.; Gardner, R. J. M.; Nellist, M.; Phillips, K.; Warner, K.; Laing, N.; Roses, A. D.: Multilocus linkage analysis in tuberous sclerosis. (Abstract) Am. J. Hum. Genet. 43: A148, 1988.

Lagos, J. C.; Gomez, M. R.: Tuberous sclerosis: reappraisal of a clinical entity. Mayo Clin. Proc. 42: 26-49, 1967.

Larbre, F.; Loire, R.; Guibaud, P.; Lauras, B.; Weill, B.: Observation clinique et anatomique d'un anevrysme de l'aorte au cours d'une sclerose tubereuse de Bourneville. Arch. Franc. Pediat. 28: 975-984, 1971.

Lygidakis, N. A.; Lindenbaum, R. H.: Pitted enamel hypoplasia in tuberous sclerosis patients and first-degree relatives. Clin. Genet. 32: 216-221, 1987.

Marshall, D.; Saul, G. B.; Sachs, E., Jr.: Tuberous sclerosis: a report of 16 cases in two family trees revealing genetic dominance. New Eng. J. Med. 261: 1102-1105, 1959.

Martin, G. I.; Kaiserman, D.; Liegler, D.; Amorosi, E. D.; Nadel, H.: Computer-assisted cranial tomography in early diagnosis of tuberous sclerosis. J.A.M.A. 235: 2323-2324, 1976.

McGrae, J. D., Jr.; Hashimoto, K.: Unilateral facial angiofibromas: a segmental form of tuberous sclerosis. Brit. J. Derm. 134: 727-730, 1996.

McWilliam, R. C.; Stephenson, J. B. P.: Depigmented hair: the earliest sign of tuberous sclerosis. Arch. Dis. Child. 132: 961-963, 1978.

Milledge, R. D.; Gerald, B. E.; Carter, W. J.: Pulmonary manifestations of tuberous sclerosis. Am. J. Roentgen. 98: 734-738, 1966.

Nagase, T.; Seki, N.; Ishikawa, K.; Ohira, M.; Kawarabayasi, Y.; Ohara, O.; Tanaka, A.; Kotani, H.; Miyajima, N.; Nomura, N.: Prediction of the coding sequences of unidentified human genes. VI. The coding sequences of 80 new genes (KIAA0201-KIAA0280) deduced by analysis of cDNA clones from cell line KG-1 and brain. DNA Res. 3: 321-329, 1996.

Nevin, N. C.; Pearce, W. G.: Diagnostic and genetical aspects of tuberous sclerosis. J. Med. Genet. 5: 273-280, 1968.

Nickel, W. R.; Reed, W. B.: Tuberous sclerosis: special reference to the microscopic alterations in the cutaneous hamartomas. Arch. Derm. 85: 209-226, 1962.

Norio, R.: Personal Communication. Helsinki, Finland 5/27/1981.

Northrup, H.; Beaudet, A. L.; O'Brien, W. E.; Herman, G. E.; Lewis, R. A.; Pollack, M. S.: Linkage of tuberous sclerosis to ABO blood group.(Letter) Lancet I: 804-805, 1987.

Northrup, H.; Kwiatkowski, D. J.; Roach, E. S.; Dobyns, W. B.; Lewis, R. A.; Herman, G. E.; Rodriguez, E., Jr.; Daiger, S. P.; Blanton, S. H.: Evidence for genetic heterogeneity in tuberous sclerosis: one locus on chromosome 9 and at least one locus elsewhere. Am. J. Hum. Genet. 51: 709-720, 1992.

O'Callaghan, F. J. K.; Clarke, A. C.; Joffe, H.; Keeton, B.; Martin, R.; Salmon, A.; Thomas, R. D.; Osborne, J. P.: Tuberous sclerosis complex and Wolff-Parkinson-White syndrome. Arch. Dis. Child. 78: 159-162, 1998.

O'Callaghan, T. J.; Edwards, J. A.; Tobin, M.; Mookerjee, B. K.: Tuberous sclerosis with striking renal involvement in a family. Arch. Intern. Med. 135: 1082-1087, 1975.

Pascual-Castroviejo, I.; Patron, M.; Gutierrez, M.; Carceller, F.; Pascual-Pascual, S. I.: Tuberous sclerosis associated with histologically confirmed ocular and cerebral tumors. Pediat. Neurol. 13: 172-174, 1995.

Povey, S.; Burley, M. W.; Attwood, J.; Benham, F.; Hunt, D.; Jeremiah, S. J.; Franklin, D.; Gillett, G.; Malas, S.; Robson, E. B.; Tippett, P.; Edwards, J. H.; Kwiatkowski, D. J.; Super, M.; Mueller, R.; Fryer, A.; Clarke, A.; Webb, D.; Osborne, J.: Two loci for tuberous sclerosis: one on 9q34 and one on 16p13. Ann. Hum. Genet. 58: 107-127, 1994.

Povey, S.; Burley, M. W.; Fryer, A. E.; Osborne, J.; Al-Gazali, L. I.; Mueller, R.: Genetic recombination between tuberous sclerosis and oncogene v-abl. (Letter) Lancet II: 279-280, 1988.

Rattan, P. K.; Knuppel, R. A.; Scerbo, J. C.; Foster, G.: Tuberous sclerosis in pregnancy. Obstet. Gynec. 62 (suppl.): 21S-22S, 1983.

Renwick, J. H.: Tuberous sclerosis and ABO. (Letter) Lancet II: 1096-1097, 1987.

Rott, H.-D.; Fahsold, R.: Tuberous sclerosis in two sibs of normal parents. Clin. Genet. 39: 306-308, 1991.

Ruggieri, M.; Carbonara, C.; Magro, G.; Migone, N.; Grasso, S.; Tine, A.; Pavone, L.; Gomez, M. R.: Tuberous sclerosis complex: neonatal deaths in three of four children of consanguineous, non-expressing parents. J. Med. Genet. 34: 256-260, 1997.

Rushton, A. R.; Shaywitz, B. A.: Tuberous sclerosis: possible modification of phenotypic expression by an unlinked dominant gene. J. Med. Genet. 16: 32-35, 1979.

Sampson, J.; Scahill, S. J.; Stephenson, J. B. P.; Connor, J. M.: Genetic aspects of tuberous sclerosis in the west of Scotland. (Abstract) J. Med. Genet. 25: 275, 1988.

Sampson, J. R.; Attwood, D.; Al Mughery, A. S.; Reid, J. S.: Pitted enamel hypoplasia in tuberous sclerosis. Clin. Genet. 42: 50-52, 1992.

Sampson, J. R.; Janssen, L. A. J.; Sandkuijl, L. A.; the Tuberous Sclerosis Collaborative Group: Linkage investigation of three putative tuberous sclerosis determining loci on chromosomes 9q, 11q, and 12q. J. Med. Genet. 29: 861-866, 1992.

Sampson, J. R.; Patel, A.; Mee, A. D.: Multifocal renal cell carcinoma in sibs from a chromosome 9 linked (TSC1) tuberous sclerosis family. J. Med. Genet. 32: 848-850, 1995.

Sampson, J. R.; Scahill, S. J.; Stephenson, J. B. P.; Mann, L.; Connor, J. M.: Genetic aspects of tuberous sclerosis in the west of Scotland. J. Med. Genet. 26: 28-31, 1989.

Sampson, J. R.; Yates, J. R. W.; Pirrit, L. A.; Fleury, P.; Winship, I.; Beighton, P.; Connor, J. M.: Evidence for genetic heterogeneity in tuberous sclerosis. J. Med. Genet. 26: 511-516, 1989.

Scappaticci, S.; Cerimele, D.; Tondi, M.; Vivarelli, R.; Fois, A.; Fraccaro, M.: Chromosome abnormalities in tuberous sclerosis. Hum. Genet. 79: 151-156, 1988.

Scheig, R. L.; Bornstein, P.: Tuberous sclerosis in the adult: an unusual case without mental deficiency or epilepsy. Arch. Intern. Med. 108: 789-795, 1961.

Schull, W. J.; Crowe, F. W.: Neurocutaneous syndromes in a kindred: a case of simultaneous occurrence of tuberous sclerosis and neurofibromatosis. Neurology 3: 904-909, 1953.

Schwartz, P. L.; Beards, J. A.; Maris, P. J. G.: Tuberous sclerosis associated with a retinal angioma. Am. J. Ophthal. 90: 485-488, 1980.

Sepp, T.; Yates, J. R. W.; Green, A. J.: Loss of heterozygosity in tuberous sclerosis hamartomas. J. Med. Genet. 33: 962-964, 1996.

Short, M. P.; Haines, J. L.; Bove, C.; Henske, E. P.; Guillemette, W.; Sitsma, M.; Amos, J.; Andermann, E.; Gusella, J. F.; Kwiatkowski, D. J.: Linkage and heterogeneity in tuberous sclerosis: linkage to chromosome 16 and resolution of old problems. (Abstract) Am. J. Hum. Genet. 51 (suppl.): A201, 1992.

Smith, H. C.; Watson, G. H.; Patel, R. G.; Super, M.: Cardiac rhabdomyomata in tuberous sclerosis: their course and diagnostic value. Arch. Dis. Child. 64: 196-200, 1989.

Smith, M.; Haines, J.; Trofatter, J.; Dumars, K.; Pandolfo, M.; Conneally, M.: Linkage studies in tuberous sclerosis. (Abstract) Cytogenet. Cell Genet. 46: 694-695, 1987.

Smith, M.; Haines, J.; Trofatter, J.; Dumars, K.; Pandolfo, M.; Conneally, M.: Linkage studies in tuberous sclerosis. (Abstract) Am. J. Hum. Genet. 41: A186, 1987.

Stapleton, F. B.; Johnson, D.; Kaplan, G. W.; Griswold, W.: The cystic renal lesion in tuberous sclerosis. J. Pediat. 97: 574-579, 1980.

Stevenson, A. C.; Fisher, O. D.: Frequency of epiloia in Northern Ireland. Brit. J. Prev. Soc. Med. 10: 134-135, 1956.

Sugita, K.; Itoh, K.; Takeuchi, Y.; Cho, H.; Kakinuma, H.; Nakajima, H.; Takase, M.: Tuberous sclerosis: report of two cases studied by computer-assisted cranial tomography within one week after birth. Brain Dev. 7: 438-443, 1985.

Sybert, V. P.; Hall, J. G.: Inheritance of tuberous sclerosis.(Letter) Lancet I: 783, 1979.

Teplick, J. G.: Tuberous sclerosis. Extensive roentgen findings without usual clinical picture: a case report. Radiology 93: 53-55, 1969.

van Baal, J. G.; Fleury, P.; Brummelkamp, W. H.: Tuberous sclerosis and the relation with renal angiomyolipoma: a genetic study on the clinical aspects. Clin. Genet. 35: 167-173, 1989.

van Slegtenhorst, M.; de Hoogt, R.; Hermans, C.; Nellist, M.; Janssen, B.; Verhoef, S.; Lindhout, D.; van den Ouweland, A.; Halley, D.; Young, J.; Burley, M.; Jeremiah, S.; and 29 others: Identification of the tuberous sclerosis gene TSC1 on chromosome 9q34. Science 277: 805-808, 1997.

van Slegtenhorst, M.; Nellist, M.; Nagelkerken, B.; Cheadle, J.; Snell, R.; van den Ouweland, A.; Reuser, A.; Sampson, J.; Halley, D.; van der Sluijs, P.: Interaction between hamartin and tuberin, the TSC1 and TSC2 gene products. Hum. Molec. Genet. 7: 1053-1057, 1998.

Verhoef, S.; Vrtel, R.; van Essen, T.; Bakker, L.; Sikkens, E.; Halley, D.; Lindhout, D.; van den Ouweland, A.: Somatic mosaicism and clinical variation in tuberous sclerosis complex. Lancet 345: 202 only, 1995.

Webb, D.; Osborne, J. P.; Clarke, A.: Pitted enamel hypoplasia in tuberous sclerosis.(Letter) Clin. Genet. 45: 269, 1994.

Webb, D. W.; Clarke, A.; Fryer, A.; Osborne, J. P.: The cutaneous features of tuberous sclerosis: a population study. Brit. J. Dermatol. 135: 1-5, 1996.

Webb, D. W.; Osborne, J. P.: Non-penetrance in tuberous sclerosis. J. Med. Genet. 28: 417-419, 1991.

Webb, D. W.; Thomas, R. D.; Osborne, J. P.: Echocardiography and genetic counselling in tuberous sclerosis. J. Med. Genet. 29: 487-489, 1992.

Wilson, J.; Carter, C.: Genetics of tuberose sclerosis.(Letter) Lancet I: 340, 1978.

TUBEROUS SCLEROSIS 2 (*191092)

JJM NOTE

No clinical distinctions are certain among the various loci for tuberous sclerosis; hence, the same tumor types should be suspected.

Using tuberous sclerosis (TSC) families in which data had excluded linkage to chromosome 9, Kandt et al. (1992) failed to detect linkage with loci on chromosomes 11 and 12. Linkage was demonstrated, however, with D16S283, the closest marker on the proximal side of the locus for polycystic kidney disease type 1 (173900), on chromosome 16p13. A lod score of 9.50 at theta = 0.02 was observed; 1 family independently presented a lod score of 4.44 at theta = 0.05. Although the cystic kidneys of PKD1 and those of tuberous sclerosis can be distinguished histologically, possible allelism of TSC4 and PKD1 is worthy of consideration. Confirmation of a tuberous sclerosis locus on chromosome 16 was provided by Pericak-Vance et al. (1992), Short et al. (1992), and Smith et al. (1992). Approximately 60% of tuberous sclerosis families probably have their disorder as a result of mutation on chromosome 16.

(With the growing consensus that there is no form of tuberous sclerosis encoded by either chromosome 11 (191090) or chromosome 12 (191091), it seemed desirable to refer to the chromosome 16 form of tuberous sclerosis as TSC2.)

Using pulsed-field gel electrophoresis, the European Chromosome 16 Tuberous Sclerosis Consortium (1993) identified 5 tuberous sclerosis-associated deletions at 16p13.3. These were mapped to a 120-kb region that was cloned in cosmids and from which 4 genes were isolated. One gene, designated TSC2, was interrupted by all 5 PFGE deletions, and closer examination revealed several intragenic mutations, including 1 de novo deletion. In this case, Northern blot analysis identified a shortened transcript, while reduced expression was observed in another TSC family, confirming TSC2 as the chromosome 16 TSC gene. The 5.5-kb TSC2 transcript was found to be widely expressed, and its protein product, designated tuberin, to have a region of homology to the GTPase-activating protein GAP3.

Using DNA markers, Green et al. (1994), found allele loss on 16p13.3 in 3 angiomyolipomas, 1 cardiac rhabdomyoma, 1 cortical tuber, and 1 giant cell astrocytoma. This led them to suggest that the TSC2 gene functions as a tumor suppressor gene, in accordance with the Knudson hypothesis. Similar evidence for a tumor suppressor function for the TSC1 gene (191100) had been adduced from studies of loss of heterozygosity (LOH).

Carbonara et al. (1996) studied LOH in both the TSC1 and TSC2 loci and 7 tumor suppressor gene-containing regions, p53 (191170), NF1 (162200), NF2 (101000), BRCA1 (113705), APC (175100), VHL (193300), and MLM (155600), in 20 hamartomas from 18 tuberous sclerosis patients. Overall, 8 angiomyolipomas, 8 giant cell astrocytomas, 1 cortical tuber, and 3 rhabdomyomas were analyzed. LOH at either TSC locus was found in a large fraction of the informative patients, both sporadic (7/14) and familial (1/4). A statistically significant preponderance of LOH of TSC2 was observed in the sporadic group (P less than 0.01). Carbonara et al. (1996) suspected that bias in the selection for TSC patients with the most severe organ impairment was responsible for the finding. According to this suggestion, a TSC2 defect may confer a greater risk for early kidney failure or, possibly, a more rapid growth of a giant cell astrocytoma. None of the 7 antioncogenes tested showed LOH, indicating that the loss of either TSC gene product may be sufficient to promote hamartomatis cell growth. The observation of LOH at different markers in an astrocytoma and in an angiomyolipoma from the same patients suggested to the authors the multifocal origin of a second-hit mutation.

Henske et al. (1996) analyzed 87 lesions from 47 TSC patients for LOH in the TSC1 and TSC2 regions. Of the 28 patients with angiomyolipomas or rhabdomyomas LOH for 16p13 was detected in in lesions from 12 (57%). LOH for 9q34 was detected only in 1 patient. The authors noted that LOH occurred only in 4% of TSC brain lesions and suggested that TSC brain lesions may result from a different pathogenetic mechanism than TSC kidney or rhabdomyoma lesions.

Sepp et al. (1996) described the spectrum of LOH in 51 hamartomas from 34 cases of tuberous sclerosis. Of 51 hamartomas analyzed, 21 (41%) showed LOH; 16 hamartomas showed LOH around TSC2 and 5 showed LOH in the vicinity of TSC1. No hamartomas showed LOH for markers around both loci. Sepp et al. (1996) reported that there did not appear to be any major differences in the frequency of LOH between the different types of hamartoma.

Bjornsson et al. (1996) studied 6 TSC-associated renal cell carcinomas (RCCs). Their findings suggested that some TSC-associated RCCs have clinical, pathologic, and genetic features which distinguish them from sporadic RCC. Clinically, the TSC-associated RCC occurred at a younger age (36 years) than sporadic tumors and occurred primarily in women (5 out of 6 cases). LOH was observed on 9q34, 16p13.3, and in 2 cases on chromosome 3p.

Hereditary renal carcinoma was described in the rat by Eker (1954). These tumors share morphologic similarities with human renal cancer. Yeung et al. (1994) localized the inherited mutation to rat chromosome 10q12 by linkage analysis. This region is known to be syntenic with human 16p13.3, the site of the TSC2 gene. Yeung et al. (1994) found a specific rearrangement of the rat homolog of TSC2 which cosegregated with carriers of the predisposing mutation. Tumors with or without LOH expressed only the mutant allele, consistent with the 2-hit hypothesis of Knudson and the role of TSC2 as a tumor suppressor gene. The mutation in the rat gave rise to an aberrant transcript that deleted the 3-prime end, which normally contains a region of homology with the catalytic domain of rap1GAP. Kobayashi et al. (1995) likewise identified a germline mutation in the TSC2 gene in the Eker rat. In a separate publication, Kobayashi et al. (1995) reported the complete cDNA and genomic structure of the rat TSC2 gene. The deduced amino acid sequence (1,743 amino acids) showed 92% identity to the human counterpart. Surprisingly, there were 41 or more coding exons with small introns spanning a total of only approximately 35 kb of genomic DNA. Two alternative splicing events were recognized.

Pilz et al. (1995) demonstrated that the mouse homolog of TSC2 maps to chromosome 17. They showed that, although it maps to the same general region as t(w18) and t(h20) (2 previously described deletions associated with the T complex), Tsc2 actually did not fall within either of these deletions. Xu et al. (1995) described alternatively spliced isoforms of TSC2, one of which lacked the 43 amino acids encoded by exon 25. A third isoform exhibited a deletion of 44 amino acids spanning codons 946-989; amino acid 989 is a serine residue encoded by the first codon of exon 26. The 2 isoforms exist in newborn and adult mouse tissues, reinforcing the potential functional importance of these alternatively spliced products. Xu et al. (1995) speculated that the distinct polypeptides encoded by the TSC2 gene may have different targets as well as functions involved in the regulation of cell growth.

The 400-Mb genome of the Japanese pufferfish, Fugu rubripes, is relatively free of repetitive DNA and contains genes with small introns at high density. Sandford et al. (1996) demonstrated that the genes that are mutant in polycystic kidney disease-1 (PKD1; 601313) and tuberous sclerosis-2 are conserved in the Fugu genome where they are tightly linked. In addition, sequences homologous to the SSTR5 gene (182455) were identified 5-prime to PKD1, defining a larger syntenic region. As in genomes of mouse and human, the Fugu TSC2 and PKD1 genes are adjacent in a tail-to-tail orientation.

Wienecke et al. (1995) generated antisera against the N-terminal and C-terminal portions of tuberin, and found that these antisera specifically recognize a 180-kD protein in immunoprecipitation and immunoblotting analyses. A wide variety of human cell lines expressed the 180-kD tuberin protein, and subcellular fractionation showed that most tuberin is found in a membrane/particulate fraction. They found that immunoprecipitates of native tuberin contain an activity that specifically stimulated the intrinsic GTPase activity of the RAS-related protein RAP1A (179520). Tuberin did not stimulate the GTPase activity of RAP2 (179540), HRAS (190020), Rac, or RHO (165370). These results suggested to the authors that the loss of tuberin leads to constitutive activation of RAP1 in tumors of patients with tuberous sclerosis.

To facilitate search for mutations in tuberin, Wilson et al. (1996) designed an RT-PCR-based assay system to scan the expressed coding region of the TSC2 gene in lymphoblasts. Using 34 overlapping PCR assays, they performed SSCP analysis of DNA from 26 apparently sporadic TSC cases, 2 TSC families uninformative for linkage, and 2 confirmed chromosome 16-linked TSC families. Of the 60 chromosomes scanned, 14 showed abnormal SSCP mobility shifts. Using direct PCR sequencing, they identified 5 missense mutations, 1 3-bp in-frame deletion, and 1 2-bp frameshift deletion, 1 missense mutation, 1 25-bp tandem duplication, and 5 silent nucleotide changes thought to be polymorphisms. There was no apparent clustering of mutations within the TSC2 gene. The authors commented that the diversity of mutation types argued that TSC2 may not act in a classic tumor suppressor fashion. In addition, they saw no specific correlation between the different mutations and clinical severity or expression.

Au et al. (1997) tested 88 TSC probands with the TSC2 cDNA by Southern blotting, searching for gross deletions, rearrangements, or insertions. They detected 2 deletions and a rare intragenic polymorphic variant.

Van Bakel et al. (1997) stated that mutations in the TSC2 gene on 16p13.3 are responsible for approximately 50% of familial tuberous sclerosis. The gene has 41 small exons spanning 45 kb of genomic DNA and encodes a 5.5 kb mRNA. Large germline deletions of TSC2 occur in less than 5% of cases, and a number of small intragenic mutations have been described. Using the protein truncation test (PTT), van Bakel et al. (1997) analyzed mRNA from 18 unrelated cases of tuberous sclerosis for TSC2 mutations. Three cases were predicted to be TSC2 mutations on the basis of linkage analysis or because a hamartoma from the patient showed LOH for 16p13.3 markers. Confirmed mutations were identified in 5 (28%) of the families studied.

Sampson et al. (1997) studied 27 unrelated patients with tuberous sclerosis and renal cystic disease. They found that 22 patients had contiguous deletions of TSC2 and PKD1. In 17 patients with constitutional deletions, cystic disease was severe, with early renal insufficiency. One patient with deletion of TSC2 and of only the 3-prime untranslated region (UTR) of PKD1 had few cysts. Four patients were somatic mosaics; the severity of their cystic disease varied considerably. Mosaicism and mild cystic disease also were demonstrated in the parents of 3 of the constitutionally deleted patients. Five patients without contiguous deletions had relatively mild cystic disease, 3 of whom had gross rearrangements of TSC2, while 2 had no identified mutation. Thus, Sampson et al. (1997) concluded that significant renal cystic disease in tuberous sclerosis usually reflects mutational involvement of the PKD1 gene, and mosaicism for large deletions of TSC2 and PKD1 occurs frequently.

Jones et al. (1997) comprehensively defined the TSC1 mutational spectrum in 171 sequentially ascertained, unrelated TSC patients by SSCP and heteroduplex analysis of all 21 coding exons, and by assaying a restriction fragment spanning the whole locus. Mutations were identified in 9 of 24 familial cases, but in only 13 of 147 sporadic cases. In contrast, a limited screen revealed TSC2 mutations in 2 of the 24 familial cases and in 45 of the 147 sporadic cases. Thus, TSC1 mutations were significantly underrepresented among sporadic cases. Both large deletions and missense mutations were common at the TSC2 locus, whereas most TSC1 mutations were small truncated lesions. Mental retardation was significantly less frequent among carriers of TSC1 mutations than TSC2 mutations (odds ratio, 5.54 for sporadic cases only; 6.78 +/- 1.54 when a single randomly selected patient per multigeneration family was also included). No correlation between mental retardation and the type of mutation was found. Jones et al. (1997) concluded that there is a reduced risk of mental retardation in TSC1 as opposed to TSC2 disease and that consequent ascertainment bias, at least in part, explains the relative paucity of TSC1 mutations in sporadic TSC.

Au et al. (1998) tested 90 patients with tuberous sclerosis complex for mutations in the TSC2 gene by means of single-strand conformation analysis (SSCA) of genomic DNA. Patients included 56 sporadic cases and 34 familial probands. All 41 exons of the TSC2 gene were studied. They identified 32 SSCA changes: 22 disease-causing mutations, and 10 polymorphic variants. Mutations were detected at a much higher frequency in the sporadic cases (32%) than in the multiplex families (9%). Among the 8 families for which linkage to the TSC2 region of chromosome 16 had been determined, only 1 mutation was found. Mutations were distributed uniformly across the gene; they included 5 deletions, 3 insertions, 10 missense mutations, 2 nonsense mutations, and 2 tandem duplications. No mutations were detected in exons 25 and 31, which are spliced out in the isoforms. No correspondence between variability of phenotype and type of mutation (missense vs early termination) was found. They commented that diagnostic testing is difficult because of the genetic heterogeneity of TSC, with at least 2 causative genes, the large size of the TSC2 gene, and the variety of mutations.

Pulmonary lymphangiomyomatosis (LAM) is a rare disease of unknown etiology, affecting women almost exclusively. It was first described by Van Stossel (1937). The average age of onset symptoms, which include shortness of breath (67%), lung collapse (25%), coughing (12%), and chest pain (10%), is 33 years (Taylor et al., 1990; Johnson et al., 1993). Chest x-rays typically show a diffuse interstitial infiltrate, with no predominance in any one lung zone. Although most LAM is pulmonary, cases with retroperitoneal, pelvic, or perirenal involvement in lymph nodes and extranodal sites have been reported (Torres et al., 1995). The disorder is usually diagnosed by open lung biopsy. Microscopically, there is cystic distortion of the normal pulmonary architecture by an atypical smooth muscle infiltrate. Most patients have a slowly declining clinical course. Lung transplantation is the only effective therapy for patients with end-stage disease (Boehler et al., 1996). Lymphangiomyomatosis can occur as an isolated disorder or in association with tuberous sclerosis. Among patients with tuberous sclerosis it is said to be the third most frequent cause of TSC-related death, after renal disease and brain tumors (Castro et al., 1995). Renal angiomyolipomas occur in approximately 50% of sporadic LAM patients and in 70% of TSC patients. Loss of heterozygosity (LOH) in the chromosomal region for the TSC2 gene occurs in 60% of TSC-associated angiomyolipomas. Because of the similar pulmonary and renal manifestations of TSC and sporadic LAM, Smolarek et al. (1998) hypothesized that LAM and TSC have a common genetic basis. They analyzed renal angiomyolipomas from 13 women with sporadic LAM for LOH in the region of the TSC1 (9q34) and TSC2 (16p13) genes. TSC2 LOH was detected in 7 (54%) of the angiomyolipomas. They also found TSC2 LOH in 4 lymph nodes from a woman with retroperitoneal LAM. No TSC1 LOH was found. The findings indicated that the TSC2 gene may be involved in the pathogenesis of sporadic LAM. They noted, however, that genetic transmission of LAM had not been reported. Women with LAM may have low-penetrance germline TSC2 mutations, or they may be mosaic, with TSC2 mutations in the lung and the kidney but not in other organs.

See also Kandt et al. (1992); Kobayashi et al. (1995)

Au, K.-S.; Rodriguez, J. A.; Finch, J. L.; Volcik, K. A.; Roach, E. S.; Delgado, M. R.; Rodriguez, E., Jr.; Northrup, H.: Germ-line mutational analysis of the TSC2 gene in 90 tuberous-sclerosis patients. Am. J. Hum. Genet. 62: 286-294, 1998.

Au, K.-S.; Rodriguez, J. A.; Rodriguez, E., Jr.; Dobyns, W. B.; Delgado, M. R.; Northrup, H.: Mutations and polymorphisms in the tuberous sclerosis complex gene on chromosome 16. Hum. Mutat. 9: 23-29, 1997.

Bjornsson, J.; Short, M. P.; Kwiatkowski, D. J.; Henske, E. P.: Tuberous sclerosis-associated renal cell carcinoma. Am. J. Path. 149: 1201-1208, 1996.

Boehler, A.; Speich, R.; Russi, E. W.; Weder, W.: Lung transplantation for lymphangioleiomyomatosis. New Eng. J. Med. 335: 1275-1280, 1996.

Bowen, P.: Achondroplasia in two sisters with normal parents. Birth Defects Orig. Art. Ser. X(12): 31-36, 1974.

Carbonara, C.; Longa, L.; Grosso, E.; Mazzucco, G.; Borrone, C.; Garre, M. L.; Brisigotti, M.; Filippi, G.; Scabar, A.; Giannotti, A.; Falzoni, P.; Monga, G.; Garini, G.; Gabrielli, M.; Riegler, P.; Danesino, C.; Ruggieri, M.; Magro, G.; Migone, N.: Apparent preferential loss of heterozygosity at TSC2 over TSC1 chromosomal region in tuberous sclerosis hamartomas. Genes Chromosomes Cancer 15: 18-25, 1996.

Castro, M.; Shepherd, C. W.; Gomez, M. R.; Lie, J. T.; Ryu, J. H.: Pulmonary tuberous sclerosis. Chest 107: 189-195, 1995.

Eker, R.: Familial renal adenomas in Wistar rats: a preliminary report. Acta Path. Microbiol. Scand. 34: 554-562, 1954.

European Chromosome 16 Tuberous Sclerosis Consortium: Identification and characterization of the tuberous sclerosis gene on chromosome 16. Cell 75: 1305-1315, 1993.

Green, A. J.; Smith, M.; Yates, J. R. W.: Loss of heterozygosity on chromosome 16p13.3 in hamartomas from tuberous sclerosis patients. Nature Genet. 6: 192-196, 1994.

Henske, E. P.; Scheithauer, B. W.; Short, M. P.; Wollmann, R.; Nahmias, J.; Hornigold, N.; van Slegtenhorst, M.; Welsh, C. T.; Kwiatkowski, D. J.: Allelic loss is frequent in tuberous sclerosis kidney lesions but rare in brain lesions. Am. J. Hum. Genet. 59: 400-406, 1996.

Johnson, S. F.; Davey, D. D.; Cibull, M. L.; Schwartz, R. W.; Strodel, W. E.: Lymphangioleiomyomatosis. Am. Surg. 59: 395-399, 1993.

Jones, A. C.; Daniells, C. E.; Snell, R. G.; Tachataki, M.; Idziaszczyk, S. A.; Krawczak, M.; Sampson, J. R.; Cheadle, J. P.: Molecular genetic and phenotypic analysis reveals differences between TSC1 and TSC2 associated familial and sporadic tuberous sclerosis. Hum. Molec. Genet. 6: 2155-2161, 1997.

Kandt, R. S.; Haines, J. L.; Smith, M.; Northrup, H.; Gardner, R. J. M.; Short, M. P.; Dumars, K.; Roach, E. S.; Steingold, S.; Wall, S.; Blanton, S. H.; Flodman, P.; Kwiatkowski, D. J.; Jewell, A.; Weber, J. L.; Roses, A. D.; Pericak-Vance, M. A.: Linkage of a major gene locus for tuberous sclerosis to a chromosome 16 marker for polycystic kidney disease. (Abstract) Am. J. Hum. Genet. 51 (suppl.): A4, 1992.

Kandt, R. S.; Haines, J. L.; Smith, M.; Northrup, H.; Gardner, R. J. M.; Short, M. P.; Dumars, K.; Roach, E. S.; Steingold, S.; Wall, S.; Blanton, S. H.; Flodman, P.; Kwiatkowski, D. J.; Jewell, A.; Weber, J. L.; Roses, A. D.; Pericak-Vance, M. A.: Linkage of an important gene locus for tuberous sclerosis to a chromosome 16 marker for polycystic kidney disease. Nature Genet. 2: 37-41, 1992.

Kobayashi, T.; Hirayama, Y.; Kobayashi, E.; Kubo, Y.; Hino, O.: A germline insertion in the tuberous sclerosis (Tsc2) gene gives rise to the Eker rat model of dominantly inherited cancer. Nature Genet. 9: 70-74, 1995.

Kobayashi, T.; Nishizawa, M.; Hirayama, Y.; Kobayashi, E.; Hino, O.: cDNA structure, alternative splicing and exon-intron organization of the predisposing tuberous sclerosis (Tsc2) gene of the Eker rat model. Nucleic Acids Res. 23: 2608-2613, 1995.
Kumar, A.; Kandt, R. S.; Wolpert, C.; Roses, A. D.; Pericak-Vance, M. A.; Gilbert, J. R.: Mutation analysis of the TSC2 gene in an African-American family. Hum. Molec. Genet. 4: 2295-2298, 1995.
Kumar, A.; Wolpert, C.; Kandt, R. S.; Segal, J.; Pufky, J.; Roses, A. D.; Pericak-Vance, M. A.; Gilbert, J. R.: A de novo frame-shift mutation in the tuberin gene. Hum. Molec. Genet. 4: 1471-1472, 1995.
Pericak-Vance, M. A.; Gardner, R. J. M.; Steingold, S.; Wall, S. L.; Carter, S.; DiMario, F. J.; Vance, J. M.; Reeders, S.; Roses, A. D.; Kandt, R. S.: Confirmation of linkage of tuberous sclerosis to chromosome 16p. (Abstract) Am. J. Hum. Genet. 51 (suppl.): A198, 1992.
Pilz, A.; Glenister, P.; Povey, S.; Lyon, M.; Abbott, C.: The mouse homologue of the human tuberous sclerosis 2 (TSC2) gene maps to chromosome 17, but does not fall within the t(w18) or t(h20) deletions. Genomics 26: 420-421, 1995.
Sampson, J. R.; Maheshwar, M. M.; Aspinwall, R.; Thompson, P.; Cheadle, J. P.; Ravine, D.; Roy, S.; Haan, E.; Bernstein, J.; Harris, P. C.: Renal cystic disease in tuberous sclerosis: role of the polycystic kidney disease 1 gene. Am. J. Hum. Genet. 61: 843-851, 1997.
Sandford, R.; Sgotto, B.; Burn, T.; Brenner, S.: The tuberin (TSC2), autosomal dominant polycystic kidney disease (PKD1), and somatostatin type V receptor (SSTR5) genes form a synteny group in the Fugu genome. Genomics 38: 84-86, 1996.
Sepp, T.; Yates, J. R. W.; Green, A. J.: Loss of heterozygosity in tuberous sclerosis hamartomas. J. Med. Genet. 33: 962-964, 1996.
Short, M. P.; Haines, J. L.; Bove, C.; Henske, E. P.; Guillemette, W.; Sitsma, M.; Amos, J.; Andermann, E.; Gusella, J. F.; Kwiatkowski, D. J.: Linkage and heterogeneity in tuberous sclerosis: linkage to chromosome 16 and resolution of old problems. (Abstract) Am. J. Hum. Genet. 51 (suppl.): A201, 1992.
Smith, M.; Handa, K.; Sokolov, G.; Postle, S.; Flodman, P.; Spence, M. A.: Further evidence for a tuberous sclerosis gene locus on chromosome 16p13. (Abstract) Am. J. Hum. Genet. 51 (suppl.): A201, 1992.
Smolarek, T. A.; Wessner, L. L.; McCormack, F. X.; Mylet, J. C.; Menon, A. G.; Henske, E. P.: Evidence that lymphangiomyomatosis is caused by TSC2 mutations: chromosome 16p13 loss of heterozygosity in angiomyolipomas and lymph nodes from women with lymphangiomyomatosis. Am. J. Hum. Genet. 62: 810-815, 1998.
Taylor, J. R.; Ryu, J.; Colby, T. V.; Raffin, T. A.: Lymphangioleiomyomatosis: clinical course in 32 patients. New Eng. J. Med. 323: 1254-1260, 1990.
Torres, V. E.; Bjornsson, J.; King, B. F.; Kumar, R.; Zincke, H. Edell, E. S.; Wilson, T. O.; Hattery, R. R.; Gomez, M. R.: Extrapulmonary lymphangioleiomyomatosis and lymphangiomatous cysts in tuberous sclerosis complex. Mayo Clin. Proc. 70: 641-648, 1995.
van Bakel, I.; Sepp, T.; Ward, S.; Yates, J. R. W.; Green, A. J.: Mutations in the TSC2 gene: analysis of the complete coding sequence using the protein truncation test (PTT). Hum. Molec. Genet. 6: 1409-1414, 1997.
Van Stossel, E.: Uber muskulare cirrhose der lunge. Beitr. Klin. Tuberk. 90: 432-442, 1937.
Verhoef, S.; Vrtel, R.; van Essen, T.; Bakker, L.; Sikkens, E.; Halley, D.; Lindhout, D.; van den Ouweland, A.: Somatic mosaicism and clinical variation in tuberous sclerosis complex. Lancet 345: 202 only, 1995.
Vrtel, R.; Verhoef, S.; Bouman, K.; Maheshwar, M. M.; Nellist, M.; van Essen, A. J.; Bakker, P. L. G.; Hermans, C. J.; Bink-Boelkens, M. T. E.; van Elburg, R. M.; Hoff, M.; Lindhout, D.; Sampson, J.; Halley, D. J. J.; van den Ouweland, A. M. W.: Identification of a nonsense mutation at the 5-prime end of the TSC2 gene in a family with a presumptive diagnosis of tuberous sclerosis complex. J. Med. Genet. 33: 47-51, 1996.
Webb, D. W.; Osborne, J. P.: Non-penetrance in tuberous sclerosis. J. Med. Genet. 28: 417-419, 1991.
Wienecke, R.; Konig, A.; DeClue, J. E.: Identification of tuberin, the tuberous sclerosis-2 product: tuberin possesses specific Rap1GAP activity. J. Biol. Chem. 270: 16409-16414, 1995.
Wilson, P. J.; Ramesh, V.; Kristiansen, A.; Bove, C.; Jozwiak, S.; Kwiatkowski, D. J.; Short, M. P.; Haines, J. L.: Novel mutations detected in the TSC2 gene from both sporadic and familial TSC patients. Hum. Molec. Genet. 5: 249-256, 1996.
Xu, L.; Sterner, C.; Maheshwar, M. M.; Wilson, P. J.; Nellist, M.; Short, P. M.; Haines, J. L.; Sampson, J. R.; Ramesh, V.: Alternative splicing of the tuberous sclerosis 2 (TSC2) gene in human and mouse tissues. Genomics 27: 475-480, 1995.
Yates, J. R. W.; van Bakel, I.; Sepp, T.; Payne, S. J.; Webb, D. W.; Nevin, N. C.; Green, A. J.: Female germline mosaicism in tuberous sclerosis confirmed by molecular genetic analysis. Hum. Molec. Genet. 6: 2265-2269, 1997.
Yeung, R. S.; Xiao, G.-H.; Jin, F.; Lee, W.-C.; Testa, J. R.; Knudson, A. G.: Predisposition to renal carcinoma in the Eker rat is determined by germ-line mutation of the tuberous sclerosis 2 (TSC2) gene. Proc. Nat. Acad. Sci. 91: 11413-11416, 1994.

TUBEROUS SCLEROSIS 3 (191091)

JJM NOTE

This entity is included for completeness.

The observation of a patient suffering from tuberous sclerosis with a de novo translocation t(3;12)(p26.3;q23.3) prompted Fahsold et al. (1991) to search for a third TSC gene on chromosome 12. Maximum lod scores of 2.3 (at theta = 0.14) and 4.3 (at theta = 4.3) with markers D12S7 and PAH (261600), respectively, provided strong indication of the location of a TSC3 gene on the long arm of chromosome 12. The PAH locus is located at 12q24.1. Short et al. (1992) found little or no evidence for a TSC locus on 12q. According to Edwards (1992), the location of a tuberous sclerosis locus on chromosome 12 has been excluded.

See also Fahsold et al. (1991)

Edwards, J. H.: Personal Communication. Oxford, England 10/25/1992.
Fahsold, R.; Rott, H.-D.; Lorenz, P.: A third gene locus for tuberous sclerosis is closely linked to the phenylalanine hydroxylase gene locus. Hum. Genet. 88: 85-90, 1991.
Fahsold, R.; Rott, H.-D.; Lorenz, P.; Claussen, U.: A new gene locus for tuberous sclerosis maps to chromosome 12. (Abstract) Cytogenet. Cell Genet. 58: 1976 only, 1991.
Short, M. P.; Haines, J. L.; Bove, C.; Henske, E. P.; Guillemette, W.; Sitsma, M.; Amos, J.; Andermann, E.; Gusella, J. F.; Kwiatkowski, D. J.: Linkage and heterogeneity in tuberous sclerosis: linkage to chromosome 16 and resolution of old problems. (Abstract) Am. J. Hum. Genet. 51 (suppl.): A201 only, 1992.

TUBEROUS SCLEROSIS 4 (191090)

JJM NOTE

This entity is included as another candidate locus.

Clark et al. (1989) described a t(11q23.3;22q11.1) translocation in a liveborn infant with tuberous sclerosis. Because the neural cell adhesion molecule NCAM (116930) maps to the same band, they raised the possibility that this is a candidate gene for a form of tuberous sclerosis. In an analysis of 9 families, Flodman et al. (1989) found that 4 families showed recombination with at least 1 of the 9q34 loci by pairwise linkage analysis; of these families, 3 showed evidence of linkage to 11q, and in the fourth family data on chromosome 11 markers were not available. Four families showed recombination with at least 1 of the 11q markers by pairwise linkage analysis; of these families, 3 showed evidence of linkage to the ABO locus at 9q34, and the fourth family was uninformative at the ABO locus. Evidence of linkage to 11q was clear in 1 family. Smith et al. (1989) found lod scores for linkage to marker loci on 11q strongly supporting assignment there. Haines et al. (1989) also presented evidence in support of linkage heterogeneity in tuberous sclerosis. In studies of 15 families, Smith et al. (1990) found evidence of linkage to a chromosome 11 marker, D11S144, which maps to 11q22-q23; maximum lod = 3.26 at theta = 0.08. The tyrosinase locus (11q14-q22) gave a maximum lod score of 2.88 at theta = 0.0 for linkage to TSC4. Homogeneity analysis provided no evidence for genetic heterogeneity (i.e., non-11-linked families) in this data set. In a South African family in which no evidence of linkage to chromosome 9 markers was found, Winship et al. (1990) found atypical clinical features, namely, prominent nuchal skin tags (multiple acrochordons), a confetti pattern of hypopigmentation of the skin of the lower legs, and absence of ungual fibromata. Kandt et al. (1992) and Short et al. (1992) found little or no evidence for a tuberous sclerosis gene on 11q.

Clark, R. D.; Smith, M.; Pandolfo, M.; Fausel, R. E.; Bustillo, A. M.: Tuberous sclerosis in a liveborn infant with trisomy due to t(11q23.3;22q11.1) translocation: is neural cell adhesion molecule a candidate gene for tuberous sclerosis?. Am. J. Hum. Genet. 43: A44, 1989.
Flodman, P.; Baumann, R.; Yoshiyama, K.; Smith, M.: Classification of tuberous sclerosis families based on linkage analysis with 9q34 and 11q22-11q23 markers. (Abstract) Am. J. Hum. Genet. 45 (suppl.): A139, 1989.
Haines, J. L.; Amos, J.; Atwood, J.; Bech-Hansen, N. T.; Burley, M.; Conneally, P. M.; Connor, J. M.; Fahsold, R.; Fryer, A.; Kandt, R. S.; Northrup, H.; Osborne, J.; Pericak-Vance, M. A.; Povey, S.; Sampson, J. R.; Short, P.; Smith, M.; Speer, M. C.; Trofatter, J. A.; Yates, J. R. W.: Linkage heterogeneity in tuberous sclerosis. (Abstract) Cytogenet. Cell Genet. 51 (suppl.): 1010, 1989.
Kandt, R. S.; Haines, J. L.; Smith, M.; Northrup, H.; Gardner, R. J. M.; Short, M. P.; Dumars, K.; Roach, E. S.; Steingold, S.; Wall, S.; Blanton, S. H.; Flodman, P.; Kwiatkowski, D. J.; Jewell, A.; Weber, J. L.; Roses, A. D.; Pericak-Vance, M. A.: Linkage of a major gene locus for tuberous sclerosis to a chromosome 16 marker for polycystic kidney disease. (Abstract) Am. J. Hum. Genet. 51 (suppl.): A4, 1992.
Short, M. P.; Haines, J. L.; Bove, C.; Henske, E. P.; Guillemette, W.; Sitsma, M.; Amos, J.; Andermann, E.; Gusella, J. F.; Kwiatkowski, D. J.: Linkage and heterogeneity in tuberous sclerosis: linkage to chromosome 16 and resolution of old problems. (Abstract) Am. J. Hum. Genet. 51 (suppl.): A201, 1992.
Smith, M.; Dumars, K.; Baumann, R.; Yoshiyama, K.; Nakamura, Y.; Julier, C.; Trofatter, J. A.; Weeks, D. E.; Pandolfo, M.; Conneally, P. M.: Evidence for genetic heterogeneity in tuberous sclerosis: one gene maps to the 9q34 region and a second gene maps in the 11q22-11q23 region. (Abstract) Cytogenet. Cell Genet. 51: 1082, 1989.
Smith, M.; Smalley, S.; Cantor, R.; Pandolfo, M.; Gomez, M. I.; Baumann, R.; Flodman, P.; Yoshiyama, K.; Nakamura, Y.; Julier, C.; Dumars, K.; Haines, J.; Trofatter, J.; Spence, M. A.; Weeks, D.; Conneally, M.: Mapping of a gene determining tuberous sclerosis to human chromosome 11q14-11q23. Genomics 6: 105-114, 1990.

Winship, I. M.; Connor, J. M.; Beighton, P. H.: Genetic heterogeneity in tuberous sclerosis: phenotypic correlations. J. Med. Genet. 27: 418-421, 1990.

Multisystem Disorders Predisposing to Multiple Tumors

ATAXIA-TELANGIECTASIA (*208900)

JJM NOTE

This condition, with perhaps more investigators than affected patients, is rightly considered a Rosetta stone of modern biology, for it has manifestations that represent immunology, neurosciences, DNA repair, altered gene expression, chromosome breakage and translocation, clonal progression, cancer predisposition including in heterozygotes, population screening, and reproductive biology. The predisposed tumor has been the mode of clinical presentation in childhood, especially under age 6 years. The tumors that occur to excess are Hodgkin and non-Hodgkin lymphoma, acute lymphocytic leukemia, and, in adults, adenocarcinoma of the stomach. The alleged predisposition of heterozygotes to breast cancer awaits confirmation, especially with specific correlation with mutations in *ATM*.

Description

Ataxia-telangiectasia (AT) is an autosomal recessive disorder characterized by cerebellar ataxia, telangiectases, immune defects, and a predisposition to malignancy. Chromosomal breakage is a feature. AT cells are abnormally sensitive to killing by ionizing radiation, and abnormally resistant to inhibition of DNA synthesis by ionizing radiation. The latter trait has been used to identify 4 complementation groups for the classical form of the disease (Jaspers et al., 1988). At least 4 of these (A, C, D, and E) map to chromosome 11q23 (Sanal et al., 1990) and are associated with mutations in the ATM gene.

Homozygotes

Patients present in early childhood with progressive cerebellar ataxia and later develop conjunctival telangiectases, other progressive neurologic degeneration, sinopulmonary infection, and malignancies. Telangiectases typically develop between 3 and 5 years of age. The earlier ataxia can be misdiagnosed as ataxic cerebral palsy before the appearance of oculocutaneous telangiectases. Gatti et al. (1991) contended that oculocutaneous telangiectases eventually occur in all patients, while Maserati et al. (1988) wrote that patients without telangiectases are not uncommon. A characteristic oculomotor apraxia, i.e., difficulty in the initiation of voluntary eye movements, frequently precedes the development of telangiectases.

Thibaut et al. (1994) reviewed cases of necrobiosis lipoidica in association with ataxia-telangiectasia.

Gonadal dysfunction in ataxia-telangiectasia was discussed by Miller and Chatten (1967) and Zadik et al. (1978), and others.

According to Boder (1985), the oldest known AT patients (as of 1985) were C.P., who died in November 1978 at age 52 years, and his sister M.P., who died in July 1979 at the age of almost 49 years. The sister was the subject of the report by Saxon et al. (1979) on T-cell leukemia in AT. The possibility of heteroalleles at the ataxia-telangiectasia loci might be suggested.

Neurologic Manifestations

AT may be the most common syndromic progressive cerebellar ataxia of early childhood. Truncal ataxia precedes appendicular ataxia (Gatti et al., 1991). Oculomotor apraxia is progressive and opticokinetic nystagmus is absent. Choreoathetosis and/or dystonia occur in 90% of patients and can be severe. Neurologic dysfunction is a clinically invariable feature in homozygotes. Woods and Taylor (1992) studied 70 affected persons in the British Isles, 29 females and 41 males with an age range of 2 to 42 years. Most presented by 3 years of age with truncal ataxia. All had ataxia, ocular motor apraxia, an impassive face, and dysarthria, although clinical immune deficiency was present only in 43 of 70 patients. Ocular telangiectases was seen in all but one. All 60 tested showed increased sensitivity to ionizing radiation, 43 of 48 had an elevated alpha-fetoprotein level, and 14 of 21 had an immunoglobulin deficiency.

Deep tendon reflexes become diminished or absent by age 8 and patients later develop diminished large-fiber sensation. Gatti et al. (1991) pointed out that 'a significant proportion of older patients in their twenties and early thirties develop progressive spinal muscular atrophy, affecting mostly hands and feet, and dystonia.' Interosseous muscular atrophy in the hands in combination with the early-onset dystonic posturing leads to striking combined flexion-extension contractures of the fingers, which they illustrated. Mental retardation is not a feature of AT, although some older patients have a severe loss of short-term memory (Gatti et al., 1991).

Malignancy

Patients with AT have a strong predisposition to malignancy. Hecht et al. (1966) observed lymphocytic leukemia in patients with AT. A nonleukemic sib and 2 unrelated patients with AT had multiple chromosomal breaks and impaired responsiveness to phytohemagglutinin. This was the first report of chromosomal breakage in AT. Leukemia and chromosomal abnormalities occur in at least 2 other mendelian disorders—Fanconi pancytopenia (FA; 227650) and Bloom syndrome (BS; 210900).

Saxon et al. (1979) demonstrated thymic origin of the neoplastic cells in a 48-year-old woman with AT and chronic lymphatic leukemia. The neoplastic cells had the specific 14q+ translocation and showed both helper and suppressor function, suggesting that the malignant transformation had occurred in an uncommitted T-lymphocyte precursor that was capable of differentiation. This is a situation comparable to chronic myeloid leukemia in which the Philadelphia chromosome occurs in a stem cell progenitor of both polymorphs and megakaryocytes.

In general, lymphomas in AT patients tend to be of B-cell origin, whereas the leukemias tend to be of the T-CLL type. Rosen and Harris (1987) discussed the case of a 30-year-old man with AT who developed a malignant lymphoma of B-cell type involving the tonsil and lungs. Haerer et al. (1969) described a black sibship of 12, of whom 5 had ataxia-telangiectasia; 2 of those affected died of mucinous adenocarcinoma of the stomach at ages 21 and 19 years. Bigbee et al. (1989) demonstrated an increased frequency of somatic cell mutation in vivo in individuals with AT. Obligate heterozygotes for the disease did not appear to have a significantly increased frequency of such mutations. The authors speculated that the predisposition to somatic cell mutation may be related to the increased susceptibility to cancer in AT homozygotes.

Other solid tumors, including medulloblastomas and gliomas, occur with increased frequency in AT (Gatti et al., 1991).

Immune Disorders

Defects of the immune mechanism and hypoplasia of the thymus have been demonstrated. Serum IgG2 or IgA levels are diminished or absent in 80% and 60% of patients, respectively (Gatti et al., 1991). IgE levels can be diminished, IgM levels diminished or normal. Peripheral lymphopenia as well as decreased cellular immunity to intradermally injected test antigens can be seen early in the disorder. Sinopulmonary infections are frequent, but their severity cannot be simply correlated with the degree of immunodeficiency.

Atypical Cases

Ying and Decoteau (1981) described a family in which a brother and sister may have had an allelic (and milder) form of AT. The proband, a 58-year-old male of Saskatchewan Mennonite origin, had spinocerebellar degeneration associated with choreiform movements beginning at about age 10 years. Despite considerable physical handicap, he was able to work as a delivery man in the family store. No telangiectases were found at age 44 (they were carefully sought because of typical AT in a niece) or on later examinations. He showed total absence of IgA in serum and concentrated saliva and low IgE in serum. He was anergic on skin testing. Glucose tolerance was markedly decreased. Serum alpha-fetoprotein was 840 ng per ml (normal, less than 10 ng per ml). Lymphocyte response to phytohemagglutinin was blunted. He died of lymphoma at age 58. He showed cytogenetic abnormalities typical of AT; 4 abnormal clones were identified, all involving chromosome 14 in some way. The proband had 4 brothers and 2 sisters. A brother died of leukemia at age 16. A sister was likewise diagnosed as having spinocerebellar degeneration with choreiform movements at age 46; she died at age 55 of breast cancer. The proband's niece with typical AT had telangiectases of the bulbar conjunctivae and earlobes noted at age 3, when she began to have recurrent and severe sinopulmonary infections. She died at age 20 of staphylococcal pneumonia superimposed on bronchiectasis. The brother and sister who died in their 50s may have been genetic compounds. Their parents denied consanguinity.

Taylor et al. (1987) described 3 patients who were atypical in terms of clinical features and cellular features as observed in vitro. One of the patients was a 45-year-old woman with onset of neurologic manifestations in her early twenties. Maserati et al. (1988) described 2 sisters, aged 9 and 11 years, with a progressive neurologic disorder similar to AT, chromosome instability with rearrangements involving chromosomes 7 and 14, but no telangiectases or immunologic anomalies typical of AT. Byrne et al. (1984) reported similar cases of ataxia without telangiectases with selective IgE deficiency but normal IgA and alpha-fetoprotein. Ziv et al. (1989) described 2 Turkish sibs with an atypically prolonged course and atypical behavior of cultured fibroblasts. See 208910 and 208920 for AT-like syndromes.

Heterozygotes

Welshimer and Swift (1982) studied families of homozygotes for AT, Fanconi anemia (FA), and xeroderma pigmentosum (XP; 278700) to test the hypothesis that heterozygotes may be predisposed to some of the same congenital malformations and developmental disabilities that are common among homozygotes. Among XP relatives, 11 of 1,100 had unexplained mental retardation, whereas only 3 of 1,439 relatives of FA and AT homozygotes showed mental retardation. Four XP relatives but no FA or AT relatives had microcephaly. Idiopathic scoliosis and vertebral anomalies occurred in excess in AT relatives, while genitourinary and distal limb malformations were found in FA families.

Swift et al. (1987) examined the cancer risk of heterozygotes for AT in 128 families, including 4 of Amish ancestry, 110 white non-Amish families, and 14 black families. They measured documented cancer incidence rather than cancer mortality based solely on death certificates and compared the cancer incidence in adult blood relatives of probands directly with that in spouse controls. The incidence rates in AT relatives were significantly elevated over those in spouse controls. In persons heterozygous for AT, the relative risk of cancer was estimated to be 2.3 for men and 3.1 for women. Breast cancer in women was the cancer most clearly associated with heterozygosity for AT. Swift et al. (1987) estimated that 8 to 18% of patients with breast cancer in the U.S. white population would be heterozygous for AT. Intuitively, it is difficult to believe that such a high proportion of breast cancer women are AT heterozygotes. Pippard et al. (1988) confirmed this observation, however. They reported an excess of breast cancer deaths in British mothers of AT patients (significant at the 5% level), but no excess mortality from malignant neoplasms in the grandparents.

Morrell et al. (1990) reported cancer incidence measured retrospectively in 574 close blood relatives of AT patients and 213 spouse controls in 44 previously unreported families. For heterozygous carriers of the AT gene, the relative risk of cancer was estimated to be 6.1 as compared with non-heterozygotes. The most frequent cancer site in the blood relatives was the female breast, with 9 cancers observed. Gatti et al. (1991) provided a review in which they noted the possibly high frequency of breast cancer in AT heterozygotes. Swift (1980) defended, from the viewpoint of not causing anxiety, the usefulness and safety of cancer risk counseling of heterozygotes for AT. Swift et al. (1991) reported the results of a prospective study of 1,599 adult blood relatives of patients with AT and 821 of their spouses distributed in 161 families. Cancer rates were significantly higher among the blood relatives than in their spouses, specifically in the subgroup of 294 blood relatives who were known to be heterozygous for the AT gene. The estimated risk of cancer of all types among heterozygotes as compared with noncarriers was 3.8 in men and 3.5 in women, and that for breast cancer in carrier women was 5.1. Among the blood relatives, women with breast cancer were more likely to have been exposed to selected sources of ionizing radiation than controls without cancer. Male and female blood relatives also had 3-fold and 2.6-fold excess mortality from all causes, respectively, from the ages of 20 through 59 years. Swift et al. (1991) suggested that diagnostic or occupational exposure to ionizing radiation increases the risk of breast cancer in women heterozygous for AT. The work of Swift et al. (1991) on the frequency of breast cancer in AT was critiqued by numerous authors, including Bridges and Arlett (1992).

Since the genes responsible for most cases of AT are located on 11q, Wooster et al. (1993) typed 5 DNA markers in the AT region in 16 breast cancer families. They found no evidence for linkage between breast cancer and these markers and concluded that the contribution of AT to familial breast cancer is likely to be minimal.

Athma et al. (1996) determined the AT gene carrier status of 776 blood relatives in 99 AT families by tracing the ATM gene in each family through tightly linked flanking DNA markers. There were 33 women with breast cancer who could be genotyped; 25 of these were AT heterozygotes, compared to an expected 14.9. For 21 breast cancers with onset before age 60, the odds ratio was 2.9 and for 12 cases with onset at age 60 or older, the odds ratio was 6.4. Thus, the breast cancer risk for AT heterozygous women is not limited to young women but appeared to be even higher at older ages. Athma et al. (1996) estimated that, of all breast cancers in the U.S., 6.6% may occur in women who are AT heterozygotes. This proportion is several times greater than the estimated proportion of carriers of BRCA1 mutations (113705) in breast cancer cases with onset at any age.

The reported increased risk for breast cancer for AT family members has been most evident among younger women, leading to an age-specific relative risk model predicting that 8% of breast cancer in women under age 40 arises in AT carriers, compared with 2% of cases between 40 and 59 years (Easton, 1994). To test this hypothesis, FitzGerald et al. (1997) undertook a germline mutational analysis of the ATM gene in a population of women with early onset of breast cancer, using a protein truncation (PTT) assay to detect chain-terminating mutations, which account for 90% of mutations identified in children with AT. They detected a heterozygous ATM mutation in 2 of 202 (1%) controls, consistent with the frequency of AT carriers predicted from epidemiologic studies. ATM mutations were present in only 2 of 401 (0.5%) women with early onset of breast cancer ($P = 0.6$). FitzGerald et al. (1997) concluded that heterozygous ATM mutations do not confer genetic predisposition to early onset of breast cancer.

The results of FitzGerald et al. (1997) are discrepant with those of Athma et al. (1996), who conducted a study 'from the other direction' by following identified AT mutations through the families of those with clinically recognized AT. Analysis of DNA markers flanking the AT gene allowed them to identify precisely which female relatives with breast cancer carried the AT mutation. On the basis of the genetic relationship between each case and the AT proband, the a priori probability that these 2 share the AT mutation was calculated. This led to an estimated relative risk of 3.8 as compared to noncarriers. This result was similar to that found by Easton (1994), who reanalyzed the previous studies of breast cancer risk in mothers (and other close relatives) of AT cases. Bishop and Hopper (1997) analyzed these 2 studies and suggested that they may not be discrepant. Indeed, they estimated that the study of FitzGerald et al. (1997) yielded an upper limit of the 95% confidence interval for the proportion of early onset breast cancer occurring in AT heterozygotes as 2.4% (assuming that their assay identified 75% of all mutations).

Biochemical Features

The nature of the basic defect is a mystery but is believed to involve one or more of the enzymes concerned with DNA repair or processing.

Shaham and Becker (1981) showed that the AT clastogenic (chromosome breaking) factor present in plasma of AT patients and in the culture medium of AT skin fibroblasts is a peptide with a molecular weight in the range of 500 to 1000. No clastogenic activity could be demonstrated in extracts of cultured AT fibroblasts.

Mutants of the mei-41 gene in Drosophila melanogaster were first identified on the basis of a defect in meiotic recombination and subsequently by their mutagen sensitivity to a wide range of mutagens, including ionizing radiation, ultraviolet radiation, methyl methanesulfonate, and hydroxyurea (Banga et al., 1986; Baker et al., 1976). Indeed, there is no overlap between the x-ray dose-kill curves for wildtype and for mei-41 mutants, and females heterozygous for mei-41 mutations display an intermediate level of mutagen sensitivity (Boyd et al., 1976; Nguyen et al., 1979). Hari et al. (1995) noted that mutations in the mei-41 gene also cause high levels of chromosome breakage and instability in mitotic cells and in the male germline. A number of gaps and breaks are enhanced following treatment with x-rays to the extent that after 220 R of irradiation, virtually all of the subsequent metaphases possess at least 1 break or rearrangement. They commented that the observation of chromatid breaks and gaps in the metaphase chromosomes of both mutagenized and unmutagenized mei-41 cells is surprising, because many organisms possess cell cycle checkpoint controls that prevent cells with damaged DNA from exiting G2 and entering the M phase. Hari et al. (1995) demonstrated that mei-41 has a similar if not identical effect on G2/M progression of X-irradiated neuroblasts in Drosophila as is observed in ataxia-telangiectasia: cells irradiated in G2 fail to display an initial block in cell cycle progression that is characteristic of normal cells. Hari et al. (1995) also showed that the Drosophila mei-41 and the human ATM gene are homologous at the level of predicted amino acid sequence. Like the ATM protein, the mei-41 protein belongs to a family of phosphatidylinositol 3-kinase (PI3K)-like proteins that include the yeast rad3 and Mec1p proteins. Hari et al. (1995) concluded that the mei-41 gene of Drosophila is a functional homolog of the human ATM gene.

Greenwell et al. (1995) showed that the ATM gene has strong homology to 2 known yeast genes, ESR1/MEC1 of Saccharomyces cerevisiae and rad3 of Schizosaccharomyces pombe, and to a yeast open reading frame, YBLO88. Greenwell et al. (1995) showed that YBLO88 encodes TEL1, a gene required for maintaining wildtype telomere length. Yeast chromosomes terminate in tracts of simple repetitive DNA, poly-G1-3-T. Mutations in the TEL1 gene result in shortened telomeres. Sequence analysis of TEL1 indicated that it encodes a very large protein (322 kD) with amino acid motifs found in phosphatidylinositol/protein kinases. The authors found that the closest homolog to TEL1 is the human ATM gene. Morrow et al. (1995) presented data indicating that TEL1 and the checkpoint gene MEC1 in S. cerevisiae are functionally related and that functions of the human ATM gene are apparently divided between at least 2 S. cerevisiae homologs. Paulovich and Hartwell (1995) likewise identified MEC1 as a yeast homolog of ATM.

Zakian (1995) provided a dendrogram indicating the relationship between ATM and the various ATM-like genes. She pointed out that 'whether or not these ATM-like genes are ATM homologs, continued inquiry in genetically tractable model organisms like yeast and Drosophila will surely provide valuable insight into the functions of this family of proteins.'

Telomeres are essential for stable maintenance of linear chromosomes in eukaryotes. As indicated earlier, the ATM family of genes, including TEL1 of budding yeast, rad3+ of fission yeast, and human ATM itself, appear to be involved in telomere length regulation. Naito et al. (1998) cloned another fission yeast ATM homolog, tel1+, and found that a tel1rad3 double mutant lost all telomeric DNA sequences. Thus, the ATM homologs are essential in telomere maintenance. The mutant grew poorly and formed irregular-shaped colonies, probably due to chromosome instability; however, during prolonged culture of double mutants, cells forming normal round-shaped colonies arose at a relatively high frequency. All 3 chromosomes in these derivative cells were circular and lacked telomeric sequences. This appeared to be the first report of eukaryotic cells whose chromosomes were all circular. Upon meiosis, these derivative cells produced few viable spores. Therefore, the exclusively circular genome lacking telomeric sequences is proficient for mitotic growth, but does not permit meiosis.

Hawley and Friend (1996) commented upon the current state of ATM research and concluded that ATM must play crucial roles in normally developing or undamaged cells, as well as the studied role in irradiated cells, in order to explain the neurologic, immune, and reproductive problems observed in AT patients. They also proposed that ATM may be intimately associated with both p53 (TP53; 191170) and the molecular machinery required for chromosomal exchange, perhaps as components of the recombination nodules.

Other Features

Waldmann and McIntire (1972) showed raised alpha-fetoprotein in the blood of patients with AT. This, they felt, suggests immaturity of the liver and is consistent with the view that the primary defect is in tissue differentiation, specifically, a defect in the interaction necessary for differentiation of gut-associated organs such as the thymus and liver. Ishiguro et al. (1986) concluded that the elevated alpha-fetoprotein in patients with AT probably originates in the liver.

On the circulating monocytes of AT patients, Bar et al. (1978) demonstrated an 80 to 85% decrease in insulin receptor affinity. This decrease was not observed in the cultured fibroblasts of AT patients or in the monocytes and fibroblasts of relatives of these patients. In addition, they found that whole plasma and immunoglobulin-enriched fractions of plasma from AT patients inhibited the normal binding of insulin to its receptors on cultured human lymphocytes and on human placental membranes. This suggested the presence of antireceptor immunoglobulins. AT and type B acanthosis nigricans have several features in common that suggest the possibility of similar causes for the insulin resistance each demonstrates.

Mohamed et al. (1987) found marked reduction of topoisomerase II (126430) in some but not all AT cell lines. DNA topoisomerases I and II are enzymes that introduce transient single- and double-strand breaks into DNA and thus are capable of interconverting various DNA conformations. The isolation of mutants of the 2 enzymes in yeast and the increased levels of DNA topoisomerase II in cells undergoing DNA synthesis provide evidence for the role of these enzymes in DNA replication and in chromosome segregation and organization.

Kojis et al. (1989) suggested that the very high frequency of lymphocyte-associated rearrangements (LARs) in peripheral blood chromosome preparations is a diagnostic criterion of the disease. They pointed out a striking difference in the types of rearrangements observed in lymphocytes and fibroblasts. LARs are not commonly observed in fibroblasts, despite the increased but random instability of chromosomes from these cells relative to lymphocytes. The region of location of the AT gene, 11q22-q23, is not involved in site-specific rearrangements in either lymphocytes or fibroblasts.

Carbonari et al. (1990) found that patients with AT have more circulating T cells bearing gamma/delta receptors characteristic of immature cells than alpha/beta receptors typical of mature cells. Normal ratios were found in the patients with other immune deficits, except for 1 child with a primary T-cell defect. Peterson and Funkhouser (1990) proposed that these findings are consistent with a defect in genetic recombination leading to the switch from gamma/delta to alpha/beta. There may also be a defect in DNA ligation or some other aspect of DNA repair. Elucidation of the molecular abnormalities of lymphocytes may demonstrate fundamental molecular mechanisms for cellular differentiation not only of lymphocytes but of other cell systems such as the nervous system.

Lipkowitz et al. (1990) showed that an abnormal V(D)J recombination, joining V segments of the T-cell receptor gamma gene (186970) with J segments of the T-cell receptor beta gene (186930), occurs in peripheral blood lymphocytes of AT patients at a frequency 50- to 100-fold higher than normal. This frequency is roughly the same as the increase in the risk for lymphoid malignancy in these individuals. There is also an increase in the frequency of the lymphocyte-specific cytogenetic abnormalities thought to be due to interlocus recombination in non-AT patients with non-Hodgkin lymphoma, further suggesting a relationship between these translocations and lymphoid malignancies. Agriculture workers occupationally exposed to pesticides used in the production and storage of grain have a high frequency of cytogenetic abnormalities in peripheral blood lymphocytes in a pattern reminiscent of those in AT patients. Furthermore, these agriculture workers have an increased risk of developing T- and D-lymphoid malignancies. Lipkowitz et al. (1992) used a PCR-based assay developed for the study of AT patients to demonstrate a 10- to 20-fold increased frequency of hybrid antigen-receptor genes in peripheral blood lymphocytes of agriculture workers with chemical exposure.

Inheritance

In a study of 47 families ascertained throughout the United Kingdom, Woods et al. (1990) found a low parental consanguinity rate; no parents were first cousins or more closely related, whereas 10% had been expected. Furthermore, the incidence of the disorder in 79 sibs of index cases was 1 in 7, rather than the expected 1 in 4.

Cytogenetics

Oxford et al. (1975) found that chromosome 14 was often involved in rearrangements in AT and that band 14q12 was a highly specific exchange point. In addition to the changes in chromosome 14, a pericentric inversion of chromosome 7 is characteristic. McCaw et al. (1975) described t(14;14)(q11;q32) translocation in T-cell malignancies of patients with AT. T cells show a t(14;14)q12q32 rearrangement in about 10% of AT patients.

Aurias et al. (1986) described a possible 'new' type of chromosome rearrangement, namely, telomere-centromere translocation (tct) followed by double duplication. This type of rearrangement was found between chromosomes 7 and 14 in cases of AT (Aurias et al., 1986). Gatti et al. (1985) and Aurias and Dutrillaux (1986) found that the sites of breaks in rearrangements (7p14, 7q35, 14q12, 14qter, 2p11, 2p12, and 22q11-q12) are those where members of the immunoglobulin superfamily are located: IGK, IGH, IGL, TCRA, TCRB, TCRG. The somatic gene rearrangement must precede expression of these genes.

Kennaugh et al. (1986) studied a patient with an inversion of 14q which had been present for many years in T cells. It was found that the breakpoint in 14q32 lay outside the IgH locus and proximal to it. The constant region gene of the T-cell receptor alpha chain (TCRA) locus was translocated to the 14q32 position. Johnson et al. (1986) found that the 14q32 breakpoint in the 14/14 translocation found in T-CLL cells and in an AT patient occurred within the immunoglobulin gene cluster. The AT patient had the characteristic chromosome 14 tandem translocation in 100% of karyotyped T cells 10 years before her death from T-cell leukemia. (This was the same patient described earlier by Saxon et al. (1979).) Stern et al. (1988) used in situ chromosomal hybridization to map the TCRA gene in 3 different nonmalignant T-cell clones derived from patients with AT. The constant region was translocated in each clone; the variable region remained in its original position in 2 clones and was deleted in 1 which lost the derivative chromosome 14.

Stern et al. (1988) mapped the 14q32.1 recurrent breakpoint of AT clones by in situ hybridization. They found that the breakpoint lay between D14S1 (107750) and PI (107400). In a t(14;14) clone they found an interstitial duplication including D14S1 and a part of the IGH locus. Studying the chromosomes by R-banding, Zhang et al. (1988) concluded that the distal breakpoint in the chromosome 14 inversion in an AT clone was different from that in the chromosome 14 inversion in a malignant T-cell line; specifically, in AT, the breakpoint was centromeric to both the immunoglobulin heavy chain locus and the D14S1 anonymous locus (107750). They suggested that this finding favors the existence of an unknown oncogene in band 14q32.1.

Russo et al. (1989) presented evidence for a cluster of breakpoints in the 14q32.1 region, the site of the putative oncogene TCL1 (186960), in cases of ataxia-telangiectasia with chronic lymphocytic leukemia. The 14q32.1 breakpoint is at least 10,000 kb centromeric to the immunoglobulin heavy chain locus. In a cell line with a translocation t(14;14)(q11;q32) from an AT patient with T-cell chronic lymphocytic leukemia, Russo et al. (1989) showed that a J(alpha) sequence from the TCRA locus (186880) was involved. This was again the patient first reported by Saxon et al. (1979). Humphreys et al. (1989) found some rearrangements involving chromosomes 7 and 14 at the usual 4 sites associated with AT—7p14, 7q35, 14q12, and 14q32—all sites of T-cell receptor genes.

Mapping

By linkage to RFLP markers, Gatti et al. (1988) localized the AT gene to 11q22-q23. They had previously excluded 171 markers, comprising approximately 35% of the genome. The most promising marker in a large Amish pedigree was found to be THY1 (188230), which is located at 11q22.3; it showed linkage with maximum lod = 1.8 at theta = 0.00. When data from the other 4 informative group A AT families were added, the maximum lod score rose to 3.63 with no observed recombinants. The maximum lod score for all 31 families studied for linkage of AT to THY1 was 4.34 at theta = 0.10. The large Amish pedigree diagrammed in their figure 1 is the kindred reported by McKusick and Cross (1966), Ginter and Tallapragada (1975), and Rary et al. (1975). By further mapping with a panel of 10 markers, Sanal et al. (1990) concluded that the AT locus is in band 11q23.

It is of interest that the site of the AT1 gene (11q22-q23) is the same as or adjacent to the region occupied by the CD3 (186790), THY1, and NCAM (116930) genes, all of which are members of the immunoglobulin-gene superfamily and therefore may be subject to the same defect that afflicts the T-cell receptor and immunoglobulin molecules in AT. Concannon et al. (1990) excluded the AT1 gene from a region extending 15 cM to either side of ETS1 (164720), which maps to 11q24. According to Gatti (1990), the gene in families from complementation groups A, C, and D, representing approximately 97% of all families, has been mapped to 11q23. Thus, a single gene may exist with various intragenic defects permitting complementation.

In studies of 35 consecutively obtained families in the British Isles, McConville et al. (1990) found support for linkage with THY1 at zero recombination. They found evidence suggesting a second AT locus on 11q, centromeric to the site previously postulated. With 3 exceptions, the families had not been assigned to complementation groups. The series of families included the only group E family described to date. They quoted Jaspers et al. (1988) as giving the proportion of group A, group C, and group D cases as approximately 56%, 28%, and 14%, respectively.

By linkage studies in a Jewish-Moroccan family with AT of the group C type, Ziv et al. (1991) found that the disorder was linked to the same region (11q22-q23) as found in group A families. McConville et al. (1990) located the AT1 gene to a 5-cM region in 11q22-q23, flanked by NCAM and DRD2 (126450) on one side and STMY1 (185250) on the other.

In a large inbred family with AT including 2 affected individuals related as first cousins, Hernandez et al. (1991) excluded linkage to markers in the region of 11q where the AT1 gene was previously assigned. They observed that the affected individuals had less severe ataxia than is usually the case, although oculocutaneous telangiectases and increased cellular radiosensitivity typical of AT were present.

On the basis of an 18-point map of the 11q23 region of 11q, derived from linkage analysis of 40 CEPH families, Foroud et al. (1991) analyzed 111 AT families from Turkey, Israel, England, Italy, and the United States, localizing the gene to an 8-cM sex-averaged interval between the markers STMY1 and D11S132/NCAM. Ziv et al. (1992) obtained results from linkage study indicating that the ATA gene in 3 large Arab families was located in 11q23. However, in a Druze family unassigned to a specific complementation group, several recombinants between AT and the same markers were observed.

Sobel et al. (1992) pointed to linkage evidence suggesting that there are 2 AT loci on 11q and that group D AT may be located distal to the site of groups A and C in the 11q23 region. Kapp et al. (1992) reported the cloning of a putative group D gene that mapped, by means of radiation hybrids, to the region between markers THY1 and D11S83; see 208905. In linkage studies of 14 Turkish families, 12 of which were consanguineous, Sanal et al. (1992) obtained results indicating that the most likely location for a single AT locus is within a 6-cM sex-averaged interval defined by STMY and the marker CJ77. However, it appeared that there are at least 2 distinct AT loci (ATA and ATD) at 11q22-q23, with perhaps a third locus, ATC, located very near the ATA gene.

Hernandez et al. (1993) described a large family in which 2 adult cousins had AT with a somewhat milder clinical course than usual. Since genetic linkage analysis did 'not provide any evidence that the gene for AT in this family is located at 11q22-23,' further locus heterogeneity was suggested.

Gatti et al. (1993) reported prenatal genotyping in this disorder. They pointed out that although at least 5 complementation groups have been defined, linkage studies of more than 160 families from various parts of the world have failed to show linkage heterogeneity. All but 2 families were linked to a 6-cM (sex-averaged) region at 11q22.3 defined by the markers STMY1 and D11S385. A further analysis of 50 British families narrowed the localization to a 4-cM (sex-averaged) region defined by D11S611 and D11S535. The demonstrated complementation groups may represent different intragenic mutations or separate ataxia-telangiectasia genes clustered within the 11q22.3 region, neither of which would challenge the validity of linkage or haplotyping studies. A possible reinterpretation of the complementation data is that the radiosensitivity of AT fibroblasts can be complemented by many genes besides the AT gene or genes. Gatti et al. (1993) used the flanking markers to show that the haplotypes in a fetus were identical to those in a previously born affected child. The parents chose to continue the pregnancy.

Matsuda et al. (1996) determined the chromosomal locations of the Atm and Acat1 (mitochondrial acetoacetyl-CoA thiolase; 203750) genes in mouse, rat, and Syrian hamster by direct R-banding fluorescence in situ hybridization. The 2 genes colocalized to mouse 9C-D, the proximal end of rat 8q24.1, and 12qa4-qa5 of Syrian hamster. The regions in the mouse and rat are homologous to human chromosome 11q. In the study of interspecific backcross mice, no recombinants were found among Atm, Npat (601448), and Acat1. NPAT is a gene identified by positional cloning in the AT region and the gene product may be transported into the nucleus because it has sequences matching the nuclear localization signals. Therefore, the gene was tentatively designated NPAT, for 'nuclear protein mapped to the ataxia-telangiectasia locus' (Imai, 1996).

Pecker et al. (1996) also mapped the mouse Atm gene to band 9C by FISH. Using a panel of 67 progeny derived from an interspecific backcross, Xia et al. (1996) demonstrated that the mouse homolog maps to chromosome 9.

Molecular Genetics

At least 2 stages in the cell cycle are regulated in response to DNA damage, the G1-S and the G2-M transitions (Hartwell, 1992). These transitions serve as checkpoints at which cells delay progress through the cell cycle to allow repair of damage before entering either S-phase, when damage would be perpetuated, or M-phase, when breaks would result in the loss of genomic material. Checkpoints are thought to consist of surveillance mechanisms that can detect DNA damage, signal transduction pathways that transmit and amplify the signal to the replication or segregation machinery, and possibly repair activities. Both the G1-S and G2-M checkpoints are known to be under genetic control, since there are mutants that abolish the arrest or delay occurring in normal cells in response to DNA damage. Painter et al. (1982) showed that the G1-S checkpoint is abolished in cells from AT patients. Kastan et al. (1992) provided strong evidence that p53 is necessary for the G1-S checkpoint. They found that the AT gene(s) is upstream of the p53 gene in a pathway that activates the G1-S checkpoint. p53 levels increase 3- to 5-fold by a posttranscriptional mechanism after gamma-irradiation, coincident with a delay of the G1-S transition (Kastan et al., 1991); the induction of p53 does not occur in AT cells (Kastan et al., 1992). Induction by ionizing radiation of the GADD45 gene (126335), an induction that is also defective in AT cells, is dependent on wildtype p53 function (Kastan et al., 1992). Thus, Kastan et al. (1992) identified 3 participants—AT gene(s), p53, and GADD45—in a signal transduction pathway that controls cell cycle arrest following DNA damage. Abnormalities in this pathway probably contribute to tumor development. Kastan et al. (1992) pointed out that lymphoid malignancies are the most common tumor seen both in AT patients and in p53-deficient mice. Lymphoid cells normally experience DNA strand breaks during gene rearrangements. The G1 checkpoint may be important in the avoidance of errors in that process. Breast cancer and other nonlymphoid cancers are increased in individuals heterozygous for germline mutations of either p53 (e.g., the Li-Fraumeni syndrome; 191170.0001) or the AT gene(s) (172,171:Swift et al., 1987, 1991).

The p53 tumor-suppressor protein is a sequence-specific DNA-binding transcription factor that induces cell cycle arrest or apoptosis in response to genotoxic stress. Activation of p53 by DNA-damaging agents is critical for eliminating cells with damaged genomic DNA and underlies the apoptotic response of human cancers treated with ionizing radiation and radiomimetic drugs. Both the levels of p53 protein and its affinity for specific DNA sequences increase in response to genotoxic stress. In vitro, the affinity of p53 for DNA is regulated by its carboxyl terminus. Waterman et al. (1998) therefore examined whether this region of p53 is targeted by DNA-damage signaling pathways in vivo. In nonirradiated cells, serines 376 and 378 of p53 were phosphorylated. IR led to dephosphorylation of ser376, creating a consensus binding site for 14-3-3 proteins (113508) and leading to association of p53 with 14-3-3. In turn, this increased the affinity of p53 for sequence-specific DNA. Consistent with the lack of p53 activation by ionizing radiation in AT, neither ser376 dephosphorylation nor the interaction of p53 with 14-3-3 proteins occurred in AT cells.

Using 2 recombination vectors to study recombination in AT and control human fibroblast lines, Meyn (1993) found that the spontaneous intrachromosomal recombination rates were 30 to 200 times higher in AT fibroblast lines than in normal cells, whereas extrachromosomal recombination frequencies were near normal. Increased recombination is thus a component of genetic instability in AT and may contribute to the cancer risk. Other evidence of in vitro and in vivo genomic instability includes increased frequencies of translocations and other chromosomal aberrations in lymphocytes and fibroblasts, micronucleus formation in epithelial cells, and loss of heterozygosity in erythrocytes. Hyperrecombination is a specific feature of the AT phenotype rather than a genetic consequence of defective DNA repair because a xeroderma pigmentosum cell line exhibited normal spontaneous recombination rates.

Croce et al. (1985) assigned the gene for the alpha subunit of the T-cell antigen receptor (186880) to the region of one of the common breakpoints in AT (14q11.2) and suggested that the oncogene TCL1 (186960) is located in the region of the other breakpoint (14q32.3). It is thought that the TCL1 gene may be activated by chromosome inversion or translocation, either of which results in juxtaposition of the TCL1 gene and the TCRA gene. In AT, circulating lymphocytes show characteristic rearrangements involving the site of the T-cell receptor gamma genes (7p15), T-cell receptor beta genes (7q35), T-cell receptor alpha genes (14q11), and immunoglobulin heavy chain genes (14q32) (McFarlin et al., 1972; Ying and Decoteau, 1981).

Jung et al. (1995) isolated cDNA that corrected the radiation sensitivity and DNA synthesis defects in fibroblasts from an AT1 group D patient by expression cloning, and showed that the cDNA encoded NFKBI, a truncated form of I-kappa-B (164008), which is an inhibitor of NFKB1,the nuclear factor kappa-B transcriptional activator (164011). The parental AT1 fibroblast expressed large amounts of the NFKBI transcript and showed constitutive activation of NFKB1. The AT1 fibroblast transfected with the truncated NFKBI expressed normal amounts of the NFKBI transcript and showed regulated activation of NFKB1. Since the NFKBI gene is located on chromosome 14 and not chromosome 11, it is probably not the site of the primary defect; Jung et al. (1995) hypothesized that its contribution to the ataxia-telangiectasia phenotype may work downstream of the gene representing the primary defect.

By the strategy of positional cloning, Savitsky et al. (1995) identified a gene that they designated ATM (for 'AT mutated'). ATM, which has a transcript of 12 kb, was found to be mutated in AT patients from all complementation groups, indicating that it is probably the sole gene responsible for this disorder. A partial ATM cDNA clone of 5.9 kb encoded a putative protein that is similar to several yeast and mammalian phosphatidylinositol 3-prime (PI-3) kinases (e.g., 171833, 171834) that are involved in mitogenic signal transduction, meiotic recombination, and cell cycle control. Savitsky et al. (1995) speculated that the discovery of ATM may allow the identification of AT heterozygotes who are at increased risk of cancer.

In order to clone ATM, Savitsky et al. (1995) constructed a YAC contig and cosmid contigs spanning the interval between D11S384 and D11S1818, known from linkage studies to contain the AT1 gene. Two complementary methods were used for identification of transcribed sequences: hybrid selection based on direct hybridization of genomic DNA with cDNAs, and exon amplification to identify putative exons in genomic DNA by their splicing capacity. One cDNA clone had an open reading frame that predicted a protein of 1,708 amino acids, beginning with the first available methionine and ending in a termination codon. Southern blot analysis revealed homozygous deletion in this gene in affected members of family ISAT9, an extended Palestinian-Arab AT family that had not been assigned to a complementation group (Ziv et al., 1992). The deletion detected by Savitsky et al. (1995) included almost the entire genomic region spanned by the original cDNA clone and segregated with the disease. This finding led to a systematic search for mutations in additional AT patients. The restriction endonuclease fingerprinting (REF) method was applied to DNA products of RT-PCR-amplified RNA derived from AT cell lines. When an abnormal REF pattern was found, the relevant portion of the transcript was directly sequenced. Most of the mutations identified in 14 patients, including 3 sib pairs, were predicted to lead to premature truncation of the protein product. Three mutations were predicted to create in-frame deletions of 1, 2, or 3 amino acid residues; see 208900.0001, 208900.0002, and 208900.0003.

Pecker et al. (1996) showed that the mouse Atm gene encodes a 3,066-amino acid predicted protein with 84% identity to the human sequence. Northern blots showed a 13-kb transcript in brain, skeletal muscle, and testis with lower amounts in other tissues. A 10.5-kb band was also seen in testis mRNA.

Brzoska et al. (1995) isolated a candidate AT gene (which they called ATDC) on the basis of its ability to complement the ionizing radiation sensitivity of AT group D fibroblasts. They found that the ATDC protein physically interacts with the intermediate-filament protein vimentin (193060), which is a protein kinase C substrate and colocalizing protein, and with an inhibitor of protein kinase C, PKCI1. Indirect immunofluorescence analysis of cultured cells transfected with a plasmid encoding an epitope-tagged ATDC protein localized the protein to vimentin filaments. Brzoska et al. (1995) suggested that the ATDC and PKCI1 proteins may be components of a single transduction pathway that is induced by ionizing radiation and mediated by protein kinase C (176960). The fact that the ATM gene encodes a protein with a putative phosphatidylinositol 3-kinase domain and functions as a lipid-mediated signaling molecule is consistent with a model in which ATDC and PKC function downstream from ATM in this pathway.

Kastan (1995) reviewed the implications of the cloning of the ATM gene.

Savitsky et al. (1995) sequenced the 3-prime half of the ATM gene; Byrd et al. (1996) reported the 1,348-amino acid sequence of the N-terminal half of the ATM gene product. No homology with other genes was found within the N-terminal half of the AT protein. They also identified 6 mutations affecting this half of the protein. One of these mutations was found to be associated with a haplotype that is common to 4 apparently unrelated families of Irish descent. All the patients so far examined for both AT alleles had been shown to be compound heterozygotes. None of these mutations affected the putative promoter region which made direct divergent transcription of both the ATM gene and a novel gene, E14 (601605). All of the mutations identified by Byrd et al. (1996) were deletions except for 1 insertion.

Savitsky et al. (1995) reported the sequence of a cDNA contig spanning the entire open reading frame of the ATM gene. The predicted protein of 3,056 amino acids showed significant sequence similarities to several large proteins in yeast, Drosophila, and mammals, all of which share the PI 3-kinase domain. Many of these proteins are involved in the detection of DNA damage and the control of cell cycle progression. Mutations in the genes confer a variety of phenotypes with features similar to those observed in human AT cells.

Uziel et al. (1996) determined the genomic organization of ATM using long distance PCR between exons. The gene contains 66 exons spanning approximately 150 kb of genomic DNA. The initiating methionine occurs in exon 4. The first 2 exons, 1a and 1b, are used differentially in alternative transcripts; the initiation codon lies within exon 4; and the final, 3.8-kb exon has about 3.6 kb of 3-prime untranslated sequence.

Gilad et al. (1996) performed mutation analysis in 55 families with AT, using RT-PCR of total RNA from cultured fibroblasts or lymphoblasts, followed by restriction endonuclease fingerprinting of PCR products. Of the 44 AT mutations identified, 39 (89%) were expected to inactivate the ATM protein by truncating it, abolishing correct initiation or termination of translation, or deleting large segments. Additional mutations included 4 smaller in-frame deletions and insertions, and 1 substitution of a highly conserved amino acid at the PI-3 kinase domain.

Wright et al. (1996) assayed 38 cell lines, including 36 from unrelated AT patients and 2 control cell lines, for ATM gene mutations. They detected 30 mutations. Twenty-five of these were distinct, and most of the patients were compound heterozygotes. The mutations included nucleotide substitutions (2 cases), insertions (1 case), and deletions of 2 to 298 nucleotides (27 cases). The most frequent variant (detected in 3 unrelated patients) was a 9-bp deletion at codon 2546 in exon 54 (208900.0007). This mutation had been previously reported in 5 different patients and constituted 8% of the reported mutations at that time. Twenty-two of the observed alterations would be predicted to lead to protein alterations. Wright et al. (1996) discussed the practicality of population screening in light of their results.

Using an antiserum developed to a peptide corresponding to the deduced amino acid sequence of ATM, Brown et al. (1997) showed that the ATM protein is a single, high-molecular-weight protein predominantly confined to the nucleus of human fibroblasts, although it is present in both nuclear and microsomal fractions from human lymphoblast cells and peripheral blood lymphocytes. ATM protein levels and localization remain constant throughout all stages of the cell cycle. Truncated ATM protein was not detected in lymphoblasts from AT patients homozygous for mutations leading to premature protein termination. Exposure of normal human cells to gamma-irradiation and the radiomimetic drug neocarzinostatin had no effect on ATM protein levels, in contrast to a noted rise in p53 levels over the same time interval. The findings of constitutive expression and nuclear localization of the ATM protein were consistent with its potential role in choreographing appropriate cellular responses to genomic damage.

Atm-null mice, as well as those null for p53, develop mainly T-cell lymphomas, supporting the view that these genes have similar roles in thymocyte development. To study the interactions of these 2 genes on an organismal level, Westphal et al. (1997) bred mice heterozygous for null alleles of both atm and p53 to produce all genotypic combinations. Mice doubly null for atm and p53 exhibited a dramatic acceleration of tumor formation relative to singly null mice, suggesting that the 2 genes collaborate to prevent tumorigenesis. With respect to their roles in apoptosis, loss of atm rendered thymocytes only partly resistant to irradiation-induced apoptosis, whereas additional loss of p53 engendered complete resistance. This implied that the irradiation-induced atm and p53 apoptotic pathways are not completely congruent. Furthermore, Westphal et al. (1997) found that atm and p53 do not appear to interact in acute radiation toxicity, suggesting a separate atm effector pathway for this DNA damage response and having implications for the prognosis and treatment of human tumors.

Zhang et al. (1997) discussed the cloning of a full-length cDNA encoding ATM and correction of multiple aspects of the radiosensitive phenotype of AT cells by transfection with this cDNA. Overexpression of ATM cDNA in AT cells enhanced their survival after radiation exposure, decreased radiation-induced chromosome aberrations, reduced radioresistant DNA synthesis, and partially corrected defective cell cycle checkpoints and induction of stress-activated protein kinase. This correction of the defects of AT cells provided further evidence of the multiplicity of effector functions of the ATM protein and suggested possible approaches to gene therapy.

Banin et al. (1998) and Canman et al. (1998) observed enhanced phosphorylation of p53 by ATM in response to DNA damage. Both found that ATM had intrinsic protein kinase activity and phosphorylated p53 on serine-15 in a manganese-dependent manner. Ionizing radiation, but not ultraviolet radiation, rapidly enhanced the p53-directed kinase activity of endogenous ATM. Phosphorylation of p53 on serine-15 in response to ionizing radiation was reduced in ataxia-telangiectasia cells.

Khanna et al. (1998) reported direct interaction between ATM and p53 involving 2 regions in ATM, one at the N terminus and the other at the C terminus, corresponding to the PI-3 kinase domain. Recombinant ATM protein phosphorylated p53 on serine 15 near the N terminus. Ectopic expression of ATM in AT cells restored normal ionizing radiation (IR)-induced phosphorylation of p53, whereas expression of ATM antisense RNA in control cells abrogated the rapid IR-induced phosphorylation of p53 on serine 15. The results demonstrated that ATM can bind p53 directly and is responsible for its serine 15 phosphorylation, thereby contributing to the activation and stabilization of p53 during the IR-induced DNA damage response.

By analyzing tumor DNA from patients with sporadic T-cell prolymphocytic leukemia (TPLL), a rare clonal malignancy with similarities to a mature T-cell leukemia seen in ataxia-telangiectasia, Vorechovsky et al. (1997) demonstrated a high frequency of ATM mutations in TPLL. In marked contrast to the ATM mutation pattern in AT, the most frequent nucleotide changes in this leukemia were missense mutations (e.g.; 208900.0009). These clustered in the region corresponding to the kinase domain, which is highly conserved in ATM-related proteins in mouse, yeast, and Drosophila. The resulting amino acid substitutions were predicted to interfere with ATMP binding or substrate recognition. Two of 17 mutated TPLL samples had a previously reported AT allele. One of these, a 9-bp deletion causing loss of a string of 3 amino acids (208900.0007), was the most frequent ATM allele reported in AT patients. The other, val2424gly (208900.0005), had been observed in a patient with atypically mild ataxia-telangiectasia. In neither of these 2 cases was any wildtype allele detectable. Vorechovsky et al. (1997) also studied B-cell non-Hodgkin lymphomas (BNHL) for ATM mutations and detected 3 missense mutations (e.g.; 208900.0010). Concannon and Gatti (1997) pointed out that, although in vitro cell fusion studies had suggested that AT is genetically heterogeneous, all AT patients studied to that time had been found to harbor mutations in the ATM gene. More than 100 ATM mutations had been documented; these were broadly distributed throughout the gene. Except for patients from families with known consanguinity, most AT patients were compound heterozygotes. More than 70% of the mutations were predicted to lead to protein truncation. Many of the reported mutations affected mRNA splicing with at least half of the coding exons (32/62) having been found to undergo exon skipping. The large size of the ATM gene (66 exons spanning approximately 150 kb of genomic DNA), together with the diversity and broad d istribution of mutations in AT patients, limited direct mutation screening as a diagnostic tool, or method of carrier identification, except where founder effect mutations are involved.

Using the yeast 2-hybrid system, Lim et al. (1998) demonstrated that the ATM protein binds to beta-adaptin (600157), one of the components of the AP-2 adaptor complex, which is involved in clathrin-mediated endocytosis of receptors. The interaction between ATM and beta-adaptin was confirmed in vitro, and coimmunoprecipitation and colocalization studies showed that the proteins also associate in vivo. The ATM protein also interacted in vitro with beta-NAP (602166), a neuronal beta-adaptin homolog that had been identified as an autoantigen in a patient with cerebellar degeneration. The finding of an association of ATM with beta-adaptin in vesicles indicated that ATM may play a role in intracellular vesicle and/or protein transport mechanisms. The interaction may help explain how mutations in the ATM gene cause the pleiotropic nature of the ataxia-telangiectasia phenotype. The large size of the ATM protein and its multiple subcellular localization suggest that ATM may have more than one function.

Heterogeneity

On the basis of complementation studies of DNA repair in cultured fibroblasts, Paterson et al. (1977) suggested the existence of 2 distinct types of ataxia-telangiectasia. By genetic complementation analysis, Jaspers and Bootsma (1982) concluded that extensive genetic heterogeneity exists in AT. Their method involved cell fusion and was based on the observation that the rate of DNA synthesis is inhibited by x-rays to a lesser extent in AT cells than in normal cells. At least 5 complementation groups have been identified (Murnane and Painter, 1982; Jaspers and Bootsma, 1982). Heterogeneity in AT has also been indicated by the clinical work of Fiorilli et al. (1983).

Jaspers et al. (1988) reported the results of complementation studies on fibroblast strains from 50 patients with AT or Nijmegen breakage syndrome (NBS; 251260), using the radioresistant DNA replication characteristic as a marker. Six different genetic complementation groups were identified. Four of these, called AB, C, D, and E (of which AB is the largest), represented patients with clinical signs of AT. (According to Gatti (1990), the frequencies of these 4 groups are approximately 55%, 28%, 14%, and 3%, respectively.) Patients having NBS fell into 2 groups, designated V1 and V2. A patient with clinical symptoms of both AT and NBS was found in group V1, indicating that the 2 disorders are closely related (Curry et al., 1989). No group-specific patterns of clinical characteristics or ethnic origin were apparent among the AT cases. In addition to the radiosensitive ATs, a separate category of patients was found, characterized by a relatively mild clinical course and weak radiosensitivity. Jaspers et al. (1988) concluded that a defect in 1 of at least 6 different genes may underlie inherited radiosensitivity in humans.

Curry et al. (1989) used the designation AT(Fresno) (208900.0014) for the V1 disorder in twin girls who had clinical features combining those of ataxia-telangiectasia and the Nijmegen breakage syndrome (251260). Complementation studies with Sendai virus-mediated fusion of fibroblast cell lines showed complementation with AT groups A, C, and E but not with the cell line from a patient with the Nijmegen breakage syndrome. Hernandez et al. (1993) cited evidence for the existence of 4 complementation groups: AB, C, D, and E. Loci for AB, C, and D have been identified on 11q. However, Komatsu et al. (1996) could demonstrate that the gene for the V2 form of Nijmegen breakage syndrome is not located on chromosome 11. They found that cells from a patient with this form were highly sensitive to radiation and that the sensitivity was unchanged after the transfer of an extra copy of normal chromosome 11.

Gatti et al. (1988) noted the existence of at least 4 clinically indistinguishable complementation groups (A, C, D, and E) among 80 affecteds (Jaspers et al., 1985; Jaspers et al., 1988). The Amish pedigree represents group A. This locus was designated ATA (HGM9). Since the Thy-1 glycoproteins are major cell surface constituents of rodent thymocytes and neurons (Tse et al., 1985), the question might be raised as to whether mutation in the THY1 gene is the basis of AT. The fact that recombination was found between THY1 and AT in the overall study may indicate that AT is not due to a defect in THY1 or it may mean that complementation group A is caused by mutation in THY1 but a mutation at another site is responsible for other forms of the disorder. When genetic linkage data from group C families are pooled, it appears that group C also may be linked to 11q22-q23 (Gatti, 1989).

The group D defect is correctable by transfer of chromosome 11 into an SV40-transformed fibroblast cell line (Komatsu et al., 1990). Ejima et al. (1990) corrected the radiosensitivity of a group D fibroblast line by introducing an 11q fragment into these cells. Lambert et al. (1991) showed by microcell-mediated chromosome transfer that immortalized AT cells from complementation group D were corrected by genetic material from region 11q22-q23. A deoxyribophosphodiesterase deficiency has been identified in cells from group E patients. Together, groups A and C encompass about 85% of AT patients. Genetic linkage studies should also clarify whether AT variant families are linked to chromosome 11q22-q23 or to group D or E defects.

As indicated earlier, Savitsky et al. (1995) identified mutations in a single gene (termed ATM by them) in ataxia-telangiectasia cases of complementation groups A, C, D, and E and in 4 other patients in whom the complementation group was not determined. Thus it appears that the complementation that is observed is intragenic and that all AT patients have mutations in a single gene.

See Concannon and Gatti (1997) for discussion of the suggested genetic heterogeneity in AT and an update of mutations in the ATM gene. They noted that most AT patients from nonconsanguineous families were compound heterozygotes.

Rare AT patients with milder manifestations of the clinical or cellular characteristics of the disease have been reported and have been designated 'AT variants.' Gilad et al. (1998) quantified ATM protein levels in 6 AT variants and searched their ATM genes for mutations. Cell lines from these patients exhibited considerable variability in radiosensitivity while showing the typical radioresistant DNA synthesis of AT cells.

Unlike classic AT patients, however, these patients exhibited 1% to 17% of the normal level of ATM. The underlying genotypes were either homozygous for mutations expected to produce mild phenotypes or compound heterozygous for a mild and a severe mutation. In an attempt to determine whether the AT(Fresno) variation correlated with ATM mutations and levels of ATM protein expression, Gilad et al. (1998) searched for ATM mutations in a cell line derived from one of the sisters studied by Curry et al. (1989). This cell line was found to be devoid of the ATM protein and homozygous for a severe ATM mutation. Gilad et al. (1998) concluded that certain AT variant phenotypes, including some of those without telangiectasia, represent ATM mutations.

Sasaki et al. (1998) presented the results of a mutation screen in 14 unrelated AT patients, most of them Japanese. They used a hierarchical strategy in which they extensively analyzed the entire coding region of the cDNA. In the first stage, point mutations were sought by PCR-SSCP in short patches. In the second and third stages, the products of medium- and long-patch PCR, each covering the entire region, were examined by agarose gel electrophoresis to search for length changes. They found a total of 15 mutations (including 12 new) and 4 polymorphisms. Abnormal splicing of ATM was frequent among Japanese, and no hotspot was obvious, suggesting no strong founder effects in that ethnic group. Eleven patients carried either 1 homozygous or 2 compound heterozygous mutations, 1 patient carried only 1 detectable heterozygous mutation, and no mutation was found in 2 patients. Overall, mutations were found in at least 75% of the different ATM alleles examined.

Diagnosis

The presence of early-onset ataxia with oculocutaneous telangiectases permits diagnosis of AT. The clinical diagnosis of AT can be problematic before the appearance of telangiectases. Oculomotor apraxia is a useful aid to early clinical diagnosis. Early-onset cerebellar ataxia and oculomotor apraxia are also typical of X-linked Pelizaeus-Merzbacher disease (312080) and can be seen in Joubert's syndrome (213300). These disorders can be distinguished by leukoencephalopathy in the former, and by profound cerebellar hypoplasia in the latter. See also 257550. Elevated levels of alpha-fetoprotein (126430) and carcinoembryonic antigen are the most useful readily available markers for confirmation of the diagnosis of AT (Gatti et al., 1991). Dysgammaglobulinemia, decreased cellular immune responses, and peripheral lymphopenia are supportive findings but are not invariable.

Henderson et al. (1985) devised a rapid diagnostic method based on the hypersensitivity of AT lymphocytes to killing by gamma irradiation. Similar studies in fibroblasts require skin biopsy and a prolonged culture time. Llerena et al. (1989) concluded that in chorionic villus sampling, gamma radiation is a reliable way of discriminating between unaffected fetuses and those with AT. The reliability of this approach is in question, however. Painter and Young (1980) suggested that the radiosensitivity of AT cells may be caused by their failure to respond to DNA damage with a delay in DNA synthesis that could give time for repair to take place.

Shiloh et al. (1989) presented evidence that the extent of chromatid damage induced in the G2 phase of the cell cycle by moderate dosage of x-rays is markedly higher in AT heterozygous cells than in normal controls. They used this as a test of heterozygosity.

Rosin and Ochs (1986) applied the exfoliated cell micronucleus test to the question of in vivo chromosomal instability in AT. This test is performed on exfoliated cells from the oral cavity collected by swabbing the mucosa with a moistened tongue depressor and also on urinary bladder cells obtained by centrifugation of freshly voided urine specimens. Micronuclei in these cells result from fragmentation of chromosomes in the dividing cells from the epithelium, resulting in acentric fragments which are excluded from the main nucleus when the cell divides. These fragments form their own membrane and can be identified as extranuclear Feulgen-positive bodies in daughter cells which migrate up through the epithelium to be exfoliated. Rosin and Ochs (1986) found that AT homozygotes had a 5- to 14-fold increase in the frequency of exfoliated cell micronuclei. Heterozygotes could be reliably identified by this method (Rosin et al., 1989).

Using X-radiation with 1 Gy on G2-phase lymphocytes from 7 AT patients, 13 obligate AT heterozygotes, and 14 normal controls, Tchirkov et al. (1997) found that both AT homozygotes and heterozygotes showed significantly increased levels of radiation-induced chromatid damage relative to that of normal controls.

Clinical Management

Patients with AT and their cultured cells are unusually sensitive to x-ray just as patients and cells with xeroderma pigmentosum are sensitive to ultraviolet. Treatment of malignancy with conventional dosages of radiation can be fatal to AT patients.

Population Genetics

On the basis of a 'vigorous case finding' in the United States in 2 time periods, Swift et al. (1986) estimated the incidence and gene frequency of AT. The highest observed incidence was in the state of Michigan for the period 1965 to 1969 when white AT patients were born at the rate of 11.3 per million births. Based on the incidence data, the minimum frequency of a single hypothetical AT gene in the U.S. white population was estimated to be 0.0017. Pedigree analysis, which estimates the gene frequency from the proportion of affected close blood relatives of homozygous probands, estimated the most likely gene frequency to be 0.007 on the assumption that AT is a single homogeneous genetic syndrome. Given that complementation analysis has demonstrated genetic heterogeneity in AT, the AT heterozygote frequency might fall between 0.68% and 7.7%, with 2.8% being a likely estimate. In the West Midlands of England, the birth frequency of AT was estimated to be about 1 in 300,000.

Stankovic et al. (1998) reported the spectrum of 59 ATM mutations observed in AT patients in the British Isles. Of the 51 ATM mutations identified in families native to the British Isles, 11 were founder mutations, and 2 of these 11 conferred a milder clinical phenotype with respect to both cerebellar degeneration and cellular features. In 2 AT families, a 7271T-G mutation of the ATM gene appeared to be associated with an increased risk of breast cancer in both homozygotes and heterozygotes, although there was a less severe AT phenotype in terms of the degree of cerebellar degeneration. This mutation was associated with expression of full-length ATM protein at a level comparable to that in unaffected individuals. In addition, Stankovic et al. (1998) studied 18 AT patients, in 15 families, who developed leukemia, lymphoma, preleukemic T-cell proliferation, or Hodgkin lymphoma, mostly in childhood. A wide variety of ATM mutation types, including missense mutations and in-frame deletions, were seen in this group of patients. The authors showed that 25% of all AT patients carried in-frame deletions or missense mutations, many of which were also associated with expression of mutant ATM protein.

Ejima and Sasaki (1998) studied 8 unrelated Japanese families with ataxia-telangiectasia for mutations in the ATM gene. Six different mutations were found on 12 of the 16 alleles examined. Two mutations, 4612del165 (208900.0014) and 7883del5, were found more frequently than the others; 7 of 16 (44%) of the mutant alleles had 1 of these 2 mutations. Microsatellite genotyping demonstrated that a common haplotype was shared by the mutant alleles for both common mutations. The authors suggested that the 2 founder mutations may be predominant among Japanese ATM mutant alleles.

Telatar et al. (1998) found that 4 mutations accounted for 86 to 93% of 41 Costa Rican AT patients studied. They suggested that the Costa Rican population might be useful for analyzing the role of ATM heterozygosity in cancer.

See monographs edited by Bridges and Harnden (1982) and Gatti and Swift (1985).

Animal Model

Barlow et al. (1996) created a murine model of ataxia-telangiectasia by disrupting the Atm locus via gene targeting. Mice homozygous for the disrupted Atm allele displayed growth retardation, neurologic dysfunction, male and female infertility secondary to the absence of mature gametes, defects in T lymphocyte maturation, and extreme sensitivity to gamma-irradiation. Most of the animals developed malignant thymic lymphomas between 2 and 4 months of age. Several chromosomal anomalies were detected in one of these tumors. Fibroblasts from these mice grew slowly and exhibited abnormal radiation-induced G1 checkpoint function. Atm-disrupted mice recapitulated the ataxia-telangiectasia phenotype in humans. The authors noted that humans also show incomplete sexual maturation in ATM (Boder, 1975).

Elson et al. (1996) generated a mouse model for ataxia-telangiectasia using gene targeting to generate mice that did not express the Atm protein. Atm-deficient mice were retarded in growth, did not produce mature sperm, and exhibited severe defects in T-cell maturation while going on to develop thymomas. Atm-deficient fibroblasts grew poorly in culture and displayed a high level of double-stranded chromosome breaks. Atm-deficient thymocytes underwent spontaneous apoptosis in vitro significantly more often than controls. Atm-deficient mice then exhibited many of the same symptoms found in ataxia-telangiectasia patients and in cells derived from them. Furthermore, Elson et al. (1996) demonstrated that the Atm protein exists as 2 discrete molecular species, and that loss of 1 or both of these can lead to the development of the disease.

Xu and Baltimore (1996) disrupted the mouse ATM gene by homologous recombination. Xu et al. (1996) reported that ATM -/- mice are viable, growth-retarded, and infertile. The infertility results from meiotic failure, as meiosis is arrested at the zygotene/pachytene stage of prophase I as a result of abnormal chromosomal synapsis and subsequent chromosome fragmentation. The cerebella of ATM -/- mice appear normal by histologic examination, and the mice have no gross behavioral abnormalities. ATM -/- mice exhibit multiple immune defects similar to those of AT patients, and most develop thymic lymphomas at 3 to 4 months of age and die of the tumors by 4 months. Xu and Baltimore (1996) showed that mouse ATM -/- cells are hypersensitive to gamma irradiation and defective in cell cycle arrest following radiation, and ATM -/- thymocytes are more resistant to apoptosis induced by gamma radiation than normal thymocytes. They also provide direct evidence that ATM acts as an upregulator of p53.

See also Al Saadi et al. (1980); Ammann et al. (1969); Amromin et al. (1979); Aurias and Dutrillaux (1986); Aurias et al. (1980); Aurias et al. (1983); Becker et al. (1989); Bender et al. (1985); Bender et al. (1985); Bernstein et al. (1981); Bochkov et al. (1974); Boder and Sedgwick (1958); Cohen et al. (1979); Cohen et al. (1975); Cornforth and Bedford (1985); Cox et al. (1978); DeLeon et al. (1976); Feigin et al. (1970); Fiorilli et al. (1985); Ford and Lavin (1981); Frais (1979); Gatti et al. (1982); Hagberg et al. (1970); Hansen et al. (1977); Harnden (1974); Hoar and Sargent (1976); Hodge et al. (1980); Huang and Sheridan (1981); Johnson et al. (1985); Korein et al. (1961); Krishna Kumar et al. (1979); Levin and Perlov (1971); Lisker and Cobo (1970); Littlefield et al. (1981); McConville et al. (1990); Oxelius et al. (1982); Paterson et al. (1976); Paterson and Smith (1979); Peterson and Funkhouser (1989); Peterson et al. (1964); Rary et al. (1974); Reye and Mosman (1960); Richkind et al. (1982); Schalch et al. (1970); Scheres et al. (1980); Sedgwick and Boder (1972); Shultz et al. (1982); Shuster et al. (1966); Sourander et al. (1966); Stern et al. (1988); Sugimoto et al. (1982); Swift et al. (1976); Tadjoedin and Fraser (1965); Taylor et al. (1975); Taylor et al. (1976); Teplitz (1978); Toledano and Lang (1980); Vincent et al. (1975); Waldmann et al. (1983); Watanabe et al. (1977); Weinstein et al. (1985); Yount (1982)

Al Saadi, A.; Palutke, M.; Krishna Kumar, G.: Evolution of chromosomal abnormalities in sequential cytogenetic studies of ataxia telangiectasia. Hum. Genet. 55: 23-29, 1980.

Ammann, A. J.; Cain, W. A.; Ishizaka, K.; Hong, R.; Good, R. A.: Immunoglobulin E deficiency in ataxia-telangiectasia. New Eng. J. Med. 281: 469-472, 1969.

Amromin, G. D.; Boder, E.; Teplitz, R.: Ataxia-telangiectasia with a 32-year survival: a clinicopathological report. J. Neuropath. Exp. Neurol. 38: 621-643, 1979.

Athma, P.; Rappaport, R.; Swift, M.: Molecular genotyping shows that ataxia-telangiectasia heterozygotes are predisposed to breast cancer. Cancer Genet. Cytogenet. 92: 130-134, 1996.

Aurias, A.; Croquette, M. F.; Nuyts, J. P.; Griscelli, C.; Dutrillaux, B.: New data on clonal anomalies of chromosome 14 in ataxia telangiectasia: tct(14;14) and inv(14). Hum. Genet. 72: 22-24, 1986.

Aurias, A.; Dutrillaux, B.: Probable involvement of immunoglobulin superfamily genes in most recurrent chromosomal rearrangements from ataxia telangiectasia. Hum. Genet. 72: 210-214, 1986.

Aurias, A.; Dutrillaux, B.: A possible new type of chromosome rearrangement: telomere-centromere translocation (tct) followed by double duplication. Hum. Genet. 72: 25-26, 1986.

Aurias, A.; Dutrillaux, B.; Buriot, D.; Lejeune, J.: High frequencies of inversions and translocations of chromosomes 7 and 14 in ataxia-telangiectasia. Mutat. Res. 69: 369-374, 1980.

Aurias, A.; Dutrillaux, B.; Griscelli, C.: Tandem translocation t(14;14) in isolated and clonal cells in ataxia telangiectasia are different. Hum. Genet. 63: 320-322, 1983.

Baker, B. S.; Boyd, J. B.; Carpenter, A. T. C.; Green, M. M.; Nguyen, T. D.; Ripoll, P.; Smith, P. D.: Genetic controls of meiotic recombination and somatic DNA metabolism in Drosophila melanogaster. Proc. Nat. Acad. Sci. 73: 4140-4144, 1976.

Banga, S. S.; Shenkar, R.; Boyd, J. B.: Hypersensitivity of Drosophila mei-41 mutants to hydroxyurea is associated with reduced mitotic chromosome stability. Mutat. Res. 163: 157-165, 1986.

Banin, S.; Moyal, L.; Shieh, S.-Y.; Taya, Y.; Anderson, C. W.; Chessa, L.; Smorodinsky, N. I.; Prives, C.; Reiss, Y.; Shiloh, Y.; Ziv, Y.: Enhanced phosphorylation of p53 by ATM in response to DNA damage. Science 281: 674-677, 1998.

Bar, R. S.; Levis, W. R.; Rechler, M. M.; Harrison, L. C.; Siebert, C.; Podskalny, J.; Roth, J.; Muggeo, M.: Extreme insulin resistance in ataxia telangiectasia: defect in affinity of insulin receptors. New Eng. J. Med. 298: 1164-1171, 1978.

Barlow, C.; Hirotsune, S.; Paylor, R.; Liyanage, M.; Eckhaus, M.; Collins, F.; Shiloh, Y.; Crawley, J. N.; Ried, T.; Tagle, D.; Wynshaw-Boris, A.: Atm-deficient mice: a paradigm of ataxia telangiectasia. Cell 86: 159-171, 1996.

Becker, Y.; Tabor, E.; Asher, Y.: Ataxia-telangiectasia fibroblasts have less fibronectin mRNA than control cells but have the same levels of integrin and beta-actin mRNA. Hum. Genet. 81: 165-170, 1989.

Bender, M. A.; Rary, J. M.; Kale, R. P.: G(2) chromosomal radiosensitivity in ataxia telangiectasia lymphocytes. Mutat.Res. 152: 39-47, 1985.

Bender, M. A.; Rary, J. M.; Kale, R. P.: G(0) chromosomal radiosensitivity in ataxia telangiectasia lymphocytes. Mutat. Res. 150: 277-282, 1985.

Bernstein, R.; Pinto, M.; Jenkins, T.: Ataxia telangiectasia with evolution of monosomy 14 and emergence of Hodgkin's disease. Cancer Genet. Cytogenet. 4: 31-37, 1981.

Bigbee, W. L.; Langlois, R. G.; Swift, M.; Jensen, R. H.: Evidence for an elevated frequency of in vivo somatic cell mutations in ataxia telangiectasia. Am. J. Hum. Genet. 44: 402-408, 1989.

Bishop, D. T.; Hopper, J.: AT-tributable risks? Nature Genet. 15: 226 only, 1997.

Bochkov, N. P.; Lopukhin, Y. M.; Kuleshov, N. P.; Kovalchuk, L. V.: Cytogenetic study of patients with ataxia-telangiectasia. Humangenetik 24: 115-128, 1974.

Boder, E.: Ataxia-telangiectasia: some historic, clinical and pathologic observations.In: Bergsma, D.: Immunodeficiency in Man and Animals. New York: National Foundation-March of Dimes (pub.) 1975. Pp. 255-300.

Boder, E.: Ataxia-telangiectasia: an overview.In: Gatti, R. A.; Swift, M.: Ataxia-telangiectasia: Genetics, Neuropathology and Immunology of a Degenerative Disease of Childhood. New York: Alan R. Liss (pub.) 1985. Pp. 1-63.

Boder, E.; Sedgwick, R. P.: Ataxia-telangiectasia: a familial syndrome of progressive cerebellar ataxia, oculocutaneous telangiectasia and frequent pulmonary infection. Pediatrics 21: 526-554, 1958.

Boyd, J. B.; Golino, M. D.; Nguyen, T. D.; Green, M. M.: Isolation and characterization of X-linked mutants of Drosophila melanogaster which are sensitive to mutagens. Genetics 84: 485-506, 1976.

Bridges, B. A.; Arlett, C. F.: Risk of breast cancer in ataxia-telangiectasia.(Letter) New Eng. J. Med. 326: 1357-1361, 1992.

Bridges, B. A.; Harnden, D. G.: Ataxia-telangiectasia: A Cellular and Molecular Link between Cancer, Neuropathology, and Immune Deficiency. New York: John Wiley (pub.) 1982.

Brown, K. D.; Ziv, Y.; Sadanandan, S. N.; Chessa, L.; Collins, F. S.; Shiloh, Y.; Tagle, D. A.: The ataxia-telangiectasia gene product, a constitutively expressed nuclear protein that is not up-regulated following genome damage. Proc. Nat. Acad. Sci. 94: 1840-1845, 1997.

Brzoska, P. M.; Chen, H.; Zhu, Y.; Levin, N. A.; Disatnik, M.-H.; Mochly-Rosen, D.; Murnane, J. P.; Christman, M. F.: The product of the ataxia-telangiectasia group D complementing gene, ATDC, interacts with a protein kinase C substrate and inhibitor. Proc. Nat. Acad. Sci. 92: 7824-7828, 1995.

Byrd, P. J.; McConville, C. M.; Cooper, P.; Parkhill, J.; Stankovic, T.; McGuire, G. M.; Thick, J. A.; Taylor, A. M. R.: Mutations revealed by sequencing the 5-prime half of the gene for ataxia telangiectasia. Hum. Molec. Genet. 5: 145-149, 1996.

Byrne, E.; Hallpike, J. F.; Manson, J. F.; Sutherland, G. R.; Thong, Y. H.: Ataxia-without-telangiectasia: progressive multisystem degeneration with IgE deficiency and chromosome instability. J. Neurol. Sci. 66: 307-317, 1984.

Canman, C. E.; Lim, D.-S.; Cimprich, K. A.; Taya, Y.; Tamai, K.; Sakaguchi, K.; Appella, E.; Kastan, M. B.; Siliciano, J. D.: Activation of the ATM kinase by ionizing radiation and phosphorylation of p53. Science 281: 677-679, 1998.

Carbonari, M.; Cherchi, M.; Paganelli, R.; Giannini, G.; Galli, E.; Gaetano, C.; Papetti, C.; Fiorilli, M.: Relative increase of T cells expressing the gamma/delta rather than the alpha/beta receptor in ataxia-telangiectasia. New Eng. J. Med. 322: 73-76, 1990.

Chen, P.; Imray, F. P.; Kidson, C.: Gene dosage and complementation analysis of ataxia telangiectasia lymphoblastoid cell lines assayed by induced chromosome aberrations. Mutat. Res. 129: 165-172, 1984.

Chou, P. Y.; Fasman, G. D.: Prediction of the secondary structure of proteins from their amino acid sequence. Adv. Enzymol. Relat. Areas Molec. Biol. 47: 45-148, 1978.

Cohen, M. M.; Sagi, M.; Ben-Zur, Z.; Schaap, T.; Voss, R.; Kohn, G.; Ben-Bassat, H.: Ataxia telangiectasia: chromosomal stability in continuous lymphoblastoid cell lines. Cytogenet. Cell Genet. 23: 44-52, 1979.

Cohen, M. M.; Shaham, M.; Dagan, J.; Shmueli, E.; Kohn, G.: Cytogenetic investigations in families with ataxia-telangiectasia. Cytogenet. Cell Genet. 15: 338-356, 1975.

Concannon, P.; Gatti, R. A.: Diversity of ATM gene mutations detected in patients with ataxia-telangiectasia. Hum. Mutat. 10: 100-107, 1997.

Concannon, P.; Malhotra, U.; Charmley, P.; Reynolds, J.; Lange, K.; Gatti, R. A.: The ataxia-telangiectasia gene (ATA) on chromosome 11 is distinct from the ETS-1 gene. Am. J. Hum. Genet. 46: 789-794, 1990.

Cornforth, M. N.; Bedford, J. S.: On the nature of a defect in cells from individuals with ataxia-telangiectasia. Science 227: 1589-1591, 1985.

Cox, R.; Hosking, G. P.; Wilson, J.: Ataxia telangiectasia: evaluation of radiosensitivity in cultured skin fibroblasts as a diagnostic test. Arch. Dis. Child. 53: 386-390, 1978.

Croce, C. M.; Isobe, M.; Palumbo, A.; Puck, J.; Ming, J.; Tweardy, D.; Erikson, J.; Davis, M.; Rovera, G.: Gene for alpha-chain of human T-cell receptor: location on chromosome 14 region involved in T-cell neoplasms. Science 227: 1044-1047, 1985.

Curry, C. J. R.; Tsai, J.; Hutchinson, H. T.; Jaspers, N. G. J.; Wara, D.; Gatti, R. A.: AT-Fresno: a phenotype linking ataxia-telangiectasia with the Nijmegen breakage syndrome. Am. J. Hum. Genet. 45: 270-275, 1989.

DeLeon, G. A.; Grover, W. D.; Huff, D. S.: Neuropathologic changes in ataxia-telangiectasia. Neurology 26: 947-951, 1976.

Easton, D. F.: Cancer risks in A-T heterozygotes. Int. J. Rad. Biol. 66: S177-S182, 1994.

Ejima, Y.; Oshimura, M.; Sasaki, M. S.: Establishment of a novel immortalized cell line from ataxia-telangiectasia fibroblasts and its use for the chromosomal assignment of radiosensitivity gene. Int. J. Radiat. Biol. 58: 989-997, 1990.

Ejima, Y.; Sasaki, M. S.: Mutations of the ATM gene detected in Japanese ataxia-telangiectasia patients: possible preponderance of the two founder mutations 4612del165 and 7883del5. Hum. Genet. 102: 403-408, 1998.

Elson, A.; Wang, Y.; Daugherty, C. J.; Morton, C. C.; Zhou, F.; Campos-Torres, J.; Leder, P.: Pleiotropic defects in ataxia-telangiectasia protein-deficient mice. Proc. Nat. Acad. Sci. 93: 13084-13089, 1996.

Feigin, R. D.; Vietti, T. J.; Wyatt, R. G.; Kaufman, D. G.; Smith, C. H.: Ataxia telangiectasia with granulocytopenia. J. Pediat. 77: 431-438, 1970.

Fiorilli, M.; Antonelli, A.; Russo, G.; Crescenzi, M.; Carbonari, M.; Petrinelli, P.: Variant of ataxia-telangiectasia with low-level radiosensitivity. Hum. Genet. 70: 274-277, 1985.

Fiorilli, M.; Businco, L.; Pandolfi, F.; Paganelli, R.; Russo, G.; Aiuti, F.: Heterogeneity of immunological abnormalities in ataxia-telangiectasia. J. Clin. Immun. 3: 135-141, 1983.

FitzGerald, M. G.; Bean, J. M.; Hedge, S. R.; Unsal, H.; MacDonald, D. J.; Harkin, D. P.; Finkelstein, D. M.; Isselbacher, K. J.; Haber, D. A.: Heterozygous ATM mutations do not contribute to early onset of breast cancer. Nature Genet. 15: 307-310, 1997.

Ford, M. D.; Lavin, M. F.: Ataxia telangiectasia: an anomaly in DNA replication after irradiation. Nucleic Acids Res. 9: 1395-1404, 1981.

Foroud, T.; Wei, S.; Ziv, Y.; Sobel, E.; Lange, E.; Chao, A.; Goradia, T.; Huo, Y.; Tolun, A.; Chessa, L.; Charmley, P.; Sanal, O.; Salman, N.; Julier, C.; Concannon, P.; McConville, C.; Taylor, A. M. R.; Shiloh, Y.; Lange, K.; Gatti, R. A.: Localization of an ataxia-telangiectasia locus to a 3-cM interval on chromosome 11q23: linkage analysis of 111 families by an international consortium. Am. J. Hum. Genet. 49: 1263-1279, 1991.

Frais, M. A.: Gastric adenocarcinoma due to ataxia-telangiectasia (Louis-Bar syndrome). J. Med. Genet. 16: 160-161, 1979.

Garnier, J.; Osguthorpe, D. J.; Robson, B.: Analysis of the accuracy and implications of simple methods for predicting the secondary structure of globular proteins. J. Molec. Biol. 120: 97-120, 1978.

Gatti, R. A.: Personal Communication. Los Angeles, Calif. 6/13/1989.

Gatti, R. A.: Personal Communication. Los Angeles, Calif. 6/1990.

Gatti, R. A.; Aurias, A.; Griscelli, C.; Sparkes, R. S.: Translocations involving chromosomes 2p and 22q in ataxia-telangiectasia. Dis. Markers 3: 169-195, 1985.

Gatti, R. A.; Berkel, I.; Boder, E.; Braedt, G.; Charmley, P.; Concannon, P.; Ersoy, R.; Foroud, T.; Jaspers, N. G. J.; Lange, K.; Lathrop, G. M.; Leppert, M.; Nakamura, Y.; O'Connell, P.; Paterson, M.; Salser, W.; Sanal, O.; Silver, J.; Sparkes, R. S.; Susi, E.; Weeks, D. E.; Wei, S.; White, R.; Yoder, F.: Localization of an ataxia-telangiectasia gene to chromosome 11q22-23. Nature 336: 577-580, 1988.

Gatti, R. A.; Bick, M.; Tam, C. F.; Medici, M. A.; Oxelius, V.-A.; Holland, M.; Goldstein, A. L.; Boder, E.: Ataxia-telangiectasia: a multiparameter analysis of eight families. Clin. Immun. Immunopath. 23: 501-516, 1982.

Gatti, R. A.; Boder, E.; Vinters, H. V.; Sparkes, R. S.; Norman, A.; Lange, K.: Ataxia-telangiectasia: an interdisciplinary approach to pathogenesis. Medicine 70: 99-117, 1991.

Gatti, R. A.; Peterson, K. L.; Novak, J.; Chen, X.; Yang-Chen, L.; Liang, T.; Lange, E.; Lange, K.: Prenatal genotyping of ataxia-telangiectasia. (Letter) Lancet 342: 376, 1993.

Gatti, R. A.; Swift, M.: Ataxia-telangiectasia: Genetics, Neuropathology, and Immunology of a Degenerative Disease of Childhood. New York: Alan R. Liss (pub.) 1985.

Gilad, S.; Bar-Shira, A.; Harnik, R.; Shkedy, D.; Ziv, Y.; Khosravi, R.; Brown, K.; Vanagaite, L.; Xu, G.; Frydman, M.; Lavin, M. F.; Hill, D.; Tagle, D. A.; Shiloh, Y.: Ataxia-telangiectasia: founder effect among North African Jews. Hum. Molec. Genet. 5: 2033-2037, 1996.

Gilad, S.; Chessa, L.; Khosravi, R.; Russell, P.; Galanty, Y.; Piane, M.; Gatti, R. A.; Jorgensen, T. J.; Shiloh, Y.; Bar-Shira, A.: Genotype-phenotype relationships in ataxia-telangiectasia and variants. Am. J. Hum. Genet. 62: 551-561, 1998.

Gilad, S.; Khosravi, R.; Shkedy, D.; Uziel, T.; Ziv, Y.; Savitsky, K.; Rotman, G.; Smith, S.; Chessa, L.; Jorgensen, T. J.; Harnik, R.; Frydman, M.; Sanal, O.; Portnoi, S.; Goldwicz, Z.; Jaspers, N. G. J.; Gatti, R. A.; Lenoir, G.; Lavin, M. F.; Tatsumi, K.; Wegner, R. D.; Shiloh, Y.; Bar-Shira, A.: Predominance of null mutations in ataxia-telangiectasia. Hum. Molec. Genet. 5: 433-439, 1996.

Ginter, D. N.; Tallapragada, R.: Ataxia-telangiectasia. Birth Defects Orig. Art. Ser. XI(2): 408-409, 1975.

Greenwell, P. W.; Kronmal, S. L.; Porter, S. E.; Gassbenhuber, J.; Obermaier, B.; Petes, T. D.: TEL1, a gene involved in controlling telomere length in S. cerevisiae, is homologous to the human ataxia telangiectasia gene. Cell 82: 823-829, 1995.

Haerer, A. F.; Jackson, J. F.; Evers, C. G.: Ataxia-telangiectasia with gastric adenocarcinoma. J.A.M.A. 210: 1884-1887, 1969.

Hagberg, A.; Hansson, O.; Liden, S.; Nilsson, K.: Familial ataxic diplegia with deficient cellular immunity: a new clinical entity. Acta Paediat. Scand. 59: 545-550, 1970.

Hansen, R. L.; Marx, J. J.; Ptacek, L. J.; Roberts, R. C.: Immunological studies on an aberrant form of ataxia-telangiectasia. Am. J. Dis. Child. 131: 518-521, 1977.

Hari, K. L.; Santerre, A.; Sekelsky, J. J.; McKim, K. S.; Boyd, J. B.; Hawley, R. S.: The mei-41 gene of D. melanogaster is a structural and functional homolog of the human ataxia telangiectasia gene. Cell 82: 815-821, 1995.

Harnden, D. G.: Ataxia-telangiectasia syndrome: cytogenetic and cancer aspects.In: German, J.: Chromosomes and Cancer. New York: Wiley (pub.) 1974. Pp. 619-636.

Hartwell, L.: Defects in a cell cycle checkpoint may be responsible for the genomic instability of cancer cells. Cell 71: 543-546, 1992.

Hawley, R. S.; Friend, S. H.: Strange bedfellows in even stranger places: the role of ATM in meiotic cells, lymphocytes, tumors, and its functional links to p53. Genes Dev. 10: 2383-2388, 1996.

Hecht, F.; Koler, R. D.; Rigas, D. A.; Dahnke, G. S.; Case, M. P.; Tisdale, V.; Miller, R. W.: Leukemia and lymphocytes in ataxia-telangiectasia. (Letter) Lancet II: 1193, 1966.

Henderson, L.; Cole, H.; Arlett, C.; James, S. E.; Cole, J.; Lehmann, A.; Rosenbloom, L.; Redmond, T.; Meller, S.: Diagnosis of ataxia-telangiectasia by T-lymphocyte cloning assay.(Letter) Lancet II: 1242, 1985.

Hernandez, D.; McConville, C. M.; Stacey, M.; Woods, C. G.; Brown, M. M.; Shutt, P.; Rysiecki, G.; Taylor, A. M. R.: A family showing no evidence of linkage between the ataxia telangiectasia gene and chromosome 11q22-23. J. Med. Genet. 30: 135-140, 1993.

Hernandez, D.; McConville, C. M.; Taylor, M. A.: Failure to map the AT gene to 11q22-23 in a large inbred family.(Abstract) Cytogenet. Cell Genet. 58: 1962-1963, 1991.

Hoar, D. I.; Sargent, P.: Chemical mutagen hypersensitivity in ataxia-telangiectasia. Nature 261: 590-592, 1976.

Hodge, S. E.; Berkel, A. I.; Gatti, R. A.; Boder, E.; Spence, M. A.: Ataxia-telangiectasia and xeroderma pigmentosum: no evidence of linkage to HLA. Tissue Antigens 15: 313-317, 1980.

Huang, P. C.; Sheridan, R. B., III: Genetic and biochemical studies with ataxia telangiectasia. Hum. Genet. 59: 1-9, 1981.

Humphreys, M. W.; Nevin, N. C.; Wooldridge, M. A. W.: Cytogenetic investigations in a family with ataxia telangiectasia. Hum. Genet. 83: 79-82, 1989.

Imai, T.: Personal Communication. Chiba, Japan 9/12/1996.

Ishiguro, T.; Taketa, K.; Gatti, R. A.: Tissue of origin of elevated alpha-fetoprotein in ataxia-telangiectasia. Dis. Markers 4: 293-297, 1986.

Jaspers, N. G. J.; Bootsma, D.: Genetic heterogeneity in ataxia-telangiectasia studied by cell fusion. Proc. Nat. Acad. Sci. 79: 2641-2644, 1982.

Jaspers, N. G. J.; Gatti, R. A.; Baan, C.; Linssen, P. C. M. L.; Bootsma, D.: Genetic complementation analysis of ataxia telangiectasia and Nijmegen breakage syndrome: a survey of 50 patients. Cytogenet. Cell Genet. 49: 259-263, 1988.

Jaspers, N. G. J.; Painter, R. B.; Paterson, M. C.; Kidson, C.; Inoue, T.: Complementation analysis of ataxia-telangiectasia.In: Gatti, R. A.; Swift, M.: Ataxia-telangiectasia: Genetics, Neuropathology and Immunology of a Degenerative Disease of Childhood. New York: Alan R. Liss (pub.) 1985. Pp. 147-162.

Johnson, J. P.; Gatti, R. A.; Sears, T. S.; White, R. L.: Inverted duplication of J(H) associated with chromosome 14 translocation and T-cell leukemia in ataxia-telangiectasia. Am. J. Hum. Genet. 39: 787-796, 1986.

Johnson, J. P.; White, R. L.; Gatti, R. A.: Rearrangement of J(H) genes in a patient with ataxia telangiectasia, chromosome 14 translocation, and T-cell leukemia.(Abstract) Am. J. Hum. Genet. 37: A100, 1985.

Jung, M.; Zhang, Y.; Lee, S.; Dritschilo, A.: Correction of radiation sensitivity in ataxia telangiectasia cells by a truncated I-kappa-B-alpha. Science 268: 1619-1621, 1995.

Kapp, L. N.; Painter, R. B.; Yu, L.-C.; van Loon, N.; Richard, C. W., III; James, M. R.; Cox, D. R.; Murnane, J. P.: Cloning of a candidate gene for ataxia-telangiectasia group D. Am. J. Hum. Genet. 51: 45-54, 1992.

Kastan, K.: Clinical implications of basic research: ataxia-telangiectasia—broad implications for a rare disorder. New Eng. J. Med. 333: 662-663, 1995.

Kastan, M. B.; Onyekwere, O.; Sidransky, D.; Vogelstein, B.; Craig, R. W.: Participation of p53 protein in the cellular response to DNA damage. Cancer Res. 51: 6304-6311, 1991.

Kastan, M. B.; Zhan, Q.; El-Deiry, W. S.; Carrier, F.; Jacks, T.; Walsh, W. V.; Plunkett, B. S.; Vogelstein, B.; Fornace, A. J., Jr.: A mammalian cell cycle checkpoint pathway utilizing p53 and GADD45 is defective in ataxia-telangiectasia. Cell 71: 587-597, 1992.

Kennaugh, A. A.; Butterworth, S. V.; Hollis, R.; Baer, R.; Rabbitts, T. H.; Taylor, A. M. R.: The chromosome breakpoint at 14q32 in an ataxia telangiectasia t(14;14) T cell clone is different from the 14q32 breakpoint in Burkitts and an inv(14) T cell lymphoma. Hum. Genet. 73: 254-259, 1986.

Khanna, K. K.; Keating, K. E.; Kozlov, S.; Scott, S.; Gatei, M.; Hobson, K.; Taya, Y.; Gabrielli, B.; Chan, D.; Lees-Miller, S. P.; Lavin, M. F.: ATM associates with and phosphorylates p53: mapping the region of interaction. Nature Genet. 20: 398-400, 1998.

Kojis, T. L.; Schreck, R. R.; Gatti, R. A.; Sparkes, R. S.: Tissue specificity of chromosomal rearrangements in ataxia-telangiectasia. Hum. Genet. 83: 347-352, 1989.

Komatsu, K.; Kodama, S.; Okumura, Y.; Koi, M.; Oshimura, M.: Restoration of radiation resistance in ataxia-telangiectasia cells by the introduction of normal human chromosome 11. Mutat. Res. 235: 59-63, 1990.

Komatsu, K.; Matsuura, S.; Tauchi, H.; Endo, S.; Kodama, S.; Smeets, D.; Weemaes, C.; Oshimura, M.: The gene for Nijmegen breakage syndrome (V2) is not located on chromosome 11.(Letter) Am. J. Hum. Genet. 58: 885-888, 1996.

Korein, J.; Steinman, P. A.; Senz, E. H.: Ataxia-telangiectasia: report of a case and review of the literature. Arch. Neurol. 4: 272-280, 1961.

Krishna Kumar, G.; Al Saadi, A.; Yang, S. S.; McCaughey, R. S.: Ataxia-telangiectasia and hepatocellular carcinoma. Am. J. Med. Sci. 278: 157-160, 1979.

Laake, K.; Telatar, M.; Geitvik, G. A.; Hansen, R. O.; Heiberg, A.; Andresen, A. M.; Gatti, R.; Borresen-Dale, A.-L.: Identical mutation in 55% of the ATM alleles in 11 Norwegian AT families: evidence for a founder effect. Europ. J. Hum. Genet. 6: 235-244, 1998.

Lambert, C.; Schultz, R. A.; Smith, M.; Wagner-McPherson, C.; McDaniel, L. D.; Donlon, T.; Stanbridge, E. J.; Friedberg, E. C.: Functional complementation of ataxia-telangiectasia group D (AT-D) cells by microcell-mediated chromosome transfer and mapping of the AT-D locus to the region 11q22-23. Proc. Nat. Acad. Sci. 88: 5907-5911, 1991.

Levin, S.; Perlov, S.: Ataxia-telangiectasia in Israel, with observations on its relationship to malignant disease. Israel J. Med. Sci. 7: 1535-1541, 1971.

Lim, D.-S.; Kirsch, D. G.; Canman, C. E.; Ahn, J.-H.; Ziv, Y.; Newman, L. S.; Darnell, R. B.; Shiloh, Y.; Kastan, M. B.: ATM binds to beta-adaptin in cytoplasmic vesicles. Proc. Nat. Acad. Sci. 95: 10146-10151, 1998.

Lipkowitz, S.; Garry, V. F.; Kirsch, I. R.: Interlocus V-J recombination measures genomic instability in agriculture workers at risk for lymphoid malignancies. Proc. Nat. Acad. Sci. 89: 5301-5305, 1992.

Lipkowitz, S.; Stern, M.-H.; Kirsch, I. R.: Hybrid T cell receptor genes formed by interlocus recombination in normal and ataxia-telangiectasia lymphocytes. J. Exp. Med. 172: 409-418, 1990.

Lisker, R.; Cobo, A.: Chromosome breakage in ataxia-telangiectasia. (Letter) Lancet I: 618, 1970.

Littlefield, L. G.; Colyer, S. P.; Joiner, E. E.; DuFrain, R. J.; Frome, E.; Cohen, M. M.: Chromosomal radiation sensitivity in ataxia telangiectasia long-term lymphoblastoid cell lines. Cytogenet. Cell Genet. 31: 203-213, 1981.

Llerena, J., Jr.; Murer-Orlando, M.; McGuire, M.; Zahed, L.; Sheridan, R. J.; Berry, A. C.; Bobrow, M.: Spontaneous and induced chromosome breakage in chorionic villus samples: a cytogenetic approach to first trimester prenatal diagnosis of ataxia telangiectasia syndrome. J. Med. Genet. 26: 174-178, 1989.

Maserati, E.; Ottolini, A.; Veggiotti, P.; Lanzi, G.; Pasquali, F.: Ataxia-without-telangiectasia in two sisters with rearrangements of chromosomes 7 and 14. Clin. Genet. 34: 283-287, 1988.

Matsuda, Y.; Imai, T.; Shiomi, T.; Saito, T.; Yamauchi, M.; Fukao, T.; Akao, Y.; Seki, N.; Ito, H.; Hori, T.: Comparative genome mapping of the ataxia-telangiectasia region in mouse, rat, and Syrian hamster. Genomics 34: 347-352, 1996.

McCaw, B. K.; Hecht, F.; Harden, D. G.; Teplitz, R. L.: Somatic rearrangement of chromosome 14 in human lymphocytes. Proc. Nat. Acad. Sci. 72: 2071-2075, 1975.
McConville, C. M.; Formstone, C. J.; Hernandez, D.; Thick, J.; Taylor, A. M.: Fine mapping of the chromosome 11q22-23 region using PFGE, linkage and haplotype analysis; localization of the gene for ataxia telangiectasia to a 5cM region flanked by NCAM/DRD2 and STMY/CJ52.75, phi 2.22. Nucleic Acids Res. 18: 4335-4343, 1990.
McConville, C. M.; Stankovic, T.; Byrd, P. J.; McGuire, G. M.; Yao, Q.-Y.; Lennox, G. G.; Taylor, A. M. R.: Mutations associated with variant phenotypes in ataxia-telangiectasia. Am. J. Hum. Genet. 59: 320-330, 1996.
McConville, C. M.; Woods, C. G.; Farrall, M.; Metcalfe, J. A.; Taylor, A. M. R.: Analysis of 7 polymorphic markers at chromosome 11q22-23 in 35 ataxia telangiectasia families; further evidence of linkage. Hum. Genet. 85: 215-220, 1990.
McFarlin, D. E.; Strober, W.; Waldmann, T. A.: Ataxia-telangiectasia. Medicine 51: 281-314, 1972.
McKusick, V. A.; Cross, H. E.: Ataxia-telangiectasia and Swiss-type agammaglobulinemia. Two genetic disorders of the immune mechanism in related Amish sibships. J.A.M.A. 195: 739-745, 1966.
Meyn, M. S.: High spontaneous intrachromosomal recombination rates in ataxia-telangiectasia. Science 260: 1327-1330, 1993.
Miller, M. E.; Chatten, J.: Ovarian changes in ataxia-telangiectasia. Acta Paediat. Scand. 56: 559-561, 1967.
Mohamed, R.; Singh, S. P.; Kumar, S.; Lavin, M. F.: A defect in DNA topoisomerase II activity in ataxia-telangiectasia cells. Biochem. Biophys. Res. Commun. 149: 233-238, 1987.
Morrell, D.; Chase, C. L.; Swift, M.: Cancers in 44 families with ataxia-telangiectasia. Cancer Genet. Cytogenet. 50: 119-123, 1990.
Morrow, D. M.; Tagle, D. A.; Shiloh, Y.; Collins, F. S.; Hieter, P.: TEL1, an S. cerevisiae homolog of the human gene mutated in ataxia telangiectasia, is functionally related to the yeast checkpoint gene MEC1. Cell 82: 831-840, 1995.
Murnane, J. P.; Painter, R. B.: Complementation of the effects in DNA synthesis in irradiated and unirradiated ataxia-telangiectasia cells. Proc. Nat. Acad. Sci. 79: 1960-1963, 1982.
Naito, T.; Matsuura, A.; Ishikawa, F.: Circular chromosome formation in a fission yeast mutant defective in two ATM homologues. Nature Genet. 20: 203-206, 1998.
Nguyen, T. D.; Boyd, J. B.; Green, M. M.: Sensitivity of Drosophila mutants to chemical carcinogens. Mutat. Res. 63: 67-77, 1979.
Oxelius, V.-A.; Berkel, A. I.; Hanson, L. A.: IgG2 deficiency in ataxia-telangiectasia. New Eng. J. Med. 306: 515-517, 1982.
Oxford, J. M.; Harnden, D. G.; Parrington, J. M.; Delhanty, J. D. A.: Specific chromosome aberrations in ataxia-telangiectasia. J. Med. Genet. 12: 251-262, 1975.
Painter, R. B.; Cramer, P.; Howard, R.; Young, B. R.: Two forms of inhibition of DNA replicon initiation in human cells.In: Harris, C. C.; Cerutti, P. C.: Mechanisms of Chemical Carcinogenesis. New York: Alan R. Liss (pub.) 1982. Pp. 383-386.
Painter, R. B.; Young, B. R.: Radiosensitivity in ataxia-telangiectasia: a new explanation. Proc. Nat. Acad. Sci. 77: 7315-7317, 1980.
Paterson, M. C.; Smith, B. P.; Knight, P. A.; Anderson, A. K.: Ataxia telangiectasia: an inherited human disease involving radiosensitivity, malignancy and defective DNA repair.In: Castellani, A.: Research in Photobiology. New York: Plenum (pub.) 1977. Pp. 207-218.
Paterson, M. C.; Smith, B. P.; Lohman, P. H. M.; Anderson, A. K.; Fishman, L.: Defective excision repair of gamma-ray-damaged DNA in human (ataxia-telangiectasia) fibroblasts. Nature 260: 444-447, 1976.
Paterson, M. C.; Smith, P. J.: Ataxia telangiectasia: an inherited human disorder involving hypersensitivity to ionizing radiation and related DNA-damaging chemicals. Ann. Rev. Genet. 13: 291-318, 1979.
Paulovich, A. G.; Hartwell, L. H.: A checkpoint regulates the rate of progression through S phase in S. cerevisiae in response to DNA damage. Cell 82: 841-847, 1995.
Pecker, I.; Avraham, K. B.; Gilbert, D. J.; Savitsky, K.; Rotman, G.; Harnik, R.; Fukao, T.; Schrock, E.; Hirotsune, S.; Tagle, D. A.; Collins, F. S.; Wynshaw-Boris, A.; Ried, T.; Copeland, N. G.; Jenkins, N. A.; Shiloh, Y.; Ziv, Y.: Identification and chromosomal localization of Atm, the mouse homolog of the ataxia-telangiectasia gene. Genomics 35: 39-45, 1996.
Peterson, R. D. A.; Funkhouser, J. D.: Speculations on ataxia-telangiectasia: defective regulation of the immunoglobulin gene superfamily. Immun. Today 10: 313-315, 1989.
Peterson, R. D. A.; Funkhouser, J. D.: Ataxia-telangiectasia: an important clue.(Editorial) New Eng. J. Med. 322: 124-125, 1990.
Peterson, R. D. A.; Kelly, W. D.; Good, R. A.: Ataxia-telangiectasia: its association with a defective thymus, immunological-deficiency disease and malignancy. Lancet I: 1189-1193, 1964.
Pippard, E. C.; Hall, A. J.; Barker, D. J. P.; Bridges, B. A.: Cancer in homozygotes and heterozygotes of ataxia-telangiectasia and xeroderma pigmentosum in Britain. Cancer Res. 48: 2929-2932, 1988.
Rary, J. M.; Bender, M. A.; Kelly, T. E.: A 14/14 marker chromosome lymphocyte clone in ataxia telangiectasis. J. Hered. 66: 33-35, 1975.
Rary, J. M.; Bender, M. A.; Kelly, T. E.: Cytogenetic studies of ataxia-telangiectasia.(Abstract) Am. J. Hum. Genet. 26: 70, 1974.
Reye, C.; Mosman, N. S. W.: Ataxia-telangiectasia. Am. J. Dis. Child. 99: 238-247, 1960.
Richkind, K. E.; Boder, E.; Teplitz, R. L.: Fetal proteins in ataxia-telangiectasia. J.A.M.A. 248: 1346-1347, 1982.
Rosen, F. S.; Harris, N. L.: Case records of the Massachusetts General Hospital: a 30-year-old man with ataxia-telangiectasia and dysphagia. New Eng. J. Med. 316: 91-100, 1987.
Rosin, M. P.; Ochs, H. D.: In vivo chromosomal instability in ataxia-telangiectasia homozygotes and heterozygotes. Hum. Genet. 74: 335-340, 1986.
Rosin, M. P.; Ochs, H. D.; Gatti, R. A.; Boder, E.: Heterogeneity of chromosomal breakage levels in epithelial tissue of ataxia-telangiectasia homozygotes and heterozygotes. Hum. Genet. 83: 133-138, 1989.
Russo, G.; Isobe, M.; Gatti, R.; Finan, J.; Batuman, O.; Huebner, K.; Nowell, P. C.; Croce, C. M.: Molecular analysis of a t(14;14) translocation in leukemic T-cells of an ataxia telangiectasia patient. Proc. Nat. Acad. Sci. 86: 602-606, 1989.
Sanal, O.; Lange, E.; Telatar, M.; Sobel, E.; Salazar-Novak, J.; Ersoy, F.; Morrison, A.; Concannon, P.; Tolun, A.; Gatti, R. A.: Ataxia-telangiectasia: linkage analysis of chromosome 11q22-23 markers in Turkish families. FASEB J. 6: 2848-2852, 1992.
Sanal, O.; Wei, S.; Foroud, T.; Malhotra, U.; Concannon, P.; Charmley, P.; Salser, W.; Lange, K.; Gatti, R. A.: Further mapping of an ataxia-telangiectasia locus to the chromosome 11q23 region. Am. J. Hum. Genet. 47: 860-866, 1990.
Sasaki, T.; Tian, H.; Kukita, Y.; Inazuka, M.; Tahira, T.; Imai, T.; Yamauchi, M.; Saito, T.; Hori, T.; Hashimoto-Tamaoki, T.; Komatsu, K.; Nikaido, O.; Hayashi, K.: ATM mutations in patients with ataxia telangiectasia screened by a hierarchical strategy. Hum. Mutat. 12: 186-195, 1998.
Savitsky, K.; Bar-Shira, A.; Gilad, S.; Rotman, G.; Ziv, Y.; Vanagaite, L.; Tagle, D. A.; Smith, S.; Uziel, T.; Sfez, S.; Ashkenazi, M.; Pecker, I.; Frydman, M.; Harnik, R.; Patanjali, S. R.; Simmons, A.; Clines, G. A.; Sartiel, A.; Gatti, R. A.; Chessa, L.; Sanal, O.; Lavin, M. F.; Jaspers, N. G. J.; Taylor, A. M. R.; Arlett, C. F.; Miki, T.; Weissman, S. M.; Lovett, M.; Collins, F. S.; Shiloh, Y.: A single ataxia telangiectasia gene with a product similar to PI-3 kinase. Science 268: 1749-1753, 1995.
Savitsky, K.; Sfez, S.; Tagle, D. A.; Ziv, Y.; Sartiel, A.; Collins, F. S.; Shiloh, Y.; Rotman, G.: The complete sequence of the coding region of the ATM gene reveals similarity to cell cycle regulators in different species. Hum. Molec. Genet. 4: 2025-2032, 1995.
Saxon, A.; Stevens, R. H.; Golde, D. W.: Helper and suppressor T-lymphocyte leukemia in ataxia-telangiectasia. New Eng. J. Med. 300: 700-704, 1979.
Schalch, D. S.; McFarlin, D. E.; Barlow, M. H.: An unusual form of diabetes mellitus in ataxia-telangiectasia. New Eng. J. Med. 282: 1396-1402, 1970.
Scheres, J. M. J. C.; Hustinx, T. W. J.; Weemaes, C. M. R.: Chromosome 7 in ataxia-telangiectasia. J. Pediat. 97: 440-441, 1980.
Sedgwick, R. P.; Boder, E.: Ataxia-telangiectasia.In: Vinken, P. J.; Bruyn, G. W.: Handbook of Clinical Neurology. Amsterdam: North-Holland Publishing Co. (pub.) 14: 1972. Pp. 267-339.
Shaham, M.; Becker, Y.: The ataxia telangiectasia clastogenic factor is a low molecular weight peptide. Hum. Genet. 58: 422-424, 1981.
Shiloh, Y.; Parshad, R.; Frydman, M.; Sanford, K. K.; Portnoi, S.; Ziv, Y.; Jones, G. M.: G(2) chromosomal radiosensitivity in families with ataxia-telangiectasia. Hum. Genet. 84: 15-18, 1989.
Shultz, L. D.; Sweet, H. O.; Davisson, M. T.; Coman, D. R.: 'Wasted,' a new mutant of the mouse with abnormalities characteristic of ataxia telangiectasia. Nature 297: 402-404, 1982.
Shuster, J.; Hart, Z.; Stimson, C. W.; Brough, A. J.; Poulik, M. D.: Ataxia-telangiectasia with cerebellar tumor. Pediatrics 37: 776-786, 1966.
Sobel, E.; Lange, E.; Jaspers, N. G. J.; Chessa, L.; Sanal, O.; Shiloh, Y.; Taylor, A. M. R.; Weemaes, C. M. A.; Lange, K.; Gatti, R. A.: Ataxia-telangiectasia: linkage evidence for genetic heterogeneity.(Letter) Am. J. Hum. Genet. 50: 1343-1348, 1992.
Sourander, P.; Bonnevier, J. O.; Olsson, Y.: A case of ataxia-telangiectasia with lesions in the spinal cord. Acta Neurol. Scand. 42: 354-366, 1966.
Stankovic, T.; Kidd, A. M. J.; Sutcliffe, A.; McGuire, G. M.; Robinson, P.; Weber, P.; Bedenham, T.; Bradwell, A. R.; Easton, D. F.; Lennox, G. G.; Haites, N.; Byrd, P. J.; Taylor, A. M. R.: ATM mutations and phenotypes in ataxia-telangiectasia families in the British Isles: expression of mutant ATM and the risk of leukemia, lymphoma, and breast cancer. Am. J. Hum. Genet. 62: 334-345, 1998.
Stern, M.-H.; Zhang, F.; Griscelli, C.; Thomas, G.; Aurias, A.: Molecular characterization of different ataxia telangiectasia T-cell clones. I. A common breakpoint at the 14q11.2 band splits the T-cell receptor alpha-chain gene. Hum. Genet. 78: 33-36, 1988.
Stern, M.-H.; Zhang, F.; Thomas, G.; Griscelli, C.; Aurias, A.: Molecular characterization of ataxia telangiectasia T cell clones. III. Mapping the 14q32.1 distal breakpoint. Hum. Genet. 81: 18-22, 1988.
Sugimoto, T.; Kidowaki, T.; Sawada, T.; Ohtsuka-Urano, T.; Kusunoki, T.: Ataxia-telangiectasia associated with non-T, non-B cell acute lymphocytic leukemia. Acta Paediat. Scand. 71: 509-510, 1982.
Swift, M.: Cancer risk counseling.(Letter) Science 210: 1074, 1980.
Swift, M.; Morrell, D.; Cromartie, E.; Chamberlin, A. R.; Skolnick, M. H.; Bishop, D. T.: The incidence and gene frequency of ataxia-telangiectasia in the United States. Am. J. Hum. Genet. 39: 573-583, 1986.
Swift, M.; Morrell, D.; Massey, R. B.; Chase, C. L.: Incidence of cancer in 161 families affected by ataxia-telangiectasia. New Eng. J. Med. 325: 1831-1836, 1991.
Swift, M.; Reitnauer, P. J.; Morrell, D.; Chase, C. L.: Breast and other cancers in families with ataxia-telangiectasia. New Eng. J. Med. 316: 1289-1294, 1987.
Swift, M. R.; Sholman, L.; Perry, M.; Chase, C.: Malignant neoplasms in the families of patients with ataxia-telangiectasia. Cancer Res. 36: 209-215, 1976.
Tadjoedin, M. K.; Fraser, F. C.: Heredity of ataxia-telangiectasia (Louis-Bar syndrome). Am. J. Dis. Child. 110: 64-68, 1965.
Taylor, A. M. R.; Flude, E.; Laher, B.; Stacey, M.; McKay, E.; Watt, J.; Green, S. H.; Harding, A. E.: Variant forms of ataxia telangiectasia. J. Med. Genet. 24: 669-677, 1987.
Taylor, A. M. R.; Harnden, D. G.; Arlett, C. F.; Harcourt, S. A.; Lehmann, A. R.; Stevens, S.; Bridges, B. A.: Ataxia-telangiectasia: a human mutation with abnormal radiation sensitivity. Nature 258: 427-429, 1975.
Taylor, A. M. R.; Metcalfe, J. A.; Oxford, J. M.; Harnden, D. G.: Is chromatid-type damage in ataxia-telangiectasia after irradiation at G(0) a consequence of defective repair? Nature 260: 441-443, 1976.

Tchirkov, A.; Bay, J.-O.; Pernin, D.; Bignon, Y.-J. Rio, P.; Grancho, M.; Kwiatkowski, F.; Giollant, M.; Malet, P.; Verrelle, P.: Detection of heterozygous carriers of the ataxia-telangiectasia (ATM) gene by G(2) phase chromosomal radiosensitivity of peripheral blood lymphocytes. Hum. Genet. 101: 312-316, 1997.
Telatar, M.; Wang, Z.; Castellvi-Bel, S.; Tai, L.-Q.; Sheikhavandi, S.; Regueiro, J. R.; Porras, O.; Gatti, R. A.: A model for ATM heterozygote identification in a large population: four founder-effect ATM mutations identify most of Costa Rican patients with ataxia telangiectasia. Molec. Genet. Metab. 64: 36-43, 1998.
Teplitz, R. L.: Ataxia-telangiectasia. Arch. Neurol. 35: 553-554, 1978.
Thibaut, S.; Sass, U.; Khoury, A.; Simonart, J.-M.: Ataxia-telangiectasia and necrobiosis lipoidica: an explainable association. Europ. J. Derm. 4: 509-513, 1994.
Toledano, S. R.; Lang, B. J.: Ataxia-telangiectasia and acute lymphoblastic leukemia. Cancer 45: 1675-1678, 1980.
Toyoshima, M.; Hara, T.; Zhang, H.; Yamamoto, T.; Akaboshi, S.; Nanba, E.; Ohno, K.; Hori, N.; Sato, K.; Takeshita, K.: Ataxia-telangiectasia without immunodeficiency: novel point mutations within and adjacent to the phosphatidylinositol 3-kinase-like domain. Am. J. Med. Genet. 75: 141-144, 1998.
Tse, A. G. D.; Barclay, A. N.; Watts, A.; Williams, A. F.: A glycophospholipid tail at the carboxyl terminus of the Thy-1 glycoprotein of neurons and thymocytes. Science 230: 1003-1008, 1985.
Uziel, T.; Savitsky, K.; Platzer, M.; Ziv, Y.; Helbitz, T.; Nehls, M.; Boehm, T.; Rosenthal, A.; Shiloh, Y.; Rotman, G.: Genomic organization of the ATM gene. Genomics 33: 317-320, 1996.
van Belzen, M. J.; Hiel, J. A. P.; Weemaes, C. M. R.; Gabreels, F. J. M.; van Engelen, B. G. M.; Smeets, D. F. C. M.; van den Heuvel, L. P. W. J.: A double missense mutation in the ATM gene of a Dutch family with ataxia telangiectasia. Hum. Genet. 102: 187-191, 1998.
Vincent, R. A., Jr.; Sheridan, R. B., III; Huang, P. C.: DNA strand breakage repair in ataxia-telangiectasia fibroblast-like cells. Mutat. Res. 33: 357-366, 1975.
Vorechovsky, I.; Luo, L.; Dyer, M. J. S.; Catovsky, D.; Amlot, P. L.; Yaxley, J. C.; Foroni, L.; Hammarstrom, L.; Webster, A. D. B.; Yuille, M. A. R.: Clustering of missense mutations in the ataxia-telangiectasia gene in a sporadic T-cell leukaemia. Nature Genet. 17: 96-99, 1997.
Vorechovsky, I.; Luo, L.; Lindblom, A.; Negrini, M.; Webster, A. D.; Croce, C. M.; Hammarstrom, L.: ATM mutations in cancer families. Cancer Res. 56: 4130-4133, 1996.
Waldmann, T. A.; McIntire, K. R.: Serum-alpha-fetoprotein levels in patients with ataxia-telangiectasia. Lancet II: 1112-1115, 1972.
Waldmann, T. A.; Misiti, J.; Nelson, D. L.; Kraemer, K. H.: Ataxia-telangiectasia: a multisystem hereditary disease with immunodeficiency, impaired organ maturation, x-ray hypersensitivity, and a high incidence of neoplasia. Ann. Intern. Med. 99: 367-379, 1983.
Watanabe, A.; Hanazono, H.; Sogawa, H.; Takaya, H.: Stomach cancer in a 14-year-old-boy with ataxia-telangiectasia. Tohoku J. Exp. Med. 121: 127-131, 1977.
Waterman, M. J. F.; Stavridi, E. S.; Waterman, J. L. F.; Halazonetis, T. D.: ATM-dependent activation of p53 involves dephosphorylation and association with 14-3-3 proteins. Nature Genet. 19: 175-178, 1998.
Weinstein, S.; Scottolini, A. G.; Loo, S. Y. T.; Caldwell, P. C.; Bhagavan, N. V.: Ataxia telangiectasia with hepatocellular carcinoma in a 15-year-old girl and studies of her kindred. Arch. Path. Lab. Med. 109: 1000-1004, 1985.
Welshimer, K.; Swift, M.: Congenital malformations and developmental disabilities in ataxia-telangiectasia, Fanconi anemia, and xeroderma pigmentosum families. Am. J. Hum. Genet. 34: 781-793, 1982.
Westphal, C. H.; Rowan, S.; Schmaltz, C.; Elson, A.; Fisher, D. E.; Leder, P.: atm and p53 cooperate in apoptosis and suppression of tumorigenesis, but not in resistance to acute radiation toxicity. Nature Genet. 16: 397-401, 1997.
Woods, C. G.; Bundey, S. E.; Taylor, A. M. R.: Unusual features in the inheritance of ataxia telangiectasia. Hum. Genet. 84: 555-562, 1990.
Woods, C. G.; Taylor, A. M. R.: Ataxia telangiectasia in the British Isles: the clinical and laboratory features of 70 affected individuals. Quart. J. Med. 82: 169-179, 1992.
Wooster, R.; Ford, D.; Mangion, J.; Ponder, B. A. J.; Peto, J.; Easton, D. F.; Stratton, M. R.: Absence of linkage to the ataxia telangiectasia locus in familial breast cancer. Hum. Genet. 92: 91-94, 1993.
Wright, J.; Teraoka, S.; Onengut, S.; Tolun, A.; Gatti, R. A.; Ochs, H. D.; Concannon, P.: A high frequency of distinct ATM gene mutations in ataxia-telangiectasia. Am. J. Hum. Genet. 59: 839-846, 1996.
Xia, Y.-R.; Welch, C. L.; Warden, C. H.; Lange, E.; Fukao, T.; Lusis, A. J.; Gatti, R. A.: Assignment of the mouse ataxia-telangiectasia gene (Atm) to mouse chromosome 9. Mammalian Genome 7: 554-555, 1996.
Xu, Y.; Ashley, T; Brainerd, E. E.; Bronson, R. T.; Meyn, M. S.; Baltimore, D.: Targeted disruption of ATM leads to growth retardation, chromosomal fragmentation during meiosis, immune defects, and thymic lymphoma. Genes Dev. 10: 2411-2422, 1996.
Xu, Y.; Baltimore, D.: Dual roles of ATM in the cellular response to radiation and in cell growth control. Genes Dev. 10: 2401-2410, 1996.
Ying, K. L.; Decoteau, W. E.: Cytogenetic anomalies in a patient with ataxia, immune deficiency, and high alpha-fetoprotein in the absence of telangiectasia. Cancer Genet. Cytogenet. 4: 311-317, 1981.
Yount, W. J.: IgG2 deficiency and ataxia-telangiectasia. (Editorial) New Eng. J. Med. 306: 541-543, 1982.
Zadik, Z.; Levin, S.; Prager-Lewin, R.; Laron, Z.: Gonadal dysfunction in patients with ataxia telangiectasia. Acta Paediat. Scand. 67: 477-479, 1978.
Zakian, V. A.: ATM-related genes: what do they tell us about functions of the human gene? Cell 82: 685-687, 1995.
Zhang, F.; Stern, M.-H.; Thomas, G.; Aurias, A.: Molecular characterization of ataxia telangiectasia T cell clones. II. The clonal inv(14) in ataxia telangiectasia differs from the inv(14) in T cell lymphoma. Hum. Genet. 78: 316-319, 1988.
Zhang, N.; Chen, P.; Khanna, K. K.; Scott, S.; Gatei, M.; Kozlov, S.; Watters, D.; Spring, K.; Yen, T.; Lavin, M. F.: Isolation of full-length ATM cDNA and correction of the ataxia-telangiectasia cellular phenotype. Proc. Nat. Acad. Sci. 94: 8021-8026, 1997.
Ziv, Y.; Amiel, A.; Jaspers, N. G. J.; Berkel, A. I.; Shiloh, Y.: Ataxia-telangiectasia: a variant with altered in vitro phenotype of fibroblast cells. Mutat. Res. 210: 211-219, 1989.
Ziv, Y.; Frydman, M.; Lange, E.; Zelnik, N.; Rotman, G.; Julier, C.; Jaspers, N. G. J.; Dagan, Y.; Abeliovicz, D.; Dar, H.; Borochowitz, Z.; Lathrop, M.; Gatti, R. A.; Shiloh, Y.: Ataxia-telangiectasia: linkage analysis in highly inbred Arab and Druze families and differentiation from an ataxia-microcephaly-cataract syndrome. Hum. Genet. 88: 619-626, 1992.
Ziv, Y.; Rotman, G.; Frydman, M.; Dagan, J.; Cohen, T.; Foroud, T.; Gatti, R. A.; Shiloh, Y.: The ATC (ataxia-telangiectasia complementation group C) locus localizes to 11q22-q23. Genomics 9: 373-375, 1991.

BLOOM SYNDROME (*210900)

JJM NOTE

This rare disorder predisposes children to acute myeloid leukemia, Hodgkin disease, Wilms tumor, and medulloblastoma; in affected adults, squamous cell carcinoma of the skin and esophagus and sigmoid carcinoma have been seen. Patients, families, and biomedical scientists interested in this syndrome are indebted to James German III for his lifelong dedication to the study of this condition.

Description

Bloom syndrome is an autosomal recessive disorder characterized by proportionate pre- and postnatal growth deficiency; sun-sensitive, telangiectatic, hypo- and hyperpigmented skin; predisposition to malignancy; and chromosomal instability.

Clinical Features

Landau et al. (1966) described a patient whose parents were second cousins and who showed low gamma-A and gamma-M serum proteins. Ferrara et al. (1967) described the disease in a 'Chinese-American'; however, the diagnosis was later (Ferrara, 1972) revised to focal dermal hypoplasia (305600).

German et al. (1984) collected information on 103 patients. German and Takebe (1989) suggested that differences in skin pigmentation in various ethnic groups may confer a degree of protection against actinic radiation and thus obscure one of the characteristic facial signs of Bloom syndrome, i.e., telangiectasia. As a result, Bloom syndrome may be underdiagnosed in some populations. Legum et al. (1991) described an affected Iranian Jewish male, possibly the first definite non-Ashkenazi Jewish patient. The patient had another unique complication, myocardopathy.

The 14 Japanese cases reported by German and Takebe (1989) differed somewhat from most cases recognized elsewhere in that dolichocephaly was a less constant feature, the facial skin lesions were less prominent, and life-threatening infections were less frequent. The characteristic predisposition to neoplasia, as well as the probable tendency to diabetes mellitus, was found, however. German (1990) stated that diabetes mellitus of maturity-onset type, developing, however, in the second or third decade, is proving to be a frequent feature. Mori et al. (1990) reported diabetes mellitus in Bloom syndrome. Kelly (1977) observed a case of Bloom syndrome in a black female. German (Szalay, 1978) confirmed the diagnosis of Bloom syndrome in the black female reported by Szalay (1972).

German (1988) stated that the longest survival known to him is that of a man who died of esophageal cancer at the age of 48 years, having survived sigmoid cancer which had occurred 10 years earlier. Almost 150 cases worldwide have been cataloged by German (1990); he has personally examined 96 of these patients. Jewish patients represent 32% of the group, all but 1 of them being Ashkenazi. Complementation studies using sister chromatid exchange (SCE) as the measure of cross-correction indicate that this is one disease. Parental consanguinity was identified in 2 of 36 Jewish cases and in 25 of 75 non-Jewish cases. Heterozygotes do not show increased sister chromatid exchanges. German (1990) has not found an increased frequency of cancer in obligatory heterozygotes.

Passarge (1991) observed 10 patients in Germany during a 20-year period. One patient died at the age of 5 years of acute leukemia, a second at the age of 18 years of pulmonary fibrosis and bronchiectasis, and a third at the age of 21 years of Hodgkin lymphoma and subsequently leukemia.

German (1992) reported that there were 132 cases in the Bloom's Syndrome Registry as of January 1, 1990. One hundred and twenty seven had survived infancy. In all, 93 were still alive. Of the 39 deceased patients, 31 had died of cancer at a mean age of 27.8; cancer had been diagnosed at ages ranging from 4 years to 46 years. Of the 46 cancer patients, 14 had more than 1 primary, 2 had more than 2 primaries, and 1 had more than 3 primaries.

Biochemical Features

Vijayalaxmi et al. (1983) found that lymphocytes from patients with Bloom syndrome showed an incidence of cell resistance to the purine analog 6-thioguanine about 8 times the normal. Cells with specific locus mutations have been reported to be present in abnormally great numbers in BS fibroblast cultures, e.g., 6-thioguanine-resistant and diphtheria toxin-resistant cells.

Seal et al. (1991) compared the uracil DNA glycosylase from 2 nontransformed cell strains derived from persons of different ethnic backgrounds, with 2 different, similarly highly purified, normal human uracil DNA glycosylases. For each of the 4, a molecular mass of 37 kD was observed. The Bloom syndrome enzymes differed substantially in their isoelectric point and were thermolabile as compared to the normal human enzymes. They displayed a different K(m) and V(max) and were strikingly insensitive to 5-fluorouracil and 5-bromouracil, pyrimidine analogs that drastically decreased the activity of the normal human enzymes. In particular, each Bloom syndrome enzyme required 10- to 100-fold higher concentrations of each analog to achieve comparable inhibition of enzyme activity.

Other Features

Langlois et al. (1989) used a glycophorin A assay to measure the frequency in persons of blood type MN of variant erythrocytes that lack the expression of 1 allelic form of the protein, presumably due to mutational or recombinational events in erythroid precursor cells. Blood from persons with Bloom syndrome showed 50- to 100-fold increases in the frequency of variants of 3 types, those with a hemizygous phenotype, those with a homozygous phenotype, and those with what appeared to be partial loss of the expression of 1 locus. The high frequency of homozygous variants, indicating altered allelic segregation, could be taken as evidence for increased somatic crossing-over in vivo.

An increased generation of functional hemizygosity and homozygosity in somatic cells may play a role in the high cancer risk of persons with Bloom syndrome. Accumulation of p53 (191170) protein is seen in the nuclei of mammalian cells following DNA damage caused by ultraviolet radiation, x-ray, or a restriction enzyme. Promoters containing p53-binding sites show a dramatic transcriptional response to DNA damage. The p53 response to x-ray is rapid, reaching a peak at 2 hours after radiation, but is very transitory and reduced in magnitude compared with that seen in response to UV. Lu and Lane (1993) found no substantive defect in the p53 response of cells from ataxia-telangiectasia (208900) or xeroderma pigmentosum complementation group A (278700) patients. In contrast, 2 out of 11 primary cultures from Bloom syndrome patients showed complete absence of p53 accumulation following UV irradiation or SV40 infection and a grossly delayed and aberrant response following x-ray.

Van Kerckhove et al. (1988) found a specific defect in the pokeweed mitogen-induced alternative pathway of lymphocyte activation in BS patients.

Inheritance

Szalay (1963) provided the first evidence of a genetic basis. He described an isolated case in the child of first-cousin parents and 2 affected sibs. Autosomal recessive inheritance was established by German (1969), who maintains a worldwide registry which he periodically reports on (e.g., German et al., 1979). Of the then-known 21 families with Bloom syndrome, 12 were Ashkenazi and in these only 1 parental couple was consanguineous. On the other hand, 6 of the 9 non-Jewish unions were consanguineous. The Jewish gene appeared to have originated in a local area of eastern Europe.

German and Takebe (1989) reported that 14 cases in 12 families had been identified in Japan. Widely separated birthplaces and a frequency of parental consanguinity greater than in the general population suggested that the mutation, although rare, is widely distributed in that country. Complementation studies indicated that the same genetic locus is involved in the Japanese cases as in the Ashkenazi Jewish cases and non-Ashkenazi Jewish cases.

Cytogenetics

Multiple seemingly nonspecific chromosomal breaks have been observed in Bloom syndrome, as in Fanconi anemia (227650), and may be related causally to the high frequency of leukemia (German et al., 1965; German, 1992). (The Bloom and Fanconi syndromes are chromosome breakage or clastogenic syndromes.)

Schroeder and German (1974) showed that chromosomal aberrations were more numerous in Fanconi cells than in Bloom cells. In Bloom syndrome most interchanges were between homologous chromosomes, i.e., sister chromatid exchanges, whereas in Fanconi syndrome they were usually between nonhomologous chromosomes. Sister chromatid exchanges represent a cytologic marker useful for diagnosis including prenatal diagnosis. No test for the carrier state is known; the frequency of sister chromatid exchanges is not abnormal in heterozygotes (German et al., 1977).

Although the nature of the basic defect is not known, the absence of a substance that is supplied by cocultivated normal cells and reduced the rate of sister chromatid exchanges in Bloom syndrome fibroblasts was suggested by the work of Rudiger et al. (1980). Spontaneous SCE, but not mutagen-induced SCE, is inhibited by the Bloom corrective factor present in normal cell-conditioned culture medium. Control cells and cells of Fanconi anemia and xeroderma pigmentosum reduced the rate of sister chromatid exchanges in Bloom cells by about 45 to 50% (Bartram et al., 1981). In contrast, Bloom heterozygous cells reduced the rate of SCE by only 16 to 18%. Bartram et al. (1981) interpreted the findings as indicative of dosage effect. They concluded that the data suggest the existence of a 'corrective factor' which is either inactive or absent in homozygous Bloom cells and reduced in heterozygotes. It may be identical with or closely related to the normal gene product of the Bloom locus.

Thompson et al. (1982) found greatly increased sister chromatid exchanges in a mutant Chinese hamster ovary (CHO) cell line (EM9) with a DNA repair deficiency (see 126340). The deficiency was complemented in human-CHO somatic cell hybrids by human chromosome 19. Is this the Bloom syndrome defect? The mapping of Bloom syndrome to chromosome 15 excludes that possibility. (See Mapping Information.)

Weksberg et al. (1988) addressed the issue of dominance or recessivity of the low-SCE Bloom syndrome phenotype. Although most cells from BS patients exhibit high SCE, lymphoid cells from some patients exhibit dimorphism for high and low SCE. A high-SCE lymphoblast line was mutagenized, and a clone carrying the markers ouabain resistance and thioguanine resistance was isolated to serve as a fusion parent. When fused with low-SCE BS lines, the hybrid was found to have low SCE levels, establishing dominance of the low-SCE phenotype. By the same methodology, Weksberg et al. (1988) did a complementation analysis using high-SCE lymphoblast cell lines derived from patients of diverse ethnic origin: Ashkenazi Jewish, French-Canadian, Mennonite, and Japanese. No correction of the high SCE characteristic of BS cells was seen in any hybrids. Thus, a single gene is responsible for the high-SCE phenotype in BS patients.

Mapping

Lander and Botstein (1987) pointed out that recessive disorders can be efficiently mapped using RFLPs in the study of the DNA of affected children from consanguineous marriages. The method, which the authors called 'homozygosity mapping,' involves detection of the disease locus by virtue of the fact that the adjacent region will preferentially be homozygous by descent in such inbred children. They showed that a single affected child of a first-cousin marriage contains the same total information about linkage as a nuclear family with 3 affected children. They presented calculations to show that it should be practical to map a recessive disease gene by studying DNA from fewer than a dozen unrelated, affected inbred children, given a complete RFLP linkage map. Bloom syndrome was pointed to by Lander and Botstein (1987) as a good candidate for this type of mapping. About 100 living affected persons are known, but there are only 8 known families with 2 living affected members and 1 with 3 affected members. This is insufficient for traditional linkage analysis. By contrast, at least 24 affected persons are children of marriages between cousins. In some more common recessive disorders such as Werdnig-Hoffmann syndrome (253300), multiplex families may be difficult to collect because affected children die at a young age.

Because of the high parental consanguinity rate in non-Ashkenazi families with Bloom syndrome, Ellis et al. (1992) were able to do homozygosity mapping. Tight linkage was found with loci on distal 15q, specifically 15q26.1 (German et al., 1994). A polymorphic tetranucleotide repeat in an intron of the FES gene (190030) was homozygous in 25 of 26 individuals with Bloom syndrome whose parents were consanguineous (German et al., 1994). The location of the BLM gene on chromosome 15 was further supported by the finding of maternal uniparental disomy for that chromosome in a patient reported by Woodage et al. (1994). The patient had features of both Bloom syndrome and Prader-Willi syndrome (176270). Meiotic recombination between the 2 chromosomes 15 derived from the mother had resulted in heterodisomy for proximal 15q and isodisomy for distal 15q. In this individual, Bloom syndrome was probably due to homozygosity for a gene located telomeric to D15S95, which is at 15q25, rather than to genetic imprinting, the mechanism responsible for the development of PWS. This report represented the first application of disomy analysis to the regional localization of a disease gene. Ellis et al. (1994) found a striking association of a specific allele at the FES locus and at the D15S127 locus, both of which are tightly linked to BLM. This linkage disequilibrium constituted strong support for a founder-effect hypothesis to account for the fact that approximately 1 in 110 Ashkenazi Jews carries the Bloom syndrome mutation.

The rarity of Bloom syndrome and the recessive nature of its inheritance limits mapping of the gene by linkage approaches. McDaniel and Schultz (1992) used Bloom syndrome cells as recipients for microcell-mediated chromosome transfer to map a locus that results in complementation of the elevated sister chromatid exchange phenotype. Studying the Bloom cell line GM08505 (Coriell Institute) with a stable frequency of SCEs 10-fold higher than control values, they demonstrated that transfer of human chromosome 15 corrected the defect.

Straughen et al. (1996) described a 2-Mb contiguous map of the 15q26.1 region constructed from P1 clones and yeast artificial chromosomes (YACs) that contains the BLM gene. They also reported a long-range restriction map of this region.

Phenotype/genotype

Willis and Lindahl (1987) and Chan et al. (1987) independently demonstrated an abnormality of DNA ligase I (126391) in Bloom syndrome. DNA ligase I and DNA polymerase alpha (312040) are enzymes that function during DNA replication; DNA ligase II and DNA polymerase beta (174760) function during DNA repair. That the primary defect resides in the structural gene for DNA ligase I was suggested by the changes in the physical properties of the enzyme, specifically, heat sensitivity (Willis and Lindahl, 1987) and altered aggregation properties (Chan et al., 1987). Experiments with a fibroblast line derived from a Japanese case of Bloom syndrome showed that DNA ligase I from that source was not obviously heat sensitive or present in reduced amounts. Chan and Becker (1988) also came to the conclusion that mutation of the DNA ligase I gene may account for the primary metabolic defect in Bloom syndrome. Their data suggested that the defect in DNA ligase I is not due to a reduction in the number of protein molecules or to inhibitory substances but rather, at least in part, to the ATP binding and/or hydrolytic activity of the enzyme.

Willis et al. (1987) found that all cell lines derived from 7 patients with Bloom syndrome contained a DNA ligase I with unusual properties. In 6 lines the enzyme activity was reduced and the residual enzyme was anomalously heat-labile. In the seventh line, they found a dimeric rather than a monomeric form of ligase I. Several cell lines representative of other inherited disorders had apparently normal DNA ligases. The data were interpreted as indicating that BS is due to a defect in the structure of DNA ligase I caused by a 'leaky' point mutation occurring at one of at least 2 alternative sites. If the primary defect lies in the structural gene for DNA ligase I, then the mutation for Bloom syndrome is on chromosome 19, which encodes DNA ligase I (Barnes et al., 1990).

Since alteration of the DNA ligase I activity is a consistent biochemical feature of Bloom syndrome cells, Petrini et al. (1991) cloned DNA ligase I cDNA from normal human cells. Human DNA ligase I cDNAs from normal and BS cells complemented an S. cerevisiae DNA ligase mutation, and protein extracts prepared from S. cerevisiae transformants expressing normal and BS cDNA contained comparable levels of DNA ligase I activity. DNA sequencing and Northern blot analysis of DNA ligase I expression in 2 BS fibroblast lines representing each of 2 aberrant DNA ligase I molecular phenotypes demonstrated that the gene was unchanged in BS cells. Thus, a factor other than mutation in the ligase I gene must be involved as the basic defect.

Nicotera et al. (1989) suggested that the major biochemical defect in Bloom syndrome is chronic overproduction of the superoxide radical anion. They thought that inefficient removal of peroxide might be responsible for high rates of sister chromatid exchange and chromosomal damage in Bloom syndrome cells. Seal et al. (1988) described a monoclonal antibody, defined by enzyme-linked immunosorbent assay (ELISA), that reacted with normal uracil DNA glycosylase (191525) of human placenta as well as with the glycosylases from normal human cell types and 13 abnormal human cell strains. On the other hand, the antibody neither recognized nor inhibited native uracil DNA glycosylase from any of 5 separate Bloom syndrome cell strains. Lack of immunoreactivity with this antibody, which the authors designated 40.10.09, was suggested as a test for the early diagnosis of Bloom syndrome.

Cairney et al. (1987) described Wilms tumor in 3 patients with Bloom syndrome. Wilms tumor was bilateral in 1 of the 3 patients. Cairney et al. (1987) postulated that the elevated somatic recombination may mediate a high rate of conversion to homozygosity. Somatic recombination leading to homozygosity in Bloom syndrome has been suggested by several findings, including 'twin spots' or areas of hyper- and hypopigmentation on the skin of affected black children (Festa et al., 1979), increased frequency of exchange between the satellite stalks of acrocentric chromosomes (Therman et al., 1981), and increased variant blood group phenotypes and red cells from a patient with Bloom syndrome who was heterozygous for the AB blood group (Ben-Sasson et al., 1985). Petrella et al. (1991) observed autosomal triple trisomy involving chromosomes 2, 8, and 11 in a pregnancy conceived by a couple at risk for an offspring with Bloom syndrome. The SCE rate suggested that the conceptus was either heterozygous for the Bloom syndrome mutation or homozygous normal. They also found the Bloom syndrome gene in a non-Ashkenazi Jew and reported medulloblastoma in a patient with Bloom syndrome.

The hypermutability of Bloom syndrome cells includes hyperrecombinability. Ellis et al. (1995) noted that, although cells from all persons with Bloom syndrome exhibit the diagnostic high SCE rate, in some persons a minor population of low SCE lymphocytes exist in the blood. Lymphoblastoid cell lines (LCLs) with low SCE rates can be developed from these low SCE lymphocytes. In multiple low SCE LCLs examined from 11 patients with BS, polymorphic loci distal to BLM on 15q had become homozygous in LCLs from 5 persons, whereas polymorphic loci proximal to BLM remained heterozygous in all low SCE LCLs. These observations supported the hypothesis that low SCE lymphocytes arose through recombination within BLM in persons with BS who had inherited paternally and maternally derived BLM alleles mutated at different sites. Such a recombination event in a precursor stem cell in these compound heterozygotes thus gave rise to a cell whose progeny had a functionally wildtype gene and phenotypically a low SCE rate (Ellis et al., 1995). Ellis et al. (1995) used the low SCE LCLs in which reduction to homozygosity had occurred for localizing BLM by an approach referred to as somatic crossover point (SCP) mapping. The precise map position of BLM was determined by comparing the genotypes of the recombinant low SCE LCLs from the 5 persons mentioned above with their constitutional genotypes at loci in the region around BLM. The strategy was to identify the most proximal polymorphic locus possible that was constitutionally heterozygous and that had been reduced to homozygosity in the low SCE LCLs, and to identify the most distal polymorphic locus possible that had remained constitutionally heterozygous in them. BLM would have to be in the short interval defined by the reduced (distal) and the unreduced (proximal) heterozygous markers. The power of this approach was limited only by the density of polymorphic loci available in the immediate vicinity of BLM. A candidate for BLM was identified by direct sele ction of a cDNA derived from a 250-kb segment of the genome to which BLM had been assigned by SCP mapping. cDNA analysis of the candidate gene identified a 4437-bp cDNA that encoded a 1417-amino acid peptide with homology to the RecQ helicases, a subfamily of DExH box-containing DNA and RNA helicases. The presence of chain-terminating mutations in the candidate gene in persons with Bloom syndrome proved that it was BLM. Mutational analysis in the first 13 unrelated persons with BS examined permitted the identification of 7 unique mutations in 10 of them. The fact that 4 of the 7 mutations resulted in premature termination of translation indicated that the cause of most Bloom syndrome is the loss of enzymatic activity of the BLM gene product. Identification of loss-of-function mutations in BLM is consistent with the autosomal recessive transmission, and the homology of BLM and RecQ suggested that BLM has enzymatic activity. In 4 persons with Jewish ancestry, a 6-bp deletion and a 7-bp insertion at nucleotide 2281 were identified, and each of the 4 persons were homozygous for the mutation. Homozygosity was predictable because linkage disequilibrium had been detected in Ashkenazi Jews with Bloom syndrome between BLM, D15S127, and FES (Ellis et al., 1994). Thus a person who carried this deletion/insertion mutation was a founder of Ashkenazi Jewish population and nearly all Ashkenazi Jews with Bloom syndrome inherited the mutation identical by descent from this common ancestor.

The RecQ gene family, of which BLM is a member, is named after the E. coli gene. RecQ is an E. coli gene that is a member of the RecF recombination pathway, a pathway of genes in which mutations abolish the conjugational recombination proficiency and ultraviolet resistance of a mutant strain. RecQL (600537) is a human gene isolated from HeLa cells, the product of which possesses DNA-dependent ATPase, DNA helicase, and 3-prime-to-5-prime single-stranded DNA translocation activities. Ellis et al. (1995) suggested that the absence of the BLM gene product probably destabilizes other enzymes that participate in DNA replication and repair, perhaps through direct interaction and through more general responses to DNA damage.

Ellis and German (1996) reported that the BLM protein has similarity to 2 other proteins that are members of the RecQ family of helicases, namely the gene product encoded by the Werner syndrome gene (WRN; 277700) and the product of the yeast gene SGS1. SGS1 was identified by a mutation that suppressed the slow-growth phenotype of mutations in the topoisomerase gene. These proteins have 42% to 44% amino acid identity across the conserved helicase motifs. In addition, the proteins are of similar length and contain highly negatively charged N-terminal regions and highly positively charged C-terminal regions. Ellis and

German (1996) noted that these similarities in overall structure have raised the possibility that the proteins play similar roles in metabolism. Since the SGS1 gene product is known to interact with the products of the yeast topoisomerase genes, they predicted that the BLM and WRN genes interact with human topoisomerases.

Sinclair et al. (1997) showed that mutation of the yeast SGS1 gene causes premature aging in yeast mother cells as demonstrated by shortened life span and the aging-induced phenotypes of sterility and redistribution of the Sir3 silencing protein from telomeres to the nucleolus. Further, in old SGS1 cells the nucleolus was enlarged and fragmented, changes that also occur in old wildtype cells. Their findings suggested a conserved mechanism of cellular aging that may be related to nucleolar structure. The similar effect of the related SGS1 and WRN genes on yeast and human aging, along with age-associated changes in rDNA content reported for several mammalian species, suggested that a common mechanism may underlie aging in eukaryotes.

Clinical Management

German (1992) commented that BS neoplasms themselves, e.g., leukemic marrow cells, demonstrate a high sister chromatid exchange rate similar to nonneoplastic cells of BS patients. A clinical difference is that leukemia in BS usually presents itself with leukopenia rather than leukocytosis. He also commented on the fact that there is pitifully little that one can do in relation to the proneness to cancer. In contrast to the situation with carcinoma, early diagnosis of leukemia is at present not known to improve the chances of curative therapy.

German (1992) advised against frequent hematologic examinations in children for fear of untoward psychologic effects. Allogeneic marrow grafting has not been carried out in BS. An argument can be made for identifying as soon as possible a potential donor of bone marrow for any person with BS, and the cryopreservation of cord-blood stem cells of HLA-matched sibs who might be born after the BS child is identified can be considered, for possible later transplantation.

See also Arase et al. (1980); Bloom (1966); Bryant et al. (1979); Cerutti et al. (1981); Chaganti et al. (1974); Ellis et al. (1994); German (1990); German (1992); German et al. (1974); German et al. (1977); Hutteroth et al. (1975); Kawashima et al. (1980); Kuhn (1981); Leroux et al. (1984); Mulcahy and French (1981); Rosin and German (1985); Sawitsky et al. (1966); Schmidt-Preuss et al. (1981); Siciliano et al. (1983); van Buul et al. (1978); Warren et al. (1981)

Arase, S.; Takahashi, O.; Ishizaki, K.; Takebe, H.: Bloom's syndrome in a Japanese boy with lymphoma. Clin. Genet. 18: 123-127, 1980.

Barnes, D. E.; Johnston, L. H.; Kodama, K.; Tomkinson, A. E.; Lasko, D. D.; Lindahl, T.: Human DNA ligase I cDNA: cloning and functional expression in Saccharomyces cerevisiae. Proc. Nat. Acad. Sci. 87: 6679-6683, 1990.

Bartram, C. R.; Rudiger, H. W.; Schmidt-Preuss, U.; Passarge, E.: Functional deficiency of fibroblasts heterozygous for Bloom syndrome as specific manifestation of the primary defect. Am. J. Hum. Genet. 33: 928-934, 1981.

Ben-Sasson, S. A.; Cohen, T.; Voss, R.: Background allelic variants in normal hemopoietic cells and Bloom's syndrome erythrocytes and the possible implication of somatic crossing-over. Cancer Genet. Cytogenet. 15: 237-242, 1985.

Bloom, D.: The syndrome of congenital telangiectatic erythema and stunted growth. J. Pediat. 68: 103-113, 1966.

Bryant, E. M.; Hoehn, H.; Martin, G. M.: Normalisation of sister chromatid exchange frequencies in Bloom's syndrome by euploid cell hybridization. Nature 279: 795-796, 1979.

Cairney, A. E. L.; Andrews, M.; Greenberg, M.; Smith, D.; Weksberg, R.: Wilms tumor in three patients with Bloom syndrome. J. Pediat. 111: 414-416, 1987.

Cerutti, P.; Emerit, I.; Hirschi, M.; Zbinden, I.: Bloom's syndrome: a deficiency in the detoxification of active oxygen species? (Abstract) Europ. J. Cancer Clin. Oncol. 17: 938, 1981.

Chaganti, R. S. K.; Schonberg, S.; German, J.: A manyfold increase in sister chromatid exchanges in Bloom's syndrome lymphocytes. Proc. Nat. Acad. Sci. 71: 4508-4512, 1974.

Chan, J. Y.-H.; Becker, F. F.: Defective DNA ligase I in Bloom's syndrome cells: simultaneous analysis using immunoblotting and the ligase-(32-P)AMP adduct assay. J. Biol. Chem. 263: 18231-18235, 1988.

Chan, J. Y. H.; Becker, F. F.; German, J.; Ray, J. H.: Altered DNA ligase I activity in Bloom's syndrome cells. Nature 325: 357-359, 1987.

Ellis, N. A.; German, J.: Molecular genetics of Bloom's syndrome. Hum. Molec. Genet. 5: 1457-1463, 1996.

Ellis, N. A.; Groden, J.; Ye, T.-Z.; Straughen, J.; Lennon, D. J.; Ciocci, S.; Proytcheva, M.; German, J.: The Bloom's syndrome gene product is homologous to RecQ helicases. Cell 83: 655-666, 1995.

Ellis, N. A.; Lennon, D. J.; Proytcheva, M.; Alhadeff, B.; Henderson, E. E.; German, J.: Somatic intragenic recombination within the mutated locus BLM can correct the high SCE phenotype of Bloom syndrome cells. Am. J. Hum. Genet. 57: 1019-1027, 1995.

Ellis, N. A.; Roe, A. M.; Kozloski, J.; Proytcheva, M.; Falk, C.; German, J.: Linkage disequilibrium between the FES, D15S127, and BLM loci in Ashkenazi Jews with Bloom syndrome. Am. J. Hum. Genet. 55: 453-460, 1994.

Ellis, N. A.; Roe, A. M.; Otterud, B.; Leppert, M.; German, J.: Homozygosity mapping of the Bloom's syndrome locus. (Abstract) Am. J. Hum. Genet. 51 (suppl.): A187, 1992.

Ellis, N. A.; Ye, T.-Z.; Patton, S.; German, J.; Goodfellow, P. N.; Weller, P.: Cloning of PBDX, an MIC2-related gene that spans the pseudoautosomal boundary on chromosome Xp. Nature Genet. 6: 394-399, 1994.

Ferrara, A.: Goltz's syndrome. (Letter) Am. J. Dis. Child. 123: 262, 1972.

Ferrara, A.; Fontana, V. J.; Numsen, G.: Bloom's syndrome in Oriental male. New York J. Med. 67: 3258-3262, 1967.

Festa, R. S.; Meadows, A. T.; Boshes, R. A.: Leukemia in a black child with Bloom syndrome. Cancer 44: 1507-1510, 1979.

German, J.: Personal Communication. New York, N. Y. 3/1990.

German, J.: Bloom's syndrome. I. Genetical and clinical observations in the first twenty-seven patients. Am. J. Hum. Genet. 21: 196-227, 1969.

German, J.: Personal Communication. New York, N. Y. 2/17/1988.

German, J.: Personal Communication. New York, N. Y. 11/12/1992.

German, J.: Personal Communication. New York, N. Y. 6/12/1990.

German, J.: Bloom's syndrome: incidence, age of onset, and types of leukemia in the Bloom's Syndrome Registry.In: Bartsocas, C. S.; Loukopoulos, D.: Genetics of Hematological Disorders. Washington, D. C.: Hemisphere Publishers (pub.) 1992. Pp. 241-258.

German, J.; Archibald, R.; Bloom, D.: Chromosomal breakage in a rare and probably genetically determined syndrome of man. Science 148: 506-507, 1965.

German, J.; Bloom, D.; Passarge, E.: Bloom's syndrome XI. Progress report for 1983. Clin. Genet. 25: 166-174, 1984.

German, J.; Bloom, D.; Passarge, E.: Bloom's syndrome. VII. Progress report for 1978. Clin. Genet. 15: 361-367, 1979.

German, J.; Bloom, D.; Passarge, E.; Fried, K.; Goodman, R. M.; Katzenellenbogen, I.; Larson, Z.; Legum, C.; Levine, S.; Wahrman, J.: Bloom's syndrome. VI. The disorder in Israel and an estimation of the gene frequency in the Ashkenazim. Am. J. Hum. Genet. 29: 553-562, 1977.

German, J.; Crippa, L. P.; Bloom, D.: Bloom's syndrome. III. Analysis of the chromosome aberration characteristic of this disorder. Chromosoma 48: 361-366, 1974.

German, J.; Roe, A. M.; Leppert, M. F.; Ellis, N. A.: Bloom syndrome: an analysis of consanguineous families assigns the locus mutated to chromosome band 15q26.1. Proc. Nat. Acad. Sci. 91: 6669-6673, 1994.

German, J.; Schonberg, S.; Louie, E.; Chaganti, R. S. K.: Bloom's syndrome. IV. Sister chromatid exchanges in lymphocytes. Am. J. Hum. Genet. 29: 248-255, 1977.

German, J.; Takebe, H.: Bloom's syndrome. XIV. The disorder in Japan. Clin. Genet. 35: 93-110, 1989.

Hutteroth, T. H.; Litwin, S. D.; German, J.: Abnormal immune responses of Bloom's syndrome lymphocytes in vitro. J. Clin. Invest. 56: 1-7, 1975.

Kawashima, H.; Sato, T.; Taniguchi, N.; Yagi, T.; Ishizaki, K.; Takebe, H.: Bloom's syndrome in a Japanese girl. Clin. Genet. 17: 143-148, 1980.

Kelly, T. E.: Personal Communication. Charlottesville, Va. 1977.

Kuhn, E. M.: A high incidence of mitotic chiasmata in endoreduplicated Bloom's syndrome cells. Hum. Genet. 58: 417-421, 1981.

Landau, J. W.; Sasaki, M. S.; Newcomer, V. D.; Norman, A.: Bloom's syndrome: the syndrome of telangiectatic erythema and growth retardation. Arch. Derm. 94: 687-694, 1966.

Lander, E. S.; Botstein, D.: Homozygosity mapping: a way to map human recessive traits with the DNA of inbred children. Science 236: 1567-1570, 1987.

Langlois, R. G.; Bigbee, W. L.; Jensen, R. H.; German, J.: Evidence for increased in vivo mutation and somatic recombination in Bloom's syndrome. Proc. Nat. Acad. Sci. 86: 670-674, 1989.

Legum, C.; Furman, N.; Diamant, S.: Bloom's syndrome in an Iranian Jewish male. Ann. Genet. 34: 198-200, 1991.

Leroux, D.; Chmara, D.; Jalbert, P.: Bloom's syndrome: in vitro correction of the sister chromatid exchange rate by normal cells. Cancer Genet. Cytogenet. 12: 139-143, 1984.

Lu, X.; Lane, D. P.: Differential induction of transcriptionally active p53 following UV or ionizing radiation: defects in chromosome instability syndromes? Cell 75: 765-778, 1993.

McDaniel, L. D.; Schultz, R. A.: Elevated sister chromatid exchange phenotype of Bloom syndrome cells is complemented by human chromosome 15. Proc. Nat. Acad. Sci. 89: 7968-7972, 1992.

Mori, S.; Kondo, N.; Motoyoshi, F.; Yamaguchi, S.; Kaneko, H.; Orii, T.: Diabetes mellitus in a young man with Bloom's syndrome. Clin. Genet. 38: 387-390, 1990.

Mulcahy, M. T.; French, M.: Pregnancy in Bloom's syndrome. Clin. Genet. 19: 156-158, 1981.

Nicotera, T. M.; Notaro, J.; Notaro, S.; Schumer, J.; Sandberg, A. A.: Elevated superoxide dismutase in Bloom's syndrome: a genetic condition of oxidative stress. Cancer Res. 49: 5239-5243, 1989.

Passarge, E.: Bloom's syndrome: the German experience. Ann. Genet. 34: 179-197, 1991.

Petrella, R.; Hirschhorn, K.; German, J.: Triple autosomal trisomy in a pregnancy at risk for Bloom's syndrome. Am. J. Med. Genet. 40: 316-318, 1991.

Petrini, J. H. J.; Huwiler, K. G.; Weaver, D. T.: A wild-type DNA ligase I gene is expressed in Bloom's syndrome cells. Proc. Nat. Acad. Sci. 88: 7615-7619, 1991.

Rosin, M. P.; German, J.: Evidence for chromosome instability in vivo in Bloom syndrome: increased numbers of micronuclei in exfoliated cells. Hum. Genet. 71: 187-191, 1985.

Rudiger, H. W.; Bartram, C. R.; Harder, W.; Passarge, E.: Rate of sister chromatid exchanges in Bloom syndrome fibroblasts reduced by co-cultivation with normal fibroblasts. Am. J. Hum. Genet. 32: 150-157, 1980.

Sawitsky, A.; Bloom, D.; German, J.: Chromosomal breakage and acute leukemia in congenital telangiectatic erythema and stunted growth. Ann. Intern. Med. 65: 487-495, 1966.

Schmidt-Preuss, U.; Maack, P.; Bartram, C. R.; Rudiger, H. W.: Mutagen-induced sister chromatid exchange rate in Bloom syndrome remains unaltered in the presence of Bloom corrective factor. Hum. Genet. 58: 432-433, 1981.

Schroeder, T. M.; German, J. B.: Bloom's syndrome and Fanconi's anemia: demonstration of two distinctive patterns of chromosome disruption and rearrangement. Humangenetik 25: 299-306, 1974.

Seal, G.; Brech, K.; Karp, S. J.; Cool, B. L.; Sirover, M. A.: Immunological lesions in human uracil DNA glycosylase: association with Bloom syndrome. Proc. Nat. Acad. Sci. 85: 2339-2343, 1988.
Seal, G.; Tallarida, R. J.; Sirover, M. A.: Purification and properties of the uracil DNA glycosylase from Bloom's syndrome. Biochim. Biophys. Acta 1097: 299-308, 1991.
Siciliano, M. J.; Stallings, R. L.; Adair, G. M.; Humphrey, R. M.; Siciliano, J.: Provisional assignment of TPI, GPI, and PEPD to Chinese hamster autosomes 8 and 9: a cytogenetic basis for functional haploidy of an autosomal linkage group in CHO cells. Cytogenet. Cell Genet. 35: 15-20, 1983.
Sinclair, D. A.; Mills, K.; Guarente, L.: Accelerated aging and nucleolar fragmentation in yeast sgs1 mutants. Science. 277: 1313-1316, 1997.
Straughen, J.; Ciocci, S.; Ye, T.-Z.; Lennon, D. N.; Proytcheva, M.; Alhadeff, B.; Goodfellow, P.; German, J.; Ellis, N. A.; Groden, J.: Physical mapping of the Bloom syndrome region by the identification of YAC and P1 clones from human chromosome 15 band q26.1. Genomics 35: 118-128, 1996.
Straughen, J. E.; Johnson, J.; McLaren, D.; Proytcheva, M.; Ellis, N.; German, J.; Groden, J.: A rapid method for detecting the predominant Ashkenazi Jewish mutation in the Bloom's syndrome gene. Hum. Mutat. 11: 175-178, 1998.
Szalay, G. C.: Dwarfism with skin manifestations. J. Pediat. 62: 686-695, 1963.
Szalay, G. C.: Questionable Bloom syndrome in a Negro girl. Am. J. Dis. Child. 124: 245-248, 1972.
Szalay, G. C.: Personal Communication. Harbor City, Calif. 5/28/1978.
Therman, E.; Otto, P. G.; Shahidi, N. T.: Mitotic recombination and segregation of satellites in Bloom's syndrome. Chromosoma 82: 627-636, 1981.
Thompson, L. H.; Brookman, K. W.; Dillehay, L. E.; Carrano, A. V.; Mazrimas, J. A.; Mooney, C. L.; Minkler, J. L.: A CHO-cell strain having hypersensitivity to mutagens, a defect in DNA strand-break repair, and an extraordinary baseline frequency of sister-chromatid exchange. Mutat. Res. 95: 427-440, 1982.
van Buul, P. P. W.; Natarajan, A. T.; Verdegaal-Immerzeel, E. A. M.: Suppression of the frequencies of sister chromatid exchanges in Bloom's syndrome fibroblasts by co-cultivation by Chinese hamster cells. Hum. Genet. 44: 187-189, 1978.
Van Kerckhove, C. W.; Ceuppens, J. L.; Vanderschueren-Lodeweyckx, M.; Eggermont, E.; Vertessen, S.; Stevens, E. A. M.: Bloom's syndrome: clinical features and immunologic abnormalities of four patients. Am. J. Dis. Child. 142: 1089-1093, 1988.
Vijayalaxmi, (NI); Evans, H. J.; Ray, J. H.; German, J.: Bloom's syndrome: evidence for an increased mutation frequency in vivo. Science 221: 851-853, 1983.
Warren, S. T.; Schultz, R. A.; Chang, C.-C.; Wade, M. H.; Trosko, J. E.: Elevated spontaneous mutation rate in Bloom syndrome fibroblasts. Proc. Nat. Acad. Sci. 78: 3133-3137, 1981.
Weksberg, R.; Smith, C.; Anson-Cartwright, L.; Maloney, K.: Bloom syndrome: a single complementation group defines patients of diverse ethnic origin. Am. J. Hum. Genet. 42: 816-824, 1988.
Willis, A. E.; Lindahl, T.: DNA ligase I deficiency in Bloom's syndrome. Nature 325: 355-357, 1987.
Willis, A. E.; Weksberg, R.; Tomlinson, S.; Lindahl, T.: Structural alterations of DNA ligase I in Bloom syndrome. Proc. Nat. Acad. Sci. 84: 8016-8020, 1987.
Woodage, T.; Prasad, M.; Dixon, J. W.; Selby, R. E.; Romain, D. R.; Columbano-Green, L. M.; Graham, D.; Rogan, P. K.; Seip, J. R.; Smith, A.; Trent, R. J.: Bloom syndrome and maternal uniparental disomy for chromosome 15. Am. J. Hum. Genet. 55: 74-80, 1994.

NEVOID BASAL CELL CARCINOMA SYNDROME (#109400)

JJM NOTE

Of 63 youths, ages 6–19, with at least one basal cell carcinoma, 13 proved to have the nevoid basal cell carcinoma syndrome (NBCCS), and another 10 had a preexisting nevus sebaceous of Jadassohn [Rahbari H, Mehregan AH: Basal cell epithelioma (carcinoma) in children and teenagers. Cancer 49:350–53, 1982]. Other malignant neoplasms that may be the first obvious manifestation of the disorder include medulloblastoma, astrocytoma, ovarian carcinoma, and cardiac fibrous histiocytoma. As one of few single-gene disorders with a predisposition to cancer and frank birth defects, the NBCCS gene may act as a tumor-suppressor gene and, at least in embryonic and fetal cells, a gene important in migration, differentiation, or programmed cell death, as articulated by Bale and associates (1995, 1996).

A number sign (#) is used with this entry because of evidence that the disorder results from mutations in PTCH (601309), the human homolog of the Drosophila 'patched' gene.

Nevoid basal cell carcinoma syndrome may be the most satisfactory designation for this disorder. Gorlin and Goltz (1960) suggested autosomal dominant inheritance. About 40% of cases represent new mutation (Gorlin, 1982). Jones et al. (1975) found evidence of paternal age effect in new mutations for this disorder. Herzberg and Wiskemann (1963) described what they termed the 'fifth phakomatosis,' basal cell nevus syndrome with medulloblastoma. A father and son had basal cell nevi. The son had medulloblastoma and congenital thoracic scoliosis. One of the patients studied by Cawson and Kerr (1964) had astrocytoma with severe hydrocephalus. The palms and soles may show pits. (Palmar pits occur also with Cowden syndrome (158350) and atypical palmar pits with Darier disease (124200).) Other features include bridging of the sella turcica, mild mandibular prognathism, lateral displacement of the inner canthi, frontal and biparietal bossing, odontogenic keratocysts of the jaws, kyphoscoliosis, bifid, missing, fused and/or splayed ribs, imperfect segmentation of cervical vertebrae, characteristic lamellar calcification of the falx cerebri, ovarian fibromata and lymphomesenteric cysts which tend to calcify, and short fourth metacarpal. The basal cell nevi occur in enormous numbers. Some may resemble seborrheic keratoses. The nevi are rarely congenital, most often appearing in increasing numbers around the time of puberty. Lip and/or palatal clefts probably occur in about 5% of cases, and mental retardation in about the same frequency.

Lile et al. (1968) observed 4 cases in 3 generations. In 2 of these patients the terminal phalanx of the thumb was short. Ovarian carcinoma has been observed (Berlin et al., 1966). Huge calcified ovarian fibromas were present in a CPC case discussed by Holmes (1976) at the Massachusetts General Hospital. The occurrence of lymphomesenteric cysts, described by Clendenning et al. (1963), was emphasized in another CPC case by Ottinger and Vickery (1986).

Schwartz (1978) pointed to hamartomatous polyps of the stomach and mesenteric cyst as features of the basal cell nevus syndrome. Totten (1980) observed a large congenital lung cyst occupying the left thoracic cavity in a patient with the basal cell nevus syndrome. Patients with BCNS are abnormally sensitive to radiotherapeutic doses of ionizing radiation; several patients so-treated have developed an unusually large number of basal cell tumors in the irradiated area a short time after exposure. Radiosensitivity could not be detected at the cellular level, however (Featherstone et al., 1983). Unilateral coloboma of the iris and glaucoma occurred in a familial case known to me (GS, P19000).

Evans et al. (1993) studied the clinical complications of this disorder in 84 cases. Basal cell carcinomas and jaw cysts occurred in more than 90% of patients by 40 years of age, but both sometimes occurred before 10 years of age. Less well-described complications included ovarian calcification or fibroma (24%), medulloblastoma (5%), cardiac fibroma (3%), cleft palate (5%), and ophthalmic abnormalities such as squint or cataract (26%).

Cramer and Niederdellmann (1983) described 9 subjects from 3 families with the syndrome of cerebral gigantism (117550); 7 of the patients also had signs of the basal cell nevus syndrome. In 1 family, a father was 193 cm tall at age 45 and his son was 197 cm tall at age 18; both had jaw cysts and other signs of basal cell nevus syndrome. Another son was 198 cm tall at age 17 years. Macrocephaly, mild hydrocephalus, intracranial calcification and EEG abnormalities were described.

The basal cell nevus syndrome has features that are compatible with the Knudson hypothesis, which has had great heuristic value in retinoblastoma (180200), Wilms tumor (194070), and other neoplasms in which tumor-suppressor genes have been found. It combines malformations with neoplasia. In 6 cases from 4 families, Gibbs et al. (1986) found no chromosomal abnormality. In 1 of 5 affected related individuals, Fletcher and Morton (1988) found a constitutional chromosomal rearrangement involving chromosomes 5 and 15. The site involved on chromosome 5 was in the same approximate region on the long arm as that involved in adenomatous polyposis coli (175100). Fletcher and Morton (1988) pointed out similarities between BCNS and Gardner syndrome and proposed that they may be genetically related.

Search for palmar pits is warranted in any patient with a basal cell carcinoma before age 30. As might be expected of an autosomal dominant disorder, a considerable proportion of persons judged to have this syndrome in the course of family studies may have no skin lesions. Jones et al. (1986) presented the case of a woman who at age 19 underwent cardiac transplantation for an unresectable fibrous histiocytoma of the left ventricle (Jamieson et al., 1981). The patient showed marfanoid build, frontal bossing with large occipitofrontal circumference, ocular hypertelorism, broad nasal root, enlarged jaw, glaucoma, long fingers, multiple odontogenic keratocysts, postaxial polydactyly of right foot, and bony bridging of right metatarsals 4 and 5. The palmar and plantar pits are like those of the form of porokeratosis described in 175850. Holubar et al. (1970) found basal cell epitheliomas in multiple palmar pits in an 8-year-old girl with BCNS. Levine et al. (1987) described subconjunctival epithelial cysts presenting a dramatic appearance of everted upper eyelids in patients with this condition.

Farndon et al. (1992) estimated that the minimum prevalence is 1 per 57,000; 1 in 200 patients with basal cell carcinomas (1 or more) had the syndrome, but the proportion is much higher (1 in 5) among those in whom a basal cell carcinoma develops before age 19. Only a few of the nevi grow and become locally invasive, and basal cell carcinomas do not develop at all in about 15% of affected persons. Radiation treatment can result in fresh crops of aggressive basal cell carcinomas and can lead to severe disfigurement.

Gorlin (1987) gave a comprehensive review. Evans et al. (1991) found abnormal ribs in 2 infants delivered preterm at 29 and 25 weeks. The finding at first was thought unimportant but subsequently was shown to indicate that some members of their families had Gorlin syndrome. In the first case, early routine chest radiographs showed an incidental finding of bifid ribs. The 25-year-old father had dislocated shoulder from birth due to Sprengel deformity (184400), pronounced frontal bossing with enlarged head, hypertelorism, calcification of the falx cerebri, pituitary fossa totally bridged by bone, and bilateral bifid ribs. A 5-year-old brother had calcification within the falx cerebri. A 4-year-old brother had been diagnosed as having arrested congenital hydrocephalus, and chest radiograph showed 2 bifid ribs. In the second index case, in addition to bifid ribs, there was possible Sprengel deformity. The baby's mother had had 5 jaw cysts removed between ages 11 and 31; she had enlarged head with pronounced frontal bossing and pits in the palms of her hands and feet. She had multiple milia on her forehead and hypertelorism. Radiographs showed scoliosis, calcified ovarian fibroma, calcification of the falx cerebri, and minor rib anomalies.

Waaler and Aarskog (1980) reported a family in which the mother had hydrocephalus, rib malformations, dysplasia of thoracic vertebrae and Sprengel anomaly, and each of her 3 daughters had 1 or more of these 4 features. The hydrocephalus (present in the mother and a daughter) was moderate and compensated spontaneously, making shunt operation unnecessary. It seems probable that the family reported by Waaler and Aarskog (1980) also suffered from Gorlin syndrome.

Goldstein et al. (1994) examined 11 African Americans from 2 families with Gorlin syndrome (which they abbreviated NBCC for nevoid basal cell carcinoma syndrome). They also reviewed the literature on this condition in African Americans. They found that reduced expression of the basal cell carcinomas but full expression of the other components of the syndrome was characteristic in African Americans. The 3 most common findings in their 11 cases were jaw cysts, palmar palmar and/or plantar pits, and calcification of the falx cerebri. Only 4 of the 11 had 1 or more confirmed basal cell carcinomas, whereas the frequency of basal cell carcinomas in whites is estimated at 90%. In a study that aimed to ascertain all the affected families in Australia, Shanley et al. (1994) identified 118 cases. The frequency of most manifestations were similar to those reported by Evans et al. (1993). A major difference, however, was that the multiple basal cell carcinomas were manifest from an earlier age in the Australian population, which probably reflects greater exposure to ultraviolet radiation. Of the 64 families ascertained, 37 represented simplex cases (that is, the only proband in the family), and accordingly the new mutation rate appeared to be high, a surprising finding in light of the lack of impact Gorlin syndrome (which they abbreviated NBCCS) on reproductive capabilities. Their data showed the occurrence of multiple BCCs or onset under 20 years of age in 90 of 118 cases (75%).

Loose linkage to Rh was suggested by Anderson (1968) and to Charcot-Marie-Tooth disease by Heimler et al. (1978). Both mutations are located on chromosome 1. Consequently, of considerable interest is the finding by Bale et al. (1985) of a suggestion of linkage to amylase-2 (104650), which is located at 1p21. McConville et al. (1987) investigated linkage of BCNS with the NRAS oncogene locus (164790). Farndon and Simmons (1987) found negative lod scores with 5 markers on chromosome 1, suggesting that BCNS may not be on that chromosome. By linkage analysis, Farndon et al. (1992) localized BCNS to 9q22.3-q31. They concluded that the most likely position was between DNA markers D9S12 and D9S53. The maximum lod scores with these 2 markers were 3.597 at theta = 0.04 and 6.457 at theta = 0.03, respectively. Reis et al. (1992) confirmed the assignment to chromosome 9. They concluded that D9S43 is centromeric to BCNS and that GSN (137350) and ASS (215700) are telomeric to BCNS in that sequence. In their collection of Australasian pedigrees, Wicking et al. (1994) further refined the localization of the gene to a site between markers D9S196 and D9S180, an interval reported to be approximately 2 cM. Farndon et al. (1994) concluded that the gene involved with NBCCS lies in a 2.6 cM interval centromeric to D9S287. Recombinants also mapped the gene for Fanconi anemia, group C (227645) to the same region; no recombinants were detected between FACC and NBCCS; maximum lod = 5.601 at theta = 0.0. The gene for epithelioma, self-healing, squamous (132800), and the gene for xeroderma pigmentosum, complementation group A (XPAC; 278700) mapped to the same region.

Johnson et al. (1996) demonstrated 2 independent sequence changes in exon 15 of the human homolog of the Drosophila patched gene (601309) in patients with Gorlin syndrome. They cloned the human PTC gene (symbol = PTCH) and mapped it by radiation hybrid analysis to band 9q22.3, a region implicated in BCNS. It was mapped very close to D9S287, which lies between D9S196 and D9S176. One 49-year-old man had an insertion of 9 bp at nucleotide 2445 of the coding sequence resulting in the insertion of 3 amino acids (pro-asn-ile) after amino acid 815 (601309.0001). This change produces a tandem duplication of 3 amino acids. A second affected individual was heterozygous for an 11-bp deletion that removes nucleotides 2442 to 2452 (601309.0002). The resulting frameshift truncates the ORF by creating a stop codon 9 amino acids after amino acid 813. Johnson et al. (1996) stated that PTC is expressed in developing sclerotome, branchial arches, limbs, and spinal cord and in vertebrate skin. They noted that the pattern of vertebrate PTC expression is consistent with the abnormalities found in BCNS. Therefore, Johnson et al. (1996) concluded that human PTC is a strong candidate gene for BCNS.

Hahn et al. (1996) isolated a human sequence with strong homology to the Drosophila segment polarity gene PTC from a YAC and cosmid contig of the NBCCS region. They carried out mutation analysis and demonstrated alteration of PTC in NBCCS patients and in tumors from patients with this condition. Of the mutations identified in unrelated patients, 4 were deletions or insertions resulting in frameshifts (e.g., 601309.0004) and 2 were point mutations leading to premature stops (601309.0003 and 601309.0005). An additional finding of Hahn et al. (1996) that confirmed the relationship between mutations in PTC and the disease was the identification of a frameshift mutation (2000insC) in a sporadic NBCCS patient and the absence of this mutation in her unaffected parents. The authors proposed that a reduction in the expression of the PTC gene can lead to the developmental abnormalities observed in this syndrome and that complete loss of PTC function contributes to transformation of certain cell types. To analyze the role of PTC in neoplasia, tumors related to the NBCCS were screened for mutations. Hahn et al. (1996) reported that 2 sporadic basal-cell carcinomas with allelic loss of the NBCCS region had inactivating mutations of the remaining allele.

Bialer et al. (1994) made the prenatal diagnosis of Gorlin syndrome in a pregnancy sired by a man with Gorlin syndrome. There were 2 other affected members in the family. Polymorphic DNA markers on chromosome 9 were used and the fetal diagnosis was confirmed by ultrasound scan which showed unilateral cleft lip, probable cleft palate, and hydrocephalus. The parents elected to terminate the pregnancy and examination of the fetus revealed aqueductal stenosis, cleft lip, and cleft palate with a prominent forehead and macrocephaly.

Gailani et al. (1992) found allelic loss in the chromosome 9q31 region in 11 of 16 sporadic basal cell carcinomas, in 2 hereditary basal cell carcinomas, and in 1 hereditary ovarian fibroma. Furthermore, in a study of 5 Gorlin syndrome kindreds, tight linkage was found with a genetic marker in this region. Loss of heterozygosity implies that the gene is homozygously inactivated and normally functions as a tumor suppressor. In contrast, hemizygous germline mutations lead to multiple congenital anomalies of the types seen in this disorder. D9S29 was the closest linked marker. Gailani et al. (1992) pointed out that the Ferguson-Smith syndrome (132800) maps to the same region and raised the possibility that Gorlin syndrome and the Ferguson-Smith syndrome may be allelic disorders. The tumors in Ferguson-Smith syndrome are variably described as self-healing squamous cell carcinomas or as keratoacanthomas. Developmental defects do not occur in that syndrome. The most important risk factor for basal cell carcinomas in persons who are not genetically predisposed is ultraviolet exposure. UV-radiation contributes to carcinogenesis through production of C-to-T transitions in DNA and not by induction of large-scale chromosome rearrangements. It is likely that in tumors not showing allelic loss both copies of the gene on chromosome 9 have undergone point mutations. Furthermore, one might predict that basal cell carcinomas related to exposure to ionizing radiation, which causes chromosome breaks, would more often show allelic loss. Chenevix-Trench et al. (1993) demonstrated that, in Australasian pedigrees, the NBCCS gene was linked to markers in the same region of chromosome 9 with no evidence of significant heterogeneity. Loss of heterozygosity (LOH) was detected in half of sporadic cases of basal cell carcinoma, a rate significantly higher than that in other skin lesions used as controls. This suggests that sporadic basal cell carcinomas may be due to mutation in the same gene.

Levanat et al. (1996) suggested a 2-hit mechanism for neoplasia in the Gorlin syndrome according to the Knudson model. The authors concluded that the causative gene probably functions as a tumor suppressor based on deletion of the relevant region of 9q found in many neoplasms occurring in the syndrome. They suggested that some of the associated developmental defects may also arise through a 2-hit mechanism. Like neoplasms in familial cancer predisposition syndromes, the jaw cysts in Gorlin syndrome are multiple and appear in a random pattern, but similar defects are seen occasionally as a isolated finding in the general population. Levanat et al. (1996) examined a series of chromosome 9 polymorphisms in abnormal and matched constitutional tissue and found that the lining of the jaw cysts lost the normal copy of the Gorlin syndrome region while retaining the mutant copy. These results suggested to them that a somatic mutation of a particular gene in an embryonic or fetal cell leads to abnormal migration, or differentiation, or perhaps failure to undergo programmed cell death, manifested later as a developmental defect.

Shimkets et al. (1996) reported cytogenetic and molecular characterization of germline deletions in one patient with a chromosome 9q22 deletion and a second patient with a deletion of 9q22-q31. Both had typical features of Gorlin syndrome plus additional findings. Shimkets et al. (1996) noted that the fact that Gorlin syndrome can be caused by null mutations (deletions) has several implications. In conjunction with previous analysis of allelic loss in tumors, this study provided evidence that associated neoplasms arise with homozygous inactivation of the gene.

(Although Gorlin has described many syndromes, several of which have been given his name, none is more intimately connected with his name than the basal cell nevus syndrome. See Gorlin (1993) for an autobiography.)

Wicking et al. (1997) screened 71 unrelated individuals with NBCCS for mutations in the PTCH exons. They identified 28 mutations that were distributed throughout the entire gene and predicted that 86% would cause protein truncation. Wicking et al. (1997) identified 3 families bearing identical genotypes with variable phenotypes. From this they concluded that phenotypic variability in NBCCS is a complex genetic event. No phenotype/genotype correlation between the position of the truncation mutations and major clinical features was evident. Wicking et al. (1997) concluded that the preponderance of truncation mutations in the germline of NBCCS patients suggests that the developmental defects associated with NBCCS are likely due to haploinsufficiency.

Bale (1997) reviewed factors contributing to the variable expressivity of PTCH mutations in NBCCS. He reported that clinical features of NBCCS syndrome differ more among families than between families. Shimkets et al. (1996) reported 2 patients with small interstitial deletions on chromosome 9q which involved the PTCH gene. Phenotypes of the 2 patients differed with respect to several key findings (e.g., occurrence of jaw cysts, palmar pits, and skeletal abnormalities). Bale (1997) noted that developmental defects may also arise through a 2-hit mechanism and he reviewed evidence for loss of the normal allele in epithelial cells lining jaw cysts. Bale (1997) noted the absence of genotype/phenotype correlation in NBCCS and concluded that modifying genes and germline variants resulting in hypomorphic or hypermorphic alleles may play an important role in determining the phenotype.

Kimonis et al. (1997) reviewed the findings in 105 individuals with Gorlin syndrome examined at the NIH since 1985. The data derived from 48 males and 57 females ranging in age from 4 months to 87 years. They found that 80% of whites (71 of 90) and 38% of African Americans (5 of 13) had at least 1 basal cell carcinoma, with the first tumor occurring at a mean age of 23 years (median = 20) and 21 years (median = 20), respectively. Excluding individuals exposed to radiation therapy, the number of BCCs ranged from 1 to more than 1000 (median = 8) and 1 to 3 (median = 2), respectively, in the 2 ethnic groups. Jaw cysts occurred in 74% (78 of 105) with the first tumor occurring in 80% by the age of 20 years. The number of total jaw cysts ranged from 1 to 28 (median = 3). Palmar pits and plantar pits were seen in 87%. Ovarian fibromas were diagnosed by ultrasound in 9 of 52 (17%) at a mean age of 30 years. Medulloblastoma occurred in 4 patients at a mean age of 2.3 years. Three patients had cleft lip or palate. Physical findings included 'coarse face' in 54%, relative macrocephaly in 50%, hypertelorism in 42%, frontal bossing in 27%, pectus deformity in 13%, and Sprengel deformity in 11%. Important radiologic signs included calcification of the falx cerebri in 65%, of the tentorium cerebelli in 20%, bridged sella in 68%, bifid ribs in 26%, hemivertebrae in 15%, fusion of the vertebral bodies in 10%, and flame-shaped lucencies of the phalanges, metacarpal, and carpal bones of the hands in 30%. Several traits previously considered components of the syndrome (including short fourth metacarpal, scoliosis, cervical ribs, and spina bifida occulta) were not found to be significantly increased in the affected individuals. Kimonis et al. (1997) tabulated major criteria and minor criteria which could be used in making the diagnosis of Gorlin syndrome by the presence of 2 major or 1 major and 2 minor criteria. They suggested that a child who is at risk for having inherited the gene should have careful physical examination at birth for pits and other minor features and radiologic evaluation with rib, skull, and spine films to look for skeletal abnormalities. In their study, the risk of developing a medulloblastoma was greatest between 2 and 3 years; however, such had been reported up to age 7 years (Evans et al., 1991). Evans et al. (1993) reported an incidence of 3% of cardiac fibroma in Gorlin syndrome. Kimonis et al. (1997) suggested that advising patients to reduce exposure to UV radiation may lessen their risk of developing BCCs. Radiation therapy should also be avoided, but this is understandably difficult when there is a need to make the diagnosis and provide for proper surveillance.

Korczak et al. (1997) reported the case of an African-American boy found at the age of 2 years to have medulloblastoma for which he underwent surgery, chemotherapy, and craniospinal irradiation. At the age of 6 years he was diagnosed with Gorlin syndrome following his presentation with a large odontogenic keratocyst of the mandible, pits of the palms and soles, and numerous BCCs in the area of the back and neck that had been irradiated previously. Examination of other relatives showed that the mother also had Gorlin syndrome but was more mildly affected; in particular, she had no BCCs. The authors stated that this case illustrates complex gene-environment interaction. Increased skin pigmentation in African-Americans is presumably protective against UV, but not ionizing, radiation. The case and similar cases in the literature illustrated the importance of considering Gorlin syndrome in the differential diagnosis of any patient who presents with a medulloblastoma, especially before the age of 5 years, and of examining other close relatives for signs of Gorlin syndrome to determine the patient's at-risk status. Korczak et al. (1997) also concluded that for individuals who are radiosensitive, protocols that utilize chemotherapy in lieu of radiotherapy should be considered.

See also Anderson and Cook (1966); Anderson et al. (1967); Bale et al. (1995); Dahl et al. (1976); Gorlin et al. (1976); Gorlin and Sedano (1971); Gundlach and Kiehn (1979); Howell and Mehregan (1970); Lorenz and Fuhrmann (1978); Satinoff and Wells (1969); Southwick and Schwartz (1979)

Anderson, D. E.: Linkage analysis of the nevoid basal cell carcinoma syndrome. Ann. Hum. Genet. 32: 113-123, 1968.

Anderson, D. E.; Cook, W. A.: Jaw cysts and basal cell nevus syndrome. J. Oral Surg. 24: 15-26, 1966.

Anderson, D. E.; Taylor, W. B.; Falls, H. F.; Davidson, R. T.: The nevoid basal cell carcinoma syndrome. Am. J. Hum. Genet. 19: 12-22, 1967.

Bale, A. E.: Variable expressivity of patched mutations in flies and humans. (Editorial) Am. J. Hum. Genet. 60: 10-12, 1997.

Bale, A. E.; Bale, S. J.; Mulvihill, J. J.: Linkage between the nevoid basal cell carcinoma syndrome (NBCCS) gene and chromosome 1 markers. (Abstract) Am. J. Hum. Genet. 37: A44 only, 1985.

Bale, A. E.; Gailani, M. R.; Leffell, D. J.: The Gorlin syndrome gene: a tumor suppressor active in basal cell carcinogenesis and embryonic development. Proc. Assoc. Am. Phys. 107: 253-257, 1995.

Berlin, N. I.; Van Scott, E. J.; Clendenning, W. E.; Archard, H. O.; Block, J. B.; Witkop, C. J., Jr.; Haynes, H. A.: Basal cell nevus syndrome. Ann. Intern. Med. 64: 403-421, 1966.

Bialer, M. G.; Gailani, M. R.; McLaughlin, J. A.; Petrikovsky, B.; Bale, A. E.: Prenatal diagnosis of Gorlin syndrome. (Letter) Lancet 344: 477 only, 1994.

Cawson, R. A.; Kerr, G. A.: The syndrome of jaw cysts, basal cell tumours and skeletal anomalies. Proc. Roy. Soc. Med. 57: 799-801, 1964.

Chenevix-Trench, G.; Wicking, C.; Berkman, J.; Sharpe, H.; Hockey, A.; Haan, E.; Oley, C.; Ravine, D.; Turner, A.; Goldgar, D.; Searle, J.; Wainwright, B.: Further localization of the gene for nevoid basal cell carcinoma syndrome (NBCCS) in 15 Australasian families: linkage and loss of heterozygosity. Am. J. Hum. Genet. 53: 760-767, 1993.

Clendenning, W. E.; Herdt, J. R.; Block, J. B.: Ovarian fibromas and mesenteric cysts: their association with hereditary basal cell cancer of the skin. Am. J. Obstet. Gynec. 87: 1008-1012, 1963.

Cramer, H.; Niederdellmann, H.: Cerebral gigantism associated with jaw cyst basal cell naevoid syndrome in two families. Arch. Psychiat. Nervenkr. 233: 111-124, 1983.

Dahl, E.; Kreiborg, S.; Jensen, B. L.: Craniofacial morphology in the nevoid basal cell carcinoma syndrome. Int. J. Oral Surg. 33: 300-303, 1976.

Evans, D. G. R.; Farndon, P. A.; Burnell, L. D.; Rao Gattamaneni, H.; Birch, J. M.: The incidence of Gorlin syndrome in 173 consecutive cases of medulloblastoma. Brit. J. Cancer 64: 959-961, 1991.

Evans, D. G. R.; Ladusans, E. J.; Rimmer, S.; Burnell, L. D.; Thakker, N.; Farndon, P. A.: Complications of the naevoid basal cell carcinoma syndrome: results of a population based study. J. Med. Genet. 30: 460-464, 1993.

Evans, D. G. R.; Sims, D. G.; Donnai, D.: Family implications of neonatal Gorlin's syndrome. Arch. Dis. Child. 66: 1162-1163, 1991.

Farndon, P. A.; Del Mastro, R. G.; Evans, D. G. R.; Kilpatrick, M. W.: Location of gene for Gorlin syndrome. Lancet 339: 581-582, 1992.

Farndon, P. A.; Morris, D. J.; Hardy, C.; McConville, C. M.; Weissenbach, J.; Kilpatrick, M. W.; Reis, A.: Analysis of 133 meioses places the genes for nevoid basal cell carcinoma (Gorlin) syndrome and Fanconi anemia group C in a 2.6-cM interval and contributes to the fine map of 9q22.3. Genomics 23: 486-489, 1994.

Farndon, P. A.; Simmons, J.: Linkage analysis of the naevoid basal cell carcinoma syndrome (NBCCS) and chromosome 1 markers. (Abstract) Cytogenet. Cell Genet. 46: 612 only, 1987.

Featherstone, T.; Taylor, A. M. R.; Harnden, D. G.: Studies on the radiosensitivity of cells from patients with basal cell naevus syndrome. Am. J. Hum. Genet. 35: 58-66, 1983.

Fletcher, J. A.; Morton, C. C.: Basal cell nevus syndrome: cytogenetic evidence for a genetic origin in common with familial adenomatous polyposis. (Abstract) Am. J. Hum. Genet. 43: A23 only, 1988.

Gailani, M. R.; Bale, S. J.; Leffell, D. J.; DiGiovanna, J. J.; Peck, G. L.; Poliak, S.; Drum, M. A.; Pastakia, B.; McBride, O. W.; Kase, R.; Greene, M.; Mulvihill, J. J.; Bale, A. E.: Developmental defects in Gorlin syndrome related to a putative tumor suppressor gene on chromosome 9. Cell 69: 111-117, 1992.

Gibbs, P. M.; Stevens, P. R.; Garson, O. M.: The multiple basal cell nevus syndrome: a cytogenetic study of six cases. Cancer Genet. Cytogenet. 20: 369-370, 1986.

Goldstein, A. M.; Pastakia, B.; DiGiovanna, J. J.; Poliak, S.; Santucci, S.; Kase, R.; Bale, A. E.; Bale, S. J.: Clinical findings in two African-American families

with nevoid basal cell carcinoma syndrome (NBCC). Am. J. Med. Genet. 50: 272-281, 1994.
Gorlin, R. J.: Personal Communication. Minneapolis, Minn. 1982.
Gorlin, R. J.: Nevoid basal-cell carcinoma syndrome. Medicine 66: 98-113, 1987.
Gorlin, R. J.: From oral pathology to craniofacial genetics. Am. J. Med. Genet. 46: 317-334, 1993.
Gorlin, R. J.; Goltz, R. W.: Multiple nevoid basal-cell epithelioma, jaw cysts and bifid rib: a syndrome. New Eng. J. Med. 262: 908-912, 1960.
Gorlin, R. J.; Pindborg, J. J.; Cohen, M. M., Jr.: Syndromes of the Head and Neck. New York: Blakiston Division, McGraw-Hill (pub.) (2nd ed.): 1976. Pp. 520-526. Note: Multiple nevoid basal cell carcinoma syndrome. . . . .
Gorlin, R. J.; Sedano, H. O.: The multiple nevoid basal cell carcinoma syndrome revisited. Birth Defects Orig. Art. Ser. VII(8): 140-148, 1971.
Gundlach, K. K. H.; Kiehn, M.: Multiple basal cell carcinoma and keratocysts—the Gorlin and Goltz syndrome. J. Maxillofac. Surg. 7: 299-307, 1979.
Hahn, H.; Wicking, C.; Zaphiropoulos, P. G.; Gailani, M. R.; Shanley, S.; Chidambaram, A.; Vorechovsky, I.; Holmberg, E.; Unden, A. B.; Gillies, S.; Negus, K.; Smyth, I.; Pressman, C.; Leffell, D. J.; Gerrard, B.; Goldstein, A. M.; Dean, M.; Toftgard, R.; Chenevix-Trench, G.; Wainwright, B.; Bale, A. E. Mutations of the human homolog of Drosophila patched in the nevoid basal cell carcinoma syndrome. Cell 85: 841-851, 1996.
Heimler, A.; Friedman, E.; Rosenthal, A.: Naevoid basal cell carcinoma syndrome and Charcot-Marie-Tooth disease. J. Med. Genet. 15: 288-291, 1978.
Herzberg, J. J.; Wiskemann, A.: Die fuenfte Phakomatose. Basalzellnaevus mit familiaerer Belastung und Medulloblastom. Dermatologica 126: 106-123, 1963.
Holmes, L. B.: Cabot case. New Eng. J. Med. 294: 772-777, 1976.
Holubar, K.; Matras, H.; Smalik, A. V.: Multiple palmar basal cell epitheliomas in basal cell nevus syndrome. Arch. Derm. 101: 679-682, 1970.
Howell, J. B.; Mehregan, A. H.: Pursuit of the pits in the nevoid basal cell carcinoma syndrome. Arch. Derm. 102: 586-597, 1970.
Jamieson, S. W.; Gaudiani, V. A.; Reitz, B. A.; Oyer, P. E.; Stinson, E. B.; Shumway, N. E.: Operative treatment of unresectable tumor of the left ventricle. J. Thorac. Cardiovasc. Surg. 81: 797-799, 1981.
Johnson, R. L.; Rothman, A. L.; Xie, J.; Goodrich, L. V.; Bare, J. W.; Bonifas, J. M.; Quinn, E. H.; Myers, R. M.; Cox, D. R.; Epstein, E. H., Jr.; Scott, M. P.: Human homolog of patched, a candidate gene for the basal cell nevus syndrome. Science 272: 1668-1671, 1996.
Jones, K. L.; Smith, D. W.; Harvey, M. A. S.; Hall, B. D.; Quan, L.: Older paternal age and fresh gene mutation: data on additional disorders. J. Pediat. 86: 84-88, 1975.
Jones, K. L.; Wolf, P. L.; Jensen, P.; Dittrich, H.; Benirschke, K.; Bloor, C.: The Gorlin syndrome: a genetically determined disorder associated with cardiac tumor. Am. Heart J. 111: 1013-1015, 1986.
Kimonis, V. E.; Goldstein, A. M.; Pastakia, B.; Yang, M. L.; Kase, R.; DiGiovanna, J. J.; Bale, A. E.; Bale, S. J.: Clinical manifestations in 105 persons with nevoid basal cell carcinoma syndrome. Am. J. Med. Genet. 69: 299-308, 1997.
Korczak, J. F.; Brahim, J. S.; DiGiovanna, J. J.; Kase, R. G.; Wexler, L. H.; Goldstein, A. M.: Nevoid basal cell carcinoma syndrome with medulloblastoma in an African-American boy: a rare case illustrating gene-environment interaction. Am. J. Med. Genet. 69: 309-314, 1997.
Levanat, S.; Gorlin, R. J.; Fallet, S.; Johnson, D. R.; Fantasia, J. E.; Bale, A. E.: A two-hit model for developmental defects in Gorlin syndrome. Nature Genet. 12: 85-87, 1996.
Levine, D. J.; Robertson, D. B.; Varma, V. A.: Familial subconjunctival epithelial cysts associated with the nevoid basal cell carcinoma syndrome. (Letter) Arch. Derm. 123: 23-24, 1987.
Lile, H. A.; Rogers, J. F.; Gerald, B.: The basal cell nevus syndrome. Am. J. Roentgen. 103: 214-217, 1968.
Lorenz, R.; Fuhrmann, W.: Familial basal cell nevus syndrome. Hum. Genet. 44: 153-163, 1978.
McConville, C. M.; Taylor, A. M. R.; Byrd, P. J.; Woolgar, J. A.; Hollis, R.: Basal cell naevus syndrome and N-ras polymorphism. (Abstract) Cytogenet. Cell Genet. 46: 660 only, 1987.
Ottinger, L. W.; Vickery, A. L., Jr.: Case records of the Massachusetts General Hospital (Case 10-1986). New Eng. J. Med. 314: 700-706, 1986.
Reis, A.; Kuster, W.; Linss, G.; Gebel, E.; Hamm, H.; Fuhrmann, W.; Wolff, G.; Groth, W.; Gustafson, G.; Kuklik, M.; Burger, J.; Wegner, R. D.; Neitzel, H.: Localisation of gene for the naevoid basal-cell carcinoma syndrome. (Letter) Lancet 339: 617 only, 1992.
Satinoff, M. I.; Wells, C.: Multiple basal cell naevus syndrome in ancient Egypt. Med. Hist. 13: 294-297, 1969.
Schwartz, R. A.: Basal-cell-nevus syndrome and gastrointestinal polyposis. (Letter) New Eng. J. Med. 299: 49 only, 1978.
Shanley, S.; Ratcliffe, J.; Hockey, A.; Haan, E.; Oley, C.; Ravine, D.; Martin, N.; Wicking, C.; Chenevix-Trench, G.: Nevoid basal cell carcinoma syndrome: review of 118 affected individuals. Am. J. Med. Genet. 50: 282-290, 1994.
Shimkets, R.; Gailani, M. R.; Siu, V. M.; Yang-Feng, T.; Pressman, C. L.; Levanat, S.; Goldstein, A.; Dean, M.; Bale, A. E.: Molecular analysis of chromosome 9q deletions in two Gorlin syndrome patients. Am. J. Hum. Genet. 59: 417-422, 1996.
Southwick, G. J.; Schwartz, R. A.: The basal cell nevus syndrome: disasters occurring among a series of 36 patients. Cancer 44: 2294-2305, 1979.
Totten, J. R.: The multiple nevoid basal cell carcinoma syndrome: report of its occurrence in four generations of a family. Cancer 46: 1456-1462, 1980.
Waaler, P. E.; Aarskog, D.: Syndrome of hydrocephalus, costovertebral dysplasia and Sprengel anomaly with autosomal dominant inheritance. Neuropediatrics 11: 291-297, 1980.
Wicking, C.; Berkman, J.; Wainwright, B.; Chenevix-Trench, G.: Fine genetic mapping of the gene for nevoid basal cell carcinoma syndrome. Genomics 22: 505-511, 1994.
Wicking, C.; Shanley, S.; Smyth, I.; Gillies, S.; Negus, K.; Graham, S.; Suthers, G.; Haites, N.; Edwards, M.; Wainwright, B.; Chenevix-Trench, G.: Most germ-line mutations in the nevoid basal cell carcinoma syndrome lead to a premature termination of the PATCHED protein, and no genotype-phenotype correlations are evident. Am. J. Hum. Genet. 60: 21-26, 1997.

HOMOLOG OF DROSOPHILA PATCHED (*601309)

JJM NOTE

This entity goes further into the nevoid basal cell carcinoma syndrome with the major association of basal cell carcinoma, medulloblastoma, and ovarian fibroma.

The Drosophila 'patched' (ptc) gene encodes a transmembrane protein which represses transcription in specific cells of genes encoding members of the TGF beta (see 190180) and Wnt (164820) families of signaling proteins. Vertebrate homologs of ptc have been identified in mice chickens and zebra fish. Johnson et al. (1996) reported the isolation and mapping of the human homolog of the Drosophila (ptc) gene. They also reported that this gene is a candidate gene for the basal cell nevus syndrome (BCNS; 109400), also called Gorlin syndrome. Johnson et al. (1996) cloned the human PTC gene by screening a human lung cDNA library with mouse ptc cDNA clones. They assembled 5.1 kb of contiguous sequence containing a 4.5-kb open reading frame that encodes a 1447-amino acid protein. The predicted amino acid sequence has 96% identity to mouse and a 40% identity to Drosophila ptc proteins. The human PTC protein is predicted to contain 12 hydrophobic membrane-spanning domains and 2 large hydrophilic extracellular loops. Johnson et al. (1996) mapped the human PTC gene to chromosome 9q22.3 by radiation hybrid analysis. PTCH is the preferred symbol.

Johnson et al. (1996) identified 2 mutations in the PTC coding sequence that were associated with BCNS (601309.0001 and 610309.0002). They also examined the DNA of 12 sporadic basal cell carcinomas (BCCs) and found a point mutation that resulted in a leu175-to-phe amino acid substitution in the predicted first extracellular loop of the protein. Leucine-175 is in exon 3 and is conserved in all reported ptc sequences of mouse, Drosophila, and chicken.

Hahn et al. (1996) likewise isolated a human sequence with strong homology to the Drosophila segment polarity gene 'patched' from a YAC and cosmid contig of the nevoid basal cell carcinoma (NBCCS) region on chromosome 9q22.3. Their mapping data placed the human PTC gene between FACC (227645) and the marker D9S287. The physical map distance between FACC and PTC is less than 650 kb and the map distance between PTC and D9S287 is less than 290 kb. Hahn et al. (1996) defined the intron exon boundaries of the PTC gene and reported that the PTC gene contains 23 exons spanning approximately 34 kb. They noted that there are at least 3 different forms of the PTC protein present in mammalian cells; the ancestral form and 2 human forms. The first in-frame methionine codon for 1 of the forms is in the third exon. The other human form of PTC contains an open reading frame that extends through to the 5-prime end and may be initiated by upstream sequences which have not yet been isolated. Hahn et al.(1996) pointed out that the identification of several potential forms of the PTC protein provides a mechanism whereby a single PTC gene could play a role in different pathways. They stressed that determination of the regulation of different splice forms of PTC mRNA may shed light on the apparent role of the gene in embryonic development and growth control in adult cells.

Hahn et al. (1996) used exon sequence and SSCP to search for mutations in the PTC gene in patients with nevoid basal cell carcinoma syndrome (NBCCS). They identified 4 different germline mutations (601309.0003, 601309.0004, 601309.0005, and 601309.0006) in unrelated familial cases of NBCCS. They also identified 2 germline mutations in sporadic cases of NBCCS (601309.0007, 601309.0008). In addition, they identified 2 somatic mutations in tumor DNA derived from basal cell carcinomas.

Using SSCP to screen human 'patched' in 37 sporadic BCCs in humans, Gailani et al. (1996) detected mutations in one third of the tumors. Direct sequencing of 2 BCCs without SSCP variants revealed mutations in those tumors as well, suggesting to the investigators that inactivation of 'patched' is probably a necessary step in BCC development. By Northern blots and RNA in situ hybridization Gailani et al. (1996) showed that 'patched' is expressed at high levels in tumor cells but not normal skin, suggesting to them that mutational inactivation of the gene leads to overexpression of mutant transcript owing to failure of a negative feedback mechanism. Nine tumors with loss of heterozygosity (LOH) had mutations of the remaining allele and 2 tumors without LOH had 2 inactivating mutations. Basal cell carcinoma is the most common cancer in humans. Epidemiologic studies show a correlation between exposure to sunlight and BCCs, but the association is less striking than that of squamous cell carcinoma of the skin and sunlight. In 15 of the 16 mutations identified in this study, the tumors were from sun-exposed sites. Seven mutations were typical of ultraviolet-B damage: C-T substitutions at dypyrimidine sites, including 2 CC-to-TT double-base mutations. Gailani et al. (1996) noted that the other 8 mutations, including deletions, transversion point mutations, and double-base substitutions other than CC-to-TT, can be caused by ultraviolet-B but are not UVB-specific.

To assess the role of Ptc in cell physiology and development, Marigo et al. (1996) expressed the chick patched gene in Xenopus laevis oocytes (oocytes do not express endogenous Ptc). Protein of the size expected for Ptc was detected 48 hours after injection. They then performed binding assays on injected, uninjected, and control-injected oocytes using the N-terminal fragment (N-Shh) of human sonic hedgehog protein (SHH; 600725). The binding assay showed that labeled N-Shh protein could bind to Ptc-injected oocytes, but not to the control oocytes. Injected oocytes bound human N-Shh produced in E. coli and mouse N-Shh produced in the baculovirus system. Marigo et al. (1996) demonstrated direct interaction between Ptc and Shh using co-immunoprecipitation studies. They also showed that the 2 extracellular loops of the Ptc protein are necessary for binding and that binding also requires that the Ptc protein be glycosylated. Marigo et al. (1996) proposed that Ptc does not carry out signaling to the cell directly but that an additional molecule is involved, namely the 7-transmembrane protein Smo (601500).

Independently and simultaneously, Stone et al. (1996) concluded that the Ptc gene encodes a candidate receptor for Shh by showing that epitope-tagged N-Shh binds specifically to human embryonic kidney 293 cells expressing mouse Ptc. Ptc also could be immunoprecipitated by N-Shh-IgG. The authors calculated a K(d) of 460 picoM for binding of N-Shh and mouse Ptc. By expression of genes in 293 cells with subsequent lysis and immunoprecipitation, Stone et al. (1996) showed that Ptc, Smo, and Shh form a physical complex in vivo and that a Smo-Shh complex does not form in the absence of Ptc. They proposed that the Hedgehog system may provide mitogenic or differentiative signals to basal cells in the skin throughout life. They also raised the possibility that BCNS and BCC might result from constitutive activation of Smo which becomes oncogenic after its release from inhibition by Ptc.

On the basis of their studies in Drosophila, Chen and Struhl (1996) presented evidence that Ptc acts as a receptor for hedgehog (Hh) proteins. They suggested a novel signal transduction mechanism in which Hh proteins bind to Ptc or to a Ptc-Smo complex and thereby induces Smo activity. Their results showed further that Ptc limits the range of Hh action and that the high levels of Ptc induced by Hh serve to sequester any free Hh and thereby create a barrier to its further movement.

Chidambaram et al. (1996) used the Jackson Laboratory Backcross DNA panel map service to map the murine PTCH homolog, Ptc, to mouse chromosome 13. Ptc maps close to the murine Facc locus (0 recombinants in 188 meioses). They noted that mouse mutations such as flexed tail (f), purkinje cell degeneration (pcd), and mesenchymal dysplasia (mes) which involve abnormal development of skeletal and neural tissues are also located in this region of chromosome 13 and may be allelic to Ptc.

Wicking et al. (1997) screened 71 unrelated individuals with NBCCS for mutations in the PTCH exons. They identified 28 mutations that were distributed throughout the entire gene and predicted that 86% would cause protein truncation. Wicking et al. (1997) identified 3 families bearing identical genotypes with variable phenotypes. From this they concluded that phenotypic variability in NBCCS is a complex genetic event. No phenotype/genotype correlation between the position of the truncation mutations and major clinical features was evident. Wicking et al. (1997) concluded that the preponderance of truncation mutations in the germline of NBCCS patients suggests that the developmental defects associated with NBCCS are likely due to haploinsufficiency. They noted that studies in Drosophila indicate that developmental pathways are particularly sensitive to dosage effects, with absolute levels of certain proteins being critical to the correct functioning of such pathways.

Bale (1997) reviewed factors contributing to the variable expressivity of PTCH mutations in NBCCS. He reported that clinical features of NBCCS syndrome differ more among families than between families. Shimkets et al. (1996) reported 2 patients with small interstitial deletions on chromosome 9q which involved the PTCH gene. Phenotypes of the 2 patients differed with respect to several key findings (e.g., occurrence of jaw cysts, palmar pits, and skeletal abnormalities). Bale (1997) noted that developmental defects may also arise through a 2-hit mechanism and he reviewed evidence for loss of the normal allele in epithelial cells lining jaw cysts. Bale (1997) noted the absence of genotype/phenotype correlation in NBCCS and concluded that modifying genes and germline variants resulting in hypomorphic or hypermorphic alleles may play an important role in determining the phenotype.

Goodrich et al. (1997) investigated the function of the ptc gene by inactivating the gene in mice by homologous recombination in ES cells. Mice homozygous for the mutation died during embryogenesis and were found to have open and overgrown neural tubes. Two sonic hedgehog (Shh) target genes, ptc itself and Gli (165220), were derepressed in the ectoderm and mesoderm but not in the endoderm. Shh targets that are, under normal conditions, transcribed ventrally were aberrantly expressed in dorsal and lateral neural tube cells. Goodrich et al. (1997) concluded that ptc is essential for repression of genes that are locally activated by Shh. Mice heterozygous for the ptch mutation were larger than normal, and a subset of them developed hind limb defects (including extra digits, syndactyly and soft tissue tumors) or cerebellar medulloblastomas, abnormalities also seen in patients with the basal cell nevus syndrome. The authors speculated that their failure to observe basal cell carcinomas in the heterozygous mice may have been because somatic inactivation of the second ptc gene is required as it is in human basal cell carcinomas.

Approximately 5% of patients with Gorlin syndrome develop medulloblastoma in the first few years of life, and 10% of patients with medulloblastoma diagnosed at age 2 years or under have Gorlin syndrome. Cowan et al. (1997) found that 1 out of 3 unrelated patients with medulloblastoma complicated by Gorlin syndrome had lost the wildtype allele on 9q, indicating that the Gorlin locus probably acts a tumor suppressor in the development of this tumor. They also confirmed this role in a basal cell carcinoma from the same individual. They suggested that Gorlin syndrome is more common than previously recognized and may not be diagnosed on clinical grounds alone even in middle life. In their Table 1 they provided diagnostic criteria for Gorlin syndrome. Five major and 6 minor criteria were listed. A positive diagnosis can be made, they suggested, on the basis of 2 major or 1 major and 2 minor criteria. Major criteria included multiple (more than 2) BCCs or 1 before age 30 years, or more than 10 basal cell nevi; any odontogenic keratocyst or polyostotic bone cyst; palmar and plantar pits; ectopic calcification; and a family history of NBCCS. Minor criteria included rib or vertebral anomalies; large head circumference with frontal bossing; cardiac or ovarian fibroma; and lymphomesenteric cysts. Falx calcification under the age of 20 years and palmar or plantar pits were among the major criteria.

Studying patients who presented with multiple odontogenic keratocysts, Lench et al. (1997) identified 5 novel germline mutations in PTCH. Four mutations caused premature stop codons and 1 resulted in an amino acid substitution toward the C terminus of the predicted protein.

Wicking et al. (1997) presented an additional 4 novel PTCH mutations in nevoid basal cell carcinoma syndrome, having previously reported 28 mutations. They identified 8 individuals who carried a de novo mutation in the PTCH gene. In 5 of these cases, clinical and radiologic examination had not unequivocally ruled out a diagnosis in one of the parents. On the basis of the findings in the parents, Wicking et al. (1997) presented the following review of diagnostic criteria for this syndrome: (1) although palmar and plantar pitting is pathognomonic of NBCCS, it can be falsely reported; (2) a caution must be exercised in using 'multiple BCC' as a diagnostic criterion, especially in areas of high sun exposure; (3) high arched palate, a minor diagnostic anomaly, is quite common in the general population; and (4) a dense calcification of the falx was not found in these parents, but is an almost invariable finding in adults with mutations in the PTCH gene.

Aszterbaum et al. (1998) screened the 23 exons of the PTCH gene for mutations by use of single-strand conformation polymorphism analysis of DNA from 86 basal cell nevus syndrome probands, 26 sporadic basal cell carcinomas, and 7 basal cell nevus syndrome-associated basal cell carcinomas. This screen identified mutations located in 8 exons in 13 of the basal cell nevus syndrome patients and in 3 of the tumors. The most common mutations were frameshifts resulting in premature chain termination. Of 26 sporadic basal cell carcinomas screened, 11 showed loss of heterozygosity at 1 or more of the polymorphic markers examined in the PTCH gene region. Of these 11, 3 tumors were found to have PTCH gene mutations, each in a different exon of the gene. One of these was predicted to result in an amino acid substitution, 1 in a premature stop codon, and 1 in a frameshift. The latter 2 mutations caused premature chain termination. These 3 mutations were not those considered to be characteristic of UV-induced changes.

Aszterbaum, M.; Rothman, A.; Johnson, R. L.; Fisher, M.; Xie, J.; Bonifas, J. M.; Zhang, X.; Scott, M. P.; Epstein, E. H., Jr. Identification of mutations in the human PATCHED gene in sporadic basal cell carcinomas and in patients with the basal cell nevus syndrome. J. Invest. Derm. 110: 885-888, 1998.

Bale, A. E.: Variable expressivity of patched mutations in flies and humans. (Editorial) Am. J. Hum. Genet. 60: 10-12, 1997.

Chen, Y.; Struhl, G.: Dual roles for Patched in sequestering and transducing Hedgehog. Cell 87: 553-563, 1996.

Chidambaram, A.; Hahn, H.; Allikmets, R.; Bale, A. E.; Dean, M.: The murine homolog (Ptc) of the Drosophila segment polarity gene patched and the human NBCCS locus maps to mouse chromosome 13. (Abstract) Am. J. Hum. Genet. 59: A300 only, 1996.

Cowan, R.; Hoban, P.; Kelsey, A.; Birch, J. M.; Gattamaneni, R.; Evans, D. G. R.: The gene for the naevoid basal cell carcinoma syndrome acts as a tumor-suppressor gene in medulloblastoma. Brit. J. Cancer 76: 141-145, 1997.
Gailani, M. R.; Stahle-Backdahl, M.; Leffell, D. J.; Glynn, M.; Zaphiropoulos, P. G.; Pressman, C.; Unden, A. B.; Dean, M.; Brash, D. E.; Bale, A. E.; Toftgard, R.: The role of the human homologue of Drosophila patched in sporadic basal cell carcinomas. Nature Genet. 14: 78-81, 1996.
Goodrich, L. V.; Milenkovic, L.; Higgins, K. M.; Scott, M. P.: Altered neural cell fates and medulloblastoma in mouse patched mutants. Science 277: 1109-1113, 1997.
Hahn, H.; Wicking, C.; Zaphiropoulos, P. G.; Gailani, M. R.; Shanley, S.; Chidambaram, A.; Vorechovsky, I.; Holmberg, E.; Unden, A. B.; Gillies, S.; Negus, K.; Smyth, I.; Pressman, C.; Leffell, D. J.; Gerrard, B.; Goldstein, A. M.; Dean, M.; Toftgard, R.; Chenevix-Trench, G.; Wainwright, B.; Bale, A. E.: Mutations of the human homolog of Drosophila patched in the nevoid basal cell carcinoma syndrome. Cell 85: 841-851, 1996.
Johnson, R. L.; Rothman, A. L.; Xie, J.; Goodrich, L. V.; Bare, J. W.; Bonifas, J. M.; Quinn, E. H.; Myers, R. M.; Cox, D. R.; Epstein, E. H., Jr.; Scott, M. P.: Human homolog of patched, a candidate gene for the basal cell nevus syndrome. Science 272: 1668-1671, 1996.
Lench, N. J.; Telford, E. A. R.; High, A. S.; Markham, A. F.; Wicking, C.; Wainwright, B. J.: Characterisation of human patched germ line mutations in naevoid basal cell carcinoma syndrome. Hum. Genet. 100: 497-502, 1997.
Marigo, V.; Davey, R. A.; Zuo, Y.; Cunningham, J. M.; Tabin, C. J.: Biochemical evidence that Patched is the Hedgehog receptor. Nature 384: 176-179, 1996.
Shimkets, R.; Gailani, M. R.; Siu, V. M.; Yang-Feng, T.; Pressman, C. L.; Levanat, S.; Goldstein, A.; Dean, M.; Bale, A. E.: Molecular analysis of chromosome 9q deletions in two Gorlin syndrome patients. Am. J. Hum. Genet. 59: 417-422, 1996.
Stone, D. M.; Hynes, M.; Armanini, M.; Swanson, T. A.; Gu, Q.; Johnson, R. L.; Scott, M. P.; Pennica, D.; Goddard, A.; Phillips, H.; Noll, M.; Hooper, J. E.; de Sauvage, F.; Rosenthal, A.: The tumour-suppressor gene patched encodes a candidate receptor for Sonic hedgehog. Nature 384: 129-134, 1996.
Wicking, C.; Gillies, S.; Smyth, I.; Shanley, S.; Fowles, L.; Ratcliffe, J.; Wainwright, B.; Chenevix-Trench, G.: De novo mutations of the patched gene in nevoid basal cell carcinoma syndrome help to define the clinical phenotype. Am. J. Med. Genet. 73: 304-307, 1997.
Wicking, C.; Shanley, S.; Smyth, I.; Gillies, S.; Negus, K.; Graham, S.; Suthers, G.; Haites, N.; Edwards, M.; Wainwright, B.; Chenevix-Trench, G.: Most germ-line mutations in the nevoid basal cell carcinoma syndrome lead to a premature termination of the PATCHED protein, and no genotype-phenotype correlations are evident. Am. J. Hum. Genet. 60: 21-26, 1997.

PROTEUS SYNDROME (176920)

JJM NOTE

Neurofibromatosis 1 has been termed the *quintessential neurocristopathy* (Bolande RP: Neurofibromatosis—the quintessential neurocristopathy: pathogenetic concepts and relationships. Adv. Neurol. 29:67–75, 1981). I think of Proteus syndrome as the quintessential phacomatosis; that is, its features are seen in one or another of all the other classic disorders considered in this heuristic group. In briefing me on the first patient I saw with the disorder, the nurse said that I had my choice of which height to use because the leg hypertrophy was so severe—the legs differed by some 6 cm in length. Telling parents of a newly diagnosed child with neurofibromatosis 1 that Elephant Man had Proteus syndrome and not neurofibromatosis 1 is part of every educational session.

Wiedemann et al. (1983) described a 'new' syndrome in 4 unrelated boys with the combination of manifestations listed. The authors considered the disorder to fall into the category of congenital hamartomatous disorders and to be 'undoubtedly genetically determined,' perhaps as an autosomal dominant. They named the syndrome for the Greek god Proteus, 'the polymorphous,' who could change his shape at will to avoid capture. The disorder might be confused with the Klippel-Trenaunay-Weber syndrome (149000) and with Ollier disease and Maffucci syndrome (166000). Wiedemann et al. (1983) suggested that the patient reported by Temtamy and Rogers (1976) and probably also the patient of Graetz (1928) may have had this disorder. Happle (1986) pointed out that the lesions follow the lines of Blaschko and suggested that the cause is a dominant lethal gene surviving by mosaicism. He suggested this mechanism also for Schimmelpenning-Feuerstein-Mims syndrome (165630) and the McCune-Albright syndrome (174800). (The hypothesis has been proved in the case of the latter condition.) Rescue of a lethal genotype by chimerism with normal embryos (Bennett, 1978) is an experimental model of this mode of inheritance. Smeets et al. (1994) reported a patient with regional manifestations of Proteus syndrome. Major findings included multiple hyperostoses of the calvaria, facial bones, and mandible. Additionally, the patient had a scleral tumor. The observations were interpreted as supporting the hypothesis of somatic mosaicism.

Goodship et al. (1991) presented a possible case of father-son transmission of Proteus syndrome. The son had cranial hemihypertrophy, a lymphangioma, a lipoma, and epidermal nevi. The father had had a large lymphangioma resected from the right side of his face as a child. The possibility that the father was mosaic was raised. Kruger et al. (1993) observed mild Proteus syndrome in a boy whose mother had very mild manifestations. The mentally normal son had mild hypertrophy of the left side of the upper lip and cheek with impaired mimic expression in this region, hypertrophy of the left arm, partial gigantism of the left middle finger, and a large subcutaneous swelling in the upper left abdomen thought clinically and sonographically to be a lipoma. The mother had facial asymmetry with hypertrophy of the right lower cheek and impaired mimic expression in that region. Both the mother and the son had 'distinct venous marking' over the upper thorax.

Costa et al. (1985) reported 2 additional cases; both had abdominal and pelvic lipomatosis. One, a 7-year-old boy, was noted at age 3 to have a conjunctival dermoid. Laparotomy at age 6 for acute abdominal pain showed right iliac fossa lipomatosis and twisted necrotic mesenteric fat as the presumed cause of pain. Some similarities to the Bannayan syndrome (153480) and linear sebaceous nevus syndrome (163200) were noted. Bialer et al. (1988) also discussed the differential diagnosis of the Proteus and Bannayan-Zonana syndromes. In their review of the literature, they found a history of consanguinity in 2 of 36 families with the Proteus syndrome. They stated that overlap among syndromes that include hamartomata as prominent features suggests they may be etiologically or pathogenetically related, perhaps involving abnormal secretion of a growth factor or abnormal tissue or tissue response to a growth factor. Costa et al. (1985) noted that mean paternal age at the time of birth of 10 of the patients was 30 (range 23 to 40), which is probably not significantly elevated. Beluffi et al. (1990) added a fifth case with pelvic lipomatosis. They demonstrated the use of CT scan for revealing pelvic lipomatosis. Burgio and Wiedemann (1984) found that the skin changes are papillomatous epidermal nevi. Viljoen et al. (1987), Clark et al. (1987) and Malamitsi-Puchner et al. (1987) reported 6 cases, 11 cases, and 1 case, respectively. All emphasized lipomatosis as a feature. Dean and Cole (1988) raised the possibility that encephalocraniocutaneous lipomatosis (ECCL) is 'a variant of Proteus syndrome.' On the other hand, McCall et al. (1992) studied 3 patients with an array of defects associated with the Proteus syndrome and related hamartoneoplastic conditions. They compared the findings in these 3 patients with those of 50 others with Proteus syndrome and 9 with ECCL reported in the literature. They concluded that Proteus syndrome and ECCL are distinct identiti es even though some clinical manifestations are shared and a few patients have manifestations of both disorders. Rizzo et al. (1993) reported a patient with manifestations of ECCL and Proteus syndrome: multiple lipomas of the head, neck, paravertebral region, and heart; hyperostosis of the skull; rugose, yellow-brown skin pigmentation; vertebral anomalies; and alopecia. From a comparison with other cases, they concluded that these disorders represent a continuum rather than distinct entities, supporting the concept of somatic mosaicism, lethal in the nonmosaic state. Cohen (1993) reported 2 unusual cases that also supported the concept of somatic mosaicism. In 1 patient, a huge connective tissue nevus covered the chest and abdomen, and hyperostoses of the calvaria were observed. In the other patient, linear verrucous epidermal nevi, epibulbar dermoids, and hyperostoses were found. No enlargement of the limbs or digits occurred, and the plantar surfaces of the feet were normal. Cohen (1993) also reviewed selective aspects of Proteus syndrome, including uncommon neoplasms, pulmonary and renal abnormalities, brain malformations, and types of abnormal growth in the craniofacial skeleton.

Viljoen et al. (1988) described the skin manifestations of the Proteus syndrome in 6 patients. All had marked hypertrophy of the skin of the soles, which the authors believed to be a unique feature of this syndrome. Large epidermal nevi and linear macular lesions with areas of depigmentation and hyperpigmentation were seen in 3 patients. Light microscopy of affected skin from the soles demonstrated elongation of the cytoplasm of basal cells. Samlaska et al. (1989) reported a typical case and reviewed 34 reported cases, all sporadic. Malamitsi-Puchner et al. (1990) provided follow-up on the severely affected child originally reported by Malamitsi-Puchner et al. (1987). Viljoen et al. (1987) mentioned that surgical removal of lymphatic, fatty, or hemangiomatous elements is difficult and results in unsightly scars and keloids. Malamitsi-Puchner et al. (1990) found that striking overgrowth of tissues occurred after surgical operations. Furthermore, the patient, a 4.5 year-old child, developed testicular malignancy. Hotamisligil and Ertogan (1990) described the case of a 9-month-old girl who, in addition to other features, had soft tissue masses in the paravertebral and gluteal areas with aggressive involvement of the spinal canal and a hyperpigmented epidermal nevus with hyperkeratosis on the left half of the body. There was macrodactyly of both feet and the left hand with syndactyly of the third and fourth left fingers. Nephrogenic diabetes insipidus was described in the Proteus syndrome for the first time. Skovby et al. (1993) reported 2 patients who illustrated the 2 ways in which spinal compromise may develop in Proteus syndrome: vertebral anomalies or tumor infiltration. In one patient, spinal stenosis resulted from an angular kyphoscoliosis; in the other, cord compression resulted from infiltration of a paraspinal, intrathoracic angiolipoma. McMullin et al. (1993) described 2 unrelated children with diffuse swelling of one cheek and hypertrophy of the underlying maxilla and mandible. Both developed verrucous pigmented streaks over the area of swelling, and had epilepsy and severe mental subnormality. One of the children had a contralateral hemiplegia, and his condition was progressive. The other child had no focal neurologic signs, and his disease seemed to be nonprogressive. Although the facial appearance suggested encephalocraniocutaneous lipomatosis (ECCL) syndrome, no lipomata could be demonstrated. McMullin et al. (1993) suggested that the 2 cases may represent part of a spectrum of conditions, including the ECCL syndrome and Proteus syndrome.

Haramoto et al. (1995) described a sporadic case of hemifacial hamartomatous hyperplasia consisting of sebaceous nevus-like skin changes, subcutaneous lipomatous mass, cranial bone hyperplasia, and bony change of the meninges. The lesion involved the anterior half of the face and cranial base and was delimited immediately by the midline. This was thought to represent somatic mosaicism for a dominant mutation, lethal in its nonmosaic state, and to be a variant of Proteus syndrome.

Gordon et al. (1995) observed 2 patients with Proteus syndrome who developed neoplasms. Patient 1 had a probable mesothelioma, although papillary carcinoma of the thyroid could not be completely ruled out; the patient, who died suddenly while sleeping at age 5 years, at autopsy had a papillary neoplasm, most likely of mesothelial origin, involving the inferior surface of the diaphragm and infiltrating into the musculature, within the omentum, in the pelvic area, within the scrotum, and within some of the mesenteric lymph nodes. Patient 2 had bilateral ovarian serous cystadenomas with nuclear atypia identified at 6 years and 3 months of age. A right ovarian oophorectomy was performed; invasion of the right fallopian tube was noted. A tabulation of uncommon neoplasms in Proteus syndrome was provided.

Lacombe and Battin (1996) described 2 unrelated children diagnosed at birth as having isolated macrodactyly (155500). Follow-up examination showed development of hemihypertrophy in both cases. Three dorsal angiomas were found in 1 child, a female, at the age of 4 years. The symptoms of both of these patients better fit the diagnostic criteria of Proteus syndrome.

Biesecker et al. (1998) described their experience with 18 patients with a referring diagnosis of Proteus syndrome. Splenic hyperplasia was found to be a manifestation. The spleen was enlarged in 2 of the 18 cases, and another patient with Proteus syndrome and asymptomatic splenomegaly was known to them. Ceelen et al. (1997) described a man with Proteus syndrome who sustained a rupture of an enlarged spleen following a fall from a horse. Biesecker et al. (1998) also observed enlargement of the thymus. Proteus syndrome was frequently confused with hemihyperplasia. Biesecker et al. (1998) described a distinct subtype of hemihyperplasia defined by static or mildly progressive hemihyperplasia and multiple lipomata.

Although Wiedemann et al. (1983) made this syndrome generally known, it had previously been delineated by Cohen and Hayden (1979). The assignment of a name with mnemonic value perhaps accounts for the greater success of Wiedemann than of Cohen in bringing the disorder to general attention. (Also, Cohen and Hayden's report was in a more obscure publication than Wiedemann's.) 10,9,8:Cohen (1987, 1988, 1988) believed that Joseph Merrick (the Elephant Man), the famous patient of Sir Frederick Treves, had Proteus syndrome, not neurofibromatosis (162200) as often thought.

See also Gorlin (1984); Lezama and Buyse (1984); Mucke et al. (1985)

Beluffi, G.; DiGiulio, G.; Fiori, P.: Pelvic lipomatosis in the Proteus syndrome: a further diagnostic sign. (Letter) Europ. J. Pediat. 149: 866 only, 1990.

Bennett, D.: Rescue of a lethal T/t locus genotype by chimaerism with normal embryos. Nature 272: 539 only, 1978.

Bialer, M. G.; Riedy, M. J.; Wilson, W. G.: Proteus syndrome versus Bannayan-Zonana syndrome: a problem in differential diagnosis. Europ. J. Pediat. 148: 122-125, 1988.

Biesecker, L. G.; Peters, K. F.; Darling, T. N.; Choyke, P.; Hill, S.; Schimke, N.; Cunningham, M.; Meltzer, P.; Cohen, M. M., Jr.: Clinical differentiation between Proteus syndrome and hemihyperplasia: description of a distinct form of hemihyperplasia. Am. J. Med. Genet. 79: 311-318, 1998.

Burgio, G. R.; Wiedemann, H. R.: Further and new details on the Proteus syndrome. Europ. J. Pediat. 143: 71-73, 1984.

Ceelen, W.; De Waele, J.; Kunnen, M.; de Hemptinne, B.: Non-operative management of a splenic laceration in a patient with the Proteus syndrome. J. Accid. Emerg. Med. 14: 111-113, 1997.

Clark, R. D.; Donnai, D.; Rogers, J.; Cooper, J.; Baraitser, M.: Proteus syndrome: an expanded phenotype. Am. J. Med. Genet. 27: 99-117, 1987.

Cohen, M. M., Jr.: Further diagnostic thoughts about the Elephant Man. Am. J. Med. Genet. 29: 777-782, 1988.

Cohen, M. M., Jr.: Understanding Proteus syndrome, unmasking the Elephant Man, and stemming elephant fever. Neurofibromatosis 1: 260-280, 1988.

Cohen, M. M., Jr.: The Elephant Man did not have neurofibromatosis. Proc. Greenwood Genet. Center 6: 187-192, 1987.

Cohen, M. M., Jr.: Proteus syndrome: clinical evidence for somatic mosaicism and selective review. Am. J. Med. Genet. 47: 645-652, 1993.

Cohen, M. M., Jr.; Hayden, P. W.: A new recognized hamartomatous syndrome. Birth Defects Orig. Art. Ser. 15(5B): 291-296, 1979.

Costa, T.; Fitch, N.; Azouz, E. M.: Proteus syndrome: report of two cases with pelvic lipomatosis. Pediatrics 76: 984-989, 1985.

Dean, J. C. S.; Cole, G. F.: When is Proteus syndrome not Proteus syndrome? (Abstract) Clin. Genet. Soc.: Aberdeen, Scotland , 9/22/1988.

Goodship, J.; Redfearn, A.; Milligan, D.; Gardner-Medwin, D.; Burn, J.: Transmission of Proteus syndrome from father to son?. J. Med. Genet. 28: 781-785, 1991.

Gordon, P. L.; Wilroy, R. S.; Lasater, O. E.; Cohen, M. M., Jr.: Neoplasms in Proteus syndrome. Am. J. Med. Genet. 57: 74-78, 1995.

Gorlin, R. J.: Proteus syndrome. J. Clin. Dysmorph. 2: 8-9, 1984.

Graetz, I.: Ueber einen Fall von sogenannter 'totaler halbseitiger Korperhypertrophie'. Z. Kinderheilk. 45: 381-403, 1928.

Happle, R.: Cutaneous manifestation of lethal genes. Hum. Genet. 72: 280 only, 1986.

Haramoto, U.; Kobayashi, S.; Ohmori, K.: Hemifacial hyperplasia with meningeal involvement: a variant of Proteus syndrome? Am. J. Med. Genet. 59: 164-167, 1995.

Hotamisligil, G. S.; Ertogan, F.: The Proteus syndrome: association with nephrogenic diabetes insipidus. Clin. Genet. 38: 139-144, 1990.

Kruger, G.; Pelz, L.; Wiedemann, H.-R.: Transmission of Proteus syndrome from mother to son? (Letter) Am. J. Med. Genet. 45: 117-118, 1993.

Lacombe, D.; Battin, J.: Isolated macrodactyly and Proteus syndrome. (Letter) Clin. Dysmorph. 5: 255-257, 1996.

Lezama, D. B.; Buyse, M. L.: The Proteus syndrome: the emergence of an entity. J. Clin. Dysmorph. 2: 10-13, 1984.

Malamitsi-Puchner, A.; Dimitriadis, D.; Bartsocas, C.; Wiedemann, H.-R.: Proteus syndrome: course of a severe case. Am. J. Med. Genet. 35: 283-285, 1990.

Malamitsi-Puchner, A.; Kitsiou, S.; Bartsocas, C. S.: Severe Proteus syndrome in an 18-month-old boy. Am. J. Med. Genet. 27: 119-125, 1987.

McCall, S.; Ramzy, M. I.; Cure, J. K.; Pai, G. S.: Encephalocraniocutaneous lipomatosis and the Proteus syndrome: distinct entities with overlapping manifestations. Am. J. Med. Genet. 43: 662-668, 1992.

McMullin, G. P.; Super, M.; Clarke, M. A.: Cranial hemihypertrophy with ipsilateral naevoid streaks, intellectual handicap and epilepsy: a report of two cases. Clin. Genet. 44: 249-253, 1993.

Mucke, J.; Willgerodt, H.; Kunzel, R.; Brock, D.: Variability in the Proteus syndrome: report of an affected child with progressive lipomatosis. Europ. J. Pediat. 143: 320-323, 1985.

Rizzo, R.; Pavone, L.; Micali, G.; Nigro, F.; Cohen, M. M., Jr.: Encephalocraniocutaneous lipomatosis, Proteus syndrome, and somatic mosaicism. Am. J. Med. Genet. 47: 653-655, 1993.

Samlaska, C. P.; Levin, S. W.; James, W. D.; Benson, P. M.; Walker, J. C.; Perlik, P. C.: Proteus syndrome. Arch. Derm. 125: 1109-1114, 1989.

Skovby, F.; Graham, J. M., Jr.; Sonne-Holm, S.; Cohen, M. M., Jr.: Compromise of the spinal canal in Proteus syndrome. Am. J. Med. Genet. 47: 656-659, 1993.

Smeets, E.; Fryns, J.-P.; Cohen, M. M., Jr.: Regional Proteus syndrome and somatic mosaicism. Am. J. Med. Genet. 51: 29-31, 1994.

Temtamy, S. A.; Rogers, J. G.: Macrodactyly, hemihypertrophy, and connective tissue nevi: report of a new syndrome and review of the literature. J. Pediat. 89: 924-927, 1976.

Viljoen, D. L.; Nelson, M. M.; de Jong, G.; Beighton, P.: Proteus syndrome in Southern Africa: natural history and clinical manifestations in six individuals. Am. J. Med. Genet. 27: 87-97, 1987.

Viljoen, D. L.; Saxe, N.; Temple-Camp, C.: Cutaneous manifestations of the Proteus syndrome. Pediat. Derm. 5: 14-21, 1988.

Wiedemann, H.-R.; Burgio, G. R.; Aldenhoff, P.; Kunze, J.; Kaufmann, H. J.; Schirg, E.: The Proteus syndrome: partial gigantism of the hands and/or feet, nevi, hemihypertrophy, subcutaneous tumors, macrocephaly or other skull anomalies and possible accelerated growth and visceral affections. Europ. J. Pediat. 140: 5-12, 1983.

COWDEN MULTIPLE HAMARTOMA SYNDROME (#158350)

JJM NOTE

This disorder should be sought in every woman with breast cancer. There are benign and malignant neoplasms of the breast, colon, and thyroid. Other malignancies include astrocytoma; non-Hodgkin lymphoma; renal cell adenocarcinoma; primary neuroendocrine carcinoma of the skin (Merkel cell carcinoma); and cancers of the endometrium, lung, bladder, and skin. My introduction to the disorder was with a patient on the Surgery Branch of the National Cancer Institute (Mulvihill and McKeen 1977). From age

41 to 57, she had had cancers of the endometrium, breast, cecum, kidney, and urinary bladder; benign tumors of the thyroid, sigmoid colon, tongue, and skin were removed as well. An inveterate cigarette smoker, she died of an adenocarcinoma of the lung at age 59 years. Material collected from this family contributed to the mapping of the gene on chromosome 10q22-23 by Nelen et al. (1996).

A number sign (#) is used with this entry because of evidence that Cowden disease is caused by germline mutations of the PTEN gene (601728).

Multiple hamartomatous lesions, especially of the skin, mucous membranes, breast and thyroid, are encountered. Verrucous skin lesions of the face and limbs, cobblestone-like papules of the gingiva and buccal mucosa, and multiple facial trichilemmomas are leading findings (Brownstein et al., 1977). Hamartomatous polyps of the colon and other intestines occur also. Only 1 case had been reported before the report of Weary et al. (1972). Weary et al. (1972) proposed the designation multiple hamartoma syndrome and identified an autosomal dominant pattern of inheritance in the family of Rachel Cowden, for whom Lloyd and Dennis (1963) had named the disorder. Affected brother and sister were observed by Gentry et al. (1974). Gentry et al. (1974) observed affected persons in 4 generations, with father-to-son transmission. Brownstein et al. (1979) reported on the dermatopathology in 19 patients. Twenty-nine of 53 facial lesions biopsied were trichilemmomas. All oral mucosal lesions were fibromas. Biopsies from the hands and feet showed benign keratosis. Ruschak et al. (1981) described a patient who at the age of 18 years, after experiencing several years of recurrent diarrhea, underwent colectomy and ileostomy for multiple colonic polyposis. Several lipomas on the trunk were also removed. The patient was unique in having deficiency of T-lymphocyte function with recurrent cellulitis and abscess formation and the eventual development of acute myelogenous leukemia. Breast cancer is an indication for prophylactic mastectomy in the view of Walton et al. (1986). Elston et al. (1986) described a 70-year-old woman in whom the diagnosis of Cowden syndrome had been made on the basis of facial trichilemmomas at age 63. Adenoid facies, high-arched palate, thickened, furrowed tongue, pectus excavatum, and scoliosis were described. Trichilemmomas were found in the sacral area. The need for close surveillance for malignancy was emphasized by the development of 3 different malignant neoplasms in this patient in a 16-year period.

Starink et al. (1986) analyzed 21 cases of Cowden syndrome in 7 families, with multiple cases in 4 of the families. They reconfirmed autosomal dominant inheritance with high penetrance in both sexes and high frequency of breast cancer in females. Craniomegaly was the most frequent extracutaneous finding. About 60% of patients had gastrointestinal polyps and 76% had cutaneous fibromas. Williard et al. (1992) described the case of a woman who presented at age 32 with breast cancer. Her mother had died of breast cancer at age 42, and 2 maternal aunts had had premenopausal breast cancer. She had multiple soft, fleshy, 3- to 5-mm papillomatous lesions consistent with acrochordons in the axillae, inframammary folds, groins, and posterior neck. She also had nodular and papillomatous lesions of the tongue and frenulum.

Carlson et al. (1986) could demonstrate no linkage with any of a battery of markers in a family with 4 affected in 3 generations. Furthermore, measurements of epidermal growth factor (131530) in body fluids yielded normal findings. Hanssen et al. (1993) described Cowden syndrome in a large 4-generation family. Anticipation was demonstrated with greater severity and earlier onset of signs and symptoms in successive generations. Macrocephaly was present in all affected individuals, was markedly progressive in 3 of 6 affected children in the fourth generation, and was associated with slight to moderate delay in psychomotor development. There was 1 instance of male-to-male transmission. In a survey of 87 reported patients, Hanssen et al. (1993) found a marked excess of affected females; the male-to-female ratio was 26 to 61.

Padberg et al. (1991) suggested that the disorder previously referred to as cerebelloparenchymal disorder VI (Lhermitte-Duclos disease) is merely part of the multiple hamartoma syndrome. Mental dullness and in some cases signs of increased intracranial pressure are features; the latter is the result of herniation of the cerebellar tonsils. The condition was first described by Lhermitte and Duclos (1920). Ambler et al. (1969), who described the disorder in mother and son, stated that a total of 35 cases had been reported. Padberg et al. (1991) observed 2 unrelated patients who had macrocephaly, seizures and mild cerebellar signs resulting from dysplastic gangliocytoma of the cerebellum (Lhermitte-Duclos disease). Both had autosomal dominant Cowden disease evidenced by facial, oral, and acral papules. In the 2 families, 9 sibs demonstrated the mucocutaneous lesions, thyroid disease, breast tumors, and ovarian tumors compatible with the diagnosis of Cowden disease. Some of them also showed various degrees of neurologic signs such as macrocephaly, mental retardation, seizures, tremors, and dysdiadochokinesia. Padberg et al. (1991) suggested that the combination of Lhermitte-Duclos disease and Cowden disease represents a new phakomatosis. Albrecht et al. (1992) described 2 patients who had both Cowden syndrome and Lhermitte-Duclos disease and concluded that these are the same entity, a hamartoma-neoplasia syndrome. The skin lesions were described as multiple trichilemmomas, a type of benign skin appendage tumor. Oral papillomatosis and cutaneous keratoses also occurred. Trichilemmomas covering the pinna and around the mouth were pictured as well as papillomatosis of the tongue and keratoses of the sole. Eng et al. (1994) described a 3-generation family with Cowden disease and Lhermitte-Duclos disease. Lhermitte-Duclos disease is believed to be a hamartomatous overgrowth of hypertrophic ganglion cells which replace the granular cell layer and Purkinje cells of the cere bellum. In the grandfather of the proband in the family reported by Eng et al. (1994), the diagnosis of Lhermitte-Duclos disease had been made by cerebellar biopsy. Eng et al. (1994) found no chromosomal abnormality in the peripheral lymphocytes of the proposita and her affected mother and by single-strand conformation polymorphism analysis found no evidence of mutation in the p53 gene. Wells et al. (1994) found reports of 7 cases of Lhermitte-Duclos syndrome occurring in adults with Cowden syndrome and reported this association in a 16-year-old girl with craniomegaly, choroidal hamartoma, right conjunctival papilloma, and a history of bilateral multinodular adenomatous goiter and cystic hygroma. Although Cowden syndrome has traditionally been defined by mucocutaneous criteria, it typically also involves hamartomas and neoplasms of internal organs, most commonly in the thyroid, breast, and female genitourinary tract. Because the mucocutaneous features may develop several decades after birth, the patient reported by Wells et al. (1994) highlighted the need for long-term follow-up of a pediatric patient with Lhermitte-Duclos syndrome because of the risk of malignancies associated with Cowden syndrome. The characteristic pathologic features of Lhermitte-Duclos syndrome are global hypertrophy of the cerebellum, coarse gyri, and the typical 'inverted cortex' pattern.

Vinchon et al. (1994) described an affected woman who was first seen at the age of 16 for ataxia and symptoms of raised intracranial pressure. A ventriculoatrial shunting relieved the symptoms. A nodule in her left breast was removed at the age of 18. At the age of 24, the symptoms of raised intracranial pressure recurred, CT scan disclosed a trigonoseptal tumor, and partial resection of the lesion showed it to be a benign astrocytoma. Hypothalamic hyperprolactinemia and goiter due to microinvasive vesicular carcinoma were other complications. Dermatologic examinations showed several trichilemmomas over the nose, subclavian area, and right elbow. Vinchon et al. (1994) found 72 reported cases of Lhermitte-Duclos disease; 26 had conditions suggesting Cowden disease and 7 were definite cases of Cowden disease.

Haibach et al. (1992) reported renal cell adenocarcinoma and primary neuroendocrine carcinoma of the skin in association with Cowden syndrome. They searched for abnormalities in the EGFR gene (131550) in kidney, liver and thyroid as well as in the tissue of the primary neuroendocrine carcinoma, but found none. Primary neuroendocrine carcinoma of the skin (PNECS) is also known as trabecular carcinoma or Merkel cell carcinoma. First described by Tang and Toker (1978), it originates from the Merkel cell, a pluripotential basal epidermal cell. Lyons et al. (1993) observed meningioma in a 41-year-old woman known to have Cowden disease. In addition to many skin lesions, she had a history of follicular adenoma of the thyroid and breast carcinoma as well as a family history of carcinoma of the colon affecting 2 previous generations.

Hanssen and Fryns (1995) indicated that progressive macrocephaly, scrotal tongue, and mild-to-moderate mental retardation are important signs of Cowden syndrome in young children. Trichilemmomas in the nasolabial folds and palmar and plantar hyperkeratotic pits usually become evident later in childhood. They are often accompanied by the appearance of subcutaneous lipomas and cutaneous hemangiomas.

Schrager et al. (1998) analyzed the clinical and pathologic features of breast disease in 19 women with Cowden disease. The 19 women showed a spectrum of benign histopathologic findings, including ductal hyperplasia, intraductal papillomatosis, adenosis, lobular atrophy, fibroadenomas, and fibrocystic change. Features suggestive of a breast hamartoma were found in 17 (89%). Malignant disease, most of which was ductal carcinoma, was found in 14 women (74%): ductal carcinoma in situ in 12, and infiltrating ductal carcinoma in 12. A common benign breast lesion in CD is a densely fibrotic hyalinized nodule.

To localize the gene for Cowden disease, Nelen et al. (1996) performed an autosomal genome scan using DNA markers. They examined a total of 12 families, and obtained a maximum lod score of 8.92 at theta = 0.02 with the marker D10S573 located on 10q22-q23. The authors stated that the neurologic and neoplastic features of Cowden disease are consistent with the possibility that the Cowden gene is a tumor suppressor gene. The high frequency of breast cancer in female

Cowden patients (approximately 30%) makes it a strong candidate for a breast cancer susceptibility gene. It should also be considered in connection with nonmedullary thyroid cancer.

The chromosomal region containing the Cowden disease gene was known to contain a tumor suppressor gene (PTEN; 601728) that had been found to be mutated in sporadic brain, breast, and prostate cancer. Liaw et al. (1997) found germline mutations in the PTEN gene in 4 of 5 families with Cowden syndrome. Missense (601728.0001) and nonsense mutations were predicted to disrupt the protein tyrosine/dual-specificity phosphatase domain of the protein. All affected individuals of the 5 families studied manifested trichilemmomas, regardless of whether their mutation was a missense or nonsense mutation. Nonsense mutations were associated with macrocephaly in 2 families. In 1 of these families, a premature stop codon at position 157 (601728.0003) was also associated with Lhermitte-Duclos disease (LDD), manifested by ataxia and dysplastic cerebellar gangliocytomatosis. In the other family, a stop codon at position 233 (601728.0002) was not associated with cerebellar manifestation. Liaw et al. (1997) speculated that the larger N-terminal truncation may be responsible for the more severe LDD phenotype. Their data indicated that PTEN is a tumor suppressor gene in the germline and that it plays a role in organizing the relationship of different cell types within an organ during development. Because germline mutations of PTEN predispose to a breast and thyroid cancer syndrome (Cowden disease) and somatic mutations are found in sporadic breast cancer, PTEN was an obvious candidate for predisposition to non-CD breast cancer.

Nelen et al. (1997) confirmed that the PTEN is indeed the gene for Cowden disease by a refined localization of the gene to the interval between D10S1761 and D10S541, which contains the PTEN gene, and by mutation analysis in 8 unrelated familial and 11 sporadic patients with Cowden disease. They detected 8 different mutations in various regions of the PTEN gene (e.g., 601728.0005). One mutation was detected twice (601728.0007). All detected changes in the gene were predicted to have a deleterious effect on the putative protein. They found no indications of a correlation between genotype and phenotype. In 10 patients, no mutation could be detected. These were patients who showed linkage to the same region, 10q22-q23; no evidence emerged from the phenotype of these patients to suggest genetic heterogeneity.

Arch et al. (1997) described an 18-month-old boy with macrocephaly, pseudopapilledema, 2 small lipomas, hamartomatous neoplasms, and polyps of the duodenum and colon who had an interstitial deletion of 10q23.2-q24.1. The initial diagnosis had been Bannayan-Riley-Ruvalcaba syndrome (BRRS; 153480). Cowden disease is relatively poorly described in young children. The overlap of clinical features of Cowden disease and BBR syndrome, the demonstration of an interstitial deletion of the region of chromosome 10 containing the Cowden disease locus, and the specific demonstration that the PTEN gene was missing from the deleted chromosome 10 in their patient led Arch et al. (1997) to suggest that these 2 disorders are allelic.

See also Burnett et al. (1975); Marano et al. (1988); Milbouw et al. (1988); Mulvihill and McKeen (1977); Thyresson and Doyle (1981); Yuasa et al. (1980)

Albrecht, S.; Haber, R. M.; Goodman, J. C.; Duvic, M.: Cowden syndrome and Lhermitte-Duclos disease. Cancer 70: 869-876, 1992.

Ambler, M.; Pogacar, S.; Sidman, R.: Lhermitte-Duclos disease (granule cell hypertrophy of the cerebellum): pathological analysis of the first familial cases. J. Neuropath. Exp. Neurol. 28: 622-647, 1969.

Arch, E. M.; Goodman, B. K.; Van Wesep, R. A.; Liaw, D.; Clarke, K.; Parsons, R.; McKusick, V. A.; Geraghty, M. T.: Deletion of PTEN in a patient with Bannayan-Riley-Ruvalcaba syndrome suggests allelism with Cowden disease. Am. J. Med. Genet. 71: 489-493, 1997.

Brownstein, M. H.; Mehregan, A. H.; Bikowski, J. B.: Trichilemmomas in Cowden's disease. (Letter) J.A.M.A. 238: 26, 1977.

Brownstein, M. H.; Mehregan, A. H.; Bikowski, J. B. B.; Lupulescu, A.; Patterson, J. C.: The dermatopathology of Cowden's syndrome. Brit. J. Derm. 100: 667-673, 1979.

Burnett, J. W.; Goldner, R.; Calton, G. J.: Cowden disease: report of two additional cases. Brit. J. Derm. 93: 329-336, 1975.

Carlson, H. E.; Burns, T. W.; Davenport, S. L.; Luger, A. M.; Spence, M. A.; Sparkes, R. S.; Orth, D. N.: Cowden disease: gene marker studies and measurements of epidermal growth factor. Am. J. Hum. Genet. 38: 908-917, 1986.

Elston, D. M.; James, W. D.; Rodman, O. G.; Graham, G. F.: Multiple hamartoma syndrome (Cowden's disease) associated with non-Hodgkin's lymphoma. Arch. Derm. 122: 572-575, 1986.

Eng, C.; Murday, V.; Seal, S.; Mohammed, S.; Hodgson, S. V.; Chaudary, M. A.; Fentiman, I. S.; Ponder, B. A. J.; Eeles, R. A.: Cowden syndrome and Lhermitte-Duclos disease in a family: a single genetic syndrome with pleiotropy? J. Med. Genet. 31: 458-461, 1994.

Gentry, W. C., Jr.; Eskritt, N. R.; Gorlin, R. J.: Multiple hamartoma syndrome (Cowden disease). Arch. Derm. 109: 521-525, 1974.

Haibach, H.; Burns, T. W.; Carlson, H. E.; Burman, K. D.; Deftos, L. J.: Multiple hamartoma syndrome (Cowden's disease) associated with renal cell carcinoma and primary neuroendocrine carcinoma of the skin (Merkel cell carcinoma). Am. J. Clin. Path. 97: 705-712, 1992.

Hanssen, A. M. N.; Fryns, J. P.: Cowden syndrome. J. Med. Genet. 32: 117-119, 1995.

Hanssen, A. M. N.; Werquin, H.; Suys, E.; Fryns, J. P.: Cowden syndrome: report of a large family with macrocephaly and increased severity of signs in subsequent generations. Clin. Genet. 44: 281-286, 1993.

Lhermitte, J.; Duclos, P.: Sur un ganglioneurome diffus du cortex du cervelet. Bull. Assoc. Franc. Cancer 9: 99-107, 1920.

Liaw, D.; Marsh, D. J.; Li, J.; Dahia, P. L. M.; Wang, S. I.; Zheng, Z.; Bose, S.; Call, K. M.; Tsou, H. C.; Peacocke, M.; Eng, C.; Parsons, R.: Germline mutations of the PTEN gene in Cowden disease, an inherited breast and thyroid cancer syndrome. Nature Genet. 16: 64-67, 1997.

Lloyd, K. M.; Dennis, M.: Cowden's disease: a possible new symptom complex with multiple system involvement. Ann. Intern. Med. 58: 136-142, 1963.

Lyons, C. J.; Wilson, C. B.; Horton, J. C.: Association between meningioma and Cowden's disease. Neurology 43: 1436-1437, 1993.

Marano, S. P.; Johnson, P. C.; Spetzler, R. F.: Recurrent Lhermitte-Duclos disease in a child. J. Neurosurg. 69: 599-603, 1988.

Milbouw, G.; Born, J. D.; Martin, D.; Collignon, J.; Hans, P.; Reznik, M.; Bonnal, J.: Clinical and radiological aspects of dysplastic gangliocytoma (Lhermitte-Duclos disease): a report of two cases with review of the literature. Neurosurgery 22: 124-128, 1988.

Mulvihill, J. J.; McKeen, E. A.: Genetics of multiple primary tumors: a clinical etiologic approach illustrated by three patients. Cancer 40: 1867-1871, 1977.

Nelen, M. R.; Padberg, G. W.; Peeters, E. A. J.; Lin, A. Y.; van den Helm, B.; Frants, R. R.; Coulon, V.; Goldstein, A. M.; van Reen, M. M. M.; Easton, D. F.; Eeles, R. A.; Hodgson, S.; Mulvihill, J. J.; Murday, V. A.; Tucker, M. A.; Mariman, E. C. M.; Starink, T. M.; Ponder, B. A. J.; Ropers, H. H.; Kremer, H.; Longy, M.; Eng, C.: Localization of the gene for Cowden disease to chromosome 10q22-23. Nature Genet. 13: 114-116, 1996.

Nelen, M. R.; van Staveren, W. C. G.; Peeters, E. A. J.; Ben Hassel, M.; Gorlin, R. J.; Hamm, H.; Lindboe, C. F.; Fryns, J.-P.; Sijmons, R. H.; Woods, D. G.; Mariman, E. C. M.; Padberg, G. W.; Kremer, H.: Germline mutations in the PTEN/MMAC1 gene in patients with Cowden disease. Hum. Molec. Genet. 6: 1383-1387, 1997.

Padberg, G. W.; Schot, J. D. L.; Vielvoye, G. J.; Bots, G. T. A. M.; de Beer, F. C.: Lhermitte-Duclos disease and Cowden disease: a single phakomatosis. Ann. Neurol. 29: 517-523, 1991.

Ruschak, P. J.; Kauh, Y. C.; Luscombe, H. A.: Cowden's disease associated with immunodeficiency. Arch. Derm. 117: 573-575, 1981.

Schrager, C. A.; Schneider, D.; Gruener, A. C.; Tsou, H. C.; Peacocke, M.: Clinical and pathological features of breast disease in Cowden's syndrome: an underrecognized syndrome with an increased risk of breast cancer. Hum. Path. 29: 47-53, 1998.

Starink, T. M.; van der Veen, J. P. W.; Arwert, F.; de Waal, L. P.; de Lange, G. G.; Gille, J. J. P.; Eriksson, A. W.: The Cowden syndrome: a clinical and genetic study in 21 patients. Clin. Genet. 29: 222-233, 1986.

Tang, C.-K ; Toker, C.: Trabecular carcinoma of the skin: an ultrastructural study. Cancer 42: 2311-2321, 1978.

Thyresson, H. N.; Doyle, J. A.: Cowden's disease (multiple hamartoma syndrome). Mayo Clin. Proc. 56: 179-184, 1981.

Vinchon, M.; Blond, S.; Lejeune, J. P.; Krivosik, I.; Fossati, P.; Assaker, R.; Christiaens, J. L.: Association of Lhermitte-Duclos and Cowden disease: report of a new case and review of the literature. J. Neurol. Neurosurg. Psychiat. 57: 699-704, 1994.

Walton, B. J.; Morain, W. D.; Baughman, R. D.; Jordan, A.; Crichlow, R. W.: Cowden's disease: a further indication for prophylactic mastectomy. Surgery 99: 82-86, 1986.

Weary, P. E.; Gorlin, R. J.; Gentry, W. C., Jr.; Comer, J. E.; Greer, K. E.: Multiple hamartoma syndrome (Cowden's disease). Arch. Derm. 106: 682-690, 1972.

Wells, G. B.; Lasner, T. M.; Yousem, D. M.; Zager, E. L.: Lhermitte-Duclos disease and Cowden's syndrome in an adolescent patient: case report. J. Neurosurg. 81: 133-136, 1994.

Williard, W.; Borgen, P.; Bol, R.; Tiwari, R.; Osborne, M.: Cowden's disease: a case report with analyses at the molecular level. Cancer 69: 2969-2974, 1992.

Yuasa, T.; Hanano, M.; Ohshima, F.; Tsubaki, T.: The association of myasthenia gravis with multiple hamartoma syndrome (Cowden disease). (Letter) Ann. Neurol. 7: 591-592, 1980.

PHOSPHATASE AND TENSIN HOMOLOG; PTEN (*601728)

JJM NOTE

Germline (i.e., hereditary) changes in this gene have been seen in the juvenile polyposis and the syndromes of Cowden, Lhermitte-Duclos, Bannayan-Zonana, and Bannayan-Riley-Ruvalcaba. Hence, a long list of cancers and tumors includes breast, glioma, colon, and thyroid. Somatic mutations, in part reflected by studies of loss of heterozygosity, have been seen in glioblastoma and cancers of the endometrium, prostate, kidney, and breast.

As tumors progress to more advanced stages, they acquire an increasing number of genetic alterations. Li et al. (1997) noted that 1 alteration which occurs at high frequency in a variety of human tumors is loss of heterozygosity (LOH) at 10q23. Although rarely seen in low-grade glial tumors and early-stage prostate cancers, LOH at 10q23 occurs in approximately 70% of glioblastomas (the most advanced form of glial tumor) and approximately 60% of advanced prostate cancers. This pattern of LOH and the finding that wildtype chromosome 10 suppresses the tumorigenicity of glioblastoma cells in mice suggested to Li et al. (1997) that 10q23 encodes a tumor suppressor gene. By mapping of homozygous deletions on 10q23, they isolated a candidate tumor suppressor gene that they called PTEN for 'phosphatase and tensin homolog deleted on chromosome ten.' Sequence analysis of the predicted 403-amino acid open reading frame (ORF) revealed a protein tyrosine phosphatase domain and a large region of homology (approximately 175 amino acids) to chicken tensin (600076; a protein that interacts with actin filaments at focal adhesions) and bovine auxilin. In preliminary screens, Li et al. (1997) detected mutations of PTEN in 31% (13 of 42) of glioblastoma cell lines and xenografts, 100% (4 of 4) of prostate cancer cell lines, 6% (4 of 65) of breast cancer cell lines and xenografts, and 17% (3 of 18) of primary glioblastomas. The homologies displayed by the structure of PTEN suggested to the investigators that it may suppress tumor cell growth by antagonizing protein tyrosine kinases and may regulate tumor cell invasion and metastasis through interactions at focal adhesions.

The same gene was independently isolated by Steck et al. (1997), who designated it MMAC1 (mutated in multiple advanced cancers 1), by use of exon trapping. They began their search for the gene from deletions involving 10q23-q24, which occur in the majority of cases of human glioblastoma multiformes. Homozygous deletions in 4 glioma cells lines further refined the location. The MMAC1 gene spans these deletions and encodes a widely expressed 5.5-kb mRNA with a 403-amino acid ORF. The predicted MMAC1 protein contains sequence motifs with significant homology to the catalytic domain of protein phosphatases and to the cytoskeletal proteins tensin and auxilin. MMAC1 coding-region mutations were observed in a number of glioma, prostate, kidney, and breast carcinoma cell lines or tumor specimens. The MMAC1 gene was mapped to 10q23.3. Steck et al. (1997) also cloned mouse and dog homologs of MMAC1. Hansen and Justice (1998) mapped the Pten gene to mouse chromosome 19.

Tamura et al. (1998) found that overexpression of PTEN inhibited cell migration, whereas antisense PTEN enhanced migration. Integrin-mediated cell spreading and the formation of focal adhesions were downregulated by wildtype PTEN but not by PTEN with an inactive phosphatase domain. PTEN interacted with the focal adhesion kinase FAK (600758) and reduced its tyrosine phosphorylation. Overexpression of FAK partially antagonized the effects of PTEN. Thus, PTEN phosphatase may function as a tumor suppressor by negatively regulating cell interactions with the extracellular matrix.

Liaw et al. (1997) identified germline mutations in 4 of 5 families with Cowden disease (CD; 158350). They found missense (601728.0001) and nonsense (601728.0002; 601728.0003) mutations predicted to disrupt the protein tyrosine/dual-specificity phosphatase domain of the protein. Nelen et al. (1997) confirmed the PTEN gene as the Cowden disease gene in 8 unrelated families and reported germline missense mutations in CD families (e.g., 601728.0005).

In a patient referred with a diagnosis of Bannayan-Riley-Ruvalcaba syndrome (BRRS; 153480), Arch et al. (1997) demonstrated an interstitial deletion of 10q23.2-q24.1. They demonstrated that the PTEN gene was missing from the deleted chromosome. Because of phenotypic overlap between the BRR syndrome and Cowden disease and because of the apparent mapping to the same chromosomal area, Arch et al. (1997) proposed that Cowden disease and BRR syndrome are allelic.

Marsh et al. (1997) identified germline mutations in PTEN in the hamartomatous polyposis syndrome, which they referred to as Bannayan-Zonana syndrome (BZS; see 153480), but which goes by a number of other designations, including Bannayan-Riley-Ruvalcaba syndrome. Hamartomas are benign, hyperplastic, disorganized growths. Characteristic features of BZS include macrocephaly, lipomas, intestinal hamartomatous polyps, vascular malformations, Hashimoto thyroiditis, and speckled pigmentation of the penis. Many patients present at birth with macrocephaly and failure to thrive. The disorder has been described particularly in young patients. Cowden disease has been described mainly in older individuals and is characterized by hamartomas in multiple organs and a risk of breast and thyroid cancer. Breast tumors are said to occur in 75% of females with Cowden disease and thyroid disease in 50 to 75%; macrocephaly (38%) and lipomas (less than 5%) are less common, and speckled penis apparently does not occur. Hamartomatous polyps occur in the Peutz-Jeghers syndrome (175200), which is associated with melanin spots of the lips, buccal mucosa, and digits. The Peutz-Jeghers syndrome maps to 19p, without genetic heterogeneity, and juvenile polyposis (174900) to 10q22. Cowden syndrome maps to a region of 10q thought to be telomeric to the juvenile polyposis region. In studies of 2 unrelated well-characterized BZS families, Marsh et al. (1997) found that affected individuals demonstrated haplotype sharing for the 10q22-q23 region. Therefore, PTEN mutations were sought, with the demonstration of germline mutations, arg233ter (R233X; exon 7; 601728.0002) in one family and ser170arg (S170R; exon 6; 601728.0004) in the second family. The R233X mutation had been found in a family with classic Cowden disease by Liaw et al. (1997). The identical mutation occurred in these 2 families with 2 distinct syndromes on 2 different 10q22-q23 haplotypes, arguing against a common ancestor or founder effect. The only common clinical features in the CD family and the BZS family with R233X were macrocephaly and thyroid disease.

Genetic heterogeneity of Cowden syndrome was suggested by the fact that Tsou et al. (1997) found no coding sequence mutations in 23 CS families for whom linkage to the PTEN locus had never been established. They reported 3 novel PTEN mutations in Cowden syndrome and demonstrated that these mutations were associated with Cowden syndrome and breast cancer. They found no PTEN mutations in a group of individuals with early-onset breast cancer, suggesting that germline mutations in the PTEN gene are not common in this group.

Tsou et al. (1998) found 3 mutations in the PTEN gene in unrelated individuals with Cowden syndrome. These included a missense mutation in exon 5, a splice site mutation in intron 7 causing exon skipping, and a missense mutation in exon 3. Tsou et al. (1998) also reported a rare polymorphism in exon 7 of the PTEN coding sequence.

Marsh et al. (1998) identified PTEN mutations in 30 of 37 (81%) Cowden disease families, including missense and nonsense point mutations, deletions, insertions, a deletion/insertion, and splice site mutations. These mutations were scattered over the entire length of PTEN, with the exception of the first, fourth, and last exons. A 'hotspot' for PTEN mutation in Cowden disease was identified in exon 5, which contains the PTPase core motif, with 13 of 30 (43%) Cowden disease mutations identified in that exon. Thus, 7 of 30 (23%) were within the core motif, the majority (5 of 7) of which were missense mutations, possibly pointing to the functional significance of this region.

Marsh et al. (1998) identified germline PTEN mutations in 4 of 7 (57%) Bannayan-Zonana syndrome families studied. None of these mutations was observed in the PTPase core motif.

To examine the role of the dual-specificity phosphatase PTEN in ontogenesis and tumor suppression, Di Cristofano et al. (1998) disrupted mouse Pten by homologous recombination. Pten inactivation resulted in early embryonic lethality. Homozygous deficient ES (embryonic stem) cells formed aberrant embryoid bodies and displayed an altered ability to differentiate into endodermal, ectodermal, and mesodermal derivatives. Heterozygous knockout mice and chimeric mice derived from heterozygous ES cells showed hyperplastic/dysplastic changes in the prostate, skin, and colon, which are characteristic of Cowden disease, Lhermitte-Duclos disease, and Bannayan-Zonana syndrome. They also spontaneously developed germ cell, gonadostromal, thyroid, and colon tumors. In addition, Pten inactivation enhanced the ability of ES cells to generate tumors in nude and syngeneic mice, due to increased anchorage-independent growth and aberrant differentiation. These results supported the notion that PTEN haploinsufficiency plays a causal role in the 3 disorders in which mutations had been found, and demonstrated that Pten is a tumor suppressor essential for embryonic development.

Stambolic et al. (1998) found that Pten-mutant mouse embryos displayed regions of increased proliferation. In contrast, Pten-deficient immortalized mouse embryonic fibroblasts exhibited decreased sensitivity to cell death in response to a number of apoptotic stimuli, accompanied by constitutively elevated activity and phosphorylation of protein kinase B (PKB)/Akt, a crucial regulator of cell survival. Expression of exogenous Pten in mutant cells restored both their sensitivity to agonist-induced apoptosis and normal pattern of PKB/Akt phosphorylation. Furthermore, Pten negatively regulated intracellular levels of phosphatidylinositol 3,4,5-trisphosphate in cells and dephosphorylated it in vitro. These results showed that PTEN may exert its role as a tumor suppressor by negatively regulating the PI3K/PKB/Akt signaling pathway.

Arch, E. M.; Goodman, B. K.; Van Wesep, R. A.; Liaw, D.; Clarke, K.; Parsons, R.; McKusick, V. A.; Geraghty, M. T.: Deletion of PTEN in a patient with Bannayan-Riley-Ruvalcaba syndrome suggests allelism with Cowden disease. Am. J. Med. Genet. 71: 489-493, 1997.

Di Cristofano, A.; Pesce, B.; Cordon-Cardo, C.; Pandolfi, P. P.: Pten is essential for embryonic development and tumour suppression. Nature Genet. 19: 348-355, 1998.

Eng, C.; Peacocke, M.: PTEN and inherited hamartoma-cancer syndromes. (Letter) Nature Genet. 19: 223 only, 1998.

Hansen, G. M.; Justice, M. J.: Pten, a candidate tumor suppressor gene, maps to mouse chromosome 19. Mammalian Genome 9: 88-90, 1998.

Kong, D.; Suzuki, A.; Zou, T.-T.; Sakurada, A.; Kemp, L. W.; Wakatsuki, S.; Yokoyama, T.; Yamakawa, H.; Furukawa, T.; Sato, M.; Ohuchi, N.; Sato, S.; Yin, J.; Wang, S.; Abraham, J. M.; Souza, R. F.; Smolinski, K. N.; Meltzer, S. J.; Horii, A.: PTEN1 is frequently mutated in primary endometrial carcinomas. (Letter) Nature Genet. 17: 143-144, 1997.

Li, J.; Yen, C.; Liaw, D.; Podsypanina, K.; Bose, S.; Wang, S. I.; Puc, J.; Miliaresis, C.; Rodgers, L.; McCombie, R.; Bigner, S. H.; Giovanella, B. C.; Ittmann, M.; Tycko, B.; Hibshoosh, H.; Wigler, M. H.; Parsons, R.: PTEN, a putative protein tyrosine phosphatase gene mutated in human brain, breast, and prostate cancer. Science 275: 1943-1946, 1997.

Liaw, D.; Marsh, D. J.; Li, J.; Dahia, P. L. M.; Wang, S. I.; Zheng, Z.; Bose, S.; Call, K. M.; Tsou, H. C.; Peacocke, M.; Eng, C.; Parsons, R.: Germline mutations of the PTEN gene in Cowden disease, an inherited breast and thyroid cancer syndrome. Nature Genet. 16: 64-67, 1997.
Marsh, D. J.; Coulon, V.; Lunetta, K. L.; Rocca-Serra, P.; Dahia, P. L. M.; Zheng, Z.; Liaw, D.; Caron, S.; Duboue, B.; Lin, A. Y.; Richardson, A.-L.; Bonnetblanc, J.-M.; and 25 others: Mutation spectrum and genotype-phenotype analyses in Cowden disease and Bannayan-Zonana syndrome, two hamartoma syndromes with germline PTEN mutation. Hum. Molec. Genet. 7: 507-515, 1998.
Marsh, D. J.; Dahia, P. L. M.; Zheng, Z.; Liaw, D.; Parsons, R.; Gorlin, R. J.; Eng, C.: Germline mutations in PTEN are present in Bannayan-Zonana syndrome. (Letter) Nature Genet. 16: 333-334, 1997.
Nelen, M. R.; Padberg, G. W.; Peeters, E. A.; Lin, A. Y.; van den Helm, B.; Frants, R. R.; Coulon, V.; Goldstein, A. M.; van Reen, M. M.; Easton, D. F.; Eeles, R. A.; Hodgsen, S.; and 10 others: Localization of the gene for Cowden disease to chromosome 10q22-23. Nature Genet. 13: 114-116, 1996.
Nelen, M. R.; van Staveren, W. C. G.; Peeters, E. A. J.; Ben Hassel, M.; Gorlin, R. J.; Hamm, H.; Lindboe, C. F.; Fryns, J.-P.; Sijmons, R. H.; Woods, D. G.; Mariman, E. C. M.; Padberg, G. W.; Kremer, H.: Germline mutations in the PTEN/MMAC1 gene in patients with Cowden disease. Hum. Molec. Genet. 6: 1383-1387, 1997.
Olschwang, S.; Serova-Sinilnikova, O. M.; Lenoir, G. M.; Thomas, G.: PTEN germ-line mutations in juvenile polyposis coli. (Letter) Nature Genet. 18: 12-14, 1998.
Stambolic, V.; Suzuki, A.; de la Pompa, J. L.; Brothers, G. M.; Mirtsos, C.; Sasaki, T.; Ruland, J.; Penninger, J. M.; Siderovski, D. P.; Mak, T. W.: Negative regulation of PKB/Akt-dependent cell survival by the tumor suppressor PTEN. Cell 95: 29-39, 1998.
Steck, P. A.; Pershouse, M. A.; Jasser, S. A.; Yung, W. K. A.; Lin, H.; Ligon, A. H.; Langford, L. A.; Baumgard, M. L.; Hattier, T.; Davis, T.; Frye, C.; Hu, R.; Swedlund, B.; Teng, D. H. F.; Tavtigian, S. V.: Identification of a candidate tumour suppressor gene, MMAC1, at chromosome 10q23.3 that is mutated in multiple advanced cancers. Nature Genet. 15: 356-362, 1997.
Tamura, M.; Gu, J.; Matsumoto, K.; Aota, S.; Parsons, R.; Yamada, K. M.: Inhibition of cell migration, spreading, and focal adhesions by tumor suppressor PTEN. Science 280: 1614-1617, 1998.
Tsou, H. C.; Ping, X. L.; Xie, X. X.; Gruener, A. C.; Zhang, H.; Nini, R.; Swisshelm, K.; Sybert, V.; Diamond, T. M.; Sutphen, R.; Peacocke, M.: The genetic basis of Cowden's syndrome: three novel mutations in PTEN/MMAC1/TEP1. Hum. Genet. 102: 467-473, 1998.
Tsou, H. C.; Teng, D. H.-F.; Ping, X. L.; Brancolini, V.; Davis, T.; Hu, R.; Xie, X. X.; Gruener, A. C.; Schrager, C. A. Christiano, A. M.; Eng, C.; Steck, P.; Ott, J.; Tavtigian, S. V.; Peacocke, M.: The role of MMAC1 mutations in early-onset breast cancer: causative in association with Cowden syndrome and excluded in BRCA1-negative cases. Am. J. Hum. Genet. 61: 1036-1043, 1997.

HEMANGIOMATOSIS, CUTANEOUS, WITH ASSOCIATED FEATURES (234800)

JJM NOTE

It would be interesting to revisit this family, update the clinical information with a current perspective on related syndromes, and try mapping the gene (e.g., against von Hippel-Lindau or Cowden disease loci).

Gluszcz et al. (1963) described 4 sibs with cutaneous hemangiomatosis, acrocyanosis, hyperflexibility of joints, and phimosis. Some showed slight abnormalities of the vertebral bodies and ocular hypertelorism. In 2 (a female aged 15 and a male aged 19), tumors resembling cerebellar angioblastoma of von Hippel-Lindau disease (193300) were removed from the cervicothoracic portion of the spinal canal. See disseminated hemangiomatosis (106070).

Gluszcz, A.; Polis, Z.; Waleszkowski, J.: Familial syndrome of general dysplasia of the connective tissue and of the vascular system associated with angioblastoma of the spinal canal. Pol. Med. J. 2: 924-936, 1963.

MACROCEPHALY, MULTIPLE LIPOMAS, AND HEMANGIOMAS (*153480)

JJM NOTE

The eponymic cacophony should not detract from recognition of a pattern of vascular malformations and neoplasms, angiolipoma, hamartomatous intestinal polyps, and, in one case, meningioma. The proband in an unreported family I followed at the NIH for more than a decade had bilateral breast cancer in her thirties, two serious arteriovenous malformations, and a brother whose head was so large that we needed to use the body CT scanner to image it.

A number sign (#) is used with this entry because of evidence that this disorder results from mutations in the PTEN gene (601728). The genetic etiology may be heterogenous since Carethers et al. (1998) could find no PTEN germline mutations in 3 sporadic cases of this disorder.

Bannayan (1971) first described this disorder in a single child observed at autopsy at the Johns Hopkins Hospital. 27,28:Zonana et al. (1975, 1976) described the disorder in a father and 2 sons. One son had overgrowth of the right index finger and involvement of the small bowel mesentery by hamartoma with angiomatous, lipomatous and lymphangiomatous components. Miles et al. (1981) documented the syndrome in 11 persons in 4 additional families. Features included high palate, scaphocephaly, lipomas of the anterior abdominal wall, thigh, perineum, scapula area, etc., hemangiomas of the anterior abdominal wall, wrist, knee, and foot, bleeding from intracranial hemangioma, and arteriovenous malformation leading to leg amputation. Computerized axial tomography showed no enlargement of the cerebral ventricles. The lipomas spontaneously regress with age. The affected persons show high birthweight and length, but growth levels off at age 6 or 7 years. Bone age is normal. Motor development is delayed and incoordination is a lifelong feature. Speech development is delayed and intelligence is usually mildly retarded. Seizures result from intracranial hemorrhage. Drooling is a problem in children. Some children have pectus excavatum. No pseudopapilledema is observed in this disorder. The disorder appears to be an autosomal dominant but about 80% of affected persons have been male. In 1 instance the disorder was transmitted by a man with a head of normal size. Higginbottom and Schultz (1982) described the syndrome in 3 generations of an American black kindred. They concluded that affected persons may have an increased risk of intracranial tumors: a woman in their family had meningothelial meningioma removed at age 28.

Riley and Smith (1960) described mother and 2 of 7 children who had macrocephaly, pseudopapilledema, and multiple hemangiomata. Two other sibs had macrocephaly and pseudopapilledema. Intellect and vision were unimpaired. Ruvalcaba et al. (1980) described as instances of the Sotos syndrome (117550) 2 unrelated patients with macrocephaly, intestinal polyposis, and pigmented macules of the penis. Smith (1982) subsequently suggested that these patients had a different disorder. DiLiberti et al. (1983) described a patient with RMSS whose mother was probably affected. The proband was a 7.5-year-old boy with macrocephaly, hamartomatous intestinal polyps, and cafe-au-lait spots on the penis. The mother had macrocephaly, a facial appearance similar to the son's, and a hamartomatous intestinal polyp. DiLiberti et al. (1983) added prominent Schwalbe lines (a frequent normal finding but a consistent feature of 'anterior chamber cleavage syndromes'), prominent corneal nerves, and lipid storage myopathy as features of the disorder. On the basis of 4 patients, DiLiberti et al. (1984) extended the description of the lipid storage myopathy associated with this disorder. The patients had delayed psychomotor development and/or hypotonia in childhood. Electromyography in 3 patients showed evidence of a myopathic process. Muscle biopsy in all 4 showed a lipid storage myopathy with increased numbers of neutral lipid droplets—predominantly in type 1 fibers. Type 2 fibers were consistently smaller than expected. Dvir et al. (1988) described a patient with macrocephaly, pseudopapilledema, enlarged penis, lipoangiomatosis, and spotted pigmentation of the glans penis. The patient's father and a brother had macrocephaly; the father had enlarged penis. The authors suggested that the Riley-Smith syndrome and the Bannayan-Zonana syndrome (BZS) are the same as the disorder first reported by Riley and Smith (1960). Dvir et al. (1988) proposed to unify the 3 conditions in 1 hereditar y syndrome and name it macrocephaly-hamartomas-papilledema.

Gorlin (1988) also suggested that RMSS is the same as the Bannayan-Zonana syndrome. He referred to the case of an 8-year-old male who had 50 or more hamartomas of the bowel through which he lost serum protein. The diagnosis of BZS had been made, but Gorlin found that the patient had speckled penis. In a 38-year-old man (R.D.T., JHH 2358085) with macrocephaly, multiple lipomas, and vascular anomalies, Pyeritz (1988) observed 'unstable angina,' dilated aortic root, and ascending aorta. Halal and Silver (1989) described an 8.5-year-old boy with slowly progressive macrocephaly, psychomotor retardation, multiple subcutaneous angiolipomas, hypertelorism, exotropia, prolonged drooling to age 5 years, cutis marmorata, telangiectases over the shoulders, atrial septal defect repaired at age 4 years, broad thumbs and great toes, and muscle wasting. The angiolipomas were bluish subcutaneous nodules scattered all over his body. The parents were second cousins. The father was thought to have minor manifestations of the disorder, i.e., mild hypertelorism and broad thumbs and great toes, as well as mild cutis marmorata over the inner aspect of the forearms and a cluster of telangiectases and dilated vessels on the anterior aspect of both legs. Halal and Silver (1989) concluded that the proband and his father may have had the BZ syndrome (with previously undescribed additional anomalies) overlapping with the syndrome of cutis marmorata telangiectatica congenita (CMTC; 219250); or, alternatively, that the disorder in the father and son represents a new syndrome of macrocephaly and hamartomas with overlapping manifestations with the BZ and CMTC syndromes. Moretti-Ferreira et al. (1989) commented on the variability of severity, which may represent different allelic mutations or genetic heterogeneity. In the sporadic cases of Bannayan (1971) and of Okumura et al. (1986), death occurred as a result of severe visceral lipomatosis at ages 3.5 years and 5.75 years, respectively. C ohen (1990) suggested the designation of Bannayan-Riley-Ruvalcaba syndrome (BRRS). DiLiberti (1990) questioned the evidence that these syndromes are the same. In an 11-year-old male with features of the Bannayan-Zonana syndrome, Israel et al. (1991) found a 19;Y translocation in circulating lymphocytes: 46X,t(Y;19)(q11;q13). They raised the possibility that a small deletion or position effect of chromosome 19q is responsible for this syndrome. DiLiberti (1992) extended the boundaries of this entry further to include benign familial macrocephaly (153470, 248000). He examined the muscle biopsy results from 14 children with macrocephaly and hypotonia/weakness and correlated them with clinical findings. Of the 14, 13 had evidence of lipid storage myopathy, either generalized or focal. All 13 had examinations consistent with either benign familial macrocephaly, Ruvalcaba-Myhre-Smith syndrome, or Bannayan-Zonana syndrome. DiLiberti (1992) suggested that all 3 of these disorders may represent phenotypic variability resulting from mutation at a single genetic locus. Gorlin et al. (1992) reported a kindred with 12 affected members. They expanded the phenotypic spectrum to include Hashimoto thyroiditis, which occurred in 7 of the patients.

Fryburg et al. (1994) suggested that a defect in long-chain fatty acid oxidation resulting from deficiency of long-chain-L-3-hydroxyacyl-CoA dehydrogenase (LCHAD; 143450) may be responsible for the lipid myopathy in the Bannayan-Riley-Ruvalcaba syndrome. Their patient had macrocephaly with prominent frontal bossing and low-set ears, hypertelorism, and hemangiomas.

Zigman et al. (1997) localized the gene for the Bannayan-Riley-Ruvalcaba syndrome to 10q23 by study of 2 patients with intestinal juvenile polyposis and karyotypic abnormalities involving 10q. Patient 1 was a boy who displayed cognitive and developmental delay, macrocephaly, subcutaneous hamartoma, and hypotonia, in addition to multiple intestinal juvenile polyps. He ultimately underwent a total abdominal colectomy for refractory anemia and failure to thrive. Patient 2 was a girl who had features consistent with Bannayan-Riley-Ruvalcaba syndrome, including hypotonia, cognitive and developmental delay, and multiple intestinal juvenile polyps. This patient also had additional features not typical of BRRS, including atrial and ventricular septal defects, dysplastic pulmonic valve, left superior vena cava draining into a dilated coronary sinus, and bilateral clubfoot. (Three other unrelated patients with the diagnosis of BRRS but without karyotypic abnormalities were also studied; 2 of the 3 patients had pigmentary changes of the penis.) Patient 1 had an unbalanced translocation between chromosomes 10 and 9; patient 2 has an interstitial deletion of 10q23.1-q24.2. In each case, DNA markers allowed localization of the defect to 10q23.

Arch et al. (1997) described an 18-month-old patient previously thought to have the BRR syndrome. They found an interstitial deletion of 10q23.2-q24.1. This is the same region as that to which Cowden disease (158350) maps, as well as the PTEN gene (601728), which is the site of causative mutations in Cowden disease. Cowden disease has been described mainly in adults. Because of the considerable phenotypic overlap between the BRR syndrome and Cowden disease and because of a demonstration in their patient that the PTEN gene was deleted on chromosome 10, Arch et al. (1997) suggested that these are allelic disorders. The 18-month-old patient had macrocephaly (head circumference more than 4 S.D. above mean for age), facial dysmorphism (prominent forehead and hypertelorism), a lipoma on the right abdomen and on the right thumb, 2 small hemangiomas on the back and shoulder, and 2 skin tags at the base of the spine and in the groin area. Psychomotor development was delayed. Ophthalmologic examination showed pseudopapilledema bilaterally. Neither parent had any features of either Cowden disease or BRR syndrome. Hematochezia prompted upper and lower gastrointestinal endoscopy, which showed multiple sessile polyps throughout the duodenum and large intestine. Histology showed that these were hamartomatous.

Marsh et al. (1997) demonstrated germline mutations in the PTEN gene in patients with Bannayan-Zonana syndrome. Thus, Cowden disease and BZS are allelic disorders. One of the mutations that Marsh et al. (1997) found associated with BZS, R233X (601728.0002), had previously been reported in association with Cowden disease. The identical mutation occurred in 2 unrelated families with 2 distinct syndromes on 2 different 10q22-q23 haplotypes, arguing against a common ancestor or a founder effect.

Carethers et al. (1998) failed to find PTEN germline mutations in 3 sporadic cases of Bannayan-Riley-Ruvalcaba syndrome. One patient, a male, presented with macrocephaly, hypotonia, cognitive and developmental delays, cutaneous lipomas, and a 2-cm intestinal metaplastic polyp located in the ascending colon. The second patient, also a male, presented with macrocephaly, multiple intestinal juvenile polyps, pigmentation of the genitalia, cutaneous and visceral lipomas, cutaneous hemangiomas, and hyporeflexia. Patient 3, another male, presented with macrocephaly, multiple rectal juvenile polyps, pigmentary spotting of the penis, cutaneous lipomas, and cognitive and developmental delay.

DiLiberti (1998) proposed a new nomenclature reflecting the unification of multiple syndromes that are now known to be caused by mutations in the PTEN gene. He proposed that it be called the PTEN MATCHS syndrome; MATCHS was derived from macrocephaly, autosomal dominant, thyroid disease, cancer, hamartomata, and skin abnormalities.

See also Miles et al. (1984)

Arch, E. M.; Goodman, B. K.; Van Wesep, R. A.; Liaw, D.; Clarke, K.; Parsons, R.; McKusick, V. A.; Geraghty, M. T.: Deletion of PTEN in a patient with Bannayan-Riley-Ruvalcaba syndrome suggests allelism with Cowden disease. Am. J. Med. Genet. 71: 489-493, 1997.

Bannayan, G. A.: Lipomatosis, angiomatosis, and macrencephalia: a previously undescribed congenital syndrome. Arch. Path. 92: 1-5, 1971.

Carethers, J. M.; Furnari, F. B.; Zigman, A. F.; Lavine, J. E.; Jones, M. C.; Graham, G. E.; Teebi, A. S.; Huang, H.-J. S.; Ha, H. T.; Chauhan, D. P.; Chang, C. L.; Cavenee, W. K.; Boland, C. R.: Absence of PTEN/MMAC1 germ-line mutations in sporadic Bannayan-Riley-Ruvalcaba syndrome. Cancer. Res. 58: 2724-2726, 1998.

Cohen, M. M., Jr.: Bannayan-Riley-Ruvalcaba syndrome: renaming three formerly recognized syndromes as one etiologic entity. (Letter) Am. J. Med. Genet. 35: 291 only, 1990.

DiLiberti, J. H.: Inherited macrocephaly-hamartoma syndromes. Am. J. Med. Genet. 79: 284-290, 1998.

DiLiberti, J. H.: Comments on Dr. Cohen's letter. (Letter) Am. J. Med. Genet. 35: 292 only, 1990.

DiLiberti, J. H.: Correlation of skeletal muscle biopsy with phenotype in the familial macrocephaly syndromes. J. Med. Genet. 29: 46-49, 1992.

DiLiberti, J. H.; D'Agostino, A. N.; Ruvalcaba, R. H. A.; Schimschock, J. R.: A new lipid storage myopathy observed in individuals with the Ruvalcaba-Myhre-Smith syndrome. Am. J. Med. Genet. 18: 163-167, 1984.

DiLiberti, J. H.; Weleber, R. G.; Budden, S.: Ruvalcaba-Myhre-Smith syndrome: a case with probable autosomal-dominant inheritance and additional manifestations. Am. J. Med. Genet. 15: 491-495, 1983.

Dvir, M.; Beer, S.; Aladjem, M.: Heredofamilial syndrome of mesodermal hamartomas, macrocephaly, and pseudopapilledema. Pediatrics 81: 287-290, 1988.

Fryburg, J. S.; Pelegano, J. P.; Bennett, M. J.; Bebin, E. M.: Long-chain 3-hydroxyacyl-coenzyme A dehydrogenase (L-CHAD) deficiency in a patient with the Bannayan-Riley-Ruvalcaba syndrome. Am. J. Med. Genet. 52: 97-102, 1994.

Gorlin, R. J.: Personal Communication. Minneapolis, Minn. 7/18/1988.

Gorlin, R. J.; Cohen, M. M., Jr.; Condon, L. M.; Burke, B. A.: Bannayan-Riley-Ruvalcaba syndrome. Am. J. Med. Genet. 44: 307-314, 1992.

Halal, F.; Silver, K.: Slowly progressive macrocephaly with hamartomas: a new syndrome?. Am. J. Med. Genet. 33: 182-185, 1989.

Higginbottom, M. C.; Schultz, P.: The Bannayan syndrome: an autosomal dominant disorder consisting of macrocephaly, lipomas, hemangiomas, and risk for intracranial tumors. Pediatrics 69: 632-634, 1982.

Israel, J.; Lessick, M.; Szego, K.; Wong, P.: Translocation 19;Y in a child with Bannayan-Zonana phenotype. J. Med. Genet. 28: 427-428, 1991.

Marsh, D. J.; Dahia, P. L. M.; Zheng, Z.; Liaw, D.; Parsons, R.; Gorlin, R. J.; Eng, C.: Germline mutations in PTEN are present in Bannayan-Zonana syndrome. (Letter) Nature Genet. 16: 333-334, 1997.

Miles, J. H.; Zonana, J.; Mcfarlane, J.; Aleck, K. A.; Bawle, E.: Macrocephaly with hamartomas: Bannayan-Zonana syndrome. Am. J. Med. Genet. 19: 225-234, 1984.

Miles, J. H.; Zonana, J.; Mcfarlane, J. P.; Aleck, K.; Bawle, E.: Familial macrocephaly with lipomas and hemangiomas. (Abstract) Am. J. Hum. Genet. 33: 86A only, 1981.

Moretti-Ferreira, D.; Koiffmann, C. P.; Souza, D. H.; Diament, A. J.; Wajntal, A.: Macrocephaly, multiple lipomas, and hemangiomata (Bannayan-Zonana syndrome): genetic heterogeneity or autosomal dominant locus with at least two different allelic forms?. Am. J. Med. Genet. 34: 548-551, 1989.

Okumura, K.; Sasaki, Y.; Ohyama, M.; Nishi, T.: Bannayan syndrome: generalized lipomatosis associated with megalencephaly and macrodactyly. Acta Path. Jpn. 36: 269-277, 1986.

Pyeritz, R. E.: Personal Communication. Baltimore, Md. 12/9/1988.

Riley, H. D., Jr.; Smith, W. R.: Macrocephaly, pseudopapilledema and multiple hemangiomata: a previously undescribed heredofamilial syndrome. Pediatrics 26: 293-300, 1960.

Ruvalcaba, R. H. A.; Myhre, S.; Smith, D. W.: Sotos syndrome with intestinal polyposis and pigmentary changes of the genitalia. Clin. Genet. 18: 413-416, 1980.

Smith, D. W.: Recognizable Patterns of Human Malformations. Philadelphia: W. B. Saunders (pub.) (3rd ed.): 1982. Pp. 387 only.

Zigman, A. F.; Lavine, J. E.; Jones, M. C.; Boland, C. R.; Carethers, J. M.: Localization of the Bannayan-Riley-Ruvalcaba syndrome gene to chromosome 10q23. Gastroenterology 113: 1433-1437, 1997.

Zonana, J.; Davis, D.; Rimoin, D. L.: Multiple lipomas, hemangiomas and macrocephaly—an autosomal dominant hamartomatous syndrome. (Abstract) Am. J. Hum. Genet. 27: 97A only, 1975.

Zonana, J.; Rimoin, D. L.; Davis, D. C.: Macrocephaly with multiple lipomas and hemangiomas. J. Pediat. 89: 600-603, 1976.

MYXOMA, SPOTTY PIGMENTATION, AND ENDOCRINE OVERACTIVITY (CARNEY SYNDROME) (*160980)

JJM NOTE

Based on some 60 cases, the Carney syndrome is diagnosed when two of these features are seen: cardiac myxoma, cutaneous myxoma, myxoid mammary fibroadenoma, spotty mucocutaneous pigmentation including lentigines and blue nevi, primary pigmented nodular adrenocortical disease, testicular tumor, growth hormone-secreting pituitary adenoma, and psammomatous melanotic schwannoma (Carney JA, Young WF: Primary pigmented nodular adrenocortical disease and its associated conditions. Endocrinologist 21:1440–44, 1992). Other malignant tumors specifically include thyroid carcinoma, large-cell calcifying Sertoli cell tumor, liposarcoma, pheochromocytoma, acoustic neuroma, and adrenal cortical carcinoma (Nwokoro et al. 1997). Two separate cardiac myxomas from our patient showed a chromosome 10q26 breakpoint in culture (Gollin SM et al.: 10q26: Consistent chromosomal breakpoint in two cardiac myxomas from a patient with Carney syndrome. Am. J. Hum. Genet. 57:343, 1995). In some, but not all families, the gene maps to chromosome 2p16 (Stratakis CA et al.: Carney complex: diagnosis and management of the complex of spotty skin pigmentation, myxomas, endocrine overactivity, and schwannomas. Am. J. Med. Genet. 80:183–85, 1998; Milunsky J et al.: Evidence for genetic heterogeneity of the Carney complex. Cancer Genet. Cytogenet. 106:173–76, 1998).

Atherton et al. (1980) described a 10-year-old boy with profuse cutaneous pigmented lesions, subcutaneous myxoid neurofibromata, and atrial myxoma. At birth, 3 pigmented lesions were noted on the neck, trunk and thigh; a large number of pigmented lesions developed in the first few weeks of life. These were always more prominent in the summer months. A myxoid liposarcoma was removed from behind the right ear at age 2 and again at age 8, and 2 others were removed from the chin and anterior chest at age 3. Both parents were heavily 'freckled,' although less so than the patient. At age 10, a left atrial myxoma arising from the interatrial septum was removed. Part of the tumor extended through the foramen ovale into the right atrium, and a smaller, separate tumor was attached to the free wall of the right atrium opposite the foramen. The boy had blue eyes and hair of a distinctive rust-red color. Atherton et al. (1980) suggested the designation NAME syndrome, an acronym for the several features: nevi, atrial myxoma, myxoid neurofibromata, ephelides. Koopman and Happle (1991) suggested that it would be more useful to use the acronym NAME for nevi, atrial myxoma, mucinosis of the skin, and endocrine overactivity. Rees et al. (1973) reported a case of this syndrome in a young man with red hair and fair skin. Autosomal dominant inheritance was suggested. Follow-up by Atherton et al. (1980) referred to a palatal tumor with characteristics of myxoid neurofibroma, as in skin tumors of the Atherton case.

Schweizer-Cagianut et al. (1980) described a brother and sister in whom, respectively, Cushing disease was diagnosed at 18 years and bilateral adrenalectomy for Cushing disease was performed at age 28 years. Significantly, hyperpigmentation did not occur in either. The brother had fibromas of the skin, leading to the diagnosis of von Recklinghausen disease, but cafe-au-lait spots were neither described nor evidenced by the published photographs. In both of the sister's pregnancies, flaccid hemiparesis developed at the end of the second month and spontaneous abortion occurred. She had documented intracranial bleed, fibromas of the eyelid, and microcalcification of the breasts. A brother had died at age 5 years of atrial myxoma; he had a hemangioma of the right groin. A sister had syndactyly, hypertelorism, and low IQ. The parents were normal and nonconsanguineous. The adrenals in the brother and sister showed the typical lesions of microadenomatosis, or primary adrenocortical nodular dysplasia, with foci of eosinophilic giant cells. The results of adrenal function tests were compatible with autonomy of the adrenals. Bilateral adrenocortical nodular dysplasia has often been described previously but only in sporadic cases with one exception. Arce et al. (1978) described the identical histologic picture in 3 sibs they interpreted as having familial hypothalamic-pituitary Cushing disease. A fourth sib had virilizing adrenocortical carcinoma. It is likely that Schweizer-Cagianut et al. (1980) described a syndrome of which adrenocortical microadenomatosis is merely one feature. This prediction, made soon after the publication of Schweizer-Cagianut et al. (1980), was quickly confirmed by the conclusion of Schweizer-Cagianut et al. (1982) that cardiac myxoma was a part of this 'peculiar familial syndrome.' A sib in the family they reported apparently did not have the adrenal disorder but died of cardiac myxoma at age 5. The sister originally reported was found at autopsy to have a cardiac myxoma; the fibroma of her eyelid was reinterpreted as a 'myxoma'; both breasts contained multiple small 'benign fibroadenomas with an unusual myxomatous and vascularized stroma' and she was noted to have had finely freckled pigmentation around the mouth and lips.' Shenoy et al. (1984) reported 4 patients with Cushing syndrome and an unusual bilateral adrenal disorder, the same as that described above but termed by them 'primary pigmented nodular adrenocortical disease.' Barlow et al. (1983) described 2 sisters with the combination of Cushing syndrome, cardiac myxomas, other myxoid tumors, and spotty facial and labial pigmentation.

Carney et al. (1985) presented evidence for the existence of a distinct syndrome consisting of spotty cutaneous pigmentation, myxomas of the heart and elsewhere, and Cushing syndrome from nodular adrenocortical dysplasia. From the report of the family history of case 14 of Carney et al. (1985), it is clear that endocrine adenomatosis is not limited to the adrenal cortex: a brother of the patient had pigmented spots of the face and lips, had had nodular and pedunculated myxomas of the skin removed, and at age 21 was found to have acromegaly caused by pituitary adenoma. Successful hypophysectomy was performed. In 2 of 40 patients reported by Carney et al. (1985), a single pigmented macule was found in the mouth. Thus, buccal spotting is uncommon, in contrast to the Peutz-Jeghers syndrome (175200) in which it is a standard feature. In addition to the face, eyelids, ears, and vermilion borders of the lips, sites of spots included conjunctiva or sclera, vulva, back of hands and fingers, anal verge, and glans penis. In 9 of 17 male patients, testicular tumors were found; they were bilateral in 7 patients and multicentric in each affected testis. The testicular tumors were large-cell calcifying Sertoli cell tumor, Leydig cell tumor, or adrenocortical rest tumor. Sexual precocity occurred with the first 2 types. Large-cell calcifying Sertoli cell tumors are ordinarily very rare; in Carney's series they occurred in more than half the males. In 2 patients an unusual tumor (in the retroperitoneum in one, in a suprapatellar pouch in the other) occurred and was termed 'calcifying pigmented neuroectodermal tumor' by the authors. Carney et al. (1985) suggested that the NAME syndrome, as well as the LAMB syndrome (lentigines, atrial myxoma, mucocutaneous myxoma, and blue nevi) reported by Rhodes et al. (1984), represent this pleiotropic syndrome of cutaneous, cardiac, and endocrine involvement. The mode of inheritance is not established. The nature of the lesions suggests autosomal domin ant inheritance but the familial cases seem to have been limited to 1 generation. However, in the case of Carney's patient 14, who had affected sibs, the mother had pigmented spots on the eyelids, lips, back of hands, and vulva, and had had myxoid tumors removed from an ear, breast, and lower eyelid. Dominant, probably autosomal dominant, inheritance was indicated by the findings of Carney et al. (1986), who have done much to delineate this syndrome of lentigines, cardiac myxoma, and endocrine overactivity. At least one manifestation of the syndrome occurred in 3 successive generations of the family. Both males and females were affected, and 5 of 11 children of affected persons had the disorder. No male-to-male transmission was found in this or in any reported family (until the report of Koopman and Happle, 1991). A brother of the proband had acromegaly due to a growth-hormone secreting tumor of the pituitary. Carney et al. (1986) followed up on a family reported by Proppe and Scully (1980) in the original description of large-cell calcifying Sertoli cell tumor of the testis. Two brothers in their family III had, in addition to the testis tumor, cardiac myxoma and nodular adrenocortical hyperplasia. Study showed typical pigmentation and left atrial myxoma in the mother.

Wilsher et al. (1986) described an affected mother and her son and daughter. The mother had left and right atrial myxomata. Her daughter had a left ventricular myxoma, subcutaneous myxoid neurofibromata, and mammary fibroadenosis. The son had no evidence of cardiac myxoma. However, the figures demonstrated that all 3 had melanin spots around the lips and over the bridge of the nose. The authors referred to these as ephelides, i.e., freckles. Vidaillet et al. (1987) found 5 cases of what they termed 'syndrome myxoma' out of a total of 75 patients with cardiac myxoma seen at the Mayo Clinic between January 1954 and December 1985. They compared 49 cases of syndrome myxoma with cases of sporadic myxoma; the former showed a younger age, a higher frequency of familial occurrence, ventricular location of tumor (13% vs 0%), multiple tumors (50% vs 1%), and recurrent tumor (18% vs 0%). Kennedy et al. (1987) emphasized the diagnostic usefulness of ophthalmologic abnormalities. Of 63 patients studied, eyelid myxomas were present in 16%, facial and eyelid lentigines in 70%, and pigmented lesions on the caruncle or conjunctival semilunar fold in 27%. Spotty pigmentation of the skin involved the vermilion border of the lips but only infrequently affected the buccal mucosa. Young et al. (1989) provided a 50-year follow-up on a patient seen at the Mayo Clinic at the age of 17. A photograph taken at that time showed spotty pigmentation involving the face and vermilion borders of the lips as well as the chest and shoulders. The patient also showed signs of Cushing syndrome, which first developed at the age of 13. The adrenals were found to be mottled brown in appearance and, histologically, to have nodules composed of large adrenocortical cells with moderately intense brown cytoplasmic pigment. She had 2 daughters, one when she was 24 and the other when she was 27. After the delivery of her second child, her hirsutism disappeared and she lost her excessive body weight. Examination at the age of 66 showed no features of Cushing syndrome. The spotty facial pigmentation had faded but was still evident. A fraternal twin of this woman had the same features of Cushing syndrome and spotty facial and labial pigmentation. His development was normal until age 13 when growth stopped and weight gain started. He remained overweight but his spotty facial pigmentation faded with age. He died suddenly at the age of 44 after elective left inguinal hernioplasty. Autopsy showed severe atherosclerotic narrowing of the left anterior descending coronary artery, but no cardiac myxoma. The adrenals showed 2- to 3-mm reddish brown bulging nodules and, histologically, coarsely granular brown pigment. Both daughters of the proband had spotty facial pigmentation, and one had primary pigmented nodular adrenal cortical disease for which bilateral total adrenalectomy was performed. One of the daughters also had a nasopharyngeal schwannoma. Several other members of the family had spotty facial pigmentation. Handley et al. (1992) described the Carney syndrome in a mother and her son and daughter. All 3 had varying degrees of centrofacial/mucosal lentigines and cutaneous myxoid tumors. The mother had myxoid mammary fibroadenomatosis and a left atrial myxoma; her daughter developed a prolactin-secreting pituitary adenoma; the son had bilateral large-cell calcified Sertoli cell testicular tumors and an axillary psammomatous melanotic schwannoma.

Carney (1995) described a fascinating search, which was eventually successful, for Harvey Cushing's case of Minnie G., who was reported in his 1912 monograph on the pituitary and its disorders. Carney (1995) postulated that this patient might have had 'his' syndrome. The patient, then 23 years old, had been referred to Cushing at The Johns Hopkins Hospital in 1910. Minnie, whose actual given name was Maita, died in 1958 at the age of 71 years. Having assembled the family pedigree, Carney (1995) could find no evidence of other affected members in the extended family.

Stratakis et al. (1995) and Stratakis et al. (1996) studied 101 patients from 11 North American kindreds with Carney complex. Skin myxomas, cardiac myxomas, Cushing syndrome, and acromegaly were present in, respectively, 62, 30, 31, and 8% of the patients. They demonstrated linkage to markers on the short arm of chromosome 2 (2p16) with a maximum 2-point lod score of 5.97 at theta = 0.03 for linkage to a (CA)n dinucleotide repeat polymorphism. Candidate genes in the proximity, including the proopiomelanocortin gene (176830) and the DNA-mismatch repair gene MSH2 (120435), were excluded.

Stratakis et al. (1996) examined 15 tumor and normal tissue specimens from 13 patients with Carney complex. DNA was extracted from peripheral blood, tumor cell lines, and frozen or paraffin-embedded tissues and subjected to PCR amplification with primers from 64 microsatellite locations covering chromosomes 1 and 3 to 22 and 14 loci on chromosome 2. The alterations detected were loss and gain of heterozygosity (LOH and GOH; 49% and 26%, respectively), deletions of both alleles (DEL; 10%), and microsatellite length instability (15%). GOH and LOH were the most frequent changes, with telomeric markers significantly overrepresented (P less than 0.05). Chromosomes 6, 11, 22, 10, and 19 demonstrated mostly LOH, GOH, or DEL in over 40% of the informative loci tested (73, 59, 47, 46, and 44%, respectively), whereas markers on chromosome 2 showed only microsatellite length instability (10%). The authors concluded that tumors and tumor cell lines from patients with Carney complex demonstrate significant genomic, but not microsatellite length, instability. Thus, the Carney complex gene or genes on chromosome 2p16 are different from the MSH2 and MSH6 genes and have a dominant, rather than recessive, tumor function. These genes appear to be involved in the regulation of genomic stability of dividing cells, in particular the structure of telomeres in replicating chromosomes and/or the function of the mitotic apparatus.

Stratakis et al. (1997) reviewed the medical records of 12 families with Carney complex (53 affected patients) for evidence of thyroid abnormalities. The authors identified 2 patients with thyroid carcinoma (1 papillary and 1 follicular; 3.8%) and 1 with follicular adenoma in 3 unrelated kindreds. They also studied 6 affected members of these kindreds for the presence of thyroid disease (familial cases) and 5 patients with Carney complex who had no affected relatives (sporadic cases). These 11 patients consisted of 5 adults and 6 children or adolescents. All had normal results of physical and biochemical examination of the thyroid gland (total and free T4, T3, and TSH levels). Thyroid ultrasonography showed hypoechoic, cystic, solid, or mixed lesions in 3 of the 5 adults (60%) and 4 of the 6 children (67%). Thyroid gland abnormalities were documented in 5 sibs and 1 parent-child pair. The authors concluded that thyroid gland pathology is common in patients with Carney complex, and includes a spectrum of abnormalities ranging from follicular hyperplasia and/or cystic changes to carcinoma. Since, additionally, thyroid gland pathology appeared to be inherited in an autosomal dominant manner, as are the other manifestations of Carney complex, the authors considered it to be a candidate component of the Carney complex syndrome.

Nwokoro et al. (1997) described a Carney syndrome family in which the proposita, a 34-year-old woman, had a history of multiple nevi, diffuse facial lentigines, and labial pigmentation, all apparent at an early age. In addition, right ventricular myxomas were resected at the age of 30. The patient had invasive follicular carcinoma of the thyroid gland, Barrett metaplasia of the esophagus, neoplastic colonic polyps, bipolar affective disorder, and atypical mesenchymal neoplasm of the uterine cervix distinct from the myxoid uterine leiomyoma usually seen in this syndrome. This patient's most aggressive manifestation was the thyroid carcinoma. Diagnosis of Carney syndrome was established also in her 9-year-old son and there was a probable diagnosis in her 12-year-old daughter. Endocrine manifestations were prominent in the family with at least 9 relatives in 3 generations affected with this aspect of Carney syndrome. Hirsutism was a feature of the affected members of the family. The mother of the proposita had hypopituitarism. Hyperparathyroidism was present in 3 affected relatives. Pituitary microadenoma and calcifying testicular tumor were present in 1 relative each.

Basson et al. (1997) presented evidence from linkage studies that the Carney complex is genetically heterogeneous. They studied a kindred with 7 affected individuals in 4 generations. All 6 living affected individuals exhibited pleomorphic spotty skin pigmentation typical of the Carney complex. All 4 of the 6 affected individuals who had reached adulthood had had left and/or right atrial myxomas. Three individuals whose atrial myxomas were resected had recurrence at least once, distant from the operative site. Three of them had histories of extracardiac myxomas, 2 in the breast and 1 in the vagina. Endocrine abnormalities were present in 3, 2 showed thyroid dysfunction, and 1 required adrenalectomy for Cushing syndrome. Linkage studies yielded lod scores of greater than -2.0 over the 10-cM interval between D2S391 and D2S393, where the Carney complex had previously been mapped.

See also Carney et al. (1986)

Arce, B.; Licea, M.; Hung, S.; Padron, R.: Familial Cushing's syndrome. Acta Endocr. 87: 139-147, 1978.

Atherton, D. J.; Pitcher, D. W.; Wells, R. S.; MacDonald, D. M.: A syndrome of various cutaneous pigmented lesions, myxoid neurofibromata and atrial myxoma: the NAME syndrome. Brit. J. Derm. 103: 421-429, 1980.

Barlow, J. F.; Abu-Gazeleh, S.; Tam, G. E.; Wirtz, P. S.; Ofstein, L. C.; O'Brien, C. P.; Woods, G. L.; Drymalski, W. G.: Myxoid tumor of the uterus and right atrial myxomas. S. Dakota J. Med. 36: 9-13, 1983.

Basson, C. T.; MacRae, C. A.; Korf, B.; Merliss, A.: Genetic heterogeneity of familial atrial myxoma syndromes (Carney complex). Am. J. Cardiol. 79: 994-995, 1997.

Carney, J. A.: The search for Harvey Cushing's patient, Minnie G., and the cause of her hypercortisolism. Am. J. Surg. Path. 19: 100-108, 1995.

Carney, J. A.; Gordon, H.; Carpenter, P. C.; Shenoy, B. V.; Go, V. L. W.: The complex of myxomas, spotty pigmentation, and endocrine overactivity. Medicine 64: 270-283, 1985.

Carney, J. A.; Headington, J. T.; Su, W. P. D.: Cutaneous myxomas: a major component of the complex of myxomas, spotty pigmentation, and endocrine overactivity. Arch. Derm. 122: 790-798, 1986.

Carney, J. A.; Hruska, H. S.; Beauchamp, G. D.; Gordon, H.: Dominant inheritance of the complex of myxomas, spotty pigmentation, and endocrine overactivity. Mayo Clin. Proc. 61: 165-172, 1986.

Handley, J.; Carson, D.; Sloan, J.; Walsh, M.; Thornton, C.; Hadden, D.; Bingham, E. A.: Multiple lentigines, myxoid tumours and endocrine overactivity; four cases of Carney's complex. Brit. J. Derm. 126: 367-371, 1992.

Kennedy, R. H.; Waller, R. R.; Carney, J. A.: Ocular pigmented spots and eyelid myxomas. Am. J. Ophthal. 104: 533-538, 1987.

Koopman, R. J. J.; Happle, R.: Autosomal dominant transmission of the NAME syndrome (nevi, atrial myxoma, mucinosis of the skin and endocrine overactivity). Hum. Genet. 86: 300-304, 1991.

Nwokoro, N. A.; Korytkowski, M. T.; Rose, S.; Gorin, M. B.; Stadler, M. P.; Witchel, S. F.; Mulvihill, J. J.: Spectrum of malignancy and premalignancy in Carney syndrome. Am. J. Med. Genet. 73: 369-377, 1997.

Proppe, K. H.; Scully, R. E.: Large-cell calcifying Sertoli cell tumor of the testis. Am. J. Clin. Path. 74: 607-619, 1980.

Rees, J. R.; Ross, F. G. M.; Keen, G.: Lentiginosis and left atrial myxoma. Brit. Heart J. 35: 874-876, 1973.

Rhodes, A. R.; Silverman, R. A.; Harrist, T. J.; Perez-Atayde, A. R.: Mucocutaneous lentigines, cardiomucocutaneous myxomas, and multiple blue nevi: the 'LAMB' syndrome. J. Am. Acad. Derm. 10: 72-82, 1984.

Schweizer-Cagianut, M.; Froesch, E. R.; Hedinger, C.: Familial Cushing's syndrome with primary adrenocortical microadenomatosis (primary adrenocortical nodular dysplasia). Acta Endocr. 94: 529-535, 1980.

Schweizer-Cagianut, M.; Salomon, F.; Hedinger, C. E.: Primary adrenocortical nodular dysplasia with Cushing's syndrome and cardiac myxomas: a peculiar familial disease. Virchows Arch. Path. Anat. 397: 183-192, 1982.

Shenoy, B. V.; Carpenter, P. C.; Carney, J. A.: Bilateral primary pigmented nodular adrenocortical disease: rare cause of the Cushing syndrome. Am. J. Surg. Path. 8: 335-344, 1984.

Stratakis, C. A.; Carney, J. A.; Lin, J.-P.; Papanicolaou, D. A.; Karl, M.; Kastner, D. L.; Pras, E.; Chrousos, G. P.: Carney complex, a familial multiple neoplasia and lentiginosis syndrome: analysis of 11 kindreds and linkage to the short arm of chromosome 2. J. Clin. Invest. 97: 699-705, 1996.
Stratakis, C. A.; Courcoutsakis, N. A.; Abati, A.; Filie, A.; Doppman, J. L.; Carney, J. A.; Shawker, T.: Thyroid gland abnormalities in patients with the syndrome of spotty skin pigmentation, myxomas, endocrine overactivity, and schwannomas (Carney complex). J. Clin. Endocr. Metab. 82: 2037-2043, 1997.
Stratakis, C. A.; Jenkins, R. B.; Pras, E.; Mitsiadis, C. S.; Raff, S. B.; Stalboerger, P. G.; Tsigos, C.; Carney, J. A.; Chrousos, G. P.: Cytogenetic and microsatellite alterations in tumors from patients with the syndrome of myxomas, spotty skin pigmentation, and endocrine overactivity (Carney complex). J. Clin. Endocr. Metab. 81: 3607-3614, 1996.
Stratakis, C. A.; Pras, E.; Lin, J.-P.; Kastner, D. L.; Carney, J. A.; Chrousos, G. P.: Carney complex, a multiple endocrine neoplasia and familial lentiginosis syndrome: clinical analysis and linkage to the D2S123 locus (chromosome 2p16). (Abstract) Am. J. Hum. Genet. 57: A54, 1995.
Vidaillet, H. J., Jr.; Seward, J. B.; Fyke, F. E., III; Su, W. P. D.; Tajik, A. J.: 'Syndrome myxoma': a subset of patients with cardiac myxoma associated with pigmented skin lesions and peripheral and endocrine neoplasms. Brit. Heart J. 57: 247-255, 1987.
Wilsher, M. L.; Synek, B. J. L.; Roche, A. H. G.; Holdaway, I. M.; Neutze, J. M.; Nicholson, G. I.: A familial syndrome of cardiac myxomas, myxoid neurofibromata, cutaneous pigmented lesions, and endocrine abnormalities. Aust. New Zeal. J. Med. 16: 393-396, 1986.
Young, W. F., Jr.; Carney, J. A.; Musa, B. U.; Wulffraat, N. M.; Lens, J. W.; Drexhage, H. A.: Familial Cushing's syndrome due to primary pigmented nodular adrenocortical disease: reinvestigation 50 years later. New Eng. J. Med. 321: 1659-1664, 1989.

NOONAN SYNDROME 1 (163950)

JJM NOTE

Besides the case reports cited by McKusick (malignant schwannoma, cervical hygromas, and cutaneous lymphangiomas), recent case reports suggest an association of Noonan syndrome with neuroblastoma (two patients), cancer in the cryptorchid testis, and vaginal rhabdomyosarcoma (Fryns JP, Moerman P: Trisomy 21 in a second trimester male hydropic fetus with "Noonan-syndrome-phenotype": nuchal and thoracic lymphangiomatosis, oedema of the dorsum of hands and feet, and bicuspid pulmonary valve. Genet. Couns. 9:59–60, 1998; Lopez-Miranda B et al.: Noonan syndrome associated with neuroblastoma: a case report. Pediatr. Radiol. 27:324–26, 1997; Cotton JL, Williams RG: Noonan syndrome and neuroblastoma. Arch. Pediatr. Adolesc. Med. 149:1280–81, 1995; Khan S et al.: Vaginal rhabdomyosarcoma in a patient with Noonan syndrome. J. Med. Genet. 32:743–45, 1995; Sriram K et al.: Noonan's syndrome with carcinoma of undescended testis. Ill. Med. J. 171:294–96, 1987). More intriguing is the emerging association with childhood acute lymphoblastic leukemia (some six cases worldwide, which could be by chance) and childhood chronic myelomonocytic leukemia (four cases in France). The latter evokes the association of that malignancy with neurofibromatosis 1, thereby strengthening the overlap between neurofibromatosis and Noonan syndrome (Bader-Meunier B et al.: Occurrence of myeloproliferative disorder in patients with Noonan syndrome. Pediatrics 130:885–89, 1997).

The disorder now known as Noonan syndrome bears similarities to the disorder described by Turner (1938) and shown by Ford et al. (1959) to have its basis in a 45,X chromosomal aberration. (With considerable justification, Ullrich's name is combined with that of Turner as the designation for the 45,X syndrome. Ullrich described the disorder 8 years before Turner. Wiedemann and Glatzl (1991) provided a follow-up of Ullrich's original patient with Ullrich-Turner syndrome and demonstrated that the restudy in 1987, when the patient was 66 years old, confirmed the 45,X chromosome constitution.)

Noonan (1968) reported 19 cases of whom 17 had pulmonary stenosis and 2 had patent ductus arteriosus. Twelve were males and 7 were females. Deformity of the sternum with precocious closure of sutures was a frequent feature. Kaplan et al. (1968) described 2 brothers with elevated alkaline phosphatase levels, one of whom also had malignant schwannoma of the forearm. Nora and Sinha (1968) observed mother-to-offspring transmission in 3 families; in 1 family, transmission was through 3 generations. They suggested X-linked dominant inheritance of either a single mutant gene or a submicroscopic deletion. Among 95 male patients with pulmonary stenosis, Celermajer et al. (1968) found the Turner phenotype in 8. In 5 of these, karyotyping was performed. In 4 the chromosomes were normal. In one an extra acrocentric chromosome was present. Abdel-Salem and Temtamy (1969) reported 2 affected sibs from a first-cousin marriage. A deceased female sib may have been affected also. They suggested autosomal recessive inheritance. Baird and De Jong (1972) described 7 cases in 3 generations. One affected woman had 5 affected children (out of 6) with 2 different husbands. Seizures and anomalous upper lateral incisors may have been coincidental. Diekmann et al. (1967) described 2 brothers and a sister, with normal and unrelated parents, who had somatic characteristics of the Noonan syndrome, particularly pterygium colli and deformed sternum, and had myocardiopathy leading to death at ages 12 and 10 in two of them. Migeon and Whitehouse (1967) described 2 families, each with 2 sibs with somatic features of the Turner syndrome. In 1 family, 2 brothers had webbing of the neck, coarctation of the aorta and cryptorchidism. In the second, a brother and sister were affected. Simpson et al. (1969) reported experiences suggesting that rubella embryopathy may result in the Turner phenotype, thereby accounting for either the male Turner syndrome or the female pseudo-Turner syndrome. A part icularly convincing pedigree for autosomal dominant inheritance was reported by Bolton et al. (1974), who found the condition in a man and 4 sons (in a sibship of 10). Four of the 5 affected persons had pulmonic stenosis. Father-to-son transmission was reported by Qazi et al. (1974).

Koretzky et al. (1969) described an unusual type of pulmonary valvular dysplasia which showed a familial tendency with either affected parent and offspring or affected sibs. Although some relatives had pulmonary valvular stenosis of the standard dome-shaped variety, the valvular dysplasia in others was characterized by the presence of three distinct cusps and no commissural fusion. The obstructive mechanism was related to markedly thickened, immobile cusps, with disorganized myxomatous tissue. Other features were retarded growth, abnormal facies (triangular face, hypertelorism, low-set ears and ptosis of the eyelids), absence of ejection click, and unusually marked right axis deviation by electrocardiogram. It now seems clear that the patients of Koretzky et al. (1969) had the Noonan syndrome. Mendez and Opitz (1985) stated that the Watson syndrome (193520) and the LEOPARD syndrome (151100) 'are essentially indistinguishable from the Noonan syndrome.' Witt et al. (1987) reviewed the occurrence of lymphedema in Noonan syndrome. When it does occur, it opens the possibility of prenatal diagnosis by imaging methods or by AFP level. Noonan syndrome was one of the causes found for posterior cervical hygroma in a series of previable fetuses studied by Kalousek and Seller (1987). The authors found, furthermore, that 45,X Turner syndrome lethal in the fetal period showed a constant association of 3 defects, posterior cervical cystic hygroma, generalized subcutaneous edema, and preductal aortic coarctation. Evans et al. (1991) found a large cutaneous lymphangioma of the right cheek and amegakaryocytic thrombocytopenia in a male infant with Noonan syndrome. Donnenfeld et al. (1991) presented a case of Noonan syndrome in which posterior nuchal cystic hygroma was diagnosed at 13 to 14 weeks of gestation by ultrasonography. The hygroma had regressed by the time of birth leaving nuchal skin fold redundancy and pterygium colli. On the basis of studies of genital tract fun ction in 11 adult males with Noonan syndrome, Elsawi et al. (1994) concluded that bilateral testicular maldescent was a main factor in contributing to impairment of fertility. Four of the 11 men had fathered children.

Thrombocytopenia occurs in some cases of the Noonan syndrome (Goldstein, 1979). Partial deficiency of factor XI was described by Kitchens and Alexander (1983). Out of 9 patients with Noonan syndrome, de Haan et al. (1988) found 4 with partial deficiency of factor XI (30-65% of normal). They reviewed the other reports of bleeding tendency associated with thrombocytopenia or with abnormal platelet function. Witt et al. (1988) described bleeding diathesis in 19 patients with Noonan syndrome. Several different defects were identified in the coagulation and platelet systems, occurring singly or in combination; for example, 2 patients had factor XI deficiency, 3 had presumptive von Willebrand disease, and 1 had thrombocytopenia. In 5 of the patients an unusually pungent odor of urine and sweat was noted by parents. One of these patients was reported by Humbert et al. (1970) as a case of trimethylaminuria (136131) and another patient was suspected of having this condition. Sharland et al. (1990) also described a variety of coagulation factor deficiencies. The most common abnormality was a partial factor XI deficiency in the heterozygote range, found in 21 of 31 patients. Of 72 patients studied (37 male, 35 female, mean age 11.4 years) by Sharland et al. (1992), 47 (65%) had a history of abnormal bruising or bleeding. In 29 patients (40%), prolonged activated partial thromboplastin time was found. In 36 patients (50%) specific abnormalities were found in the intrinsic pathway of coagulation, i.e., partial deficiency of factor XI:C, XII:C, and VIII:C. Multiple abnormalities among these 36 patients included combined deficiencies of factors XI and XII (4 patients), of factors XI and VIII (4 patients), and of factors VIII, XI, and XII (1 patient). In 5 families, similar coagulation-factor deficiencies were present in first-degree relatives. Sharland et al. (1992)

suggested that because of the involvement of several factors, either singly or in combination, there ar e likely to be regulatory factors that control the intrinsic (contact activation) system; that these factors are under chromosomal genetic control; and that abnormalities of this regulation occur in Noonan syndrome.

Allanson et al. (1985) studied the changes in facial appearance with age. They pointed out that the manifestations may be subtle in adults. Ranke et al. (1988) analyzed the clinical features of 144 patients from 2 West German centers. The size at birth was normal in both sexes. In both males and females, the mean height followed along the 3rd percentile until puberty, but decreased transiently due to an approximately 2-year delay in onset of puberty. Final height approaches the lower limits of normal at the end of the second decade of life. The mean adult height was 162.5 cm in males and 152.7 cm in females, respectively. Allanson (1987) provided a useful review. The fetal primidone syndrome, occurring in the offspring of mothers taking this anticonvulsant, closely simulates the Noonan syndrome. Baraitser and Patton (1986) reported 4 unrelated children (2 boys, 2 girls) with a Noonan-like syndrome associated with sparse hair as a conspicuous feature.

Leichtman (1996) reported a family suggesting that CFC syndrome (115150) is a variable expression of Noonan syndrome. He described a 4-year-old girl who had all of the manifestations of CFC syndrome (characteristic facial and cardiac anomalies, developmental delay, hypotrichosis, eczematic eruption with resistance to treatment), whose mother had typical characteristics of Noonan syndrome. Lorenzetti and Fryns (1996) reported a 13-year-old boy with Noonan syndrome and retinitis pigmentosa. Because similar eye defects are found in CFC syndrome, the authors suggested that CFC and Noonan syndromes might be variable manifestations of the same entity. However, Neri and Zollino (1996) noted distinctions between the patient reported by Lorenzetti and Fryns (1996) and CFC syndrome, and stated that similarity of eye defects is not enough to conclude that CFC and Noonan syndromes are the same condition.

Using a number of probes at the neurofibromatosis type I locus (NF1; 162200) in the study of 11 families with Noonan syndrome in 2 or 3 generations, Sharland et al. (1992) excluded proximal 17q as the location of the gene. Studying six 2-generation families with classic Noonan syndrome, Flintoff et al. (1993) could find no evidence of linkage of this disorder to NF1 on 17q or to NF2 (101000) on 22q. Edman Ahlbom et al. (1995) likewise could find no evidence of linkage between NF1 and Noonan syndrome with cafe-au-lait spots.

By means of a genomewide linkage analysis in a large Dutch kindred with autosomal dominant Noonan syndrome, Jamieson et al. (1994) localized the gene to chromosome 12; maximum lod = 4.04 at theta = 0.0. Linkage analysis using chromosome 12 markers in 20 smaller, 2-generation families gave a maximum lod of 2.89 at theta = 0.07, but haplotype analysis showed nonlinkage in 1 family. These data suggested that a gene for Noonan syndrome is located in the 12q22-qter region between markers D12S84 and D12S366. Clinical studies in this kindred were reported by van der Burgt et al. (1994).

Brady et al. (1997) further analyzed the 3-generation Dutch family studied by Jamieson et al. (1994) using newly isolated CA-repeat markers derived from the interval between D12S84 and D12S366. In this way they were able to reduce the localization to an interval bounded by markers D12S105 and NOS1 (163731), which has been mapped to 12q24.2-q24.31.

Legius et al. (1998) performed linkage analysis in a 4-generation Belgian family with Noonan syndrome in some individuals and CFC syndrome in others. Clinical and linkage data in this family indicated that the 2 syndromes result from variable expression of the same genetic defect. They found a maximum lod score of 4.43 at zero recombination for marker D12S84 in 12q24. A crossover in this pedigree narrowed the candidate gene region to a 5-cM interval between D12S84 and D12S1341. A remarkable feature of the family studied by Legius et al. (1998) was the presence of 3 dizygotic twins in the offspring of 2 affected females. A dizygotic twin pair was observed in the offspring of an affected female in the family in which linkage was studied by Jamieson et al. (1994). It is possible that an increased frequency of dizygotic twinning is associated with NS1/CFC linked to 12q24. The fragile X syndrome (309550) is another mendelian disorder with a possibly increased frequency of dizygotic twinning (Partington et al., 1996; Schwartz et al., 1994).

Robin et al. (1995) described 6 patients with Noonan syndrome who underwent molecular evaluation for submicroscopic deletion of chromosome 22q11. None of these patients presented with conotruncal heart defects. Evidence for 22q11 hemizygosity was demonstrated in only 1 patient. This patient had Noonan-like manifestations without clinical manifestations of DiGeorge or velocardiofacial syndromes. Digilio et al. (1996) studied 4 patients with Noonan syndrome and tetralogy of Fallot. Chromosome analysis was normal in all 4 patients. DNA analysis showed no hemizygosity for the 22q11 region in any of the patients.

History

Cole (1980) pointed out that the blacksmith in the famous painting 'Among Those Left' by Ivan Le Lorraine Albright appears to have had Noonan syndrome. The contour of the sternum, the low-set ears, and the short stature are suggestive. Genetic confirmation was provided by studies of a great-grandson with general features of the Noonan syndrome and cardiac abnormalities consistent with that diagnosis (pulmonic stenosis and regurgitation, abnormal architecture of the left ventricular musculature). Opitz and Pallister (1979) reproduced the first published illustration of the Noonan syndrome by Kobylinski (1883), and Opitz (1985) republished the photograph of Rickey E., the first patient with 'her' syndrome studied at the State University of Iowa by Jacqueline A. Noonan.

See also Alslev and Reinwein (1958); Char et al. (1972); Duncan et al. (1981); Fisher et al. (1982); Golabi et al. (1985); Hall et al. (1982); Levy et al. (1970); Linde et al. (1973); Miller and Motulsky (1978); Nora et al. (1974); Pierini and Pierini (1979); Sharland et al. (1992); Wiedemann (1991); Witt et al. (1985)

Abdel-Salem, E.; Temtamy, S. A.: Familial Turner phenotype. J. Pediat. 74: 67-72, 1969.

Allanson, J. E.: Noonan syndrome. J. Med. Genet. 24: 9-13, 1987.

Allanson, J. E.; Hall, J. G.; Hughes, H. E.; Preus, M.; Witt, R. D.: Noonan syndrome: the changing phenotype. Am. J. Med. Genet. 21: 507-514, 1985.

Alslev, J.; Reinwein, H.: Ueber das familiaere Vorkommen des sogenannten Ullrich-Turner-Syndromes und das Vorhandensein eines pterygium colli, eines Kryptorchismus und des Meige-Syndromes bei zwei Bruedern mit kongenitalen Vitien. Dtsch. Med. Wschr. 83: 601-604, 1958.

Baird, P. A.; De Jong, B. P.: Noonan's syndrome (XX and XY Turner phenotype) in three generations of a family. J. Pediat. 80: 110-114, 1972.

Baraitser, M.; Patton, M. A.: A Noonan-like short stature syndrome with sparse hair. J. Med. Genet. 23: 161-164, 1986.

Bolton, M. R.; Pugh, D. M.; Mattioli, L. F.; Dunn, M. I.; Schimke, R. N.: The Noonan syndrome: a family study. Ann. Intern. Med. 80: 626-629, 1974.

Brady, A. F.; Jamieson, C. R.; van der Burgt, I.; Crosby, A.; van Reen, M.; Kremer, H.; Mariman, E.; Patton, M. A.; Jeffery, S.: Further delineation of the critical region for Noonan syndrome on the long arm of chromosome 12. Europ. J. Hum. Genet. 5: 336-337, 1997.

Celermajer, J. M.; Bowdler, J. D.; Cohen, D. H.: Pulmonary stenosis in patients with the Turner phenotype in the male. Am. J. Dis. Child. 116: 351-358, 1968.

Char, F.; Rodriquez-Fernandez, H. L.; Scott, C. I., Jr.; Borgaonkar, D. S.; Bell, B. B.; Rowe, R. D.: The Noonan syndrome—a clinical study of forty-five cases. Birth Defects Orig. Art. Ser. VIII(5): 110-118, 1972.

Cole, R. B.: Noonan's syndrome: a historical perspective. Pediatrics 66: 468-469, 1980.

de Haan, M.; van de Kamp, J. J. P.; Briet, E.; Dubbeldam, J.: Noonan syndrome: partial factor XI deficiency. Am. J. Med. Genet. 29: 277-282, 1988.

Diekmann, L.; Pfeiffer, R. A.; Hilgenberg, F.; Bender, F.; Reploh, H. D.: Familiaere Kardiomyopathie mit Pterygium colli. Muench. Med. Wschr. 109: 2638-2645, 1967.

Digilio, M. C.; Marino, B.; Giannotti, A.; Dallapiccola, B.: Exclusion of 22q11 deletion in Noonan syndrome with tetralogy of Fallot. (Letter) Am. J. Med. Genet. 62: 413-414, 1996.

Donnenfeld, A. E.; Nazir, M. A.; Sindoni, F.; Librizzi, R. J.: Prenatal sonographic documentation of cystic hygroma regression in Noonan syndrome. Am. J. Med. Genet. 39: 461-465, 1991.

Duncan, W. J.; Fowler, R. S.; Farkas, L. G.; Ross, R. B.; Wright, A. W.; Bloom, K. R.; Huot, D. J.; Sondheimer, H. M.; Rowe, R. D.: A comprehensive scoring system for evaluating Noonan syndrome. Am. J. Med. Genet. 10: 37-50, 1981.

Edman Ahlbom, B.; Dahl, N.; Zetterqvist, P.; Anneren, G.: Noonan syndrome with cafe-au-lait spots and multiple lentigines syndrome are not linked to the neurofibromatosis type 1 locus. Clin. Genet. 48: 85-89, 1995.

Elsawi, M. M.; Pryor, J. P.; Klufio, G.; Barnes, C.; Patton, M. A.: Genital tract function in men with Noonan syndrome. J. Med. Genet. 31: 468-470, 1994.

Evans, D. G. R.; Lonsdale, R. N.; Patton, M. A.: Cutaneous lymphangioma and amegakaryocytic thrombocytopenia in Noonan syndrome. Clin. Genet. 39: 228-232, 1991.

Fisher, E.; Weiss, E. B.; Michals, K.; DuBrow, I. W.; Hastrieter, A. R.; Matalon, R.: Spontaneous chylothorax in Noonan's syndrome. Europ. J. Pediat. 138: 282-284, 1982.

Flintoff, W. F.; Bahuau, M.; Lyonnet, S.; Gilgenkrantz, S.; Lacombe, D.; Marcon, F.; Levilliers, J.; Kachaner, J.; Munnich, A.; Le Merrer, M.: No evidence for linkage to the type 1 or type 2 neurofibromatosis loci in Noonan syndrome families. Am. J. Med. Genet. 46: 700-705, 1993.

Ford, C. E.; Miller, O. J.; Polani, P. E.; de Almeida, J. C.; Briggs, J. H.: A sex-chromosome anomaly in a case of gonadal dysgenesis (Turner's syndrome). Lancet 1: 711-713, 1959.

Golabi, M.; Hall, B. D.; Clericuzio, C.; Johnston, K.: Fetal primidone syndrome: a distinct pattern of malformations. (Abstract) Am. J. Hum. Genet. 37: A54 only, 1985.

Goldstein, J. L.: Personal Communication. Dallas, Tex. 1/25/1979.

Hall, J. G.; Reed, S. D.; Greene, G.: The distal arthrogryposes: delineation of new entities—review and nosologic discussion. Am. J. Med. Genet. 11: 185-239, 1982.

Humbert, J. R.; Hammond, K. B.; Hathaway, W. E.; Marcoux, J. G.; O'Brien, D.: Trimethylaminuria: the fish-odour syndrome. (Letter) Lancet II: 770-771, 1970.

Jamieson, C. R.; van der Burgt, I.; Brady, A. F.; van Reen, M.; Elsawi, M. M.; Hol, F.; Jeffery, S.; Patton, M. A.; Mariman, E.: Mapping a gene for Noonan syndrome to the long arm of chromosome 12. Nature Genet. 8: 357-360, 1994.

Kalousek, D. K.; Seller, M. J.: Differential diagnosis of posterior cervical hygroma in previable fetuses. Am. J. Med. Genet. Suppl. 3: 83-92, 1987.
Kaplan, M. S.; Opitz, J. M.; Gosset, F. R.: Noonan's syndrome. A case with elevated serum alkaline phosphatase levels and malignant schwannoma of the left forearm. Am. J. Dis. Child. 116: 359-366, 1968.
Kitchens, C. S.; Alexander, J. A.: Partial deficiency of coagulation factor XI as a newly recognized feature of Noonan syndrome. J. Pediat. 102: 224-227, 1983.
Kobylinski, O.: Ueber eine flughautahnliche Ausbreitung am Halse. Arch. Anthrop. 14: 343-348, 1883.
Koretzky, E. D.; Moller, J. H.; Korns, M. E.; Schwartz, C. J.; Edwards, J. E.: Congenital pulmonary stenosis resulting from dysplasia of valve. Circulation 40: 43-53, 1969.
Legius, E.; Schollen, E.; Matthijs, G.; Fryns, J.-P.: Fine mapping of Noonan/cardio-facio cutaneous syndrome in a large family. Europ. J. Hum. Genet. 6: 32-37, 1998.
Leichtman, L. G.: Are cardio-facio-cutaneous syndrome and Noonan syndrome distinct? A case of CFC offspring of a mother with Noonan syndrome. Clin. Dysmorph. 5: 61-64, 1996.
Levy, E. P.; Pashayan, H.; Fraser, F. C.; Pinsky, L.: XX and XY Turner phenotypes in a family. Am. J. Dis. Child. 120: 36-43, 1970.
Linde, L. M.; Turner, S. W.; Sparkes, R. S.: Pulmonary valvular dysplasia, a cardiofacial syndrome. Brit. Heart J. 35: 301-304, 1973.
Lorenzetti, M.-E.; Fryns, J.-P.: Retinitis pigmentosa in a young man with Noonan syndrome: further evidence that Noonan syndrome (NS) and the cardio-facio-cutaneous syndrome (CFC) are variable manifestations of the same entity? Am. J. Med. Genet. 65: 97-99, 1996.
Mendez, H. M. M.; Opitz, J. M.: Noonan syndrome: a review. Am. J. Med. Genet. 21: 493-506, 1985.
Migeon, B. R.; Whitehouse, D.: Familial occurrence of the somatic phenotype of Turner's syndrome. Johns Hopkins Med. J. 120: 78-80, 1967.
Miller, M.; Motulsky, A.: Noonan syndrome in an adult family presenting with chronic lymphedema. Am. J. Med. 65: 379-383, 1978.
Neri, G.; Zollino, M.: More on the Noonan-CFC Controversy (Editorial) Am. J. Med. Genet. 65: 100-(only), 1996.
Noonan, J. A.: Hypertelorism with Turner phenotype. A new syndrome with associated congenital heart disease. Am. J. Dis. Child. 116: 373-380, 1968.
Nora, J. J.; Nora, A. H.; Sinha, A. K.; Spangler, R. D.; Lubs, H. A., Jr.: The Ullrich-Noonan syndrome (Turner phenotype). Am. J. Dis. Child. 127: 48-55, 1974.
Nora, J. J.; Sinha, A. K.: Direct familial transmission of the Turner phenotype. Am. J. Dis. Child. 116: 343-350, 1968.
Opitz, J. M.: The Noonan syndrome. (Editorial) Am. J. Med. Genet. 21: 515-518, 1985.
Opitz, J. M.; Pallister, P. D.: Brief historical note: the concept of 'gonadal dysgenesis.'. Am. J. Med. Genet. 4: 333-343, 1979.
Partington, M. W.; Moore, D. Y.; Turner, G. M.: Confirmation of early menopause in fragile X carriers. Am. J. Med. Genet. 64: 370-372, 1996.
Pierini, D. O.; Pierini, A. M.: Keratosis pilaris atrophicans faciei (ulerythema ophryogenes) a cutaneous marker in the Noonan syndrome. Brit. J. Derm. 100: 409-416, 1979.
Qazi, Q. H.; Arnon, R. G.; Paydar, M. H.; Mapa, H. C.: Familial occurrence of Noonan syndrome. Am. J. Dis. Child. 127: 696-698, 1974.
Ranke, M. B.; Heidemann, P.; Knupfer, C.; Enders, H.; Schmaltz, A. A.; Bierich, J. R.: Noonan syndrome: growth and clinical manifestations in 144 cases. Europ. J. Pediat. 148: 220-227, 1988.
Robin, N. H.; Sellinger, B.; McDonald-McGinn, D.; Zackai, E. H.; Emanuel, B. S.; Driscoll, D. A.: Classical Noonan syndrome is not associated with deletions of 22q11. Am. J. Med. Genet. 56: 94-96, 1995.
Schwartz, C. E.; Dean, J.; Howard-Peebles, P. N.; Bugge, M.; Mikkelson, M.; Tommerup, N.; Hull, C.; Hagerman, R.; Holden, J. J.; Stevenson, R. E.: Obstetrical and gynecological complications in fragile X carriers: a multicenter study. Am. J. Med. Genet. 51: 400-402, 1994.
Sharland, M.; Patton, M.; Chittolie, A.; Talbot, S.; Bevan, D.: Coagulation factor abnormalities in Noonan syndrome. (Abstract) J. Med. Genet. 27: 646 only, 1990.
Sharland, M.; Patton, M. A.; Talbot, S.; Chitolie, A.; Bevan, D. H.: Coagulation-factor deficiencies and abnormal bleeding in Noonan's syndrome. Lancet 339: 19-21, 1992.
Sharland, M.; Taylor, R.; Patton, M. A.; Jeffery, S.: Absence of linkage of Noonan syndrome to the neurofibromatosis type 1 locus. J. Med. Genet. 29: 188-190, 1992.
Simpson, J. W.; Nora, J. J.; Singer, D. B.; McNamara, D. G.: Multiple valvular sclerosis in Turner phenotypes and rubella syndrome. Am. J. Cardiol. 23: 94-97, 1969.
Turner, H. H.: A syndrome of infantilism, congenital webbed neck and cubitus valgus. Endocrinology 23: 566-574, 1938.
van der Burgt, I.; Berends, E.; Lommen, E.; van Beersum, S.; Hamel, B.; Mariman, E.: Clinical and molecular studies in a large Dutch family with Noonan syndrome. Am. J. Med. Genet. 53: 187-191, 1994.
Wiedemann, H.-R.: Otto Ullrich and his syndromes. Am. J. Med. Genet. 41: 128-133, 1991.
Wiedemann, H.-R.; Glatzl, J.: Follow-up of Ullrich's original patient with 'Ullrich-Turner' syndrome. Am. J. Med. Genet. 41: 134-136, 1991.
Witt, D.; Allanson, J.; Wadsworth, L.; Hall, J. G.: Bleeding disorders in 7 cases of Noonan syndrome; further evidence of heterogeneity. (Abstract) Am. J. Hum. Genet. 37: A83 only, 1985.
Witt, D. R.; Hoyme, E.; Zonana, J.; Manchester, D. K.; Fryns, J. P.; Stevenson, J. G.; Curry, C. J. R.; Hall, J. G.: Lymphedema in Noonan syndrome: clues to pathogenesis and prenatal diagnosis and review of the literature. Am. J. Med. Genet. 27: 841-856, 1987.
Witt, D. R.; McGillivray, B. C.; Allanson, J. E.; Hughes, H. E.; Hathaway, W. E.; Zipursky, A.; Hall, J. G.: Bleeding diathesis in Noonan syndrome: a common association. Am. J. Med. Genet. 31: 305-317, 1988.

BLUE RUBBER BLEB NEVUS (BEAN SYNDROME) (*112200)

JJM NOTE

Given the rarity of the disorder, the associated cases with medulloblastoma, chronic lymphocytic leukemia, and renal cell carcinoma may be coincidental.

This is a bladderlike variety of hemangioma found particularly on the trunk and upper arms. Nocturnal pain and regional hyperhidrosis are features. Bleeding hemangiomas of the gastrointestinal tract are an important complication. Berlyne and Berlyne (1960) demonstrated transmission through 5 generations. Other cases have been sporadic, perhaps new dominant mutations. Fretzin and Potter (1965) described a particularly dramatic case with involvement of the skin and gastrointestinal tract and angiomatous gigantism of the right arm requiring amputation in infancy. In a single case in a Japanese woman, Sakurane et al. (1967) described cavernous hemangiomas characteristic of blue rubber bleb nevi over the entire surface of the body and in the mucosa of the oropharynx, esophagus, distal ileum and anus. In addition the patient had multiple enchondromatosis. This, then, had many of the features of Maffucci syndrome (see 166000). Two families with affected persons in 3 and 5 successive generations, supporting autosomal dominant inheritance, were reported by Walshe et al. (1966). Bean (1958) gave the name to this condition which, furthermore, he was mainly instrumental in delineating. Rice and Fischer (1962) observed the association of cerebellar medulloblastoma. They illustrated the extraordinary appearance of the skin lesions. Intestinal hemangiomas were found at autopsy. Munkvad (1983) reported a family with 7 affected persons in 3 generations, including father-to-son transmission. The skin tumors are rubber-like nipples, easily compressible and promptly refilling after compression. They vary in color, size, shape and number and may be tender. The affected persons in Munkvad's pedigree had no evidence of visceral abnormality. Satya-Murti et al. (1986) described a 19-year-old man with extensive central nervous involvement who had a chronic, slowly progressive, and nonfatal course.

See 600195 for a familial venous malformation syndrome (VMCM) that may be the same as the blue rubber bleb nevus syndrome of Bean (1958), which maps to 9p. Gallione et al. (1995) suggested that VMCM is identical to the Bean syndrome. Several members of the family they studied had gastrointestinal bleeding from vascular lesions just as did the family originally described by Bean (1958).

See also Fine et al. (1961); Hoffman et al. (1978); McCauley et al. (1979); Morris et al. (1978); Nakagawara et al. (1977); Talbot and Wyatt (1970)

Bean, W. B.: Vascular Spiders and Related Lesions of the Skin. Springfield, Ill.: Charles C Thomas (pub.) 1958. Pp. 178-185.
Berlyne, G. M.; Berlyne, N.: Anaemia due to 'blue-rubber-bleb' naevus disease. Lancet II: 1275-1277, 1960.
Fine, R. M.; Derbes, V. J.; Clark, W. H.: Blue rubber bleb nevus. Arch. Derm. 84: 802-805, 1961.
Fretzin, D. F.; Potter, B.: Blue rubber bleb nevus. Arch. Intern. Med. 116: 924-929, 1965.
Gallione, C. J.; Pasyk, K. A.; Boon, L. M.; Lennon, F.; Johnson, D. W.; Helmbold, E. A.; Markel, D. S.; Vikkula, M.; Mulliken, J. B.; Warman, M. L.; Pericak-Vance, M. A.; Marchuk, D. A.: A gene for familial venous malformations maps to chromosome 9p in a second large kindred. J. Med. Genet. 32: 197-199, 1995.
Hoffman, T.; Chasko, S.; Safai, B.: Association of blue rubber bleb nevus syndrome with chronic lymphocytic leukemia and hypernephroma. Johns Hopkins Med. J. 142: 91-94, 1978.
McCauley, R. G. K.; Leonidas, J. C.; Bartoshesky, L. E.: Blue rubber bleb nevus syndrome. Radiology 133: 375-377, 1979.
Morris, S. J.; Kaplan, S. R.; Ballan, K.; Tedesco, F. J.: Blue rubber-bleb nevus syndrome. J.A.M.A. 239: 1887 only, 1978.
Munkvad, M.: Blue rubber bleb nevus syndrome. Dermatologica 167: 307-309, 1983.
Nakagawara, G.; Asano, E.; Kimura, S.; Akimoto, R.; Miyazaki, I.: Blue rubber bleb nevus syndrome: report of a case. Dis. Colon Rectum 20: 421-427, 1977.
Rice, J. S.; Fischer, D. S.: Blue rubber bleb nevus syndrome. Arch. Derm. 86: 503-511, 1962.
Sakurane, H. F.; Sugai, T.; Saito, T.: The association of blue rubber bleb nevus and Maffucci's syndrome. Arch. Derm. 95: 28-36, 1967.
Satya-Murti, S.; Navada, S.; Eames, F.: Central nervous system involvement in blue-rubber-bleb-nevus syndrome. Arch. Neurol. 43: 1184-1186, 1986.
Talbot, S.; Wyatt, E. H.: Blue rubber bleb naevi (report of a family in which only males were affected). Brit. J. Derm. 82: 37-39, 1970.
Walshe, M. M.; Evans, C. D.; Warin, R. P.: Blue rubber bleb naevus. Brit. Med. J. 2: 931-932, 1966.

BECKWITH-WIEDEMANN EMG SYNDROME (*130650)

JJM NOTE

This disorder is one of several that predispose, perhaps by dysregulation of growth processes, to several embryonal tumors of the abdomen (Wilms tumor, hepatoblastoma, neuroblastoma), as well as to adrenal cortical benign and malignant neoplasms, and perhaps rhabdomyosarcoma.

A number sign (#) is used with this entry because of evidence that mutation of the p57(KIP2) gene (CDKN1C; 600856) causes the disorder.

Description

This disorder, which is now most often referred to as Beckwith-Wiedemann syndrome (BWS), has as its cardinal features in the neonate exomphalos, macroglossia, and gigantism. This was the origin of the initialism EMG syndrome, used earlier as the preferred designation. BWS patients are at increased risk of developing specific tumors. BWS syndrome is caused by mutation in the chromosome 11p15.5 region. Duplication in this region appears to be involved in the pathogenesis and imprinting results in anomalous patterns of transmission. Most cases are sporadic.

Clinical Features

The enlarged tongue, together with omphalocele or other umbilical abnormalities, permits recognition of the disorder at birth. Because many of the affected infants have hypoglycemia in the first days of life, anticipation of this complication can prevent serious neurologic sequelae. Visceromegaly, adrenocortical cytomegaly, and dysplasia of the renal medulla are conspicuous features. Adrenal carcinoma, nephroblastoma, hepatoblastoma, and rhabdomyosarcoma occur with increased frequency.

42,43:Irving (1967, 1970) initially described the 'typical linear indentations of the lobe' that have become one of the diagnostic criteria, also well documented by Best and Hoekstra (1981). Peculiar posterior helical ear pits were first described in the BWS by Kosseff et al. (1972) and later by many others (see Best, 1991).

Two reported patients had hearing loss due to fixation of the stapes (Paulsen, 1973 and Daugbjerg and Everberg, 1984). In 3 patients, BWS and type III polycystic kidney disease occurred simultaneously (Mulvihill et al., 1989). An adult woman developed a progressive virilization due to her androgen-secreting adrenal carcinoma (Clouston et al., 1989).

A review of 31 patients with BWS and malignant tumors showed that 18 had Wilms tumor (Sotelo-Avila et al., 1980). Wiedemann (1983) reported that of 388 children, 29 developed 32 neoplasms. Of these tumors, 26 were intra-abdominal, 14 being Wilms tumors and 5 adrenocortical carcinoma. Hemihypertrophy, partial or complete, was noted in 12.5% of the cases but in more than 49% of the children with neoplasms.

Wiedemann (1989) commented on overgrowth of the external genitalia in both males and females with BWS. Sippell et al. (1989) reported longitudinal data on height, bone maturation, weight, and pubertal development in 7 children with BWS. The children reached an average height of 2.5 SD above the mean at or after puberty. Growth velocity was above the ninetieth percentile until 4 to 6 years of age, and normal thereafter. Bone age was significantly advanced in all patients studied. One of the patients had latent hypothyroidism. The association of BWS and thyroid disorders may be more than coincidental (Leung, 1985 and Leung and McArthur, 1989). Emery et al. (1983) reported 2 affected sibs, one with thoracic neuroblastoma and the other who died at age 2 months of cardiomyopathy and respiratory failure.

A 'new' aspect of the natural history of BWS was reported by Chitayat et al. (1990) who observed 2 infants who were apparently normal at birth but later developed characteristics of the disorder. Both had hypoglycemia neonatally and gradually developed coarse facial changes, umbilical hernia, and macroglossia. Renal sonography done after the macroglossia developed showed large kidneys in both. The placenta was carefully examined in each case but findings described as typical of BWS were found only in one. Chitayat et al. (1990) postulated that the cellular hyperplasia and hypertrophy characteristic of BWS is caused by persistent rests of embryonal cells that secrete paracrine and/or endocrine growth factors and that the effects may not become evident until postnatal life. Neuroblastoma is another form of embryonal neoplasm that occurs in BWS (Chitayat et al., 1990). Falik-Borenstein et al. (1991) described an affected infant with congenital gastric teratoma.

In a study of 53 affected children, Carlin et al. (1990) suggested that this disorder may be milder in many cases than one would guess from published descriptions. In 11 families (21%), more than one child had BWS, including 2 sets of twins, one monozygotic and one dizygotic. Additionally, 24 families had one or sometimes both parents, and/or other relatives, affected with one or more signs of BWS. They suggested that hemihypertrophy is an underappreciated diagnostic clue for BWS in the relatives of probands. Knight et al. (1980) and Watanabe and Yamanaka (1990) described prune belly syndrome (100100) in association with BWS.

Mental retardation was documented in 6 out of 39 observed by Martinez-y-Martinez et al. (1992), one being related to neonatal hypoglycemia.

Elliott et al. (1994) observed 76 patients with Beckwith-Wiedemann syndrome. The criteria for diagnosis were the presence of 3 major findings (macroglossia, pre- or postnatal growth greater than the 90th centile, and abdominal wall defects) or 2 major findings plus minor manifestations. In this preselected group, macroglossia was found in 97% of the patients, overgrowth in 88%, and abdominal wall defects in 80%. Hypoglycemia occurred in 48 patients and neoplasias in 3. Intellectual development was normal in all. Congenital heart defects were reported in 5 patients. Three patients had postaxial polydactyly of the foot. In one family, the mother of the index case had an ear pit and macroglossia as a child. In one family, 2 first cousins were affected. In 2 other families, 2 sibs were affected. Of 68 apparently sporadic cases, 15 had a relative with minor features of the syndrome. Elliott et al. (1994) suggested that incomplete penetrance may lead to underdiagnosis of familial cases.

Weng et al. (1995) reported the results of a follow-up study on 15 patients with WBS. They found that the pregnancies in these cases tended to have polyhydramnios with large placentas that were almost twice the normal placental weight. The large fetal size and polyhydramnios often resulted in early delivery with occasional perinatal mortality (observed in 3 cases). Excessive umbilical cord length was a manifestation of the increased placental size and was a useful sign in suspecting WBS before delivery. Abdominal wall defects and/or macroglossia helped confirm the diagnosis at birth. The newborn patients were almost 2 standard deviations above the expected mean length and weight for gestational age. The trend to increased size continued through early childhood and became less dramatic with increasing age. No cytogenetic abnormality was detected in 9 patients studied and the only tumor detected was a gastric teratoma evident in one infant at birth. Four of 15 patients had surgery for macroglossia. The findings were compared with those of Pettenati et al. (1986), who studied 22 patients.

Drut and Drut (1996) studied affected members of a family in which 4 members had WBS as a result of trisomy 11p15. Clinical examination showed nonimmune hydrops and placentomegaly in 2 sibs and multiple phenotypic abnormalities consistent with WBS in the 2 other relatives.

Inheritance

The mode of inheritance is complex. Possible patterns include autosomal dominant inheritance with variable expressivity, contiguous gene duplication at 11p15, and genomic imprinting resulting from a defective or absent copy of the maternally derived gene.

Wiedemann (1964) reported 3 affected sibs, and Irving (1967) observed a family with 2 affected sibs and an affected second cousin. Autosomal dominant inheritance was suggested by Kosseff et al. (1972), Forrester (1973), and Lubinsky et al. (1974). Mausuura et al. (1975) presented a pedigree in which each of 3 sibships related as second cousins had 1 case. Kosseff et al. (1976) reviewed the pedigrees of this syndrome and invoked premutation (a special form of dominant inheritance) to explain the findings in some pedigrees. Chemke (1976) described 8 cases in 2 sibships of an inbred kindred. Seven of those affected died in the neonatal period. Piussan et al. (1980) described a family in which all 6 infants (including twins) appeared to have had BWS. Five who died in the neonatal period had congenital omphalocele. The surviving twin presented only minimal umbilical hernia, which reduced spontaneously. Autopsy in the last-born confirmed the diagnosis of BWS. The parents were normal and not related. Sommer et al. (1977) reported a kindred in which 3 normal sisters gave birth to 8 infants with BWS. Autosomal dominant inheritance with the purported phenomenon of 'delayed mutation' was proposed. Best and Hoekstra (1981) described BWS in a mother, her brother, and 2 of her children by different fathers. Nivelon-Chevallier et al. (1983) described a family in which 4 offspring of 3 normal sisters were affected. In 2 cases, antenatal diagnosis was established by ultrasonography which showed exomphalos. In one of these cases, histologic examination of the abortus confirmed the diagnosis.

Niikawa et al. (1986) analyzed 5 unrelated kindreds with 18 affected persons and 19 families from the literature, each with more than 1 affected person. The clinical findings were highly variable, tending to become less distinctive with age. The syndrome was transmitted directly and vertically through 3 generations in 4 families and through 2 generations in 7 families. Male-to-male transmission was noted once.

The sex ratio was not significantly different from 1. The segregation ratio among sibs of the proband was 0.571 +/- 0.066.

Pettenati et al. (1986) examined 22 cases of BWS clinically and cytogenetically and compared the findings with those in 226 previously reported cases. All 22 patients were chromosomally normal. The authors thought that transmission was most consistent with autosomal dominant inheritance with incomplete penetrance. There were 2 affected sisters in one family and 3 affected sisters in another; in many families, there were isolated manifestations such as macroglossia, ear crease, or omphalocele.

Olney et al. (1988) reported 3 pairs of monozygotic (MZ) twins and reviewed 3 previously reported MZ twin pairs, in which one twin showed typical BWS with minimal or no expression of the condition in the cotwin. Indeed, phenotypic concordance in MZ twin pairs had not been reported. Olney et al. (1988) concluded that the most likely mode of inheritance is autosomal dominant. Clayton-Smith et al. (1992) brought to 11 the total number of sets of monozygotic twins with BWS. Ten of these were female. Since the likelihood of this occurring due to chance alone was less than 1 in 200, the authors thought that this supported an association between MZ twinning in females and the BWS locus. Lubinsky and Hall (1991) suggested that the process of X inactivation may be responsible for imprinting at the autosomal locus for BWS as a result of a 'spillover' effect; thus would discordance for BWS in monozygotic twins be explicable, particularly in light of the fact that the so-called unaffected twin may have minor anomalies such as earlobe creases or mild macroglossia. Orstavik et al. (1995) described 13-year-old MZ female twins who were discordant for WBS. They used PCR analysis at the androgen receptor locus to determine the pattern of X inactivation. Only the paternal allele was active in all cells of the affected child. In her sister, the pattern of X inactivation was moderately skewed in the same direction, whereas in their mother, the pattern was random. Orstavik et al. (1995) suggested that nonrandom X inactivation may be related to expression of the autosomal locus for WBS.

Aleck and Hadro (1989) reported a 4-generation family in which affected children, both male and female, occurred among the offspring of each of 4 sisters, 2 of whom were monozygotic twins. Affected individuals also occurred among the children of 2 affected females. No microsigns of BWS were found in presumptive carrier women. Aleck and Hadro (1989) concluded that their findings supported the hypothesis of Lubinsky et al. (1974) that BWS is an irregular dominant resulting from the segregation of a stable premutation passed through males and females but with expression (telomutation) only in offspring of female carriers of the premutation. They suggested that an ovum-mediated sex-associated factor may be involved in the process of telomutation. One child with BWS in full-blown form had juvenile fibromatosis (220600) of the face. In addition to earlobe creases, some affected members of the family had 'dents in their ears,' punched-out lesions on the posterior aspect of the pinnae; see also Best and Hoekstra (1981). Meckel diverticulum and bicornuate uterus were also noted in one fully affected woman.

Koufos et al. (1989) discussed the peculiarities of genetic transmission of familial BWS. Whether a pleiotropic mutation at 11p15.5 or a variety of allelic mutations at 11p15.5 underlies the pathogenesis of BWS as well as of the related tumors (rhabdomyosarcoma, hepatoblastoma, adrenal tumors) or, alternatively, whether these diseases may be due to defects at closely linked but separate loci, was unclear.

Henry et al. (1991) used 11p15.5 markers to determine the parental origin of chromosome 11 in 8 sporadic cases of BWS. Probands in 3 informative families had uniparental paternal disomy for region 11p15.5. Furthermore, an overall greatly increased frequency of homozygosity for several 11p15.5 markers in 21 sporadic BWS patients suggested that isodisomy probably accounts for an even higher proportion of BWS sporadic cases. The findings were compared with the lack of paternal alleles in the Prader-Willi syndrome (176270) and the lack of maternal alleles in the Angelman syndrome (105830). Paternal duplication in trisomic BWS patients, retention of paternal alleles in tumors, and higher penetrance in individuals born to female carriers in familial BWS (Lubinsky et al., 1974; Brown et al., 1990) corroborate the involvement of maternal genomic imprinting. An unbalanced dosage of maternal and paternal alleles may be the common factor in the different etiologic forms of BWS and associated tumors.

On the basis of an analysis of 19 published pedigrees suggesting autosomal dominant inheritance, Moutou et al. (1992) confirmed an excess of transmitting females and showed that there were 2 reasons for this excess: reduced fecundity in affected males compared to affected females in a ratio of 1 to 4.6; and a smaller risk of being affected, in a ratio of 1 to 3, for subjects who inherited the gene from their father. The latter finding suggested genomic imprinting. These results, together with the occurrence of uniparental disomy in some sporadic cases and the fact that all cases trisomic for the region 11p15.5 have had the duplicated region derived from the father, support the proposal that overgrowth in BWS patients and malignant proliferation in associated tumors reflect an imbalance between paternal and maternal alleles.

Viljoen and Ramesar (1992) also examined the case for imprinting in BWS. They presented a new pedigree that appeared to support paternal imprinting. An unaffected male had affected grandchildren through unaffected daughters born of different wives. Under the postulate of paternal imprinting, the phenotypically normal progenitor was a carrier for the imprinted gene and had a 50% chance of passing it to progeny, all of whom would lack stigmata of BWS. According to the hypothesis, change of the imprint status would occur in the germline of female carriers, and their children, both male and female, would have a 50% risk of manifesting the disorder, that is, 100% of those carrying the gene would be affected. The pedigree was consistent with this interpretation. An affected granddaughter of the progenitor had an affected daughter. Viljoen and Ramesar (1992) also reviewed 27 previously published pedigrees with 2 or more affected persons and in all but 4 concluded that paternal imprinting would explain the nonmendelian inheritance of BWS.

Elliott and Maher (1994) reviewed the subject of BWS. They pointed out that it is useful to consider the complications of the syndrome by the age of the subject. Multiple births are more common in BWS, with an excess of both monozygotic and dizygotic twins. Twin pairs are invariably discordant for BWS, although the second twin may occasionally show minor features. An excess of female monozygotic twin pairs (13 female, 1 male) has been observed among twin pairs with normal chromosomes.

Cytogenetics

In 2 unrelated children with features of the Beckwith-Wiedemann syndrome, Waziri et al. (1983) found partial duplication of 11p. They reviewed 6 other reported cases of partial duplication of 11p and found description of features suggesting BWS. Their first patient had deletion of 11q23.33-qter and duplication of 11p13-p15. In the second case, duplication of 11p15 was suspected. Since the duplicated region presumably contains the insulin locus (INS; 176730) and perhaps also the locus for insulin-like growth factor-2 (IGF2; 147470), the finding suggested that the EMG syndrome may be 'caused' by excess of one or both of these.

Pueschel and Padre-Mendoza (1984) described a child with this syndrome and a balanced 11/22 translocation: 46,XX,t(11p;22q). The phenotypically normal mother had the same balanced translocation. Turleau et al. (1984) found trisomy 11p15 in 2 cases of Beckwith-Wiedemann syndrome. One was an instance of de novo duplication of 11p15; the other was the result of t(4;11)(q33;p14)pat.

Turleau and de Grouchy (1985) found no clear evidence of phenotypic differences between patients with and those without chromosome abnormality. Okano et al. (1986) described an infant with partial trisomy of the terminal part of 11p. The father had a balanced translocation between chromosomes 4 and 11 with a breakpoint at 11p13. The proband had duplication of the 11p13-pter segment. Okano et al. (1986) analyzed the clinical features of this and 14 other reported cases; findings in 13 patients with duplication of 11p15 resembled those of BWS.

Henry et al. (1989) reported the first 2 cases of BWS with dup11p15 and adrenocortical carcinoma (ADCC). Together with evidence for somatic chromosomal events leading to loss of 11p15.5 alleles in familial ADCC cases, they hypothesized that a gene involved in predisposition to ADCC maps to region 11p15.5.

Norman et al. (1992) observed a phenotypically normal mother and 2 offspring with BWS, all 3 of whom carried the same paracentric inversion, inv(11)(p11.2p15.5). The phenotypically normal mother was a carrier of the translocation; both of her parents had normal chromosomes. The breakpoint on 11p was mapped by in situ hybridization to a site proximal to the insulin and insulin-growth factor 2 loci, and distal to D11S12. That the translocation t(9;11) had originated in the grandfather was proven because the derivative translocation chromosome 9 had a large block of methyl green/DAPI-positive heterochromatin on the long arm which could be traced back to him. Testing for loci on chromosome 11 likewise indicated that the derivative chromosome 11 originated from the grandfather. Thus, BWS associated with balanced chromosome translocations is transmitted in the same sex-dependent pattern as are noncytogenetic forms of familial BWS. Tommerup et al. (1993) reported the case of a patient with BWS and a reciprocal translocation t(9;11)(p11;p15.5).

Fryns et al. (1993) observed a patient with BWS and duplication 4q/deficiency 18p as the result of an unbalanced paternal translocation. The authors suggest that other contiguous gene duplication/deletions exist in BWS.

Drut and Drut (1996) used fluorescence in situ hybridization for interphase analysis to demonstrate trisomy 11p15 in interphase nuclei of formalin-fixed paraffin-embedded placenta from a stillborn fetus and in peripheral blood lymphocytes from 2 liveborn female relatives with

WBS. Karyotype of the father of 2 stillborn sibs demonstrated a balanced reciprocal translocation: 46,XY,t(10;11)(q26;p15).

Slavotinek et al. (1997) described a 3-generation family in which a father and son had a balanced reciprocal translocation t(5;11)(p15.3;p15.3). Two family members with unbalanced products of this translocation resulting in trisomy for chromosome 11p15.3-pter had features of BWS, with prenatal overgrowth, macroglossia, coarse facial features, and broad hands. Slavotinek et al. (1997) reviewed the clinical features of BWS patients with paternally derived duplications of 11p15.5 compared to those with normal chromosomes and found that patients with duplications had a decreased incidence of hypoglycemia, facial nevus flammeus, and hemihypertrophy, but an increased incidence of moderate-to-severe learning difficulties. They also noted persistent physical differences, including prominent occiput, prominent forehead, round face with full cheeks, deep-set eyes with epicanthic folds, hypertelorism, broad, flat nasal bridge, and micrognathia.

Mapping

In a family with multiple cases of BWS in multiple sibships and at least 2 generations (Aleck and Hadro, 1989), Koufos et al. (1989) demonstrated tight linkage of the mutation causing the syndrome to markers located at 11p15.5. Stimulated by this observation, Koufos et al. (1989) studied Wilms tumors of sporadic nature, not associated with BWS, and identified a subset of these tumors which had attained somatic homozygosity through mitotic recombination, with the smallest shared region of overlap being distal to the beta-globin complex at 11p15.5. The data suggested that there is a second locus, distinct from that involved in the WAGR syndrome, that plays a role in the association of Wilms tumor with BWS. See also Henry et al. (1989).

In 2 families, Ping et al. (1989) found linkage of BWS to the insulin gene, which is located at 11p15.5 (maximum lod = 3.6; theta = 0.00). Weksberg et al. (1990) found molecular evidence consistent with a duplication of a portion of 11p in 2 patients. In one, the duplicated segment was inherited; in the second, it had originated de novo and contained a portion of 11p15. Weksberg et al. (1990) showed that the duplicated segment included the genes for insulin-like growth factor-2, beta-globin (141900), calcitonin, and parathyroid hormone (168450), but did not include the gene for myogenic differentiation factor (MYOD1; 159970). These data place the BWS gene and the other genes mentioned distal to MYOD1 on 11p15.

Molecular Genetics

Somatic development of homozygosity for a mutant gene at a locus on chromosome 11 is a common pathogenetic mechanism in these disparate, rare tumors—hepatoblastoma, rhabdomyosarcoma, adrenal carcinoma, and Wilms tumor—that are frequent in BWS. The evidence was obtained by studying polymorphic DNA probes that map to 11p and consisted of the finding of homozygosity in the tumor from a person with heterozygosity.

Jeanpierre et al. (1985) found no obligate or consistent duplication of any 11p markers in BWS and concluded that duplication of INS (176730), HRAS1 (190020), and IGF2 (147470) cannot be directly responsible for the hyperinsulinism, predisposition to neoplasm, or gigantism in this disorder. Spritz et al. (1986) studied 7 patients and found no evidence of extra dosage of the insulin or IGF2 genes. Mannens et al. (1987) found that a tumor from a BWS patient had loss of heterozygosity for all 11p markers tested and for one 11q marker, PGA (169700); APOA1 (107680) remained heterozygous.

Hayward et al. (1988) found loss of heterozygosity at the HRAS1 locus (190020) on 11p in an adrenal adenoma from a 45-year-old woman with BWS. Little et al. (1988) found homozygosity for a region of 11p defined by the calcitonin (114130) and insulin (176730) genes in a hepatoblastoma from a child with BWS.

Litz et al. (1988) described monozygotic twins of whom only one had BWS. No cytogenetic or molecular abnormality of 11p could be detected in either normal or affected tissues obtained from the BWS twin. Using cloned DNA fragments homologous to 4 genes located on 11p, namely, catalase, parathyroid hormone, insulin-like growth factor II, and HRAS, Schofield et al. (1989) could find no evidence of large-scale deletions or amplifications in this chromosome region in 4 patients with BWS.

Since the insulin-like growth factor II gene is parentally imprinted in the mouse, it has been suggested that, in the human, duplication of the nonimprinted locus might lead to diploid expression of the gene and consequent general hyperplasia. However, using RFLPs for 4 linked markers on 11p and genomic clones internal to the IGF2 locus, Nystrom et al. (1992) found no evidence of alteration or amplification in 11 patients. In one patient who developed Wilms tumor, they found no evidence for loss of material on 11p. The possibility of mutation in unknown transacting factors affecting the expression of IGF2 remained. In a related study, Nystrom et al. (1992) found that in 1 of 14 sporadic cases of BWS, both copies of chromosome 11 were derived from the father, indicating paternal isodisomy. Schinzel (1993) commented that BWS was, to that date, the only example of partial parental disomy, which he referred to as 'mosaic-segmental uniparental disomy.' The paternal isodisomy suggests that the BWS gene is maternally imprinted. The IGF2 gene is also maternally imprinted, functions as a fetal growth factor, and is overexpressed in Wilms tumor.

Weksberg et al. (1993) studied allele-specific expression of the IGF2 gene, using an ApaI polymorphism in the 3-prime untranslated region of IGF2. Control skin fibroblasts were shown to maintain monoallelic expression of the paternal IGF2 allele, whereas skin fibroblasts from 3 out of 5 patients with BWS demonstrated biallelic IGF2 expression. In a sixth BWS patient, fresh tongue tissue as well as the fibroblasts derived from this tissue demonstrated biallelic expression, whereas the tongue tissue obtained from a control showed monoallelic expression. Weksberg et al. (1993) excluded paternal heterodisomy, using RFLPs in the IGF2, insulin, and tyrosine hydroxylase (191290) genes. They concluded that biallelic expression reflects disruption of IGF2 imprinting and that the BWS phenotype can result from the loss of normal suppression of the maternally inherited IGF2 gene. In contrast to previous reports in which imprinting of IGF2 has been invoked as the mechanism to explain sporadic cases of BWS (especially in situations where uniparental disomy and trisomy of the 11p15.5 region has occurred), Ramesar et al. (1993) suggested that paternal imprinting of a growth suppressor gene, e.g., H19 (103280), may be one of the causes of familial BWS.

Brown et al. (1996) described evidence of alteration of imprinting in a BWS family with an inversion, inv(11)(p11.2;p15.5). The carrier mother had no BWS manifestations, and had apparently inherited the inversion from her father. The 2 children who inherited the inversion from the mother were affected. The maternally inherited inversion was located approximately 300 kb centromeric to the IGF2 gene. Allele-specific expression analysis revealed that in the affected children the IGF2 was expressed from both parental alleles. Brown et al. (1996) demonstrated that the inversion led to biallelic expression of IGF2 and altered DNA replication patterns in the IGF2 region. The H19 imprinting in affected individuals was normal, suggesting an H19-independent pathway to biallelic IGF2 transcription. DNA methylation in IGF2 remained monoallelic, suggesting to the investigators that the mutation caused by the translocation had uncoupled allele-specific methylation from gene expression.

Working from the proposal that the paternally derived gene(s) at 11p15.5 are selectively expressed in BWS while the maternally transmitted gene(s) are inactive, Kubota et al. (1994) examined 18 patients for the parental origin of their 11p15 regions. Two patients had duplications of 11p15 from their respective fathers and 1 from the mother, indicating the transmission of an excessive dosage of the paternal gene in each. In the series of 12 sporadic cases, uniparental paternal disomy, either total constitutional or segmental, was not observed.

Zhang et al. (1997) produced targeted disruption of the p57(KIP2) gene (CDKN1C; 600856) in mice and demonstrated that they have altered cell proliferation and differentiation, leading to abdominal muscle defects; cleft palate; endochondral bone ossification defects with incomplete differentiation of hypertrophic chondrocytes; renal medullary dysplasia; adrenal cortical hyperplasia and cytomegaly; and lens cell hyperproliferation and apoptosis. Since many of these phenotypes are observed in patients with BWS, Zhang et al. (1997) suggested that the observations support a loss of p57(KIP2) expression as having a role in that disorder. Omphalocele was a feature of the mutant mice; mutant embryos showed umbilical abnormalities as early as E16.5. Neonatal lethality was due to defects in the closure of the secondary palate, with aspiration of milk and swallowing of air causing inflation and stretching of the stomach and intestines. Renal medullary dysplasia caused enlargement of the kidneys. Zhang et al. (1997) noted that type X collagen (120110) is expressed in hypertrophic chondrocytes and has been implicated in proper bone development. In mutant mice, expression of type X collagen was significantly reduced in the mutant hypertrophic zone. Thus, the investigators concluded that p57(KIP2) is required for expression of collagen X, and perhaps for other genes that facilitate the ossification of chondrocytes. Expression of p57(KIP2) is restricted to the fetal adrenal cortex and presumably plays a role in controlling cell proliferation; its absence leads to adrenal cortex hyperplasia and cytomegaly. The adrenal gland is among the most consistently enlarged organs in BWS patients. Some other manifestations of BWS are not so easily explained by the loss of control by this cyclin-dependent kinase inhibitory protein, e.g., the defects in kidney development and formation of the secondary palate.

Heterogeneity

Since different patterns of inheritance have been documented, true genetic heterogeneity is not yet proved.

Pathogenesis

Gardner (1973) pointed out some similarities between the Beckwith-Wiedemann syndrome and the disorder in the offspring of diabetic mothers. The similarities suggest that fetal hyperinsulinism may be involved in the latter condition, inasmuch as the insulin gene (INS; 176730) is located in the 11p15 region. However, see the work of Jeanpierre et al. (1985) and Spritz et al. (1986) outlined earlier.

The finding of loss of IGF2 imprinting in sporadic Wilms tumor (Rainier et al., 1993) further strengthens the view that IGF2 overexpression plays an important role in somatic overgrowth and the development of embryonal tumors.

In connection with studies of the methylation of the H19 (103280) and IGF2 (147470) genes in BWS, Reik et al. (1994) outlined the following distinctive features of the disorder. In approximately 15% of patients, BWS is inherited as an autosomal dominant with variable expressivity and incomplete penetrance. The penetrance is affected by parental transmission with an increase on maternal and a decrease on paternal transmission. In familial cases, the BWS gene maps to 11p15. In some sporadic cases, there are 11p15 chromosomal abnormalities, including partial trisomy, where the duplicated segment is always of paternal origin, balanced translocations and inversions that are maternally inherited, and paternal isodisomy for 11p, which occurs in approximately 20% of sporadic BWS patients. These observations suggest that genes for BWS are located on 11p15 and that they are either maternally imprinted genes with growth enhancing activity or paternally imprinted genes with growth suppressing activity. Reik et al. (1994) analyzed the allele-specific methylation patterns in the IGF2 gene and in the neighboring and reciprocally imprinted H19 gene in 42 BWS patients, 10 of whom represented mosaic uniparental disomy (UPD) cases. They found that allelic methylation of both genes was normal in all non-UPD cases, with the paternal allele being methylated, and was increased in UPD cases in proportion with the disomic lineage. These findings suggested to Reik et al. (1994) that sporadic BWS is not associated with a general alteration of methylation imprinting of the IGF2 and H19 genes. The methylation assay used in this study thus offers a simple and reliable diagnostic test of UPD for 11p15.5.

Reik et al. (1995) identified BWS patients who had inherited a normal biparental chromosome complement of the 11p15.5 region, where IGF2 and H19 reside, but had an altered pattern of allelic methylation of both genes, with the maternal chromosome carrying a paternal imprinting pattern. In fibroblasts, IGF2 was expressed from both parental alleles and H19 was not expressed, precisely as predicted from the altered pattern of allelic methylation. DNA replication patterns in the 11p15.5 region remained asynchronous as in controls, however. Reik et al. (1995) stated that the results provided the first example of a dissociation of regional control of DNA replication from regional control of allelic methylation and expression in imprinting. The authors suggested that the altered pattern of allelic methylation and expression arises in the germline or in the early embryo from defects in resetting or setting of imprinting in the maternal germline. Potential candidate regions for mutations include the previously identified translocation breakpoint clusters and the H19 gene itself. The finding of possible 'imprinting mutations' in BWS raises the prospect of identifying genetic factors that control imprinting in this region.

Hoovers et al. (1995) isolated YACs, and from the YACs cosmid libraries, representing the region of loss of heterozygosity in embryonal tumors associated with BWS. They isolated 5 germline balanced chromosomal rearrangement breakpoint sites from BWS patients, as well as a balanced chromosomal translocation breakpoint from a rhabdoid tumor, within a 295- to 320-kb cluster defined by a complete cosmid contig crossing these breakpoints. This breakpoint cluster terminated approximately 100 kb centromeric to the imprinted gene IGF2 (147470) and 100 kb telomeric to p57(KIP2) (600856), an inhibitor of cyclin-dependent kinases, and was located within subchromosomal transferable fragments that suppressed the growth of embryonal tumor cells in genetic complementation experiments. They identified 11 transcribed sequences in this BWS/tumor suppressor coincident region of 11p15, one of which corresponded to p57(KIP2). However, 3 additional BWS breakpoints were more than 4 megabases centromeric to the other 5 breakpoints and were excluded from the tumor suppressor region defined by subchromosomal transferable fragments. Thus, multiple genetic loci defined BWS and tumor suppression on 11p15. The authors speculated that, most likely, a group of cancer-related genes falls within a several megabase region, similar to 1p, 3p, and 9p.

Studying DNA samples from 9 unrelated Japanese patients with BWS, Hatada et al. (1996) analyzed the entire coding region of p57(KIP2) (600856), including intron/exon boundaries, by direct PCR using 5 PCR primer sets. They detected mutations in 2 patients (e.g., 600856.0001). In one other patient, Hatada et al. (1996) demonstrated reduced expression of the p57(KIP2) gene in adrenal gland. They concluded that the studies provided evidence for a new mechanism for producing a phenotype with dominant transmission and little or no gene product: one allele with an inactive product is expressed and the other allele is repressed by genomic imprinting. Hatada et al. (1996) commented that other loci may possibly be involved in BWS, since there are 3 other known balanced translocations leading to BWS which map several megabases from the p57(KIP2) region.

Several lines of evidence suggest that BWS may be caused by relative overexpression of the maternally imprinted IGF2 gene. Although cytogenetic and molecular findings are normal in most cases, a few show paternally inherited duplication of 11p15, segmental or mosaic paternal uniparental disomy for 11p15, or a maternal translocation with one breakpoint at 11p15. In familial cases, furthermore, transmission is always through the mother. These observations suggest that 2 paternal copies of IGF2 result from either paternal duplication or paternal UPD, that deletion or disruption associated with maternal translocation results in activation of the maternal gene, and that the mutations which are maternally transmitted activate the maternal IGF2 gene. Another overgrowth syndrome, Simpson-Golabi-Behmel syndrome (SGBS; 312870), has been shown to be due to mutations in an extracellular proteoglycan, glypican-3 (300037), that is inferred to play an important role in growth control in embryonic mesodermal tissues in which it is selectively expressed. It forms a complex with IGF2 and probably modulates IGF2 action. Thus there may be a commonality in the pathogenesis of BWS and SGBS.

Further evidence of the role of IGF2 in BWS was presented by Sun et al. (1997). They introduced Igf2 transgenes into the mouse genome by using embryonic stem (ES) cells and thereby caused transactivation of the endogenous Igf2 gene. The consequent overexpression of Igf2 resulted in most of the symptoms of Beckwith-Wiedemann syndrome, including prenatal overgrowth, polyhydramnios, fetal and neonatal lethality, disproportionate organ overgrowth including tongue enlargement, and skeletal abnormalities. This was presented as evidence that IGF2 overexpression is a key determinant of BWS. The Igf2 dosage-dependent phenotypes seen in the transgenic mouse models BWS closely. These phenotypes overlap with other fetal overgrowth syndromes attributed to increased IGF2 levels, such as the IGF2 overgrowth syndrome (Morison et al., 1996), in which symptoms are less severe, and the Simpson-Golabi-Behmel syndrome, in which symptoms are more severe, than in BWS. Sun et al. (1997) commented that the transgenic Igf2 mouse is a better model than the H19 mouse, which shows overgrowth only, or the Igf2r mouse (147280), which shows disproportionate overgrowth of the heart but not other organs, and an overall increase in body weight to 130%. Exomphalos or omphalocele, enlargement of the adrenal cortex, and renal medullary dysplasia were observed in the p57(KIP2) knockout mouse together with some other phenotypes that might constitute minor features of BWS. Overexpression of IGF2 and underexpression of p57(KIP2) account for almost all features of BWS, with the possible exception of tumors and hypoglycemia.

Catchpoole et al. (1997) studied the molecular pathology of 106 sporadic BWS cases. Fourteen of 83 informative cases had segmental paternal isodisomy for a region of 11p15.5 bounded by D11S861 and D11S2071. Isodisomy for 11q was detected in 3 cases, but only as far telomerically as 11q13-11q21. The allele-specific methylation status of the H19 gene (103280) was determined in 80 sporadic BWS cases. Thirteen had UPD and showed H19 hypermethylation. Of 63 cases with biparental inheritance, 5 showed H19 hypermethylation consistent with an imprinting center mutation or imprinting error lesion. The phenotype of these 5 patients overlapped with that of sporadic BWS cases without UPD and with normal H19 methylation; however, exomphalos was more common (p less than 0.05) in the latter group. Catchpoole et al. (1997) concluded that the expression of imprinted genes in this region may be dependent on the precise molecular pathology and that H19 methylation is useful in diagnosis of UPD or altered imprinting in BWS.

Diagnosis

Diagnosis is based on clinical findings. A careful cytogenetic analysis of the 11p15 region is recommended. Prenatal diagnosis by ultrasonography is possible (Nivelon-Chevallier et al., 1983; Winter et al., 1986; Cobelis et al., 1988). When the pregnancy is not terminated, the prenatal diagnosis helps to prevent neonatal complications (Viljoen et al., 1991).

Clinical Management

Since neonatal hypoglycemia is frequent (1 in 3 cases) and potentially deleterious for the CNS, Martinez-y-Martinez et al. (1992) proposed monitoring the glycemia in BWS newborns every 6 hours during the first 3 days in order to correct blood glucose levels below 2.6 mmol/l (46.8 mg/dl).

Adrenal carcinoma, nephroblastoma, hepatoblastoma, and rhabdomyosarcoma occur with increased frequency and justify biannual abdominal ultrasound examinations (Azouz et al., 1990). Wiedemann (1983) recommended that children with this syndrome be examined with renal sonography: first, at 3-month intervals, and after the third year of life, at 6-month intervals. Although less frequent, thoracic neuroblastoma occurs. A periodic chest radiograph is necessary (Sirinelli et al., 1989).

Population Genetics

Thorburn et al. (1970) described 6 cases in Jamaican blacks and estimated an incidence of 1 in 13,700 births.

Animal Model

In human and mouse, most imprinted genes are arranged in chromosomal clusters. Their linked organization suggests coordinated mechanisms controlling imprinting and gene expression. Identification of local and regional elements responsible for the epigenetic control of imprinted gene expression is important for understanding the molecular basis of disorders associated with imprinting such as BWS. Paulsen et al. (1998) established a complete contig of clones along the murine imprinting cluster on distal chromosome 7 syntenic with the human imprinting region at 11p15.5 associated with BWS. The cluster comprises approximately 1 Mb of DNA, contains at least 8 imprinted genes, and is demarcated by the 2 maternally expressed genes Ipl (602131) and H19 (103280), which are directly flanked by the nonimprinted genes Nap1l4 (601651) and L23mrp (600789), respectively. Paulsen et al. (1998) also localized Kvlqt1 (192500) and Tapa1 (186845) between Cdkn1c (600856) and Mash2 (601886). The mouse Kvlqt1 gene was maternally expressed in most fetal tissues but biallelically transcribed in most neonatal tissues, suggesting relaxation of imprinting during development.

History

For biographical and autobiographical accounts of Hans-Rudolf Wiedemann, see Opitz and Mullen (1992) and Wiedemann (1992), respectively.

See also Balcom et al. (1985); Beckwith (1969); Ben-Galim et al. (1977); Berry et al. (1980); Bose et al. (1985); Chitayat et al. (1990); Engstrom et al. (1988); Estabrooks et al. (1989); Filippi and McKusick (1970); Francheschini et al. (1993); Gerver et al. (1991); Haas et al. (1986); Hecht and Sandberg (1986); Henry et al. (1989); Journel et al. (1985); Koh et al. (1986); Koufos et al. (1985); Leung et al. (1979); Martinez et al. (1985); Martinez-y-Martinez et al. (1985); McCowan and Becroft (1994); Nystrom et al. (1992); Reddy et al. (1972); Schmutz (1986); Vidailhet et al. (1971); Weksberg et al. (1993); Wiedemann (1969); Wiedemann et al. (1968); Wilfong et al. (1992); Wockel et al. (1981); Wojciechowski and Pritchard (1981)

Aleck, K. A.; Hadro, T. A.: Dominant inheritance of Wiedemann-Beckwith syndrome: further evidence for transmission of 'unstable premutation' through carrier women. Am. J. Med. Genet. 33: 155-160, 1989.

Azouz, E. M.; Larson, E. J.; Patel, J.; Gyepes, M. T.: Beckwith-Wiedemann syndrome: development of nephroblastoma during the surveillance period. Pediat. Radiol. 20: 550-552, 1990.

Balcom, R. J.; Hakanson, D. O.; Werner, A.; Gordon, L. P.: Massive thymic hyperplasia in an infant with Beckwith-Wiedemann syndrome. Arch. Path. Lab. Med. 109: 153-155, 1985.

Beckwith, J. B.: Macroglossia, omphalocele, adrenal cytomegaly, gigantism, and hyperplastic visceromegaly. Birth Defects Orig. Art. Ser. V(2): 188-196, 1969.

Ben-Galim, E.; Gross-Kieselstein, E.; Abrahamov, A.: Beckwith-Wiedemann syndrome in a mother and her son. Am. J. Dis. Child. 131: 801-803, 1977.

Berry, A. C.; Belton, E. M.; Chantler, C.: Monozygotic twins discordant for Wiedemann-Beckwith syndrome and the implications for genetic counselling. J. Med. Genet. 17: 136-138, 1980.

Best, L. G.: Familial posterior helical ear pits and Wiedemann-Beckwith syndrome. Am. J. Med. Genet. 40: 188-195, 1991.

Best, L. G.; Hoekstra, R. E.: Wiedemann-Beckwith syndrome: autosomal-dominant inheritance in a family. Am. J. Med. Genet. 9: 291-299, 1981.

Bose, B.; Wilkie, R. A.; Madlom, M.; Forsyth, J. S.; Faed, M. J.: Wiedemann-Beckwith syndrome in one of monozygotic twins. Arch. Dis. Child. 60: 1191-1192, 1985.

Brown, K. W.; Villar, A. J.; Bickmore, W.; Clayton-Smith, J.; Catchpoole, D.; Maher, E. R.; Reik, W.: Imprinting mutation in the Beckwith-Wiedemann syndrome leads to biallelic IGF2 expression through an H19-independent pathway. Hum. Molec. Genet. 5: 2027-2032, 1996.

Brown, K. W.; Williams, J. C.; Maitland, N. J.; Mott, M. G.: Genomic imprinting and the Beckwith-Wiedemann syndrome. (Letter) Am. J. Hum. Genet. 46: 1000-1001, 1990.

Carlin, M. E.; Escobar, L. F.; Ward, R. E.; Wielgus, T. Y.: A reassessment of Beckwith-Wiedemann syndrome (BWS). (Abstract) Am. J. Hum. Genet. 47 (suppl.): A50, 1990.

Catchpoole, D.; Lam, W. W. K.; Valler, D.; Temple, I. K.; Joyce, J. A.; Reik, W.; Schofield, P. N.; Maher, E. R.: Epigenetic modification and uniparental inheritance of H19 in Beckwith-Wiedemann syndrome. J. Med. Genet. 34: 353-359, 1997.

Chemke, J.: Familial macroglossia-omphalocele syndrome. J. Genet. Hum. 24: 271-279, 1976.

Chitayat, D.; Friedman, J. M.; Dimmick, J. E.: Neuroblastoma in a child with Wiedemann-Beckwith syndrome. Am. J. Med. Genet. 35: 433-436, 1990.

Chitayat, D.; Rothchild, A.; Ling, E.; Friedman, J. M.; Couch, R. M.; Yong, S.-L.; Baldwin, V. J.; Hall, J. G.: Apparent postnatal onset of some manifestations of the Wiedemann-Beckwith syndrome. Am. J. Med. Genet. 36: 434-439, 1990.

Clayton-Smith, J.; Read, A. P.; Donnai, D.: Monozygotic twinning and Wiedemann-Beckwith syndrome. Am. J. Med. Genet. 42: 633-637, 1992.

Clouston, W. M.; Cannel, G. C.; Fryar, B. G.; Searle, J. W.; Martin, N. I.; Mortimer, R. H.: Virilizing adrenal adenoma in an adult with the Beckwith-Wiedemann syndrome: paradoxical response to dexamethasone. Clin. Endocr. 31: 467-473, 1989.

Cobelis, G.; Iannoto, P.; Stabile, M.; Lonardo, F.; Della Bruna, M.; Caliendo, E.; Ventruto, V.: Prenatal ultrasound diagnosis of macroglossia in the Wiedemann-Beckwith syndrome. Prenatal Diag. 8: 79-81, 1988.

Daugbjerg, P.; Everberg, G.: A case of Beckwith-Wiedemann syndrome with conductive hearing loss. Acta Paediat. Scand. 73: 408-410, 1984.

Drut, R. M.; Drut, R.: Nonimmune fetal hydrops and placentomegaly: diagnosis of familial Wiedemann-Beckwith syndrome with trisomy 11p15 using FISH. Am. J. Med. Genet. 62: 145-149, 1996.

Elliott, M.; Bayly, R.; Cole, T.; Temple, I. K.; Maher, E. R.: Clinical features and natural history of Beckwith-Wiedemann syndrome: presentation of 74 new cases. Clin. Genet. 46: 168-174, 1994.

Elliott, M.; Maher, E. R.: Beckwith-Wiedemann syndrome. J. Med. Genet. 31: 560-564, 1994.

Emery, L. G.; Shields, M.; Shah, N. R.; Garbes, A.: Neuroblastoma associated with Beckwith-Wiedemann syndrome. Cancer 52: 176-179, 1983.

Engstrom, W.; Lindham, S.; Schofield, P.: Wiedemann-Beckwith syndrome. Europ. J. Pediat. 147: 450-457, 1988.

Estabrooks, L. L.; Lamb, A. N.; Kirkmann, H.; Boyer, S.; Wiley, J.; Callahan, N. D.; Rao, K. W.: Beckwith-Wiedemann syndrome in twins with a duplication of chromosome (15q11.2-q13)mat. Am. J. Hum. Genet. 14: A75, 1989.

Falik-Borenstein, T. C.; Korenberg, J. R.; Davos, I.; Platt, L. D.; Gans, S.; Goodman, B.; Schreck, R.; Graham, J. M., Jr.: Congenital gastric teratoma in Wiedemann-Beckwith syndrome. Am. J. Med. Genet. 38: 52-57, 1991.

Filippi, G.; McKusick, V. A.: The Beckwith-Wiedemann syndrome (the exomphalos-macroglossia-gigantism syndrome): report of two cases and review of the literature. Medicine 49: 279-298, 1970.

Forrester, R. M.: Wiedemann-Beckwith syndrome. (Letter) Lancet II: 47, 1973.

Francheschini, P.; Guala, A.; Vardeu, M. P.; Franceschini, D.: Monozygotic twinning and Wiedemann-Beckwith syndrome. (Letter) Am. J. Med. Genet. 46: 353-354, 1993.

Fryns, J. P.; Kleczkowska, A.; Devriendt, K.; Devliegher, H.; Van den Berghe, H.: Wiedemann-Beckwith syndrome and chromosomal duplication 4q/deficiency 18p. Genet. Counsel. 4: 37-41, 1993.

Gardner, L. I.: Pseudo-Beckwith-Wiedemann syndrome: interaction with maternal diabetes. (Letter) Lancet II: 911-912, 1973.

Gerver, W. J. M.; Menheere, P. P. C. A.; Schaap, C.; Degraeuwe, P.: The effects of a somatostatin analogue on the metabolism of an infant with Beckwith-Wiedemann syndrome and hyperinsulinaemia. Europ. J. Pediat. 150: 634-637, 1991.

Haas, O. A.; Zoubek, A.; Grumayer, E. R.; Gadner, H.: Constitutional interstitial deletion of 11p11 and pericentric inversion of chromosome 9 in a patient with Wiedemann-Beckwith syndrome and hepatoblastoma. Cancer Genet. Cytogenet. 23: 95-104, 1986.

Hatada, I.; Ohashi, H.; Fukushima, Y.; Kaneko, Y.; Inoue, M.; Komoto, Y.; Okada, A.; Ohishi, S.; Nabetani, A.; Morisaki, H.; Nakayama, M.; Niikawa, M,; Mukai, T.: An imprinted gene p57(KIP2) is mutated in Beckwith-Wiedemann syndrome. Nature Genet. 14: 171-173, 1996.

Hayward, N. K.; Little, M. H.; Mortimer, R. H.; Clouston, W. M.; Smith, P. J.: Generation of homozygosity at the c-Ha-ras-1 locus of chromosome 11p in an adrenal adenoma from an adult with Wiedemann-Beckwith syndrome. Cancer Genet. Cytogenet. 30: 127-132, 1988.

Hecht, F.; Sandberg, A. A.: Wiedemann-Beckwith syndrome: cancer predisposition and chromosome 11. Cancer Genet. Cytogenet. 23: 159-161, 1986.

Henry, I.; Bonaiti-Pellie, C.; Chehensse, V.; Beldjord, C.; Schwartz, C.; Utermann, G.; Junien, C.: Uniparental paternal disomy in a genetic cancer-predisposing syndrome. Nature 351: 665-667, 1991.

Henry, I.; Grandjouan, S.; Couillin, P.; Barichard, F.; Huerre-Jeanpierre, C.; Glaser, T.; Philip, T.; Lenoir, G.; Chaussain, J. L.; Junien, C.: Tumor-specific loss of 11p15.5 alleles in del11p13 Wilms tumor and in familial adrenocortical carcinoma. Proc. Nat. Acad. Sci. 86: 3247-3251, 1989.

Henry, I.; Jeanpierre, M.; Couillin, P.; Barichard, F.; Serre, J. L.; Journel, H.; Lamouroux, A.; Turleau, C.; de Grouchy, J.; Junien, C.: Molecular definition of the 11p15.5 region involved in Beckwith-Wiedemann syndrome and probably in predisposition to adrenocortical carcinoma. Hum. Genet. 81: 273-277, 1989.

Hoovers, J. M. N.; Kalikin, L. M.; Johnson, L. A.; Alders, M.; Redeker, B.; Law, D. J.; Bliek, J.; Steenman, M.; Benedict, M.; Wiegant, J.; Lengauer, C.; Taillon-Miller, P.; Schlessinger, D.; Edwards, M. C.; Elledge, S. J.; Ivens, A.; Westerveld, A.; Little, P.; Mannens, M.; Feinberg, A. P.: Multiple genetic loci within 11p15 defined by Beckwith-Wiedemann syndrome rearrangement breakpoints and subchromosomal transferable fragments. Proc. Nat. Acad. Sci. 92: 12456-12460, 1995.

Irving, I. M.: Exomphalos with macroglossia: a study of eleven cases. J. Pediat. Surg. 2: 499-507, 1967.

Irving, I. M.: The 'E.M.G.' syndrome (exomphalos, macroglossia, gigantism). Prog. Pediat. Surg. 1: 1-16, 1970.

Jeanpierre, M.; Henry, I.; Turleau, C.; Ullrich, A.; Mallet, J.; de Grouchy, J.; Junien, C.: Beckwith-Wiedemann syndrome (BWS) and 11p15 markers. (Abstract) Cytogenet. Cell Genet. 40: 661, 1985.

Journel, H.; Lucas, J.; Allaire, C.; Le Mee, F.; Defawe, G.; Lecornu, M.; Jouan, H.; Roussey, M.; Le Marec, B.: Trisomy 11p15 and Beckwith-Wiedemann syndrome: report of two new cases. Ann. Genet. 28: 97-101, 1985.

Knight, J. A.; Palmer, W. M.; Gardner, A. Y.; Bryden, W. L.: Association of the Beckwith-Wiedemann and prune belly syndromes. Clin. Pediat. 19: 485-488, 1980.
Koh, T. H. H. G.; Cooper, J. E.; Newman, C. L.; Walker, T. M.; Kiely, E. M.; Hoffmann, E. B.: Pancreatoblastoma in a neonate with Wiedemann-Beckwith syndrome. Europ. J. Pediat. 145: 435-438, 1986.
Kosseff, A. L.; Herrmann, J.; Gilbert, E. F.; Viseskul, C.; Lubinsky, M.; Opitz, J. M.: The Wiedemann-Beckwith syndrome. Europ. J. Pediat. 123: 139-166, 1976.
Kosseff, A. L.; Herrmann, J.; Opitz, J. M.: The Wiedemann-Beckwith syndrome: genetic consideration and a diagnostic sign. (Letter) Lancet I: 844, 1972.
Koufos, A.; Grundy, P.; Morgan, K.; Aleck, K. A.; Hadro, T.; Lampkin, B. C.; Kalbakji, A.; Cavenee, W. K.: Familial Wiedemann-Beckwith syndrome and a second Wilms tumor locus both map to 11p15.5. Am. J. Hum. Genet. 44: 711-719, 1989.
Koufos, A.; Hansen, M. F.; Copeland, N. G.; Jenkins, N. A.; Lampkin, B. C.; Cavenee, W. K.: Loss of heterozygosity in three embryonal tumours suggests a common pathogenetic mechanism. Nature 316: 330-334, 1985.
Kubota, T.; Saitoh, S.; Matsumoto, T.; Narahara, K.; Fukushima, Y.; Jinno, Y.; Niikawa, N.: Excess functional copy of allele at chromosomal region 11p15 may cause Wiedemann-Beckwith (EMG) syndrome. Am. J. Med. Genet. 49: 378-383, 1994.
Leung, A. K. C.: Wiedemann-Beckwith syndrome and hypothyroidism. (Letter) Europ. J. Pediat. 144: 295, 1985.
Leung, A. K. C.; McArthur, R. G.: EMG-syndrome and hypothyroidism. (Letter) Clin. Genet. 36: 269, 1989.
Leung, A. K. C.; McArthur, R. G.; Ross, S. A.; McMillan, D. D.; Sauve, R. S.: Thyroxine-binding globulin deficiency in Beckwith syndrome. J. Pediat. 95: 752-754, 1979.
Little, M. H.; Thomson, D. B.; Hayward, N. K.; Smith, P. J.: Loss of alleles on the short arm of chromosome 11 in a hepatoblastoma from a child with Beckwith-Wiedemann syndrome. Hum. Genet. 79: 186-189, 1988.
Litz, C. E.; Taylor, K. A.; Qiu, J. S.; Pescovitz, O. H.; de Martinville, B.: Absence of detectable chromosomal and molecular abnormalities in monozygotic twins discordant for the Wiedemann-Beckwith syndrome. Am. J. Med. Genet. 30: 821-833, 1988.
Lubinsky, M.; Herrmann, J.; Kosseff, A. L.; Opitz, J. M.: Autosomal dominant sex-dependent transmission of the Wiedemann-Beckwith syndrome. (Letter) Lancet I: 932, 1974.
Lubinsky, M. S.; Hall, J. G.: Genomic imprinting, monozygous twinning, and X inactivation. (Letter) Lancet 337: 1288, 1991.
Mannens, M.; Slater, R. M.; Heyting, C.; Bliek, J.; Hoovers, J.; Bleeker-Wagemakers, E. M.; Voute, P. A.; Coad, N.; Frants, R. R.; Pearson, P. L.: Chromosome 11, Wilms' tumour and associated congenital diseases. (Abstract) Cytogenet. Cell Genet. 46: 655, 1987.
Martinez, R.; Ocampo-Campos, R.; Perez-Arroyo, R.; Corona-Rivera, E.; Cantu, J. M.: The Wiedemann-Beckwith syndrome in four sibs including one with associated congenital hypothyroidism. Europ. J. Pediat. 143: 233-235, 1985.
Martinez-y-Martinez, R.; Martinez-Carboney, R.; Ocampo-Campos, R.; Rivera, H.; Gomez Plascencia y Castillo, J.; Cuevas, A.; Martin Manrique, M. C.: Wiedemann-Beckwith syndrome: clinical, cytogenetical and radiological observations in 39 new cases. Genet. Counsel. 3: 67-76, 1992.
Martinez-y-Martinez, R.; Ocampo-Campos, R.; Perez-Arroyo, R.; Corona-Rivera, E.; Cantu, J. M.: The Wiedemann-Beckwith syndrome in four sibs including one with associated congenital hypothyroidism. Europ. J. Pediat. 143: 233-235, 1985.
Mausuura, N.; Endo, M.; Okayasu, T.; Okuno, A.: Wiedemann-Beckwith syndrome. (Letter) Lancet II: 508, 1975.
McCowan, L. M. E.; Becroft, D. M. O.: Beckwith-Wiedemann syndrome, placental abnormalities and gestational proteinuric hypertension. Obstet. Gynecol . 83: 813-817, 1994.
Morison, I. M.; Becroft, D. M.; Taniguchi, T.; Woods, C. G.; Reeve, A. E.: Somatic overgrowth associated with overexpression of insulin-like growth factor II. Nature Med. 2: 311-316, 1996.
Moutou, C.; Junien, C.; Henry, I.; Bonaiti-Pellie, C.: Beckwith-Wiedemann syndrome: a demonstration of the mechanisms responsible for the excess of transmitting females. J. Med. Genet. 29: 217-220, 1992.
Mulvihill, D. M.; Mercado, M. G.; Boineau, F. G.: Beckwith-Wiedemann syndrome and its association with type III polycystic kidney disease. Pediat. Nephrol. 3: 286-289, 1989.
Niikawa, N.; Ishikiriyama, S.; Takahashi, S.; Inagawa, A.; Tonoki, H.; Ohta, Y.; Hase, N.; Kamei, T.; Kajii, T.: The Wiedemann-Beckwith syndrome: pedigree studies on five families with evidence for autosomal dominant inheritance with variable expressivity. Am. J. Med. Genet. 24: 41-55, 1986.
Nivelon-Chevallier, A.; Mavel, A.; Michiels, R.; Bethenod, M.: Syndrome de Wiedemann-Beckwith familial: diagnostic antenatal echographique et confirmation histologique. J. Genet. Hum. 31: 397-402, 1983.
Norman, A. M.; Read, A. P.; Clayton-Smith, J.; Andrews, T.; Donnai, D.: Recurrent Wiedemann-Beckwith syndrome with inversion of chromosome (11)(p11.2p15.5). Am. J. Med. Genet. 42: 638-641, 1992.
Nystrom, A.; Cheetham, J. E.; Engstrom, W.; Schofield, P. N.: Molecular analysis of patients with Wiedemann-Beckwith syndrome. II. Paternally derived disomies of chromosome 11. Europ. J. Pediat. 151: 511-514, 1992.
Nystrom, A.; Engstrom, W.; Cheetham, J.; Schofield, P. N.: Molecular analysis of patients with Wiedemann-Beckwith syndrome. I. Gene dosage on the short arm of chromosome 11. Europ. J. Pediat. 151: 504-510, 1992.
Okano, Y.; Osasa, Y.; Yamamoto, H.; Hase, Y.; Tsuruhara, T.; Fujita, H.: An infant with Beckwith-Wiedemann syndrome and chromosomal duplication 11p13-pter: correlation of symptoms between 11p trisomy and Beckwith-Wiedemann syndrome. Jpn. J. Hum. Genet. 31: 365-372, 1986.
Olney, A. H.; Buehler, B. A.; Waziri, M.: Wiedemann-Beckwith syndrome in apparently discordant monozygotic twins. Am. J. Med. Genet. 29: 491-499, 1988.
Opitz, J. M.; Mullen, P.: Hans-Rudolf Wiedemann: an appreciation. Am. J. Med. Genet. 43: 737-739, 1992.
Orstavik, R. E.; Tommerup, N.; Eiklid, K.; Orstavik, K. H.: Non-random X chromosome inactivation in an affected twin in a monozygotic twin pair discordant for Wiedemann-Beckwith syndrome. Am. J. Med. Genet. 56: 210-214, 1995.
Paulsen, K.: Otological features in exomphalos-macro-glossia-gigantism syndrome (Wiedemann's syndrome). Z. Laryngol. Rhinol. Otol. 52: 793-798, 1973.
Paulsen, M.; Davies, K. R.; Bowden, L. M.; Villar, A. J.; Franck, O.; Fuermann, M.; Dean, W. L.; Moore, T. F.; Rodrigues, N.; Davies, K. E.; Hu, R.-J.; Feinberg, A. P.; Maher, E. R.; Reik, W.; Walter, J.: Syntenic organization of the mouse distal chromosome 7 imprinting cluster and the Beckwith-Wiedemann syndrome region in chromosome 11p15.5. Hum. Molec. Genet. 7: 1149-1159, 1998.
Pettenati, M. J.; Haines, J. L.; Higgins, R. R.; Wappner, R. S.; Palmer, C. G.; Weaver, D. D.: Wiedemann-Beckwith syndrome: presentation of clinical and cytogenetic data on 22 new cases and review of the literature. Hum. Genet. 74: 143-154, 1986.
Ping, A. J.; Reeve, A. E.; Law, D. J.; Young, M. R.; Boehnke, M.; Feinberg, A. P.: Genetic linkage of Beckwith-Wiedemann syndrome to 11p15. Am. J. Hum. Genet. 44: 720-723, 1989.
Piussan, C.; Risbourg, B.; Lenaerts, C.; Delvallez, N.; Gontier, M. F.; Vitse, M.: Syndrome de Wiedemann et Beckwith: une nouvelle observation familiale. J. Genet. Hum. 28: 281-291, 1980.
Pueschel, S. M.; Padre-Mendoza, T.: Chromosome 11 and Beckwith-Wiedemann syndrome. (Letter) J. Pediat. 104: 484-485, 1984.
Rainier, S.; Johnson, L. A.; Dobry, C. J.; Ping, A. J.; Grundy, P. E.; Feinberg, A. P.: Relaxation of imprinted genes in human cancer. Nature 362: 747-749, 1993.
Ramesar, R.; Babaya, M.; Viljoen, D.: Molecular investigation of familial Beckwith-Wiedemann syndrome: a model for paternal imprinting. Europ. J. Hum. Genet. 1: 109-113, 1993.
Reddy, J. K.; Schimke, R. N.; Chang, C. H. J.; Svoboda, D. J.; Slaven, J.: Beckwith-Wiedemann syndrome: Wilms' tumor, cardiac hamartoma, persistent visceromegaly, and glomeruloneogenesis in a 2-year-old boy. Arch. Path. 94: 523-532, 1972.
Reik, W.; Brown, K. W.; Schneid, H.; Le Bouc, Y.; Bickmore, W.; Maher, E. R.: Imprinting mutations in the Beckwith-Wiedemann syndrome suggested by an altered imprinting pattern in the IGF2-H19 domain. Hum. Molec. Genet. 4: 2379-2385, 1995.
Reik, W.; Brown, K. W.; Slatter, R. E.; Sartori, P.; Elliott, M.; Maher, E. R.: Allelic methylation of H19 and IGF2 in the Beckwith-Wiedemann syndrome. Hum. Molec. Genet. 3: 1297-1301, 1994.
Schinzel, A.: Genomic imprinting: consequences of uniparental disomy for human disease. Am. J. Med. Genet. 46: 683-684, 1993.
Schmutz, S. M.: Deletion of chromosome 11 (p11p13) in a patient with Beckwith-Wiedemann syndrome. Clin. Genet. 30: 154-156, 1986.
Schofield, P. N.; Lindham, S.; Engstrom, W.: Analysis of gene dosage on chromosome 11 in children suffering from Beckwith-Wiedemann syndrome. Europ. J. Pediat. 148: 320-324, 1989.
Sippell, W. G.; Partsch, C.-J.; Wiedemann, H. R.: Growth, bone maturation and pubertal development in children with the EMG-syndrome. Clin. Genet. 35: 20-28, 1989.
Sirinelli, D.; Silberman, B.; Baudon, J. J.; Sinassamy, P.; Gruner, M.; Montagne, J. P.: Beckwith-Wiedemann syndrome and neural crest tumors. Pediat. Radiol. 19: 242-245, 1989.
Slavotinek, A.; Gaunt, L.; Donnai, D.: Paternally inherited duplications of 11p15.5 and Beckwith-Wiedemann syndrome. J. Med. Genet. 34: 819-826, 1997.
Sommer, A.; Cutler, E. A.; Cohen, B. L.; Harper, D.; Backes, C.: Familial occurrence of the Wiedemann-Beckwith syndrome and persistent fontanel. Am. J. Med. Genet. 1: 59-63, 1977.
Sotelo-Avila, C.; Gonzalez-Crussi, F.; Fowler, J. W.: Complete and incomplete forms of Beckwith-Wiedemann syndrome: their oncogenic potential. J. Pediat. 96: 47-50, 1980.
Spritz, R. A.; Mager, D.; Pauli, R. M.; Laxova, R.: Normal dosage of the insulin and insulin-like growth factor II genes in patients with the Beckwith-Wiedemann syndrome. Am. J. Hum. Genet. 39: 265-273, 1986.
Sun, F.-L.; Dean, W. L.; Kelsey, G.; Allen, N. D.; Reik, W.: Transactivation of IGF2 in a mouse model of Beckwith-Wiedemann syndrome. Nature 389: 809-815, 1997.
Thorburn, M. J.; Wright, E. S.; Miller, C. G.; Smith-Read, E. H. M.: Exomphalos-macroglossia-gigantism syndrome in Jamaican infants. Am. J. Dis. Child. 119: 316-321, 1970.
Tommerup, N.; Brandt, C. A.; Pedersen, S.; Bolund, L.; Kamper, J.: Sex dependent transmission of Beckwith-Wiedemann syndrome associated with a reciprocal translocation t(9;11)(p11.2;p15.5). J. Med. Genet. 30: 958-961, 1993.
Turleau, C.; de Grouchy, J.: Beckwith-Wiedemann syndrome: clinical comparison between patients with and without 11p15 trisomy. Ann. Genet. 28: 93-96, 1985.
Turleau, C.; de Grouchy, J.; Chavin-Colin, F.; Martelli, H.; Voyer, M.; Charlas, R.: Trisomy 11p15 and Beckwith-Wiedemann syndrome: a report of two cases. Hum. Genet. 67: 219-221, 1984.
Vidailhet, M.; Rauber, G.; Grogon, G.; Ratajczak, F.; Neimann, N.; Pierson, M.: Syndrome de Beckwith-Wiedemann. Etude de quatre observations, dont une avec nephroblastome. Arch. Franc. Ped. 28: 10-11, 1971.
Viljoen, D.; Ramesar, R.: Evidence for paternal imprinting in familial Beckwith-Wiedemann syndrome. J. Med. Genet. 29: 221-225, 1992.
Viljoen, D. L.; Jaquire, Z.; Woods, D. L.: Prenatal diagnosis in autosomal dominant Beckwith-Wiedemann syndrome. Pren. Diagn. 11: 167-175, 1991.
Watanabe, H.; Yamanaka, T.: A possible relationship between Beckwith-Wiedemann syndrome, urinary tract anomaly and prune belly syndrome. Clin. Genet. 38: 410-414, 1990.

Waziri, M.; Patil, S. R.; Hanson, J. W.; Bartley, J. A.: Abnormality of chromosome 11 in patients with features of Beckwith-Wiedemann syndrome. J. Pediat. 102: 873-876, 1983.
Weksberg, R.; Glaves, M.; Teshima, I.; Waziri, M.; Patil, S.; Williams, B. R. G.: Molecular characterization of Beckwith-Wiedemann syndrome (BWS) patients with partial duplication of chromosome 11p excludes the gene MYOD1 from the BWS region. Genomics 8: 693-698, 1990.
Weksberg, R.; Shen, D. R.; Fei, Y.-L.; Song, Q. L.; Squire, J.: Disruption of insulin-like growth factor 2 imprinting in Beckwith-Wiedemann syndrome. (Abstract) Am. J. Hum. Genet. 53 (suppl.): A17, 1993.
Weksberg, R.; Shen, D. R.; Fei, Y.-L.; Song, Q. L.; Squire, J.: Disruption of insulin-like growth factor 2 imprinting in Beckwith-Wiedemann syndrome. Nature Genet. 5: 143-150, 1993.
Weng, E. Y.; Moeschler, J. B.; Graham, J. M., Jr.: Longitudinal observations on 15 children with Wiedemann-Beckwith syndrome. Am. J. Med. Genet. 56: 366-373, 1995.
Wiedemann, H.-R.: Hans-Rudolf Wiedemann in a half-century of German pediatric genetics. Am. J. Med. Genet. 43: 740-746, 1992.
Wiedemann, H.-R.: Complexe malformatif familial avec hernie ombilicale et macroglossie—un 'syndrome nouveau'?. J. Genet. Hum. 13: 223-232, 1964.
Wiedemann, H.-R.: Das EMG-Syndrome: Exomphalos, Makroglossie, Gigantismus und Kohlenhydratstoffwechselstoerung. Z. Kinderheilk. 106: 171-185, 1969.
Wiedemann, H.-R.: Tumours and hemihypertrophy associated with Wiedemann-Beckwith syndrome. (Letter) Europ. J. Pediat. 141: 129, 1983.
Wiedemann, H.-R.: Genital overgrowth in the EMG syndrome. (Letter) Am. J. Med. Genet. 32: 255-256, 1989.
Wiedemann, H.-R.; Spranger, J. W.; Mogharei, M.; Kubler, W.; Tolksdorf, M.; Bontemps, M.; Drescher, J.; Gunschera, H.: Ueber das Syndrom Exomphalos-Makroglossie-Gigantismus, ueber generalisierte Muskelhypertrophie, progressive Lipodystrophie und Miescher-Syndrom im Sinne diencephaler Syndrome. Z. Kinderheilk. 102: 1-36, 1968.
Wilfong, A. A.; Parke, J. T.; McCrary, J. A.: Opsoclonus-Myoclonus with Beckwith-Wiedemann syndrome and hepatoblastoma. Pediat. Radiol. 8: 77-79, 1992.
Winter, S. C.; Curry, C. J. R.; Smith, J. C.; Kassel, S.; Miller, L.; Andrea, J.: Prenatal diagnosis of the Beckwith-Wiedemann syndrome. Am. J. Med. Genet. 24: 137-141, 1986.
Wockel, W.; Scheibner, K.; Lageman, A.: A variant of the Wiedemann-Beckwith syndrome. Europ. J. Pediat. 135: 319-324, 1981.
Wojciechowski, A. H.; Pritchard, J.: Beckwith-Wiedemann (exomphalos-macroglossia-gigantism—EMG) syndrome and malignant lymphoma. Europ. J. Pediat. 137: 317-321, 1981.
Zhang, P.; Leigeois, N. J.; Wong, C.; Finegold, M.; Hou, H.; Thompson, J. C.; Silverman, A.; Harper, J. W.; DePinho, R. A.; Elledge, S. J.: Altered cell differentiation and proliferation in mice lacking p57(KIP2) indicates a role in Beckwith-Wiedemann syndrome. Nature 387: 151-158, 1997.

CEREBRAL GIGANTISM (SOTOS SYNDROME) (*117550)

JJM NOTE

Malignant neoplasms that complicate this overgrowth syndrome include Wilms tumor, hepatocellular carcinoma, vaginal carcinoma, neuroectodermal tumor, mixed parotid tumor, small cell lung carcinoma, neuroblastoma, and osteochondroma.

Sotos et al. (1964) described 5 children with a disorder characterized by excessively rapid growth, acromegalic features, and a nonprogressive cerebral disorder with mental retardation. High-arched palate and prominent jaw were noted in several of them. Birth length was between the 90th and 97th centiles in all. Bone age was advanced in most of them.

Except for a concordant set of identical twins (Hook and Reynolds, 1967), most cases have been sporadic. (I observed the case of an affected boy whose father, not available for study, was described as having similar features.) The reported cases may represent new dominant mutations. Large hands and feet are present from birth. Growth is rapid in the first years of life but final height may not be excessive. Bone age is advanced. The skull is large with moderate prognathism. Mild dilation of the cerebral ventricles, nonspecific EEG changes, and seizures have been observed. Poor coordination and mental retardation are features. The differential diagnosis should include the XYY syndrome. In 2 patients, Bejar et al. (1970) found abnormal dermatoglyphics, normal growth hormone levels, and high levels of valine, isoleucine and leucine in the blood. The glycine-to-valine ratio seemed particularly useful in distinguishing patients from controls. Hooft et al. (1968) described cerebral gigantism in 2 first cousins. Nevo et al. (1974) described affected brother and sister and their affected double first cousin, in an inbred Arab family in Israel. Two of the 3 showed generalized edema and flexion contractures of the feet at birth. This may represent a distinct disorder; see Nevo syndrome (601451).

Hansen and Friis (1976) described affected mother and child. Zonana et al. (1976) described affected mother and 2 children (male and female). The mother's father may have been affected. Rosenbaum (1977) showed me mother and infant daughter with cerebral gigantism. The mother had a master's degree in education, exostoses of the alveolar ridges, and size 11 shoes. Both mother and daughter showed early eruption of teeth. Zonana et al. (1977) reported 3 families showing vertical transmission and equal severity in males and females; no male-to-male transmission was observed. As an addendum, they commented on a fourth instance of affected mother and son.

Ruvalcaba et al. (1980) found hamartomatous polyps of the intestine and melanin spots of the penis in 2 males with the Sotos syndrome. Smith et al. (1981) observed affected mother and daughter—the presumed fifth instance of dominant inheritance. The mother had primary hypothyroidism due to Hashimoto disease. Halal (1982) reported a family in which the father and 2 of his sons were affected. She knew of no other instance of documented male-to-male transmission. Halal (1983) reported that the older of the boys she reported with cerebral gigantism had pigmented spots on the genitalia and that the father had been found to have a rectal polyp—findings like those in 2 unrelated adult males reported by Ruvalcaba et al. (1980). Presumed Sotos syndrome was described in a mother and 2 daughters by Bale et al. (1985). They suggested that instances of seemingly autosomal recessive inheritance may be examples of incomplete penetrance, gonadal mosaicism, or genetic heterogeneity.

In a study of the metacarpophalangeal pattern profile in Sotos syndrome, Butler et al. (1985) found no evidence of heterogeneity and developed a diagnostic tool they suggested may be useful. Winship (1985) described a 'Cape Coloured' family with affected father and 4 children by 2 different, unrelated wives. Kaneko et al. (1987) found congenital heart defects in 5 of 10 patients with typical Sotos syndrome.

Noreau et al. (1998) reported on 3 patients with congenital heart defects out of 14 Sotos syndrome patients. Review of reports showed another 17 patients with variable cardiac defects, mostly closure defects, making an overall incidence of approximately 8%.

Goldstein et al. (1988) described 2 unrelated children with macrocephaly, excessive growth, strabismus, hypotonia and developmental delay, and improvement with age. Minor changes in the mother of each infant suggested dominant inheritance of a Sotos sequence. Fryns (1988) referred to cases of the fragile X syndrome (309550) in which Sotos syndrome had been diagnosed; he therefore suggested that this disorder be designated the Sotos sequence or the mental retardation-overgrowth sequence. Nance et al. (1990) described a 15-month-old child with Sotos syndrome and a paraspinal neuroblastoma. From this and other evidence, they concluded that children with this disorder may be at an increased risk for developing tumors.

In a review, Cole and Hughes (1990) emphasized that the handicaps in Sotos syndrome are fewer than previously believed and tend to improve with age. The latter feature makes identification of affected adults difficult. Cole and Hughes (1994) clinically assessed 79 patients with a provisional diagnosis of Sotos syndrome and evaluated their photographs between ages 1 and 6 years. These photographs, together with photographs of first-degree relatives, also at ages 1 to 6 years, were reviewed by 4 clinical geneticists. In 41 probands, but no first-degree relatives, the facial gestalt was thought to be characteristic of Sotos syndrome. Comparison of anthropometric measurements, bone age, and developmental delay in these 41 probands showed marked differences between them and the remaining 38 probands. Length was identified as the most significantly increased prenatal parameter. In childhood, occipitofrontal head circumference (OFC), height, and weight were all increased. OFC remained above the 97th percentile in all but one case throughout childhood and adulthood, whereas height and weight had a tendency to return toward the mean. This 'normalization' was more pronounced in females and was probably related to their early puberty. Early developmental delay and an advanced bone age were seen in 100% and 84% of cases, respectively. Cole and Hughes (1994) suggested that facial gestalt, growth pattern, bone age, and developmental delay are the major diagnostic criteria. Using these criteria, no affected first-degree relatives were identified.

Scarpa et al. (1994) described a sister and brother with macrocrania and coarse face (frontal bossing, highly arched palate, prognathism, pointed chin, large ears). Psychomotor development of the sister, who also had advanced osseous maturation, improved significantly at the age of 7 years. Accelerated growth with normal bone age, optic atrophy, renal agenesis with contralateral double kidney, and significant mental retardation (IQ, 45) were shown in the brother at 3.5 years of age. The father of these children was tall, with macrocrania and large hands and feet. He had had learning difficulties in school and was a manual laborer. Scarpa et al. (1994) suggested that these children and their father showed different manifestations of Sotos syndrome.

Schrander-Stumpel et al. (1990) described a 6-year-old boy with Sotos syndrome who also had a de novo, apparently balanced translocation, t(3;6)(p21;p21). They suggested that the autosomal dominant gene for the Sotos syndrome may be located either at 3p21 or 6p21. Tsukahara and Kajii (1991) could find no abnormality in high resolution-banded chromosomes from 5 patients. Involvement of genes at 3p21 was also suggested by the case reported by Cole et al. (1992); a 22-year-old female with Sotos syndrome, a nonsmoker, died of small cell lung carcinoma (182280) for which genetic determinants in the 3p21 region are suggested by loss-of-heterozygosity studies. Maroun et al. (1994) reported the case of a 4-year-old girl with Sotos phenotype and a de novo balanced translocation between 5q and 15q: 46,XX,t(5,15)(q35;q22). They thus suggested 5q35 or 15q22 as the site of an autosomal dominant gene determining Sotos syndrome.

Allanson and Cole (1996) presented anthropometric evaluation of the head in 45 patients with Sotos syndrome between age 1 and 25 years. With increasing age, the face lengthens and the chin becomes more striking.

The possibility of uniparental disomy in Sotos syndrome was investigated by Smith et al. (1997). Using 112 dinucleotide repeat DNA polymorphisms, they examined parental inheritance of all autosomal pairs, except chromosome 15, in 29 patients with Sotos syndrome. All informative cases showed biparental inheritance and no cases of UPD were found.

Opitz et al. (1998) discussed the differentiation of 2 overgrowth syndromes, that of Sotos and that of Weaver, and the question of whether the similarities are sufficient to consider them 1 entity. The apparent differences between them may be artificial as they were between the G and BBB syndromes, which were listed as separate entities, even with different mode of inheritance, autosomal dominant and X-linked recessive, respectively, from the fourth (1975) through the ninth (1990) print editions of this catalog (McKusick, 1998). As noted by Opitz et al. (1998) vertebrate development is constrained into only a very few final or common developmental pathways; therefore, no developmental anomaly seen in humans is unique to ('pathognomonic of') one syndrome.

Possible phenotypic differences between the syndromes of Sotos and Weaver pointed out by Opitz et al. (1998) were the following: the Sotos syndrome may be a cancer syndrome, whereas the Weaver syndrome is not (although a neuroblastoma had been reported in the latter disorder). In Sotos syndrome there is remarkably advanced dental maturation; this is rarely commented on in Weaver syndrome. In Weaver syndrome, there are more conspicuous contractures and a facial appearance that experts find convincingly different from that in Sotos syndrome. Opitz et al. (1998) favored allelic heterogeneity as the explanation for the similarities between Sotos and Weaver syndromes. They suggested that mapping and isolation of the causative gene or genes would settle the issue.

Opitz et al. (1998) reported affected mother and daughter. The mother was described as a large infant and 'as tall as her teacher in school.' Her adult height was 185.4 cm, and she had mandibular prognathism and a prominent pointed chin. The daughter showed a prominent forehead with sparseness of frontal hair and a 'ruddy' or flushed complexion, especially of the nose and perioral area. She had prominent features of the congenital hypotonia/lymphedema sequence with hypermobile joints, especially at the knees and ankles, lymphedema nails (especially toenails), and a high total ridge count (TRC) of the fingertip dermatoglyphics. The mother also had a high TRC and a receding frontal hairline.

Brown et al. (1998) described a pair of 5-year-old male monozygotic twins who were discordant for Sotos syndrome.

See also Boman and Nilsson (1980); Dodge et al. (1983); Stephenson et al. (1968)

Allanson, J. E.; Cole, T. R. P.: Sotos syndrome: evolution of facial phenotype subjective and objective assessment. Am. J. Med. Genet. 65: 13-20, 1996.
Bale, A. E.; Drum, M. A.; Parry, D. M.; Mulvihill, J. J.: Familial Sotos syndrome (cerebral gigantism): craniofacial and psychological characteristics. Am. J. Med. Genet. 20: 613-624, 1985.
Bejar, R. L.; Smith, G. F.; Park, S.; Spellacy, W. N.; Wolfson, S. L.; Nyhan, W. L.: Cerebral gigantism: concentrations of amino acids in plasma and muscle. J. Pediat. 76: 105-111, 1970.
Boman, H.; Nilsson, D.: Sotos syndrome in two brothers. Clin. Genet. 18: 421-427, 1980.
Brown, W. T.; Wisniewski, K. E.; Sudhalter, V.; Keogh, M.; Tsiouris, J.; Miezejeski, C.; Schaefer, G. B.: Identical twins discordant for Sotos syndrome. Am. J. Med. Genet. 79: 329-333, 1998.
Butler, M. G.; Meaney, F. J.; Kittur, S.; Hersh, J. H.; Hornstein, L.: Metacarpophalangeal pattern profile analysis in Sotos syndrome. Am. J. Med. Genet. 20: 625-629, 1985.
Cole, T. R. P.; Hughes, H. E.: Sotos syndrome: a study of the diagnostic criteria and natural history. J. Med. Genet. 31: 20-32, 1994.
Cole, T. R. P.; Hughes, H. E.: Sotos syndrome. J. Med. Genet. 27: 571-576, 1990.
Cole, T. R. P.; Hughes, H. E.; Jeffreys, M. J.; Williams, G. T.; Arnold, M. M.: Small cell lung carcinoma in a patient with Sotos syndrome: are genes at 3p21 involved in both conditions?. J. Med. Genet. 29: 338-341, 1992.
Dodge, P. R.; Homes, S. J.; Sotos, J. F.: Cerebral gigantism. Dev. Med. Child Neurol. 25: 248-252, 1983.
Fryns, J. P.: The Prader-Willi syndrome and the Sotos syndrome: syndromes or sequences? (Letter) Clin. Genet. 33: 457-458, 1988.
Goldstein, D. J.; Ward, R. E.; Moore, E.; Fremion, A. S.; Wappner, R. S.: Overgrowth, congenital hypotonia, nystagmus, strabismus, and mental retardation: variant of dominantly inherited Sotos sequence?. Am. J. Med. Genet. 29: 783-792, 1988.
Halal, F.: Cerebral gigantism, intestinal polyposis, and pigmentary spotting of the genitalia. (Letter) Am. J. Med. Genet. 15: 161, 1983.
Halal, F.: Male to male transmission of cerebral gigantism. Am. J. Med. Genet. 12: 411-419, 1982.
Hansen, F. J.; Friis, B.: Familial occurrence of cerebral gigantism, Sotos' syndrome. Acta Paediat. Scand. 65: 387-389, 1976.
Hooft, C.; Schotte, H.; Van Hooren, G.: Familial cerebral gigantism. Acta Paediat. Belg. 22: 173-186, 1968.
Hook, E. B.; Reynolds, J. W.: Cerebral gigantism: endocrinological and clinical observations of six patients including a congenital giant, concordant monozygotic twins, and a child who achieved adult gigantic size. J. Pediat. 70: 900-914, 1967.
Kaneko, H.; Tsukahara, M.; Tachibana, H.; Kurashige, H.; Kuwano, A.; Kajii, T.: Congenital heart defects in Sotos sequence. Am. J. Med. Genet. 26: 569-576, 1987.
Maroun, C.; Schmerler, S.; Hutcheon, R. G.: Child with Sotos phenotype and a 5:15 translocation. Am. J. Med. Genet. 50: 291-293, 1994.
McKusick, V. A.: Mendelian Inheritance in Man: A Catalog of Human Genes and Genetic Disorders. Baltimore: Johns Hopkins University Press (12th ed.), 1998.
Nance, M. A.; Neglia, J. P.; Talwar, D.; Berry, S. A.: Neuroblastoma in a patient with Sotos' syndrome. J. Med. Genet. 27: 130-132, 1990.
Nevo, S.; Zeltzer, M.; Benderly, A.; Levy, J.: Evidence for autosomal recessive inheritance in cerebral gigantism. J. Med. Genet. 11: 158-165, 1974.
Noreau, D. R.; Al-Ata, J.; Jutras, L.; Teebi, A. S.: Congenital heart defects in Sotos syndrome. Am. J. Med. Genet. 79: 327-328, 1998.
Opitz, J. M.; Weaver, D. W.; Reynolds, J. F., Jr.: The syndromes of Sotos and Weaver: reports and review. Am. J. Med. Genet. 79: 294-304, 1998.
Rosenbaum, K. N.: Personal Communication. Baltimore, Md. 1977.
Ruvalcaba, R. H. A.; Myhre, S.; Smith, D. W.: Sotos syndrome with intestinal polyposis and pigmentary changes of the genitalia. Clin. Genet. 18: 413-416, 1980.
Scarpa, P.; Faggioli, R.; Voghenzi, A.: Familial Sotos syndrome: longitudinal study of two additional cases. Genet. Counsel. 5: 155-159, 1994.
Schrander-Stumpel, C. T. R. M.; Fryns, J. P.; Hamers, G. G.: Sotos syndrome and de novo balanced autosomal translocation (t(3;6)(p21;p21)). Clin. Genet. 37: 226-229, 1990.
Smith, A.; Farrar, J. R.; Silink, M.; Judzewitsch, R.: Investigations in dominant Sotos syndrome. Ann. Genet. 24: 226-228, 1981.
Smith, M.; Fullwood, P.; Qi, Y.; Palmer, S.; Upadhyaya, M.; Cole, T.: No evidence for uniparental disomy as a common cause of Sotos syndrome. J. Med. Genet. 34: 10-12, 1997.
Sotos, J. F.; Dodge, P. R.; Muirhead, D.; Crawford, J. D.; Talbot, N. B.: Cerebral gigantism in childhood: a syndrome of excessively rapid growth with acromegalic features and a nonprogressive neurologic disorder. New Eng. J. Med. 271: 109-116, 1964.
Stephenson, J. N.; Mellinger, R. C.; Manson, G.: Cerebral gigantism. Pediatrics 41: 130-138, 1968.
Tsukahara, M.; Kajii, T.: High resolution-banded chromosomes from patients with Sotos syndrome. (Letter) Clin. Genet. 39: 313-314, 1991.
Winship, I. M.: Sotos syndrome—autosomal dominant inheritance substantiated. Clin. Genet. 28: 243-246, 1985.
Zonana, J.; Rimoin, D. L.; Fisher, D. A.: Cerebral gigantism—apparent dominant inheritance. Birth Defects Orig. Art. Ser. XII(6): 63-69, 1976.
Zonana, J.; Sotos, J. F.; Romshe, C. A.; Fisher, D. A.; Elders, M. J.; Rimoin, D. L.: Dominant inheritance of cerebral gigantism. J. Pediat. 91: 251-256, 1977.

METAPHYSEAL CHONDRODYSPLASIA (CARTILAGE-HAIR HYPOPLASIA) (*250250)

JJM NOTE

The list of associated neoplasms is long and diverse: pulmonary lymphoma, intestinal lymphosarcoma, hepatoma, duodenal carcinoma, skin carcinoma, Hodgkin disease, and testicular progonoma. Yet, formal analysis is needed to estimate which might be truly occurring to excess. Given the bone marrow hypoplasia and immune defect, one might predict leukemia, in addition to lymphoma.

This disorder was first recognized (McKusick et al., 1965) as a distinct entity in the Old Order Amish, a religious isolate (McKusick, 1978), but was later identified in other groups. CHH is a form of short-limbed dwarfism due to skeletal dysplasia. Incomplete extension at the elbows, anterolateral chest deformity like Harrison grooves, genu varum, and excessively long fibula distally relative to the tibia are specific skeletal features. The fingers are usually loose-jointed and the fingernails foreshortened. By x-ray the changes are of the type called in the past metaphyseal dysostosis by radiologists. Biopsy shows hypoplasia of cartilage to be the nature of the skeletal abnormality. The hair is fine, sparse and light-colored. Microscopically, it has an abnormally small caliber. Many patients with CHH have a major immunologic defect that is manifested, for example, as an unusual susceptibility to chickenpox; in the original study of the Amish, 6 cases of fatal varicella pneumonia were noted (McKusick et al., 1965). In addition to lymphopenia, some patients have anemia and neutropenia. Aganglionic megacolon is found in some patients. Autosomal recessive inheritance seems quite certain although penetrance is reduced (when dwarfism is taken as the phenotype for ascertainment).

Kaitila and Perheentupa (1980) referred to 33 cases in 28 families in Finland. Leukocyte interferon might be useful in children with CHH and varicella, judging from the apparent benefit in immunosuppressed children with cancer (Arvin et al., 1982). The relationship of the syndrome described by Burgert et al. (1965), with aregenerative anemia and celiac syndrome, is unclear. Unexplained features present in some cases of CHH include anemia, malabsorption, Hirschsprung disease, and susceptibility to chickenpox. Some of these features resemble those of the patients with pancreatic insufficiency and neutropenia reported by Burke et al. (1967) and some of their patients had metaphyseal changes. See also the report of Theodorou and Adams (1963). Levin (1978) characterized a possible variant of CHH. Two sibs and one isolated case, in addition to being shorter than most patients with CHH, had dental abnormalities consisting of microdontia, a single notch in the center of the incisal edge of each permanent incisor, and doubling of the lingual cusps of the lower premolars.

Polmar and Pierce (1986) described marked impairment of T-cell function due to an intrinsic defect in cell proliferation. Since defective proliferation was found also in B cells and fibroblasts from CHH persons, they concluded that the defect in cell proliferation is generalized. They stated that, as might be expected, proliferation-independent natural killer activity was normal. They thought this explained that 'an increased incidence of malignancy was not observed in CHH patients.' This is not our experience; the malignancy experience of CHH patients seems excessive (Francomano, 1984). Lymphoma appears to have an increased frequency. Primary liver carcinoma and duodenal carcinoma were diagnoses in other cases; several patients have succumbed to virulent viral hepatitis. Skin carcinoma in exposed areas may be related to the hypopigmentation of the skin. Herpes labialis is extraordinarily severe in some patients and may be accompanied by unusual malignancy of the mouth. The literature on 63 non-Amish cases was summarized by van der Burgt et al. (1991). They concluded that there seems to be no increased susceptibility to malignancy in persons with CHH. However, Gorlin (1992) referred to a patient treated for Hodgkin disease at the age of about 33 years. Even among the Amish, where the self-same gene must be responsible for all the manifestations, there is a remarkable degree of variability in all features of the syndrome. The hair may be seemingly completely unaffected; some patients suffer no apparent immunologic or hematologic defect; Hirschsprung disease of clinically evident severity occurs in a minority. Patients may be relatively tall; at least 1 Amish woman, the daughter of 2 affected persons, is 55 inches tall. She was thought to be unaffected until about age 9 when x-rays showed characteristic changes. At the other end of the spectrum of severity is severe involvement in all aspects of the disorder, including extensive aganglionosis of the bowel leading to early demise . In 1 such patient the fingers showed distal arthrogryposis; stiffness of the interphalangeal joints prevented bending of the fingers, which were characteristically short as well, and there were dimples over the metacarpophalangeal joints. Le Merrer and Maroteaux (1991) illustrated the difficulties in making the diagnosis in the first year or so of life.

Makitie (1992) presented an epidemiologic and genetic study of 107 (46 males and 61 females) CHH patients distributed in 85 families in Finland. Of the 107, 18 had died, 7 before the age of 1 year. The living patients ranged in age from 1 to 51 years (median 21 years). The incidence was estimated to be 1:23,000 live births. Consanguinity was found in 2 families. The geographic distribution of the birthplaces of the patients and their great-grandparents showed accumulation in a small area in western Finland, and regional clusters were seen in other parts of the country as well. Segregation analysis was consistent with recessive inheritance with reduced penetrance, as had been suggested by McKusick et al. (1965). In another report, Makitie and Kaitila (1993) reviewed the clinical manifestations of 108 patients. They emphasized the fact that there is not only short-limb dwarfism, hypoplastic hair, and defective immunity, but also defective erythrogenesis. Six patients had normal hair. Childhood anemia had occurred in 79% of the patients; it was usually mild but was severe in 14 patients. Hirschsprung disease was observed in 8, anal stenosis in 1, and esophageal atresia in 1. Intrafamilial variation, as studied in 16 sibships, was considerable. According to Makitie et al. (1992), macrocytosis accompanied the anemia or occurred independent of the anemia during childhood. The reticulocyte index was always low in relation to the anemia. Anemia, which was most prevalent and severe during infancy, underwent spontaneous remission before adulthood in all patients except 3 infants with fatal hypoplastic anemia. Lymphopenia was present in 62% of 88 patients studied, and neutropenia in 24%. Anemia correlated significantly with the severity of the immunodeficiency and growth failure and the presence of neutropenia. Makitie et al. (1992) found that the stature at birth showed a mean relative length of -3.0 SD. The median adult height was 131.1 cm for 15 males (range, 110.7-149.0 cm) and 122.5 cm (range, 103.7-137.4 cm) for 20 females. Adult height showed no correlation with midparent height. Makitie et al. (1992) analyzed 149 skeletal radiographic surveys of 82 Finnish patients. Skeletal age was markedly reduced in 14% of patients. Mild scoliosis was observed in one-fourth of patients. Excessive distal length of the fibula and irregularity and flaring of the ribs at the costochondral junction were noted. Six of 108 patients had malignancy: pulmonary lymphoma leading to death at age 10; intestinal lymphosarcoma with death at age 22; malignant testis tumor (progonoma) removed surgically at the age of 6 months; and skin carcinoma in 3 patients.

McKusick et al. (1965) pointed to what appeared to be a deficiency of affected individuals in the CHH families among the Amish. The deficiency appeared to be statistically significant and reduced penetrance was suggested as a possible explanation. Makitie (1992) found a similar deficiency in Finnish families, but it did not reach statistical significance. In linkage studies of CHH, Sulisalo et al. (1994) found no evidence of healthy sibs with homozygosity for 'affected' haplotypes. Sulisalo (1994) studied all available families in Finland with only 1 affected child and found 2 examples of probands with uniparental disomy for chromosome 9; both were maternal. UPD can, of course, explain a deficiency of affected individuals in recessively inherited disorders; however, 2 out of 54 Finnish uniplex patients would be inadequate to explain the deficiency of affected individuals in the Finnish pedigrees (de la Chapelle, 1994). Among 12 Amish uniplex families, Sulisalo (1994) found that 1 of the affected children was a UPD case. (In 1 Amish family with only 1 affected out of 12 children, the results were inconclusive because of homozygosity of markers close to CHH due probably to the relatively high degree of inbreeding in the Amish.)

Kaitila et al. (1975) and Ochs (1975) reported normal red cell adenosine deaminase in CHH. In a boy with CHH, the son of first-cousin parents, Sanchez-Corona et al. (1990) found a 4-fold increase in red blood cell adenosine deaminase activity.

Acyclovir should have a useful place in the treatment of varicella in patients with CHH (Dunkle et al., 1991; Gnann, 1994; Wood, 1994), and the use of varicella vaccine (Hardy et al., 1991) for prophylaxis will be worthy of consideration. An attenuated live polio vaccine might have risks of severe disease in a patient with CHH.

By linkage analysis, Sulisalo et al. (1993) assigned the CHH gene to chromosome 9, between D9S43 and D9S50; multipoint lod score = 9.94. Based on strong linkage disequilibrium, the closest marker, D9S50, was thought to be less than 1 cM from the gene. No heterogeneity was observed in 14 Finnish families, and there was no evidence of reduced penetrance. Sulisalo et al. (1993) were 'reluctant to assign the CHH locus to either side of the centromere although the evidence probably favours 9p.' Sulisalo et al. (1995) stated that the CHH gene is on 9p close to D9S163 within an interval of 1.5 cM flanked by D9S165 and D9S50. In a study of 9 families with CHH and no genealogic connections to either Amish or Finns, Sulisalo et al. (1995) found no recombinants between the CHH gene and any of the 3 closest marker loci, suggesting that CHH in these families results from mutation(s) at the same locus as in the Amish and Finnish families.

On 'closer inspection of the pedigrees' reported by Sulisalo et al. (1995), Krawczak (1995) concluded that their power to disclose linkage heterogeneity was limited and that a substantial level of heterogeneity might have gone unnoticed.

Ridanpaa et al. (1995) refined the localization of the CHH gene to 9p13 using in situ hybridization for the localization of markers involved in linkage and linkage disequilibrium analyses. They concluded that CHH is approximately 0.3 cM on the proximal side of the microsatellite marker D9S163. PAX5 (167414) also maps to 9p13 and is located between D9S163 and another microsatellite marker within an interval of no more than 56 kb. The following order was proposed for loci on 9p13: cen—D9S50—CHH—D9S163—PAX5—tel.

In the initial study of CHH in the Amish by McKusick et al. (1965), a deficiency of affected members of sibships resulted in a lower than expected segregation ratio. A similar deficiency of affected members was not found in analysis of 83 Finnish families by Makitie (1992). In the Amish series, reduced penetrance was suggested as the most likely mechanism of the deficiency, but after the CHH locus was assigned to proximal 9q and several closely linked polymorphic markers were identified, it became possible to search for reduced penetrance by determining the haplotypes of unaffected sibs in multiplex families. Sulisalo et al. (1994) studied 66 clinically unaffected sibs from Finnish and Amish families and found none to be haploidentical with affected sibs. Thus reduced penetrance or underdiagnosis of mildly manifesting individuals appeared to be ruled out in both populations. Sulisalo et al. (1997) initiated a repeat segregation analysis in an extended series of Finnish CHH families and analyzed uniplex families in order to look for cases of uniparental disomy, 1 such case having been observed in a family counseled for prenatal diagnosis. Although segregation analysis of 101 families again failed to give evidence of a statistically significant deficiency of affected members, the study of 54 uniplex families revealed 1 additional patient with CHH and uniparental disomy for chromosome 9. In Finland, therefore, 2 of 54 uniplex families showed CHH as a result of chromosome 9 uniparental disomy.

Makitie et al. (1995) gave an extensive review of CHH.

See also Coupe and Lowry (1970); Halle et al. (1970); Harris et al. (1981); Hong et al. (1972); Lowry et al. (1970); Lux et al. (1970); Makitie et al. (1992); Makitie et al. (1992); Mancini and Morabito (1974); Pierce and Polmar (1982); Saulsbury et al. (1975); Seige (1980); Sulisalo et al. (1994); Virolainen et al. (1978); Wilson et al. (1978)

Arvin, A. M.; Kushner, J. H.; Feldman, S; Baehner, R. L.; Hammond, D.; Merigan, T. C.: Human leukocyte interferon from the treatment of varicella in children with cancer. New Eng. J. Med. 306: 761-765, 1982.

Burgert, E. O., Jr.; Dower, J. C.; Tauxe, W. N.: A new syndrome—aregenerative anemia, malabsorption (celiac), dyschondroplasia and hyperphosphatemia. (Abstract) J. Pediat. 67: 711-712, 1965.

Burke, V.; Colebatch, J. H.; Anderson, C. M.; Simons, M. J.: Association of pancreatic insufficiency and chronic neutropenia in childhood. Arch. Dis. Child. 42: 147-157, 1967.

Coupe, R. L.; Lowry, R. B.: Abnormality of the hair in cartilage-hair hypoplasia. Dermatologica 141: 329-334, 1970.

de la Chapelle, A.: Personal Communication. Helsinki, Finland 12/14/1994.

Dunkle, L. M.; Arvin, A. M.; Whitley, R. J.; Rotbart, H. A.; Feder, H. M., Jr.; Feldman, S.; Gershon, A. A.; Levy, M. L.; Hayden, G. F.; McGuirt, P. V.; Harris, J.; Balfour, H. H., Jr.: A controlled trial of acyclovir for chickenpox in normal children. New Eng. J. Med. 325: 1539-1544, 1991.

Francomano, C.: Personal Communication. Baltimore, Md. 1984.

Gnann, J. W., Jr.: New antivirals with activity against varicella-zoster virus. Ann. Neurol. 34: S69-S72, 1994.

Gorlin, R. J.: Cartilage-hair-hypoplasia and Hodgkin disease. (Letter) Am. J. Med. Genet. 44: 539 only, 1992.

Halle, M. A.; Collipp, P. J.; Roginsky, M.: Cartilage-hair hypoplasia in childhood. New York J. Med. 70: 2705-2708, 1970.

Hardy, I.; Gershon, A. A.; Steinberg, S. P.; LaRussa, P.; the Varicella Vaccine Collaborative Study Group: The incidence of zoster after immunization with live attenuated varicella vaccine: a study in children with leukemia. New Eng. J. Med. 325: 1545-1550, 1991.

Harris, R. E.; Baehner, R. L.; Gleiser, S.; Weaver, D. D.; Hodes, M. E.: Cartilage-hair hypoplasia, defective T-cell function, and Diamond-Blackfan anemia in an Amish child. Am. J. Med. Genet. 8: 291-297, 1981.

Hong, R.; Ammann, J.; Haung, S. W.; Levy, R. L.; Davenport, G.; Bach, M. L.; Bach, F. H.; Bortin, M. M.; Kay, H. E. M.: Cartilage-hair hypoplasia: effect of thymus transplants. Clin. Immun. Immunopath. 1: 15-25, 1972.

Kaitila, I.; Perheentupa, J.: Cartilage-hair hypoplasia (CHH).In: Eriksson, A. W.; Forsius, H. R.; Nevanlinna, H. R.; Workman, P. L.; Norio, R. K.: Population Structure and Genetic Disorders. New York: Academic Press (pub.) 1980. Pp. 588-591.

Kaitila, I. J.; Tanaka, K. R.; Rimoin, D. L.: Normal red cell adenosine deaminase activity in cartilage-hair hypoplasia. J. Pediat. 87: 153-154, 1975.

Krawczak, M.: Locus homogeneity for cartilage-hair hypoplasia proven?. (Letter) Hum. Genet. 96: 371-372, 1995.

Le Merrer, M.; Maroteaux, P.: Cartilage hair hypoplasia in infancy: a misleading chondrodysplasia. Europ. J. Pediat. 150: 847-851, 1991.

Levin, L. S.: Personal Communication. Baltimore, Md. 1978.

Lowry, R. B.; Wood, B. J.; Birkbeck, J. A.; Padwick, P. H.: Cartilage-hair hypoplasia: a rare and recessive cause of dwarfism. Clin. Pediat. 9: 44-46, 1970.

Lux, S. E.; Johnston, R. B., Jr.; August, C. S.; Say, B.; Penchaszadeh, V. B.; Rosen, F. S.; McKusick, V. A.: Neutropenia and abnormal cellular immunity in cartilage-hair hypoplasia. New Eng. J. Med. 282: 234-236, 1970.

Makitie, O.: Cartilage-hair hypoplasia in Finland: epidemiological and genetic aspects of 107 patients. J. Med. Genet. 29: 652-655, 1992.

Makitie, O.; Kaitila, I.: Cartilage-hair hypoplasia—clinical manifestations in 108 Finnish patients. Europ. J. Pediat. 152: 211-217, 1993.

Makitie, O.; Marttinen, E.; Kaitila, I.: Skeletal growth in cartilage-hair hypoplasia—a radiological study of 82 patients. Pediat. Radiol. 22: 434-439, 1992.

Makitie, O.; Perheentupa, J.; Kaitila, I.: Growth in cartilage-hair hypoplasia. Pediat. Res. 31: 176-180, 1992.

Makitie, O.; Rajantie, J.; Kaitila, I.: Anaemia and macrocytosis—unrecognized features in cartilage-hair hypoplasia. Acta Paediat. 81: 1026-1029, 1992.

Makitie, O.; Sulisalo, T.; de la Chapelle, A.; Kaitila, I.: Cartilage-hair hypoplasia. J. Med. Genet. 32: 39-43, 1995.

Mancini, R.; Morabito, F.: La condrodisplasia metafisaria, tipo McKusick, o 'cartilage-hair hypoplasia': studio di due casi. Acta Med. Auxol. 6: 131-148, 1974.

McKusick, V. A.: Medical Genetic Studies of the Amish: Selected Papers. Baltimore: Johns Hopkins Univ. Press (pub.) 1978.

McKusick, V. A.; Eldridge, R.; Hostetler, J. A.; Egeland, J. A.; Ruangwit, U.: Dwarfism in the Amish. II. Cartilage-hair hypoplasia. Bull. Johns Hopkins Hosp. 116: 285-326, 1965.

Ochs, H. D.: Normal red cell adenosine deaminase activity in cartilage-hair hypoplasia—reply. (Letter) J. Pediat. 87: 154 only, 1975.

Pierce, G. F.; Polmar, S. H.: Lymphocyte dysfunction in cartilage-hair hypoplasia: evidence for an intrinsic defect in cellular proliferation. J. Immun. 129: 570-575, 1982.

Polmar, S. H.; Pierce, G. F.: Cartilage hair hypoplasia: immunological aspects and their clinical implications. Clin. Immun. Immunopath. 40: 87-93, 1986.

Ridanpaa, M.; Sulisalo, T.; de la Chapelle, A.; Kaitila, I.: Genetic and physical mapping of the cartilage-hair hypoplasia locus on 9p13. (Abstract) Am. J. Hum. Genet. 57: A201 only, 1995.

Sanchez-Corona, J.; Garcia-Cruz, D.; Medina, C.; Cantu, J. M.; Ramos-Zepeda, R.; Rivas, F.: Increased adenosine deaminase activity in a patient with cartilage-hair hypoplasia. Ann. Genet. 33: 99-102, 1990.

Saulsbury, F. T.; Winkelstein, J. A.; Davis, L. E.; Hsu, S.; Souza, B. J.; Gutcher, G. R.; Butler, I. J.: Combined immunodeficiency and vaccine-related poliomyelitis in a child with cartilage-hair hypoplasia. J. Pediat. 86: 868-872, 1975.

Seige, M.: Metaphysare Chondrodysplasie vom Typ McKusick (Knorpel-Haar Hypoplasie). Mschr. Kinderheilk. 128: 157-159, 1980.

Sulisalo, T.: Genetic Mapping of the gene for cartilage-hair hypoplasia. Thesis: Univ. Helsinki (pub.) 1994. Pp. 35-38.

Sulisalo, T.; Francomano, C. A.; Sistonen, P.; Maher, J. F.; McKusick, V. A.; de la Chapelle, A.; Kaitila, I.: High-resolution genetic mapping of the cartilage-hair hypoplasia (CHH) gene in Amish and Finnish families. Genomics 20: 347-353, 1994.

Sulisalo, T.; Klockars, J.; Makitie, O.; Francomano, C. A.; de la Chapelle, A.; Kaitila, I.; Sistonen, P.: High-resolution linkage-disequilibrium mapping of the cartilage-hair hypoplasia gene. Am. J. Hum. Genet. 55: 937-945, 1994.

Sulisalo, T.; Makitie, O.; Sistonen, P.; Ridanpaa, M.; El-Rifai, W.; Ruuskanen, O.; de la Chapelle, A.; Kaitila, I.: Uniparental disomy in cartilage-hair hypoplasia. Europ. J. Hum. Genet. 5: 35-42, 1997.

Sulisalo, T.; Sistonen, P.; Hastbacka, J.; Wadelius, C.; Makitie, O.; de la Chapelle, A.; Kaitila, I.: Cartilage-hair hypoplasia gene assigned to chromosome 9 by linkage analysis. Nature Genet. 3: 338-341, 1993.

Sulisalo, T.; van der Burgt, I.; Rimoin, D. L.; Bonaventure, J.; Sillence, D.; Campbell, J. B.; Chitayat, D.; Scott, C. I.; de la Chapelle, A.; Sistonen, P.; Kaitila, I.: Genetic homogeneity of cartilage-hair hypoplasia. Hum. Genet. 95: 157-160, 1995.

Theodorou, S. D.; Adams, J.: An unusual case of metaphyseal dysplasia. J. Bone Joint Surg. 45B: 364-369, 1963.

van der Burgt, I.; Haraldsson, A.; Oosterwijk, J. C.; van Essen, A. J.; Weemaes, C.; Hamel, B.: Cartilage hair hypoplasia, metaphyseal chondrodysplasia type McKusick: description of seven patients and review of the literature. Am. J. Med. Genet. 41: 371-380, 1991.

Virolainen, M.; Savilahti, E.; Kaitila, I.; Perheentupa, J.: Cellular and humoral immunity in cartilage-hair hypoplasia. Pediat. Res. 12: 961-966, 1978.

Wilson, W. G.; Aylsworth, A. S.; Folds, J. D.; Whisnant, J. K.: Cartilage-hair hypoplasia (metaphyseal chondrodysplasia, type McKusick) with combined immune deficiency: variable expression and development of immunologic functions in sibs. Birth Defects Orig. Art. Ser. XIV(6A): 117-129, 1978.

Wood, M. J.: Current experience with antiviral therapy for acute herpes zoster. Ann. Neurol. 35: S65-S68, 1994.

RUBINSTEIN-TAYBI SYNDROME (#180849)

JJM NOTE

A comprehensive literature reveiw, through 1994, revealed 36 tumors in patients with Rubinstein-Taybi syndrome (Miller RW, Rubinstein JH: Tumors in Rubinstein-Taybi syndrome. Am. J. Med. Genet. 56:112-15, 1995). The authors were impressed with the excess of neural and developmental tumors and keloids. Cell types were leukemia, oligodendroglioma, medulloblastoma, neuroblastoma, benign meningioma, pheochromocytoma, nasopharyngeal rhabdomyosarcoma, leiomyosarcoma, seminoma, embryonal carcinoma, odontoma, choristoma, dermoid cyst, and pilomatrixomas. Subsequent case reports strengthen some of the associations and add a granular cell tumor of the cervical spinal cord (Burton BJ et al.: Granular cell tumour of the spinal cord in a patient with Rubenstein-Taybi syndrome. Br. J. Neurosurg. 11:257-59, 1997; Masuno M et al.: Pilomatrixomas in Rubinstein-Taybi syndrome. Am. J. Med. Genet. 77:81-82, 1998; Skousen GJ et al.: Medulloblastoma in patient with Rubinstein-Taybi syndrome. Am. J. Med. Genet. 66:367, 1996).

A number sign (#) is used with this entry because of evidence that the characteristic disorder is caused by mutation in the gene encoding the transcriptional coactivator CREB-binding protein (600140).

In addition to the anomalies listed above, pulmonary stenosis, keloid formation in surgical scars, large foramen magnum, and vertebral and sternal anomalies should be mentioned. The dermatoglyphic changes described by Giroux and Miller (1967) suggest a chromosomal abnormality. Wulfsberg et al. (1983) could demonstrate no abnormality by high resolution cytogenetics, but Berry (1987) reviewed the etiogenetic basis and concluded that microdeletion is most likely. A figure of 0.1% for recurrence risk in sibs was stated as appropriate for genetic counseling purposes. Padfield et al. (1968) studied 17 cases and found none of 50 sibs affected. The frequency of Rubinstein syndrome was estimated to be 1 per 300-500 institutionalized persons with mental retardation over age 5 years. Father-daughter incest produced another case (Padfield et al., 1968). Rubinstein (1969) found parental age to be about average. Der Kaloustian et al. (1972) described affected brother and sister from consanguineous parents. However, whereas the facies was characteristic, broad first digits were absent clinically and questionable radiographically. Pfeiffer (1968) described the syndrome in only 1 of presumably monozygotic twins. Baraitser and Preece (1983), on the other hand, reported the Rubinstein syndrome in all 4 members of 2 pairs of monozygotic twins. Gillies and Roussounis (1985) reported 2 families: in one, 2 sibs were affected; in the other, the uncle of the index case was affected and other members of the family were judged to show varying degrees of expression of the disorder. Hennekam et al. (1989) described this disorder in mother and son. The mother's IQ was estimated to be about 65. Marion et al. (1993) presented another instance of parent-to-child transmission. Mother and daughter were affected. Having been labeled 'learning disabled' in elementary school, the mother had attended special education classes and had dropped out of school in the eleventh grade. In a Japanese boy who died at the age of 20 months, Kimura et al. (1993) described the apparent association of RSTS and the DiGeorge syndrome (188400). Bonioli et al. (1993) described the association of RSTS with slipped capital femoral epiphysis in a 10-year-old girl. Levy (1976) described juvenile glaucoma in RSTS and McKusick (1968) observed congenital glaucoma. Bonioli and Bellini (1992) reported the association of RSTS with pheochromocytoma in a 7-year-old girl.

Stevens and Bhakta (1995) evaluated cardiac abnormalities with a questionnaire survey. Of 138 patients in the study, 45 (32.6%) had a known cardiac abnormality; 27 patients had single defects, including atrial septal defect, ventricular septal defect, patent ductus arteriosus, coarctation of the aorta, pulmonic stenosis, or bicuspid aortic valve. In 8 of these individuals the problem resolved spontaneously, while 8 required surgery. Complex congenital heart defects of 2 or more abnormalities were present in 16 patients; 2 of these patients had spontaneous resolution, while 7 required surgery. Surgery was planned in 5 additional patients. It is noteworthy that pulmonic stenosis was present in only 1 patient as an isolated finding.

Stevens (1997) described 11 patients with RSTS and patellar dislocation. The age at diagnosis of patellar dislocation ranged from birth to 16 years. Chronic dislocations were present in 10 patients, and 8 of 11 had bilateral patellar dislocation. Surgical stabilization of the patella was required in 8 patients; most achieved a good outcome with surgical repair. All families reported that the patellar dislocations impaired developmental skills, which improved after surgery. Other joint abnormalities, including congenital dislocations and laxity of joints, were described in 7 of the 11 patients.

It was difficult to accept the suggestion of multifactorial inheritance offered by Roy et al. (1968). Multifactorial inheritance seemed unlikely in the case of such rare entities. One might expect a graded severity among cases and in close relatives, since there is no obvious mechanism for a threshold effect. The second family of Gillies and Roussounis (1985) may demonstrate just that. Stirt (1982) warned of the risk of cardiac arrhythmia with use of succinylcholine in the Rubinstein-Taybi syndrome. Hennekam et al. (1990) reported that among 45 patients in the Netherlands, all had broad halluces but only 39 had broad thumbs. Persistent fetal pads of the fingers, shawl scrotum, and frequent fractures were found in several patients. Constipation was a problem, and easily collapsible laryngeal walls caused difficulties in sleep and anesthesia. Stevens et al. (1990) found no second cases among the 91 sibs of 50 probands. Hennekam and Van Doorne (1990) commented on short upper lip and pouting lower lip, a feature documented in many photographs by Hennekam et al. (1990). A high, slit-like palate was also noted. Hennekam et al. (1990) reviewed data on 502 cases. In 12 of 13 proven or possible monozygotic twins, both children were affected. Two patients had reproduced, with 1 affected and 2 normal offspring. They found 1 recurrence among 708 sibs of 502 probands. From this information and the scarcity of affected sibs reported in the literature, they suggested that the recurrence risk figure for sibs is on the order of 0.1%, lower than the figure of 1.0% suggested by Berry (1987) for use in genetic counseling. The recurrence risk for offspring of affected persons may be 50%. Hennekam et al. (1990) favored an autosomal dominant mutation as the most likely cause. Robinson et al. (1993) reported another set of monozygotic twins concordant for RSTS. Preis and Majewski (1995) reported monozygotic twin sisters concordant for RSTS diagnosed at the age of 10 weeks. They commented that the typical features of the syndrome increasingly developed in early infancy toward the total 'Gestalt' by the age of 2 years.

Garcia et al. (1992) reported the cases of mother and daughter with RSTS. Lowry (1990) discussed the phenotypic overlap with the Saethre-Chotzen syndrome (101400). Guion-Almeida and Richieri-Costa (1992) described agenesis of the corpus callosum, iris coloboma, and megacolon in a boy with RSTS. Kanjilal et al. (1992) described pulmonary hypertension, mitral valve regurgitation, and patent ductus arteriosus as well as hypoplastic right kidney in a 3-month-old child with delayed motor and mental development corresponding to that of a 1-month-old infant. Shashi and Fryburg (1995) described mediastinal vascular ring causing tracheoesophageal obstruction with respiratory symptoms and dysphagia in a child with RSTS. The patient had a remarkable improvement in swallowing ability after surgery and some decrease in the frequency of respiratory infections. Chun et al. (1992) had reported 1 case of RSTS among their large series of cases of vascular rings at Johns Hopkins.

Imaizumi and Kuroki (1991) observed a sporadic case of Rubinstein-Taybi syndrome with de novo reciprocal translocation t(2;16)(p13.3;p13.3). Bazacliu et al. (1973) reported a patient with this condition and a mosaic chromosomal aberration including a deleted chromosome 2p. They suggested that the RSTS gene may be at 2p13.3 rather than at 16p13.3. During a systematic chromosomal survey of 7 unrelated patients with Rubinstein-Taybi syndrome, 45,44:Tommerup et al. (1991, 1992) found an apparently balanced de novo reciprocal translocation, t(7;16)(q34;p13.3), in an affected boy. The breakpoint in chromosome 16 involved the same subband p13.3 as was observed in a 2;16 translocation by Imaizumi and Kuroki (1991). Lacombe et al. (1992) provided confirmation for the assignment of the RSTS gene to 16p13.3. A 2-month-old girl with typical features showed a de novo pericentric inversion of one chromosome 16; her karyotype was 46,XX,inv(16)(p13.3;q13). In a large collaborative study, Breuning et al. (1993) investigated the region of 16p13.3 where 2 distinct reciprocal translocations occurred in patients with a clinical diagnosis of RSTS. The breakpoints lay between the cosmids N2 and RT1. Using 2-color fluorescence in situ hybridization, Breuning et al. (1993) demonstrated that the signal from RT1 was missing from 1 chromosome 16 in 6 of 24 patients with RSTS. The parents of 5 of these patients did not show a deletion of RT1, indicating a de novo rearrangement. They estimated that RSTS is caused by submicroscopic interstitial deletions within 16p13.3 in approximately one-fourth of patients. By molecular studies, Hennekam et al. (1993) found a copy of chromosome 16 from each parent in all 19 patients with RSTS studied. Uniparental disomy was also excluded for 5 other chromosome arms known to be imprinted in mice. Clinical features were essentially the same in patients with or without visible deletions, with a possible exception for the incidence of microcephaly, angulation of thumbs and halluces, and partial duplication of the halluces ('big toes'). Masuno et al. (1994) screened 25 Japanese patients with RSTS for microdeletions using high-resolution GTG banding and fluorescence in situ hybridization with a cosmid probe (RT1, D16S237). In 1 patient, a microdeletion was demonstrated by in situ hybridization, but none was detected by high-resolution banding.

Petrij et al. (1995) showed that the breakpoints at 16p13.3 that have been demonstrated in patients with RSTS are all restricted to a region that contains the gene for the human CREB-binding protein (600140), a nuclear protein participating as a co-activator in cyclic AMP-regulated gene expression. They showed, furthermore, that RSTS can result also from point mutations in the CREBBP gene. Because the patients are heterozygous for the mutations, Petrij et al. (1995) proposed that the loss of one functional copy of the CBP gene underlies the developmental abnormalities in RSTS and possibly the propensity for malignancy. They suggested that the somatic translocation t(8;16), associated with acute nonlymphocytic leukemia type M5b, whose breakpoint was located by Wessels et al. (1991) to be in the same area as the RSTS breakpoints, may cause leukemia by joining CBP to an unknown gene on chromosome 8. The unusual incidence of neoplasms, as well as the propensity to form keloids, may be explained by the role proposed for CBP in cAMP-regulated cell immortalization. The X-linked alpha-thalassemia/mental retardation syndrome (301040) is another example of multiple congenital malformations with mental retardation caused by a generalized dysregulation of gene expression. In that case, the mutation is located in the gene encoding the human homolog of yeast RAD54, or X-linked helicase-2 (300032). Since CREB-binding protein is a critical coactivator for thyroid hormone receptors, Olson and Koenig (1997) hypothesized that thyroid hormone resistance might occur in RSTS. To assess the function of the thyroid axis in RSTS, they measured free thyroxine (T4) and thyroid-stimulating hormone (TSH; see 188530) in 12 affected subjects. Free T4 and TSH levels were normal in all 12 subjects, indicating that overt thyroid hormone resistance is not a typical feature of RSTS.

Wallerstein et al. (1997) used the RT1 probe to screen 64 patients with clinical evidence of RSTS; 7 (11%) had a deletion. Another patient had a translocation involving the region without evidence of deletion. The features of coloboma, growth retardation, nevus flammeus, and hypotonia had a positive predictive value for the presence of an RT1 deletion. The authors commented that, because of the relatively low frequency of deletions in RSTS, the RT1 probe is useful in diagnostic confirmation but has limited use as a screening tool.

Nomenclature: Although the acronym RTS is sometimes used for Rubinstein-Taybi syndrome, the use of this acronym for 2 other syndromes, Rothmund-Thomson syndrome (268400) and Rett syndrome (312750), may lead to confusion; hence, RSTS, the HGM symbol, is recommended.

See also Coffin (1964); Hennekam et al. (1990); Kajii et al. (1981); Rubinstein (1974); Rubinstein and Taybi (1963)

Baraitser, M.; Preece, M. A.: The Rubinstein-Taybi syndrome: occurrence in two sets of identical twins. Clin. Genet. 23: 318-320, 1983.
Bazacliu, E.; Tonceanu, S.; Carp, G.; Ghisoiu, V.; Rosca, G. H.; Rosca, S.: Rubinstein-Taybi syndrome with karyotype changes and recurring pneumopathy (original in Hungarian). Ftziologia 22: 645-650, 1973.
Berry, A. C.: Rubinstein-Taybi syndrome. J. Med. Genet. 24: 562-566, 1987.
Bonioli, E.; Bellini, C.: Rubinstein-Taybi syndrome and pheochromocytoma. (Letter) Am J. Med. Genet. 44: 386 only, 1992.
Bonioli, E.; Bellini, C.; Senes, F. M.; Palmieri, A.; Di Stadio, M.; Pinelli, G.: Slipped capital femoral epiphysis associated with Rubinstein-Taybi syndrome. Clin. Genet. 44: 79-81, 1993.
Breuning, M. H.; Dauwerse, H. G.; Fugazza, G.; Saris, J. J.; Spruit, L.; Wijnen, H.; Tommerup, N.; van der Hagen, C. B.; Imaizumi, K.; Kuroki, Y.; van den Boogaard, M.-J.; de Pater, J. M.; Mariman, E. C. M.; Hamel, B. C. J.; Himmelbauer, H.; Frischauf, A.-M.; Stallings, R. L.; Beverstock, G. C.; van Ommen, G.-J. B.; Hennekam, R. C. M.: Rubinstein-Taybi syndrome caused by submicroscopic deletions within 16p13.3. Am. J. Hum. Genet. 52: 249-254, 1993.
Chun, K.; Colombani, P. M.; Dudgeon, D. L.; Haller, J. A., Jr.: Diagnosis and management of congenital vascular rings: a 22-year experience. Ann. Thorac. Surg. 53: 597-603, 1992.
Coffin, G. S.: Brachydactyly, peculiar facies and mental retardation. Am. J. Dis. Child. 108: 351-359, 1964.
Der Kaloustian, V. M.; Afifi, A. K.; Sinno, A. A.; Mire, J.: The Rubinstein-Taybi syndrome: clinical and muscle electron microscopic study. Am. J. Dis. Child. 124: 897-902, 1972.
Garcia, D.; Karasik, J.; Marion, R.: Apparent dominant transmission of Rubinstein-Taybi syndrome (RTS). (Abstract) Am. J. Hum. Genet. 51 (suppl.): A95, 1992.
Gillies, D. R. N.; Roussounis, S. H.: Rubinstein-Taybi syndrome: further evidence of a genetic aetiology. Dev. Med. Child Neurol. 27: 751-755, 1985.
Giroux, J.; Miller, J. R.: Dermatoglyphics of the broad thumb and great toe syndrome. Am. J. Dis. Child. 113: 207-209, 1967.
Guion-Almeida, M. L.; Richieri-Costa, A.: Callosal agenesis, iris coloboma, and megacolon in a Brazilian boy with Rubinstein-Taybi syndrome. Am. J. Med. Genet. 43: 929-931, 1992.
Hennekam, R. C. M.; Lommen, E. J. P.; Strengers, J. L. M.; Van Spijker, H. G.; Jansen-Kokx, T. M. G.: Rubinstein-Taybi syndrome in a mother and son. Europ. J. Pediat. 148: 439-441, 1989.
Hennekam, R. C. M.; Stevens, C. A.; Van de Kamp, J. J. P.: Etiology and recurrence risk in Rubinstein-Taybi syndrome. Am. J. Med. Genet. Suppl. 6: 56-64, 1990.
Hennekam, R. C. M.; Tilanus, M.; Hamel, B. C. J.; Voshart-van Heeren, H.; Mariman, E. C. M.; van Beersum, S. E. C.; van den Boogaard, M.-J. H.; Breuning, M. H.: Deletion at chromosome 16p13.3 as a cause of Rubinstein-Taybi syndrome: clinical aspects. Am. J. Hum. Genet. 52: 255-262, 1993.
Hennekam, R. C. M.; Van den Boogaard, M.-J.; Sibbles, B. J.; Van Spijker, H. G.: Rubinstein-Taybi syndrome in the Netherlands. Am. J. Med. Genet. Suppl. 6: 17-29, 1990.
Hennekam, R. C. M.; Van Doorne, J. M.: Oral aspects of Rubinstein-Taybi syndrome. Am. J. Med. Genet. Suppl. 6: 42-47, 1990.
Imaizumi, K.; Kuroki, Y.: Rubinstein-Taybi syndrome with de novo reciprocal translocation t(2;16)(p13.3;p13.3). Am. J. Med. Genet. 38: 636-639, 1991.
Kajii, T.; Hagiwara, K.; Tsukahara, M.; Nakajima, H.; Fukuda, Y.: Monozygotic twins discordant for Rubinstein-Taybi syndrome. J. Med. Genet. 18: 312-314, 1981.
Kanjilal, D.; Basir, M. A.; Verma, R. S.; Rajegowda, B. K.; Lala, R.; Nagaraj, A.: New dysmorphic features in Rubinstein-Taybi syndrome. J. Med. Genet. 29: 669-670, 1992.
Kimura, H.; Ito, Y.; Koda, Y.; Hase, Y.: Rubinstein-Taybi syndrome with thymic hypoplasia. Am. J. Med. Genet. 46: 293-296, 1993.
Lacombe, D.; Saura, R.; Taine, L.; Battin, J.: Confirmation of assignment of a locus for Rubinstein-Taybi syndrome gene to 16p13.3. Am. J. Med. Genet. 44: 126-128, 1992.
Levy, N. S.: Juvenile glaucoma in the Rubinstein-Taybi syndrome. J. Pediat. Ophthal. 13: 141-143, 1976.
Lowry, R. B.: Overlap between Rubinstein-Taybi and Saethre-Chotzen syndromes: a case report. Am. J. Med. Genet. Suppl. 6: 73-76, 1990.
Marion, R. W.; Garcia, D. M.; Karasik, J. B.: Apparent dominant transmission of the Rubinstein-Taybi syndrome. Am. J. Med. Genet. 46: 284-287, 1993.
Masuno, M.; Imaizumi, K.; Kurosawa, K.; Makita, Y.; Petrij, F.; Dauwerse, H. G.; Breuning, M. H.; Kuroki, Y.: Submicroscopic deletion of chromosome region 16p13.3 in a Japanese patient with Rubinstein-Taybi syndrome. Am. J. Med. Genet. 53: 352-354, 1994.
McKusick, V. A.: Personal Communication. Baltimore: Md. 1968.
Olson, D. P.; Koenig, R. J.: Thyroid function in Rubinstein-Taybi syndrome. J. Clin. Endocr. Metab. 82: 3264-3266, 1997.
Padfield, C. J.; Partington, M. W.; Simpson, N. E.: The Rubinstein-Taybi syndrome. Arch. Dis. Child. 43: 94-101, 1968.
Petrij, F.; Giles, R. H.; Dauwerse, H. G.; Saris, J. J.; Hennekam, R. C. M.; Masuno, M.; Tommerup, N.; van Ommen, G.-J. B.; Goodman, R. H.; Peters, D. J. M.; Breuning, M. H.: Rubinstein-Taybi syndrome caused by mutations in the transcriptional co-activator CBP. Nature 376: 348-351, 1995.
Pfeiffer, R. A.: Rubinstein-Taybi-Syndrom bei wahrscheinlich eineiigen Zwillingen. Humangenetik 6: 84-87, 1968.
Preis, S.; Majewski, F.: Monozygotic twins concordant for Rubinstein-Taybi syndrome: changing phenotype during infancy. Clin. Genet. 48: 72-75, 1995.
Robinson, T. W.; Stewart, D. L.; Hersh, J. H.: Monozygotic twins concordant for Rubinstein-Taybi syndrome and implications for genetic counseling. Am. J. Med. Genet. 45: 671-673, 1993.
Roy, F. H.; Summitt, R. L.; Hiatt, R. L.; Hughes, J. G.: Ocular manifestations of the Rubinstein-Taybi syndrome: case report and review of the literature. Arch. Ophthal. 79: 272-278, 1968.
Rubinstein, J. H.: The broad thumb syndrome—progress report 1968. Birth Defects Orig. Art. Ser. V(2): 25-41, 1969.
Rubinstein, J. H.: Fatherhood of the so-called Rubinstein-Taybi-syndrome. Am. J. Dis. Child. 128: 424, 1974.
Rubinstein, J. H.; Taybi, H.: Broad thumbs and toes and facial abnormalities. Am. J. Dis. Child. 105: 588-608, 1963.
Shashi, V.; Fryburg, J. S.: Vascular ring leading to tracheoesophageal compression in a patient with Rubinstein-Taybi syndrome. Clin. Genet. 48: 324-327, 1995.
Stevens, C. A.: Patellar dislocation in Rubinstein-Taybi syndrome. Am. J. Med. Genet. 72: 188-190, 1997.
Stevens, C. A.; Bhakta, M. G.: Cardiac abnormalities in the Rubinstein-Taybi syndrome. Am. J. Med. Genet. 59: 346-348, 1995.
Stevens, C. A.; Carey, J. C.; Blackburn, B. L.: Rubinstein-Taybi syndrome: a natural history study. Am. J. Med. Genet. Suppl. 6: 30-37, 1990.
Stirt, J. A.: Succinylcholine in Rubinstein-Taybi syndrome. (Letter) Anesthesiology 57: 429, 1982.
Tommerup, N.; van der Hagen, C. B.; Heiberg, A.: Tentative assignment of a locus for Rubinstein-Taybi syndrome to 16p13.3 by a de novo reciprocal translocation, t(7;16)(q34;p13.3). Am. J. Med. Genet. 44: 237-241, 1992.
Tommerup, N.; van der Hagen, C. B.; Heiberg, A.: Tentative assignment of a locus for Rubinstein-Taybi syndrome to 16p13.3 by a de novo reciprocal translocation, t(7;16)(q34;p13.3). (Abstract) Cytogenet. Cell Genet. 58: 2002-2003, 1991.
Wallerstein, R.; Anderson, C. E.; Hay, B.; Gupta, P.; Gibas, L.; Ansari, K.; Cowchock, F. S.; Weinblatt, V.; Reid, C.; Levitas, A.; Jackson, L.: Submicroscopic deletions at 16p13.3 in Rubinstein-Taybi syndrome: frequency and clinical manifestations in a North American population. J. Med. Genet. 34: 203-206, 1997.
Wessels, J. W.; Mollevanger, P.; Dauwerse, J. G.; Cluitmans, F. H. M.; Breuning, M. H.; Beverstock, G. C.: Two distinct loci on the short arm of chromosome 16 are involved in myeloid leukemia. Blood 77: 1555-1559, 1991.
Wulfsberg, E. A.; Klisak, I. J.; Sparkes, R. S.: High resolution chromosome banding in the Rubinstein-Taybi syndrome. Clin. Genet. 23: 35-37, 1983.

CREB-BINDING PROTEIN; CREBBP (*600140)

JJM NOTE

Mutation of this gene underlies the Rubinstein-Taybi syndrome, in the entry above. Also, the gene is involved in a translocation breakpoint in the M4/M5 subtype of acute myeloid leukemia.

When cellular levels of cAMP increase, a cascade of events leads to the induction of genes that contain cis-regulatory elements called cAMP-response elements (CREs). Elevated cAMP levels cause stimulation and nuclear translocation of protein kinase A (PKA; see 176911), which activates the transcription factor CREB (CRE-binding protein; see 123810) by phosphorylating it at a single residue, serine 133 (Gonzalez and Montminy, 1989). The precise molecular mechanisms by which CREB induces gene expression are not known. Chrivia et al. (1993) reported the discovery of a nuclear transcriptional coactivator protein, CREB-binding protein (CBP), that binds specifically to the PKA-phosphorylated form of the CREB protein. CBP is a large protein (relative mass = 250,000) that contains a bromodomain, i.e., a conserved structural unit important for protein-protein interactions. In Drosophila and yeast, this domain is found in coactivator proteins involved in signal-dependent, but not basal, transcription (Nordheim, 1994). Kwok et al. (1994) used fluorescence anisotropy measurements to define the equilibrium binding parameters of the phosphoCREB:CBP interaction and reported that CBP can activate transcription through a region in its C terminus. The activation domain of CBP interacts directly with the basal transcription factor TFIIB (189963). As TFIIB interacts with the TATA-box-binding protein TBP1 (186852), a continuous chain of physical contacts is established linking the stimulus-activated phospho-CREB, bound distal to the promoter, with the polymerase II complex that initiates transcription (Nordheim, 1994). Acting as a coactivator, CBP augments the activity of phosphorylated CREB to activate transcription of cAMP-responsive genes.

Arias et al. (1994) microinjected an anti-CBP antiserum into fibroblasts and determined that transcription from a cAMP promoter can be inhibited. Arias et al. (1994) also reported that CBP cooperates with upstream activators, such as JUN (165160), that are involved in mitogen-responsive transcription. Studies indicated that when JUN was phosphorylated at the transcriptionally stimulatory sites ser73 and ser63, it bound CBP with comparable affinity to CREB. They proposed that CBP is recruited to the promoter through interaction with certain phosphorylated factors and that CBP may thus play a critical role in the transmission of inductive signals from cell surface receptors to the transcriptional apparatus.

Sterol regulatory element binding proteins (SREBPs; e.g., 184756) activate transcription of genes whose products are involved in the cellular uptake and synthesis of cholesterol. Oliner et al. (1996) showed that the putative activation domain of SREBP binds specifically to N-terminal domains of recombinant CBP and p300 (a CREBBP-related protein). Transfection studies demonstrated that CBP enhances the ability of SREBP to activate transcription of reporter genes in HeLa cells, thus suggesting that CBP may serve as a coactivator for SREBP.

CBP binds to the ser133 phosphorylated region of CREB via a domain called KIX (Parker et al., 1996). The phosphorylated domain of CREB was termed KID for kinase-inducible domain. Radhakrishnan et al. (1997) used nuclear magnetic resonance spectroscopy to study the molecular interactions of the KIX:KID domains of CBP and CREB, respectively. The KIX domain of CBP comprises amino acid residues 586 to 666. The KID domain of CREB comprises amino acid residues 101 to 160. The KID undergoes a coil-to-helix folding transition upon binding to KIX, forming 2 alpha helices. The amphipathic helix alpha-B of KID interacts with a hydrophobic groove defined by helices alpha-1 and alpha-3 of KIX. The other KID helix, alpha-A, contacts a different face of the alpha-3 helix. The phosphate group of the critical phosphoserine residue of KID forms a hydrogen bond to the side chain of tyr658 of KIX. The structure provides a model for interactions between other transactivation domains and their targets.

Petrij et al. (1995) found that all the breakpoints and microdeletions in Rubinstein-Taybi syndrome (RTS; 180849) are located in 16p13.3 in a region they showed contains the gene for the CREB binding protein. They showed, furthermore, that RTS results not only from gross chromosomal rearrangements of 16p but also from point mutations in the CBP gene itself. Because the patients are heterozygous for the mutations, they proposed that the loss of one functional copy of the CBP gene underlies the developmental abnormalities in RTS and possibly the propensity for malignancy. Keloids are unusually frequent and they pointed to a case of somatic translocation t(8;16) associated with acute nonlymphocytic leukemia of the M5b type, whose breakpoint they located in the same region as the RTS breakpoints (Wessels et al., 1991). They proposed that the translocation may have joined CBP to an unknown gene on chromosome 8, leading to the malignancy. Fluorescence in situ hybridization (FISH) was used to situate all RTS breakpoints in an area of 150 kb; microdeletions were found to range from 130 to more than 650 kb in size. Complementary DNAs from this area showed very high sequence homology with mouse CBP. Petrij et al. (1995) used FISH to show that mouse CBP hybridized to human 16p13.3. Further studies indicated that the human CBP gene spans at least the 150-kb genomic area containing the RTS breakpoints. Southern and Northern hybridization uncovered cases with aberrant genomic fragments, as well as truncated mRNA.

By positional cloning on chromosome 16, Borrow et al. (1996) implicated the CREB-binding protein (600140), a transcriptional adaptor/coactivator protein, in the M4/M5 subtype of acute myeloid leukemia. Borrow et al. (1996) also used positional cloning to identify a novel gene at the chromosome 8 breakpoint, which they termed MOZ (601408) for monocytic leukemia zinc finger protein.

Chen and Korenberg (1995) also mapped CREBBP to 16p13.3 using a cDNA probe for the mouse gene in fluorescence in situ hybridization. Using a human genomic clone for FISH, Wydner et al. (1995) mapped CREBBP to 16p13.3-p13.2.

To search for mutations in RTS patients without rearrangements, Petrij et al. (1995) used the protein truncation test (PTT). In 2 of 16 patients, a truncated protein product was found; both patients, who had classic RTS phenotype, had a change of a glutamine codon to a stop codon by a C-to-T substitution (see 600140.0001 and 600140.0002).

The recurring translocation t(11;16)(q23;p13.3) has been observed only in cases of acute leukemia or myelodysplasia secondary to therapy with drugs targeting DNA topoisomerase II (126430). Sobulo et al. (1997) showed that in 2 patients the MLL gene was fused in-frame to a different exon of the CREBBP gene, producing chimeric proteins containing the AT-hooks, methyltransferase homology domain, and transcriptional repression domain of MLL fused to the CREB binding domain or to the bromodomain of CREBBP. Both fusion products retained the histone acetyltransferase domain of CREBBP and may lead to leukemia by promoting histone acetylation of genomic regions targeted by the MLL AT-hooks, leading to transcriptional deregulation via aberrant chromatin organization.

Taine et al. (1998) found microdeletion of the RT1 probe (D16S237) in 3 of 30 French patients with RTS. By pooling data from their study and previous series, they found the cumulative frequency of the 16p13.3 microdeletion to be 11.9% (19 in 159).

While mapping and cloning the human CREBBP gene on 16p13.3, Giles et al. (1997) noticed an emerging pattern of relationship between this chromosome band and a region of chromosome 22q. CBP exhibits extensive homology to the adenovirus E1A-associated protein p300, whose gene (EP300; 602700) maps to 22q13 (Eckner et al., 1994; Lundblad et al., 1995). They noted that the heme oxygenase-1 gene (HMOX1; 141250) maps to 22q12 and the heme oxygenase-2 gene (HMOX2; 141251) to 16p13.3 Furthermore, the gene for phosphomannomutase-2 (PMM2; 601785) is on 16p13 and that for phosphomannomutase-1 (PMM1; 601786) is on 22q13. Sequence comparison at the amino acid level revealed that homology between these paralogous proteins is high: 63% between CBP and p300, 66% between PMM1 and PMM2, and 74% between HMOX1 and HMOX2. By examination of genome databases, e.g., OMIM, Giles et al. (1998) found 6 additional sets of paralogs mapping to 16p13 and 22q11-q13, although the extent of homology between these paralogous sets was not known. These pairs were SSTR5 (182455) and SSTR3 (182453) on 16 and 22, respectively; CSNK2A2 (115442) and CSNK1E (600863); UBE2I (601661) and UB2L3; MYH11 (160745) and MYH9 (160775); CRYM (123740) and a cluster of crystallin genes on chromosome 22, e.g., CRYBB1 (600929); IL4RA (147781) and IL2RB (146710). Giles et al. (1998) postulated that these bands show homology because of an ancestral event of the type proposed by Ohno (1993). Arguing from the fact that chromosome 14q carries SSTR1 (182451), UBE2L1 (600012), MYH6 (160710), and MYH7 (160760), whereas SSTR2 (182452), CSNK1D (600864), and CRYBA1 (123610) map to 17q, combined with the location on leukemia breakpoints, strongly suggested to Giles et al. (1998) that these gene families arose by tetraploidization resulting in members on 14q, 16p, 17q, and 22q. Genetic redundancy is potentially of great relevance to organismal evolution since it may protect against potentia lly harmful mutations and may provide a pool of diverse yet functionally similar proteins for further evolution.

Arias, J.; Alberts, A. S.; Brindle, P.; Claret, F. X.; Smeal, T.; Karin, M.; Feramisco, J.; Montminy, M.: Activation of cAMP and mitogen responsive genes relies on a common nuclear factor. Nature 370: 226-229, 1994.

Borrow, J.; Stanton, V. P., Jr.; Andresen, J. M.; Becher, R.; Behm, F. G.; Chaganti, R. S. K.; Civin, C. I.; Disteche, C.; Dube, I.; Frischauf, A. M.; Horsman, D.; Mitelman, F.; Volinia, S.; Watmore, A. E.; Housman, D. E.: The translocation t(8;16)(p11;p13) of acute myeloid leukaemia fuses a putative acetyltransferase to the CREB-binding protein. Nature Genet. 14: 33-41, 1996.

Chen, X.-N.; Korenberg, J. R.: Localization of human CREBBP (CREB binding protein) to 16p13.3 by fluorescence in situ hybridization. Cytogenet. Cell Genet. 71: 56-57, 1995.

Chrivia, J. C.; Kwok, R. P. S.; Lamb, N.; Hagiwara, M.; Montminy, M. R.; Goodman, R. H.: Phosphorylated CREB binds specifically to the nuclear protein CBP. Nature 365: 855-859, 1993.

Eckner, R.; Ewen, M. E.; Newsome, D.; Gerdes, M.; DeCaprio, J. A.; Lawrence, J. B.; Livingston, D. M.: Molecular cloning and functional analysis of the adenovirus E1A-associated 300-kD protein (p300) reveals a protein with properties of a transcriptional adaptor. Genes Dev. 15: 869-884, 1994.

Giles, R. H.; Dauwerse, H. G.; van Ommen, G.-J. B.; Breuning, M. H.: Do human chromosomal bands 16p13 and 22q11-13 share ancestral origins? (Letter) Am. J. Hum. Genet. 63: 1240-1242, 1998.

Giles, R. H.; Petrij, F.; Dauwerse, H. G.; den Hollander, A. I.; Lushnikova, T.; van Ommen, G.-J. B.; Goodman, R. H.; Deaven, L. L.; Doggett, N. A.; Peters, D. J.; Breuning, M. H.: Construction of a 1.2-Mb contig surrounding, and molecular analysis of, the human CREB-binding protein (CBP/CREBBP) gene on chromosome 16p13.3. Genomics 42: 96-114, 1997.

Gonzalez, G. A.; Montminy, M. R.: Cyclic AMP stimulates somatostatin gene transcription by phosphorylation of CREB at serine 133. Cell 59: 675-680, 1989.

Kwok, R. P. S.; Lundblad, J. R.; Chrivia, J. C.; Richards, J. P.; Bachinger, H. P.; Brennan, R. G.; Roberts, S. G. E.; Green, M. R.; Goodman, R. H.: Nuclear protein CBP is a coactivator for the transcription factor CREB. Nature 370: 223-226, 1994.

Lundblad, J. R.; Kwok, R. P. S.; Laurance, M. E.; Harter, M. L.; Goodman, R. H.: Adenoviral E1A-associated protein p300 as a functional homologue of the transcriptional co-activator CBP. Nature 374: 85-88, 1995.

Nordheim, A.: CREB takes CBP to tango. Nature 370: 177-178, 1994.

Ohno, S.: Patterns in genome evolution. Curr. Opin. Genet. Dev. 3: 911-914, 1993.

Oliner, J. D.; Andresen, J. M.; Hansen, S. K.; Zhou, S.; Tjian, R.: SREBP transcriptional activity is mediated through an interaction with the CREB-binding protein. Genes Dev. 10: 2903-2911, 1996.

Parker, D.; Ferreri, K.; Nakajima, T.; LaMorte, V. J.; Evans, R.; Koerber, S. C.; Hoeger, C.; Montminy, M. R.: Phosphorylation of CREB at ser-133 induces complex formation with CREB-binding protein via a direct mechanism. Molec. Cell. Biol. 16: 694-703, 1996.

Petrij, F.; Giles, R. H.; Dauwerse, H. G.; Saris, J. J.; Hennekam, R. C. M.; Masuno, M.; Tommerup, N.; van Ommen, G.-J. B.; Goodman, R. H.; Peters, D. J. M.; Breuning, M. H.: Rubinstein-Taybi syndrome caused by mutations in the transcriptional co-activator CBP. Nature 376: 348-351, 1995.

Radhakrishnan, I.; Perez-Alvarado, G. C.; Parker, D.; Dyson, H. J.; Montminy, M. R.; Wright, P. E.: Solution structure of the KIX domain of CBP bound to the transactivation domain of CREB: a model for activator:coactivator interactions. Cell 91: 741-752, 1997.

Sobulo, O. M.; Borrow, J.; Tomek, R.; Reshmi, S.; Harden, A.; Schlegelberger, B.; Housman, D.; Doggett, N. A.; Rowley, J. D.; Zeleznik-Le, N. J.: MLL is fused to CBP, a histone acetyltransferase, in therapy-related acute myeloid leukemia with a t(11;16)(q23;p13.3). Proc. Nat. Acad. Sci. 94: 8732-8737, 1997.

Taine, L.; Goizet, C.; Wen, Z. Q.; Petrij, F.; Breuning, M. H.; Ayme, S.; Saura, R.; Arveiler, B.; Lacombe, D.: Submicroscopic deletion of chromosome 16p13.3 in patients with Rubinstein-Taybi syndrome. Am. J. Med. Genet. 78: 267-270, 1998.

Wessels, J. W.; Mollevanger, P.; Dauwerse, J. G.; Cluitmans, F. H. M.; Breuning, M. H.; Beverstock, G. C.: Two distinct loci on the short arm of chromosome 16 are involved in myeloid leukemia. Blood 77: 1555-1559, 1991.

Wydner, K. L.; Bhattacharya, S.; Eckner, R.; Lawrence, J. B.; Livingston, D. M.: Localization of human CREB-binding protein gene (CREBBP) to 16p13.2-p13.3 by fluorescence in situ hybridization. Genomics 30: 395-396, 1995.

2. NERVOUS SYSTEM

Ophthalmic Tumors

RETINOBLASTOMA (*180200)

JJM NOTE

Just when we are tempted to think the last chapter has been written on what retinoblastoma can teach us about cancer genetics, another chapter begins. It continues to serve as a model system for investigating the most common cancers of human beings. Steps include recognizing familial aggregation, associating birth defects and additional primary cancers, discovering multifocality and bilaterality, discovering germline and somatic cytogenetic defects, gene mapping, isolation, mutation detection, roles in other cancers, application to clinical management, and, eventually, genetically engineered or gene transfer therapy.

Description

Retinoblastoma is an embryonic neoplasm of retinal origin. It almost always presents in early childhood and is often bilateral. Spontaneous regression ('cure') occurs in some cases.

Clinical Features

Connolly et al. (1983) reported a 4-generation family with 3 patterns of expression of the retinoblastoma gene: frank retinoblastoma, unilateral or bilateral; retinoma; and no visible retinal pathology except for 'normal degeneration' with age. ('Paving stone degeneration' of the type observed in 2 of 3 RB carriers, aged 49 and 59, is said by Duane (1980) to occur in about 20% of the adult population.)

Retinoma

Gallie and Phillips (1982) described benign lesions in the retina in retinoblastoma patients. The distinctive characteristics of these lesions, referred to by the authors as retinomas, included a translucent, grayish retinal mass protruding into the vitreous, 'cottage-cheese' calcification in 75%, and retinal pigment epithelial migration and proliferation in 60%. They suggested that retinomas represent not the heterozygous state postulated by the Knudson 2-stage model of carcinogenesis but rather the homozygous state occurring in differentiated cell(s). Gallie et al. (1982) suggested that retinomas represent either spontaneous regression of a retinoblastoma or a benign manifestation of the RB gene.

Pinealoma

Brownstein et al. (1984) described 3 children with bilateral retinoblastoma and a morphologically similar neoplasm in the region of the pineal. They referred to this as trilateral retinoblastoma. The pineal gland has sometimes been referred to as 'the third eye.' Lueder et al. (1991) described a fourth case of pinealoma associated with bilateral retinoblastoma. The patient was one of 56 with heritable RB. Amoaku et al. (1996) reported 5 patients with trilateral retinoblastoma (including 2 previously reported), diagnosed among 146 consecutive patients with retinoblastoma in the West Midlands Health Authority Region in England between 1957 and 1994. This represented an incidence of 3%. There were 4 patients with pineoblastoma, only 1 of whom had a positive family history. The mean age at diagnosis of RB in the entire series was 6 months, whereas the patients with pineoblastoma were diagnosed at a mean age of 2 years. The tumors were not evident on the initial computed tomographic (CT) scans. One child presented with a calcified suprasellar mass 13 months before bilateral sporadic retinoblastoma was identified. Death occurred within 1 month of diagnosis of the intracranial tumor in 3 patients who received no treatment. In the other 2 patients, who were treated, death occurred at 15 months and 2 years, respectively, after diagnosis of intracranial tumor.

Osteogenic Sarcoma

The risk of osteogenic sarcoma is increased 500-fold in bilateral retinoblastoma patients, the bone malignancy being at sites removed from those exposed to radiation treatment of the eye tumor (Abramson et al., 1976). Francois (1977) concluded that there is a special predisposition to osteogenic sarcoma, both radiogenic and nonradiogenic, in retinoblastoma patients and possibly in their relatives. Matsunaga (1980) estimated that the relative risk of development of nonradiogenic osteosarcoma in persons with the retinoblastoma gene is 230. That osteosarcoma is a direct effect of the genomic change that underlies retinoblastoma is indicated by the cases of osteosarcoma without retinoblastoma but with genomic changes like those of retinoblastoma.

Chromosome Deletion-associated Features

Motegi et al. (1983) suggested that patients with retinoblastoma from interstitial deletion of 13q have a characteristic appearance of the midface: prominent eyebrows, broad nasal bridge, bulbous tip of the nose, large mouth with thin upper lip, and long philtrum. Fukushima et al. (1987) described a mother and son with retinoblastoma due to deletion of 13q14. They both had peculiar facies with high forehead, low and broad nasal root, short and bulbous nose, and open mouth with thin upper lip, as well as prominent earlobes. The son had enucleation of the right eye at age 13 months and no evidence of involvement of the left eye at 21 months. The mother had enucleation of the right eye at age 6 months and prophylactic x-ray therapy to the left eye.

Other Features

Bonaiti-Pellie et al. (1975) found an increased frequency of malformations, especially cleft palate, in association with retinoblastoma and proposed that this argued for germinal mutation rather than somatic mutation. Friend et al. (1987) found that the same DNA segment that they had isolated from 13q14 and shown to have attributes of the retinoblastoma gene is additionally the target of somatic mutations in mesenchymal tumors among patients having no apparent predisposition to retinoblastoma. Sequence analysis of this cDNA provided little insight into its normal functional role. Almost two-thirds of the secondary tumors arising in patients with retinoblastoma are mesenchymal in origin. Over 60% of the mesenchymal tumors are osteosarcomas; the soft tissue sarcomas include fibrosarcoma, leiomyosarcoma, liposarcoma, and others. Friend et al. (1987) specifically demonstrated homozygous deletions of the RB1 locus in sporadic cases of leiomyosarcoma, malignant fibrous histiocytoma, and undifferentiated sarcoma in the absence of any history of retinoblastoma. Friend et al. (1988) reviewed the subject of tumor-suppressing genes in retinoblastoma and other disorders. Henson et al. (1994) found loss of heterozygosity for the RB1 protein in 16 of 54 informative high-grade astrocytomas but not in 12 low-grade gliomas. Deletion mapping with ranking markers revealed that the retinoblastoma locus was preferentially targeted by the deletions. SSCP analysis and direct DNA sequencing demonstrated mutations in the remaining retinoblastoma allele. This evidence suggested to the authors that, whereas mutation of the p53 tumor suppressor gene is an early event in the formation of many astrocytomas, mutation of the retinoblastoma gene is associated with progression of the astrocytoma into high grade astrocytoma or glioblastoma multiforme. Previous work had demonstrated frequent loss of 9p, 10, 13q, 17p, 19q, and 22q in astrocytomas. Mutations in the RB1 gene were described. Cance et al. (1990) found that leiomyosarcomas and other soft tissue sarcomas in which expression of the RB gene product was decreased were more aggressive than tumors in which this protein was expressed by nearly all cells. Gibbons et al. (1995) presented a patient in whom sporadic unilateral retinoblastoma occurred at an unusually early age (4 months) in the presence of Fanconi anemia (227650). In a second child, they found retinoblastoma in association with Bloom syndrome (210900), another chromosome breakage syndrome. Fearon (1997) reviewed more than 20 hereditary cancer syndromes that had been defined and attributed to specific germline mutations in inherited cancer genes. All but 4 of these behaved as autosomal dominants. Retinoblastoma was one of the first to be defined at the molecular level.

Inheritance

The genetic nature of retinoblastoma and its dominant inheritance came to light early because prompt recognition and treatment of individual cases permitted survival. The genetic basis of other embryonic tumors, such as Wilms tumor, neuroblastoma and medulloblastoma, was later to be appreciated.

Smith and Sorsby (1958) concluded that bilateral cases are most often familial. In their opinion, estimates of mutation rate of 2.3 x 10(-5) as given by Falls and Neel (1951) were too high. Many unilateral cases may be sporadic with a low risk (empirically, about 4%) to subsequent children or to offspring of the proband. Knudson (1971) proposed that a 2-mutation model best fits the data. In this view, a fraction of cases are nonhereditary and result from 2 somatic mutational events in one cell. The remainder are hereditary cases, occurring in persons susceptible by reason of having inherited one of the mutational events. (See review of Knudson (1986) on the 2-mutation model and other aspects of the genetics of human cancer.) Matsunaga (1982) suggested that the almost synchronous appearance of bilateral retinoblastoma argues against the 2-mutation model, which assumes that in the gene carriers the eyes acquire tumors independently. Age-specific incidence rates for 96 New

Zealand patients with sporadic retinoblastoma peaked earlier for bilateral cases than for unilateral ones (Fitzgerald et al., 1983). The cumulative log survival until diagnosis for bilateral and unilateral patients followed linear and quadratic curves, respectively, thus supporting the 2-hit hypothesis. A germ cell mutation rate of 9.3 x 10(-6) to 10.9 x 10(-6) was estimated. Interestingly, retinoblastoma, which behaves as a dominant in pedigrees, results from a gene that is expressed only in the homozygote, i.e., is recessive. Since retinoblastoma results from the homozygous or hemizygous state of a gene at 13q14, it should perhaps be stated that the susceptibility is dominant. The total experience with retinoblastoma has been that 5 to 10% of cases have been inherited; 20 to 30% have been new germinal mutations; and 60 to 70% have been sporadic, i.e., somatic mutations.

Dryja et al. (1989) and Zhu et al. (1989) found that in bilateral retinoblastoma there is a preferential retention of the paternal chromosome in the process of loss of heterozygosity (LOH). This may indicate that either (1) mutation of RB1 is more common during spermatogenesis than oogenesis as a result of differences between male and female meiosis, DNA methylation or environmental exposure; or (2) the paternal chromosome in the early embryo is more at risk for mutation, or deficient in DNA repair. Dryja et al. (1989) analyzed the parental origin of mutations at the retinoblastoma locus using RFLPs known to map at chromosomal band 13q14. Ten of 10 new germline mutations were derived from the father's chromosome 13, while 4 of 7 somatic mutations occurred in the maternal retinoblastoma allele. The authors suggested that new germline mutations at the retinoblastoma locus arise more frequently during spermatogenesis than during oogenesis, but that genomic imprinting does not play an important role in the development of somatic mutations leading to this malignancy. Matsunaga et al. (1990) found no parental age effect in 225 sporadic cases of bilateral retinoblastoma and in 10 sporadic cases of chromosome deletion or translocation involving 13q14 that was identified as of paternal origin. Parental exposure to ionizing radiation or chemical mutagens, the effect of which accumulates with age, does not seem to play a major role in the production of germinal mutations at the RB1 locus. Since analysis of month of birth of 753 children with sporadic unilateral retinoblastoma showed no significant deviation from controls or a cyclic trend, nonheritable retinoblastoma is not likely to be associated with viruses whose activity varies markedly with season. DerKinderen et al. (1990) found that fathers 50 years of age or older had a relative risk of 5.0 to have a child with sporadic hereditary retinoblastoma compared with fathers in the population in general and that mothers 35 years of age or older had a relative risk of 1.7 compared with mothers in the population in general. Sporadic retinoblastoma was defined as occurrence without a family history of the disease. These cases were further classified as sporadic hereditary retinoblastoma when the patient was bilaterally affected or when a patient with unilateral retinoblastoma without family history later had the birth of a child with retinoblastoma; or as sporadic nonhereditary retinoblastoma when there was no family history, only unilateral retinoblastoma, and no occurrence of a relative in whom retinoblastoma was subsequently diagnosed. In connection with retinoblastoma, there is no parent of origin effect in the inherited disorder to suggest imprinting; however, there is evidence of imprinting in relation to osteosarcoma (Hall, 1993). The gene is more often transmitted from the father in the case of RB-related osteosarcoma. New mutations in the retinoblastoma gene most often occur in the father. Munier et al. (1992) found evidence of segregation distortion in a study of 8 kindreds with hereditary retinoblastoma by concomitant ophthalmologic examination and determination of 7 intragenic RFLPs. There was preferential transmission of the mutant allele from fathers; there was no difference from a 1:1 segregation ratio among the children of female carriers.

Naumova and Sapienza (1994) presented epidemiologic and genetic analyses of sporadic and familial retinoblastoma suggesting the existence of an X-linked gene (308290) involved in the genesis of a significant fraction of new bilateral cases of the disease. From the finding of both sex-ratio distortion in favor of males and transmission-ratio distortion in favor of affecteds among the offspring of males with bilateral sporadic disease, they proposed the existence of a defective imprinting gene on the X chromosome. Sporadic cases of Wilms tumor and embryonal rhabdomyosarcoma exhibit preferential loss of maternal alleles at loci linked to the putative tumor suppressor genes on 11p; sporadic cases of osteosarcoma and bilateral retinoblastoma similarly show preferential loss of maternal alleles at loci linked to the tumor suppressor gene on 13q. These observations may be explained by preferential germline mutation of the father's tumor suppressor gene or by genome imprinting. Naumova et al. (1994) examined 74 cases of sporadic retinoblastoma for tumors in which at least 2 genetic events had occurred: loss of heterozygosity for 13q markers and formation of an isochromosome 6p. They found 16 cases containing both events. In 13 of 16 such tumors, the chromosome 13q that was lost and the chromosome 6p that was duplicated were derived from the same parent. Because the formation of an isochromosome 6p is thought to be a somatic event and perhaps related to tumor progression (Horsthemke et al., 1989), the germline mutation model does not predict a relationship between the 2 events. The genome imprinting model, on the other hand, assumes that the original cell that gave rise to the tumor bore a genome imprint. Because the imprinting process affects loci on many different chromosomes, genetic events involving 2 unlinked loci would be expected to be related with respect to parent of origin, if both loci are imprinted. Toguchida et al. (1989) examined 13 sporadic osteosarcomas and in 12 found evidence indicating that the initial mutation was in the paternal gene. The finding suggests the involvement of germinal imprinting in producing differential susceptibility of the 2 genes to mutation. There was a growing body of data indicating a difference in behavior of maternally and paternally derived autosomal genes. Germinal imprinting may be mediated by some epigenetic process such as de novo DNA methylation and be carried over to postzygotic stages.

Sippel et al. (1998) evaluated 156 families with retinoblastoma in which the initial oncogenic mutation in the RB1 gene had been identified. In 15 of the families (approximately 10%) they were able to document mosaicism for the initial mutation in the retinoblastoma gene, either in the proband or in one of the proband's parents. Sippel et al. (1998) stated that the true incidence of mosaicism in this group was probably higher than 10%; in some additional families mosaicism was likely but could not be proven, because somatic or germline DNA from key family members was unavailable. In one mosaic father the mutation was detected in both sperm and leukocyte DNA; in a second, the mutation was detected only in sperm. The possibility of mosaicism should always be considered during genetic counseling of newly identified families with retinoblastoma and other disorders in which a high proportion of cases represent new mutations. Genetic tests of germline DNA can provide valuable information that is not available through analysis of somatic (leukocyte) DNA.

Penetrance

Macklin (1959) demonstrated irregularities in the inheritance, suggesting incomplete penetrance. In 10.5% of cases, affected persons were identified in collateral lines. Examples included: (1) a bilateral case, his unilaterally affected brother and a bilaterally affected daughter of the latter person; (2) six bilaterally affected offspring of a woman who had one microphthalmic eye but refused examination; (3) several instances of two or more affected sibs with normal parents. Reduced penetrance was reported by Connolly et al. (1983) and Onadim et al. (1992) among others. A striking difference in penetrance between 2 generations of the family described by Connolly et al. (1983) suggested the segregation of an additional epistatic, host-resistance gene. Bundey and Morten (1981) reported a rather similar pattern of intergenerational difference in penetrance. Huang et al. (1988) demonstrated suppression of the neoplastic phenotype in cultured retinoblastoma or osteosarcoma cells by introduction of a cloned RB gene via retrovirus-mediated gene transfer. This is compelling evidence that the inactivation of the endogenous RB gene in these cells was a crucial step in tumorigenesis. Scheffer et al. (1989) identified nonpenetrance of the RB gene by the use of linkage markers in family studies.

Dryja et al. (1993) examined the molecular basis of reduced penetrance in some retinoblastoma families. In 1 family, a germline deletion was shared by affected and unaffected obligate carriers. The deletion encompassed exon 4 of the RB gene and corresponded to a mutant protein without residues 127-166. In another pedigree, the phenomenon was only 'pseudo-low penetrance' because the appearance of reduced penetrance was created by the fact that 2 distant relatives had independently derived mutations. Munier et al. (1993) reported 2 families with 'pseudo low penetrance' resulting from familial aggregation of sporadic cases caused by independently derived mutations.

Bia and Cowell (1995) noted that rare families show evidence of incomplete penetrance where individuals transmit the mutant gene without being affected themselves. Formal proof of incomplete penetrance requires identification of the predisposing mutation. They reported an extraordinary family in which different mutations were identified in first cousins with bilateral disease. One cousin carried a C-to-T transition in exon 8, which changed CGA (arg) at codon 123 to TGA (stop). This mutation was present also in his affected mother. The other cousin carried an 8-bp deletion in exon 20 that resulted in the generation of a downstream stop codon. This mutation was not present in his mother who was a half-sister of the affected mother of his affected cousin. Thus, this was an example not of reduced penetrance but of independent constitutional germline mutations.

Population Genetics

Macklin (1960) stated that in the U.S. the frequency of retinoblastoma is about 1 in 23,000 live births. Jensen and Miller (1971) found that at ages 2 to 3 years a peak of mortality occurred which was 2.5 times greater in blacks than in whites. Whether this reflects a truly high frequency in blacks or some other factor such as higher mortality from delayed diagnosis is not clear.

Pendergrass and Davis (1980) found an incidence of 3.58 cases among each million children under age 15 years. Over 90% were diagnosed before age 5 years. No difference was found between whites and blacks, but other non-whites had rates more than 4 times greater than those of whites. Bilateral disease occurred in 20%. No nonhereditary retinoblastomas (which represent 55-65% of all retinoblastoma cases) are bilateral. Bilateral and unilateral hereditary retinoblastoma represent, respectively, about 25-30% and 10-15% of all cases.

Cytogenetics

The earliest example of a cytogenetic change in a solid tumor was the description of partial deletion of a D group chromosome in retinoblastoma by Stallard (1962) and Lele et al. (1963).

In 12 reported patients with a deletion of the long arm of a D chromosome, 7 had retinoblastoma, which in 3 instances was bilateral (Taylor, 1970; Gey, 1970). Cytogenetic evidence suggested that a locus for retinoblastoma is on the long arm of chromosome 13. In the patients of Orye et al. (1971) and that of Wilson et al. (1969), in which a 14q- karyotype was found, no clinical features of the type usually associated with 13q- were present. Orye et al. (1971) found deletion of a distal part of the long arm of one chromosome 13 in a case of bilateral retinoblastoma. The broadest of the three Giemsa bands normally present on the long arm was missing. Grace et al. (1971) described a patient with typical 13q- syndrome plus retinoblastoma. The karyotype contained a ring D chromosome. Wilson et al. (1973) restudied their case of bilateral retinoblastoma with new banding techniques and concluded that it, like all the other deleted D-chromosome cases, was an instance of 13q-. Orye et al. (1974) suggested that deletion of 13q21 is mainly responsible for retinoblastoma. With the advent of banding techniques, the chromosome involved was identified as chromosome 13 and the critical segment common to all deletions as band 13q14 (Francke, 1976).

Deletion of 13q22 was found by Riccardi et al. (1979), who reviewed published cases of retinoblastoma with abnormality of chromosome 13. They noted that, contrariwise, duplication of this segment has only mildly deleterious consequences. In 1 of 8 patients with retinoblastoma, Davison et al. (1979) found a reciprocal translocation of chromosomes 1 and 13. The breakpoint in chromosome 13 was at band q12, suggesting that the retinoblastoma locus is more proximal than thought from other data. 159,161:Sparkes et al. (1979, 1980) found that retinoblastoma and esterase D (ESD; 133280) map to the same band, 13q14. They observed that quantitative and qualitative expression of esterase D in 5 persons with partial deletions or duplications of chromosome 13 supported localization of the gene to 13q14. The same band had been found deleted in cases of retinoblastoma. They suggested that linkage of familial retinoblastoma and esterase D should be sought, to check on whether familial retinoblastoma represents mutation at the same locus as that deleted in the chromosomal form of the disorder. Rivera et al. (1981) concluded that the retinoblastoma and esterase D loci are in the proximal half of the 13q14 band. Benedict et al. (1983) studied a patient who had ESD activity 50% of normal but no deletion of 13q14 at the 550-band level. However, in 2 stem lines identified in a retinoblastoma from this patient, they found a missing chromosome 13 and no detectable ESD activity was found in the tumor. Therefore, in the tumor, the patient had total loss of genetic information at the location of the retinoblastoma gene. Thus, homozygosity appears to underlie this tumor.

Knight et al. (1980) studied linkage of familial retinoblastoma with fluorescent markers of chromosome 13. Instances of discordant segregation were attributed to crossing-over. Nichols et al. (1980) studied a patient with a 13q;Xp translocation and retinoblastoma. The 13q14 band was translocated intact to the X chromosome rather than being the breakpoint of the translocation. Genetic inactivation of the derivative X chromosome shown by late labeling and other findings with resulting functional monosomy of the 13q14 band was considered likely. About 20 cases of abnormality involving the 13q14 band, as an aberration in all somatic cells, had been described by 1981 (Balaban-Malenbaum et al., 1981).

Mosaicism for a 13q- cell line was found in 2 of 3 patients in a series of 42 that had abnormal karyotypes (Motegi, 1981). All 3 had bilateral sporadic retinoblastoma and the 2 mosaic cases had an apparently normal phenotype except for the eye tumors.

Comparison of the structural changes in the tumor cells with those in fibroblasts of some patients supported Knudson's 2-hit hypothesis. Using 2 probes, Horsthemke et al. (1987) found deletions in nearly 20% of patients with bilateral or multifocal unilateral retinoblastoma. One probe also detected a RFLP useful as a genetic marker in some families. Three of 8 deletions did not include the esterase D locus and were undetectable by conventional cytogenetic analysis. Sequence analysis of the RB cDNA clones demonstrated a long open reading frame encoding a hypothetical protein with features suggestive of a DNA-binding function.

Ejima et al. (1988) showed that cytogenetically visible germline mutations resulting in retinoblastoma are usually in the paternally derived gene. Such a bias would not be expected for sporadic (nonfamilial) tumors, where both mutations occur in somatic tissue, but there had been some indication of a bias toward initial somatic mutation in the paternally derived gene on chromosome 11 in sporadic Wilms tumor (Shroeder et al., 1987). Lemieux et al. (1989) used a method of minute band analysis to locate the RB locus in subband 13q14.11. The method involved high resolution G banding by BrdU using antibodies and gold. The banding was visualized by electron microscopy. Lemieux et al. (1989) suggested that the gain in resolution afforded by immunochemical banding, coupled with electron microscopy, would prove highly useful in detecting small chromosome anomalies in clinical syndromes and neoplasms. Greger et al. (1990) studied a family in which the father and 2 of his children had bilateral retinoblastoma. The father was found to be a mosaic for 2 different deletions, one of which showed a complex rearrangement. The 2 deletions shared 1 breakpoint but extended in opposite directions.

Mapping

Sparkes et al. (1983) showed that the locus for the nondeletion form of retinoblastoma is closely linked to the ESD locus. Tight linkage to ESD was established by Connolly et al. (1983). Dryja et al. (1983) found quantitatively normal esterase D in all of 51 patients with retinoblastoma and no known chromosomal abnormality. Cowell et al. (1986) surveyed 200 retinoblastoma patients (75% bilateral and 25% familial) for dosage evidence of deletion of the esterase D locus. Nine patients with about half-normal levels were found; 5 had not been previously recognized as carriers of deletion.

Duncan et al. (1987) found that a cDNA probe hybridized most strongly to 13q14.2 and 13q14.3. They restudied an individual in whom a deletion was said to have separated the closely linked ESD and RB1 loci, placing ESD proximal to RB1. Quantitative in situ hybridization studies of this deletion showed, however, that ESD was missing from the deleted chromosome 13 and duplicated on a normal homolog. This led them to conclude that the deletion in this individual cannot be used to determine the orientation or sublocalization of ESD and RB1 within the 13q14 region.

In the mother and son with retinoblastoma due to deletion of 13q14 described by Fukushima et al. (1987), the esterase D activity of red blood cells was about 50% of normal and the esterase D haplotype was 1-0 in the mother and 2-0 in the son. Cowell et al. (1988) found a 50% level of esterase D activity and a small deletion in region 13q14 in a patient with retinoblastoma. The mother showed the same 50% of normal esterase D level and the same deletion, but had no retinal abnormality. Cowell et al. (1988) stated that this was the first instance of direct transmission of the deletion and also the first instance in which the deletion did not predispose to tumor formation.

Stone et al. (1989) mapped the murine homolog of the human retinoblastoma gene, symbolized Rb-1, to mouse chromosome 14 by analysis of somatic cell hybrids. In recombinant inbred strains, the findings suggested close linkage of Rb-1 and Es-10, which appears to be the murine homolog of ESD. By in situ hybridization, Ono and Yoshida (1993) assigned the RB1 gene to mouse 14D3 and the rat homolog to 15q12. A unique sequence human RB1 cosmid DNA probe was used by Verma et al. (1996) to localize the RB1 homolog in chimpanzee, gorilla, and orangutan to chromosome 14 by fluorescence in situ hybridization.

Higgins et al. (1989) used field inversion gel electrophoresis (FIGE) to construct a restriction map of approximately 1,000 kb of DNA surrounding the RB1 locus and to detect the translocation breakpoints in 3 retinoblastoma patients. The large size of the presumed RB1 gene, approximately 200 kb, and its multiple dispersed exons, complicate molecular screening for prenatal and presymptomatic diagnosis and for carrier detection. By FIGE, Higgins et al. (1989) could detect and map the translocation breakpoints of all 3 retinoblastoma patients within the putative RB1 gene, thus substantiating the authenticity of this candidate sequence and demonstrating the utility of FIGE in detecting chromosomal rearrangements affecting this locus.

In the course of constructing a long-range restriction map around the RB gene by means of PFGE analysis, Blanquet et al. (1991) found evidence of possible genomic imprinting: the NruI restriction pattern differed according to the parental origin of the deletion or other rearrangement.

Molecular Genetics

Fung et al. (1987) used a cDNA probe to determine the lesion in retinoblastomas. In 16 of 40 retinoblastomas studied with a cDNA probe by Fung et al. (1987), a structural change in the RB gene was identifiable, including, in some cases, homozygous internal deletions with corresponding truncated transcripts. An osteosarcoma also had a homozygous internal deletion with a truncated transcript. Possible hotspots for deletion were identified within the RB genomic locus. Among those tumors with no identifiable structural change, there was either absence of an RB transcript or abnormal expression of the RB transcript.

Hensel et al. (1988) found that all 3 patients with retinoblastoma whose DNA was heterozygous for a RFLP detected by an RB1 probe showed loss of 1 allele in DNA from small-cell lung cancer (182280) tissue. Two markers more distal to the RB1 locus, at 13q22-q31 and 13q33, showed loss in 9 of 12 and 3 of 8 SCLC cases, respectively. Harbour et al. (1988) found structural abnormalities of the RB gene in 1 of 8 primary small-cell lung cancer tumors, in 4 of 22 small-cell lung cancer cell lines, and in 1 of 4 pulmonary carcinoid lines. RB mRNA was absent in 60% of SCLC lines and in 75% of pulmonary carcinoid lines, including all samples with DNA abnormalities. In contrast, RB transcripts were found in 90% of non-SCLC lines and in all normal human lung. It is of interest that both SCLC and pulmonary carcinoids are neuroendocrine tumors.

Bookstein et al. (1988) identified at least 20 exons in genomic clones of the RB gene and provisionally numbered them. With a unique sequence probe from intron 1, they detected heterozygous deletions in genomic DNA from 3 retinoblastoma cell lines and genomic rearrangements in fibroblasts from 2 hereditary retinoblastoma patients, indicating that intron 1 includes a frequent site for mutations conferring predisposition to retinoblastoma. Demonstration of a DNA deletion of exons 2-6 from 1 RB allele, as well as the demonstration of other deletions, explains the origin of shortened RB mRNA transcripts. The retinoblastoma candidate gene, 4.7R, does not show gross deletion or rearrangement in most retinoblastomas. Dunn et al. (1988) searched for more subtle mutations using the ribonuclease protection method for analysis of 4.7R mRNA from retinoblastomas. In the ribonuclease (RNase) protection assay, RNases A and T1 cleave single-stranded RNA at basepair mismatches in RNA:RNA or RNA:DNA hybrids. The test identifies only about 50% of single basepair mutations. The sequence of the 4.7R cDNA, as reported by Lee et al. (1987), gives a coding region of 2,784 bp and a polypeptide of 928 amino acids. Dunn et al. (1988) found that 5 of 11 RB tumors, which exhibited normal 4.7R DNA and normal-sized RNA transcripts, showed abnormal ribonuclease cleavage patterns. Three of the 5 mutations affected the same region of the mRNA, consistent with an effect on splicing involving an as yet unidentified 5-prime exon. Canning and Dryja (1989) found deletions in the retinoblastoma gene in 12 of 49 tumors from patients with retinoblastoma or osteosarcoma. Mapping of the deletion breakpoints revealed coincidence of no 2 breakpoints. Thus they could not support the conclusion of others regarding the existence of a 'hotspot' for deletion breakpoints in this gene. In 4 tumors, they sequenced 200 basepairs surrounding each deletion breakpoint. Three deletions had termini within pairs of short, direct repeats ranging in size from 4 to 7 basepairs. They interpreted this as indicating that 'slipped mispairing' may predominate in the generation of deletions at the RB locus. Short, direct repeats were incriminated in 15 of 20 deletions in the beta-globin locus. In other loci, Alu sequences at breakpoints are frequently found, e.g., in the LDL receptor gene (143890), the ADA gene (102700) and the beta-hexosaminidase gene (272800). Efstratiadis et al. (1980) argued that the repeats found at the breakpoints of deletions of the beta-globin locus are not long enough to mediate unequal crossing over via homologous recombination, thus requiring the alternative model, 'slipped mispairing' during DNA replication.

Dunn et al. (1989) extended the characterization of mutations in RB1 using RNase protection of RB1 transcripts to locate probable mutations, followed by polymerase chain reaction (PCR) to amplify and sequence the mutant allele. Mutations were identified in 15 of 21 RB tumors; in 8 tumors, the precise error in nucleotide sequence was characterized. Each of 4 germline mutations involved a small deletion or duplication while 3 somatic mutations were point mutations leading to splice alterations and loss of an exon from the mature RB1 mRNA. By PCR techniques, Yandell et al. (1989) demonstrated single nucleotide changes in tumors from 7 patients with simplex retinoblastoma (with no family history of the disease). In 4 patients, the mutation involved only the tumor cells, and in 3 it involved normal somatic cells as well as tumor cells but was not found in either parent. Thus, these 3 represent new germinal mutations. All 3 were C-to-T transitions in the coding strand in the retinoblastoma gene. Two of the 3 occurred at CpG pairs. Since new germinal mutations of the retinoblastoma gene are more likely to occur on the paternal allele (Dryja et al., 1989), the overrepresentation of C-to-T transitions is probably the result of processes that occur during male gametogenesis. Direct analysis of disease-causing mutations are particularly applicable to diseases characterized by a high proportion of new mutations. This approach had been used successfully with Lesch-Nyhan syndrome (380000) and Duchenne muscular dystrophy (310200), which are associated with high frequencies of new point mutations and new deletions, respectively. Analysis of genomic DNA obviates the difficulty of having an RNA transcript or protein gene product for analysis and allows the detection of mutations that may occur at splice sites or other sequences that are excluded from the RNA transcript. Of the 10 mutations (7 retinoblastoma tumors plus 3 others), 5 occurred in the exon 21-24 region which represents only 15% o f the coding sequence of the retinoblastoma gene. It has been found that aberrant proteins from which amino acids coded by these exons have been deleted lack binding activity for the adenovirus E1A dominant transforming protein.

Blanquet et al. (1995) performed a mutation survey of the RB1 gene in 232 patients with hereditary or nonhereditary retinoblastoma. They systematically explored all 27 exons and flanking sequences, as well as the promoter. All types of point mutations were represented and found to be unequally distributed along the RB1 gene sequence. In the population studied, exons 3, 8, 18, and 19 were preferentially altered. Correlations between the phenotypic expression and molecular alterations were difficult to discern, at least for the missense and in-frame mutations. However, Blanquet et al. (1995) observed that some patients carrying mutations in exon 19 also developed nonocular tumors such as pineoblastoma, fibrosarcoma, or osteosarcoma. Germline mutations were detected in 36% of 25 familial cases, in 20.5% of bilateral sporadic or unilateral multifocal cases, and in 7.1% of unilateral sporadic cases. Because of the low level of detection of germline mutations in hereditary cases, they reasoned that other mechanisms of inactivation of RB1 must be involved.

Lohmann et al. (1996) found no correlation between the location of frameshift or nonsense mutations and phenotypic features, including age at diagnosis, the number of tumor foci, and the manifestations of nonocular tumors.

Lohmann et al. (1997) investigated the frequency and nature of constitutional RB1-gene mutations in patients with isolated unilateral retinoblastoma. A total of 45 mutations were detected in tumors from 36 patients. Thirty-nine of the mutations—including 34 small mutations, 2 large structural alterations, and hypermethylation in 3 tumors—were not detected in the corresponding peripheral blood DNA. In 6 (17%) of the 36 patients, a mutation was detected in constitutional DNA, and 1 of these mutations was known to be associated with reduced expressivity. The presence of a constitutional mutation was not associated with an early age at treatment. In 1 patient, somatic mosaicism was demonstrated by molecular analysis of DNA and RNA from peripheral blood. In 2 patients without a detectable mutation in peripheral blood, mosaicism was suggested because 1 of the patients showed multifocal tumors and the other later developed bilateral retinoblastoma. Lohmann et al. (1996) studied 119 patients with hereditary retinoblastoma for germline RB1 mutations. Southern blot hybridization and PCR fragment-length analysis revealed mutations in 48 patients. In the remaining 71 patients, they detected mutations in 51 (72%) by applying heteroduplex analysis, nonisotopic SSCP, and direct sequencing. Rare sequence variants were also found in 4 patients. No region of the RB1 gene was preferentially involved in single base substitutions. Recurrent transitions were observed at most of the 14 CGA codons within the RB1 gene. No mutation was observed in exons 25-27, although this region contains 2 CGA codons. This suggested to the authors that mutations within the 3-prime terminal region of the RB1 gene may not be oncogenic. For the entire series of 119 patients, mutations were identified in 99 (83%). The spectrum comprised 15% large deletions, 26% small length alterations, and 42% base substitutions.

Pathogenesis

Murphree and Benedict (1984) suggested that the retinoblastoma gene is a model for a class of recessive human cancer genes that have a 'suppressor' or 'regulatory' function. The primary mechanism in the development of retinoblastoma is loss or inactivation of both alleles of this gene. This mechanism contrasts with that of putative human oncogenes which are thought to induce cancers following activation or alteration. The high incidence of second primary tumors among patients who inherit one retinoblastoma gene suggests that this cancer gene plays a key role in the etiology of several other primary malignancies. In some retinoblastomas, extra nonrandom copies of specific chromosomal regions occur, suggesting that an 'expressor' gene (possibly an oncogene) may be involved. Dryja et al. (1984) concluded that about half of retinoblastomas show homozygosity of part or all of 13q in tumor cells. The conclusion was based on the finding of homozygosity for ESD (133280) and for cloned DNA fragments that are homologous to loci on chromosome 13 and show RFLPs. These tumors came from patients heterozygous for these loci in circulating leukocytes. Discrepant homozygosity was not found in other chromosomes. Homozygosity was found in both multifocal (hereditary) and unifocal (presumably sporadic) cases. Somatic recombinations or abnormal chromosomal segregation in mitosis was postulated. The authors pointed out the usefulness of this approach for identifying persons at risk for retinoblastoma. If the tumor in a proband with hereditary tumor has deletion in a 13q, then the submicroscopic change must be in the other chromosome (which can be traced by the linkage principle). Results of Lee et al. (1987) imply that the RB gene product regulates other genes within the cell. On morphologic criteria, retinoblastoma has long been considered a malignancy of the photoreceptor cell lineage (Bogenmann et al., 1988). Horowitz et al. (1990) found that inactivation of the retinoblastoma protein , p105-Rb, which is universal in retinoblastoma cells, is present in most small-cell lung cancers and in one-third of bladder cancers but is infrequent in other human tumors.

Shiio et al. (1992) found evidence that wildtype p53 (191170) suppresses transcription of the RB gene. From deletion and mutagenesis experiments, a cis-acting element (GGAAGTGA) susceptible to regulation by p53 was mapped within the RB promoter. Benedict et al. (1990) suggested that retinoblastoma may be unique in that functional loss of both RB alleles suffices for malignant development. In osteosarcoma, additional change of the p53 locus may be necessary, and in small-cell lung carcinoma, changes in both p53 and the short arm of chromosome 3 may be necessary.

The results of many experiments indicate that the normal function of the RB gene product is a negative regulator of cellular proliferation which is achieved by sequestering a variety of nuclear proteins involved in cellular growth. DNA tumor virus oncoproteins transform cells, at least in part, by releasing these cellular proteins from inactive complex. Two of the cellular proteins that are bound, E2F (189971) and the MYC protein (NMYC; 164840), are transcription factors. Bandara et al. (1991) reported a naturally occurring loss-of-function RB allele encoding a protein that fails to complex with DRTF1 (189902). They also showed that cyclin A (CCNA; 123835) also complexes with DRTF1 and facilitates the efficient assembly of the RB protein into the complex. Goodrich et al. (1991) found that injection into cells of either full-length or a truncated form of the RB protein containing the T antigen-binding region inhibited progression from G1 into the S phase. Coinjection of anti-RB antibodies antagonized this effect. The results indicated that RB regulates cell proliferation by restricting cell cycle progression at a specific point in G1 and established a biologic assay for RB activity. Coinjection of RB with a T-antigen peptide or injection into cells expressing T antigen was accompanied by no inhibition of progression into S phase. This was interpreted as indicating that the transforming proteins of some DNA tumor viruses, including SV40 T antigen and adenovirus E1A, may promote cell growth, at least in part, by binding and inactivating RB. Mancini et al. (1994) demonstrated by immunoblotting and immunolabeling that a significant portion of hypophosphorylated Rb (retinoblastoma gene product) associates with the nuclear matrix during the early G1 phase. They suggested that Rb interactions with a nuclear matrix may be important for its ability to regulate cell cycle progression. Mutant Rb in tumor cells did not associate with the matrix, whereas Rb-reconstituted cells contained abundant matrix-bound Rb.

Luo et al. (1998) demonstrated that Rb can repress transcription of endogenous cell cycle genes containing E2F sites through recruitment of histone deacetylase, which deacetylates histones on the promoter, thereby promoting formation of nucleosomes that inhibit transcription.

Sakai et al. (1991) studied the methylation pattern at the 5-prime end of the RB gene, including its promoter region and exon 1, in DNA purified from 56 primary retinoblastomas. The purpose was to investigate the possibility that hypermethylation of the promoter region might be responsible for loss of function of the gene giving rise to tumor. In 5 of the tumors, they found evidence for hypermethylation; all tumors were from unilateral, 'simplex' patients. No methylation abnormality was detected in DNA purified from the leukocytes of these patients. In 1 of the tumors, the hypermethylation was confined to 1 allele. No mutations to account for the allele-specific hypermethylation was found in a 1,306-bp sequence including the hypermethylated region. If hypermethylation indeed is the cause of the tumor, therapeutic agents that interfere with methylation of DNA might be effective.

A significant proportion of disease-causing mutations in the RB1 gene result in the premature termination of protein synthesis, and most of these mutations occur as C-to-T transitions at CpG dinucleotides. Mancini et al. (1997) presented evidence confirming the view that such recurring CpG mutations are the result of the deamination of 5-methylcytosine within these CpGs. They used the sodium-bisulfite conversion method to detect cytosine methylation in representative exons of RB1. They analyzed DNA from a variety of tissues and specifically targeted CGA codons in RB1, where recurrent termination mutations had been reported. They found that DNA methylation within RB1 exons 8, 14, 25, and 27 appeared to be restricted to CpGs, including 6 CGA codons. Other codons containing methylated cytosines had not been reported to be mutated. Therefore, disease-causing mutations at CpGs in RB1 appear to be determined by several factors, including the constitutive presence of DNA methylation at cytosines within CpGs, the specific codon within which the methylated cytosine is located, and the particular region of the gene within which that codon resides.

Weinberg (1995) reviewed the role of the retinoblastoma protein in the control of the cell cycle. He pointed out that it 'operates in the midst of the cell cycle clock apparatus.' Its main role is to act as a signal transducer connecting the cell cycle clock with the transcriptional machinery. He suggested that despite the possible roles of the RB protein in normal cell physiology that have been revealed, major enigmas remain: how can most of mouse development proceed in an essentially normal fashion in its absence? Why does loss of the RB protein in humans lead specifically to retinal tumors and not to malignancies in other tissues of young children? Why do humans, unlike all other mammals, suffer retinal tumors when the RB protein is lost?

Gallie (1997) commented that the 'biology of the protein product of RB1 has ignited the field of cell-cycle regulation, so that much more literature now refers to the retinoblastoma protein than to the disease.' The main point of her editorial, however, was to emphasize the need for a new kind of partnership between research, innovative entrepreneurial activities, and clinical management to get the most out of the investment that has been made in identifying human disease genes such as RB1.

Diagnosis and Counseling

Ophthalmoscopic examination typically shows a white 'cat's eye' reflex and a retinal tumor in one or both eyes, usually by age 3 years.

159,161:Sparkes et al. (1979, 1980) suggested that if linkage between retinoblastoma and esterase D were found, this would provide a means of genetic counseling and early diagnosis, including prenatal diagnosis. Turleau et al. (1983) added a fourth family to those with retinoblastoma due to deletion of the critical portion of 13q in the offspring of a parent with a balanced insertional translocation (Riccardi et al., 1979; Rivera et al., 1981; Strong et al., 1981). They pointed out that without karyotyping the recurrence risk after the birth of 1 case from normal parents might be thought to be virtually nil whereas in fact it is 25% if one of the normal parents is a carrier of an insertional translocation. The same risk, 25%, applies to occurrence of a trisomic offspring. The RFLPs found by Dryja et al. (1984) should be linked to the RB locus and perhaps close enough to be useful in genetic counseling in some families. Cavenee et al. (1985) showed that the chromosome 13 remaining in tumors from 2 hereditary retinoblastoma cases was derived from the affected parents. The ability to identify which chromosome of an affected parent carries the mutation predisposing to retinoblastoma would have obvious usefulness in genetic counseling. Cavenee et al. (1986) demonstrated the usefulness of multiple RFLP and isozymic markers flanking the RB1 locus in prenatal and postnatal prediction of susceptibility to retinoblastoma. Wiggs et al. (1988) used cloned fragments of a 'retinoblastoma gene' that detected RFLPs and tested the usefulness of these RFLPs in predicting cancer in 20 families with hereditary retinoblastoma. Predictions were possible in 19 of the 20 families; in 18 of the 19, the marker RFLPs showed a consistent association with the mutation predisposing to retinoblastoma. In the 19th kindred, there was a lack of cosegregation, which, however, may have been due to inaccuracy of the clinical diagnosis of the retinal lesion in the key member of the kin dred. Greger et al. (1988) demonstrated the usefulness of linkage analysis with DNA markers in genetic counseling of families with hereditary retinoblastoma.

Maat-Kievit et al. (1993) detected an enormous retinoblastoma at 21 weeks of gestation by means of ultrasound.

Since about 75% of cases of retinoblastoma due to constitutional mutations represent new mutations, Janson and Nordenskjold (1994) recognized a need for methods to identify carriers of such germline mutations so that informed genetic counseling can be offered. Using pulsed field gel electrophoresis in a screening of 20 unrelated cases with bilateral retinoblastoma, 1 constitutional mutation was detected and found to be caused by a balanced translocation t(4;13), with the breakpoint within intron 17 of the retinoblastoma gene.

Screening

Noorani et al. (1996) compared the direct health care costs of molecular and conventional screening of relatives of individuals affected with retinoblastoma. With variables set at the most likely values (baseline), the expected cost (in 1994 Canadian dollars) of conventional screening was $31,430 for a prototype family consisting of 7 at-risk relatives. The cost included 3 clinic examinations and 8 examinations under anesthetic over the first 3 years of life for each relative. Using baseline variables, the molecular strategy consisted of the screening of a prototype family of 1 proband and 7 at-risk relatives at a cost of $8,674, including identification of the RB1 mutation in the proband, subsequent testing of the relevant relatives for that mutation, and clinical follow-up similar to the conventional strategy for relatives with the mutation. Sensitivity analysis over the range of values for each variable revealed a significant saving of health care dollars by the molecular route, indicating the benefit of redirecting economic resources to molecular diagnosis in retinoblastoma.

Animal Model

Windle et al. (1990) created transgenic mice by microinjecting fertilized ova with a chimeric gene containing the protein coding region of the SV40 T antigen (Tag) driven by the promoter of the luteinizing hormone beta-subunit gene. One of the male founders developed bilateral retinoblastomas at about age 5 months. The phenotype was heritable with complete penetrance in transgenic offspring in whom the tumors were first observed at about 2 months. Windle et al. (1990) demonstrated specific association between p105(Rb) and T antigen in mouse retinoblastoma tumor cells. Thus, evidence is provided for oncogenesis due to the ocular-specific expression of an Rb-binding oncoprotein that can functionally inactivate the Rb protein.

Cloning

Dryja et al. (1984) cloned DNA fragments from chromosome 13. Three of these identified RFLPs; they were from region 13q12-13q22, which contains the retinoblastoma 'locus.' Using chromosome 13-specific RFLPs, Dryja et al. (1984) studied osteogenic sarcomas from 15 patients without retinoblastoma. In 3, homozygosity was found at every chromosome 13 locus tested. Constitutional DNA was available on only 1 of the 3; heterozygosity was found.

Friend et al. (1986) isolated a cDNA that detects a chromosomal segment having the properties of the gene at the retinoblastoma locus. The gene was found to be expressed in many tumor types, but no RNA transcript was found in retinoblastomas or osteosarcomas. The locus spanned at least 70 kb of DNA. Friend et al. (1986) started with a 1.5 kb DNA sequence which could detect deletions involving 13q14 in 3 of 37 retinoblastomas. They then used chromosome walking techniques to isolate and map 30 kb of surrounding genomic DNA. One of the single copy fragments recognized a DNA sequence in the mouse genome and also in human chromosome 13. The conservation of this DNA sequence between mouse and humans suggested that the cloned fragment contained a coding exon of a gene. Therefore, they tested the ability of this fragment to hybridize to RNA derived from retinoblastoma cells and from human retinal cells. They found that indeed it recognized a 4.7-kb RNA transcript in the retinal cell line but that this transcript was not detectable in 4 retinoblastomas. As outlined by Cavenee (1986), when the cDNA described by Friend et al. (1986) was used as a probe to screen RNA samples from different tumor types, it was shown to hybridize to all tested except retinoblastomas and retinoblastoma-associated osteosarcomas. Furthermore, use of this cDNA to analyze the genomic structure of its homologous locus in 50 retinoblastomas or associated osteosarcomas showed that about 30% had somatically altered genomic loci. These alterations took the form of fragments of altered mobility (suggesting gene rearrangements), underrepresented fragments (suggesting heterozygous deletions), and missing fragments (suggesting homozygous deletions). Since one of the homozygous deletions was entirely contained within the genomic locus homologous to the cDNA probe, it may well be that this expressed gene is indeed the RB1 locus. Dryja et al. (1986) isolated a cDNA fragment derived from human retinal mRNA that detect s a locus within 13q14 which is often deleted in retinoblastoma. The gene identified by Dryja et al. (1986) may be the same as that which has been found by others and named RB-MYC.

Toguchida et al. (1993) reported the complete genomic sequence of the RB1 gene, which was contained in a 180,388-bp contig. The 27 exons of the gene are known to span over 180 kb of genomic DNA. The gene produces a 4.7-kb transcript that encodes a nuclear phosphoprotein consisting of 928 amino acids. Analysis of the sequence indicated a high ratio of (A+T)/(G+C) and a high density of Line-1 (L1) repeat sequences, suggesting that the locus maps to G-bands 13q14.12 or 13q14.2. Hong et al. (1989) demonstrated that the RB transcript is encoded in 27 exons dispersed over about 200 kb of genomic DNA. The length of individual exons ranges from 31 to 1,889 bp. The largest intron spans more than 60 kb and the smallest one has only 80 bp. Deletion of exons 13-17 is frequently observed in various types of tumors, including retinoblastoma, breast cancer, and osteosarcoma, and the presence of a potential 'hotspot' for recombination in the region was predicted. A putative 'leucine-zipper' motif is exclusively encoded by exon 20. Transcription of RB is initiated at multiple positions and the sequences surrounding the initiation sites have a high G+C content. Several features of the RB promoter are reminiscent of those associated with many so-called housekeeping genes, consistent with the ubiquitous expression of the RB gene.

Function

Lee et al. (1987) prepared a rabbit antiserum against the RB protein studied by Horsthemke et al. (1987) and showed that it was present in all cell lines expressing normal RB mRNA but was not detected in 5 retinoblastoma cell lines. The RB protein can be metabolically labeled with (32)P-phosphoric acid, indicating that it is a phosphoprotein. Biochemical fractionation and immunofluorescence studies demonstrated that most of the protein is located in the nucleus. Furthermore, the protein was retained by and could be eluted from DNA-cellulose columns, suggesting a DNA binding activity. A gene encoding a messenger RNA of 4.6 kb, located in the proximity of esterase D, was identified by Lee et al. (1987) as the retinoblastoma susceptibility gene on the basis of chromosomal location, homozygous deletion, and tumor-specific alterations in expression. Transcription of the gene was abnormal in all of 6 retinoblastomas examined: in 2, mRNA was not detectable, while 4 others expressed variable quantities of the mRNA with decreased molecular size of about 4.0 kb. In contrast, full-length RB mRNA was present in human fetal retina and placenta, and in other tumors such as neuroblastoma and medulloblastoma. The sequence of cDNA clones indicated a hypothetical protein of 816 amino acids. A computer-assisted search of protein sequences revealed no closely related proteins. Kimchi et al. (1988) demonstrated that retinoblastoma cells lack receptors for transforming growth factor, beta-type (TGFB; 190180). Whereas in control cells, in this case retinal cells from fetuses, treatment with TGF-beta inhibited cell division, treatment of retinoblastoma cells with the same substance had no effect on DNA synthesis, cell division, or cell morphology. Bogenmann et al. (1988) showed that this tumor grown in vitro expresses highly specific photoreceptor cell genes: transducin alpha subunit, which is specific to the cone cell, and transcripts for red or green cone cell photopigment. Bogenmann et al. (1988) found no marker genes specific to rod cells. Yokota et al. (1988) found markedly reduced amounts of RB transcript in some small-cell carcinomas. Weichselbaum et al. (1988) presented evidence suggesting that in soft tissue sarcomas as well as in osteosarcomas, transcriptional inactivation or posttranscriptional downregulation of the RB gene may be important in etiology. They used hybridization with a cDNA probe for RB mRNA to analyze 3 soft tissue sarcomas and 4 osteosarcomas in patients without retinoblastoma. Most of the tumors studied did not express detectable levels of RB mRNA, whereas normal cells and epithelial tumor cells did. One osteosarcoma expressed a 2.4-kb transcript in addition to a normal 4.7-kb species. Whyte et al. (1988) demonstrated that a 105,000-MW cellular protein, which is one of the cellular targets implicated in the process of transformation by the adenovirus E1A proteins, is in fact the product of the retinoblastoma gene. This interaction with the formation of a stable protein/protein complex was the first demonstration of a physical link between an oncogene and an antioncogene. A similar case can be made for numerous other disorders, many of which are more common. Gallie (1997) stated that 'At last our resources and efforts will deliver 'health,' through prevention and early management of disease, which will replace the much more expensive treatment of end-stage illness.'

Fearon (1997) provided a schematic representation of the cellular localization and presumed functions of the proteins encoded by inherited cancer genes.

DeCaprio et al. (1989), Buchkovich et al. (1989), and Chen et al. (1989) demonstrated that the RB1 gene product has the properties of a cell cycle regulatory element and that its function is modulated by a phosphorylation/dephosphorylation mechanism during cell proliferation and differentiation. In G0/G1 cells, virtually all the RB protein is unphosphorylated, whereas during S and G2 phases, it is largely, if not exclusively, phosphorylated.

See also Aherne and Roberts (1975); Bader et al. (1982); Bonaiti-Pellie and Briard-Guillemot (1981); Briard-Guillemot et al. (1974); Bundey (1980); Carlson and Desnick (1979); Cavenee et al. (1983); de Grouchy et al. (1980); Dryja et al. (1984); Dryja et al. (1989); Dryja et al. (1986); Dryja et al. (1984); Ejima et al. (1982); Eldridge et al. (1972); Francois (1968); Francois et al. (1975); Godbout et al. (1983); Green and Wyke (1985); Hoegerman (1979); Kitchin and Ellsworth (1974); Knudson (1985); Knudson et al. (1975); Knudson et al. (1976); Lee et al. (1987); MacKay et al. (1984); Manchester (1961); Matsunaga (1978); Matsunaga (1981); Matsunaga (1978); Michalova et al. (1982); Motegi (1982); Motegi et al. (1983); Motegi et al. (1982); Nirankari et al. (1962); Nussbaum and Puck (1976); Orye et al. (1982); Sakai et al. (1991); Schappert-Kimmijser et al. (1966); Schimke et al. (1974); Sparkes (1985); Sparkes et al. (1979); Squire et al. (1985); Squire et al. (1984); Turleau et al. (1985); Vogel (1968); Vogel (1969); Vogel (1979); Warburg (1974); Weichselbaum et al. (1977); Wilson et al. (1977)

Abramson, D. H.; Ellsworth, R. M.; Zimmerman, L. E.: Monocular cancer in retinoblastoma survivors. Trans. Am. Acad. Ophthal. Otolaryng. 81: 454-457, 1976.

Aherne, G. E. S.; Roberts, D. F.: Retinoblastoma—a clinical survey and its genetic implications. Clin. Genet. 8: 275-290, 1975.

Amoaku, W. M. K.; Willshaw, H. E.; Parkes, S. E.; Shah, K. J.; Mann, J. R.: Trilateral retinoblastoma: a report of five patients. Cancer 78: 858-863, 1996.

Bader, J. L.; Meadows, A. T.; Zimmerman, L. E.; Rorke, L. B.; Voute, P. A.; Champion, L. A. A.; Miller, R. W.: Bilateral retinoblastoma with ectopic intracranial retinoblastoma: trilateral retinoblastoma. Cancer Genet. Cytogenet. 5: 203-213, 1982.

Balaban-Malenbaum, G.; Gilbert, F.; Nichols, W. W.; Hill, R.; Shields, J.; Meadows, A. T.: A deleted chromosome no. 13 in human retinoblastoma cells: relevance to tumorigenesis. Cancer Genet. Cytogenet. 3: 243-250, 1981.

Bandara, L. R.; Adamczewski, J. P.; Hunt, T.; La Thangue, N. B.: Cyclin A and the retinoblastoma gene product complex with a common transcription factor. Nature 352: 249-251, 1991.

Benedict, W. F.; Murphree, A. L.; Banerjee, A.; Spina, C. A.; Sparkes, M. C.; Sparkes, R. S.: Patient with 13 chromosome deletion: evidence that the retinoblastoma gene is a recessive cancer gene. Science 219: 973-975, 1983.

Benedict, W. F.; Xu, H.-J.; Hu, S.-X.; Takahashi, R.: Role of the retinoblastoma gene in the initiation and progression of human cancer. J. Clin. Invest. 85: 988-993, 1990.

Bia, B.; Cowell, J. K.: Independent constitutional germline mutations occurring in the RB1 gene in cousins with bilateral retinoblastoma. Oncogene 11: 977-979, 1995.

Blanquet, V.; Turleau, C.; de Grouchy, J.; Creau-Goldberg, N.: Physical map around the retinoblastoma gene: possible genomic imprinting suggested by NruI digestion. Genomics 10: 350-355, 1991.

Blanquet, V.; Turleau, C.; Gross-Morand, M. S.; Senamaud-Beaufort, C.; Doz, F.; Besmond, C.: Spectrum of germline mutations in the RB1 gene: a study of 232 patients with hereditary and non hereditary retinoblastoma. Hum. Molec. Genet. 4: 383-388, 1995.

Bogenmann, E.; Lochrie, M. A.; Simon, M. I.: Cone cell-specific genes expressed in retinoblastoma. Science 240: 76-78, 1988.

Bonaiti-Pellie, C.; Briard-Guillemot, M. L.: Segregation analysis in hereditary retinoblastoma. Hum. Genet. 57: 411-419, 1981.

Bonaiti-Pellie, C.; Briard-Guillemot, M. L.; Feingold, J.; Frezal, J.: Associated congenital malformations in retinoblastoma. Clin. Genet. 7: 37-39, 1975.

Bookstein, R.; Lee, E. Y.-H. P.; To, H.; Young, L.-J.; Sery, T. W.; Hayes, R. C.; Friedmann, T.; Lee, W.-H.: Human retinoblastoma susceptibility gene: genomic organization and analysis of heterozygous intragenic deletion mutants. Proc. Nat. Acad. Sci. 85: 2210-2214, 1988.

Briard-Guillemot, M. L.; Bonaiti-Pellie, C.; Feingold, J.; Frezal, J.: Etude genetique du retinoblastome. Humangenetik 24: 271-284, 1974.

Brownstein, S.; de Chadarevian, J.-P.; Little, J. M.: Trilateral retinoblastoma: report of two cases. Arch. Ophthal. 102: 257-262, 1984.

Buchkovich, K.; Duffy, L. A.; Harlow, E.: The retinoblastoma protein is phosphorylated during specific phases of the cell cycle. Cell 58: 1097-1105, 1989.

Bundey, S.: Recent views on genetic factors in retinoblastoma. (Abstract) J. Med. Genet. 17: 386-387, 1980.

Bundey, S.; Morten, J. E. N.: An unusual pedigree with retinoblastoma. Does it shed light on the delayed mutation and host resistance theories?. Hum. Genet. 59: 434-436, 1981.

Cance, W. G.; Brennan, M. F.; Dudas, M. E.; Huang, C.-M.; Cordon-Cardo, C.: Altered expression of the retinoblastoma gene product in human sarcomas. New Eng. J. Med. 323: 1457-1462, 1990.

Canning, S.; Dryja, T. P.: Short, direct repeats at the breakpoints of deletions of the retinoblastoma gene. Proc. Nat. Acad. Sci. 86: 5044-5048, 1989.

Carlson, E. A.; Desnick, R. J.: Mutational mosaicism and genetic counseling in retinoblastoma. Am. J. Med. Genet. 4: 365-381, 1979.

Cavenee, W. K.: The genetic basis of neoplasia: the retinoblastoma paradigm. Trends Genet. 2: 299-300, 1986.

Cavenee, W. K.; Dryja, T. P.; Phillips, R. A.; Benedict, W. F.; Godbout, R.; Gallie, B. L.; Murphree, A. L.; Strong, L. C.; White, R. L.: Expression of recessive alleles by chromosomal mechanisms in retinoblastoma. Nature 305: 779-784, 1983.

Cavenee, W. K.; Hansen, M. F.; Nordenskjold, M.; Kock, E.; Maumenee, I.; Squire, J. A.; Phillips, R. A.; Gallie, B. L.: Genetic origin of mutations predisposing to retinoblastoma. Science 228: 501-503, 1985.

Cavenee, W. K.; Murphree, A. L.; Shull, M. M.; Benedict, W. F.; Sparkes, R. S.; Kock, E.; Nordenskjold, M.: Prediction of familial predisposition to retinoblastoma. New Eng. J. Med. 314: 1201-1207, 1986.

Chen, P.-L.; Scully, P.; Shew, J.-Y.; Wang, J. Y. J.; Lee, W.-H.: Phosphorylation of the retinoblastoma gene product is modulated during the cell cycle and cellular differentiation. Cell 58: 1193-1198, 1989.

Connolly, M. J.; Payne, R. H.; Johnson, G.; Gallie, B. L.; Allderdice, P. W.; Marshall, W. H.; Lawton, R. D.: Familial, EsD-linked, retinoblastoma with reduced penetrance and variable expressivity. Hum. Genet. 65: 122-124, 1983.

Cowell, J. K.; Bia, B.: A novel missense mutation in patients from a retinoblastoma pedigree showing only mild expression of the tumor phenotype. Oncogene 16: 3211-3213, 1998.

Cowell, J. K.; Rutland, P.; Hungerford, J.; Jay, M.: Deletion of chromosome region 13q14 is transmissible and does not always predispose to retinoblastoma. Hum. Genet. 80: 43-45, 1988.

Cowell, J. K.; Rutland, P.; Jay, M.; Hungerford, J.: Deletions of the esterase D locus from a survey of 200 retinoblastoma patients. Hum. Genet. 72: 164-167, 1986.

Cowell, J. K.; Smith, T.; Bia, B.: Frequent constitutional C to T mutations in CGA-arginine codons in the RB1 gene produce premature stop codons in patients with bilateral (hereditary) retinoblastoma. Europ. J. Hum. Genet. 2: 281-290, 1994.

Davison, E. V.; Gibbons, B.; Aherne, G. E. S.; Roberts, D. F.: Chromosomes in retinoblastoma patients. Clin. Genet. 15: 505-508, 1979.

DeCaprio, J. A.; Ludlow, J. W.; Lynch, D.; Furukawa, Y.; Griffin, J.; Piwnica-Worms, H.; Huang, C.-M.; Livingston, D. M.: The product of the retinoblastoma susceptibility gene has properties of a cell cycle regulatory element. Cell 58: 1085-1095, 1989.

de Grouchy, J.; Turleau, C.; Cabanis, M. O.; Richardet, J. M.: Retinoblastome et deletion intercalaire du chromosome 13. Arch. Franc. Pediat. 37: 531-535, 1980.

DerKinderen, D. J.; Koten, J. W.; Tan, K. E. W. P.; Beemer, F. A.; Van Romunde, L. K. J.; Den Otter, W.: Parental age in sporadic hereditary retinoblastoma. Am. J. Ophthal. 110: 605-609, 1990.

Dryja, T.; Cavenee, W.; Epstein, J.; Rapaport, J.; Goorin, A.; Koufos, A.: Chromosome 13 homozygosity in osteogenic sarcoma without retinoblastoma. (Abstract) Am. J. Hum. Genet. 36: 28S, 1984.

Dryja, T. P.; Bruns, G. A. P.; Gallie, B.; Petersen, R.; Green, W.; Rapaport, J. M.; Albert, D. M.; Gerald, P. S.: Low incidence of deletion of the esterase D locus in retinoblastoma patients. Hum. Genet. 64: 151-155, 1983.

Dryja, T. P.; Cavenee, W.; White, R.; Rapaport, J. M.; Petersen, R.; Albert, D. M.; Bruns, G. A. P.: Homozygosity of chromosome 13 in retinoblastoma. New Eng. J. Med. 310: 550-553, 1984.

Dryja, T. P.; Friend, S.; Weinberg, R. A.: Isolation of a cDNA fragment derived from human retina mRNA which detects a locus within 13q14 often deleted in retinoblastomas. (Abstract) Am. J. Hum. Genet. 39: A29, 1986.

Dryja, T. P.; Mukai, S.; Petersen, R.; Rapaport, J. M.; Walton, D.; Yandell, D. W.: Parental origin of mutations of the retinoblastoma gene. Nature 339: 556-558, 1989.

Dryja, T. P.; Mukai, S.; Rapaport, J. M.; Yandell, D. W.: Parental origin of mutations of the retinoblastoma gene. (Abstract) Am. J. Hum. Genet. 45 (suppl.): A19, 1989.

Dryja, T. P.; Rapaport, J.; McGee, T. L.; Nork, T. M.; Schwartz, T. L.: Molecular etiology of low-penetrance retinoblastoma in two pedigrees. Am. J. Hum. Genet. 52: 1122-1128, 1993.

Dryja, T. P.; Rapaport, J. M.; Joyce, J. M.; Petersen, R. A.: Molecular detection of deletions involving band q14 of chromosome 13 in retinoblastomas. Proc. Nat. Acad. Sci. 83: 7391-7394, 1986.

Dryja, T. P.; Rapaport, J. M.; Weichselbaum, R.; Bruns, G. A. P.: Chromosome 13 restriction fragment length polymorphisms. Hum. Genet. 65: 320-324, 1984.

Duane, T. B.: Clinical Ophthalmology. Hagerstown: Harper and Row (pub.) 3: 1980. Pp. 13 only.

Duncan, A. M. V.; Morgan, C.; Gallie, B. L.; Phillips, R. A.; Squire, J.: Re-evaluation of the sublocalization of esterase D and its relation to the retinoblastoma locus by in situ hybridization. Cytogenet. Cell Genet. 44: 153-157, 1987.

Dunn, J. M.; Phillips, R. A.; Becker, A. J.; Gallie, B. L.: Identification of germline and somatic mutations affecting the retinoblastoma gene. Science 241: 1797-1800, 1988.

Dunn, J. M.; Phillips, R. A.; Zhu, X.; Becker, A.; Gallie, B. L.: Mutations in the RB1 gene and their effects on transcription. Molec. Cell. Biol. 9: 4596-4604, 1989.

Efstratiadis, A.; Posakony, J. W.; Maniatis, T.; Lawn, R. M.; O'Connell, C.; Spritz, R. A.; DeRiel, J. K.; Forget, B. G.; Weissman, S. M.; Slightom, J. L.; Blechl, A. E.; Smithies, O.; Baralle, F. E.; Shoulders, C. C.; Proudfoot, N. J.: The structure and evolution of the human beta-globin gene family. Cell 21: 653-668, 1980.

Ejima, Y.; Sasaki, M. S.; Kaneko, A.; Tanooka, H.: Types, rates, origin and expressivity of chromosome mutations involving 13q14 in retinoblastoma patients. Hum. Genet. 79: 118-123, 1988.

Ejima, Y.; Sasaki, M. S.; Kaneko, A.; Tanooka, H.; Hara, Y.; Hida, T.; Kinoshita, Y.: Possible inactivation of part of chromosome 13 due to 13qXp translocation associated with retinoblastoma. Clin. Genet. 21: 357-361, 1982.

Eldridge, R.; O'Meara, K.; Kitchin, D.: Superior intelligence in sighted retinoblastoma patients and their families. J. Med. Genet. 9: 331-335, 1972.

Falls, H. F.; Neel, J. V.: Genetics of retinoblastoma. Arch. Ophthal. 46: 367-389, 1951.

Fearon, E. R.: Human cancer syndromes: clues to the origin and nature of cancer. Science 278: 1043-1050, 1997.

Fitzgerald, P. H.; Stewart, J.; Suckling, R. D.: Retinoblastoma mutation rate in New Zealand and support for the two-hit model. Hum. Genet. 64: 128-130, 1983.

Francke, U.: Retinoblastoma and chromosome 13. Cytogenet. Cell Genet. 14: 131-134, 1976.

Francois, J.: Hereditary malignant tumor of the eye. Congenital Anomalies of The Eye. St. Louis: C. V. Mosby Co. (pub.) 1968. Pp. 205-246.

Francois, J.: Retinoblastoma and osteogenic sarcoma. Ophthalmologica 175: 185-191, 1977.

Francois, J.; Matton, M. T.; De Bie, S.; Tanaka, Y.; Vandenbulcke, D.: Genesis and genetics of retinoblastoma. Ophthalmologica 170: 405-425, 1975.

Friend, S. H.; Bernards, R.; Rogelj, S.; Weinberg, R. A.; Rapaport, J. M.; Albert, D. M.; Dryja, T. P.: A human DNA segment with properties of the gene that predisposes to retinoblastoma and osteosarcoma. Nature 323: 643-646, 1986.

Friend, S. H.; Dryja, T. P.; Weinberg, R. A.: Oncogenes and tumor-suppressing genes. New Eng. J. Med. 318: 618-622, 1988.

Friend, S. H.; Horowitz, J. M.; Gerber, M. R.; Wang, X.-F.; Bogenmann, E.; Li, F. P.; Weinberg, R. A.: Deletions of a DNA sequence in retinoblastomas and mesenchymal tumors: organization of the sequence and its encoded protein. Proc. Nat. Acad. Sci. 84: 9059-9063, 1987. Note: Correction: Proc. Nat. Acad. Sci. 85: 2234 only, 1988..

Fukushima, Y.; Kuroki, Y.; Ito, T.; Kondo, I.; Nishigaki, I.: Familial retinoblastoma (mother and son) with 13q14 deletion. Hum. Genet. 77: 104-107, 1987.

Fung, Y.-K. T.; Murphree, A. L.; T'Ang, A.; Qian, J.; Hinrichs, S. H.; Benedict, W. F.: Structural evidence for the authenticity of the human retinoblastoma gene. Science 236: 1657-1661, 1987.

Gallie, B. L.: Predictive testing for retinoblastoma comes of age. (Editorial) Am. J. Hum. Genet. 61: 279-281, 1997.

Gallie, B. L.; Ellsworth, R. M.; Abramson, D. M.; Phillips, R. A.: Retinoma: spontaneous regression of retinoblastoma or benign manifestation of the mutation?. Brit. J. Cancer 45: 513-521, 1982.

Gallie, B. L.; Phillips, R. A.: Multiple manifestations of the retinoblastoma gene. Birth Defects Orig. Art. Ser. 18(6): 689-701, 1982.

Gey, W.: Dq-, multiple Missbildungen und Retinoblastom. Humangenetik 10: 362-365, 1970.

Gibbons, B.; Scott, D.; Hungerford, J. L.; Cheung, K. L.; Harrison, C.; Attard-Montalto, S.; Evans, M.; Birch, J. M.; Kingston, J. E.: Retinoblastoma in association with the chromosome breakage syndromes Fanconi's anaemia and

Bloom's syndrome: clinical and cytogenetic findings. Clin. Genet. 47: 311-317, 1995.
Godbout, R.; Dryja, T. P.; Squire, J.; Gallie, B. L.; Phillips, R. A.: Somatic inactivation of genes on chromosome 13 is a common event in retinoblastoma. Nature 304: 451-453, 1983.
Goodrich, D. W.; Wang, N. P.; Qian, Y.-W.; Lee, E. Y.-H. P.; Lee, W.-H.: The retinoblastoma gene product regulates progression through the G1 phase of the cell cycle. Cell 67: 293-302, 1991.
Grace, E.; Drennan, J.; Colver, D.; Gordon, R. R.: The 13q deletion syndrome. J. Med. Genet. 8: 351-357, 1971.
Green, A. R.; Wyke, J. A.: Anti-oncogenes: a subset of regulatory genes involved in carcinogenesis?. Lancet II: 475-477, 1985.
Greger, V.; Kerst, S.; Messmer, E.; Hopping, W.; Passarge, E.; Horsthemke, B.: Application of linkage analysis to genetic counselling in families with hereditary retinoblastoma. J. Med. Genet. 25: 217-221, 1988.
Greger, V.; Passarge, E.; Horsthemke, B.: Somatic mosaicism in a patient with bilateral retinoblastoma. Am. J. Hum. Genet. 46: 1187-1193, 1990.
Hall, J. G.: Personal Communication. Vancouver, British Columbia, Canada 5/29/1993.
Harbour, J. W.; Lai, S.-L.; Whang-Peng, J.; Gazdar, A. F.; Minna, J. D.; Kaye, F. J.: Abnormalities in structure and expression of the human retinoblastoma gene in SCLC. Science 241: 353-357, 1988.
Hensel, C.; Hsieh, C.-L.; Lee, W.-H.; Pam-Lee, E.; Gazdar, A.; Sakaguchi, A. Y.; Naylor, S. L.: Allele loss and lack of expression of the RB-1 locus in small cell lung cancer. (Abstract) Am. J. Hum. Genet. 43: A25, 1988.
Henson, J. W.; Schnitker, B. L.; Correa, K. M.; von Diemling, A.; Fassbender, F.; Xu, H.-J.; Benedict, W. F.; Yandell, D. W.; Louis, D. N.: The retinoblastoma gene is involved in malignant progression of astrocytomas. Ann. Neurol. 36: 714-721, 1994.
Higgins, M. J.; Hansen, M. F.; Cavenee, W. K.; Lalande, M.: Molecular detection of chromosomal translocations that disrupt the putative retinoblastoma susceptibility locus. Molec. Cell. Biol. 9: 1-5, 1989.
Hoegerman, S. F.: Chromosome 13 long arm interstitial deletion may result from maternal inverted insertion. Science 205: 1035-1036, 1979.
Hogg, A.; Bia, B.; Onadim, Z.; Cowell, J. K.: Molecular mechanisms of oncogenic mutations in tumours from patients with bilateral and unilateral retinoblastoma. Proc. Nat. Acad. Sci. 90: 7351-7355, 1993.
Hong, F. D.; Huang, H.-J. S.; To, H.; Young, L.-J. S.; Oro, A.; Bookstein, R.; Lee, E. Y.-H. P.; Lee, W.-H.: Structure of the human retinoblastoma gene. Proc. Nat. Acad. Sci. 86: 5502-5506, 1989.
Horowitz, J. M.; Park, S.-H.; Bogenmann, E.; Cheng, J.-C.; Yandell, D. W.; Kaye, F. J.; Minna, J. D.; Dryja, T. P.; Weinberg, R. A.: Frequent inactivation of the retinoblastoma anti-oncogene is restricted to a subset of human tumor cells. Proc. Nat. Acad. Sci. 87: 2775-2779, 1990.
Horowitz, J. M.; Park, S. H.; Yandell, D. W.; Weinberg, R. A.: Involvement of the retinoblastoma gene in the genesis of various human tumors: In: Kavenee, W.; Hastie, N.; Stanbridge, E.: Recessive Oncogenes and Tumor Suppression: Current Communications in Molecular Biology. Cold Spring Harbor, New York: Cold Spring Harbor Laboratory Press (pub.) 1989. Pp. 101-108.
Horsthemke, B.; Greger, V.; Barnert, H. J.; Hopping, W.; Passarge, E.: Detection of submicroscopic deletions and a DNA polymorphism at the retinoblastoma locus. Hum. Genet. 76: 257-261, 1987.
Horsthemke, B.; Greger, V.; Becher, R.; Passarge, E.: Mechanism of i(6p) formation in retinoblastoma tumor cells. Cancer Genet. Cytogenet. 37: 95-102, 1989.
Huang, H.-J. S.; Yee, J.-K.; Shew, J.-Y.; Chen, P.-L.; Bookstein, R.; Friedmann, T.; Lee, E. Y.-H. P.; Lee, W.-H.: Suppression of the neoplastic phenotype by replacement of the RB gene in human cancer cells. Science 242: 1563-1566, 1988.
Janson, M.; Nordenskjold, M.: A constitutional mutation within the retinoblastoma gene detected by PFGE. Clin. Genet. 45: 5-10, 1994.
Jensen, R. D.; Miller, R. W.: Retinoblastoma: epidemiologic characteristics. New Eng. J. Med. 285: 307-311, 1971.
Kimchi, A.; Wang, X.-F.; Weinberg, R. A.; Cheifetz, S.; Massague, J.: Absence of TGF-beta receptors and growth inhibitory responses in retinoblastoma cells. Science 240: 196-199, 1988.
Kitchin, F. D.; Ellsworth, R. M.: Pleiotropic effects of the gene for retinoblastoma. J. Med. Genet. 11: 244-246, 1974.
Knight, L. A.; Gardner, H. A.; Gallie, B. L.: Familial retinoblastoma: segregation of chromosome 13 in four families. Am. J. Hum. Genet. 32: 194-201, 1980.
Knudson, A. G.: Hereditary cancer, oncogenes and anti-oncogenes. Cancer Res. 45: 1437-1443, 1985.
Knudson, A. G., Jr.: Mutation and cancer: statistical study of retinoblastoma. Proc. Nat. Acad. Sci. 68: 820-823, 1971.
Knudson, A. G., Jr.: Genetics of human cancer. Annu. Rev. Genet. 20: 231-251, 1986.
Knudson, A. G., Jr.; Hethcote, H. W.; Brown, B. W.: Mutation and childhood cancer: a probabilistic model for the incidence of retinoblastoma. Proc. Nat. Acad. Sci. 72: 5116-5120, 1975.
Knudson, A. G., Jr.; Meadows, A. T.; Nichols, W. W.; Hill, R.: Chromosomal deletion and retinoblastoma. New Eng. J. Med. 295: 1120-1123, 1976.
Lee, W.-H.; Bookstein, R.; Hong, F.; Young, L.-J.; Shew, J.-Y.; Lee, E. Y.-H. P.: Human retinoblastoma susceptibility gene: cloning, identification, and sequence. Science 235: 1394-1399, 1987.
Lee, W.-H.; Shew, J.-Y.; Hong, F. D.; Sery, T. W.; Donoso, L. A.; Young, L.-J.; Bookstein, R.; Lee, E. Y.-H. P.: The retinoblastoma susceptibility gene encodes a nuclear phosphoprotein associated with DNA binding activity. Nature 329: 642-645, 1987.
Lele, K. P.; Penrose, L. S.; Stallard, H. B.: Chromosome deletion in a case of retinoblastoma. Ann. Hum. Genet. 27: 171-174, 1963.
Lemieux, N.; Messier, P. E.; Jacob, J. L.; Milot, J.; Richer, C. L.: Precise cytogenetic localization of the Rb locus at subband 13q14.11 by ultrastructural detection after immunochemical chromosome banding. (Abstract) Am. J. Hum. Genet. 45 (suppl.): A27, 1989.
Liu, Z.; Song, Y.; Bia, B.; Cowell, J. K.: Germline mutations in the RB1 gene in patients with hereditary retinoblastoma. Genes Chromosomes Cancer 14: 277-284, 1995.
Lohmann, D. R.; Brandt, B.; Hopping, W.; Passarge, E.; Horsthemke, B.: Spectrum of small length germline mutations in the RB1 gene. Hum. Molec. Genet. 3: 2187-2193, 1994.
Lohmann, D. R.; Brandt, B.; Hopping, W.; Passarge, E.; Horsthemke, B.: The spectrum of RB1 germ-line mutations in hereditary retinoblastoma. Am. J. Hum. Genet. 58: 940-949, 1996.
Lohmann, D. R.; Gerick, M.; Brandt, B.; Oelschlager, U.; Lorenz, B.; Passarge, E.; Horsthemke, B.: Constitutional RB1-gene mutations in patients with isolated unilateral retinoblastoma. Am. J. Hum. Genet. 61: 282-294, 1997.
Lueder, G. T.; Judisch, G. F.; Wen, B.-C.: Heritable retinoblastoma and pinealoma. Arch. Ophthal. 109: 1707-1709, 1991.
Luo, R. X.; Postigo, A. A.; Dean, D. C.: Rb interacts with histone deacetylase to repress transcription. Cell 92: 463-473, 1998.
Maat-Kievit, J. A.; Oepkes, D.; Hartwig, N. G.; Vermeij-Keers, C.; van Kamp, I. L.; van de Kamp, J. J. P.: A large retinoblastoma detected in a fetus at 21 weeks of gestation. Prenatal Diag. 13: 377-384, 1993.
MacKay, C. J.; Abramson, D. H.; Ellsworth, R. M.: Metastatic patterns of retinoblastoma. Arch. Ophthal. 102: 391-396, 1984.
Macklin, M. T.: Inheritance of retinoblastoma in Ohio. Arch. Ophthal. 62: 842-851, 1959.
Macklin, M. T.: A study of retinoblastoma in Ohio. Am. J. Hum. Genet. 12: 1-43, 1960.
Manchester, P. T., Jr.: Retinoblastoma among offspring of adult survivors. Arch. Ophthal. 65: 546-549, 1961.
Mancini, D.; Singh, S.; Ainsworth, P.; Rodenhiser, D.: Constitutively methylated CpG dinucleotides as mutation hot spots in the retinoblastoma gene (RB1). Am. J. Hum. Genet. 61: 80-87, 1997.
Mancini, M. A.; Shan, B.; Nickerson, J. A.; Penman, S.; Lee, W.-H.: The retinoblastoma gene product is a cell cycle-dependent, nuclear matrix-associated protein. Proc. Nat. Acad. Sci. 91: 418-422, 1994.
Matsunaga, E.: Recurrence risks to relatives of patients with retinoblastoma. Jpn. J. Ophthal. 22: 313-319, 1978.
Matsunaga, E.: Almost synchronous appearance of bilateral retinoblastomas. (Letter) Am. J. Med. Genet. 11: 485-487, 1982.
Matsunaga, E.: Retinoblastoma: mutational mosaicism or host resistance?. Am. J. Med. Genet. 8: 375-387, 1981.
Matsunaga, E.: Hereditary retinoblastoma: host resistance and second primary tumors. J. Nat. Cancer Inst. 65: 47-51, 1980.
Matsunaga, E.: Hereditary retinoblastoma: delayed mutation or host resistance?. Am. J. Hum. Genet. 30: 406-425, 1978.
Matsunaga, E.; Minoda, K.; Sasaki, M. S.: Parental age and seasonal variation in the births of children with sporadic retinoblastoma: a mutation-epidemiologic study. Hum. Genet. 84: 155-158, 1990.
Michalova, K.; Kloucek, F.; Musilova, J.: Deletion of 13q in two patients with retinoblastoma, one probably due to 13q- mosaicism in the mother. Hum. Genet. 61: 264-266, 1982.
Motegi, T.: High rate of detection of 13q14 deletion mosaicism among retinoblastoma patients (using more extensive methods). Hum. Genet. 61: 95-97, 1982.
Motegi, T.: Lymphocyte chromosome survey in 42 patients with retinoblastoma: effort to detect 13q14 deletion mosaicism. Hum. Genet. 58: 168-173, 1981.
Motegi, T.; Kaga, M.; Yanagawa, Y.; Kadowaki, H.; Watanabe, K.; Inoue, A.; Komatsu, M.; Minoda, K.: A recognizable pattern of the midface of retinoblastoma patients with interstitial deletion of 13q. Hum. Genet. 64: 160-162, 1983.
Motegi, T.; Komatsu, M.; Minoda, K.: Is the interstitial deletion of 13q in retinoblastoma patients not transmissible?. (Letter) Hum. Genet. 64: 205, 1983.
Motegi, T.; Komatsu, M.; Nakazato, Y.; Ohuchi, M.; Minoda, K.: Retinoblastoma in a boy with a de novo mutation of a 13/18 translocation: the assumption that the retinoblastoma locus is at 13q141, particularly at the distal portion of it. Hum. Genet. 60: 193-195, 1982.
Munier, F.; Spence, M. A.; Pescia, G.; Balmer, A.; Gailloud, C.; Thonney, F.; van Melle, G.; Rutz, H. P.: Paternal selection favoring mutant alleles of the retinoblastoma susceptibility gene. Hum. Genet. 89: 508-512, 1992.
Munier, F. L.; Wang, M. X.; Spence, M. A.; Thonney, F.; Balmer, A.; Pescia, G.; Donoso, L. A.; Murphree, A. L.: Pseudo low penetrance in retinoblastoma: fortuitous familial aggregation of sporadic cases caused by independently derived mutations in two large pedigrees. Arch. Ophthal. 111: 1507-1511, 1993.
Murphree, A. L.; Benedict, W. F.: Retinoblastoma: clues to human oncogenesis. Science 223: 1028-1033, 1984.
Naumova, A.; Hansen, M.; Strong, L.; Jones, P. A.; Hadjistilianou, D.; Mastrangelo, D.; Griegel, S.; Rajewsky, M. F.; Shields, J.; Donoso, L.; Wang, M.; Sapienza, C.: Concordance between parental origin of chromosome 13q loss and chromosome 6p duplication in sporadic retinoblastoma. Am. J. Hum. Genet. 54: 274-281, 1994.
Naumova, A.; Sapienza, C.: The genetics of retinoblastoma, revisited. Am. J. Hum. Genet. 54: 264-273, 1994.
Nichols, W. W.; Miller, R. C.; Sobel, M.; Hoffman, E.; Sparkes, R. S.; Mohandas, T.; Veomett, I.; Davis, J. R.: Further observations on a 13qXp translocation associated with retinoblastoma. Am. J. Ophthal. 89: 621-627, 1980.
Nirankari, M. S.; Gulati, G. C.; Chaddah, M. R.: Retinoblastoma: genetics and report of a family. Am. J. Ophthal. 53: 523-532, 1962.
Noorani, H. Z.; Khan, H. N.; Gallie, B. L.; Detsky, A. S.: Cost comparison of molecular versus conventional screening of relatives at risk for retinoblastoma. Am. J. Hum. Genet. 59: 301-307, 1996.
Nussbaum, R.; Puck, J.: Recurrence risks for retinoblastoma: a model for autosomal dominant disorders with complex inheritance. J. Pediat. Ophthal. 13: 89-98, 1976.
Onadim, Z.; Hogg, A.; Baird, P. N.; Cowell, J. K.: Oncogenic point mutations in exon 20 of the RB1 gene in families showing incomplete penetrance and mild expression of the retinoblastoma phenotype. Proc. Nat. Acad. Sci. 89: 6177-6181, 1992.

Onadim, Z.; Woolford, A. J.; Kingston, J. E.; Hungerford, J. L.: The RB1 gene mutation in a child with ectopic intracranial retinoblastoma. Brit. J. Cancer 76: 1405-1409, 1997.
Ono, T.; Yoshida, M. C.: Chromosomal assignment of retinoblastoma 1 gene (RB1) to mouse 14D3 and rat 15q12 by fluorescence in situ hybridization. Jpn. J. Genet. 68: 617-621, 1993.
Orye, E.; Benoit, Y.; Coppieters, R.; Jeannin, P.; Vercruysse, C.; Delaey, J.; Delbeke, M.-J.: A case of retinoblastoma, associated with histiocytosis-X and mosaicism of a deleted D-group chromosome (13q14-q31). Clin. Genet. 22: 37-39, 1982.
Orye, E.; Delbeke, M. J.; Vandenabeele, B.: Retinoblastoma and long arm deletion of chromosome 13. Attempts to define the deleted segment. Clin. Genet. 5: 457-464, 1974.
Orye, E.; Delbeke, M. J.; Vandenabeele, B.: Retinoblastoma and D-chromosome deletions. (Letter) Lancet II: 1376, 1971.
Pendergrass, T. W.; Davis, S.: Incidence of retinoblastoma in the United States. Arch. Ophthal. 98: 1204-1210, 1980.
Riccardi, V. M.; Hittner, H. M.; Francke, U.; Pippin, S.; Holmquist, G. P.; Kretzer, F. L.; Ferrell, R.: Partial triplication and deletion of 13q: study of a family presenting with bilateral retinoblastomas. Clin. Genet. 15: 332-345, 1979.
Rivera, H.; Turleau, C.; de Grouchy, J.; Junien, C.; Despoisse, S.; Zucker, J.-M.: Retinoblastoma-del(13q14): report of two patients, one with a trisomic sib due to maternal insertion; gene-dosage effect for esterase D. Hum. Genet. 59: 211-214, 1981.
Sakai, T.; Ohtani, N.; McGee, T. L.; Robbins, P. D.; Dryja, T. P.: Oncogenic germ-line mutations in Sp1 and ATF sites in the human retinoblastoma gene. Nature 353: 83-86, 1991.
Sakai, T.; Toguchida, J.; Ohtani, N.; Yandell, D. W.; Rapaport, J. M.; Dryja, T. P.: Allele-specific hypermethylation of the retinoblastoma tumor-suppressor gene. Am. J. Hum. Genet. 48: 880-888, 1991.
Schappert-Kimmijser, J.; Hemmes, G. D.; Nijland, R.: The heredity of retinoblastoma. Ophthalmologica 151: 197-213, 1966.
Scheffer, H.; te Meerman, G. J.; Kruize, Y. C. M.; van den Berg, A. H. M.; Penninga, D. P.; Tan, K. E. W. P.; der Kinderen, D. J.; Buys, C. H. C. M.: Linkage analysis of families with hereditary retinoblastoma: nonpenetrance of mutation, revealed by combined use of markers within and flanking the RB1 gene. Am. J. Hum. Genet. 45: 252-260, 1989.
Schimke, R. N.; Lowman, J.; Cowan, G.: Retinoblastoma and osteogenic sarcoma in sibs. Cancer 34: 2077-2079, 1974.
Schubert, E. L.; Strong, L. C.; Hansen, M. F.: A splicing mutation in RB1 in low penetrance retinoblastoma. Hum. Genet. 100: 557-563, 1997.
Shiio, Y.; Yamamoto, T.; Yamaguchi, N.: Negative regulation of Rb expression by the p53 gene product. Proc. Nat. Acad. Sci. 89: 5206-5210, 1992.
Shroeder, W. T.; Chao, L.-Y.; Dao, D. D.; Strong, L. C.; Pathak, S.; Riccardi, V.; Lewis, W. H.; Saunders, G. F.: Nonrandom loss of maternal chromosome 11 alleles in Wilms tumors. Am. J. Hum. Genet. 40: 413-420, 1987.
Sippel, K. C.; Fraioli, R. E.; Smith, G. D.; Schalkoff, M. E.; Sutherland, J.; Gallie, B. L.; Dryja, T. P.: Frequency of somatic and germ-line mosaicism in retinoblastoma: implications for genetic counseling. Am. J. Hum. Genet. 62: 610-619, 1998.
Smith, S. M.; Sorsby, A.: Retinoblastoma: some genetic aspects. Ann. Hum. Genet. 23: 50-58, 1958.
Sparkes, R. S.: The genetics of retinoblastoma. Biochim. Biophys. Acta 780: 95-118, 1985.
Sparkes, R. S.; Muller, H.; Klisak, I.; Abram, J. A.: Retinoblastoma with 13q; chromosomal deletion associated with maternal paracentric inversion of 13q. Science 203: 1027-1029, 1979.
Sparkes, R. S.; Murphree, A. L.; Lingua, R. W.; Sparkes, M. C.; Field, L. L.; Funderburk, S. J.; Benedict, W. F.: Gene for hereditary retinoblastoma assigned to human chromosome 13 by linkage to esterase D. Science 219: 971-973, 1983.
Sparkes, R. S.; Sparkes, M. C.; Wilson, M. G.; Towner, J. W.; Benedict, W.; Murphree, A. L.; Yunis, J. J.: Regional assignment of genes for human esterase D and retinoblastoma to chromosome band 13q14. Science 208: 1042-1044, 1980.
Sparkes, R. S.; Sparkes, M. C.; Wilson, M. G.; Towner, J. W.; Benedict, W.; Murphree, A. L.; Yunis, J. J.: Regional assignment of genes for human esterase D and retinoblastoma to chromosome band 13q14. (Abstract) Cytogenet. Cell Genet. 25: 209, 1979.
Squire, J.; Gallie, B. L.; Phillips, R. A.: A detailed analysis of chromosomal changes in heritable and non-heritable retinoblastoma. Hum. Genet. 70: 291-301, 1985.
Squire, J.; Phillips, R. A.; Boyce, S.; Godbout, R.; Rogers, B.; Gallie, B. L.: Isochromosome 6p, a unique chromosomal abnormality in retinoblastoma: verification by standard staining techniques, new densitometric methods, and somatic cell hybridization. Hum. Genet. 66: 46-53, 1984.
Stallard, H. B.: The conservation treatment of retinoblastoma. Trans. Ophthal. Soc. 82: 473, 1962.
Stone, J. C.; Crosby, J. L.; Kozak, C. A.; Schievella, A. R.; Bernards, R.; Nadeau, J. H.: The murine retinoblastoma homolog maps to chromosome 14 near Es-10. Genomics 5: 70-75, 1989.
Strong, L. C.; Riccardi, V. M.; Ferrell, R. E.; Sparkes, R. S.: Familial retinoblastoma and chromosome 13 deletion transmitted via an insertional translocation. Science 213: 1501-1503, 1981.
Taylor, A. I.: Dq-, Dr and retinoblastoma. Humangenetik 10: 209-217, 1970.
Toguchida, J.; Ishizaki, K.; Sasaki, M. S.; Nakamura, Y.; Ikenaga, M.; Kato, M.; Sugimot, M.; Kotoura, Y.; Yamamuro, T.: Preferential mutation of paternally derived RB gene as the initial event in sporadic osteosarcoma. Nature 338: 156-158, 1989.
Toguchida, J.; McGee, T. L.; Paterson, J. C.; Eagle, J. R.; Tucker, S.; Yandell, D. W.; Dryja, T. P.: Complete genomic sequence of the human retinoblastoma susceptibility gene. Genomics 17: 535-543, 1993.
Turleau, C.; de Grouchy, J.; Chavin-Colin, F.; Despoisses, S.; Leblanc, A.: Two cases of del(13q)-retinoblastoma and two cases of partial trisomy due to a familial insertion. Ann. Genet. 26: 158-160, 1983.
Turleau, C.; de Grouchy, J.; Chavin-Colin, F.; Junien, C.; Seger, J.; Schlienger, P.; Leblanc, A.; Haye, C.: Cytogenetic forms of retinoblastoma: their incidence in a survey of 66 patients. Cancer Genet. Cytogenet. 16: 321-334, 1985.
Verma, R. S.; Ramesh, K. H.; Samonte, R. V.; Conte, R. A.: Mapping the homolog of the human Rb1 gene to chromosome 14 of higher primates. Mammalian Genome 7: 591-592, 1996.
Vogel, F.: Genetics of retinoblastoma. Modern Trends in Ophthalmology. (pub.) 1968.
Vogel, F.: Genetics of retinoblastoma.In: Genetic Counseling. Heidelberg University, Science Library. Trans. by Sabine Kurth. New York: Springer Verlag (pub.) 1969.
Vogel, F.: The genetics of retinoblastoma. Hum. Genet. 52: 1-54, 1979.
Warburg, M.: Retinoblastoma.In: Goldberg, M. F.: Genetic and Metabolic Eye Disease. Boston: Little, Brown and Co. (pub.) 1974. Pp. 447-461.
Weichselbaum, R. R.; Beckett, M.; Diamond, A.: Some retinoblastomas, osteosarcomas, and soft tissue sarcomas may share a common etiology. Proc. Nat. Acad. Sci. 85: 2106-2109, 1988.
Weichselbaum, R. R.; Nove, J.; Little, J. B.: Fibroblasts from a D-deletion type retinoblastoma patient are abnormally x-ray sensitive. Nature 266: 726-727, 1977.
Weinberg, R. A.: The retinoblastoma protein and cell cycle control. Cell 81: 323-330, 1995.
Whyte, P.; Buchkovich, K. J.; Horowitz, J. M.; Friend, S. H.; Raybuck, M.; Weinberg, R. A.; Harlow, E.: Association between an oncogene and an anti-oncogene: the adenovirus E1A proteins bind to the retinoblastoma gene product. Nature 334: 124-129, 1988.
Wiggs, J.; Nordenskjold, M.; Yandell, D.; Rapaport, J.; Grondin, V.; Janson, M.; Werelius, B.; Petersen, R.; Craft, A.; Riedel, K.; Liberfarb, R.; Walton, D.; Wilson, W.; Dryja, T. P.: Prediction of the risk of hereditary retinoblastoma, using DNA polymorphisms within the retinoblastoma gene. New Eng. J. Med. 318: 151-157, 1988.
Wilson, M. G.; Ebbin, A. J.; Towner, J. W.; Spencer, W. H.: Chromosomal anomalies in patients with retinoblastoma. Clin. Genet. 12: 1-8, 1977.
Wilson, M. G.; Melnyk, J.; Towner, J. W. J.: Retinoblastoma and deletion D(14) syndrome. J. Med. Genet. 6: 322-327, 1969.
Wilson, M. G.; Towner, J. W.; Fujimoto, A.: Retinoblastoma and D-chromosome deletions. Am. J. Hum. Genet. 25: 57-61, 1973.
Windle, J. J.; Albert, D. M.; O'Brien, J. M.; Marcus, D. M.; Disteche, C. M.; Bernards, R.; Mellon, P. L.: Retinoblastoma in transgenic mice. Nature 343: 665-669, 1990.
Yandell, D. W.; Campbell, T. A.; Dayton, S. H.; Petersen, R.; Walton, D.; Little, J. B.; McConkie-Rosell, A.; Buckley, E.; Dryja, T.: Oncogenic point mutations in the human retinoblastoma gene: their application to genetic counseling. New Eng. J. Med. 321: 1689-1695, 1989.
Yokota, J.; Akiyama, T.; Fung, Y.-K. T.; Benedict, W. F.; Namba, Y.; Hanaoka, M.; Wada, M.; Terasaki, T.; Shimosato, Y.; Sugimura, T.; Terada, M.: Altered expression of the retinoblastoma (RB) gene in small-cell carcinoma of the lung. Oncogene 3: 471-475, 1988.
Zhu, X.; Dunn, J. M.; Phillips, R. A.; Goddard, A. D.; Paton, K. E.; Becker, A.; Gallie, B. L.: Preferential germline mutation of the paternal allele in retinoblastoma. Nature 340: 312-313, 1989.

RETINOBLASTOMA-LIKE 2 (*180203)

JJM NOTE

It is a leap to implicate this homologous *RB* gene in cancers of the breast, ovary, liver, and prostate because of the cytogenetic finding of 16q deletions in such tumors.

Mayol et al. (1993) cloned a retinoblastoma-related human gene, referred to as RB2, on the basis of sequence homology of the E1A-binding domain of the retinoblastoma gene (RB1; 180200). Structural homology with RB1 suggested a possible function of RB2 as a tumor suppressor gene. Yeung et al. (1993) mapped the gene to human chromosome 16q12.2 and rat chromosome 19, using fluorescence in situ hybridization and somatic hybrid cell analysis, respectively. Based on known syntenic relationships among human, rat and mouse, the data suggested that the mouse homolog resides on chromosome 8. Deletions of chromosome 16q have been found in several human neoplasms, including breast, ovarian, hepatic, and prostatic cancers, which supports the involvement of RB2 in human cancer as a tumor suppressor gene.

This locus is symbolized RBL2 because it was identified after the gene on chromosome 20, which is symbolized RBL1 (116957). RBL1 has a molecular weight of 107 kD; RBL2 has a molecular weight of about 120 kD.

Baldi et al. (1996) characterized the organization and 5-prime flanking region of the Rb2, or p130, gene. They determined that the gene contains 22 exons and spans over 50 kb.

Baldi, A.; Boccia, V.; Claudio, P. P.; De Luca, A.; Giordano, A.: Genomic structure of the human retinoblastoma-related Rb2/p130 gene. Proc. Nat. Acad. Sci. 93: 4629-4632, 1996.

Mayol, X.; Grana, X.; Baldi, A.; Sang, N.; Hu, Q.; Giordano, A.: Cloning of a new member of the retinoblastoma gene family (pRb2) which binds to the E1A transforming domain. Oncogene 8: 2561-2566, 1993.
Yeung, R. S.; Bell, D. W.; Testa, J. R.; Mayol, X.; Baldi, A.; Grana, X.; Klinga-Levan, K.; Knudson, A. G.; Giordano, A.: The retinoblastoma-related gene, RB2, maps to human chromosome 16q12 and rat chromosome 19. Oncogene 8: 3465-3468, 1993.

CHOROIDAL OSTEOMA, BILATERAL (118865)

JJM NOTE

Although this entity is histopathologically benign, the threat to vision is severe. The bilaterality and embryologic origins of this tumor are flags of its genetic origins.

Choroidal osteoma is a benign tumor found in the peripapillary region, most characteristically in one eye of otherwise healthy, young females. Histopathologic changes include bone formation with trabeculae, blood-filled cavernous spaces, and cells typical of bone formation, i.e., osteoblasts, osteocytes, and osteoclasts. In each of 2 families, choroidal osteoma has been observed in 2 successive generations. Noble (1990) described a family in which a girl and her monozygotic twin brothers had bilateral choroidal osteomas. The tumors showed growth between ages 11 and 13 years in the sister. The twin brothers' tumors remained stable between 9 and 11 years except for a new, isolated lesion in one eye of one of them. Their mother had a yellow mottling situated nasal to the disk in each eye resembling in appearance that in one eye in one of the twins. She showed no calcium on ultrasonography. Cuhna (1984) observed an affected mother and daughter: the 33-year-old mother had visual acuity reduced to the finger counting range in each eye since age 11 years and had bilateral peripapillary atrophic chorioretinal lesions. Her 5-year-old daughter had bilateral yellow-white lesions associated with bilateral macular hemorrhages. The familial occurrence as well as the bilaterality is consistent with the origin of the osteomas in a choristoma. A choristoma is a benign tumefaction of a chorista, which is an embryonic tissue rest composed of cellular and tissue elements not normally present at the affected site.

Cuhna, S. L.: Osseous choristoma of the choroid: a familial disease. Arch. Ophthal. 102: 1052-1054, 1984.
Noble, K. G.: Bilateral choroidal osteoma in three siblings. Am. J. Ophthal. 109: 656-660, 1990.

Meningiomas

MENINGIOMA (*156100)

JJM NOTE

In a little-known monograph, a definitive literature review of familial brain tumors summarizes the associations with meningioma (Tijssen CC et al.: *Familial Brain Tumors: Developments in Oncology*, vol. 9. The Hague: Nijhoff, 1982). Meningioma has been associated with exposure to ionizing radiation and to certain occupations and, in case reports, with Werner syndrome, Cowden multiple hamartoma syndrome, and Bannayan-Riley-Ruvalcaba syndrome. When familial aggregation is seen, however, neurofibromatosis 2 is the major known explanation, although a hereditary translocation involving chromosome 22 should be ruled out.

Some meningiomas result from mutation in the NF2 tumor suppressor gene (101000) and some appear to arise from mutations in the SIS oncogene (see 190040.0001); both of these loci are on chromosome 22. There is evidence from linkage studies (see later) that some familial multiple meningioma is determined by a locus distinct from either NF2 or SIS.

A few reports have described familial occurrence of meningioma. In addition to reports of multiple affected first-degree relatives (Joynt and Perret, 1961; Memon, 1980), affected monozygotic twins were reported by Sedzimir et al. (1973). Delleman et al. (1978) observed meningioma in 5 members of 2 generations, with male-to-male transmission. One of the 5 had acoustic neurinoma. A member of the second generation, not available for examination, was said to have multiple cafe-au-lait spots. The authors viewed the kindred as one of neurofibromatosis with wide variability. The suspicion that neurofibromatosis (162200) is frequently complicated by meningioma and, hence, that this could account for familial meningioma is not upheld by the review of Hope and Mulvihill (1981). The tumor tissue of sporadic meningiomas often shows absence of one chromosome 22 or, less frequently, absence of distal 22q (Sandberg, 1980; Zang, 1982). Maltby et al. (1988) studied the chromosomes from 50 meningiomas. Twenty-three of these tissue specimens had a normal diploid karyotype. Twenty-five had monosomy for chromosome 22, with or without a normal diploid line and/or additional aneuploidy. Cell lines monosomic for chromosome 22 also frequently had random chromosome loss. (Ewing sarcoma and the related peripheral neuroepithelioma (133450) likewise show abnormality involving chromosome 22.) Bolger et al. (1985) described a family in which a father and 3 of his 8 children had meningioma with clinical onset at ages ranging from 35 to 65 years. A fourth offspring died of multiple neoplasms arising at 29 years. No one in the family had signs of von Recklinghausen neurofibromatosis. The 3 sibs with meningioma carried a constitutional Robertsonian 14/22 translocation, t(14;22)(14qter-cen-22qter), in circulating leukocytes. Three of these sibs without meningioma had a normal karyotype. In the next generation, 4 carriers of the translocation were asymptomatic, except for one with bre ast cancer. Both living sibs with meningioma were found to have a variant of the SIS oncogene in peripheral leukocyte DNA. Family studies showed that the SIS variant segregated with the normal chromosome 22. Later studies by Smidt et al. (1990) showed that the affected members of this family had deletion of Alu sequences in the fifth intron of the SIS gene (see 190040.0001).

Arinami et al. (1986) described an institutionalized mentally retarded male with a constitutional ring chromosome 22 discovered during a cytogenetic survey. There appeared to be deletion of 22q13.3-qter. He died at age 27 and was found to have meningeal hypoplasia and dysplasia, including meningiomatosis and an unusual cyst containing meningocytic elements. Pressure of the cyst on the brain stem was thought to have caused death. Casalone et al. (1987) confirmed the high frequency of abnormalities involving chromosome 22 in studies of 31 cases of meningiomas. They suggested that bands 22q11-q12 are critical. Seizinger et al. (1987) contributed to the evidence that chromosome 22 is involved in meningioma. In 17 of 40 constitutionally heterozygous patients (43%), hemizygosity for the corresponding DNA marker was found in their tumors. Meese et al. (1987) did similar studies; 3 of 12 patients showed loss of heterozygosity (LOH) of markers on chromosome 22. Lekanne Deprez et al. (1995) found partial or complete loss of chromosome 22 in 66 of 93 meningiomas. Dumanski et al. (1987) studied LOH for chromosome 22 loci in tumor and normal tissue from 35 unrelated patients. Sixteen tumors retained the constitutional genotype along chromosome 22, while 14 tumors (40%) showed loss of 1 constitutional allele at all informative loci, consistent with monosomy 22 in the tumor DNA. The remaining 5 tumors (14%) showed LOH in the tumor DNA at 1 or more chromosome 22 loci and retained heterozygosity at other loci, consistent with other variable terminal deletions of 1 chromosome 22 in the tumor DNA. The results in these cases suggested that a meningioma locus is located distal to the myoglobin locus, within 22q12.3-qter. Multiple loci on other chromosomes also were studied, and 12 of the 19 tumors with losses of chromosome 22 alleles showed additional losses of heterozygosity at loci on 1 to 3 other chromosomes. All tumors that retained the constitutional genotype on chromosome 22 also re tained heterozygosity on all informative loci on other chromosomes analyzed, suggesting that the rearrangement of chromosome 22 was the primary event in the tumorigenesis of meningioma. Using DNA markers, Okazaki et al. (1988) found loss of constitutional heterozygosity in 3 of 11 informative cases of meningioma. In a larger series that incorporated the earler one, Dumanski et al. (1990) compared constitutional and tumor tissue genotypes from 81 unrelated patients with meningioma, studying 25 polymorphic loci on chromosome 22. In 30 tumors (37%) constitutional genotype was retained, consistent with no detectable aberrations on chromosome 22. In 42 tumors (52%) there was loss of 1 allele at all informative loci, consistent with monosomy 22 in the tumor DNA. The remaining 9 tumors (11%) retained constitutional heterozygosity in the tumor DNA at 1 or more centromeric loci and showed a loss of heterozygosity at other telomeric loci, consistent with variable terminal deletions of 1 chromosome 22q in the tumor DNA. The localization of breakpoints in these 9 cases with deletions suggested that a meningioma locus is located distal to the myoglobin locus, within 22q12.3-qter. They also concluded that the meningioma locus is distinct from the NF2 locus (acoustic neurinoma; 101000) inasmuch as they appear to be on opposite sides of the DNA marker D22S28.

Fontaine et al. (1990) found that LOH in sporadic meningiomas involved the maternally derived chromosome 22 in 5 tumors and the paternally derived chromosome 22 in 4. Thus, in contrast to embryonal tumors (retinoblastoma and Wilms tumor), imprinting is apparently not a factor. Sanson et al. (1990) also found both maternal and paternal origin of the lost chromosome 22 in meningioma. Chung and Seizinger (1992) stated that although similar deletions of chromosome 22 have been found in acoustic neuroma and meningioma, it remained unclear whether distinct genetic loci were involved. 26,27:Pulst et al. (1992, 1993) presented linkage data suggesting that a meningioma locus is separate from the NF2 locus. They described multiple meningiomas and ependymomas in 2 generations of a family. Multipoint linkage analysis resulted in location scores less than

See also Gaist and Piazza (1959); Joynt and Perret (1965); Sahar (1965)

Arinami, T.; Kondo, I.; Hamaguchi, H.; Nakajima, S.: Multifocal meningiomas in a patient with a constitutional ring chromosome 22. J. Med. Genet. 23: 178-179, 1986.
Bolger, G. B.; Stamberg, J.; Kirsch, I. R.; Hollis, G. F.; Schwarz, D. F.; Thomas, G. H.: Chromosome translocation t(14;22) and oncogene (c-sis) variant in a pedigree with familial meningioma. New Eng. J. Med. 312: 564-567, 1985.
Buijs, A.; Sherr, S.; van Baal, S.; van Bezouw, S.; van der Plas, D.; Geurts van Kessel, A.; Riegman, P.; Lekanne Deprez, R.; Zwarthoff, E.; Hagemeijer, A.; Grosveld, G.: Translocation (12;22)(p13;q11) in myeloproliferative disorders results in fusion of the ETS-like TEL gene on 12q13 to the MN1 gene on 22q11. Oncogene 10: 1511-1519, 1995.
Casalone, R.; Granata, P.; Simi, P.; Tarantino, E.; Butti, G.; Buonaguidi, R.; Faggionato, F.; Knerich, R.; Solero, L.: Recessive cancer genes in meningiomas? An analysis of 31 cases. Cancer Genet. Cytogenet. 27: 145-159, 1987.
Chung, R. Y.; Seizinger, B. R.: Molecular genetics of neurological tumours. J. Med. Genet. 29: 361-367, 1992.
Delleman, J. W.; De Jong, J. G. Y.; Bleeker, G. M.: Meningiomas in five members of a family over two generations, in one member simultaneously with acoustic neurinomas. Neurology 28: 567-570, 1978.
Dumanski, J. P.; Carlbom, E.; Collins, V. P.; Nordenskjold, M.: Deletion mapping of a locus on human chromosome 22 involved in the oncogenesis of meningioma. Proc. Nat. Acad. Sci. 84: 9275-9279, 1987.
Dumanski, J. P.; Rouleau, G. A.; Nordenskjold, M.; Collins, V. P.: Molecular genetic analysis of chromosome 22 in 81 cases of meningioma. Cancer Res. 50: 5863-5867, 1990.
Fontaine, B.; Rouleau, G. A.; Seizinger, B.; Jewell, A. F.; Hanson, M. P.; Martuza, R. L.; Gusella, J. F.: Equal parental origin of chromosome 22 losses in human sporadic meningioma: no evidence for genomic imprinting. Am. J. Hum. Genet. 47: 823-827, 1990.
Gaist, G.; Piazza, G.: Meningiomas in two members of the same family (with no evidence of neurofibromatosis). J. Neurosurg. 16: 110-113, 1959.
Hope, D. G.; Mulvihill, J. J.: Malignancy in neurofibromatosis.In: Riccardi, V. M.; Mulvihill, J. J.: Neurofibromatosis (von Recklinghausen Disease). New York: Raven Press (pub.) 1981. Note: Alternate: Adv. Neurol. 29: 33-56, 1981.
Joynt, R. J.; Perret, G. E.: Familial meningiomas. J. Neurol. Neurosurg. Psychiat. 28: 163-164, 1965.
Joynt, R. J.; Perret, G. E.: Meningiomas in a mother and daughter: cases without evidence of neurofibromatosis. Neurology 11: 164-165, 1961.
Kinzler, K. W.; Vogelstein, B.: A gene for neurofibromatosis 2. Nature 363: 495-496, 1993.
Lekanne Deprez, R. H.; Groen, N. A.; van Biezen, N. A.; Hagemeijer, A.; van Drunen, E.; Koper, J. W.; Avezaat, C. J. J.; Bootsma, D.; Zwarthoff, E. C.: A t(4;22) in a meningioma points to the localization of a putative tumor-suppressor gene. Am. J. Hum. Genet. 48: 783-790, 1991.
Lekanne Deprez, R. H.; Riegman, P. H.; van Drunen, E.; Warringa, U. L.; Groen, N. A.; Stefanko, S. Z.; Koper, J. W.; Avezaat, C. J. J.; Mulder, P. G. H.; Zwarthoff, E. C.; Hagemeijer, A.: Cytogenetic, molecular genetic and pathological analyses in 126 meningiomas. J. Neuropath. Exp. Neurol. 54: 224-235, 1995.
Lekanne Deprez, R. H.; Riegman, P. H. J.; Groen, N. A.; Warringa, U. L.; van Biezen, N. A.; Molijn, A. C.; Bootsma, D.; de Jong, P. J.; Menon, A. G.; Kley, N. A.; Seizinger, B. R.; Zwarthoff, E. C.: Cloning and characterization of MN1, a gene from chromosome 22q11, which is disrupted by a balanced translocation in a meningioma. Oncogene 10: 1521-1528, 1995.
Lele, K. P.; Penrose, L. S.; Stallard, H. B.: Chromosome deletion in a case of retinoblastoma. Ann. Hum. Genet. 27: 171-174, 1963.
Maltby, E. L.; Ironside, J. W.; Battersby, R. D. E.: Cytogenetic studies in 50 meningiomas. Cancer Genet. Cytogenet. 31: 199-210, 1988.
Maxwell, M.; Shih, S. D.; Galanopoulos, T.; Hedley-Whyte, E. T.; Cosgrove, G. R.: Familial meningioma: analysis of expression of neurofibromatosis 2 protein merlin: report of two cases. J. Neurosurg. 88: 562-569, 1998.
Meese, E.; Blin, N.; Zang, K. D.: Loss of heterozygosity and the origin of meningioma. Hum. Genet. 77: 349-351, 1987.
Memon, M. Y.: Multiple and familial meningiomas without evidence of neurofibromatosis. Neurosurgery 7: 262-264, 1980.
Okazaki, M.; Nishisho, I.; Tateishi, H.; Motomura, K.; Yamamoto, M.; Miki, T.; Hayakawa, T.; Takai, S.; Honjo, T.; Mori, T.: Loss of genes on the long arm of chromosome 22 in human meningiomas. Molec. Biol. Med. 5: 15-22, 1988.
Petrella, R.; Levine, S.; Wilmot, P. L.; Ashar, K. D.; Casamassima, A. C.; Shapiro, L. R.: Multiple meningiomas in a patient with constitutional ring chromosome 22. Am. J. Med. Genet. 47: 184-186, 1993.
Peyrard, M.; Fransson, I.; Xie, Y.-G.; Han, F.-Y.; Ruttledge, M. H.; Swahn, S.; Collins, J. E.; Dunham, I.; Collins, V. P.; Dumanski, J. P.: Characterization of a new member of the human beta-adaptin gene family from chromosome 22q12, a candidate meningioma gene. Hum. Molec. Genet. 3: 1393-1399, 1994.
Pulst, S.-M.; Rouleau, G.; Fain, P.; Sieb, J. P.: Familial meningioma is not allelic to neurofibromatosis 2. (Abstract) Am. J. Hum. Genet. 51 (suppl.): A106 only, 1992.
Pulst, S.-M.; Rouleau, G. A.; Marineau, C.; Fain, P.; Sieb, J. P.: Familial meningioma is not allelic to neurofibromatosis 2. Neurology 43: 2096-2098, 1993.
Ruttledge, M. H.; Sarrazin, J.; Rangaratnam, S.; Phelan, C. M.; Twist, E.; Merel, P.; Delattre, O.; Thomas, G.; Nordenskjold, M.; Collins, V. P.; Dumanski, J. P.; Rouleau, G. A.: Evidence for the complete inactivation of the NF2 gene in the majority of sporadic meningiomas. Nature Genet. 6: 180-184, 1994.
Sahar, A.: Familial occurrence of meningiomas: case report. J. Neurosurg. 23: 444-445, 1965.
Sandberg, A. A.: The Chromosomes in Human Cancer and Leukemia. New York: Elsevier (pub.) 1980. Pp. 535-543.
Sanson, M.; Delattre, O.; Couturier, J.; Philippon, J.; Cophignon, J.; Derome, P.; Rouleau, G. A.; Thomas, G.: Parental origin of chromosome 22 alleles lost in meningioma. (Letter) Am. J. Hum. Genet. 47: 877-880, 1990.
Sanson, M.; Marineau, C.; Desmaze, C.; Lutchman, M.; Ruttledge, M.; Baron, C.; Narod, S.; Delattre, O.; Lenoir, G.; Thomas, G.; Aurias, A.; Rouleau, G. A.: Germline deletion in a neurofibromatosis type 2 kindred inactivates the NF2 gene and a candidate meningioma locus. Hum. Molec. Genet. 2: 1215-1220, 1993.
Sedzimir, C. B.; Frazer, A. K.; Roberts, J. R.: Cranial and spinal meningiomas in a pair of identical twin boys. J. Neurol. Neurosurg. Psychiat. 36: 368-376, 1973.
Seizinger, B. R.; de la Monte, S.; Atkins, L.; Gusella, J. F.; Martuza, R. L.: Molecular genetic approach to human meningioma: loss of genes on chromosome 22. Proc. Nat. Acad. Sci. 84: 5419-5423, 1987.
Smidt, M.; Kirsch, I.; Ratner, L.: Deletion of Alu sequences in the fifth c-sis intron in individuals with meningiomas. J. Clin. Invest. 86: 1151-1157, 1990.
Stallard, H. B.: The conservative treatment of retinoblastoma. Trans. Ophthal. Soc. 82: 473-534, 1962.
Wellenreuther, R.; Kraus, J. A.; Lenartz, D.; Menon, A. G.; Schramm, J.; Louis, D. N.; Ramesh, V.; Gusella, J. F.; Wiestler, O. D.; von Deimling, A.: Analysis of the neurofibromatosis 2 gene reveals molecular variants of meningioma. Am. J. Path. 146: 827-832, 1995.
Zang, K. D.: Cytological and cytogenetical studies on human meningioma. Cancer Genet. Cytogenet. 6: 249-274, 1982.
Zang, K. D.; Singer, H.: Chromosomal constitution of meningiomas. Nature 216: 84-85, 1967.

TRANSCOBALAMIN DEFICIENCY 2 (*275350)

JJM NOTE

This entity is selected only because meningioma was seen in heterozygotes (used for gene mapping), a possibly coincidental association (Arwert et al. 1986).

Transcobalamin II (TC II), a plasma globulin, is the primary transport protein for vitamin B12. Genetic absence leads to severe megaloblastic anemia in early infancy. Hakami et al. (1971) described macrocytic anemia and other manifestations of vitamin B12 deficiency in 2 infant sibs who had normal levels of serum B12. Deficiency in TC II was demonstrated. A partial deficiency in both parents and other hematologically normal relatives indicated autosomal recessive inheritance. Decreased intestinal absorption of B12, uncorrected by intrinsic factor, suggested that transcobalamin II is involved in B12 absorption. B12-responsive megaloblastic anemias of the pediatric age group include a form due to lack of intrinsic factor (261000) and a form due to a defect in intestinal absorption, with associated proteinuria (261100). Scott et al. (1972) concluded that no defect in homocysteine methyltransferase or methylmalonyl CoA mutase occurs in patients with TC II deficiency and that TC II is normally necessary mainly for delivery of the cobalamin molecule to the hematopoietic system. Immunodeficiency occurs in some families with deficiency of TC II. The patient with TC deficiency reported by Hitzig et al. (1974) came from a Moroccan family in which 2 sons had died of severe infections in early infancy. The diagnosis of Kostmann infantile genetic agranulocytosis had been made. The proband was delivered by cesarean section and maintained in a sterile unit for 115 days, in anticipation of bone marrow transplantation at a later stage. Agammaglobulinemia and lack of antibody response to strong antigenic stimuli were found. Cellular immune reactions were normal. After removal from isolation, he developed severe diarrhea and an upper respiratory infection. Severe atrophy of the intestinal mucosa and deficiency of disaccharidases were found. At 7 months of age he developed macromegaloblastic anemia with low reticulocyte counts, leukopenia with granulocytopenia, and thrombocytopenia with severe h emorrhagic diathesis. While in the sterile environment he had received vitamin B12 and folic acid; vitamin B12 quickly restored him to normal health. The blood, the intestine and the immune system returned to normal. With B12 therapy, an anomalous B12-binding protein, possibly 'transcobalamin III,' or fetal vitamin B12-binding globulin, appeared in the patient's serum.

Arrabal et al. (1988) reported normal growth and mental development of a Spanish brother and sister with TC II deficiency after 10 years of vitamin B12 therapy. Two sibs had died at 1.5 and 2 months of age and were apparently affected. Seligman et al. (1980) described a patient who represented a compound heterozygote for absent TC II and defective TC II that failed to bind cobalamin. The father was heterozygous for the absence of TC II and the mother was heterozygous for the defective form. The proband had 2 children: a heterozygote for absence of TC II and a heterozygote for abnormal TC II. The patient had megaloblastic anemia. Tested only by immunoreactive TC II, abnormality would not be recognized. Continued treatment with folate, to which the anemia responded, would lead to serious neurologic abnormality. Hypogammaglobulinemia (Hitzig and Kenny, 1975) and disturbed phagocytic function of leukocytes (Seger et al., 1980) are corrected by appropriate therapy. Thomas et al. (1982) described a child of Maltese ancestry who presented in early infancy with megaloblastic anemia and was treated with folinic acid from 6 weeks of age when the diagnosis of dihydrofolate reductase deficiency was made (Case 2, Tauro et al., 1976). TC II deficiency was not recognized until age 2 years by which time he had severe mental retardation, ataxia, and a pyramidal deficit in the limbs. Following treatment with intramuscular hydroxycobalamin, his condition slowly improved, but at age 7 years he was left with severe neurologic deficit. Hoffbrand et al. (1984) reported that this patient had no serum transcobalamin II binding capacity, whereas his parents and sisters had values about 50% of normal. Immunoreactive TC II was present in the patient's serum but at about 39% of normal. The patient reported by Kaikov et al. (1991) presented at 6 weeks of age with failure to thrive, diarrhea, macrocytic anemia, and decreased IgG. Both parents had intermediate levels of transcobalamin II consistent with the heterozygous state.

Hall (1981) gave a clinically oriented review of congenital defects of vitamin B12 transport, and Frater-Schroder (1983) gave a genetically oriented review. Because of the evidence presented by Porck et al. (1983) that the TC II of cord blood is of fetal origin, cord blood can be used in the neonatal diagnosis of deficiency. Mayes et al. (1987) studied the production of transcobalamin II by amniotic cells in a pregnancy at risk for deficiency. They predicted heterozygosity which was confirmed in the live-born child. Sacher et al. (1983) described a boy, born healthy, who developed diarrhea, vomiting and ulcerative stomatitis together with megaloblastic anemia and thrombocytopenia and neutropenia at the age of 5 weeks. Serum apo-TC2 was not detectable. They could show that cultured skin fibroblasts failed to secrete functioning TC II. They demonstrated heterozygosity for a silent TCN2 allele in both parents and the maternal grandfather. All 3 were asymptomatic. The level of serum cobalamin is usually normal in cases of TC II deficiency because the bulk of serum cobalamin normally circulates attached to R binder, whose function is unknown, rather than to TC II. However, Meyers and Carmel (1984) observed a case of TC II deficiency with subnormal serum cobalamin. Barshop et al. (1990) reported the case of a black infant with very severe TC II deficiency. The infant had not only megaloblastic anemia but also thrombocytopenia and neutropenia, i.e., pancytopenia. Methylmalonic acid and homocystine were found in the urine. Vance et al. (1993) reported the cytogenetic findings of bone marrow aspirates and identified fragile site expression in the hematopoietic cells of a patient.

Barshop et al. (1990) presented evidence that TC II as well as intrinsic factor is required for transport of cobalamin from the intestine to the blood. TC II is immunologically, biochemically and functionally distinct from the R binder protein (193090). The protein deficient in the cases referred to earlier is the same as that found to be polymorphic electrophoretically by Daiger et al. (1975), who in a study of about 100 Caucasians and in subsequent family studies found evidence for 4 alleles. The first demonstration of polymorphism in vitamin B12 binding (Daiger et al., 1975) did not identify which of the 2 protein classes was responsible for the genetic variation. Frater-Schroder and Hitzig (1977) showed that TC II is the site of the variation. Genetic heterogeneity of the TC II protein was demonstrated by Frater-Schroder et al. (1979), who applied their analytic methods to the improved detection of carriers (of silent alleles, for example). At least 5 alleles were identified.

Cavalli-Sforza et al. (1979) found a suggestion of linkage of the TCN2 locus and adenylate kinase (AK1; 103000); lod score = 1.78 at theta = 0.139. Linkage studies by Ott and Frater-Schroder (1981) excluded TCN2 versus ABO linkage, however. Yang et al. (1981) could find no evidence of linkage with several loci, including ABO, AK1, ADA, GLO1, PI, and HLA. Arwert et al. (1983) reported that transcobalamin II (symbolized Tcn-2 in the mouse) is linked to the alpha-globin locus and the esterase-3 (Es-3) locus on mouse chromosome 11. In the mouse, classic linkage studies could not confirm linkage to Es-3. The recombination frequency between Hba and Tcn-2 was 10.2% in one series of experiments and in other experiments with 5 different inbred strains the mean was 14% (range, 7% to 24%) (Acklin et al., 1984). In mice, Frater-Schroder et al. (1985) used somatic cell hybrids, recombinant inbred (RI) mouse strains, and backcross breeding experiments to map the Tcn-2 locus to chromosome 11, linked to the alpha-globin locus (recombination frequency = 19.2%). Arwert et al. (1984) presented evidence from studies of human-rodent somatic cell hybrids that appeared to exclude chromosome 16 as the site of TCN2 in man. Eiberg et al. (1986) showed that TCN2 and blood group P are linked; the maximum lod score was 7.91 at theta = 0.14 for males and theta = 0.20 for females. Arwert et al. (1986) demonstrated that TC II is synthesized by cells in culture and secreted into the medium. Thus, they were able to map the TCN2 gene to chromosome 22 by somatic cell hybridization. They also showed that meningioma cells obtained from patients heterozygous for TCN2 showed a concomitant loss of one chromosome 22 and one of the two TCN2 alleles, which strongly supported the assignment to chromosome 22.

The human TCN2 gene, referred to also as TC2, was cloned by Platica et al. (1991) and Li et al. (1993) to characterize the molecular basis of transcobalamin II deficiency. Li et al. (1994) studied skin fibroblasts since TC2 synthesis had been demonstrated in cultured skin fibroblasts from normal patients. They found that the affected child was a compound heterozygote with a gross deletion of the maternally derived allele and a 4-bp deletion in the coding region of the paternally derived allele. Both of these deletions caused markedly reduced levels of TC2 mRNA and protein. Li et al. (1995) isolated the TC2 gene and showed that it is composed of 9 exons and 8 introns spanning about 20 kb. The organization of the gene with respect to the number, size, and position of the exons is similar to that of the other cobalamin-binding proteins human gastric intrinsic factor (GIF; 261000) and TC1 (189905). Unlike IF and TC1, analysis of the promoter indicated that TC2 is a housekeeping gene. Li et al. (1995) mapped the genomic clone for TC2 to 22q12-q13 by fluorescence in situ hybridization.

Regec et al. (1995) also cloned TCN2 and showed that the TCN2 gene spans a minimum of 18 kb and contains 9 exons and 8 introns, with a polyadenylation signal sequence located 509 bp downstream from the termination codon and a transcription initiation site beginning 158-bp upstream from the ATG translation start site. The 5-prime flanking DNA does not have a TATA or CCAAT regulatory element, but a 34-nucleotide stretch beginning just upstream of the CAP site contains 4 tandemly organized 5-prime-CCCC-3-prime tetramers. This sequence is a motif for a trans-active transcription factor (ETF) that regulates expression of the epidermal growth factor receptor gene (EGFR; 131550), which also lacks TATA and CCAAT regulatory elements (Kageyama et al., 1989). Regec et al. (1995) stated that a number of the exon/intron splice junctions of human TCN2, TCN1 (189905), and intrinsic factor (GIF; 261000) genes are located in homologous regions, providing evidence that they evolved by duplication of an ancestral gene.

Data on gene frequencies of allelic variants were tabulated by Roychoudhury and Nei (1988).

See also Burman et al. (1979); Chanarin et al. (1978); Daiger et al. (1978); Frater-Schroder et al. (1982); Gallmann et al. (1986); Masina et al. (1979); Niebrugge et al. (1982); Porck et al. (1984); Quadros et al. (1986); Rosenblatt et al. (1987); Zeitlin et al. (1985)

Acklin, M.; Frater-Schroder, M.; Haller, O.; Lundin, L. G.; Prochazka, M.; Skow, L. C.: Localization of transcobalamin II (Tcn-2) on chromosome 11: linkage to waved-2 (wa-2) and the hemoglobin alpha-chain locus (Hba). Mouse Newsletter , 1984.

Arrabal, M. C.; Villegas, A.; Mariscal, E.; Lentijo, B.; Luck, W.; Hoffbrand, A. V.: Transcobalamin II deficiency: long-term follow-up of two cases. Acta Haemat. 80: 162-166, 1988.

Arwert, F.; Bruderer, S.; Frater-Schroder, M.; Haller, O.; Hilgers, J.; Hilkens, J.; Porck, H. J.; Skow, L.: Tentative assignment of transcobalamin II (Tcn-2) to chromosome 11. Mouse Newsletter 69: July only, 1983.

Arwert, F.; Geurts van Kessel, A.; Porck, H.; Frater-Schroder, M.; Westerveld, A.; Meera Khan, P.: Human transcobalamin II (TC2) in man-rodent somatic cell hybrids. (Abstract) Cytogenet. Cell Genet. 37: 404 only, 1984.

Arwert, F.; Porck, H. J.; Frater-Schroder, M.; Brahe, C.; Geurts van Kessel, A.; Westerveld, A.; Meera Khan, P.; Zang, K.; Frants, R. R.; Kortbeek, H. T.; Eriksson, A. W.: Assignment of human transcobalamin II (TC2) to chromosome 22 using somatic cell hybrids and monosomic meningioma cells. Hum. Genet. 74: 378-381, 1986.

Barshop, B. A.; Wolff, J.; Nyhan, A. L.; Yu, A.; Prodanos, C.; Jones, G.; Sweetman, L.; Leslie, J.; Holm, J.; Green, R.; Jacobsen, D. W.; Cooper, B. A.; Rosenblatt, D.: Transcobalamin II deficiency presenting with methylmalonic aciduria and homocystinuria and abnormal absorption of cobalamin. Am. J. Med. Genet. 35: 222-228, 1990.

Burman, J. F.; Mollin, D. L.; Sourial, N. A.; Sladden, R. A.: Inherited lack of transcobalamin II in serum and megaloblastic anaemia: a further patient. Brit. J. Haemat. 43: 27-38, 1979.

Cavalli-Sforza, L. L.; King, M. C.; Go, R. C. P.; Namboodiri, K. K.; Lynch, H. T.; Wong, L.; Kaplan, E. B.; Elston, R. C.: Possible linkage between transcobalamin II (TC II) and adenylate kinase (AK). (Abstract) Cytogenet. Cell Genet. 25: 140-141, 1979.

Chanarin, I.; Muir, M.; Hughes, A.; Hoffbrand, A. V.: Evidence for intestinal origin of transcobalamin II during vitamin B12 absorption. Brit. Med. J. 1: 1453-1455, 1978.

Daiger, S. P.; Labowe, M. L.; Cavalli-Sforza, L. L.: Polymorphic electrophoretic variants of vitamin B12 binding proteins in human plasma. (Abstract) Am. J. Hum. Genet. 27: 31A only, 1975.

Daiger, S. P.; Labowe, M. L.; Parsons, M.; Wang, L.; Cavalli-Sforza, L. L.: Detection of genetic variation with radioactive ligands. III. Genetic polymorphism of transcobalamin II in human plasma. Am. J. Hum. Genet. 30: 202-214, 1978.

Eiberg, H.; Moller, N.; Mohr, J.; Nielsen, L. S.: Linkage of transcobalamin II (TC2) to the P blood group system and assignment to chromosome 22. Clin. Genet. 29: 354-359, 1986.
Frater-Schroder, M.: Genetic patterns of transcobalamin II and the relationships with congenital defects. Molec. Cell. Biochem. 56: 5-31, 1983.
Frater-Schroder, M.; Hitzig, W. H.: The transcobalamin II isoprotein pattern. (Abstract) Experientia 33: 791 only, 1977.
Frater-Schroder, M.; Hitzig, W. H.; Butler, R.: Studies on transcobalamin: detection of TC II isoproteins in human serum. Blood 53: 193-203, 1979.
Frater-Schroder, M.; Porck, H. J.; Eriksson, A. W.; Daiger, S. P.; Cavalli-Sforza, L. L.: Standardization of nomenclature for transcobalamin II variants. Hum. Genet. 61: 165-166, 1982.
Frater-Schroder, M.; Prochazka, M.; Haller, O.; Arwert, F.; Porck, H. J.; Skow, L. C.; Lundin, L.-G.; Hilkens, J.; Hilgers, J.: Localization of the gene for the vitamin B12 binding protein, transcobalamin II, near the centromere on mouse chromosome 11, linked with the hemoglobin alpha-chain locus. Biochem. Genet. 23: 139-153, 1985.
Gallmann, M.; Frater-Schroder, M.; Scheffrahn, W.; Ott, J.; Schmid, B.; Butler, E.; Biedermann, V.; Kierat, L.: Indication against genetic localisation of the human transcobalamin II gene (TC2) on chromosome 16. Clin. Genet. 29: 349-353, 1986.
Hakami, N.; Neiman, P. E.; Canellos, G. P.; Lazerson, J.: Neonatal megaloblastic anemia due to inherited transcobalamin II deficiency in two siblings. New Eng. J. Med. 285: 1163-1170, 1971.
Hall, C. A.: Congenital disorders of vitamin B12 transport and their contributions to concepts. II. Yale J. Biol. Med. 54: 485-495, 1981.
Hitzig, W. H.; Dohmann, U.; Pluss, H. J.; Vischer, D.: Hereditary transcobalamin-II deficiency—clinical findings in a new family. J. Pediat. 85: 622-628, 1974.
Hitzig, W. H.; Kenny, A. B.: The role of vitamin B12 and its transport globulins in the production of antibodies. Clin. Exp. Immun. 20: 105-111, 1975.
Hoffbrand, A. V.; Tripp, E.; Jackson, B. F. A.; Luck, W. E.; Frater-Schroder, M.: Hereditary abnormal transcobalamin II previously diagnosed as congenital dihydrofolate reductase deficiency. (Letter) New Eng. J. Med. 310: 789-790, 1984.
Kageyama, R.; Merlino, G. T.; Pastan, I.: Nuclear factor ETF specifically stimulates transcription from promoters without a TATA box. J. Biol. Chem. 264: 15508-15514, 1989.
Kaikov, Y.; Wadsworth, L. D.; Hall, C. A.; Rogers, P. C. J.: Transcobalamin II deficiency: case report and review of the literature. Europ. J. Pediat. 150: 841-843, 1991.
Li, N.; Rosenblatt, D. S.; Kamen, B. A.; Seetharam, S.; Seetharam, B.: Identification of two mutant alleles of transcobalamin II in an affected family. Hum. Molec. Genet. 3: 1835-1840, 1994.
Li, N.; Seetharam, S.; Lindemans, J.; Alpers, D. H.; Arwert, F.; Seetharam, B.: Isolation and sequence analysis of variant forms of human transcobalamin II. Biochim. Biophys. Acta 1172: 21-30, 1993.
Li, N.; Seetharam, S.; Seetharam, B.: Genomic structure of human transcobalamin II: comparison to human intrinsic factor and transcobalamin I. Biochem. Biophys. Res. Commun. 208: 756-764, 1995.
Masina, P.; Ramunno, L.; Iannelli, D.: Evidence for 15 genetically determined electrophoretic variants of transcobalamin II in rabbit serum. Biochem. Genet. 17: 757-767, 1979.
Mayes, J. S.; Say, B.; Marcus, D. L.: Prenatal studies in a family with transcobalamin II deficiency. (Letter) Am. J. Hum. Genet. 41: 686-687, 1987.
Meyers, P. A.; Carmel, R.: Hereditary transcobalamin II deficiency with subnormal serum cobalamin levels. Pediatrics 74: 866-871, 1984.
Niebrugge, D. J.; Benjamin, D. R.; Christie, D.; Scott, C. R.: Hereditary transcobalamin II deficiency presenting as red cell hypoplasia. J. Pediat. 101: 732-735, 1982.
Ott, J.; Frater-Schroder, M.: Absence of linkage between transcobalamin II and ABO. Hum. Genet. 59: 164-165, 1981.
Platica, O.; Janeczko, R.; Quadros, E. V.; Regec, A.; Romain, R.; Rothenberg, S. P.: The cDNA sequence and the deduced amino acid sequence of human transcobalamin II show homology with rat intrinsic factor and human transcobalamin I. J. Biol. Chem. 266: 7860-7863, 1991.
Porck, H. J.; Fleming, A. F.; Frants, R. R.: Distribution of genetic variants of transcobalamin II in Nigerian black populations. Am. J. Hum. Genet. 36: 710-717, 1984.
Porck, H. J.; Frater-Schroder, M.; Frants, R. R.; Kierat, L.; Eriksson, A. W.: Genetic evidence for fetal origin of transcobalamin II in human cord blood. Blood 62: 234-237, 1983.
Quadros, E. V.; Rothenberg, S. P.; Pan, Y.-C. E.; Stein, S.: Purification and molecular characterization of human transcobalamin II. J. Biol. Chem. 261: 15455-15460, 1986.
Regec, A.; Quadros, E. V.; Platica, O.; Rothenberg, S. P.: The cloning and characterization of the human transcobalamin II gene. Blood 85: 2711-2719, 1995.
Rosenblatt, D. S.; Hosack, A.; Matiaszuk, N.: Expression of transcobalamin II by amniocytes. Prenatal Diag. 7: 35-39, 1987.
Roychoudhury, A. K.; Nei, M.: Human Polymorphic Genes: World Distribution. New York: Oxford Univ. Press (pub.) 1988.
Sacher, M.; Paky, F.; Frater-Schroder, M.: Vererbter Transcobalamin-II-Mangel: klinische, genetische Untersuchungen und Diagnose in der Fibroblastenkultur. Helv. Paediat. Acta 38: 549-558, 1983.
Scott, C. R.; Hakami, N.; Teng, C. C.; Sagerson, R. N.: Hereditary transcobalamin II deficiency: the role of transcobalamin II in vitamin B12 dependent reactions in man. J. Pediat. 81: 1106-1111, 1972.
Seger, R.; Wildfeuer, A.; Frater-Schroder, M.; Linnel, J.; Hitzig, W. H.: Granulocyte dysfunction in transcobalamin II deficiency responding to leucovorin or hydroxocobalamin-plasma transfusion. J. Inherit. Metab. Dis. 3: 3-9, 1980.
Seligman, P. A.; Steiner, L. L.; Allen, R. H.: Studies of a patient with megaloblastic anemia and an abnormal transcobalamin II. New Eng. J. Med. 303: 1209-1212, 1980.
Tauro, G. P.; Danks, D. M.; Rowe, P. B.; Van der Weyden, M. B.; Schwarz, M. A.; Collins, V. L.; Neal, B. W.: Dihydrofolate reductase deficiency causing megaloblastic anemia in two families. New Eng. J. Med. 294: 466-470, 1976.
Thomas, P. K.; Hoffbrand, A. V.; Smith, I. S.: Neurological involvement in hereditary transcobalamin II deficiency. J. Neurol. Neurosurg. Psychiat. 45: 74-77, 1982.
Vance, G. H.; Moncino, M.; Heerema, N. A.: Cytogenetic findings of a child with transcobalamin II deficiency. Am. J. Med. Genet. 46: 615-619, 1993.
Yang, S. Y.; Coleman, P.; Ochs, H. D.; Dupont, B.: Inheritance and genetic linkage of transcobalamin II. Hum. Genet. 57: 307-311, 1981.
Zeitlin, H. C.; Sheppard, K.; Baum, J. D.; Bolton, F. G.; Hall, C. A.: Homozygous transcobalamin II deficiency maintained on oral hydroxocobalamin. Blood 66: 1022-1027, 1985.

Gliomas

MELANOMA-ASTROCYTOMA SYNDROME (155755)

JJM NOTE

In Connecticut and Denmark, survivors of brain tumors (not necessarily astrocytomas) have an excess of melanoma (NCI Monograph 68, 1985), so the familial association is not unexpected. The few brain tumors in families with the hereditary dysplastic nevus syndrome may not exceed expectations.

Kaufman et al. (1993) described a family in which cutaneous malignant melanoma (CMM) or cerebral astrocytoma, or both, developed in 8 members over 3 generations. In 2 patients with both malignant melanoma and astrocytoma, the brain tumor followed the diagnosis of melanoma by a period of 2 and 10 years, respectively, and was the primary cause of morbidity and mortality. Kaufman et al. (1993) suggested that this may be a newly described genetic disorder. There were instances of male-to-male transmission. Liu et al. (1995) analyzed the family reported by Kaufman et al. (1993) for germline mutations in the p15 gene (CDKN2B; 600431), the p16 gene (CDKN2A; 600160), including exon 1-beta, and the CDK4 gene (123829); no mutations were found.

Azizi et al. (1995) specifically examined the question whether tumors of the nervous system have an increased likelihood in families of patients with cutaneous melanoma. They examined the occurrence of nervous system tumors as second cancers in a series of 904 patients with cutaneous melanoma and/or their family members. In 15 families with 17 members with cutaneous melanoma, one or more additional relatives had tumors of the nervous system, including astrocytoma, medulloblastoma, glioblastoma multiforme, ependymoma, glioma, meningioma, and acoustic neurilemmoma. Another subgroup of 10 patients with cutaneous melanoma had either a meningioma (n = 9) or acoustic neurilemmoma (n = 1) as a second tumor. In 20 of 25 of the cutaneous melanoma patients in this series and in additional first-degree relatives of 9 of 11 of the affected families, atypical melanocytic nevi had occurred as are found in the familial atypical multiple mole-melanoma syndrome (155600).

Bahuau et al. (1997) also described a family with a cancer syndrome of CMM and nervous system tumors. Pathologic findings of CMM, dysplastic nevi, astrocytoma, neurofibroma, schwannoma, and meningioma indicate a strong overlap with the NF1 (162200) tumor spectrum. However, mutation screening of the NF1 gene and segregation analyses excluded allelism of this familial syndrome with NF1. The families reported by Bahuau et al. (1997) and Kaufman et al. (1993) were the basis of a study by Bahuau et al. (1998) which showed that both families were linked to deletions of 9p. In the family of Bahuau et al. (1997), a large germline deletion ablated the entire p16, p19, and p15 gene cluster, whereas a more circumscribed molecular lesion disrupting p16 and p19 but leaving p15 unaltered segregated with the melanoma-astrocytoma syndrome family.

Azizi, E.; Friedman, J.; Pavlovsky, F.; Iscovich, J.; Bornstein, A.; Shafir, R.; Trau, H.; Brenner, H.; Nass, D.: Familial cutaneous malignant melanoma and tumors of the nervous system. Cancer 76: 1571-1578, 1995.
Bahuau, M.; Vidaud, D.; Jenkins, R. B.; Bieche, I.; Kimmel, D. W.; Assouline, B.; Smith, J. S.; Alderete, B.; Cayuela, J.-M.; Harpey, J.-P.; Caille, B.; Vidaud, M.: Germ-line deletion involving the INK4 locus in familial proneness to melanoma and nervous system tumours. Cancer Res. 58: 2298-2303, 1998.
Bahuau, M.; Vidaud, D.; Kujas, M.; Palangie, A.; Assouline, B.; Chaignaud-Lebreton, M.; Prieur, M.; Vidaud, M.; Harpey, J.-P.; Lafourcade, J.; Caille, B.: Familial aggregation of malignant melanoma/dysplastic naevi and tumors of the nervous system: an original syndrome of tumour proneness. Ann. Genet. 40: 78-91, 1997.
Kaufman, D. K.; Kimmel, D. W.; Parisi, J. E.; Michels, V. V.: A familial syndrome with cutaneous malignant melanoma and cerebral astrocytoma. Neurology 43: 1728-1731, 1993.

Liu, L.; Lassam, N. J.; Slingerland, J. M.; Bailey, D.; Cole, D.; Jenkins, R.; Hogg, D.: Germline p16(INK4A) mutation and protein dysfunction in a family with inherited melanoma. Oncogene 11: 405-412, 1995.

GLIOMA OF BRAIN (*137800)

JJM NOTE

Of course, glioblastoma multiforme and astrocytoma are included in this common group of brain tumors. The clinician who has a patient with a glioma has a long list of single-gene traits to think of: neurofibromatosis 1 and 2, tuberous sclerosis, von Hippel-Lindau disease, Turcot syndrome, Li-Fraumeni cancer family syndrome, ataxia-telangiectasia, and the nevoid basal cell carcinoma syndrome. The best-substantiated environmental factor is exposure to vinyl chloride. Significant sibship aggregation in childhood was confirmed in Britain by Draper et al. (Occurrence of childhood cancers among sibs and estimation of familial risks. J. Med. Genet. 14:81–90, 1977) and in the Connecticut population-based registry by Farwell and Flannery (Cancer in relatives of children with central-nervous-system neoplasms. N. Engl. J. Med. 311:749–53, 1984). The definitive literature review is Tijssen CC et al.: *Familial Brain Tumors. Developments in Oncology,* vol. 9. The Hague: Nijhoff, 1982.

King and Eisinger (1966) described glioma multiforme of the frontal lobes in father and daughter with development of symptoms at age 50 and 34 years, respectively. Schianchi and Kraus-Ruppert (1980) described affected father and son. Others have reported multiple affected sibs or other relatives. Armstrong and Hanson (1969) described 3 sibs who died of brain glioma in adulthood. Thuwe et al. (1979) observed 6 cases of brain glioma and a possible seventh on an isolated Swedish coastal island. The affected persons were related as cousins, all in different sibships. One instance of parental consanguinity, the lack of parent-child transmission, and the longtime isolation of the population suggest recessive inheritance. In further studies in this island community, Thuwe (1984) reported 4 closely related cases of brain tumor. It was found that 30 probands with brain tumor were more often the product of a consanguineous marriage than were controls and a higher proportion could be traced to a common ancestor living in the 1600s. It was concluded that genetic factors play a role, although a single major gene seems unlikely. In a study of cancer mortality during childhood in sibs, Miller (1971) found 8 pairs of nontwin sibs with brain tumor versus 0.9 expected. There were 8 other families versus 0.9 expected in which 1 child died of brain tumor and another died of cancer of bone or muscle. In a highly inbred Arab family in Israel, Chemke et al. (1985) observed 5 cases of glioblastoma multiforme in 2 sibships. Curiously, all were male and in all the tumor was located on the right side of the brain. The ages of presentation ranged from 4 to 11 years. 'Astrocytoma type 3' was the histologic diagnosis. Glioblastoma multiforme is often associated with neurofibromatosis; such was not the case in this kindred. Leblanc et al. (1986) described father and son operated on at ages 26 and 37 years, respectively, for mixed oligodendrocytic-astrocytic glioma. Heuch and Blom (19 86) reported glioblastoma multiforme in 2 brothers, aged 65 and 68 years, and in their paternal aunt, aged 81. The father had died of tuberculosis before age 40. True multicentric origin, consistent with a hereditary basis, was observed in 1 of the 3 cases. Tijssen (1987) referred to an international register of familial brain tumors maintained in Leyden.

Vieregge et al. (1987) gave an extensive review of reported cases of familial glioma with or without other malformations. They reported a family in which members of several generations had one or another abnormality: father and son had glioma; another man and his daughter had colonic polyps; and skeletal abnormalities in the form of short stature and exostoses were present in some members. Munoz et al. (1988) described a brother and sister without a history of phacomatosis or cerebral tumors who developed malignant tumors with ependymal and choroidal differentiation. The girl presented at 28 months with a tumor of the posterior fossa, and the boy presented at 15 months with a tumor of the left cerebral hemisphere. Duhaime et al. (1989) reported histologically identical glioblastoma multiforme in 2 sibs, aged 2 and 5 years old, whose symptoms developed simultaneously. The case of prenatal diagnosis reported by Geraghty et al. (1989) illustrated the occurrence of glioblastoma multiforme in the fetus.

The most striking and consistent chromosomal finding in a series of human glioblastoma cell lines was an increase in copy number of chromosome 7. Henn et al. (1986) found that in all of the cell lines ERBB-specific mRNA was increased to levels even higher than expected from the number of chromosomes 7 present. These changes were not found in benign astrocytomas. Bigner et al. (1988) found chromosome abnormalities in 12 of 54 malignant gliomas. Structural abnormalities of 9p and 19q were increased to a statistically significant degree. Chromosome 10 was implicated in glioblastoma multiforme by Fujimoto et al. (1989), who found loss of constitutional heterozygosity in tumor samples from 10 of 13 patients in whom paired tumor and lymphocyte DNA samples were screened. Bigner et al. (1988) concluded that the most frequent chromosomal changes in malignant gliomas are gains of chromosome 7 and losses of chromosome 10. Double minute chromosomes, indicating the presence of gene amplification, are found in about 50% of malignant gliomas. Most tumors with double minute chromosomes contain 1 of 5 amplified genes, most often the EGF receptor gene (EGFR; 131550). Following up on the observation that the EGFR gene is amplified in 40% of malignant gliomas, Wong et al. (1992) characterized the rearrangements in 5 malignant gliomas. In one they found deletion of most of the extracytoplasmic domain of the receptor. The 4 other tumors had internal deletions of the gene. El-Azouzi et al. (1989) described loss of constitutional heterozygosity for markers on the short arm of chromosome 17 in both low grade and high grade malignant astrocytomas, suggesting that this region may contain a tumor suppressor gene associated with the early events in tumorigenesis. See 165220 for discussion of oncogene GLI1, which was identified in cerebral glioma.

Olopade et al. (1992) found molecular evidence of deletion in 9p in 10 of 15 glioma-derived cell lines and 13 of 35 primary gliomas. The shortest region of overlap of these deletions mapped to the interval between the centromeric end of the interferon gene cluster and the methylthioadenosine phosphorylase locus (156540).

To evaluate whether loss of chromosome 19 alleles is common in glial tumors of different types and grades, von Deimling et al. (1992) performed Southern blot RFLP analysis for multiple chromosome 19 loci in 122 gliomas from 116 patients. In 29 tumors, loss of constitutional heterozygosity of 19q was demonstrated; 4 tumors had partial deletion of 19q. The results were interpreted as indicating the presence of a glial tumor suppressor gene on 19q.

In a search for submicroscopic deletions in chromosome 10, Fults and Pedone (1993) performed a RFLP analysis in 30 patients, using markers that had been mapped accurately on chromosome 10 by genetic linkage studies. Loss of heterozygosity (LOH) at one or more loci was found in 15 of the 30 patients. In 7 cases, LOH was found at every informative locus. LOH was confined to a portion of the long arm in 6 patients; the smallest region of overlap among these 6 deletions was flanked by markers D10S12 proximally and D10S6 distally, a 33.4-cM region mapped physically near the telomere, 10q25.1-qter.

Karlbom et al. (1993) analyzed a panel of glial tumors consisting of 11 low grade gliomas, 9 anaplastic gliomas, and 29 glioblastomas for loss of heterozygosity by examining at least one locus for each chromosome. The frequency of allele loss was highest among the glioblastomas, suggesting that genetic alterations accumulate during glial tumor development. The most common genetic alteration was found to involve allele losses of chromosome 10 loci, these being found in all glioblastomas and in 3 anaplastic tumors. Deletion mapping analysis revealed partial loss of chromosome 10 in 8 glioblastomas and 2 anaplastic tumors in 3 distinct regions: one telomeric region on 1p and both telomeric and centromeric locations on 10q. These data suggested to Karlbom et al. (1993) the existence of multiple chromosome 10 tumor suppressor gene loci whose inactivation is involved in the malignant progression of glioma. In studies of 20 gliomas with microsatellite markers from chromosome 10, the locus that exhibited the most loss (69%) was the region bordered by D10S249 and D10S558 and inclusive of D10S594, with a linkage distance of 3 cM (Kimmelman et al., 1996). This region was known to be deleted in various grades of tumor, including low- and high-grade tumors. These results suggested to Kimmelman et al. (1996) that chromosome region 10p15 is involved in human gliomas of diverse grades and that this region may harbor genes important in the development of and progression to the malignant phenotype. (As indicated earlier, regions of 10q had been implicated in other studies.) See the DMBT1 gene (601969), so designated for 'deleted in malignant brain tumors,' for a discussion of a gene on 10q25.3-q26.1 that showed intragenic homozygous deletions in medulloblastoma and glioblastoma multiforme tumor tissue and in brain tumor cell lines (Mollenhauer et al., 1997).

Hamilton et al. (1995) found that glioblastoma multiforme is a feature of the form of Turcot syndrome (276300) due to defects in the mismatch repair genes MLH1 (120436) or PMS2 (600259). Bogler et al. (1995) reviewed the role of the p53 tumor protein gene (191170) in the initiation and progression of human gliomas.

Von Deimling et al. (1995) proposed a simplified model for the pathogenesis of human gliomas. Reviewing their work and that of others, they suggested 3 distinct pathways. The first pathway, which leads to pilocytic astrocytomas (WHO grade I), is caused by loss of heterozygosity for chromosome 17q, presumably unmasking mutations in the NF1 (162200) gene. The second pathway begins with LOH at 17p, unmasking mutations in p53 and leading to astrocytoma (WHO grade II). Further LOH at 13q, 19q and 9p, unmasking mutations in RB1, p16 and p15, provides a further step on the second pathway, giving rise to grade III astrocytomas. The final step on the second pathway, LOH on chromosome 10 and perhaps other chromosomes, culminates in glioblastoma multiforme type 1 (WHO grade IV). The third independent pathway begins with LOH at chromosomes 10 and 9p, followed by gene amplification of EGFR, CDK4 (123829), MDM2 (164785), and SAS (181035), and culminating in glioblastoma multiforme type 2 (WHO grade IV).

Ueki et al. (1997) studied gliomas for tumor-specific alterations in the ANOVA gene (601991), which they cloned from the glioma candidate region at 19q13.3. They found no alterations of ANOVA in gliomas by Southern blotting and SSCP analysis, suggesting that ANOVA is not the chromosome 19q glioma tumor suppressor gene.

See also Bigner et al. (1988); de Tribolet et al. (1979); Isamat et al. (1974); Kjellin et al. (1960); Koch and Waldbaur (1981); Parkinson and Hall (1962); Reese et al. (1944); Von Motz et al. (1977)

Armstrong, R. M.; Hanson, C. W.: Familial gliomas. Neurology 19: 1061-1063, 1969.
Bigner, S. H.; Mark, J.; Burger, P. C.; Mahaley, M. S., Jr.; Bullard, D. E.; Muhlbaier, L. H.; Bigner, D. D.: Specific chromosomal abnormalities in malignant human gliomas. Cancer Res. 88: 405-411, 1988.
Bigner, S. H.; Vogelstein, B.; Bigner, D. D.: Chromosomal abnormalities and gene amplification in malignant gliomas. Atlas Sci. Biochem. 333-336, 1988.
Bogler, O.; Huang, H.-J. S.; Kleihues, P.; Cavenee, W. K.: The p53 gene and its role in human brain tumors. GLIA 15: 308-327, 1995.
Chemke, J.; Katznelson, D.; Zucker, G.: Familial glioblastoma multiforme without neurofibromatosis. Am. J. Med. Genet. 21: 731-735, 1985.
de Tribolet, N.; Deruaz, J.-P.; Zander, E.: Familial gliomas. Neurochirgia 22: 225-228, 1979.
Duhaime, A. C.; Bunin, G.; Sutton, L.; Rorke, L. B.; Packer, R. J.: Simultaneous presentation of glioblastoma multiforme in siblings two and five years old: case report. Neurosurgery 24: 434-439, 1989.
El-Azouzi, M.; Chung, R. Y.; Farmer, G. E.; Martuza, R. L.; Black, P. M.; Rouleau, G. A.; Hettlich, C.; Hedley-Whyte, E. T.; Zervas, N. T.; Panagopoulos, K.; Nakamura, Y.; Gusella, J. F.; Seizinger, B. R.: Loss of distinct regions on the short arm of chromosome 17 associated with tumorigenesis of human astrocytomas. Proc. Nat. Acad. Sci. 86: 7186-7190, 1989.
Fujimoto, M.; Fults, D. W.; Thomas, G. A.; Nakamura, Y.; Heilbrun, M. P.; White, R.; Story, J. L.; Naylor, S. L.; Kagan-Hallet, K. S.; Sheridan, P. J.: Loss of heterozygosity on chromosome 10 in human glioblastoma multiforme. Genomics 4: 210-214, 1989.
Fults, D.; Pedone, C.: Deletion mapping of the long arm of chromosome 10 in glioblastoma multiforme. Genes Chromosomes Cancer 7: 173-177, 1993.
Geraghty, A. V.; Knott, P. D.; Hanna, H. M.: Prenatal diagnosis of fetal glioblastoma multiforme. Prenatal Diag. 9: 613-616, 1989.
Hamilton, S. R.; Liu, B.; Parsons, R. E.; Papadopoulos, N.; Jen, J.; Powell, S. M.; Krush, A. J.; Berk, T.; Cohen, Z.; Tetu, B.; Burger, P. C.; Wood, P. A.; Taqi, F.; Booker, S. V.; Petersen, G. M.; Offerhaus, G. J. A.; Tersmette, A. C.; Giardiello, F. M.; Vogelstein, B.; Kinzler, K. W.: The molecular basis of Turcot's syndrome. New Eng. J. Med. 332: 839-847, 1995.
Henn, W.; Blin, N.; Zang, K. D.: Polysomy of chromosome 7 is correlated with overexpression of the erbB oncogene in human glioblastoma cell lines. Hum. Genet. 74: 104-106, 1986.
Heuch, I.; Blom, G. P.: Glioblastoma multiforme in three family members, including a case of true multicentricity. J. Neurol. 233: 142-144, 1986.
Isamat, F.; Miranda, A. M.; Bartumeus, F.; Prat, J.: Genetic implications of familial brain tumors. J. Neurosurg. 41: 573-575, 1974.
Karlbom, A. E.; James, C. D.; Boethius, J.; Cavenee, W. K.; Collins, V. P.; Nordenskjold, M.; Larsson, C.: Loss of heterozygosity in malignant gliomas involves at least three distinct regions on chromosome 10. Hum. Genet. 92: 169-174, 1993.
Kimmelman, A. C.; Ross, D. A.; Liang, B. C.: Loss of heterozygosity of chromosome 10p in human gliomas. Genomics 34: 250-254, 1996.
King, A. B.; Eisinger, G.: May glioma multiforme be hereditary? Guthrie Clin. Bull. 35: 169-175, 1966.
Kjellin, K.; Muller, R.; Astrom, K. E.: The occurrence of brain tumor in several members of a family. J. Neuropath. Exp. Neurol. 19: 528-537, 1960.
Koch, G.; Waldbaur, H.: Erbliche Hirngeschwuelste (brain tumour family syndrome). Z. Allg. Med. 57: 1219-1224, 1981.
Leblanc, R.; Lozano, A.; Robitaille, Y.: Familial mixed oligodendrocytic-astrocytic gliomas. Neurosurgery 18: 480-482, 1986.
Miller, R. W.: Deaths from childhood leukemia and solid tumors among twins and other sibs in the United States, 1960-67. J. Nat. Cancer Inst. 46: 203-209, 1971.
Mollenhauer, J.; Wiemann, S.; Scheurlen, W.; Korn, B.; Hayashi, Y.; Wilgenbus, K. K.; van Deimling, A.; Poustka, A.: DMBT1, a new member of the SRCR superfamily, on chromosome 10q25.3-26.1 is deleted in malignant brain tumours. Nature Genet. 17: 32-39, 1997.
Munoz, D. G.; Griebel, R.; Rozdilsky, B.; George, D.: Cerebral malignant tumors with ependymal and choroidal differentiation in two siblings. Neurosurgery 22: 928-933, 1988.
Olopade, O. I.; Jenkins, R. B.; Ransom, D. T.; Malik, K.; Pomykala, H.; Nobori, T.; Cowan, J. M.: Rowley, J. D. and Diaz, M. O.: Molecular analysis of deletions of the short arm of chromosome 9 in human gliomas. Cancer Res. 52: 2523-2529, 1992.
Parkinson, D.; Hall, C. W.: Oligodendrogliomas: simultaneous appearance in frontal lobes in siblings. J. Neurosurg. 19: 424-426, 1962.
Reese, W.; Meredith, J. M.; Zfass, I. S.: Cerebral glioma in siblings. Sth. Med. J. 37: 424-428, 1944.
Schianchi, P.; Kraus-Ruppert, R.: Familial brain tumors: rhombencephalon-astrocytoma grade I in father and son. Acta Neuropath. 52: 153-155, 1980.
Thuwe, I.: Glioma cerebri in an island community. Sweden: St Jorgen's Hospital, Univ. of Goteberg (pub.) 1984.
Thuwe, I.; Lundstrom, B.; Walinder, J.: Familial brain tumour.(Letter) Lancet I: 504 only, 1979.
Tijssen, C. C.: Familial mixed oligodendrocytic-astrocytic gliomas.(Letter) Neurosurgery 20: 62 only, 1987.
Ueki, K.; Ramaswamy, S.; Billings, S. J.; Mohrenweiser, H. W.; Louis, D. N.: ANOVA, a putative astrocytic RNA-binding protein gene that maps to chromosome 19q13.3. Neurogenetics 1: 31-36, 1997.
Vieregge, P.; Gerhard, L.; Nahser, H. C.: Familial glioma: occurrence within the 'familial cancer syndrome' and systemic malformations. J. Neurol. 234: 220-232, 1987.
von Deimling, A.; Louis, D. N.; von Ammon, K.; Petersen, I.; Wiestler, O. D.; Seizinger, B. R.: Evidence for a tumor suppressor gene on chromosome 19q associated with human astrocytomas, oligodendrogliomas, and mixed gliomas. Cancer Res. 52: 4277-4279, 1992.
von Deimling, A.; Louis, D. N.; Wiestler, O. D.: Molecular pathways in the formation of gliomas. GLIA 15: 328-338, 1995.
Von Motz, I. P.; Bots, G. T. A. M.; Endtz, L. J.: Astrocytoma in three sisters. Neurology 27: 1038-1041, 1977.
Wong, A. J.; Ruppert, J. M.; Bigner, S. H.; Grzeschik, C. H.; Humphrey, P. A.; Bigner, D. S.; Vogelstein, B.: Structural alterations of the epidermal growth factor receptor gene in human gliomas. Proc. Nat. Acad. Sci. 89: 2965-2969, 1992.

NORRIE DISEASE (*310600)

JJM NOTE

This is another pseudoglioma condition that needs further clinical delineation.

Warburg (1961) reported 7 cases of a hereditary degenerative disease in 7 generations of a Danish family. The proband was a 12-month-old boy who was normal except for lens opacities found at initial examination at 3 months of age. The irides were atrophic. The fundus was filled with a proliferating retrolental yellowish mass. At 8 months of age the left eye was enucleated on suspicion of retinoblastoma. Histologic examination showed a hemorrhagic necrotic mass in the posterior chamber, surrounded by undifferentiated glial tissue. Histologic diagnosis was pseudotumor of the retina, retinal hyperplasia, hyperplasia of retinal, ciliary, and iris pigment epithelium, hypoplasia and necrosis of the inner layer of the retina, cataract, and phthisis bulbi. Six relatives had a similar ocular disease. In 5 of these 7 cases deafness developed in later years, and in 4 of the 7 cases the mental capacity was low. Warburg (1961) found 48 similar cases in 9 families described in the literature under different categories which she believed belonged to this disease. She suggested that the disorder be named for Gordon Norrie (1933). Warburg (1963) presented 2 new families with 11 patients suffering from this disease. Patients examined varied from 2 months to 58 years of age. At earliest examination, pseudoglioma, synechiae and atrophy of the iris were observed. Blindness was found during the first month of life. By 8 months cataract was observed, and at 10 years the eyes were atrophic with band-shaped corneal degeneration and dense cataract. By the age of 50 years the atrophy had advanced to opaque white cornea, obliterated anterior chamber, atrophic white iris, and cataractous lens. Though some afflicted had normal intelligence, many were mentally deficient. The mental retardation is a deterioration inasmuch as the affected infants seem to be normal for the first 1 to 2 years. Five of 9 in 1 family were hard of hearing and 2 of these 5 had diabetes. The mode of inheritance in both families wa s X-chromosomal recessive. Whitnall and Norman (1940) reported the neuropathology of a case. The optic nerves and lateral geniculate bodies were small. Taylor et al. (1959) reported a Greek family with this condition living in Episkopi in Cyprus. The condition was popularly known as Episkopi blindness. The published pedigree showed 16 affected males in 5 generations. All affected males were retarded. Woodruff et al. (1993) described the case of a 2-year-old girl who showed severe visual impairment resulting from cataract and total retinal detachment in the right eye with a vascularized mass behind the lens, and in the left eye a retinal fold and traction retinal detachment in the temporal periphery. She had a brother and 2 maternal uncles with Norrie disease. Woodruff et al. (1993) suggested that she was an example of a manifesting heterozygote.

As discussed under microphthalmia (309700), 'pseudoglioma,' microphthalmos, and Norrie disease are confused in the literature. Pseudoglioma is a nonspecific term for any condition more or less mimicking retinoblastoma. Thus pseudoglioma can have as diverse causes as inflammation, hemorrhage, trauma, neoplasia or congenital malformation. Many of the causes lead only to unilateral involvement. ND is a form of bilateral and congenital pseudoglioma. It should be evident from the above discussion that pseudoglioma is merely any condition of the eye liable to be mistaken for true glioma and is therefore not an acceptable clinical or pathologic diagnosis (Duke-Elder, 1958). In his system of ophthalmology, Duke-Elder (1958) mistakenly classified the disorder as band-shaped keratopathy. In the family reported by Forssman (1960), 'pseudoglioma' was combined with mental deficiency present from infancy and apparently of progressive nature. Forssman's patients (1960) were first described by Dahlberg-Parrow (1956). Three of the blind boys were reexamined by Warburg (1966), who concluded that the histories and ocular findings were typical of ND. In the extensive pedigree from a Canadian Indian group reported by Wilson (1949), histologic changes were similar to those of ND. Clarke (1898) described possible homozygous affected females. A man blind from probable bilateral 'pseudoglioma' married his first cousin. Of their 6 children, 2 girls and 1 boy had unilateral or bilateral 'pseudoglioma.'

Moreira-Filho and Neustein (1979) described 6 brothers with what they viewed as a variant of ND, because microcephaly was present in all. (In some ways the patients resembled those reported by Goldberg and McKusick as discussed in entry 309800.) The pedigree was informative for linkage with Xg; negative lod scores were obtained. Warburg et al. (1965) had demonstrated no linkage with the Xg blood groups. Johnston et al. (1982) described 2 families with 8 affected males—the first families reported from Ireland. Harendra de Silva and de Silva (1988) described an extensively affected family in Sri Lanka. Gal et al. (1985) found close linkage of Norrie disease to the L1.28/TaqI RFLP, DXS7 (maximum lod = 3.50 at theta = 0.00). Thus, ND may be in or slightly proximal to band Xp11.3 and near the retinitis pigmentosa locus (312600), which is also linked to DXS7. Gal et al. (1985) found a peak lod score of 4.1 at theta 0.00 for linkage with DXS7; no recombination was found. DXS7 has been localized to Xp11.3 (or Xp11.3-p11.2). See Bleeker-Wagemakers et al. (1985) for the full data. 25,28:Gal et al. (1985, 1986) also described a 14-year-old boy with a complex syndrome dominated by Norrie disease who appeared to have a small deletion involving DXS7 as well; seemingly, the deletion had been transmitted through 3 generations. Other features of the complex were severe mental retardation, hypogonadism, growth disturbances, and increased susceptibility to infections. De la Chapelle et al. (1985) found a deletion defined by DXS7 in 4 affected members of a family. Using probe L1.28 in the study of a chorion villus sample, they could show that the male fetus of a carrier woman was unaffected. Kivlin et al. (1987) presented further linkage data. They stated that no recombination had been identified between ND and the DNA marker L1.28; with their data, the total lod score became 5.42. Ohba and Yamashita (1986) presented evidence suggesting that the Norrie diseas e locus may be on Xp at band Xp11. A female infant with typical clinical and histopathologic features of vitreoretinal dysplasia was found to have a reciprocal translocation at t(X;10)(p11;p14). Her parents and sibs had normal karyotypes. Donnai et al. (1988) found that the DXS7 locus was deleted in 2 affected brothers; DXS7 is located in Xp11.3. OTC (311250), located at Xp21.1, and DXS84, also located at Xp21.1, were intact. Ngo et al. (1988) and Katayama et al. (1988) found the first recombinant between Norrie disease and the DXS7 locus. The addition of their family brought a total of published informative families to 7, with a maximum lod score of 7.58 at a recombination fraction of 0.038. They stated the assignment of the DXS7 locus (defined by probe L1.28) as Xp11.3-p11.2. Ngo et al. (1989) pointed out that a single recombination event had been reported twice (Ngo et al., 1988; Katayama et al., 1988); otherwise, a distorted impression of the distance between the marker and Norrie disease might be given. Gal et al. (1988) described prenatal exclusion with flanking DNA markers. In an addendum, they stated that 3 families with Norrie disease and DXS7 deletion had been reported, bringing the compiled lod score for NDP vs DXS7 linkage to 11.18 at theta = 0.00. Using a RFLP detected by the ornithine amino transferase (OAT)-related DNA sequences that map to Xp (see 258870), Ngo et al. (1989) found a suggestion of linkage to the Norrie disease locus. In 3 generations of a family with 4 affected males in 3 sibships of 2 generations, Zhu et al. (1989) demonstrated deletion of 2 loci, DXS7 and DXS77. DNA studies of chorion villus biopsy material from the fetus of an obligatory carrier indicated that the fetus had inherited the normal allele from the carrier mother. This prediction was confirmed on eye examination at age 5 months. Deletions at the DXS7 locus have been detected in 3 other families (de la Chapelle et al. , 1985; Gal et al., 1985; Donnai et al., 1988). Diergaarde et al. (1989) further refined the localization of the deletion in a Dutch case of ND. Sims et al. (1989) demonstrated that the Norrie disease gene is distinct from the monoamine oxidase genes, although some males with atypical Norrie disease who have a submicroscopic deletion in the region of the DXS7 locus have been shown to have disruption of the MAOA (309850) and MAOB (309860) genes. The authors studied genomic DNA from 19 males in 9 families affected with Norrie disease. No deletions or rearrangements in the region of DSX7 or MAOA were observed in the DNA of these patients. Linkage analysis between the NDP gene and the DSX7 or MAOA loci showed no recombination, with a lod score of 2.80 and 2.58 at a theta of 0.0 for MAOA and DSX7, respectively. MAO activities in fibroblasts and platelets were normal. Although the MAO and NDP loci appear to be distinct, the high level of polymorphism at the MAO locus should prove useful in the molecular diagnosis of the disease. Collins et al. (1992) reported a male with Norrie disease and 2 obligate heterozygous females who were shown to have a submicroscopic deletion involving the Norrie disease locus and the loci for MAOA and MAOB. The propositus was a profoundly retarded, blind male; he also had neurologic abnormalities including myoclonus and stereotypy-habit disorder (persistent stereotypic and self-injurious behavior with a deleterious effect on the patient's adaptation to home and school environments). Both obligate carriers had a normal IQ. In the propositus, MAO activity was undetectable; in the female heterozygotes, the levels were reduced to the range observed in patients receiving MAO-inhibiting antidepressants. One of the carriers, the mother of the propositus, met diagnostic criteria for 'chronic hypomania and schizotypal features.' Lindsay et al. (1992) did linkage studies using a highly informative microsatellite marker, DXS426, which maps proximal to DXS7 in the interval Xp11.4-p11.23. A multiply informative crossover localized the NDP gene proximal to DXS7. In conjunction with information from 2 NDP patients who had a deletion for DXS7 but not for DXS426, their data indicated that the NDP gene is between DXS7 and DXS426 on proximal Xp. Wolff et al. (1992) restudied the family with Episkopi blindness originally studied by Taylor et al. (1959). DNA studies revealed no deletion of any of the probes from proximal Xq. Linkage analysis yielded positive lod scores for all informative markers; e.g., with DXS255, maximum lod = 6.54 at theta = 0.0. The findings confirmed that Episkopi blindness and Norrie disease are the same entity.

By positional cloning, Berger et al. (1992) and Chen et al. (1992) isolated a likely candidate gene for the site of the mutation in Norrie disease. Berger et al. (1992) made use of the submicroscopic deletion reported by Bergen et al. (1993), using differential Alu PCR fingerprint cloning. (Positional cloning papers get published more promptly than do descriptions of chromosomal aberrations.) The gene was found to be expressed in retina, choroid, and fetal brain by Berger et al. (1992) and in fetal and adult brain by Chen et al. (1992). It was evolutionarily conserved and encoded a predicted protein of 133 amino acids. The genomic equivalent of the cDNA spanned a maximum of 50 kb (Chen et al., 1992) and was partly deleted in several typical Norrie disease patients. Chen et al. (1992) identified an EcoRI polymorphism with a calculated heterozygosity value of 43% in the gene. Berger et al. (1992) found no homologous sequences in DNA and protein databases, indicating that the cDNA they found is part of a gene encoding a 'pioneer' protein, i.e., a new settler in the territories of the databases. Sims et al. (1992) narrowed the mapping of the NDP gene to a 150-kb region defined by a recombination and by the smallest submicroscopic chromosomal deletion associated with Norrie disease identified to date. They concluded that the order of loci is as follows: pter—DXS7—MAOA—MAOB—NDP—cen.

Chen et al. (1992) studied the end point of the deletion in a Norrie disease patient who had been shown to lack both DXS7 and MAO coding sequences, these being closely situated telomeric to the NDP locus. The pattern of retention of subclones within the deletion patient placed the end point of the deletion within 30 to 130 kb of the proximal end of a 650-kb YAC containing DXS7 and the MAO genes. They concluded that the NDP gene lay in whole or in part in the interval of approximately 250 kb within the YAC. As a prerequisite for identifying point mutations in the NDP gene, Berger et al. (1992) established the exon-intron structure of the gene in 12 of 17 unrelated patients. Altered PCR fragments were detected using PCR products derived from the promoter region, exons 1 and 2, as well as the coding part of exon 3, in single-strand conformation polymorphism (SSCP) analyses. Direct sequencing revealed 11 different mutations, all but one of which were likely to give rise to significant structural changes in the predicted protein. (The exception was a leu61-to-phe mutation.) The diagnosis of Norrie disease is often difficult. Although specific ocular symptoms, progressive deafness, and mental disturbance are considered classic features, more than half of the patients are neither deaf nor mentally retarded. The ability to test for specific mutations will permit clarification of diagnosis. The fact that only one of the point mutations was detected twice is in keeping with the high proportion of new mutations expected for an X-linked disorder with greatly reduced male reproductive fitness. Because of the small size of the ND mRNA and the coding sequence, mutation detection in Norrie disease is particularly simple and fast.

Although Berger et al. (1992) and Chen et al. (1992) could identify no strong homologies with the candidate gene they identified, by studying the number and spacing of cysteine residues, Meindl et al. (1992) later detected homologies between the Norrie disease gene product and a C-terminal domain that is common to a group of proteins including mucins. Furthermore, they characterized 3 missense mutations, replacing evolutionarily conserved cysteines or creating new cysteine codons, emphasizing the functional importance of these sites. These findings and the clinical features of Norrie disease suggested a possible role for the NDP gene in a neuroectodermal cell-cell interaction. Only exons 2 and 3 of the NDP gene are translated. Exon 2 contains the first 58 codons of the open reading frame. The intron that follows it is roughly 14.5 kb. Exon 3 is the largest exon and contains residues 59-133 of the open reading frame and a 917-bp untranslated 3-prime-region. Chen et al. (1993) isolated genomic DNA clones encompassing the NDP gene which they found spans 28 kb and consists of 3 exons, the first of which is entirely contained within the 5-prime untranslated region. By PCR analysis, they found that expression of the NDP gene is not confined to the eye or to the brain. They found homology with cysteine-rich protein-binding domains of intermediate-early genes implicated in the regulation of cell proliferation. This led them to propose that the NDP molecule likewise may be involved in the pathway that regulates neural cell differentiation and proliferation. Meitinger et al. (1993) reported that sequence pattern searches and 3-dimensional modeling suggested that the Norrie disease protein (NDP) has a tertiary structure similar to that of a transforming growth factor beta (e.g., 190180). The model identified NDP as a member of an emerging family of growth factors containing a cystine knot motif, with direct implications for the physiologic role of NDP.

To elucidate the cellular and molecular processes involved in Norrie disease, Berger et al. (1996) used gene targeting technology to generate ND mutant mice. The murine homolog of the ND gene was cloned and shown to encode a polypeptide that shares 94% amino acid sequence identity with its human counterpart. RNA in situ hybridization revealed expression in retina, brain, and olfactory bulb and epithelium of 2-week-old mice. Hemizygous mice carrying a replacement mutation in exon 2 of the ND gene developed retrolental structures in the vitreous body and showed an overall disorganization of the retinal ganglion cell layer. The outer plexiform layer disappeared occasionally, resulting in a juxtaposed inner and outer nuclear layer. The ocular findings were consistent with observations in ND patients, and in the process of the work Berger et al. (1996) demonstrated that the murine gene is, like the human gene, on the X chromosome.

Schuback et al. (1995) reported their experience to date in mutation identification in Norrie disease. They had carried out mutational analysis in 26 kindreds. All coding exons, as well as noncoding regions of exons 1 and 2, 636 nucleotides in the noncoding region of exon 3, and 197 nucleotides of the 5-prime flanking sequence were analyzed for single-strand conformation polymorphisms (SSCPs) by PCR amplification of genomic DNA. DNA fragments that showed altered SSCP band mobilities were sequenced to locate these specific mutations. In addition to 3 previously described submicroscopic deletions encompassing the entire ND gene, they identified 6 intragenic deletions, 8 missense mutations (7 point mutations, 1 9-bp deletion), 6 nonsense mutations (3 point mutations, 3 single bp deletions/frameshift), and 1 10-bp insertion that created an expanded repeat in the 5-prime noncoding region of exon 1. Mutations had been identified in 24 of 26 of the kindreds they had studied. With the exception of 2 different mutations, each found in 2 apparently unrelated kindreds, these mutations were unique. Location of most point mutations at or near cysteine residues, potentially critical in protein tertiary structure, supported a previous protein model for norrin as a member of a cystine knot growth factor family (Meitinger et al., 1993). Genotype-phenotype correlations were not evident with the limited clinical data available, except in the cases of larger submicroscopic deletions associated with a more severe neurologic syndrome. Given the remarkable intra- and interfamilial variability in hearing and brain dysfunction in this disease, Schuback et al. (1995) predicted that other phenotypic expressions of the disease gene that do not match the classic Norrie phenotype may be identified in the future.

Isashiki et al. (1995) tabulated known mutations in the NDP gene together with the clinical manifestations and described a new cys95-to-arg mutation.

Walker et al. (1997) described 2 new mutations in exon 3 of the NDP gene: ser73ter and ser101phe, associated with severe and less severe ocular phenotype, respectively. They referred to the Norrie disease protein as norrin. The affected individuals in both families presented with neither the sensorineural deafness nor the mental retardation that often accompanies Norrie disease. Whereas the ocular features of the S73X mutation were fairly typical of Norrie disease, the amino acid change S101F within the C-terminal region of norrin was associated with milder features. The S101F mutation was the most C-terminal reported to that time. Others (Chen et al., 1993; Meindl et al., 1995) had previously made the observation that mutations in the C-terminal portion of the gene product appear to result in less severe Norrie disease.

Shastry et al. (1997) reported 4 novel missense mutations in the NDP gene associated with 1 X-linked and 4 sporadic cases of familial exudative vitreoretinopathy (see 310600.0013).

See also Anderson and Warburg (1961); Berger et al. (1992); Gal et al. (1985); Holmes (1971); Isashiki et al. (1995); Nance et al. (1969); Ngo et al. (1989); Phillips et al. (1986); Sims et al. (1989); Warburg (1966)

Anderson, S. R.; Warburg, M.: Norrie's disease. Arch. Ophthal. 66: 614-618, 1961.

Bergen, A. A. B.; Wapenaar, M. C.; Schuurman, E. J. M.; Diergaarde, P. J.; Lerach, H.; Monaco, A. P.; Bakker, E.; Bleeker-Wagemakers, E. M.; van Ommen, G. J. B.: Detection of a new submicroscopic Norrie disease deletion interval with a novel DNA probe isolated by differential Alu PCR fingerprint cloning. Cytogenet. Cell Genet. 62: 231-235, 1993.

Berger, W.; Meindl, A.; van de Pol, T. J. R.; Cremers, F. P. M.; Ropers, H. H.; Doerner, C.; Monaco, A.; Bergen, A. A. B.; Lebo, R.; Warburg, M.; Zergollern, L.; Lorenz, B.; Gal, A.; Bleeker-Wagemakers, E. M.; Meitinger, T.: Isolation of a candidate gene for Norrie disease by positional cloning. Nature Genet. 1: 199-203, 1992.

Berger, W.; van de Pol, D.; Bachner, D.; Oerlemans, F.; Winkens, H.; Hameister, H.; Wieringa, B.; Hendriks, W.; Ropers, H.-H.: An animal model for Norrie disease (ND): gene targeting of the mouse ND gene. Hum. Molec. Genet. 5: 51-59, 1996.

Berger, W.; van de Pol, D.; Warburg, M.; Gal, A.; Bleeker-Wagemakers, L.; de Silva, H.; Meindl, A.; Meitinger, T.; Cremers, F.; Ropers, H.-H.: Mutations in the candidate gene for Norrie disease. Hum. Molec. Genet. 1: 461-465, 1992.

Bleeker-Wagemakers, L. M.; Friedrich, U.; Gal, A.; Wienker, T. F.; Warburg, M.; Ropers, H.-H.: Close linkage between Norrie disease, a cloned DNA sequence from the proximal short arm, and the centromere of the X chromosome. Hum. Genet. 71: 211-214, 1985.

Chen, Z.-Y.; Battinelli, E. M.; Fielder, A.; Bundey, S.; Sims, K.; Breakefield, X. O.; Craig, I. W.: A mutation in the Norrie disease gene (NDP) associated with X-linked familial exudative vitreoretinopathy. Nature Genet. 5: 180-183, 1993.

Chen, Z.-Y.; Battinelli, E. M.; Hendriks, R. W.; Powell, J. F.; Middleton-Price, H.; Sims, K. B.; Breakefield, X. O.; Craig, I. W.: Norrie disease gene: characterization of deletions and possible function. Genomics 16: 533-535, 1993.

Chen, Z.-Y.; Battinelli, E. M.; Woodruff, G.; Young, I.; Breakefield, X. O.; Craig, I. W.: Characterization of a mutation within the NDP gene in a family with a manifesting female carrier. Hum. Molec. Genet. 2: 1727-1729, 1993.

Chen, Z.-Y.; Hendriks, R. W.; Jobling, M. A.; Powell, J. F.; Breakefield, X. O.; Sims, K. B.; Craig, I. W.: Isolation and characterization of a candidate gene for Norrie disease. Nature Genet. 1: 204-208, 1992.

Chen, Z.-Y.; Sims, K. B.; Coleman, M.; Donnai, D.; Monaco, A.; Breakefield, X. O.; Davies, K. E.; Craig, I. W.: Characterization of a YAC containing part or all of the Norrie disease locus. Hum. Molec. Genet. 1: 161-164, 1992.

Chynn, E. W.; Walton, D. S.; Hahn, L. B.; Dryja, T. P.: Norrie disease: diagnosis of a simplex case by DNA analysis. Arch. Ophthal. 114: 1136-1138, 1996.

Clarke, E.: 'Pseudo-glioma' in both eyes. Trans. Ophthal. Soc. U.K. 18: 136-138, 1898.

Collins, F. A.; Murphy, D. L.; Reiss, A. L.; Sims, K. B.; Lewis, J. G.; Freund, L.; Karoum, F.; Zhu, D.; Maumenee, I. H.; Antonarakis, S. E.: Clinical, biochemical, and neuropsychiatric evaluation of a patient with a contiguous gene syndrome due to a microdeletion Xp11.3 including the Norrie disease locus and monoamine oxidase (MAOA and MAOB) genes. Am. J. Med. Genet. 42: 127-134, 1992.

Dahlberg-Parrow, R.: Congenital sex-linked pseudoglioma and grave mental deficiency. Acta Ophthal. 34: 250-254, 1956.

de la Chapelle, A.; Sankila, E.-M.; Lindlof, M.; Aula, P.; Norio, R.: Norrie disease caused by a gene deletion allowing carrier detection and prenatal diagnosis. Clin. Genet. 28: 317-320, 1985.

Diergaarde, P. J.; Wieringa, B.; Bleeker-Wagemakers, E. M.; Sims, K. B.; Breakefield, X. O.; Ropers, H.-H.: Physical fine-mapping of a deletion spanning the Norrie gene. Hum. Genet. 84: 22-26, 1989.

Donnai, D.; Mountford, R. C.; Read, A. P.: Norrie disease resulting from a gene deletion: clinical features and DNA studies. J. Med. Genet. 25: 73-78, 1988.

Dudgeon, J.: Familial exudative vitreo-retinopathy. Trans. Ophthal. Soc. U.K. 99: 45-49, 1979.

Duke-Elder, J. R.: Pseudoglioma in children: aspects of clinical and pathological diagnosis. Sth. Med. J. 51: 754-759, 1958.

Forssman, H.: Mental deficiency and pseudoglioma, a syndrome inherited as an X-linked recessive. Am. J. Ment. Defic. 64: 984-987, 1960.

Fuchs, S.; Kellner, U.; Wedemann, H.; Gal, A.: Missense mutation (Arg121Trp) in the Norrie disease gene associated with X-linked exudative vitreoretinopathy. Hum. Mutat. 6: 257-259, 1995.

Fuchs, S.; Xu, S. Y.; Caballero, M.; Salcedo, M.; La O, A.; Wedemann, H.; Gal, A.: A missense point mutation (Leu13Arg) of the Norrie disease gene in a large Cuban kindred with Norrie disease. Hum. Molec. Genet. 3: 655-656, 1994.

Fullwood, P.; Jones, J.; Bundey, S.; Dudgeon, J.; Fielder, A. R.; Kilpatrick, M. W.: X linked exudative vitreoretinopathy: clinical features and genetic linkage analysis. Brit. J. Ophthal. 77: 168-170, 1993.

Gal, A.; Bleeker-Wagemakers, L.; Wienker, T. F.; Warburg, M.; Ropers, H.-H.: Localization of the gene for Norrie disease by linkage to the DXS7 locus. (Abstract) Cytogenet. Cell Genet. 40: 633, 1985.

Gal, A.; Stolzenberger, C.; Wienker, T.; Wieacker, P.; Ropers, H.-H.; Friedrich, U.; Bleeker-Wagemakers, L.; Pearson, P.; Warburg, M.: Norrie's disease: close linkage with genetic markers from the proximal short arm of the X chromosome. Clin. Genet. 27: 282-283, 1985.

Gal, A.; Uhlhaas, S.; Glaser, D.; Grimm, T.: Prenatal exclusion of Norrie disease with flanking DNA markers. Am. J. Med. Genet. 31: 449-453, 1988.

Gal, A.; Wieringa, B.; Smeets, D. F. C. M.; Bleeker-Wagemakers, L.; Ropers, H. H.: Submicroscopic interstitial deletion of the X chromosome explains a complex genetic syndrome dominated by Norrie disease. Cytogenet. Cell Genet. 42: 219-224, 1986.
Harendra de Silva, D. G.; de Silva, D. B. K.: Norrie's disease in an Asian family. Brit. J. Ophthal. 72: 62-64, 1988.
Harmon, D. L.; Gardner-Medwin, D.; Stirling, J. L.: Two new mutations in a late infantile Tay-Sachs patient are both in exon 1 of the beta-hexosaminidase alpha subunit gene. J. Med. Genet. 30: 123-128, 1993.
Hattori, Y.; Yamashiro, Y.; Ohba, Y.; Miyaji, T.; Morishita, M.; Yamamoto, K.; Yamamoto, K.; Narai, S.; Kimura, A.: A new beta-thalassemia mutation (initiation codon ATG-to-GTG) found in the Japanese population. Hemoglobin 15: 317-325, 1991.
Holmes, L. B.: Norrie's disease—an X-linked syndrome of retinal malformation, mental retardation and deafness. New Eng. J. Med. 284: 367-368, 1971.
Isashiki, Y.; Ohba, N.; Yanagita, T.; Hokita, N.; Doi, N.; Nakagawa, M.; Ozawa, M.; Kuroda, N.: Novel mutation at the initiation codon in the Norrie disease gene in two Japanese families. Hum. Genet. 95: 105-108, 1995.
Isashiki, Y.; Ohba, N.; Yanagita, T.; Hokita, N.; Hotta, Y.; Hayakawa, M.; Fujiki, K.; Tanabe, U.: Mutations in the Norrie disease gene: a new mutation in a Japanese family. (Letter) Brit. J. Ophthal. 79: 703-708, 1995.
Johnson, K.; Mintz-Hittner, H. A.; Conley, Y. P.; Ferrell, R. E.: X-linked exudative vitreoretinopathy caused by an arginine to leucine substitution (R121L) in the Norrie disease protein. Clin. Genet. 50: 113-115, 1996.
Johnston, S. S.; Hanna, J. E.; Nevin, N. C.; Bryars, J. H.: Norrie's disease. Birth Defects Orig. Art. Ser. 18(6): 729-738, 1982.
Katayama, S.; Wohlferd, M.; Golbus, M. S.: First demonstration of recombination between the gene for Norrie disease and probe L1.28. Am. J. Med. Genet. 30: 967-970, 1988.
Kivlin, J. D.; Sanborn, G. E.; Wright, E.; Cannon, L.; Carey, J.: Further linkage data on Norrie disease. Am. J. Med. Genet. 26: 733-736, 1987.
Lindsay, S.; Thiselton, D. L.; Bateman, J. B.; Ngo, J. T.; Sparkes, R. S.; Coleman, M.; Davies, K. E.; Bhattacharya, S. S.: Localisation of the gene for Norrie disease to between DXS7 and DXS426 on Xp. Hum. Genet. 88: 349-350, 1992.
Meindl, A.; Berger, W.; Meitinger, T.; van de Pol, D.; Achatz, H.; Dorner, C.; Haasemann, M.; Hellebrand, H.; Gal, A.; Cremers, F.; Ropers, H.-H.: Norrie disease is caused by mutations in an extracellular protein resembling C-terminal globular domain of mucins. Nature Genet. 2: 139-143, 1992.
Meindl, A.; Lorenz, B.; Achatz, H.; Hellebrand, H.; Schmitz-Valckenberg, P.; Meitinger, T.: Missense mutations in the NDP gene in patients with a less severe course of Norrie disease. Hum. Molec. Genet. 4: 489-490, 1995.
Meitinger, T.; Meindl, A.; Bork, P.; Rost, B.; Sander, C.; Haasemann, M.; Murken, J.: Molecular modeling of the Norrie disease protein predicts a cystine knot growth factor tertiary structure. Nature Genet. 5: 376-380, 1993.
Moreira-Filho, C. A.; Neustein, I.: A presumptive new variant of Norrie's disease. J. Med. Genet. 16: 125-128, 1979.
Nance, W. E.; Hara, S.; Hansen, A.; Elliott, J.; Lewis, M.; Chown, B.: Genetic linkage studies in a Negro kindred with Norrie's disease. Am. J. Hum. Genet. 21: 423-429, 1969.
Ngo, J.; Spence, M. A.; Cortessis, V.; Bateman, J. B.; Sparkes, R. S.: Duplicate report crossing over in Norrie disease family. (Letter) Am. J. Med. Genet. 33: 286, 1989.
Ngo, J. T.; Bateman, J. B.; Cortessis, V.; Sparkes, R. S.; Mohandas, T.; Inana, G.; Spence, M. A.: Norrie disease: linkage analysis using a 4.2-kb RFLP detected by a human ornithine aminotransferase cDNA probe. Genomics 4: 539-545, 1989.
Ngo, J. T.; Spence, M. A.; Cortessis, V.; Sparkes, R. S.; Bateman, J. B.: Recombinational event between Norrie disease and DXS7 loci. Clin. Genet. 34: 43-47, 1988.
Norrie, G.: Nogle Blindhedsaarsager: en oversigt. Hospitalstidende 76: 141-147, 1933.
Ohba, N.; Yamashita, T.: Primary vitreoretinal dysplasia resembling Norrie's disease in a female: association with X autosome chromosomal translocation. Brit. J. Ophthal. 70: 64-71, 1986.
Patten, J. L.; Johns, D. R.; Valle, D.; Eil, C.; Gruppuso, P. A.; Steele, G.; Smallwood, P. M.; Levine, M. A.: Mutation in the gene encoding the stimulatory G protein of adenylate cyclase in Albright's hereditary osteodystrophy. New Eng. J. Med. 322: 1412-1419, 1990.
Phillips, C. I.; Newton, M.; Duvall, J.; Holloway, S.; Levy, A. M.: Probably Norrie's disease due to mutation: two sporadic sibships of two males each, a necropsy of one case, and, given Norrie's disease, a calculation of the gene mutation frequency. Brit. J. Ophthal. 70: 305-313, 1986.
Rehm, H. L.; Gutierrez-Espeleta, G. A.; Garcia, R.; Jimenez, G.; Khetarpal, U.; Priest, J. M.; Sims, K. B.; Keats, B. J. B.; Morton, C. C.: Norrie disease gene mutation in a large Costa Rican kindred with a novel phenotype including venous insufficiency. Hum. Mutat. 9: 402-408, 1997.
Schuback, D. E.; Chen, Z. Y.; Craig, I. W.; Breakefield, X. O.; Sims, K. B.: Mutations in the Norrie disease gene. Hum. Mutat. 5: 285-292, 1995.
Shastry, B. S.; Hejtmancik, J. F.; Plager, D. A.; Hartzer, M. K.; Trese, M. T.: Linkage and candidate gene analysis in X-linked familial exudative vitreoretinopathy. Genomics 27: 341-344, 1995.
Shastry, B. S.; Hejtmancik, J. F.; Trese, M. T.: Identification of novel missense mutations in the Norrie disease gene associated with one X-linked and four sporadic cases of familial exudative vitreoretinopathy. Hum. Mutat. 9: 396-401, 1997.
Sims, K. B.; de la Chapelle, A.; Norio, R.; Sankila, E.-M.; Hsu, Y.-P. P.; Rinehart, W. B.; Corey, T. J.; Ozelius, L.; Powell, J. F.; Bruns, G.; Gusella, J. F.; Murphy, D. L.; Breakefield, X. O.: Monoamine oxidase deficiency in males with an X chromosome deletion. Neuron 2: 1069-1076, 1989.
Sims, K. B.; Lebo, R. V.; Benson, G.; Shalish, C.; Schuback, D.; Chen, Z. Y.; Bruns, G.; Craig, I. W.; Golbus, M. S.; Breakefield, X. O.: The Norrie disease gene maps to a 150 kb region on chromosome Xp11.3. Hum. Molec. Genet. 1: 83-89, 1992.
Sims, K. B.; Ozelius, L.; Corey, T.; Rinehart, W. B.; Liberfarb, R.; Haines, J.; Chen, W. J.; Norio, R.; Sankila, E.; de la Chapelle, A.; Murphy, D. L.; Gusella, J.; Breakefield, X. O.: Norrie disease gene is distinct from the monoamine oxidase genes. Am. J. Hum. Genet. 45: 424-434, 1989.
Taylor, P. J.; Coates, T.; Newhouse, M. L.: Episkopi blindness: hereditary blindness in a Greek Cypriot family. Brit. J. Ophthal. 43: 340-344, 1959.
Torrente, I.; Mangino, M.; Gennarelli, M.; Novelli, G.; Giannotti, A.; Vadala, P.; Dallapiccola, B.: Two new missense mutations (A105T and C110G) in the Norrin gene in two Italian families with Norrie disease and familial exudative vitreoretinopathy. (Letter) Am. J. Med. Genet. 72: 242-244, 1997.
Walker, J. L.; Dixon, J.; Fenton, C. R.; Hungerford, J.; Lynch, S. A.; Stenhouses, S. A. R.; Christian, A.; Craig, I. W.: Two new mutations in exon 3 of the NDP gene: S73X and S101F associated with severe and less severe ocular phenotype, respectively. Hum. Mutat. 9: 53-56, 1997.
Warburg, M.: Norrie's disease: a new hereditary bilateral pseudotumour of the retina. Acta Ophthal. (Copenh) 39: 757-772, 1961.
Warburg, M.: Norrie's disease (atrofia bulborum hereditaria). Acta Ophthal. 41: 134-146, 1963.
Warburg, M.: Personal Communication. Copenhagen, Denmark 1966.
Warburg, M.: Norrie's disease, a congenital progressive oculo-acoustico-cerebral degeneration. Acta Ophthal. 89 (suppl.): 1-147, 1966.
Warburg, M.; Hauge, M.; Sanger, R.: Norrie's disease and the Xg blood group system: linkage data. Acta Genet. Statist. Med. 15: 103-115, 1965.
Whitnall, S. E.; Norman, R. M.: Microphthalmia and the visual pathways: a case associated with blindness and imbecility, and sex-linked. Brit. J. Ophthal. 24: 229-244, 1940.
Wilson, W. M. G.: Congenital blindness (pseudoglioma) occurring as a sex-linked developmental anomaly. Canad. Med. Assoc. J. 60: 580-584, 1949.
Wolff, G.; Mayerova, A.; Wienker, T. F.; Atalianis, P.; Ioannou, P.; Warburg, M.: Clinical reinvestigation and linkage analysis in the family with Episkopi blindness (Norrie disease). J. Med. Genet. 29: 816-819, 1992.
Wong, F.; Goldberg, M. F.; Hao, Y.: Identification of a nonsense mutation at codon 128 of the Norrie's disease gene in a male infant. Arch. Ophthal. 111: 1553-1557, 1993.
Woodruff, G.; Newbury-Ecob, R.; Plaha, D. S.; Young, I. D.: Manifesting heterozygosity in Norrie's disease?. Brit. J. Ophthal. 77: 813-814, 1993.
Zhu, D.; Antonarakis, S. E.; Schmeckpeper, B. J.; Diergaarde, P. J.; Greb, A. E.; Maumenee, I. H.: Microdeletion in the X-chromosome and prenatal diagnosis in a family with Norrie disease. Am. J. Med. Genet. 33: 485-488, 1989.

OSTEOPOROSIS-PSEUDOGLIOMA SYNDROME (*259770)

JJM NOTE

This entity is included because of the mass effect; it is not a true neoplasm.

Bianchine et al. (1972) described 3 families. One of the patients, who had earlier been reported by Bianchine and Murdoch (1969), had had many fractures, suggesting osteogenesis imperfecta. In addition, at the age of a few weeks, presumed retinoblastoma was discovered in each eye. Enucleation was performed after preparatory irradiation. Histology showed pseudoglioma. Opitz (1972) observed the syndrome in 2 brothers and a sister who were mentally retarded. Briard and Frezal (1976) suggested that muscular hypotonia and ligamentous laxity are integral features and that intelligence is normal in most cases. Cases such as that of Meyer (1955) may have been instances of this disorder. A review of published cases suggested to Frontali et al. (1985) that OPS may be more frequent in Mediterranean countries. Beighton et al. (1985) gave the designation ocular form of osteogenesis imperfecta to the disorder they observed in 4 brothers and a nephew and niece of theirs (related as first cousins). In addition to severe osteogenesis imperfecta, they had blindness due to hyperplasia of the vitreous, corneal opacity, and secondary glaucoma. Robinow (1985), Brude (1986), and Superti-Furga et al. (1986) suggested that this is really the osteoporosis-pseudoglioma syndrome; earlier when I wrote to Beighton with the same suggestion, he agreed that the latter diagnosis was likely. Two obligate heterozygotes in the family described by Superti-Furga et al. (1986) had had 3 fractures each and had osteoporosis, suggesting the possibility of clinical expression of this recessive gene in heterozygotes (Superti-Furga, 1989). Frontali and Dallapiccola (1986) likewise concluded that Beighton's ocular OI is osteoporosis-pseudoglioma syndrome, and Beighton (1986) acknowledged the diagnosis. Teebi et al. (1988) described OPS in 2 brothers and a sister whose parents were phenotypically normal first cousins. All 3 sibs had the additional feature of ventricular septa l defect. Somer et al. (1988) described affected brother and sister from consanguineous parents in a Finnish kindred. De Paepe et al. (1993) reported the cases of 2 unrelated patients who were single offspring of nonconsanguineous, healthy parents. One, a 17-year-old girl, had serious visual impairment since birth. Retrolental masses were demonstrated echographically. The eye abnormality had been diagnosed as Reese retinal dysplasia (266400). She was severely dwarfed and had major skeletal deformities resulting in inability to walk after age 2 years. The second patient was an 18-year-old girl with unilateral neonatal blindness, short stature, and deformities, mainly of pelvis and lower limbs. Visual acuity was decreased in the right eye and even light perception was completely absent on the left. At the age of 16 years, vision in the right eye worsened as a result of a vitreous hemorrhage and persistence of the primary vitreous was noted. Furthermore, a retrolental dense mass was documented ultrasonographically in the right eye. Skeletal x-rays demonstrated the clinical variability in this disorder. Collagen studies yielded normal results in both patients, thus differentiating it from severe osteogenesis imperfecta, which it resembled.

Gong et al. (1996) analyzed 16 DNA samples (7 affected individuals) from 3 different consanguineous kindreds with OPS. Using a combination of traditional linkage analysis and homozygosity mapping, they assigned the OPS locus to 11q12-q13. Studies in 10 additional families of varying ehtnic background narrowed the candidate region to a 13-cM interval between D11S1298 and D11S971. The maximum combined lod score was 5.99 at theta = 0 with marker D11S987. Gong et al. (1996) found that the most likely location of the OPS gene is in a 3-cM region between GSTP1 (138370) and D11S1296. Their data supported locus homogeneity and provided no evidence of a founder effect.

See also Neuhauser et al. (1976); Saraux et al. (1967)

Beighton, P.: Osteoporosis-pseudoglioma syndrome. (Letter) Clin. Genet. 29: 263 only, 1986.
Beighton, P.; Winship, I.; Behari, D.: The ocular form of osteogenesis imperfecta: a new autosomal recessive syndrome. Clin. Genet. 28: 69-75, 1985.
Bianchine, J. W.; Briard-Guillemot, M. L.; Maroteaux, P.; Frezal, J.; Harrison, H. E.: Generalized osteoporosis with bilateral pseudoglioma—an autosomal recessive disorder of connective tissue: report of three families—review of the literature. (Abstract) Am. J. Hum. Genet. 24: 34A only, 1972.
Bianchine, J. W.; Murdoch, J. L.: Juvenile osteoporosis (?) in a boy with bilateral enucleation of the eyes for pseudoglioma. Birth Defects Orig. Art. Ser. V(4): 225-226, 1969.
Briard, M. L.; Frezal, J.: Le pseudo-gliome bilateral avec osteoporose generalisee: une affection recessive autosomique. J. Genet. Hum. 24 (suppl.): 65-74, 1976.
Brude, E.: Ocular osteogenesis imperfecta. (Letter) Clin. Genet. 29: 187 only, 1986.
De Paepe, A.; Leroy, J. G.; Nuytinck, L.; Meire, F.; Capoen, J.: Osteoporosis-pseudoglioma syndrome. Am. J. Med. Genet. 45: 30-37, 1993.
Frontali, M.; Dallapiccola, B.: Osteoporosis-pseudoglioma syndrome and the ocular form of osteogenesis imperfecta. (Letter) Clin. Genet. 29: 262 only, 1986.
Frontali, M.; Stomeo, C.; Dallapiccola, B.: Osteoporosis-pseudoglioma syndrome: report of three affected sibs and an overview. Am. J. Med. Genet. 22: 35-47, 1985.
Gong, Y.; Vikkula, M.; Boon, L.; Beighton, P.; Ramesar, R.; Peltonen, L.; Somer, H.; Hirose, T.; Dallapiccola, B.; De Paepe, A.; Swoboda, W.; Zabel, B.; Superti-Furga, A.; Steinmann, B.; Brunner, H. G.; Jans, A.; Boles, R. G.; Adkins, W.; van den Boogaard, M.-J.; Olsen, B. R.; Warman, M. L.: Osteoporosis-pseudoglioma syndrome, a disorder affecting skeletal strength and vision, is assigned to chromosome region 11q12-13. Am. J. Hum. Genet. 59: 146-151, 1996.
Meyer, H. J.: Atypical osteogenesis imperfecta: Lobstein's disease. Arch. Pediat. 72: 182-186, 1955.
Neuhauser, G.; Kaveggia, E. G.; Opitz, J. M.: Autosomal recessive syndrome of pseudogliomatous blindness, osteoporosis and mild mental retardation. Clin. Genet. 9: 324-332, 1976.
Opitz, J. M.: Personal Communication. Madison, Wis. 1972.
Robinow, M.: Osteoporosis-pseudoglioma syndrome?. (Letter) Clin. Genet. 28: 359 only, 1985.
Saraux, H.; Frezal, J.; Roy, C.; Aron, J. J.; Hayat, B.; Lamy, M.: Pseudo-gliome et fragilite osseuse hereditaire a transmission autosomal recessive. Ann. Oculist. 200: 1241-1252, 1967.
Somer, H.; Palotie, A.; Somer, M.; Hoikka, V.; Peltonen, L.: Osteoporosis-pseudoglioma syndrome: clinical, morphological, and biochemical studies. J. Med. Genet. 25: 543-549, 1988.
Superti-Furga, A.: Personal Communication. Zurich, Switzerland 4/12/1989.
Superti-Furga, A.; Steinmann, B.; Perfumo, F.: Osteoporosis-pseudoglioma or osteogenesis imperfecta?. (Letter) Clin. Genet. 29: 184-185, 1986.
Teebi, A. S.; Al-Awadi, S. A.; Marafie, M. J.; Bushnaq, R. A.; Satyanath, S.: Osteoporosis-pseudoglioma syndrome with congenital heart disease: a new association. J. Med. Genet. 25: 32-36, 1988.

AMAUROTIC IDIOCY, ADULT-TYPE (*204300)

JJM NOTE

This entity is included because of one case of astrocytoma complicating the disorder (which is also listed next as 162350).

The case of neuronal ceroid lipofuscinosis (NCL or CLN) reported by Kufs (1925) had onset at age 26 and death at age 38. Fine et al. (1960) found reports of 18 complete histologic descriptions. Cases with the anatomic characteristics listed for the juvenile form may have late onset, making it possible that the adult and juvenile forms are in fact one entity (Zeman and Hoffman, 1962). Chou and Thompson (1970) reported the morphologic changes in a man who was well until age 17 and died at age 32. A sister was said to have died of a similar clinical picture (seizures, intellectual deterioration, lack of motor control, development of athetoid movements). The parents were well and were related as first cousins.

A kindred with presumed autosomal dominant inheritance was reported by Boehme et al. (1971). Armstrong et al. (1974) reported deficiency of leukocyte peroxidase in patients with the late infantile and juvenile forms of ceroid lipofuscinosis (204200, 204500). Armstrong et al. (1974) later studied 3 sibs from Boehme's family and found that all 3 had low peroxidase although only 2 were clinically affected. Brodner and Noh (1976) studied a 24-year-old man from the family with presumed dominant Kufs disease reported by Boehme et al. (1971). Cortical biopsy at the time of craniotomy for removal of astrocytoma showed changes indicative of Kufs disease. Age of onset, which is usually uniform within families, was 31 in this man's kindred. (The family of Boehme et al. (1971) is presented elsewhere (162350) as a distinct dominant disorder.)

Dom et al. (1979) described 2 brothers with Kufs disease. In both, the first manifestation took the form of generalized epileptic seizures, at ages 30 and 32, followed by a cerebellar syndrome with myoclonic jerks and extrapyramidal symptoms. Autopsy showed extensive storage of ceroid lipofuscin in the central nervous system (curvilinear bodies), in hepatocytes, in heart muscle and in the retina. Muscle biopsy is a diagnostic tool; the surviving younger brother showed typical accumulation on biopsy of peroneal muscle. The eye grounds were normal.

See also Armstrong et al. (1974)

Armstrong, D.; Dimmitt, S.; Boehme, D. H.: Leukocyte peroxidase deficiency in a family with a dominant form of Kufs' disease. Science 186: 155-156, 1974.
Armstrong, D.; Dimmitt, S.; VanWormer, D. E.: Studies in Batten disease. Arch. Neurol. 30: 144-152, 1974.
Boehme, D. H.; Cottrell, J. C.; Leonberg, S. C.; Zeman, W.: A dominant form of neuronal ceroid-lipofuscinosis. Brain 94: 745-760, 1971.
Brodner, R. A.; Noh, J. M.: Early diagnosis of Kufs' disease (Letter) Lancet II: 1024 only, 1976.
Chou, S. M.; Thompson, H. G.: Electron microscopy of storage cytosomes in Kufs' disease. Arch. Neurol. 23: 489-501, 1970.
Dom, R.; Brucher, J. M.; Ceuterick, C.; Carton, H.; Martin, J. J.: Adult ceroid-lipofuscinosis (Kufs' disease) in two brothers. Retinal and visceral storage in one; diagnostic muscle biopsy in the other. Acta Neuropath. 45: 67-72, 1979.
Fine, D. I.; Barron, K. D.; Hirano, A.: Central nervous system lipidosis in an adult with atrophy of the cerebellar granular layer. A case report. J. Neurol. 19: 355-369, 1960.
Kufs, H.: Uber eine Spatform der amaurotischen Idiotie und ihre heredofamiliaren Grundlagen. Z. Ges. Neurol. Psychiat. 95: 169-188, 1925.
Zeman, W.; Hoffman, J.: Juvenile and late forms of amaurotic idiocy in one family. J. Neurol. Neurosurg. Psychiat. 25: 352-362, 1962.

NEURONAL CEROID-LIPOFUSCINOSIS, DOMINANT (PARRY) TYPE (*162350)

JJM NOTE

This entity, like the previous, is included because of one case complicated by astrocytoma.

The strikingly consistent clinical picture has onset at about age 31 years and presents as a cerebellar syndrome. Major fits, myoclonic jerks, progressive dementia, and hypertension are other features. Boehme et al. (1971) reported 11 cases in 4 generations of a family named Parry from the southern part of New Jersey. Zeman and Dyken (1969) abandoned the age-dependent classification of 'amaurotic familial idiocies' and divided them into the gangliosidoses and the neuronal ceroid lipofuscinoses. The latter group, of which the biochemical lesions remain obscure, includes amaurotic idiocy, late infantile type (204500), amaurotic idiocy, juvenile type (204200) and amaurotic idiocy, adult type (204300). All of these are recessive. The disorder in the Parry kindred was clearly dominant. Several similar reported families were found, but none could confidently be said to be identical. Premorbid diagnosis was achieved by brain biopsy in a member of the kindred of Boehme et al. (1971) who was undergoing surgery for astrocytoma (Brodner and Noh, 1976).

See also Armstrong et al. (1974); Brodner et al. (1976)

Armstrong, D.; Dimmitt, S.; Boehme, D. H.: Leukocyte perioxidase deficiency in a family with a dominant form of Kufs' disease. Science 186: 155-156, 1974.
Boehme, D. H.; Cottrell, J. C.; Leonberg, S. C.; Zeman, W.: A dominant form of neuronal ceroid-lipofuscinosis. Brain 94: 745-760, 1971.
Brodner, R. A.; Noh, J. M.: Early diagnosis of Kufs' disease. (Letter) Lancet II: 1024 only, 1976.
Brodner, R. A.; Noh, J. M.; Fine, E. J.: A dominant form of adult neuronal ceroid-lipofuscinosis (Kufs' disease) with an associated occipital astrocytoma: early

diagnosis by cortical biopsy. J. Neurol. Neurosurg. Psychiat. 39: 231-238, 1976.
Zeman, W.; Dyken, P.: Neuronal ceroid-lipofuscinosis (Batten's disease). Relationship to amaurotic familial idiocy?. Pediatrics 44: 570-583, 1969.

SEIP SYNDROME (*269700)

JJM NOTE

This entity is included because of the possibly hamartomatous nature of the brain lesion seen at autopsy.

The features are generalized lipodystrophy, hyperlipemia, hepatomegaly, acanthosis nigricans, elevated basal metabolic rate, and nonketotic insulin-resistant diabetes mellitus. Studies of pituitary and adrenal function including growth hormone assays have been normal. Polycystic ovaries, muscular hypertrophy, and mental retardation have occurred in some cases. Two affected sibs have been reported in each of 5 families and in 4 other families the parents were consanguineous (Brunzell et al., 1968). Seip (1959) described affected brother and sister. Lipodystrophic muscular hypertrophy (Senior, 1961) may be the same entity. Substances with insulin-antagonizing and fat-mobilizing properties have been found in the urine (Hamwi et al., 1966). Leprechaunism (246200) has some similar features. See Prader-Willi syndrome (176270) for a condition in which abundant fat, muscle hypotonia and small hands and feet are the opposite of the findings in this syndrome. See systemic cystic angiomatosis and Seip syndrome (272500) for discussion of what may be the same entity. Mabry and Hollingsworth (1971) presented evidence for abnormal pituitary function with secretion of an abnormal hormone with melanotrophic and growth hormone properties. In 1 case surgical hypophysectomy was followed by marked improvement. Seip (1971) reviewed published cases. Berge et al. (1976) reported postmortem findings in a case. Specifically they found hypothalamic lesions judged to be of a malformative or hamartomatous nature. They pointed out that Russell's emaciation syndrome (Russell, 1951), which does not appear to be mendelian, has been shown to be due to a hypothalamic lesion, usually glioma, and that cerebral gigantism may likewise be of diencephalic origin. Huseman et al. (1978) reported 3 black sibs. Huseman et al. (1979) pointed out the association of polycystic ovarian disease in 3 black sisters. Labial hypertrophy, sexual precocity, and oligomenorrhea were noted previously in patients with congenital lipodystrophy (e.g., Brunzell et al., 1968). McLean and Hoefnagel (1980) observed partial lipodystrophy (affecting the face, arms and upper torso) in a 16-year-old girl with familial C3 deficiency. This may be the first indication of an immunologic basis of a form of lipodystrophy. Dorasamy (1980) reported a case of an affected female child with first-cousin parents.

Seip and Trygstad (1996) presented information on their 7 patients with congenital generalized lipodystrophy followed over a period of almost 40 years and one patient with acquired generalized lipodystrophy, a disorder that was first reported by Ziegler (1928) and later by Lawrence (1946), and may represent an autoimmune disorder. In congenital lipodystrophy, insulin resistance is present from birth, resulting in hyperinsulinemia, dyslipidemia, and insulin-resistant diabetes with an anabolic syndrome worsened by a voracious appetite. Growth velocity is increased in preschool age children, and organomegaly is observed with hypertrophic cardiomyopathy that can be lethal in early adulthood. Three of the 4 patients of Seip and Trygstad (1996) died at the ages of 24, 32, and 37 years. The fourth, alive at age 39 years, suffered from stenocardia (angina pectoris). Their first patient was born in 1952 of second-cousin parents. She had a healthy twin brother and an affected younger brother. She suffered from severe hyperhidrosis with moist and warm hands, attributable to an increased energy metabolism. Insulin-resistant diabetes developed at age 12 years and diabetic nephropathy and neuropathy were evident by age 16 years. She died at age 32. The affected brother was 192 cm tall at the age of 17 years. He was married with 3 healthy children.

Seip and Trygstad (1996) urged that patients with congenital lipodystrophy avoid excessive food intake. They should have 4 regular meals a day and avoid large meals because they have limited ability to store energy as fat, lacking the buffer capacity of a normal adipose organ. They considered it unimportant whether calories are given as carbohydrate or fat. Easily digestible carbohydrates should be restricted, and dietary fiber is important.

Gedde-Dahl et al. (1996) stated that 5 of the 6 families studied by Seip and Trygstad (1996) came from adjacent rural 'municipalities' of southwestern Norway. Six patients from this area were born between 1951 and 1973, and none between 1974 and 1995. The absence of new cases might be explained by a decrease in the intraregion marriage rate and inbreeding. Genealogic investigation showed that the mutation must have occurred at least 400 years ago. The sixth family was clinically different and geographically sporadic from a Finnish-descent rural east Norwegian population. In general, males appeared to be less severely affected than females. The patient born in the Finnish-derived population of Norway was patient no. 3 of Seip and Trygstad (1996). In infancy he was described as 'not as athletic as the other patients' and pneumoencephalography showed more extensive changes than in the other patients with dilatation of both lateral ventricles, the third ventricle, and the basal cisterns, and he was somewhat more mentally retarded. He died of heart failure at the age of 35 years. Bjornstad et al. (1996) reported that all 8 patients of Seip and Trygstad (1996) had hypertrophic hearts, mostly with deranged diastolic, but also systolic, function. One had pulmonary hypertension. Conspicuous acanthosis nigricans was illustrated in 2 patients aged 8 and 9 years (Seip and Trygstad, 1996).

Uzun et al. (1997) described 3 patients with multiple peripheral pulmonary artery stenoses in association with congenital generalized lipodystrophy. Two were a brother and sister, aged 2 and 6 years, respectively; the third was a 14-year-old girl.

Lie (1996) gave a tribute to Seip on his seventy-fifth birthday.

See also Afifi et al. (1976); Berardinelli (1954); Oseid et al. (1977); Reed et al. (1965); Seip and Trygstad (1963)

Afifi, A. K.; Mire-Salman, J.; Najjar, S.: The myopathology of congenital generalized lipodystrophy: light and electron microscopic observations. Johns Hopkins Med. J. 139: 61-68, 1976.
Berardinelli, W.: An undiagnosed endocrinometabolic syndrome: report of two cases. J. Clin. Endocr. 14: 193-204, 1954.
Berge, T.; Brun, A.; Hansing, B.; Kjellman, B.: Congenital generalized lipodystrophy: report on one case, with special reference to postmortem findings. Acta Path. Microbiol. Scand. 84A: 47-54, 1976.
Bjornstad, P. G.; Foerster, A.; Ihlen, H.: Cardiac findings in generalized lipodystrophy. Acta Paediat. Suppl. 413: 39-43, 1996.
Brunzell, J. D.; Shankle, S. W.; Bethune, J. E.: Congenital generalized lipodystrophy and systemic cystic angiomatosis: the simultaneous occurrence of two unusual syndromes in a single family. Ann. Intern. Med. 69: 501-516, 1968.
Dorasamy, D. S.: Congenital lipodystrophy: a case report. S. Afr. Med. J. 58: 417-420, 1980.
Gedde-Dahl, T., Jr.; Trygstad, O.; Van Maldergem, L.; Magre, J.; van der Hagen, C. B.; Olaisen, B.; Stenersen, M.; Mevag, B.; Berardinelli-Seip Study Group Genetics of the Berardinelli-Seip syndrome (congenital generalized lipodystrophy) in Norway: epidemiology and gene mapping. Acta Paediat. Suppl. 413: 52-58, 1996.
Hamwi, G. J.; Kruger, F. A.; Eymontt, M. J.; Scarpelli, D. G.; Gwinup, G.; Byron, R.: Lipoatrophic diabetes. Diabetes 15: 262-268, 1966.
Huseman, C.; Johanson, A.; Varma, M.; Blizzard, R. M.: Congenital lipodystrophy: an endocrine study in three siblings. I. Disorders of carbohydrate metabolism. J. Pediat. 93: 221-226, 1978.
Huseman, C. A.; Johanson, A. J.; Blizzard, R. M.: Congenital lipodystrophy. II. Association with polycystic ovarian disease. J. Pediat. 95: 72-74, 1979.
Lawrence, R. D.: Lipodystrophy and hepatomegaly with diabetes, lipaemia, and other metabolic disturbances: a case throwing new light on the action of insulin. Lancet I: 724-731 and 773-775, 1946.
Lie, S. O.: A tribute to Martin Seip. Acta Paediat. Suppl. 413: 1 only, 1996.
Mabry, C. C.; Hollingsworth, D. R.: Generalized lipodystrophy (lipoatrophic diabetes): evidence for abnormal pituitary function. (Abstract) Pediat. Res. 5 (suppl.): 397-398, 1971.
McLean, R. H.; Hoefnagel, D.: Partial lipodystrophy and familial C3 deficiency. Hum. Hered. 30: 149-154, 1980.
Oseid, S.; Beck-Nielsen, H.; Pedersen, O.; Sovik, O.: Decreased binding of insulin to its receptor in patients with congenital generalized lipodystrophy. New Eng. J. Med. 296: 245-248, 1977.
Reed, W. B.; Dexter, R. A.; Corley, C. C.; Fish, C.: Congenital lipodystrophic diabetes with acanthosis nigricans: the Seip-Lawrence syndrome. Arch. Derm. 91: 326-334, 1965.
Russell, A.: A diencephalic syndrome of emaciation in infancy and childhood. Arch. Dis. Child. 26: 274 only, 1951.
Seip, M.: Generalized lipodystrophy. Ergeb. Inn. Med. Kinderheilk. 31: 59-95, 1971.
Seip, M.: Lipodystrophy and gigantism with associated endocrine manifestation: a new diencephalic syndrome?. Acta Paediat. 48: 555-574, 1959.
Seip, M.; Trygstad, O.: Generalized lipodystrophy, congenital and acquired (lipoatrophy). Acta Paediat. Suppl. 413: 2-28, 1996.
Seip, M.; Trygstad, O.: Generalized lipodystrophy. Arch. Dis. Child. 38: 447-453, 1963.
Senior, B.: Lipodystrophic muscular hypertrophy. Arch. Dis. Child. 36: 426-431, 1961.
Uzun, O.; Blackburn, M. E. C.; Gibbs, J. L.: Congenital total lipodystrophy and peripheral pulmonary artery stenosis. Arch. Dis. Child. 76: 456-457, 1997.
Ziegler, L. H.: Lipodystrophies: report of seven cases. Brain 51: 147-167, 1928.

Pinealomas

PINEAL HYPERPLASIA, INSULIN-RESISTANT DIABETES MELLITUS, AND SOMATIC ABNORMALITIES (#262190)

JJM NOTE

Although the original case had an ovarian tumor, this entity is included in nervous system disorders because pineal enlargement is a defining feature.

A number sign (#) is used with this entry because of the evidence that the causative mutation(s) reside in the insulin receptor gene (147670).

Rabson and Mendenhall (1956) described 3 sibs (2 girls, 1 boy) who initially presented with dental and skin abnormalities, abdominal distension, and phallic enlargement. The children demonstrated early dentition, a coarse, senile-appearing facies, and striking hirsutism. An 'adult growth of hair of head' at 5 years of age was pictured in the case of one of the girls. In the older girl the genitalia were large enough at the age of 6 months to permit vaginal examination for diagnosis of a left ovarian tumor which was removed soon afterward. The children were mentally precocious. Prognathism and very thick fingernails as well as acanthosis nigricans were also described. Insulin-resistant diabetes developed, and the patients died during childhood of ketoacidosis and intercurrent infections. At necropsy pineal hyperplasia was found in all 3. Rabson and Mendenhall (1956) stated that 'the histories. . .gain in value because the family physician (E.N.M.) brought them and their siblings into the world, and cared for them until they died. . .' Cases probably of the same condition in 2 sisters, offspring of first-cousin parents, were described by Barnes et al. (1974). They did not have abnormality of the teeth and nails and were living at ages 23 and 11 years, when reported. (Accili et al. (1989) demonstrated that these 2 sisters had a point mutation of the insulin receptor gene (147670.0011).) West et al. (1975) described brother and sister with the syndrome. They showed unusual facies, dry skin, acanthosis nigricans, thickened nails, hirsutism, dental precocity and dysplasia, abdominal protuberance, and phallic enlargement. Both had diabetic ketoacidosis with insulin resistance. The elder, a girl, had recurrent sepsis and died at the age of 7.8 years. The pineal body was hyperplastic at autopsy. West and Leonard (1980) gave follow-up information on the surviving brother, aged 12 years. Monocyte-binding studies suggested that the condition is a genetic deficiency of insulin recept ors.

In a 6-year-old boy with clinical features of Rabson-Mendenhall syndrome, Takata et al. (1986) found decreased insulin binding in erythrocytes, cultured fibroblasts, and transformed lymphocytes. This was taken as an indication of a primary defect in insulin receptors. In a patient with the Rabson-Mendenhall syndrome, Moncada et al. (1986) found a 90% decrease in the number of insulin receptors. Since previous studies had shown that the marked decrease is not due to accelerated receptor degradation, they investigated the possibility that reduced synthesis might be responsible. Their experiments suggested that the receptor precursor is synthesized at a normal rate and that once the receptor is inserted into the membrane, it is degraded at a normal rate. Therefore they proposed that there may be a defect between synthesis and insertion into the membrane. Taylor et al. (1983) found residual insulin binding by cultured fibroblasts even though cultured lymphocytes showed markedly reduced insulin binding. It appeared that the residual binding in the case of fibroblasts had a different affinity of insulin binding. Taylor et al. (1983) suggested that the low level of residual binding might involve primarily receptors for insulin-like growth factors rather than insulin itself. Rittey et al. (1988) documented elevated melatonin metabolite excretion in the urine of a patient with Mendenhall syndrome presenting with neonatal hypoglycemia. The secretion of melatonin from the pineal gland is linked to the light-dark cycle, being greater at night in all species, including humans, and it is pulsatile (de Leiva et al., 1990; Lewy et al., 1980). Some patients with hypothalamic hypogonadism have unusually high plasma melatonin concentrations, suggesting that increased pineal activity may be involved in the pathogenesis of this condition (Puig-Domingo et al., 1992).

Accili, D.; Frapier, C.; Mosthaf, L.; McKeon, C.; Elbein, S. C.; Permutt, M. A.; Ramos, E.; Lander, E.; Ullrich, A.; Taylor, S. I.: A mutation in the insulin receptor gene that impairs transport of the receptor to the plasma membrane and causes insulin-resistant diabetes. EMBO J. 8: 2509-2517, 1989.

Barnes, N. D.; Palumbo, P. J.; Hayles, A. B.; Folgar, H.: Insulin resistance, skin changes and virilization: a recessively inherited syndrome possibly due to pineal gland dysfunction. Diabetologia 10: 285-289, 1974.

de Leiva, A.; Tortosa, F.; Peinado, M. A.; Serrano, J.; Rodriguez-Espinosa, J.; Puig-Domingo, M.: Episodic nyctohemeral secretion of melatonin in adult humans: lack of relation with LH pulsatile pattern. Acta Endocr. 122: 76-82, 1990.

Lewy, A. J.; Wehr, T. A.; Goodwin, F. K.; Newsome, D. A.; Markey, S. P.: Light suppresses melatonin secretion in humans. Science 210: 1267-1269, 1980.

Moncada, V. Y.; Hedo, J. A.; Serrano-Rios, M.; Taylor, S. I.: Insulin-receptor biosynthesis in cultured lymphocytes from an insulin-resistant patient (Rabson-Mendenhall syndrome): evidence for defect before insertion of receptor into plasma membrane. Diabetes 35: 802-807, 1986.

Puig-Domingo, M.; Webb, S. M.; Serrano, J.; Peinado, M.-A.; Corcoy, R.; Ruscalleda, J.; Reiter, R. J.; de Leiva, A.: Melatonin-related hypogonadotropic hypogonadism. New Eng. J. Med. 327: 1356-1359, 1992.

Rabson, S. M.; Mendenhall, E. N.: Familial hypertrophy of pineal body, hyperplasia of adrenal cortex and diabetes mellitus. Am. J. Clin. Path. 26: 283-290, 1956.

Rittey, C. D. C.; Evans, T. J.; Gray, C. E.; Paton, R. D.; Bojkowski, C.: Melatonin state in Mendenhall's syndrome. Arch. Dis. Child. 63: 852-854, 1988.

Takata, Y.; Kobayashi, M.; Maegawa, H.; Watanabe, N.; Ishibashi, O.; Shigeta, Y.; Fujinami, A.: A primary defect in insulin receptor in a young male patient with insulin resistance. Metabolism 35: 950-955, 1986.

Taylor, S. I.; Underhill, L. H.; Hedo, J. A.; Roth, J.; Serrano Rios, M.; Blizzard, R. M.: Decreased insulin binding to cultured cells from a patient with the Rabson-Mendenhall syndrome: dichotomy between studies with cultured lymphocytes and cultured fibroblasts. J. Clin. Endocr. Metab. 56: 856-861, 1983.

West, R. J.; Leonard, J. V.: Familial insulin resistance with pineal hyperplasia: metabolic studies and effect of hypophysectomy. Arch. Dis. Child. 55: 619-621, 1980.

West, R. J.; Lloyd, J. K.; Turner, W. M. L.: Familial insulin-resistant diabetes, multiple somatic anomalies, and pineal hyperplasia. Arch. Dis. Child. 50: 703-708, 1975.

TERATOMA, PINEAL (273120)

JJM NOTE

In general, childhood pinealoma seems to occur more frequently in Japan than in the United States.

From Japan, Wakai et al. (1980) described 2 teenage brothers with pineal teratoma. The parents were not related. Both showed upward gaze palsy and horizontal nystagmus with lateral gaze. One showed abnormal intracranial calcification on plain skull film. One presented at age 13 years with a 7-month history of polydipsia, polyuria and loss of appetite. The second presented at age 17 years with progressive right-sided hemiparesis and a 2-year history of diplopia. The familial occurrence of presacral teratoma is noted elsewhere (176450). As noted by Wakai et al. (1980), medulloblastoma has been reported in sibs on a number of occasions and occurs as part of the basal cell nevus syndrome (109400). Schimke (1983) described a paraventricular germinoma in a 22-year-old male whose paternal uncle had a benign cystic teratoma removed from his mediastinum at the age of 18 years.

Schimke, R. N.: Familial extragonadal germ cell tumors. (Letter) Am. J. Med. Genet. 15: 175-176, 1983.

Wakai, S.; Segawa, H.; Kitahara, S.; Asano, T.; Sano, K.; Ogihara, R.; Tomita, S.: Teratoma in the pineal region in two brothers: case reports. J. Neurosurg. 53: 239-243, 1980.

Medulloblastomas

MEDULLOBLASTOMA (*155255)

JJM NOTE

Medulloblastoma can be the presenting feature of several Mendelian disorders: the nevoid basal cell carcinoma syndrome, Turcot syndrome, Li-Fraumeni syndrome, and blue rubber bleb nevus.

Isochromosome 17q has been observed in high frequency in cytogenetic studies of medulloblastoma, the most common posterior fossa neoplasm in children. By studies using restriction fragment length polymorphisms, Cogen et al. (1990) showed loss of heterozygosity for 17p sequences in 45% of medulloblastomas. The finding was predictive of a poor clinical response to treatment. Furthermore, a deletion could be mapped to 17p13.1-p12, the same chromosomal region for which loss of alleles has been shown in tumor specimens from patients with colon cancer, and the same region to which the p53 gene (191170) has been mapped. However, using denaturing gradient gel electrophoresis and direct sequencing, Cogen et al. (1992) detected p53 mutations in only 2 of 20 medulloblastoma specimens. Moreover, additional RFLP studies of these 20 specimens showed loss of heterozygosity at a more distal and distinct site, 17p13.3.

Cerebellar medulloblastoma is a feature of the basal cell nevus syndrome (109400), the Von Hippel-Lindau syndrome (193300), and the form of Turcot syndrome (276300) due to mutations in the APC gene (e.g., 175100.0022). In a formal risk analysis for brain tumors in familial adenomatous polyposis, Hamilton et al. (1995) found that the relative risk of cerebellar medulloblastoma in patients with familial adenomatous polyposis was 92 times that for the general population (95% confidence interval, 29 to 269; P less than 0.001).

See the DMBT1 gene (601969), so named for 'deleted in malignant brain tumors,' which Mollenhauer et al. (1997) found to be the site of homozygous intragenic deletions at chromosome 10q25.3-q26.1 in medulloblastoma and glioblastoma multiforme tumor tissue, as well as in brain tumor cell lines.

Cogen, P. H.; Daneshvar, L.; Metzger, A. K.; Duyk, G.; Edwards, M. S. B.; Sheffield, V. C.: Involvement of multiple chromosome 17p loci in medulloblastoma tumorigenesis. Am. J. Hum. Genet. 50: 584-589, 1992.

Cogen, P. H.; Daneshvar, L.; Metzger, A. K.; Edwards, M. S. B.: Deletion mapping of the medulloblastoma locus on chromosome 17p. Genomics 8: 279-285, 1990.

Hamilton, S. R.; Liu, B.; Parsons, R. E.; Papadopoulos, N.; Jen, J.; Powell, S. M.; Krush, A. J.; Berk, T.; Cohen, Z.; Tetu, B.; Burger, P. C.; Wood, P. A.; Taqi, F.; Booker, S. V.; Petersen, G. M.; Offerhaus, G. J. A.; Tersmette, A. C.; Giardiello, F. M.; Vogelstein, B.; Kinzler, K. W.: The molecular basis of Turcot's syndrome. New Eng. J. Med. 332: 839-847, 1995.

Mollenhauer, J.; Wiemann, S.; Scheurlen, W.; Korn, B.; Hayashi, Y.; Wilgenbus, K. K.; van Deimling, A.; Poustka, A.: DMBT1, a new member of the SRCR superfamily, on chromosome 10q25.3-26.1 is deleted in malignant brain tumours. Nature Genet. 17: 32-39, 1997.

C SYNDROME (*211750)

JJM NOTE

This entity is included because of one patient with medulloblastoma.

Opitz et al. (1969) described a brother and sister with a malformation syndrome that included unusual facies, polydactyly, cardiac abnormality and, in the boy, cryptorchidism. Preus et al. (1975) described 2 similar patients who were unrelated. Oberklaid and Danks (1975) described a patient and suggested that the disorder be called the Opitz trigonocephaly syndrome. They were dubious that the cases of Preus et al. (1975) were the same. The peculiar shape of the skull, the unusual facies and the bizarre conformation of the palate were illustrated and described. Flexion deformity of the elbows, wrists and fingers were seen. The child died at 2 weeks of age. About half the patients die in the first year. Antley et al. (1981) brought the total number of cases to 11 and pointed (in an addendum) to affected brother and sister reported earlier under another designation. Normal karyotype, normal parents with multiple affected offspring, equal sex ratio of affected persons, and parental consanguinity make autosomal recessive inheritance highly likely.

Sargent et al. (1985) presented 12 cases of trigonocephaly of which 6 were associated with other malformations. Partial or complete obliteration of the metopic suture is characteristic. The forehead is narrow and pointed, often associated with biparietal widening and a triangular shape of the skull when viewed from above. Trigonocephaly has been observed as part of several chromosomal syndromes. The cases of Sargent et al. (1985) included an example of first-cousin parents and a pair of affected sibs. Isolated trigonocephaly is usually a trivial anomaly. Complex trigonocephaly, even after chromosomal aberrations are excluded, may be heterogeneous and the risk of recurrence for the group as a whole is probably on the order of 10% rather than 25% (Sargent et al., 1985).

Lalatta et al. (1990) reported 3 unrelated patients. One had a large omphalocele. Cabral de Almeida et al. (1992) reported another example of large omphalocele with the C syndrome.

Stratton et al. (1990) reported a presumed case with apparently normal development, and Camera et al. (1990) reported 2 further cases with typical characteristics. Haaf et al. (1991) described this syndrome in the daughter of consanguineous parents. The child showed striking upslanting of the palpebral fissures, small nose with broad root, abnormally modeled ears, short neck with loose skin, polysyndactyly, and prominent clitoris and labia majora. The patient also had Eisenmenger syndrome (ventricular septal defect with pulmonary hypertension and right-to-left shunt through a persistent ductus arteriosus) and was mentally retarded. Glickstein et al. (1995) reported a girl who, in addition to typical manifestations of the C syndrome, had tetralogy of Fallot and agenesis of the corpus callosum. These defects had previously been described in only 1 or 2 patients with this syndrome.

De Koster et al. (1990) described a patient who also had pseudohypoaldosteronism (264350). Since this disorder may represent a defect in the mineralocorticoid receptor, which has been assigned to chromosome 4, De Koster et al. (1990) raised the possibility that the association is due to a cytogenetically undetectable microdeletion in chromosome 4. Schaap et al. (1992) reiterated the suggestion that this disorder may be a microdeletion syndrome which is cytogenetically undetectable. They described 2 cases of trigonocephaly, only one of which appeared to have the Opitz-C syndrome, and reviewed 22 cases from the literature.

Omran et al. (1997) described a case of probable Opitz trigonocephaly syndrome (which they abbreviated OTS) who presented at the age of 12 years with symptoms of raised intercranial pressure due to medulloblastoma. Both feet featured postaxial hexadactyly. Addor et al. (1995) reported a 6-year-old girl with C-trigonocephaly syndrome and diaphragmatic hernia. Two other examples of this complication of the syndrome were included in their report. Their first patient showed retrognathia, high arched palate, broad alveolar ridges, and abnormal oral frenula. Their other 2 cases showed similar oral facial findings.

See also Flatz et al. (1984)

Addor, M.-C.; Stefanutti, D.; Farron, F.; Meinecke, P.; Lacombe, D.; Sarlangue, J.; Prescia, G.; Schorderet, D. F.: C trigonocephaly syndrome with diaphragmnatic hernia. Genet. Counsel. 6: 113-120, 1995.

Antley, R. M.; Hwang, D. S.; Theopold, W.; Gorlin, R. J.; Steeper, T.; Pitt, D.; Danks, D. M.; McPherson, E.; Bartels, H.; Wiedemann, H.-R.; Opitz, J. M.: Further delineation of the C (trigonocephaly) syndrome. Am. J. Med. Genet. 9: 147-163, 1981.

Cabral de Almeida, J. C.; Llerena, J. C., Jr.; Vargas, F. R.; Alonso, M. R.: C syndrome and omphalocele: another example. (Letter) Am. J. Med. Genet. 43: 385 only, 1992.

Camera, G.; Serra, G.; Selicorni, A.: 'C' trigonocephaly syndrome: two additional cases. (Letter) Am. J. Med. Genet. 37: 463-464, 1990.

De Koster, J.; Legius, E.; de Zegher, F.; Devlieger, H.; Fryns, J.-P.; Eggermont, E.: Opitz C syndrome and pseudohypoaldosteronism. Am. J. Med. Genet. 37: 457-459, 1990.

Flatz, S. D.; Schinzel, A.; Doehring, E.; Kamran, D.; Eilers, E.: Opitz trigonocephaly syndrome: report of two cases. Europ. J. Pediat. 141: 183-185, 1984.

Glickstein, J.; Karasik, J.; Garcia Caride, D.; Marion, R. W.: 'C' trigonocephaly syndrome: report of a child with agenesis of the corpus callosum and tetralogy of Fallot, and review. Am. J. Med. Genet. 56: 215-218, 1995.

Haaf, T.; Hofmann, R.; Schmid, M.: Opitz trigonocephaly syndrome. Am. J. Med. Genet. 40: 444-446, 1991.

Lalatta, F.; Clerici Bagozzi, D.; Salmoiraghi, M. G.; Tagliabue, P.; Tischer, C.; Zollino, M.; Di Rocco, C.; Neri, G.; Opitz, J. M.: 'C' trigonocephaly syndrome: clinical variability and possibility of surgical treatment. Am. J. Med. Genet. 37: 451-456, 1990.

Oberklaid, F.; Danks, D. M.: The Opitz trigonocephaly syndrome: a case report. Am. J. Dis. Child. 129: 1348-1349, 1975.

Omran, H.; Hildebrandt, F.; Korinthenberg, R.; Brandis, M.: Probable Opitz trigonocephaly C syndrome with medulloblastoma. Am. J. Med. Genet. 69: 395-399, 1997.

Opitz, J. M.; Johnson, R. C.; McCreadie, S. R.; Smith, D. W.: The C syndrome of multiple congenital anomalies. Birth Defects Orig. Art. Ser. V(2): 161-166, 1969.

Preus, M.; Alexander, W. J.; Fraser, F. C.: The C syndrome. Birth Defects Orig. Art. Ser. XI(2): 58-62, 1975.

Sargent, C.; Burn, J.; Baraitser, M.; Pembrey, M. E.: Trigonocephaly and the Opitz C syndrome. J. Med. Genet. 22: 39-45, 1985.

Schaap, C.; Schrander-Stumpel, C. T. R. M.; Fryns, J. P.: Opitz-C syndrome: on the nosology of mental retardation and trigonocephaly. Genetic Counseling 3: 209-215, 1992.

Stratton, R. F.; Sykes, N. J.; Hassler, T. W.: C syndrome with apparently normal development. Am. J. Med. Genet. 37: 460-462, 1990.

DIHYDROPYRIMIDINE DEHYDROGENASE (*274270)

JJM NOTE

This entity is included because of one case report of a child with medulloblastoma (Berglund et al., 1979).

Dihydropyrimidine dehydrogenase (DPD, or DPYD; EC 1.3.1.2) is the initial and rate-limiting enzyme in the 3-step pathway of uracil and thymidine catabolism and in the pathway leading to the formation of beta-alanine. Berger et al. (1984) presented findings in 3 unrelated patients (2 boys and 1 girl) with a nonspecific clinical picture of cerebral dysfunction and persistent urinary excretion of excessive amounts of uracil, thymine, and 5-hydroxymethyluracil. The excretory pattern suggested deficiency of DPD. Autosomal recessive inheritance was supported by the finding that the parents of 1 patient were first cousins. DPD is also the principal enzyme involved in the degradation of the chemotherapeutic drug 5-fluorouracil (5-FU), which acts by inhibiting thymidylate synthase (188350).

Wadman et al. (1984) postulated deficiency of DPD as the cause of the thymine-uraciluria they observed in a child with autism. This is an example of a pharmacogenetic condition of the pharmacokinetic variety, pharmacodynamic being the other main category.

Tuchman et al. (1985) described a 27-year-old woman who suffered an unusually severe reaction to 5-FU given in limited dosage on a weekly schedule. Symptoms included stomatitis, leukopenia, thrombocytopenia, hair loss, diarrhea, fever, marked weight loss, cerebellar ataxia, and neurologic symptoms, progressing to semicoma. High levels of uracil and thymine were found in the urine of the patient and of one brother; both had very high plasma and urinary concentrations of pyrimidine bases. Serum levels and urinary excretion of uric acid were normal in all members of the family, and the patient's white cell thymidine kinase was normal. The mother's urine showed a small amount of uracil but no thymine. A second brother and a sister showed none of the abnormalities. The authors suggested that the defect may be in dihydropyrimidine dehydrogenase which is involved in pyrimidine base degradation. The defect would not be expected to be apparent clinically unless the subject is given a pyrimidine-base analog. Deficiency of DPD in fibroblasts from patients with thymine-uraciluria was described by Berglund et al. (1979) in a child with medulloblastoma and by Bakkeren et al. (1984).

Diasio et al. (1988) established autosomal recessive inheritance by describing consanguinity in the parents of a homozygote and partial deficiency of DPD in both of her children. The proband, a previously completely healthy 40-year-old woman, suffered severe neurotoxicity due to 5-fluorouracil given for breast cancer. Diasio et al. (1988) found that after administration of a 'test' dose of 5-fluorouracil to this patient, a markedly prolonged elimination half-life (159 min) was observed with no evidence of catabolites of the drug in plasma or cerebrospinal fluid and with 89.7% of the administered dose being excreted into the urine as unchanged drug. Partial deficiency of the enzyme was found also in the patient's father; the mother was deceased. The enzyme was assayed in peripheral blood mononuclear cells.

Brockstedt et al. (1990) commented that 7 of the 8 previously reported pediatric patients were Dutch. They reported a ninth pediatric case. Van Gennip et al. (1994) described a Dutch family in which 1 individual was hospitalized at the age of 25 months with bilateral microphthalmia, coloboma of the iris and choroid, nystagmus, and a gradually increasing psychomotor retardation. The patient was found to be homozygous for a deletion of an exon of 165 nucleotides (274270.0001), whereas both parents and 1 sib were heterozygous.

DPD deficiency is a clinically heterogeneous disorder. Vreken et al. (1997) stated that patients with a nearly complete enzyme defect show convulsive disorders in about 50% of cases, whereas patients experiencing acute 5-fluorouracil toxicity usually show DPD enzymatic activities in the heterozygous range. Vreken et al. (1997) found a 4-bp deletion in the DPYD gene (274270.0003) in a Dutch consanguineous family with DPD deficiency.

To evaluate the prevalence of dihydropyrimidinuria (DHPuria), Sumi et al. (1998) analyzed urine samples from 21,200 healthy Japanese infants and found 2 cases of DHPuria without clinical symptoms. Based on this result, they estimated the prevalence to be approximately 1 in 10,000 births in Japan. In addition, they analyzed pyrimidine catabolism on a previously reported family with an adult DHPuria case (Sumi et al., 1996; Hayashi et al., 1996). In this patient, Hayashi et al. (1996) had demonstrated metabolic changes predicting a risk of severe 5-fluorouracil toxicity. On restudy of the family of this patient, a sister of the proband was found to excrete large amounts of dihydrouracil and dihydrothymine. The parents and the child of the propositus showed slight increases of dihydrouracil and dihydrothymine. This was the first reported instance of a family with 2 cases of DHPuria. To determine the inheritance of DHPuria in this family and to examine the risk of 5-FU toxicity, a uracil loading test was performed on the parents. Urinary dihydrouracil concentrations in the parents after the loading was several times higher than those in normal control persons, the finding being consistent with DHPuria heterozygotes. Sumi et al. (1998) concluded that DHPuria is an autosomal recessive condition. Sumi et al. (1998) suggested that homozygotes may have a high risk of 5-FU toxicity, while the risk may be relatively low in heterozygotes.

Yokota et al. (1994) cloned and sequenced pig and human DPD cDNAs. The pig and human enzymes contain 1,025 amino acids and have a calculated molecular mass of 111,416 and 111,398, respectively. The sequence of the gene suggested that DPD has at least 3 distinct domains. Using somatic cell hybrid strategies, Yokota et al. (1994) mapped the gene for DPD, symbolized DPYD, to the centromeric region of human chromosome 1 between 1p22 and 1q21. By fluorescence in situ hybridization, Takai et al. (1994) assigned the DPYD gene to 1p22.

See also Wadman et al. (1985); Wilcken et al. (1985)

Albin, N.; Johnson, M. R.; Shahinian, H.; Wang, K.; Diasio, R. B.: Initial characterization of the molecular defect in human dihydropyrimidine dehydrogenase deficiency. Proc. Am. Assoc. Cancer Res. (Abstract) 36: 211 only, 1995.

Bakkeren, J. A. J. M.; DeAbreu, R. A.; Sengers, R. C. A.; Gabreels, F. J. M.; Maas, J. M.; Renier, W. O.: Elevated urine, blood and cerebrospinal fluid levels of uracil and thymine in a child with dihydrothymine dehydrogenase deficiency. Clin. Chim. Acta 140: 247-256, 1984.

Berger, R.; Stoker-de Vries, S. A.; Wadman, S. K.; Duran, M.; Beemer, F. A.; de Bree, P. K.; Weits-Binnerts, J. J.; Penders, T. J.; van der Woude, J. K.: Dihydropyrimidine dehydrogenase deficiency leading to thymine-uraciluria: an inborn error of pyrimidine metabolism. Clin. Chim. Acta 141: 227-234, 1984.

Berglund, G.; Greter, J.; Lindstedt, S.; Steen, G.; Waldenstrom, J.; Wass, U.: Urinary excretion of thymine and uracil on a two-year-old child with a malignant tumor of the brain. Clin. Chem. 25: 1325-1328, 1979.

Brockstedt, M.; Jakobs, C.; Smit, L. M. E.; van Gennip, A. H.; Berger, R.: A new case of dihydropyrimidine dehydrogenase deficiency. J. Inherit. Metab. Dis. 13: 121-124, 1990.

Diasio, R. B.; Beavers, T. L.; Carpenter, J. T.: Familial deficiency of dihydropyrimidine dehydrogenase: biochemical basis for familial pyrimidinemia and severe 5-fluorouracil-induced toxicity. J. Clin. Invest. 81: 47-51, 1988.

Harris, B. E.; Carpenter, J. T.; Diasio, R. B.: Severe 5-fluorouracil toxicity secondary to dihydropyrimidine dehydrogenase deficiency: a potentially more common pharmacogenetic syndrome. Cancer 68: 499-501, 1991.

Hayashi, K.; Kidouchi, K.; Sumi, S.; Mizokami, M.; Orito, E.; Kumada, K.; Ueda, R.; Wada, Y.: Possible prediction of adverse reactions to pyrimidine chemotherapy from urinary pyrimidine levels and a case of asymptomatic adult dihydropyrimidinuria. Clin. Cancer Res. 2: 1937-1941, 1996.

Meinsma, R.; Fernandez-Salguero, P.; Van Kuilenburg, A. B. P.; van Gennip A. H.; Gonzalez, F. J.: Human polymorphism in drug metabolism: mutation in the dihydropyrimidine dehydrogenase gene results in exon skipping and thymine uracilurea (sic). DNA Cell Biol. 14: 1-6, 1995.

Ridge, S. A.; Brown, O.; McMurrough, J.; Fernandez-Salguero, P.; Evans, W. E.; Gonzalez, F. J.; McLeod, H. L.: Mutations at codon 974 of the DPYD gene are a rare event. Brit. J. Cancer 75: 178-179, 1997.

Ridge, S. A.; Brown, O.; McMurrough, J.; Fernandez-Salguero, P.; Evans, W. E.; Gonzalez, F. J.; McLeod, H. L.: Mutations at codon 974 of the DPYD gene are a rare event. Brit. J. Cancer 75: 178-179, 1997.

Sumi, S.; Imaeda, M.; Kidouchi, K.; Ohba, S.; Hamajima, N.; Kodama, K.; Togari, H.; Wada, Y.: Population and family studies of dihydropyrimidinuria: prevalence, inheritance mode, and risk of fluorouracil toxicity. Am. J. Med. Genet. 78: 336-340, 1998.

Sumi, S.; Kidouchi, K.; Hayashi, K.; Ohba, S.; Wada, Y.: Dihydropyrimidinuria without clinical symptoms. J. Inherit. Metab. Dis. 19: 701-702, 1996.

Takai, S.; Fernandez-Salguero, P.; Kimura, S.; Gonzalez, F. J.; Yamada, K.: Assignment of the human dihydropyrimidine dehydrogenase gene (DPYD) to chromosome region 1p22 by fluorescence in situ hybridization. Genomics 24: 613-614, 1994.

Tuchman, M.; Stoeckeler, J. S.; Kiang, D. T.; O'Dea, R. F.; Ramnaraine, M. L.; Mirkin, B. L.: Familial pyrimidinemia and pyrimidinuria associated with severe fluorouracil toxicity. New Eng. J. Med. 313: 245-249, 1985.

van Gennip, A. H.; Abeling, N. G. G. M.; Stroomer, A. E. M.; van Lenthe, H.; Bakker, H. D.: Clinical and biochemical findings in six patients with pyrimidine degradation defects. J. Inherit. Metab. Dis. 17: 130-132, 1994.

Van Kuilenburg, A. B. P.; Vreken, P.; Beex, L. V. A. M.; Meinsma, R.; Van Lenthe, H.; De Abreu, R. A.; Van Gennip, A. H.: Heterozygosity for a point mutation in an invariant splice donor site of dihydropyrimidine dehydrogenase and severe 5-fluorouracil related toxicity. Europ. J. Cancer 33: 2258-2264, 1997.

Vreken, P.; van Kuilenburg, A. B. P.; Meinsma, R.; Beemer, F. A.; Duran, M.; van Gennip, A. H.: Dihydropyrimidine dehydrogenase deficiency: a novel mutation and expression of missense mutations in E. coli. J. Inherit. Metab. Dis. 21: 276-279, 1998.

Vreken, P.; Van Kuilenburg, A. B. P.; Meinsma, R.; De Abreu, R. A.; Van Gennip, A. H.: Identification of a four-base deletion (delTCAT296-299) in the dihydropyrimidine dehydrogenase gene with variable clinical expression. Hum. Genet. 100: 263-265, 1997.

Vreken, P.; Van Kuilenburg, A. B. P.; Meinsma, R.; Smit, G. P. A.; Bakker, H. D.; De Abreu, R. A.; van Gennip, A. H.: A point mutation in an invariant splice donor site leads to exon skipping in two unrelated Dutch patients with dihydropyrimidine dehydrogenase deficiency. J. Inherit. Metab. Dis. 19: 645-654, 1996.

Vreken, P.; Van Kuilenburg, A. B. P.; Meinsma, R.; van Gennip, A. H.: Identification of novel point mutations in the dihydropyrimidine dehydrogenase gene. J. Inherit. Metab. Dis. 20: 335-338, 1997.

Vreken, P.; Van Kuilenburg, A. B. P.; Meinsma, R.; van Gennip, A. H.: Dihydropyrimidine dehydrogenase (DPD) deficiency: identification and expression of missense mutations C29R, R886H and R235W. Hum. Genet. 101: 333-338, 1997.

Wadman, S. K.; Beemer, F. A.; de Bree, P. K.; Duran, M.; van Gennip, A. H.; Ketting, D.; van Sprang, F. J.: New defects of pyrimidine metabolism. Adv. Exp. Med. Biol. 165A: 109-114, 1984.

Wadman, S. K.; Berger, R.; Duran, M.; de Bree, P. K.; Stoker-de Vries, S. A.; Beemer, F. A.; Weits-Binnerts, J. J.; Penders, T. J.; van der Woude, J. K.: Dihydropyrimidine dehydrogenase deficiency leading to thymine-uraciluria: an inborn error of pyrimidine metabolism. J. Inherit. Metab. Dis. 8 (suppl. 2): 113-114, 1985.

Wei, X.; McLeod, H. L.; McMurrough, J.; Gonzalez, F. J.; Fernandez-Salguero, P.: Molecular basis of the human dihydropyrimidine dehydrogenase deficiency and 5-fluorouracil toxicity. J. Clin. Invest. 98: 610-615, 1996.

Wilcken, B.; Hammond, J.; Berger, R.; Wise, G.; James, C.: Dihydropyrimidine dehydrogenase deficiency—a further case. J. Inherit. Metab. Dis. 8 (suppl. 2): 115-116, 1985.

Yokota, H.; Fernandez-Salguero, P.; Furuya, H.; Lin, K.; McBride, O. W.; Podschun, B.; Schnackerz, K. D.; Gonzalez, F. J.: cDNA cloning and chromosome mapping of human dihydropyrimidine dehydrogenase, an enzyme associated with 5-fluorouracil toxicity and congenital thymine uraciluria. J. Biol. Chem. 269: 23192-23196, 1994.

POLYSYNDACTYLY WITH PECULIAR SKULL SHAPE (#175700)

JJM NOTE

This entity is included because of one affected young female who also had medulloblastoma.

A number sign is used with this entry because of the evidence that this disorder is caused by mutation in the GLI3 gene (165240).

Greig (1928) described digital malformations and peculiar skull shape in mother and daughter. The mother had syndactyly of both hands. The daughter, of above average intelligence, had polysyndactyly and a peculiar skull shape in the form of expanded cranial vault leading to high forehead and bregma, with no evidence of precocious closure of cranial sutures. The thumbs and great toes had bifid terminal phalanges. Hootnick and Holmes (1972) presented a family and suggested that the 'new' disorder designated frontodigital syndrome by Marshall and Smith (1970) was in fact this condition. Temtamy and McKusick (1978) studied a particularly instructive family in which 10 members of 4 generations in 6 sibships were affected in the pattern of a fully penetrant autosomal dominant trait. Merlob et al. (1981) reported the case of a female infant with postaxial polydactyly of the hands and preaxial polydactyly of the feet, with syndactyly, and craniofacial dysmorphism characterized by frontal bossing. X-ray examination revealed markedly advanced bone age. There was also bilateral hip dislocation. The father of the infant had a high forehead and mild hypertelorism. Fryns et al. (1981) described the disorder in dizygotic 4-month-old twin brothers and in their father; the twins had severe affection, the father mild. Fryns (1982) documented the variability and autosomal dominant inheritance on the basis of 7 cases. In 1 family, a mother and son were affected. Gollop and Fontes (1985) described affected mother and 2 of her 3 sons. Chudley and Houston (1982) described the syndrome in 3 generations of a family and perhaps by implication in a fourth. They commented on phenotypic overlap with the acrocallosal syndrome (200990). Baraitser et al. (1983) reported 13 affected persons in 3 kindreds with, curiously, no male-to-male transmission. They also commented on similarity to the acrocallosal syndrome. The main clinical distinction is mental retardation (which has agene sis of the corpus callosum). Legius et al. (1985) also proposed that the acrocallosal syndrome is the same as Greig syndrome. Baraitser et al. (1983) observed that the facial features of Greig syndrome can be so mild as to be indistinguishable from the normal. Therefore they suggested that type IV preaxial polydactyly, or uncomplicated polysyndactyly (174700), as delineated by Temtamy and McKusick (1978) may be Greig syndrome.

In an analysis of reported cases, Baccichetti et al. (1982) suggested that deletion of part of the 7p21 may be critical in GCPS. Tommerup and Nielsen (1983) described a translocation t(3;7)(p21.1;p13) segregating through 4 generations in invariable association with GCPS. High resolution cytogenetic analysis using G and R banding did not uncover any imbalance of the affected chromosomes, nor were the late replicating patterns changed. A girl with GCPS died with medulloblastoma. See craniosynostosis (123100), which is also thought to be determined by mutation in a gene on 7p. Motegi et al. (1985) reported the case of a boy with a tiny deletion of 7p15.3-7p21.3. From comparison with other cases of 7p deletion, with or without craniosynostosis, they suggested that the critical segment for craniosynostosis may be at 7p21.2 or the proximal part of 7p21.3. Sage et al. (1987) subjected the breakpoints on chromosome 3 and chromosome 7 to molecular genetic analysis.

In 7 GCPS pedigrees with no chromosome abnormality, Brueton et al. (1988) found linkage with epidermal growth factor receptor (EGFR; 131550); maximum lod score = 3.17 at theta = 0.0 (confidence limits = 0-12 cM). EGFR is located at 7p13-p11. Drabkin et al. (1989) identified 2 very closely linked DNA sequences that flanked the translocation breakpoint; no recombination between the disorder and these sequences was found. A pulsed field analysis showed that the disorder is also linked to the TCRG locus, but by this approach Drabkin et al. (1989) found no evidence of linkage to EGFR (131550). The family reported by Ridler et al. (1977) as an example of type II syndactyly (186000) was in fact a family with Greig syndrome, as established by Winter (1989), who revisited the family.

In a patient with Greig cephalopolysyndactyly syndrome and deletion of 7p13-p11.2, Rosenkranz et al. (1989) found molecular evidence of deletion of the EGFR (131550) gene. In a second patient with deletion of 7p14.2-p12.3, they found that the EGFR genes were intact. From the data available, they concluded that the EGFR gene is probably in band 7p12.3-p12.1 and the GCPS gene more distally situated in 7p13-p12.3. 19,18:Pettigrew et al. (1989, 1991) confirmed the assignment to 7p13 by study of the sporadic case of an 11-month-old infant with typical features including macrocephaly, frontal bossing, syndactyly, postaxial polydactyly of the hands, and preaxial polydactyly of the feet. High resolution chromosome analysis showed a 46,XX,del(7)(p13p14)pat chromosome pattern. This was the first report of an interstitial deletion associated with Greig syndrome. Cytogenetic analysis of polymorphisms of the heterochromatin in the pericentromeric region suggested that the deleted chromosome was of paternal origin. Review of clinical features and published reports of patients with a deletion involving 7p13 showed a number to have features overlapping with Greig syndrome. Kruger et al. (1989) reported on cases of Greig syndrome segregating in a large kindred over 4 generations. The disorder was due to reciprocal translocation t(6;7)(q27;p13). One patient in this pedigree had a severe malformation syndrome due to duplication 7p13-pter. Wagner et al. (1990) studied 2 patients with GCPS and a cytogenetically visible microdeletion of 7p with gene probes that had been assigned close to the proposed Greig locus. One patient showed loss of the TCRG gene cluster and both showed hemizygosity for PGAM2 (261670). On the other hand, HOX 1.4 (142950) and IFNB2 (147620) showed normal gene dosage. This suggested that PGAM2 and GCPS are in 7p13-p12.3; TCRG in the distal part of 7p14.2-p13; and HOX 1.4 and IFNB2 distal to 7p14.2. The findings excluded the HOX 1.4 gene from involvement in the pathogene sis of GCPS.

Vortkamp et al. (1991) used a candidate gene approach to test the possible implication of the GLI3 gene in this disorder inasmuch as the GLI3 gene had been mapped to 7p13. Vortkamp et al. (1991) demonstrated that 2 of 3 translocations found to be associated with GCPS interrupt the GLI3 gene. The breakpoints were within the first third of the coding sequence. In the third translocation, chromosome 7 was broken at about 10 kb downstream of the 3-prime end of GLI3.

Winter and Huson (1988) called attention to the evidence that, on both morphologic and comparative gene mapping grounds, Greig cephalopolysyndactyly syndrome is homologous to the mouse mutant 'extra toes' (Xt) on mouse chromosome 13. The pattern of polydactyly in the 2 species is very similar and both conditions probably map close to the T-gamma receptor locus (TCRG; 186970). Vortkamp et al. (1992) reported deletion in the 5-prime end of the Gli-3 gene in an Xt mutant, and Schimmang et al. (1992) reported that expression of Gli-3 is reduced in this mutant. Hui and Joyner (1993) described the molecular characteristics of the Xt mutation. They found that deficiency of expression of Gli-3 in the mutant mouse is due to a deletion within the 3-prime end of the gene. Furthermore, structures affected in the mouse mutant and in the human syndrome were found to correlate with expression domains of Gli-3 in the mouse. These findings strongly supported the suggestion that deficiency of GLI3 function leads to human GCPS.

D. M. Grieg, a Scot, pronounced his name 'Gregg' (Ferguson-Smith, 1996).

Baccichetti, C.; Artifoni, L.; Zanardo, V.: Deletions of the short arm of chromosome 7 without craniosynostosis. (Letter) Clin. Genet. 21: 348-349, 1982.

Baraitser, M.; Winter, R. M.; Brett, E. M.: Greig cephalopolysyndactyly: report of 13 affected individuals in three families. Clin. Genet. 24: 257-265, 1983.

Brueton, L.; Huson, S. M.; Winter, R. B.; Williamson, R.: Chromosomal localisation of a developmental gene in man: direct DNA analysis demonstrates that Greig cephalopolysyndactyly maps to 7p13. Am. J. Med. Genet. 31: 799-804, 1988.

Chudley, A. E.; Houston, C. S.: The Greig cephalopolysyndactyly syndrome in a Canadian family. Am. J. Med. Genet. 13: 269-276, 1982.

Drabkin, H.; Sage, M.; Helms, C.; Green, P.; Gemmill, R.; Smith, D.; Erickson, P.; Hart, I.; Ferguson-Smith, A.; Ruddle, F.; Tommerup, N.: Regional and physical mapping studies characterizing the Greig polysyndactyly 3;7 chromosome translocation, t(3;7)(p21.1;p13). Genomics 4: 518-529, 1989.

Ferguson-Smith, M. A.: Personal Communication. Cambridge, England 12/6/1996.

Fryns, J. P.: Le syndrome de Greig: une polysyndactylie variable associee a une dysmorphie craniofaciale distincte. J. Genet. Hum. 30 (suppl. 5): 403-408, 1982.

Fryns, J. P.; Van Noyen, G.; Van Den Berghe, H.: The Greig polysyndactyly craniofacial dysmorphism syndrome: variable expression in a family. Europ. J. Pediat. 136: 217-220, 1981.

Gollop, T. R.; Fontes, L. R.: The Greig cephalopolysyndactyly syndrome: report of a family and review of the literature. Am. J. Med. Genet. 22: 59-68, 1985.

Greig, D. M.: Oxycephaly. Edinburgh Med. J. 33: 189-218, 1928.

Hootnick, D.; Holmes, L. B.: Familial polysyndactyly and craniofacial anomalies. Clin. Genet. 3: 128-134, 1972.

Hui, C.; Joyner, A. L.: A mouse model of Greig cephalopolysyndactyly syndrome: the extra-toes(J) mutation contains an intragenic deletion of the Gli3 gene. Nature Genet. 3: 241-245, 1993.

Kruger, G.; Gotz, J.; Kvist, U.; Dunker, H.; Erfurth, F.; Pelz, L.; Zech, L.: Greig syndrome in a large kindred due to reciprocal chromosome translocation t(6;7)(q27;p13). Am. J. Med. Genet. 32: 411-416, 1989.

Legius, E.; Fryns, J. P.; Casaer, P.; Boel, M.; Eggermont, E.: Schinzel acrocallosal syndrome: a variant example of the Greig syndrome?. Ann. Genet. 28: 239-240, 1985.

Marshall, R. E.; Smith, D. W.: Frontodigital syndrome: a dominantly inherited disorder with normal intelligence. J. Pediat. 77: 129-133, 1970.

Merlob, P.; Grunebaum, M.; Reisner, S. H.: A newborn infant with craniofacial dysmorphism and polysyndactyly (Greig's syndrome). Acta Paediat. Scand. 70: 275-277, 1981.

Motegi, T.; Ohuchi, M.; Ohtaki, C.; Fujiwara, K.; Enomoto, S.; Hasegawa, T.; Kishi, K.; Hayakawa, H.: A craniosynostosis in a boy with a

del(7)(p15.3p21.3): assignment by deletion mapping of the critical segment for craniosynostosis to the mid-portion of 7p21. Hum. Genet. 71: 160-162, 1985.
Pettigrew, A. L.; Greenberg, F.; Caskey, C. T.; Ledbetter, D. H.: Greig syndrome associated with an interstitial deletion of 7p: confirmation of the localization of Greig syndrome to 7p13. Hum. Genet. 87: 452-456, 1991.
Pettigrew, A. L.; Greenberg, F.; Ledbetter, D. H.; Caskey, C. T.: Greig syndrome associated with an interstitial deletion of 7p: confirmation of the localization of Greig syndrome to 7p13. (Abstract) Am. J. Hum. Genet. 45 (suppl.): A87 only, 1989.
Ridler, M. A. C.; Laxova, R.; Dewhurst, K.; Saldana-Garcia, P.: A family with syndactyly type II (synpolydactyly). Clin. Genet. 12: 213-220, 1977.
Rosenkranz, W.; Kroisel, P. M.; Wagner, K.: Deletion of EGFR-gene in one of two patients with Greig cephalopolysyndactyly syndrome and microdeletion of chromosome 7p. (Abstract) Cytogenet. Cell Genet. 51: 1069 only, 1989.
Sage, M.; Hart, I.; Green, P.; Tommerup, N.; Drabkin, H.: Isolation and analysis of the Greig polysyndactyly-craniofacial anomalies syndrome 3;7 translocation. (Abstract) Am. J. Hum. Genet. 41: A106 only, 1987.
Schimmang, T.; Lemaistre, M.; Vortkamp, A.; Ruther, U.: Expression of the zinc finger gene Gli3 is affected in the morphogenetic mouse mutant extra-toes (Xt). Development 116: 799-804, 1992.
Temtamy, S. A.; McKusick, V. A.: The Genetics of Hand Malformations. New York: Alan R. Liss (pub.) 1978.
Tommerup, N.; Nielsen, F.: A familial reciprocal translocation t(3;7)(p21.1;p13) associated with the Greig polysyndactyly-craniofacial anomalies syndrome. Am. J. Med. Genet. 16: 313-321, 1983.
Vortkamp, A.; Franz, T.; Gessler, M.; Grzeschik, K. H.: Deletion of GLI3 supports the homology of the human Greig cephalopolysyndactyly syndrome (GCPS) and the mouse mutant extra toes (Xt). Mammalian Genome 3: 461-463, 1992.
Vortkamp, A.; Gessler, M.; Grzeschik, K.-H.: GLI3 zinc-finger gene interrupted by translocations in Greig syndrome families. Nature 352: 539-540, 1991.
Wagner, K.; Kroisel, P. M.; Rosenkranz, W.: Molecular and cytogenetic analysis in two patients with microdeletions of 7p and Greig syndrome: hemizygosity for PGAM2 and TCRG genes. Genomics 8: 487-491, 1990.
Winter, R. M.: Personal Communication. Harrow, England 4/17/1989.
Winter, R. M.; Huson, S. M.: Greig cephalopolysyndactyly syndrome: a possible mouse homologue (Xt-extra toes). Am. J. Med. Genet. 31: 793-798, 1988.

Periventricular Tumors

PAPILLOMA OF CHOROID PLEXUS (260500)

JJM NOTE

Besides these case reports, choroid plexus papilloma has been seen in individuals with the Aicardi syndrome (304050) and hypomelanosis of Ito (146150), the latter having a constitutional translocation (X;17). I have heard of cases in the Li-Fraumeni cancer family syndrome.

Komminoth et al. (1965) observed intraventricular papilloma of the choroid plexus in a 2-year-old boy and his 4-year-old sister. Coons et al. (1989) described malignant degeneration of choroid plexus papillomas in 2 sibs. Zwetsloot et al. (1991) observed papillomas of the choroid plexus in a son and daughter of first-cousin parents.

Coons, S.; Johnson, P. C.; Dickman, C. A.; Rekate, H.: Choroid plexus carcinoma in siblings: a study by light and electron microscopy with Ki-67 immunocytochemistry. J. Neuropath. Exp. Neurol. 48: 483-493, 1989.
Komminoth, R.; Woringer, E.; Baumgartner, J.; Braun, J. P.; Le Maistre, D.: Papillome intraventriculaire familial: caracteristiques angiographiques. Neurochirurgie 11: 267-272, 1965.
Zwetsloot, C. P.; Kros, J. M.; Paz y Geuze, H. D.: Familial occurrence of tumours of the choroid plexus. J. Med. Genet. 28: 492-494, 1991.

HYPOTHALAMIC HAMARTOBLASTOMA, HYPOPITUITARISM, IMPERFORATE ANUS, AND POSTAXIAL POLYDACTYLY (PALLISTER-HALL SYNDROME) (#146510)

JJM NOTE

Until additional clinical history is documented, this disorder must be suspected in any person with a hypothalamic hamartoma or hamartoblastoma, in part because of its autosomal dominant inheritance and resulting implications for other family members.

A number sign (#) is used with this entry because of evidence that Pallister-Hall syndrome is caused by frameshift mutations in the GLI3 gene (165240).

Hall et al. (1980) reported 6 infants with a neonatally lethal malformation syndrome of hypothalamic hamartoblastoma, postaxial polydactyly, and imperforate anus. Some had laryngeal cleft, abnormal lung lobation, renal agenesis or dysplasia, short 4th metacarpals, nail dysplasia, multiple buccal frenula, hypoadrenalism, microphallus, congenital heart defect, and intrauterine growth retardation. All cases were sporadic and chromosomes were apparently normal. The parents were nonconsanguineous. No environmental exposure was common to all cases. The ages of the fathers were 21, 25, 25, 29, 43, and unknown. The anterior pituitary was not found in any patient. The hypothalamic tumor was apparent on the inferior surface of the cerebrum and extended from the optic chiasma to the interpeduncular fossa. The tumor replaced the hypothalamus and other nuclei that originate in the embryonic hypothalamic plate. It was composed mainly of cells resembling primitive, undifferentiated germinal cells.

Graham et al. (1983) described an infant with abnormal auricles, short nose with flattened bridge, microglossia, micrognathia, cleft palate, short limbs, dislocated hips, and 4-limb postaxial polydactyly. The infant died at 2 hours of age and autopsy showed hypothalamic hamartoblastoma. A sister of the mother died at 17 hours of age and showed 4-limb polydactyly, recessed mandible, and small tongue; autopsy was not done. See 241800.

Donnai et al. (1987) suggested that the Pallister-Hall syndrome and severe Smith-Lemli-Opitz syndrome (270400) are the same disorder. Iafolla et al. (1989) reported that nail dysplasia accompanies the postaxial polydactyly. They pointed out that magnetic resonance imaging is the most valuable diagnostic tool; CT scan has been reported to miss the tumor. Pallister et al. (1989) described 3 additional cases. One patient had imperforate anus and right hydronephrosis and hydroureter with absent left kidney; another had imperforate anus without the renal anomalies. A teratogen has been suspected because of the fact that cases were not recognized before February 1978. Pallister et al. (1989) suggested, however, that autopsy of cases of imperforate anus with associated CNS lesions should be reviewed.

Finnigan et al. (1991) described 2 unrelated patients with the usual features of Pallister-Hall syndrome, including diencephalic anomalies, but without hamartoblastomas. Kuller et al. (1992) reported 3 cases. Two of them were affected brothers, one of whom had a chromosome abnormality: the father was a balanced carrier for the translocation t(3;7)(p25.3;q36); the offspring had an unbalanced karyotype with a der(7) chromosome ('7q+'). The parents, of Filipino extraction, were not known to be related. The infant, who died at age 5 weeks of necrotizing enterocolitis, had cleft lip and palate, agenesis of the nasal septum, hypopituitarism by endocrine evaluation, and fused thalami and a hypothalamic hamartomatous lesion by MRI. Postaxial polydactyly was not present and the anus was patent. The older brother died at the age of 3 years. A nonbanded karyotype performed in the Philippines was said to have been normal. Micropenis was present in both brothers. Kuller et al. (1992) suggested that the gene for this disorder may lie on either chromosome 3 or chromosome 7.

Autosomal dominant inheritance was supported by the findings of Topf et al. (1993) in a 9-year-old boy and his 34-year-old father. The son had precocious puberty, imperforate anus, postaxial polydactyly, hypospadias, a hypothalamic mass, and a displaced pituitary gland; the father had polydactyly, a hypothalamic mass, and a flattened pituitary gland. In the son imperforate anus, postaxial polydactyly, and hypospadias, were surgically corrected early in life. His subsequent growth and development were normal. He had minor craniofacial dysmorphism. At the age of 8 years, he presented because of precocious puberty and was found by MRI scan to have a large hypothalamic mass and displaced pituitary. The father's polydactyly consisted of an accessory digit, apparently originating from the third metacarpal. His facial features resembled those of the son. He did not have a history of imperforate anus or hypospadias. Kletter and Biesecker (1992) interpreted the findings to indicate that the Pallister-Hall syndrome is due to an autosomal dominant gene, that most cases are sporadic and the result of new mutation, and that the gene manifests variable expressivity. Penman Splitt et al. (1994) also reported an instance of male-to-male transmission: a 9-year-old boy had hamartoblastoma, midaxial and postaxial polydactyly with brachydactyly, hypospadias, imperforate anus, and precocious puberty. The 34-year-old father had central and postaxial polydactyly, macrocephaly, and a hypothalamic mass. Both were of normal intelligence.

Thomas et al. (1994) reported a definite sib recurrence of Pallister-Hall syndrome in a family without a cytogenetically visible chromosome abnormality. The father of the 2 affected sibs was born with nearly identical digital abnormalities and could represent either mild expression or mosaicism for a dominant gene. The first-born infant, a female, was noted at birth to have choanal atresia, bifid epiglottis, and cleft upper larynx with posterior web in the subglottic area. Her nose was short, with depressed nasal bridge and anteverted nares. There was bilateral hexadactyly with osseous 2/3 syndactyly of the right hand that was thought to represent an insertional type of polydactyly. Subsequently she developed signs of hypopituitarism, had recurrent problems with upper airway obstruction eventually necessitating tracheostomy, and required gastrostomy because of feeding difficulties. She died at 12 months of age after a respiratory infection. At necropsy, a tracheal diverticulum and abnormal lobation of the lungs were found. The pituitary gland was absent with a rudimentary stalk. A 2 x 2 cm soft nodular mass completely replaced the hypothalamus. The male sib was born at 30 weeks' gestation and required respiratory support after birth for upper airway obstruction. He had 4/5 syndactyly of the right hand with postaxial polydactyly of both hands and short fourth and fifth digits. He also had micropenis. Laryngoscopy and bronchoscopy showed a bifid uvula and epiglottis with cleft larynx. He subsequently was shown to have hypopituitarism and a cerebral MRI scan showed a 2 x 2 cm mass in the hypothalamic region with signal characteristics identical to those of normal brain tissue. He died suddenly at 9.5 months of age after a respiratory infection. The father was born with nearly complete 4/5 cutaneous syndactyly of the right hand and postaxial polydactyly of both hands. He had no other abnormalities. Both parents also had normal chromosomes.

Sama et al. (1994) reported an affected newborn female with a large suprasellar mass with a posterior cystic component, bilateral choanal atresia, renal hypoplasia, exomphalos, postaxial polydactyly, and underdevelopment of the 4th metacarpal. Her father was 38 years old.

Verloes et al. (1992) commented that 'Some years ago, syndromologists and clinical geneticists were pleasantly divided into splitters. . .and lumpers. . .' With the increasing volume of reported congenital anomalies, they suggested that 2 new categories may be delineated: the 'stretchers,' who are mainly preoccupied with extending the limits of a given phenotype by adding milder or unusually severe variants, and the 'cut-and-pasters,' who displace the same atypical or borderline cases from one syndrome to another. Verloes et al. (1992) reviewed the differential diagnosis of Pallister-Hall syndrome and of syndromal hypothalamic hamartoblastoma in general. The conditions that they considered as related included Smith-Lemli-Opitz syndrome type II (268670), the pseudotrisomy 13 syndrome or holoprosencephaly-polydactyly syndrome (264480), orofaciodigital syndrome type VI or Varadi-Papp syndrome (277170), and the hydrolethalus syndrome (236680). Because of the insolvable overlap, Verloes et al. (1992) suggested the creation of a phenotypic classification called the multiplex syndrome, defined as a clinical frame encompassing an unknown number of genetic and/or nongenetic multiple congenital anomaly (MCA) syndromes, for which differential diagnosis cannot be performed unequivocally. They suggested calling this the cerebroacrovisceral early lethality (CAVE) multiplex syndrome.

Verloes et al. (1995) described a 24-week-old fetus with agenesis of the corpus callosum, arhinencephaly, hypothalamic hamartoblastoma, absence of the right thumb, hypoplastic left thumb, hypoplastic lungs, intestinal malrotation, microgastria, asplenia, 'inverted' horseshoe kidney, blind vagina, and absence of internal genitalia. Underdevelopment of preaxial structures had not previously been reported in the Pallister-Hall syndrome, in which postaxial polydactyly is a cardinal feature. Although some features suggested the the 'microgastria-limb reduction' complex (156810), patients with that complex are usually not so severely affected and have never had associated hypothalamic hamartoblastoma.

Sills et al. (1993) reported Pallister-Hall syndrome in a male infant and his female sib fetus. Sills et al. (1994) reported that the parents elected to terminate their third pregnancy because prenatal ultrasonographic findings suggested PHS. At 12 weeks, ultrasound demonstrated an abnormal-appearing fetus with large head and a large midline anterior fluid-filled structure, possibly representing a dilated, malformed ventricle. Repeat examination at 13 weeks showed evidence of holoprosencephaly, enlarged cisterna magna, short umbilical cord, polydactyly, and possible syndactyly. The limbs appeared short. At termination at 15 weeks, chromosome studies of chorionic villus cells showed a normal 46,XX karyotype.

Verloes et al. (1995) reported the cases of 2 unrelated, long surviving patients, ages 2 and 17 years, with Pallister-Hall syndrome. In addition to hypothalamic hamartoblastoma, both showed mild facial dysmorphism (downward slanted palpebral fissures, ptosis, microretrognathia), cleft epiglottis, and developmental delay. The younger child had stenosis of the pulmonary arteries, complex urogenital malformations, and anal atresia. The older patient had precocious puberty caused by the hamartoma, combined with complete growth hormone deficiency. Both patients showed bony anomalies of the limbs: variable proximal synostosis between the second to fourth metacarpals or intercalary polydactyly with generalized brachydactyly, severe brachytelephalangism, syndactyly, and nail hypoplasia.

Using multivariate analysis and numerical taxonomy, Verloes et al. (1995) concluded that on review, 'most overlapping cases (and, in fact, most cases reported as Pallister-Hall, including some from the original report)' could be unambiguously classified as Smith-Lemli-Opitz syndrome, type II (268670), orofaciodigital syndrome type VI (277170), or holoprosencephaly-polydactyly syndrome (264480). Together with the absence of anomalies of cholesterol metabolism, a combination of oral frenula, laryngeal malformations, digestive abnormalities, intercalary polysyndactyly, generalized brachytelephalangism, and nail hypoplasia should allow the delineation of Pallister-Hall syndrome, even when a CNS tumor is absent (Verloes, 1995). The radiologic abnormalities in the hand are helpful in differentiating Pallister-Hall syndrome from other syndromes in which hypothalamic hamartoblastoma is observed. This may be of major importance for genetic counseling, since Pallister-Hall syndrome may be an autosomal dominant disorder, thus contrasting with most of the other disorders with the CAVE phenotype, which are recessively inherited.

Unsinn et al. (1995) described a patient who had hydrocolpos and postaxial polydactyly as well as hypothalamic hamartoblastoma, raising the possibility that the Pallister-Hall syndrome and the McKusick-Kaufman syndrome (236700) are one entity. Lurie (1995) questioned whether the Kaufmann-McKusick syndrome and the Pallister-Hall syndrome can be considered one entity. Lurie and Wulfsberg (1994) found in a survey of 43 reported familial cases neither preaxial nor central forms of polydactyly. Lurie (1995) stated that he also did not remember vaginourethral fistula as occurring in these patients.

The range of phenotypic variability in the Pallister-Hall syndrome may be demonstrated by the mother and son reported by Low et al. (1995). A 53-year-old woman and her 20-year-old son both presented with polysyndactyly but without other external malformations or mental retardation. MRI revealed, as an incidental finding, asymptomatic hypothalamic hamartomas in both patients. The sibs of both the mother and the son were unaffected.

After conducting a genomewide search with highly polymorphic markers in 4 affected families including 32 affected and 36 unaffected persons, Kang et al. (1996) found linkage initially to D7S672. Physical mapping data from this region resulted in refinement of marker order. Genotyping with additional markers resulted in the maximum 2-point lod score between PHS and D7S691 of 8.04 at theta = 0.0. Multipoint analysis showed that the phenotype mapped between D7S521 and D7S678 with a peak lod score of 13.6 at D7S691. This region contains at least 2 candidate genes, GLI3 (165240) and inhibin beta-1 (147290). These candidate genes had previously been mapped to the 7p15-p13 region. See later for evidence that the GLI3 gene is the site of the mutations causing PHS.

An international workshop on Pallister-Hall syndrome (Biesecker et al., 1996) developed minimal diagnostic criteria for this entity. The index case in a family must have both hypothalamic hamartoma and central polydactyly to meet the diagnostic criteria. First degree relatives of the index case must have either hypothalamic hamartoma or polydactyly (central or postaxial) and show inheritance in an autosomal dominant pattern or in a manner consistent with gonadal mosaicism. Recommendations for clinical evaluation of suspected cases were presented. Biesecker et al. (1996) concluded that hypothalamic hamartoma is not specific to PHS. Because of a possible relationship of PHS to Smith-Lemli-Opitz syndrome, as noted earlier, Biesecker et al. (1996) analyzed levels of cholesterol and intermediate metabolites of the later stages of cholesterol biosynthesis and found no evidence of a generalized disorder of cholesterol biosynthesis in patients with familial PHS. On genetic and biochemical grounds, they concluded that PHS and Smith-Lemli-Opitz syndrome are not allelic variants of a single locus.

Kang et al. (1997) noted that large deletions or translocations resulting in haploinsufficiency of the GLI3 gene have been associated with Greig cephalopolysyndactyly syndrome GCPS; (175700), which maps to the same region of chromosome 7p, although no mutations in GLI3 were identified in GCPS patients with normal karyotypes. Both PHS and GCPS have polysyndactyly and abnormal craniofacial features and are inherited in an autosomal dominant pattern, but are clinically distinct. The polydactyly of GCPS is commonly preaxial and that of PHS is typically central or postaxial. No reported cases of GCPS had hypothalamic hamartoma and PHS does not cause hypertelorism or broadening of the nasal root or forehead. Because of the colocalization of the loci for PHS and GCPS, Kang et al. (1997) investigated GLI3 as a candidate gene for PHS. They reported 2 PHS families with frameshift mutations in GLI3 (165240.0002; 165240.0003) located 3-prime of the zinc finger-encoding domains, including 1 family with a de novo mutation.

Kang et al. (1997) described the clinical characteristics of a family with 22 affected members and the genetic mapping of the locus. Clinical, radiographic, and endoscopic evaluations showed that the disorder was fully penetrant with variable expressivity and low morbidity. Using anonymous STRP markers, they established linkage to 7p13 by 2-point analysis with D7S691, resulting in a lod score of 7.0 at theta = 0.0, near the GLI3 locus. The linkage data reported by Kang et al. (1997) suggested that the larger, mildly affected families may also have mutations in GLI3. The proband in their first family studied was a 21-month-old child who was noted to have ptosis and polydactyly at the time of birth. Cranial MRI showed a hypothalamic mass compatible with hamartoma.

The family history showed that she had 21 relatives with polydactyly in an autosomal dominant pattern. The pedigree included 9 obligate heterozygotes, all of whom had some manifestation of the disorder. The ratio of males to females was 6 to 16. Several members of the family had central polydactyly (also referred to as mesoaxial or insertional polydactyly). This form typically includes partial osseous syndactyly with proximal fusion of the metacarpals and is characteristic of only a few syndromes.

See also Clarren et al. (1980); Huff and Fernandes (1982)

Biesecker, L. G.; Abbott, M.; Allen, J.; Clericuzio, C.; Feuillan, P.; Graham, J. M., Jr.; Hall, J.; Kang, S.; Olney, A. H.; Lefton, D.; Neri, G.; Peters, K.; Verloes, A.: Report from the workshop on Pallister-Hall syndrome and related phenotypes. Am. J. Med. Genet. 65: 76-81, 1996.
Biesecker, L. G.; Kang, S.; Schaffer, A. A.; Abbott, M.; Kelley, R. I.; Allen, J. C.; Clericuzio, C.; Grebe, T.; Olney, A.; Graham, J. M., Jr.: Exclusion of candidate loci and cholesterol biosynthetic abnormalities in familial Pallister-Hall syndrome. J. Med. Genet. 33: 947-951, 1996.
Clarren, S. K.; Alvord, E. C., Jr.; Hall, J. G.: Congenital hypothalamic hamartoblastoma, hypopituitarism, imperforate anus, and postaxial polydactyly—a new syndrome? Part II: neuropathological considerations. Am. J. Med. Genet. 7: 75-83, 1980.
Donnai, D.; Burn, J.; Hughes, H.: Smith-Lemli-Opitz syndromes: do they include the Pallister-Hall syndrome? (Letter) Am. J. Med. Genet. 28: 741-743, 1987.
Finnigan, D. P.; Clarren, S. K.; Haas, J. E.: Extending the Pallister-Hall syndrome to include other central nervous system malformations. Am. J. Med. Genet. 40: 395-400, 1991.
Graham, J. M.; Perl, D.; O'Keefe, T.; Rawnsley, E.; Little, G. A.: Apparent familial recurrence of hypothalamic hamartoblastoma syndrome. (Abstract) Proc. Greenwood Genet. Center 2: 117-118, 1983.
Hall, J. G.; Pallister, P. D.; Clarren, S. K.; Beckwith, J. B.; Wiglesworth, F. W.; Fraser, F. C.; Cho, S.; Benke, P. J.; Reed, S. D.: Congenital hypothalamic hamartoblastoma, hypopituitarism, imperforate anus, and postaxial polydactyly—a new syndrome? Part I: clinical, causal, and pathogenetic considerations. Am. J. Med. Genet. 7: 47-74, 1980.
Huff, D. S.; Fernandes, M.: Two cases of congenital hypothalamic hamartoblastoma, polydactyly, and other congenital anomalies (Pallister-Hall syndrome). (Letter) New Eng. J. Med. 306: 430-431, 1982.
Iafolla, K.; Fratkin, J. D.; Spiegel, P. K.; Cohen, M. M., Jr.; Graham, J. M., Jr.: Case report and delineation of the congenital hypothalamic hamartoblastoma syndrome (Pallister-Hall syndrome). Am. J. Med. Genet. 33: 489-499, 1989.
Kang, S.; Allen, J.; Graham, J. M., Jr.; Grebe, T.; Clericuzio, C.; Patronas, N.; Ondrey, F.; Green, E.; Schaffer, A.; Abbott, M.; Biesecker, L. G.: Linkage mapping and phenotypic analysis of autosomal dominant Pallister-Hall syndrome. J. Med. Genet. 34: 441-446, 1997.
Kang, S.; Graham, J. M., Jr.; Abbott, M.; Schaffer, A.; Green, E. D.; Rosenberg, M.; Allen, J.; Clericuzio, C.; Grebe, T.; Haskins-Olney, A.; Biesecker, L. G.: Autosomal dominant Pallister-Hall syndrome maps to 7p13. (Abstract) Am. J. Hum. Genet. 59 (suppl.): A17 only, 1996.
Kang, S.; Graham, J. M., Jr.; Olney, A. H.; Biesecker, L. G.: GLI3 frameshift mutations cause autosomal dominant Pallister-Hall syndrome. Nature Genet. 15: 266-268, 1997.
Kletter, G. B.; Biesecker, L. G.: Male-to-male transmission of the Pallister-Hall syndrome. (Abstract) Am. J. Hum. Genet. 51 (suppl.): A100 only, 1992.
Kuller, J. A.; Cox, V. A.; Schonberg, S. A.; Golabi, M.: Pallister-Hall syndrome associated with an unbalanced chromosome translocation. Am. J. Med. Genet. 43: 647-650, 1992.
Low, M.; Moringlane, J. R.; Reif, J.; Barbier, D.; Beige, G.; Kolles, H.; Kujat, C.; Zang, K. D.; Henn, W.: Polysyndactyly and asymptomatic hypothalamic hamartoma in mother and son: a variant of Pallister-Hall syndrome. Clin. Genet. 48: 209-212, 1995.
Lurie, I. W.: Pallister-Hall and McKusick-Kaufmann syndromes. (Letter) J. Med. Genet. 32: 668-672, 1995.
Lurie, I. W.; Wulfsberg, E. A.: The McKusick-Kaufmann syndrome: phenotypic variation observed in familial cases as a clue for the evaluation of sporadic cases. Genet. Counsel. 5: 275-281, 1994.
Pallister, P. D.; Hecht, F.; Herrman, J.: Three additional cases of the congenital hypothalamic 'hamartoblastoma' (Pallister-Hall) syndrome. (Letter) Am. J. Med. Genet. 33: 500-501, 1989.
Penman Splitt, M.; Wright, C.; Perry, R.; Burn, J.: Autosomal dominant transmission of Pallister-Hall syndrome. Clin. Dysmorph. 3: 301-308, 1994.
Sama, A.; Mason, J. D. T.; Gibbin, K. P.; Young, I. D.; Hewitt, M.: The Pallister-Hall syndrome. (Letter) J. Med. Genet. 31: 740 only, 1994.
Sills, I. N.; Rapaport, R.; Desposito, F.; Lieber, C.: Familial Pallister-Hall syndrome: three affected offspring. (Letter) Am. J. Med. Genet. 52: 251 only, 1994.
Sills, I. N.; Rapaport, R.; Robinson, L. P.; Lieber, C.; Shih, L. Y.; Horlick, M. N. B.; Schwartz, M.; Desposito, F.: Familial Pallister-Hall syndrome: case report and hormonal evaluation. Am. J. Med. Genet. 47: 321-325, 1993.
Thomas, H. M.; Todd, P. J.; Heaf, D.; Fryer, A. E.: Recurrence of Pallister-Hall syndrome in two sibs. J. Med. Genet. 31: 145-147, 1994.
Topf, K. F.; Kletter, G. B.; Kelch, R. P.; Brunberg, J. A.; Biesecker, L. G.: Autosomal dominant transmission of the Pallister-Hall syndrome. J. Pediat. 123: 943-946, 1993.
Unsinn, K. M.; Neu, N.; Krejci, A.; Posch, A.; Menardi, G.; Gassner, I.: Pallister-Hall syndrome and McKusick-Kaufmann (sic) syndrome: one entity?. J. Med. Genet. 32: 125-128, 1995.
Verloes, A.: Numerical syndromology: a mathematical approach to the nosology of complex phenotypes. Am. J. Med. Genet. 55: 433-443, 1995.
Verloes, A.; David, A.; Ngo, L.; Bottani, A.: Stringent delineation of Pallister-Hall syndrome in two long surviving patients: importance of radiological anomalies of the hands. J. Med. Genet. 32: 605-611, 1995.
Verloes, A.; Gillerot, Y.; Langhendries, J.-P.; Fryns, J.-P.; Koulischer, L.: Variability versus heterogeneity in syndromal hypothalamic hamartoblastoma and related disorders: review and delineation of the cerebro-acro-visceral early lethality (CAVE) multiplex syndrome. Am. J. Med. Genet. 43: 669-677, 1992.
Verloes, A.; Narcy, F.; Fallet-Bianco, C.: Syndromal hypothalamic hamartoblastoma with holoprosencephaly sequence, microphthalmia, pulmonary malformations, radial hypoplasia and Mullerian regression: further delineation of a new syndrome? Clin. Dysmorph. 4: 33-37, 1995.

GLI-KRUPPEL FAMILY MEMBER 3; GLI3 (*165240)

JJM NOTE

Mutations in this gene rise to the previous entry (Pallister-Hall syndrome), of which hypothalamic hamartoblastoma is a cardinal feature, and the Greig cephalopolysyndactyly syndrome (175700) (also in this *Catalog*) because one affected patient also had medulloblastoma.

Like GLI2 (165230), the GLI3 gene was isolated by virtue of its cross-hybridization to the zinc finger gene GLI (165220), which is amplified in certain glioblastomas and represents a potential sequence-specific DNA-binding transcription factor from the GLI-Kruppel gene family. The GLI3 gene is expressed in tissues such as testis, myometrium, placenta, and lung and as an 8.5-kb mRNA and the protein product (relative molecular mass, 190,000) shows sequence-specific DNA binding. The GLI3 gene was mapped to chromosome 7 by analysis of DNA from human-rodent hybrid panels (Ruppert et al., 1988) and to 7p13 by in situ hybridization (Ruppert et al., 1990). The gene encodes a predicted 1,596-amino acid polypeptide.

See 165240.0001 for evidence that GLI3 is the site of the mutation responsible for Greig cephalopolysyndactyly syndrome (GCPS; 175700). Although GLI1 (165220), located on 12q13, has been associated with glioma, GLI3 has not been related to glioma or other form of neoplasia (Vogelstein, 1994).

Vortkamp et al. (1994) isolated a YAC contig of more than 1,000 kb, including the GLI3 gene. In this contig the gene itself spanned at least 200 to 250 kb. A CpG island was located in the vicinity of the 5-prime region of the GLI3 cDNA, implying a potential promoter region. Kang et al. (1997) demonstrated that the coding region of GLI3 is composed of 14 exons, including a large exon of more than 2,500 bp. In addition, the gene contains 2 intragenic dinucleotide repeats, and 4 single-bp polymorphisms in the coding region. Kang et al. (1997) used these coding region polymorphisms to design an allele-specific expression study useful for investigating patients with Greig syndrome.

Large deletions or translocations resulting in haploinsufficiency of the GLI3 gene have been found in association with GCPS (Pettigrew et al., 1991; Vortkamp et al., 1991; Brueton et al., 1988), although no mutations were identified in GCPS patients with normal karyotypes. Kang et al. (1997) identified frameshift mutations in the GLI3 gene (e.g., 165240.0002) in Pallister-Hall syndrome (PHS; 146510).

Postaxial polydactyly type A1 (PAPA1; 174200) is an autosomal trait characterized by an extra digit in the ulnar and/or fibular side of the upper and/or lower extremities. The extra digit is usually functional, since it is well formed and articulates with the fifth or extrametacarpal/metatarsal. The gene responsible for this trait was mapped to the centromeric region of 7p by Radhakrishna et al. (1997). Since the GLI3 gene maps to the same region, Radhakrishna et al. (1997) performed a mutation search in GLI3 cDNA from an Indian family with PAPA1 and demonstrated a mutation at codon 764 in 1 GLI3 allele. The mutation occurred 3-prime to the conserved domain termed Pzf1 (post zinc finger 1). Haploinsufficiency GLI3 is associated with some cases of GCPS, while frameshift mutations 3-prime to the zinc finger domains but upstream of the Pzf1 domain cause the Pallister-Hall syndrome. Thus, the coding of presumably different truncated GLI3 proteins are associated with different clinical syndromes, suggesting specific roles for the variant protein domains during development.

For a review of the role of this gene in limb development, see Johnson and Tabin (1997).

Brueton, L.; Huson, S. M.; Winter, R. M.; Williamson, R.: Chromosomal localisation of a developmental gene in man: direct DNA analysis demonstrates that Greig cephalopolysyndactyly maps to 7p13. Am. J. Med. Genet. 31: 799-804, 1988.
Johnson, R. L.; Tabin, C. J.: Molecular models for vertebrate limb development. Cell 90: 979-990, 1997.
Kang, S.; Graham, J. M., Jr.; Olney, A. H.; Biesecker, L. G.: GLI3 frameshift mutations cause autosomal dominant Pallister-Hall syndrome. Nature Genet. 15: 266-268, 1997.
Kang, S.; Rosenberg, M.; Ko, V. D.; Biesecker, L. G.: Gene structure and allelic expression assay of the human GLI3 gene. Hum. Genet. 101: 154-157, 1997.

Pettigrew, A. L.; Greenberg, F.; Caskey, C. T.; Ledbetter, D. H.: Greig syndrome associated with an interstitial deletion of 7p: confirmation of the localization of Greig syndrome to 7p13. Hum. Genet. 87: 452-456, 1991.
Radhakrishna, U.; Blouin, J.-L.; Mehenni, H.; Patel, U. C.; Patel, M. N.; Solanki, J. V.; Antonarakis, S. E.: Mapping one form of autosomal dominant postaxial polydactyly type A to chromosome 7p15-q11.23 by linkage analysis. Am. J. Hum. Genet. 60: 597-604, 1997.
Radhakrishna, U.; Wild, A.; Grzeschik, K.-H.; Antonarakis, S. E.: GLI3 mutations in postaxial polydactyly type A. (Abstract) Am. J. Hum. Genet. 61 (suppl.): A48 only, 1997.
Ruppert, J. M.; Kinzler, K. W.; Wong, A. J.; Bigner, S. H.; Kao, F.-T.; Law, M. L.; Seuanez, H. N.; O'Brien, S. J.; Vogelstein, B.: The GLI-Kruppel family of human genes. Molec. Cell. Biol. 8: 3104-3113, 1988.
Ruppert, J. M.; Vogelstein, B.; Arheden, K.; Kinzler, K. W.: GLI3 encodes a 190-kilodalton protein with multiple regions of GLI similarity. Molec. Cell. Biol. 10: 5408-5415, 1990.
Vogelstein, B.: Personal Communication. Baltimore, Md. 5/9/1994.
Vortkamp, A.; Gessler, M.; Grzeschik, K.-H.: GLI3 zinc-finger gene interrupted by translocations in Greig syndrome families. Nature 352: 539-540, 1991.
Vortkamp, A.; Gessler, M.; Le Paslier, D.; Elaswarapu, R.; Smith, S.; Grzeschik, K.-H.: Isolation of a yeast artificial chromosome contig spanning the Greig cephalopolysyndactyly syndrome (GCPS) gene region. Genomics 22: 563-568, 1994.

HYPOTHALAMIC HAMARTOMAS (241800)

JJM NOTE

This entity may well overlap with the above syndrome of hypothalamic hamartoblastomas (146510).

In a case of hypothalamic hamartoma, Marcuse et al. (1953) reported that 1 sib had internal hydrocephalus and died at age 6 months, within a day after operation. Another sib had a similar clinical course and died at age 2 months; autopsy was not performed. A double first cousin of the proband likewise had a similar clinical course; postmortem at age 5 months showed 'mature glioma of the brain stem.' See 146510.

Encha-Razavi et al. (1992) described 3 unrelated fetuses (one was actually a term neonate who had died a few minutes after birth) in whom, at autopsy, congenital hypothalamic hamartomas were found. Two were associated with skeletal dysplasia. A third was associated with malformations suggestive of Meckel syndrome (249000): heart defect, pulmonary hypoplasia, renal dysplasia, and posterior encephalocele. The first family reported by Encha-Razavi et al. (1992) had, in addition to the fully studied female newborn, an apparently identically affected brother who also died in the newborn period but was not studied neuropathologically. That boy showed micropenis, suggesting hypopituitarism, and had micromelia, short ribs, midline cleft of the upper lip, short nose, alveolar frenula, and macrocephaly. The parents were young, healthy, and unrelated. Neither child had postaxial polydactyly. Encha-Razavi et al. (1992) suggested the designation congenital hypothalamic hamartoma syndrome (CHHS) for a possibly familial disorder that combines orofacial abnormalities and skeletal dysplasia with hypothalamic hamartomas.

The familial case reported by Graham et al. (1983) has some similarities. The infant showed abnormal auricles, short nose with flattened bridge, microglossia, micrognathia, cleft palate, short limbs, dislocated hips, and 4-limb postaxial polydactyly. The infant died at 2 hours of age and autopsy showed hypothalamic hamartoblastoma. A sister of the mother died at 17 hours of age and showed 4-limb polydactyly, recessed mandible, and small tongue; autopsy was not done.

Encha-Razavi, F.; Larroche, J. C.; Roume, J.; Migne, G.; Delezoide, A. L.; Gonzales, M.; Mulliez, N.: Congenital hypothalamic hamartoma syndrome: nosological discussion and minimum diagnostic criteria of a possibly familial form. Am. J. Med. Genet. 42: 44-50, 1992.
Graham, J. M.; Perl, D.; O'Keefe, T.; Rawnsley, E.; Little, G. A.: Apparent familial recurrence of hypothalamic hamartoblastoma syndrome. (Abstract) Proc. Greenwood Genet. Center 2: 117-118, 1983.
Marcuse, P. M.; Burger, R. A.; Salmon, G. W.: Hamartoma of the hypothalamus: report of two cases with associated developmental defects. J. Pediat. 43: 301-308, 1953.

CEREBRAL SARCOMA (117600)

JJM NOTE

I wonder if these families have been tested for constitutional mutation in *TP53*, as in the Li-Fraumeni cancer family syndrome.

In 2 families Gainer et al. (1975) observed 4 cases of cerebral fibrosarcoma (father and daughter; 2 sisters).

Gainer, J. V., Jr.; Chou, S. M.; Chadduck, W. M.: Familial cerebral sarcomas. Arch. Neurol. 32: 665-669, 1975.

Spinal Cord Tumors

CHORDOMA (215400)

JJM NOTE

Although rare, chordoma is, in a sense, the epitome of neoplasia in an embryonic remanent and might be investigated in the same manner as other embryonal tumors.

Chordomas are rare, clinically malignant tumors derived from notochordal remnants. They occur along the length of the spinal axis, predominantly in the sphenooccipital, vertebral, and sacrococcygeal regions. They are characterized by slow growth, local destruction of bone, extension into adjacent soft tissues, and, rarely, distant metastatic spread.

Foote et al. (1958) described middle-aged brother and sister with sacrococcygeal chordoma. Recurrence and metastases occurred in both.

Stepanek et al. (1998) described chordoma in 4 members of a family in a pattern they considered most consistent with autosomal dominant inheritance with reduced penetrance. Two females and 2 males were affected. The index case was in a 20-year-old woman who had a chordoma in the region of the pituitary gland. A paternal aunt of this woman presented at 39 years with persistent nasal stuffiness and was found to have a chordoma abutting the posterior aspect of the nasal septum. A brother of that woman was found, at the age of 28 years, to have a sacral chordoma. A first cousin of that brother and sister had a chordoma of the nasopharynx.

Foote, R. F.; Ablin, G.; Hall, W.: Chordoma in siblings. Calif. Med. 88: 383-386, 1958.
Stepanek, J.; Cataldo, S. A.; Ebersold, M. J.; Lindor, N. M.; Jenkins, R. B.; Unni, K.; Weinshenker, B. G.; Rubenstein, R. L.: Familial chordoma with probable autosomal dominant inheritance. (Letter) Am. J. Med. Genet. 75: 335-336, 1998.

SPINAL EXTRADURAL CYST (271100)

JJM NOTE

This entity is included because of the spinal mass effect.

Chynn (1967) described spinal extradural cyst in a black brother and sister, aged 12 and 10 at the time of diagnosis. Progressive weakness in the legs was the main symptom. Another sib may also have been affected. In all 3 sibs, congenital lymphedema of the leg and double rows of eyelashes (distichiasis) were present. Bergland (1968) found 3 out of 4 sibs affected but this appears to be the same family as that described by Chynn (1967). Spinal extradural cysts are very rare. Spinal anomalies of this type were present in affected persons with lymphedema and distichiasis reported by Robinow et al. (1970) but were asymptomatic. It is likely that 'spinal extradural cyst' is not a separate genetic entity but merely part of the lymphedema-distichiasis syndrome, a dominant (153400). These so-called cysts are in fact often arachnoid diverticula (Cilluffo et al., 1981). They have been reported with the Marfan syndrome (154700).

Bergland, R. M.: Congenital intraspinal extradural cyst: report of three cases in one family. J. Neurosurg. 28: 495-499, 1968.
Chynn, K.-Y.: Congenital spinal extradural cyst in two siblings. Am. J. Roentgen. 101: 204-215, 1967.
Cilluffo, J. M.; Gomez, M. R.; Reese, D. F.; Onofrio, B. M.; Miller, R. H.: Idiopathic ('congenital') spinal arachnoid diverticula: clinical diagnosis and surgical results. Mayo Clin. Proc. 56: 93-101, 1981.
Robinow, M.; Johnson, G. F.; Verhagen, A. D.: Distichiasis-lymphedema: a hereditary syndrome of multiple congenital defects. Am. J. Dis. Child. 119: 343-347, 1970.

EPENDYMOMA, FAMILIAL (131445)

JJM NOTE

Monosomy 22 here raises the issue of meningioma, not ependymoma, as the cell type. Ependymoma is a feature of the Li-Fraumeni cancer family syndrome (151623, 191170), tuberous sclerosis (191092, 191100), and the melanoma-astrocytoma syndrome (155755).

Gilchrist and Savard (1989) described the familial occurrence of ependymoma in 2 sisters and a maternal male cousin. Karyotypic analysis of the tumor from 1 sister showed mosaicism for the loss of one chromosome 22.

Gilchrist, D. M.; Savard, M. L.: Ependymomas in two sisters and a maternal male cousin. (Abstract) Am. J. Hum. Genet. 45 (suppl.): A22 only, 1989.

SUBEPENDYMOMA (600139)

JJM NOTE

A rare cell type might show genetic origins more obviously than a common one.

Subependymomas are unusual tumors believed to arise from the bipotential subependymal cell, which normally differentiates into either ependymal cells or astrocytes. They were characterized as a distinct entity by Scheinker (1945). They tend to be slow-growing, noninvasive, and located in the ventricular system, septum pellucidum, cerebral aqueduct, or proximal spinal cord. Clarenbach et al. (1979) described the simultaneous occurrence of fourth ventricular subependymomas in monozygotic twins, both of whom became symptomatic at 22 years of age. Honan et al. (1987) described subependymomas in 3 out of 11 sibs. Ryken et al. (1994) reported the occurrence of fourth ventricular subependymomas in a father and son.

See 131445 for a discussion of ependymoma.

Clarenbach, P.; Kleihues, P.; Metzel, E.; Dichgans, J.: Simultaneous clinical manifestation of subependymoma of the fourth ventricle in identical twins. J. Neurosurg. 50: 655-659, 1979.

Honan, W. P.; Anderson, M.; Carey, M. P.; Williams, B.: Familial subependymomas. Br. J. Neurosurg. 1: 317-321, 1987.

Ryken, T. C.; Robinson, R. A.; VanGilder, J. C.: Familial occurrence of subependymoma: report of two cases. J. Neurosurg. 80: 1108-1111, 1994.

Scheinker, I. M.: Subependymoma: a newly recognized tumor of subependymal derivation. J. Neurosurg. 2: 232-240, 1945.

3. ENDOCRINE SYSTEM

Multiple Endocrine Tumors

MULTIPLE ENDOCRINE NEOPLASIA 1 (*131100)

JJM NOTE

The memory aid for this syndrome is PPP—*p*arathyroid, *p*ancreatic islet cell, and *p*ituitary—with manifestations in that order of frequency. Progress in understanding multiple endocrine neoplasia 1 holds several lessons, as a prototypic genetic neoplastic syndrome. First, the clinical delineation of the disorder continues with insightful reports of cases and small and large kindreds. Second, interdisciplinary and international collaborations have used linkage analysis, loss of heterozygosity studies, and cooperative teamwork to speed research and get the findings back to family members. For example, the comparison of biochemical detection to DNA diagnostics proved the latter to be superior, not simply because it can reduce the priority risk of some relatives from 50% to close to 0%. Third, with the mapping of *MEN* genes, the old "overlap" families can be assigned to the *MEN1* locus on chromosome 11 or the *MEN2* locus on 10. As both genes are now isolated, questions of genotype-phenotype correlations will be addressed.

Description

Multiple endocrine neoplasia is an autosomal dominant disorder characterized by a high frequency of peptic ulcer disease and primary endocrine abnormalities involving the pituitary, parathyroid, and pancreas.

Clinical Features

Underwood and Jacobs (1963) found father, son, and daughter affected. Hypoglycemia was the presenting manifestation in all 3. In addition to islet cell adenomas, the father had bronchial carcinoma and hyperparathyroidism from parathyroid adenomas. The son and daughter had been followed from childhood as cases of idiopathic epilepsy unresponsive to anticonvulsive therapy. Guida et al. (1966) described pituitary adenoma and duodenal carcinoid in patients with this condition. Bronchial carcinoid (Williams and Celestin, 1962) also occurs as a feature of endocrine adenomatosis. Kipnis et al. (1969) described a patient with MEA who succumbed to metastatic schwannoma.

On the basis of studies of 8 affected members of a family, Vance et al. (1972) suggested that the primary genetic lesion in endocrine adenomatosis is one that leads to neoplasia and hyperfunction of the islets of Langerhans and that the other endocrine tumors arise as secondary effects of hypersecretion of islet hormones. There are families that have only hyperparathyroidism. Either the families with only hyperparathyroidism represent a distinct entity or the theory of Vance et al. (1972) is not valid.

A notable feature of this disease is intrafamilial uniformity. Some kindreds (e.g., Ballard et al., 1964; Wermer, 1954) have a high frequency of severe peptic ulcer disease with islet cell tumors, whereas other kindreds (e.g., Johnson et al., 1967) are devoid of peptic disease.

Bilateral pheochromocytomas occur in MEN2 and MEN3 and pancreatic islet cell tumor in MEN1. Tateishi et al. (1978) described a patient with both forms of endocrine neoplasia. They also reviewed 14 reported cases of MEA with features overlapping MEA I and II. For example, 7 patients with acromegaly due to pituitary adenoma had pheochromocytoma, 2 with Sipple syndrome (MEN2) had pituitary adenoma, and so on. See 171400 for discussion of MEA II (MEN2) and 162300 for MEA III (MEN3).

Prosser et al. (1979) found 4 patients in 3 unrelated families who had prolactin-secreting pituitary adenomas. Farid (1980) observed 4 kindreds in Newfoundland. Although their ancestors came from the same small community in the British Isles to the Burin Peninsula of Newfoundland, Farid (1980) could identify no genealogic connections. All 4 families had hyperparathyroidism and prolactinoma, but no pancreatic tumors were documented. Two kindreds had carcinoid tumors at unusual sites, either thymus or peripheral lung parenchyma. In contrast to the benign course of the prolactinomas and the primary hyperparathyroidism, 2 persons with thymic carcinoid died from metastatic disease. A suggestion of linkage with HLA was found. This syndrome was referred to as MEN1-Burin by Bear et al. (1985). Hershon et al. (1983) described a phenotypically similar but unrelated kindred from the Pacific Northwest, in which 6 of 7 living affected members had prolactinomas and none had pancreatic islet tumors. Farid (1994) reported evidence indicating that the 4 Burin Peninsula families are in fact related. All affected members of the 4 families carried a PYGM (232600) allele and all 4 families had ancestors who lived in the Harbor Breton region 1 century earlier. Petty et al. (1994) demonstrated by linkage studies that the gene in both the Newfoundland kindreds and the kindred from the Pacific Northwest mapped to 11q in the same region as the MEN1 gene. No recombinants were seen with PYGM in either kindred, but the PYGM allele associated with the disease was different in the 2 kindreds.

Olufemi et al. (1998) demonstrated that the Burin variant of MEN1 is due to an arg460-to-ter nonsense mutation (131100.0016) in the MEN1 gene. Furthermore, they showed by haplotype analysis that affected individuals in all 4 families shared a common haplotype over a 2.5-Mb region, thus providing convincing evidence of a common founder.

The Zollinger-Ellison syndrome may present purely as hyperparathyroidism. For example, 1 member of the family described as having hereditary hyperparathyroidism by Cutler et al. (1964) was later reported to have a malignant schwannoma, pituitary adenomas, multiple pancreatic islet cell adenomas, and multiple adrenocortical adenomas. Snyder et al. (1972) reported 5 families and noted the previously described association of lipomas.

The Zollinger-Ellison syndrome is merely hypergastrinism and may have causes other than MEA I. For example, Long et al. (1980) reported the Z-E syndrome with ectopic production of gastrin by a mucinous cystadenoma of the ovary. McCarthy (1982) distinguished 2 common forms of the Zollinger-Ellison syndrome: the sporadic and usually malignant type, seen most often in later life , and the genetic variety that occurs as part of MEN I.

Stacpoole et al. (1981) observed a family in which 3 persons had A-cell pancreatic tumors (glucagonomas) as part of MEN I. Two had the classic glucagonoma syndrome with skin rash, glucose intolerance, and hypoaminoacidemia. Administered secretin and somatostatin gave anomalous metabolic responses.

Bahn et al. (1986) reported 25-year-old monozygotic twins with MEN I who, according to the authors, had impressive differences in expression of the disorder. One had epigastric pain and diarrhea at presentation; was found to have primary hyperparathyroidism, Zollinger-Ellison syndrome, Cushing disease, and hyperprolactinemia; and underwent hypophysectomy. The second twin was asymptomatic but had primary hyperparathyroidism and hyperprolactinemia. A large, histologically benign pituitary adenoma 'that invaded dura and bone' was removed by a transphenoidal approach 2 days after parathyroidectomy.

Maton et al. (1986) suggested that the Cushing syndrome is more common in patients with the Zollinger-Ellison syndrome than previously reported, occurring in 8% of all cases. Three of 16 patients with the Zollinger-Ellison syndrome and MEN1 had the Cushing syndrome due to pituitary overproduction of ACTH. In all sporadic cases of the Z-E syndrome, Cushing syndrome was due to ectopic production of ACTH by the gastrinoma. Although it is theoretically possible that ectopic production of corticotropin-releasing factor is the mechanism in some cases, there is no evidence for such. Gaitan et al. (1993) described mother and daughter who, in addition to other manifestations of MEN1, had Cushing disease due to ACTH-secreting tumors. Wolfe and Jensen (1987) reviewed diagnosis and treatment of the Zollinger-Ellison syndrome.

Brandi et al. (1986) concluded that primary hyperparathyroidism in familial MEN type I may have a humoral cause. Since the mitogenic factor persists after parathyroidectomy, it evidently is not secreted by the hyperplastic glands themselves. It did not seem to be identical to any of the well-recognized circulating growth factors, nor did it have similar mitogenic effects on pancreatic or pituitary cells in vitro, despite the presence of islet cell and pituitary tumors in some of the same patients. Curiously, Brandi et al. (1986) did not detect a parathyroid mitogenic factor in patients with MEN II who also had hyperparathyroidism. Schimke (1986) suggested: 'Considered within the framework of a 2-step model, the general event in the multiple endocrine neoplasia syndromes may be an abnormality of a plasma-membrane receptor in the affected endocrine glands. The somatic mutation may involve derepression of a primitive gene coding for a protein that promotes the growth of endocrine glands.' The 'primitive gene' might be an oncogene. This would represent a rather different 2-mutation theory than the one that applies to retinoblastoma (180200) and Wilms tumor (194070). In this case the mutations are presumably at different loci.

Skogseid et al. (1992) reviewed adrenocortical lesions in 31 MEN I patients. In 12 (37%), they found adrenal enlargement, which was bilateral in 7. One person developed unilateral adrenocortical carcinoma manifested by rapid adrenal expansion, feminization, and an abnormal urinary steroid profile after 4 years of observation for bilateral minor adrenal enlargement. In the other patients, adrenal enlargement was not associated with ascertainable biochemical disturbances in the hypothalamic-pituitary-adrenocortical axis. Pancreatic endocrine tumors were significantly overrepresented in the patients with adrenal lesions, being present in all 12. In agreement with findings in sporadic cases, the MEN1 adrenocortical carcinoma showed loss of constitutional heterozygosity for alleles at 17p, 13q, 11p, and 11q. The benign adrenal lesions retained heterozygosity for the MEN1 locus at 11q13. Skogseid et al. (1992) concluded that the pituitary-independent adrenocortical proliferation is not the result of the primary lesion in MEN I but may represent a secondary phenomenon, perhaps related to the pancreatic endocrine tumor.

Courseaux et al. (1996) used a combination of methods to refine maps of the approximately 5-Mb region of 11q13 that includes MEN1. They proposed the following gene order: cen—PGA—FTH1—UGB—AHNAK—ROM1—MDU1—CHRM1—COX8—EMK1—FKBP2—PLCB3—[PYGM, ZFM1]—FAU—CAPN1—[MLK3, RELA]—FOSL1—SEA—CFL1—tel. The location of MEN1 was narrowed to a 2-Mb region beginning centromeric to COX8 and extending to approximately CAPN1.

Biochemical Features

Brandi et al. (1986) cultured bovine parathyroid cells to test for mitogenic activity in plasma from patients with type I MEN. Whereas normal plasma stimulated incorporation of labeled thymidine to the same extent as did plasma-free culture medium, the plasma from the patients increased mitogenic activity 2400% over the control value. Bovine parathyroid cells were stimulated to proliferation, whereas plasma from normal subjects inhibited proliferation of the bovine parathyroid cells. The mitogenic activity had an apparent molecular weight of 50,000 to 55,000.

Inheritance

MEN I may be the only hereditary form of prolactinoma in humans.

Mapping

The study of Farid (1980) suggested linkage with HLA. 8,5:Bale et al. (1987, 1989) studied linkage with multiple markers in a single large kindred. INT2 (164950), which is located at 11q13, was found to be closely linked to MEN1. This raised the possibility that MEN1 may be due to mutation in the INT2 gene.

In studies of 3 families, Thakker et al. (1989) established linkage with INT2; peak lod score = 3.30 at theta = 0.00. Larsson et al. (1988) mapped the MEN1 locus to chromosome 11 by demonstrating linkage to a DNA probe derived from the PYGM locus (232600), which, in turn, has been mapped to 11q13-qter.

By comparing constitutional and tumor tissue genotypes of insulinomas from 2 brothers who had inherited the disorder from their mother, Larsson et al. (1988) demonstrated loss of the linked locus in tumor tissue. Tumors in both showed loss of 1 constitutional allele at all informative loci on chromosome 11. The informative markers extended from HRAS1 (190020), located at 11p15.5, to APOA1 (107680), located at 11q13. (Larsson et al. (1988) quoted a paper in press indicating that the APOA1 locus is in band q23.) In 3 families, the authors observed a total maximum lod score of 4.37 at theta = 0.00. Larsson et al. (1989) and Nordenskjold et al. (1989) both provided linkage data for markers at 11q13 with multiple endocrine neoplasia type I.

Nakamura et al. (1989) identified 6 closely linked markers in the vicinity of 11q13 which are useful for identification of carriers in MEN I families. The target region containing the gene was narrowed to about 12 cM. Friedman et al. (1989) and Thakker et al. (1989) demonstrated loss of heterozygosity for chromosome 11 alleles in parathyroid tumors from patients with MEN I. Friedman et al. (1989) found such loss in 10 of 16 tumors from 14 patients. In 7 of 10 tumors, the subregion of loss was less than the full length of chromosome 11 but always included 1 copy of the MEN I locus.

Bale et al. (1989) studied the DNA from 66 tumors removed from patients with MEN I. They demonstrated allelic loss of chromosome 11 RFLPs in 10 of 16 parathyroid tumors from MEN I patients and in 9 of 34 sporadic parathyroid adenomas. The smallest consistent region of loss was between INT2 and D11S149. Bystrom et al. (1990) showed that the pathogenesis of MEN1-associated parathyroid lesions involves unmasking of a recessive mutation at the disease locus and that sporadic primary hyperparathyroidism shares the same mechanisms. By examination of allele losses in MEN1-associated lesions, they defined deletions of chromosome 11 and mapped the MEN1 locus to a small region within 11q13, telomeric to PYGM.

Richard et al. (1991) created a high-resolution radiation hybrid map of the proximal long arm of chromosome 11 containing the MEN1 and BCL1 (151400) gene loci. By statistical analysis of the cosegregation of markers in radiation hybrids, they arrived at the following most likely order of loci: C1NH (106100)—OSBP (167040)—CD5 (153340)/CD20(112210)—PGA (169710)—FTH1 (134770)—COX8 (123870)—PYGM—SEA (165110)—KRN1 (148021)—HSTF1 (164980)/INT2—GST3 (138370)—PPP1A (176875). They suggested that the localization of the protooncogene SEA between PYGM and INT2, 2 markers that flank MEN1, makes SEA a potential candidate for the MEN1 locus. The positions of PPP1A and GST3 are such that neither is likely to be directly involved in BCL1 or MEN1. Fujimori et al. (1992) placed the MEN1 locus within an 8-cM region between D11S480 and D11S546.

Larsson et al. (1992) found that 13 marker systems tested with 17 DNA probes were located within a region on chromosome 11 spanning a 14% meiotic recombination region, with the MEN1 locus in the middle: based on meiotic crossovers, 4 systems were on the telomeric side and 4 on the centromeric side of MEN1. The remaining 5 were closely linked to MEN1, with no crossovers in their set of 6 families including 59 affected persons. The calculated accuracy of prediction of MEN1 was more than 99.5% when 3 of the marker systems were informative.

Sawicki et al. (1992) reported loss of heterozygosity (LOH) on chromosome 11 in 5 of 11 sporadic gastrinomas. Four of these tumors had LOH for markers flanking the MEN1 region. LOH on chromosome 11 had previously been found in 3 types of tumors that occur in patients with the MEN I syndrome: sporadic pituitary adenomas, sporadic and familial parathyroid neoplasms, and pancreatic endocrine tumors including insulinoma, glucagonoma, vasointestinal peptide tumor, and nonfunctional tumors. Although gastrinomas account for the majority of both sporadic and familial pancreatic endocrine tumors, LOH had not previously been demonstrated for this form.

From a comparative mapping analysis of 10q and the pericentric region of 11q in the human and mouse chromosome 19, Rochelle et al. (1992) concluded that the murine homolog of the MEN1 locus may lie in the proximal segment of chromosome 19. MDU1 (158070) lies in that region and is a candidate gene for MEN1.

Guru et al. (1997) mapped and sequenced the MEN1 (131100) genomic region. They produced a precisely ordered map of 33 transcribed genes within this 2-Mb region.

Molecular Genetics

Chandrasekharappa et al. (1997) identified several MEN1 candidate genes in a previously identified minimal interval on 11q13. One of the genes contained 12 different frameshift, nonsense, missense, and in-frame deletion mutations in 14 probands from 15 families (e.g., 131100.0001). The MEN1 gene contains 10 exons and encodes a ubiquitously expressed 2.8-kb transcript. The predicted 610-amino acid protein product, termed menin by them, exhibited no apparent similarities to any previously known proteins. They commented that the identification of the MEN1 gene should enable improved understanding of the mechanism of endocrine tumorigenesis and should facilitate early diagnosis.

To identify additional candidate genes in the segment of less than 300 kb where the MEN1 locus is situated, Lemmens et al. (1997) used a BAC to isolate cDNAs from a bovine parathyroid cDNA library by direct selection. One of the novel genes that they identified, called SCG2 (for 'suppressor candidate gene 2') by them, proved to be identical to the MEN1 gene reported by Chandrasekharappa et al. (1997). The SCG2 transcript was 2.9 kb in all tissues with an additional 4.2 kb transcript also being present in the pancreas and thymus. A human SCG2 cDNA clone, covering 2.3 kb at the 3-prime end of the gene, was isolated by hybridization screening. Northern blot analysis with this human sequence gave results identical to those from the bovine sequence. Mutational analysis of human SCG2 in 10 unrelated MEN1 families identified 1 polymorphism and 9 different heterozygous mutations (1 missense, 4 nonsense, 1 insertional, and 3 deletional frameshifts) that segregated with the disease, hence providing an independent confirmation for the identification of the MEN1 gene.

Agarwal et al. (1997) failed to find germline mutations of the MEN1 gene in 5 kindreds with familial hyperparathyroidism. Heppner et al. (1997) found somatic mutation of the MEN1 gene in 21% of parathyroid tumors not associated with MEN1, representing 54% of parathyroid tumors with 11q13 LOH. The authors suggested that parathyroid tumor formation in kindreds with somatic mutation of MEN1 may be initiated by germline mutation of an unidentified tumor suppressor gene or oncogene. The finding of somatic mutation (131100.0013) in a single tumor from a member of such a kindred indicated that somatic MEN1 gene mutation may also contribute to tumorigenesis in such individuals. Previous studies had found frequent 11q13 LOH in sporadic tumors as follows: gastrinoma (45%), insulinoma (19%), anterior pituitary gland tumors (3 to 30%), carcinoid tumors (78%), thyroid follicular tumors (15%), and aldosteronomas (36%). Heppner et al. (1997) suggested that many of these tumors likewise may have MEN1 somatic mutations. Whether somatic mutations of the MEN1 gene are confined to tumors of endocrine tissue or whether somatic mutation of the MEN1 gene is operative in tumor formation of nonendocrine tissues remained to be established.

Agarwal et al. (1997) extended their mutation analysis to 34 more unrelated familial MEN1 probands (to a total of 50 kindreds) and to 2 related disorders, sporadic MEN1 and familial hyperparathyroidism (145000). In 8 of 11 cases of sporadic MEN1, they found heterozygous germline MEN1 mutations (e.g., 131100.0014); such mutations were found in 47 of 50 familial MEN1 probands. They proved that the mutation was new in 1 case of sporadic MEN1. Among the familial MEN1 cases, 8 were observed more than once. In all, 40 different mutations (32 familial and 8 sporadic) were distributed across the MEN1 gene. A predicted loss of function of the encoded menin protein supported the prediction that MEN1 is a tumor suppressor gene. No MEN1 germline mutations were found in 5 probands with familial hyperparathyroidism, suggesting that this disorder is often caused by mutation in another gene. Agarwal et al. (1997) suggested that further work must be done to establish the prevalence of MEN1 mutations among patients with other sporadic disorders. The most obvious sporadic disorders to investigate are Zollinger-Ellison syndrome, prolactinoma, and foregut carcinoid.

Prezant et al. (1998) screened the complete coding sequence of the MEN1 gene for mutations in 45 sporadic anterior pituitary tumors, including 14 hormone-secreting tumors and 31 nonsecreting tumors, by dideoxy fingerprinting and sequence analysis. No pathogenic sequence changes were found in the MEN1 coding region. The MEN1 gene was expressed in 43 of these tumors with sufficient RNA, including 1 tumor with LOH for several polymorphic markers on chromosomal region 11q13. Also, both alleles were expressed in 19 tumors in which the constitutional DNA was heterozygous for intragenic polymorphisms. The authors concluded that inactivation of the MEN1 tumor suppressor gene, by mutation or by imprinting, does not appear to play a prominent role in sporadic pituitary adenoma pathogenesis.

The amino acid sequence of menin offers no clue to the function or subcellular location of this putative tumor suppressor. Based on immunofluorescence, Western blotting of subcellular fractions, and epitope tagging with enhanced green fluorescent protein, Guru et al. (1998) demonstrated that menin is located primarily in the nucleus. They identified at least 2 independent nuclear localizations signals (NLSs), both located in the C-terminal fourth of the protein. They pointed out that among the 68 known independent disease-associated mutations, none of the 22 missense and 3 in-frame deletions affect either of the putative NLS sequences. However, if expressed, none of the truncated menin proteins resulting from the 43 known frameshift/nonsense mutations would retain both the NLSs. The precise role(s) of menin in the nucleus remained to be understood.

Bassett et al. (1998) investigated 63 unrelated MEN1 kindreds (195 affected and 396 unaffected members) for mutations in the 2,790-bp coding region and splice sites, by SSCP and DNA sequence analysis. They identified 47 mutations (12 nonsense mutations, 21 deletions, 7 insertions, 1 donor splice site mutation, and 6 missense mutations) that were scattered throughout the coding region, together with 6 polymorphisms that had heterozygosity frequencies of 2 to 44%. More than 10% of the mutations arose de novo, and 4 mutation hotspots accounted for more than 25% of the mutations. SSCP was found to be a sensitive and specific mutational screening method that detected more than 85% of the mutations. MEN1 mutant-gene carrier status was detected in 201 individuals (155 affected and 46 unaffected). By analysis of these cases, they defined the age-related penetrance of MEN1 as 7%, 52%, 87%, 98%, 99%, and 100% at 10, 20, 30, 40, 50, and 60 years of age, respectively. The number of disease alleles and the frequent occurrence of de novo mutations, often at hotspots with short repeat sequences, suggested that haplotype analysis is of limited use for the diagnosis of MEN1.

Lipomatous tumors are known to occur in a relatively high proportion of patients with MEN1. By fluorescence in situ hybridization analysis of lipomas from 2 patients with MEN1, Vortmeyer et al. (1998) demonstrated deletion of 1 MEN1 allele in 53% of cells examined from case 1 and in 63% of cells examined from case 2. In both cases, both MEN1 gene copies were visualized in normal cellular constituents. To investigate the role of the MEN1 gene in sporadic lipomas, Vortmeyer et al. (1998) analyzed 6 sporadic tumors. In 1 case, SSCP analysis and subsequent sequencing revealed a 4-bp deletion in exon 2 (131100.0017). This deletion was present only in the tumor tissue, and not in the normal tissue from the same patient.

Giraud et al. (1998) studied a total of 84 families and/or isolated patients with either MEN1 or MEN1-related inherited endocrine tumors. They screened for MEN1 germline mutations by heteroduplex and sequence analysis of the gene-coding region of the MEN1 gene and its untranslated exon 1. Germline MEN1 alterations were identified in 47 of 54 (87%) MEN1 families, in 9 of 11 (82%) isolated MEN1 patients, and in only 6 of 19 (31.5%) atypical MEN1-related inherited cases. They characterized 52 distinct mutations in a total of 62 MEN1 germline alterations. Truncating mutations, frameshifts and nonsense mutations, accounted for 35 of the 52 alterations. No genotype/phenotype correlation could be made. Age-related penetrance was estimated to be more than 95% over age 30 years. No MEN1 germline mutations were found in 7 of 54 (13%) MEN1 families.

Animal Model

Chedid et al. (1988) described hereditary pituitary prolactinomas in the rat. It was thought to be an autosomal dominant characteristic with incomplete penetrance and a greater incidence in males.

History

Wermer first reported 'his' syndrome in 1954, and Zollinger and Ellison 'theirs' in 1955 (see Wermer, 1954; Zollinger and Ellison, 1955). The Zollinger-Ellison syndrome of intractable peptic ulcer with pancreatic islet adenoma is a facet of multiple endocrine adenomatosis. Recognition that the 2 eponymic syndromes are one subsequently occurred (Lulu et al., 1968), with multiple endocrine neoplasia type I (MEN1) as the preferred designation.

See also Aach and Kissane (1969); Bale et al. (1989); Bale et al. (1989); Betts et al. (1980); Buchta and Kaplan (1971); Cocco and Conway (1975); Deveney et al. (1978); Ellison and Wilson (1964); Ellison and Wilson (1967); Friesen et al. (1972); Groussin et al. (1980); Jones et al. (1970); Lamers and Froeling (1979); Lamers et al. (1978); Mallette et al. (1974); McCarthy et al. (1977); Mee et al. (1983); Regan and Malagelada (1978); Reimer and Singh (1981); Schimke (1976); Schimke (1984); Snyder et al. (1974); Stadil et al. (1976); Straus et al. (1977); Vance et al. (1969); Way et al. (1968)

Aach, R.; Kissane, J. M.: Clinicopathologic conference: multiple endocrine adenomatosis. Am. J. Med. 47: 608-618, 1969.

Agarwal, S. K.; Kester, M. B.; Debelenko, L. V.; Heppner, C.; Emmert-Buck, M. R.; Skarulis, M. C.; Doppman, J. L.; Kim, Y. S.; Lubensky, I. A.; Zhuang, Z.; Green, J. S.; Guru, S. C.; Manickam, P.; Olufemi, S.-E.; Liotta, L. A.; Chandrasekharappa, S. C.; Collins, F. S.; Spiegel, A. M.; Burns, A. L.; Marx, S. J.: Germline mutations of the MEN1 gene in familial multiple endocrine neoplasia type 1 and related states. Hum. Molec. Genet. 6: 1169-1175, 1997.

Agarwal, S. K.; Kester, M. B.; Debelenko, L. V.; Heppner, C.; Emmert-Buck, M. R.; Skarulis, M. C.; Doppman, J. L.; Kim, Y. S.; Lubensky, I. A.; Zhuang, Z.; Green, J. S.; Guru, S. C.; Manickam, P.; Olufemi, S.-E.; Liotta, L. A.; Chandrasekharappa, S. C.; Collins, F. S.; Spiegel, A. M.; Burns, A. L.; Marx, S. J.: Germline mutations of the MEN1 gene in familial multiple endocrine neoplasia type 1 and related states. Hum. Molec. Genet. 7: 1169-1175, 1997.

Bahn, R. S.; Scheithauer, B. W.; van Heerden, J. A.; Laws, E. R., Jr.; Horvath, E.; Gharib, H.: Nonidentical expressions of multiple endocrine neoplasia, type I, in identical twins. Mayo Clin. Proc. 61: 689-696, 1986.

Bale, A. E.; Friedman, E.; Sakaguchi, K.; Nakamura, Y.; McBride, O. W.; Spiegel, A. M.; Aurbach, G. D.; Marx, S. J.: Allelic loss in hereditary and sporadic tumors associated with MEN1. (Abstract) Am. J. Hum. Genet. 45 (suppl.): A15, 1989.

Bale, S. J.; Bale, A. E.; Marx, S. J.: Mapping of multiple endocrine neoplasia type 1 with markers on chromosome 11q. (Abstract) Cytogenet. Cell Genet. 51: 956, 1989.

Bale, S. J.; Bale, A. E.; Stewart, K.; Dachowski, L.; McBride, O. W.; Glaser, T.; Green, J. E., III; Mulvihill, J. J.; Brandi, M. L.; Sakaguchi, K.; Aurbach, G. D.; Marx, S. J.: Linkage analysis of multiple endocrine neoplasia type 1 with INT2 and other markers on chromosome 11. Genomics 4: 320-322, 1989.

Bale, S. J.; Marx, S. J.; Langfield, D.; Bale, A. E.; Mulvihill, J.; Green, J., III: Multiple endocrine neoplasia type 1 (MEN1): genetic linkage studies. (Abstract) Cytogenet. Cell Genet. 46: 575, 1987.

Ballard, H. S.; Frame, B.; Hartsock, R. J.: Familial multiple endocrine adenoma-peptic ulcer complex. Medicine 43: 481-516, 1964.

Bassett, J. H. D.; Forbes, S. A.; Pannett, A. A. J.; Lloyd, S. E.; Christie, P. T.; Wooding, C.; Harding, B.; Besser, G. M.; Edwards, C. R.; Monson, J. P.; Sampson, J.; Wass, J. A. H.; Wheeler, M. H.; Thakker, R. V.: Characterization of mutations in patients with multiple endocrine neoplasia type 1. Am. J. Hum. Genet. 62: 232-244, 1998.

Bear, J. C.; Briones-Urbina, R.; Fahey, J. F.; Farid, N. R.: Variant multiple endocrine neoplasia I (MEN I-Burin): further studies and non-linkage to HLA. Hum. Hered. 35: 15-20, 1985.

Betts, J. B.; O'Malley, B. P.; Rosenthal, F. D.: Hyperparathyroidism: a prerequisite for Zollinger-Ellison syndrome in multiple endocrine adenomatosis type 1—report of a further family and a review of the literature. Quart. J. Med. 49: 69-76, 1980.

Boni, R.; Vortmeyer, A. O.; Pack, S.; Park, W.-S.; Burg, G.; Hofbauer, G.; Darling, T.; Liotta, L.; Zhuang, Z.: Somatic mutations of the MEN1 tumor suppressor gene detected in sporadic angiofibromas. (Letter) J. Invest. Derm. 111: 539-540, 1998.

Brandi, M. L.; Aurbach, G. D.; Fitzpatrick, L. A.; Quarto, R.; Spiegel, A. M.; Bliziotes, M. M.; Norton, J. A.; Doppman, J. L.; Marx, S. J.: Parathyroid mitogenic activity in plasma from patients with familial multiple endocrine neoplasia type 1. New Eng. J. Med. 314: 1287-1293, 1986.

Buchta, R. M.; Kaplan, J. M.: Zollinger-Ellison syndrome in a nine-year-old child: a case report and review of the entity in childhood. Pediatrics 47: 594-598, 1971.

Bystrom, C.; Larsson, C.; Blomberg, C.; Sandelin, K.; Falkmer, U.; Skogseid, B.; Oberg, K.; Werner, S.; Nordenskjold, M.: Localization of the MEN1 gene to a small region within chromosome 11q13 by deletion mapping in tumors. Proc. Nat. Acad. Sci. 87: 1968-1972, 1990.

Chandrasekharappa, S. C.; Guru, S. C.; Manickam, P.; Olufemi, S.-E.; Collins, F. S.; Emmert-Buck, M. R.; Debelenko, L. V.; Zhuang, Z.; Lubensky, I. A.; Liotta, L. A.; Crabtree, J. S.; Wang, Y.; Roe, B. A.; Weisemann, J.; Boguski, M. S.; Agarwal, S. K.; Kester, M. B.; Kim, Y. S.; Heppner, C.; Dong, Q.; Spiegel, A. M.; Burns, A. L.; Marx, S. J.: Positional cloning of the gene for multiple endocrine neoplasia-type 1. Science 276: 404-406, 1997.

Chedid, A.; Hoversland, R. C.; Chejfec, G.; McCune, S. A.; Jurin, R. R.; Evans, E. E.: Hereditary pituitary prolactinomas: a new rat model. Acta Endocr. 119: 535-542, 1988.

Cocco, A. E.; Conway, S. J.: Zollinger-Ellison syndrome associated with ovarian mucinous cystadenocarcinoma. New Eng. J. Med. 293: 485-486, 1975.

Courseaux, A.; Grosgeorge, J.; Gaudray, P.; Pannett, A. A. J.; Forbes, S. A.; Williamson, C.; Bassett, D.; Thakker, R. V.; Teh, B. T.; Farnebo, F.; Shepherd, J.; Skogseid, B.; Larsson, C.; Giraud, S.; Zhang, C. X.; Salandre, J.; Calender, A.: Definition of the minimal MEN1 candidate area based on a 5-Mb integrated map of proximal 11q13. Genomics 37: 354-365, 1996.

Cutler, R. E.; Reiss, E.; Ackerman, L. V.: Familial hyperparathyroidism: a kindred involving eleven cases, with a discussion of primary chief-cell hyperplasia. New Eng. J. Med. 270: 859-865, 1964.

Darling, T. N.; Skarulis, M. C.; Steinberg, S. M.; Marx, S. J.; Spiegel, A. M.; Turner, M.: Multiple facial angiofibromas and collagenomas in patients with multiple endocrine neoplasia type 1. Arch. Derm. 133: 853-857, 1997.

Debelenko, L. V.; Brambilla, E.; Agarwal, S. K.; Swalwell, J. I.; Kester, M. B.; Lubensky, I. A.; Zhuang, Z.; Guru, S. C.; Manickam, P.; Olufemi, S.-E.; Chandrasekharappa, S. C.; Crabtree, J. S.; Kim, Y. S.; Heppner, C.; Burns, A. L.; Spiegel, A. M.; Marx, S. J.; Liotta, L. A.; Collins, F. S.; Travis, W. D.; Emmert-Buck, M. R.: Identification of MEN1 gene mutations in sporadic carcinoid tumors of the lung. Hum. Molec. Genet. 6: 2285-2290, 1997.

Deveney, C. W.; Deveney, K. S.; Way, L. W.: The Zollinger-Ellison syndrome—23 years later. Ann. Surg. 188: 384-391, 1978.

Ellison, E. H.; Wilson, S. D.: The Zollinger-Ellison syndrome. Re-appraisal and evaluation of 260 registered cases. Ann. Surg. 160: 512-530, 1964.

Ellison, E. H.; Wilson, S. D.: The Zollinger-Ellison syndrome updated. Surg. Clin. N. Am. 47: 1115-1124, 1967.

Farid, N. R.: Personal Communication. Riyadh, Saudi Arabia 1/11/1994.

Farid, N. R.: Personal Communication. St. John's, Newfoundland, Canada 7/28/1980.

Farid, N. R.; Buehler, S.; Russell, N. A.; Maroun, F. B.; Allerdice, P.; Smyth, H. S.: Prolactinomas in familial multiple endocrine neoplasia syndrome type I: relationship to HLA and carcinoid tumors. Am. J. Med. 69: 874-880, 1980.

Friedman, E.; Sakaguchi, K.; Bale, A. E.; Falchetti, A.; Streeten, E.; Zimering, M. B.; Weinstein, L. S.; McBride, W. O.; Nakamura, Y.; Brandi, M.-L.; Norton, J. A.; Aurbach, G. D.; Spiegel, A. M.; Marx, S. J.: Clonality of parathyroid tumors in familial multiple endocrine neoplasia type 1. New Eng. J. Med. 321: 213-218, 1989. Note: Erratum: New Eng. J. Med. 321: 1057 only, 1989. . .

Friesen, S. R.; Schimke, R. N.; Pearse, A. G. E.: Genetic aspects of the Z-E syndrome: prospective studies in two kindred: antral gastrin cell hyperplasia. Ann. Surg. 176: 370-383, 1972.

Fujimori, M.; Wells, S. A., Jr.; Nakamura, Y.: Fine-scale mapping of the gene responsible for multiple endocrine neoplasia type 1 (MEN1). Am. J. Hum. Genet. 50: 399-403, 1992.

Gaitan, D.; Loosen, P. T.; Orth, D. N.: Two patients with Cushing's disease in a kindred with multiple endocrine neoplasia type I. J. Clin. Endocr. Metab. 76: 1580-1582, 1993.

Giraud, S.; Zhang, C. X.; Serova-Sinilnikova, O.; Wautot, V.; Salandre, J.; Buisson, N.; Waterlot, C.; Bauters, C.; Porchet, N.; Aubert, J.-P.; Emy, P.; Cadiot, G.; and 29 others: Germ-line mutation analysis in patients with multiple endocrine neoplasia type 1 and related disorders. Am. J. Hum. Genet. 63: 455-467, 1998.

Groussin, P.; Renard, J.-P.; Chavanne, D.; Bertrand, G.; Avril, J.; Simier, J.-L.: Antigens HLA d'une famille ayant une adenomatose polyendocrienne de type I. (Letter) Presse Med. 9: 1033-1034, 1980.

Guida, P. M.; Todd, J. E.; Moore, S. W.; Beal, J. M.: Zollinger-Ellison syndrome with interesting variations: report of twelve cases including one of carcinoid of the duodenum. Am. J. Surg. 112: 807-817, 1966.

Guru, S. C.; Agarwal, S. K.; Manickam, P.; Olufemi, S.-E.; Crabtree, J. S.; Weisemann, J. M.; Kester, M. B.; Kim, Y. S.; Wang, Y.; Emmert-Buck, M. R.; Liotta, L. A.; Spiegel, A. M.; Boguski, M. S.; Roe, B. A.; Collins, F. S.; Marx, S. J.; Burns, L.; Chandrasekharappa, S. C.: A transcript map for the 2.8-Mb region containing the multiple endocrine neoplasia type 1 locus. Genome Res. 7: 725-735, 1997.

Guru, S. C.; Goldsmith, P. K.; Burns, A. L.; Marx, S. J.; Spiegel, A. M.; Collins, F. S.; Chandrasekharappa, S. C.: Menin, the product of the MEN1 gene, is a nuclear protein. Proc. Nat. Acad. Sci. 95: 1630-1634, 1998.

Heppner, C.; Kester, M. B.; Agarwal, S. K.; Debelenko, L. V.; Emmert-Buck, M. R.; Guru, S. C.; Manickam, P.; Olufemi, S.-E.; Skarulis, M. C.; Doppman, J. L.; Alexander, R. H.; Kim, Y. S.; Saggar, S. K.; Lubensky, I. A.; Zhuang, Z.; Liotta, L. A.; Chandrasekharappa, S. C.; Collins, F. S.; Spiegel, A. M.; Burns, A. L.; Marx, S. J.: Somatic mutation of the MEN1 gene in parathyroid tumours. Nature Genet. 16: 375-378, 1997.

Hershon, K. S.; Kelly, W. A.; Shaw, C. M.; Schwartz, R.; Bierman, E. L.: Prolactinomas as part of the multiple endocrine neoplastic syndrome type 1. Am. J. Med. 74: 713-720, 1983.

Johnson, G. J.; Summerskill, W. H. J.; Anderson, V. E.; Keating, F. R.: Clinical and genetic investigation of a large kindred with multiple endocrine adenomatosis. New Eng. J. Med. 277: 1379-1385, 1967.

Jones, B. S.; O'Hagan, J. J.; Phear, D. N.; Sheville, E.: A case of the Zollinger-Ellison syndrome associated with hyperplasia of salivary and Brunner's glands. Gut 11: 837-839, 1970.

Lamers, C. B. H. W.; Froeling, P. G. A. M.: Clinical significance of hyperparathyroidism in familial multiple endocrine adenomatosis. Am. J. Med. 66: 422-424, 1979.

Lamers, C. B. H. W.; Stadil, F.; van Tongeren, J. H.: Prevalence of endocrine abnormalities in patients with the Zollinger-Ellison syndrome in their families. Am. J. Med. 64: 607-612, 1978.

Larsson, C.; Shepherd, J.; Nakamura, Y.; Blomberg, C.; Weber, G.; Werelius, B.; Hayward, N.; Teh, B.; Tokino, T.; Seizinger, B.; Skogseid, B.; Oberg, K.; Nordenskjold, M.: Predictive testing for multiple endocrine neoplasia type 1 using DNA polymorphisms. J. Clin. Invest. 89: 1344-1349, 1992.

Larsson, C.; Skogseid, B.; Oberg, K.; Nakamura, Y.; Nordenskjold, M.: Multiple endocrine neoplasia type 1 gene maps to chromosome 11 and is lost in insulinoma. Nature 332: 85-87, 1988.

Larsson, C.; Skogseid, B.; Oberg, K.; Nordenskjold, M.: Linkage of multiple endocrine neoplasia type 1 (MEN1) to markers at 11q13. (Abstract) Cytogenet. Cell Genet. 51: 1029, 1989.

Lemmens, I.; Van de Ven, W. J. M.; Kas, K.; Zhang, C. X.; Giraud, S.; Wautot, V.; Buisson, N.; De Witte, K.; Salandre, J.; Lenoir, G.; Pugeat, M.; Calender, A.; and 19 others: Identification of the multiple endocrine neoplasia type 1 (MEN1) gene. Hum. Molec. Genet. 6: 1177-1183, 1997.

Long, T. T.; Barton, T. K.; Draffin, R.; Reeves, W. J.; McCarty, K. S., Jr.: Conservative management of the Zollinger-Ellison syndrome: ectopic gastrin production by an ovarian cystadenoma. J.A.M.A. 243: 1837-1839, 1980.

Lulu, D. J.; Corcoran, T. E.; Andre, M.: Familial endocrine adenomatosis with associated Zollinger-Ellison syndrome: Wermer's syndrome. Am. J. Surg. 115: 695-701, 1968.

Mallette, L. E.; Bilezikian, J. P.; Ketcham, A. S.; Aurbach, G. D.: Parathyroid carcinoma in familial hyperparathyroidism. Am. J. Med. 57: 642-648, 1974.

Maton, P. N.; Gardner, J. D.; Jensen, R. T.: Cushing's syndrome in patients with the Zollinger-Ellison syndrome. New Eng. J. Med. 315: 1-5, 1986.

McCarthy, D. M.: Zollinger-Ellison syndrome. Ann. Rev. Med. 33: 197-215, 1982.

McCarthy, D. M.; Olinger, E. J.; May, R. J.; Long, B. W.; Gardner, J. D.: H2-histamine receptor blocking agents in the Zollinger-Ellison syndrome: experience in seven cases and implications for long-term therapy. Ann. Intern. Med. 87: 668-675, 1977.

Mee, A. S.; Ismail, S.; Bornman, P. C.; Marks, I. N.: Changing concepts in the presentation, diagnosis and management of the Zollinger-Ellison syndrome. Quart. J. Med. 206: 256-267, 1983.

Nakamura, Y.; Larsson, C.; Julier, C.; Bystrom, C.; Skogseid, B.; Wells, S.; Oberg, K.; Carlson, M.; Taggart, T.; O'Connell, P.; Leppert, M.; Lalouel, J.-M.; Nordenskjold, M.; White, R.: Localization of the genetic defect in multiple endocrine neoplasia type I within a small region of chromosome 11. Am. J. Hum. Genet. 44: 751-755, 1989.

Nordenskjold, M.; Jones, C. A.; Glaser, T.; Werelius, B.; Larsson, C.; Jansson, M.: Mapping of chromosome 11q probes on somatic cell hybrids. (Abstract) Cytogenet. Cell Genet. 51: 1054, 1989.

Olufemi, S.-E.; Green, J. S.; Manickam, P.; Guru, S. C.; Agarwal, S. K.; Kester, M. B.; Dong, Q.; Burns, A. L.; Spiegel, A. M.; Marx, S. J.; Collins, F. S.; Chandrasekharappa, S. C.: Common ancestral mutation in the MEN1 gene is likely responsible for the prolactinoma variant of MEN1 (MEN1-Burin) in four kindreds from Newfoundland. Hum. Mutat. 11: 264-269, 1998.

Petty, E. M.; Green, J. S.; Marx, S. J.; Taggart, R. T.; Farid, N.; Bale, A. E.: Mapping the gene for hereditary hyperparathyroidism and prolactinoma (MEN1-Burin) to chromosome 11q: evidence for a founder effect in patients from Newfoundland. Am. J. Hum. Genet. 54: 1060-1066, 1994.

Prezant, T. R.; Levine, J.; Melmed, S.: Molecular characterization of the Men1 tumor suppressor gene in sporadic pituitary tumors. J. Clin. Endocr. Metab. 83: 1388-1391, 1998.

Prosser, P. R.; Karam, J. H.; Townsend, J. J.; Frosham, P. H.: Prolactin-secreting pituitary adenomas in multiple endocrine adenomatosis, type I. Ann. Intern. Med. 91: 41-44, 1979.

Regan, P. T.; Malagelada, J.-R.: A reappraisal of clinical, roentgenographic, and endoscopic features of the Zollinger-Ellison syndrome. Mayo Clin. Proc. 53: 19-23, 1978.

Reimer, D.; Singh, S. M.: A kindred with 5 cases of multiple endocrine adenomatosis type I. Hum. Hered. 31: 84-88, 1981.

Richard, C. W., III; Withers, D. A.; Meeker, T. C.; Maurer, S.; Evans, G. A.; Myers, R. M.; Cox, D. R.: A radiation hybrid map of the proximal long arm of human chromosome 11 containing the multiple endocrine neoplasia type 1 (MEN-1) and bcl-1 disease loci. Am. J. Hum. Genet. 49: 1189-1196, 1991.

Rochelle, J. M.; Watson, M. L.; Oakey, R. J.; Seldin, M. F.: A linkage map of mouse chromosome 19:definition of comparative mapping relationships with human chromosomes 10 and 11 including the MEN1 locus. Genomics 14: 26-31, 1992.

Sawicki, M. P.; Wan, Y.-J. Y.; Johnson, C. L.; Berenson, J.; Gatti, R.; Passaro, E., Jr.: Loss of heterozygosity on chromosome 11 in sporadic gastrinomas. Hum. Genet. 89: 445-449, 1992.

Schimke, R. N.: Multiple endocrine neoplasia: search for the oncogenic trigger. (Editorial) New Eng. J. Med. 314: 1315-1316, 1986.

Schimke, R. N.: Multiple endocrine adenomatosis syndromes. Adv. Intern. Med. 21: 249-265, 1976.

Schimke, R. N.: Genetic aspects of multiple endocrine neoplasia. Ann. Rev. Med. 35: 25-31, 1984.

Skogseid, B.; Larsson, C.; Lindgren, P.-G.; Kvanta, E.; Rastad, J.; Theodorsson, E.; Wide, L.; Wilander, E.; Oberg, K.: Clinical and genetic features of adrenocortical lesions in multiple endocrine neoplasia type 1. J. Clin. Endocr. Metab. 75: 76-81, 1992.

Snyder, N.; Scurry, M.; Hughes, W.: Hypergastrinemia in familial multiple endocrine adenomatosis. Ann. Intern. Med. 80: 321-325, 1974.

Snyder, N., III; Scurry, M. T.; Deiss, W. P., Jr.: Five families with multiple endocrine adenomatosis. Ann. Intern. Med. 76: 53-58, 1972.

Stacpoole, P. W.; Jaspan, J.; Kasselberg, A. G.; Halter, S. A.; Polonsky, K.; Gluck, F. W.; Liljenquist, J. E.; Rabin, D.: A familial glucagonoma syndrome: genetic, clinical and biochemical features. Am. J. Med. 70: 1017-1026, 1981.

Stadil, F.; Stage, G.; Rehfeld, J. F.; Efsen, F.; Fischerman, K.: Treatment of Zollinger-Ellison syndrome with streptozotocin. New Eng. J. Med. 294: 1440-1442, 1976.
Straus, E.; Johnson, G. F.; Yalow, R. S.: Canine Zollinger-Ellison syndrome. Gastroenterology 72: 380-381, 1977.
Tateishi, R.; Wada, A.; Ishiguro, S.; Ehara, M.; Sakamoto, H.; Miki, T.; Mori, Y.; Matsui, Y.; Ishikawa, O.: Coexistence of bilateral pheochromocytoma and pancreatic islet cell tumor: report of a case and review of the literature. Cancer 42: 2928-2934, 1978.
Thakker, R. V.; Bouloux, P.; Wooding, C.; Chotai, K.; Broad, P. M.; Spurr, N. K.; Besser, G. M.; O'Riordan, J. L. H.: Association of parathyroid tumors in multiple endocrine neoplasia type I with loss of alleles on chromosome 11. New Eng. J. Med. 321: 218-224, 1989.
Underwood, L. E.; Jacobs, N. M.: Familial endocrine adenomatosis. Am. J. Dis. Child. 106: 218-223, 1963.
Vance, J. E.; Stoll, R. W.; Kitabchi, A. E.; Buchanan, K. D.; Hollander, D.; Williams, R. H.: Familial nesidioblastosis as the predominant manifestation of multiple endocrine adenomatosis. Am. J. Med. 52: 211-227, 1972.
Vance, J. E.; Stoll, R. W.; Kitabchi, A. E.; Williams, R. H.; Wood, F. C., Jr.: Nesidioblastosis in familial endocrine adenomatosis. J.A.M.A. 207: 1679-1682, 1969.
Vortmeyer, A. O.; Boni, R.; Pak, E.; Pack, S.; Zhuang, Z.: Multiple endocrine neoplasia 1 gene alterations in MEN1-associated and sporadic lipomas. (Letter) J. Nat. Cancer Inst. 90: 398-399, 1998.
Way, L.; Goldman, L.; Dunphy, J. E.: Zollinger-Ellison syndrome: an analysis of twenty-five cases. Am. J. Surg. 116: 293-304, 1968.
Wermer, P.: Genetic aspects of adenomatosis of endocrine glands. Am. J. Med. 16: 363-371, 1954.
Williams, E. D.; Celestin, L. R.: The association of bronchial carcinoid and pluriglandular adenomatosis. Thorax 17: 120-127, 1962.
Wolfe, M. M.; Jensen, R. T.: Zollinger-Ellison syndrome: current concepts in diagnosis and management. New Eng. J. Med. 317: 1200-1209, 1987.
Zollinger, R. M.; Ellison, E. H.: Primary peptic ulcerations of the jejunum associated with islet cell tumors of the pancreas. Ann. Surg. 142: 709-728, 1955.

MULTIPLE ENDOCRINE NEOPLASIA, TYPE II; MEN2 (#171400)

JJM NOTE

Multiple endocrine neoplasia 2, MEN2, is the Sipple syndrome, "simply" pheochromocytoma and medullary thyroid carcinoma. Another memory aid (parallel to MEN1 as PPP) is MMM: *m*edullary thyroid and *m*edullary adrenal *m*alignancy. The gene locus is on chromosome 10 and the isolated gene is the oncogene, *RET*, a receptor tyrosine kinase, with a protein domain homologous to the torso gene of drosophila. The association of gene changes elsewhere than chromosome 10 and the occurrence of overlapping syndromes suggest the necessary involvement of other genes in *MEN2*. All families should be referred for evaluation by a medical geneticist. After proper counseling, the results of genetic testing permit half the individuals at risk to escape lifelong biochemical screening, because they did not receive the mutant gene (Ledger GA et al.: Genetic testing in the diagnosis and management of multiple endocrine neoplasia type II. Ann. Intern. Med. 122:118–24, 1995).

Description

A number sign (#) is used with this entry because of evidence indicating that MEN2A results from mutation in the RET oncogene (164761).

Multiple endocrine neoplasia, type IIA, is an autosomal dominant syndrome of multiple endocrine neoplasms, including medullary thyroid carcinoma, pheochromocytoma, and parathyroid adenomas.

Clinical Features

Schimke and Hartmann (1965) described a syndrome of pheochromocytoma and medullary thyroid carcinoma with abundant amyloid stroma. A similar but distinct condition is described under neuromata, mucosal, with endocrine tumors (MEN2B; MEN3; 162300). Steiner et al. (1968) described a family with 11 cases in successive generations.

The pheochromocytomas were bilateral, parathyroid adenoma was present in several, and one patient had Cushing syndrome.

Steiner et al. (1968) referred to this disorder as 'multiple endocrine neoplasia, type II' to distinguish it from the multiple endocrine adenomatosis described by Wermer (MEN I; 131100) and called type I by Steiner et al. (1968).

Urbanski (1967) found parathyroid adenoma to be part of the syndrome also.

Meyer and Abdel-Bari (1968) presented observations consistent with the view that medullary carcinoma is a thyrocalcitonin-producing neoplasm of parafollicular cells of the thyroid.

Parathyroid hyperplasia or adenomas in some of these patients may be secondary to hypocalcemic effects of thyrocalcitonin.

Johnston et al. (1970), as well as others, have shown calcitonin-secretion by medullary thyroid carcinoma.

Kaplan et al. (1970) showed that the adrenal medulla produces a calcitonin-like material indistinguishable from that of the thyroid by bio- and radioimmunoassay.

They suggested that the parafollicular cells of the thyroid are of neural crest origin.

The finding that medullary carcinoma of the thyroid arises from parafollicular cells and that, like the cell of origin, it sometimes produces thyrocalcitonin may account for the association of parathyroid hyperplasia and perhaps parathyroid adenoma.

Poloyan et al. (1970) was impressed with the histologic similarity between the medullary thyroid cancer and pheochromocytoma metastases. Keiser et al. (1973) pointed out that histaminase is useful in the identification of metastases of medullary carcinoma.

In their opinion parathyroid adenomas are a primary feature of the disorder.

Pearson et al. (1973) studied 21 members of a kindred with surgically confirmed multiple endocrine neoplasms.

All 21 had medullary carcinoma of the thyroid.

Adrenal pheochromocytomas were present in 10 and were bilateral in 6.

Three had one or more parathyroid glands showing adenomatous hyperplasia and 10 showed chief cell hyperplasia.

The thyroid cancer metastasized to other areas including the liver, lungs, and bone in several of the patients.

All patients had elevated peripheral thyrocalcitonin.

Peripheral parathyroid hormone was elevated in only 2; however, parathyroid hormone was elevated in the inferior thyroid vein of all patients examined.

Hamilton et al. (1978) suggested that an increased urinary epinephrine fraction is a sensitive and reliable screening test for pheochromocytoma in MEN II, comparable to the calcitonin radioimmunoassay for medullary carcinoma of the thyroid.

Carney et al. (1975) found bilateral adrenal medullary hyperplasia in an asymptomatic 12-year-old girl. She had bilateral thyroid carcinoma and hyperparathyroidism. The adrenals were explored because of elevated urinary levels of vanillylmandelic acid.

Migrating neural crest cells are able to decarboxylate and store precursors of aromatic amines that fluoresce after exposure to formaldehyde vapor. The last is a method for identifying neural crest origin of enterochromaffin, argyrophil cells of the bronchi, islets of Langerhans, and parafollicular cells of the thyroid, among others.

These are collectively termed the amine precursor uptake and decarboxylase (APUD) system.

Tischler et al. (1976) extended the evidence of neural origin by demonstrating that cultured cells from medullary carcinoma of the thyroid, bronchial carcinoid, and pheochromocytoma display all-or-nothing action potentials of short duration.

Le Marec et al. (1980) reported congenital megacolon with plexus hyperplasia in a family with Sipple syndrome.

Megacolon of this type seems to be more usual in MEN III than in MEN II.

Cameron et al. (1978) described the Zollinger-Ellison syndrome with type II MEA, a first. These families represent an overlap of phenotypic features in the 3 forms of MEN.

Gagel et al. (1989), Nunziata et al. (1989), and Kousseff et al. (1991) observed primary localized cutaneous amyloidosis (PLCA) in association with MEN2A.

Gagel et al. (1989) and Kousseff et al. (1991) referred to it as cutaneous lichen amyloidosis.

In a family with 6 affected members in 5 generations, a mother and her daughter had interscapular cutaneous pruritic lesions (Kousseff et al., 1991).

Kousseff (1992) provided a pedigree and photographs of the skin lesions. Cutaneous lichen amyloidosis as an apparently independent autosomal dominant trait has also been described (105250).

The skin deposits of amyloid associated with pruritus in the interscapular region represents a form of 'friction amyloidosis' (Wong and Lin, 1988).

It is related to notalgia paresthetica, an inherited neuropathy of the posterior dorsal nerve rami.

('Notalgia' means 'back pain.') The neuropathy hypothesis was supported by the finding of mutations in the RET protooncogene which is expressed in the peripheral and central nervous system.

Ceccherini et al. (1994) demonstrated a specific cys634-to-tyr missense mutation (164761.0004) in affected members of a pedigree in which MEN2A was combined with localized cutaneous lichen amyloidosis.

Easton et al. (1989) estimated on the basis of clinical history that 41% of gene carriers are asymptomatic at age 70.

Screening by the standard tests for detecting the earliest manifestations of the syndrome increased the penetrance to an estimated 93% by age 31.

There was some suggestion of an earlier onset of medullary thyroid cancer in female gene carriers, and of a tendency for pheochromocytoma to cluster in families.

Multiple endocrine neoplasia type 2A can be subclassified as: (1) medullary thyroid carcinoma with pheochromocytomas and parathyroid tumors; (2) medullary thyroid carcinoma with pheochromocytomas alone; or (3) medullary thyroid carcinoma with parathyroid tumors alone.

Hereditary medullary thyroid carcinoma alone (see 155240) may be a fourth phenotype resulting, in some instances, from mutation at the MEN2 locus.

Cytogenetics

In affected members of 7 families, Van Dyke et al. (1982) found a small deletion of chromosome 20p12.2 segregating with MEN II.

Van Dyke et al. (1981) demonstrated the 20p deletion in 16 cases of MEN II in 7 families, and in 1 case of MEN III.

Hsu et al. (1981) found a higher frequency of metaphases with chromosome and chromatid abnormalities (average, 11.0%) in cases of Sipple syndrome than in controls (average, 3.8%).

In one pair of sibs, they failed to find the same deletion in chromosome 20.

Jackson (1982) stated that his group had correctly identified 18 persons in blinded studies.

A small deletion was found in 18 members of 8 families with Sipple syndrome and also in 2 unrelated patients with MEN III.

See Babu et al. (1982).

Although, as they pointed out, their experience is not proof of the absence of deletion, Gustavson et al. (1983) could not demonstrate such by high resolution banding in either MEN I or MEN II.

High resolution cytogenetics in 5 persons studied by Emmertsen et al. (1983) showed no deletion in band 20p12.2.

In both MEN2A and MEN2B, Babu et al. (1984) reported the finding of an interstitial deletion in band 20p12.2.

In a double-blind study, 2 of 13 controls were thought to have the deletion; all 9 blood samples from 8 affected members of 4 MEN2A families were found to have the deletion; from 3 MEN2B families, 5 blood samples from 4 affected members showed the deletion, whereas 3 did not.

The authors suggested that these 2 entities are genetically closely related; that the dominant expression of the mutation at 20p12.2 is hyperplasia of thyroid C cells and adrenal medullary cells; that in accordance with Knudson's 2-mutation-event theory and in analogy to retinoblastoma, thyroid cancer and pheochromocytoma are recessive manifestations.

Zatterale et al. (1984) could not detect a 20p12.2 deletion in prometaphase banding studies.

Babu et al. (1987) reported on an expanded double-blind study of chromosome 20 deletion in MEN2A and MEN2B.

A deletion in 20p12.2 was found in 15 of 21 MEN2A patients and in 4 of 8 MEN2B patients.

These findings await explanation since the locus for these disorders maps elsewhere.

Mapping

Jackson et al. (1976) found a suggestion of linkage to P blood group (111400) but not to HLA.

Simpson et al. (1979) found 23.1% recombination between MEN II and HLA, but this linkage was rendered highly unlikely by further studies by Simpson and Falk (1982).

Jackson et al. (1979) concluded that medullary thyroid cancer fits a 2-mutation theory.

They suggested that C-cell hyperplasia is the gene-determined first mutational event and cancer the second.

Emmertsen et al. (1983) found no significantly positive lod scores between MEN II and 25 different genetic markers.

Simpson et al. (1984) assigned the calcitonin gene to chromosome 11 by use of a cDNA clone isolated from medullary thyroid carcinoma and a somatic cell hybrid panel.

With a TaqI RFLP detected by this probe, they studied linkage of the calcitonin locus and MEN2; negative lod scores were found at all recombination values.

Goodfellow et al. (1984) studied linkage between MEN2 and DNA probes assigned to 20p12.2 by in situ hybridization.

Negative lod scores were obtained.

In 2 large MEN2A pedigrees, Goodfellow et al. (1985) studied linkage with 2 RFLPs found in an anonymous DNA segment D20S5, which had been isolated from a chromosome 19/20 flow-sorted library and shown by in situ hybridization to be located at 20p12.

Linkage was excluded at theta equal to or less than 0.13.

In studies of a single large kindred, Kruger et al. (1986) excluded close linkage with several markers and found no statistically significant linkage with any marker.

Low positive lod scores were obtained with GC (139200), GPT (138200) and HP (140100).

Perrier et al. (1987) found no linkage of MEN2 and HLA. By multipoint linkage analysis, Farrer et al. (1987) excluded a large portion of chromosome 13 as the site of the MEN2A locus.

Farrer et al. (1987) could find no linkage of MEN2A to 3 DNA markers that mapped to chromosome 20.

Simpson et al. (1987) performed linkage studies in 5 families, each with at least 1 member found to have a deletion at 20p12.2.

In these families, the MEN2 locus was found to be linked to a DNA marker on chromosome 10, D10S5, which by in situ hybridization maps to 10q21.1.

The maximal lod score was 3.58 for a recombination fraction of 0.19. With 2 RFLPs recognized by an RBP3 probe (180290), Simpson et al. (1987) found a maximal lod score of 8.0 at theta = 0.11.

In studies of 5 Japanese families, Yamamoto et al. (1989) found that MEN2A is closely linked to RBP3; the maximum lod score was 5.19 at a recombination fraction of 0.00.

In 2 kindreds, Mathew et al. (1987) found a maximal lod score of 3.88 at theta = 0.04 for linkage with RBP3, which has been positioned at 10p11.2-q11.2.

Sobol et al. (1989) reported on linkage studies in 35 families with medullary carcinoma of the thyroid, with or without pheochromocytoma. Their results suggest that a susceptibility gene for hereditary medullary carcinoma of the thyroid may be located at the same locus on chromosome 10 as that of MEN2A.

Narod et al. (1989) did linkage studies in 18 families, 9 with MEN2A and 9 with medullary carcinoma of the thyroid without pheochromocytoma, with probes specific for the pericentromeric region of chromosome 10 and concluded that the mutations for the 2 presentations are closely situated.

Genetic heterogeneity of the susceptibility locus was not seen among these 18 families.

The genetic mutation for medullary carcinoma was in disequilibrium with alleles at 2 closely linked markers.

Using a panel of markers from the pericentric region of chromosome 10, Lairmore et al. (1991) performed linkage studies in 2 large families with medullary thyroid carcinoma (MTC1) and 6 families with MEN2B.

The maximum lod score between MTC1 and marker D10Z1 was 5.88 with 0% recombination.

MEN2B showed similarly tight linkage to D10Z1, with a maximum lod score of 3.58 at 0% recombination.

The multipoint lod score for MEN2B at D10Z1 was 4.08.

Linkage studies in a single large MEN2A kindred showed tight linkage to D10Z1 in this condition as well, with a maximum multipoint lod score of 7.04 at a recombination fraction of 0.

The highest lod score obtained was between MEN2A and the haplotyped locus RBP3 with a lod score of 11.33 at a recombination fraction of 0.

Linkage data between MEN2A and 3 additional markers on 10q, as well as between MEN2A and the FNRB locus (135630) on 10p11.2, were also presented.

Lairmore et al. (1991) found no evidence for genetic heterogeneity among families with MEN2A, MEN2B, and MTC.

In 2 families with medullary thyroid carcinoma with parathyroid tumors alone, Carson et al. (1990) demonstrated that the disorder was closely linked to 2 markers in the vicinity of the centromere of chromosome 10, namely, RBP3 and D10Z1.

Using a centromeric marker at the D10Z1 locus in 30 families with MEN2A, Narod et al. (1991) demonstrated tight linkage with the centromere.

Brooks-Wilson et al. (1992) characterized a dense cluster of CpG islands at D10S94 in proximal 10q11.2.

No recombinants between D10S94 and MEN2A had been identified. They generated a 570-kb restriction map by pulsed field gel electrophoresis. Six CpG islands were clustered within a 180-kb region.

They suggested that these CpG islands may represent the 5-prime ends of candidate genes for MEN2A, MEN2B, and/or MTC1.

McDonald et al. (1992) further identified an MEN2A candidate gene by use of an evolutionarily conserved sequence from D10S94. The gene spanned 11 kb and had an unmethylated CpG island at its 5-prime end.

It encoded a putative 415-amino acid polypeptide similar in sequence to nucleolin (162090), an abundant nucleolar protein.

In a patient with MEN2A, they found no difference in the candidate gene, termed mcs94-1, from the MEN2A chromosome or its wildtype homolog.

By genetic linkage analysis, Gardner et al. (1993) demonstrated that the MEN2A gene is located in a small region of 10q11.2 flanked by D10S141 proximally and D10S94 distally, these two markers being separated by a sex-average genetic distance of 0.55 cM. Mole et al. (1993) constructed a YAC contig spanning 1.1 Mb of band 10q11.2 which must include MEN2A because it encompassed 3 markers, D10S141, RET, and D10S94.

A 480-kb region separated D10S141 and D10S94.

Janson et al. (1991) constructed a physical map of the MEN2 region by combining data from pulsed field gel electrophoresis (PFGE) with those generated from a panel of radiation-reduced somatic cell hybrids. Comparison of the physical map with the linkage map showed a recombination rate higher than expected: thus, for the closest pair of linked markers on the centromeric side of MEN2, 1 centimorgan corresponded to approximately 300 kb, and for markers on the telomeric side, 1 centimorgan corresponded to approximately 350 to 600 kb.

There is evidence from other sources that the 11q12-q13 region is unusual in having a high G-C content, suggesting a high concentration of genes and other characteristics including increased meiotic recombination usually associated with telomeric regions (Saccone et al., 1992).

Molecular Genetics

Mathew et al. (1987) found deletion of a hypervariable region of DNA on 1p in 7 of 14 tumors (pheochromocytomas and medullary carcinomas) developing in patients with MEN2.

In 1 of 2 families examined, the deleted chromosome was that inherited from the affected parent.

Thus, the site of deletion presumably does not represent the location of the inherited gene.

The deleted region was distal to the breakpoint commonly detected in neuroblastomas (256700), which share with the tumors of MEN2 embryologic origin from neuroectoderm.

The most frequent breakpoint involved in neuroblastomas is 1p32, whereas the genes deleted in the tumors studied by Mathew et al. (1987) were located at 1p35-p33.

In an analysis of tumor DNA from 42 patients with MEN2A, Landsvater et al. (1989) showed that markers on chromosome 10 were lost in only 1 tumor, a result that contrasts with studies in other tumors for which both familial and sporadic cases are known.

That MEN2A is genetically heterogeneous is suggested by the linkage in some families to markers at the 10q11.2 region and the lack of linkage in other families.

The basis of the above subclassification, whether different mutations in one gene or mutations in adjacent genes in the 10q11.2 region, is not clear (Simpson, 1992).

The subclasses do seem to 'breed true' in different families.

In a panel of 34 families with MEN2A, Narod et al. (1992) found no evidence of genetic heterogeneity. No recombination was observed between MEN2A and any of 4 DNA marker loci.

Narod et al. (1992) constructed haplotypes for 11 polymorphisms in the MEN2A region for mutation-bearing chromosomes in 24 French families and for 100 spouse controls.

One haplotype was present in 4 MEN2A families but was not observed in any control (P = less than 0.01).

Two additional families shared a core segment of this haplotype near the MEN2A gene.

Narod et al. (1992) suggested that these 6 families had a common affected ancestor.

Because the incidence of pheochromocytoma among carriers varied from 0.0 to 74% in these 6 families, they suggested that additional factors modify the expression of the gene.

Curiously, the most consistent molecular genetic abnormality that has been found in pheochromocytomas and medullary thyroid cancers, either sporadic or part of MEN2, is loss of heterozygosity (LOH) on 1p.

Using RFLP analysis, Moley et al. (1992) identified loss of all or a portion of 1p in 12 of 18 pheochromocytomas.

LOH of 1p was found in all 9 pheochromocytomas in MEN2A and MEN2B patients, compared with only 2 of 7 sporadic pheochromocytomas.

They also found 1p LOH in the pheochromocytoma of 1 of 2 von Hippel-Lindau patients (193300).

LOH on 1p was noted in only 3 of 24 informative medullary thyroid carcinomas, and these were from patients with MEN2A.

Mulligan et al. (1993) identified missense mutations in the RET protooncogene in 20 of 23 apparently distinct MEN2A families, but not in 23 normal controls.

Of these 20 mutations, 19 affected the same conserved cysteine residue at the boundary of the RET extracellular and intracellular domains.

Diagnosis

In the Netherlands, Vasen et al. (1987) demonstrated the usefulness of screening and a central registry for the long-term follow-up of cases.

In an 18-year study of a large kindred, Gagel et al. (1988) found that prospective screening and early treatment of manifestations of multiple endocrine neoplasia can prevent metastasis of medullary thyroid carcinoma and the morbidity and mortality of pheochromocytoma. Medullary carcinoma of the thyroid is the most consistent single manifestation of this disorder and occurs in almost all cases by age 40.

Before age 40 in particular, it is necessary to use a provocative test of the combined calcitonin secretagogues enterogastrone and calcium in order to detect the disorder since the basal levels are not elevated (Baylin, 1989).

As part of a French national program, Sobol et al. (1989) used DNA probes in a genetic linkage study of 130 members of 11 families of European and North African origin who were ascertained through members with MEN2A.

No recombination was found between the mutation causing MEN2A and 2 of 3 markers used.

All 11 families were informative for at least 1 of the markers and linkage information was adequate to permit genetic counseling in 8 families.

Sobol et al. (1989) concluded that RFLP analysis is more useful in predicting the carrier state than conventional endocrine challenge, especially in younger persons, but accuracy is maximal when both methods are used.

Mathew et al. (1991), including 23 members of the MEN2A International Collaborative Group, described 4 markers from the pericentric region of chromosome 10 that are tightly linked to MEN2A and are useful for testing for carrier status in individuals genetically at risk but showing a negative biochemical screening test for thyroid C-cell hyperplasia.

The tests were also accurate for prenatal diagnosis.

Calmettes et al. (1992) reported the consensus on biochemical and genetic screening formulated by the European Community Concerted Action on the subject of medullary thyroid carcinoma.

For biochemical screening, measurement of the basal and pentagastrin- and/or calcium-stimulated serum levels of calcitonin by radioimmunoassay was considered essential starting at the age of 3 and continuing annually until the age of 35.

Furthermore, annual screening for pheochromocytoma by measurement of urinary excretion of catecholamines and for hyperparathyroidism by serum calcium determination was considered indicated.

Biochemical screening can be reserved for gene carriers in some families; genetic screening using genetic markers can be done with 95% accuracy in informative families whenever DNA is available from at least 2 family members proven to be affected.

Total thyroidectomy at an early stage usually cures the patient with medullary thyroid carcinoma.

On the basis of studies in a very large kindred, Landsvater et al. (1993) found 7 individuals with abnormal calcitonin test results. Five of these people were thyroidectomized, and C-cell hyperplasia was diagnosed.

Four were the offspring of a mother at risk for the development of MEN2A who showed, however, normal calcitonin test results throughout the years, whereas the father, who was not at risk, had abnormal test results over a period of 10 years, without evidence of progressive elevation.

None of the 7 individuals developed other manifestations of MEN2A. DNA analysis using markers linked to the MEN2A gene demonstrated, with more than 99% likelihood, that none of the persons who could be genotyped was a gene carrier.

Thus, C-cell hyperplasia due to some mechanism other than the presence of the MEN2A gene may occur in MEN2A kindreds.

Schuffenecker et al. (1997) reported that 5.6% to 9% of cases of MEN2A/MTC are de novo cases with no family history.

They reported further that new mutations in the RET oncogene in these cases were demonstrated exclusively on the paternal allele.

Retrospective analysis on 274 MEN2A cases revealed that in 40.2% of patients pheochromocytoma occurred 2 to 11 years subsequent to MTC. Schuffenecker et al. (1997) concluded that all apparently sporadic MTC patients should be examined for de novo RET mutations.

Sporadic medullary thyroid carcinoma has usually been found to result from a mutational event occurring at the single-cell level, indicating that they are monoclonal.

By clonality assay of medullary carcinoma of the thyroid in MEN type 2, Ferraris et al. (1997) showed the carcinomas they studied to be polyclonal in most instances.

They used a polymorphic trinucleotide repeat of the X-linked human androgen receptor gene (313700) to demonstrate that 10 out of 11 MTCs expressed a polyclonal pattern of X inactivation; furthermore, a significant percentage of cases clinically defined as sporadic showed a polyclonal pattern.

See also Anderson et al. (1971); Baylin et al. (1978); Block et al. (1967); Cerny et al. (1982); Cushman (1962); Farrer et al. (1987); Francke (1985); Gagel et al. (1982); Graze (1978); Li et al. (1974); Lima and Smith (1971); Lips et al. (1978); Lips et al. (1981); Mathew et al. (1987); O'Dorisio et al. (1982); Pearson (1985); Ponder (1993); Sarosi and Doe (1968); Simpson et al. (1987); Sipple (1961); Sobol et al. (1989); Tashjian and Melvin (1968); Valk et al. (1981); Wood (1979)

Anderson, T. E.; Spackman, T. J.; Schwartz, S. S.: Roentgen findings in intestinal ganglioneuromatosis: its association with medullary thyroid carcinoma and pheochromocytoma. Radiology 101: 93-96, 1971.
Babu, V. R.; Jackson, C. E.; Van Dyke, D. L.: Chromosome 20 deletion in multiple endocrine neoplasia syndrome types 2A and 2B. (Abstract) Clin. Res. 30: 489A only, 1982.
Babu, V. R.; Van Dyke, D. L.; Flejter, W. L.; Jackson, C. E.: Chromosome 20 deletion in multiple endocrine neoplasia type 2: expanded double-blind studies. Am. J. Med. Genet. 27: 739-748, 1987.
Babu, V. R.; Van Dyke, D. L.; Jackson, C. E.: Chromosome 20 deletion in human multiple endocrine neoplasia types 2A and 2B: a double-blind study. Proc. Nat. Acad. Sci. 81: 2525-2528, 1984.
Baylin, S. B.: Personal Communication. Baltimore, Md. 4/22/1989.
Baylin, S. B.; Hsu, S. H.; Gann, D. S.; Smallridge, R. C.; Wells, S. A., Jr.: Inherited medullary thyroid carcinoma: a final monoclonal mutation in one of multiple clones of susceptible cells. Science 199: 429-431, 1978.
Block, M. A.; Horn, R. C., Jr.; Miller, J. M.; Barrett, J. L.; Brush, B. E.: Familial medullary carcinoma of the thyroid. Ann. Surg. 166: 403-412, 1967.
Brooks-Wilson, A. R.; Smailus, D. E.; Goodfellow, P. J.: A cluster of CpG islands at D10S94, near the locus responsible for multiple endocrine neoplasia type 2A (MEN2A). Genomics 13: 339-343, 1992.
Calmettes, C.; Ponder, B. A. J.; Fischer, J. A.; Raue, F.; the members of the European Community Concerted Action: Medullary Thyroid Carcinoma: Early diagnosis of the multiple endocrine neoplasia type 2 syndrome: consensus statement. Europ. J. Clin. Invest. 22: 755-760, 1992.
Cameron, D.; Spiro, H. M.; Landsberg, L.: Zollinger-Ellison syndrome with multiple endocrine adenomatosis type II. (Letter) New Eng. J. Med. 299: 152-153, 1978.
Carney, J. A.; Sizemore, G. W.; Tyce, G. M.: Bilateral adrenal medullary hyperplasia in multiple endocrine neoplasia, type II. The precursor of bilateral pheochromocytoma. Mayo Clin. Proc. 50: 3-10, 1975.
Carson, N. L.; Wu, J.; Jackson, C. E.; Kidd, K. K.; Simpson, N. E.: The mutation for medullary thyroid carcinoma with parathyroid tumors (MTC with PTs) is closely linked to the centromeric region of chromosome 10. Am. J. Hum. Genet. 47: 946-951, 1990.
Ceccherini, I.; Romei, C.; Barone, V.; Pacini, F.; Martino, E.; Loviselli, A.; Pinchera, A.; Romeo, G.: Identification of the cys634-to-tyr mutation of the RET proto-oncogene in a pedigree with multiple endocrine neoplasia type 2A and localized cutaneous lichen amyloidosis. J. Endocr. Invest. 17: 201-204, 1994.
Cerny, J. C.; Jackson, C. E.; Talpos, G. B.; Yott, J. B.; Lee, M. W.: Pheochromocytoma in multiple endocrine neoplasia type II: an example of the two-hit theory of neoplasia. Surgery 92: 849-852, 1982.
Cushman, P., Jr.: Familial endocrine tumors: report of two unrelated kindreds affected with pheochromocytomas, one also with multiple thyroid carcinomas. Am. J. Med. 32: 352-360, 1962.
Easton, D. F.; Ponder, M. A.; Cummings, T.; Gagel, R. F.; Hansen, H. H.; Reichlin, S.; Tashjian, A. H., Jr.; Telenius-Berg, M.; Ponder, B. A. J.; the Cancer Research Campaign Medullary Thyroid Group: The clinical and screening age-at-onset distribution for the MEN-2 syndrome. Am. J. Hum. Genet. 44: 208-215, 1989.
Emmertsen, K.; Lamm, L. U.; Rasmussen, K. Z.; Elbrond, O.; Hansen, H. H.; Henningsen, K.; Jorgensen, J.; Petersen, G. B.: Linkage and chromosome study of multiple endocrine neoplasia IIa. Cancer Genet. Cytogenet. 9: 251-259, 1983.
Farrer, L. A.; Goodfellow, P. J.; Lamarche, C. M.; Franjkovic, I.; Myers, S.; White, B. N.; Holden, J. J. A.; Kidd, J. R.; Simpson, N. E.; Kidd, K. K.: An efficient strategy for gene mapping using multipoint linkage analysis: exclusion of the multiple endocrine neoplasia 2A (MEN2A) locus from chromosome 13. Am. J. Hum. Genet. 40: 329-337, 1987.
Farrer, L. A.; Goodfellow, P. J.; White, B. N.; Holden, J. J. A.; Kidd, J. R.; Simpson, N. E.; Kidd, K. K.: Linkage analysis of multiple endocrine neoplasia type 2A (MEN-2A) and three DNA markers on chromosome 20: evidence against synteny. Cancer Genet. Cytogenet. 27: 327-334, 1987.
Ferraris, A. M.; Mangerini, R.; Gaetani, G. F.; Romei, C.; Pinchera, A.; Pacini, F.: Polyclonal origin of medullary carcinoma of the thyroid in multiple endocrine neoplasia type 2. Hum. Genet. 99: 202-205, 1997.
Francke, U.: Personal Communication. New Haven, Conn. 1985.
Gagel, R. F.; Jackson, C. E.; Block, M. A.; Feldman, Z. T.; Reichlin, S.; Hamilton, B. P.; Tashjian, A. H., Jr.: Age-related probability of development of hereditary medullary thyroid carcinoma. J. Pediat. 101: 941-946, 1982.
Gagel, R. F.; Levy, M. L.; Donovan, D. T.; Alford, B. R.; Wheeler, T.; Tschen, J. A.: Multiple endocrine neoplasia type 2a associated with cutaneous lichen amyloidosis. Ann. Intern. Med. 111: 802-806, 1989.
Gagel, R. F.; Tashjian, A. H., Jr.; Cummings, T.; Papathanasopoulos, N.; Kaplan, M. M.; DeLellis, R. A.; Wolfe, H. J.; Reichlin, S.: The clinical outcome of prospective screening for multiple endocrine neoplasia type 2a: an 18-year experience. New Eng. J. Med. 318: 478-484, 1988.
Gardner, E.; Papi, L.; Easton, D. F.; Cummings, T.; Jackson, C. E.; Kaplan, M.; Love, D. R.; Mole, S. E.; Moore, J. K.; Mulligan, L. M.; Norum, R. A.; Ponder, M. A.; Reichlin, S.; Stall, G.; Telenius, H.; Telenius-Berg, M.; Tunnacliffe, A.; Ponder, B. A. J.: Genetic linkage studies map the multiple endocrine neoplasia type 2 loci to a small interval on chromosome 10q11.2. Hum. Molec. Genet. 2: 241-246, 1993.
Goodfellow, P. J.; White, B. N.; Holden, J. J. A.; Duncan, A. M. V.; Sears, E. V. P.; Wang, H.-S.; Berlin, L.; Kidd, K. K.; Simpson, N. E.: Linkage analysis of a DNA marker localized to 20p12 and multiple endocrine neoplasia type 2A. Am. J. Hum. Genet. 37: 890-897, 1985.
Goodfellow, P. J.; White, B. N.; Holden, J. J. A.; Duncan, A. M. V.; Wang, H.-S.; Greenberg, C. R.; Sears, E. V. P.; Ghent, W. R.; Simpson, N. E.: Linkage studies in multiple endocrine neoplasia type-2 using chromosome 20 markers. (Abstract) Am. J. Hum. Genet. 36: 138S only, 1984.
Graze, K.: Natural history of a familial medullary thyroid carcinoma. New Eng. J. Med. 299: 980-985, 1978.
Gustavson, K.-H.; Jansson, R.; Oberg, K.: Chromosomal breakage in multiple endocrine adenomatosis (types I and II). Clin. Genet. 23: 143-149, 1983.
Hamilton, B. P.; Landsberg, L.; Levine, R. J.: Measurement of urinary epinephrine in screening for pheochromocytoma in multiple endocrine neoplasia type II. Am. J. Med. 65: 1027-1032, 1978.
Hsu, T. C.; Pathak, S.; Samaan, N.; Hickey, R. C.: Chromosome instability in patients with medullary carcinoma of the thyroid. J.A.M.A. 246: 2046-2048, 1981.
Jackson, C. E.: Personal Communication. Detroit, Mich. 4/14/1982.
Jackson, C. E.; Block, M. A.; Greenawalt, K. A.; Tashjian, A. H., Jr.: The two-mutational-event theory in medullary thyroid cancer. Am. J. Hum. Genet. 31: 704-710, 1979.
Jackson, C. E.; Conneally, P. M.; Sizemore, G. W.; Tashjian, A. H., Jr.: Possible linear order of genes for endocrine neoplasia type 2, the P red cell antigen and HL-A on chromosome 6. Birth Defects Orig. Art. Ser. 12(1): 159-164, 1976.
Janson, M.; Larsson, C.; Werelius, B.; Jones, C.; Glaser, T.; Nakamura, Y.; Jones, C. P.; Nordenskjold, M.: Detailed physical map of human chromosomal region 11q12-13 shows high meiotic recombination rate around the MEN1 locus. Proc. Nat. Acad. Sci. 88: 10609-10613, 1991.
Johnston, C. I.; Martin, T. J.; Riddell, J.: Medullary thyroid carcinoma: a functional peptide secreting tumor. Aust. Ann. Med. 19: 50-53, 1970.
Kaplan, E. L.; Arnaud, C. D.; Hill, B. J.; Peskin, G. W.: Adrenal medullary calcitonin-like factor: a key to multiple endocrine neoplasia, type 2?. Surgery 68: 146-149, 1970.
Keiser, H. R.; Beaven, M. A.; Doppman, J.; Wells, S. A., Jr.; Buja, L. M.: Sipple's syndrome: medullary thyroid carcinoma, pheochromocytoma, and parathyroid disease. Ann. Intern. Med. 78: 561-579, 1973.
Kousseff, B. G.: Sipple syndrome with lichen amyloidosis as a paracrinopathy. (Letter) Am. J. Med. Genet. 42: 751-753, 1992.
Kousseff, B. G.; Espinoza, C.; Zamore, G. A.: Sipple syndrome with lichen amyloidosis as a paracrinopathy: pleiotropy, heterogeneity, or a contiguous gene?. J. Am. Acad. Derm. 25: 651-657, 1991.
Kruger, S. D.; Gertner, J. M.; Sparkes, R. S.; Haedt, L. E.; Crist, M.; Sparkes, M. C.; Genel, M.; Kidd, K. K.: Linkage analyses of multiple endocrine neoplasia, type 2 (MEN-2) with 23 classical genetic polymorphisms. Hum. Hered. 36: 6-11, 1986.
Lairmore, T. C.; Howe, J. R.; Korte, J. A.; Dilley, W. G.; Aine, L.; Aine, E.; Wells, S. A., Jr.; Donis-Keller, H.: Familial medullary thyroid carcinoma and multiple endocrine neoplasia type 2B map to the same region of chromosome 10 as multiple endocrine neoplasia type 2A. Genomics 9: 181-192, 1991.
Landsvater, R. M.; Mathew, C. G. P.; Smith, B. A.; Marcus, E. M.; te Meerman, G. J.; Lips, C. J. M.; Geerdink, R. A.; Nakamura, Y.; Ponder, B. A. J.; Buys, C. H. C. M.: Development of multiple endocrine neoplasia type 2A does not involve substantial deletions of chromosome 10. Genomics 4: 246-250, 1989.
Landsvater, R. M.; Rombouts, A. G. M.; te Meerman, G. J.; Schillhorn-van Veen, J. M. J.; Berends, M. J. H.; Geerdink, R. A.; Struyvenberg, A.; Buys, C. H. C. M.; Lips, C. J. M.: The clinical implications of a positive calcitonin test for C-cell hyperplasia in genetically unaffected members of an MEN2A kindred. Am. J. Hum. Genet. 52: 335-342, 1993.
Le Marec, B.; Roussey, M.; Cornec, A.; Calmettes, C.; Kerisit, J.; Allanic, H.: Cancer de la thyroide a stroma amyloide, syndrome de Sipple, megacolon congenital avec hyperplasie des plexus: une seule et meme affection autosomique dominante a penetrance complete. J. Genet. Hum. 28: 169-174, 1980.
Li, F. P.; Melvin, K. E. W.; Tashjian, A. H., Jr.; Levine, P. H.; Fraumeni, J. F., Jr.: Familial medullary thyroid carcinoma and pheochromocytoma: epidemiologic investigations. J. Nat. Cancer Inst. 52: 285-287, 1974.
Lima, J. B.; Smith, P. D.: Sipple's syndrome (pheochromocytoma and thyroid carcinoma) with bilateral breast carcinoma. Am. J. Surg. 121: 732-735, 1971.
Lips, C. J. M.; Minder, W. H.; Leo, J. R.; Alleman, A.; Hackeng, W. H. L.: Evidence of multicentric origin of the multiple endocrine neoplasia syndrome type 2A (Sipple's syndrome) in a large family in the Netherlands: diagnostic and therapeutic implications. Am. J. Med. 64: 569-578, 1978.
Lips, C. J. M.; Van der Sluys Veer, J.; Struyvenberg, A.; Alleman, A.; Leo, J. R.; Wittebol, P.; Minder, W. H.; Kooiker, C. J.; Geerdink, R. A.; Van Waes, P. F. G. M.; Hackeng, W. H. L.: Bilateral occurrence of pheochromocytoma in patients with the multiple endocrine neoplasia syndrome type 2A (Sipple's syndrome). Am. J. Med. 70: 1051-1060, 1981.
Mathew, C. G. P.; Chin, K. S.; Easton, D. F.; Thorpe, K.; Carter, C.; Liou, G. I.; Fong, S.-L.; Bridges, C. D. B.; Haak, H.; Nieuwenhuijzen Kruseman, A. C.; Schifter, S.; Hansen, H. H.; Telenius, H.; Telenius-Berg, M.; Ponder, B. A. J.: A linked genetic marker for multiple endocrine neoplasia type 2A on chromosome 10. Nature 328: 527-528, 1987.
Mathew, C. G. P.; Easton, D. F.; Nakamura, Y.; Ponder, B. A. J.; the MEN 2A International Collaborative Group: Presymptomatic screening for multiple endocrine neoplasia type 2A with linked DNA markers. Lancet 337: 7-11, 1991.
Mathew, C. G. P.; Smith, B. A.; Thorpe, K.; Wong, Z.; Royle, N. J.; Jeffreys, A. J.; Ponder, B. A. J.: Deletion of genes on chromosome 1 in endocrine neoplasia. Nature 328: 524-526, 1987.
McDonald, H.; Smailus, D.; Jenkins, H.; Adams, K.; Simpson, N. E.; Goodfellow, P. J.: Identification and characterization of a gene at D10S94 in the MEN2A region. Genomics 13: 344-348, 1992.

Meyer, J. S.; Abdel-Bari, W.: Granules and thyrocalcitonin-like activity in medullary carcinoma of the thyroid gland. New Eng. J. Med. 278: 523-529, 1968.
Mole, S. E.; Mulligan, L. M.; Healey, C. S.; Ponder, B. A. J.; Tunnacliffe, A.: Localisation of the gene for multiple endocrine neoplasia type 2A to a 480 kb region in chromosome band 10q11.2. Hum. Molec. Genet. 2: 247-252, 1993.
Moley, J. F.; Brother, M. B.; Fong, C. T.; White, P. S.; Baylin, S. B.; Nelkin, B.; Wells, S. A.; Brodeur, G. M.: Consistent association of 1p loss of heterozygosity with pheochromocytomas from patients with multiple endocrine neoplasia type 2 syndromes. Cancer Res. 52: 770-774, 1992.
Mulligan, L. M.; Kwok, J. B. J.; Healey, C. S.; Elsdon, M. J.; Eng, C.; Gardner, E.; Love, D. R.; Mole, S. E.; Moore, J. K.; Papi, L.; Ponder, M. A.; Telenius, H.; Tunnacliffe, A.; Ponder, B. A. J. : Germ-line mutations of the RET protooncogene in multiple endocrine neoplasia type 2A. Nature 363: 458-460, 1993.
Narod, S. A.; Lavoue, M.-F.; Morgan, K.; Calmettes, C.; Sobol, H.; Goodfellow, P. J.; Lenoir, G. M.: Genetic analysis of 24 French families with multiple endocrine neoplasia type 2A. Am. J. Hum. Genet. 51: 469-477, 1992.
Narod, S. A.; Sobol, H.; Nakamura, Y.; Calmettes, C.; Baulieu, J. L.; Bigorgne, J. C.; Chabrier, G.; Couette, J.; de Gennes, J. L.; Duprey, J.; et al.: Linkage analysis of hereditary thyroid carcinoma with and without pheochromocytoma. Hum. Genet. 83: 353-358, 1989.
Narod, S. A.; Sobol, H.; Schuffenecker, I.; Lavoue, M.-F.; Lenoir, G. M.: The gene for MEN 2A is tightly linked to the centromere of chromosome 10. Hum. Genet. 86: 529-530, 1991.
Nunziata, V.; di Giovanni, G.; Lettera, A. M.; D'Armiento, M.; Mancini, M.: Cutaneous lichen amyloidosis associated with multiple endocrine neoplasia type 2A. Henry Ford Hosp. Med. J. 37: 144-146, 1989.
O'Dorisio, T. M.; Falko, J. M.; Cataland, S.; Almoney, R. W.; George, J. M.; Mazzaferri, E. L.; Reynolds, J. C.: Gastrin responses to a test meal in patients with familial medullary thyroid carcinoma. J. Clin. Endocr. Metab. 54: 798-802, 1982.
Pearson, K. D.; Wells, S. A., Jr.; Keiser, H. R.: Familial medullary carcinoma of the thyroid, adrenal pheochromocytoma and parathyroid hyperplasia. A syndrome of multiple endocrine neoplasia. Radiology 107: 249-256, 1973.
Pearson, P. L.: Personal Communication. Leiden, The Netherlands 1985.
Perrier, P.; Thomas, J. L.; Bonaiti-Pellie, C.; Gay, G.; Leclere, J.; Streiff, F.: Lack of linkage between HLA and multiple endocrine neoplasia type 2 in a French family. Tissue Antigens 29: 13-17, 1987.
Poloyan, E.; Scanu, A.; Straus, F. H.; Pickleman, J. R.; Poloyan, D.: Familial pheochromocytoma, medullary thyroid carcinoma, and parathyroid adenomas. J.A.M.A. 214: 1443-1447, 1970.
Ponder, B. A. J.: Personal Communication. Cambridge, England 5/30/1993.
Saccone, S.; De Sario, A.; Della Valle, G.; Bernardi, G.: The highest gene concentrations in the human genome are in telomeric bands of metaphase chromosomes. Proc. Nat. Acad. Sci. 89: 4913-4917, 1992.
Sarosi, G.; Doe, R. P.: Familial occurrence of parathyroid adenomas, pheochromocytoma, and medullary carcinoma of the thyroid with amyloid stroma (Sipple's syndrome). Ann. Intern. Med. 68: 1305-1309, 1968.
Schimke, R. N.; Hartmann, W. H.: Familial amyloid-producing medullary thyroid carcinoma and pheochromocytoma, a distinct genetic entity. Ann. Intern. Med. 63: 1027-1039, 1965.
Schuffenecker, I.; Ginet, N.; Goldgar, D.; Eng, C.; Chambe, B.; Boneu, A.; Houdent, C.; Pallo, D.; Schlumberger, M.; Thivolet, C.; Lenoir, G. M.; Le Groupe d'Etude des Tumeurs a Calcitonine: Prevalence and parental origin of de novo RET mutations in multiple endocrine neoplasia type 2A and familial medullary thyroid carcinoma. (Letter) Am. J. Hum. Genet. 60: 233-237, 1997.
Simpson, N. E.: Personal Communication. Kingston, Ontario, Canada 1/14/1992.
Simpson, N. E.; Falk, J.: Exclusion of linkage between the loci for multiple endocrine neoplasia type-2 (MEN-2) and HLA. Hum. Genet. 60: 157 only, 1982.
Simpson, N. E.; Falk, J.; Forster-Gibson, C. J.; Goodall, J.; Sears, E.; Partington, M. W.; Ghent, W.: Possible linkage between the loci for multiple endocrine neoplasias type II (MEN2) and HLA. (Abstract) Cytogenet. Cell Genet. 25: 204 only, 1979.
Simpson, N. E.; Goodfellow, P. J.; Riddell, D. C.; Hamerton, J. L.; Holden, J. J. A.; White, B. N.: Assignment of the calcitonin gene to chromosome 11 and probable exclusion of linkage between the gene and the locus for multiple endocrine neoplasia type 2. (Abstract) Am. J. Hum. Genet. 36: 153S only, 1984.
Simpson, N. E.; Kidd, K. K.; Goodfellow, P. J.; McDermid, H.; Myers, S.; Kidd, J. R.; Jackson, C. E.; Duncan, A. M. V.; Farrer, L. A.; Brasch, K.; Castiglione, C.; Genel, M.; Gertner, J.; Greenberg, C. R.; Gusella, J. F.; Holden, J. J. A.; White, B. N.: Assignment of multiple endocrine neoplasia type 2A to chromosome 10 by linkage. Nature 328: 528-530, 1987.
Simpson, N. E.; Kidd, K. K.; Goodfellow, P. J.; McDermid, H.; Myers, S.; Kidd, J. R.; Jackson, C. E.; Duncan, A. M. V.; Farrer, L. A.; Brasch, K.; Castiglione, C.; Genel, M.; Gertner, J.; Greenberg, C. R.; Gusella, J. F.; Holden, J. J. A.; White, B. N.: Linkage of multiple endocrine neoplasia type 2A (MEN2A) and a new DNA marker (D10S5) at 10q21.1. (Abstract) Cytogenet. Cell Genet. 46: 693 only, 1987.
Sipple, J. H.: The association of pheochromocytoma with carcinoma of the thyroid gland. Am. J. Med. 31: 163-166, 1961.
Sobol, H.; Narod, S. A.; Lenoir, G. M.: Linkage of medullary thyroid carcinoma with and without pheochromocytoma. (Abstract) Am. J. Hum. Genet. 45 (suppl.): A34 only, 1989.
Sobol, H.; Narod, S. A.; Nakamura, Y.; Boneu, A.; Calmettes, C.; Chadenas, D.; Charpentier, G.; Chatal, J. F.; Delepine, N.; Delisle, M. J.; et al.: Screening for multiple endocrine neoplasia type 2a with DNA-polymorphism analysis. New Eng. J. Med. 321: 996-1001, 1989.
Steiner, A. L.; Goodman, A. D.; Powers, S. R., Jr.: Study of a kindred with pheochromocytoma, medullary thyroid carcinoma, hyperparathyroidism and Cushing's disease: multiple endocrine neoplasia, type II. Medicine 47: 371-409, 1968.
Tashjian, A. H., Jr.; Melvin, K. E. W.: Medullary carcinoma of the thyroid: thyrocalcitonin in plasma and tumor. New Eng. J. Med. 279: 279-283, 1968.
Tischler, A. S.; Dichter, M. A.; Biales, B.; DeLellis, R. A.; Wolfe, H.: Neural properties of cultured human endocrine tumor cells of proposed neural crest origin. Science 192: 902-904, 1976.
Urbanski, F. X.: Medullary thyroid carcinoma, parathyroid adenoma, and bilateral pheochromocytoma. An unusual triad of endocrine tumors. J. Chronic Dis. 20: 627-636, 1967.
Valk, T. W.; Frager, M. S.; Gross, M. D.; Sisson, J. C.; Wieland, D. M.; Swanson, D. P.; Mangner, T. J.; Beierwaltes, W. H.: Spectrum of pheochromocytoma in multiple endocrine neoplasia: a scintigraphic portrayal using (131)I-metaiodobenzylguanidine. Ann. Intern. Med. 94: 762-767, 1981.
Van Dyke, D. L.; Jackson, C. E.; Babu, V. R.: Multiple endocrine neoplasia type 2 (MEN-2): an autosomal dominant syndrome with a possible chromosome 20 deletion. (Abstract) Am. J. Hum. Genet. 33: 69A only, 1981.
Van Dyke, D. L.; Jackson, C. E.; Babu, V. R.: Localization of autosomal dominant multiple endocrine neoplasia 2 syndrome (MEN2) to 20p12.2. (Abstract) Cytogenet. Cell Genet. 32: 324 only, 1982.
Vasen, H. F. A.; Nieuwenhuijzen Kruseman, A. C.; Berkel, H.; Beukers, E. K. M.; Delprat, C. C.; Van Doorn, R. G.; Geerdink, R. A.; Haak, H. R.; Hackeng, W. H. L.; Koppeschaar, H. P. F.; Krenning, E. P.; Lamberts, S. W. J.; Lekkerkerker, F. J. F.; Michels, R. P. J.; Moers, A. M. J.; Pieters, G. F. F. M.; Wiersinga, W. M.; Lips, C. J. M.: Multiple endocrine neoplasia syndrome type 2: the value of screening and central registration: a study of 15 kindreds in The Netherlands. Am. J. Med. 83: 847-852, 1987.
Wong, C.-K.; Lin, C.-S.: Friction amyloidosis. Int. J. Dermat. 27: 302-307, 1988.
Wood, D.: Multiple endocrine neoplasia II at Hartford Hospital. Hartford Hosp. Bull. 32: 121-134, 1979.
Yamamoto, M.; Takai, S.; Miki, T.; Motomura, K.; Okazaki, M.; Nishisho, I.; Tateishi, H.; Miyauchi, A.; Honjo, T.; Pakstis, A. J.; Mori, T.: Close linkage of MEN2A with RBP3 locus in Japanese kindreds. Hum. Genet. 82: 287-288, 1989.
Zatterale, A.; Stabile, M.; Nunziata, V.; Di Giovanni, G.; Vecchione, R.; Ventruto, V.: Multiple endocrine neoplasia type 2 (Sipple's syndrome): clinical and cytogenetic analysis of a kindred. J. Med. Genet. 21: 108-111, 1984.

MULTIPLE MUCOSAL NEUROMA SYNDROME (MEN2B) (#162300)

JJM NOTE

Genetic linkage has clarified that the multiple mucosal neuroma syndrome (*MEN2B*) is a different mutation at the same gene locus as the Sipple syndrome (*MEN2A*) rather than being a distinct genetic disease. Therefore, the designation of MEN3 is defunct. The clinical pattern evokes the image of neurofibromatosis 1 and Marfan syndrome.

A number sign (#) is used with this entry because of evidence that the MEN2B phenotype is determined by mutation at the same locus, i.e., the RET gene (164761), as that involved in MEN2A (171400). However, Mulligan et al. (1994) found no mutations in exons 10 and 11 of the RET gene in 22 families with MEN2B. The causative mutations might, of course, be located elsewhere in the gene. Indeed, in all 9 unrelated MEN2B patients studied by Hofstra et al. (1994), a mutation in codon 918 was demonstrated, causing the substitution of a threonine for a methionine in the tyrosine kinase domain of the RET protein (164761.0013). (In the previous numbering convention, this mutation was designated met664-to-thr.) Since MEN2A and MEN2B are allelic disorders, this nomenclature is justified; use of the designation MEN III can be discontinued. The met918-to-thr mutation in the tyrosine kinase domain is the only RET mutation identified in cases of MEN2B (Bongarzone et al., 1998).

Williams and Pollock (1966) described 2 unrelated patients with multiple true neuromas, pheochromocytoma and thyroid carcinoma. The thyroid cancer was of the medullary type as in the PTC syndrome (171400). Although the association of pheochromocytoma with neurofibromatosis is well known, the nervous tumor is a true neuroma, i.e., consists mainly of nerve cells, in this condition. The patients sometimes have cafe-au-lait spots. The neuromas occur as pedunculated nodules on the eyelid margins, lips and tongue. The lips are diffusely hypertrophied with a 'Negroid' appearance. The father of one of Williams and Pollock's cases had very thick lips and eyelid and tongue lesions as did his daughters. He had a medullary thyroid cancer and died at age 38 after an abdominal operation, having had symptoms suggestive of pheochromocytoma. Schimke et al. (1968) also reported cases. Cunliffe et al. (1968) demonstrated calcitonin-secretion in a medullary carcinoma of the thyroid. The patient was a 19-year-old girl with acne, features of Marfan syndrome, neuromas of tongue and eyelid, prominent lips, nodular goiter, pigmentation of hands, feet and circumoral area, proximal myopathy, loose motions, and flushing attacks. (The first patient I saw with this syndrome was referred to me as possible Marfan syndrome.) The features suggesting Marfan syndrome were high arched palate, pectus excavatum, bilateral pes cavus, high patella and scoliosis. Marfanoid habitus and pes cavus are striking features in most. Megacolon with plexus hyperplasia is a feature (Carney and Hayles, 1977). Colonic diverticula also occur. Mucosal neuromas involve the lips, anterior tongue, conjunctiva and nasal and laryngeal mucosa. Medullated nerve fibers traverse the cornea. Bartlett et al. (1968) described affected persons in 6 generations. Prophylactic thyroidectomy should be performed when the phenotype is recognized (Wolfe et al., 1973). In addition to histaminase, DOPA decarboxylase is high in medullary carcinoma of the thyroid (Atkins et al., 1973). The latter enzyme is found in pheochromocytomas also.

Carney et al. (1976) and Dyck et al. (1979) described a kindred with 7 affected persons in 3 generations. In contrast to Sipple syndrome (171400), no parathyroid disease was found in any of them. The authors proposed the designation 'multiple endocrine neoplasia, type 2b.' Carney et al. (1980) confirmed the lack of clinically important involvement of the parathyroids in this disorder. Kullberg and Nieuwenhuijzen Kruseman (1987) described a patient who was known to have had liver metastases of medullary carcinoma of the thyroid for at least 22 years. Other features illustrating an indolent course of the disorder were displayed by affected members of this family. Derangement of esophageal motility with an achalasia-type clinical picture may be a leading problem (Baylin, 1989). Jackson et al. (1988) found evidence suggesting that MEN3 (MEN2B) maps to chromosome 10, probably in the same region as MEN2A (171400). It is possible that MEN2A and MEN2B are allelic disorders; if indeed such is the case, there would be every justification for calling them MEN2A and MEN2B rather than MEN2 and MEN3. Norum (1989) presented compelling evidence that MEN3 is allelic to MEN2: both mapped to chromosome 10 near the centromere and are on the same side of the RBP3 gene (180290). Lairmore et al. (1991) presented evidence that the MEN2B locus maps in the pericentric region of chromosome 10. They found no evidence for linkage genetic heterogeneity among families with medullary thyroid carcinoma, MEN2B, or MEN2A. By genetic linkage analysis, Gardner et al. (1993) demonstrated that the MEN2B locus is flanked by D10S141 and RBP3, proximally and distally, respectively; both markers are located in 10q11.2.

Fryns and Chrzanowska (1988) gave a review. Mutation in a gene controlling a paracrine or an autocrine growth factor seemed a possibility in this disorder.

Carlson et al. (1994) estimated that 50% of cases of MEN2B arise de novo. A single identical point mutation in the catalytic core of the tyrosine kinase domain of RET has been found to be associated with both inherited and de novo MEN2B. Carlson et al. (1994) analyzed 25 cases of de novo MEN2B in order to determine the parental origin of the mutated RET allele. In all cases, the new mutation was of paternal origin. They observed a distortion of sex ratio in both de novo MEN2B patients and the affected offspring of MEN2B-transmitting males. Specifically, as pointed out by Sapienza (1994), the report by Carlson et al. (1994) and that of another group includes 45 females and 24 males with de novo MEN2B. These results suggested a differential susceptibility of RET mutation in paternally and maternally derived DNA and a possible role for imprinting of RET during development.

Enlarged nerves of the gastrointestinal tract (ganglioneuromatosis), often with megacolon, is a feature of MEN2B; corneal nerve thickening (medullated corneal nerve fibers) is a comparable feature, as are mucosal neuromas. Kane et al. (1995) described an unusual kindred in which medullary thyroid carcinoma was associated with corneal nerve thickening without other aspects of the multiple endocrine neoplasia syndromes. The grandfather and 1 daughter, both of whom had corneal nerve thickening, had medullary thyroid carcinoma, and a daughter of that daughter had both corneal nerve thickening and C-cell hyperplasia. Three adults in the middle generation had corneal nerve thickening only. DNA sequence analysis revealed no mutations in exon 10 or 11, or in exon 16 of the RET gene, regions where mutations had been described for MEN2A and MEN2B, respectively.

Morrison and Nevin (1996) presented a review of this syndrome. They used the term Wagenmann-Froboese syndrome, because Wagenmann (1922) and Froboese (1923) initially described this disorder. Thyroid carcinoma and multiple mucosal neuromata occur in virtually all affected persons. Neoplasias of enteric ganglia were found in more than 40% of patients; pheochromocytoma in 50%; and marfanoid habitus in 75%. The absence of mucosal neuromata in persons with MEN2A is the main criterion for differential diagnosis.

Prophylactic total thyroidectomy before the age of 4 years was proposed for children with this disorder (Morrison et al., 1991; Utiger, 1994).

See also Baum and Adler (1972); Baylin et al. (1970); Braley (1954); Carlson et al. (1994); Carney et al. (1981); Carney et al. (1976); DeSchryver-Kecskemeti et al. (1983); Gorlin and Mirkin (1972); Gorlin et al. (1968); Moyes and Alexander (1977)

Atkins, F. L.; Beaven, M. A.; Keiser, H. R.: DOPA decarboxylase in medullary carcinoma of the thyroid. New Eng. J. Med. 289: 545-548, 1973.

Bartlett, R. C.; Bean, L. R.; Mandelstam, P.: Hereditary study of neuroendocrine dysplasia in six generations. (Abstract) Int. Assoc. Dent. Res., San Francisco 36, 3/1968.

Baum, J. L.; Adler, M. E.: Pheochromocytoma, medullary thyroid carcinoma, multiple mucosal neuroma: a variant of the syndrome. Arch. Ophthal. 87: 574-584, 1972.

Baylin, S. B.: Personal Communication. Baltimore, Md. 4/22/1989.

Baylin, S. B.; Beaven, M. A.; Engelman, K.; Sjoerdsma, A.: Elevated histaminase activity in medullary carcinoma of the thyroid gland. New Eng. J. Med. 283: 1239-1244, 1970.

Bongarzone, I.; Vigano, E.; Alberti, L.; Borrello, M. G.; Pasini, B.; Greco, A.; Mondellini, P.; Smith, D. P.; Ponder, B. A. J.; Romeo, G.; Pierotti, M. A.: Full activation of MEN2B mutant RET by an additional MEN2A mutation or by ligand GDNF stimulation. Oncogene 16: 2295-2301, 1998.

Braley, A. E.: Medullated corneal nerves and plexiform neuroma associated with pheochromocytoma. Trans. Am. Ophthal. Soc. 52: 189-197, 1954.

Carlson, K. M.; Bracamontes, J.; Jackson, C. E.; Clark, R.; Lacroix, A.; Wells, S. A., Jr.; Goodfellow, P. J.: Parent-of-origin effects in multiple endocrine neoplasia type 2B. Am. J. Hum. Genet. 55: 1076-1082, 1994.

Carlson, K. M.; Dou, S.; Chi, D.; Scavarda, N.; Toshima, K.; Jackson, C. E.; Wells, S. A., Jr.; Goodfellow, P. J.; Donis-Keller, H.: Single missense mutation in the tyrosine kinase catalytic domain of the RET protooncogene is associated with multiple endocrine neoplasia type 2B. Proc. Nat. Acad. Sci. 91: 1579-1583, 1994.

Carney, J. A.; Go, V. L. W.; Sizemore, G. W.; Hayles, A. B.: Alimentary-tract ganglioneuromatosis: a major component of the syndrome of multiple endocrine neoplasia, type 2b. New Eng. J. Med. 295: 1287-1291, 1976.

Carney, J. A.; Hayles, A. B.: Alimentary tract manifestations of multiple endocrine neoplasia, type 2b. Mayo Clin. Proc. 52: 543-548, 1977.

Carney, J. A.; Hayles, A. B.; Pearse, A. G. E.; Perry, H. O.; Sizemore, G. W.: Abnormal cutaneous innervation in multiple endocrine neoplasia, type 2b. Ann. Intern. Med. 94: 362-363, 1981.

Carney, J. A.; Roth, S. I.; Heath, H., III; Sizemore, G. W.; Hayles, A. B.: The parathyroid glands in multiple endocrine neoplasia type 2b. Am. J. Path. 99: 387-398, 1980.

Carney, J. A.; Sizemore, G. W.; Lovestedt, S. A.: Mucosal ganglioneuromatosis, medullary thyroid carcinoma, and pheochromocytoma: multiple endocrine neoplasia, type 2b. Oral Surg. 41: 739-752, 1976.

Cunliffe, W. J.; Black, M. M.; Hall, R.; Johnston, I. D. A.; Hudgson, P.; Shuster, S.; Gudmundsson, T. V.; Joplin, G. F.; Williams, E. D.; Woodhouse, N. J. Y.; Galante, L.; MacIntyre, I.: A calcitonin-secreting thyroid carcinoma. Lancet II: 63-66, 1968.

DeSchryver-Kecskemeti, K.; Clouse, R. E.; Goldstein, M. N.; Gersell, D.; O'Neal, L.: Intestinal ganglioneuromatosis: a manifestation of overproduction of nerve growth factor?. New Eng. J. Med. 308: 635-639, 1983.

Dyck, P. J.; Carney, J. A.; Sizemore, G. W.; Okazaki, H.; Brimijoin, W. S.; Lambert, E. H.: Multiple endocrine neoplasia, type 2b: phenotype recognition; neurological features and their pathological basis. Ann. Neurol. 6: 302-314, 1979.

Froboese, C.: Das aus markhaltigen nervenfascern bestehende gangliezellenlose echte neurom in rankenformzugleich ein beitrag zu den nervosen Geschwulsten der zunge und des augenlides. Virchows Arch. Pathol. Anat. 240: 312-327, 1923.

Fryns, J. P.; Chrzanowska, K.: Mucosal neuromata syndrome (MEN type IIb (III)). J. Med. Genet. 25: 703-706, 1988.

Gardner, E.; Papi, L.; Easton, D. F.; Cummings, T.; Jackson, C. E.; Kaplan, M.; Love, D. R.; Mole, S. E.; Moore, J. K.; Mulligan, L. M.; Norum, R. A.; Ponder, M. A.; Reichlin, S.; Stall, G.; Telenius, H.; Telenius-Berg, M.; Tunnacliffe, A.; Ponder, B. A. J.: Genetic linkage studies map the multiple endocrine neoplasia type 2 loci to a small interval on chromosome 10q11.2. Hum. Molec. Genet. 2: 241-246, 1993.

Gorlin, R. J.; Mirkin, B. L.: Multiple mucosal neuromas, pheochromocytoma, medullary carcinoma of the thyroid and marfanoid body build with muscle wasting: syndrome of hyperplasia and neoplasia of neural crest derivatives; a unitarian concept. Z. Kinderheilk. 113: 313-321, 1972.

Gorlin, R. J.; Sedano, H. O.; Vickers, R. A.; Cervenka, J.: Multiple mucosal neuromas, pheochromocytoma and medullary carcinoma of the thyroid—a syndrome. Cancer 22: 293-299, 1968.

Hofstra, R. M. W.; Landsvater, R. M.; Ceccherini, I.; Stulp, R. P.; Stelwagen, T.; Luo, Y.; Pasini, B.; Hoppener, J. W. M.; Ploos van Amstel, H. K.; Romeo, G.; Lips, C. J. M.; Buys, C. H. C. M.: A mutation in the RET protooncogene associated with multiple endocrine neoplasia type 2B and sporadic medullary thyroid carcinoma. Nature 367: 375-376, 1994.

Jackson, C. E.; Norum, R. A.; O'Neal, L. W.; Nikolai, T. F.; Delaney, J. P.: Linkage between MEN2B and chromosome 10 markers linked to MEN2A. (Abstract) Am. J. Hum. Genet. 43: A147, 1988.

Kane, L. A.; Tsai, M. S.; Gharib, H.; Khosla, S.; Robertson, D. M.; Schaid, D. J.; Honchel, R.; Thibodeau, S. N.: Familial medullary thyroid cancer and prominent corneal nerves: clinical and genetic analysis. J. Clin. Endocr. Metab. 80: 289-293, 1995.

Kullberg, B. J.; Nieuwenhuijzen Kruseman, A. C.: Multiple endocrine neoplasia type 2b with a good prognosis. Arch. Intern. Med. 147: 1125-1127, 1987.

Lairmore, T. C.; Howe, J. R.; Korte, J. A.; Dilley, W. G.; Aine, L.; Aine, E.; Wells, S. A., Jr.; Donis-Keller, H.: Familial medullary thyroid carcinoma and multiple endocrine neoplasia type 2B map to the same region of chromosome 10 as multiple endocrine neoplasia type 2A. Genomics 9: 181-192, 1991.

Morrison, P. J.; Hadden, D. R.; Hughes, A. E.; Russell, C. F. J.; Nevin, N. C.: Gene probe analysis in an informative family with the MEN2A syndrome. Improvement in carrier risk estimations. Quart. J. Med. 78: 597-603, 1991.

Morrison, P. J.; Nevin, N. C.: Multiple endocrine neoplasia type 2B (mucosal neuroma syndrome, Wagenmann-Froboese syndrome). J. Med. Genet. 33: 779-782, 1996.

Moyes, C. D.; Alexander, F. W.: Mucosal neuroma syndrome presenting in a neonate. Dev. Med. Child Neurol. 19: 518-521, 1977.

Mulligan, L. M.; Eng, C.; Healey, C. S.; Clayton, D.; Kwok, J. B. J.; Gardner, E.; Ponder, M. A.; Frilling, A.; Jackson, C. E.; Lehnert, H.; Neumann, H. P. H.; Thibodeau, S. N.; Ponder, B. A. J.: Specific mutations of the RET proto-oncogene are related to disease phenotype in MEN 2A and FMTC. Nature Genet. 6: 70-74, 1994.
Norum, R. A.: Personal Communication. Detroit, Mich. 5/1989.
Sapienza, C.: Parental origin effects, genome imprinting, and sex-ratio distortion: double or nothing? (Editorial) Am. J. Hum. Genet. 55: 1073-1075, 1994.
Schimke, R. N.; Hartmann, W. H.; Prout, T. E.; Rimoin, D. L.: Pheochromocytoma, medullary thyroid carcinoma and multiple neuromas. New Eng. J. Med. 279: 1-7, 1968.
Utiger, R. D.: Medullary thyroid carcinoma, genes and the prevention of cancer. New Eng. J. Med. 331: 870-871, 1994.
Wagenmann, A.: Multiple neurome des auges und der Zunge. Ber. Dtsch. Ophthal. 43: 282-285, 1922.
Williams, E. D.; Pollock, D. J.: Multiple mucosal neuromata with endocrine tumours: a syndrome allied to von Recklinghausen's disease. J. Path. Bact. 91: 71-80, 1966.
Wolfe, H. J.; Melvin, K. E. W.; Cervi-Skinner, S. J.; Al Saadi, A. A.; Juliar, J. F.; Jackson, C. E.; Tashjian, A. H., Jr.: C-cell hyperplasia preceding medullary thyroid carcinoma. New Eng. J. Med. 289: 437-441, 1973.

MEDULLARY THYROID CARCINOMA, FAMILIAL (#155240)

JJM NOTE

This entity seems infrequent, and the onus for the clinician is to be sure that MEN2A or MEN2B is *not* present, before settling for pure familial medullary thyroid carcinoma in one family member. In the next family member, the concern about MEN2 has to be raised, until genotype and phenotype correlations become stronger than they are at present. A few sporadic patients with medullay thyroid carcinoma have germline (hereditary) mutations *RET*,

A number sign (#) is used with this entry because of evidence that this form of familial MTC occurs from mutation in the RET gene (164761) on chromosome 10.

About 75% of medullary thyroid carcinoma is sporadic; these cases are unilateral. Bilateral multifocal medullary carcinoma is a cardinal feature of autosomal dominant multiple endocrine neoplasia type II (MEN2A). In addition, there are cases of familial medullary thyroid carcinoma in which there are no extra-thyroid manifestations of multiple endocrine neoplasia; otherwise the behavior of the tumor is the same as that in the hereditary disease. Such families were observed by Farndon et al. (1986). The linkage studies of Narod et al. (1989) appear to indicate conclusively that familial medullary carcinoma of the thyroid (without pheochromocytoma) is caused by an allele in the same gene that is the site of the mutation in MEN2. They studied 18 families, 9 with MEN2 and 9 with medullary carcinoma of the thyroid without pheochromocytoma, with probes specific for the pericentromeric region of chromosome 10 and found close linkage in both cases. Genetic heterogeneity of the susceptibility locus was not observed. The genetic mutation for medullary carcinoma was in disequilibrium with alleles of 2 closely linked marker loci. In 2 large families with medullary thyroid carcinoma (MTC1), Lairmore et al. (1991) showed that the maximum lod score between the neoplasia and marker D10Z1 was 5.88 with 0% recombination. They found no evidence for genetic heterogeneity among families with medullary thyroid carcinoma, MEN2B (162300), or MEN2A. The mutations causing these disorders thus appear to be related to each other as alleles. On the other hand, Carson et al. (1991) reported 2 families in which the MTC mutation appeared not to be linked to the pericentromeric markers on chromosome 10.

Primary localized cutaneous amyloidosis (PLCA), which has been observed with MEN2A, was reported also in a family in which multiple affected members had an association only with medullary thyroid carcinoma (Ferrer et al., 1991).

Simpson (1991) proposed a branching classification of the forms of medullary thyroid carcinoma (MTC). MTC has 2 forms: MTC1 (with no other primary tumors) and MEN2. MEN2 has 2 forms: MEN2A and MEN2B. MEN2A has 3 forms: MEN2A-1 (MTC with pheochromocytoma and parathyroid tumor), MEN2A-2 (MTC with pheochromocytoma), and MEN2A-3 (MTC with parathyroid tumor).

Rakover et al. (1994) described 2 sibs in whom isolated familial medullary carcinoma of the thyroid was diagnosed at the ages of 16 and 19 years. Hirschsprung disease was identified at the age of 1 year in both of them. Twelve other members of the family had medullary carcinoma of the thyroid. The authors stated that, although this was the first report of an association between Hirschsprung disease and isolated familial medullary carcinoma of the thyroid, the association should not be surprising because of the known association of both with mutations in the RET gene.

Carson, N. L.; Simpson, N. E.; Myers, S.; Anderson, L.; Clarke, C.; Simpson, W. J.; Bridge, P. J.: Genetic heterogeneity in families with familial medullary thyroid carcinoma. (Abstract) Cytogenet. Cell Genet. 58: 1948 only, 1991.
Farndon, J. R.; Leight, G. S.; Dilley, W. G.; Baylin, S. B.; Smallridge, R. C.; Harrison, T. S.; Wells, S. A., Jr.: Familial medullary thyroid carcinoma without associated endocrinopathies: a distinct clinical entity. Brit. J. Surg. 73: 278-281, 1986.
Ferrer, J. P.; Halperin, I.; Conget, J. I.; Alsina, M.; Martinez-Osaba, M. J.; Palou, J.; Bombi, J. A.; Vilardell, E.: Primary localized cutaneous amyloidosis and familial medullary thyroid carcinoma. Clin. Endocr. 34: 435-439, 1991.
Lairmore, T. C.; Howe, J. R.; Korte, J. A.; Dilley, W. G.; Aine, L.; Aine, E.; Wells, S. A., Jr.; Donis-Keller, H.: Familial medullary thyroid carcinoma and multiple endocrine neoplasia type 2B map to the same region of chromosome 10 as multiple endocrine neoplasia type 2A. Genomics 9: 181-192, 1991.
Narod, S. A.; Sobol, H.; Nakamura, Y.; Calmettes, C.; Baulieu, J.-L.; Bigorgne, J.-C.; Chabrier, G.; Couette, J.; de Gennes, J. L.; Duprey, J.; Gardet, P.; Guillausseau, P.-J.; Guilloteau, D.; Houdent, C.; Lefebvre, J.; Modigliani, E.; Parmentier, C.; Pugeat, M.; Siame, C.; Tourniaire, J.; Vandroux, J.-C.; Vinot, J.-M.; Lenoir, G. M.: Linkage analysis of hereditary thyroid carcinoma with and without pheochromocytoma. Hum. Genet. 83: 353-358, 1989.
Rakover, Y.; Dharan, M.; Luboshitsky, R.: Hirschsprung's disease associated with isolated familial medullary carcinoma of the thyroid. J. Pediat. Endocr. 7: 373-377, 1994.
Simpson, N. E.: Personal Communication. Kingston, Ontario, Canada 9/3/1991.

RET PROTO-ONCOGENE; RET (*164761)

JJM NOTE

Mutations in differing parts of this gene give rise to vastly different disorders: multiple endocrine neoplasia 2A, multiple endocrine neoplasia 2B (otherwise known as the multiple mucosal neuroma syndrome), and Hirschsprung disease.

Description

Mutations in the RET gene are associated with the disorders multiple endocrine neoplasia, type IIA (MEN2A; 171400), multiple endocrine neoplasia, type IIB (MEN2B; 162300), Hirschsprung disease (HSCR; aganglionic megacolon; 142623), and medullary thyroid carcinoma (MTC; 155240).

RET (REarranged during Transfection), as its acronym suggests, was cloned as a chimeric oncogene during a classic NIH 3T3 transformation assay (Takahashi et al., 1985).

The RET protooncogene is one of the receptor tyrosine kinases, which are cell-surface molecules that transduce signals for cell growth and differentiation. The RET gene was defined as an oncogene by a classical transfection assay. Later it was shown (Grieco et al., 1990) that RET can undergo oncogenic activation in vivo and in vitro by cytogenetic rearrangement. Attie et al. (1995) studied the 20 exons of the RET gene by a combination of denaturing gradient gel electrophoresis and SSCP in 45 sporadic cases and 35 familial cases of Hirschsprung disease. They found mutations of the RET gene in 50% of familial HSCR, regardless of the length of the aganglionic segment. The mean penetrance of the mutant allele in familial HSCR was significantly higher in males (72%) than in females (51%). Mutations at the RET locus were scattered along the length of the gene and accounted for at least one-third of sporadic HSCR cases in this series. Among the mutations identified in sporadic cases (16/45), 7 proved to be de novo mutations. Taken together, the low penetrance of the mutant gene, the lack of genotype/phenotype correlation, the sex-dependent effect of RET mutations, and the variable clinical expression of the disease support the existence of one or more modifier genes in familial HSCR.

Bolk et al. (1996) stated that 16% of children with congenital central hypoventilation syndrome (CCHS; 209880) have Hirschsprung disease. Because RET mutations have been found in Hirschsprung disease, Bolk et al. (1996) used SSCP analysis to study mutations of the RET gene in 14 patients with CCHS. All detected nucleotide changes in the RET gene were classified as polymorphic variants. On the other hand, Amiel et al. (1998) found a mutation in the RET gene in 1 of 7 children with isolated CCHS.

Eng et al. (1995) screened 7 exons of the RET oncogene for mutations in 48 sporadic pheochromocytomas and found a RET mutation in exons 10, 11, and 16 in 5 of the 48. Of these, 1 was proven to be germline and 2 were proven to be somatic mutations. Mutations of the VHL gene were found in 4 of the 48 sporadic pheochromocytomas; these 4 included the 2 bilateral cases in the series, of which 1 was proven to be a germline mutation, and 2 others, of which 1 was proven to be somatic.

Pasini et al. (1995) cloned the entire RET genomic sequence in a contig of cosmids and established the position of the 20 exons of the RET gene with respect to a detailed restriction map based on 8 endonucleases. A new highly polymorphic CA repeat sequence was identified within intron 5. The estimated size of the gene is 55 kb. Intron 1 accounts for approximately 24 kb, while exons 2 to 20 are contained within a region of 31 kb. This overall gene structure of a large first intron with small exons interspersed at the 5-prime half and more clustered at the 3-prime half is reminiscent of that of PDGFRB (173410) and KIT (164920), genes that also encode tyrosine kinase receptors. No evidence of RET-related genes or RET pseudogenes in 10q11.2 or elsewhere in the genome was found. They could demonstrate that the orientation of the RET gene on 10q11.2 is 5-prime centromeric/3-prime telomeric.

Angrist et al. (1995) analyzed the RET gene in 80 HSCR probands by PCR and identified 8 putative mutations.

Iwashita et al. (1996) introduced 5 HSCR mutations into the extracellular domain of human RET cDNA. These mutations were introduced with or without a MEN2A mutation (cys634 to arg; 164761.0011). The 5 mutations in the RET extracellular domain inhibited the transport of the RET protein to the plasma membrane. Introduction of the extracellular domain RET mutation along with the MEN2A mutation led to significant reduction of the transforming activity of MEN2A-RET, for which cell surface expression is required. Iwashita et al. (1996) demonstrated that with the 5 HSCR extracellular domain RET mutations cell surface expression is low. The authors concluded that sufficient levels of RET expression on the cell surface are required for ganglia migration toward the distal portion of the colon or for full differentiation.

Using the approach of SSCP analysis established for all 20 exons of the RET gene, Seri et al. (1997) identified 7 additional mutations among 39 sporadic and familial cases of Hirschsprung disease (detection rate 18%). They considered that the relatively low efficiency of detecting mutations of RET in Hirschsprung patients cannot be accounted for by genetic heterogeneity, which is not supported by the results of linkage analysis in pedigrees analyzed to date. Almost 74% of the point mutations in their series, as well as in other patient series, were identified among long-segment patients, who represented only 25% of the patient population. Seri et al. (1997) found a C620R substitution in a patient affected with total colonic aganglionosis; the same mutation had been found in medullary thyroid carcinoma. An R313Q mutation (164761.0026) was identified in homozygous state in a child born of consanguineous parents and was associated with the most severe Hirschsprung phenotype, namely, a total colonic aganglionosis with small bowel involvement.

Familial adenomatous polyposis (FAP) is caused by germline mutations of the adenomatous polyposis coli (APC) gene, and it is associated with an increased risk of developing papillary thyroid carcinomas. A significant fraction of sporadic human papillary thyroid carcinomas have RET protooncogene rearrangements. These rearrangements generate chimeric transforming oncogenes designated RET/PTC. Cetta et al. (1998) used an immunohistochemical and RT-PCR approach to analyze for RET/PTC activation in papillary thyroid carcinomas in 2 FAP kindreds, both showing typical APC gene mutations. Kindred 1 had 7 members affected by FAP, and among these, 3 patients had papillary thyroid carcinomas. Kindred 2 had 2 patients, mother and daughter, who were affected by colonic polyposis; the daughter also had a papillary carcinoma. Cetta et al. (1998) found RET/PTC1 oncogene activation in 2 of 3 papillary carcinomas of FAP kindred 1 and in the papillary carcinoma of FAP kindred 2. These findings showed that loss of function of APC coexists with gain of function of RET in some papillary thyroid carcinomas, suggesting that RET/PTC1 oncogene activation could be a progression step in the development of FAP-associated thyroid tumors.

Shirahama et al. (1998) investigated the spectrum of RET mutations among Japanese patients by screening the RET gene in 71 patients with thyroid carcinoma. They found mutations in 33 of 34 MEN2A patients and in 5 of 6 FMTC families studied. The met918-to-thr mutation (164761.0013) was found in 4 patients with MEN2B and in 2 of the 22 patients with sporadic medullary thyroid carcinoma. A total of 5 germline mutations were found among the 22 sporadic cases studied, 4 of which were found to be de novo mutations. The authors commented that the high frequency of germline mutations among patients with sporadic medullary thyroid carcinoma has important implications for the clinical management of family members of any patient with this malignancy.

Mapping

By fluorescence in situ hybridization, Ishizaka et al. (1989) assigned the RET oncogene to 10q11.2. Because of the location, they suggested that this might be a candidate gene for multiple endocrine neoplasia type IIA. Lairmore et al. (1993) developed a 1.5-Mb YAC contig containing 3 loci closely linked to the MEN2A locus. The orientation of the contig and order of the 3 markers were cen—RET—D10S94—D10S102—tel. A critical crossover event placed the MEN2A locus centromeric to D10S102. Lairmore et al. (1993) pointed out that no recombination events had been reported between MEN2A and either D10S94 or RET. Mulligan et al. (1993) and Donis-Keller et al. (1993) demonstrated mutations in the RET oncogene that are associated with MEN2A and medullary thyroid carcinoma.

Reviews

Eng (1996) reviewed the role of the RET protooncogene in multiple endocrine neoplasia type II and in Hirschsprung disease. Hoppener and Lips (1996) also reviewed RET gene mutations from the point of view of the molecular biology and the clinical aspects. Eng and Mulligan (1997) tabulated mutations of the RET gene in MEN2, the related sporadic tumors medullary thyroid carcinoma and pheochromocytoma, and familial and sporadic Hirschsprung disease. Germline mutations in 1 of 8 codons within RET cause the 3 subtypes of MEN2, namely, MEN2A, MEN2B, and familial medullary thyroid carcinoma. They stated that a somatic M918T mutation (164761.0013) accounts for the largest proportion of RET mutations detected in medullary thyroid carcinomas, most series showing a 30% to 50% range. It appeared that pheochromocytomas have a wider range of RET mutations. In contrast to MEN2, approximately 25% of patients with Hirschsprung disease have germline mutations scattered throughout the length of RET.

Fearon (1997) reviewed more than 20 different hereditary cancer syndromes that had been defined and attributed to specific germline mutations in various inherited cancer genes. In a useful diagram, he illustrated the roles of allelic variation ('1 gene - different syndromes') and genetic heterogeneity ('different genes - 1 syndrome') in inherited cancer syndromes. For example, some missense mutations, e.g., in codon 609 cause MEN2A and a familial medullary thyroid carcinoma; others, e.g., missense mutations in codon 918, cause MEN2B; yet other mutations cause Hirschsprung disease.

Decker et al. (1998) found that Hirschsprung disease (HSCR1; 142623) cosegregated with MEN2A in 7 (16%) of 44 families ascertained through MEN2A. The predisposing RET mutations in all 7 families had previously been reported in MEN2A or FMTC and occurred in exon 10 at codons 609, 618, or 620: cys609-to-tyr (164761.0029), cys618-to-ser (164761.0008), cys620-to-arg (164761.0009), and cys620-to-trp (164761.0032).

Pelet et al. (1998) investigated the effect on RET function of 7 HSCR-related missense mutations by introducing them into either a 114-amino acid wildtype RET isoform (RET51) or a constitutively activated form of RET51 (RET-MEN2A). Pelet et al. (1998) reported that 1 mutation affecting the extracytoplasmic cadherin domain (arg231 to his; R231H) and 2 mutations located in the tyrosine kinase domain (lys907 to glu, K907E; or glu921 to lys, E921K) impaired the biologic activity of RET-MEN2A when tested in cultured fibroblast and pheochromocytoma cells. However, the mechanisms resulting in RET inactivation differed since the receptor bearing the R231H extracellular mutation results in an absent RET protein at the cell surface, while the E921K mutation located within the catalytic domain abolished its enzymatic activity. In contrast, 3 mutations mapping to the intracytoplasmic domain neither modified the transforming capacity of RET-MEN2A nor stimulated the catalytic activity of RET in a ligand-independent system (ser767 to arg, pro1039 to leu, met1064 to thr). Finally, the cys609-to-trp HSCR mutation exerted a dual effect on RET since it led to a decrease of the receptor at the cell surface and converted RET51 into a constitutively activated kinase due to the formation of disulfide-linked homodimers. The data demonstrated that allelic heterogeneity at the RET locus in HSCR is associated with various molecular mechanisms responsible for RET dysfunction.

See also Eng et al. (1995); Eng et al. (1995); Ikeda et al. (1990); Pachnis et al. (1993); Pierotti et al. (1992); Rodrigues and Park (1993); Santoro et al. (1992); van Heyningen (1994)

Amiel, J.; Salomon, R.; Attie, T.; Pelet, A.; Trang, H.; Mokhtari, M.; Gaultier, C.; Munnich, A.; Lyonnet, S.: Mutations of the RET-GDNF signaling pathway in Ondine's curse. (Letter) Am. J. Hum. Genet. 62: 715-717, 1998.

Angrist, M.; Bolk, S.; Thiel, B.; Puffenberger, E. G.; Hofstra, R. M.; Buys, C. H. C. M.; Cass, D. T.; Chakravarti, A.: Mutation analysis of the RET receptor tyrosine kinase in Hirschsprung disease. Hum. Molec. Genet. 4: 821-830, 1995.

Attie, T.; Pelet, A.; Edery, P.; Eng, C.; Mulligan, L. M.; Amiel, J.; Boutrand, L.; Beldjord, C.; Nihoul-Fekete, C.; Munnich, A.; Ponder, B. A. J.; Lyonnet, S.: Diversity of RET proto-oncogene mutations in familial and sporadic Hirschsprung disease. Hum. Molec. Genet. 4: 1381-1386, 1995.

Berndt, I.; Reuter, M.; Saller, B.; Frank-Raue, K.; Groth, P.: Grubendorf, M.; Raue, F.; Ritter, M. M.; Hoppner, W.: A new hot spot for mutations in the ret protooncogene causing familial medullary thyroid carcinoma and multiple endocrine neoplasia type 2A. J. Clin. Endocr. Metab. 83: 770-774, 1998.

Boccia, L. M.; Green, J. S.; Joyce, C.; Eng, C.; Taylor, S. A. M.; Mulligan, L. M.: Mutation of RET codon 768 is associated with the FMTC phenotype. Clin. Genet. 51: 81-85, 1997.

Bolino, A.; Schuffenecker, I.; Luo, Y.; Seri, M.; Silengo, M.; Tocco, T.; Chabrier, G.; Houdent, C.; Murat, A.; Schlumberger, M.; Tourniaire, J.; Lenoir, G. M.; Romeo, G.: RET mutations in exons 13 and 14 of FMTC patients. Oncogene 10: 2415-2419, 1995.
Bolk, S.; Angrist, M.; Schwartz, S.; Silvestri, J. M.; Weese-Mayer, D. E.; Chakravarti, A.: Congenital central hypoventilation syndrome: mutation analysis of the receptor tyrosine kinase RET. Am. J. Med. Genet. 63: 603-609, 1996.
Bongarzone, I.; Vigano, E.; Alberti, L.; Borrello, M. G.; Pasini, B.; Greco, A.; Mondellini, P.; Smith, D. P.; Ponder, B. A. J.; Romeo, G.; Pierotti, M. A.: Full activation of MEN2B mutant RET by an additional MEN2A mutation or by ligand GDNF stimulation. Oncogene 16: 2295-2301, 1998.
Carlson, K. M.; Bracamontes, J.; Jackson, C. E.; Clark, R.; Lacroix, A.; Wells, S. A., Jr.; Goodfellow, P. J.: Parent-of-origin effects in multiple endocrine neoplasia type 2B. Am. J. Hum. Genet. 55: 1076-1082, 1994.
Carlson, K. M.; Dou, S.; Chi, D.; Scavarda, N.; Toshima, K.; Jackson, C. E.; Wells, S. A., Jr.; Goodfellow, P. J.; Donis-Keller, H.: Single missense mutation in the tyrosine kinase catalytic domain of the RET protooncogene is associated with multiple endocrine neoplasia type 2B. Proc. Nat. Acad. Sci. 91: 1579-1583, 1994.
Ceccherini, I.; Romei, C.; Barone, V.; Pacini, F.; Martino, E.; Loviselli, A.; Pinchera, A.; Romeo, G.: Identification of the cys634-to-tyr mutation of the RET proto-oncogene in a pedigree with multiple endocrine neoplasia type 2A and localized cutaneous lichen amyloidosis. J. Endocr. Invest. 17: 201-204, 1994.
Cetta, F.; Chiappetta, G.; Melillo, R. M.; Petracci, M.; Montalto, G.; Santoro, M.; Fusco, A.: The ret/ptc1 oncogene is activated in familial adenomatous polyposis-associated thyroid papillary carcinomas. J. Clin. Endocr. Metab. 83: 1003-1006, 1998.
Decker, R. A.; Peacock, M. L.; Watson, P.: Hirschsprung disease in MEN 2A: increased spectrum of RET exon 10 genotypes and strong genotype-phenotype correlation. Hum. Molec. Genet. 7: 129-134, 1998.
Donis-Keller, H.; Dou, S.; Chi, D.; Carlson, K. M.; Toshima, K.; Lairmore, T. C.; Howe, J. R.; Moley, J. F.; Goodfellow, P.; Wells, S. A., Jr.: Mutations in the RET proto-oncogene are associated with MEN 2A and FMTC. Hum. Molec. Genet. 2: 851-856, 1993.
Doray, B.; Salomon, R.; Amiel, J.; Pelet, A.; Touraine, R.; Billaud, M.; Attie, T.; Bachy, B.; Munnich, A.; Lyonnet, S.: Mutation of the RET ligand, neurturin, supports multigenic inheritance in Hirschsprung disease. Hum. Molec. Genet. 7: 1449-1452, 1998.
Edery, P.; Lyonnet, S.; Mulligan, L. M.; Pelet, A.; Dow, E.; Abel, L.; Holder, S.; Nihoul-Fekete, C.; Ponder, B. A. J.; Munnich, A.: Mutations of the RET proto-oncogene in Hirschsprung's disease. Nature 367: 378-380, 1994.
Eng, C.: The RET proto-oncogene in multiple endocrine neoplasia type 2 and Hirschsprung's disease. New Eng. J. Med. 335: 943-951, 1996.
Eng, C.; Crossey, P. A.; Mulligan, L. M.; Healey, C. S.; Houghton, C.; Prowse, A.; Chew, S. L.; Dahia, P. L. M.; O'Riordan, J. L. H.; Toledo, S. P. A.; Smith, D. P.; Maher, E. R.; Ponder, B. A. J.: Mutations in the RET proto-oncogene and the von Hippel-Lindau disease tumour suppressor gene in sporadic and syndromic phaeochromocytomas. J. Clin. Genet. 32: 934-937, 1995.
Eng, C.; Mulligan, L. M.: Mutations of the RET proto-oncogene in the multiple endocrine neoplasia type 2 syndromes, related sporadic tumours, and Hirschsprung disease. Hum. Mutat. 9: 97-109, 1997.
Eng, C.; Mulligan, L. M.; Smith, D. P.; Healey, C. S.; Frilling, A.; Raue, F.; Neumann, H. P. H.; Pfragner, R.; Behmel, A.; Lorenzo, M. J.; Stonehouse, T. J.; Ponder, M. A.; Ponder, B. A. J.: Mutation of the RET protooncogene in sporadic medullary thyroid carcinoma. Genes Chromosomes Cancer 12: 209-212, 1995.
Eng, C.; Smith, D. P.; Mulligan, L. M.; Healey, C. S.; Zvelebil, M. J.; Stonehouse, T. J.; Ponder, M. A.; Jackson, C. E.; Waterfield, M. D.; Ponder, B. A. J.: A novel point mutation in the tyrosine kinase domain of the RET proto-oncogene in sporadic medullary thyroid carcinoma and in a family with FMTC. Oncogene 10: 509-513, 1995.
Eng, C.; Smith, D. P.; Mulligan, L. M.; Nagai, M. A.; Healey, C. S.; Ponder, M. A.; Gardner, E.; Scheumann, G. F. W.; Jackson, C. E.; Tunnacliffe, A.; Ponder, B. A. J.: Point mutation within the tyrosine kinase domain of the RET proto-oncogene in multiple endocrine neoplasia type 2B and related sporadic tumors. Hum. Molec. Genet. 3: 237-241, 1994.
Fearon, E. R.: Human cancer syndromes: clues to the origin and nature of cancer. Science 278: 1043-1050, 1997.
Gardner, E.; Mulligan, L. M.; Eng, C.; Healey, C. S.; Kwok, J. B. J.; Ponder, M. A.; Ponder, B. A. J.: Haplotype analysis of MEN 2 mutations. Hum. Molec. Genet. 3: 1771-1774, 1994.
Grieco, M.; Santoro, M.; Berlingieri, M. T.; Melillo, R. M.; Donghi, R.; Bongarzone, I.; Pierotti, M. A.; Della Porta, G.; Fusco, A.; Vecchio, G.: PTC is a novel rearranged form of the ret proto-oncogene and is frequently detected in vivo in human thyroid papillary carcinomas. Cell 60: 557-563, 1990.
Hofstra, R. M. W.; Landsvater, R. M.; Ceccherini, I.; Stulp, R. P.; Stelwagen, T.; Luo, Y.; Pasini, B.; Hoppener, J. W. M.; Ploos van Amstel, H. K.; Romeo, G.; Lips, C. J. M.; Buys, C. H. C. M.: A mutation in the RET proto-oncogene associated with multiple endocrine neoplasia type 2B and sporadic medullary thyroid carcinoma. Nature 367: 375-376, 1994.
Hofstra, R. M. W.; Sijmons, R. H.; Stelwagen, T.; Stulp, R. P.; Kousseff, B. G.; Lips, C. J. M.; Steijlen, P. M.; Van Voorst Vader, P. C.; Buys, C. H. C. M.: RET mutation screening in familial cutaneous lichen amyloidosis and in skin amyloidosis associated with multiple endocrine neoplasia. J. Invest. Derm. 107: 215-218, 1996.
Hoppener, J. W. M.; Lips, C. J. M.: RET receptor tyrosine kinase gene mutations: molecular biological, physiological and clinical aspects. Europ. J. Clin. Invest. 26: 613-624, 1996.
Hoppner, W.; Ritter, M. M.: A duplication of 12 bp in the critical cysteine rich domain of the RET proto-oncogene results in a distinct phenotype of multiple endocrine neoplasia type 2A. Hum. Molec. Genet. 6: 587-590, 1997.
Ikeda, I.; Ishizaka, Y.; Tahira, T.; Suzuki, T.; Onda, M.; Sugimura, T.; Nagao, M.: Specific expression of the ret proto-oncogene in human neuroblastoma cell lines. Oncogene 5: 1291-1296, 1990.
Ishizaka, Y.; Itoh, F.; Tahira, T.; Ikeda, I.; Sugimura, T.; Tucker, J.; Fertitta, A.; Carrano, A. V.; Nagao, M.: Human ret proto-oncogene mapped to chromosome 10q11.2. Oncogene 4: 1519-1521, 1989.
Iwashita, T.; Murakami, H.; Asai, N.; Takahashi, M.: Mechanism of Ret dysfunction by Hirschsprung mutations affecting its extracellular domain. Hum. Molec. Genet. 5: 1577-1580, 1996.
Lairmore, T. C.; Dou, S.; Howe, J. R.; Chi, D.; Carlson, K.; Veile, R.; Mishra, S. K.; Wells, S. A., Jr.; Donis-Keller, H.: A 1.5-megabase yeast artificial chromosome contig from human chromosome 10q11.2 connecting three genetic loci (RET, D10S94, and D10S102) closely linked to the MEN2A locus. Proc. Nat. Acad. Sci. 90: 492-496, 1993.
Mulligan, L. M.; Eng, C.; Healey, C. S.; Clayton, D.; Kwok, J. B. J.; Gardner, E.; Ponder, M. A.; Frilling, A.; Jackson, C. E.; Lehnert, H.; Neumann, H. P. H.; Thibodeau, S. N.; Ponder, B. A. J.: Specific mutations of the RET proto-oncogene are related to disease phenotype in MEN 2A and FMTC. Nature Genet. 6: 70-74, 1994.
Mulligan, L. M.; Kwok, J. B. J.; Healey, C. S.; Elsdon, M. J.; Eng, C.; Gardner, E.; Love, D. R.; Mole, S. E.; Moore, J. K.; Papi, L.; Ponder, M. A.; Telenius, H.; Tunnacliffe, A.; Ponder, B. A. J.: Germ-line mutations of the RET proto-oncogene in multiple endocrine neoplasia type 2A. Nature 363: 458-460, 1993.
Pachnis, V.; Mankoo, B.; Costantini, F.: Expression of the c-ret proto-oncogene during mouse embryogenesis. Development 119: 1005-1017, 1993.
Pasini, B.; Hofstra, R. M. W.; Yin, L.; Bocciardi, R.; Santamaria, G.; Grootscholten, P. M.; Ceccherini, I.; Patrone, G.; Priolo, M.; Buys, C. H. C. M.; Romeo, G.: The physical map of the human RET proto-oncogene. Oncogene 11: 1737-1743, 1995.
Pelet, A.; Geneste, O.; Edery, P.; Pasini, A.; Chappuis, S.; Attie, T.; Munnich, A.; Lenoir, G.; Lyonnet, S.; Billaud, M.: Various mechanisms cause RET-mediated signaling defects in Hirschsprung's disease. J. Clin. Invest. 101: 1415-1423, 1998.
Pierotti, M. A.; Santoro, M.; Jenkins, R. B.; Sozzi, G.; Bongarzone, I.; Grieco, M.; Monzini, N.; Miozzo, M.; Herrmann, M. A.; Fusco, A.; Hay, I. D.; Della Porta, G.; Vecchio, G.: Characterization of an inversion on the long arm of chromosome 10 juxtaposing D10S170 and RET and creating the oncogenic sequence RET/PTC. Proc. Nat. Acad. Sci. 89: 1616-1620, 1992.
Robinson, M. F.; Cote, G. J.; Nunziata, V.; Brandi, M. L.; Ferrer, J. P.; Martins Bugalho, M. J. G.; Almeida Ruas, M. M.; Chik, C.; Colantuoni, V.; Gagel, R. F.: Mutation of a specific codon of the RET proto-oncogene in the multiple endocrine neoplasia type 2A/cutaneous lichen amyloidosis syndrome. (Abstract) Fifth International Workshop on Multiple Endocrine Neoplasia, Stockholm, Archipelago , 1994.
Rodrigues, G. A.; Park, M.: Dimerization mediated through a leucine zipper activates the oncogenic potential of the met receptor tyrosine kinase. Molec. Cell. Biol. 13: 6711-6722, 1993.
Romeo, G.; Ronchetto, P.; Luo, Y.; Barone, V.; Seri, M.; Ceccherini, I.; Pasini, B.; Bocciardi, R.; Lerone, M.; Kaariainen, H.; Martucciello, G.: Point mutations affecting the tyrosine kinase domain of the RET proto-oncogene in Hirschsprung's disease. Nature 367: 377-378, 1994.
Santoro, M.; Carlomagno, F.; Hay, I. D.; Herrmann, M. A.; Grieco, M.; Melillo, R.; Pierotti, M. A.; Bongarzone, I.; Della Porta, G.; Berger, N.; Peix, J. L.; Paulin, C.; Fabien, N.; Vecchio, G.; Jenkins, R. B.; Fusco, A.: Ret oncogene activation in human thyroid neoplasms is restricted to the papillary cancer subtype. J. Clin. Invest. 89: 1517-1522, 1992.
Santoro, M.; Carlomagno, F.; Romano, A.; Bottaro, D. P.; Dathan, N. A.; Grieco, M.; Fusco, A.; Vecchio, G.; Matoskova, B.; Kraus, M. H.; Di Fiore, P. P.: Activation of RET as a dominant transforming gene by germline mutations of MEN2A and MEN2B. Science 267: 381-383, 1995.
Schuchardt, A.; D'Agati, V.; Larsson-Blomberg, L.; Costantini, F.; Pachnis, V.: Defects in the kidney and enteric nervous system of mice lacking the tyrosine kinase receptor Ret. Nature 367: 380-383, 1994.
Seri, M.; Celli, I.; Betsos, N.; Claudiani, F.; Camera, G.; Romeo, G.: A Cys634Gly substitution of the RET proto-oncogene in a family with recurrence of multiple endocrine neoplasia type 2A and cutaneous lichen amyloidosis. Clin. Genet. 51: 86-90, 1997.
Seri, M.; Yin, L.; Barone, A.; Bolino, A.; Celli, I.; Bocciardi, R.; Pasini, B.; Ceccherini, I.; Lerone, M.; Kristoffersson, U.; Larsson, L. T.; Casasa, J. M.; Cass, D. T.; Abramowicz, M. J.; Vanderwinden, J.-M.; Kravcenkiene, I.; Baric, I.; Silengo, M.; Martucciello, G.; Romeo, G.: Frequency of RET mutations in long- and short-segment Hirschsprung disease. Hum. Mutat. 9: 243-249, 1997.
Shirahama, S.; Ogura, K.; Takami, H.; Ito, K.; Tohsen, T.; Miyauchi, A.; Nakamura, Y.: Mutational analysis of the RET proto-oncogene in 71 Japanese patients with medullary thyroid carcinoma. J. Hum. Genet. 43: 101-106, 1998.
Svensson, P.-J.; Anvret, M.; Molander, M.-L.; Nordenskjold, A.: Phenotypic variation in a family with mutations in two Hirschsprung-related genes (RET and endothelin receptor B). Hum. Genet. 103: 145-148, 1998.
Takahashi, M.; Buma, Y.; Hiai, H.: Isolation of ret proto-oncogene cDNA with an amino-terminal signal sequence. Oncogene 4: 805-806, 1989.
Takahashi, M.; Buma, Y.; Iwamoto, T.; Inaguma, Y.; Ikeda, H.; Hiai, H.: Cloning and expression of the ret proto-oncogene encoding a tyrosine kinase with two potential transmembrane domains. Oncogene 3: 571-578, 1988.
Takahashi, M.; Ritz, J.; Cooper, G. M.: Activation of a novel human transforming gene, ret, by DNA rearrangement. Cell 42: 581-588, 1985.
van Heyningen, V.: One gene—four syndromes. Nature 367: 319-320, 1994.
Xue, F.; Yu, H.; Maurer, L. H.; Memoli, V. A.; Nutile-McMenemey, N.; Schuster, M. K.; Bowden, D. W.; Mao, J.; Noll, W. W.: Germline RET mutations in MEN 2A and FMTC and their detection by simple DNA diagnostic tests. Hum. Molec. Genet. 3: 635-638, 1994.
Yin, L.; Ceccherini, I.; Pasini, B.; Matera, I.; Bicocchi, M. P.; Barone, V.; Bocciardi, R.; Kaariainen, H.; Weber, D.; Devoto, M.; Romeo, G.: Close linkage with the RET protooncogene and boundaries of deletion mutations in

autosomal dominant Hirschsprung disease. Hum. Molec. Genet. 2: 1803-1808, 1993.

CHROMOGRANIN A (*118910)

JJM NOTE

This tumor marker is included with the endocrine neoplasias rather than with other tumor markers, since it seems so specific for peptide-producing endocrine tumors.

Chromogranin A is a protein costored and coreleased with catecholamines from storage granules in the adrenal medulla. Secretory protein I (parathyroid secretory protein; PSP) is a protein costored and coreleased with parathyroid hormone (PTH) from storage granules in the parathyroid gland. Its function is unknown. Like PTH (168450), its secretion is inversely proportional to extracellular calcium concentration. Bhargava et al. (1983) showed that the degree of phosphorylation of PSP is also inversely proportional to serum calcium, and that PSP is the major phosphorylated protein released by the parathyroid gland. Cohn et al. (1982) demonstrated a close similarity of these 2 proteins in amino acid composition, physical properties, and immunologic crossreactivity. Kruggel et al. (1985) determined the N-terminal sequences of bovine and human adrenal medullary chromogranin A; the sequences are identical to each other and also to the published sequence of secretory protein I. O'Connor and Deftos (1986) showed that chromogranin A is secreted by a great variety of peptide-producing endocrine neoplasms: pheochromocytoma, parathyroid adenoma, medullary thyroid carcinoma, carcinoids, oat-cell lung cancer, pancreatic islet-cell tumors, and aortic-body tumor. Deftos et al. (1986) cloned a cDNA for CHGA using mRNA from CHGA-producing medullary thyroid carcinoma cells in an expression vector, lamba gt11. Konecki et al. (1987) isolated a full-length clone encoding human chromogranin A from a lambda-gt10 cDNA library of a human pheochromocytoma. The nucleotide sequence showed that human chromogranin A is a 439-residue protein preceded by an 18-residue signal peptide. Sequence findings suggested that pancreastatin is derived from chromogranin A itself rather than from a protein that is only similar to chromogranin A. The pancreastatin sequence contained in human chromogranin A is flanked by sites for proteolytic processing. In man, pancreastatin may be important for the physiologic hom eostasis of blood insulin levels as well as pathologic aberrations such as diabetes mellitus. Wu et al. (1991) found that the chromogranin A gene has 8 exons and 7 introns spanning about 11 kb.

Using immunohistochemistry on plastic sections, Cetin et al. (1993) investigated the occurrence and cellular distribution of CHGA, pancreastatin, and chromostatin (CST), a CHGA-derived bioactive peptide, in human endocrine pancreas of healthy and disease states and in the adrenal medulla. In the normal and diabetic pancreas, CST immunoreactivity was localized exclusively in beta cells, which were mostly unreactive for PST and CHGA. Both latter peptides were confined mainly to glucagon (alpha) cells. Insulinoma cells displayed strong insulin, PST, and CHGA immunoreactivities, but they were faintly immunoreactive for CST or unreactive. Adrenal chromaffin cells exhibited strong immunoreactivity for CHGA but lacked CST and PST immunoreactivities. Based on the peculiar distribution pattern of CST, PST, and CHGA, Cetin et al. (1993) suggested that CHGA is differentially processed in chromaffin and islet tissues and in insulinoma cells. The unique cellular localization of CST in the endocrine pancreas of normal and pathologic conditions may indicate that CST is involved in beta-cell function.

Murray et al. (1987) mapped CHGA to chromosome 14 by probing DNA from a hybrid cell panel with specific cDNA. Using a cDNA clone for the chromogranin A gene, Modi et al. (1989) mapped the gene to 14q32 by Southern blot analysis of human-rodent somatic cell hybrid DNAs and by in situ hybridization. Simon-Chazottes et al. (1993) demonstrated that the chromogranin A gene is present in single dose in both the mouse and rat. Analysis of the allele distribution in an interspecific mouse backcross by single-strand conformation polymorphism positioned the Chga locus on mouse chromosome 12. By study of a rat/mouse somatic cell hybrid panel, they determined that the corresponding gene is on rat chromosome 6. In each case (mouse, rat, and human), chromogranin A is encoded in a conserved region with nearby markers, including the immunoglobulin heavy chain locus. Nobels et al. (1997) evaluated the clinical usefulness of chromogranin A, termed CgA by the authors, as a neuroendocrine serum marker. Serum levels of CgA, neuron-specific enolase (NSE), and the alpha-subunit of glycoprotein hormone (alpha-SU) were determined in 211 patients with neuroendocrine tumors and 180 control subjects with nonendocrine tumors. The concentrations of CgA, NSE, and alpha-SU were elevated in 50%, 43% and 24% of patients with neuroendocrine tumors, respectively. Serum CgA was most frequently increased in subjects with gastrinomas (100%), pheochromocytomas (89%), carcinoid tumors (80%), nonfunctioning tumors of the endocrine pancreas (69%), and medullary thyroid carcinomas (50%). The highest levels were observed in subjects with carcinoid tumors; NSE was most frequently elevated in patients with small cell lung carcinoma (74%) and alpha-SU was most frequently elevated in patients with carcinoid tumors (39%). A significant positive relationship was demonstrated between the tumor load and serum CgA levels (P less than 0.01, by X2 test). Elevated concentrations of CgA, NSE, and alpha-SU were present in, respective ly, 7%, 35%, and 15% of control subjects. Markedly elevated serum levels of CgA, exceeding 300 mu-g/L, were observed in only 2% of control patients (n = 3) compared to 40% of patients with neuroendocrine tumors (n = 76). The authors concluded that CgA is the best general neuroendocrine serum marker available at that time.

See also Angeletti (1986); Hagn et al. (1986); Modi et al. (1989)

Angeletti, R. H.: Chromogranins and neuroendocrine secretion. (Editorial) Lab. Invest. 55: 387-390, 1986.
Bhargava, G.; Russell, J.; Sherwood, L. M.: Phosphorylation of parathyroid secretory protein. Proc. Nat. Acad. Sci. 80: 878-881, 1983.
Cetin, Y.; Aunis, D.; Bader, M.-F.; Galindo, E.; Jorns, A.; Bargsten, G.; Grube, D.: Chromostatin, a chromogranin A-derived bioactive peptide, is present in human pancreatic insulin (beta) cells. Proc. Nat. Acad. Sci. 90: 2360-2364, 1993.
Cohn, D. V.; Zangerle, R.; Fischer-Colbrie, R. R.; Chu, L. L. H.; Elting, J. J.; Hamilton, J. W.; Winkler, H.: Similarity of secretory protein I from parathyroid gland to chromogranin A from the adrenal medulla. Proc. Nat. Acad. Sci. 79: 6056-6059, 1982.
Deftos, L. J.; Murray, S. S.; Burton, D. W.; Parmer, R. J.; O'Connor, D. T.; Delegeane, A. M.; Mellon, P. L.: A cloned chromogranin A (CgA) cDNA detects a 2.3kb mRNA in diverse neuroendocrine tissues. Biochem. Biophys. Res. Commun. 137: 418-423, 1986.
Hagn, C.; Schmid, K. W.; Fischer-Colbrie, R.; Winkler, H.: Chromogranin A, B, and C in human adrenal medulla and endocrine tissues. Lab. Invest. 55: 405-411, 1986.
Konecki, D. S.; Benedum, U. M.; Gerdes, H.-H.; Huttner, W. B.: The primary structure of human chromogranin A and pancreastatin. J. Biol. Chem. 262: 17026-17030, 1987.
Kruggel, W.; O'Connor, D. T.; Lewis, R. V.: The amino terminal sequences of bovine and human chromogranin A and secretory protein I are identical. Biochem. Biophys. Res. Commun. 127: 380-383, 1985.
Modi, W. S.; Levine, M. A.; Dean, M.; Seuanez, H.; O'Brien, S. J.: The chromogranin A gene: chromosome assignment and RFLP analysis. (Abstract) Cytogenet. Cell Genet. 51: 1046 only, 1989.
Modi, W. S.; Levine, M. A.; Seuanez, H. N.; Dean, M.; O'Brien, S. J.: The human chromogranin A gene: chromosome assignment and RFLP analysis. Am. J. Hum. Genet. 45: 814-818, 1989.
Murray, S. S.; Deaven, L. L.; Burton, D. W.; O'Connor, D. T.; Mellon, P. L.; Deftos, L. J.: The gene for human chromogranin A (CgA) is located on chromosome 14. Biochem. Biophys. Res. Commun. 142: 141-146, 1987.
Nobels, F. R. E.; Kwekkeboom, D. J.; Coopmans, W.; Schoenmakers, C. H. H.; Lindemans, J.; De Herder, W. W.; Krenning, E. P.; Bouillon, R.; Lamberts, S. W. J.: Chromogranin A as serum marker for neuroendocrine neoplasia: comparison with neuron-specific enolase and the alpha-subunit of glycoprotein hormones. J. Clin. Endocr. Metab. 82: 2622-2628, 1997.
O'Connor, D. T.; Deftos, L. J.: Secretion of chromogranin A by peptide-producing endocrine neoplasms. New Eng. J. Med. 314: 1145-1151, 1986.
Simon-Chazottes, D.; Wu, H.; Parmer, R. J.; Rozansky, D. J.; Szpirer, J.; Levan, G.; Kurtz, T. W.; Szpirer, C.; Guenet, J. L.; O'Connor, D. T.: Assignment of the chromogranin A (Chga) locus to homologous regions on mouse chromosome 12 and rat chromosome 6. Genomics 17: 252-255, 1993.
Wu, H.-J.; Rozansky, D. J.; Parmer, R. J.; Gill, B. M.; O'Connor, D. T.: Structure and function of the chromogranin A gene: clues to evolution and tissue-specific expression. J. Biol. Chem. 266: 13130-13134, 1991.

CALCITONIN (*114130)

JJM NOTE

This protein is included here rather than with other tumor markers because it is an exclusive product of the C cell of the thyroid; hence, it is a clinically useful marker of neoplasia of those cells, as seen, for example, in multiple endocrine neoplasia 2.

Calcitonin is a peptide hormone synthesized by the parafollicular cells of the thyroid. It causes reduction in serum calcium—an effect opposite to that of parathyroid hormone (PTH; 168450). Human calcitonin contains 32 amino acids and has a molecular weight of 3,421. See Dayhoff (1972) for amino acid sequence data. Multiple calcitonin polypeptides are encoded in a single messenger RNA (Jacobs et al., 1981). Rosenfeld et al. (1982) presented evidence that alternative RNA splicing of the transcripts of the calcitonin gene is responsible for the production of different polypeptide products. See Amara et al. (1982). Genomic mapping results are consistent with the existence of a single calcitonin gene (Rosenfeld et al., 1982).

Katacalcin (kata-, Gr. down) was the name given by A. P. Waterson to a 21-amino acid peptide that flanks calcitonin on its C-terminal side in the large precursor polyprotein from which calcitonin is cleaved. Its concentration is higher in males than in females and approximately equimolar with calcitonin; doubles within 5 min of calcium infusion; and is markedly raised in cases of medullary thyroid carcinoma. Katacalcin is a new hormone discovered by use of recombinant DNA technology rather than by traditional techniques of tissue extraction and purification based on biologic assay; like calcitonin, it may be involved in both plasma calcium regulation and skeletal maintenance (Hillyard et al., 1983). Rosenfeld et al. (1983) showed that alternative processing of the RNA transcribed from the calcitonin gene results in the production of an mRNA in neural tissue distinct from that in thyroidal 'C' cells. The novel neuropeptide was referred to as calcitonin gene-related peptide (CGRP). The distribution of CGRP-producing cells and pathways in the brain and other tissues suggests functions for CGRP in nociception, ingestive behavior, and modulation of the autonomic and endocrine systems. The approach described here has general applicability, viz., the use of recombinant DNA technology to analyze complex neurobiologic systems in the absence of prior structural or biologic information. CGRP-containing neurons were detected particularly in association with heart and blood vessels. CGRP was shown to have potent vasodilator action and probably is an important regulator of vascular tone and blood flow (Tippins, 1986). Tschopp et al. (1985) determined the location of CGRP and its binding sites in the CNS and pituitary. Goltzman and Mitchell (1985) identified discrete receptors for CT and CGRP in the nervous system and in peripheral tissues. Tiller-Borcich et al. (1988) found that CGRP is concentrated in the locus caeruleus in the human. CGRP has very potent hemodynamic activity, an d the locus caeruleus is the main source of noradrenergic neurotransmission in the CNS.

Using a molecular probe containing a 584-basepair sequence corresponding to part of the human calcitonin mRNA in the study of somatic cell hybrids, Hoppener et al. (1984) assigned the calcitonin gene to 11p14-qter. The calcitonin gene was found to contain a polymorphic site for restriction endonuclease TaqI. Przepiorka et al. (1984) mapped the calcitonin gene to 11p by molecular hybridization of a human calcitonin cDNA probe to DNA from human-rodent hybrid cells. In situ hybridization narrowed the assignment to 11p15-p13. In a cell line derived from a particular virulent medullary carcinoma of the thyroid, Testa (1984) found a chromosomal rearrangement affecting 11p. Simpson et al. (1984) assigned the calcitonin gene to chromosome 11 by use of a cDNA clone isolated from medullary thyroid carcinoma and a somatic cell hybrid panel. With a TaqI RFLP detected by this probe, they studied linkage of the calcitonin locus and MEN2; negative lod scores were found at all recombination values. In 2 tumors with mitotic deletions, Henry et al. (1989) found that CALCA must lie distal to PTH in 11p15.5; in both tumors the CALCA locus was lost and the PTH locus retained. It may be of functional significance that the PTH and calcitonin genes are close together, since they are the yin and the yang of the control of calcium metabolism. In the mouse, both PTH and calcitonin are coded by chromosome 7 (Lalley et al., 1987). Hoppener et al. (1988) described a calcitonin pseudogene and reviewed information on the CALC genes. The CALC1 gene produces calcitonin (encoded by exon 4) or calcitonin gene-related peptide (encoded by exon 5) in a tissue-specific fashion. The CALC2 gene (CALCB; 114160) produces a second calcitonin gene-related peptide, but probably not a second calcitonin. The presumed pseudogene CALC3 does not seem to encode either peptide. Like the other 2 CALC genes, the CALC3 gene was found to be located on human chromosome 11 (Hoppener et al., 1988). Hoovers et al. (1993) used fluorescence in situ hybridization to prometaphase chromosomes, pulsed field gel electrophoresis analysis, and 2-color in situ hybridization to interphase nuclei to map CALCA, CALCB, and the pseudogene CALC3 to a 220-kb SacII fragment on 11p15.2-p15.1. The related islet amyloid polypeptide (IAPP; 147940) gene was assigned to 12p12.3-p12.1 by the same methods. The results supported the evolutionary relationship between the calcitonin/CGRP genes and the IAPP gene and between parts of human chromosomes 11 and 12.

Breimer et al. (1988) reviewed the organization, expression, and splicing of the calcitonin genes and the structure and function of the peptides they encode. The alpha-calcitonin/CGRP gene stretches over approximately 6.5 kb and consists of 6 exons. The first 3 exons are present in both calcitonin and CGRP mRNA, although exon 1 is not translated. Exon 4 contains the calcitonin-coding sequence. Exon 5 encodes the CGRP sequence. The organization of the beta gene is similar to that of the alpha gene, and it is likely that the 2 arose by duplication. There is no evidence for the production of calcitonin mRNA from the beta gene. Perhaps the beta gene allows the expression of CGRP at sites or in circumstances where the inadvertent production of calcitonin could be deleterious. Mathe et al. (1994) examined the concentration of calcitonin gene-related peptide immunoreactivity in the cerebrospinal fluid of 63 patients with major depression (cf., 125480) with that found in cerebrospinal fluid of 28 patients with schizophrenia (e.g., 181510) and 20 controls. Patients with all forms of major depression had higher levels of this peptide in the spinal fluid than did patients with schizophrenia or controls. The authors suggested that the increased concentration of CGRP may be a marker trait of major depressive disorder.

See also Edbrooke et al. (1985); Girgis et al. (1985); Jonas et al. (1985); Kittur et al. (1985); MacIntyre et al. (1982); Neher et al. (1968); New and Mudge (1986); Struthers et al. (1986)

Alevizaki, M.; Stevenson, J. C.; Girgis, S. I.; MacIntyre, I.; Legon, S.: Altered calcitonin gene in a young patient with osteoporosis. Brit. Med. J. 298: 1215-1216, 1989.

Amara, S. G.; Jonas, V.; Rosenfeld, M. G.; Ong, E. S.; Evans, R. M.: Alternative RNA processing in calcitonin gene expression generates mRNAs encoding different polypeptide products. Nature 298: 240-244, 1982.

Breimer, L. H.; MacIntyre, I.; Zaidi, M.: Peptides from the calcitonin genes: molecular genetics, structure and function. Biochem. J. 255: 377-390, 1988.

Dayhoff, M. O.: Atlas of Protein Sequence and Structure. Hormones, active peptides and toxins. Washington: National Biomedical Research Foundation (pub.) 5: 1972. Pp. D205 only.

Edbrooke, M. R.; Parker, D.; McVey, J. H.; Riley, J. H.; Sorenson, G. D.; Pettengill, O. S.; Craig, R. K.: Expression of the human calcitonin/CGRP gene in lung and thyroid carcinoma. EMBO J. 4: 715-724, 1985.

Girgis, S. I.; Macdonald, D. W. R.; Stevenson, J. C.; Bevis, P. J. R.; Lynch, C.; Wimalawansa, S. J.; Self, C. H.; Morris, H. R.; MacIntyre, I.: Calcitonin gene-related peptide: potent vasodilator and major product of calcitonin gene. Lancet II: 14-16, 1985.

Goltzman, D.; Mitchell, J.: Interaction of calcitonin and calcitonin gene-related peptide at receptor sites in target tissues. Science 227: 1343-1345, 1985.

Henry, I.; Grandjouan, S.; Barichard, F.; Huerre-Jeanpierre, C.; Junien, C.: Mitotic deletions of 11p15.5 in two different tumors indicate that the CALCA locus is distal to the PTH locus. Cytogenet. Cell Genet. 50: 155-157, 1989.

Hillyard, C. J.; Myers, C.; Abeyasekera, G.; Stevenson, J. C.; Craig, R. K.; MacIntyre, I.: Katacalcin: a new plasma calcium-lowering hormone. Lancet I: 846-848, 1983.

Hoovers, J. M. N.; Redeker, E.; Speleman, F.; Hoppener, J. W. M.; Bhola, S.; Bliek, J.; van Roy, N.; Leschot, N. J.; Westerveld, A.; Mannens, M.: High-resolution chromosomal localization of the human calcitonin/CGRP/IAPP gene family members. Genomics 15: 525-529, 1993.

Hoppener, J. W. M.; Steenbergh, P. H.; Zandberg, J.; Adema, G. J.; Geurts van Kessel, A. H. M.; Lips, C. J. M.; Jansz, H. S.: A third human CALC (pseudo)gene on chromosome 11. FEBS Lett. 233: 57-63, 1988.

Hoppener, J. W. M.; Steenbergh, P. H.; Zandberg, J.; Bakker, E.; Pearson, P. L.; Geurts van Kessel, A. H. M.; Jansz, H. S.; Lips, C. J. M.: Localization of the polymorphic human calcitonin gene on chromosome 11. Hum. Genet. 66: 309-312, 1984.

Jacobs, J. W.; Goodman, R. H.; Chin, W. W.; Dee, P. C.; Habener, J. F.; Bell, N. H.; Potts, J. T., Jr.: Calcitonin messenger RNA encodes multiple polypeptides in a single precursor. Science 213: 457-459, 1981.

Jonas, V.; Lin, C. R.; Kawashima, E.; Semon, D.; Swanson, L. W.; Mermod, J.-J.; Evans, R. M.; Rosenfeld, M. G.: Alternative RNA processing events in human calcitonin/calcitonin gene-related peptide gene expression. Proc. Nat. Acad. Sci. 82: 1994-1998, 1985.

Kittur, S. D.; Hoppener, J. W. M.; Antonarakis, S. E.; Daniels, J. D. J.; Meyers, D. A.; Maestri, N. E.; Jansen, M.; Korneluk, R. G.; Nelkin, B. D.; Kazazian, H. H., Jr.: Linkage map of the short arm of human chromosome 11: location of the genes for catalase, calcitonin, and insulin-like growth factor II. Proc. Nat. Acad. Sci. 82: 5064-5067, 1985.

Lalley, P. A.; Sakaguchi, A. Y.; Eddy, R. L.; Honey, N. H.; Bell, G. I.; Shen, L.-P.; Rutter, W. J.; Jacobs, J. W.; Heinrich, G.; Chin, W. W.; Naylor, S. L.: Mapping polypeptide hormone genes in the mouse: somatostatin, glucagon, calcitonin, and parathyroid hormone. Cytogenet. Cell Genet. 44: 92-97, 1987.

MacIntyre, I.; Hillyard, C. J.; Murphy, P. K.; Reynolds, J. J.; Gaines-Das, R. E.; Craig, R. K.: A second plasma calcium-lowering peptide from the human calcitonin precursor. Nature 300: 460-462, 1982.

Mathe, A. A.; Agren, H.; Lindstrom, L.; Theodorsson, E.: Increased concentration of calcitonin gene-related peptide in cerebrospinal fluid of depressed patients: a possible trait marker of major depressive disorder. Neurosci. Lett. 182: 138-142, 1994.

Neher, R.; Riniker, B.; Rittel, W.; Zuber, H.: Thyrocalcitonin. II. Struktur von alpha-Thyrocalcitonin. Helv. Chim. Acta 51: 917-924, 1968.

New, H. V.; Mudge, A. W.: Calcitonin gene-related peptide regulates muscle acetylcholine receptor synthesis. Nature 323: 809-811, 1986.

Przepiorka, D.; Baylin, S. B.; McBride, D. W.; Testa, J. R.; de Bustros, A.; Nelkin, B. D.: The human calcitonin gene is located on the short arm of chromosome 11. Biochem. Biophys. Res. Commun. 120: 493-499, 1984.

Rosenfeld, M. G.; Lin, C. R.; Amara, S. G.; Stolarsky, L.; Roos, B. A.; Ong, E. S.; Evans, R. M.: Calcitonin mRNA polymorphism: peptide switching associated with alternative RNA splicing events. Proc. Nat. Acad. Sci. 79: 1717-1721, 1982.

Rosenfeld, M. G.; Mermod, J.-J.; Amara, S. G.; Swanson, L. W.; Sawchenko, P. E.; Rivier, J.; Vale, W. W.; Evans, R. M.: Production of a novel neuropeptide encoded by the calcitonin gene via tissue-specific RNA processing. Nature 304: 129-135, 1983.

Simpson, N. E.; Goodfellow, P. J.; Riddell, D. C.; Hamerton, J. L.; Holden, J. J. A.; White, B. N.: Assignment of the calcitonin gene to chromosome 11 and probable exclusion of linkage between the gene and the locus for multiple endocrine neoplasia type 2. (Abstract) Am. J. Hum. Genet. 36: 153S, 1984.

Struthers, A. D.; Brown, M. J.; Macdonald, D. W. R.; Beacham, J. L.; Stevenson, J. C.; Morris, H. R.; MacIntyre, I.: Human calcitonin gene related peptide: a potent endogenous vasodilator in man. Clin. Sci. 70: 389-393, 1986.
Testa, J. R.: Personal Communication. Baltimore, Md. 3/1984.
Tiller-Borcich, J. K.; Capili, H.; Gordan, G. S.: Human brain calcitonin gene-related peptide (CGRP) is concentrated in the locus caeruleus. Neuropeptides 11: 55-61, 1988.
Tippins, J. R.: CGRP: a novel neuropeptide from the calcitonin gene is the most potent vasodilator known. J. Hypertension 4 (suppl. 5): S102-S105, 1986.
Tschopp, F. A.; Henke, H.; Petermann, J. B.; Tobler, P. H.; Janzer, R.; Hokfelt, T.; Lundberg, J. M.; Cuello, C.; Fischer, J. A.: Calcitonin gene-related peptide and its binding sites in the human central nervous system and pituitary. Proc. Nat. Acad. Sci. 82: 248-252, 1985.

Pituitary Tumors

ACROMEGALY (*102200)

JJM NOTE

With the cloning of the gene for multiple endocrine neoplasia 1 (*MEN1*), the burden of proof will be to rule out localization or allelism in these or future examples of mendelizing acromegaly. Another gene predisposing to growth hormone-secreting pituitary adenoma is *GNAS1* (139320).

Koch and Tiwisina (1959) reviewed 8 examples of affected persons in 2 successive generations, including 4 instances of father and 1 or more sons affected. Some reported instances of familial acromegaly may in fact be pachydermoperiostosis (167100), acromegaloid-cutis verticis gyrata-leukoma syndrome (102100), or cerebral gigantism (117550). Furthermore, familial acromegaly can be a partial expression of the multiple endocrine adenomatosis syndrome, specifically multiple endocrine neoplasia type I (MEN1; 131100). Pituitary growth hormone (GH)-secreting tumors (somatotrophinomas) occur also as part of McCune-Albright syndrome (174800) due to mutations in the GNAS1 gene (139320). Levin et al. (1974) reported the cases of 2 brothers with acromegaly confirmed by elevated GH levels. Both had acanthosis nigricans and pituitary tumors. Pestell et al. (1989) described a family in which 5 members over 3 generations had isolated functional pituitary adenomas. In 4 cases this was associated with acromegaly, and in the fifth galactorrhea from prolactin excess was the presenting feature. The tumors were histologically of either atypical mixed cell or undifferentiated cell type. No parent-child transmission was observed. The 5 individuals were related as uncle and nephew or uncle and niece or as second cousins. There were no consanguineous marriages in the family. Autosomal dominant inheritance with reduced penetrance was proposed. Pestell et al. (1989) considered the disorder in this family to be distinct from MEN1. Jones et al. (1984), Abbassioun et al. (1986), and McCarthy et al. (1990) also reported cases of familial acromegaly.

Growth hormone secreting pituitary adenomas (somatotrophinomas) occur in families either as an isolated autosomal dominant endocrinopathy (as illustrated by the examples cited above) or as part of MEN1. Thakker et al. (1993) compared DNA in somatotrophinomas and peripheral leukocytes obtained from 13 patients with acromegaly; one patient also suffered from MEN1. Five DNA probes identifying RFLPs from 11q demonstrated allele loss in pituitary tumors from 5 patients, 4 non-MEN1 and 1 MEN1. Deletion mapping revealed that the region of allele loss common to the somatotrophinomas involved 11q13. Similar allelic deletions at 12 other loci distributed through the genome did not reveal generalized allele loss in the somatotrophinomas. Thakker et al. (1993) interpreted these results as indicating that a recessive oncogene on 11q13 is specifically involved in the monoclonal development of somatotrophinomas; 11q13 is also the site of the gene for MEN1 in which somatotrophinomas are a feature. (It is a well known phenomenon that tumors that occur as a component of a familial neoplasia syndrome also occur as sporadic tumors on the basis of somatic mutation. Is it not possible that the findings of Thakker et al. (1993) have the same basis as sporadic meningioma due to mutation in the NF2 gene (e.g., 101000.0003), cerebellar hemangioblastoma, sporadic cerebellar hemangioblastoma, or sporadic renal carcinoma due to mutation in the gene for von Hippel-Lindau syndrome (e.g., 193300.0002 and 193300.0007, respectively)? VAM.)

In addition, Thakker et al. (1993) found mutations in the GNAS1 gene (139320) in 2 non-MEN1 somatotrophinomas, one of which also demonstrated allele loss of chromosome 11. (The authors referred to GNAS1 as GSP.)

Barlier et al. (1998) examined adenomas from 30 acromegalic patients (8 gsp+ and 22 gsp-). The gsp+ adenomas appeared to secrete significantly more growth hormone when the ratio of basal GH level/tumor size was considered. A better octreotide sensitivity of mutated adenomas was seen under in vivo (short- and long-term) and in vitro conditions. During the acute octreotide test, the GH nadir was significantly lower in the gsp+ adenomas (85% of maximal inhibition vs 52%). Eighteen patients were treated with octreotide for at least 3 months before surgery; the percent inhibition of GH hypersecretion was higher in gsp+ adenomas (76% vs 47%). In cell culture, the octreotide-induced inhibition of GH release was significantly higher in gsp+ adenomas (71% vs 30%). During 2 years of postoperative follow-up, GH hypersecretion was controlled in all patients with gsp mutations, even in those in whom tumor tissue remained after surgery. On the contrary, in the gsp- group, octreotide treatment did not control hypersecretion in 4 patients bearing tumor remnants. Gs-alpha mutations could, therefore, be a new marker to foresee the susceptibility of the tumor to be controlled by somatostatin analogs, which improves prognosis.

Because phosphodiesterase (PDE) expression and activity are controlled by cAMP, Lania et al. (1998) investigated whether activating Gs-alpha mutations that occur in human GH-secreting adenomas are associated with increased PDE activity. They studied 10 adenomas with wildtype Gs-alpha (gsp-) and 8 with mutant Gs-alpha (gsp+) phenotypes. Although, in the absence of PDE inhibitors, intracellular cAMP levels were similar in gsp+ and gsp- adenomas, PDE blockade with 3-isobutyl-1-methylxanthine induced a marked increase in cAMP in all but 1 gsp+ adenoma (77 to 2900% increase) and a slight rise in only 2 gsp- adenomas. In vitro GH release was significantly higher in gsp+ than in gsp- adenomas (315 +/- 158 vs 82 +/- 53 micro g/well; P less than 0.01), and PDE blockade caused a further increase in 3 of 5 gsp+ adenomas but not in gsp- tumors. By direct measurement, PDE activity was about 7-fold higher in gsp+ than in gsp- adenomas (320 +/- 213 vs 48 +/- 23 pmol/min mg protein; P less than 0.05). The authors concluded that activating mutations of the Gs-alpha gene that naturally occur in pituitary adenomas are associated with an increased PDE activity that might, at least partially, counteract the constitutive activation of the cAMP-dependent pathway.

Shimon and Melmed (1997) noted that multiple molecular events can initiate pituitary adenomas. These events include early chromosomal mutations (11q13, 13q14 LOH) and possibly expression of pituitary-specific protooncogenes and/or growth factors including GSP, CREB (see CREB1; 123810 and CREB2; 123811), HST (see FGF4; 164980) and TGF alpha (see TGFA; 190170). Subsequent permissive factors allowing clonal expansion of the transformed cell include hypothalamic hormone receptor signals, paracrine growth factor signals, and disordered cell cycle regulation. Shimon and Melmed (1997) conclude that this cascade of events results in unrestrained hormone transcription and secretion, as well as cell proliferation, which are hallmarks of the pituitary adenoma. Since only monoclonality and GSP mutations have reproducibly been shown to be operative in sporadic human pituitary adenomas, the fundamental intrinsic defect leading to adenoma formation remains elusive.

To evaluate the relative contributions of past or present growth hormone hypersecretion and of hypertension to the cardiac abnormalities present in acromegaly, Lopez-Velasco et al. (1997) studied the serum GH and insulin-like growth factor I concentrations, systolic and diastolic blood pressures, and morphological and functional cardiac indexes. As measured by echocardiography-Doppler, 39 patients with active or cured acromegaly, 16 hypertensive controls, and 17 normotensive controls were screened. Hypertension was present in approximately 43% of patients with active acromegaly and in 28% of patients in whom acromegaly was cured. The authors concluded that asymptomatic cardiac abnormalities present in acromegalic patients are independently related to acromegaly and hypertension, suggesting the existence of a specific acromegalic myocardiopathy that might be aggravated by the coexistence of hypertension.

See also Koch (1949)

Abbassioun, K.; Fatourehchi, V.; Amirjamshidi, A.; Meibodi, N. A.: Familial acromegaly with pituitary adenoma: report of three affected siblings. J. Neurosurg. 64: 510-512, 1986.
Barlier, A.; Gunz, G.; Zamora, A. J.; Morange-Ramos, I.; Figarella-Branger, D.; Dufour, H.; Enjalbert, A.; Jaquet, P.: Pronostic (sic) and therapeutic consequences of Gs-alpha mutations in somatotroph adenomas. J. Clin. Endocr. Metab. 83: 1604-1610, 1998.
Jones, M. K.; Evans, P. J.; Jones, I. R.; Thomas, J. P.: Familial acromegaly. Clin. Endocr. 20: 355-358, 1984.
Koch, G.: Erbliche Hirngeschwuelste. Z. Menschl. Vererb. Konstitutionsl. 29: 400-423, 1949.
Koch, G.; Tiwisina, T.: Beitrag zur Erblichkeit der Akromegalie und der Hyperostosis generalisata mit Pachydermie. Aerztl. Forsch. 13: 489-504, 1959.
Lania, A.; Persani, L.; Ballare, E.; Mantovani, S.; Losa, M.; Spada, A.: Constitutively active Gs-alpha is associated with an increased phosphodiesterase activity in human growth hormone-secreting adenomas. J. Clin. Endocr. Metab. 83: 1624-1628, 1998.
Levin, S. R.; Hafeldt, F. D.; Becker, N.; Wilson, C. B.; Seymour, R.; Forsham, P. H.: Hypersomatotropism and acanthosis nigricans in two brothers. Arch. Intern. Med. 134: 365-367, 1974.

Lopez-Velasco, R.; Escobar-Morreale, H. F.; Vega, B.; Villa, E.; Sancho, J. M.; Moya-Mur, J. L.; Garcia-Robles, R.: Cardiac involvement in acromegaly: specific myocardiopathy or consequence of systemic hypertension? J. Clin. Endocr. Metab. 82: 1047-1053, 1997.

McCarthy, M. I.; Noonan, K.; Wass, J. A. H.; Monson, J. P.: Familial acromegaly: studies in three families. Clin. Endocr. 32: 719-728, 1990.

Pestell, R. G.; Alford, F. P.; Best, J. D.: Familial acromegaly. Acta Endocr. 121: 286-289, 1989.

Shimon, I.; Melmed, S.: Genetic basis of endocrine disease. Pituitary tumor pathogenesis. J. Clin. Endocr. Metab. 82: 1675-1681, 1997.

Thakker, R. V.; Pook, M. A.; Wooding, C.; Boscaro, M.; Scanarini, M.; Clayton, R. N.: Association of somatotrophinomas with loss of alleles on chromosome 11 and with gsp mutations. J. Clin. Invest. 91: 2815-2821, 1993.

PROLACTINOMA, FAMILIAL (600634)

JJM NOTE

With *MEN1* identified, probands should be examined for mutants within the gene.

Prolactin-secreting pituitary adenoma (prolactinoma) is the most common type of hormonally active pituitary adenoma. These tumors are a feature of multiple endocrine neoplasia, type I (MEN1; 131100). Berezin and Karasik (1995) studied 4 families in each of which more than one member was found to have prolactinoma. They concluded that there is a familial tendency to prolactinoma independent of its association with MEN-1.

Berezin, M.; Karasik, A.: Familial prolactinoma. Clin. Endocr. 42: 483-486, 1995.

AMENORRHEA-GALACTORRHEA SYNDROME (104600)

JJM NOTE

This disorder is included because of pituitary adenoma and the possibility that it overlaps with multiple endocrine neoplasia 1.

The association of secondary amenorrhea and galactorrhea is generally thought to occur in 2 distinct syndromes: the Forbes-Albright syndrome, in which amenorrhea and galactorrhea are accompanied by a pituitary tumor, with or without prior pregnancy, and the Chiari-Frommel syndrome, in which amenorrhea and galactorrhea commence after pregnancy, without associated pituitary tumor. This distinction may be artificial (Rimoin and Schimke, 1971), because the pituitary adenoma may be too small to identify clinically and progression from the benign to the neoplastic syndrome has been documented (Young et al., 1967). Linquette et al. (1967) described mother and daughter with amenorrhea-galactorrhea associated with pituitary adenoma. The mother first developed clinical signs after a pregnancy, whereas the daughter was never pregnant and amenorrhea followed emotional trauma. The sella turcica was enlarged in both and tumor was confirmed by craniotomy. The tumors resembled chromophobe adenomas, but there was fine eosinophilic granulation on tetrachrome staining, as seen in prolactin cells. Since the amenorrhea-galactorrhea syndrome has been described as a part of a multiple endocrine adenomatosis syndrome, it is not certain that the ailment in the mother and daughter reported by Linquette et al. (1967) represented a distinct entity.

Linquette, M.; Herlant, M.; Laine, E.; Fossati, P.; Dupont-Lecompte, M.: Adenome prolactive chez une jeune fille dont la mere etait porteuse d'un adenome hypophysaire avec amenorrhee-galactorrhee. Ann. Endocr. 28: 773-780, 1967.

Rimoin, D. L.; Schimke, R. N.: Genetic Disorders of the Endocrine Glands. St. Louis: C. V. Mosby (pub.) 1971.

Young, R. L.; Bradley, E. M.; Goldzieher, J. W.; Myers, P. W.; Lecocq, F. R.: Spectrum of nonpuerperal galactorrhea: report of two cases evolving through the various syndromes. J. Clin. Endocr. 27: 461-466, 1967.

PITUITARY DWARFISM WITH LARGE SELLA TURCICA (262710)

JJM NOTE

It would be interesting to have a follow-up of this family, given advances in imaging and genetics. Could it, too, be a variant of multiple endocrine neoplasia 1?

Parks et al. (1978) described this combination in 2 sisters and a brother, the only children of nonconsanguineous parents. Basal thyrotropin levels were low despite hypothyroidism, and increased little after injection of thyrotropin-releasing hormone. Stimulated growth hormone levels were less than 5 nanograms per milliliter. Treatment with both thyroxine and growth hormone was necessary for rapid growth. The findings were judged compatible with either familial neoplasia of the anterior pituitary or a regulatory defect promoting hyperplasia and inhibiting hormone release.

Parks, J. S.; Tenore, A.; Bongiovanni, A. M.; Kirkland, R. F.: Familial hypopituitarism with large sella turcica. New Eng. J. Med. 298: 698-702, 1978.

CUSHING DISEASE, PITUITARY (219090)

JJM NOTE

Follow-up of these families with clinical and *MEN1* gene testing needs to be done.

Swinburn and Holdaway (1988) described 2 teenaged sisters with pituitary-based Cushing syndrome. In 1 patient, the disorder was cured by pituitary radiation; in the other, biopsy showed pituitary adenoma. The familial association of pituitary-based Cushing disease had seemingly not been reported previously, although familial Cushing syndrome has been described secondary to nodular adrenocortical hyperplasia (see 160980). Swinburn and Holdaway (1988) granted that 'despite the negative family history, these patients could still represent the first generation of a MEN I family.' See 131100. Maton et al. (1986) suggested that the Cushing syndrome is more common in patients with the Zollinger-Ellison syndrome and MEN I than previously reported. They demonstrated Cushing syndrome due to pituitary overproduction of ACTH in 3 of 16 patients with MEN I. Salti and Mufarrij (1981) described Cushing disease in a 28-year-old woman and her niece, aged 25 years. The parents of the niece were first cousins; thus, the aunt and the niece were related to each other also as first cousins once removed. Endocrine studies established pituitary ACTH excess. The sella was normal-sized in the aunt and enlarged in the niece; both received external pituitary irradiation which in the aunt had produced clinical and hormonal remission for 9 years. Gardner et al. (1989) reported the cases of black sisters, aged 38 and 40 years, with Cushing disease due to ACTH-immunostaining pituitary adenoma demonstrated at transsphenoidal pituitary surgery. Gardner et al. (1989) considered it unlikely that these sisters had MEN I, because usually hyperparathyroidism is present at the time of diagnosis of MEN I. Serum calcium levels were normal on repeated occasions in these patients. Furthermore, there was no clinical evidence of pancreatic islet cell tumor in either patient.

Gardner, D. F.; Barlascini, C. O., Jr.; Downs, R. W., Jr.; Sahni, K. S.: Case report: Cushing's disease in two sisters. Am. J. Med. Sci. 297: 387-389, 1989.

Maton, P. N.; Gardner, J. D.; Jensen, R. T.: Cushing's syndrome in patients with the Zollinger-Ellison syndrome. New Eng. J. Med. 315: 1-5, 1986.

Salti, I. S.; Mufarrij, I. S.: Familial Cushing disease. Am. J. Med. Genet. 8: 91-94, 1981.

Swinburn, B. A.; Holdaway, I. M.: Familial Cushing's disease. Aust. New Zeal. J. Med. 18: 169-171, 1988.

Parathyroid Tumors

HYPERPARATHYROIDISM, FAMILIAL PRIMARY (#145000)

JJM NOTE

With any patient who has hyperparathyroidism, it is crucial to rule out familial hypocalciuric hypercalcemia before operation. Then, the possibility of a familial pattern should be addressed.

A number sign (#) is used with this entry because of evidence that familial isolated hyperparathyroidism is probably caused by the same gene that causes the hyperparathyroidism-jaw tumor syndrome (HRPT2; 145001).

Familial hyperparathyroidism is usually part of endocrine adenomatosis (131100). Primary chief cell hyperplasia is the usual histologic change in the hyperparathyroidism of that condition and also in the type which occurs in families without evidence of other endocrine disease. This suggests that discovery of this histologic change should prompt study for other endocrine disease in the patient and for other cases in the family. Cases of apparently isolated familial hyperparathyroidism have been reported by Cameron et al. (1966), Cutler et al. (1964), and Peters et al. (1966), among others. The pedigrees have usually been consistent with autosomal dominant inheritance. Families with multiple cases of parathyroid adenoma have been described by Cassidy and Anderson (1960), Jackson et al. (1960) and others. Such families may have multiple endocrine adenomatosis, although the possibility of a separate and distinct dominantly inherited entity cannot be excluded (Cutler et al., 1964). Cutler et al. (1964) among others have emphasized that chief cell hyperplasia rather than adenoma may be the characteristic histologic change in the familial cases of hyperparathyroidism. The two types of changes are often distinguished with difficulty. Jackson and Boonstra (1967) studied 8 families, each with multiple cases (55 in all) of parathyroid adenoma. In at least 3 kindreds some individuals had other endocrine adenomata. Marx et al. (1977) concluded that primary parathyroid hyperplasia is separate from MEA I (131100). In the former disorder, close to half the offspring of affected persons show hypercalcemia in the first 2 decades, nephrolithiasis and peptic ulcer disease are unusual, and concentrations of peptic hormones other than parathyroid hormone are normal. The distinctness of 'pure' hyperparathyroidism is supported by the findings of Goldsmith et al. (1976) in an extensively affected kindred. Boey et al. (1975) found associated endocrine disease indicat ive of multiple endocrine adenomatosis in 21 of 119 cases of primary hyperparathyroidism. The clinical pattern of hypercalcemia was the same in the 21 as in the others, but MEA was found more often in patients with several diseased parathyroid glands. Most hyperparathyroidism is nonfamilial. The frequency of familial primary hyperparathyroidism was estimated to be 0.14 per thousand by Jackson and Boonstra (1967) and 0.13 per thousand by Christensson (1976). Thymic and nonthymic carcinoid tumors have been reported with MEA I and with hereditary hyperparathyroidism (Lokich and Li, 1978). In addition to MEA I, familial hypocalciuric hypercalcemia (145980) can be confused with familial primary hyperparathyroidism. A severe neonatal form of hyperparathyroidism, reported in a single informative family (see 239200), is probably autosomal recessive. Law et al. (1983) reported a family with 2 affected sisters and an affected brother with 10 other unaffected sibs and with normal parents and 2 offspring of one of the affected sisters and many unaffected paternal and maternal aunts and uncles. The parents were unrelated. Autosomal recessive inheritance, proposed by the authors for this apparently unique family, is much less likely in my opinion than gonadal mosaicism of one of the parents. The 3 affected persons had adenomatous hyperparathyroidism which became symptomatic in their twenties. Arnold et al. (1988) presented evidence for the clonal origin of parathyroid adenomas. They found tumor-cell-specific restriction fragment-length alterations involving the parathyroid hormone gene (168450) in 2 parathyroid adenomas. Furthermore, using a RFLP related to HPRT (308000), they found that of 8 adenomas in women, 6 had the DNA hybridization pattern indicative of monoclonality, and the other 2 had an equivocal pattern. None of 5 hyperplastic parathyroid glands showed a monoclonal pattern. Familial primary hyperparathyroidism with multiple ossifyin g jaw fibromas is a genetically distinct disorder; see 145001.

To establish whether familial isolated hyperparathyroidism is a distinct disease entity or a variant of one of the multiple endocrine neoplasia syndromes, Wassif et al. (1993) tested 19 members of a large family in which the disease had been transmitted through 4 generations in an autosomal dominant fashion. Location of the causative mutation on chromosome 10 in the vicinity of the MEN1 locus (131100) or on chromosome 11 in the vicinity of the MEN2A locus (171400) was excluded by linkage study, as was also linkage to the parathyroid hormone gene (PTH; 168450). In 1 member of the family, a parathyroid carcinoma was found after postsurgical recurrence of hypercalcemia. Because parathyroid carcinoma has been reported in other families with primary hyperparathyroidism (Dinnen et al., 1977; Frayha et al., 1972; Leborgne et al., 1975; Streeten et al., 1992), malignant hyperparathyroid disease is clearly a risk in these families.

Contrary results were reported by Kassem et al. (1994), who suggested that familial primary hyperparathyroidism can exist as a separate entity, judging from findings in 1 family, and may be caused by mutation in a gene located in the MEN1 region on 11q13, possibly the MEN1 locus. They studied a large family in which 7 patients had been previously operated upon for PHPT. Through family studies 4 new patients with PHPT and 2 with probable PHPT were found. No clinical or biochemical evidence of MEN syndromes could be detected. However, DNA marker D11S97, located at 11q13, gave a maximum 2-point lod score of 2.12 at a recombination fraction of 0.05.

Stefenelli et al. (1997) found a high incidence of left ventricular hypertrophy (82%) and aortic and/or mitral valve calcifications (46 and 39%, respectively) in patients with primary hyperparathyroidism. Follow-up at 1 year and at 41 months after successful parathyroidectomy showed regression of hypertrophy. While no comments were made regarding inclusion of familial cases in their study, their findings indicate that parathyroid hormone per se may play an important role in myocardial hypertrophy and that these cardiac complications may also occur in familial cases.

In connection with a study of germline mutations in the MEN1 gene in familial multiple endocrine neoplasia type I, Agarwal et al. (1997) studied the DNA of the probands from 5 families with familial hyperparathyroidism. No MEN1 germline mutation was found in any of them, suggesting that familial hyperparathyroidism is often caused by mutation in another gene or genes.

Tanaka et al. (1998) studied germline mutations of the MEN1 gene in 5 cases of familial and 4 cases of sporadic multiple endocrine neoplasia type I, 6 cases in 3 independent pedigrees of familial pituitary adenoma without MEN1, and 3 cases of familial primary hyperparathyroidism in Japanese patients. Eight different types of germline mutations in all 9 MEN1 cases were distributed in exons 2, 3, 7, and 10 and intron 7 of the MEN1 gene. Loss of heterozygosity (LOH) on 11q13 was detected in all 9 tumors of the cases with microsatellite analysis. No germline mutation of the MEN1 gene was detected in 3 pedigrees of familial pituitary adenoma and 3 cases of familial primary hyperparathyroidism. LOH on 11q13 was detected in 2 cases in 1 pedigree of familial pituitary adenoma, and 1 of them showed a heterozygous somatic mutation of the MEN1 gene. No LOH on 11q13 was detected in 3 cases of familial primary hyperparathyroidism. The authors concluded that the loss of function of MEN1 causes familial or sporadic MEN1, but does not cause familial primary hyperparathyroidism or most familial pituitary adenoma without MEN1.

Teh et al. (1998) reported 3 families with familial primary hyperparathyroidism and evaluated their clinical, pathologic, and genetic profiles. In 2 families with a total of 10 affected cases and 3 female obligate carriers, there was no evidence of jaw or renal lesions despite careful radiologic investigations. In both families, the disease was linked to the 1q21-q32 region with lod scores of 3.10 and 3.43 for markers D1S222 and D1S249, respectively, at recombination fractions of 0. In 1 family, 2 types of parathyroid pathology were found: 3 of chief cell type and 1 of oxyphil/oncocytic cell type. Two chief cell tumors and 1 oxyphil tumor were found to have LOH involving loss of the wildtype alleles for chromosome 1q markers. In the third family, with 4 affected sibs, a parathyroid carcinoma and 2 cases of polycystic kidney disease were found. The parathyroid carcinoma also showed LOH in the 1q region. Teh et al. (1998) interpreted these findings as indicating that a subset of familial isolated hyperparathyroidism families represent a variant of the hyperparathyroidism-jaw tumor syndrome and that the gene involved is a tumor suppressor gene.

See also Graber and Jacobs (1968); Grevsten et al. (1974); Mallette et al. (1974); Sandler and Moncrieff (1980)

Agarwal, S. K.; Kester, M. B.; Debelenko, L. V.; Heppner, C.; Emmert-Buck, M. R.; Skarulis, M. C.; Doppman, J. L.; Kim, Y. S.; Lubensky, I. A.; Zhuang, Z.; Green, J. S.; Guru, S. C.; Manickam, P.; Olufemi, S.-E.; Liotta, L. A.; Chandrasekharappa, S. C.; Collins, F. S.; Spiegel, A. M.; Burns, A. L.; Marx, S. J.: Germline mutations of the MEN1 gene in familial multiple endocrine neoplasia type 1 and related states. Hum. Molec. Genet. 6: 1169-1175, 1997.

Arnold, A.; Staunton, C. E.; Kim, H. G.; Gaz, R. D.; Kronenberg, H. M.: Monoclonality and abnormal parathyroid hormone genes in parathyroid adenomas. New Eng. J. Med. 318: 658-662, 1988.

Boey, J. H.; Cooke, T. J. C.; Gilbert, G. M.; Sweeney, E. C.; Taylor, S.: Occurrence of other endocrine tumours in primary hyperparathyroidism. Lancet II: 781-784, 1975.

Cameron, K. M.; Ogg, C. S.; Harrison, A. R.: Familial hyperparathyroidism. Lancet II: 1006-1007, 1966.

Cassidy, C. E.; Anderson, A. S.: A familial occurrence of hyperparathyroidism caused by multiple parathyroid adenomas. Metabolism 9: 1152-1158, 1960.

Christensson, T.: Familial hyperparathyroidism. (Letter) Ann. Intern. Med. 85: 614-615, 1976.

Cutler, R. E.; Reiss, E.; Ackerman, L. V.: Familial hyperparathyroidism: a kindred involving eleven cases, with a discussion of primary chief-cell hyperplasia. New Eng. J. Med. 270: 859-865, 1964.

Dinnen, J. S.; Greenwood, R. H.; Jones, J. H.; Walker, D. A.; Williams, E. D.: Parathyroid carcinoma in familial hyperparathyroidism. J. Clin. Path. 30: 966-975, 1977.

Frayha, R. A.; Nassar, V. H.; Dagher, F.; Salti, I. S.: Familial parathyroid carcinoma. Leban. Med. J. 25: 299-309, 1972.

Goldsmith, R. E.; Sizemore, G. W.; Chen, I.-W.; Zalme, E.; Altemeier, W. A.: Familial hyperparathyroidism: description of a large kindred with physiologic observations and a review of the literature. Ann. Intern. Med. 84: 36-43, 1976.

Graber, A. L.; Jacobs, K.: Familial hyperparathyroidism. Medical and surgical considerations. J.A.M.A. 204: 542-544, 1968.

Grevsten, S.; Grimelius, L.; Thoren, L.: Familial hyperparathyroidism. Upsala J. Med. Sci. 79: 109-115, 1974.

Jackson, C. E.; Boonstra, C. E.: The relationship of hereditary hyperparathyroidism to endocrine adenomatosis. Am. J. Med. 43: 727-734, 1967.

Jackson, C. E.; Talbert, P. C.; Taylor, H. D.: Hereditary hyperparathyroidism. J. Indiana Med. Assoc. 53: 1313-1316, 1960.

Kassem, M.; Zhang, X.; Brask, S.; Eriksen, E. F.; Mosekilde, L.; Kruse, T. A.: Familial isolated primary hyperparathyroidism. Clin. Endocr. 41: 415-420, 1994.

Law, W. M., Jr.; Hodgson, S. F.; Heath, H., III: Autosomal recessive inheritance of familial hyperparathyroidism. New Eng. J. Med. 309: 650-653, 1983.

Leborgne, J.; Le Neel, J.-C.; Buzelin, F.; Malvy, P.: Cancer familial des parathyroides: interet de l'angiographie dans le diagnostic des recidives loco-regionales—considerations a propos de deux cas. J. Chir. (Paris) 109: 315-326, 1975.
Lokich, J. J.; Li, F. P.: Carcinoid of the thymus with hereditary hyperparathyroidism. Ann. Intern. Med. 89: 364-365, 1978.
Mallette, L. E.; Bilezikian, J. P.; Ketcham, A. S.; Aurbach, G. D.: Parathyroid carcinoma in familial hyperparathyroidism. Am. J. Med. 57: 642-648, 1974.
Marx, S. J.; Spiegel, A. M.; Brown, E. M.; Aurbach, G. D.: Family studies in patients with primary parathyroid hyperplasia. Am. J. Med. 62: 698-706, 1977.
Peters, N.; Chalmers, T. M.; Rack, J. H.; Truscott, B. M.; Adams, P. H.: Familial hyperparathyroidism. Postgrad. Med. J. 42: 228-233, 1966.
Sandler, L. M.; Moncrieff, M. W.: Familial hyperparathyroidism. Arch. Dis. Child. 55: 146-157, 1980.
Stefenelli, T.; Abela, C.; Frank, H.; Koller-Strametz, J.; Globits, S.; Bergler-Klein, J.; Niederle, B.: Cardiac abnormalities in patients with primary hyperparathyroidism: implications for follow-up. J. Clin. Endocr. Metab. 82: 106-112, 1997.
Streeten, E. A.; Weinstein, L. S.; Norton, J. A.; Mulvihill, J. J.; White, B. J.; Friedman, E.; Jaffe, G.; Brandi, M. L.; Stewart, K.; Zimering, M. B.; Spiegel, A. M.; Aurbach, G. D.; Marx, S. J.: Studies in a kindred with parathyroid carcinoma. J. Clin. Endocr. Metab. 75: 362-366, 1992.
Tanaka, C.; Yoshimoto, K.; Yamada, S.; Nishioka, H.; Ii, S.; Moritani, M.; Yamaoka, T.; Itakura, M.: Absence of germ-line mutations of the multiple endocrine neoplasia type 1 (MEN1) gene in familial pituitary adenoma in contrast to MEN1 in Japanese. J. Clin. Endocr. Metab. 83: 960-965, 1998.
Teh, B. T.; Farnebo, F.; Twigg, S.; Hoog, A.; Kytola, S.; Korpi-Hyovalti, E.; Wong, F. K.; Nordenstrom, J.; Grimelius, L.; Sandelin, K.; Robinson, B.; Farnebo, L.-O.; Larsson, C.: Familial isolated hyperparathyroidism maps to the hyperparathyroidism-jaw tumor locus in 1q21-q32 in a subset of families. J. Clin. Endocr. Metab. 83: 2114-2120, 1998.
Wassif, W. S.; Moniz, C. F.; Friedman, E.; Wong, S.; Weber, G.; Nordenskjold, M.; Peters, T. J.; Larsson, C.: Familial isolated hyperparathyroidism: a distinct genetic entity with an increased risk of parathyroid cancer. J. Clin. Endocr. Metab. 77: 1485-1489, 1993.

PARATHYROID ADENOMATOSIS 1 (*168461)

JJM NOTE

The principal association of somatic mutations in this gene is with parathyroid adenomas, but glioblastoma and breast cancer tissues have abnormalities of it as well.

Arnold et al. (1989) found 2 parathyroid adenomas bearing clonal restriction fragment abnormalities involving the parathyroid hormone locus (168450) and showed in one of these tumors that a DNA rearrangement occurred at the PTH locus; that the rearrangement separated the PTH gene's 5-prime flanking region from its coding exons, perhaps placing a newly adjacent gene under the influence of PTH regulatory elements; that the DNA that combined with PTH normally maps to 11q13, the known chromosomal localization of several oncogenes (e.g., 164980; 164950; 151400) and the gene for multiple endocrine neoplasia type I (131100); and that the rearrangement was a reciprocal, conservative recombination of the locus on 11q13 (initially designated D11S287) with PTH (on 11p15). Gailani et al. (1991) performed pulsed field gel electrophoresis (PFGE) analysis of the proximal region of 11q. The maximal distance between BCL1 and INT2 was less than or equal to 160 kb. BCL1 and PRAD1 were less than or equal to 48 kb from each other. Richard et al. (1991) described a high resolution radiation hybrid map of the 11q12-q13 region.

PRAD1 encodes a cyclin. Inaba et al. (1992) used murine cDNA clones for 3 cyclin D genes that are normally expressed during the G1 phase of the cell cycle to clone the cognate human genes. They assigned the CCND1 gene to 11q13 by combination of somatic cell hybrid studies and fluorescence in situ hybridization.

From a human glioblastoma cell line, Xiong et al. (1991) identified a gene capable of 'rescuing' a mitosis-defective strain of budding yeast. Transcription of cyclin D1 gave rise to 2 major transcripts through alternative polyadenylation. The transcript and its 34-kD product were both abundant in the glioblastoma cell line of origin.

PRAD1 is dramatically overexpressed in a subset of parathyroid adenomas. Rosenberg et al. (1991) found that in some parathyroid adenomas a somatic mutation places the candidate gene PRAD1 under the promoter of the parathyroid hormone gene (168450). PRAD1 may also be important in the pathogenesis of a subset of breast and squamous cell cancers that contain DNA amplification in 11q13. Motokura et al. (1991) found that the cloned placental PRAD1 gene encodes a protein of 295 amino acids with sequence similarities to the cyclins. Cyclins can form a complex with and activate p34(cdc2) protein kinase, thereby regulating progress through the cell cycle. PRAD1 mRNA levels were found to vary dramatically across the cell cycle in HeLa cells.

Rosenberg et al. (1991) speculated that PRAD1 was probably the BCL1 oncogene (151400): in addition to being no more than 130 kb apart from BCL1, as demonstrated by PFGE, PRAD1 mRNA was abundantly expressed in 7 of 7 centrocytic lymphomas (Kiel classification), in contrast to 13 closely related but noncentrocytic lymphomas. In 3 of the 7 centrocytic lymphomas, BCL1 DNA rearrangement was detectable. Also, 2 unusual cases of chronic lymphocytic leukemia (CLL) with BCL1 rearrangement overexpressed PRAD1, in contrast to 5 CLL controls. Komatsu et al. (1994) presented findings in a variant chromosomal translocation at 11q13 which strongly supported the conclusion that the BCL1 gene is the same as PRAD1. The patient they studied showed juxtaposition of the lambda immunoglobulin light chain gene (147220) to the PRAD1 gene at its 3-prime end, resulting in overexpression of PRAD1 mRNA.

Hinds et al. (1994) presented evidence that the cyclin D1 gene can function as an oncogene. In cultured cells, a cDNA clone of the CCND1 gene contributed to cell transformation by complementing a defective adenovirus E1A oncogene. The product of the retinoblastoma susceptibility gene (180200) acts as a tumor suppressor and loss of its function is involved in the development of various types of cancer. DNA tumor viruses are supposed to disturb the normal regulation of the cell cycle by inactivating the RB gene product. Muller et al. (1994) demonstrated that the cell cycle-dependent expression of cyclin D1 is dependent on the presence of a functional RB protein. RB-deficient tumor cell lines, as well as cells expressing viral oncoproteins, had low or barely detectable levels of cyclin D1. Expression of cyclin D1, but not of cyclins A and E, was induced by transfection of the RB gene into RB-deficient tumor cells.

Dysregulation of oncogenes by translocation to the IgH locus on 14q32 is a seminal event in the pathogenesis of B-cell tumors. In multiple myeloma (254500), translocation to the IgH locus has been reported at an incidence of 20 to 60%. For most translocations, the partner chromosome is unknown; for the others, a diverse array of chromosomal partners have been demonstrated, with 11q13 (where cyclin D1 is located) being the only chromosome that is frequently involved (Chesi et al., 1996; Bergsagel et al., 1996; Chesi et al., 1997).

Animal Model

Wang et al. (1994) produced transgenic mice containing the MMTV-LTR (mouse mammary tumor virus long terminal repeat) as a promoter sequence upstream of cyclin D1. Overexpression of the transgene resulted in abnormal proliferation of mammary cells and the development of mammary adenocarcinomas. The authors proposed that reports of amplification of DNA from 11q13, the location of human cyclin D1, in 15-20% of breast cancers, may be accounted for by this gene.

Sicinski et al. (1995) used gene targeting in ES cells to generate mice lacking cyclin D1. Cyclin D1-null mutant mice developed to term but showed reduced body size, reduced viability, and symptoms of neurologic impairment. Their retinas displayed a striking reduction in cell number due to proliferative failure during embryonic development. In situ hybridization studies of normal mouse embryos demonstrated an extremely high level of cyclin D1 in the retina, suggesting to the authors a special dependence of this tissue on cyclin D1. In adult mutant females, the breast epithelial compartment failed to undergo the massive proliferative changes associated with pregnancy despite normal levels of ovarian steroid hormones. Thus, the authors speculated that steroid-induced proliferation of mammary epithelium during pregnancy may be driven through cyclin D1.

Ma et al. (1998) studied cyclin D1-deficient mice, which have small eyes with thin retinas, and observed that there was a lower level of retinal cell proliferation and a unique pattern of photoreceptor cell death. Death was first observed in scattered clusters of cells in the retina. It then appeared to spread from these few cells to nearby photoreceptors, eventually producing extensive holes in the photoreceptor layer. These holes appeared to be filled with interneurons from the inner nuclear layer. The death occurred mainly during the second to fourth postnatal weeks. Other models of photoreceptor degeneration in rodents differed in that they occur more uniformly across the retina, with death proceeding over a longer period of time until all, or nearly all, of the photoreceptors degenerate. Ma et al. (1998) found that expression of a bcl2 transgene could not prevent the death.

See also Richard et al. (1991)

Arnold, A.; Kim, H. G.; Gaz, R. D.; Eddy, R. L.; Fukushima, Y.; Byers, M. G.; Shows, T. B.; Kronenberg, H. M.: Molecular cloning and chromosomal mapping of DNA rearranged with the parathyroid hormone gene in a parathyroid adenoma. J. Clin. Invest. 83: 2034-2040, 1989.
Bergsagel, P. L.; Chesi, M.; Nardini, E.; Brents, L. A.; Kirby, S. L.; Kuehl, W. M.: Promiscuous translocations into immunoglobulin heavy chain switch regions in multiple myeloma. Proc. Nat. Acad. Sci. 93: 13931-13936, 1996.
Chesi, M.; Bergsagel, P. L.; Brents, L. A.; Smith, C. M.; Gerhard, D. S.; Kuehl, W. M.: Dysregulation of cyclin D1 by translocation into an IgH gamma switch region in two multiple myeloma cell lines. Blood 88: 674-681, 1996.
Chesi, M.; Nardini, E.; Brents, L. A.; Schrock, E.; Ried, T.; Kuehl, W. M.; Bergsagel, P. L.: Frequent translocation t(4;14)(p16.3;q32.3) in multiple mye-

loma is associated with increased expression and activating mutations of fibroblast growth factor receptor 3. Nature Genet. 16: 260-264, 1997.
Gailani, M. R.; Petty, E. M.; Horsthemke, B.; Arnold, A.; Marx, S. J.; Bale, A. E.: Physical mapping of chromosome 11q12-13 by pulsed field gel electrophoresis (PFGE). (Abstract) Cytogenet. Cell Genet. 58: 1959, 1991.
Hinds, P. W.; Dowdy, S. F.; Eaton, E. N.; Arnold, A.; Weinberg, R. A.: Function of a human cyclin gene as an oncogene. Proc. Nat. Acad. Sci. 91: 709-713, 1994.
Inaba, T.; Matsushime, H.; Valentine, M.; Roussel, M. F.; Sherr, C. J.; Look, A. T.: Genomic organization, chromosomal localization, and independent expression of human cyclin D genes. Genomics 13: 565-574, 1992.
Komatsu, H.; Iida, S.; Yamamoto, K.; Mikuni, C.; Nitta, M.; Takahashi, T.; Ueda, R.; Seto, M.: A variant chromosome translocation at 11q13 identifying PRAD1/cyclin D1 as the BCL-1 gene . Blood 84: 1226-1231, 1994.
Ma, C.; Papermaster, D.; Cepko, C. L.: A unique pattern of photoreceptor degeneration in cyclin D1 mutant mice. Proc. Nat. Acad. Sci. 95: 9938-9943, 1998.
Motokura, T.; Bloom, T.; Kim, H. G.; Juppner, H.; Ruderman, J. V.; Kronenberg, H. M.; Arnold, A.: A novel cyclin encoded by a bcl1-linked candidate oncogene. Nature 350: 512-515, 1991.
Muller, H.; Lukas, J.; Schneider, A.; Warthoe, P.; Bartek, J.; Eilers, M.; Strauss, M.: Cyclin D1 expression is regulated by the retinoblastoma protein. Proc. Nat. Acad. Sci. 91: 2945-2949, 1994.
Richard, C. W.; Withers, D. A.; Meeker, T. C.; Myers, R. M.: A radiation hybrid map of the proximal long arm of human chromosome 11 containing the MEN-1 and bcl-1 disease locus. (Abstract) Cytogenet. Cell Genet. 58: 1970, 1991.
Richard, C. W., III; Withers, D. A.; Meeker, T. C.; Maurer, S.; Evans, G. A.; Myers, R. M.; Cox, D. R.: A radiation hybrid map of the proximal long arm of human chromosome 11 containing the multiple endocrine neoplasia type I (MEN-1) and bcl-1 disease loci. Am. J. Hum. Genet. 49: 1189-1196, 1991.
Rosenberg, C. L.; Kim, H. G.; Shows, T. B.; Kronenberg, H. M.; Arnold, A.: Rearrangement and overexpression of D11S287E, a candidate oncogene on chromosome 11q13 in benign parathyroid tumors. Oncogene 6: 449-453, 1991.
Rosenberg, C. L.; Wong, E.; Petty, E. M.; Bale, A. E.; Tsujimoto, Y.; Harris, N. L.; Arnold, A.: PRAD1, a candidate BCL1 oncogene: mapping and expression in centrocytic lymphoma. Proc. Nat. Acad. Sci. 88: 9638-9642, 1991.
Sicinski, P.; Donaher, J. L.; Parker, S. B.; Li, T.; Fazeli, A.; Gardner, H.; Haslam, S. Z.; Bronson, R. T.; Elledge, S. J.; Weinberg, R. A.: Cyclin D1 provides a link between development and oncogenesis in the retina and breast. Cell 82: 621-630, 1995.
Wang, T. C.; Cardiff, R. D.; Zukerberg, L.; Lees, E.; Arnold, A.; Schmidt, E. V.: Mammary hyperplasia and carcinoma in MMTV-cyclin D1 transgenic mice. Nature 369: 669-671, 1994.
Xiong, Y.; Connelly, T.; Futcher, B.; Beach, D.: Human D-type cyclin. Cell 65: 691-699, 1991.

HYPERPARATHYROIDISM, FAMILIAL PRIMARY, WITH JAW FIBROMAS (*145001)

JJM NOTE

Finally, confirmatory reports of this rare syndrome have appeared (e.g., Jackson et al. 1990).

Although the most common familial form of primary hyperparathyroidism is parathyroid hyperplasia (HRPT1; 145000), a few families have manifested parathyroid adenomas. Mallette et al. (1987) described a family in which 4 members developed cystic parathyroid adenomas. Although calcium levels returned to normal after resection of the adenoma, a second adenoma often developed several years later; thus, Mallette et al. (1987) termed the condition adenomatosis. Each adenoma had a cystic histologic appearance, and 3 of the 4 normal-sized parathyroid glands also contained many cysts. No other endocrine tumors developed, but in 3 of the patients the hyperparathyroidism was complicated by fibrous maxillary or mandibular tumors that resembled ossifying fibromas rather than the brown tumors generally found in patients with hyperparathyroidism. Each patient with an adenoma was hypercalciuric, but 2 were considered obligate carriers of hypocalciuric hypercalcemia (145980). The adenomatosis occurred in a father and 3 sons. A brother and sister of the father had hypercalciuric hypercalcemia, as did the son of 1 of the 3 'sons.'

Jackson (1958) reported a family with hereditary hyperparathyroidism in which 4 of 5 of the affected members of the first generation had jaw tumors. Because 3 affected members of the third generation developed similar jaw tumors which progressed even after surgical correction of their hypercalcemia, the family was reinvestigated. The maxillary and mandibular tumors could be differentiated from the 'brown tumors' of hyperparathyroidism; they were histologically distinct fibroosseous lesions without giant cells. As far as is known, ossified fibrous tumors do not occur in areas other than the jaw, such as the knees or ribs which can be the site of 'brown tumors.' Parathyroid enlargement was mostly uniglandular with multiple tumors found only occasionally. Linkage studies in this and a second family with the same association demonstrated linkage to neither chromosome 11 markers (the site of the gene for MEN1; see 131100) nor markers on chromosome 10 (the site of the gene for MEN2; see 171400). (According to Jackson (1994), linkage studies in 5 families exclude the locus for this disorder, symbolized HRPT2, from the sites of MEN1 and MEN2.) Thus, hereditary hyperparathyroidism with multiple ossifying jaw fibromas may be a genetically distinct disorder. The family of Mallette et al. (1987) was included in the study of Jackson et al. (1990). Other families were reported by Kennett and Pollick (1971), Rosen and Palmer (1981), and Warnakulasuriya et al. (1985). Inoue et al. (1995) reported a 53-year-old Japanese woman diagnosed as having primary hyperparathyroidism caused by hyperplasia of the parathyroid glands and causing renal stones and hypercalcemia. One year after she underwent total parathyroidectomy and implantation of parathyroid tissue, she underwent surgery for a cementifying fibroma. A 19-year-old nephew was found to have elevated serum calcium levels and levels of serum parathyroid hormone and a parathyroid adenoma was removed at surgery.

Szabo et al. (1995) performed genetic linkage studies in 5 families containing a total of 20 individuals with the hereditary hyperparathyroidism-jaw tumor syndrome (HPT-JT). They mapped the HRPT2 locus to 1q21-q31 and found a maximum lod score of 6.10 at theta = 0.0 with marker D1S212.

Szabo et al. (1995) found no instance of parathyroid carcinoma in their families; 1 case of this malignancy had been reported by Dinnen et al. (1977). The occurrence of Wilms tumor in 2 female members of unrelated families in their study raised the possibility that Wilms tumor may be a component of the HPT-JT syndrome. Further evidence that parathyroid carcinoma and Wilms tumor are part of the HPT-JT syndrome came from a report by Kakinuma et al. (1994) in which one sib had parathyroid carcinoma, a second had parathyroid adenoma plus Wilms tumor, and a third had parathyroid adenoma plus jaw tumor. Szabo et al. (1995) studied 6 hereditary Wilms tumor families, including 29 affected members, and found no linkage to 1q markers closely linked with HRPT2. Furthermore, 9 parathyroid adenomas and one Wilms tumor from 9 members of 3 HPT-JT families showed no loss of heterozygosity at the HRPT2-linked loci.

Teh et al. (1996) reported 2 families with HPT-JT syndrome in which adult renal hamartomas or cystic kidney disease were prominent associated features, possibly representing a new phenotypic variant of the HPT-JT syndrome. In the first family, renal lesions were present in 5 of 6 affected individuals, whereas HPT and jaw tumors (JT) were seen in 4 and 2 cases, respectively. In the second family, JT was found in 3 of the 5 affected individuals, and 2 affected members also exhibited polycystic kidney disease. The possibility of the latter cosegregating as a separate autosomal dominant gene can not be ruled out. A sex-dependent penetrance of primary HPT, resulting in predominantly male-affected cases was evident in the 2 families. The disease in these 2 kindreds was linked to 5 markers in the 1q21-q32 region (lod scores: 3.2-4.2), whereas linkage to the MEN1 and MEN2 regions was excluded. Meiotic recombinations detected in affected individuals placed the locus telomeric of D1S215, thus narrowing the HRPT2 region from over 60 to approximately 34 cM. Loss of heterozygosity (LOH) was studied in 7 renal hamartomas from 2 affected individuals in the first family, as well as in a jaw tumor and a parathyroid tumor from the second family. All renal hamartomas showed LOH in the 1q21-q32 region. All losses involved the wildtype allele derived from the unaffected parent, suggesting the inactivation of a tumor suppressor gene in this region.

Pidwirny et al. (1995) found that the proband in the Canadian family reported by Kennett and Pollick (1971) had died of parathyroid carcinoma, that the family was a branch of a kindred reported by Rosen and Palmer (1981), and that a member of the second branch had also died of parathyroid cancer. This and other experiences established parathyroid cancer as part of the hyperparathyroidism-jaw tumor syndrome occurring in at least 1 patient in 5 (42%) of the 12 known families.

Fujikawa et al. (1998) described 2 sisters and a brother, young adults, with hyperparathyroidism due to multiple parathyroid adenomas without evidence of other endocrinologic abnormalities. A 22-year-old woman had 2 parathyroid adenomas complicated by multiple ossifying jaw fibromas. Her sister, aged 29, also suffered from primary hyperparathyroidism associated with 2 parathyroid adenomas, 1 of which was also suspected to be a carcinoma. These 2 woman had unusual multiple small uterine polyps, which were diagnosed as adenomyomatous polyps. Their brother, aged 17, had 2 parathyroid adenomas complicated by urolithiasis. Fujikawa et al. (1998) tabulated the findings in previously reported families. The tabulation indicated that parathyroid lesions tend to be malignant in familial idiopathic hyperparathyroidism.

Dinnen, J. S.; Greenwood, R. H.; Jones, J. H.; Walker, D. A.; Williams, E. D.: Parathyroid carcinoma in familial hyperparathyroidism. J. Clin. Path. 30: 966-975, 1977.
Fujikawa, M.; Okamura, K.; Sato, K.; Mizokami, T.; Tamaki, K.; Yanagida, T.; Fujishima, M.: Familial isolated hyperparathyroidism due to multiple adenomas associated with ossifying jaw fibroma and multiple uterine adenomyomatous polyps. Europ. J. Endocr. 138: 557-561, 1998.
Inoue, H.; Miki, H.; Oshimo, K.; Tanaka, K.; Monden, Y.; Yamamoto, A.; Kagawa, S.; Sano, N.; Hayashi, E.; Nagayama, M.; Hayashi, Y.: Familial

hyperparathyroidism associated with jaw fibroma: case report and literature review. Clin. Endocr. 43: 225-229, 1995.
Jackson, C. E.: Hereditary hyperparathyroidism associated with recurrent pancreatitis. Ann. Intern. Med. 49: 829-836, 1958.
Jackson, C. E.: Personal Communication. Detroit, Mich. 1994.
Jackson, C. E.; Norum, R. A.; Boyd, S. B.; Talpos, G. B.; Wilson, S. D.; Taggert, R. T.; Mallette, L. E.: Hereditary hyperparathyroidism and multiple ossifying jaw fibromas: a clinically and genetically distinct syndrome. Surgery 108: 1006-1013, 1990.
Kakinuma, A.; Morimoto, I.; Nakano, Y.; Fujimoto, R.; Ishida, O.; Okada, Y.; Inokuchi, N.; Fujihara, T.; Eto, S.: Familial primary hyperparathyroidism complicated with Wilms' tumor. Intern Med. 33: 123-126, 1994.
Kennett, S.; Pollick, H.: Jaw lesions in familial hyperparathyroidism. Oral Surg. Oral Med. Oral Path. 31: 502-510, 1971.
Mallette, L. E.; Malini, S.; Rappaport, M. P.; Kirkland, J. L.: Familial cystic parathyroid adenomatosis. Ann. Intern. Med. 107: 54-60, 1987.
Pidwirny, G. N.; Szabo, J.; Hobbs, M.; Heath, H.; Jackson, C. E.: Followup of two hyperparathyroidism-jaw tumor syndrome families reported in 1971 and 1981 reveals that they are related and that parathyroid cancer is a part of the syndrome. (Abstract) Am. J. Hum. Genet. 57: A75 only, 1995.
Rosen, I. B.; Palmer, J. A.: Fibroosseous tumors of the facial skeleton in association with primary hyperparathyroidism: an endocrine syndrome or coincidence?. Am. J. Surg. 142: 494-498, 1981.
Szabo, J.; Heath, B.; Hill, V. M.; Jackson, C. E.; Zarbo, R. J.; Mallette, L. E.; Chew, S. L.; Besser, G. M.; Thakker, R. V.; Huff, V.; Leppert, M. F.; Heath, H., III: Hereditary hyperparathyroidism-jaw tumor syndrome: the endocrine tumor gene HRPT2 maps to chromosome 1q21-q31. Am. J. Hum. Genet. 56: 944-950, 1995.
Teh, B. T.; Farnebo, F.; Kristoffersson, U.; Sundelin, B.; Cardinal, J.; Axelson, R.; Yap, A.; Epstein, M.; Heath, H., III; Cameron, D.; Larsson, C.: Autosomal dominant primary hyperparathyroidism and jaw tumor syndrome associated with renal hamartomas and cystic kidney disease: linkage to 1q21-q32 and loss of the wild type allele in renal hamartomas. J. Clin. Endocr. Metab. 81: 4204-4211, 1996.
Warnakulasuriya, S.; Markwell, B. D.; Williams, D. M.: Familial hyperparathyroidism associated with cementifying fibromas of the jaws in two siblings. Oral Surg. Oral Med. Oral Path. 59: 269-274, 1985.

HYPERPARATHYROIDISM, PRIMARY (600166)

JJM NOTE

Perhaps haplotype analysis of this rare trait could localize the gene, using the previously cited genes as candidate areas with which to start.

Primary hyperparathyroidism due to water clear cell hyperplasia (WCCH) shows a strong association with blood group O (Tisell et al., 1981). Hedbaeck and Oden (1994) compared the blood groups of 32 cases of WCCH with those of 2 control groups, one with primary hyperparathyroidism due to other causes and the other with the population in a geographically defined area of Sweden. The blood group distribution differed between the 2 control groups, but the findings in the patients with WCCH differed with high significance ($P = 0.00040$). This association was thought to be by far the strongest association with the ABO system demonstrated to date.

Hedbaeck, G.; Oden, A.: Parathyroid water clear cell hyperplasia, an O-allele associated condition. Hum. Genet. 94: 195-197, 1994.
Tisell, L.-E.; Hedman, I.; Hansson, G.: Clinical characteristics and surgical results in hyperparathyroidism caused by water-clear cell hyperplasia. World J. Surg. 5: 565-571, 1981.

HYPOCALCIURIC HYPERCALCEMIA, FAMILIAL (*145980)

JJM NOTE

This is included because of parathyroid hyperplasia and the important *lack* of parathyroid adenoma. Lipomas may also occur to excess.

A number sign (#) is used with this entry because of evidence that this phenotype is caused by mutation in the gene encoding the calcium-sensing receptor (CASR; 601199). Gain-of-function mutations in the CASR gene are responsible for familial hypercalciuric hypercalcemia (HHC1) and for neonatal severe hyperparathyroidism (NSHPT; 239200). Loss-of-function mutations result in autosomal dominant hypocalcemia (601198).

From studies of the families of 25 index patients with primary parathyroid hyperplasia, Marx et al. (1977) identified two autosomal dominant disorders, type I multiple endocrine adenomatosis (131100) and one they termed familial hypocalciuric hypercalcemia. The latter was present in the families of 2 of the patients. Among offspring of affected persons in the kindreds with FHH, as distinct from MEA I, the prevalence of hypercalcemia approached the expected 50% during the first 2 decades. Nephrolithiasis and peptic ulcer were uncommon. Moderate hypercalcemia occurred without hypercalciuria. Subtotal parathyroidectomy did not abolish hypercalcemia. Concentrations of peptide hormones other than parathyroid hormones were common in patients with FHH.

Marx et al. (1978) and Marx et al. (1981) contrasted FHH with primary hyperparathyroidism (HP). Patients with FHH had higher creatinine clearance values than HP patients but higher serum magnesium levels than both normals and HP patients. Elevated magnesium level was proportional to elevated calcium level in FHH but was inversely related in HP. Urinary excretion of both calcium and magnesium was significantly lower in FHH than in HP. Abnormal serum protein binding of calcium and magnesium in FHH was excluded.

Attie et al. (1980) stated that familial hypocalciuric hypercalcemia, which was first reported by Foley et al. (1972) as familial benign hypercalcemia, is the first-to-be-described parathormone-independent renal tubular defect in calcium reabsorption. Menko et al. (1984) presented the hypothesis that the abnormality may involve the 'setting of the parathyroid gland,' a process that seems to occur in the perinatal period, and that the fundamental defect may be in renal calcium handling. Among 67 patients referred after unsuccessful surgery for presumed primary hyperparathyroidism, Marx et al. (1980) found that 6 were members of kindreds with familial hypocalciuric hypercalcemia. This disorder achieves greater practical importance as routine biochemical screening becomes widely practiced. Marx (1980) estimated that about 25 patients with this disorder undergo unsuccessful parathyroidectomy in the United States each year. Furthermore, their hypercalcemic relatives are usually not recognized or informed of the mild nature of their disorder. Unlike primary hyperparathyroidism, hypercalcemia of this origin begins before age 10 years and is not accompanied by urinary stone or renal damage. The only complications attributable to the hypercalcemia are pancreatitis and chondrocalcinosis. Parathyroid hyperplasia is found in most cases, but hypercalcemia usually persists after parathyroidectomy. Both the kidneys and the parathyroid glands seem insensitive to chronic hypercalcemia. In some cases circulating parathormone levels are elevated and can lead to neonatal severe 'primary hyperparathyroidism' (239200) in offspring of affected women. A simple diagnostic test is the ratio of renal calcium clearance to creatinine clearance; a value below 0.01 suggests familial hypocalciuric hypercalcemia. The finding of hypercalcemia in first-degree relatives supports the diagnosis, particularly when found in children under age 10 years. Lipomas may be a pleiotropic effect of the FHH gene (22 :Levine, 1980).

Paterson and Gunn (1981) found this disorder in at least 10 members of 4 generations of a large kindred. Parathyroid exploration had been performed in 3 members (twice in 1) before it was realized that they did not have primary hyperparathyroidism. The relation to neonatal severe primary hyperparathyroidism was discussed further by Marx et al. (1982). In some instances, NSPH may represent the homozygous state of FHH. Menko et al. (1983) identified 27 hypercalcemic persons in 3 generations of a large kindred. Five had had parathyroid surgery. The patients tend to have hypermagnesemia as opposed to the hypomagnesemia of hyperparathyroidism. Increased renal tubular calcium reabsorption and persistent normal functioning of the parathyroid glands in the face of hypercalcemia remain the sole definite abnormalities of the syndrome. Steinmann et al. (1984) and Marx et al. (1985) presented evidence that FHH can show only intermittent and very mild hypercalcemia in heterozygotes and that in the homozygous state the gene can cause neonatal severe primary hyperparathyroidism. This hypothesis was proven by Pollak et al. (1994). The kindred on which Marx et al. (1985) based this conclusion was first reported by Hillman et al. (1964) as an instance of autosomal recessive neonatal severe primary hyperparathyroidism. Two offspring of first-cousin parents were affected. Only later was FHH described and was it realized that most cases of neonatal severe primary hyperparathyroidism occur in families with FHH.

Marx et al. (1985) concluded that of 22 reported cases of NSPH, 9 were in kindreds with definite or probable FHH. In 3 kindreds, because of normocalcemia in both parents and, in 2 of them, parental consanguinity, autosomal recessive inheritance was suggested. It was one of these 3 kindreds that Marx et al. (1985) restudied. The mild and intermittent nature of hypercalcemia in heterozygotes was responsible for the earlier misinterpretation. The frequency of gallstones is increased; indeed, this is the only discernible increase in medical problems. Skeletal mass is normal and fractures do not occur with increased frequency (Law and Heath, 1985).

Mapping

In a linkage study in an extensively affected Dutch family, Menko et al. (1984) excluded linkage with several markers; low positive lod scores were observed with Duffy. On the basis of linkage studies, Heath et al. (1992) concluded that HHC is not related to the multiple endocrine neoplasia syndromes (131100, 171400, 162300). They also excluded basic fibroblast growth factor (134920), parathyroid hormone (168450), and several other candidate loci. In each of 4 unrelated families with FHH, Chou et al. (1992) demonstrated linkage of the disease phenotype with DNA markers on 3q; combined maximum multipoint lod score = 20.67. The FHH locus lies within 15 cM of RHO (180380) (99% confidence interval). The FHH gene presumably lies in the 3q21-q24 region. Chou et al. (1992) suggested that the gene defect perturbs the ability of the parathyroid gland and kidney to recognize and/or respond to changes in extracellular calcium concentration. In a further study of 5 families, Heath et al. (1993) found that HHC mapped to 3q in 4, but in a fifth kindred it mapped to 19p13.3 (145981), thus indicating locus heterogeneity.

By Southern analysis of hamster-human hybrid cell DNAs containing only human chromosome 3, Pollak et al. (1993) demonstrated that the human homolog of the bovine gene is located on chromosome 3. The human receptor is encoded by 6 exons that span more than 20 kb.

Finegold et al. (1994) presented evidence for linkage of a form of autosomal dominant hypoparathyroidism to a region of 3q13 flanking marker D3S1303 and suggested that it may be caused in this family by an inactivating mutation in the Ca(2+)-sensing receptor suppressing PTH secretion and lowering the 'set point' for serum calcium levels.

Janicic et al. (1995) mapped the CASR gene to 3q13.3-q21 by fluorescence in situ hybridization. The localization to chromosome 3 was confirmed by somatic cell hybrid analysis. By interspecific backcross analysis, they found that the Casr gene segregated with D16Mit4 on mouse chromosome 16. The corresponding gene was found to be on rat chromosome 11.

Molecular Genetics

Parathyroid cells respond to decreases in extracellular calcium concentration by means of the calcium sensing receptor (601199), a cell surface receptor that alters phosphatidylinositol turnover and intracellular calcium, ultimately effecting an increase in PTH secretion. The 'set point' of parathyroid cells is defined as that calcium concentration at which PTH secretion is half-maximal. Parathyroid glands from FHH patients have an increase in this set point, and in vitro studies of parathyroid tissue from neonatal severe hyperparathyroidism patients show a still greater increase in this set point. Calcium handling by the kidney is also abnormal in individuals with FHH, who fail to show a hypercalciuric response to hypercalcemia. Brown et al. (1993) identified a putative bovine parathyroid cell Ca(2+)-sensing receptor cDNA by expression cloning in Xenopus laevis oocytes. The cDNA encoded a predicted 120-kD polypeptide containing a large extracellular domain and 7 membrane-spanning regions characteristic of G protein-coupled cell surface receptors. In addition to parathyroid tissue, the receptor was also expressed in regions of the kidney involved in Ca(2+)-regulated Ca(2+) and Mg(2+) reabsorption.

The Ca(2+)-sensing receptor belongs to the superfamily of 7-membrane-spanning G protein-coupled receptors. Pollak et al. (1993) demonstrated that mutations in the human Ca(2+)-sensing receptor gene cause both familial hypocalciuric hypercalcemia and neonatal severe hyperparathyroidism. They discovered 3 nonconservative missense mutations, 2 in the extracellular N-terminal domain of the receptor (601199.0002 and 601199.0003) and 1 in the final intracellular loop (601199.0002). The wildtype receptor expressed in Xenopus laevis oocytes elicited large inward currents in response to perfused polyvalent cations; in contrast, a markedly attenuated response was observed with the protein expressed by 1 of the mutations.

Clapham (1993) pointed out that familial hypocalciuric hypercalcemia joins the list of disorders due to defective G protein receptors, others being defects in the thyrotropin receptor (275200), the luteinizing hormone receptor (152790), the V2 vasopressin receptor (304800), rhodopsin (180380), the ACTH receptor (202200), and the cone opsin receptors (303800, 303900). Diseases have been related to defects in G protein itself in the case of the alpha subunit of Gs (139320) and to mutations in the alpha subunit of Gi found in pituitary, adrenal cortex, ovary, and thyroid tumors (Lyons et al., 1990)—the GIP oncogene.

In addition to familial hypocalciuric hypercalcemia and neonatal severe hyperparathyroidism, mutation in the CASR gene can cause autosomal dominant hypocalcemia. Pollak et al. (1994) hypothesized that, in contrast to familial hypocalciuric hypercalcemia in which mild hypercalcemia is caused by mutations that reduce the activity of the Ca(2+)-sensing receptor, mild hypocalcemia might be caused by a mutation that inappropriately activates the receptor at subnormal Ca(2+) levels. Such activating mutations have been described in other G protein-coupled receptors. In 1 of 2 probands with hypocalcemia, they indeed found a missense mutation (glu128ala; see 601199.0004) in the CASR gene, which they symbolized PCAR1.

Chou et al. (1995) reported 5 novel mutations (see 601199.0022) in the CASR gene (called PCAR1 by them) in FHH or neonatal severe hyperparathyroidism: arg228gln, thr139met, gly144glu, arg63met, and arg67cys. Each resulted in a nonconservative amino acid alteration and each was predicted to be in the large extracellular domain of the Ca(2+)-sensing receptor.

In the case of the probands from 3 other families with FHH linked to 3q, no mutations were identified in PCAR1. In a Japanese FHH family, Aida et al. (1995) identified an HHC1 mutation by PCR and SSCP. Nucleotide sequencing showed a G-to-C transversion at nucleotide 118 that resulted in a pro40-to-ala amino acid substitution. The proband was homozygous and the consanguineous parents were heterozygous for the mutation. The parents showed borderline elevations of serum calcium.

Pearce et al. (1995) did a mutation search of the CASR gene in 9 unrelated kindreds with a total of 39 affected members with familial benign hypercalcemia and in 3 unrelated children with sporadic NHPT. In 6 of 9 FBH kindreds, heterozygosity for a novel mutation (1 missense and 5 missense) were found; in the 3 children with NHPT, 2 de novo heterozygous missense mutations and 1 homozygous frameshift mutation were identified (see 601199.0006, 601199.0007, and 601199.0008). SSCP analysis was found by the authors to be a sensitive and specific mutational screening method that detected more than 85% of these CASR gene mutations. Pearce et al. (1995) noted that the identification of CASR mutations may help distinguish FBH from mild primary hyperparathyroidism which otherwise can be clinically difficult. These results indicated that NHPT is not exclusively the result of homozygosity for a mutation that causes familial benign hypercalcemia in the heterozygous state but rather can be due to heterozygosity for mutations at the CASR locus. Indeed, the parents and sibs of the 3 children with NHPT were normocalcemic. All 3 children with NHPT presented with neonatal hypercalcemia that was associated with marked bony undermineralization. Parathyroidectomy and histologic examination revealed T-cell hyperplasia of all 4 parathyroid glands in the 3 NHPT children, who all became hypocalcemic and required vitamin D replacement postoperatively. The clinical features of 2 of the cases had been previously reported by Meeran et al. (1994) and Dezateux et al. (1984).

Janicic et al. (1995) studied family members of a Nova Scotian deme in which both FHH and NSHPT were segregating and found, by PCR amplification of CASR exons, that FHH individuals were heterozygous and NSHPT individuals were homozygous for an abnormally long exon 7. This was due to an insertion at codon 877 of an Alu-repetitive element of the predicted-variant/human-specific-1 subfamily. The Alu insertion was in the opposite orientation to the PCAR1 gene and contained an exceptionally long poly(A) tract. Stop signals were found in all reading frames within the Alu sequence, leading to a predicted shortening of the Ca(2+)-sensing receptor protein. Janicic et al. (1995) observed that the loss of most of the carboxy-terminal intracellular domain of the protein would dramatically impair its signal transduction capability. Identification of the specific mutation in this community will allow rapid testing of at-risk individuals. Clinical features of affected members of the kindred had previously been reported by Pratt et al. (1947), Goldbloom et al. (1972), and Cole et al. (1990). This was a common ancestry that dated back at least 11 generations to settlement of the area by New England fishing families in the mid-1700s. Bai et al. (1997) demonstrated that insertion of this Alu element resulted in the production of a nonfunctional protein of molecular weight 30 kD less than wildtype with decreased cell surface expression. They also showed that transcription of the Alu-containing CASR produced both a full-length product and a product that was truncated due to stalling at the poly(T) tract. Subsequent in vitro translation produced 3 truncated proteins due to termination in all reading frames as predicted.

Bai et al. (1997) characterized the in vivo, cellular and molecular pathophysiology of a case of NHPT resulting from a de novo, heterozygous missense mutation in the CASR gene. The female neonate was admitted to the hospital for suspected osteogenesis imperfecta. She presented with markedly undermineralized bones, multiple metaphyseal fractures, but moderately severe hypercalcemia. Subtotal parathyroidectomy was performed at 6 weeks; hypercalcemia recurred rapidly, but the bone disease improved gradually with reversion to an asymptomatic state resembling FHH. Dispersed parathyroid cells from the resected tissue showed a set-point (the level of Ca(2+) half maximally inhibiting PTH secretion) substantially higher than for normal human parathyroid cells (1.8 vs 1.0 mM Ca(2+), respectively). A similar increase in the calcium set-point was observed in vivo (serum calcium 3.2 vs 2.4 mM). The proband's CASR gene showed a missense mutation (R185Q) at codon 185 (601199.0003). Her normocalcemic parents were homozygous for the wildtype CASR sequence. While cotransfection of normal and mutant receptors showed a higher Ca(2+) level than for wildtype (6.3 vs 4.6 mM, respectively) for eliciting a half-maximal increase in inositol phosphates, transient expression of the mutant R185Q CASR in human embryonic kidney cells revealed a substantially attenuated Ca(2+)-evoked accumulation of total inositol phosphates, Bai et al. (1997) concluded that this de novo, heterozygous CASR mutation exerts a dominant-negative action on the normal CASR, producing NHPT and more severe hypercalcemia than typically seen in FHH. Moreover, the authors presented evidence that normal maternal calcium homeostasis prompted additional secondary hyperparathyroidism in the fetus, thus contributing to the severity of the NHPT in this patient with FHH. Of interest, the same R185Q mutation (601199.0003) had been described previously by Pollak et al. (1993) in a U.S. kindred (family A) reported by Marx et al. (1982). Affected family members had a degree of hypercalcemia (a mean of 3.08 mM with a range of 2.72 to 3.43 mM) that is similar to that of the proband described by Bai et al. (1997); 2 neonates in one branch of this family presented with NHPT and one of them, patient A-26, inherited the abnormal CASR from her father.

Animal Model

To examine the receptor's role in calcium homeostasis and to elucidate the mechanism by which inherited human CASR gene defects cause disease, Ho et al. (1995) created mice in which the Casr gene was disrupted by standard methods of homologous recombination. They found that the phenotype of heterozygous mice mimicked familial hypocalciuric hypercalcemia and homozygous deficient mice exhibited the phenotype of neonatal severe parathyroidism. The findings suggested to the authors that human CASR mutations cause these disorders by reducing the number of functional receptor molecules on the cell surface.

See also Arnaud (1984); Davies et al. (1984)

Aida, K.; Koishi, S.; Inoue, M.; Nakazato, M.; Tawata, M.; Onaya, T.: Familial hypocalciuric hypercalcemia associated with mutation in the human Ca(2+)-sensing receptor gene. J. Clin. Endocr. Metab. 80: 2594-2598, 1995.

Arnaud, C. D.: Familial benign hypercalcemia: nature's solution to neonatal hyperparathyroidism?. (Editorial) Mayo Clin. Proc. 59: 864-865, 1984.

Attie, M. F.; Gill, J. R., Jr.; Stock, J. L.; Spiegel, A. M.; Downs, R. W., Jr.; Levine, M. A.; Marx, S. J.: Parathyroid hormone (PTH) independent abnormality of renal tubular transport of calcium in familial hypocalciuric hypercalcemia. (Abstract) Clin. Res. 28: 384A, 1980.

Bai, M.; Janicic, N.; Trivedi, S.; Quinn, S. J.; Cole, D. E. C.; Brown, E. M.; Hendy, G. N.: Markedly reduced activity of mutant calcium-sensing receptor with an inserted Alu element from a kindred with familial hypocalciuric hypercalcemia and neonatal severe hyperparathyroidism. J. Clin. Invest. 99: 1917-1925, 1997.

Bai, M.; Pearce, S. H. S.; Kifor, O.; Trivedi, S.; Stauffer, U. G.; Thakker, R. V.; Brown, E. M.; Steinmann, B.: In vivo and in vitro characterization of neonatal hyperparathyroidism resulting from a de novo, heterozygous mutation in the Ca(2+)-sensing receptor gene: Normal maternal calcium homeostasis as a cause of secondary hyperparathyroidism in familial benign hypocalciuric hypercalcemia. J. Clin. Invest. 99: 88-96, 1997.

Brown, E. M.; Gamba, G.; Riccardi, D.; Lombardi, M.; Butters, R.; Kifor, O.; Sun, A.; Hediger, M.; Lytton, M. A.; Hebert, S. C.: Cloning and characterization of an extracellular Ca(2+)-sensing receptor from bovine parathyroid. Nature 366: 575-580, 1993.

Chou, Y.-H. W.; Brown, E. M.; Levi, T.; Crowe, G.; Atkinson, A. B.; Arnqvist, H. J.; Toss, G.; Fuleihan, G. E.-H.; Seidman, J. G.; Seidman, C. E.: The gene responsible for familial hypocalciuric hypercalcemia maps to chromosome 3q in four unrelated families. Nature Genet. 1: 295-300, 1992.

Chou, Y.-H. W.; Pollak, M. R.; Brandi, M. L.; Toss, G.; Arnqvist, H.; Atkinson, A. B.; Papapoulos, S. E.; Marx, S.; Brown, E. M.; Seidman, J. G.; Seidman, C. E.: Mutations in the human Ca(2+)-sensing-receptor gene that cause familial hypocalciuric hypercalcemia. Am. J. Hum. Genet. 56: 1075-1079, 1995.

Clapham, D. E.: Mutations in G protein-linked receptors: novel insights on disease. Cell 75: 1237-1239, 1993.

Cole, D. E. C.; Forsythe, C. R.; Dooley, J. M.; Grantmyre, E. B.; Salisbury, S. R.: Primary neonatal hyperparathyroidism: a devastating neurodevelopmental disorder if left untreated. J. Craniofacial Genet. Dev. Biol. 10: 205-214, 1990.

Davies, M.; Adams, P. H.; Lumb, G. A.; Berry, J. L.; Loveridge, N.: Familial hypocalciuric hypercalcaemia: evidence for continued enhanced renal tubular reabsorption of calcium following total parathyroidectomy. Acta Endocr. 106: 499-504, 1984.

Dezateux, C. A.; Hyde, J. C.; Hoey, H. M. C. V.; O'Riordan, J. L. H.; Spitz, L.; Taylor, G. W.; Grant, D. B.: Neonatal hyperparathyroidism. Europ. J. Pediat. 142: 135-136, 1984.

Finegold, D. N.; Armitage, M. M.; Galiani, M.; Matise, T. C.; Pandian, M. R.; Perry, Y. M.; Deka, R.; Ferrell, R. E.: Preliminary localization of a gene for autosomal dominant hypoparathyroidism to chromosome 3q13. Pediat. Res. 36: 414-417, 1994.

Foley, T. P.; Harrison, H. C.; Arnaud, C. D.; Harrison, H. E.: Familial benign hypercalcemia. J. Pediat. 81: 1060-1067, 1972.

Goldbloom, R. B.; Gillis, D. A.; Prasad, M.: Hereditary parathyroid hyperplasia: a surgical emergency of early infancy. Pediatrics 49: 514-523, 1972.

Heath, H., III; Jackson, C. E.; Otterud, B.; Leppert, M. F.: Genetic linkage analysis in familial benign (hypocalciuric) hypercalcemia: evidence for locus heterogeneity. Am. J. Hum. Genet. 53: 193-200, 1993.

Heath, H., III; Leppert, M. F.; Lifton, R. P.; Penniston, J. T.; Edens, M.; Jerominski, L.; Laakso, K. J.; Nelson, L.; Otterud, B.; Stauffer, D.; Walker, D.: Genetic linkage analysis in familial benign hypercalcemia using a candidate gene strategy. I. Studies in four families. J. Clin. Endocr. Metab. 75: 846-851, 1992.

Hillman, D. A.; Scriver, C. R.; Pedvis, S.; Shragovitch, I.: Neonatal familial primary hyperparathyroidism. New Eng. J. Med. 270: 483-490, 1964.

Ho, C.; Conner, D. A.; Pollak, M. R.; Ladd, D. J.; Kifor, O.; Warren, H. B.; Brown, E. M.; Seidman, J. G.; Seidman, C. E.: A mouse model of human familial hypocalciuric hypercalcemia and neonatal severe hyperparathyroidism. Nature Genet. 11: 389-394, 1995.

Janicic, N.; Soliman, E.; Pausova, Z.; Seldin, M. F.; Riviere, M.; Szpirer, J.; Szpirer, C.; Hendy, G. N.: Mapping of the calcium-sensing gene (CASR) to human chromosome 3q13.3-21 by fluorescence in situ hybridization, and localization to rat chromosome 11 and mouse chromosome 16. Mammalian Genome 6: 798-801, 1995.

Law, W. M., Jr.; Heath, H., III: Familial benign hypercalcemia (hypocalciuric hypercalcemia): clinical and pathogenetic studies in 21 families. Ann. Intern. Med. 102: 511-519, 1985.

Levine, M. A.: Personal Communication. Bethesda, Md. 11/1980.

Lyons, J.; Landis, C. A.; Harsh, G.; Vallar, L.; Grunewald, K.; Feichtinger, H.; Duh, Q.-Y.; Clark, O. H.; Kawasaki, E.; Bourne, H. R.; McCormick, F.: Two G protein oncogenes in human endocrine tumors. Science 249: 655-659, 1990.

Marx, S. J.: Familial hypocalciuric hypercalcemia. (Editorial) New Eng. J. Med. 303: 810-811, 1980.

Marx, S. J.; Attie, M. F.; Levine, M. A.; Spiegel, A. M.; Downs, R. W., Jr.; Lasker, R. D.: The hypocalciuric or benign variant of familial hypercalcemia: clinical and biochemical features in fifteen kindreds. Medicine 60: 397-412, 1981.

Marx, S. J.; Attie, M. F.; Spiegel, A. M.; Levine, M. A.; Lasker, R. D.; Fox, M.: An association between neonatal severe primary hyperparathyroidism and familial hypocalciuric hypercalcemia in three kindreds. New Eng. J. Med. 306: 257-264, 1982.

Marx, S. J.; Fraser, D.; Rapoport, A.: Familial hypocalciuric hypercalcemia: mild expression of the gene in heterozygotes and severe expression in homozygotes. Am. J. Med. 78: 15-22, 1985.

Marx, S. J.; Spiegel, A. M.; Brown, E. M.; Aurbach, G. D.: Family studies in patients with primary parathyroid hyperplasia. Am. J. Med. 62: 698-706, 1977.

Marx, S. J.; Spiegel, A. M.; Brown, E. M.; Koehler, J. O.; Gardner, D. G.; Brennan, M. F.; Aurbach, G. D.: Divalent cation metabolism; familial hypercalciuric hypercalcemia versus typical primary hyperparathyroidism. Am. J. Med. 65: 235-242, 1978.

Marx, S. J.; Spiegel, A. M.; Levine, M. A.; Rizzoli, R. E.; Lasker, R. D.; Santora, A. C.; Downs, R. W., Jr.; Aurbach, G. D.: Familial hypocalciuric hypercalcemia: the relation to primary parathyroid hyperplasia. New Eng. J. Med. 307: 416-426, 1982.

Marx, S. J.; Stock, J. L.; Attie, M. F.; Downs, R. W., Jr.; Gardner, D. G.; Brown, E. M.; Spiegel, A. M.; Doppman, J. L.; Brennan, M. F.: Familial hypocalciuric hypercalcemia: recognition among patients referred after unsuccessful parathyroid exploration. Ann. Intern. Med. 92: 351-356, 1980.

Meeran, K.; Husain, M.; Puccini, M.; Scott, H.; Dionisi-Vici, D.; Harvey, D. R.; Lynn, J.; Thakker, R. V.: Neonatal primary hyperparathyroidism masked by vitamin D deficiency. Clin. Endocr. 41: 531-534, 1994.

Menko, F. H.; Bijvoet, O. L. M.; Fronen, J. L. H. H.; Sandler, L. M.; Adami, S.; O'Riordan, J. L. H.; Schopman, W.; Heynen, G.; Blomen-Kuneken, W.: Familial benign hypercalcaemia: study of a large family. Quart. J. Med. 206: 120-140, 1983.

Menko, F. H.; Bijvoet, O. L. M.; Meera Khan, P.; Nijenhuis, L. E.; van Loghem, E.; Schreuder, I.; Bernini, L. F.; Pronk, J. C.; Madan, K.; Went, L. N.: Familial benign hypercalcaemia (FBH; McK. no. 14598, 1983): linkage studies in a large Dutch family. Hum. Genet. 67: 452-454, 1984.

Paterson, C. R.; Gunn, A.: Familial benign hypercalcaemia. Lancet I: 61-63, 1981.

Pearce, S. H. S.; Trump, D.; Wooding, C.; Besser, G. M.; Chew, S. L.; Grant, D. B.; Heath, D. A.; Hughes, I. A.; Paterson, C. R.; Whyte, M. P.; Thakker, R. V.: Calcium-sensing receptor mutations in familial benign hypercalcemia and neonatal hyperparathyroidism. J. Clin. Invest. 96: 2683-2692, 1995.

Pollak, M. R.; Brown, E. M.; Chou, Y.-H. W.; Hebert, S. C.; Marx, S. J.; Steinmann, B.; Levi, T.; Seidman, C. E.; Seidman, J. G.: Mutations in the human Ca(2+)-sensing receptor gene cause familial hypocalciuric hypercalcemia and neonatal severe hyperparathyroidism. Cell 75: 1297-1303, 1993.

Pollak, M. R.; Brown, E. M.; Estep, H. L.; McLaine, P. N.; Kifor, O.; Park, J.; Hebert, S. C.; Seidman, C. E.; Seidman, J. G.: Autosomal dominant hypocalcaemia caused by a Ca(2+)-sensing receptor gene mutation. Nature Genet. 8: 303-307, 1994.

Pollak, M. R.; Chou, Y.-H. W.; Marx, S. J.; Steinmann, B.; Cole, D. E. C.; Brandi, M. L.; Papapoulos, S. E.; Menko, F. H.; Hendy, G. N.; Brown, E. M.; Seidman, C. E.; Seidman, J. G.: Familial hypocalciuric hypercalcemia and neonatal severe hyperparathyroidism: effects of mutant gene dosage on phenotype. J. Clin. Invest. 93: 1108-1112, 1994.

Pratt, E. L.; Geren, B. B.; Neuhauser, E. B. D.: Hypercalcemia and idiopathic hyperplasia of the parathyroid glands in an infant. J. Pediat. 30: 388-399, 1947.

Steinmann, B.; Gnehm, H. E.; Rao, V. H.; Kind, H. P.; Prader, A.: Neonatal severe primary hyperparathyroidism and alkaptonuria in a boy born to related parents with familial hypocalciuric hypercalcemia. Helv. Paediat. Acta 39: 171-186, 1984.

HYPOPHOSPHATEMIA, X-LINKED (*307800)

JJM NOTE

Old literature associated this disorder with parathyroid adenoma (Thomas WC Jr, Fry RM: Parathyroid adenomas in chronic rickets. Am. J. Med. 49:404–07, 1970).

The description of hereditary hypophosphatemia with vitamin D-resistant rickets and delineation of its X-linked inheritance is attributable mainly to Winters et al. (1958). See the commentary by Scriver et al. (1991) with a survey of progress since 1958. In a large North Carolina family of English-Scottish extraction, Winters et al. (1958) noted that hypophosphatemia was the most dependable marker of the presence of the gene. The degree of depression of serum phosphate was the same in males and females, although the severity of bone disease was much less severe in females. There were no instances of male-to-male transmission of either bone disease or hypophosphatemia, and all daughters of hypophosphatemic males were themselves hypophosphatemic. In their literature review, the authors noted a report of male-to-male transmission of 'vitamin D resistant rickets,' which may have been an instance of hypophosphatemic bone disease (146350). Affected persons show a reduction in renal phosphate Tm to about 50% of normal. Males and females are not significantly different in this respect.

Stickler (1969) concluded that hypophosphatemia is already present in the neonatal period, that alkaline phosphatase is elevated at 1 month of age, and that early treatment with high doses of vitamin D does not prevent growth failure. The concept of vitamin D resistance has shortcomings because the mimicry of nutritional rickets is not close. The X-linked disorder never shows myopathy, tetany, or hypocalcemia. Furthermore, complete healing with vitamin D in high dosage and restoration of normal growth is difficult, if not impossible.

Ponchon et al. (1969) concluded that the liver is the major, if not the only, physiologic site of hydroxylation of vitamin D3 (cholecalciferol) to its biologically active metabolite 25-hydroxycholecalciferol. Earp et al. (1970) found 25-HCC ineffective in 5 patients. Negative results with 25-HCC (see also Cohanim et al., 1972) made it unlikely that the basic defect is in the conversion of vitamin D to the active form. On the basis of a follow-up study, McNair and Stickler (1969) questioned whether vitamin D therapy has any beneficial effect on growth. Stickler and Morgenstern (1989) analyzed heights and symptoms of 52 patients, aged at least 18 years, with this disorder. They could find no evidence that any form of treatment had any effect on adult height, symptoms, or alkaline phosphatase levels. There was a negative relation between adult height and the number of osteotomies undergone. Complications of treatment, such as renal failure, which appeared secondary to vitamin D intoxication in 3 patients in their twenties, may outweigh any possible benefits. Glorieux and Scriver (1972) suggested that the defect in this condition resides in the parathyroid-hormone sensitive component of phosphate transport in kidney cells. Since calcium promotes phosphate reabsorption, the authors suggested that the beneficial effect of vitamin D therapy is secondary to the effects on calcium metabolism. Glorieux et al. (1972) found restoration of growth when inorganic phosphate salt supplement and vitamin D2 were administered. They interpreted this to support their conclusion that the defect is primarily due to loss of phosphate at the level of the renal tubule. They also showed a direct correlation between the level of serum inorganic phosphate and whole blood oxygen pressure at 50% oxygen saturation. They speculated that low Pi may inhibit synthesis of 2,3-diphosphoglycerate in red cells with resulting inhibition of release of oxygen to tissues. They suggested that this might be the mechanism of growth retardation. Because of the disputed value of phosphate and vitamin D therapy and the recognized complication of nephrocalcinosis, Verge et al. (1991), in Sydney, Australia, studied therapy in 9 boys and 15 girls and compared the results with those reported in 1971 in 16 untreated Australian patients. The 19 patients treated for at least 2 years before the onset of puberty had a mean height standard-deviation score of -1.08, as compared with

See also Albertsen et al. (1987); Archard and Witkop (1966); Blackard et al. (1962); Burnett et al. (1964); Chesney et al. (1983); Cole and Scriver (1984); Condon et al. (1971); Drezner et al. (1979); Lobaugh and Drezner (1983); Meyer et al. (1979); Rasmussen et al. (1981); Tenenhouse and Scriver (1975)

Adams, J. E.; Davies, M.: Intra-spinal new bone formation and spinal cord compression in familial hypophosphataemic vitamin D resistant osteomalacia. Quart. J. Med. 61: 1117-1129, 1986.

Albertsen, H. M.; Ahrens, P.; Frey, D.; Machler, M.; Kruse, T. A.: Close linkage between X-linked hypophosphatemia and DXS207 defined by the DNA probe pPA4B. (Abstract) Cytogenet. Cell Genet. 46: 569, 1987.

Archard, H. O.; Witkop, C. J., Jr.: Hereditary hypophosphatemia (vitamin D-resistant rickets) presenting primary dental manifestations. Oral Surg. 22: 184-193, 1966.

Beck, L.; Soumounou, Y.; Martel, J.; Krishnamurthy, G.; Gauthier, C.; Goodyer, C. G.; Tenenhouse, H. S.: Pex/PEX tissue distribution and evidence for a deletion in the 3-prime region of the Pex gene in X-linked hypophosphatemic mice. J. Clin. Invest. 99: 1200-1209, 1997.

Bell, C. L.; Tenenhouse, H. S.; Scriver, C. R.: Primary cultures of renal epithelial cells from X-linked hypophosphatemic (Hyp) mice express defects in phosphate transport and vitamin D metabolism. Am. J. Hum. Genet. 43: 293-303, 1988.

Blackard, W. G.; Robinson, R. R.; White, J. E.: Familial hypophosphatemia: report of a case, with observations regarding pathogenesis. New Eng. J. Med. 266: 899-905, 1962.

Buckle, V. J.; Edwards, J. H.; Evans, E. P.; Jonasson, J. A.; Lyon, M. F.; Peters, J.; Searle, A. G.: Comparative maps of human and mouse X chromosomes. (Abstract) Cytogenet. Cell Genet. 40: 594-595, 1985.

Burnett, C. H.; Dent, C. E.; Harper, C.; Warland, B. J.: Vitamin D-resistant rickets: analysis of twenty-four pedigrees with hereditary and sporadic cases. Am. J. Med. 36: 222-232, 1964.

Carpenter, T. O.; Keller, M.; Schwartz, D.; Mitnick, M.; Smith, C.; Ellison, A.; Carey, D.; Comite, F.; Horst, R.; Travers, R.; Glorieux, F. H.; Gundberg, C. M.; Poole, A. R.; Insogna, K. L.: 24,25 Dihydroxyvitamin D supplementation corrects hyperparathyroidism and improves skeletal abnormalities in X-linked hypophosphatemic rickets—a clinical research center study. J. Clin. Endocr. Metab. 81: 2381-2388, 1996.

Carpenter, T. O.; Mitnick, M. A.; Ellison, A.; Smith, C.; Insogna, K. L.: Nocturnal hyperparathyroidism: a frequent feature of X-linked hypophosphatemia. J. Clin. Endocr. Metab. 78: 1378-1383, 1994.

Chesney, R. W.; Mazess, R. B.; Rose, P.; Hamstra, A. J.; DeLuca, H. F.; Breed, A. L.: Long-term influence of calcitriol (1,25-dihydroxyvitamin D) and supplemental phosphate in X-linked hypophosphatemic rickets. Pediatrics 71: 559-567, 1983.

Cohanim, M.; DeLuca, H. F.; Yendt, E. R.: Effects of prolonged treatment with 25-hydroxycholecalciferol in hypophosphatemic (vitamin D refractory) rickets and osteomalacia. Johns Hopkins Med. J. 131: 118-132, 1972.

Cole, D. E. C.; Scriver, C. R.: The effects of mendelian mutation on renal sulfate and phosphate transport in man and mouse. Pediat. Res. 18: 25-29, 1984.

Condon, J. R.; Nassim, J. R.; Rutter, A.: Defective intestinal phosphate absorption in familial and non-familial hypophosphatemia. Brit. Med. J. 3: 138-141, 1970.

Condon, J. R.; Nassim, J. R.; Rutter, A.: Pathogenesis of rickets and osteomalacia in familial hypophosphataemia. Arch. Dis. Child. 46: 269-272, 1971.

Davies, M.; Kane, R.; Valentine, J.: Impaired hearing in X-linked hypophosphataemic (vitamin-D-resistant) osteomalacia. Ann. Intern. Med. 100: 230-232, 1984.

Dennis, V. W.; Bello-Reuss, E.; Robinson, R. R.: Response of phosphate transport to parathyroid hormone in segments of rabbit nephron. Am. J. Physiol. 233: F29-F38, 1977.

Drezner, M. K.; Haussler, M. R.; Lyles, K. W.; Harrelson, J. M.: The role of 1,25-dihydroxycholecalciferol (DHCC) in the pathogenesis and treatment of X-linked hypophosphatemic rickets (XHL). (Abstract) Clin. Res. 27: 365A, 1979.

Du, L.; Desbarats, M.; Viel, J.; Glorieux, F. H.; Cawthorn, C.; Ecarot, B.: cDNA cloning of the murine Pex gene implicated in X-linked hypophosphatemia and evidence for expression in bone. Genomics 36: 22-28, 1996.

Earp, H. S.; Ney, R. L.; Gitelman, H. J.; Richman, R.; DeLuca, H. F.: Effects of 25-hydroxycholecalciferol in patients with familial hypophosphatemia and vitamin-D-resistant rickets. New Eng. J. Med. 283: 627-630, 1970.

Ecarot-Charrier, B.; Glorieux, F. H.; Travers, R.; Desbarats, M.; Bouchard, F.; Hinek, A.: Defective bone formation by transplanted Hyp mouse bone cells into normal mice. Endocrinology 123: 768-773, 1988.

Econs, M. J.; Barker, D. F.; Speer, M. C.; Pericak-Vance, M. A.; Fain, P. R.; Drezner, M. K.: A linkage map of the human X-linked hypophosphatemic rickets gene locus. (Abstract) Clin. Res. 39: 330A only, 1991.

Econs, M. J.; Pericak-Vance, M. A.; Betz, H.; Bartlett, R. J.; Speer, M. C.; Drezner, M. K.: The human glycine receptor: a new probe that is linked to the X-linked hypophosphatemic rickets gene. Genomics 7: 439-441, 1990.

Eicher, E. M.; Southard, J. L.; Scriver, C. R.; Glorieux, F. H.: Hypophosphatemia: mouse model for human familial hypophosphatemic (vitamin D-resistant) rickets. Proc. Nat. Acad. Sci. 73: 4667-4671, 1976.

Glorieux, F. H.: Rickets, the continuing challenge. (Editorial) New Eng. J. Med. 325: 1875-1877, 1991.

Glorieux, F. H.; Scriver, C. R.: Loss of a parathyroid hormone-sensitive component of phosphate transport in X-linked hypophosphatemia. Science 175: 997-1000, 1972.

Glorieux, F. H.; Scriver, C. R.; Reade, T. M.; Goldman, H.; Rosenborough, A.: Use of phosphate and vitamin D to prevent dwarfism and rickets in X-linked hypophosphatemia. New Eng. J. Med. 287: 481-487, 1972.

Grieff, M.; Mumm, S.; Waeltz, P.; Mazzarella, R.; Whyte, M. P.; Thakker, R. V.; Schlessinger, D.: Expression and cloning of the human X-linked hypophosphatemia gene cDNA. Biochem. Biophys. Res. Commun. 231: 635-639, 1997.

Hardy, D. C.; Murphy, W. A.; Siegel, B. A.; Reid, I. R.; Whyte, M. P.: X-linked hypophosphatemia in adults: prevalence of skeletal radiographic and scintigraphic features. Radiology 171: 403-414, 1989.

Harrell, R. M.; Lyles, K. W.; Harrelson, J. M.; Friedman, N. E.; Drezner, M. K.: Healing of bone disease in X-linked hypophosphatemic rickets/osteomalacia; induction and maintenance with phosphorus and calcitriol. J. Clin. Invest. 75: 1858-1868, 1985.

Harrison, H. E.; Harrison, H. C.; Lifshitz, F.; Johnson, A. D.: Growth disturbance in hereditary hypophosphatemia. Am. J. Dis. Child. 112: 290-297, 1966.

Highman, J. H.; Sanderson, P. H.; Sutcliffe, M. M. L.: Vitamin-D-resistant osteomalacia as a cause of spinal cord compression. Quart. J. Med. 39: 529-537, 1970.

Holm, I. A.; Huang, X.; Kunkel, L. M.: Mutational analysis of the PEX gene in patients with X-linked hypophosphatemic rickets. Am. J. Hum. Genet. 60: 790-797, 1997.
HYP Consortium: A gene (HYP) with homologies to endopeptidases is mutated in patients with X-linked hypophosphatemic rickets. Nature Genet. 11: 130-136, 1995.
Lobaugh, B.; Drezner, M. K.: Abnormal regulation of renal 25-hydroxyvitamin D-1-alpha-hydroxylase activity in the X-linked hypophosphatemic mouse. J. Clin. Invest. 71: 400-403, 1983.
Machler, M.; Frey, D.; Gal, A.; Orth, U.; Wienker, T. F.; Fanconi, A.; Schmid, W.: X-linked dominant hypophosphatemia is closely linked to DNA markers DXS41 and DXS43 at Xp22. Hum. Genet. 73: 271-275, 1986.
McNair, S. L.; Stickler, G. B.: Growth in familial hypophosphatemic vitamin-D-resistant rickets. New Eng. J. Med. 281: 511-516, 1969.
Meyer, R. A., Jr.; Jowsey, J.; Meyer, M. H.: Osteomalacia and altered magnesium metabolism in the X-linked hypophosphatemic mouse. Calcif. Tissue Int. 27: 19-26, 1979.
Meyer, R. A., Jr.; Meyer, M. H.; Gray, R. W.: Parabiosis suggests a humoral factor is involved in X-linked hypophosphatemia in mice. J. Bone Miner. Res. 4: 493-500, 1989.
Moser, C. R.; Fessel, W. J.: Rheumatic manifestations of hypophosphatemia. Arch. Intern. Med. 134: 674-678, 1974.
Nehgme, R.; Fahey, J. T.; Smith, C.; Carpenter, T. O.: Cardiovascular abnormalities in patients with X-linked hypophosphatemia. J. Clin. Endocr. Metab. 82: 2450-2454, 1997.
Nesbitt, T.; Coffman, T. M.; Griffiths, R.; Drezner, M. K.: Crosstransplantation of kidneys in normal and Hyp mice: evidence that the Hyp mouse phenotype is unrelated to an intrinsic renal defect. J. Clin. Invest. 89: 1453-1459, 1992.
Nesbitt, T.; Lobaugh, B.; Drezner, M. K.: Calcitonin stimulation of renal 25-hydroxyvitamin D-1(alpha)-hydroxylase activity in hypophosphatemic mice: evidence that the regulation of calcitriol production is not universally abnormal in X-linked hypophosphatemia. J. Clin. Invest. 79: 15-19, 1987.
Petersen, D. J.; Boniface, A. M.; Schranck, F. W.; Rupich, R. C.; Whyte, M. P.: X-linked hypophosphatemic rickets: a study (with literature review) of linear growth response to calcitriol and phosphate therapy. J. Bone Miner. Res. 7: 583-597, 1992.
Polisson, R. P.; Martinez, S.; Khoury, M.; Harrell, R. M.; Lyles, K. W.; Friedman, N.; Harrelson, J. M.; Reisner, E.; Drezner, M. K.: Calcification of entheses associated with X-linked hypophosphatemic osteomalacia. New Eng. J. Med. 313: 1-6, 1985.
Ponchon, G.; Kennan, A. L.; DeLuca, H. F.: 'Activation' of vitamin D by the liver. J. Clin. Invest. 48: 2032-2037, 1969.
Rasmussen, H.; Pechet, M.; Anast, C.; Mazur, A.; Gertner, J.; Broadus, A. E.: Long-term treatment of familial hypophosphatemic rickets with oral phosphate and 1-alpha-hydroxyvitamin D(3). J. Pediat. 99: 16-25, 1981.
Read, A. P.; Thakker, R. V.; Davies, K. E.; Mountford, R. C.; Brenton, D. P.; Davies, M.; Glorieux, F.; Harris, R.; Hendy, G. N.; King, A.; McGlade, S.; Peacock, C. J.; Smith, R.; O'Riordan, J. L. H.: Mapping of human X-linked hypophosphataemic rickets by multilocus linkage analysis. Hum. Genet. 73: 267-270, 1986.
Reitz, R. E.; Weinstein, R. L.: Parathyroid hormone secretion in familial vitamin-D-resistant rickets. New Eng. J. Med. 289: 941-945, 1973.
Rowe, P. S. N.; Read, A. P.; Mountford, R.; Benham, F.; Kruse, T. A.; Camerino, G.; Davies, K. E.; O'Riordan, J. L. H.: Three DNA markers for hypophosphataemic rickets. Hum. Genet. 89: 539-542, 1992.
Scriver, C. R.; Tenenhouse, H. S.; Glorieux, F. H.: X-linked hypophosphatemia: an appreciation of a classic paper and a survey of progress since 1958. Medicine 70: 218-228, 1991.
Seikaly, M. G.; Brown, R.; Baum, M.: The effect of recombinant human growth hormone in children with X-linked hypophosphatemia. Pediatrics 100: 879-884, 1997.
Shields, E. D.; Scriver, C. R.; Reade, T.; Fujiwara, T. M.; Morgan, K.; Ciampi, A.; Schwartz, S.: X-linked hypophosphatemia: the mutant gene is expressed in teeth as well as in kidney. Am. J. Hum. Genet. 46: 434-442, 1990.
Short, E. M.; Binder, J. H.; Rosenberg, L. E.: Familial hypophosphatemic rickets: defective transport of inorganic phosphate by intestinal mucosa. Science 179: 700-702, 1973.
Short, E. M.; Sebastian, A.; Spencer, M.; Morris, R. C., Jr.: Hyperresponsiveness to the phosphaturic effect of parathyroid hormone in X-linked hypophosphatemic vitamin D-resistant rickets (FHR). (Abstract) J. Clin. Invest. 53: 75A, 1974.
Stickler, G. B.: Familial hypophosphatemic vitamin D resistant rickets: the neonatal period and infancy. Acta Paediat. Scand. 58: 213-219, 1969.
Stickler, G. B.; Morgenstern, B. Z.: Hypophosphataemic rickets: final height and clinical symptoms in adults. Lancet II: 902-905, 1989.
Strom, T. M.; Francis, F.; Lorenz, B.; Boddrich, A.; Econs, M. J.; Lehrach, H.; Meitinger, T.: Pex gene deletions in Gy and Hyp mice provide mouse models for X-linked hypophosphatemia. Hum. Molec. Genet. 6: 165-171, 1997.
Tenenhouse, H. S.; Scriver, C. R.: Orthophosphate transport in the erythrocyte of normal subjects and of patients with X-linked hypophosphatemia. J. Clin. Invest. 55: 644-654, 1975.
Tenenhouse, H. S.; Werner, A.; Biber, J.; Ma, S.; Martel, J.; Roy, S.; Murer, H.: Renal Na(+)-phosphate cotransport in murine X-linked hypophosphatemic rickets: molecular characterization. J. Clin. Invest. 93: 671-676, 1994.
Thakker, R. V.; Davies, K. E.; Read, A. P.; Tippett, P.; Wooding, C.; Flint, T.; Wood, S.; Kruse, T. A.; Whyte, M. P.; O'Riordan, J. L. H.: Linkage analysis of two cloned DNA sequences, DXS197 and DXS207, in hypophosphatemic rickets families. Genomics 8: 189-193, 1990.
Thakker, R. V.; Read, A. P.; Davies, K. E.; Whyte, M. P.; Weksberg, R.; Glorieux, F.; Davies, M.; Mountford, R. C.; Harris, R.; King, A.; Kim, G. S.; Fraser, D.; Kooh, S. W.; O'Riordan, J. L. H.: Bridging markers defining the map position of X linked hypophosphataemic rickets. J. Med. Genet. 24: 756-760, 1987.
Verge, C. F.; Lam, A.; Simpson, J. M.; Cowell, C. T.; Howard, N. J.; Silink, M.: Effects of therapy in X-linked hypophosphatemic rickets. New Eng. J. Med. 325: 1843-1848, 1991.
Winters, R. W.; Graham, J. B.; Williams, T. F.; McFalls, V. W.; Burnett, C. H.: A genetic study of familial hypophosphatemia and vitamin D-resistant rickets with a review of the literature. Medicine 37: 97-142, 1958.

Thyroid Tumors

THYROID CARCINOMA, PAPILLARY (#188550)

JJM NOTE

Although papillary thyroid carcinoma is not classically part of the Sipple syndrome (multiple endocrine neoplasia 2A), it is curious that the oncogene, *RET*, discovered in human *papillary* thyroid carcinoma, is the gene for *MEN2* and is seen as mutants in sporadic medullary thyroid carcinomas. To further complicate the clinical phenotype, our Genetic Endocrine Neoplasia Clinic at the University of Pittsburgh saw an adult male from a *MEN1* family who also had papillary thyroid carcinoma.

A number sign (#) is used with this entry because of evidence that papillary thyroid carcinoma is caused by a number of different genetic changes, particularly by chimeric oncogenes formed by fusion of the tyrosine kinase domain of the RET protooncogene (164761) to the 5-prime-terminal region of another gene. In the case of the chimeric gene PTC1, the fusion partner is called H4 (D10S170; 601985), which, like RET, is located on chromosome 10 and becomes fused with RET through an intrachromosomal rearrangement. PTC3 results from a structural rearrangement between the ELE1 (601984) and RET genes. PTC2 is generated through fusion of RET with RI-alpha (188830), situated on chromosome 17. An oncogenic rearrangement of the RET gene is involved in about 35% of cases of papillary thyroid carcinoma; in about 15% of cases, an oncogenic rearrangement has occurred in the NTRK1 protooncogene (191315), which is activated through fusion with neighboring genes TPM3 (191030) and TPR (189940) on 1q, and TFG (602498), located on chromosome 3.

Lote et al. (1980) identified 2 kindreds with 7 and 4 cases of papillary carcinoma in otherwise healthy, nonirradiated subjects. All grew up in 1 of 2 small fishing villages in northern Norway. The familial cases showed an earlier mean age at diagnosis (37.6 years) than did sporadic cases from the same region (52.8 years). Multiple endocrine adenomatosis, Gardner syndrome (175100), and arrhenoblastoma (107950) were excluded. Phade et al. (1981) described 3 affected sibs, of normal parents, with discovery of cancer at ages 12, 7 and 20 years. The authors found 1 other report of familial papillary carcinoma without polyposis coli, in a father and daughter, aged 40 and 12, respectively, at discovery (Lacour et al., 1973). The young age at occurrence and frequent bilateral involvement are characteristic of hereditary cancers. 18,19:Stoffer et al. (1985, 1986) presented evidence for the existence of a familial form of papillary carcinoma of the thyroid, possibly inherited as an autosomal dominant. Four parents of patients with familial PACT had colon cancer and 5 other family members died of intraabdominal malignancy that was not further defined. Perkel et al. (1988) presented evidence suggesting a familial susceptibility factor in radiation-induced thyroid neoplasms.

From 5 thyroid papillary carcinomas and 2 of their respective lymph node metastases, Fusco et al. (1987) isolated a new oncogene by demonstrating transforming activity using DNA transfection analysis on NIH 3T3 cells. Radice et al. (1988) found a TaqI polymorphism in a probe prepared from a genomic lambda phage library constructed with a tertiary NIH 3T3 transfectant obtained by transfection with a human thyroid papillary carcinoma DNA (Fusco et al., 1987). The gene was localized to 10q through the use of a panel of rodent-human cell hybrid DNA and by in situ hybridization. (It is of interest that MEN2 (171400) maps to 10q; another form of thyroid carcinoma, medullary carcinoma, is a component of that disorder.) Radice et al. (1989) refined the assignment to 10q11-q12 by in situ hybridization. They also identified RFLPs of the PACT gene that can be used for study of linkage to MEN2A. Such studies may resolve the question as to whether the PACT gene is the site of the mutation in MEN2A. Grieco et al. (1990) demonstrated that the gene, called by them PTC, is a novel, rearranged form of the RET protooncogene (164761). Herrmann et al. (1991) found clonal abnormalities on cytogenetic analysis in 9 out of 26 papillary thyroid cancers and 5 follicular thyroid cancers. In the former group, the abnormalities included loss of the Y chromosome, addition of an extra chromosome 5, or inversion in chromosome 10, inv(10)(q11.2q21.2). Using DNA probes specific for chromosomes 1, 3, 10, 16, and 17, they carried out RFLP analyses of 12 papillary cancers. No loss of heterozygosity (LOH) was observed for loci mapped to chromosome 10. Jenkins et al. (1990) likewise found the inv(10)(q11.2q21) with breakpoints where RET and another sequence of unknown function, D10S170 (601985), are located. Among 18 cases of PTC, Pierotti et al. (1992) identified 5 with the identical abnormality. They reported the cytogenetic and molecular characterization of 4 of these tumors and demonstrated that the cytogenetic inversion provided the structural basis for the D10S170/RET fusion, leading to the generation of the chimeric transforming sequence which they referred to as RET/PTC. Santoro et al. (1992) found the activated form of the RET oncogene in 33 (19%) of 177 papillary carcinomas and in none of 109 thyroid tumors of other histotypes.

Grossman et al. (1995) identified 13 families with 30 individuals affected by familial nonmedullary thyroid cancer. In 14 of these affected individuals whom they personally treated, 13 had multifocal tumors, and 6 of these were bilateral. The incidence of lymph node metastasis was 57%, as was the incidence of local invasion. Recurrences occurred in 7 patients during follow-up. The histologic diagnosis was papillary thyroid carcinoma in 13 of the 14 patients; in 1 patient it was Hurthle cell carcinoma. Grossman et al. (1995) used the abbreviation FNMTC. Takami et al. (1996) identified 34 families in Japan with 72 individuals affected by nonmedullary thyroid cancer: 17 men and 55 women. Pathologic diagnosis was papillary carcinoma in 64 patients, follicular carcinoma in 6, and anaplastic carcinoma in 2. From the findings in their study they concluded that familial nonmedullary thyroid cancer behaves more aggressively than sporadic nonmedullary thyroid cancer.

Tumor specific rearrangements of the RET gene are frequently detected in papillary thyroid carcinomas. These rearrangements result in the formation of chimeric genes showing the tyrosine kinase domain of RET fused with the 5-prime end sequences of different genes. Bongarzone et al. (1994) examined tumors from a series of 52 patients with papillary thyroid carcinomas and identified 10 cases of RET fusion with the D10S170 locus (also known as H4) resulting in the generation of the RET/PTC1 oncogene, 2 cases with the gene encoding the regulatory subunit RI-alpha of PRKAR1A (188830), and 6 cases with a newly discovered gene they called ELE1 (601984) located on chromosome 10 and leading to the formation of the RET/PTC3 oncogene. The RET/PTC3 hybrid gene was expressed in all 6 cases and was associated with the synthesis of 2 constitutively phosphorylated isoforms of the oncoprotein (p75 and p80). The chromosome 10 localization of both RET and ELE1 and the detection, in all cases, of a sequence reciprocal to that generating the oncogenic rearrangements, strongly suggested that RET/PTC3 formation is a consequence of an intrachromosomal inversion of chromosome 10. The RET/PTC3 hybrid oncogene was observed in both sporadic and radiation-associated post-Chernobyl papillary thyroid carcinomas. Bongarzone et al. (1997) examined the genomic regions containing the ELE1/RET breakpoints in 6 sporadic and 3 post-Chernobyl tumors in 2 papillary carcinomas of different origins. Notably, in all sporadic tumors and in 1 post-Chernobyl tumor, the ELE1/RET recombination corresponded with short sequences of homology (3 to 7 bp) between the 2 rearranging genes. In addition, they observed an interesting distribution of the post-Chernobyl breakpoints in the ELE1 break cluster region (bcr) located within an Alu element, or between 2 closely situated elements, and always in AT-rich regions. Pierotti et al. (1996) indicated that oncogenic rearrangements of the RET gene are found in about 35% of cases of papi llary thyroid carcinoma; rearrangements involving the NTRK1 gene (191315) on chromosome 1 are involved in about 15% of cases. The RET and NTRK1 genes encode membrane receptor-like proteins with tyrosine kinase activity. Their expression in strictly regulated and confined to subsets of neural crest-derived cells. The oncogenic rearrangements cause deletion of the N-terminal domain and fusion of the remaining TK domain of the receptor genes with the 5-prime end of different unrelated genes, designated activating genes. Since all the activating genes are ubiquitously expressed and also contain a dimerization domain, each RET and NTRK1 rearrangement produces chimeric mRNAs and proteins in the thyroid cells in which rearrangements occur. Moreover, the fusion products express an intrinsic and constitutive tyrosine kinase activity. In the case of NTRK1, fusions occur with TPM3 (191030), TPR (189940), and TFG genes. Tong et al. (1997) found that the PTC1 oncoprotein forms a dimer in vivo, and the leucine zipper in the N-terminal region of H4 is responsible for this dimerization. Dimerization is essential for tyrosine hyperphosphorylation and the transforming activity of the PTC1 oncogene.

See also Flannigan et al. (1983)

Bongarzone, I.; Butti, M. G.; Fugazzola, L.; Pacini, F.; Pinchera, A.; Vorontsova, T. V.; Demidchik, E. P.; Pierotti, M. A.: Comparison of the breakpoint regions of ELE1 and RET genes involved in the generation of RET/PTC3 oncogene in sporadic and in radiation-associated papillary thyroid carcinomas. Genomics 42: 252-259, 1997.

Bongarzone, L.; Butti, M. G.; Coronelli, S.; Borrello, M. G.; Santoro, M.; Mondellini, P.; Pilotti, S.; Fusco, A.; Della Porta, G.; Pierotti, M. A.: Frequent activation of ret protooncogene by fusion with a new activating gene in papillary thyroid carcinomas. Cancer Res. 54: 2979-2985, 1994.

Flannigan, G. M.; Clifford, R. P.; Winslet, M.; Lawrence, D. A. S.; Fiddian, R. V.: Simultaneous presentation of papillary carcinoma of thyroid in a father and son. Brit. J. Surg. 70: 181-182, 1983.

Fusco, A.; Grieco, M.; Santoro, M.; Berlingieri, M. T.; Pilotti, S.; Pierotti, M. A.; Della Porta, G.; Vecchio, G.: A new oncogene in human thyroid papillary carcinomas and their lymph-nodal metastases. Nature 328: 170-172, 1987.

Grieco, M.; Santoro, M.; Berlingieri, M. T.; Melillo, R. M.; Donghi, R.; Bongarzone, I.; Pierotti, M. A.; Della Porta, G.; Fusco, A.; Vecchio, G.: PTC is a novel rearranged form of the ret proto-oncogene and is frequently detected in vivo in human thyroid papillary carcinomas. Cell 60: 557-563, 1990.

Grossman, R. F.; Tu, S.-H.; Duh, Q.-Y.; Siperstein, A. E.; Novosolov, F.; Clark, O. H.: Familial nonmedullary thyroid cancer: an emerging entity that warrants aggressive treatment. Arch. Surg. 130: 892-899, 1995.

Herrmann, M. A.; Hay, I. D.; Bartelt, D. H., Jr.; Ritland, S. R.; Dahl, R. J.; Grant, C. S.; Jenkins, R. B.: Cytogenetic and molecular genetic studies of follicular and papillary thyroid cancers. J. Clin. Invest. 88: 1596-1604, 1991.

Jenkins, R. B.; Hay, I. D.; Herath, J. F.; Schultz, C. G.; Spurbeck, J. L.; Grant, C. S.; Goellner, J. R.; Dewald, G. W.: Frequent occurrence of cytogenetic abnormalities in sporadic nonmedullary thyroid carcinoma. Cancer 66: 1213-1220, 1990.

Lacour, J.; Vignalou, J.; Perez, R.; Gerard-Marchant, R.: Epithelioma papillaire du corps thyroide; a propos de deux cas familiaux. Nouv. Presse Med. 2: 2249-2252, 1973.

Lote, K.; Andersen, K.; Nordal, E.; Brennhovd, I. O.: Familial occurrence of papillary thyroid carcinoma. Cancer 46: 1291-1297, 1980.

Perkel, V. S.; Gail, M. H.; Lubin, J.; Pee, D. Y.; Weinstein, R.; Shore-Freedman, E.; Schneider, A. B.: Radiation-induced thyroid neoplasms: evidence for familial susceptibility factors. J. Clin. Endocr. Metab. 66: 1316-1322, 1988.

Phade, V. R.; Lawrence, W. R.; Max, M. H.: Familial papillary carcinoma of the thyroid. Arch. Surg. 116: 836-837, 1981.

Pierotti, M. A.; Bongarzone, I.; Borello, M. G.; Greco, A.; Pilotti, S.; Sozzi, G.: Cytogenetics and molecular genetics of the carcinomas arising from thyroid epithelial follicular cells. Genes Chromosomes Cancer 16: 1-14, 1996.

Pierotti, M. A.; Santoro, M.; Jenkins, R. B.; Sozzi, G.; Bongarzone, I.; Grieco, M.; Monzini, N.; Miozzo, M.; Herrmann, M. A.; Fusco, A.; Hay, I. D.; Della Porta, G.; Vecchio, G.: Characterization of an inversion on the long arm of chromosome 10 juxtaposing D10S170 and RET and creating the oncogenic sequence RET/PTC. Proc. Nat. Acad. Sci. 89: 1616-1620, 1992.

Radice, P.; Donghi, R.; Pierotti, M. A.; Longoni, A.; Fusco, A.; Grieco, M.; Santoro, M.; Vecchio, G.; Della Porta, G.: RFLP for TaqI of the human thyroid papillary carcinoma (PTC) oncogene. Nucleic Acids Res. 16: 9062 only, 1988.

Radice, P.; Donghi, R.; Sozzi, G.; Pierotti, M. A.; Miozzo, M.; Longoni, A.; Mondini, P.; Fusco, A.; Grieco, M.; Santoro, M.; Vecchio, G.; Spurr, N. K.; Della Porta, G.: DNA sequences linked to the human papillary thyroid carcinoma (PTC) oncogene map to chromosome 10q11-q12 and identify TaqI and HincII RFLPs. (Abstract) Cytogenet. Cell Genet. 51: 1062 only, 1989.

Santoro, M.; Carlomagno, F.; Hay, I. D.; Herrmann, M. A.; Grieco, M.; Melillo, R.; Pierotti, M. A.; Bongarzone, I.; Della Porta, G.; Berger, N.; Peix, J. L.; Paulin, C.; Fabien, N.; Vecchio, G.; Jenkins, R. B.; Fusco, A.: Ret oncogene activation in human thyroid neoplasms is restricted to the papillary cancer subtype. J. Clin. Invest. 89: 1517-1522, 1992.

Stoffer, S. S.; Bach, J. V.; Van Dyke, D. L.; Szpunar, W.; Weiss, L.: Familial papillary carcinoma of the thyroid (FPCT): is it autosomal dominant?. (Abstract) Am. J. Hum. Genet. 37: A40 only, 1985.

Stoffer, S. S.; Van Dyke, D. L.; Bach, J. V.; Szpunar, W.; Weiss, L.: Familial papillary carcinoma of the thyroid. Am. J. Med. Genet. 25: 775-782, 1986.

Takami, H.; Ozaki, O.; Ito, K.: Familial nonmedullary thyroid cancer: an emerging entity that warrants aggressive treatment. (Letter) Arch. Surg. 131: 676 only, 1996.

Tong, Q.; Xing, S.; Jhiang, S. M.: Leucine zipper-mediated dimerization is essential for the PTC1 oncogenic activity. J. Biol. Chem. 272: 9043-9047, 1997.

RET-ACTIVATING GENE; ELE1 (*601984)

JJM NOTE

This gene could also be listing among translocation breakpoints (chapter 11), but is here because of its role in papillary thyroid carcinoma, especially that associated with radiation exposure from Chernobyl.

The PTC3 oncogene (RET/PTC3) is an activated form of the RET protooncogene (164761), which is frequently rearranged in papillary thyroid carcinoma (188550). RET/PTC3 results from a structural rearrangement between the ELE1 and RET genes (Bongarzone et al., 1994), and has been observed in both sporadic and radiation-associated post-Chernobyl papillary thyroid carcinoma. To understand the molecular basis that predisposes RET and ELE1 genes to be recurrent targets of 'illegitimate' recombination, Bongarzone et al. (1997) examined the genomic regions containing the ELE1/RET breakpoints in 6 Italian sporadic, RET/PTC3-positive tumors and 3 radiation-associated tumors from children living in areas contaminated by the Chernobyl accident; the latter tumors also expressed RET/PTC3 oncogene. They found that the breakpoints in both genes clustered in regions that they designated ELE1-bcr (1.8 kb) and RET-bcr (1.9 kb). In all sporadic tumors and in 1 post-Chernobyl tumor, the ELE1/RET recombination corresponded with short sequences of homology (3 to 7 bp) between the 2 rearranging genes. In addition, Bongarzone et al. (1997) observed an interesting distribution of the post-Chernobyl breakpoints in ELE1-bcr located within an Alu element, or between 2 close Alu elements, and always in AT-rich regions. In the case of the ELE1/RET fusion gene, the 5-prime end of ELE1 is fused to the tyrosine kinase portion of the RET gene. See 601985 and 188830 for discussion of the PTC1 and PTC3 chimeric oncogenes, respectively.

Bongarzone, I.; Butti, M. G.; Fugazzola, L.; Pacini, F.; Pinchera, A.; Vorontsova, T. V.; Demidchik, E. P.; Pierotti, M. A.: Comparison of the breakpoint regions of ELE1 and RET genes involved in the generation of RET/PTC3 oncogene in sporadic and in radiation-associated papillary thyroid carcinomas. Genomics 42: 252-259, 1997.

Bongarzone, L.; Butti, M. G.; Coronelli, S.; Borrello, M. G.; Santoro, M.; Mondellini, P.; Pilotti, S.; Fusco, A.; Della Porta, G.; Pierotti, M. A.: Frequent activation of ret protooncogene by fusion with a new activating gene in papillary thyroid carcinomas. Cancer Res. 54: 2979-2985, 1994.

GOITER, MULTINODULAR, WITH INTRATHYROIDAL CALCIFICATION (#138800)

JJM NOTE

This disorder is included because of goiter, which, in this entity, seems to be multiple adenomas and possibly low-grade papillary carcinoma.

Murray et al. (1966) described a family in which members of 5 generations had nontoxic goiter appearing in the early teens. Calcification and firm, nodular consistency were unusual features. None of the known defects in thyroid hormonogenesis could be demonstrated. Radioactive iodine studies showed increased thyroid avidity and rapid turnover. No certain male-to-male transmission was observed. Couch et al. (1986) observed a family in which 18 members had a form of euthyroid adolescent multinodular goiter. Histologic study showed multiple adenomata with areas of epithelial hyperplasia, hemorrhage, and calcification. In 2, there were focal areas of epithelial hyperplasia reminiscent of low-grade papillary carcinoma. The pattern of inheritance was consistent with autosomal dominance, with diminished penetrance in males; in 3 instances, unaffected males had daughters with multinodular goiter. One example of male-to-male transmission was noted. Couch et al. (1986) suggested that although the patients were euthyroid, the likely basis for the disorder is an abnormality in thyroglobulin structure and function. If so, the mutation in this case may be located in the structural gene for thyroglobulin on chromosome 8. Cooper et al. (1981) reported a similar family. Baas et al. (1984) presented evidence supporting autosomal dominance of multinodular euthyroidal goiter in association with a defective thyroglobulin allele.

2,3:Bignell et al. (1997, 1997) performed a genomewide linkage scan in a large Canadian family with 18 cases of nontoxic multinodular goiter, including 2 individuals who also had papillary lesions highly suggestive of papillary carcinoma. They identified a locus on chromosome 14q, designated MNG1 for 'multinodular goiter 1,' with a maximum 2-point lod score of 3.8 at D14S1030 and a multipoint lod score of 4.88, defined by D14S1062 (proximally) and D14S267 (distally). The gene encoding thyroid stimulating hormone receptor (TSHR; 275200), which is located at 14q31, was considered a possible candidate gene but was found to be outside the linked region. TSHR appeared to be excluded as a candidate gene for multinodular goiter by recombination in 5 individuals from the Canadian family. Moreover, examination of TSHR in 2 affected individuals from this family did not reveal any variants in the coding sequence. To determine the role of the MNG1 gene in familial nonmedullary thyroid cancer (NMTC), Bignell et al. (1997) studied 37 smaller pedigrees, each containing at least 2 cases of NMTC. Analysis by both parametric and nonparametric methods indicated that only a very small proportion of familial NMTC is attributable to MNG1.

Baas, F.; Bikker, H.; Van Ommen, G.-J. B.; de Vijlder, J. J. M.: Unusual scarcity of restriction site polymorphism in the human thyroglobulin gene: a linkage study suggesting autosomal dominance of a defective thyroglobulin allele. Hum. Genet. 67: 301-305, 1984.

Bignell, G. R.; Canzian, F.; Shayeghi, M.; Stark, M.; Shugart, Y. Y.; Biggs, P.; Mangion, J.; Hamoudi, R.; Rosenblatt, J.; Buu, P.; Sun, S.; Stoffer, S. S.; Goldgar, D. E.; Romeo, G.; Houlston, R. S.; Narod, S. A.; Stratton, M. R.; Foulkes, W. D.: Familial nontoxic multinodular thyroid goiter locus maps to chromosome 14q but does not account for familial nonmedullary thyroid cancer. Am. J. Hum. Genet. 61: 1123-1130, 1997.

Bignell, G. R.; Canzian, F.; Shayeghi, M.; Stark, M.; Shugart, Y. Y.; Biggs, P.; Mangion, J.; Hamoudi, R.; Rosenblatt, J.; Buu, P.; Sun, S.; Stoffer, S. S.; Goldgar, D. E.; Romeo, G.; Houlston, R. S.; Narod, S. A.; Stratton, M. R.; Foulkes, W. D.: A familial non-toxic multinodular thyroid goiter locus maps to chromosome 14q but does not account for familial non-medullary thyroid cancer. (Abstract) Am. J. Hum. Genet. 61 (suppl.): A268 only, 1997.

Cooper, D. S.; Axelrod, L.; deGroot, L. J.; Vickery, A.; Maloof, F.: Congenital goiter and the development of metastatic follicular carcinoma with evidence for a leak of nonhormonal iodide: clinical, pathological, kinetic and biochemical studies and a review of the literature. J. Clin. Endocr. Metab. 52: 294-306, 1981.

Couch, R. M.; Hughes, I. A.; DeSa, D. J.; Schiffrin, A.; Guyda, H.; Winter, J. S. D.: An autosomal dominant form of adolescent multinodular goiter. Am. J. Hum. Genet. 39: 811-816, 1986.

Murray, I. P.; Thomson, J. A.; McGirr, E. M.; MacDonald, E. M.; Kennedy, J. S.; McLennan, I.: Unusual familial goiter associated with intrathyroidal calcification. J. Clin. Endocr. 26: 1039-1040, 1966.

THYROID HORMONE PLASMA MEMBRANE TRANSPORT DEFECT (188560)

JJM NOTE

This entity is included because of mention of goiter, despite no histopathologic documentation.

Wortsman et al. (1983) described eumetabolic hyperthyroxinemia without alteration in serum protein binding of T4 in a 74-year-old woman. The basic defect appeared to be a selective defect in transport of T4 across the plasma membrane as shown by in vitro studies of erythrocytes. A 53-year-old daughter and a 23-year-old grandson had similar findings. The proband and her daughter had enlarged thyroid glands located mostly substernally. A deceased sister of the proband had had a goiter. Two other children of the proband and 7 other grandchildren were unaffected.

Wortsman, J.; Premachandra, B. N.; Williams, K.; Burman, K. D.; Hay, I. D.; Davis, P. J.: Familial resistance to thyroid hormone associated with decreased transport across the plasma membrane. Ann. Intern. Med. 98: 904-909, 1983.

THYROID HORMONE RESISTANCE (#188570)

JJM NOTE

This disorder is included because of goiter.

A number sign (#) is used with this entry because the causative mutation is in the thyroid hormone receptor gene (THR1; 190160).

Brooks et al. (1981) found thyroid hormone resistance in 8 persons in 4 generations of a kindred. All were clinically euthyroid but all had goiters and markedly increased serum thyroid hormone levels. Serum thyrotropin (TSH) levels were normal or slightly elevated and responded normally to the administration of thyrotropin-releasing hormone (TRH) and L-triiodothyronine. The kindred had 3 instances of male-to-male transmission. Dominant inheritance has also been noted by Lamberg et al. (1975), Elewaut et al. (1976), and Maxon et al. (1980). A recessive form of thyroid hormone resistance is also well substantiated (see 274300). Gharib and Klee (1985) observed a family with 6 affected persons in 4 sibships of 3 generations of a family. Of the 6, 5 had goiter and all had increased concentrations of triiodothyronine and free thyroxine without symptoms or signs of hyperthyroidism. Basal serum levels of TSH were normal in all 6; in the 4 tested, these levels responded normally to TRH. This is, then, a form of familial euthyroid hyperthyroxinemia not due to abnormality in the binding proteins, albumin ('dysalbuminemic hyperthyroxinemia'; 103600) or prealbumin (176300). Hopwood et al. (1986) described 2 affected families with involvement of 2 or more generations and examples of male-to-male transmission in each. They emphasized the probable frequency of this condition, particularly in children. Thyrotoxicosis is often suspected because of the goiter and the elevated thyroxine and triiodothyronine levels. Misdiagnosis can have obvious ill effects. Sakurai et al. (1989) stated that about 200 cases of generalized resistance to thyroid hormone have been described. They demonstrated the defect in father and son. The proband was 6.5 years when first suspected of having this disorder. Delayed verbal expression led to a diagnosis of attention-deficit/hyperactivity disorder at age 3. At age 5.5, thyroid gland enlargement and elevated serum thyroxine and triiodothyronine were i nterpreted as hyperthyroidism but specific therapy for this condition was not successful. Indeed, normalization of the thyroid hormone levels in blood slowed his growth rate and strikingly increased TSH concentration. His verbal IQ was 85% but his performance IQ was 120% of the standard score. The father also had a history of delayed speech development, hyperactivity, and learning disability but later obtained higher education including a master's degree. Physical examination was normal except for minimal thyroid gland enlargement. In studies of 18 families with 49 persons with generalized resistance to thyroid hormone, Hauser et al. (1993) found a high frequency of attention deficit/hyperactivity disorder in the affected individuals as children.

One form of familial thyroid hormone resistance may have as its basis a mutation in the thyroid hormone receptor, which has been identified with oncogene ERBA2 (190160). The locus maps to human chromosome 3. Bale et al. (1988) found no recombination between generalized thyroid hormone resistance and RFLPs of the ERBA2 gene; the maximum lod score was 3.91 at theta = 0.0 (Usala et al., 1988). The family studied was kindred A of Magner et al. (1986). A mutation in the ERBA2 gene appears to result in abnormalities of thyroid hormone action in many tissues. In addition to inappropriate secretion of TSH from the pituitary, affected members of the kindred showed delayed bone age and short stature. Most also showed mild mental retardation, diminished or inappropriately normal basal metabolic rates, and hepatic resistance to thyroid hormone (relatively low sex hormone binding globulin and relatively high cholesterol). Usala et al. (1990) also showed linkage between GTHR and ERBA2 in 2 other families; the combined maximum lod score for the 3 kindreds was 5.77 at a recombination fraction of 0.0.

To date, all individuals with the GTHR phenotype have had mutations in the thyroid hormone receptor beta (TR-beta) gene that impair T3-mediated function. Weiss et al. (1996) reported a family in which dominantly inherited GTHR is not associated with abnormalities in the TR-alpha or TR-beta genes, as determined by sequencing and linkage analysis. Affected family members manifest a severe form of GTHR, with reduced responses of thyrotrophs and peripheral tissues requiring 8 to 10 fold the normal replacement doses of L-T4 and L-T3. No other endocrine abnormalities were detected. The defect developed de novo in the proposita and was transmitted to her 2 children from unrelated fathers. An abnormal cofactor with a specific function in the regulation of thyroid hormone action is probably involved in the expression of the GTHR phenotype in this family.

Generalized thyroid hormone resistance is a notable example of a phenotype which can be either completely recessive or completely dominant even though the causative mutations are in the same gene, THR1. The dominant forms represent dominant-negative mutations in which the mutant gene product interferes with the function of the normal receptors; the completely recessive forms are due to a deletion of the THR1 gene (see 190160.0003).

See the review of the syndromes of resistance to thyroid hormone by Refetoff et al. (1993).

See also Bantle et al. (1982)

Bale, A. E.; Usala, S. J.; Gesundheit, N.; Weinberger, C.; Lash, R. W.; Wondisford, F. E.; McBride, O. W.; Weintraub, B. D.: The gene for generalized thyroid hormone resistance (GTHR) is tightly linked to and may be identical with the proto-oncogene, c-erbA-beta. (Abstract) Am. J. Hum. Genet. 43: A77 only, 1988.

Bantle, J. P.; Seeling, S.; Mariash, C. N.; Ulstrom, R. A.; Oppenheimer, J. H.: Resistance to thyroid hormones: a disorder frequently confused with Graves' disease. Arch. Intern. Med. 142: 1867-1871, 1982.

Brooks, M. H.; Barbato, A. L.; Collins, S.; Garbincius, J.; Neidballa, R. G.; Hoffman, D.: Familial thyroid hormone resistance. Am. J. Med. 71: 414-421, 1981.

Elewaut, A.; Mussche, M.; Vermeulen, A.: Familial partial target organ resistance to thyroid hormones. J. Clin. Endocr. Metab. 43: 575-580, 1976.

Gharib, H.; Klee, G. G.: Familial euthyroid hyperthyroxinemia secondary to pituitary and peripheral resistance to thyroid hormones. Mayo Clin. Proc. 60: 9-15, 1985.

Hauser, P.; Zametkin, A. J.; Martinez, P.; Vitiello, B.; Matochik, J. A.; Mixson, A. J.; Weintraub, B. D.: Attention deficit-hyperactivity disorder in people with generalized resistance to thyroid hormone. New Eng. J. Med. 328: 997-1001, 1993.

Hopwood, N. J.; Sauder, S. E.; Shapiro, B.; Sisson, J. C.: Familial partial peripheral and pituitary resistance to thyroid hormone: a frequently missed diagnosis?. Pediatrics 78: 1114-1122, 1986.

Lamberg, B. A.; Sandstrom, R.; Rosegard, S.; Saarinen, P.; Evered, D. C.: Sporadic and familial partial peripheral resistance to thyroid hormones.In: Harland, W. A.; Orr, J. S.: Thyroid Hormone Metabolism. London: Academic Press (pub.) 1975. Pp. 139-161.

Magner, J. A.; Petrick, P.; Menezes-Ferreira, M. M.; Stelling, M.; Weintraub, B. D.: Familial generalized resistance to thyroid hormones: report of three kindreds and correlation of patterns of affected tissues with the binding of (125-I) triiodothyronine to fibroblast nuclei. J. Endocr. Invest. 9: 459-469, 1986.

Maxon, H. R.; Burman, K. D.; Premachandra, B. N.: Euthyroid familial hyperthyroxinemia. (Letter) New Eng. J. Med. 302: 1263 only, 1980.

Refetoff, S.; Weiss, R. E.; Usala, S. J.: The syndromes of resistance to thyroid hormone. Endocr. Rev. 14: 348-399, 1993.

Sakurai, A.; Takeda, K.; Ain, K.; Ceccarelli, P.; Nakai, A.; Seino, S.; Bell, G. I.; Refetoff, S.; DeGroot, L. J.: Generalized resistance to thyroid hormone associated with a mutation in the ligand-binding domain of the human thyroid hormone receptor beta. Proc. Nat. Acad. Sci. 86: 8977-8981, 1989.

Usala, S. J.; Bale, A. E.; Gesundheit, N.; Weinberger, C.; Lash, R. W.; Wondisford, F. E.; McBride, O. W.; Weintraub, B. D.: Tight linkage between the syndrome of generalized thyroid hormone resistance and the human c-erbA(beta) gene. Molec. Endocr. 2: 1217-1220, 1988.

Usala, S. J.; Tennyson, G. E.; Bale, A. E.; Lash, R. W.; Gesundheit, N.; Wondisford, F. E.; Accili, D.; Hauser, P.; Weintraub, B. D.: A base mutation of the C-erbA-beta thyroid hormone receptor in a kindred with generalized thyroid hormone resistance: molecular heterogeneity in two other kindreds. J. Clin. Invest. 85: 93-100, 1990.

Weiss, R. E.; Hayashi, Y., Nagaya, T.; Petty, K. J.; Murata, Y.; Tunca, H., Seo, H.; Refetoff, S.: Dominant inheritance of resistance to thyroid hormone not linked to defects in the thyroid hormone receptor alpha or beta genes may be due to a defective cofactor. J. Clin. Endocr. Metab. 81: 4196-4203, 1996.

THYROID HORMONOGENESIS, GENETIC DEFECT IN, 2A (*274500)

JJM NOTE

This disorder is included because of goiter.

This represents a group of defects. The common denominator is the discharge of a significant percentage of labeled iodide from the thyroid upon administration of thiocyanate or perchlorate, indicating a defect in converting accumulated iodide to organically bound iodine. Discharge may be partial or complete; therefore, organification may be partially or completely deficient. Pommier et al. (1976) summarized the in vitro kinetics and proposed mechanisms for the 3 peroxidase reactions: (1) iodide oxidation, (2) thyroglobulin iodination, and (3) iodothyronine coupling. Haddad and Sidbury (1959) first demonstrated an in vitro deficiency of thyroid peroxidase activity from one of these patients. A peroxide generating system did not improve activity. Hagen et al. (1971) described an intelligent, euthyroid child of normal stature with recurrent goiter. She and her similarly affected sister had normal hearing. Like patients with Pendred syndrome (274600), she discharged 50% of the thyroidal iodide after perchlorate. Her thyroid tissues showed no iodide peroxidation or tyrosine iodination activity. Addition of excessive hematin, the prosthetic group of peroxidase, restored tyrosine iodination.

Evidence excluding the thyroid peroxidase gene as a candidate gene for Pendred syndrome was reported by Gausden et al. (1996). They used a highly informative VNTR 1.5-kb downstream of exon 10 of the thyroid peroxidase gene in linkage analysis from 7 kindreds from the UK and in 1 large inbred Israeli kindred. In 6 of 9 sibships they found evidence of recombination between the thyroid peroxidase gene and Pendred syndrome.

Niepomniszcze et al. (1975) called this the 'apo enzyme-prothetic group defect' and pointed out that an organification defect may be produced by a defective or deficient iodide acceptor (i.e., thyroglobulin). Niepomniszcze et al. (1973) described a cretinous child with a goiter who completely discharged radioiodide after administration of perchlorate. The total in vitro peroxidase deficiency was not improved by peroxide, hematin or enzyme solubilization. Pommier et al. (1974) found that tissue from a euthyroid woman with a recurrent goiter and partial iodide discharge had normal iodide peroxidation but deficient thyroglobulin iodination. Partial solubilization of the enzyme resulted in a 3-fold increase in thyroglobulin iodination activity. Though the allelic nature of these defects is in question, Perez-Cuvit et al. (1977) described partial iodide discharge in the euthyroid, identical twin grandnieces of the 4 sibs first described with the complete organification defect. The twins' hearing was normal and their parents were unrelated. The findings were interpreted as indicating a partial peroxidase defect resulting from compound heterogeneity for two different abnormal alleles. Medeiros-Neto et al. (1982) described thyroid peroxidase deficiency in a congenitally goitrous, mentally retarded, hypothyroid child, whose parents were first cousins. Both parents showed a thyroid abnormality. Wolff (1983) stated that only 22 persons with this abnormality had been reported. Couch et al. (1985) reported a Hutterite kindred with 9 affected persons including identical twins.

In collaboration with others, Dumont (1987) assigned the thyroid peroxidase gene to 2pter-p12 by probing of the DNA of a panel of somatic cell hybrids. Kimura et al. (1987) concluded that 2 thyroid peroxidase proteins are generated through alternate splicing of the same gene. By analysis of DNA from somatic cell hybrids, they mapped the thyroid peroxidase gene to the short arm of chromosome 2. De Vijlder et al. (1988) mapped the TPO gene to 2pter-p12 by Southern blot analysis of genomic DNA from somatic cell hybrids. By in situ hybridization using a biotinylated cDNA probe, Barnett et al. (1990) refined the localization to 2p13. Cross-hybridization was found with the gene for myeloperoxidase on chromosome 17, and a small peak at 2p24-p23 was thought to represent cross-hybridization with another related peroxidase. Endo et al. (1995) used fluorescence in situ hybridization to confirm the localization of human and mouse TPO genes to human chromosome 2 and mouse chromosome 12 and to assign them regionally to 2p25 and 12C, respectively. They commented on the fact that Barnett et al. (1990) had assigned the TPO gene to 2p13. They observed no signal in that region and observed that 2p13 is homologous to mouse chromosome 6 or 11 and not to chromosome 12, where the mouse Tpo gene is located.

On the basis of DNA sequence homology, Seto et al. (1987) concluded that the thyroid microsomal antigen involved in autoimmune thyroid disease is, at least in part, TPO. Kimura et al. (1989) demonstrated that the TPO gene contains 17 exons and covers at least 150 kb of chromosome 2. They found that the positions of the third through eleventh exon-intron junctions in TPO coincide exactly with those of the second through eleventh exon-intron junctions in myeloperoxidase (MPO; 254600), except the seventh myeloperoxidase junction, which does not have any counterpart in thyroid peroxidase. Four exons unique to TPO were found at the 3-prime end of the gene (exons 13-16), each of which encompasses a different protein module. Three of these modules, representing exons 13, 14, and 15, bear significant similarities to C4b/beta-2-glycoprotein, the EGF/LDL receptor, and a typical transmembrane domain, respectively. The genes coding for these modules were probably fused to an ancestral peroxidase gene to generate the present TPO gene. Thus, intron loss and/or insertion, and exon shuffling have played important roles in the evolution of the TPO gene. Also see Kimura (1990). In 5 families, which included 9 goitrous subjects with complete or partial TPO deficiency, Mangklabruks et al. (1991) found a lod score (2.08) compatible with linkage of the disorder to a RFLP in the TPO gene in 1 family with inbreeding. In 2 other families, the lod score was inconsistent with linkage between the disease and the TPO gene. In a fourth family, partial deletion of the TPO gene was suggested by Southern blotting. Anker et al. (1992) identified a tetranucleotide repeat (AATG) in intron 10 of the TPO gene. By study of CEPH pedigrees, they found a heterozygosity of 67% and a PIC value of 0.61. By linkage studies, they subregionalized TPO to 2pter-p23.

Thyroid peroxidase plays a central role in thyroid gland function. The enzyme catalyzes 2 important reactions of thyroid hormone synthesis: the iodination of tyrosine residues in thyroglobulin and phenoxy-ester formation between pairs of iodinated tyrosines to generate the thyroid hormones, thyroxin and triiodothyronine.

Total iodide organification defect (TIOD), in which iodide taken up by the thyroid gland cannot be oxidized and bound to protein, was found by Bikker et al. (1995) to be the most common hereditary inborn error causing congenital hypothyroidism in The Netherlands. They found 7 different mutations in 9 apparently unrelated families with TIOD by screening for mutations in the TPO gene, using denaturing gradient gel electrophoresis of PCR-amplified genomic DNA. Two duplications and 1 insertion resulted in frameshift mutations; 4 single nucleotide substitutions were identified. Nine affected members of 5 families were compound heterozygotes and 6 patients from 4 families were homozygous for 1 of the 7 mutations.

See also Burrow et al. (1973); Leszynsky (1964); Libert et al. (1987); Ljunggren et al. (1973); Parker and Beierwaltes (1961); Stanbury (1963); Stanbury (1978); Valenta et al. (1971)

Abramowicz, M. J.; Targovnik, H. M.; Varela, V.; Cochaux, P.; Krawiec, L.; Pisarev, M. A.; Propato, F. V. E.; Juvenal, G.; Chester, H. A.; Vassart, G.: Identification of a mutation in the coding sequence of the human thyroid peroxidase gene causing congenital goiter. J. Clin. Invest. 90: 1200-1204, 1992.

Anker, R.; Steinbrueck, T.; Donis-Keller, H.: Tetranucleotide repeat polymorphism at the human thyroid peroxidase (hTPO) locus. Hum. Molec. Genet. 1: 137 only, 1992.

Barnett, P. S.; Bhatt, B.; Pagliuca, A.; Mulfti, G.; McGregor, A. M.; Banga, J. P.: The thyroid peroxidase gene: accurate localisation by non-isotopic in situ hybridisation to chromosome 2p13.In: Carayon, J.; Ruf, J.: Thyroperoxidase and Thyroid Autoimmunity. London: INSERM, John Libbey Eurotext Ltd. (pub.) 207: 1990. Pp. 55 only.

Bikker, H.; Baas, F.; De Vijlder, J. J. M.: Molecular analysis of mutated thyroid peroxidase detected in patients with total iodide organification defects. J. Clin. Endocr. Metab. 82: 649-653, 1997.

Bikker, H.; den Hartog, M. T.; Baas, F.; Gons, M. H.; Vulsma, T.; de Vijlder, J. J. M.: A 20-basepair duplication in the human thyroid peroxidase gene results in a total iodide organification defect and congenital hypothyroidism. J. Clin. Endocr. Metab. 79: 248-252, 1994.

Bikker, H.; Vulsma, T.; Baas, F.; de Vijlder, J. J. M.: Identification of five novel inactivating mutations in the human thyroid peroxidase gene by denaturing gradient gel electrophoresis. Hum. Mutat. 6: 9-16, 1995.

Bikker, H.; Waelkens, J. J. J.; Bravenboer, B.; de Vijlder, J. J. M.: Congenital hypothyroidism caused by a premature termination signal in exon 10 of the human thyroid peroxidase gene. J. Clin. Endocr. Metab. 81: 2076-2079, 1996.

Burrow, G. N.; Spaulding, S. W.; Alexander, N. M.; Bower, B. F.: Normal peroxidase activity in Pendred's syndrome. J. Clin. Endocr. 36: 522-530, 1973.

Couch, R. M.; Dean, H. J.; Winter, J. S. D.: Congenital hypothyroidism caused by defective iodide transport. J. Pediat. 106: 950-953, 1985.

de Vijlder, J. J. M.; Dinsart, C.; Libert, F.; Geurts van Kessel, A.; Bikker, H.; Bolhuis, P. A.; Vassart, G.: Regional localization of the thyroid peroxidase gene to human chromosome 2pter-p12. Cytogenet. Cell Genet. 47: 170-172, 1988.

Dumont, J. E.: Personal Communication. Brussels, Belgium 8/12/1987.

Endo, Y.; Onogi, S.; Umeki, K.; Yamamoto, I.; Kotani, T.; Ohtaki, S.; Fujita, T.: Regional localization of the gene for thyroid peroxidase to human chromosome 2p25 and mouse chromosome 12C. Genomics 25: 760-761, 1995.

Gausden, E.; Armour, J. A. L.; Coyle, B.; Coffey, R.; Hochberg, Z.; Pembrey, M.; Britton, K. E.; Grossman, A.; Reardon, W.; Trembath, R.: Thyroid peroxidase: evidence for disease gene exclusion in Pendred's syndrome. Clin. Endocr. 44: 441-446, 1996.

Haddad, H. M.; Sidbury, J. B., Jr.: Defect of the iodinating system in congenital goitrous cretinism: report of a case with biochemical studies. J. Clin. Endocr. 19: 1446-1457, 1959.

Hagen, G. A.; Niepomniszcze, H.; Haibach, H.; Bigazzi, M.; Hati, R.; Rapoport, B.; Jimenez, C.; DeGroot, L. J.; Frawley, T. F.: Peroxidase deficiency in familial goiter with iodide organification defect. New Eng. J. Med. 285: 1394-1398, 1971.

Kimura, S.: Structure and regulation of the human thyroid peroxidase gene.In: Carayon, P.; Ruf, J.: Thyroperoxidase and Thyroid Autoimmunity. London: INSERM, John Libbey Eurotext Ltd. (pub.) 207: 1990. Pp. 3-10.

Kimura, S.; Hong, Y.-S.; Kotani, T.; Ohtaki, S.; Kikkawa, F.: Structure of the human thyroid peroxidase gene: comparison and relationship to the human myeloperoxidase gene. Biochemistry 28: 4481-4489, 1989.

Kimura, S.; Kotani, T.; McBride, O. W.; Umeki, K.; Hirai, K.; Nakayama, T.; Ohtaki, S.: Human thyroid peroxidase: complete cDNA and protein sequence, chromosome mapping, and identification of two alternately spliced mRNAs. Proc. Nat. Acad. Sci. 84: 5555-5559, 1987.

Leszynsky, H. E.: Genetic studies in familial goitrous cretinism. (Abstract) Acta Endocr. 46: 103-110, 1964.

Libert, F.; Ruel, J.; Ludgate, M.; Swillens, S.; Alexander, N.; Vassart, G.; Dinsart, C.: Complete nucleotide sequence of the human thyroperoxidase-microsomal antigen cDNA. Nucleic Acids Res. 15: 6735 only, 1987.

Ljunggren, J. G.; Lindstrom, H.; Hjern, B.: The concentration of peroxidase in normal and adenomatous human thyroid tissue with special reference to patients with Pendred's syndrome. Acta Endocr. 72: 272-278, 1973.

Mangklabruks, A.; Billerbeck, A. E.; Wajchenberg, B.; Knobel, M.; Cox, N. J.; De Groot, L. J.; Medeiros-Neto, G.: Genetic linkage studies of thyroid peroxidase (TPO) gene in families with TPO deficiency. J. Clin. Endocr. Metab. 72: 471-476, 1991.

Medeiros-Neto, G. A.; Okamura, K.; Cavaliere, H.; Taurog, A.; Knobel, M.; Bisi, H.; Kallas, W. G.; Mattar, E.: Familial thyroid peroxidase defect. Clin. Endocr. 17: 1-14, 1982.

Niepomniszcze, H.; Castells, S.; DeGroot, L. J.; Refetoff, S.; Kim, O. S.; Rapoport, B.; Hati, R.: Peroxidase defect in congenital goiter with complete organification block. J. Clin. Endocr. 36: 347-357, 1973.

Niepomniszcze, H.; Rosenbloom, A. L.; Degroot, L. J.; Shimaoka, K.; Refetoff, S.; Yamamoto, K.: Differentiation of two abnormalities in thyroid peroxidase causing organification defect and goitrous hypothyroidism. Metabolism 24: 57-67, 1975.

Parker, R. H.; Beierwaltes, W. H.: Inheritance of defective organification of iodine in familial goitrous cretinism. J. Clin. Endocr. 21: 21-30, 1961.

Perez-Cuvit, E.; Crigler, J. F., Jr.; Stanbury, J. B.: Partial and total iodide organification defect in different sibships in a kindred. Am. J. Hum. Genet. 29: 142-148, 1977.

Pommier, J.; Tourniaire, J.; Deme, D.; Chalendar, D.; Bornet, H.; Nunez, J.: A defective thyroid peroxidase solubilized from a familial goiter with iodine organification defect. J. Clin. Endocr. 39: 69-80, 1974.

Pommier, J.; Tourniaire, J.; Rahmoun, B.; Deme, D.; Pallo, D.; Bornet, H.; Nuzer, J.: Thyroid iodine organification defects. J. Clin. Endocr. 42: 319-329, 1976.

Seto, P.; Hirayu, H.; Magnusson, R. P.; Gestautas, J.; Portmann, L.; DeGroot, L. J.; Rapoport, B.: Isolation of a complementary DNA clone for thyroid microsomal antigen: homology with the gene for thyroid peroxidase. J. Clin. Invest. 80: 1205-1208, 1987.

Stanbury, J. B.: The metabolic errors in certain types of familial goiter. Recent Prog. Horm. Res. 19: 547-577, 1963.

Stanbury, J. B.: Familial goiter.In: Stanbury, J. B.; Wyngaarden, J. B.; Fredrickson, D. S.: The Metabolic Basis of Inherited Disease. New York: McGraw-Hill (pub.) (4th ed.): 1978. Pp. 206-239.

Valenta, L.; Bode, H. H.; Vickery, A. L.; Maloof, F.: Lack of thyroid peroxidase activity: a cause of congenital goitrous hypothyroidism. (Abstract) J. Clin. Invest. 50: 94A-95A, 1971.

Wolff, J.: Congenital goiter with defective iodide transport. Endocr. Rev. 4: 240-254, 1983.

THYROID HORMONOGENESIS, GENETIC DEFECT IN, 2B (*274600)

JJM NOTE

This disorder is included because of goiter, as well as thyroid carcinoma.

A mild type of organification defect is associated with congenital deafness. Patients show only partial discharge of iodide (25 to 50%) when thiocyanate or perchlorate is given (Fraser et al., 1960). Their thyroids are moderately enlarged from childhood. Patients are usually euthyroidal, though an exaggerated response to thyrotropin-releasing hormone (TRH; 275120) suggests a compensated hypothyroidism (Gomez-Pan et al., 1974); mental retardation has been reported (Thompson et al., 1970). Thyroid carcinoma has been observed (Thieme, 1957; Elman, 1958; Milutinovic et al., 1969); because of the characteristically 'wild' histology, malignancy may be incorrectly diagnosed. The deafness is neurosensory in type and sometimes associated with defective vestibular function. Vestibular disturbance was a striking feature in cases seen in Glasgow in 1969 with Professor J. H. Hutchinson. The deafness may be present at birth or develop in early childhood. Batsakis and Nishiyama (1962) estimated that Pendred syndrome accounts for 1 to 10% of hereditary deafness.

Illum et al. (1972) reported 15 cases. They showed a pedigree in which 8 proven cases and several presumed cases occurred in 3 generations of a family in a pseudodominant pattern (same family as that of Johnsen, 1958). In 1 patient, histologic examination showed a Mondini type malformation of the cochlea (i.e., only the basal cochlear turn was retained while the apical turns formed a common cavity). In 6 and perhaps 7 of the other 14 cases, the same defect was demonstrated by tomography of the temporal bones in the axial-pyramidal projection. The authors suggested that peroxidase deficiency may be responsible for the cochlear lesion as well as the thyroid defect. I noted a possible relationship between progression of deafness and the occurrence of trauma (JHH 141-02-07; JHH 147-02-08). Lesions in the organ of Corti have been produced in the chick and rat by administration of propylthiouracil during embryogenesis. The lesion did not occur when thyroxine was given with the antithyroid drug (Bargman and Gardner, 1967). Fraser (1965) estimated the frequency in the British Isles to be about 0.000075.

Cremers et al. (1998) described a boy whose bilateral sensorineural hearing loss progressed rapidly from about 50 to 60 decibels at the age of 3 years and 3 months to more than 100 decibels at the age of 4 years and 4 months. A search for the cause of the progressive hearing loss led to discovery of dysplasia of the cochlea and a widened vestibular aqueduct. Thyroid function tests were normal, but thyroglobulin was elevated. The diagnosis of Pendred syndrome was confirmed by the positive results of a potassium perchlorate test, indicating defective organic binding of the iodine in the thyroid gland. Aside from the branchiootorenal syndrome (113650), Pendred syndrome is the only known genetic disorder with a widened vestibular aqueduct.

Phelps et al. (1998) presented data on the radiologic malformations of the ear in 40 patients, all complying with strict diagnostic criteria for Pendred syndrome. Deficiency of the interscalar septum in the distal coils of the cochlea (Mondini deformity) was found to be common but probably not a constant feature of Pendred syndrome. On the other hand, enlargement of the endolymphatic sac and duct in association with a large vestibular aqueduct was present in all 20 patients examined by MRI. Phelps et al. (1998) concluded that thin section high resolution MRI on a T2 protocol in the axial and sagittal planes is the imaging investigation of choice in this disorder. They concluded that 'if there is uncertainty about cochlear malformation being a constant feature of Pendred syndrome, there can be little doubt as to the importance of enlarged vestibular aqueduct, endolymphatic sac, and endolymphatic duct, which were almost constant features' in their series.

Peroxidase activity is normal in Pendred syndrome (Burrow et al., 1973; Ljunggren et al., 1973; Cave and Dunn, 1975). There appear to be different varieties of Pendred syndrome, because Hollander et al. (1964) found a defect involving, apparently, an abnormal condensation of iodotyrosines to form iodothyronines, rather than an inadequate iodination of tyrosine. Milutinovic et al. (1969) found about 80% of the protein-bound radioiodine in the normal 19S human thyroglobulin fraction from a Pendred gland. On the other hand, Medeiros-Neto et al. (1968) found less than 15% in the 19S fraction. Desai et al. (1974) found a 15.2-16.8S radioiodinated thyroidal protein with immunologic properties of normal thyroglobulin. Fraser (1967) raised the question of whether the organification defect without deafness, as described by Stanbury and Hedge (1950) and later by Furth et al. (1967), was different from the organification defect with deafness as described by Pendred. He proposed that variability in severity of the same defect may be involved and supported this contention with a description of a patient with unilateral deafness whose sister had Pendred syndrome and bilateral deafness. Also, cases of the full syndrome and cases with near-normal hearing occurred in the same family.

Van Wouwe et al. (1986) found Pendred syndrome in a severely retarded girl with duplication-deficiency: duplication in 10p and deficiency in distal 8q. The father carried a de novo balanced translocation between 8q and 10p: t(8;10)(q24;p11). The authors raised the possibility that Pendred syndrome maps to 8q24-qter. It should be noted that the structural gene for thyroglobulin is in this segment. Although the proposita had deafness and hypothyroidism consistent with Pendred syndrome, as well as a marked reduction of (123)I uptake on perchlorate test, is it possible she has a thyroglobulin synthesis defect such as that discussed in 274900 and that the deafness has other cause?

By linkage, Gausden et al. (1996) excluded the thyroid peroxidase gene (274500) on chromosome 2 as the site of the mutation in their families with Pendred syndrome. This confirmed that at least 1 further, as yet unidentified step is required for complete organification of iodide within the thyroid. Observations that when both parents are affected there was no complementation for the disease phenotype suggested homogeneity. Coyle et al. (1996) excluded at least 8 autosomal recessive loci for nonsyndromic deafness that had been reported, although up to that time none had been cloned. However, they did find significant linkage to the DFNB4 locus (600791) located on 7q31. Twelve families with 2 or more affected individuals with Pendred syndrome were studied.

Sheffield et al. (1996) used a DNA-pooling strategy in an inbred kindred with 5 affected sibs and 2 other affected members to perform a genomewide linkage search for the Pendred syndrome disease locus. They mapped the locus to an approximately 9-cM interval on chromosome 7 in the 7q21-q34 region. Very few in vitro studies of thyroid tissue from Pendred patients had previously been reported, and no biochemical abnormality had been found consistently. Therefore, the possibility of a circulating inhibitor of organification had not been previously excluded. By measuring the major steps of thyroid hormonogenesis simultaneously in cryopreserved cultured cells from Pendred patients, Sheffield et al. (1996) conclusively showed the presence of a post-cAMP and post-iodine uptake defect. Furthermore, they found that the magnitude of the organification defects was similar to the decrease in T3 secretion, suggesting in Pendred patients iodide organification may be the rate-limiting step in thyroid hormone secretion.

Gausden et al. (1997), who concluded that the PDS locus is situated at 7q31 where a form of nonsyndromic deafness, DNFB4, has also been mapped, studied 5 kindreds and narrowed the locus to an interval between D7S501 and D7S525, separated by a genetic distance estimated to be 2.5 cM.

Coucke et al. (1997) concluded that the candidate region of approximately 1.7 cM is flanked by markers D7S501 and D7S692.

The perchlorate discharge test, the gold-standard investigation for Pendred syndrome, is nonspecific, and in the absence of alternative means of confirming the diagnosis, its sensitivity is unknown. Reardon et al. (1997) used the mapping of the Pendred syndrome gene to 7q to identify pedigrees, and assessed the prevalence of clinical parameters of disease in affected individuals. Cosegregation between disease and the locus on 7q was found in 36 familial cases. Clinical and investigative findings were compared in index cases (n = 18) versus affected sibs (n = 18). The overall prevalence of goiter was 73%, higher in index cases (94%) than in sibs (56%), many of whom had not previously been considered to have the condition. One perchlorate discharge test was false-negative (2.9%). Radiologic malformations of the cochlea were identified in 86% of cases. Reardon et al. (1997) concluded that securing a diagnosis of Pendred syndrome may be difficult, especially in the single case. They noted that the perchlorate discharge test, although valuable, is difficult to undertake in the younger patient, and radiology may assist in diagnosing such patients.

Everett et al. (1997) used a positional cloning strategy to identify the PDS gene mutated in Pendred syndrome. The PDS gene produces a transcript of approximately 5 kb that was found to be expressed at significant levels only in the thyroid. The predicted protein, called pendrin by the authors, is closely related to a number of sulfate transporters. Everett et al. (1997) found that PDS encodes a 780-amino acid (86-kD) protein. Sequences resembling that of human pendrin were found across a large taxonomic span, including animals, plants, and yeast, although the 2 closest relatives were the proteins encoded by the human DRA (126650) and DTD (222600) genes. The proteins within this family are highly hydrophobic; for example, 57% of the amino acids within pendrin are hydrophobic. In PDS families, Everett et al. (1997) found 3 apparently deleterious mutations, each segregating with the disease in the respective families in which they occurred (e.g., 274600.0001). Implicating loss-of-function mutations in PDS as a direct cause of thyroid disease in Pendred syndrome is relatively straightforward, as the gene is heavily expressed in thyroid, a tissue known to produce sulfated proteins. In particular, thyroglobulin (188450), the major secretory product of thyroid follicular cells, is sulfated on both tyrosine residues and asparagine-linked oligosaccharides. Sulfation of thyroglobulin is ubiquitous in organisms ranging from humans to lower eukaryotes, suggesting that this form of posttranslational modification serves an intrinsic role in thyroglobulin function. It had been suggested that thyroglobulin sulfation can influence subsequent iodination of the protein. The role of pendrin in cochlear development is less obvious. Everett et al. (1997) speculated that Pendred syndrome may be more common than previously thought. They pointed out that another recessive locus for deafness, designated DFNB4 (600791), maps to 7q31, the same region as the PDS gene. They considered it likel y that the DFNB4 individuals reported actually have Pendred syndrome, rather than mutations in another gene.

Van Hauwe et al. (1998) performed mutation analysis of the PDS gene in 14 Pendred families originating from 7 countries and identified all mutations. The mutations included 3 single base deletions, 1 splice site mutation, and 10 missense mutations. One missense mutation (leu236 to pro; 274600.0005) was found in homozygous state in 2 consanguineous families and in heterozygous state in 5 additional nonconsanguineous families, thus being present in half the families. Another missense mutation (thr416 to pro; 274600.0006) was found in homozygous state in 1 family and in heterozygous state in 4 families. Pendred patients in 3 nonconsanguineous families were shown to be compound heterozygotes for these 2 frequent mutations. In total, 1 or both of these mutations were found in 9 of the 14 families analyzed. The findings should facilitate molecular diagnosis of Pendred syndrome. Coyle et al. (1998) analyzed 56 kindreds, all with features suggestive of a diagnosis of Pendred syndrome. Molecular analysis of the PDS gene identified 47 of the 60 (78%) mutant alleles in 31 families; the families included 3 homozygous consanguineous kindreds and 1 extended family segregating 3 mutant alleles. Four recurrent mutations accounted for 35 (74%) of PDS disease chromosomes detected and haplotype analysis favored common founders rather than mutational hotspots within the PDS gene. A donor splice mutation (1001+1G-A) was observed in 10 families, in 5 of which descendants could be traced to the northeast of England, the region in which the family described in 1896 by Pendred resided (Reardon et al., 1997). Analysis of parental DNA failed to reveal any de novo mutations in this study cohort, either within the familial or the isolated cases of Pendred syndrome.

In a study of the structure of the DRA gene, (126650), which is mutant in congenital chloride diarrhea (CLD; 214700), Haila et al. (1998) found that the PDS gene is highly homologous to the DRA gene, with a similar genomic structure, and that it was situated 3-prime of the DRA gene in the same BAC clone.

History

This syndrome was described by Vaughan Pendred (1896). It was exactly a century later that Coyle et al. (1996) and Sheffield et al. (1996) showed that the disorder maps to chromosome 7q.

See also Baldwin et al. (1995)

Baldwin, C. T.; Weiss, S.; Farrer, L.; De Stefano, A.; Adair, R.; Franklyn, B.; Kidd, K. K.; Korostishevsky, M.; Bonne-Tamir, B.: Linkage of congenital, recessive deafness (DFNB4) to chromosome 7q31 and evidence for genetic heterogeneity in the Middle Eastern Druze population. Hum. Molec. Genet. 4: 1637-1642, 1995.
Bargman, G. J.; Gardner, L. I.: Otic lesions and congenital hypothyroidism in the developing chick. J. Clin. Invest. 46: 1828-1839, 1967.
Batsakis, J. G.; Nishiyama, R. H.: Deafness with sporadic goiter: Pendred's syndrome. Arch. Otolaryng. 76: 401-406, 1962.
Burrow, G. N.; Spaulding, S. W.; Alexander, N. M.; Bower, B. F.: Normal peroxidase activity in Pendred's syndrome. J. Clin. Endocr. 36: 522-530, 1973.
Cave, W. T., Jr.; Dunn, J. T.: Studies on the thyroidal defect in an atypical form of Pendred's syndrome. J. Clin. Endocr. Metab. 41: 590-599, 1975.
Coucke, P.; Van Camp, G.; Demirhan, O.; Kabakkaya, Y.; Balemans, W.; Van Hauwe, P.; Van Agtmael, T.; Smith, R. J. H.; Parving, A.; Bolder, C. H. H. M.; Cremers, C. W. R. J.; Willems, P. J.: The gene for Pendred syndrome is located between D7S501 and D7S692 in a 1.7-cM region on chromosome 7q. Genomics 40: 48-54, 1997.
Coyle, B.; Coffey, R.; Armour, J. A. L.; Gausden, E.; Hochberg, Z.; Grossman, E.; Britton, K.; Pembrey, M.; Reardon, W.; Trembath, R.: Pendred syndrome (goitre and sensorineural hearing loss) maps to chromosome 7 in the region containing the nonsyndromic deafness gene DFNB4. Nature Genet. 12: 421-423, 1996.
Coyle, B.; Reardon, W.; Herbrick, J.-A.; Tsui, L.-C.; Gausden, E.; Lee, J.; Coffey, R.; Grueters, A.; Grossman, A.; Phelps, P. D.; Luxon, L.; Kendall-Taylor, P.; Scherer, S. W.; Trembath, R. C.: Molecular analysis of the PDS gene in Pendred syndrome (sensorineural hearing loss and goitre). Hum. Molec. Genet. 7: 1105-1112, 1998.
Cremers, C. W. R. J.; Bolder, C.; Admiraal, R. J. C.; Everett, L. A.; Joosten, F. B. M.; van Hauwe, P.; Green, E. D.; Otten, B. J.: Progressive sensorineural hearing loss and a widened vestibular aqueduct in Pendred syndrome. Arch. Otolaryng. Head Neck Surg. 124: 501-505, 1998.
Desai, K. B.; Metita, M. N.; Patel, M. C.; Ramanna, L.; Ganatra, R. D.: Thyroidal iodoproteins in Pendred syndrome. J. Endocr. 63: 409-410, 1974.
Elman, D. S.: Familial association of nerve deafness with nodular goiter and thyroid carcinoma. New Eng. J. Med. 259: 219-223, 1958.
Everett, L. A.; Glaser, B.; Beck, J. C.; Idol, J. R.; Buchs, A.; Heyman, M.; Adawi, F.; Hazani, E.; Nassir, E.; Baxevanis, A. D.; Sheffield, V. C.; Green, E. D.: Pendred syndrome is caused by mutations in a putative sulphate transporter gene (PDS). Nature Genet. 17: 411-422, 1997.
Fraser, G. R.: Association of congenital deafness with goiter (Pendred's syndrome): a study of 207 families. Ann. Hum. Genet. 28: 201-249, 1965.
Fraser, G. R.: Personal Communication. Adelaide, Australia 1967.
Fraser, G. R.; Morgans, M. E.; Trotter, W. R.: The syndrome of sporadic goiter and congenital deafness. Quart. J. Med. 29: 279-295, 1960.
Furth, E. D.; Carvalho, M.; Vianna, B.: Familial goiter due to an organification defect in euthyroid siblings. J. Clin. Endocr. 27: 1137-1140, 1967.
Gausden, E.; Armour, J. A. L.; Coyle, B.; Coffey, R.; Hochberg, Z.; Pembrey, M.; Britton, K. E.; Grossman, A.; Reardon, W.; Trembath, R.: Thyroid peroxidase: evidence for disease gene exclusion in Pendred's syndrome. Clin. Endocr. 44: 441-446, 1996.
Gausden, E.; Coyle, B.; Armour, J. A. L.; Coffey, R.; Grossman, A.; Fraser, G. R.; Winter, R. M.; Pembrey, M. E.; Kendall-Taylor, P.; Stephens, D.; Luxon, L. M.; Phelps, P. D.; Reardon, W.; Trembath, R.: Pendred syndrome: evidence for genetic homogeneity and further refinement of linkage. J. Med. Genet. 34: 126-129, 1997.
Gomez-Pan, A.; Evered, D. C.; Hall, R.: Pituitary-thyroid function in Pendred syndrome. Brit. Med. J. 2: 152-153, 1974.
Haila, S.; Hoglund, P.; Scherer, S. W.; Lee, J. R.; Kristo, P.; Coyle, B.; Trembath, R.; Holmberg, C.; de la Chapelle, A.; Kere, J.: Genomic structure of the human congenital chloride diarrhea (CLD) gene. Gene 214: 87-93, 1998.
Hollander, C. S.; Prout, T. E.; Rienhoff, M.; Ruben, R. J.; Asper, S. P., Jr.: Congenital deafness and goiter: studies of a patient with a cochlear defect and inadequate formation of iodothyronines. Am. J. Med. 37: 630-637, 1964.
Illum, P.; Kiaer, H. W.; Hvidberg-Hansen, J.; Sondergaard, G.: Fifteen cases of Pendred's syndrome. Arch. Otolaryng. 96: 297-304, 1972.
Johnsen, S.: Familial deafness and goiter in persons with a low level of protein-bound iodine. Acta Otolaryng. 140 (suppl.): 168-177, 1958.
Li, X. C.; Everett, L. A.; Lalwani, A. K.; Desmukh, D.; Friedman, T. B.; Green, E. D.; Wilcox, E. R.: A mutation in PDS causes non-syndromic recessive deafness. (Letter) Nature Genet. 18: 215-217, 1998.
Ljunggren, J. G.; Lindstrom, H.; Ajern, B.: The concentration of peroxidase in normal and adenomatous human thyroid tissue with special reference to patients with Pendred syndrome. Acta Endocr. 72: 272-278, 1973.
Medeiros-Neto, G. A.; Nicolau, W.; Kieffer, J.; Ulhoa Cintra, A. B.: Thyroidal iodoproteins in Pendred's syndrome. J. Clin. Endocr. 28: 1205-1213, 1968.
Milutinovic, P. S.; Stanbury, J. B.; Wicken, J. V.: Thyroid function in a family with the Pendred syndrome. J. Clin. Endocr. 29: 962-969, 1969.
Pendred, V.: Deaf-mutism and goitre. Lancet II: 532, 1896.
Phelps, P. D.; Coffey, R. A.; Trembath, R. C.; Luxon, L. M.; Grossman, A. B.; Britton, K. E.; Kendall-Taylor, P.; Graham, J. M.; Cadge, B. C.; Stephens, S. G. D.; Pembrey, M. E.; Reardon, W.: Radiological malformations of the ear in Pendred syndrome. Clin. Radiol. 53: 268-273, 1998.
Reardon, W.; Coffey, R.; Phelps, P. D.; Luxon, L. M.; Stephens, D.; Kendall-Taylor, P.; Britton, K. E.; Grossman, A.; Trembath, R.: Pendred syndrome—100 years of underascertainment? Quart. J. Med. 90: 443-447, 1997.
Sheffield, V. C.; Kraiem, Z.; Beck, J. C.; Nishimura, D.; Stone, E. M.; Salameh, M.; Sadeh, O.; Glaser, B.: Pendred syndrome maps to chromosome 7q21-34 and is caused by an intrinsic defect in thyroid iodine organification. Nature Genet. 12: 424-426, 1996.
Stanbury, J. B.; Hedge, A. N.: A study of a family of goitrous cretins. J. Clin. Endocr. 10: 1471-1484, 1950.
Thieme, E. T.: A report of the occurrence of deaf-mutism and goiter in four of six siblings of a North American family. Ann. Surg. 146: 941-948, 1957.
Thompson, J.; Maguire, W. C.; Hurwitz, L. J.: A family with deafness, goiter, epilepsy and low intelligence segregating independently. Irish J. Med. Sci. 3: 427-431, 1970.
Van Hauwe, P.; Everett, L. A.; Coucke, P.; Scott, D. A.; Kraft, M. L.; Ris-Stalpers, C.; Bolder, C.; Otten, B.; de Vijlder, J. J. M.; Dietrich, N. L.; Ramesh, A.; Srisailapathy, S. C. R.; Parving, A.; Cremers, C. W. R. J.; Willems, P. J.; Smith, R. J. H.; Green, E. D.; Van Camp, G.: Two frequent missense mutations in Pendred syndrome. Hum. Molec. Genet. 7: 1099-1104, 1998.
van Wouwe, J. P.; Wijnands, M. C.; Mourad-Baars, P. E. C.; Geraedts, J. P. M.; Beverstock, G. C.; van de Kamp, J. J. P.: A patient with dup(10p)del(8q) and Pendred syndrome. Am. J. Med. Genet. 24: 211-217, 1986.

THYROID HORMONOGENESIS, GENETIC DEFECT IN, 3 (*274700)

JJM NOTE

This disorder is included because of goiter.

Nunez et al. (1975) showed that thyroid peroxidase catalyzes three different reactions and exists in two interchangeable forms, A and B. Form A catalyzes iodide oxidation and high-rate thyroglobulin iodination, whereas form B catalyzes low-rate thyroglobulin iodination and iodotyrosyl coupling. Pommier et al. (1974) studied thyroid tissue from a euthyroid patient with a childhood goiter in whom iodide oxidation was normal and thyroglobulin iodination was only slightly reduced, yet coupling of the iodotyrosines was markedly reduced. They proposed that this defect was secondary to a lack of conformational change from form A to form B. Both a defect in the third peroxidase reaction (primary coupling defect) and alteration of amino acid sequence within thyroglobulin, changing the total number or the intramolecular position of the iodotyrosines (secondary coupling defect), could result in the same phenotype. Some patients have been cretinous while others only had goiters; therefore, heterogeneity may exist in this group of patients. Buhler et al. (1964) reported an early case of 18p- syndrome. In addition to the typical manifestations of that syndrome, the girl had an abnormality of thyroxine synthesis, most likely a coupling defect (Buhler, 1983). Since the proband was the youngest of 7 sibs and none of the other sibs had a thyroxine defect, Buhler (1983) suggested that the deletion may have 'uncovered' a heterozygous mutation inherited from 1 parent and that the gene locus is on the short arm of chromosome 18.

Stanbury and Dumont (1983) indicated that the coupling defects represent a 'poorly defined group, which is almost surely heterogeneous.'

See also Alexander and Burrow (1970); Morris (1964); Nunez et al. (1976)

Alexander, N. M.; Burrow, G. N.: Thyroxine biosynthesis in human goitrous cretinism. J. Clin. Endocr. 30: 308-315, 1970.
Buhler, E. M.: Unmasking of heterozygosity by inherited balanced translocations: implications for prenatal diagnosis and gene mapping. Ann. Genet. 26: 133-137, 1983.
Buhler, E. M.; Buhler, U. K.; Stalder, G. R.: Partial monosomy 18 and anomaly of thyroxine synthesis. Lancet I: 170-171, 1964.
Morris, J. H.: Defective coupling of iodotyrosine in familial goiters: report of two patients. Arch. Intern. Med. 114: 417-423, 1964.
Nunez, J.; Pommier, J.; Dominici, R.; Rahmoun, B.; Deme, D.; Tourniaire, J.: Peroxidases and thyroglobulins from different goiters. In: Robbins, J.; Braverman, L. E.: Proc. 7th Int. Thyroid Conf., Boston, 1975. Amsterdam: Excerpta Medica (pub.) 1976. Pp. Abstract only.
Pommier, J.; Tourniaire, J.; Deme, D.; Chalendar, D.; Bornet, H.; Nunez, J.: A defective thyroid peroxidase solubilized from a familial goiter with iodine organification defect. J. Clin. Endocr. 39: 69-80, 1974.
Stanbury, J. B.; Dumont, J. E.: Familial goiter and related disorders. In: Stanbury, J. B.; Wyngaarden, J. B.; Fredrickson, D. S.; Goldstein, J. L.; Brown, M. S.: The Metabolic Basis of Inherited Disease. New York: McGraw-Hill (pub.) (5th ed.): 1983. Pp. 231-269.

THYROID HORMONOGENESIS, GENETIC DEFECT IN, 4 (*274800)

JJM NOTE

This disorder is included because of goiter.

The deiodinases are a group of membrane bound, NADPH-dependent, FAD-enhanced isoenzymes found in the thyroid, kidney, liver, and other organs. Patients with this defect lack the ability to deiodinate radiolabeled monoiodotyrosine (MIT) and diiodotyrosine (DIT). This results in continuous urinary loss of iodine and tyrosine from the body. Most of the earlier patients had severe congenital hypothyroidism. In a classic study, 2,3:Hutchison and McGirr (1954, 1956) studied this disorder in inbred, itinerant tinkers from western Scotland. Werdnig-Hoffman disease occurred also in this family, apparently as an independently inherited disorder. Kusakabe and Miyake (1963) described a mother and daughter plus 2 other unrelated individuals who showed defective peripheral iodotyrosine deiodination but essentially normal in vitro thyroidal activity. The same authors (1964) reported 3 affected sisters, from a sibship of 7 born to healthy, first-cousin parents, whose thyroidal tissue lacked deiodinase activity, but whose peripheral tissues deiodinated radiolabeled MIT and DIT normally. The patients with these limited tissue defects had normal growth and development. Ismail-Beigi and Rahimifar (1977) found a mild variant of the defect in 3 children born to unaffected first cousins. The hypothyroidism in this disorder results not from dyshormonogenesis, but from simple iodine depletion (Stanbury, 1978). This has been well demonstrated by Codaccioni et al. (1970), who maintained euthyroidism in 5 patients with the complete defect by giving supplemental iodide alone. Rochiccioli and Dutau (1974) showed reduced deiodination of radiolabeled DIT after loading obligated carriers with cold DIT, thereby making heterozygote detection possible.

See also Kusakabe and Miyake (1964)

Codaccioni, J. L.; Pierron, H.; Rouault, F.; Aquaron, R.; Jaquet, P.: Hypothyroidie infantile par defaut d-iodotyrosine-des-halogenase. II Resultants du traitment par l'iode de 5 cas. Ann. Endocr. 31: 1174-1182, 1970.
Hutchison, J. H.; McGirr, E. M.: Hypothyroidism as an inborn error of metabolism. J. Clin. Endocr. Metab. 14: 869-886, 1954.
Hutchison, J. H.; McGirr, E. M.: Sporadic non-endemic goitrous cretinism. Hereditary transmission. Lancet 1: 1035-1037, 1956.
Ismail-Beigi, F.; Rahimifar, M.: A variant of iodotyrosine-dihalogenase deficiency. J. Clin. Endocr. 44: 499-506, 1977.
Kusakabe, T.; Miyake, T.: Defective deiodination of I-131-labeled L-diiodo-tyrosine in patients with simple goiter. J. Clin. Endocr. 23: 132-139, 1963.
Kusakabe, T.; Miyake, T.: Thyroidal deiodination defect in three sisters with simple goiter. J. Clin. Endocr. 24: 456-459, 1964.
Rochiccioli, P.; Dutau, G.: Trouble de l'hormonosynthese thyroidienne par deficit en iodotyrosine-deshalogenase. Arch. Franc. Pediat. 31: 25-36, 1974.
Stanbury, J. B.: Familial goiter.In: Stanbury, J. B.; Wyngaarden, J. B.; Fredrickson, D. S.: The Metabolic Basis of Inherited Disease. New York: McGraw-Hill (pub.) (4th ed.): 1978. Pp. 206-239.

THYROID HORMONOGENESIS, GENETIC DEFECT IN, 5 (274900)

JJM NOTE

This disorder is included because of goiter.

Thyroglobulin is a large glycoprotein with 2 identical subunits. Iodination occurs in the extracellular colloid space. Numerous defects could produce the same phenotype. Reports have varied, with few patients having severe hypothyroidism. Some have a low PBI commensurate with their low serum T4; others have a normal or elevated PBI with a low or low-normal serum T4. The thyroid may contain very small amounts of normally iodinated thyroglobulin or ample amounts of immunologically identifiable thyroglobulin with abnormal physical properties such as solubility (Michel et al., 1964), protease digestibility (McGirr et al., 1960), and iodinatibility (Kusakabe, 1972). Riddick et al. (1969) reported on 3 goitrous members of a sibship of 4. These patients had hypothyroidism or compensated hypothyroidism, and had normal or high uptake of radioiodine; biochemical measurements on removed thyroid tissue showed absence of thyroglobulin with the appearance of abnormal light iodoproteins. (Any block in thyroglobulin synthesis is associated with increased iodination of serum albumin that diffuses into the gland.) Lissitzky et al. (1975) found a marked reduction of the carbohydrate moieties, supposedly necessary for secretion, in the thyroglobulin from a congenital goiter. Electron micrographs showed scarcity of colloid in the follicular lumen and overdistended, protein-filled endoplasmic reticulum. Increased metabolic activity of the gland or any block in thyroglobulin synthesis results in iodination of the serum albumin that diffuses into the hyperplastic thyroid (Stanbury, 1978). Patients previously classified as having familial plasma iodoprotein defects can now be categorized as having some type of thyroglobulin abnormality or a nonmendelian familial abnormality of the thyroid, such as Hashimoto struma (140300) or Graves disease (275000). In an excellent review of all inherited disorders of thyroid metabolism, Lever et al. (1983) discussed the heterogeneous group of disorder s of hormonogenesis related to abnormal TG formation or secretion. Hypothyroidism in the goat and sheep has been related to a defect in mRNA for TG (Van Voorthuizen et al., 1978; Falconer et al., 1970). Baas et al. (1984) suggested that hereditary congenital hypothyroidism due to a defect in the synthesis and structure of thyroglobulin may be autosomal dominant. They found an RFLP involving the TG gene that cosegregated with the clinical phenotype in a kindred (see 188450). Baas et al. (1984) suggested that since the tertiary and quaternary structure of TG is very important for hormone formation, a change in one of the two subunits (heterozygosity) may lead to severely impaired hormonogenesis of the heterodimeric TG and thus autosomal dominant inheritance of this disorder characterized by relatively high levels of abnormal TG.

See also Lissitzky et al. (1967)

Baas, F.; Bikker, H.; van Ommen, G.-J. B.; de Vijlder, J. J. M.: Unusual scarcity of restriction site polymorphism in the human thyroglobulin gene: a linkage study suggesting autosomal dominance of a defective thyroglobulin allele. Hum. Genet. 67: 301-305, 1984.
Falconer, I. R.; Roitt, I. M.; Seamark, R. F.; Torrigiani, G.: Studies of the congenitally goitrous sheep: iodoproteins of the goitre. Biochem. J. 117: 417-424, 1970.
Kusakabe, T.: A goitrous subject with a structural abnormality of thyroglobulin. J. Clin. Endocr. 35: 785-794, 1972.
Lever, E. G.; Medeiros-Neto, G. A.; DeGroot, L. J.: Inherited disorders of thyroid metabolism. Endocr. Rev. 4: 213-239, 1983.
Lissitzky, S.; Codaccioni, J. L.; Bismuth, J.; Depieds, R.: Congenital goiter with hypothyroidism and iodo-serum albumin replacing thyroglobulin. J. Clin. Endocr. 27: 185-196, 1967.
Lissitzky, S.; Torresani, J.; Burrow, G. N.; Bouchilloux, S.; Ghabaud, O.: Defective thyroglobulin export as a cause of congenital goitre. Clin. Endocr. 4: 363-392, 1975.
McGirr, E. M.; Hutchison, J. H.; Clement, W. E.; Kennedy, J. S.; Currie, A. R.: Goitre and cretinism due to the production of an abnormal iodinated thyroid compound. Scot. Med. J. 5: 189-203, 1960.
Michel, R.; Rall, J. E.; Roche, J.; Tubiana, M.: Thyroidal iodoproteins in patients with goitrous hypothyroidism. J. Clin. Endocr. 24: 352-358, 1964.
Riddick, F. A., Jr.; Desai, K. B.; Murison, P. J.; Stanbury, J. B.: Familial goiter with diminished synthesis of thyroglobulin. J. Exp. Med. 150: 203-212, 1969.
Stanbury, J. B.: Familial goiter. In: Stanbury, J. B.; Wyngaarden, J. B.; Fredrickson, D. S.: The Metabolic Basis of Inherited Disease. New York: McGraw-Hill (pub.) (4th ed.): 1978. Pp. 206-239.
Van Voorthuizen, W. F.; Dinsart, C.; Flavell, R. A.; DeVijlder, J. J. M.; Vassart, G.: Abnormal cellular localization of thyroglobulin mRNA associated with hereditary congenital goiter and thyroglobulin deficiency. Proc. Nat. Acad. Sci. 75: 74-78, 1978.

THYROTOXICOSIS (GRAVES DISEASE) (275000)

JJM NOTE

This disorder is included because of goiter.

Bartels (1941) claimed that this disorder is inherited as a simple autosomal recessive with relative sex limitation to females and reduced penetrance (70-80%) in homozygotes. Martin and Fisher (1945) also postulated a recessive factor predisposing to exophthalmic goiter. In contrast, these workers (1951) could find no evidence of hereditary basis of toxic nodular goiter. Ingbar et al. (1956) found abnormalities of thyroid metabolism in euthyroid relatives of a thyrotoxic patient. Levit, in early studies in the U.S.S.R., was more inclined toward dominant inheritance (Fraser, 1967). Neither the recessive nor the dominant hypothesis has satisfactory proof. Impressively extensive involvement occurs in some families. Skillern (1972) favored polygenic inheritance for hyperthyroidism and for Hashimoto thyroiditis (140300) which may accompany it. Hollingsworth and Mabry (1976) focused on congenital Graves disease, which usually has been described as a transient disorder in the newborn offspring of women who have or have had hyperthyroidism. The authors concluded that long-acting thyroid stimulator (LATS), an immunoglobulin, is not the cause because it has a half-life of only 6 days and the thyroid problem in the patients they studied persisted for months or years. They favored autosomal dominant inheritance with female predilection. They quoted an experience of Dr. A. M. DiGeorge (Philadelphia) with affected father and son. The father had symptoms from age 3 years and the son from birth; both required antithyroid medication. On the other hand, Wit et al. (1986) held that neonatal thyrotoxicosis, which they observed in sibs, is caused by transplacental passage of thyroid-stimulating immunoglobulins (TSI). The first infant was treated with exchange transfusion and potassium iodide; the second with intrauterine prophylthiouracil (given to the mother) to control fetal heart rate, followed by potassium iodide. The mother was euthyroid as a result of subtotal thyroidectomy. This is a disorder that simulates mendelism. Mortality from neonatal thyrotoxicosis as well as intellectual impairment in later life is considerable. Horwitz and Refetoff (1977) concluded that Graves disease has an increased frequency in patients with familial deficiency of TBG (314200).

Sasazuki (1981) presented data he interpreted as indicating that two major genes, one linked to human leukocyte antigen, or HLA (on chromosome 6) and one linked to Gm (on chromosome 14), control susceptibility to Graves disease. He studied 30 Japanese families with 2 or more affected first-degree relatives. All affected sibs, except for 1 female, shared both HLA and Gm haplotypes. The exceptional case was explicable, he thought, on the basis of recombination between HLA and an HLA-linked susceptibility gene for Graves disease. The HLA-linked gene appeared to be in strong linkage disequilibrium with HLA-DR5, but no significant association was noted between Graves disease and a particular immunoglobulin allotype. From statistical analysis, using the estimated incidence of Graves disease in Japanese as 0.008, he concluded that both the HLA- and the Gm-linked genes appear to be recessive and that their gene frequencies are 0.18 and 0.20, respectively, with a penetrance between 0.45 and 1.0. Uno et al. (1981) presented evidence that one of these genes is closely linked to HLA-DR and that the other is linked to the gene coding for the Gm allotype. Payami et al. (1989) confirmed the positive association of Graves disease with HLA-DR3. In addition, they found a negative association between DR5 and Graves disease; this was not a result of the increase in DR3. When patients with or without eye disease were considered separately, a significant increase in the frequency of DR4 was found in those without eye disease. When the authors applied the sib-pair method to the data, they found frequencies of haplotype sharing that deviated significantly from the expected one-fourth, one-half, one-fourth distribution that is expected if the disease is not linked to HLA. This indicated the presence of a disease gene in or near the HLA region in both the Caucasian and the Chinese populations studied. The autoimmunogen in Graves disease is thyrotropin receptor (TSHR; 275200); see 275200.0001.

Donner et al. (1997) determined the frequency of the cytotoxic T-lymphocyte-associated serine esterase-4 (CTLA4; 123890) position 49 polymorphism (G/A; Ala/Thr) in 305 patients with Graves disease, 293 patients with insulin-dependent diabetes mellitus (IDDM; 222100), and 325 controls. Patients with Graves disease had significantly more Ala-17 alleles than controls, both as homozygotes (21 vs 13%) and as heterozygotes (53 vs 46%), and fewer Thr-17 alleles as homozygotes (26 vs 42%; P less than 2 x 10-4). The phenotypic frequency of Ala-positive patients (73%) was significantly higher than of controls (58%; P = 10-4; relative risk = 2). Patients with IDDM also had significantly more Ala alleles as homozygotes (19%) or heterozygotes (50%; P = 0.01). Wilkin (1990) reviewed endocrine disorders of hormone excess and hormone deficiency resulting from receptor autoimmunity.

To identify susceptibility genes, Tomer et al. (1997) studied DNA markers in the regions of 8 candidate genes: the HLA region, TSHR, thyroid peroxidase, thyroglobulin (188450), IDDM4 (600319), IDDM5 (600320), immunoglobulin heavy chain gene (147100), and CTLA-4. In patients with Graves disease, negative lod scores were obtained for all the candidate genes, except for markers in the TSHR region on chromosome 14q31. Marker D14S81 gave the highest score (maximum lod = 2.05, theta = 0.01) assuming a dominant mode of inheritance and a penetrance of 0.8. These data confirm the authors' previous observations of a lack of a necessary disease locus for Graves disease in the HLA gene region and suggested the presence of an important susceptibility gene on 14q31, but at a considerable distance from the TSHR gene.

Congenital hyperthyroidism has been shown in some cases to be due to de novo mutations in the thyrotropin receptor gene (TSHR; 275200.0004) that result in a constitutive activation of the receptor. The TSHR gene is located on 14q31. Furthermore, familial nonautoimmune hyperthyroidism (Thomas et al., 1982) was shown to be due to germline mutations in the TSHR gene (Duprez et al., 1994). Patients with autosomal dominant nonautoimmune hyperthyroidism do not have the clinical manifestations of Graves disease, such as thyroid-associated ophthalmopathy, pretibial myxedema, or lymphocytic infiltration of the thyroid, nor do they have thyroid antibodies (Paschke and Ludgate, 1997). The thyroid gland is enlarged in most patients. Hyperthyroidism can occur at any time from the neonatal period to adulthood. This variability in age at onset is probably the result of other genetic components and environmental factors, such as iodine intake and dietary goitrogens.

See also Adams et al. (1983); Farid et al. (1979); Green (1976); Hollingsworth et al. (1972); Martin and Fisher (1951)

Adams, D. D.; Adams, Y. J.; Knight, J. G.; McCall, J.; White, P.; Parkinson, R.; Horrocks, R.; van Loghem, E.: On the nature of the genes influencing the prevalence of Graves' disease. Life Sci. 31: 3-13, 1983.

Bartels, E. D.: Heredity in Graves' Disease. Copenhagen: Munksgaard (pub.) 1941.

Donner, H.; Rau, H.; Walfish, P. G.; Braun, J.; Siegmund, T.; Finke, R.; Herwig, J.; Usadel, K. H.; Badenhoop, K.: CTLA4 alanine-17 confers genetic susceptibility to Graves' disease and to type 1 diabetes mellitus. J. Clin. Endocr. Metab. 82: 143-146, 1997.

Duprez, L.; Parma, J.; Van Sande, J.; Allgeier, A.; Leclere, J.; Schvartz, C.; Delisle, M.-J.; Decoulx, M.; Orgiazzi, J.; Dumont, J.; Vassart, G.: Germline mutations in the thyrotropin receptor gene cause non-autoimmune autosomal dominant hyperthyroidism. Nature Genet. 7: 396-401, 1994.

Farid, N. R.; Sampson, L.; Noel, E. P.; Barnard, J. M.; Mandeville, R.; Larsen, B.; Marshall, W. H.: A study of human leukocyte D locus related antigens in Graves' disease. J. Clin. Invest. 63: 108-113, 1979.

Fraser, G. R.: Personal Communication. Adelaide, Australia 1967.

Green, W. L.: Humoral and genetic factors in thyrotoxic Graves disease and neonatal thyrotoxicosis. J.A.M.A. 235: 1449-1450, 1976.

Hollingsworth, D. R.; Mabry, C. C.: Congenital Graves disease: four familial cases with long-term follow-up and perspective. Am. J. Dis. Child. 130: 148-155, 1976.

Hollingsworth, D. R.; Mabry, C. C.; Eckerd, J. M.: Hereditary aspects of Graves' disease in infancy and childhood. J. Pediat. 81: 446-459, 1972.

Horwitz, D. L.; Refetoff, S.: Graves' disease associated with familial deficiency of thyroxine-binding globulin. J. Clin. Endocr. Metab. 44: 242-247, 1977.

Ingbar, S. H.; Freinkel, N.; Dowling, J. T.; Kumagai, L. F.: Abnormalities of iodine metabolism in euthyroid relatives of patients with Graves' disease. (Abstract) J. Clin. Invest. 35: 714 only, 1956.

Martin, L.; Fisher, R. A.: The hereditary and familial aspects of toxic nodular goitre (secondary thyrotoxicosis). Quart. J. Med. 20: 293-297, 1951.

Martin, L.; Fisher, R. A.: The hereditary and familial aspects of exophthalmic goiter and nodular goitre. Quart. J. Med. 14: 207-219, 1945.

Paschke, R.; Ludgate, M.: The thyrotropin receptor in thyroid diseases. New Eng. J. Med. 337: 1675-1681, 1997.

Payami, H.; Joe, S.; Farid, N. R.; Stenszky, V.; Chan, S. H.; Yeo, P. P. B.; Cheah, J. S.; Thomson, G.: Relative predispositional effects (RPEs) of marker alleles with disease: HLA-DR alleles and Graves disease. Am. J. Hum. Genet. 45: 541-546, 1989.

Sasazuki, T.: Two major genes, linked to HLA and Gm, control the susceptibility to Graves' disease. (Abstract) Sixth Int. Cong. Hum. Genet., Jerusalem 210 only, 1981.

Skillern, P. G.: Genetics of Graves' disease. Mayo Clin. Proc. 47: 848-849, 1972.

Thomas, J. L.; Leclere, J.; Hartemann, P.; Duheille, J.; Orgiazzi, J.; Petersen, M.; Janot, C.; Guedenet, J.-C.: Familial hyperthyroidism without evidence of autoimmunity. Acta Endocr. 100: 512-518, 1982.

Tomer, Y.; Barbesino, G.; Keddache, M.; Greenberg, D. A.; Davies, T. F.: Mapping of a major susceptibility locus for Graves' Disease (GD-1) to chromosome 14q31. J. Clin. Endocr. Metab. 82: 1645-1648, 1997.

Uno, H.; Sasazuki, T.; Tamai, H.; Matsumoto, H.: Two major genes, linked to HLA and Gm, control susceptibility to Graves' disease. Nature 292: 768-770, 1981.

Wilkin, T. J.: Receptor autoimmunity in endocrine disorders. New Eng. J. Med. 323: 1318-1324, 1990.

Wit, J. M.; Rees-Smith, B.; Creagh, F. M.; Bruinse, H. W.; van der Heide, D.; Docter, R.; Gerards, L. J.: Thyroid-stimulating immunoglobulins and thyroid function tests in two siblings with neonatal thyrotoxicosis. Europ. J. Pediat. 145: 143-147, 1986.

Pancreatic Islet Tumors

NESIDIOBLASTOSIS OF PANCREAS (#256450)

JJM NOTE

Many molecular genetic approaches could be explored when these rare patients undergo pancreatectomy: loss of heterozygosity and cytogenetics studies using the candidate genes (e.g., multiple endocrine neoplasia 1 and Beckwith-Wiedemann); clonal origins (e.g., using X-linked probes in females).

A number sign (#) is used with this entry because of evidence that the disorder is caused by mutations in the beta-cell high-affinity sulfonylurea receptor gene (SUR1; 600509) or in the inwardly rectifying potassium channel gene KCNJ11 (600937). Both of these genes are located on 11p15 where the phenotype of persistent hyperinsulinemic hypoglycemia of infancy (PHHI) has been mapped.

Nesidioblastosis is a diffuse abnormality of the pancreas in which there is extensive, often disorganized formation of new islets. The term nesidioblastosis (meaning neoformation of islets of Langerhans from pancreatic duct epithelium) was coined by Laidlaw (1938). 'Nesidio' comes from a Greek word for islet. Yakovac et al. (1971) was first to report nesidioblastosis in a series of infants with intractable hypoglycemia. Woo et al. (1976) reported sibs. Schwartz et al. (1979) observed the disorder in 5 children of both sexes from 2 families with unaffected parents. The disorder presented as severe neonatal hypoglycemia. Pancreatectomy was required. Dahms et al. (1980) recognized 2 histologic groups of nesidioblastosis among cases of hyperinsulinemic hypoglycemia: group I had diffuse hyperplasia of the islets of Langerhans as well as nesidioblastosis; group II had more subtle nesidioblastosis alone. Group I patients were 8 months old or younger. Group II patients ranged in age from 3 to 15 years. Four of the group I patients had the Beckwith EMG syndrome (130650). Wuthrich et al. (1986) described affected sibs. Moreno et al. (1989) described the disorder in a brother and sister whose parents were consanguineous. In addition, a brother and sister who died in the neonatal period were probably affected. Glaser et al. (1990) reported a total of 21 cases in 7 pedigrees, which included a large Bedouin family, an Arab family, and 5 Jewish families of Eastern European origin (Ashkenazi). Consanguinity was evident in the Bedouin and Arab families. Woolf et al. (1991) found consanguineous parentage in 5 of 28 families. Segregation analysis showed that the ratio of subsequent affected to unaffected sibs was similar to that expected of an autosomal recessive condition. The sex ratio was close to 1.

Of 26 families seen over a period of 15 years, Thornton et al. (1991) found that 5 had more than 1 affected child (19%). There was no apparent difference clinically, biochemically, or histologically between patients in the 5 multiplex and 21 simplex families. The segregation ratio, 0.254, agreed closely with that expected for an autosomal recessive disorder. Pancreatic histologic data were available for 22 of the 25 patients who underwent pancreatectomy. Differing histologic findings were found in 3 infants in 1 family and 2 in another family; both within and among multiplex families, there was no consistency of the histologic change. Four of the simplex cases had an adenoma; all were cured by its removal. None of these patients had additional histologic abnormalities. Thornton et al. (1991) commented that multiple cases of adenoma had not been reported in sibships. On the basis of their large pathologic and clinical review, Rahier et al. (1984) also suggested that adenomas constitute a separate cause of persistent neonatal hypoglycemia. The variable histologic changes suggested that it is not the specific histologic changes that are genetically determined but rather an abnormality of regulation of the beta cells. Rahier (1989) discussed the relevance of pancreatic histologic changes in patients with hyperinsulinism. Horev et al. (1991) described a family in which 4 of 13 children had hyperinsulinism with variable outcomes. One sib required pancreatectomy and 3 were treated successfully with diazoxide and were able to discontinue this therapy between ages 1 and 2 years.

Burman et al. (1992) described a brother and sister, aged 42 and 34, respectively, with recurrent syncope due to severe hyperinsulinemic hypoglycemia. Exploratory laparotomy in the brother showed a grossly normal pancreas, but histologic examination showed islet cell hyperplasia. Neither sib had any evidence of multiple endocrine neoplasia type I (131100) or any other family history to suggest this diagnosis. In adults, islet cell hyperplasia occurs almost exclusively in MEN I. Because the brother related his symptoms to exercise, Burman et al. (1992) used treadmill exercise in both patients to diagnose hyperinsulinism and to observe its response to therapy.

Bianchi et al. (1992) described the prenatal diagnosis of nesidioblastosis on the basis of high levels of insulin and C-peptide and low values of glucose in the amniotic fluid. The diagnosis was suspected because of the previous unexplained death of twins in utero at 8 months' gestation followed by the birth of a macrosomic female (4200 g) who showed persistent neonatal hyperinsulinemic hypoglycemia, treated by subtotal pancreatectomy, on the 21st day of life. The histology confirmed the presence of nesidioblastosis. During the pregnancy in question, an accelerated growth rate was noticed on echo examinations. Because of the findings in the amniotic fluid at the 30th and 35th week, prompt therapy could be instituted at the birth of the child. Aparicio et al. (1993) likewise achieved prenatal diagnosis of familial neonatal hyperinsulinemia. Three sibs were affected. Infant A was macrosomatic and stillborn. Infant B was macrosomatic at birth following a pregnancy uncomplicated by maternal diabetes. Following diagnosis of hyperinsulinemic hypoglycemia, the patient was treated with oral diazoxide with continuation of therapy until hyperinsulinemia was resolved by age 2 years. The pregnancy with infant C was closely monitored with ultrasonography and amniocentesis. Based on these results, infant C was delivered immediately upon obtaining evidence of lung maturation. Neonatal hyperinsulinemia was confirmed by a markedly increased cord plasma insulin concentration.

Bruining (1990) estimated that the incidence of PHHI is 1/50,000 live births in a randomly mating population. In a Saudi Arabian population in which 51% of births occurred to parents who were first or second cousins, Mathew et al. (1988) established the incidence as 1/2,675 live births.

Glaser et al. (1994) mapped PHHI to the interval between markers D11S926 and D11S928 in 15 nuclear families of Ashkenazi-Jewish origin. Thomas et al. (1995) used the homozygosity gene-mapping strategy to localize the mutation for this disorder to a region of 11p between markers D11S1334 and D11S899 in 5 consanguineous families of Saudi Arabian origin; maximum lod score = 5.02 at theta = 0.0 for marker D11S926. Thomas et al. (1995) found that the PHHI locus is more than 13 cM centromeric to the beta-globin locus (HBB; 141900), thus excluding it as a candidate gene for Beckwith-Wiedemann syndrome (BWS; 130650).

Fantes et al. (1995) mapped markers linked to PHHI to a precise region of 11p15.1-p14. In evaluation of 6 additional families and 5 new markers, Glaser et al. (1995) localized the gene between D11S419 and D11S1310. The region was estimated to be 0.8 cM in length. Significant linkage disequilibrium between markers and the PHHI gene was observed over a region of 10.3 cM for Ashkenazi Jewish chromosomes. Haplotype analysis showed that 12 of 36 PHHI chromosomes (versus 1 of 36 non-PHHI chromosomes) bore a specific haplotype for D11S419—D11S902—D11S921 (p less than 0.0007), strongly suggesting a founder effect in this ethnic group.

The beta-cell high-affinity sulfonylurea receptor gene (SUR; 600509) maps to the same chromosomal region as PHHI and is the site of mutations in PHHI patients. 24,25:Thomas et al. (1995, 1996) reported mutations in the SUR gene in PHHI patients. Nestorowicz et al. (1996) reported mutations of the SUR gene in 25 probands from Ashkenazi Jewish families with PHHI.

Thomas et al. (1996) screened genomic DNA from infants with PHHI for mutations in the inwardly rectifying potassium channel referred to by them as Kir6.2 (KCNJ11; 600937). They identified a homozygous KCNJ11 gene mutation (L147P; 600937.0001) in an affected individual with severe PHHI. The finding of a mutation in the inward rectifier gene therefore indicated to them that there is genetic heterogeneity in PHHI, since mutations of the SUR gene also occur in cases of PHHI.

Kukuvitis et al. (1997) described an autosomal dominant form of PHHI in a French-Canadian family and excluded linkage to SUR and KCNJ11 on 11p15.

Two types of histopathologic lesions are associated with PHHI despite their similar clinical presentation, a focal form and a diffuse form. The focal form, present in approximately 30% of cases, is characterized by focal hypoplasia of islet-like cells, including hypertrophied insulin cells with giant nuclei. In the diffuse form, all the islets of Langerhans throughout the pancreas are irregular in shape and contain distinctly hypertrophied insulin cells. These 2 forms can be distinguished by pancreatic venous sampling, and peroperative extemporaneous histologic examination can determine whether subtotal or partial pancreatectomy is required (Dubois et al., 1995). Sixteen infants with sporadic PHHI resistant to diazoxide, who had undergone pancreatectomy, were investigated by de Lonlay et al. (1997). Venous sampling and histologic studies during surgery allowed identification of 10 cases of the focal form and 6 cases of the diffuse form. De Lonlay et al. (1997) showed that in cases of the focal form, but not those of the diffuse form, there was specific loss of maternal alleles of the imprinted chromosome region 11p15 in cells of the hyperplastic area of the pancreas but not in normal pancreatic cells. This somatic event was consistent with a proliferative monoclonal lesion. It involves disruption of the balance between monoallelic expression of several maternally and paternally expressed genes. Thus, they provided the first molecular explanation for the heterogeneity of sporadic forms of PHHI such that it is possible to perform only partial pancreatectomy, limited to the focal somatic lesion, so as to avoid iatrogenic diabetes in patients with focal adenomatous hyperplasia.

See also Aynsley-Green et al. (1981)

Aparicio, L.; Carpenter, M. W.; Schwartz, R.; Gruppuso, P. A.: Prenatal diagnosis of familial neonatal hyperinsulinemia. Acta Paediat. 82: 683-686, 1993.
Aynsley-Green, A.; Polak, J. M.; Bloom, S. R.; Gough, M. H.; Keeling, J.; Ashcroft, S. J. H.; Turner, R. C.; Baum, J. D.: Nesidioblastosis of the pancreas: definition of the syndrome and the management of the severe neonatal hyperinsulinaemic hypoglycaemia. Arch. Dis. Child. 56: 496-508, 1981.
Bianchi, C.; Corbella, E.; Beccaria, L.; Bolla, P.; Chiumello, G.: A case of familial nesidioblastosis: prenatal diagnosis of foetal hyperinsulinism. Acta Paediat. 81: 853-855, 1992.
Bruining, G. J.: Recent advances in hyperinsulinism and the pathogenesis of diabetes mellitus. Curr. Opin. Pediat. 2: 758-765, 1990.
Burman, W. J.; McDermott, M. T.; Bornemann, M.: Familial hyperinsulinism presenting in adults. Arch. Intern. Med. 152: 2125-2127, 1992.
Dahms, B. B.; Landing, B. H.; Blaskovics, M.; Roe, T. F.: Nesidioblastosis and other islet cell abnormalities in hyperinsulinemic hypoglycemia of childhood. Hum. Path. 11: 641-649, 1980.
de Lonlay, P.; Fournet, J.-C.; Rahier, J.; Gross-Morand, M.-S.; Poggi-Travert, F.; Foussier, V.; Bonnefont, J.-P.; Brusset, M.-C.; Brunelle, F.; Robert, J.-J.; Nihoul-Fekete, C.; Saudubray, J.-M.; Junien, C.: Somatic deletion of the imprinted 11p15 region in sporadic persistent hyperinsulinemic hypoglycemia of infancy is specific of focal adenomatous hyperplasia and endorses partial pancreatectomy. J. Clin. Invest. 100: 802-807, 1997.
Dubois, J.; Brunelle, F.; Touati, G.; Sebag, G.; Nuttin, C.; Thach, T.; Nikoul-Fekete, C.; Rahier, J.; Saudubray, J. M.: Hyperinsulinism in children: diagnostic value of pancreatic venous sampling correlated with clinical, pathological and surgical outcome in 25 cases. Pediat. Radiol. 25: 512-516, 1995.
Fantes, J. A.; Oghene, K.; Boyle, S.; Danes, S.; Fletcher, J. M.; Bruford, E. A.; Williamson, K.; Seawright, A.; Schedl, A.; Hanson, I.; Zehetner, G.; Bhogal, R.; Lehrach, H.; Gregory, S.; Williams, J.; Little, P. F. R.; Sellar, G. C.; Hoovers, J.; Mannens, M.; Weissenbach, J.; Junien, C.; van Heyningen, V.; Bickmore, W. A.: A high-resolution integrated physical, cytogenetic, and genetic map of human chromosome 11: distal p13 to proximal p15.1. Genomics 25: 447-461, 1995.
Glaser, B.; Chiu, K. C.; Anker, R.; Nestorowicz, A.; Landau, H.; Ben-Bassat, H.; Shlomai, Z.; Kaiser, N.; Thornton, P. S.; Stanley, C. A.; Spielman, R. S.; Gogolin-Ewens, K.; Cerasi, E.; Baker, L.; Rice, J.; Donis-Keller, H.; Permutt, M. A.: Familial hyperinsulinism maps to chromosome 11p14-15.1, 30 cM centromeric to the insulin gene. Nature Genet. 7: 185-188, 1994.
Glaser, B.; Chiu, K. C.; Liu, L.; Anker, R.; Nestorowicz, A.; Cox, N. J.; Landau, H.; Kaiser, N.; Thornton, P. S.; Stanley, C. A.; Cerasi, E.; Baker, L.; Donis-Keller, H.; Permutt, M. A.: Recombinant mapping of the familial hyperinsulinism gene to an 0.8 cM region on chromosome 11p15.1 and demonstration of a founder effect in Ashkenazi Jews. Hum. Molec. Genet. 4: 879-886, 1995.
Glaser, B.; Phillip, M.; Carmi, R.; Lieberman, E.; Landau, H.: Persistent hyperinsulinemic hypoglycemia of infancy ('nesidioblastosis'): autosomal recessive inheritance in 7 pedigrees. Am. J. Med. Genet. 37: 511-515, 1990.
Horev, Z.; Ipp, M.; Levey, P.; Daneman, D.: Familial hyperinsulinism: successful conservative management. J. Pediat. 119: 717-720, 1991.
Kukuvitis, A.; Deal, C.; Arbour, L.; Polychronakos, C.: An autosomal dominant form of familial persistent hyperinsulinemic hypoglycemia of infancy, not linked to the sulfonylurea receptor locus. J. Clin. Endocr. Metab. 82: 1192-1194, 1997.
Laidlaw, G. F.: Nesidioblastoma, the islet tumor of the pancreas. Am. J. Path. 14: 125-134, 1938.
Mathew, P. M.; Young, J. M.; Abu-Osba, Y. K.; Mulhern, B. D.; Hammoudi, S.; Hamdan, J. A.; Sa'di, A. R.: Persistent neonatal hyperinsulinism. Clin. Pediat. 27: 148-151, 1988.
Moreno, L. A.; Turck, D.; Gottrand, F.; Fabre, M.; Manouvrier-Hanu, S.; Farriaux, J.: Familial hyperinsulinism with nesidioblastosis of the pancreas: further evidence for autosomal recessive inheritance. Am. J. Med. Genet. 34: 584-586, 1989.
Nestorowicz, A.; Wilson, B. A.; Schoor, K. P.; Inoue, H.; Glaser, B.; Landau, H.; Stanley, C. A.; Thornton, P. S.; Clement, J. P., IV; Bryan, J.; Aguilar-Bryan, L.; Permutt, M. A.: Mutations in the sulfonylurea receptor gene are associated with familial hyperinsulinism in Ashkenazi Jews. Hum. Molec. Genet. 5: 1813-1822, 1996.
Rahier, J.: Relevance of endocrine pancreas nesidioblastosis to hyperinsulinemic hypoglycemia. Diabetes Care 12: 164-166, 1989.
Rahier, J.; Falt, K.; Muntefering, H.; Becker, K.; Gepts, W.; Falkmer, S.: The basic structural lesion of persistent neonatal hypoglycaemia with hyperinsulinism: deficiency of pancreatic D cells or hyperactivity of B cells?. Diabetologia 26: 282-289, 1984.
Schwartz, S. S.; Rich, B. H.; Lucky, A. W.; Straus, F. H., II; Gonen, B.; Wolfsdorf, J.; Thorp, F. W.; Burrington, J. D.; Madden, J. D.; Rubenstein, A. H.; Rosenfield, R. L.: Familial nesidioblastosis: severe neonatal hypoglycemia in two families. J. Pediat. 95: 44-53, 1979.
Thomas, P.; Ye, Y.; Lightner, E.: Mutation of the pancreatic islet inward rectifier Kir6.2 also leads to familial persistent hyperinsulinemic hypoglycemia of infancy. Hum. Molec. Genet. 5: 1809-1812, 1996.
Thomas, P. M.; Cote, G. J.; Hallman, D. M.; Mathew, P. M.: Homozygosity mapping, to chromosome 11p, of the gene for familial persistent hyperinsulinemic hypoglycemia of infancy. Am. J. Hum. Genet. 56: 416-421, 1995.
Thomas, P. M.; Cote, G. J.; Wohllk, N.; Haddad, B.; Mathew, P. M.; Rabl, W.; Aguilar-Bryan, L.; Gagel, R. F.; Bryan, J.: Mutations in the sulfonylurea receptor gene in familial persistent hyperinsulinemic hypoglycemia of infancy. Science 268: 426-429, 1995.
Thomas, P. M.; Wohllk, N.; Huang, E.; Kuhnle, U.; Rabl, W.; Gagel, R. F.; Cote, G. J.: Inactivation of the first nucleotide-binding fold of the sulfonylurea receptor, and familial persistent hyperinsulinemic hypoglycemia of infancy. Am. J. Hum. Genet. 59: 510-518, 1996.
Thornton, P. S.; Sumner, A. E.; Ruchelli, E. D.; Spielman, R. S.; Baker, L.; Stanley, C. A.: Familial and sporadic hyperinsulinism: histopathologic findings and segregation analysis support a single autosomal recessive disorder. J. Pediat. 119: 721-724, 1991.
Woo, D.; Scopes, J. W.; Polak, J. M.: Idiopathic hypoglycaemia in sibs with morphological evidence of nesidioblastosis of the pancreas. Arch. Dis. Child. 51: 528-531, 1976.
Woolf, D. A.; Leonard, J. V.; Trembath, R. C.; Pembrey, M. E.; Grant, D. B.: Nesidioblastosis: evidence for autosomal recessive inheritance. Arch. Dis. Child. 66: 529-530, 1991.
Wuthrich, C. H.; Schubiger, G.; Zuppinger, K.: Persistent neonatal hyperinsulinemic hypoglycemia in two siblings successfully treated with diazoxide. Helv. Paediat. Acta 41: 455-459, 1986.
Yakovac, W. C.; Baker, L.; Hummeler, K.: Beta cell nesidioblastosis in idiopathic hypoglycemia of infancy. J. Pediat. 79: 226-231, 1971.

SULFONYLUREA RECEPTOR (*600509)

JJM NOTE

Several mutations in this gene cause the clinical phenotype given above, otherwise known as persistent hyperinsulinemic hypoglycemia of infancy due to focal adenomatous hyperplasia.

Sulfonylureas are a class of drugs widely used as oral hypoglycemics to promote insulin secretion in the treatment of noninsulin-dependent diabetes mellitus. These drugs interact with the sulfonylurea receptor of pancreatic beta cells and inhibit the conductance of adenosine triphosphate (ATP)-dependent potassium channels. Aguilar-Bryan et al. (1995) cloned the cDNA for the high-affinity sulfonylurea receptor. Analysis of the predicted amino acid sequence indicated that the gene is a member of the ATP-binding cassette or traffic ATPase superfamily with multiple membrane-spanning domains and 2 nucleotide-binding folds. The results suggested that the sulfonylurea receptor may sense changes in ATP and ADP concentration, affect K(ATP) channel activity, and thereby modulate insulin release. Biochemical studies by Aguilar-Bryan et al. (1995) indicated that SUR is a large membrane protein (140 to 170 kD).

Familial persistent hyperinsulinemic hypoglycemia of infancy (PHHI; 256450), an autosomal recessive disorder characterized by unregulated insulin secretion, was mapped to 11p15.1-14 by linkage analysis (Glaser et al., 1994; Thomas et al., 1995; Fantes et al., 1995). This localization excluded as candidate genes for PHHI such previously mapped genes involved in beta-cell function as glucokinase (138079), islet glucose transporter (138160), and glucagon-like peptide-1 receptor (138032). The SUR protein was thought to be a likely candidate since it is a putative subunit of the beta-cell ATP-sensitive potassium channel, K(ATP), a modulator of insulin secretion. By fluorescence in situ hybridization, Thomas et al. (1995) showed that the SUR gene maps to 11p15.1. Mutation screening uncovered 2 separate SUR gene splice site mutations that segregated with the disease phenotype in affected individuals from 9 different families (600509.0001 and 600509.0002). Both mutations resulted in aberrant processing of the RNA sequence and disruption of the putative second nucleotide-binding domain (NBF-2) of the SUR protein. Philipson and Steiner (1995) discussed the significance of the work on SUR. They compared the SUR protein with other members of the family of ATP-binding cassette proteins, including the multidrug resistance (MDR) proteins (e.g., 171050), which cause chemotherapeutic drug resistance when overexpressed, and the CFTR protein (602421), which is mutant in cystic fibrosis patients.

In further studies on the role of the SUR nucleotide-binding folds in PHHI, Thomas et al. (1996) demonstrated that disruption of the first nucleotide-binding fold (NBF-1) likewise leads to PHHI (600509.0003). Their data indicated that both nucleotide-binding-fold regions of the sulfonylurea receptor are required for normal regulation of beta-cell ATP-dependent potassium channel activity and insulin secretion.

Inagaki et al. (1995) cloned a member of the inwardly rectifying potassium channel family, which they called BIR (600937). The gene was expressed in large amounts in rat pancreatic islets and glucose-responsive insulin-secreting cell lines. Coexpression with SUR reconstituted an inwardly rectifying potassium conductance of 76 picosiemens that was sensitive to ATP, inhibited by sulfonylureas, and activated by diazoxide. Inagaki et al. (1995) presented data that these pancreatic beta-cell potassium channels are a complex composed of at least 2 subunits: BIR and SUR. By fluorescence in situ hybridization, they mapped the BIR gene to 11p15.1. One lambda clone from the 3-prime end of the SUR gene contained sequence from the 5-prime end of the BIR gene. Thus, they concluded that the 2 genes are clustered at 11p15.1, with the BIR gene immediately 3-prime of the SUR gene.

An isoform of SUR, designated as sulfonylurea receptor-2 (SUR2), was cloned by Inagaki et al. (1996) from a rat brain cDNA library. The 5300-bp cDNA sequence encodes a protein of 1545 amino acids that shares 68% identity with SUR. Although the structural features of SUR and SUR2 are similar, the ATP sensitivity and pharmacological properties of the K(ATP) channel reconstituted from SUR2 and BIR are distinct from those obtained from the SUR/BIR channel.

Nestorowicz et al. (1996) screened 25 probands in Ashkenazi Jewish families with PHHI for mutations in the SUR gene by SSCP analysis of genomic DNA and subsequent sequence analysis. They identified 2 common mutations. One mutation was a novel in-frame deletion of the codon for F1388 (600509.0006). This mutation creates a unique BseR1 restriction site in exon 34 allowing detection of noncarriers, homozygotes, and heterozygotes. The second mutation was a previously described G-to-A transition at position -9 of the 3-prime splice site of intron 32 (600509.0002). Extended haplotype analysis in the D11S1901-D11S1310 region revealed that these 2 mutations were associated with the H1 haplotype in 88% of patients studied. The deletion mutation was associated with a specific haplotype H1, indicating a founder effect. Nestorowicz et al. (1996) analyzed the functional consequences of the deletion mutation on K(ATP) channel activity by examining 86Rb (+) efflux in COSm6 cells coexpressing Kir6.2 (600937) and either wildtype or F1388 deletion SUR transfected DNA. These studies revealed that the F1388-deleted SUR is unable to form a functional K(ATP) channel with Kir6.2.

Genomic mismatch scanning (GMS) is a technique that enriches for regions of identity by descent (IBD) between 2 individuals without the need for genotyping or sequencing. Regions of IBD selected by GMS are mapped by hybridization to a microarray containing ordered clones of genomic DNA from chromosomes of interest. Cheung et al. (1998) demonstrated the feasibility and efficacy of this form of linkage mapping, using congenital hyperinsulinism (256450), an autosomal recessive disorder, whose relatively high frequency in Ashkenazi Jews suggests a founder effect. Cheung et al. (1998) showed that the combination of GMS and hybridization of IBD products to a chromosome-11 microarray correctly maps the gene for congenital hyperinsulinemia, SUR, to a 2-Mb region, thereby demonstrating linkage disequilibrium mapping without genotyping.

(The method of genomic mismatch scanning used by Cheung et al. (1998) in humans had already been successfully applied in yeast, where GMS products are hybridized to a DNA microarray containing every open reading frame from Saccharomyces cerevisiae, yielding definitive genotypes for over 85% of the 6,400 ORFs in a single hybridization (Kruglyak, 1997).)

Nestorowicz et al. (1998) screened 45 familial hyperinsulinism probands of various ethnic origins for mutations in the SUR gene. Single-strand conformation polymorphism (SSCP) and nucleotide sequence analyses of genomic DNA revealed a total of 17 novel and 3 previously described mutations in SUR. The novel mutations comprised 1 nonsense and 10 missense mutations, 2 deletions, 3 mutations in consensus splice site sequences, and an in-frame insertion of 6 nucleotides. One mutation occurred in the first nucleotide-binding domain (NBF-1) of the SUR molecule and another 8 mutations were located in the second nucleotide-binding domain (NBF-2), including 2 at highly conserved amino acid residues within the Walker A sequence motif. Most of the other mutations were distributed throughout the 3 putative transmembrane domains of the SUR protein. With the exception of the 3993-9G-to-A mutation, which was detected on 4.5% (4 of 88) disease chromosomes, allelic frequencies for the identified mutations varied between 1.1% and 2.3%. The clinical manifestations of familial hyperinsulinism in those patients homozygous for mutations in the SUR gene were described. In contrast to the allelic homogeneity of familial hyperinsulinism found in Ashkenazi Jewish patients (Nestorowicz et al., 1996), the findings of the wider study suggested that a large degree of allelic heterogeneity at the SUR locus exists in non-Ashkenazi patients with familial hyperinsulinism.

From a morphologic standpoint, there are 2 types of histopathologic lesions underlying PHHI: a focal adenomatous hyperplasia of islet cells of the pancreas in approximately 30% of sporadic cases, and a diffuse form. In sporadic focal forms, specific losses of maternal alleles (LOH) of the imprinted chromosomal region 11p15, restricted to the hyperplastic area of the pancreas, were detected by de Lonlay et al. (1997). Similar mechanisms are observed in embryonal tumors and in the Beckwith-Wiedemann syndrome (BWS; 130650); the latter condition is also associated with neonatal but transient hyperinsulinism. The same chromosomal region, 11p15.1, includes the SUR gene and the KCNJ11 gene (600937) which code for the 2 subunits of the beta cell K(+)-ATP channel. Recessive mutations in these genes cause recessive familial forms of PHHI, but appear not to be imprinted. Although the parental bias in loss of maternal alleles did not argue in favor of direct involvement of the SUR or KCNJ11 genes, the LOH may unmask a recessive mutation leading to persistent hyperinsulinism. Verkarre et al. (1998) reported somatic reduction to hemizygosity or homozygosity of a paternal SUR constitutional heterozygous mutation in 4 patients with a focal form of PHHI. Thus, this somatic event, which leads to both beta cell proliferation and hyperinsulinism, can be considered as the somatic equivalent, restricted to a microscopic focal lesion, of constitutional uniparental disomy associated with unmasking of a heterozygous parental mutation leading to a somatic recessive disorder.

See also Thomas et al. (1995)

Aguilar-Bryan, L.; Nichols, C. G.; Wechsler, S. W.; Clement, J. P., IV; Boyd, A. E., III; Gonzalez, G.; Herrera-Sosa, H.; Nguy, K.; Bryan, J.; Nelson, D. A.: Cloning of the beta cell high-affinity sulfonylurea receptor: a regulator of insulin secretion. Science 268: 423-426, 1995.

Cheung, V. G.; Gregg, J. P.; Gogolin-Ewens, K. J.; Bandong, J.; Stanley, C. A.; Baker, L.; Higgins, M. J.; Nowak, N. J.; Shows, T. B.; Ewens, W. J.; Nelson, S. F.; Spielman, R. S.: Linkage-disequilibrium mapping without genotyping. Nature Genet. 18: 225-230, 1998.

de Lonlay, P.; Fournet, J.-C.; Rahier, J.; Gross-Morand, M.-S.; Poggi-Travert, F.; Foussier, V.; Bonnefont, J.-P.; Brusset, M.-C.; Brunelle, F.; Robert, J.-J.; Nihoul-Fekete, C.; Saudubray, J.-M.; Junien, C.: Somatic deletion of the imprinted 11p15 region in sporadic persistent hyperinsulinemic hypoglycemia of infancy is specific of focal adenomatous hyperplasia and endorses partial pancreatectomy. J. Clin. Invest. 100: 802-807, 1997.

Dunne, M. J.; Kane, C.; Shepherd, R. M.; Sanchez, J. A.; James, R. F. L.; Johnson, P. R. V.; Aynsley-Green, A.; Lu, S.; Clement, J. P., IV; Lindley, K. J.; Seino, S.; Aguilar-Bryan, L.: Familial persistent hyperinsulinemic hypoglycemia of infancy and mutations in the sulfonylurea receptor. New Eng. J. Med. 336: 703-706, 1997.

Fantes, J. A.; Oghene, K.; Boyle, S.; Danes, S.; Fletcher, J. M.; Bruford, E. A.; Williamson, K.; Seawright, A.; Schedl, A.; Hanson, I.; Zehetner, G.; Bhogal, R.; Lehrach, H.; Gregory, S.; Williams, J.; Little, P. F. R.; Sellar, G. C.; Hoovers, J.; Mannens, M.; Weissenbach, J.; Junien, C.; van Heyningen, V.; Bickmore, W. A.: A high-resolution integrated physical, cytogenetic, and genetic map of human chromosome 11: distal p13 to proximal p15.1. Genomics 25: 447-461, 1995.

Glaser, B.; Chiu, K. C.; Anker, R.; Nestorowicz, A.; Landau, H.; Ben-Bassat, H.; Shlomai, Z.; Kaiser, N.; Thornton, P. S.; Stanley, C. A.; Spielman, R. S.; Gogolin-Ewens, K.; Cerasi, E.; Baker, L.; Rice, J.; Donis-Keller, H.; Permutt, M. A.: Familial hyperinsulinism maps to chromosome 11p14-15.1, 30 cM centromeric to the insulin gene. Nature Genet. 7: 185-188, 1994.

Inagaki, N.; Gonoi, T.; Clement, J. P., IV; Wang, C.-Z.; Aguilar-Bryan, L.; Bryan, J.; Seino, S.: A family of sulfonylurea receptors determines the pharmacological properties of ATP-sensitive K(+) channels. Neuron 16: 1011-1017, 1996.

Inagaki, N.; Gonoi, T.; Clement, J. P., IV.; Namba, N.; Inazawa, J.; Gonzalez, G.; Aguilar-Bryan, L.; Seino, S.; Bryan, J.: Reconstitution of I(KATP): an inward rectifier subunit plus the sulfonylurea receptor. Science 270: 1166-1169, 1995.

Kruglyak, L.: The use of a genetic map of biallelic markers in linkage studies. Nature Genet. 17: 21-24, 1997.

Nestorowicz, A.; Glaser, B.; Wilson, B. A.; Shyng, S.-L.; Nichols, C. G.; Stanley, C. A.; Thornton, P. S.; Permutt, M. A.: Genetic heterogeneity in familial hyperinsulinism. Hum. Molec. Genet. 7: 1119-1128, 1998.

Nestorowicz, A.; Wilson, B. A.; Schoor, K. P.; Inoue, H.; Glaser, B.; Landau, H.; Stanley, C. A.; Thornton, P. S.; Clement, J. P., IV; Bryan, J.; Aguilar-Bryan, L.; Permutt, M. A.: Mutations in the sulfonylurea receptor gene are associated with familial hyperinsulinism in Ashkenazi Jews. Hum. Molec. Genet. 5: 1813-1822, 1996.

Philipson, L. H.; Steiner, D. F.: Pas de deux or more: the sulfonylurea receptor and K+ channels. Science 268: 372-373, 1995.

Thomas, P. M.; Cote, G. J.; Hallman, D. M.; Mathew, P. M.: Homozygosity mapping, to chromosome 11p, of the gene for familial persistent hyperinsulinemic hypoglycemia of infancy. Am. J. Hum. Genet. 56: 416-421, 1995.

Thomas, P. M.; Cote, G. J.; Wohllk, N.; Haddad, B.; Mathew, P. M.; Rabl, W.; Aguilar-Bryan, L.; Gagel, R. F.; Bryan, J.: Mutations in the sulfonylurea receptor gene in familial persistent hyperinsulinemic hypoglycemia of infancy. Science 268: 426-429, 1995.

Thomas, P. M.; Wohllk, N.; Huang, E.; Kuhnle, U.; Rabl, W.; Gagel, R. F.; Cote, G. J.: Inactivation of the first nucleotide-binding fold of the sulfonylurea receptor, and familial persistent hyperinsulinemic hypoglycemia of infancy. Am. J. Hum. Genet. 59: 510-518, 1996.

Verkarre, V.; Fournet, J.-C.; de Lonlay, P.; Gross-Morand, M.-S.; Devillers, M.; Rahier, J.; Brunelle, F.; Robert, J.-J.; Nihoul-Fekete, C.; Saudubray, J.-M.; Junien, C.: Paternal mutation of the sulfonylurea receptor (SUR1) gene and maternal loss of 11p15 imprinted genes lead to persistent hyperinsulinism in focal adenomatous hyperplasia. J. Clin. Invest. 102: 1286-1291, 1998.

POTASSIUM CHANNEL, INWARDLY RECTIFYING, SUBFAMILY J MEMBER 11 (*600937)

JJM NOTE

Mutations in this gene also cause the clinical phenotype called nesidioblastosis.

ATP-sensitive potassium currents, I(KATP), were discovered in cardiac muscle and later found in pancreatic beta cells, pituitary tissue, skeletal muscle, brain, and vascular and nonvascular smooth muscle. I(KATP) functions in secretion and muscle contraction by coupling metabolic activity to membrane potential. In pancreatic beta cells, ATP-potassium channels are crucial for the regulation of glucose-induced insulin secretion and are the target for the sulfonylureas, oral hypoglycemic agents widely used in the treatment of non-insulin-dependent diabetes mellitus (NIDDM), and for diazoxide, a potassium channel opener. The sulfonylurea receptor SUR (600509) is a member of the ATP-binding cassette superfamily with multiple transmembrane-spanning domains and 2 potential nucleotide-binding folds. Truncation of SUR that removes the second nucleotide-binding fold causes familial persistent hyperinsulinemic hypoglycemia of infancy (see, for example, 600509.0001), a rare disorder of glucose homeostasis characterized by unregulated insulin secretion despite severe hypoglycemia. Although these observations imply that SUR is closely associated with, or even a subunit of, K(ATP) channels, expression of SUR alone had not produced a measurable I(KATP). Inagaki et al. (1995) cloned a member of the inwardly rectifying potassium channel family, which they called BIR or Kir6.2, in the nomenclature of Chandy and Gutman (1993). The channel was expressed in large amounts in rat pancreatic islets and glucose-responsive insulin-secreting cell lines. The sequence showed a single open reading frame encoding a 390-amino acid protein with 2 putative transmembrane segments. The mouse homolog also had a single open reading frame encoding a 390-amino acid protein with 96% amino acid identity with human BIR, thus confirming that the gene encoding human BIR is intronless in the protein-coding region. Several other genes encoding inward rectifiers lack introns. By fluorescence in situ hybridization, Inagaki et al . (1995) mapped the BIR gene to 11p15.1. The sequence obtained from 1 lambda clone at the 3-prime end of the SUR gene matched a part of the gene encoding BIR; with a sense primer near the 3-prime end of the SUR gene and an antisense primer near the 5-prime end of the BIR gene they PCR-amplified an approximately 4.5-kb fragment. Thus, the authors determined that the 2 genes are clustered at 11p15.1, with the BIR gene immediately 3-prime of the SUR gene. The SUR gene had been previously mapped to 11p15.1 by fluorescence in situ hybridization (Thomas et al., 1995). (This gene is also symbolized KCNJ11.)

Inagaki et al. (1995) demonstrated that coexpression of BIR with the sulfonylurea receptor SUR reconstituted an inwardly rectifying potassium conductance of 76 picosiemens that was sensitive to ATP and was inhibited by sulfonylureas and activated by diazoxide. The data indicated to the authors that these pancreatic beta-cell potassium channels are a complex composed of at least 2 subunits: BIR and SUR.

An isoform of SUR, designated SUR2 (601439), was cloned by Inagaki et al. (1996) from a rat brain cDNA library. The 5300-bp cDNA sequence encodes a polypeptide of 1545 amino acids that shares 68% identity with SUR. Coexpression of SUR2 and BIR in COS1 cells reconstitutes the properties of K(ATP) channels described in cardiac and skeletal muscle. However, the investigators found that the SUR2/BIR channel is less sensitive than the SUR/BIR channel both to ATP and to the sulfonylurea glibenclamide, and is activated by the cardiac KATP channel openers cromakalim and pinacidil but not by diazoxide. The affinity of SUR2 for sulfonylureas is 500 times lower than that of SUR.

Thomas et al. (1996) screened genomic DNA from infants with persistent hyperinsulinemic hypoglycemia of infancy (PHHI; 256450) for mutations in the inwardly rectifying potassium channel referred to by them as Kir6.2. They discovered a Kir6.2 gene mutation in an affected individual with severe PHHI. This individual was the product of a consanguineous mating and was homozygous for a 649T-C point mutation of the genomic sequence (600937.0001). Her parents were heterozygous for this mutation. Thomas et al. (1996) noted that Kir6.2 lacks introns within its coding region, which is less than 1.5 kb in length. Genomic DNA may therefore be used as a template for PCR amplifications of coding sequence. The finding of a mutation in this inward rectifier gene therefore indicated to the authors that there is genetic heterogeneity in PHHI, since mutations of the SUR gene also occur in cases of PHHI.

Chandy, K. G.; Gutman, G. A.: Nomenclature for mammalian potassium channel genes. Trends Pharm. Sci. 14: 434, 1993.

Inagaki, N.; Gonoi, T.; Clement, J. P.; Namba, N.; Inazawa, J.; Gonzalez, G.; Aguilar-Bryan, L.; Seino, S.; Bryan, J.: Reconstitution of I(KATP): an inward rectifier subunit plus the sulfonylurea receptor. Science 270: 1166-1170, 1995.

Inagaki, N.; Gonoi, T.; Clement, J. P., IV; Wang, C.-Z.; Aguilar-Bryan, L.; Bryan, J.; Seino, S.: A family of sulfonylurea receptors determines the pharmacological properties of ATP-sensitive K(+) channels. Neuron 16: 1011-1017, 1996.

Thomas, P.; Ye, Y.; Lightner, E.: Mutation of the pancreatic islet inward rectifier Kir6.2 also leads to familial persistent hyperinsulinemic hypoglycemia of infancy. Hum. Molec. Genet. 5: 1809-1812, 1996.

Thomas, P. M.; Cote, G. J.; Hallman, D. M.; Mathew, P. M.: Homozygosity mapping, to chromosome 11p, of the gene for familial persistent hyperinsulinemic hypoglycemia of infancy. Am. J. Hum. Genet. 56: 416-421, 1995.

HYPOGLYCEMIA, LEUCINE-INDUCED (240800)

JJM NOTE

This is a trait that should be tested against *MEN1* markers and that could be sought in individuals with sporadic islet adenomas.

Several types of familial infantile hypoglycemia have been reported (see 256450 240600, 602485, 130650), including those precipitated by leucine (Cochrane et al., 1956; DiGeorge and Auerbach, 1960). Ebbin et al. (1967) observed symptomatic hypoglycemia with leucine sensitivity in a mother and daughter. Other cases of adults with leucine sensitivity have had islet adenomas, which apparently were not present in this case.

See also McQuarrie (1954)

Cochrane, W. A.; Payne, W. W.; Simpkiss, M. J.; Woolf, L. I.: Familial hypoglycemia precipitated by amino acids. J. Clin. Invest. 35: 411-422, 1956.

DiGeorge, A. M.; Auerbach, V. H.: Leucine-induced hypoglycemia: a review and speculations. Am. J. Med. Sci. 240: 792-801, 1960.

Ebbin, A. J.; Huntley, C.; Tranquada, R. E.: Symptomatic leucine sensitivity in a mother and daughter. Metabolism 16: 926-932, 1967.

McQuarrie, I.: Idiopathic spontaneously occurring hypoglycemia in infants. Clinical significance of problems and treatment. Am. J. Dis. Child. 87: 399-428, 1954.

Adrenal Medullary Tumors

PHEOCHROMOCYTOMA (*171300)

JJM NOTE

The clinician with a patient who has pheochromocytoma should ask, Why? The answer may be an underlying Mendelian trait, such as von Hippel-Lindau disease, neurofibromatosis 1, or multiple endocrine neoplasia 2 (Sipple syndrome). Or, family history may be positive. In any case, the familiality of pheochromocytoma is high enough (although low) and the consequences of a missed diagnosis serious enough that relatives should be alerted to their risk, and appropriate evaluation or screening should be offered. In three decades of follow-up, the kindred reported by Tisherman et al. (1962) revealed eight additional relatives with pheochromocytoma (at an average age of 19 years) and others with recurrent or additional pheochromocytomas, meningioma, paraganglioma, and a glomus vagale. The report itself was a family affair, with three Tishermans and a daughter-in-law (Tisherman SE et al.: Three decade investigation of familial pheochromocytoma: An allele of von Hippel-Lindau disease? Arch. Intern. Med. 154:2550–56, 1993). Our subtitle, in the form of a question, has proved correct, as Zbar and Ferrell found a missense mutation in the pheochromocytoma hot spot of the von Hippel-Lindau gene (Mulvihill JJ et al.: Familial pheochromocytoma due to mutant von Hippel-Lindau disease gene. Arch. Intern. Med. 157:1390–91, 1997).

A number sign (#) is used with this entry because of evidence that this tumor occurs not only as a part of several syndromes, e.g., neurofibromatosis I (NF1; 162200), von Hippel-Lindau syndrome (VHL; 193300), and multiple endocrine neoplasia types IIA (MEN2A; 171400) and IIB (MEN2B; 162300), but also as an isolated disorder due to mutation in the RET gene (164761) or the MEN2 gene.

Adrenal medullary tumor as part of a syndrome or as an isolated defect is probably inherited as a simple dominant in some families. The relation of the condition in the family reported by Hadorn (1963) is uncertain. Three sibs had adrenal tumors. A brother and sister suffered from tachycardia, sweating, hypertension, and albuminuria. The sister had advanced hypertensive retinopathy and the brother had congestive heart failure. At autopsy the sister showed cerebral hemorrhage and bilateral adrenocortical tumors. A surviving sib developed similar symptoms. Pheochromocytoma was tentatively diagnosed. The Regitine test was strongly positive, the urine contained large amounts of norepinephrine, and pneumoperitoneum demonstrated an enlarged right adrenal. At operation a mixed tumor containing hypernephromatous and paraganglion tissue was found.

The large kindred studied by Tisherman et al. (1962) had at least 7 patients with pheochromocytoma. One or more cafe-au-lait spots (in 22 persons), extensive hemangiomas (in 2 persons) and angiomatosis retinae (in 2 persons) were discovered in members of the family. Pheochromocytoma was associated with congenital cataracts in 1 patient and with renal artery stenosis in another. Tisherman et al. (1993) provided a 30-year follow-up of this family and suggested that the mutant gene is allelic to von Hippel-Lindau disease. Zbar et al. (1996) confirmed this hypothesis and demonstrated that the mutation is tyr112 to his in the VHL gene (193300.0012). There were 22 affected members of the kindred; 19 had pheochromocytoma and none had renal carcinoma. The form of von Hippel-Lindau syndrome with pheochromocytoma but not renal carcinoma was referred to by Zbar et al. (1996) as VHL type 2A.

Pheochromocytoma occurred in at least 4 members, including father and son, of a family studied by Swinton et al. (1972). They pointed out that hypercalcemia, corrected by adrenalectomy, can be associated with pheochromocytoma. This may be due to secretion of a calcitonin-like substance. The difficulty in discriminating this from multiple endocrine adenomatosis is obvious. Familial pheochromocytoma is usually bilateral and the patients are likely to show resistance to the vasopressor effects of tyramine (Engelman et al., 1968). Knudson and Strong (1972) applied Knudson's 2-mutation theory to pheochromocytoma (see 180200) and concluded that it fits. Familial pheochromocytoma was first reported by Calkins and Howard (1947). Seven documented and other possible cases occurred in the family reported by Kaufman and Franklin (1979). Fairchild et al. (1979) described a 29-year-old woman who had neuroblastoma during infancy, developed an extraadrenal pheochromocytoma at age 16 years, with subsequent hepatic recurrence, and was found to have multifocal renal cell carcinoma. Renal cell carcinoma and pheochromocytoma are combined in the von Hippel-Lindau syndrome (for which there was no evidence in this patient or her family). The association of pheochromocytoma and neuroblastoma had, it seemed, not been previously noted.

Bolande (1974) introduced the designation, and popularized the concept, of the neurocristopathies. Pearse (1969) defined and named the APUD cell system. Bolande (1974) identified 'simple neurocristopathies' such as pheochromocytoma and medullary carcinoma of the thyroid, and complex neurocristopathies and neurocristopathic syndromes such as von Recklinghausen disease and Sipple syndrome. Ohno et al. (1982) observed pheochromocytoma in 2 sisters whose father had pheochromocytoma. One of the sisters had aniridia and her pheochromocytoma was malignant. In 34 sporadic and 7 familial instances of pheochromocytoma, Khosla et al. (1991) found evidence of loss of heterozygosity (LOH) at multiple sites: 1p in 42%, 3p in 16%, 17p in 24%, and 22q in 31%. They also noted a correlation between LOH on 1p and urinary excretion of metanephrine by these patients (P = 0.02). LOH on 1p, 3p, and 17p also appeared to be associated with increased tumor volume. They suggested that tumor formation and/or progression in pheochromocytoma might involve multiple genes, analogous with the model proposed for colon carcinoma (Fearon and Vogelstein, 1990). The findings of Moley et al. (1992) suggested that LOH on 1p is particularly frequent in pheochromocytoma, being found by them in all 9 pheochromocytomas in MEN2A and MEN2B, in 2 of 7 sporadic pheochromocytomas, and in 1 of 2 pheochromocytomas in von Hippel-Lindau patients.

Among 82 unselected patients with pheochromocytoma, Neumann et al. (1993) found that 3 had MEN2 and 16 had von Hippel-Lindau disease. In the latter 16 patients, pheochromocytoma was the first symptomatic lesion. In a prospective study, 36 of 79 subjects at risk for pheochromocytoma were found to have unsuspected pheochromocytomas. Overall, there were 130 patients with 185 pheochromocytomas; 43 had von Hippel-Lindau disease, 24 had MEN2, and 63 had sporadic tumors. The patients with familial pheochromocytoma were younger, had multifocal localization much more often, and had cancer more frequently; the frequency of extraadrenal tumors was lower in the familial cases, however. Pheochromocytoma was the only clinical manifestation of their syndrome in 38% of carriers of von Hippel-Lindau disease and 24% of carriers of MEN2. Neumann et al. (1993) concluded that all patients with pheochromocytoma should be screened for MEN2 and von Hippel-Lindau disease, and that all patients in families with MEN2 or von Hippel-Lindau disease should be screened for pheochromocytoma, even if they are asymptomatic.

Eng et al. (1995) screened 7 exons of the RET oncogene (164761) for mutations in 48 sporadic pheochromocytomas and found a RET mutation in exons 10, 11, and 16 in 5 of the 48. Of these, 1 was proven to be germline and 2 were proven to be somatic mutations. Mutations of the VHL gene were found in 4 of the 48 sporadic pheochromocytomas; these 4 included the 2 bilateral cases in the series, of which 1 was proven to be a germline mutation, and 2 others, of which 1 was proven to be somatic. Conversely, Crossey et al. (1995) suggested that patients with familial, multiple, or early-onset pheochromocytoma should be investigated for germline VHL mutations. In 2 kindreds with no clinical evidence of VHL disease, they found 2 missense VHL gene mutations: V155L and R238W (193300.0003).

Woodward et al. (1997) undertook an analysis of 3 groups of patients with no evidence of von Hippel Lindau disease, multiple endocrine neoplasia type 2, or neurofibromatosis, all conditions in which pheochromocytoma can occur: group A, 8 kindreds with familial pheochromocytoma; group B, 2 patients with isolated bilateral pheochromocytoma; and group C, 6 cases of multiple extraadrenal pheochromocytoma or adrenal pheochromocytoma with a family history for neuroectodermal tumors. Germline missense VHL mutations were identified in 3 of 8 kindreds with familial pheochromocytoma. A germline VHL mutation was also characterized in 1 of the 2 patients with bilateral pheochromocytoma. In the group of patients with multiple extraadrenal pheochromocytoma or adrenal pheochromocytoma with a family history of neuroectodermal tumors, no mutations were found in VHL or in RET (the site of mutations in MEN2). The absence of germline VHL and RET gene mutations in many of these families suggested that other pheochromocytoma susceptibility loci may exist. Glial cell line-derived neurotrophic factor (GDNF; 600837) is a natural ligand for RET. Woodward et al. (1997) searched for germline mutations in GDNF in 16 cases of familial pheochromocytoma from groups A, B, and C and looked for evidence of somatic change in GDNF in 28 sporadic pheochromocytomas, 12 MEN2 pheochromocytomas, and 5 VHL pheochromocytomas. No GDNF mutations were identified in patients with familial pheochromocytoma, but a 277C-T (R93W) sequence variant was identified in 1 of the 28 sporadic tumors. This candidate mutation was identified in the germline and tumor tissue but was not present in 104 control GDNF alleles. GDNF sequence variants, including R93W, had been suggested previously to represent low penetrance susceptibility mutations for Hirschsprung disease and the R93W was not identified in 376 control alleles studied by others (see 600837). These findings suggested that although GDNF mutations do not have a major role in the patho genesis of familial or sporadic pheochromocytomas, allelic variation at the GDNF locus may modify pheochromocytoma susceptibility.

Brauch et al. (1997) and Bar et al. (1997) investigated the question of the role of germline mutations in cases of sporadic pheochromocytoma. Brauch et al. (1997) investigated 62 German patients with pheochromocytoma without a history of hereditary disease; 2 patients (3%) showed VHL mutations. No mutations were detected in the RET gene. Bar et al. (1997) studied 27 sporadic cases. Only 1 patient, with an atypical clinical presentation, had a VHL germline mutation, and none had a RET mutation. Both groups concluded that sporadic pheochromocytomas are rarely associated with germline mutations in either of these genes. Brauch et al. (1997) concluded that, because the costs of genetic screening are relatively low and the study of index cases allows optimal care for family members, molecular testing may be cost effective, however.

In 68 patients who had been operated on for pheochromocytomas, van der Harst et al. (1998) retrospectively screened for VHL mutations. Variants were found on SSCP analysis of the VHL gene in 8 of the 68 patients. Two of the patients were related (an uncle and his nephew) and they carried the same missense mutation: R64P. In 4 other patients, missense mutations, P25L, L63P, G144Q, and I147T, were identified. All of the mutations were novel, and 3 of them (P25L, L63P, and R64P) were located closer to the N terminus of the VHL protein than any previously reported VHL mutations. In the remaining 2 cases, the mutations were localized not in the coding sequence, but in the intronic sequence (but not within splice sites) adjacent to the exon, so they were probably not related to the disease.

See also Carman and Brashear (1960); Cook et al. (1960); Ho et al. (1978); Melicow (1977); Strunge et al. (1972); Tradec et al. (1961); Von Doepp (1962)

Bar, M.; Friedman, E.; Jakobovitz, O.; Leibowitz, G.; Lerer, I.; Abeliovich, D.; Gross, D. J.: Sporadic phaeochromocytomas are rarely associated with germline mutations in the von Hippel-Lindau and RET genes. Clin. Endocr. 47: 707-712, 1997.

Bolande, R. P.: The neurocristopathies: a unifying concept of disease arising in neural crest maldevelopment. Hum. Path. 5: 409-429, 1974.

Brauch, H.; Hoeppner, W.; Jahnig, H.; Wohl, T.; Engelhardt, D.; Spelsberg, F.; Ritter, M. M.: Sporadic pheochromocytomas are rarely associated with germline mutations in the vhl tumor suppressor gene or the ret protooncogene. J. Clin. Endocr. Metab. 82: 4101-4104, 1997.

Calkins, E.; Howard, J. E.: Bilateral familial pheochromocytoma with paroxysmal hypertension: successful surgical removal of tumors in 2 cases, with discussion of certain diagnostic and physiological considerations. J. Clin. Endocr. 7: 475-492, 1947.

Carman, C. T.; Brashear, R. E.: Pheochromocytoma as an inherited abnormality: report of the tenth affected kindred and review of the literature. New Eng. J. Med. 263: 419-423, 1960.

Cook, J. E.; Urich, R. W.; Sample, H. G., Jr.; Fawcett, N. W.: Peculiar familial and malignant pheochromocytomas of the organs of Zuckerkandl. Ann. Intern. Med. 52: 126-133, 1960.
Crossey, P. A.; Eng, C.; Ginalska-Malinowska, M.; Lennard, T. W. J.; Wheeler, D. C.; Ponder, B. A. J.; Maher, E. R.: Molecular genetic diagnosis of von Hippel-Lindau disease in familial phaeochromocytoma. J. Med. Genet. 32: 885-886, 1995.
Eng, C.; Crossey, P. A.; Mulligan, L. M.; Healey, C. S.; Houghton, C.; Prowse, A.; Chew, S. L.; Dahia, P. L. M.; O'Riordan, J. L. H.; Toledo, S. P. A.; Smith, D. P.; Maher, E. R.; Ponder, B. A. J.: Mutations in the RET proto-oncogene and the von Hippel-Lindau disease tumour suppressor gene in sporadic and syndromic phaeochromocytomas. J. Med. Genet. 32: 934-937, 1995.
Engelman, K.; Horwitz, D.; Ambrose, I. M.; Sjoerdsma, A.: Further evaluation of the tyramine test for pheochromocytoma. New Eng. J. Med. 278: 705-709, 1968.
Fairchild, R. S.; Kyner, J. L.; Hermreck, A.; Schimke, R. N.: Neuroblastoma, pheochromocytoma, and renal cell carcinoma: occurrence in a single patient. J.A.M.A. 242: 2210-2211, 1979.
Fearon, E. R.; Vogelstein, B.: A genetic model for colorectal tumorigenesis. Cell 61: 759-767, 1990.
Hadorn, W.: Maligne Hypernephroide und paraganglionaere Mischgeschwuelste der Nebenniere bei drei Geschwistern. Helv. Med. Acta 30: 291-296, 1963.
Ho, A. D.; Feurle, G.; Gless, K.-H.; Brandeis, W. E.: Normotensive familial phaeochromocytoma with predominant noradrenaline secretion. Brit. Med. J. 1: 81-82, 1978.
Kaufman, J. J.; Franklin, S.: Familial pheochromocytoma: a report of 2 cases in a kindred. J. Urol. 121: 801-804, 1979.
Khosla, S.; Patel, V. M.; Hay, I. D.; Schaid, D. J.; Grant, C. S.; van Heerden, J. A.; Thibodeau, S. N.: Loss of heterozygosity suggests multiple genetic alterations in pheochromocytomas and medullary thyroid carcinomas. J. Clin. Invest. 87: 1691-1699, 1991.
Knudson, A. G., Jr.; Strong, L. C.: Mutation and cancer: neuroblastoma and pheochromocytoma. Am. J. Hum. Genet. 24: 514-532, 1972.
Melicow, M. M.: One hundred cases of pheochromocytoma (107 tumors) at the Columbia-Presbyterian Medical Center, 1926-1976: a clinicopathological analysis. Cancer 40: 1987-2004, 1977.
Moley, J. F.; Brother, M. B.; Fong, C. T.; White, P. S.; Baylin, S. B.; Nelkin, B.; Wells, S. A.; Brodeur, G. M.: Consistent association of 1p loss of heterozygosity with pheochromocytomas from patients with multiple endocrine neoplasia type 2 syndromes. Cancer Res. 52: 770-774, 1992.
Neumann, H. P. H.; Berger, D. P.; Sigmund, G.; Blum, U.; Schmidt, D.; Parmer, R. J.; Volk, B.; Kirste, G.: Pheochromocytomas, multiple endocrine neoplasia type 2, and von Hippel-Lindau disease. New Eng. J. Med. 329: 1531-1538, 1993.
Ohno, F.; Yamano, T.; Kataoka, K.: A case of congenital aniridia and familial pheochromocytoma—with special reference to aniridia-Wilms' tumor syndrome. Jpn. J. Hum. Genet. 27: 335-340, 1982.
Pearse, A. G. E.: The cytochemistry and ultrastructure of polypeptide hormone-producing cells of the APUD series and the embryologic, physiologic, and pathologic implications of the concept. J. Histochem. Cytochem. 17: 303-313, 1969.
Strunge, P.; Ingsrup, H. M.; Lochte, J. J.; Zimmermann-Nielsen, C.: Bilateral phaeochromocytoma in two brothers. Acta Paediat. Scand. 61: 729-732, 1972.
Swinton, N. W.; Clerkin, E. P.; Flint, L. D.: Hypercalcemia and familial pheochromocytoma. Correction after adrenalectomy. Ann. Intern. Med. 76: 455-457, 1972.
Tisherman, S. E.; Gregg, F. J.; Danowski, T. S.: Familial pheochromocytoma. J.A.M.A. 182: 152-156, 1962.
Tisherman, S. E.; Tisherman, B. G.; Tisherman, S. A.; Dunmire, S.; Levey, G. S.; Mulvihill, J. J.: Three-decade investigation of familial pheochromocytoma: an allele of von Hippel-Lindau disease? Arch. Int. Med. 153: 2550-2556, 1993.
Tradec, E.; Maratka, Z.; Palecrova, M.: Le pheochromocytome avec caractere familial. J. Chir. 81: 479, 1961.
van der Harst, E.; de Krijger, R. R.; Dinjens, W. N. M.; Weeks, L. E.; Bonjer, H. J.; Bruining, H. A.; Lamberts, S. W. J.; Koper, J. W.: Germline mutations in the vhl gene in patients presenting with phaeochromocytomas. Int. J. Cancer. 77: 337-340, 1998.
Von Doepp, C. E.: Das Phaeochromocytom als dominant vererbbare dysgenetische Geschwulst. Virchows Arch. Path. Anat. 335: 231-239, 1962.
Woodward, E. R.; Eng, C.; McMahon, R.; Voutilainen, R.; Affara, N. A.; Ponder, B. A. J.; Maher, E. R.: Genetic predisposition to phaeochromocytoma: analysis of candidate genes GDNF, RET and VHL. Hum. Molec. Genet. 6: 1051-1056, 1997.
Zbar, B.; Kishida, T.; Chen, F.; Schmidt, L.; Maher, E. R.; Richards, F. M.; Crossey, P. A.; Webster, A. R.; Affara, N. A.; Ferguson-Smith, M. A.; Brauch, H.; Glavac, D.; and 14 others: Germline mutations in the Von Hippel-Lindau disease (VHL) gene in families from North America, Europe, and Japan. Hum. Mutat. 8: 348-357, 1996.

PHEOCHROMOCYTOMA, FAMILIAL EXTRA-ADRENAL (171350)

JJM NOTE

Additional clinical reports or a molecular insight are needed to verify the distinctiveness of this entity.

Glowniack et al. (1985) suggested that primary extraadrenal pheochromocytoma may be a distinct genetic entity and that within families similarities may occur. These conclusions were based on the report by Cook et al. (1960) of a brother and sister with pheochromocytoma of the organs of Zuckerkandl (masses of chromaffin tissue found at the aortic bifurcation), on the report by Spring and Palubinskas (1977) of mother and son with pheochromocytoma involving the lower urinary tract, and particularly on the family they reported in detail in which 3 persons in as many successive generations has pheochromocytoma of the right renal hilum. In the 2 previous generations, pheochromocytoma was suspected in 3 persons; 1 was a hypertensive girl who died suddenly at age 16 years. The pedigree reported by Glowniack et al. (1985) contained an obligatory carrier (according to the hypothesis of autosomal dominant inheritance) who at age 52 years showed no signs of pheochromocytoma by any test including CT scans and scintigraphy with radioiodine-tagged metalodobenzylguanidine.

Cook, J. E.; Urich, R. W.; Sample, H. G., Jr.; Fawcett, N. W.: Peculiar familial and malignant pheochromocytomas of the organs of Zuckerkandl. Ann. Intern. Med. 52: 126-133, 1960.
Glowniack, J. V.; Shapiro, B.; Sisson, J. C.; Thompson, N. W.; Coran, A. G.; Lloyd, R.; Kelsch, R. C.; Beierwaltes, W. H.: Familial extra-adrenal pheochromocytoma: a new syndrome. Arch. Intern. Med. 145: 257-261, 1985.
Spring, D. B.; Palubinskas, A. J.: Familial pheochromocytoma: a rare case of hydronephrosis and hydroureter in two generations. Brit. J. Radiol. 50: 596-599, 1977.

PHEOCHROMOCYTOMA—ISLET CELL TUMOR SYNDROME (171420)

JJM NOTE

Surely this disorder will be clarified when the molecular genetic markers for the relevant genes are applied.

Carney et al. (1980) suggested that pheochromocytoma and/or islet cell tumor is an autosomal dominant endocrine adenomatosis distinct from MEA I, II and III. They reported 3 families. Among 11 affected patients (aged 5 to 53 years), 10 had pheochromocytoma (bilateral in 6), 4 had islet cell tumor (multicentric in 1), and 3 had both tumors. Clinical presentation was due to pheochromocytoma in 10 (symptoms or signs beginning before age 10 years in 3) and islet cell tumor in 1. Two patients died from pheochromocytoma and 2 from islet cell carcinoma. One of their probands had numerous cafe-au-lait spots up to 4 cm in diameter and axillary freckling. There was, however, no family history of von Recklinghausen disease. Most of the other families in the literature with this combination of endocrine tumors had von Hippel-Lindau disease (VHL; 193300). For example, Hull et al. (1979) reported sibs with VHL syndrome who had both pheochromocytoma and islet cell adenomas. Thus, the case for a separate pheochromocytoma-islet cell tumor cannot be considered proved. Zeller et al. (1982) reported the eleventh case of this association.

Carney, J. A.; Go, V. L. W.; Gordon, H.; Northcutt, R. C.; Pearse, A. G. E.; Sheps, S. G.: Familial pheochromocytoma and islet cell tumor of the pancreas. Am. J. Med. 68: 515-521, 1980.
Hull, M. T.; Warfel, K. A.; Muller, J.; Higgins, J. T.: Familial islet cell tumors in von Hippel-Lindau disease. Cancer 44: 1523-1526, 1979.
Zeller, J. R.; Kauffman, H. M.; Komorowski, R. A.; Itskovitz, H. D.: Bilateral pheochromocytoma and islet cell adenoma of the pancreas. Arch. Surg. 117: 827-830, 1982.

NEUROBLASTOMA (*256700)

JJM NOTE

The chromosomal and oncogene findings in neuroblastoma (as well as its embryonal histology) suggest major genetic determinants for this relatively common childhood tumor. It is ironic that the familial aggregation is infrequent and that no Mendelian disorder has been associated with it—not even neurofibromatosis, the quintessential neurocristopathy.

Dodge and Benner (1945) reported a brother and sister with neuroblastoma of the adrenal medulla. The father and 3 of his 5 sibs in the report of Chatten and Voorhees (1967) had cafe-au-lait spots; the 3 affected sibs were female. Griffin and Bolande (1969) described 2 sisters with congenital disseminated neuroblastoma. In both, regression of the retroperitoneal tumors to fibrocalcific residues and maturation to ganglioneuroma were observed. In one of them, metastatic nodules in the skin matured to ganglioneuromas and by progressive loss of ganglion cells came to resemble neurofibromas closely. A 15-year-old sister showed, by x-ray, a small focus of adrenal calcification. Wong et al. (1971) described an affected brother and sister, each of whom was diagnosed at the age of 5.5 months. The father showed increased amounts of vanillylmandelic acid in the urine. In the family of Zimmerman (1951), the father had had a mediastinal ganglioneuroma removed at age 10 years. Helson et al. (1969) found elevated catecholamines in sibs of children with overt neuroblastomas. Neuroblastoma may be much more frequent than clinical detection suggests (Beckwith and Perrin, 1963). Hardy and Nesbit (1972) reported neuroblastoma in a brother and sister and a male first cousin. Wagget et al. (1973) described 2 sib pairs of which all 4 died with metastatic neuroblastoma. There was no evidence of tumor or neurofibromatosis in sibs or parents. Gerson et al. (1974) gave a follow-up on the family reported by Chatten and Voorhees (1967). The mother of 4 sibs with neuroblastoma had persistently elevated urinary catecholamines, but was asymptomatic. She was subsequently found to have a posterior mediastinal mass which in retrospective review of radiographs was found to have been present and of constant size for at least 16 years. Knudson and Strong (1972) concluded that Knudson's 2-mutation theory of cancer can be applied to neuroblastoma. Two other conditions to which the theory had been applied (see 171300 and 180200) are 'dominant.' Neuroblastoma at times shows spontaneous regression. Particularly dramatic are cases of disseminated disease involving the liver, skin, and bone marrow, which are labeled Stage IV-S to differentiate them from the more usual, fatal disseminated form, Stage IV. Knudson and Meadows (1980) postulated that neuroblastoma IV-S represents an hereditary one-hit neoplastic disorder of multicellular origin. Thus, they would place it in the same category as von Recklinghausen neurofibromatosis, C-cell hyperplasia of the thyroid, and adrenal medullary hyperplasia. The 2-hit tumors (monoclonal in origin) relating to these 4 conditions are neuroblastoma, neurofibrosarcoma, medullary carcinoma of the thyroid, and pheochromocytoma. The theory, if true, may have implications for the management of Stage IV. Should efforts be made to reduce the population of cells that might be transformed by a second hit into true neuroblastoma? Should mutagenic therapeutic agents be avoided? The best agent would be a nonmutagen that stimulates differentiation. In the view of Knudson and Meadows (1980), neuroblastoma IV-S represents a collection of nonmalignant neural crest cells bearing a mutation that interferes with their normal differentiation. Delayed maturation ultimately transforms them into ganglioneuromas or neurofibromas or causes them to die. The neurofibromatosis mutation interfered with development at a later stage of maturation; hence, the tumors contain Schwann cells, nerve fibers and even ganglion cells.

Fairchild et al. (1979) described a 29-year-old woman who had neuroblastoma during infancy, developed an extra adrenal pheochromocytoma at age 16 years, with subsequent hepatic recurrence, and was found to have multifocal renal cell carcinoma. Renal cell carcinoma and pheochromocytoma are combined in the von Hippel-Lindau syndrome (for which there was no evidence in this patient or her family). The association of pheochromocytoma and neuroblastoma had not, it seemed, been noted previously.

Morell et al. (1977) noted association between neuroblastoma and an uncommon Gm phenotype.

Balaban-Malenbaum and Gilbert (1977) described characteristic chromosomal changes in human neuroblastoma cell lines: a long nonbanding homogeneously staining region (HSR) and double minute chromosomes. The two never coexisted in a cell. Different cell lines had the HSR on different chromosomes. Gilbert et al. (1981) suggested that chromosome 1p34 contains 1 or more genes responsible for the control of neuroblast proliferation and that the loss of activity of said genes through deletion, rearrangements or point mutations is involved in tumorigenesis in neuroblastoma. This conclusion was based on the finding of chromosomal abnormalities at this site in direct preparations from neuroblastomas or from cell lines. Earlier, Brodeur et al. (1977) found similar changes. Perucho et al. (1981) found that DNA from a neuroblastoma cell line contained a transforming element, i.e., one that would transform mouse fibroblasts which became thereby tumorigenic in nude mice. Pegelow et al. (1975) reported a remarkable family with 3 instances of neuroblastoma. The proposita had neuroblastoma at birth and both parents had had children, by previous matings, who had died of neuroblastoma. Hecht et al. (1982) reported further information on the family. The proposita was well at age 8.5 years, after receiving chemotherapy early in life. By the previous marriage, the father of the proposita had a healthy son who fathered a child with congenital metastatic neuroblastoma. Two chromosomal variants were segregating in the family (paracentric inversion of chromosome 11, with breaks at 11q21 and 11q23, and 21p-), but neither correlated with neuroblastoma or the presumed carrier status. One might suggest that heterozygotes for a 'neuroblastoma gene' have only neuroblastoma in situ; microscopic foci of neuroblastoma in the adrenals have an incidence of about 1 in 200 during infancy (Beckwith and Perrin, 1963). Shimizu et al. (1983) demonstrated that DNA from a human neuroblastoma cell line is capable of inducing foci of transformed NIH 3T3 cells after DNA-mediated gene transfer. Human sequences responsible for the transformation were isolated from the mouse cells and were shown to be present in all human cells. No gross rearrangement was demonstrable in the sequences in the neuroblastoma cell line. Although clearly distinct from 2 other human transforming genes present in bladder, lung, and colon carcinoma cell lines, all 3 may be related members of the ras gene family. Preliminary results of experiments in Ruddle's laboratory suggest that the SK-N-SH transforming gene reported by Wigler's group in neuroblastoma is situated on chromosome 1. Shimizu et al. (1983) demonstrated a transforming gene in a neuroblastoma cell line that is related to both v-K-ras and v-H-ras and probably codes for an immunologically crossreactive and structurally related protein. In a human neuroblastoma cell line, Cowell and Rupniak (1983) found a consistent abnormality of 1p: the region distal to 1p31 had homogeneously staining characteristics. Double minutes were also present as a second constitutive feature. Unexpectedly, Suzuki et al. (1989) found frequent loss of heterozygosity on chromosome 14q but not on chromosome 1p.

Fong et al. (1989) found that somatic loss of heterozygosity occurred most consistently between 1p36.1 and 1p36.3 in neuroblastomas. They found a correlation between loss of heterozygosity on 1p and amplification of MYCN (164840), which is on chromosome 2. Stating that at least 70% of neuroblastomas show cytogenetically visible aberrations in 1p, Weith et al. (1989) presented the results of studies of loss of heterozygosity in 9 different tumors and the corresponding normal tissue. Most of the probes used to detect polymorphic DNA loci were derived from a library of microdissected distal 1p chromosome fragments. Allelic loss was observed with at least 2 probes in 8 of 9 neuroblastomas. The consensus deletion in all 8 tumors included the segment 1p36.2-p36.1. The study of Weith et al. (1989) provided no support for a correlation between 1p deletion and amplification of the MYCN gene in this disorder. Ritke et al. (1989) concluded that the data suggesting that the breakpoints are often located in bands other than 1p32 are in error and that the evidence points to the involvement of specific DNA sequences within 1p32 as important to the development of neuroblastoma. Noting that tumor cells from about 70% of neuroblastoma patients show deletion of part of 1p, they concluded that the abnormality is most likely due to complex translocation and deletion mechanisms. For example, the MYCL locus (164850) is often moved from the altered chromosome 1 to another chromosome. One of the patients in their study had a breakpoint at 1p32 located between JUN (165160) and MYCL, thus establishing the order as centromere—JUN—MYCL—telomere. Hunt and Tereba (1990) used a panel of 20 cloned sequences and 2 isozymes to determine the location of the breakpoints in 7 neuroblastoma cell lines. They concluded that the most distal deletion breakpoint occurred between MYCL1 and D1S57. They proposed that the neuroblastoma susceptibility gene is located distal to MYCL1, since in 3 of the 5 lines in which MYCL1 was deleted from a chromosome 1, the gene had been translocated to another chromosome, thus retaining the diploid complement. They concluded that another gene linked to MYCL1 may be involved in this neoplasm. Martinsson et al. (1989) isolated a set of microclones and mapped them on the short arm of chromosome 1. Using some of these, they showed that the breakpoint in a neuroblastoma cell line carrying a translocation involving 1p had a breakpoint at 1p36.1. On the basis of loss of heterozygosity (LOH), Takayama et al. (1992) concluded that the commonly deleted region in neuroblastoma is distal to the D1S112 locus, i.e., on 1pter-p36.1. They also found LOH for markers on 14q in 10 out of 25 informative cases (40%). The commonly deleted regions were distal to the D14S13 locus, i.e., on 14q32-qter. Biegel et al. (1993) reported the case of a child with dysmorphic features, as well as developmental and growth delay, who developed neuroblastoma at 5 months of age. Cytogenetic analysis of blood lymphocytes demonstrated an interstitial deletion of 1p36.2-p36.1, which was apparent only with high-resolution banding. The interstitial deletion involving subbands of 1p36 was corroborated by molecular analysis with a collection of polymorphic DNA probes for 1p. Caron et al. (1993) and Cheng et al. (1993) obtained seemingly contradictory results. Caron et al. (1993) reported that allelic loss of 1p36 was found in 15 of 53 neuroblastomas; in 13 of the 15, the lost allele was of maternal origin. Cheng et al. (1993) found that the paternal allele was lost from 1p in 6 of 10 cases, consistent with a random distribution. Cheng et al. (1993) also found that the paternal allele of N-myc (MYCN; 164840) was preferentially amplified in 12 of 13 cases of neuroblastoma.

Bader et al. (1991) demonstrated that the reintroduction of chromosome 1 into a human neuroblastoma cell line resulted in differentiation and cell death, consistent with the idea that mutation of a 1p36 gene allows unrestricted growth. The et al. (1993) pursued the observation that the introduction of human chromosome 17 suppresses the tumorigenicity of the same neuroblastoma cell line in the absence of any effect on in vitro growth. They concluded that the neurofibromatosis type 1 (NF1; 162200) gene may be responsible. In 4 out of 10 human neuroblastoma lines, there was little or no neurofibromin expression, and 2 of these lines showed evidence of NF1 mutations.

Caron et al. (1994) described 3 neuroblastoma tumors and 2 cell lines in which LOH in 1p resulted from an unbalanced translocation between the short arm of chromosome 1 and the long arm of chromosome 17. Southern blot and cytogenetic analyses showed that in all cases the chromosome 17 homolog from which the 1;17 translocation was derived was still present and intact. This suggested a model in which a translocation between the short arm of chromosome 1 and the long arm of chromosome 17 takes place in the S/G2 phase of the cell cycle and results in LOH 1p. Nonhomologous mitotic recombination in the S/G2 phase had not previously been observed as a mechanism of LOH. Caron et al. (1995) examined the question of genomic imprinting of the neuroblastoma tumor suppressor gene on 1p36; previously reported results had been conflicting. They studied the parental origin of 1p36 alleles lost in 47 neuroblastomas. The results were remarkably different for tumors with and without amplification of the N-myc oncogene. In the N-myc single copy tumors, they found that the lost 1p36 alleles were preferentially of maternal origin (16 of 17 cases) and that the commonly deleted region mapped to 1p36.3-p36.2. In contrast, all N-myc-amplified neuroblastomas had larger 1p deletions, extending from the telomere to at least 1p36.1-p35. These deletions were of random parental origin (18 of 30 maternal LOH). This strongly suggested that different suppressor genes on 1p are inactivated in these 2 types of neuroblastoma. Deletion of a more proximal suppressor gene is associated with N-myc amplification, while a distal, probably imprinted, suppressor can be deleted in N-myc single copy cases.

The human DAN gene (600613), formerly known as the N03 gene, was suggested by Enomoto et al. (1994) to be the tumor suppressor gene of neuroblastoma. The gene has growth suppressing properties in cultured cells and the human gene was mapped to 1p36.13-p36.11, the same area as that involved in neuroblastoma.

To characterize the region of consistent deletion of 1p in neuroblastoma, White et al. (1995) performed LOH studies on 122 neuroblastoma tumor samples using 30 distal chromosome 1p polymorphisms. LOH was detected in 32 of the 122 tumors (26%). A single region of LOH, marked distally by the D1Z2 and proximally by D1S228, was deleted in all tumors demonstrating loss. Also, cells from a patient with a constitutional deletion of 1p36 and from a neuroblastoma cell line with a small 1p36 deletion were analyzed by fluorescence in situ hybridization. Cells from both sources had interstitial deletions of 1p36.3-p36.2 that overlapped the consensus region of LOH defined by the tumors. Four proposed candidate genes—DAN, ID3 (600277), CDC2L1 (176873), and TNFR2 (191191)—were shown to lie outside the consensus region of allelic LOH, as defined by the above deletions. These results more precisely defined the location of the neuroblastoma suppressor gene within 1p36.3-p36.2, eliminating 33 cM of proximal 1p36 from consideration. Furthermore, a consensus region of loss that excluded the 4 leading candidate genes was found in all tumors with 1p36 LOH.

Plon (1997) reported a family in which 2 sibs and a second cousin had the diagnosis of neuroblastoma at less than 1 year of age. The father of the sibs had had Wilms tumor in childhood, but neither the grandparents nor the great-grandparents of the affected children were known to be affected. There was no history of consanguinity. Maris et al. (1996) described a family with 2 affected sibs and an affected second cousin. In the classic study of the genetics of neuroblastoma, Knudson and Strong (1972) found in a review of 529 cases 13 instances of multiple occurrences within a family and 1 case with 2 second cousins affected. The remainder of the 12 families had multiple sibs affected, including 1 half-sib pair. Taken together, these reports represented 5 families in which unaffected adults may have transmitted a predisposition to their offspring. This is consistent with the Knudson and Strong (1972) conclusion that the age at onset of neuroblastoma in familial versus sporadic cases is consistent with a 2-hit mechanism and that autosomal recessive inheritance is unlikely.

See also Arenson et al. (1976); Brodeur et al. (1981); Gilbert et al. (1982); Klein et al. (1975); Montgomery et al. (1983); Roberts and Lee (1975); Schwab et al. (1983); Schwab et al. (1984); Shimizu et al. (1983)

Arenson, E. B.; Hutter, J. J., Jr.; Restuccia, R. D.; Holton, C. P.: Neuroblastoma in father and son. J.A.M.A. 235: 727-729, 1976.

Bader, S. A.; Fasching, C.; Brodeur, G. M.; Stanbridge, E. J.: Dissociation of suppression of tumorigenicity and differentiation in vitro effected by transfer of single human chromosomes into human neuroblastoma cells. Cell Growth Differ. 2: 245-255, 1991.

Balaban-Malenbaum, G.; Gilbert, F.: Double minute chromosomes and the homogeneously staining regions in chromosomes of a human neuroblastoma cell line. Science 198: 739-741, 1977.

Beckwith, J. B.; Perrin, E. V.: In situ neuroblastomas: a contribution to the natural history of neural crest tumors. Am. J. Path. 43: 1089-1104, 1963.

Biegel, J. A.; White, P. S.; Marshall, H. N.; Fujimori, M.; Zackai, E. H.; Scher, C. D.; Brodeur, G. M.; Emanuel, B. S.: Constitutional 1p36 deletion in a child with neuroblastoma. Am. J. Hum. Genet. 52: 176-182, 1993.

Brodeur, G. M.; Green, A. A.; Hayes, F. A.; Williams, K. J.; Williams, D. L.; Tsiatis, A. A.: Cytogenetic features of human neuroblastomas and cell lines. Cancer Res. 41: 4678-4686, 1981.

Brodeur, G. M.; Sekhon, G. S.; Goldstein, M. N.: Chromosomal aberrations in human neuroblastomas. Cancer 40: 2256-2263, 1977.

Caron, H.; Peter, M.; van Sluis, P.; Speleman, F.; de Kraker, J.; Laureys, G.; Michon, J.; Brugieres, L.; Voute, P. A.; Westerveld, A.; Slater, R.; Delattre, O.; Versteeg, R.: Evidence for two tumour suppressor loci on chromosomal bands 1p35-36 involved in neuroblastoma: one probably imprinted, another associated with N-myc amplification. Hum. Molec. Genet. 4: 535-539, 1995.

Caron, H.; van Sluis, P.; van Hoeve, M.; de Kraker, J.; Bras, J.; Slater, R.; Mannens, M.; Voute, P. A.; Westerveld, A.; Versteeg, R.: Allelic loss of chromosome 1p36 in neuroblastoma is of preferential maternal origin and correlates with N-myc amplification. Nature Genet. 4: 187-190, 1993.

Caron, H.; van Sluis, P.; van Roy, N.; de Kraker, J.; Speleman, F.; Voute, P. A.; Westerveld, A.; Slater, R.; Versteeg, R.: Recurrent 1;17 translocations in human neuroblastoma reveal nonhomologous mitotic recombination during the S/G2 phase as a novel mechanism for loss of heterozygosity. Am. J. Hum. Genet. 55: 341-347, 1994.

Chatten, J.; Voorhees, M. L.: Familial neuroblastoma. Report of a kindred with multiple disorders, including neuroblastomas in four siblings. New Eng. J. Med. 277: 1230-1236, 1967.

Cheng, J. M.; Hiemstra, J. L.; Schneider, S. S.; Naumova, A.; Cheung, N.-K.; Cohn, S. L.; Diller, L.; Sapienza, C.; Brodeur, G. M.: Preferential amplification of the paternal allele of the N-myc gene in human neuroblastomas. Nature Genet. 4: 191-194, 1993.

Cowell, J. K.; Rupniak, H. T.: Chromosome analysis of human neuroblastoma cell line TR14 showing double minutes and an aberration involving chromosome 1. Cancer Genet. Cytogenet. 9: 273-280, 1983.

Dodge, H. J.; Benner, M. C.: Neuroblastoma of the adrenal medulla in siblings. Rocky Mt. Med. J. 42: 35-38, 1945.

Enomoto, H.; Ozaki, T.; Takahashi, E.; Nomura, N.; Tabata, S.; Takahashi, H.; Ohnuma, N.; Tanabe, M.; Iwai, J.; Yoshida, H.; Matsunaga, T.; Sakiyama, S.: Identification of human DAN gene, mapping to the putative neuroblastoma tumor suppressor gene. Oncogene 9: 2785-2791, 1994.

Fairchild, R. S.; Kyner, J. L.; Hermreck, A.; Schimke, R. N.: Neuroblastoma, pheochromocytoma, and renal cell carcinoma: occurrence in a single patient. J.A.M.A. 242: 2210-2211, 1979.

Fong, C.-T.; Dracopoli, N. C.; White, P. S.; Merrill, P. T.; Griffith, R. C.; Housman, D. E.; Brodeur, G. M.: Loss of heterozygosity for the short arm of chromosome 1 in human neuroblastomas: correlation with N-myc amplification. Proc. Nat. Acad. Sci. 86: 3753-3757, 1989.

Gerson, J. M.; Chatten, J.; Eisman, S.: Familial neuroblastoma—a follow-up. (Letter) New Eng. J. Med. 290: 1487 only, 1974.

Gilbert, F.; Balaban, G.; Moorhead, P.: Chromosome 1p deletion-rearrangements in neuroblastoma: relevance to tumorigenesis. (Abstract) Am. J. Hum. Genet. 33: 104A only, 1981.

Gilbert, F.; Balaban, G.; Moorhead, P.; Bianchi, D.; Schlesinger, H.: Abnormalities of chromosome 1p in human neuroblastoma tumors and cell lines. Cancer Genet. Cytogenet. 7: 33-42, 1982.

Griffin, M. E.; Bolande, R. P.: Familial neuroblastoma with regression and maturation to ganglioneurofibroma. Pediatrics 43: 377-382, 1969.

Hardy, P. C.; Nesbit, M. E., Jr.: Familial neuroblastoma: report of a kindred with a high incidence of infantile tumors. J. Pediat. 80: 74-77, 1972.

Hecht, F.; Hecht, B. K.; Northrup, J. C.; Trachtenberg, N.; Wood, T. S.; Cohen, J. D.: Genetics of familial neuroblastoma: long-range studies. Cancer Genet. Cytogenet. 7: 227-230, 1982.

Helson, L.; Blasco, P.; Murphy, M. L.: Familial neuroblastoma. (Abstract) Clin. Res. 17: 614 only, 1969.

Hunt, J. D.; Tereba, A.: Molecular evaluation of abnormalities of the short arm of chromosome 1 in neuroblastoma. Genes Chromosomes Cancer 2: 137-146, 1990.

Klein, H.; Plochl, E.; Lampert, F.: Familial neuroblastoma: cytogenetic investigation of the peripheral blood. Humangenetik 28: 217-220, 1975.

Knudson, A. G., Jr.; Meadows, A. T.: Regression of neuroblastoma IV-S: a genetic hypothesis. New Eng. J. Med. 302: 1254-1256, 1980.

Knudson, A. G., Jr.; Strong, L. C.: Mutation and cancer: neuroblastoma and pheochromocytoma. Am. J. Hum. Genet. 24: 514-532, 1972.

Maris, J. M.; Kyemba, S. M.; Rebbeck, T. R.; White, P. S.; Sulman, E. P.; Jensen, S. J.; Allen, C.; Biegel, J. A.; Yanofsky, R. A.; Feldman, G. L.; Brodeur, G. M.: Familial predisposition to neuroblastoma does not map to chromosome band 1p36. Cancer Res. 56: 3421-3425, 1996.

Martinsson, T.; Weith, A.; Cziepluch, C.; Schwab, M.: Chromosome 1 deletions in human neuroblastomas: generation and fine mapping of microclones from the distal 1p region. Genes Chromosomes Cancer 1: 67-78, 1989.

Montgomery, K. T.; Biedler, J. L.; Spengler, B. A.; Melera, P. W.: Specific DNA sequence amplification in human neuroblastoma cells. Proc. Nat. Acad. Sci. 80: 5724-5728, 1983.

Morell, A.; Scherz, R.; Kaser, H.; Skvaril, F.: Evidence for an association between uncommon Gm phenotypes and neuroblastoma. Lancet I: 23-24, 1977.

Pegelow, C. H.; Ebbin, A. J.; Powars, D.; Turner, J. W.: Familial neuroblastoma. J. Pediat. 87: 763-765, 1975.

Perucho, M.; Goldfarb, M.; Shimizu, K.; Lama, C.; Fogh, J.; Wigler, M.: Human-tumor-derived cell lines contain common and different transforming genes. Cell 27: 467-476, 1981.

Plon, S. E.: Anticipation of pediatric malignancies. (Letter) Am. J. Hum. Genet. 60: 1256-1257, 1997.

Ritke, M. K.; Shah, R.; Valentine, M.; Douglass, E. C.; Tereba, A.: Molecular analysis of chromosome 1 abnormalities in neuroblastoma. Cytogenet. Cell Genet. 50: 84-90, 1989.

Roberts, F. F.; Lee, K. R.: Familial neuroblastoma presenting as multiple tumors. Radiology 116: 133-136, 1975.

Schwab, M.; Alitalo, K.; Klempnauer, K.-H.; Varmus, H. E.; Bishop, J. M.; Gilbert, F.; Brodeur, G.; Goldstein, M.; Trent, J.: Amplified DNA with limited

homology to myc cellular oncogene is shared by human neuroblastoma cell lines and a neuroblastoma tumour. Nature 305: 245-248, 1983.
Schwab, M.; Varmus, H. E.; Bishop, J. M.; Grzeschik, K.-H.; Naylor, S. L.; Sakaguchi, A. Y.; Brodeur, G.; Trent, J.: Chromosome localization in normal human cells and neuroblastomas of a gene related to c-myc. Nature 308: 288-291, 1984.
Shimizu, K.; Goldfarb, M.; Perucho, M.; Wigler, M.: Isolation and preliminary characterization of the transforming gene of a human neuroblastoma cell line. Proc. Nat. Acad. Sci. 80: 383-387, 1983.
Shimizu, K.; Goldfarb, M.; Suard, Y.; Perucho, M.; Li, Y.; Kamata, T.; Feramisco, J.; Stavnezer, E.; Fogh, J.; Wigler, M. H.: Three human transforming genes are related to the viral ras oncogenes. Proc. Nat. Acad. Sci. 80: 2112-2116, 1983.
Suzuki, T.; Yokota, J.; Mugishima, H.; Okabe, I.; Ookuni, M.; Sugimura, T.; Terada, M.: Frequent loss of heterozygosity on chromosome 14q in neuroblastoma. Cancer Res. 49: 1095-1098, 1989.
Takayama, H.; Suzuki, T.; Mugishima, H.; Fujisawa, T.; Ookuni, M.; Schwab, M.; Gehring, M.; Nakamura, Y.; Sugimura, T.; Terada, M.; Yokota, J.: Deletion mapping of chromosomes 14q and 1p in human neuroblastoma. Oncogene 7: 1185-1189, 1992.
The, I.; Murthy, A. E.; Hannigan, G. E.; Jacoby, L. B.; Menon, A. G.; Gusella, J. F.; Bernards, A.: Neurofibromatosis type 1 gene mutations in neuroblastoma. Nature Genet. 3: 62-66, 1993.
Wagget, J.; Aherne, G.; Aherne, W.: Familial neuroblastoma: report of two sib pairs. Arch. Dis. Child. 48: 63-66, 1973.
Weith, A.; Martinsson, T.; Cziepluch, C.; Bruderlein, S.; Amler, L. C.; Berthold, F.; Schwab, M.: Neuroblastoma consensus deletion maps to 1p36.1-2. Genes Chromosomes Cancer 1: 159-166, 1989.
White, P. S.; Maris, J. M.; Beltinger, C.; Sulman, E.; Marshall, H. N.; Fujimori, M.; Kaufman, B. A.; Biegel, J. A.; Allen, C.; Hilliard, C.; Valentine, M. B.; Look, A. T.; Enomoto, H.; Sakiyama, S.; Brodeur, G. M.: A region of consistent deletion in neuroblastoma maps within human chromosome 1p36.2-36.3. Proc. Nat. Acad. Sci. 92: 5520-5524, 1995.
Wong, K. Y.; Hanenson, I. B.; Lampkin, B. C.: Familial neuroblastoma. Am. J. Dis. Child. 121: 415-416, 1971.
Zimmerman, J.: Ganglioneuroblastome als erbliche Systemerkrankung des Sympathicus. Beitr. Path. Anat. 111: 355-372, 1951.

MEGALENCEPHALY (*155350)

JJM NOTE

This entity is included because of one case with ganglioneuroblastoma, a possible coincidence.

DeMyer (1972) reported instances of apparently autosomal dominant megalencephaly, with male-to-male transmission in some cases. In a family with 3 affected generations, the proband also had mediastinal ganglioneuroblastoma. In the group of megalencephalics as a whole, males predominated. It is of interest, therefore, that X-linked megalencephaly has been suggested (see 248000). Most persons with megalencephaly are mentally retarded. DeMyer (1972) and Hall (1974) gave me information on other families showing autosomal dominant inheritance of megalencephaly. A distinction should be made between megalencephaly and macrocephaly (153470), depending, as the etymology suggests, on whether large brain size is present. The distinction is not always easy but is aided by recent imaging methods. Schreier et al. (1974) found 10 persons in 3 generations of a family with presumed megalencephaly, indicating the possibility of autosomal dominant inheritance. An additional 2 members were considered to have true hydrocephalus. Among 557 children who presented a diagnostic problem of a large head, Lorber and Priestley (1981) found 109 with megalencephaly as the primary diagnosis. At least half the cases had a familial incidence of large head. Males predominated 4 to 1. Only 7 of the 109 were retarded and these showed also a variety of neurologic and other somatic abnormalities; some had superior intelligence.

See also DeMyer (1972)

DeMyer, W. E.: Personal Communication. Indianapolis, Ind. 12/1972.
DeMyer, W. E.: Megalencephaly in children: clinical syndromes, genetic patterns, and differential diagnosis from other causes of megalocephaly. Neurology 22: 634-643, 1972.
Hall, J. G.: Personal Communication. Seattle, Wash. 1974.
Lorber, J.; Priestley, B. L.: Children with large heads: a practical approach to diagnosis in 557 children, with special reference to 109 children with megalencephaly. Dev. Med. Child Neurol. 23: 494-504, 1981.
Schreier, H.; Rapin, I.; Davis, J.: Familial megalencephaly or hydrocephalus?. Neurology 24: 232-236, 1974.

CUSHING DISEASE, ADRENAL (219080)

JJM NOTE

To date, most patients with Cushing disease seem to be sporadic, but I wonder what would be the results of a detailed search for predisposing conditions, such as Carney syndrome, Li-Fraumeni cancer family syndrome, and various multiple endocrine neoplasias.

Cushing syndrome has been reported in type II multiple endocrine neoplasia (Steiner et al., 1968; see 171400); in these cases, the adrenal tumors may be primary. Maton et al. (1986) documented Cushing disease in 3 of 16 patients with the Zollinger-Ellison syndrome and MEN I (131100), but in this situation the Cushing disease was due to ACTH-secreting pituitary adenoma. Cushing disease due to primary nodular hyperplasia of the adrenals occurs also as part of the Carney syndrome (160980). Other features of the Carney syndrome include cutaneous and cardiac myxomas, spotty pigmentation of the skin, and endocrine overactivity that extends also to sexual precocity in the presence of Sertoli-Leydig cell tumors. Families such as that of Acre et al. (1978) may represent Carney syndrome; of 4 sibs with Cushing syndrome, 3 had adenomatous hyperplasia of the adrenal glands and a fourth had a virilizing adrenal carcinoma. It seems, however, that there are cases of familial Cushing disease that cannot be attributed to either of these 2 syndromes and may be inherited as an autosomal recessive, unlike these 2 syndromes which are autosomal dominants. Donaldson et al. (1981) described a brother and sister with bilateral nodular adrenal hyperplasia manifesting already at birth. A picture showing features of Cushing syndrome at the age of 1 week was published. Teding Van Berkhout et al. (1989) described Cushing syndrome due to nodular adrenocortical dysplasia in the 2 sisters. The disorder was successfully treated by complete adrenalectomy. No evidence of associated disorders suggesting Carney syndrome was found in the 2 sisters or their first-degree relatives.

However, the serum of both girls and their mother contained immunoglobulins capable of stimulating adrenal DNA synthesis and cortisol production in vitro. The results were interpreted as indicating that the disorder is an inherited disease of immunologic origin (Wulffraat et al., 1988). Findlay et al. (1993) reported the cases of a black mother and daughter who each presented with clinical features of Cushing syndrome at age 38 years and were found to have ACTH-independent macronodular adrenal hyperplasia. Hodge and Froesch (1988) described 2 sisters with primary micronodular adrenocortical dysplasia leading to Cushing syndrome. Bilateral adrenalectomy was performed at ages 14 and 30 years, respectively. The patients appeared to have a disease characterized by autonomous overactivity of nests of abnormal adrenal cells with suppression of endogenous corticotropin.

Lacroix et al. (1997) studied a 36-year-old woman with Cushing syndrome, ACTH-independent macronodular adrenal hyperplasia, and orthostatic hypotension. During upright posture, the patient's cortisol and aldosterone levels were stimulated despite suppression of ACTH and renin. Arginine vasopressin (AVP, 10 U im), under dexamethasone suppression, increased plasma cortisol (3.4-fold), aldosterone (67-fold), and androgens in this patient but not in controls. Following adrenalectomy, orthostatic hypotension persisted; a prolonged vasoconstrictive response to AVP was found in vitro in the patient's small arteries. Lacroix et al. (1997) proposed that altered adrenal and vascular responses of the V1-AVP receptor-effector pathway underlie this syndrome.

Cushing syndrome due to ACTH-producing pituitary adenoma can be caused by mutations in the GNAS1 gene (e.g., 139320.0012), which is the site of the causative mutations in McCune-Albright syndrome (174800) and Albright hereditary osteodystrophy (103580), and in the GNAI2 gene (e.g., 139360.0003).

Acre, B.; Licea, M.; Hung, S.; Padron, R.: Familial Cushing's syndrome. Acta Endocr. 87: 139-147, 1978.
Donaldson, M. D. C.; Grant, D. B.; O'Hare, M. J.; Shackleton, C. H. L.: Familial congenital Cushing's syndrome due to bilateral nodular adrenal hyperplasia. Clin. Endocr. 14: 519-526, 1981.
Findlay, J. C.; Sheeler, L. R.; Engeland, W. C.; Aron, D. C.: Familial adrenocorticotropin-independent Cushing's syndrome with bilateral macronodular adrenal hyperplasia. J. Clin. Endocr. Metab. 76: 189-191, 1993.
Hodge, B. O.; Froesch, T. A.: Familial Cushing's syndrome: micronodular adrenocortical dysplasia. Arch. Intern. Med. 148: 1133-1136, 1988.
Lacroix, A.; Tremblay, J.; Touyz, R. M.; Deng, L. Y.; Lariviere, R.; Cusson, J. R.; Schiffrin, E. L.; Hamet, P.: Abnormal adrenal and vascular responses to vasopressin mediated by a V1-vasopressin receptor in a patient with adrenocorticotropin-independent macronodular adrenal hyperplasia, Cushing's syndrome, and orthostatic hypotension. J. Clin. Endocr. Metab. 82: 2414-2422, 1997.

Maton, P. N.; Gardner, J. D.; Jensen, R. T.: Cushing's syndrome in patients with the Zollinger-Ellison syndrome. New Eng. J. Med. 315: 1-5, 1986.
Steiner, A. L.; Goodman, A. D.; Powers, S. R., Jr.: Study of a kindred with pheochromocytoma, medullary thyroid carcinoma, hyperparathyroidism and Cushing's disease: multiple endocrine neoplasia, type II. Medicine 47: 371-409, 1968.
Teding Van Berkhout, F.; Croughs, R. J. M.; Wulffraat, N. M.; Drexhage, H. A.: Familial Cushing's syndrome due to nodular adrenocortical dysplasia is an inherited disease of immunological origin. Clin. Endocr. 31: 185-191, 1989.
Wulffraat, N. M.; Drexhage, H. A.; Wiersinga, W. M.; van der Gaag, R. D.; Jencken, P.; Mol, J. A.: Immunoglobulins of patients with Cushing's syndrome due to pigmented adrenocortical micronodular dysplasia stimulate in-vitro steroidogenesis. J. Clin. Endocr. Metab. 66: 301-307, 1988.

ADRENOCORTICAL CARCINOMA, HEREDITARY (#202300)

JJM NOTE

Despite its rarity, adrenocortical carcinoma (and adenoma) have some conspicuous genetic and congenital determinants. Beckwith-Wiedemann syndrome, urogenital anomalies (preceding the tumor), and hemihypertrophy seem to occur to excess in children and adolescents with adrenal tumors. Most striking, however, is the inclusion of adrenal tumors in the spectrum of malignancies in the Li-Fraumeni cancer family syndrome (151623, 191170) (Li FP et al.: A cancer family syndrome in twenty-four kindreds. Cancer Res. 48:5358–62, 1988). Congenital adrenal hyperplasia of 21-hydroxylase deficiency (201910) may have an excess of benign and malignant adrenal cortical tumors (Lack EE et al.: Adrenal cortical neoplasms in the pediatric and adolescent age group. Pathol. Ann. 27:1–52, 1992).

A number sign (#) is used with this entry because of evidence that it represents a part of the 'multiple tumor associated chromosome region-1' (MTACR1; 194071).

Adrenocortical carcinoma (ADCC) is a rare childhood tumor, representing about 0.4% of childhood tumors, with a high incidence of associated tumors. ADCC occurs with increased frequency in patients with the Beckwith-Wiedemann syndrome (130650). Fraumeni and Miller (1967) mentioned affected sibs. Mahloudji et al. (1971) observed affected brother and sister who were products of a consanguineous union. Nichols (1968) also described affected brother and sister. Henry et al. (1987) described a family in which 2 brothers developed ADCC with signs of virilism. Studies with 11p markers indicated loss of heterozygosity (LOH) in the tumor cells of both brothers; furthermore, the chromosome 11 region that remained in the tumor was the same and was inherited from the same parent. 3,4:Henry et al. (1987, 1989) interpreted this as indicating the existence of a recessive oncogene responsible for hereditary ADCC and located on 11p, probably 11p15.5.

Gicquel et al. (1997) evaluated the frequency and distribution of 11p15 LOH and IGF2 gene (147470) overexpression in a series of 82 sporadic adrenocortical tumors. They screened for pathologic functional imprinting of the 11p15 region in tumors not exhibiting LOH and evaluated the expression of the H19 gene (103280) in these tumors. Abnormalities of the 11p15 region, including LOH with the loss of the maternal allele and duplication of the paternal allele and/or IGF2 gene overexpression, are frequent features of the malignant state; these abnormalities were found in 27 of 29 (93%) of the malignant tumors but in only 3 of 35 (9%) of the benign tumors. Tumors without abnormality of the 11p15 region, mainly benign tumors, did not exhibit pathologic functional imprinting. The authors concluded that dysregulation of the 11p15 region occurs in late steps of adrenocortical tumorigenesis and suggested that LOH and IGF2 overexpression are new molecular markers for better diagnostic and prognostic evaluations of adrenocortical tumors.

See also Werder et al. (1975)

Fraumeni, J. F., Jr.; Miller, R. W.: Adrenocortical neoplasms with hemihypertrophy, brain tumors, and other disorders. J. Pediat. 70: 129-138, 1967.
Gicquel, C.; Raffin-Sanson, M.-L.; Gaston, V.; Bertagna, X.; Plouin, P.-F.; Schlumberger, M.; Louvel, A.; Luton, J.-P.; Le Bouc, Y.: Structural and functional abnormalities at 11p15 are associated with the malignant phenotype in sporadic adrenocortical tumors: study on a series of 82 tumors. J. Clin. Endocr. Metab. 82: 2559-2565, 1997.
Henry, I.; Huerre-Jeanpierre, C.; Azoulay, M.; Chaussain, J. L.; Junien, C.: A recessive oncogene for familial adrenocortical carcinoma (ADCC) maps to 11p. (Abstract) Cytogenet. Cell Genet. 46: 629 only, 1987.
Henry, I.; Jeanpierre, M.; Couillin, P.; Barichard, F.; Serre, J.-L.; Journel, H.; Lamouroux, A.; Turleau, C.; de Grouchy, J.; Junien, C.: Molecular definition of the 11p15.5 region involved in Beckwith-Wiedemann syndrome and probably in predisposition to adrenocortical carcinoma. Hum. Genet. 81: 273-277, 1989.
Mahloudji, M.; Ronaghi, H.; Dutz, W.: Virilizing adrenal carcinoma in two sibs. J. Med. Genet. 8: 160-163, 1971.
Nichols, J.: Adrenal cortex.In: Bloodworth, J. M. B.: Endocrine Pathology. Baltimore: Williams and Wilkins 1968. Pp. 224-255.
Werder, E. A.; Haller, R.; Vetter, W.; Zachmann, M.; Siebenmann, R.: Isolated glucocorticoid insufficiency. Helv. Paediat. Acta 30: 175-183, 1975.

THYROTOXIC PERIODIC PARALYSIS (188580)

JJM NOTE

Although many fascinating associations occur with this entity, it is included because of the apparent excess of aldosterone-secreting adrenal adenomas seen in China.

Periodic paralysis with hypokalemia occurs with thyrotoxicosis with a clinical picture virtually indistinguishable from the hereditary form without hyperthyroidism (170400). The thyrotoxicosis may be mild. Attacks can be precipitated by high carbohydrate intake. Spontaneous or induced attacks do not occur in persons whose hyperthyroidism has been corrected. A specific genetic basis is suggested by the fact that although occasional cases occur in Caucasians (e.g., Ali, 1975; Shah et al., 1979), the disorder is seen predominantly in Orientals. In 1,366 consecutive Southern Chinese patients with thyrotoxicosis, 25 had a history of 1 or more attacks of periodic paralysis (McFadzean and Yeung, 1967). This is predominantly a disease of males; 23 of the 25 patients of McFadzean and Yeung were male. I was told (Hsu, 1978) that Chinese males with thyrotoxicosis who have HLA types BW22 and BW17 and do not have BW46 are most susceptible to periodic paralysis. Thus, thyrotoxic periodic paralysis may be due to a genetic peculiarity of muscle membranes. Multiple cases in families have been observed (Yeung, 1981) but the mode of inheritance is not clear. Layzer and Goldfield (1974) observed periodic paralysis in a Japanese-American male who 'abused' thyroid hormone. Leung (1985) observed thyrotoxicosis and periodic paralysis in 4 adult members of a Chinese family: mother and 3 children (2 male, 1 female). Two had 'hashitoxicosis' (Graves disease and Hashimoto thyroiditis) as evidenced by the presence of thyroid antibodies and elevated thyroxine levels. The other 2 were not available for study. In a report of 50 consecutive Chinese patients with primary hyperaldosteronism due to adrenal cortical adenoma, Ma et al. (1986) found that periodic paralysis was a presenting feature in 42% of the patients. This was quite unusual. Other notable symptoms were palpitations (30%) and syncope (12%). Mean serum potassium level at presentation was significantly lower and plasma aldosteron e higher in patients with periodic paralysis. Hypokalemic paralysis occurs in Orientals with barium poisoning, and as a side-effect of gossypol, a fertility regulating agent used in China. Thus it is likely that the Oriental periodic paralysis may be precipitated by any one of many factors. Raskin et al. (1981) and Pun et al. (1989) observed periodic paralysis with Sjogren syndrome (270150), which can be complicated by distal renal tubular acidosis.

Kilpatrick et al. (1994) found 6 reports of thyrotoxic hypokalemic periodic paralysis in African-Americans and described 4 additional cases, all in males. They concluded that the disorder may be more frequent in blacks than previously suspected and should be considered when patients present with unexplained hypokalemia, muscular weakness, and rhabdomyolysis.

See also Au and Yeung (1972); Bernard et al. (1972); Kusakabe et al. (1976); Ramsay (1974); Yeung and Tse (1974)

Ali, K.: Hypokalaemic periodic paralysis complicating thyrotoxicosis. Brit. Med. J. 2: 503-504, 1975.
Au, K.-S.; Yeung, R. T. T.: Thyrotoxic periodic paralysis: periodic variation in the muscle calcium pump activity. Arch. Neurol. 26: 543-546, 1972.
Bernard, J. D.; Larson, M. A.; Norris, F. H., Jr.: Thyrotoxic periodic paralysis in Californians of Mexican and Filipino ancestry. Calif. Med. 116: 70-74, 1972.
Hsu, T. H.: Personal Communication. Baltimore, Md. 2/27/1978.
Kilpatrick, R. E.; Seiler-Smith, S.; Levine, S. N.: Thyrotoxic hypokalemic periodic paralysis: report of four cases in black American males. Thyroid 4: 441-445, 1994.
Kusakabe, T.; Yoshida, M.; Nishikawa, M.: Thyrotoxic periodic paralysis: a peculiar case with unusual dystonic behavior and variable relations of paralysis to serum potassium levels. J. Clin. Endocr. Metab. 43: 730-740, 1976.
Layzer, R. B.; Goldfield, E.: Periodic paralysis caused by abuse of thyroid hormone. Neurology 24: 949-952, 1974.
Leung, A. K. C.: Familial 'hashitoxic' periodic paralysis. J. Roy. Soc. Med. 78: 638-640, 1985.
Ma, J. T. C.; Wang, C.; Lam, K. S. L.; Yeung, R. T. T.; Chan, F. L.; Boey, J.; Cheung, P. S. Y.; Coghlan, J. P.; Scoggins, B. A.; Stockigt, J. R.: Fifty cases of primary hyperaldosteronism in Hong Kong Chinese with a high frequency of

periodic paralysis; evaluation of techniques for tumour localisation. Quart. J. Med. 61: 1021-1037, 1986.
McFadzean, A. J. S.; Yeung, R.: Periodic paralysis complicating thyrotoxicosis in Chinese. Brit. Med. J. 1: 451-455, 1967.
Pun, K.-K.; Wong, C.-K.; Tsui, E. Y.-L.; Tam, S. C.-F.; Kung, A. W.-C.; Wang, C. C.-L.: Hypokalemic periodic paralysis due to the Sjogren syndrome in Chinese patients. Ann. Intern. Med. 110: 405-406, 1989.
Ramsay, I. D.: Thyrotoxic periodic paralysis. In: Thyroid Disease and Muscle Dysfunction. Chicago: Year Book Med. Publ. (pub.) 1974. Pp. 96-125.
Raskin, R. J.; Tesar, J. T.; Lawless, O. J.: Hypokalemic periodic paralysis in Sjogren's syndrome. Arch. Intern. Med. 141: 1671-1673, 1981.
Shah, N.; Kussman, M. J.; Tulgan, H.: Familial periodic paralysis and hyperthyroidism. New York J. Med. 79: 1770-1771, 1979.
Yeung, R. T. T.: Personal Communication. Hong Kong 10/9/1981.
Yeung, R. T. T.; Tse, T. F.: Thyrotoxic periodic paralysis: effect of propranolol. Am. J. Med. 57: 584-590, 1974.

ALDOSTERONISM, SENSITIVE TO DEXAMETHASONE (#103900)

JJM NOTE

This entity is one very treatable explanation for multiple adrenal adenomas.

A number sign (#) is used with this entry because of evidence that glucocorticoid-remediable aldosteronism (GRA) is the result of an anti-Lepore-type fusion of the CYP11B2 (124080) and CYP11B1 (202010) genes. (The various hemoglobins Lepore (e.g., 142000.0019) have a fusion beta-type subunit that is delta globin at the NH2 end and beta globin at the COOH end. This chimeric structure results from nonhomologous pairing and unequal crossing-over between the contiguous delta and beta globin genes. The hemoglobins Lepore result from delta-beta fusion because the delta globin gene (142000) is located upstream from the beta globin gene (141900). The hemoglobins anti-Lepore, e.g., Hb Miyada (141900.0179) and Hb P(Nilotic) (141900.0215), are the reciprocal product of nonhomologous pairing and unequal crossing-over between the HBD and HBB genes; they are beta-delta fusion globins. In GRA, the 5-prime portion of the downstream gene is the 5-prime portion of the fusion gene; hence, it is an anti-Lepore fusion.)

Sutherland et al. (1966) and Salti et al. (1969) described a father and son with hypertension, low plasma renin activity, and increased aldosterone secretion responsive to dexamethasone. Growth and sexual development were normal. At laparotomy the father was found to have multiple adrenocortical adenomas. This appears to be distinct from Conn syndrome (primary aldosteronism) which is not sensitive to dexamethasone. New and Peterson (1967) described 2 cases in a family. Giebink et al. (1973) studied 2 brothers and their mother who had glucocorticoid-remediable aldosteronism. Ganguly et al. (1981) showed that the paradoxic decline in plasma aldosterone when the patient is in the upright posture, usually observed in aldosterone-producing adenoma, is also seen in GSH. Thus, in patients with primary aldosteronism in whom GSH is suspected on the basis of young age and family history and a postural decline in plasma aldosterone is demonstrated, treatment with glucocorticoid should be given for 4 to 6 weeks before localization procedures are begun. Ganguly et al. (1981) studied 2 families, each with 3 affected persons. The diagnosis of hyperaldosteronism was established by failure of saline infusion to suppress plasma aldosterone normally and by the failure of furosemide or a low sodium diet to stimulate plasma renin activity. One family had basal serum potassium levels below 3.5 mmol per liter, whereas values were normal in the second family. Although primary aldosteronism is rare (about 2% of hypertensives have it), it has been subdivided into 3 types: aldosterone-producing adenoma (50-90% of cases), idiopathic form thought to be due to bilateral adrenal hyperplasia, and GSH (the rarest form). Mulrow (1981) speculated that the primary defect in GSH resides in the anterior pituitary gland. Experiments in animals have hinted at the existence of another aldosterone-regulating hormone, possibly originating in the pituitary. Mulrow (1981) asked: 'Is it possible that in t he familial disorder of glucocorticoid-suppressible hyperaldosteronism, the pituitary gland is synthesizing or processing a more potent form of (a fragment of proopiomelanocortin) that enhances the response of the adrenal glomerulosa cell to normal concentrations of ACTH?' If the answer is 'yes,' GSH might appropriately be discussed in entry 176830. This hypothesis proved untrue, however.

Aldosterone synthase (124080), like steroid 11-beta-hydroxylase (202010), is expressed in both adrenal fasciculata and glomerulosa; they are 95% identical (Mornet et al., 1989) and lie on chromosome 8q (Mornet et al., 1989; Chua et al., 1987). That they are immediately adjacent is indicated by the fact that a chimeric, anti-Lepore-like gene has been identified as the cause of glucocorticoid-remediable aldosteronism. In glucocorticoid-remediable aldosteronism (GRA, an alternative acronym for GSH) there are high levels of the abnormal adrenal steroids 18-oxocortisol and 18-hydroxycortisol. The hypertension, variable hyperaldosteronism, and abnormal steroid production are all under the control of ACTH and suppressible by glucocorticoids. The fusion gene has the promoter and some other 5-prime parts of the CYP11B2 gene. As is the practice with other hybrid genes, the details are given as an allelic variant of the gene that contributes the 5-prime portion; therefore, see 202010.0002.

Glucocorticoid suppressible hyperaldosteronism is the result of CYP11B2 activity under the control of ACTH (which normally regulates CYP11B1) and results from a unequal crossing-over involving the CYP11B1 and CYP11B2 genes. Normally, these genes are in the following orientation: 5-prime—CYP11B2—CYP11B1–3-prime; the hybrid anti-Lepore gene lies between CYP11B2 and CYP11B1 and has B1 sequence at its 5-prime end and B2 sequence at its 3-prime end. The breakpoints of the various hybrid genes that have been studied have been found to be 5-prime of intron 4. Pascoe et al. (1992) demonstrated that hybrid cDNAs containing 5-prime sequences from CYP11B1 and 3-prime sequences from CYP11B2, when transfected into COS-1 cells, resulted in aldosterone synthesis at near normal levels when the constructs contained up to the first 3 exons of CYP11B1, while those with 5 or more exons from CYP11B1 produced no detectable aldosterone.

Gordon (1995) stated that 'in a study on approximately 1,000 descendants of an English convict transported to Australia in 1837 for highway robbery in Northamptonshire,' his colleagues and he had confirmed, in 21 affected members thus far identified, the extraordinary phenotypic heterogeneity in glucocorticoid-remediable aldosteronism. The affected members were often normokalemic, and some remained normotensive until late in life. This disorder, which he referred to as familial hyperaldosteronism type 1, is associated with hybrid genes showing somewhat different crossover points linking the CYP11B1 and CYP11B2 portions. To that extent, the disorder shows genetic heterogeneity; however, no other gene has been implicated in the syndrome of ACTH-dependent hyperaldosteronism.

Pascoe et al. (1995) studied a French kindred in which 7 members had GSH; of the 7, 2 also had adrenal tumors and 2 other members of the family had micronodular adrenal hyperplasia. One of the adrenal tumors and the surrounding adrenal tissue had been removed, giving a rare opportunity to study the regulation and action of the hybrid CYP11B1/CYP11B2 gene causing the disease. The hybrid gene was demonstrated to be expressed at higher levels than either CYP11B1 or CYP11B2 in the cortex of the adrenal by RT-PCR and Northern blot analysis. In situ hybridization showed that both CYP11B1 and the hybrid chain were expressed in all 3 zones of the cortex. In cell culture experiments, hybrid gene expression was stimulated by ACTH, leading to increased production of aldosterone and the hybrid steroids characteristic of GSH. The genetic basis of the tumors and hyperplasia in this family was not known but may have been related to the duplication causing the hyperaldosteronism.

Gates et al. (1996) described 2 large pedigrees with many subjects who had the abnormal chimeric gene associated with glucocorticoid remediable aldosteronism. Most of the affected members, who had only mild hypertension and normal biochemistry, were clinically indistinguishable from patients with essential hypertension. This suggested to the authors that GRA is an underdiagnosed condition.

See also Ganguly et al. (1981); Grim and Weinberger (1980)

Chua, S. C.; Szabo, P.; Vitek, A.; Grzeschik, K.-H.; John, M.; White, P. C.: Cloning of cDNA encoding steroid 11-beta-hydroxylase (P450C11). Proc. Nat. Acad. Sci. 84: 7193-7197, 1987.
Ganguly, A.; Grim, C. E.; Bergstein, J.; Brown, R. D.; Weinberger, M. H.: Genetic and pathophysiologic studies of a new kindred with glucocorticoid-suppressible hyperaldosteronism manifest in three generations. J. Clin. Endocr. Metab. 53: 1040-1046, 1981.
Ganguly, A.; Grim, C. E.; Weinberger, M. H.: Anomalous postural aldosterone response in glucocorticoid-suppressible hyperaldosteronism. New Eng. J. Med. 305: 991-993, 1981.
Gates, L. J.; MacConnachie, A. A.; Lifton, R. P.; Haites, N. E.; Benjamin, N.: Variation of phenotype in patients with glucocorticoid remediable aldosteronism. J. Med. Genet. 33: 25-28, 1996.
Giebink, G. S.; Gotlin, R. W.; Biglieri, E. G.; Katz, F. H.: A kindred with familial glucocorticoid-suppressible aldosteronism. J. Clin. Endocr. 36: 715-723, 1973.
Gordon, R. D.: Heterogeneous hypertension. Nature Genet. 11: 6-9, 1995.
Grim, C. E.; Weinberger, M. H.: Familial, dexamethasone-suppressible, normokalemic hyperaldosteronism. Pediatrics 65: 597-604, 1980.
Mornet, E.; Dupont, J.; Vitek, A.; White, P. C.: Characterization of two genes encoding human steroid 11-beta-hydroxylase (P-45011-beta). J. Biol. Chem. 264: 20961-20967, 1989.
Mulrow, P. J.: Glucocorticoid-suppressible hyperaldosteronism: a clue to the missing hormone?. (Editorial) New Eng. J. Med. 305: 1013-1014, 1981.
New, M. I.; Peterson, R. E.: A new form of congenital adrenal hyperplasia. J. Clin. Endocr. 27: 300-305, 1967.
Pascoe, L.; Curnow, K. M.; White, P. C.: Mutations in the CYP11B1 (11-beta-hydroxylase) and CYP11B2 (aldosterone synthase) genes causing CMOII deficiency, 11-hydroxylase deficiency and glucocorticoid suppressible hyperaldosteronism. (Abstract) Am. J. Hum. Genet. 51 (suppl.): A28, 1992.

Pascoe, L.; Jeunemaitre, X.; Lebrethon, M.-C.; Curnow, K. M.; Gomez-Sanchez, C. E.; Gasc, J.-M.; Saez, J. M.; Corvol, P.: Glucocorticoid-suppressible hyperaldosteronism and adrenal tumors occurring in a single French pedigree. J. Clin. Invest. 96: 2236-2246, 1995.
Salti, I. S.; Stiefel, M.; Ruse, J. L.; Laidlaw, J. C.: Non-tumorous 'primary' aldosteronism. I. Type relieved by glucocorticoid (glucocorticoid-remediable aldosteronism). Canad. Med. Assoc. J. 101: 1-10, 1969.
Sutherland, D. J.; Ruse, J. L.; Laidlaw, J. C.: Hypertension, increased aldosterone secretion and low plasma renin activity relieved by dexamethasone. Canad. Med. Assoc. J. 95: 1109-1119, 1966.

MACROSOMIA ADIPOSA CONGENITA (248100)

JJM NOTE

This entity is included because of the association with adrenocortical adenomas. Given the follow-up capabilities in Denmark, I wonder if the family's experience over the last six decades could be documented and interpreted with current knowledge.

Christiansen (1929) described a Danish kindred in which 7 infants (6 females, 1 male), the offspring of sisters by presumably unrelated husbands, developed gross obesity beginning soon after birth. Precocious skeletal development was evident in the ossification centers and teeth. Marked voracity was a feature. Relative eosinophilia and low vitality with death of 5 of the children in the first year were noted. Adrenocortical adenomas were found at autopsy. The nature of the disorder is obscure.

Christiansen, T.: Macrosomia adiposa congenita. A new dysendocrine syndrome of familial occurrence. Endocrinology 13: 149-163, 1929.

Paragangliomas

PARAGANGLIOMAS (*168000)

JJM NOTE

Much clinical and laboratory research remains to be conducted to sort out the genetics of these neoplasms. For now, the clinician encountering a patient with a paraganglioma or related neoplasm should take a definitive family history and address the possibility of neurofibromatosis 1, tumors at other sites in the body, including pheochromocytoma, chondroma, gastric leiomyoma, and von Hippel-Lindau disease. I mention the last because of one large pheochromocytoma family with a von Hippel-Lindau gene mutation that has two cases of paragangliomas (Tisherman SE et al.: Three decade investigation of familial pheochromocytoma: An allele of von Hippel-Lindau disease? Arch. Intern. Med. 153:2550–56, 1993).

Kroll et al. (1964) found carotid body tumors in 12 members of a family in an autosomal dominant pattern of inheritance. Carotid body tumors and glomus jugulare tumors are considered to be chemodectomas, this being a term for tumors arising in chemoreceptor structures. Some would question the appropriateness of calling these paragangliomas. Resler et al. (1966) described a patient with bilateral carotid body tumors and a glomus jugulare tumor. They commented that familial carotid body tumors tend to be multiple. Familial glomus jugulare tumors are probably rare; one of the few reported families may be that with 3 affected sisters described in 1937 by Goekoop (cited by Rosen, 1952). Also see Ladenheim and Sachs (1961). (The glomus jugulare, of which tumors may occur as an expression of this same gene, was first discovered by Stacy R. Guild at Johns Hopkins in 1941 (12,13:Guild, 1941, 1953).)

Bartels (1949) found carotid body tumors in members of 3 successive generations. Wilson (1970) reviewed the familial reports and described a family with male-to-male transmission and a 'skipped generation.' Pratt (1973) reviewed the literature and reported 8 new cases of either unilateral or bilateral carotid body tumors in 4 generations of a kindred. In 1 generation of this family, 4 sisters had bilateral tumors and 1 brother had unilateral tumors. None of the 8 tumors reported were malignant. Chedid and Jao (1974) found carotid body tumors in 6 members of 2 consecutive generations of a family. Four also had chronic obstructive pulmonary disease with persistently high arterial pCO(2) and low pO(2). They theorized that the tumors start as hyperplasia secondary to the stimulus of these altered blood gases.

Nissenblatt (1978) pointed out the connection between high altitude living, emphysema and cyanotic congenital heart disease on the one hand, and carotid body tumors on the other.

Van Baars (1980) found 26 affected persons in a Dutch kindred, of whom 13 were not known to be affected before the study. The disorder was traced through 6 generations. Grufferman et al. (1980) reviewed reports of 88 familial and 835 nonfamilial CBT1 cases. The sex ratio was 1.0 in familial cases. Autosomal dominant inheritance was supported. Bilateral disease occurred in 31.8% of familial cases and 4.4% of sporadic cases. In 6% of cases, second primary tumors, mostly other paragangliomas, occurred. Parkin (1981) reported the familial association of multiple paragangliomata and pheochromocytomas. He referred to this entity as 'multiple glomus tumors.' Although indeed the tumors in some instances involve the glomus tympanicum and glomus jugulare, these are not multiple glomus tumors in the same sense as the term is used in entry 138000. The occurrence of pheochromocytoma and multiple paragangliomas in neurofibromatosis (162200) was described by DeAngelis et al. (1987).

Parry et al. (1982) sought genetic factors in CBT1 by reviewing the records of 222 histologically diagnosed cases: 146 females, 76 males; mean age of development of tumors 44.7 years. In 16 patients who had other extraadrenal paragangliomas, suggesting a multiple primary tumor syndrome, CBT1 was diagnosed earlier (mean, 35.4 years; p less than 0.01). Familial CBT1 was more often bilateral and diagnosed slightly earlier. In a sporadic case, Karasov et al. (1982) described a girl who had a record number of 21 paragangliomas removed between the ages of 13 and 17 years, with evidence of remaining tumors. The tumors were catecholamine-producing.

Khafagi et al. (1987) reported the cases of 2 unrelated adults, both of whom had a positive family history of paraganglioma. Both had malignant nonfunctional paraganglioma which could be detected by uptake of iodine-131 metaiodobenzylguanidine. This agent proved to be of therapeutic value in the management of unresectable metastases. This led me to raise the possibility that the gene for carotid body tumor may be situated in 13q34 since genes for the coagulation factors VII (227500) and X (227600) were known from deletion mapping to be situated on 13q34 (Pfeiffer et al., 1982) and hereditary tumors are often associated with small deletions (witness retinoblastoma and Wilms tumor).

Kroll et al. (1964) observed an association of combined deficiency of factors VII and X with carotid body tumor in a large kindred. Morton (1984) provided follow-up information on the family of Kroll et al. (1964). In the proband M.L., chromosome analysis of metaphase chromosomes after methotrexate synchronization showed a normal 46,XX karyotype with no evidence of a deletion of chromosome 13. Examination of M.L. in 1984 showed a 1-cm palpable mass on the left side of the neck as was reported by Kroll et al. (1964). In no member of the kindred had the tumor been removed because it gave them no cosmetic or other trouble.

From a study of 15 pedigrees, van der Mey et al. (1989) concluded that the phenotype multiple paragangliomata (referred to be them as 'hereditary glomus tumors') is inherited almost exclusively via the paternal line, a finding inconsistent with simple autosomal dominant transmission. They suggested that the results can be explained in terms of genomic imprinting: the maternally derived gene is inactivated during female oogenesis and can be reactivated only during spermatogenesis. Heutink et al. (1992) stated that all individuals with hereditary paragangliomas have inherited the disease gene from their father; expression of the phenotype is not observed in the offspring of an affected female until subsequent transmission of the gene through a male carrier. They stated that this finding strongly suggests involvement of genomic imprinting.

In a large Dutch pedigree with hereditary paragangliomas, Heutink et al. (1992) found linkage with D11S147 located in the region 11q23-qter; peak lod score = 6.0 at theta = 0.0. Likelihood calculations yielded a very high odds ratio in favor of genomic imprinting versus absence of genomic imprinting. Devilee et al. (1992) performed haplotyping in this large Dutch family, using 15 markers in the region 11q13-q23. Two recombination events placed the PGL locus distal to STMY (185260) and proximal to D11S836, thus excluding the oncogenes INT2 (164950) and ETS1 (164720) as the site of the mutation. In 2 distantly related pedigrees with the same disorder, Mariman et al. (1993) confirmed the location of the PGL gene on 11q. A maximum lod score of 5.4 at theta = 0.0 was obtained for linkage between the disorder and the chromosomal region defined by 2 markers, INT2 and D11S527. They concluded that the gene is proximal to the tyrosinase locus (TYR; 203100), which is located at 11q14-q21. This places the gene more proximal than the previously indicated site. INT2 is located at 11q13. The reason for the discrepancy in mapping was not clear. That INT2 is the site of the mutation was still considered possible.

Devilee et al. (1994) found loss of heterozygosity (LOH) only on chromosome 11, with a marked clustering on the distal half of the long arm. In all 8 cases in which they could determine the parental origin, the allele undergoing loss was maternally derived.

The possibility that paraganglioma is genetically heterogeneous, with 2 different genes on 11q being able to act as predisposing factors, was suggested by the mapping of the gene to 11q13 by Mariman et al. (1993) and to 11q22-q23 by Heutink et al. (1992). Heutink et al. (1994) narrowed the assignment of the distal locus to 11q22.3-q23.3 between STMY and CD3D (186790). By linkage analysis and haplotyping of a single large family in which the pattern of inheritance was consistent with genomic imprinting, Mariman et al. (1995) mapped the proximal gene locus to a 5-cM region of 11q13.1 between D11S956 and PYGM (232600). A maximum lod score of 7.62 at theta = 0.0 was obtained for D11S480. The interval did not overlap with the locus for glomus tumors in other families: 11q22.3-q23.3. In an analysis of a second family showing the imprinting phenomenon, Mariman et al. (1995) excluded the 5-cM region as the location of the disease gene, whereas indication of linkage was obtained with markers from the distal locus. These observations suggested the presence of 2 distinct imprinted genes for glomus tumors on 11q. Assuming that both genes code for functionally equivalent tumor suppressors, the Knudson hypothesis (Knudson, 1986) would predict that both copies of the 2 genes have to be turned off before glomus tumors can develop. Genomic imprinting would account for the silencing of the 2 maternal copies, whereas a paternally inherited mutation in 1 of the genes would represent the third hit. A single somatic mutation in the remaining active paternal copy would subsequently suffice to initiate tumor growth. Maternally inherited mutations would only very rarely give rise to tumor development because it would require the inactivation of the 2 paternally transmitted gene copies in a single cell. Thus, paraganglioma might represent an example of a digenic disorder comparable to 1 form of retinitis pigmentosa (180721.0001).

Oosterwijk et al. (1996) offered genetic counseling on the basis of DNA linkage diagnosis in an extended Dutch pedigree. Presymptomatic testing of the paternal allele was performed in 16 cases; 4 of these appeared to have the at-risk haplotype and in 2 of the 4 a glomus tumor was subsequently detected on MRI. In 1 case linkage results were inconclusive because of recombination and 1 person did not want to learn his test result. The maternal allele was tested for carrier status in 4 cases of which 1 appeared to be a carrier. Oosterwijk et al. (1996) gave the location of the PGL gene as 11q22.3-q23.

Baysal et al. (1997) reported linkage of paraganglioma (PGL) to chromosome 11q23 in 3 out of 6 North American families. Recombinants narrowed the critical region to a 4.5-Mb interval flanked by the markers D11S1647 and D11S622. Partial allelic loss of strictly maternal origin was detected in 5 out of 19 tumors. Baysal et al. (1997) noted that the age of onset of symptoms was significantly different between fathers and children: affected children had an earlier age of onset in 39 of 57 father-child pairs. Milunsky et al. (1997) confirmed linkage to 11q23 in studies of 3 families. The inheritance pattern was consistent with genetic imprinting; the disorder was transmitted only by males.

Another 10 families with hereditary head and neck paragangliomas were ascertained by van Schothorst et al. (1998) from the same geographic region as that from which the large PGL1-linked family originated. They determined the disease-linked haplotype, as defined by 13 markers encompassing a large interval on 11q21-q23, in these families and showed that alleles were identical for 6 contiguous markers, spanning a genetic distance of 6 cM and containing PGL1. Despite this strong indication of a common ancestor, no kinship between the families could be demonstrated through genealogic surveys going back to 1800.

(The triad of gastric leiomyosarcoma, pulmonary chondroma, and extraadrenal paraganglioma constitutes a syndrome that occurs mainly in young women and is known as the Carney triad (Carney, 1983). At least 28 cases had been identified by the time of the report by Margulies and Sheps (1988). There have, it seems, been no familial cases. This triad is not to be confused with the other Carney syndrome of myxoma, spotty pigmentation and endocrinopathy (160980).)

See also Chase (1933); Desai and Patel (1961); Gruber and Metson (1980); Herrmann (1977); Katz (1964); Lee et al. (1977); van Baars et al. (1982); van Baars et al. (1981); van Baars et al. (1981)

Bartels, J.: De tumoren van het glomus jugulare. Thesis: Groningen (pub.) 1949.

Baysal, B. E.; Farr, J. E.; Rubinstein, W. S.; Galus, R. A.; Johnson, K. A.; Aston, C. E.; Myers, E. N.; Johnson, J. T.; Carrau, R.; Kirkpatrick, S. J.; Myssiorek, D.; Singh, D.; Saha, S.; Gollin, S. M.; Evans, G. A.; James, M. R.; Richard, C. W., III: Fine mapping of an imprinted gene for familial nonchromaffin paragangliomas, on chromosome 11q23. Am. J. Hum. Genet. 60: 121-132, 1997.

Carney, J. A.: The triad of gastric epithelioid leiomyosarcoma, pulmonary chondroma, and functioning extra-adrenal paraganglioma: a five-year review. Medicine 62: 159-169, 1983.

Chase, W. H.: Familial and bilateral tumours of the carotid body. J. Path. Bact. 36: 1-12, 1933.

Chedid, A.; Jao, W.: Hereditary tumors of the carotid bodies and chronic obstructive pulmonary disease. Cancer 33: 1635-1641, 1974.

DeAngelis, L. M.; Kelleher, M. B.; Post, K. D.; Fetell, M. R.: Multiple paragangliomas in neurofibromatosis: a new neuroendocrine neoplasia. Neurology 37: 129-133, 1987.

Desai, M. G.; Patel, C. C.: Heredo-familial carotid body tumours. Clin. Radiol. 12: 214-217, 1961.

Devilee, P.; Heutink, P.; van der Mey, A. G. L.; Oostra, B. A.; van Ommen, G. J.-B.; Cornelisse, C. J.: Mapping the genomically imprinted gene responsible for hereditary paragangliomas on 11q. (Abstract) Am. J. Hum. Genet. 51 (suppl.): A61 only, 1992.

Devilee, P.; van Schothorst, E. M.; Bardoel, A. F. J.; Bonsing, B.; Kuipers-Dijkshoorn, N.; James, M. R.; Fleuren, G.; van der Mey, A. G. L.; Cornelisse, C. J.: Allelotype of head and neck paragangliomas: allelic imbalance is confined to the long arm of chromosome 11, the site of the predisposing locus PGL. Genes Chromosomes Cancer 11: 71-78, 1994.

Gruber, H.; Metson, R.: Carotid body paraganglioma regression with relief of hypoxemia. Ann. Intern. Med. 92: 800-802, 1980.

Grufferman, S.; Gillman, M. W.; Pasternak, L. R.; Peterson, C. L.; Young, W. G., Jr.: Familial carotid body tumors: case report and epidemiologic review. Cancer 46: 2116-2122, 1980.

Guild, S. R.: A hitherto unrecognized structure, the glomus jugularis in man. (Abstract) Anat. Rec. 79 (suppl. 2): 28 only, 1941.

Guild, S. R.: The glomus jugulare, a nonchromaffin paraganglion, in man. Ann. Otol. Rhinol. Laryng. 62: 1045-1071, 1953.

Herrmann, J.: Delayed mutation model: carotid body tumors and retinoblastoma.In: Mulvihill, J. J.; Miller, R. W.; Fraumeni, J. F., Jr.: Genetics of Human Cancer. New York: Raven Press (pub.) 1977. Pp. 417-438.

Heutink, P.; van der Mey, A. G. L.; Sandkuijl, L. A.; van Gils, A. P. G.; Bardoel, A.; Breedveld, G. J.; van Vliet, M.; van Ommen, G.-J. B.; Cornelisse, C. J.; Oostra, B. A.; Weber, J. L.; Devilee, P.: A gene subject to genomic imprinting and responsible for hereditary paragangliomas maps to chromosome 11q23-qter. Hum. Molec. Genet. 1: 7-10, 1992.

Heutink, P.; van Schothorst, E. M.; van der Mey, A. G. L.; Bardoel, A.; Breedveld, G.; Pertijs, J.; Sandkuijl, L. A.; van Ommen, G.-J. B.; Cornelisse, C. J.; Oostra, B. A.; Devilee, P.: Further localization of the gene for hereditary paragangliomas and evidence for linkage in unrelated families. Europ. J. Hum. Genet. 2: 148-158, 1994.

Karasov, R. S.; Sheps, S. G.; Carney, J. A.; van Heerden, J. A.; DeQuattro, V.: Paragangliomatosis with numerous catecholamine-producing tumors. Mayo Clin. Proc. 57: 590-595, 1982.

Katz, A. D.: Carotid body tumors in a large family group. Am. J. Surg. 108: 570-573, 1964.

Khafagi, F.; Egerton-Vernon, J.; van Doorn, T.; Foster, W.; McPhee, I. B.; Allison, R. W. G.: Localization and treatment of familial malignant nonfunctional paraganglioma with iodine-131 MIBG: report of two cases. J. Nucl. Med. 28: 528-531, 1987.

Knudson, A. G., Jr.: Genetics of human cancer. Annu. Rev. Genet. 20: 231-251, 1986.

Kroll, A. J.; Alexander, B.; Cochios, F.; Pechet, L.: Hereditary deficiencies of clotting factors VII and X associated with carotid-body tumors. New Eng. J. Med. 270: 6-13, 1964.

Ladenheim, J. C.; Sachs, E., Jr.: Familial tumors of the 'glomus jugulare': report of 2 cases, with a note on the vascular origin of the glomera. Neurology 11: 303-309, 1961.

Lee, S. P.; Nicholson, G. I.; Hitchcock, G. C.: Familial abdominal chemodectomas with associated cutaneous angiolipomas. Pathology 9: 173-177, 1977.

Margulies, K. B.; Sheps, S. G.: Carney's triad: guidelines for management. Mayo Clin. Proc. 63: 496-502, 1988.

Mariman, E. C. M.; van Beersum, S. E. C.; Cremers, C. W. R. J.; Struycken, P. M.; Ropers, H. H.: Fine mapping of a putatively imprinted gene for familial non-chromaffin paragangliomas to chromosome 11q13.1: evidence for genetic heterogeneity. Hum. Genet. 95: 56-62, 1995.

Mariman, E. C. M.; van Beersum, S. E. C.; Cremers, C. W. R. J.; van Baars, F. M.; Ropers, H. H.: Analysis of a second family with hereditary non-chromaffin paragangliomas locates the underlying gene at the proximal region of chromosome 11q. Hum. Genet. 91: 357-361, 1993.

Milunsky, J.; DeStefano, A. L.; Huang, X.-L.; Baldwin, C. T.; Michels, V. V.; Jako, G.; Milunsky, A.: Familial paragangliomas: linkage to chromosome 11q23 and clinical implications. Am. J. Med. Genet. 72: 66-70, 1997.

Morton, C. C.: Personal Communication. Boston, Mass. 5/23/1984.

Nissenblatt, M. J.: Cyanotic heart disease: 'low altitude' risk for carotid body tumor?. Johns Hopkins Med. J. 142: 18-22, 1978.

Oosterwijk, J. C.; Jansen, J. C.; van Schothorst, E. M.; Oosterhof, A. W.; Devilee, P.; Bakker, E.; Zoeteweij, M. W.; van der Mey, A. G. L.: First experiences with genetic counselling based on predictive DNA diagnosis in hereditary glomus tumours (paragangliomas). J. Med. Genet. 33: 379-383, 1996.

Parkin, J. L.: Familial multiple glomus tumors and pheochromocytomas. Ann. Otol. Rhinol. Laryng. 90: 60-63, 1981.

Parry, D. M.; Li, F. P.; Strong, L. C.; Carney, J. A.; Schottenfeld, D.; Reimer, R. R.; Grufferman, S.: Carotid body tumors in humans: genetics and epidemiology. J. Nat. Cancer Inst. 68: 573-578, 1982.

Pfeiffer, R. A.; Ott, R.; Gilgenkrantz, S.; Alexandre, P.: Deficiency of coagulation factors VII and X associated with deletion of a chromosome 13 (q34): evidence from two cases with 46,XY,t(13;Y)(q11;q34). Hum. Genet. 62: 358-360, 1982.

Pratt, L. W.: Familial carotid body tumors. Arch. Otolaryng. 97: 334-336, 1973.

Resler, D. R.; Snow, J. B.; Williams, G. R.: Multiplicity and familial incidence of carotid body and glomus jugulare tumors. Ann. Otolaryng. 75: 114-122, 1966.

Rosen, S.: Glomus jugulare tumor of the middle ear with normal drum, improved biopsy technique. Ann. Otolaryng. 61: 448-451, 1952.

van Baars, F.; Cremers, C.; van den Broek, P.; Geerts, S.; Veldman, J.: Genetic aspects of nonchromaffin paraganglioma. Hum. Genet. 60: 305-309, 1982.

van Baars, F.; van den Broek, P.; Cremers, C.; Veldman, J.: Familial non-chromaffinic paragangliomas (glomus tumors): clinical aspects. Laryngoscope 91: 988-996, 1981.

van Baars, F. M.: Glomustumoren et herediteit. Thesis: Catholic Univ. of Nijmegen (pub.) 1980.
van Baars, F. M.; Cremers, C. W. R. J.; van den Broek, P.; Veldman, J. E.: Familial non-chromaffinic paragangliomas (glomus tumors): clinical and genetic aspects. Acta Otolaryng. 91: 589-593, 1981.
van der Mey, A. G.; Maaswinkel-Mooy, P. D.; Cornelisse, C. J.; Schmidt, P. H.; van de Kamp, J. J.: Genomic imprinting in hereditary glomus tumours: evidence for new genetic theory. Lancet II: 1291-1294, 1989.
van Schothorst, E. M.; Jansen, J. C.; Grooters, E.; Prins, D. E. M.; Wiersinga, J. J.; van der Mey, A. G. L.; van Ommen, G.-J. B.; Devilee, P.; Cornelisse, C. J.: Founder effect at PGL1 in hereditary head and neck paraganglioma families from The Netherlands. Am. J. Hum. Genet. 63: 468-473, 1998.
Wilson, H.: Carotid body tumors: familial and bilateral. Ann. Surg. 171: 843-848, 1970.

PARAGANGLIOMAS, FAMILIAL NONCHROMAFFIN, 2 (*601650)

JJM NOTE

Because of late clinical diagnosis, usual benign course, and imprinting, family history could well be negative at first unless deep probing is done.

As reviewed in PGL1 168000, 3,4:Heutink et al. (1992, 1994) mapped familial nonchromaffin paragangliomas to 11q22.3-q23.3, between markers STMY1 (185250) and CD3D (186790). A single large Dutch pedigree gave a lod score of more than 3 for 11q23 markers; 5 other Dutch families provided additional support for linkage to 11q23. Linkage analysis in another unrelated Dutch pedigree revealed the presence of a second, more proximal locus (PGL2) on 11q13, in a 5-cM region flanked by markers D11S956 and PYGM (232600) (6,5:Mariman et al., 1993, 1995). Both the 11q23 PGL1 and the 11q13.1 PGL2 Dutch pedigrees demonstrated maternal imprinting (inactivation) of the mutant PGL gene. Baysal et al. (1997) ascertained a total of 11 North American PGL families and confirmed maternal imprinting. In 3 of 6 families, linkage analysis provided evidence of linkage to the PGL1 locus at 11q23. Recombinants narrowed the critical region to an approximately 4.5-Mb interval flanked by markers D11S1647 and D11S622. Partial allelic loss of strictly maternal origin was detected in 5 of 19 tumors. The greatest degree of imbalance was detected at 11q23, distal to D11S1327 and proximal to CD3D. Age at onset of symptoms was significantly different between fathers and children. Affected children had earlier age at onset of symptoms in 39 of 57 father-child pairs. However, a more conservative comparison of the number of pairs in which the child had 5 or more years earlier age at onset vis-a-vis that of complementary pairs revealed no significant difference. Baysal et al. (1997) stated that whether these data represented genetic anticipation or ascertainment bias could be addressed only by analysis of a larger number of father-child pairs.

To assess whether the PGL gene is subject to a 2-hit inactivation mechanism of Knudson, Devilee et al. (1994) assayed loss of heterozygosity (LOH) in head and neck PGLs. LOH was found only on chromosome 11, with marked clustering on the distal half of the q arm. Furthermore, the maternal allele was lost in all 8 cases in which parental origin of the mutation could be determined. Attempts to define precisely the smallest region of overlap were hampered by the discovery of only partial allele loss, suggesting either tumor heterogeneity or polyclonality.

Baysal, B. E.; Farr, J. E.; Rubinstein, W. S.; Galus, R. A.; Johnson, K. A.; Aston, C. E.; Myers, E. N.; Johnson, J. T.; Carrau, R.; Kirkpatrick, S. J.; Myssiorek, D.; Singh, D.; Saha, S.; Gollin, S. M.; Evans, G. A.; James, M. R.; Richard, C. W., III: Fine mapping of an imprinted gene for familial nonchromaffin paragangliomas, on chromosome 11q23. Am. J. Hum. Genet. 60: 121-132, 1997.
Devilee, P.; van Schothorst, E. M.; Bardoel, A. F. J.; Bonsing, B.; Kuipers-Dijkshoorn, N.; James, M. R.; Fleuren, G.; van der Mey, A. G. L.; Cornelisse, C. J.: Allelotype of head and neck paragangliomas: allelic imbalance is confined to the long arm of chromosome 11, the site of the predisposing locus PGL. Genes Chromosomes Cancer 11: 71-78, 1994.
Heutink, P.; van der Mey, A. G. L.; Sandkuijl, L. A.; van Gils, A. P. G.; Bardoel, A.; Breedveld, G. J.; van Vliet, M.; van Ommen, G.-J. B.; Cornelisse, C. J.; Oostra, B. A.; Weber, J. L.; Devilee, P.: A gene subject to genomic imprinting and responsible for hereditary paragangliomas maps to chromosome 11q23-qter. Hum. Molec. Genet. 1: 7-10, 1992.
Heutink, P.; van Schothorst, E. M.; van der Mey, A. G. L.; Bardoel, A.; Breedveld, G.; Pertijs, J.; Sandkuijl, L. A.; van Ommen, G.-J. B.; Cornelisse, C. J.; Oostra, B. A.; Devilee, P.: Further localization of the gene for hereditary paragangliomas and evidence for linkage in unrelated families. Europ. J. Hum. Genet. 2: 148-158, 1994.
Mariman, E. C. M.; van Beersum, S. E. C.; Cremers, C. W. R. J.; Struycken, P. M.; Ropers, H. H.: Fine mapping of a putatively imprinted gene for familial non-chromaffin paragangliomas to chromosome 11q13.1: evidence for genetic heterogeneity. Hum. Genet. 95: 56-62, 1995.
Mariman, E. C. M.; van Beersum, S. E. C.; Cremers, C. W. R. J.; van Baars, F. M.; Ropers, H. H.: Analysis of a second family with hereditary non-chromaffin paragangliomas locates the underlying gene at the proximal region of chromosome 11q. Hum. Genet. 91: 357-361, 1993.

FACTOR X DEFICIENCY (*227600)

JJM NOTE

This entity is included because of the family of Kroll et al. (1964), updated by Morton (1984), that segregates both deficiencies of factors VII and X and carotid body tumors.

The bleeding tendency is manifested by prolonged nasal and mucosal hemorrhage, menorrhagia, hematuria, and occasionally hemarthrosis. Mr. Stuart and Miss Prower were the first persons shown to have this abnormality. Mr. Stuart was the product of an aunt-nephew mating. Girolami et al. (1970) described a hemorrhagic disorder that may be the result of an abnormal factor X. Polymorphism of factors IX and X was suggested by the findings of Lester et al. (1972). Enfield et al. (1975) showed that the amino-terminal sequence of the light chain of bovine Stuart factor is homologous with the amino-terminal region of bovine prothrombin and, like the latter, appears to contain several residues of the unusual amino acid, gamma-carboxyglutamic acid. The heavy chains of bovine factors X and IX are homologous with trypsin, thrombin, and other mammalian serine proteases. Furie et al. (1977) presented evidence that the acquired deficiency of factor X associated with systemic amyloidosis is caused by binding to amyloid. In an editorial on variants of vitamin K-dependent coagulation factors, Bertina et al. (1979) stated that 9 defective variants of factor II, 5 variants of factor X, and many variants (about 180 pedigrees) of factor IX have been identified. At least 1 variant of factor VII (Padua) is also known. See factor X, quantitative variation in (134530). Endo (1981) observed spontaneously developing hematomyelia with incomplete transverse paralysis in a 17-year-old patient with factor X deficiency.

Pregnancy in women with congenital deficiencies of coagulation factors such as factor X is often associated with adverse fetal outcomes. Recurrent spontaneous abortions, placental abruptions, and premature births are reported. Kumar and Mehta (1994) reviewed the outcome of 4 pregnancies in a patient with factor X deficiency. The first 2 pregnancies resulted in the birth of premature babies at 21 and 25 weeks of gestation, both of whom died in the neonatal period. The patient had been treated with fresh frozen plasma for acute bleeding episodes during these pregnancies. In addition, during the second conception she was given factor IX complex prophylactically during the latter half of her pregnancy. During her next 2 pregnancies, she was treated early in pregnancy with prophylactic replacement of factor X. She delivered healthy babies at 34 and 32 weeks of gestation and both babies thrived.

Stoll and Roth (1980) described a girl with a duplication-deficiency syndrome involving chromosomes 4 and 13. The mother had a balanced translocation t(4;13)(q26;q34). The child had partial trisomy of 4q and partial monosomy of 13q. Factor X level was half normal. Pfeiffer et al. (1982) presented evidence suggesting that factors VII and X may be coded by 13q34. They found deficiency of the 2 factors in 2 cases with 46,XY,t(13;Y)(q11;q34) including probable deletion of a terminal segment of 13q. A prolonged prothrombin time was found before surgery in the first case, leading to studies of coagulation; neither patient had clinical abnormality of coagulation. In 1 case, factor VII was measured as 42, 40, and 45% and factor X as 59, 44, and 60% of normal, in 2 different laboratories; in the second case, factor VII was 55 and 54% of normal and factor X was 25 and 62%. These values were normal in all 4 parents.

Synthesis of factors VII and X, as well as factors II and IX, takes place in the liver and requires vitamin K. Structural homologies of these factors, which are precursors of serine proteases, have been shown (Zur and Nemerson, 1981). Ratnoff and Bennett (1973) stated: 'Although a persuasive argument can be made that (they) are derived from a single molecule most evidence suggests that they are distinctive proteins.' Kroll et al. (1964) found combined deficiency of factors VII and X in association with carotid body tumor in a large kindred (see 168000). Because hereditary tumors are so often associated with small deletions, such should be sought in the family of Kroll et al. (1964) and others. Factor X is composed of a light chain of M(r) 16,200 and a heavy chain held together by a disulfide bond. Factor X is synthesized in the liver as a single-chain molecule. Vitamin K is involved in the biosynthesis and is required for the carboxylation of the first 11 glutamic acid residues in the amino-terminal portion of the human molecule. Leytus et al. (1984) isolated and characterized a cDNA coding for human factor X. The DNA sequence coding for the active site showed a high degree of homology with prothrombin and factor IX, 2 other vitamin K-dependent serine proteases that participate in blood coagulation. Perhaps the cDNA probes could be used to corroborate the assignment of the gene locus to chromosome 13. Royle et al. (1985) confirmed the assignment by Southern analysis of somatic cell hybrids. 42,41:Rocchi et al. (1985, 1986) confirmed assignment of F10 to chromosome 13 by means of blotting studies of DNA from somatic cell hybrids using a cDNA probe. Fung et al. (1985) predicted the amino acid factor X sequence of plasma factor X from the cDNA. Factor X is synthesized as a single polypeptide chain precursor in which the light and heavy chains are linked by the tripeptide arg-lys-arg. Scambler and Williamson (1985) studied a female monosomic for 13q34 and deficient in clotting factors VII and X, as well as her brother, who was trisomic for 13q34 and had elevated levels of these factors. (These persons suffered the effects of segregation from a reciprocal translocation in the mother involving the tip of chromosome 13 (13q34) and 6q24-6q26.) DNA dosage studies with a cloned human factor X gene showed that the low levels of factor X expression was due to absence of one copy of the factor X structural gene. By in situ hybridization, Royle et al. (1986) mapped factor X to 13q32-qter. By dosage effects in 7 cases of abnormal chromosome 13, Gilgenkrantz et al. (1986) mapped the factor VII and factor X genes to 13q34. The clotting factors were increased in trisomy of this segment and decreased in monosomy. In constructing the CEPH consortium linkage map of chromosome 13, Bowcock et al. (1993) concluded that F10 is the most telomeric marker on that chromosome. Using interspecific and intersubspecific mapping panels, Koizumi et al. (1995) mapped the mouse homolog, Cf10, to the centromeric region of mouse chromosome 8.

De Stefano et al. (1988) identified a new variant, which they called factor X Roma. It showed normal antigen level that was responsive by Russell viper venom (RVV), but severely impaired in activation via the intrinsic pathway and partially defective in activation via the extrinsic pathway. Bernardi et al. (1989) found that a patient with factor X deficiency was a genetic compound. From her mother she had received a partially deleted gene missing the 3-prime portion coding for the catalytic domain of the factor.

Other vitamin K-dependent proteins known to be involved in blood coagulation are factor VII (227500), factor IX (306900), prothrombin (176930), protein S (176880), and protein C (176860). Leytus et al. (1986) reported that the F10 gene spans approximately 25 kb of genomic DNA. It comprises 7 introns and 8 exons and shows considerable structural homology with the genes encoding other vitamin K-dependent clotting factors. Exon 1 codes for the signal peptide, exon 2 for the propeptide and the gamma-carboxyglutamic acid-rich domain, exon 3 for the short aromatic acid-rich stack, exon 4 for the first epidermal growth factor domain, exon 5 for the second epidermal growth factor domain, exon 6 for the activation peptide, and exons 7 and 8 for the catalytic domain. The splice junction sites follow the GT/AG rule and delineate 7 introns, ranging in size from 950 to 7,400 nucleotides. Human F10 mRNA comprises approximately 1,500 nucleotides, including a coding region of 1,475 bases and a short 3-prime-untranslated region of 10 nucleotides. The polyadenylation signal (ATTAAA) is located in the coding sequence and precedes the stop codon by 1 nucleotide. It appears that the vitamin K-dependent proteins evolved from a single common gene and that this ancestral gene arose through a process involving the assembly of small protein coding units. Wieland et al. (1991) identified an instance of germline mosaicism for deletion of exons 7 and 8. One offspring had this deletion and a different deletion inherited from the mother, i.e., was a compound heterozygote.

Factor X is the zymogen of factor Xa, a serine protease that occupies a pivotal position in the clotting process. It is activated either by the contact-activated (intrinsic) pathway or by the tissue factor (extrinsic) pathway. Factor Xa, in combination with factor V, then activates prothrombin to form the effector enzyme of the coagulation cascade. Cooper et al. (1997) reviewed the molecular genetics of factor X and of factor X deficiency. These included CRM-negative factor X deficiency due to gene deletions and CRM-positive dysfunctional variants due to missense mutations. They tabulated 18 missense mutations.

See also Bachmann (1958); de Grouchy et al. (1984); Dodds (1973); Furie et al. (1981); Girolami et al. (1983); Girolami et al. (1971); Girolami et al. (1975); Girolami et al. (1970); Girolami et al. (1973); Girolami et al. (1985); Greipp et al. (1979); Jaye et al. (1985); Kaul et al. (1986); Mori et al. (1981); Pfeiffer et al. (1985); Roos and Huizinga (1959); Sumer et al. (1986)

Bachmann, F.: Familienuntersuchungen beim kongenitalen Stuart-Prower-Factor Mangel. Arch. Klaus Stift. Vererbungsforsch. 33: 27-78, 1958.

Bernardi, F.; Marchetti, G.; Patracchini, P.; Volinia, S.; Gemmati, D.; Simioni, P.; Girolami, A.: Partial gene deletion in a family with factor X deficiency. Blood 73: 2123-2127, 1989.

Bertina, R. M.; Briet, E.; Veltkamp, J. J.: Variants of vitamin K-dependent coagulation factors. (Editorial) Acta Haemat. 62: 1-3, 1979.

Bowcock, A. M.; Gerken, S. C.; Barnes, R. I.; Shiang, R.; Jabs, E. W.; Warren, A. C.; Antonarakis, S.; Retief, A. E.; Vergnaud, G.; Leppert, M.; Lalouel, J.-M.; White, R. L.; Cavalli-Sforza, L. L.: The CEPH consortium linkage map of human chromosome 13. Genomics 16: 486-496, 1993.

Cooper, D. N.; Millar, D. S.; Wacey, A.; Pemberton, S.; Tuddenham, E. G. D.: Inherited factor X deficiency: molecular genetics and pathophysiology. Thromb. Haemost. 78: 161-172, 1997.

de Grouchy, J.; Dautzenberg, M.-D.; Turleau, C.; Beguin, S.; Chavin-Colin, F.: Regional mapping of clotting factors VII and X to 13q34: expression of factor VII through chromosome 8. Hum. Genet. 66: 230-233, 1984.

De Stefano, V.; Leone, G.; Ferrelli, R.; Hassan, H. J.; Macioce, G.; Bizzi, B.: Factor X Roma: a congenital factor X variant defective at different degrees in the intrinsic and the extrinsic activation. Brit. J. Haemat. 69: 387-391, 1988.

Dodds, W. J.: Canine factor X (Stuart-Prower factor) deficiency. J. Lab. Clin. Med. 82: 560-566, 1973.

Endo, Y.: Congenital factor X deficiency and incomplete transverse paralysis. J.A.M.A. 246: 1708, 1981.

Enfield, D. L.; Ericsson, L. H.; Walsh, K. A.; Neurath, H.; Titani, K.: Bovine factor X (Stuart factor): primary structure of the light chain. Proc. Nat. Acad. Sci. 72: 16-19, 1975.

Fung, M. R.; Hay, C. W.; MacGillivray, R. T. A.: Characterization of an almost full-length cDNA coding for human blood coagulation factor X. Proc. Nat. Acad. Sci. 82: 3591-3595, 1985.

Furie, B.; Greene, E.; Furie, B. C.: Syndrome of acquired factor X deficiency and systemic amyloidosis: in vivo studies of the metabolic fate of factor X. New Eng. J. Med. 297: 81-85, 1977.

Furie, B.; Voo, L.; McAdam, K. P. W. J.; Furie, B. C.: Mechanism of factor X deficiency in systemic amyloidosis. New Eng. J. Med. 304: 827-830, 1981.

Gilgenkrantz, S.; Briquel, M.-E.; Andre, E.; Alexandre, P.; Jalbert, P.; Le Marec, B.; Pouzol, P.; Pommereuil, M.: Structural genes of coagulation factors VII and X located on 13q34. Ann. Genet. 29: 32-35, 1986.

Girolami, A.; Lazzarin, M.; Procidano, M.; Luzzatto, G.: A family with heterozygous factor X Friuli defect outside Friuli. Blut 46: 149-154, 1983.

Girolami, A.; Lazzarin, M.; Scarpa, R.; Brunetti, A.: Further studies on the abnormal factor X (factor X Friuli) coagulation disorder: a report of another family. Blood 37: 534-541, 1971.

Girolami, A.; Molaro, G.; Falomo, R.: Factor X Friuli coagulation disorder: the demise of the index patient. Acta Haemat. 54: 120-125, 1975.

Girolami, A.; Molaro, G.; Lazzarin, M.; Scarpa, R.; Brunetti, A.: Congenital haemorrhagic condition similar but not identical to factor X deficiency: a haemorrhagic state due to an abnormal factor X?. Scand. J. Haemat. 7: 91-99, 1970.

Girolami, A.; Molaro, G.; Lazzarin, M.; Scarpa, R.; Brunetti, A.: A 'new' congenital haemorrhagic condition due to the presence of an abnormal factor X (factor X Friuli): study of a large kindred. Brit. J. Haemat. 19: 179-192, 1970.

Girolami, A.; Nicolini, R.; Furlani, E.; Bareggi, G.: Abnormal factor X (factor X Friuli) coagulation disorder. Acta Haemat. 49: 114-122, 1973.

Girolami, A.; Vicarioto, M.; Ruzza, G.; Cappellato, G.; Vergolani, A.: Factor X Padua: a 'new' congenital factor X abnormality with a defect only in the extrinsic system. Acta Haemat. 73: 31-36, 1985.

Greipp, P. R.; Kyle, R. A.; Bowie, E. J. W.: Factor X deficiency in primary amyloidosis: resolution after splenectomy. New Eng. J. Med. 301: 1050-1051, 1979.

James, H. L.; Girolami, A.; Fair, D. S.: Molecular defect in coagulation factor X(Friuli) results from a substitution of serine for proline at position 343. Blood 77: 317-323, 1991.

Jaye, M.; Ricca, G.; Kaplan, R.; Howk, R.; Mudd, R.; Ngo, K. Y.; Fair, D. S.; Drohan, W.: Polymorphism associated with the human coagulation factor X (F10) gene. Nucleic Acids Res. 13: 8286, 1985.

Kaul, R. K.; Hildebrand, B.; Roberts, S.; Jagadeeswaran, P.: Isolation and characterization of human blood-coagulation factor X cDNA. Gene 41: 311-314, 1986.

Kim, D. J.; Thompson, A. R.; James, H. L.: Factor X(Ketchikan): a variant molecule in which gly replaces a gla residue at position 14 in the light chain. Hun. Genet. 95: 212-214, 1995.

Koizumi, T.; Hendel, E.; Lalley, P. A.; Tchetgen, M.-B. N.; Nadeau, J. H.: Homologs of genes and anonymous loci on human chromosome 13 map to mouse chromosomes 8 and 14. Mammalian Genome 6: 263-268, 1995.

Kroll, A. J.; Alexander, B.; Cochios, F.; Pechet, L.: Hereditary deficiencies of clotting factors VII and X associated with carotid-body tumors. New Eng. J. Med. 270: 6-13, 1964.

Kumar, M.; Mehta, P.: Congenital coagulopathies and pregnancy: report of four pregnancies in a factor X-deficient woman. Am. J. Hemat. 46: 241-244, 1994.

Lester, R. H.; Elston, R. C.; Graham, J. B.: Variations in levels of blood clotting factors IX and X in a population of normal men: possible genetic polymorphisms. Am. J. Hum. Genet. 24: 168-180, 1972.

Leytus, S. P.; Chung, D. W.; Kisiel, W.; Kurachi, K.; Davie, E. W.: Characterization of a cDNA coding for human factor X. Proc. Nat. Acad. Sci. 81: 3699-3702, 1984.

Leytus, S. P.; Foster, D. C.; Kurachi, K.; Davie, E. W.: Gene for human factor X: a blood coagulation factor whose gene organization is essentially identical with that of factor IX and protein C. Biochemistry 25: 5098-5102, 1986.

Marchetti, G.; Castaman, G.; Pinotti, M.; Lunghi, B.; di Iasio, M. G.; Ruggieri, M.; Rodeghiero, F.: Molecular bases of CRM(+) factor X deficiency: a frequent mutation (Ser334Pro) in the catalytic domain and a substitution (Glu102Lys) in the second EGF-like domain. Brit. J. Haemat. 90: 910-915, 1995.

Messier, T. L.; Wong, C. Y.; Bovill, E. G.; Long, G. L.; Church, W. R.: Factor X Stockton: a mild bleeding diathesis associated with an active site mutation in factor X. Blood Coagulation Fibrinolysis 7: 5-14, 1996.

Mori, K.; Sakai, H.; Nakano, N.; Suzuki, S.; Sugal, K.; Hisa, S.; Goto, Y.: Congenital factor X deficiency in Japan. Tohoku J. Exp. Med. 133: 1-19, 1981.

Pfeiffer, R. A.; Ott, R.; Gilgenkrantz, S.; Alexandre, P.: Deficiency of coagulation factors VII and X associated with deletion of a chromosome 13 (q34): evidence from two cases with 46,XY,t(13;Y)(q11;q34). Hum. Genet. 62: 358-360, 1982.

Pfeiffer, R. A.; Ott, R.; Taben, K. D.: Clotting factors VII and X as useful markers of terminal deletion of chromosome 13. (Letter) Hum. Genet. 69: 192, 1985.
Racchi, M.; Watzke, H. H.; High, K. A.; Lively, M. O.: Human coagulation factor X deficiency caused by a mutant signal peptide that blocks cleavage by signal peptidase but not targeting and translocation to the endoplasmic reticulum. J. Biol. Chem. 268: 5735-5740, 1993.
Ratnoff, O. D.; Bennett, B.: The genetics of hereditary disorders of blood coagulation. Science 179: 1291-1298, 1973.
Reddy, S. V.; Zhou, Z.-Q.; Rao, K. J.; Scott, J. P.; Watzke, H.; High, K. A.; Jagadeeswaran, P.: Molecular characterization of human factor X(San Antonio). Blood 74: 1486-1490, 1989.
Rocchi, M.; Roncuzzi, L.; Santamaria, R.; Archidiacono, N.; Dente, L.; Romeo, G.: Mapping through somatic cell hybrids and cDNA probes of protein C to chromosome 2, factor X to chromosome 13, and alpha-1-acid glycoprotein to chromosome 9. Hum. Genet. 74: 30-33, 1986.
Rocchi, M.; Roncuzzi, L.; Santamaria, R.; Sbarra, D.; Mochi, M.; Archidiacono, N.; Covone, A.; Cortese, R.; Romeo, G.: Mapping of coagulation factors protein C and factor X on chromosome 2 and 13 respectively. (Abstract) Cytogenet. Cell Genet. 40: 734, 1985.
Roos, J.; Huizinga, J.: Genetic investigation of the Stuart coagulation defect. Acta Genet. Statist. Med. 9: 115-122, 1959.
Royle, N. J.; Fung, M. R.; MacGillivray, R.; Hamerton, J. L.: The localization of the factor X gene for clotting to chromosome 13. (Abstract) Cytogenet. Cell Genet. 40: 736, 1985.
Royle, N. J.; Fung, M. R.; MacGillivray, R. T. A.; Hamerton, J. L.: The gene for clotting factor 10 is mapped to 13q32-qter. Cytogenet. Cell Genet. 41: 185-188, 1986.
Rudolph, A. E.; Mullane, M. P.; Porche-Sorbet, R.; Tsuda, S.; Miletich, J. P.: Identification of a glycine substitution at residue 7 and characterization of the recombinant protein. J. Biol. Chem. 271: 28601-28606, 1996.
Scambler, P. J.; Williamson, R.: The structural gene for human coagulation factor X is located on chromosome 13q34. Cytogenet. Cell Genet. 39: 231-233, 1985.
Stoll, C.; Roth, M. P.: Partial 4q duplication due to inherited der(13), t(4;13)(q26;q34)mat in a girl with a deficiency of factor X. Hum. Genet. 53: 303-304, 1980.
Sumer, T.; Ahmad, M.; Sumer, N. K.; Al-Mouzan, M. I.: Severe congenital factor X deficiency with intracranial haemorrhage. Europ. J. Pediat. 145: 119-120, 1986.
Watzke, H. H.; Lechner, K.; Roberts, H. R.; Reddy, S. V.; Welsch, D. J.; Friedman, P.; Mahr, G.; Jagadeeswaran, P.; Monroe, D. M.; High, K. A.: Molecular defect (gla(+14)-to-lys) and its functional consequences in a hereditary factor X deficiency (factor X 'Vorarlberg'). J. Biol. Chem. 265: 11982-11989, 1990.
Watzke, H. H.; Wallmark, A.; Hamaguchi, N.; Giardina, P.; Stafford, D. W.; High, K. A.: Factor X (Santo Domingo): evidence that the severe clinical phenotype arises from a mutation blocking secretion. J. Clin. Invest. 88: 1685-1689, 1991.
Wieland, K.; Millar, D. S.; Grundy, C. B.; Mibashan, R. S.; Kakkar, V. V.; Cooper, D. N.: Molecular genetic analysis of factor X deficiency: gene deletion and germline mosaicism. Hum. Genet. 86: 273-278, 1991.
Zur, M.; Nemerson, Y.: Tissue factor pathways of blood coagulation.In: Bloom, A. L.; Thomas, D. P.: Haemostasis and thrombosis. Edinburgh: Churchill Livingstone (pub.) 1981. Pp. 124-139.

CHEMODECTOMA, INTRA-ABDOMINAL, WITH CUTANEOUS ANGIOLIPOMAS (118350)

JJM NOTE

It would be of interest to follow up on this family.

Lee et al. (1977) described 2 brothers with cutaneous angiolipomas and retroperitoneal chemodectomas. Both died of malignant dissemination of the chemodectomas. Two other brothers died of tumors before age 45, and one of them also had skin lumps. Thus, they may have been affected also. See paragangliomata (168000).

Lee, S. P.; Nicholson, G. I.; Hitchcock, G.: Familial abdominal chemodectomas with associated cutaneous angiolipomas. Pathology 9: 173-177, 1977.

CARCINOID, INTESTINAL (114900)

JJM NOTE

Follow-up of these three kindreds could be illuminating, especially if thorough clinical and molecular characterization could be done.

Anderson (1966) observed appendiceal carcinoid in father and daughter. Eschbach and Rinaldo (1962) reported fatal malignant carcinoid of the ileum in brother and sister. Duodenal carcinoid is described with multiple endocrine neoplasia (131100, 162300, 171400). Debelenko et al. (1997) described somatic mutations in the MEN1 gene (131100.0015) in sporadic carcinoid tumors of lung.

See also Moertel and Dockerty (1973)

Anderson, R. E.: A familial instance of appendiceal carcinoid. Am. J. Surg. 111: 738-740, 1966.
Debelenko, L. V.; Brambilla, E.; Agarwal, S. K.; Swalwell, J. I.; Kester, M. B.; Lubensky, I. A.; Zhuang, Z.; Guru, S. C.; Manickam, P.; Olufemi, S.-E.; Chandrasekharappa, S. C.; Crabtree, J. S.; Kim, Y. S.; Heppner, C.; Burns, A. L.; Spiegel, A. M.; Marx, S. J.; Liotta, L. A.; Collins, F. S.; Travis, W. D.; Emmert-Buck, M. R.: Identification of MEN1 gene mutations in sporadic carcinoid tumors of the lung. Hum. Molec. Genet. 6: 2285-2290, 1997.
Eschbach, J. W.; Rinaldo, J. A., Jr.: Metastatic carcinoid: a familial occurrence. Ann. Intern. Med. 57: 647-650, 1962.
Moertel, C. G.; Dockerty, M. B.: Familial occurrence of metastasizing carcinoid tumors. Ann. Intern. Med. 78: 389-390, 1973.

4. MESODERMAL DISORDERS

Malignant Soft-Tissue Sarcomas

LI-FRAUMENI CANCER FAMILY SYNDROME (#151623)

JJM NOTE

For their tenacity and success in defining this cancer family syndrome, mapping and isolating its gene, and applying those finding to populations, Fred Li and Joe Fraumeni won the General Motors prize in 1995. While researchers are clarifying the details of its genetic basis and routes to biologic control, the first priority for oncologists is to take a relevant cancer family history, suspect the occasional family, and refer them to a cancer genetics program for further evaluation, which might include genetic testing, currently in a research context. The mnemonic for the tumor types is ***SBLA:*** ***S***arcoma, ***B***reast, ***B***one, ***B***rain, ***L***ung, ***L***aryngeal, ***L***eukemia, and ***A***drenal cortical neoplasm.

A number sign (#) is used with this entry because mutations in the TP53 gene (191170) have been found in this disorder.

In reviewing medical records and death certificates of 648 childhood rhabdomyosarcoma patients, Li and Fraumeni (1969) identified 4 families in which sibs or cousins had a childhood sarcoma. These 4 families also had striking histories of breast cancer and other neoplasms, suggesting a new familial cancer syndrome of diverse tumors. Subsequent prospective studies confirmed the high risk in family members of the tumor types that comprise LFS (Li and Fraumeni, 1982). Studies in other geographic and ethnic groups by 2,1:Birch et al. (1984, 1990) corroborated the syndrome. The spectrum of cancers in the syndrome was shown to include, in addition to breast cancer and soft tissue sarcomas, brain tumors, osteosarcoma, leukemia, and adrenocortical carcinoma.

Fraumeni et al. (1975) described a kindred in which in 1 sibship of 9 adults, 4 died of lymphocytic or histiocytic lymphomas and one, a male, of Waldenstrom macroglobulinemia complicated by adenocarcinoma of the lung. In the next generation, 1 person died of Hodgkin disease; 4 of 9 healthy persons had impaired lymphocyte transformation with phytohemagglutinin and 3 of these had polyclonal elevation of IgM. Subsequent to the studies, adenocarcinoma of the lung developed in one of those with an immune defect, a woman, and her 3-year-old grandson developed lymphocytic leukemia. This is the first suggestion of a genetic or immunologic basis of lung adenocarcinoma. Pearson et al. (1982) reported 2 families resembling that reported by Li and Fraumeni (1969). In 1, the mother had breast cancer and 3 of her 4 children had adrenocortical carcinoma, medulloblastoma and rhabdomyosarcoma; in the other, the mother had breast cancer and 2 of her 3 children had adrenocortical carcinoma and rhabdomyosarcoma. Mulvihill (1982) used the designation sarcoma family syndrome of Li and Fraumeni for the familial association of breast cancer, soft tissue sarcoma and other tumors. Hartley et al. (1987) referred to this as the SBLA syndrome, a designation derived from the tumors that occur in this cancer family syndrome: sarcoma, breast and brain tumors, leukemia, laryngeal and lung cancer, and adrenal cortical carcinoma (Lynch et al., 1978).

Possible component tumors of LFS are melanoma, gonadal germ cell tumors, and carcinomas of the lung, pancreas, and prostate (Strong et al., 1987; Li et al., 1988). The diverse tumor types in family members characteristically developed at unusually early ages, and multiple primary tumors are frequent. By segregation analysis, Strong et al. (1987) demonstrated that the observed cancer distribution in families best fit a rare autosomal dominant gene model. The model also predicted that in families at risk the probability of developing any invasive cancer (excluding carcinomas of the skin) reaches almost 50% by age 30, when only 1% of the general population has developed cancer. More than 90% of the gene carriers would develop cancer by age 70 (review by Malkin et al., 1990). Because tumor suppressor genes had been found to be associated with familial neoplasms, Malkin et al. (1990) suspected mutation in this type of gene in LFS. The retinoblastoma gene (RB1; 180200) was an unlikely candidate for a germline mutation in LFS because retinoblastoma had not been observed in these families. On the other hand, the TP53 gene was a more likely candidate because inactivating mutations therein had been associated with sporadic osteosarcomas, soft tissue sarcomas, brain tumors, leukemias, and carcinomas of the lung and breast. Furthermore, transgenic mice carrying a mutant p53 gene have an increased incidence of osteosarcomas, soft tissue sarcomas, adenocarcinomas of the lung, and adrenal and lymphoid tumors—all tumors that occur as part of LFS. Such was the basis for the successful search for p53 mutations in this disorder (see 191170.0001).

To determine the frequency and distribution of germline p53 mutations in LFS families, Frebourg et al. (1995) sequenced the 10 coding exons of TP53 in lymphocytes and fibroblast cell lines derived from 15 families with the syndrome. Germline mutations were observed in 8 such families; of these, 6 were missense mutations located between exons 5 and 8. One mutation was a nonsense mutation in exon 6 and one was a splicing mutation in intron 4; each of these mutations generated aberrantly short p53 RNAs. The study indicated that most germline p53 mutations in LFS are located between exons 5 and 8 and that approximately 50% of patients with LFS have no germline mutations in the coding region of the p53 gene. Significantly, in 3 families, a mutation of the p53 gene was observed in a fibroblast cell line derived from the proband, but the mutation was not found in affected relatives in 2 families or in the blood from 1 of the probands. This indicated that the mutation probably occurred during cell culture; thus, it is necessary that analysis for germline p53 mutations be performed on cells that have not been grown in vitro.

Mutations in p53 have not been detected in approximately 30% of LFS families. To address the possibility either that TP53 mutations were missed or that another predisposing gene is altered in LFS, Evans et al. (1998) used a variety of methods to determine the TP53 status in a large LFS kindred. A transcriptional activation assay on exons 4 to 10 of TP53 excluded a mutation within the DNA-binding domain. SSCP analysis, using intronic primers and sequencing of all the coding exons and intron/exon junctions, also yielded no mutations. Finally, linkage analysis excluded potential mutations in the noncoding regions of TP53. The family had been ascertained through systematic surveys of cancer in relatives of 382 childhood osteosarcoma patients. Classic LFS was defined as the diagnosis of a sarcoma in an individual before 45 years of age, having 2 first-degree relatives with cancer before the age of 45 years. In the LFS kindred studied, in addition to osteosarcoma in the proband, a sister had osteosarcoma and the mother had breast cancer before the age of 45 years. There was also lung cancer in the family as well as cancers of the ovary, thyroid, tongue, and kidney. Thirteen members were affected in 7 sibships in 3 generations.

Hisada et al. (1998) quantified the incidence of second and third primary cancers in individuals from 24 LFS families originally diagnosed with cancer between 1968 and 1986. Among 200 LFS family members diagnosed with cancer, 30 (15%) developed a second cancer. Eight individuals (4%) had a third cancer, while 4 (2%) eventually developed a fourth cancer. Overall, the relative risk of occurrence of a second cancer was 5.3, with a cumulated probability of second cancer occurrence of 57% at 30 years after diagnosis of a first cancer. Relative risks of second cancers occurring in families with this syndrome were 83.0, 9.7, and 1.5 for individuals with a first cancer at ages 0 to 19 years, 20 to 44 years, and 45 years or more, respectively. Thirty (71%) of 42 subsequent cancers in this group were component cancers of LFS.

See also Li and Fraumeni (1969); Li and Fraumeni (1975); Lynch et al. (1985)

Birch, J. M.; Hartley, A. L.; Blair, V.; Kelsey, A. M.; Harris, M.; Teare, M. D.; Jones, P. H. M.: Identification of factors associated with high breast cancer risk in the mothers of children with soft tissue sarcoma. J. Clin. Oncol. 8: 583-590, 1990.

Birch, J. M.; Hartley, A. L.; Marsden, H. B.; Harris, M.; Swindell, R.: Excess risk of breast cancer in the mothers of children with soft tissue sarcomas. Brit. J. Cancer 49: 325-331, 1984.

Evans, S. C.; Mims, B.; McMasters, K. M.; Foster, C. J.; deAndrade, M.; Amos, C. I.; Strong, L. C.; Lozano, G.: Exclusion of a p53 germline mutation in a classic Li-Fraumeni syndrome family. Hum. Genet. 102: 681-686, 1998.

Fraumeni, J. F., Jr.; Wertelecki, W.; Blattner, W. A.; Jensen, R. D.; Leventhal, B. G.: Varied manifestations of a familial lymphoproliferative disorder. Am. J. Med. 59: 145-151, 1975.

Frebourg, T.; Barbier, N.; Yan, Y.; Garber, J. E.; Dreyfus, M.; Fraumeni, J., Jr.; Li, F. P.; Friend, S. H.: Germ-line p53 mutations in 15 families with Li-Fraumeni syndrome. Am. J. Hum. Genet. 56: 608-615, 1995.

Hartley, A. L.; Birch, J. M.; Marsden, H. B.; Harris, M.: Malignant melanoma in families of children with osteosarcoma, chondrosarcoma, and adrenal cortical carcinoma. J. Med. Genet. 24: 664-668, 1987.

Hisada, M.; Garber, J. E.; Fung, C. Y.; Fraumeni, J. F., Jr.; Li, F. P.: Multiple primary cancers in families with Li-Fraumeni syndrome. J. Nat. Cancer Inst. 90: 606-611, 1998.

Li, F. P.; Fraumeni, J. F.: Rhabdomyosarcoma in children: an epidemiologic study and identification of a familial cancer syndrome. J. Nat. Cancer Inst. 43: 1364-1373, 1969.

Li, F. P.; Fraumeni, J. F., Jr.: Soft-tissue sarcomas, breast cancer, and other neoplasms: a familial syndrome? Ann. Intern. Med. 71: 747-752, 1969.

Li, F. P.; Fraumeni, J. F., Jr.: Familial breast cancer, soft-tissue sarcomas, and other neoplasms. (Letter) Ann. Intern. Med. 83: 833-834, 1975.

Li, F. P.; Fraumeni, J. F., Jr.: Prospective study of a family cancer syndrome. J.A.M.A. 247: 2692-2694, 1982.
Li, F. P.; Fraumeni, J. F., Jr.; Mulvihill, J. J.; Blattner, W. A.; Dreyfus, M. G.; Tucker, M. A.; Miller, R. W.: A cancer family syndrome in twenty-four kindreds. Cancer Res. 48: 5358 only, 1988.
Lynch, H. T.; Katz, D. A.; Bogard, P. J.; Lynch, J. F.: The sarcoma, breast cancer, lung cancer, and adrenocortical carcinoma syndrome revisited. Am. J. Dis. Child. 139: 134-136, 1985.
Lynch, H. T.; Mulcahy, G. M.; Harris, R. E.; Guirgis, H. A.; Lynch, J. F.: Genetic and pathologic findings in a kindred with hereditary sarcoma breast cancer, brain tumors, leukemia, lung, laryngeal, and adrenal cortical carcinoma. Cancer 41: 2055-2064, 1978.
Malkin, D.; Li, F. P.; Strong, L. C.; Fraumeni, J. F., Jr.; Nelson, C. E.; Kim, D. H.; Kassel, J.; Gryka, M. A.; Bischoff, F. Z.; Tainsky, M. A.; Friend, S. H.: Germ line p53 mutations in a familial syndrome of breast cancer, sarcomas, and other neoplasms. Science 250: 1233-1238, 1990.
Mulvihill, J. J.: Personal Communication. Bethesda, Md. 6/11/1982.
Pearson, A. D. J.; Craft, A. W.; Ratcliffe, J. M.; Birch, J. M.; Morris-Jones, P.; Roberts, D. F.: Two families with the Li-Fraumeni cancer family syndrome. J. Med. Genet. 19: 362-365, 1982.
Strong, L. C.; Stine, M.; Norsted, T. L.: Cancer in survivors of childhood soft tissue sarcoma and their relatives. J. Nat. Cancer Inst. 79: 1213-1220, 1987.

TUMOR PROTEIN P53 (*191170)

JJM NOTE

The loosely adhered to convention is that the gene is called *TP53* and that the resultant protein is p53. It is important to distinguish the broad spectrum of tumor types that have *somatic cell* mutations in the *TP53* gene from the narrower but still diverse range of tumors seen in germline mutations, most of which fall into the classic or nonclassic form of the Li-Fraumeni cancer family syndrome. To date, the Li-Fraumeni tumors with known *TP53* mutations include various sarcomas (osteosarcoma, rhabdomyosarcoma, chondrosarcoma, leiomyosarcoma), breast cancer, medulloblastoma, ependymoma, thyroid carcinoma, and non-Hodgkin lymphoma. The logistics and counseling issues concerning the clinical use of *TP53* gene analysis suggest the need for practicing oncologists to refer members from such families to cancer genetics clinics.

The p53 tumor antigen is found in increased amounts in a wide variety of transformed cells. The protein is also detectable in many actively proliferating, nontransformed cells, but it is undetectable or present at low levels in resting cells. Several lines of evidence suggest its involvement in cell cycle regulation. The synthesis of p53 increases in resting cells after stimulation with mitogen, and stimulated cells can be blocked from entering S phase by microinjections of p53-specific antibody. The P53 gene has structural and functional similarities to the MYC family of protooncogenes. Since tumor antigen p53 can complement activated RAS genes in the transformation of rat fibroblasts, the gene encoding p53 may act as an oncogene. As reviewed by Levine et al. (1991), alteration or inactivation of p53 by mutation, or the interaction of p53 with oncogene products of DNA tumor viruses, can lead to cancer. These mutations seem to be the most common genetic change in human cancers. The mutations are particularly localized in 4 regions of the protein that are highly conserved among several different species. There are at least 3 mutation 'hotspots,' affecting residues 175, 248, and 273. The highest percentage of mutations (13%) were found at position 273 (Levine et al., 1991). Wildtype p53, like the retinoblastoma gene product, seems to regulate cell growth and division negatively; unlike Rb, some mutant forms of p53 gain a function so as to stimulate cell division. It is perhaps because of the gain-of-function mutations, which can act in the heterozygous state, that p53 mutations are so common in human cancers.

Using a cDNA clone in the analysis of man-rodent hybrid cells, 79,78:McBride et al. (1985, 1986) assigned the P53 gene to chromosome 17; through study of hybrids with a chromosome 17 translocation and by in situ hybridization to metaphase spreads, they regionalized the locus to the terminal band of the short arm, 17p13. Benchimol et al. (1985) also assigned the P53 gene to the short arm of chromosome 17. By in situ hybridization with a mouse DNA probe, Le Beau et al. (1985) mapped the P53 gene to 17q21-q22. (When the work was repeated with a human probe, a considerable number of grains were found on the short arm.) In situ hybridization of TP53 and ERBA probes to malignant cells from 3 acute promyelocytic leukemia (APL) patients indicated that the TP53 locus is translocated to chromosome 15, whereas ERBA stays with chromosome 17. Subsequently, this group concluded that indeed the gene is on the short arm; the group studied the nature of the homologous long arm sequences that are translocated in APL (Rowley, 1986). By Southern filter hybridization of DNAs from human-rodent hybrids, Miller et al. (1986) also localized the P53 gene to 17p. They suggested that the use of the mouse gene as the probe in the work of Le Beau et al. (1985) may have been responsible for the inaccurate results; the murine and human genes are not completely homologous. Isobe et al. (1986) also assigned TP53 to 17p13. Somatic cell hybrid studies by vanTuinen and Ledbetter (1987) narrowed the assignment to 17p13.105-p12.

The open reading frame of p53 is 393 amino acids long, with the central region (consisting of amino acids from about 100 to 300) containing the DNA-binding domain. This proteolysis-resistant core is flanked by a C-terminal end mediating oligomerization and an N-terminal end containing a strong transcription activation signal (Vogelstein and Kinzler, 1994). Cho et al. (1994) succeeded in co-crystallizing the core domain of p53 bound to DNA. The structure of p53 appears to be unique, consisting of a large beta-sandwich that acts as a scaffold for 3 loop-based elements. The sandwich is composed of 2 anti-parallel beta-sheets containing 4 and 5 beta-strands, respectively. The first loop binds to DNA within the major groove, the second loop binds to DNA within the minor groove, and the third loop packs against the second loop to stabilize it. Vogelstein and Kinzler (1994) pointed out that one of the most notable features of the structure is its correlation with the data on mutations. The residues most frequently mutated in cancers are all at or near the protein-DNA interface, and over two-thirds of the missense mutations are in 1 of the 3 DNA loops.

Vogelstein and Kinzler (1992) proposed a model in which p53 binds as a tetramer to a p53-binding site (PBS) and activates the expression of downstream genes that inhibit growth and/or invasion . Pavletich et al. (1993) stated that tetramerization occurs by interactions between the p53 monomers through a C-terminal domain comprising amino acid residues 325-356. Jeffrey et al. (1995) reported the crystal structure of this tetramerization domain at 1.7-angstrom resolution and described the physical properties of the tetrameric interaction.

Masuda et al. (1987) surveyed 134 human carcinomas, sarcomas, leukemias, and lymphomas obtained at surgery or from peripheral blood and found rearrangements of the P53 gene only in osteogenic sarcomas (259500). The change was identified in 3 of 6 osteogenic sarcomas examined. Normal tissue from one of these patients had an unrearranged gene, indicating that the genetic abnormality in the tumor was acquired. Two of the sarcomas with rearranged genes expressed levels of p53 protein that were elevated relative to other tumors. Alterations of the P53 gene were also found in 3 human osteogenic sarcoma cell lines. Romano et al. (1989) described a mutation in the human nuclear phosphoprotein P53 gene in the DNA of an osteosarcoma cell line. A G-to-C point mutation in codon 156 results in an arginine-to-proline amino acid substitution. Ara et al. (1990) demonstrated the polymorphic nature of the arg72-to-pro variation. The polymorphism is associated with a polymorphic AccII restriction site in the fourth exon. On the basis of 50 unrelated persons, the following frequencies were estimated: A1 = 0.36; A2 = 0.64. In the course of study of 241 patients with various neoplasms, Mulligan et al. (1990) found particular changes in p53 in 31 and 29 cases, respectively, of rhabdomyosarcomas and osteosarcomas. The alterations of p53 in rhabdomyosarcomas included complete deletion of both p53 alleles, complete deletion of one allele with or without point mutation of the remaining allele, and absence of detectable RNA. In osteosarcomas, Mulligan et al. (1990) detected homozygous deletion and lack of expression of p53 RNA or aberrant expression of p53 protein. The fact that other, primary mutations have been defined in these tumors suggests that the change in p53 plays a progressional role in tumorigenesis. In 26 retinoblastoma tumors, no change in the P53 gene was found, despite the frequency with which such alterations were found in the clinically associated tumor, osteosarcoma. Thus, retinoblastoma and osteosarcoma may have common requirements for an initiating mutation, whereas different progressional mutations, isochromosome 6p in retinoblastoma, are involved in progression.

Multifocal osteogenic sarcoma (OS) is diagnosed when bone tumors occur at 2 or more sites that are commonly associated with 'primary disease.' These tumors arise as synchronous or metachronous lesions with or without metastatic visceral involvement. The most common location of OS is the metaphyseal region of bones in the upper and lower limbs. The lung is the most frequent site of metastasis from unifocal OS. Osteogenic sarcoma is known to occur in patients with the Li-Fraumeni syndrome and in transgenic mice carrying a mutant p53 allele. Using single-strand conformation polymorphism (SSCP) analysis, Iavarone et al. (1992) found p53 point mutations in tumor DNA from 4 patients with multifocal osteogenic sarcoma without familial histories of increased tumor predisposition. Furthermore, a germline p53 mutation was detected in one of the patients, whose tumor tissue showed a further rearrangement of the residual wildtype allele.

Reisman et al. (1988) used the chloramphenicol acetyltransferase (CAT) assay to characterize the regulatory regions of the human P53 gene. They identified 2 promoters: the first located 100 to 250 bp upstream of the noncoding first exon, and the second, stronger promoter, within the first intron. Baker et al. (1989) concluded that TP53 mutations may be involved in colorectal cancer, perhaps through inactivation of a tumor suppressor function of the wildtype gene.

Since lung cancer frequently shows loss of heterozygosity on 17p, Takahashi et al. (1989) sought changes in the P53 gene which is located there. They found that the gene was frequently mutated or inactivated in all types of human lung cancer. The genetic abnormalities included gross changes such as homozygous deletions and abnormally sized mRNAs, along with a variety of point or small mutations, which map to the p53 open reading frame and change amino acid sequence in a region highly conserved between man and mouse. Very low or absent expression of p53 mRNA in lung cancer cell lines compared to normal lung was seen. In all of 10 small cell lung cancer cell lines and 9 non-SCLC cell lines, they found coexistence of 3 abnormalities involving 3p, RB (180200) on chromosome 13, and P53. The authors stated that studies addressing when these lesions develop and whether their correction reverses a malignant phenotype will be of great interest.

Iggo et al. (1990) suggested that P53 is the protooncogene that most commonly undergoes mutation in lung cancer. They found several mutations that were G-to-T transversions resulting in missense changes in amino acids highly conserved in evolution. Chung et al. (1990) found that human glioblastomas in which P53 gene mutations were found had an earlier age of onset than did tumors without P53 gene mutations as detected by direct sequencing of primer directed DNA amplified product. The average postoperative survival among patients with demonstrable P53 mutations was considerably longer than that of the group without such mutations.

Schiffer et al. (1995) used SSCP analysis of exons 5 to 8 as well as direct sequence analysis to search for p53 mutations in 30 cases of childhood astrocytic tumors. Mutations were found in 2 of 8 glioblastomas and 1 out of 9 anaplastic astrocytomas, but in none of the more benign pilocytic astrocytomas.

The role of cytosine methylation in CpG dinucleotides in germline mutations that cause mendelian disorders was pointed out by Cooper and Youssoufian (1988); Rideout et al. (1990) emphasized the contribution of 5-methylcytosine to somatic mutation resulting in human disease, particularly tumorigenesis. They suggested that as many as 43% (9 of 21) of P53 somatic mutations may be due to the presence of 5-methylcytosine. There are 82 CpGs in the 2,362 nucleotides of the P53 double-stranded coding sequence. Thus, about 3.5% of the sequence may be contributing to as much as 43% of the point mutations, each of which is a transition from 5-methylcytosine to thymine (or a corresponding transition from guanine to adenine). Methylation of cytosine seems to increase the potential for mutation at cytosine residues by a factor of at least 10. Halevy et al. (1990) demonstrated that P53 mutants involved in the neoplastic process display various properties including gain of function. Takahashi et al. (1990) identified intronic point mutations as a mechanism for inactivation of the tumor suppressor function of the P53 gene. They demonstrated point mutations at the splice acceptor site in the third intron and in the splice donor site in the seventh intron accounting for abnormal mRNA splicing. In 1 patient the same intronic point mutation was found in the tumor cell line derived from a bone marrow metastasis and in multiple liver metastases, but not in normal DNA. By immunohistologic staining of primary colorectal carcinomas with antibodies specific for p53, Rodrigues et al. (1990) demonstrated gross overexpression of the protein in about 50% of cases. Benign adenomas were all negative for p53 overexpression. By direct sequencing and chemical-mismatch-cleavage analysis of p53 cDNA using PCR in 6 cell lines expressing high levels of p53, they showed that all were synthesizing mRNAs that encoded mutant p53 proteins. In 2 of 4 cell lines in which p53 expression was lower, point mutations we re still detected. An arg273-to-his mutation was found in 4 of the 7 instances in which specific point mutations were identified. Using a human osteosarcoma cell line, Saos-2, which completely lacked endogenous p53, Chen et al. (1990) studied the effects of producing single copies of exogenous P53 genes containing either point-mutated or wildtype versions of the P53 cDNA sequence, when introduced by infecting the cells with recombinant retroviruses. Expression of wildtype p53 suppressed the neoplastic phenotype. In a 2-allele configuration, wildtype P53 was phenotypically dominant to mutated p53. In studies of 4 human esophageal carcinoma cell lines and 14 human esophageal squamous cell carcinomas, Hollstein et al. (1990) found a mutated p53 allele (1 frameshift and 6 missense mutations) in 2 cell lines and 5 of the tumor specimens. All missense mutations detected occurred at G:C basepairs in codons at or adjacent to mutations previously reported in other cancers. In 11 of 18 invasive bladder cancers, Sidransky et al. (1991) found alterations of p53. Point mutations resulting in single amino acid substitutions were present in 10 and a 24-bp deletion was found in 1. In all but 1, the mutations were associated with 17p allelic deletions, leaving the cells with only mutant forms of the P53 gene product. Using the PCR and oligomer-specific hybridization, P53 mutations were identified in 1 to 7% of the cells in the urine sediment from each of 3 patients tested. This was the first genetic alteration demonstrated in a high proportion of primary invasive bladder cancers (109800). The possibility of demonstrating the mutation in cells shed into the urine is obviously of great practical importance.

Hollstein et al. (1991) reviewed the repertoire of mutations in the evolutionarily conserved codons of the P53 tumor suppressor gene found in diverse types of human cancer. Transitions predominated in colon, brain, and lymphoid malignancies whereas G:C to T:A transversions were the most frequent substitutions observed in cancers of the lung and liver. Mutations at A:T basepairs were seen more frequently in esophageal carcinomas than in other solid tumors. Most transitions in colorectal carcinomas, brain tumors, leukemias, and lymphomas were at CpG dinucleotide mutational hotspots. G-to-T transversions in lung, breast, and esophageal carcinomas were dispersed among numerous codons. In liver tumors in persons from geographic areas in which both aflatoxin B1 (AFB1) and hepatitis B virus (HBV) are cancer risk factors, most mutations were at one nucleotide pair of codon 249.

As is well known, epidemiologic data etiologically associate cigarette smoke with lung cancer, aflatoxin exposure with hepatoma, and sunlight exposure with skin cancer. When the P53 gene was used as a 'mutagen test,' the observed patterns of mutation differed dramatically in each case and followed patterns predicted by laboratory-based studies; e.g., the strong excess of G-to-T transversions found in lung cancers are associated with benzo(alpha)pyrene and conjoiners (Chiba et al., 1990). Hepatomas in large regions of Africa and China, but not in regions without exposure to aflatoxins, show predominantly transversion mutations at amino acid 249 in the P53 gene. In a series of New England and Swedish patients, Brash et al. (1991) found that 14 of 24 (58%) invasive squamous cell carcinomas of the skin contained mutations in the P53 gene, each altering the amino acid sequence. Involvement of UV light in these p53 mutations was indicated by the presence in 3 of the tumors of a CC-to-TT double-base change, which is only known to be induced by UV. UV was also implicated by a UV-like occurrence of mutations exclusively at dipyrimidine sites, including a high frequency of C-to-T substitutions. In internal malignancies, p53 mutations do not show these UV-specific mutations. Ziegler et al. (1994) pointed out that squamous cell carcinoma of the skin can progress by stages: sun-damaged epidermis, with individual disordered keratinocytes; actinic keratosis, spontaneously regressing keratinized patches having aberrant cell differentiation and proliferation; carcinoma in situ; squamous cell carcinoma of the skin, and metastasis. They showed that the p53 mutations induced by ultraviolet radiation and found in more than 90% of human squamous cell carcinomas were present in actinic keratoses. Inactivating p53 in mouse skin reduced the appearance of sunburn cells (apoptotic keratinocytes generated by overexposure to ultraviolet). Skin thus appears to possess a p53-dependent 'guardian-of-the-tissue' response to DNA damage which aborts precancerous cells. If this response is reduced in a single cell by a prior p53 mutation, sunburn can select for clonal expansion of the p53-mutated cell into an actinic keratosis. Thus, sunlight can act twice: as tumor initiator and tumor promoter.

Borresen et al. (1991) devised a modification of denaturing gradient gel electrophoresis (DGGE) termed 'constant denaturant gel electrophoresis' (CDGE) and applied it for the rapid and sensitive screening of 4 conserved regions of the P53 gene where most of the mutations have been found. CDGE detected P53 mutations in 11 of 32 breast carcinomas. By sequencing 18 human genomic clones that bind to p53 in vitro, El-Deiry et al. (1992) identified the consensus binding site with striking internal symmetry, consisting of 2 copies of a 10-basepair motif separated by 0-13 bp. One copy of the motif was insufficient for binding, and subtle alterations of the motif, even when present in multiple copies, resulted in loss of affinity for p53. Mutants of p53, representing each of the 4 'hotspots' found to be altered frequently in human cancers, failed to bind to the consensus dimer. Toguchida et al. (1992) identified a germline mutation in 8 of 15 patients with sarcoma selected (out of a total of 196 patients with sarcoma) because they had had multiple primary cancers or had a family history of cancer. In 3 of the patients there was no known family history of cancer; in the other 5 there was an unusual personal or family history of cancer. The tumor was osteosarcoma in 7 of the patients and malignant fibrous histiocytoma in 1. Four mutations caused amino acid substitutions and 4 caused stop codons. The group of patients with cancer who carry germline mutations of the P53 gene is more diverse than suggested by the clinical delineation of the Li-Fraumeni syndrome. Identification of carriers might have considerable clinical importance. Many of the seemingly sporadic cases may in fact have germline mutations which are not identified if the patient does not survive. In the 8 patients studied, the tumor was osteosarcoma in 7 and malignant fibrous histiocytoma in the eighth. Malkin et al. (1992) likewise found germline mutations in the p53 gene despite the fact that the family histories were not indicative of the Li-Fraumeni syndrome.

The development of cervical and anal cancers has been linked with infection by human papillomaviruses (HPVs), most commonly HPV16 or HPV18 (167960). The virally encoded oncoproteins E6 and E7 form a complex with cell-encoded protein products of tumor suppressor genes, E6 binding to the cell p53 protein and E7 to the retinoblastoma gene product (RB). It is likely that these viral-host protein interactions result in loss of the negative growth control normally exerted by p53 and RB. E6 can direct the rapid proteolytic degradation of p53, and E6 expression following HPV infection is likely to be a loss of functional wildtype p53 protein within the cell. Crook et al. (1992) presented evidence that HPV-negative cervical cancers have point mutations in the TP53 gene. They also presented data indicating that loss of wildtype p53 function is a critical event in the pathology of anogenital cancer, and that in the absence of a virally encoded E6 protein that mediates p53 degradation, this loss of function occurs via somatic mutation. The prognosis of HPV-negative cervical cancers is worse than that of HPV-positive cancers. Kaelbling et al. (1992) likewise found loss of heterozygosity on chromosome 17p and mutant p53 in HPV-negative cervical carcinomas.

Using a monoclonal antibody, Moll et al. (1992) demonstrated that 10 of 27 cases of breast cancer (37%) showed a pattern of p53 protein staining, indicating that the protein was limited to the cytoplasm and absent from the nucleus. In 8 cases (30%) the nucleus of cancer cells showed a high level of P53; in 9 cases (33%) there was complete lack of detectable staining. Nucleotide sequencing of P53 cDNAs derived from the samples with cytoplasmic staining revealed only wildtype p53 alleles in 6 out of 7 cases. An eighth case was determined to be wildtype by SSCP. In contrast, the samples containing nuclear p53 contained a variety of missense mutations and a nonsense mutation. Tumors that lacked detectable p53 staining had wildtype nucleotide sequences. A case of normal lactating breast tissue also showed intense cytoplasmic staining with nuclear sparing. Moll et al. (1992) interpreted the data as indicating that some breast cancers inactivate the tumor-suppressing activity of p53 by sequestering the protein in the cytoplasm, away from its site of action in the cell nucleus. The detection of cytoplasmic p53 in normal lactating breast tissue suggested that this mechanism may be employed in specific physiologic situations to permit transient cell proliferation. Moll et al. (1992) referred to the phenomenon as nuclear exclusion.

Felix et al. (1992) used SSCP analysis to study the P53 gene in 10 families with multiple members affected with leukemia. The diagnoses included acute and chronic leukemias and Hodgkin disease. The experience suggested that p53 mutations are not a primary event responsible for hereditary susceptibility to familial leukemia.

Vogelstein and Kinzler (1992) provided a review of function and dysfunction in the p53 gene. They outlined 5 mechanisms for p53 inactivation. P53 is postulated to bind as a tetramer to a p53-binding site (PBS) and to activate the expression of adjacent genes that inhibit growth and/or invasion. Deletion of one or both p53 alleles reduces the expression of tetramers, resulting in decreased expression of the growth inhibitory genes. This mechanism is found in occasional tumors of several types. Nonsense or splice site mutations that result in truncation of the protein do not allow oligomerization, thus resulting in a similar reduction of p53 tetramers. Mutations of this type are fairly common in lung, esophagus, and other cancers. A third mechanism involves missense mutations resulting in dominant-negative effects with an even greater reduction of functionally active tetramers. Such missense mutations are common in colon, brain, lung, breast, skin, bladder, and other cancers. A fourth mechanism by which p53 is involved in oncogenesis is common in cervical cancers where the expression of the E6 gene of human papillomavirus (HPV) results in the functional inactivation of p53 through binding and degradation. The p53 pathway may also be disrupted by alteration of a cellular gene, MDM2 (164785). This gene was originally identified by virtue of its amplification in a spontaneously transformed mouse cell line (Fakharzadeh et al., 1991). The MDM2 gene is amplified in the significant fraction of the most common human sarcomas and the consequent overexpression of MDM2 is likely to interfere with p53 activity (Oliner et al., 1992). Shieh et al. (1997) showed that, upon DNA damage, p53 is phosphorylated at ser15 and that this event leads to reduced interaction of p53 with its negative regulator, the oncoprotein MDM2. Furthermore, these investigators demonstrated that the phosphorylation of p53 at ser15 and ser37 by purified DNA-dependent protein kinase (600899) impairs the ability of MDM2 to inhibit p53-dependent transactivation. They presented evidence that these effects are most likely due to a conformational change induced by phosphorylation of p53. Shieh et al. (1997) proposed the following potential mechanism for the involvement of p53 in DNA damage: under normal unstressed conditions, p53 associates with its inhibitor, and p53-dependent transactivation is repressed. Upon DNA damage, p53 is phosphorylated at ser15, which in turn induces a conformational change that makes MDM2 unable to bind p53 and results in the relief of the inhibitory effect of MDM2 on p53.

In a review of germline P53 mutations, Frebourg and Friend (1992) collected information on 18 separate germline mutations of the P53 gene. These were widely distributed over the P53 gene between amino acid residues 72 and 325.

See Fields and Jang (1990), Unger et al. (1992), and Chumakov et al. (1993) for discussion of the DNA-binding properties of wildtype and mutant p53 and their roles in transcriptional transactivation.

It has been suggested that wildtype p53 may play a role in DNA repair and that expression of mutant forms of p53 may alter cellular resistance to the DNA damage caused by gamma-radiation. Furthermore, p53 had been thought to function as a cell cycle checkpoint after irradiation, also suggesting that mutant p53 might change the cellular proliferative response to radiation. Lee and Bernstein (1993) used transgenic mice expressing 1 of 2 mutant alleles of p53 to test this prediction. They found that expression of either mutant significantly increased the cellular resistance of a variety of hematopoietic cell lineages to gamma-radiation. Lee and Bernstein (1993) speculated that wildtype P53 may serve as a 'guardian of the genome,' preventing proliferation of a cell that has sustained genetic damage. If this is the case, cells lacking wildtype p53 protein as the result of a dominant-negative action of mutant p53 might not undergo radiation-induced cell death, thereby increasing radiation resistance. The ability of p53 to activate transcription from specific sequences suggests that genes induced by p53 may mediate its biologic role as a tumor suppressor. El-Deiry et al. (1993) identified a gene called WAF1 (gene symbol=CDKN1A; 116899), whose induction was associated with wildtype but not mutant p53 gene expression in a human brain tumor cell line. WAF1 is identical to a gene called CIP1 (Harper et al., 1993) whose product binds to cyclin complexes and inhibits the function of cyclin-dependent kinases. El-Deiry et al. (1993) suggested the following model for p53 function: p53 is not required for normal development, but in certain cellular environments (DNA damage, cellular stress), p53 expression is stimulated. In turn, p53 binds to WAF1 regulatory elements and transcriptionally activates its expression. The WAF1 protein subsequently binds to and inhibits cyclin-dependent kinase activity, preventing phosphorylation of critical cyclin-dependent kinase substrates and blocking cell cycle progression. In tumor cells with inactive p53, this pathway would thereby be defective, permitting unregulated growth. See also the report of a tumor-associated polymorphism in CDKN1A (116899.0001).

Donehower et al. (1992) showed that although development was normal in mice lacking one or both alleles of p53 as a result of gene targeting of ES cells, spontaneous tumors, specifically lymphomas and sarcomas, occurred in high frequency. Kemp et al. (1994) reported that a single dose of 4 Gy radiation dramatically decreased the latency for tumor development in p53 heterozygous mice. The pattern of genetic alterations at the remaining wildtype allele in these tumors differed substantially from that in spontaneous tumors from similar mice, indicating that p53 itself may have been a target for radiation-induced alterations. Radiation at a lower dose, 1 Gy, of preweanling p53 null mice also significantly decreased tumor latency, suggesting that there are additional genetic targets involved in radiation-induced malignancy. The p53-deficient mice provide a sensitive system for studies of the consequences of radiation exposure.

Caelles et al. (1994) investigated p53 function by developing immortalized somatotropic progenitor cells that express a temperature-sensitive mutation in the p53 gene. In these cells, induction of apoptosis (programmed cell death) by DNA damage depends strictly on p53 function. Temperature-shift experiments indicated that the extent of apoptotic DNA cleavage is directly proportional to the period during which p53 is functional. A shift to the permissive temperature triggers apoptosis following DNA damage, but this is independent of new RNA or protein synthesis. Caelles et al. (1994) suggested, therefore, that p53 is not an activator of apoptosis-mediator genes, but rather that p53 either represses genes necessary for cell survival or is a component of the enzymatic machinery for apoptotic cleavage or repair of DNA.

Osteosarcomas are among the tumors most frequently associated with mutations in the TP53 gene; this is reflected by the allelic variants reported here, which also indicate that germline TP53 mutations occur in osteosarcoma patients with Li-Fraumeni syndrome. Osteosarcomas are also among the tumors occurring in highest incidence in transgenic mice carrying mutated TP53 genes (Lavigueur et al., 1989). Smith-Sorensen et al. (1993) used constant denaturant gel electrophoresis (CDGE) followed by direct sequencing to identify mutations in the conserved domains II-V. Of 28 osteosarcomas, 6 were shown to have a TP53 mutation. Two of these had previously been identified in osteosarcomas; ser241-to-phe (191170.0013) and arg282-to-trp (191170.0018).

Harris and Hollstein (1993) gave an extensive review of the clinical implications of the TP53 gene. They tabulated the characteristics of 7 tumor-suppressor genes known to be involved in human cancers: TP53, RB1, APC, WT1, NF1, NF2, and VHL. Their figure 2 presented graphically the difference in spectrum of TP53 mutations causing liver cancer in high-incidence areas as opposed to low-incidence areas. Eighty percent of TP53 mutations are missense mutations; this is in contrast to the findings in other tumor-suppressor genes, e.g., in the APC gene in adenomatous polyposis, in which only about 9% of mutations are of the missense type.

Dumaz et al. (1993) used reverse transcription-PCR and SSCP to analyze more than 40 skin tumors (mainly basal and squamous cell carcinomas) from patients with xeroderma pigmentosum (278700). They found that 17 out of 43 contained at least one point mutation on the TP53 gene. All the mutations were located at dipyrimidine sites, essentially at CC sequences, which are hotspots for UV-induced DNA lesions. In 14 out of 19 mutations at CC sites, a tandem double base CC-to-TT transition, considered to be unique to UV-induced lesions, was found; these mutations are not observed in internal human tumors. The mutations were considered to be due to translesion synthesis of unrepaired dipyrimidine lesions left on the untranscribed strand.

Science magazine designated p53 the 'molecule of the year' for 1993. Culotta and Koshland (1993) and Harris (1993) gave an extensive account of its discovery and elucidation of function, as well as the use of P53 in cancer risk assessment.

Mutations in the p53 gene are found in only about 25 to 33% of hepatocellular carcinoma cases. However, the majority of cases of hepatocellular carcinoma are associated with chronic hepatitis B (HBV) infections. In order to investigate the roles of HBV and p53 in the initiation of hepatocellular carcinoma, Ueda et al. (1995) developed a transgenic mouse model in which expression of a single HBV gene product, the HBx transcriptional trans-activator protein, leads to progressive neoplastic changes in the liver. They demonstrated that tumor development correlates precisely with p53 binding to HBx in the cytoplasm and complete blockage of p53 entry into the nucleus. Analysis of tumor DNA revealed p53 mutations in only a small proportion of advanced tumors, strongly suggesting that p53 mutations are not the cause of tumors but may contribute to tumor progression.

Krawczak et al. (1995) compared the spectrum of somatic TP53 single basepair substitutions detected in 955 cancers with that of 2,224 different germline mutations in 279 different human genes (other than TP53), reported as the cause of inherited disease. The comparison demonstrated that, disregarding a relatively small subset (12%) of TP53 mutations that probably resulted from the action of exogenous mutagens, both the relative rates and the nearest-neighbor spectra of single basepair substitutions are similar in the 2 datasets. This resemblance suggested that a substantial proportion of cancer-associated somatic TP53 mutations result from endogenous cellular mechanisms. The likelihood of clinical observation of a particular mutation type differs, however, between tumors and genetic diseases, when the chemical properties of the resulting amino acid substitutions are considered. Together with a 6-fold higher observation likelihood for mutations at evolutionarily conserved residues, this finding argues that selection is a critical factor in determining which TP53 mutations are found to be associated with human cancer.

Transgenic mice lacking the p53 tumor suppressor gene are indistinguishable from their wildtype litter mates except for their dramatically early onset of tumor formation. It has been noted, however, that the frequency of recovery of these homozygous mutant animals is slightly less than anticipated. Sah et al. (1995) described a developmental analysis demonstrating that a variable percentage of p53-deficient animals exhibit mid-brain exencephaly, a neural tube defect that is incompatible with postnatal survival.

DNA damage is thought to initiate the teratogenicity caused by numerous drugs and environmental chemicals, collectively termed xenobiotics. Nicol et al. (1995) hypothesized that p53 deficiencies may enhance susceptibility to chemical teratogenesis. To test this hypothesis, they chose benzo[a]pyrene, a known teratogen, and tested its effects on pregnant heterozygous p53-deficient mice. Such mice exhibited between 2- and 4-fold higher embryo toxicity and teratogenicity than normal p53 controls. Fetal resorptions reflecting in utero death were genotyped using PCR and found to be increased 2.6-fold and 3.6-fold, respectively, with heterozygous and homozygous p53-deficient embryos. The results were interpreted as evidence that p53 is an important teratologic suppressor gene that protects the embryo from DNA-damaging chemicals and developmental oxidative stress.

Restoration of p53 function has been proposed as a means of counteracting oncogenesis. Such therapy requires, however, prior knowledge of the proper balance of p53 function in a given target tissue. Nakamura et al. (1995) addressed the problem by targeting expression of the wildtype human p53 gene to the lens, a tissue entirely composed of epithelial cells that differentiate into elongated fiber cells. Transgenic mice expressing wildtype human p53 develop microphthalmia as a result of a defect in fiber formation that sets in shortly after birth. Apoptotic cells were observed that failed to undergo proper differentiation. Nakamura et al. (1995) also generated mice expressing a mutant human p53 allele that lacks wildtype function. In normal lens, phenotype was restored in double transgenic animals that carry both wildtype and mutant human p53 alleles. This study highlighted the difficulties that can arise if p53 levels are improperly balanced in a differentiating tissue.

Brennan et al. (1995) presented results pointing to the molecular target of tobacco and alcohol which epidemiologic studies demonstrate to be associated with squamous cell carcinoma of the head and neck. They found P53 mutations in 42% of patients (54/129); in 58% of patients who smoked cigarettes and used alcohol (37/64); in 33% of patients who smoked but abstained from alcohol (13/39); and 17% of patients who neither smoked nor drank alcohol (4/24). All the mutations in the patients who neither drank nor smoked occurred at sites containing CpG dinucleotides (potentially representing endogenous mutations) within the P53 gene, whereas only 23% of those in cigarette smokers consisted of such changes.

Since cigarette smoke carcinogens such as benzo(a)pyrene are implicated in the development of lung cancer, Denissenko et al. (1996) investigated the relationship between benzo(a)pyrene diol epoxide (BPDE) adduct formation and P53 mutations. (BPDE is the ultimate carcinogenic metabolite of benzo(a)pyrene.) HeLa cells and normal bronchial epithelial cells were treated with BPDE, and DNA was isolated and cleaved at the sites of modified bases with UvrABC nuclease complex from E. coli. They mapped the distribution of BPDE adducts along the P53 gene using a modification of ligation-mediated PCR (LMPCR) with P53 oligonucleotide primers. They also examined isolated genomic DNA. Denissenko et al. (1996) demonstrated strong and selective adduct formation at guanine positions in codons 157, 248, and 273. They noted that these positions are the major mutational hotspots in human lung cancers and stated that the coincidence of mutational hotspots and adduct hotspots suggests that benzo(a)pyrene metabolites or structurally related compounds are involved in transformation of human lung tissue. Denissenko et al. (1996) concluded that their study provided a direct link between a defined cigarette smoke carcinogen and human cancer mutations.

To investigate possible mechanisms underlying the selectivity of BPDE binding, Denissenko et al. (1997) mapped the adducts in plasmid DNA containing genomic P53 sequences. They found that when cytosines in CpG sequences were converted to 5-methylcytosines by the CpG-specific methylase, SssI, and the DNA was subsequently treated with BPDE, adduct hotspots were created that were similar to those seen in genomic DNA when all CpGs were methylated. A strong positive effect of 5-methylcytosine on BPDE adduct formation at CpG sites was also documented with sequences of the PGK1 gene (311800) derived from an active or inactive human X chromosome and having differential methylation patterns. These results showed that methylated CpG dinucleotides, in addition to being an endogenous promutagenic factor, may represent a preferential target for exogenous chemical carcinogens. The data open new avenues concerning the reasons that the majority of mutational hotspots in human genes are at CpGs.

The centrosome plays a vital role in mitotic fidelity, ensuring establishment of bipolar spindles and balanced chromosome segregation. Centrosome duplication occurs only once during the cell cycle and is therefore highly regulated. Fukasawa et al. (1996) showed that in mouse embryonic fibroblasts lacking the p53 tumor suppressor protein, multiple copies of functionally competent centrosomes are generated during a single cell cycle. In contrast, mouse embryonic fibroblasts prepared from normal mice or mice deficient in the retinal blastoma tumor suppressor gene product did not display these abnormalities. The abnormally amplified centrosomes profoundly affected mitotic fidelity, resulting in unequal segregation of chromosomes. These observations implicated p53 in the regulation of centrosome duplication and suggested to the authors a possible mechanism by which loss of p53 may cause genetic instability.

Hollstein et al. (1996) gave an update on a listing of point mutations in the p53 gene of human tumors and cell lines compiled from the published literature and made available electronically through the file server at the EMBL Data Library. In July 1995, the database contained records on almost 4,500 published mutations. In their figure 1, they mapped the growth of mutations in the database since 1989. The data were made available from the European Bioinformatics Institute (EBI) network server through several means. Beroud et al. (1996) also described software and database for p53 gene mutations available for both MS-DOS and Macintosh platforms. Their database contained over 4,200 mutations as of September 1995. De Vries et al. (1996) described a database of p53 mutations. Hainaut et al. (1997) reported further on the p53 mutation database maintained at the International Agency for Research on Cancer (IARC) in Lyon, France and reported on earlier by Hollstein et al. (1996). Updated twice annually, the current version contained records on 5,091 published mutations and was expected to surpass the 6,000 mark in the January 1997 release. Availability and searchability were described.

Aas et al. (1996) presented data linking specific mutations in the P53 gene to primary resistance to doxorubicin therapy and early relapse in breast cancer patients. The L2 domain (codons 163-195) and the L3 domain (codons 236-256) contain zinc-finger sequences and are involved with p53's DNA binding function. Of 63 patients studied, 11 had mutations affecting L2 and/or L3. Four out of 11 patients with mutations affecting L2/L3 became progressively worse during doxorubicin treatment versus 2 out of 52 patients without mutations or with mutations not affecting the L2/L3 domain. P53 mutations in patients with progressive disease all affected the L3 domain. Of 8 patients with L3 mutations, 4 expressed primary resistance to anthracycline therapy. Aas et al. (1996) reported that patients with P53 mutations affecting the zinc binding domain who responded to primary chemotherapy initially in most cases had a relapse of their disease within 3 months.

Roth et al. (1996) cited evidence that, despite the multiplicity of genetic lesions within a single cancer cell, the correction of a single critical genetic lesion is sufficient to abrogate tumorigenicity in human cancer cells. Furthermore, tumor suppressor genes undergo homozygous loss of function by mutation, deletion, methylation, or a combination of these within cancer cells. Roth et al. (1996) investigated the effects of injection into tumors (non-small-cell lung carcinomas) of a retroviral vector containing the wildtype p53 gene under control of a beta-actin promoter. Following injection of the p53-containing viral vector, tumor regression occurred in 3 patients and tumor growth stabilized in 3 other patients.

Two promoters have been demonstrated at the 5-prime end of the human p53 gene. The first is located upstream of the first exon and is responsible for transcription of the major p53 mRNA species. The second (p53p2) is a stronger promoter than the first and is located within the 10,738-bp first intron, approximately 1,000 bp downstream of exon 1. Reisman et al. (1996) stated that mRNA transcripts initiated from the second promoter had been identified in HL-60 cells by primer extension analysis and were observed to increase in abundance during differentiation of HL-60 cells to granulocytes. By screening a cDNA library with a probe derived from sequences downstream of the p53p2 start site, Reisman et al. (1996) cloned and characterized a cDNA that represented an mRNA that appeared to have been initiated from the p53p2 promoter. The authors designated the gene encoding the transcript Hp53int1; the GDB designation was D17S2179E. The cDNA is 1,125 bp in length and is polyadenylated downstream from a consensus poly(A) addition site. The entire 1,125 bp were derived from intron 1 of the p53 gene, with no introns having been removed. Reisman et al. (1996) concluded that Hp53int1 may be a pseudogene for a gene that may have encoded a DNA-binding protein. Alternatively, the transcript may have a function, since RNA transcripts of this gene are present in a number of human cells and their levels are induced during terminal differentiation of myeloid leukemia cells.

The p53 tumor suppressor is found to be mutated and abundant in a wide variety of tumors. In tumors showing rapid growth, the type II isoform of hexokinase (HK2; 601125) is also highly expressed to facilitate high rates of glucose catabolism, which in turn promote rapid tumor proliferation. Mathupala et al. (1997) cloned p53 from the AS-30D rat hepatoma cell line and identified 2 point mutations at the periphery of the p53 core DNA binding domain. Using coexpression studies they showed that overexpressed mutant p53 significantly and reproducibly activated the HK2 promoter and increased gene expression. The authors noted that their novel findings are consistent with reports describing the transactivating effects of p53 on various genes (Unger et al., 1992; Chumakov et al., 1993; Zhang et al., 1993). Their findings contrast with previous reports that mutant p53 functions in tumor cells only to prevent wildtype p53 from transactivating genes involved in suppressing cell proliferation (Fields and Jang, 1990; Farmer et al., 1992).

Levine (1997) reviewed all aspects of p53, which he referred to as the cellular gatekeeper for growth and division.

To determine the mechanism of p53-induced apoptosis, Polyak et al. (1997) examined in detail the transcripts induced by p53 expression before the onset of apoptosis. Of 7,202 transcripts identified, only 14 (0.19%) were found to be markedly increased in p53-expressing cells compared with controls. Strikingly, many of these genes were predicted to encode proteins that could generate or respond to oxidative stress, including 1 that is implicated in apoptosis in plant meristems. Additional biochemical and pharmacologic experiments suggested that p53 results in apoptosis through a 3-step process: (1) the transcriptional induction of redox-related genes; (2) the formation of reactive oxygen species; and (3) the oxidative degradation of mitochondrial components, culminating in cell death.

Individuals with 1 malignancy of the aerodigestive tract have a high incidence of second primary aerodigestive tumors. The mechanism for this field effect had not been determined. Franklin et al. (1997) studied an individual with widespread dysplastic changes in the respiratory epithelium but no overt carcinoma. The entire tracheobronchial tree obtained at autopsy was embedded in paraffin, and bronchial epithelial cells were isolated by microdissection. DNA extracted from the microdissected cells was analyzed for point mutations in the p53 gene. A single, identical point mutation consisting of a G:C to T:A transversion in codon 245 was identified in bronchial epithelium from 7 of 10 sites in both lungs. Epithelium at sites containing the p53 mutation was morphologically abnormal, exhibiting squamous metaplasia and mild-to-moderate atypia. No invasive tumor was found in the tracheobronchial tree or any other location. Cells from the peripheral blood, kidney, liver, and lymph node exhibited no abnormality in the p53 gene. Widespread presence of a single somatic p53 point mutation in the bronchi of a smoker suggested to Franklin et al. (1997) that a single progenitor bronchial epithelial clone may expand to populate broad areas of the bronchial mucosa—a novel mechanism for field carcinogenesis in the respiratory epithelium that may be important in assessing individuals for risk of a second primary tumor, as well as in devising effective strategies for chemoprevention of lung cancer. The patient was a 66-year-old man with a 50-pack-per-year history of smoking and chronic obstructive pulmonary disease. His sputum cytology had shown moderate atypia. He died unexpectedly after laparotomy for small bowel obstruction.

Conseiller et al. (1998) constructed a chimeric tumor suppressor gene from the wildtype p53 gene for possible gene therapy of cancer based on the tumor suppressor function of p53. In so-called chimeric tumor suppressor 1 (CTS1), the domains that mediate the inactivation of p53 were removed. The authors found that CTS1 enhanced transcriptional activity, was resistant to inactivation by the p53-binding protein MDM2 (164785), had the ability to suppress cell growth, and showed faster induction of apoptosis. Thus, CTS1 was considered an alternative for use in gene therapy for wildtype p53-resistant human tumors.

The E6 oncoprotein derived from tumor-associated human papillomaviruses (HPVs) binds to and induces the degradation of the cellular tumor-suppressor protein p53. Storey et al. (1998) investigated the effect of the pro72-to-arg polymorphism (191170.0005) on the susceptibility of p53 to E6-mediated degradation and found that the arginine form of p53 is significantly more susceptible than the proline form. Moreover, allelic analysis of patients with HPV-associated tumors revealed a striking overrepresentation of homozygous arginine-72 p53 compared with the normal population, which indicated that individuals homozygous for arginine-72 are about 7 times more susceptible to HPV-associated tumorigenesis than heterozygotes.

See also Crawford (1985); Felix et al. (1992); Harlow et al. (1985); Jenkins et al. (1985); Lamb and Crawford (1986); Oda et al. (1992)

Aas, T.; Borresen, A.-L.; Geisler, S.; Smith-Sorenson, B.; Johnsen, H.; Varhaug, J. E.; Akslen, L. A.; Lonning, P. E.: Specific P53 mutations are associated with de novo resistance to doxorubicin in breast cancer patients. Nature Med. 2: 811-814, 1996.

Aguilar, F.; Harris, C. C.; Sun, T.; Hollstein, M.; Cerutti, P.: Geographic variation of p53 mutational profile in nonmalignant human liver. Science 264: 1317-1319, 1994.

Aguilar, F.; Hussain, S. P.; Cerutti, P.: Aflatoxin B1 induces the transversion of G-to-T in codon 249 of the p53 tumor suppressor gene in human hepatocytes. Proc. Nat. Acad. Sci. 90: 8586-8590, 1993.

Ara, S.; Lee, P. S. Y.; Hansen, M. F.; Saya, H.: Codon 72 polymorphism of the TP53 gene. Nucleic Acids Res. 18: 4961, 1990.

Baker, S. J.; Fearon, E. R.; Nigro, J. M.; Hamilton, S. R.; Preisinger, A. C.; Jessup, J. M.; vanTuinen, P.; Ledbetter, D. H.; Barker, D. F.; Nakamura, Y.; White, R.; Vogelstein, B.: Chromosome 17 deletions and p53 gene mutations in colorectal carcinomas. Science 244: 217-221, 1989.

Benchimol, S.; Lamb, P.; Crawford, L. V.; Sheer, D.; Shows, T. B.; Bruns, G. A. P.; Peacock, J.: Transformation associated p53 protein is encoded by a gene on human chromosome 17. Somat. Cell Molec. Genet. 11: 505-509, 1985.

Beroud, C.; Verdier, F.; Soussi, T.: p53 gene mutation: software and database. Nucleic Acids Res. 24: 147-150, 1996.

Borresen, A.-L.; Hovig, E.; Smith-Sorensen, B.; Malkin, D.; Lystad, S.; Andersen, T. I.; Nesland, J. M.; Isselbacher, K. J.; Friend, S. H.: Constant denaturant gel electrophoresis as a rapid screening technique for p53 mutations. Proc. Nat. Acad. Sci. 88: 8405-8409, 1991.

Borresen, A. L.; Andersen, T. I.; Garber, J.; Barbier-Piraux, N.; Thorlacius, S.; Eyfjord, J.; Ottestad, L.; Smith-Sorensen, B.; Hovig, E.; Malkin, D.; Friend, S. H.: Screening for germ line TP53 mutations in breast cancer patients. Cancer Res. 52: 3234-3236, 1992.

Brash, D. E.; Rudolph, J. A.; Simon, J. A.; Lin, A.; McKenna, G. J.; Baden, H. P.; Halperin, A. J.; Ponten, J.: A role for sunlight in skin cancer: UV-induced p53 mutations in squamous cell carcinoma. Proc. Nat. Acad. Sci. 88: 10124-10128, 1991.

Brennan, J. A.; Boyle, J. O.; Koch, W. M.; Goodman, S. N.; Hruban, R. H.; Eby, Y. J.; Couch, M. J.; Forastiere, A. A.; Sidransky, D.: Association between cigarette smoking and mutation of the p53 gene in squamous-cell carcinoma of the head and neck. New Eng. J. Med. 332: 712-717, 1995.

Bressac, B.; Kew, M.; Wands, J.; Ozturk, M.: Selective G to T mutations of p53 gene in hepatocellular carcinoma from southern Africa. Nature 350: 429-431, 1991.

Buetow, K. H.; Sheffield, V. C.; Zhu, M.; Zhou, T.; Shen, F.; Hino, O.; Smith, M.; McMahon, B. J.; Lanier, A. P.; London, W. T.; Redeker, A. G.; Govindarajan, S.: Low frequency of p53 mutations observed in a diverse collection of primary hepatocellular carcinomas. Proc. Nat. Acad. Sci. 89: 9622-9626, 1992.

Caelles, C.; Helmberg, A.; Karin, M.: p53-dependent apoptosis in the absence of transcriptional activation of p53-target genes. Nature 370: 220-223, 1994.
Carrere, N.; Leblanc, R. M.; Begueret, J.; Blouin, P.; Cheyrou, A.: A new mutation of exon 5 of the P53 gene in breast cancer. Hum. Molec. Genet. 2: 1075, 1993.
Casey, G.; Yamanaka, Y.; Freiss, H.; Kobrin, M. S.; Lopez, M. E.; Buchler, M.; Beger, H. G.; Korc, M.: p53 mutations are common in pancreatic cancer and are absent in chronic pancreatitis. Cancer Lett. 69: 151-160, 1993.
Chakrani, F.; Armand, J.-P.; Lenoir, G.; Ju, L.; Liang, J.-P.; May, E.; May, P.: Mutations clustered in exon 5 of the p53 gene in primary nasopharyngeal carcinomas from southeastern Asia. Int. J. Cancer 61: 316-320, 1995.
Chen, L.-C.; Neubauer, A.; Kurisu, W.; Waldman, F. M.; Ljung, B.-M.; Goodson, W., III; Goldman, E. S.; Moore, D., II; Balazs, M.; Liu, E.; Mayall, B. H.; Smith, H. S.: Loss of heterozygosity on the short arm of chromosome 17 is associated with high proliferative capacity and DNA aneuploidy in primary human breast cancer. Proc. Nat. Acad. Sci. 88: 3847-3851, 1991.
Chen, P.-L.; Chen, Y.; Bookstein, R.; Lee, W.-H.: Genetic mechanisms of tumor suppression by the human p53 gene. Science 250: 1576-1580, 1990.
Chiba, I.; Takahashi, T.; Nau, M. M.; D'Amico, D.; Curiel, D. T.; Mitsudomi, T.; Buchhagen, D. L.; Carbone, D.; Piantadosi, S.; Koga, H.; Reissman, P.; Slamon, D. J.; Holmes, E. C.; Minna, J. D.: Mutations in the p53 gene are frequent in primary, resected non-small cell lung cancer. Oncogene 5: 1603-1610, 1990.
Cho, Y.; Gorina, S.; Jeffrey, P. D.; Pavletich, N. P.: Crystal structure of a p53 tumor suppressor-DNA complex: understanding tumorigenic mutations. Science 265: 346-355, 1994.
Chumakov, A. M.; Miller, C. W.; Chen, D. L.; Koeffler, H. P.: Analysis of p53 transactivation through high-affinity binding sites. Oncogene 8: 3005-3011, 1993.
Chung, R. Y.; Whaley, J. M.; Anderson, K. M.; Freiman, R. N.; Menon, A. G.; Seizinger, B. R.: p53 gene mutations in human glioblastomas associated with early age onset and better survival. (Abstract) Am. J. Hum. Genet. 47 (suppl.): A4, 1990.
Conseiller, E.; Debussche, L.; Landais, D.; Venot, C.; Maratrat, M.; Sierra, V.; Tocque, B.; Bracco, L.: CTS1: a p53-derived chimeric tumor suppressor gene with enhanced in vitro apoptotic properties. J. Clin. Invest. 101: 120-127, 1998.
Cooper, D. N.; Youssoufian, H.: The CpG dinucleotide and human genetic disease. Hum. Genet. 78: 151-155, 1988.
Crawford, L.: Human p53 and human tumours. BioEssays 3: 117-120, 1985.
Crook, T.; Wrede, D.; Tidy, J. A.; Mason, W. P.; Evans, D. J.; Vousden, K. H.: Clonal p53 mutation in primary cervical cancer: association with human-papillomavirus-negative tumours. Lancet 339: 1070-1073, 1992.
Culotta, E.; Koshland, D. E., Jr.: p53 sweeps through cancer research. Science 262: 1958-1959, 1993.
Denissenko, M. F.; Chen, J. X.; Tang, M.; Pfeifer, G. P.: Cytosine methylation determines hot spots of DNA damage in the human P53 gene. Proc. Nat. Acad. Sci. 94: 3893-3898, 1997.
Denissenko, M. F.; Pao, A.; Tang, M.; Pfeifer, G. P.: Preferential formation of benzo(a)pyrene adducts at lung cancer mutational hot spots in P53. Science 274: 430-434, 1996.
De Vries, E. M. G.; Ricke, D. O.; De Vries, T. N.; Hartmann, A.; Blaszyk, H.; Liao, D.; Soussi, T.; Kovach, J. S.; Sommer, S. S.: Database of mutations in the p53 and APC tumor suppressor genes designed to facilitate molecular epidemiological analyses. Hum. Mutat. 7: 202-213, 1996.
Donehower, L. A.; Harvey, M.; Slagle, B. L.; McArthur, M. J.; Montgomery, C. A., Jr.; Butel, J. S.; Bradley, A.: Mice deficient for p53 are developmentally normal but susceptible to spontaneous tumours. Nature 356: 215-221, 1992.
Dumaz, N.; Drougard, C.; Sarasin, A.; Daya-Grosjean, L.: Specific UV-induced mutation spectrum in the p53 gene of skin tumors from DNA-repair-deficient xeroderma pigmentosum patients. Proc. Nat. Acad. Sci. 90: 10529-10533, 1993.
El-Deiry, W. S.; Kern, S. E.; Pietenpol, J. A.; Kinzler, K. W.; Vogelstein, B.: Definition of a consensus binding site for p53. Nature Genet. 1: 45-49, 1992.
El-Deiry, W. S.; Tokino, T.; Velculescu, V. E.; Levy, D. B.; Parsons, R.; Trent, J. M.; Lin, D.; Mercer, E.; Kinzler, K. W.; Vogelstein, B.: WAF1, a potential mediator of p53 tumor suppression. Cell 75: 817-825, 1993.
Fagin, J. A.; Matsuo, K.; Karmakar, A.; Chen, D. L.; Tang, S.-H.; Koeffler, H. P.: High prevalence of mutations of the p53 gene in poorly differentiated human thyroid carcinomas. J. Clin. Invest. 91: 179-184, 1993.
Fakharzadeh, S.; Trusko, S. P.; George, D. L.: Tumorigenic potential associated with enhanced expression of a gene that is amplified in a mouse tumor cell line. EMBO J. 10: 1565-1569, 1991.
Farmer, G.; Bargonetti, J.; Zhu, H.; Friedman, P.; Prywes, R.; Prives, C.: Wild-type p53 activates transcription in vitro. Nature 358: 83-86, 1992.
Felix, C. A.; D'Amico, D.; Mitsudomi, T.; Nau, M. M.; Li, F. P.; Fraumeni, J. F., Jr.; Cole, D. E.; McCalla, J.; Reaman, G. H.; Whang-Peng, J.; Knutsen, T.; Minna, J. D.; Poplack, D. G.: Absence of hereditary p53 mutations in 10 familial leukemia pedigrees. J. Clin. Invest. 90: 653-658, 1992.
Felix, C. A.; Nau, M. M.; Takahashi, T.; Mitsudomi, T.; Chiba, I.; Poplack, D. G.; Reaman, G. H.; Cole, D. E.; Letterio, J. J.; Whang-Peng, J.; Knutsen, T.; Minna, J. D.: Hereditary and acquired p53 gene mutations in childhood acute lymphoblastic leukemia. J. Clin. Invest. 89: 640-647, 1992.
Fields, S.; Jang, S. K.: Presence of a potent transcription activating sequence in the p53 protein. Science 249: 1046-1049, 1990.
Foster, P. L.; Eisenstadt, E.; Miller, J. H.: Base substitution mutations induced by metabolically activated aflatoxin B1. Proc. Nat. Acad. Sci. 80: 2695-2698, 1983.
Franklin, W. A.; Gazdar, A. F.; Haney, J.; Wistuba, I. I.; La Rosa, F. G.; Kennedy, T.; Ritchey, D. M.; Miller, Y. E.: Widely dispersed p53 mutation in respiratory epithelium: a novel mechanism for field carcinogenesis. J. Clin. Invest. 100: 2133-2137, 1997.
Frebourg, T.; Friend, S. H.: Cancer risks from germline P53 mutations. J. Clin. Invest. 90: 1637-1641, 1992.
Frebourg, T.; Kassel, J.; Lam, K. T.; Gryka, M. A.; Barbier, N.; Andersen, T. I.; Borresen, A.-L.; Friend, S. H.: Germ-line mutations of the p53 tumor suppressor gene in patients with high risk for cancer inactivate the p53 protein. Proc. Nat. Acad. Sci. 89: 6413-6417, 1992.
Fukasawa, K.; Choi, T.; Kuriyama, R.; Rulong, S.; Vande Woude, G. F.: Abnormal centrosome amplification in the absence of p53. Science 271: 1744-1747, 1996.
Hainaut, P.; Soussi, T.; Shomer, B.; Hollstein, M.; Greenblatt, M.; Hovig, E.; Harris, C. C.; Montesano, R.: Database of p53 gene somatic mutations in human tumors and cell lines: updated compilation and future prospects. Nucleic Acids Res. 25: 151-157, 1997.
Halevy, O.; Michalovitz, D.; Oren, M.: Different tumor-derived p53 mutants exhibit distinct biological activities. Science 250: 113-116, 1990.
Harlow, E.; Williamson, N. M.; Ralston, R.; Helfman, D. M.; Adams, T. E.: Molecular cloning and in vitro expression of a cDNA clone for human cellular tumor antigen p53. Molec. Cell. Biol. 5: 1601-1610, 1985.
Harper, J. W.; Adami, G. R.; Wei, N.; Keyomarsi, K.; Elledge, S. J.: The p21 Cdk-interacting protein Cip1 is a potent inhibitor of G1 cyclin-dependent kinases. Cell 75: 805-816, 1993.
Harris, C. C.: p53: at the crossroads of molecular carcinogenesis and risk assessment. Science 262: 1980-1981, 1993.
Harris, C. C.; Hollstein, M.: Clinical implications of the p53 tumor-suppressor gene. New Eng. J. Med. 329: 1318-1327, 1993.
Herrmann, M. A.; Hay, I. D.; Bartelt, D. H.; Ritland, S. R., Jr.; Dahl, R. J.; Grant, C. S.; Jenkins, R. B.: Cytogenetic and molecular genetic studies of follicular and papillary thyroid cancers. J. Clin. Invest. 88: 1596-1604, 1991.
Hollstein, M.; Shomer, B.; Greenblatt, M.; Soussi, T.; Hovig, E.; Montesano, R.; Harris, C. C.: Somatic point mutations in the p53 gene of human tumors and cell lines: updated compilation. Nucleic Acids Res. 24: 141-146, 1996.
Hollstein, M.; Sidransky, D.; Vogelstein, B.; Harris, C. C.: p53 mutations in human cancers. Science 253: 49-53, 1991.
Hollstein, M. C.; Metcalf, R. A.; Welsh, J. A.; Montesano, R.; Harris, C. C.: Frequent mutation of the p53 gene in human esophageal cancer. Proc. Nat. Acad. Sci. 87: 9958-9961, 1990.
Hsu, I. C.; Metcalf, R. A.; Sun, T.; Welsh, J. A.; Wang, N. J.; Harris, C. C.: Mutational hotspot in the p53 gene in human hepatocellular carcinomas. Nature 350: 427-428, 1991.
Iavarone, A.; Matthay, K. K.; Steinkirchner, T. M.; Israel, M. A.: Germ-line and somatic p53 gene mutations in multifocal osteogenic sarcoma. Proc. Nat. Acad. Sci. 89: 4207-4209, 1992.
Iggo, R.; Gatter, K.; Bartek, J.; Lane, D.; Harris, A. L.: Increased expression of mutant forms of p53 oncogene in primary lung cancer. Lancet 335: 675-679, 1990.
Isobe, M.; Emanuel, B. S.; Givol, D.; Oren, M.; Croce, C. M.: Localization of gene for human p53 tumour antigen to band 17p13. Nature 320: 84-85, 1986.
Jeffrey, P. D.; Gorina, S.; Pavletich, N. P.: Crystal structure of the tetramerization domain of the p53 tumor suppressor at 1.7 angstroms. Science 267: 1498-1502, 1995.
Jenkins, J. R.; Rudge, K.; Chumakov, P.; Currie, G. A.: The cellular oncogene p53 can be activated by mutagenesis. Nature 317: 816-818, 1985.
Kaelbling, M.; Burk, R. D.; Atkin, N. B.; Johnson, A. B.; Klinger, H. P.: Loss of heterozygosity on chromosome 17p and mutant p53 in HPV-negative cervical carcinomas. Lancet 340: 140-142, 1992.
Kemp, C. J.; Wheldon, T.; Balmain, A.: p53-deficient mice are extremely susceptible to radiation-induced tumorigenesis. Nature Genet. 8: 66-69, 1994.
Krawczak, M.; Smith-Sorensen, B.; Schmidtke, J.; Kakkar, V. V.; Cooper, D. N.; Hovig, E.: Somatic spectrum of cancer-associated single basepair substitutions in the TP53 gene is determined mainly by endogenous mechanisms of mutation and by selection. Hum. Mutat. 5: 48-57, 1995.
Lamb, P.; Crawford, L.: Characterization of the human p53 gene. Molec. Cell. Biol. 6: 1379-1385, 1986.
Lavigueur, A.; Maltby, V.; Mock, D.; Rossant, J.; Pawson, T.; Bernstein, A.: High incidence of lung, bone, and lymphoid tumors in transgenic mice overexpressing mutant alleles of the p53 oncogene. Molec. Cell. Biol. 9: 3982-3991, 1989.
Law, J. C.; Strong, L. C.; Chidambaram, A.; Ferrell, R. E.: A germ line mutation in exon 5 of the p53 gene in an extended cancer family. Cancer Res. 51: 6385-6387, 1991.
Le Beau, M. M.; Westbrook, C. A.; Diaz, M. O.; Rowley, J. D.; Oren, M.: Translocation of the p53 gene in t(15;17) in acute promyelocytic leukaemia. Nature 316: 826-828, 1985.
Lee, J. M.; Bernstein, A.: p53 mutations increase resistance to ionizing radiation. Proc. Nat. Acad. Sci. 90: 5742-5746, 1993.
Levine, A. J.: p53, the cellular gatekeeper for growth and division. Cell 88: 323-331, 1997.
Levine, A. J.; Momand, J.; Finlay, C. A.: The p53 tumour suppressor gene. Nature 351: 453-456, 1991.
Malkin, D.; Jolly, K. W.; Barbier, N.; Look, A. T.; Friend, S. H.; Gebhardt, M. C.; Andersen, T. I.; Borresen, A.-L.; Li, F. P.; Garber, J.; Strong, L. C.: Germline mutations of the p53 tumor-suppressor gene in children and young adults with second malignant neoplasms. New Eng. J. Med. 326: 1309-1315, 1992.
Malkin, D.; Li, F. P.; Strong, L. C.; Fraumeni, J. F., Jr.; Nelson, C. E.; Kim, D. H.; Kassel, J.; Gryka, M. A.; Bischoff, F. Z.; Tainsky, M. A.; Friend, S. H.: Germ line p53 mutations in a familial syndrome of breast cancer, sarcomas, and other neoplasms. Science 250: 1233-1238, 1990.
Masuda, H.; Miller, C.; Koeffler, H. P.; Battifora, H.; Cline, M. J.: Rearrangement of the p53 gene in human osteogenic sarcomas. Proc. Nat. Acad. Sci. 84: 7716-7719, 1987.
Mathupala, S. P.; Heese, C.; Pedersen, P. L.: Glucose catabolism in cancer cells: the type II hexokinase promoter contains functionally active response elements for the tumor suppressor p53. J. Biol. Chem. 272: 22776-22780, 1997.
Mazoyer, S.; Lalle, P.; Moyret-Lalle, C.; Marcais, C.; Schraub, S.; Frappaz, D.; Sobol, H.; Ozturk, M.: Two germ-line mutations affecting the same nucleotide

at codon 257 of p53 gene, a rare site for mutations. Oncogene 9: 1237-1239, 1994.
McBride, O. W.; Merry, D.; Givol, D.: The gene for human p53 cellular tumor antigen is located on chromosome 17 short arm (17p13). Proc. Nat. Acad. Sci. 83: 130-134, 1986.
McBride, O. W.; Merry, D. E.; Oren, M.; Givol, D.: Human p53 cellular tumor antigen is on chromosome 17p13. (Abstract) Cytogenet. Cell Genet. 40: 694-695, 1985.
Metzger, A. K.; Sheffield, V. C.; Duyk, G.; Daneshvar, L.; Edwards, M. S. B.; Cogen, P. H.: Identification of a germ-line mutation in the p53 gene in a patient with an intracranial ependymoma. Proc. Nat. Acad. Sci. 88: 7825-7829, 1991.
Miller, C.; Mohandas, T.; Wolf, D.; Prokocimer, M.; Rotter, V.; Koeffler, H. P.: Human p53 gene localized to short arm of chromosome 17. Nature 319: 783-784, 1986.
Moll, U. M.; Riou, G.; Levine, A. J.: Two distinct mechanisms alter p53 in breast cancer: mutation and nuclear exclusion. Proc. Nat. Acad. Sci. 89: 7262-7266, 1992.
Mulligan, L. M.; Matlashewski, G. J.; Scrable, H. J.; Cavenee, W. K.: Mechanisms of p53 loss in human sarcomas. Proc. Nat. Acad. Sci. 87: 5863-5867, 1990.
Nakamura, T.; Pichel, J. G.; Williams-Simons, L.; Westphal, H.: An apoptotic defect in lens differentiation caused by human p53 is rescued by a mutant allele. Proc. Nat. Acad. Sci. 92: 6142-6146, 1995.
Nicol, C. J.; Harrison, M. L.; Laposa, R. R.; Gimelshtein, I. L.; Wells, P. G.: A teratologic suppressor role for p53 in benzo[a]pyrene-treated transgenic p53-deficient mice. Nature Genet. 10: 181-187, 1995.
Nigro, J. M.; Baker, S. J.; Preisinger, A. C.; Jessup, J. M.; Hostetter, R.; Cleary, K.; Bigner, S. H.; Davidson, N.; Baylin, S.; Devilee, P.; Glover, T.; Collins, F. S.; Weston, A.; Modali, R.; Harris, C. C.; Vogelstein, B.: Mutations in the p53 gene occur in diverse human tumour types. Nature 342: 705-708, 1989.
Oda, T.; Tsuda, H.; Scarpa, A.; Sakamoto, M.; Hirohashi, S.: p53 gene mutation spectrum in hepatocellular carcinoma. Cancer Res. 52: 6358-6364, 1992.
Oliner, J. D.; Kinzler, K. W.; Meltzer, P. S.; George, D. L.; Vogelstein, B.: Amplification of a gene encoding a p53-associated protein in human sarcomas. Nature 358: 80-83, 1992.
Olschwang, S.; Laurent-Puig, P.; Vassal, A.; Salmon, R.-J.; Thomas, G.: Characterization of a frequent polymorphism in the coding sequence of the Tp53 gene in colonic cancer patients and a control population. Hum. Genet. 86: 369-370, 1991.
Patel, P.; Stephenson, J.; Scheuer, P. J.; Francis, G. E.: p53 codon 249-ser mutations in hepatocellular carcinoma patients with low aflatoxin exposure. (Letter) Lancet 339: 881, 1992.
Pavletich, N. P.; Chambers, K. A.; Pabo, C. O.: The DNA-binding domain of p53 contains the four conserved regions and the major mutation hot spots. Genes Dev. 7: 2556-2564, 1993.
Polyak, K.; Xia, Y.; Zweler, J. L.; Kinzler, K. W.; Vogelstein, B.: A model for p53-induced apoptosis. Nature 389: 300-305, 1997.
Reisman, D.; Balint, E.; Loging, W. T.; Rotter, V.; Almon, E.: A novel transcript encoded within the 10-kb first intron of the human p53 tumor suppressor gene (D17S2179E) is induced during differentiation of myeloid leukemia cells. Genomics 38: 364-370, 1996.
Reisman, D.; Greenberg, M.; Rotter, V.: Human p53 oncogene contains one promoter upstream of exon 1 and a second, stronger promoter within intron 1. Proc. Nat. Acad. Sci. 85: 5146-5150, 1988.
Rideout, W. M., III; Coetzee, G. A.; Olumi, A. F.; Jones, P. A.: 5-Methylcytosine as an endogenous mutagen in the human LDL receptor and p53 genes. Science 249: 1288-1290, 1990.
Rodrigues, N. R.; Rowan, A.; Smith, M. E. F.; Kerr, I. B.; Bodmer, W. F.; Gannon, J. V.; Lane, D. P.: p53 mutations in colorectal cancer. Proc. Nat. Acad. Sci. 87: 7555-7559, 1990.
Romano, J. W.; Ehrhart, J. C.; Duthu, A.; Kim, C. M.; Appella, E.; May, P.: A mutation in the p53 gene of a human osteosarcoma cell line. (Abstract) Am. J. Hum. Genet. 45 (suppl.): A32, 1989.
Roth, J. A.; Nguyen, D.; Lawrence, D. D.; Kemp, B. L.; Carrasco, C. H.; Ferson, D. Z.; Hong, W. K.; Komaki, R.; Lee, J. J.; Nesbitt, J. C.; Pisters, K. M. W.; Putnam, J. B.; Schea, R.; Shin, D. M.; Walsh, G. L.; Dolormente, M. M.; Han, C.-I.; Martin, F. D.; Yen, N.; Xu, K.; Stephens, L. C.; McDonnell, T. J.; Mukhopadhyay, T.; Cai, D.: Retrovirus-mediated wild-type p53 gene transfer to tumors of patients with lung cancer. Nature Med. 2: 985-990, 1996.
Rowley, J. D.: Personal Communication. Chicago, Ill. 1/3/1986.
Sah, V. P.; Arradi, L. D.; Mulligan, G. J.; Williams, B. O.; Bronson, R. T.; Jacks, T.: A subset of p53-deficient embryos exhibit exencephaly. Nature Genet. 10: 175-180, 1995.
Santibanez-Koref, M. F.; Birch, J. M.; Hartley, A. L.; Morris Jones, P. H.; Craft, A. W.; Eden, T.; Crowther, D.; Kelsey, A. M.; Harris, M.: p53 germline mutations in Li-Fraumeni syndrome. Lancet 338: 1490-1491, 1991.
Schiffer, D.; Cavalla, P.; Di Sapio, A.; Giordana, M. T.; Mauro, A.: Mutations and immunohistochemistry of p53 and proliferation markers in astrocytic tumors of childhood. Child's Nerv. Syst. 11: 517-522, 1995.
Sedlacek, Z.; Kodet, R.; Kriz, V.; Seemanova, E.; Vodvarka, P.; Wilgenbus, P.; Mares, J.; Poustka, A.; Goetz, P.: Two Li-Fraumeni syndrome families with novel germline p53 mutations: loss of the wild-type p53 allele in only 50% of tumours. Brit. J. Cancer 77: 1034-1039, 1998.
Shieh, S.-Y.; Ikeda, M.; Taya, Y.; Prives, C.: DNA damage-induced phosphorylation of p53 alleviates inhibition by MDM2. Cell 91: 325-334, 1997.
Sidransky, D.; Von Eschenbach, A.; Tsai, Y. C.; Jones, P.; Summerhayes, I.; Marshall, F.; Paul, M.; Green, P.; Hamilton, S. R.; Frost, P.; Vogelstein, B.: Identification of p53 gene mutations in bladder cancers and urine samples. Science 252: 706-709, 1991.
Smith, H. S.: Personal Communication. San Francisco, Calif. 11/16/1993.
Smith-Sorensen, B.; Gebhardt, M. C.; Kloen, P.; McIntyre, J.; Aguilar, F.; Cerutti, P.; Borresen, A.-L.: Screening for TP53 mutations in osteosarcomas using constant denaturant gel electrophoresis (CDGE). Hum. Mutat. 2: 274-285, 1993.
Srivastava, S.; Zou, Z.; Pirollo, K.; Blattner, W.; Chang, E. H.: Germ-line transmission of a mutated p53 gene in a cancer-prone family with Li-Fraumeni syndrome. Nature 348: 747-749, 1990.
Storey, A.; Thomas, M.; Kalita, A.; Harwood, C.; Gardiol, D.; Mantovani, F.; Breuer, J.; Leigh, I. M.; Matlashewski, G.; Banks, L.: Role of a p53 polymorphism in the development of human papilloma-virus-associated cancer. Nature 393: 229-234, 1998.
Sun, Y.; Hegamyer, G.; Cheng, Y.-J.; Hildesheim, A.; Chen, J.-Y.; Chen, I.-H.; Cao, Y.; Yao, K.-T.; Colburn, N. H.: An infrequent point mutation of the p53 gene in human nasopharyngeal carcinoma. Proc. Nat. Acad. Sci. 89: 6516-6520, 1992.
Takahashi, T.; D'Amico, D.; Chiba, I.; Buchhagen, D. L.; Minna, J. D.: Identification of intronic point mutations as an alternative mechanism for p53 inactivation in lung cancer. J. Clin. Invest. 86: 363-369, 1990.
Takahashi, T.; Nau, M. M.; Chiba, I.; Birrer, M. J.; Rosenberg, R. K.; Vinocour, M.; Levitt, M.; Pass, H.; Gazdar, A. F.; Minna, J. D.: p53: a frequent target for genetic abnormalities in lung cancer. Science 246: 491-494, 1989.
Toguchida, J.; Yamaguchi, T.; Dayton, S. H.; Beauchamp, R. L.; Herrera, G. E.; Ishizaki, K.; Yamamuro, T.; Meyers, P. A.; Little, J. B.; Sasaki, M. S.; Weichselbaum, R. R.; Yandell, D. W.: Prevalence and spectrum of germline mutations of the p53 gene among patients with sarcoma. New Eng. J. Med. 326: 1301-1308, 1992.
Ueda, H.; Ullrich, S. J.; Gangemi, J. D.; Kappel, C. A.; Ngo, L.; Feitelson, M. A.; Jay, G.: Functional inactivation but not structural mutation of p53 causes liver cancer. Nature Genet. 9: 41-47, 1995.
Unger, T.; Nau, M. M.; Segal, S.; Minna, J. D.: p53: a transdominant regulator of transcription whose function is ablated by mutations occurring in human cancer. EMBO J. 11: 1383-1390, 1992.
vanTuinen, P.; Ledbetter, D. H.: Construction and utilization of a detailed somatic cell hybrid mapping panel for human chromosome 17: localization of an anonymous clone to the critical region of Miller-Dieker syndrome, deletion 17p13. (Abstract) Cytogenet. Cell Genet. 46: 708-709, 1987.
Varley, J. M.; McGowan, G.; Thorncroft, M.; Tricker, K. J.; Teare,, M. D.; Santibanez-Koref, M. F.; Houlston, R. S.; Martin, J.; Birch, J. M.; Evans, D. G. R.: An extended Li-Fraumeni kindred with gastric carcinoma and a codon 175 mutation in TP53. J. Clin. Genet. 32: 942-945, 1995.
Varley, J. M.; McGown, G.; Thorncroft, M.; Cochrane, S.; Morrison, P.; Woll, P.; Kelsey, A. M.; Mitchell, E. L. D.; Boyle, J.; Birch, J. M.; Evans, D. G. R.: A previously undescribed mutation within the tetramerisation domain of TP53 in a family with Li-Fraumeni sydrome (sic). Oncogene 12: 2437-2442, 1996.
Vogelstein, B.; Kinzler, K. W.: p53 function and dysfunction. Cell 70: 523-526, 1992.
Vogelstein, B.; Kinzler, K. W.: X-rays strike p53 again. Nature 370: 174-175, 1994.
Zhang, W.; Funk, W. D.; Wright, W. E.; Shay, J. W.; Deisseroth, A. B.: Novel DNA binding of p53 mutants and their role in transcriptional activation. Oncogene 8: 2555-2559, 1993.
Ziegler, A.; Jonason, A. S.; Leffell, D. J.; Simon, J. A.; Sharma, H. W.; Kimmelman, J.; Remington, L.; Jacks, T.; Brash, D. E.: Sunburn and p53 in the onset of skin cancer. Nature 372: 773-776, 1994.

CANCER, FAMILIAL (114450)

JJM NOTE

At the time this observation was made, the interpretation was that the radioresistance allowed the survival of mutated cells that otherwise would have died to progress to clinical cancer. Many biological assays were conducted on Li-Fraumeni cells; radio resistance has been newly found to make sense, now that the function of p53 is known. This family has a proved *TP53* mutation.

Bech-Hansen et al. (1981) studied a family in which members had had a diversity of neoplasms over 6 generations (originally reported by Blattner et al., 1979). Two members had neoplasms of possible radiogenic origin. Gamma-irradiation survival studies of cultured skin fibroblasts in these 2 patients and in 3 other relatives, but not their spouses, over 3 generations demonstrated resistance to cell killing. Radioresistance, as well as radiosensitivity (e.g., in ataxia-telangiectasia and xeroderma pigmentosum to gamma- and UV-irradiation, respectively), measured in vitro may be a marker for increased cancer risk. Thus, one subset of 'cancer families,' such as that described by Li and Fraumeni (1969), may represent this category.

Bech-Hansen, N. T.; Blattner, W. A.; Sell, B. M.; McKeen, E. A.; Lampkin, B. C.; Fraumeni, J. F., Jr.; Paterson, M. C.: Transmission of in-vitro radioresistance in a cancer-prone family. Lancet I: 1335-1337, 1981.
Blattner, W. A.; McGuire, D. B.; Mulvihill, J. J.; Lampkin, B. C.; Hananian, J.; Fraumeni, J. F., Jr.: Genealogy of cancer in a family. J.A.M.A. 241: 259-261, 1979.
Li, F. P.; Fraumeni, J. F., Jr.: Rhabdomyosarcoma in children: epidemiologic study and identification of a familial cancer syndrome. J. Nat. Cancer Inst. 43: 1365-1373, 1969.

RHABDOMYOSARCOMA 1 (#268210)

JJM NOTE

As a possible genetic etiology, Li-Fraumeni syndrome should be thought of first when a rhabdomyosarcoma is diagnosed. I have never been convinced that the tumor occurs to excess in Beckwith-Wiedemann and WAGR syndromes; rather, the presence of one or two possibly coincidental cases was convenient support for the chromosome 11 data described in this entry.

A number sign (#) is used with this entry because of evidence that rhabdomyosarcoma, which maps to 11p15.5, is caused by mutation in the Beckwith-Wiedemann region 1A gene (BWR1A; 602631).

Scrable et al. (1987) generalized the hypothesis of somatic chromosomal interchanges at mitosis as the basis of retinoblastoma (180200). To the study of other tumors, the proposed model suggests that predisposing recessive mutations are revealed by exchanges that result in homozygous (or hemizygous) defect at the tumor locus. This would suggest that for tumors with no known cytogenetic aberrations, one could find the location of these genes by delineating the smallest overlapping region of genetic homozygosity shared among tumors of the same type. They applied this rationale to rhabdomyosarcoma, a 'pediatric tumor.' Koufos et al. (1985) had suggested that a locus on chromosome 11 was involved. By the use of multiple markers on chromosome 11 in 4 tumors (comparing the tumor DNA with constitutional DNA in each case), Scrable et al. (1987) mapped the rhabdomyosarcoma locus to 11pter-p15.5. They suggested that the clinical association of rhabdomyosarcoma with the Beckwith-Wiedemann syndrome (130650) may have its basis in involvement of this locus; likewise, this locus may be involved in those cases of the WAGR syndrome (194070) in which rhabdomyosarcoma occurs. Stenman and Sager (1987) reviewed extensive prior data from chromosome studies of tumorigenic and tumor-derived Chinese hamster cells suggesting the presence of a tumor-suppressor gene on hamster 3p. They showed that 6 genes on human 11p (INS, CAT, HBBC, CALC, PTH, and HRAS) can be localized by in situ hybridization to Chinese hamster chromosome 3. INS and CAT were located close to the centromere on 3p, whereas HBBC, CALC, and PTH were at 3q3-q4 and HRAS at 3q4. Two tumor-suppressor genes have been described on 11p: one that determines Wilms tumor, which is linked to CAT (115500), and one that determines rhabdomyosarcoma, which is linked to INS (176730). Stenman and Sager (1987) suggested that the Chinese hamster-suppressor gene may be closely linked to INS or CAT in that species. The striking histologic resemblance of rhabdomyosarcoma to fetal striated muscle extends to the genes they express. MYOD1 (159970) is one of these. Scrable et al. (1990) demonstrated an evolutionarily conserved muscle-specific linkage group comprising MYOD1 and LDHA (150000), which is located close to the RMSCR locus.

Scrable et al. (1989) examined the role of genome imprinting in the development of embryonal rhabdomyosarcoma. Imprinting is an epigenetic, gamete-of-origin-dependent, allele-inactivation process. Pursuing the observation that these tumors arise from cells that are clonally isodisomic for loci on chromosome 11p, they demonstrated that the isodisomic chromosome 11p alleles were always of paternal origin in both familial and sporadic cases. If a tumor-suppressor allele was subject either to mutation or to epigenetic inactivation by passage through the male germ line, then any cell that contained such an allele would be functionally hemizygous at the affected locus. Thus, in either a genetically or an epigenetically caused hemizygote, only 1 additional event would be required to produce a null phenotype. In the classic Knudson model, tumor-suppressor alleles that carry alterations in nucleotide sequence will be inherited as the predisposing mutation. In such families, the disease will be genetically linked to markers on the chromosome that carries the tumor-suppressor allele, and the sex of the parent from whom the defective allele is inherited is not predicted to have an effect. This appears to be the case for families with retinoblastoma (180200), adenomatous polyposis coli (175100), and bilateral acoustic neuroma (101000). In a second class of familial tumors, an epigenetically inactivated tumor-suppressor allele will be inherited as the predisposing mutation. Because the inactivation of this allele need not be dependent on the allele itself but will reflect the activity of the gene or genes involved in generating or maintaining the genome imprint, the inheritance of the tumor phenotype will not be linked to the tumor-suppressor locus.

The findings of Pedone et al. (1994) suggest that acquisition of a double dosage of the gene for insulinlike growth factor-2 (IGF2; 147470) may be an important step before the initiation or progression of rhabdomyosarcoma. The IGF2 gene is imprinted, the maternally derived allele being inactive in normal tissues. Pedone et al. (1994) found that whereas monoallelic expression of IGF2 was conserved in normal adult muscle tissue, 2 or more copies of active IGF2 alleles, arising by either relaxation of imprinting or duplication of the active allele, were found in 9 out of 11 (82%) rhabdomyosarcomas retaining heterozygosity at 11p15, regardless of the histologic subtype. The findings did not extend to leiomyosarcomas; they detected only 1 case of partial reactivation of the maternal IGF2 allele out of 7 tumors of that subtype tested.

In a rhabdomyosarcoma cell line, Schwienbacher et al. (1998) found a G-to-A transition at nucleotide 688 that introduced an arginine in place of a cysteine in the product of the BWR1A gene (602631.0002). The change was present in homozygous state, suggesting that loss of the normal allele occurred during tumorigenesis.

Koufos, A.; Hansen, M. F.; Copeland, N. G.; Jenkins, N. A.; Lampkin, B. C.; Cavenee, W. K.: Loss of heterozygosity in three embryonal tumours suggests a common pathogenetic mechanism. Nature 316: 330-334, 1985.

Pedone, P. V.; Tirabosco, R.; Cavazzana, A. O.; Ungaro, P.; Basso, G.; Luksch, R.; Carli, M.; Bruni, C. B.; Frunzio, R.; Riccio, A.: Mono- and bi-allelic expression of insulin-like growth factor II gene in human muscle tumors. Hum. Molec. Genet. 3: 1117-1121, 1994.

Schwienbacher, C.; Sabbioni, S.; Campi, M.; Veronese, A.; Bernardi, G.; Menegatti, A.; Hatada, I.; Mukai, T.; Ohashi, H.; Barbanti-Brodano, G.; Croce, C. M.; Negrini, M.: Transcriptional map of 170-kb region at chromosome 11p15.5: identification and mutational analysis of the BWR1A gene reveals the presence of mutations in tumor samples. Proc. Nat. Acad. Sci. 95: 3873-3878, 1998.

Scrable, H.; Cavenee, W.; Ghavimi, F.; Lovell, M.; Morgan, K.; Sapienza, C.: A model for embryonal rhabdomyosarcoma tumorigenesis that involves genome imprinting. Proc. Nat. Acad. Sci. 86: 7480-7484, 1989.

Scrable, H. J.; Johnson, D. K.; Rinchik, E. M.; Cavenee, W. K.: Rhabdomyosarcoma-associated locus and MYOD1 are syntenic but separate loci on the short arm of human chromosome 11. Proc. Nat. Acad. Sci. 87: 2182-2186, 1990.

Scrable, H. J.; Witte, D. P.; Lampkin, B. C.; Cavenee, W. K.: Chromosomal localization of the human rhabdomyosarcoma locus by mitotic recombination mapping. Nature 329: 645-647, 1987.

Stenman, G.; Sager, R.: Genetic analysis of tumorigenesis: a conserved region in the human and Chinese hamster genomes contains genetically identified tumor-suppressor genes. Proc. Nat. Acad. Sci. 84: 9099-9102, 1987.

FORK HEAD IN RHABDOMYOSARCOMA (*136533)

JJM NOTE

One could speculate that, eventually, a familial or even sporadic instance of rhabdomyosarcoma could be due to a constitutional translocation of this gene (if it were not lethal to a fetus).

In alveolar rhabdomyosarcoma, the translocation t(2;13)(q35;q14) is frequently found. Barr et al. (1993) determined that PAX3 (193500), which had previously been found to be mutated in Waardenburg syndrome, was affected by this t(2;13) in alveolar rhabdomyosarcoma. Galili et al. (1993) isolated the chromosome 13 gene that is fused with PAX3. The rearrangement breakpoints occurred within an intron downstream of the paired box and homeodomain-encoding regions. and identified it as a member of the forkhead domain family, which encodes transcription factors containing a conserved DNA-binding motif related to the Drosophila region-specific homeotic gene 'forkhead.' The distal half of the forkhead and the C-terminal region of the FKHR gene are involved in the chimeric transcript and fusion protein. (Because of the homology to 'forkhead,' the gene was symbolized FKHR, for 'forkhead' in rhabdomyosarcoma.) See human T-cell leukemia virus enhancer factor (143089) and interleukin enhancer binding factor (147685) for other members of the forkhead domain family of transcription factors.

Fredericks et al. (1995) demonstrated expression of a 97-kD PAX3-FKHR fusion protein in a t(2;13)-positive rhabdomyosarcoma cell line and verified that a single polypeptide contained epitopes derived from each protein. The fusion protein was localized to the nucleus in these cells, as was wildtype PAX3 in cells lacking the translocation. They found that DNA binding of the fusion protein was significantly impaired relative to that of PAX3 despite the fact that the 2 proteins had identical PAX DNA-binding domains. However, the fusion protein was a much more potent transcriptional activator than PAX3. Thus, the fusion protein may function as an oncogenic transcription factor by enhancing activation of normal PAX3 target genes.

Sublett et al. (1995) found that the PAX3/FKHR hybrid protein binds DNA in vitro in a sequence specific manner and transactivates the expression of artificial reporter genes, suggesting that its aberrant expression could subvert the transcriptional programs that normally control the growth, differentiation, and survival of primitive myogenic precursors in vivo.

In a review of 28 published cases of alveolar rhabdomyosarcoma with cytogenetic studies, Whang-Peng et al. (1992) found the characteristic t(2;13) translocation in 64%; in 18% of cases they found a variant t(1;13)(p36;q14) translocation which was subsequently shown to result in fusion of the FKHR gene with the PAX7 gene (167410) on chromosome 1.

Using a retroviral vector, Scheidler et al. (1996) introduced the PAX3/FKHR fusion gene into chicken embryo fibroblasts. Expression of the PAX3/FKHR protein in these cells led to transformation: the cells became enlarged, grew tightly packed and in multiple layers, and acquired the ability for anchorage-independent growth. Davis and Barr (1997) demonstrated that in both the t(2;13) and the variant t(1;13) translocations, which generate PAX3/FKHR and PAX7/FKHR fusion proteins, respectively, there is overexpression of the fusion product in addition to functional alterations. In the t(2;13) translocation, transcription of PAX3/FKHR is increased relative to wildtype PAX3 by a copy number-independent process. In contrast, PAX7/FKHR overexpression results from fusion gene amplification.

Anderson et al. (1998) identified a processed pseudogene that is similar to FKHR and maps to 5q35.2-q35.3.

Anderson, M. J.; Viars, C. S.; Czekay, S.; Cavenee, W. K.; Arden, K. C.: Cloning and characterization of three human forkhead genes that comprise an FKHR-like gene subfamily. Genomics 47: 187-199, 1998.

Barr, F. G.; Galili, N.; Holick, J.; Biegel, J. A.; Rovera, G.; Emanuel, B. S.: Rearrangement of the PAX3 paired box gene in the paediatric solid tumor alveolar rhabdomyosarcoma. Nature Genet. 3: 113-117, 1993.

Davis, R. J.; Barr, F. G.: Fusion genes resulting from alternative chromosomal translocations are overexpressed by gene-specific mechanisms in alveolar rhabdomyosarcoma. Proc. Nat. Acad. Sci. 94: 8047-8051, 1997.

Fredericks, W. J.; Galili, N.; Mukhopadhyay, S.; Rovera, G.; Bennicelli, J.; Barr, F. G.; Rauscher, F. J., III: The PAX3-FKHR fusion protein created by the t(2;13) translocation in alveolar rhabdomyosarcomas is a more potent transcriptional activator than PAX3. Molec. Cell. Biol. 15: 1522-1535, 1995.

Galili, N.; Davis, R. J.; Fredericks, W. J.; Mukhopadhyay, S.; Rauscher, F. J., III; Emanuel, B. S.; Rovera, G.; Barr, F. G.: Fusion of a fork head domain gene to PAX3 in the solid tumour alveolar rhabdomyosarcoma. Nature Genet. 5: 230-235, 1993.

Scheidler, S.; Fredericks, W. J.; Rauscher, F. J., III; Barr, F. G.; Vogt, P. K.: The hybrid PAX3-FKHR fusion protein of alveolar rhabdomyosarcoma transforms fibroblasts in culture. Proc. Nat. Acad. Sci. 93: 9805-9809, 1996.

Sublett, J. E.; Jeon, I-S.; Shapiro, D. N.: The alveolar rhabdomyosarcoma PAX3/FKHR fusion protein is a transcriptional activator. Oncogene 11: 545-552, 1995.

Whang-Peng, J.; Knutsen, T.; Theil, K.; Horowitz, M. E.; Triche, T.: Cytogenetic studies in subgroups of rhabdomyosarcoma. Genes Chromosomes Cancer 5: 299-310, 1992.

SNF5, YEAST, HOMOLOG OF (*601607)

JJM NOTE

This entity underlies a tumor of uncertain origin called extrarenal malignant rhabdoid tumor. Germline mutation has not been described.

Rhabdoid tumors are a highly malignant group of neoplasms that usually occur in children less than 2 years of age. Malignant rhabdoid tumors (MRTs) of the kidney were first described as a sarcomatous variant of Wilms tumors (Beckwith and Palmer, 1978). Later, extrarenal rhabdoid tumor was reported in numerous locations, including the central nervous system (Parham et al., 1994). Classification has been difficult because of considerable variation in the histologic and immunologic characteristics within and between rhabdoid tumors of the liver, soft tissues, and CNS. Bonnin et al. (1984) described 7 patients with rhabdoid tumors of the kidney who had CNS tumors that differed in their histology. Weeks et al. (1989) reported on a series of 111 cases of which 13.5% with renal rhabdoid tumors had a CNS malignancy. The cell of origin for rhabdoid tumors is not known, and the question of whether rhabdoid tumors in various locations are related is moot. In the CNS, rhabdoid tumors may be pure rhabdoid tumors or a variant that has been designated atypical teratoid tumor. Biegel et al. (1990) described monosomy 22 in 3 rhabdoid tumors of the CNS, and Biegel et al. (1992) reported a rhabdoid tumor with an unbalanced 9;22 translocation leading to loss of 22q11.2-qter. Douglass et al. (1990) reported a CNS tumor with monosomy 22; Muller et al. (1995) reported on a rhabdoid tumor of the pineal region with monosomy 22, and Yu et al. (1996) presented a series of CNS rhabdoid tumors with monosomy 22, which was identified by fluorescence in situ hybridization. Biegel et al. (1996) sublocalized a rhabdoid tumor locus to the region between the constant region genes of the immunoglobulin lambda locus (146770) and BCR (151410) in a 500-kb span of 22q11.

Since cytogenetic and molecular analyses showed that deletion of 22q11.2 is a recurrent genetic characteristic of MRTs, this locus may encode a tumor suppressor gene. Versteege et al. (1998) mapped the most frequently deleted part of 22q11.2 from a panel of 13 MRT cell lines. They observed 6 homozygous deletions that delineated the smallest region of overlap. The region was found to fall in the SNF5/INI1 gene, which encodes a member of the chromatin-remodeling SWI/SNF multiprotein complexes. Versteege et al. (1998) analyzed the sequence of SNF5/INI1 and found frameshift or nonsense mutations of this gene in 6 other cells lines. These truncating mutations of 1 allele were associated with a loss of the other allele. Identical alterations were observed in corresponding primary tumor DNAs but not in matched constitutional DNAs, indicating that they had been acquired somatically.

Versteege et al. (1998) found that the SNF5/INI1 gene contains 9 exons and spans approximately 50 kb. They concluded that biallelic, somatic alteration of SNF5/INI1, arising through a combination of complete deletions of the gene, frameshift alterations, and nonsense mutations, is a frequent, if not constant, feature of MRTs. This was considered fully consistent with the paradigm of the 2-hit recessive model of oncogenesis and supported the hypothesis that SNF5/INI1 is the MRT tumor suppressor gene. Although constitutional mutations were not detected in this study, which was based on sporadic tumors, previous reports of multifocal and/or familial MRTs indicated that constitutional inactivation of 1 allele of SNF5/INI1 could be responsible for hereditary cases. Given the low incidence of MRT and its aggressive evolution, which frequently leads to death at a young age, familial cases, if existing, would be expected to be rare. Altered chromatin organization at specific DNA sites may be crucial in the process of oncogenesis. The SWI/SNF complexes, which have been identified in organisms from yeast to humans, are thought to be important in the remodeling of chromatin structure.

Beckwith, J. B.; Palmer, N. F.: Histopathology and prognosis of Wilms tumors: results from the First National Wilms' Tumor Study. Cancer 41: 1937-1948, 1978.

Biegel, J. A.; Allen, C. S.; Kawasaki, K.; Shimizu, N.; Budarf, M. L.; Bell, C. J.: Narrowing the critical region for a rhabdoid tumor locus in 22q11. Genes Chromosomes Cancer 16: 94-105, 1996.

Biegel, J. A.; Burk, C. D.; Parmiter, A. H.; Emanuel, B. S.: Molecular analysis of a partial deletion of 22q in a central nervous system rhabdoid tumor. Genes Chromosomes Cancer 5: 104-108, 1992.

Biegel, J. A.; Rorke, L. B.; Packer, R. J.; Emanuel, B. S.: Monosomy 22 in rhabdoid or atypical teratoid tumors of the brain. J. Neurosurg. 73: 710-714, 1990.

Bonnin, J. M.; Rubinstein, L. J.; Palmer, N. F.; Beckwith, J. B.: The association of embryonal tumors originating in the kidney and in the brain: a report of seven cases. Cancer 54: 2137-2146, 1984.

Douglass, E. C.; Valentine, M.; Rowe, S. T.; Parham, D. M.; Wilimas, J. A.; Sanders, J. M.; Houghton, P. J.: Malignant rhabdoid tumor: a highly malignant childhood tumor with minimal karyotypic changes. Genes Chromosomes Cancer 2: 210-216, 1990.

Muller, M.; Hubbard, S. L.; Fukuyama, K.; Dirks, P.; Matsuzawa, K.; Rutka, J. T.: Characterization of a pineal region malignant rhabdoid tumor. Pediat. Neurosurg. 22: 204-209, 1995.

Parham, D. M.; Weeks, D. A.; Beckwith, J. B.: The clinicopathologic spectrum of putative extrarenal rhabdoid tumors. Am. J. Surg. Path. 18: 1010-1029, 1994.

Versteege, I.; Sevenet, N.; Lange, J.; Rousseau-Merck, M.-F.; Ambros, P.; Handgretinger, R.; Aurias, A.; Delattre, O.: Truncating mutations of hSNF5/INI1 in aggressive paediatric cancer. Nature 394: 203-206, 1998.

Weeks, D. A.; Beckwith, J.; Mierau, G. W.; Luckey, D. W.: Rhabdoid tumor of kidney: a report of 111 cases from the National Wilms' Tumor Study Pathology Center. Am. J. Surg. Path. 13: 439-458, 1989.

Yu, I. T.; Burger, P. C.; Griffin, C. A.; Perlman, E. J.: Monosomy for chromosome 22 in rhabdoid tumors of the brain by fluorescence in situ hybridization. Cancer Genet. Cytogenet. (in-press), 1996.

SYNOVIAL, SARCOMA, X BREAKPOINT 1 (*312820)

JJM NOTE

This tumor is a fine example of how tumor cytogenetics helps the histopathologist make a specific diagnosis from the large group of sarcomas. I am not aware of familial aggregation of this specific tumor, but it has been seen, I believe, in Li-Fraumeni syndrome.

Synovial sarcoma, according to the experience of Enzinger and Weiss (1983), is the fourth most common type of soft tissue sarcoma. It usually develops in adolescents and young adults, is more common in males than in females, and has no racial predilection. Turc-Carel et al. (1987) found a translocation involving chromosome X (band p11.2) and chromosome 18 (band q11.2) in short-term cultures of cells from 5 synovial sarcomas and 1 malignant fibrous histiocytoma. In 4 of the tumors, the translocation t(X;18)(p11.2;q11.2) was reciprocal. The 2 other tumors had complex translocations, which, however, always involved chromosomes X and 18 at the 2 sites mentioned. The X;18 translocation was not detected in other histologic types of soft tissue sarcoma. This was the first description of a sex chromosome abnormality in a human solid tumor. Griffin and Emanuel (1987) confirmed the original findings of Turc-Carel et al. (1987). The karyotype of the tumor in their patient was 46,XX,t(X;18)(p11;q11). It may be significant that the ARAF1 oncogene (311010) maps to the same region. Inactivation of the normal X chromosome in cells carrying the X/autosome translocation may result in loss of expression of the normal allele and allow expression of the altered gene at the breakpoint on the X chromosome. This phenomenon permitted localization of the gene for Duchenne muscular dystrophy (310200) and a number of other autosomal recessive disorders that have been observed in females with X/autosome translocations. Smith et al. (1987) identified the translocation t(X;18)(p11.2;q11.2) in every cell analyzed from each of 3 synovial sarcomas. Karakousis et al. (1987) found a specific translocation between the X chromosome and chromosome 18 in 6 cases of synovial sarcoma. Wang-Wuu et al. (1987) likewise found t(X;18) in a case of synovial sarcoma. In a 13-year-old Japanese girl, Ueda et al. (1988) found t(X;18) in a synovial sarcoma, together with an insertion of chromosome 11q material into 15q. Knight et al. (1989) noted that the oncogene ARAF1 is not directly involved in the X;18 translocation. Miozzo et al. (1992) reported the instructive case of a patient with the Turner syndrome in whom the only X chromosome was involved in a translocation of typical form: t(X;18)(p11;q11). Knight et al. (1992) examined a hybrid cell line by Southern analysis using 13 additional markers located at Xp11.3-cen. The location of the breakpoint was further confirmed by fluorescence in situ hybridization. Two YAC probes of 300 kb and 450 kb, containing the OATL2 locus (311241), hybridized to both derivative chromosomes, indicating that these YACs span the translocation breakpoint. Similar results were obtained with 2 independent cell lines carrying the translocation. Sinke et al. (1993) concluded that the translocation breakpoint in Xp11.2 that is associated with synovial sarcoma is different from the translocation breakpoint in Xp11.2 associated with renal adenocarcinoma (312390).

By screening a synovial sarcoma cDNA library with a YAC spanning the X chromosome breakpoint, Clark et al. (1994) identified a hybrid transcript that contained 5-prime sequences (designated SYT by them; see SSXT; 600192) mapping to chromosome 18 and 3-prime sequences (designated SSX by them) mapping to the X chromosome. An SYT probe detected genomic rearrangements in 10 of 13 synovial sarcomas. Sequencing of cDNA clones showed that the normal SYT gene encodes a protein rich in glutamine, proline, and glycine and that in synovial sarcoma, rearrangement of the SYT gene results in the formation of an SYT-SSX fusion protein. Both SYT and SSX failed to exhibit significant homology to known gene sequences.

Nomenclature: SYT is the symbol for synaptotagmin (185605), an unrelated protein. Clark et al. (1994) used SSX for 'synovial sarcoma X chromosome breakpoint;' they used SYT for 'synovial sarcoma translocation.' SSXT, another proposed symbol, avoids the confusion with the gene to which the symbol SYT was previously assigned.

Crew et al. (1995) demonstrated that the t(X;18)(p11.2;q11.2) found in human synovial sarcoma results in the fusion of the chromosome 18 SSXT gene (which they symbolized SYT) to either of 2 distinct genes, SSX1 or SSX2 (600677), at Xp11.2. The SSX1 and SSX2 genes encode closely related proteins (81% identity) of 188 amino acids that are rich in charged amino acids. The N-terminal portion of each SSX protein exhibits homology to the Kruppel-associated box (KRAB), a transcriptional repressor domain previously found only in Kruppel-type zinc finger proteins, e.g., zinc finger protein-117 (ZNF117; 194624) and ZNF83 (194558). PCR analysis demonstrated the presence of SYT-SSX1 or SYT-SSX2 fusion transcripts in 29 of 32 synovial sarcomas examined, indicating that the detection of these hybrid transcripts by PCR may represent a useful diagnostic method. Sequence analysis demonstrated further heterogeneity in the fusion transcripts with the formation of 2 distinct SYT-SSX1 fusion junctions and 2 distinct SYT-SSX2 fusion junctions. Both the SYT-SSX1 and the SYT-SSX2 hybrid transcripts encode fusion proteins in which the C-terminal 8 amino acids of the normal SYT protein have been replaced by 78 amino acids encoded by an SSX gene. The SSX1 and SSX2 protein sequences present in these 2 fusion products have 66 of 78 amino acids in common.

Chand et al. (1995) found that at least 5 copies of the SSX1 locus are clustered on the X chromosome in a region that also contains several OAT-like sequences (311240). It remains to be seen how many of these are functional genes.

See also Cannizzaro et al. (1989); Gilgenkrantz et al. (1989)

Cannizzaro, L. A.; Dal Cin, P.; Mendez, M. J.; Sandberg, A. A.: Sublocalization of Xp breakpoint in the X;18 chromosome translocation of synovial sarcoma. (Abstract) Cytogenet. Cell Genet. 51: 974, 1989.

Chand, A.; Clark, J.; Cooper, C. S.; Craig, I. W.: Long-range organization of reiterated sequences, including the SSX1 cDNA, at the OATL1 cluster in Xp11.23. Genomics 30: 545-552, 1995.

Clark, J.; Rocques, P. J.; Crew, A. J.; Gill, S.; Shipley, J.; Chan, A. M.-L.; Gusterson, B. A.; Cooper, C. S.: Identification of novel genes, SYT and SSX, involved in the t(X;18)(p11.2;q11.2) translocation found in human synovial sarcoma. Nature Genet. 7: 502-508, 1994.

Crew, A. J.; Clark, J.; Fisher, C.; Gill, S.; Grimer, R.; Chand, A.; Shipley, J.; Gusterson, B. A.; Cooper, C. S.: Fusion of SYT to two genes, SSX1 and SSX2, encoding proteins with homology to the Kruppel-associated box in human synovial sarcoma. EMBO J. 14: 2333-2340, 1995.

Enzinger, F. M.; Weiss, S. W.: Soft Tissue Tumors. St. Louis: C. V. Mosby (pub.) 1983. Pp. 519-549.

Gilgenkrantz, S.; Mujica, P.; Chery, M.; Gregoire, M.-J.; Leotard, B.; Mandel, J.-L.: Mapping the breakpoint at Xp11.2 in synovial sarcoma. (Abstract) Cytogenet. Cell Genet. 51: 1004, 1989.

Griffin, C. A.; Emanuel, B. S.: Translocation (X;18) in a synovial sarcoma. (Letter) Cancer Genet. Cytogenet. 26: 181-183, 1987.

Karakousis, C. P.; Dal Cin, P.; Turc-Carel, C.; Limon, J.; Sandberg, A. A.: Chromosomal changes in soft-tissue sarcomas: a new diagnostic parameter. Arch. Surg. 122: 1257-1260, 1987.

Knight, J. C.; Martin, C. A.; Warren, W.; Smith, S.; Westbury, G.; Fisher, C.; Gusterson, B. A.; Cooper, C. S.; Reeves, B. R.: Mapping the translocation t(X;18)(p11.2;q11.2) in human synovial sarcoma. (Abstract) Cytogenet. Cell Genet. 51: 1023, 1989.

Knight, J. C.; Reeves, B. R.; Kearney, L.; Monaco, A. P.; Lehrach, H.; Cooper, C. S.: Localization of the synovial sarcoma t(X;18)(p11.2;q11.2) breakpoint by fluorescence in situ hybridization. Hum. Molec. Genet. 1: 633-637, 1992.

Miozzo, M.; Sozzi, G.; Cariani, C. T.; Di Palma, S.; Luksch, R.; Azzarelli, A.; Pierotti, M. A.; Della Porta, G.: A synovial sarcoma with t(X;18)(p11;q11) in a patient with Turner's syndrome. Genes Chromosomes Cancer 4: 192-193, 1992.

Sinke, R. J.; de Leeuw, B.; Janssen, H. A. P.; Olde Weghuis, D.; Suijkerbuijk, R. F.; Meloni, A. M.; Gilgenkrantz, S.; Berger, W.; Ropers, H. H.; Sandberg, A. A.; Geurts van Kessel, A.: Localization of X chromosome short arm markers relative to synovial sarcoma- and renal adenocarcinoma-associated translocation breakpoints. Hum. Genet. 92: 305-308, 1993.

Smith, S.; Reeves, B. R.; Wong, L.; Fisher, C.: A consistent chromosome translocation in synovial sarcoma. (Letter) Cancer Genet. Cytogenet. 26: 179-180, 1987.

Turc-Carel, C.; Dal Cin, P.; Limon, J.; Rao, U.; Li, F. P.; Corson, J. M.; Zimmerman, R.; Parry, D. M.; Cowan, J. M.; Sandberg, A. A.: Involvement of chromosome X in primary cytogenetic change in human neoplasia: nonrandom translocation in synovial sarcoma. Proc. Nat. Acad. Sci. 84: 1981-1985, 1987.

Ueda, T.; Aozasa, K.; Tsujimoto, M.; Uchida, A.; Taniwaki, M.; Abe, T.: Translocation X;18 and insertion 15;11 in a case of synovial sarcoma. Cancer Genet. Cytogenet. 30: 183-185, 1988.

Wang-Wuu, S.; Soukup, S. W.; Lange, B. J.: Another synovial sarcoma with t(X;18). Cancer Genet. Cytogenet. 29: 179-181, 1987.

SYNOVIAL, SARCOMA, X BREAKPOINT 2 (*600677)

JJM NOTE

A candidate gene to be studied further in synovial sarcomas.

See SSX1 (312820) for the discussion of 2 genes located in Xp11.2, SSX1 and SSX2, one or the other of which is fused with the SSXT gene (600192) in the translocation t(X;18) (Crew et al., 1995).

Crew, A. J.; Clark, J.; Fisher, C.; Gill, S.; Grimer, R.; Chand, A.; Shipley, J.; Gusterson, B. A.; Cooper, C. S.: Fusion of SYT to two genes, SSX1 and SSX2, encoding proteins with homology to the Kruppel-associated box in human synovial sarcoma. EMBO J. 14: 2333-2340, 1995.

SYNOVIAL, SARCOMA, TRANSLOCATED TO X CHROMOSOME (*600192)

JJM NOTE

This entity is included to complete the preceeding entry.

Human synovial sarcomas contain a recurrent and specific chromosomal translocation t(X;18)(p11.2;q11.2). By screening a synovial sarcoma cDNA library with a YAC spanning the X chromosome breakpoint, Clark et al. (1994) identified a hybrid transcript that contained 5-prime sequences mapping to chromosome 18 and 3-prime sequences mapping to the X chromosome (see SSX1; 312820). A probe from the chromosome 18 gene sequence (symbolized SSXT) detected genomic rearrangements in 10 of 13 synovial sarcomas. The chromosome 18 gene was symbolized SYT by Clark et al. (1994); since this symbol had previously been assigned to a separate gene, that for synaptotagmin (185605), the symbol SSXT was used by the HGM nomenclature committee. Sequencing of cDNA clones showed that the normal SSXT gene encodes a protein rich in glutamine, proline, and glycine, and that in synovial sarcoma, rearrangement of the SSXT gene results in the formation of a fusion protein. Both the chromosome 18 and the X chromosome components failed to exhibit significant homology to known gene sequences.

Crew et al. (1995) demonstrated that the t(X;18)(p11.2;q11.2) found in human synovial sarcoma results in the fusion of the chromosome 18 SSXT gene (which they symbolized SYT) to either of 2 distinct genes, SSX1 or SSX2 (600677), at Xp11.2. The SSX1 and SSX2 genes encode closely related proteins (81% identity) of 188 amino acids that are rich in charged amino acids. The N-terminal portion of each SSX protein exhibits homology to the Kruppel-associated box (KRAB), a transcriptional repressor domain previously found only in Kruppel-type zinc finger proteins, e.g., zinc finger protein-117 (ZNF117; 194624) and ZNF83 (194558). PCR analysis demonstrated the presence of SYT-SSX1 or SYT-SSX2 fusion transcripts in 29 of 32 synovial sarcomas examined, indicating that the detection of these hybrid transcripts by PCR may represent a useful diagnostic method. Sequence analysis demonstrated further heterogeneity in the fusion transcripts with the formation of 2 distinct SYT-SSX1 fusion junctions and 2 distinct SYT-SSX2 fusion junctions. Both the SYT-SSX1 and the SYT-SSX2 hybrid transcripts encode fusion proteins in which the C-terminal 8 amino acids of the normal SYT protein have been replaced by 78 amino acids encoded by an SSX gene. The SSX1 and SSX2 protein sequences present in these 2 fusion products have 66 of 78 amino acids in common.

The mouse homolog of SSXT was isolated and sequenced in full by de Bruijn et al. (1996) (who referred to the gene as SYT). It was found to have been conserved during evolution and to be part of a region of synteny between human and mouse chromosomes 18. In early embryogenesis, mouse Syt is ubiquitously expressed. In later stages, the expression becomes confined to cartilage tissues, specific neuronal cells, and some epithelium-derived tissues. In mature testis, expression was specifically observed in primary spermatocytes.

Clark, J.; Rocques, P. J.; Crew, A. J.; Gill, S.; Shipley, J.; Chan, A. M.-L.; Gusterson, B. A.; Cooper, C. S.: Identification of novel genes, SYT and SSX, involved in the t(X;18)(p11.2;q11.2) translocation found in human synovial sarcoma. Nature Genet. 7: 502-508, 1994.

Crew, A. J.; Clark, J.; Fisher, C.; Gill, S.; Grimer, R.; Chand, A.; Shipley, J.; Gusterson, B. A.; Cooper, C. S.: Fusion of SYT to two genes, SSX1 and SSX2, encoding proteins with homology to the Kruppel-associated box in human synovial sarcoma. EMBO J. 14: 2333-2340, 1995.

de Bruijn, D. R. H.; Baats, E.; Zechner, U.; de Leeuw, B.; Balemans, M.; Olde Weghuis, D.; Hirning-Folz, U.; Geurts van Kessel, A.: Isolation and characterization of the mouse homolog of SYT, a gene implicated in the development of human synovial sarcomas. Oncogene 13: 643-648, 1996.

DERMOID CYSTS, FAMILIAL FRONTONASAL (600679)

JJM NOTE

This fascinating syndrome is included not for frank malignancy but because of mass lesion.

Plewes and Jacobson (1971) described a father, his son, and 2 daughters who had frontonasal dermoid cysts. The clinical x-ray findings resembled those of a slow-growing intracranial space-occupying lesion. Plain x-ray films showed a widened nasal septum, and contrast radiology helped outline an avascular mass arising from the cribriform plate. Resection, utilizing a synchronous nasal and frontal operation, proved effective. The father had been seen at the age of 33 years with a 10-year history of epilepsy. A 'nasal cyst' had been removed some years previously, but no precise information was available on the nature of the lesion. The epileptiform attacks were nocturnal with hallucinations and postictal dysphasia. There was bilateral anosmia and early papilledema. Skull x-ray films showed translucency in the right frontal region together with erosion of the anterior clinoids. The eldest offspring, a 13-year-old boy, was first seen for complaints of recurrent episodes of nasal obstruction. A submucous resection of a septal swelling had been carried out 4 years previously when deviation of the septum to the right with a high left-sided cystic swelling was noted. X-ray films showed a widened nasal septum, maldevelopment of the right frontal sinuses, and incomplete fusion of the frontonasal process. The elder sister was seen at the age of 12 years with recent personality changes as the main reason for hospital referral. There was hypertelorism with a widened nasal bridge and small dermal sinuses just below the glabella, which had shown a white discharge intermittently throughout childhood. Apart from anosmia, no neurologic abnormality was detected. Widening of the nasal septum was easily visualized. Plain x-ray films and tomograms of the frontonasal region confirmed the presence of a wide nasal septum and showed a small defect in the midline frontally, as well as a large defect in the cribiform plate. Personality changes and memory deficits noted before surgery resolved, and anosmia remained as th e only deficit. The third sib was a 9-year-old girl who was asymptomatic at the time of the report by Plewes and Jacobson (1971), but examination showed an anosmia, widening of the nasal septum, and radiologic changes similar in most respects to those of her 2 sibs and father.

Samuel (1995) provided follow-up on the family reported by Plewes and Jacobson (1971).

Plewes, J. L.; Jacobson, I.: Familial frontonasal dermoid cysts: report of four cases. J. Neurosurg. 34: 683-686, 1971.

Samuel, D.: Personal Communication. Bethesda, Md. 6/29/1995.

EWING SARCOMA (*133450)

JJM NOTE

Two case reports document sib aggregation of Ewing sarcoma (Hutter RBP et al.: Ewing's sarcoma in siblings. Am. J. Surg. 107:598–603, 1964; Huntington RW et al.: Malignant bone tumors in siblings. J. Bone Jt. Surg. 42A:1065–75, 1960). A possible excess of urogenital anomalies was seen in a review of 154 NIH patients with Ewing sarcoma (McKeen EA et al.: Birth defects in Ewing's sarcoma. N. Engl. J. Med. 309:1522, 1983). A large case-control study of 208 patients revealed a possible association with umbilical and inguinal hernias, but in comparison only with randomly selected telephone controls in the same geographic region (not with sibling controls). Family history was likewise negative: the case families included no relatives with the tumor (Winn DM et al.: A case-control study of the etiology of Ewing's sarcoma. Cancer Epidemiol. Biomark. Preven. 1:525–32, 1992).

Although Ewing sarcoma and its cognate neoplasms have apparently not been reported in parent and child and there is no other evidence of heritability, the findings listed below indicate the existence of a determinant in a specific region on the long arm of chromosome 22. Thus, the listing here, with asterisk, is deemed warranted.

The sarcoma described by Ewing (1921) was considered by him to be a diffuse endothelioma of bone. It affects children and adolescents. In a study of 5 Ewing sarcoma cell lines established from 4 patients, Turc-Carel et al. (1984) found a consistent reciprocal translocation t(11;22)(q24;q12). In 4 patients, Aurias et al. (1984) studied fresh tumor cells derived by biopsy of primary or metastatic tumors. Abnormal karyotypes with translocations involving 22q12 were found in all. In 2 cases, t(11;22)(q24;q12) was found. Histologic differentiation of ES from several other childhood tumors is often difficult; the marker chromosome may be very useful to precise diagnosis.

Peripheral neuroepithelioma (peripheral neuroblastoma) is an uncommon malignant tumor of the peripheral nervous system with a histologic appearance similar to that of classic childhood neuroblastoma. In contrast to neuroblastoma, however, it typically presents in older children and young adults, can be associated with peripheral nerves, and spares the adrenal glands and sympathetic ganglia. In both tumor specimens and tumor cell lines from 2 patients with PNE, Whang-Peng et al. (1984) found a reciprocal 11;22 translocation indistinguishable from that reported in Ewing sarcoma (Turc-Carel et al., 1983). The findings support the hypothesis that these 2 neoplasms have a common histogenesis and that the malignant transformation leading to these 2 tumors has a common basis. In a case of secondary ES presenting 11 years after treatment of acute lymphoblastic leukemia in childhood, Tilly et al. (1984) found the same t(11;22)(q24;q12) rearrangement described in cases of primary ES. In somatic cell hybrids obtained by fusion of translocation (11;22)-positive Ewing sarcoma cells and Chinese hamster fibroblasts, Geurts van Kessel et al. (1985) showed that the SIS oncogene (190040) was translocated to chromosome 11 and that the lambda constant genes (147220) and the Philadelphia chromosome 'breakpoint cluster region' are not translocated. These results confirmed the cytogenetic observation that the Ewing sarcoma-associated breakpoint in chromosome 22 is distal to that in translocation (8;22)-positive Burkitt lymphoma and that in translocation (9;22)-positive chronic myeloid leukemia.

Askin tumor, a malignancy of the thoracopulmonary region, also has been found to have t(11;22) (Whang-Peng et al., 1986) and is likely a neuroepithelioma of the chest wall (Triche and Askin, 1983). Balanced reciprocal translocation t(11;22)(q23;q11) is a relatively frequent constitutional finding. Indeed, Budarf et al. (1989) stated that it is the most common non-Robertsonian constitutional translocation in humans. Among 13 cases, Douglass et al. (1986) found that 9 had t(11;22) and that 2 additional cases had only a deleted chromosome 22. Griffin et al. (1986) could distinguish the cytologically indistinguishable tumor-related t(11;22) by doing in situ hybridization with probes for the constant region of the lambda light chain located at 22q11 and the ETS1 oncogene (164720) located at 11q23.3-q24. In a study of 23 cases, Turc-Carel et al. (1987) found that all had reciprocal translocations involving

11q24 and 22q12. Mugneret et al. (1987) found a consistent secondary chromosomal change in 4 of 23 Ewing sarcomas. The translocation involves chromosomes 1 and 16. The derivative 16 was in place of 1 of the normal chromosomes 16. Nonrandom secondary changes are rare in hematologic malignancies. This was presumably the first observation of such a change in a solid tumor.

Esthesioneuroblastoma is an exceedingly rare malignancy that originates in the neuroectodermal stem cells of the olfactory epithelium. In a malignant lesion from 1 such patient, Whang-Peng et al. (1987) found a reciprocal translocation, t(11;22)(q24;q12), indistinguishable from the one that has been reported in Ewing sarcoma and Askin tumor. Turc-Carel et al. (1988) corroborated the remarkable consistency of t(11;22)(q24;q12) in all but 1 of their own 31 cases and in the majority of 54 cases reported by other investigators. The translocation mentioned was present in 83% of the total series. In 4% of cases, variant translocations involved 22q12, but with a chromosome other than number 11. In summary, the breakpoint on chromosome 22q12 was found in 92% of cases, whereas the breakpoint at 11q24 was observed in 88% of cases. Among 43 cases of Ewing sarcoma analyzed for numerical changes, trisomy 8 was observed consistently in half; an unbalanced t(1;16) was found in 18% of 82 cases analyzed for structural changes (Mugneret et al., 1988). Budarf et al. (1989) used DNA markers to distinguish the location of the breakpoints in the 11;22 translocation of Ewing sarcoma from those in the frequent 'normal' constitutional translocation. By both in situ hybridization and somatic cell hybrids, Zhang et al. (1990) showed that the recurrent translocation breakpoint on chromosome 22 of neuroepithelioma lies between 2 probes, D22S1 and D22S15. The 2 probes were genetically linked with a lod score of 5.3 at theta = 0.0. Furthermore, they were unaffected by a partial deletion of 22q occurring with meningioma (156100), showing that the meningioma locus is distal to that of neuroepithelioma. Selleri et al. (1991) demonstrated that the gene encoding the leukemia inhibitory factor (LIF; 159540) is translocated to the ES and PNE derivative chromosome 11 in the immediate vicinity of the most centromeric marker flanking the breakpoint on chromosome 11. The physical distance between flanki ng cosmid markers on chromosome 11 was determined to be approximately 1 megabase. Chromosomal in situ suppression hybridization was used to localize the ES and PNE breakpoints between 2 closely spaced DNA markers. The observations confirmed the localization of the breakpoint and of the LIF gene to 22q12. Since LIF has been shown to suppress in vitro proliferation of myeloid leukemia cell lines and to prevent differentiation of embryonic cells in culture, a chromosomal translocation in the vicinity of this gene might induce oncogenesis. Pulsed field gel electrophoresis demonstrated no abnormalities in a 650-kb region surrounding the LIF locus, however. The ES/PNE breakpoint is located telomeric to NCAM (116930), THY1 (188230), CD3 (186790), APOA1 (107680), and ETS1. The breakpoints in the t(4;11), t(9;11), and t(11;19) translocations occurring in leukemias mapped to more centromeric sites on 11q.

Phylogenetically conserved restriction fragments in the vicinity of the breakpoints on chromosome 22 and chromosome 11 allowed identification of transcribed sequences from these regions and indicated that a hybrid transcript might be generated by the t(11;22)(q24;q12) translocation. Delattre et al. (1992) used these conserved fragments to screen cDNA libraries to show that the translocation alters the open reading frame of an expressed gene on chromosome 22 by substituting a sequence encoding a putative RNA-binding domain for that of the DNA-binding domain of the human homolog of murine Fli-1. The protein encoded by chromosome 22, termed EWS, had 656 amino acids. The fragment from chromosome 11 predicted an amino acid sequence with 97% identity with the mouse Fli-1 gene (Ben-David et al., 1991). The retroviral insertion site activating murine Fli-1 resides on mouse chromosome 9 in a region with homology of synteny to human chromosome 11 in the proximity of the ETS1 gene. May et al. (1993) showed that the 11;22 translocation of Ewing sarcoma produces a chimeric transcription factor that requires the DNA-binding domain encoded by FLI1 (193067) for transformation.

Just as the fusion of the 5-prime portion of the EWS gene to the FLI1 gene results in Ewing sarcoma, fusion of the same part of the EWS gene to the ATF1 gene (123803) results in malignant melanoma of soft parts (Zucman et al., 1993).

Plougastel et al. (1993) demonstrated that the EWS gene spans about 40 kb of DNA and has 17 exons. The first 7 exons encode the N-terminal domain, which consists of a repeated degenerated polypeptide of 7 to 12 residues rich in tyrosine, serine, threonine, glycine, and glutamine. Exons 11, 12, and 13 encode the putative RNA binding domain. Of 19 Ewing tumors, the chromosome 22 breakpoint was localized in intron 7 or 8 in 18 cases and in intron 10 in 1 case. Plougastel et al. (1994) found that the cDNA sequence of the mouse Ews gene demonstrates 93% nucleotide homology with the human cDNA and that the translation product shares 98% homology with the human EWS protein. They found that the murine Ews locus lies within a segment of chromosome 11 that shows conserved synteny with human 22q12.

Desmoplastic small round cell tumor (DSRCT) is associated with a recurrent chromosomal translocation, t(11;22)(p13;q12). DSRCT is characterized by a predilection for young males, abdominal serosal involvement, poor prognosis, and a primitive histologic appearance. Typically there is a prominent desmoplastic stromal reaction and simultaneous multi-immunophenotypic differentiation with neoplastic cells expressing proteins associated with epithelial, muscular, and neural differentiation. Because the chromosome translocation breakpoints are located cytogenetically in the regions of 2 tumor-associated genes, WT1 (194070) on chromosome 11 and EWS on chromosome 22, Gerald et al. (1995) investigated these genes as potential translocation partners in DSRCT and showed that they are consistently rearranged in genomic DNA isolated from tumor tissue. The breakpoints involved the intron between EWS exons 7 and 8 and the intron between WT1 exons 7 and 8. Chimeric transcripts corresponding to the fusion gene were detected in 4 of 6 cases studied. Analyses of these transcripts showed an in-frame fusion of RNA encoding the N-terminal domain of EWS to both alternatively spliced forms of the last 3 zinc fingers of the DNA-binding domain of WT1. Thus, DSRCT represents the third tumor type associated with translocation of EWS; it is the first tumor associated with a consistent translocation involving WT1. The chimeric products were predicted to modulate transcription at WT1 target sites and contribute to the development of this unique tumor.

Ohno et al. (1994) demonstrated differential splicing of the EWS protein involving 2 exons encoding 72 amino acids. Both alternatively spliced transcripts are expressed in a variety of cells. One of the forms binds to RNA in vitro, and specifically to poly-G and poly-U. The RNA-binding activity was localized to the carboxy-terminal 86 amino acids, which constitute an RGG box. Thus, the N-terminal domain of EWS, which is involved in chromosome translocation, may regulate the specificity of RNA-binding activity of EWS. By mutational analysis of the EWS-ERG chimeric protein, Ohno et al. (1994) found that the N-terminal EWS functions as a regulatory domain for the transcriptional activation properties of the chimeric protein.

Aman et al. (1996) reported that the exon/intron structures of FUS (137070) and EWS show extensive similarities in the RNP regions and suggested that the 2 genes may have been derived from a common ancestor. They noted that although the N-terminal ends of FUS and EWS are different, they share extensive homology and they are distinct from the N-terminal regions of other RNP-carrying proteins. They proposed that FUS and EWS may be regarded as the first members of a new family of RNA binding proteins. Aman et al. (1996) stressed that the nature of the promoter regions of FUS and EWS is important for understanding the control of the fusion genes involving FUS and EWS in tumors. All of the tumor-specific translocations lead to the formation of fusion genes with FUS or EWS promoter regions and 5-prime coding regions linked to transcription factor genes. Aman et al. (1996) concluded that the transcriptional control of the fusion genes is most likely dominated by the FUS and EWS promoters.

In a case of myxoid chondrosarcoma, Gill et al. (1995) demonstrated that a translocation t(9;22)(q22-31;q11-12) led to fusion of the EWS gene with a DNA segment from 9q22-q31. This particular t(9;22) is a recurrent cytogenetic finding in the myxoid variant of chondrosarcoma, which is the fourth type of human sarcoma in which the EWS gene was shown to be involved. The other 3 are: fusion with ERG (165080) on chromosome 21 in some cases of Ewing sarcoma; fusion with FLI1 (193067) on chromosome 11 in a larger proportion of cases of Ewing sarcoma; and fusion with ATF1 (123803) on chromosome 12 in cases of malignant melanoma of soft parts, MMSP (also known as soft tissue clear cell sarcoma). The fusion of EWS with the WT1 gene in desmoplastic small round cell tumors is another example.

Jeon et al. (1995) identified a variant type of Ewing sarcoma translocation, t(7;22)(p22;q12), that fused EWS to the human homolog of the murine ETS gene ER81. This gene, designated ETV1 (for ETS translocation variant; 600541), was located on 7p22 as indicated by the site of the breakpoints in the translocation, which was confirmed by fluorescence in situ hybridization.

Using a competitive PCR technique, Tanaka et al. (1997) showed that there might be a correlation between the expression levels of the EWS-FLI1 fusion gene and the proliferative activity of Ewing sarcoma and primitive neuroectodermal tumor cells. Furthermore, when the EWS-FLI1 expression was inhibited by antisense oligodeoxynucleotides against the fusion RNA, the growth of tumor cells was significantly reduced both in vitro and in vivo. Their data further indicated that the growth inhibition of the cells by the antisense sequence might be mediated by G0/G1 block in the cell cycle progression. Burchill et al. (1997) used RT-PCR for evaluation of EWS-FLI1 fusion transcripts in 18 neurally derived small round cell tumors. These included 6 tumors of the Ewing family and 12 neuroblastomas. EWS-FLI1 fusion transcripts were identified in all 6 Ewing tumors, but also in 2 of the 12 neuroblastomas.

See also Maletz et al. (1986)

Aman, P.; Panagopoulos, I.; Lassen, C.; Fioretos, T.; Mencinger, M.; Toresson, H.; Hoglund, M.; Forster, A.; Rabbitts, T. H.; Ron, D.; Mandahl, N.; Mitelman, F.: Expression patterns of the human sarcoma-associated genes FUS and EWS and the genomic structure of FUS. Genomics 37: 1-8, 1996.

Aurias, A.; Rimbaut, C.; Buffe, D.; Zucker, J.-M.; Mazabraud, A.: Translocation involving chromosome 22 in Ewing's sarcoma: a cytogenetic study of four fresh tumors. Cancer Genet. Cytogenet. 12: 21-25, 1984.

Ben-David, Y.; Giddens, E. B.; Letwin, K.; Bernstein, A.: Erythroleukemia induction by Friend murine leukemia virus: insertional activation of a new member of the ets gene family, Fli-1, closely linked to c-ets-1. Genes Dev. 5: 908-918, 1991.

Budarf, M.; Sellinger, B.; Griffin, C.; Emanuel, B. S.: Comparative mapping of the constitutional and tumor-associated 11;22 translocations. Am. J. Hum. Genet. 45: 128-139, 1989.

Burchill, S. A.; Wheeldon, J.; Cullinane, C.; Lewis, I. J.: EWS-FLI1 fusion transcripts identified in patients with typical neuroblastoma. Europ. J. Cancer 33: 239-243, 1997.

Delattre, O.; Zucman, J.; Plougastel, B.; Desmaze, C.; Melot, T.; Peter, M.; Kovar, H.; Joubert, I.; de Jong, P.; Rouleau, G.; Aurias, A.; Thomas, G.: Gene fusion with an ETS DNA-binding domain caused by chromosome translocation in human tumours. Nature 359: 162-165, 1992.

Douglass, E. C.; Valentine, M.; Green, A. A.; Hayes, F. A.; Thompson, E. I.: t(11;22) and other chromosomal rearrangements in Ewing's sarcoma. J. Nat. Cancer Inst. 77: 1211-1213, 1986.

Ewing, J.: Diffuse endothelioma of bone. Proc. N. Y. Path. Soc. 21: 17-24, 1921.

Gerald, W. L.; Rosai, J.; Ladanyi, M.: Characterization of the genomic breakpoint and chimeric transcripts in the EWS-WT1 gene fusion of desmoplastic small round cell tumor. Proc. Nat. Acad. Sci. 92: 1028-1032, 1995.

Geurts van Kessel, A.; Turc-Carel, C.; de Klein, A.; Grosveld, G.; Lenoir, G.; Bootsma, D.: Translocation of oncogene c-sis from chromosome 22 to chromosome 11 in a Ewing sarcoma-derived cell line. Molec. Cell. Biol. 5: 427-429, 1985.

Gill, S.; McManus, A. P.; Crew, A. J.; Benjamin, H.; Sheer, D.; Gusterson, B. A.; Pinkerton, C. R.; Patel, K.; Cooper, C. S.; Shipley, J. M.: Fusion of the EWS gene to a DNA segment from 9q22-31 in a human myxoid chondrosarcoma. Genes Chromosomes Cancer 12: 307-310, 1995.

Griffin, C. A.; McKeon, C.; Israel, M. A.; Gegonne, A.; Ghysdael, J.; Stehelin, D.; Douglass, E. C.; Green, A. A.; Emanuel, B. S.: Comparison of constitutional and tumor-associated 11;22 translocations: nonidentical breakpoints on chromosomes 11 and 22. Proc. Nat. Acad. Sci. 83: 6122-6126, 1986.

Jeon, I.-S.; Davis, J. N.; Braun, B. S.; Sublett, J. E.; Roussel, M. F.; Denny, C. T.; Shapiro, D. N.: A variant Ewing's sarcoma translocation (7;22) fuses the EWS gene to the ETS gene ETV1. Oncogene 10: 1229-1234, 1995.

Maletz, N.; McMorrow, L. E.; Greco, M. A.; Wolman, S. R.: Ewing's sarcoma: pathology, tissue culture, and cytogenetics. Cancer 58: 252-257, 1986.

May, W. A.; Gishizky, M. L.; Lessnick, S. L.; Lunsford, L. B.; Lewis, B. C.; Delattre, O.; Zucman, J.; Thomas, G.; Denny, C. T.: Ewing sarcoma 11;22 translocation produces a chimeric transcription factor that requires the DNA-binding domain encoded by FLI1 for transformation. Proc. Nat. Acad. Sci. 90: 5752-5756, 1993.

Mugneret, F.; Aurias, A.; Lizard, S.; Turc-Carel, C.: Der(16)t(1;16)(q11;q11.1) is a consistent secondary chromosome change in Ewing's sarcoma. (Abstract) Cytogenet. Cell Genet. 46: 665 only, 1987.

Mugneret, F.; Lizard, S.; Aurias, A.; Turc-Carel, C.: Chromosomes in Ewing's sarcoma. II. Nonrandom additional changes, trisomy 8 and der(16)t(1;16). Cancer Genet. Cytogenet. 32: 239-245, 1988.

Ohno, T.; Ouchida, M.; Lee, L.; Gatalica, Z.; Rao, V. N.; Reddy, E. S. P.: The EWS gene, involved in Ewing family of tumors, malignant melanoma of soft parts and desmoplastic small round cell tumors, codes for an RNA binding protein with novel regulatory domains. Oncogene 9: 3087-3097, 1994.

Plougastel, B.; Mattei, M.-G.; Thomas, G.; Delattre, O.: Cloning and chromosome localization of the mouse Ews gene. Genomics 23: 278-281, 1994.

Plougastel, B.; Zucman, J.; Peter, M.; Thomas, G.; Delattre, O.: Genomic structure of the EWS gene and its relationship to EWSR1, a site of tumor-associated chromosome translocation. Genomics 18: 609-615, 1993.

Selleri, L.; Hermanson, G. G.; Eubanks, J. H.; Lewis, K. A.; Evans, G. A.: Molecular localization of the t(11;22)(q24;q12) translocation of Ewing sarcoma by chromosomal in situ suppression hybridization. Proc. Nat. Acad. Sci. 88: 887-891, 1991.

Tanaka, K.; Iwakuma, T.; Harimaya, K.; Sato, H.; Iwamoto, Y.: EWS-Fli1 antisense oligodeoxynucleotide inhibits proliferation of human Ewing's sarcoma and primitive neuroectodermal tumor cells. J. Clin. Invest. 99: 239-247, 1997.

Tilly, H.; Bastard, C.; Chevallier, B.; Halkin, E.; Monconduit, M.: Chromosomal abnormalities in secondary Ewing's sarcoma. (Letter) Lancet II: 812 only, 1984.

Triche, T. J.; Askin, F. B.: Neuroblastoma and the differential diagnosis of small-, round-, blue-cell tumors. Hum. Path. 14: 569-595, 1983.

Turc-Carel, C.; Aurias, A.; Mugneret, F.; Lizard, S.; Sidaner, I.; Philip, I.; Philip, T.; Lenoir, G. M.; Mazabraud, A.: Outstanding consistency of the chromosomal breakpoints 11q24 and 22q12 in Ewing's sarcoma. (Abstract) Cytogenet. Cell Genet. 46: 706 only, 1987.

Turc-Carel, C.; Aurias, A.; Mugneret, F.; Lizard, S.; Sidaner, I.; Volk, C.; Thiery, J. P.; Olschwang, S.; Philip, I.; Berger, M. P.; Philip, T.; Lenoir, G. M.; Mazabraud, A.: Chromosomes in Ewing's sarcoma. I. An evaluation of 85 cases and remarkable consistency of t(11;22)(q24;q12). Cancer Genet. Cytogenet. 32: 229-238, 1988.

Turc-Carel, C.; Philip, I.; Berger, M.-P.; Philip, T.; Lenoir, G. M.: Chromosomal translocations in Ewing's sarcoma. New Eng. J. Med. 309: 497-498, 1983.

Turc-Carel, C.; Philip, I.; Berger, M.-P.; Philip, T.; Lenoir, G. M.: Chromosome study of Ewing's sarcoma (ES) cell lines: consistency of a reciprocal translocation t(11;22)(q24;q12). Cancer Genet. Cytogenet. 12: 1-19, 1984.

Whang-Peng, J.; Freter, C. E.; Knutsen, T.; Nanfro, J. J.; Gazdar, A.: Translocation t(11;22) in esthesioneuroblastoma. Cancer Genet. Cytogenet. 29: 155-157, 1987.

Whang-Peng, J.; Triche, T. J.; Knutsen, T.; Miser, J.; Douglass, E. C.; Israel, M. A.: Chromosome translocation in peripheral neuroepithelioma. New Eng. J. Med. 311: 584-585, 1984.

Whang-Peng, J.; Triche, T. J.; Knutsen, T.; Miser, J.; Kao-Shan, S.; Tsai, S.; Israel, M. A.: Cytogenetic characterization of selected small round cell tumors of childhood. Cancer Genet. Cytogenet. 21: 185-208, 1986.

Zhang, F. R.; Delattre, O.; Rouleau, G.; Couturier, J.; Lefrancois, D.; Thomas, G.; Aurias, A.: The neuroepithelioma breakpoint on chromosome 22 is proximal to the meningioma locus. Genomics 6: 174-177, 1990.

Zucman, J.; Delattre, O.; Desmaze, C.; Epstein, A. L.; Stenman, G.; Speleman, F.; Fletchers, C. D. M.; Aurias, A.; Thomas, G.: EWS and ATF-1 gene fusion induced by t(12;22) translocation in malignant melanoma of soft parts. Nature Genet. 4: 341-345, 1993.

CHONDROSARCOMA, MYXOID EXTRASKELETAL, FUSED TO EWS (*600542)

JJM NOTE

Once again, a rare tumor studied cytogenetically contributes to understanding various genes and tumors.

A recurrent translocation t(9;22)(q22-31;q11-12) was observed in extraskeletal myxoid chondrosarcomas by Hinrichs et al. (1985), Turc-Carel et al. (1988), and Orndal et al. (1991). Gill et al. (1995) demonstrated that the EWS gene on chromosome 22 (133450) is rearranged in such cases and defines the segment on chromosome 9 that is fused to the amino-terminal region of the EWS gene as a result of the reciprocal translocation.

EMCs are soft tissue tumors of chondroblastic origin that occur primarily in muscle. Despite their relatively low-grade malignancy, recurrence and metastasis may appear many years after the initial diagnosis. Histologic tissue section examination reveals a nodular pattern and the presence of mucoid material. EMC patients are generally older than 35 years; however, a few cases have been described in children and adolescents. Labelle et al. (1995) noted that the reports of recurrent t(9;22)(q22;q12) translocation in EMC (Orndal et al., 1991; Stenman et al., 1995) are consistent with the involvement of the Ewing sarcoma gene (133450) located on 22q. In the Ewing family of tumors, EWS can be fused with 3 different ETS transcription factors: FLI1 (193067) on 11q, ERG (165080) on 21q, and ETV1 (600541) on 7p. In malignant melanoma of soft parts, EWS is fused to ATF1 (123803) on 12p and in desmoplasmic small round cell tumors. It is fused to the Wilms tumor gene (194070) on 11p. Although the exact function of the intact EWS protein remains unknown, it possesses a conserved RNA-recognition motif and can bind RNA in vitro, suggesting that it may be involved in RNA metabolism. In all known EWS fusion proteins, the RNA-recognition motif of EWS is replaced by the DNA-binding domain of the corresponding transcription factor.

Labelle et al. (1995) demonstrated by fluorescence in situ hybridization that in 1 EMC tumor the chromosome 22 breakpoint occurred in the EWS gene. Northern blot analysis revealed an aberrant transcript that was cloned by a modified RT-PCR procedure. This transcript consisted of an in-frame fusion of the 5-prime end of EWS to a previously unidentified gene that was named TEC. This fusion transcript was detected in 6 of 8 EMCs studied, and 3 different junction types between the 2 genes were found. EWS linked to the entire TEC protein. Homology analysis showed that the predicted TEC protein contains a DNA-binding domain characteristic of nuclear receptors. The highest identity scores were observed with the NURR1 family of orphan nuclear receptors (see 601828). (This situation is reminiscent of the FUS/CHOP fusion protein in which the entire CHOP (126337) protein is linked to the amino-terminal domain of FUS (137070) by an additional 26 amino acid sequence.) Labelle et al. (1995) stated that the EWS/TEC gene fusion is the second example of the oncogenic conversion of a nuclear receptor in human tumorigenesis, the first being the PML/RARA gene fusion generated by the t(15;17) translocation in acute promyelocytic leukemia. In all, EWS has been shown to be rearranged with 6 different partners, each located on a different chromosomal segment, and each resulting in a different type of tumor. The only example of a gene having more fusion partners generated by chromosomal translocation is the MLL gene (159555), which can be fused to at least 7 different partners in a variety of childhood acute leukemias.

Gill et al. (1995) described the fusion of the EWS gene with the 9q gene in a patient with skeletal myxoid chondrosarcoma. Furthermore,

Tarkkanen et al. (1994) observed it in the dedifferentiated chondrosarcoma with no evidence of myxoid component.

The CSMF gene appears to be the human homolog of the rat gene NOR1, which had been identified as a sequence overexpressed in rat brain cells undergoing apoptosis (Clark et al., 1996). It is, furthermore, a member of the steroid/thyroid receptor gene superfamily. Clark et al. (1996) (who symbolized the CSMF gene as CHN) found that the chimeric EWS-CHN gene encodes an EWS-CHN fusion protein in which the C-terminal RNA-binding domain of EWS is replaced by the entire CHN protein, comprising a long N-terminal domain, a central DNA binding domain, and a C-terminal ligand-binding/dimerization domain.

Nomenclature: It is to be noted that the designation TEC was assigned to a quite distinct entity (600583) and this gene, involved in the fusion protein in EMC, had been previously designated CSMF. Clark et al. (1996) symbolized the gene as CHN. This designation should not be confused with that for chimerin (118423), which is located on chromosome 2q31.

Clark, J.; Benjamin, H.; Gill, S.; Sidhar, S.; Goodwin, G.; Crew, J.; Gusterson, B. A.; Shipley, J.; Cooper, C. S.: Fusion of the EWS gene to CHN, a member of the steroid/thyroid receptor gene superfamily, in a human myxoid chondrosarcoma. Oncogene 12: 229-235, 1996.

Gill, S.; McManus, A. P.; Crew, A. J.; Benjamin, H.; Sheer, D.; Gusterson, B. A.; Pinkerton, C. R.; Patel, K.; Cooper, C. S.; Shipley, J. M.: Fusion of the EWS gene to a DNA segment from 9q22-31 in a human myxoid chondrosarcoma. Genes Chromosomes Cancer 12: 307-310, 1995.

Hinrichs, S. H.; Jaramillo, M. A.; Gumerlock, P. H.; Gardner, M. B.; Lewis, J. P.; Freeman, A. E.: Myxoid chondrosarcoma with a translocation involving chromosomes 9 and 22. Cancer Genet. Cytogenet. 14: 219-226, 1985.

Labelle, Y.; Zucman, J.; Stenman, G.; Kindblom, L.-G.; Knight, J.; Turc-Carel, C.; Dockhorn-Dworniczak, B.; Mandahl, N.; Desmaze, C.; Peter, M.; Aurias, A.; Delattre, O.; Thomas, G.: Oncogenic conversion of a novel orphan nuclear receptor by chromosome translocation. Hum. Molec. Genet. 4: 2219-2226, 1995.

Orndal, C.; Carlen, B.; Akerman, M.; Willen, H.; Mandahl, N.; Heim, S.; Rydholm, A.; Mitelman, F.: Chromosomal abnormality t(9;22)(q22;q12) in an extraskeletal myxoid chondrosarcoma characterized by fine needle aspiration cytology, electron microscopy, immunohistochemistry and DNA flow cytometry. Cytopathology 2: 261-270, 1991.

Stenman, G.; Andersson, H.; Mandahl, N.; Meis-Kindblom, J. M.; Kindblom, L.-G.: Translocation t(9;22)(q22;q12) is a primary cytogenetic abnormality in extraskeletal myxoid chondrosarcoma. Int. J. Cancer 62: 398-402, 1995.

Tarkkanen, M.; Wiklund, T.; Virolainen, M.; Elomaa, I.; Knuutila, S.: Dedifferentiated chondrosarcoma with t(9;22)(q34;q11-12). Genes Chromosomes Cancer 9: 136-140, 1994.

Turc-Carel, C.; Dal Cin, P.; Rao, U.; Karakousis, C.; Sandberg, A. A.: Recurrent breakpoints at 9q31 and 22q12.2 in extraskeletal myxoid chondrosarcoma. Cancer Genet. Cytogenet. 30: 145-150, 1988.

ROTHMUND-THOMSON SYNDROME (*268400)

JJM NOTE

Heuristically, this disorder is usually considered a syndrome of premature aging. It is included with the mesenchymal tumors because of the apparent excess of sarcomas. An excess of skin cancers also seems clear, and the tumor predisposition might include rare chemoreceptor tumors. My colleagues and I noted a 37-year-old female with Rothmund-Thomson syndrome and multiple glomus body tumors that caused cranial nerve palsies (Dinulos MB, Mulvihill JJ: Bilateral glomus body tumors in a patient with Rothmund-Thomson syndrome. Am. J. Hum. Genet. 51:A274, 1992).

This is a hereditary dermatosis characterized by atrophy, pigmentation, and telangiectasia and frequently accompanied by juvenile cataract, saddle nose, congenital bone defects, disturbances of hair growth, and hypogonadism. Prognosis for survival is fairly good. The family described by Rothmund (1868) was further investigated by Siemens (cited by Waardenburg et al., 1961). It is possible that the condition described by Thomson (1936) is a different recessive disorder from that described by Rothmund. Saddle nose was not present and cataract did not occur. An excellent color illustration was provided by Thomson (1936). Starr et al. (1985) reported 2 new cases and emphasized the less well-known nondermatologic features, namely, hypodontia, soft tissue contractures, proportionate short stature, hypogonadism, anemia, and osteogenic sarcoma. Birth weight in the 2 cases was 3 kg and 2.83 kg. Drouin et al. (1993) described osteosarcoma of the distal femoral metaphysis in an 11-year-old French-Canadian boy with RTS. Several such cases had previously been reported.

Ying et al. (1990) described trisomy 8 mosaicism in association with Rothmund-Thomson syndrome. Der Kaloustian et al. (1990) found mosaic trisomy 8 and mosaic supernumerary i(2q) in fibroblasts from normal skin; lymphocytes, however, had a normal karyotype. Orstavik et al. (1994) likewise found instability of lymphocyte chromosomes in a young girl who had severe skeletal abnormalities of the upper limbs with absence of both radii, short dysmorphic ulnae, a rudimentary right thumb, and aplasia of the left thumb. She also had anal atresia with a rectovaginal fistula. Poikiloderma developed on the face and extensor surfaces of the limbs beginning at the age of 3 months. Mental development was normal. Lindor et al. (1996) reported a brother and sister with typical clinical manifestations of RTS and tibial sarcoma. Three cell lines [46,XY; 47,XY,+8; 47,XY,+i(8q)] were found in lymphocyte cultures of the brother and 2 cell lines [46,XX; 47,XX,+i(8q)] were detected in the sister. FISH studies with a probe specific for the chromosome 8 centromere showed 3 chromosome 8 signals in 4 to 16% of uncultivated lymphocytes and buccal cells of these patients, suggesting the in vivo presence of abnormal cell lines. RTS may be associated with clonal rearrangements causing acquired somatic mosaicism.

See 270240 for a skeletal dysplasia with telangiectases and mesodermal dysgenesis of the iris thought to represent Rothmund-Thomson syndrome.

See also Blinstrub et al. (1964); Block and Stauffer (1929); Cole et al. (1945); Dechenne et al. (1983); Franceschetti (1953); Hall et al. (1980); Kraus et al. (1970); Sexton (1954); Taylor (1957)

Blinstrub, R. S.; Lehman, R.; Steinberg, T. H.: Poikiloderma congenitale. Report of two cases. Arch. Derm. 89: 659-664, 1964.

Block, B.; Stauffer, H.: Skin diseases of endocrine system (dyshormonal dermatoses). Poikiloderma-like changes in connection with under development of the sexual glands and dystrophia adiposogenitalis. Arch. Derm. Syph. 19: 22-34, 1929.

Cole, H. N.; Giffen, H. K.; Simmons, J. T.; Stroud, G. M., III : Congenital cataracts in sisters with congenital ectodermal dysplasia. J.A.M.A. 129: 723-728, 1945.

Dechenne, C.; Chantraine, J. M.; Davin, J. C.: A Rothmund-Thomson case with hypertension. Clin. Genet. 24: 266-272, 1983.

Der Kaloustian, V. M.; McGill, J. J.; Vekemans, M.; Kopelman, H. R.: Clonal lines of aneuploid cells in Rothmund-Thomson syndrome. Am. J. Med. Genet. 37: 336-339, 1990.

Drouin, C. A.; Mongrain, E.; Sasseville, D.; Bouchard, H.-L.; Drouin, M.: Rothmund-Thomson syndrome with osteosarcoma. J. Am. Acad. Derm. 28: 301-305, 1993.

Franceschetti, A.: Les dysplasies ectodermiques et les syndromes hereditaires apparentes. Dermatologica 106: 129-156, 1953.

Hall, J. C.; Pagon, R. A.; Wilson, K. M.: Rothmund-Thomson syndrome with severe dwarfism. Am. J. Dis. Child. 134: 165-169, 1980.

Kraus, B. S.; Gottlieb, M. A.; Meliton, H. R.: The dentition in Rothmund's syndrome. J. Am. Dent. Assoc. 81: 894-915, 1970.

Lindor, N. M.; Devries, E. M. G.; Michels, V. V.; Schad, C. R.; Jalal, S. M.; Donovan, K. M.; Smithson, W. A.; Kvols, L. K.; Thibodeau, S. N.; Dewald, G. W.: Rothmund-Thomson syndrome in siblings: evidence for acquired in vivo mosaicism. Clin. Genet. 49: 124-129, 1996.

Orstavik, K. H.; McFadden, N.; Hagelsteen, J.; Ormerod, E.; van der Hagen, C. B.: Instability of lymphocyte chromosomes in a girl with Rothmund-Thomson syndrome. J. Med. Genet. 31: 570-572, 1994.

Rothmund, A.: Ueber Cataracte in Verbindung mit einer eigenthuemlichen Hautdegeneration. Albrecht von Graefes Arch. Klin. Exp. Ophthal. 14: 159-182, 1868.

Sexton, G. B.: Thomson's syndrome (poikiloderma congenitale). Canad. Med. Assoc. J. 70: 662-665, 1954.

Starr, D. G.; McClure, J. P.; Connor, J. M.: Non-dermatological complications and genetic aspects of the Rothmund-Thomson syndrome. Clin. Genet. 27: 102-104, 1985.

Taylor, W. B.: Rothmund's syndrome—Thomson's syndrome. Arch. Derm. 75: 236-244, 1957.

Thomson, M. S.: Poikiloderma congenitale. Brit. J. Derm. 48: 221-234, 1936.

Waardenburg, P. J.; Franceschetti, A.; Klein, D.: Genetics and Ophthalmology. Springfield, Ill.: Charles C Thomas (pub.) 1: 1961. Pp. 896 only.

Ying, K. L.; Oizumi, J.; Curry, C. J. R.: Rothmund-Thomson syndrome associated with trisomy 8 mosaicism. J. Med. Genet. 27: 258-260, 1990.

WERNER SYNDROME (*277700)

JJM NOTE

The neoplasms associated with Werner syndrome have not been critically tallied since the early 1980s (Salk 1982; Usui M et al.: The occurrence of soft tissue sarcomas in three siblings with Werner's syndrome. Cancer 54:2580–86, 1984). Of some 300 patients known by then, 58 had a malignant neoplasm and at least 23 had a benign tumor, including 9 with meningioma. At least 18 sarcomas occurred, in addition to 4 leukemias and 5 cutaneous malignant melanomas, mostly of the foot. There is a paucity of "typical" cancers, associated with old age in humans, despite the fact that Werner syndrome is considered a model of aging. A patient with scleroderma and cancer should be suspected of having Werner syndrome.

The features of Werner syndrome are scleroderma-like skin changes, especially in the extremities, cataract, subcutaneous calcification, premature arteriosclerosis, diabetes mellitus, and a wizened and prematurely aged facies. A particularly instructive pedigree was reported by McKusick et al. (1963). The habitus is characteristic, with short stature, slender limbs, and stocky trunk. The nose is beaked. Epstein et al. (1966) studied a Japanese patient living in Seattle. Goto et al. (1981) studied 42 Japanese families containing 80 affected persons. Autosomal recessive inheritance was confirmed. Malignancy was frequent in the patients and in the families generally. HLA was not linked. The frequency of Werner syndrome in Japan was estimated to be about 3 per million persons. The origin of the grandparents of the cases would be of interest. Khraishi et al. (1992) described a 47-year-old woman who had been misdiagnosed as having progressive systemic sclerosis with metastatic calcification for 12 years and then developed a painful, distal femoral, osteoblastic cortical juxtaarticular lesion with exuberant soft tissue calcification. This lesion proved to be an osteosarcoma requiring amputation. Ruprecht (1989) reported that in 10 of 18 eyes from 9 patients with Werner syndrome, cataract surgery was complicated by wound dehiscence and its consequences. Additionally, corneal endothelial decompensation occurred in 8 eyes. In view of the reduced growth potential of fibroblasts, he suggested small surgical incisions and other modifications of the usual procedures of cataract surgery, including no local or systemic use of cortisone.

'Variegated translocation mosaicism' is the designation proposed by W. W. Nichols (Hoehn et al., 1975) for a phenomenon he and others have observed in Werner syndrome: skin fibroblast cell lines from such patients are usually composed of one or several clones, each marked by a distinctive, apparently balanced translocation. In fibroblast cell lines and lymphoblastoid cell lines made from circulating B-lymphocytes in 2 brothers born of first-cousin parents, Schonberg et al. (1984) demonstrated variegated translocation mosaicism as well as the abbreviated life span characteristic of cell lines from these patients. Scappaticci et al. (1990) found multiple numerical and structural chromosomal abnormalities in cultured lymphocytes of 4 patients with Werner syndrome; several of the changes were clonal. Bauer et al. (1986) found that fibroblasts from a patient with the Werner syndrome had a markedly attenuated mitogenic response to platelet-derived growth factor (173430, 190040) and fibroblast growth factor (131220, 134920). Both PDGF binding and receptor numbers per cell were unaltered. The Werner syndrome cells expressed high constitutive levels of collagenase in vitro. In studies with clastogens, Gebhart et al. (1988) concluded that Werner syndrome cells demonstrate differences clearly distinguishing them from those of classic chromosome instability syndromes. Fukuchi et al. (1990) found an 8-fold higher average frequency of 6-thioguanine-resistant lymphocytes in Werner syndrome patients than in sex- and age-matched normal controls. Fukuchi et al. (1989) had found that the 'mutator phenotype' of Werner syndrome is characterized by extensive deletions. Monnat et al. (1992) determined the junction region sequences of deletions in the HPRT gene from thioguanine-resistant Werner syndrome fibroblasts. Given the potential for homologous recombination between copies of repeated DNA sequences that constitute approximately a third of the human HPRT gene, they were surprised to discover that all the deletions were generated by nonhomologous recombination of donor DNA duplexes that share little nucleotide sequence identity. No difference in structure or complexity was observed between deletions isolated from Werner syndrome fibroblasts or from myeloid leukemia cells. This suggested to Monnat et al. (1992) that the Werner syndrome deletion mutator uses deletion mutagenesis pathways that are similar or identical to those used in other human somatic cells.

In a study of 21 Japanese families originating in 16 different prefectures, Goto et al. (1992) did linkage studies demonstrating close linkage of WRN to a group of markers on chromosome 8. At least 3 of the 4 following major signs were required for the diagnosis: characteristic habitus and stature, premature senescence, scleroderma-like skin changes, and endocrine abnormalities. The first suggestion of linkage was increased homozygosity for ankyrin (182900) and D8S87. The ANK1 locus is located at 8p11.2. The Werner syndrome showed a maximum lod score of 2.89 at theta = 0.058 for linkage with ANK1. A multipoint lod score of 9.92 was obtained for the linkage of Werner syndrome with 3 markers. No linkage was found with lipoprotein lipase (238600) and other evidence suggested that this locus lies closer to 8pter than does the Werner syndrome locus. A likely location for the WRN gene appeared to be 8p12-p11. Schellenberg et al. (1992) confirmed the assignment by homozygosity mapping, i.e., linkage analysis using affected individuals from first- or second-cousin marriages. A peak lod score of 5.58 at a recombination fraction of 0.03 was obtained with D8S87. By linkage studies, Thomas et al. (1993) determined that the heregulin locus (142445) is distal to WRN and that ANK1 and PLAT (173370) are in that order on the centromeric side of WRN. Nakura et al. (1994) studied 27 Werner syndrome kindreds of various ethnic origins, 26 of which were consanguineous. In 24 of these families, the affected subject was given the diagnosis of definite Werner syndrome and affected subjects in the remaining 3 pedigrees were given the diagnosis of probable Werner syndrome. With 2-point linkage analysis using 13 short tandem repeat polymorphic sites on 8p, Nakura et al. (1994) found that the locus yielding a maximum lod score at the smallest recombination fraction was D8S339. Lod scores in excess of 3.0 were obtained with this marker for both Japanese and Caucasian families. Multipoint ana lysis of the markers yielded a maximum lod score of 17.05 at a distance of approximately 0.6 cM from D8S339. Combined with the analysis of homozygosity in subjects from inbred pedigrees, the data indicated that the WRN locus is between D8S131 and D8S87, in an 8.3-cM interval containing D8S339.

The finite replicative life span of human cells in vitro, the Hayflick phenomenon (Hayflick, 1965), is due to the stochastic loss of replicative ability in a continuously increasing fraction of newborn cells at every generation. Normal human fibroblasts achieve approximately 60 population doublings in culture, while Werner syndrome cells usually achieve only about 20 population doublings. There are 2 alternative kinetic explanations for the decreased life span of Werner syndrome cells. First, the initial fraction of cycling cells in a fresh explant may be approximately the same as in an explant derived from a normal subject, but the rate of loss of reproductive ability may be much higher in Werner syndrome cells. Second, when freshly explanted, the Werner syndrome cells may contain a much smaller fraction of cycling cells, which lose their reproductive ability at a normal rate. Of course, a combination of the 2 mechanisms is possible. To distinguish between the 2 main hypotheses, Faragher et al. (1993) studied cells from an obligate heterozygote, determining the fraction of cells in S phase throughout the life span of cultures. They found that the cells in these cultures usually exited, apparently irreversibly, from the cell cycle at a faster rate than did normal cells, although for the most part they started off with good replicative ability. They proposed that the Werner syndrome gene is a 'counting' gene controlling the number of times that human cells are able to divide before terminal differentiation. Thweatt and Goldstein (1993) arrived at a similar hypothesis. They pointed out that several overexpressed gene sequences isolated from a Werner syndrome fibroblast cDNA library possessed the capacity to inhibit DNA synthesis and disrupt many normal biochemical processes. Because a similar constellation of genes is overexpressed in senescent normal fibroblasts, the findings suggested a common molecular genetic pathway for replicative senescence in the 2 types of cells. Thweatt and Goldstein (1993) proposed that the primary defect in WRN is a mutation in a gene for a trans-acting repressor protein that reduces its binding affinity for shared regulatory regions of several genes, including those that encode inhibitors of DNA synthesis. The mutant WRN repressor gene triggers a sequence of premature expression of inhibitors of DNA synthesis and other genes, with resulting inhibition of DNA synthesis and early cellular senescence, events that occur much later in normal cells.

Because DNA polymerase beta (174760) maps to the same region, 8p12-p11, as the Werner syndrome phenotype, Sadakane et al. (1994) focused on this gene as the possible site of mutation causing the disorder. In blood samples from Werner syndrome patients, they found a novel DNA polymerase beta cDNA species that contained 107-bp insertions or 87-bp deletions in the portion of the gene encoding the catalytic domain of the protein. Since the mutations change the structure of DNA polymerase beta, the capacity of the DNA repair system would be impaired. The elevated rates of mutations, especially deletions, of the HPRT gene (308000), as demonstrated by Fukuchi et al. (1989), may be explained by a defect in DNA polymerase beta. The 107-bp insertion was found in 2 independent Werner syndrome patients and in the carrier mother of 1 of the patients but not in an unaffected sister or in a healthy population. Since Salk (1982) had found that somatic cells from Werner syndrome patients reveal a propensity to develop chromosomal aberrations, including translocations, inversions, and deletions, further study was required to establish whether the changes in the POLB gene are primary or secondary. Chang et al. (1994) presented several lines of evidence suggesting that POLB is not the Werner syndrome gene. Activity gels showed normal enzyme activity and electrophoretic mobility. Nucleotide sequence analysis of the entire coding region failed to demonstrate mutations (although mistakes in the published sequence for POLB were discovered). Single-strand conformation polymorphism (SSCP) and heteroduplex analyses failed to reveal evidence of mutations in the promoter region. A newly discerned polymorphism failed to reveal homozygosity by descent in a consanguineous patient. Fluorescence in situ hybridization placed the POLB gene centromeric to D8S135 at 8p11.2, beyond the region of peak lod scores for Werner syndrome.

Yu et al. (1994) used linkage disequilibrium in an attempt to narrow down the location of the WRN gene. They found that D8S339 and 2 polymorphisms at the glutathione reductase locus (138300) showed strong statistically significant evidence of disequilibrium with WRN in the Japanese population but not in the Caucasian population. In addition, they showed that a limited number of haplotypes are associated with the disease in both populations and that these haplotypes define clusters of apparently related haplotypes that may identify as many as 8 or 9 independent WRN mutations in these 2 populations.

Ye et al. (1995) used homozygosity mapping with markers derived from an 8p22-p12 microdissection library to type members of Japanese families with WRN. One marker, MS8-134 (D8S1055), showed a lod score of over 20 at theta = 0.00.

Yu et al. (1996) identified the WRN gene by positional cloning. The predicted protein is 1,432 amino acids long and shows significant similarity to DNA helicases. Matsumoto et al. (1997) defined the genomic structure of the gene and determined intron-exon boundaries for 35 exons (each 68 to 767 bp in size). The region most highly homologous to RecQ-type helicases was contained in exons 14 through 21. Yu et al. (1997) determined the intron-exon structure of the WRN gene. A total of 35 exons were defined, with the coding sequence beginning in the second exon.

The protein encoded by the WRN gene has 1,432 amino acids. Bloom syndrome (BS), which has features quite different from those of Werner syndrome, is caused by defects in another human RecQ-like helicase, BLM (210900).

Yu et al. (1996) identified 4 mutations in patients with Werner syndrome. Two of the mutations (277700.0003 and 277700.0004) were splice-junction mutations with the predicted result being the exclusion of exons from the final messenger RNA. One of these mutations (277700.0004), which resulted in a frameshift and a predicted truncated protein, was found in the homozygous state in 60% of Japanese Werner syndrome patients examined. The other 2 mutations were nonsense mutations (277700.0001 and 277700.0002). The identification of a mutated putative helicase as the gene product of the WRN gene suggested to Yu et al. (1996) that defective DNA metabolism is involved in a complex process of aging in Werner syndrome patients.

Oshima et al. (1996) reported 9 new WRN mutations in 10 Werner syndrome patients, including 4 Japanese patients and 6 Caucasian patients. These mutations were located at different sites across the coding region. Oshima et al. (1996) noted that all of the WRN mutations found to date either create a stop codon or cause frameshifts that lead to premature terminations. They noted that the WRN protein is partially homologous to RecQ helicases and that it contains 7 helicase motifs, 2 of which have been found in all ATP-binding proteins. Oshima et al. (1996) briefly reviewed the functions of helicases and reported that DNA helicases have been implicated in a number of molecular processes, including unwinding of DNA during replication, DNA repair, and accurate chromosomal segregation.

Goto et al. (1997) studied the helicase gene mutations previously described by Yu et al. (1996) in 89 Japanese Werner syndrome patients. Thirty-five (39.3%) were homozygous for mutation 4 (277700.0004); 1 (1.1%) was homozygous for mutation 1 (277700.0001); 6 (6.7%) were positive for both mutations 1 and 4; 1 was homozygous for a new mutation, which they designated mutation 5 (277700.0005); 13 (14.6%) had a single copy of mutation 4; 3 (3.4%) had a single copy of mutation 1; and the remaining 30 (33.8%) were negative for all 5 mutations. Of the 178 chromosomes in the 89 patients, 89 (50%) carried mutation 4, 11 (6.2%) carried mutation 1, and 2 (1.1%) carried mutation 5. In 76 chromosomes (42.7%), no mutation was identified.

Yu et al. (1997) screened Werner syndrome subjects for mutations and identified 5 new ones. Four of these new mutations either partially disrupted the helicase domain region or resulted in predicted protein products lacking the entire helicase region. Their results confirmed that mutations in the WRN gene are responsible for Werner syndrome. In addition, the location of the mutations indicated that the presence or absence of the helicase domain does not influence the Werner syndrome phenotype, suggesting that this syndrome is the result of complete loss of function of the WRN gene product.

Matsumoto et al. (1997) presented evidence that the helicase that is defective in Werner syndrome is missing the nuclear localization signal (NLS) and that this leads to impaired nuclear import as a major contributing factor in the molecular pathology of the disorder. The finding helped to explain the enigma that most Werner syndrome patients have similar clinical phenotypes no matter how different their mutations. The role the Werner syndrome helicase plays in the nucleus in preventing premature aging remained to be clarified.

Gray et al. (1997) pointed out that the homology of the WRN protein to members of the RecQ family of DNA helicases does not necessarily mean that the WRN gene encodes an active helicase. For example, the Saccharomyces cerevisiae RAD26 gene protein and the human transcription-repair coupling factor mutant in Cockayne syndrome B (133540) are highly homologous to known helicases, yet neither encodes an active helicase. Gray et al. (1997) reported work indicating that the Werner syndrome protein does indeed catalyze DNA unwinding.

The yeast sgs1 gene encodes a DNA helicase with similarity to the WRN and Bloom syndrome (BLM; 210900) genes. Sinclair et al. (1997) showed that mutation of sgs1 causes premature aging in yeast mother cells as demonstrated by shortened life span and the aging-induced phenotypes of sterility and redistribution of the Sir3 silencing protein from telomeres to the nucleolus. Furthermore, in old sgs1 cells the nucleolus was enlarged and fragmented, changes that also occur in old wildtype cells. Their findings suggested a conserved mechanism of cellular aging that may be related to nucleolar structure. The similar effect of the related sgs1 and WRN genes on yeast and human aging, along with age-associated changes in rDNA content reported for several mammalian species, suggested that a common mechanism may underlie aging in eukaryotes.

Both Werner syndrome and Bloom syndrome show an increased rate of chromosomal aberration. Yamagata et al. (1998) showed that yeast sgs1 helicase acts as a suppressor of illegitimate recombination through homologous recombination and that human BLM and WRN helicases can suppress the increased homologous and illegitimate recombinations in the S. cerevisiae sgs1 mutant. The results implied a role of BLM and WRN helicases in the control of genomic stability in human cells. Similar to sgs1 helicase, BLM helicase suppressed the cell growth in the top3 sgs1 mutation background and restored the increased sensitivity of the sgs1 mutant to hydroxyurea, but the WRN helicase did not. Thus, BLM and WRN helicases may have different roles in human cells.

To define better the function of WRN protein, Marciniak et al. (1998) determined its subcellular localization. By indirect immunofluorescence, using polyclonal anti-human WRN, they showed a predominant nucleolar localization. Studies of WRN mutant cell lines confirmed the specificity of antibody recognition. No difference was seen in the subcellular localization of the WRN protein in a variety of normal and transformed human cell lines, including both carcinomas and sarcomas. The nucleolar localization of human WRN protein was supported by the finding that upon biochemical subcellular fractionation, WRN protein was present in an increased concentration in a subnuclear fraction enriched for nucleolar proteins. Marciniak et al. (1998) also determined the subcellular localization of the mouse WRN homolog. In contrast to human WRN protein, mouse WRN protein was present diffusely throughout the nucleus. Imamura et al. (1997) cloned the mouse homolog of the human WRN gene and assigned it to 8A4 by fluorescence in situ hybridization.

Sinclair and Guarente (1997) found that mutants for sgs1 accumulate extrachromosomal rDNA circles (ERCs) more rapidly than cells with wildtype sgs1 leading to premature aging and a shorter life span. These ERCs are nucleic acid circles that excise from the yeast rDNA locus and replicate via the autonomously replicating sequence present in each rDNA repeat.

The initiation of DNA replication involves a minimum of 4 factors: a specific DNA sequence (origin), an initiator protein that binds to the origin, a helicase that unwinds the origin, and a protein that binds single-stranded DNA and stabilizes the unwound origin. In eukaryotic cells, the origin recognition complex (ORC) is the initiator protein, and replication protein A (RPA1; 179835) is the single-stranded DNA-binding protein. However, the helicase has not been identified and the nature of origins remains elusive, except in the case of S. cerevisiae. A unique feature of eukaryotic DNA replication is that it occurs at a few-hundred discrete foci. Thus, it has been proposed that a real origin must contain a specific DNA sequence and must be attached to replication foci. Using Xenopus laevis egg extracts, Yan and Newport (1995) identified and purified a 170-kD protein, called focus-forming activity 1 (FFA1), that is required for the formation of replication foci. Yan et al. (1998) reported that FFA1 has DNA-helicase activity. Moreover, it is a homolog of the human Werner syndrome gene product WRN, a protein associated with premature aging in humans.

Martin (1997) gave a thoughtful review of the question of whether the Werner mutation is a bonafide reflection of mechanisms of 'normal aging.'

Because insulin resistance in Werner syndrome may be due to defective signaling distal to the insulin receptor, Izumino et al. (1997) analyzed the metabolic effects of troglitazone (TRO), an antidiabetic drug that sensitizes insulin action, in 5 patients with Werner syndrome. Each patient was treated with 400 mg/day of TRO for 4 weeks and underwent a 75-g oral glucose tolerance test (OGTT) and frequently sampled iv glucose tolerance tests. Treatment reduced the area under the curve of glucose and insulin in the OGTT by 26% and 43%, respectively. Glucose tolerance, expressed as the glucose disappearance rate, improved significantly (1.36 +/- 0.16 to 1.94 +/- 0.30%/min; P less than 0.005). The authors found that TRO ameliorates glucose intolerance mediated by increased insulin sensitivity as well as glucose effectiveness, as assessed by minimal analysis, in Werner syndrome patients.

Ogburn et al. (1997) found that immortalized B lymphocytes from Werner syndrome subjects are hypersensitive to 4-nitro-quinoline-1-oxide (4NQO), supporting earlier work on T lymphocytes. They also showed that B cell lines from clinically normal heterozygous carriers exhibit sensitivities to this genotoxic agent, which were intermediate to those of homozygous wildtype homozygous mutants; 4NQO was shown to induce an apoptotic response. The data raised the question of an enhanced sensitivity of the relatively prevalent heterozygous carriers to certain environmental genotoxic agents. Ogburn et al. (1997) pointed out that most or all of the mutations identified to that time through the clinical diagnosis of Werner syndrome had been null mutations; not a single missense mutation had been identified. This would predict that levels of the putative Werner helicase activity in heterozygous carriers is approximately 50% of wildtype. Since the prevalence of carriers is as high as 1 in 150 to 1 in 200, a deleterious phenotype associated with a carrier state could have potential public health concern.

See also Boyd and Grant (1959); Cerimele et al. (1982); Gebhart et al. (1985); Goto et al. (1978); Motulsky et al. (1962); Nordenson (1977); Rabbiosi and Borroni (1979); Salk et al. (1981); Salk et al. (1981); Samantray et al. (1977); Scappaticci et al. (1982); Tri and Combs (1978)

Bauer, E. A.; Silverman, N.; Busiek, D. F.; Kronberger, A.; Deuel, T. F.: Diminished response of Werner's syndrome fibroblasts to growth factors PDGF and FGF. Science 234: 1240-1243, 1986.

Boyd, M. W. J.; Grant, A. P.: Werner's syndrome (progeria of the adult): further pathological and biochemical observations. Brit. Med. J. 2: 920-925, 1959.

Cerimele, D.; Cottoni, F.; Scappaticci, S.; Rabbiosi, G.; Borroni, G.; Sanna, E.; Zei, G.; Fraccaro, M.: High prevalence of Werner's syndrome in Sardinia: description of six patients and estimate of the gene frequency. Hum. Genet. 62: 25-30, 1982.

Chang, M.; Burmer, G. C.; Sweasy, J.; Loeb, L. A.; Edelhoff, S.; Disteche, C. M.; Yu, C.-E.; Anderson, L.; Oshima, J.; Nakura, J.; Miki, T.; Kamino, K.; Ogihara, T.; Schellenberg, G. D.; Martin, G. M.: Evidence against DNA polymerase beta as a candidate gene for Werner syndrome. Hum. Genet. 93: 507-512, 1994.

Epstein, C. J.; Martin, G. M.; Schultz, A. L.; Motulsky, A. G.: Werner's syndrome: a review of its symptomatology, natural history, pathologic features, genetics and relationship to the natural aging process. Medicine 45: 177-222, 1966.

Faragher, R. G. A.; Kill, I. R.; Hunter, J. A. A.; Pope, F. M.; Tannock, C.; Shall, S.: The gene responsible for Werner syndrome may be a cell division 'counting' gene. Proc. Nat. Acad. Sci. 90: 12030-12034, 1993.

Fukuchi, K.; Martin, G. M.; Monnat, R. J., Jr.: Mutator phenotype of Werner syndrome is characterized by extensive deletions. Proc. Nat. Acad. Sci. 86: 5893-5897, 1989.

Fukuchi, K.; Tanaka, K.; Kumahara, Y.; Marumo, K.; Pride, M. B.; Martin, G. M.; Monnat, R. J., Jr.: Increased frequency of 6-thioguanine-resistant peripheral blood lymphocytes in Werner syndrome patients. Hum. Genet. 84: 249-252, 1990.

Gebhart, E.; Bauer, R.; Raub, U.; Schinzel, M.; Ruprecht, K. W.; Jonas, J. B.: Spontaneous and induced chromosomal instability in Werner syndrome. Hum. Genet. 80: 135-139, 1988.

Gebhart, E.; Schinzel, M.; Ruprecht, K. W.: Cytogenetic studies using various clastogens in two patients with Werner syndrome and control individuals. Hum. Genet. 70: 324-327, 1985.

Goto, M.; Hariuchi, Y.; Tanimoto, K.; Ishii, T.; Nakashima, H.: Werner's syndrome: analysis of 15 cases with a review of the Japanese literature. J. Am. Geriat. Soc. 26: 341-347, 1978.

Goto, M.; Imamura, O.; Kuromitsu, J.; Matsumoto, T.; Yamabe, Y.; Tokutake, Y.; Suzuki, N.; Mason, B.; Drayna, D.; Sugawara, M.; Sugimoto, M.; Furuichi, Y.: Analysis of helicase gene mutations in Japanese Werner's syndrome patients. Hum. Genet. 99: 191-193, 1997.

Goto, M.; Rubenstein, M.; Weber, J.; Woods, K.; Drayna, D.: Genetic linkage of Werner's syndrome to five markers on chromosome 8. Nature 355: 735-738, 1992.

Goto, M.; Tanimoto, K.; Horiuchi, Y.; Sasazuki, T.: Family analysis of Werner's syndrome: a survey of 42 Japanese families with a review of the literature. Clin. Genet. 19: 8-15, 1981.

Gray, M. D.; Shen, J.-C.; Kamath-Loeb, A. S.; Blank, A.; Sopher, B. L.; Martin, G. M.; Oshima, J.; Loeb, L. A.: The Werner syndrome protein is a DNA helicase. Nature Genet. 17: 100-103, 1997.

Hayflick, L.: The limited in vitro lifetime of human diploid cell strains. Exp. Cell Res. 37: 614-636, 1965.

Hoehn, H.; Bryant, E. M.; Au, K.; Norwood, T. H.; Boman, H.; Martin, G. M.: Variegated translocation mosaicism in human skin fibroblast cultures. Cytogenet. Cell Genet. 15: 282-298, 1975.

Huang, S.; Li, B.; Gray, M. D.; Oshima, J.; Mian, I. S.; Campisi, J.: The premature ageing syndrome protein, WRN, is a 3-prime-5-prime exonuclease. Nature Genet. 20-114-115, 1998.

Imamura, O.; Ichikawa, K.; Yamabe, Y.; Goto, M.; Sugawara, M.; Furuichi, Y.: Cloning of a mouse homologue of the human Werner syndrome gene and assignment to 8A4 by fluorescence in situ hybridization. Genomics 41: 298-300, 1997.

Izumino, K.; Sakamaki, H.; Ishibashi, M.; Takino, H.; Yamasaki, H.; Yamaguchi, Y.; Chikuba, N.; Matsumoto, K.; Akazawa, S.; Tokuyama, K.; Nagataki, S.: Troglitazone ameliorates insulin resistance in patients with Werner's syndrome. J. Clin. Endocr. Metab. 82: 2391-2395, 1997.

Khraishi, M.; Howard, B.; Little, H.: A patient with Werner's syndrome and osteosarcoma presenting as scleroderma. J. Rheum. 19: 810-813, 1992.

Marciniak, R. A.; Lombard, D. B.; Johnson, F. B.; Guarente, L.: Nucleolar localization of the Werner syndrome protein in human cells. Proc. Nat. Acad. Sci. 95: 6887-6892, 1998.

Martin, G. M.: The Werner mutation: does it lead to a 'public' or 'private' mechanism of aging? Molec. Med. 3: 356-358, 1997.

Matsumoto, T.; Imamura, O.; Yamabe, Y.; Kuromitsu, J.; Tokutake, Y.; Shimamoto, A.; Suzuki, N.; Satoh, M.; Kitao, S.; Ichikawa, K.; Kataoka, H.; Sugawara, K.; Thomas, W.; Mason, B.; Tsuchihashi, Z.; Drayna, D.; Sugawara, M.; Sugimoto, M.; Furuichi, Y.; Goto, M.: Mutation and haplotype analyses of the Werner's syndrome gene based on its genomic structure: genetic epidemiology in the Japanese population. Hum. Genet. 100: 123-130, 1997.

Matsumoto, T.; Shimamoto, A.; Goto, M.; Furuichi, Y.: Impaired nuclear localization of defective DNA helicases in Werner's syndrome. (Letter) Nature Genet. 16: 335-336, 1997.

McKusick, V. A.; Abbey, H.; Bartalos, M.; Bowen, P.; Boyer, S. H., IV; Cohen, B. H.; Danks, D. M.; Duchastel, Y.; Emery, A. E. H.; Epstein, E. J.; Fainer, D. C.; Finn, R.; and 18 others: Medical genetics 1962. J. Chronic Dis. 16: 457-634, 1963.

Meisslitzer, C.; Ruppitsch, W.; Weirich-Schwaiger, H.; Weirich, H. G.; Jabkowsky, J.; Klein, G.; Schweiger, M.; Hirsch-Kauffmann, M.: Werner syndrome: characterization of mutations in the WRN gene in an affected family. Europ. J. Hum. Genet. 5: 364-370, 1997.

Monnat, R. J., Jr.; Hackmann, A. F. M.; Chiaverotti, T. A.: Nucleotide sequence analysis of human hypoxanthine phosphoribosyltransferase (HPRT) gene deletions. Genomics 13: 777-787, 1992.

Motulsky, A. G.; Schultz, A.; Priest, J. H.: Werner's syndrome: chromosomes, genes, and the ageing process. Lancet I: 160-161, 1962.

Nakura, J.; Wijsman, E. M.; Miki, T.; Kamino, K.; Yu, C.-E.; Oshima, J.; Fukuchi, K.; Weber, J. L.; Piussan, C.; Melaragno, M. I.; Epstein, C. J.; Scappaticci, S.; Fraccaro, M.; Matsumura, T.; Murano, S.; Yoshida, S.; Fujiwara, Y.; Saida, T.; Ogihara, T.; Martin, G. M.; Schellenberg, G. D.: Homozygosity mapping of the Werner syndrome locus (WRN). Genomics 23: 600-608, 1994.

Nordenson, I.: Chromosome breaks in Werner's syndrome and their prevention in vitro by radical-scavenging enzymes. Hereditas 87: 151-154, 1977.

Ogburn, C. E.; Oshima, J.; Poot, M.; Chen, R.; Hunt, K. E.; Gollahon, K. A.; Rabinovitch, P. S.; Martin, G. M.: An apoptosis-inducing genotoxin differentiates heterozygotic carriers for Werner helicase mutations from wild-type and homozygous mutants. Hum. Genet. 101: 121-125, 1997.

Oshima, J.; Yu, C.-E.; Piussan, C.; Klein, G.; Jabkowski, J.; Balci, S.; Miki, T.; Nakura, J.; Ogihara, T.; Ells, J.; Smith, M. A. C.; Melaragno, M. I.; Fraccaro, M.; Scappaticci, S.; Matthews, J.; Ouais, S.; Jarzebowicz, A.; Schellenberg, G. D.; Martin, G. M.: Homozygous and compound heterozygous mutations at the Werner syndrome locus. Hum. Molec. Genet. 5: 1909-1913, 1996.

Rabbiosi, G.; Borroni, G.: Werner's syndrome: seven cases in one family. Dermatologica 158: 355-360, 1979.

Ruprecht, K. W.: Ophthalmological aspects in patients with Werner's syndrome. Arch. Gerontol. Geriat. 9: 263-270, 1989.

Sadakane, Y.; Maeda, K.; Kuroda, Y.; Hori, K.: Identification of mutations in DNA polymerase beta mRNAs from patients with Werner syndrome. Biochem. Biophys. Res. Commun. 200: 219-225, 1994.

Salk, D.: Werner's syndrome: a review of recent research with an analysis of connective tissue metabolism, growth control of cultured cells, and chromosomal aberrations. Hum. Genet. 62: 1-15, 1982.

Salk, D.; Au, K.; Hoehn, H.; Martin, G. M.: Effects of radical-scavenging enzymes and reduced oxygen exposure on growth and chromosome abnormalities of Werner syndrome cultured skin fibroblasts. Hum. Genet. 57: 269-275, 1981.

Salk, D.; Bryant, E.; Au, K.; Hoehn, H.; Martin, G. M.: Systematic growth studies, cocultivation, and cell hybridization studies of Werner syndrome cultured skin fibroblasts. Hum. Genet. 58: 310-316, 1981.

Samantray, S. K.; Samantray, S.; Johnson, S. C.; Bhaktaviziam, A.: Werner syndrome. Aust. New Zeal. J. Med. 7: 309-311, 1977.

Scappaticci, S.; Cerimele, D.; Fraccaro, M.: Clonal structural chromosomal rearrangements in primary fibroblast cultures and in lymphocytes of patients with Werner's syndrome. Hum. Genet. 62: 16-24, 1982.

Scappaticci, S.; Forabosco, A.; Borroni, G.; Orecchia, G.; Fraccaro, M.: Clonal structural chromosomal rearrangements in lymphocytes of four patients with Werner's syndrome. Ann. Genet. 33: 5-8, 1990.

Schellenberg, G. D.; Martin, G. M.; Wijsman, E. M.; Nakura, J.; Miki, T.; Ogihara, T.: Homozygosity mapping and Werner's syndrome. (Letter) Lancet 339: 1002, 1992.

Schonberg, S.; Niermeijer, M. F.; Bootsma, D.; Henderson, E.; German, J.: Werner's syndrome: proliferation in vitro of clones of cells bearing chromosome translocations. Am. J. Hum. Genet. 36: 387-397, 1984.

Sinclair, D. A.; Guarente, L.: Extrachromosomal rDNA circles—a cause of aging in yeast. Cell 91: 1033-1042, 1997.

Sinclair, D. A.; Mills, K.; Guarente, L.: Accelerated aging and nucleolar fragmentation in yeast sgs1 mutants. Science. 277: 1313-1316, 1997.

Thomas, W.; Rubenstein, M.; Goto, M.; Drayna, D.: A genetic analysis of the Werner syndrome region on human chromosome 8p. Genomics 16: 685-690, 1993.

Thweatt, R.; Goldstein, S.: Werner syndrome and biological ageing: a molecular genetic hypothesis. BioEssays 15: 421-426, 1993.

Tri, T. B.; Combs, D. T.: Congestive cardiomyopathy in Werner's syndrome. (Letter) Lancet I: 1052-1053, 1978.

Yamagata, K.; Kato, J.; Shimamoto, A.; Goto, M.; Furuichi, Y.; Ikeda, H.: Bloom's and Werner's syndrome genes suppress hyperrecombination in yeast sgs1 mutant: implication for genomic instability in human diseases. Proc. Nat. Acad. Sci. 95: 8733-8738, 1998.

Yan, H.; Chen, C.-Y.; Kobayashi, R.; Newport, J.: Replication focus-forming activity 1 and the Werner syndrome gene product. Nature Genet. 19: 375-378, 1998.

Yan, H.; Newport, J.: FFA-1, a protein that promotes the formation of replication centers within nuclei. Science 269: 1883-1885, 1995.
Ye, L.; Nakura, J.; Mitsuda, N.; Fujioka, Y.; Kamino, K.; Ohta, T.; Jinno, Y.; Niikawa, N.; Miki, T.; Ogihara, T.: Genetic association between chromosome 8 microsatellite (MS8-134) and Werner syndrome (WRN): chromosome microdissection and homozygosity mapping. Genomics 28: 566-569, 1995.
Yu, C.-E.; Oshima, J.; Fu, Y.-H.; Wijsman, E. M.; Hisama, F.; Alisch, R.; Matthews, S.; Nakura, J.; Miki, T.; Quais, S.; Martin, G. M.; Mulligan, J.; Schellenberg, G. D.: Positional cloning of the Werner's syndrome gene. Science 272: 258-262, 1996.
Yu, C.-E.; Oshima, J.; Goddard, K. A. B.; Miki, T.; Nakura, J.; Ogihara, T.: Poot, M.; Hoehn, H.; Fraccaro, M.; Piussan, C.; Martin, G. M.; Schellenberg, G. D.; Wijsman, E. M.: Linkage disequilibrium and haplotype studies of chromosome 8p11.1-21.1 markers and Werner syndrome. Am. J. Hum. Genet. 55: 356-364, 1994.
Yu, C.-E.; Oshima, J.; Wijsman, E. M.; Nakura, J.; Miki, T.; Piussan, C., Matthews, S.; Fu, Y.-H.; Mulligan, J.; Martin, G. M.; Schellenberg, G. D.; Werner's Syndrome Collaborative Group: Mutations in the consensus helicase domains of the Werner syndrome gene. Am. J. Hum. Genet. 60: 330-341, 1997.

LYMPHEDEMA, HEREDITARY 2 (*153200)

JJM NOTE

One prior case has been published, a 59-year-old male with congenital lymphedema of the left leg and a rapidly progressing keratotic lymphangioma that concealed or contained a malignant angiosarcoma (Wendt T et al.: Progressives Lymphangiokeratom und Angiosarkom (Stewart-Treves-Syndrom) bei kongenitalem Lymphoedem. Hautarzt 39:155–60, 1988). The mechanism of carcinogenesis probably lies more in the resulting stasis than in the hypoplasia of lymph vessels produced by the mutant gene. After all, Stewart-Treves syndrome (sarcoma after radical mastectomy and lymph node excision) is acquired and not hereditary nor congenital.

Edema, particularly severe below the waist, develops about the time of puberty. Meige (1898) described 8 cases in 4 generations without male-to-male transmission. Goodman (1962) reported the condition in 2 sisters and a brother with presumed normal parents who were not known to be related. Herbert and Bowen (1983) described a kindred with many cases of lymphedema of postpubertal onset. Involvement of the upper limbs (as well as the lower limbs), face, and larynx and, in one, a persistent pleural effusion were notable features. Scintilymphangiography indicated paucity or absence of lymph nodes in the axillae and above the inguinal ligaments. The authors pointed to the difficulties of nosology. For example, since 'yellow nail disease' (153300) has yellow or dystrophic nails as a variable feature, this could be the same disorder. They pointed also to the association of late-onset lymphedema with deafness (Emberger et al., 1979) and with primary pulmonary hypertension and cerebrovascular malformations (152900; Avasthey and Roy, 1968). In their family (Herbert and Bowen, 1983), chronic facial swelling resulted in a characteristic appearance of affected members including puffiness, shiny skin, deep creases, and, in some, excessive wrinkling. Emerson (1966) noted similar facial features and remarked on the possible erroneous diagnosis of myxedema. Figueroa et al. (1983) reported the association of cleft palate. In their family, the mother, with only lymphedema praecox of the legs, gave birth to 5 sons, 3 of whom had both lymphedema of the legs and cleft palate. A mild form of lymphedema affecting mainly the medial aspect of both ankles in a 21-year-old son was pictured.

Andersson et al. (1995) described a family in which 3 individuals, a grandmother, her son and her grandson, had onset of lymphedema in their mid-20s or 30s. The grandson was 23 years old when he had his first episode of lymphedema, which was thought to be due to thrombophlebitis. During the ensuing decade, he had episodic waxing and waning of lymphedema of both lower limbs and was treated with anticoagulant therapy. At the age of 35, he developed lymphangiosarcoma on the inner right thigh and died of metastases some months later. Lymphangiosarcoma, usually associated with postmastectomy lymphedema, had not been described previously in late-onset hereditary lymphedema. Andersson et al. (1995) raised the question of whether a genetic predisposition to malignancy combined with the lymphedema was etiologically significant. There seemed to be an unusually high frequency of cancer (uterine, colon, lung, prostate, breast, and bone) in the proband's family.

See also Juchems (1963); Osterland (1961); Wheeler et al. (1981)

Andersson, H. C.; Parry, D. M.; Mulvihill, J. J.: Lymphangiosarcoma in late-onset hereditary lymphedema: case report and nosological implications. Am. J. Med. Genet. 56: 72-75, 1995.
Avasthey, P.; Roy, S. B.: Primary pulmonary hypertension, cerebrovascular malformation, and lymphoedema of the feet in a family. Brit. Heart J. 30: 769-775, 1968.
Emberger, J. M.; Navarro, M.; Dejean, M.; Izarn, P.: Surdi-mutite, lymphoedeme des membres inferieurs et anomalies hematologiques (leucose aigue cytopenies) a transmission autosomique dominante. J. Genet. Hum. 27: 237-245, 1979.
Emerson, P. A.: Yellow nails, lymphoedema, and pleural effusions. Thorax 21: 247-253, 1966.
Figueroa, A. A.; Pruzansky, S.; Rollnick, B. R.: Meige disease (familial lymphedema praecox) and cleft palate: report of a family and review of the literature. Cleft Palate J. 20: 151-157, 1983.
Goodman, R. M.: Familial lymphedema of the Meige's type. Am. J. Med. 32: 651-656, 1962.
Herbert, F. A.; Bowen, P. A.: Hereditary late-onset lymphedema with pleural effusion and laryngeal edema. Arch. Intern. Med. 143: 913-915, 1983.
Juchems, R.: Das hereditaere Lymphoedem, Typ Meige. Klin. Wschr. 41: 328-332, 1963.
Meige, H.: Dystrophie oedemateuse hereditaire. Presse Med. 6: 341-343, 1898.
Osterland, G.: Beobachtungen zum Nonne-Milroy-Meige-Syndrom. Z. Menschl. Vererb. Konstitutionsl. 36: 108-117, 1961.
Wheeler, E. S.; Chan, V.; Wassman, R.; Rimoin, D. L.; Lesavoy, M. A.: Familial lymphedema praecox: Meige's disease. Plast. Reconst. Surg. 67: 362-364, 1981.

Benign Soft-Tissue Tumors

LIPOMATOSIS, MULTIPLE (*151900)

JJM NOTE

At least one patient with multiple lipomatosis has developed liposarcoma. Distinct changes on chromosomes 12 and 16 seem to be necessary for such malignant progression. In general, of course, there is no reason to suggest that a solitary lipoma has any malignant potential.

Stephens and Isaacson (1959) observed 17 cases in 3 generations. Usually the condition did not become evident until the age of about 35 years, although in 1 case lipomas were present at age 9. The gastrointestinal tract may be involved (Lang et al., 1959). Mandahl et al. (1987) studied 10 consecutive lipomas; supernumerary ring chromosomes were found in 3; balanced rearrangements in 6; and a normal karyotype in 1. Chromosome 12 was involved in 5 of the balanced rearrangements and probably in all the ring chromosomes, with breakpoints localized to 12q13 or 12q14. They suggested that the development of myxoid liposarcoma requires the recombination of 2 specific chromosome bands, 12q13 and 16p11, whereas for some types of benign lipogenic tumors structural changes in 12q13-q14 may be sufficient for neoplastic growth. In a follow-up study of 25 additional solitary lipomas, Mandahl et al. (1988) found supernumerary ring chromosomes in 2 cases; balanced rearrangements involving 12q13-q14 in 8; hypodiploid or diploid chromosomes involving aberrations other than rings or rearrangements of 12q13-q14 in 7; and normal karyotypes in 8. Dal Cin et al. (1988) reported on the cytogenetic analysis of 3 different lipomas from a patient with multiple subcutaneous lipomas. Two distinct reciprocal translocations were observed, both with breakpoints at 12q14. The reciprocal translocation t(3;12)(q28;q14) was found in 2 lipomas from this patient; t(1;12)(q34.2;q14) was seen in the third. The authors noted that 2 previously described translocations involving chromosome 12q14 also had breakpoints in the region of 3q27-q28. It would be of great interest to determine whether familial multiple lipomatosis shows linkage to markers on 12q. Turc-Carel et al. (1988) concluded that the breakpoint on chromosome 12 may be found at either 12q13 or 12q14. Heim et al. (1988) concluded, on the basis of cytogenetic analysis of tumor cells from 50 lipomas, that 4 main cytogenetic subtypes can be recog nized: (1) tumors with normal karyotype (18 cases); (2) tumors with rearrangements of 12q13-q14 (18 cases); (3) tumors with ring chromosomes (6 cases); and (4) tumors with other clonal changes (8 cases). All 6 tumors with ring marker chromosomes were histopathologically classified as atypical lipomas. All 7 multiple lipomas were karyotypically normal. In 9 different solitary subcutaneous lipomas, all with clonal abnormalities affecting 12q13-q15, Arheden et al. (1989) found no rearrangement or amplification of the INT1 gene (164820), which maps to the same region.

Limon et al. (1986) and Turc-Carel et al. (1986) described a specific chromosome change, t(12;16)(q13;p11), in the tumors of all 4 cases of myxoid liposarcoma studied. One patient had a history of multiple lipomatosis. A relationship to the INT1 oncogene on chromosome 12 or to a fragile site on that chromosome was suggested. Turc-Carel et al. (1987) found, however, by Southern blot analysis of DNA from 2 myxoid liposarcomas carrying the translocation t(12;16)(q13;p11), that the translocation did not disrupt the INT1 gene. Karakousis et al. (1987) found the specific translocation between chromosomes 12 and 16 in 5 cases of myxoid liposarcoma. Walter et al. (1988) and Bridge et al. (1988) reported 12;16 translocation in myxoid liposarcoma. See also Smith et al. (1987) and Mertens et al. (1987). Sreekantaiah et al. (1991) found abnormalities of chromosome 12 in 34 of 92 patients with lipomas. Eneroth et al. (1990) narrowed the localization of the chromosome breaks in t(12;16) of myxoid liposarcomas to 12q13.3 and 16p11.2.

Aman et al. (1992) used the Southern blot technique to demonstrate that CHOP (126337), a gene of the CCAAT/enhancer binding protein (C/EBP) family which maps to 12q13 and is assumed to be involved in adipocyte differentiation, is rearranged in myxoid liposarcomas. Rabbitts et al. (1993) demonstrated that the translocation t(12;16)(q13;p11) in malignant liposarcoma can result in a fusion of the CHOP dominant-negative transcription factor gene with a novel gene called FUS (137070). The distinctness of benign lipomas, including multiple lipomatosis, from myxoid liposarcoma is indicated by the findings of Mrozek et al. (1993) that the chromosome 12 breakpoint in myxoid liposarcoma is at 12q13.3, as reported by many others, but that the breakpoint in benign lipoma, including multiple lipomatosis, is more distally situated, at 12q15. They suggested that the more distal location is identical to that found in other benign tumors such as uterine leiomyoma (150699) and pleomorphic adenoma of the salivary gland (181030).

In a note added in proof to their report of the HMGIC gene (600698) of chromosome 12 involved in soft tissue tumors with translocations of 12q15, Schoenmakers et al. (1995) referred to the isolation and sequencing of the lipoma-preferred-partner gene (LPP; 600700). Chromosome 3 is the preferred partner in lipomas resulting from translocation involving 12q15. Other translocation partners of 12q15 include chromosomes 1, 2, 4, 5, 6, 7, 10, 11, 13, 15, 17, 21, and X (Sreekantaiah et al., 1991).

Golsch and Worret (1995) described a family with 9 cases of familial multiple lipomatosis in 3 generations. The oldest living member of the family suffered from a nonalcoholic peripheral neuropathy. Polyneuropathy has been described in association with multiple symmetric lipomatosis (151800) and may be an intrinsic part of that disorder. Its relationship to familial multiple lipomatosis is less clear.

See also Heim et al. (1986); Humphrey and Kingsley (1938); Krabble and Bartels (1944); Kurzweg and Spencer (1951); Rabbiosi et al. (1977); Shanks et al. (1957); Turc-Carel et al. (1986); Weinberg et al. (1980)

Aman, P.; Ron, D.; Mandahl, N.; Fioretos, T.; Heim, S.; Arheden, K.; Willen, H.; Rydholm, A.; Mitelman, F.: Rearrangement of the transcription factor gene CHOP in myxoid liposarcomas with t(12;16)(q13;p11). Genes Chromosomes Cancer 5: 278-285, 1992.

Arheden, K.; Mandahl, N.; Heim, S.; Mitelman, F.: The INT1 oncogene is not rearranged or amplified in lipomas with structural chromosomal abnormalities of 12q13-15. Cancer Genet. Cytogenet. 42: 143-146, 1989.

Bridge, J. A.; Shaffer, B.; Neff, J. R.; Sanger, W. G.; Moran, J.: A complex translocation involving chromosomes 12 and 16 in a metastatic myxoid liposarcoma. (Letter) Cancer Genet. Cytogenet. 34: 119-120, 1988.

Dal Cin, P.; Turc-Carel, C.; Sandberg, A. A.: Consistent involvement of band 12q14 in two different translocations in three lipomas from the same patient. Cancer Genet. Cytogenet. 31: 237-240, 1988.

Eneroth, M.; Mandahl, N.; Heim, S.; Willen, H.; Rydholm, A.; Alberts, K. A.; Mitelman, F.: Localization of the chromosomal breakpoints of the t(12;16) in liposarcoma to subbands 12q13.3 and 16p11.2. Cancer Genet. Cytogenet. 48: 101-107, 1990.

Golsch, S.; Worret, W.-I.: Familial multiple lipomatosis with polyneuropathy. Europ. J. Derm. 5: 283-285, 1995.

Heim, S.; Mandahl, N.; Kristoffersson, U.; Mitelman, F.; Roser, B.; Rydholm, A.; Willen, H.: Reciprocal translocation t(3;12)(q27;q13) in lipoma. Cancer Genet. Cytogenet. 23: 301-304, 1986.

Heim, S.; Mandahl, N.; Rydholm, A.; Willen, H.; Mitelman, F.: Different karyotypic features characterize different clinicopathologic subgroups of benign lipogenic tumors. Int. J. Cancer 42: 863-867, 1988.

Humphrey, A. A.; Kingsley, P. C.: Familial multiple lipomas: report of a family. Arch. Derm. Syph. 37: 30-34, 1938.

Karakousis, C. P.; Dal Cin, P.; Turc-Carel, C.; Limon, J.; Sandberg, A. A.: Chromosomal changes in soft-tissue sarcomas: a new diagnostic parameter. Arch. Surg. 122: 1257-1260, 1987.

Krabble, K. H.; Bartels, E. D.: La lipomatose circonscripte multiple. Copenhagen: Munksgaard (pub.) 1944.

Kurzweg, F. T.; Spencer, R.: Familial multiple lipomatosis. Am. J. Surg. 82: 762-765, 1951.

Lang, C. S.; Leagus, C.; Stahlgren, L. H.: Intestinal lipomatosis. Surgery 46: 1054-1059, 1959.

Limon, J.; Turc-Carel, C.; Dal Cin, P.; Rao, U.; Sandberg, A. A.: Recurrent chromosome translocations in liposarcoma. (Letter) Cancer Genet. Cytogenet. 22: 93-94, 1986.

Mandahl, N.; Heim, S.; Arheden, K.; Rydholm, A.; Willen, H.; Mitelman, F.: Three major cytogenetic subgroups can be identified among chromosomally abnormal solitary lipomas. Hum. Genet. 79: 203-208, 1988.

Mandahl, N.; Heim, S.; Johansson, B.; Bennet, K.; Mertens, F.; Olsson, G.; Rooser, B.; Rydholm, A.; Willen, H.; Mitelman, F.: Lipomas have characteristic structural chromosomal rearrangements of 12q13-q14. Int. J. Cancer 39: 685-688, 1987.

Mertens, F.; Johansson, B.; Mandahl, N.; Heim, S.; Bennett, K.; Rydholm, A.; Willen, H.; Mitelman, F.: Clonal chromosome abnormalities in two liposarcomas. Cancer Genet. Cytogenet. 28: 137-144, 1987.

Mrozek, K.; Karakousis, C. P.; Bloomfield, C. D.: Chromosome 12 breakpoints are cytogenetically different in benign and malignant lipogenic tumors: localization of breakpoints in lipoma to 12q15 and in myxoid liposarcoma to 12q13.3. Cancer Res. 53: 1670-1675, 1993.

Rabbiosi, G.; Borroni, G.; Scuderi, N.: Familial multiple lipomatosis. Acta Derm. Venerol. 57: 265-267, 1977.

Rabbitts, T. H.; Forster, A.; Larson, R.; Nathan, P.: Fusion of the dominant negative transcription regulator CHOP with a novel gene FUS by translocation t(12;16) in malignant liposarcoma. Nature Genet. 4: 175-180, 1993.

Schoenmakers, E. P. P. M.; Wanschura, S.; Mols, R.; Bullerdiek, J.; Van den Berghe, H.; Van de Ven, W. J. M.: Recurrent rearrangements in the high mobility group protein gene, HMGI-C, in benign mesenchymal tumours. Nature Genet. 10: 436-444, 1995.

Shanks, J. A.; Paranchych, W.; Tuba, J.: Familial multiple lipomatosis. Canad. Med. Assoc. J. 77: 881-884, 1957.

Smith, S.; Reeves, B. R.; Wong, L.: Translocation t(12;16) in a case of myxoid liposarcoma. Cancer Genet. Cytogenet. 26: 185-186, 1987.

Sreekantaiah, C.; Leong, S. P. L.; Karakousis, C. P.; McGee, D. L.; Rappaport, W. D.; Villar, H. V.; Neal, D.; Fleming, S.; Wankel, A.; Herrington, P. N.; Carmona, R.; Sandberg, A. A.: Cytogenetic profile of 109 lipomas. Cancer Res. 51: 422-433, 1991.

Stephens, F. E.; Isaacson, A.: Hereditary multiple lipomatosis. J. Hered. 50: 51-53, 1959.

Turc-Carel, C.; Dal Cin, P.; Boghosian, L.; Leong, S. P. L.; Sandberg, A. A.: Breakpoints in benign lipoma may be at 12q13 or 12q14. Cancer Genet. Cytogenet. 36: 131-135, 1988.

Turc-Carel, C.; Dal Cin, P.; Rao, U.; Karakousis, C.; Sandberg, A. A.: Cytogenetic studies of adipose tissue tumors. I. A benign lipoma with reciprocal translocation t(3;12)(q28;q14). Cancer Genet. Cytogenet. 23: 283-289, 1986.

Turc-Carel, C.; Limon, J.; Dal Cin, P.; Rao, U.; Karakousis, C.; Sandberg, A. A.: Cytogenetic studies of adipose tissue tumors. II. Recurrent reciprocal translocation t(12;16)(q13;p11) in myxoid liposarcomas. Cancer Genet. Cytogenet. 23: 291-299, 1986.

Turc-Carel, C.; Pietrzak, E.; Kakati, S.; Kinniburgh, A. J.; Sandberg, A. A.: The human int-1 gene is located at chromosome region 12q12-12q13 and is not rearranged in myxoid liposarcoma with t(12;16)(q13;p11). Oncogene Res. 1: 397-405, 1987.

Walter, T. A.; Decker, H. H.; Leong, S. P. L.; Sandberg, A. A.: A case of myxoid liposarcoma with translocation t(12;16) as the only abnormality. Cancer Genet. Cytogenet. 34: 117-118, 1988.

Weinberg, J. B.; Hasstedt, S. J.; Skolnick, M. H.; Kimberling, W. J.; Baty, B.: Analysis of a large pedigree with elliptocytosis, multiple lipomatosis, and biological false-positive serological tests for syphilis. Am. J. Med. Genet. 5: 57-67, 1980.

LIPOMATOSIS, FAMILIAL BENIGN CERVICAL (151800)

JJM NOTE

This entity is included because of benign neoplasia, apart from the report of malignant degeneration by Tizian et al. (1983).

McKusick et al. (1962) described 3 brothers with a collar of fat around the neck in the submandibular area and involving the nape of the neck. The age of onset was said to be 45, 39 and 29 years in the 3 patients. The mother was said to be definitely unaffected, having died at age 61, but 2 sisters and a maternal aunt were also affected. In advanced stages the process extended into the upper mediastinum. In the 3 brothers lipomata of conventional type were present in the epitrochlear area, back, axillae, internal aspect of forearm, etc. Brodie (1846) is said to have first described diffuse symmetrical lipomatosis with predilection for the neck. It was called 'fat neck' (Fetthals) by Madelung (1888). In cretinism we have seen not only the supraclavicular fossa but also the axillae filled with fat (Cross et al., 1968). Familial cretinism may raise a suspicion of a separate recessively inherited supraclavicular lipomatosis. Cervical lipomatosis was associated with gout and hyperlipoproteinemia type IV in the sisters reported by Greene et al. (1970). Oligomenorrhea, muscle cramps, pes cavus, and extensor plantar reflexes were also described. See 151660. Lyon (1910) reported a striking case which was familial. Michon and Rose (1935) observed familial cases. Taylor et al. (1961) described surgical procedures adopted in a case similar to those reported by McKusick et al. (1962). Since fat cells are smaller than normal, Enzi et al. (1977) concluded that lipomata are attributable to neoformation of adipocytes. In their studies of 10 affected males, reduced glucose tolerance and hyperlipoproteinemia were no more frequent than in controls. In lipomatous tissue but not in normal fat tissue from these subjects, in vitro insensitivity to the lipolytic effect of catecholamines was demonstrated. The block appeared to be proximal to cyclic AMP formation because theophylline induced a prompt and significant decrease in intracellular ATP in lipomatous tissue. Enzi (1984) had studied 34 patients of whom 3 had other affected family members: a brother in 2 instances and a brother and father in the third. The other patients declared that none of their sibs (34 brothers, 28 sisters) or parents was affected. Enzi et al. (1985) suggested that MSL is not rare in Mediterranean areas and that the frequency in Italy is about 1 per 25,000 males. Enzi et al. (1985) documented the high frequency of somatic and autonomic neuropathies. In 28 of 33 male patients changes varying from vibratory sensory loss to incapacitating trophic ulcers or Charcot arthropathy were found. HDL was increased, consistent with the diagnosis of hyperalphalipoproteinemia, and LDL fractions were reduced with a marked enhancement of lipoprotein lipase activity in adipose tissue.

Pollock et al. (1988) pointed out that, with increasing age, peripheral neuropathy becomes more common in multiple symmetric lipomatosis and is a principal cause of severe disability. The peripheral neuropathy is often attributed to alcoholism, but the pathologic findings of Pollock et al. (1988) led them to conclude that the neuropathy is in fact an integral part of the syndrome. Biochemical observations suggested a defect in catecholamine-stimulated lipolysis at the level of cell membranes. Chalk et al. (1990) also felt that alcoholism could be excluded. Sural nerve biopsy in 1 patient showed nerve fiber loss, predominantly affecting large myelinated fibers. The relationship between myelin sheath thickness and axon diameter was normal, arguing that this neuropathy is not due to primary axonal atrophy. Chalk et al. (1990) described coexisting peripheral neuropathy and multiple symmetric lipomatosis in 4 of 7 sibs, 3 female and 1 male. They favored autosomal recessive inheritance because of absence of either condition in 3 other generations of this family. Zancanaro et al. (1990) presented studies suggesting that multiple symmetric lipomatosis may be a neoplastic disease that originates in brown fat.

Williams et al. (1993) described the findings on magnetic resonance imaging in 2 unrelated women, aged 27 and 48 years. They commented that the standard treatment is surgical debulking, but prognosis is guarded because of frequent recurrences. Tizian et al. (1983) described malignant degeneration within the lipomatosis tissue. This must be very rare inasmuch as Williams et al. (1993) could find no other report of this complication.

Klopstock et al. (1994) pointed out that ragged-red fibers are occasionally found in muscle of patients with multiple symmetric lipomatosis, suggesting a mitochondrial abnormality. They studied 11 unrelated patients with this disorder by means of neurophysiology, muscle morphology, muscle biochemistry, Southern blot, and PCR analysis of mitochondrial DNA. All patients were men, aged 41 to 63 years. Clinical or electrophysiologic signs of sensorimotor polyneuropathy were present in 9 patients, 8 of whom had a history of alcoholism. In muscle biopsy specimens, the most prominent feature was pathologic subsarcolemmal accumulations of mitochondria. Biochemical analysis of respiratory chain enzymes revealed a moderate but significant decrease of cytochrome c oxidase activity as compared with age-matched controls. In 1 patient, Southern blot analysis showed multiple deletions of mitochondrial DNA.

See also Enzi (1984)

Brodie, B. C.: Clinical Lectures on Surgery, Delivered at St. George's Hospital. Philadelphia: Lea and Blanchard (pub.) 1846. Pp. 201-202.
Chalk, C. H.; Mills, K. R.; Jacobs, J. M.; Donaghy, M.: Familial multiple symmetric lipomatosis with peripheral neuropathy. Neurology 40: 1246-1250, 1990.
Cross, H. E.; Hollander, C. S.; Rimoin, D. L.; McKusick, V. A.: Familial agoitrous cretinism accompanied by muscular hypertrophy. Pediatrics 41: 413-420, 1968.
Enzi, G.: Multiple symmetric lipomatosis: an updated clinical report. Medicine 63: 56-64, 1984.
Enzi, G.: Personal Communication. Padua, Italy 7/11/1984.
Enzi, G.; Angelini, C.; Negrin, P.; Armani, M.; Pierobon, S.; Fedele, D.: Sensory, motor, and autonomic neuropathy in patients with multiple symmetric lipomatosis. Medicine 64: 388-393, 1985.
Enzi, G.; Inelmen, E. M.; Baritussio, A.; Dorigo, P.; Prosdocimi, M.; Mazzoleni, F.: Multiple symmetric-lipomatosis: a defect in adrenergic-stimulated lipolysis. J. Clin. Invest. 60: 1221-1229, 1977.
Greene, M. L.; Glueck, C. J.; Fujimoto, W. Y.; Seegmiller, J. E.: Benign symmetric lipomatosis (Launois-Bensaude adenolipomatosis) with gout and hyperlipoproteinemia. Am. J. Med. 48: 239-246, 1970.
Klopstock, T.; Naumann, M.; Schalke, B.; Bischof, F.; Seibel, P.; Kottlors, M.; Eckert, P.; Reiners, K.; Toyka, K. V.; Reichmann, H.: Multiple symmetric lipomatosis: abnormalities in complex IV and multiple deletions in mitochondrial DNA. Neurology 44: 862-866, 1994.
Lyon, I. P.: Adiposis and lipomatosis: considered in reference to their constitutional relations and symptomatology. Arch. Intern. Med. 6: 28-120, 1910.
Madelung, (NI): Ueber den Fetthals (diffuses Lipom des Halses). Arch. Klin. Chir. 37: 106-130, 1888.
McKusick, V. A.; Abbey, H.; Bowen, P.; Boyer, S. H., IV; Cohen, B. H.; Danks, D. M.; Emery, A. E. H.; Finn, R.; Ferguson-Smith, M. A.; Goodman, R. M.; Handmaker, S. D.; Harris, W. S.; and 15 others : Medical genetics 1961. J. Chronic Dis. 15: 417-572, 1962. Fig. 24.
Michon, P.; Rose, F.: Adenolipomatose symetrique familiale. Bull. Soc. Franc. Derm. Syph. 42: 1005-1007, 1935.
Pollock, M.; Nicholson, G. I.; Nukada, H.; Cameron, S.; Frankish, P.: Neuropathy in multiple symmetric lipomatosis: Madelung's disease. Brain 111: 1157-1171, 1988.
Taylor, L. M.; Beahrs, O. H.; Fontana, R. S.: Benign symmetric lipomatosis. Proc. Staff Meet. Mayo Clinic 36: 96-100, 1961.
Tizian, C.; Berger, A.; Vykoupil, K. F.: Malignant degeneration in Madelung's disease (benign lipomatosis of the neck): case report. Brit. J. Plast. Surg. 36: 187-189, 1983.
Williams, D. W., III; Ginsberg, L. E.; Moody, D. M.; McCain, B. L.: Madelung disease: MR findings. Am. J. Neuroradiol. 14: 1070-1073, 1993.
Zancanaro, C.; Sbarbati, A.; Morroni, M.; Carraro, R.; Cigolini, M.; Enzi, G.; Cinti, S.: Multiple symmetric lipomatosis: ultrastructural investigation of the tissue and preadipocytes in primary culture. Lab. Invest. 63: 253-258, 1990.

ADIPOSIS DOLOROSA (DERCUM DISEASE) (103200)

JJM NOTE

This disease is included because of benign, not malignant neoplasia.

This disorder, which was first described by Dercum (1892), is characterized by painful subcutaneous lipomas in a background of obesity. It is about 5 times more frequent in females than in males. Onset of symptoms is generally in middle age. The fatty tumors are most often located on the trunk and limbs with sparing of the face and hands. Severe asthenia has been emphasized as a feature by some (Wohl and Pastor, 1938). Lynch and Harlan (1963) observed the disease in 4 members of 3 generations of 1 family and in 2, possibly 4, persons in 2 generations of a second family.

See also Cantu et al. (1973)

Cantu, J. M.; Ruiz-Barquin, E.; Jimenez, M.; Castillo, L.; Macotela-Ruiz, E.: Autosomal dominant inheritance in adiposis dolorosa (Dercum's disease). Humangenetik 18: 89-91, 1973.
Dercum, F. X.: Three cases of a hitherto unclassified affection resembling in its grosser aspects obesity, but associated with special nervous symptoms: adiposis dolorosa. Am. J. Med. Sci. 104: 521-535, 1892.
Lynch, H. T.; Harlan, W. L.: Hereditary factors in adiposis dolorosa (Dercum's disease). Am. J. Hum. Genet. 15: 184-190, 1963.
Wohl, M. G.; Pastor, N.: Adipositas dolorosa (Dercum's disease). J.A.M.A. 110: 1261-1264, 1938.

FRONTOFACIONASAL DYSOSTOSIS (*229400)

JJM NOTE

This entity, with three reported persons, includes one young male with lipomas and a sister with a corneal dermoid.

Gollop (1981) suggested that a disorder observed in a brother and sister whose parents were first cousins once removed represented a 'new' autosomal recessive disorder, which he called frontofacionasal dysostosis. The girl showed brachycephaly, bilateral blepharophimosis and ptosis, S-shaped palpebral fissures, bilateral lower lid lagophthalmos ('hare eye'; inability to close the eye completely), limbic dermoid of the left eye, coloboma of the upper lid, bilateral cleft lip and palate, deformed nostrils, and severe midface hypoplasia. The brother had, in addition, prefrontal lipomata, cranium bifidum occultum, and ankyloblepharon filiforme on the right. Gollop et al. (1984) reported a 2-month-old girl whose parents were first cousins. Features were encephalocele, hypertelorism, midface hypoplasia, hypoplasia of frontal bone on the left side, malformed left eye, absent inner eyelashes, irregular S-shaped palpebral fissures, deformed nostrils, hypoplastic right nasal wing and cleft lip, and clefts of premaxilla, palate, and uvula. White et al. (1991) reported a fourth case of this disorder.

Gollop, T. R.: Fronto-facio-nasal dysostosis—a new autosomal recessive syndrome. (Letter) Am. J. Med. Genet. 10: 409-412, 1981.
Gollop, T. R.; Kiota, M. M.; Martins, R. M. M.; Lucchesi, E. A.; Alvarenga, E.: Frontofacionasal dysplasia: evidence for autosomal recessive inheritance. Am. J. Med. Genet. 19: 301-305, 1984.
White, E. W.; Figueroa, R.; Flannery, D. B.: Frontofacionasal dysplasia. Am. J. Med. Genet. 40: 338-340, 1991.

DUPUYTREN CONTRACTURE (*126900)

JJM NOTE

To the extent that the pathologic process is neoplasia or hyperplasia, this common disorder is included. Molecular studies are needed to address vigorously the rationale.

Dupuytren contracture is a connective tissue disorder which is mainly confined to Caucasians and is particularly prevalent among persons of northern European extraction. It may be the most common heritable disorder of connective tissue. Manson (1931) described affected father and 3 sons with contractures of fingers. Autosomal dominance with variable penetrance is likely. There are certain fundamental similarities to Peyronie disease (171000) and the two are associated more frequently than chance alone would dictate. Knuckle pads (149100) are also associated frequently.

Under the designation 'familial fibromatosis,' Young and Fortt (1981) described a family with at least 5 affected members. At the age of 4 months, the proband was first observed to have several small lumps on his trunk. At 10 months, discrete, painless, smooth, nonfluctuant lumps were noted over the left scapula, right axilla, and lower anterior abdomen. These subsequently subsided and, at age 3 years, he showed only 2 small swellings on the back of the neck. The histologic changes were those of fibromatosis. The father had thickening of the palmar fascia bilaterally from at least age 10 years, but no contracture. A paternal uncle of the proband had extensive palmar and plantar fibrosis dating from childhood with contracture of the right fifth finger. His 15-year-old daughter had thickening in the right palm. The proband's paternal grandfather had thickening of the right palmar fascia from age 11 years, resulting in a contracture of the fifth finger by age 45. This family illustrates nosologic uncertainty; see congenital generalized fibromatosis (228550) and juvenile fibromatosis (228600). In a review of disorders associated with knuckle pads, Ramer et al. (1994) referred to the condition in the family of Young and Fortt (1981) as Touraine polyfibromatosis.

Plantar fibromas are also observed with familial Dupuytren contractures (Warren, 1985). Wurster-Hill et al. (1988) performed chromosome studies on the nodular growth in 8 patients. All showed chromosome abnormalities that included numerical and structural clones, random numerical and structural aberrations, and premature centromere separation. Transverse fascial tissue unexpectedly showed the same types of abnormalities as the nodular tissue. In a 3-generation family, Ranta et al. (1990) observed hypodontia associated with Dupuytren contracture. Contracture was observed in 5 individuals, 4 of whom had hypodontia.

Nagai et al. (1996) described plantar fibromatosis in 4 of 7 Japanese sibs. The proband was a 68-year-old man who had noted subcutaneous plantar nodules for 5 years. The nodules recurred after surgical excision; a 58-year-old brother had noted nodules in his left sole 20 years earlier and had 2 operations but the nodules recurred each time. A 53-year-old brother had nodules from the age of 20 years. A 48-year-old sister had noted nodules at the age of 20 years. None of them had diabetes mellitus, epilepsy, alcoholism, or Dupuytren contracture. They cited the report of recurrent plantar fibromatosis in twins (Schindler and Schindler, 1986).

See also Atasu and Ozdemir (1979); Bazin et al. (1980); Bowser-Riley et al. (1975); Hueston (1963); Kipikasa (1972); Kostia (1957); Ling (1963); Lygonis (1966); Matthews (1979); Maza and Goodman (1968); Pierce (1974); Skoog (1948)

Atasu, M.; Ozdemir, N.: Dupuytren's contracture in two families. Hacettepe Bull. Med. Surg. 12: 35-41, 1979.

Bazin, S.; LeLous, M.; Duance, V. C.; Sims, T. J.; Bailey, A. J.; Gabbiani, G.; D'Andiran, G.; Pizzolato, G.; Browski, A.; Nicoletis, C.; Delaunay, A.: Biochemistry and histology of the connective tissue of Dupuytren's disease lesions. Europ. J. Clin. Invest. 10: 9-16, 1980.

Bowser-Riley, S.; Bain, A. D.; Nobel, J.; Lamb, D. W.: Chromosome abnormalities in Dupuytren's disease. Lancet II: 1282-1283, 1975.

Hueston, J. T.: Dupuytren's Contracture. Baltimore: Williams and Wilkins (pub.) 1963.

Kipikasa, A.: The share of heredity in the formation of Dupuytren's contracture. Acta Chir. Plast. 14: 52-59, 1972.

Kostia, J.: A Dupuytren's contracture family. Ann. Chir. Gynaec. Fenn. 46: 351-358, 1957.

Ling, R. S. M.: The genetic factor in Dupuytren's disease. J. Bone Joint Surg. 45-B: 709-718, 1963.

Lygonis, C. S.: Familial Dupuytren's contracture. Hereditas 56: 142-143, 1966.

Manson, J. S.: Heredity and Dupuytren's contracture. Brit. Med. J. 2: 11 only, 1931.

Matthews, P.: Familial Dupuytren's contracture with predominantly female expression. Brit. J. Plast. Surg. 32: 120-123, 1979.

Maza, R. K.; Goodman, R. M.: A family with Dupuytren's contracture. J. Hered. 59: 155-156, 1968.

Nagai, Y.; Okugi, Y.; Ishikawa, O.: Familial plantar fibromatosis. (Letter.) Europ. J. Derm. 6: 463 only, 1996.

Pierce, E. R.: Dupuytren's contractures in three successive generations. Birth Defects Orig. Art. Ser. X(5): 206-207, 1974.

Ramer, J. C.; Vasily, D. B.; Ladda, R. L.: Familial leuconychia, knuckle pads, hearing loss, and palmoplantar hyperkeratosis: an additional family with Bart-Pumphrey syndrome. J. Med. Genet. 31: 68-71, 1994.

Ranta, H.; Knif, J.; Ranta, K.: Familial hypodontia associated with Dupuytren's disease. Scand. J. Dent. Res. 98: 457-460, 1990.

Schindler, S. L.; Schindler, E. I.: Recurrent plantar fibromatosis in twins. (Letter) J. Am. Podiat. Med. Assoc. 76: 654 only, 1986.

Skoog, T.: Dupuytren's contraction with special reference to aetiology and improved surgical treatment: its occurrence in epileptics: note on knuckle-pads. Acta Chir. Scand. 96 (suppl. 139): 1-190, 1948.

Warren, W. D.: Personal Communication. Atlanta, Ga. 5/3/1985.

Wurster-Hill, D. H.; Brown, F.; Park, J. P.; Gibson, S. H.: Cytogenetic studies in Dupuytren contracture. Am. J. Hum. Genet. 43: 285-292, 1988.

Young, I. D.; Fortt, R. W.: Familial fibromatosis. Clin. Genet. 20: 211-216, 1981.

FIBROMATOSIS, CONGENITAL GENERALIZED (228550)

JJM NOTE

This entity illustrates the lethality (malignancy) of histopathologically benign neoplasia. In the absence of other candidate gene loci for this disorder, the locus for neurofibromatosis 1 might be assessed when another family is encountered. Other candidate genes might include cell-cycle proteins.

Congenital generalized fibromatosis is a rare disorder characterized by multiple fibroblastic tumors involving skin, striated muscles, bones and viscera. The tumors are present at birth or develop during the first weeks of life. This disorder was described by Stout (1954), who distinguished it from other forms of juvenile fibromatosis. The radiologic findings are similar to those of Ollier disease (166000). Multiple cystic lesions involve the metaphyses. Multiple soft tissue nodules occur (Shnitka et al., 1958), as in multiple neurofibromatosis (162200), but a cutaneous pigmentary anomaly is not a feature. Hower et al. (1971) described affected half sisters with the same mother. Affected sibs have also been observed by McAdams (as cited by Stout et al., 1961). Bartlett et al. (1961) observed 4 cases among first cousins. The mother of affected brother and sister and the father of another affected brother-sister pair were sibs. Familusi et al. (1976) observed gingival hypertrophy and ankylosis of many joints in a Nigerian infant. Baer and Radkowski (1973) distinguished congenital multiple fibromatosis which carries a better prognosis because viscera are not affected. According to their tabulation of reported cases, Bartlett et al. (1961) and Kauffman and Stout (1965) reported patients of both categories. Stout (1954) and Teng et al. (1963) dealt with congenital generalized fibromatosis and Heiple et al. (1972) reported on congenital multiple fibromatosis (despite the title of his paper). Some consider the tumors a form of hamartomatosis since they often contain smooth muscle and vascular channels in addition to fibrous tissue.

Chung and Enzinger (1981) preferred the term juvenile myofibromatosis because of the presumed myofibroblastic origin of the cells. The prognosis is poor when several internal organs are affected; 80% of such infants are said to die in the first 4 months of life. On the other hand, complete spontaneous regression has been observed (Teng et al., 1963). These characteristics are reminiscent of those of neuroblastoma (256700). Venencie et al. (1987) described infant brothers with cutaneous and skeletal myofibromas. The cutaneous lesions healed spontaneously and completely. Although this and other reports suggest autosomal recessive inheritance, such cannot explain the affected half sisters reported by Hower et al. (1971) and the cases of this disorder in a female infant and her father reported by Jennings et al. (1984). Castro et al. (1987) reported a case of multicentric fibromatosis in a 35-year-old woman who had first onset of a cutaneous tumor at age 5 years. Although the parents came from Jewish lineages that had resided in the same city in Morocco for many generations, no consanguinity was known. One aunt and a female cousin had similar subcutaneous tissues, but the sibs, parents, and grandparents were apparently free of tumors. Bracko et al. (1992) described 2 brothers with multicentric infantile myofibromatosis. In both, tumors were present at birth; the tumors regressed spontaneously but new lesions developed throughout the follow-up periods of 15 and 8 years. Bracko et al. (1992) reviewed 9 additional families with this disorder in more than one member. The occurrence of the disorder in 8 sets of sibs, with consanguinity in 2, favored autosomal recessive inheritance; as noted previously, however, the occurrence in half-sisters and in successive generations favored autosomal dominant inheritance.

See also Altemani et al. (1985); Baird and Worth (1976); Modi (1982); Touraine and Ruel (1945)

Altemani, A. M.; Amstalden, E. I.; Filho, J. M.: Congenital generalized fibromatosis causing spinal cord compression. Hum. Path. 16: 1063-1065, 1985.

Baer, J. W.; Radkowski, M. A.: Congenital multiple fibromatosis: a case report with review of the world literature. Am. J. Roentgen. 118: 200-205, 1973.

Baird, P. A.; Worth, A. J.: Congenital generalized fibromatosis: an autosomal recessive condition?. Clin. Genet. 9: 488-494, 1976.

Bartlett, R. C.; Otis, R. D.; Laakso, A. O.: Multiple congenital neoplasms of soft tissue. Cancer 14: 913-920, 1961.

Bracko, M.; Cindro, L.; Golouh, R.: Familial occurrence of infantile myofibromatosis. Cancer 69: 1294-1299, 1992.

Castro, D. J.; Hoover, L.; Lufkin, R. B.; Castro, D. J.; Fu, T.-S.: Multicentric fibromatosis of familial inheritance. Arch. Path. Lab. Med. 111: 867-869, 1987.

Chung, E. B.; Enzinger, F. M.: Infantile myofibromatosis. Cancer 48: 1807-1818, 1981.

Familusi, J. B.; Nottidge, V. A.; Antia, A. U.; Attah, E. B.: Congenital generalized fibromatosis: an African case with gingival hypertrophy and other unusual features. Am. J. Dis. Child. 130: 1215-1217, 1976.

Heiple, K. G.; Perrin, E.; Aikawa, M.: Congenital generalized fibromatosis. A case limited to osseous lesions. J. Bone Joint Surg. 54A: 663-669, 1972.

Hower, J.; Gobel, F. J.; Ruttner, J. R.; Wurster, K.: Familiaere kongenitale generalisierte Fibromatose bei zwei Halbschwestern. Schweiz. Med. Wschr. 101: 1381-1385, 1971.

Jennings, T. A.; Duray, P. H.; Collins, F. S.; Sabetta, J.; Enzinger, F. M.: Infantile myofibromatosis: evidence for an autosomal dominant disorder. Am. J. Surg. Path. 8: 529-538, 1984.

Kauffman, S. L.; Stout, A. P.: Congenital mesenchymal tumors. Cancer 18: 460-476, 1965.

Modi, N.: Congenital generalised fibromatosis. Arch. Dis. Child. 57: 881-882, 1982.

Shnitka, T. K.; Asp, D. M.; Horner, R. H.: Congenital generalized fibromatosis. Cancer 11: 627-639, 1958.

Stout, A. P.: Juvenile fibromatoses. Cancer 7: 953-978, 1954.

Teng, P.; Warden, M. J.; Cohn, W. L.: Congenital generalized fibromatosis (renal and skeletal) with complete spontaneous regression. J. Pediat. 62: 748-753, 1963.

Touraine, A.; Ruel, H.: La polyfibromatose hereditaire. Ann. Derm. Syph. 29: 1-5, 1945.

Venencie, P. Y.; Bigel, P.; Desgruelles, C.; Lortat-Jacob, S.; Dufier, J. L.; Saurat, J. H.: Infantile myofibromatosis: report of two cases in one family. Brit. J. Derm. 117: 255-259, 1987.

DESMOID DISEASE, HEREDITARY (#135290)

JJM NOTE

This "benign" neoplasm of the mesentery can be life threatening. Although the few known families have had mutations in the *APC* gene for familial polyposis of the colon, the sole feature is the infiltrating fibromatosis without intestinal polyps.

A number sign (#) is used with this entry because of evidence that the defect, at least in some cases, resides in the APC gene (175100).

Maher et al. (1992) described a mother and son with marked infiltrative fibromatosis of the mesentery of the type that occurs as part of the Gardner syndrome, which is caused by a mutation in the APC gene. However, colonic polyps, osteomas, sebaceous cysts, and congenital hypertrophy of the retinal pigment epithelium typical of Gardner syndrome were not observed. In the mother, the diagnosis of infiltrative fibromatosis of the mesentery was made at age 30. Barium enema was negative for colonic polyps. She died at 32 years of age of bowel perforation and intractable abdominal sepsis. No colonic polyps were discovered postmortem. The son was found to have an abdominal mass at age 14 years. Colonoscopic examination at age 15 and search for the extrabowel features of Gardner syndrome were negative on study at age 15. Studies of the family showed that the grandmother, mother, and an aunt and uncle of the 'mother' had died, according to firm or circumstantial information, of nonpolyposis colon cancer syndrome. The mother of the 'mother' also had a breast tumor removed. Maher et al. (1992) recognized the fact that affected members of this family may have had a mutation at the APC locus that predisposed them to the development of desmoid tumor and perhaps colon cancer but not multiple polyposis. It would be of interest to determine whether a mutation of the APC gene is present in this family since somatic mutations of the APC gene have been described in the desmoid tumors that occur in familial adenomatous polyposis (Sen-Gupta et al., 1993).

Eccles et al. (1996) observed a germline mutation in the APC gene segregating with hereditary desmoid disease but without colonic polyposis (see 175100.0025).

Scott et al. (1996) identified a germline APC mutation in the original FIF kindred which they described (Maher et al., 1992). Direct sequencing of genomic DNA revealed a 4-bp deletion at nucleotides 5844-5847 (codon 1962) of the APC sequence (175100.0026). The same mutation occurred in 2 other apparently unrelated families with desmoid tumors. Haplotype analysis suggested a common origin for the APC mutation in the 3 families. From their findings, Scott et al. (1996) concluded that familial adenomatous polyposis and FIF are allelic and that APC mutations that truncate the APC protein distal to the beta-catenin-binding domain are associated with desmoid tumors, absence of congenital hypertrophy of the retinal pigment epithelium (CHRPE), and variable but attenuated polyposis expression.

Eccles, D. M.; van der Luijt, R.; Breukel, C.; Bullman, H.; Bunyan, D.; Fisher, A.; Barber, J.; du Boulay, C.; Primrose, J.; Burn, J.; Fodde, R.: Hereditary desmoid disease due to a frameshift mutation at codon 1924 of the APC gene. Am. J. Hum. Genet. 59: 1193-1201, 1996.

Maher, E. R.; Morson, B.; Beach, R.; Hodgson, S. V.: Phenotypic variation in hereditary nonpolyposis colon cancer syndrome: association with infiltrative fibromatosis (desmoid tumor). Cancer 69: 2049-2051, 1992.

Scott, R. J.; Froggart, N. J.; Trembath, R. C.; Evans, D. G. R.; Hodgson, S. V.; Maher, E. R.: Familial infiltrative fibromatosis (desmoid tumours) (MIM135290) caused by a recurrent 3-prime APC gene mutation. Hum. Molec. Genet. 5: 1921-1924, 1996.

Sen-Gupta, S.; Van der Luijt, R. B.; Bowles, L. V.; Meera Khan, P.; Delhanty, J. D. A.: Somatic mutation of APC gene in desmoid tumour in familial adenomatous polyposis. (Letter) Lancet 342: 552-553, 1993.

FIBROMATOSIS, JUVENILE HYALINE (*228600)

JJM NOTE

This entity is included not because of malignant potential, but because of the benign, but progressive neoplastic nature of the pathologic process(es).

Both males and females are affected and in at least 2 instances (Drescher et al., 1967; Enjoji et al., 1968) 2 affected sibs have been reported. The disorder is characterized by multiple subcutaneous tumors, particularly of the scalp, appearing at about age 2 years and slowly growing, causing deformities. Gingival fibromatosis is associated. The tumors recur after removal. Histologically they demonstrate an abundance of homogeneous, amorphous, acidophilic ground substance in which spindle-shaped cells form minute streaks. Murray (1873) and Whitfield and Robinson (1903) reported 3 affected sibs whose unaffected parents were first cousins. Microscopic and ultrastructural features have been described. Kitano (1976) reported a case in the son of a consanguineous marriage. Large tumors were found on the scalp and whitish nodules on the nape and sides of the neck. Hypertrophic gingivae and tumors at both commissures of the lips were illustrated. Histopathologically, tumor cells were embedded in an amorphous eosinophilic ground substance. X-ray films showed numerous osteolytic and osteoclastic lesions of the skeleton. Fayad et al. (1987) reviewed 15 previously reported cases; the great majority of these had a positive family history and several had parental consanguinity. Aldred and Crawford (1987) gave a comprehensive review of 23 cases from 17 families. Long-term follow-up information is badly needed.

Puretic et al. (1962) described an apparently 'new' form of connective tissue disorder. In addition to the proband, a brother and sister were apparently affected, having died in infancy with painful flexural contractures of the elbows, shoulder joints and knees which developed at about 3 months of age. In addition to contractures, the proband showed deformity of the face and skull; stunted growth; osteolysis of terminal phalanges; multiple large subcutaneous nodes, some calcified; dysseborrheic, sclerodermiform and atrophic changes of the skin; recurrent suppurative infections of the skin, eyes, nose and ears; and gingival fibromatosis. Kitano et al. (1972) reported the disorder in 2 sibs. Ishikawa and Hori (1964) described a 2.5-year-old Japanese infant whose sib had died at 8 months, probably of the same condition. 'Systemic hyalinosis' was suggested as a designation. Cells store a metachromatic substance (Kitano et al., 1972). Suschke and Kunze (1971) considered the condition to be a mucopolysaccharidosis. Landing and Nadorra (1986) suggested that this is the same as juvenile hyaline fibromatosis but distinct from infantile systemic hyalinosis (236490). Gorlin et al. (1990) agreed with this classification. They presented photographs of typical cases (their Fig. 26-2) and emphasized the occurrence of flexion contractures and gingival fibromatosis. Bedford et al. (1991) described 2 severely affected unrelated children with painful flexion contractures of all the large joints, perianal granulomas, and recurrent infections from which both died at 19 months and 2 years. The parents of 1 of the children were consanguineous.

Breier et al. (1997) described a 14-month-old girl with this disorder. Compared with controls, synthesis and degradation of type I collagen (120150, 120160) was found increased in fibroblasts derived from the patient; in contrast, overall metabolism of type III collagen was reduced by 36%.

See also Roggli et al. (1980); Woyke et al. (1970)

Aldred, M. J.; Crawford, P. J. M.: Juvenile hyaline fibromatosis. Oral Surg. Oral Med. Oral Path. 63: 71-77, 1987.
Bedford, C. D.; Sills, J. A.; Sommelet-Olive, D.; Boman, F.; Beltramo, F.; Cornu, G.: Juvenile hyaline fibromatosis: a report of two severe cases. J. Pediat. 119: 404-410, 1991.
Breier, F.; Fang-Kircher, S.; Wolff, K.; Jurecka, W.: Juvenile hyaline fibromatosis: impaired collagen metabolism in human skin fibroblasts. Arch. Dis. Child. 77: 436-440, 1997.
Drescher, E.; Woyke, S.; Markiewicz, C.; Tegi, S.: Juvenile fibromatosis in siblings (fibromatosis hyalinica multiplex juvenilis). J. Pediat. Surg. 2: 427-430, 1967.
Enjoji, M.; Kato, N.; Kamikarzuru, K.; Arima, E.: Juvenile fibromatosis of the scalp in siblings. Acta Med. Univ. Kagoshima Suppl. 10: 145-151, 1968.
Fayad, M. N.; Yacoub, A.; Salman, S.; Khudr, A.; Der Kaloustian, V. M.: Juvenile hyaline fibromatosis: two new patients and review of the literature. Am. J. Med. Genet. 26: 123-131, 1987.
Gorlin, R. J.; Cohen, M. M., Jr.; Levin, L. S.: Syndromes of the Head and Neck. New York: Oxford Univ. Press (pub.) (3rd ed.) : 1990. Pp. 849-850.
Ishikawa, H.; Hori, Y.: Systematisierte Hyalinose in Zusammenhang mit Epidermolysis bullosa polydystrophica und hyalinosis cutis et mucosae. Arch. Klin. Exp. Derm. 218: 30-51, 1964.
Kitano, Y.: Juvenile hyalin fibromatosis. Arch. Derm. 112: 86-88, 1976.
Kitano, Y.; Horiki, M.; Aoki, T.; Sagami, S.: Two cases of juvenile hyalin fibromatosis: some histological, electron microscopic, and tissue culture observations. Arch. Derm. 106: 877-883, 1972.
Landing, B. H.; Nadorra, R.: Infantile systemic hyalinosis: report of four cases of a disease, fatal in infancy, apparently different from juvenile systemic hyalinosis. Pediat. Path. 6: 55-79, 1986.
Murray, J.: On three peculiar cases of molluscum fibrosum in one family. Med. Chir. Trans. London 56: 235-238, 1873.
Puretic, S.; Puretic, B.; Fiser-Herman, M.; Adamcic, M.: A unique form of mesenchymal dysplasia. Brit. J. Derm. 74: 8-19, 1962.
Roggli, V. L.; Kim, H.-S.; Hawkins, E.: Congenital generalized fibromatosis with visceral involvement: a case report. Cancer 45: 954-960, 1980.
Suschke, J.; Kunze, D.: Ein neuer Mucopolysaccharidose. Dtsch. Med. Wschr. 96: 1941-1943, 1971.
Whitfield, A.; Robinson, A. H.: A further report on the remarkable series of cases of molluscum fibrosum in children communicated to the society by Dr. John Murray in 1873. Med.-Chir. Trans. London 86: 293 only, 1903.
Woyke, S.; Wenancjusz, D.; Olszewski, W.: Ultrastructure of a fibromatosis hyalinica multiplex juvenilis. Cancer 26: 1157-1168, 1970.

RAMON SYNDROME (266270)

JJM NOTE

Although not histopathologically malignant, the curious pathogenic mechanism of this syndrome seems to be both dysplasia and multiple neoplasia.

Ramon et al. (1967) described 2 sibs with cherubism (maxillary fibrous dysplasia; see 118400), gingival fibromatosis (see 135300), epilepsy, mental deficiency, hypertrichosis, and stunted growth. De Pina-Neto et al. (1986) described the same disorder in 4 individuals in a Brazilian kindred. A male and 2 females were the offspring of a first-cousin marriage; the fourth case was a male related to them as a second cousin. The features were identical to those of the cases reported by Ramon et al. (1967) except that 3 of the 4 patients also had juvenile rheumatoid arthritis, which de Pina-Neto et al. (1986) suggested should be considered a part of the syndrome. De Pina-Neto et al. (1998) provided information on the clinical evolution of the disorder in the Brazilian family. Three members had developed pigmentary changes in the retina and paleness of the optic disk. Another had developed giant hypertrophy of the labia minora that, when examined histopathologically, was found to be due to neoplastic fibroblast and epithelial proliferation caused by a fibromatous process similar to that reported in the gingivae of patients with this disorder.

Some of the features of Ramon syndrome are found in patients treated with phenytoin. Gingival hyperplasia also occurs in the Rutherfurd syndrome (180900), the Laband syndrome (135500), and the Jones syndrome (135550).

de Pina-Neto, J. M.; Moreno, A. F. C.; Silva, L. R.; Velludo, M. A. S. L.; Patean, E. B. L.; Ribeiro, M. V. M.; Athayde-Junior, L.; Voltarelli, J. C.: Cherubism, gingival fibromatosis, epilepsy, and mental deficiency (Ramon syndrome) with juvenile rheumatoid arthritis. Am. J. Med. Genet. 25: 433-441, 1986.
de Pina-Neto, J. M.; Vieira de Souza, N.; Velludo, M. A. S. L.; Perosa, G. B. D.; de Freitas, M. M. S.; Colafemina, J. F.: Retinal changes and tumorigenesis in Ramon syndrome: follow-up of a Brazilian family. Am. J. Med. Genet. 77: 43-46, 1998.
Ramon, Y.; Berman, W.; Bubis, J. J.: Gingival fibromatosis combined with cherubism. Oral Surg. 24: 436-448, 1967.

FIBROMATOSIS, GINGIVAL (135300)

JJM NOTE

This entity, as well as related ones, needs to be revisited with current genetic techniques, such as segregation and linkage analysis, cytogenetics, and assays of clonality and of loss of heterozygosity.

A mutation for 'gingival fibromatosis' may not be separate from that for 'gingival fibromatosis with hypertrichosis' (135400). Although Zackin and Weisberger (1961) stated that there was 'slight hypertrichosis in all members' of the Italian family they studied, they did not clearly state whether persons without fibromatosis as well as those with it were hirsute. Emerson (1965) reported 13 affected individuals in 4 generations. Becker et al. (1967) described gingival fibromatosis without other features in mother, son and daughter. Ramon et al. (1967) described 2 brothers with features of gingival fibromatosis and cherubism; the parents, who were first-cousin Sephardic Jews, and 6 sibs were healthy (see 266270). Witkop (1971) described an extensively affected kindred in which none of 13 examined cases and none of many others reportedly with the disorder had hypertrichosis. Gorlin (1977) suggested that the disorder described by Chatterjee and Mazumder (1967) as 'fibroosseous dysplasia of the jaws' in a man and his 2 sons may have been gingival fibromatosis in an unusually neglected form. The tumorous involvement reached amazing proportions as shown in the published photographs. The father had progressive swelling of the upper jaw from childhood; involvement of the lower jaw was later in onset. Jorgenson and Cocker (1974) opined that there are both dominant and recessive forms, with generalized and focal types being variable expression of the same disorder.

Both genetic and pharmacologically induced forms of gingival fibromatosis are known. The most common genetic form, hereditary gingival fibromatosis, is usually transmitted as an autosomal dominant trait, although sporadic cases are common and autosomal recessive inheritance has been reported. Most of the recessively inherited gingival fibromatosis occurs as a part of a syndrome, e.g., 228560 and 266270. Syndromal forms of gingival fibromatosis, inherited as autosomal dominant, are represented by 135500 and 135400. Hart et al. (1998) performed a genomewide search for linkage to polymorphic markers in an extended family with an autosomal dominant form of isolated gingival fibromatosis. They found a maximum lod score of 5.05 at theta = 0.00 for the 2p21 region, bounded by loci D2S1788 and D2S441.

Becker, W.; Collings, C. K.; Zimmerman, E. R.; De La Rosa, M.; Singdahlsen, D.: Hereditary gingival fibromatosis. A report on a family in which three members were affected with fibromatosis of the gingiva. Oral Surg. 24: 313-318, 1967.
Chatterjee, S. K.; Mazumder, J. K.: Massive fibro-osseous dysplasia of the jaws in two generations. Brit. J. Surg. 54: 335-340, 1967.
Emerson, T. G.: Hereditary gingival hyperplasia: a family pedigree of four generations. Oral Surg. 19: 1-9, 1965.
Gorlin, R. J.: Personal Communication. Minneapolis, Minn. 1977.
Hart, T. C.; Pallos, D.; Bowden, D. W.; Bolyard, J.; Pettenati, M. J.; Cortelli, J. R.: Genetic linkage of hereditary gingival fibromatosis to chromosome 2p21. Am. J. Hum. Genet. 62: 876-883, 1998.
Jorgenson, R. J.; Cocker, M. E.: Variation in the inheritance and expression of gingival fibromatosis. J. Periodont. 45: 472-477, 1974.
Ramon, Y.; Berman, W.; Bubis, J. J.: Gingival fibromatosis combined with cherubism. Oral Surg. 24: 435-448, 1967.
Witkop, C. J., Jr.: Heterogeneity in gingival fibromatosis. Birth Defects Orig. Art. Ser. VII(7): 210-221, 1971.
Zackin, S. J.; Weisberger, D.: Hereditary gingival fibromatosis. Report of a family. Oral Surg. 14: 828-836, 1961.

FIBROMATOSIS, GINGIVAL, WITH HYPERTRICHOSIS (135400)

JJM NOTE

The same considerations apply to this disorder as are expressed elsewhere in this section.

Extreme hirsutism with gingival fibromatosis follows a dominant pattern of inheritance (Weski, 1920; Garn and Hatch, 1950). I have seen a sporadic case of a severely retarded child who had muscular hypotonia in addition to hypertrichosis and gingival hyperplasia. There is no necessary relationship between the age of development of the gingival changes and the hypertrichosis. The latter may be present at birth but often appears at puberty (Anderson et al., 1969). Horning et al. (1985) described a case of gingival fibromatosis with hypertrichosis. Dilantin produces a phenocopy (Vogel, 1977). Bondeson and Miles (1993) concluded that Julia Pastrana (1834-1860), who was widely exhibited as a 'freak' in America and Europe, had this condition. ('Nondescript' was a term used at the time for strange animals and monsters from remote parts of the world.) Bondeson and Miles (1993) presented historical evidence and x-rays to support the conclusion that gingival hyperplasia was present. This individual and her son, who was also affected, had been thought to have hypertrichosis lanuginosa congenita (145700), but histologic studies demonstrated that Julia Pastrana had hypertrichosis terminalis, the hairy growth being unmistakably terminal in character. Facial deformity gave her the description 'monkey-faced.'

Anderson, J.; Cunliffe, W. J.; Roberts, D. F.; Close, H.: Hereditary gingival fibromatosis. Brit. Med. J. 3: 218-219, 1969.
Bondeson, J.; Miles, A. E. W.: Julia Pastrana, the nondescript: an example of congenital, generalized hypertrichosis terminalis with gingival hyperplasia. Am. J. Med. Genet. 47: 198-212, 1993.
Garn, S. M.; Hatch, C. E.: Hereditary general gingival hyperplasia. J. Hered. 41: 41-42, 1950.
Horning, G. M.; Fisher, J. G.; Barker, B. F.; Killoy, W. J.; Lowe, J. W.: Gingival fibromatosis with hypertrichosis: a case report. J. Periodont. 56: 344-347, 1985.
Vogel, R. I.: Gingival hyperplasia and folic acid deficiency from anticonvulsive drug therapy: a theoretical relationship. J. Theor. Biol. 67: 269-278, 1977.
Weski, H.: Elephantiasis gingivae hereditaria. Dtsch. Mschr. Zahnheilk. 38: 557-584, 1920.

FIBROMATOSIS, GINGIVAL, WITH ABNORMAL FINGERS, FINGERNAILS, NOSE, AND EARS, AND SPLENOMEGALY (*135500)

JJM NOTE

This entity is another variant to be sorted out with current clinical and laboratory tools as the rare families become available.

Gorlin (1967) called my attention to reports of 2 Asiatic Indian families (one living in the Caribbean and one in India) in which gingival fibromatosis occurred in association with 'whittling' of the terminal phalanges and absence or dysplasia of the fingernails. One report (Laband et al., 1964) described the disorder in a 38-year-old Trinidad woman and 5 of her 7 children. The mother showed large, soft ears, hypertension, hyperextensibility of metacarpophalangeal joints, and splenomegaly. The affected children had soft tissue enlargement of the nose and ears, splenomegaly, skeletal abnormalities, obscure or reduced size of toenails and thumbnails, short terminal phalanges, and hypermobility of several joints. In the second report (Alavandar, 1965), 5 persons in 3 generations showed associated features of thickening of the soft tissues of the nose and ear with softness of the cartilages, hyperextensible joints, and hepatomegaly. Chodirker et al. (1986) reported a case of this syndrome with profound mental retardation. Pina-Neto et al. (1988) also reported a case with mental retardation.

This syndrome has also been called Zimmermann-Laband syndrome (ZLS). According to the review by Van Buggenhout et al. (1995) 23 cases, including 11 patients from 2 families, had been reported. Most patients had normal intelligence, although some were mildly retarded. They reported a patient with ZLS and severe mental retardation and concluded that severe mental retardation can be a feature of the syndrome.

Robertson et al. (1998) reported on a 4-decade follow-up of a male with ZLS who developed a cardiomyopathy and dilatation of the aortic root and arch, anomalies previously undescribed in this disorder. From birth he had had a coarse face, protruding tongue, hirsutism, dry thick skin, hypotonia, umbilical hernia, and hepatosplenomegaly. Congestive heart failure due to patent ductus arteriosus resolved after spontaneous closure of the duct. At the age of 8 years, myopia and poorly developed teeth with hypertropic gingival margins were additional features. Extensive areas of long downy body hair had required shaving from age 4 years. Liver, skin, and skeletal muscle histologic findings were all normal after extensive attempts to demonstrate storage material. He progressed to manual employment in a sheltered position, but did not achieve skills sufficient for independent living. A pericardial effusion of uncertain cause was detected during the second decade of life, and increasing dyspnea and reduced exercise tolerance necessitated the formation of a pleuropericardial window at age 24 years. Histologic findings of the pericardium were normal; no evidence for tuberculosis was found. Echocardiography demonstrated left atrial dilatation and progressive dilatation of the aortic arch from the aortic root to the descending aorta. At the age of 37 years his height was 167 cm and head circumference was 60 cm. Gingival fibromatosis was encroaching on the dental crowns and extending posteriorly along the palatal shelves. The thumbs and great toes were small and the toenails and fingernails hypoplastic.

Alavandar, G.: Elephantiasis gingivae: report of an affected family with associated hepatomegaly, soft tissue and skeletal abnormalities. J. All India Dent. Assoc. 37: 349-353, 1965.
Chodirker, B. N.; Chudley, A. E.; Toffler, M. A.; Reed, M. H.: Zimmerman-Laband syndrome and profound mental retardation. Am. J. Med. Genet. 25: 543-547, 1986.
Gorlin, R. J.: Personal Communication. Minneapolis, Minn. 1967.
Laband, P. F.; Habib, G.; Humphreys, G. S.: Hereditary gingival fibromatosis. Report of an affected family with associated splenomegaly and skeletal and soft-tissue abnormalities. Oral Surg. 17: 339-351, 1964.
Pina-Neto, J. M.; Martelli-Soares, L. R.; Oliveira-Souza, A. H.; Lopes-Petean, E. B.; Velludo, M. A. S. L.; Campos de Freitas, A.; Ribas, J. P.: A new case of Zimmerman-Laband syndrome with mild mental retardation, asymmetry of limbs, and hypertrichosis. (Letter) Am. J. Med. Genet. 31: 691-695, 1988.
Robertson, S. P.; Lipp, H.; Bankier, A.: Zimmermann-Laband syndrome in an adult. Long-term follow-up of a patient with vascular and cardiac complications. Am. J. Med. Genet. 78: 160-164, 1998.
Van Buggenhout, G. J. C. M.; Brunner, H. G.; Trommelen, J. C. M.; Hamel, B. C. J.: Zimmermann-Laband syndrome in a patient with severe mental retardation. Genet. Counsel. 6: 321-327, 1995.
#splenomegaly

FIBROMATOSIS, GINGIVAL, WITH PROGRESSIVE DEAFNESS (*135550)

JJM NOTE

The same considerations apply to this entity as with this entire series.

Jones et al. (1977) described a family with gingival fibromatosis associated with progressive sensorineural hearing loss in 5 generations. While 9 persons had GF without demonstrated or reported deafness, all 16 persons with a hearing loss had GF. The proband with the full syndrome was 10 years old. Hartsfield et al. (1985) reported 5 cases of GFD in 3 generations with male-to-male transmission.

Hartsfield, J. K., Jr.; Bixler, D.; Hazen, R. H.: Gingival fibromatosis with sensorineural hearing loss: an autosomal dominant trait. Am. J. Med. Genet. 22: 623-627, 1985.
Jones, G.; Wilroy, R. S., Jr.; McHaney, V.: Familial gingival fibromatosis associated with progressive deafness in five generations of a family. Birth Defects Orig. Art. Ser. XIII(3B): 195-201, 1977.

FIBROMATOSIS, GINGIVAL, WITH DISTINCTIVE FACIES (228560)

JJM NOTE

This is another syndrome with gingival fibromatosis that needs further clinical and genetic characterization.

Goldblatt and Singer (1992) described a brother and sister with gingival fibromatosis associated with craniofacial dysmorphism: relative macrocephaly, bushy eyebrows with synophrys, hypertelorism and downslanting palpebral fissures, flattened nasal bridge and hypoplastic nares, cupid-bow mouth, and highly arched palate. Autosomal recessive inheritance has been suggested for isolated gingival fibromatosis (Jorgenson and Cocker, 1974) and for a disorder with associated cherubism and other abnormalities (266270).

Goldblatt, J.; Singer, S. L.: Autosomal recessive gingival fibromatosis with distinctive facies. Clin. Genet. 42: 306-308, 1992.
Jorgenson, R. J.; Cocker, M. E.: Variation in the inheritance and expression of gingival fibromatosis. J. Periodont. 45: 472-477, 1974.

LIP, MEDIAN NODULE OF UPPER (*151630)

JJM NOTE

This entity is included as a curious mass, even though it is not known to be neoplastic.

Prystowsky and Rogers (1976) observed a nodule in the middle of the upper lip in a mother and all of her 4 children (by 3 husbands). All 9 of the mother's sibs and her father were reported to be affected. Prystowsky and Rogers (1976) proposed that the grandfather may have been homozygous for an autosomal dominant trait. Tsukahara et al. (1994) described median nodule of the upper lip in 18 persons in 3 families in a pattern entirely consistent with autosomal dominant inheritance.

Prystowsky, S. D.; Rogers, J. G.: A median nodule of the upper lip as a familial trait. Oral Surg. Oral Med. Oral Path. 42: 653-655, 1976.
Tsukahara, M.; Fernandez, I.; Sugio, Y.; Shiota, K.: Median nodule of the upper lip: an autosomal dominant trait. Am. J. Med. Genet. 51: 13-15, 1994.

DERMOIDS OF CORNEA (*304730)

JJM NOTE

The fact that this single gene, on chromosome X, is "used" to control mesoblastic proliferation only on the cornea means that thousands of other genes are yet to be discovered as micromanagers of normal structure and function.

Henkind et al. (1973) described 2 male cousins of Puerto Rican extraction who were born with bilateral opaque corneas with no other ocular or systemic abnormality. Their mothers were sisters. The unaffected sister of one of the affected males (Henkind's III-3) gave birth to an affected male (Nitowsky, 1978). Histopathologic study showed that opacifications were due to corneal dermoids, i.e., abnormal mesoblastic tissue covered by epithelium. The authors reviewed other forms of neonatal corneal opacities. Congenital hereditary corneal dystrophy (217700) and sclerocornea (269400) are most likely to be confused with corneal dermoid. Ring dermoid of the cornea (180550) is presumably a distinct entity. The condition described by Guizar-Vazquez et al. (1981) may have been autosomal recessive and, therefore, distinct. Because corneal dermoids represent tissue elements not normally found in the cornea, they are considered to be chistomas. Igbal et al. (1987) performed linkage analysis in a 3-generation family demonstrating typical X-linked recessive inheritance. No recombination was found between the CND gene and DXS43, which maps to Xp22.2-p22.1 (theta = 0.00; lod = 2.4).

Guizar-Vazquez, J.; Luengas-Munoz, F. J.; Antillon, F.: Corneal dermoids and short stature in brother and sister—a new syndrome?. Am. J. Med. Genet. 8: 229-234, 1981.
Henkind, P.; Marinoff, G.; Manas, A.; Freidman, A.: Bilateral corneal dermoids. Am. J. Ophthal. 76: 972-977, 1973.
Igbal, M. A.; Chitayat, D.; Hahm, S. Y. E.; Nitowsky, H. M.: Linkage of gene for corneal dermoids with the DXS43 (Xp22.2-p22.1) locus. (Abstract) Am. J. Hum. Genet. 41: A171 only, 1987.
Nitowsky, H. M.: Personal Communication. New York, N. Y. 1978.

RING DERMOID OF CORNEA (180550)

JJM NOTE

Does this curious tumor represent still another mechanism of tumorigenesis?

Mattos et al. (1980) reported a Peruvian family in which the grandmother, all 3 of her children (daughters), and 1 of the 2 sons of a daughter had bilateral annular limbal dermoids with corneal and conjunctival extensions. There were no associated extraocular anomalies. The choristomas (mass of tissue histologically normal for another site) involved the limbus for 360 degrees, and extended anteriorly onto the cornea and posteriorly about 5 mm. (Conjunctival and limbal dermoids occur with the Goldenhar syndrome (164210).) Each patient had hairs growing from the tumor mass.

Mattos, J.; Contreras, F.; O'Donnell, F. E., Jr.: Ring dermoid syndrome: a new syndrome of autosomal-dominantly inherited, bilateral, annular limbal dermoids with corneal and conjunctival extension. Arch. Ophthal. 98: 1059-1061, 1980.

PALPEBRAL COLOBOMA-LIPOMA SYNDROME (*167730)

JJM NOTE

This syndrome represents a strange union of midfacial dysplasia, malformation, and neoplasia.

Penchaszadeh et al. (1980, 1982) described a Venezuelan family in which 8 persons in 3 generations showed bilateral coloboma of the upper and lower lids at the junction between their middle and inner thirds; prominent circumscribed, rounded deposits of fat in the medial region of both upper lids; aplasia or malposition of the lacrimal puncta; ocular hypertelorism; broad nasal bridge and fatty accumulations on the nasal bridge and nasolabial areas. Male-to-male transmission was observed. Penetrance appeared to be complete. One affected male in the second generation had affected children by each of 2 wives and only normal children by a third. This syndrome was validated by the family reported by Akarsu and Sayli (1991). The family contained 2 instances of male-to-male transmission. The syndrome is easily distinguished from frontonasal dysplasia (136760) and from craniofrontonasal dysplasia (304110), which it only superficially resembles. The appearance as pictured from both the Venezuelan and the Turkish family was striking.

Akarsu, A. N.; Sayli, B. S.: Nasopalpebral lipoma-coloboma syndrome. Clin. Genet. 40: 342-344, 1991.
Penchaszadeh, V. B.; Velasquez, D.; Arrivillaga, R.: Bilateral upper and lower palpebral colobomata and associated facial anomalies: a new autosomal dominant condition. (Abstract) Am. J. Hum. Genet. 32: 122A only, 1980.
Penchaszadeh, V. B.; Velasquez, D.; Arrivillaga, R.: The nasopalpebral lipoma-coloboma syndrome: a new autosomal dominant dysplasia-malformation syndrome with congenital nasopalpebral lipomas, eyelid colobomas, telecanthus, and maxillary hypoplasia. Am. J. Med. Genet. 11: 397-410, 1982.

LIPOMA OF THE CONJUNCTIVA (151700)

JJM NOTE

What has happened to the next two generations of this family?

Saebo (1948) described 3 persons in 3 successive generations: grandfather, mother, and daughter. The tumor is distinct from the conjunctival lipodermoid of the Goldenhar syndrome (164210).

Saebo, J.: Lipoma conjunctivae in three generations. Acta Ophthal. 26: 447-450, 1948.

CARDIOMYOPATHY-HYPOGONADISM-COLLAGENOMA SYNDROME (115250)

JJM NOTE

Could each of these families represent a different entity? The various locations for "cardiomyopathy genes" might be a place to start sorting out these families or the next one that is identified.

Sacks et al. (1980) described a 48-year-old man with tricuspid regurgitation and, at autopsy, a cardiomyopathy involving both ventricles but with predominant involvement of the right ventricle. He also had primary testicular failure and a distinctive type of cutaneous collagenoma. The patient's 2 brothers were found to have similar collagenomas and testicular failure, as well as signs of a mild to moderate degree of cardiomyopathy. The father was 68 years old at death. For several years he had cardiomegaly with atrial fibrillation and chronic congestive heart failure. From birth he had a posterior occipital scalp lesion devoid of hair (this also being the description of the lesion in his 3 sons).

Henderson et al. (1968) described 3 brothers with numerous skin nodules on the back. These consisted of thickened dermis due to increased collagenous tissue. One brother had idiopathic myocardiopathy, a second had atrophy of the left iris and severe high frequency sensorineural hearing loss, and the third had recurrent vasculitis. Thus, the cutaneous abnormality may be merely part of a systemic disorder.

Uitto et al. (1979) reported an American black family with 7 affected in 3 generations, including 1 instance of male-to-male transmission. The asymptomatic skin nodules were mainly on the back and chest. Individual lesions varied from a few millimeters to several centimeters in size, were indurated, and showed minimal epidermal changes. Histologically, they were characterized by excessive accumulations of dense, coarse collagen fibers in the dermis. Onset was in the teens and the number of lesions increased during pregnancy. Hormonal influence is suggested.

Henderson, R. R.; Wheeler, C. E., Jr.; Abele, D. C.: Familial cutaneous collagenoma. Arch. Derm. 98: 23-27, 1968.
Sacks, H. N.; Crawley, I. S.; Ward, J. A.; Fine, R. M.: Familial cardiomyopathy, hypogonadism, and collagenoma. Ann. Intern. Med. 93: 813-817, 1980.
Uitto, J.; Santa-Cruz, D. J.; Eisen, A. Z.: Familial cutaneous collagenoma: genetic studies on a family. Brit. J. Derm. 101: 185-195, 1979.

5. ALIMENTARY SYSTEM

Colorectal Cancer Syndromes without Polyposis

CANCER FAMILY SYNDROME OF LYNCH 1 (#114500)

JJM NOTE

Nearly two dozen entries concern colorectal carcinoma. This entry gives a general overview of colorectal cancer and specifically addresses one phenotype—the cancer family syndrome of Lynch 1—that has been associated with mutations at four loci: *MSH2, MLH1, PMSL1,* and *PMSL2.* The term *cancer family syndrome of Lynch 1* is synonymous with *site-specific familial colon cancer* and *hereditary nonpolypotic colorectal cancer.* For now, I prefer the term *Burt syndrome* for familial colorectal cancer with only a few polyps or adenomas (Burt et al. 1985). It is a mutation at the same locus (*APC*) as familial polyposis coli, but has a much different clinical problem. The research and synthesis of Vogelstein and his group (e.g., Kinzler KW, Vogelstein B: Landscaping the cancer terrain. Science 280:1036–37, 1998) have fleshed out a sturdy model of multistage carcinogenesis that proves to be simplistic, but that emphasizes the sequential acquisition of somatic genetic changes that correlate with the histopathologic pathway from hyperplasia to frank malignancy and metastases. The observations champion an ecogenetic interpretation of colon carcinogenesis, thereby defusing the previous tension between environmental and genetic theories.

A number sign (#) is used with this entry because it is clear that more than one gene locus can be involved alone or in combination in the production of the phenotype. Kinzler and Vogelstein (1996) gave a review of hereditary colorectal cancer (HCC) and the multistep process of carcinogenesis which typically develops over decades and appears to require at least 7 genetic events for completion. Inheritance of a single altered gene can result in a marked predisposition to colorectal cancer in 2 distinct syndromes, familial adenomatous polyposis (FAP; 175100) and hereditary nonpolyposis colorectal cancer. The genetic defect in FAP involves the rate of tumor initiation by targeting the gatekeeper function of the APC gene (175100). In contrast, the defect in HNPCC largely affects tumor aggression by targeting the genome guardian function of DNA repair. Studies of these syndromes have provided unique insights into both inherited and sporadic forms of human tumors.

Colon cancer is a well-known feature of familial polyposis coli (APC; 175100). Cancer of the colon occurred in 7 members of 4 successive generations of the family reported by Kluge (1964), leading him to suggest a simple genetic basis for colonic cancer independent of polyposis. Morson (1973) studied a similar family. The combination of colonic and endometrial cancer has been observed in many families (e.g., Williams, 1978). Sivak et al. (1981) studied a kindred with the familial cancer syndrome in which every confirmed affected member had at least 1 primary carcinoma of the colon. The average age at which cancer appeared was 38 years. Multiple primary neoplasms occurred in 23% of cancer patients. Linkage studies with HLA showed 3 crossovers out of 17 opportunities. The lod score was 1.06 at a recombination fraction of 0.20. Budd and Fink (1981) reported a family with a high frequency of mucoid colonic carcinoma. Since endometrial carcinoma, atypical endometrial hyperplasia, uterine leiomyosarcoma, bladder transitional carcinoma, and renal cell carcinoma also occurred in the family, this may be the same disorder as the cancer family syndrome of Lynch, type II (114400). Lynch and Lynch (1979) pointed out that cancer of the right colon is particularly characteristic of the cancer-family syndrome. In the DNA from 1 colon and 2 lung carcinoma cell lines, Perucho et al. (1981) demonstrated the same or closely related transforming elements. By DNA-mediated gene transfer, mouse fibroblasts could be morphologically transformed and rendered tumorigenic in nude mice. Bamezai et al. (1984) reported an Indian Sikh kindred in which 8 persons suffered from cancer of the cecum, not associated with polyposis. Burt et al. (1985) studied a large Utah kindred called to attention because of occurrence of colorectal cancer in a brother, a sister, and a nephew. No clear inheritance pattern was discernible until systematic screening was undertaken for colonic polyps using flexible proctosigmoidoscopy. One or more adenomatous polyps were found in 41 of 191 family members (21%) and 12 of 132 controls (9%)—p less than 0.005. Pedigree analysis showed best fit with autosomal dominant inheritance. Cannon-Albright et al. (1988) extended the studies with investigations of 33 additional kindreds. The kindreds were selected through either a single person with an adenomatous polyp or a cluster of relatives with colonic cancer. The kindreds all had common colorectal cancers, not the rare inherited condition of familial polyposis coli or nonpolyposis inherited colorectal cancer. Likelihood analysis strongly supported dominant inheritance of a susceptibility to colorectal adenomas and cancers, with a gene frequency of 19%. According to the most likely genetic model, adenomatous polyps and colorectal cancers occur only in genetically susceptible persons; however, the 95% confidence interval for this proportion was 53 to 100%. Ponz de Leon et al. (1992) analyzed data on 605 families of probands with colorectal cancer in the province of Modena in Italy. Among the 577 presumed nonpolyposis cases, both parents had colorectal cancer in 11, one parent in 130, and neither parent in 436. Segregation was compatible with dominant transmission of susceptibility to cancer. In preliminary observations, Pathak and Goodacre (1986) found deletion of 12p in colorectal cancer specimens.

Fearon et al. (1987) studied the clonal composition of human colorectal tumors. Using X-linked RFLPs, they showed that all 50 tumors from females showed a monoclonal pattern of X-chromosome inactivation; these tumors included 20 carcinomas and 30 adenomas of either familial or spontaneous type. In over 75% of carcinomas examined, somatic loss of chromosome 17p sequences was found; such loss was rare in adenomas. Fearon et al. (1987) suggested that a gene on the short arm of chromosome 17 may be associated with progression from the benign to the malignant state. Mecklin (1987) investigated the frequency of hereditary colorectal cancer among all colorectal cancer patients diagnosed in 1 Finnish county during the 1970s. The cancer family syndrome type of hereditary nonpolyposis colorectal carcinoma emerged as the most common verifiable risk factor, involving between 3.8 and 5.5% of all colorectal cancer patients. The frequencies of familial adenomatosis and ulcerative colitis were 0.2% and 0.6%, respectively. The observed frequency is probably an underestimate. The patients with cancer family syndrome were young, accounting for 29 to 39% of the patients under 50 years of age, and their tumors were located predominantly (65%) in the right hemicolon. By a combination of DNA hybridization analyses and tissue sectioning techniques, Bos et al. (1987) demonstrated that RAS gene mutations occur in over a third of colorectal cancers, that most of the mutations are at codon 12 of the KRAS gene (190070), and that the mutations usually precede the development of malignancy. In 38 tumors from 25 patients with familial polyposis coli, and in 20 sporadic colon carcinomas, Okamoto et al. (1988) found frequent occurrence of allele loss on chromosome 22, with some additional losses on chromosomes 5, 6, 12q, and 15. The DNA probe C11p11, which has been found to be linked to familial polyposis coli, also detected frequent allele loss in both familial and sporadic colon carcinomas but not in b enign adenomas. In a more extensive study, Vogelstein et al. (1988) studied the interrelationships of the 4 alterations demonstrated in colorectal cancer (RAS gene mutations and deletions of chromosome 5, 17 and 18 sequences) and determined their occurrence with respect to different stages of colorectal tumorigenesis. They found RAS gene mutations frequently in adenomas, this being the first demonstration of such in benign human tumors. In adenomas greater than 1 cm in size, the prevalence was similar to that observed in carcinomas (58% and 47%, respectively). Sequences on chromosome 5 that are linked to familial adenomatous polyposis were seldom lost in adenomas from such patients. Therefore, the Knudson model is unlikely to be applicable to the adenoma/carcinoma sequence in this disorder. Chromosome 18 sequences were lost frequently in colon carcinomas (73%) and in advanced adenomas (47%), but only occasionally in earlier stage adenomas (11-13%); see 120470. Chromosome 17 sequences were usually lost only in carcinomas (75%). The results suggested a model wherein the steps required for malignancy involve the activation of a dominantly acting oncogene coupled with the loss of several genes that normally suppress tumorigenesis; see 120460.

Wildrick and Boman (1988) found deletion of the glucocorticoid receptor locus (138040), located on 5q, in colorectal cancers. Law et al. (1988) examined the question of whether the gene for familial polyposis coli on chromosome 5 may be the site of changes leading to colorectal cancer in the general population, analogous to recessive tumor genes in retinoblastoma and Wilms tumor. To avoid error in interpretation of allelic loss from a study of nonhomogeneous samples, tumor cell populations were first microdissected from 24 colorectal carcinomas, an additional 9 cancers were engrafted in nude mice, and nuclei were flow-sorted in an additional 2. Of 31 cancers informative for chromosome 5 markers, only 6 (19%) showed loss of heterozygosity of chromosome 5 alleles, compared to 19 of 34 (56%) on chromosome 17, and 17 of 33 (52%) on chromosome 18. Law et al. (1988) concluded that FPC is a true dominant for adenomatosis but not a common recessive gene for colon cancer, and that simple mendelian models involving loss of alleles at a single locus may be inappropriate for understanding common human solid tumors. Vogelstein et al. (1989) examined the extent and variation of allelic loss for polymorphic DNA markers in every nonacrocentric autosomal arm in 56 paired colorectal carcinoma and adjacent normal colonic mucosa specimens. They referred to the analysis as an allelotype, in analogy with a karyotype. Three major conclusions were drawn from the study: (1) Allelic deletions are remarkably common; 1 of the alleles of each polymorphic marker tested was lost in at least some tumors, and some tumors lost more than half of their parental alleles. (2) In addition to allelic deletions, new DNA fragments not present in normal tissue were identified in 5 carcinomas; these new fragments contained repeated sequences (of the variable-number-of-tandem-repeat type). (3) Patients with more than the median percentage of allelic deletions had a considerably worse prognosis than did the other p atients, although the stage and size of the primary tumors were very similar in the 2 groups.

Delattre et al. (1989) reviewed the 3 general types of genetic alterations in colorectal cancer: (1) change in DNA content of the malignant cells as monitored by flow cytometry; (2) specific loss of genetic material, i.e., a complete loss of chromosome 18 and a structural rearrangement of chromosome 17 leading most often to the loss of 1 short arm, and loss of part of 5q as demonstrated by loss of heterozygosity; and (3) in nearly 40% of tumors, activation by point mutation of RAS oncogenes (never HRAS, rarely NRAS, and most frequently KRAS). In KRAS, with 1 exception, the activation has always occurred by a change in the coding properties of the 12th or 13th codon. In studies of the multiple genetic alterations in colorectal cancer, Delattre et al. (1989) found that deletions and mitotic abnormalities occurred more frequently in distal than in proximal tumors. The frequency of KRAS mutations did not differ between proximal and distal cancers. In studies of 15 colorectal tumors, Konstantinova et al. (1991) found rearrangements of the short arm of chromosome 17, leading to deletion of this arm or part of it in 12; in 2 others, one of the homologs of pair 17 was lost. One chromosome 18 was lost in 12 out of 13 cases with fully identified numerical abnormalities; chromosome 5, in 6 tumors; and other chromosomes in lesser numbers of cases. See 120470 for a discussion of a gene on chromosome 18 called DCC ('deleted in colorectal cancer') that shows mutations, including point mutations, in colorectal tumor tissue; also see 164790 for a discussion of a mutation in the NRAS oncogene in colorectal cancer. Kikuchi-Yanoshita et al. (1992) presented evidence that genetic changes in both alleles of the TP53 gene through mutation and LOH, which result in abnormal protein accumulation, are involved in the conversion of adenoma to early carcinoma in both familial adenomatous polyposis and in nonfamilial polyposis cases. On the basis of complex segregation analysis of a published series of consecutive pedigrees ascertained through patients undergoing treatment for colorectal cancer, Houlston et al. (1992) concluded that a dominant gene (or genes) with a frequency of 0.006 with a lifetime penetrance of 0.63 is likely. The gene was thought to account for 81% of colorectal cancer in patients under 35 years of age; however, by age 65, about 85% appeared to be phenocopies.

Fearon and Vogelstein (1990) reviewed the evidence supporting their multistep genetic model for colorectal tumorigenesis. They suggested that multiple mutations lead to a progression from normal epithelium to metastatic carcinoma through hyperplastic epithelium—early adenoma—intermediate adenoma—late adenoma—and carcinoma. The genes in which mutations occur at steps in this process include APC on chromosome 5, KRAS on chromosome 12, TP53 (191170) on 17p, and DCC on chromosome 18. Hereditary nonpolyposis colorectal carcinoma (HNPCC) comprises about 5% of all colorectal carcinomas (Lynch, 1986; Mecklin, 1987). The minimum criterion of HNPCC is that colorectal carcinoma is diagnosed and histologically verified in at least 3 relatives belonging to 2 or more successive generations. Moreover, the age of onset should be less than 50 years in at least 1 patient. In addition to the colon (most often the right side), organs commonly affected with cancer include the endometrium, stomach, biliary and pancreatic system, and urinary tract (Mecklin and Jarvinen, 1991). Peltomaki et al. (1992) performed linkage studies in 9 Finnish families, demonstrating that HNPCC is not linked to the MCC (mutated in colon cancer; 159350)/APC region on 5q21; combined maximum lod score = -22.57 at a recombination fraction of 0.00. The report demonstrated the feasibility of studying DNA not only from blood samples from living family members but also from formaldehyde-fixed archival pathology specimens from deceased individuals.

Other genes that have been demonstrated or suspected of involvement in colorectal cancer include MSH2 (120435) on chromosome 2 and the DRA candidate colon tumor-suppressor gene (126650) on chromosome 7.

In addition, the state of DNA methylation appears to play a role in genetic instability in colorectal cancer cells. Lengauer et al. (1997) noted that DNA methylation is essential in prokaryotes, dispensable in lower eukaryotes (such as Saccharomyces cerevisiae) yet present and presumably important in mammals. Many cancers have been shown to have a global hypomethylation of DNA compared with normal tissues. Treatment of cells or animals with 5-azacytidine (5-aza-C), a demethylating agent that irreversibly inactivates methyltransferase (see 156569), is oncogenic in vitro and in vivo. Conversely, other studies showed that hypermethylation of specific sequences found in some tumors can be associated with the inactivation of tumor suppressor gene expression. Mice genetically deficient in methyltransferase are resistant to colorectal tumorigenesis initiated by mutation of the APC tumor suppressor gene, and treatment of these mice with 5-aza-C enhances the resistance (Laird et al., 1995).

Lengauer et al. (1997) reported a striking difference in the expression of exogenously introduced retroviral genes in various colorectal cancer cell lines. Extinguished expression was associated with DNA methylation and could be reversed by treatment with the demethylating agent 5-aza-C. A striking correlation between genetic instability and methylation capacity suggested that methylation abnormalities may play a role in the chromosome segregation processes in cancer cells. It has been speculated that genetic instability is necessary for a tumor to accumulate the numerous genetic alterations that accompany carcinogenesis. There appeared to exist 2 pathways of genetic instability in colorectal cancer. The first is found in about 15% of tumors and involves point mutations, microdeletions, and microinsertions associated with deficiency of mismatch repair (MMR). The second is found in MMR-proficient cells and involves gains and losses of whole chromosomes. Lengauer et al. (1997) suggested that methylation abnormalities are intrinsically and directly involved in the generation of the second type of instability, thus allowing for the selection of methylation-negative cells during the clonal evolution of tumors. The hypothesis was supported by the observation that demethylation is associated with chromosomal aberrations, including mitotic dysfunction and translocation, and was consistent with the hypothesis relating methylation and aneuploidy put forward by Thomas (1995). Jones and Gonzalgo (1997) commented on altered DNA methylation and genome instability as a new pathway to cancer.

In a second report, Lengauer et al. (1997) showed that tumors without microsatellite instability exhibit a striking defect in chromosome segregation, resulting in gains or losses in excess of 10(-2) per chromosome per generation. This form of chromosomal instability reflected a continuing cellular defect that persisted throughout the lifetime of the tumor cell and was not simply related to chromosome number. While microsatellite instability is a recessive trait, chromosomal instability appeared to be dominant. The data indicated that persistent genetic instability may be critical for the development of all colorectal cancers, and that this instability can arise through 2 distinct pathways.

See also Lovett (1976); Lovett (1976); Mathis (1962)

Bamezai, R.; Singh, G.; Khanna, N. N.; Singh, S.: Genetics of site specific colon cancer: a family study. Clin. Genet. 26: 129-132, 1984.

Bos, J. L.; Fearon, E. R.; Hamilton, S. R.; Verlaan-de Vries, M.; van Boom, J. H.; van der Eb, A. J.; Vogelstein, B.: Prevalence of ras gene mutations in human colorectal cancers. Nature 327: 293-297, 1987.

Budd, D. C.; Fink, D. L.: Mucoid colonic carcinoma as an autosomal-dominant inherited syndrome. Arch. Surg. 116: 901-905, 1981.

Burt, R. W.; Bishop, D. T.; Cannon, L. A.; Dowdle, M. A.; Lee, R. G.; Skolnick, M. H.: Dominant inheritance of adenomatous colonic polyps and colorectal cancer. New Eng. J. Med. 312: 1540-1544, 1985.

Cannon-Albright, L. A.; Skolnick, M. H.; Bishop, T.; Lee, R. G.; Burt, R. W.: Common inheritance of susceptibility to colonic adenomatous polyps and associated colorectal cancers. New Eng. J. Med. 319: 533-537, 1988.

Delattre, O.; Olschwang, S.; Law, D. J.; Melot, T.; Remvikos, Y.; Salmon, R. J.; Sastre, X.; Validire, P.; Feinberg, A. P.; Thomas, G.: Multiple genetic alterations in distal and proximal colorectal cancer. Lancet II: 353-356, 1989.

Fearon, E. R.; Hamilton, S. R.; Vogelstein, B.: Clonal analysis of human colorectal tumors. Science 238: 193-197, 1987.

Fearon, E. R.; Vogelstein, B.: A genetic model for colorectal tumorigenesis. Cell 61: 759-767, 1990.

Houlston, R. S.; Collins, A.; Slack, J.; Morton, N. E.: Dominant genes for colorectal cancer are not rare. Ann. Hum. Genet. 56: 99-103, 1992.

Jones, P. A.; Gonzalgo, M. L.: Altered DNA methylation and genome instability: a new pathway to cancer? Proc. Nat. Acad. Sci. 94: 2103-2105, 1997.

Kikuchi-Yanoshita, R.; Konishi, M.; Ito, S.; Seki, M.; Tanaka, K.; Maeda, Y.; Iino, H.; Fukayama, M.; Koike, M.; Mori, T.; Sakuraba, H.; Fukunari, H.; Iwama, T.; Miyaki, M.: Genetic changes of both p53 alleles associated with the conversion from colorectal adenoma to early carcinoma in familial adenomatous polyposis and non-familial adenomatous polyposis patients. Cancer Res. 52: 3965-3971, 1992.

Kinzler, K. W.; Vogelstein, B.: Lessons from hereditary colorectal cancer. Cell 87: 159-170, 1996.

Kluge, T.: Familial cancer of the colon. Acta Chir. Scand. 127: 392-398, 1964.

Konstantinova, L. N.; Fleischman, E. W.; Knisch, V. I.; Perevozchikov, A. G.; Kopnin, B. P.: Karyotype pecularities (sic) of human colorectal adenocarcinomas. Hum. Genet. 86: 491-496, 1991.

Laird, P. W.; Jackson-Grusby, L.; Fazeli, A.; Dickinson, S. L.; Jung, W. E.; Li, E.; Weinberg, R. A.; Jaenisch, R.: Suppression of intestinal neoplasia by DNA hypomethylation. Cell 81: 197-205, 1995.
Law, D. J.; Olschwang, S.; Monpezat, J. P.; Lefrancois, D.; Jagelman, D.; Petrelli, N. J.; Thomas, G.; Feinberg, A. P.: Concerted nonsyntenic allelic loss in human colorectal carcinoma. Science 241: 961-965, 1988.
Lengauer, C.; Kinzler, K. W.; Vogelstein, B.: Genetic instability in colorectal cancers. Nature 386: 623-627, 1997.
Lengauer, C.; Kinzler, K. W.; Vogelstein, B.: DNA methylation and genetic instability in colorectal cancer cells. Proc. Nat. Acad. Sci. 94: 2545-2550, 1997.
Lovett, E.: Familial cancer of the gastro-intestinal tract. Brit. J. Surg. 63: 19-22, 1976.
Lovett, E.: Family studies in cancer of the colon and rectum. Brit. J. Surg. 63: 13-18, 1976.
Lynch, H. T.: Frequency of hereditary nonpolyposis colorectal carcinoma (Lynch syndromes I and II). Gastroenterology 90: 486-492, 1986.
Lynch, H. T.; Lynch, P.: The cancer-family syndrome: a pragmatic basis for syndrome identification. Dis. Colon Rectum 22: 106-110, 1979.
Mathis, V. M.: Familiaeres Colon Karzinom. Ein Stammbaum aus dem Kanton Aargau. Schweiz. Med. Wschr. 92: 1673-1678, 1962.
Mecklin, J.-P.: Frequency of hereditary colorectal carcinoma. Gastroenterology 93: 1021-1025, 1987.
Mecklin, J.-P.; Jarvinen, H. J.: Tumor spectrum in cancer family syndrome (hereditary nonpolyposis colorectal cancer). Cancer 68: 1109-1112, 1991.
Morson, B. C.: Personal Communication. London 1973.
Okamoto, M.; Sasaki, M.; Sugio, K.; Sato, C.; Iwama, T.; Ikeuchi, T.; Tonomura, A.; Sasazuki, T.; Miyaki, M.: Loss of constitutional heterozygosity in colon carcinoma from patients with familial polyposis coli. Nature 331: 273-277, 1988.
Pathak, S.; Goodacre, A.: Specific chromosome anomalies and predisposition to human breast, renal cell, and colorectal carcinoma. Cancer Genet. Cytogenet. 19: 29-36, 1986.
Peltomaki, P.; Sistonen, P.; Mecklin, J.-P.; Pylkkanen, L.; Aaltonen, L.; Nordling, S.; Kere, J.; Jarvinen, H.; Hamilton, S. R.; Petersen, G.; Kinzler, K. W.; Vogelstein, B.; de la Chapelle, A.: Evidence that the MCC-APC gene region in 5q21 is not the site for susceptibility to hereditary nonpolyposis colorectal carcinoma. Cancer Res. 52: 4530-4533, 1992.
Perucho, M.; Goldfarb, M.; Shimizu, K.; Lama, C.; Fogh, J.; Wigler, M.: Human-tumor-derived cell lines contain common and different transforming genes. Cell 27: 467-476, 1981.
Ponz de Leon, M.; Scapoli, C.; Zanghieri, G.; Sassatelli, R.; Sacchetti, C.; Barrai, I.: Genetic transmission of colorectal cancer: exploratory data analysis from a population based registry. J. Med. Genet. 29: 531-538, 1992.
Sivak, M. V., Jr.; Sivak, D. S.; Braun, W. A.; Sullivan, B. H., Jr.: A linkage study of HLA and inherited adenocarcinoma of the colon. Cancer 48: 76-81, 1981.
Thomas, J. H.: Genomic imprinting proposed as a surveillance mechanism for chromosome loss. Proc. Nat. Acad. Sci. 92: 480-482, 1995.
Vogelstein, B.; Fearon, E. R.; Hamilton, S. R.; Kern, S. E.; Preisinger, A. C.; Leppert, M.; Nakamura, Y.; White, R.; Smits, A. M. M.; Bos, J. L.: Genetic alterations during colorectal-tumor development. New Eng. J. Med. 319: 525-532, 1988.
Vogelstein, B.; Fearon, E. R.; Kern, S. E.; Hamilton, S. R.; Preisinger, A. C.; Nakamura, Y.; White, R.: Allelotype of colorectal carcinomas. Science 244: 207-211, 1989.
Wildrick, D. M.; Boman, B. M.: Chromosome 5 allele loss at the glucocorticoid receptor locus in human colorectal carcinomas. Biochem. Biophys. Res. Commun. 150: 591-598, 1988.
Williams, C.: Management of malignancy in 'cancer families.'. Lancet I: 198-199, 1978.

CANCER FAMILY SYNDROME OF LYNCH 2 (#114400)

JJM NOTE

This description is of historic interest because the molecular basis of the Lynch cancer family syndromes points to at least five loci that contribute to the two clinically distinct syndromes.

A number sign (#) is used with this entry because it is likely that germinal mutations in any one of several different genes can be responsible for familial cancer and because the Lynch cancer family syndrome II appears to be due to mutation in a gene on chromosome 2 (MSH2; 120435) and perhaps to mutation in other genes, including one on chromosome 18 (see later).

Using the fourth edition (1975) of these catalogs, Mulvihill et al. (1977) counted 200 entries in which neoplasia was a regular or occasional feature. Some, such as von Recklinghausen neurofibromatosis (162200), tylosis (148500), the several types of intestinal polyposis (e.g., 175100), von Hippel-Lindau syndrome (193300), and the basal cell nevus syndrome (109400), are listed in the dominant catalog. Xeroderma pigmentosum (e.g., 278700), a recessive, is complicated by skin malignancy unless exposure to ultraviolet light is stringently avoided. In addition, notable instances of 'cancer families' are on record. For example, Lynch et al. (1966) reported 2 large 'cancer families.' In 1 family, 9 of 11 sibs had histologically confirmed cancers, with 4 of these showing multiple primary tumors. In the second family, 7 of 13 sibs showed histologically proven cancers, with multiple primary malignant neoplasm in 4. The 2 families contained 6 instances of the ordinarily rare combination of primary colonic and endometrial carcinoma. The 'cancer family' of Warthin is another notable example (Hauser and Weller, 1936).

Lynch et al. (1966) and Lynch and Krush (1967) suggested the existence of a syndrome, which they called the cancer-family syndrome, that is characterized by (1) increased occurrence of endometrial carcinoma and adenocarcinoma of the colon as well as multiple primary malignant neoplasms, and (2) autosomal dominant inheritance. Lynch and Lynch (1979) pointed out that cancer of the right colon is particularly characteristic of the cancer-family syndrome. Familial aggregation alone may be chance inasmuch as about 1 in 4 Americans develops cancer in a lifetime. Features of cancers with a genetic origin include early age of onset, bilaterality or multifocality, multiplicity of primary cancers, and, of course, familiality. Biochemical insight into familial susceptibility to cancer is beginning. Lynch et al. (1973) suggested that among families with breast cancer some have an excess of ovarian cancer, others are prone to sarcoma, brain tumors and leukemia, whereas yet others have associated gastrointestinal cancer. Genetic differences in inducibility of aryl hydrocarbon hydroxylase (see 108340) may underlie susceptibility to lung cancer and colon cancer.

Hereditary nonpolyposis colorectal cancer (HNPCC) is subdivided into (1) Lynch syndrome I (site-specific colonic cancer; see 114500) and (2) Lynch syndrome II (colonic cancer in association with other forms of cancer, particularly carcinoma of the endometrium and ovary, but also cancer of the pancreas) (Lynch et al., 1985). Both HNPCC disorders show a proclivity to early onset, predominant proximal location of colon cancer, a dominant pattern of inheritance, an excess of multiple primary cancers, and significantly improved survival when compared stage for stage with the American College of Surgeons Audit Series. Use of these features in surveillance and management programs mandates periodic colonoscopy or double air-contrast barium enema, because of the proximal location, and a lifelong watchfulness with attention to other tumors integral to the syndrome.

Warthin's original description of 'Family G' (Warthin, 1913) showed an excess of gastric cancer. The decline in incidence of gastric carcinoma and increase in colonic cancer in recent years was found to have its parallel in Family G on update by Lynch and Krush (1971). Cristofaro et al. (1987) and Guanti et al. (1990) described an extensively affected Italian family with the characteristic features of the Lynch cancer family syndrome II: early age of onset of tumors, increased frequency of adenocarcinomas of the colon, mainly with proximal location, and high occurrence of gastric, endometrial, and multiple primary malignancies. Unique pathologic findings included chronic atrophic gastritis and an excess of macrophages in association with atrophy of crypts in the colonic mucosa. Abusamra et al. (1987) described a family in which 8 cancers (6 colonic and 2 endometrial) occurred in 7 members of 3 generations. The colonic cancer was diagnosed in 5 of the 6 affected patients at an unusually young age, had a predilection for the proximal colon, and was of the mucinous type in 4 patients. No polyposis was found.

Lynch et al. (1991) estimated that hereditary nonpolyposis colorectal cancer accounts for about 4 to 6% of colorectal cancer.

Lynch et al. (1985) described linkage of the cancer family syndrome (Lynch syndrome II) to Kidd blood group (111000); a lod score of 3.19 was obtained. The Kidd blood group locus has been assigned to 18q11-q12; colorectal cancer-related sequences have been identified in the region 18q23.3-qter (120470).

See also Blattner et al. (1979); Brisman et al. (1967); Dubosson (1977); Dunstone and Knaggs (1972); Fielding (1969); Lynch (1967); Lynch et al. (1985); Maack and Rudiger (1983)

Abusamra, H.; Maximova, S.; Bar-Meir, S.; Krispin, M.; Rotmensch, H. H.: Cancer family syndrome of Lynch. Am. J. Med. 83: 981-983, 1987.
Blattner, W. A.; McGuire, D. B.; Mulvihill, J. J.; Lampkin, B. C.; Hananian, J.; Fraumeni, J. F., Jr.: Genealogy of cancer in a family. J.A.M.A. 241: 259-261, 1979.
Brisman, R.; Baker, R. R.; Elkins, R.; Hartmann, W. H.: Carcinoma of lung in four siblings. Cancer 20: 2048-2053, 1967.
Cristofaro, G.; Lynch, H. T.; Caruso, M. L.; Attolini, A.; DiMatteo, G.; Giorgio, P.; Senatore, S.; Argentieri, A.; Sbano, E.; Guanti, G.; Fusaro, R.; Giorgio, I.: New phenotypic aspects in a family with Lynch syndrome II. Cancer 60: 51-58, 1987.
Dubosson, J.-D.: Adenocarcinomatose hereditaire dans quatre generations d'une famille Valaissanne. J. Genet. Hum. 25: 233-278, 1977.
Dunstone, G. H.; Knaggs, T. W. L.: Familial cancer of the colon and rectum. J. Med. Genet. 9: 451-456, 1972.
Fielding, J. F.: Familial non-polypotic carcinoma of the colon. Brit. Med. J. 1: 512-513, 1969.
Guanti, G.; Susca, F.; Cristofaro, G.; Caruso, M. L.; Massari, S.; Porsia, R.; Stella, A.; Giorgio, I.: Cancer family syndrome: cytogenetic investigations, in vitro tetraploidy, and biomarker studies in a large family. J. Med. Genet. 27: 441-445, 1990.

Hauser, I. J.; Weller, C. V.: A further report on the cancer family of Warthin. Am. J. Cancer 27: 434-449, 1936.
Lynch, H. T.: Hereditary Factors in Carcinoma. Berlin and New York: Springer (pub.) 1967.
Lynch, H. T.; Krush, A. J.: Heredity and adenocarcinoma of the colon. Gastroenterology 53: 517-527, 1967.
Lynch, H. T.; Krush, A. J.: Cancer family 'G' revisited: 1895-1970. Cancer 27: 1505-1511, 1971.
Lynch, H. T.; Krush, A. J.; Guirgis, H.: Genetic factors in families with combined gastrointestinal and breast cancer. Am. J. Gastroent. 59: 31-40, 1973.
Lynch, H. T.; Lanspa, S.; Smyrk, T.; Boman, B.; Watson, P.; Lynch, J.: Hereditary nonpolyposis colorectal cancer (Lynch syndromes I & II): genetics, pathology, natural history, and cancer control. Part I. Cancer Genet. Cytogenet. 53: 143-160, 1991.
Lynch, H. T.; Lynch, P.: The cancer-family syndrome: a pragmatic basis for syndrome identification. Dis. Colon Rectum 22: 106-110, 1979.
Lynch, H. T.; Schuelke, G. S.; Kimberling, W. J.; Albano, W. A.; Lynch, J. F.; Biscone, K. A.; Lipkin, M. L.; Deschner, E. E.; Mikol, Y. B.; Sandberg, A. A.; Elston, R. C.; Bailey-Wilson, J. E.; Danes, B. S.: Hereditary nonpolyposis colorectal cancer (Lynch syndromes I and II). II. Biomarker studies. Cancer 56: 939-951, 1985.
Lynch, H. T.; Shaw, M. W.; Magnuson, C. W.; Larsen, A. L.; Krush, A. J.: Hereditary factors in cancer: study of two large midwestern kindreds. Arch. Intern. Med. 117: 206-212, 1966.
Lynch, H. T.; Voorhees, G. J.; Lanspa, S. J.; McGreevy, P. S.; Lynch, J. F.: Pancreatic carcinoma and hereditary nonpolyposis colorectal cancer: a family study. Brit. J. Cancer 52: 271-273, 1985.
Maack, P.; Rudiger, H. W.: Familial cancer or cancer family syndrome: report on a cancer family and consideration of genetic mechanisms. Clin. Genet. 24: 36-40, 1983.
Mulvihill, J. J.; Miller, R. W.; Fraumeni, J. F.: Genetics of Human Cancer. New York: Raven Press (pub.) 1977.
Warthin, A. S.: Heredity with reference to carcinoma. Arch. Intern. Med. 12: 546-555, 1913.

COLORECTAL CANCER, HEREDITARY NONPOLYPOSIS (HNPCC); hMSH2 (*120435)

JJM NOTE

This gene locus is one of four that account for the occasional family that meets the criteria for the Lynch cancer family syndrome. Since the syndrome was rigidly defined, by the so-called Amsterdam criteria, for purposes of gene-mapping research, it remains to be seen how much colorectal cancer is accounted for by somatic and germ-cell mutation in the four genes. My colleagues and I encountered a family in which only a mother and her daughter had colon cancer and yet both had a germline mutation of *MSH2* (Mulvihill JJ et al.: Inadvertent off-protocol testing for colon cancer-predisposing gene hMSH2. Am. J. Hum. Genet. 57:A72, 1995). Eventually, all colon cancers might be assayed for the RER phenotype. When an abnormal RER phenotype is seen, the patient and family might be referred for genetics consultation to rule out the Lynch cancer family syndrome.

Although one of the earliest descriptions of 'inherited cancer' involved adenocarcinoma of the colon in a large family described by Warthin (1913), no definitive proof that the cancers in these families were due primarily to hereditary factors has been forthcoming. One problem in establishing such proof is that colon cancers are so common that it is difficult to rule out chance clustering and other nonhereditary factors. Moreover, the environment, notably diet, has been shown to play a substantial role in colon cancer; members of families are likely to share similar environments, thus complicating definitive analysis. Peltomaki et al. (1993) and Aaltonen et al. (1993) searched for evidence of a genetic component through linkage analysis. In studies of 2 large kindreds, many individuals who had colon cancer with or without endometrial cancer were found to show linkage to an anonymous microsatellite marker on chromosome 2, D2S123, which means that a gene is in the region of 2p16-p15. The residence of the 2 kindreds on 2 different continents made an environmental explanation unlikely. See also familial nonpolyposis colon cancer type 2 (120436).

Aaltonen et al. (1993) studied an additional 14 smaller kindreds. Linkage could be excluded in 3 families by lod scores less than -2.0, whereas the remaining 11 smaller families displayed varying degrees of positive and negative lod scores, suggesting genetic heterogeneity. Aaltonen et al. (1993) found, furthermore, no loss of heterozygosity for the D2S123 or other chromosome 2 markers in either familial or sporadic cases of FCC and the incidence of mutations in KRAS, P53, and APC was similar in the 2 groups of tumors. They found, however, that most of the familial cancers had widespread alteration in short repeated DNA sequences, (CA)n dinucleotide repeat fragments, suggesting that numerous replication errors had occurred in the sequences during tumor development. In 13% of sporadic cancers, identical abnormalities were found and these cancers shared biologic properties with the familial cases, namely, location on the right side of the colon and preservation of diploidy or near-diploidy. Aaltonen et al. (1993) proposed that these findings reflect the existence on chromosome 2 of a gene which is neither an oncogene nor a tumor suppressor gene but rather a gene leading to genomic instability. Thibodeau et al. (1993) examined colorectal tumor DNA for somatic instability at (CA)n repeats on 5q, 15q, 17p, and 18q. Differences between tumor and normal DNA were detected in 25 of the 90 tumors studied. The instability appeared as either a substantial change in repeat length (often heterogeneous in nature) or a minor change (typically 2 bp). There was a significant correlation between microsatellite instability and location of the tumor in the proximal colon, i.e., the right colon, and with increased patient survival; instability was correlated inversely with loss of heterozygosity for 5q, 17p, and 18q.

Ionov et al. (1993) found that 12% of colorectal carcinomas (CRCs) carry somatic deletions in poly(dA/dT) sequences and other simple repeats. They estimated that cells from these tumors can carry more than 100,000 such mutations. They concluded that these mutations reflect a previously undescribed form of carcinogenesis in the colon mediated by a mutation in a DNA replication factor resulting in reduced fidelity for replication or repair—a 'mutator mutation.'

The subset of sporadic colorectal tumors and most tumors developing in hereditary nonpolyposis colorectal cancer (HNPCC) patients, containing alterations in microsatellite sequences, are thought to manifest replication errors and are referred to as RER(+) for 'replication errors.' Using genetic criteria, Parsons et al. (1993) demonstrated that the mutation rate of (CA)n repeats in RER(+) tumor cells is at least 100-fold that in RER(-) tumor cells and affects extrachromosomal as well as endogenous genomic sequences. Moreover, using in vitro assays, they showed that the mutability of RER(+) cells is associated with a profound defect in strand-specific mismatch repair. This deficiency was observed with microsatellite heteroduplexes as well as with heteroduplexes containing single base-base mismatches and affected an early step in the repair pathway. Thus, a true mutator phenotype exists in a subset of human tumors. The responsible defect is likely to cause transitions and transversions in addition to microsatellite alterations.

Fishel et al. (1993) studied human homologs of the mismatch repair system in Escherichia coli referred to as the MutHLS pathway. The pathway promotes a long patch (approximately 2 kb) excision repair reaction that is dependent on the products of the MutH, MutL, MutS, and MutU genes. Genetic analysis suggested that Saccharomyces cerevisiae has a mismatch repair system similar to the bacterial MutHLS system. The S. cerevisiae pathway has a MutS homolog, MSH2. In both bacteria and S. cerevisiae, mismatch repair plays a role in maintaining the genetic stability of DNA. In S. cerevisiae, Msh2 mutants exhibit increased rates of expansion and contraction of dinucleotide repeat sequences. Fishel et al. (1993) cloned and characterized a human MutS homolog, MSH2, and demonstrated that the gene maps to 2p22-p21 by study of a mapping panel of somatic cell hybrid DNAs using PCR. A T-to-C transition mutation was detected in the -6 position of a splice acceptor site in sporadic colon tumors and as a constitutional change in affected members of 2 small families with HNPCC. Fishel et al. (1993) were prompted to study the human homolog MSH2 following the report by Aaltonen et al. (1993) that the mutation in nonpolyposis colon cancer that maps to 2p behaves like a defect in DNA repair of the MutHLS type which they had previously been studying.

Leach et al. (1993) used chromosome microdissection to obtain highly polymorphic markers from 2p16. These and other markers were ordered in a panel of somatic cell hybrids and used to define a 0.8-Mb interval containing the locus for HNPCC. Candidate genes were mapped with respect to this locus, and one gene was found to lie within the 0.8-Mb interval. This gene was homologous to a prokaryotic gene, MutS, that participates in mismatch repair. (Disruption of the MutL and MutS mismatch repair genes produces microsatellite instability in bacteria and yeast (Levinson and Gutman, 1987; Strand et al., 1993).) Using the sequence of cDNA clones of the gene, they demonstrated the existence of germline mutations that substantially altered the predicted gene product and cosegregated with disease in the HNPCC kindreds. The highest homology was to the yeast Msh-2 gene in the helix-turn-helix domain, perhaps responsible for MutS binding to DNA. The yeast and human Msh-2 proteins were 77% identical between codons 615 and 788. There were 10 other blocks of similar amino acids distributed throughout the length of the 2 proteins. Furthermore, Leach et al. (1993) succeeded in identifying specific germline mutations in each of the 2 kindreds that originally established linkage to chromosome 2 (Peltomaki et al., 1993). It is noteworthy that both the candidate gene approach and positional cloning, the 2 main methods of map-based cloning, were used in identifying the MSH2 gene.

Lishanski et al. (1994) developed an experimental strategy for detecting heterozygosity in genomic DNA based on preferential binding of Escherichia coli MutS protein to DNA molecules containing mismatched bases. The binding was detected by a gel mobility-shift assay. The approach was tested by using as a model the most commonly occurring mutations within the cystic fibrosis gene (CFTR; 602421).

Using a panel of microsatellite polymorphisms in the vicinity of D2S123, Green et al. (1994) tested 7 Canadian HNPCC families. Whereas 1 family was clearly linked to the COCA1 locus (lod = 4.21) and a second family was probably linked (lod = 0.92), linkage was excluded in 3 families. In the remaining 2 families, the data were inconclusive. In the definitely linked family, individuals with cancer of the endometrium or ureter shared a common haplotype with 12 family members with colorectal cancer. This supported the suspected association between these extracolonic neoplasms and the HNPCC syndrome. In addition, 5 of the 6 persons with adenomatous polyps, but no colorectal cancer, had the same haplotype as the affected persons, while the sixth carried a recombination. One individual with colorectal cancer carried a recombination that placed the COCA1 locus telomeric to D2S123.

Aquilina et al. (1994) detected a mismatch binding defect leading to a mutator phenotype in LoVo, a human colorectal carcinoma cell line. Umar et al. (1994) described a deletion in the MSH2 gene in LoVo cells together with a defect in mismatch repair by LoVo cell extracts.

The microsatellite DNA instability that is associated with alteration in the MSH2 gene in hereditary nonpolyposis colon cancer and several forms of sporadic cancer is thought to arise from defective repair of DNA replication errors that create insertion-deletion loop-type (IDL) mismatched nucleotides. Fishel et al. (1994) showed that purified MSH2 protein efficiently and specifically binds DNA containing IDL mismatches of up to 14 nucleotides. The findings supported a direct role for MSH2 in mutation avoidance and microsatellite stability in human cells.

Kolodner et al. (1994) found that the genomic MSH2 locus covers approximately 73 kb and contains 16 exons. The sequence of all the intron-exon junctions was determined and used to develop methods for analyzing each MSH2 exon for mutations. These methods were used to analyze 2 large HNPCC kindreds exhibiting features of the Muir-Torre syndrome (158320) and to demonstrate that cancer susceptibility was due to the inheritance of a frameshift mutation in the MSH2 gene in 1 family and a nonsense mutation in the MSH2 gene in the other family. Linkage of the cancer phenotype to chromosome 2p had been described in these families by Hall et al. (1994).

Orth et al. (1994) found that 5 of 10 ovarian tumor cell lines were genetically unstable at most microsatellite loci analyzed. In clones and subclones derived serially from 1 of these cell lines (serous cystadenocarcinoma), a very high proportion of microsatellites distributed in many different regions of the genome changed their size in a mercurial fashion. In 1 ovarian tumor, they identified the source of the genetic instability as a point mutation (R524P) in the MSH2 gene. The patient was a 38-year-old heterozygote for this mutation and her normal tissue carried both mutant and wildtype alleles of the MSH2 gene. However, the wildtype allele was lost at some point early during tumorigenesis so that DNA isolated either from the patients ovarian tumor or from the cell line carried only the mutant MSH2 allele. The genetic instability observed in the tumor and cell line DNA, together with the germline mutation in a mismatch-repair gene, suggested that MSH2 is involved in the onset and/or progression in a subset of ovarian cancer.

Using denaturing gradient gel electrophoresis (DGGE) to screen for mutations in all 16 exons of the MSH2 gene in 34 unrelated HNPCC kindreds, Wijnen et al. (1995) found 7 novel pathogenic germline mutations resulting in stop codons, either directly or through frameshifts. Four nonpathogenic variations, including 1 useful polymorphism, were also identified. MSH2 mutations were found in 21% of the families. They could not establish any correlation between the site of the individual mutations and the spectrum of tumor types.

To investigate the role of the MSH2 gene in genome stability and tumorigenesis, de Wind et al. (1995) generated cells and mice deficient for the gene. Msh2-deficient mouse embryonic stem cell lines were found to have lost mismatch binding and acquired microsatellite instability, a mutator phenotype, and tolerance to methylation agents. Moreover, in these cells, homologous recombination had lost dependence on complete identity between interacting DNA sequences, suggesting that Msh2 is involved in safeguarding the genome from promiscuous recombination. MSH2-deficient mice displayed no major abnormalities, but a significant fraction developed lymphomas at an early age.

Reitmair et al. (1995) described a mouse strain homozygous for a 'knockout' mutation at the MSH2 locus. Surprisingly, these mice were found to be viable, produced offspring in a mendelian ratio, and bred through at least 2 generations. Starting at 2 months of age, homozygous MSH2-deficient mice began to develop lymphoid tumors with high frequency that contained microsatellite instabilities. These data established a direct link between MSH2 deficiency and the pathogenesis of cancer.

Maliaka et al. (1996) identified 6 different new mutations in the MLH1 and MSH2 genes in Russian and Moldavian HNPCC families. Three of these mutations occurred in CpG dinucleotides and led to a premature stop codon, splicing defect, or an amino acid substitution in evolutionarily conserved residues. Analysis of a compilation of published mutations including the new data suggested to the authors that CpG dinucleotides within the coding regions of the MSH2 and MLH1 genes are hotspots for single basepair substitutions.

The penetrance of mutations in the DNA mismatch repair (MMR) genes is unknown except in classic HNPCC kindreds because the families studied to date have specifically been selected for research purposes. Using a population-based strategy, Dunlop et al. (1997) calculated the lifetime cancer risk associated with germline MMR gene mutations, irrespective of their family history. They identified 67 gene carriers whose risk to age 70 for all cancers was 91% for males and 69% for females. Risk of developing colorectal cancer was significantly greater for males than for females (74% vs 30%, P = 0.006). The risk of uterine cancer (42%) exceeded that for colorectal cancer in females, emphasizing the need for uterine screening. Their findings gave further insight into the biologic effect of defective DNA MMR. Dunlop et al. (1997) demonstrated a systematic approach to identifying individuals with high risk of cancer who may not be part of classic HNPCC families. The risk estimates derived from these analyses provided a rational basis on which to guide genetic counseling and tailor clinical surveillance.

Mice carrying a targeted germline disruption of the MSH2 gene are viable and susceptible to lymphoid tumors; however, defects in this gene had not been identified in human lymphomas. To determine if the lymphomas these mice develop are related to a particular subtype of human lymphoma, Lowsky et al. (1997) evaluated 20 clinically ill homozygous MSH2 -/- mice ranging in age from 2 to 13 months. The murine tumors comprised a single histopathologic entity representing the malignant counterpart of precursor thymic T cells and closely resembling human precursor T-cell lymphoblastic lymphoma (LBL). Evaluation of the expression of 3 T-cell malignancy-associated genes showed that rhombotin-2 (RBTN2; 180385), TAL1 (187040), and HOX11 (186770) were expressed in 100, 40, and 0% of the murine tumors, respectively. The MSH2 -/- murine model of precursor T-cell LBL was substantiated by the finding of a newly identical expression pattern of RBTM2, TAL1, and HOX11 in 10 well-characterized cases of human LBL. Direct evidence for MSH2 abnormalities in human LBL was established by sequence analysis of exon 13 of human MSH2, which revealed coding region mutations in 2 of 10 cases. The findings of Lowsky et al. (1997) implicated defects in the mismatch repair system with the aberrant expression of T-cell specific protooncogenes and defined a new pathway of human lymphomagenesis.

Pearson et al. (1997) studied the interaction of the human mismatch repair protein MSH2 with slipped-strand structures formed from a triplet repeat sequence in order to address the possible role of MSH2 in trinucleotide expansion, which is associated with several neurodegenerative diseases such as myotonic dystrophy (DM; 160900). Genomic clones of the myotonic dystrophy locus containing disease-relevant lengths of (CTG)n(CAG)n triplet repeats were examined. They found that the affinity of MSH2 increased with the length of the repeat sequence. Furthermore, MHS bound preferentially to looped-out CAG repeat sequences, implicating a strand asymmetry in MSH2 recognition. Pearson et al. (1997) suggested that MSH2 may participate in trinucleotide repeat expansion via its role in repair and/or recombination.

All homologs of the MutS proteins contain a highly conserved region of approximately 150 amino acids that encompasses a helix-turn-helix domain associated with an adenine nucleotide and magnesium binding motif, termed Walker-A motif. This part of the molecule has ATPase activity. Gradia et al. (1997) found that this ATPase activity and the associated adenine nucleotide-binding domain functions to regulate mismatch binding as a molecular switch. The MSH2/MSH6 complex is 'on' (binds mismatched nucleotides) in the ADP-bound form and 'off' in the ATP-bound form. Hydrolysis of ATP results in the recovery of mismatch binding, while ADP-to-ATP exchange results in mismatch dissociation. These results suggested to Gradia et al. (1997) a new model for the function of MutS proteins during mismatch repair in which the switch determines the timing of downstream events.

Pensotti et al. (1997) screened 14 Italian families affected with HNPCC for germline mutations of the MSH2, MLH1, and GTBP genes. DNA alterations were observed in 6 of the families, involving MSH2 or MLH1. No mutations were detected in the G/T binding protein (GTBP; 600678). (GTBP codes for a 160-kD protein that interacts with the product of MSH2. Although mutations of the GTBP gene had been observed in colon cancer cell lines, germline alterations had not been reported in HNPCC patients.) A finding of note by Pensotti et al. (1997) was that the kindreds with demonstrated mutations had a mean age of colorectal cancer onset of 43 years versus an average age of 53 years for the families without mutations.

From findings in the Danish HNPCC Register, Myrhoj et al. (1997) concluded that colorectal cancer in HNPCC behaves differently from colorectal cancer in general. The mean age at diagnosis of primary CRC was 41 years as compared with 70 years for all Danish CRC. In 68%, the colon cancer was located proximal to the splenic flexure as compared with 49% in Danish CRC in general; the numbers for rectal location were 20% for HNPCC as compared with the general experience of 43%. In 7% of HNPCC, more than 1 colorectal cancer was found at the first operation, versus 1% in the general group. Metastatic tendency was less than in sporadic CRC and survival was better.

Fearon (1997) reviewed more than 20 different hereditary cancer syndromes that had been defined and attributed to specific germline mutations in various inherited cancer genes. In a useful illustration, he diagrammed the roles of allelic variation ('one gene - different syndromes') and genetic heterogeneity ('different genes - one syndrome') in inherited cancer syndromes. The example he used of the latter was HNPCC resulting from mutations in various mismatch repair genes.

Genuardi et al. (1998) reported the existence of alternative splicing in the MSH2 gene. Coupled RT-PCR of various tissue samples from normal individuals and hereditary nonpolyposis colon cancer patients identified MSH2 gene products lacking exons 5, 13, 2-7, and 2-8. The levels of expression varied among different samples. All isoforms were found in 43 to 100% of the mononuclear blood samples, as well as in other tissues. The authors cautioned that knowledge of the existence of multiple alternative splicing events not caused by genomic DNA changes is important for the evaluation of the results of molecular diagnostic tests based on RNA analysis.

To evaluate the feasibility of molecular screening for HNPCC, Aaltonen et al. (1998) prospectively screened tumor specimens obtained from 509 consecutive Finnish patients with colorectal adenocarcinomas for DNA replication errors, which are characteristic of hereditary colorectal cancers. These replication errors were detected through microsatellite-marker analyses of tumor DNA. DNA from normal tissue from the patients with replication errors were screened for germline mutations in MLH1 and MSH2. Among the 509 patients, 63 (12%) had replication errors. Specimens of normal tissue from 10 of these 63 patients had a germline mutation of MLH1 or MSH2. Of these 10 patients (2% of the 509 patients), 9 had a first-degree relative with endometrial or colorectal cancer, 7 were under 50 years of age, and 4 had had colorectal or endometrial cancer previously. Aaltonen et al. (1998) recommended testing for replication errors in all patients with colorectal cancer who meet one or more of the following criteria: a family history of colorectal or endometrial cancer, an age of less than 50 years, and a history of multiple colorectal or endometrial cancers. Patients found to have replication errors should undergo further analysis for germline mutations in DNA mismatch-repair genes.

Lynch and Smyrk (1998) noted that several groups have proposed less restrictive criteria, including the Bethesda criteria (Rodriguez-Bigas et al., 1997) and the Japanese criteria (Nakahara et al., 1997). While applauding the trend toward less restrictive criteria, they doubted that in actual practice clinicians will find any algorithm particularly useful in deciding on which patients to test.

Wijnen et al. (1998) assessed the prevalence of MSH2 and MLH1 mutations in families suspected of having hereditary nonpolyposis colorectal cancer and evaluated whether clinical findings can predict the outcome of genetic testing. They studied 184 kindreds; mutations of one or the other gene were found in 47 (26%). Clinical factors associated with these mutations were early age at diagnosis of colorectal cancer, the occurrence in the kindred of endometrial cancer or tumors of the small intestine, a higher number of family members with colorectal or endometrial cancer, the presence of multiple colorectal cancers or both colorectal and endometrial cancers in a single family member, and fulfillment of the Amsterdam criteria for the diagnosis of hereditary nonpolyposis colorectal cancer (at least 3 family members in 2 or more successive generations must have colorectal cancer, 1 of whom is a first-degree relative of the other 2; cancer must be diagnosed before the age of 50 in at least 1 family member; and familial adenomatous polyposis must be ruled out). Multivariate analysis showed that a younger age at diagnosis of colorectal cancer, fulfillment of the Amsterdam criteria, and the presence of endometrial cancer in the kindred were independent predictors of germline mutations of MSH2 or MLH1. Wijnen et al. (1998) used these results to devise a logistic model for estimating the likelihood of a mutation in MSH2 and MLH1.

Nonsteroidal antiinflammatory drugs (NSAIDs) have a cancer preventive action that has been attributed largely to their antiproliferative and apoptosis-inducing activities. Ruschoff et al. (1998) demonstrated that microsatellite instability (MSI) in colorectal cancer cells deficient for the mismatch repair (MMR) genes MLH1, MSH2, and MSH6, is markedly reduced during exposure to aspirin or sulindac (or Clinoril, which is chemically related to indomethacin, trade name Indocin). This effect was reversible, time and concentration dependent, and apparently independent of proliferation rate and cylcooxygenase function. In contrast, the MSI phenotype of an endometrial cancer cell line deficient in PMS2 (600259) was unaffected by aspirin/sulindac. Ruschoff et al. (1998) showed that the MSI reduction in the susceptible MMR-deficient cells was confined to nonapoptotic cells, whereas apoptotic cells remained unstable and were eliminated from the growing population. These results suggested that aspirin/sulindac induces a genetic selection for microsatellite stability in a subset of MMR-deficient cells and may provide an effective prophylactic therapy for hereditary nonpolyposis colorectal cancer kindreds where alteration of the MSH2 and MLH1 genes are associated with most cases of cancer susceptibility.

Chronic oxidative stress may play a critical role in the pathogenesis of many human cancers. DeWeese et al. (1998) reported that mouse embryonic stem (ES) cells from mice carrying either 1 or 2 disrupted Msh2 alleles displayed an increased survival following protracted exposures to low-level ionizing radiation as compared with wildtype ES cells. The increases in survival exhibited by ES cells deficient in DNA mismatch repair appeared to have resulted from a failure to execute cell death (apoptosis) efficiently in response to radiation exposure. For each of the ES cell types, prolonged low-level radiation treatment generated oxidative genome damage that manifested as an accumulation of oxidized bases in genomic DNA. However, ES cells from Msh2 +/- and Msh2 -/- mice accumulated more oxidized bases as a consequence of low-level radiation exposure than did ES cells from Msh2 +/+ mice. The propensity for normal cells with mismatch repair enzyme deficiencies, including cells heterozygous for inactivating mismatch repair enzyme gene mutations, to survive promutagenic genome insults accompanying stresses may contribute to the increased cancer risk characteristic of the hereditary nonpolyposis colorectal cancer syndrome.

Aaltonen, L. A.; Peltomaki, P.; Leach, F. S.; Sistonen, P.; Pylkkanen, L.; Mecklin, J.-P.; Jarvinen, H.; Powell, S. M.; Jen, J.; Hamilton, S. R.; Petersen, G. M.; Kinzler, K. W.; Vogelstein, B.; de la Chapelle, A.: Clues to the pathogenesis of familial colorectal cancer. Science 260: 812-816, 1993.

Aaltonen, L. A.; Salovaara, R.; Kristo, P.; Canzian, F.; Hemminki, A.; Peltomaki, P.; Chadwick, R. B.; Kaariainen, H.; Eskelinen, M.; Jarvinen, H.; Mecklin, J.-P.; de la Chapelle, A.: Incidence of hereditary nonpolyposis colorectal cancer and the feasibility of molecular screening for the disease. New Eng. J. Med. 338: 1481-1487, 1998.

Aquilina, G.; Hess, P.; Branch, P.; MacGeoch, C.; Casciano, I.; Karran, P.; Bignami, M.: A mismatch recognition defect in colon carcinoma confers DNA microsatellite instability and a mutator phenotype. Proc. Nat. Acad. Sci. 91: 8905-8909, 1994.

DeWeese, T. L.; Shipman, J. M.; Larrier, N. A.; Buckley, N. M.; Kidd, L. R.; Groopman, J. D.; Cutler, R. G.; te Riele, H.; Nelson, W. G.: Mouse embryonic stem cells carrying one or two defective Msh2 alleles respond abnormally to oxidative stress inflicted by low-level radiation. Proc. Nat. Acad. Sci. 95: 11915-11920, 1998.

de Wind, N.; Dekker, M.; Berns, A.; Radman, M.; te Riele, H.: Inactivation of the mouse Msh2 gene results in mismatch repair deficiency, methylation tolerance, hyperrecombination, and predisposition to cancer. Cell 82: 321-330, 1995.

Dunlop, M. G.; Farrington, S. M.; Carothers, A. D.; Wyllie, A. H.; Sharp, L.; Burn, J.; Liu, B.; Kinzler, K. W.; Vogelstein, B.: Cancer risk associated with germline DNA mismatch repair gene mutations. Hum. Molec. Genet. 6: 105-110, 1997.

Esche, C.; Kruse, R.; Lamberti, C.; Friedl, W.; Propping, P.; Lehmann, P.; Ruzicka, T.: Muir-Torre syndrome: clinical features and molecular genetic analysis. Brit. J. Derm. 136: 913-917, 1997.

Fearon, E. R.: Human cancer syndromes: clues to the origin and nature of cancer. Science 278: 1043-1050, 1997.

Fishel, R.; Ewel, A.; Lee, S.; Lescoe, M. K.; Griffith, J.: Binding of mismatched microsatellite DNA sequences by the human MSH2 protein. Science 266: 1403-1405, 1994.

Fishel, R.; Lescoe, M. K.; Rao, M. R. S.; Copeland, N. G.; Jenkins, N. A.; Garber, J.; Kane, M.; Kolodner, R.: The human mutator gene homolog MSH2 and its association with hereditary nonpolyposis colon cancer. Cell 75: 1027-1038, 1993.

Genuardi, M.; Viel, A.; Bonora, D.; Capozzi, E.; Bellacosa, A.; Leonardi, F.; Valle, R.; Ventura, A.; Pedroni, M.; Boiocchi, M.; Neri, G.: Characterization of MLH1 and MSH2 alternative splicing and its relevance to molecular testing of colorectal cancer susceptibility. Hum. Genet. 102: 15-20, 1998.

Gradia, S.; Acharya, S.; Fishel, R.: The human mismatch recognition complex hMSH2-hMSH6 functions as a novel molecular switch. Cell 91: 995-1005, 1997.

Green, R. C.; Narod, S. A.; Morasse, J.; Young, T.-L.; Cox, J.; Fitzgerald, G. W. N.; Tonin, P.; Ginsburg, O.; Miller, S.; Jothy, S.; Poitras, P.; Laframboise, R.; Routhier, G.; Plante, M.; Morissette, J.; Weissenbach, J.; Khandjian, E. W.; Rousseau, F.: Hereditary nonpolyposis colon cancer: analysis of linkage to 2p15-16 places the COCA1 locus telomeric to D2S123 and reveals genetic heterogeneity in seven Canadian families. Am. J. Hum. Genet. 54: 1067-1077, 1994.

Hall, N. R.; Murday, V. A.; Chapman, P.; Williams, M. A.; Burn, J.; Finan, P. J.; Bishop, D. T.: Genetic linkage in Muir-Torre syndrome to the same chromosomal site as cancer family syndrome. Europ. J. Cancer 30A: 180-182, 1994.

Ionov, Y.; Peinado, M. A.; Malkhosyan, S.; Shibata, D.; Perucho, M.: Ubiquitous somatic mutations in simple repeated sequences reveal a new mechanism for colonic carcinogenesis. Nature 363: 558-561, 1993.

Jeon, H. M.; Lynch, P. M.; Howard, L.; Ajani, J.; Levin, B.; Frazier, M. L.: Mutation of the hMSH2 gene in two families with hereditary nonpolyposis colorectal cancer. Hum. Mutat. 7: 327-333, 1996.
Kolodner, R. D.; Hall, N. R.; Lipford, J.; Kane, M. F.; Rao, M. R. S.; Morrison, P.; Wirth, L.; Finan, P. J.; Burn, J.; Chapman, P.; Earabino, C.; Merchant, E.; Bishops, D. T.: Structure of the human MSH2 locus and analysis of two Muir-Torre kindreds for msh2 mutations. Genomics 24: 516-526, 1994.
Leach, F. S.; Nicolaides, N. C.; Papadopoulos, N.; Liu, B.; Jen, J.; Parsons, R.; Peltomaki, P.; Sistonen, P.; Aaltonen, L. A.; Nystrom-Lahti, M.; Guan, X.-Y.; Zhang, J.; Meltzer, P. S.; Yu, J.-W.; Kao, F.-T.; Chen, D. J.; Cerosaletti, K. M.; Fournier, R. E. K.; Todd, S.; Lewis, T.; Leach, R. J.; Naylor, S. L.; Weissenbach, J.; Mecklin, J.-P.; Jarvinen, H.; Petersen, G. M.; Hamilton, S. R.; Green, J.; Jass, J.; Watson, P.; Lynch, H. T.; Trent, J. M.; de la Chapelle, A.; Kinzler, K. W.; Vogelstein, B.: Mutations of a MutS homolog in hereditary non-polyposis colorectal cancer. Cell 75: 1215-1225, 1993.
Levinson, G.; Gutman, G. A.: High frequencies of short frameshifts in poly-CA/TG tandem repeats borne by bacteriophage M13 in Escherichia coli K-12. Nucleic Acids Res. 15: 5323-5338, 1987.
Lishanski, A.; Ostrander, E. A.; Rine, J.: Mutation detection by mismatch binding protein, MutS, in amplified DNA: application to the cystic fibrosis gene. Proc. Nat. Acad. Sci. 91: 2674-2678, 1994.
Lowsky, R.; DeCoteau, J. F.; Reitmair, A. H.; Ichinohasama, R.; Dong, W.-F.; Xu, Y.; Mak, T. W.; Kadin, M. E.; Minden, M. D.: Defects of the mismatch repair gene MSH2 are implicated in the development of murine and human lymphoblastic lymphomas and are associated with the aberrant expression of rhombotin-2 (Lmo-2) and Tal-1 (SCL). Blood 89: 2276-2282, 1997.
Lynch, H. T.; Smyrk, T. C.: Identifying hereditary nonpolyposis colorectal cancer. (Editorial) New Eng. J. Med. 338: 1537-1538, 1998.
Maliaka, Y. K.; Chudina, A. P.; Belev, N. F.; Alday, P.; Bochkov, N. P.; Buerstedde, J.-M.: CpG dinucleotides in the hMSH2 and hMLH1 genes are hotspots for HNPCC mutations. Hum. Genet. 97: 251-255, 1996.
Mary, J.-L.; Bishop, T.; Kolodner, R.; Lipford, J. R.; Kane, M.; Weber, W.; Torhorst, J.; Muller, H.; Spycher, M.; Scott, R. J.: Mutational analysis of the hMSH2 gene reveals a three base pair deletion in a family predisposed to colorectal cancer development. Hum. Molec. Genet. 3: 2067-2069, 1994.
Myrhoj, T.; Bisgaard, M. L.; Bernstein, I.; Svendsen, L. B.; Sondergaard, J. O.; Bulow, S.: Hereditary non-polyposis colorectal cancer: clinical features and survival. Scand. J. Gastroenterol. 32: 572-576, 1997.
Nakahara, M.; Yokozaki, H.; Yasui, W.; Dohl, K.; Tahara, E.: Identification of concurrent germ-line mutations in hMSH2 and/or hMLH1 in Japanese hereditary nonpolyposis colorectal cancer kindreds. Cancer Epidemiol. Biomarkers Prev. 6: 1057-1064, 1997.
Orth, K.; Hung, J.; Gazdar, A.; Bowcock, A.; Mathis, J. M.; Sambrook, J.: Genetic instability in human ovarian cancer cell lines. Proc. Nat. Acad. Sci. 91: 9495-9499, 1994.
Parsons, R.; Li, G.-M.; Longley, M. J.; Fang, W. H.; Papadopoulos, N.; Jen, J.; de la Chapelle, A.; Kinzler, K. W.; Vogelstein, B.; Modrich, P.: Hypermutability and mismatch repair deficiency in RER(+) tumor cells. Cell 75: 1227-1236, 1993.
Pearson, C. E.; Ewel, A.; Acharya, S.; Fishel, R. A.; Sinden, R. R.: Human MSH2 binds to trinucleotide repeat DNA structures associated with neurodegenerative diseases. Hum. Molec. Genet. 6: 1117-1123, 1997.
Peltomaki, P.; Aaltonen, L. A.; Sistonen, P.; Pylkkanen, L.; Mecklin, J.-P.; Jarvinen, H.; Green, J. S.; Jass, J. R.; Weber, J. L.; Leach, F. S.; Petersen, G. M.; Hamilton, S. R.; de la Chapelle, A.; Vogelstein, B.: Genetic mapping of a locus predisposing to human colorectal cancer. Science 260: 810-812, 1993.
Pensotti, V.; Radice, P.; Presciuttini, S.; Calistri, D.; Gazzoli, I.; Perez, A. P. G.; Mondini, P.; Buonsanti, G.; Sala, P.; Rossetti, C.; Ranzani. G. N.; Bertario, L.; Pierotti, M. A.: Mean age of tumor onset in hereditary nonpolyposis colorectal cancer (HNPCC) families correlates with the presence of mutations in DNA mismatch repair genes. Genes Chromosomes Cancer 19: 135-142, 1997.
Reitmair, A. H.; Schmits, R.; Ewel, A.; Bapat, B.; Redston, M.; Mitri, A.; Waterhouse, P.; Mittrucker, H.-W.; Wakeham, A.; Liu, B.; Thomason, A.; Griesser, H.; Gallinger, S.; Ballhausen, W. G.; Fishel, R.; Mak, T. W.: MSH2 deficient mice are viable and susceptible to lymphoid tumours. Nature Genet. 11: 64-70, 1995.
Rodriguez-Bigas, M. A.; Boland, C. R.; Hamilton, S. R.; Henson, D. E.; Jass, J. R.; Khan, P. M.; Lynch, H.; Perucho, M.; Smyrk, T.; Sobin, L.; Srivastava, S.: A National Cancer Institute Workshop on Hereditary Nonpolyposis Colorectal Cancer Syndrome: meeting highlights and Bethesda guidelines. J. Nat. Cancer Inst. 89: 1758-1762, 1997.
Ruschoff, J.; Wallinger, S.; Dietmaier, W.; Bocker, T.; Brockhoff, G.; Hofstadter, F.; Fishel, R.: Aspirin suppresses the mutator phenotype associated with hereditary nonpolyposis colorectal cancer by genetic selection. Proc. Nat. Acad. Sci. 95: 11301-11306, 1998.
Strand, M.; Prolla, T. A.; Liskay, R. M.; Petes, T. D.: Destabilization of tracts of simple repetitive DNA in yeast by mutations affecting DNA mismatch repair. Nature 365: 274-276, 1993.
Thibodeau, S. N.; Bren, G.; Schaid, D.: Microsatellite instability in cancer of the proximal colon. Lancet 260: 816-819, 1993.
Umar, A.; Boyer, J. C.; Thomas, D. C.; Nguyen, D. C.; Risinger, J. I.; Boyd, J.; Ionov, Y.; Perucho, M.; Kunkel, T. A.: Defective mismatch repair in extracts of colorectal and endometrial cancer cell lines exhibiting microsatellite instability. J. Biol. Chem. 269: 14367-14370, 1994.
Warthin, A. S.: Heredity with reference to carcinoma. Arch. Intern. Med. 12: 546-555, 1913.
Wijnen, J.; Vasen, H.; Khan, P. M.; Menko, F. H.; van der Klift, H.; van Leeuwen, C.; van den Broek, M.; van Leeuwen-Cornelisse, I.; Nagengast, F.; Meijers-Heijboer, A.; Lindhout, D.; Griffioen, G.; Cats, A.; Kleibeuker, J.; Varesco, L.; Bertario, L.; Bisgaard, M. L.; Mohr, J.; Fodde, R.: Seven new mutations in hMSH2, an HNPCC gene, identified by denaturing gradient-gel electrophoresis. Am. J. Hum. Genet. 56: 1060-1066, 1995.
Wijnen, J. T.; Vasen, H. F. A.; Meera Khan, P.; Zwinderman, A. H.; van der Klift, H.; Mulder, A.; Tops, C.; Moller, P.; Fodde, R.: Clinical findings with implications for genetic testing in families with clustering of colorectal cancer. New Eng. J. Med. 339: 511-518, 1998.

COLORECTAL CANCER, HEREDITARY NONPOLYPOSIS (HNPCC); hMLH1 (*120436)

JJM NOTE

hMLH1 is the second major gene locus associated with the Lynch cancer family syndrome (but also with Turcot syndrome).

Using RFLPs and microsatellite markers for linkage analysis in 3 hereditary nonpolyposis colon cancer families, Lindblom et al. (1993) demonstrated linkage to 3p23-p21. Tumor DNA from 1 tumor in each family was included in the study to look for rearrangements related to tumor development. None of the colon tumors showed loss of heterozygosity (LOH) for any of the informative markers used on 20 different chromosomes. However, after they had detected linkage to 3p, Lindblom et al. (1993) observed a gain of bands for several dinucleotide markers located on 3p. A gain of bands was observed with markers on many chromosomes.

After human homologs of the mutS gene of bacteria and yeast were found to have mutations responsible for hereditary nonpolyposis colorectal cancer (120435), Papadopoulos et al. (1994) searched for other human mismatch repair (MMR) genes. A survey of a large database of expressed sequenced tags (ESTs) derived from random cDNA clones revealed 3 additional human MMR genes, all related to the bacterial mutL gene. Papadopoulos et al. (1994) mapped one of these genes (MLH1) to 3p21.3 by fluorescence in situ hybridization. (The other 2 genes had a slightly greater similarity to the yeast mutL homolog PMS1 and were therefore denoted PMS1 and PMS2, respectively.) The mapping of MLH1 to 3p21 was of interest because markers in that area had been linked to hereditary nonpolyposis colon cancer in several families (Lindblom et al., 1993). Searching for mutations in the MLH1 gene, Papadopoulos et al. (1994) performed RT-PCR analyses of lymphoblastoid cell RNA and directly sequenced the coding region of the gene in 10 HNPCC kindreds linked to 3p markers. Remarkably, all affected individuals from 7 Finnish kindreds exhibited a heterozygous deletion of codons 578 to 632. The derivation of 5 of these 7 kindreds could be traced to a common ancestor, and the presence of the same presumptive defect in 2 other kindreds supported a 'founder effect' for many cases of HNPCC in the Finnish population. Codons 578 to 632 were found to constitute a single exon that was deleted from 1 allele in the 7 kindreds. This exon encodes several highly conserved amino acids found at identical positions in yeast MLH1. In another 3p-linked family, Papadopoulos et al. (1994) observed a 4-nucleotide deletion beginning at the first position of codon 727 and producing a frameshift with a new stop codon located 166 nucleotides downstream. As a result, the COOH-terminal 19 amino acids of MLH1 were substituted with 53 different amino acids, some encoded by nucleotides normally in the 3-prime untranslated region. Anot her kindred displayed a 4-nucleotide insertion between codons 755 and 756. This insertion resulted in a frameshift and extension of the open reading frame to include 99 nucleotides downstream of the normal stop codon. One cell line showed a transversion from TCA to TAA in codon 252, resulting in conversion of a serine to a stop (120436.0001).

Simultaneously and independently, Bronner et al. (1994) likewise implicated the human MutL homolog, MLH1, in the form of HNPCC that maps to 3p. They mapped the MLH1 gene to the same region, 3p23-p21.3, by fluorescence in situ hybridization. Furthermore, in 1 chromosome 3-linked HNPCC family, they demonstrated a missense mutation in affected individuals (120436.0002). Using 19 dinucleotide markers and haplotype analysis in 2 families in which the disease was linked to 3p23-p21, Tannergard et al. (1994) also localized the gene specifically to 3p23-p21.3.

Since defects in the MSH2 gene (120435) on chromosome 2 may account for as many as 60% of cases of hereditary nonpolyposis colon cancer and the MLH1 gene on chromosome 3 may play a role in up to 30%, defects in these 2 genes can account for the vast majority of cases.

Between 1984 and 1994, extensive clinical and genealogic studies in Finland had identified approximately 40 hereditary nonpolyposis colorectal families that met the internationally accepted criteria for the disorder. Nystrom-Lahti et al. (1994) focused on 18 of these families. Since convincing evidence of 2p linkage had not been found in Finnish families, the role of the proposed locus on 3p was investigated. Of 18 apparently unrelated families living in different parts of Finland, 11 could be traced genealogically to a common ancestry dating at least 13 generations back in a small geographic area. Linkage studies were possible in 9 families, revealing conclusive or probable linkage to markers on 3p in 8. Of the 8, 5 were among those having shared ancestry. By analysis of recombinations in the 'linked' families, this second HNPCC locus was assigned to the 1-cM interval between marker loci D3S1561 and D3S1298. A haplotype encompassing 10 cM around the HNPCC locus was conserved in 5 of the pedigrees with shared ancestry and was present in 2 further families in which linkage analysis was not possible. The results suggested the presence of widespread single ancestral founding mutation. Studies in vitro indicate that heterozygosity of mutations in DNA MMR genes, unlike homozygosity, does not affect mismatch repair. Hemminki et al. (1994) demonstrated that loss of heterozygosity of markers within or adjacent to the MLH1 gene on 3p occurs nonrandomly in tumors from members of families in which the disease phenotype cosegregates with MLH1. In every informative case, the loss affected the wildtype allele. These results suggested that DNA MMR genes resemble tumor suppressor genes in that 2 hits are required to cause a phenotypic effect.

Han et al. (1995) reported that the human MLH1 gene consists of 19 coding exons spanning approximately 100 kb, and that exons 1 to 7 contain a region that is highly conserved in the MLH1 and PMS1 (600258) genes of yeast. Using PCR-SSCP analysis and DNA sequencing to examine the entire coding region of the MLH1 gene in DNAs of 34 unrelated cancer patients from HNPCC pedigrees, they found germline mutations in 8 (24%): 4 missense mutations, 1 intron mutation that would affect splicing, and 3 frameshift mutations resulting in truncation of the gene product downstream of the mutation site.

Hypermutable H6 colorectal tumor cells are defective in strand-specific mismatch repair and bear defects in both alleles of the human MLH1 gene. Li and Modrich (1995) purified to near homogeneity an activity from HeLa cells that complemented H6 nuclear extracts to restore repair proficiency on a set of heteroduplex DNAs representing the 8 base-base mismatches as well as a number of slipped-strand, insertion/deletion mispairs. The activity behaved as a single species during fractionation and copurified with proteins of 85 and 100 kD. Microsequence analysis demonstrated both of these proteins to be homologs of bacterial MutL, with the former corresponding to the human MLH1 product and the latter to the product of human PMS2 (600259) or a closely related gene. The 1:1 molar stoichiometry of the 2 polypeptides and their hydrodynamic behavior indicated formation of a heterodimer. These observations indicated that interactions between members of the family of the human MutL homologs may be restricted.

The Turcot syndrome (276300) is the association of colonic polyposis with brain tumor. Hamilton et al. (1995) demonstrated that a majority of such cases (at least 10 out of 15) have mutations in the APC gene (175100) and that the brain tumor is usually medulloblastoma. Some of the families with Turcot syndrome have mutations in the MMR genes MLH1 or PMS2; in these cases, the type of brain tumor is glioblastoma multiforme.

Liu et al. (1996) evaluated tumors from 74 HNPCC kindreds for genomic instability characteristic of a mismatch repair deficiency and found such instability in 68 (92%) of the kindreds. The entire coding regions of the 5 known human mismatch repair genes were evaluated in 48 kindreds with instability, and mutations were identified in 70%: mutations in the MSH2 gene in 15 (31%), in the MLH1 gene in 16 (33%), in the PMS1 gene in 1 (2%), in the PMS2 gene, in 2 (4%), and in the GTBP gene (600678) in none. The study was interpreted as demonstrating that a combination of techniques can be used for genetic diagnosis of tumor susceptibility in most HNPCC kindreds and lays the foundation for genetic testing of this relatively common disease. Plummer and Casey (1996) pointed out that one of challenges of genetic testing for HNPCC is the development of a standard set of protocols that can be applied to the analysis of multiple candidate genes. In families meeting the strict International Collaborative Group (ICG) definition of HNPCC, the youngest living affected family member should initially be screened for germline mutations in the 2 most commonly mutated mismatch repair genes (MSH2 and MLH1). In individuals who do not meet the strict ICG definition but appear to have a familial predisposition to colon cancer reminiscent of HNPCC, Liu et al. (1996) proposed that mutation analysis be performed only if RER indicating genomic instability characteristic of a mismatch repair deficiency is identified in the tumor of an affected individual. They noted that difficulty is that it is often impossible to obtain blood samples from living affected relatives. In fact, in the study by Liu et al. (1996), blood samples for germline analysis could be obtained from only 48 of the 68 kindreds showing the RER phenotype. Maliaka et al. (1996) identified 6 different new mutations in the MLH1 and MSH2 genes in Russian and Moldavian HNPCC families. Three of these mutations occurred in CpG dinucleotides and led to a premature stop codon, splicing defect, or an amino acid substitution in evolutionarily conserved residues. Analysis of a compilation of published mutations including the new data suggested to the authors that CpG dinucleotides within the coding regions of the MSH2 and MLH1 genes are hotspots for single basepair substitutions.

From a study of unrelated HNPCC families, Wijnen et al. (1996) commented that, whereas the spectrum of mutations at the MSH2 gene is heterogeneous, a cluster of MLH1 mutations were found in the region encompassing exons 15 and 16, which accounts for 50% of all the independent MLH1 mutations described to date. They stated that their finding has great practical value in the design of clinical genetic services.

By screening members of Finnish families displaying HNPCC for predisposing germline mutations in MSH2 and MLH1, Nystrom-Lahti et al. (1995) showed that 2 mutations in MLH1 together account for 63% (19/30) of kindreds meeting international diagnostic criteria. One mutation, originally detected as a 165-bp deletion in MLH1 cDNA comprising exon 16, was shown to represent a 3.5-kb genomic deletion most likely resulting from Alu-mediated recombination (120436.0004). The second mutation destroyed the splice acceptor site of exon 6 (120436.0005). They commented that this was the first report of Alu-mediated recombination causing a prevalent, dominantly inherited predisposition to cancer. Nystrom-Lahti et al. (1995) designed a simple diagnostic test based on PCR for both mutations. Thus 2 ancestral founding mutations account for most Finnish HNPCC kindreds.

Sasaki et al. (1996) studied 43 tumors and corresponding normal tissues from 23 Japanese patients with multiple primary cancers. They found no germline mutations of the MLH1 gene and detected only 2 somatic missense mutations among the 43 tumors examined. These 2 tumors had each shown increased replication error (RER(+)) at more than 1 of the 5 microsatellite loci examined. Only the second of these 2 mutations occurred in an evolutionarily conserved domain of the protein.

Baker et al. (1996) generated mice with a null mutation of the Mlh1 gene. They reported that in addition to compromising replication fidelity, Mlh1 deficiency appeared to cause both male and female sterility associated with reduced levels of chiasmata. Mlh1-deficient spermatocytes exhibited high levels of prematurely separated chromosomes and cell-cycle arrest occurred in the first division of meiosis. Baker et al. (1996) also carried out analysis of the Mlh1 protein in spermatocytes and oocytes using immunostaining. They demonstrated that Mlh1 localizes at chiasma sites on meiotic chromosomes. They concluded that Mlh1 in the mouse is involved in both DNA mismatch repair and meiotic crossing over.

Bellacosa et al. (1996) reviewed genetic counseling aspects of HNPCC against the background of the clinical and molecular genetics.

Depending on the presence or absence of extracolonic tumors, HNPCC has been divided into Lynch syndrome I and Lynch syndrome II. Jager et al. (1997) reported studies based on the Danish HNPCC register comprising 28 families that fulfilled the Amsterdam criteria (Vasen et al., 1991): i.e., (1) families should exhibit 3 histologically verified cases of colorectal cancer, of which at least 1 should be diagnosed before the age of 50 years; (2) there should be affected individuals in 2 generations, and 1 of these individuals should be a first-degree relative to the other 2; and (3) familial adenomatous polyposis should be excluded. In this study, Jager et al. (1997) found an intron 14 founder mutation in the MLH1 gene (120436.0007) in approximately 25% of the kindreds and showed that it was associated with an attenuated HNPCC phenotype characterized by a highly reduced frequency of extracolonic tumors. The mutation was a combined 7-bp deletion and 4-bp insertion that 'silenced the mutated allele,' i.e., it was not expressed. Tumors exhibited microsatellite instability, and loss of the wildtype MLH1 allele was prevalent. Jager et al. (1997) proposed that the mutation resulted in a milder phenotype because the mutated MLH1 protein was prevented from exerting a dominant-negative effect on the concerted action of the mismatch repair system.

Genuardi et al. (1998) characterized the normal alternative splicing of the MLH1 gene and reported a number of splice variants that exist in various tissue types. They observed splice variants lacking exons 6/9, 9, 9/10, 9/10/11, 10/11, 12, 16, and 17. The level of expression varied among different samples. All isoforms were found in 43 to 100% of the mononuclear blood cell samples, as well as in other tissues. The authors cautioned that knowledge of existence of multiple alternative splicing events not caused by genomic DNA changes is important for the evaluation of the results of molecular diagnostic tests based on RNA analysis.

Herman et al. (1998) reported that hypermethylation of the 5-prime CpG island of the MLH1 gene is found in most sporadic primary colorectal cancers with microsatellite instability (MSI) and that this methylation was often, but not invariably, associated with loss of MLH1 protein expression. Such methylation also occurred, but was less prominent, in MSI-negative tumors, as well as in MSI-positive tumors with known mutations of a mismatch repair gene. No hypermethylation of MSH2 (120435) was found. Hypermethylation of colorectal cancer cell lines with microsatellite instability also was frequently observed, and in such cases, reversal of the methylation with 5-aza-2-prime-deoxycytidine not only resulted in reexpression of MLH1 protein, but also in restoration of the mismatch repair capacity in MMR-deficient cell lines. The results suggested that microsatellite instability in sporadic colorectal cancer often results from epigenetic inactivation of MLH1 in association with DNA methylation.

Shimodaira et al. (1998) described a new method for detecting mutations in MLH1 HNPCC using a dominant mutator effect of MLH1 cDNA expressed in Saccharomyces cerevisiae. Most MLH1 missense mutations identified in HNPCC patients abolish the dominant mutator effect. Furthermore, PCR amplification of MLH1 cDNA from mRNA of an HNPCC patient, followed by in vivo recombination into a gap expression vector, allowed detection of a heterozygous loss-of-function missense mutation in MLH1 using this method. This functional assay offers a simple method for detecting and evaluating pathogenic mutations in MLH1.

Germline defects in DNA mismatch repair genes account for the inherited familial cancer syndrome of hereditary nonpolyposis colon cancers in which affected individuals show accelerated development of cancers of the proximal colon, endometrium, and stomach. These cancers typically demonstrate inactivation of the residual wildtype MMR allele inherited opposite the germline mutant, absence of DNA MMR activity in in vitro assays, and acquisition of an in vivo mutator phenotype showing up to 1,000-fold increased gene mutation rates. Additionally, these cancers display an associated instability of genomic MSI. MSI is similarly found in approximately 15 to 20% of sporadic colon cancers that arise in individuals without any family history of colon cancer. Like HNPCC-associated colon cancers, sporadic MSI colon cancers arise predominantly in the proximal colon and show a high rate of frameshift mutations at a mutational hotspot in the transforming growth factor-beta type II receptor tumor suppressor gene (TGFBR2; 190182). Familial and sporadic MSI colon cancers thus appear to share a common carcinogenic pathway. Liu et al. (1995) established that MMR gene inactivation via somatic mutation was the cause of some cases of sporadic MSI colon cancers. However, unexpectedly, in many sporadic MSI colon cancers, MMR genes were found to remain wildtype. MMR coding sequences were similarly reported to be wildtype in many sporadic MSI endometrial cancers (Katabuchi et al., 1995). Kane et al. (1997) described methylation of the MLH1 promoter region in some MSI tumors. Veigl et al. (1998) investigated a group of MSI cancer cell lines, most of which were documented as established from antecedent MSI-positive malignant tumors. In 5 of 6 such cases, they found that MLH1 protein was absent, even though MHL1-coding sequences were wildtype. In each case, absence of MLH1 protein was associated with the methylation of the MLH1 gene promoter. Furthermore, in each case, treatment with the demethylating a gent 5-azacytidine induced expression of the absent MLH1 protein. Moreover, in single cell clones, MLH1 expression could be turned on, off, and on again by 5-azacytidine exposure, washout, and reexposure. This epigenetic inactivation of MLH1 additionally accounted for the silencing of both maternal and paternal tumor MLH1 alleles, both of which could be reactivated by 5-azacytidine. Thus, substantial numbers of human MSI cancers appear to arise by MLH1 silencing via an epigenetic mechanism that can inactivate both of the MLH1 alleles. Promoter methylation is intimately associated with this epigenetic silencing mechanism.

Ban and Yang (1998) determined the crystal structure of a 40-kD N-terminal fragment of E. coli MutL that retains all of the conserved residues in the MutL family. The structure of MutL is homologous to that of an ATPase-containing fragment of DNA gyrase. The authors demonstrated that MutL binds and hydrolyzes ATP to ADP and Pi. Mutations in the MutL family that cause deficiencies in DNA mismatch repair and a predisposition to cancer mainly occur in the putative ATP-binding site. Ban and Yang (1998) also provided evidence that the flexible, yet conserved loops surrounding this ATP-binding site undergo conformational changes upon ATP hydrolysis, thereby modulating interactions between MutL and other components of the repair machinery.

Baker, S. M.; Plug, A. W.; Prolla, T. A.; Bronner, C. E.; Harris, A. C.; Yao, X.; Christie, D.-M.; Monell, C.; Arnheim, N.; Bradley, A.; Ashley, T.; Liskay, R. M.: Involvement of mouse Mlh1 in DNA mismatch repair and meiotic crossing over. Nature Genet. 336-342, 1996.

Ban, C.; Yang, W.: Crystal structure and ATPase activity of MutL: implications for DNA repair and mutagenesis. Cell 95: 541-552, 1998.

Bapat, B.; Xia, L.; Madlensky, L.; Mitri, A.; Tonin, P.; Narod, S. A.; Gallinger, S.: The genetic basis of Muir-Torre syndrome includes the hMLH1 locus. (Letter) Am. J. Hum. Genet. 59: 736-739, 1996.

Bellacosa, A.; Genuardi, M.; Anti, M.; Viel, A.; Ponz de Leon, M.: Hereditary nonpolyposis colorectal cancer: review of clinical, molecular genetics, and counseling aspects. Am. J. Med. Genet. 62: 353-364, 1996.

Bronner, C. E.; Baker, S. M.; Morrison, P. T.; Warren, G.; Smith, L. G.; Lescoe, M. K.; Kane, M.; Earabino, C.; Lipford, J.; Lindblom, A.; Tannergard, P.; Bollag, R. J.; Godwin, A. R.; Ward, D. C.; Nordenskjold, M.; Fishel, R.; Kolodner, R.; Liskay, R. M.: Mutation in the DNA mismatch repair gene homologue hMLH1 is associated with hereditary non-polyposis colon cancer. Nature 368: 258-261, 1994.

Genuardi, M.; Viel, A.; Bonora, D.; Capozzi, E.; Bellacosa, A.; Leonardi, F.; Valle, R.; Ventura, A.; Pedroni, M.; Boiocchi, M.; Neri, G.: Characterization of MLH1 and MSH2 alternative splicing and its relevance to molecular testing of colorectal cancer susceptibility. Hum. Genet. 102: 15-20, 1998.

Green, R. C.; Narod, S. A.; Morasse, J.; Young, T. L.; Cox, J.; Fitzgerald, G. W. N.; Tonin, P.; Ginsburg, O.; Miller, S.; Poitras, P.; Laframboise, R.; Routhier, G.; Plante, M.; Morissette, J.; Weissenbach, J.: Khandjian, E. W.; Rousseau, F.: Hereditary nonpolyposis colon cancer: analysis of linkage to 2p15-16 places the COCA1 locus telomeric to D2S123 and reveals genetic heterogeneity in seven Canadian families. Am. J. Hum. Genet. 54: 1067-1077, 1994.

Hamilton, S. R.; Liu, B.; Parsons, R. E.; Papadopoulos, N.; Jen, J.; Powell, S. M.; Krush, A. J.; Berk, T.; Cohen, Z.; Tetu, B.; Burger, P. C.; Wood, P. A.; Taqi, F.; Booker, S. V.; Petersen, G. M.; Offerhaus, G. J. A.; Tersmette, A. C.; Giardiello, F. M.; Vogelstein, B.; Kinzler, K. W.: The molecular basis of Turcot's syndrome. New Eng. J. Med. 332: 839-847, 1995.

Han, H.-J.; Maruyama, M.; Baba, S.; Park, J.-G.; Nakamura, Y.: Genomic structure of human mismatch repair gene, hMLH1, and its mutation analysis in patients with hereditary non-polyposis colorectal cancer (HNPCC). Hum. Molec. Genet. 4: 237-242, 1995.

Hemminki, A.; Peltomaki, P.; Mecklin, J.-P.; Jarvinen, H.; Salovaara, R.; Nystrom-Lahti, M.; de la Chapelle, A.; Aaltonen, L. A.: Loss of the wild type MLH1 gene is a feature of hereditary nonpolyposis colorectal cancer. Nature Genet. 8: 405-410, 1994.

Herman, J. G.; Umar, A.; Polyak, K.; Graff, J. R.; Ahuja, N.; Issa, J.-P. J.; Markowitz, S.; Willson, J. K. V.; Hamilton, S. R.; Kinzler, K. W.; Kane, M. F.; Kolodner, R. D.; Vogelstein, B.; Kunkel, T. A.; Baylin, S. B.: Incidence and functional consequences of hMLH1 promoter hypermethylation in colorectal carcinoma. Proc. Nat. Acad. Sci. 95: 6870-6875, 1998.

Jager, A. C.; Bisgaard, M. L.; Myrhoj, T.; Bernstein, I.; Rehfeld, J. F.; Nielsen, F. C.: Reduced frequency of extracolonic cancers in hereditary nonpolyposis colorectal cancer families with monoallelic hMLH1 expression. Am. J. Hum. Genet. 61: 129-138, 1997.

Kane, M. F.; Loda, M.; Gaida, G. M.; Lipman, J.; Mishra, R.; Goldman, H.; Jessup, J. M.; Kolodner, R.: Methylation of the hMLH1 promoter correlates with lack of expression of hMLH1 in sporadic colon tumors and mismatch repair-defective human tumor cell lines. Cancer Res. 57: 808-811, 1997.

Katabuchi, H.; van Rees, B.; Lambers, A. R.; Ronnett, B. M.; Blazes, M. S.; Leach, F. S.; Cho, K. R.; Hedrick, L.: Mutations in DNA mismatch repair genes are not responsible for microsatellite instability in most sporadic endometrial carcinomas. Cancer Res. 55: 5556-5560, 1995.

Li, G.-M.; Modrich, P.: Restoration of mismatch repair to nuclear extracts of H6 colorectal tumor cells by a heterodimer of human MutL homologs. Proc. Nat. Acad. Sci. 92: 1950-1954, 1995.

Lindblom, A.; Tannergard, P.; Werelius, B.; Nordenskjold, M.: Genetic mapping of a second locus predisposing to hereditary non-polyposis colon cancer. Nature Genet. 5: 279-282, 1993.

Liu, B.; Nicolaides, N. C.; Markowitz, S.; Willson, J. K. V.; Parsons, R. E.; Jen, J.; de la Chapelle, A.; Hamilton, S. R.; Kinzler, K. W.; Vogelstein, B.: Mismatch repair gene defects in sporadic colorectal cancers with microsatellite instability. Nature Genet. 9: 48-55, 1995.

Liu, B.; Parsons, R.; Papadopoulos, N.; Nicolaides, N. C.; Lynch, H. T.; Watson, P.; Jass, J. R.; Dunlop, M.; Wyllie, A.; Peltomaki, P.; de la Chapelle, A.; Hamilton, S. R.; Vogelstein, B.; Kinzler, K. W.: Analysis of mismatch repair genes in hereditary non-polyposis colorectal cancer patients. Nature Med. 2: 169-174, 1996.

Lynch, H. T.; Fusaro, R. M.; Roberts, L.; Voorhees, G. J.; Lynch, J. F.: Muir-Torre syndrome in several members of a family with a variant of the cancer family syndrome. Brit. J. Derm. 113: 295-301, 1985.

Maliaka, Y. K.; Chudina, A. P.; Belev, N. F.; Alday, P.; Bochkov, N. P.; Buerstedde, J.-M.: CpG dinucleotides in the hMSH2 and hMLH1 genes are hotspots for HNPCC mutations. Hum. Genet. 97: 251-255, 1996.

Moisio, A.-L.; Sistonen, P.; Weissenbach, J.; de la Chapelle, A.; Peltomaki, P.: Age and origin of two common MLH1 mutations predisposing to hereditary colon cancer. Am. J. Hum. Genet. 59: 1243-1251, 1996.

Nystrom-Lahti, M.; Kristo, P.; Nicolaides, N. C.; Chang, S.-Y.; Aaltonen, L. A.; Moisio, A.-L.; Jarvinen, H. J.; Mecklin, J.-P.; Kinzler, K. W.; Vogelstein, B.; de la Chapelle, A.; Peltomaki, P.: Founding mutations and Alu-mediated recombination in hereditary colon cancer. Nature Med. 1: 1203-1206, 1995.

Nystrom-Lahti, M.; Sistonen, P.; Mecklin, J.-P.; Pylkkanen, L.; Aaltonen, L. A.; Jarvinen, H.; Weissenbach, J.; de la Chapelle, A.; Peltomaki, P.: Close linkage to chromosome 3p and conservation of ancestral founding haplotype in hereditary nonpolyposis colorectal cancer families. Proc. Nat. Acad. Sci. 91: 6054-6058, 1994.

Papadopoulos, N.; Nicolaides, N. C.; Wei, Y.-F.; Ruben, S. M.; Carter, K. C.; Rosen, C. A.; Haseltine, W. A.; Fleischmann, R. D.; Fraser, C. M.; Adams, M. D.; Venter, J. C.; Hamilton, S. R.; Petersen, G. M.; Watson, P.; Lynch, H. T.; Peltomaki, P.; Mecklin, J.-P.; de la Chapelle, A.; Kinzler, K. W.; Vogelstein, B.: Mutation of a mutL homolog in hereditary colon cancer. Science 263: 1625-1629, 1994.

Paraf, F.; Sasseville, D.; Watters, A. K.; Narod, S.; Ginsburg, O.; Shibata, H.; Jothy, S.: Clinicopathological relevance of the association between gastrointestinal and sebaceous neoplasms: the Muir-Torre syndrome. Hum. Pathol. 26: 422-427, 1995.

Plummer, S. J.; Casey, G.: Are we any closer to genetic testing for common malignancies? Nature Med. 2: 156-158, 1996.

Sasaki, S.; Horii, A.; Shimada, M.; Han, H.-J.; Yanagisawa, A.; Muto, T.; Nakamura, Y.: Somatic mutations of a human mismatch repair gene, hMLH1, in tumors from patients with multiple primary cancers. Hum. Mutat. 7: 275-278, 1996.

Shimodaira, H.; Filosi, N.; Shibata, H.; Suzuki, T.; Radice, P.; Kanamaru, R.; Friend, S. H.; Kolodner, R. D.; Ishioka, C.: Functional analysis of human MLH1 mutations in Saccharomyces cerevisiae. Nature Genet. 19: 384-389, 1998.

Tannergard, P.; Zabarovsky, E.; Stanbridge, E.; Nordenskjold, M.; Lindblom, A.: Sublocalization of a locus at 3p21.3-23 predisposing to hereditary nonpolyposis colon cancer. Hum. Genet. 94: 210-214, 1994.
Vasen, H. F.; Mecklin, J. P.; Khan, P. M.; Lynch, H. T.: The International Collaborative Group on Hereditary Non-Polyposis Colorectal Cancer (ICG-HNPCC). Dis. Colon Rectum 34: 424-425, 1991.
Veigl, M. L.; Kasturi, L.; Olechnowicz, J.; Ma, A.; Lutterbaugh, J. D.; Periyasamy, S.; Li, G.-M.; Drummond, J.; Modrich, P. L.; Sedwick, W. D.; Markowitz, S. D.: Biallelic inactivation of hMLH1 by epigenetic gene silencing, a novel mechanism causing human MSI cancers. Proc. Nat. Acad. Sci. 95: 8698-8702, 1998.
Wang, Y.; Friedl, W.; Lamberti, C.; Ruelfs, C.; Kruse, R.; Propping, P.: Hereditary nonpolyposis colorectal cancer: causative role of a germline missense mutation in the hMLH1 gene confirmed by the independent occurrence of the same somatic mutation in tumour tissue. Hum. Genet. 100: 362-364, 1997.
Wijnen, J.; Khan, P. M.; Vasen, H.; Menko, F.; van der Klift, H.; van den Broek, M.; van Leeuwen-Cornelisse, I.; Nagengast, F.; Meijers-Heijboer, E. J.; Lindhout, D.; Griffioen, G.; Cats, A.; Kleibeuker, J.; Varesco, L.; Bertario, L.; Bisgaard, M.-L.; Mohr, J.; Kolodner, R.; Fodde, R.: Majority of hMLH1 mutations responsible for hereditary nonpolyposis colorectal cancer cluster at the exonic region 15-16. Am. J. Hum. Genet. 58: 300-307, 1996.

MUIR-TORRE SYNDROME (#158320)

JJM NOTE

Full phenotypic characterization of this syndrome awaits further cases and families, in conjunction with genotyping for at least the MSH2 locus. When President Ronald Reagan and his brother had colectomies for cancer within a month of each other, the possibility of a familial syndrome could be raised. In fact, as a federal government physician-scientist, I sent a memorandum through channels, raising the possibility of Muir-Torre syndrome because of the president's multiple skin cancers (if they were not due to the sun). My suggestion was stopped cold and returned with a note, "If the White House wanted to hear about Muir-Torre syndrome, they would have asked."

A number sign (#) is used with this entry because of evidence that the Muir-Torre syndrome is part of the Lynch cancer family syndrome II (114400), which has been related to mutation in the MSH2 gene (120435) located on 2p. Mutations in the MLH1 gene (e.g., 120436.0006), located on 3p, also cause the syndrome.

Muir-Torre syndrome represents the association of sebaceous skin tumors with internal malignancy. The Gardner and Peutz-Jeghers syndromes are examples of skin-polyposis syndromes. Polyps of the stomach have been reported with the basal cell nevus syndrome (109400). Muir et al. (1967) described a Maltese male with multiple primary carcinomata of the colon, duodenum and larynx in association with keratoacanthomata of the face. Although he was one of 22 sibs, including 4 sets of twins, no family history of malignancy was obtained. Torre (1968) emphasized the occurrence of multiple sebaceous tumors, e.g., sebaceous adenoma, like that of tuberous sclerosis, in a patient who had a primary carcinoma of the ampulla of Vater resected at age 50 years and a primary carcinoma of the colon resected at age 53. Stewart et al. (1977) reported a woman who had 11 keratoacanthomata removed over a 21-year period and who also had Bowen disease of the vulva and a carcinoma of the rectum. The cancers, although multiple, are usually relatively indolent. Reiffers et al. (1976) presented evidence for autosomal dominant inheritance including male-to-male transmission. The families have characteristics of the Lynch II cancer family syndrome. Lynch et al. (1981) observed sebaceous neoplasms in affected members of cancer families and postulated that the purportedly distinct Muir-Torre syndrome is actually one mode of manifestation of the cancer family syndrome. Lynch et al. (1981) expressed doubts that intestinal polyposis is a part of this syndrome. See also Lynch et al. (1985), who reported continuing uncertainty as to the nosologic place of the Muir-Torre syndrome. That the Muir-Torre phenotype is an expression of the Lynch II cancer family syndrome was supported by the fact that one of the cases described by Lynch et al. (1981) was found to be a descendant of Warthin's family G which is considered to be the first description of the familial cancer syndrome now termed 'Lynch II' (Finan and Connolly, 1984). Rothenberg et al. (1990) described 2 cases suggesting that a single sebaceous adenoma can be the sole tip-off to the presence of colon cancer in the Muir-Torre syndrome. A 68-year-old woman had a solitary, 2-mm tumor of the forehead. Both she and her mother had had colonic resections for carcinomas. A 50-year-old male physician with a solitary, 6-mm sebaceous adenoma removed from the scalp was investigated for colonic abnormalities and found to have a fungating low-grade adenocarcinoma at the splenic flexure. The man's father had had a colon carcinoma removed at the age of 29 years. Guitart et al. (1991) described a case of a 44-year-old man with a keratoacanthoma of the left nasolabial fold; his mother died of gastric carcinoma and his sister of colonic cancer. He and his brother had alpha-1-antitrypsin deficiency (107400). The keratoacanthoma developed after immunosuppressive therapy for necrotizing vasculitis. Stone et al. (1986) reported a patient with Muir-Torre syndrome in whom exacerbation of the cutaneous lesions occurred during immunosuppression after heart transplant. Cohen (1992) used the label Muir-Torre syndrome in describing a man with Hodgkin lymphoma who subsequently developed a sebaceous carcinoma of the upper eyelid. Rodenas et al. (1993) reported a family in which multiple members had the Muir-Torre syndrome and hyperlipidemia.

Approximately 15% of female patients with Muir-Torre syndrome develop endometrial cancer (Cohen et al., 1991).

Hall et al. (1994) reported a 5-generation family in which at least 2 persons displayed the Muir-Torre phenotype, while many other family members had tumors consistent with cancer family syndrome. Most of the tumors were gastrointestinal, gynecologic, and urologic, with several persons having multiple synchronous or metachronous primaries. The prognosis appeared to be better than might be expected. Hall et al. (1994) found that in this and a second Muir-Torre family, the disease locus mapped to 2p in the same region as the Lynch II syndrome, namely to the MSH2 locus. Haplotype analysis of the pedigree showed that the persons with colorectal or endometrial cancer (with or without sebaceous lesions of the skin) shared the same segment of 2p at this locus.

Schwartz and Torre (1995) reviewed the Muir-Torre syndrome defined as the association of certain types of sebaceous neoplasms of the skin, with or without keratoacanthomas, with one or more low-grade visceral malignancies in the absence of other predisposing factors. The sebaceous tumors were relatively uncommon or rare types: sebaceous adenoma, sebaceous epithelioma, basal cell epithelioma with sebaceous differentiation, and sebaceous carcinoma. They suggested that the visceral cancers are often multiple and usually indolent, permitting prolonged survival. Even metastatic disease may respond well to aggressive surgical treatment. The sebaceous cancers in this syndrome, like the visceral malignancies, are less aggressive than their counterparts unassociated with this syndrome.

Bapat et al. (1996) demonstrated that mutations in the MLH1 locus can also underlie the Muir-Torre syndrome.

Kruse et al. (1996) identified germline mutations in the MSH2 DNA mismatch repair gene in 2 unrelated Muir-Torre syndrome patients ascertained because of their skin tumors. They suggested that the results, together with published cases of Muir-Torre syndrome, supported the hypothesis that this disorder with its characteristic skin lesions is confined to mutations in the MSH2 gene. They tabulated 4 reported cases in addition to the 2 of their own.

About half of MTS cases, defined as the coincidence of at least 1 sebaceous skin tumor and 1 internal malignancy, are affected by colorectal cancer. In a subgroup of MTS patients, the disease has an underlying DNA mismatch-repair (MMR) defect and thus is allelic to hereditary nonpolyposis colorectal cancer (HNPCC). In 16 MTS patients with sebaceous skin tumors and colorectal cancer, Kruse et al. (1998) found that all exhibited high genomic instability in at least 1 tumor from the skin or colon. A search for germline mutations in the MSH2 and MLH1 genes in 13 of the patients revealed truncating mutations in 9 (69%): 8 mutations in the MSH2 gene and 1 in the MLH1 gene. Thus, MTS patients exhibited significantly more mutations in the MHS2 gene than in the MLH1 gene. The subpopulation of MTS patients who were also affected by colorectal cancer, irrespective of family history and age at onset of tumors are likely to have an underlying DNA MMR defect similar to that for patients with a family history fulfilling the strict clinical criteria for HNPCC.

See also Bitran and Pellettiere (1974); Fahmy et al. (1982); Hall et al. (1994); Lynne-Davies and Brown (1974); Rulon and Helwig (1973); Schwartz et al. (1980); Tschang et al. (1976)

Bapat, B.; Xia, L.; Madlensky, L.; Mitri, A.; Tonin, P.; Narod, S. A.; Gallinger, S.: The genetic basis of Muir-Torre syndrome includes the hMLH1 locus. (Letter) Am. J. Hum. Genet. 59: 736-739, 1996.
Bitran, J.; Pellettiere, E. V.: Multiple sebaceous gland tumors and internal carcinoma: Torre's syndrome. Cancer 33: 835-836, 1974.
Cohen, P. R.: Muir-Torre syndrome in patients with hematologic malignancies. Am. J. Hemat. 40: 64-65, 1992.
Cohen, P. R.; Kohn, S. R.; Kurzrock, R.: Association of sebaceous gland tumors and internal malignancy: the Muir-Torre syndrome. Am. J. Med. 90: 606-613, 1991.
Fahmy, A.; Burgdorf, W. H. C.; Schosser, R. H.; Pitha, J.: Muir-Torre syndrome: report of a case and reevaluation of the dermatopathologic features. Cancer 49: 1898-1903, 1982.
Finan, M. C.; Connolly, S. M.: Sebaceous gland tumors and systemic disease: a clinicopathologic analysis. Medicine 63: 232-242, 1984.

Guitart, J.; McGillis, S. T.; Bergfeld, W. F.; Tuthill, R. J.; Bailin, P. L.; Camisa, C.: Muir-Torre syndrome associated with alpha-1-antitrypsin deficiency and cutaneous vasculitis: report of a case with exacerbation of a cutaneous neoplasm during immunosuppressive therapy. J. Am. Acad. Derm. 24: 875-877, 1991.
Hall, N. R.; Angela, M.; Williams, T.; Murday, V. A.; Newton, J. A.; Bishop, D. T.: Muir-Torre syndrome: a variant of the cancer family syndrome. J. Med. Genet. 31: 627-631, 1994.
Hall, N. R.; Murday, V. A.; Chapman, P.; Williams, M. A.; Burn, J.; Finan, P. J.; Bishop, D. T.: Genetic linkage in Muir-Torre syndrome to the same chromosomal site as cancer family syndrome. Europ. J. Cancer 30A: 180-182, 1994.
Kruse, R.; Lamberti, C.; Wang, Y.; Ruelfs, C.; Bruns, A.; Esche, C.; Lehmann, P.; Ruzicka, T.; Rutten, A.; Friedl, W.; Propping, P.: Is the mismatch repair deficient type of Muir-Torre syndrome confined to mutations in the hMSH2 gene? Hum. Genet. 98: 747-750, 1996.
Kruse, R.; Rutten, A.; Lamberti, C.; Hosseiny-Malayeri, H. R.; Wang, Y.; Ruelfs, C.; Jungck, M.; Mathiak, M.; Ruzicka, T.; Hartschuh, W.; Bisceglia, M.; Friedl, W.; Propping, P.: Muir-Torre phenotype has a frequency of DNA mismatch-repair-gene mutations similar to that in hereditary nonpolyposis colorectal cancer families defined by the Amsterdam criteria. Am. J. Hum. Genet. 63: 63-70, 1998.
Lynch, H. T.; Fusaro, R. M.; Roberts, L.; Voorhees, G. J.; Lynch, J. F.: Muir-Torre syndrome in several members of a family with a variant of the cancer family syndrome. Brit. J. Derm. 113: 295-301, 1985.
Lynch, H. T.; Lynch, P. M.; Pester, J.; Fusaro, R. M.: The cancer family syndrome: rare cutaneous phenotypic linkage of Torre's syndrome. J.A.M.A. 141: 607-611, 1981.
Lynne-Davies, G.; Brown, J.: Multiple sebaceous gland tumours associated with polyposis of the colon and bony abnormalities. Canad. Med. Assoc. J. 110: 1377-1379, 1974.
Muir, G. G.; Bell, A. Y.; Barlow, K. A.: Multiple primary carcinomata of the colon, duodenum, and larynx associated with keratoacanthomata of the face. Brit. J. Surg. 54: 191-195, 1967.
Reiffers, J.; Laugier, P.; Hunziker, N.: Hyperplasies sebacees, kerato-acanthomes, epitheliomas du visage et cancer du colon: une nouvelle entite?. Dermatologica 153: 23-33, 1976.
Rodenas, J. M.; Herranz, T.; Tercedor, J.; Lopez, B.; Naranjo, R.; Delgado, V.: Muir-Torre syndrome associated with a family history of hyperlipidemia. J. Am. Acad. Derm. 28: 285-288, 1993.
Rothenberg, J.; Lambert, W. C.; Vail, J. T., Jr.; Nemlick, A. S.; Schwartz, R. A.: The Muir-Torre (Torre's) syndrome: the significance of a solitary sebaceous tumor. J. Am. Acad. Derm. 23: 638-640, 1990.
Rulon, D. B.; Helwig, E. B.: Multiple sebaceous neoplasms of the skin: an association with multiple visceral carcinomas, especially of the colon. Am. J. Clin. Path. 60: 745-752, 1973.
Schwartz, R. A.; Flieger, D. N.; Saied, N. K.: The Torre syndrome with gastrointestinal polyposis. Arch. Derm. 116: 312-314, 1980.
Schwartz, R. A.; Torre, D. P.: The Muir-Torre syndrome: a 25-year retrospect. J. Am. Acad. Dermat. 33: 90-104, 1995.
Stewart, W. M.; Lauret, P.; Hemet, J.; Thomine, E.; Gueville, R. M.: Kerato-acanthomes multiples et carcinomes visceraux: syndrome de Torre. Ann. Derm. Venerol. 104: 622-626, 1977.
Stone, M. S.; Duncan, W. C.; McGavran, M. H.: Torre's syndrome: exacerbation of cutaneous manifestations with immunosuppression. J. Am. Acad. Derm. 15: 1101-1103, 1986.
Torre, D.: Multiple sebaceous tumors. Arch. Derm. 98: 549-551, 1968.
Tschang, T.-P.; Poulos, E.; Ho, C.-K.; Kuo, T.-T.: Multiple sebaceous adenomas and internal malignant disease: a case report with chromosomal analysis. Hum. Path. 7: 589-594, 1976.

MISMATCH REPAIR GENE 1; hPMS1 (*600258)

JJM NOTE

This gene, *PMS1,* is one of the infrequent causes of the Lynch cancer family syndrome, but would in theory be detected by a survey assay of colon cancer tissue for replication error repair (RER) phenomenon or microsatellite instability.

Genomic instability in tumor DNA is a feature of hereditary nonpolyposis colon cancer (HNPCC). Two types of HNPCC had been tied to mismatch repair of dinucleotide and trinucleotide repeat sequences: MSH2 (120435) on 2p and MLH1 (120436) on 3p. Two other genes were implicated in the mismatch repair complex: PMSL1 on chromosome 2 and PMSL2 (600259) on chromosome 7 (Papadopoulos et al., 1994). Nicolaides et al. (1994) regionalized the PMSL2 gene to 2q31-q33 by fluorescence in situ hybridization. Horii et al. (1994) isolated and analyzed the human counterparts of yeast PMSL genes. DNA sequencing analyses indicated that human PMSL genes constitute a multiple gene family. Some of the family members were mapped to 7q11.23 and 7q22 by fluorescence in situ hybridization. None of the cosmid clones studied by Horii et al. (1994) mapped to chromosome 2.

MLH1 is a homolog of the bacterial DNA mismatch repair gene mutL. In screening a database of human genes identified by the expressed sequence tag (EST) method (Adams et al., 1991), Papadopoulos et al. (1994) identified 2 other ESTs with homology to bacterial and yeast mutL genes. The 2 had greatest homology to the yeast PMS1 gene and therefore the human genes were designated PMS1 and PMS2 (600259). Sequence analysis of human PMS1 cDNA identified an open reading frame (ORF) of 2,795 bp thought to encode a 932-residue protein with 27% identity to yeast PMS1. Sequence analysis of the human PMS2 clone identified a 2,586-bp ORF thought to encode an 862-residue protein with 32% identity to yeast PMS1. To evaluate the role of PMS1 and PMS2 in HNPCC, Nicolaides et al. (1994) studied 22 patients with a family history of HNPCC who did not have demonstrable mutations in MSH2 or MLH1. To identify potential mutations, they used an in vitro synthesized protein (IVSP) assay, which detects deletions, insertions, frameshifts, and nonsense mutations that result in an altered size of the protein product. Three overlapping segments of PMS1 were amplified using RT-PCR and the products were transcribed and translated in vitro. A reproducible truncated product was detected in a segment containing codons 1 to 500 from 1 patient (C.W.). Sequence analysis of RT-PCR products revealed an in-frame deletion that removed codons 195 to 233. Sequence analysis of a PMS1 genomic clone identified an exon that corresponded exactly to the boundaries of the deletion. They then amplified the skipped exon from the genomic DNA of patient C.W.; while no alterations of the splicing consensus sites were identified, a single C-to-T transition at codon 233, which converted a glutamine to a stop codon (600258.0001), was identified. This nonsense codon appeared to be responsible for the exon skipping. (Dietz et al. (1993) had demonstrated that exons containing nonsense codons can be selectively skipped; see 258870.0036.)

Adams, M. D.; Kelley, J. M.; Gocayne, J. D.; Dubnick, M.; Polymeropoulos, M. H.; Xiao, H.; Merril, C. R.; Wu, A.; Olde, B.; Moreno, R. F.; Kerlavage, A. R.; McCombie, W. R.; Venter, J. C.: Complementary DNA sequencing: expressed sequence tags and human genome project. Science 252: 1651-1656, 1991.
Dietz, H. C.; Valle, D.; Francomano, C. A.; Kendzior, R. J., Jr.; Pyeritz, R. E.; Cutting, G. R.: The skipping of constitutive exons in vivo induced by nonsense mutations. Science 259: 680-683, 1993.
Horii, A.; Han, H.-J.; Sasaki, S.; Shimada, M.; Nakamura, Y.: Cloning, characterization and chromosomal assignment of the human genes homologous to yeast PMS1, a member of mismatch repair genes. Biochem. Biophys. Res. Commun. 204: 1257-1264, 1994.
Nicolaides, N. C.; Papadopoulos, N.; Liu, B.; Wei, Y.-F.; Carter, K. C.; Ruben, S. M.; Rosen, C. A.; Haseltine, W. A.; Fleischmann, R. D.; Fraser, C. M.; Adams, M. D.; Venter, J. C.; Dunlop, M. G.; Hamilton, S. R.; Petersen, G. M.; de la Chapelle, A.; Vogelstein, B.; Kinzler, K. W.: Mutations of two PMS homologues in hereditary nonpolyposis colon cancer. Nature 371: 75-80, 1994.
Papadopoulos, N.; Nicolaides, N. C.; Wei, Y.-F.; Ruben, S. M.; Carter, K. C.; Rosen, C. A.; Haseltine, W. A.; Fleischmann, R. D.; Fraser, C. M.; Adams, M. D.; Venter, J. C.; Hamilton, S. R.; Petersen, G. M.; Watson, P.; Lynch, H. T.; Peltomaki, P.; Mecklin, J.-P.; de la Chapelle, A.; Kinzler, K. W.; Vogelstein, B.: Mutation of a mutL homolog in hereditary colon cancer. Science 263: 1625-1629, 1994.

MISMATCH REPAIR GENE 2; hPMS2 (*600259)

JJM NOTE

This is the fourth gene associated with the Lynch cancer family syndrome and the other gene locus associated with Turcot syndrome.

See PMS1 (600258). The PMS2 gene encodes a protein involved in the mismatch repair (MMR) complex. It was found to map to chromosome 7 by analysis of somatic cell hybrids and to 7p22 by fluorescence in situ hybridization (Nicolaides et al., 1994).

For mutational analysis of the PMS2 gene, Nicolaides et al. (1994) studied 22 colon cancer samples from patients with a family history of hereditary nonpolyposis colorectal cancer (HNPCC). They used cDNA in an in vitro synthesized protein (IVSP) assay. In one case, G.C., analysis demonstrated a truncated PMS2 protein. Analysis of RNA from normal fibroblasts revealed an altered transcript due to a 1,230-bp in-frame deletion of codons 268 to 669 as well as the expected normal transcript. In contrast, analysis of the tumor identified a second transcript due to a somatic out-of-frame deletion that removed codons 301 to 381. The germline deletion of codons 268 to 669 was also observed in the tumor, but no normal transcript could be identified. During the sequence analysis of PMS2 cDNA, Nicolaides et al. (1994) noted a G-to-A transition in codon 20 that resulted in an arginine to glutamine change. The change was observed in 3 of 18 HNPCC patients and in 5 of 77 controls. They concluded that the change represents a polymorphism rather than a functional mutation.

Turcot syndrome (276300) is characterized clinically by the concurrence of a primary brain tumor and multiple colorectal adenomas. Hamilton et al. (1995) demonstrated that this association can result from 2 distinct types of germline defects: mutation of the APC gene (175100), which also causes familial polyposis of the colon, or mutation of the mismatch repair gene PMS2.

The form of familial nonpolyposis colon cancer due to mutations in the PMS2 gene might be referred to as type 3, type 1 being that due to mutation in the MSH2 gene (120435) and type 2 being that due to mutation in the MLH1 gene (120436).

Parsons et al. (1995) found that HNPCC patients with germline mutations of either PMS2 or MLH1 had widespread mutations not only in their tumors but also in their non-neoplastic cells. Although these patients had numerous mutations in all the tissues examined, they had very few tumors. The hypermutability was associated with a profound defect in MMR at the biochemical level. These results had implications for the relationship between mutagenesis and carcinogenesis, and they suggested that MMR deficiency is compatible with normal human development. Furthermore, drugs designed to be lethal to MMR-deficient tumor cells would likely be toxic to the non-neoplastic cells of the MMR-deficient patients studied in this report. Careful evaluation of the germline defects in HNPCC patients would therefore seem critical once such therapeutic agents are developed. The findings suggested their strategy for making any cell MMR-deficient by the transfer of genes encoding the mutant proteins found in patients of this study. The expression of the altered gene products could be driven by cell type-specific promoters to create highly mutable cell types in transgenic organisms. This strategy might be useful for a variety of experimental purposes.

Using gene targeting in embryonic stem cells, Baker et al. (1995) generated a line of mice with a null mutation in the Pms2 gene. They observed microsatellite instability in the male germline and in tumor DNA of Pms2-deficient animals. They concluded, therefore, that Pms2 is involved in DNA mismatch repair in a variety of tissues. The deficient animals appeared prone to sarcomas and lymphomas. Pms2-deficient males were infertile, producing only abnormal spermatozoa. Analysis of axial element and synaptonemal complex formation during prophase of meiosis I indicated abnormalities in chromosome synapsis. These observations suggested links among MMR, genetic recombination, and chromosome synapsis in meiosis.

Nicolaides et al. (1995) found that the PMS2 genomic locus encompasses 16 kb and consists of 15 exons. They also identified a family of PMS2-related genes located on chromosome 7 at 7p13-p12, 7q11, and 7q22. Exons 1 through 5 of these homologs share a high degree of identity with PMS2. Nicolaides et al. (1995) presented the sequence of 7 novel genes representing the PMS2-related gene family. The similarity and number of these genes made specific amplification of PMS2 problematic, but knowledge of them aided the successful design of oligonucleotides for this purpose.

In the process of characterizing PMS2 cDNA and its promoter region, Nicolaides et al. (1995) discovered another gene, provisionally called JTV1 (600859), that is transcribed from the opposite DNA strand and overlaps with PMS2. The genes lie in a head-to-head arrangement and both are ubiquitously expressed.

To determine the effect of Pms2 inactivation on genomic integrity in vivo, Narayanan et al. (1997) constructed hybrid transgenic mice that carry targeted disruptions at the Pms2 locus along with a chromosomally integrated mutation reporter gene. In the absence of any mutagenic treatment, mice nullizygous for Pms2 showed a 100-fold elevation in mutation frequency in all tissues examined compared with both wild-type and heterozygous littermates. The mutation pattern in the nullizygotes was notable for frequent 1-bp deletions and insertions within mononucleotide repeat sequences, consistent with an essential role for PMS2 in the repair of replication slippage errors. Further, the results demonstrated that high rates of mutagenesis in multiple tissues are compatible with normal development and life and are not necessarily associated with accelerated aging. Also, the finding of genetic instability in all tissues tested contrasted with the limited tissue distribution of cancers in the animals, raising important questions regarding the role of mutagenesis in carcinogenesis.

Miyaki et al. (1997) stated that up to that time 2 PMS2 germline mutations had been identified in hereditary nonpolyposis colorectal cancer families.

Nomenclature: It is proposed here to refer to HNPCC families with mutation of the PMS2 gene as type 4 (HNPCC4); types 1, 2, 3, and 5 have mutations in MSH2 (120435), MLH1 (120436), PMS1 (600258), and MSH6 (600678), respectively.

Baker, S. M.; Bronner, C. E.; Zhang, L.; Plug, A. W.; Robatzek, M.; Warren, G.; Elliott, E. A.; Yu, J.; Ashley, T.; Arnheim, N.; Flavell, R. A.; Liskay, R. M.: Male mice defective in the DNA mismatch repair gene PMS2 exhibit abnormal chromosome synapsis in meiosis. Cell 82: 309-319, 1995.

Hamilton, S. R.; Liu, B.; Parsons, R. E.; Papadopoulos, N.; Jen, J.; Powell, S. M.; Krush, A. J.; Berk, T.; Cohen, Z.; Tetu, B.; Burger, P. C.; Wood, P. A.; Taqi, F.; Booker, S. V.; Petersen, G. M.; Offerhaus, G. J. A.; Tersmette, A. C.; Giardiello, F. M.; Vogelstein, B.; Kinzler, K. W.: The molecular basis of Turcot's syndrome. New Eng. J. Med. 332: 839-847, 1995.

Miyaki, M.; Konishi, M.; Tanaka, K.; Kikuchi-Yanoshita, R.; Muraoka, M.; Yasuno, M.; Igari, T.; Koike, M.; Chiba, M.; Mori, T.: Germline mutation of MSH6 as the cause of hereditary nonpolyposis colorectal cancer. (Letter) Nature Genet. 17: 271-272, 1997.

Narayanan, L.; Fritzell, J. A.; Baker, S. M.; Liskay, R. M.; Glazer, P. M.: Elevated levels of mutation in multiple tissues of mice deficient in the DNA mismatch repair gene Pms2. Proc. Nat. Acad. Sci. 94: 3122-3127, 1997.

Nicolaides, N. C.; Carter, K. C.; Shell, B. K.; Papadopoulos, N.; Vogelstein, B.; Kinzler, K. W.: Genomic organization of the human PMS2 gene family. Genomics 30: 195-206, 1995.

Nicolaides, N. C.; Kinzler, K. W.; Vogelstein, B.: Analysis of the 5-prime region of PMS2 reveals heterogeneous transcripts and a novel overlapping gene. Genomics 29: 329-334, 1995.

Nicolaides, N. C.; Papadopoulos, N.; Liu, B.; Wei, Y.-F.; Carter, K. C.; Ruben, S. M.; Rosen, C. A.; Haseltine, W. A.; Fleischmann, R. D.; Fraser, C. M.; Adams, M. D.; Venter, J. C.; Dunlop, M. G.; Hamilton, S. R.; Petersen, G. M.; de la Chapelle, A.; Vogelstein, B.; Kinzler, K. W.: Mutations of two PMS homologues in hereditary nonpolyposis colon cancer. Nature 371: 75-80, 1994.

Parsons, R.; Li, G.-M.; Longley, M.; Modrich, P.; Liu, B.; Berk, T.; Hamilton, S. R.; Kinzler, K. W.; Vogelstein, B.: Mismatch repair deficiency in phenotypically normal human cells. Science 268: 738-740, 1995.

COLORECTAL CANCER–RELATED CHROMOSOME SEQUENCE-17 (120460)

JJM NOTE

This gene is a major step in the multistage sequence of genetic changes leading to metastatic colon cancer, but must be distinguished from the *TP53* gene that is constitutionally mutated or absent in the Li-Fraumeni cancer family syndrome. One wonders about the effects of a hereditary (germline) mutation of this gene.

Fearon et al. (1987) used X-linked RFLPs to examine the pattern of X-chromosome inactivation in colorectal tumors of females. All 50 examined showed monoclonal patterns of X-chromosome inactivation; these tumors included 20 carcinomas and 30 adenomas of either familial or spontaneous type. Fearon et al. (1987) also used RFLPs of autosomes as clonal markers to detect somatic loss or gain of specific chromosomal sequences in colorectal tumors. They found that somatic loss of chromosome 17p sequences occurred in over 75% of the carcinomas examined, but that such loss was rare in adenomas. Monpezat et al. (1988) also found loss of alleles located on 17p, loss of alleles from chromosome 18 (120470), or both. By the study of 20 chromosome 17p markers, Baker et al. (1989) localized the common region of deletion in colorectal carcinomas to 17p13.3-p12. This region contains the gene for the transformation-associated protein p53 (TP53; 191170). In 2 tumors in which the TP53 gene was deleted on 1 chromosome, they could show that the remaining allele was mutant; alanine substituted for valine at codon 143 in 1 tumor and histidine substituted for arginine at codon 175 of the second tumor. Both mutations were in a highly conserved region of the TP53 gene that had previously been found to be mutated in murine p53 oncogenes. Baker et al. (1989) concluded that TP53 mutations may be involved in colorectal cancer (114500), perhaps through inactivation of a tumor suppressor function of the wildtype gene. Evidence suggests that the p53 gene is usually the relevant target in cases of 17p deletion (Vogelstein, 1990).

Baker, S. J.; Fearon, E. R.; Nigro, J. M.; Hamilton, S. R.; Preisinger, A. C.; Jessup, J. M.; vanTuinen, P.; Ledbetter, D. H.; Barker, D. F.; Nakamura, Y.; White, R.; Vogelstein, B.: Chromosome 17 deletions and p53 gene mutations in colorectal carcinomas. Science 244: 217-221, 1989.

Fearon, E. R.; Hamilton, S. R.; Vogelstein, B.: Clonal analysis of human colorectal tumors. Science 238: 193-197, 1987.

Monpezat, J. P.; Delattre, O.; Bernard, A.; Grunwald, D.; Remvikos, Y.; Muleris, M.; Salmon, R. J.; Frelat, G.; Dutrillaux, B.; Thomas, G.: Loss of alleles on chromosome 18 and on the short arm of chromosome 17 in polyploid colorectal carcinomas. Int. J. Cancer 41: 404-408, 1988.

Vogelstein, B.: Personal Communication. Baltimore, Md. 7/12/1990.

DELETED IN COLORECTAL CARCINOMA; DCC (*120470)

JJM NOTE

This gene is another major step in multistage colorectal carcinogenesis, and seems to have a role in the pathogenesis of pancreatic cancer. Again, germline mutations might be sought in certain cancer families not linked to loci associated with the cancer family syndrome of Lynch.

Vogelstein et al. (1988) found that chromosome 18 sequences were lost frequently in colorectal carcinomas (73%) and in advanced adenomas (47%), but only occasionally in earlier-stage adenomas (11 to 13%). Taken in connection with other findings of changes in chromosome 17, as well as chromosome 5, these findings suggest a model wherein the steps required for malignancy often involve the activation of a dominantly acting oncogene coupled with the loss of several genes that normally suppress tumorigenesis. The critical area in chromosome 18 appeared to reside between 18q21.3 and the telomere. It is of interest that Lynch et al. (1985) found a lod score of 3.19 for linkage between a familial cancer syndrome (Lynch syndrome II; 114400) and Kidd blood group (JK; 111000); the Kidd blood group has been assigned to 18q11-q12. Boman et al. (1988) found tumor-specific allele loss at the D18S6 locus on chromosome 18 in 2 patients with familial adenomatous polyposis and in 2 patients with sporadic colon cancer. D18S6 is closely linked to LCFS2 and JK. Fearon et al. (1990) cloned a contiguous stretch of DNA, comprising 370 kb, from the region of 18q suspected to contain the tumor suppressor gene. Potential exons in the 370-kb region were defined by human-rodent sequence identities, and the expression of potential exons was assessed by an 'exon-connection' strategy based on the polymerase chain reaction. Expressed exons were used as probes for screening of cDNA to obtain clones that encoded a gene the authors termed DCC ('deleted in colorectal carcinomas'). This cDNA was encoded by at least 8 exons. The predicted amino acid sequence specified a protein with sequence similarity to neural cell adhesion molecules (116930) and related cell surface glycoproteins. While the DCC gene was expressed in most normal tissues, including colonic mucosa, its expression was greatly reduced or absent in most colorectal carcinomas tested. Somatic mutations within the DCC gene observed in colorectal can cers included a homozygous deletion of the 5-prime end, a point mutation within one of the introns, and 10 examples of DNA insertions within a 0.17-kb fragment immediately downstream of 1 of the exons.

Thiagalingam et al. (1996) evaluated sporadic colon cancer tumors for allelic deletions. They defined a minimally lost region (MLR) on chromosome 18q21 which extended between markers D18S535 and D18S858. It encompassed 16 cM between D18S535 and 20CO3 and included 2 candidate tumor suppressor genes: DPC4 (600993) and DCC. DPC4 was deleted in up to one-third of cases and DCC or a neighboring gene was deleted in the remaining tumors.

Tanaka et al. (1991) demonstrated that transfer of a normal human chromosome 18 into a colon carcinoma cell line through microcell hybridization severely reduced the cloning efficiency of the hybrid cells in soft agar and completely suppressed tumorigenicity in athymic nude mice. Similar results were obtained when a normal chromosome 5, which carries the locus for adenomatous polyposis coli (175100), was transferred into the cells, but the growth properties of the hybrid cells were unchanged when chromosome 11 was introduced. Nigro et al. (1991) observed a curious phenomenon of scrambled exons in the DCC gene in a variety of normal and neoplastic cells of rodent and human origin. Abnormally spliced transcripts showed that exons were joined accurately at consensus splice sites, but in an order different from that present in the primary transcript. Thus, a novel type of RNA product resulted. Using molecular markers in an interspecific backcross between C57BL/6J and Mus spretus, Justice et al. (1992) mapped the corresponding gene to mouse chromosome 18. In a study of 28 cases of surgically resected gastric cancer, excluding the diffuse type, Uchino et al. (1992) concluded that loss of heterozygosity (LOH) on chromosome 18q occurs at an earlier stage than LOH on chromosome 17p and that tumor suppressor genes located on these 2 chromosome arms are critically involved in the development of most gastric cancers. Involvement of DCC may be rather selective for gastrointestinal cancers. Hohne et al. (1992) presented evidence that loss of DCC gene expression is an important factor in the development or progress of pancreatic adenocarcinoma. In 8 of 11 pancreatic carcinoma cell lines and in 4 of 8 primary ductal adenocarcinomas of the pancreas, a complete extinction of DCC gene expression was observed, whereas the KRAS gene (190070) was mutated at codon 12 in 7 of the 8 primary tumors. Reduced or absent DCC expression tended to be associated with undifferentiated pancreatic tumor cell lines, whereas in the more differentiated ones, DCC expression was conserved.

Cho et al. (1994) commented that the DCC gene encodes a protein with sequence similarity to cell adhesion molecules such as N-CAM (116930). Studying a YAC contig containing the entire DCC coding region, they showed that the DCC gene spans approximately 1.4 Mb. They used lambda phage clones to demonstrate the existence of 29 DCC exons, and the sequences of the exon-intron boundaries were determined. In a panel of primary colorectal tumors, they found that most had lost the region containing DCC.

The DCC protein has structural features in common with certain types of cell-adhesion molecules and may participate with other proteins in cell-cell and cell-matrix interactions. Zetter (1993) found that expression of the DCC gene was absent in most colorectal cancers that were metastatic to the liver, but was lost only in a minority of nonmetastatic cancers. Furthermore, Jen et al. (1994) found that allelic loss of 18q in the region occupied by the DCC gene carried a worse prognosis than that in cases with no loss of chromosome 18q. They developed procedures to examine the status of 18q with microsatellite markers and PCR-amplified DNA from formalin-fixed, paraffin-embedded tumors. Normal tissue and tumor tissue could be examined on the same microscopic slide. Allelic loss of 18q was assessed in 145 consecutively resected stage II or III colorectal carcinomas. The prognosis in patients with stage II cancer (Dukes stage B; tumor extending through the bowel wall, without lymph-node metastasis) was similar to that in patients with stage III cancer, who were thought to benefit from adjuvant therapy. In contrast, patients with stage II disease who did not have chromosome 18q allelic loss in their tumor had a survival rate similar to that of patients with stage I disease and might not require additional therapy.

Shibata et al. (1996) reported findings that extended the observations of Jen et al. (1994), who had found that allelic loss of 18q predicted a poor outcome in patients with stage II colorectal cancer. They studied the DCC gene as a possible specific prognostic marker. Expression of DCC was evaluated immunohistochemically in 132 paraffin-embedded samples from patients with curatively resected stage II or stage III colorectal carcinomas. They found that expression of DCC was a strong positive predictive factor for survival in both stage II and stage III colorectal carcinomas. In patients with stage II disease whose tumors expressed DCC, the 5-year survival rate was 94.3%, whereas in patients with DCC-negative tumors, the survival rate was 61.6%. In patients with stage III disease, the respective survival rates were 59.3% and 33.2%.

Vogelstein (1995) stated that the precise location of the DCC gene was thought to be 18q21.3.

Maesawa et al. (1996) screened tumor specimens from 111 patients with esophageal squamous cell carcinoma for LOH at the DCC locus and observed LOH in 10 of 61 informative cases (16%). No statistically significant correlation was observed between DCC-LOH and lymph node metastasis, histopathologic grade, or tumor stage. Survivorship of DCC-LOH patients was not statistically different from that of patients without LOH. These results suggested to Maesawa et al. (1996) that LOH at the DCC locus is not related to the acquisition of metastatic potential or the state of tumor cell differentiation in esophageal squamous cell carcinoma.

Keino-Masu et al. (1996) noted that the establishment of neuronal connections involves the accurate guidance of developing axons to their targets through the combined actions of attractive and repulsive guidance cues in the extracellular environment. Diffusible chemoattractants secreted by target cells are involved, as well as diffusible chemorepellents secreted by nontarget cells which generate exclusion zones that axons avoid (Keynes and Cook, 1995). Two recently identified families of guidance molecules, the netrins (see netrin-1, 601614) and semaphorins (see 601281 and 601124), are proteins that can function as diffusible attractants or repellents for developing neurons, but the receptors and signal transduction mechanisms through which they produce their effects are poorly understood. Netrins are chemoattractants for commissural axons in the vertebral spinal cord. Keino-Masu et al. (1996) showed that DCC, a transmembrane protein of the immunoglobulin superfamily, is expressed on spinal commissural axons and possesses netrin-1-binding activity. Moreover, an antibody to DCC selectively blocked the netrin-1-dependent outgrowth of commissural axons in vitro. These results indicated that DCC is a receptor or a component of a receptor that mediates the effects of netrin-1 on commissural axons, and they complement genetic evidence for interactions between DCC and netrin homologs in C. elegans (UNC-40; see Chan et al., 1996) and Drosophila (frazzled; see Kolodziej et al., 1996).

To help elucidate the functions of the DCC gene and to test the suggestion that it functions as a receptor for the axonal chemoattractant netrin-1, Fazeli et al. (1997) inactivated the DCC homolog in the mouse genome through use of homologous recombination and studied the effects of this inactivation on both the intestine and the developing nervous system. They found defects in axonal projections that were similar to those observed in netrin-1-deficient mice, but there was no effect on growth, differentiation, morphogenesis, or tumorigenesis in mouse intestine. These observations failed to support a tumor-suppressor function for DCC in the mouse, but were consistent with the hypothesis that DCC is a component of the receptor for netrin-1.

Gotley et al. (1996) examined the expression of the DCC gene in colorectal cancers and metastases. Using RT-PCR, they detected DCC mRNA in all colonic tissue specimens through all stages of tumor development. Using monoclonal antibodies against DCC, they found DCC protein is abundant in normal human bladder and is detectable in colon, pancreas and kidney, but not in liver. DCC protein could be detected in varying abundance in all specimens of normal colonic mucosa analyzed as well as in all specimens of adenomatous polyps, colorectal carcinoma and colorectal liver metastases. In some patients, tumor tissue contained less DCC protein than the adjacent normal mucosa. Gotley et al. (1996) found no cases of complete loss of DCC mRNA or protein in colon cancers or metastases.

The mechanism of action of DCC was unknown. Mehlen et al. (1998) showed that DCC induces apoptosis in the absence of ligand binding, but blocks apoptosis when engaged by netrin-1. Furthermore, DCC is a caspase substrate, and mutation of the site at which caspase-3 (CASP3; 600636) cleaves DCC suppresses the proapoptotic effect of DCC completely. These results indicated the DCC may function as a tumor-suppressor protein by inducing apoptosis in settings in which ligand is unavailable (for example, during metastasis or tumor growth beyond local blood supply) through functional caspase cascades by a mechanism that requires cleavage of DCC at asp1290.

Boman, B. M.; Wildrick, D. M.; Alfaro, S. R.: Chromosome 18 allele loss at the D18S6 locus in human colorectal carcinomas. Biochem. Biophys. Res. Commun. 155: 463-469, 1988.

Chan, S. S.-Y.; Zheng, H.; Su, M.-W.; Wilk, R.; Killeen, M. T.; Hedgecock, E. M.; Culotti, J. G.: UNC-40, a C. elegans homolog of the DCC (deleted in colorectal cancer), is required in motile cells responding to UNC-6 netrin cues. Cell 87: 187-195, 1996.

Cho, K. R.; Oliner, J. D.; Simons, J. W.; Hedrick, L.; Fearon, E. R.; Preisinger, A. C.; Hedge, P.; Silverman, G. A.; Vogelstein, B.: The DCC gene: structural analysis and mutations in colorectal carcinomas. Genomics 19: 525-531, 1994.

Fazeli, A.; Dickinson, S. L.; Hermiston, M. L.; Tighe, R. V.; Steen, R. G.; Small, C. G.; Stoeckli, E. T.; Keino-Masu, K.; Masu, M.; Rayburn, H.; Simons, J.; Bronson, R. T.; Gordon, J. I.; Tessier-Lavigne, M.; Weinberg, R. A.: Phenotype of mice lacking functional Deleted in colorectal cancer (Dcc) gene. Nature 386: 796-804, 1997.

Fearon, E. R.; Cho, K. R.; Nigro, J. M.; Kern, S. E.; Simons, J. W.; Ruppert, J. M.; Hamilton, S. R.; Preisinger, A. C.; Thomas, G.; Kinzler, K. W.; Vogelstein, B.: Identification of a chromosome 18q gene that is altered in colorectal cancers. Science 247: 49-56, 1990.

Gotley, D. C.; Reeder, J. A.; Fawcett, J.; Walsh, M. D.; Bates, P.; Simmons, D. L.; Antalis, T. M.: The deleted in colon cancer (DCC) gene is consistently expressed in colorectal cancers and metastases. Oncogene 13: 787-795, 1996.

Hohne, M. W.; Halatsch, M.-E.; Kahl, G. F.; Weinel, R. J.: Frequent loss of expression of the potential tumor suppressor gene DCC in ductal pancreatic adenocarcinoma. Cancer Res. 52: 2616-2619, 1992.

Jen, J.; Kim, H.; Piantadosi, S.; Liu, Z.-F.; Levitt, R. C.; Sistonen, P.; Kinzler, K. W.; Vogelstein, B.; Hamilton, S. R.: Allelic loss of chromosome 18q and prognosis in colorectal cancer. New Eng. J. Med. 331: 213-221, 1994.

Justice, M. J.; Gilbert, D. J.; Kinzler, K. W.; Vogelstein, B.; Buchberg, A. M.; Ceci, J. D.; Matsuda, Y.; Chapman, V. M.; Patriotis, C.; Makris, A.; Tsichlis, P. N.; Jenkins, N. A.; Copeland, N. G.: A molecular genetic linkage map of mouse chromosome 18 reveals extensive linkage conservation with human chromosomes 5 and 18. Genomics 13: 1281-1288, 1992.

Keino-Masu, K.; Masu, M.; Hinck, L.; Leonardo, E. D.; Chan, S. S.-Y.; Culotti, J. G.; Tessier-Lavigne, M.: Deleted in colorectal cancer (DCC) encodes a netrin receptor. Cell 87: 175-185, 1996.

Keynes, R.; Cook, G. M. W.: Axon guidance molecules. Cell 83: 161-169, 1995.

Kolodziej, P. A.; Timpe, L. C.; Mitchell, K. J.; Fried, S. R.; Goodman, C. S.; Jan, L. Y.; Jan, Y. N.: Frazzled encodes a Drosophila member of the DCC immunoglobulin subfamily and is required for CNS and motor axon guidance. Cell 87: 197-204, 1996.

Lynch, H. T.; Schuelke, G. S.; Kimberling, W. J.; Albano, W. A.; Lynch, J. F.; Biscone, K. A.; Lipkin, M. L.; Deschner, E. E.; Mikol, Y. B.; Sandberg, A. A.; Elston, R. C.; Bailey-Wilson, J. E.; Danes, B. S.: Hereditary nonpolyposis colorectal cancer (Lynch syndromes I and II). II. Biomarker studies. Cancer 56: 939-951, 1985.

Maesawa, C.; Tamura, G.; Ogasawara, S.; Suzuki, Y.; Sakata, K.; Sugimura, J.; Nishizuka, S.; Sato, N.; Ishida, K.; Saito, K.; Satodate, R.: Loss of heterozygosity at the DCC gene locus is not crucial for the acquisition of metastatic potential in oesophageal squamous cell carcinoma. Europ. J. Cancer. 32A: 896-898, 1996.

Mehlen, P.; Rabizadeh, S.; Snipas, S. J.; Assa-Munt, N.; Salvesen, G. S.; Bredesen, D. E.: The DCC gene produce induces apoptosis by a mechanism requiring receptor proteolysis. Nature 395: 801-804, 1998.

Miyake, S.; Nagai, K.; Yoshino, K.; Oto, M.; Endo, M.; Yuasa, Y.: Point mutations and allelic deletion of tumor suppressor gene DCC in human esophageal squamous cell carcinomas and their relation to metastasis. Cancer Res. 54: 3007-3010, 1994.

Nigro, J. M.; Cho, K. R.; Fearon, E. R.; Kern, S. E.; Ruppert, J. M.; Oliner, J. D.; Kinzler, K. W.; Vogelstein, B.: Scrambled exons. Cell 64: 607-613, 1991.

Shibata, D.; Reale, M. A.; Lavin, P.; Silverman, M.; Fearon, E. R.; Steele, G., Jr.; Jessup, J. M.; Loda, M.; Summerhayes, I. C.: The DCC protein and prognosis in colorectal cancer. New Eng. J. Med. 335: 1727-1732, 1996.

Tanaka, K.; Oshimura, M.; Kikuchi, R.; Seki, M.; Hayashi, T.; Miyaki, M.: Suppression of tumorigenicity in human colon carcinoma cells by introduction of normal chromosome 5 or 18. Nature 349: 340-342, 1991.

Thiagalingam, S.; Lengauer, C.; Leach, F. S.; Schutte, M.; Hahn, S. A.; Overhauser, J.; Willson, J. K. V.; Markowitz, S.; Hamilton, S. R.; Kern, S. E.; Kinzler, K. W.; Vogelstein, B.: Evaluation of candidate tumour suppressor genes on chromosome 18 in colorectal cancers. Nature Genet. 13: 343-346, 1996.

Uchino, S.; Tsuda, H.; Noguchi, M.; Yokota, J.; Terada, M.; Saito, T.; Kobayashi, M.; Sugimura, T.; Hirohashi, S.: Frequent loss of heterozygosity at the DCC locus in gastric cancer. Cancer Res. 52: 3099-3102, 1992.

Vogelstein, B.: Personal Communication. Baltimore, Md. 11/30/1995.

Vogelstein, B.; Fearon, E. R.; Hamilton, S. R.; Kern, S. E.; Preisinger, A. C.; Leppert, M.; Nakamura, Y.; White, R.; Smits, A. M. M.; Bos, J. L.: Genetic alterations during colorectal-tumor development. New Eng. J. Med. 319: 525-532, 1988.

Zetter, B. R.: Adhesion molecules in tumor metastasis. Semin. Cancer Biol. 4: 219-229, 1993.

TRANSFORMING GROWTH FACTOR-BETA RECEPTOR, TYPE 2 (*190182)

JJM NOTE

A single sibship had late-onset colon cancer without polyps or signs of the Lynch syndrome (Lu et al., 1998). In addition, somatic gene changes have been seen in colon cancer lines and a gastic cancer.

Growth factor-induced protein phosphorylation plays a key role in the signal transduction that leads to mitogenic responses. Most growth factor receptors are transmembrane tyrosine kinases or are associated with cytoplasmic tyrosine kinases. However, another class of transmembrane receptors is predicted to function as serine-threonine kinases. The type II activin receptor (102576) and the types I (190181) and II TGF-beta receptors (Lin et al., 1992) belong to the serine-threonine kinase family. The many activities of TGF-beta in regulating cell proliferation and differentiation and extracellular matrix production are mediated through these receptors. With the use of cells overexpressing truncated type II receptors as dominant-negative mutants to block type II receptor signaling selectively, Chen et al. (1993) demonstrated the existence of 2 receptor pathways. The type II receptors, possibly in conjunction with type I receptors, mediate the induction of growth inhibition and hypophosphorylation of the retinoblastoma gene product. The type I receptors are responsible for effects on extracellular matrix, such as the induction of fibronectin and plasminogen activator inhibitor I, and for increased JUNB (165161) expression. Selective inactivation of the type II receptors alters the response to TGF-beta in a manner similar to the functional inactivation of the RB protein, suggesting a role for RB in the type II, but not the type I, receptor pathway.

Using a full-length cDNA and a genomic probe in Southern blot analysis of a human/rodent somatic cell hybrid panel and by direct fluorescence in situ hybridization to normal metaphase chromosomes, Mathew et al. (1994) showed that the TGFBR2 gene maps to 3p22. Bonyadi et al. (1996) mapped the mouse Tgfbr2 gene to distal mouse chromosome 9 within a region of synteny with human 3p22-p21. The mapping was done by linkage studies.

Markowitz et al. (1995) stated that transforming growth factor-beta (190180) inhibits the growth of multiple epithelial cell types, and loss of this negative regulation is thought to contribute to tumor development. The TGFB growth inhibitory signal is transduced through 2 receptors, type I (TGFBR1) and type II (TGFBR2), which function as a heteromeric complex. Markowitz et al. (1995) investigated whether inactivation of TGF-beta receptors is a mechanism by which human colon cancer cells lose responsiveness to TGF-beta. They found that the TGFBR2 gene was inactivated in a subset of colon cancer cell lines (referred to as RER(+), for 'replication errors') exhibiting microsatellite instability, but not in RER(-) cells. Eight such examples, due to 3 different mutations, were identified. The mutations (190182.0001) were clustered within small repeated sequences in the TGFBR2 gene and were accompanied by the absence of cell surface receptors. Markowitz et al. (1995) stated that TGFBR2 mutation, by inducing the escape of cells from TGF-beta-mediated growth control, links DNA repair defects (120435) with a specific pathway of tumor progression. The small repeat sequences in the TGFBR2 gene make it a favorable target for RER(+)-associated mutator mechanisms. Once generated, the proliferative advantage of cells with inactivated type II receptor would drive colon tumor progression. This pathway may also be operative in other human malignancies in which the RER(+) phenotype has been detected (reviewed by Eshleman and Markowitz, 1995).

Whereas TGF-beta (TGFB) inhibits the growth of many epithelial cell types including nontransformed colon epithelial cells, colon and many other cancer cell lines are resistant to suppression of growth by TGF-beta. Parsons et al. (1995) confirmed the presence of TGFBR2 mutations in 100 of 111 cases of RER+ colon cancers. As in the report of Markowitz et al. (1995), in each of these cases, a frameshift mutation was detected within a small adenine mononucleotide repeat at nucleotides 709-718 of the TGFBR2 cDNA. Insertions or deletions of adenines within this repeat produce -1,

Bonyadi, M.; Cui, W.; Nagase, H.; Akhurst, R. J.: The TGF-beta type II receptor, Tgfbr2, maps to distal mouse chromosome 9. Genomics 33: 328-329, 1996.

Chen, R.-H.; Ebner, R.; Derynck, R.: Inactivation of the type II receptor reveals two receptor pathways for the diverse TGF-beta activities. Science 260: 1335-1338, 1993.

Eshleman, J. R.; Markowitz, S. D.: Microsatellite instability in inherited and sporadic neoplasms. Curr. Opin. Oncol. 7: 83-89, 1995.

Lin, H. Y.; Wang, X.-F.; Ng-Eaton, E.; Weinberg, R. A.; Lodish, H. F.: Expression cloning of the TGF-beta type II receptor, a functional transmembrane serine/threonine kinase. Cell 68: 775-785, 1992.

Lu, S.-L.; Kawabata, M.; Imamura, T.; Akiyama, Y.; Nomizu, T.; Miyazono, K.; Yuasa, Y.: HNPCC associated with germline mutation in the TGF-beta type II receptor gene. (Letter) Nature Genet. 19: 17-18, 1998.

Markowitz, S.; Wang, J.; Myeroff, L.; Parsons, R.; Sun, L.; Lutterbaugh, J.; Fan, R. S.; Zborowska, E.; Kinzler, K. W.; Vogelstein, B.; Brattain, M.; Willson, J. K. V.: Inactivation of the type II TGF-beta receptor in colon cancer cells with microsatellite instability. Science 268: 1336-1338, 1995.

Mathew, S.; Murty, V. V. V. S.; Cheifetz, S.; George, D.; Massague, J.; Chaganti, R. S. K.: Transforming growth factor receptor gene TGFBR2 maps to human chromosome band 3p22. Genomics 20: 114-115, 1994.

Myeroff, L. L.; Parsons, R.; Kim, S.-J.; Hedrick, L.; Cho, K. R.; Orth, K.; Mathis, M.; Kinzler, K. W.; Lutterbaugh, J.; Park, K.; Bang, Y.-J.; Lee, H. Y.; Park, J.-G.; Lynch, H. T.; Roberts, A. B.; Vogelstein, B.; Markowitz, S. D.: A transforming growth factor beta receptor type II gene mutation common in colon and gastric but rare in endometrial cancers with microsatellite instability. Cancer Res. 55: 5545-5547, 1995.

Parsons, R.; Myeroff, L. L.; Liu, B.; Willson, J. K. V.; Markowitz, S. D.; Kinzler, K. W.; Vogelstein, B.: Microsatellite instability and mutations of the transforming growth factor beta type II receptor gene in colorectal cancer. Cancer Res. 55: 5548-5550, 1995.

Souza, R. F.; Appel, R.; Yin, J.; Wang, S.; Smolinski, K. N.; Abraham, J. M.; Zou, T.-T.; Shi, Y.-Q.; Lei, J.; Cottrell, J.; Cymes, K.; Biden, K.; Simms, L.; Leggett, B.; Lynch, P. M.; Frazier, M.; Powell, S. M.; Harpaz, N.; Sugimura, H.; Young, J.; Meltzer, S. J.: Microsatellite instability in the insulin-like growth factor II receptor gene in gastrointestinal tumours. (Letter) Nature Genet. 14: 255-257, 1996.

Tannergard, P.; Liu, T.; Weger, A.; Nordenskjold, M.; Lindbloom, A.: Tumorigenesis in colorectal tumors from patients with hereditary non-polyposis colorectal cancer. Hum. Genet. 101: 51-55, 1997.

PHOSPHOLIPASE A2 (*172411)

JJM NOTE

Seen as a germline mutation in only one 80-year-old male with apparently sporadic colon cancer, this gene may turn out to epitomize the much postulated but poorly documented class of "modifier" genes—genes that influence the expression of the "major" cancer genes, namely, those associated with infrequent syndromes of familial cancer. Such genes, hard to map and discover by family studies, may account for much human cancer.

Elevated levels of synovial fluid and plasma phospholipase A2 correlate with the activity of inflammatory arthritis. Seilhamer et al. (1989) characterized multiple forms of phospholipase A2 in arthritic synovial fluid. Johnson et al. (1990) purified and cloned the human gene encoding the major form of PLA2 expressed in rheumatoid arthritis synovial fluid. The gene product is a member of the type II PLA2 family, which includes the inflammatory PLA2 found in the venom of the viperid class of snakes. Johnson et al. (1990) showed that there is diversity of phospholipase A2 in synovial fluid which is apparently related to alternative splicing of the synovial phospholipase A2 gene, which they mapped to chromosome 1p35. Pancreatic PLA2, which is a member of the type I PLA2 family, is encoded by a gene on chromosome 12 (172410).

Masharani et al. (1988) described RFLPs of a gene homologous to the PPLA2 gene. 17,16:Seilhamer et al. (1988, 1989) assigned this gene (symbolized PLA2L by them) to chromosome 1 by hybridization to DNA from human-rodent hybrids. Seilhamer et al. (1989) called the PLA2L gene 'nonpancreatic,' or npPLA2, since the potential role(s) of its gene product were unknown. They described its DNA sequence in human, rat, and pig. Subsequent work by Johnson et al. (1990) indicated that the phospholipase A2 that appears in synovial fluid in rheumatoid arthritis is coded by a gene on chromosome 1; hence, it is this gene that was previously called PLA2L. (The preferred symbol is PLA2G2A; see 601658 for the gene phospholipase A2-like (PLA2L).)

Moser et al. (1990) identified a dominantly transmitted, fully penetrant mouse mutation named Min (multiple intestinal neoplasia) that causes a phenotype closely resembling familial adenomatous polyposis (FAP; 175100). Heterozygotes for the Min mutation developed numerous intestinal and colonic adenomas similar in morphology to the adenomas seen in FAP. If left untreated, the adenomas eventually become locally invasive (Moser et al., 1992). In addition, about 10% of Min/+ females on a hybrid background develop spontaneous mammary adenoacanthomas or adenocarcinomas (Moser et al., 1993). Homozygotes for Min die in utero. The Min mice carry a nonsense mutation in the mouse homolog of the FAP gene (Su et al., 1992). In the laboratory of William Dove at the University of Wisconsin, where the Min mutation was discovered and characterized, it was noted that the number of intestinal tumors in Min/+ mice was strongly affected by genetic background: C57BL/6J-Min/+ mice had an average of 29 tumors, whereas their Min/+ F1 progeny with AKR mice showed an average of only 6 tumors. This finding suggested that the AKR strain carries alleles that act in a dominant fashion to modify the tumorigenic effect of Min. Dietrich et al. (1993) mapped the postulated modifying gene, called Mom-1 (modifier of Min-1), to the distal portion of mouse chromosome 4. Allelic variation in Mom-1 appeared to control about 50% of the genetic variation in tumor number in 2 different intraspecific backcrosses. A cumulative lod score exceeding 14 supported the mapping. Mom-1 was found to lie in a region of synteny conservation with human chromosome 1p36-p35, a region of frequent somatic loss of heterozygosity in a variety of human tumors, including colon tumors.

MacPhee et al. (1995) identified a candidate gene for Mom1 in the mouse. The gene for secretory type II phospholipase A2 (Pla2s) was shown to map to the same region as Mom1 and displayed 100% concordance between allele type and tumor susceptibility. Expression and sequence analysis revealed that Mom1-susceptible strains are most likely null for Pla2s activity. MacPhee et al. (1995) interpreted the results as indicating that Pla2s acts as a novel modifier of polyp number by altering the cellular microenvironment within the intestinal crypt. The homologous gene in the human, PLA2G2A, maps to 1p35 in a region of homology to distal mouse chromosome 4. The authors stated that it is unclear how loss of PLA2G2A in tumor cells would contribute to neoplasia in the human. One possibility is that PLA2G2A involvement is reflected by the variable numbers of adenomas identified among FAP family members inheriting the same mutation. The variation in tumor burden could be attributable to the segregation of different PLA2G2A alleles. Population surveys would be useful for determining the extent of PLA2G2A allelic variation, as well as for identifying individuals at risk for developing intestinal cancer. The identification of mouse Pla2s as a candidate for a major modifier of intestinal polyp formation may provide a missing link between high fat diets and increased incidence of colon cancer.

In a large familial adenomatous polyposis kindred in which patients harbored the same germline mutation but showed markedly different disease characteristics, Dobbie et al. (1997) investigated the region of human chromosome 1p36-p35 that is homologous to the region of murine chromosome 4 containing the Mom1 gene. They looked for linkage between each of 14 microsatellite markers and the development of extracolonic symptoms in polyposis coli patients. Depending on the mode of inheritance of the modifier locus, a maximum lod score was observed for markers D1S211 and D1S197, reaching 2.08 and 1.77, respectively. The observed values obtained within a large FAP family were considered supportive of a phenotype-modifying locus within this chromosomal region.

To test the hypothesis of the correlation between PLA2G2A gene alterations and human tumor development of the sort demonstrated in the mouse model of FAP, Nimmrich et al. (1997) screened 14 patients with FAP and 20 patients with sporadic colorectal cancer for germline and somatic PLA2G2A gene mutations. None of the individuals with FAP showed PLA2G2A germline alterations. However, a germline mutation (172411.0001) was observed in 1 patient with an apparently sporadic colorectal tumor; the wildtype allele was somatically lost in the tumor of this patient.

Cormier et al. (1997) provided further evidence on the role of PLA2G2A in resistance to intestinal tumorigenesis. They reported that a cosmid transgene overexpressing Pla2g2a in mice caused a reduction in tumor multiplicity and size, comparable to that conferred by a single copy of the resistance allele of Mom1. Thus, they concluded that this secretory phospholipase can provide active tumor resistance. The association of Pla2g2a with Mom1 thus withstood a strong functional test. The work probably represents the successful identification of a polymorphic quantitative trait locus (QTL) in mammals.

Dragani and Manenti (1997) reviewed other work in rodents defining counteractive cancer susceptibility and cancer resistance genes. This phenomenon is observed in mouse models of lung cancer, where the tumor-promoting effect of the Pas1 locus is downregulated by the presence of the Par resistance loci. In rats, tumors presumed to arise as a consequence of the hepatocarcinogen sensitivity locus, Hcs, are discouraged from growing by coincident presence of resistance (Hcr) loci.

A role for PLA2G2A in the pathogenesis of ulcerative colitis (191390) was postulated by Haapamaki et al. (1997), who demonstrated expression of the PLA2G2A gene in metaplastic Paneth cells and columnar epithelial cells in inflamed colonic mucosa from patients with ulcerative colitis. No expression was detected in other tissues from the same patients or, by Northern blot analysis, in colonic biopsies from disease-free controls. Haapamaki et al. (1997) hypothesized that intraluminal secretion of PLA2G2A during the active phase of ulcerative colitis is a host defense mechanism.

See also Johnson et al. (1990)

Cormier, R. T.; Hong, K. H.; Halberg, R. B.; Hawkins, T. L.; Richardson, P.; Mulherkar, R.; Dove, W. F.; Lander, E. S.: Secretory phospholipase Pla2g2a confers resistance to intestinal tumorigenesis. Nature Genet. 17: 88-91, 1997.
Dietrich, W. F.; Lander, E. S.; Smith, J. S.; Moser, A. R.; Gould, K. A.; Luongo, C.; Borenstein, N.; Dove, W.: Genetic identification of Mom-1, a major modifier locus affecting Min-induced intestinal neoplasia in the mouse. Cell 75: 631-639, 1993.
Dobbie, Z.; Heinimann, K.; Bishop, D. T.; Muller, H.; Scott, R. J.: Identification of a modifier gene locus on chromosome 1p35-36 in familial adenomatous polyposis. Hum. Genet. 99: 653-657, 1997.
Dragani, T. A.; Manenti, G.: Mom1 leads the pack. Nature Genet. 17: 7-8, 1997.
Haapamaki, M. M.; Gronroos, J. M.; Nurmi, H.; Alanen, K.; Kallajoki, M.; Nevalainen, T. J.: Gene expression of group II phospholipase A2 in intestine in ulcerative colitis. Gut 40: 95-101, 1997.
Johnson, L. K.; Frank, S.; Vades, P.; Pruzanski, W.; Lusis, A. J.; Seilhamer, J. J.: Localization and evolution of two human phospholipase A2 genes and two related genetic elements.In: Wong, P. Y.-K.; Dennis, E. A.: Phospholipase A2. New York: Plenum Press (pub.) 1990. Pp. 17-34.
Johnson, L. K.; Seilhamer, J. J.; Frank, S.; Lusis, A.; Vadas, P.; Pruzanski, W.: Synovial fluid phospholipase A(2): chromosomal co-localization with homologous genes may provide disease-related PLA(2) diversity. (Abstract) Arthritis Rheum. 33 (suppl.): S79, 1990.
MacPhee, M.; Chepenik, K. P.; Liddell, R. A.; Nelson, K. K.; Siracusa, L. D.; Buchberg, A. M.: The secretory phospholipase A2 gene is a candidate for the Mom1 locus, a major modifier of Apc(Min)-induced intestinal neoplasia. Cell 81: 957-966, 1995.
Masharani, U.; Coleman, R. T.; Johnson, L. K.; Seilhamer, J. J.: EcoRI and NsiI RFLPs at a human PLA2 gene on chromosome 1. Nucleic Acids Res. 16: 9073, 1988.
Moser, A. R.; Dove, W. F.; Roth, K. A.; Gordon, J. I.: The Min (multiple intestinal neoplasia) mutation: its effect on gut epithelial cell differentiation and interaction with a modifier system. J. Cell Biol. 116: 1517-1526, 1992.
Moser, A. R.; Mattes, E. M.; Dove, W. F.; Lindstrom, M. J.; Haag, J. D.; Gould, M. N.: Apc(Min), a mutation in the murine Apc gene, predisposes to mammary carcinomas and focal alveolar hyperplasias. Proc. Nat. Acad. Sci. 90: 8977-8981, 1993.
Moser, A. R.; Pitot, H. C.; Dove, W. F.: A dominant mutation that predisposes to multiple intestinal neoplasia in the mouse. Science 247: 322-324, 1990.
Nimmrich, I.; Friedl, W.; Kruse, R.; Pietsch, S.; Hentsch, S.; Deuter, R.; Winde, G.; Muller, O.: Loss of the PLA2G2A gene in a sporadic colorectal tumor of a patient with a PLA2G2A germline mutation and absence of PLA2G2A germline alterations in patients with FAP. Hum. Genet. 100: 345-349, 1997.
Seilhamer, J. J.; Plant, S.; Pruzanski, W.; Schilling, J.; Stefanski, E.; Vadas, P.; Johnson, L. K.: Multiple forms of phospholipase A(2) in arthritic synovial fluid. J. Biochem. 106: 38-42, 1989.
Seilhamer, J. J.; Pruzanski, W.; Vadas, P.; Plant, S.; Miller, J. A.; Kloss, J.; Johnson, L. K.: Cloning and recombinant expression of phospholipase A(2) present in rheumatoid arthritic synovial fluid. J. Biol. Chem. 264: 5335-5338, 1989.
Seilhamer, J. J.; Randall, T. L.; Johnson, L. K.; Heinzmann, C.; Klisak, I.; Sparkes, R. S.; Lusis, A. J.: Novel gene exon homologous to pancreatic phospholipase A(2): sequence and chromosomal mapping of both human genes. J. Cell. Biochem. 39: 327-337, 1989.
Seilhamer, J. J.; Randall, T. L.; Johnson, L. K.; Lusis, A.; Sparkes, R. S.; Heinzman, C.: Chromosomal mapping of human pancreatic PLA2 and a homologous PLA2 exon. (Abstract) J. Cell. Biochem. (Suppl. 12E): 55 only, 1988.
Su, L.-K.; Kinzler, K. W.; Vogelstein, B.; Preisinger, A. C.; Moser, A. R.; Luongo, C.; Gould, K. A.; Dove, W. F.: Multiple intestinal neoplasia caused by a mutation in the murine homolog of the APC gene. Science 256: 668-670, 1992.

NAIL-PATELLA SYNDROME (#161200)

JJM NOTE

This entry is included because of one report of familial colon cancer in the syndrome (Gilula and Kantor 1975).

A number sign (#) is used with this entry because of evidence that the disorder is caused by mutations in the LIM-homeodomain protein LMX1B (602575).

Dysplasia of the nails and absent or hypoplastic patellae are the cardinal features but others are iliac horns, abnormality of the elbows interfering with pronation and supination, and in some cases nephropathy. Nephropathy was an associated abnormality in the family of Hawkins and Smith (1950). The renal change resembles glomerulonephritis. It is relatively benign although fatality at a young age from this complication has been described (Leahy, 1966). The renal disorder in the case of Simila et al. (1970) took the appearance of congenital nephrosis; 8 persons in the family had nail-patella syndrome, of whom 5 also had renal disease. The seeming familial aggregation of the renal complications suggests 2 separate genes, one for a nephropathic form and one for a nonnephropathic form. They might be allelic since no heterogeneity has been detected in the linkage with the ABO locus. As demonstrated by electron microscopy by Morita et al. (1973), among others, many collagen fibrils are present in the thickened basement membranes and in mesangial matrix of otherwise normal glomeruli. Abnormalities of collagen at this site have also been demonstrated in Alport syndrome (104200). Both of these conditions may be special forms of heritable disorders of connective tissue. Gilula and Kantor (1975) found colon cancer in association with the nail-patella syndrome. Sabnis et al. (1980) reported 3 patients with collagenation of glomerular basement membrane like that of the nail-patella syndrome. However, typical bone and nail changes were said to be absent. It is not clear how thoroughly the changes were sought or whether minor changes were present. An 8-year-old boy, son of first-cousin Palestinian Arabs, presented with the nephrotic syndrome. A sister and brother had died of renal disease at ages 6.5 and 9 years, respectively. A 13-year-old girl presented with recurrent urinary tract infections, proteinuria, and edema. No family information was provided. During evaluation of aortic regurgitation, a 27-year-old asymptomatic woman was noted to have proteinuria and renal insufficiency. A sister had undergone renal transplant (diagnosis unknown) and a brother had nephritis. The father's autopsy report stated 'severe interstitial nephritis and hypernephroma.' Questions include: (1) Do these cases represent variable expression of the classic nail-patella gene? (2) Is there a genetic form of glomerular basement membrane collagenosis that lacks the bone and nail involvement of the nail-patella syndrome, and is, perhaps, in light of case 1, inherited as an autosomal recessive? The range of variability of manifestation in a large series of cases ascertained through family studies has not, to my knowledge, been determined.

The Goodpasture antigen, an autoantibody (233450), is directed against type IV collagen of basement membrane (120090, 120130) or some element closely associated with it. The occurrence of Goodpasture syndrome in a patient with NPS (Curtis et al., 1976) may be more than coincidence.

This condition is sometimes called Fong disease for the physician who discovered it (Fong, 1946) in a patient on whom he performed intravenous pyelography while investigating hypertension and albuminuria related to pregnancy. No similar abnormality had been seen by many persons who reviewed the patient's films. No comments were made concerning nail dysplasia or absence of patellae. The name of Turner (1933) is associated with the disorder because of his description of 2 extensively affected families. (The designation Turner syndrome, however, leads to confusion with the XO syndrome.) He described the absence of the patellas and the inability to extend the elbows completely, as well as nail dysplasia. He also showed an x-ray of the pelvis which demonstrated the iliac horns clearly, but he made no reference to this finding in either the legend to the figure or in the text. Aschner (1934) reported that in a study on the genetics of the human skeletal system, she had found in the literature 8 families in which defect or hypoplasia of the patella was inherited through several generations with 'a hereditary defect of the thumbnails in all the affected members' of 3 of the families. Citing her mentor Julius Bauer as the authority, Aschner (1934) suggested that the association of nail dysplasia and absence of the patella was the consequence of close linkage of 2 mutant genes. She was sharply critical of the interpretation given by Oesterreicher (1930), who described affected members in 5 generations and interpreted the association as representing 'polyphenia' (which we would call pleiotropy, or pleiotropism). Aschner (1934) wrote: 'It is rather curious, therefore, that even Oesterreicher, who is well acquainted with the related literature, accepts the hypothesis of polyphenia. In my opinion this may be understood by the fact that this author, a neurologist, did not feel sufficiently secure in genetics to form an opinion of his own in regard to the genotypic connections of the observed symptoms. He therefore sought the advice of Paula Hertwig, the biologist. Hertwig, on the other hand, was naturally not familiar enough with the clinical facts, which would make a polyphenia impossible.' Both iliac horns and renal dysplasia were clearly identified as features of this syndrome by Hawkins and Smith (1950). The deformity of the pelvis characteristic of trisomy 8 suggests somewhat the presence of iliac horns but the appearance is quite different from that in the nail-patella syndrome (Giedion, 1990).

Taguchi et al. (1988) demonstrated that characteristic ultrastructural changes in the glomerulus can be present even in patients without apparent clinical renal involvement. They reported a patient observed for 3 years, between ages 8 and 11. In a patient who had proteinuria before the age of 2 years, Browning et al. (1988) found characteristic changes in the glomeruli at age 27 months. Strong immunofluorescent staining, particularly for IgM, raised the possibility of superimposed immune complex disease. Richieri-Costa (1991) reported on 4 affected persons in a 3-generation Brazilian family with NPS. Three affected individuals had antecubital pterygium, 2 had proteinuria, and 1 had cleft lip and palate. Rizzo et al. (1993) described an Italian family in which bilateral antecubital pterygia, labeled arthrogryposis, was the presenting sign. Renal involvement was also severe in this family. From study of a large family with 30 patients with NPS (which the authors referred to as HOOD, hereditary osteoonychodysplasia), Looij et al. (1988) concluded that a person with NPS has a risk of about 1 in 4 of having a child with NPS nephropathy and a risk of about 1 in 10 of having a child in whom renal failure will develop. (It should be noted that they calculated their risk figures as a percentage in the article, i.e., 24% for the first risk, but stated the risk as a ratio, i.e., 1:4 in the abstract; these are not the same.) In the kidneys of an 18-week spontaneously aborted fetus of a mother with NPS, Drut et al. (1992) found changes which they suggested might be the basis of prenatal diagnosis by intrauterine kidney biopsy.

The nail-patella locus and the ABO blood group locus (110300) are linked. The recombination fraction is about 10% but is higher in females than in males (Renwick and Schulze, 1965). Ferguson-Smith et al. (1976) assigned the ABO-NPS1-AK1 linkage group to 9q34 by regional assignment of AK1 (103000).

Ghiggeri et al. (1993) described a patient with an unbalanced translocation resulting in monosomy of 9q32-qter. The patient had dysplastic nails but normal patellas. She was found to have heterozygous deletion of COL5A1 (120215) and underexpression of alpha-1 chains of type V collagen by fibroblasts. Ghiggeri et al. (1993) suggested that mutations in the COL5A1 gene may be involved in the pathogenesis of the nail-patella syndrome. The patient also had skin and bone manifestations resembling those of Goltz syndrome (305600), an X-linked disorder. The patient had previously been described by Zuffardi et al. (1989), who used the deletion to map the AK1 and ORM1 (138600) loci.

Duba et al. (1998) identified a balanced t(9;17)(q34.1;q25) associated with NPS. By fluorescence in situ hybridization with probes from 9q, they narrowed the breakpoint region to a 17-cM interval between D9S262 and ABL (189980). The translocation was thought to have resulted from a break within or near the NPS gene, causing defective expression. Duba et al. (1998) suggested that the translocation may aid in the identification of the gene.

By linkage analysis in 3 families, using highly informative dinucleotide repeat polymorphisms on 9q33-q34, Campeau et al. (1995) confirmed the assignment of NPS1 to that region and localized the gene to an interval on 9q34.1 distal to D9S60 and proximal to the argininosuccinate synthetase gene (ASS; 215700).

McIntosh et al. (1997) further reduced the NPS1 candidate interval to 1-2 cM. They used 13 polymorphic markers in 5 families to position NPS1 between D9S60 and AK1. Maximum LOD scores of 27.0 and 22.0 were obtained with D9S112 and D9S315, respectively, at zero recombination. The order cen—D9S60—(D9S112/D9S315/NPS1)—AK1—tel was determined by linkage analysis.

The LIM-homeodomain protein Lmx1b plays a central role in dorsal/ventral patterning of the vertebrate limb. Targeted disruption of Lmx1b results in skeletal defects, including hypoplastic nails, absent patellae, and a unique form of renal dysplasia (Chen et al., 1998). Dreyer et al. (1998) showed that the LMX1B gene maps to 9q in the same region as the NPS locus by fluorescence in situ hybridization. Furthermore, they demonstrated that 3 unrelated NPS patients carried de novo heterozygous mutations in this gene. Functional studies showed that one of these mutations disrupted sequence-specific DNA binding, while the other 2 mutations resulted in premature termination of translation. These were the first described mutations in a LIM-homeodomain protein that accounted for an inherited form of abnormal skeletal patterning and renal failure.

Lichter et al. (1997) and McIntosh et al. (1997) described cosegregation of open angle glaucoma (137750) and NPS. In 2 families, Lichter et al. (1997) found linkage results indicating that the cosegregation was the result of a pleiotropic effect of the NPS1 gene at 9q34. In a further study of 29 additional families, 28 of them ascertained on the basis of NPS, glaucoma was present in 9 (31%). In 1 family an individual with open angle glaucoma but no nail-patella syndrome was found; there may have been another explanation for the open angle glaucoma. That the glaucoma is a genuine pleiotropic effect of the NPS1 gene is supported by the finding of anterior segment ocular abnormalities in mice with targeted disruption of the Lmx1b gene (McIntosh, 1998). Vollrath et al. (1998) demonstrated mutations in the LMX1B gene in 4 families with combined NPS and open angle glaucoma.

See also Bennett et al. (1973); Cottereill and Jacobs (1961); Daniel et al. (1980); Darlington and Hawkins (1967); Eisenberg et al. (1972); Mark et al. (1983); Myers et al. (1980); Pillay (1965); Renwick and Lawler (1955); Schroeder (1961); Vernier et al. (1974); Von Knorre (1961); Westerveld et al. (1976); Zimmerman (1961)

Aschner, B.: A typical hereditary syndrome: dystrophy of the nails, congenital defect of the patella and congenital defect of the head of the radius. J.A.M.A. 102: 2017-2020, 1934.

Bennett, W. M.; Musgrave, J. E.; Campbell, R. A.; Elliot, D.; Cox, R.; Brooks, R. E.; Lovrien, E. W.; Beals, R. K.; Porter, G. A.: The nephropathy of the nail-patella syndrome: clinicopathologic analysis of 11 kindreds. Am. J. Med. 54: 304-319, 1973.

Browning, M. C.; Weidner, N.; Lorentz, W. B., Jr.: Renal histopathology of the nail-patella syndrome in a two-year-old boy. Clin. Nephrol. 29: 210-213, 1988.

Campeau, E.; Watkins, D.; Rouleau, G. A.; Babul, R.; Buchanan, J. A.; Meschino, W.; Der Kaloustian, V. M.: Linkage analysis of the nail-patella syndrome. Am. J. Hum. Genet. 56: 243-247, 1995.

Chen, H.; Lun, Y.; Ovchinnikov, D.; Kokubo, H.; Oberg, K. C.; Pepicelli, C. V.; Gan, L.; Lee, B.; Johnson, R. L.: Limb and kidney defects in Lmx1b mutant mice suggest an involvement of LMX1B in human nail patella syndrome. Nature Genet. 19: 51-55, 1998.

Cottereill, C. P.; Jacobs, P.: Hereditary arthro-osteo-onychodysplasia associated with iliac horns. Brit. J. Clin. Pract. 15: 933-941, 1961.

Curtis, J. J.; Bhathena, D.; Leach, R. P.; Galla, J. H.; Lucas, B. A.; Luke, R. G.: Goodpasture's syndrome in a patient with the nail-patella syndrome. Am. J. Med. 61: 401-406, 1976.

Daniel, C. R., III; Osment, L. S.; Noojin, R. O.: Triangular lunulae: a clue to nail-patella syndrome. Arch. Derm. 116: 448-449, 1980.

Darlington, D.; Hawkins, C. F.: Nail patella syndrome with iliac horns and hereditary nephropathy: necropsy report and anatomical dissection. J. Bone Joint Surg. 49B: 164-174, 1967.

Dreyer, S. D.; Zhou, G.; Baldini, A.; Winterpacht, A.; Zabel, B.; Cole, W.; Johnson, R. L.; Lee, B.: Mutations in LMX1B cause abnormal skeletal patterning and renal dysplasia in nail patella syndrome. Nature Genet. 19: 47-50, 1998.

Drut, R. M.; Chandra, S.; Latorraca, R.; Gilbert-Barness, E.: Nail-patella syndrome in a spontaneously aborted 18-week fetus: ultrastructural and immunofluorescent study of the kidneys. Am. J. Med. Genet. 43: 693-696, 1992.

Duba, H.-C.; Erdel, M.; Loffler, J.; Wirth, J.; Utermann, B.; Utermann, G.: Nail patella syndrome in a cytogenetically balanced t(9;17)(q34.1;q25) carrier. Europ. J. Hum. Genet. 6: 75-79, 1998.

Eisenberg, K. S.; Potter, D. E.; Bovill, E. G., Jr.: Osteo-onychodystrophy with nephropathy and renal osteodystrophy: a case report. J. Bone Joint Surg. 54: 1301-1305, 1972.

Ferguson-Smith, M. A.; Aitken, D. A.; Turleau, C.; de Grouchy, J.: Localisation of the human ABO: Np-1:AK-1 linkage group by regional assignment of AK-1 to 9q34. Hum. Genet. 34: 35-43, 1976.

Fong, E. E.: 'Iliac horns' (symmetrical bilateral central posterior iliac processes): a case report. Radiology 47: 517-518, 1946.

Ghiggeri, G. M.; Caridi, G.; Altieri, P.; Pezzolo, A.; Gimelli, G.; Zuffardi, O.: Are the nail-patella syndrome and the autosomal Goltz-like syndrome the phenotypic expressions of different alleles at the COL5A1 locus?. Hum. Genet. 91: 175-177, 1993.

Giedion, A.: Personal Communication. Zurich, Switzerland 4/26/1990.

Gilula, L. A.; Kantor, O. S.: Familial colon carcinoma in nail-patella syndrome. Am. J. Roentgen. 123: 783-790, 1975.

Hawkins, C. F.; Smith, O. E.: Renal dysplasia in a family with multiple hereditary abnormalities including iliac horns. Lancet I: 803-808, 1950.

Leahy, M. S.: The hereditary nephropathy of osteo-onychodysplasia (nail patella syndrome). Am. J. Dis. Child. 112: 237-241, 1966.

Lichter, P. R.; Richards, J. E.; Downs, C. A.; Stringham, H. M.; Boehnke, M.; Farley, F. A.: Cosegregation of open-angle glaucoma and the nail-patella syndrome. Am. J. Ophthal. 124: 506-515, 1997.

Looij, B. J., Jr.; Te Slaa, R. L.; Hogewind, B. L.; van de Kamp, J. J. P.: Genetic counselling in hereditary osteo-onychodysplasia (HOOD, nail-patella syndrome) with nephropathy. J. Med. Genet. 25: 682-686, 1988.

Mark, T. M.; Rywlin, A. M.; Unger, H.: Cystic adventitial degeneration of the popliteal artery: its occurrence in a patient with the nail-patella syndrome. Arch. Path. Lab. Med. 107: 186-188, 1983.

McIntosh, I.: Personal Communication. Baltimore, Md. 4/28/1998.

McIntosh, I.; Clough, M. V.; Schaffer, A. A.; Puffenberger, E. G.; Horton, V. K.; Peters, K.; Abbott, M. H.; Roig, C. M.; Cutone, S.; Ozelius, L.; Kwiatkowski, D. J.; Pyeritz, R. E.; Brown, L. J.; Pauli, R. M.; McCormick, M. K.; Francomano, C. A.: Fine mapping of the nail-patella syndrome locus at 9q34. Am. J. Hum. Genet. 60: 133-142, 1997.

McIntosh, I.; Lichter, P. R.; Clough, M. V.; Roig, C. M.; Wong, D.; Scott, C.; Downs, C.; Stringham, H.; Boehnke, M.; Francomano, C. A.; Richards, J. E.: Glaucoma and nail patella syndrome (NPS) co-segregate: one gene or two? (Abstract) Am. J. Hum. Genet. 61 (suppl.): A13 only, 1997.

Morita, T.; Laughlin, O.; Kawano, K.; Kimmelstiel, P.: Nail-patella syndrome. Arch. Intern. Med. 131: 271-277, 1973.

Myers, H. S.; Gregory, M.; Beighton, P.: Clinical pathologic conference: renal failure in a 44-year-old female. Urol. Radiol. 1: 251-253, 1980.

Oesterreicher, W.: Gemeinsame Vererbung von Anonychie bzw. Onychatrophie, Patellardefekt und Luxatio radii. Dominantes Auftreten in 5 Generationen. Ztschr. Konstitutionslehre 15: 465-476, 1930.

Pillay, V. K.: Onycho-osteodysplasia (nail-patella syndrome): study of a Chinese family with this condition. Ann. Hum. Genet. 28: 301-307, 1965.

Renwick, J. H.; Lawler, S. D.: Genetical linkage between the ABO and nail-patella loci. Ann. Hum. Genet. 19: 312-331, 1955.

Renwick, J. H.; Schulze, J.: Male and female recombination fractions for the nail patella: ABO linkage in man. Ann. Hum. Genet. 28: 379-392, 1965.

Richieri-Costa, A.: Antecubital pterygium and cleft lip/palate presenting as signs of the nail-patella syndrome: report of a Brazilian family. Am. J. Med. Genet. 38: 9-12, 1991.

Rizzo, R.; Pavone, L.; Micali, G.; Hall, J. G.: Familial bilateral antecubital pterygia with severe renal involvement in nail-patella syndrome. Clin. Genet. 44: 1-7, 1993.

Sabnis, S. G.; Antonovych, T. T.; Argy, W. P.; Rakowski, T. A.; Gandy, D. R.; Salcedo, J. R.: Nail-patella syndrome. Clin. Nephrol. 14: 148-153, 1980.

Schroeder, G.: Osteo-onycho-dysplasia hereditaria. Z. Menschl. Vererb. Konstitutionsl. 36: 42-73, 1961.

Simila, S.; Vesa, L.; Wasz-Hockert, O.: Hereditary onycho-osteodysplasia (the nail-patella syndrome) with nephrosis-like renal disease in a newborn boy. Pediatrics 46: 61-65, 1970.

Taguchi, T.; Takebayashi, S.; Nishimura, M.; Tsuru, N.: Nephropathy of nail-patella syndrome. Ultrastruct. Path. 12: 175-183, 1988.

Turner, J. W.: An hereditary arthrodysplasia associated with hereditary dystrophy of the nails. J.A.M.A. 100: 882-884, 1933.

Vernier, R. L.; Hoyer, J. R.; Michael, A. F.: The nail-patella syndrome—pathogenesis of the kidney lesion. Birth Defects Orig. Art. Ser. X(4): 57-59, 1974.

Vollrath, D.; Jaramillo-Babb, V. L.; Clough, M. V.; McIntosh, I.; Scott, K. M.; Lichter, P. R.; Richards, J. E.: Loss-of-function mutations in the LIM-homeodomain gene, LMX1B, in nail-patella syndrome. Hum. Molec. Genet. 7: 1091-1098, 1998.

Von Knorre, G. V.: Ueber die hereditaere Arthro-osteo-onycho-dysplasie (Turner-Kieser-syndrom). Z. Menschl. Vererb. Konstitutionsl. 36: 118-129, 1961.

Westerveld, A.; Jongsma, A. P. M.; Meera Khan, P.; Van Someren, H.; Bootsma, D.: Assignment of the AK(1): Np: ABO linkage group to human chromosome 9. Proc. Nat. Acad. Sci. 73: 895-899, 1976.

Zimmerman, C.: Iliac horns: a pathognomonic roentgen sign of familial onycho-osteodysplasia. Am. J. Roentgen. 86: 478-483, 1961.

Zuffardi, O.; Caiulo, A.; Maraschio, P.; Tupler, R.; Bianchi, E.; Amisano, P.; Beluffi, G.; Moratti, R.; Liguri, G.: Regional assignment of the loci for adenylate kinase to 9q32 and for alpha-1-acid glycoprotein to 9q31-q32: a locus for Goltz syndrome in region 9q32-qter?. Hum. Genet. 82: 17-19, 1989.

MILIA, MULTIPLE ERUPTIVE (*157400)

JJM NOTE

This entity is included because one adult male with this condition also had colon cancer, most likely a coincidence.

Thies and Schwarz (1961) described this condition. Heard et al. (1971) described a seemingly identical situation in a man who had onset of the abnormality in childhood and whose father had the same condition. (The father died at age 72 of carcinoma of the colon.) The case of Thies and Schwarz was not familial and was late in onset.

Heard, M. G.; Horton, W. A.; Hambrick, G. W., Jr.: Multiple eruptive milia. Birth Defects Orig. Art. Ser. VII(8): 333-337, 1971.

Thies, W.; Schwarz, E.: Multiple eruptive milia—an organoid follicle hamartoma. Arch. Klin. Exp. Derm. 214: 21-34, 1961.

METHANE PRODUCTION (250650)

JJM NOTE

Far be it for me to suggest that this trait is not the Rosetta stone for understanding colon carcinogenesis. A related observation is the allegation that bovine flatus is a larger contributor to atmospheric hydrocarbons than fossil fuel combustion from vehicles and power plants; also, the consumption of red meat is positively correlated with the risk for colon cancer.

Bond et al. (1970) made the following observations: Methane (CH4) in man is derived solely from the metabolism of the colonic flora. Respiratory CH4 excretion is a simple but reliable indicator of intestinal CH4 production. In the adult population about one-third excrete large amounts of CH4 whereas the others excrete very little. No adult changed his excretion status over a period of 1 year. Children below the age of 3 excrete no CH4. If both parents excrete CH4 all offspring over age 7 excrete CH4. The concordance between marriage partners was random. Eleven of 12 identical twins and 14 of 16 fraternal twins were concordant. The genetics of this trait is unclear. Levitt and Duane (1972) noted that floating of stools is related more to gas content, especially methane, than to fat. Engel (1973) never found enteric CH4 production in the first month of life. He suspected that babies acquire methane-producing bacteria from their mothers, since there was a 5-fold difference in frequency of CH4 production depending on whether the mother was or was not a producer. Because of a lack of correlation with fathers and because an adult can convert from consistently negative to consistently positive, Engel doubted a genetic basis. Haines et al. (1977) observed that 80% of colon cancer patients had detectable methane in their breath, compared with 39% of nonmalignant colonic disease patients and 40% of persons without colon disease. This suggested a difference in anaerobic intestinal flora in colon cancer. Whether the difference antedated or followed the development of cancer was unclear.

See also Bond et al. (1971); Levitt and Bond (1980); Levitt and Bond (1970)

Bond, J. H.; Engel, R. R.; Levitt, M. D.: Methane production in man. (Abstract) Gastroenterology 58: 1035 only, 1970.

Bond, J. H., Jr.; Engel, R. R.; Levitt, M. D.: Factor influencing pulmonary methane excretion in man: an indirect method of studying the in situ metabolism of the methane-producing colonic bacteria. J. Exp. Med. 133: 572-588, 1971.

Engel, R. R.: Personal Communication. Minneapolis, Minn. 10/8/1973.

Haines, A.; Metz, G.; Delwari, J.; Blendis, L. M.; Wiggins, H.: Breath-methane in patients with cancer of the large bowel. Lancet II: 481-483, 1977.

Levitt, M. D.; Bond, J. H.: Flatulence. Ann. Rev. Med. 31: 127-137, 1980.

Levitt, M. D.; Bond, J. H., Jr.: Volume, composition, and source of intestinal gas. Gastroenterology 59: 921-929, 1970.

Levitt, M. D.; Duane, W. C.: Floating stools—flatus versus fat. New Eng. J. Med. 286: 973-975, 1972.

Colorectal Cancer Syndromes with Polyposis

ADENOMATOUS POLYPOSIS OF THE COLON (175100)

JJM NOTE

According to H. J. R. Bussey (*Familial Polyposis Coli.* Baltimore, Johns Hopkins University Press, 1975), the first authentic reported case dates to 1847 (Corvisart L: Hypertrophie partielle de la muqueuse intestinale. Bull. Soc. Anat. 22:400, 1847). Among the lessons taught by the interdisciplinary research on this condition are (1) the value of a disease registry, started in 1925 with three families by Lockhart-Mummery; (2) the concept of ecogenetics (e.g., the disease has a different picture in Japan, with frequent gastric polyps, perhaps owing to the same factors that cause so much stomach cancer in Japan); and (3) the role of karyotyping individuals with a single-gene disorder with atypical features, especially growth and mental retardation.

Description

Familial adenomatous polyposis (FAP) is an autosomal dominant disorder which typically presents with colorectal cancer (CRC) in early adult life secondary to extensive adenomatous polyps of the colon. Polyps also develop in the upper gastrointestinal tract and malignancies may occur in other sites including the brain and the thyroid. Helpful diagnostic features include pigmented retinal lesions known as congenital hypertrophy of the retinal pigment, jaw cysts, sebaceous cysts, and osteomata. The APC gene at 5q21 is mutant in FAP.

Nomenclature

Multiple polyposis of the colon, hereditary polyposis coli, familial multiple polyposis, and familial polyposis of the colon (FPC) were early terms for this disorder. The designation familial adenomatous polyposis (FAP) is most often used today, particularly in Great Britain, based in part on the appreciation that the polyps are not confined to the colon. FAP has also been used as an acronym for familial amyloid polyneuropathy (176300) and for fibroblast activation protein (600403).

The specific gene on chromosome 5 that is the site of the mutation in this disorder is symbolized APC, for adenomatous polyposis coli, and the same designation and acronym is used by many for the disorder as well. Here we use FAP and FPC interchangeably for the clinical disorder and APC for the gene.

Gardner syndrome, colonic polyposis with extrabowel tumors, especially osteomas, and a rather characteristic retinal lesion, is now known to be a phenotypic variant of FAP, caused by mutation in the APC gene.

Clinical Features

Familial adenomatous polyposis (FAP) is characterized by adenomatous polyps of the colon and rectum; in extreme cases the bowel is carpeted with a myriad of polyps. This is a viciously premalignant disease with one or more polyps progressing through dysplasia to malignancy in untreated gene carriers with a median age at diagnosis of 40 years. Carcinoma may arise at any age from late childhood through the seventh decade. The presenting features are usually those of malignancy, such as weight loss and inanition, bowel obstruction, or bloody diarrhea. Cases of new mutation still present in these ways but in areas with well organized registers most other gene carriers are detected by bowel examination while still asymptomatic. Occasionally, the extracolonic features of the condition lead to presentation.

In the past, patients with extracolonic features were treated as a distinct phenotype labeled Gardner syndrome. Detailed evaluation has shown that a majority of FAP patients have one or more extracolonic features (Krush, 1988). Gardner syndrome and FAP may occur in sibships, map to the same chromosomal location (Bodmer et al., 1987; Nakamura et al., 1988), and may be associated with identical pathological mutations in the APC gene (Nishisho et al., 1991), so the terms are regarded as synonymous. The terms are used interchangeably although there is some evidence to suggest that some extracolonic features correlate with the nature of the mutation. The best evidence of distinctness was provided by large kindreds such as that of Asman and Pierce (1970) in which no extraintestinal features were found and that of Gardner (1962) in which association of extrabowel features was consistently found. Furthermore, restudy of an earlier reported kindred (Kelly and McKinnon, 1961) showed that the disorder is in fact Gardner syndrome with about 60 affected persons (Pierce et al., 1970). Danes and Gardner (1978) claimed that in some kindreds, including kindred 109, the Mormon family in which Gardner first described this syndrome, only the extrabowel lesions occur in some branches, whereas in other branches the full syndrome (including colonic polyps and cancer) is observed.

Extracolonic features may be examined in 3 groups: adenomatous polyps in the upper gastrointestinal tract, which have become a major determinant of long-term morbidity; ocular, cutaneous, and skeletal features, which are important diagnostic features but are usually benign; and malignancy in organs outside the gastrointestinal tract.

Upper Gastrointestinal Tract

Periampullary cancer is now a well-recognized feature of FAP (Harned and Williams, 1982; Jones and Nance, 1977). The clustering of polyps around the ampulla of Vater is strong evidence for bile being implicated in the pathological process (Pauli et al., 1980). Offerhaus et al. (1992) found that in the families in the Johns Hopkins Polyposis Registry, there was a greatly increased relative risk of duodenal adenocarcinoma and ampullary adenocarcinoma. No significant increased risk was found for gastric or nonduodenal small intestinal cancer. The demonstration by Bapat et al. (1993) of mutations in the APC gene in periampullary adenomas (175100.0019, 175100.0020) is in keeping with the periampullary tumors being an extension of the same pathological process.

Iida et al. (1988) reviewed the natural history of gastric adenomas in adenomatous polyposis of the colon. Thirteen of 26 patients were found to have gastric adenomas; during a 6.8-year follow-up, 6 of the 13 patients developed additional gastric adenomas. Gastric cancer is much more common in Japan than in Western countries which is likely to be associated with the greater frequency of this type of malignancy in Japanese carriers of the APC gene defect (Burt et al., 1984; Sivak and Jagelman, 1984.) Offerhaus et al. (1992) commented on the fact that in Japan gastric cancer is more common than duodenal cancer in patients with FAP and that gastric adenomas develop in 50% of Japanese patients with FAP whereas Jagelman et al. (1988) reported that duodenal cancer was much more common than stomach cancer in Western gene carriers.

Erbe and Welch (1978) presented a case with polyps of the small bowel and 2 adenocarcinomas of the jejunum. Jejunal or ileal carcinoma is rare.

Ocular, Cutaneous, and Skeletal Features

Blair and Trempe (1980) pointed out that congenital hypertrophy of the retinal pigment epithelium (CHRPE) is a frequent finding in Gardner syndrome and can be a valuable clue to the presence of the gene in persons who have not yet developed other manifestations. The pigmented fundus lesion may be mistaken for malignant melanoma. Lewis et al. (1984) described multiple and bilateral patches of CHRPE in affected members of 3 families with Gardner syndrome. In 4 other families, a total of 8 patients did not show CHRPE. Genetic heterogeneity was suggested. Most CHRPE lesions are unilateral, solitary, nonfamilial, and not known to be associated with other ocular or systemic disorders. The patches may be 1 or 2 disc diameters in size with a surrounding area of depigmentation. They are often referred to as 'pigmented scars.' The center of the lesion may show chorioretinal atrophy and the peripheral hyperpigmentation. Bull et al. (1985) also reported observations on CHRPE in the Gardner syndrome.

Traboulsi et al. (1987) examined 134 members of 16 families with Gardner syndrome for pigmented ocular fundus lesions. Of 41 patients with documented Gardner syndrome, 37 (90.2%) had such lesions. The lesions were bilateral in 32 of the patients and in 2 of 42 controls. Twenty (46.5%) of 43 first-degree relatives at 50% risk for Gardner syndrome had bilateral pigmented fundus lesions indicating that they probably had inherited the abnormal gene. The presence of bilateral lesions, multiple lesions (more than 4), or both appeared to be a specific (specificity = 0.952) and sensitive (sensitivity = 0.780) clinical marker for Gardner syndrome. Since the lesions were observed in a 3-month-old baby at risk, they are probably congenital. Color photographs were provided.

Diaz-Llopis and Menezo (1988) suggested that CHRPE may be a useful marker to detect patients at risk for FAP in families with the gene. Combining eye examination for CHRPE with data on age of onset and linked DNA markers appeared to be highly effective in carrier exclusion. Lyons et al. (1988) concluded that the CHRPE phenotype is a more powerful marker than other phenotypic features of Gardner syndrome. Baker et al. (1988) claimed that CHRPE is not specific for Gardner syndrome as opposed to FPC. When ophthalmic examinations were performed on 56 patients with 'validated' familial adenomatous polyposis, 8 patients were found to have the retinal lesions without any of the extracolonic features of Gardner syndrome. It is possible, of course, that the eye lesion may be the only extracolonic feature of Gardner syndrome. Chapman et al. (1989) searched for CHRPE in 40 patients representing all 25 pedigrees with familial polyposis identified in the 'Northern region' of the U.K. All had multiple lesions, ranging in number from 2 to more than 40. None of 35 controls had more than 2 lesions.

Houlston et al. (1992) suggested that CHRPE is not exclusively a manifestation of mutation at the APC locus. They described 3 patients who had 4 or more patches with no other extracolonic manifestations of adenomatous polyposis coli and all having fewer than 5 adenomatous polyps detected by colonoscopy. In the families of the 3 patients, a parent and the proband in each case had colorectal cancer. In 2 families, there was cancer of other types. Houlston et al. (1992) suggested that CHRPE can occur with cancer family syndromes (114400 and 114500). It should be pointed out, however, that no search for mutations of the APC gene was made in these cases; the range of phenotype associated with such mutations might include the picture observed in these cases. Patients expressing CHRPE tend to cluster within specific polyposis families. See later for the findings of Olschwang et al. (1993) for evidence that CHRPE lesions are absent if the protein-truncating mutation in the APC gene occurs before exon 9 but are consistently present if it occurs after this exon.

In Gardner syndrome, polyps of the colon and sometimes of the stomach and small intestine are associated with osseous and soft tissue tumors. Globoid osteomata of the mandible with overlying fibromata are characteristic. Osteomatous changes in the calvaria with associated fibromas (of the forehead, for example) are also observed. Sebaceous or epidermoid cysts occur on the back.

Fader et al. (1962) first reported dental anomalies in Gardner syndrome. These include impacted teeth, supernumerary teeth, congenitally missing teeth, and abnormally long and pointed roots on the posterior teeth (Carl and Herrera, 1987). Jarvinen et al. (1982) found dental anomalies in 18% of patients but jaw osteomata were very frequent. Utsunomiya and Nakamura (1975) recorded jaw osteomata, which appear as radiopaque lesions without a translucent halo, in 95% of FAP patients although interpretation of the orthopantomograms is difficult and limits this as a diagnostic investigation.

Tumors and Malignancy Outside the Gastrointestinal Tract

Mesenteric fibromatosis may develop, especially after surgery (Simpson et al., 1964). Also known as desmoid tumors, these slowly growing lesions are locally invasive and can reach enormous proportions. Though typically associated with the abdominal cavity they may develop anywhere in the body. Fraumeni et al. (1968) described a family in which the father and a daughter had a malignant mesenchymal tumor, a son had polyposis coli, and another son had both polyposis coli and malignant mesenchymal tumor. Klemmer et al. (1987) found an increased frequency of desmoids in persons with familial adenomatous polyposis of the colon. The crude frequency was about 6%, but the risk was dependent on age and sex. The lifetime risk was estimated to be 8% for males and 13% for females.

Heimann et al. (1987) described a male patient who presented at 25 months of age with precocious puberty and an abdominal mass that was found to be a virilizing hepatoblastoma. Liver transplant was performed, and 7 months later a metastatic lesion in the upper lobe of the left lung was resected without complication. Shneider et al. (1992) reported that the patient remained disease-free and in good health without additional therapy for 53 months following transplantation. Then, at the age of 8 years, he developed painless rectal bleeding. Colonoscopy revealed multiple adenomatous polyps of the colon and a strong maternal family history of polyposis and colon cancer was obtained. Ophthalmologic examination revealed CHRPE. Total colectomy and ileoanal reconstruction was performed when he was 10 years of age.

Several groups noted the association of hepatoblastoma with polyposis coli (e.g., Kingston et al., 1982; Li et al., 1987; Krush et al., 1988). Li et al. (1987) observed hepatoblastoma in 4 unrelated children who had a family history of polyposis coli and found this association in 10 other kindreds in the literature. One child who survived hepatoblastoma showed multiple colonic adenomas at 7 years of age. She and 8 affected maternal relatives also had CHRPE. Krush et al. (1988) reported hepatoblastoma in 4 children from unrelated families. One child, 19-years-old at the time of the report, survived after a resection of a hepatoblastoma in infancy and had recently been found to have Gardner syndrome. He, like many others in these 4 families, both affected and at risk, had osteomatous jaw lesions and pigmented ocular fundus lesions. Hepatoblastoma has also been observed in type I glycogen storage disease (232200) and in Beckwith-Wiedemann syndrome (130650).

In a worldwide collaborative study, Garber et al. (1988) identified 11 children with hepatoblastoma and a family history of adenomatous polyposis; 14 additional instances of the association were collected from the literature. Among the 11 survivors of hepatoblastoma in the combined series, adenomatous lesions of the colon had been sought in 7 and detected in 6 patients at ages 7 to 25 years. Five of these patients also had CHRPE. Giardiello et al. (1991) studied the frequency of hepatoblastoma in the families registered in the familial polyposis registry maintained at Johns Hopkins since 1973. Seven members of these families had hepatoblastoma diagnosed at ages varying from 1 month to 4.5 years. Six of them were from Gardner syndrome families and 1 was from a polyposis family without extrabowel manifestations. Giardiello et al. (1991) calculated the relative risk of hepatoblastoma in persons with the APC gene from birth through age 4 as being 3.3 per 1,000 person/years.

In a retrospective review of their family history data, Hughes and Michels (1992) found that 2 of 470 (0.42%) children born to 241 patients with FAP had hepatoblastoma. This figure is significantly higher than the 1 in 100,000 incidence of hepatoblastoma in the general population. However, for genetic counseling purposes, an empiric risk of less than 1% for hepatoblastoma can be cited to persons with FAP for their children.

To elucidate the possible roles of APC gene alterations in sporadic hepatoblastomas, Oda et al. (1996) examined LOH at the APC and MCC loci and sequenced part of the APC gene, including the mutation cluster region, in 13 hepatoblastomas of patients without familial adenomatous polyposis. LOH at the APC and/or MCC loci was observed in 4 of 7 informative cases (57%). Of the 13 cases, somatic mutations were detected in 8 (61.5%), with 9 cases (69%) showing genetic alterations in the APC gene as LOH or somatic mutations. In 2 cases, double mutations were demonstrated. Furthermore, the nature of the somatic mutations observed in this study was unusual because 9 of the 10 mutations were missense, with only 1 case featuring a frameshift mutation due to an insertion. By contrast, more than 90% of mutations in the APC gene in colorectal tumors result in a truncated APC protein due to either frameshift or nonsense mutations.

Marshall et al. (1967) described a case of Gardner syndrome with adrenal cortical carcinoma with Cushing syndrome. Camiel et al. (1968) described thyroid carcinoma in 2 sisters who also had Gardner syndrome, which was probably present in at least 3 generations of the family. Smith (1968) also described patients with the association of colonic polyps and papillary carcinoma of the thyroid. Herve et al. (1995) reported a case of papillary carcinoma in a 16-year-old girl with Gardner syndrome. They reviewed the literature and estimated that the incidence of thyroid carcinoma in patients with Gardner syndrome approached 100 times that of the general population.

Hoffmann and Brooke (1970) described a family in which 6 persons in 3 generations had FAP and a mother and son had sarcoma of bone leading to death from metastases at 28 and 13 years of age, respectively. No evidence of polyposis was found in either but special studies including autopsies were not done.

Capps et al. (1968) described a family with 4 generations of carcinoma of the colon and polyposis of the colon. A brother of the proband died of brain tumor at age 9 years and had colonic polyposis. The proband, aged 14 years at first presentation, had carcinoma of the colon, ampulla of Vater, and urinary bladder.

In a member of the original Utah kindred with Gardner syndrome, Naylor and Gardner (1981) observed bilateral adrenal adenomas. They found reports of 6 cases of adrenal adenoma and 1 of primary adrenal carcinoma. They reviewed 15 reported cases of thyroid tumors in Gardner syndrome. Bell and Mazzaferri (1993) reported what they alleged to represent the 37th report of the association of Gardner syndrome with papillary thyroid carcinoma. They pointed out that 94.3% of the patients have been women.

In a 26-year-old woman with Gardner syndrome, Walsh et al. (1987) found multifocal adenomatous change with severe dysplasia in the gallbladder. They referred to observations of others on bile duct cancer and carcinoma in situ of the gallbladder in patients with this form of hereditary polyposis.

Turcot syndrome (276300) is characterized clinically by the concurrence of a primary brain tumor and multiple colorectal adenomas. Hamilton et al. (1995) demonstrated that this association can result from at least 2 distinct types of germline defects: mutation in the APC gene, which is responsible for adenomatous polyposis coli, or a mutation in mismatch repair gene PMS2 (600259) or MLH1 (120436). Medulloblastoma was the brain tumor in most cases with APC mutations; glioblastoma was the brain tumor in most cases with mismatch repair gene mutations.

Shull and Fitts (1974) reported a family in which the father and 2 sons had both adenomatous and lymphoid polyps. Venkitachalam et al. (1978) pointed out that lymphoid polyposis has several times been reported in families with polyposis coli.

Hodgson et al. (1993) reported 2 cases of deletion of 5q associated with FAP. They noted that 5 such cases had previously been reported. In their 2 cases, as well as in one of those previously reported, macroscopic polyposis was confined to the proximal colon in patients in their thirties, although microscopic adenomatosis was shown in the more distal colon with occasional single polyps. Both of their subjects had dermoid cysts, and CHRPE was seen in one. The latter patient, who had the more extensive deletion, also showed Caroli disease (dilatation of distal intrahepatic bile ducts with intrahepatic stone formation); see 263200. Hodgson et al. (1994) suggested that heterozygous deletion of the entire APC gene may be associated with a form of FAP characterized by more proximal distribution of adenomas than usual, of which some are sessile and some may be nonpolypoid or flat. They postulated that in the usual type of FAP where the mutation results in a truncated protein, this protein may interfere with the function of the protein product of the normal allele to cause a more severe disease than seen in their patients. They pointed to the large kindreds reported by Leppert et al. (1990) and Lynch et al. (1992) as possible examples of this particular phenotype. Samowitz et al. (1995) pointed out that this seemingly different phenotype was referred to by Lynch et al. (1992) as 'hereditary flat adenoma syndrome.' Later, when it was found that the family reported by Leppert et al. (1990) and the families of Lynch et al. (1992) had characteristic mutations in the 5-prime end of the APC gene, the syndrome was renamed 'attenuated adenomatous polyposis coli' (AAPC).

Eccles et al. (1996) described a family in which members of 3 generations showed hereditary desmoid disease characterized by multifocal fibromatosis of the paraspinal muscles, breast, occiput, arms, lower ribs, abdominal wall, and mesentery. Osteomas, epidermal cysts, and congenital features such as imperforate anus were observed. One patient developed carcinoma of the head of the pancreas (ampulla of Vater); however, no member of the family had polyposis of the colon or stomach. The affected family was found to carry a germline chain-terminating mutation at the 3-prime end of the APC gene (codon 1924) with somatic loss of the wildtype allele leading to tumor development at multiple sites (175100.0025).

Other Features

Hsu et al. (1983) found that the polyps of Gardner syndrome are multiclonal in origin, as are the tumors in neurofibromatosis (162200) and trichoepithelioma (132700). Rasheed et al. (1983) demonstrated that skin fibroblasts from patients with Gardner syndrome and familial polyposis coli showed increased susceptibility to retrovirus-induced transformation and chromosomal aneuploidy. Chen et al. (1989) concluded that the in vitro life span of cultured skin fibroblasts from individuals with FPC is markedly extended when compared with that of normal individuals.

Luk and Baylin (1984) concluded that the activity of ornithine decarboxylase (EC 4.1.1.17) may be a useful marker for the genotype of familial polyposis. This rate-limiting enzyme in the polyamine biosynthetic pathway is essential for intestinal mucosal proliferation. High levels of activity were found in normal-appearing colonic mucosa from 11 of 13 patients with familial polyposis and in all polyps biopsied from these same patients. Mucosa from dysplastic polyps showed higher mean ornithine decarboxylase activity than mucosa from polyps that were not dysplastic. Among clinically unaffected first-degree relatives of patients with familial polyposis, a bimodal distribution of ornithine decarboxylase activity was observed; one peak at the mean of normal controls and the other at the mean for normal-appearing mucosa from affected patients. In 40 Danish patients with familial polyposis coli, Pandey et al. (1986) found an increased frequency of Gm3;5;13. The 2 conditions could not be distinguished by this approach.

Smith et al. (1993) produced monoclonal and polyclonal antibodies to APC for characterizing the protein in normal and tumor cells. They found that 81% of colon tumor lines were totally devoid of the normal, full-length protein, whereas 40 cell lines derived from sporadic tumors of other organs had only full-length APC. Immunohistochemical analysis of APC in normal colonic mucosa demonstrated cytoplasmic staining with more intense staining in the basolateral margins of the epithelial cells. The staining was markedly increased in the upper portions of the crypts, suggesting an increased level of expression with maturation.

Midgley et al. (1997) developed antisera to APC N- and C-terminal epitopes. They found that the APC protein was expressed in epithelial and mesenchymal cells in many tissues. In epithelium of bladder, small and large intestine, esophagus, stomach, and epidermis, APC expression was restricted to regions where cell replication has ceased and terminal differentiation is established. APC staining was often diffusely cytoplasmic; however, in surface cells there was accentuated expression in the subapical regions and along the lateral margins. Midgley et al. (1997) concluded that this distribution is compatible with APC function related to signaling at the adherens junction and indicates that APC plays a role in cells committed to terminal differentiation.

Inheritance

FAP behaves as an autosomal dominant trait with almost complete penetrance but striking variation in expression.

Cytogenetics

Herrera et al. (1986) found a constitutional interstitial deletion of 5q in a man with possible Gardner syndrome; the large bowel was 'carpeted' with more than 100 adenomatous polyps and contained a well-differentiated carcinoma of the rectum, a similar carcinoma of the ascending colon, and melanosis coli. A small mesenteric neurofibroma was also found. In addition, the patient had severe mental retardation, horseshoe kidney, absence of the left lobe of the liver, and agenesis of the gallbladder. This finding of deletion prompted the search for linkage with RFLP markers on 5q; positive results indicated that the mutation determining Gardner syndrome is located on 5q, probably near bands 5q21-q22. Like Herrera et al. (1986), Kobayashi et al. (1991) described an interstitial deletion of 5q in a boy with Gardner syndrome who had mental retardation and multiple minor anomalies. The deletion involved q22.1-q31.1.

Hockey et al. (1989) described an interstitial deletion of 5q15-q22 in 2 intellectually handicapped brothers with familial adenomatous polyposis. Their retarded mother died of an inoperable carcinoma of the colon with extensive polyposis.

Cross et al. (1992) reported a male patient and his maternal aunt with familial adenomatous polyposis coli associated with epidermoid cysts, osteomata, and areas of congenital hypertrophy of the retinal pigment epithelium, and mild-to-moderate mental handicap. Both affected persons had an interstitial deletion of the long arm of chromosome 5: del(5)(q22q23.2). Two other unaffected family members had the direct insertion of chromosome 5 which was responsible for the deletion. Molecular analyses indicated loss of one allele at the MCC and APC loci. Barber et al. (1994) described an institutionalized adult woman who was referred for chromosome studies because of autistic behavior. A high risk of colorectal cancer was predicted when an interstitial deletion of 5q was found in lymphocytes and deletion of the MCC and APC gene confirmed by molecular analysis. Adenomatous polyposis coli and carcinoma of the rectum were subsequently diagnosed in the patient. The del(5)(q15q22.3) arose as a result of recombination within the small insertion loop formed at meiosis by the direct insertion found in the patient's mother.

Other cytogenetic observations include that of Danes and Gardner (1978) who described in vitro tetraploidy in skin cultures containing epithelioid cells only from persons in the branch of kindred 109 with full expression of the syndrome.

Gardner et al. (1982) observed an excessive random loss and gain of single chromosomes in lymphocytes and fibroblasts cultured from patients with Gardner syndrome and familial polyposis coli and from children at risk for multiple adenomas in the colorectum. A consistent heteromorphism of chromosome 2, tentatively identified as a deletion, was observed in 17 patients with multiple colonic polyps and in 2 persons, aged 6 and 13 years, at risk for Gardner syndrome but as yet without colorectal polyps. The heteromorphism was not found in 2 patients with occasional discrete colorectal adenomas or in 18 controls without Gardner syndrome or familial polyposis coli. The portion of chromosome 2 affected was 2q14.3-q21.3. Fineman et al. (1984) did high resolution cytogenetic studies of mitotic chromosomes in peripheral blood of 2 patients with Gardner syndrome and 2 with familial polyposis; no deletion was found in chromosome 2. Kasukawa et al. (1983) also could find no abnormality of chromosome 2.

In a Dutch FAP family, van der Luijt et al. (1995) detected a germline rearrangement in the form of a constitutional reciprocal translocation t(5;10)(q22;q25), resulting in the disruption of the APC gene. The translocation apparently resulted in a null allele. Clinically the patients exhibited atypical features, namely a slightly delayed age of onset of colorectal cancer and a reduced number of colorectal polyps. The latter were mainly sessile and were located predominantly in the proximal colon. This was thought to be the first description of a reciprocal translation disrupting the APC gene.

Mapping

The identification of the gene mutation(s) in FAP, including Gardner syndrome, started with the observation of a patient with a presumed 'contiguous gene syndrome' that included multiple polyposis of the colon (Herrera et al., 1986).

Bodmer et al. (1987) found a maximum lod score of 3.26 at theta = 0.0 for linkage to a marker called C11p11, which was known to map to chromosome 5. Solomon et al. (1987) used a highly polymorphic 'minisatellite' probe to demonstrate that at least 20% of sporadic colorectal adenocarcinomas lose one of the alleles present in normal tissue of the host. In the study of Bodmer et al. (1987), the disease phenotype was labeled as familial adenomatous polyposis. That many and possibly the majority of the families studied were instances of Gardner syndrome is suggested by the findings in the Johns Hopkins Polyposis Registry, as mentioned earlier. The findings of Bodmer et al. (1987) suggest that Gardner syndrome and perhaps even familial colorectal cancer may be allelic disorders; all appear to be linked to markers on 5q.

Simultaneously and independently, Leppert et al. (1987) demonstrated linkage to a chromosome 5 marker in the region of 5q22. In 2 of the families, affected individuals had lesions typical of Gardner syndrome; in 3 persons, it was typical of familial polyposis coli. Almost all of the linkage information was provided by the families with the familial polyposis coli phenotype. By use of multilocus analysis, Leppert et al. (1987) found a maximum lod score of 5.0 for mapping of the familial polyposis locus to 5q. This is another indication that Gardner syndrome and familial polyposis coli are allelic disorders. Like Bodmer et al. (1987), Leppert et al. (1987) had been prompted to look at chromosome 5 because of the case reported by Herrera et al. (1986).

Through studies in 6 families, Nakamura et al. (1988) refined the genetic localization of the polyposis locus to a position about 17 cM distal to the DNA probe C11p11 at 5q21-q22. Three of the families had the familial polyposis coli phenotype and 3 had Gardner syndrome. The data were fully consistent with the idea that these are allelic disorders.

In Dutch kindreds, Meera Khan et al. (1988) found that the RFLP used by Bodmer et al. (1987) and Leppert et al. (1987) was minimally informative because of its low heterozygosity. On the other hand, another RFLP related to D5S37 and previously localized on 5q21 showed close linkage to FPC (lod = 7.85 at theta = 0.048 with 95% probability limits 0.005-0.145). The results were interpreted as indicating that the most likely location of FPC is in the band 5q22 very close to 5q21 or perhaps even in the transitional zone between these 2 bands.

Endo and Kasukawa (1987) could find no evidence of deletion in 5q in patients studied by them. Using cells from patients with deletions in the 5q15-q21 region, Varesco et al. (1989) constructed human/hamster hybrid cell lines that retained either the normal or the deleted chromosome 5. This allowed localization of probes to the deleted region by a single Southern blot. These lines were used to identify markers from the region of the polyposis gene obtained by cloning the ends of 0.5- to 2-megabase BssHII rare-cutting fragments purified by pulsed field gel electrophoresis. Three markers derived from CpG-rich islands mapped within the deletions and are therefore close to the APC gene.

Cross et al. (1992) indicated that familial direct insertions should be considered as a cause of recurrent microdeletion syndromes. Molecular studies in patients with a cytogenetically evident deletion of 5q21 indicated that the MCC locus is situated there (Lindgren et al., 1992). Since APC is about 150 kb distal to MCC, APC may be located in 5q21 or, as suggested by others, in the region between 5q21 and 5q22.

Tops et al. (1993) described a family with seemingly typical autosomal dominant colonic polyposis but not linked to the APC locus. The gene for Turcot syndrome (276300) had previously been shown to be nonallelic with the APC locus by the same group (Tops et al., 1992).

Stella et al. (1993) reported studies in 2 kindreds with a variant form of familial adenomatous polyposis: the number of colonic polyps was low and variable (from 5 to 100) and the disease showed a slower evolution than in the usual cases of FAP, with colon cancer occurring at a more advanced age in spite of early onset of intestinal manifestations. The APC gene was excluded as the site of the mutation by linkage studies using intragenic and tightly linked markers.

Molecular Genetics

Chen et al. (1989) hypothesized that the recessive FPC growth-control gene is either the same as or closely linked to a recessive gene responsible for the programmed senescence of normal human skin fibroblasts.

As indicated in 114500, loss of constitutional heterozygosity (LOH) for markers on chromosome 5 is a frequent finding in colon cancers. The observations such as those of Ashton-Rickardt et al. (1989) indicate that it is loss of the APC locus that is responsible in the main for contributions to the malignant process. Using APC-linked probes, as well as those mapping to the outlying regions of chromosome 5, Okamoto et al. (1990) analyzed 51 colorectal tumors and 7 desmoids from 19 cases of familial polyposis coli and 5 Gardner syndrome patients, as well as 15 sporadic colon cancers. They found a high incidence of allelic deletion in hereditary colon cancers, as well as in sporadic colon cancers, with a peak at the APC locus. APC loss resulted primarily from interstitial deletion or mitotic recombination. Combined tumor and pedigree analysis in a Gardner syndrome family showed loss of normal 5q alleles in 3 tumors, including a desmoid tumor, which suggested the involvement of hemizygosity or homozygosity of the defective APC gene in colon carcinogenesis and possibly in extracolonic neoplasms associated with APC.

Simultaneously and independently, the group of Bert Vogelstein in Baltimore, in collaboration with the group of Yusuke Nakamura in Tokyo (Kinzler et al., 1991; Nishisho et al., 1991) and the group of Ray White in Salt Lake City (Groden et al., 1991; Joslyn et al., 1991), announced the identification and characterization of the APC gene together with identification of mutations in the gene. In 61 unrelated patients with adenomatous polyposis coli, Groden et al. (1991) sought mutations in 3 genes (DP1, SRP19, and DP2.5) located within a 100-kb region deleted in 2 patients with APC (Joslyn et al., 1991). The intron-exon boundary sequences were defined for each of these genes, and single-strand conformation polymorphism (SSCP) analysis of exons from DP2.5 identified 4 mutations specific to APC patients. Each of 2 aberrant alleles contained a base substitution changing an amino acid to a stop codon in the predicted peptide; the other mutations were small deletions leading to frameshifts. Analysis of DNA from parents of one of these patients showed that his 2-bp deletion was a new mutation; furthermore, the mutation was transmitted to 2 of his children. Groden et al. (1991) found that the DP2.5 gene consists of 15 exons. Use of the same method of SSCP analysis to examine 12 of the 17 exons of MCC (159350) failed to show any variants unique to APC patients.

In 2 small, nested deletions of 100 to 260 kb identified by Joslyn et al. (1991), they found that the MCC gene was located outside the deleted region. One of the new genes found in the deleted area had a sequence identical to that of SRP19 (182175), the gene coding for the 19-kD component of the ribosomal signal recognition particle. The second cDNA, provisionally designated DP1 ('deleted in polyposis' 1; 125265), was found to be transcribed in the same orientation as MCC. The other 2 cDNAs, DP2 and DP3, were found to overlap, forming a single gene, DP2.5, that was transcribed in the same orientation as SRP19. (In SSCP analysis, sequence alterations are detected as shifts in electrophoretic mobility of single-stranded DNA on denaturing acrylamide gels; the 2 complementary strands of a DNA segment usually resolve as 2 SSCP conformers (so called) of distinct mobilities. However, if the sample is from an individual heterozygous for a basepair variant within the amplified segment, often 3 or more bands are seen. In some cases, even the sample from a homozygous individual will show multiple bands. Basepair-change variants are identified by differences in pattern of conformers in DNAs from various individuals.)

Kinzler et al. (1991) isolated 6 contiguous stretches of sequence (contigs) containing approximately 5.5 Mb of DNA from the region tightly linked to APC on 5q21. The region was found to contain 6 genes, all of which were expressed in normal colonic mucosa. (These were FER (176942), MCC, SRP19, TB1, and TB2, in addition to the APC gene itself. TB2 is the gene called DP1 by Groden et al. (1991) and Kinzler (1991).) The APC gene, identified in a contig initiated from the MCC gene, was found to encode an unusually large protein. Both genes encode proteins predicted to contain coiled-coil regions and both are expressed in a wide variety of tissues.

Nishisho et al. (1991) demonstrated point mutations in the APC gene in the germline of APC and GS patients. Nishisho et al. (1991) screened for mutations using the ribonuclease (RNase) protection assay on PCR-amplified DNA. In addition, in some of the families, the PCR products were cloned and sequenced to search for mutations not detected by RNase protection. Five variants were detected among 103 kindreds in whom 3 of the APC gene exons were analyzed. In addition to germline mutations, Nishisho et al. (1991) found a somatic APC mutation in a sporadic colorectal cancer (see 175100.0009).

In a patient with familial adenomatous polyposis, Seki et al. (1992) found that an adrenocortical carcinoma had loss of heterozygosity at the APC locus, i.e., had lost the normal APC gene. Sen-Gupta et al. (1993) reported a somatic mutation of the APC gene in desmoid tissue in a patient reported by Hodgson et al. (1993) as having familial adenomatous polyposis caused by a constitutional chromosome 5q22 deletion. The somatic mutation consisted of a 107-bp deletion between codons 1438 and 1473, causing a frameshift that would lead to a truncated gene product.

Lambertz and Ballhausen (1993) isolated cDNA clones representing transcripts expressed in human fetal brain and coding for the 5-prime end of the APC gene. Sequence analyses revealed an alternative 5-prime untranslated region (UTR) comprising at least 103 bp. This finding suggested that 2 APC-specific promoter elements exist, giving rise to 2 different untranslated regions. Within the alternative UTR, Lambertz and Ballhausen (1993) identified 3 additional AUG codons, located 5-prime to the intrinsic APC initiation site. These codons may be relevant for the translational regulation of APC gene expression.

Horii et al. (1993) noted that transcriptional initiation of APC occurs at 3 sites in 2 distinct non-translated exons at the 5-prime end of the gene. Studies of transcripts from human colorectal tumor cell lines suggested the presence of mutations in the transcriptional control region. Horii et al. (1993) also detected at least 5 different forms of 5-prime noncoding sequences which were generated by alternative splicing. They stated that the splicing mechanism appeared to be regulated in a tissue-specific fashion, and 1 transcript, expressed exclusively in brain, contained an extra exon.

Xia et al. (1995) described an alternatively spliced APC transcript which had not been reported previously. Within this transcript, they found an evolutionarily conserved but previously unidentified exon between the known exons 10 and 11. The exon contains a heptad repeat motif.

Using a ribonuclease protection analysis coupled with PCR amplification of genomic DNA, Miyoshi et al. (1992) searched for germline mutations of APC in 79 unrelated patients with familial adenomatous polyposis. Mutations were found in 53 patients (67%); 28 of the mutations were small deletions and 2 were insertions of 1 or 2 bp; 19 were point mutations resulting in stop codons, and only 4 were missense point mutations. Thus, 92% of the mutations were predicted to result in truncation of the APC protein. More than two-thirds (68%) of the mutations were clustered in the 5-prime half of the last exon, and nearly two-fifths of the total mutations occurred at 1 of 5 positions. This information has significant implications for understanding the role of APC mutation in colorectal neoplasia and for designing effective methods for genetic counseling and presymptomatic diagnosis.

In 22 unrelated APC patients, Nagase et al. (1992) found that germline mutations between codons 1250 and 1464 were associated with profuse polyps (10 or more), whereas mutations in other regions of the APC gene were associated with sparse polyps (fewer than 10). Nagase et al. (1992) summarized the results of screening for germline mutations in 150 unrelated patients with familial adenomatous polyposis. They could find no indication that extracolonic manifestations such as osteomas and desmoid tumors correlate with the type or intragenic location of a particular germline mutation.

Reporting on studies of somatic mutations of the APC gene in 16 adenomas and 47 carcinomas developed in FPC and non-FPC patients, Miyoshi et al. (1992) described loss of heterozygosity (LOH) at the APC locus in 30 tumors and 43 other somatic mutations. In 41 of 43 mutations (95%), truncation of the APC protein resulted. Over 60% of the somatic mutations were clustered within a small region of exon 15 designated as MCR (for 'mutation cluster region'), which accounted for less than 10% of the coding region. Combining all the data, more than 80% of tumors had at least one mutation in the APC gene, of which more than 60% had 2 mutations. The results strongly suggested that somatic mutations of the APC gene are associated with the development of the great majority of colorectal tumors.

Fodde et al. (1992) confirmed the fact that FPC is caused by a highly heterogeneous spectrum of point mutations. All the mutations found in their study were chain terminating. Denaturing gradient gel electrophoresis (DGGE) proved to be a rapid and sensitive technique for detection of mutations in the unusually large APC gene. Fodde et al. (1992) also found a polymorphism, a neutral single nucleotide substitution at codon 545 (GCA to GCG), that did not alter the amino acid residue alanine. On the basis of data from 98 unrelated individuals, they estimated the minor allele frequency to be 0.378 and derived the corresponding polymorphism information content (PIC) to be 0.36 (Botstein et al., 1980). Hampton et al. (1992) isolated 2 overlapping YACs containing the MCC gene; one of the YACs also included the complete APC gene.

Powell et al. (1992) presented evidence that APC mutations occur early during colorectal tumorigenesis. Sequence analysis of 41 colorectal tumors revealed that most carcinomas (60%) and adenomas (63%) contained a mutated APC gene. Furthermore, mutations of the APC gene were found in the earliest tumors that could be analyzed, including adenomas as small as 0.5 cm in diameter. The frequency of such mutations remained constant as tumors progressed from benign to malignant stages. This finding provided support for the multistage process of colorectal carcinogenesis with the APC gene at or near the initial step (Fearon and Vogelstein, 1990).

Boland et al. (1995) performed microallelotyping of many regions from individual colorectal tumors to determine the sequence and tempo of allelic loss at 5q, 17p, and 18q during neoplastic progression. No allelic losses were found in normal tissues surrounding colorectal neoplasms, but losses occurred abruptly on 5q at the transition from normal colonic epithelium to benign adenoma, and on 17p at the transition from adenoma to carcinoma, indicating an essential role for these losses in tumor progression. Allelic losses were uniform throughout extensively microdissected benign adenomas and carcinomas. However, substantial allelic heterogeneity was found in high-grade dysplasia, the transition lesion between adenoma and carcinoma. Thus, Boland et al. (1995) concluded that 5q (presumably APC) and 17p (presumably p53) are associated with abrupt waves of clonal neoplastic expansion.

Using SSCP analysis, Cottrell et al. (1992) also identified APC mutations in sporadic colon cancers. The mutations they found were either point mutations or small deletions or insertions causing frameshifts, and all generated stop codons. Their results suggested that highly localized short sequences, such as runs that code for adenine and thymine, may account for up to 20% of all observed APC mutations. One 5-bp deletion was found in a sporadic colon tumor and in 4 unrelated APC patients. The APC gene contains an 8,538-bp open reading frame and is predicted to encode a 2,843-amino acid polypeptide with few homologies to other proteins (Groden et al., 1991; Kinzler et al., 1991). The large size of the APC coding region makes identification of mutations labor intensive and costly. However, because most of the mutations predict truncation of the APC protein, the mutations might be more conveniently detected through analysis of the APC protein.

Nagase and Nakamura (1993) summarized the germline APC mutations in 174 FAP patients and in somatic APC mutations in 103 colorectal tumors, as well as mutations in tumors arising in organs other than the colon and rectum. They concluded that inactivation of both alleles at the APC locus is required for development of most tumors in the colon and rectum. The great majority of the known mutations would result in truncation of the APC product. Almost all known mutations had been located within the 5-prime half of the coding region, although somatic mutations in colorectal tumors tended to cluster in the MCR (mutation cluster region), which represents only 8% of the coding sequence (codons 1286-1513). Most identified point mutations in the APC gene are changes from cytosine to other nucleotides. It was suggested that the location of germline mutations is correlated with the number of colorectal polyps in FAP patients.

Spirio et al. (1993) determined the APC mutations in 7 families with attenuated adenomatous polyposis coli, i.e., familial adenomatous polyposis coli with relatively few colonic polyps but still a significant risk of colon cancer. Four distinct mutations in the APC gene were identified in the 7 AAPC families. These mutations that predicted truncation products, either by single basepair changes or frameshifts, were similar to mutations identified in families with classic APC. However, they differ in that the 4 mutated sites were located very close to one another and nearer the 5-prime end of the APC gene than any base substitutions or small deletions previously discovered in patients with classic APC. Another correlation between genotype and phenotype was provided by Olschwang et al. (1993), who found that the extent of congenital hypertrophy of the retinal pigment epithelium correlated with the position of the protein-truncating mutation in APC. CHRPE lesions were almost always absent if the mutation occurred before exon 9, but were consistently present if it occurred after this exon. The same finding was recorded by Bunyan et al. (1995), who also noted that a more distal mutation site was associated with an earlier age of onset of symptoms and a larger number of colonic polyps.

Clues to the normal function of the APC gene product were obtained by immunoprecipitation experiments performed by Rubinfeld et al. (1993), who showed that antibodies to APC precipitated a 95-kD protein identified as beta-catenin (CTNNB1; 116806), and by Su et al. (1993), who found association with 2 cellular proteins, alpha-catenin and beta-catenin. Both proteins bind to the cell adhesion molecule E-cadherin (192090). The results suggested that APC is involved in cell adhesion. Peifer (1993) reviewed the role of the catenins in a homolog in Drosophila. One possibility is that the APC complex regulates transmission of the contact inhibition signal into the cell. This hypothesis is consistent with the observation that APC mutations are associated with the development of hyperplasia, an early event in tumorigenesis. A second possibility is that the APC-catenin complex regulates adhesion. Although this idea is supported by evidence that loss of cadherin-mediated adhesion can contribute to metastasis, it is less consistent with the evidence that APC acts early in tumorigenesis.

Papadopoulos et al. (1995) observed that 'detection of germline mutations in a sensitive and specific manner presents a continuing challenge.' In dominantly inherited diseases, since the mutation occurs in only 1 allele, it is often masked by the normal allele. To counter this problem, they reported the development of a sensitive and specific diagnostic strategy based on somatic cell hybridization termed monoallelic mutation analysis (MAMA). This simple and ingenious method involves the use of hamster/human somatic cell hybrids, which could be expected in many cases to have only 1 of the 2 alleles present. To show that single alleles were isolated in the clones, microsatellite markers proximal and distal to the gene of interest were assessed. They demonstrated the utility of this strategy in FAP and in hereditary nonpolyposis cancer.

Sulekova and Ballhausen (1995) identified a novel coding exon of the APC gene. To that point, the 54-bp exon (exon 10A) was the smallest coding exon in the gene, and was located 1.6-kb downstream from exon 10. It is alternatively spliced and inserted in-frame into mature transcripts; it gives an APC protein with an additional 18 amino acids. This highly conserved exon is expressed in a tissue-independent fashion. APC exon 10A flanking sequences were presented so that this exon could be included in mutation screening procedures.

Unlike some other tumor suppressor genes, loss or mutation in the wildtype gene is not essential to the development of intestinal polyps (Fearon and Vogelstein, 1990).

Studies to identify proteins that interact with APC were undertaken in search of clues to the mechanism by which APC acts in tumor initiation. One such protein is the E-cadherin associated protein beta-catenin (116806). Miyashiro et al. (1995) discussed findings that suggested an important link between the role of APC in tumor initiation and the process of cellular adhesion. Miyashiro et al. (1995) examined the distribution and localization of the APC protein and alpha-catenin (116805) in the normal mouse intestine by light and immunoelectron microscopy using specific antibodies. The results suggested to them that a portion of the APC protein is localized in the lateral cytoplasm of intestinal epithelial cells and functions in cooperation with catenins, whereas the APC protein in microvilli and in the apical cytoplasm has other functions independent of catenins.

Matsumine et al. (1996) showed that the APC-beta-catenin complex binds to DLG (see 601014), the human homolog of the Drosophila discs large tumor suppressor protein. This interaction required the carboxyl-terminal region of APC and the DLG homology repeat region of DLG. APC colocalized with DLG at the lateral cytoplasm in rat colon epithelial cells and at the synapse in cultured hippocampal neurons. These results suggested to Matsumine et al. (1996) that the APC-DLG complex may participate in regulation of both cell cycle progression and neuronal function. Rubinfeld et al. (1996) showed that when beta-catenin is present in excess, APC binds to a human homolog of 'Armadillo' (glycogen synthase kinase 3-beta), another component of the Wingless signaling pathway in Drosophila development.

It is widely accepted that tumors are monoclonal in origin, arising from a mutation or series of mutations in a single cell and its descendants (Fialkow, 1979). Contradictory findings were reported by Novelli et al. (1996) who used direct in situ hybridization with Y chromosome probes to examine the clonal origin of colonic adenomas and uninvolved intestinal mucosa from an XO/XY mosaic individual with FAP. In this patient, the crypts of the small and large intestine were clonal, but at least 76% of the microadenomas were polyclonal in origin. Although other interpretations were considered, such as collision of separate tumors, Novelli et al. (1996) felt the results most strongly favored a true polyclonal nature of the polyps.

In screening the entire coding region of the APC gene for mutations in an unselected series of 105 Dutch FAP kindreds, van der Luijt et al. (1997) found 67 germline mutations. They used GC-clamped denaturing gradient gel electrophoresis (DGGE) for exons 1-14, and the protein truncation test (PTT), for the large exon 15. By this approach they identified 65 pathogenic mutations in these apparently unrelated FAP families. The mutations were predominantly either frameshifts (39/65) or single base substitutions (18/65), resulting in premature stop codons. Mutations that would predict abnormal RNA splicing were identified in 7 cases. In only 1 of the families did a nonconservative amino acid change segregate with the disease. In spite of the large number of APC mutations that had previously been reported, 27 of the mutations were novel, thus again emphasizing the great heterogeneity of the mutational spectrum in FAP. In addition to the point mutations, structural rearrangements of APC were found in 2 pedigrees by Southern blot analysis. In about 35% of the FAP kindreds, a detailed molecular analysis of the coding region did not reveal pathogenic mutations.

De Vries et al. (1996) described an on-line database of mutations in the APC gene. Beroud and Soussi (1996) described a database of germline and somatic mutations in the APC gene in human tumors and cell lines. The database held 737 mutations of which 332 were germline and 402 somatic. Almost all led to truncation of the APC protein either by a nonsense mutation (30%) or by a frameshift mutation (68%). Most of the mutations occurred in the first half of the coding region. Germline mutations were scattered throughout the 5-prime half of the gene, whereas somatic mutations (60%) were concentrated in a region called the mutation cluster region (MCR). In germline mutations, 2 hotspot codons were identified, one at position 1061 and the second at position 1309. In somatic mutations, 2 hotspots occurred at positions 1309 and 1450. The authors noted that the concentration of mutations in the 5-prime part of the gene was believed to be involved in a dominant effect of the N-terminus of the APC protein. This APC region contains a dimerization domain and it has been demonstrated that wildtype and mutant APC are associated in vivo. Although the precise biologic function of APC is not known, it had been shown that APC protein associates with cations or with microtubules. They stated that the database was available in both IBM and Macintosh format on floppy discs.

Neufeld and White (1997) used a combination of immunofluorescence microscopy and biochemical fractionation to determine the location of APC protein in epithelial cells. The former method placed full-length APC protein in both the nucleus and the cytoplasm. The nuclear APC protein was concentrated in discrete subnuclear regions, including the nucleoli, whereas the cytoplasmic APC protein concentrated at the leading edge of migrating cells. Colocalization of APC protein with rRNA confirmed a nucleolar localization. These immunocytochemical findings were supported by cell fractionation, which demonstrated that full length APC protein is located in both the membrane/cytoskeletal and the nuclear fractions.

Analyzing 2 sets of data, Marshall et al. (1997) commented on the finding that 1- to 2-bp deletions and 1-bp insertions were much more commonly found among somatic mutations of the APC gene than among germline mutations of the APC gene. On the other hand, deletions of more than 2 bp were much more frequent among the germline mutations than among somatic mutations. Insertions of more than 1 bp were more frequent in somatic mutations than in germline mutations. Possible explanations for these differences were discussed.

Van der Luijt et al. (1996) noted that mutations in the 3-prime region of APC are infrequent in FAP. They examined 2 families having APC mutations in the 3-prime region of the gene (exon 15), 1 at codon 1987 (175100.0027) and 1 at codons 1860 to 1862 (175100.0028). In both cases, the mutation caused a frameshift and premature stop codon. Both families had the attenuated form of FAP (AAPC). No truncated polypeptides were detected in the cells of patients in these families. Van der Luijt et al. (1996) speculated that these 3-prime mutations represent null alleles and that the AAPC phenotype is the result of a dosage effect; patients with such mutations make less than a critical quantity of the normal gene product.

Laken et al. (1997) described a novel mechanism for predisposition to colorectal cancer (CRC) involving a 3920T-A point mutation in APC (175100.0029). The change converted the sequence AAATAAAA to (A)8 and may result in failure of the cellular transcriptional or translational machinery. They found the mutation in 6% of Ashkenazi Jews and about 28% of Ashkenazim with a family history of CRC.

Fearon (1997) reviewed more than 20 different hereditary cancer syndromes that had been defined and attributed to specific germline mutations in various inherited cancer genes. In a useful diagram he illustrated how APC protein regulates beta-catenin (CTNNB1; 116806) levels in normal cells, and how mutations in APC or CTNNB1 in cancer cell genes deregulate cell growth via T-cell transcription factor 4 (TCF4; 602228).

As outlined earlier, inactivating mutations of APC, which are found in most colorectal cancers, cause aberrant accumulation of beta-catenin, which then binds TCF4, causing increased transcriptional activation of unknown genes. He et al. (1998) identified the c-myc (190080) oncogene as a target gene in this signaling pathway. Expression of MYC was shown to be repressed by wildtype APC and activated by beta-catenin, and these effects were mediated through TCF4 binding sites in the MYC promoter. These results provided a molecular framework for understanding the previously enigmatic overexpression of MYC in colorectal cancers.

Homfray et al. (1998) examined the question of whether APC mutations or mutations leading to genomic instability are the initiating events in colonic tumorigenesis. If defective mismatch repair (MMR) precedes APC mutations, the APC mutations of colon tumors with defective MMR and hence replication errors (RER+) should differ from those of RER- tumors, in at least 3 specific ways: (1) a higher frequency of allele loss at APC in RER- tumors; (2) more frameshift than nonsense mutations in RER+ tumors; and (3) APC mutations in simple repeat sequences in RER+ tumors. Homfray et al. (1998) found no evidence that sporadic RER+ and RER- colon cancers (including cell lines) differ in any of these 3 ways. Although it remained possible that MMR is abnormal in tumors from HNPCC families before APC mutations occur, it is likely that in sporadic colon tumors, APC mutations, rather than genomic instability, are the initiating events in tumorigenesis.

As already noted, a subset of APC mutations, the so-called attenuated or APC(AP) alleles, predispose to only a few colon polyps (Spirio et al., 1993; see 175100.0021). This leads to the hypothesis that if mutation of the inherited normal allele is rate limiting in polyp development (according to the Knudson 2-hit model), the increased number of polyps associated with the usual APC alleles indicates that the frequency of mutations that can lead to polyp formation is higher among carriers of such mutations than among APC(AP) carriers. Spirio et al. (1998) suggested that the APC protein may modulate the frequency of mutations, such as loss of heterozygosity (LOH), necessary for colon polyp formation. Spirio et al. (1998) thus reasoned that tumors from patients who carry an APC(AP) allele may show a reduced frequency of LOH compared with tumors from patients who carry an APC allele of the conventional type. Indeed, they found, in screening of tumors from APC(AP) carriers, a reduction of LOH compared with that of an unselected group of polyposis patients. In fact, no loss of the inherited normal APC allele was observed, although sequencing showed that the inherited normal APC allele had frequently undergone point mutations and small deletions in the tumors. A low frequency loss of the inherited APC(AP) allele was seen. These findings supported the suggestion that the APC(AP) allele has residual gene activity and that this activity modulates the spectrum and frequency of mutations that lead to adenoma formation.

Diagnosis

Thakker et al. (1995) presented as a diagnostic tool a weighted scoring system for changes on dental panoramic radiographs called the Dental Panoramic Radiographs Score (DPRS). The score took into account the nature, extent, and sight of osseous and dental changes, as well as the incidence of the anomaly in the general population. Using the highest threshold, a specificity of 100% and sensitivity of approximately 68% were obtained. If all positive findings were considered as significant, sensitivity was increased to approximately 82%, but the specificity was reduced to approximately 88%. Overall, approximately 68% of the affected subjects had significant changes, and approximately 18% had normal appearance on DPR, with the remainder having changes classified as minimal or equivocal.

Tops et al. (1989) showed that 2 linked polymorphic DNA markers closely flank the APC locus, one on either side. They estimated that use of these markers allows prenatal and presymptomatic diagnosis with more than 99.9% reliability in most families. Dunlop et al. (1990) described 6 DNA markers flanking the APC gene that are useful for presymptomatic diagnosis. In 1991, Dunlop et al. performed presymptomatic analysis of DNA from 41 individuals at risk for FPC. Of these, 28 individuals were informative, and 14 whose probe-derived risk was greater than 0.93 were subsequently demonstrated to be affected by clinical screening. The authors suggested that an integrated risk analysis, including genotypic, colonic, and ophthalmologic evaluation for the presence of CHRPE, should be used in FPC screening programs. Cachon-Gonzalez et al. (1991) concluded on the basis of linkage studies using 4 DNA probes that presymptomatic diagnosis can be given with only 90% probability based on DNA typing alone. This level of risk proved acceptable to one couple requesting prenatal diagnosis where the initial request had been for sterilization.

Petersen et al. (1993) demonstrated the feasibility of presymptomatic direct detection of APC mutations in each of 4 families. No change in the conventional FAP colon screening regimen was recommended for children found to have a mutation. In contrast, when direct tests indicated that an individual did not have the mutation, they recommended that screening be decreased. Three of the mutations were nonsense mutations and one was a frameshift mutation due to insertion of 1 nucleotide. In an evaluation of molecular genetic diagnosis in the management of familial polyposis, Maher et al. (1993) concluded that intragenic and closely linked DNA markers are informative in most families and that, in addition to the clinical benefits of presymptomatic diagnosis, the reduction in screening for low-risk relatives means that molecular genetic diagnosis is a cost-effective procedure. In their population based study, they estimated a minimum heterozygote prevalence of 1/26,000. Of 33 probands, 8 (24%) represented new mutations. Ophthalmologic assessment showed more than 3 CHRPEs in 27 of 43 (63%) affected patients and high-risk relatives and none of 18 low-risk relatives. Interfamilial variation in CHRPE expression was evident.

Caspari et al. (1994) and Gayther et al. (1994) found that patients with the 5-bp deletion at codon 1309 had gastrointestinal symptoms and death from colorectal cancer that occurred about 10 years earlier than in patients with other mutations. In contrast, mutations near the 5-prime end of the APC gene resulted in an attenuated form of FAP; these patients had relatively few colonic polyps and a later onset of symptoms (Wallis et al., 1994). Spirio et al. (1993) and Olschwang et al. (1993) found that the patients with mutations in codons 136 to 302 of the APC gene did not develop CHRPE, whereas those with mutations in a region beyond exon 9 and up to codon 1387 of the gene presented with ophthalmic lesions. Patients with mutations in exon 9 either did or did not have CHRPE. Caspari et al. (1995) demonstrated that patients with a mutation between codons 1445 and 1578 do not express CHRPE; however, these patients developed severe desmoid tumors. Their conclusion was based on observations in 36 patients from 20 families. With the exception of 3 prepubertal children, all patients with mutations in codons 1445-1578 developed desmoid tumors.

Davies et al. (1995) found that families with mutations 3-prime of codon 1444 had significantly more lesions on dental panoramic radiographs (P less than 0.001) and appeared to have a higher incidence of desmoid tumors than did families with mutations at the 5-prime end. All 7 families except one with mutations 5-prime of exon 9 did not express CHRPE. All of 38 individuals from 16 families with mutations between exon 9 and codon 1444 expressed CHRPE. The 11 individuals from 4 families with mutations 3-prime of codon 1444 did not express CHRPE. These results suggested that the severity of some of the features of Gardner syndrome may correlate with genotype in FAP.

Evans et al. (1993) likewise reported families with an attenuated form of FAP. In 1 family, a 59-year-old patient showed no abnormality; late onset of polyps was discovered by endoscopy and biopsy in other members of that family and in 2 other families. Mutation analysis in these families was not reported.

Soravia et al. (1998) defined attenuated adenomatous polyposis coli by the occurrence of less than 100 colonic adenomas and a later onset of colorectal cancer (age greater than 40 years). By protein-truncation test (PTT) assay, they screened the entire coding region of the APC gene in affected individuals from 11 AAPC kindreds, and their phenotypic differences were examined. They identified 5 novel germline APC mutations in 7 kindreds. The mutations were located in 3 different regions of the APC gene: (1) at the 5-prime end spanning exons 4 and 5, (2) within exon 9, and (3) at the 3-prime distal end of the gene. Variability in the number of colorectal adenomas was most apparent in individuals with mutations in region 1, and upper-gastrointestinal manifestations were more severe in them. In individuals with mutations in either region 2 or region 3, the average number of adenomas tended to be lower than that in individuals with mutations in region 1, although age at diagnosis was similar. In all AAPC kindreds, a predominance of right-sided colorectal adenomas and rectal polyp sparing was observed. No desmoid tumors were found in these kindreds. A rarity of rectal polyps in AAPC was noted; however, further clinical studies with longer follow-up were considered necessary to prove that the risk of rectal cancer is lower in AAPC than in classic FAP. The results were considered relevant for the clinical management of AAPC kindreds: colonoscopy as opposed to sigmoidoscopy should be advised for endoscopic surveillance because of the right-sided location of colorectal adenomas; upper GI endoscopic surveillance is warranted for detection of premalignant gastric or duodenal tumors; and individuals affected with AAPC may require total colectomy with ileorectal anastomosis only when prophylactic colectomy is advised. Lynch and Smyrk (1998) stated that multiple fundic gland polyps had preceded the finding of pathology in the colon in a relatively large number of their patients with attenuated FAP.

The widespread distribution of many different mutations within the very large APC gene represents a problem for molecular genetic diagnosis. Powell et al. (1993) presented a novel approach for the rapid and sensitive detection of APC mutations. Because most mutations in patients with familial adenomatous polyposis result in truncation of the APC gene product due to frameshift, nonsense mutation, or splice site changes, they developed a method based on the examination of APC proteins synthesized in vitro and study of endogenous APC transcripts. In 62 unrelated patients from the Johns Hopkins Familial Adenomatous Polyposis Registry, primary screening was accomplished by analysis of protein synthesized in vitro from surrogate APC genes. In addition, the relative amount of transcript from each APC allele was determined with an allele-specific expression assay. The first of these assays revealed truncated protein in 51 of the 62 patients (82%). In 3 of the 11 remaining patients, the allele-specific expression assay demonstrated significantly reduced expression of one allele of the APC gene. Use of the 2 assays in combination successfully identified germline APC mutations in 87% of the 62 patients. A so-called protein truncation test, based on the in vitro transcription and translation of genomic PCR products, was developed also by van der Luijt et al. (1994).

Friedl et al. (1996) reported 2 families with a rather mild phenotype of FAP due to a frameshift mutation at codon 1597. They suggested that these families may represent a clue for defining a 5-prime border for the occurrence of a second region of attenuated FAP that is located in the 3-prime part of the APC gene.

The use of commercially available tests for genes linked to familial cancer is a source of concern about the possible adverse impact on patients. Giardiello et al. (1997) assessed indications for APC gene testing, whether informed consent was obtained and genetic counseling was offered before testing, and the interpretation of the results, through telephone interviews with physicians and genetic counselors in a nationwide sample of 177 patients from 125 patients who underwent testing during 1995. Of the 177 patients tested, 83% had clinical features of FAP or were at risk for the disease—both valid indications for testing. Only 18.6% (33 of 177) received genetic counseling before the tests, and only 16.9% (28 of 166) provided written informed consent. In 31.6% of the cases, the physicians misinterpreted the test results. Among the patients with unconventional indications for testing, the rate of positive results was only 2.3% (1 of 44). Giardiello et al. (1997) concluded that physicians should be prepared to offer genetic counseling if they order genetic tests.

Since an alteration of the APC gene occurs early in most colorectal tumors, detection of APC mutations in fecal tumor DNA could be a powerful tool for the diagnosis of noninvasive cancer. Deuter and Muller (1998) described a highly sensitive and nonradioactive heteroduplex-PCR method (HD-PCR) for detecting APC mutations in stool DNA.

Clinical Management

Petersen et al. (1989) demonstrated how one could use linkage information to modify the standard recommendations for follow-up. For example, in the family of an affected 36-year-old man with a positive family history of APC, there were 4 asymptomatic children under the age of 10 years. Before linkage analysis, all children had a 50% risk. Screening protocols would call for annual sigmoidoscopy in all beginning at age 12 years. With the linkage information, one could state to the family with 98% confidence that 3 of the children did not inherit the gene and that 1 child did. That child could be screened annually; the others would have screening every 3 years beginning at ages 12 or 13 and continuing until age 35.

Morton et al. (1992) demonstrated that DNA extracted from preserved tissue of dead relatives could be used to extend informativeness in APC families.

Waddell and Loughry (1983) were the first to make a connection between nonsteroidal antiinflammatory drugs (NSAIDs) and colon cancer. Astutely, they observed the disappearance of rectal polyps in a patient with Gardner syndrome and correctly attributed this disappearance to treatment with sulindac, an NSAID that was given for unrelated reasons. In a random, double-blind, placebo-controlled study of 22 patients with familial adenomatous polyposis, including 18 who had not undergone colectomy, Giardiello et al. (1993) found that sulindac, at a dose of 150 mg orally twice a day for 9 months, reduced the number and size of colorectal adenomas. The effect was incomplete, however, leading the authors to conclude that it is unlikely to replace colectomy as primary therapy. However, the effect has been confirmed in numerous clinical trials (Marcus, 1995) and the interpretation that NSAIDs promote regression of colon polyps by inhibiting prostaglandin synthesis has been supported by genetic studies in mice (Oshima et al., 1996).

It had been reported by Giovannucci et al. (1994) and others that patients who take aspirin or other NSAIDs on a regular basis have a 40 to 50% lower relative risk of colorectal cancer when compared with persons not regularly taking these medications. In studying the molecular basis of this observation, Tsujii and DuBois (1995) showed that overexpression of cyclooxygenase 2 (COX2; 600262) in rat intestinal epithelial cells resulted in increased adhesion to extracellular matrix and resistance to butyrate-induced apoptosis. These phenotypic changes that might enhance tumorigenic potential were reversed by sulindac sulfide, a COX inhibitor.

Schnitzler et al. (1996) demonstrated in an in vitro study that sulindac inhibits the growth of carcinoma cells and causes an increase in APC mRNA. They suggested that the effect of these agents on colonic carcinogenesis is not mediated entirely by means of an inhibition of prostaglandin biosynthesis. Boolbol et al. (1996) found that histologically normal-appearing small bowel from mice carrying the Min mutation showed increased amounts of Cox-2 and prostaglandin E(2) compared to +/+ littermates. Using 2 different in situ techniques, terminal transferase-mediated dUTP nick end labeling and a direct immunoperoxidase method, Min animals also demonstrated a 27%-47% decrease in enterocyte apoptosis compared to +/+ animals. Treatment with sulindac not only inhibited tumor formation but decreased small bowel Cox2 and prostaglandin E(2) to baseline and restored normal levels of apoptosis.

Prescott and White (1996) reviewed the intimate connections between APC and prostaglandin H synthase-2 (PTGS2; 600262). Oshima et al. (1996) provided key evidence on the role of COX2 (which is encoded by the Ptgs2 gene) in colon carcinogenesis. They bred mice carrying an APC mutation with a mouse with a disrupted Ptgs2 gene. All of the animals were heterozygous at the Apc locus; if homozygous for wildtype COX2, they developed an average of 652 polyps at 10 weeks, while heterozygotes had 224 polyps and homozygously deficient mice had only 93 polyps. This experiment provided definitive genetic evidence that induction of COX2 is an early, rate-limiting step for adenoma formation. As supporting evidence, a drug that inhibited COX2 but not COX1 (176805) also markedly reduced the number of polyps. Thus, Oshima et al. (1996) added the PTGS2 gene to the list of genes involved in colon neoplasia. The findings suggested to Prescott and White (1996) that drugs which inhibit COX2 should be broadly effective in chemoprevention of colon cancer. Prescott and White (1996) noted that among the questions remaining are the following: Will specific inhibitors of COX2 be more effective than those that inhibit both isoforms? Will inhibition of COX-2 be as effective in sporadic polyps and hereditary nonpolyposis colon cancer as in FAP patients? How should chemoprevention with COX inhibitors be integrated with current surveillance and intervention protocols?

In connection with the effectiveness and cost of periodic screening in persons at risk for familial polyposis coli, the study of Maule (1994) is of interest: he found that nurses could carry out screening by flexible sigmoidoscopy as accurately and safely as experienced gastroenterologists. (Screening in the case of type I familial nonpolyposis colorectal cancer (120435) will require full-length fiberoptic colonoscopy because of the predominance of lesions in the right colon.) Kassirer (1994), while accepting the usefulness of many more nurse practitioners in collaborative practice arrangements, doubted that their function as entirely independent practitioners of primary care would be either cost-effective or safe.

Westbrook et al. (1994) explored the potential for gene therapy by studying the transient expression of the APC gene in normal rat colonic epithelium, using liposomal gene delivery by rectal catheter infusion. Expression of a beta-galactosidase reporter gene and of the human APC gene under a constitutive promoter was demonstrated. Close to 100% of epithelial cells expressed the introduced gene. Expression was transient and did not persist beyond 4 days, consistent with the normal turnover time of gut epithelium, but it could be maintained by repeated treatments. Human APC was expressed for 3 weeks under these conditions at approximately one-tenth the level of the endogenous APC gene, and no toxicity was observed beyond that attributed to repeated rectal enemas.

Population Genetics

In the Johns Hopkins Hospital Colon Polyposis Registry, established in 1973 and covering 6 states and the District of Columbia, 98 Gardner syndrome kindreds and 47 APC kindreds were recorded by April 1988. (The Peutz-Jeghers syndrome (175200) was registered in 19 kindreds.)

Burn et al. (1991) estimated the prevalence of APC as 2.29 x 10(-5) in the northern region of England. Bisgaard et al. (1994) reported results based on a nationwide Danish polyposis registry that included all known Danish cases of FAP and their relatives. By identifying all FAP patients born between 1920 and 1949, they found the frequency of the disease to be 1 in 13,528. By comparing the number of affected and nonaffected offspring born to affected parents during the same period, they concluded that penetrance of the disease for inherited cases was close to 100% by age 40 years. The mutation rate found by the direct method was 9 mutations per million gametes per generation, and the proportion of new mutants was estimated to be 25%. Fitness for patients between 15 and 29 years was found to be close to 1, while for patients older than 30 the fitness was reduced. Fitness increased over the 3 decades from date of birth (from 0.44 to 0.71), probably because treatment became more widespread and effective. When Bisgaard et al. (1994) used the overall fitness in the period, 0.87, to estimate the mutation rate by the indirect method, they found a lower value than by the direct method, namely, 5 mutations per million gametes per generation.

Animal Model

Lander (1991) used his microsatellite map to locate the Min (multiple intestinal neoplasia) gene to mouse chromosome 18. The extensive homology of synteny between mouse chromosome 18 and human chromosome 5 suggests that Min corresponds to APC in the human. Luongo et al. (1993) showed that Min, a mutant allele of Apc, maps to proximal mouse chromosome 18. Thus, Apc and Mcc (159350) are syntenic in both mouse and human, although the gene order in the interval between the 2 genes is different in the 2 species. Su et al. (1992) showed specifically that the murine homolog of the APC gene (Apc) is tightly linked to the phenotypically defined Min locus. Furthermore, the authenticity of Min as a model of adenomatous polyposis of the colon was clinched by the demonstration of a nonsense mutation in the gene: a transversion from T to A at nucleotide 2549, converting codon 850 from leucine (TTG) to a stop (TAG) codon. Fodde et al. (1994) introduced a chain-termination mutation in the fifteenth exon of the mouse Apc gene and used it to modify the endogenous gene by homologous recombination in embryonic stem cells. Mice heterozygous for the Apc gene modification progressively developed intestinal tumors in a manner similar to that observed in patients with FAP and in mice carrying the Min mutation.

Moser et al. (1993) showed that female mice carrying the Min mutation are also prone to develop mammary tumors. Min/+ mammary glands are more sensitive to chemical carcinogenesis than are +/+ mammary glands. Transplantation of mammary cells from Min/+ or +/+ donors into +/+ hosts demonstrated that the propensity to develop mammary tumors is intrinsic to the Min/+ mammary cells.

Dietrich et al. (1993) reported the genetic mapping of a locus that strongly modifies tumor number in Min/+ animals. This gene, Mom-1 ('modifier of Min-1'), mapped to distal chromosome 4 and controlled about 50% of genetic variation in tumor number in 2 intraspecific backcrosses. It was found to lie in the region of synteny conservation with human chromosome 1p36-p35, a region of frequent somatic loss of heterozygosity in a variety of human tumors, including colon tumors (see 172411).

To investigate the molecular mechanism of polyp formation followed by carcinogenesis in the digestive tract of Min mice, Oshima et al. (1995) used homologous recombination in embryonic stem (ES) cells and generated mice containing a mutant gene encoding a product truncated at residue 716 of the 2845 amino acid wildtype gene product. Mendelian transmission of the gene caused most homozygous mice to die in utero before day 8 of gestation. The heterozygotes developed multiple polyps throughout the intestinal tract, mostly in the small intestine. The earliest polyps arose multifocally during the third week after birth, and new polyps continued to appear thereafter. Surprisingly, every nascent polyp consisted of a microadenoma covered with a layer of the normal villous epithelium. These microadenomas originated from single crypts by forming abnormal outpockets into the inner (lacteal) side of the neighboring villi. Oshima et al. (1995) dissected each microadenoma from the nascent polyps by peeling off the normal epithelium and determined their genotype by PCR. They found that all microadenomas had already lost the wildtype Apc allele, whereas the mutant allele remained unchanged. These results indicated the loss of heterozygosity followed by formation of intravillous microadenomas is responsible for polyposis in the intestinal mucosa. It is unlikely that the truncated product interacts with the wildtype protein and causes the microadenomas by a dominant-negative mechanism. Earlier studies had suggested that polyps could arise as a result of heterozygosity for an APC mutation without the need for a further genetic event, and it had been suggested that this was the earliest event in the development of all, or at least most, colon cancer.

Takaku et al. (1998) inactivated the mouse Dpc4 (Smad4) gene, which is homologous to human SMAD4 (600993). The homozygous mutants were embryonic lethal, whereas the heterozygotes showed no abnormality. These investigators then introduced the Dpc4 mutation into the knockout mice for the mouse Apc-delta716 gene. Because both Apc and Dpc4 are located on mouse chromosome 18, they constructed compound heterozygotes carrying both mutations on the same chromosome by meiotic recombination. In such mice, intestinal polyps developed into more malignant tumors than those in the simple Apc-delta716 heterozygotes, showing an extensive stromal cell proliferation, submucosal invasion, cell type heterogeneity, and in vivo transplantability. Takaku et al. (1998) suggested that mutations in DPC4 (SMAD4) play a significant role in the malignant progression of colorectal tumors.

Differences in colon cancer incidence worldwide led to the hypothesis that this variation can be explained largely by environmental, and notably dietary, influences. Wasan et al. (1997) stated that, although a positive correlation between dietary fat intake and incidence was suggested by some human epidemiologic and rodent carcinogenic studies, a direct association remained to be established. Using the multiple intestinal neoplasia (MIN) mouse which undergoes the development of colon cancer spontaneously, Wasan et al. (1997) demonstrated that there is a general increase in tumor counts in both the large and the small bowel with higher dietary fat. Increasing dietary fat also increased polyp size in the small bowel. These changes appeared independent of total calorific intake as assessed by body weights. Halving the crude fiber intake together with an increase in dietary fat from 3% to 10% did not have as marked an effect on tumor counts as did an increase of fat alone to 15%, which also decreased survival.

History

Gardner (1972) gave a personal account of the discovery of the Gardner syndrome. He was introduced to the large Utah family with colonic polyposis by a premedical student who was in his course in genetics in 1947. This family was the basis of the report on polyposis by Gardner (1951). He also studied multiple osteomas and described the pattern of autosomal dominant inheritance and stated that 'as a working hypothesis the same gene is postulated to influence both abnormalities,' polyposis and osteomas (Gardner and Plenk, 1952). In the period 1950 to 1953, 4 kinds of abnormal growths (multiple intestinal polyposis, osteomas, fibromas, and sebaceous cysts) were observed in the same family members. Desmoid tumors, dental abnormalities, carcinoma of the ampulla of Vater, and thyroid carcinoma were later described. Smith (1958) observed desmoid tumors and postoperative scars in polyposis patients and initiated the designation Gardner syndrome.

De Vries et al. (1996) described an on-line database of mutations in the APC gene. Beroud and Soussi (1996) described a database of germline and somatic mutations in the APC gene in human tumors and cell lines. The database held 737 mutations of which 332 were germline and 402 somatic.

See also Benecke (1931); Berk et al. (1981); Butson (1983); Chang et al. (1968); Danes (1975); Danes (1976); Denzler et al. (1979); Duhamel et al. (1960); Dunlop et al. (1991); Gorlin and Chaudhry (1960); Greer et al. (1977); Haggitt and Booth (1970); Hyson and Burrell (1976); Ingram and Oldfield (1937); Kaplan et al. (1982); Leffall et al. (1977); Leppard and Bussey (1975); Leppert et al. (1987); Lewis and Mitchell (1971); MacDonald et al. (1967); McKusick (1962); Miyoshi et al. (1992); Murphy et al. (1980); Nagase et al. (1992); Naylor and Gardner (1977); Naylor and Lebenthal (1980); Pavlides et al. (1977); Phillips (1981); Pierce (1972); Rider et al. (1986); Sanchez et al. (1979); Savage (1964); Schneider et al. (1983); Schnur et al. (1973); Shemesh (1985); Sugihara et al. (1982); Thompson et al. (1983); Vanhoutte (1970); Veale (1960); Watanabe et al. (1978); Yonemoto et al. (1969)

Armstrong, J. G.; Davies, D. R.; Guy, S. P.; Frayling, I. M.; Evans, D. G. R.: APC mutations in familial adenomatous polyposis families in the northwest of England. Hum. Mutat. 10: 376-380, 1997.

Ashton-Rickardt, P. G.; Dunlop, M. G.; Nakamura, Y.; Morris, R. G.; Purdie, C. A.; Steel, C. M.; Evans, H. J.; Bird, C. C.; Wyllie, A. H.: High frequency of APC loss in sporadic colorectal carcinoma due to breaks clustered in 5q21-22. Oncogene 4: 1169-1174, 1989.

Asman, H. B.; Pierce, E. R.: Familial multiple polyposis: a statistical study of a large Kentucky kindred. Cancer 25: 972-981, 1970.

Baker, R. H.; Heinemann, M.-H.; Miller, H. H.; DeCosse, J. J.: Hyperpigmented lesions of the retinal pigment epithelium in familial adenomatous polyposis. Am. J. Med. Genet. 31: 427-435, 1988.
Bapat, B.; Odze, R.; Mitri, A.; Berk, T.; Ward, M.; Gallinger, S.: Identification of somatic APC gene mutations in periampullary adenomas in a patient with familial adenomatous polyposis (FAP). Hum. Molec. Genet. 2: 1957-1959, 1993.
Barber, J. C. K.; Ellis, K. H.; Bowles, L. V.; Delhanty, J. D. A.; Ede, R. F.; Male, B. M.; Eccles, D. M.: Adenomatous polyposis coli and a cytogenetic deletion of chromosome 5 resulting from a maternal intrachromosomal insertion. J. Med. Genet. 31: 312-316, 1994.
Bell, B.; Mazzaferri, E. L.: Familial adenomatous polyposis (Gardner's syndrome) and thyroid carcinoma: a case report and review of the literature. Digest. Dis. Sci. 38: 185-190, 1993.
Benecke, E.: Ueber Epitheliome auf Atheromen (Epidermoide) und Dermoidcysten der Haut. Frankf. Z. Pathol. 42: 502-515, 1931.
Berk, T.; Cohen, Z.; Cullen, J. B.: Familial polyposis and the role of the preventive registry. Canad. Med. Assoc. J. 124: 1427-1428, 1981.
Beroud, C.; Soussi, T.: APC gene: database of germline and somatic mutations in human tumors and cell lines. Nucleic Acids Res. 24: 121-124, 1996.
Bisgaard, M. L.; Fenger, K.; Bulow, S.; Niebuhr, E.; Mohr, J.: Familial adenomatous polyposis (FAP): frequency, penetrance, and mutation rate. Hum. Mutat. 3: 121-125, 1994.
Blair, N. P.; Trempe, C. L.: Hypertrophy of the retinal pigment epithelium associated with Gardner's syndrome. Am. J. Ophthal. 90: 661-667, 1980.
Bodmer, W. F.; Bailey, C. J.; Bodmer, J.; Bussey, H. J. R.; Ellis, A.; Gorman, P.; Lucibello, F. C.; Murday, V. A.; Rider, S. H.; Scambler, P.; Sheer, D.; Solomon, E.; Spurr, N. K.: Localization of the gene for familial adenomatous polyposis on chromosome 5. Nature 328: 614-616, 1987.
Boland, C. R.; Sato, J.; Appelman, H. D.; Bresalier, R. S.; Feinberg, A. P.: Microallelotyping defines the sequence and tempo of allelic losses at tumour suppressor gene loci during colorectal cancer progression. Nature Med. 1: 902-909, 1995.
Boolbol, S. K.; Dannenberg, A. J.; Chadburn, A.; Martucci, C.; Guo, X.-J.; Ramonetti, J. T.; Abreu-Goris, M.; Newmark, H. L.; Lipkin, M. L.; DeCosse, J. J.; Bertagnolli, M. M.: Cyclooxygenase-2 overexpression and tumor formation are blocked by sulindac in a murine model of familial adenomatous polyposis. Cancer Res. 56: 2556-2560, 1996.
Botstein, D.; White, R. L.; Skolnick, M.; Davis, R. W.: Construction of a genetic linkage map in man using restriction fragment length polymorphisms. Am. J. Hum. Genet. 32: 314-331, 1980.
Bull, M. J.; Ellis, F. D.; Sato, S.; Weaver, D. D.: Hypertrophy of retinal pigment epithelium in Gardner syndrome. (Abstract) Proc. Greenwood Genet. Center 4: 136 only, 1985.
Bunyan, D. J.; Shea-Simonds, J.; Reck, A. C.; Finnis, D.; Eccles, D. M.: Genotype-phenotype correlations of new causative APC gene mutations in patients with familial adenomatous polyposis. J. Med. Genet. 32: 728-731, 1995.
Burn, J.; Chapman, P.; Delhanty, J.; Wood, C.; Lalloo, F.; Cachon-Gonzalez, M. B.; Tsioupra, K.; Church, W.; Rhodes, M.; Gunn, A.: The UK northern region genetic register for familial adenomatous polyposis coli: use of age of onset, congenital hypertrophy of the retinal pigment epithelium, and DNA markers in risk calculations. J. Med. Genet. 28: 289-296, 1991.
Burt, R. W.; Berenson, M. M.; Lee, R. G.; Tolman, K. G.; Freston, J. W.; Gardner, E. J.: Upper gastrointestinal polyps in Gardner's syndrome. Gastroenterology 86: 295-301, 1984.
Butson, A. R. C.: Familial multiple polyposis coli with multiple associated tumors. Dis. Colon Rectum 26: 578-582, 1983.
Cachon-Gonzalez, M. B.; Delhanty, J. D. A.; Burn, J.; Tsioupra, K.; Davis, M. B.; Attwood, J.; Chapman, P.: Linkage analysis in adenomatous polyposis coli: the use of four closely linked DNA probes in 20 UK families. J. Med. Genet. 28: 681-685, 1991.
Cama, A.; Esposito, D. L.; Palmirotta, R.; Curia, M. C.; Ranieri, A.; Ficari, F.; Valanzano, R.; Modesti, A.; Battista, P.; Tonelli, F.; Mariani-Costantini, R.: A novel mutation at the splice junction of exon 9 of the APC gene in familial adenomatous polyposis. Hum. Mutat. 3: 305-308, 1994.
Camiel, M. R.; Mule, J. E.; Alexander, L. L.; Benninghoff, D. L.: Association of thyroid carcinoma with Gardner's syndrome in siblings. New Eng. J. Med. 278: 1056-1058, 1968.
Capps, W. F., Jr.; Lewis, M. I.; Gazzaniga, D. A.: Carcinoma of the colon, ampulla of Vater and urinary bladder associated with familial multiple polyposis: a case report. Dis. Colon Rectum 11: 298-304, 1968.
Carl, W.; Herrera, L.: Dental and bone abnormalities in patients with familial polyposis coli. Sem. Surg. Oncol 3: 77-83, 1987.
Caspari, R.; Friedl, W.; Mandl, M.; Moslein, G.; Kadmon, M.; Knapp, M.; Jacobasch, K.-H.; Ecker, K.-W.; Kreissler-Haag, D.; Timmermanns, G.; Propping, P.: Familial adenomatous polyposis: mutation at codon 1309 and early onset of colon cancer. Lancet 343: 629-632, 1994.
Caspari, R.; Olschwang, S.; Friedl, W.; Mandl, M.; Boisson, C.; Boker, T.; Augustin, A.; Kadmon, M.; Moslein, G.; Thomas, G.; Propping, P.: Familial adenomatous polyposis: desmoid tumours and lack of ophthalmic lesions (CHRPE) associated with APC mutations beyond codon 1444. Hum. Molec. Genet. 4: 337-340, 1995.
Chang, C. H.; Platt, E. D.; Thomas, K. E.; Watne, A. L.: Bone abnormalities in Gardner's syndrome. Am. J. Roentgen. 103: 645-652, 1968.
Chapman, P. D.; Church, W.; Burn, J.; Gunn, A.: Congenital hypertrophy of retinal pigment epithelium: a sign of familial adenomatous polyposis. Brit. Med. J. 298: 353-354, 1989.
Chen, S.; Kazim, D.; Kraveka, J.; Pollack, R. E.: Skin fibroblasts from individuals hemizygous for the familial adenopolyposis susceptibility gene show delayed crisis in vitro. Proc. Nat. Acad. Sci. 86: 2008-2012, 1989.
Cottrell, S.; Bicknell, D.; Kaklamanis, L.; Bodmer, W. F.: Molecular analysis of APC mutations in familial adenomatous polyposis and sporadic colon carcinomas. Lancet 340: 626-630, 1992.
Cross, I.; Delhanty, J.; Chapman, P.; Bowles, L. V.; Griffin, D.; Wolstenholme, J.; Bradburn, M.; Brown, J.; Wood, C.; Gunn, A.; Burn, J.: An intrachromosomal insertion causing 5q22 deletion and familial adenomatous polyposis coli in two generations. J. Med. Genet. 29: 175-179, 1992.
Curia, M. C.; Esposito, D. L.; Aceto, G.; Palmirotta, R.; Crognale, S.; Valanzano, R.; Ficari, F.; Tonelli, F.; Battista, P.; Mariani-Costantini, R.; Cama, A.: Transcript dosage effect in familial adenomatous polyposis: model offered by two kindreds with exon 9 APC gene mutations. Hum. Mutat. 11: 197-201, 1998.
Danes, B. S.: The Gardner syndrome. Cancer 36: 2327-2333, 1975.
Danes, B. S.: The Gardner syndrome: increased tetraploidy in cultured skin fibroblast. J. Med. Genet. 13: 52-56, 1976.
Danes, B. S.; Gardner, E. J.: The Gardner syndrome: a cell culture study of kindred 109. J. Med. Genet. 15: 346-551, 1978.
Davies, D. R.; Armstrong, J. G.; Thakker, N.; Horner, K.; Guy, S. P.; Clancy, T.; Sloan, P.; Blair, V.; Dodd, C.; Warnes, T. W.; Harris, R.; Evans, D. G. R.: Severe Gardner syndrome in families with mutations restricted to a specific region of the APC gene. Am. J. Hum. Genet. 57: 1151-1158, 1995.
Denzler, T. B.; Harned, R. K.; Pergam, C. J.: Gastric polyps in familial polyposis coli. Radiology 130: 63-66, 1979.
Deuter, R.; Muller, O.: Detection of APC mutations in stool DNA of patients with colorectal cancer by HD-PCR. Hum. Mutat. 11: 84-89, 1998.
De Vries, E. M. G.; Ricke, D. O.; De Vries, T. N.; Hartmann, A.; Blaszyk, H.; Liao, D.; Soussi, T.; Kovach, J. S.; Sommer, S. S.: Database of mutations in the p53 and APC tumor suppressor genes designed to facilitate molecular epidemiological analyses. Hum. Mutat. 7: 202-213, 1996.
Diaz-Llopis, M.; Menezo, J. L.: Congenital hypertrophy of the retinal pigment epithelium in familial adenomatous polyposis. Arch. Ophthal. 106: 412-413, 1988.
Dietrich, W. F.; Lander, E. S.; Smith, J. S.; Moser, A. R.; Gould, K. A.; Luongo, C.; Borenstein, N.; Dove, W.: Genetic identification of Mom-1, a major modifier locus affecting Min-induced intestinal neoplasia in the mouse. Cell 75: 631-639, 1993.
Distante, S.; Nasioulas, S.; Somers, G. R.; Cameron, D. J. S.; Young, M. A.; Forrest, S. M.; Gardner, R. J. M.: Familial adenomatous polyposis in a 5 year old child: a clinical, pathological, and molecular genetic study. J. Med. Genet. 33: 157-160, 1996.
Duhamel, J.; Berthon, G.; Dubarry, J. J.: Etude mathematique de l'heredite de la polypose recto-colique. J. Genet. Hum. 9: 65-77, 1960.
Dunlop, M. G.; Wyllie, A. H.; Nakamura, Y.; Steel, C. M.; Evans, H. J.; White, R. L.; Bird, C. C.: Genetic linkage map of six polymorphic DNA markers around the gene for familial adenomatous polyposis on chromosome 5. Am. J. Hum. Genet. 47: 982-987, 1990.
Dunlop, M. G.; Wyllie, A. H.; Steel, C. M.; Piris, J.; Evans, H. J.: Linked DNA markers for presymptomatic diagnosis of familial adenomatous polyposis. Lancet 337: 313-316, 1991.
Eccles, D. M.; Lunt, P. W.; Wallis, Y.; Griffiths, M.; Sandhu, B.; McKay, S.; Morton, D.; Shea-Simonds, J.; Macdonald, F.: An unusually severe phenotype for familial adenomatous polyposis. Arch. Dis. Child. 77: 431-435, 1997.
Eccles, D. M.; van der Luijt, R.; Breukel, C.; Bullman, H.; Bunyan, D.; Fisher, A.; Barber, J.; du Boulay, C.; Primrose, J.; Burn, J.; Fodde, R.: Hereditary desmoid disease due to a frameshift mutation at codon 1924 of the APC gene. Am. J. Hum. Genet. 59: 1193-1201, 1996.
Endo, A.; Kasukawa, T.: Gardner syndrome and interstitial chromosome deletion. (Letter) Am. J. Med. Genet. 28: 511-512, 1987.
Erbe, R. W.; Welch, W. R.: Case records of the Massachusetts General Hospital. Weekly clinicopathological exercises. Case 47-1978. New Eng. J. Med. 299: 1237-1245, 1978.
Evans, D. G. R.; Guy, S. P.; Thakker, N.; Armstrong, J. G.; Dodd, C.; Davies, D. R.; Babbs, C.; Clancy, T.; Warnes, T.; Sloan, P.; Taylor, T. V.; Harris, R.: Non-penetrance and late appearance of polyps in families with familial adenomatous polyposis. Gut 34: 1389-1393, 1993.
Fader, M.; Kline, S. N.; Spatz, S. S.; Zubrow, H. J.: Gardner's syndrome (intestinal polyposis, osteomas, sebaceous cysts) and a new dental discovery. Oral Surg. 15: 153-172, 1962.
Fearon, E. R.: Human cancer syndromes: clues to the origin and nature of cancer. Science 278: 1043-1050, 1997.
Fearon, E. R.; Vogelstein, B.: A genetic model for colorectal tumorigenesis. Cell 61: 759-767, 1990.
Fialkow, P. J.: Clonal origin of human tumors. Ann. Rev. Med. 30: 135-143, 1979.
Fineman, R. M.; Morgan, M.; Burt, R. W.; Gardner, E. J.: Failure to demonstrate a chromosome 2 deletion in adenomatous colorectal polyposis patients. Cancer 53: 317-318, 1984.
Fodde, R.; Edelmann, W.; Yang, K.; van Leeuwen, C.; Carlson, C.; Renault, B.; Breukel, C.; Alt, E.; Lipkin, M.; Meera Khan, P.; Kucherlapati, R.: A targeted chain-termination mutation in the mouse Apc gene results in multiple intestinal tumors. Proc. Nat. Acad. Sci. 91: 8969-8973, 1994.
Fodde, R.; van der Luijt, R.; Wijnen, J.; Tops, C.; van der Klift, H.; van Leeuwen-Cornelisse, I.; Griffioen, G.; Vasen, H.; Meera Khan, P.: Eight novel inactivating germ line mutations at the APC gene identified by denaturing gradient gel electrophoresis. Genomics 13: 1162-1168, 1992.
Fraumeni, J. F., Jr.; Vogel, C. L.; Easton, J. M.: Sarcomas and multiple polyposis in a kindred. A genetic variety of hereditary polyposis? Arch. Intern. Med. 121: 57-61, 1968.
Frayling, I. M.; Beck, N. E.; Ilyas, M.; Dove-Edwin, I.; Goodman, P.; Pack, K.; Bell, J. A.; Williams, C. B.; Hodgson, S. V.; Thomas, H. J. W.; Talbot, I. C.; Bodmer, W. F.; Tomlinson, I. P. M.: The APC variants I1307K and E1317Q are associated with colorectal tumors, but not always with a family history. Proc. Nat. Acad. Sci. 95: 10722-10727, 1998.
Friedl, W.; Meuschel, S.; Caspari, R.; Lamberti, C.; Krieger, S.; Sengteller, M.; Propping, P.: Attenuated familial adenomatous polyposis due to a mutation in the 3-prime part of the APC gene: a clue for understanding the function of the APC protein. Hum. Genet. 97: 579-584, 1996.

Garber, J. E.; Li, F. P.; Kingston, J. E.; Krush, A. J.; Strong, L. C.; Finegold, M. J.; Bertario, L.; Bulow, S.; Filippone, A., Jr.; Gedde-Dahl, T., Jr.; Jarvinen, H. J.: Hepatoblastoma and familial adenomatous polyposis. J. Nat. Cancer Inst. 80: 1626-1628, 1988.
Gardner, E. J.: Follow-up study of a family group exhibiting dominant inheritance for a syndrome including intestinal polyps, osteomas, fibromas and epidermal cysts. Am. J. Hum. Genet. 14: 376-390, 1962.
Gardner, E. J.: A genetic and clinical study of intestinal polyposis, a predisposing factor for carcinoma of the colon and rectum. Am. J. Hum. Genet. 3: 167-176, 1951.
Gardner, E. J.: Discovery of the Gardner syndrome. Birth Defects Orig. Art. Ser. VIII(2): 48-51, 1972.
Gardner, E. J.; Plenk, H. P.: Hereditary pattern for multiple osteomas in a family group. Am. J. Hum. Genet. 4: 31-36, 1952.
Gardner, E. J.; Rogers, S. W.; Woodward, S.: Numerical and structural chromosome aberrations in cultured lymphocytes and cutaneous fibroblasts of patients with multiple adenomas of the colorectum. Cancer 49: 1413-1419, 1982.
Gayther, S. A.; Wells, D.; SenGupta, S. B.; Chapman, P.; Neale, K.; Tsioupra, K.; Delhanty, J. D. A.: Regionally clustered APC mutations are associated with a severe phenotype and occur at a high frequency in new mutation cases of adenomatous polyposis coli. Hum. Mol. Genet. 3: 53-56, 1994.
Giardiello, F. M.; Brensinger, J. D.; Petersen, G. M.; Luce, M. C.; Hylind, L. M.; Bacon, J. A.; Booker, S. V.; Parker, R. D.; Hamilton, S. R.: The use and interpretation of commercial APC gene testing for familial adenomatous polyposis. New Eng. J. Med. 336: 823-827, 1997.
Giardiello, F. M.; Hamilton, S. R.; Krush, A. J.; Piantadosi, S.; Hylind, L. M.; Celano, P.; Booker, S. V.; Robinson, C. R.; Offerhaus, G. J. A.: Treatment of colonic and rectal adenomas with sulindac in familial adenomatous polyposis. New Eng. J. Med. 328: 1313-1316, 1993.
Giardiello, F. M.; Offerhaus, G. J. A.; Krush, A. J.; Booker, S. V.; Tersmette, A. C.; Mulder, J.-W. R.; Kelley, C. N.; Hamilton, S. R.: Risk of hepatoblastoma in familial adenomatous polyposis. J. Pediat. 119: 766-768, 1991.
Giovannucci, E.; Rimm, E. B.; Stampfer, M. J.; Colditz, G. A.; Ascherio, A.; Willett, W. C.: Aspirin use and the risk for colorectal cancer and adenoma in male health professionals. Ann. Intern. Med. 121: 241-246, 1994.
Gorlin, R. J.; Chaudhry, A. P.: Multiple osteomatosis, fibromas, lipomas and fibrosarcomas of the skin and mesentery, epidermoid inclusion cysts of the skin, leiomyomas and multiple intestinal polyposis: an heritable disorder of connective tissue. New Eng. J. Med. 263: 1151-1158, 1960.
Greer, J. A., Jr.; Devine, K. D.; Dahlin, D. C.: Gardner's syndrome and chondrosarcoma of the hyoid bone. Arch. Otolaryng. 103: 425-427, 1977.
Groden, J.; Thliveris, A.; Samowitz, W.; Carlson, M.; Gelbert, L.; Albertsen, H.; Joslyn, G.; Stevens, J.; Spirio, L.; Robertson, M.; Sargeant, L.; Krapcho, K.; Wolff, E.; Burt, R.; Hughes, J. P.; Warrington, J.; McPherson, J.; Wasmuth, J.; Le Paslier, D.; Abderrahim, H.; Cohen, D.; Leppert, M.; White, R.: Identification and characterization of the familial adenomatous polyposis coli gene. Cell 66: 589-600, 1991.
Haggitt, R. C.; Booth, J. L.: Bilateral fibromatosis of the breast in Gardner's syndrome. Cancer 25: 161-166, 1970.
Hamilton, S. R.; Liu, B.; Parsons, R. E.; Papadopoulos, N.; Jen, J.; Powell, S. M.; Krush, A. J.; Berk, T.; Cohen, Z.; Tetu, B.; Burger, P. C.; Wood, P. A.; Taqi, F.; Booker, S. V.; Petersen, G. M.; Offerhaus, G. J. A.; Tersmette, A. C.; Giardiello, F. M.; Vogelstein, B.; Kinzler, K. W.: The molecular basis of Turcot's syndrome. New Eng. J. Med. 332: 839-847, 1995.
Hampton, G. M.; Ward, J. R. T. J.; Cottrell, S.; Howe, K.; Thomas, H. J. W.; Ballhausen, W. G.; Jones, T.; Sheer, D.; Solomon, E.; Frischauf, A.-M.; Bodmer, W. F.: Yeast artificial chromosomes for the molecular analysis of the familial polyposis APC gene region. Proc. Nat. Acad. Sci. 89: 8249-8253, 1992.
Harned, R. K.; Williams, S. M.: Familial polyposis coli and periampullary malignancy. Dis. Colon Rectum 25: 227-229, 1982.
He, T.-C.; Sparks, A. B.; Rago, C.; Hermeking, H.; Zawel, L.; da Costa, L. T.; Morin, P. J.; Vogelstein, B.; Kinzler, K. W.: Identification of c-MYC as a target of the APC pathway. Science 281: 1509-1512, 1998.
Heimann, A.; White, P. F.; Riely, C. A.; Ritchey, A. K.; Flye, M. W.; Barwick, K. W.: Hepatoblastoma presenting as isosexual precocity: the clinical importance of histologic and serologic parameters. J. Clin. Gastroent. 9: 105-110, 1987.
Herrera, L.; Kakati, S.; Gibas, L.; Pietrzak, E.; Sandberg, A. A.: Gardner syndrome in a man with an interstitial deletion of 5q. Am. J. Med. Genet. 25: 473-476, 1986.
Herve, R.; Farret, O.; Mayaudon, H.; Helie, C.; Denee, J. M.; Bauduceau, B.; Molinie, C.: Association syndrome de Gardner et carcinome thyroidien. (Letter) Presse Med. 24: 415 only, 1995.
Hockey, K. A.; Mulcahy, M. T.; Montgomery, P.; Levitt, S.: Deletion of chromosome 5q and familial adenomatous polyposis. J. Med. Genet. 26: 61-62, 1989.
Hodgson, S. V.; Coonar, A. S.; Hanson, P. J. V.; Cottrell, S.; Scriven, P. N.; Jones, T.; Hawley, P. R.; Wilkinson, M. L.: Two cases of 5q deletions in patients with familial adenomatous polyposis: possible link with Caroli's disease. J. Med. Genet. 30: 369-375, 1993.
Hodgson, S. V.; Fagg, N. L. K.; Talbot, I. C.; Wilkinson, M.: Deletions of the entire APC gene are associated with sessile colonic adenomas. (Letter) J. Med. Genet. 31: 426 only, 1994.
Hoffmann, D. C.; Brooke, B. N.: Familial sarcoma of bone in a polyposis coli family. Dis. Colon Rectum 13: 119-120, 1970.
Homfray, T. F. R.; Cottrell, S. E.; Ilyas, M.; Rowan, A.; Talbot, I. C.; Bodmer, W. F.; Tomlinson, I. P. M.: Defects in mismatch repair occur after APC mutations in the pathogenesis of sporadic colorectal tumours. Hum. Mutat. 11: 114-120, 1998.
Horii, A.; Nakatsuru, S.; Ichii, S.; Nagase, H.; Nakamura, Y.: Multiple forms of the APC gene transcripts and their tissue-specific expression. Hum. Molec. Genet. 2: 283-287, 1993.
Horii, A.; Nakatsuru, S.; Miyoshi, Y.; Ichii, S.; Nagase, H.; Kato, Y.; Yanagisawa, A.; Nakamura, Y.: The APC gene, responsible for familial adenomatous polyposis, is mutated in human gastric cancer. Cancer Res. 52: 3231-3233, 1992.
Houlston, R. S.; Fallon, T.; Harocopos, C.; Williams, C. B.; Davey, C.; Slack, J.: Congenital hypertrophy of retinal pigment epithelium in patients with colonic polyps associated with cancer family syndrome. Clin. Genet. 42: 16-18, 1992.
Hsu, S. H.; Luk, G. D.; Krush, A. J.; Hamilton, S. R.; Hoover, H. H., Jr.: Multiclonal origin of polyps in Gardner syndrome. Science 221: 951-953, 1983.
Hughes, L. J.; Michels, V. V.: Risk of hepatoblastoma in familial adenomatous polyposis. Am. J. Med. Genet. 43: 1023-1025, 1992.
Hyson, E. A.; Burrell, M.: Familial gastric polyposis (cont.) (Letter) New Eng. J. Med. 295: 905 only, 1976.
Iida, M.; Yao, T.; Itoh, H.; Watanabe, H.; Matsui, T.; Iwashita, A.; Fujishima, M.: Natural history of gastric adenomas in patients with familial adenomatosis coli/Gardner's syndrome. Cancer 61: 605-611, 1988.
Ingram, J. T.; Oldfield, M. C.: Hereditary sebaceous cysts. Brit. Med. J. 1: 960-963, 1937.
Jagelman, D. G.; DeCosse, J. J.; Bussey, H. J. R.; the Leeds Castle Polyposis Group: Upper gastrointestinal cancer in familial adenomatous polyposis. Lancet I: 1149-1151, 1988.
Jarvinen, H. J.; Peltokallio, P.; Landtman, M.; Wolf, J.: Gardner's stigmas in patients with familial adenomatosis coli. Brit. J. Surg. 69: 718-721, 1982.
Jones, T. R.; Nance, F. C.: Periampullary malignancy in Gardner's syndrome. Ann. Surg. 185: 565-573, 1977.
Joslyn, G.; Carlson, M.; Thliveris, A.; Albertsen, H.; Gelbert, L.; Samowitz, W.; Groden, J.; Stevens, J.; Spirio, L.; Robertson, M.; Sargeant, L.; Krapcho, K.; Wolff, E.; Burt, R.; Hughes, J. P.; Warrington, J.; McPherson, J.; Wasmuth, J.; Le Paslier, D.; Abderrahim, H.; Cohen, D.; Leppert, M.; White, R.: Identification of deletion mutations and three new genes at the familial polyposis locus. Cell 66: 601-613, 1991.
Kaplan, J.; Cushing, B.; Chang, C.-H.; Poland, R.; Roscamp, J.; Perrin, E.; Bhaya, N.: Familial T-cell lymphoblastic lymphoma: association with von Recklinghausen neurofibromatosis and Gardner syndrome. Am. J. Hemat. 12: 247-250, 1982.
Kassirer, J. P.: What role for nurse practitioner in primary care? (Editorial) New Eng. J. Med. 330: 204-205, 1994.
Kasukawa, T.; Watanabe, T.; Endo, A.: Absence of heteromorphism of chromosome number 2 homologues in patients with hereditary adenomatosis of the colon and rectum. Cancer Genet. Cytogenet. 9: 283-295, 1983.
Kelly, P. B.; McKinnon, D. A.: Familial multiple polyposis of the colon: review and description of a large kindred. McGill Med. J. 30: 67-85, 1961.
Kingston, J. E.; Draper, G. J.; Mann, J. R.: Hepatoblastoma and polyposis coli. (Letter) Lancet I: 457 only, 1982.
Kinzler, K. W.: Personal Communication. Baltimore, Md. 9/30/1991.
Kinzler, K. W.; Nilbert, M. C.; Su, L.-K.; Vogelstein, B.; Bryan, T. M.; Levy, D. B.; Smith, K. J.; Preisinger, A. C.; Hedge, P.; McKechnie, D.; Finniear, R.; Markham, A.; Groffen, J.; Boguski, M. S.; Altschul, S. F.; Horii, A.; Ando, H.; Miyoshi, Y.; Miki, Y.; Nishisho, I.; Nakamura, Y.: Identification of FAP locus genes from chromosome 5q21. Science 253: 661-665, 1991.
Klemmer, S.; Pascoe, L.; DeCosse, J.: Occurrence of desmoids in patients with familial adenomatous polyposis of the colon. Am. J. Med. Genet. 28: 385-392, 1987.
Kobayashi, T.; Narahara, K.; Yokoyama, Y.; Ueyama, S.; Mohri, O.; Fujii, T.; Fujimoto, M.; Ohtsuki, S.; Tsuji, K.; Seino, Y.: Gardner syndrome in a boy with interstitial deletion of the long arm of chromosome 5. Am. J. Med. Genet. 41: 460-463, 1991.
Krush, A. J.: Personal Communication. Baltimore, Md. 4/1/1988.
Krush, A. J.; Traboulsi, E. I.; Offerhaus, G. J. A.; Maumenee, I. H.; Yardley, J. H.; Levin, L. S.: Hepatoblastoma, pigmented ocular fundus lesions and jaw lesions in Gardner syndrome. Am. J. Med. Genet. 29: 323-332, 1988.
Laken, S. J.; Petersen, G. M.; Gruber, S. B.; Oddoux, C.; Ostrer, H.; Giardiello, F. M.; Hamilton, S. R.; Hampel, H.; Markowitz, A.; Klimstra, D.; Jhanwar, S.; Winawer, S.; Offit, K.; Luce, M. C.; Kinzler, K. W.; Vogelstein, B.: Familial colorectal cancer in Ashkenazim due to a hypermutable tract in APC. Nature Genet. 17: 79-83, 1997.
Lambertz, S.; Ballhausen, W. G.: Identification of an alternative 5-prime untranslated region of the adenomatous polyposis coli gene. Hum. Genet. 90: 650-652, 1993.
Lander, E.: Personal Communication. Cambridge, Mass. 10/1991.
Leffall, L. D.; Chung, E. B.; Dewitty, R. L.; Cornwell, E. E.; Blakey, T. M.: Familial polyposis coli in Black patients. Ann. Surg. 186: 324-333, 1977.
Leppard, B.; Bussey, H. J. R.: Epidermoid cysts, polyposis coli and Gardner's syndrome. Brit. J. Surg. 62: 387-393, 1975.
Leppert, M.; Burt, R.; Hughes, J. P.; Samowitz, W.; Nakamura, Y.; Woodward, S.; Gardner, E.; Lalouel, J.-M.; White, R.: Genetic analysis of an inherited predisposition to colon cancer in a family with a variable number of adenomatous polyps. New Eng. J. Med. 322: 904-908, 1990.
Leppert, M.; Dobbs, M.; Scambler, P.; O'Connell, P.; Nakamura, Y.; Stauffer, D.; Woodward, S.; Burt, R.; Hughes, J.; Gardner, E.; Lathrop, M.; Wasmuth, J.; Lalouel, J.-M.; White, R.: The gene for familial polyposis coli maps to the long arm of chromosome 5. Science 238: 1411-1413, 1987.
Leppert, M.; Dobbs, M.; Scambler, P.; O'Connell, P.; Nakamura, Y.; Stauffer, D.; Woodward, S.; Burt, R.; Hughes, J. P.; Gardner, E.; Lathrop, M.; Wasmuth, J.; Lalouel, J. M.; White, R.: The gene for familial polyposis maps to the long arm of chromosome 5. (Abstract) Cytogenet. Cell Genet. 46: 647 only, 1987.
Lewis, R. A.; Crowder, W. E.; Eierman, L. A.; Nussbaum, R. L.; Ferrell, R. E.: The Gardner syndrome: significance of ocular features. Ophthalmology 91: 916-925, 1984.
Lewis, R. J.; Mitchell, J. C.: Basal cell carcinoma in Gardner's syndrome. Acta Derm. Venerol. 51: 67-68, 1971.
Li, F. P.; Thurber, W. A.; Seddon, J.; Holmes, G. E.: Hepatoblastoma in families with polyposis coli. J.A.M.A. 257: 2475-2477, 1987.
Lindgren, V.; Bryke, C. R.; Ozcelik, T.; Yang-Feng, T. L.; Francke, U.: Phenotypic, cytogenetic, and molecular studies of three patients with constitutional

deletions of chromosome 5 in the region of the gene for familial adenomatous polyposis. Am. J. Hum. Genet. 50: 988-997, 1992.
Luk, G. D.; Baylin, S. B.: Ornithine decarboxylase as a biologic marker in familial colonic polyposis. New Eng. J. Med. 311: 80-83, 1984.
Luongo, C.; Gould, K. A.; Su, L.-K.; Kinzler, K. W.; Vogelstein, B.; Dietrich, W.; Lander, E. S.; Moser, A. R.: Mapping of multiple intestinal neoplasia (Min) to proximal chromosome 18 of the mouse. Genomics 15: 3-8, 1993.
Lynch, H. T.: Desmoid tumors: genotype-phenotype differences in familial adenomatous polyposis: a nosological dilemma. (Editorial) Am. J. Hum. Genet. 59: 1184-1185, 1996.
Lynch, H. T.; Smyrk, T. C.: Classification of familial adenomatous polyposis: a diagnostic nightmare. (Editorial) Am. J. Hum. Genet. 62: 1288-1289, 1998.
Lynch, H. T.; Smyrk, T. C.; Watson, P.; Lanspa, S. J.; Lynch, P. M.; Jenkins, J. X.; Rouse, J.; Cavalieri, J.; Howard, L.; Lynch, J.: Hereditary flat adenoma syndrome: a variant of familial adenomatous polyposis? Dis. Colon Rectum 35: 411-421, 1992.
Lyons, L. A.; Lewis, R. A.; Strong, L. C.; Zuckerbrod, S.; Ferrell, R. E.: A genetic study of Gardner syndrome and congenital hypertrophy of the retinal pigment epithelium. Am. J. Hum. Genet. 42: 290-296, 1988.
MacDonald, J. M.; Davis, W. C.; Crago, H. R.; Berk, A. D.: Gardner's syndrome and periampullary malignancy. Am. J. Surg. 113: 425-430, 1967.
Maher, E. R.; Barton, D. E.; Slatter, R.; Koch, D. J.; Jones, M. H.; Nagase, H.; Payne, S. J.; Charles, S. J.; Moore, A. T.; Nakamura, Y.; Ferguson-Smith, M. A.: Evaluation of molecular genetic diagnosis in the management of familial adenomatous polyposis coli: a population based study. J. Med. Genet. 30: 675-678, 1993.
Maher, E. R.; Morson, B.; Beach, R.; Hodgson, S. V.: Phenotypic variation in hereditary nonpolyposis colon cancer syndrome: association with infiltrative fibromatosis (desmoid tumor). Cancer 69: 2049-2051, 1992.
Marcus, A. J.: Aspirin as prophylaxis against colorectal cancer. New. Eng. J. Med. 333: 656-658, 1995.
Marshall, B.; Isidro, G.; Carvalhas, R.; Boavida, M. G.: Germline versus somatic mutations of the APC gene: evidence for mechanistic differences. (Letter) Hum. Mutat. 9: 286-288, 1997.
Marshall, W. H.; Martin, F. I. R.; MacKay, I. R.: Gardner's syndrome with adrenal carcinoma. Aust. Ann. Med. 16: 242-244, 1967.
Matsumine, A.; Ogai, A.; Senda, T.; Okumura, N.; Satoh, K.; Baeg, G.-H.; Kawahara, T.; Kobayashi, S.; Okada, M.; Toyoshima, K.; Akiyama, T.: Binding of APC to the human homolog of the Drosophila discs large tumor suppressor protein. Science 272: 1020-1023, 1996.
Maule, W. F.: Screening for colorectal cancer by nurse endoscopists. New Eng. J. Med. 330: 183-187, 1994.
McKusick, V. A.: Genetic factors in intestinal polyposis. J.A.M.A. 182: 271-277, 1962.
Meera Khan, P.; Tops, C. M. J.; van den Broek, M.; Breukel, C.; Wijnen, J. T.; Oldenburg, M.; v.d. Bos, J.; van Leeuwen-Cornelisse, I. S. J.; Vasen, H. F. A.; Griffioen, G.; Verspaget, H. M.; den Hartog Jager, F. C. A.; Lamers, C. B. H. W.: Close linkage of a highly polymorphic marker (D5S37) to familial adenomatous polyposis (FAP) and confirmation of FAP localization on chromosome 5q21-q22. Hum. Genet. 79: 183-185, 1988.
Midgley, C. A.; White, S.; Howitt, R.; Save, V.; Dunlop, M. G.; Hall, P. A.; Lane, D. P.; Wyllie, A. H.; Bubb, V. J.: APC expression in normal human tissues. J. Path. 181: 426-433, 1997.
Miyashiro, I.; Senda, T.; Matsumine, A.; Baeg, G.-H.; Kuroda, T.; Shimano, T.; Miura, S.; Noda, T.; Kobayashi, S.; Monden, M.; Toyoshima, K.; Akiyama, T.: Subcellular localization of the APC protein: immunoelectron microscopic study of the association of the APC protein with catenin. Oncogene 11: 89-96, 1995.
Miyoshi, Y.; Ando, H.; Nagase, H.; Nishisho, I.; Horii, A.; Miki, Y.; Mori, T.; Utsunomiya, J.; Baba, S.; Petersen, G.; Hamilton, S. R.; Kinzler, K. W.; Vogelstein, B.; Nakamura, Y.: Germ-line mutations of the APC gene in 53 familial adenomatous polyposis patients. Proc. Nat. Acad. Sci. 89: 4452-4456, 1992.
Miyoshi, Y.; Nagase, H.; Ando, H.; Horii, A.; Ichii, S.; Nakatsuru, S.; Aoki, T.; Miki, Y.; Mori, T.; Nakamura, Y.: Somatic mutations of the APC gene in colorectal tumors: mutation cluster region in the APC gene. Hum. Molec. Genet. 1: 229-233, 1992.
Morton, D. G.; Macdonald, F.; Cachon-Gonzales, M. B.; Rindl, P. M.; Neoptolemos, J. P.; Keighley, M. R. B.; Delhanty, J. D. A.; McKeown, C. M.; Kilpatrick, M.; Hulten, M. A.: The use of DNA from paraffin wax preserved tissue for predictive diagnosis in familial adenomatous polyposis. J. Med. Genet. 29: 571-573, 1992.
Moser, A. R.; Mattes, E. M.; Dove, W. F.; Lindstrom, M. J.; Haag, J. D.; Gould, M. N.: Apc(Min), a mutation in the murine Apc gene, predisposes to mammary carcinomas and focal alveolar hyperplasias. Proc. Nat. Acad. Sci. 90: 8977-8981, 1993.
Murphy, E. A.; Krush, A. J.; Dietz, M.; Rohde, C. A.: Hereditary polyposis coli. III. Genetic and evolutionary fitness. Am. J. Hum. Genet. 32: 700-713, 1980.
Nagase, H.; Miyoshi, Y.; Horii, A.; Aoki, T.; Ogawa, M.; Utsunomiya, J.; Baba, S.; Sasazuki, T.; Nakamura, Y.: Correlation between the location of germ-line mutations in the APC gene and the number of colorectal polyps in familial adenomatous polyposis patients. Cancer Res. 52: 4055-4057, 1992.
Nagase, H.; Miyoshi, Y.; Horii, A.; Aoki, T.; Petersen, G. M.; Vogelstein, B.; Maher, E.; Ogawa, M.; Maruyama, M.; Utsunomiya, J.; Baba, S.; Nakamura, Y.: Screening for germ-line mutations in familial adenomatous polyposis patients: 61 new patients and a summary of 150 unrelated patients. Hum. Mutat. 1: 467-473, 1992.
Nagase, H.; Nakamura, Y.: Mutations of the APC (adenomatous polyposis coli) gene. Hum. Mutat. 2: 425-434, 1993.
Nakamura, Y.; Lathrop, M.; Leppert, M.; Dobbs, M.; Wasmuth, J.; Wolff, E.; Carlson, M.; Fujimoto, E.; Krapcho, K.; Sears, T.; Woodward, S.; Hughes, J.; Burt, R.; Gardner, E.; Lalouel, J.-M.; White, R.: Localization of the genetic defect in familial adenomatous polyposis within a small region of chromosome 5. Am. J. Hum. Genet. 43: 638-644, 1988.
Naylor, E. W.; Gardner, E. J.: Adrenal adenomas in a patient with Gardner's syndrome. Clin. Genet. 20: 67-73, 1981.
Naylor, E. W.; Gardner, E. J.: Penetrance and expressivity of the gene responsible for the Gardner syndrome. Clin. Genet. 11: 381-393, 1977.
Naylor, E. W.; Lebenthal, E.: Gardner's syndrome: recent developments in research and management. Digest. Dis. Sci. 25: 945-959, 1980.
Neufeld, K. L.; White, R. L.: Nuclear and cytoplasmic localizations of the adenomatous polyposis coli protein. Proc. Nat. Acad. Sci. 94: 3034-3039, 1997.
Nishisho, I.; Nakamura, Y.; Miyoshi, Y.; Miki, Y.; Ando, H.; Horii, A.; Koyama, K.; Utsunomiya, J.; Baba, S.; Hedge, P.; Markham, A.; Krush, A. J.; Petersen, G.; Hamilton, S. R.; Nilbert, M. C.; Levy, D. B.; Bryan, T. M.; Preisinger, A. C.; Smith, K. J.; Su, L.-K.; Kinzler, K. W.; Vogelstein, B.: Mutations of chromosome 5q21 genes in FAP and colorectal cancer patients. Science 253: 665-669, 1991.
Novelli, M. R.; Williamson, J. A.; Tomlinson, I. P. M.; Elia, G.; Hodgson, S. V.; Talbot, I. C.; Bodmer, W. F.; Wright, N. A.: Polyclonal origin of colonic adenomas in an XO/XY patient with FAP. Science 272: 1187-1190, 1996.
Oda, H.; Imai, Y.; Nakatsuru, Y.; Hata, J.; Ishikawa, T.: Somatic mutations of the APC gene in sporadic hepatoblastomas. Cancer Res. 56: 3320-3323, 1996.
Offerhaus, G. J. A.; Giardiello, F. M.; Krush, A. J.; Booker, S. V.; Tersmette, A. C.; Kelley, N. C.; Hamilton, S. R.: The risk of upper gastrointestinal cancer in familial adenomatous polyposis. Gastroenterology 102: 1980-1982, 1992.
Okamoto, M.; Sato, C.; Kohno, Y.; Mori, T.; Iwama, T.; Tonomura, A.; Miki, Y.; Utsunomiya, J.; Nakamura, Y.; White, R.; Miyaki, M.: Molecular nature of chromosome 5q loss in colorectal tumors and desmoids from patients with familial adenomatous polyposis. Hum. Genet. 85: 595-599, 1990.
Olschwang, S.; Tiret, A.; Laurent-Puig, P.; Muleris, M.; Parc, R.; Thomas, G.: Restriction of ocular fundus lesions to a specific subgroup of APC mutations in adenomatous polyposis coli patients. Cell 75: 959-968, 1993.
Oshima, M.; Dinchuk, J. E.; Kargman, S. L.; Oshima, H.; Hancock, B.; Kwong, E.; Trzaskos, J. M.; Evans, J. F.; Taketo, M. M.: Suppression of intestinal polyposis in Apc(delta-716) knockout mice by inhibition of cyclooxygenase 2 (COX-2). Cell 87: 803-809, 1996.
Oshima, M.; Oshima, H.; Kitagawa, K.; Kobayashi, M.; Itakura, C.; Taketo, M.: Loss of Apc heterozygosity and abnormal tissue building in nascent intestinal polyps in mice carrying a truncated Apc gene. Proc. Nat. Acad. Sci. 92: 4482-4486, 1995.
Pandey, J. P.; Ebbesen, P.; Bulow, S.; Svendsen, L. B.; Fudenberg, H. H.: IgG heavy-chain (Gm) allotypes in familial polyposis coli. Am. J. Hum. Genet. 39: 133-136, 1986.
Papadopoulos, N.; Leach, F. S.; Kinzler, K. W.; Vogelstein, B.: Monoallelic mutation analysis (MAMA) for identifying germline mutations. Nature Genet. 11: 99-102, 1995.
Pauli, R. M.; Pauli, M. E.; Hall, J. G.: Gardner syndrome and periampullary malignancy. Am. J. Med. Genet. 6: 205-219, 1980.
Pavlides, G. P.; Milligan, F. D.; Clark, D. N.; Cohen, S. B.; Wennstrom, C. J.; Burbige, E. J.; Krush, A. J.; Murphy, E. A.: Hereditary polyposis coli. I. The diagnostic value of colonoscopy, barium enema, and fecal occult blood. Cancer 40: 2632-2639, 1977.
Peifer, M.: Cancer, catenins, and cuticle pattern: a complex connection. Science 262: 1667-1668, 1993.
Petersen, G. M.; Francomano, C.; Kinzler, K.; Nakamura, Y.: Presymptomatic direct detection of adenomatous polyposis coli (APC) gene mutations in familial adenomatous polyposis. Hum. Genet. 91: 307-311, 1993.
Petersen, G. M.; Parmigiani, G.; Thomas, D.: Missense mutations in disease genes: a Bayesian approach to evaluate causality. Am. J. Hum. Genet. 62: 1516-1524, 1998.
Petersen, G. M.; Shohat, T.; Brown, J.; Nakamura, Y.: Genetic counseling for familial adenomatous polyposis (FAP) with chromosome 5q linkage information. (Abstract) Am. J. Hum. Genet. 45 (suppl.): A125 only, 1989.
Phillips, L. G., Jr.: Polyposis and carcinoma of the small bowel and familial colonic polyposis. Dis. Colon Rectum 24: 478-481, 1981.
Pierce, E. R.: Pleiotropism and heterogeneity in hereditary intestinal polyposis. Birth Defects Orig. Art. Ser. VII(2): 52-62, 1972.
Pierce, E. R.; Weisbord, T.; McKusick, V. A.: Gardner's syndrome: formal genetics and statistical analysis of a large Canadian kindred. Clin. Genet. 1: 65-80, 1970.
Powell, S. M.; Petersen, G. M.; Krush, A. J.; Booker, S.; Jen, J.; Giardiello, F. M.; Hamilton, S. R.; Vogelstein, B.; Kinzler, K. W.: Molecular diagnosis of familial adenomatous polyposis. New Eng. J. Med. 329: 1982-1987, 1993.
Powell, S. M.; Zilz, N.; Beazer-Barclay, Y.; Bryan, T. M.; Hamilton, S. R.; Thibodeau, S. N.; Vogelstein, B.; Kinzler, K. W.: APC mutations occur early during colorectal tumorigenesis. Nature 359: 235-237, 1992.
Prescott, S. M.; White, R. L.: Self-promotion? Intimate connections between APC and prostaglandin H synthase-2. Cell 87: 783-786, 1996.
Rasheed, S.; Rhim, J. S.; Gardner, E. J.: Inherited susceptibility to retrovirus-induced transformation of Gardner syndrome cells. Am. J. Hum. Genet. 35: 919-931, 1983.
Redston, M.; Nathanson, K. L.; Yuan, Z. Q.; Neuhausen, S. L.; Satagopan, J.; Wong, N.; Yang, D.; Nafa, D.; Abrahamson, J.; Ozcelik, H.; Antin-Ozerkis, D.; Andrulis, I.; and 13 others: The APC I1307K allele and breast cancer risk. Nature Genet. 20: 13-14, 1998.
Rider, S. H.; Mazzullo, H. A.; Davis, M. B.; Delhanty, J. D. A.: Familial polyposis coli: growth characteristics of karyotypically variable cultured fibroblasts, response to epidermal growth factor and the tumour promoter 12-0-tetradecanoyl phorbol-13-acetate. J. Med. Genet. 23: 131-144, 1986.
Rubinfeld, B.; Albert, I.; Porfiri, E.; Fiol, C.; Munemitsu, S.; Polakis, P.: Binding of GSK3-beta to the APC-beta-catenin complex and regulation of complex assembly. Science 272: 1023-1026, 1996.
Rubinfeld, B.; Souza, B.; Albert, I.; Muller, O.; Chamberlain, S. H.; Masiarz, F. R.; Munemitsu, S.; Polakis, P.: Association of the APC gene product with beta-catenin. Science 262: 1731-1734, 1993.
Samowitz, W. S.; Burt, R. W.; Leppert, M.: Gastrointestinal polyposis and non-polyposis syndromes. (Letter) New Eng. J. Med. 332: 1518 only, 1995.

Sanchez, M. A.; Zali, M. R.; Khalil, A. A.; Ponce, R.; Font, R. G.: Be aware of Gardner's syndrome: a review of the literature. Am. J. Gastroent. 71: 68-73, 1979.
Savage, P. T.: Polyposis coli associated with multiple tumours in other parts of the body (Gardner's syndrome). Proc. Roy. Soc. Med. 57: 402-403, 1964.
Schneider, N. R.; Cubilla, A. L.; Chaganti, R. S. K.: Association of endocrine neoplasia with multiple polyposis of the colon. Cancer 51: 1171-1175, 1983.
Schnitzler, M.; Dwight, T.; Robinson, B. G.: Sulindac increases the expression of APC mRNA in malignant colonic epithelial cells: an in vitro study. Gut 38: 707-713, 1996.
Schnur, P. L.; David, E.; Brown, P. W., Jr.; Beahrs, O. H.; Remine, W. H.; Harrison, E. G., Jr.: Duodenal cancer and Gardner syndrome. J.A.M.A. 223: 1229-1232, 1973.
Scott, R. J.; Froggart, N. J.; Trembath, R. C.; Evans, D. G. R.; Hodgson, S. V.; Maher, E. R.: Familial infiltrative fibromatosis (desmoid tumours) (MIM135290) caused by a recurrent 3-prime APC gene mutation. Hum. Molec. Genet. 5: 1921-1924, 1996.
Seki, M.; Tanaka, K.; Kikuchi-Yanoshita, R.; Konishi, M.; Fukunari, H.; Iwama, T.; Miyaki, M.: Loss of normal allele of the APC gene in an adrenocortical carcinoma from a patient with familial adenomatous polyposis. Hum. Genet. 89: 298-300, 1992.
Sen-Gupta, S.; Van der Luijt, R. B.; Bowles, L. V.; Meera Khan, P.; Delhanty, J. D. A.: Somatic mutation of APC gene in desmoid tumour in familial adenomatous polyposis. (Letter) Lancet 342: 552-553, 1993.
Shemesh, E.: Adenomatous polyp of the common bile duct in familial polyposis coli. Israel J. Med. Sci. 21: 701-702, 1985.
Shneider, B. L.; Haque, S.; van Hoff, J.; Touloukian, R. J.; West, A. B.: Familial adenomatous polyposis following liver transplantation for a virilizing hepatoblastoma. J. Pediat. Gastroent. Nutr. 15: 198-201, 1992.
Shull, L. N., Jr.; Fitts, C. T.: Lymphoid polyposis associated with familial polyposis and Gardner's syndrome. Ann. Surg. 180: 319-322, 1974.
Simpson, R. D.; Harrison, E. G., Jr.; Mayo, C. W.: Mesenteric fibromatosis in familial polyposis: a variant of Gardner's syndrome. Cancer 17: 526-534, 1964.
Sivak, M. V.; Jagelman, D. G.: Upper gastrointestinal endoscopy in polyposis syndromes: familial polyposis coli and Gardner's syndrome. Gastrointest. Endosc. 30: 102-104, 1984.
Smith, K. J.; Johnson, K. A.; Bryan, T. M.; Hill, D. E.; Markowitz, S.; Willson, J. K. V.; Paraskeva, C.; Petersen, G. M.; Hamilton, S. R.; Vogelstein, B.; Kinzler, K. W.: The APC gene product in normal and tumor cells. Proc. Nat. Acad. Sci. 90: 2846-2850, 1993.
Smith, W. G.: Multiple polyposis, Gardner's syndrome and desmoid tumors. Dis. Colon Rectum 1: 323-332, 1958.
Smith, W. G.: Familial multiple polyposis: research tool for investigating the etiology of carcinoma of the colon? Dis. Colon Rectum 11: 17-31, 1968.
Solomon, E.; Voss, R.; Hall, V.; Bodmer, W. F.; Jass, J. R.; Jeffreys, A. J.; Lucibello, F. C.; Patel, I.; Rider, S. H.: Chromosome 5 allele loss in human colorectal carcinomas. Nature 328: 616-619, 1987.
Soravia, C.; Berk, T.; Madlensky, L.; Mitri, A.; Cheng, H.; Gallinger, S.; Cohen, Z.; Bapat, B.: Genotype-phenotype correlations in attenuated adenomatous polyposis coli. Am. J. Hum. Genet. 62: 1290-1301, 1998.
Spirio, L.; Olschwang, S.; Groden, J.; Robertson, M.; Samowitz, W.; Joslyn, G.; Gelbert, L.; Thliveris, A.; Carlson, M.; Otterud, B.; Lynch, H.; Watson, P.; Lynch, P.; Laurent-Puig, P.; Burt, R.; Hughes, J. P.; Thomas, G.; Leppert, M.; White, R.: Alleles of the APC gene: an attenuated form of familial polyposis. Cell 75: 951-957, 1993.
Spirio, L. N.; Samowitz, W.; Robertson, J.; Robertson, M.; Burt, R. W.; Leppert, M.; White, R.: Alleles of APC modulate the frequency and classes of mutations that lead to colon polyps. Nature Genet. 20: 385-388, 1998.
Stella, A.; Resta, N.; Gentile, M.; Susca, F.; Mareni, C.; Montera, M. P.; Guanti, G.: Exclusion of the APC gene as the cause of a variant form of familial adenomatous polyposis (FAP). Am. J. Hum. Genet. 53: 1031-1037, 1993.
Su, L.-K.; Kinzler, K. W.; Vogelstein, B.; Preisinger, A. C.; Moser, A. R.; Luongo, C.; Gould, K. A.; Dove, W. F.: Multiple intestinal neoplasia caused by a mutation in the murine homolog of the APC gene. Science 256: 668-670, 1992.
Su, L.-K.; Vogelstein, B.; Kinzler, K. W.: Association of the APC tumor suppressor protein with catenins. Science 262: 1734-1737, 1993.
Sugihara, K.; Muto, T.; Kamiya, J.; Konishi, F.; Sawada, T.; Morioka, Y.: Gardner's syndrome associated with periampullary carcinoma, duodenal and gastric adenomatosis: report of a case. Dis. Colon Rectum 25: 766-771, 1982.
Sulekova, Z.; Ballhausen, W. G.: A novel coding exon of the human adenomatous polyposis coli gene. Hum. Genet. 96: 469-471, 1995.
Takaku, K.; Oshima, M.; Miyoshi, H.; Matsui, M.; Seldin, M. F.; Taketo, M. M.: Intestinal tumorigenesis in compound mutant mice of both Dpc4 (Smad4) and Apc genes. Cell 92: 645-656, 1998.
Thakker, N.; Davies, R.; Horner, K.; Armstrong, J.; Clancy, T.; Guy, S.; Harris, R.; Sloan, P.; Evans, G.: The dental phenotype in familial adenomatous polyposis: diagnostic application of a weighted scoring system for changes on dental panoramic radiographs. J. Med. Genet. 32: 458-464, 1995.
Thompson, J. S.; Harned, R. K.; Anderson, J. C.; Hodgson, P. E.: Papillary carcinoma of the thyroid and familial polyposis coli. Dis. Colon Rectum 26: 583-585, 1983.
Tops, C. M. J.; van der Klift, H. M.; van der Luijt, R. B.; Griffioen, G.; Taal, B. G.; Vasen, H. F. A.; Khan, P. M.: Non-allelic heterogeneity of familial adenomatous polyposis. Am. J. Med. Genet. 47: 563-567, 1993.
Tops, C. M. J.; Vasen, H. F. A.; van Berge Henegouwen, G.; Simoons, P. P.; van de Klift, H. M.; van Leeuwen, I. S. J.; Breukel, C.; Fodde, R.; den Hartog Jager, F. C. A.; Nagengast, F. M.; Griffioen, G.; Meera Khan, P.: Genetic evidence that Turcot syndrome is not allelic to familial adenomatous polyposis. Am. J. Med. Genet. 43: 888-893, 1992.
Tops, C. M. J.; Wijnen, J. T.; Griffioen, G.; van Leeuwen, I. S. J.; Vasen, H. F. A.; den Hartog Jager, F. C. A.; Breukel, C.; Nagengast, F. M.; van der Klift, H. M.; Lamers, C. B. H. W.; Meera Khan, P.: Presymptomatic diagnosis of familial adenomatous polyposis by bridging DNA markers. Lancet II: 1361-1363, 1989.
Traboulsi, E. I.; Krush, A. J.; Gardner, E. J.; Booker, S. V.; Offerhaus, G. J. A.; Yardley, J. H.; Hamilton, S. R.; Luk, G. D.; Giardiello, F. M.; Welsh, S. B.; Hughes, J. P.; Maumenee, I. H.: Prevalence and importance of pigmented ocular fundus lesions in Gardner's syndrome. New Eng. J. Med. 316: 661-667, 1987.
Tsujii, M.; DuBois, R. N.: Alterations in cellular adhesion and apoptosis in epithelial cells overexpressing prostaglandin endoperoxide synthase 2. Cell 83: 493-501, 1995.
Turcot, J.; Despres, J. P.; St. Pierre, F.: Malignant tumors of the central nervous system associated with familial polyposis of the colon: report of two cases. Dis. Colon Rectum 2: 465-468, 1959.
Utsunomiya, J.; Nakamura, T.: The occult osteomatous changes in the mandible in patients with familial polyposis coli.. Brit. J. Surg. 62: 45-51, 1975.
van der Luijt, R.; Meera Khan, P.; Vasen, H.; van Leeuwen, C.; Tops, C.; Roest, P.; den Dunnen, J.; Fodde, R.: Rapid detection of translation-terminating mutations at the adenomatous polyposis coli (APC) gene by direct protein truncation test. Genomics 20: 1-4, 1994.
van der Luijt, R. B.; Khan, P. M.; Vasen, H. F. A.; Breukel, C.; Tops, C. M. J.; Scott, R. J.; Fodde, R.: Germline mutations in the 3-prime part of APC exon 15 do not result in truncated proteins and are associated with attenuated adenomatous polyposis coli. Hum. Genet. 98: 727-734, 1996.
van der Luijt, R. B.; Khan, P. M.; Vasen, H. F. A.; Tops, C. M. J.; van Leeuwen-Cornelisse, I. S. J.; Wijnen, J. T.; van der Klift, H. M.; Plug, R. J.; Griffioen, G.; Fodde, R.: Molecular analysis of the APC gene in 105 Dutch kindreds with familial adenomatous polyposis: 67 germline mutations identified by DGGE, PTT, and Southern analysis. Hum. Mutat. 9: 7-16, 1997.
van der Luijt, R. B.; Tops, C. M. J.; Khan, P. M.; van der Klift, H. M.; Breukel, C.; van Leeuwen-Cornelisse, I. S. J.; Dauwerse, H. G.; Beverstock, G. C.; van Noort, E.; Snel, P.; Slors, F. J. M.; Vasen, H. F. A.; Fodde, R.: Molecular, cytogenetic, and phenotypic studies of a constitutional reciprocal translocation t(5;10)(q22;q25) responsible for familial adenomatous polyposis in a Dutch pedigree. Genes Chromosomes Cancer 13: 192-202, 1995.
Vanhoutte, J. J.: Polypoid lymphoid hyperplasia of the terminal ileum in patients with familial polyposis coli and with Gardner's syndrome. Am. J. Roentgen. 110: 340-342, 1970.
Varesco, L.; Thomas, H. J. W.; Cottrell, S.; Murday, V.; Fennell, S. J.; Williams, S.; Searle, S.; Sheer, D.; Bodmer, W. F.; Frischauf, A.-M.; Solomon, E.: CpG island clones from a deletion encompassing the gene for adenomatous polyposis coli. Proc. Nat. Acad. Sci. 86: 10118-10122, 1989.
Veale, A. M.: Clinical and genetic problems in familial intestinal polyposis. Gut 1: 285-290, 1960.
Venkitachalam, P. S.; Hirsch, E.; Elguezabal, A.; Littman, L.: Multiple lymphoid polyposis and familial polyposis of the colon: a genetic relationship. Dis. Colon Rectum 21: 336-341, 1978.
Waddell, W. R.; Loughry, R. W.: Sulindac for polyposis of the colon. J. Surg. Oncol. 24: 83-87, 1983.
Wallis, Y. L.; Macdonald, F.; Hulten, M.; Morton, J. E. V.; McKeown, C. M.; Neoptolemos, J. P.; Keighley, M.; Morton, D. G.: Genotype-phenotype correlation between position of constitutional APC gene mutation and CHRPE expression in familial adenomatous polyposis. Hum. Genet. 94: 543-548, 1994.
Walsh, N.; Qizilbash, A.; Banerjee, R.; Waugh, G. A.: Biliary neoplasia in Gardner's syndrome. Arch. Path. Lab. Med. 111: 76-77, 1987.
Wasan, H. S.; Novelli, M.; Bee, J.; Bodmer, W. F.: Dietary fat influences on polyp phenotype in multiple intestinal neoplasia mice. Proc. Nat. Acad. Sci. 94: 3308-3313, 1997.
Watanabe, H.; Enjoji, M.; Yao, T.; Ohsato, K.: Gastric lesions in familial adenomatosis coli: their incidence and histologic analysis. Hum. Path. 9: 269-283, 1978.
Westbrook, C. A.; Chmura, S. J.; Arenas, R. B.; Kim, S. Y.; Otto, G.: Human APC gene expression in rodent colonic epithelium in vivo using liposomal gene delivery. Hum. Molec. Genet. 3: 2005-2010, 1994.
Woodage, T.; King, S. M.; Wacholder, S.; Hartge, P.; Struewing, J. P.; McAdams, M.; Laken, S. J.; Tucker, M. A.; Brody, L. C.: The APC I1307K allele and cancer risk in a community-based study of Ashkenazi Jews. Nature Genet. 20: 62-65, 1998.
Xia, L.; St. Denis, K. A.; Bapat, B.: Evidence for a novel exon in the coding region of the adenomatous polyposis coli (APC) gene. Genomics 28: 589-591, 1995.
Yonemoto, R. H.; Slayback, J. B.; Byron, R. L., Jr.; Rosen, R. B.: Familial polyposis of the entire gastrointestinal tract. Arch. Surg. 99: 427-434, 1969.
Young, J.; Simms, L. A.; Tarish, J.; Buttenshaw, R.; Knight, N.; Anderson, G. J.; Bell, A.; Leggett, B.: A family with attenuated familial adenomatous polyposis due to a mutation in the alternatively spliced region of APC exon 9. Hum. Mutat. 11: 450-455, 1998.

BCL2-RELATED PROTEIN A1 (*601056)

JJM NOTE

This gene could be sought in familial gastric cancer.

Programmed cell death (apoptosis) plays an important role in embryonic development, deletion of autoreactive T lymphocytes and homeostasis. Genes regulating apoptosis included P53 (191170), a tumor suppressor gene, MYC (190080), a protooncogene, and BCL2 (151430). Lin et al. (1993) isolated a novel mouse cDNA sequence, designated BCL2-related protein A1 (Bcl2a1) by them, and identified it as a member of the BCL2 family of apoptosis regulators by the predicted protein sequence. Lin et al. (1996) demonstrated that the A1 protein, although regulated differently from BCL2, has similar anti-apoptotic activity. Lin et al. (1997) sequenced a portion of the A1 gene, and using the A1 probe on mouse and human DNA panels, demonstrated that the mouse gene is situated on chromosome 9 and the human homolog is on chromosome 15. The BCL2A1 gene has also been referred to as BCL2L5, BFL1, and GRS.

Choi et al. (1995) isolated a BCL2-related gene from human fetal liver. Homology to the BH1 and BH2 domains of BCL2 was striking. BFL1 is abundantly expressed in bone marrow and at a low level in some other tissues. A correlation was noted between the expression level of BFL1 and the development of stomach cancer in 8 sets of clinical samples. Choi et al. (1995) speculated that BFL1 is involved in the promotion of cell survival during development or progression of stomach cancer. Choi et al. (1997) showed that BFL1 is the human homolog of murine A1. By fluorescence in situ hybridization analysis, they mapped the human gene to 15q24.3.

D'Sa-Eipper et al. (1996) showed that the BFL1 protein suppresses apoptosis induced by the p53 tumor suppressor protein in a manner similar to other BCL2 family members. The BFL1 gene showed a dominant cooperating oncogenic activity with the E1A oncogene in transformation of primary rodent epithelial cells.

Choi, S. S.; Park, I.-C.; Yun, J. W.; Sung, Y. C.; Hong, S.-I.; Shin, H.-S.: A novel Bcl-2 related gene, Bfl-1, is overexpressed in stomach cancer and preferentially expressed in bone marrow. Oncogene 11: 1693-1698, 1995.

Choi, S. S.; Park, S. H.; Kim, U.-J.; Shin, H.-S.: Bfl-1, a Bcl-2-related gene, is the human homolog of the murine A1, and maps to chromosome 15q24.3. Mammalian Genome 8: 781-782, 1997.

D'Sa-Eipper, C.; Subramanian, T.; Chinnadurai, G.: Bfl-1, a bcl-2 homologue, suppresses p53-induced apoptosis and exhibits potent cooperative transforming activity. Cancer Res. 56: 3879-3882, 1996.

Lin, E. Y.; Kozak, C. A.; Orlofsky, A.; Prystowsky, M. B.: The bcl-2 family member, Bcl2a1, maps to mouse chromosome 9 and human chromosome 15. Mammalian Genome 8: 293-294, 1997.

Lin, E. Y.; Orlofsky, A.; Berger, M. S.; Prystowsky, M. B.: Characterization of A1, a novel hemopoietic-specific early-response gene with sequence similarity to bcl-2. J. Immun. 151: 1979-1988, 1993.

Lin, E. Y.; Orlofsky, A.; Wang, H.-G.; Reed, J. C.; Prystowsky, M. B.: A1, a Bcl-2 family member, prolongs cell survival and permits myeloid differentiation. Blood 87: 983-992, 1996.

TURCOT SYNDROME (#276300)

JJM NOTE

A new level of understanding of the clinical and molecular genetics of this phenotype is emerging, but questions remain on the unusual families described here. (The terminal *t* is silent in Turcot.)

A number sign (#) is used with this entry because of evidence that the disorder is due to mutations in either the adenomatous polyposis coli gene (APC; 175100) or in the mismatch repair genes MLH1 (120436) or PMS2 (600259).

Turcot et al. (1959) described affected brother and sister. The parents were third cousins (Turcot, 1961). Because of the association of colonic polyps with tumors of many types in the Gardner syndrome (APC; 175100), it is possible that the sibs reported by Turcot et al. (1959) had that condition. The possibility is strengthened by the description by Yaffee (1964) of a patient with Gardner syndrome whose 'uncle died of Turcot syndrome.' This might suggest that Turcot syndrome is merely an unusual mode of presentation of the Gardner syndrome. The contrary view, that there exists a genuine syndrome of glioma and polyposis inherited as a recessive, is supported by the family reported by Baughman et al. (1969). A brother and 2 sisters had the full syndrome and another brother may have been affected. The parents were healthy and unrelated. Everson and Fraumeni (1976) described 2 sibs who died from glioblastoma multiforme associated with focal nodular hyperplasia of the liver and cafe-au-lait spots. One sib had four adenomatous sigmoid polyps removed at age 22. At autopsy no further colonic polyps were found. Itoh et al. (1979) described 2 sisters, of first-cousin parents, with malignant cerebral neoplasm and colonic polyposis. One sister presented at age 19 with multiple (about 100) polyps of the colon for which colectomy was performed; at the age of 22 she presented with a grade 3 astrocytoma of the left frontal lobe, from which she eventually died. She had several cafe-au-lait spots. The younger sister was found at age 17 to have about 80 adenomas of the colon (adenocarcinoma was found in two large polyps) and 14 small primary gastric cancers, all of the signet-ring-cell type. She had total gastrectomy and total colectomy. She had several cafe-au-lait spots and, on the left thigh, three small lipomas. At the age of 21, she was found to have a grade 3 astrocytoma of the right temporal lobe, from which she eventually expired. Panoramic radiographs of the jaws and ra diologic survey of the skeleton showed no abnormality suggestive of the Gardner syndrome. From study of their own cases and those in the literature, they concluded that in the Turcot syndrome polyps are somewhat fewer in number than in familial polyposis coli (APC), but are in general larger in size. The ratio of polyps over 3 cm in diameter to all polyps was frequently more than 1% in Turcot syndrome, but usually less than 0.2% in FPC. They were able to collect reports of 12 families plus several nonfamilial cases of the Turcot syndrome.

Bussey (1975) thought he could define a recessive adenomatous polyposis. Itoh et al. (1979) observed cases of this apparent type and found that the polyposis was of the Turcot type in terms of number and size. Hence, the Bussey polyposis may represent incomplete Turcot syndrome. Michels and Stevens (1982) reported a 22-year-old female with multiple polyposis of the colon requiring colectomy at age 17, multiple basal cell carcinoma in the scalp first presenting at age 18, pontine glioma presenting at age 19, and a tumor of the left posterior parietal region, either a second primary tumor or metastatic adenocarcinoma of the colon (invasive adenocarcinoma had been found in the colon specimen). A sister died of cerebral glioma at age 8. Li et al. (1983) reported the case of a woman who developed colonic polyposis and carcinoma at the age of 31 years, and astrocytoma at age 37. Her brother and sister had died of astrocytoma at ages 18 and 33 years, respectively. Progressive neutropenia developing 3 months after radiotherapy for the brain tumor and acute myelomonocytic leukemia developing 19 months after treatment suggested radiosensitivity. Studies of cultured skin fibroblasts in 3 laboratories showed slight but significant radiosensitivity in an early passage subculture (after 6 to 10 doublings), but no abnormality in later subculture (after 21 to 29 doublings). Selective in vitro loss of radiosensitive cells may account for the normality of later subcultures.

Itoh and Ohsato (1985) suggested that there may be 3 types of polyposis coli occurring in combination with CNS tumors. Type 1, which is that seen in Turcot's original cases, is characterized by multiple colonic polyps numbering between 20 and 100, some of which may exceed 3 cm in diameter; the polyps frequently show malignant transformation in the second and third decades of life. In several of the type 1 cases, parental consanguinity was described and the patients were found as solitary cases or as multiple cases within 1 sibship. In type 2, the number of colonic polyps is small, usually less than 10. The mode of inheritance is uncertain. In type 3, there are numerous small colonic polyps similar to those of classic familial polyposis coli. The mode of inheritance in this group may be autosomal dominant with respect to colonic lesions; indeed this may be standard familial adenomatous polyposis with the chance association of CNS tumor. Tops et al. (1992) presented evidence that the gene responsible for Turcot syndrome is not allelic to that responsible for APC: a brother and sister with Turcot syndrome had completely different haplotypes for RFLPs from the 5q21-q22 region; furthermore, an unaffected sister had the same haplotypes as an affected brother. Kikuchi et al. (1993) found no constitutional abnormality of the conserved regions of the TP53 (191170) gene (exons 5 to 9) in 2 patients with the glioma-polyposis syndrome but did find independent alterations in the TP53 gene in the tumors, suggesting that TP53 may play a role in progression but not in initiation of the disease.

Lasser et al. (1994) studied a 2-generation, 12-member family in which 3 individuals, a father and a daughter and son by different mothers, had Gardner syndrome and 2 sons by the second wife had Turcot syndrome. The brain tumors were medulloblastomas in both of the Turcot syndrome patients. Linkage data using marker D5S346, located 30 to 70 kb from the APC locus, showed evidence highly suggestive of linkage to the disease phenotype (lod score = 1.92).

Hamilton et al. (1995) studied 14 families with Turcot syndrome identified in 2 registries and the family originally described by Turcot et al. (1959). Genetic abnormalities were identified in 13 of the 14 registry families. Germline APC mutations were detected in 10. The predominant brain tumor in these 10 families was medulloblastoma (11 of 14 patients, or 79%), and the relative risk of cerebellar medulloblastoma in patients with familial adenomatous polyposis was 92 times that found in the general population according to a formal risk analysis for brain tumors in familial adenomatous polyposis in one of the registry databases. In contrast, the type of brain tumor in the other 4 families was glioblastoma multiforme. The glioblastomas and colorectal tumors in 3 of these families and in the original family studied by Turcot et al. (1959) had replication errors characteristic of hereditary nonpolyposis colorectal cancer. In addition, germline mutations in the mismatch repair gene MLH1 (120436) or PMS2 (600259) were found in 2 families.

It is noteworthy that several of the patients with Turcot syndrome, shown by Hamilton et al. (1995) to have APC mutations, also had pigmented ocular fundus lesions, epidermal inclusion cysts, or osteosclerotic jaw lesions consistent with Gardner syndrome. One of the patients with a truncating point mutation in the PMS2 gene (600259.0001) had cafe-au-lait spots.

Dupuis and Verellen-Dumoulin (1995) suggested that CNS tumors occur as an extracolonic feature with increased frequency in autosomal dominant FAP and that separate from this association is autosomal recessive Turcot syndrome, with central nervous system tumors as primary features. In the latter condition, the colonic polyps occur earlier, are less numerous and larger, and undergo an even earlier malignant transformation than in FAP. Furthermore, the CNS tumors in autosomal recessive Turcot syndrome are mainly glioblastomas. They pointed to the report by Tops et al. (1993) in which linkage to the APC gene was excluded. Mastronardi et al. (1991) reviewed the reported cases of Turcot syndrome and concluded that there are 2 distinct syndromes: medulloblastoma in young patients in kindreds affected by FAP, and glioma in older patients in kindreds with autosomal recessive inheritance, including the original kindred described by Turcot et al. (1959).

Pronunciation: Turcot, a French Canadian, pronounced his name with a silent terminal 't.'

See also Bigorgne et al. (1983); Chowdhary et al. (1985); Lewis et al. (1983); Rothman et al. (1975)

Baughman, F. A., Jr.; List, C. F.; Williams, J. R.; Muldoon, J. P.; Segarra, J. M.; Volkel, J. S.: The glioma-polyposis syndrome. New Eng. J. Med. 281: 1345-1346, 1969.

Bigorgne, J.-C.; Fanello, S.; Rohmer, V.; Ben Bouali, A.; Guy, G.; Ronceray, J.: Craniopharyngiome associe a une polypose rectocolique: syndrome de Turcot?. (Letter) Gastroenterol. Clin. Biol. 7: 1047 only, 1983.

Bussey, H. J. R.: Familial Polyposis Coli. Baltimore: Johns Hopkins Univ. Press (pub.) 1975.

Chowdhary, U. M.; Boehme, D. H.; Al-Jishi, M.: Turcot syndrome (glioma polyposis): case report. J. Neurosurg. 63: 804-807, 1985.

Dupuis, M. J. M.; Verellen-Dumoulin, C.: Gastrointestinal polyposis and nonpolyposis syndromes. (Letter) New Eng. J. Med. 332: 1518 only, 1995.

Everson, R. B.; Fraumeni, J. F., Jr.: Familial glioblastoma with hepatic focal nodular hyperplasia. Cancer 38: 310-313, 1976.

Hamilton, S. R.; Liu, B.; Parsons, R. E.; Papadopoulos, N.; Jen, J.; Powell, S. M.; Krush, A. J.; Berk, T.; Cohen, Z.; Tetu, B.; Burger, P. C.; Wood, P. A.; Taqi, F.; Booker, S. V.; Petersen, G. M.; Offerhaus, G. J. A.; Tersmette, A. C.; Giardiello, F. M.; Vogelstein, B.; Kinzler, K. W.: The molecular basis of Turcot's syndrome. New Eng. J. Med. 332: 839-847, 1995.

Itoh, H.; Ohsato, K.: Turcot syndrome and its characteristic colonic manifestations. Dis. Colon Rectum 28: 399-402, 1985.

Itoh, H.; Ohsato, K.; Yao, T.; Iida, M.; Watanabe, H.: Turcot's syndrome and its mode of inheritance. Gut 20: 414-419, 1979.

Kikuchi, T.; Rempel, S. A.; Rutz, H.-P.; de Tribolet, N.; Mulligan, L.; Cavenee, W. K.; Jothy, S.; Leduy, L.; Van Meir, E. G.: Turcot's syndrome of glioma and polyposis occurs in the absence of germ line mutations of exons 5 to 9 of the p53 gene. Cancer Res. 53: 957-961, 1993.

Lasser, D. M.; DeVivo, D. C.; Garvin, J.; Wilhelmsen, K. C.: Turcot's syndrome: evidence for linkage to the adenomatous polyposis coli (APC) locus. Neurology 44: 1083-1086, 1994.

Lewis, J. H.; Ginsberg, A. L.; Toomey, K. E.: Turcot's syndrome: evidence for autosomal dominant inheritance. Cancer 51: 524-528, 1983.

Li, F. P.; Little, J. B.; Bech-Hansen, N. T.; Paterson, M. C.; Arlett, C.; Garnick, M. B.; Mayer, R. J.: Acute leukemia after radiotherapy in a patient with Turcot's syndrome: impaired colony formation in skin fibroblast cultures after irradiation. Am. J. Med. 74: 343-348, 1983.

Mastronardi, L.; Ferrante, L.; Lunardi, P.; Cervoni, L.; Fortuna, A.: Association between neuroepithelial tumor and multiple intestinal polyposis (Turcot's syndrome): report of a case and critical analysis of the literature. Neurosurgery 28: 449-452, 1991.

Michels, V. V.; Stevens, J. C.: Basal cell carcinoma in a patient with intestinal polyposis. Clin. Genet. 22: 80-82, 1982.

Rothman, D.; Su, C. P.; Kendall, A. B.: Dilemma in a case of Turcot's (glioma-polyposis) syndrome: report of a case. Dis. Colon Rectum 18: 514-515, 1975.

Tops, C. M. J.; van der Klift, H. M.; van der Luijt, R. B.; Griffioen, G.; Taal, B. G.; Vasen, H. F. A.; Khan, P. M.: Non-allelic heterogeneity of familial adenomatous polyposis. Am. J. Med. Genet. 47: 563-567, 1993.

Tops, C. M. J.; Vasen, H. F. A.; van Berge Henegouwen, G.; Simoons, P. P.; van de Klift, H. M.; van Leeuwen, I. S. J.; Breukel, C.; Fodde, R.; den Hartog Jager, F. C. A.; Nagengast, F. M.; Griffioen, G.; Meera Khan, P.: Genetic evidence that Turcot syndrome is not allelic to familial adenomatous polyposis. Am. J. Med. Genet. 43: 888-893, 1992.

Turcot, J.: Personal Communication. Quebec City, Quebec, Canada 1961.

Turcot, J.; Despres, J. P.; St. Pierre, F.: Malignant tumors of the central nervous system associated with familial polyposis of the colon: report of two cases. Dis. Colon Rectum 2: 465-468, 1959.

Yaffee, H. S.: Gastric polyposis and soft tissue tumors: a variant of Gardner's syndrome. Arch. Derm. 89: 806-808, 1964.

PEUTZ-JEGHERS SYNDROME (#175200)

JJM NOTE

This polyposis is the outlier: the polyps are hamartomatous, not adenomatous nor preneoplastic, but certain otherwise rare gastrointestinal cancers occur to excess (such as duodenal carcinoma); other tumor types are unpredictable, such as gonadal tumors and perhaps breast and pancreatic cancer. The pigmentation can be subtle. Some features suggest an overlap with the Carney syndrome (LAMB or NAME, 160980). Of historic note, this was Dr. McKusick's first published foray into genetics (Jeghers et al. 1949), and cancer genetics, at that.

A number sign (#) is used with this entry because of evidence that the disorder is due to mutations in the serine/threonine kinase STK11 gene (602216).

In the syndrome named for Peutz (1921) and Jeghers (Jeghers et al., 1949), polyps may occur in any part of the gastrointestinal tract but jejunal polyps are a consistent feature. Intussusception and bleeding are the usual symptoms. Melanin spots of the lips, buccal mucosa, and digits represent the second part of the syndrome. Malignant degeneration of the small intestinal polyps is rare. Metastases from a malignant polyp in Peutz-Jeghers syndrome was reported by Williams and Knudsen (1965). Dodds et al. (1972) found 15 cases of gastrointestinal carcinoma in Peutz-Jeghers syndrome: 5 in colon, 4 in duodenum, 4 in stomach, 1 in ileum, and 1 in both jejunum and stomach. In the family reported by Farmer et al. (1963), the father had only polyps, the son apparently only pigmentation, and the daughter both polyps and pigmentation. Kieselstein et al. (1969), who found polycystic kidney disease in the same family, also noted a dissociation of signs. Brigg et al. (1976) observed a case of presumed Peutz-Jeghers syndrome without spots or positive family history. Hamartomatous polyps were limited to the jejunum and caused bleeding. Griffith and Bisset (1980) reported 3 cases. In 2 of them, the family history was negative; in the third, the father and a paternal uncle had melanin spots of the lips but no history of intestinal disorder. Sommerhaug and Mason (1970) added the ureter to the sites of polyps described in the Peutz-Jeghers syndrome. Previously described extraintestinal sites include esophagus, bladder, renal pelvis, bronchus and nose. Burdick and Prior (1982) reported nonresectable adenocarcinoma of the jejunum arising in a Peutz-Jeghers polyp and accompanied by metastases in mesenteric lymph nodes. Two developed breast carcinoma of which 1 arose in a fibroadenoma. Three had benign ovarian tumors, 1 had a benign breast tumor and 1 had a benign colloid thyroid nodule. One of my first 5 cases (case 7 in Jeghers et al., 1949) died of pancreatic ca ncer. Bowlby (1986) reported pancreatic cancer in an adolescent boy with PJS.

Affected females are prone to develop ovarian tumor, especially granulosa cell tumor (Christian et al., 1964). Wilson et al. (1986) described gynecomastia and multifocal and bilateral testicular tumors in a 6-year-old boy with PJS. The testicular tumors appear to be of Sertoli cell origin and most are calcifying. Two previously reported cases were found. Coen et al. (1991) reported the case of a 4-year-old boy with Peutz-Jeghers syndrome and bilateral sex-cord testicular tumors resulting in gynecomastia. Studies led to the conclusion that increase in aromatase activity (107910) in the gonadal tumors was responsible for estrogen excess and gynecomastia. Three other reported male patients with Peutz-Jeghers syndrome and gonadal tumors had presented with gynecomastia between birth and 6 years of age. They pointed out that multifocal sex-cord tumors were found in palpably normal testes. The occurrence of ovarian tumors far exceeds that of testicular tumors in this disorder. The production of estrogen by ovarian tumors is indicated by the reported appearance of isosexual precocity in girls with PJS (Solh et al., 1983). Young et al. (1995) reported 2 black boys, aged 3.5 and 5.5 years, who were evaluated for gynecomastia and found to have multicentric Sertoli cell testicular tumors responsible for their feminization. Both had rapid growth and advanced bone age, and serum levels of estradiol were markedly elevated.

In a patient with both psoriasis and Peutz-Jeghers syndrome (sine polyps), Banse-Kupin and Douglass (1986) described a peculiar phenomenon: the development of characteristic pigmented macules within preexisting psoriatic plaques in sites highly unusual for PJS, e.g., on the elbow, back of the neck and occipital scalp, buttocks, and legs. Sommerhaug and Mason (1970) suggested that patients with PJS develop polyps in areas of frequent trauma. Banse-Kupin and Douglass (1986) proposed that pigmented macules may likewise be located in areas of frequent trauma or areas of inflammation. Inflammation may induce blockage of pigment transfer from melanocyte to keratinocyte, resulting in a macule. As the inflammation or trauma subsides, so may the blockage and the lesion may fade. Histologically, the oral mucosal lesions resemble lentigo simplex, but the acral lesions are distinctive (Yamada et al., 1981). There is an increased number of melanocytes with long dendrites filled with melanosomes but few melanosomes in keratocytes, suggesting a pigment block.

Giardiello et al. (1987) investigated the occurrence of cancer in 31 patients with PJS followed from 1973 to 1985. Gastrointestinal carcinoma developed in 4, nongastrointestinal carcinoma in 10, and multiple myeloma in 1. Adenomatous polyps of the stomach and colon occurred in 3 other patients. There were 4 cases of pancreatic cancer. Foley et al. (1988) provided a 49-year follow-up of the 'Harrisburg family,' 3 affected members of which were reported by Jeghers et al. (1949). The family had also been studied earlier by Bartholomew et al. (1962). In all, 12 affected members have been identified, making this the largest PJS kindred reported. One member of the family had developed a duodenal carcinoma and a hamartoma with adenomatous changes. Another member developed short bowel syndrome. In the follow-up of 72 patients with PJS in the St. Mark's Polyposis Registry, Spigelman et al. (1989) found that malignant tumors had developed in 16 (22%), of whom all but 1 had died. There were 9 gastrointestinal and 7 nongastrointestinal tumors. The chance of dying of cancer by age 57 was 48%.

Unusually early age of onset was observed by Fernandez Seara et al. (1995) in a 15-day-old girl who was found to have generalized gastrointestinal polyposis manifested by abdominal distention, hematemesis, bloody diarrhea, and edema. At 15 days of age, ileocecal intussusception causing intestinal obstruction was diagnosed radiologically and reduced by hydrostatic enema; ileocecal surgical resection was required, however. Rectal prolapse due to a large polyp occurred at one month of age. Esophagogastroscopy showed polyps in the stomach; one in the antrum partially obstructed the lumen. No hyperpigmentation of the lips or oral mucosa was observed at any time and none were present in her relatives. The histologic appearance of the polyps removed during life and at autopsy was consistent with Peutz-Jeghers syndrome.

Studying 2 extended families, Bali et al. (1995) found positive evidence for linkage with several microsatellite markers on chromosome 1. Seldin (1997) reported that addition of more family members in the 2 largest families decreased the lod scores substantially as did the addition of more markers in the region. Indeed, in the original study, the maximum 2-point lod was below 2.0. Multipoint linkage analysis yielded a maximum lod score of 4.00 at D1S220. This is located in the distal region of 1p, where the human homolog of the putative modifier of multiple intestinal neoplasias (172411) had been previously mapped.

In a patient with Peutz-Jeghers syndrome, Markie et al. (1996) demonstrated a pericentric inversion in chromosome 6. Using fluorescence in situ hybridization with YAC clones selected to contain genetic markers from chromosome 6 and with a probe for the centromeric alphoid array, they located 1 inversion breakpoint within the alphoid repeat array, in a 1-cM interval between D6S257 and D6S402, and the other in a 4-cM interval between D6S403 and D6S311.

To localize the susceptibility locus for Peutz-Jeghers syndrome, Hemminki et al. (1997) used comparative genomic hybridization (CGH) and targeted linkage analysis, combined with loss of heterozygosity (LOH) study. They demonstrated a high-penetrance locus in distal 19p with a multipoint lod score of 7.00 at marker D19S886 without evidence of genetic heterogeneity. The study demonstrated the power of CGH combined with LOH analysis in identifying putative tumor suppressor loci. In comparative genomic hybridization, a single hybridization allows DNA copy number changes in the whole genome of a tumor to be assessed in comparison with normal tissue DNA (Kallioniemi et al., 1992). The findings of Hemminki et al. (1997) suggested that in most or all of the families they studied, the PJS was caused by a defect in a single locus on 19p. That the Peutz-Jeghers syndrome is genetically homogeneous required, however, confirmation by linkage analysis in further families. Amos et al. (1997) confirmed the mapping of PJS to the telomeric region of 19p. In the 5 families examined, there were no recombinants with the marker D19S886. The multipoint lod score at D19S886 was 7.52, and they found no evidence for genetic heterogeneity or of reduced penetrance.

Mehenni et al. (1997) performed a genomewide linkage analysis, using DNA polymorphisms in 6 families (2 from Spain, 2 from India, 1 from the U.S., and 1 from Portugal), including 39 affected individuals and 6 individuals of unknown status. Marker D19S886 yielded a maximum lod score of 4.74 at a recombination fraction of 0.45; multipoint linkage analysis resulted in a lod score of 7.51 for the interval between D19S886 and 19pter. However, markers on 19q13.4 also showed significant evidence for linkage. For example, D19S880 resulted in a maximum lod score of 3.8 at theta = 0.13. Most of this positive linkage was contributed by a single family. Thus, the results confirmed the mapping of a common PJS locus on 19p13.3, but also suggested the existence, in a minority of families, of a potential PJS locus on 19q13.4.

Within a distance of 190 kb proximal to D19S886, the marker with the highest lod score in the study of Hemminki et al. (1997), Jenne et al. (1998) identified and characterized a novel human gene encoding the serine/threonine kinase STK11. In a 3-generation PJS family, they found an STK11 allele with a deletion of exons 4 and 5 and an inversion of exons 6 and 7 (602216.0001) segregating with the disease. Sequence analysis of STK11 exons in 4 unrelated PJS patients identified 3 nonsense mutations (602216.0002, 602216.0003, 602216.0004) and 1 acceptor splice site mutation (602216.0005). All 5 germline mutations were predicted to disrupt the function of the kinase domain. Jenne et al. (1998) concluded that germline mutations in STK11, probably in conjunction with acquired genetic defects of the second allele in somatic cells, caused the manifestations of PJS.

Independently and simultaneously, Hemminki et al. (1998), the group that identified the linkage of PJS to chromosome 19, demonstrated mutations in the serine/threonine kinase gene in 11 of 12 unrelated patients with PJS.

Jenne (1998) speculated that cellular context between melanocytes and keratinocytes are regulated by STK11 activity. He pointed to the wide tissue distribution of STK11 and suggested that effects in melanocytes may be observed preferentially at sites of mechanical and physical stress.

See also Andre et al. (1966); Cantu et al. (1980); Cochet et al. (1979); Humphries et al. (1966); Joishy et al. (1979); Keen and Murray (1962); Lehur et al. (1984); Lin et al. (1977); Matuchansky et al. (1979); McAllister et al. (1967); McAllister and Richards (1977); McKittrick et al. (1971); Michalany and Ferraz (1962); Parker and Knight (1983); Peloquin et al. (1981); Riley and Swift (1980); Scully (1970); Sheward (1962); Tweedie and McCann (1984)

Amos, C. I.; Bali, D.; Thiel, T. J.; Anderson, J. P.; Gourley, I.; Frazier, M. L.; Lynch, P. M.; Luchtefeld, M. A.; Young, A.; McGarrity, T. J.; Seldin, M. F.: Fine mapping of a genetic locus for Peutz-Jeghers syndrome on chromosome 19p. Cancer Res. 57: 3653-3656, 1997.

Andre, R.; Duhamel, G.; Bruaire, M.; Tiollais, P.: Syndrome de Peutz-Jeghers avec polypose oesophagienne. Bull. Mem. Soc. Med. Hop. Paris 117: 505-510, 1966.

Bali, D.; Gourley, I. S.; McGarrity, T. J.; Spencer, C. A.; Howard, L.; Frazier, M. L.; Lynch, P. M.; Seldin, M. F.; Amos, C. I.: Peutz-Jegher's syndrome maps to chromosome 1p. (Abstract) Am. J. Hum. Genet. 57 (supp.): A186 only, 1995.

Banse-Kupin, L. A.; Douglass, M. C.: Localization of Peutz-Jeghers macules to psoriatic plaques. Arch. Derm. 122: 679-683, 1986.

Bartholomew, L. G.; Moore, C.; Dahlin, D. C.; Waugh, J. M.: Intestinal polyposis associated with mucocutaneous pigmentation. Surg. Gynec. Obstet. 115: 1-11, 1962.

Bowlby, L. S.: Pancreatic adenocarcinoma in an adolescent male with Peutz-Jeghers syndrome. Hum. Path. 17: 97-99, 1986.

Brigg, J. K.; Taylor, T. V.; Torrance, H. B.: Unusual manifestations of the Peutz-Jeghers syndrome. Brit. Med. J. 2: 853, 1976.

Burdick, D.; Prior, J. T.: Peutz-Jeghers syndrome: a clinicopathologic study of a large family with a 27-year follow-up. Cancer 50: 2139-2146, 1982.

Cantu, J. M.; Rivera, H.; Ocampo-Campos, R.; Bedolla, N.; Cortes-Gallegos, V.; Gonzalez-Mendoza, A.; Diaz, M.; Hernandez, A.: Peutz-Jeghers syndrome with feminizing Sertoli cell tumor. Cancer 46: 223-228, 1980.

Christian, C. D.; McLoughlin, T. G.; Cathcart, E. S.; Eisenberg, M. M.: Peutz-Jeghers syndrome associated with functioning ovarian tumor. J.A.M.A. 190: 935-938, 1964.

Cochet, B.; Carrel, J.; Desbaillets, L.; Widgren, S.: Peutz-Jeghers syndrome associated with gastrointestinal carcinoma: report of two cases in a family. Gut 20: 169-175, 1979.

Coen, P.; Kulin, H.; Ballantine, T.; Zaino, R.; Frauenhoffer, E.; Boal, D.; Inkster, S.; Brodie, A.; Santen, R.: An aromatase-producing sex-cord tumor resulting in prepubertal gynecomastia. New Eng. J. Med. 324: 317-322, 1991.

Dodds, W. J.; Schulte, W. J.; Hensley, G. T.; Hogan, W. J.: Peutz-Jeghers syndrome and gastrointestinal malignancy. Am. J. Roentgen. 115: 374-377, 1972.

Farmer, R. G.; Hawks, W. A.; Turnbull, R. B.: The spectrum of the Peutz-Jeghers syndrome: report of 3 cases. Am. J. Dig. Dis. 8: 953-961, 1963.

Fernandez Seara, M. J.; Martinez Soto, M. I.; Fernandez Lorenzo, J. R.; Trabazo, S.; Gamborino, E.; Forteza Vila, J.: Peutz-Jeghers syndrome in a neonate. J. Pediat. 126: 965-967, 1995.

Foley, T. R.; McGarrity, T. J.; Abt, A. B.: Peutz-Jeghers syndrome: a clinicopathologic survey of the 'Harrisburg family' with a 49-year follow-up. Gastroenterology 95: 1535-1540, 1988.

Giardiello, F. M.; Welsh, S. B.; Hamilton, S. R.; Offerhaus, G. J. A.; Gittelsohn, A. M.; Booker, S. V.; Krush, A. J.; Yardley, J. H.; Luk, G. D.: Increased risk of cancer in the Peutz-Jeghers syndrome. New Eng. J. Med. 316: 1511-1514, 1987.

Griffith, C. D. M.; Bisset, W. H.: Peutz-Jeghers syndrome. Arch. Dis. Child. 55: 866-869, 1980.

Hemminki, A.; Markie, D.; Tomlinson, I.; Avizienyte, E.; Roth, S.; Loukola, A.; Bignell, G.; Warren, W.; Aminoff, M.; Hoglund, P.; Jarvinen, H.; Kristo, P.; Pelin, K.; Ridanpaa, M.; Salovaara, R.; Toro, T.; Bodmer, W.; Olschwang, S.; Olsen, A. S.; Stratton, M. R.; de la Chapelle, A.; Aaltonen, L. A.: A serine/threonine kinase gene defective in Peutz-Jeghers syndrome. Nature 391: 184-187, 1998.

Hemminki, A.; Tomlinson, I.; Markie, D.; Jarvinen, H.; Sistonen, P.; Bjorkqvist, A.-M.; Knuutila, S.; Salovaara, R.; Bodmer, W.; Shibata, D.; de la Chapelle, A.; Aaltonen, L. A.: Localization of a susceptibility locus for Peutz-Jeghers syndrome to 19p using comparative genomic hybridization and targeted linkage analysis. Nature Genet. 15: 87-90, 1997.

Humphries, A. L., Jr.; Shepherd, M. H.; Peters, H. J.: Peutz-Jeghers syndrome with colonic adenocarcinoma and ovarian tumors. J.A.M.A. 197: 296-298, 1966.

Jeghers, H.; McKusick, V. A.; Katz, K. H.: Generalized intestinal polyposis and melanin spots of the oral mucosa, lips and digits. New Eng. J. Med. 241: 993-1005 and 1031-1036, 1949.

Jenne, D. E.: Personal Communication. Munich, Germany 1/8/1998.

Jenne, D. E.; Reimann, H.; Nezu, J.; Friedel, W.; Loff, S.; Jeschke, R.; Muller, O.; Back, W.; Zimmer, M.: Peutz-Jeghers syndrome is caused by mutations in a novel serine threonine kinase. Nature Genet. 18: 38-43, 1998.

Joishy, S. K.; Leela, M. P.; Balasegaram, M.: Peutz-Jeghers syndrome and its complications: first case report from Malaysia with review of literature. Am. J. Surg. 138: 716-720, 1979.
Kallioniemi, A.; Kallioniemi, O. P.; Sudar, D.; Rutovitz, D.; Gray, J. W.; Waldmann, F.; Pinkel, D.: Comparative genomic hybridization for molecular cytogenetic analysis of solid tumors. Science 258: 818-821, 1992.
Keen, G.; Murray, M. A.: Peutz-Jeghers syndrome: a further family history. Brit. Med. J. 1: 923-924, 1962.
Kieselstein, M.; Herman, G.; Wahrman, J.; Voss, R.; Gitelson, S.; Feuchtwanger, M.; Kadar, S.: Mucocutaneous pigmentation and intestinal polyposis (Peutz-Jeghers syndrome) in a family of Iraqi Jews with polycystic kidney disease, with a chromosome study. Israel J. Med. Sci. 5: 81-90, 1969.
Lehur, P.-A.; Madarnas, P.; Devroede, G.; Perey, B. J.; Menard, D. B.; Hamade, N.: Peutz-Jeghers syndrome: association of duodenal and bilateral breast cancers in the same patient. Digest. Dis. Sci. 29: 178-182, 1984.
Lin, J. I.; Caracta, P. F.; Lidner, A.; Gutzman, L. G.: Peutz-Jeghers polyposis with metastasizing duodenal carcinoma. Sth. Med. J. 70: 882-884, 1977.
Markie, D.; Huson, S.; Maher, E.; Davies, A.; Tomlinson, I.; Bodmer, W. F.: A pericentric inversion of chromosome six in a patient with Peutz-Jeghers' syndrome and the use of FISH to localise the breakpoints on a genetic map. Hum. Genet. 98: 125-128, 1996.
Matuchansky, C.; Babin, P.; Coutrot, S.; Druart, F.; Barbier, J.; Maire, P.: Peutz-Jeghers syndrome with metastasizing carcinoma arising from a jejunal hamartoma. Gastroenterology 77: 1311-1315, 1979.
McAllister, A. J.; Hicken, N. F.; Latimer, R. G.; Condon, V. R.: Seventeen patients with Peutz-Jeghers syndrome in four generations. Am. J. Surg. 114: 839-843, 1967.
McAllister, A. J.; Richards, K. F.: Peutz-Jeghers syndrome: experience with twenty patients in five generations. Am. J. Surg. 134: 717-720, 1977.
McKittrick, J. E.; Lewis, W. M.; Doane, W. A.; Gerwig, W. H.: The Peutz-Jeghers syndrome: report of two cases, one with 30-year follow-up. Arch. Surg. 103: 57-62, 1971.
Mehenni, H.; Blouin, J.-L.; Radhakrishna, U.; Bhardwaj, S. S.; Bhardwaj, K.; Dixit, V. B.; Richards, K. F.; Bermejo-Fenoll, A.; Leal, A. S.; Raval, R. C.; Antonarakis, S. E.: Peutz-Jeghers syndrome: confirmation of linkage to chromosome 19p13.3 and identification of a potential second locus, on 19p13.4. Am. J. Hum. Genet. 61: 1327-1334, 1997.
Michalany, J.; Ferraz, M. D.: Peutz syndrome in a mulatto family with special reference to the histological structure of the intestinal polyps. Gastroenterology 97: 119-129, 1962.
Parker, M. C. O.; Knight, M.: Peutz-Jeghers syndrome causing obstructive jaundice due to polyp in common bile duct. J. Roy. Soc. Med. 76: 701-703, 1983.
Peloquin, A. B.; Lauze, S.; Band, P.; Queeneville, G.: Syndrome de Peutz-Jeghers cancerise avec metastases disseminees. Canad. J. Surg. 24: 90-94, 1981.
Peutz, J. L. A.: Very remarkable case of familial polyposis of mucous membrane of intestinal tract and nasopharynx accompanied by peculiar pigmentations of skin and mucous membrane. (Dutch). Nederl. Maandschr. Geneesk. 10: 134-146, 1921.
Riley, E.; Swift, M.: A family with Peutz-Jeghers syndrome and bilateral breast cancer. Cancer 46: 815-817, 1980.
Scully, R. E.: Sex cord tumors with annular tubules—a distinctive ovarian tumor of the Peutz-Jeghers syndrome. Cancer 25: 1107-1121, 1970.
Seldin, M. F.: Personal Communication. Davis, Calif. 1/18/1997.
Sheward, J. D.: Peutz-Jeghers syndrome in childhood: unusual radiological features. Brit. Med. J. 1: 921-923, 1962.
Solh, H. M.; Azoury, R. S.; Najjar, S. S.: Peutz-Jeghers syndrome associated with precocious puberty. J. Pediat. 103: 593-595, 1983.
Sommerhaug, R. G.; Mason, T.: Peutz-Jeghers syndrome and ureteral polyposis. J.A.M.A. 211: 120-122, 1970.
Spigelman, A. D.; Murday, V.; Phillips, R. K. S.: Cancer and the Peutz-Jeghers syndrome. Gut 30: 1588-1590, 1989.
Tweedie, J. H.; McCann, B. G.: Peutz-Jeghers syndrome and metastasising colonic adenocarcinoma. Gut 25: 1118-1123, 1984.
Williams, J. P.; Knudsen, A.: Peutz-Jeghers syndrome with metastasizing duodenal carcinoma. Gut 6: 179-184, 1965.
Wilson, D. M.; Pitts, W. C.; Hintz, R. L.; Rosenfeld, R. G.: Testicular tumors with Peutz-Jeghers syndrome. Cancer 57: 2238-2240, 1986.
Yamada, K.; Matsukawa, A.; Hori, Y.; Kukita, A.: Ultrastructural studies on pigmented macules of the Peutz-Jeghers syndrome. J. Derm. 8: 367-377, 1981.
Young, S.; Gooneratne, S.; Straus, F. H., II; Zeller, W. P.; Bulun, S. E.; Rosenthal, I. M.: Feminizing Sertoli cell tumors in boys with Peutz-Jeghers syndrome. Am. J. Surg. Path. 19: 50-58, 1995.

POLYPOSIS, INTESTINAL, SCATTERED AND DISCRETE (175400)

JJM NOTE

It is time to revisit this family with current clinical tools and DNA markers to resolve whether or not this entity is distinct. I wonder if it represents Burt syndrome (familial colon cancer with few adenomas).

Woolf et al. (1955) suggested that some families have scattered polyps as a dominant trait distinct from multiple polyposis of the colon. The kindred of Lindberg and Kock (1975) had these features. However, studies of polyposis I families (175100) show such wide variability in the number of polyps that it is difficult to accept the idea that a separate mutation exists. The evidence is, to say the least, inconclusive.

Lindberg, B.; Kock, N. G.: A family with atypical colonic polyposis and gastric cancer: a three-decade follow-up. Cancer 35: 255-259, 1975.
Woolf, C. M.; Richards, R. C.; Gardner, E. J.: Occasional discrete polyps of the colon and rectum showing inherited tendency in a kindred. Cancer 8: 403-408, 1955.

POLYPOSIS, GENERALIZED JUVENILE, WITH PULMONARY ARTERIOVENOUS MALFORMATION (175050)

JJM NOTE

The recognition of two families validates this syndrome, but both families should be restudied with available probes for the two components.

Cox et al. (1980) described a 28-year-old woman and her 10-year-old daughter with this combination. Both showed severe digital clubbing. Polyps were discovered in the colon of the daughter at age 5 years. At the age of 8 years, a density was discovered in the left lower lobe on chest x-ray and shown by pulmonary arteriography to be an AV malformation. The mother had had a pulmonary AV malformation resected at the age of 10 years, and at the age of 12, partial colectomy was performed for multiple polyposis, leading to severe rectal bleeding. At age 16 years, 16 cm of the distal ileum was resected. Although this may represent the coincidence of autosomal dominant juvenile polyposis (174900) and Osler-Rendu-Weber disease (187300)— the loci linked and both involved in a minute chromosomal aberration in this family— this may be a unique syndrome due to mutation in a single gene. Conte et al. (1982) described an autosomal dominant syndrome of juvenile gastrointestinal polyposis, cutaneous telangiectasia, and pulmonary arteriovenous malformations in a father and his son and daughter. The father died from colon cancer at the age of 36. The brother and sister presented with recurrent rectal bleeding at an early age. All 3 had radiologically demonstrated pulmonary AV malformations with clubbing of the digits and hypertrophic pulmonary osteoarthropathy prompting arthritis clinic care. The sister additionally had repeated episodes of subarachnoid hemorrhage from cerebral AV malformations. Juvenile polyps are characteristically smooth surfaced but cystic on section. Adenomatous polyps are lobulated without cystic dilatation of glands. Microscopically, juvenile polyps are hamartomas. Baert et al. (1983) described a 15-year-old girl with juvenile polyps, mild digital clubbing, subperiosteal thickening of metacarpals, and AV malformations of the lung. Simpson and Dalinka (1985) noted a 14-year-old girl with juvenile polyps of the colon, clubbing of digits, and hypertrophic osteoarthropathy of long bones, but no AV malformations of the lungs.

Baert, A. L.; Casteels-Van Daele, M.; Broeckx, J.; Wijndaele, L.; Wilms, G.; Eggermont, E.: Generalized juvenile polyposis with pulmonary arteriovenous malformations and hypertrophic osteoarthropathy. Am. J. Roentgen. 141: 661-662, 1983.
Conte, W. J.; Rotter, J. I.; Schwartz, A. G.; Congleton, J. E.: Hereditary generalized juvenile polyposis, arteriovenous malformations and colonic carcinoma. (Abstract) Clin. Res. 30: 93A only, 1982.
Cox, K. L.; Frates, R. C., Jr.; Wong, A.; Gandhi, G.: Hereditary generalized juvenile polyposis associated with pulmonary arteriovenous malformation. Gastroenterology 78: 1566-1570, 1980.
Simpson, E. L.; Dalinka, M. K.: Association of hypertrophic osteoarthropathy with gastrointestinal polyposis. Am. J. Roentgen. 144: 983-984, 1985.

JUVENILE INTESTINAL POLYPOSIS (#174900)

JJM NOTE

This entry describes a multifacted clinical picture that needs to be clarified with additional observations, since mutations in various genes have been associated with the disorder(s). The most recent family I saw in clinic came with three questions: When should the next colon examination take place? Will you map the gene in our family? Can we start talking about gene thereapy for our family?

A number sign (#) is used with this entry because of evidence that mutations in the SMAD4/DPC4 gene (600993) located on 18q21.1 result in juvenile polyposis. It has also been suggested that juvenile polyposis can be caused by mutations in the PTEN gene (601728), the same gene that is mutant in Cowden disease (158350) and in Bannayan-Zonana syndrome (153480). A variant of the juvenile polyposis syndrome (JPS), termed hereditary mixed polyposis syndrome (MPSH; 601228), is characterized by atypical juvenile polyps, with mixed features of hamartomas and adenomas. The MPSH locus maps to 6q16.

Veale et al. (1966) investigated the families of 11 patients. Juvenile polyps may be isolated or multiple, even very numerous. The histology and natural history of these polyps suggest that they are hamartomas. Per se they are probably not precancerous. In 4 families, multiple polyposis and/or colonic carcinoma occurred in relatives. For example, the father of an affected brother and sister had colonic cancer. In 2 instances a parent of a case of juvenile polyposis had colonic cancer and multiple polyposis. Smilow et al. (1966) described a 7-year-old boy with juvenile polyposis, his mother who at 10 years of age had noted a prolapsed polyp during defecation, and his maternal grandfather who at age 60 had surgery for adenocarcinoma of the colon. In the grandfather, various polyps were present, some resembling adenomatous polyps and others resembling the juvenile polyps found in the proband and his mother. In the proband's mother the lesions were so numerous that total colectomy and ileostomy were performed. Veale et al. (1966) described juvenile polyps in 2 sisters and their mother. Haggitt and Pitcock (1970) described a girl who had onset of intermittent bright red rectal bleeding at age 3 years. Her father, an aunt and an uncle had 'well-differentiated adenocarcinoma with invasion of the submucosa.' The grandfather died at age 42 of colonic cancer. Gathright and Cofer (1974) described the disorder in a mother and 5 sons. Three brothers and a sister in another family were affected. Rozen and Baratz (1982) found multiple juvenile polyps of the colon in mother and son. Later a metastatic adenocarcinoma of the colon developed in the mother. In both patients, histology of the polyps showed no adenomas, but some of the juvenile polyps contained adenoma-like elements. Because of the mother's history and the adenomatous features of some of the son's polyps, he underwent subtotal colectomy.

Grosfeld and West (1986) presented 5 isolated cases. Clinical findings included abdominal pain, weakness, rectal bleeding, diarrhea, rectal prolapse, intussusception, digital clubbing, and failure to thrive. Laboratory findings included anemia, hypoalbuminemia, hypokalemia, and skin test anergy. The authors insisted that hamartomatous is an inappropriate adjective; the histology is quite different from that of the Peutz-Jeghers polyp. Soper (1986), in discussing the paper of Grosfeld and West (1986), stated that 'internists at the University of Iowa have studied a family through 4 generations, in which 15 members had juvenile polyposis. . .' GI malignancy was present in 11 of the 15 but no cancer arose in polyps. Walpole and Cullity (1989) described a patient with first presentation of polyposis in the second year of life. The patient had a large head, bilateral inguinal hernias and cryptorchidism, an umbilical hernia, and finger clubbing. Death occurred due to adenocarcinoma of the pancreas at age 19. Sharma et al. (1995) described the clinical and pathologic features of affected members in a large family with juvenile polyposis. Five living affected members had typical juvenile colonic polyps as well as atypical lobulated polyps with adenomatous change in the rectosigmoid area. Four members of the family had died of colonic malignancy between ages 30 and 55.

In a discussion of juvenile polyps in relation to Bannayan-Riley-Ruvalcaba syndrome (153480), Gorlin et al. (1992) defined juvenile polyps as retention polyps usually presenting before 10 years of age, although at least 15% are found in adults. Rectal bleeding is the most common presentation. Grossly and histologically, juvenile polyps are distinctive. They are usually pedunculated and spherical with a smooth surface. A cut section demonstrates numerous large cystic spaces of variable size, filled with a grayish or yellowish mucus surrounded by copious reddish stroma; hence the term 'retention polyp.' In contrast to the hamartomatous polyps of Peutz-Jeghers syndrome (175200), muscle fibers are not present in the stroma.

Although some cases of familial polyposis of the entire gastrointestinal tract represent juvenile polyposis, some seem to be adenomatous polyposis. Yonemoto et al. (1969) described a family with multiple cases consistent with dominant inheritance. One patient had a desmoid tumor of the abdominal wall. Whether the condition is distinct from Gardner syndrome (175100) or familial polyposis of the colon is unclear. Early development of symptoms is typical. The case described by Ravitch (1948) proved to be juvenile polyposis (Ravitch, 1974).

Familial juvenile polyposis of the stomach (Watanabe et al., 1979) is presumably the same disorder. Gastric cancer was reported in association with juvenile polyposis of the stomach by Watanabe et al. (1979) and Sassatelli et al. (1993).

Subramony et al. (1994) described an extensively affected kindred showing preponderance of right-sided polyps. Subtotal colectomy with ileorectal anastomosis resulted in prompt recurrence of polyps. It was followed by prompt recurrence of polyps in the retained rectum. Further investigations of the family were reported by Scott-Conner et al. (1995). Of 34 living members of the kindred, 15 were investigated, and histologically typical juvenile polyps were found in 11. In each instance, polyps were most numerous in the right colon, with few polyps in the descending colon and none in the rectum. Eight patients had had subtotal colectomies with ileorectal anastamoses; the other 3 patients were managed by polypectomy (with 1 recurrence after 10 years). In addition to juvenile polyps, polyps with adenomatous or villous elements were identified in 3 patients. Invasive adenocarcinoma in a large mixed polyp of the cecum was found in 1 of these patients. Coexisting carcinoma of the stomach was found in 2 patients with polyps. Polyps had recurred in the rectal remnants of 3 patients at a mean of 36 months after subtotal colectomy. Two patients had undergone conversion to total proctocolectomy with ileoanal anastomosis and J pouch; 1 patient was found to have juvenile polyps in the pouch 40 months after surgery. Scott-Conner et al. (1995) concluded that despite the preponderance of right-sided polyps at initial diagnosis, the rapid recurrence of polyps after subtotal colectomy argues in favor of performing proctocolectomy with preservation of anal sphincter function (restorative proctocolectomy) at the time of initial surgery. Patients with a small number of polyps may choose instead to undergo periodic colonoscopy with colonoscopic polypectomy. Histologically juvenile polyps are comprised primarily of stromal elements, often overlayed with a thin layer of normal-appearing epithelium that is easily traumatized. The proliferative changes seen in adenomas are notably absent. The gross appearan ce of the surface is smooth, and cystic spaces containing retained mucus may be seen when the polyp is sectioned. The stock is thin and many of these polyps slough spontaneously.

Juvenile polyps, with normal epithelium but hypertrophic lamina propria, are histologically distinct from adenomatous polyps that have mutations in the APC gene (175100) on chromosome 5 as their basis. Furthermore, hereditary juvenile polyposis is not linked to APC or to the MCC gene (159350), which is also located on chromosome 5 (Leggett et al., 1993). Jacoby et al. (1997) reported a case that suggests that a tumor suppressor gene associated with juvenile polyposis may be located at 10q22.3-q24.1. Cytogenetic studies of the patient, who had juvenile polyposis and multiple congenital anomalies of the head, limbs, and abdomen, revealed a de novo interstitial deletion of chromosome 10 as the only defect. The deleted area was between bands 10q22.3 and 10q24.1. Other syndromes associated with hamartomatous polyps similar to those observed in this patient include the Ruvalcaba-Myhre-Smith syndrome (153480) and Cowden disease (158350). The latter condition maps to the same region of 10q and has been shown to be due to mutation in the PTEN gene (601728). The abnormalities other than the juvenile polyps included short stature, broad nasal tip and long philtrum, widely spaced canthi, hypoplastic ears, small head, and redundant neck skin. There were also a small umbilical hernia, hypoplastic oblique muscles with bilateral abdominal bulging laterally, and prominent venous patterning on thorax and abdomen. The hands and feet were plump dorsally. Both the hands and feet were short and broad with redundant skin over the hands. Motor and language skills were developmentally delayed. Marsh et al. (1997) commented that the JPS locus is located probably between markers D10S219 and D10S1696 and that Cowden disease maps between markers D10S215 and D10S564, a region telomeric of the JPS region.

Olschwang et al. (1998) demonstrated mutations in the PTEN gene in 3 patients with juvenile polyposis coli. One of them was a 14-year-old boy who underwent colonoscopy that revealed juvenile polyposis (see 601728.0009). He had no personal or family history that could be related to Cowden disease or Bannayan-Zonana syndrome. The second patient was a 74-year-old man with anemia and hypoalbuminemia in whom gastroscopy and colonoscopy showed polyps throughout the digestive tract, which were classified as juvenile by histology (see 601728.0010). The third patient was found to have juvenile polyps throughout the stomach, duodenum, and colon when gastroscopy and colonoscopy were performed at the age of 7 years, after a 3-year history of intermittent rectal bleeding (see 601728.0011). Eng and Peacocke (1998) questioned the role of PTEN mutations in the juvenile polyposis syndrome. They pointed out that 3 groups (Marsh et al., 1997; Riggins et al., 1997; Howe et al., 1998) had found no evidence of germline PTEN mutations in 21 JPS families and 16 sporadic cases, and power calculations indicated that if 10% of JPS cases were due to germline PTEN mutations, there should have been a 0.99 likelihood of detecting at least 1 mutation among these 37 cases. They suggested further that the 3 patients found by Olschwang et al. (1998) to have germline PTEN mutations may, in fact, have had either Cowden disease or Bannayan-Zonana syndrome; the 74-year-old man had manifestations they interpreted as suggestive of Cowden disease, and the 2 children may not have yet demonstrated features of Cowden disease, which has a penetrance well below 10% under 15 years of age (Nelen et al., 1996).

In cases of juvenile intestinal polyposis, Howe et al. (1998) demonstrated mutations in the gene variously known as SMAD4 and DPC4 (600993), located on 18q21.1. Houlston et al. (1998) analyzed 8 families in which members had the juvenile polyposis syndrome, looking for linkage to DPC4. Overall, there was no evidence for linkage to DPC4; linkage could be excluded in 2 of the 8 pedigrees and was unlikely in 2 others. They then tested these 8 families and 13 familial and sporadic JPS cases for germline mutations in DPC4. Only 1 germline DPC4 mutation was found (in a familial JPS patient with a pedigree unsuitable for linkage analysis). Although they confirmed DPC4 as a cause of JPS, they concluded that this gene is the site of the causative mutation in a minority of cases.

Tsuchiya et al. (1998) described the case of a 6-year-old African American boy with mental retardation, dysmorphic features, and juvenile polyposis coli. A cytogenetically visible interstitial deletion of 10q23 was identified and characterized by fluorescence in situ hybridization. Five YACs that span an 11- to 15-cM region within the deletion were identified. The patient's deletion contained the putative locus for Cowden syndrome and the PTEN gene. The African-American patient first had rectal bleeding at 2 years of age. Polyps were identified extending from the duodenum to the rectum. Dysmorphic features included macrocephaly with a head circumference of 53.6 cm, frontal bossing, hypertelorism, flat nasal bridge, hyperpigmentation at the corners of the mouth, thickened gingiva, bilateral preauricular pits, and a flat midface. He had a large penis with a hyperpigmented macule on the dorsal shaft. Multiple other hyperpigmented macules as well as a large area of hypopigmentation were identified. The distal phalanges of all fingers and toes were strikingly bulbous and nail clubbing was noted. This was clearly a case of the Bannayan-Riley-Ruvalcaba syndrome.

Nomenclature: Because the polyps are not limited to the colon, juvenile intestinal polyposis appears to be a better designation than polyposis coli, juvenile type.

See also Marsh et al. (1997)

Eng, C.; Peacocke, M.: PTEN and inherited hamartoma-cancer syndromes. (Letter) Nature Genet. 19: 223 only, 1998.
Gathright, J. B., Jr.; Cofer, T. W., Jr.: Familial incidence of juvenile polyposis coli. Surg. Gynec. Obstet. 138: 185-188, 1974.
Gorlin, R. J.; Cohen, M. M., Jr.; Condon, L. M.; Burke, B. A.: Bannayan-Riley-Ruvalcaba syndrome. Am. J. Med. Genet 44: 307-314, 1992.
Grosfeld, J. L.; West, K. W.: Generalized juvenile polyposis coli: clinical management based on long-term observations. Arch. Surg. 121: 530-534, 1986.
Haggitt, R. C.; Pitcock, J. A.: Familial juvenile polyposis of the colon. Cancer 26: 1232-1238, 1970.
Houlston, R.; Bevan, S.; Williams, A.; Young, J.; Dunlop, M.; Rozen, P.; Eng, C.; Markie, D.; Woodford-Richens, K.; Rodriguez-Bigas, M. A.; Leggett, B.; Neale, K.; Phillips, R.; Sheridan, E.; Hodgson, S.; Iwama, T.; Eccles, D.; Bodmer, W.; Tomlinson, I.: Mutations in DPC4 (SMAD4) cause juvenile polyposis syndrome, but only account for a minority of cases. Hum. Molec. Genet. 7: 1907-1912, 1998.
Howe, J. R.; Ringold, J. C.; Summers, R. W.; Mitros, F. A.; Nishimura, D. Y.; Stone, E. M.: A gene for familial juvenile polyposis maps to chromosome 18q21.1. Am. J. Hum. Genet. 62: 1129-1136, 1998.
Howe, J. R.; Roth, S.; Ringold, J. C.; Summers, R. W.; Jarvinen, H. J.; Sistonen, P.; Tomlinson, I. P. M.; Houlston, R. S.; Bevan, S.; Mitros, F. A.; Stone, E. M.; Aaltonen, L. A.: Mutations in the SMAD4/DPC4 gene in juvenile polyposis. Science 280: 1086-1088, 1998.
Jacoby, R. F.; Schlack, S.; Sekhon, G.; Laxova, R.: Del(10)(q22.3q24.1) associated with juvenile polyposis. Am. J. Med Genet. 70: 361-364, 1997.
Leggett, B. A.; Thomas, L. R.; Knight, N.; Healey, S.; Chenevix-Trench, G.; Searle, J.: Exclusion of APC and MCC as the gene defect in one family with familial juvenile polyposis. Gastroenterology 105: 1313-1316, 1993.
Marsh, D. J.; Dahia, P. L. M.; Zheng, Z.; Liaw, D.; Parsons, R.; Gorlin, R. J.; Eng, C.: Germline mutations in PTEN are present in Bannayan-Zonana syndrome. (Letter) Nature Genet. 16: 333-334, 1997.
Marsh, D. J.; Roth, S.; Lunetta, K. L.; Hemminki, A.; Dahia, P. L.; Sistonen, P.; Zheng, Z.; Caron, S.; van Orsouw, N. J.; Bodmer, W. F.; Cottrell, S. E.; Dunlop, M. G.; and 13 others: Exclusion of PTEN and 10q22-24 as the susceptibility locus for juvenile polyposis syndrome. Cancer Res. 57: 5017-5021, 1997.
Nelen, M. R.; Padberg, G. W.; Peeters, E. A.; Lin, A. Y.; van den Helm, B.; Frants, R. R.; Coulon, V.; Goldstein, A. M.; van Reen, M. M.; Easton, D. F.; Eeles, R. A.; Hodgsen, S.; and 10 others: Localization of the gene for Cowden disease to chromosome 10q22-23. Nature Genet. 13: 114-116, 1996.
Olschwang, S.; Serova-Sinilnikova, O. M.; Lenoir, G. M.; Thomas, G.: PTEN germ-line mutations in juvenile polyposis coli. (Letter) Nature Genet. 18: 12-14, 1998.
Ravitch, M. M.: Polypoid adenomatosis of entire gastro-intestinal tract. Ann. Surg. 128: 283-298, 1948.
Ravitch, M. M.: Personal Communication. Pittsburgh, Pa. 1974.
Riggins, G. J.; Hamilton, S. R.; Kinzler, K. W.; Vogelstein, B.: Normal PTEN gene in juvenile polyposis. J. Neg. Observ. Genet. Oncol. 1: 1, 1997. Note: Electronic Article.
Rozen, P.; Baratz, M.: Familial juvenile colonic polyposis with associated colon cancer. Cancer 49: 1500-1503, 1982.
Sassatelli, R.; Bertoni, G.; Serra, L.; Bedogni, G.; Ponz de Leon, M.: Generalized juvenile polyposis with mixed pattern and gastric cancer. Gastroenterology 104: 910-915, 1993.
Scott-Conner, C. E. H.; Hausmann, M.; Hall, T. J.; Skelton, D. S.; Anglin, B. L.; Subramony, C.: Familial juvenile polyposis: patterns of recurrence and implications for surgical management. J. Am. Coll. Surg. 181: 407-413, 1995.
Sharma, A. K.; Sharma, S. S.; Mathur, P.: Familial juvenile polyposis with adenomatous-carcinomatous change. J. Gastroenterol. Hepatol. 10: 131-134, 1995.
Smilow, P. C.; Pryor, C. A., Jr.; Swinton, N. W.: Juvenile polyposis coli: a report of three patients in three generations of one family. Dis. Colon Rectum 9: 248-254, 1966.
Soper, R. T.: Discussion. Arch. Surg. 121: 534, 1986.
Subramony, C.; Scott-Conner, C. E. H.; Skelton, D. S.; Hall, T. J.: Familial juvenile polyposis. A study of a kindred: evolution of polyps and relationship to gastrointestinal carcinoma. Am. J. Clin. Pathol. 102: 91-97, 1994.
Tsuchiya, K. D.; Wiesner, G.; Cassidy, S. B.; Limwongse, C.; Boyle, J. T.; Schwartz, S.: Deletion 10q23.2-q23.33 in a patient with gastrointestinal juvenile polyposis and other features of a Cowden-like syndrome. Genes Chromosomes Cancer 21: 113-118, 1998.
Veale, A. M.; McColl, I.; Bussey, H. J. R.; Morson, B. C.: Juvenile polyposis coli. J. Med. Genet. 3: 5-16, 1966.
Walpole, I. R.; Cullity, G.: Juvenile polyposis: a case with early presentation and death attributable to adenocarcinoma of the pancreas. Am. J. Med. Genet. 32: 1-8, 1989.
Watanabe, A.; Nagashima, H.; Motoi, M.; Ogawa, K.: Familial juvenile polyposis of the stomach. Gastroenterology 77: 148-151, 1979.
Yonemoto, R. H.; Slayback, J. B.; Byron, R. L., Jr.; Rosen, R. B.: Familial polyposis of the entire gastrointestinal tract. Arch. Surg. 99: 427-434, 1969.

POLYPOSIS, SKIN PIGMENTATION, ALOPECIA, AND FINGERNAIL CHANGES (CRONKITE-CANADA SYNDROME) (175500)

JJM NOTE

This disorder is another atypical syndrome in the heuristic category of polyposis. It needs to be distinguished from the more common and more likely hereditary ones.

This syndrome was first described by Cronkhite and Canada (1955) and later by Jarnum and Jensen (1966). Manousos and Webster (1966) reported the fourth patient. All cases have been sporadic and all have been adults. The prognosis is poor. The etiology is unknown but it appears to be nongenetic. The pigmentation is diffuse rather than spotted as in the Peutz-Jeghers syndrome (175200).

See also Dacruz (1967)

Cronkhite, L. W., Jr.; Canada, W. J.: Generalized gastrointestinal polyposis: an unusual syndrome of polyposis, pigmentation, alopecia and onychotrophia. New Eng. J. Med. 252: 1011-1015, 1955.
Dacruz, G. M. G.: Generalized gastrointestinal polyposis: an unusual syndrome of adenomatous polyposis, alopecia, onychorotrophia. Am. J. Gastroent. 47: 504-510, 1967.
Jarnum, S.; Jensen, H.: Diffuse gastrointestinal polyposis with ectodermal changes: a case with severe malabsorption and enteric loss of plasma proteins and electrolytes. Gastroenterology 50: 107-118, 1966.
Manousos, O.; Webster, C. V.: Diffuse gastrointestinal polyposis with ectodermal changes. Gut 7: 375-378, 1966.

POLYPOSIS, INTESTINAL, WITH MULTIPLE EXOSTOSES (175450)

JJM NOTE

A follow-up investigation might clarify this family's exact disorder.

Fuchs (1975) described multiple exostoses (133700) in a 41-year-old man and his 2 sons, aged 11 and 15 years. The older son had extensive polyposis of the sigmoid colon and gastric antrum. The younger son had x-ray changes suggestive of sigmoid polyposis, whereas x-ray studies in the father were negative for signs of polyposis.

Fuchs, G. A.: Multiple kartilaginaere Exostosen bie Kolon- und Magen-polypose: Mittelungen einer neuen, vom Gardner-syndrome abweichenden erblichen Kombinationserkrankung. Dtsch. Med. Wschr. 100: 2316-2319, 1975.

LEUKEMIA, ACUTE MYELOCYTIC, WITH POLYPOSIS COLI AND COLON CANCER (246470)

JJM NOTE

The final tumor type, in only two brothers, was urinary bladder tumors; clinical and laboratory follow-up, if possible, would be fascinating.

In an Italian-American family with first-cousin parents, Greenberg et al. (1981) described 2 brothers, in a sibship of 5, who had this combination. In 1, colectomy with ileoproctostomy was performed for multiple adenomatous polyps of the colon at age 35. Because of multiple polyps and adenocarcinoma in situ in the residual rectum, abdominoperineal resection and ileostomy were done at age 40. Leukemia of acute myeloblastic type was discovered at age 45 and the patient died after a 2-month course. The second brother had fulguration of 2 papillary tumors of the bladder at age 25. In the next 3 years, he had multiple colonic polypectomies and abdominoperineal resection for Duke's stage C-1 carcinoma of the rectum. At age 34 he was discovered to have acute myeloblastic leukemia and died after a very short course.

Greenberg, M. S.; Anderson, K. C.; Marchetto, D. J.; Li, F. P.: Acute myelocytic leukemia in two brothers with polyposis coli and carcinoma of the colon. Ann. Intern. Med. 95: 702-703, 1981.

POLYPOSIS, GASTRIC (*175020)

JJM NOTE

This is another family that could be clarified by molecular techniques.

Dos Santos and de Magalhaes (1980) described a family in which 10 members of 3 generations had multiple polyposis, with adenocarcinomatous propensities, limited to the stomach. No male-to-male transmission was observed. Seruca et al. (1991) restudied the family reported by dos Santos and de Magalhaes (1980) and added an observation of male-to-male transmission. Severe psoriasis was present in some members of the family as a probable coincidental disorder.

dos Santos, J. G.; de Magalhaes, J.: Familial gastric polyposis: a new entity. J. Genet. Hum. 28: 293-297, 1980.
Seruca, R.; Carneiro, F.; Castedo, S.; David, L.; Lopes, C.; Sobrinho-Simoes, M.: Familial gastric polyposis revisited: autosomal dominant inheritance confirmed. Cancer Genet. Cytogenet. 53: 97-100, 1991.

HEREDITY MIXED POLYPOSIS SYNDROME (*601228)

JJM NOTE

Although the clinical features are not especially characteristic, a single family led to mapping the responsible gene, distinct from other polyposis genes.

The hereditary mixed polyposis syndrome (HMPS) is characterized by atypical juvenile polyps (174900), colonic adenomas, and colorectal carcinomas. Thomas et al. (1996) noted that HMPS patients may also have an increased propensity to develop inflammatory and metaplastic polyps. The authors stated that in a large HMPS family identified at St. Mark's Hospital, London, HMPS appeared to be inherited as an autosomal dominant. Most older individuals presented with colorectal carcinoma; younger individuals, most of whom had undergone screening by colonoscopy, tended to present with benign colonic lesions, often the characteristic polyp of the hereditary mixed polyposis syndrome. Linkage analysis data did not support linkage to the APC locus (175100) on chromosome 5 or to any of the loci for hereditary nonpolyposis colorectal cancer (120435, 120436) known at that time. Thomas et al. (1996) found evidence that the HMPS locus lies on 6q in significant 2-point lod scores for linkage between the phenotype and the D6S283 locus identified by a microsatellite marker (CA repeat). Analysis of recombinants and multipoint linkage analysis suggested that the HMPS locus lies in a 4-cM interval containing the D6S283 locus and flanked by markers D6S468 and D6S301.

It is unclear whether HMPS is a variant of juvenile polyposis or a distinct disease (Murday and Slack (1989)). Thomas et al. (1996) stated that it will be of interest to study linkage between the disease and the putative HMPS locus in juvenile polyposis families. They speculated that if HMPS is indeed a variant of juvenile polyposis, then the distinctive phenotype of the St. Mark's family linked to chromosome 6 (SM96) may result from a specific mutation in a juvenile polyposis gene, from environmental effects, or from the effects of modifier loci. The SM96 family was distributed across the world and subject to many different environments, yet distant relatives produced the characteristic HMPS polyp. The possibility of modifier loci is suggested by the mom1 modifier locus in the mouse, for which a homolog has been identified in humans (172411).

Murday, V.; Slack, J.: Inherited disorders associated with colorectal cancer. Cancer Surv. 8: 139-157, 1989.
Thomas, H. J. W.; Whitelaw, S. C.; Cottrell, S. E.; Murday, V. A.; Tomlinson, I. P. M.; Markie, D.; Jones, T.; Bishop, D. T.; Hodgson, S. V.; Sheer, D.; Northover, J. M. A.; Talbot, I. C.; Solomon, E.; Bodmer, W. F.: Genetic mapping of the hereditary mixed polyposis syndrome to chromosome 6q. Am. J. Hum. Genet. 58: 770-776, 1996.

Oral Tumors

CHEILITIS GLANDULARIS (118330)

JJM NOTE

This condition illustrates, in a small way, the multiple determinants of a rare cancer and the concept of ecogenetics.

Cheilitis glandularis is characterized by enlargement and eversion of the lower lip associated with hypertrophy of the labial mucous glands, dilatation of the excretory ducts, and variable inflammation. In whites, it is associated with a relatively high incidence of squamous cell carcinoma of the lower lip, presumably due to actinic exposure of the mucosa. Weir and Johnson (1971) described the disorder in a black man and his son and daughter.

See also Rada et al. (1985)

Rada, D. C.; Koranda, F. C.; Katz, F. S.: Cheilitis glandularis-a disorder of ductal ectasia. J. Derm. Surg. Oncol. 11: 372-375, 1985.
Weir, T. W.; Johnson, W. C.: Cheilitis glandularis. Arch. Derm. 103: 433-437, 1971.

LIP, HAMARTOMATOUS (151640)

JJM NOTE

Another familial lip tumor, but the observer has, unfortunately, passed away.

Levin (1986) told me of a family with this anomaly in an autosomal dominant pedigree pattern.

Levin, L. S.: Personal Communication. Baltimore, Md. 4/10/1986.

ORAL-FACIAL-DIGITAL SYNDROME 3 (*258850)

JJM NOTE

This entity is included because of the harmartomas of the tongue.

Sugarman et al. (1971) reported a new form of oral-facial-digital syndrome in 2 sisters. Features were mental retardation, eye abnormalities, lobulated hamartomatous tongue, dental abnormalities, bifid uvula, postaxial hexadactyly of hands and feet, pectus excavatum, short sternum, and kyphosis. One of the sibs showed ceaseless 'see-saw winking' of the eyes. The parents were not related. We have observed a family in which 3 of 4 sibs (2 males, 1 female) were affected. None had 'see-saw' winking. However, all had myoclonic jerks, affected lids, extraocular muscles, arms, etc. Hypertelorism, exotropia, irregular teeth, hamartomatous tongue, postaxial polydactyly, and profound mental retardation were features strikingly like those in Sugarman's cases. The sibs showed severe spasticity. One had a macular red spot, leading some to classify the disorder as cerebromacular degeneration (Ford, 1960). Others had suggested Biedl-Bardet syndrome (209900) as the diagnosis, as often happens when the combination of mental retardation and polydactyly is encountered. Also see OFD syndrome type I (311200) and type II (252100). Smith and Gardner-Medwin (1993) described a brother and sister with mental retardation, malformations of the cerebellar vermis, characteristic 'metronome eye movements,' lingual hamartomas, and postaxial polydactyly. Some of the features overlapped with those of OFD syndrome VI (277170) and Joubert syndrome (213300).

Ford, F. R.: Diseases of the Nervous System in Infancy, Childhood and Adolescence. Springfield, Ill.: Charles C Thomas (pub.) (4th ed.): 1960. Pp. 811 only.
Smith, R. A.; Gardner-Medwin, D.: Orofaciodigital syndrome type III in two sibs. J. Med. Genet. 30: 870-872, 1993.
Sugarman, G. I.; Katakia, M.; Menkes, J. H.: See-saw winking in a familial oral-facial-digital syndrome. Clin. Genet. 2: 248-254, 1971.

ORAL-FACIAL-DIGITAL SYNDROME 4 (258860)

JJM NOTE

This entity is included because of the harmartomas of the tongue.

Baraitser (1986) suggested the existence of a fourth type of orofaciodigital syndrome but recognized that considerable overlap of the features of the various forms gives rise to difficulties in precise clinical differentiation. The mode of inheritance and the presence of severe tibial dysplasia differentiate type IV from type I (311200). Burn et al. (1984) described 2 sisters, the children of first-cousin Pakistani Moslem parents, who had unusual facies, tongue hamartomata, pre- and postaxial polydactyly, severe talipes, and mesomelic limb shortening associated with tibial dysplasia. Many of the features resembled those of the Majewski type of short rib-polydactyly syndrome (263520). Indeed, the older of the 2 sibs reported by Burn et al. (1984) had been reported as having either that syndrome (Cooper and Hall, 1982) or the Mohr-Majewski compound (Baraitser et al., 1983). Temtamy and McKusick (1978) had described a patient that they felt showed features of both syndromes. Nevin and Thomas (1989) reported a fifth patient. The clinical characteristics included lobulated tongue, pseudo-cleft of the lip, pre- and postaxial polydactyly of the hands and feet, severe talipes equinovarus, mesomelic limb shortness associated with tibial hypoplasia, and severe bilateral deafness. Meinecke and Hayek (1990) described a severely affected infant. Nevin et al. (1992) added to the genetic understanding of this disorder by describing an affected second cousin of the patient reported by Nevin and Thomas (1989). Ades et al. (1994) entertained a severe form of OFD IV as the diagnosis in a child of first-cousin Lebanese parents who was spontaneously delivered stillborn at 27 weeks' gestation. The fetus showed occipitoschisis, polydactyly, campomelia, cleft palate, laryngeal dysplasia, ocular colobomata, hepatic fibrosis with intrahepatic cyst, ambiguous genitalia, cystic dysplastic kidneys, and brain malformation.

Digilio et al. (1995) described a female patient with OFD IV who, in addition to typical manifestations of the syndrome, had previously undescribed malformations and deformations, including cerebral and renal anomalies, anal atresia, and dislocation of elbows and knees. They stated that this was the tenth reported patient. Toriello et al. (1997) described 6 children diagnosed as having OFD syndrome with tibial defects. They concluded that the patients probably had different conditions, although all had oral, facial, digital, and tibial defects.

Ades, L. C.; Clapton, W. K.; Morphett, A.; Morris, L. L.; Haan, E. A.: Polydactyly, campomelia, ambiguous genitalia, cystic dysplastic kidneys, and cerebral malformation in a fetus of consanguineous parents: a new multiple malformation syndrome, or a severe form of oral-facial-digital syndrome type IV?. Am. J. Med. Genet. 49: 211-217, 1994.

Baraitser, M.: The orofaciodigital (OFD) syndromes. J. Med. Genet. 23: 116-119, 1986.

Baraitser, M.; Burn, J.; Fixsen, J.: A female infant with features of Mohr and Majewski syndromes: variable expression, a genetic compound, or a distinct entity?. J. Med. Genet. 20: 65-67, 1983.

Burn, J.; Dezateux, C.; Hall, C. M.; Baraitser, M.: Orofaciodigital syndrome with mesomelic limb shortening. J. Med. Genet. 21: 189-192, 1984.

Cooper, C. P.; Hall, C. M.: Lethal short-rib polydactyly syndrome of the Majewski type: a report of three cases. Radiology 144: 513-517, 1982.

Digilio, M. C.; Giannotti, A.; Pagnotta, G.; Mingarelli, R.; Dallapiccola, B.: Joint dislocation and cerebral anomalies are consistently associated with oral-facial-digital syndrome type IV. Clin. Genet. 48: 156-159, 1995.

Meinecke, P.; Hayek, H.: Orofaciodigital syndrome type IV (Mohr-Majewski syndrome) with severe expression expanding the known spectrum of anomalies. J. Med. Genet. 27: 200-202, 1990.

Nevin, N. C.; Magee, A. C.; Mudenda, V.; Thomas, P. S.: Orofaciodigital syndrome type IV: report of a patient. (Letter) Am. J. Med. Genet. 43: 902-904, 1992.

Nevin, N. C.; Thomas, P. S.: Orofaciodigital syndrome type IV: report of a patient. Am. J. Med. Genet. 32: 151-154, 1989.

Temtamy, S. A.; McKusick, V. A.: The Genetics of Hand Malformations. New York: Alan R. Liss (pub.) 1978.

Toriello, H. V.; Carey, J. C.; Suslak, E.; Desposito, F. R.; Leonard, B.; Lipson, M.; Friedman, B. D.; Hoyme, H. E.: Six patients with oral-facial-digital syndrome IV: the case for heterogeneity. Am. J. Med. Genet. 69: 250-260, 1997.

BRANCHIAL CLEFTS WITH CHARACTERISTIC FACIES, GROWTH RETARDATION, IMPERFORATE NASOLACRIMAL DUCT, AND PREMATURE AGING (*113620)

JJM NOTE

This entity is included because of the hemangiomatous branchial clefts and periorbital cysts, as well as the subcutaneous scalp cysts.

Lee et al. (1982) described a 38-year-old woman and her 8-year-old son who had low birth weight for dates and retarded postnatal growth, bilateral branchial cleft sinuses, congenital strabismus, obstructed nasolacrimal ducts, broad nasal bridge, protruding upper lip, and carp mouth. Graying of the mother's hair occurred at age 18. Intelligence was normal. The same disorder may have been reported by Hall et al. (1983) and Fujimoto et al. (1987). Hall et al. (1983) described 2 unrelated children (1 male, 1 female) with hemangiomatous branchial clefts and pseudocleft of the upper lip (resembling a surgically repaired cleft or a fused cleft). They found reports of 2 additional patients who, they suspected, also represented sporadic cases of this syndrome. In several persons in 3 families, Fujimoto et al. (1987) observed an autosomal dominant disorder of abnormal upper lip, which resembled a poorly repaired median cleft lip, malformed nose with broad bridge and flattened tip, lacrimal duct obstruction, malformed ears, and branchial cleft sinuses and/or linear skin lesions behind the ears. In each of the 3 families an affected parent had at least 1 affected child, and father-to-son transmission was observed in 1. Other anomalies included coloboma, microphthalmia, auricular pits, lip pits, highly arched palate, dental anomalies, and subcutaneous cysts of the scalp. Premature greying of hair occurred in affected adults. The abnormality of the upper lip might be described as an unusually broad and prominent philtrum.

Mazzone et al. (1992) reported a patient who, in addition to typical features of BOFS, had partial agenesis of the cerebellar vermis. Lin et al. (1992) concluded that the father and son reported by Legius et al. (1990) had the BOF syndrome and that this additional finding of male-to-male transmission confirmed autosomal dominant inheritance. Fielding and Fryer (1992) described 2 sibs with this syndrome, each of whom also had orbital hemangiomatous cysts. Both parents were clinically normal and unrelated. Thus this may have represented an autosomal recessive form of the disorder or germline mosaicism for the dominant gene. Schmerler et al. (1992) reviewed the development of an affected child over a 12-year period of observation. Normal intelligence, regular class placement, hypernasal speech, and continued growth along the third centile were noted. The infant had been referred at the age of 5 months for evaluation of his facial appearance and 'burn-like' lesions behind both ears. McCool and Weaver (1994) observed the BOF syndrome in a mother and her son who lacked the ocular and branchial abnormalities but had bilateral supraauricular sinuses and hearing loss. The son had bilateral cleft lip and right alveolar cleft; the mother had asymmetric nostrils and upper lip. The supraauricular sinuses were thought to represent persistence of the otic vesicle sinus tract.

Lin et al. (1995) described 15 new observations of the BOF syndrome and reviewed previously reported cases (28 with typical and 5 with atypical manifestations) in detail. Postauricular cervical branchial defects were found in 40 of 43 patients, and supraauricular defects were found in 6. Pathologic findings of the excised branchial defects showed thymic remnants in several cases. Colobomata were found in 16 of 35 patients, cataracts in 8 of 33, deafness in 14 of 38, scalp cysts in 4 of 38, and premature graying of hair in 9 of 38. Pseudoclefts were observed in 23 patients, and cleft lip and/or palate in 20. Urologic examination of 19 patients revealed kidney abnormalities (agenesis, cysts, hydronephrosis) in 7. Autosomal dominant inheritance of the BOF syndrome is supported by a 3-generation German family, 2 instances of father-to-son transmission, and 7 other parent-offspring families (Fujimoto et al., 1987; Lin et al., 1995).

Richardson et al. (1996) described a boy with cleft lip and palate, microphthalmos, colobomata of optic nerves and irides, and cystic dysplasia of the left kidney. His mother had similar ocular abnormalities (plus polycoria), obstruction of nasolacrimal ducts, bifid nasal tip, abnormal philtrum, hypodontia, and premature graying of the hair. His maternal grandmother had the same facial defects and nasolacrimal duct obstruction, but normal eyes. The spectrum of abnormalities in this family fits the BOF syndrome, although cervical hemangiomata or branchial sinuses were not found in affected persons in this family.

See also Legius and Fryns (1992)

Fielding, D. W.; Fryer, A. E.: Recurrence of orbital cysts in the branchio-oculo-facial syndrome. J. Med. Genet. 29: 430-431, 1992.

Fujimoto, A.; Lipson, M.; Lacro, R. V.; Shinno, N. W.; Boelter, W. D.; Jones, K. L.; Wilson, M. G.: New autosomal dominant branchio-oculo-facial syndrome. Am. J. Med. Genet. 27: 943-951, 1987.
Hall, B. D.; deLorimier, A.; Foster, L. H.: A new syndrome of hemangiomatous branchial clefts, lip pseudoclefts, and unusual facial appearance. Am. J. Med. Genet. 14: 135-138, 1983.
Lee, W. K.; Root, A. W.; Fenske, N.: Bilateral branchial cleft sinuses associated with intrauterine and postnatal growth retardation, premature aging, and unusual facial appearance: a new syndrome with dominant transmission. Am. J. Med. Genet. 11: 345-352, 1982.
Legius, E.; Fryns, J.-P.: Reply to Dr. Lin. (Letter) Clin. Genet. 41: 223 only, 1992.
Legius, E.; Fryns, J. P.; Van Den Berghe, H.: Dominant branchial cleft syndrome with characteristics of both branchio-oto-renal and branchio-oculo-facial syndrome. Clin. Genet. 37: 347-350, 1990.
Lin, A. E.; Doherty, R.; Lea, D.: Branchio-oculo-facial and branchio-oto-renal syndromes are distinct entities. (Letter) Clin. Genet. 41: 221-222, 1992.
Lin, A. E.; Gorlin, R. J.; Lurie, I. W.; Brunner, H. G.; van der Burgt, I.; Naumchik, I. V.; Rumyantseva, N. V.; Stengel-Rutkowski, S.; Rosenbaum, K.; Meinecke, P.; Muller, D.: Further delineation of the branchio-oculo-facial syndrome. Am. J. Med. Genet. 56: 42-59, 1995.
Mazzone, D.; Milana, A.; Carpinato, C.: Branchio-oculo-facial syndrome: report of a new case with agenesis of cerebellar vermis. Europ. J. Pediat. 151: 312 only, 1992.
McCool, M.; Weaver, D. D.: Branchio-oculo-facial syndrome: broadening the spectrum. Am. J. Med. Genet. 49: 414-421, 1994.
Richardson, E.; Davison, C.; Moore, A.: Colobomatous microphthalmia with midfacial clefting: part of the spectrum of branchio-oculo-facial syndrome? Ophthal. Genet. 17: 59-65, 1996.
Schmerler, S.; Kushnick, T.; Desposito, F.: Long-term evaluation of a child with the branchio-oculo-facial syndrome. Am. J. Med. Genet. 44: 177-178, 1992.
duct . . .

LICHEN PLANUS, FAMILIAL (151620)

JJM NOTE

This entity is included (as a stretch) because of the premalignant nature of lichen planus in the mouth and because the features in families, as described, evoke the characteristics of a neoplasia of genetic origins (early onset, atypical nature).

Sodaify and Vollum (1978) described a Jewish family in Shiraz, Iran, in which the mother and 4 sons and a daughter from a sibship of 9 had lichen planus. Copeman et al. (1978) reported 5 families, each with 2 affected members—mother-offspring in four and 2 brothers in the fifth. They found an increased frequency of HLA-B7 (in 8 of the 10). They suggested that familial lichen planus affects younger persons and recurs and erupts more acutely, extensively and severely, attacking also nails and mucous membranes, than is the case with nonfamilial lichen planus. They concluded: 'We hint that their genotype might have rendered them susceptible to a pathogen that precipitated their disease.' Mahood (1983) reported 9 affected persons in 4 families. The disease appeared to differ from the nonfamilial form in an earlier age of onset, tendency to chronicity, increased severity, and the frequent presence of 'atypical' disease. Valsecchi et al. (1990) described an Italian family in which the 68-year-old father and his monozygotic 34-year-old twin sons were affected. The father had his first attack of lichen planus at the age of 54 years with involvement of the entire trunk. There was no involvement of the mucous membranes. The eruption was controlled with topical steroid therapy but recurred after a remission of 1 year. Systemic prednisolone was administered. The twins developed typical lesions within 1 year of each other and both were treated with steroids, either topical or systemic.

See also Saffron (1940)

Copeman, P. W. M.; Tan, R. S.-H.; Timlin, D.; Samman, P. D.: Familial lichen planus: another disease or a distinct people?. Brit. J. Derm. 98: 573-577, 1978.
Mahood, J. M.: Familial lichen planus: a report of nine cases from four families with a brief review of the literature. Arch. Derm. 119: 292-294, 1983.
Saffron, M. H.: Familial lichen planus: a report of four cases of lichen planus in one family, with a brief review of the literature. Arch. Derm. Syph. 42: 653-655, 1940.
Sodaify, M.; Vollum, D. I.: Familial lichen planus: a case report. Brit. J. Derm. 98: 579-581, 1978.
Valsecchi, R.; Bontempelli, M.; di Landro, A.; Barcella, A.; Lainelli, T.: Familial lichen planus. (Letter) Acta Derm. Venerol. 70: 272-273, 1990.

FOCAL EPITHELIAL HYPERPLASIA OF THE ORAL MUCOSA (136400)

JJM NOTE

This hyperplasia is chosen for its ethnic concentration, although environmental determinants may well be sufficient explanation.

Most cases of this rare lesion have been nonfamilial, although Schock (1969) described the disorder in an Indian woman and 3 of her daughters. The majority of cases have been in Eskimos and in American Indians, from the Warm Springs Indians of Oregon to the Chavante Indians of Brazil. The disorder may be viral rather than genetic.

Schock, R. K.: Familial focal epithelial hyperplasia. Report of a case. Oral Surg. 28: 598-602, 1969.

Esophageal Tumors

GASTROINTESTINAL STROMAL TUMORS (*164920)

JJM NOTE

These tumors seem to arise from cells of Cajal, imbedded in intestinal trait musculature and essential for peristalsis. Sporadic tumors of various gastrointestinal sites have had somatic mutations in the gene, and one familial aggregation had a germline mutation (Nishida et al. 1998). The gene is the oncogene *kit* and the name of this entry in OMIM is "*v-kit* Hardy-Zuckerman 4 feline sarcoma viral oncogene homolog." The same tumor type has complicated systemic mast-cell disease. Other malignancies with somatic mutation in *kit* are mast cell leukemia or mastocytosis and acute myeloid leukemia of the M5 subtype.

The provirus of the Hardy-Zuckerman 4 feline sarcoma virus was molecularly cloned. A segment from the middle of the provirus, showing homology to mammalian genomic DNA, was termed v-kit. Its human homolog was assigned to chromosome 4 by Barker et al. (1985), using human-mouse somatic cell hybrids. By in situ hybridization, Mattei et al. (1987) mapped the KIT gene to 4q11-q12, the largest number of grains being in the q12 band; see d'Auriol et al. (1988). By the same method, Yarden et al. (1987) assigned the KIT locus to 4cen-q21. They demonstrated that the homologous gene is on chromosome 5 in the mouse.

Geissler et al. (1988) showed that in the mouse the kit gene is disrupted in 2 spontaneous mutant W alleles. It was previously known that mutations at the W locus in the mouse (producing changes, including white-coat color, sterility, and anemia, attributable to failure of stem cell populations to migrate and/or proliferate effectively during development) map in the same region as kit. Thus, the KIT protooncogene, which encodes a putative tyrosine kinase receptor, was a candidate for the W locus. The protooncogene KIT encodes a transmembrane tyrosine kinase. A strong structural homology of KIT to the CSF1 receptor (164770) and the platelet-derived growth factor receptor (173490) suggests a model for the action of the KIT gene product during differentiation. The phenotypes of W mutants suggest that 3 cell populations in which KIT function is critical are the pluripotent hematopoietic stem cell, the migrating melanoblast during early embryonic development, and the primordial germ cell during this same period of development. Identification of the ligand for KIT will help in the understanding of its function. A genetic clue to its nature may be found through characterization of the Sl ('steel') locus of the mouse. The phenotypes of mutants at this locus closely resemble mutants at the W locus; however, unlike W, the defect in Sl is not intrinsic to the progenitor stem cells of the affected tissues, but rather lies in the environment in which melanoblast, germ cell, and hematopoietic progenitors differentiate and proliferate. Chabot et al. (1988) also related kit to a W mutation in the mouse. These findings will prompt search for comparable changes in disorders such as Fanconi anemia (227650). In the dominant W(42) spotting phenotype in the mouse, Tan et al. (1990) demonstrated an asp790-to-asparagine mutation in the KIT protein product. Aspartic acid-790 is a conserved residue in all protein kinases. Nocka et al. (1990) identified mutations in other alleles at the W locus: W(37), E-to-K at position 582; W(v), T-to-M at position 660; and W(41), V-to-M at position 831. The W mutation is the result of a 78-amino acid deletion that includes the transmembrane of the KIT protein. Nocka et al. (1990) detected a 125-kD KIT protein in homozygous W/W mast cells which lacked kinase activity and did not express KIT on the cell surface. Thus, in mice, the c-kit receptor tyrosine kinase is the gene product of the W locus, whereas Sl encodes the ligand for this growth factor receptor. Microphthalmia (mi/mi) in mice also shows deficiency in melanocytes and mast cells. In addition, whereas W and Sl mutants can be anemic and sterile, 'mi' mice are osteopetrotic due to a monocyte/macrophage defect. Dubreuil et al. (1991) found that the fms gene (164770) complements the mitogenic defect in mast cells of mutant W mice but not of mi/mi mice. Brannan et al. (1991) detected a HaeIII polymorphism in the KIT gene. This polymorphism was linked to other 4q markers. Poduslo et al. (1991) described a strategy for identifying polymorphic genetic markers in the 3-prime untranslated regions of coding loci. The method combined PCR and single-strand conformation polymorphism (SSCP) analysis.

Vandenbark et al. (1992) demonstrated that the KIT gene spans more than 70 kb of DNA and includes 21 exons. The longest transcript is 5,230 bp and is alternatively spliced. Its overall gene structure closely resembles that of the CSF1R gene. Using pulsed field gel electrophoresis, Vandenbark et al. (1992) demonstrated that the KIT gene and the PDGFRA gene (173490) reside on the same 700 kb BssHI fragment. Spritz et al. (1992) found deletion of both the KIT gene and the PDGRFA gene in a patient with piebaldism, mental retardation, and multiple congenital anomalies associated with a 46,XY,del(4)(q12q21.1) karyotype. The patient was hemizygous for the 2 deleted genes. This might suggest that the KIT gene is located in 4q12; the PDGFRA gene is considered to be in 4q12.

The pacemaker activity in the mammalian gut is responsible for generating anally propagating phasic contractions. Although the cellular basis for this intrinsic activity is unknown, the smooth muscle cells of the external muscle layers and the innervated cellular network of interstitial cells of Cajal (ICC), which is closely associated with the external muscle layers, have both been proposed to stimulate pacemaker activity. The interstitial cells of Cajal were identified in the 19th century but their developmental origin and function remained unclear until the studies of Huizinga et al. (1995). Injection of antibodies directed against the extracellular domain of kit into newborn mice leads to changes in the in vitro contraction patterns in the small intestine and absence of kit mRNA in the myenteric plexus area. Huizinga et al. (1995) found that mice with W mutations at the kit locus lacked the network of ICC in the myenteric plexus region. Using a polyclonal antibody directed against the intracellular domain of the kit receptor tyrosine kinase, they demonstrated that wildtype and heterozygous mice showed high levels of Kit expression between the longitudinal and circular muscle layers at the level of the Auerbach plexus. By contrast, no kit immunoreactivity was found in or between the muscle layers of homozygous W/W mice. Huizinga et al. (1995) suggested that functional gut abnormalities and megacolon observed in patients with piebaldism (Bolognia and Pawelek, 1988) reflects an identical function of the KIT signaling pathway in the development of the interstitial cells of Cajal in humans; see 172800 for a discussion of megacolon in association with piebald trait. Mutations in RET (164761), another member of the receptor tyrosine kinase family that is expressed in neural crest-derived ganglion cells, are also associated with megacolon.

Gastrointestinal stromal tumors (GISTs) are the most common mesenchymal neoplasms in the human digestive tract. Hirota et al. (1998) investigated the mutational status of KIT in 58 mesenchymal tumors that developed in the GI wall (4 in the esophagus, 36 in the stomach, 14 in the small intestine, and 4 in the large intestine). KIT expression was examined by immunohistochemistry. Eight authentic glial leiomyomas and an authentic schwannoma did not express KIT. The remaining 49 mesenchymal tumors were diagnosed as GISTs, and 94% (46 of 49) of these expressed KIT. Examination of these tumors for expression of CD34 (142230), which is a reliable marker for GISTs, revealed that 82% (40 of 49) were CD34-positive, and 78% (38 of 49) were positive for both KIT and CD34. Three of 5 KIT-negative GISTs were also CD34-negative. Hirota et al. (1998) compared the immunohistochemical characteristic of GISTs with those of interstitial cells of Cajal (ICCs), which regulate autonomous contraction of the GI tract. They found that ICCs are double-positive for KIT and CD34. In 5 GISTs, they found mutations in the region between the transmembrane and tyrosine kinase domains (e.g., 164920.0011). All of the corresponding mutant KIT proteins were constitutively activated without the KIT ligand, stem cell factor (SCF; 184745). Stable transfection of the mutant KIT cDNAs induced malignant transformation of murine lymphoid cells, suggesting that the mutations contribute to tumor development. Hirota et al. (1998) suggested that GISTs may originate from the interstitial cells of Cajal (ICCs) because the development of ICCs is dependent on the SCF-KIT interaction and because, like GISTs, these cells express both KIT and CD34. It is noteworthy that the 5 mutations identified in GISTs lay between codons 550 and codon 559. One of the 2 missense mutations found in patients with mast cell leukemia or mastocytosis (164920.0009) was a val560-to-gly mutation.

As outlined earlier, networks of interstitial cells of Cajal embedded in the musculature of the gastrointestinal tract are involved in the generation of electrical pacemaker activity for gastrointestinal motility. This pacemaker activity manifests itself as rhythmic slow waves in membrane potential, and controls the frequency and propagation characteristics of gut contractile activity. Mice that lack a functional Kit receptor failed to develop the network of interstitial cells of Cajal. Thomsen et al. (1998) provided direct evidence that a single Cajal cell generates spontaneous contractions and a rhythmic inward current that is insensitive to L-type calcium channel blockers.

Most GISTs are solitary and the gain-of-function mutations found by Hirota et al. (1998) were somatic. Nishida et al. (1998) described a family with multiple GISTs. Affected members all had a KIT mutation occurring between the transmembrane and tyrosine kinase domains, which is also the region where mutations had been demonstrated in solitary GISTs. The KIT mutation in this family was detected not only in tumors but also in leukocytes, indicating that GISTs constitute a familial cancer syndrome. Seven individuals in 5 sibships of 4 generations of the family were affected by either benign and/or malignant GISTs. Nishida et al. (1998) pointed to the woman reported by El-Omar et al. (1994) to be a probable case of benign and malignant GISTs associated with hyperpigmentation of the perineal skin. Marshall et al. (1990) reported a family in which several members suffered from benign GISTs associated with either urticaria pigmentosa or systemic mast-cell disease. Two members of the family reported by Nishida et al. (1998) were reported to have hyperpigmentation of the perineum.

Comparative mapping data suggested that the dominant white coat color in pigs may be due to a mutation in KIT. Moller et al. (1996) reported that dominant white pigs lack melanocytes in the skin, as would be anticipated for a KIT mutation. They found, furthermore, a complete association between the dominant white mutation and a duplication of the KIT gene, or part of it, in samples of unrelated pigs representing 6 different breeds. Duplication was revealed by SSCP analysis and subsequent sequence analysis showing that white pigs transmitted 2 nonallelic KIT sequences. The presence of a gene duplication in white pigs was confirmed with quantitative Southern blot, quantitative PCR, and fluorescence in situ hybridization (FISH) analyses. FISH analysis showed that KIT and the very closely linked gene encoding platelet-derived growth factor receptor are located on pig 8p12. The results of Moller et al. (1996) demonstrated an extremely low rate of recombination in the centromeric region of this chromosome since the closely linked (0.5 cM) serum albumin (103600) locus had previously been mapped by FISH to 8q12. The authors noted that pig chromosome 8 shares extensive conserved synteny with human chromosome 4, but the gene order is rearranged.

Marklund et al. (1998) reported that the dominant white phenotype in domestic pigs is caused by 2 mutations in the KIT gene, 1 gene duplication associated with a partially dominant phenotype and a splice mutation in 1 of the copies leading to the fully dominant allele. The splice mutation is a G-to-A substitution in the first nucleotide of intron 17, leading to a skipping of exon 17. The duplication is most likely a regulatory mutation affecting KIT expression, whereas the splice mutation is expected to cause a receptor with impaired or absent tyrosine kinase activity. Immunocytochemistry showed that this variant form is expressed in 17- to 19-day-old pig embryos. Hundreds of millions of white pigs around the world are assumed to be heterozygous or homozygous for the 2 mutations.

See also Spritz et al. (1992); Spritz et al. (1992)

Barker, P. E.; Besmer, P.; Ruddle, F. H.: Human c-kit oncogene on human chromosome 4. (Abstract) Am. J. Hum. Genet. 37: A143, 1985.

Beghini, A.; Larizza, L.; Cairoli, R.; Morra, E.: c-kit activating mutations and mast cell proliferation in human leukemia. (Letter) Blood 92: 701-702, 1998.

Bolognia, J. L.; Pawelek, J. M.: Biology of hypopigmentation. J. Am. Acad. Derm. 19: 217-255, 1988.

Brannan, C. I.; Lyman, S. D.; Williams, D. E.; Eisenman, J.; Anderson, D. M.; Cosman, D.; Bedell, M. A.; Jenkins, N. A.; Copeland, N. G.: Steel-Dickie mutation encodes a c-kit ligand lacking transmembrane and cytoplasmic domains. Proc. Nat. Acad. Sci. 88: 4671-4674, 1991.

Chabot, B.; Stephenson, D. A.; Chapman, V. M.; Besmer, P.; Bernstein, A.: The proto-oncogene c-kit encoding a transmembrane tyrosine kinase receptor maps to the mouse W locus. Nature 335: 88-89, 1988.

d'Auriol, L.; Mattei, M.-G.; Andre, C.; Galibert, F.: Localization of the human c-kit protooncogene on the q11-q12 region of chromosome 4. Hum. Genet. 78: 374-376, 1988.

Dubreuil, P.; Forrester, L.; Rottapel, R.; Reedijk, M.; Fujita, J.; Bernstein, A.: The c-fms gene complements the mitogenic defect in mast cells derived from mutant W mice but not mi (microphthalmia) mice. Proc. Nat. Acad. Sci. 88: 2341-2345, 1991.

el-Omar, M.; Davies, J.; Gupta, S.; Ross, H.; Thompson, R.: Leiomyosarcoma in leiomyomatosis of the small intestine. Postgrad. Med. J. 70: 661-664, 1994.

Fleischman, R. A.: Human piebald trait resulting from a dominant negative mutant allele of the c-kit membrane receptor gene. J. Clin. Invest. 89: 1713-1717, 1992.

Fleischman, R. A.; Saltman, D. L.; Stastny, V.; Zneimer, S.: Deletion of the c-kit protooncogene in the human developmental defect piebald trait. Proc. Nat. Acad. Sci. 88: 10885-10889, 1991.
Furitsu, T.; Tsujimura, T.; Tono, T.; Ikeda, H.; Kitayama, H.; Koshimizu, U.; Sugahara, H.; Butterfield, J. H.; Ashman, L. K.; Kanayama, Y.; Matsuzawa, Y.; Kitamura, Y.; Kanakura, Y.: Identification of mutations in the coding sequence of the proto-oncogene c-kit in a human mast cell leukemia cell line causing ligand-independent activation of c-kit product. J. Clin. Invest. 92: 1736-1744, 1993.
Geissler, E. N.; Ryan, M. A.; Housman, D. E.: The dominant-white spotting (W) locus of the mouse encodes the c-kit proto-oncogene. Cell 55: 185-192, 1988.
Giebel, L. B.; Spritz, R. A.: Mutation of the KIT (mast/stem cell growth factor receptor) protooncogene in human piebaldism. Proc. Nat. Acad. Sci. 88: 8696-8699, 1991.
Hirota, S.; Isozaki, K.; Moriyama, Y.; Hashimoto, K.; Nishida, T.; Ishiguro, S.; Kawano, K.; Hanada, M.; Kurata, A.; Takeda, M.; Tunio, G. M.; Matsuzawa, Y.; Kanakura, Y.; Shinomura, Y.; Kitamura, Y.: Gain-of-function mutations of c-kit in human gastrointestinal stromal tumors. Science 279: 577-580, 1998.
Huizinga, J. D.; Thuneberg, L.; Kluppel, M.; Malysz, J.; Mikkelsen, H. B.; Bernstein, A.: W/kit gene required for interstitial cells of Cajal and for intestinal pacemaker activity. Nature 373: 347-349, 1995.
Longley, B. J.; Tyrrell, L.; Lu, S.-Z.; Ma, Y.-S.; Langley, K.; Ding, T.; Duffy, T.; Jacobs, P.; Tang, L. H.; Modlin, I.: Somatic c-KIT activating mutation in urticaria pigmentosa and aggressive mastocytosis: establishment of clonality in a human mast cell neoplasm. Nature Genet. 12: 312-314, 1996.
Marklund, S.; Kijas, J.; Rodriguez-Martinez, H.; Ronnstrand, L.; Funa, K.; Moller, M.; Lange, D.; Edfors-Lilja, I.; Andersson, L.: Molecular basis for the dominant white phenotype in the domestic pig. Genome Res. 8: 826-833, 1998.
Marshall, J. B.; Diaz-Arias, A. A.; Bochna, G. S.; Vogele, K. A.: Achalasia due to diffuse esophageal leiomyomatosis and inherited as an autosomal dominant disorder. Gastroenterology 98: 1358-1365, 1990.
Mattei, M. G.; d'Auriol, L.; Andre, C.; Passage, E.; Mattei, J. F.; Galibert, F.: Assignment of the human c-kit proto-oncogene to the q11-q12 region of chromosome 4, using in situ hybridization. (Abstract) Cytogenet. Cell Genet. 46: 657, 1987.
Moller, M. J.; Chaudhary, R.; Hellmen, E.; Hoyheim, B.; Chowdhary, B.; Andersson, L.: Pigs with the dominant white coat color phenotype carry a duplication of the KIT gene encoding the mast/stem cell growth factor receptor. Mammalian Genome 7: 822-830, 1996.
Nagata, H.; Worobec, A. S.; Oh, C. K.; Chowdhury, B. A.; Tannenbaum, S.; Suzuki, Y.; Metcalfe, D. D.: Identification of a point mutation in the catalytic domain of the protooncogene c-kit in peripheral blood mononuclear cells of patients who have mastocytosis with an associated hematologic disorder. Proc. Nat. Acad. Sci. 92: 10560-10564, 1995.
Nishida, T.; Hirota, S.; Taniguchi, M.; Hashimoto, K.; Isozaki, K.; Nakamura, H.; Kanakura, Y.; Tanaka, T.; Takabayashi, A.; Matsuda, H.; Kitamura, Y.: Familial gastrointestinal stromal tumours with germline mutation of the KIT gene. (Letter) Nature Genet. 19: 323-324, 1998.
Nocka, K.; Tan, J. C.; Chiu, E.; Chu, T. Y.; Ray, P.; Traktman, P.; Besmer, P.: Molecular bases of dominant negative and loss of function mutations at the murine c-kit/white spotting locus: W-37, W-v, W-41 and W. EMBO J. 9: 1805-1813, 1990.
Pignon, J.-M.; Giraudier, S.; Duquesnoy, P.; Jouault, H.; Imbert, M.; Vainchenker, W.; Vernant, J.-P.; Tulliez, M.: A new c-kit mutation in a case of aggressive mast cell disease. Brit. J. Haemat. 96: 374-376, 1997.
Poduslo, S. E.; Dean, M.; Kolch, U.; O'Brien, S. J.: Detecting high-resolution polymorphisms in human coding loci by combining PCR and single-strand conformation polymorphism (SSCP) analysis. Am. J. Hum. Genet. 49: 106-111, 1991.
Selmanowitz, V. J.; Rabinowitz, A. D.; Orentreich, N.; Wenk, E.: Pigmentary correction of piebaldism by autografts. I. Procedures and clinical findings. J. Derm. Surg. Oncol. 3: 615-622, 1977.
Spritz, R. A.; Beighton, P.: Piebaldism with deafness: molecular evidence for an expanded syndrome. Am. J. Med. Genet. 75: 101-103, 1998.
Spritz, R. A.; Droetto, S.; Fukushima, Y.: Deletion of the KIT and PDGFRA genes in a patient with piebaldism. Am. J. Med. Genet. 44: 492-495, 1992.
Spritz, R. A.; Giebel, L. B.: Mutation of the c-kit (mast/stem cell growth factor receptor) proto-oncogene in human piebaldism. (Abstract) Am. J. Hum. Genet. 49 (suppl.): 38, 1991.
Spritz, R. A.; Giebel, L. B.; Holmes, S. A.: Dominant negative and loss of function mutations of the c-kit (mast/stem cell growth factor receptor) proto-oncogene in human piebaldism. Am. J. Hum. Genet. 50: 261-269, 1992.
Spritz, R. A.; Holmes, S. A.; Ramesar, R.; Greenberg, J.; Curtis, D.; Beighton, P.: Mutations of the KIT (mast/stem cell growth factor receptor) proto-oncogene account for a continuous range of phenotypes in human piebaldism. Am. J. Hum. Genet. 51: 1058-1065, 1992.
Tan, J. C.; Nocka, K.; Ray, P.; Traktman, P.; Besmer, P.: The dominant W42 spotting phenotype results from a missense mutation in the c-kit receptor kinase. Science 247: 209-212, 1990.
Thomsen, L.; Robinson, T. L.; Lee, J. C. F.; Farraway, L. A.; Hughes, M. J. G.; Andrews, D. W.; Huizinga, J. D.: Interstitial cells of Cajal generate a rhythmic pacemaker current. Nature Med. 4: 848-851, 1998.
Tsujimura, T.; Furitsu, T.; Morimoto, M.; Isozaki, K.; Nomura, S.; Matsuzawa, Y.; Kitamura, Y.; Kanakura, Y.: Ligand-independent activation of c-kit receptor tyrosine kinase in a murine mastocytoma cell line P-815 generated by a point mutation. Blood 83: 2619-2626, 1994.
Vandenbark, G. R.; deCastro, C. M.; Taylor, H.; Dew-Knight, S.; Kaufman, R. E.: Cloning and structural analysis of the human c-kit gene. Oncogene 7: 1259-1266, 1992.
Yarden, Y.; Kuang, W.-J.; Yang-Feng, T.; Coussens, L.; Munemitsu, S.; Dull, T. J.; Chen, E.; Schlessinger, J.; Francke, U.; Ullrich, A.: Human proto-oncogene c-kit: a new cell surface receptor tyrosine kinase for an unidentified ligand. EMBO J. 6: 3341-3351, 1987.

BARRETT ESOPHAGUS (109350)

JJM NOTE

This disorder is another candidate ecogenetic trait associated with cancer. My colleagues and I reported a 33-year-old female with Carney syndrome (160980) complicated by Barrett esophageal metaplasia (Stadler MP et al.: The broadened spectrum of malignancy and premalignancy in the Carney complex. Am. J. Hum. Genet. 53[Suppl.]:369, 1993).

Barrett (1950) described a patient with chronic ulcerating esophagitis in which columnar rather than squamous epithelium surrounded the ulcers. Allison and Johnstone (1953), followed by many others, showed that the columnar epithelium-lined intrathoracic structure is anatomically and functionally esophagus. The proximal esophagus usually retains its normal squamous epithelium. The Barrett esophagus is a complication of gastroesophageal reflux. Why it develops only in some patients is not clear; Sjogren and Johnson (1983) suggested that it 'may be congenitally determined in part.' Familial occurrence was reported by Borrie and Goldwater (1976). Adenocarcinoma of the esophagus has an incidence of about 10% in the Barrett esophagus. Adenocarcinoma constitutes a minority of esophageal cancers but most of these originate in a Barrett esophagus. Gelfand (1983) reported Barrett esophagus in identical twins. 7,6:Everhart et al. (1978, 1983) described Barrett esophagus in 3 persons in 2 generations of a family. Crabb et al. (1985) described a family in which the proband had both GER and Barrett esophagus, 3 of 5 children also had both, the other 2 children had only GER, and 2 grandchildren had GER. One of the children with both developed adenocarcinoma of the esophagus. We have seen adenocarcinoma of the esophagus in a man (G.D., 1474651) with the Barrett anomaly and a brother who died of esophageal cancer. Prior and Whorwell (1986) observed 2 sisters, each of whom presented at age 66. Jochem et al. (1992) described a family with 6 cases of Barrett esophagus, all in males, in 3 successive generations. In 3 of the patients there was associated adenocarcinoma. Cameron (1992) reviewed the literature.

See also Mossberg (1966)

Allison, P. R.; Johnstone, A. S.: The esophagus lined with gastric mucus membrane. Thorax 8: 87-101, 1953.
Barrett, N. R.: Chronic peptic ulcer of the oesophagus and esophagitis. Brit. J. Surg. 38: 175-182, 1950.
Borrie, J.; Goldwater, L.: Columnar cell-lined esophagus: assessment of etiology and treatment: a 22-year experience. J. Thorac. Cardiovasc. Surg. 71: 825-834, 1976.
Cameron, A. J.: Barrett's esophagus and adenocarcinoma: from the family to the gene. (Editorial) Gastroenterology 102: 1421-1424, 1992.
Crabb, D. W.; Berk, M. A.; Hall, T. R.; Conneally, P. M.; Biegel, A. A.; Lehman, G. A.: Familial gastroesophageal reflux and development of Barrett's esophagus. Ann. Intern. Med. 103: 52-54, 1985.
Everhart, C. W.; Holtzapple, P. G.; Humphries, T. J.: Barrett's esophagus: inherited epithelium or inherited reflex?. (Editorial) J. Clin. Gastroent. 5: 357-358, 1983.
Everhart, C. W., Jr.; Holtzapple, P. G.; Humphries, T. J.: Occurrence of Barrett's esophagus in three members of the same family: first report of familial incidence. (Abstract) Gastroenterology 74: 1032 only, 1978.
Gelfand, M. D.: Barrett's esophagus in sexagenarian identical twins. J. Clin. Gastroent. 5: 251-253, 1983.
Jochem, V. J.; Fuerst, P. A.; Fromkes, J. J.: Familial Barrett's esophagus associated with adenocarcinoma. Gastroenterology 102: 1400-1402, 1992.
Mossberg, S. M.: The columnar-lined esophagus (Barrett syndrome)—an acquired condition?. Gastroenterology 50: 671-676, 1966.
Prior, A.; Whorwell, P. J.: Familial Barrett's oesophagus?. Hepatogastroenterology 33: 86-87, 1986.
Sjogren, R. W., Jr.; Johnson, L. F.: Barrett's esophagus: a review. Am. J. Med. 74: 313-321, 1983.

LEIOMYOMA OF VULVA AND ESOPHAGUS (150700)

JJM NOTE

The cytogenetic and molecular approach of Surti et al. to the common leiomyoma of the uterus might yield insight into the clonality and origins of these rare tumors (Lynch RA et al.: Genomic and functional map of the chromosome 14 t(12;14) breakpoint cluster region in uterine leiomyoma. Genomics 52:17-26, 1998; Ishwad CS et al.: Two discrete regions of deletion at 7q in uterine leiomyomas. Genes Chromosomes Cancer 19:156-60, 1997).

Wahlen and Astedt (1965) described this combination in mother and daughter. The esophageal tumor was an obstructing lesion in the lower portion. In both women the presenting complaint with reference to the vulval lesions was enlargement of the clitoris due to growth of the tumor at its base. Chromosome and endocrinologic studies showed nothing abnormal. The authors emphasized that leiomyoma of the vulva should prompt x-ray studies of the esophagus, and leiomyoma of the esophagus should prompt search for vulval leiomyoma. Schapiro and Sandrock (1973) reported 2 unrelated women in their 20s who for years had been thought to have achalasia. It is possible that the disorder in these 2 reports was in fact the leiomyomatosis nephropathy syndrome (308940), an entity in which expression is milder in females, and which appears to be a contiguous gene deletion syndrome (Antignac et al., 1992) that in many cases also includes congenital cataract.

Antignac, C.; Gros, F.; Geffriaud, C.; Kashtan, C.; Proesmans, W.; Tryggvason, K.; Gubler, M. C.: Alport syndrome (AS) and diffuse esophageal leiomyomatosis (DL): deletions in the 5-prime end and upstream region of the COL4A5 collagen gene. (Abstract) Am. J. Hum. Genet. 51 (suppl.): A7 only, 1992.

Schapiro, R. L.; Sandrock, A. R.: Esophagogastric and vulvar leiomyomatosis: a new radiologic syndrome. J. Canad. Assoc. Radiol. 24: 184-187, 1973.

Wahlen, T.; Astedt, B.: Familial occurrence of coexisting leiomyoma of vulva and oesophagus. Acta Obstet. Gynec. Scand. 44: 197-203, 1965.

KERATOSIS PALMARIS ET PLANTARIS WITH ESOPHAGEAL CANCER (*148500)

JJM NOTE

This is the best candidate gene for familial esophageal cancer, which is quite infrequent. Have the findings of the significant but low rates of familial aggregation in various parts of China (Chang-Claude J et al.: Familial aggregation of oesophageal cancer in a high incidence area in China. Int. J. Epidemiol. 26:1159–65, 1997; Hu N et al.: Familial aggregation of oesophageal cancer in Yangcheng County, Shanxi Province, China. Int. J. Epidemiol. 21:877–82, 1992) been adequately scrutinized for this gene?

Palmoplantar keratoderma (PPK) is a complex group of hereditary syndromes that have been classified into diffuse, punctate, and focal forms according to the pattern of hyperkeratosis on the palms and soles (Lucker et al., 1994). Diffuse palmoplantar keratodermas can be further subdivided by the presence of epidermolysis on histopathology into epidermolytic PPK (EPPK) and nonepidermolytic PPK (NEPPK or tylosis).

The same disorder as that described in entry 144200 was associated with esophageal cancer in the 2 kindreds (which perhaps are related) studied in Liverpool by Howel-Evans et al. (1958); the association was first reported by Clarke and McConnell (1954). The disorder is apparently distinct. Whether allelic with the other form is unknown. From Oxford, Shine and Allison (1966) described another family in which multiple members showed the association. The authors suggested that 'probably a different allele in this pedigree is involved than in the pedigrees of Howel-Evans et al. (1958).' Esophageal cancer was later in onset (average 61 years as compared with average 45 in the Howel-Evans families), a sliding hiatal hernia was present, and the lower esophagus was lined by gastric mucosa. Harper et al. (1970) gave further information on the Liverpool families and added 2 families, each with 1 case of esophageal cancer with tylosis. Age of onset of the tylosis appears to be a feature distinguishing the cancer-prone from the nonprone form. Tylosis is late in onset in the form with esophageal cancer. Tyldesley (1974) pointed out that oral leukoplakia is also a feature of these cases. Ellis et al. (1994) reviewed the largest of the Liverpool families of tylosis with esophageal cancer. Tylosis had been diagnosed in 89 of 345 persons and 57 of the 89 were still alive. Of the 32 tylotic members of the family who had died, 21 succumbed to cancer of the esophagus and 11 from other causes. In this family, tylosis had a late onset and was inherited as an autosomal dominant with complete penetrance by puberty, and involved other anatomic sites including oral leukoplakia and follicular changes. The epidermis from biopsies of the skin of the soles showed gross acanthosis, hyperkeratosis (without parakeratosis) and hypergranulosis, but there was no evidence of intracellular edema or lysis of any keratinocytes and no spongiosis.

In mapping the genetic locus for keratosis palmaris et plantaris (without associated features) to chromosome 17 near the keratin type I genes, Rogaev et al. (1993) raised the possibility that the association with cancer may represent a contiguous gene syndrome. Blanchet-Bardon et al. (1987) observed hereditary epidermolytic palmoplantar keratoderma in association with breast and ovarian cancer in a large kindred. Rogaev et al. (1993) mapped the tylosis gene to the same region as that which carries the breast cancer locus (BRCA1; 113705). The possibility of a contiguous gene syndrome on chromosome 17 was disproved by Reis (1994) who showed that palmoplantar keratoderma with esophageal cancer was not linked to either the keratin cluster on chromosome 17 or that on chromosome 12. Thus, it is distinct from the 2 forms of palmoplantar keratoderma represented by entries 139350 and 144200.

In the largest of the original Liverpool families, Risk et al. (1994) excluded chromosome 12 as the site of the mutation but found linkage to markers on chromosome 17, indicating that the tylosis with esophageal cancer gene (TOC) lies within a 4-cM region between D17S785 and D17S937. It is located in the 17q23-qter region, telomeric to the type I keratin gene cluster. Hennies et al. (1995) likewise mapped the TOC gene to 17q, distal to the type I keratin gene cluster. Kinship studies were performed in a 4-generation family from a formerly isolated Black Forest Valley, in which 19 of 43 members were affected by palmoplantar keratoderma. Plantar hyperkeratosis became evident between ages 8 and 15 years, whereas manifestation of palmoplantar keratosis occurred later, between 14 and 25 years of age. Two members of the family had died from esophageal cancer, first diagnosed at 56 and 48 years of age, respectively. Further features of the disorder in this family were congenital oral and esophageal leukoplakia and a follicular keratosis, especially on the upper arms and thighs.

Kelsell et al. (1996) used haplotype analysis for fine mapping of the TOC locus. Five representative samples from each of the 2 large pedigrees in Liverpool and in the U.S. and from 1 smaller pedigree in Germany were used to investigate the possibility of a common ancestral mutation. Based on these studies, Kelsell et al. (1996) reduced the genetic interval containing the TOC locus to approximately 1 cM. The Liverpool and German families appeared to share a common haplotype in the disease gene region despite a difference in incidence of esophageal cancer between the pedigrees. The U.S. and German pedigrees appeared to be genetically unrelated in the region of the disease locus.

Stevens et al. (1996) performed linkage analysis in an extensively affected American pedigree originally reported by Marger and Marger (1993). They identified 125 affected individuals in 7 generations, with squamous cell carcinoma of the esophagus occurring in 8. The family was part of a larger pedigree with affected individuals still residing in Germany. The PPK was inherited as an autosomal dominant with full penetrance and usually developed between 6 and 12 years of age. Predominantly, the thickening involved the pressure areas of the sole and calluses disappeared completely with prolonged bedrest or inactivity. The palms were not involved except for exaggerated calluses in manual workers. Frictional hyperkeratosis developed in areas other than the hands and feet at sites of recurrent physical trauma, such as the dorsal aspects of the proximal interphalangeal joints, elbows, knees, and outer aspect of the arm in an individual who repeatedly used the site to operate a machine. As in the Liverpool family, linkage to 17q4 was demonstrated. Stevens et al. (1996) stated that the clinical phenotype of the American family clearly identified it as a form of focal nonepidermolytic PPK. Review of the data on the Howell-Evans Liverpool pedigree supported the reassessment of the clinical phenotype as follows: the thickened skin is focal in nature, related to physical activity with sparing of the hands, and regresses completely on bedrest; the onset at 7 to 8 years of age is that of a focal not a diffuse PPK; and the presence of oral and follicular hyperkeratosis is typical. Stevens et al. (1996) referred to the Howell-Evans syndrome as palmoplantar ectodermal dysplasia type III. They provided an extensive classification of the palmoplantar keratodermas. It seemed quite clear in the family reported by Stevens et al. (1996) that the occurrence of esophageal cancer was related to the use of tobacco. In general, there was a low incidence of smoking on the part of affected and nonaffected members, but of the 8 patients with esophageal cancer, 7 had either smoked heavily or chewed tobacco.

Nomenclature: Risk et al. (1994) used TOCG as a symbol for 'tylosis oesophageal cancer gene.' The gene was also symbolized TOC, dropping the 'G.' Although TEC would appear to be a more appropriate symbol in American English, which does not use the diphthong, the symbol TEC has already been used for a protein tyrosine kinase (600058).

See also Yesudian et al. (1980)

Blanchet-Bardon, C.; Nazzaro, V.; Chevrant-Breton, J.; Espie, M.; Kerbrat, P.; Le Marec, B.: Hereditary epidermolytic palmoplantar keratoderma associated with breast and ovarian cancer in a large kindred. Brit. J. Derm. 117: 363-370, 1987.

Clarke, C. A.; McConnell, R. B.: Six cases of carcinoma of the oesophagus occurring in one family. Brit. Med. J. 2: 1137-1138, 1954.

Ellis, A.; Field, J. K.; Field, E. A.; Friedmann, P. S.; Fryer, A.; Howard, P.; Leigh, I. M.; Risk, J.; Shaw, J. M.; Whittaker, J.: Tylosis associated with carcinoma of the oesophagus and oral leukoplakia in a large Liverpool family—a review of six generations. Europ. J. Cancer 30B: 102-112, 1994.

Harper, P. S.; Harper, R. M. J.; Howel-Evans, A. W.: Carcinoma of the oesophagus with tylosis. Quart. J. Med. 39: 317-333, 1970.

Hennies, H.-C.; Hagedorn, M.; Reis, A.: Palmoplantar keratoderma in association with carcinoma of the esophagus maps to chromosome 17q distal to the keratin gene cluster. Genomics 29: 537-540, 1995.

Howel-Evans, W.; McConnell, R. B.; Clarke, C. A.; Sheppard, P. M.: Carcinoma of the oesophagus with keratosis palmaris et plantaris (tylosis): a study of two families. Quart. J. Med. 27: 413-429, 1958.
Kelsell, D. P.; Risk, J. M.; Leigh, I. M.; Stevens, H. P.; Ellis, A.; Hennies, H. C.; Reis, A.; Weissenbach, J.; Bishop, D. T.; Spurr, N. K.; Field, J. K.: Close mapping of the focal non-epidermolytic palmoplantar keratoderma (PPK) locus associated with oesophageal cancer (TOC). Hum. Molec. Genet. 5: 857-860, 1996.
Lucker, G. P. H.; van de Kerkhof, P. C. M.; Steijlen, P. M.: The hereditary palmoplantar keratoses: an updated review and classification. Brit. J. Derm. 131: 1-14, 1994.
Marger, R. S.; Marger, D.: Carcinoma of the esophagus and tylosis: a lethal genetic combination. Cancer 72: 17-19, 1993.
Reis, A.: Personal Communication. Berlin, Germany 3/28/1994.
Risk, J. M.; Field, E. A.; Field, J. K.; Whittaker, J.; Fryer, A.; Ellis, A.; Shaw, J. M.; Friedmann, P. S.; Bishop, D. T.; Bodmer, J.; Leigh, I. M.: Tylosis oesophageal cancer mapped. (Letter) Nature Genet. 8: 319-321, 1994.
Rogaev, E. I.; Rogaeva, E. A.; Ginter, E. K.; Korovaitseva, G. I.; Farrer, L. A.; Shlensky, A. B.; Pritkov, A. N.; Mordovtsev, V. N.; St. George-Hyslop, P. H.: Identification of the genetic locus for keratosis palmaris et plantaris on chromosome 17 near the RARA and keratin type I genes. Nature Genet. 5: 158-162, 1993.
Shine, I.; Allison, P. R.: Carcinoma of the esophagus with tylosis. Lancet I: 951-953, 1966.
Stevens, H. P.; Kelsell, D. P.; Bryant, S. P.; Bishop, D. T.; Spurr, N. K.; Weissenbach, J.; Marger, D.; Marger, R. S.; Leigh, I. M.: Linkage of an American pedigree with palmoplantar keratoderma and malignancy (palmoplantar ectodermal dysplasia type III) to 17q24: literature survey and proposed updated classification of the keratodermas. Arch. Derm. 132: 640-651, 1996.
Tyldesley, W. R.: Oral leukoplakia associated with tylosis and esophageal carcinoma. J. Oral Path. 3: 62-70, 1974.
Yesudian, P.; Premalatha, S.; Thambiah, A. S.: Genetic tylosis with malignancy: a study of a South Indian pedigree. Brit. J. Derm. 102: 597-600, 1980.

Pancreatic Acinar Tumors

PANCREATIC CARCINOMA (#260350)

JJM NOTE

A preliminary note has been made of my Registry of Familial Pancreatic Cancer (Mulvihill JJ et al.: Familial adenocarcinoma of the exocrine pancreas. Am. J. Hum. Genet. 47:A13, 1990). Forty-three nuclear families met an operating definition of two first-degree relatives with pancreatic cancer or two affected second-degree relatives with the connective relative having any cancer. Both vertical and horizontal aggregations were seen, as well as a possible excess of breast cancer. Lynch et al., likewise, reported a clinical series (Lynch HT et al.: Familial pancreatic cancer: clinicopathologic study of 18 nuclear families. Am. J. Gastroenterol. 85:54–60, 1990). Mendelian traits that seem to have an excess of pancreatic cancer include hereditary pancreatitis, hereditary dysplastic nevus syndrome, Lynch cancer family syndrome type 2, Gardner polyposis, tuberous sclerosis, von Hippel-Lindau syndrome, Li-Fraumeni cancer family syndrome, and Peutz-Jeghers syndrome.

A number sign (#) is used with this entry because a number of genes have been found to carry mutations associated with pancreatic carcinoma, either familial or sporadic. These include CDKN2 (600160) and MADH4 (600993).

Friedman and Fialkow (1976) observed cancer of the pancreas in 4 brothers from a sibship of 6. Diagnosis was made between ages 66 and 75 years. None had a history of pancreatitis or tumors at other sites. Reimer et al. (1977) reported pancreatic cancer in father and son. Pancreatic carcinoma occurs in the von Hippel-Lindau syndrome (193300), hereditary pancreatitis (Appel, 1974), and ataxia-telangiectasia (Swift et al., 1976). In a family of European Jewish ancestry, Ehrenthal et al. (1987) identified pancreatic adenocarcinoma in women of 3 successive generations. The diagnosis was histologically confirmed in each case. The youngest of the 3 women presented at age 29, her mother at age 42, and her grandmother at age 76. When Billy Carter, brother of President Jimmy Carter, died of pancreatic cancer in 1988, the news media reported that his sister Jean Carter Stapleton, evangelist and faith healer, had died of pancreatic cancer at age 54 and their mother of pancreatic, bone, and breast cancer at age 85. In the New York Times of December 1, 1989, it was announced that President Carter's last surviving sib, Gloria Carter Spann, was found to be suffering from pancreatic cancer, 'the same disease that killed their father, sister, and brother, and contributed to the death of their mother.' The possibility of environmental causation might be raised, e.g., an agricultural insecticide or an oncogenic agent such as aflatoxin.

Lynch et al. (1995) studied 8 families in which 2 or more first-degree relatives had pancreatic cancer. In the 8 families, 25 pancreatic cancers were found. The mean age of occurrence was 62.8 years, with a range from 45 to 90 years. Parent and child occurrence was observed in 4 of the 8 families. In one family, the progenitor had an affected brother and 5 of his 8 children developed cancer; 2 had pancreatic cancer, 2 had breast cancer (one of these also had pancreatic cancer), and 1 daughter had ovarian cancer. Of the 7 carcinomas available for pathologic review, 6 were typical ductal adenocarcinomas and the seventh was a giant cell variant of ductal carcinoma.

Caldas et al. (1994) quoted Boring et al. (1993) as indicating that pancreatic ductal adenocarcinoma is the fourth leading cause of cancer death in either sex in the United States, with an estimated incidence in 1993 of nearly 27,700 and a mortality of 24,500. Although KRAS mutations (190070) are found in 85% or more of cases and p53 mutations (191170) in at least of 50% of cases, little is known of the genetic changes that occur in pancreatic carcinogenesis. Caldas et al. (1994) described experiments suggesting that mutation in the p16 gene (CDKN2; 600160), which they referred to as MTS1, is a frequent cause of pancreatic adenocarcinoma. Liu et al. (1995) found that 50% of pancreatic cancer cell lines had deletions of both exons 1 and 2 of the CDKN2 gene; furthermore, an additional 30% of pancreatic cancer cell lines harbored point mutations or microdeletions based on DNA sequencing.

Evans et al. (1995) described a large pedigree in which pancreatic cancer was inherited as an autosomal dominant. Diabetes and exocrine insufficiency were observed in all family members who eventually developed pancreatic cancer. The presence of diabetes, often years before the diagnosis of cancer, allowed identification of those people who had inherited the predisposing allele. The germline mutation that was responsible for the predispositioned pancreatic cancer was unknown. However, since the KRAS gene (KRAS2; 190070) is mutated in 75 to 90% of cases of sporadic pancreatic cancer (Almoguera et al., 1988), Evans et al. (1995) analyzed this gene in pancreatic tumors from patients in this family. Pancreatic tissue from all 3 persons analyzed showed KRAS point mutations and in all 3 the mutations were located in codon 13. The observation of several different KRAS mutations in pancreatic tumors from this family and the presence of only wildtype KRAS in normal tissue clearly showed that the mutant KRAS gene was not the germline event underlying pancreatic cancer in this family. The lack of attacks of abdominal pain seemingly distinguishes the affected members of this family from hereditary pancreatitis (167800) in which diabetes mellitus and pancreatic cancer occur, but the distinction is by no means clear because it was stated that 6 of the 9 persons who developed pancreatic cancer in this family also had a history clinically compatible with pancreatic insufficiency before diagnosis of cancer: weight loss, fatty and foul smelling stools, increased fecal fat, and relief with exogenous pancreatic enzyme supplementation. All 9 members of the family with pancreatic cancer were males.

Schutte et al. (1995) observed that, although pancreatic adenocarcinoma is relatively frequent, description of the genetic alterations has been hampered for several reasons. Because of the aggressive clinical course, the number of resected specimens, especially at early stages of the cancer, have been somewhat limited. Furthermore, pancreatic adenocarcinomas characteristically exhibit an exuberant host desmoplastic response, resulting in a high mixture of nonneoplastic cells and hampering the molecular genetic analysis of primary tumor samples. Finally, familial patterns of pancreatic adenocarcinoma usually do not involve young ages of onset, high penetrance, or extensive pedigrees. Schutte et al. (1995) used the method of representational difference analysis (RDA), as described by Lisitsyn et al. (1995), to demonstrate a homozygous deletion mapping to a 1-cM region at 13q12.3. The deletion was flanked by the markers D13S171 and D13S260 and lay within the 6-cM region identified as containing the BRCA2 (600185) locus of heritable breast cancer susceptibility. This suggests that the same gene may be involved in multiple tumor types and that its function is that of a tumor suppressor rather than that of a dominant oncogene. RDA is a means for isolating DNA fragments that are present in only 1 of 2 nearly identical complex genomes. It utilizes a subtractive hybridization method but differs from conventional methods by using 'representations' of the genomes that have a reduction in complexity. Representations are generated by a PCR-based size selection applied to the restriction fragments of both genomes. Moreover, RDA takes advantage of both subtractive hybridization and DNA reassociation kinetics to favor the reiterated PCR amplification of the the difference between the 2 genomes. RDA can enrich difference products over a millionfold after 3 rounds of selection. The patient in whose pancreatic cancer the homozygous deletion was demonstrated was an 84-year-old woman who had success ful resection of a right-sided colon carcinoma at the age of 61. Her mother had had an adenocarcinoma of the colon resected and died of breast carcinoma at age 80. Her mother's sister died of breast carcinoma at age 94. Her mother's brother died of 'stomach' cancer in his 80s. The patient's brother died of colorectal cancer at the age of 52. Whether the members of this kindred were hemizygous for the deletion in the 13q12.3 region was not determined; Schutte et al. (1995) stated that 'it will be of interest to determine whether the individuals of the presently reported familial cluster are hemizygous in (this) region.'

Further evidence of the role of CDKN2 in pancreatic tumorigenesis was provided by Whelan et al. (1995), who described a kindred with an increased risk of pancreatic cancers, melanomas, and possibly additional types of tumors cosegregating with a CDKN2 mutation (600160.0005).

See also MacDermott and Kramer (1973)

Almoguera, C.; Shibata, D.; Forrester, K.; Martin, J.; Arnheim, N.; Perucho, M.: Most human carcinomas of the exocrine pancreas contain mutant c-K-ras genes. Cell 53: 549-554, 1988.

Appel, M. F.: Hereditary pancreatitis: review and presentation of an additional kindred. Arch. Surg. 108: 63-65, 1974.

Boring, C. C.; Squires, T. S.; Tong, T.: Cancer statistics, 1993. CA Cancer J. Clin. 43: 7-26, 1993.

Caldas, C.; Hahn, S. A.; da Costa, L. T.; Redston, M. S.; Schutte, M.; Seymour, A. B.; Weinstein, C. L.; Hruban, R. H.; Yeo, C. J.; Kern, S. E.: Frequent somatic mutations and homozygous deletions of the p16 (MTS1) gene in pancreatic adenocarcinoma. Nature Genet. 8: 27-32, 1994.

Ehrenthal, D.; Haeger, L.; Griffin, T.; Compton, C.: Familial pancreatic adenocarcinoma in three generations: a case report and a review of the literature. Cancer 59: 1661-1664, 1987.

Evans, J. P.; Burke, W.; Chen, R.; Bennett, R. L.; Schmidt, R. A.; Patchen Dellinger, E.; Kimmey, M.; Crispin, D.; Brentnall, T. A.; Byrd, D. R.: Familial pancreatic adenocarcinoma: association with diabetes and early molecular diagnosis. J. Med. Genet. 32: 330-335, 1995.

Friedman, J. M.; Fialkow, P. J.: Familial carcinoma of the pancreas. Clin. Genet. 9: 463-469, 1976.

Lisitsyn, N. A.; Lisitsina, N. M.; Dalbagni, G.; Barker, P.; Sanchez, C. A.; Gnarra, J.; Linehan, W. M.; Reid, B. J.; Wigler, M. H.: Comparative genomic analysis of tumors: detection of DNA losses and amplification. Proc. Nat. Acad. Sci. 92: 151-155, 1995.

Liu, Q.; Yan, Y.-X.; McClure, M.; Nakagawa, H.; Fujimura, F.; Rustgi, A. K.: MTS-1 (CDKN2) tumor suppressor gene deletions are a frequent event in esophagus squamous cancer and pancreatic adenocarcinoma cell lines. Oncogene 10: 619-622, 1995.

Lynch, H. T.; Fusaro,L.; Smyrk, T. C.; Watson, P.; Lanspa, S.; Lynch, J. F.: Medical genetic study of eight pancreatic cancer-prone families. Cancer Invest. 13: 141-149, 1995.

MacDermott, R. P.; Kramer, P.: Adenocarcinoma of the pancreas in 4 siblings. Gastroenterology 65: 137-139, 1973.

Reimer, R. R.; Fraumeni, J. F., Jr.; Ozols, R. F.; Bender, R.: Pancreatic cancer in father and son. (Letter) Lancet I: 911, 1977.

Schutte, M.; da Costa, L. T.; Hahn, S. A.; Moskaluk, C.; Hoque, A. T. M. S.; Rozenblum, E.; Weinstein, C. L.; Bittner, M.; Meltzer, P. S.; Trent, J. M.; Yeo, C. J.; Hruban, R. H.; Kern, S. E.: Identification by representational difference analysis of a homozygous deletion in pancreatic carcinoma that lies with the BRCA2 region. Proc. Nat. Acad. Sci. 92: 5950-5954, 1995.

Swift, M.; Sholman, L.; Perry, M.; Chase, C.: Malignant neoplasms in the families of patients with ataxia-telangiectasia. Cancer Res. 36: 209-215, 1976.

Whelan, A. J.; Bartsch, D.; Goodfellow, P. J.: Brief report: a familial syndrome of pancreatic cancer and melanoma with a mutation in the CDKN2 tumor-suppressor gene. New Eng. J. Med. 333: 975-977, 1995.

PANCREATITIS, HEREDITARY (#167800)

JJM NOTE

A family history of symptoms possibly related to pancreatitis should be specifically elicited in patients with pancreatic cancer. Genetic mapping might be a route to understanding the origins of pancreatic cancer.

A number sign (#) is used with this entry because of evidence that the disorder can be caused by mutation in the cationic trypsinogen gene PRSS1 (276000). Furthermore, idiopathic pancreatitis has been found to be associated with mutations in the cystic fibrosis gene (CFTR; 602421).

Gross et al. (1962) described a kindred with affected persons in 4 generations. Four other families had been reported from the Mayo Clinic, including the first reported example by Comfort and Steinberg (1952). A puzzling feature is the urinary excretion of lysine and cystine by about half the members of affected kindreds (with or without pancreatitis). Cystine urinary stones have not been observed. Singer and Cohen (1966) reported onset at about age 20 in a man whose younger sister and a cousin were similarly affected. The attacks were characterized by severe abdominal pains, fever, and marked elevation of serum amylase. Except for the last symptom, differentiation from familial Mediterranean fever (249100), also called 'familial paroxysmal peritonitis,' might be difficult. The aminoaciduria was almost certainly an incidental finding since family members without pancreatitis showed it and because other families with pancreatitis have not had this feature (Davidson et al., 1968). Robechek (1967) observed a family with 5 affected persons. He suggested that hypertrophy of the sphincter of Oddi together with a common ampulla of the biliary and pancreatic ducts may be the inherited factor. Mann and Rubin (1969) described a 17-month-old boy with steatorrhea whose 26-year-old brother and mother had steatorrhea and pancreatic calcification. Hereditary pancreatitis occurs with hyperparathyroidism in the multiple endocrine adenomatosis syndrome (131100). McElroy and Christiansen (1972) described a family in which 10 persons had definite pancreatitis and 16 others may have been affected. They pointed out that thrombosis in the portal or splenic vein occurs with significant frequency. Sibert (1978) identified 72 patients in 7 families in England and Wales. Penetrance was about 80%. The mean age of onset was 13.6 years. There were 2 peaks, one at 5 years and one at 17 years. The second peak was thought to represent genetically susceptible persons with symptoms preci pitated by alcohol, rather than genetic heterogeneity. In 5 of the families, members with both childhood and adult onset were identified. In most cases the attacks were of nuisance value only. Only 4 of the 72 patients had life-threatening disease. Pancreatic insufficiency (5.5%), diabetes mellitus (12.5%), pseudocysts (5.5%) and hemorrhagic pleural effusion were observed. Portal vein thrombosis occurred in 2 and was suspected in 3 others. Patients seemed to improve later in life. Attacks were precipitated by emotional upset, alcohol, or high fat intake. Sarles et al. (1982) pointed out that chronic calcifying pancreatitis is characterized by pancreatic stones in the ducts and acini. They had shown that 'stone protein' (see 167770) inhibits in vitro calcium carbonate nucleation and decreases the rate of crystal growth, suggesting that it acts as a physiologic inhibitor of spontaneous calcium carbonate formation in supersaturated pancreatic juice. (A similar function has been suggested for statherin in human saliva (Schlesinger and Hay, 1977).) Sarles et al. (1982) found absence of stone protein in the pancreatic stones in a case of calcific pancreatitis and interpreted this as indicating that the protein was not secreted into the pancreatic juice. Freud et al. (1992) described the cases of monozygotic twin girls of Ashkenazi origin who were admitted to hospital at the age of 9 years because of recurrent attacks of pancreatitis. Dalton-Clarke et al. (1985) found 10 definite and 4 suspected cases of pancreatitis in an English family. Lewis and Gazet (1993) reported pancreatitis in members of 4 successive generations of a second English family. A male in each of the first generations had a combination of calcific pancreatitis and pancreatic carcinoma.

Rumenapf et al. (1994) stated that more than 50 families of hereditary pancreatitis had been reported since the first description by Comfort and Steinberg (1952). They reported on the case of a 26-year-old man from a family in which 6 of 34 members had confirmed pancreatitis and an additional 3 members had suspected pancreatitis. A great uncle had died of pancreatic cancer after suffering from pancreatitis for years. Numerous pancreatic calculi were removed surgically, and a side-to-side pancreaticojejunostomy with a Roux-Y loop was performed. Rumenapf et al. (1994) suggested that surgery may be superior to endoscopic drainage. Lowenfels et al. (1997) assembled records on 246 patients (125 males and 121 females) thought to have hereditary pancreatitis. In 218 patients the diagnosis appeared to be highly probable and in 28 patients it was thought to be less certain. The mean age of onset of symptoms of pancreatitis was 13.9 +/- 12.2 years. Compared with an expected number of 0.150, 8 pancreatic adenocarcinomas developed during 8,531 person-years of follow-up. The mean age at diagnosis of pancreatic cancer was 56.9 +/- 11.2 years. Frequency of other tumors was not increased. Eight of 20 reported deaths in the cohort were from pancreatic cancer. Thirty members of the cohort had been tested and all were found to have a mutated copy of the trypsinogen gene. The estimated cumulative risk of pancreatic cancer to age 70 years in patients with hereditary pancreatitis approached 40%. For patients with a paternal inheritance pattern, the cumulative risk of pancreatic cancer was approximately 75%.

Sarles et al. (1996) reported 11 families with hereditary pancreatitis characterized by the presence of calculi in pancreatic ducts. The disorder in 1 family with 5 cases was classified as calcic lithiasis because the calculi were composed of more than 95% calcium salts. Protein lithiasis was present in the other 10 families, the calculi being composed of degraded amorphous residues of lithostathine (167770), the pancreatic secretory protein that inhibits salt crystallization. Average age at clinical onset of symptoms was 15 years. Clinical progression seemed to be less severe than that in alcoholic chronic pancreatitis (alcoholic calcic lithiasis).

Le Bodic et al. (1996) analyzed the genomic segregation of highly informative microsatellite markers in a French family of 147 individuals, 47 of whom had hereditary pancreatitis. Linkage was found between HPC and 6 chromosome 7q markers. The marker D7S661 was linked to HPC with a lod score of 8.58 at theta = 0.077. Multipoint linkage analysis indicated that the HPC gene is most likely located in the region encompassed by markers D7S661 and D7S676 on 7q33-qter. Le Bodic et al. (1996) noted that the gene encoding carboxypeptidase A1 (CPA1; 114850), which is a pancreatic exopeptidase, has been mapped to the same region. Whitcomb et al. (1996) performed a genomewide linkage analysis on a family extensively affected with hereditary pancreatitis centered in eastern Kentucky and western Virginia. Using microsatellite markers, they established linkage between the hereditary pancreatitis phenotype and 7q. A maximal lod score of 4.73 at a recombination fraction of 0.0 was obtained with D7S684 located in the 7q35 region. Using 3 large HP families located in Virginia, West Virginia, and Tennessee, Pandya et al. (1996) confirmed the tight linkage of HP to marker D7S684. They placed the HP locus within a 16-cM interval between markers D7S495 and D7S688.

Dasouki et al. (1998) reported on the results of linkage and direct mutational analysis for the common arg117-to-his (R117H) mutation (276000.0001) in 8 unrelated families with hereditary pancreatitis. By 2-point linkage analysis with the 7q35 marker D7S676, done initially in 4 families, positive lod scores were found in 2, a negative lod score in 1, and a weakly positive lod score in 1. Direct mutational analysis of exon 3 of the cationic trypsinogen gene in 6 families showed that all symptomatic individuals tested were heterozygous for the R117H mutation. Also, several asymptomatic but at-risk relatives were found to be heterozygous for this mutation. Affected individuals in the remaining 2 families did not have the mutation. Radiation hybrid mapping assigned the gene to 7q35 between 2 specific markers. The negative linkage and absence of the trypsinogen mutation in 2 of 8 families suggested locus heterogeneity in hereditary pancreatitis.

Several genes previously mapped to 7q were considered candidates for HPC because they were known to be expressed in the exocrine pancreas and to encode enzymes that could potentially activate digestive enzymes within the pancreas. The hypothesis that pancreatitis results from inappropriate activation of pancreatic proenzymes was first promulgated by Chiara (1896) and subsequently demonstrated to be an experimental model for pancreatitis (Steer and Meldolesi, 1987). Although Le Bodic et al. (1996) considered the CPA1 to be the primary candidate, this gene mapped centromeric to the HPC locus as defined by obligate recombinations in their study. However, at least 8 trypsinogen genes are located on 7q35 between markers D7S495 and D7S498 and within the V and D-C segments of the complex T-cell receptor beta-chain gene (186930). Trypsinogen is an inactive proenzyme for trypsin, which becomes active when an 8-amino acid N-terminal peptide is removed. Of the 8 trypsinogen-like genes sequenced and identified within the TCRB locus by Rowen et al. (1996), 3 were determined by sequence analysis to be pseudogenes. Another group of 5 trypsinogen genes, including the cationic and anionic pancreatic trypsinogen genes, were found to be in a cluster located between 2 elements near the 3-prime end of the TCRB locus. A sixth trypsinogen-like gene may also be functional, but no corresponding cDNAs had been identified. Whitcomb et al. (1996) noted that the 5 trypsinogen genes are highly homologous, each residing within a tandemly duplicated 10-kb segment and each composed of 5 exons. Mutational screening analyses for each of the exons from the cationic and anionic trypsinogen genes in multiple affected and unaffected family members allowed Whitcomb et al. (1996) to identify a single G-to-A transition in the third exon of cationic trypsinogen in all affected members and obligate carriers in 1 family (276000.0001). The mutation was predicted to result in an arg (CGC)-to-his (CAC) substitution at amino acid residue 105 of trypsin (codon number 117 in the more common chymotrypsin numbering system). The finding of identical mutations in 5 separate kindreds raised the possibility that these families might be distantly related and the mutation centuries old. Although no genealogic link could be found through 8 generations, subsequent haplotyping revealed that all 4 of the American families had the same high-risk haplotype over a 4-cM region encompassing 7 STR markers, confirming the likelihood that these kindreds share a common ancestor. A fifth family from Naples, Italy, displayed a unique haplotype indicating that the same mutation had occurred on at least 2 occasions. The R117H mutation created a novel restriction enzyme recognition site for AflIII that permitted facile screening for the mutation in the general population. The mutation was not found in any of 140 unrelated control individuals. X-ray crystal structure analysis, molecular modeling, and protein digest data indicated that the arg117 residue is a trypsin-sensitive site. Whitcomb et al. (1996) provided a diagram of a model of the trypsin self-destruct mechanism designed to prevent pancreatic autodigestion. Active trypsin is inhibited normally by a limited supply of trypsin inhibitor (167790). If trypsin activity exceeds the inhibitory capacity of PSTI, then proenzymes, including mesotrypsin and enzyme Y, are activated. The activation of these enzymes is postulated to be part of a feedback mechanism for inactivating wildtype trypsinogen, trypsin, and other zymogens. When the arg117 cleavage site for mesotrypsin, enzyme Y, and trypsin is replaced by histidine, trypsin continues to activate trypsinogen and other zymogens unabated, leading to autodigestion of the pancreas and pancreatitis.

Sharer et al. (1998) and Cohn et al. (1998) demonstrated that mutations in the cystic fibrosis gene can cause idiopathic pancreatitis when present in heterozygous state in association with the variable number of thymidines in intron 8 of the CFTR gene, specifically the 5T allele.

History

The 'S.' family used by Whitcomb et al. (1996) in their map-based gene discovery in hereditary pancreatitis had the name Slone, according to an editorial accompanying the paper of Whitcomb et al. (1996) (Anonymous, 1996). This family had been reported by McElroy and Christiansen (1972). Whitcomb (1997) indicated that the family with the R117H mutation in the PRSS1 gene (276000.0001) was identified as the S-family, but the 'S.' did not stand for Slone.

See also Carey and Fitzgerald (1968); Freeman et al. (1976); Girard and Archambault (1980); Gross et al. (1964); Kattwinkel et al. (1973); Makela and Aarimaa (1985); Riccardi et al. (1975); Sato and Saitoh (1974); Sibert (1973)

Anonymous: It takes a family. (Editorial) Nature Genet. 14: 117-118, 1996.

Carey, M. C.; Fitzgerald, O.: Hyperparathyroidism associated with chronic pancreatitis in a family. Gut 9: 700-703, 1968.

Chiara, H.: Ueber Selbstverdauung des menschlichen Pankreas. Ztschr. Heilkunde 17: 70-96, 1896.

Cohn, J. A.; Friedman, K. J.; Noone, P. G.; Knowles, M. R.; Silverman, L. M.; Jowell, P. S.: Relation between mutations of the cystic fibrosis gene and idiopathic pancreatitis. New Eng. J. Med. 339: 653-658, 1998.

Comfort, M. W.; Steinberg, A. G.: Pedigree of a family with hereditary chronic relapsing pancreatitis. Gastroenterology 21: 54-63, 1952.

Dalton-Clarke, H. J.; Lewis, M. H.; Levi, A. J.; Blumgart, L. H.: Familial chronic calcific pancreatitis: a family study. Brit. J. Surg. 72: 307-308, 1985.

Dasouki, M. J.; Cogan, J.; Summar, M. L.; Neblitt, W., III; Foroud, T.; Koller, D.; Phillips, J. A., III: Heterogeneity in hereditary pancreatitis. Am. J. Med. Genet. 77: 47-53, 1998.

Davidson, P.; Costanza, D.; Swieconek, J. A.; Harris, J. B.: Hereditary pancreatitis: a kindred without gross aminoaciduria. Ann. Intern. Med. 68: 88-96, 1968.

Freeman, H. J.; Weinstein, W. M.; Shnitka, T. K.; Crockford, P. M.; Herbert, F. A.: Alpha-1-antitrypsin deficiency and pancreatic fibrosis. Ann. Intern. Med. 85: 73-76, 1976.

Freud, E.; Barak, R.; Ziv, N.; Leiser, A.; Dinari, G.; Mor, C.; Zer, M.: Familial chronic recurrent pancreatitis in identical twins: case report and review of the literature. Arch. Surg. 127: 1125-1128, 1992.

Girard, R. M.; Archambault, A.: Hereditary chronic pancreatitis. (Letter) New Eng. J. Med. 303: 286-287, 1980.

Gross, J. B.; Gambill, E. E.; Ulrich, J. A.: Hereditary pancreatitis. Description of a fifth kindred and summary of clinical features. Am. J. Med. 33: 358-364, 1962.

Gross, J. B.; Ulrich, J. A.; Jones, J. D.: Urinary excretion of amino acids in a kindred with hereditary pancreatitis and aminoaciduria. Gastroenterology 47: 41-48, 1964.

Kattwinkel, J.; Lapey, A.; Di Sant'Agnese, P. A.; Edwards, W. A.; Huffy, M. P.: Hereditary pancreatitis: three new kindreds and a critical review of the literature. Pediatrics 51: 55-69, 1973.

Le Bodic, L.; Bignon, J.-D.; Raguenes, O.; Mercier, B.; Georgelin, T.; Schnee, M.; Soulard, F.; Gagne, K.; Bonneville, F.; Muller, J.-Y.; Bachner, L.; Ferec, C.: The hereditary pancreatitis gene maps to long arm of chromosome 7. Hum. Molec. Genet. 5: 549-554, 1996.

Lewis, M. P. N.; Gazet, J.-C.: Hereditary calcific pancreatitis in an English family. Brit. J. Surg. 80: 487-488, 1993.

Lowenfels, A. B.; Maisonneuve, P.; DiMagno, E. P.; Elitsur, Y.; Gates, L. K., Jr.; Perrault, J.; Whitcomb, D. C.; The International Hereditary Pancreatitis Study Group: Hereditary pancreatitis and the risk of pancreatic cancer. J. Nat. Cancer Inst. 89: 442-446, 1997.

Makela, P.; Aarimaa, M.: Pancreatography in a family with hereditary pancreatitis. Acta Radiol. 26: 63-66, 1985.

Mann, T. P.; Rubin, J.: Familial pancreatic exocrine dysfunction with pancreatic calcification. Proc. Roy. Soc. Med. 62: 326, 1969.

McElroy, R.; Christiansen, P. A.: Hereditary pancreatitis in a kinship associated with portal vein thrombosis. Am. J. Med. 52: 228-241, 1972.

Pandya, A.; Blanton, S. H.; Landa, B.; Javaheri, R.; Melvin, E.; Nance, W. E.; Markello, T.: Linkage studies in a large kindred with hereditary pancreatitis confirms mapping of the gene to a 16-cM region on 7q. Genomics 38: 227-230, 1996.

Riccardi, V. M.; Shih, V. E.; Holmes, L. B.; Nardi, G. L.: Hereditary pancreatitis—nonspecificity of aminoaciduria and diagnosis of occult disease. Arch. Intern. Med. 135: 822-825, 1975.

Robechek, P. J.: Hereditary chronic relapsing pancreatitis: a clue to pancreatitis in general?. Am. J. Surg. 113: 819-824, 1967.

Rowen, L.; Koop, B. F.; Hood, L.: The complete 685-kilobase DNA sequence of the human beta T cell receptor locus. Science 272: 1755-1762, 1996.

Rumenapf, G.; Kamm, M.; Rupprecht, H.; Scheele, J.: Surgical management of hereditary pancreatitis: report of a case and presentation of a new family. (Letter) Pancreas 9: 398-399, 1994.

Sarles, H.; Camarena, J.; Bernard, J. P.; Sahel, J.; Laugier, R.: Two forms of hereditary chronic pancreatitis. Pancreas 12: 138-141, 1996.

Sarles, H.; De Caro, A.; Multigner, L.; Martin, E.: Giant pancreatic stones in teetotal women due to absence of the 'stone protein'?. (Letter) Lancet II: 714-715, 1982.

Sato, T.; Saitoh, Y.: Familial chronic pancreatitis associated with pancreatic lithiasis. Am. J. Surg. 127: 511-517, 1974.

Schlesinger, D. H.; Hay, D. I.: Complete covalent structure of statherin, a tyrosine-rich acidic peptide which inhibits calcium phosphate precipitation from human parotid saliva. J. Biol. Chem. 252: 1689-1695, 1977.
Sharer, N.; Schwarz, M.; Malone, G.; Howarth, A.; Painter, J.; Super, M.; Braganza, J.: Mutations of the cystic fibrosis gene in patients with chronic pancreatitis. New Eng. J. Med. 339: 645-652, 1998.
Sibert, J. R.: Hereditary pancreatitis in England and Wales. J. Med. Genet. 15: 189-201, 1978.
Sibert, J. R.: Hereditary pancreatitis in a Newcastle family. Arch. Dis. Child. 48: 618-621, 1973.
Singer, M.; Cohen, F. B.: Hereditary chronic relapsing pancreatitis. J. Newark Beth Israel Hosp. 21: 121-126, 1966.
Steer, M. L.; Meldolesi, J.: The cell biology of experimental pancreatitis. New Eng. J. Med. 316: 144-150, 1987.
Whitcomb, D. C.: Personal Communication. Pittsburgh, Pa. 12/9/1997.
Whitcomb, D. C.; Gorry, M. C.; Preston, R. A.; Furey, W.; Sossenheimer, M. J.; Ulrich, C. D.; Martin, S. P.; Gates, L. K., Jr.; Amann, S. T.; Toskes, P. P.; Liddle, R.; McGrath, K.; Uomo, G.; Post, J. C.; Ehrlich, G. D.: Hereditary pancreatitis is caused by a mutation in the cationic trypsinogen gene. Nature Genet. 14: 141-145, 1996.
Whitcomb, D. C.; Preston, R. A.; Aston, C. E.; Sossenheimer, M. J.; Barua, P. S.; Zhang, Y.; Wong-Chong, A.; White, G. J.; Wood, P. G.; Gates, L. K., Jr.; Ulrich, C.; Martin, S. P.; Post, J. C.; Ehrlich, G. D.: A gene for hereditary pancreatitis maps to chromosome 7q35. Gastroenterology 110: 1975-1980, 1996.

Small Bowel Tumors

CYSTIC FIBROSIS (#219700)

JJM NOTE

The 1985 Thanksgiving weekend on which word got out that the cystic fibrosis gene was mapped by linkage to the *met* oncogene had those of us at the National Cancer Institute searching for a cancer risk in cystic fibrosis. A decade later, analysis of 28,511 patients showed no general excess of cancer, but a specific excess of digestive tract cancers (Neglia et al. 1995). Case reports continue to trickle out, and, by virtue of greater longevity in current patients compared with those of the prenous era, more coincidental cases with cancer will occur. Hence, the reported patient with cystic fibrosis and colon cancer might be due to random association, apart from the patient's age at cancer diagnosis—27 years (Kuster P et al.: Cancer of the colon in a patient with cystic fibrosis. Monats. Kind. 142:794–97, 1994). The rarity of ileal adenocarcinoma in the general population suggests that case reports correctly point to a huge relative risk for that cancer (Siraganian et al. 1987; Samuels LE et al.: Adenocarcinoma of the ileum in a patient with cystic fibrosis. Digest. Dis. Sci. 40:107–8, 1995). It remains to be seen whether or not pancreatic cancer and acute leukemia occur to excess in children with cystic fibrosis (Tsongalis GL et al.: Association of pancreatic adenocarcinoma, mild lung disease, and delta F508 mutation in a cystic fibrosis patient. Clin. Chem. 40:1972–74, 1994; Gururangan S et al.: Successful treatment of acute lymphoblastic leukemia in a child with cystic fibrosis. Med. Pediatr. Oncol. 22:414–16, 1994). The latter association resembles Shwachman-Diamond syndrome (260400).

Description

A number sign (#) is used with this entry because the disorder is caused by mutations in the cystic fibrosis conductance regulator gene (CFTR; 602421), located on chromosome 7.

Formerly known as cystic fibrosis of the pancreas, this entity has increasingly been labeled simply 'cystic fibrosis.' Manifestations relate not only to the disruption of exocrine function of the pancreas but also to intestinal glands (meconium ileus), biliary tree (biliary cirrhosis), bronchial glands (chronic bronchopulmonary infection with emphysema), and sweat glands (high sweat electrolyte with depletion in a hot environment). Infertility occurs in males and females.

Clinical Features

The mildest extreme of CF is represented by patients not diagnosed until middle age (Scully et al., 1977). The phenotypic variability in CF was analyzed by Sing et al. (1982). In an inbred kindred in North Carolina, a mild form of cystic fibrosis was described by Knowles et al. (1989). There was 1 instance of mother-daughter involvement, the mother being related to her husband. One of the presumed homozygotes was a 62-year-old woman. Another was her 52-year-old sister, the mother of the affected proposita. The daughter was an intensive care nurse, the mother of a normal daughter. Manifestations in the family were predominantly pulmonary; pancreatic exocrine insufficiency was not a conspicuous feature, especially in the older patients.

The 2 subgroups defined by the A and C haplotypes of polymorphisms closely linked to the CF locus on chromosome 7, reported by Estivill et al. (1987), have clinical differences in terms of the frequency of meconium ileus, pseudomonas infections, and pancreatic disease (Woo, 1988).

Gasparini et al. (1990) described a RFLP DNA marker closely linked to the CF locus which showed an allelic correlation with severity of the disorder: the genotype 2/2 was associated with severe disease; the genotype 1/2 was overrepresented in patients with very mild clinical manifestations, including pancreatic insufficiency, absence of meconium ileus, and absence of Pseudomonas colonization.

Meconium Ileus

Allan et al. (1981) showed that sibs tend to show recurrence of meconium ileus as a feature of cystic fibrosis. The distal intestinal obstruction syndrome is a 'meconium ileus equivalent' that occurs in adolescents and adults with CF. It is the consequence of the abnormally viscid mucofeculant material in the terminal ileum and right colon, where the fecal stream is normally liquid.Typical features are recurrent episodes of RLQ pain with palpable mass in the right iliac fossa. Symptoms are exacerbated by eating.

Mornet et al. (1988) determined the haplotype associated with cystic fibrosis in 41 families using 4 DNA probes, all of which are tightly linked to the CF gene. In 17 of the families an affected child had meconium ileus, and in the other 24 families there was a child without meconium ileus. A different haplotype was associated with the 2 types of families, suggesting that multiple allelism, i.e., different mutations at the same locus, accounts for CF with or without meconium ileus.

Liver Disease

Gaskin et al. (1988) found that 96% of patients with cystic fibrosis and evidence of liver disease had biliary tract obstruction, usually a stricture of the distal common bile duct. All patients without liver disease had normal intrahepatic and common-duct excretion of tracer.

Bilton et al. (1990) described a case of cystic fibrosis complicated by common bile duct stenosis.

Pancreatic Insufficiency

Approximately 15% of CF patients do not have pancreatic insufficiency, i.e., are 'pancreatic sufficient.' Kerem et al. (1989) performed linkage disequilibrium and haplotype association studies of patients in 2 clinical subgroups, one pancreatic insufficient (PI) and the other pancreatic sufficient (PS). Significant differences were found in allelic and haplotype distributions in the 2 groups. The data suggested that most of the CF-PI patients were descendants of a single mutational event at the CF locus, whereas the CF-PS patients resulted from multiple, different mutations. Corey et al. (1989) commented on the intrafamilial concordance for pancreatic insufficiency in CF.

Devoto et al. (1989) studied the allele and haplotype frequencies of 5 polymorphic DNA markers near the CF locus in 355 CF patients from Belgium, the German Democratic Republic, Greece, and Italy who were divided into 2 groups according to whether or not they were taking supplementary pancreatic enzymes. The distributions of alleles and haplotypes revealed by 2 of the probes were always different in patients with or without pancreatic insufficiency in all the populations studied. In the case of 1 haplotype that was present in 73% of all the CF chromosomes in their sample, they found homozygosity in only 28% of patients without pancreatic insufficiency as contrasted with 64% who were homozygous and had pancreatic insufficiency. Like other workers, they concluded that this indicated that pancreatic insufficiency and sufficiency are associated with different mutations at the CF locus.

Ferrari et al. (1990) studied the distribution of haplotypes based on 8 polymorphic DNA markers linked to CF in 163 Italian patients and correlated the findings with clinical presentation. Among 19 pancreatic sufficient patients, 6 (31.6%) showed at least 1 copy of a rare phenotype which was present in only 16 of 138 patients (11.6%) with pancreatic insufficiency. In addition, only 5 pancreatic sufficient patients were homozygous for the common 2,1 haplotype as compared with 88 patients (63.8%) with pancreatic insufficiency. Kristidis et al. (1992) likewise found intrafamilial consistency of the pancreatic phenotype, whether pancreatic sufficient or insufficient. Furthermore, the PS phenotype occurred in patients who had 1 or 2 mild CFTR mutations, such as arg117-to-his (602421.0005), arg334-to-trp (602421.0034), arg347-to-pro (602421.0006), ala455-to-glu (602421.0007), and pro574-to-his (602421.0018), whereas the PI phenotype occurred in patients with 2 severe alleles, such as phe508-to-del (602421.0001), ile507-to-del (602421.0002), gln493-to-ter (602421.0003), gly542-to-ter

(602421.0009), arg553-to-ter (602421.0014), and trp1282-to-ter (602421.0022).

Borgo et al. (1993) commented on the phenotypic intrafamilial heterogeneity displayed by an Italian family in which 3 sibs, 2 of whom were dizygotic twins, were compound heterozygotes for the delF508 (602421.0001) and the 1717,-1,G-A splicing mutation (602421.0008). While close intrafamilial concordance was found for exocrine pancreatic phenotype, the pulmonary phenotype varied widely. They suggested that interaction of the CFTR protein with tissue-specific proteins or the action of modifier loci (which may be operationally identical possibilities) plays a role in intrafamilial variability.

Barreto et al. (1991) concluded that the father of a girl with severe CF also had CF but was mildly affected. The child was homozygous for the delta-F508 mutation associated with haplotype B; the father was a compound heterozygote for this mutation and a second CF mutation associated with haplotype C. Perhaps it should not be surprising that some patients with cystic fibrosis have no pancreatic lesions (Oppenheimer, 1972).

Sharer et al. (1998) and Cohn et al. (1998) demonstrated that heterozygosity for CFTR mutations can lead to 'idiopathic' chronic pancreatitis, especially when the mutation is associated with the 5T allele of the variable number of thymidines in intron 8 of the CFTR gene.

Pulmonary Disease

Pier et al. (1996) provided an experimental explanation for the susceptibility of CF patients to chronic Pseudomonas aeruginosa lung infections. They found that cultured human airway epithelial cells expressing the delta-F508 allele of the CFTR gene were defective in uptake of P. aeruginosa compared with cells expressing the wildtype allele. P. aeruginosa lipopolysaccharide-core oligosaccharide was identified as the bacterial ligand for epithelial cell ingestion; exogenous oligosaccharide inhibited bacterial ingestion in a neonatal mouse model, resulting in increased amounts of bacteria in the lungs. The authors concluded that CFTR may normally contribute to a host-defense mechanism that is important for clearance of P. aeruginosa from the respiratory tract.

Infertility

Oppenheimer et al. (1970) suggested that characteristics of cervical mucus may account for infertility in females with cystic fibrosis. Congenital bilateral absence of the vas deferens (CBAVD; 277180) is a usual cause of male infertility in cystic fibrosis. It also occurs with CFTR mutations in heterozygous state, especially when associated with the polymorphic number of thymidines in intron 8, specifically the 5T allele.

Carcinoma

Siraganian et al. (1987) pointed to adenocarcinoma of the ileum in 3 males with cystic fibrosis. The diagnosis was made between ages 29 and 34 years.

From a pancreatic adenocarcinoma developing in a 26-year-old patient with cystic fibrosis due to the phenylalanine-508 deletion, Schoumacher et al. (1990) established a cell line in which the cells showed morphologic and chemical characteristics typical of pancreatic duct cells and showed physiologic properties of CF cells. Schoumacher et al. (1990) suggested that the cell line, which had been stable through more than 80 passages over a 2-year period, could serve as a continuous cell line for studies of the CF defect. Bradbury et al. (1992) demonstrated that the CFTR protein is involved in cAMP-dependent regulation of endocytosis and exocytosis. In a study of pancreatic cancer cells derived from a CF patient, they found that plasma membrane recycling did not occur until normal CFTR was provided.

Neglia et al. (1995) performed a retrospective cohort study of the occurrence of cancer in 28,511 patients with cystic fibrosis from 1985 through 1992 in the United States and Canada. The number of cases observed was compared with the number expected, calculated from population-based data on the incidence of cancer. They also analyzed proportional incidence ratios to assess the association between specific cancers and cystic fibrosis in Europe. The final results indicated that although the overall risk of cancer in patients with cystic fibrosis is similar to that of the general population, there is an increased risk of digestive tract cancers. They recommended that persistent or unexplained gastrointestinal symptoms in CF patients should be carefully investigated.

Inheritance

Recessive inheritance of cystic fibrosis was first shown clearly by Lowe et al. (1949). Roberts (1960) collected family data which appeared to him inconsistent with the quarter ratio expected of a recessive trait. Bulmer (1961) pointed out, however, that when proper correction is made for ascertainment bias, the observed proportions may agree with those expected for a recessive trait.

Rather than estimating the frequency of the CF gene from the square root of the incidence figure, Danks et al. (1983) used the frequency of CF in first cousins. The estimate of gene frequency was 0.0281 as contrasted with 0.0198 (based on direct count). Danks et al. (1983) suggested that the disparity between the 2 estimates might be the existence of 2 gene loci, each with a frequency of 0.0140 for the CF gene and a heterozygote frequency of 1 in 36. Thus, in Victoria, Australia, 1 in 18 persons might be heterozygous at one or the other locus. Later, however, the authors published a retraction and concluded that they had no evidence of more than 1 locus.

For risk analysis in cystic fibrosis, Edwards and Miciak (1990) proposed a simple procedure called the 'slash sheet.' They pointed out that the various methods of estimating genetic risk fall into 2 main groups: first, enumerating all possibilities and excluding those inconsistent with the tests, a simple procedure in small families, and second, using conditional arguments. The latter approach uses Bayes theorem. The former approach, Edwards and Miciak (1990) pointed out, follows a procedure advanced in 1654 by Pascal, following correspondence with Fermat, on the problem of the Chevalier de Mere, now known as the 'problem of points.' Two noblemen were gambling, and, while one was winning, the other was called away and the game was abandoned. How should the stakes be divided? Edwards and Miciak (1990) noted that 'genetic risk is merely an unfinished game of chance.'

Cytogenetics

Park et al. (1987) concluded that CF is distal to and on the 5-prime side of MET. They determined this by in situ hybridization on metaphase and prometaphase chromosomes of normal lymphocytes as well as lymphoblastoid cells containing a t(5;7)(q35;q22). Normal cells showed clustering of MET grains to 7q31. Furthermore, in the lymphoblastoid cell line, there was significant labeling within the 5q+ chromosome, confirming that MET is located distal to 7q22 with most grains clustered at 7q31. Somatic cell hybrids containing the derivative 7 showed on Southern analysis that the 3-prime portion of the MET gene, but not the 5-prime portion, was located there; thus, MET is at the translocation breakpoint. Studies in another cell line with a 7q32 translocation breakpoint indicated that MET is located at or proximal to 7q32. A break at this site was accompanied by loss of 3 markers within 1 cM of CF, suggesting that if MET is at the breakpoint on 7q31, CF is located distally.

In the course of studying a case of cystic fibrosis, Spence et al. (1988) discovered what appeared to be a case of uniparental disomy: the father did not contribute alleles to the propositus for markers near the CF locus or for centromeric markers on chromosome 7. High-resolution cytogenetic analysis was normal, and the result could not be explained by nonpaternity or a submicroscopic deletion. Uniparental disomy could be explained by various mechanisms such as monosomic conception with subsequent chromosome gain, trisomic conception followed by chromosome loss, postfertilization error, or gamete complementation. Patients with more than one genetic disorder might be suspected of having isodisomy, which should also be suspected in cases of an apparent new mutation leading to a recessive disorder when only 1 parent is heterozygous, and in cases of females affected with X-linked recessive disorders. Engel (1980) appears to have originated the concept of uniparental disomy and resulting isodisomy. 205,204:Voss et al. (1988, 1989) also demonstrated uniparental disomy for chromosome 7 in a patient with cystic fibrosis.

Mapping

Mayo et al. (1980) attempted to map the cystic fibrosis gene by study of CF x mouse cell hybrids and examination for production of the cystic fibrosis mucociliary inhibitor. The strongest chance of assignment was for chromosome 4. Scambler et al. (1985) found that the albumin locus labeled by a DNA clone did not segregate with CF or with any of 6 other chromosome 4 markers. They estimated that about half the length of chromosome 4 was accounted for by the markers used. Eiberg et al. (1984) found a hint of linkage to F13B (134580); the maximum lod score was 1.71 at a recombination fraction of 0.05 for males and females combined. Linkage with 56 other genetic markers was negative (Eiberg et al., 1984). Eiberg et al. (1985) showed that cystic fibrosis and paraoxonase (PON; 168820) are linked; the maximum lod score was 3.70 at theta = 0.07 in males and 0.00 in females.

Tsui et al. (1985) found that the CF locus is linked to that of a DNA marker which is also linked to the PON locus, which in turn by independent evidence is linked to CF, thus closing the circle. The DNA marker was provisionally called D0CRI-917. The interval between the marker and PON was about 5 cM and the interval between it and CF about 15 cM. Whether the order is marker—PON—CF or PON—marker—CF was not certain; the former order was favored by 9:5 odds. Knowlton et al. (1985) reported that the anonymous probe D0CRI-917, linked to CF with about 15% recombination, is located on chromosome 7. White et al. (1985) showed very tight linkage to the MET oncogene

(164860), which was assigned to the midportion of 7q. Wainwright et al. (1985) reported tight linkage also to the gene for another anonymous DNA probe, pJ3.11, which was assigned to 7cen-7q22. The closely linked probes pJ3.11 and MET are sufficiently informative to permit carrier detection in 80% of families in which there is a living CF child and unaffected sibs (Farrall et al., 1986). Scambler et al. (1985) showed that the COL1A2 gene (120160) is linked to CF (maximum lod for the sexes combined = 3.27 at a male recombination fraction of 0.08 and a female recombination fraction of 0.15.) PON and CF show recombination frequency of about 10%. CF is about 10 cM from both TCRB (186930) and COL1A2. TCRB and COL1A2 are not closely linked; thus, CF lies between them in the proximal part of 7q22. Wainwright et al. (1986) presented linkage data for COL1A2 versus CF (lod = 3.58 at theta = 0.10), TCRB versus CF (lod = 2.20 at theta = 0.15) and TCRB versus PON (all lods negative). Based on combined linkage data from 50 informative 2-generation families, Buchwald et al. (1986) concluded that CF is 19 cM from COL1A2, which is located at 7q21.3-q22.1. COL1A2 is closely linked to D7S15 and to PON. The probable order is COL1A2—D7S15—PON—CF. The regional localization of CF is 7q22.3-q23.1. Linkage of cystic fibrosis to various DNA markers and/or classical markers was reported in a series of articles by Beaudet et al. (1986), White et al. (1986), Bowcock et al. (1986), Farrall et al. (1986), Tsui et al. (1986), Spence et al. (1986), and Watkins et al. (1986). In Amish/Mennonite/Hutterite kindreds, Klinger et al. (1986) and Watkins et al. (1986) found close linkage with markers on chromosome 7, consistent with locus homogeneity for the defect causing CF in the populations that had been examined to date.

Estivill et al. (1987) identified a candidate for the cystic fibrosis locus by using a 'rare-cutter cosmid library.' They found a genomic region with the characteristics of an HTF island in high linkage disequilibrium with CF. The fact that the sequence was conserved throughout mammalian evolution strengthens the view that this is the CF gene. HTF islands, standing for HpaII tiny fragments, have a sequence length of between 500 and 1000 bp and often include the first exons as well as upstream sequences 5-prime to coding genes (Bird, 1986; Brown and Bird, 1986). These HTF islands are regions of DNA rich in the nonmethylated dinucleotide CpG and contain clusters of sites for CpG-methylation-sensitive restriction enzymes. (There are about 30,000 HTF islands in the human genome.) Estivill et al. (1987) stated that 94% of the chromosomes are of haplotype B, which is present in only 34% of the chromosomes in the general population. In 127 Italian families, Estivill et al. (1988) studied linkage disequilibrium of markers at the locus containing the CpG-enriched methylation-free island designated D7S23. In a search for deletions by means of field inversion gel electrophoresis (FIGE), Morreau et al. (1988) analyzed DNA from 10 cystic fibrosis patients representing 19 different CF chromosomes. No differences were detected after digestion of the samples with 2 different restriction enzymes and hybridization with 4 different probes. The authors estimated that the percentage of deletions occurring within the CF region is less than 15.2% (95% confidence interval, N = 19). The fact that no patient with a combination of cystic fibrosis and a genetic syndrome due to a second affected locus in close vicinity to the CF locus has been described suggests that deletions are rare. Beaudet et al. (1989) found strong linkage disequilibrium between the CF locus and closely linked markers on chromosome 7. By in situ hybridization Duncan et al. (1988) mapped 2 DNA sequences closely linked to the CF locus to 7q31.3-q32. This is a more distal location than had been inferred from previous data.

Using cystic fibrosis and published CF haplotypes as the test bed, Collins and Morton (1998) illustrated how allelic association can be efficiently combined with linkage evidence to identify a region for positional cloning of a disease gene.

Molecular Genetics

For an extensive discussion of the molecular genetics of cystic fibrosis and a listing of allelic variants of the CFTR gene, see (602421).

Collins (1992) gave an update concerning the molecular biology of CF and the therapeutic implications thereof.

Heterogeneity

Vitale et al. (1986) found close linkage of the CF gene and the MET locus in 12 unrelated Italian cystic fibrosis families, thus supporting their hypothesis of genetic homogeneity based on the analysis of consanguineous marriages among 624 couples of CF parents. Lander and Botstein (1986) and Romeo et al. (1986) discussed further the consanguinity method for studying heterogeneity in cystic fibrosis. Estivill et al. (1987) used their haplotype data to argue against genetic heterogeneity at the CF locus. They proposed that the great majority of CF mutations found in the population arose from an original mutational event which occurred in the Caucasian population after racial divergence in man.

Pathogenesis

Frizzell (1987) pointed out that cystic fibrosis is of interest to neuroscientists because it appears to be a disease of ion channels. It is apparently not the conduction properties of ion channels that are affected, but rather their gating by chemical agonists. These conductance pathways appear to be unique to epithelial cells in which salt and water transport rates are governed by cyclic AMP and calcium-dependent regulatory processes.

Decrease in fluid and salt secretion is responsible for the blockage of exocrine outflow from the pancreas and the accumulation of heavy dehydrated mucous in the airways. In sweat glands, salt reabsorption is defective. This is the basis of the folkloric anecdote that the midwife would lick the forehead of the newborn and, if the sweat tasted abnormally salty, predict that the infant was destined to die of pulmonary congestion and its side effects. Quinton (1983) and Knowles et al. (1983) first suggested that the primary defect of cystic fibrosis may be in chloride transport. Widdicombe et al. (1985) demonstrated a cyclic AMP-dependent transepithelial chloride current in normal but not CF epithelia. The pathophysiology of cystic fibrosis, specifically the impermeability of epithelia to chloride ion, was reviewed by Welsh and Fick (1987).

Landry et al. (1989) purified several proteins from kidney and trachea that exhibit chloride channel activity when they are reconstituted into artificial phospholipid bilayer membranes. One or more of these proteins may turn out to be all or part of the secretory chloride channel that is defective in CF. Using antibodies against CFTR peptides, Marino et al. (1991) demonstrated that the CFTR molecule is located in and confined to the apical domain of pancreatic centroacinar and intralobular duct cells. From this they concluded that the proximal duct epithelial cells play a key role in the early events leading to pancreatic insufficiency in CF and that apical chloride transport by these cells is essential for normal pancreatic secretory function. Jetten et al. (1989) created a stable human airway epithelial cell line by retroviral transformation of CF airway epithelium. They found that it maintains the defect in the secretory chloride channel. Rich et al. (1990) expressed the CFTR gene in cultured cystic fibrosis airway epithelial cells and assessed chloride ion channel activation in single cells by means of a fluorescence microscopic assay and a patch-clamp technique. In cells from patients with CF, expression of the CFTR gene but not of the mutant form corrected the chloride ion channel defect. Since there is no animal model for CF, the authors viewed the cell line as very important in studies of the basic defect and for screening of candidate genes which would complement the defect and thus identify the site of the mutation. Bradbury et al. (1992) raised the question as to whether there may be more to the pathogenesis of cystic fibrosis than merely a defect in chloride passage across cell membranes and the concomitant defect in secretion of water.

Diagnosis

Boue et al. (1986) reported on prenatal diagnostic studies in 200 pregnancies with a presumed 1-in-4 risk of recurrence of cystic fibrosis. The method involved measurement of total enzymes and isoenzymes of gamma-glutamyl-transpeptidase, aminopeptidase M, and alkaline phosphatase in amniotic fluid in the second trimester. The recurrence rate of cystic fibrosis was 22.5% in 147 cases in which the index case had cystic fibrosis without meconium ileus at birth but was 47.5% when the index case had meconium ileus. The authors speculated on the mechanism of the 50% recurrence rate and favored the view that 1 parent was in fact a homozygote for a mild allele. With use of their method, the authors suggested 98% accuracy in prenatal diagnosis of cystic fibrosis. Allan et al. (1981), Super (1987), and Boue et al. (1986) found that in families in which a CF child did not have meconium ileus the observed recurrence rate agreed with the expected 1-in-4 risk, but that in families with a history of meconium ileus in the index case the recurrence rate was much higher, 43.7% in the study of Boue et al. (1986). Mornet et al. (1989) found different haplotype associations in the 2 types of families. A distortion of the segregation ratio was suggested to explain the high recurrence rate. Estivill et al. (1987) pointed out that individuals with haplotypes A and C as determined by their cosmid library, whether homozygous or heterozygous, have a considerably reduced risk of being carriers as compared to the 1 in 20 average risk in the British population. On the other hand, a homozygote for haplotype B had a risk of about 1 in 7 of being a carrier. It appears that about 85% of cases of CF in northern Europeans have 1 particular haplotype and the rest a second haplotype. CF with or without meconium ileus may be different entities. Baxter et al. (1988) stated that the meconium ileus form of CF is often lethal so that families with this form are underrepresented in linkage studies. On the other hand, couples who seek prenatal diagnosis often have had children with this problem. Harris et al. (1988) found that 30 of 37 British CF families were sufficiently informative with 3 RFLP probes to enable prenatal diagnosis. They also used linkage analysis to exclude CF in 2 cases in which diagnosis of the disease was equivocal in the sib of an affected child.

Strain et al. (1988), Krawczak et al. (1988), and Beaudet et al. (1989) discussed the use of linkage disequilibrium between CF and DNA markers in genetic risk calculation. Handyside et al. (1992) achieved preimplantation diagnosis. In vitro fertilization techniques were used to recover oocytes from each of 3 women and fertilize them with the husband's sperm. Both members of the 3 couples carried the delF508 mutation. Three days after insemination, embryos in the cleavage stage underwent biopsy with removal of 1 or 2 cells for DNA amplification and analysis. In 2 of the women the oocytes produced noncarrier, carrier, and affected embryos. Both couples chose to have 1 noncarrier embryo and 1 carrier embryo transferred. One woman became pregnant and gave birth to a girl free of the deletion in both chromosomes. Curnow (1994) used cystic fibrosis to illustrate how, in genetic counseling, one can calculate carrier risk for recessive diseases when not all the mutant alleles are detectable. Dean (1995) reviewed the 5 main methods used for detecting mutations at the time.

Savov et al. (1995) demonstrated the presence of 2 different mutations carried by the same CF allele in 4 out of 44 Bulgarian CF patients during a systematic search of the entire coding sequence of the CFTR gene. Two of the double mutant alleles include 1 nonsense and 1 missense mutation, and although the nonsense mutation could be considered to be the main defect, the amino acid substitutions are candidates for disease-causing mutations as well. Savov et al. (1995) suggested that double mutant alleles may be more common than expected and could account for some of the problems in phenotype-genotype correlations. Stern (1997) reviewed the diagnosis of cystic fibrosis. He presented a table of conditions, all readily distinguishable from cystic fibrosis, that can cause moderately elevated sweat electrolytes. With mutation analysis, in approximately 1% of cases no abnormal gene can be found and in about 18% more only 1 abnormal gene will be identified. Stern (1997) pointed out, however, that even if both genes were abnormal, the patient could have an ameliorating or neutralizing second mutation elsewhere. For example, patients homozygous for delF508 (602421.0001) have normal sweat electrolyte concentrations if a second mutation, R553Q, is also present.

Screening

Under the chairmanship of Beaudet and Kazazian (1990), a workshop at the National Institutes of Health laid down guidelines concerning screening for the cystic fibrosis gene. The following points were emphasized: screening should be voluntary, and confidentiality must be assured; screening requires informed consent; providers of screening services have an obligation to ensure adequate education and counseling; quality control in all aspects of testing is required; and there should be equal access to testing.

Newborn babies with CF have abnormally high levels of immunoreactive trypsin (IRT) in serum, which has been the basis for a screening test. Hammond et al. (1991) reported on the results of a Colorado statewide test of the feasibility and efficacy of measuring immunoreactive trypsinogen in blood spots to screen for neonatal cystic fibrosis. They found an incidence of cystic fibrosis of 1 in 3,827 (0.26 per 1,000), with 3.2 newborns per 1,000 requiring repeat measurements. When adjusted for race and compliance with testing, the incidence among the white infants (1 in 2,521) was close to the expected incidence. They concluded that screening was feasible and could be implemented with acceptable rates of repeat testing and false positive and false negative results. Laroche and Travert (1991) found 9 F508 deletion heterozygotes among 149 infants with neonatal transitory mild hypertrypsinemia. Dumur et al. (1990) found an increased frequency of heterozygosity for the same mutation in adults with chronic bronchial hypersecretion. Observing that many patients with cystic fibrosis are malnourished by the time the diagnosis is made, Farrell et al. (1997) sought out to determine whether newborn screening and early treatment might prevent the development of nutritional deficiency. A total of 650,341 newborn infants were screened by measuring immunoreactive trypsinogen on dried blood spots (from April 1985 through June 1991) or by combining the trypsinogen test with DNA analysis (from July 1991 through June 1994). Of 325,171 infants assigned to an early-diagnosis group, cystic fibrosis was diagnosed in 74 infants, including 5 with negative screening tests. Excluding infants with meconium ileus, Farrell et al. (1997) evaluated nutritional status for up to 10 years by anthropometric and biochemical methods in 56 of the infants who received an early diagnosis and in 40 of the infants in whom the diagnosis was made by standard methods (the control group). Pancreatic insufficiency was managed with nutritional interventions that included high-calorie diets, pancreatic enzyme therapy, and fat-soluble vitamin supplements. The diagnosis of cystic fibrosis was confirmed by a positive sweat test at a younger age in the early-diagnosis group than in the control group (mean age, 12 vs 72 weeks). At the time of diagnosis, the early-diagnosis group had significantly higher height and weight percentiles and a higher head circumference percentile. The early-diagnosis group also had significantly higher anthropometric indices during the follow-up, especially the children of pancreatic insufficiency and those who were homozygous for the delta-F508 mutation. Dankert-Roelse and te Meerman (1997) raised the question of whether the time has not arrived for adoption of routine neonatal screening for cystic fibrosis.

Macek et al. (1997) reported a large-scale study for mutation identification in African-American CF patients. The entire coding and flanking intronic sequence of the CFTR gene was analyzed by denaturing gradient-gel electrophoresis (DGGE) and sequencing in 82 African-American CF chromosomes. One novel mutation, 3120+1G-A (602421.0120), occurred with a frequency of 12.3% and was also detected in a native African patient. To establish gene frequencies, an additional group of 66 African-American CF chromosomes were screened for mutations identified in 2 or more African-American patients. Screening for 16 'common Caucasian' mutations identified 52% of CF alleles in African-Americans, while screening for 8 'common African' mutations accounted for an additional 23%. The combined detection rate of 75% was comparable to the sensitivity of mutation analysis in Caucasian CF patients. These results indicated that African-Americans have their own set of 'common' CF mutations that originated from the native African population.

To examine whether the 3120+1G-A mutation has a common origin in the diverse populations in which it has been observed or whether its widespread distribution is the result of recurrent mutational events, Dork et al. (1998) analyzed DNA samples obtained from 17 unrelated CF patients in 4 different populations and from 8 unrelated African CF carriers. They found identical extended CFTR haplotypes for the 3120+1G-A alleles in Arab, African, and African American patients, strongly suggesting that the mutation had a common origin. This finding was not surprising in the case of Africans and African Americans; it was not as easy to explain the presence of the 3120+1G-A mutation in African and Saudi Arab patients. Recent ethnic admixture accounts for a few percent of Africans in Saudi Arabia; however, this was considered an unlikely explanation of the finding, since none of the Saudi families with the mutation had any anthropomorphologic signs of an African descent. In the past, a continuous gene flow between Arab and African populations probably persisted for many centuries, in association with trading and with the spread of the Islamic religion. Thus far, the Greeks are the only Caucasian population in which the 3120+1G-A mutation has been identified. A recurrent mutational event seems to be unlikely, because the Greek haplotype differs from the others in only minor respects. Historical contacts, e.g., under Alexander the Great or during the ancient Minoan civilization, may provide an explanation for the common ancestry of the disease mutation in these ethnically diverse populations. Dork et al. (1998) concluded that 3120+1G-A is an ancient mutation that may be more common than previously thought in populations of the tropical and subtropical belt, where CF probably is an underdiagnosed disorder.

Clinical Management

Cleghorn et al. (1986) obtained good results from oral administration of a balanced solution rendered nonabsorbable by addition of polyethylene glycol.

Hubbard et al. (1992) reported on the use of human deoxyribonuclease I produced by recombinant DNA techniques for cleaving DNA in the sputum of patients with cystic fibrosis and thereby reducing sputum viscosity. Improvement of lung function was documented.

Rosenfeld et al. (1992) evaluated the direct transfer of the normal CFTR gene to airway epithelium using a replication-deficient recombinant adenovirus (Ad) vector containing normal human CFTR cDNA (Ad-CFTR). Two days after in vivo intratracheal introduction of Ad-CFTR in cotton rats, in situ analysis demonstrated human CFTR gene expression in lung epithelium. Northern analysis of lung RNA revealed human CFTR transcripts for up to 6 weeks. Human CFTR protein was detected in epithelial cells using anti-human CFTR antibody 11 to 14 days after infection. While the safety and effectiveness remained to be demonstrated, these observations suggested the feasibility of in vivo CFTR gene transfer as therapy for the pulmonary manifestations of CF.

Hyde et al. (1993) illustrated the feasibility of gene therapy for the pulmonary aspects of CF in humans. They used liposomes to deliver a CFTR expression plasmid to epithelia of the airway and to alveoli deep in the lung, leading to the correction of the ion conductance defects found in the trachea of transgenic (cf/cf) mice. Yang et al. (1993) described a similar approach to the treatment of hepatobiliary disease of cystic fibrosis. In situ hybridization and immunocytochemical analysis of rat liver sections indicated that the endogenous CFTR gene is primarily expressed in the intrahepatic biliary epithelial cells. To target recombinant genes specifically to the biliary epithelium in vivo, Yang et al. (1993) infused recombinant adenoviruses expressing lacZ or human CFTR into the biliary tract through the common bile duct. Conditions were established for achieving recombinant gene expression in virtually all cells of the intrahepatic bile ducts in vivo. Expression persisted in the smaller bile ducts for the duration of the experiment, which was 21 days.

Crystal et al. (1994) administered a recombinant adenovirus vector containing the normal human CFTR cDNA to the nasal and bronchial epithelium of 4 individuals with CF. They found that the vector can express the CFTR cDNA in the CF respiratory epithelium in vivo. With doses up to 2 x 10(9) pfu, there was no recombination/complementation or shedding of the vector or rise of neutralizing antibody titers. At 2 x 10(9) pfu, a transient systemic and pulmonary syndrome was observed. The syndrome was thought to have been caused by vector-induced inflammation of the lower respiratory tract and was possibly induced by interleukin-6, which was found at high levels in the serum of a patient. Follow-up at 6 to 12 months demonstrated no long-term adverse effects. Crystal et al. (1994) concluded that correction of the CF phenotype in the airway epithelium might be achieved with this approach.

A controlled study of adenoviral-vector-mediated gene transfer in the nasal epithelium of patients with cystic fibrosis by Knowles et al. (1995) yielded less encouraging results than those predicted by Crystal et al. (1994). Knowles et al. (1995) did not succeed in correcting the functional defects in nasal epithelium and local inflammatory responses limited the dose of adenovirus that could be administered to overcome the inefficiency of gene transfer. Wilson (1995) reviewed gene therapy for cystic fibrosis. Transplantation of ex vivo manipulated stem cells was the concept of gene therapy used in ADA deficiency (102700). Wide distribution of possible cellular targets for gene therapy in the CF lung and the absence of a known lung epithelial stem cell suggested that an ex vivo approach to gene therapy would not be feasible. Therefore research focused on in vivo approaches for gene transfer that could conveniently be delivered into the airway via aerosols.

Smyth et al. (1994) described colonic strictures, later referred to as fibrosing colonopathy, in children with cystic fibrosis. The patients presented with intestinal obstruction and required surgical resection of a thickened and narrowed area of the colon. The only aspect of these children's management that had changed was a switch to new 'high strength' pancreatic enzyme preparations about 12 months previously. It was not clear whether the preparation was responsible for the problem or whether this was a part of the pathology of cystic fibrosis. In some instances, the clinical and radiologic features were suggestive of Crohn disease or an inflammatory colitis, but the histologic findings were strikingly different (Smyth, 1996). The stenoses, which are frequently long segment, result from submucosal thickening by fibrous connective tissue. This leads to intraluminal narrowing which occurs without a significant reduction in the external diameter of the colon. The epithelium is generally intact with very little inflammatory change in the affected areas. FitzSimmons et al. (1997) studied 29 patients (mean age, 5.0 years) with fibrosing colonopathy that required colectomy for colonic strictures and compared the patients with 105 controls (mean age 5.2 years) who were other patients with cystic fibrosis matched for age at the time of surgery and who did not have fibrosing colonopathy. They found that the relative risk of fibrosing colonopathy that was associated with a dose of 24,001 to 50,000 units of lipase per kilogram per day, as compared with the dose of 0 to 24,000 units per kilogram per day, was 10.9, and relative risk associated with a dose of more than 50,000 units per kilogram per day was 199.5. The findings were considered to support the recommendation that the daily dose of pancreatic enzymes for most patients should remain below 10,000 units of lipase per kilogram.

Population Genetics

Attempting total ascertainment of cases in white children born alive in Ohio during the years 1950 through 1953, Steinberg and Brown (1960) estimated the phenotype frequency to be about 1 in 3,700, a value only about one-fourth that of some earlier estimates. Cystic fibrosis even at this lower estimate is the most frequent lethal genetic disease of childhood. The gene frequency was estimated to be about 0.016, and about 3% of white persons are heterozygotes. In Connecticut, Honeyman and Siker (1965) arrived at higher phenotype frequency estimates of 1 in 489 (maximal) and 1 in 1,863 (minimal). Bois et al. (1978) reported a frequency of at least 1 in 377 births in an area of Brittany.

Klinger (1983) found an incidence of 1/569 among 10,816 live births in the Old Order Amish of Holmes County, Ohio. The gene frequency was estimated to be at least 0.042. On the other hand, not a single case was found among 4,448 live births in the Geauga County, Ohio, Amish. In Italy, to estimate the incidence of CF, Romeo et al. (1985) used the increase in first- and second-cousin parentage, as compared with the general level of consanguinity indicated by the archive of consanguineous marriages maintained by the Catholic Church. The incidence was estimated to be about 1/2000. The data were consistent with a single gene locus; consanguinity would have been higher if more than one were present. The segregation ratio in 624 CF sibships was 0.252.

In Hutterite families with cystic fibrosis, Ober et al. (1987) found close linkage to chromosome 7 markers as in non-Hutterite families. Because 3 different chromosome 7 haplotypes carried the CF mutation in these families, they suggested that the CF gene may have been introduced into the Hutterite population by as many as 3 different ancestors. Fujiwara et al. (1989) confirmed these observations.

From studies in Caucasian families in Utah, Jorde and Lathrop (1988) concluded that fertility differences are unlikely to account for the observed Caucasian CF gene frequency. They compared 143 grandparent couples of Utah CF cases with 20 replicate sets of matched control couples drawn from the Utah Genealogical Database. Before ascertainment correction was applied, CF carriers appeared to manifest a significant fertility advantage over controls. After the correction formula (not used in previous studies) was applied, this difference disappeared. Also, no differences were found between carriers and controls in the length of intervals between births.

In the Hutterites, Klinger et al. (1990) demonstrated that 1 of the 3 previously identified CF haplotypes carries the phe508 deletion. The other 2 Hutterite CF haplotypes are generally rare in Caucasian populations and must carry different CF mutations. Thus, there must have been at least 3 original carriers of CF mutations among the founders of the Hutterite population. They found 1 Hutterite CF patient who had both of the haplotypes that do not carry the phe508 deletion. From a study in Northern Ireland, Hill et al. (1989) concluded that the CF locus is in strong linkage disequilibrium with KM19 and Xv-2C, as it is in other Caucasian populations. These findings indicate that CF in northern European populations may have resulted from a single ancestral mutation. A further finding was preferential inheritance of the paternal CF allele (22 of 28) as opposed to the maternal CF allele (6 of 28) with no significant difference in the sex of the children inheriting these alleles. Cutting et al. (1989) concluded from the analysis of closely linked DNA marker haplotypes that the majority of CF mutations in the Caucasian population arose from a single mutational event. Similar analysis in American black families suggested that multiple mutant alleles are found in this population.

Although CF had been thought to be very rare in Arabs, Nazer et al. (1989) documented CF in 13 children in Saudi Arabia. El-Harith et al. (1998) reported that 6 mutations, detectable by PCR with subsequent restriction enzyme digestion, would allow detection of 70% of Saudi CFTR mutations.

Estivill et al. (1989) reported that in Spanish and Italian populations, deletion of phe508 is present in only 46.2% of CF chromosomes. In all cases, it occurred with haplotype 2, which accounts for about 75% of southern European CF chromosomes; thus, at least 2 independent mutations must have occurred on this haplotype. McIntosh et al. (1989) found a frequency of 74.4% for the phe508 deletion in Scotland. Colten (1990) indicated that one-third of the more than 15,000 patients listed in the registry of the North American National Cystic Fibrosis Foundation are older than 21 years.

Using PCR and hybridization with allele-specific oligonucleotides, Lemna et al. (1990) found the phe508 deletion in 75.8% of 439 cystic fibrosis chromosomes. The 3-base deletion was found in only 30.3% of cystic fibrosis chromosomes from Ashkenazi families. In 5 southern European populations (Albanian, Greek, Italian, Spanish, and Yugoslavian), Nunes et al. (1991) found that, apart from delF508, the most frequent mutations were G542X (602421.0009), 6.04%; R1162X (602421.0033), 3.61%; and N1303K (602421.0032), 3.24%. Of the 14 mutations tested, 7 others had frequencies of less than 1% and 4 mutations were not found at all.

Ten Kate et al. (1991) demonstrated that consanguinity, even if present, may be irrelevant: a family with 2 brothers with cystic fibrosis whose parents were consanguineous, being members of an isolated religious group, were found to have inherited different mutations from the parents. They presented a diagram relating the likelihood of 'autozygosity', depending on gene frequency with consanguinity of various degrees.

In a systematic study of 365 CF chromosomes in the Celtic population in Brittany, Ferec et al. (1992) identified more than 98% of the cystic fibrosis gene mutations. By use of the denaturing gradient gel electrophoresis (DGGE) method, they detected 19 different CFTR mutations located in 9 exons. Nine new mutations were found.

Kerem et al. (1995) reported that the incidence of CF and the frequency of disease-causing mutations varies considerably among the Jewish ethnic subgroups in Israel. Among Ashkenazi Jews, the frequency of CF is 1:3300, which is similar to the frequency in most Caucasian populations. Among non-Ashkenazi Jews, the disease occurs at a similar frequency among Jews from Libya (1:2700), Georgia (1:2700), Greece and Bulgaria (1:2400), but is rare in Jews from Yemen (1:8800), Morocco, (1:15000), Iraq (1:32000), and Iran (1:39000). To that point, only 12 mutations had been identified in Israeli Jews, and this enabled the identification of 91% of the CF chromosomes in the entire Jewish CF population. However, in each Jewish ethnic group, the disease is caused by a different repertoire of mutations.

In a study in the Netherlands, de Vries et al. (1997) tested for the carrier frequency of the delta-F508 mutation by analyzing mouthwashes and matched blood samples from 11,654 blood donors from all over the country. They detected a delF508 carrier frequency of 1 in 42 (95% CI 1/37-1/47). By assuming that the relative frequency of the delF508 mutation among carriers and patients is comparable in the Netherlands, they estimated the overall CF carrier frequency as 1 in 32, significantly less than 1 in 25, the usual figure cited. An increase in carrier frequency with increasing distance from the northeastern region of the country was observed, thus confirming that there is a gradient in gene frequency with low frequencies in the northeastern part of the country and high frequencies in the southern part.

Brock et al. (1998) studied a total of 27,161 women enrolled in prenatal clinics in Scotland between 1990 and 1997. All 27,161 were screened for the delta-F508 (602421.0001), G551D (602421.0013), and G542X (602421.0009) mutations. In 14,360 women R117H was also measured. In addition, 183 patients with cystic fibrosis were studied for the presence of these mutations. Based on their data, the authors estimated that the incidence of CF in the Scottish population is 1 in 1984, with 95% confidence intervals between 1 in 1,692 to 1 in 2,336.

Evolution

Hansson (1988) speculated that if the defect in the control of apical membrane chloride ion channels in CF extends to the intestine, a resistance to bacterial toxin-mediated diarrhea might confer a selective advantage on carriers for the CF gene. Baxter et al. (1988) presented actual observations indicating that intestine in CF homozygotes fails to exhibit a secretory response on exposure to bacterial toxins that would normally induce a secretory diarrhea. They were proceeding to investigate intestinal secretory responses of heterozygotes. The high frequency of the CF gene might be explained by this mechanism. Romeo et al. (1989) also suggested that a selective advantage consisting of high resistance to chloride-ion-secreting diarrheas might have favored, in the past, survival of infants heterozygous for the CF gene.

McMillan et al. (1989) demonstrated an apparent association between heterozygosity at the cystic fibrosis locus and heterozygosity for a RFLP near the constant region of the T-cell receptor beta chain (186930). They suggested that this previously unreported disease association might indicate some form of epistatic interaction between the CF gene and the TCRB gene such that the double heterozygote is immunologically advantaged.

Rodman and Zamudio (1991) suggested that resistance to cholera may have been the environmental factor that selected CF heterozygotes over their 'normal' homozygote cohort. This suggestion obtained experimental support from the observations of Gabriel et al. (1994). In a study of the CFTR(-/-) mouse, created by disruption by the CFTR gene at exon 10 by insertion of an in-frame stop codon to replace ser489, they found that transgenic mice that expressed no CFTR protein did not secrete fluid in response to cholera toxin. Heterozygotes expressed 50% of the normal amount of CFTR protein in the intestinal epithelium and secreted 50% of the normal fluid and chloride ion in response to cholera toxin. The findings suggested that CF heterozygotes may possess a selective advantage of resistance to cholera.

Pier et al. (1998) investigated whether increased resistance to typhoid fever in the heterozygote could be a factor in maintaining mutant CFTR alleles at high levels in selected populations. Typhoid fever is initiated when Salmonella typhi enters gastrointestinal epithelial cells for submucosal translocation. They found that S. typhi, but not the related murine pathogen S. typhimurium, uses CFTR for entry into epithelial cells. Cells expressing wildtype CFTR internalized more S. typhi than isogenic cells expressing the most common CFTR mutation, delta-F508 (602421.0001). Monoclonal antibodies and synthetic peptides containing a sequence corresponding to the first predicted extracellular domain of CFTR inhibited uptake of S. typhi. Heterozygous delta-F508 Cftr mice translocated 86% fewer S. typhi into the gastrointestinal submucosa than did wildtype Cftr mice; no translocation occurred in delta-F508 Cftr homozygous mice. The Cftr genotype had no effect on the translocation of S. typhimurium. Immunoelectron microscopy revealed that more CFTR bound S. typhi in the submucosa of Cftr wildtype mice than in delta-F508 heterozygous mice. Pier et al. (1998) concluded that diminished levels of CFTR in heterozygotes decreases susceptibility to typhoid fever.

Genetic Variability

Wine (1992) pointed out that CFTR mutations associated with pancreatic sufficiency, milder pulmonary disease, and improved sweat gland function are associated with residual CFTR chloride-ion channel function. He questioned the disruptive effects proposed for the delta-F508 mutation because variation in homozygotes for this mutation is very large. At the same time, those homozygous for stop codons have been severely affected, showing pancreatic insufficiency and pulmonary function values (FEV1) in the same range as those of delta-F508 subjects. Disruptive effects of delta-F508 would be expected to give rise to a dominant pattern of inheritance. Wine (1992) concluded that the observations are consistent with the recessive nature of CF and with the likelihood that gene or protein replacement therapy for CF will be effective on their own, without requiring concomitant silencing of the delta-F508 gene. Sheppard et al. (1993) found that some CFTR mutations, such as delF508, which disrupt normal processing and hence are missing from the apical membrane, generate no chloride current and are associated with severe disease. Other mutants, such as R117H (602421.0005), R334W (602421.0034), and R347P (602421.0006), which are correctly processed and retain significant apical chloride channel function, are associated with a milder form of the disease. Thus, the CF genotype determines the biochemical abnormality, which determines the clinical phenotype. Because these 3 'mild' mutants have normal regulation, interventions designed to increase the activity of mutant CFTR may have therapeutic efficacy in patients with these mutations. Studying 267 children and adolescents with CF who were regularly seen at the same center, Kubesch et al. (1993) found that the age-specific colonization rates with Pseudomonas aeruginosa were significantly lower in pancreatic sufficient than in pancreatic insufficient patients. The missense and splice site mutations that were 'mild' CF alleles with respect to exocrine pancreatic function were also 'low risk' alleles for the acquisition of P. aeruginosa. On the other hand, the proportion of P. aeruginosa-positive patients increased most rapidly in the pancreatic insufficient delF508 compound heterozygotes who were carrying a termination mutation in the nucleotide binding fold-encoding exons.

Animal Model

Because the pulmonary complications of CF are the most morbid aspects of the disease, a potential therapeutic strategy is to reconstitute expression of the normal CFTR gene in airway epithelia by somatic gene transfer. Engelhardt et al. (1992) developed an animal model of the human airway, using bronchial xenografts engrafted on rat tracheas and implanted into nude mice, and tested the efficiency of in vivo retroviral gene transfer. They found that in undifferentiated regenerating epithelium, 5 to 10% retroviral gene transfer was obtained, whereas in fully differentiated epithelium, no gene transfer was noted. These findings suggested that retroviral-mediated gene transfer to the airways in vivo may be feasible if the proper regenerative state can be induced.

Several groups succeeded in constructing a transgenic mouse model of cystic fibrosis (Clarke et al., 1992; Colledge et al., 1992; Dorin et al., 1992; Snouwaert et al., 1992). Unlike the HPRT-deficient mouse, constructed as a model for the Lesch-Nyhan syndrome (308000), the CFTR-deficient homozygote showed measurable defects in ion permeability of airway and intestinal epithelia, similar to those demonstrable in human CF tissues. Furthermore, most of the deficient mice had intestinal pathology similar to that of meconium ileus. Also, there appeared to be no prenatal loss from litters produced from crosses between heterozygotes. Most of the mice, however, died soon after birth as a consequence of intestinal blockage. Unlike the human male, the homozygous mouse male in at least one instance was fertile. In a transgenic mouse model of CF created by Dorin et al. (1994) through insertion in exon 10, only a low incidence of meconium ileus was observed. In contrast to the very high level of fatal intestinal obstruction in 3 other CF mouse models, they showed that the partial duplication consequent upon insertional gene targeting allowed exon skipping and aberrant splicing to produce normal Cftr mRNA, but at levels greatly reduced compared with wildtype mice. Instead of the predicted mutant Cftr transcript, a novel mRNA was produced that utilized cryptic splice sites in the disrupting plasmid sequence. Although residual wildtype mRNA in the exon 10 insertional mutant mouse seems to ameliorate the severity of the intestinal phenotype observed in the absolute 'null' CF mice, the presence of low-level residual wildtype Cftr mRNA does not correct the CF ion transport defect. The long-term survival of this insertional mutant mouse provides the opportunity to address factors important in the development of lung disease. To correct the lethal intestinal abnormalities that occur in the transgenic CFTR-null mouse, Zhou et al. (1994) used the human CFTR gene under the control of the rat intestinal fatty acid binding protein (134640) gene promoter. The mice survived and showed functional correction of ileal goblet cell and crypt cell hyperplasia and cAMP-stimulated chloride secretion. The results supported the concept that transfer of the human CFTR gene may be a useful strategy for correcting physiologic defects in patients with CF.

Mice homozygous for disruption of the Cftr gene, unlike the human disease, fail to show any gross lung pathology (Rozmahel et al. (1996)). It was proposed that a calcium-activated Cl(-) conductance could compensate for the lack of the Cftr-encoded Cl(-) channel function in these mice. The absence of this alternative chloride transport mechanism in the intestinal epithelial cells was believed to be responsible for the severe intestinal pathology observed in the same mice. Prolonged survival in these mice was demonstrated among backcross and intercross progeny with different inbred strains, suggesting that modulation of disease severity was genetically determined. A genome scan showed that the major modifier locus mapped near the centromere of mouse chromosome 7 in a region of conserved synteny with human chromosome 19q13. Candidate genes in that region include the gamma-subunit of protein kinase C (176980), the alpha-3 subunit of the type 1 Na(+)/K(+) exchanging ATPase (182350), and the sodium channel, type 1, beta-polypeptide (600235).

In connection with the design of a large-animal model for cystic fibrosis, Tebbutt et al. (1995) cloned and sequenced the CFTR cDNA of sheep. It showed a high degree of conservation at the DNA coding and predicted polypeptide levels with human CFTR; at the nucleotide level, there was a 90% conservation (compared with 80% between human and mouse). At the polypeptide level, the degree of similarity was 95% (compared with 88% between human and mouse). Northern blot analysis and reverse transcription-PCR showed that the patterns of expression of the ovine CFTR gene are very similar to those seen in humans. Further, the developmental expression of CFTR in the sheep is equivalent to that observed in humans.

Harris (1997) pointed out that the generation of cloned sheep (Campbell et al., 1996; Wilmut et al., 1997) establishes the practicality of creating an ovine model of CF. The failure of mice with disruption of the Cftr gene to reproduce the pulmonary and pancreatic features of CF may be due, in the case of the lung at least, in part to considerable differences in submucosal gland distribution in mouse and human. Mice have very few of these glands and they are restricted to the tracheal submucosa. The CFTR chloride ion channel is not expressed at high levels in the mouse pancreas, in contrast to humans where the pancreas is the site of most abundant CFTR expression. Sheep and human CFTR show greater identity and similarity than do human and mouse. Furthermore, Harris (1997) noted that the tissue-specific pattern of expression of the ovine CFTR gene and the developmental expression of CFTR in the sheep are very similar to that in humans. CF pathology commences in utero; for example, obstruction of CF pancreatic ducts by deposits of secreted material commences in the midtrimester of human gestation and by term the pancreas is structurally and functionally destroyed. Thus, in utero therapy might be necessary.

Kent et al. (1997) described the phenotype of an inbred congenic strain of CFTR-knockout mouse that developed spontaneous and progressive lung disease of early onset. The major features of the lung disease included failure of effective mucociliary transport, postbronchiolar overinflation of alveoli, and parenchymal interstitial thickening, with evidence of fibrosis and inflammatory cell recruitment. Kent et al. (1997) speculated that the basis for development of lung disease in the congenic CFTR-knockout mice is their observed lack of a non-CFTR chloride channel normally found in CFTR-knockout mice of mixed genetic background.

History

Spock et al. (1967) observed that patients have a factor in serum that inhibits the action of cilia in explants of rabbit tracheal mucosa. Serum from heterozygotes contained an amount of the factor intermediate between none (the normal situation) and the level in patients.

Smith et al. (1968) found cystic fibrosis in a child with cri-du-chat syndrome. Only the mother was heterozygous by Spock test. They suggested that loss of part of the short arm of the chromosome 5 derived from the father had occurred and that the deleted portion carried the cystic fibrosis locus. Danes and Bearn (1968) found vesicular metachromasia in the fibroblasts of both parents suggesting that the reported experience cannot be taken as evidence of localization of the CF gene on the short arm of chromosome 5.

Edwards et al. (1984) reported a family in which deficiency at the tip of 13q was associated with cystic fibrosis. Weak evidence supporting assignment to 13q was provided by a boy with both cystic fibrosis and hemophilia A; no translocation was visualized but the authors postulated a telomeric translocation that disrupted both loci at the tip of the X chromosome and chromosome 13. They cited 2 other observations of cystic fibrosis with chromosome 13 abnormality.

Williamson (1984) excluded cystic fibrosis from chromosome 13; none of the DNA probes that were monosomic in the case of Edwards et al. (1984) were linked to cystic fibrosis in studies of affected sibs.

In skin fibroblasts from both homozygotes and heterozygotes, Danes and Bearn (1968) found cytoplasmic intravesicular metachromasia of a type readily distinguished from that of mucopolysaccharidoses. Danes and Bearn (1969) described a morphologic change in the fibroblasts and furthermore suggested that homozygosity at either of two different loci can produce cystic fibrosis. In type I, the fibroblasts show discrete metachromatic cytoplasmic vesicles and normal mucopolysaccharide content. In type II, fibroblast metachromasia is present in both vesicles and granules and is evenly distributed through the cytoplasm; mucopolysaccharide content of the cells is markedly increased. On the basis of cell culture phenotype, Danes et al. (1978) identified 3 classes of cystic fibrosis and concluded that there is a prognostic difference between classes. They also suggested that their Class III represents the genetic compound. A deficiency of arginine esterase has been suggested by Rao and Nadler (1974), who reported absence of 1 of 3 isozymes in various cases of cystic fibrosis. Their hypothesis is that the ciliary factor and related substances are present because of failure of degradation when the enzyme is deficient. Stern et al. (1978) described a cystic fibrosis variant with little pancreatic abnormality. Hosli and Vogt (1979) claimed the successful discrimination of cystic fibrosis patients, obligatory heterozygotes (parents), and normal controls by heat inactivation of acid phosphatase and alpha-mannosidase in plasma. In this test, normals retain 80 to 100% activity, heterozygotes 40 to 60%, and CF patients almost none. There was no overlap between groups. Katznelson et al. (1983) did a stringently blinded trial of the reliability of the test of Hosli and Vogt (1979), submitting doubly coded samples to Dr. Hosli. The genotype was correctly identified in each of 45 cases. Shapiro and Lam (1982) found that the usual increase in intracellular calcium in fi broblasts with successive time (passages) in culture is exaggerated in cystic fibrosis fibroblasts. In kidney specimens obtained at autopsy from patients with cystic fibrosis, Katz et al. (1988) documented microscopic nephrocalcinosis in 35 of 38 specimens. Hypercalciuria was present in 5 of 14 patients studied. The presence of microscopic nephrocalcinosis in 3 patients less than 1 year of age suggested to these authors that the mutation in cystic fibrosis involves a primary abnormality of renal calcium metabolism. Shapiro et al. (1982) reported anomalous kinetics of mitochondrial NADH dehydrogenase in cystic fibrosis homozygotes and heterozygotes. Studying white cells, Sanguinetti-Briceno and Brock (1982) could not identify a correlation between NADH dehydrogenase and CF genotype. Shepherd et al. (1988) found that cystic fibrosis infants had 25% higher rates of total energy expenditure compared to healthy infants matched for age and body weight. The authors suggested that the data point to an energy-requiring basic defect.

See also Abeliovich et al. (1992); Baylin et al. (1980); Blanck and Mendoza (1976); Breslow et al. (1978); Breslow et al. (1981); Brock et al. (1982); Brusilow (1970); Bullock et al. (1982); Cattanach and Kirk (1985); Cavenee et al. (1984); Cheadle et al. (1992); Cheadle et al. (1992); Colledge et al. (1995); Curtis et al. (1991); Cutting et al. (1992); Danes and Bearn (1969); Danes and Bearn (1968); Danes and Bearn (1969); Danes et al. (1977); Danks et al. (1965); Devoto et al. (1991); Di Sant'Agnese and Davis (1979); Di Sant'Agnese and Davis (1976); Di Sant'Agnese and Talamo (1967); Dork et al. (1991); Eiberg et al. (1985); Eiberg et al. (1985); European Working Group on CF Genetics (1990); Fanen et al. (1992); Farrall et al. (1986); Farrall et al. (1986); Farrall et al. (1986); Frydman (1979); Goodchild et al. (1976); Harris and Riley (1968); Horn et al. (1986); Katznelson and Ben-Yishay (1978); Kerem et al. (1989); Levitan (1989); Mangos and McSherry (1968); Manson and Brock (1980); Naylor et al. (1986); Oppenheimer and Esterly (1969); Oppenheimer and Esterly (1970); Rao and Nadler (1975); Rao et al. (1972); Scambler et al. (1985); Scambler et al. (1986); Searle et al. (1987); Shier (1979); Shwachman et al. (1977); Stern et al. (1982); The Cystic Fibrosis Genotype-Phenotype Consortium (1993); Warner et al. (1976); Wilson et al. (1975); Wright and Morton (1968); Yang et al. (1993)

Abeliovich, D.; Lavon, I. P.; Lerer, I.; Cohen, T.; Springer, C.; Avital, A.; Cutting, G. R.: Screening for five mutations detects 97% of cystic fibrosis (CF) chromosomes and predicts a carrier frequency of 1:29 in the Jewish Ashkenazi population. Am. J. Hum. Genet. 51: 951-956, 1992.

Allan, J. L.; Robbie, M.; Phelan, P. D.; Danks, D. M.: Familial occurrence of meconium ileus. Europ. J. Pediat. 135: 291-292, 1981.

Barreto, C.; Marques Pinto, L.; Duarte, A.; Lavinha, J.; Ramsay, M.: A fertile male with cystic fibrosis: molecular genetic analysis. J. Med. Genet. 28: 420-421, 1991.

Baxter, P. S.; Goldhill, J.; Hardcastle, J.; Hardcastle, P. T.; Taylor, C. J.: Accounting for cystic fibrosis. (Letter) Nature 335: 211 only, 1988.

Baylin, S. B.; Rosenstein, B. J.; Marton, L. J.; Lockwood, D. H.: Age-related abnormalities of circulating polyamines and diamine oxidase activity in cystic fibrosis heterozygotes and homozygotes. Pediat. Res. 14: 921-925, 1980.

Beaudet, A.; Bowcock, A.; Buchwald, M.; Cavalli-Sforza, L.; Farrall, M.; King, M.-C.; Klinger, K.; Lalouel, J.-M.; Lathrop, G.; Naylor, S.; Ott, J.; Tsui, L.-C.; Wainwright, B.; Watkins, P.; White, R.; Williamson, R.: Linkage of cystic fibrosis to two tightly linked DNA markers: joint report from a collaborative study. Am. J. Hum. Genet. 39: 681-693, 1986.

Beaudet, A. L.; Feldman, G. L.; Fernbach, S. D.; Buffone, G. J.; O'Brien, W. E.: Linkage disequilibrium, cystic fibrosis, and genetic counseling. Am. J. Hum. Genet. 44: 319-326, 1989.

Beaudet, A. L.; Kazazian, H. H., Jr.: Statement from the National Institutes of Health Workshop on Population Screening for the Cystic Fibrosis Gene. New Eng. J. Med. 323: 70-71, 1990.

Bilton, D.; Fox, R.; Webb, A. K.; Lawler, W.; McMahon, R. F. T.; Howat, J. M. T.: Pathology of common bile duct stenosis in cystic fibrosis. Gut 31: 236-238, 1990.

Bird, A. P.: CpG-rich islands and the function of DNA methylation. Nature 321: 209-213, 1986.

Blanck, R. R.; Mendoza, E. M.: Fertility in a man with cystic fibrosis. J.A.M.A. 235: 1364, 1976.

Bois, E.; Feingold, J.; Demenais, F.; Runavot, Y.; Jehanne, M.; Toudic, L.: Cluster of cystic fibrosis cases in a limited area of Brittany (France). Clin. Genet. 14: 73-76, 1978.
Borgo, G.; Cabrini, G.; Mastella, G.; Ronchetto, P.; Devoto, M.; Romeo, G.: Phenotypic intrafamilial heterogeneity in cystic fibrosis. (Letter) Clin. Genet. 44: 48-49, 1993.
Boue, A.; Muller, F.; Nezelof, C.; Oury, J. F.; Duchatel, F.; Dumez, Y.; Aubry, M. C.; Boue, J.: Prenatal diagnosis in 200 pregnancies with a 1-in-4 risk of cystic fibrosis. Hum. Genet. 74: 288-297, 1986.
Bowcock, A. M.; Crandall, J.; Daneshvar, L.; Lee, G. M.; Young, B.; Zunzunegui, V.; Craik, C.; Cavalli-Sforza, L. L.; King, M.-C.: Genetic analysis of cystic fibrosis: linkage of DNA and classical markers in multiplex families. Am. J. Hum. Genet. 39: 699-706, 1986.
Bradbury, N. A.; Jilling, T.; Berta, G.; Sorscher, E. J.; Bridges, R. J.; Kirk, K. L.: Regulation of plasma membrane recycling by CFTR. Science 256: 530-532, 1992.
Breslow, J. L.; Epstein, J.; Fontaine, J. H.; Forbes, G. B.: Enhanced dexamethasone resistance in cystic fibrosis cells: potential use for heterozygote detection and prenatal diagnosis. Science 201: 180-182, 1978.
Breslow, J. L.; McPherson, J.; Epstein, J.: Distinguishing homozygous and heterozygous cystic fibrosis fibroblasts from normal cells by differences in sodium transport. New Eng. J. Med. 304: 1-5, 1981.
Brock, D. J. H.; Gilfillan, A.; Holloway, S.: The incidence of cystic fibrosis in Scotland calculated from heterozygote frequencies. Clin. Genet. 53: 47-49, 1998.
Brock, D. J. H.; Hayward, C.; Super, M.: Controlled trial of serum isoelectric focusing in the detection of the cystic fibrosis gene. Hum. Genet. 60: 30-31, 1982.
Brown, W. R.; Bird, A. P.: Long-range restriction site mapping of mammalian genomic DNA. Nature 322: 447-481, 1986.
Brusilow, S. W.: Cystic fibrosis in adults. Annu. Rev. Med. 21: 99-104, 1970.
Buchwald, M.; Zsiga, M.; Markiewicz, D.; Plavsic, N.; Kennedy, D.; Zengerling, S.; Willard, H. F.; Tsipouras, P.; Schmiegelow, K.; Schwartz, M.; Eiberg, H.; Mohr, J.; Barker, D.; Donis-Keller, H.; Tsui, L.-C.: Linkage of cystic fibrosis to the pro-alpha2(I) collagen gene, COL1A2, on chromosome 7. Cytogenet. Cell Genet. 41: 234-239, 1986.
Bullock, S.; Hayward, C.; Manson, J.; Brock, D. J. H.; Raeburn, J. A.: Quantitative immunoassays for diagnosis and carrier detection in cystic fibrosis. Clin. Genet. 21: 336-341, 1982.
Bulmer, M. G.: Fibrocystic disease of the pancreas: a comment. Ann. Hum. Genet. 25: 163-164, 1961.
Campbell, K. H. S.; McWhir, J.; Ritchie, W. A.; Wilmut, I.: Sheep cloned by nuclear transfer from a cultured cell line. Nature 380: 64-66, 1996.
Cattanach, B. M.; Kirk, M.: Differential activity of maternally and paternally derived chromosome regions in mice. Nature 315: 496-498, 1985.
Cavenee, W.; Leach, R.; Mohandas, T.; Pearson, P.; White, R.: Isolation and regional localization of DNA segments revealing polymorphic loci from human chromosome 13. Am. J. Hum. Genet. 36: 10-24, 1984.
Cheadle, J.; Al-Jader, L.; Goodchild, M.; Meredith, A. L.: Mild pulmonary disease in a cystic fibrosis child homozygous for R553X. J. Med. Genet. 29: 597, 1992.
Cheadle, J. P.; Meredith, A. L.; Al-Jader, L. N.: A new missense mutation (R1283M) in exon 20 of the cystic fibrosis transmembrane conductance regulator gene. Hum. Molec. Genet. 1: 123-125, 1992.
Clarke, L. L.; Grubb, B. R.; Gabriel, S. E.; Smithies, O.; Koller, B. H.; Boucher, R. C.: Defective epithelial chloride transport in a gene-targeted mouse model of cystic fibrosis. Science 257: 1125-1128, 1992.
Cleghorn, G. J.; Stringer, D. A.; Forstner, G. G.; Durie, P. R.: Treatment of distal intestinal obstruction syndrome in cystic fibrosis with a balanced intestinal lavage solution. Lancet I: 8-11, 1986.
Cohn, J. A.; Friedman, K. J.; Noone, P. G.; Knowles, M. R.; Silverman, L. M.; Jowell, P. S.: Relation between mutations of the cystic fibrosis gene and idiopathic pancreatitis. New Eng. J. Med. 339: 653-658, 1998.
Colledge, W. H.; Abella, B. S.; Southern, K. W.; Ratcliff, R.; Jiang, C.; Cheng, S. H.; MacVinish, L. J.; Anderson, J. R.; Cuthbert, A. W.; Evans, M. J.: Generation and characterisation of a delta-F508 cystic fibrosis mouse model. Nature Genet. 10: 445-452, 1995.
Colledge, W. H.; Ratcliff, R.; Foster, D.; Williamson, R.; Evans, M. J.: Cystic fibrosis mouse with intestinal obstruction. (Letter) Lancet 340: 680 only, 1992.
Collins, A.; Morton, N. E.: Mapping a disease locus by allelic association. Proc. Nat. Acad. Sci. 95: 1741-1745, 1998.
Collins, F. S.: Cystic fibrosis: molecular biology and therapeutic implications. Science 256: 774-779, 1992.
Colten, H. R.: Screening for cystic fibrosis.(Editorial) New Eng. J. Med. 322: 328-329, 1990.
Corey, M.; Durie, P.; Moore, D.; Forstner, G.; Levison, H.: Familial concordance of pancreatic function in cystic fibrosis. J. Pediat. 115: 274-277, 1989.
Crystal, R. G.; McElvaney, N. G.; Rosenfeld, M. A.; Chu, C.-S.; Mastrangeli, A.; Hay, J. G.; Brody, S. L.; Jaffe, H. A.; Eissa, N. T.; Danel, C.: Administration of an adenovirus containing the human CFTR cDNA to the respiratory tract of individuals with cystic fibrosis. Nature Genet. 8: 42-51, 1994.
Curnow, R. N.: Carrier risk calculations for recessive diseases when not all the mutant alleles are detectable. Am. J. Med. Genet. 52: 108-114, 1994.
Curtis, A.; Nelson, R.; Porteous, M.; Burn, J.; Bhattacharya, S. S.: Association of less common cystic fibrosis mutations with a mild phenotype. J. Med. Genet. 28: 34-37, 1991.
Cutting, G. R.; Antonarakis, S. E.; Buetow, K. H.; Kasch, L. M.; Rosenstein, B. J.; Kazazian, H. H., Jr.: Analysis of DNA polymorphism haplotypes linked to the cystic fibrosis locus in North American black and Caucasian families supports the existence of multiple mutations of the cystic fibrosis gene. Am. J. Hum. Genet. 44: 307-318, 1989.
Cutting, G. R.; Curristin, S. M.; Nash, E.; Rosenstein, B. J.; Lerer, I.; Abeliovich, D.; Hill, A.; Graham, C.: Analysis of four diverse population groups indicates that a subset of cystic fibrosis mutations occur in common among Caucasians. Am. J. Hum. Genet. 50: 1185-1194, 1992.
Danes, B. S.; Bearn, A. G.: A genetic cell marker in cystic fibrosis of the pancreas. Lancet I: 1061-1063, 1968.
Danes, B. S.; Bearn, A. G.: Cystic fibrosis: an improved method for studying white blood-cells in culture.(Letter) Lancet II: 437 only, 1969.
Danes, B. S.; Bearn, A. G.: Cystic fibrosis: distribution of mucopolysaccharides in fibroblast cultures. Biochem. Biophys. Res. Commun. 36: 919-924, 1969.
Danes, B. S.; Bearn, A. G.: Localisation of the cystic-fibrosis gene.(Letter) Lancet II: 1303 only, 1968.
Danes, B. S.; Bearn, A. G.: Cystic fibrosis of the pancreas. A study in cell culture. J. Exp. Med. 129: 775-794, 1969.
Danes, B. S.; Beck, B.; Flensborg, E. W.: Cystic fibrosis: cell culture classes in a Danish population. Clin. Genet. 13: 327-334, 1978.
Danes, B. S.; Hodson, M. E.; Batten, J.: Cystic fibrosis: evidence for a genetic compound from a family study in cell culture. Clin. Genet. 11: 83-90, 1977.
Dankert-Roelse, J. E.; te Meerman, G. J.: Screening for cystic fibrosis—time to change our position? (Editorial) New Eng. J. Med. 337: 997-999, 1997.
Danks, D. M.; Allan, J.; Anderson, C. M.: A genetic study of fibrocystic disease of the pancreas. Ann. Hum. Genet. 28: 323-356, 1965.
Danks, D. M.; Allan, J.; Phelan, P. D.; Chapman, C.: Mutations at more than one locus may be involved in cystic fibrosis—evidence based on first-cousin data and direct counting of cases. Am. J. Hum. Genet. 35: 838-844, 1983. Note: Retraction: Am. J. Hum. Genet. 36: 1401-1402, 1984.
Dean, M.: Resolving DNA mutations. Nature Genet. 9: 103-104, 1995.
Devoto, M.; De Benedetti, L.; Seia, M.; Piceni Sereni, L.; Ferrari, M.; Bonduelle, M. L.; Malfroot, A.; Lissens, W.; Balassopoulou, A.; Adam, G.; Loukopoulos, D.; Cochaux, P.; Vassart, G.; Szibor, R.; Hein, J.; Grade, K.; Berger, W.; Wainwright, B.; Romeo, G.: Haplotypes in cystic fibrosis patients with or without pancreatic insufficiency from four European populations. Genomics 5: 894-898, 1989.
Devoto, M.; Ronchetto, P.; Fanen, P.; Orriols, J. J. T.; Romeo, G.; Goossens, M.; Ferrari, M.; Magnani, C.; Seia, M.; Cremonesi, L.: Screening for non-delta-F508 mutations in five exons of the cystic fibrosis transmembrane conductance regulator (CFTR) gene in Italy. Am. J. Hum. Genet. 48: 1127-1132, 1991.
de Vries, H. G.; Collee, J. M.; de Walle, H. E. K.; van Veldhuizen, M. H. R.; Smit Sibinga, C. T.; Scheffer, H.; ten Kate, L. P.: Prevalence of delta-F508 cystic fibrosis carriers in The Netherlands: logistic regression on sex, age, region of residence and number of offspring. Hum. Genet. 99: 74-79, 1997.
Di Sant'Agnese, P. A.; Davis, P. B.: Cystic fibrosis in adults: 75 cases and a review of 232 cases in the literature. Am. J. Med. 66: 121-132, 1979.
Di Sant'Agnese, P. A.; Davis, P. B.: Research in cystic fibrosis. New Eng. J. Med. 295: 481-485, and 534-541, and 597-602, 1976.
Di Sant'Agnese, P. A.; Talamo, R. C.: Pathogenesis and physiopathology of cystic fibrosis of the pancreas: fibrocystic disease of the pancreas (muco-viscidosis). New Eng. J. Med. 277: 1287-1294 and 1344-1352, 1967.
Dorin, J. R.; Dickinson, P.; Alton, E. W. F. W.; Smith, S. N.; Geddes, D. M.; Stevenson, B. J.; Kimber, W. L.; Fleming, S.; Clarke, A. R.; Hooper, M. L.; Anderson, L.; Beddington, R. S. P.; Porteous, D. J.: Cystic fibrosis in the mouse by targeted insertional mutagenesis. Nature 359: 211-215, 1992.
Dorin, J. R.; Stevenson, B. J.; Fleming, S.; Alton, E. W. F. W.; Dickinson, P.; Porteous, D. J.: Long-term survival of the exon 10 insertional cystic fibrosis mutant mouse is a consequence of low level residual wild-type Cftr gene expression. Mammalian Genome 5: 465-472, 1994.
Dork, T.; El-Harith, E.-H. A.; Stuhrmann, M.; Macek, M., Jr.; Egan, M.; Cutting, G. R.; Tzetis, M.; Kanavakis, E.; Carles, S.; Claustres, M.; Padoa, C.; Ramsay, M.; Schmidtke, J.: Evidence for a common ethnic origin of cystic fibrosis mutation 3120+1G-to-A in diverse populations. (Letter) Am. J. Hum. Genet. 63: 656-662, 1998.
Dork, T.; Wulbrand, U.; Richter, T.; Neumann, T.; Wolfes, H.; Wulf, B.; Maass, G.; Tummler, B.: Cystic fibrosis with three mutations in the cystic fibrosis transmembrane conductance regulator gene. Hum. Genet. 87: 441-446, 1991.
Dumur, V.; Lafitte, J. J.; Gervais, R.; Debaecker, D.; Kesteloot, M.; Lalau, G.; Roussel, P.: Abnormal distribution of cystic fibrosis delta-F508 allele in adults with chronic bronchial hypersecretion. Lancet 335: 1340, 1990.
Duncan, A. M. V.; Buchwald, M.; Tsui, L.-C.: In situ hybridization of two cloned chromosome 7 sequences tightly linked to the cystic fibrosis locus. Cytogenet. Cell Genet. 49: 309-310, 1988.
Edwards, J. H.; Jonasson, J. A.; Blackwell, N. L.: Locus for cystic fibrosis.(Letter) Lancet I: 1020 only, 1984.
Edwards, J. H.; Miciak, A.: The slash sheet: a simple procedure for risk analysis in cystic fibrosis.(Letter) Am. J. Hum. Genet. 47: 1024-1028, 1990.
Eiberg, H.; Mohr, J.; Nielsen, L. S.: Linkage relationships of human coagulation factor XIIIB. (Abstract) Cytogenet. Cell Genet. 37: 463 only, 1984.
Eiberg, H.; Mohr, J.; Schmiegelow, K.; Nielsen, L. S.; Williamson, R.: Linkage relationships of paraoxonase (PON) with other markers: indication of PON-cystic fibrosis synteny. Clin. Genet. 28: 265-271, 1985.
Eiberg, H.; Schmiegelow, K.; Koch, C.; Mohr, J.; Schwartz, M.; Niebuhr, E.: Cystic fibrosis; hint of linkage with F13B. Clin. Genet. 27: 206, 1985.
Eiberg, H.; Schmiegelow, K.; Tsui, L.-C.; Buchwald, M.; Niebuhr, E.; Phelan, P. D.; Williamson, R.; Warwick, W.; Koch, C.; Mohr, J.: Cystic fibrosis, linkage with PON. (Abstract) Cytogenet. Cell Genet. 40: 623, 1985.
El-Harith, E.-H. A.; Stuhrmann, M.; Dork, T.; Eskandarani, H. A.; Schmidtke, J.: PCR-based analysis of cystic fibrosis mutations specific for Saudi patients. Saudi Med. J. 19: 148-152, 1998.
Engel, E.: A new genetic concept: uniparental disomy and its potential effect, isodisomy. Am. J. Med. Genet. 6: 137-143, 1980.
Engelhardt, J. F.; Yankaskas, J. R.; Wilson, J. M.: In vivo retroviral gene transfer into human bronchial epithelia of xenografts. J. Clin. Invest. 90: 2598-2607, 1992.
Estivill, X.; Chillon, M.; Casals, T.; Bosch, A.; Morral, N.; Nunes, V.; Gasparini, P.; Seia, A.; Pignatti, P. F.; Novelli, G.; Dallapiccola, B.; Fernandez, E.; Benitez, J.; Williamson, R.: Delta-F508 gene deletion in cystic fibrosis in Southern Europe.(Letter) Lancet II: 1404-1405, 1989.

Estivill, X.; Farrall, M.; Scambler, P. J.; Bell, G. M.; Hawley, K. M. F.; Lench, N. J.; Bates, G. P.; Kruyer, H. C.; Frederick, P. A.; Stanier, P.; Watson, E. K.; Williamson, R.; Wainwright, B. J.: A candidate for the cystic fibrosis locus isolated by selection for methylation-free islands. Nature 326: 840-845, 1987.

Estivill, X.; Farrall, M.; Williamson, R.; Ferrari, M.; Seia, M.; Giunta, A. M.; Novelli, G.; Potenza, L.; Dallapicolla, B.; Borgo, G.; Gasparini, P.; Pignatti, P. F.; De Benedetti, L.; Vitale, E.; Devoto, M.; Romeo, G.: Linkage disequilibrium between cystic fibrosis and linked DNA polymorphisms in Italian families: a collaborative study. Am. J. Hum. Genet. 43: 23-28, 1988.

European Working Group on CF Genetics: Gradient of distribution in Europe of the major CF mutation and of its associated haplotype. Hum. Genet. 85: 436-445, 1990.

Fanen, P.; Ghanem, N.; Vidaud, M.; Besmond, C.; Martin, J.; Costes, B.; Plassa, F.; Goossens, M.: Molecular characterization of cystic fibrosis: 16 novel mutations identified by analysis of the whole cystic fibrosis conductance transmembrane regulator (CFTR) coding regions and splice site junctions. Genomics 13: 770-776, 1992.

Farrall, M.; Law, H.-Y.; Rodeck, C. H.; Warren, R.; Stanier, P.; Super, M.; Lissens, W.; Scambler, P.; Watson, E.; Wainwright, B.; Williamson, R.: First-trimester prenatal diagnosis of cystic fibrosis with linked DNA probes. Lancet I: 1402-1405, 1986.

Farrall, M.; Scambler, P.; Klinger, K. W.; Davies, K.; Worrall, C.; Williamson, R.; Wainwright, B.: Cystic fibrosis carrier detection using a linked gene probe. J. Med. Genet. 23: 295-299, 1986.

Farrall, M.; Scambler, P.; North, P.; Williamson, R.: The analysis of multiple polymorphic loci on a single human chromosome to exclude linkage to inherited disease: cystic fibrosis and chromosome 4. Am. J. Hum. Genet. 38: 75-83, 1986.

Farrall, M.; Watson, E.; Bates, G.; Bell, G.; Bell, J.; Davies, K. A.; Estivill, X.; Kruyer, H.; Law, H.-Y.; Lench, N.; Lissens, W.; Simon, P.; Scambler, P.; Stanier, P.; Vassart, G.; Worrall, C.; Williamson, R.; Wainwright, B. J.: Further data supporting linkage between cystic fibrosis and the met oncogene and haplotype analysis with met and pJ3.11. Am. J. Hum. Genet. 39: 713-719, 1986.

Farrell, P. M.; Kosorok, M. R.; Laxova, A.; Shen, G.; Koscik, R. E.; Bruns, W. T.; Splaingard, M.; Mischler, E. H.: Nutritional benefits of neonatal screening for cystic fibrosis. New Eng. J. Med. 337: 963-969, 1997.

Ferec, C.; Audrezet, M. P.; Mercier, B.; Guillermit, H.; Moullier, P.; Quere, I.; Verlingue, C.: Detection of over 98% cystic fibrosis mutations in a Celtic population. Nature Genet. 1: 188-191, 1992.

Ferrari, M.; Antonelli, M.; Bellini, F.; Borgo, G.; Castiglione, O.; Curcio, L.; Dallapiccola, B.; Devoto, M.; Estivill, X.; Gasparini, P.; Giunta, A.; Marianelli, L.; Mastella, G.; Novelli, G.; Pignatti, P.; Romano, L.; Romeo, G.; Seia, M.; Williamson, R.: Genetic differences in cystic fibrosis patients with and without pancreatic insufficiency: an Italian collaborative study. Hum. Genet. 84: 435-438, 1990.

FitzSimmons, S. C.; Burkhart, G. A.; Borowitz, D.; Grand, R. J.; Hammerstrom, T.; Durie, P. R.; Lloyd-Still, J. D.; Lowenfels, A. B.: High-dose pancreatic-enzyme supplements and fibrosing colonopathy in children with cystic fibrosis. New Eng. J. Med. 336: 1283-1289, 1997.

Frizzell, R. A.: Cystic fibrosis: a disease of ion channels? Trends Neurosci. 10: 190-193, 1987.

Frydman, M. I.: Epidemiology of cystic fibrosis: a review. J. Chronic Dis. 32: 211-219, 1979.

Fujiwara, T. M.; Morgan, K.; Schwartz, R. H.; Doherty, R. A.; Miller, S. R.; Klinger, K.; Stanislovitis, P.; Stuart, N.; Watkins, P. C.: Genealogical analysis of cystic fibrosis families and chromosome 7q RFLP haplotypes in the Hutterite brethren. Am. J. Hum. Genet. 44: 327-337, 1989.

Gabriel, S. E.; Brigman, K. N.; Koller, B. H.; Boucher, R. C.; Stutts, M. J.: Cystic fibrosis heterozygote resistance to cholera toxin in cystic fibrosis mouse model. Science 266: 107-109, 1994.

Gaskin, K. J.; Waters, D. L. M.; Howman-Giles, R.; de Silva, M.; Earl, J. W.; Martin, H. C. O.; Kan, A. E.; Brown, J. M.; Dorney, S. F. A.: Liver disease and common-bile-duct stenosis in cystic fibrosis. New Eng. J. Med. 318: 340-346, 1988.

Gasparini, P.; Novelli, G.; Estivill, X.; Olivieri, D.; Savoia, A.; Ruzzo, A.; Nunes, V.; Borgo, G.; Antonelli, M.; Williamson, R.; Pignatti, P. F.; Dallapiccola, B.: The genotype of a new linked DNA marker, MP6d-9, is related to the clinical course of cystic fibrosis. J. Med. Genet. 27: 17-20, 1990.

Goodchild, M. C.; Edwards, J. H.; Glenn, K. P.; Grindey, C.; Harris, R.; Mackintosh, P.; Wentzel, L.: A search for linkage in cystic fibrosis. J. Med. Genet. 13: 417-419, 1976.

Hammond, K. B.; Abman, S. H.; Sokol, R. J.; Accurso, F. J.: Efficacy of statewide neonatal screening for cystic fibrosis by assay of trypsinogen concentrations. New Eng. J. Med. 325: 769-774, 1991.

Handyside, A. H.; Lesko, J. G.; Tarin, J. J.; Winston, R. M. L.; Hughes, M. R.: Birth of a normal girl after in vitro fertilization and preimplantation diagnostic testing for cystic fibrosis. New Eng. J. Med. 327: 905-909, 1992.

Hansson, G. C.: Cystic fibrosis and chloride-secreting diarrhoea. (Letter) Nature 333: 711, 1988.

Harris, A.: Towards an ovine model of cystic fibrosis. Hum. Molec. Genet. 6: 2191-2193, 1997.

Harris, A.; Quinlan, C.; Bobrow, M.: Cystic fibrosis typing with DNA probes: experience of a screening laboratory. Hum. Genet. 79: 76-79, 1988.

Harris, R. L.; Riley, H. D., Jr.: Cystic fibrosis in the American Indian. Pediatrics 41: 733-738, 1968.

Hill, A. J. M.; Graham, C. A.; Kelly, E. D.; Morrison, P. J.; Nevin, N. C.: Linkage disequilibrium and CF allele segregation analysis in cystic fibrosis families in Northern Ireland. Hum. Genet. 83: 391-394, 1989.

Honeyman, M. S.; Siker, E.: Cystic fibrosis of the pancreas: an estimate of the incidence. Am. J. Hum. Genet. 17: 461-465, 1965.

Horn, S. D.; Horn, R. A.; Sharkey, P. D.; Beall, R. J.; Hoff, J. S.; Rosenstein, B. J.: Misclassification problems in diagnosis-related groups: cystic fibrosis as an example. New Eng. J. Med. 314: 484-487, 1986.

Hosli, P.; Vogt, E.: Detection of cystic fibrosis homozygotes and heterozygotes with plasma. Lancet I: 543-546, 1979.

Hubbard, R. C.; McElvaney, N. G.; Birrer, P.; Shak, S.; Robinson, W. W.; Jolley, C.; Wu, M.; Chernick, M. S.; Crystal, R. G.: A preliminary study of aerosolized recombinant human deoxyribonuclease I in the treatment of cystic fibrosis. New Eng. J. Med. 326: 812-815, 1992.

Hyde, S. C.; Gill, D. R.; Higgins, C. F.; Trezise, A. E. O.; MacVinish, L. J.; Cuthbert, A. W.; Ratcliff, R.; Evans, M. J.; Colledge, W. H.: Correction of the ion transport defect in cystic fibrosis transgenic mice by gene therapy. Nature 362: 250-255, 1993.

Jetten, A. M.; Yankaskas, J. R.; Stutts, M. J.; Willumsen, N. J.; Boucher, R. C.: Persistence of abnormal chloride conductance regulation in transformed cystic fibrosis epithelia. Science 244: 1472-1475, 1989.

Jorde, L. B.; Lathrop, G. M.: A test of the heterozygote-advantage hypothesis in cystic fibrosis carriers. Am. J. Hum. Genet. 42: 808-815, 1988.

Katz, S. M.; Krueger, L. J.; Falkner, B.: Microscopic nephrocalcinosis in cystic fibrosis. New Eng. J. Med. 319: 263-266, 1988.

Katznelson, D.; Ben-Yishay, M.: Cystic fibrosis in Israel: clinical and genetic aspects. Israel J. Med. Sci. 14: 204-211, 1978.

Katznelson, D.; Blau, H.; Sack, J.: Detection of cystic-fibrosis genotypes.(Letter) Lancet II: 622 only, 1983.

Kent, G.; Iles, R.; Bear, C. E.; Huan, L.-J.; Griesenbach, U.; McKerlie, C.; Frndova, H.; Ackerley, C.; Gosselin, D.; Radzioch, D.; O'Brodovich, H.; Tsui, L.-C.; Buchwald, M.; Tanswell, A. K.: Lung disease in mice with cystic fibrosis. J. Clin. Invest. 100: 3060-3069, 1997.

Kerem, B.; Buchanan, J. A.; Durie, P.; Corey, M. L.; Levison, H.; Rommens, J. M.; Buchwald, M.; Tsui, L.-C.: DNA marker haplotype association with pancreatic sufficiency in cystic fibrosis. Am. J. Hum. Genet. 44: 827-834, 1989.

Kerem, B.; Rommens, J. M.; Buchanan, J. A.; Markiewicz, D.; Cox, T. K.; Chakravarti, A.; Buchwald, M.; Tsui, L.-C.: Identification of the cystic fibrosis gene: genetic analysis. Science 245: 1073-1080, 1989.

Kerem, E.; Kalman, Y. M.; Yahav, Y.; Shoshani, T.; Abeliovich, D.; Szeinberg, A.; Rivlin, J.; Blau, H.; Tal, A.; Ben-Tur, L.; Springer, C.; Augarten, A.; Godfrey, S.; Lerer, I.; Branski, D.; Friedman, M.; Kerem, B.: Highly variable incidence of cystic fibrosis and different mutation distribution among different Jewish ethnic groups in Israel. Hum. Genet. 96: 193-197, 1995.

Klinger, K.; Horn, G. T.; Stanislovitis, P.; Schwartz, R. H.; Fujiwara, T. M.; Morgan, K.: Cystic fibrosis mutations in the Hutterite brethren. Am. J. Hum. Genet. 46: 983-987, 1990.

Klinger, K.; Stanislovitis, P.; Hoffman, N.; Watkins, P. C.; Schwartz, R.; Doherty, R.; Scambler, P.; Farrall, M.; Williamson, R.; Wainwright, B.: Genetic homogeneity of cystic fibrosis. Nucleic Acids Res. 14: 8681-8686, 1986.

Klinger, K. W.: Cystic fibrosis in the Ohio Amish: gene frequency and founder effect. Hum. Genet. 65: 94-98, 1983.

Knowles, M.; Gatzy, J.; Boucher, R.: Relative ion permeability of normal and cystic fibrosis nasal epithelium. J. Clin. Invest. 71: 1410-1417, 1983.

Knowles, M. R.; Barnett, T. B.; McConkie-Rosell, A.; Sawyer, C.; Kahler, S. G.: Mild cystic fibrosis in a consanguineous family. Ann. Intern. Med. 110: 599-605, 1989.

Knowles, M. R.; Hohneker, K. W.; Zhou, Z.; Olsen, J. C.; Noah, T. L.; Hu, P.-C.; Leigh, M. W.; Engelhardt, J. F.; Edwards, L. J.; Jones, K. R.; Grossman, M.; Wilson, J. M.; Johnson, L. G.; Boucher, R. C.: A controlled study of adenoviral-vector-mediated gene transfer in the nasal epithelium of patients with cystic fibrosis. New Eng. J. Med. 333: 823-831, 1995.

Knowlton, R. G.; Cohen-Haguenauer, O.; Van Cong, N.; Frezal, J.; Brown, V. A.; Barker, D.; Braman, J. C.; Schumm, J. W.; Tsui, L.-C.; Buchwald, M.; Donis-Keller, H.: A polymorphic DNA marker linked to cystic fibrosis is located on chromosome 7. Nature 318: 380-382, 1985.

Krawczak, M.; Konecki, D. S.; Schmidtke, J.; Duck, M.; Engel, W.; Nutzenadel, W.; Trefz, F. K.: Allelic association of the cystic fibrosis locus and two DNA markers, XV2c and KM19, in 55 German families. Hum. Genet. 80: 78-80, 1988.

Kristidis, P.; Bozon, D.; Corey, M.; Markiewicz, D.; Rommens, J.; Tsui, L.-C.; Durie, P.: Genetic determination of exocrine pancreatic function in cystic fibrosis. Am. J. Hum. Genet. 50: 1178-1184, 1992.

Kubesch, P.; Dork, T.; Wulbrand, U.; Kalin, N.; Neumann, T.; Wulf, B.; Geerlings, H.; Weissbrodt, H.; von der Hardt, H.; Tummler, B.: Genetic determinants of airways' colonisation with Pseudomonas aeruginosa in cystic fibrosis. Lancet 341: 189-193, 1993.

Lander, E. S.; Botstein, D.: Consanguinity and heterogeneity: cystic fibrosis need not be homogeneous in Italy.(Letter) Am. J. Hum. Genet. 39: 282-283, 1986.

Landry, D. W.; Akabas, M. H.; Redhead, C.; Edelman, A.; Cragoe, E. J., Jr.; Al-Awqati, Q.: Purification and reconstitution of chloride channels from kidney and trachea. Science 244: 1469-1472, 1989.

Laroche, D.; Travert, G.: Abnormal frequency of delta-F(508) mutation in neonatal transitory hypertrypsinaemia.(Letter) Lancet 337: 55 only, 1991.

Lemna, W. K.; Feldman, G. L.; Kerem, B.; Fernbach, S. D.; Zevkovich, E. P.; O'Brien, W. E.; Riordan, J. R.; Collins, F. S.; Tsui, L.-C.; Beaudet, A. L.: Mutation analysis for heterozygote detection and the prenatal diagnosis of cystic fibrosis. New Eng. J. Med. 322: 291-296, 1990.

Levitan, I. B.: The basic defect in cystic fibrosis. Science 244: 1423, 1989.

Lowe, C. U.; May, C. D.; Reed, S. C.: Fibrosis of the pancreas in infants and children: a statistical study of clinical and hereditary features. Am. J. Dis. Child. 78: 349-374, 1949.

Macek, M., Jr.; Mackova, A.; Hamosh, A.; Hilman, B. C.; Selden, R. F.; Lucotte, G.; Friedman, K. J.; Knowles, M. R.; Rosenstein, B. J.; Cutting, G. R.: Identification of common cystic fibrosis mutations in African-Americans with cystic fibrosis increases the detection rate to 75%. Am. J. Hum. Genet. 60: 1122-1127, 1997.

Mangos, J. A.; McSherry, N. R.: Studies on the mechanism of inhibition of sodium transport in cystic fibrosis of the pancreas. Pediat. Res. 2: 378-384, 1968.

Manson, J. C.; Brock, D. J. H.: Development of a quantitative immunoassay for the cystic fibrosis gene. Lancet I: 330-331, 1980.

Marino, C. R.; Matovcik, L. M.; Gorelick, F. S.; Cohn, J. A.: Localization of the cystic fibrosis transmembrane conductance regulator in pancreas. J. Clin. Invest. 88: 712-716, 1991.
Mayo, B. J.; Klebe, R. J.; Barnett, D. R.; Lankford, B. J.; Bowman, B. H.: Somatic cell genetic studies of the cystic fibrosis mucociliary inhibitor. Clin. Genet. 18: 379-386, 1980.
McIntosh, I.; Lorenzo, M.-L.; Brock, D. J. H.: Frequency of delta-F508 mutation on cystic fibrosis chromosomes in UK.(Letter) Lancet II: 1404 only, 1989.
McMillan, S. A.; Hill, A. J. M.; Graham, C. A.; Nevin, N. C.; Fay, A. C.: T cell receptor beta chain polymorphisms are associated with cystic fibrosis. J. Med. Genet. 26: 431-433, 1989.
Mornet, E.; Simon-Bouy, B.; Serre, J. L.; Estivill, X.; Farrall, M.; Boue, J.; Williamson, R.; Boue, J.; Boue, A.: Genetic differences between cystic fibrosis with and without meconium ileus. Lancet I: 376-378, 1988.
Mornet, E.; Simon-Bouy, B.; Serre, J. L.; Muller, F.; Taillandier, A.; Martinez, M.; Boue, J.; Boue, A.: Genetic heterogeneity between two clinical forms of cystic fibrosis evidenced by familial analysis and linked DNA probes. Clin. Genet. 35: 81-87, 1989.
Morreau, J.; Sinaasappel, M.; Oostra, B. A.; Halley, D. J. J.: Cystic fibrosis: screening for a DNA deletion by field inversion gel electrophoresis. Hum. Genet. 79: 64-67, 1988.
Naylor, S. L.; Barnett, D. R.; Buchanan, J. M.; Latimer, J.; Wieder, K.; Marshall, S.; Gardner, J.; Denning, C. R.; Gluckson, M.; Pinero, R.; Rendon, H.; Miranda, L. I.; Kammerer, C.; Zansky, S. M.; King, R. H.; Bowman, B. H.; MacCluer, J. W.: Linkage of cystic fibrosis locus and polymorphic DNA markers in 14 families. Am. J. Hum. Genet. 39: 707-712, 1986.
Nazer, H.; Riff, E.; Sakati, N.; Mathew, R.; Majeed-Saidan, M. A.; Harfi, H.: Cystic fibrosis in Saudi Arabia. Europ. J. Pediat. 148: 330-332, 1989.
Neglia, J. P.; FitzSimmons, S. C.; Maisonneuve, P.; Schoni, M. H.; Schoni-Affolter, F.; Corey, M.; Lowenfels, A. B.; Boyle, P.; Dozor, A. J.; Durie, P.: The risk of cancer among patients with cystic fibrosis. New Eng. J. Med. 332: 494-499, 1995.
Nunes, V.; Gasparini, P.; Novelli, G.; Gaona, A.; Bonizzato, A.; Sangiuolo, F.; Balassopoulou, A.; Gimenez, F. J.; Dognini, M.; Ravnik-Glavac, M.; Cikuli, M.; Mokini, V.; Komel, R.; Dallapiccola, B.; Pignatti, P. F.; Loukopoulos, D.; Casals, T.; Estivill, X.: Analysis of 14 cystic fibrosis mutations in five south European populations. Hum. Genet. 87: 737-738, 1991.
Ober, C.; Bombard, A.; Dhaliwal, R.; Elias, S.; Fagan, J.; Laffler, T. G.; Martin, A. O.; Rosinsky, B.: Studies of cystic fibrosis in Hutterite families by using linked DNA probes. Am. J. Hum. Genet. 41: 1145-1151, 1987.
Oppenheimer, E. H.: Absence of pancreatic lesions in cystic fibrosis. Birth Defects Orig. Art. Ser. VIII(2): 108-113, 1972.
Oppenheimer, E. H.; Case, A. L.; Esterly, J. R.; Rothberg, R. M.: Cervical mucus in cystic fibrosis: a possible cause of infertility. Am. J. Obstet. Gynec. 108: 673-674, 1970.
Oppenheimer, E. H.; Esterly, J. R.: Observations on cystic fibrosis of the pancreas. V. Developmental changes in the male genital system. J. Pediat. 75: 806-811, 1969.
Oppenheimer, E. H.; Esterly, J. R.: Observations on cystic fibrosis of the pancreas. VI. The uterine cervix. J. Pediat. 77: 991-995, 1970.
Park, M.; Testa, J. R.; Blair, D. G.; Dean, M.; Parsa, N. Z.; Vande Woude, G. F.: The CF locus is distal to and upstream from the met protooncogene transcription unit which is located at 7q31-32. Cytogenet. Cell Genet. 46: 674-675, 1987.
Pier, G. B.; Grout, M.; Zaidi, T.; Meluleni, G.; Mueschenborn, S. S.; Banting, G.; Ratcliff, R.; Evans, M. J.; Colledge, W. H.: Salmonella typhi uses CFTR to enter intestinal epithelial cells. Nature 393: 79-82, 1998.
Pier, G. B.; Grout, M.; Zaldi, T. S.; Olsen, J. C.; Johnson, L. G.; Yankaskas, J. R.; Goldberg, J. B.: Role of mutant CFTR in hypersusceptibility of cystic fibrosis patients to lung infections. Science 271: 63-67, 1996.
Quinton, P. M.: Chloride impermeability in cystic fibrosis. Nature 301: 421-422, 1983.
Rao, G. J. S.; Nadler, H. L.: Arginine esterase in cystic fibrosis of the pancreas. Pediat. Res. 8: 684-686, 1974.
Rao, G. J. S.; Nadler, H. L.: Deficiency of arginine esterase in cystic fibrosis of pancreas—demonstration of proteolytic nature of activity. Pediat. Res. 9: 739-741, 1975.
Rao, G. J. S.; Posner, L. A.; Nadler, H. L.: Deficiency of kallikrein activity in plasma of patients with cystic fibrosis. Science 177: 610-611, 1972.
Rich, D. P.; Anderson, M. P.; Gregory, R. J.; Cheng, S. H.; Paul, S.; Jefferson, D. M.; McCann, J. D.; Klinger, K. W.; Smith, A. E.; Welsh, M. J.: Expression of cystic fibrosis transmembrane conductance regulator corrects defective chloride channel regulation in cystic fibrosis airway epithelial cells. Nature 347: 358-363, 1990.
Roberts, G. B. S.: Familial incidence of fibrocystic disease of the pancreas. Ann. Hum. Genet. 24: 127-135, 1960.
Rodman, D. M.; Zamudio, S.: The cystic fibrosis heterozygote—advantage in surviving cholera? Med. Hypotheses 36: 253-258, 1991.
Romeo, G.; Bianco, M.; Devoto, M.; Menozzi, P.; Mastella, G.; Giunta, A. M.; Micalizzi, C.; Antonelli, M.; Battistini, A.; Santamaria, F.; Castello, D.; Marianelli, A.; Marchi, A. G.; Manca, A.; Miano, A.: Incidence in Italy, genetic heterogeneity, and segregation analysis of cystic fibrosis. Am. J. Hum. Genet. 37: 338-349, 1985.
Romeo, G.; Devoto, M.; Bianco, M.: Homogeneity vs. heterogeneity of cystic fibrosis in Italy.(Letter) Am. J. Hum. Genet. 39: 283-284, 1986.
Romeo, G.; Devoto, M.; Galietta, L. J. V.: Why is the cystic fibrosis gene so frequent? Hum. Genet. 84: 1-5, 1989.
Rosenfeld, M. A.; Yoshimura, K.; Trapnell, B. C.; Yoneyama, K.; Rosenthal, E. R.; Dalemans, W.; Fukayama, M.; Bargon, J.; Stier, L. E.; Stratford-Perricaudet, L.; Perricaudet, M.; Guggino, W. B.; Pavirani, A.; Lecocq, J.-P.; Crystal, R. G.: In vivo transfer of the human cystic fibrosis transmembrane conductance regulator gene to the airway epithelium. Cell 68: 143-155, 1992.
Rozmahel, R.; Wilschanski, M.; Matin, A.; Plyte, S.; Oliver, M.; Auerbach, W.; Moore, A.; Forstner, J.; Durie, P.; Nadeau, J.; Bear, C.; Tsui, L.-C.: Modulation of disease severity in cystic fibrosis transmembrane conductance regulator deficient mice by a secondary genetic factor. Nature Genet. 12: 280-287, 1996.
Sanguinetti-Briceno, N. R.; Brock, D. J. H.: NADH dehydrogenase in cystic fibrosis. Clin. Genet. 22: 308-311, 1982.
Savov, A.; Angelicheva, D.; Balassopoulou, A.; Jordanova, A.; Noussia-Arvanitakis, S.; Kalaydjieva, L.: Double mutant alleles: are they rare? Hum. Molec. Genet. 4: 1169-1171, 1995.
Scambler, P.; Robbins, T.; Gilliam, C.; Boylston, A.; Tippett, P.; Williamson, R.; Davies, K. E.: Linkage studies between polymorphic markers on chromosome 4 and cystic fibrosis. Hum. Genet. 69: 250-254, 1985.
Scambler, P. J.; Wainwright, B. J.; Farrall, M.; Bell, J.; Stanier, P.; Lench, N. J.; Bell, G.; Kruyer, H.; Ramirez, F.; Williamson, R.: Linkage of COL1A2 collagen gene to cystic fibrosis, and its clinical implications.(Letter) Lancet II: 1241-1242, 1985.
Scambler, P. J.; Wainwright, B. J.; Watson, E.; Bates, G.; Bell, G.; Williamson, R.; Farrall, M.: Isolation of a further anonymous informative DNA sequence from chromosome seven closely linked to cystic fibrosis. Nucleic Acids Res. 14: 1951-1956, 1986.
Schoumacher, R. A.; Ram, J.; Iannuzzi, M. C.; Bradbury, N. A.; Wallace, R. W.; Tom Hon, C.; Kelly, D. R.; Schmid, S. M.; Gelder, F. B.; Rado, T. A.; Frizzell, R. A.: A cystic fibrosis pancreatic adenocarcinoma cell line. Proc. Nat. Acad. Sci. 87: 4012-4016, 1990.
Scully, R. E.; Goldabini, J. J.; McNeely, B. V.: Case reports of the Massachusetts General Hospital (CPCs). New Eng. J. Med. 296: 1519-1526, 1977.
Searle, A. G.; Peters, J.; Lyon, M. F.; Evans, E. P.; Edwards, J. H.; Bauckle, B. J.: Chromosome maps of man and mouse, III. Genomics 1: 3-18, 1987.
Shapiro, B. L.; Lam, L. F.-H.: Calcium and age in fibroblasts from control subjects and patients with cystic fibrosis. Science 216: 417-419, 1982.
Shapiro, B. L.; Lam, L. F.-H.; Feigal, R. J.: Mitochondrial NADH dehydrogenase in cystic fibrosis: enzyme kinetics in cultured fibroblasts. Am. J. Hum. Genet. 34: 846-852, 1982.
Sharer, N.; Schwarz, M.; Malone, G.; Howarth, A.; Painter, J.; Super, M.; Braganza, J.: Mutations of the cystic fibrosis gene in patients with chronic pancreatitis. New Eng. J. Med. 339: 645-652, 1998.
Shepherd, R. W.; Holt, T. L.; Vasques-Velasquez, L.; Coward, W. A.; Prentice, A.; Lucas, A.: Increased energy expenditure in young children with cystic fibrosis. Lancet I: 1300-1303, 1988.
Sheppard, D. N.; Rich, D. P.; Ostedgaard, L. S.; Gregory, R. J.; Smith, A. E.; Welsh, M. J.: Mutations in CFTR associated with mild-disease form Cl- channels with altered pore properties. Nature 362: 160-164, 1993.
Shier, W. T.: Increased resistance to influenza as a possible source of heterozygote advantage in cystic fibrosis. Med. Hypotheses 5: 661-667, 1979.
Shwachman, H.; Kowalski, M.; Khaw, K.-T.: Cystic fibrosis: a new outlook: 70 patients above 25 years of age. Medicine 56: 129-149, 1977.
Sing, C. F.; Risser, D. R.; Howatt, W. F.; Erickson, R. P.: Phenotypic heterogeneity in cystic fibrosis. Am. J. Med. Genet. 13: 179-195, 1982.
Siraganian, P. A.; Miller, R. W.; Swender, P. T.: Cystic fibrosis and ileal carcinoma.(Letter) Lancet II: 1158 only, 1987.
Smith, D. W.; Docter, J. M.; Ferrier, P. E.; Frias, J. L.; Spock, A.: Possible localisation of the gene for cystic fibrosis of the pancreas to the short arm of chromosome 5. Lancet II: 309-312, 1968.
Smyth, R. L.: Fibrosing colonopathy in cystic fibrosis. Arch. Child. Dis. 74: 464-466, 1996.
Smyth, R. L.; van Velzen, D.; Smyth, A. R.; Lloyd, D. A.; Heaf, D. P.: Strictures of ascending colon in cystic fibrosis and high-strength pancreatic enzymes. Lancet 343: 85-86, 1994.
Snouwaert, J. N.; Brigman, K. K.; Latour, A. M.; Malouf, N. N.; Boucher, R. C.; Smithies, O.; Koller, B. H.: An animal model for cystic fibrosis made by gene targeting. Science 257: 1083-1088, 1992.
Spence, J. E.; Perciaccante, R. G.; Greig, G. M.; Willard, H. F.; Ledbetter, D. H.; Hejtmancik, J. F.; Pollack, M. S.; O'Brien, W. E.; Beaudet, A. L.: Uniparental disomy as a mechanism for human genetic disease. Am. J. Hum. Genet. 42: 217-226, 1988.
Spence, J. E.; Rosenbloom, C. L.; O'Brien, W. E.; Seilheimer, D. K.; Cole, S.; Ferrell, R. E.; Stern, R. C.; Beaudet, A. L.: Linkage of DNA markers to cystic fibrosis in 26 families. Am. J. Hum. Genet. 39: 729-734, 1986.
Spock, A.; Heick, H. M. C.; Cress, H.; Logan, W. S.: Abnormal serum factor in patients with cystic fibrosis of the pancreas. Pediat. Res. 1: 173-177, 1967.
Steinberg, A. G.; Brown, D. C.: On the incidence of cystic fibrosis of the pancreas. Am. J. Hum. Genet. 12: 416-424, 1960.
Stern, R. C.: The diagnosis of cystic fibrosis. New Eng. J. Med. 336: 487-491, 1997.
Stern, R. C.; Boat, T. F.; Abramowsky, C. R.; Matthews, L. W.; Wood, R. E.; Daershuk, C. F.: Intermediate-range sweat chloride concentration and Pseudomonas bronchitis: a cystic fibrosis variant with preservation of exocrine pancreatic function. J.A.M.A. 239: 2676-2680, 1978.
Stern, R. C.; Boat, T. F.; Doershuk, C. F.: Obstructive azoospermia as a diagnostic criterion for the cystic fibrosis syndrome. Lancet I: 1401-1404, 1982.
Strain, L.; Curtis, A.; Mennie, M.; Holloway, S.; Brock, D. J. H.: Use of linkage disequilibrium data in prenatal diagnosis of cystic fibrosis. Hum. Genet. 80: 75-77, 1988.
Super, M.: Genetic counselling and antenatal diagnosis of cystic fibrosis. J. Roy. Soc. Med. 80: 13-15, 1987.
Tebbutt, S. J.; Wardle, C. J. C.; Hill, D. F.; Harris, A.: Molecular analysis of the ovine cystic fibrosis transmembrane conductance regulator gene. Proc. Nat. Acad. Sci. 92: 2293-2297, 1995.
Ten Kate, L. P.; Scheffer, H.; Cornel, M. C.; vanLookeren Campagne, J. G.: Consanguinity sans reproche. Hum. Genet. 86: 295-296, 1991.
The Cystic Fibrosis Genotype-Phenotype Consortium: Correlation between genotype and phenotype in patients with cystic fibrosis. New Eng. J. Med. 329: 1308-1313, 1993.
Tsui, L.-C.; Buchwald, M.; Barker, D.; Braman, J. C.; Knowlton, R.; Schumm, J. W.; Eiberg, H.; Mohr, J.; Kennedy, D.; Plavsic, N.; Zsiga, M.; Markiewicz, D.; Akots, G.; Brown, V.; Helms, C.; Gravius, T.; Parker, C.; Rediker, K.; Donis-

Keller, H.: Cystic fibrosis locus defined by a genetically linked polymorphic DNA marker. Science 230: 1054-1057, 1985.
Tsui, L.-C.; Buetow, K.; Buchwald, M.: Genetic analysis of cystic fibrosis using linked DNA markers. Am. J. Hum. Genet. 39: 720-728, 1986.
Vitale, E.; Devoto, M.; Mastella, G.; Romeo, G.: Homogeneity of cystic fibrosis in Italy. Am. J. Hum. Genet. 39: 832-836, 1986.
Voss, R.; Ben-Simon, E.; Avital, A.; Godfrey, S.; Zlotogora, J.; Dagan, J.; Tikochinski, T.; Hillel, J.: Isodisomy of chromosome 7 in a patient with cystic fibrosis: could uniparental disomy be common in humans? Am. J. Hum. Genet. 45: 373-380, 1989.
Voss, R.; Ben-Simon, E.; Zlotogora, Y.; Dagan, J.; Godfry, S.; Haberfeld, A.; Hillel, Y.: Uniparental disomy for chromosome 7—cause for homozygosity at the cystic fibrosis locus. (Abstract) Am. J. Hum. Genet. 43: A73 only, 1988.
Wainwright, B.; Scambler, P.; Farrall, M.; Schwartz, M.; Williamson, R.: Linkage between the cystic fibrosis locus and markers on chromosome 7q. Cytogenet. Cell Genet. 41: 191-192, 1986.
Wainwright, B. J.; Scambler, P. J.; Schmidtke, J.; Watson, E. A.; Law, H.-Y.; Farrall, M.; Cooke, H. J.; Eiberg, H.; Williamson, R.: Localization of cystic fibrosis locus to human chromosome 7cen-q22. Nature 318: 384-385, 1985.
Warner, J. O.; Norman, A. P.; Soothill, J. F.: Cystic fibrosis heterozygosity in the pathogenesis of allergy. Lancet I: 990-991, 1976.
Watkins, P. C.; Schwartz, R.; Hoffman, N.; Stanislovitis, P.; Doherty, R.; Klinger, K.: A linkage study of cystic fibrosis is extended multigenerational pedigrees. Am. J. Hum. Genet. 39: 735-743, 1986.
Welsh, M. J.; Fick, R. B.: Cystic fibrosis. J. Clin. Invest. 80: 1523-1526, 1987.
White, R.; Leppert, M.; O'Connell, P.; Nakamura, Y.; Woodward, S.; Hoff, M.; Herbst, J.; Dean, M.; Vande Woude, G.; Lathrop, G. M.; Lalouel, J.-M.: Further linkage data on cystic fibrosis: the Utah study. Am. J. Hum. Genet. 39: 694-698, 1986.
White, R.; Woodward, S.; Leppert, M.; O'Connell, P.; Hoff, M.; Herbst, J.; Lalouel, J.-M.; Dean, M.; Vande Woude, G.: A closely linked genetic marker for cystic fibrosis. Nature 318: 382-384, 1985.
Widdicombe, J. H.; Welsh, M. J.; Finkbeiner, W. E.: Cystic fibrosis decreases the apical membrane chloride permeability of monolayers cultured from cells of tracheal epithelium. Proc. Nat. Acad. Sci. 82: 6167-6171, 1985.
Williamson, R.: Personal Communication. London, England 11/22/1984.
Wilmut, I.; Schnieke, A. E.; McWhir, J.; Kind, A. J.; Campbell, K. H. S.: Viable offspring derived from fetal and adult mammalian cells. Nature 385: 810-813, 1997.
Wilson, G. B.; Fudenberg, H. H.; Jahn, T. L.: Studies on cystic fibrosis using isoelectric focusing. I. An assay for detection of cystic fibrosis homozygotes and heterozygote carriers from serum. Pediat. Res. 9: 635-640, 1975.
Wilson, J. M.: Gene therapy for cystic fibrosis: challenges and future directions. J. Clin. Invest. 96: 2547-2554, 1995.
Wine, J. J.: No CFTR: are CF symptoms milder? (Letter) Nature Genet. 1: 10 only, 1992.
Woo, S.: Personal Communication. Houston, Tex. 1988.
Wright, S. W.; Morton, N. E.: Genetic studies on cystic fibrosis in Hawaii. Am. J. Hum. Genet. 20: 157-162, 1968.
Yang, Y.; Devor, D. C.; Engelhardt, J. F.; Ernst, S. A.; Strong, T. V.; Collins, F. S.; Cohn, J. A.; Frizzell, R. A.; Wilson, J. M.: Molecular basis of defective anion transport in L cells expressing recombinant forms of CFTR. Hum. Molec. Genet. 2: 1253-1261, 1993.
Yang, Y.; Raper, S. E.; Cohn, J. A.; Engelhardt, J. F.; Wilson, J. M.: An approach for treating the hepatobiliary disease of cystic fibrosis by somatic gene transfer. Proc. Nat. Acad. Sci. 90: 4601-4605, 1993.
Zhou, L.; Dey, C. R.; Wert, S. E.; DuVall, M. D.; Frizzell, R. A.; Whitsett, J. A.: Correction of lethal intestinal defect in a mouse model of cystic fibrosis by human CFTR. Science 266: 1705-1708, 1994.

HEMANGIOMAS OF SMALL INTESTINE (140900)

JJM NOTE

Localized hemangiomas seem to occur in many organ systems.

Bandler (1960) reported a family in 3 generations of which there were 3 proved and 2 possible instances of cavernous hemangioma involving almost the entire small intestine. One patient had mucocutaneous pigment spots precisely like those of the Peutz-Jeghers syndrome. See blue rubber nevus syndrome (112200).

Bandler, M.: Hemangiomas of the small intestine associated with mucocutaneous pigmentation. Gastroenterology 38: 641-645, 1960.

Anal Tumors

ANAL CANAL CARCINOMA (105580)

JJM NOTE

There may be interactions with HIV, given the excess of this cancer in HIV-positive individuals.

Squamous carcinoma of the anal canal is a relatively uncommon tumor. Most of these carcinomas develop from squamous epithelium, but some arise from the transitional zone between the columnar epithelium of the rectum and the squamous epithelium of the anal canal. The latter type is called cloacogenic or transitional carcinoma. In a cytogenetic study of 8 cases of anal canal cancer (1 cloacogenic and 7 squamous cell), Muleris et al. (1987) found that all had chromosomal abnormalities. A rearrangement involving the long arm of chromosome 11 was found in all instances. Rearrangements of chromosome 3, detected in 6 tumors, led to a deletion of the short arm in 5 cases. The smallest common deleted segments were 11q22-qter and 3p22. The authors pointed out that this may be a situation comparable to that in retinoblastoma (180200), Wilms tumor (194070), osteosarcoma (259500), and acoustic neurinoma (101000).

Muleris, M.; Salmon, R.-J.; Girodet, J.; Zafrani, B.; Dutrillaux, B.: Recurrent deletions of chromosomes 11q and 3p in anal canal carcinoma. Int. J. Cancer 39: 595-598, 1987.

BLOOD GROUP—LEWIS SYSTEM (*111100)

JJM NOTE

Koprowski et al. (1982) suggested a nonrandom association of Lewis-negative status with a risk for certain gastrointestinal cancers. I believe the tumor marker CA 19-9 emerged from the initial work.

The Lewis system involves genetically variable antigens in the body fluids and only secondarily are the antigens absorbed to red cells. Grollman et al. (1969) showed that Lewis-negative women lack a specific fucosyltransferase which is present in the milk of Lewis-positive women. The enzyme is apparently required for synthesis of the structural determinants of both Lewis (a) and Lewis (b) specificity. The same enzyme is involved in the synthesis of milk oligosaccharides, because 2 oligosaccharides containing the relevant linkage were absent from the milk of Lewis-negative women. Grubb (1953) provided the ingenious interpretation of the interactions between the Les locus determining presence/absence of Lewis substance in the saliva and on red cells and the Se locus (182100) determining secretion of ABH blood group substances in the saliva and Le(a) or Le(b) expression in red cells.

Weitkamp et al. (1974) presented evidence that the Lewis blood group locus and the C3 locus are linked. The assignment of C3 to chromosome 19 (see 120700) indicated that Lewis blood group is also on that chromosome. Gedde-Dahl et al. (1984) used Les as the symbol for this locus.

From linkage studies using microsatellite markers, Reguigne-Arnould et al. (1995) concluded that FUT3 is in a cluster of loci with FUT6 (136836) and FUT5 (136835) on 19p13.3. The following gene order was deduced: 19pter—D19S216—FUT6—FUT3—FUT5—D19S567—cen.

In transfusion medicine, it has been found that some individuals who Sheinfeld et al. (1989) found an increased frequency of Lewis blood group nonsecretor—Le(a+1pb-)—and of recessive Le(a-b-) phenotypes among women with recurrent urinary tract infections. No significant difference was found in the distribution of ABO or P phenotypes between a group of 49 white women with histories of recurrent urinary tract infections and 49 healthy control women without recurrent urinary tract infections.

The last step in the biosynthesis of Lewis antigen, the addition of a fucose to precursor polysaccharides, can be catalyzed by at least several types of alpha-3-fucosyltransferases: FUT1, or Bombay (211100), FUT2, or secretor (182100), FUT3, or Lewis, and FUT4, which is the myeloid form of alpha-3-fucosyltransferase (104230).

Nishihara et al. (1994) found that all le alleles had a T59G mutation, whereas none of the Le alleles did. The le alleles were divided into 2 subtypes, le1, having a G508A mutation, and le2, having a T1067A mutation. The T1067A mutation reduced the enzyme activity less than 10%, whereas the G508A mutation in the catalytic domain made the enzyme completely inactive. The frequency of Le, le1, and le2 in the Japanese population was found to be 66%, 30%, and 4%, respectively.

Yazawa et al. (1996) demonstrated that genotyping of Le genes by PCR-RFLP methods could be used for determining Lewis blood type on human hairs and blood stains, as well as in paternity testing.

type as Lewis-positive on erythrocytes can change their erythrocyte phenotype to Lewis-negative during diseases or during pregnancy. Orntoft et al. (1996) noted that these patients have been named non-genuine Lewis-negative individuals as they have alpha-1-4 fucosyltransferase activity in saliva. Due to this phenomenon, the Lewis-negative phenotype is more common among cancer patients (approximately 20%) than among healthy individuals (approximately 8%). Orntoft et al. (1996) examined the mutational spectrum of the Lewis gene in Denmark and found 6 different mutations. Five, 59T-G (L20R; 111100.0001), 202T-C (W68R), 314C-T (T105M), 508G-A (G170S; 111100.0001), and 1067T-A (I356K), were frequent, and 1, 445C-A (L146M), was only detected in 1 of 40 individuals. The authors demonstrated that the nucleotide 202 and 314 mutations were located on the same allele. COS-7 cells transfected with an allele having the 202/314 mutations lacked enzyme activity. Lewis-negative patients, whose erythrocytes converted from Lewis-positive to Lewis-negative during their disease, showed FUT3 heterozygosity significantly more often than did others (p less than 0.05). Pang et al. (1998) identified 5 novel missense mutations in the FUT3 gene in African (Xhosa) and Caucasian subjects in South Africa.

Data on gene frequencies of allelic variants were tabulated by Roychoudhury and Nei (1988).

See also Koprowski et al. (1982)

Gedde-Dahl, T., Jr.; Olaisen, B.; Teisberg, P.; Wilhelmy, M. C.; Mevag, B.; Helland, R.: The locus for apolipoprotein E (apoE) is close to the Lutheran (Lu) blood group locus on chromosome 19. Hum. Genet. 67: 178-182, 1984.

Grollman, E. F.; Kobata, A.; Ginsburg, V.: An enzymatic basis for Lewis blood types in man. J. Clin. Invest. 48: 1489-1494, 1969.

Grubb, R.: Zur Genetik des Lewis-Systems. Naturwissenschaften 21: 560-561, 1953.

Koda, Y.; Kimura, H.; Mekada, E.: Analysis of Lewis fucosyltransferase genes from the human gastric mucosa of Lewis-positive and -negative individuals. Blood 82: 2915-2919, 1993.

Koprowski, H.; Blaszczyk, M.; Steplewski, Z.; Brockhaus, M.; Magnani, J.; Ginsburg, V.: Lewis blood-type may affect the incidence of gastrointestinal cancer. Lancet I: 1332-1333, 1982.

Nishihara, S.; Narimatsu, H.; Iwasaki, H.; Yazawa, S.; Akamatsu, S.; Ando, T.; Seno, T.; Narimatsu, I.: Molecular genetic analysis of the human Lewis histo-blood group system. J. Biol. Chem. 269: 29271-29278, 1994.

Orntoft, T. F.; Vestergaard, E. M.; Holmes, E.; Jakobsen, J. S.; Grunnet, N.; Mortensen, M.; Johnson, P.; Bross, P.; Gregersen, N.; Skorstengaard, K.; Jensen, U. B.; Bolund, L.; Wolf, H.: Influence of Lewis alpha-1-3/4-L-fucosyltransferase (FUT3) gene mutations on enzyme activity, erythrocyte phenotyping, and circulating tumor marker sialyl-Lewis a levels. J. Biol. Chem. 271: 32260-32268, 1996.

Pang, H.; Liu, Y.; Koda, Y.; Soejima, M.; Jia, J.; Schlaphoff, T.; du Toit, E. D.; Kimura, H.: Five novel missense mutations of the Lewis gene (FUT3) in African (Xhosa) and Caucasian populations in South Africa. Hum. Genet. 102: 675-680, 1998.

Reguigne-Arnould, I.; Couillin, P.; Mollicone, R.; Faure, S.; Fletcher, A.; Kelly, R. J.; Lowe, J. B.; Oriol, R.: Relative positions of two clusters of human alpha-L-fucosyltransferases in 19q (FUT1-FUT2) and 19p (FUT6-FUT3-FUT5) within the microsatellite genetic map of chromosome 19. Cytogenet. Cell Genet. 71: 158-162, 1995.

Roychoudhury, A. K.; Nei, M.: Human Polymorphic Genes: World Distribution. New York: Oxford Univ. Press (pub.) 1988.

Sheinfeld, J.; Schaeffer, A. J.; Cordon-Cardo, C.; Rogatko, A.; Fair, W. R.: Association of the Lewis blood-group phenotype with recurrent urinary tract infections in women. New Eng. J. Med. 320: 773-777, 1989.

Weitkamp, L. R.; Johnston, E.; Guttormsen, S. A.: Probable genetic linkage between the loci for the Lewis blood group and complement C3. Cytogenet. Cell Genet. 13: 183-184, 1974.

Yazawa, S.; Oh-Kawara, H.; Nakajima, T.; Hosomi, O.; Akamatsu, S.; Kishi, K.: Histo-blood group Lewis genotyping from human hairs and blood. Jpn. J. Hum. Genet. 41: 177-188, 1996.

Liver Tumors

HEPATOCELLULAR CARCINOMA 1 (*114550)

JJM NOTE

Despite its relatively low incidence in North America, hepatocellular carcinoma is a leading cancer, if not the first worldwide. It clearly has environmental determinants, such as aflatoxin, hepatitis B, and others. And, a few major gene traits predispose to or are complicated by hepatocellular carcinoma, sometimes, but not always, with the common intermediate of cirrhosis, such as alpha-1-antitrypsin deficiency, various types of glycogen storage disease, various porphyrias, Wilson disease, tyrosinemia type 1, cartilage-hair hypoplasia, ataxia-telangiectasia, and Russell-Silver syndrome. As details emerge, however, hepatocellular carcinoma seems to be a premier example of ecogenetics in carcinogenesis. One or several integration sites of hepatitis B virus may alter susceptibility to the virus. Various xenobiotic metabolizing enzymes (such as alcohol dehydrogenase, epoxide hydrolase, and glutathione *S*-transferase) seem to interact with a clear environmental determinant to determine the final risk.

Primary cancer of the liver in 3 brothers was described by Kaplan and Cole (1965) and by Hagstrom and Baker (1968). In these patients there was no recognized preexisting liver disease. Denison et al. (1971) described 2 adult brothers who died of primary hepatocellular carcinoma. Both had micronodular cirrhosis with features of subacute progressive viral hepatitis. Australia antigen was demonstrated in the brother in whom it was sought. Their father had died much earlier of hepatocellular carcinoma. See 231100 for description of liver cancer as a complication of giant cell hepatitis of infancy. Familial LCC might also have its explanation in alpha-1-antitrypsin deficiency (107400), hemochromatosis (235200), and tyrosinemia (276700). Integration of the hepatitis B virus (HBV) into cellular DNA occurs during long-term persistent infection in man. Hepatocellular carcinomas isolated from carriers of virus often contain clonally propagated viral DNA. Shen et al. (1991) presented evidence for the interaction of inherited susceptibility and hepatitis B viral infection in cases of primary hepatocellular carcinoma in eastern China. Complex segregation analysis of 490 extended families supported the existence of a recessive allele with population frequency approximately 0.25, which results in a lifetime risk of HCC in the presence of both HBV infection and genetic susceptibility, of 0.84 for males and 0.46 for females. The model further predicted that, in the absence of genetic susceptibility, lifetime risk of HCC is 0.09 for HBV-infected males and 0.01 for HBV-infected females and that regardless of genotype the risk is virtually zero for uninfected persons.

The finding of small deletions in retinoblastoma and Wilms tumor prompted Rogler et al. (1985) to look for the same in association with HBV integration in hepatocellular carcinoma. They demonstrated a deletion of at least 13.5 kb of cellular sequences in a liver cancer. The HBV integration and the deletion occurred on the short arm of chromosome 11 at location 11p14-p13. The deleted sequences were lost in tumor cells leaving only a single copy. Clones of the DNA flanking the deleted segment were used for the mapping of the deletion in somatic cell hybrids and by in situ hybridization. Cellular sequences homologous to the deleted region were cloned and used to exclude the possibility that this DNA had been moved to other positions in the genome. Fisher et al. (1987) extended the observations of Rogler et al. (1985). Using somatic cell hybrids that contained defined 11p deletions, 2 cloned DNA sequences that flank the deletion generated by a hepatocellular carcinoma (as a consequence of hepatitis B virus integration) were mapped to 11p13. Wilms tumor (194070) and the tumors of Beckwith-Wiedemann syndrome (130650) are also determined by changes on 11p. See 142330 for familial hepatic adenoma, sometimes associated with hepatocellular carcinoma.

Henderson et al. (1988) found that unique cellular DNA to the left of an HBV DNA integration site cloned from a primary tumor mapped to chromosome 18q (18q11.1-q11.2), whereas right-hand flanking DNA mapped to chromosome 17 at a subterminal region of the long arm. In a hepatoma specimen from Shanghai, Zhou et al. (1988) identified integration of hepatitis B virus into 17p12-p11.2, which is near the human protooncogene p53 (191170). Furthermore, the sequence of flanking cellular DNA showed highly significant homology with a conserved region of a number of functional mammalian DNAs, including the human autonomously replicated sequence-1 (ARS1; 109110). ARS1 is a sequence of human DNA that allows replication of Saccharomyces cerevisiae integrative plasmids as autonomously replicating elements in S. cerevisiae cells. Since integration of viral DNA is not a required step in the replicative cycle of the hepatitis virus, the presence of integrated HBV sequences in many human hepatocellular carcinomas suggests a causal relationship. Since any 1 of several integration sites may lead to the same result, the crucial cellular targets involved in triggering liver cell malignant transformation may differ from tumor to tumor. Smith et al. (1989) gave evidence for microdeletions of chromosome 4q involving the alcohol dehydrogenase isoenzyme gene ADH3 and hepatomas from 3 of 5 individuals heterozygous for an XbaI RFLP detectable by the ADH probe. Two of 7 individuals heterozygous for an epidermal growth factor RFLP had lost 1 EGF allele in their hepatoma tissue.

Agarwal et al. (1998) reported a case of severe gynecomastia in a 17 and one-half-year-old boy due to high levels of aromatase expression in a large fibrolamellar hepatocellular carcinoma, which caused extremely elevated serum levels of estrone (1200 pg/mL) and estradiol-17 (312 pg/mL) that suppressed follicle-stimulating hormone (FSH) and luteinizing hormone (LH) (1.3 and 2.8 IU/L, respectively) and consequently testosterone (1.53 ng/mL). After removal of the 1.5-kg tumor, gynecomastia partially regressed, and normal hormone levels were restored. By immunohistochemistry, diffuse intracytoplasmic aromatase expression was detected in the liver cancer cells. Northern blot analysis showed P450 aromatase transcripts in total RNA from the hepatocellular cancer but not in the adjacent liver nor in disease-free adult liver samples.

Promoters I.3 and II were used for P450 aromatase transcription in the cancer.

Primary hepatocellular carcinoma occurs at high frequencies in east Asia and sub-Saharan Africa. In these areas of the world, chronic infection with the hepatitis B virus is the best documented risk factor; however, only 20 to 25% of HBV carriers develop HCC. Exposure to the fungal toxin aflatoxin B1 (AFB1) has been suggested to increase HCC risk, in part because in vitro experiments demonstrated that AFB1 mutagenic metabolites bind to DNA and are capable of inducing G-to-T transversions. In certain areas of the HCC endemic regions, a mutational hotspot has been reported in the p53 tumor suppressor gene (TP53; 191170): an AGG-to-AGT transversion (arginine to serine) of codon 249 in exon 7 (191170.0006). Microsomal epoxide hydrolase (EPHX; 132810) and glutathione-S-transferase M1 (GSTM1; 138350) are both involved in AFB1 detoxification in hepatocytes. Polymorphism of both genes has been identified. In Ghana and China, McGlynn et al. (1995) conducted studies to determine whether mutant alleles at one or both of these loci are associated with increased levels of serum AFB1-albumin adducts, with HCC, and with mutations at codon 249 of p53. In a cross-sectional study, they found that mutant alleles at both loci were significantly overrepresented in individuals with serum AFB1 albumin adducts. Additionally, in a case-control study, mutant alleles of EPHX were significantly overrepresented in persons with HCC. The relationship of EPHX to HCC varied by hepatitis B surface antigen status, indicating that a synergistic effect may exist. Mutations at codon 249 of p53 were observed only among HCC patients with one or both high-risk genotypes. These findings by McGlynn et al. (1995) supported the existence of genetic susceptibility in humans to the environmental carcinogen AFB1 and indicated that there is a synergistic increase in risk of HCC with the combination of hepatitis B virus infection and susceptible genotype.

See also Chang et al. (1984); Lynch et al. (1984)

Agarwal, V. R.; Takayama, K.; Van Wyk, J. J.; Sasano, H.; Simpson, E. R.; Bulun, S. E.: Molecular basis of severe gynecomastia associated with aromatase expression in a fibrolamellar hepatocellular carcinoma. J. Clin. Endocr. Metab. 83: 1797-1800, 1998.

Chang, M.-H.; Hsu, H.-C.; Lee, C.-Y.; Chen, D.-S.; Lee, C.-H.; Lin, K.-S.: Fraternal hepatocellular carcinoma in young children in two families. Cancer 53: 1807-1810, 1984.

Denison, E. K.; Peters, R. L.; Reynolds, T. B.: Familial hepatoma with hepatitis-associated antigen. Ann. Intern. Med. 74: 391-394, 1971.

Fisher, J. H.; Scoggin, C. H.; Rogler, C. E.: Sequences which flank an 11p deletion observed in an hepatocellular carcinoma map to 11p13. Hum. Genet. 75: 66-69, 1987.

Hagstrom, R. M.; Baker, T. D.: Primary hepatocellular carcinoma in three male siblings. Cancer 22: 142-150, 1968.

Henderson, A. S.; Ripley, S.; Hino, O.; Rogler, C. E.: Identification of a chromosomal aberration associated with a hepatitis B DNA integration site in human cells. Cancer Genet. Cytogenet. 30: 269-275, 1988.

Kaplan, L.; Cole, L.: Fraternal primary hepatocellular carcinoma in three male, adult siblings. Am. J. Med. 39: 305-311, 1965.

Lynch, H. T.; Srivatanskul, P.; Phornthutkul, K.; Lynch, J. F.: Familial hepatocellular carcinoma in an endemic area of Thailand. Cancer Genet. Cytogenet. 11: 11-18, 1984.

McGlynn, K. A.; Rosvold, E. A.; Lustbader, E. D.; Hu, Y.; Clapper, M. L.; Zhou, T.; Wild, C. P.; Xia, X.-L.; Baffoe-Bonnie, A.; Ofori-Adjei, D.; Chen, G.-C.; London, W. T.; Shen, F.-M.; Buetow, K. H.: Susceptibility to hepatocellular carcinoma is associated with genetic variation in the enzymatic detoxification of aflatoxin B1. Proc. Nat. Acad. Sci. 92: 2384-2387, 1995.

Rogler, C. E.; Sherman, M.; Su, C. Y.; Shafritz, D. A.; Summers, J.; Shows, T. B.; Henderson, A.; Kew, M.: Deletion in chromosome 11p associated with a hepatitis B integration site in hepatocellular carcinoma. Science 230: 319-322, 1985.

Shen, F.-M.; Lee, M. K.; Gong, H.-M.; Cai, X.-Q.; King, M.-C.: Complex segregation analysis of primary hepatocellular carcinoma in Chinese families: interaction of inherited susceptibility and hepatitis B viral infection. Am. J. Hum. Genet. 49: 88-93, 1991.

Smith, M.; Yoshiyama, K.; Kew, M.: Evidence for chromosome 4q deletions in human hepatomas. (Abstract) Cytogenet. Cell Genet. 51: 1081 only, 1989.

Zhou, Y.-Z.; Slagle, B. L.; Donehower, L. A.; vanTuinen, P.; Ledbetter, D. H.; Butel, J. S.: Structural analysis of a hepatitis B virus genome integrated into chromosome 17p of a human hepatocellular carcinoma. J. Virol. 62: 4224-4231, 1988.

HEPATOCELLULAR CARCINOMA 2 (*142380)

JJM NOTE

One wonders if germline mutations might be seen in families with apparent excesses of hepatocellular carcinoma.

Integration of DNA from the hepatitis B virus has been shown to occur frequently in human hepatocellular carcinomas. Recombinant DNA probes have been isolated from such a tumor. These probes detected rearrangements of the corresponding DNA domain in 10% of liver tumors regardless of whether they were HBV-related or not (Pasquinelli et al., 1988). 1,2:Blanquet et al. (1987, 1988) cloned the normal allele and used it for mapping the gene by somatic cell hybrid studies and by in situ hybridization. These experiments showed that the gene is located in the area 4q32.1. The Human Gene Mapping Workshops have used HVBS6 as the designation of the integration site mapped to 4q32.1 (McAlpine, 1989). Buetow et al. (1989) found that 7 of 11 primary liver tumors tested against a panel of RFLPs demonstrated loss of constitutional heterozygosity for markers on chromosome 4, particularly 4q. Buetow et al. (1989) suggested that chronic hepatitis B virus infection and other environmental agents may operate through genetic events leading to loss of a tumor suppressor locus (anti-oncogene) on chromosome 4. The findings were thought to be consistent with those of Pasquinelli et al. (1988), which placed the critical region in the vicinity of 4q32. Wang and Rogler (1988), on the other hand, found loss of heterozygosity in 11p and 13q.

Blanquet, V.; Garreau, F.; Brechot, C.; Turleau, C.: Regional mapping by in situ hybridization of a DNA segment rearranged in human liver tumors to 4q32.1. (Abstract) Cytogenet. Cell Genet. 46: 582 only, 1987.

Blanquet, V.; Garreau, F.; Chenivesse, X.; Brechot, C.; Turleau, C.: Regional mapping to 4q32.1 by in situ hybridization of a DNA domain rearranged in human liver cancer. Hum. Genet. 80: 274-276, 1988.

Buetow, K. H.; Murray, J. C.; Israel, J. L.; London, W. T.; Smith, M.; Kew, M.; Blanquet, V.; Brechot, C.; Redeker, A.; Govindarajah, S.: Loss of heterozygosity suggests tumor suppressor gene responsible for primary hepatocellular carcinoma. Proc. Nat. Acad. Sci. 86: 8852-8856, 1989.

McAlpine, P. J.: Personal Communication. Winnipeg, Manitoba, Canada 1989.

Pasquinelli, C.; Garreau, F.; Bougueleret, L.; Cariani, E.; Grzeschik, K. H.; Thiers, V.; Croissant, O.; Hadchouel, M.; Tiollais, P.; Brechot, C.: Rearrangement of a common cellular DNA domain on chromosome 4 in human primary liver tumors. J. Virol. 62: 629-632, 1988.

Wang, H. P.; Rogler, C. E.: Deletions in human chromosome arms 11p and 13q in primary hepatocellular carcinomas. Cytogenet. Cell Genet. 48: 72-78, 1988.

HEMOCHROMATOSIS (*235200)

JJM NOTE

A hematocrit and iron studies are ordered for many reasons. In one sense, they are tests for genetic predisposition to liver cancer, since hemochromatosis might be diagnosed. The scenario resembles testing for other cancer-predisposing genes, such as *BRCA1* or *hMSH2,* but without the professional pressure for extensive pretest counseling. Imagine if the consequences of diagnosing hemochromatosis had to be explained before every blood count was done!

Description

The features of hemochromatosis include cirrhosis of the liver, diabetes, hypermelanotic pigmentation of the skin, and heart failure. Primary hepatocellular carcinoma (HCC; 114550), complicating cirrhosis, is responsible for about one-third of deaths in affected homozygotes. Since hemochromatosis is a relatively easily treated disorder if diagnosed, this is a form of preventable cancer.

Nomenclature

Mercier et al. (1997) urged strongly that the symbol HFE be used for the hemochromatosis gene rather than 'HLA-H' as used by Feder et al. (1996). The designation HLA-H was used also for a presumed pseudogene in the HLA class I region; see 142925. Similarly, Bodmer et al. (1997) argued that 'HLA-H' is an undesirable designation and pointed to the accepted authority of the WHO Nomenclature Committee for Factors of the HLA System in determining symbols of genes in this region.

Clinical Features

Muir et al. (1984) recognized 4 different 'types of disease expression' which 'bred true' in families and may indicate that more than one genetic lesion in iron metabolism can lead to hereditary hemochromatosis. The 4 groups were as follows: group I—classic form with elevated transferrin saturation, serum ferritin levels, and liver iron content; group II—severe iron overload, accelerated disease manifesting at an early age; group III—elevated total body iron stores, normal transferrin saturation and serum ferritin levels; and group IV—markedly elevated findings on serum biochemical tests, i.e., transferrin saturation and serum ferritin, with minimal elevation in total body iron stores.

Diamond et al. (1989) studied the prevalence and pathogenesis of osteopenia in 22 men with hemochromatosis. They concluded that a significant decrease in bone density is observed in this condition, particularly when hypogonadism is present. They speculated that low serum free-testosterone concentrations, rather than calciotrophic hormones, determine bone mass in this disorder. Milman et al. (1992) found no relationship between genetic subtypes of transferrin and the expression of disease in hemochromatosis patients.

Deugnier et al. (1993) analyzed the occurrence of primary liver cancer in hemochromatosis; there was 1 instance of cholangiocarcinoma and 53 instances of hepatocellular carcinoma. Of the 54 patients, 32 were untreated and 22 had been 'de-ironed.' Three of the patients had HCC in noncirrhotic but only fibrotic liver. Chronic alcoholism and tobacco smoking was higher in patients with hepatocellular carcinoma than in matched hemochromatosis patients without HCC.

Roldan et al. (1998) described acute liver failure after iron supplementation in a 29-year-old woman with unrecognized hemochromatosis.

Edwards et al. (1980) identified 35 homozygotes through pedigree studies, using the close linkage to HLA-A in the identification. Thirteen were asymptomatic. Arthropathy was present in 20, hepatomegaly in 19, transaminasemia in 16, skin pigmentation in 15, splenomegaly in 14, cirrhosis in 14, hypogonadism in 6, and diabetes in 2. None had congestive heart failure. Only 1 had the triad of hepatomegaly, hyperpigmentation, and diabetes. Serum iron was increased in 30 of 35, transferrin saturation was increased in all 35, serum ferritin in 23 of 32, urinary iron excretion after deferoxamine in 28 of 33, hepatic parenchymal cell stainable iron in 32 of 33, and hepatic iron in 27 of 27. Iron loading was 2.7 times greater in men than in women. No female had hepatic cirrhosis. These studies illustrate the usefulness of pedigree studies in revealing the full range of clinical expression of a genetic disease.

Since 10% of whites are heterozygous for the HLA-linked hemochromatosis mutation, Bulaj et al. (1996) sought to define the phenotype in these individuals. They performed a cross-sectional analysis of 1,058 genotyped heterozygotes to define the effects of age and sex. The study covered 545 males and 553 female members of 202 pedigrees, each with an HLA-typed homozygous proband. They measured serum iron, transferrin saturation, and ferritin in all heterozygotes and in 321 genetically normal subjects (unaffected family members or spouses of family members). Liver biopsies were performed in a subgroup of heterozygotes. The mean serum iron concentrations and transferrin-saturation values were higher in heterozygotes than in normal subjects and did not increase with age. Initial transferrin-saturation levels exceeding the threshold associated with the homozygous genotype were found in 4% of males and 8% of female heterozygotes. The geometric mean serum ferritin concentration was higher in heterozygotes than in normal subjects and increased with age. Higher-than-normal values were found in 20% of males and 8% of female heterozygotes. The clinical and biochemical expression of hemochromatosis was more marked in heterozygotes with paternally transmitted mutations than in those with maternally transmitted mutations. Liver biopsy abnormalities were generally associated with alcohol abuse, hepatitis, or porphyria cutanea tarda. Bulaj et al. (1996) concluded that complications due to iron overload alone in hemochromatosis heterozygotes are 'extremely rare.' This was the first description of parent-of-origin effects in hemochromatosis.

Other Features

Murphy (1987) cited the following example of the ill-effects of hemochromatosis: A considerable proportion of the patients who develop Vibrio vulnificus septicemia are persons with hemochromatosis. This organism thrives in an environment with abundant iron. It is found in brackish waters, especially estuaries, where it is concentrated by 'filter feeders' such as oysters.

Anand et al. (1983) and Eriksson et al. (1986) described cases suggesting a possible relationship between alpha-1-antitrypsin deficiency (107400) and hemochromatosis. In a series of 15 patients referred to a liver transplantation center in the U.S., Rabinovitz et al. (1992) found a significant correlation between heterozygous PiZ alpha-1-antitrypsin deficiency and hemochromatosis. Other studies, however, failed to show a relationship between the 2 inborn errors of metabolism. To investigate the matter further, Elzouki et al. (1995) used a monoclonal antibody against the PiZ variant of AAT in 67 consecutive patients with genetic hemochromatosis seen in 2 Swedish hospitals. In 3 of the patients with hemochromatosis, homozygosity for the PiZ variant was found. Liver biopsy was performed in 65 of the 67 patients; 2 of the 3 PiZ homozygotes were found to have cirrhosis, compared to 10% (6 of 59) of the noncarriers of the PiZ variant. None of the homozygous or heterozygous AAT-deficient patients had developed hepatocellular carcinoma compared with 2 of 59 of the non-PiZ gene carriers. Severe emphysema developed in 2 of the patients with the homozygous phenotype. Elzouki et al. (1995) concluded that the data suggested that the presence of the PiZ allele in double dose when associated with genetic hemochromatosis contributes to the earlier onset of cirrhosis, although it may not increase the risk of hepatocellular carcinoma.

Inheritance

Debre et al. (1958) concluded that the biochemical defect of idiopathic hemochromatosis is present in heterozygotes and that whether the disease develops is dependent on other influences on iron metabolism. They suggested that juvenile hemochromatosis resulting from consanguineous marriages may represent the homozygous state of the gene. See later for a description of the work of Camaschella et al. (1997) indicating that juvenile hemochromatosis (JH; 602390) is a genetic disorder distinct from adult hemochromatosis. Bothwell et al. (1959), Debre et al. (1958), and several others concluded that one form of hemochromatosis is inherited as an autosomal dominant with incomplete penetrance in females because of loss of blood in menstruation and pregnancy. Saddi and Feingold (1974) reported a study of 96 pedigrees which, they concluded, supported autosomal recessive inheritance. Consanguinity was increased among the parents. No parent or offspring was affected. Segregation analysis was consistent with autosomal recessive inheritance if reduced penetrance in females was assumed.

Simon et al. (1977) concluded that idiopathic hemochromatosis is recessive, although polygenic (probably oligogenic) inheritance could not be excluded.

Bassett et al. (1982) provided evidence that clarifies some of the previous confusion of whether hemochromatosis is a recessive or a dominant. They observed 5 families with hemochromatosis in 2 successive generations. HLA typing of the subjects indicated that a homozygous-heterozygous mating almost certainly had occurred in 4 of the 5 families, resulting in homozygous offspring. Powell et al. (1987) restudied a family reported by Bassett et al. (1982) in which 2 children apparently homozygous for hemochromatosis did not manifest overt disease; alternative explanations such as dominant inheritance were postulated. Subsequent studies provided the correct explanation (pseudodominant inheritance) and added further evidence for the tight linkage of HFE to HLA-A.

Mapping

Simon et al. (1976) found HLA-A3 in 78.4% of cases and 27% of controls; HLA-B14 in 25.5% of cases and 3.4% of controls. Among sibs with hemochromatosis, Simon et al. (1977) found a highly significant association between hemochromatosis and possession of the same two haplotypes. For 6 families a lod score of 2.239 at a recombination fraction of 0.005 supported linkage of HLA and hemochromatosis. Stevens et al. (1977) concluded that a gene for hemochromatosis may be on chromosome 6 close to the HLA-A locus in linkage disequilibrium with high frequency of A3 in patients with hemochromatosis. Cartwright et al. (1978) obtained lod scores well above the 3.0 generally taken as indicating linkage, for the HLA-hemochromatosis linkage. That the high lod score is not an artifact due to A3, B7 and B14 associations was supported by the finding of a lod score of 4.14 at theta 0.00 in 5 pedigrees in which these antigens were not present in the probands (Dadone et al., 1982). Skolnick (1983) contended that linkage disequilibrium cannot explain the HLA-hemochromatosis association because the association is with a haplotype, either A3-B7 or A3-B14. 46,47:Edwards et al. (1985, 1986) presented the first known example of recombination between the HLA-A and hemochromatosis loci and proposed that the (or at least a) hemochromatosis locus lies between the HLA-A and HLA-B loci. 35,36:David et al. (1986, 1987) studied an exceptional recombinant family with 3 HLA-identical sibs: one (the proband) had hemochromatosis, while the other 2 were free of any clinical or biologic signs of the disease. The study of restriction patterns using 2 MHC class I probes showed 2 differences between the proband and his sibs which were attributed to an unbalanced crossover or a genetic conversion. The absence of a 7.7-kb HIND III fragment in the proband suggested that this segment is the location of at least part of the hemochromatosis gene. Furthermore, it appears that the hemochromatosis gene lies te lomeric to the HLA-A locus. Lucotte and Coulondre (1986) found that a specific PvuII restriction fragment correlates absolutely with the HLA-A3 serologic allele and with the hemochromatosis allele.

In situ hybridization studies suggest that a ferritin heavy chain gene (FTH; see 134770) lies in the region 6p21.3-p12 (Cragg et al., 1985; McGill et al., 1987). David et al. (1989) studied restriction polymorphism of the FTH gene in 83 hemochromatosis patients and 84 controls as well as in 19 nuclear families. No significant difference was found with 10 restriction enzymes used; hence, the genomic abnormality responsible for this disorder is not a major deletion of the FTH gene. The possibility that the locus on chromosome 6 represents a second functional FTH gene in addition to the one on chromosome 11 is suggested by the fact that 2 H-type ferritin subunits have been isolated from porcine spleen and 2 are known in tadpoles; furthermore, 2 isoelectric forms of the H subunit have been found by in vitro translation of HeLa cell mRNA (reviewed by David et al., 1989). Dugast et al. (1990) found that 2 human ferritin heavy chain genes lie near the hemochromatosis locus on 6p. One of these was shown to be a processed pseudogene. Comparison of its sequence with those of other FTH pseudogenes indicated that these pseudogenes may have derived from a functional FTH gene other than that on chromosome 11. This raises the possibility that the other gene on 6p may be functional and may be the site of the mutation in hemochromatosis. Zappone et al. (1991) found no major deletions or alterations in the region of 6p containing these 2 ferritin H genes in patients with hemochromatosis. They also described a polymorphism in one of the genes that they had previously shown to be a processed pseudogene. This polymorphism did not correlate with the presence of HFE, however. The PIC value of the polymorphism was calculated as 0.49. Using a somatic cell hybrid regional mapping panel for the short arm of chromosome 6, as well as linkage analysis in hemochromatosis families and a population study of hemochromatosis patients and normal individuals, Summers et al. (1991) concluded t hat the FTH pseudogene sequence on chromosome 6, described by Dugast et al. (1990), maps to 6p, centromeric to the glyoxalase (138750) locus and distant from the hemochromatosis locus. Thus, it is not a candidate gene for hemochromatosis.

In patients with hemochromatosis, Lord et al. (1990) used pulsed field gel electrophoresis in conjunction with probes that map within, or in the vicinity of, the HLA class I region to search for structural lesions that might help them spot the HFE locus. However, in members of 3 pedigrees affected with hemochromatosis, and in 6 unrelated patients with the disorder, no disease-specific differences were detected using 4 restriction enzymes and 8 probes. From these data, they concluded that the lesion responsible for hereditary hemochromatosis lies beyond the resolution of this technique and does not involve large structural deletions or extensive rearrangements.

Jazwinska et al. (1993) found a maximum lod score of 9.90 at theta = 0.0 for HLA-A and 8.26 at theta = 0.0 for a microsatellite marker at D6S105. No recombination was observed with either marker. Other markers were separated from the HFE locus by recombination, thereby defining the centromeric and telomeric limits for the HFE gene as HLA-B and D6S109, respectively. A multipoint map indicated that HFE is located in a region less than 1 cM proximal to HLA-A and less than 1 cM telomeric of HLA-A.

Boretto et al. (1992) reported linkage studies with restriction polymorphisms which were consistent with location of the HFE gene either less than 100 kb centromeric to the HLA-A locus or on its telomeric side.

El Kahloun et al. (1993) used a yeast artificial chromosome with a 320-kb insert of genomic DNA which included the HLA-A gene to screen a cDNA library of human duodenal mucosa. They isolated 7 cDNA clones that corresponded to 7 new non-class I structural genes. Since these genes were located within a region that may contain the gene that is mutant in hemochromatosis, they referred to the genes as HCG (hemochromatosis candidate gene) I-VII. They concluded that HCG I, III, V, and VI are probably single copy genes situated 180, 155, 140, and 230 kb, respectively, centromeric to HLA-A. There were several copies of the other 3 genes. Each of the genes was associated with a CpG/HTF island. Goei et al. (1994) likewise isolated novel gene fragments from a cDNA library derived from human duodenum, using cDNA hybridization selection with a 320-kb YAC containing the HLA-A gene. Also in search of the HFE gene, Yaouang et al. (1994) identified a zone of linkage disequilibrium which suggested that the HFE gene may reside within a 400-kb expanse of DNA between the locus they referred to as i97 and HLA-F (143110).

In a single family with hemochromatosis, Calandro et al. (1995) identified 2 recombinant individuals and confirmed recombination by analysis of 16 polymorphic markers located near HLA-A and D6S105. One of the recombinants provided evidence for the location of the hemochromatosis gene telomeric to the 5-prime end of the HLA-F locus. The HLA-F locus was placed approximately 0.027 cM distal to HLA-A, which in turn was 0.01 cM distal of HLA-B. Raha-Chowdhury et al. (1996) showed that a highly polymorphic polypurine tract in the 5-prime untranslated region of HLA-F is as strongly associated with hemochromatosis as HLA-A3 or D6S105-8. The observed frequency of heterozygosity at the HLA-F polymorphism was 95% and the locus was found to be informative in pedigrees that are not informative at HLA-A and D6S105.

Totaro et al. (1996) generated a detailed 1.2-Mb physical and transcription map of 6p spanning the HLA class I region from HLA-E to approximately 500 kb telomeric of HLA-F. The localization of known genes was refined, and a new gene from the RNA helicase superfamily was identified. Overall, 14 transcription units in addition to the HLA genes were detected and integrated into the map. Thirteen cDNA fragments showed no similarity with known sequences, and could be candidates for the disease hemochromatosis. In a study of hereditary hemochromatosis (HH) patients, Feder et al. (1996) used linkage disequilibrium and full haplotype analysis to identify a 250-kb region more than 3 Mb telomeric of the major histocompatibility complex on chromosome 6 that is identical by descent in 85% of patient chromosomes. Within this region, they identified a gene related to the MHC class I family (termed HLA-H by them) containing 2 missense alterations. One mutation (235200.0001) was predicted to inactivate this class of proteins and was homozygous in 83% of 178 patients. The role of this gene in hemochromatosis was supported by the frequency and nature of the major mutation and by prior studies implicating MHC class I-like proteins in iron metabolism.

Lonjou et al. (1998) contrasted linkage and allelic association. Recombination acts on the genetic map, not on the physical map. On the other hand, the physical map is usually more accurate. Choice of the genetic or physical map for positional cloning by allelic association depends on the goodness of fit of data to each map under an established model. Huntington disease illustrates the usual case in which the greater reliability of physical data outweighs recombinational heterogeneity. Hemochromatosis represents an exceptional case in which unrecognized recombinational heterogeneity retarded positional cloning for a decade. In this case recombinational heterogeneity was demonstrated by the fact that the ratio of physical to genetic distance was 0.97 distally and 6.14 proximally. The power of allelic association was limited by scarcity of markers until microsatellites were introduced and subsequently by failure to recognize that 1 cM corresponds to several Mb in the region telomeric to HLA-A. Finally HFE was shown to lie more distally than earlier researchers had assumed, but the preferred marker D6S105 was still nearly 2 Mb from HFE. Allowance for nonuniform recombination would have saved a decade of fruitless search near HLA-A, 4.6 Mb from HFE.

Heterogeneity

Edwards et al. (1981) suggested that 2 families reported by Wands et al. (1976) and Rowe et al. (1977) may have a rare distinct form of hemochromatosis. In these families neither serum ferritin concentration nor transferrin saturation was a reliable indicator of hepatic siderosis and fibrosis. Hepatic fibrosis was observed in some individuals with a very modest increase in hepatic iron and in a few individuals with normal hepatic iron content. The disorder appeared to be transmitted as an autosomal dominant. Since no HLA data were reported in these families, it is impossible to say whether the mutation is on chromosome 6 or not.

In Australia, Jazwinska et al. (1996) found that all patients of northern European origin with hemochromatosis were homozygous for the cys282-to-tyr mutation (C282Y; 235200.0001). The frequency was greater than 90% in Brittany (Jouanolle et al., 1996). However, in Italy, Carella et al. (1997) performed mutation analysis on the HLA-H gene in patients from families with the 6p-linked disease but without the C282Y mutation and failed to find nucleotide abnormalities in coding sequences and intron/exon boundaries that could account for the disorder. The negative findings of RNA-SSCP were supported by the absence of mutations in the HLA-H gene by direct sequencing. Major deletions or rearrangements of the gene were excluded by Southern blotting. Thus, Carella et al. (1997) concluded that hemochromatosis in Italy appears to be more heterogeneous than reported in northern Europe. Abnormalities in unexplored portions of introns, RNA untranslated regions, or regulatory elements are possibilities. Also possible is the existence of another tightly linked locus in which a mutation can cause the same phenotype.

Juvenile Hemochromatosis

Juvenile hemochromatosis (602390) has clinical features similar to those of HFE but its clinical course is more severe, is characterized by an earlier onset of cardiac symptoms and endocrine dysfunctions, and does not show a male preponderance.

Escobar et al. (1987) established the diagnosis of hemochromatosis in a 7-year-old boy and his 29-month-old brother. These were said to be the youngest children with primary hemochromatosis reported to that time. They were members of a family in which 3 generations had affected individuals. Data from the literature on values of serum iron, serum ferritin, transferrin saturation, and hepatic iron were reviewed.

Kaikov et al. (1992) described hemochromatosis in asymptomatic sibs in whom the diagnosis was made after an unexpected finding of elevated serum iron concentrations. The sibs were 7, 6 and 4 years of age. Elevated red cell mean corpuscular volume (MCV) was elevated in all 3, at 90 to 92 fL. In their review of the literature, they found 16 cases of symptomatic homozygous children of ages ranging from 4 to 19 years at the time of diagnosis. They suggested that normalization of the MCV may be an additional indirect index of adequate phlebotomy.

Camaschella et al. (1997) described 7 Italian patients from 5 unrelated families with features typical of juvenile hemochromatosis. In 4 of 5 families, the parents were consanguineous. Analysis of HFE gene mutations in all cases and nucleotide sequencing of the HFE gene in 1 case excluded this gene as responsible for juvenile hemochromatosis. Furthermore, segregation analysis of 6p markers closely associated with HFE in families with consanguineous parents clearly showed that JH is unlinked to 6p and thus genetically distinct from HFE. See hemochromatosis type 2 (HFE2; 602390). The cases of Escobar et al. (1987) and Kaikov et al. (1992) may have been instances of classic HFE of unusually early detection. The work of Camaschella et al. (1997) establishes the existence of a separate entity, hemochromatosis type 2.

Biochemical Features

Sargent et al. (1979) suggested that chromium may be an essential trace mineral in man as it is in the rat, being required for normal function of insulin. Chromium is transported bound to transferrin and competes with iron for that binding. Sargent et al. (1979) showed that less chromium is retained in patients with hemochromatosis than in normals and suggested that the diabetes of hemochromatosis may be due, at least in part, to chromium deficiency.

Barton et al. (1994) demonstrated that hemochromatosis homozygotes and, to a lesser extent, heterozygotes, both male and female, have increased levels of blood lead. The mean blood lead of subjects with transfusion-induced iron overload did not differ significantly from that of normal controls. The findings in homozygotes could not be related to age, presence or absence of iron loading, or the extent of therapeutic phlebotomy. Increased absorption of iron and cobalt had previously been documented in homozygotes; the new findings were interpreted as indicating increased absorption of lead as well. The absorptive pathway for the 2 metals may be the same. The findings suggested that patients with hemochromatosis, like children with iron deficiency, are more susceptible to lead poisoning.

Dadone et al. (1982) found saturation of transferrin above 62% to be the best simply measured indicator of genotype: homozygosity was accurately predicted in 92% of cases. The logarithmic transform of serum ferritin concentration was only 71% accurate. The frequency of the hemochromatosis gene was estimated at 0.069 +/- 0.020, corresponding to a heterozygote frequency of 0.13 and a homozygote frequency of 0.005.

Borecki et al. (1989) performed a segregation analysis on 147 pedigrees from Brittany, France, with hereditary hemochromatosis as indexed by the measurement of latent capacity of transferrin. No evidence for heterozygous expression was observed, either in the biochemical domain of latent capacity of transferrin, or in increased liability to overt disease. The analysis allowed clear resolution of the recessive single gene inheritance pattern in these families. Borecki et al. (1990) concluded that the hemochromatosis gene is completely recessive with respect to both clinical manifestations and serum iron abnormalities, with significant differences in expression by sex. Clinical manifestations were present in all male homozygotes, suggesting that the recessive hemochromatosis genotype is fully penetrant at all ages in males. This was not the case for younger females, however.

Molecular Genetics

Hardy et al. (1990) sought evidence for major rearrangements or deletions at the HLA class I region and at 3 loci directly involved in iron metabolism (H- and L-apoferritin and the transferrin receptor). They studied 5 probands with neonatal hemochromatosis and 14 first-degree relatives in a total of 6 kindreds. They also sought out HLA associations by collating the results of HLA serotyping in these 19 persons and in 17 members of 7 additional kindreds in which neonatal hemochromatosis had occurred, including 5 probands and 12 first-degree relatives. They found no evidence for major rearrangements or deletions in the genes studied and found no evidence for linkage of neonatal hemochromatosis to HLA serotypes. They concluded that while hereditary hemochromatosis and neonatal hemochromatosis are similar in their patterns of iron loading, they are not genetically related.

As a first step in understanding the function of the HLA-H gene product, Parkkila et al. (1997) generated an antibody to a C-terminal peptide and used it for immunolocalization of the HLA-H protein in the gastrointestinal tract of Finnish and American subjects presumed not to have hereditary hemochromatosis. Although staining for the HLA-H protein was seen in some epithelial cells in every segment of the alimentary canal, its cellular and subcellular expression in the small intestine was distinct from that in other segments. In contrast to the stomach and colon, where staining is polarized and restricted to the basal lateral surfaces, and in contrast to the epithelial cells of the esophagus and submucosal leukocytes, which showed nonpolarized staining around the entire plasma membrane, the staining in the small intestine was mainly intracellular and perinuclear, limited to cells in deep crypts. Parkkila et al. (1997) concluded that the unique subcellular localization in the crypts of the small intestine in proximity to the presumed sites of iron absorption supported the implication of this protein in the molecular basis of hemochromatosis. Studies in cultured cells (Feder et al., 1997) showed that the C282Y mutation abrogates the binding of recombinant HFE protein to B2M and disrupts its transport to the cell surface. The HFE protein was shown by immunohistochemistry to be expressed in certain epithelial cells throughout the human alimentary tract and to have a unique localization in the cryptal cells of small intestine, where signals to regulate iron absorption are received from the body (Parkkila et al., 1997). Parkkila et al. (1997) demonstrated by immunohistochemistry that the HFE protein is expressed in human placenta in the apical plasma membrane of the syncytiotrophoblasts, where the transferrin-bound iron is normally transported to the fetus via receptor-mediated endocytosis. Western blot analyses showed that the HFE protein is associated with B2M in placent al membranes. Unexpectedly, the transferrin receptor (TFR; 190010) was also found to be associated with the HFE protein/B2M complex. These studies placed the normal HFE protein at the site of contact with the maternal circulation where its association with transferrin receptor raises the possibility that the HFE protein plays some role in determining maternal/fetal iron homeostasis. These findings also raised the possibility that mutations in the HFE gene can disrupt this association and contribute to some forms of neonatal iron overload; see idiopathic neonatal hemochromatosis (231100).

The C282Y mutation is found in 85% of the chromosomes 6 from patients with hemochromatosis (Feder et al., 1998). The H63D mutation, which has no effect on B2M association, was found in 8 of 9 patients heterozygous for the C282Y mutant. In cultured cells overexpressing wildtype or mutant HFE proteins, Feder et al. (1998) demonstrated that both the wildtype and H63D HFE proteins form stable complexes with the transferrin receptor. The C282Y mutation nearly completely prevented the association of the mutant HFE protein with TFR. Studies on cell-associated transferrin at 37 degrees C suggested that the overexpressed wildtype HFE protein decreases the affinity of TFR for transferrin. The overexpressed H63D protein did not have this effect, providing the first direct evidence for a functional consequence of the H63D mutation. These results established a molecular link between HFE and a key protein involved in iron transport and raised the possibility that alterations in this regulatory mechanism may play a role in the pathogenesis of hereditary hemochromatosis.

Lebron et al. (1998) stated that HFE binds to transferrin receptor and reduces its affinity for iron-loaded transferrin, implicating HFE in iron metabolism. The 2.6-angstrom crystal structure of HFE revealed the locations of hemochromatosis mutations and a patch of histidines that could be involved in pH-dependent interactions. Lebron et al. (1998) demonstrated that soluble TFR and HFE bind tightly at the basic pH of the cell surface, but not at the acidic pH of intracellular vesicles. TFR:HFE stoichiometry (2:1) differs from TFR:transferrin stoichiometry (2:2), implying a different mode of binding for HFE and transferrin to TFR, consistent with the demonstration that HFE, transferrin, and TFR form a ternary complex.

Robson et al. (1997) reviewed the identification of the probable gene mutant in hemochromatosis. They pointed out that Simon et al. (1976) first reported the association between specific HLA antigens and hemochromatosis and that it took 20 years to identify the strongest candidate gene to that time. They emphasized that formal proof from functional studies was still required to prove that mutations in this gene cause hemochromatosis. They also discussed why the gene has proved so elusive.

Beutler et al. (1997) pointed out that calreticulin (CALR; 109091), like beta-2-microglobulin (B2M; 109700), associates with class I HLA proteins and appears to be identical to mobilferrin, a putative iron transport protein. Thus these 2 proteins were considered candidates for mutations in patients with hemochromatosis. The investigators sequenced the coding region and parts of introns of the HFE gene (called by them HLA-H), the B2M gene, and the CALR gene in 10, 7, and 5 hemochromatosis patients, respectively, selecting those who were not homozygous for the common C282Y mutation. No additional mutations were found in the HLA-H gene and no disease related mutations in the other 2 genes. The basis for hemochromatosis in more than 10% of European patients and in most Asian patients awaits explanation. Beutler et al. (1997) speculated that the finding of some effects in heterozygotes (Bulaj et al., 1996) and the rarity of mutations other than C282Y and his63 to asp (H63D; 235200.0002) may point to a gain-of-function consequence of these mutations, similar, they suggested, to sickle cell anemia, which is caused by only 1 type of mutation (see 141900.0038) and represents in effect a gain-of-function mutation. The unique mutation causing achondroplasia, gly380 to arg (G380R; 134934.0001), might also be cited.

Population Genetics

The frequency of the hemochromatosis gene in Utah was placed at 5.6% (Cartwright et al., 1979). Homozygotes had a frequency of 0.3% and heterozygotes a frequency of 10.6% . A similar gene frequency was estimated for Brittany (Beaumont et al., 1979). Krikker (1982) described the newly established Hemochromatosis Research Foundation, Inc. As justification for its existence, Krikker wrote as follows: 'The incidence of heterozygosity for the hemochromatosis allele in the white population is approximately 10%. The expected incidence of homozygosity is about 2 to 3 per 1000, an estimate supported by the finding of homozygosity in 1 in 333 residents of Utah (Cartwright et al., 1979), 1 in 400 Bretons (Beaumont et al., 1979), and in an autopsy study 1 in 500 Scots (MacSween and Scott, 1973).' In an extensive study of hemochromatosis in Brittany, Lalouel et al. (1985) confirmed the Salt Lake City data (Cartwright et al., 1979; Kravitz et al., 1979).

In the county of Jamtland in central Sweden, an area known in the past for a high prevalence of iron deficiency, 106,105:Olsson et al. (1983, 1984) screened for iron overload by a laboratory routine that automatically included determination of serum iron and transferrin saturation. They found a prevalence of 0.5% for genetic iron overload which means that 12.8% of the population are gene carriers.

Meyer et al. (1987) used serum ferritin concentration as a screening test for iron overload in 599 Afrikaners living in the South Western Cape, South Africa. Sixteen subjects, all males from different families, had concentrations greater than 400 micrograms/liter. Reevaluations 3 and 5 years later included remeasurement of serum ferritin, assessment of alcohol intake, measurements of serum gamma-glutamyltransferase, percentage saturation of transferrin, and HLA typing. The serum ferritin concentration is significantly raised after excessive alcohol consumption; however, the measurement of serum gamma-glutamyltransferase helps resolve the confusion because a serum ferritin concentration above 300 micrograms/liter is very unlikely to be the result of alcohol-induced hepatic damage if the gamma-glutamyltransferase is less than 50 units per liter. Of the 16 index persons, 4 were diagnosed as homozygous for the HLA-linked iron-loading gene. Six appeared to be heterozygotes, 3 were heterozygotes who were also abusing alcohol, and 2 did not fit into any of the diagnostic groups. The calculated gene frequency was 0.082, with an expectant heterozygote frequency of 0.148. The fact that no females were identified in the study suggested to the authors that their criteria for homozygosity were set too high. When the data were recalculated for the 300 males, the gene frequency became 0.115 and the heterozygote frequency became 0.204. Simon et al. (1987) presented findings they interpreted as fitting well with the hypothesis that 'the hemochromatosis mutation was a rare if not unique event that produced an ancestral HLA marking that was subsequently modified by recombinations and geographical scattering due to migrations.'

Among 11,065 presumably healthy blood donors (5,840 men and 5,225 women), Edwards et al. (1988) found that transferrin saturation of 62% or more after an overnight fast had a frequency of 0.008 in men and 0.003 in women. Detailed studies were performed in 38 persons with values higher than 62%; 35 underwent liver biopsy. Liver iron stores ranged from normal to markedly increased. Twelve sibs with an identical HLA match to a proband underwent liver biopsy, and 11 had increased liver iron stores. Analysis of pedigrees led to the conclusion that 26 of the 38 probands were homozygotes and 12 were heterozygotes. Basing the estimate of the frequency of homozygosity on the data in men, Edwards et al. (1988) arrived at an estimate of 0.0045, corresponding to a gene frequency of 0.067. By means of a screening using transferrin saturation followed by repeat transferrin saturation and serum ferritin, clinical examination, and laparoscopy, Karlsson et al. (1988) concluded that the prevalence of hemochromatosis in Finland is about 5 per 10,000.

Jazwinska et al. (1995) found that hemochromatosis shows a very strong founder effect in Australia, with the majority of patients being of Celtic (Scottish/Irish) origin. By analyzing chromosomes from 26 multiply affected hemochromatosis pedigrees for linkage disequilibrium and genetic heterogeneity, they were able to assign hemochromatosis status unambiguously to 107 chromosomes—64 as affected and 43 as unaffected. With the serologic marker HLA-A and 4 microsatellite markers, highly significant allelic association with hemochromatosis was found. One predominant ancestral haplotype was present in 33% of 64 affected chromosomes and was associated exclusively with hemochromatosis (haplotype relative risk 903). No other common haplotype was significantly associated with hemochromatosis. Thus, the common mutation probably underlies hemochromatosis in Australian patients, having been introduced into this population on an ancestral haplotype. Furthermore, the candidate HFE region extends between and includes D6S248 and D6S105.

Milman et al. (1990) studied 1 Faroese and 4 Danish kindreds with hemochromatosis. Milman (1991) analyzed 179 patients ascertained in Denmark between 1950 and 1985, as well as 13 preclinical subjects ascertained through family studies or high serum transferrin saturation values. The high frequency of the HFE gene may account, through the mechanism of pseudodominance, for the simulation of dominant inheritance and the consequent debates in the past as to the mode of inheritance of hemochromatosis. Dokal et al. (1991) reported on a family with affected members in 2 generations in a pseudodominant pedigree pattern. The affected father was deceased. The heterozygous mother and all 6 children (3 homozygotes, 3 heterozygotes) were HLA identical (A1B8/A3B14). Affected sibs were recognized in the precirrhotic stage of hemochromatosis by analysis of serum parameters of iron status in combination with magnetic resonance imaging. In the Saguenay-Lac St-Jean region of northeastern Quebec, De Braekeleer (1993) estimated the prevalence of hereditary hemochromatosis to be 0.014, giving a heterozygote frequency of 0.21. These were among the highest frequencies found to that date in white populations. Fertility studies showed that carriers of the gene tended to have more children than noncarriers. However, since the differences were not statistically significant, genetic drift could not be excluded.

In an analysis of 82 unrelated HFE patients and 82 unrelated healthy controls, Jazwinska et al. (1993) found that allele 8 at the D6S105 locus was present in 93% of patients and only 21% of controls, giving an approximate relative risk for this allele of 48.4. HLA-A3 was present in 62% of patients and 26% of controls, giving an approximate relative risk for A3 of 4.8. They concluded that the microsatellite marker D6105 was the closest marker to HFE reported to that date.

Using a relative risk of 1.0 for the C282Y homozygote, Risch (1997) calculated the risk of the C282Y/H63D compound heterozygote to be 0.00525 and the relative risk of other genotypes to be 0.00015. There appeared to be a modestly increased risk (about 4-fold) associated with homozygosity for H63D. Great haplotype diversity on non-C282Y chromosomes has been observed in patients. This was not surprising, as the disequilibrium on H63D chromosomes spans a much shorter distance (700 kb) than on C828Y chromosomes (more than 7 Mb), consistent with the higher frequency and likely older origin for H63D. Indeed, the disequilibrium analysis of H63D chromosomes provided compelling evidence both for the implication of H63D in hemochromatosis and that HFE is the hemochromatosis gene. Beckman et al. (1997) found that the C282Y mutation is rare or absent in Asiatic (Indian, Chinese) populations. The highest allele frequency they found was in Swedes (7.5%).

Parkkila et al. (1997) suggested a selective advantage of the C282Y mutation on the basis of improved survival during infancy, childhood, and pregnancy in times past, by leading to increased iron absorption and accumulation of larger body iron stores. Although this selection could operate at the level of increased dietary iron absorption, such mutations might also lead to enhanced maternal/fetal iron transport. Such an effect might confer a selective advantage on the fetus under conditions of maternal iron deprivation.

HLA Association

In 50 unselected and unrelated patients with hemochromatosis, Ritter et al. (1984) found a high association with the HLA haplotype A3B14 (relative risk 23.4). For one large group of homozygotes for this haplotype they identified a common ancestor living in the early 1700s. They suggested that the high frequency of the hemochromatosis gene might be the result of a selective advantage under conditions of iron deficiency: homozygotes do not lose reproductive capacity from testicular effects, and females, homozygous and perhaps heterozygous as well, are better prepared to meet the increased iron demands of pregnancy. Simon et al. (1988) suggested that a single, ancient mutation of a gene involved in iron homeostasis resulted in the present-day hemochromatosis allele. This mutation was thought to have occurred on a chromosome 6 carrying HLA-A3 and HLA-B7. Over the years recombination events between the HLA-A and HLA-B loci presumably led to the observed association with other HLA-B alleles on haplotypes carrying HLA-A3, and recombinations between HLA-A and the hemochromatosis locus produced associations with other HLA-A alleles and haplotypes. The original mutation should be progressing toward equilibrium with the HLA alleles, with the residual association resulting either because there has been insufficient time to reach equilibrium or because the association confers a selective advantage (Kushner et al., 1988). A recent recombination event or perhaps a new mutation has placed a hemochromatosis allele on an HLA-A2,B12 chromosome in a population that made a major contribution to the present-day Australian gene pool. Because of the predominant origin of the present-day Australian population, Summers et al. (1989) suggested that this chromosome originated in England or perhaps, in view of the family names of many of the patients, Ireland.

Jouanolle et al. (1990) studied RFLPs from the HLA-A region and identified a significantly high frequency of a particular EcoRI fragment among the hemochromatosis patients who were HLA-A3 in tissue type.

In Denmark, Milman et al. (1988) found the pattern of HLA-antigens associated with hemochromatosis to be similar to those reported both in Germany, where HLA-A3,B7 dominated, and in Brittany, Great Britain, and central Sweden, where HLA-A3,B14 dominated. In South Wales, Cragg et al. (1988) found that 80% of 15 unrelated patients had HLA-A3 compared with 24% of 600 unrelated and unaffected persons. The most common haplotype was HLA-A3,B7. They could find no evidence in support of the possibility that either the ferritin heavy chain gene (134770) or HLA class I genes are candidates for the gene mutant in hemochromatosis. In studies of 24 Australian families, Summers et al. (1989) found linkage to HLA in at least 23. The evidence was interpreted as indicating the involvement of a single genetic locus in most (probably all) cases of familial hemochromatosis in Australia. As in all other populations reported, an association of HLA-A3 and HLA-B7 with the disease was found in the Australian cases. In addition, HLA-A2 and HLA-B12 were in significant linkage disequilibrium in patients but not in controls, which might indicate a new mutation or recent recombination between HLA-A and hemochromatosis either in the Australian patient population or in the founding population. In 74 Danish patients with hemochromatosis and 21 homozygous relatives, Milman et al. (1992) found atypical frequencies of HLA type: A3 was present in 53.6% as compared to 15.1% in the general population. B7 was present in 33.1% as compared with 15.6% in the general population. The 2 most frequent haplotypes were A3,B5 (10.3% vs 0.3%) and A3,B7 (25.6% vs 6.6%). In a review of 57 families with hereditary hemochromatosis, Adams (1992) found 3 pairs of HLA-identical, sex-matched sibs in which the younger sib demonstrated considerably more iron loading than the older sib. In 19 pairs of HLA-identical, sex-matched sibs homozygous for hemochromatosis, the iron loading was more marked in the older sib. There was no evidence of blood loss, difference in alcohol consumption, or dietary iron loading to explain the increased iron loading in the younger sibs. Yaouanq et al. (1992) used 5 biallelic polymorphisms located in the HLA class I region to test 198 HLA-typed subjects from the families of 22 hemochromatosis patients. The 5 polymorphisms provided sufficient information to identify unequivocally extended restriction haplotypes in all families. The restriction haplotypes cosegregated with the HFE allele and enabled identification of genotypically identical sibs in all families studied. The method avoids the disadvantages of HLA serologic typing and should be useful for genetic counseling in HFE families.

Diagnosis

Early diagnosis by clinical features is difficult. Unexplained elevation of transferrin saturation should prompt study for hemochromatosis. Diagnosis in asymptomatic stages is important because organ damage can be prevented by early therapy. Hepatic iron is the most sensitive index of preclinical disease; of noninvasive tests, serum ferritin is unreliable, whereas transferrin saturation correlates with hepatic iron content (Rowe et al., 1977; Edwards et al., 1977). Elevated serum iron is a diagnostically valuable finding which can be sought in relatives of full-blown cases.

On the basis of data generated by an ongoing study of hemochromatosis in Brittany, France, Borecki et al. (1990) concluded that percent transferrin saturation is a reliable indicator of the homozygous state but that, contrary to previous studies, there is no evidence for partial expression of this value in heterozygotes.

Kaikov et al. (1992) described hemochromatosis in asymptomatic sibs in whom the diagnosis was made after an unexpected finding of elevated serum iron concentrations. The sibs were 7, 6 and 4 years of age. Elevated red cell mean corpuscular volume (MCV) was elevated in all 3, at 90 to 92 fL. In their review of the literature, they found 16 cases of symptomatic homozygous children of ages ranging from 4 to 19 years at the time of diagnosis. They suggested that normalization of the MCV may be an additional indirect index of adequate phlebotomy.

Clinical Management

Prophylactic venesection is indicated. Niederau et al. (1985) concluded that patients diagnosed in the precirrhotic stage and treated with venesection have a normal life expectancy, whereas cirrhotic patients have a shortened life expectancy and a high risk of liver cancer even when complete iron depletion has been achieved. Siemons and Mahler (1987) found that phlebotomy conducted over a 16-month period restored fertility and normal endocrinologic findings in a 37-year-old man with severe hypogonadotropic hypogonadism due to hemochromatosis. From a screening of 1,968 employees of 2 large corporations, Leggett et al. (1990) concluded that the prevalence of significant iron overload due to homozygous hemochromatosis warranting treatment is approximately 1 in 300 among Australians (predominantly Caucasians). They suggested that transferrin saturation should be included in adult health screening programs. Worwood et al. (1991) urged that a regular program be instituted for identifying homozygotes for hemochromatosis on the basis of ferritin concentrations and inviting these individuals to donate frequently to keep the ferritin concentration toward the lower end of the normal range. Such a program would be beneficial both to persons with this common disease and to the blood supply.

Feder et al. (1996) viewed hemochromatosis as a model disorder for genetic testing since it is a frequent disorder and effective intervention, namely therapeutic phlebotomy, is available for the prevention of the disorder. The findings of Feder et al. (1996) were achieved by scientists at a biotechnology company. Cox (1996) discussed the importance of their simple PCR-based test to detect homozygosity for the mutant hemochromatosis gene.

Animal Model

De Sousa et al. (1994) reported a comparative histologic and quantitative analysis of iron distribution in the tissues of mice homozygous and heterozygous for knockout of the beta-2-microglobulin gene. Progressive hepatic iron overload, indistinguishable from that observed in human hemochromatosis, was found only in mice homozygous for the mutated B2M gene. (B2M is complexed with HLA class I molecules and is essential to their function.)

Rothenberg and Voland (1996) identified a multigene system in the murine major histocompatibility complex that contains excellent candidates for the murine equivalent of the human HFE locus and implicates nonclassic class I genes in the control of iron absorption. This gene system is characterized by multiple copies of 2 head-to-head genes encoded on opposite strands and driven by a common regulatory motif. This regulatory motif has striking homology to the promoter region of the beta-globin gene (141900), a gene obviously involved in iron metabolism, and hence termed beta-globin analogous promoter, beta-GAP or BGAP. Upstream of the BGAP sequence are nonclassic class I genes. At least 1 of these nonclassic class I genes, Q2, is expressed in the gastrointestinal tract, the primary site of iron absorption. Also expressed in the gastrointestinal tract and downstream of the BGAP motif is a second set of putative genes, termed Hephaestus (HEPH). Based on these observations, Rothenberg and Voland (1996) hypothesized that the genes that seemed to be controlled by BGAP regulatory motifs would be responsible for the control of iron absorption. As a test of this hypothesis, they predicted that mice with altered expression of class I gene products, the beta-2-microglobulin knockout mice, would develop iron overload. This prediction was confirmed, and these results indicated to the authors that B2M-associated proteins are involved in the control of intestinal iron absorption and are strong candidates for the site of the mutation in hemochromatosis. The most frequent mutation in the HFE gene responsible for hemochromatosis, cys282 to tyr (235200.0001), interferes with the binding of beta-2-microglobulin to the HFE gene product.

Searching for new human MHC class I related genes, Hashimoto et al. (1995) found MHC-related protein-1 (MR1; 600764) and a second MHC-related protein they designated MR2. MR2 was located at 6p22 by fluorescence in situ hybridization analysis. The HLA-H gene reported by Feder et al. (1996) as a candidate gene for hereditary hemochromatosis turned out to be identical to the MR2 gene of Hashimoto et al. (1995). Hashimoto et al. (1997) isolated the murine homolog of this gene. It was found to be similar to its human counterpart with an overall predicted amino acid sequence similarity of approximately 66% and expression in various tissues as in human. An extra 8 amino acid residues between the alpha-1 and the alpha-2 domains in the mouse molecule compared to the human counterpart could be explained by the creation of the additional coding sequence from the intron. They showed that while the human gene is located telomeric to the MHC region on human 6p, the mouse homolog was translocated from the site telomeric to MHC on mouse chromosome 17 to chromosome 13 along with other genes. Hashimoto et al. (1997) commented that this mouse gene should be important in verifying a possible role of MR2 in an iron metabolism in that experimental species.

To test the hypothesis that the HFE gene is involved in regulation of iron homeostasis, Zhou et al. (1998) studied the effects of a targeted disruption of the murine homolog of the HFE gene. The HFE-deficient mice showed profound differences in parameters of iron homeostasis. Even on a standard diet, by 10 weeks of age, fasting transferrin saturation was significantly elevated compared with normal littermates, and hepatic iron concentration was 8-fold higher than that of wildtype littermates. Stainable hepatic iron in the HFE mutant mice was predominantly in hepatocytes in a periportal distribution. Iron concentrations in spleen, heart, and kidney were not significantly different from that in littermates. Erythroid parameters were normal, indicating that the anemia did not contribute to the increased iron storage. The study showed that HFE protein is involved in the regulation of iron homeostasis and that mutations in the gene are responsible for hereditary hemochromatosis. The knockout mouse model should be useful for determining the pathogenesis of increased iron accumulation and for evaluating therapeutic strategies for prevention or correction of iron overload (Beutler, 1998).

History

The first description of hemochromatosis is attributed to Trousseau (1865). His first patient was a 28-year-old man with severe diabetes. Trousseau wrote: 'From the time this man came into the hospital, I was struck by the almost bronzed appearance of his countenance, and the blackish color of his penis.' At autopsy the liver was found to be very large. 'The entire surface of the organ was granular; it was of a uniform grayish-yellow color; it was very dense, resisting pressure so much as to prevent penetration by the finger. It creaked under the scalpel, and the surface of the cut was granular in place of being smooth.'

The hereditary nature of hemochromatosis was emphasized particularly by Sheldon (1935). In his classic monograph entitled 'Haemochromatosis,' Sheldon (1935) reviewed references to a familial or hereditary basis of the disease made by 14 authors and stated: 'Further evidence is greatly desirable on this aspect of the disease, since the fact of an occasional familial incidence must obviously be taken into account in any theory regarding the origin of the disease.'

The pedigree of Nussbaumer et al. (1952) was reproduced by Sorsby (1953).

See also Balcerzak et al. (1966); Beaumont et al. (1980); Borecki et al. (1990); Cazzola et al. (1983); Charlton et al. (1967); Cox and Peters (1978); Cutler et al. (1980); Dadone et al. (1982); Feller et al. (1977); Felts et al. (1967); Gordeuk et al. (1992); Halliday et al. (1977); Johnson and Frey (1962); Kidd (1979); Kuhnl et al. (1978); Lamon et al. (1979); Le Mignon et al. (1983); Lipinski et al. (1978); Maddrey et al. (1979); Milman et al. (1992); Muir et al. (1978); Perkins et al. (1965); Pollycove (1972); Simon et al. (1977); Simon et al. (1977); Simon et al. (1980); Trousseau (1882); Valberg et al. (1980)

Adams, P. C.: Intrafamilial variation in hereditary hemochromatosis. Digest. Dis. Sci. 37: 361-363, 1992.

Anand, S.; Schade, R. R.; Bendetti, C.; Kelly, R.; Rabin, B. S.; Krause, J.; Starzl, T. E.; Iwatsuki, S. I.; Van Thiel, D. H.: Idiopathic alpha-hemochromotosis (sic) and 1-antitrypsin deficiency: coexistence in a family with progressive liver disease in the proband. Hepatology 3: 714-718, 1983.

Balcerzak, S. P.; Westerman, M. P.; Lee, R. E.; Doyle, A. P.: Idiopathic hemochromatosis: a study of three families. Am. J. Med. 40: 857-873, 1966.

Barton, J. C.; Patton, M. A.; Edwards, C. Q.; Griffen, L. M.; Kushner, J. P.; Meeks, R. G.; Leggett, R. W.: Blood lead concentrations in hereditary hemochromatosis. J. Lab. Clin. Med. 124: 193-198, 1994.

Bassett, M. L.; Doran, T. J.; Halliday, J. W.; Bashir, H. V.; Powell, L. W.: Idiopathic hemochromatosis: demonstration of homozygous-heterozygous mating by HLA typing of families. Hum. Genet. 60: 352-356, 1982.

Beaumont, C.; Simon, M.; Fauchet, R.; Hespel, J.-P.; Brissot, P.; Genetet, B.; Bourel, M.: Serum ferritin as a possible marker of the hemochromatosis allele. New Eng. J. Med. 301: 169-174, 1979.

Beaumont, C.; Simon, M.; Smith, P. M.; Worwood, M.: Hepatic and serum ferritin concentrations in patients with idiopathic hemochromatosis. Gastroenterology 79: 877-883, 1980.

Beckman, L. E.; Saha, N.; Spitsyn, V.; Van Landeghem, G.; Beckman, L.: Ethnic differences in the HFE codon 282 (Cys/Tyr) polymorphism. Hum. Hered. 47: 263-267, 1997.

Beutler, E.: Targeted disruption of the HFE gene. Proc. Nat. Acad. Sci. 95: 2033-2034, 1998.

Beutler, E.; Gelbart, T.; West, C.; Lee, P.; Adams, M.; Blackstone, R.; Pockros, P.; Kosty, M.; Venditti, C. P.; Phatak, P. D.; Seese, N. K.; Chorney, K. A.; Ten Elshof, A. E.; Gerhard, G. S.; Chorney, M.: Mutation analysis in hereditary hemochromatosis. Blood Cells Molec. Dis. 22: 187-194, 1996.

Beutler, E.; West, C.; Gelbart, T.: HLA-H and associated proteins in patients with hemochromatosis. Molec. Med. 3: 397-402, 1997.

Bodmer, J. G.; Parham, P.; Albert, E. D.; Marsh, S. G. E.: Putting a hold on 'HLA-H.' (Letter) Nature Genet. 15: 234-235, 1997.

Borecki, I. B.; Lathrop, G. M.; Bonney, G. E.; Yaouanq, J.; Rao, D. C.: Combined segregation and linkage analysis of genetic hemochromatosis using affection status, serum iron, and HLA. Am. J. Hum. Genet. 47: 542-550, 1990.

Borecki, I. B.; Rao, D. C.; Yaouanq, J.; Lalouel, J. M.: Percent transferrin saturation in segregating hemochromatosis. Am. J. Med. Genet. 36: 301-305, 1990.

Borecki, I. B.; Rao, D. C.; Yaouanq, J.; Lalouel, J. M.: Segregation of genetic hemochromatosis indexed by latent capacity of transferrin. Am. J. Hum. Genet. 45: 465-470, 1989.

Boretto, J.; Jouanolle, A.-M.; Yaouanq, J.; El Kahloun, A.; Mauvieux, V.; Blayau, M.; Perichon, M.; Le Treut, A.; Clayton, J.; Borot, N.; Le Gall, J.-Y.; Pontarotti, P.; David, V.: Anonymous markers located on chromosome 6 in the HLA-A class I region: allelic distribution in genetic haemochromatosis. Hum. Genet. 89: 33-36, 1992.

Bothwell, T. H.; Cohen, I.; Abrahams, O. L.; Perold, S. M.: A familial study in idiopathic hemochromatosis. Am. J. Med. 27: 730-738, 1959.

Bulaj, Z. J.; Griffen, L. M.; Jorde, L. B.; Edwards, C. Q.; Kushner, J. P.: Clinical and biochemical abnormalities in people heterozygous for hemochromatosis. New Eng. J. Med. 335: 1799-1805, 1996.

Calandro, L. M.; Baer, D. M.; Sensabaugh, G. F.: Characterization of a recombinant that locates the hereditary hemochromatosis gene telomeric to HLA-F. Hum. Genet. 96: 339-342, 1995.

Camaschella, C.; Roetto, A.; Cicilano, M.; Pasquero, P.; Bosio, S.; Gubetta, L.; Di Vito, F.; Girelli, D.; Totaro, A.; Carella, M.; Grifa, A.; Gasparini, P.: Juvenile and adult hemochromatosis are distinct genetic disorders. Europ. J. Hum. Genet. 5: 371-375, 1997.

Carella, M.; D'Ambrosio, L.; Totaro, A.; Grifa, A.; Valentino, M. A.; Piperno, A.; Girelli, D.; Roetto, A.; Franco, B.; Gasparini, P.; Camaschella, C.: Mutation analysis of the HLA-H gene in Italian hemochromatosis patients. Am. J. Hum. Genet. 60: 828-832, 1997.

Cartwright, G. E.; Edwards, C. Q.; Kravitz, K.; Skolnick, M.; Amos, D. B.; Johnson, A.; Bushjaer, L.: Hereditary hemochromatosis: phenotypic expression of the disease. New Eng. J. Med. 301: 175-179, 1979.

Cartwright, G. E.; Skolnick, M.; Amos, D. B.; Edwards, C. Q.; Kravitz, K.; Johnson, A.: Inheritance of hemochromatosis: linkage to HLA. Trans. Assoc. Am. Phys. 91: 273-281, 1978.

Cazzola, M.; Ascari, E.; Barosi, G.; Claudiani, G.; Dacco, M.; Kaltwasser, J. P.; Panaiotopoulos, N.; Schalk, K. P.; Werner, E. E.: Juvenile idiopathic haemochromatosis: a life-threatening disorder presenting as hypogonadotropic hypogonadism. Hum. Genet. 65: 149-154, 1983.

Charlton, R. W.; Abrahams, C.; Bothwell, T. H.: Idiopathic hemochromatosis in young subjects. Arch. Path. 83: 132-140, 1967.

Cox, T.: Haemochromatosis: strike while the iron is hot. Nature Genet. 13: 386-388, 1996.

Cox, T. M.; Peters, T. J.: Uptake of iron by duodenal biopsy specimens from patients with iron-deficiency anaemia and primary haemochromatosis. Lancet I: 123-124, 1978.

Cragg, S. J.; Darke, C.; Worwood, M.: HLA class I and H ferritin gene polymorphisms in normal subjects and patients with haemochromatosis. Hum. Genet. 80: 63-68, 1988.

Cragg, S. J.; Drysdale, J.; Worwood, M.: Genes for the 'H' subunit of human ferritin are present on a number of human chromosomes. Hum. Genet. 71: 108-112, 1985.

Cullen, L. M.; Gao, X.; Easteal, S.; Jazwinska, E. C.: The hemochromatosis 845 G-to-A and 187 C-to-G mutations: prevalence in non-Caucasian populations. Am. J. Hum. Genet. 62: 1403-1407, 1998.

Cutler, D. J.; Isner, J. M.; Bracey, A. W.; Hufnagel, C. A.; Conrad, P. W.; Roberts, W. C.; Kerwin, D. M.; Weintraub, A. M.: Hemochromatosis heart disease: an unemphasized cause of potentially reversible restrictive cardiomyopathy. Am. J. Med. 69: 923-928, 1980.

Dadone, M.; Skolnick, M.; Edwards, C.: Linkage between hereditary hemochromatosis and HLA. (Abstract) Cytogenet. Cell Genet. 32: 261-262, 1982.

Dadone, M. M.; Kushner, J. P.; Edwards, C. Q.; Bishop, D. T.; Skolnick, M. H.: Hereditary hemochromatosis: analysis of laboratory expression of the disease by genotype in 18 pedigrees. Am. J. Clin. Path. 78: 196-207, 1982.

David, V.; Papadopoulos, P.; Yaouanq, J.; Blayau, M.; Abel, L.; Zappone, E.; Perichon, M.; Drysdale, J.; Le Gall, J.-Y.; Simon, M.: Ferritin H gene polymorphism in idiopathic hemochromatosis. Hum. Genet. 81: 123-126, 1989.

David, V.; Paul, P.; Simon, M.; Le Gall, J.-Y.; Fauchet, R.; Gicquel, I.; Dugast, I.; Le Mignon, L.; Yaouanq, J.; Cohen, D.; Bourel, M.: DNA polymorphism related to the idiopathic hemochromatosis gene: evidence in a recombinant family. Hum. Genet. 74: 113-120, 1986.

David, V.; Paul, P.; Yaouanq, J.; Blayau, M.; Fauchet, R.; Cohen, D.; Le Gall, J. Y.; Simon, M.: Molecular genetic approach to the hemochromatosis gene. (Abstract) Cytogenet. Cell Genet. 46: 604, 1987.

De Braekeleer, M.: A prevalence and fertility study of haemochromatosis in Saguenay-Lac-Saint-Jean. Ann. Hum. Biol. 20: 501-505, 1993.

Debre, R.; Dreyfus, J.-C.; Frezal, J.; Labie, D.; Lamy, M.; Maroteaux, P.; Schapira, F.; Schapira, G.: Genetics of haemochromatosis. Ann. Hum. Genet. 23: 16-30, 1958.

de Sousa, M.; Reimao, R.; Lacerda, R.; Hugo, P.; Kaufmann, S. H. E.; Porto, G.: Iron overload in beta-2-microglobulin-deficient mice. Immun. Lett. 39: 105-111, 1994.

Deugnier, Y. M.; Guyader, D.; Crantock, L.; Lopez, J.-M.; Turlin, B.; Yaouanq, J.; Jouanolle, H.; Campion, J.-P.; Launois, B.; Halliday, J. W.; Powell, L. W.; Brissot, P.: Primary liver cancer in genetic hemochromatosis: a clinical, pathological, and pathogenetic study of 54 cases. Gastroenterology 104: 228-234, 1993.

Diamond, T.; Stiel, D.; Posen, S.: Osteoporosis in hemochromatosis: iron excess, gonadal deficiency, or other factors?. Ann. Intern. Med. 110: 430-436, 1989.

Dokal, I.; Lord, D.; Rhodes, D.; Bydder, G.; Cox, T.: Detection of hereditary haemochromatosis in an HLA-identical pedigree showing discordance between HLA class I genes and the disease locus. Hum. Genet. 88: 209-214, 1991.

Dugast, I. J.; Papadopoulos, P.; Zappone, E.; Jones, C.; Theriault, K.; Handelman, G. J.; Benarous, R.; Drysdale, J. W.: Identification of two human ferritin H genes on the short arm of chromosome 6. Genomics 6: 204-211, 1990.

Edwards, C. Q.; Carroll, M.; Bray, P. F.; Cartwright, G. E.: Hereditary hemochromatosis: diagnosis in siblings and children. New Eng. J. Med. 297: 7-13, 1977.

Edwards, C. Q.; Cartwright, G. E.; Skolnick, M. H.; Amos, D. B.: Homozygosity for hemochromatosis: clinical manifestations. Ann. Intern. Med. 93: 519-525, 1980.

Edwards, C. Q.; Griffen, L. M.; Dadone, M. M.; Skolnick, M. H.; Kushner, J. P.: The locus for hereditary hemochromatosis maps between HLA-A and HLA-B. (Abstract) Cytogenet. Cell Genet. 40: 620, 1985.

Edwards, C. Q.; Griffen, L. M.; Dadone, M. M.; Skolnick, M. H.; Kushner, J. P.: Mapping the locus for hereditary hemochromatosis: localization between HLA-B and HLA-A. Am. J. Hum. Genet. 38: 805-811, 1986.

Edwards, C. Q.; Griffen, L. M.; Goldgar, D.; Drummond, C.; Skolnick, M. H.; Kushner, J. P.: Prevalence of hemochromatosis among 11,065 presumably healthy blood donors. New Eng. J. Med. 318: 1355-1362, 1988.

Edwards, C. Q.; Skolnick, M. H.; Kushner, J. P.: Hereditary hemochromatosis: contributions of genetic analyses. Prog. Hemat. 12: 43-71, 1981.

El Kahloun, A.; Chauvel, B.; Mauvieux, V.; Dorval, I.; Jouanolle, A.-M.; Gicquel, I.; Le Gall, J.-Y.; David, V.: Localization of seven new genes around the HLA-A locus. Hum. Molec. Genet. 2: 55-60, 1993.

Elzouki, A.-N. Y.; Hultcrantz, R.; Stal, P.; Befrits, R.; Eriksson, S.: Increased PiZ gene frequency for alpha(1) antitrypsin in patients with genetic haemochromatosis. Gut 36: 922-926, 1995.

Eriksson, S.; Lindmark, B.; Hanik, L.: A Swedish family with alpha-1-antitrypsin deficiency, haemochromatosis, haemoglobinopathy D and early death in liver cirrhosis. J. Hepatol. 2: 65-72, 1986.

Escobar, G. J.; Heyman, M. B.; Smith, W. B.; Thaler, M. M.: Primary hemochromatosis in childhood. Pediatrics 80: 549-554, 1987.

Feder, J. N.; Gnirke, A.; Thomas, W.; Tsuchihashi, Z.; Ruddy, D. A.; Basava, A.; Dormishian, F.; Domingo, R., Jr.; Ellis, M. C.; Fullan, A.; Hinton, L. M.; Jones, N. L.; Kimmel, B. E.; Kronmal, G. S.; Lauer, P.; Lee, V. K.; Loeb, D. B.; Mapa, F. A.; McClelland, E.; Meyer, N. C.; Mintier, G. A.; Moeller, N.; Moore, T.; Morikang, E.; Prass, C. E.; Quintana, L.; Starnes, S. M.; Schatzman, R. C.; Brunke, K. J.; Drayna, D. T.; Risch, N. J.; Bacon, B. R.; Wolff, R.

K.: A novel MHC class I-like gene is mutated in patients with hereditary haemochromatosis. Nature Genet. 13: 399-408, 1996.
Feder, J. N.; Penny, D. M.; Irrinki, A.; Lee, V. K.; Lebron, J. A.; Watson, N.; Tsuchihashi, Z.; Sigal, E.; Bjorkman, P. J.; Schatzman, R. C.: The hemochromatosis gene product complexes with the transferrin receptor and lowers its affinity for ligand binding. Proc. Nat. Acad. Sci. 95: 1472-1477, 1998.
Feder, J. N.; Tsuchihashi, Z.; Irrinki, A.; Lee, V. K.; Mapa, F. A.; Morikang, E.; Prass, C. E.; Starnes, S. M.; Wolff, R. K.; Parkkila, S.; Sly, W. S.; Schatzman, R. C.:J. Biol. Chem. 272: 14025-14028, 1997.
Feller, E. R.; Pont, A.; Wands, J. R.; Carter, E. A.; Foster, G.; Kourides, I. A.; Isselbacher, K. J.: Familial hemochromatosis: physiologic studies in the precirrhotic stage of the disease. New Eng. J. Med. 296: 1422-1426, 1977.
Felts, J. H.; Nelson, J. R.; Herndon, C. N.; Spurr, C. L.: Hemochromatosis in two young sisters: case studies and a family survey. Ann. Intern. Med. 67: 117-123, 1967.
Goei, V. L.; Parimoo, S.; Capossela, A.; Chu, T. W.; Gruen, J. R.: Isolation of novel non-HLA gene fragments from the hemochromatosis region (6p21.3) by cDNA hybridization selection. Am. J. Hum. Genet. 54: 244-251, 1994.
Gordeuk, V.; Mukiibi, J.; Hasstedt, S.; Samowitz, W.; Edwards, C. Q.; West, G.; Ndambire, S.; Emmamial, J.; Nkanza, N.; Chapanduka, Z.; Randall, M.; Boone, P.; Romano, P.; Martell, R. W.; Yamashita, T.; Effler, P.; Brittenham, G.:New Eng. J. Med. 326: 95-100, 1992.
Halliday, J. W.; Cowlishaw, J. L.; Russo, A. M.; Powell, L. W.: Serum-ferritin in diagnosis of haemochromatosis: a study of 43 families. Lancet II: 621-624, 1977.
Hardy, L.; Hansen, J. L.; Kushner, J. P.; Knisely, A. S.: Neonatal hemochromatosis: genetic analysis of transferrin-receptor, H-apoferritin, and L-apoferritin loci and of the human leukocyte antigen class I region. Am. J. Path. 137: 149-153, 1990.
Hashimoto, K.; Hirai, M.; Kurosawa, Y.: Identification of a mouse homolog for the human hereditary haemochromatosis candidate gene. Biochem. Biophys. Res. Commun. 230: 35-39, 1997.
Hashimoto, K.; Hirai, M.; Kurosawa, Y.: A gene outside the human MHC related to classical HLA class I genes. Science 269: 693-695, 1995.
Jazwinska, E. C.; Cullen, L. M.; Busfield, F.; Pyper, W. R.; Webb, S. I.; Powell, L. W.; Morris, C. P.; Walsh T. P.: Haemochromatosis and HLA-H. (Letter) Nature Genet. 14: 249-251, 1996.
Jazwinska, E. C.; Lee, S. C.; Webb, S. I.; Halliday, J. W.; Powell, L. W.: Localization of the hemochromatosis gene close to D6S105. Am. J. Hum. Genet. 53: 347-352, 1993.
Jazwinska, E. C.; Pyper, W. R.; Burt, M. J.; Francis, J. L.; Goldwurm, S.; Webb, S. I.; Lee, S. C.; Halliday, J. W.; Powell, L. W.: Haplotype analysis in Australian hemochromatosis patients: evidence for a predominant ancestral haplotype exclusively associated with hemochromatosis. Am. J. Hum. Genet. 56: 428-433, 1995.
Johnson, G. B., Jr.; Frey, W. G., III: Familial aspects of idiopathic hemochromatosis. J.A.M.A. 179: 747-751, 1962.
Jouanolle, A.-M.; Yaouanq, J.; Blayau, M.; Perichon, M.; Fauchet, R.; Font, M.-P.; Le Gall, J.-Y.; David, V.: HLA class I gene polymorphism in genetic hemochromatosis. Hum. Genet. 85: 279-282, 1990.
Jouanolle, A. M.; Fergelot, P.; Gandon, G.; Yaouanq, J.; Le Gall, J. Y.; David, V.: A candidate gene for hemochromatosis: frequency of the C282Y and H63D mutations. Hum. Genet. 100: 544-547, 1997.
Jouanolle, A. M.; Gandon, G.; Jezequel, P.; Blayau, M.; Campion, M. L.; Yaouanq, J.; Mosser, J.; Fergelot, P.; Chauvel, B.; Bouric, P.; Carn, G.; Andrieux, N.; Gicquel, I.; Le Gall, J.-Y.; David, V.: Haemochromatosis and HLA-H. (Letter) Nature Genet. 14: 251-252, 1996.
Kaikov, Y.; Wadsworth, L. D.; Hassall, E.; Dimmick, J. E.; Rogers, P. C. J.: Primary hemochromatosis in children: report of three newly diagnosed cases and review of the pediatric literature. Pediatrics 90: 37-42, 1992.
Karlsson, M.; Ikkala, E.; Reunanen, A.; Takkunen, H.; Vuori, E.; Makinen, J.: Prevalence of hemochromatosis in Finland. Acta Med. Scand. 224: 385-390, 1988.
Kidd, K. K.: Genetic linkage and hemochromatosis. (Editorial) New Eng. J. Med. 301: 209-210, 1979.
Kravitz, K.; Skolnick, M.; Cannings, C.; Carmelli, D.; Baty, B.; Amos, B.; Johnson, A.; Mendell, N.; Edwards, C.; Cartwright, G.: Genetic linkage between hereditary hemochromatosis and HLA. Am. J. Hum. Genet. 31: 601-619, 1979.
Krikker, M. A.: A foundation for hemochromatosis. (Letter) Ann. Intern. Med. 97: 782-783, 1982.
Kuhnl, P.; Kaltwasser, J. P.; Seidl, S.: HLA antigens in patients with idiopathic hemochromatosis (IH). Tissue Antigens 12: 398-401, 1978.
Kushner, J. P.; Skolnick, M. H.; Edwards, C. Q.; Goldgar, D.; Griffen, L. M.; Drummond, C.: Advantageous hemochromatosis. (Letter) New Eng. J. Med. 319: 1156, 1988.
Lalouel, J. M.; Le Mignon, L.; Simon, M.; Fauchet, R.; Bourel, M.; Rao, D. C.; Morton, N. E.: Genetic analysis of idiopathic hemochromatosis using both qualitative (disease status) and quantitative (serum iron) information. Am. J. Hum. Genet. 37: 700-718, 1985.
Lamon, J. M.; Marynick, S. P.; Roseblatt, R.; Donnelly, S.: Idiopathic hemochromatosis in a young female: a case study and review of the syndrome in young people. Gastroenterology 76: 178-183, 1979.
Lebron, J. A.; Bennett, M. J.; Vaughn, D. E.; Chirino, A. J.; Snow, P. M.; Mintier, G. A.; Feder, J. N.; Bjorkman, P. J.: Crystal structure of the hemochromatosis protein HFE and characterization of its interaction with transferrin receptor. Cell 93: 111-123, 1998.
Leggett, B. A.; Halliday, J. W.; Brown, N. N.; Bryant, S.; Powell, L. W.: Prevalence of haemochromatosis amongst asymptomatic Australians. Brit. J. Haemat. 74: 525-530, 1990.
Le Mignon, L.; Simon, M.; Fauchet, R.; Edan, G.; Le Reun, M.; Brissot, P.; Genetet, B.; Bourel, M.: An HLA-All association with the hemochromatosis allele?. Clin. Genet. 24: 171-176, 1983.
Lipinski, M.; Hors, J.; Saleun, J.-P.; Saddi, R.; Passa, P.; Lafaurie, S.; Feingold, N.; Dausset, J.: Idiopathic hemochromatosis: linkage with HLA. Tissue Antigens 11: 471-474, 1978.
Lonjou, C.; Collins, A.; Ajioka, R. S.; Jorde, L. B.; Kushner, J. P.; Morton, N. E.: Allelic association under map error and recombinational heterogeneity: a tale of two sites. Proc. Nat. Acad. Sci. 95: 11366-11370, 1998.
Lord, D. K.; Dunham, I.; Campbell, R. D.; Bomford, A.; Strachan, T.; Cox, T. M.: Molecular analysis of the human MHC class I region in hereditary haemochromatosis: a study by pulsed-field gel electrophoresis. Hum. Genet. 85: 531-536, 1990.
Lucotte, G.; Coulondre, C.: Association between a 10 kb PvuII restriction fragment of genomic DNA with the hemochromatosis gene. Exp. Clin. Immunogenet. 3: 219-223, 1986.
MacSween, R. N. M.; Scott, A. R.: Hepatic cirrhosis: a clinico-pathological review of 520 cases. J. Clin. Path. 26: 936-942, 1973.
Maddrey, W. C.; Hamilton, S. R.; Belitsos, N. J.: Familial hemochromatosis. Johns Hopkins Med. J. 144: 66-69, 1979.
McGill, J. R.; Naylor, S. L.; Sakaguchi, A. Y.; Moore, C. M.; Boyd, D.; Barrett, K. J.; Shows, T. B.; Drysdale, J. W.: Human ferritin H and L sequences lie on ten different chromosomes. Hum. Genet. 76: 66-72, 1987.
Mercier, B.; Mura, C.; Ferec, C.: Putting a hold on 'HLA-H.' (Letter) Nature Genet. 15: 234 only, 1997.
Merryweather-Clarke, A. T.; Pointon, J. J.; Shearman, J. D.; Robson, K. J. H.: Global prevalence of putative haemochromatosis mutations. J. Med. Genet. 34: 275-278, 1997.
Meyer, T. E.; Ballot, D.; Bothwell, T. H.; Green, A.; Derman, D. P.; Baynes, R. D.; Jenkins, T.; Jooste, P. L.; du Toit, E. D.; Jacobs, P. J.: The HLA linked iron loading gene in an Afrikaner population. J. Med. Genet. 24: 348-356, 1987.
Milman, N.: Hereditary haemochromatosis in Denmark 1950-1985: clinical, biochemical and histological features in 179 patients and 13 preclinical cases. Danish Med. Bull. 38: 385-393, 1991.
Milman, N.; Eiberg, H.; Thymann, M.; Fenger, K.: Transferrin subtypes in 51 Danish patients with hereditary haemochromatosis and in 847 normal subjects. Hum. Genet. 88: 475-476, 1992.
Milman, N.; Graudal, N.; Nielsen, L. S.; Fenger, K.: An HLA study in 74 Danish haemochromatosis patients and in 21 of their families. Clin. Genet. 41: 6-11, 1992.
Milman, N.; Graudal, N.; Nielsen, L. S.; Fenger, K.: HLA determinants in 70 Danish patients with idiopathic haemochromatosis. Clin. Genet. 33: 286-292, 1988.
Milman, N.; Graudal, N.; Nielsen, L. S.; Mathiassen, B.; Tauris, P.; Lund, B.; Kristensen, J. S.; Fenger, K.: Family studies of hereditary hemochromatosis in Denmark and the Faroe Islands. Hum. Genet. 85: 228-232, 1990.
Muir, W. A.; McLaren, G. D.; Braun, W.; Askari, A.: Evidence for heterogeneity in hereditary hemochromatosis: evaluation of 174 persons in nine families. Am. J. Med. 76: 806-814, 1984.
Muir, W. A.; McLaren, G. D.; Braun, W. E.; Askari, A. K.: Hereditary hemochromatosis: dominant inheritance and probable linkage to HLA. (Abstract) Am. J. Hum. Genet. 30: 61A, 1978.
Mura, C.; Nousbaum, J.-B.; Verger, P.; Moalic, M.-T.; Raguenes, O.; Mercier, A.-Y.; Ferec, C.: Phenotype-genotype correlation in haemochromatosis subjects. Hum. Genet. 101: 271-276, 1997.
Murphy, P. A.: Personal Communication. Baltimore, Md. 12/12/1987.
Niederau, C.; Fischer, R.; Sonnenberg, A.; Stremmel, W.; Trampisch, H. J.; Strohmeyer, G.: Survival and causes of death in cirrhotic and in noncirrhotic patients with primary hemochromatosis. New Eng. J. Med. 313: 1256-1262, 1985.
Nussbaumer, T.; Plattner, H. C.; Rywlin, A. M.: Hemochromatose juvenile chez trois soeurs et un frere avec consanguinite des parents: etude anatomoclinique et genetique du syndrome endocrinohepato-myocardique. J. Genet. Hum. 1: 53-59, 1952.
Olsson, K. S.; Eriksson, K.; Ritter, B.; Heedman, P. A.: Screening for iron overload using transferrin saturation. Acta Med. Scand. 215: 105-112, 1984.
Olsson, K. S.; Ritter, B.; Rosen, U.; Heedman, P. A.; Staugard, F.: Prevalence of iron overload in central Sweden. Acta Med. Scand. 213: 145-150, 1983.
Parkkila, S.; Waheed, A.; Britton, R. S.; Bacon, B. R.; Zhou, X. Y.; Tomatsu, S.; Fleming, R. E.; Sly, W. S.: Association of the transferrin receptor in human placenta with HFE, the protein defective in hereditary hemochromatosis. Proc. Nat. Acad. Sci. 94: 13198-13202, 1997.
Parkkila, S.; Waheed, A.; Britton, R. S.; Feder, J. N.; Tsuchihashi, Z.; Schatzman, R. C.; Bacon, B. R.; Sly, W. S.: Immunohistochemistry of HLA-H, the protein defective in patients with hereditary hemochromatosis, reveals unique pattern of expression in gastrointestinal tract. Proc. Nat. Acad. Sci. 94: 2534-2539, 1997.
Perkins, K. W.; McInnes, I. W. S.; Blackburn, C. R. B.; Beal, R. W.: Idiopathic hemochromatosis in children: report of a family. Am. J. Med. 39: 118-126, 1965.
Pollycove, M.: Hemochromatosis.In: Stanbury, J. B.; Wyngaarden, J. B.; Fredrickson, D. S.: The Metabolic Basis of Inherited Disease. New York: McGraw-Hill (pub.) (3rd ed.): 1972. Pp. 1051-1084.
Powell, L. W.; Ferluga, J.; Halliday, J. W.; Bassett, M. L.; Kohonen-Corish, M.; Serjeantson, S.: Genetic hemochromatosis and HLA linkage. Hum. Genet. 77: 55-56, 1987.
Rabinovitz, M.; Gavaler, J. S.; Kelly, R. H.; Van Thiel, D. H.: Association between heterozygous alpha-1-antitrypsin deficiency and genetic hemochromatosis. Hepatology 16: 145-148, 1992.
Raha-Chowdhury, R.; Bowen, D. J.; Worwood, M.: A new highly polymorphic marker in the 5-prime untranslated region of HLA-F shows strong allelic association with haemochromatosis. Hum. Genet. 97: 228-231, 1996.
Rhodes, D. A.; Raha-Chowdhury, R.; Cox, T. M.; Trowsdale, J.: Homozygosity for the predominant Cys282Tyr mutation and absence of disease expression in hereditary haemochromatosis. J. Med. Genet. 34: 761-764, 1997.
Risch, N.: Haemochromatosis, HFE and genetic complexity. (Letter) Nature Genet. 17: 375-376, 1997.

Ritter, B.; Safwenberg, J.; Olsson, K. S.: HLA as a marker of the hemochromatosis gene in Sweden. Hum. Genet. 68: 62-66, 1984.
Robson, K. J. H.; Shearman, J. D.; Merryweather-Clarke, A. T.; Pointon, J. J.; Rosenberg, W. M.; Walker, A. P.; Dooley, J. S.; Bomford, A.; Raha-Chowdhury, R.; Worwood, M.: Haemochromatosis: a gene at last? J. Med. Genet. 34: 148-151, 1997.
Roldan, F. P.; Echenagusia, A. A.; Carro, P. G.: Hemochromatosis presenting as acute liver failure after iron supplementation. (Letter) New Eng. J. Med. 339: 269-270, 1998.
Roth, M.-P.; Giraldo, P.; Hariti, G.; Poloni, E. S.; Sanchez-Mazas, A.; De Stefano, G. F.; Dugoujon, J.-M.; Coppin, H.: Absence of the hemochromatosis gene Cys282Tyr mutation in three ethnic groups from Algeria (Mzab), Ethiopia, and Senegal. Immunogenetics 46: 222-225, 1997.
Rothenberg, B. E.; Voland, J. R.: Beta-2 knockout mice develop parenchymal iron overload: a putative role for class I genes of the major histocompatibility complex in iron metabolism. Proc. Nat. Acad. Sci. 93: 1529-1534, 1996.
Rowe, J. W.; Wands, J. R.; Mezey, S. E.; Waterbury, L. A.; Wright, J. R.; Tobin, J.; Andres, R.: Familial hemochromatosis: characteristics of the precirrhotic stage in a large kindred. Medicine 56: 197-211, 1977.
Saddi, R.; Feingold, J.: Idiopathic haemochromatosis: an autosomal recessive disease. Clin. Genet. 5: 234-241, 1974.
Sargent, T., III; Lim, T. H.; Jenson, R. L.: Reduced chromium retention in patients with hemochromatosis, a possible basis of hemochromatotic diabetes. Metabolism 28: 70-79, 1979.
Sheldon, J. H.: Haemochromatosis. London: Oxford Univ. Press (pub.) 1935. Pp. 19 only.
Siemons, L. J.; Mahler, C.: Hypogonadotropic hypogonadism in hemochromatosis: recovery of reproductive function after iron depletion. J. Clin. Endocr. Metab. 65: 585-587, 1987.
Simon, M.; Alexandre, J. L.; Bourel, M.; Le Marec, B.; Scordia, C.: Heredity of idiopathic haemochromatosis: a study of 106 families. Clin. Genet. 11: 327-341, 1977.
Simon, M.; Bourel, M.; Fauchet, R.; Genetet, B.: Association of HLA-A3 and HLA-B14 antigens with idiopathic haemochromatosis. Gut 17: 332-334, 1976.
Simon, M.; Bourel, M.; Genetet, B.; Fauchet, R.: Idiopathic hemochromatosis: demonstration of recessive transmission and early detection by family HLA typing. New Eng. J. Med. 297: 1017-1021, 1977.
Simon, M.; Bourel, M.; Genetet, B.; Fauchet, R.: Heredity of idiopathic haemochromatosis. (Letter) Lancet I: 706, 1977.
Simon, M.; Fauchet, R.; Hespel, J. P.; Beaumont, C.; Brissot, P.; Hary, B.; De Nercy, H. Y. H.; Genetet, B.; Bourel, M.: Idiopathic hemochromatosis: a study of biochemical expression in 247 heterozygous members of 63 families: evidence for a single major HLA-linked gene. Gastroenterology 78: 703-708, 1980.
Simon, M.; Fauchet, R.; Le Gall, J. Y.; Brissot, P.; Bourel, M.: Immunogenetics of idiopathic hemochromatosis and secondary iron overload.In: Farid, N. R.: Immunogenetics of Endocrine Disorders. New York: Alan R. Liss (pub.) 1988. Pp. 345-371.
Simon, M.; Le Mignon, L.; Fauchet, R.; Yaouanq, J.; David, V.; Edan, G.; Bourel, M.: A study of 609 HLA haplotypes marking for the hemochromatosis gene: (1) mapping of the gene near the HLA-A locus and characters required to define a heterozygous population and (2) hypothesis concerning the underlying cause of hemochromatosis-HLA association. Am. J. Hum. Genet. 41: 89-105, 1987.
Skolnick, M.: Personal Communication. Salt Lake City, Utah 9/27/1983.
Sorsby, A.: Clinical Genetics. St. Louis: C. V. Mosby (pub.) 1953. Pp. 206 only.
Stevens, F. M.; Walters, J. M.; Watt, D. W.; McCarthy, C. F.: Inheritance of idiopathic haemochromatosis. (Letter) Lancet I: 1107, 1977.
Summers, K. M.; Tam, K. S.; Bartley, P. B.; Drysdale, J.; Zoghbi, H. Y.; Halliday, J. W.; Powell, L. W.: Fine mapping of a human chromosome 6 ferritin heavy chain pseudogene: relevance to haemochromatosis. Hum. Genet. 88: 175-178, 1991.
Summers, K. M.; Tam, K. S.; Halliday, J. W.; Powell, L. W.: HLA determinants in an Australian population of hemochromatosis patients and their families. Am. J. Hum. Genet. 45: 41-48, 1989.
The UK Haemochromatosis Consortium: A simple genetic test identifies 90% of UK patients with haemochromatosis. Gut 41: 841-844, 1997.
Totaro, A.; Rommens, J. M.; Grifa, A.; Lunardi, C.; Carella, M.; Huizenga, J. J.; Roetto, A.; Camaschella, C.; De Sandre, G.; Gasparini, P.: Hereditary hemochromatosis: generation of a transcription map within a refined and extended map of the HLA class I region. Genomics 31: 319-326, 1996.
Trousseau, A.: Clinique Med. de l'Hotel de Paris. (pub.) II: 1865. Pp. 663-698.
Trousseau, A.: Clinical Medicine Lectures Delivered at the Hotel-Dieu Paris by A. Trousseau. Philadelphia: Blakiston (pub.) II: 1882. Pp. 313-315. Note: Translation: 3rd revised and enlarged edition by Sir John Rose Cormack and P. Victor Bazire. . . . ...
Valberg, L. S.; Lloyd, D. A.; Ghent, C. N.; Flanagan, P. R.; Sinclair, N. R.; Stiller, C. R.; Chamberlain, M. J.: Clinical and biochemical expression of the genetic abnormality in idiopathic hemochromatosis. Gastroenterology 79: 884-892, 1980.
Waheed, A.; Parkkila, S.; Zhou, X. Y.; Tomatsu, S.; Tsuchihashi, Z.; Feder, J. N.; Schatzman, R. C.; Britton, R. S.; Bacon, B. R.; Sly, W. S.: Hereditary hemochromatosis: effects of C282Y and H63D mutations on association with beta-2-microglobulin, intracellular processing, and cell surface expression of the HFE protein in COS-7 cells. Proc. Nat. Acad. Sci. 94: 12384-12389, 1997.
Wands, J. R.; Rowe, J. A.; Mezey, S. E.; Waterbury, L. A.; Wright, J. R.; Halliday, J. W.; Isselbacher, K. J.; Powell, L. W.: Normal serum ferritin concentrations in precirrhotic hemochromatosis. New Eng. J. Med. 294: 302-305, 1976.
Worwood, M.; Darke, C.; Trenchard, P.: Hereditary haemochromatosis and blood donation. (Letter) Brit. Med. J. 302: 593, 1991.
Yaouang, J.; Perichon, M.; Chorney, M.; Pontarotti, P.; Le Treut, A.; El Kahloun, A.; Mauvieux, V.; Blayau, M.; Jouanolle, A. M.; Chauvel, B.; Moirand, R.; Nouel, O.; Le Gall, J. Y.; Feingold, J.; David, V.: Anonymous marker loci within 400 kb of HLA-A generate haplotypes in linkage disequilibrium with the hemochromatosis gene (HFE). Am. J. Hum. Genet. 54: 252-263, 1994.
Yaouanq, J.; El Kahloun, A.; Chorney, M.; Jouanolle, A. M.; Mauvieux, V.; Perichon, M.; Blayau, M.; Pontarotti, P.; Le Gall, J. Y.; David, V.: Familial screening for genetic haemochromatosis by means of DNA markers. J. Med. Genet. 29: 320-322, 1992.
Zappone, E.; Dugast, I.; Papadopoulos, P.; Theriault, K.; David, V.; LeGall, J.-Y.; Summers, K.; Powell, L.; Drysdale, J.: Polymorphism in a ferritin H gene from chromosome 6p. Hum. Genet. 86: 557-561, 1991.
Zhou, X. Y.; Tomatsu, S.; Fleming, R. E.; Parkkila, S.; Waheed, A.; Jiang, J.; Fei, Y.; Brunt, E. M.; Ruddy, D. A.; Prass, C. E.; Schatzman, R. C.; O'Neill, R.; Britton, R. S.; Bacon, B. R.; Sly, W. S.: HFE gene knockout produces mouse model of hereditary hemochromatosis. Proc. Nat. Acad. Sci. 95: 2492-2497, 1998.

HEPATOCELLULAR CARCINOMA COMPLICATING HEMOCHROMATOSIS (*602254)

JJM NOTE

Like the mouse *Mim* gene that modifies expression of the gene for familial polyposis, this gene may modify the risk of liver cancer in hemochromatosis.

Hemochromatosis (HC; 235200) is an iron storage disorder caused by an inherited defect in iron metabolism. Iron accumulates in the tissues, particularly in hepatocytes, leading to cellular toxicity and liver fibrosis and eventually to cirrhosis and hepatocellular carcinoma.

Kedar et al. (1996) used differential display reverse transcriptase polymerase chain reaction (DDRT-PCR) to identify a partial mRNA, termed HC/HCC by them, that was strongly expressed in the liver of a patient with HC and in tumor tissue from a subject with hepatocellular carcinoma complicating HC. The HC/HCC cDNA, encoding a 9.5-kb transcript, consists of 342 amino acid residues. By Northern blot analysis, Kedar et al. (1996) showed that the message of this gene is present at very low levels in normal human liver, and is also expressed in lymphocytes of HC patients and in MOLT-4 human T-lymphoid cells.

Kedar, V.; Fletcher, L.; Powell, L.: Molecular cloning of a novel mRNA highly expressed in haemochromatotic human liver and proliferating cells. Biochem. Biophys. Res. Commun. 226: 461-466, 1996.

HEPATIC ADENOMAS, FAMILIAL (142330)

JJM NOTE

This entity is a candidate example of ecogenetics in cancer origins. It is surprising that a confirming report has not appeared.

Foster et al. (1978) observed a family in which the mother and 3 of 5 children, a son and 2 daughters, all teenaged, had liver-cell adenomas. All 4 plus several other members of the kindred had insulin-dependent maturity-onset diabetes of the young (MODY; 125850). Sclerocystic ovaries were present in the 2 daughters. The mother's father and paternal grandfather had histologically confirmed hepatocellular carcinoma (see 114550). The family came to attention when the 18-year-old daughter developed sudden abdominal pain and was laparotomized. The tumors were highly vascular and hemorrhage into the tumor was a probable cause of pain. The authors noted the occurrence of discrete liver-cell adenomas in type 1 glycogen storage disease (232200) and in patients on oral contraceptives. The liver tumors in this family may have been related in some way to a peculiar metabolic defect that led also to MODY. Nonetheless, a small chromosomal deletion or other abnormality should be sought.

Foster, J. H.; Donohue, T. A.; Berman, M. M.: Familial liver-cell adenomas and diabetes mellitus. New Eng. J. Med. 299: 239-241, 1978.

ANTITRYPSIN DEFICIENCY (*107400)

JJM NOTE

The role of alpha-1-antitrypsin deficiency in the origins of hepatocellular carcinoma has not been pursued in recent years. It seems clear that homozygotes, at least for the Z allele, are at high risk for liver cancer. A question with important public health consequences is, What is the cancer risk of the much more common heterozygote? Early evidence was largely based on the presence of characteristic nonglycogen glycoprotein cytoplasmic globules in normal liver adjacent to hepatomas (Berg and Eriksson 1972; Brunt PW: Progress report: Antitrypsin and the liver. Gut 15:573–80, 1974.)

Description

Alpha-1-antitrypsin is a protease inhibitor, deficiency of which is associated with emphysema and liver disease. The protein is encoded by a gene (PI) located on the distal long arm of chromosome 14.

Clinical Features

Deficiency of protease inhibitor activity is associated with several of the electrophoretic variants of serum alpha-1-antitrypsin; Axelsson and Laurell (1965) first suggested that the genes for electrophoretic variants are allelic with the deficiency gene. Laurell and Eriksson (1963) described absence of alpha-1-antitrypsin from the plasma in patients with degenerative lung disease leading to death in middle life. (Eriksson (1989) gave an interesting historical account including the pedigree of his first family (Eriksson, 1965).) Emphysematous changes involve primarily the lower lung fields (Bell, 1970).

Many electrophoretic variants of serum alpha-1-antitrypsin have been described, beginning with those reported by Axelsson and Laurell (1965). Some of these variants are associated with reduced protease activity and occasionally with clinical consequences. Kueppers and Bearn (1967) studied an Italian family with multiple members heterozygous for an electrophoretic variant that could not be distinguished from that which Axelsson and Laurell (1965) found in a Swedish family. The polymorphism of prealbumin described by Fagerhol and Braend (1965) was shown by Fagerhol and Laurell (1967) to be the same as the alpha-1-antitrypsin polymorphism. Fagerhol (1968) suggested that the system be called Pi for protease inhibitor.

Gans et al. (1969) described familial infantile liver cirrhosis in presumed homozygotes for alpha-1-antitrypsin deficiency. Udall et al. (1982) speculated that a factor in the pathogenesis of infantile cirrhosis may be lack of protease inhibitor to counteract the effects of proteases that cross the intestinal barrier in the neonate. Lake-Bakaar and Dooley (1982) found that alpha-1-antitrypsin is an important proteolytic inhibitor in bile, thus providing support of the pathogenetic theory of Udall et al. (1982). Aagenaes et al. (1972) described the clinical picture in children with the ZZ genotype as neonatal cholestasis. Five such cases were described.

Morin et al. (1975) concluded that heterozygotes are not at increased risk of alcoholic cirrhosis. Eriksson et al. (1986) concluded that the risk of cirrhosis and liver cancer is increased for males homozygous for alpha-1-antitrypsin deficiency but not for females. The finding suggested additive effects of exogenous factors. Geddes et al. (1977) found that the frequency of non-M phenotypes was increased to a significant extent in patients with sclerosing alveolitis with or without rheumatoid arthritis. Fargion et al. (1981) found an increased frequency of non-M phenotypes in patients with hepatocellular carcinoma. Furthermore, patients with liver cancer and a non-M phenotype had a lower average age than those with an M phenotype.

Phenotypic correlations indicate that there are 3 types of mutations: those producing deficiency of enzyme action, null mutations, and mutations causing altered function of the gene product. PI Z and PI M(Malton) are deficiency mutations that are associated with both emphysema and liver disease. PI S, M(Heerlen), M(Mineral Springs), M(Procida), M(Nichinan), I, and P(Lowell) are deficiency mutations with an increased risk of emphysema only. The null mutations are associated with emphysema only. PI Pittsburgh is an example of a mutation leading to altered function of the gene product with a bleeding disorder as the clinical presentation. Crystal (1990) gave a comprehensive review of the pathogenetic relationship between AAT deficiency and emphysema and liver disease, including a detailed listing of the various mutations that have been identified and a discussion of the possibilities for therapy.

An adult with antitrypsin deficiency and combined liver and lung disease was reported by Gherardi (1971). See the study of 12 cases of combined disease by Berg and Eriksson (1972). Contrary to the usual view that liver disease, while a risk in children, is not a great risk to adults with alpha-1-antitrypsin deficiency, Cox and Smyth (1983) found a relatively high risk in men between 51 and 60 years. A low concentration of serum prealbumin was a sensitive indicator of impaired liver function.

Wiebicke et al. (1996) confirmed the absence of pulmonary function abnormalities in the vast majority of children with homozygous alpha-1-PI deficiency. Rodriguez-Cintron et al. (1995) suggested that bronchiectasis should be considered part of the spectrum of pulmonary pathology that may be encountered in individuals with AAT deficiency. They described a 21-year-old man with massive hemoptysis and homozygous ZZ AAT deficiency. Neither panlobular emphysema nor cirrhosis of the liver was present.

Biochemical Features

About 30 variants of alpha-1-antitrypsin had been described by 1981 (Hug et al., 1981). The alleles have been given symbols according to the relative electrophoretic mobility of the allele product. Cox (1978) reported the recommendations of a workshop on Pi nomenclature. Several reports (Bell and Carroll, 1973; Kuhlenschmidt et al., 1974; Eriksson and Larsson, 1975) have suggested that the defect may be in a sialyltransferase and that deficiency of antitrypsin in the blood is the result of impaired secretion from hepatocytes, increased clearance of the undersialated protein, or both. It is difficult to see how this could cause codominant inheritance or account for the different types that appear to be the products of at least 30 different alleles, unless an amino acid substitution interferes with sialidation.

Yoshida et al. (1976) studied a variant protein from a ZZ homozygote and showed 2 amino acid substitutions, glutamic acid to lysine and glutamic acid to glutamine. The sialic acid content of the variant protein was reduced, presumably as a result of change in configuration of the protein since none of the carbohydrate-binding amino acids were substituted.

The leukocytes of chronic granulomatous disease have a defect in inactivation of antitrypsin. In their experiments, George et al. (1984) found that addition of azide or catalase enhanced the effectiveness of the mutant inhibitor. AT, like other serine protease inhibitors such as antithrombin III, alpha-2-antiplasmin, and C1-inhibitor, has a single inhibitor-specific reactive site peptide bond that is formed between adjacent amino acid residues termed P(1) and P-prime(1). The reactivity of these inhibitors with proteolytic enzymes depends heavily on the nature of the residue at position P(1), the central position of the reactive center.

Harrison et al. (1990) described an improved method for detecting what they termed 'low Z expressor' phenotype in MZ heterozygotes. An obligate carrier mother who was being typed as part of a family study appeared to be a PI(M)/PI(null) heterozygote. By routine isoelectric focusing, she was typed as M, her affected child as Z, and her husband as MZ. Atypically low concentrations of circulating Z peptides were demonstrated by the improved method.

Weitz et al. (1992) demonstrated a correlation between plasma levels of elastase-specific fibrinopeptides and PI genotype. The levels of these peptides were highest in ZZ homozygotes and intermediate in MZ heterozygotes. This was interpreted as evidence that unopposed human neutrophil elastase is responsible for emphysema in patients with alpha-1-proteinase inhibitor deficiency.

Inheritance

Family studies indicate recessive inheritance of antitrypsin deficiency. In early studies, heterozygotes, who can be detected chemically, were unaffected clinically; later studies suggested that heterozygosity may predispose to lung disease (Lieberman, 1969). For example, of 12 patients with obstructive lung disease present before age 40 years, 2 were judged by Tarkoff et al. (1968) to be homozygous for the deficiency and 1 heterozygous. Among 103 patients, Kueppers et al. (1969) found 5 homozygotes and 25 heterozygotes for the deficiency gene. They suggested that, especially in males, heterozygosity may predispose to chronic obstructive lung disease. Stevens et al. (1971) concluded that heterozygotes may develop emphysema qualitatively like that in homozygotes, but at a later age. The importance of prompt treatment of respiratory infections and avoidance of proteolytic aerosols, smoking and employment entailing exposure to respiratory irritants are important preventive measures in these families.

Lieberman et al. (1979) found an increased frequency of heterozygosity for antitrypsin deficiency in twins and parents of twins. They concluded that 'increased' fertility and twinning may be heterozygous advantages for antitrypsin deficiency. Clark and Martin (1982) found that the frequency of the S allele in mothers of dizygotic twins (0.088) was double that in controls (0.044). The frequency of S in the parents of monozygotic twins and in fathers of DZ twins was no higher than in controls. Normal frequencies were observed for the Z allele. No fertility indices other than twinning itself were available. To study relationships between Pi types, fertility, and twinning, Boomsma et al. (1992) studied 90 DZ and 70 MZ Dutch twin pairs and their parents. They found that mothers of dizygotic twins had frequencies of the S and Z alleles that were 3 times higher than those in a control sample. Mothers of identical twins also had a higher frequency of S than controls. The S allele may thus both increase ovulation rate and enhance the success of multiple pregnancies.

Mapping

A possible heterogeneity in recombination frequency between Pi variants believed to be allelic was reported by Gedde-Dahl et al. (1972): Pi(Z) had less recombination with Gm than Pi(non-Z). Gedde-Dahl et al. (1975) gave further data on the Gm-Pi linkage. They considered heterogeneity of recombination fraction among males of different Pi types to be likely. The major difference seemed to be between the Pi(Z) and other alleles. Possible explanations included a chromosomal deletion, inversion or locus regulating recombination in linkage disequilibrium with the Pi locus. Gedde-Dahl et al. (1981) showed that the allele-specific heterogeneity of Gm-Pi linkage is attributable to 'reduced' recombination in Z-allele heterozygotes. They found an equal sex ratio for Pi 'non-Z' variants, as opposed to a 1:2 male-female ratio for 'Z' families.

The location of Gm and Pi on 6p was excluded by Bender et al. (1979). By studying hybrids of mouse or rat hepatoma cells with human lymphocytes, Darlington et al. (1982) and Pearson et al. (1981) achieved direct assignment of the Pi locus to chromosome 14. From study of 2 families with abnormalities of the long arm of chromosome 14, Cox et al. (1982) localized GM to 14q32.3 and PI to a more proximal position between 14q24.3 and 14q32.1. The immunoglobulin genes are in a chromosome region noted for its high frequency of breaks associated with chromosome rearrangement, occurring both spontaneously in cultured lymphocytes and in certain malignancies.

By study of human-Chinese hamster somatic cell hybrids, both genes were assigned to chromosome 14. By in situ hybridization, Schroeder et al. (1985) narrowed the assignment of the PI locus to 14q31-q32. Turleau et al. (1984) studied a patient with an interstitial deletion of 14q and assigned the PI locus to 14q32.1 by exclusion mapping. In a similar patient with an interstitial deletion of 14q, Yamamoto et al. (1986) confirmed the assignment to 14q32.1. By the dosage principle, the level of alpha-1-antitrypsin in the patient was only about half of that in his parents and in controls.

Molecular Genetics

Lai et al. (1983) cloned the alpha-1-antitrypsin gene and showed that it contains 3 introns in the peptide-coding region. All persons showed 2 distinct bands (9.6 kb and 8.5 kb long) when the cloned gene was used as a hybridization probe to analyze EcoRI-digested genomic DNA. Analysis using only intronic DNA as probe showed that the authentic gene resides in the 9.6-kb fragment. The other gene may be a pseudogene. See 107410.

Long et al. (1984) provided the complete cDNA sequence of the PI gene. The genomic length of the gene is 10.2 kb with a 1,434-bp coding region. The gene has 4 introns; exon 1, the 5-prime portion of exon 2, and the 3-prime portion of exon 5 are noncoding regions. The first intron, 5.3 kb long, contains a 143-amino acid open reading frame (which does not appear to be an actual protein coding region), an Alu family sequence, and a pseudotranscription initiation region. Carrell (1986) cited evidence for the existence of 2 genes coding for alpha-1-antitrypsin, although the plasma findings are compatible with expression of the alleles at a single locus. Sefton et al. (1989) used pulsed field gel electrophoresis to demonstrate that the genes encoding alpha-1-antitrypsin and alpha-1-antichymotrypsin (AACT) are approximately 220 kb apart and oriented in opposite directions. By in situ hybridization, Ledbetter et al. (1987) localized the AAT locus to mouse chromosome 12.

Molecular studies of a ring chromosome 14 showed that the IGH and D14S1 loci were missing, whereas the PI locus was present (Keyeux et al., 1989). Thus, PI is proximal to the other 2 loci, a conclusion that was supported by much earlier data. A noncoding alpha-1-antitrypsin-like gene (PIL; 107410) is located 12 kb 3-prime of the AAT gene. Billingsley et al. (1989) found that this gene and the PI and AACT genes are carried by a single 550-kb NarI fragment. Also see Billingsley et al. (1993).

Nukiwa et al. (1986) found in the Z gene a second substitution in addition to that which is responsible for the change from glutamic acid to lysine at amino acid 342. The latter mutation is located in exon 5; the second mutation, GTG to GCG, was located in exon 3 and led to a predicted substitution of alanine for valine as amino acid residue number 213. The valine-to-alanine change at 213 was found in all of 40 Z haplotypes, using synthetic oligonucleotide gene probes directed toward the mutated exon 3 sequences in the Z gene. Furthermore, the exon 3 mutation eliminated a BstEII restriction endonuclease site, allowing rapid identification of the change in genomic DNA. Surprisingly, only 23% of the M1 haplotypes were found to be BstEII site negative. The new form of M1, i.e., M1(ala-213), is identical to M1 but has an isoelectric focusing 'silent' amino acid substitution. M1 has a frequency of 68 to 76%; M2, 14 to 20%; and M3, 10 to 12%. The Z gene represents 1 to 2% of all alpha-1-antitrypsin haplotypes.

Garver et al. (1986) investigated the molecular basis of the Pi null-null alpha-1-antitrypsin phenotype. The gene appeared to be intact without discernible deletion or other structural abnormality, yet there was no detectable mRNA produced. The 5-prime promoter region also appeared to be normal. No evidence of hypermethylation of cytosine nucleotides in the promoter region was detected. The defect may be comparable to that in some forms of thalassemia in which a change, at a splicing site, for example, may lead to greatly reduced mRNA production. The null-null phenotype is accompanied by emphysema as is the ZZ and SZ phenotypes but an important difference is that cirrhosis and liver cancer do not occur with the null-null phenotype; there is no abnormal antitrypsin produced that is excreted with difficulty from the cells of synthesis.

Nukiwa et al. (1987) identified a null form of alpha-1-antitrypsin resulting from a frameshift causing a stop codon to be formed approximately 44% from the N terminus of the precursor protein. Although the molecular basis of antitrypsin deficiency was quite different from that in the Z haplotype, the phenotypic consequences were similar: severe deficiency associated with high risk of emphysema. Perlino et al. (1987) found that the AAT gene in macrophages is transcribed from a macrophage-specific promoter located about 2000 bp upstream of the hepatocyte-specific promoter. Transcription from the 2 AAT promoters is mutually exclusive; the macrophage promoter is silent in hepatocytes and the hepatocyte promoter is silent in macrophages. In macrophages, 2 distinct mRNAs are generated by alternative splicing.

Cox et al. (1987) studied RFLPs associated with the AAT gene. They gave information on extensive variability expressed by the polymorphic information content (PIC) as proposed by Botstein et al. (1980). PI types and M subtypes tended to be associated with specific RFLP haplotypes. Bamforth and Kalsheker (1988) discussed a rare Pi null allele that in homozygous state leads to pulmonary emphysema at an early age. In 3 families, all the affected individuals presented in early childhood with recurrent chest infections and wheezing, presumably related to passive smoking. Even though there was no detectable AAT, no partial or complete deletion of the gene could be identified.

By means of isoelectric focusing, Weber and Weidinger (1988) found a PI variant that they called PI S (Cologne). A father and daughter were heterozygous. Alpha-1-antitrypsin concentrations were within the normal range.

Nukiwa et al. (1988) indicated that approximately 75 AAT alleles have been identified at the protein and/or gene level. The most common alleles are the 2 forms of M1, that with valine at position 213 and that with alanine at position 213 (called here M1A and M1V).

Hafeez et al. (1992) demonstrated that the AAT gene has 3 macrophage-specific transcriptional initiation sites upstream from a single hepatocyte-specific transcriptional initiation site. Macrophages use these sites during basal and modulated expression. Hepatoma cells use the hepatocyte-specific transcriptional initiation site during basal and modulated expression but also switch on transcription from the upstream macrophage transcriptional initiation sites during modulation by the acute phase mediator interleukin-6 (IL6; 147620).

Pathogenesis

Lomas et al. (1992) presented an explanation for the accumulation of insoluble intracellular inclusions in the ZZ homozygote. Only about 15% of the AAT protein is secreted in the plasma in ZZ homozygotes. The 85% that is not secreted accumulates in the endoplasmic reticulum (ER) of the hepatocyte; much of it is degraded but the remainder aggregates to form insoluble intracellular inclusions. About 10% of newborn ZZ homozygotes develop liver disease that often leads to fatal childhood cirrhosis. Lomas et al. (1992) demonstrated the molecular pathology underlying this accumulation and described how the Z mutation in antitrypsin results in a unique molecular interaction between the reactive center loop of one molecule and the gap in the A-sheet of another. This loop-sheet polymerization of Z antitrypsin occurs spontaneously at 37 degrees C and is completely blocked by the insertion of a specific peptide into the A-sheet of the antitrypsin molecule. The loop-sheet polymerization is concentration- and temperature-dependent. At times of stress, the formation of inclusions in the hepatocyte will likely overwhelm the degradative mechanisms. Antitrypsin is an acute phase protein and as such undergoes a manifold increase in association with temperature elevations during bouts of inflammation. Control of inflammation and pyrexia in ZZ homozygote infants is important. In the long-term, more specific interventions may be possible, e.g., the delivery to the hepatocyte of engineered loop peptides specific to alpha-1-antitrypsin.

Liver injury in individuals with the ZZ genotype presumably results from toxic effects of the abnormal AAT molecule accumulating within the endoplasmic reticulum of liver cells; however, only 12 to 15% of individuals with this genotype develop liver disease. Therefore, Wu et al. (1994) predicted that other genetic factors determine susceptibility to liver disease. To examine this hypothesis, they transduced skin fibroblasts from ZZ individuals with liver disease and from ZZ individuals without liver disease with amphotropic recombinant retroviral particles designed to express the mutant AAT*Z gene under direction of a constitutive viral promoter. Expression of the AAT gene was conferred on each fibroblast cell line. Compared to the same cell line transduced with the wildtype gene, there was selective intracellular accumulation of the mutant protein in each case. However, there was a marked delay in degradation of the mutant protein after it accumulated in the fibroblasts from ZZ individuals with liver disease ('susceptible hosts') as compared to those without liver disease ('protected hosts'). Appropriate disease controls showed that the lag in degradation in susceptible hosts is specific for the combination of the ZZ genotype and liver disease. Biochemical characteristics of the ATT*Z degradation in the protected hosts was found to be similar to those of a common ER degradation pathway previously described for T-cell receptor alpha subunits and asialoglycoprotein receptor subunits, therefore raising the possibility that the lag in degradation in the susceptible host is a defect in this common ER degradation pathway.

As reviewed by Lomas (1996), inclusions in the most frequent cause of antitrypsin deficiency, the C mutation (glu342lys; 107400.0011), is accompanied by accumulation of protein in the endoplasmic reticulum of the liver. These hepatic inclusions in turn result from a protein-protein interaction between the reactive center loop of 1 molecule and the beta-pleated sheet of a second. This loop-sheet polymerization is the basis of deficiencies associated also with mutations of C1-inhibitor (106100), antithrombin III (107300), and alpha-1-antichymotrypsin (107280), all of which are serine proteinase inhibitors (serpins).

Sigsgaard et al. (1992) showed that in cotton workers the airborne concentration of respirable endotoxin was associated with byssinosis. Endotoxin might induce byssinosis through the release of biochemical mediators at the bronchoalveolar surface. Alpha-1-antitrypsin, which neutralizes enzymes released by granulocytes, might have a counteracting role. Sigsgaard et al. (1994) found that the MZ phenotype was associated with an increased prevalence of byssinosis compared with the MM phenotype: 3/8 (38%) and 25/187 (13%). An association between the MZ phenotype and familial allergy was also found, although the association was somewhat weaker.

Diagnosis

Kidd et al. (1983) used a chemically synthesized specific oligonucleotide probe (19-mer) as a sensitive and direct test for the presence or absence of the Z gene (glu342 to lys; GAG to AAG). Kidd et al. (1984) reported the use of such probes in the prenatal diagnosis of the deficiency syndrome. George et al. (1984) showed that replacement of this methionine-358 with valine in a genetically engineered mutant of human alpha-1-antitrypsin resulted in an inhibitor of connective tissue breakdown when tested in a model of inflammation. Degradation of basement membrane collagen was efficiently inhibited by a concentration of the mutant substance that was tenfold lower than that of the normal antitrypsin.

Clinical Management

Wewers et al. (1987) reported on treatment of patients with alpha-1-antitrypsin deficiency with intravenous plasma-derived enzyme once a week. Although granting that a completely rigorous study was impossible, the authors concluded that infusions of enzyme are safe and can reverse the biochemical abnormalities in serum and lung fluid and, further, that lifetime avoidance of cigarette smoking together with such replacement may be a logical approach to long-term therapy.

The liver represents an excellent organ for gene therapy since many genetic disorders result from deficiency of liver-specific gene products. Kay et al. (1992) demonstrated the autologous transplantation of canine hepatocytes transduced with a retroviral vector containing the human alpha-1-antitrypsin cDNA under transcriptional control of the cytomegalovirus promoter. At least 1 billion hepatocytes or 5% of the liver mass could be transplanted by the portal vasculature. Human alpha-1-antitrypsin was demonstrable in the serum of 2 dogs for 1 month. Although the serum levels of the human enzyme eventually fell due to inactivation of the cytomegalovirus promoter, PCR analysis demonstrated that a significant fraction of the transduced hepatocytes migrated to the liver and continued to survive in vivo.

As a model for gene therapy, Garver et al. (1987) used a retroviral vector to insert human alpha-1-antitrypsin cDNA into the genome of mouse fibroblasts. After demonstrating that the clone produced human antitrypsin after more than 100 population doublings in the absence of selection pressure, they transplanted the clone into the peritoneal cavities of nude mice. When the animals were evaluated 4 weeks later, human antitrypsin was detected in both sera and the epithelial surface of the lungs. Lemarchand et al. (1992) reported experiments supporting the feasibility of in vivo human gene transfer of recombinant human AAT cDNA to endothelial cells by means of replication-deficiency adenovirus vectors.

Population Genetics

Roychoudhury and Nei (1988) tabulated worldwide gene frequencies for allelic variants M (M1, M2, M3, M4), S, Z, F, I, and V. Cox (1989) and Crystal (1989) reviewed the variants, 'normal' and pathologic, of the PI gene.

DeCroo et al. (1991) studied the frequency of alpha-1-antitrypsin alleles in U.S. whites, U.S. blacks, and African blacks (living in Nigeria). While the PI*S allele was present at a polymorphic level in U.S. whites, it was present only sporadically in U.S. blacks and completely absent in African blacks. The PI*Z allele was not detected in the black populations tested. DeCroo et al. (1991) used the PI allele frequency data to calculate white admixture in U.S. blacks. The average white admixture estimate in U.S. blacks, based on all PI alleles, was about 13%. This value was about 24% when only the S and Z alleles were used.

Studies of the distribution of the Pi types S and Z in Europe demonstrated that they occur mainly among those of European stock. Hutchison (1998) found that the frequency of the gene for Pi type Z is highest on the northwestern seaboard of the continent and that the mutation seems to have arisen in southern Scandinavia. The distribution of type S is quite different: the gene frequency is highest in the Iberian peninsula and the mutation is likely to have arisen in that region.

Animal Model

Kurachi et al. (1981) found more than 96% homology of cDNA and amino acid sequences between the alpha-1-antitrypsin of man and baboon. Comparison of baboon alpha-1-antitrypsin, human antithrombin III, and chicken ovalbumin indicated about 30% homology of amino acid sequence.

The pallid (pa) mouse develops emphysema late in life. Martorana et al. (1993) demonstrated that pallid mice have markedly reduced levels of serum alpha-1-antitrypsin associated with severe deficiency in serum elastase inhibitory capacity. However, they have normal alpha-1-antitrypsin mRNA levels in the liver.

See also Arnaud et al. (1977); Arnaud et al. (1978); Carrell et al. (1982); Chan et al. (1978); Chapuis-Cellier et al. (1981); Cox (1975); Cox (1980); Cox (1981); Curiel et al. (1989); Curiel et al. (1989); Faber et al. (1994); Faber et al. (1989); Fagerhol and Cox (1981); Fagerhol and Gedde-Dahl (1969); Fagerhol and Hauge (1968); Fagerhol and Laurell (1970); Fagerhol and Tenfjord (1968); Falk and Briscoe (1970); Falk and Briscoe (1970); Fraizer et al. (1989); Frants and Eriksson (1980); Freeman et al. (1976); Gedde-Dahl et al. (1975); Graham et al. (1990); Guenter et al. (1971); Hall et al. (1976); Hepper et al. (1969); Hodges et al. (1981); Holmes et al. (1990); Holmes et al. (1990); Hug et al. (1980); Iammarino et al. (1979); Jeppsson et al. (1975); Kew et al. (1978); Kramps et al. (1981); Kueppers et al. (1964); Kueppers and Christopherson (1978); Langley et al. (1979); Lieberman et al. (1971); Lopez et al. (1964); Meisen et al. (1988); Morse (1978); Neumann et al. (1976); Nukiwa et al. (1986); Owen et al. (1976); Perrault et al. (1979); Pierce et al. (1969); Rodriguez-Soriano et al. (1978); Rosenthal et al. (1979); Schmitt et al. (1975); Sharp et al. (1969); Starzl et al. (1983); Stockley (1979); Talamo et al. (1968); Talamo and Feingold (1973); Townley et al. (1970); Weitkamp et al. (1978); Welch et al. (1980); Yoshida et al. (1979)

Aagenaes, O.; Matlary, A.; Elgjo, K.; Munthe, E.; Fagerhol, M.: Neonatal cholestasis in alpha-1-antitrypsin deficient children: clinical, genetic, histological and immunohistochemical findings. Acta Paediat. Scand. 61: 632-642, 1972.

Abbott, C. M.; McMahon, C. J.; Whitehouse, D. B.; Povey, S.: Prenatal diagnosis of alpha-1-antitrypsin deficiency using polymerase chain reaction. (Letter) Lancet I: 763-764, 1988.

Arnaud, P.; Galbraith, R. M.; Faulk, W. P.: Increased frequency of the MZ phenotype of alpha-1-protease inhibitor in juvenile chronic polyarthritis. J. Clin. Invest. 60: 1442-1444, 1977.

Arnaud, P.; Galbraith, R. M.; Galbraith, G. M. P.; Allen, R. C.; Fudenberg, H. H.: A new allele of human alpha-1-antitrypsin: Pi (N Hampton). Am. J. Hum. Genet. 30: 653-659, 1978.

Axelsson, U.; Laurell, C. B.: Hereditary variants of serum alpha-1-antitrypsin. Am. J. Hum. Genet. 17: 466-472, 1965.

Babron, M. C.; Constans, J.; Dugoujon, J. M.; Cambon-Thomsen, A.; Bonaiti-Pellie, C.: The Gm-Pi linkage in 843 French families: effect of the alleles Pi Z and Pi S. Ann. Hum. Genet. 54: 107-113, 1990.

Bamforth, F. J.; Kalsheker, N. A.: Alpha-1 antitrypsin deficiency due to Pi null: clinical presentation and evidence for molecular heterogeneity. J. Med. Genet. 25: 83-87, 1988.

Bell, O. F.; Carroll, R. W.: Basis of the defect in alpha-1-antitrypsin deficiency. Nature 243: 410-411, 1973.

Bell, R. S.: The radiographic manifestations of alpha-1 antitrypsin deficiency: an important recognizable pattern of chronic obstructive pulmonary disease (COPD). Radiology 95: 19-24, 1970.

Bender, K.; Muller, C. R.; Schmidt, A.; Strohmaier, U.; Wienker, T. F.: Linkage studies on the human Pi, Gm, GLO, and HLA genes. Hum. Genet. 49: 159-166, 1979.

Berg, N. O.; Eriksson, S.: Liver disease in adults with alpha-1-antitrypsin deficiency. New Eng. J. Med. 287: 1264-1267, 1972.

Billingsley, G. D.; Walter, M. A.; Cox, D. W.: Physical linkage of alpha-1-antitrypsin and alpha-1-antichymotrypsin by pulsed field gel electrophoresis. (Abstract) Am. J. Hum. Genet. 45 (suppl.): A130, 1989.
Billingsley, G. D.; Walter, M. A.; Hammond, G. L.; Cox, D. W.: Physical mapping of four serpin genes: alpha-1-antitrypsin, alpha-1-antichymotrypsin, corticosteroid-binding globulin, and protein C inhibitor, within a 280-kb region on chromosome 14q32.1. Am. J. Hum. Genet. 52: 343-353, 1993.
Boomsma, D. I.; Frants, R. R.; Bank, R. A.; Martin, N. G.: Protease inhibitor (Pi) locus, fertility and twinning. Hum. Genet. 89: 329-332, 1992.
Botstein, D.; White, R. L.; Skolnick, M.; Davis, R. W.: Construction of a genetic linkage map in man using restriction fragment length polymorphisms. Am. J. Hum. Genet. 32: 314-331, 1980.
Brantly, M.; Courtney, M.; Crystal, R. G.: Repair of the secretion defect in the Z form of alpha-1-antitrypsin by addition of a second mutation. Science 242: 1700-1702, 1988.
Brennan, S. O.; Carrell, R. W.: Alpha-1-antitrypsin Christchurch, 363 glu-to-lys: mutation at the P-prime-5 position does not affect inhibitory activity. Biochim. Biophys. Acta 873: 13-19, 1986.
Carrell, R. W.: Alpha-1-antitrypsin: molecular pathology, leukocytes, and tissue damage. J. Clin. Invest. 78: 1427-1431, 1986.
Carrell, R. W.; Jeppsson, J.-O.; Laurell, C.-B.; Brennan, S. O.; Owen, M. C.; Vaughan, L.; Boswell, D. R.: Structure and variation of human alpha-1-antitrypsin. Nature 298: 329-334, 1982.
Chan, C. H.; Steer, C. J.; Vergalla, J.; Jones, E. A.: Alpha-1-antitrypsin deficiency with cirrhosis associated with the protease inhibitor phenotype SZ. Am. J. Med. 65: 978-986, 1978.
Chapuis-Cellier, C.; Verdier, M.; Lepetit, J. C.; Fudenberg, H. H.; Arnaud, P.: Pi-Gm linkage: evidence for linkage in males but not in females and for an effect of the S allele of the Pi system. J. Immunogenet. 8: 257-262, 1981.
Clark, P.; Breit, S. N.; Dawkins, R. L.; Penny, R.: Genetic study of a family with two members with Weber-Christian disease (panniculitis) and alpha-1-antitrypsin deficiency. Am. J. Med. Genet. 13: 57-62, 1982.
Clark, P.; Martin, N. G.: An excess of the Pi(s) allele in dizygotic twins and their mothers. Hum. Genet. 61: 171-174, 1982.
Cox, D. W.: Alpha-1-antitrypsin deficiency.In: Scriver, C. R.; Beaudet, A. L.; Sly, W. S.; Valle, D.: The Metabolic Basis of Inherited Disease. New York: McGraw-Hill (pub.) (6th ed.): 1989.
Cox, D. W.: The effect of neuraminidase on genetic variants of alpha-1-antitrypsin. Am. J. Hum. Genet. 27: 165-177, 1975.
Cox, D. W.: Genetic variation in alpha-1-antitrypsin. (Editorial) Am. J. Hum. Genet. 30: 660-662, 1978.
Cox, D. W.: Transmission of Z allele from heterozygotes for alpha-1-antitrypsin deficiency. (Letter) Am. J. Hum. Genet. 32: 455-457, 1980.
Cox, D. W.: New variants of alpha-1-antitrypsin: comparison of Pi typing techniques. Am. J. Hum. Genet. 33: 354-365, 1981.
Cox, D. W.; Billingsley, G. D.; Mansfield, T.: DNA restriction-site polymorphisms associated with the alpha-1-antitrypsin gene. Am. J. Hum. Genet. 41: 891-906, 1987.
Cox, D. W.; Markovic, V. D.; Teshima, I. E.: Genes for immunoglobulin heavy chains and for alpha-1-antitrypsin are localized to specific regions of chromosome 14q. Nature 297: 428-430, 1982.
Cox, D. W.; Smyth, S.: Risk for liver disease in adults with alpha-1-antitrypsin deficiency. Am. J. Med. 74: 221-227, 1983.
Cox, D. W.; Woo, S. L. C.; Mansfield, T.: DNA restriction fragments associated with alpha-1-antitrypsin indicate a single origin for deficiency allele PI Z. Nature 316: 79-81, 1985.
Crystal, R. G.: Alpha-1-antitrypsin deficiency, emphysema, and liver disease: genetic basis and strategies for therapy. J. Clin. Invest. 85: 1343-1352, 1990.
Crystal, R. G.: The alpha-1-antitrypsin gene and its deficiency states. Trends Genet. 5: 411-417, 1989.
Curiel, D.; Brantly, M.; Curiel, E.; Stier, L.; Crystal, R. G.: Alpha-1-antitrypsin deficiency caused by the alpha-1-antitrypsin null(Mattawa) gene: an insertion mutation rendering the alpha-1-antitrypsin gene incapable of producing alpha-1-antitrypsin. J. Clin. Invest. 83: 1144-1152, 1989.
Curiel, D. T.; Chytil, A.; Courtney, M.; Crystal, R. G.: Serum alpha-1-antitrypsin deficiency associated with the common S-type (glu264-to-val) mutation results from intracellular degradation of alpha-1-antitrypsin prior to secretion. J. Biol. Chem. 264: 10477-10486, 1989.
Curiel, D. T.; Holmes, M. D.; Okayama, H.; Brantly, M. L.; Vogelmeier, C.; Travis, W. D.: Stier, L. E.; Perks, W. H. and Crystal, R. G.: Molecular basis of the liver and lung disease associated with the alpha-1-antitrypsin deficiency allele M(Malton). J. Biol. Chem. 264: 13938-13945, 1989.
Curiel, D. T.; Vogelmeier, C.; Hubbard, R. C.; Stier, L. E.; Crystal, R. G.: Molecular basis of alpha-1-antitrypsin deficiency and emphysema associated with the alpha-1-antitrypsin M(Mineral Springs) allele. Molec. Cell. Biol. 10: 47-56, 1990.
Darlington, G. J.; Astrin, K. H.; Muirhead, S. P.; Desnick, R. J.; Smith, M.: Assignment of human alpha-1-antitrypsin to chromosome 14 by somatic cell hybrid analysis. Proc. Nat. Acad. Sci. 79: 870-873, 1982.
DeCroo, S.; Kamboh, M. I.; Ferrell, R. E.: Population genetics of alpha-1-antitrypsin polymorphism in US whites, US blacks and African blacks. Hum. Hered. 41: 215-221, 1991.
Dry, P. J.: Rapid detection of alpha-1-antitrypsin deficiency by analysis of a PCR-induced TaqI restriction site. Hum. Genet. 87: 742-744, 1991.
Dycaico, M. J.; Grant, S. G. N.; Felts, K.; Nichols, W. S.; Geller, S. A.; Hager, J. H.; Pollard, A. J.; Kohler, S. W.; Short, H. P.; Jirik, F. R.; Hanahan, D.; Sorge, J. A.: Neonatal hepatitis induced by alpha-1-antitrypsin: a transgenic mouse model. Science 242: 1409-1415, 1988.
Emmerich, J.; Alhenc-Gelas, M.; Gandrille, S.; Guichet, C.; Fiessinger, J.-N.; Aiach, M.: Mechanism of protein C deficiency in a patient with arginine 358 alpha(1)-antitrypsin (Pittsburgh mutation): role in the maintenance of hemostatic balance. J. Lab. Clin. Med. 125: 531-539, 1995.
Eriksson, S.: Studies in alpha 1-antitrypsin deficiency. Acta Med. Scand. 177 (suppl. 432): 1-85, 1965.
Eriksson, S.: Alpha-1-antitrypsin deficiency: lessons learned from the bedside to the gene and back again. Chest 95: 181-189, 1989.
Eriksson, S.; Carlson, J.; Velez, R.: Risk of cirrhosis and primary liver cancer in alpha-1-antitrypsin deficiency. New Eng. J. Med. 314: 736-739, 1986.
Eriksson, S.; Larsson, C.: Purification and partial characterization of PAS-positive inclusion bodies from the liver in alpha-1-antitrypsin deficiency. New Eng. J. Med. 292: 176-180, 1975.
Faber, J.-P.; Weidinger, S.; Olek, K.: Sequence data of the rare deficient alpha-1-antitrypsin variant PI Z(augsburg). Am. J. Hum. Genet. 46: 1158-1162, 1990.
Faber, J. P.; Poller, W.; Weidinger, S.; Kirchgesser, M.; Schwaab, R.; Bidlingmaier, S.; Olek, K.: Identification and DNA sequence analysis of 15 new alpha-1-antitrypsin variants, including two Pi* Q alleles and one deficient Pi* M allele. Am. J. Hum. Genet. 55: 1113-1121, 1994.
Faber, J. P.; Weidinger, S.; Goedde, H.-W.; Olek, K.: The deficient alpha-1-antitrypsin phenotype Pi P is associated with an A-to-T transversion in exon III of the gene. (Letter) Am. J. Hum. Genet. 45: 161-163, 1989.
Faber, J. P.; Weidinger, S.; Olek, K.: Sequence data of two rare deficient alpha-1-antitrypsin phenotypes. (Abstract) Cytogenet. Cell Genet. 51: 995-996, 1989.
Fagerhol, M. K.: The Pi system. Genetic variants of serum alpha-1-antitrypsin. Ser. Haematol. 1: 153-161, 1968.
Fagerhol, M. K.; Braend, M.: Serum prealbumin: polymorphism in man. Science 149: 986-987, 1965.
Fagerhol, M. K.; Cox, D. W.: The Pi polymorphism: genetic, biochemical, and clinical aspects of human alpha-1-antitrypsin. Adv. Hum. Genet. 11: 1-62, 1981.
Fagerhol, M. K.; Gedde-Dahl, T., Jr.: Genetics of the Pi serum types: family studies of the inherited variants of serum alpha-1-antitrypsin. Hum. Hered. 19: 354-359, 1969.
Fagerhol, M. K.; Hauge, H. E.: The Pi phenotype MP: discovery of a ninth allele belonging to the system of inherited variants of serum alpha-1-antitrypsin. Vox Sang. 15: 396-400, 1968.
Fagerhol, M. K.; Laurell, C. B.: The Pi system—inherited variants of serum alpha-1-antitrypsin. Prog. Med. Genet. 7: 96-111, 1970.
Fagerhol, M. K.; Laurell, C. B.: The polymorphism of 'prealbumins' and alpha-1-antitrypsin in human sera. Clin. Chim. Acta 16: 199-203, 1967.
Fagerhol, M. K.; Tenfjord, O. W.: Serum Pi types in some European, American, Asian and African populations. Acta Path. Microbiol. Scand. 72: 601-608, 1968.
Falk, G. A.; Briscoe, W. A.: Chronic obstructive pulmonary disease and heterozygous alpha-1-antitrypsin deficiency. (Editorial) Ann. Intern. Med. 72: 595-596, 1970.
Falk, G. A.; Briscoe, W. A.: Alpha-1-antitrypsin deficiency in chronic obstructive pulmonary disease. (Editorial) Ann. Intern. Med. 72: 427-429, 1970.
Fargion, S.; Klasen, E. C.; Lalatta, F.; Sangalli, G.; Tommasini, M.; Fiorelli, G.: Alpha-1-antitrypsin in patients with carcinoma and chronic active hepatitis. Clin. Genet. 19: 134-139, 1981.
Fortin, P. R.; Fraser, R. S.; Watts, C. S.; Esdaile, J. M.: Alpha-1 antitrypsin deficiency and systemic necrotizing vasculitis. J. Rheum. 18: 1613-1616, 1991.
Fraizer, G. C.; Harrold, T. R.; Hofker, M. H.; Cox, D. W.: In-frame single codon deletion in the M(Malton) deficiency allele of alpha-1-antitrypsin. Am. J. Hum. Genet. 44: 894-902, 1989.
Fraizer, G. C.; Siewertsen, M.; Harold, T. R.; Cox, D. W.: Deletion/frameshift mutation in the alpha-1-antitrypsin null allele, PI*QO(Bolton). Hum. Genet. 83: 377-382, 1989.
Fraizer, G. C.; Siewertsen, M. A.; Hofker, M. H.; Brubacher, M. G.; Cox, D. W.: A null deficiency allele of alpha-1-antitrypsin, QOludwigshafen, with altered tertiary structure. J. Clin. Invest. 86: 1878-1884, 1990.
Frants, R.; Eriksson, A. W.: A new unstable Pi M variant of alpha-1-antitrypsin in a Finnish isolate. Hum. Hered. 30: 333-342, 1980.
Freeman, H. J.; Weinstein, W. M.; Shnitka, T. K.; Crockford, P. M.; Herbert, F. A.: Alpha-1-antitrypsin deficiency and pancreatic fibrosis. Ann. Intern. Med. 85: 73-76, 1976.
Gans, H.; Sharp, H. L.; Tan, B. H.: Antiprotease deficiency and familial infantile liver cirrhosis. Surg. Gynec. Obstet. 129: 289-299, 1969.
Garver, R. I., Jr.; Chytil, A.; Courtney, M.; Crystal, R. G.: Clonal gene therapy: transplanted mouse fibroblast clones express human alpha-1-antitrypsin gene in vivo. Science 237: 762-764, 1987.
Garver, R. I., Jr.; Mornex, J.-F.; Nukiwa, T.; Brantly, M.; Courtney, M.; LeCocq, J.-P.; Crystal, R. G.: Alpha-1-antitrypsin deficiency and emphysema caused by homozygous inheritance of non-expressing alpha-1-antitrypsin genes. New Eng. J. Med. 314: 762-766, 1986.
Gedde-Dahl, T., Jr.; Cook, P. J. L.; Fagerhol, M. K.; Pierce, J. A.: The Gm-Pi linkage: a summary estimate. Birth Defects Orig. Art. Ser. XI(3): 157-158, 1975. Note: Alternate: Cytogenet. Cell Genet. 14: 327-328, 1975..
Gedde-Dahl, T., Jr.; Cook, P. J. L.; Fagerhol, M. K.; Pierce, J. A.: Improved estimate of the Gm-Pi linkage. Ann. Hum. Genet. 39: 43-50, 1975.
Gedde-Dahl, T., Jr.; Fagerhol, M. K.; Cook, P. J. L.; Noades, J.: Autosomal linkage between the Gm and Pi loci in man. Ann. Hum. Genet. 35: 393-400, 1972.
Gedde-Dahl, T., Jr.; Frants, R.; Olaisen, B.; Eriksson, A. W.; van Loghem, E.; Lamm, L.: The Gm-Pi linkage heterogeneity in view of Pi M subtypes. Ann. Hum. Genet. 45: 143-153, 1981.
Geddes, D. M.; Webley, M.; Brewerton, D. A.; Turton, C. W.; Turner-Warwick, M.; Murphy, A. H.; Ward, A. M.: Alpha-1-antitrypsin phenotypes in fibrosing alveolitis and rheumatoid arthritis. Lancet II: 1049-1053, 1977.
George, P. M.; Vissers, M. C. M.; Travis, J.; Winterbourn, C. C.; Carrell, R. W.: A genetically engineered mutant of alpha-1-antitrypsin protects connective tissue from neutrophil damage and may be useful in lung disease. Lancet II: 1426-1428, 1984.
Gherardi, G. J.: Alpha-1-antitrypsin deficiency and its effect on the liver. Hum. Path. 2: 173-175, 1971.
Graham, A.; Hayes, K.; Weidinger, S.; Newton, C. R.; Markham, A. F.; Kalsheker, N. A.: Characterisation of the alpha-1-antitrypsin M3 gene, a normal variant. Hum. Genet. 85: 381-382, 1990.

Graham, A.; Kalsheker, N. A.; Bamforth, F. J.; Newton, C. R.; Markham, A. F.: Molecular characterization of two alpha-1-antitrypsin deficiency variants: proteinase inhibitor (Pi) null(Newport) (gly(115)-to-ser) and (Pi) Z Wrexham (ser(-19)-to-leu). Hum. Genet. 85: 537-540, 1990.

Graham, A.; Kalsheker, N. A.; Newton, C. R.; Bamforth, F. J.; Powell, S. J.; Markham, A. F.: Molecular characterisation of three alpha-1-antitrypsin deficiency variants: proteinase inhibitor (Pi) Null (Cardiff) (asp256-to-val), Pi M(Malton) (phe51-deletion) and Pi I (arg39-to-cys). Hum. Genet. 84: 55-58, 1989.

Guenter, C. A.; Welch, M. H.; Hammarsten, J. F.: Alpha-1-antitrypsin deficiency and pulmonary emphysema. Annu. Rev. Med. 22: 283-292, 1971.

Hafeez, W.; Ciliberto, G.; Perlmutter, D. H.: Constitutive and modulated expression of the human alpha-1 antitrypsin gene: different transcriptional initiation sites used in three different cell types. J. Clin. Invest. 89: 1214-1222, 1992.

Hall, W. J.; Hyde, R. W.; Schwartz, R. H.; Mudholkar, G. S.; Webb, D. R.; Chaubey, Y. P.; Townes, P. L.: Pulmonary abnormalities in intermediate alpha-1-antitrypsin deficiency. J. Clin. Invest. 57: 1069-1077, 1976.

Harrison, H. H.; Miller, K. L.; Whitington, P. F.: Improved detection, via ISO-DALT two-dimensional electrophoresis, of 'low Z expressor' individuals with alpha-1-antitrypsin MZ phenotype. Clin. Chem. 36: 1850-1851, 1990.

Hejtmancik, J. F.; Sifers, R. N.; Ward, P. A.; Harris, S.; Mansfield, T.; Cox, D. W.: Prenatal diagnosis of alpha-1-antitrypsin deficiency by restriction fragment length polymorphisms, and comparison with oligonucleotide probe analysis. Lancet II: 767-769, 1986.

Hendrick, S. J.; Silverman, A. K.; Solomon, A. R.; Headington, J. T.: Alpha-1-antitrypsin deficiency associated with panniculitis. J. Am. Acad. Derm. 18: 684-692, 1988.

Hepper, N. G.; Black, L. F.; Gleich, G. J.; Kueppers, F.: The prevalence of alpha-1-antitrypsin deficiency in selected groups of patients with chronic obstructive lung disease. Mayo Clin. Proc. 44: 697-710, 1969.

Hildesheim, J.; Kinsley, G.; Bissell, M.; Pierce, J.; Brantly, M.: Genetic diversity from a limited repertoire of mutations on different common allelic backgrounds: alpha-1-antitrypsin deficiency variant P(Duarte). Hum. Mutat. 2: 221-228, 1993.

Hodges, J. R.; Millward-Sadler, G. H.; Barbatis, C.; Wright, R.: Heterozygous MZ alpha-1-antitrypsin deficiency in adults with chronic active hepatitis and cryptogenic cirrhosis. New Eng. J. Med. 304: 557-560, 1981.

Hofker, M. H.; Nelen, M.; Klasen, E. C.; Nukiwa, T.; Curiel, D.; Crystal, R. G.; Frants, R. R.: Cloning and characterization of an alpha-1-antitrypsin like gene 12 kb downstream of the genuine alpha-1-antitrypsin gene. Biochem. Biophys. Res. Commun. 155: 634-642, 1988.

Hofker, M. H.; Nukiwa, T.; van Paassen, H. M. B.; Nelen, M.; Kramps, J. A.; Klasen, E. C.; Frants, R. R.; Crystal, R. G.: A pro-to-leu substitution in codon 369 of the alpha-1-antitrypsin deficiency variant PI MHeerlen. Hum. Genet. 81: 264-268, 1989.

Holmes, M. D.; Brantly, M. L.; Crystal, R. G.: Molecular analysis of the heterogeneity among the P-family of alpha-1-antitrypsin alleles. Am. Rev. Resp. Dis. 142: 1185-1192, 1990.

Holmes, M. D.; Brantly, M. L.; Curiel, D. T.; Weidinger, S.; Crystal, R. G.: Characterization of the normal alpha-1-antitrypsin allele V(Munich): a variant associated with a unique protein isoelectric focusing pattern. Am. J. Hum. Genet. 46: 810-816, 1990.

Holmes, M. D.; Brantly, M. L.; Fells, G. A.; Crystal, R. G.: Alpha-1-antitrypsin W(Bethesda): molecular basis of an unusual alpha-1-antitrypsin deficiency variant. Biochem. Biophys. Res. Commun. 170: 1013-1020, 1990.

Hug, G.; Chuck, G.; Fagerhol, M. K.: Pi(P-Clifton): a new alpha(1) antitrypsin allele in an American Negro family. J. Med. Genet. 18: 43-45, 1981.

Hug, G.; Chuck, G.; Slemmer, T. M.; Fagerhol, M. K.: Pi (Ecincinnati): a new alpha-1-antitrypsin allele in three Negro families. Hum. Genet. 54: 361-364, 1980.

Hutchison, D. C. S.: Alpha-1-antitrypsin deficiency in Europe: geographical distribution of Pi types S and Z. Resp. Med. 92: 367-377, 1998.

Iammarino, R. M.; Wagener, D. K.; Allen, R. C.: Segregation distortion of the alpha-1-antitrypsin Pi Z allele. Am. J. Hum. Genet. 31: 508-517, 1979.

Jeppsson, J.-O.; Larsson, C.; Eriksson, S.: Characterization of alpha-1-antitrypsin in the inclusion bodies from the liver in alpha-1-antitrypsin deficiency. New Eng. J. Med. 293: 576-579, 1975.

Kalsheker, N.; Hayes, K.; Weidinger, S.; Graham, A.: What is Pi (proteinase inhibitor) null or PiQO?: a problem highlighted by the alpha-1-antitrypsin M(heerlen) mutation. J. Med. Genet. 29: 27-29, 1992.

Kalsheker, N. A.; Hodgson, I. J.; Watkins, G. L.; White, J. P.; Morrison, H. M.; Stockley, R. A.: Deoxyribonucleic acid (DNA) polymorphism of the alpha-1-antitrypsin gene in chronic lung disease. Brit. Med. J. 294: 1511-1514, 1987.

Kawakami, Y.; Irie, T.; Kishi, F.; Asanuma, Y.; Shida, A.; Yoshikawa, T.; Kamishima, K.; Hasegawa, H.; Murao, M.: Familial aggregation of abnormal ventilatory control and pulmonary function in chronic obstructive pulmonary disease. Europ. J. Resp. Dis. 62: 56-64, 1981.

Kay, M. A.; Baley, P.; Rothenberg, S.; Leland, F.; Fleming, L.; Ponder, K. P.; Liu, T.; Finegold, M.; Darlington, G.; Pokorny, W.; Woo, S. L. C.: Expression of human alpha-1-antitrypsin in dogs after autologous transplantation of retroviral transduced hepatocytes. Proc. Nat. Acad. Sci. 89: 89-93, 1992.

Kennedy, M.; Brett, W.: Monozygotic twins with alpha-1-antitrypsin deficiency. (Letter) Lancet I: 527-528, 1985.

Kew, M. C.; Turnbull, R.; Prinsloo, I.: Alpha-1-antitrypsin deficiency and hepatocellular cancer. Brit. J. Cancer 37: 635-638, 1978.

Keyeux, G.; Gilgenkrantz, S.; Lefranc, G.; Lefranc, M.-P.: Molecular characterization of a ring chromosome 14 showing that the PI locus is centromeric to the D14S1 and IGH loci. Hum. Genet. 82: 219-222, 1989.

Kidd, V. J.; Golbus, M. S.; Wallace, R. B.; Itakura, K.; Woo, S. L. C.: Prenatal diagnosis of alpha-1-antitrypsin deficiency by direct analysis of the mutation site in the gene. New Eng. J. Med. 310: 639-642, 1984.

Kidd, V. J.; Wallace, R. B.; Itakura, K.; Woo, S. L. C.: Alpha-1-antitrypsin deficiency detection by direct analysis of the mutation in the gene. Nature 304: 230-234, 1983.

Klasen, E. C.; Biemond, I.; Laros, C. D.: Alpha-1-antitrypsin deficiency and the flaccid lung syndrome: the heterozygote controversy. Clin. Genet. 29: 211-215, 1986.

Kramps, J. A.; Brouwers, J. W.; Maesen, F.; Dijkman, J. H.: Pi(M-Heerlen), a Pi(M) allele resulting in very low alpha-1-antitrypsin serum levels. Hum. Genet. 59: 104-107, 1981.

Kueppers, F.; Bearn, A. G.: An inherited alpha-1-antitrypsin variant. Humangenetik 4: 217-220, 1967.

Kueppers, F.; Briscoe, W. A.; Bearn, A. G.: Hereditary deficiency of serum alpha 1-antitrypsin. Science 146: 1678-1679, 1964.

Kueppers, F.; Christopherson, M. J.: Alpha-1-antitrypsin: further genetic heterogeneity revealed by isoelectric focusing. Am. J. Hum. Genet. 30: 359-365, 1978.

Kueppers, F.; Fallat, R.; Larson, R. K.: Obstructive lung diseases and alpha-antitrypsin deficiency gene heterozygosity. Science 165: 899-901, 1969.

Kuhlenschmidt, M. S.; Yunis, E. J.; Iammarino, R. M.; Turco, S. J.; Peters, S. P.; Glew, R. H.: Demonstration of sialyltransferase deficiency in the serum of a patient with alpha-1-antitrypsin deficiency and hepatic cirrhosis. Lab. Invest. 31: 413-419, 1974.

Kurachi, K.; Chandra, T.; Degen, S. J. F.; White, T. T.; Marchioro, T. L.; Woo, S. L. C.; Davie, E. W.: Cloning and sequence of cDNA coding for alpha-1-antitrypsin. Proc. Nat. Acad. Sci. 78: 6826-6830, 1981.

Lai, E. C.; Kao, F.-T.; Law, M. L.; Woo, S. L. C.: Assignment of the alpha-1-antitrypsin gene and a sequence-related gene to human chromosome 14 by molecular hybridization. Am. J. Hum. Genet. 35: 385-392, 1983.

Lake-Bakaar, G.; Dooley, J. S.: Alpha-1-antitrypsin deficiency and liver disease. (Letter) Lancet II: 159, 1982.

Langley, C. E.; Berninger, R. W.; Wolfson, S. L.; Talamo, R. C.: An unusual type of alpha-1-antitrypsin deficiency in a child. Johns Hopkins Med. J. 144: 161-165, 1979.

Laubach, V. E.; Ryan, W. J.; Brantly, M.: Characterization of a human alpha-1-antitrypsin null allele involving aberrant mRNA splicing. Hum. Molec. Genet. 2: 1001-1005, 1993.

Laurell, C.-B.; Eriksson, S.: The electrophoretic alpha-1-globulin pattern of serum in alpha-1-antitrypsin deficiency. Scand. J. Clin. Lab. Invest. 15: 132-140, 1963.

Ledbetter, S. A.; Ledbetter, D. H.; Ledley, F. D.; Woo, S.: Localization of phenylalanine hydroxylase (PAH) and alpha-1 antitrypsin (AAT) loci in mouse genome by synteny and in situ hybridization. (Abstract) Am. J. Hum. Genet. 41: A173, 1987.

Lemarchand, P.; Jaffe, H. A.; Danel, C.; Cid, M. C.; Kleinman, H. K.; Stratford-Perricaudet, L. D.; Perricaudet, M.; Pavirani, A.; Lecocq, J.-P.; Crystal, R. G.: Adenovirus-mediated transfer of a recombinant human alpha-1-antitrypsin cDNA to human endothelial cells. Proc. Nat. Acad. Sci. 89: 6482-6486, 1992.

Lewis, J. H.; Iammarino, R. M.; Spero, J. A.; Hasiba, U.: Antithrombin Pittsburgh: an alpha-1-antitrypsin variant causing hemorrhagic disease. Blood 51: 129-137, 1978.

Lieberman, J.: Heterozygous and homozygous alpha-1-antitrypsin deficiency in patients with pulmonary emphysema. New Eng. J. Med. 281: 279-284, 1969.

Lieberman, J.; Borhani, N. O.; Feinleib, M.: Alpha-1-antitrypsin deficiency in twins and parents-of-twins. Clin. Genet. 15: 29-36, 1979.

Lieberman, J.; Mittman, C.; Kent, J. R.: Screening for heterozygous alpha-1-antitrypsin deficiency. III. A provocative test with diethylstilbestrol and effect of oral contraceptives. J.A.M.A. 217: 1198-1206, 1971.

Lomas, D. A.: New insights into the structural basis of alpha-1-antitrypsin deficiency. Quart. J. Med. 89: 807-812, 1996.

Lomas, D. A.; Evans, D. L.; Finch, J. T.; Carrell, R. W.: The mechanism of Z alpha-1-antitrypsin accumulation in the liver. Nature 357: 605-607, 1992.

Long, G. L.; Chandra, T.; Woo, S. L. C.; Davie, E. W.; Kurachi, K.: Complete sequence of the cDNA for human alpha-1-antitrypsin and the gene for the S variant. Biochemistry 23: 4828-4837, 1984.

Lopez, V.; Oetliker, O.; Colombo, J. P.; Butler, R.: Ein Fall von familiaerem alpha-1-Antitrypsinmangel. Helv. Paediat. Acta 19: 296-303, 1964.

Lovegrove, J. U.; Jeremiah, S.; Gillett, G. T.; Temple, I. K.; Povey, S.; Whitehouse, D. B.: A new alpha 1-antitrypsin mutation, thr-met 85, (PI Z-bristol) associated with novel electrophoresis properties. Ann. Hum. Genet. 67: 385-391, 1997.

Martorana, P. A.; Brand, T.; Gardi, C.; van Even, P.; de Santi, M. M.; Calzoni, P.; Marcolongo, P.; Lungarella, G.: The pallid mouse: a model of genetic alpha-1-antitrypsin deficiency. Lab. Invest. 68: 233-241, 1993.

Matsunaga, E.; Shiokawa, S.; Nakamura, H.; Maruyama, T.; Tsuda, K.; Fukumaki, Y.: Molecular analysis of the gene of the alpha-1-antitrypsin deficiency variant, Mnichinan. Am. J. Hum. Genet. 46: 602-612, 1990.

Meisen, C.; Higuchi, M.; Brautigam, S.; Driesel, A. J.; Blandfort, M.; Olek, K.: Prenatal diagnosis of alpha-1-antitrypsin deficiency using oligonucleotide probe analysis. Hum. Genet. 79: 190-192, 1988.

Morgan, K.; Scobie, G.· Kalsheker, N. A.: Point mutation in a 3-prime flanking sequence of the alpha-1-antitrypsin gene associated with chronic respiratory disease occurs in a regulatory sequence. Hum. Molec. Genet. 2: 253-257, 1993.

Morin, T.; Martin, J.-P.; Feldmann, G.; Rueff, B.; Benhamou, J.-P.; Ropartz, C.: Heterozygous alpha (1)-antitrypsin deficiency and cirrhosis in adults, a fortuitous association. Lancet I: 250-251, 1975.

Morse, J. O.: Alpha-1-antitrypsin deficiency. New Eng. J. Med. 299: 1045-1048 and 1099-1105, 1978.

Nakamura, H.; Ogawa, A.; Hisano, S.; Fukuma, M.; Tachibana, N.; Tsuda, K.: A family with a new deficient variant of alpha-1-antitrypsin PiM(Nichinan): with special reference to diastase-resistant, periodic acid-Schiff positive globules in the liver cells. J. Jpn. Soc. Intern. Med. 69: 47-54, 1980.

Neumann, F.; Meirom, R.; Rattner, D.; Trainin, Z.; Klopfer, U.; Nobel, T. A.: Animal model of human disease: alpha-1-antitrypsin deficiency. Am. J. Path. 84: 427-430, 1976.

Nukiwa, T.; Brantly, M.; Garver, R.; Paul, L.; Courtney, M.; LeCocq, J.-P.; Crystal, R. G.: Evaluation of 'at risk' alpha-1-antitrypsin genotype SZ with synthetic oligonucleotide gene probes. J. Clin. Invest. 77: 528-537, 1986.

Nukiwa, T.; Brantly, M. L.; Ogushi, F.; Fells, G. A.; Crystal, R. G.: Characterization of the gene and protein of the common alpha-1-antitrypsin normal M2 allele. Am. J. Hum. Genet. 43: 322-330, 1988.
Nukiwa, T.; Satoh, K.; Brantly, M. L.; Ogushi, F.; Fells, G. A.; Courtney, M.; Crystal, R. G.: Identification of a second mutation in the protein-coding sequence of the Z type alpha-1-antitrypsin gene. J. Biol. Chem. 261: 15989-15994, 1986.
Nukiwa, T.; Takahashi, H.; Brantly, M.; Courtney, M.; Crystal, R. G.: Alpha-1-antitrypsin null (Granite Falls), a nonexpressing alpha-1-antitrypsin gene associated with a frameshift to stop mutation in a coding exon. J. Biol. Chem. 262: 11999-12004, 1987.
Okayama, H.; Brantly, M.; Holmes, M.; Crystal, R. G.: Characterization of the molecular (sic) basis of the alpha-1-antitrypsin F allele. Am. J. Hum. Genet. 48: 1154-1158, 1991.
Owen, M. C.; Brennan, S. O.; Lewis, J. H.; Carrell, R. W.: Mutation of antitrypsin to antithrombin: alpha-1-antitrypsin Pittsburgh (358 met-to-arg), a fatal bleeding disorder. New Eng. J. Med. 309: 694-698, 1983.
Owen, M. C.; Carrell, R. W.: Alpha-1-antitrypsin: molecular abnormality of S variant. Brit. Med. J. 1: 130-131, 1976.
Owen, M. C.; Carrell, R. W.; Brennan, S. O.: The abnormality of the S variant of human alpha-1-antitrypsin. Biochim. Biophys. Acta 453: 257-261, 1976.
Pearson, S.; Tetri, P.; George, D. L.; Francke, U.: Alpha-1-antitrypsin (PI) expression in rat hepatoma-human somatic cell hybrids: evidence for PI locus on chromosome 14 and for regulatory locus on the X chromosome. (Abstract) Am. J. Hum. Genet. 33: 148A, 1981.
Perlino, E.; Cortese, R.; Ciliberto, G.: The human alpha-1-antitrypsin gene is transcribed from two different promoters in macrophages and hepatocytes. EMBO J. 6: 2767-2771, 1987.
Perrault, J. L.; Malo, J.-L.; Bake, B.; Renzi, G.; Grassino, A.: Alpha-1-antitrypsin deficiency: genetic, clinical and functional correlations in a three generation family. Respiration 37: 291-300, 1979.
Pierce, J. A.; Eisen, A. Z.; Dhingra, H. K.: Relationship of antitrypsin deficiency to the pathogenesis of emphysema. Trans. Assoc. Am. Phys. 82: 87-97, 1969.
Poller, W.; Faber, J.-P.; Weidinger, S.; Olek, K.: DNA polymorphisms associated with a new alpha-1-antitrypsin PI Q0 variant (PI Q0-Riedenburg). Hum. Genet. 86: 522-524, 1991.
Poller, W.; Meisen, C.; Olek, K.: DNA polymorphisms of the alpha-1-antitrypsin gene region in patients with chronic obstructive pulmonary disease. Europ. J. Clin. Invest. 20: 1-7, 1990.
Rodriguez-Cintron, W.; Guntupalli, K.; Fraire, A. E.: Bronchiectasis and homozygous (P1ZZ) alpha1-antitrypsin deficiency in a young man. Thorax 50: 424-425, 1995.
Rodriguez-Soriano, J.; Fidalgo, I.; Camarero, C.; Vallo, A.; Oliveros, R.: Juvenile cirrhosis and membranous glomerulonephritis in a child with alpha-1-antitrypsin deficiency PiSZ. Acta Paediat. Scand. 67: 793-796, 1978.
Rosenthal, P.; Liebman, W. M.; Thaler, M. M.: Alpha-1-antitrypsin deficiency and severe infantile liver disease. Am. J. Dis. Child. 133: 1195-1196, 1979.
Roychoudhury, A. K.; Nei, M.: Human Polymorphic Genes: World Distribution. New York: Oxford Univ. Press (pub.) 1988. Pp. 132-135.
Satoh, K.; Nukiwa, T.; Brantly, M.; Garver, R. I., Jr.; Hofker, M.; Courtney, M.; Crystal, R. G.: Emphysema associated with complete absence of alpha-1-antitrypsin of a stop codon in an alpha-1-antitrypsin-coding exon. Am. J. Hum. Genet. 42: 77-83, 1988.
Schapira, M.; Ramus, M.-A.; Jallat, S.; Carvallo, D.; Courtney, M.: Recombinant alpha-1-antitrypsin Pittsburgh (met-358 to arg) is a potent inhibitor of plasma kallikrein and activated factor XII fragment. J. Clin. Invest. 77: 635-637, 1986.
Schmitt, M. G., Jr.; Phillips, R. B.; Matzen, R. N.; Rodey, G.: Alpha-1-antitrypsin deficiency: a study of the relationship between the Pi system and genetic markers. Am. J. Hum. Genet. 27: 315-321, 1975.
Schroeder, W. T.; Miller, M. F.; Woo, S. L. C.; Saunders, G. F.: Chromosomal localization of the human alpha-antitrypsin gene (PI) to 14q31-32. Am. J. Hum. Genet. 37: 868-872, 1985.
Scott, C. F.; Carrell, R. W.; Glaser, C. B.; Kueppers, F.; Lewis, J. H.; Colman, R. W.: Alpha-1-antitrypsin-Pittsburgh: a potent inhibitor of human plasma factor XIa, kallikrein, and factor XII. J. Clin. Invest. 77: 631-634, 1986.
Sefton, L.; Kearney, P.; Kelsey, G.; Povey, S.; Wolfe, J.: Physical linkage of the genes PI and AACT. (Abstract) Cytogenet. Cell Genet. 51: 1076, 1989.
Seri, M.; Magi, B.; Cellesi, C.; Olia, P. M.; Renieri, A.; De Marchi, M.: Molecular characterization of the P and I variants of alpha-1-antitrypsin. Int. J. Clin. Lab. Res. 22: 119-121, 1992.
Sharp, H. L.; Bridges, R. A.; Krivit, W.; Freier, E. F.: Cirrhosis associated with alpha-1-antitrypsin deficiency: a previously unrecognized inherited disorder. J. Lab. Clin. Med. 73: 934-939, 1969.
Sifers, R. N.; Brashears-Macatee, S.; Kidd, V. J.; Muensch, H.; Woo, S. L. C.: A frameshift mutation results in a truncated alpha-1-antitrypsin that is retained within the rough endoplasmic reticulum. J. Biol. Chem. 263: 7330-7335, 1988.
Sigsgaard, T.; Brandslund, I.; Rasmussen, J. B.; Lund, E. D.; Varming, H.: Low normal alpha-1-antitrypsin serum concentrations and MZ-phenotype are associated with byssinosis and familial allergy in cotton mill workers. Pharmacogenetics 4: 135-141, 1994.
Sigsgaard, T.; Pedersen, O. F.; Juul, S.; Gravesen, S.: Respiratory disorders and atopy in cotton, wool and other textile mill workers in Denmark. Am. J. Ind. Med. 22: 163-184, 1992.
Starzl, T. E.; Porter, K. A.; Francavilla, A.; Iwatsuki, S.: Reversal of hepatic alpha-1-antitrypsin deposition after portacaval shunt. Lancet II: 424-426, 1983.
Stevens, P. M.; Hnilica, V.; Johnson, P. C.; Bell, R. L.: Pathophysiology of hereditary emphysema. Ann. Intern. Med. 74: 672-680, 1971.
Stockley, R. A.: Alpha-1-antitrypsin phenotypes in cor pulmonale due to chronic obstructive airways disease. Quart. J. Med. 48: 419-428, 1979.
Takahashi, H.; Crystal, R. G.: Alpha-1-antitrypsin Null(isola di procida): an alpha-1-antitrypsin deficiency allele caused by deletion of all alpha-1-antitrypsin coding exons. Am. J. Hum. Genet. 47: 403-413, 1990.
Takahashi, H.; Nukiwa, T.; Satoh, K.; Ogushi, F.; Brantly, M.; Fells, G.; Stier, L.; Courtney, M.; Crystal, R. G.: Characterization of the gene and protein of the alpha-1-antitrypsin 'deficiency' allele M(procida). J. Biol. Chem. 263: 15528-15534, 1988.
Talamo, R. C.; Allen, J. D.; Kahan, M. G.; Austen, K. F.: Hereditary alpha-1-antitrypsin deficiency. New Eng. J. Med. 278: 345-351, 1968.
Talamo, R. C.; Feingold, M.: Infantile cirrhosis with hereditary alpha-1-antitrypsin deficiency. Am. J. Dis. Child. 125: 845-849, 1973.
Tarkoff, M. P.; Kueppers, F.; Miller, W. F.: Pulmonary emphysema and alpha-1-antitrypsin deficiency. Am. J. Med. 45: 220-228, 1968.
Townley, R. G.; Ryning, F.; Lynch, H. T.; Brody, A. W.: Obstructive lung disease in hereditary alpha-1-antitrypsin deficiency. J.A.M.A. 214: 325-331, 1970.
Turleau, C.; de Grouchy, J.; Chavin-Colin, F.; Dore, F.; Seger, J.; Dautzenberg, M.; Arthuis, M.; Jeanson, C.: Two patients with interstitial del(14q), one with features of Holt-Oram syndrome: exclusion mapping of PI (alpha-1-antitrypsin). Ann. Genet. 27: 237-240, 1984.
Udall, J. N.; Bloch, K. J.; Walker, W. A.: Transport of proteases across neonatal intestine and development of liver disease in infants with alpha-1-antitrypsin deficiency. Lancet I: 1441-1443, 1982.
Vidaud, D.; Emmerich, J.; Alhenc-Gelas, M.; Yvart, J.; Fiessinger, J. N.; Aiach, M.: Met358-to-arg mutation of alpha-1-antitrypsin associated with protein C deficiency in a patient with mild bleeding tendency. J. Clin. Invest. 89: 1537-1543, 1992.
Weber, W.; Weidinger, S.: PI Scologne: a new variant in the alpha-1-antitrypsin system. Hum. Genet. 80: 102, 1988.
Weidinger, S.; Jahn, W.; Cujnik, F.; Schwarzfischer, F.: Alpha-1-antitrypsin: evidence for a fifth PI M subtype and a new deficiency allele PI*Z(Augsburg). Hum. Genet. 71: 27-29, 1985.
Weitkamp, L. R.; Cox, D.; Guttormsen, S.; Johnston, E.; Hempfling, S.: Allelic specific heterogeneity in the Pi-Gm linkage group. Cytogenet. Cell Genet. 22: 647-650, 1978.
Weitz, J. I.; Silverman, E. K.; Thong, B.; Campbell, E. J.: Plasma levels of elastase-specific fibrinopeptides correlate with proteinase inhibitor phenotype: evidence for increased elastase activity in subjects with homozygous and heterozygous deficiency of alpha-1-proteinase inhibitor. J. Clin. Invest. 89: 766-773, 1992.
Welch, S. G.; McGregor, I. A.; Williams, K.: Alpha-1-antitrypsin (Pi) phenotypes in a village population from the Gambia, West Africa. Hum. Genet. 53: 233-235, 1980.
Wewers, M. D.; Casolaro, A.; Sellers, S.; Swayze, S. C.; McPhaul, K. M.; Wittes, J. T.; Crystal, R. G.: Replacement therapy for alpha-1-antitrypsin deficiency associated with emphysema. New Eng. J. Med. 316: 1055-1062, 1987.
Whitehouse, D. B.; Abbott, C. M.; Lovegrove, J. U.; McIntosh, I.; McMahon, C. J.; Mieli-Vergani, G.; Mowat, A. P.; Hopkinson, D. A.: Genetic studies on a new deficiency gene (PI*Z-Tun) at the PI locus. J. Med. Genet. 26: 744-749, 1989.
Wiebicke, W.; Niggemann, B.; Fischer, A.: Pulmonary function in children with homozygous alpha-1-protease inhibitor deficiency. Europ. J. Pediat. 155: 603-607, 1996.
Wilkie, A. O. M.: The molecular basis of genetic dominance. J. Med. Genet. 31: 89-98, 1994.
Wu, Y.; Whitman, I.; Molmenti, E.; Moore, K.; Hippenmeyer, P.; Perlmutter, D. H.: A lag in intracellular degradation of mutant alpha-1-antitrypsin correlates with the liver disease phenotype in homozygous PiZZ alpha-1-antitrypsin deficiency. Proc. Nat. Acad. Sci. 91: 9014-9018, 1994.
Yamamoto, Y.; Sawa, R.; Okamoto, N.; Matsui, A.; Yanagisawa, M.; Ikemoto, S.: Deletion 14q(q24.3 to q32.1) syndrome: significance of peculiar facial appearance in its diagnosis, and deletion mapping of Pi (alpha-1-antitrypsin). Hum. Genet. 74: 190-192, 1986.
Yoshida, A.; Chillar, R.; Taylor, J. C.: An alpha-1-antitrypsin variant, PiB Alhambra (lys-to-asp, glu-to-asp), with rapid anodal electrophoretic mobility. Am. J. Hum. Genet. 31: 555-563, 1979.
Yoshida, A.; Ewing, C.; Wessels, M.; Lieberman, J.; Gaidulis, L.: Molecular abnormality of Pi S variant of human alpha-1-antitrypsin. Am. J. Hum. Genet. 29: 233-239, 1977.
Yoshida, A.; Lieberman, J.; Gaidulis, L.; Ewing, C.: Molecular abnormality of human alpha-1-antitrypsin variant (Pi-SZ) associated with plasma activity deficiency. Proc. Nat. Acad. Sci. 73: 1324-1328, 1976.
Yoshida, A.; Taylor, J. C.; Van den Brock, W. G. M.: Structural difference between the normal PiM(1) and the common PiM(2) variant of human alpha-1-antitrypsin. Am. J. Hum. Genet. 31: 564-568, 1979.
Yuasa, I.; Sugimoto, Y.; Ichinose, M.; Matsumoto, Y.; Fukumaki, Y.; Sasaki, T.; Okada, K.: PI*S(iiyama), a deficiency gene of alpha-1-antitrypsin: evidence for the occurrence in western Japan. Jpn. J. Hum. Genet. 38: 185-191, 1993.

PORPHYRIA, ACUTE INTERMITTENT (*176000)

JJM NOTE

It seems reasonable to expect some excess risk of liver cancer, and it might be surprising that so few cases of the association have been reported (Kaczynski J et al.: Erythrocyte porphobilinogen deaminase activity and primary liver-cancer. J. Int. Med. 237:309–13, 1995).

Description

Acute intermittent porphyria is inherited as an autosomal dominant disorder and is characterized by recurrent attacks of abdominal pain, gastrointestinal dysfunction, neurologic disturbances, and excessive amounts of aminolevulinic acid and porphobilinogen in the urine. AIP results from an error in pyrrole metabolism due to deficiency of porphobilinogen deaminase (EC 4.3.1.8).

Clinical Features

Acute intermittent porphyria (AIP) is characterized by acute episodes of a variety of neuropathic symptoms; between episodes, the patient is healthy. Abdominal pain is the most common symptom, sometimes with constipation and urinary retention; paraesthesias and paralysis also occur, and death may result from respiratory paralysis (Goldberg, 1959; Stein and Tschudy, 1970; Becker and Kramer, 1977). Many other phenomena, including seizures, psychotic episodes, and hypertension occur in acute attacks.

Acute attacks rarely occur before puberty; they may be precipitated by porphyrinogenic drugs such as barbiturates and sulfonamides (for list, see Tschudy et al., 1975), some of which are known to induce the earlier rate-controlling step in heme synthesis, delta-aminolevulinic acid (ALA) synthesis. Other known precipitants are alcohol, infection, starvation, and hormonal changes; attacks are more common in women.

From a survey of AIP cases in the west of Scotland, Yeung Laiwah et al. (1983) observed an association with early-onset chronic renal failure. Porphyria-induced hypertension was considered the most likely causal factor, but enhanced susceptibility to analgesic nephropathy and nephrotoxic effects of porphyrins and their precursors were mentioned as possibilities.

Biochemical Features

Reduction in the level of activity of the heme-synthesizing enzyme uroporphyrinogen synthase (URO-S or UPS, EC 4.3.1.8) is the basic biochemical defect. The preferred name for the enzyme is porphobilinogen (PBG) deaminase (Desnick, 1981) because it catalyzes the formation of the linear tetrapyrrole hydroxymethyl bilane, which is then nonenzymatically cyclized to form uroporphyrinogen I. The enzyme is also referred to as hydroxymethylbilane synthase. Enzyme activity reduced to approximately half the normal level was demonstrated first in liver (Strand et al., 1970) and subsequently in cultured fibroblasts and red blood cells (Meyer et al., 1972; Strand et al., 1972; Sassa et al., 1974; Kreimer-Birnbaum, 1975).

In family studies, most individuals can be characterized as having either clearly normal or clearly reduced (approximately 50%) levels of heme-synthesizing enzyme UPS activity, but intermediate values are sometimes found (Sassa et al., 1974; Astrup, 1978; Kreimer-Birnbaum et al., 1980). Many latent AIP subjects never have acute attacks, but some intermittently excrete excess porphyrin precursors in urine without having symptoms (Waldenstrom, 1956).

The role of abnormalities described in steroid metabolism in AIP patients is not completely clear (Anderson et al., 1979). The essential biochemical finding in acute attacks is increased excretion of the precursors ALA and PBG in the urine; the latter is the basis for the Watson-Schwartz test (Watson and Bossenmaier, 1964). Porphyrins may be formed in the urine from the precursors.

By the method of isoelectric focusing, Meisler and Carter (1980) identified structural variants of PBG-deaminase. Louie et al. (1992) defined the 3-domain structure of PBGD by x-ray analysis. Two of the domains structurally resembled the transferrins. The x-ray structure and results from site-directed mutagenesis provided evidence for a single catalytic site.

The findings of Verma et al. (1987) indicating that porphyrins are endogenous ligands for the mitochondrial benzodiazepine receptor may have relevance as far as the neuropsychiatric aspects of porphyria are concerned. The anxiolytic effects of benzodiazepines are mediated by a 'central' benzodiazepine receptor, located primarily in the brain. There is also a 'peripheral-type' receptor which is found in numerous organs and binds benzodiazepines with high affinity. The peripheral-type receptor is selectively associated with the outer membrane of mitochondria; hence, it may be properly referred to as the mitochondrial benzodiazepine receptor. Its drug specificity differs from that of the central-type receptor. (See 125950.)

Mustajoki (1981) suggested that in one variant of acute intermittent porphyria the enzyme defect is not expressed in red cells. He studied a large kindred in which 10 members had AIP with normal erythrocyte PBG-deaminase activity. Forty-nine other Finnish patients with AIP who were unrelated to this kindred had the usual low activity of PBG-deaminase. Generally, however, reduced activity of this enzyme in red cells is a consistent finding during or between acute attacks, and characterizes latent AIP that is inherited as an autosomal dominant trait (Sassa et al., 1974; Kreimer-Birnbaum, 1975). Most enzymopathies are recessively inherited because few enzymes are so rate-limiting as to cause a serious reduction in the rate of a metabolic pathway when the enzyme has 40 to 60% normal activity.

Anderson et al. (1981) demonstrated heterogeneity in AIP. All 7 heterozygotes from 1 family showed crossreactive immunologic material (abbreviated CRM or CRIM) about 1.6 times that detected in normal red cell lysates. In the other 21 families studied, the amount of CRM showed no difference from that in normal persons. Another difference was in the pattern of the 5 stable enzyme-substrate intermediates (A, B, C, D, E) of PBG-deaminase separated by anion-exchange chromatography of erythrocyte lysates. The normal pattern shows decreasing amounts of activity A to E, with A and B intermediates representing more than 75% of total recovered activity. In most patients, the elution profile was similar to the normal with each intermediate reduced about 50%. A second profile in which the C intermediate had disproportionately higher activity than the A or B intermediates was observed in some heterozygotes.

In plasma, during acute attacks of AIP, heme precursors, delta-aminolevulinic acid, and porphobilinogen are increased. Puy et al. (1996) noted that the pathophysiology of this disorder is poorly understood. There is evidence in a porphyric rat model of increased plasma concentration and brain uptake of tryptophan and of increased synthesis of serotonin in the nervous system. The increased concentration of tryptophan and serotonin can be partly due to the hepatic heme deficiency decreasing the activity of the liver cytosolic enzyme heme-dependent tryptophan pyrolase. Women with AIP demonstrate a rise in serotonin and plasma tryptophan during the attacks, whereas both daytime and nighttime melatonin concentrations are dramatically decreased, although melatonin is produced from tryptophan. Injection of heme lowered heme precursors, tryptophan, and serotonin to normal levels but did not increase melatonin. Puy et al. (1996) used the rat experimental model and reported results suggesting that delta-aminolevulinic acid is responsible for decreased production of melatonin by the pineal gland.

Inheritance

AIP and several other genetic porphyrias (121300, 176100, 176200) are unusual among enzyme deficiency states in being manifested in the heterozygous, single gene dose (Meyer and Schmid, 1978). Doss (1989) described 4 unrelated individuals with coexistent deficiency of porphobilinogen deaminase and uroporphyrinogen decarboxylase (176100). In 2 the diagnosis was considered to be acute intermittent porphyria, and in the other 2 porphyria cutanea tarda. Family studies showed that in 1 case the dually affected father transmitted both deficiencies to 1 son and only 1 deficiency to a second son. This is consistent with the nonlinkage and nonallelism of the 2 genes, which is indeed the case: PBGD is coded by a gene on chromosome 11 and UROD by a gene on chromosome 1. The gene encoding PBGD has also been symbolized HMBS (hydroxymethylbilane synthase).

Few homozygotes for porphobilinogen deaminase deficiency have been reported. Homozygosity for null alleles in the heme biosynthesis pathway might be lethal because of the essentiality of heme not only in hemoglobin but in cytochrome P450 enzymes, catalase, etc. In those conditions that do represent the homozygous state, such as congenital erythropoietic porphyria (263700) and acute hepatic porphyria (125270), and in those dominantly inherited porphyrias for which the homozygous state has been observed also (176100, 121300), the mutations may be 'leaky.'

Beukeveld et al. (1990) described a presumed homozygous case of AIP. The clinical picture was characterized by porencephaly and severe developmental retardation. The child consistently excreted excessive amounts of delta-aminolevulinic acid, porphobilinogen, and uroporphyrin in her urine from early childhood and died at the age of 8 years. Her mother suffered from AIP; her father, although he had never had attacks, was shown by studies of blood and urine to be affected also. In fact, the presumed homozygote was shown to represent a compound heterozygote. Using allele-specific oligonucleotides, Picat et al. (1990) demonstrated that the father had the arg167-to-gln mutation (176000.0005) and the mother had the arg173-to-gln mutation (176000.0006).

Mapping

By study of mouse-human hybrid clones, 51,50:Meisler et al. (1980, 1981) showed that PBG-deaminase is determined by a gene on chromosome 11; Wang et al. (1981) assigned the locus to the long arm in the segment 11q23-qter. In 3 children with trisomy of 11qter, de Verneuil et al. (1982) studied expression of uroporphyrinogen I synthase. Dosage effect supported assignment to the region 11q23.2-qter. By in situ hybridization and by gene dosage studies in patients with monosomy or trisomy of the terminal portion of 11q, Namba et al. (1991) refined the assignment of PBGD to 11q24.1-q24.2. Llewellyn et al. (1987) found that a common 2-allele MspI RFLP of the PBGD gene was tightly linked (maximum lod = 3.14 with no recombinants) to the locus for AIP. The frequency of the polymorphism in the AIP patients did not differ significantly from that in controls. In 30 of 33 unrelated patients, no crossreacting immunologic material was produced by the mutant allele. The MspI polymorphism enabled each PBGD allele to be distinguished in subjects heterozygous for the polymorphism; thus, a major gene deletion was excluded as the cause of the CRM-negative mutation in all 18 families that contained an affected CRM-negative individual heterozygous for the polymorphism.

Tunnacliffe and McGuire (1990) constructed a long-range restriction map extending over 1.8 Mb of 11q23.3 using pulsed field gel electrophoresis and concluded that PBGD is situated in the following relation to 5 other genes: cen—CD3E—CD3D—CD3G—PBGD—CBL2—THY1—qter. They determined that the CD3G (186740) gene and PBGD are separated by 750 kb.

Molecular Genetics

Mustajoki and Desnick (1985) determined the nature of the mutation in 68 patients from 33 unrelated families in Finland. Four different porphobilinogen mutant types were found. Most (about 80%) were CRM-negative; these fell into 2 types: those with half-normal PBGD levels and those with normal PBGD levels in red cells of affected persons. The CRM-positive group was also divisible into 2 types: those with CRM:enzyme activity ratio of 1.6 and those with a ratio of 5.6 in affected persons. Mustajoki and Desnick (1985) suggested that the CRM-positive patients of the second type have milder disease. Goldberg (1985) wrote that this heterogeneity may be a factor 'among others, such as drugs, diet, and endogenous hormones, which determine whether the latent state in one patient may continue without incident or, in another, may be shattered by a painful and crippling attack.'

Major gene deletions are unlikely to be present in more than a small proportion of the commonest type of AIP, the CRM-negative form. Porphobilinogen deaminase is encoded by 2 distinct mRNA species expressed in a tissue-specific manner. The 2 mRNAs are transcribed from 2 promoters and differ only in their first exon. By PCR, Lee et al. (1990) amplified polymorphic sites in the PBGD gene that could be used for linkage analysis in AIP families. Scobie et al. (1990) identified three 2-allele RFLPs in the PBGD gene, each with allele frequencies close to 0.50. Marked linkage disequilibrium limited the number of observed haplotypes to 4, of which 1 was uncommon. In a study of 47 unrelated patients with AIP, no association was detected between any haplotype and AIP. In 3 Finnish families, each with a different subtype of AIP (CRIM-negative with low red cell enzyme activity; CRIM-positive with low enzyme activity; CRIM-negative with normal enzyme activity), Kauppinen et al. (1990) found evidence of tight linkage to PBGD RFLPs. Among 62 family members tested, 30 had the disease-associated haplotype; in 5 of them, conventional tests for AIP were normal, and in 1, uncertain. RFLP analysis can therefore be used to detect gene carriers and to help in the diagnosis of persons with uncertain results in other tests.

Grandchamp et al. (1990) used PCR amplification of patients' mRNA and sequencing of the PCR product to identify 7 point mutations causing AIP in 11 different families.

Heterogeneity

In molecular studies of 165 AIP heterozygotes from 92 unrelated families, Desnick et al. (1985) demonstrated different allelic mutations in the structural gene for PBG-deaminase and documented molecular genetic heterogeneity in AIP. Lee et al. (1991) concluded that there is heterogeneity in AIP in northern Sweden. Three different haplotypes were identified. In 10 of 28 families originating from northern Sweden, the disease segregated with the haplotype designated as 2/1/1/2. All 28 families were screened for 3 known point mutations; only 1 family was found to carry 1 of the 3 mutations.

Astrin and Desnick (1994) reviewed the 26 mutations in the hydroxymethylbilane synthase gene identified to that time. Most caused the classic form of AIP. In 28 Finnish families representing 72% of all AIP families in the Finnish population of 5 million, Kauppinen et al. (1995) found 19 separate mutations in HMBS: 13 novel mutations, including 1 de novo event, and 6 previously characterized mutations.

Diagnosis

Mustajoki and Desnick (1985) provided a useful illustration of the site (in the heme-synthesis pathway) of the mutation in each of 6 forms of porphyria. The defect is expressed in cultured fibroblasts and amniotic cells, so that prenatal diagnosis is possible (Sassa et al., 1975). The enzyme can be induced and the defect demonstrated in mitogen-stimulated lymphocytes (Sassa et al., 1978).

In suitable families, the MspI polymorphism provides a more certain way of identifying carriers of the AIP gene than current enzymatic methods.

Clinical Management

Most acute attacks, if correctly recognized, settle with supportive treatment; dextrose infusion and high carbohydrate intake may be helpful (Stein and Tschudy, 1970). Successful treatment by infusion of hematin, which is a specific feedback inhibitor of heme synthesis, has been repeatedly reported (McColl et al., 1978; Lamon et al., 1979), but hematin is neither readily available nor very soluble and its use may carry a risk of renal damage (Dhar et al., 1978).

Goetsch and Bissell (1986) described a 22-year-old woman who had had more than 15 acute attacks requiring hospitalization. Hematin was the mainstay of the patient's therapy, eliciting a well-defined clinical and biochemical response after more than 200 infusions. On one occasion, however, the patient did not respond to a batch of hematin that, in retrospect, was found to have deteriorated. Goetsch and Bissell (1986) quantified the instability of hematin and showed that the decay product (or products) is ineffective in regulating porphyrin production. The decayed material, furthermore, was found to have anticoagulant effects, thus explaining one of the complications of hematin therapy.

An experience reported by Anderson et al. (1984) suggests that in women with premenstrual exacerbation of AIP, a long-acting agonist of luteinizing hormone-releasing hormone may be an effective preventive measure. Srugo et al. (1987) described acute intermittent porphyria as the cause of 'surgical abdomen' in a 15-year-old boy. Because hypertension, tachycardia, and paralytic ileus were present, suggesting sympathetic overactivity, the beta-adrenergic blocking agent propranolol was administered in high doses with apparent dramatic improvement.

Population Genetics

High prevalence (1 in 1000) is known in northern Sweden where Waldenstrom's classic observations were made (Waldenstrom, 1956). AIP occurs with very low prevalence, perhaps 1 in 50,000, probably in all ethnic groups (Tschudy et al., 1975), including blacks (Kreimer-Birnbaum et al., 1980), but figures for prevalence based on manifest AIP, i.e., acute attacks, greatly underestimate the number of persons with latent AIP.

Puy et al. (1997) performed molecular analysis of the PBGD gene by denaturing gradient gel electrophoresis followed by direct sequencing in 405 subjects from 121 unrelated French-Caucasian AIP families that had not been screened previously at the DNA level. PBGD mutations were identified in 109 families, but only 78 were of different type, and each of these had a prevalence rate of less than 5%. Among these mutations, 33 had not been published previously. Sixty percent of the 78 mutations were located in 3 exons (exons 10, 12, and 14); 44% were missense, 18% were splice defect, 19% were frameshift, and 16% were nonsense. In addition, 2 de novo mutational events were characterized. Puy et al. (1997) found that the standard PBGD enzymatic screening method for gene-carrier detection has 95% concordancy with the molecular-based diagnosis.

Animal Model

During study of the pathogenesis of the neurologic symptoms of AIP, Lindberg et al. (1996) generated Pbgd-deficient mice by gene targeting. These mice exhibited typical biochemical characteristics of human AIP, notably, decreased hepatic Pbgd activity, increased delta-aminolevulinic acid synthase activity, and massively increased urinary excretion of the heme precursor delta-aminolevulinic acid after treatment with drugs such as phenobarbital. Behavioral tests revealed decreased motor function and histopathologic findings, including axonal neuropathy and neurologic muscle atrophy.

History

Loftus and Arnold (1991) suggested that Vincent van Gogh had suffered from attacks of acute intermittent porphyria, exacerbated by malnutrition and absinthe abuse. They suggested that this would best explain the abruptness of the onset and recovery from attacks. (It has been suggested that George III had porphyria variegata (176200).)

See also Bosch et al. (1977); Grandchamp et al. (1989); Grandchamp et al. (1989); Lamon et al. (1979); Lee et al. (1990); McColl et al. (1982); Morris et al. (1981); Raich et al. (1986); Tishler et al. (1985); Waldenstrom and Haeger-Aronsen (1963)

Anderson, K. E.; Bradlow, H. L.; Sassa, S.; Kappas, A.: Studies in porphyria. VIII. Relationship of the 5-alpha-reductive metabolism of steroid hormone to clinical expression of the genetic defect in acute intermittent porphyria. Am. J. Med. 66: 644-650, 1979.

Anderson, K. E.; Spitz, I. M.; Sassa, S.; Bardin, C. W.; Kappas, A.: Prevention of cyclical attacks of acute intermittent porphyria with a long-acting agonist of luteinizing hormone-releasing hormone. New Eng. J. Med. 311: 643-645, 1984.

Anderson, P. M.; Reddy, R. M.; Anderson, K. E.; Desnick, R. J.: Characterization of the porphobilinogen deaminase deficiency in acute intermittent porphyria: immunologic evidence for heterogeneity of the genetic defect. J. Clin. Invest. 68: 1-12, 1981.

Astrin, K. H.; Desnick, R. J.: Molecular basis of acute intermittent porphyria: mutations and polymorphisms in the human hydroxymethylbilane synthase gene. Hum. Mutat. 4: 243-252, 1994.

Astrup, E. G.: Family studies on the activity of uroporphyrinogen I synthase in diagnosis of acute intermittent porphyria. Clin. Sci. 54: 251-256, 1978.

Becker, D. M.; Kramer, S.: The neurological manifestations of porphyria: a review. Medicine 56: 411-423, 1977.
Beukeveld, G. J. J.; Wolthers, B. G.; Nordmann, Y.; Deybach, J. C.; Grandchamp, B.; Wadman, S. K.: A retrospective study of a patient with homozygous form of acute intermittent porphyria. J. Inherit. Metab. Dis. 13: 673-683, 1990.
Bosch, E. P.; Pierach, C. A.; Bossenmaier, I.; Cardinal, R.; Thornson, M.: Effect of hematin in porphyric neuropathy. Neurology 27: 1053-1056, 1977.
Daimon, M.; Yamatani, K.; Igarashi, M.; Fukase, N.; Ogawa, A.; Tominaga, M.; Sasaki, H.: Acute intermittent porphyria caused by a G to C mutation in exon 12 of the porphobilinogen deaminase gene that results in exon skipping. Hum. Genet. 92: 549-553, 1993.
Delfau, M. H.; Picat, C.; De Rooij, F.; Voortman, G.; Deybach, J. C.; Nordmann, Y.; Grandchamp, B.: Molecular heterogeneity of acute intermittent porphyria: identification of four additional mutations resulting in the CRIM-negative subtype of the disease. Am. J. Hum. Genet. 49: 421-428, 1991.
Delfau, M. H.; Picat, C.; de Rooij, F. W. M.; Hamer, K.; Bogard, M.; Wilson, J. H. P.; Deybach, J. C.; Nordmann, Y.; Grandchamp, B.: Two different point G to A mutations in exon 10 of the porphobilinogen deaminase gene are responsible for acute intermittent porphyria. J. Clin. Invest. 86: 1511-1516, 1990.
Desnick, R. J.: Personal Communication. New York, N. Y. 5/5/1981.
Desnick, R. J.; Ostasiewicz, L. T.; Tishler, P. A.; Mustajoki, P.: Acute intermittent porphyria: characterization of a novel mutation in the structural gene for porphobilinogen deaminase. Demonstration of noncatalytic enzyme intermediates stabilized by bound substrate. J. Clin. Invest. 76: 865-874, 1985.
de Verneuil, H.; Phung, N.; Nordmann, Y.; Allard, D.; Leprince, F.; Jerome, H.; Aurias, A.; Rethore, M. O.: Assignment of human uroporphyrinogen I synthase locus to region 11qter by gene dosage effect. Hum. Genet. 60: 212-213, 1982.
Dhar, G.; Bossenmaier, I.; Cardinal, R.; Petryka, Z. J.; Watson, C. J.: Transitory renal failure following rapid administration of a relatively large amount of hematin in a patient with acute intermittent porphyria in clinical remission. Acta Med. Scand. 203: 437-443, 1978.
Doss, M. O.: Dual porphyria in double heterozygotes with porphobilinogen deaminase and uroporphyrinogen decarboxylase deficiencies. Clin. Genet. 35: 146-151, 1989.
Goetsch, C. A.; Bissell, D. M.: Instability of hematin used in the treatment of acute hepatic porphyria. New Eng. J. Med. 315: 235-238, 1986.
Goldberg, A.: Acute intermittent porphyria. Quart. J. Med. 28: 183-209, 1959.
Goldberg, A.: Molecular genetics of acute intermittent porphyria. (Editorial) Brit. Med. J. 291: 499-500, 1985.
Grandchamp, B.; Delfau, M. H.; Picat, C.; de Rooij, F. W. M.; Nordmann, Y.: Heterogeneity of the molecular defects in acute intermittent porphyria. (Abstract) Am. J. Hum. Genet. 47 (suppl.): A156 only, 1990.
Grandchamp, B.; Picat, C.; de Rooij, F.; Beaumont, C.; Wilson, P.; Deybach, J. C.; Nordmann, Y.: A point mutation G-to-A in exon 12 of the porphobilinogen deaminase gene results in exon skipping and is responsible for acute intermittent porphyria. Nucleic Acids Res. 17: 6637-6649, 1989.
Grandchamp, B.; Picat, C.; Kauppinen, R.; Mignotte, V.; Peltonen, L.; Mustajoki, P.; Romeo, P. H.; Goossens, M.; Nordmann, Y.: Molecular analysis of acute intermittent porphyria in a Finnish family with normal erythrocyte porphobilinogen deaminase. Europ. J. Clin. Invest. 19: 415-418, 1989.
Grandchamp, B.; Picat, C.; Mignotte, V.; Wilson, J. H. P.; Te Velde, K.; Sandkuyl, L.; Romeo, P. H.; Goossens, M.; Nordmann, Y.: Tissue-specific splicing mutation in acute intermittent porphyria. Proc. Nat. Acad. Sci. 86: 661-664, 1989.
Gu, X.-F.; de Rooij, F.; Lee, J. S.; TeVelde, K.; Deybach, J.-C.; Nordmann, Y.; Grandchamp, B.: High prevalence of a point mutation in the porphobilinogen deaminase gene in Dutch patients with acute intermittent porphyria. Hum. Genet. 91: 128-130, 1993.
Gu, X.-F.; de Rooij, F.; Voortman, G.; Te Velde, K.; Nordmann, Y.; Grandchamp, B.: High frequency of mutations in exon 10 of the porphobilinogen deaminase gene in patients with a CRIM-positive subtype of acute intermittent porphyria. Am. J. Hum. Genet. 51: 660-665, 1992.
Gu, X.-F.; de Rooij, F.; Voortman, G.; Velde, K. T.; Deybach, J. C.; Nordmann, Y.; Grandchamp, B.: Detection of eleven mutations causing acute intermittent porphyria using denaturing gradient gel electrophoresis. Hum. Genet. 93: 47-52, 1994.
Kauppinen, R.; Mustajoki, S.; Pihlaja, H.; Peltonen, L.; Mustajoki, P.: Acute intermittent porphyria in Finland: 19 mutations in the porphobilinogen deaminase gene. Hum. Molec. Genet. 4: 215-222, 1995.
Kauppinen, R.; Peltonen, L.; Palotie, A.; Mustajoki, P.: RFLP analysis of three different types of acute intermittent porphyria. Hum. Genet. 85: 160-164, 1990.
Kauppinen, R.; Peltonen, L.; Pihlaja, H.; Mustajoki, P.: CRIM-positive mutations of acute intermittent porphyria in Finland. Hum. Mutat. 1: 392-396, 1992.
Kreimer-Birnbaum, M.: Uroporphyrinogen synthase in human blood: developmental studies. Life Sci. 17: 1473-1478, 1975.
Kreimer-Birnbaum, M.; Bannerman, R. M.; El Khatib, M.; Franco-Saenz, R.: Afro-Americans and acute intermittent porphyria. Int. J. Biochem. 12: 795-799, 1980.
Lamon, J. M.; Frykholm, B. C.; Hess, R. A.; Tschudy, D. P.: Hematin therapy for acute porphyria. Medicine 58: 252-269, 1979.
Lamon, J. M.; Frykholm, B. C.; Tschudy, D. P.: Family evaluations in acute intermittent porphyria using red cell uroporphyrinogen I synthetase. J. Med. Genet. 16: 134-139, 1979.
Lander, M.; Pitt, A. R.; Alefounder, P. R.; Bardy, D.; Abell, C.; Battersby, A. R.: Studies on the mechanism of hydroxymethyl bilane synthase concerning the role of arginine residues in substrate binding. Biochem. J. 275: 447-452, 1991.
Lee, J.-S.; Anvret, M.: Identification of the most common mutation within the porphobilinogen deaminase gene in Swedish patients with acute intermittent porphyria. Proc. Nat. Acad. Sci. 88: 10912-10915, 1991.
Lee, J.-S.; Grandchamp, B.; Anvret, M.: A point mutation of the human porphobilinogen deaminase gene in a Swedish family with acute intermittent porphyria. (Abstract) Am. J. Hum. Genet. 47 (suppl.): A162, 1990.
Lee, J.-S.; Lindsten, J.; Anvret, M.: Haplotyping of the human porphobilinogen deaminase gene in acute intermittent porphyria by polymerase chain reaction. Hum. Genet. 84: 241-243, 1990.
Lee, J.-S.; Lundin, G.; Lannfelt, L.; Forsell, L.; Picat, C.; Grandchamp, B.; Anvret, M.: Genetic heterogeneity of the porphobilinogen deaminase gene in Swedish families with acute intermittent porphyria. Hum. Genet. 87: 484-488, 1991.
Lindberg, R. L. P.; Porcher, C.; Grandchamp, B.; Ledermann, B.; Burki, K.; Brandner, S.; Aguzzi, A.; Meyer, U. A.: Porphobilinogen deaminase deficiency in mice causes a neuropathy resembling that of human hepatic porphyria. Nature Genet. 12: 195-199, 1996.
Llewellyn, D. H.; Elder, G. H.; Kalsheker, N. A.; Marsh, O. W. M.; Harrison, P. R.; Grandchamp, B.; Picat, C.; Nordmann, Y.; Romeo, P. H.; Goossens, M.: DNA polymorphism of human porphobilinogen deaminase gene in acute intermittent porphyria. Lancet II: 706-708, 1987.
Llewellyn, D. H.; Smyth, S. J.; Elder, G. H.; Hutchesson, A. C.; Rattenbury, J. M.; Smith, M. F.: Homozygous acute intermittent porphyria: compound heterozygosity for adjacent base transitions in the same codon of the porphobilinogen deaminase gene. Hum. Genet. 89: 97-98, 1992.
Llewellyn, D. H.; Whatley, S.; Elder, G. H.: Acute intermittent porphyria caused by an arginine to histidine substitution (R26H) in the cofactor-binding cleft of porphobilinogen deaminase. Hum. Molec. Genet. 2: 1315-1316, 1993.
Loftus, L. S.; Arnold, W. N.: Vincent van Gogh's illness: acute intermittent porphyria? Brit. Med. J. 303: 1589-1591, 1991.
Louie, G. V.; Brownlie, P. D.; Lambert, R.; Cooper, J. B.; Blundell, T. L.; Wood, S. P.; Warren, M. J.; Woodcock, S. C.; Jordan, P. M.: Structure of porphobilinogen deaminase reveals a flexible multidomain polymerase with a single catalytic site. Nature 359: 33-39, 1992.
Lundin, G.; Lee, J.-S.; Thunell, S.; Anvret, M.: Genetic investigation of the porphobilinogen deaminase gene in Swedish acute intermittent porphyria families. Hum. Genet. 100: 63-66, 1997.
Lundin, G.; Wedell, A.; Thunell, S.; Anvret, M.: Two new mutations in the porphobilinogen deaminase gene and a screening method using PCR amplification of specific alleles. Hum. Genet. 93: 59-62, 1994.
McColl, K. E. L.; Moore, M. R.; Thompson, G. G.; Goldberg, A.: Screening for latent acute intermittent porphyria: the value of measuring both leucocyte delta-aminolaevulinic acid synthase and erythrocyte uroporphyrinogen-1-synthase activities. J. Med. Genet. 19: 271-276, 1982.
McColl, K. E. L.; Thompson, G. T.; Moore, M. R.; Goldberg, A.: Haematin therapy and leucocyte gamma-aminolaevulinic-acid-synthase activity in prolonged attack of acute porphyria. Lancet I: 133-134, 1978.
Meisler, M. H.; Carter, M. L. C.: Rare structural variants of human and murine uroporphyrinogen I synthase. Proc. Nat. Acad. Sci. 77: 2848-2852, 1980.
Meisler, M. H.; Wanner, L.; Kao, F. T.; Jones, C.: Localization of the uroporphyrinogen I synthase locus to human chromosome region 11q13-qter and interconversion of enzyme isomers. Cytogenet. Cell Genet. 31: 124-128, 1981.
Meisler, M. H.; Wanner, L. A.; Eddy, R. E.; Shows, T. H.: Uroporphyrinogen I synthase: chromosomal linkage and isozyme expression in human-mouse hybrid cells. (Abstract) Am. J. Hum. Genet. 32: 47A only, 1980.
Meyer, U. A.; Schmid, R.: The porphyrias.In: Stanbury, J. B.; Wyngaarden, J. B.; Fredrickson, D. S.: The Metabolic Basis of Inherited Disease. New York: McGraw-Hill (pub.) (4th ed.): 1978. Pp. 1166-1220.
Meyer, U. A.; Strand, L. J.; Doss, M.; Rees, A. C.; Marver, H. S.: Intermittent acute porphyria—demonstration of a genetic defect in porphobilinogen metabolism. New Eng. J. Med. 286: 1277-1282, 1972.
Mgone, C. S.; Lanyon, W. G.; Moore, M. R.; Connor, J. M.: Detection of seven point mutations in the porphobilinogen deaminase gene in patients with acute intermittent porphyria, by direct sequencing of in vitro amplified cDNA. Hum. Genet. 90: 12-16, 1992.
Mgone, C. S.; Lanyon, W. G.; Moore, M. R.; Louie, G. V.; Conner, J. M.: Detection of a high mutation frequency in exon 12 of the porphobiliogen deaminase gene in patients with acute intermittent porphyria. Hum. Genet. 92: 619-622, 1993.
Morris, D. L.; Dudley, M. D.; Pearson, R. D.: Coagulopathy associated with hematin treatment for acute intermittent porphyria. Ann. Intern. Med. 95: 700-701, 1981.
Mustajoki, P.: Normal erythrocyte uroporphyrinogen I synthase in a kindred with acute intermittent porphyria. Ann. Intern. Med. 95: 162-166, 1981.
Mustajoki, P.; Desnick, R. J.: Genetic heterogeneity in acute intermittent porphyria: characterisation and frequency of porphobilinogen deaminase mutations in Finland. Brit. Med. J. 291: 505-509, 1985.
Namba, H.; Narahara, K.; Tsuji, K.; Yokoyama, Y.; Seino, Y.: Assignment of human porphobilinogen deaminase to 11q24.1-q24.2 by in situ hybridization and gene dosage studies. Cytogenet. Cell Genet. 57: 105-108, 1991.
Picat, C.; Delfau, M. H.; de Rooij, F. W. M.; Beukeveld, G. J. J.; Wolthers, B. G.; Wadman, S. K.; Nordmann, Y.; Grandchamp, B.: Identification of the mutations in the parents of a patient with a putative compound heterozygosity for acute intermittent porphyria. J. Inherit. Metab. Dis. 13: 684-686, 1990.
Puy, H.; Deybach, J.-C.; Bogdan, A.; Callebert, J.; Baumgartner, M.; Voisin, P.; Nordmann, Y.; Touitou, Y.: Increased delta-aminolevulinic acid and decreased pineal melatonin production: a common event in acute porphyria studies in the rat. J. Clin. Invest. 97: 104-110, 1996.
Puy, H.; Deybach, J. C.; Lamoril, J.; Robreau, A. M.; Da Silva, V.; Gouya, L.; Grandchamp, B.; Nordmann, Y.: Molecular epidemiology and diagnosis of PBG deaminase gene defects in acute intermittent porphyria. Am. J. Hum. Genet. 60: 1373-1383, 1997.
Raich, N.; Romeo, P. H.; Dubart, A.; Beaupain, D.; Cohen-Solal, M.; Goossens, M.: Molecular cloning and complete primary sequence of human erythrocyte porphobilinogen deaminase. Nucleic Acids Res. 14: 5955-5968, 1986.
Sassa, S.; Granick, S.; Bickers, D. R.; Bradlow, H. L.; Kappas, A.: A microassay for uroporphyrinogen I synthase, one of three abnormal enzyme activities in acute intermittent porphyria, and its application to the study of the genetics of this disease. Proc. Nat. Acad. Sci. 71: 732-736, 1974.

Sassa, S.; Solish, G.; Levere, R. D.; Kappas, A.: Studies in porphyria. IV. Expression of the gene defect of acute intermittent porphyria in cultured human skin fibroblasts and amniotic cells: prenatal diagnosis of the porphyric trait. J. Exp. Med. 142: 722-731, 1975.

Sassa, S.; Zalar, G. L.; Kappas, A.: Studies in porphyria: VII. Induction of uroporphyrinogen-I synthase and expression of the gene defect of acute intermittent porphyria in mitogen-stimulated human lymphocytes. J. Clin. Invest. 61: 499-508, 1978.

Scobie, G. A.; Llewellyn, D. H.; Urquhart, A. J.; Smyth, S. J.; Kalsheker, N. A.; Harrison, P. R.; Elder, G. H.: Acute intermittent porphyria caused by a C-to-T mutation that produces a stop codon in the porphobilinogen deaminase gene. Hum. Genet. 85: 631-634, 1990.

Scobie, G. A.; Urquhart, A. J.; Elder, G. H.; Kalsheker, N. A.; Llewellyn, D. H.; Smyth, J.; Harrison, P. R.: Linkage disequilibrium between DNA polymorphisms within the porphobilinogen deaminase gene. Hum. Genet. 85: 157-159, 1990.

Srugo, I.; Said, E.; Korman, S.; Jaffe, M.: Acute intermittent porphyria—an unusual cause of 'surgical' abdomen: response to propranolol therapy. Europ. J. Pediat. 146: 305-308, 1987.

Stein, J. A.; Tschudy, D. P.: Acute intermittent porphyria: a clinical and biochemical study of 46 patients. Medicine 49: 1-16, 1970.

Strand, L. J.; Felsher, B. F.; Redeker, A. G.; Marver, H. S.: Heme biosynthesis in intermittent acute porphyria: decreased hepatic conversion of porphobilinogen to porphyrins and increased delta aminolevulinic acid synthetase activity. Proc. Nat. Acad. Sci. 67: 1315-1320, 1970.

Strand, L. J.; Meyer, U. A.; Felsher, B. F.; Redeker, A. G.; Marver, H. S.: Decreased red cell uroporphyrinogen I synthetase activity in intermittent acute porphyria. J. Clin. Invest. 51: 2530-2536, 1972.

Tishler, P. V.; Woodward, B.; O'Connor, J.; Holbrook, D. A.; Seidman, L. J.; Hallett, M.; Knighton, D. J.: High prevalence of intermittent acute porphyria in a psychiatric patient population. Am. J. Psychiat. 142: 1320-1436, 1985.

Tschudy, D. P.; Valsamis, M.; Magnussen, C. R.: Acute intermittent porphyria: clinical and selected research aspects. Ann. Intern. Med. 83: 851-864, 1975.

Tunnacliffe, A.; McGuire, R. S.: A physical linkage group in human chromosome band 11q23 covering a region implicated in leukocyte neoplasia. Genomics 8: 447-453, 1990.

Verma, A.; Nye, J. S.; Snyder, S. H.: Porphyrins are endogenous ligands for the mitochondrial (peripheral-type) benzodiazepine receptor. Proc. Nat. Acad. Sci. 84: 2256-2260, 1987.

Vidaud, M.; Gattoni, R.; Stevenin, J.; Vidaud, D.; Amselem, S.; Chibani, J.; Rosa, J.; Goossens, M.: A 5-prime splice-region G-to-C mutation in exon 1 of the human beta-globin gene inhibits pre-mRNA splicing: a mechanism for beta(+)-thalassemia. Proc. Nat. Acad. Sci. 86: 1041-1045, 1989.

Waldenstrom, J.: Studies on the incidence and heredity of acute porphyria in Sweden. Acta Genet. Statist. Med. 6: 122-131, 1956.

Waldenstrom, J.; Haeger-Aronsen, B.: Different patterns of human porphyria. Brit. Med. J. 2: 272-276, 1963.

Wang, A.-L.; Arredondo-Vega, F. X.; Giampietro, P. F.; Smith, M.; Anderson, W. F.; Desnick, R. J.: Regional gene assignment of human porphobilinogen deaminase and esterase A(4) to chromosome 11q23-11qter. Proc. Nat. Acad. Sci. 78: 5734-5738, 1981.

Watson, C. J.; Bossenmaier, I.: Present status of the Ehrlich aldehyde reaction for urinary porphobilinogen. J.A.M.A. 190: 501-504, 1964.

Yeung Laiwah, A. A. C.; Mactier, R.; McColl, K. E. L.; Moore, M. R.; Goldberg, A.: Early-onset chronic renal failure as a complication of acute intermittent porphyria. Quart. J. Med. 52: 92-98, 1983.

PORPHYRIA, CUTANEA TARDA (*176100)

JJM NOTE

Several studies associate hepatocellular carcinoma with this disorder (Hardell L et al.: Aetiological aspects on primary liver cancer with special regard to alcohol, organic solvents and acute intermittent porphyria: an epidemiological investigation. Br. J. Cancer 50:389–97, 1984). A case report suggests that this entity is another example of ecogenetics in liver carcinogenesis (McConnell R et al.: Angiosarcoma, porphyria-tarda, and probable chloracne in a worker exposed to waste oil contaminated with 2,3,7,8-tetrachlorodibenzo-p-dioxin. Br. J. Ind. Med. 50:699–703, 1993).

Description

Porphyria cutanea tarda is an autosomal dominant disorder characterized by light-sensitive dermatitis and associated with the excretion of large amounts of uroporphyrin in urine.

Clinical Features

Onset of light-sensitive dermatitis in later adult life, associated with the excretion of large amounts of uroporphyrin in urine, characterizes this form of porphyria, which was so named by Waldenstrom (1937). On areas of skin exposed to sunlight, especially the face, ears and backs of the hands, chronic ulcerating lesions commence as blisters, and the skin may also be mechanically fragile (Grossman et al., 1979). Hyperpigmentation and hypertrichosis also occur. Acute neuropathic episodes do not occur in this form of porphyria. Onset is often associated with alcoholism, and occasionally with exposure to other agents, such as estrogens. Iron overload is frequently present, and may be associated, coincidentally or causally, with varying degrees of liver damage or fibrosis; liver histology may be characteristic (Cortes et al., 1980). On biopsy, liver parenchyma cells are also loaded with porphyrins and fluoresce bright red in ultraviolet light. The skin lesions are distinctly related to circulating porphyrins (Holti et al., 1958).

A similar syndrome, a 'phenocopy,' is caused by toxic exposure to certain organic chemicals such as hexachlorobenzene, as in the epidemic caused by contaminated seed wheat in Turkey (Cam and Nigogosyan, 1963; Dean, 1972) and by occupational exposure to chlorinated hydrocarbons (Bleiberg et al., 1964).

Hepatoerythropoietic porphyria (HEP) is a severe form of cutaneous porphyria that presents in infancy and is characterized biochemically by excessive excretion of acetate-substituted porphyrins and accumulation of protoporphyrin in erythrocytes (Hofstad et al., 1973; Simon et al., 1977; Czarnecki, 1980). As in porphyria cutanea tarda, uroporphyrinogen decarboxylase is deficient. However, the enzyme level is very low (7-8%) in erythrocytes and cultured skin fibroblasts, leading Elder et al. (1981) to propose that HEP is the homozygous state for porphyria cutanea tarda. Felsher et al. (1982) concluded that reduced hepatic uroporphyrinogen decarboxylase activity is a specific and intrinsic hepatic defect in PCT, but modulation of uroporphyrinogen synthesis by extrinsic factors is required for full biochemical expression of the disease.

De Verneuil et al. (1984) brought to 9 the number of known cases of HEP and confirmed that these patients are homozygous for the gene that causes PCT. The patients of de Verneuil et al. (1984) were twin daughters of a Tunisian couple related as second cousins. Both parents, although asymptomatic, showed intermediate levels of enzymatic and immunoreactive URO decarboxylase. The twins were CRM-negative, in contrast to the previously reported patients.

Toback et al. (1987) described a man with relatively mild hepatoerythropoietic porphyria and concluded that the man was a homozygote since both of his parents and his 3 children, all of whom were asymptomatic, showed moderate deficiency of UROD. They concluded that the relative mildness of the clinical symptoms in the proband was probably related to the level of residual enzyme activity and that the genetic defect in UROD in this disorder can be heterogeneous.

Malina and Lim (1988) described a 29-year-old woman who first presented with blisters and erosions on the dorsum of the fingers and hands bilaterally 3 weeks after delivery of her second child. The diagnosis of PCT was established enzymatically and by porphyrin studies. Reduced red cell UROD activity was found also in the newborn child and in the patient's mother.

Fujimoto and Brazil (1992) reported a 23-year-old woman thought to represent the 18th instance of HEP reported worldwide. She had photosensitive skin of early onset, hypertrichosis, and severe scleroderma-like lesions of the hands.

Classic congenital erythropoietic porphyria (263700) is due to deficiency of uroporphyrinogen III cosynthase. Kushner et al. (1982) described a remarkable 51-year-old man with congenital erythropoietic porphyria (Gunther disease), first manifested in infancy with eventual development of mutilating skin photosensitivity. The morphologic features of dyserythropoietic bone marrow cells, studied by light and electron microscopy, were identical to those found in congenital dyserythropoietic anemia type I (224120); such had been described before in Gunther disease. A red-orange nuclear fluorescence is not seen in type I dyserythropoietic anemia. The patient of Kushner et al. (1982) showed massive porphyrinuria, but the pattern of porphyrin excretion was atypical for classic Gunther disease: hepta-carboxyl (7-COOH) porphyrin was the major urine porphyrin, much uroporphyrin was present, and both were predominantly of the isomer III type. Erythrocyte uroporphyrinogen III cosynthase activity was normal, but uroporphyrinogen decarboxylase activity was 50% of normal. Two sons showed equally subnormal uroporphyrinogen decarboxylase activity. It was the opinion of the authors that their 51-year-old patient had 2 genetic diseases—uroporphyrinogen decarboxylase deficiency (a heterozygous state) and type I congenital dyserythropoietic anemia (a presumably homozygous state). With coexisting hepatic siderosis, heterozygous uroporphyrinogen decarboxylase deficiency leads to porphyria cutanea tarda. Homozygosity for a deficiency gene leads to hepatoerythropoietic porphyria. Thus, Gunther disease can have more than one cause. Two other reported patients with clinically typical congenital erythropoietic porphyria, but with a pattern of urinary porphyrin excretion similar to porphyria cutanea tarda, were referenced by Kushner et al. (1982).

Biochemical Features

Uroporphyrinogen decarboxylase (UROD; EC 4.1.1.37) is a cytosolic enzyme involved in the biosynthesis of heme. It catalyzes the sequential removal of 4 of the carboxymethyl side chains of uroporphyrinogen to yield coproporphyrinogen. Sassa et al. (1983) purified UROD to homogeneity. A single enzyme is involved in the 4 successive decarboxylations.

De Verneuil et al. (1984) and others distinguished 2 forms of PCT: type I, or the 'sporadic' type, associated with 50% level of URO decarboxylase in the liver only, and type II, or the 'familial' type, in which a 50% level of enzyme is found more generally, e.g., in red cells and liver and presumably in other cells as well. Is the type I genetically determined and, if so, is it determined by mutation at the same locus as that of type II? Citing Elder et al. (1978), Kappas et al. (1983) wrote: 'It is conceivable that the liver and red cell enzymes are in some manner under separate genetic control, and that the decarboxylase deficiency in liver in some sporadic PCT cases is inherited.' Most cases of familial PCT are heterozygotes. As noted earlier, homozygous cases are known (so-called hepatoerythropoietic porphyria) and both CRM-positive and CRM-negative homozygotes have been identified.

Reduced liver and red cell uroporphyrinogen decarboxylase activity has been reported in familial (Kushner et al., 1976; Lehr and Doss, 1981) and sporadic cases of porphyria cutanea tarda (Elder et al., 1978; Felsher et al., 1978). Impaired activity of this enzyme step in heme synthesis in liver could possibly explain resulting 'overflow' of uroporphyrin. Hepatic uroporphyrinogen decarboxylase activity was reduced to approximately 50% of normal levels in 17 cases of porphyria cutanea tarda and reduced levels persisted after hepatic iron overload was relieved by phlebotomy (Felsher et al., 1982). Elder et al. (1978) found normal levels of enzyme in red cells and fibroblasts. In assays of UROD activity in red cells, de Verneuil et al. (1978) found 50% levels of uroporphyrinogen decarboxylase in persons with familial porphyria cutanea, but normal enzyme levels in sporadic cases.

In hemolysates from 7 unrelated patients with familial PCT, Elder et al. (1983) found that immunoreactive uroporphyrinogen decarboxylase was decreased (average 51% of normal) to the same extent as catalytic activity (average 56% of normal), whereas in 6 sporadic cases both measurements were normal. The failure to find evidence of CRM+ mutations among the familial cases suggested to Elder et al. (1983) that a simple immunoelectrophoretic method can be used for routine diagnosis.

Inheritance

Most cases are sporadic and are more common in men than women, but familial cases have been described frequently, and apparent autosomal dominant segregation of the disorder has been reported (Holti et al., 1958; Ziprkowski et al., 1966; Topi and Gandalfo, 1977; Benedetto et al., 1978).

Although it is unusual for an enzyme deficiency to produce symptoms in the heterozygous state, i.e., in single gene dose, this is also the pattern in other types of genetic porphyrias (e.g., 121300, 176000, 176200, 177000). It seems likely that a reduced level of activity of uroporphyrinogen decarboxylase may segregate as an autosomal dominant trait, but that additional environmental factors are required for manifestation of the disorder; iron overload may have a direct metabolic role (Kushner et al., 1972; Kushner, 1982). Many possible carriers of the mutant allele may never develop symptoms. This could explain sporadic cases; alternatively, this may be a more heterogeneous syndrome and it is not yet fully elucidated (Meyer and Schmid, 1978; Romeo, 1977). Blekkenhorst et al. (1979) suggested that two forms of PCT exist: a rare familial form and a relatively common idiosyncratic form occurring sporadically as an unusual accompaniment of common hepatic disorders such as alcohol-associated liver disease.

Mapping

De Verneuil et al. (1984) assigned the locus for uroporphyrinogen decarboxylase to chromosome 1 by somatic cell hybridization and specific enzyme assay. This was the fourth enzyme of the heme biosynthetic pathway to be mapped; the other 3 are CPRD (121300) on 9, PBGD (176000) on 11, and ALAD (125270) on 9. Mattei et al. (1985) used a cDNA clone to localize UROD to 1p34 by in situ hybridization. McLellan et al. (1985) arrived at a different location, 1pter-1p21, by somatic cell hybridization using cell lines with rearranged chromosomes. Using a cDNA probe in both somatic cell and in situ hybridization, Dubart et al. (1986) confirmed the assignment to 1p34.

Molecular Genetics

Using a UROD cDNA probe in Northern blot analysis, Hansen et al. (1988) found no difference in the levels of UROD mRNA between affected individuals and their normal relatives.

Romana et al. (1987) demonstrated that the UROD gene has 10 exons spread over 3 kb. Using hybridization probes for the UROD gene in the study of genomic DNA from patients with familial PCT, Hansen et al. (1988) could not identify any major deletions, rearrangements, or restriction fragment length polymorphisms. In the UROD cDNA from a patient with familial PCT, Garey et al. (1989) demonstrated a gly-to-val substitution at amino acid position 281 (see 176100.0001). The mutation was not detected in affected persons from 7 other PCT pedigrees with an autosomal dominant pattern. They showed that the URO-D protein in the patient with the identified mutation had a greatly shortened half-life, both in vitro and in vivo (assuming, as these workers did, that one can call the findings in cultured lymphocytes an 'in vivo' observation). Hepatoerythropoietic porphyria results from a different nucleotide substitution in the same codon (see 176100.0002). The UROD protein resulting from that mutation, gly281-to-glu, also has a decreased half-life, but not so severely decreased as in the case of the gly-to-val mutation. This is an instructive instance in which 1 mutation at a particular codon causes a dominant disorder, and another mutation at the same codon causes a recessive disorder. Garey et al. (1989) suggested that the former mutation may be so severe in the homozygous state that it is lethal to the embryo; PCT can result in the heterozygote for the first mutation, but only the homozygote for the milder mutation expresses itself (as HEP). The population genetics is consistent with this suggestion. Garey et al. (1989) pointed out that familial PCT is relatively common, but only 16 cases of HEP have been described to date.

Using a cDNA clone for the UROD gene, de Verneuil et al. (1986) studied DNA from 2 homozygous patients, offspring of consanguineous parents, who suffered from HEP. They could detect neither deletions nor rearrangements in the UROD gene. Synthesis, processing, and cell-free translation of the specific transcripts appeared to be normal. The half-life of the abnormal protein was 12 times shorter than that of the normal enzyme. Thus, rapid degradation in vivo is the probable basis of the enzyme deficiency. Study of homozygous patients avoided the difficulties of studying the enzyme defect in the heterozygous PCT where both normal and abnormal protein is present. The authors suggested that use of oligonucleotide probes complementary to the normal and mutant sequences could allow them to determine if the mutation in familial PCT is the same as that in HEP; in other words, whether HEP is indeed the homozygous state of PCT.

An association between PCT and HLA-linked hereditary hemochromatosis (235200) was suggested by Kushner et al. (1985), but disputed by Beaumont et al. (1986). Santos et al. (1997) assessed the role of HLA-H mutations in PCT by an allelic-association study between PTC and the mutations identified in hemochromatosis. They studied 15 unselected, unrelated patients with PTC being treated with regular phlebotomy. The controls were 23 anonymous blood donors and 71 patients with hereditary hemochromatosis. The cys282tyr mutation (235200.0001) was found in 83% of 142 hereditary hemochromatosis chromosomes, 47% of 30 PCT chromosomes, and 9% of 46 normal blood donor chromosomes. Santos et al. (1997) concluded that the hemochromatosis gene contributes to the pathogenesis of PCT. They suggested that all first-degree relatives of patients with PCT should be screened for hereditary hemochromatosis. PCT can be viewed as having a digenic basis.

Clinical Management

Treatment is directed first to reducing iron overload by regular phlebotomy, as in the management of hemochromatosis (Epstein and Redeker, 1968; Ramsay et al., 1974; Grossman et al., 1979). Porphyrin excretion diminishes, and in many patients skin lesions disappear. When this is ineffective or when a more rapid effect is desired, oral chloroquine therapy usually induces rapid remission (Taljaard et al., 1972; Kowertz, 1973). It may also cause a transient increase in porphyrin excretion, sometimes associated with evidence of acute liver damage (Vogler et al., 1970). Remission is sustained while chloroquine is continued in regular low doses.

Several cases of porphyria cutanea tarda have been described in patients on maintenance hemodialysis for chronic renal failure (e.g., Poh-Fitzpatrick et al., 1978). The cause is thought to be insufficient removal of porphyrins through the hemodialysis membrane which leads to markedly increased levels of plasma porphyrins with resulting severe and mutilating skin lesions. The treatment of the disorder is very difficult because chloroquine is ineffective and the anemia that accompanies chronic renal failure contraindicates venesection therapy. Praga et al. (1987) found that deferoxamine was effective therapy in a patient in whom there was evidence of iron overload due to multiple blood transfusions.

Population Genetics

Porphyria cutanea tarda is the most common form of porphyria, occurring worldwide, but no figure for prevalence is available; it is particularly common in the Bantu in South Africa.

Animal Model

Bahary et al. (1991) assigned the homologous gene to chromosome 4 of the mouse.

The zebrafish mutant 'yquem' shows a photosensitive porphyria syndrome. Wang et al. (1998) showed that the porphyric phenotype is due to an inherited homozygous mutation in the UROD gene. Thus, the zebrafish mutant represented the first genetically 'accurate' model of hepatoerythropoietic porphyria; Wang et al. (1998) suggested that the model would be useful for studying the pathogenesis of UROD deficiency and evaluating gene therapy vectors. Wang et al. (1998) rescued the mutant phenotype by transient and germline expression of the wild-type allele.

See also Day et al. (1982); de Verneuil et al. (1984); de Verneuil et al. (1986); de Verneuil et al. (1988); Garey et al. (1988); Hansen et al. (1988)

Bahary, N.; Zorich, G.; Pachter, J. E.; Leibel, R. L.; Friedman, J. M.: Molecular genetic linkage maps of mouse chromosomes 4 and 6. Genomics 11: 33-47, 1991.

Beaumont, C.; Nordmann, Y.; Fauchet, R.: HLA-linked hemochromatosis and sporadic porphyria cutanea tarda. (Letter) Gastroenterology 90: 800 only, 1986.

Benedetto, A. V.; Kushner, J. P.; Taylor, J. S.: Porphyria cutanea tarda in three generations of a single family. New Eng. J. Med. 298: 358-361, 1978.

Bleiberg, J.; Wallen, M.; Brodkin, R.; Applebaum, I.: Industrially acquired porphyria. Arch. Derm. 89: 793-797, 1964.

Blekkenhorst, G. H.; Day, R. S.; Eales, L.: Two forms of porphyria cutanea tarda?. (Letter) New Eng. J. Med. 300: 93, 1979.

Cam, C.; Nigogosyan, G.: Acquired toxic porphyria cutanea tarda due to hexachlorobenzene. J.A.M.A. 183: 88-91, 1963.

Cortes, J. M.; Oliva, H.; Paradinas, F. J.; Hernandez-Guio, C.: The pathology of the liver in porphyria cutanea tarda. Histopathology 4: 471-485, 1980.

Czarnecki, D. B.: Hepatoerythropoietic porphyria. Arch. Derm. 116: 307-311, 1980.

Day, R. S.; Eales, L.; Meissner, D.: Coexistent variegate porphyria and porphyria cutanea tarda. New Eng. J. Med. 307: 36-41, 1982.

Dean, G.: The Porphyrias. A Story of Inheritance and Environment. Philadelphia: J. B. Lippincott (pub.) (2nd ed.): 1972.

de Verneuil, H.; Aitken, G.; Nordmann, Y.: Familial and sporadic porphyria cutanea: two different diseases. Hum. Genet. 44: 145-151, 1978.

de Verneuil, H.; Beaumont, C.; Deybach, J.-C.; Nordmann, Y.; Sfar, Z.; Kastally, R.: Enzymatic and immunological studies of uroporphyrinogen decarboxylase in familial porphyria cutanea tarda and hepatoerythropoietic porphyria. Am. J. Hum. Genet. 36: 613-622, 1984.

de Verneuil, H.; Bourgeois, F.; de Rooij, F.; Siersema, P. D.; Wilson, J. H. P.; Grandchamp, B.; Nordmann, Y.: Characterization of a new mutation (R292G) and a deletion at the human uroporphyrinogen decarboxylase locus in two patients with hepatoerythropoietic porphyria. Hum. Genet. 89: 548-552, 1992.

de Verneuil, H.; Grandchamp, B.; Beaumont, C.; Picat, C.; Nordmann, Y.: Uroporphyrinogen decarboxylase structural mutant (gly-281-to-glu) in a case of porphyria. Science 234: 732-734, 1986.

de Verneuil, H.; Grandchamp, B.; Foubert, C.; Weil, D.; Van Cong, N.; Gross, M.-S.; Sassa, S.; Nordmann, Y.: Assignment of the gene for uroporphyrinogen decarboxylase to human chromosome 1 by somatic cell hybridization and specific enzyme immunoassay. Hum. Genet. 66: 202-205, 1984.

de Verneuil, H.; Grandchamp, B.; Romeo, P. H.; Raich, N.; Beaumont, C.; Goossens, M.; Nicolas, H.; Nordmann, Y.: Molecular analysis of uroporphyrinogen decarboxylase deficiency in a family with two cases of hepatoerythropoietic porphyria. J. Clin. Invest. 77: 431-435, 1986.

de Verneuil, H.; Hansen, J.; Picat, C.; Grandchamp, B.; Kushner, J.; Roberts, A.; Elder, G.; Nordmann, Y.: Prevalence of the 281 (gly-to-glu) mutation in hepatoerythropoietic porphyria and porphyria cutanea tarda. Hum. Genet. 78: 101-102, 1988.

de Verneuil, H.; Hansen, J. L.; Picat, C.; Grandchamp, B.; Kushner, J.; Roberts, A.; Elder, G.; Nordmann, Y.: Prevalence of the 281 (gly-to-glu) mutation in hepatoerythropoietic porphyria and porphyria cutanea tarda. Hum. Genet. 78: 101-102, 1988.

Dubart, A.; Mattei, M. G.; Raich, N.; Beaupain, D.; Romeo, P. H.; Mattei, J. F.; Goossens, M.: Assignment of human uroporphyrinogen decarboxylase (UROD) to the p34 band of chromosome 1. Hum. Genet. 73: 277-279, 1986.

Elder, G. H.; Lee, G. B.; Tovey, J. A.: Decreased activity of hepatic uroporphyrinogen decarboxylase in sporadic porphyria cutanea tarda. New Eng. J. Med. 299: 274-278, 1978.

Elder, G. H.; Sheppard, D. M.; Tovey, J. A.; Urquhart, A. J.: Immunoreactive uroporphyrinogen decarboxylase in porphyria cutanea tarda. Lancet I: 1301-1304, 1983.

Elder, G. H.; Smith, S. G.; Herrero, C.; Mascaro, J. M.; Lecha, M.; Muniesa, A. M.; Czarnecki, D. B.; Brenan, J.; Poulos, V.; de Salamanca, R. E.: Hepatoerythropoietic porphyria: a new uroporphyrinogen decarboxylase defect or homozygous porphyria cutanea tarda?. Lancet I: 916-919, 1981.

Epstein, J. H.; Redeker, A. G.: Porphyria cutanea tarda: a study of the effect of phlebotomy. New Eng. J. Med. 279: 1301-1304, 1968.

Felsher, B. F.; Carpio, N. M.; Engleking, D. W.; Nunn, A. T.: Decreased hepatic uroporphyrinogen decarboxylase activity in porphyria cutanea tarda. New Eng. J. Med. 306: 766-769, 1982.

Felsher, B. F.; Norris, M. E.; Shih, J. C.: Red-cell uroporphyrinogen decarboxylase activity in porphyria cutanea tarda and in other forms of porphyria. New Eng. J. Med. 299: 1095-1098, 1978.

Fujimoto, A.; Brazil, J. L.: Hepatoerythropoietic porphyria in a woman with short stature and deformed hands. Am. J. Med. Genet. 44: 496-499, 1992.

Garey, J. R.; Hansen, J. L.; Harrison, L. M.; Kennedy, J. B.; Kushner, J. P.: A point mutation in the coding region of uroporphyrinogen decarboxylase associated with familial porphyria cutanea tarda. Blood 73: 892-895, 1989.

Garey, J. R.; Hansen, J. L.; Kushner, J. P.: The molecular basis of familial porphyria cutanea tarda (F-PCT). (Abstract) Clin. Res. 36: 612A, 1988.

Garey, J. R.; Harrison, L. M.; Franklin, K. F.; Metcalf, K. M.; Radisky, E. S.; Kushner, J. P.: Uroporphyrinogen decarboxylase: a splice site mutation causes the deletion of exon 6 in multiple families with porphyria cutanea tarda. J. Clin. Invest. 86: 1416-1422, 1990.

Grossman, M. E.; Bickers, D. R.; Poh-Fitzpatrick, M. B.; Delco, V. A.; Harber, L. C.: Porphyria cutanea tarda: clinical features and laboratory findings in 40 patients. Am. J. Med. 67: 277-286, 1979.

Hansen, J. L.; O'Connell, P. O.; Romana, M.; Romeo, P.-H.; Kushner, J. P.: Familial porphyria cutanea tarda: hybridization analysis of the uroporphyrinogen decarboxylase locus. Hum. Hered. 38: 283-286, 1988.

Hansen, J. L.; Pryor, M. A.; Kennedy, J. B.; Kushner, J. P.: Steady-state levels of uroporphyrinogen decarboxylase mRNA in lymphoblastoid cell lines from patients with familial porphyria cutanea tarda and their relatives. Am. J. Hum. Genet. 42: 847-853, 1988.

Hofstad, F.; Seip, M.; Eriksen, L.: Congenital erythropoietic porphyria with a hitherto undescribed porphyrin pattern. Acta Paediat. Scand. 62: 380-384, 1973.

Holti, G.; Rimington, C.; Tate, B. C.; Thomas, G.: An investigation of 'porphyria cutanea tarda.'. Quart. J. Med. 27: 1-19, 1958.

Kappas, A.; Sassa, S.; Anderson, K. E.: The porphyrias.In: Stanbury, J. B.; Wyngaarden, J. B.; Fredrickson, D. S.; Goldstein, J. L.; Brown, M. S.: The Metabolic Basis of Inherited Disease. New York: McGraw-Hill (pub.) (5th ed.): 1983. Pp. 1360 only.

Kowertz, M. J.: The therapeutic effect of chloroquine. J.A.M.A. 223: 515-519, 1973.

Kushner, J. P.: The enzymatic defect in porphyria cutanea tarda. (Editorial) New Eng. J. Med. 306: 799-800, 1982.

Kushner, J. P.; Barbuto, A. J.; Lee, G. R.: An inherited enzymatic defect in porphyria cutanea tarda: decreased uroporphyrinogen decarboxylase activity. J. Clin. Invest. 58: 1089-1097, 1976.

Kushner, J. P.; Edwards, C. Q.; Dadone, M. M.; Skolnick, M. H.: Heterozygosity for HLA-linked hemochromatosis as a likely cause of the hepatic siderosis associated with sporadic porphyria cutanea tarda. Gastroenterology 88: 1232-1238, 1985.

Kushner, J. P.; Lee, G. R.; Nacht, S.: The role of iron in the pathogenesis of porphyria cutanea tarda: an in vitro model. J. Clin. Invest. 51: 3044-3051, 1972.

Kushner, J. P.; Pimstone, N. R.; Kjeldsberg, C. R.; Pryor, M. A.; Huntley, A.: Congenital erythropoietic porphyria, diminished activity of uroporphyrinogen decarboxylase and dyserythropoiesis. Blood 59: 725-737, 1982.

Lehr, P. A.; Doss, M.: Chronishe hepatische porphyrie mit Uroporphyrinogen-Decarboxylase-Defekt in vier Generationen. Dtsch. Med. Wschr. 106: 241-245, 1981.

Malina, L.; Lim, C. K.: Manifestation of familial porphyria cutanea tarda after childbirth. Brit. J. Derm. 118: 243-245, 1988.

Mattei, M. G.; Dubart, A.; Beaupain, D.; Goossens, M.; Mattei, J. F.: Localization of the uroporphyrinogen decarboxylase gene to 1p34 band, by in situ hybridization. (Abstract) Cytogenet. Cell Genet. 40: 692, 1985.

McLellan, T.; Pryor, M. A.; Kushner, J. P.; Eddy, R. L.; Shows, T. B.: Assignment of uroporphyrinogen decarboxylase (UROD) to the pter-p21 region of human chromosome 1. Cytogenet. Cell Genet. 39: 224-227, 1985.

Meyer, U. A.; Schmid, R.: The porphyrias.In: Stanbury, J. B.; Wyngaarden, J. B.; Fredrickson, D. S. (eds.): The Metabolic Basis of Inherited Disease. New York: McGraw-Hill (pub.) (4th ed.) : 1978. Pp. 1166-1220.

Moran-Jimenez, M. J.; Ged, C.; Romana, M.; Enriquez de Salamanca, R.; Taieb, A.; Topi, G.; D'Alessandro, L.; de Verneuil, H.: Uroporphyrinogen decarboxylase: complete human gene sequence and molecular study of three families with hepatoerythropoietic porphyria. Am. J. Hum. Genet. 58: 712-721, 1996.

Poh-Fitzpatrick, M. B.; Bellet, N.; DeLeo, V. A.; Grossman, M. E.; Bickers, D. R.: Porphyria cutanea tarda in two patients treated with hemodialysis for chronic renal failure. New Eng. J. Med. 299: 292-294, 1978.

Praga, M.; de Salamanca, R. E.; Andres, A.; Nieto, J.; Oliet, A.; Perpina, J.; Morales, J. M.: Treatment of hemodialysis-related porphyria cutanea tarda with deferoxamine. (Letter) New Eng. J. Med. 316: 547-548, 1987.

Ramsay, C. A.; Magnus, I. A.; Turnbull, A.; Baker, H.: The treatment of porphyria cutanea tarda by venesection. Quart. J. Med. 43: 1-24, 1974.

Roberts, A. G.; Elder, G. H.; De Salamanca, R. E.; Herrero, C.; Lecha, M.; Mascaro, J. M.: A mutation (G281E) of the human uroporphyrinogen decarboxylase gene causes both hepatoerythropoietic porphyria and overt familial porphyria cutanea tarda: biochemical and genetic studies on Spanish patients. J. Invest. Derm. 104: 500-502, 1995.

Romana, M.; Dubart, A.; Beaupain, D.; Chabret, C.; Goossens, M.; Romeo, P.-H.: Structure of the gene for human uroporphyrinogen decarboxylase. Nucleic Acids Res. 15: 7343-7356, 1987.

Romana, M.; Grandchamp, B.; Dubart, A.; Amselem, S.; Chabret, C.; Nordmann, Y.; Goossens, M.; Romeo, P.-H.: Identification of a new mutation responsible for hepatoerythropoietic porphyria. Europ. J. Clin. Invest. 21: 225-229, 1991.

Romeo, G.: Enzymatic defects of hereditary porphyrias. Hum. Genet. 39: 261-276, 1977.

Santos, M.; Clevers, H. C.; Marx, J. J. M.: Mutations of the hereditary hemochromatosis candidate gene HLA-H in porphyria cutanea tarda. (Letter) New Eng. J. Med. 336: 1327-1328, 1997.

Sassa, S.; de Verneuil, H.; Anderson, K. E.; Kappas, A.: Purification and properties of human erythrocyte uroporphyrinogen decarboxylase: immunological demonstration of the enzyme defect in porphyria cutanea tarda. Trans. Assoc. Am. Phys. 96: 65-75, 1983.

Simon, N.; Berko, G.; Schneider, I.: Hepato-erythropoietic porphyria presenting as scleroderma and acrosclerosis in a sibling pair. Brit. J. Derm. 96: 663-668, 1977.

Taljaard, J. F.; Shanley, B. C.; Stewart-Wynne, E.; Deppe, W. M.; Joubert, S. M.: Studies on low dose chloroquine therapy and the action of chloroquine in symptomatic porphyria. Brit. J. Derm. 87: 261-269, 1972.

Toback, A. C.; Sassa, S.; Poh-Fitzpatrick, M. B.; Schechter, J.; Zaider, E.; Harber, L. C.; Kappas, A.: Hepatoerythropoietic porphyria: clinical, biochemical, and enzymatic studies in a three-generation family lineage. New Eng. J. Med. 316: 645-650, 1987.

Topi, G.; Gandalfo, L. D.: Inheritance of porphyria cutanea tarda: analysis of 14 cases in 5 families. Brit. J. Derm. 97: 617-627, 1977.
Vogler, W. R.; Galambos, J. T.; Olansky, S.: Biochemical effects of chloroquine therapy in porphyria cutanea tarda. Am. J. Med. 49: 316-321, 1970.
Waldenstrom, J.: Studien ueber Porphyrie. Acta Med. Scand. 82 (suppl.): 1-254, 1937.
Wang, H.; Long, Q.; Marty, S. D.; Sassa, S.; Lin, S.: A zebrafish model for hepatoerythropoietic porphyria. Nature Genet. 20: 239-243, 1998.
Ziprkowski, L.; Krakowski, A.; Crispin, M.; Szeinberg, A.: Porphyria cutanea tarda hereditaria. Israel J. Med. Sci. 2: 338-343, 1966.

PORPHYRIA VARIEGATA (#176200)

JJM NOTE

Only two patients with porphyria variegata and hepatocellular carcinoma could be identified. Perhaps the association is by chance, despite expectations (Germanaud J et al.: Case of association between hepatocellular carcinoma and porphyria variegata. Scand. J. Gastroenterol. 7:671–72, 1994).

A number sign (#) is used with this entry because of evidence that the disorder is caused by mutation in the gene for protoporphyrinogen oxidase (PPOX; 600923), which maps to 1q.

Affected adults show a very variable picture of skin symptoms, including hyperpigmentation and hypertrichosis, associated with acute attacks, like those of acute intermittent porphyria. Attacks may be protracted and followed by prolonged disability. Attacks are often drug-induced and show the classical neuropathic symptoms and signs, such as abdominal pain, constipation, tachycardia and hypertension, and muscular paralysis and sensory disturbances; disorientation and frank psychosis may be conspicuous features. The condition is characterized by elevated fecal levels of protoporphyrin and coproporphyrin at all times, with increased urine porphyrins at times, and an increase in urinary levels of the porphyrin precursors porphobilinogen (PBG) and delta-aminolevulinic acid (ALA) during the acute attack (Dean, 1972; Mustajoki, 1978; Meyer and Schmid, 1978).

Dean (1972) has described in an engaging manner his studies of porphyria in South Africa and comparative studies in Sweden, Holland, Turkey, and elsewhere. The high frequency of the gene for porphyria variegata in South Africa is a cardinal example of founder effect. Dean (1972) estimated that about 8,000 persons in South Africa were suffering from porphyria inherited from either Gerrit Jansz, a Dutch settler in the Cape, or his wife, Ariaantje Jacobs, who was 1 of 8 sent from an orphanage in Rotterdam to provide wives for Dutch settlers in the Cape. He estimated, furthermore, that 1 million of 3 million whites are descendants of 40 original settlers and their wives, a 12,000-fold increase. Stine and Smith (1990) calculated a coefficient of selection between 0.07 and 0.02 for the porphyria variegata gene in the Afrikaner population of South Africa. Jenkins (1990) gave a comprehensive review of medical genetics in South Africa; porphyria variegata was included in a listing of inherited conditions of unusual prevalence among some southern African populations. (See 600923.0003 for the probable founder gene defect associated causally with VP in South Africa.)

Porphyria variegata is also frequent in Finland (Mustajoki, 1980), where prevalence was estimated to be 1.3 per 100,000. Of 57 patients in 9 families, 18 had had acute attacks, but the longevity of gene carriers did not differ from that of the general population. Skin fragility was usually mild. Porphyria variegata was observed in 3 families in Sweden by Hamnstrom et al. (1967). No genealogic connection with any of the 600 known cases of acute intermittent porphyria could be shown. Cochrane and Goldberg (1968) reported studies of an extensive kindred of which the first author was a member. Children and asymptomatic adults showed the chemical features without manifest disease, but allowed the demonstration that porphyria variegata segregates is an autosomal dominant trait, with manifestation in about one half of affected adults (Cochrane and Goldberg, 1968; Dean, 1972; Hamnstrom et al., 1967; Fromke et al., 1978; Husquinet et al., 1978). The basic defect appears to be reduction in the activity of the enzyme protoporphyrinogen oxidase to approximately 50% of the normal level as determined in skin fibroblasts (Brenner and Bloomer, 1980) and presumably in other tissues, especially liver. This enzymopathy is unusual in being the effect of a single gene dose, rather than requiring a double dose for its expression. Several other genetic porphyrias share this effect; see 121300, 176000, and 176100. Biochemical confirmation of the diagnosis of variegate porphyria can be difficult, particularly in patients without neurologic dysfunction at the time of testing. Levels of porphyrin in the stool may be normal because food and bacterial metabolism in the intestine contribute to the porphyrin content of the feces and may obscure the differences in levels between normal and affected persons. To overcome this shortcoming, Logan et al. (1991) measured porphyrin levels in the bile and found considerably greater differences in bile levels than in fecal levels in pat ients and normals. Kushner (1991) reviewed the laboratory diagnosis of the porphyrias. Management consists primarily of avoidance of porphyrinogenic agents and protection for photosensitive skin.

Homozygous variegate porphyria was reported in 2 Czech sibs by 14,15:Korda et al. (1984, 1985) and in 2 unrelated English patients by Murphy et al. (1986). In the family of Korda et al. (1984) and the first case of Murphy et al. (1986), both parents were, by the biochemical evidence, heterozygotes. Photosensitivity dating from the first year or so of life, growth retardation, and raised red cell protoporphyrin concentration with 60 to 70% zinc chelation were the seemingly characteristic features. The Czech sibs also had mental retardation and nystagmus. Mustajoki et al. (1987) described a child with apparent homozygous variegate porphyria. Lymphocyte protoporphyrinogen oxidase was very low in the patient and half normal in both parents (who were first cousins). The proband developed a severe bullous skin disease a few days after birth, followed by increased fragility of exposed skin in spring and summer. High concentration of protoporphyrin in red cells was a feature. Homozygous variegate porphyria was also described by Norris et al. (1990) and by Gandolfo et al. (1991). Coakley et al. (1990) described an unusual form of variegate porphyria in a young girl with epilepsy, mental retardation, and premature adrenarche. Symptoms of porphyria commenced about the age of 12 years and death occurred about 18 months later. The patient had a very low level of protoporphyrinogen oxidase activity in her cultured fibroblasts, whereas both parents had half normal activity of the enzyme in lymphocytes. Coakley et al. (1990) suggested the patient was homozygous. Two of the subjects with homozygous VP described by Korda et al. (1984) were similar to this patient in that they had severe neurologic symptoms which included convulsions and mental retardation. The ratio of 5-beta to 5-alpha steroids in urine suggested a defect in hepatic 5-alpha-reductase activity in the patient of Coakley et al. (1990). This presumed defect as well as anticonvulsant therapy may have c ontributed to the severity of the patient's condition.

Mapping

Bissbort et al. (1988) demonstrated linkage between VP and alpha-1-antitrypsin (107400); the maximal lod score from male meioses was 4.33 at theta = 0.04, and from both sexes combined, 3.56 at theta = 0.12. In pedigrees triply informative for VP, PI and the Gm polymorphism, VP and PI cosegregated separately from Gm. These findings argued in favor of either gene order VP—PI—IGHC or PI—VP—IGHC.

Taketani et al. (1995) showed by Southern blotting of human genomic DNA that there is a single copy of the PPOX gene, and by fluorescence in situ hybridization they mapped the gene to 1q22 (see also 600923). The reason for the discrepancy between linkage mapping of variegate porphyria to chromosome 14 by Bissbort et al. (1988) and this location of the gene by fluorescence in situ hybridization is unexplained; however, Taketani et al. (1995) suggested that it is possible that an additional gene product on chromosome 14 may interact with PPOX protein to maintain the normal enzyme activity or to affect expression of the PPOX gene (thus accounting for the findings of Bissbort et al. (1988) referred to earlier).

Roberts et al. (1995) likewise mapped the PPOX gene to chromosome 1 (1q23) and showed that the clinical disorder variegate porphyria is linked to microsatellite and other markers in the region 1q21-q23. They concluded that the evidence spoke against locus heterogeneity in VP.

Molecular Genetics

The enzyme responsible for porphyria variegata, protoporphyrinogen oxidase (PPOX; EC 1.3.3.4), catalyzes the penultimate step in heme biosynthesis. In eukaryocytes, PPOX is located in the inner membrane of mitochondria and requires oxygen in order to catalyze the conversion of protoporphyrinogen IX to protoporphyrin IX. Nishimura et al. (1995) isolated the human PPOX gene (referred to as PPO by them) by complementation of a hemG mutant E. coli strain deficient in PPOX with a clone from a human placental cDNA library. The human PPOX cDNA encodes a 477-amino acid protein with sequence similarity to the hemY gene of Bacillus subtilis. The amino terminal portion contains a conserved region that forms the dinucleotide-binding site seen in many flavin-containing proteins. Northern blots detected a 1.8-kb mRNA in human K562 and HepG2 cells. When expressed in COS-1 monkey cells, the human PPOX cDNA produced much higher levels of PPOX activity which was inhibited by acifluorfen, a specific inhibitor of PPOX. The human protein could also be detected in the mitochondria of the transfected cells despite the fact that the protein lacks the apparent transport-specific leader sequence seen in a number of other mitochondrial proteins.

Taketani et al. (1995) showed that the PPOX gene has 13 exons and spans about 8 kb on 1q22. The exon/intron boundary sequences conformed to consensus acceptor (GT) and donor (AG) sequences, and exons in the gene appeared to encode functional protein domains. Primer extension analysis revealed 2 major transcriptional initiation sites in a region with sequence motifs characteristic of a promoter. They suggested that a gene product on chromosome 14 may interact with the PPOX protein to affect either normal enzyme activity or expression of the gene.

Deybach et al. (1996) investigated the molecular defect responsible for VP by sequencing the coding portions of the PPOX gene in 4 patients in 3 unrelated families of French Caucasian origin. In 1 patient, insertion of G at position 1022 of the cDNA produced a frameshift resulting in a premature stop codon (600923.0001). In 3 other patients from 2 unrelated families, they found a missense mutation leading to substitution of arginine for glycine (G232R) in exon 7 (600923.0002). In 1 VP family, they observed the cosegregation of the G232R missense mutation and the deficient PPOX activity. This was the first report of mutations in patients with VP and supported the conclusion that PPOX gene defects cause variegate porphyria.

Meissner et al. (1996) described an arg59-to-trp mutation in the PPOX gene (600923.0003) in 43 of 45 patients with VP from 26 of 27 South African families investigated, but not in 34 unaffected relatives or 9 unrelated British patients with protoporphyrinogen oxidase deficiency. Since at least one of these families could be shown to be descended from the founder of South African VP, this defect may represent the founder gene defect associated causally with VP in South Africa.

Warnich et al. (1996) identified the R59W mutation in 15 out of 17 South African patients with variegate porphyria. This mutation was shown to create a StyI restriction site. It was also shown to be associated with C(26)-C(150), 1 of 4 potential haplotypes defined by 2 polymorphisms in exon 1 of the PPOX gene. Warnich et al. (1996) reported that this data supported the founder hypothesis for variegate porphyria in South Africa. In 1 out of 17 patients there was an exon 2 mutation H20P (600923.0005). This mutation was associated with the same exon 1 polymorphism haplotype as the R59W mutation. An exon 6 mutation R168C (600923.0004) was identified in 1 patient; this mutation abolished a DsaI restriction site in genomic DNA of affected individuals and was shown to be associated with a different haplotype for the exon 1 polymorphism, A(26)-C(150).

History

Macalpine et al. (1968) suggested that George III suffered from porphyria and that the disease can be traced back to Mary Queen of Scots, thus starting a spirited controversy. Many authorities do not accept the speculation. Although the malady of George III is indistinguishable retrospectively from acute intermittent porphyria, supposed dermatologic and other manifestations in members of the family suggest that the royal porphyria, if any, was the variegate type.

Jenkins (1996) discussed the madness of King George III and also the history of 'the South African malady.' Alan Bennett's highly successful play and film 'The Madness of King George' reawakened debate on whether the sovereign suffered from porphyria. Ida Macalpine and Richard Hunter, who first formulated the retrospective diagnosis (Macalpine and Hunter, 1966), were eminent psychiatrists and were authors of a history of psychiatry (Hunter and Macalpine, 1963). Warren et al. (1996) reviewed the history of the suggested diagnosis and its possible influence on history through the members of the royal family.

Bissbort, S.; Hitzeroth, H. W.; du Wentzel, D. P.; Van den Berg, C. W.; Senff, H.; Wienker, T. F.; Bender, K.: Linkage between the variegate porphyria (VP) and the alpha-1-antitrypsin (PI) genes on human chromosome 14. Hum. Genet. 79: 289-290, 1988.

Brenner, D. A.; Bloomer, J. R.: The enzymatic defect in variegate porphyria: studies with human cultured skin fibroblasts. New Eng. J. Med. 302: 765-769, 1980.

Coakley, J.; Blake, D.; Hawkins, R.; Nordmann, Y.; Crinis, N.; Sloan, L.; McManus, J.; Connelly, J.: An unusual case of variegate porphyria with possible homozygous inheritance. Aust. New Zeal. J. Med. 20: 587-589, 1990.

Cochrane, A. L.; Goldberg, A.: A study of faecal porphyrin levels in a large family. Ann. Hum. Genet. 32: 195-208, 1968.

Dean, G.: The Porphyrias. A Story of Inheritance and Environment. Philadelphia: J. B. Lippincott (pub.) (2nd ed.): 1972.

Deybach, J.-C.; Puy, H.; Robreau, A.-M.; Lamoril, J.; Da Silva, V.; Grandchamp, B.; Nordmann, Y.: Mutations in the protoporphyrinogen oxidase gene in patients with variegate porphyria. Hum. Molec. Genet. 5: 407-410, 1996.

Fromke, V. L.; Bossenmaier, I.; Cardinal, R.; Watson, C. J.: Porphyria variegata: study of a large kindred in the United States. Am. J. Med. 65: 80-88, 1978.

Gandolfo, L. D.; Macri, A.; Biolcati, G.; Griso, D.; Phung, L. N.; Deybach, J. C.; Da Silva, V.; Nordmann, Y.; Topi, G. C.: Homozygous variegate porphyria: revision of a diagnostic error. (Letter) Brit. J. Derm. 124: 211, 1991.

Hamnstrom, B.; Haeger-Aronsen, B.; Waldenstrom, J.; Hysing, B.; Molander, J.: Three Swedish families with porphyria variegata. Brit. Med. J. 2: 449-453, 1967.

Hunter, R.; Macalpine, I.: Three Hundred Years of Psychiatry: 1538-1860. A History Presented in Selected English Texts. London, England: Oxford Univ. Press , 1963.

Husquinet, H.; Noirfalse, A.; Parent, M.-T.: Porphyria variegata: etude d'une grande famille. J. Genet. Hum. 26: 367-383, 1978.

Jenkins, T.: The South African malady. Nature Genet. 13: 7-9, 1996.

Jenkins, T.: Medical genetics in South Africa. J. Med. Genet. 27: 760-779, 1990.

Korda, V.; Deybach, J. C.; Martasek, P.; Zeman, J.; da Silva, V.; Nordmann, Y.; Houstkova, H.; Rubin, A.; Holub, J.: Homozygous variegate porphyria. (Letter) Lancet I: 851, 1984.

Korda, V.; Martasek, P.; Zeman, J.; Rubin, A.: Increased erythrocyte protoporphyrin in homozygous variegate porphyria. Photodermatology 2: 257-259, 1985.

Kushner, J. P.: Laboratory diagnosis of the porphyrias. (Editorial) New Eng. J. Med. 324: 1432-1434, 1991.

Logan, G. M.; Weimer, M. K.; Ellefson, M.; Pierach, C. A.; Bloomer, J. R.: Bile porphyrin analysis in the evaluation of variegate porphyria. New Eng. J. Med. 324: 1408-1411, 1991.

Macalpine, I.; Hunter, R.: The 'insanity' of King George III: a classic case of porphyria. Brit. Med. J. 1: 65-71, 1966.

Macalpine, I.; Hunter, R.; Rimington, C.: Porphyria in the royal houses of Stuart, Hanover and Prussia: a follow-up study of George III's illness. Brit. Med. J. 1: 7-17, 1968.

Meissner, P. N.; Dailey, T. A.; Hift, R. J.; Ziman, M.; Corrigall, A. V.; Roberts, A. G.; Meissner, D. M.; Kirsch, R. E.; Dailey, H. A.: A R59W mutation in human protoporphyrinogen oxidase results in decreased enzyme activity and is prevalent in South Africans with variegate porphyria. Nature Genet. 13: 95-97, 1996.

Meyer, U. A.; Schmid, R.: The porphyrias.In: Stanbury, J. B.; Wyngaarden, J. B.; Fredrickson, D. S.: The Metabolic Basis of Inherited Disease. New York: McGraw-Hill (pub.) (4th ed.): 1978. Pp. 1166-1220.

Murphy, G. M.; Hawk, J. L. M.; Magnus, I. A.; Barrett, D. F.; Elder, G. H.; Smith, S. G.: Homozygous variegate porphyria: two similar cases in unrelated families. J. Roy. Soc. Med. 79: 361-363, 1986.

Mustajoki, P.: Variegate porphyria: twelve years' experience in Finland. Quart. J. Med. 49: 191-203, 1980.

Mustajoki, P.: Variegate porphyria. Ann. Intern. Med. 89: 238-244, 1978.

Mustajoki, P.; Tenhunen, R.; Niemi, K. M.; Nordmann, Y.; Kaariainen, H.; Norio, R.: Homozygous variegate porphyria: a severe skin disease of infancy. Clin. Genet. 32: 300-305, 1987.

Nishimura, K.; Taketani, S.; Inokuchi, H.: Cloning of a human cDNA for protoporphyrinogen oxidase by complementation in vivo of a hemG mutant of Escherichia coli. J. Biol. Chem. 270: 8076-8080, 1995.

Norris, P. G.; Elder, G. H.; Hawk, J. L. M.: Homozygous variegate porphyria: a case report. Brit. J. Derm. 122: 253-257, 1990.

Roberts, A. G.; Whatley, S. D.; Daniels, J.; Holmans, P.; Fenton, I.; Owen, M. J.; Thompson, P.; Long, C.; Elder, G. H.: Partial characterization and assignment of the gene for protoporphyrinogen oxidase and variegate porphyria to human chromosome 1q23. Hum. Molec. Genet. 4: 2387-2390, 1995.

Stine, O. C.; Smith, K. D.: The estimation of selection coefficients in Afrikaners: Huntington disease, porphyria variegata, and lipoid proteinosis. Am. J. Hum. Genet. 46: 452-458, 1990.

Taketani, S.; Inazawa, J.; Abe, T.; Furukawa, T.; Kohno, H.; Tokunaga, R.; Nishimura, K.; Inokuchi, H.: The human protoporphyrinogen oxidase gene (PPOX): organization and location to chromosome 1. Genomics 29: 698-703, 1995.

Warnich, L.; Kotze, M. J.; Groenewald, I. M.; Groenewald, J. Z.; van Brakel, M. G.; van Heerden, C. J.; de Villiers, J. N. P.; van de Ven, W. J. M.; Schoenmakers, E. F. P. M.; Taketani, S.; Retief, A. E.: Identification of three mutations and associated haplotypes in the protoporphyrinogen oxidase gene in South African families with variegate porphyria. Hum. Molec. Genet. 5: 981-984, 1996.

Warren, M. J.; Jay, M.; Hunt, D. M.; Elder, G. H.; Rohl, J. C. G.: The maddening business of King George III and porphyria. Trends Biochem. Sci. 21: 229-234, 1996.

GLYCOGEN STORAGE DISEASE 1 (*232200)

JJM NOTE

Liver masses are a frequent problem in this disorder. Although usually adenomas, there is a definite risk of hepatocellular carcinoma and, at a younger age, hepatoblastoma. The latter, an embryonal tumor, has a very peculiar array of associated syndromes, ones that at first glance seem unrelated to each other: adenomatous polyposis of the colon, Beckwith-Wiedemann syndrome, glycogen storage disease 1 and 8, and the Schinzel-Giedion midface-retraction syndrome.

The basic defect resides in glucose-6-phosphatase (EC 3.1.3.9). The liver and kidney are involved, and hypoglycemia is a major problem. Lipidemia also occurs and may lead to xanthoma formation. Survival to adulthood, previously rare, is now the usual situation. Hyperuricemia has been observed in a considerable number of patients and in some clinical gout has occurred. Inhibited tubular secretion of uric acid due to hyperlacticacidemia and ketonemia, and overproduction of uric acid have been postulated. Senior and Loridan (1968) found that the effects of glycerol administered by mouth on levels of glucose and of lactate, together with the response to epinephrine or glucagon, permitted differentiation of the several types of hepatic glycogenosis (I, II, III and IV). The glycogen storage diseases are notable examples of genetic heterogeneity. Glycogenosis I in particular illustrates pleiotropism with simulation of primary gout and xanthomatosis. Liver adenomas are often present (Howell et al., 1978) and may undergo malignant transformation (Zangeneh et al., 1969).

Stevenson et al. (1984) described hepatocellular carcinoma developing in a 29-year-old man with GSD I who had been recognized 2 years before to have multiple liver adenomas. He was relatively asymptomatic in childhood and adolescence. His growth lagged behind that of his peers and puberty was delayed to age 17. He was active in competitive sports in high school, however, and was inducted into military service at age 27. Hepatoblastoma has been described in sibs (Fraumeni et al., 1969; Napole and Campbell, 1977; Ito et al., 1987) and has been found in association with Wilms tumor (194070), polyposis coli (175100), and the EMG syndrome (130650). Burchell et al. (1987) reported 2 women, aged 51 and 22 years, with partial GSD type Ia, and a 54-year-old man with complete GSD type Ia. The patient with complete type Ia had had unexplained hepatomegaly and a bleeding diathesis since the age of 7 years and gout since age 38. He had spider angiomas, xanthomas, gouty tophi, severe hypoglycemia, and compensated metabolic acidosis. The diagnosis was not made until he presented with hepatocellular carcinoma. The 2 patients with partial type Ia had low or absent blood-glucose response to glucagon. Their hypoglycemic symptoms occurred with exercise, suggesting that they were unable to respond by increasing their hepatic glucose production above a certain level. In both cases, symptoms resolved after the introduction of frequent meals high in cornstarch, a treatment proposed by Chen et al. (1984). Of 2 sisters with type Ia GSD, both of whom had multiple hepatic adenomata, malignancy developed in 1 at the age of 20 years. AFP levels were normal throughout the entire course of this patient, whereas the younger sister had elevated levels despite the absence of malignant lesions.

Chen et al. (1988) found that of the 38 patients with type I glycogen storage disease under their care, the 18 children under age 10 years had normal renal function. Fourteen of the 20 older patients (aged 13 to 47 years) had disturbed renal function manifested by persistent proteinuria; many also had hypertension, hematuria, or altered creatinine clearance. Progressive renal insufficiency developed in 6 of these 14 patients, leading to death from renal failure in 3. At the onset of proteinuria, creatinine clearance was increased in 7 patients. Renal biopsies performed after an average of 10 years of proteinuria demonstrateded focal segmental glomerulosclerosis. In studies of 11 patients, Restaino et al. (1993) found that 5 had renal calculi, nephrocalcinosis or both, and 5 had hypercalciuria. All 9 who were tested had evidence of impaired acid excretion. Restaino et al. (1993) interpreted the findings as indicating the presence of an incomplete form of distal renal tubular acidosis, which may be the cause of hypercalciuria and nephrocalcinosis. Obara et al. (1993) described the renal histology in 2 adult patients with type I glycogen storage disease, a 37-year-old woman and a 28-year-old man.

Smit (1993) studied retrospectively 41 patients over 10 years of age from 5 different European centers. Height was below the third percentile in 19. Hypoglycemia was still reported in 6. Hepatomegaly was present in 39 of the 40 and was marked in at least 11 of these. Adenomas were detected in 11 of 39 patients. Blood cholesterol concentration was elevated in 31 of 38 patients; blood triglycerides were elevated in 29 of 34 patients. Blood uric acid concentration was elevated in 19 of 35 patients, of whom 12 were being treated with allopurinol. Ullrich and Smit (1993) reviewed the clinical aspects of type I GSD as discussed in a symposium. Pancreatitis in association with hypertriglyceridemia or severe metabolic acidosis was reported in isolated cases. No progression of hepatic adenoma after pregnancy was observed. A huge adenoma was successfully removed surgically in a female patient. The mechanism of renal hyperfiltration manifested by elevated glomerular filtration rate remained unclear. It had been found in children below the age of 1 year.

Bianchi (1993) found 50 published cases of hepatocellular adenoma in GSD I and 10 cases of hepatocellular carcinoma. Reitsma-Bierens (1993) pointed out that adult patients may have chronic renal disease. Gout, nephropathy, and renal stones are not the only complications; after a period of 'silent' hyperfiltration, renal damage develops with proteinuria, hypertension and renal dysfunction. Biopsies of such patients show focal glomerulosclerosis. In a multicenter study in the United States and Canada, Talente et al. (1994) reviewed data from 37 patients with type Ia GSD, 5 patients with GSD type Ib, and 9 patients with GSD type III, all of whom were 18 years of age or older. In the patients with GSD type Ia, problems included short stature (90%), hepatomegaly (100%), hepatic adenomas (75%), anemia (81%), proteinuria or microalbuminuria (67%), kidney calcifications (65%), osteopenia or fractures, or both (27%), increased alkaline phosphatase (61%) and gamma-glutamyltransferase (93%) activities, and increased serum cholesterol (76%) and triglyceride (100%) levels. Hyperuricemia was present in 89%. Talente et al. (1994) suggested that the hyperuricemia and pyelonephritis should be treated to prevent nephrocalcinosis and additional renal damage.

Pizzo (1980) and Furukawa et al. (1990) described vasoconstrictive pulmonary hypertension associated with GSD I. In both patients there was no evidence of portal hypertension; both patients developed pulmonary hypertension in their second decade. Furukawa et al. (1990) suggested that this rare complication, if not a coincidence, could be due to response of the pulmonary vascular bed to a circulating agent that could not be removed or inactivated by the damaged liver, or alternatively to nocturnal oxygen desaturation caused by abnormal breathing during sleep. Michels and Beaudet (1980) and Kikuchi et al. (1991), among others, reported chronic pancreatitis as a complication of the hyperlipidemia of this disorder.

Emmett and Narins (1978) found no improvement with renal transplantation. However, Selby et al. (1993) reported therapeutically successful orthotopic liver transplantation in a 16.5-year-old girl with GSD I.

Chen et al. (1984) presented experience with raw cornstarch diet as a substitute for continuous nocturnal infusions as a measure to counteract hypoglycemia, which is the common denominator in the pathogenesis of the main manifestations of the disorder. In infants with low levels of pancreatic activity, the therapy was ineffective. Chen et al. (1990) found that indicators of proximal renal tubular dysfunction improved in patients who were given dietary therapy such as total parenteral nutrition, nocturnal nasogastric infusion of glucose, or frequent oral administration of uncooked cornstarch. The indexes used were urinary excretion of amino acids, phosphate, and beta-2-microglobulin (109700). In 14 children with GSD Ia and GSD Ib who ranged in age from 4 to 16 years, Lee et al. (1996) found that the use of uncooked cornstarch achieved satisfactory glycemia lasting only a median of 4.25 hours (range 2.5 to 6).

Lei et al. (1993) reported the molecular and biochemical characteristics of human D-glucose-6-phosphatase cDNA, its gene, and the expressed protein, which is indistinguishable from human microsomal G6Pase. Furthermore, they identified several mutations in the G6PT gene that completely inactivate the enzyme in persons with type Ia glycogen storage disease. The gene contains 5 exons and spans approximately 12.5 kb. The same group mapped the G6PT gene to chromosome 17 by analysis of somatic cell hybrids (Chou, 1993; Lei et al., 1994). The gene had previously eluded molecular characterization due primarily to its tight association with the endoplasmic reticulum and nuclear membranes.

Shelly et al. (1993) isolated cDNAs encoding G6Pase by taking advantage of an albino deletion mutant mouse that is known to express markedly reduced levels of G6Pase activity. The primary defect of this mutant mouse is a loss of the fumarylacetoacetate hydrolase gene (276700) located near the albino locus on chromosome 7. Fumarylacetoacetate hydrolase (or fumarylacetoacetase) is the final enzyme in the tyrosine degradation pathway, and a deficiency of this enzyme leads to the accumulation of toxic tyrosine metabolites, resulting in reduced expression of a group of liver-specific proteins, including G6Pase. Newborn homozygous deletion mice develop hypoglycemia shortly after birth, correlating with undetectable levels of G6Pase activity. Shelly et al. (1993) isolated a full-length cDNA encoding murine liver microsomal G6Pase by screening a normal mouse liver cDNA library differentially with probes representing mRNA populations from the normal and the albino deletion mutant mouse. They showed that the murine G6Pase transcription unit spans approximately 10 kb and consists of 5 exons.

In the course of constructing a transcription map of approximately 600 kb of genomic DNA surrounding the BRCA1 gene (113705), Brody et al. (1995) identified the G6PT gene, thus regionalizing its assignment to 17q21.

Kajihara et al. (1995) found a 91-nucleotide deletion in exon 5 of the G6PT cDNA from the liver of a Japanese patient with GSD type Ia (232200.0006). Another 8 unrelated Japanese families with a total of 9 affected individuals were found to have the same mutation, thus representing 91% of patients and carriers of GSD Ia in Japan.

Demonstration of the deficiency of glucose-6-phosphatase in type I glycogen storage disease by Cori and Cori (1952) is often pointed to as the first specific enzymopathy identified in a hereditary disorder. Although it was perhaps the first disorder characterized that can be viewed as a true garrodian inborn error of metabolism, enzyme deficiency was identified earlier in methemoglobinemia (250800) by Gibson (1948).

Ryan et al. (1994) described 3 successive pregnancies in a Chinese woman with GSD type IA. The diagnosis had been made in childhood. At the age of 10 years, she was seen with hepatomegaly, renomegaly, anemia, fasting hypoglycemia, hyperuricemia, hypertriglyceridemia, frequent epistaxis, and prolonged bleeding time. An open liver biopsy revealed only trace amounts of glucose-6-phosphatase activities (1 to 2% of normal microsomal activity). She was started on a regimen of frequent day feedings, which maintained adequate nutritional support for growth and development. However, puberty was delayed, with menarche at age 14. When she was 23 years old, 2 liver masses consistent with hepatic adenomas were noted. Diet was augmented with nocturnal nasogastric feedings of Polycose (glucose polymers derived from controlled hydrolysis of cornstarch) at 90 gm/8 hours. This resulted in a reduction in total liver size but not in the size of the adenomas. The patient's first pregnancy, at age 29, ended in an unexpected fetal death at 33 weeks 5 days of gestation. At autopsy the fetus had no gross abnormalities that could explain the cause of death. An unrecognized hypoglycemic episode in the mother was suggested as a possibility. Two subsequent pregnancies were monitored in hospital after the thirty-third and thirty-fourth weeks, respectively. In the second pregnancy, cesarean section was performed at 35 weeks 4 days with delivery of a girl. A repeat cesarean section at 35 weeks 2 days was performed for the third pregnancy with delivery of a boy. Both infants were healthy and appeared to be unaffected by von Gierke disease. Hepatic adenomas did not enlarge during the pregnancies.

GSD type IaSP is clinically indistinguishable from type Ia, but was proposed to be caused by a defect in a 21-kD stabilizing protein, SP, purified on the basis of its ability to stabilize the G6Pase catalytic unit in vitro. Only one patient had been described with the type IaSP disorder (Burchell and Waddell, 1990). Lei et al. (1995) demonstrated a mutation in exon 2 of the G6PT gene that converted an arg at codon 83 to a cys (R83C) in both alleles of the type IaSP patient. The R83C mutation (232200.0002) was also demonstrated in homozygous form in 1 and heterozygous form in 5 GSD type Ia patients, indicating that so-called type IaSP is a misclassification of GSD type Ia.

Marcolongo et al. (1998) studied the transport of glucose-6-phosphate (G6P), glucose, and orthophosphate into liver microsomes isolated from 6 patients with various subtypes of GSD I using a light-scattering method. They found that G6P, glucose, and phosphate could all cross the microsomal membrane in 4 cases of GSD type Ia. In contrast, liver microsomal transport of G6P and phosphate was deficient in the GSD Ib and Ic patients, respectively. Marcolongo et al. (1998) stated that these results supported the involvement of multiple proteins (and genes) in GSD I. Since the results obtained with the light-scattering method were in accordance with conventional kinetic analysis of the microsomal glucose-6-phosphatase system, Marcolongo et al. (1998) suggested that this technique could be used to diagnose directly GSD types Ib and Ic.

Lei et al. (1995) used SSCP analysis and DNA sequencing to characterize the G6PT gene of 70 unrelated patients with enzymatically confirmed diagnosis of type Ia GSD and detected mutations in all except 17 alleles (88%). They uncovered 16 mutations that were shown by expression to abolish or greatly reduce G6Pase activity and that, therefore, were responsible for the clinical disorder. R83C (232200.0002) and Q347X (232200.0004) were the most prevalent mutations found in Caucasians; 130X (232200.0001) and R83C were most prevalent in Hispanics; R83H was most prevalent in Chinese. The Q347X mutation was identified only in Caucasians, and the 130X mutation was identified only in Hispanic patients.

Chevalier-Porst et al. (1996) sequenced both alleles of 24 French GSD type Ia patients; 14 different mutations were found, allowing the identification of complete genotypes for all the patients. These included 9 new mutations. Five mutations, Q347X, R83C, D38V (232200.0006), G188R (232200.0012), and 158Cdel, accounted for 75% of the mutated alleles.

Parvari et al. (1997) reported the biochemical and clinical characteristics as well as mutational analyses of 12 Israeli GSD Ia patients of different families, who represent most of the GSD Ia patients in Israel. All 9 Jewish patients, as well as a Muslim Arab patient, were found to have the R83C mutation (232200.0002). Two Muslim Arab patients had the val166-to-gly (V166G) mutation (232200.0004), which had not been found in other populations.

The use of diazoxide in patients with GSD type Ia was first described by Rennert and Mukhopadhyay (1968) for improvement of glucose homeostasis. The treatment was abandoned because of skin rashes and later forgotten. Unaware of this early observation, Nuoffer et al. (1997) treated 2 originally prepubertal girls with GSD type Ia and short stature with low-dose diazoxide (3 to 4.8 mg/kg per day) for 7 and 4 years, respectively. Both showed an impressive catch-up growth. This appeared to be due to prolongation of normoglycemia after meals and reduction of fasting lacticacidosis by the drug. Nuoffer et al. (1997) speculated that blood lacticacidosis is a major cause of growth retardation in this disorder. The mode of action of diazoxide seems to be linked to K(+)-ATP channel activation. The resulting hyperpolarization of membranes decreases insulin release. The drug was given at the beginning of meals and before going to bed.

Animal Model

Lei et al. (1996) created a G6Pase knockout mouse that mimics the pathophysiology of human GSD Ia patients. In the knockout mouse, both the GTPase enzymatic activity and the glucose-6-P transport activity were destroyed. By examining G6Pase in liver and kidney, the primary gluconeogenic tissues, they demonstrated that glucose-6-P transport and hydrolysis are performed by separate proteins that are tightly coupled. They proposed a modified translocase catalytic unit model for G6Pase catalysis.

See also Cohen et al. (1985); Fine et al. (1968); Greene et al. (1976); Howell (1965); Howell (1978); Limmer et al. (1988); Malatack et al. (1983); Senior and Loridan (1968); Sidbury (1965); Spencer-Peet et al. (1971); Stamm and Webb (1975)

Bianchi, L.: Glycogen storage disease I and hepatocellular tumours. Europ. J. Pediat. 152 (suppl. 1): S63-S70, 1993.

Brody, L. C.; Abel, K. J.; Castilla, L. H.; Couch, F. J.; McKinley, D. R.; Yin, G.-Y.; Ho, P. P.; Merajver, S.; Chandrasekharappa, S. C.; Xu, J.; Cole, J. L.; Struewing, J. P.; Valdes, J. M.; Collins, F. S.; Weber, B. L.: Construction of a transcription map surrounding the BRCA1 locus of human chromosome 17. Genomics 25: 238-247, 1995.

Burchell, A.; Jung, R. T.; Lang, C. C.; Bennet, W.; Shepherd, A. N.: Diagnosis of type 1a and type 1c glycogen storage diseases in adults. Lancet I: 1059-1062, 1987.

Burchell, A.; Waddell, I. D.: Diagnosis of a novel glycogen storage disease: type 1aSP. J. Inherit. Metab. Dis. 13: 247-249, 1990.

Chen, Y.-T.; Coleman, R. A.; Scheinman, J. I.; Kolbeck, P. C.; Sidbury, J. B.: Renal disease in type I glycogen storage disease. New Eng. J. Med. 318: 7-11, 1988.

Chen, Y.-T.; Cornblath, M.; Sidbury, J. B.: Cornstarch therapy in type I glycogen-storage disease. New Eng. J. Med. 310: 171-175, 1984.

Chen, Y.-T.; Scheinman, J. I.; Park, H. K.; Coleman, R. A.; Roe, C. R.: Amelioration of proximal renal tubular dysfunction in type I glycogen storage disease with dietary therapy. New Eng. J. Med. 323: 590-593, 1990.

Chevalier-Porst, F.; Bozon, D.; Bonardot, A.-M.; Bruni, N.; Mithieux, G.; Mathieu, M.; Maire, I.: Mutation analysis in 24 French patients with glycogen storage disease type 1a. J. Med. Genet. 33: 358-360, 1996.

Chou, J. Y.: Personal Communication. Bethesda, Md. 11/4/1993.

Cohen, J. L.; Vinik, A.; Faller, J.; Fox, I. H.: Hyperuricemia in glycogen storage disease type I: contributions by hypoglycemia and hyperglucagonemia to increased urate production. J. Clin. Invest. 75: 251-257, 1985.

Cori, G. T.; Cori, C. F.: Glucose-6-phosphatase of the liver in glycogen storage disease. J. Biol. Chem. 199: 661-667, 1952.

Emmett, M.; Narins, R. G.: Renal transplantation in type I glycogenosis: failure to improve glucose metabolism. J.A.M.A. 239: 1642-1644, 1978.

Fine, R. N.; Wilson, W. A.; Donnell, G. N.: Retinal changes in glycogen storage disease type I. Am. J. Dis. Child. 115: 328-331, 1968.

Fraumeni, J. F.; Rosen, P. J.; Hull, E. W.; Barth, R. F.; Shapiro, S. R.; O'Connor, J. F.: Hepatoblastoma in infant sisters. Cancer 24: 1086-1090, 1969.

Furukawa, N.; Kinugasa, A.; Inoue, F.; Imashuku, S.; Takamatsu, T.; Sawada, T.: Type I glycogen storage disease with vasoconstrictive pulmonary hypertension. J. Inherit. Metab. Dis. 13: 102-107, 1990.

Gibson, Q. H.: The reduction of methaemoglobin in red blood cells and studies on the cause of idiopathic methaemoglobinaemia. Biochem. J. 42: 13-23, 1948.

Greene, H. L.; Slonim, A. E.; O'Neill, J. A., Jr.; Burr, I. M.: Continuous nocturnal intragastric feeding for management of type 1 glycogen-storage disease. New Eng. J. Med. 294: 423-425, 1976.

Howell, R. R.: The interrelationship of glycogen storage disease and gout. Arthritis Rheum. 8: 780-785, 1965.

Howell, R. R.: The glycogen storage diseases.In: Stanbury, J. B.; Wyngaarden, J. B.; Fredrickson, D. S.: The Metabolic Basis of Inherited Disease. New York: McGraw-Hill (pub.) (4th ed.): 1978. Pp. 137-159.

Howell, R. R.; Stevenson, R. E.; Ben-Menachem, Y.; Phyliky, R. L.; Berry, D. H.: Hepatic adenomata with type I glycogen storage disease. J. Nucl. Med. 19: 354-358, 1978.

Ito, E.; Sato, Y.; Kawauchi, K.; Munakata, H.; Kamata, Y.; Yodono, H.; Yokoyama, M.: Type 1a glycogen storage disease with hepatoblastoma in siblings. Cancer 59: 1776-1780, 1987.

Kajihara, S.; Matsuhashi, S.; Yamamoto, K.; Kido, K.; Tsuji, K.; Tanae, A.; Fujiyama, S.; Itoh, T.; Tanigawa, K.; Uchida, M.; Setoguchi, Y.; Motomura, M.; Mizuta, T.; Sakai, T.: Exon redefinition by a point mutation within exon 5 of the glucose-6-phosphatase gene is the major cause of glycogen storage disease type 1a in Japan. Am. J. Hum. Genet. 57: 549-555, 1995.

Kikuchi, M.; Hasegawa, K.; Handa, I.; Watabe, M.; Narisawa, K.; Tada, K.: Chronic pancreatitis in a child with glycogen storage disease type 1. Europ. J. Pediat. 150: 852-853, 1991.

Lee, P. J.; Dixon, M. A.; Leonard, J. V.: Uncooked cornstarch—efficacy in type I glycogenosis. Arch. Dis. Child. 74: 546-547, 1996.

Lee, W.-J.; Lee, H.-M.; Chi, C.-S.; Shu, S.-G.; Lin, L.-Y.; Lin, W.-H.: Genetic analysis of the glucose-6-phosphatase mutation of type 1a glycogen storage disease in a Chinese family. Clin. Genet. 50: 206-211, 1996.

Lei, K.-J.; Chen, H.; Pan, C.-J.; Ward, J. M.; Mosinger, B., Jr.; Lee, E. J.; Westphal, H.; Mansfield, B. C.; Chou, J. Y.: Glucose-6-phosphatase dependent substrate transport in the glycogen storage disease type-1a mouse. Nature Genet. 13: 203-209, 1996.

Lei, K.-J.; Chen, Y.-T.; Chen, H.; Wong, L.-J. C.; Liu, J.-L.; McConkie-Rosell, A.; Van Hove, J. L. K.; Ou, H. C.-Y.; Yeh, N. J.; Pan, L. Y.; Chou, J. Y.: Genetic basis of glycogen storage disease type 1a: prevalent mutations at the glucose-6-phosphatase locus. Am. J. Hum. Genet. 57: 766-771, 1995.

Lei, K.-J.; Pan, C.-J.; Shelly, L. L.; Liu, J.-L.; Chou, J. Y.: Identification of mutations in the gene for glucose-6-phosphatase, the enzyme deficient in glycogen storage disease type 1A. J. Clin. Invest. 93: 1994-1999, 1994.
Lei, K.-J.; Shelly, L. L.; Lin, B.; Sidbury, J. B.; Chen, Y.-T.; Nordlie, R. C.; Chou, J. Y.: Mutations in the glucose-6-phosphatase gene are associated with glycogen storage disease types 1a and 1aSP but not 1b and 1c. J. Clin. Invest. 95: 234-240, 1995.
Lei, K.-J.; Shelly, L. L.; Pan, C.-J.; Sidbury, J. B.; Chou, J. Y.: Mutations in the glucose-6-phosphatase gene that cause glycogen storage disease type 1a. Science 262: 580-583, 1993.
Limmer, J.; Fleig, W. E.; Leupold, D.; Bittner, R.; Ditschuneit, H.; Beger, H.-G.: Hepatocellular carcinoma in type I glycogen storage disease. Hepatology 8: 531-537, 1988.
Malatack, J. J.; Finegold, D. N.; Iwatsuki, S.; Shaw, B. W., Jr.; Gartner, J. C.; Zitelli, B. J.; Roe, T.; Starzl, T. E.: Liver transplantation for type I glycogen storage disease. Lancet I: 1073-1075, 1983.
Marcolongo, P.; Banhegyi, G.; Benedetti, A.; Hinds, C. J.; Burchell, A.: Liver microsomal transport of glucose-6-phosphate, glucose, and phosphate in type 1 glycogen storage diseases. J. Clin. Endocr. Metab. 83: 224-229, 1998.
Michels, V. V.; Beaudet, A. L.: Hemorrhagic pancreatitis in a patient with glycogen storage disease type I. Clin. Genet. 17: 220-222, 1980.
Nakai, K.; Sakamoto, H.: Construction of a novel database containing aberrant splicing mutations of mammalian genes. Gene 141: 171-177, 1994.
Napole, V. M.; Campbell, W. G., Jr.: Hepatoblastoma in infant sister and brother. Cancer 39: 2647-2650, 1977.
Nuoffer, J. M.; Mullis, P. E.; Wiesmann, U. N.: Treatment with low-dose diazoxide in two growth-retarded prepubertal girls with glycogen storage disease type Ia resulted in catch-up growth. J. Inherit. Metab. Dis. 20: 790-798, 1997.
Obara, K.; Saito, T.; Sato, H.; Ogawa, M.; Igarashi, Y.; Yoshinaga, K.: Renal histology in two adult patients with type I glycogen storage disease. Clin. Nephrol. 39: 59-64, 1993.
Parvari, R.; Lei, K.-J.; Bashan, N.; Hershkovitz, E.; Korman, S. H.; Barash, V.; Lerman-Sagie, T.; Mandel, H.; Chou, J. Y.; Moses, S. W.: Glycogen storage disease type 1a in Israel: biochemical, clinical, and mutational studies. Am. J. Med. Genet. 72: 286-290, 1997.
Pizzo, C. J.: Type I glycogen storage disease with focal nodular hyperplasia of the liver and vasoconstrictive pulmonary hypertension. Pediatrics 65: 341-343, 1980.
Qu, Y.; Abdenur, J. E.; Eng, C. M.; Desnick, R. J.: Molecular prenatal diagnosis of glycogen storage disease type Ia. Prenatal Diag. 15: 333-336, 1996.
Reitsma-Bierens, W. C. C.: Renal complications in glycogen storage disease type I. Europ. J. Pediat. 152 (suppl. 1): S60-S62, 1993.
Rennert, O. M.; Mukhopadhyay, D.: Diazoxide in Von Gierke's disease. Arch. Dis. Child. 43: 358-361, 1968.
Restaino, I.; Kaplan, B. S.; Stanley, C.; Baker, L.: Nephrolithiasis, hypocitraturia, and a distal renal tubular acidification defect in type 1 glycogen storage disease. J. Pediat. 122: 392-396, 1993.
Ryan, I. P.; Havel, R. J.; Laros, R. K., Jr.: Three consecutive pregnancies in a patient with glycogen storage disease type IA (von Gierke's disease). Am. J. Obstet. Gynec. 170: 1687-1691, 1994.
Selby, R.; Starzl, T. E.; Yunis, E.; Todo, S.; Tzakis, A. G.; Brown, B. I.; Kendall, R. S.: Liver transplantation for type I and type IV glycogen storage disease. Europ. J. Pediat. 152 (suppl. 1): S71-S76, 1993.
Senior, B.; Loridan, L.: Functional differentiation of glycogenoses of the liver with respect to the use of glycerol. New Eng. J. Med. 279: 965-970, 1968.
Senior, B.; Loridan, L.: Liver glycogenoses: metabolism of intravenously administered glycerol. New Eng. J. Med. 279: 958-965, 1968.
Shelly, L. L.; Lei, K.-J.; Pan, C.-J.; Sakata, S. F.; Ruppert, S.; Schutz, G.; Chou, J. Y.: Isolation of the gene for murine glucose-6-phosphatase, the enzyme deficient in glycogen storage disease type 1A. J. Biol. Chem. 268: 21482-21485, 1993.
Sidbury, J. B., Jr.: The genetics of the glycogen storage disease.In: Steinberg, A. G.; Bearn, A. G.: Progress in Medical Genetics. New York: Grune and Stratton (pub.) 4: 1965. Pp. 32-58.
Smit, G. P. A.: The long-term outcome of patients with glycogen storage disease type Ia. Europ. J. Pediat. 152 (suppl. 1): S52-S55, 1993.
Spencer-Peet, J.; Norman, M. E.; Lake, B. D.; McNamara, J.; Patrick, A. D.: Hepatic glycogen storage disease. Clinical and laboratory findings in 23 cases. Quart. J. Med. 40: 95-114, 1971.
Stamm, W. E.; Webb, D. I.: Partial deficiency of hepatic glucose-6-phosphatase in an adult patient. Arch. Intern. Med. 135: 1107-1109, 1975.
Stevenson, R. E.; Ben-Menachem, Y.; Dudrick, S.; Howell, R. R.: Hepatocellular carcinoma in type 1 glycogen storage disease. Proc. Greenwood Genet. Center 3: 39-46, 1984.
Talente, G. M.; Coleman, R. A.; Alter, C.; Baker, L.; Brown, B. I.; Cannon, R. A.; Chen, Y.-T.; Crigler, J. F., Jr.; Ferreira, P.; Haworth, J. C.; Herman, G. E.; Issenman, R. M.; Keating, J. P.; Linde, R.; Roe, T. F.; Senior, B.; Wolfsdorf, J. I.: Glycogen storage disease in adults. Ann. Intern. Med. 120: 218-226, 1994.
Ullrich, K.; Smit, G. P. A.: Clinical aspects of glycogen storage disease type I: summary of the discussions. Europ. J. Pediat. 152 (suppl. 1): S87-S88, 1993.
Zangeneh, F.; Limbeck, G. A.; Brown, B. I.; Emch, J. R.; Arcasoy, M. M.; Goldenberg, V. E.; Kelley, V. C.: Hepatorenal glycogenosis (type I glycogenosis) and carcinoma of the liver. J. Pediat. 74: 73-83, 1969.

GLYCOGEN STORAGE DISEASE 1B (#232220)

JJM NOTE

Despite intriguing clinical differences between the two glycogen storage diseases 1, the risk for adenomas seems similar.

A number sign (#) is used with this entry because of evidence that type Ib glycogen storage disease can result from mutation in the glucose-6-phosphate transporter gene (602671).

Senior and Loridan (1968) proposed the existence of a second type of von Gierke disease in which, although glucose-6-phosphatase activity is present on in vitro assay, glucose is not liberated from glucose-6-phosphate in vivo. They referred to this as 'functional deficiency of G6P.' They pointed out that some mutants in Neurospora show impaired enzyme function in the intact fungus despite normal activity in homogenates. Arion et al. (1975) concluded that G6Pase activity requires 2 components of the microsomal membrane: (1) a glucose-6-phosphate specific transport system that shuttles G6P from the cytoplasm to the lumen of the endoplasmic reticulum (a G6P translocase), and (2) an enzyme, glucose-6-phosphate phosphohydrolase, bound to the luminal surface of the membrane. Narisawa et al. (1978) described a patient who appeared to have a defect in the transport system. In liver without detergent, enzyme activity was very low but normal activity was obtained by addition of detergent. Kuzuya et al. (1983) reported a 25-year-old patient. Protuberant abdomen and diarrhea were noted at age 1 or 2 years, and short stature and hepatomegaly at age 4 years. At age 18, yellowish-red spots appeared on her legs and hypertension was detected. At age 20, she was 138 cm tall. Eruptive xanthoma and hyperlipidemia were present. Liver scintography suggested the presence of adenomas.

Recurrent infections and neutropenia have been recognized as distinctive features of GSD Ib. Corbeel et al. (1983) provided a 6-year follow-up on the hematologic effects of termino-lateral portacaval anastomosis. Granulocyte counts returned to normal and recurrent infections ceased after the shunt. Platelet dysfunction, evident before surgery, was also corrected. Marked hypochromic anemia, probably caused by sequestration of iron in the spleen and resistant to therapy, was a persistent feature in this patient. The mechanism of the granulocyte defect in this disorder was discussed. Roe et al. (1986) observed Crohn disease in 2 unrelated boys with GSD Ib. Their neutrophils showed severe chronic neutropenia and markedly deficient chemotactic response, whereas the leukocytes were normal in 4 patients with GSD Ia (232200). Thus, chronic inflammatory bowel disease (IBD) appears to be an integral part of GSD Ib and the abnormality of leukocytes is probably involved in the pathogenesis of the IBD. Oral lesions and perianal abscesses are common in this disorder (Ambruso et al., 1985). Ueno et al. (1986) found that neutrophils were defective in both motility and respiratory burst, whereas monocytes showed a defect only in respiratory burst. Bashan et al. (1988) showed that the rate of 2-deoxyglucose transport into GSD Ib polymorphonuclear leukocytes was 30% of that into cells of normal controls. Transport was normal in GSD Ib lymphocytes and in GSD Ia polymorphonuclear leukocytes and lymphocytes. The striking limitation of glucose transport across the cell membrane of polymorphonuclear leukocytes probably accounts for the impairment of leukocyte function that is characteristic of GSD Ib but not GSD Ia. Schroten et al. (1991) used granulocyte colony-stimulating factor (CSF3; 138970) to treat successfully the neutropenia in 2 patients with GSD Ib associated with recurrent bacterial infections. Roe et al. (1992) administered granulocyte-macrophage colony stimulating facto r (CSF2; 138960) to the 2 adolescent boys whom they had reported in 1986 (Roe et al., 1986). They observed a prompt increase in neutrophil counts to normal, complete relief from abdominal symptoms, and an increase in appetite, energy, and weight, and a feeling of well being. There was radiologic evidence of bowel healing and a decrease in the erythrocyte sedimentation rate. Both patients remained free of oral and anal lesions over a period of 10 and 12 months of treatment. One patient was switched to G-CSF (CSF3) because of a presumed allergic reaction to GM-CSF. In a multicenter study in the United States and Canada, Talente et al. (1994) identified 5 patients with GSD type Ib who were 18 years of age or older. Severe recurrent bacterial infections and gingivitis were present. One patient, a 22-year-old college student, was described in detail. She had severe recurrent stomatitis, recurrent otitis media and externa, perianal and perirectal abscesses, and, at the age of 12 years, 2 brain abscesses due to Staphylococcus aureus. At 18 years of age, she was as tall as an 8-year-old and had not undergone any pubertal changes.

In a patient with GSD Ib, Heyne and Henke-Wolter (1989) found a change in the oligosaccharide side chains of the alpha-1-antitrypsin (107400) glycoprotein suggesting effects of the limited availability of glucose or glucose derivatives for the synthesis of N-glycosidic glycoproteins. Kikuchi et al. (1990) found secondary amyloidosis in a 12-year-old girl with GSD Ib.

In 14 children (aged 4 to 16 years) with GSD Ia and GSD Ib, Lee et al. (1996) found that the use of uncooked cornstarch loads resulted in satisfactory glycemia lasting only a median of 4.25 hours (range 2.5 to 6).

Annabi et al. (1998) reported linkage of the GSD Ib locus to genetic markers spanning a 3-cM region on 11q23. The region is located between D11S939 centromerically and D11S4129 telomerically and includes the IL10R (146933), ATP1G1 (601814), and ALL1 (159555) genes. The authors studied 8 consanguineous families and 1 nonconsanguineous family of various ethnic origins. The assignment to chromosome 11 was confirmed by Kure et al. (1998), who showed that the translocase gene that is mutated in this disorder maps to chromosome 11 by study of somatic cell hybrids.

In 2 female patients with GSD Ib, Gerin et al. (1997) found 2 point mutations in the glucose-6-phosphate translocase gene (602671.0001 and 602671.0002). Kure et al. (1998) identified 3 additional mutations, one of which, W118R (602671.0003), may be unusually frequent among Japanese patients with GSD Ib.

See also Buchino et al. (1983); Heyne et al. (1984); Kamoun (1980); Narisawa et al. (1986); Sann et al. (1980); Schaub et al. (1981); Schaub and Heyne (1983); Seger et al. (1984)

Ambruso, D. R.; McCabe, E. R. B.; Anderson, D.; Beaudet, A.; Ballas, L. M.; Brandt, I. K.; Brown, B.; Coleman, R.; Dunger, D. B.; Falletta, J. M.; Friedman, H. S.; Haymond, M. W.; Keating, J. P.; Kinney, T. R.; Leonard, J. V.; Mahoney, D. H., Jr.; Matalon, R.; Roe, T. F.; Simmons, P.; Slonim, A. E.: Infectious and bleeding complications in patients with glycogenosis Ib. Am. J. Dis. Child. 139: 691-697, 1985.

Annabi, B.; Hiraiwa, H.; Mansfield, B. C.; Lei, K.-J.; Ubagai, T.; Polymeropoulos, M. H.; Moses, S. W.; Parvari, R.; Hershkovitz, E.; Mandel, H.; Fryman, M.; Chou, J. Y.: The gene for glycogen-storage disease type 1b maps to chromosome 11q23. Am. J. Hum. Genet. 62: 400-405, 1998.

Arion, W. J.; Wallin, B. K.; Lange, A. J.; Ballas, L. M.: On the involvement of a glucose-6-phosphate transport system in the function of microsomal glucose-6-phosphate. Molec. Cell. Biochem. 6: 75-83, 1975.

Bashan, N.; Hagai, Y.; Potashnik, R.; Moses, S. W.: Impaired carbohydrate metabolism of polymorphonuclear leukocytes in glycogen storage disease Ib. J. Clin. Invest. 81: 1317-1322, 1988.

Buchino, J. J.; Brown, B. I.; Volk, D. M.: Glycogen storage disease type Ib. Arch. Path. Lab. Med. 107: 283-285, 1983.

Corbeel, L.; Boogaerts, M.; Van den Berghe, G.; Everaerts, M. C.; Marchal, G.; Eeckels, R.: Haematological findings in type Ib glycogen storage disease before and after portacaval shunt. Europ. J. Pediat. 140: 273-275, 1983.

Gerin, I.; Veiga-da-Cunha, M.; Achouri, Y.; Collet, J.-F.; Van Schaftingen, E.: Sequence of a putative glucose 6-phosphate translocase, mutated in glycogen storage disease type Ib. FEBS Lett. 419: 235-238, 1997.

Heyne, K.; Henke-Wolter, J.: Glycogen storage disease Ib: modification of alpha-1-antitrypsin glycoprotein microheterogeneity. Europ. J. Pediat. 148: 341-343, 1989.

Heyne, K.; Hosenfeld, D.; Grote, W.; Schaub, J.: Glycogen storage disease type Ib: familial bleeding tendency. Europ. J. Pediat. 143: 7-9, 1984.

Kamoun, P. P.: Is type 1b glycogenosis related to an anomeric preference for glucose-6-phosphate uptake by hepatic microsomes?. Med. Hypotheses 6: 1135-1139, 1980.

Kikuchi, M.; Haginoya, K.; Miyabayashi, S.; Igarashi, Y.; Narisawa, K.; Tada, K.: Secondary amyloidosis in glycogen storage disease type Ib. Europ. J. Pediat. 149: 344-345, 1990.

Kure, S.; Suzuki, Y.; Matsubara, Y.; Sakamoto, O.; Shintaku, H.; Isshiki, G.; Hoshida, C.; Izumi, I.; Sakura, N.; Narisawa, K.: Molecular analysis of glycogen storage disease type Ib: identification of a prevalent mutation among Japanese patients and assignment of a putative glucose-6-phosphate translocase gene to chromosome 11. Biochem. Biophys. Res. Commun. 248: 426-431, 1998.

Kuzuya, T.; Matsuda, A.; Yoshida, S.; Narisawa, K.; Tada, K.; Saito, T.; Matsushita, M.: An adult case of type Ib glycogen-storage disease: enzymatic and histochemical studies. New Eng. J. Med. 308: 566-569, 1983.

Lee, P. J.; Dixon, M. A.; Leonard, J. V.: Uncooked cornstarch—efficacy in type I glycogenosis. Arch. Dis. Child. 74: 546-547, 1996.

Narisawa, K.; Igarashi, Y.; Otomo, H.; Tada, K.: A new variant of glycogen storage disease type I probably due to a defect in the glucose-6-phosphate transport system. Biochem. Biophys. Res. Commun. 83: 1360-1364, 1978.

Narisawa, K.; Ishizawa, S.; Okumura, H.; Tada, K.; Kuzuya, T.: Neutrophil metabolic dysfunction in genetically heterogeneous patients with glycogen storage disease type 1b. J. Inherit. Metab. Dis. 9: 297-300, 1986.

Roe, T. F.; Coates, T. D.; Thomas, D. W.; Miller, J. H.; Gilsanz, V.: Treatment of chronic inflammatory bowel disease in glycogen storage disease type Ib with colony-stimulating factors. New Eng. J. Med. 326: 1666-1669, 1992.

Roe, T. F.; Thomas, D. W.; Gilsanz, V.; Isaacs, H., Jr.; Atkinson, J. B.: Inflammatory bowel disease in glycogen storage disease type Ib. J. Pediat. 109: 55-59, 1986.

Sann, L.; Mathieu, M.; Bourgeous, J.; Bienvenu, J.; Bethenod, M.: In vivo evidence for defective activity of glucose-6-phosphatase in type I B glycogenosis. J. Pediat. 96: 691-694, 1980.

Schaub, J.; Bartholome, K.; Feist, D.; Schmidt, H.: Glycogenosis type Ib: further evidence for a membrane disease. (Letter) Europ. J. Pediat. 135: 325 only, 1981.

Schaub, J.; Heyne, K.: Glycogen storage disease type Ib. Europ. J. Pediat. 140: 283-288, 1983.

Schroten, H.; Roesler, J.; Breidenbach, T.; Wendel, U.; Elsner, J.; Schweitzer, S.; Zeidler, C.; Burdach, S.; Lohmann-Matthes, M.-L.; Wahn, V.; Welte, K.: Granulocyte and granulocyte-macrophage colony-stimulating factors for treatment of neutropenia in glycogen storage disease type Ib. J. Pediat. 119: 748-754, 1991.

Seger, R.; Steinmann, B.; Tiefenauer, L.; Matsunaga, T.; Gitzelmann, R.: Glycogenosis Ib: neutrophil microbicidal defects due to impaired hexose monophosphate shunt. Pediat. Res. 18: 297-299, 1984.

Senior, B.; Loridan, L.: Functional differentiation of glycogenoses of the liver with respect to the use of glycerol. New Eng. J. Med. 279: 965-970, 1968.

Talente, G. M.; Coleman, R. A.; Alter, C.; Baker, L.; Brown, B. I.; Cannon, R. A.; Chen, Y.-T.; Crigler, J. F., Jr.; Ferreira, P.; Haworth, J. C.; Herman, G. E.; Issenman, R. M.; Keating, J. P.; Linde, R.; Roe, T. F.; Senior, B.; Wolfsdorf, J. I.: Glycogen storage disease in adults. Ann. Intern. Med. 120: 218-226, 1994.

Ueno, N.; Tomita, M.; Ariga, T.; Ohkawa, M.; Nagano, S.; Takahashi, Y.; Arashima, S.; Matsumoto, S.: Impaired monocyte function in glycogen storage disease type Ib. Europ. J. Pediat. 145: 312-314, 1986.

GLYCOGEN STORAGE DISEASE 8 (*306000)

JJM NOTE

Although glycogen storage disease 1 is well known to be complicated by liver tumors, not all types predispose equally.

Deficiency of liver phosphorylase kinase (PHK; ATP:phosphotransferase; EC 2.7.1.38) produces one of the mildest of the glycogenoses of man. The clinical symptoms include hepatomegaly, growth retardation, elevation of glutamate-pyruvate transaminase and glutamate-oxaloacetate transaminase, hypercholesterolemia, hypertriglyceridemia, and fasting hyperketosis (Schimke et al., 1973; Willems et al., 1990). With age, these clinical and biochemical abnormalities gradually disappear and most adult patients are asymptomatic (Willems et al., 1990). Nagai et al. (1988) described renal tubular acidosis in a 2.5-year-old Japanese boy with phosphorylase kinase deficiency.

(The classification, particularly the numerology, of the glycogenoses was long a matter of dispute. For example, Huijing (1970) referred to this disorder as glycogen storage disease type VIA; Hug (1974) assigned number VIII to a presumedly recessive form of phosphorylase deficiency with brain involvement and number IX to phosphorylase kinase deficiency (see Schimke et al., 1973). McAdams et al. (1974) presented information on classification and morphology of the glycogenoses. Hers et al. (1989) recommended that one call this disorder simply 'X-linked phosphorylase b kinase deficiency and omit a numerical designation,' because confusingly different number systems have been used by different authors. Similarly, they would have us use the designation 'autosomal recessive phosphorylase kinase b deficiency' (261750), without numerical designation. The clinical manifestations of these 2 disorders and of liver phosphorylase deficiency (232700), called here glycogen storage disease VI, are all those of a mild form of hepatomegalic glycogenosis, without hyperlacticacidemia or hyperuricemia.)

X-linked phosphorylase b kinase deficiency is similar to glycogen storage disease VI in that both have low phosphorylase activity in the absence of adenosine monophosphate (AMP); it differs in that phosphorylase kinase activity is low (Huijing, 1967). Williams and Field (1961) found low leukocyte phosphorylase activity in 2 affected brothers and normal activity in an unaffected brother and in the father. An intermediately low level in the mother, together with affected males, suggested X-linked inheritance. Wallis et al. (1966) restudied the family and with new methods found support for X-linkage. Huijing and Fernandez (1969) studied 2 kindreds, one of which had 6 affected plus 2 possibly affected males. The other had 20 affected males, 2 affected females, and 7 probably affected males. Huijing and Fernandez (1970) suggested that the affected females studied by Hug et al. (1969) were heterozygotes. Since phosphorylase kinase is known to be enzymatically activated (Krebs et al., 1964), it is possible that an activating enzyme is controlled by the X chromosome. By cloning cells of heterozygotes, Migeon and Huijing (1974) demonstrated some fibroblasts with enzymatic levels like those of affected hemizygotes. This was presented as proof of X-linkage and X-inactivation of the phosphorylase kinase locus. An X-linked codominant electrophoretic polymorphism of phosphorylase kinase is known in the mouse (Lyon et al., 1967). Huijing (1970) pointed out similarities and differences of the human and murine defects. Malthus et al. (1980) described deficiency of liver phosphorylase kinase in rats and concluded that it is autosomal recessive. This may not be a true exception to Ohno's law of the evolutionary conservatism of the X chromosome; see comment of Hug (1974) above. Schneider et al. (1993) reviewed the animal mutants that result in PHK-linked glycogenoses. Two different X-linked disorders are known, as well as an autosomal recessive PHK deficiency affecting the liver and most other tissues but not muscle, in the rat.

In 4 boys with PHK-deficient glycogenosis, aged 29 months to 43 months, Garibaldi et al. (1978) found that dextrothyroxine (D-T4) had dramatic effects: the liver, previously greatly enlarged, returned to normal size; serum GOT, GPT, and triglycerides fell to normal; and hypoglycemia was corrected.

Protein phosphorylation is a major mechanism of signal transduction in eukaryotic cells. About 100 protein kinases have been described.

Features of primary structure are shared by all protein kinases and there are additional ones that discriminate the serine/threonine kinases from the tyrosine kinases, so-named for the residues in the proteins that undergo phosphorylation.

There appear to be 2 PHK loci on the X chromosome: one for the alpha subunit of muscle PHK (PHKA1; 311870), and one for the alpha subunit of liver PHK. Davisson (1987) considered human X-linked hepatic glycogenosis due to deficient phosphorylase kinase activity to be the human homolog of phosphorylase kinase deficiency in the I-strain mouse. The activity of phosphorylase kinase is severely deficient in skeletal muscle of these mice. Both the Phk mutation of I-strain mice and the wildtype muscle Phka gene were mapped close to the structural gene encoding phosphoglycerate kinase (Pgk-1) on the mouse X chromosome. Furthermore, severely reduced alpha subunit mRNA levels were found in these mutant mice compared to normal mice. Thus, the muscle PHKA gene is probably the site of the mutation responsible for the myopathy in the I-strain mice. However, because human hepatic glycogenosis is not due to mutation in muscle PHKA, this disease is not the human homolog of myopathy in I-strain mice. There appears to be a liver-specific alpha subunit whose mRNA does not hybridize with the cDNA of the muscle alpha subunit (Davidson et al., 1992).

Zander et al. (1988) reported the primary structure of the alpha subunit of fast-twitch skeletal muscle phosphorylase kinase, and Kilimann et al. (1988) presented the sequence of the beta subunit (PHKB; 172490). The phosphorylase kinase of liver and muscle has four dissimilar subunits of descending order of molecular weight: alpha, beta, gamma (172470), and delta (114180). Ryder-Cook et al. (1989) pointed out that the beta, gamma, and delta subunits are autosomal. Davidson et al. (1992) cloned cDNA encoding the liver isoform of the alpha subunit of phosphorylase kinase. Comparison of its sequence with that of the previously characterized muscle isoform (PHKA1; 311870) indicated a pattern of highly conserved and variable domains and demonstrated that the 2 isoforms are the products of distinct genes. In contrast to the gene for PHKA1, the gene for PHKA2 was found to be expressed predominantly in liver and other nonmuscle tissues. In X-linked phosphorylase kinase deficiency, the enzyme is lacking in liver but normal in muscle. In an autosomal recessive form (261750), the enzyme is lacking in both liver and muscle.

Willems et al. (1991) performed linkage analysis with X-chromosomal polymorphic DNA markers in 2 families with X-linked liver glycogenosis. Multipoint linkage analysis excluded the muscle PHKA region (Xq12-q13) as the site of the mutation, and indicated that the mutation responsible for X-linked liver glycogenosis is located in Xp22 between DXS143 and DXS41. By in situ hybridization, Wauters et al. (1992) demonstrated that the PHKA2 gene is located in the distal part of Xp in the same region as the mutation for X-linked liver glycogenosis. Davidson et al. (1992) mapped the PHKA2 gene to Xp22.2-p22.1 by mapping with hybrid lines that contained defined regions of the X chromosome. It is noteworthy that PHKA1 and PHKA2 are located on Xq and Xp, respectively; PHKB is located on chromosome 16. 10,7:Hendrickx et al. (1992, 1993) found a combined multipoint lod score of 16.79 for linkage of X-linked liver glycogenosis with Xp22. The corresponding gene in the rabbit was located to a comparable chromosomal region. Furthermore, a human cDNA clone for PHKA2, isolated from a human hepatoma lambda-gt11 cDNA library, showed a high degree of homology to the rabbit cDNA at the nucleotide level (89%) and amino acid level (93.5%). Furthermore, fluorescence in situ hybridization mapped the human gene to Xp22. Although Southern blot analysis in 6 unrelated X-linked liver glycogenosis patients did not reveal major gene rearrangements, the mutation in this disorder, probably small deletions or point mutations, probably reside therein.

In addition to the classic form of X-linked glycogenosis, a second form has been described by Keating et al. (1985) and others. The patients show clinical and biochemical abnormalities very similar to those of classic XLG, including growth retardation, hepatomegaly, and elevation of liver enzymes GOT and GPT. These symptoms gradually disappear with age. PHK activity is low in liver but normal in erythrocytes and leukocytes. Hendrickx et al. (1994) performed linkage analysis in 4 families with this form of X-linked liver glycogenosis and assigned the disease gene to Xp22. As the genes for both the classic X-linked liver glycogenosis and the liver alpha subunit of PHK (PHKA2) also map to Xp22, Hendrickx et al. (1994) concluded that the variant type may be allelic to the classic type and that both diseases may be caused by mutations in PHKA2. They proposed to classify X-linked glycogenosis as XLG type I (XLG1; the classic type) and XLG type II (XLG2; the variant type). Type II is characterized mainly by enlarged liver and growth retardation. In contrast to XLG1 patients, XLG2 patients do not show in vitro enzymatic deficiency of phosphorylase kinase. Both forms result from mutations in the PHKA2 gene that cause different biochemical effects (see below).

In a Dutch boy who was an isolated case of PHK deficiency type I in a nonconsanguineous family and in 4 patients from a large Dutch family with type I X-linked liver PHK deficiency extensively studied by Huijing and Fernandez (1969) and Willems et al. (1990), van den Berg et al. (1995) confirmed the occurrence of mutations in the PHKA2 gene (e.g., 306000.0005). Hendrickx et al. (1995) reported mutations in PHKA2 that cause XLG1 (e.g., 306000.0001).

In a study of the PHKA2 gene in 4 unrelated XLG2 patients, Hendrickx et al. (1996) identified 4 different mutations in the open reading frame (e.g., 306000.0011). In XLG2 the mutations lead to minor structural abnormalities in the primary structure of the liver alpha-subunit of PHK. These mutations are found in a conserved RXX(X)T motif, resembling known phosphorylation sites that may be involved in the regulation of PHK. The authors postulated that PHK activity may be regulated by phosphorylation of these sites and that XLG2 may be due to impaired activation of PHK activity. The findings may explain why the in vitro PHK enzymatic activity is not deficient in XLG2, whereas it is in XLG1. Burwinkel et al. (1996) identified mutations in the PHKA2 gene in 5 cases of XLG2 (e.g., 306000.0007). The mutations appeared to cluster in limited sequence regions. Burwinkel et al. (1996) stressed that the clustering of XLG2 mutations will further facilitate analysis by RT-PCR of blood cell mRNA and thus help avoid liver biopsy in the diagnosis of XLG2.

Burwinkel et al. (1998) described 8 new mutations and phenotypic consequences in patients with X-linked liver glycogenosis. One of the patients reported by Burwinkel et al. (1998) had low Phk activity in the liver but elevated levels in erythrocytes, typical of the XLG type 2. This patient had a lys189-to-glu missense mutation (306000.0015). The authors noted that this observation adds to the growing body of evidence that the XLG phenotype is associated with missense mutations clustering at a few sites in the PHKA2 gene.

See also Goji et al. (1985); Hers (1959); Huijing (1970); Varsanyi et al. (1980)

Burwinkel, B.; Amat, L.; Gray, R. G. F.; Matsuo, N.; Muroya, K.; Narisawa, K.; Sokol, R. J.; Vilaseca, M. A.; Kilimann, M. W.: Variability of biochemical and clinical phenotype in X-linked liver glycogenosis with mutations in the phosphorylase kinase PHKA2 gene. Hum. Genet. 102: 423-429, 1998.

Burwinkel, B.; Shin, Y. S.; Bakker, H. D.; Deutsch, J.; Lozano, M. J.: Maire, I.; Kilimann, M. W.: Mutation hotspots in the PHKA2 gene in X-linked liver glycogenosis due to phosphorylase kinase deficiency with atypical activity in blood cells (XLG2). Hum. Molec. Genet. 5: 653-658, 1996.

Davidson, J. J.; Ozcelik, T.; Hamacher, C.; Willems, P. J.; Francke, U.; Kilimann, M. W.: cDNA cloning of a liver isoform of the phosphorylase kinase alpha subunit and mapping of the gene to Xp22.2-p22.1, the region of human X-linked liver glycogenosis. Proc. Nat. Acad. Sci. 89: 2096-2100, 1992.

Davisson, M. T.: X-linked genetic homologies between mouse and man. Genomics 1: 213-227, 1987.

Garibaldi, L. R.; Borrone, C.; De Martini, I.; Battistini, E.: Dextrothyroxine treatment of phosphorylase-kinase deficiency glycogenosis in four boys. Helv. Paediat. Acta 33: 435-444, 1978.

Goji, K.; Morishita, Y.; Kodama, S.; Takahashi, T.; Matsuo, T.: Lymphocyte phosphorylase kinase activities in the sex-linked form of liver phosphorylase kinase deficiency. Europ. J. Pediat. 143: 179-182, 1985.

Hendrickx, J.; Coucke, P.; Bossuyt, P.; Wauters, J.; Raeymaekers, P.; Marchau, F.; Smit, G. P. A.; Stolte, I.; Sardharwalla, I. B.; Berthelot, J.; Van den Bergh, I.; Berger, R.; Van Broeckhoven, C.; Baussan, C.; Wapenaar, M.; Fernandes, J.; Willems, P. J.: X-linked liver glycogenosis: localization and isolation of a candidate gene. Hum. Molec. Genet. 2: 583-589, 1993.

Hendrickx, J.; Coucke, P.; Dams, E.; Lee, P.; Odievre, M.; Corbeel, L.; Fernandes, J. F.; Willems, P. J.: Mutations in the phosphorylase kinase gene PHKA2 are responsible for X-linked liver glycogen storage disease. Hum. Molec. Genet. 4: 77-83, 1995.

Hendrickx, J.; Coucke, P.; Hors-Cayla, M.-C.; Smit, G. P. A.; Shin, Y. S.; Deutsch, J.; Smeitink, J.; Berger, R.; Lee, P.; Fernandes, J.; Willems, P. J.: Localization of a new type of X-linked liver glycogenosis to the chromosomal region Xp22 containing the liver alpha-subunit of phosphorylase kinase (PHKA2). Genomics 21: 620-625, 1994.

Hendrickx, J.; Coucke, P.; Raeymaekers, P.; Willems, P. J.: X-linked liver glycogenosis: localization and isolation of a strong candidate gene. (Abstract) Am. J. Hum. Genet. 51 (suppl.): A190 only, 1992.

Hendrickx, J.; Dams, E.; Coucke, P.; Lee, P.; Fernandes, J.; Willems, P. J.: X-linked liver glycogenosis type II (XLG II) is caused by mutations in PHKA2, the gene encoding the liver alpha subunit of phosphorylase kinase. Hum. Molec. Genet. 5: 649-652, 1996.

Hers, H.-G.; Van Hoof, F.; de Barsy, T.: Glycogen storage diseases.In: Scriver, C. R.; Beaudet, A. L.; Sly, W. S.; Valle, D.: The Metabolic Basis of Inherited Disease. New York: McGraw-Hill (pub.) (6th ed.) I: 1989. Pp. 425-452.

Hers, H. G.: Etudes enzymatiques sur fragments hepatiques: application a la classification des glycogenoses. Rev. Int. Hepat. 9: 35-55, 1959.

Hug, G.: Personal Communication. Cincinnati, Ohio 1974.

Hug, G.; Schubert, W. K.; Chuck, G.: Deficient activity of dephosphophosphorylase kinase and accumulation of glycogen in the liver. J. Clin. Invest. 48: 704-715, 1969.

Huijing, F.: Phosphorylase kinase in leucocytes of normal subjects and of patients with glycogen-storage disease. Biochim. Biophys. Acta 148: 601-603, 1967.

Huijing, F.: Glycogen-storage disease type VIa: low phosphorylase kinase activity caused by a low enzyme-substrate affinity. Biochim. Biophys. Acta 206: 199-201, 1970.

Huijing, F.: Phosphorylase kinase deficiency. Biochem. Genet. 4: 187-194, 1970.

Huijing, F.; Fernandez, J.: X-chromosomal inheritance of liver glycogenosis with phosphorylase kinase deficiency. Am. J. Hum. Genet. 21: 275-284, 1969.
Huijing, F.; Fernandez, J.: Liver glycogenosis and phosphorylase kinase deficiency. (Letter) Am. J. Hum. Genet. 22: 484-485, 1970.
Keating, J. P.; Brown, B. I.; White, N. H.; DiMauro, S.: X-linked glycogen storage disease: a cause of hypotonia, hyperuricemia, and growth retardation. Am. J. Dis. Child. 139: 609-613, 1985.
Kilimann, M. W.; Zander, N. F.; Kuhn, C. C.; Crabb, J. W.; Meyer, H. E.; Heilmeyer, L. M. G., Jr.: The alpha and beta subunits of phosphorylase kinase are homologous: cDNA cloning and primary structure of the beta subunit. Proc. Nat. Acad. Sci. 85: 9381-9385, 1988.
Krebs, E. G.; Love, D. S.; Bratvold, G. E.; Trayser, K. A.; Meyer, W. L.; Fischer, E. H.: Purification and properties of rabbit skeletal muscle phosphorylase B kinase. Biochemistry 3: 1022-1033, 1964.
Lyon, J. B., Jr.; Porter, J.; Robertson, M.: Phosphorylase B kinase inheritance in mice. Science 155: 1550-1551, 1967.
Malthus, R.; Clark, D. G.; Watts, C.; Sneyd, J. G. T.: Glycogen-storage disease in rats, a genetically determined deficiency of liver phosphorylase kinase. Biochem. J. 188: 99-106, 1980.
McAdams, A. J.; Hug, G.; Bove, K. E.: Glycogen storage disease, type I to X: criteria for morphologic diagnosis. Hum. Path. 5: 463-487, 1974.
Migeon, B. R.; Huijing, F.: Glycogen-storage disease associated with phosphorylase kinase deficiency: evidence for X inactivation. Am. J. Hum. Genet. 26: 360-368, 1974.
Nagai, T.; Matsuo, N.; Tsuchiya, Y.; Cho, H.; Hasegawa, Y.; Igarashi, Y.: Proximal renal tubular acidosis associated with glycogen storage disease, type 9. Acta Paediat. Scand. 77: 460-463, 1988.
Ryder-Cook, A. S.; Derry, J. M. J.; Barnard, P. J.: Localization of the phosphorylase kinase alpha subunit gene on the mouse X chromosome. (Abstract) Cytogenet. Cell Genet. 51: 1071-1072, 1989.
Schimke, R. N.; Zakheim, R. M.; Corder, R. C.; Hug, G.: Glycogen storage disease type IX: benign glycogenosis of liver and hepatic phosphorylase kinase deficiency. J. Pediat. 83: 1031-1034, 1973.
Schneider, A.; Davidson, J. J.; Wullrich, A.; Kilimann, M. W.: Phosphorylase kinase deficiency in I-strain mice is associated with a frameshift mutation in the alpha-subunit muscle isoform. Nature Genet. 5: 381-385, 1993.
van den Berg, I. E. T.; van Beurden, E. A. C. M.; Malingre, H. E. M.; Ploos van Amstel, H. K.; Poll-The, B. T.; Smeitink, J. A. M.; Lamers, W. H.; Berger, R.: X-linked liver phosphorylase kinase deficiency is associated with mutations in the human liver phosphorylase kinase alpha subunit. Am. J. Hum. Genet. 56: 381-387, 1995.
Varsanyi, M.; Vrbica, A.; Heilmeyer, L. M. G., Jr.: X-linked dominant inheritance of partial phosphorylase kinase deficiency in mice. Biochem. Genet. 18: 247-261, 1980.
Wallis, P. G.; Sidbury, J. B., Jr.; Harris, R. C.: Hepatic phosphorylase defect. Studies on peripheral blood. Am. J. Dis. Child. 111: 278-282, 1966.
Wauters, J. G.; Bossuyt, P. J.; Davidson, J.; Hendrickx, J.; Kilimann, M. W.; Willems, P. J.: Regional mapping of a liver alpha-subunit gene of phosphorylase kinase (PHKA) to the distal region of human chromosome Xp. Cytogenet. Cell Genet. 60: 194-196, 1992.
Willems, P. J.; Gerver, W. J. M.; Berger, R.; Fernandes, J.: The natural history of liver glycogenosis due to phosphorylase kinase deficiency: a longitudinal study of 41 patients. Europ. J. Pediat. 149: 258-271, 1990.
Willems, P. J.; Hendrickx, J.; Van der Auwera, B. J.; Vits, L.; Raeymaekers, P.; Coucke, P. J.; Van den Bergh, I.; Berger, R.; Smit, G. P. A.; Van Broeckhoven, C.; Kilimann, M. W.; Van Elsen, A. F.; Fernandes, J. F.: Mapping of the gene for X-linked liver glycogenosis due to phosphorylase kinase deficiency to human chromosome region Xp22. Genomics 9: 565-569, 1991.
Williams, H. E.; Field, J. B.: Low leukocyte phosphorylase in hepatic phosphorylase deficient glycogen storage disease. J. Clin. Invest. 40: 1841-1845, 1961.
Zander, N. F.; Meyer, H. E.; Hoffmann-Posorske, E.; Crabb, J. W.; Heilmeyer, L. M. G., Jr.; Kilimann, M. W.: cDNA cloning and complete primary structure of skeletal muscle phosphorylase kinase (alpha subunit). Proc. Nat. Acad. Sci. 85: 2929-2933, 1988.

TYROSINEMIA (*276700)

JJM NOTE

Although the ecogenetic model of carcinogenesis could cite the excess tyrosine interacting with liver cells, the liver damage may occur prenatally when maternal metabolism may regulate tyrosine well. Still, the chance of liver malignancy is so great that anticipatory liver transplantation is standard. The experimental drug therapy is intriguing, but not curative.

Among the children of first-cousin parents, Lelong et al. (1963) observed 2 sons with cirrhosis, Fanconi renotubular syndrome, and marked increase in plasma tyrosine. In the sib most extensively observed, hepatosplenomegaly was discovered at 3 months of age and rickets at 18 months. Malignant changes developed in the liver, and death from pulmonary metastases occurred shortly before his 5th birthday. The author suggested that the basic defect concerns an enzyme involved with tyrosine metabolism. Earlier, Himsworth (1950) described a similar case. Zetterstrom (1963) studied 7 cases coming from an isolated area of southwestern Sweden. Perry et al. (1965) described 3 sibs (2 females and a male) in 1 sibship who died in the third month after an illness characterized by irritability and progressive somnolence, and terminally by a tendency to bleed and hypoglycemia. A peculiar odor was noted. Pathologic changes included hepatic cirrhosis, renal tubular dilatation, and pancreatic islet hypertrophy. Biochemical studies showed generalized aminoaciduria, marked elevation of methionine in the serum, and a disproportionately high urinary excretion of methionine. Alpha-keto-gamma-methiolbutyric acid was present in the urine and may account for the peculiar odor. The hypertrophy of the islets of Langerhans was probably due to stimulation by methionine or one of its metabolites. It seems likely that the disorder in the patients of Perry et al. (1965) was tyrosinemia since hypermethioninemia occurs secondary to liver failure in that condition (Scriver et al., 1967; Gaull et al., 1970). Gentz et al. (1965) described 7 patients in 4 families with multiple renal tubular defects like those of the de Toni-Debre-Fanconi syndrome, nodular cirrhosis of the liver and impaired tyrosine metabolism. In the urine, p-hydroxyphenyllactic acid was excreted in unusually large amounts. A total lack of liver p-hydroxyphenylpyruvate oxidase activity was demonstrated. Tyrosine-alpha-ketoglut arate transaminase was normal. Halvorsen et al. (1966) gave details on 6 cases from Norway. Scriver et al. (1967) identified the disease in 35 French-Canadian infants, of whom 16 were sibs (i.e., 2 or more in each of several families). Marked tyrosinemia and tyrosyluria were present. The urine contained parahydroxyphenylpyruvic acid (PHPPA) and lactic and acetic derivatives. Loading test with tyrosine and with PHPPA suggested deficient p-hydroxyphenylpyruvate oxidase activity, which was confirmed by assay of liver biopsy samples. In stage I, infants exhibit hepatic necrosis and hypermethioninemia. In stage II, nodular cirrhosis and chronic hepatic insufficiency without hypermethioninemia are found. In stage III, renal tubular damage (Baber syndrome), often with hypophosphatemic rickets, appears. Low tyrosine diet arrested progression of the disease. Lindblad et al. (1987) suggested that cardiomyopathy, usually subclinical, is a frequent finding. Mitchell et al. (1990) pointed out the significance of neurologic crises in this disorder. They found that of 48 children with tyrosinemia identified on neonatal screening since 1970, 20 (42%) had neurologic crises that began at the mean age of 1 year and led to 104 hospital admissions. These abrupt episodes of peripheral neuropathy were characterized by severe pain with extensor hypertonia (in 75%), vomiting or paralytic ileus (69%), muscle weakness (29%), and self-mutilation (8%). In 8 children, mechanical ventilation was required because of paralysis and 14 of the 20 children died. Between crises, most survivors regained normal function. They could identify no reliable biochemical marker for the crises. Urinary excretion of delta-aminolevulinic acid, a neurotoxic intermediate of porphyrin biosynthesis, was elevated during both crises and asymptomatic periods. Electrophysiologic studies and neuromuscular biopsies showed axonal degeneration and secondary demyelination. Thus, they demonstrated that episodes of acute, severe, peripheral neuropathy are common in this disorder and resemble the crises of the neuropathic porphyrias.

La Du and Gjessing (1972) discussed evidence against the hypothesis that tyrosinemia is a p-hydroxyphenylpyruvic acid oxidase deficiency and suggested that further investigation is needed to explain the clinical and pathologic features of tyrosinemia. Lindblad et al. (1977) suggested that the primary defect is in fumarylacetoacetase (EC 3.7.1.2). This leads to accumulation of succinylacetone and succinylacetoacetate. Porphobilinogen synthetase is inhibited by these substances and the authors suggested that the severe liver and kidney damage of tyrosinemia is caused by accumulation of tyrosine metabolites. A puzzling feature of hereditary tyrosinemia has been episodes similar to acute hepatic porphyria, with excretion of 5-aminolevulinic acid in the urine. The inhibition of porphobilinogen synthase explains this feature. Fumarylacetoacetase is the enzyme primarily deficient; deficiency of parahydroxyphenylpyruvate oxidase is secondary (Scriver, 1982). Kvittingen et al. (1985) described a family that may have had a pseudodeficiency gene. Presumed homozygotes for this gene had levels of fumarylacetoacetase activity only slightly higher than those in patients with tyrosinemia. No clinical abnormalities were observed. Kvittingen et al. (1992) studied a healthy 41-year-old female homozygous for the pseudodeficiency gene and 3 tyrosinemia families in which one or both parents were compound heterozygotes for the tyrosinemia and pseudodeficiency genes. Only 2 of 7 patients with typical chronic tyrosinemia had definite immunoreactivity in fibroblasts when bovine fumarylacetoacetase antibodies were used; none of the patients with the acute type had detectable immunoreactive protein in fibroblast extracts. Twenty-eight patients with hereditary tyrosinemia of various clinical phenotypes were tested. The pseudodeficiency gene product gave almost no detectable immunoreactivity in fibroblasts.

Malpuech et al. (1981) described tyrosinemia in a child with partial monosomy 4p-. The parents were not consanguineous and were chromosomally normal. It had been postulated that the severe liver damage in tyrosinemia is the result of defective degradation of tyrosine. Hostetter et al. (1983) showed, however, that liver damage is prenatal in onset (as indicated by greatly elevated alpha-fetoprotein in cord blood) and that hypertyrosinemia developed only postnatally. Thus, therapy aimed at reduction of the elevated tyrosine level is unlikely to be of fundamental value.

Prenatal diagnosis is possible either by the detection of succinylacetone in the amniotic fluid (Gagne et al., 1982) or measurement of fumarylacetoacetase in cultured amniotic cells (Kvittingen et al., 1983). Holme et al. (1985) demonstrated the feasibility of enzymatic diagnosis in chorionic villus material. Also, they showed that normal red cells have fumarylacetoacetase activity. They proposed that studies of red cells permit rapid diagnosis and recognition of heterozygotes and that enzyme replacement by blood transfusion may help patients over acute metabolic crises and until such time as definitive therapy by orthotopic liver transplantation (Fisch et al., 1978; Gartner et al., 1984) can be performed. Dehner et al. (1989) reviewed the pathologic findings in the liver on the basis of the findings in children undergoing liver transplant. They concluded that to preclude hepatocellular carcinoma, a liver replacement is necessary before the age of 2 years. In the view of Van Spronsen et al. (1989) also, orthotopic liver transplantation is the only definitive therapy for both the metabolic and the oncologic problem in this disorder. Sokal et al. (1992) recommended orthotopic liver transplantation at an early stage. The procedure was performed in 4 children under 1 year of age, within 5 months of presentation and diagnosis. During the pretransplant period, intensive medical support and restriction of dietary tyrosine was initiated to improve the patient's condition and promote weight gain.

As an alternative to liver transplantation, Lindstedt et al. (1992) treated patients with type I tyrosinemia with a potent inhibitor of 4-hydroxyphenylpyruvate dioxygenase (EC 1.13.11.27) to prevent the formation of maleylacetoacetate and fumarylacetoacetate and their saturated derivatives. The agent used in 1 acute and 4 subacute/chronic cases was 2-(2-nitro-4-trifluoromethylbenzoyl)-1,3-cyclohexanedione (NTBC). Signs of improvement included decrease in several metabolites, correction of the almost complete inhibition of porphobilinogen synthase in erythrocytes, decrease in alpha-fetoprotein, improved liver and renotubular function, and regression of hepatic abnormalities by computed tomography. No side effects were encountered. Inhibition of 4-hydroxyphenylpyruvate dioxygenase may prevent the development of liver cirrhosis and abolish or diminish the risk of liver cancer. Furthermore, normalization of porphyrin synthesis should eliminate the risk of porphyric crises. As an aid to early diagnosis for early institution of drug therapy, Holme and Lindstedt (1992) suggested a neonatal screening test based on the measurement of porphobilinogen synthase activity. Porphobilinogen synthase activity is always low in patients with tyrosinemia type I. Holme and Lindstedt (1992) were not aware of any drug used neonatally or of conditions that would interfere with the test or mimic porphobilinogen synthase activity to result in a false-normal test. Specificity of the test is not absolute because homozygous porphobilinogen synthase deficiency (125270) would be detected; in this disorder also, early diagnosis would presumably benefit the patients.

Mice homozygous for an FAH gene disruption have a neonatal lethal phenotype caused by liver dysfunction. Grompe et al. (1995) demonstrated that treatment of affected animals with NTBC abolished neonatal lethality, corrected liver function, and partially normalized the altered expression pattern of hepatic mRNAs. The prolonged life span of affected animals resulted in a phenotype analogous to human tyrosinemia type I, including hepatocellular carcinoma. These animals will serve as a useful model for studies of the pathophysiology and treatment of hereditary tyrosinemia type I as well as hepatic cancer.

De Braekeleer and Larochelle (1990) estimated the prevalence of hereditary tyrosinemia at birth as 1/1,846 liveborn and the carrier rate as 1/20 inhabitants in the Saguenay-Lac Saint Jean region. The mean coefficient of inbreeding was only slightly elevated in the tyrosinemic group compared to a control group and was due to remote consanguinity. The mean kinship coefficient was 2.3 times higher in the tyrosinemic group than in the control group. This was interpreted as indicating founder effect.

Grompe et al. (1994) found that 100% of patients from the Saguenay-Lac St. Jean region of Quebec and 28% of patients from other regions of the world carry a splice donor site mutation in intron 12. Of 25 patients from the Saguenay-Lac St. Jean region, 20 were homozygous. The frequency of carrier status, based on screening of blood spots from newborns, was about 1 per 25 in that region of Quebec and about 1 per 66 overall in Quebec. Using cDNA probes for the FAH gene, Demers et al. (1994) identified 10 haplotypes with 5 RFLPs in 118 normal chromosomes from the French-Canadian population. Among 29 children with hereditary tyrosinemia, haplotype 6 was found to be strongly associated with disease, at a frequency of 90% as compared with approximately 18% in 35 control individuals. This frequency increased to 96% in the 24 patients originating from the Saguenay-Lac St. Jean region. Most patients were found to be homozygous for a specific haplotype in this population. Analysis of 24 tyrosinemia patients from 9 countries gave a frequency of approximately 52% for haplotype 6, suggesting a relatively high association worldwide.

Tanguay et al. (1990) concluded that the acute form of hereditary tyrosinemia has absence of FAH enzyme protein, whereas the chronic form has presence of immunoreactive enzyme protein. They quoted the work of others supporting these findings. Laberge et al. (1990) described an enzyme-linked immunosorbent assay (ELISA) to measure the deficient enzyme in dried blood spots in this disorder. As mean levels of blood tyrosine in newborn specimens have declined, probably as a result of dietary changes and early discharge from nurseries, the traditional approach to screening for tyrosinemia, which was based on the fluorometric determination of tyrosine on the first dried blood spot received by neonatal screening programs, has required replacement. Russo and O'Regan (1990) reviewed the pathologic findings in the liver and kidney. In the Hopital Sainte-Justine in Montreal, 16 patients had been evaluated for liver transplantation. Renal involvement was found to be 'more abnormal than expected.' The liver was transplanted in 7 patients of whom 2 also received kidney transplantation. Hepatocarcinoma was detected in 2 of 8 patients in whom the whole liver was examined. Of the 9 patients who did not receive transplants, 5 died; of the 7 transplant patients, 1 died in an instance of combined liver-kidney transplantation. The 6 patients who survived had normal liver function, normal growth, and no recurrence of neurologic crises on a normal diet. Laine et al. (1995) studied renal function after orthotopic liver transplantation and found that the patients had normal glomerular filtration rates but showed signs of tubular dysfunction 18 to 36 months after operation.

In type I tyrosinemia, the defect in FAH, the last enzyme in the tyrosine catabolism pathway, results in accumulation of succinylacetone (SA) that reacts with amino acids and proteins to form stable adducts via Schiff base formation, lysine being the most reactive amino acid. Patients with this disorder surviving beyond infancy are at considerable risk for the development of hepatocellular carcinoma, and a high level of chromosomal breakage is observed in tyrosinemia cells, suggesting a defect in the processing of DNA. Prieto-Alamo and Laval (1998) showed that the overall DNA-ligase activity is low in tyrosinemia cells (about 20% of normal) and that Okazaki fragments are rejoined at a reduced rate compared with normal fibroblasts. No mutation was found by sequencing the ligase I cDNA (LIG1; 126391) from tyrosinemia cells, and the level of expression of the ligase I mRNA was similar in normal and tyrosinemia fibroblasts, suggesting the presence of a ligase inhibitor. SA was shown to inhibit in vitro the overall DNA-ligase activity present in normal cell extracts. The activity of purified T4 DNA-ligase, whose active site is also a lysine residue, was inhibited by SA in a dose-dependent manner. These results suggested that accumulation of SA reduces the overall ligase activity in tyrosinemia cells and indicated that metabolic errors may play a role in regulating enzymatic activities involved in DNA replication and repair.

By in situ hybridization, Berube et al. (1989) assigned the FAH gene to 15q23-q25. Using in situ hybridization, Tanguay et al. (1990) confirmed the assignment to chromosome 15 with a cDNA encoding FAH from human liver. They analyzed the FAH in livers of unrelated patients using mRNA levels, immunoreactive protein, and enzyme activity. The results suggest molecular heterogeneity of mutations causing tyrosinemia. A missense mutation in the FAH gene was demonstrated in the cDNA from 1 patient with normal FAH mRNA but without immunoreactive protein or enzymatic activity. Tanguay et al. (1990) also identified RFLPs for 4 restriction sites within the FAH gene and proposed the development of a carrier detection test by linkage analysis. Phaneuf et al. (1991) isolated human FAH cDNA clones by screening a liver cDNA expression library using specific antibodies and plaque hybridization with a rat FAH cDNA probe. From transient expression in transfected mammalian cells, a single polypeptide chain encoded by the FAH gene appeared to contain all the genetic information required for functional activity, indicating that the dimer found in vivo is a homodimer. By study of somatic cell hybrids and by in situ hybridization using the FAH cDNA, Phaneuf et al. (1991) demonstrated that the gene maps to 15q23-q25.

Kvittingen et al. (1994) demonstrated a mosaic pattern of immunoreactive FAH protein in liver tissue from 15 of 18 tyrosinemia type I patients of various ethnic origins. One additional patient had variable levels of FAH enzyme activity in liver tissue. In 4 patients exhibiting mosaicism of FAH protein, analysis for the tyrosinemia-causing mutations was performed in immunonegative and immunopositive areas of liver tissue by restriction digestion analysis and direct DNA sequencing. In all 4 patients, the immunonegative liver tissue contained the FAH mutations demonstrated in fibroblasts of the patients. In the immunopositive nodules of regenerating liver tissue, one of the mutated alleles apparently had reverted to the normal genotype. This genetic correction was observed for 3 different tyrosinemia-causing mutations. In each case, a mutant AT nucleotide pair was reverted to a normal GC pair. One of the mutations that showed reversion was the splice site mutation described in 276700.0003. Another was the glu357-to-ter mutation due to a G-to-T transversion at nucleotide 1069, which is described in 276700.0004. In a compound heterozygous patient, the same mutation was reverted to wildtype in all 4 nodules investigated. A gene conversion event or mitotic recombination between homologous chromosomes could theoretically explain the appearance of a normal allele in a compound heterozygote. Two of the patients with reverted mutations, however, were homozygous for their mutations, and no pseudogenes for FAH, for contribution of wildtype sequences, are known. Early embryonic mutation with selective growth of the mutated cells could account for the mosaicism, but a high incidence of such an event would indicate a precipitating factor. Chemical mutagenesis, reverting the disease-causing mutation, could result from the metabolites accumulating in tyrosinemia. Even if the metabolites are not direct mutagens, the compounds are toxic and induce cell necrosis with a subsequent accelerated regeneration o f hepatocytes. Rapidly replicating cells are generally prone to mutations. Reversion of the genetic defect resulting from accelerated cell regeneration should be sought in other genetic diseases in tissues with an induced, or naturally high, rate of cell replication.

In mice deficient in FAH through targeted disruption of the Fah gene, Overturf et al. (1996) found that as few as 1,000 transplanted wildtype hepatocytes were able to repopulate mutant liver, demonstrating their strong competitive growth advantage. Mutant hepatocytes corrected in situ by retroviral gene transfer were also positively selected. In mutant animals treated by multiple retrovirus injections, more than 90% of hepatocytes became FAH positive and liver function was restored to normal. These studies were prompted by a number of observations including the finding that the livers of patients with hereditary tyrosinemia frequently contained discrete nodules with FAH enzyme activity, due to a somatic reversion event (Kvittingen et al., 1993). Wilson (1996) commented on the significance of these results for the liver gene therapy for genetic diseases in general. He stated that, based on the encouraging data in the mouse model, it would seem reasonable to evaluate this approach in patients with hereditary tyrosinemia. A similar approach might be considered for other liver metabolic diseases in which genetically corrected hepatocytes would have a selective advantage over degenerating mutant cells. Wilson (1996) suggested that a useful extension of this approach might be to introduce into the vector a gene that confers upon the hepatocyte a selective advantage such as resistance to a hepatotoxic drug. This concept was being developed in bone marrow using the multidrug resistance (MDR) gene (171050).

Overturf et al. (1997) injected Fah-deficient mice with a first-generation adenoviral vector expressing the human FAH gene and followed them for up to 9 months. Nontreated FAH mutant control mice died within 6 weeks from fulminant liver failure, whereas FAH adenovirus-infected animals survived until sacrifice at 2 to 9 months. Hepatocellular cancer developed in 9 of 13 virus-treated animals. Immunohistochemical analysis revealed a mosaic of FAH-deficient and FAH-positive cells in all animals and liver function tests were improved compared to controls. Even mice harvested 9 months after viral infection had more than 50% FAH-positive cells. These results demonstrated a strong selective advantage of FAH-expressing cells in an FAH-deficient liver but also illustrated the danger of carcinomas arising from FAH-deficient hepatocytes in this disorder.

The 'albino lethal' mouse, first described by Gluecksohn-Waelsch (1979), has a large deletion on chromosome 7, including the albino locus and the Fah gene. Another Fah-deficient mouse was generated by targeted disruption of the Fah gene (Grompe et al., 1995). Endo et al. (1997) generated mice with disruption of both the Fah gene and the Hpd gene, which encodes 4-hydroxyphenylpyruvate dioxygenase at a step earlier in the metabolic pathway. This doubly mutant tyrosinemic mouse model showed apoptosis of hepatocytyes and acute onset of liver failure after administration of homogentisic acid (HGA), the intermediate metabolite between the enzymes HPD and FAH (Kubo et al., 1998). Cytochrome c was released from mitochondria prior to liver failure in the double-mutant mice after administration of HGA. In a cell-free system, the addition of fumarylacetoacetate induced release of cytochrome c from the mitochondria. Kubo et al. (1998) also found that caspase inhibitors were highly effective in preventing the liver failure induced by HGA in the double-mutant mice. Therefore, fumarylacetoacetate apparently induces the release of cytochrome c, which in turn triggers activation of the caspase cascade in hepatocytes of subjects with hereditary tyrosinemia type I.

Rootwelt et al. (1994) presented evidence for the existence of a 'pseudodeficiency' FAH allele. In an individual homozygous for pseudodeficiency of FAH and in 3 hereditary tyrosinemia type I families also carrying the pseudodeficiency allele, Western blotting of fibroblast extracts showed that the pseudodeficiency allele gave very little immunoreactive FAH protein, whereas Northern analysis revealed a normal amount of FAH mRNA. All the pseudodeficiency alleles were found to carry a C-to-T transition in nucleotide 1021, predicting an arg341-to-trp substitution. Site-directed mutagenesis and expression in a rabbit reticulocyte lysate system demonstrated that the arg341-to-trp mutation gave reduced FAH activity and reduced amounts of the full-length protein. The normal and the mutated sequences could be distinguished by BsiEI restriction digestion of PCR products. Among 516 healthy volunteers of Norwegian origin, the arg341-to-trp mutation was found in 2.2% of alleles. Testing for this specific mutation may solve the problem of prenatal diagnosis and carrier detection in families with compound heterozygote genotypes for type I tyrosinemia and pseudodeficiency.

Hahn et al. (1995) reviewed 7 previously reported mutations in tyrosinemia type I and added 2 more identified in a compound heterozygote.

Timmers and Grompe (1996) reported 6 new mutations in the FAH gene in patients with hereditary tyrosinemia type I: 2 splice mutations, 3 missense mutations, and 1 nonsense mutation.

Rootwelt et al. (1996) classified 62 hereditary tyrosinemia type 1 patients of various ethnic origins clinically into acute, chronic, or intermediate phenotypes and screened for the 14 published causal mutations in the FAH gene. Restriction analysis of PCR-amplified genomic DNA identified 74% of the mutated alleles. The IVS12,G-A,+5 mutation (276700.0003), which is predominant in the French-Canadian tyrosinemia type I patients, was the most common mutation being present in 32 alleles in patients from Europe, Pakistan, Turkey, and the United States. The IVS6,G-T,-1 transversion (276700.0010), encountered in 14 alleles, was common in Central and Western Europe. There was an apparent 'Scandinavian' 1009G-to-A combined splice and missense mutation (12 alleles), a 'Pakistani' 192G-to-T splice mutation (11 alleles), a 'Turkish' D233V mutation (6 alleles), and a 'Finnish' or northern European W262X (276700.0009) mutation (7 alleles). Rootwelt et al. (1996) commented that some of the mutations seemed to predispose for acute and others for more chronic forms of tyrosinemia type I, although no clear-cut genotype/phenotype correlation could be established.

According to the review of St-Louis and Tanguay (1997), 26 mutations in the FAH gene had been reported in type I tyrosinemia. All consisted of single-base substitutions resulting in 16 amino acid replacements, 1 silent mutation causing a splicing defect, 5 nonsense codons, and 4 putative splicing defects. The mutations were spread over the entire FAH gene, with a particular clustering between amino acid residues 230 and 250.

See also Berger et al. (1981); Fritzell et al. (1964); Gaull et al. (1968); Halvorsen and Gjessing (1964); Kang and Gerald (1970); Kvittingen et al. (1986); Kvittingen et al. (1981); Kvittingen et al. (1986); Laberge (1969); La Du (1967); Paradis et al. (1990); Pettit et al. (1985); Scriver et al. (1967); Tanguay et al. (1990); Tuchman et al. (1985); Weinberg et al. (1976); Whelan and Zannoni (1974)

Berger, R.; Smit, G. P. A.; Stoker-de Vries, S. A.; Duran, M.; Ketting, P.; Wadman, S. K.: Deficiency of fumarylacetoacetase in a patient with hereditary tyrosinemia. Clin. Chim. Acta 114: 37-44, 1981.

Berube, D.; Phaneuf, D.; Tanguay, R. M.; Gagne, R.: Assignment of the fumarylacetoacetate hydrolase gene to chromosome 15q23-15q25. (Abstract) Cytogenet. Cell Genet. 51: 962 only, 1989.

De Braekeleer, M.; Larochelle, J.: Genetic epidemiology of hereditary tyrosinemia in Quebec and in Saguenay-Lac-St-Jean. Am. J. Hum. Genet. 47: 302-307, 1990.

Dehner, L. P.; Snover, D. C.; Sharp, H. L.; Ascher, N.; Nakhleh, R.; Day, D. L.: Hereditary tyrosinemia type I (chronic form): pathologic findings in the liver. Hum. Path. 20: 149-158, 1989.

Demers, S. I.; Phaneuf, D.; Tanguay, R. M.: Hereditary tyrosinemia type I: strong association with haplotype 6 in French Canadians permits simple carrier detection and prenatal diagnosis. Am. J. Hum. Genet. 55: 327-333, 1994.

Endo, F.; Kubo, S.; Awata, H.; Kiwaki, K.; Katoh, H.; Kanegae, Y.; Saito, I.; Miyazaki, J.; Yamamoto, T.; Jakobs, C.; Hattori, S.; Matsuda, I.: Complete rescue of lethal albino c14o5 mice by null mutation of 4-hydroxyphenylpyruvate-dioxygenase and induction of apoptosis of hepatocytes in these mice in vivo retrieval of the tyrosine catabolic pathway. J. Biol. Chem. 272: 24426-24432, 1997.

Fisch, R. O.; McCabe, E. R. B.; Doeden, D.; Koep, L. J.; Kohlhoff, J. G.; Silverman, A.; Starzl, T. E.: Homotransplantation of the liver in a patient with hepatoma and hereditary tyrosinemia. J. Pediat. 93: 592-596, 1978.

Fritzell, S.; Jagenburg, O. R.; Schnurer, L. B.: Familial cirrhosis of the liver, renal tubular defects with rickets and impaired tyrosine metabolism. Acta Paediat. 53: 18-32, 1964.

Gagne, R.; Lescault, A.; Grenier, A.; Laberge, C.; Melancon, S. B.; Dallaire, L.: Prenatal diagnosis of hereditary tyrosinaemia: measurement of succinylacetone in amniotic fluid. Prenatal Diag. 2: 185-188, 1982.

Gartner, J. C.; Zitelli, B. J.; Malatack, J. J.; Shaw, B. W.; Iwatsuki, S.; Starzl, T. E.: Orthotopic liver transplantation in children: two-year experience with 47 patients. Pediatrics 74: 140-145, 1984.

Gaull, G. E.; Rassin, D. K.; Solomon, G. E.; Harris, R. C.; Sturman, J. A.: Biochemical observations on so-called hereditary tyrosinemia. Pediat. Res. 4: 337-344, 1970.

Gaull, G. E.; Rassin, D. K.; Sturman, J. A.: Significance of hypermethioninaemia in acute tyrosinosis. (Letter) Lancet I: 1318-1319, 1968.

Gentz, J.; Jagenburg, R.; Zetterstrom, R.: Tyrosinemia. J. Pediat. 66: 670-696, 1965.

Gluecksohn-Waelsch, S.: Genetic control of morphogenetic and biochemical differentiation: lethal albino deletions in the mouse. Cell 16: 225-237, 1979.

Grompe, M.; Al-Dhalimy, M.: Mutations of the fumarylacetoacetate hydrolase gene in four patients with tyrosinemia, type I. Hum. Mutat. 2: 85-93, 1993.

Grompe, M.; Lindstedt, S.; Al-Dhalimy, M.; Kennaway, N. G.; Papaconstantinou, J.; Torres-Ramos, C. A.; Ou, C.-N.; Finegold, M.: Pharmacological correction of neonatal lethal hepatic dysfunction in a murine model of hereditary tyrosinaemia type I. Nature Genet. 10: 453-460, 1995.

Grompe, M.; St-Louis, M.; Demers, S. I.; Al-Dhalimy, M.; Leclerc, B.; Tanguay, R. M.: A single mutation of the fumarylacetoacetate hydrolase gene in French Canadians with hereditary tyrosinemia type I. New. Eng. J. Med. 331: 353-357, 1994.

Hahn, S. H.; Krasnewich, D.; Brantly, M.; Kvittingen, E. A.; Gahl, W. A.: Heterozygosity for an exon 12 splicing mutation and a W234G missense mutation in an American child with chronic tyrosinemia type 1. Hum. Mutat. 6: 66-73, 1995.

Halvorsen, S.; Gjessing, L. R.: Studies of tyrosinosis. I. Effect of low-tyrosine and low-phenylalanine diet. Brit. Med. J. 2: 1171-1173, 1964.

Halvorsen, S.; Pande, H.; Loken, A. C.; Gjessing, L. R.: Tyrosinosis: a study of 6 cases. Arch. Dis. Child. 41: 238-249, 1966.

Himsworth, H. P.: Lectures on the Liver and its Diseases. Oxford: Blackwell (pub.) (2nd ed.): 1950.

Holme, E.; Lindblad, B.; Lindstedt, S.: Possibilities for treatment and for early prenatal diagnosis of hereditary tyrosinaemia. (Letter) Lancet I: 527 only, 1985.

Holme, E.; Lindstedt, S.: Neonatal screen for hereditary tyrosinaemia type I. (Letter) Lancet 340: 850 only, 1992.

Hostetter, M. K.; Levy, H. L.; Winter, H. S.; Knight, G. J.; Haddow, J. E.: Evidence for liver disease preceding amino acid abnormalities in hereditary tyrosinemia. New Eng. J. Med. 308: 1265-1267, 1983.

Kang, E. S.; Gerald, P. S.: Hereditary tyrosinemia and abnormal pyrrole metabolism: a patient with hereditary tyrosinemia is described who developed metabolic and clinical changes compatible with acute intermittent porphyria. J. Pediat. 77: 397-406, 1970.

Kubo, S.; Sun, M.; Miyahara, M.; Umeyama, K.; Urakami, K.; Yamamoto, T.; Jakobs, C.; Matsuda, I.; Endo, F.: Hepatocyte injury in tyrosinemia type 1 is induced by fumarylacetoacetate and is inhibited by caspase inhibitors. Proc. Nat. Acad. Sci. 95: 9552-9557, 1998.

Kvittingen, E. A.; Borresen, A. L.; Stokke, O.; van der Hagen, C. B.; Lie, S. O.: Deficiency of fumarylacetoacetase without hereditary tyrosinemia. Clin. Genet. 27: 550-554, 1985.

Kvittingen, E. A.; Guibaud, P. P.; Divry, P.; Mandon, G.; Rolland, M. O.; Domenichini, Y.; Jakobs, C.; Christensen, E.: Prenatal diagnosis of hereditary tyrosinaemia type I by determination of fumarylacetoacetase in chorionic villus material. (Letter) Europ. J. Pediat. 144: 597-598, 1986.

Kvittingen, E. A.; Halvorsen, S.; Jellum, E.: Deficient fumarylacetoacetate fumarylhydrolase activity in lymphocytes and fibroblasts from patients with hereditary tyrosinemia. Pediat. Res. 17: 541-544, 1983.

Kvittingen, E. A.; Jellum, E.; Stokke, O.: Assay of fumarylacetoacetate fumarylhydrolase in human liver: deficient activity in a case of hereditary tyrosinemia. Clin. Chim. Acta 115: 311-319, 1981.

Kvittingen, E. A.; Jellum, E.; Stokke, O.; Flatmark, A.; Bergan, A.; Sodal, G.; Halvorsen, S.; Schrumpf, E.; Gjone, E.: Liver transplantation in a 23-year-old tyrosinaemia patient: effects on the renal tubular dysfunction. J. Inherit. Metab. Dis. 9: 216-224, 1986.

Kvittingen, E. A.; Rootwelt, H.; Berger, R.; Brandtzaeg, P.: Self-induced correction of the genetic defect in tyrosinemia type I. J. Clin. Invest. 94: 1657-1661, 1994.

Kvittingen, E. A.; Rootwelt, H.; Brandtzaeg, P.; Bergan, A.; Berger, R.: Hereditary tyrosinemia type I: self-induced correction of the fumarylacetoacetase defect. J. Clin. Invest. 91: 1816-1821, 1993.

Kvittingen, E. A.; Rootwelt, H.; van Dam, T.; van Faassen, H.; Berger, R.: Hereditary tyrosinemia type I: lack of correlation between clinical findings and amount of immunoreactive fumarylacetoacetase protein. Pediat. Res. 31: 43-46, 1992.

Labelle, Y.; Phaneuf, D.; Leclerc, B.; Tanguay, R. M.: Characterization of the human fumarylacetoacetate hydrolase gene and identification of a missense mutation abolishing enzymatic activity. Hum. Molec. Genet. 2: 941-946, 1993.

Laberge, C.: Hereditary tyrosinemia in a French-Canadian isolate. Am. J. Hum. Genet. 21: 36-45, 1969.

Laberge, C.; Grenier, A.; Valet, J. P.; Morissette, J.: Fumarylacetoacetase measurement as a mass-screening procedure for hereditary tyrosinemia type I. Am. J. Hum. Genet. 47: 325-328, 1990.

La Du, B. N.: The enzymatic deficiency in tyrosinemia. Am. J. Dis. Child. 113: 54-57, 1967.

La Du, B. N.; Gjessing, L. R.: Tyrosinosis and tyrosinemia.In: Stanbury, J. B.; Wyngaarden, J. B.; Fredrickson, D. S.: The Metabolic Basis of Inherited Disease. New York: McGraw-Hill (pub.) (3rd ed.): 1972. Pp. 256-267.

Laine, J.; Salo, M. K.; Krogerus, L.; Karkkainen, J.; Wahlroos, O.; Holmberg, C.: The nephropathy of type I tyrosinemia after liver transplantation. Pediat. Res. 37: 640-645, 1995.

Lelong, M.; Alagille, D.; Gentil, C. I.; Colin, J.; Le Tan, V.; Gabilan, J. C.: Cirrhose congenitale et familiale avec diabete phospho-gluco-amine, rachitisme vitamin D-resistant et tyrosinurie massive. Rev. Franc. Etud. Clin. Biol. 8: 37-50, 1963.

Lindblad, B.; Fallstrom, S. P.; Hoyer, S.; Nordborg, C.; Solymar, L.; Velander, H.: Cardiomyopathy in fumarylacetoacetase deficiency (hereditary tyrosinaemia): a new feature of the disease. J. Inherit. Metab. Dis. 10 (suppl. 2): 319-322, 1987.

Lindblad, B.; Lindstedt, S.; Steen, G.: On the enzymic defects in hereditary tyrosinemia. Proc. Nat. Acad. Sci. 74: 4641-4645, 1977.

Lindstedt, S.; Holme, E.; Lock, E. A.; Hjalmarson, O.; Strandvik, B.: Treatment of hereditary tyrosinaemia type I by inhibition of 4-hydroxyphenylpyruvate dioxygenase. Lancet 340: 813-817, 1992.

Malpuech, G.; Mattei, J. F.; Gaulme, J.; Palcoux, J. B.; Lesec, G.; Vanlieferinghen, P.: Association, chez le meme sujet, d'une deletion du bras court du chromosome 4 (4p-) et d'un deficit complet en parahydroxyphenylpyruvate oxydase hepatique (tyrosinose). J. Genet. Hum. 29: 455-461, 1981.

Mitchell, G.; Larochelle, J.; Lambert, M.; Michaud, J.; Grenier, A.; Ogier, H.; Gautheir, M.; Lacroix, J.; Vanasse, M.; Larbrisseau, A.; Paradis, K.; Weber, A.; Lefevre, Y.; Melancon, S.; Dallaire, L.: Neurologic crises in hereditary tyrosinemia. New Eng. J. Med. 322: 432-437, 1990.

Overturf, K.; Al-Dhalimy, M.; Ou, C. N.; Finegold, M.; Tanguay, R.; Lieber, A.; Kay, M.; Grompe, M.: Adenovirus-mediated gene therapy in a mouse model of hereditary tyrosinemia type I. Hum. Gene Therapy 8: 513-521, 1997.

Overturf, K.; Al-Dhalimy, M.; Tanguay, R.; Brantly, M.; Ou, C.-N.; Finegold, M.; Grompe, M.: Hepatocytes corrected by gene therapy are selected in vivo in a murine model of hereditary tyrosinaemia type I. Nature Genet. 12: 266-273, 1996.

Paradis, K.; Weber, A.; Seidman, E. G.; Larochelle, J.; Garel, L.; Lenaerts, C.; Roy, C. C.: Liver transplantation for hereditary tyrosinemia: the Quebec experience. Am. J. Hum. Genet. 47: 338-342, 1990.

Perry, T. L.; Hardwick, D. F.; Dixon, G. H.; Dolman, C. L.; Hansen, S.: Hypermethioninemia: a metabolic disorder associated with cirrhosis, islet cell hyperplasia, and renal tubular degeneration. Pediatrics 36: 236-250, 1965.

Pettit, B. R.; Kvittingen, E. A.; Leonard, J. V.: Early prenatal diagnosis of hereditary tyrosinaemia. (Letter) Lancet I: 1038 only, 1985.

Phaneuf, D.; Labelle, Y.; Berube, D.; Arden, K.; Cavenee, W.; Gagne, R.; Tanguay, R. M.: Cloning and expression of the cDNA encoding human fumarylacetoacetate hydrolase, the enzyme deficient in hereditary tyrosinemia: assignment of the gene to chromosome 15. Am. J. Hum. Genet. 48: 525-535, 1991.

Phaneuf, D.; Lambert, M.; Laframboise, R.; Mitchell, G.; Lettre, F.; Tanguay, R. M.: Type 1 hereditary tyrosinemia: evidence for molecular heterogeneity and identification of a causal mutation in a French Canadian patient. J. Clin. Invest. 90: 1185-1192, 1992.

Poudrier, J.; Lettre, F.; Scriver, C. R.; Larochelle, J.; Tanguay, R. M.: Different clinical forms of hereditary tyrosinemia (type I) in patients with identical genotypes. Molec. Genet. Metab. 64: 119-125, 1998.

Prieto-Alamo, M. J.; Laval, F.: Deficient DNA-ligase activity in the metabolic disease tyrosinemia type I. Proc. Nat. Acad. Sci. 95: 12614-12618, 1998.

Rootwelt, H.; Brodtkorb, E.; Kvittingen, E. A.: Identification of a frequent pseudodeficiency mutation in the fumarylacetoacetase gene, with implications for diagnosis of tyrosinemia type I. Am. J. Hum. Genet. 55: 1122-1127, 1994.

Rootwelt, H.; Hoie, K.; Berger, R.; Kvittingen, E. A.: Fumarylacetoacetase mutations in tyrosinaemia type I. Hum. Mutat. 7: 239-243, 1996.

Russo, P.; O'Regan, S.: Visceral pathology of hereditary tyrosinemia type I. Am. J. Hum. Genet. 47: 317-324, 1990.

Scriver, C. R.: Personal Communication. Montreal, Quebec, Canada 2/15/1982.

Scriver, C. R.; Larochelle, J.; Silverberg, M.: Hereditary tyrosinemia and tyrosyluria in a French-Canadian geographic isolate. Am. J. Dis. Child. 113: 41-46, 1967.

Scriver, C. R.; Partington, M. W.; Sass-Kortsak, A.: Conference on hereditary tyrosinemia held at the Hospital for Sick Children. Canad. Med. Assoc. J. 97: 1045-1100, 1967.

Sokal, E. M.; Bustos, R.; Van Hoof, F.; Otte, J. B.: Liver transplantation for hereditary tyrosinemia—early transplantation following the patient's stabilization. Transplantation 54: 937-939, 1992.

St-Louis, M.; Leclerc, B.; Laine, J.; Salo, M. K.; Holmberg, C.; Tanguay, R. M.: Identification of a stop mutation in five Finnish patients suffering from hereditary tyrosinemia type I. Hum. Molec. Genet. 3: 69-72, 1994.

St-Louis, M.; Poudrier, J.; Phaneuf, D.; Leclerc, B.; Laframboise, R.; Tanguay, R. M.: Two novel mutations involved in hereditary tyrosinemia type I. Hum. Molec. Genet. 4: 319-320, 1995.

St-Louis, M.; Poudrier, J.; Tanguay, R. M.: Simple Detection of a (Finnish) hereditary tyrosinemia type 1 mutation. (Letter) Hum. Mutat. 7: 379-380, 1996.

St-Louis, M.; Tanguay, R. M.: Mutations in the fumarylacetoacetate hydrolase gene causing hereditary tyrosinemia type I: overview. Hum. Mutat. 9: 291-299, 1997.

Tanguay, R. M.; Phaneuf, D.; Labelle, Y.; Demers, S.: Molecular cloning and expression of the c-DNA encoding the enzyme deficient in hereditary tyrosinemia: evidence for molecular heterogeneity. (Abstract) Am. J. Hum. Genet. 47 (suppl.): A168 only, 1990.

Tanguay, R. M.; Valet, J. P.; Lescault, A.; Duband, J. L.; Laberge, C.; Lettre, F.; Plante, M.: Different molecular basis for fumarylacetoacetate hydrolase deficiency in the two clinical forms of hereditary tyrosinemia (type I). Am. J. Hum. Genet. 47: 308-316, 1990.

Timmers, C.; Grompe, M.: Six novel mutations in the fumarylacetoacetate hydrolase gene of patients with hereditary tyrosinemia type I. Hum. Mutat. 7: 367-369, 1996.

Tuchman, M.; Freese, D. K.; Sharp, H. L.; Whitley, C. B.; Ramnaraine, M. L.; Ulstrom, R. A.; Najarian, J. S.; Ascher, N.; Buist, N. R. M.; Terry, A. B.: Persistent succinylacetone excretion after liver transplantation in a patient with hereditary tyrosinaemia type I. J. Inherit. Metab. Dis. 8: 21-24, 1985.

van Spronsen, F. J.; Berger, R.; Smit, G. P. A.; de Klerk, J. B. C.; Duran, M.; Bijleveld, C. M. A.; van Faassen, H.; Slooff, M. J. H.; Heymans, H. S. A.: Tyrosinaemia type I: orthotopic liver transplantation as the only definitive answer to a metabolic as well as an oncological problem. J. Inherit. Metab. Dis. 12: 339-342, 1989.

Weinberg, A. G.; Mize, C. E.; Vorthen, H. G.: Occurrence of hepatoma in chronic form of hereditary tyrosinemia. J. Pediat. 88: 434-438, 1976.

Whelan, D. T.; Zannoni, V. G.: Microassay of tyrosine-amino transferase and p-hydroxyphenylpyruvic acid oxidase in mammalian liver and patients with hereditary tyrosinemia. Biochem. Med. 9: 19-31, 1974.

Wilson, J. M.: Round two for liver gene therapy. Nature Genet. 12: 232-233, 1996.

Zetterstrom, R.: Tyrosinosis. Ann. N.Y. Acad. Sci. 111: 220-226, 1963.

DUBIN-JOHNSON SYNDROME (#237500)

JJM NOTE

Whether this case of hepatoma is coincidental or not awaits further case reports.

A number sign (#) is used with this entry because of evidence that Dubin-Johnson syndrome is caused by mutations in the canalicular multispecific organic anion transporter (CMOAT; 601107).

The usefulness of inbred groups for the study of rare recessives is nicely illustrated by this disorder which occurs with a minimal frequency of 1 per 1,300 among Iranian Jews (Shani et al., 1970). The characteristics of the disorder are hyperbilirubinemia, deposition of melanin-like pigment in otherwise normal liver cells, in some cases hepatomegaly and abdominal pain, prolonged retention of sulfobromophthalein (which may show a higher concentration at 60 to 90 minutes than at 45 minutes), and otherwise normal liver function. Shani et al. (1970) studied 101 patients with the Dubin-Johnson syndrome ascertained in Israel between 1955 and 1969. Age at onset of jaundice varied from 10 weeks to 56 years. Penetrance is reduced in females. Sixty-four of the cases were Iranian Jews. Parents of affected sibships were consanguineous in 45% of cases as compared with a frequency of 26% among Iranian Jews generally. Segregation analysis yielded results consistent with autosomal recessive inheritance with reduced penetrance. The authors suggested that minor abnormalities may occur in heterozygotes. This suggestion is supported by the findings of Butt et al. (1966), who performed an extensive family study with liver biopsies and other examinations in many relatives. Before the report of Shani et al. (1970), the Dubin-Johnson syndrome had been described twice in mother and son (Beker and Read, 1958; Wolf et al., 1960), but at least 7 instances of multiple affected sibs with normal parents were on record (Du and Rogers, 1967). Furthermore, Calderon and Goldgraber (1961) described a Peruvian patient whose parents were first cousins. Du and Rogers (1967) described 3 affected sisters whose clinically normal parents were first cousins once removed. Three offspring of 1 affected sister had normal livers (by inspection at laparotomy in 2 and by autopsy in the third). In the Israel group of cases of DJS, Seligsohn et al. (1970) reported a striking association with deficien cy of factor VII (227500). The association was limited to Iranian Jews. I wonder if this may indicate that the DJS and factor VII loci are closely linked and that the 2 mutant genes are in coupling in a majority of the Iranian Jewish group, not having yet attained equilibrium of linkage phase. Adam (1972) questioned the linkage interpretation (although he had no alternative explanation) because the association has been seen not only in Iranian Jews but also in Iraqi and Moroccan Jews, in Ashkenazim and perhaps in other Europeans. Furthermore, he pointed out that the family (no. 55) reported by Shani et al. (1970) showed several recombinants among the children of a woman who must have been doubly heterozygous in coupling because her mother was doubly homozygous. Arias (1971) gave a useful review of all hereditary hyperbilirubinemias.

Wolkoff et al. (1973) found that urinary coproporphyrin I is a good indicator of the homozygote and heterozygote states in the Dubin-Johnson syndrome. Normals excreted 24.8% of urinary coproporphyrin as coproporphyrin I, whereas homozygotes and heterozygotes excreted 88.9 and 31.6%, respectively. The standard errors of these means were 1.3%, 1.3%, and 1.2%, respectively. In an isolated area of Japan, Kondo et al. (1974) found 40 cases of definite and 8 cases of probable DJS, distributed in 25 sibships of 22 families. Parental consanguinity was found in 52%. Kondo (1980) found no instance of factor VII deficiency in Japanese cases of DJS. Nakata et al. (1979) reported a Japanese case diagnosed in the neonatal period. Both parents and both grandfathers showed ratios of coproporphyrin isomer I to total urinary coproporphyrin in excess of 40%, characteristic of heterozygotes. The proband's ratio was 97%. Swartz et al. (1987) used electron spin resonance to analyze the hepatic pigment from a hepatoma in a patient with Dubin-Johnson syndrome. They demonstrated that the pigment associated with Dubin-Johnson syndrome has no free radical in the absence of light, thus distinguishing the pigment from typical melanin or its closely related polymers.

The human CMOAT gene (601107) maps to 10q24. Van Kuijck et al. (1997) noted that, if Dubin-Johnson syndrome were found to map to 10q24, CMOAT would become a very strong candidate for the site of the mutation causing that disorder.

Consistent with findings of defects in the homologous cmoat gene in 2 rat models of hyperbilirubinemia (Paulusma et al., 1996; Ito et al., 1997), Wada et al. (1998) reported 2 deletions and a missense mutation in the active transport family signature region of the CMOAT gene in patients with DJS.

Animal Model

The identification of the transport-deficient mutant rat strain, TR(-), which shows chronic conjugated hyperbilirubinemia as in human Dubin-Johnson syndrome, contributed to the functional characterization of CMOAT. Paulusma et al. (1996) noted that detoxification of many endogenous and xenobiotic lipophilic compounds in the liver is accomplished by transferase-mediated conjugation with glutathione, glucuronide, or sulfate moieties, resulting in negatively charged, amphiphilic compounds that are efficiently secreted into bile or urine. Hepatobiliary excretion of these conjugates is mediated by an ATP-dependent transport system, called CMOAT (canalicular multispecific organic anion transporter), located in the apical, or canalicular, membrane of the hepatocyte. Paulusma et al. (1996) identified a cDNA for rat CMOAT and also presented evidence that, while multidrug-resistance-associated protein-1 (MRP; 158343) shows a basolateral location in hepatocytes, CMOAT shows an exclusively canalicular localization. Their observation implied that CMOAT and not MRP is involved in biliary organic anion transport. They presented sequence data indicating that the 2 have different sequences, making it unlikely that they are derived from a single gene by differential splicing and suggesting instead that MRP and CMOAT are encoded by 2 different genes. Paulusma et al. (1996) found that a 1-bp deletion in the CMOAT gene, which resulted in reduced mRNA levels and a truncated nonviable protein, was responsible for the impaired transport of organic compounds from liver to bile in the TR(-) rat.

In the hepatocytes of a patient with Dubin-Johnson syndrome, Kartenbeck et al. (1996) demonstrated selected absence of the canalicular isoform of MRP, using double-label immunofluorescence and confocal laser scanning microscopy with antibodies directed against MRP and dipeptidylpeptidase IV (102720). Another isoform of MRP was detected, however, in the lateral hepatocyte membrane of the patient. Moreover, MRP was present on immunoblots of erythrocyte membranes from Dubin-Johnson syndrome and normal humans. These findings were analogous to their previous observations on the localization of the rat homolog of MRP and its canalicular isoform in normal and transport-deficient Wistar rat liver (Mayer et al., 1995).

See also Edwards (1975)

Adam, A.: Personal Communication. Tel-Aviv, Israel 1972.

Arias, I. M.: Inheritable and congenital hyperbilirubinemia: models for the study of drug metabolism. New Eng. J. Med. 285: 1416-1421, 1971.

Beker, S.; Read, A. E.: Familial Dubin-Johnson syndrome. Gastroenterology 35: 387-389, 1958.

Butt, H. R.; Anderson, E.; Foulk, W. T.; Baggenstoss, A. H.; Schoenfield, L. J.; Dickson, E. R.: Studies of chronic idiopathic jaundice (Dubin-Johnson syndrome). II. Evaluation of a large family with the trait. Gastroenterology 51: 619-630, 1966.

Calderon, A.; Goldgraber, M. B.: Chronic idiopathic jaundice: a case report. Gastroenterology 40: 244-247, 1961.

Du, J. N. H.; Rogers, A. G.: Dubin-Johnson syndrome: a family with three affected sisters. Canad. Med. Assoc. J. 97: 1225-1226, 1967.

Edwards, R. H.: Inheritance of the Dubin-Johnson-Sprinz syndrome. Gastroenterology 68: 734-749, 1975.

Ito, K.; Suzuki, H.; Hirohashi, T.; Kume, K.; Shimizu, T.; Sugiyama, Y.: Molecular cloning of canalicular multispecific organic anion transporter defective in EHBR. Am. J. Physiol. 272: G16-G22, 1997.

Kartenbeck, J.; Leuschner, U.; Mayer, R.; Keppler, D.: Absence of the canalicular isoform of the MRP gene-encoded conjugate export pump from the hepatocytes in Dubin-Johnson syndrome. Hepatology 23: 1061-1066, 1996.

Kondo, T.: Personal Communication. Uwajima City, Japan 1980.

Kondo, T.; Kuchiba, K.; Ohtsuka, Y.; Yanagisawa, W.; Shiomura, T.; Taminato, T.: Clinical and genetic studies on Dubin-Johnson syndrome in a cluster area in Japan. Jpn. J. Hum. Genet. 18: 378-392, 1974.

Mayer, R.; Kartenbeck, J.; Buchler, M.; Jedlitschky, G.; Leier, I.; Keppler, D.: Expression of the MRP gene-encoded conjugate export pump in liver and its selective absence from the canalicular membrane in transport-deficient mutant hepatocytes. J. Cell. Biol. 131: 137-150, 1995.

Nakata, F.; Oyanagi, K.; Fujiwara, M.; Sogawa, H.; Minimi, R.; Horino, K.; Nakao, T.; Kondo, T.: Dubin-Johnson syndrome in a neonate. Europ. J. Pediat. 132: 299-301, 1979.

Paulusma, C. C.; Bosma, P. J.; Zaman, G. J. R.; Bakker, C. T. M.; Otter, M.; Scheffer, G. L.; Scheper, R. J.; Borst, P.; Oude Elferink, R. P. J.: Congenital jaundice in rats with a mutation in a multidrug resistance-associated protein gene. Science 271: 1126-1128, 1996.

Paulusma, C. C.; Bosma, P. J.; Zaman, G. J. R.; Bakker, C. T. M.; Otter, M.; Scheffer, G. L.; Scheper, R. J.; Borst, P.; Oude Elferink, R. P. J.: Congenital jaundice in rats with a mutation in a multidrug resistance-associated protein gene. Science 271: 1126-1128, 1996.

Seligsohn, U.; Shani, M.; Ramot, B.; Adam, A.; Sheba, C.: Dubin-Johnson syndrome in Israel. II. Association with factor-VII deficiency. Quart. J. Med. 39: 569-584, 1970.

Shani, M.; Seligsohn, U.; Gilon, E.; Sheba, C.; Adam, A.: Dubin-Johnson syndrome in Israel. I. Clinical, laboratory, and genetic aspects of 101 cases. Quart. J. Med. 39: 549-567, 1970.

Swartz, H. M.; Chen, K.; Roth, J. A.: Further evidence that the pigment in the Dubin-Johnson syndrome is not melanin. Pigment Cell Res. 1: 69-75, 1987.

van Kuijck, M. A.; Kool, M.; Merkx, G. F. M.; Geurts van Kessel, A.; Bindels, R. J. M.; Deen, P. M. T.; van Os, C. H.: Assignment of the canalicular multispe-

cific organic anion transporter gene (CMOAT) to human chromosome 10q24 and mouse chromosome 19D2 by fluorescent in situ hybridization. Cytogenet. Cell Genet. 77: 285-287, 1997.
Wada, M.; Toh, S.; Taniguchi, K.; Nakamura, T.; Uchiumi, T.; Kohno, K.; Yoshida, I.; Kimura, A.; Sakisaka, S.; Adachi, Y.; Kuwano, M.: Mutations in the canalicular multispecific organic anion transporter (cMOAT) gene, a novel ABC transporter, in patients with hyperbilirubinemia II/Dubin-Johnson syndrome. Hum. Molec. Genet. 7: 203-207, 1998.
Wolf, R. L.; Pizette, M.; Richman, A.; Dreiling, D. A.; Jacobs, W.; Fernandez, O.; Popper, H.: Chronic idiopathic jaundice: a study of two afflicted families. Am. J. Med. 28: 32-41, 1960.
Wolkoff, A. W.; Cohen, L. E.; Arias, I. M.: Inheritance of the Dubin-Johnson syndrome. New Eng. J. Med. 288: 113-117, 1973.

GIANT-CELL HEPATITIS, NEONATAL (#231100)

JJM NOTE

This entity is included because of the case of liver cancer as a complication (Sandor et al. 1976).

A number sign (#) is used with this entry because of evidence that the disorder can be caused by any one of a number of genetic defects in bile acid synthesis, including a defect in 3-beta-hydroxy-delta-5 steroid dehydrogenase/isomerase, delta(4)-3-oxysteroid 5-beta-reductase, and oxysterol 7-alpha-hydroxylase (CYP7A1; 118455).

Increased parental consanguinity suggested autosomal recessive inheritance (Danks and Bodian, 1963). However, only 12 of 71 sibs of index cases were also affected. It was suggested that in some with the appropriate genotype the disease is manifested so severely or so mildly that the diagnosis is not made. An apparent excess of affected males may be further evidence of failure of manifestation of the genotype. Alternatively, the cases analyzed may include more than one disease. Fienberg (1960) reported 2 pairs of male sibs, and Laurendeau et al. (1961) observed 2 affected sisters. The disorder is sometimes loosely labeled 'neonatal hepatitis.' Fawaz et al. (1975) described 2 successively born sibs who died at the age of a few months of massive hepatic necrosis from hepatitis B. The mother was an asymptomatic carrier. Obviously this sort of situation must be considered in the differential diagnosis of neonatal giant cell hepatitis. Sandor et al. (1976) described brother and sister with this disorder. The male died of a rare complication—primary hepatic cancer. The female died of cirrhosis and hepatic coma. Clayton et al. (1987) described a sibship of 5 children in which 3 had progressive liver disease starting in the neonatal period. The Saudi Arabian parents were first cousins. The jaundice in these infants was cholestatic in nature and biopsy showed aggressive hepatitis with giant cells and bridging fibrosis. Chenodeoxycholic acid and cholic acid were absent from the urine and plasma. Furthermore, 2 unusual bile acid precursors were found in the urine. Consequently, the authors suggested that the infants suffered from a defect in 3-beta-hydroxy-delta(5) steroid dehydrogenase/isomerase. In the family of Clayton et al. (1987), Buchmann et al. (1990) found that in addition to the excretion of abnormal substances in the urine, 7-alpha-hydroxy-cholesterol accumulated in the circulation. Furthermore, cultured fibroblasts from the patient were completely devoid of 3- beta-hydroxy-delta 5-C(27)-steroid dehydrogenase/isomerase activity. Fibroblasts from his parents had reduced activity, compatible with the heterozygous genotype. Knisely et al. (1987) reviewed reported cases of neonatal hemochromatosis, applying rigid criteria as follows: a rapidly progressive clinical course with death in utero or in the early neonatal period; increased tissue iron deposition in multiple sites, particularly in the liver, pancreas, heart, and endocrine glands, with the extrahepatic reticuloendothelial system relatively unaffected; and no evidence for hemolytic disease, syndromes associated with hemosiderosis, or exogenous iron overload from transfusions. Evidence for abnormal iron metabolism in heterozygotes is inconclusive. Both parents of 1 patient reported by Knisely et al. (1987) had high levels of serum iron and total iron-binding capacity. Knisely et al. (1987) concluded that the available data argue against the suggestion that neonatal hemochromatosis is caused by primary placental hyperabsorption of iron. A bleeding diathesis is often observed. It seems highly probable that giant cell hepatitis is a heterogeneous category and that one or more entities appropriately termed neonatal hemochromatosis may be causative.

Driscoll et al. (1988) reported 4 cases in 2 families as evidence of autosomal recessive inheritance. In all, presentation was at birth. The clinical course was marked by hypoglycemia, hemorrhagic diathesis, and fatal renal and hepatic failure. At autopsy there was hepatic fibrosis with a distinctive pattern of iron distribution in the hepatocytes, pancreatic acinar cells, and myocardium. In 1 family both parents showed normal serum iron levels, iron-binding capacity, and transferrin and ferritin levels, but had persistently abnormal liver function tests. Neither parent nor their 1 infant had the HLA types associated with hereditary hemochromatosis. Driscoll et al. (1988) suggested that the combination of expressed maternal heterozygosity and fetal homozygosity produces enhanced iron transport across the placenta, resulting in a marked increase in iron deposition in certain fetal tissues. The reticuloendothelial system is relatively spared in this form of hemochromatosis. The clinical course and pathologic findings are distinct from those of Zellweger syndrome (214100), hereditary tyrosinemia (276700), and leprechaunism (246200), in which hepatic siderosis is also seen. Dalhoj et al. (1990) provided a 30-year follow-up of the Danish family reported by Kler and Olesen (1956) in which 6 of 9 sibs had died in utero or neonatally as a result of this disorder. No evidence of hereditary hemochromatosis (235200) or other iron storage disease was found in the parents or surviving sibs. Neonatal hemochromatosis also occurs as part of the neonatal lupus erythematosus syndrome, associated with maternal anti-Ro/SS-A (109092) and anti-Ro/SS-B (109090) autoantibodies.

Verloes et al. (1996) reported observations that throw some doubt on the notion that neonatal hemochromatosis is an autosomal recessive. They reported 2 families in which neonatal hemochromatosis was observed in half-sibs. In the first family, 2 successive girls were born of different fathers. In a second family, an affected brother and sister were followed by an affected half-brother born after donor insemination. These observations, as well as a previous abstract describing 2 affected half-sisters, revived the debate over the inheritance of neonatal hemochromatosis. Incomplete penetrance or gonadal mosaicism for a dominant disorder, a maternal 'environmental factor,' or a mitochondrial defect would appear to be more suitable explanations than autosomal recessive inheritance. Verloes et al. (1996) tabulated the reported familial cases of typical neonatal hemochromatosis. They commented that parental consanguinity was never mentioned.

See also Cassady et al. (1964)

Buchmann, M. S.; Kvittingen, E. A.; Nazer, H.; Gunasekaran, T.; Clayton, P. T.; Sjovall, J.; Bjorkhem, I.: Lack of 3-beta-hydroxy-delta(5)-C(27)-steroid dehydrogenase/isomerase in fibroblasts from a child with urinary excretion of 3-beta-hydroxy-delta(5)-bile acids: a new inborn error of metabolism. J. Clin. Invest. 86: 2034-2037, 1990.
Cassady, G.; Morrison, A. B.; Cohen, M. M.: Familial 'giant-cell hepatitis' in infancy: clinical, pathologic, and genetic studies on a large family. Am. J. Dis. Child. 107: 456-469, 1964.
Clayton, P. T.; Leonard, J. V.; Lawson, A. M.; Setchell, K. D. R.; Andersson, S.; Egestad, B.; Sjovall, J.: Familial giant cell hepatitis associated with synthesis of 3-beta,7-alpha-dihydroxy- and 3-beta,7-alpha,12-alpha-trihydroxy-5-cholenoic acids. J. Clin. Invest. 79: 1031-1038, 1987.
Dalhoj, J.; Kiaer, H.; Wiggers, P.; Grady, R. W.; Jones, R. L.; Knisely, A. S.: Iron storage disease in parents and sibs of infants with neonatal hemochromatosis: 30-year follow-up. Am. J. Med. Genet. 37: 342-345, 1990.
Danks, D.; Bodian, M.: A genetic study of neonatal obstructive jaundice. Arch. Dis. Child. 38: 378-390, 1963.
Driscoll, S. G.; Hayes, A. M.; Levy, H. L.: Neonatal hemochromatosis: evidence for autosomal recessive transmission. (Abstract) Am. J. Hum. Genet. 43: A232 only, 1988.
Fawaz, J. A.; Grady, G. F.; Kaplan, M. M.; Gellis, S. S.: Repetitive maternal-fetal transmission of fetal hepatitis B. New Eng. J. Med. 293: 1357-1359, 1975.
Fienberg, R.: Perinatal idiopathic hemochromatosis: giant cell hepatitis interpreted as an inborn error of metabolism. Am. J. Clin. Path. 33: 480-491, 1960.
Kler, W.; Olesen, M.: Hepatitis foetalis med perinatal exitus letalis hos 4 soskende. Ugeskr. Laeger 118: 868-872, 1956.
Knisely, A. S.; Magid, M. S.; Dische, M. R.; Cutz, E.: Neonatal hemochromatosis. Birth Defects Orig. Art. Ser. XXIII: 75-102, 1987.
Laurendeau, T.; Hill, J. E.; Manning, G. B.: Idiopathic neonatal hemochromatosis in siblings: an inborn error of metabolism. Arch. Path. 72: 410-423, 1961.
Sandor, T.; Surinya, M.; Monus, Z.: Familial occurrence of giant cell hepatitis in infancy. Acta Hepato-Gastroent. 23: 101-104, 1976.
Verloes, A.; Temple, I. K.; Hubert, A.-F.; Hope, P.; Gould, S.; Debauche, C.; Verellen, G.; Deville, J.-L.; Koulischer, L.; Sokal, E. M.: Recurrence of neonatal haemochromatosis in half sibs born of unaffected mothers. J. Med. Genet. 33: 444-449, 1996.

6. UROGENITAL SYSTEM

Kidney Tumors of Adults

RENAL CELL CARCINOMA (#144700)

JJM NOTE

Renal cell carcinoma, like most adult cancers, was not thought to have familial or genetic determinants until relatively recently. Yet, several classic case-control analytic epidemiologic studies had identified no major environmental agent. Then, a single family with an excess of renal cell carcinoma and a constitutional chromosomal translocation pointed to a major gene factor, at least in that one family (Cohen et al. 1979). At that time, our National Cancer Institute team had agreed to karyotype every new cancer family; by standards, karyotyping was not indicated. A gene called *HRCA1* (hereditary renal cancer associated 1) has been isolated from the family's DNA (Boldog et al., 1993). The association of renal cell carcinoma with accessory breasts (polymastia) or multiple nipples (polythelia) further documented the overlap of teratogenesis and carcinogenesis, even for an adult-onset cancer (Goedert JJ et al.: Polymastia and renal adenocarcinoma. Ann. Intern. Med. 95:182–85, 1981). J. Hall (Vancouver, British Columbia) quipped at a meeting that it was the von Hippel–multiple nipple association. Susceptibility to renal cell carcinoma in the rat has been localized to rat chromosome 10, which is homologous to human chromosome 16, not previously incriminated with kidney cancer (Yeung RS et al.: Susceptibility to renal cell carcinoma in the Eker rate involves a tumor suppressor gene on chromosome 10. Proc. Natl. Acad. Sci. USA 90:8038–42, 1993).

A number sign (#) is used with this entry because mutations in several genes have been demonstrated to lead to renal cell carcinoma. These include the von Hippel-Lindau (VHL; 193300) gene and the TRC8 gene (603046). Hereditary papillary renal cell carcinoma has been related to mutations in the MET gene (164860). Furthermore, an X-linked form of papillary renal cell carcinoma (312390) has been identified.

Familial renal cell carcinoma (RCC) is relatively rare. Reports (e.g., Franksson et al., 1972; Goldman et al., 1979) suggest an early average age at diagnosis and frequent bilateral or multiple primary tumors in familial cases. Rusche (1953) observed hypernephroma in 2 brothers. Both had distant metastasis as the first manifestation and both were in their early 30s at the time of diagnosis. Brinton (1960) described a family in which 2 brothers and a sister had hypernephroma. The father had died of kidney tumor and the mother of cancer, site unstated. One of the patients had polycythemia, a known accompaniment of hypernephroma on occasion. It should be noted that hypernephroma and cerebellar hemangioblastoma, which histologically resembles hypernephroma, are features of von Hippel-Lindau disease. Polycythemia also occurs with cerebellar hemangioblastoma. Fairchild et al. (1979) described a 29-year-old woman who had neuroblastoma during infancy, developed an extraadrenal pheochromocytoma at age 16 years, with subsequent hepatic recurrence, and was found to have multifocal renal cell carcinoma. Renal cell carcinoma and pheochromocytoma are combined in the von Hippel-Lindau syndrome, for which there was no evidence in this patient or her family. The association of pheochromocytoma and neuroblastoma had, it seemed, not been previously noted. Jakesz and Wuketich (1978) reported an instructive family in which 3 brothers had bilateral renal cell carcinoma. The index case also had cerebellar hemangioblastoma. The authors suggested that von Hippel-Lindau disease was the fundamental problem. Braun et al. (1975) studied 3 families, each with multiple cases of renal cell carcinoma. There appeared to be an association with HLA W17 tissue type.

Cohen et al. (1979) described a family in which members with an inherited chromosomal translocation, t(3;8)(p21;q24), were predisposed to renal cancer. In 1 patient, cancer developed in residual renal tissue after partial nephrectomy. One patient had polycythemia. In the family reported by Cohen et al. (1979), Wang and Perkins (1984) used high resolution prometaphase G-banding analysis to demonstrate that the breakpoints occurred at the subbands 3p14.2 (not 3p21) and 8q24.1.

Because of the findings of the role of the MYC gene (190080), located at 8q24, in Burkitt lymphoma with reciprocal translocation between 8q24 and 14q32, the possibility exists that a similar mechanism is involved in the translocation between 8q24 and 3p21 in the family reported by Cohen et al. (1979). All 10 members of the family that developed renal cancer carried the translocation, whereas no member with a normal karyotype had renal cancer. Another oncogene, RAF1 (164760), has been assigned to chromosome 3 (3p25), but its possible role in RCC (or in small cell cancer of the lung, 182280, which also is related to chromosome 3p) is unknown. Harris et al. (1986) sorted the chromosome from a cell line established from a subject with the t(3;8)(p14.2;q24.1) translocation. They found that the RAF1 gene was translocated to the derivative chromosome 8 and, conversely, that the MYC gene was translocated to the derivative chromosome 3. By in situ hybridization with an RAF1 probe, Teyssier et al. (1986) showed that in 2 renal cancers with 3p deletion, the oncogene had shifted from 3p25 to 3p14. By the use of cosmid cloning, Harris et al. (1986) showed that there was no rearrangement 31 kb 5-prime or 19 kb 3-prime to the translocated MYC gene. Gemmill et al. (1989) found that a 1.5-Mb segment surrounding MYC did not include the translocation breakpoint in familial RCC with t(3;8); thus, again, MYC seems not to be implicated in the development of RCC. Li et al. (1993) followed up on the family reported by Cohen et al. (1979). Renal carcinoma had recurred in all 5 patients carrying the 3;8 translocation at 1 to 16 years after resection. Cancer was the cause of death in 3 patients; 2 were in a second remission. The tumors from 3 family members consistently showed loss of the entire derivative chromosome 8, which bore the segment 3pter-p14. In contrast, no genetic change was detected in the derivative chromosome 3 or in normal chromosomes 3 and 8. One interpretation from these findings is that the breakpoints in chromosomes 3 and 8 per se had no relevance to the renal carcinoma. Instead, the renal carcinoma was related to the tendency toward loss of the terminal part of 3p containing the von Hippel-Lindau gene (VHL; 193300) in the tumors of the affected patients.

Boldog et al. (1993) reported positional cloning of the 3p14.2 breakpoint in the family of Cohen et al. (1979). The 3p14.2 region also contains FRA3B, the most sensitive fragile site induced by aphidicolin. A chromosome 3 probe derived from a radiation-reduced hybrid was positioned between the t(3;8) breakpoint and an aphidicolin-induced 3p14 breakpoint. A yeast artificial chromosome (YAC) contig containing this probe was developed that crossed the aphidicolin-induced breakpoint. A subsequent chromosome walk identified a YAC that crossed the t(3;8) translocation breakpoint. A lambda sublibrary allowed isolation of clones spanning the rearrangement. Unique and evolutionarily conserved DNA sequences were used to screen a kidney cDNA library. They identified a gene, referred to as HRCA1 (hereditary renal cancer associated 1), that mapped immediately adjacent to the breakpoint. On the basis of its chromosomal position, they considered HRCA1 to be a candidate tumor suppressor gene.

Ohta et al. (1996) pointed out that a 200- to 300-kb region of 3p14.2, including the fragile site locus FRA3B, is homozygously deleted in multiple tumor-derived cell lines. By exon amplification from cosmids covering this deleted region, they identified the human gene called FHIT (601153) for 'fragile histidine triad gene.' The gene encodes a protein with 69% similarity to an enzyme of Schizosaccharomyces pombe, diadenosine 5-prime,5-triple prime-P(1),P(4)-tetraphosphate, asymmetric hydrolase. They found that the FHIT locus is composed of 10 exons distributed over at least 50 kb, with 3 5-prime untranslated exons centromeric to the RCC-associated 3p14.2 breakpoint, the remaining exons telomeric to this translocation breakpoint, and exon 5 within the homozygously deleted fragile region. Aberrant transcripts of the FHIT locus were found in approximately 50% of esophageal, stomach, and colon carcinomas. The first 3 exons of the gene mapped centromeric to the t(3;8) break and between the break and the 5-prime end of the PTPRG gene (176886).

Gnarra et al. (1994) demonstrated that the von Hippel-Lindau gene was mutated in the germline of individuals with familial renal carcinoma reported by Cohen et al. (1979); furthermore, a renal tumor from one of the affected individuals carrying the constitutional translocation, t(3;8)(p14;q24) had loss of the VHL gene on the other chromosome. Thus the findings adhere to the Knudson 2-hit hypothesis as for many other tumors related to tumor suppressor genes.

Pathak et al. (1982) reported an acquired balanced 3;11 translocation in tumor cells from a patient with a normal constitutional karyotype and a history of renal cell carcinoma in the paternal grandfather and a paternal uncle. Li et al. (1982) reviewed 9 families in which 2 or more members had renal carcinoma. Multiple generations were affected in 5, sibs in 4. The median age at diagnosis was a decade earlier than usual, and individual patients had bilateral or multifocal lesions; these are features of hereditary forms of diverse cancers. No patient had von Hippel-Lindau disease and none had 3;8 translocation.

Pathak and Goodacre (1986) found reciprocal translocations involving 3p14-p13 and chromosomes 6, 8, 11 and 16. Yoshida et al. (1986)

found rearrangement of chromosome 3 in 8 of 12 nonfamilial renal cell carcinomas. Using RFLPs, Zbar et al. (1987) examined tumors from 18 patients with nonhereditary renal cell carcinomas and found loss of alleles at loci on the short arm of chromosome 3 in all 11 of the patients who could be evaluated. Szucs et al. (1987) described deletion of 3p as the only chromosome loss in a nonfamilial renal cell carcinoma. The deletion was thought to be at 3p11. Carroll et al. (1987) found clonal abnormalities affecting the short arm of chromosome 3 in the 3p21-p12 region in 5 of 6 clear cell renal carcinomas. In the remaining case, of 15 karyotyped metaphases suitable for interpretation, 1 showed a deletion in 3p. In tumor cells from a patient with renal cell carcinoma and von Hippel-Lindau disease, King et al. (1987) found a deletion in proximal 3p, thus providing support for the idea that genetic material in this area is critical to the development of RCC. In 22 of 25 primary renal cell carcinomas, Kovacs et al. (1987) found an aberration of chromosome 3, deletion of 3p, or translocation of chromosome segments to the deleted chromosome 3. The breakpoints in rearrangements of chromosome 3 clustered in the region of 3p13-p11.2. In 8 of the 25 RCCs, the rearrangement of chromosome 3 was the only karyotype change. Kovacs and Hoene (1988) studied a patient with renal cancer and a constitutional reciprocal translocation t(3;12)(q13.2;q24.1). They found that tumor cells were characterized by loss of the derivative chromosome 3, supporting the hypothesis that loss of a specific 3p segment is associated with the development of renal cancer. Kovacs et al. (1988) examined renal cell carcinoma and normal kidney tissues from 34 patients with sporadic, nonhereditary RCC. Of the 21 cytogenetically examined tumors, 18 had a detectable anomaly of 3p distal to band 3p11.2-p13, combined with the nonreciprocal translocation of a segment from another chromosome or monosomy 3. RFLP analysis showed loss of heterozygosity (LOH). The authors suggested that RCC may arise from the deletion of a 'recessive cancer gene.' Van der Hout et al. (1988) compared chromosome 3 markers in normal and tumorous nephrectomy specimens from 7 RCC patients. Three patients were not informative because of homozygosity at all loci studied. One patient showing heterozygosity at 3q in normal tissue had a tumor that remained heterozygous. In 3 patients the tumor showed LOH for a short arm marker at 3p21; in 1 of them, heterozygosity for a second short arm marker was also lost. A second of these 3 patients retained heterozygosity for the second short arm marker, as well as for a long arm marker, suggesting a chromosomal breakpoint between the loci for the 2 short arm markers. By studying LOH in 41 matched tumor/normal kidney tissue pairs, van der Hout et al. (1991) limited the commonly deleted part of 3p to the region between THRB (190160) in 3p24.3 and D3S2 in 3p21. The regions on 3p thought to be involved in the von Hippel-Lindau syndrome and in hereditary renal cell carcinoma are both outside this smallest region of overlapping deletions. Unlike some other tumor types, RCC does not show multiple allelic losses on a number of chromosomes; LOH is restricted mainly to chromosome 3.

Miles et al. (1988) found trisomy or tetrasomy of chromosome 7 as the most frequent abnormality in nonfamilial renal cell carcinoma, being present in tumors from 15 of 19 patients with cytogenetically abnormal neoplasms. An abnormal chromosome 3 was found in 10 of the cases: 2 were trisomic for chromosome 3; 2 were monosomic; 3 were hyperdiploid; and 3 had interstitial deletions with breakpoints clustered from p11 to p25. Kovacs and Brusa (1988) could find no correlation between breakpoints on 3p or 5q and fragile sites on those chromosome arms.

Erlandsson et al. (1988) compared the age distribution of 51 hereditary and 56 sporadic cases of RCC sampled from the literature. The age-incidence curve of the hereditary RCC was compatible with a single event, whereas the sporadic tumors arose as predicted from a 2-hit curve. Thus, in analogy with Knudson's original prediction for retinoblastoma and Wilms tumor, RCC appears to arise by the loss of a recessive cancer gene, probably located in band 3p14.2. Linehan et al. (1989) demonstrated that removal of leukocytes from disaggregated renal cell carcinomas improved the detection of allele loss in these carcinomas. The technique should be useful in allele deletion analysis of other solid tumors that are contaminated with host leukocytes. Anglard et al. (1992) found LOH on 3p in 25 cell lines derived from 28 informative nonpapillary forms of RCC. Deletion-mapping analysis showed retention of the distal locus D3S18 in one of the RCC lines, further localizing the putative tumor suppressor gene. Although deletions on chromosome 3 had been suggested to be specific for the clear cell type, Anglard et al. (1992) could find no correlation between LOH and clear or granular cell types. Ogawa et al. (1992) found an inverse relationship between allelic losses at chromosome 17p in renal cell carcinoma and allelic losses at 3p: none of the 5 informative RCCs with allelic losses at 17p showed allelic losses at 3p. Conversely, 17 of 25 informative RCCs with retention of 17p alleles lost alleles at 3p. They demonstrated that allelic losses at 17p were infrequent in the clear cell type of RCC, whereas allelic losses at 17p were significantly more frequent in the granular cell type of RCC. Ogawa et al. (1991) had previously reported that allelic losses at 3p were specific to the clear cell type of RCC.

Increasingly, the introduction of genes into cultured cells that are deficient in some activity or show malignant characteristics is being used as proof that the given gene is mutant and the cause of the abnormality. Shimizu et al. (1990) introduced a single chromosome containing the short arm of chromosome 3 into a human renal cell carcinoma cell line via microcell fusion. They observed suppression of tumorigenicity in nude mice or modulation of tumor-growth rate in vitro.

Levinson et al. (1990) reported that, since 1961, 28 families with multiple cases of renal cell carcinoma had been reported, with an abnormality in the constitutional karyotype having been found in only 1 family. They identified 5 more families in which a total of 12 relatives had renal cell carcinoma; peripheral blood karyotypes from 7 patients and 5 unaffected relatives showed no significant abnormalities. They suggested that members of families with multiple cases of renal cell carcinoma be screened with renal ultrasound initially at age 30, with repeat examinations every 2 or 3 years. The recommendations are similar to those for von Hippel-Lindau disease.

Everitt et al. (1992) identified a single gene mutation in rats that predisposes them to bilateral multicentric renal cell carcinoma. The disorder bore similarities to von Hippel-Lindau disease. Splenic vascular proliferative lesions, including hemangiosarcoma, were seen in 23% of 14-month-old rats of both sexes that had renal tumors. At that age, 62% of female rats with renal cell tumors had sarcomas of the lower reproductive tract of probable smooth muscle origin. In this rat model of human renal carcinoma, Walker et al. (1992) identified a germline mutation at a tumor susceptibility locus that caused a 70-fold increase in susceptibility to chemical carcinogenesis. A carcinogen that targeted both renal epithelial and mesenchymal cells caused an increase in tumors of epithelial origin in susceptible animals but no increase in carcinogen-induced mesenchymal tumors. Rats that are heterozygous for the so-called Eker mutation develop spontaneous RCCs between 4 and 12 months of age. When homozygous, the mutation is lethal prenatally at 9 to 10 days of gestation. At the histologic level, renal carcinomas in the Eker rat develop through multiple stages from early preneoplastic lesions (e.g., atypical tubules) to adenomas in virtually all heterozygotes by the age of 1 year. The homozygous mutant condition is lethal at about 10 days of fetal life. Hino et al. (1993) demonstrated that ionizing radiation induces additional tumors in a linear dose-response relationship, suggesting that in heterozygotes 2 events (one inherited, one somatic) are necessary to produce tumors, and that the predisposing gene is a tumor suppressor gene. No genetic linkage was found between the Eker mutation and rat DNA sequences homologous to those in human chromosome 3p, the presumed site of the putative tumor suppressor gene responsible for human renal cell carcinoma. Nonrandom loss of rat chromosome 5 in RCC-derived cell lines was sometimes associated with homozygous deletion of the interferon gene loci at rat c hromosome bands 5q31-q33. Since this locus is not linked to the predisposing inherited gene in the Eker rat, it probably represented a second tumor suppressor gene involved in tumor progression. Kobayashi et al. (1995) demonstrated that a germline mutation in the tuberous sclerosis-2 gene (191092), caused by the insertion of a DNA fragment approximately 5 kb long, is responsible for aberrant RNA expression from the mutant allele in the Eker rat. Except for the occurrence of renal tumors, the phenotype of tuberous sclerosis in human differs from that of the Eker rat.

Motzer et al.(1996) reviewed all aspects of renal cell carcinoma in detail, including the molecular genetic abnormalities and the evidence for a locus on 3p distinct from the von Hippel-Lindau gene; the VHL gene is involved in the great majority of cases of renal cell carcinomas of the clear-cell type.

As outlined earlier, the 3;8 chromosomal translocation in the family reported by Cohen et al. (1979) suggested the existence of a locus on 3p underlying clear-cell renal carcinoma. The frequent 3p loss of heterozygosity in sporadic RCC further led to the assumption that a critical tumor suppressor gene would be located at 3p14. Identification of the von Hippel-Lindau gene (VHL; 193300) at 3p25 provided an alternative explanation for at least some observed 3p loss of heterozygosity. Furthermore, van den Berg and Buys (1997) reported that region 3p21 may be involved in the malignant progression of renal tumors. Within 3p14, Ohta et al. (1996) identified the FHIT gene (601153), which was interrupted in its 5-prime untranslated region by the 3;8 translocation. However, a number of findings suggested that FHIT was an unlikely causative gene in the hereditary t(3;8) family. The fact that FHIT in a case of parathyroid adenoma underwent fusion with the high mobility group protein gene HMGIC (600698), the causative gene in a variety of benign tumors (Geurts et al., 1997) suggested to Gemmill et al. (1998) that FHIT might be a bystander in the fusion with an alternative candidate gene on chromosome 8. By use of 5-prime rapid amplification of cDNA ends (RACE), Gemmill et al. (1998) identified a gene, which they called TRC8 (603046), with characteristics compatible with oncogenic properties. In addition, they identified a TRC8 mutation in a sporadic renal carcinoma.

See also Lyons et al. (1977); Neel et al. (1982)

Anglard, P.; Trahan, E.; Liu, S.; Latif, F.; Merino, M. J.; Lerman, M. I.; Zbar, B.; Linehan, W. M.: Molecular and cellular characterization of human renal cell carcinoma cell lines. Cancer Res. 52: 348-356, 1992.

Boldog, F. L.; Gemmill, R. M.; Wilke, C. M.; Glover, T. W.; Nilsson, A.-S.; Chandrasekharappa, S. C.; Brown, R. S.; Li, F. P.; Drabkin, H. A.: Positional cloning of the hereditary renal carcinoma 3;8 chromosome translocation breakpoint. Proc. Nat. Acad. Sci. 90: 8509-8513, 1993.

Braun, W. E.; Strimlan, C. V.; Negron, A. G.; Straffon, R. A.; Zachary, A. A.; Bartee, S. L.; Grecek, D. R.: The association of W17 with familial renal cell carcinoma. Tissue Antigens 6: 101-104, 1975.

Brinton, L. F.: Hypernephroma—familial occurrence in one family. J.A.M.A. 173: 888-890, 1960.

Carroll, P. R.; Murty, V. V. S.; Reuter, V.; Jhanwar, S.; Fair, W. R.; Whitmore, W. F.; Chaganti, R. S. K.: Abnormalities at chromosome region 3p12-14 characterize clear cell renal carcinoma. Cancer Genet. Cytogenet. 26: 253-259, 1987.

Cohen, A. J.; Li, F. P.; Berg, S.; Marchetto, D. J.; Tsai, S.; Jacobs, S. C.; Brown, R. S.: Hereditary renal-cell carcinoma associated with chromosomal translocation. New Eng. J. Med. 301: 592-595, 1979.

Erlandsson, R.; Boldog, F.; Sumegi, J.; Klein, G.: Do human renal cell carcinomas arise by a double-loss mechanism?. Cancer Genet. Cytogenet. 36: 197-202, 1988.

Everitt, J. I.; Goldsworthy, T. L.; Wolf, D. C.; Walker, C. L.: Hereditary renal cell carcinoma in the Eker rat: a rodent familial cancer syndrome. J. Urol. 148: 1932-1936, 1992.

Fairchild, R. S.; Kyner, J. L.; Hermreck, A.; Schimke, R. N.: Neuroblastoma, pheochromocytoma, and renal cell carcinoma: occurrence in a single patient. J.A.M.A. 242: 2210-2211, 1979.

Franksson, C.; Bergstrand, A.; Ljungdahl, I.; Magnusson, G.; Nordenstam, H.: Renal carcinoma (hypernephroma) occurring in 5 siblings. J. Urol. 108: 58-61, 1972.

Gemmill, R. M.; Coyle-Morris, J.; Ware-Uribe, L.; Pearson, N.; Hecht, F.; Brown, R. S.; Li, F. P.; Drabkin, H. A.: A 1.5-megabase restriction map surrounding MYC does not include the translocation breakpoint in familial renal cell carcinoma. Genomics 4: 28-35, 1989.

Gemmill, R. M.; West, J. D.; Boldog, F.; Tanaka, N.; Robinson, L. J.; Smith, D. I.; Li, F.; Drabkin, H. A.: The hereditary renal cell carcinoma 3;8 translocation fuses FHIT to a patched-related gene, TRC8. Proc. Nat. Acad. Sci. 95: 9572-9577, 1998.

Geurts, J. M.; Schoenmakers, E. F.; Roijer, E.; Stenman, G.; Van de Ven, W. J. M.: Expression of reciprocal hybrid transcripts of HMGIC and FHIT in a pleomorphic adenoma of the parotid gland. Cancer Res. 57: 13-17, 1997.

Gnarra, J. R.; Tory, K.; Weng, Y.; Schmidt, L.; Wei, M. H.; Li, H.; Latif, F.; Liu, S.; Chen, F.; Duh, F.-M.; Lubensky, I.; Duan, D. R.; Florence, C.; Pozzatti, R.; Walther, M. M.; Bander, N. H.; Grossman, H. B.; Brauch, H.; Pomer, S.; Brooks, J. D.; Isaacs, W. B.; Lerman, M. I.; Zbar, B.; Linehan, W. M.: Mutations of the VHL tumour suppressor gene in renal carcinoma. Nature Genet. 7: 85-90, 1994.

Goldman, S. M.; Fishman, E. K.; Abeshouse, G.; Cohen, J. H.: Renal cell carcinoma diagnosed in three generations of a single family. Sth. Med. J. 72: 1457-1459, 1979.

Harris, P.; Morton, C. C.; Guglielmi, P.; Li, F.; Kelly, K.; Latt, S. A.: Mapping by chromosome sorting of several gene probes, including c-myc, to the derivative chromosomes of a 3;8 translocation associated with familial renal cancer. Cytometry 7: 589-594, 1986.

Hino, O.; Klein-Szanto, A. J. P.; Freed, J. J.; Testa, J. R.; Brown, D. Q.; Vilensky, M.; Yeung, R. S.; Tartof, K. D.; Knudson, A. G.: Spontaneous and radiation-induced renal tumors in the Eker rat model of dominantly inherited cancer. Proc. Nat. Acad. Sci. 90: 327-331, 1993.

Jakesz, R.; Wuketich, S.: Familiaeres bilaterales Nierenzellkarzinom und zerebellares Haemangioblastom. Dtsch. Med. Wschr. 103: 2040-2045, 1978.

King, C. R.; Schimke, R. N.; Arthur, T.; Davoren, B.; Collins, D.: Proximal 3p deletion in renal cell carcinoma cells from a patient with von Hippel-Lindau disease. Cancer Genet. Cytogenet. 27: 345-348, 1987.

Kobayashi, T.; Hirayama, Y.; Kobayashi, E.; Kubo, Y.; Hino, O.: A germline insertion in the tuberous sclerosis (Tsc2) gene gives rise to the Eker rat model of dominantly inherited cancer. Nature Genet. 9: 70-74, 1995.

Kovacs, G.; Brusa, P.: Recurrent genomic rearrangements are not at the fragile sites on chromosomes 3 and 5 in human renal cell carcinomas. Hum. Genet. 80: 99-101, 1988.

Kovacs, G.; Erlandsson, R.; Boldog, F.; Ingvarsson, S.; Muller-Brechlin, R.; Klein, G.; Sumegi, J.: Consistent chromosome 3p deletion and loss of heterozygosity in renal cell carcinoma. Proc. Nat. Acad. Sci. 85: 1571-1575, 1988.

Kovacs, G.; Hoene, E.: Loss of der(3) in renal carcinoma cells of a patient with constitutional t(3;12). Hum. Genet. 78: 148-150, 1988.

Kovacs, G.; Szucs, S.; De Riese, W.; Baumgartel, H.: Specific chromosome aberration in human renal cell carcinoma. Int. J. Cancer 40: 171-178, 1987.

Levinson, A. K.; Johnson, D. E.; Strong, L. C.; Pathak, S.; Huff, V.; Saunders, G. F.: Familial renal cell carcinoma: hereditary or coincidental? J. Urol. 144: 849-851, 1990.

Li, F. P.; Decker, H.-J. H.; Zbar, B.; Stanton, V. P., Jr.; Kovacs, G.; Seizinger, B. R.; Aburatani, H.; Sandberg, A. A.; Berg, S.; Hosoe, S.; Brown, R. S.: Clinical and genetic studies of renal cell carcinomas in a family with a constitutional chromosome 3;8 translocation: genetics of familial renal carcinoma. Ann. Intern. Med. 118: 106-111, 1993.

Li, F. P.; Marchetto, D. J.; Brown, R. S.: Familial renal carcinoma. Cancer Genet. Cytogenet. 7: 271-275, 1982.

Linehan, M.; Miller, E.; Anglard, P.; Merino, M.; Zbar, B.: Improved detection of allele loss in renal cell carcinomas after removal of leukocytes by immunologic selection. J. Nat. Cancer Inst. 81: 287-290, 1989.

Lyons, A. R.; Logan, H.; Johnston, G. W.: Hypernephroma in two brothers. Brit. Med. J. 1: 816-817, 1977.

Miles, J.; Michalski, K.; Kouba, M.; Weaver, D. J.: Genomic defects in nonfamilial renal cell carcinoma: possible specific chromosome change. Cancer Genet. Cytogenet. 34: 135-142, 1988.

Motzer, R. J.; Bander, N. H.; Nanus, D. M.: Renal-cell carcinoma. New Engl. J. Med. 335: 865-875, 1996.

Neel, B. G.; Jhanwar, S. C.; Chaganti, R. S. K.; Hayward, W. S.: Two human c-onc genes are located on the long arm of chromosome 8. Proc. Nat. Acad. Sci. 79: 7842-7846, 1982.

Ogawa, O.; Habuchi, T.; Kakehi, Y.; Koshiba, M.; Sugiyama, T.; Yoshida, O.: Allelic losses at chromosome 17p in human renal cell carcinoma are inversely related to allelic losses at chromosome 3p. Cancer Res. 52: 1881-1885, 1992.

Ogawa, O.; Kakehi, Y.; Ogawa, K.; Koshiba, M.; Sugiyama, T.; Yoshida, O.: Allelic loss at chromosome 3p characterizes clear cell phenotype of renal cell carcinoma. Cancer Res. 51: 949-953, 1991.

Ohta, M.; Inoue, H.; Cotticelli, M. G.; Kastury, K.; Baffa, R.; Palazzo, J.; Siprashvili, Z.; Mori, M.; McCue, P.; Druck, T.; Croce, C. M.; Huebner, K.: The FHIT gene, spanning the chromosome 3p14.2 fragile site and renal carcinoma-associated t(3;8) breakpoint, is abnormal in digestive tract cancers. Cell 84: 587-597, 1996.

Ohta, M.; Inoue, H.; Cotticelli, M. G.; Kastury, K.; Baffa, R.; Palazzo, J.; Siprashvili, Z.; Mori, M.; McCue, P.; Druck, T.; Croce, C. M.; Huebner, K.: The FHIT gene, spanning the chromosome 3p14.2 fragile site and renal carcinoma-associated t(3;8) breakpoint, is abnormal in digestive tract cancers. Cell 84: 587-597, 1996.

Pathak, S.; Goodacre, A.: Specific chromosome anomalies and predisposition to human breast, renal cell, and colorectal carcinoma. Cancer Genet. Cytogenet. 19: 29-36, 1986.

Pathak, S.; Strong, L. C.; Ferrell, R. E.; Trindade, A.: Familial renal cell carcinoma with a 3;11 chromosome translocation limited to tumor cells. Science 217: 939-941, 1982.

Rusche, C.: Silent adenocarcinoma of the kidneys with solitary metastases occurring in brothers. J. Urol. 70: 146-151, 1953.

Shimizu, M.; Yokota, J.; Mori, N.; Shuin, T.; Shinoda, M.; Terada, M.; Oshimura, M.: Introduction of normal chromosome 3p modulates the tumorigenicity of a human renal cell carcinoma cell line YCR. Oncogene 5: 185-194, 1990.

Szucs, S.; Muller-Brechlin, R.; DeRiese, W.; Kovacs, G.: Deletion 3p: the only chromosome loss in a primary renal cell carcinoma. Cancer Genet. Cytogenet. 26: 369-373, 1987.

Teyssier, J. R.; Henry, I.; Dozier, C.; Ferre, D.; Adnet, J. J.; Pluot, M.: Recurrent deletion of the short arm of chromosome 3 in human renal cell carcinoma: shift of the c-raf 1 locus. J. Nat. Cancer Inst. 77: 1187-1191, 1986.

van den Berg, A.; Buys, C. H.: Involvement of multiple loci on chromosome 3 in renal cell cancer development. Genes Chromosomes Cancer 19: 59-76, 1997.

van der Hout, A. H.; Kok, K.; van den Berg, A.; Oosterhuis, J. W.; Carritt, B.; Buys, C. H. C. M.: Direct molecular analysis of a deletion of 3p in tumors from patients with sporadic renal cell carcinoma. Cancer Genet. Cytogenet. 32: 281-285, 1988.

van der Hout, A. H.; van der Vlies, P.; Wijmenga, C.; Li, F. P.; Oosterhuis, J. W.; Buys, C. H. C. M.: The region of common allelic losses in sporadic renal cell carcinoma is bordered by the loci D3S2 and THRB. Genomics 11: 537-542, 1991.

Walker, C.; Goldsworthy, T. L.; Wolf, D. C.; Everitt, J.: Predisposition to renal cell carcinoma due to alteration of a cancer susceptibility gene. Science 255: 1693-1695, 1992.

Wang, N.; Perkins, K. L.: Involvement of band 3p14 in t(3;8) hereditary renal carcinoma. Cancer Genet. Cytogenet. 11: 479-481, 1984.

Yoshida, M. A.; Ohyashiki, K.; Ochi, H.; Gibas, Z.; Prout, G. R., Jr.; Pontes, E. J.; Huben, R.; Sandberg, A. A.: Rearrangement of chromosome 3 in renal cell carcinoma. Cancer Genet. Cytogenet. 19: 351-354, 1986.

Zbar, B.; Brauch, H.; Talmadge, C.; Linehan, M.: Loss of alleles of loci on the short arm of chromosome 3 in renal cell carcinoma. Nature 327: 721-724, 1987.

RENAL CELL CARCINOMA, PAPILLARY (179755)

JJM NOTE

Additional families are needed, but some progress on gene identification could take place even with apparently sporadic cases. The clinic lesson is to be alert to rare cell types and to document the histopathology of reported tumors in relatives.

Zbar et al. (1994) described members in 3 generations of a family who had papillary renal cell carcinoma, an uncommon histologic type. Multiple tumors of varying size were present in both kidneys of affected family members. The family contained several instances of male-to-male transmission. The disorder was not linked to polymorphic markers on chromosome 3p, and there was no loss of heterozygosity at loci on 3p identified in the renal tumors. Occurrence of papillary renal cell carcinoma in association with a translocation t(X;1)(p11;q21) (Meloni et al., 1993) suggests that there may be a gene for this type of tumor on both chromosome 1 and the X chromosome; see RCCP2 (312390).

Zbar et al. (1995) sought other families with papillary renal cell carcinoma to determine whether the large family reported by Zbar et al. (1994) represented a distinct class of inherited cancer. They identified an additional 9 families with multiple affected members. Including the affected members of the original family (designated family 150), there were 29 affected male and 12 affected female subjects. Papillary renal cell carcinomas were often detected incidentally in asymptomatic individuals or during screening of asymptomatic members of renal carcinomal cell families. The penetrance, i.e., the proportion of obligate gene carriers that showed medical evidence of the disease, was reduced. The median survival of affected individuals was 52 years. Several instances of male-to-male transmission were observed in these families. The carcinomas were often multiple and bilateral in these cases.

Sidhar et al. (1996) demonstrated that the specific chromosomal translocation t(X;1)(p11.2;q21.2) observed in human papillary renal cell carcinoma (PRCC) results in the fusion of a gene at 1q21.2, designated PRCC, to the TFE3 gene (314310) at Xp11.2. The authors noted that papillary renal cell tumors, which account for 15-20% of renal carcinomas, do not exhibit mutation of the von Hippel-Lindau locus (193300) or loss of chromosome 3p, which frequently occurs in other forms of renal carcinoma including the clear cell, granular cell, and sarcomatoid types. In their studies they demonstrated disruption of the TFE3 gene in papillary carcinomas carrying the t(X;1) translocation. Further characterization of the TFE3 gene in these tumors led to the discovery that it was fused with the PRCC gene. By Southern blot analysis using a probe corresponding to the 5-prime end of the TFE3 gene, Sidhar et al. (1996) detected rearrangements in 3 papillary renal tumor cell lines that carried the t(X:1)(p11.2:q21.2) translocation. To determine whether the TFE3 transcripts in a tumor cell line were altered at their 5-prime end, 5-prime RACE products were generated from the tumor cell line. Sequence analysis revealed that these 5-prime RACE products contained unique sequences. A probe generated from this unique sequence was used to isolate clones from human fetal and monocyte cDNA libraries. These clones hybridized to chromosome 1q21.2. Sidhar et al. (1996) reported that the gene represented by the cDNAs encodes a 491-amino acid polypeptide. The predicted PRCC protein has an N-terminal domain rich in proline, leucine, and glycine but failed to exhibit homology to other known proteins. Sidhar et al. (1996) demonstrated the presence of a PRCC-TFE3 hybrid transcript in all 3 papillary renal tumors by RT-PCR using 5-prime PRCC and 3-prime TFE3 primers. In 2 of the tumors the breakpoint could be assigned to TFE3 intron 1. Sidhar et al. (1996) concluded that the PRCC-TFE3 fusion protein apparently r epresents an N-terminal transcriptional activation domain adjacent to a TFE3 DNA binding domain. They noted that a frequent theme for translocations in sarcomas is the fusion of a transcription factor activation domain to a transcription factor DNA binding element.

Independently, Weterman et al. (1996) identified the 2 genes involved in the t(X;9)(p11;q21) translocation in human papillary renal cell carcinomas. Through the fusion of the TFE3 gene and PRCC on chromosome 1, reciprocal translocation products are formed, which are both expressed in papillary renal cell carcinomas. Weterman et al. (1996) cloned the PRCC gene from a human cDNA library. PRCC is ubiquitously expressed in normal adult and fetal tissues and encodes a putative protein of 491 amino acids with a relatively high content of proline. No relevant homologies with known sequences at either the DNA or protein level were found.

Meloni, A. M.; Dobbs, R. M.; Pontes, J. E.; Sandberg, A. A.: Translocation (X;1) in papillary renal cell carcinoma: a new cytogenetic subtype. Cancer Genet. Cytogenet. 65: 1-6, 1993.

Sidhar, S. K.; Clark. J.; Gill, S.; Hamoudi, R.; Crew, A. J.; Gwilliam, R.; Ross, M.; Linehan, W. M.; Birdsall, S.; Shipley, J.; Cooper, C. S.: The t(X;1)(p11.2;q21.2) translocation in papillary renal cell carcinoma fuses a novel gene PRCC to the TFE3 transcription factor gene. Hum. Molec. Genet. 5: 1333-1338, 1996.

Weterman, M. A. J.; Wilbrink, M.; Geurts van Kessel, A.: Fusion of the transcription factor TFE3 gene to a novel gene, PRCC, in t(X;1)(p11;q21)-positive papillary renal cell carcinomas. Proc. Nat. Acad. Sci. 93: 15294-15298, 1996.

Zbar, B.; Glenn, G.; Lubensky, I.; Choyke, P.; Walther, M. M.; Magnusson, G.; Bergerheim, U. S. R.; Pettersson, S.; Amin, M.; Hurley, K.; Linehan, W. M.: Hereditary papillary renal cell carcinoma: clinical studies in 10 families. J. Urol. 153: 907-912, 1995.

Zbar, B.; Tory, K.; Merino, M.; Schmidt, L.; Glenn, G.; Choyke, P.; Walther, M. M.; Lerman, M.; Linehan, W. M.: Hereditary papillary renal cell carcinoma. J. Urol. 151: 561-566, 1994.

MET PROTO-ONCOGENE (164860)

JJM NOTE

This long-known oncogene was discovered in an osteosarcoma cell line. Although it could not have been predicted, germline and somatic mutations of this "old" oncogene have been found in familial and sporadic papillary renal carcinoma.

Cooper et al. (1984) cloned a transforming gene from a chemically transformed human osteosarcoma-derived cell line and mapped it to 7p11.4-qter. Identity to all previously known oncogenes except ERBB (131550) was ruled out by the fact that they are encoded by other chromosomes; identity to ERBB is probably excluded by failure of direct hybridizations of the 2 probes. MET was the designation suggested by Cooper et al. (1984). Dean et al. (1985) showed that MET is in the tyrosine kinase family of oncogenes. It appeared to be most closely related in sequence to the human insulin receptor and ABL oncogene. By in situ hybridization, they assigned the MET locus to 7q21-q31, where it is tightly linked to CF (219700) which is thought to be located in the proximal part of 7q22. It is about 10 cM from TCRB (186930) and is in a region that is associated with nonrandom chromosomal deletions in some patients with acute nonlymphocytic leukemia. From the sequence of MET cDNA, Park et al. (1987) concluded that this oncogene is a cell-surface receptor for a then unknown ligand. The c-MET protooncogene product is a receptor-like tyrosine kinase comprised of disulfide-linked subunits of 50 kD (alpha) and 145 kD (beta). In the fully processed c-MET product, the alpha subunit is extracellular, and the beta subunit has extracellular, transmembrane, and tyrosine kinase domains as well as sites of tyrosine phosphorylation. Bottaro et al. (1991) demonstrated that the beta-subunit of the c-MET protooncogene product is the cell-surface receptor for hepatocyte growth factor (142409). Cooper (1992) reviewed the development of understanding of the MET oncogene in an article subtitled 'From detection by transfection to transmembrane receptor for hepatocyte growth factor.'

Dean et al. (1987) demonstrated that the MET oncogene is located on chromosome 6 in the mouse, where it is tightly linked to an obesity mutation ('ob'). Is there an obesity mutation on human chromosome 7? See 164160.

Maina et al. (1996) reported that hepatocyte growth factor and its receptor, the Met tyrosine kinase, are determinants of placenta, liver, and muscle development. They demonstrated that Met function in vivo requires signaling via 2 C-terminal tyrosines. Maina et al. (1996) demonstrated that mutation of both of these residues in the mouse genome caused embryonal death with placental liver and limb muscle defects, mimicking the phenotype of Met-null mutants. They disrupted the consensus sequences for Grb2 (600180) binding and reported that development proceeded to term without affecting placenta and liver but caused a striking reduction in limb muscle coupled to a generalized deficit of secondary fibers.

Hereditary papillary renal carcinoma (PRC; see 179755) is characterized by the development of multiple, bilateral papillary renal tumors (Zbar et al., 1995; Bernues et al., 1995). The pattern of inheritance is consistent with autosomal dominant transmission with reduced penetrance. This disorder is histologically and genetically distinct from 2 other causes of inherited renal carcinoma, von Hippel-Lindau disease (VHL; 193300), and the chromosome translocation t(3;8), as described by Cohen et al. (1979). Malignant papillary renal carcinomas are characterized by trisomy of chromosomes 7, 16, and 17, and in men, by loss of the Y chromosome (Kovacs, 1993). Inherited and sporadic clear cell renal carcinomas are characterized by inactivation of both copies of the VHL gene by mutation and/or by hypermethylation. By linkage analysis, Schmidt et al. (1997) mapped the PRC gene to 7q31.1-q34 in a 27-cM interval between D7S496 and D7S1837, the region containing the MET gene. Schmidt et al. (1997) found missense mutations in the tyrosine kinase domain of the MET gene (e.g., 164860.0004) in the germline of affected members of PRC families and in a subset of sporadic papillary renal carcinomas. Three mutations in the MET gene were located in codons that are homologous to those in KIT (164920) and RET (164761), protooncogenes that are targets of naturally occurring mutations. Schmidt et al. (1997) suggested that missense mutations located in the MET protooncogene lead to constitutive activation of the MET protein and papillary renal carcinomas. Comprising approximately 14% of all renal cell neoplasms, papillary renal carcinomas are recognized histologically by the presence of vascularized connective tissue stalks surrounded by neoplastic cells.

Jeffers et al. (1997) introduced into MET cDNA mutations that had been identified in the gene in both hereditary and sporadic forms of papillary renal carcinoma and examined the effect of each mutation in biochemical and biologic assays. They found that the MET mutants exhibited increased levels of tyrosine phosphorylation and enhanced kinase activity toward an exogenous substrate when compared with wildtype MET. Moreover, NIH 3T3 cells expressing mutant MET molecules formed foci in vitro and were tumorigenic in nude mice. A strong correlation was found between the enzymatic and biologic activity of the mutants, suggesting that tumorigenesis by MET is quantitatively related to its level of activation. Jeffers et al. (1997) raised the possibility that activating MET mutations may contribute to other human malignancies.

MET is overexpressed in a significant percentage of human cancers and is amplified during the transition between primary tumors and metastasis. To investigate whether this oncogene is directly responsible for the acquisition of the metastatic phenotype, Giordano et al. (1997) exploited a single-hit oncogenic version of MET that was able to transform and to confer invasive and metastatic properties to nontumorigenic cells, both in vitro and in nude mice. They found a point mutation in the signal transducer docking site of MET that increased the transforming ability of the oncogene, but abolished its metastatic potential. They concluded that the metastatic potential of the MET oncogene relies on the properties of its multifunctional docking site, and that a single point mutation affecting signal transduction can dissociate neoplastic transformation from metastasis.

Tumors from patients with papillary renal carcinoma and germline mutations of MET commonly show trisomy of chromosome 7 when analyzed by cytogenetic studies and comparative genomic hybridization (CGH). Zhuang et al. (1998) studied 16 renal tumors from 2 patients with documented germline mutations in exon 16 of MET. Fluorescence in situ hybridization analysis showed trisomy 7 in all tumors. To determine whether it was the mutant or the wildtype MET gene that was duplicated, Zhuang et al. (1998) performed duplex PCR and phosphoimage densitometry using polymorphic microsatellite markers D7S1801 and D7S1822, which are linked to the disease gene locus, and D1S1646 as an internal control. They determined the parental origin of chromosome alleles by genotyping parental DNA. In all 16 tumors there was an increased signal intensity (2:1 ratio) of the microsatellite allele from a chromosome bearing the mutant MET compared with the allele from the chromosome bearing the wildtype MET. The study demonstrated a nonrandom duplication of the chromosome bearing the mutated MET in HPRC and implicated this event in tumorigenesis. The 2 patients from whom tumors were studied had the identical mutation: his1112 to arg (164860.0007).

Bernues, M.; Casadevall, C.; Miro, R.; Caballin, M. R.; Villavicencio, H.; Salvador, J.; Zamarron, A.; Egozcue, J.: Cytogenetic characterization of a familial papillary renal cell carcinoma. Cancer Genet. Cytogenet. 84: 123-127, 1995.

Bottaro, D. P.; Rubin, J. S.; Faletto, D. L.; Chan, A. M.-L.; Kmiecik, T. E.; Vande Woude, G. F.; Aaronson, S. A.: Identification of the hepatocyte growth factor receptor as the c-met proto-oncogene product. Science 251: 802-804, 1991.

Cohen, A. J.; Li, F. P.; Berg, S.; Marchetto, D. J.; Tsai, S.; Jacobs, S. C.; Brown, R. S.: Hereditary renal-cell carcinoma associated with chromosomal translocation. New Eng. J. Med. 301: 592-595, 1979.

Cooper, C. S.: The met oncogene: from detection by transfection to transmembrane receptor for hepatocyte growth factor. Oncogene 7: 3-7, 1992.

Cooper, C. S.; Park, M.; Blair, D. G.; Tainsky, M. A.; Huebner, K.; Croce, C. M.; Vande Woude, G. F.: Molecular cloning of a new transforming gene from a chemically transformed human cell line. Nature 311: 29-33, 1984.

Dean, M.; Kozak, C.; Robbins, J.; Callahan, R.; O'Brien, S.; Vande Woude, G. F.: Chromosomal localization of the met proto-oncogene in the mouse and cat genome. Genomics 1: 167-173, 1987.

Dean, M.; Park, M.; Le Beau, M. M.; Robins, T. S.; Diaz, M. O.; Rowley, J. D.; Blair, D. G.; Vande Woude, G. F.: The human met oncogene is related to the tyrosine kinase oncogenes. Nature 318: 385-388, 1985.

Giordano, S.; Bardelli, A.; Zhen, Z.; Menard, S.; Ponzetto, C.; Comoglio, P. M.: A point mutation in the MET oncogene abrogates metastasis without affecting transformation. Proc. Nat. Acad. Sci. 94: 13868-13872, 1997.

Jeffers, M.; Schmidt, L.; Nakaigawa, N.; Webb, C. P.; Weirich, G.; Kishida, T.; Zbar, B.; Vande Woude, G. F.: Activating mutations for the Met tyrosine receptor in human cancer. Proc. Nat. Acad. Sci. 94: 11445-11450, 1997.

Kovacs, G.: Molecular cytogenetics of renal cell tumors. Adv. Cancer Res. 62 89-124, 1993.

Maina, F.; Casagranda, F.; Audero, E.; Simeone, A.; Comoglio, P. M.; Klein, R.; Ponzetto, C.: Uncoupling of Grb2 from the Met receptor in vivo reveals complex roles in muscle development. Cell 87: 531-542, 1996.

Park, M.; Dean, M.; Kaul, K.; Braun, M. J.; Gonda, M. A.; Vande Woude, G.: Sequence of MET protooncogene cDNA has features characteristic of the tyrosine kinase family of growth-factor receptors. Proc. Nat. Acad. Sci. 84: 6379-6383, 1987.

Schmidt, L.; Duh, F.-M.; Chen, F.; Kishida, T.; Glenn, G.; Choyke, P.; Scherer, S. W.; Zhuang, Z.; Lubensky, I.; Dean, M.; Allikmets, R.; Chidambaram, A.; and 23 others: Germline and somatic mutations in the tyrosine kinase domain of the MET proto-oncogene in papillary renal carcinoma. Nature Genet. 16: 68-73, 1997.

Zbar, B.; Glenn, G.; Lubensky, I.; Choyke, P.; Walther, M. M.; Magnusson, G.; Bergerheim, U. S. R.; Pettersson, S.; Amin, M.; Hurley, K.; Linehan, W. M.: Hereditary papillary renal cell carcinoma: clinical studies in 10 families. J. Urol. 153: 907-912, 1995.

Zhuang, Z.; Park, W.-S.; Pack, S.; Schmidt, L.; Vortmeyer, A. O.; Pak, E.; Pham, T.; Weil, R. J.; Candidus, S.; Lubensky, I. A.; Linehan, W. M.; Zbar, B.; Weirich, G.: Trisomy 7-harbouring non-random duplication of the mutant MET allele in hereditary papillary renal carcinomas. Nature Genet. 20: 66-69, 1998.

RENAL CELL CARCINOMA, PAPILLARY, X-LINKED (*312390)

JJM NOTE

It might be quite hard to spot X-linked inheritance of a rare tumor of adult onset, but such a family would have great research interest and could, in turn, benefit from the knowledge derived from the research.

Renal cell carcinoma (RCC), the most common cancer of the kidney, can be divided into papillary and non-papillary clear cell tumor subgroups (Shipley et al., 1995). The clear-cell carcinomas are characterized in cytogenetic and molecular studies by frequent abnormalities of chromosome 3p (144700). Clear-cell renal carcinomas, as well as tumors elsewhere, occur in the von Hippel-Lindau syndrome (193300), which maps also to 3p. After the cloning of the VHL gene, mutations in the gene were found in a high proportion of sporadic clear-cell renal carcinomas, e.g., 193300.0002. Papillary renal tumors, which account for 15-20% of renal carcinomas, also occur in both sporadic and familial forms (Mancilla-Jimenez et al., 1976; Zbar et al., 1994). Loss of heterozygosity and other abnormalities of 3p are not observed in these tumors but recurrent numerical abnormalities of other chromosomes have been detected. A specific reciprocal translocation between chromosomes X and 1, t(X;1)(p11.2;q21.2), was described in a case of trabecular papillary renal adenocarcinoma by de Jong et al. (1986). The same rearrangement was subsequently found in 4 additional cases of papillary renal tumors and this translocation may be a characteristic of a specific subset of papillary RCC (Meloni et al., 1993).

Translocations involving Xp11.2 have been found in synovial sarcoma (312820) and in a subgroup of renal adenocarcinomas, namely, papillary renal cell carcinoma (Meloni et al., 1993). Of particular interest is the similar localization of the breakpoints within Xp11 despite the distinct histologic origin. Sinke et al. (1993) undertook to define the precise location of the translocations in the 2 cases, making use of tumor-derived interspecies somatic cell hybrids in conjunction with Southern blot analysis. The resulting mapping data made it possible to distinguish between the 2 translocations, to review the physical map of the proximal Xp region, and to provide a basis for the eventual isolation of the breakpoints and the genes involved in the development of these 2 types of solid tumors. The chromosomal break in papillary renal cell carcinoma was found to lie between PFC (312060), OATL1 (311240), TIMP (305370), SYP (313475), and TFE3 (314310), distally; and OATL2 (311241), together with a number of DNA markers (DXS226, DXS146, DXS255, DXS14), proximally. The specific translocation reported by Meloni et al. (1993) was between the X chromosome and chromosome 1: t(X;1)(p11;q21).

A t(X;17)(p11.2;q25) and a del(X)(p11) have been described. All of these tumors with a rearrangement at Xp11 had occurred in males. Shipley et al. (1995) studied 2 RCC cell lines carrying a t(X;1)(p11.2;q21.2) translocation; 1 was derived from the first female case to be reported. By fluorescence in situ hybridization used to position YACs along contigs in the Xp11.2 region relative to the breakpoint and by correlation with mapping information from the region, they concluded that the breakpoint is situated centromeric to SYP (131475) and telomeric to DXS255 and CLCN5 (300008).

de Jong, B.; Molenaar, I. M.; Leeuw, J. A.; Idenberg, V. J.; Oosterhuis, J. W.: Cytogenetics of a renal adenocarcinoma in a 2-year-old child. Cancer Genet. Cytogenet. 21: 165-169, 1986.

Mancilla-Jimenez, R.; Stanley, R. J.; Blath, R. A.: Papillary renal cell carcinoma: a clinical, radiologic, and pathologic study of 34 cases. Cancer 38: 2469-2480, 1976.

Meloni, A. M.; Dobbs, R. M.; Pontes, J. E.; Sandberg, A. A.: Translocation (X;1) in papillary renal cell carcinoma: a new cytogenetic subtype. Cancer Genet. Cytogenet. 65: 1-6, 1993.

Shipley, J. M.; Birdsall, S.; Clark, J.; Crew, J.; Gill, S.; Linehan, M.; Gnarra, J.; Gisher, S.; Craig, I. W.; Cooper, C. S.: Mapping the X chromosome breakpoint in two papillary renal cell carcinoma cell lines with a t(X;1)(p11.2;q21.2) and the first report of a female case. Cytogenet. Cell. Genet. 71: 280-284, 1995.

Sinke, R. J.; de Leeuw, B.; Janssen, H. A. P.; Olde Weghuis, D.; Suijkerbuijk, R. F.; Meloni, A. M.; Gilgenkrantz, S.; Berger, W.; Ropers, H. H.; Sandberg, A. A.; Geurts van Kessel, A.: Localization of X chromosome short arm markers relative to synovial sarcoma- and renal adenocarcinoma-associated translocation breakpoints. Hum. Genet. 92: 305-308, 1993.

Zbar, B.; Tory, K.; Merino, M.; Schmidt, L.; Glenn, G.; Choyke, P.; Walther, M. M.; Lerman, M.; Linehan, W. M.: Hereditary papillary renal cell carcinoma. J. Urol. 151: 561-566, 1994.

TRANSLOCATION IN RENAL CARCINOMA ON CHROMOSOME 8 (*603046)

JJM NOTE

This provides an additional insight from the original and sole kidney cancer family with a 3;8 chromosomal translocation (Cohen 1979).

Cohen et al. (1979) described a family in which a 3;8 chromosomal translocation, t(3;8)(p14.2;q24.1), was associated with classic features of hereditary renal cell carcinoma (144700), i.e., autosomal dominant inheritance, early onset, and bilateral disease. Follow-up analysis found the occurrence of thyroid cancer in 2 translocation carriers with RCC (Li et al., 1993). Although studies demonstrated that the 3p14.2 breakpoint interrupted the fragile histidine triad gene (FHIT; 601153) in its 5-prime noncoding region (Ohta et al., 1996), several reasons made it unlikely that FHIT is causally related to renal or other malignancies. Gemmill et al. (1998) showed that the 8q24.1 breakpoint region encodes a 664-amino acid multiple membrane-spanning protein, which they called TRC8, for 'translocation in renal carcinoma, chromosome 8 gene.' The gene showed similarity to the hereditary basal cell carcinoma/segment polarity gene, 'patched.' This similarity involved 2 regions of 'patched,' the putative sterol-sensing domain and the second extracellular loop that participates in the binding of sonic hedgehog (SHH; 600725). In the 3;8 translocation, TRC8 was found to be fused to FHIT and disrupted within the sterol-sensing domain. In contrast, the FHIT coding region was maintained and expressed. In a series of sporadic renal carcinomas, an acquired TRC8 mutation was identified. By analogy to patched, TRC8 might function as a signaling receptor, and other pathway members, to be defined, are mutation candidates in malignant diseases involving the kidney and thyroid.

Cohen, A. J.; Li, F. P.; Berg, S.; Marchetto, D. J.; Tsai, S.; Jacobs, S. C.; Brown, R. S.: Hereditary renal-cell carcinoma associated with chromosomal translocation. New Eng. J. Med. 301: 592-595, 1979.
Gemmill, R. M.; West, J. D.; Boldog, F.; Tanaka, N.; Robinson, L. J.; Smith, D. I.; Li, F.; Drabkin, H. A.: The hereditary renal cell carcinoma 3;8 translocation fuses FHIT to a patched-related gene, TRC8. Proc. Nat. Acad. Sci. 95: 9572-9577, 1998.
Li, F. P.; Decker, H.-J. H.; Zbar, B.; Stanton, V. P., Jr.; Kovacs, G.; Seizinger, B. R.; Aburatani, H.; Sandberg, A. A.; Berg, S.; Hosoe, S.; Brown, R. S.: Clinical and genetic studies of renal cell carcinomas in a family with a constitutional chromosome 3;8 translocation: genetics of familial renal carcinoma. Ann. Intern. Med. 118: 106-111, 1993.
Ohta, M.; Inoue, H.; Cotticelli, M. G.; Kastury, K.; Baffa, R.; Palazzo, J.; Siprashvili, Z.; Mori, M.; McCue, P.; Druck, T.; Croce, C. M.; Huebner, K.: The FHIT gene, spanning the chromosome 3p14.2 fragile site and renal carcinoma-associated t(3;8) breakpoint, is abnormal in digestive tract cancers. Cell 84: 587-597, 1996.

RENAL CELL CARCINOMA 4 (*602773)

JJM NOTE

This other family with a hereditary chromosomal translocation, also involving chromosome 3, but joining to chromosome 2, allowed the suggestion of an alternative pathway to renal cell carcinoma.

Koolen et al. (1998) identified a familial case of renal cell carcinoma. Four patients over 3 generations developed nonpapillary RCCs, and 1 patient was diagnosed with squamous bladder carcinoma. Cytogenetic analysis showed that all these patients (and several unaffected members) carried a balanced t(2;3)(q35;q21) translocation. To elucidate the role of this novel chromosome 3 translocation in RCC development, Bodmer et al. (1998) performed allele segregation, loheterozygosity, and mutation analyses of various normal tissues and primary tumor samples. They demonstrated loss of the translocation-derivative chromosome 3 in 5 independent renal cell tumors of the clear-cell type, obtained from 3 members of the family carrying the constitutional translocation. In addition, analysis of the von Hippel-Lindau gene (VHL; 193300) revealed distinct insertion, deletion, and substitution mutations in 4 of the 5 tumors tested. On the basis of these findings, Bodmer et al. (1998) concluded that, in this familial case, an alternative route for RCC development was implied. In contrast to the first hit in the generally accepted 2-hit tumor-suppressor model proposed by Knudson, the familial translocation in this case may act as a primary oncogenic event, leading to nondisjunctional loss of the derivative chromosome 3 harboring the VHL tumor-suppressor gene. Mutations in the VHL gene on the other chromosome would result in subsequent tumor formation. The risk of developing renal cell carcinoma may be correlated directly with the extent of somatic (kidney) mosaicism resulting from the loss of the der(3) chromosome.

Bodmer, D.; Eleveld, M. J.; Ligtenberg, M. J. L.; Weterman, M. A. J.; Janssen, B. A. P.; Smeets, D. F. C. M.; de Wit, P. E. J.; van den Berg, A.; van den Berg, E.; Koolen, M. I.; Geurts van Kessel, A.: An alternative route for multistep tumorigenesis in a novel case of hereditary renal cell cancer and a t(2;3)(q35;q21) chromosome translocation. Am. J. Hum. Genet. 62: 1475-1483, 1998.
Koolen, M. I.; van der Meyden, A. P. M.; Bodmer, D.; Eleveld, M.; Van der Looij, E.; Brunner, H.; Smits, A.: Van den Berg, E.; Smeets, D.; Geurts van Kessel, A.: A familial case of renal cell carcinoma and a t(2;3) chromosome translocation. Kidney Int. 53: 273-275, 1998.

ONCOCYTOMA (#553000)

JJM NOTE

The tumor in the patient reported by Kadesky and Fulgham (1993) showed multifocality and bilaterality, flags of cancer of possible genetic orgins. Perhaps an interinstitutional protocol should focus on this likely informative neoplasia.

Oncocytomas are usually benign tumors that occur in various organs but particularly in the kidneys. Histologic evaluation of renal oncocytomas shows that they are composed entirely of peculiar epithelial cells with granular eosinophilic cytoplasm. Ultrastructural characterization exhibits dense packing of the cells with mitochondria that show morphologic differences from those in normal cells. On the average they are larger than those in renal carcinoma cells and their shape is abnormal. Using 5 restriction endonucleases, Welter et al. (1989) performed mitochondrial DNA analysis of 6 renal cell oncocytomas and adjacent renal tissue. In the HinfI restriction pattern, they found an additional 40-bp band in all oncocytomas but in none of the corresponding renal tissue or in renal carcinomas. By reisolating and hybridizing this band to the mitochondrial genome, they localized the sequence within the cytochrome c oxidase subunit I gene (516030). Welter et al. (1989) suggested that the alteration of mitochondrial DNA is specific for oncocytomas and that it represented the first mitochondrial change observed in a solid tumor.

Brauch et al. (1990) showed that, unlike other forms of renal neoplasia, renal oncocytomas do not show loss of alleles at loci on 3p.

Kadesky and Fulgham (1993) reported the unusual case of a 65-year-old man found to have bilateral multifocal renal oncocytomas. He underwent enucleation of 4 tumors from each kidney.

Neuhaus et al. (1997) reviewed the reported cytogenetic abnormalities that had been observed in renal oncocytoma. Loss of chromosomes 1 and Y appeared to represent at least 1 subset of oncocytomas. The frequent finding of rearrangements involving 11q13 appeared to characterize another subset. Neuhaus et al. (1997) reported the case of a 72-year-old woman with a translocation t(9;11) involving 11q13. They pointed out that the large, well-differentiated eosinophilic granular cells of renal oncocytomas express electron microscopic and immunohistochemical characteristics of the intercalated cell type A of the collecting duct.

Brauch, H.; Tory, K.; Linehan, W. M.; Weaver, D. J.; Lovell, M. A.; Zbar, B.: Molecular analysis of the short arm of chromosome 3 in five renal oncocytomas. J. Urol. 143: 622-624, 1990.
Kadesky, K. T.; Fulgham, P. F.: Bilateral multifocal renal oncocytoma: case report and review of the literature. J. Urol. 150: 1227-1228, 1993.
Neuhaus, C.; Dijkhuizen, T.; van dan Berg, E.; Storkel, S.; Stockle, M.; Mensch, B.; Huber, C.; Decker, H.-J.: Involvement of the chromosomal region 11q13 in renal oncocytoma: case report and literature review. Cancer Genet. Cytogenet. 94: 95-98, 1997.
Welter, C.; Kovacs, G.; Seitz, G.; Blin, N.: Alteration of mitochondrial DNA in human oncocytomas. Genes Chromosomes Cancer 1: 79-82, 1989.

Kidney Tumors of Children

WILMS TUMOR (*194070)

JJM NOTE

From clinical and molecular genetic observations, a number of human genes predispose to Wilms tumor. When a patient presents with Wilms tumor, it is advisable to be sure that none of the predisposing clinical syndromes or signs is present. Most Wilms tumors occur sporadically and unilaterally. The common associations are the WAGR (Wilms tumor, aniridia, genitourinary anomalies, and retardation) or Miller syndrome, Drash syndrome, hemihypertropy, Beckwith-Wiedemann syndrome, and Perlman syndrome. Wilms tumors have been seen in patients with hyperparathyroidism 2 (145001), tuberous sclerosis 1 (19110), and Bloom syndrome (210900). Mesothelioma is included because a single hamartomatous multicystic mesothelioma had a mutation in the *WT1* gene, which is expressed in murine mesoderm. There is no consensus on the schedule of screening individuals at high risk of Wilms tumor apart from a rational schedule of abdominal palpation, ultrasound examination, and urinalysis. Possibly, caretakers of high-risk infants should be taught abdominal palpation (Clericuzio CL: Clinical phenotypes and Wilms tumor. Med. Pediatr. Oncol. 21:182–87, 1993).

Rather numerous instances of multiple sibs with Wilms tumor have been described (Fitzgerald and Hardin, 1955). Strom (1957) described a family with 5 cases in 3 generations. A healthy male had 2 affected children (out of 5) by 1 wife and 1 affected child by another wife. A sister and an aunt of his had died in infancy or early childhood of abdominal tumor. Jolles (1973) described Wilms tumor in a 30-month-old girl and hypernephroma in her 67-year-old paternal grandmother. Brown et al. (1972) reported the occurrence of Wilms tumor in 4 members of 3 successive generations of a family: the proband, a girl, her mother, aunt and grandfather. The presence of Wilms tumor was histopathologically confirmed in 3 of the 4 cases. The right kidney was affected first in all. The aunt eventually developed Wilms tumor of the left kidney leading to her death at age 7. Matsunaga (1981) concluded that inheritance in familial cases, 'which constitute less than 1% of all' cases, is autosomal dominant with variable penetrance and expressivity. About 20% of familial cases are bilateral; about 3% of sporadic cases are bilateral. Bilateral cases may always be familial. Matsunaga (1981) further concluded that his 'host resistance model' fit the data.

Knudson and Strong (1972) reviewed and summarized data on 58 familial cases of Wilms tumor. They concluded that bilateral tumors are more likely to be familial, that familial tumors result from 2 mutations, 1 germinal and 1 somatic, and that sporadic tumors result from 2 somatic mutations. Work of Fearon et al. (1984), Koufos et al. (1984), Orkin et al. (1984), and Reeve et al. (1984) demonstrated that homozygosity of 11p change is present in Wilms tumor, thus providing support for the Knudson hypothesis.

The syndrome of aniridia, hemihypertrophy and other congenital anomalies with Wilms tumor, subsequently known as the WAGR syndrome, was first described by Miller et al. (1964). Meadows et al. (1974) described a family in which the mother had congenital hemihypertrophy and 3 of her children had Wilms tumor. A fourth child had a urinary tract anomaly. In 1 of the children the Wilms tumor was bilateral and in a second it was multicentric. Bond (1975) found associated congenital anomalies in 5 of 11 cases of bilateral Wilms tumor and in only 3 of 76 cases of unilateral Wilms tumor. See 235000 for discussion of familial hemihypertrophy. Riccardi et al. (1978) observed a triad of aniridia, ambiguous genitalia and mental retardation (AGR triad) in 3 patients with an interstitial deletion of the short arm of chromosome 11. One patient also had Wilms tumor. Among 6 cases, Francke et al. (1978) showed that Wilms tumor was not present in all cases, although aniridia was. For example, between monozygotic twins with an identical deletion of 11p, aniridia and mental retardation were present in both and Wilms tumor in only 1. Only 1 of the other 4 patients had Wilms tumor. The deleted segment common to all was the distal half of 11p13. Anderson et al. (1978) described aniridia, cataract and gonadoblastoma in a mentally retarded girl with an interstitial deletion of the short arm of chromosome 11. Gonadoblastoma occurs as part of the WAGR complex (Junien et al., 1980; Turleau et al., 1981). Because of the importance of awareness of this feature, the 'G' should stand for gonadoblastoma. (The 'G' of WAGR has variously meant to authors 'ambiguous genitalia' (Riccardi et al., 1978), 'genitourinary abnormalities,' or 'gonadoblastoma' (Anderson et al., 1978).) The WAGR complex qualifies as a 'contiguous gene syndrome' (Schmickel, 1986); see 194072.

Apparent close linkage of the region determining the WAGR syndrome to the catalase locus (CAT; 115500) means that assay of catalase activity can usefully indicate those cases of new-mutation aniridia that should have surveillance for the development of renal or gonadal tumors (Junien et al., 1980). In a report that focused on the aniridia component of the WAGR syndrome, Gilgenkrantz et al. (1982) analyzed the reported cases of aniridia with interstitial del(11)p. They reported a unique observation of hypertrophic cardiomyopathy in association with aniridia and catalase deficiency in a patient with del(11)(p15.1p12). Using high resolution chromosome banding, Marshall et al. (1982) studied 14 patients with aniridia. Seven were familial and had normal chromosomes; of 7 sporadic cases, 1 showed normal chromosomes and 6 had interstitial deletion of 11p of various lengths. Band 11p13 was included in the deletion in all 6 cases. In the cells of a Wilms tumor, unassociated with the WAGR syndrome and with normal constitutional chromosomes, Kaneko et al. (1981) found an interstitial deletion involving the region 11p14-p13.

Mapping studies by 17,16:de Martinville and Francke (1983, 1984) appeared to rule out a close physical association between HRAS1 and the region responsible for predisposition to Wilms tumor. Of course, deletion may bring them together. They placed HRAS1, HBB and insulin in the 11p15-p14.2 segment. By somatic cell hybridization, Junien et al. (1984) found that HRAS1 maps to 11p15.5-p15.1. In 4 cases of deletion of 11p13 with WAGR, they found that the restriction enzyme digestion patterns typical of HRAS1 were present. Thus, HRAS1 is not deleted in WAGR, a finding consistent with the difference in mapping. Reeve et al. (1984) demonstrated loss of HRAS (190020) in a sporadic case of Wilms tumor. Pointing out the conflicting evidence on the location of HRAS, they concluded that until the chromosomal location of HRAS has been determined with certainty, one cannot exclude a possible functional involvement of this oncogene in Wilms tumor development. From gene dosage studies, Narahara et al. (1984) concluded that both the CAT and the WAGR loci are in the chromosome segment 11p1306-p1305, with CAT distal to WAGR. Nakagome et al. (1984) concluded that deletion of the middle part of the 11p13 band is crucial to the WAGR syndrome; others had suggested that the distal half is of critical importance. Turleau et al. (1984) reviewed a total of 42 cases. Turleau et al. (1984) suggested that the determinant of aniridia may be separate from that for nephroblastoma, on the basis of a boy with deletion of most of 11p13, low catalase, nephroblastoma, chordee and cryptorchidism but normal irides and no mental retardation. The authors pointed out that in all published cases with aniridia the distal half of 11p13 is deleted whereas in their presently reported case there was 'a tiny residual distal segment.' The observation might suggest the order: cen—CAT—WILMS—aniridia—tel; however, Narahara et al. (1984) placed the catalase locus distal to the WAGR locus.

Riccardi et al. (1982) reported a patient with Wilms tumor and iris dysplasia, not aniridia. In the UK, Shannon et al. (1982) found the incidence of aniridia in cases of Wilms tumor to be 1 in 43. A survey detected 8 living and 3 dead children with Wilms tumor and aniridia. All 8 living children had deletion of 11p13. A high incidence of bilateral Wilms tumor (36%), male sex, early presentation, and advanced maternal age were features of the combined cases. By molecular genetic studies of cells from a patient with aniridia-Wilms tumor, Michalopoulos et al. (1985) concluded that a deletion visible cytologically in 11p13 deleted the catalase loci but not the LDHA locus, which is proximal, nor insulin, gamma-globin loci, HRAS1 and calcitonin, which are located distally. Among 49 children with Wilms tumor without aniridia, only 1 had bilateral renal tumors. By use of RFLPs that map to 11p, Raizis et al. (1985) detected mitotic recombination as the mechanism of homozygosity in a Wilms tumor. Their findings showed that insulin and beta-globin had come to homozygosity in the tumor but PTH remained heterozygous. Thus, PTH must be proximal to 11p13, the cytologically determined site of the Wilms tumor 'gene.' Scoggin et al. (1985) showed that E7-associated cell-surface antigen encoded by chromosome 11 and defined by a monoclonal antibody is deleted in cases of WAGR. This antigen is probably the same as that previously called 'a1' (151250). The studies in cases of WAGR with small deletions of 11p permitted regionalization of the assignment of antigen a1 to 11p13.

Comings (1973) proposed that dominantly inherited tumors may arise through inactivation or loss of a pair of regulatory genes that normally suppress the expression of a structural transforming gene. In 4 cases of Wilms tumor, Reeve et al. (1985) found that transcripts of insulin-like growth factor II (IGF2; 147470) were highly elevated as compared with adjacent normal kidney. Furthermore, by in situ hybridization, they mapped the IGF2 gene to 11p14.1, close to the WAGR locus. They proposed that IGF2 is the (or a) transforming gene in Wilms tumor. (Their positioning of the IGF2 locus is inconsistent with that of others who place it in a somewhat more distal band.) Scott et al. (1985) pointed out that Wilms tumor is histologically indistinguishable from the early stages of kidney development. In 12 sporadic cases of Wilms tumor, Scott et al. (1985), like Reeve et al. (1985) found that expression of the IGF2 gene was markedly increased relative to adult tissues, but was comparable to the level of expression in several fetal tissues including kidney, liver, adrenal, and striated muscle. Although this may merely reflect the stage of tumor differentiation, the possibility that IGF2 is involved in the transformation process was raised. Francke (1990) pointed out that the Wilms tumor site is close to that of IGF2, which is a candidate gene for Wilms tumor at that site. Van Heyningen et al. (1985) studied 5 persons with constitutional deletions of 11p. All had aniridia; 2 had had a Wilms tumor removed. Using a cDNA probe for catalase, they showed that the CAT locus was deleted in 4 of 5 and that it must be proximal to the Wilms and aniridia loci. HBB and CALC were deleted in none; therefore, these loci are likely to be outside the region 11p15.4-p12. A region of 11p associated with Wilms tumor has also been tied to rhabdomyosarcoma and hepatoblastoma (Koufos et al., 1985); see WT2 (194071). All 3 of these rare embryonal cancers occur in the Beckwith-Wiedema nn syndrome (130650).

Kozman et al. (1989) found loss of alleles from 11p in a Wilms tumor in a 37-year-old male. The finding indicated a common pathogenesis of childhood and adult types and suggested that molecular genetic studies may be useful in the differentiation of Wilms tumor from renal cell carcinoma or sarcoma when the histologic findings are unclear. Weissman et al. (1987) explored the role of the 11p13 deletion in Wilms tumor by introducing a normal human chromosome 11 into a Wilms tumor cell line by means of the microcell transfer technique. The ability of the cells containing the normal chromosome 11 to form tumors in nude mice was completely suppressed.

Schroeder et al. (1987) found that in 5 patients with Wilms tumor and 2 others previously reported, there was a loss of chromosome 11 alleles and that these alleles in all 7 cases were of maternal origin. All of these tumors were sporadic. The authors concluded that the initial mutation, either germinal or somatic, must have occurred on the paternal chromosome. There was no occupational history pointing to an increased risk of mutation in the fathers and, on the average, paternal age was not increased. They stated that the probability of all 7 patients losing the maternal allele in their Wilms tumor tissues, if the loss is indeed random, is less than 1%. By means of RFLPs, Huff et al. (1990) demonstrated that 7 of 8 de novo deletions of band 11p13 were of paternal origin. The 1 case of maternal origin was unremarkable in terms of the size or extent of the deletion, and the child developed Wilms tumor. Transmission of 11p13 deletions by both maternal and paternal carriers of balanced translocations has been reported, although maternal inheritance predominates. These data, in addition to the general predominance of paternally derived, de novo mutations at other loci, suggested that increased frequency of paternal deletions is due to an increased germinal mutation rate in males. Dao et al. (1987) examined the karyotype and chromosome 11 genotype of normal and tumor tissues from 13 childhood renal tumor patients. Tumors of 8 of the 12 Wilms tumor patients showed molecular evidence of loss of 11p DNA sequences by somatic recombination (4 cases), chromosome loss (2 cases), and recombination (2 cases) or chromosome loss and duplication. One malignant rhabdoid tumor in a patient heterozygous for multiple 11p markers did not show any tumor-specific 11p alteration. One of the patients had Perlman syndrome (renal hamartomas, nephroblastomatosis, and fetal gigantism; 267000).

Lewis and Yeger (1987) mapped the Wilms tumor region with 4 clones that were derived from the area of deletion in 11p, together with somatic cell hybrids containing chromosome 11 from leukemic T cells with translocation t(11;14), from fibroblasts from a familial aniridia patient with translocation t(4;11), and from lymphoblasts from a patient with Wilms tumor and deletion of 11p but no aniridia. The following order was deduced: centromere—CAT—T-cell break—aniridia break—FSHB—telomere.

Porteous et al. (1987) used chromosome-mediated gene transfer to provide an enriched source of DNA markers for 11p. They defined 10 distinct regions of 11p, 5 of which subdivided band 11p13. They also mapped 2 independent 11p13 translocation breakpoints to within the smallest region of overlap defined by WAGR deletions. The first came from a patient with familial aniridia, and the second was found in a neonate with the clinical features of Potter facies and the pathologic features of genitourinary dysplasia, with urethral and ureteral atresia and bilateral undescended testes. Porteous et al. (1987) raised the question of whether Wilms tumor and genitourinary dysplasia are alternative manifestations of mutation at the same locus. Kumar et al. (1987) demonstrated deletion of 11p14-p12 in a Wilms tumor removed from a 9-month-old boy with aniridia. Whereas morphologic transformation of normal human cells by BK virus (BKV) and by BKV DNA and its subgenomic fragments occurs in very low frequency, de Ronde et al. (1988) found that 4 individuals with various deletions in the short arm of one chromosome 11 were unusually susceptible to morphologic transformation. They suggested that the susceptibility might be explained by a 'transformation suppressor' locus situated within the deleted region. The deleted region included that of WAGR; the 'transformation suppressor' locus may be identical to the Wilms tumor locus. Using the fluorescence-activated cell sorter to select a series of somatic cell hybrids with deleted translocated chromosome 11 segregated from its normal homolog, Seawright et al. (1988) analyzed these cell hybrids with gene-specific probes and for cell-surface marker expression to order the markers and find an SRO for WAGR. They found that FSHB maps distal to WAGR and CAT maps proximal to it. Two translocation breakpoints in 11p13 (1 associated with familial aniridia and 1 with a sporadic case of congenital renal dysfunction resulting from urethral and ureteral atre sia) mapped within this SRO. Puissant et al. (1988) reported a patient with WAGR and a de novo reciprocal translocation 46,XY,t(5;11)(q11;p13). On Southern blot analysis, the gene encoding catalase had been deleted, but the gene encoding the beta subunit of follicle-stimulating hormone (FSHB) was intact.

Using a range of probes for chromosome 11, Mannens et al. (1988) demonstrated that loss of heterozygosity in Wilms tumors may not necessarily involve the proposed Wilms tumor locus at 11p13 and may be limited to 11p15.5. Jeanpierre et al. (1990) found loss of maternal alleles from the 11p15 area of the maternal chromosome in Wilms tumor tissue and a constitutional deletion of 11p13 of the maternal chromosome. There have been other instances in which the 11p region involved in loss of heterozygosity (11p15) is different from the region involved in hereditary predisposition (11p13). See 194071.

A separate gene coding for genitourinary dysplasia (symbolized GUD) was suggested by Bonetta et al. (1989), who found that the deletion breakpoint of a translocation t(11;2)(p13;p11) in a patient with Potter facies and genitourinary dysplasia mapped to the same 225-kb pulsed field gel electrophoresis fragment as did the fragment deleted in Wilms tumor. Van Heyningen et al. (1990) suggested that the Wilms tumor gene itself may be responsible for abnormalities of genitourinary development in WAGR as a pleiotropic effect. The suggestion was based on the observations that the tumor predisposition and the genitourinary malformations map to precisely the same area and that the WT candidate gene shows expression in both the developing kidney and gonads. See 137357 for a critique of the separate-gene hypothesis.

To localize a candidate for the Wilms tumor gene, Rose et al. (1990) developed a physical map of the 11p13 region by a combination of pulsed field gel electrophoresis and irradiation-reduced somatic cell hybrids of the Goss-Harris type. Restriction fragments contained in 11p13 were visualized directly using interspersed repeated DNA sequences as hybridization probes. The Wilms tumor locus was narrowed down to a region of less than 345 kb, and a transcript was identified with many of the characteristics expected for the Wilms tumor gene: a GC-rich region mapped to the 5-prime end of a transcription unit encoding a zinc finger protein. Call et al. (1990) reported further on these characteristics. The transcription unit spanned approximately 50 kb and encoded an mRNA approximately 3 kb long. The mRNA was expressed predominantly in the kidney and a subset of hematopoietic cells. The polypeptide had features suggesting a role in transcriptional regulation: the presence of 4 zinc finger domains and a region rich in proline and glutamine. The amino acid sequence of the predicted polypeptide showed significant homology to EGR1 (128990) and EGR2 (129010). Gessler et al. (1990) likewise isolated a cDNA clone derived from an RNA highly expressed in fetal kidney which is predicted to encode a Kruppel-like zinc finger protein that is probably a transcription factor. Francke (1990) commented on the significance of this putative gene for Wilms tumor. She compared the finding with that in retinoblastoma where a single gene locus has been found to be responsible. She reviewed the evidence for at least 3 genes capable of producing Wilms tumor: one in 11p13, one in 11p15.5 (194071) and at least one other not situated in either of these regions (194090). In the case of Wilms tumor, it is possible that changes at several sites are collaborative, or perhaps more likely that changes at several alternative sites result in the same tumor. A third model is that of hierarchical gene interacti on. If the function of the gene at 11p13 is to turn off the gene at 11p15.5, then loss of 11p13 expression would have the same effect as mutation or allele loss at 11p15.5.

The zinc finger protein that is likely to be a transcription factor and was isolated by Call et al. (1990) and Gessler et al. (1990) as the likely 'cause' of Wilms tumor was used by Pritchard-Jones et al. (1990) to study its role in normal development. This was done by in situ mRNA hybridization on sections of human embryos. The candidate Wilms tumor gene was expressed specifically in the condensed mesenchyme, renal vesicle, and glomerular epithelium of the developing kidney, in the related mesonephric glomeruli, and in cells approximating these structures in tumors. The other main sites of expression were the genital ridge, fetal gonad, and mesothelium. This was interpreted as indicating that the anomalies of the urinary tract and genitalia, which are frequent in both sporadic and syndrome-associated Wilms tumors, are a pleiotropic effect of the WT1 gene. Huff et al. (1991) made observations that appeared to differentiate between WT33 and LK15, 2 similar candidate Wilms tumor cDNA clones that were identified on the basis of their expression in fetal kidney and their location within the smallest region of overlap of somatic 11p13 deletions in Wilms tumors. However, Gessler (1991) concluded that LK15 and WT33 are identical, with differences due to alternative splicing at 2 exons. Huff et al. (1991) reported a patient with bilateral Wilms tumor who was heterozygous for a small germinal mutation within the WT1 gene (as identified by the WT33 clone). DNA from both tumors was homozygous for this intragenic deletion, which was predicted to encode a protein truncated by 180 amino acids.

Haber et al. (1990) described a sporadic, unilateral Wilms tumor in which 1 allele of the WT1 candidate gene contained a 25-bp deletion spanning an exon-intron junction and leading to aberrant mRNA splicing and loss of 1 of the 4 zinc finger consensus domains in the protein. The mutation was absent in the affected person's germline, consistent with the somatic inactivation of a tumor suppressor gene. In addition to the intragenic deletion affecting 1 allele, loss of heterozygosity at loci along the entire chromosome 11 indicated an earlier chromosomal nondisjunction and reduplication. Haber et al. (1992) presented evidence that this mutation of the WT1 gene behaves as a dominant negative, suppressing the function of the wildtype protein by a trans-dominant mechanism. They suspected this because the mutated allele was found to be coexpressed with the wildtype allele in a sporadic Wilms tumor. They therefore tested the ability of this mutant WT1 allele, containing an in-frame deletion within the DNA-binding domain, to transform primary baby rat kidney cells. The mutant WT1 gene was found to cooperate with the adenoviral E1A gene in transforming baby rat kidney cells. The wildtype WT1 gene in all of its alternatively spliced forms neither suppressed E1A-induced focus formation nor cooperated with E1A. Hastie (1992) also reviewed the evidence showing that mutations in the WT1 gene behave as dominant negatives, specifically in relation to causation of the Denys-Drash syndrome (DDS). This is proof that a tumor-suppressor gene plays a crucial role in normal genitourinary development. It seems quite clear that there is not a separate gene for the 'G' part of the WAGR syndrome although there is a separate gene for aniridia and almost certainly for mental retardation. Remarkably, 12 out of 25 patients from a total of 4 studies had the arg394-to-trp mutation (194070.0003) in heterozygous form as the cause of the Denys-Drash syndrome. One reason for the preponderance of this mutation is th at it represents a C-to-T transition at a CpG dimer. The same can be said for the arg-to-his change (194070.0004). This, however, appeared to be only part of the story since there are other equally vulnerable sites; Hastie (1992) suggested that the zinc fingers carrying these 2 mutations may play a particularly important role in establishing stable binding.

Mutations causing the Denys-Drash syndrome are clustered in the zinc finger encoding exons, particularly the exons encoding ZF2 and ZF3. Little et al. (1993) concluded that WT mutations resulting in the Denys-Drash syndrome may operate in a dominant-negative fashion; they observed an arg362-to-ter mutation predicted to result in a protein product lacking zinc fingers 2, 3 and 4, and therefore presumably incapable of binding DNA. Little et al. (1995) found WT1 fusion constructs containing the different classes of DDS-causing mutations. They demonstrated that the DDS mutations, indeed, disrupt DNA binding. They interpreted this as compatible with a dominant-negative mode of action, perhaps through dimerization between different WT1 isoforms. Little et al. (1995) noted that another mechanism by which the loss of DNA-binding could elicit the DDS phenotype would be a disturbed isoform dosage balance.

Nordenskjold et al. (1995) screened 27 cases of 46,XY females with gonadal dysgenesis who had previously been screened for and found not to carry SRY gene mutations (480000) to determine whether isolated gonadal dysgenesis might be due to WT mutations. Using denaturing gradient gel electrophoresis, they found a heterozygous point mutation in exon 8 in 1 of these patients: arg366-to-his, which had previously been described in a case of Denys-Drash syndrome (194070.0004). Reevaluation of the clinical data confirmed the diagnosis of Drash syndrome. Based on these results, Nordenskjold et al. (1995) concluded that isolated gonadal dysgenesis is not caused by mutations in the WT1 gene.

Ton et al. (1991) demonstrated that the smallest region of overlap between deletions causing Wilms tumor was a 16-kb segment of DNA encompassing one or more of the 5-prime exons of the zinc finger gene located on 11p13, together with an associated CpG island. This supported the authenticity of the zinc finger gene as the disease locus. Kakati et al. (1991) described a family in which a son had bilateral WT and an extra ring chromosome in the lymphocytes and in kidney tissue. The size of the ring varied considerably from cell to cell. A daughter had unilateral WT and an abnormal clone containing a small ring chromosome in PHA-stimulated and EBV-transformed lymphocytes. The mother, who was unaffected, had a karyotype similar to that of the daughter with WT. Kakati et al. (1991) hypothesized that the son's large ring chromosome was an amplified form of the small ring inherited from the mother. Chromosome 11 was cytogenetically normal in all cells examined in the affected children and the unaffected mother.

The WT1 gene comprises 10 exons, encoding a complex pattern of mRNA species. Four distinct transcripts are expressed, reflecting the presence or absence of 2 alternative splices (Haber et al., 1991). The conservation in structure and relative levels of the 4 WT1 mRNA species suggests that each encoded polypeptide makes a significant contribution to normal gene function. The control of cellular proliferation and differentiation exerted by the WT1 gene products may involve interactions between the 4 polypeptides with distinct targets and functions. Using polyclonal antibodies, Dressler and Douglass (1992) detected high levels of Pax-2 (PAX2) expression in the epithelial cells of human Wilms tumors. In the mouse they showed by immunocytochemistry that expression of the Pax-2 gene was localized to the nuclei of condensing mesenchyme cells and their epithelial derivatives in the developing kidney. The data suggested that Pax-2 is a transcription factor that is active during the mesenchyme-to-epithelium transition in early kidney development and in Wilms tumor. Pax-2 is a member of the family of genes identified in the mouse on the basis of a common protein coding domain, the paired-box, first described in the Drosophila segmentation genes 'paired' and 'gooseberry.' The Pax genes are expressed during embryogenesis in a tissue-restricted manner. The gene mutant in Waardenburg syndrome (193500) is the human homolog of the Pax-3 gene and the gene mutant in aniridia (106210) is the human homolog of the Pax-6 gene. To facilitate the search for small deletions and point mutations in the WT1 gene, Gessler et al. (1992) established the genomic organization of the gene and determined the sequence of all 10 exons and the flanking intron DNA. The pattern of alternative splicing in 2 regions was characterized in detail.

In 2 cases in which a Wilms tumor contained a somatic WT1 mutation, Park et al. (1993) found that nephrogenic rests in the same kidneys had the identical mutation. Thus, nephrogenic rests and Wilms tumor are topographically distinct lesions that are clonally derived from an early renal stem cell. Inactivation of WT1 appears to be an early genetic event that can lead to the formation of nephrogenic rests, enhancing the probability that additional genetic hits will lead to Wilms tumor. In the case of both P53 (191170) and retinoblastoma (180200), characterization of the tumor suppressor genes was provided by the dramatic growth-suppressing properties when the genes were reintroduced into cells containing inactivated endogenous genes. Similar studies in Wilms tumors had been complicated by the existence of multiple genetic loci implicated in different subsets of tumors and by the unavailability of appropriate target cell lines. To obtain an appropriate cell line for studying WT1 function, Haber et al. (1993) inoculated minced human Wilms tumors subcutaneously into nude mice and then adapted the tumor explants to growth in vitro. They found that 1 cell line could be propagated indefinitely in tissue culture without loss of tumorigenic potential. Transfection of each of 4 wildtype WT1 isoforms suppressed the growth of these cells. The endogenous WT1 transcript in these cells was devoid of exon 2 sequences, a splicing alteration that was also detected in varying amounts in all Wilms tumors tested but not in normal kidney. Production of this abnormal transcript, which encodes a functionally altered protein, may represent a distinct mechanism for inactivating WT1 in Wilms tumors.

Varanasi et al. (1994) analyzed the structural integrity of the entire WT1 gene in 98 sporadic Wilms tumors. By PCR-SSCP, they found that mutations in the WT1 gene are rare, occurring in only 6 tumors analyzed. In 1 sample, 2 independent intragenic mutations inactivated both WT1 alleles, providing a singular example of 2 different somatic alterations restricted to the WT1 gene. The data, together with the previously ascertained occurrence of large deletions/insertions in WT1, defined the frequency at which the WT1 gene is altered in sporadic tumors.

By gene targeting in embryonic stem cells, Kreidberg et al. (1993) introduced a mutation into the murine WT1 tumor suppressor gene. The mutation resulted in embryonic lethality in homozygotes, and examination of mutant embryos demonstrated a failure of kidney and gonad development. Specifically, at day 11 of gestation, the cells of the metanephric blastema underwent apoptosis, the ureteric bud failed to grow out from the Wolffian duct, and the inductive events that lead to formation of the metanephric kidney did not occur. In addition, the mutation caused abnormal development of the mesothelium, heart, and lungs. The results established a crucial role for WT1 in early urogenital development.

Gerald et al. (1995) reported the first example of a specific tumor associated with consistent translocation involving WT1. Desmoplastic small round cell tumor (DSRCT) is associated with a recurrent chromosomal translocation, t(11;22)(p13;q12). DSRCT is characterized by a predilection for young males, abdominal serosal involvement, poor prognosis, and a primitive histologic appearance. Gerald et al. (1995) found that the chromosome translocation breakpoints involved the intron between WT1 exons 7 and 8 and the intron between EWS (133450) exons 7 and 8. Chimeric transcripts corresponding to the fusion gene were detected in 4 of 6 cases of DSRCT. Analyses of these transcripts showed an in-frame fusion of RNA encoding the amino-terminal domain of EWS to both alternatively spliced forms of the last 3 zinc fingers of the DNA-binding domain of WT1. The chimeric products were predicted to modulate transcription at WT1 target sites and contribute to development of this unique tumor.

WT1 encodes a zinc finger protein expressed as distinct isoforms. Both constitutional and somatic mutations disrupting the DNA-binding domain of WT1 result in a potentially dominant-negative phenotype. In generating inducible cell lines expressing wildtype isoforms of WT1 as well as WT1 mutants, Englert et al. (1995) observed dramatic differences in the subnuclear localization of the induced proteins. The WT1 isoform that binds with high affinity to a defined DNA target, WT1(-KTS), was diffusely localized throughout the nucleus. In contrast, expression of an alternative splicing variant with reduced DNA binding affinity, WT1(+KTS), or WT1 mutants with a disrupted zinc finger domain resulted in a speckled pattern of expression within the nucleus. Though similar in appearance, the localization of WT1 variants to subnuclear clusters was clearly distinct from that of the essential splicing factor SC35, suggesting that WT1 is not directly involved in pre-mRNA splicing. Localization to subnuclear clusters required the M terminus of WT1 and coexpression of a truncated WT1 mutant and wildtype WT1(-KTS) resulted in a physical association, the redistribution of WT1(-KTS) from a diffuse to a speckled pattern, and the inhibition of its transactivational activity. These observations suggested to the authors that different WT1 isoforms and WT1 mutants have distinct subnuclear compartments. Dominant-negative WT1 proteins physically associate with wildtype WT1 in vivo and may result in its sequestration within subnuclear structures.

Products of the steroidogenic factor-1 (SF1; 184757) and WT1 genes are essential for mammalian gonadogenesis prior to sexual differentiation. In males, SF1 participates in sexual development by regulating expression of the polypeptide hormone Mullerian inhibiting substance (MIS; 600957). Nachtigal et al. (1998) showed that WT1-KTS isoforms associate and synergize with SF1 to promote MIS expression. In contrast, WT1 missense mutations, associated with male pseudohermaphroditism in Denys-Drash syndrome, fail to synergize with SF1. Additionally, the X-linked, candidate dosage-sensitive sex-reversal (DSS; 300018) gene, DAX1 (300200), antagonizes synergy between SF1 and WT1, most likely through a direct interaction with SF1. Nachtigal et al. (1998) proposed that WT1 and DAX1 functionally oppose each other in testis development by modulating SF1-mediated transactivation.

Little and Wells (1997) pointed out that only 5% of sporadic Wilms tumors have intragenic WT1 mutations, but more than 90% of patients with the Denys-Drash syndrome carry constitutional intragenic WT1 mutations. WT mutations have also been reported in juvenile granulosa-cell tumor, non-asbestos-related mesothelioma Park et al. (1993), desmoplastic small round cell tumor, and acute myeloid leukemia.

Schumacher et al. (1997) identified 19 hemizygous WT1 gene mutations/deletions in tissue samples from 64 patients. The histology of the tumors with mutations was stromal-predominant in 15, triphasic in 3, blastemal-predominant in 1, and unknown in 2 cases. Among 21 patients with stromal-predominant tumors, 15 had WT1 mutations and 10 of these were present in the germline. Of the patients with germline alterations, 6 had associated genitourinary (GU) tract malformations and a unilateral tumor, 2 had a bilateral tumor and normal GU tracts, and 2 had a unilateral tumor and normal GU tracts. Three mutations were tumor-specific and were found in patients with unilateral tumors without genital tract abnormalities. These data demonstrated the correlation of WT1 mutations with stromal-predominant histology, suggesting that a germline mutation in WT1 predisposes to the development of tumors with this histology. Twelve mutations were nonsense mutations resulting in truncation at different positions in the WT1 protein, and only 2 were missense mutations. Of the stromal-predominant tumors, 67% showed loss of heterozygosity, and in 1 tumor a different somatic mutation in addition to the germline mutation was identified. Thus, in a large proportion of a histopathologically distinct subset of Wilms tumors, the classic 2-hit inactivation model, with loss of a functional WT1 protein, is the underlying cause of tumor development.

Little et al. (1992) demonstrated that each parental allele of WT1 is equivalently expressed in normal fetal kidneys and Wilms tumors. On the other hand, Jinno et al. (1994) identified imprinting of WT1, with maternal expression in about half of preterm placental villus and fetal brain tissue. Further extensive studies showed that maternal monoallelic expression was observed in 39% of the samples, while the expression in other samples was biallelic. Mitsuya et al. (1997) studied the allele-specific expression of WT1 as well as of IGF2 and H19 in fibroblasts and lymphocytes. The expression profiles of IGF2 and H19 were constant and consistent with those in other tissues. The unexpected finding was paternal or biallelic expression of WT1 in fibroblasts and lymphocytes. This, together with the previous findings of maternal or biallelic expression in placenta and brain, suggested that the allele-specific regulatory system of WT1 is unique and may be controlled by a putative tissue- and individual-specific modifier.

Miyagawa et al. (1998) focused on the ectopic formation of skeletal muscle in Wilms tumors. They presented evidence supporting a negative regulatory role for WT1 in myogenesis. Their findings suggested that the metanephric-mesenchymal stem cells of the kidney may have the capacity to differentiate into skeletal muscle cells as well as epithelial cells. Normally, the expression of WT1 appears to prevent this ectopic differentiation program from being activated. In vitro studies suggested that WT1 may play a direct role in suppressing the formation of skeletal muscle.

Jeanpierre et al. (1998) identified WT1 mutations in patients with isolated diffuse mesangial sclerosis (256370), i.e., patients without pseudohermaphroditism and/or Wilms tumor which represent the other features of the Denys-Drash syndrome (DDS; 194080). In 4 of 10 patients, they found heterozygous mutations in the WT1 gene. Two mutations were different from those described in DDS patients. An analysis of genotype/phenotype correlation, on the basis of a WT1 mutation database of 84 germline mutations, demonstrated an association between mutations in exons 8 and 9 and DMS; among patients with DMS, a higher frequency of exon 8 mutations among 46,XY patients with female phenotype than among 46,XY patients with sexual ambiguity or male phenotype; and statistically significant evidence that mutations in exons 8 and 9 preferentially affect amino acids with different functions.

Beckwith (1998) provided useful data on the age at diagnosis of the first Wilms tumor in cases of syndrome-associated WT. Among 121 cases of Wiedemann-Beckwith syndrome, 96% were diagnosed by age 8 years; the oldest WBS patient had WT detected at 10 years, 2 months. Among 203 patients with hemihyperplasia, 94% were detected by age 8; the oldest HH patient had WT detected at 12 years, 4 months. Among 61 WAGR patients, Wilms tumor was detected in 98% by age 6 years; the oldest WAGR patient had WT detected at 7 years, 3 months. Among 52 patients with Denys-Drash syndrome, WT was detected in 96% by age 5 years; the oldest DDS patient had WT detected at 6 years of age.

See also Babaian et al. (1980); Cordero et al. (1980); DiGeorge and Harley (1966); Francke et al. (1979); Fraumeni and Glass (1968); Huerre et al. (1983); Juberg et al. (1975); Kaufman et al. (1973); Knudson and Strong (1975); Kolata (1980); Kontras and Newton (1974); Ladda et al. (1974); Neidhardt (1972); Pelletier et al. (1991); Slater and de Kraker (1982); Turleau et al. (1984); Yunis and Ramsay (1980)

Anderson, S. R.; Geertinger, P.; Larsen, H.-W.; Mikkelsen, M.; Parving, A.; Vestermark, S.; Warburg, M.: Aniridia, cataract and gonadoblastoma in a mentally retarded girl with deletion of chromosome 11: a clinicopathological case report. Ophthalmologica 176: 171-177, 1978.

Babaian, R. J.; Skinner, D. G.; Waisman, J.: Wilms' tumor in the adult patient: diagnosis, management, and review of the world medical literature. Cancer 45: 1713-1719, 1980.

Baird, P. N.; Santos, A.; Groves, N.; Jadresic, L.; Cowell, J. K.: Constitutional mutations in the WT1 gene in patients with Denys-Drash syndrome. Hum. Molec. Genet. 1: 301-305, 1992.

Barbaux, S.; Niaudet, P.; Gubler, M.-C.; Grunfeld, J.-P.; Jaubert, F.; Kuttenn, F.; Fekete, C. N.; Souleyreau-Therville, N.; Thibaud, E.; Fellous, M.; McElreavey, K.: Donor splice-site mutations in WT1 are responsible for Frasier syndrome. Nature Genet. 17: 467-470, 1997.

Beckwith, J. B.: Nephrogenic rests and the pathogenesis of Wilms tumor: developmental and clinical considerations. Am. J. Med. Genet. 79: 268-273, 1998.

Blanchet, P.; Daloze, P.; Lesage, R.; Papas, S.; van Campenhout, J.: XY gonadal dysgenesis with gonadoblastoma discovered after kidney transplantation. Am. J. Obstet. Gynec. 129: 221-222, 1977.

Bond, J. V.: Bilateral Wilms' tumor: age at diagnosis, associated congenital anomalies, and possible pattern of inheritance. Lancet II: 482-484, 1975.

Bonetta, L.; Huang, A.; Gregoris, M.; Yeger, H.; Williams, B. R. G.: Characterization of a homozygous deletion mapping to the Wilms tumor region on 11p13. (Abstract) Cytogenet. Cell Genet. 51: 965, 1989.

Brown, W. T.; Puranik, S. R.; Altman, D. H.; Hardin, H. C., Jr.: Wilms' tumor in three successive generations. Surgery 72: 756-761, 1972.

Bruening, W.; Bardeesy, N.; Silverman, B. L.; Cohn, R. A.; Machin, G. A.; Aronson, A. J.; Housman, D.; Pelletier, J.: Germline intronic and exonic mutations in the Wilms' tumour gene (WT1) affecting urogenital development. Nature Genet. 1: 144-148, 1992.

Call, K. M.; Glaser, T.; Ito, C. Y.; Buckler, A. J.; Pelletier, J.; Haber, D. A.; Rose, E. A.; Kral, A.; Yeger, H.; Lewis, W. H.; Jones, C.; Housman, D. E.: Isolation and characterization of a zinc finger polypeptide gene at the human chromosome 11 Wilms' tumor locus. Cell 60: 509-520, 1990.

Comings, D. E.: A general theory of carcinogenesis. Proc. Nat. Acad. Sci. 70: 3324-3328, 1973.

Coppes, M. J.; Liefers, G. J.; Higuchi, M.; Zinn, A. B.; Balfe, J. W.; Williams, B. R. G.: Inherited WT1 mutation in Denys-Drash syndrome. Cancer Res. 52: 6125-6128, 1992.

Cordero, J. F.; Li, F. P.; Holmes, L. B.; Gerald, P. S.: Wilms tumor in five cousins. Pediatrics 66: 716-719, 1980.

Dao, D. D.; Schroeder, W. T.; Chao, L.-Y.; Kikuchi, H.; Strong, L. C.; Riccardi, V. M.; Pathak, S.; Nichols, W. W.; Lewis, W. H.; Saunders, G. F.: Genetic mechanisms of tumor-specific loss of 11p DNA sequences in Wilms tumor. Am. J. Hum. Genet. 41: 202-217, 1987.

de Martinville, B.; Francke, U.: HRAS1, insulin and beta-globin map outside of 11p11.2-p14.1. (Abstract) Cytogenet. Cell Genet. 37: 530, 1984.

de Martinville, B.; Francke, U.: The c-Ha-ras1, insulin and beta-globin loci map outside the deletion associated with aniridia-Wilms' tumour. Nature 305: 641-643, 1983.

Denys, P.; Malvaux, P.; van den Berghe, H.; Tanghe, W.; Proesmans, W.: Association d'un syndrome anatomo-pathologique de pseudohermaphrodisme masculin, d'une tumeur de Wilms, d'une nephropathie parenchymateuse et d'un mosaicisme XX/XY. Arch. Franc. Pediat. 24: 729-739, 1967.

de Ronde, A.; Mannens, M.; Slater, R. M.; Hoovers, J.; Heyting, C.; Bleeker-Wagemakers, E. M.; Leschot, N. J.; Van Strien, A.; Sol, C. J. A.; Ter Schegget, J.; Van der Noordaa, J.: Morphological transformation by early region human

polyomavirus BK DNA of human fibroblasts with deletions in the short arm of one chromosome 11. J. Gen. Virol. 69: 467-471, 1988.
DiGeorge, A. M.; Harley, R. D. L.: The association of aniridia, Wilms tumor and genital abnormalities. Arch. Ophthal. 75: 796-798, 1966.
Dressler, G. R.; Douglass, E. C.: Pax-2 is a DNA-binding protein expressed in embryonic kidney and Wilms tumor. Proc. Nat. Acad. Sci. 89: 1179-1183, 1992.
Englert, C.; Vidal, M.; Maheswaran, S.; Ge, Y.; Ezzell, R. M.; Isselbacher, K. J.; Haber, D. A.: Truncated WT1 mutants alter the subnuclear localization of the wild-type protein. Proc. Nat. Acad. Sci. 92: 11960-11964, 1995.
Fearon, E. R.; Vogelstein, B.; Feinberg, A. P.: Somatic deletion and duplication of genes on chromosome 11 in Wilms' tumours. Nature 309: 176-178, 1984.
Fitzgerald, W. L.; Hardin, H. C., Jr.: Bilateral Wilms' tumor in a Wilms tumor family: case report. J. Urol. 73: 468-474, 1955.
Francke, U.: A gene for Wilms tumour?. Nature 343: 692-694, 1990.
Francke, U.; Holmes, L. B.; Atkins, L.; Riccardi, V. M.: Aniridia-Wilms' tumor association: evidence for specific deletion of 11p13. Cytogenet. Cell Genet. 24: 185-192, 1979.
Francke, U.; Riccardi, V. M.; Hittner, H. M.; Borges, W.: Interstitial del(11p) as a cause of the aniridia-Wilms tumor association: band localization and a heritable basis. (Abstract) Am. J. Hum. Genet. 30: 81A, 1978.
Frasier, S. D.; Bashore, R. A.; Mosier, H. D.: Gonadoblastoma associated with pure gonadal dysgenesis in monozygotic twins. J. Pediat. 64: 740-745, 1964.
Fraumeni, J. F., Jr.; Glass, A. G.: Wilms' tumor and congenital aniridia. J.A.M.A. 206: 825-828, 1968.
Gerald, W. L.; Rosai, J.; Ladanyi, M.: Characterization of the genomic breakpoint and chimeric transcripts in the EWS-WT1 gene fusion of desmoplastic small round cell tumor. Proc. Nat. Acad. Sci. 92: 1028-1032, 1995.
Gessler, M.: Personal Communication. Marburg, Germany 10/17/1991.
Gessler, M.; Konig, A.; Bruns, G. A. P.: The genomic organization and expression of the WT1 gene. Genomics 12: 807-813, 1992.
Gessler, M.; Poustka, A.; Cavenee, W.; Neve, R. L.; Orkin, S. H.; Bruns, G. A. P.: Homozygous deletion in Wilms tumours of a zinc-finger gene idey chromosome jumping. Nature 343: 774-778, 1990.
Gilgenkrantz, S.; Vigneron, C.; Gregoire, M. J.; Pernot, C.; Raspiller, A.: Association of del(11)(p15.1p12), aniridia, catalase deficiency, and cardiomyopathy. Am. J. Med. Genet. 13: 39-49, 1982.
Haber, D. A.; Buckler, A. J.; Glaser, T.; Call, K. M.; Pelletier, J.; Sohn, R. L.; Douglass, E. C.; Housman, D. E.: An internal deletion within an 11p13 zinc finger gene contributes to the development of Wilms' tumor. Cell 61: 1257-1269, 1990.
Haber, D. A.; Park, S.; Maheswaran, S.; Englert, C.; Re, G. G.; Hazen-Martin, D. J.; Sens, D. A.; Garvin, A. J.: WT1-mediated growth suppression of Wilms tumor cells expressing a WT1 splicing variant. Science 262: 2057-2059, 1993.
Haber, D. A.; Sohn, R. L.; Buckler, A. J.; Pelletier, J.; Call, K. M.; Housman, D. E.: Alternative splicing and genomic structure of the Wilms tumor gene WT1. Proc. Nat. Acad. Sci. 88: 9618-9622, 1991.
Haber, D. A.; Timmers, H. T. M.; Pelletier, J.; Sharp, P. A.; Housman, D. E.: A dominant mutation in the Wilms tumor gene WT1 cooperates with the viral oncogene E1A in transformation of primary kidney cells. Proc. Nat. Acad. Sci. 89: 6010-6014, 1992.
Haning, R. V., Jr.; Chesney, R. W.; Moorthy, A. V.; Gilbert, E. F.: A syndrome of chronic renal failure and XY gonadal dysgenesis in young phenotypic females without genital ambiguity. Am. J. Kidney Dis. 6: 40-48, 1985.
Hastie, N. D.: Dominant negative mutations in the Wilms tumour (WT1) gene cause Denys-Drash syndrome—proof that a tumour-suppressor gene plays a crucial role in normal genitourinary development. Hum. Molec. Genet. 1: 293-295, 1992.
Huerre, C.; Despoisse, S.; Gilgenkrantz, S.; Lenoir, G. M.; Junien, C.: c-Ha-ras1 is not deleted in aniridia-Wilms' tumour association. Nature 305: 638-641, 1983.
Huff, V.; Meadows, A.; Riccardi, V. M.; Strong, L. C.; Saunders, G. F.: Parental origin of de novo constitutional deletions of chromosomal band 11p13. Am. J. Hum. Genet. 47: 155-160, 1990.
Huff, V.; Miwa, H.; Haber, D. A.; Call, K. M.; Housman, D.; Strong, L. C.; Saunders, G. F.: Evidence for WT1 as a Wilms tumor (WT) gene: intragenic germinal deletion in bilateral WT. Am. J. Hum. Genet. 48: 997-1003, 1991.
Jeanpierre, C.; Antignac, C.; Beroud, C.; Lavedan, C.; Henry, I.; Saunders, G.; Williams, B.; Glaser, T.; Junien, C.: Constitutional and somatic deletions of two different regions of maternal chromosome 11 in Wilms tumor. Genomics 7: 434-438, 1990.
Jeanpierre, C.; Denamur, E.; Henry, I.; Cabanis, M.-O.; Luce, S.; Cecille, A.; Elion, J.; Peuchmaur, M.; Loirat, C.; Niaudet, P.; Gubler, M.-C.; Junien, C.: Identification of constitutional WT1 mutations, in patients with isolated diffuse mesangial sclerosis, and analysis of genotype/phenotype correlations by use of a computerized mutation database. Am. J. Hum. Genet. 62: 824-833, 1998.
Jinno, Y.; Yun, K.; Nishiwaki, K.; Kubota, T.; Ogawa, O.; Reeve, A. E.; Niikawa, N.: Mosaic and polymorphic imprinting of the WT1 gene in humans. Nature Genet. 6: 305-309, 1994.
Jolles, B.: Wilms' tumor in father and son. Lancet I: 207, 1973.
Juberg, R. C.; St. Martin, E. C.; Hundley, J. R.: Familial occurrence of Wilms's tumor: nephroblastoma in one of monozygous twins and in another sibling. Am. J. Hum. Genet. 27: 155-169, 1975.
Junien, C.; Huerre, C.; Despoisse, S.; Gilgenkrantz, S.; Lenoir, G. M.: c-Ha-ras1 is not deleted in del (11p13) Wilms' tumor (WAGR) and maps to 11p15.1-11p15.5. (Abstract) Cytogenet. Cell Genet. 37: 503, 1984.
Junien, C.; Turleau, C.; de Grouchy, J.; Said, R.; Rethore, M.-O.; Tenconi, R.; Dufier, J. L.: Regional assignment of catalase (CAT) gene to band 11p13: association with the aniridia-Wilms' tumor-gonadoblastoma (WAGR) complex. Ann. Genet. 23: 165-168, 1980.
Kakati, S.; Xiao, H.; Siddiqui, S. Y.; Sreekantaiah, C.; Weier, H.-U. G.; Green, D. M.; Fisher, J. E.; Allen, J. E.: Constitutional extra chromosomal element in a family with Wilms' tumor. Hum. Genet. 87: 183-188, 1991.
Kaneko, Y.; Celida Egues, M.; Rowley, J. D.: Interstitial deletion of short arm of chromosome 11 limited to Wilms' tumor cells in a patient without aniridia. Cancer Res. 41: 4577-4578, 1981.
Kaplinsky, C.; Ghahremani, M.; Frishberg, Y.; Rechavi, G.; Pelletier, J.: Familial Wilms' tumor associated with a WT1 zinc finger mutation. Genomics 38: 451-453, 1996.
Kaufman, R. L.; Vietti, T. J.; Wabner, C. I.: Wilms' tumor in father and son. Lancet I: 43, 1973.
Kinberg, J. A.; Angle, C. R.; Wilson, R. B.: Nephropathy-gonadal dysgenesis, type 2: renal failure in three siblings with XY dysgenesis in one. Am. J. Kidney Dis. 9: 507-510, 1987.
Klamt, B.; Koziell, A.; Poulat, F.; Wieacker, P.; Scambler, P.; Berta, P.; Gessler, M.: Frasier syndrome is caused by defective alternative splicing of WT1 leading to an altered ratio of WT1 +/-KTS splice isoforms. Hum. Molec. Genet. 7: 709-714, 1998.
Knudson, A. G., Jr.; Strong, L. C.: Familial Wilms's tumor. (Letter) Am. J. Hum. Genet. 27: 809-810, 1975.
Knudson, A. G., Jr.; Strong, L. C.: Mutation and cancer: a model for Wilms' tumor of the kidney. J. Nat. Cancer Inst. 48: 313-324, 1972.
Kolata, G. B.: Genes and cancer: the story of Wilms tumor. Science 207: 970-971, 1980.
Kontras, S. B.; Newton, W. A., Jr.: Familial Wilms' tumor. Birth Defects Orig. Art. Ser. X(4): 187-188, 1974.
Koufos, A.; Hansen, M. F.; Copeland, N. G.; Jenkins, N. A.; Lampkin, B. C.; Cavenee, W. K.: Loss of heterozygosity in three embryonal tumours suggests a common pathogenetic mechanism. Nature 316: 330-334, 1985.
Koufos, A.; Hansen, M. F.; Lampkin, B. C.; Workman, M. L.; Copeland, N. G.; Jenkins, N. A.; Cavenee, W. K.: Loss of alleles at loci on human chromosome 11 during genesis of Wilms' tumour. Nature 309: 170-172, 1984.
Kozman, H. M.; Clarke, J. M.; Little, M. H.; Smith, P. J.: Molecular genetic evidence for common pathogenesis of childhood and adult Wilms' tumor. Cancer Genet. Cytogenet. 38: 121-125, 1989.
Kreidberg, J. A.; Sariola, H.; Loring, J. M.; Maeda, M.; Pelletier, J.; Housman, D.; Jaenisch, R.: WT-1 is required for early kidney development. Cell 74: 679-691, 1993.
Kumar, S.; Harrison, C. J.; Heighway, J.; Marsden, H. B.; West, D. C.; Jones, P. M.: A cell line from Wilms' tumour with deletion in short arm of chromosome 11. Int. J. Cancer 40: 499-504, 1987.
Ladda, R. L.; Atkins, L.; Littlefield, J.: Computer-assisted analysis of chromosomal abnormalities: detection of a deletion in aniridia Wilms' tumor syndrome. Science 185: 784-787, 1974.
Lewis, W. H.; Yeger, H.: Characterization of the aniridia-Wilms tumor association region of chromosome 11. (Abstract) Cytogenet. Cell Genet. 46: 650, 1987.
Little, M.; Holmes, G.; Bickmore, W.; van Heyningen, V.; Hastie, N.; Wainwright, B.: DNA binding capacity of the WT1 protein is abolished by Denys-Drash syndrome WT1 point mutations. Hum. Molec. Genet. 4: 351-358, 1995.
Little, M.; Wells, C.: A clinical overview of WT1 gene mutations. Hum. Mutat. 9: 209-225, 1997.
Little, M. H.; Dunn, R.; Byrne, J. A.; Seawright, A.; Smith, P. J.; Pritchard-Jones, K.; van Heyningen, V.; Hastie, H. D.: Equivalent expression of paternally and maternally inherited WT1 alleles in normal fetal tissue and Wilms' tumours. Oncogene 7: 635-641, 1992.
Little, M. H.; Prosser, J.; Condie, A.; Smith, P. J.; Van Heyningen, V.; Hastie, N. D.: Zinc finger point mutations within the WT1 gene in Wilms tumor patients. Proc. Nat. Acad. Sci. 89: 4791-4795, 1992.
Little, M. H.; Williamson, K. A.; Mannens, M.; Kelsey, A.; Gosden, C.; Hastie, N. D.; van Heyningen, V.: Evidence that WT1 mutations in Denys-Drash syndrome patients may act in a dominant-negative fashion. Hum. Molec. Genet. 2: 259-264, 1993.
Mannens, M.; Slater, R. M.; Heyting, C.; Bliek, J.; de Kraker, J.; Coad, N.; de Pagter-Holthuizen, P.; Pearson, P. L.: Molecular nature of genetic changes resulting in loss of heterozygosity of chromosome 11 in Wilms' tumours. Hum. Genet. 81: 41-48, 1988.
Marshall, L. S.; Qureshi, A. R.; DiGeorge, A. M.; Kistenmacher, M. L.; Punnett, H. H.: Aniridia and the 11p13 deletion. (Abstract) Am. J. Hum. Genet. 34: 74A, 1982.
Matsunaga, E.: Genetics of Wilms' tumor. Hum. Genet. 57: 231-246, 1981.
McCoy, F. E., Jr.; Franklin, W. A.; Aronson, A. J.; Spargo, B. H.: Glomerulonephritis associated with male pseudohermaphroditism and nephroblastoma. Am. J. Surg. Path. 7: 387-395, 1983.
Meadows, A. T.; Lichtenfeld, J. L.; Koop, C. E.: Wilms' tumor in three children of a woman with congenital hemihypertrophy. New Eng. J. Med. 291: 23-24, 1974.
Michalopoulos, E. E.; Bevilacqua, P. J.; Stokoe, N.; Powers, V. E.; Willard, H. F.; Lewis, W. H.: Molecular analysis of gene deletion in aniridia—Wilms tumor association. Hum. Genet. 70: 157-162, 1985.
Miller, R. W.; Fraumen., J. F., Jr.; Manning, M. D.: Association of Wilms' tumor with aniridia, hemihypertrophy and other congenital malformations. New Eng. J. Med. 270: 922-927, 1964.
Mitsuya, K.; Sui, H.; Meguro, M.; Kugoh, H.; Jinno, Y.; Niikawa, N.; Oshimura, M.: Paternal expression of WT1 in human fibroblasts and lymphocytes. Hum. Molec. Genet. 6: 2243-2246, 1997.
Miyagawa, K.; Kent, J.; Moore, A.; Charlieu, J.-P.; Little, M. H.; Williamson, K. A.; Kelsey, A.; Brown, K. W.; Hassam, S.; Briner, J.; Hayashi, Y.; Hirai, H.; Yazaki, Y.; van Heyningen, V.; Hastie, N. D.: Loss of WT1 function leads to ectopic myogenesis in Wilms' tumour. (Letter) Nature Genet. 18: 15-17, 1998.
Nachtigal, M. W.; Hirokawa, Y.; Enyeart-VanHouten, D. L.; Flanagan, J. N.; Hammer, G. D.; Ingraham, H. A.: Wilms' tumor 1 and Dax-1 modulate the orphan nuclear receptor SF-1 in sex-specific gene expression. Cell 93: 445-454, 1998.
Nakagome, Y.; Ise, T.; Sakurai, M.; Nakajo, T.; Okamoto, E.; Takano, T.; Nakahori, Y.; Tsuchida, Y.; Nagahara, N.; Takada, Y.; Ohsawa, Y.; Sawaguchi, S.; Toyosaka, A.; Kobayashi, N.; Matsunaga, E.; Saito, S.: High-resolution studies

in patients with aniridia-Wilms tumor association, Wilms tumor or related congenital abnormalities. Hum. Genet. 67: 245-248, 1984.
Narahara, K.; Kikkawa, K.; Kimira, S.; Kimoto, H.; Ogata, M.; Kasai, R.; Hamawaki, M.; Matsuoka, K.: Regional mapping of catalase and Wilms tumor—aniridia, genitourinary abnormalities, and mental retardation triad loci to the chromosome segment 11p1305-p1306. Hum. Genet. 66: 181-185, 1984.
Neidhardt, M.: Wilms-tumor und Aniridie—ein genetisch fixiertes Syndrome?. Klin. Paediat. 184: 312-316, 1972.
Nordenskjold, A.; Fricke, G.; Anvret, M.: Absence of mutations in the WT1 gene in patients with XY gonadal dysgenesis. Hum. Genet. 96: 102-104, 1995.
Orkin, S. H.; Goldman, D. S.; Sallan, S. E.: Development of homozygosity for chromosome 11p markers in Wilms' tumour. Nature 309: 172-174, 1984.
Park, S.; Bernard, A.; Bove, K. E.; Sens, D. A.; Hazen-Martin, D. J.; Garvin, A. J.; Haber, D. A.: Inactivation of WT1 in nephrogenic rests, genetic precursors to Wilms' tumour. Nature Genet. 5: 363-367, 1993.
Park, S.; Schalling, M.; Bernard, A.; Maheswaran, S.; Shipley, G. C.; Roberts, D.; Fletcher, J.; Shipman, R.; Rheinwald, J.; Demetri, G.; Griffin, J.; Minden, M.; Housman, D. E.; Haber, D. A.: The Wilms tumour gene WT1 is expressed in murine mesoderm-derived tissues and mutated in a human mesothelioma. Nature Genet. 4: 415-420, 1993.
Pelletier, J.; Bruening, W.; Kashtan, C. E.; Mauer, S. M.; Manivel, J. C.; Striegel, J. E.; Houghton, D. C.; Junien, C.; Habib, R.; Fouser, L.; Fine, R. N.; Silverman, B. L.; Haber, D. A.; Housman, D.: Germline mutations in the Wilms' tumor suppressor gene are associated with abnormal urogenital development in Denys-Drash syndrome. Cell 67: 437-447, 1991.
Pelletier, J.; Bruening, W.; Li, F. P.; Haber, D. A.; Glaser, T.; Housman, D. E.: WT1 mutations contribute to abnormal genital system development and hereditary Wilms' tumour. Nature 353: 431-434, 1991.
Porteous, D. J.; Bickmore, W.; Christie, S.; Boyd, P. A.; Cranston, G.; Fletcher, J. M.; Gosden, J. R.; Rout, D.; Seawright, A.; Simola, K. O. J.; van Heyningen, V.; Hastie, N. D.: HRAS1-selected chromosome transfer generates markers that colocalize aniridia- and genitourinary dysplasia-associated translocation breakpoints and the Wilms tumor gene within band 11p13. Proc. Nat. Acad. Sci. 84: 5355-5359, 1987.
Pritchard-Jones, K.; Fleming, S.; Davidson, D.; Bickmore, W.; Porteous, D.; Gosden, C.; Bard, J.; Buckler, A.; Pelletier, J.; Housman, D.; van Heyningen, V.; Hastie, N.: The candidate Wilms' tumour gene is involved in genitourinary development. Nature 346: 194-197, 1990.
Puissant, H.; Azoulay, M.; Serre, J.-L.; Piet, L.; Junien, C.: Molecular analysis of a reciprocal translocation t(5;11)(q11;p13) in a WAGR patient. Hum. Genet. 79: 280-282, 1988.
Raizis, A. M.; Becroft, D. M.; Shaw, R. L.; Reeve, A. E.: A mitotic recombination in Wilms tumor occurs between the parathyroid hormone locus and 11p13. Hum. Genet. 70: 344-346, 1985.
Reeve, A. E.; Eccles, M. R.; Wilkins, R. J.; Bell, G. I.; Millow, L. J.: Expression of insulin-like growth factor-II transcripts in Wilms' tumour. Nature 317: 258-260, 1985.
Reeve, A. E.; Housiaux, P. J.; Gardner, R. J. M.; Chewings, W. E.; Grindley, R. M.; Millow, L. J.: Loss of a Harvey ras allele in sporadic Wilms' tumour. Nature 309: 174-176, 1984.
Riccardi, V. M.; Hittner, H. M.; Strong, L. C.; Fernbach, D. J.; Lebo, R.; Ferrell, R. E.: Wilms tumor with aniridia/iris dysplasia and apparently normal chromosomes. J. Pediat. 100: 574-577, 1982.
Riccardi, V. M.; Sujansky, E.; Smith, A. C.; Francke, U.: Chromosomal imbalance in the aniridia—Wilms' tumor association: 11p interstitial deletion. Pediatrics 61: 604-610, 1978.
Rose, E. A.; Glaser, T.; Jones, C.; Smith, C. L.; Lewis, W. H.; Call, K. M.; Minden, M.; Champagne, E.; Bonetta, L.; Yeger, H.; Housman, D. E.: Complete physical map of the WAGR region of 11p13 localizes a candidate Wilms' tumor gene. Cell 60: 405-508, 1990.
Salazar, H.; Kanbour, A.; Burgess, F.: Ultrastructure and observations on the histogenesis of mesotheliomas 'adenomatoid tumors' of the female genital tract. Cancer 29: 141-152, 1972.
Schmickel, R. D.: Chromosomal deletions and enzyme deficiencies. J. Pediat. 108: 244-246, 1986.
Schroeder, W. T.; Chao, L.-Y.; Dao, D. D.; Strong, L. C.; Pathak, S.; Riccardi, V.; Lewis, W. H.; Saunders, G. F.: Nonrandom loss of maternal chromosome 11 alleles in Wilms tumors. Am. J. Hum. Genet. 40: 413-420, 1987.
Schumacher, V.; Schneider, S.; Figge, A.; Wildhardt, G.; Harms, D.; Schmidt, D.; Weirich, A.; Ludwig, R.; Royer-Pokora, B.: Correlation of germ-line mutations and two-hit inactivation of the WT1 gene with Wilms tumors of stromal-predominant histology. Proc. Nat. Acad. Sci. 94: 3972-3977, 1997.
Scoggin, C. H.; Fisher, J. H.; Shoemaker, S. A.; Morse, H.; Leigh, T.; Riccardi, V. M.: The E7-associated cell-surface antigen: a marker for the 11p13 chromosomal deletion associated with aniridia-Wilms tumor. Am. J. Hum. Genet. 37: 883-889, 1985.
Scott, J.; Cowell, J.; Robertson, M. E.; Priestley, L. M.; Wadey, R.; Hopkins, B.; Pritchard, J.; Bell, G. I.; Rall, L. B.; Graham, C. F.; Knott, T. J.: Insulin-like growth factor-II gene expression in Wilms' tumour and embryonic tissues. Nature 317: 260-262, 1985.
Seawright, A.; Fletcher, J. M.; Fantes, J. A.; Morrison, H.; Porteous, D. J.; Li, S. S.-L.; Hastie, N. D.; Van Heyningen, V.: Analysis of WAGR deletions and related translocations with gene-specific DNA probes, using FACS-selected cell hybrids. Somat. Cell Molec. Genet. 14: 21-30, 1988.
Shannon, R. S.; Mann, J. R.; Harper, E.; Harnden, D. G.; Morten, J. E. N.; Herbert, A.: Wilms's tumour and aniridia: clinical and cytogenetic features. Arch. Dis. Child. 57: 685-690, 1982.
Slater, R. M.; de Kraker, J.: Chromosome number 11 and Wilms' tumor. Cancer Genet. Cytogenet. 5: 237-245, 1982.
Strom, T.: A Wilms' tumor family. Acta Paediat. 46: 601-604, 1957.
Ton, C. C. T.; Huff, V.; Call, K. M.; Cohn, S.; Strong, L. C.; Housman, D. E.; Saunders, G. F.: Smallest region of overlap in Wilms tumor deletions uniquely implicates an 11p13 zinc finger gene as the disease locus. Genomics 10: 293-297, 1991.
Turleau, C.; de Grouchy, J.; Dufier, J. L.; Phuc, L. H.; Schmelck, P. H.; Rappaport, R.; Nihoul-Fekete, C.; Diebold, N.: Aniridia, male pseudohermaphroditism, gonadoblastoma, mental retardation, and del 11p13. Hum. Genet. 57: 300-306, 1981.
Turleau, C.; de Grouchy, J.; Nihoul-Fekete, C.; Dufier, J. L.; Chavin-Colin, F.; Junien, C.: Del11p13/nephroblastoma without aniridia. Hum. Genet. 67: 455-456, 1984.
Turleau, C.; de Grouchy, J.; Tournade, M.-F.; Gagnadoux, M.-F.; Junien, C.: Del 11p/aniridia complex: report of three patients and review of 37 observations from the literature. Clin. Genet. 26: 356-362, 1984.
van Heyningen, V.; Bickmore, W. A.; Seawright, A.; Fletcher, J. M.; Maule, J.; Fekete, G.; Gessler, M.; Bruns, G. A. P.; Huerre-Jeanpierre, C.; Junien, C.; Williams, B. R. G.; Hastie, N. D.: Role for the Wilms tumor gene in genital development?. Proc. Nat. Acad. Sci. 87: 5383-5386, 1990.
van Heyningen, V.; Boyd, P. A.; Seawright, A.; Fletcher, J. M.; Fantes, J. A.; Buckton, K. E.; Spowart, G.; Porteous, D. J.; Hill, R. E.; Newton, M. S.; Hastie, N. D.: Molecular analysis of chromosome 11 deletions in aniridia-Wilms tumor syndrome. Proc. Nat. Acad. Sci. 82: 8592-8596, 1985.
Varanasi, R.; Bardeesy, N.; Ghahremani, M.; Petruzzi, M.-J.; Nowak, N.; Adam, M. A.; Grundy, P.; Shows, T. B.; Pelletier, J.: Fine structure analysis of the WT1 gene in sporadic Wilms tumor. Proc. Nat. Acad. Sci. 91: 3554-3558, 1994.
Weissman, B. E.; Saxon, P. J.; Pasquale, S. R.; Jones, G. R.; Geiser, A. G.; Stanbridge, E. J.: Introduction of normal human chromosome into a Wilms' tumor cell line controls its tumorigenic expression. Science 236: 175-180, 1987.
Yunis, J. J.; Ramsay, K. C.: Familial occurrence of the aniridia-Wilms tumor syndrome with deletion 11p13-14.1. J. Pediat. 96: 1027-1030, 1980.

WAGR SYNDROME (#194072)

JJM NOTE

I never heard Bob Miller mention the lack of eponymic recognition, and now there is another Miller syndrome of craniofacial and digital malformations. Still, his leadership in using the simple tool of informed chart review is known and is a valid route to identifying new associations. The process of looking "sideward" at patients with one disease to discover candidate syndromes and genes was termed *laterospective clinical epidemiology.*

A number sign (#) is used with this entry because the disorder is a contiguous gene syndrome due to deletion, either microscopic or submicroscopic, and the component genes are discussed in other entries.

Miller et al. (1964) first described the association of aniridia, hemihypertrophy, and other congenital anomalies with Wilms tumor. The syndrome subsequently became known as the WAGR syndrome. The G refers to 'ambiguous genitalia' (Riccardi et al., 1978), 'genitourinary abnormalities,' or 'gonadoblastoma' (Anderson et al., 1978). The R refers to mental retardation. The WAGR syndrome is one of the best-studied 'contiguous gene syndromes'; as defined by Schmickel (1986), aniridia (AN2; 106210), Wilms tumor (WT1; 194070), and probably mental retardation are determined by separate genes. Genitourinary abnormalities are probably determined by mutation in the WT1 gene; see 137357 for a critique.

Anderson, S. R.; Geertinger, P.; Larsen, H.-W.; Mikkelsen, M.; Parving, A.; Vestermark, S.; Warburg, M.: Aniridia, cataract and gonadoblastoma in a mentally retarded girl with deletion of chromosome 11: a clinicopathological case report. Ophthalmologica 176: 171-177, 1978.
Miller, R. W.; Fraumeni, J. F., Jr.; Manning, M. D.: Association of Wilms' tumor with aniridia, hemihypertrophy and other congenital malformations. New Eng. J. Med. 270: 922-927, 1964.
Riccardi, V. M.; Sujansky, E.; Smith, A. C.; Francke, U.: Chromosomal imbalance in the aniridia—Wilms' tumor association: 11p interstitial deletion. Pediatrics 61: 604-610, 1978.
Schmickel, R. D.: Chromosomal deletions and enzyme deficiencies. J. Pediat. 108: 244-246, 1986.

GENITOURINARY DYSPLASIA COMPONENT OF WAGR SYNDROME (#137357)

JJM NOTE

This is included for historical reasons.

A number sign (#) is used with this entry because of evidence that genitourinary dysplasia, which is a component of the WAGR contiguous gene syndrome (194070), is due to mutations in the same gene that is responsible for Wilms tumor (WT1; 194070).

In the WAGR syndrome, the G stands for genitourinary dysplasia, the other features being aniridia and mental retardation. The evidence from studies of balanced translocations and other observations suggested that the genitourinary dysplasia, like aniridia, is due to a separate gene in close proximity to the WT1 gene in band 11p13. Aniridia (AN2; 106210) is distal to WT1 and GUD proximal (Bickmore et al., 1989; Glaser et al., 1989). By HRAS1-selected chromosome transfer, Porteous et al. (1987) defined 10 distinct regions of the short arm of chromosome 11, 5 of which subdivided band 11p13. They also mapped 2 independent 11p13 translocation breakpoints to within the smallest region of overlap defined by the WAGR deletions. One came from a patient with familial aniridia and the second from a patient with Potter facies and genitourinary dysplasia (urethral and ureteral atresia and bilateral cryptorchidism). Porteous et al. (1987) raised the question of whether Wilms tumor and genitourinary dysplasia are alternative manifestations of mutations at the same locus. A separate gene coding for genitourinary dysplasia (symbolized GUD) was also suggested by Bonetta et al. (1989), who found that the deletion breakpoint of a translocation t(11;2)(p13;p11) in a patient with Potter facies and genitourinary dysplasia mapped to the same 225-kb pulsed field gel electrophoresis fragment as did the fragment deleted in Wilms tumor. On the other hand, van Heyningen et al. (1990) suggested that the Wilms tumor gene itself may be responsible for abnormalities of genitourinary development in WAGR as a pleiotropic effect. The suggestion was based on the observations that the tumor predisposition and the genitourinary malformations map to precisely the same area and that the WT candidate gene shows expression in both the developing kidney and gonads. That there is no GUD gene separate from the WT1 gene is supported by the fact that the Denys-Drash syndrome (nephropathy, Wilms tumor, and genital an omalies; 194080) is caused by specific point mutations in the WT1 gene (e.g., 194070.0003).

Bickmore, W. A.; Porteous, D. J.; Christie, S.; Seawright, A.; Fletcher, J. M.; Maule, J. C.; Couillin, P.; Junien, C.; Hastie, N. D.; van Heyningen, V.: CpG islands surround a DNA segment located between translocation breakpoints associated with genitourinary dysplasia and aniridia. Genomics 5: 685-693, 1989.

Bonetta, L.; Huang, A.; Gregoris, M.; Yeger, H.; Williams, B. R. G.: Characterization of a homozygous deletion mapping to the Wilms tumor region on 11p13. (Abstract) Cytogenet. Cell Genet. 51: 965 only, 1989.

Glaser, T.; Jones, C. A.; Lewis, W. H.; Call, K.; Rose, E. A.; Buckler, A.; Ito, C.; Housman, D. E.: The ultrafine structure of the WAGR gene complex. (Abstract) Cytogenet. Cell Genet. 51: 1005 only, 1989.

Porteous, D. J.; Bickmore, W.; Christie, S.; Boyd, P. A.; Cranston, G.; Fletcher, J. M.; Gosden, J. R.; Rout, D.; Seawright, A.; Simola, K. O. J.; van Heyningen, V.; Hastie, N. D.: HRAS1-selected chromosome transfer generates markers that colocalize aniridia- and genitourinary dysplasia-associated translocation breakpoints and the Wilms tumor gene within band 11p13. Proc. Nat. Acad. Sci. 84: 5355-5359, 1987.

van Heyningen, V.; Bickmore, W. A.; Seawright, A.; Fletcher, J. M.; Maule, J.; Fekete, G.; Gessler, M.; Bruns, G. A. P.; Huerre-Jeanpierre, C.; Junien, C.; Williams, B. R. G.; Hastie, N. D.: Role for the Wilms tumor gene in genital development?. Proc. Nat. Acad. Sci. 87: 5383-5386, 1990.

HEMIHYPERTROPHY (235000)

JJM NOTE

This entity is another of the major growth abnormalities, here an overgrowth of an entire body region, that is associated with Wilms tumor.

Fraumeni et al. (1967) described affected brother and sister and recorded that their maternal uncle was said to have had one leg longer than the other since childhood. They reviewed 6 other examples of familial occurrence. These included instances of successive generations affected. See 194070 for an account of Wilms tumor in children of a woman with hemihypertrophy. Frota-Pessoa (1979) reported a nonconsanguineous Polish-Brazilian family with possible involvement in grandfather, mother, and mother's sister and grandson. Nudleman et al. (1984) delineated a subtype of hemihypertrophy, designated by them the hemi-3 syndrome, which also showed hemihypesthesia, hemiareflexia and scoliosis. In the enlarged part, muscles were increased in size and strength; bones were increased in thickness but not length. Hypertrophy involved one side or quadrant, on the left in all 3, sparing the face. The neurologic defect was stationery but the scoliosis, which was convex to the left (the side of hemihypertrophy), was progressive, requiring treatment. (Classic hemihypertrophy involves more often the right side and is more frequent in males.) They observed the disorder in 3 unrelated girls. One patient had a lumbar myelomeningocele and all 3 had a family history of other neural tube defects. The authors proposed that the hemi-3 syndrome is one manifestation of neural tube defects, part of a spectrum of genetically and embryologically related CNS malformation with multifactorial inheritance. Genetic counseling, they suggested, should indicate the same order of risk as in the relatives of probands with classic neural tube defects. On the basis of 11 new cases, Viljoen et al. (1984) pointed out that bone age in the hypertrophied and normal parts do not differ and that relative body proportions remain the same during growth. No genetic or other etiologic factors could be recognized in this series. Mannens et al. (1987) studied a 3-generation family with hemihypertrophy and a case of Wilms tumor. Using 6 different 11p RFLPs, they found that a combination of alleles cosegregated with hemihypertrophy in this family.

Hoyme et al. (1998) reported on a prospective multicenter study of isolated hemihyperplasia (as they preferred to call hemihypertrophy). They looked particularly at the risk for tumor development in children with isolated hemihyperplasia. Among 168 patients ascertained, a total of 10 tumors developed in 9 patients, for an overall incidence of 5.9%. Tumors were of embryonal origin (similar to those noted in other overgrowth disorders): 4 unilateral Wilms tumors, 2 bilateral Wilms tumors, 2 adrenal cell carcinomas, 1 hepatoblastoma, and 1 leiomyosarcoma of the small bowel.

Hoyme et al. (1998) provided an anatomic classification of hemihyperplasia: complex hemihyperplasia is involvement of half of the body (at least 1 arm and 1 leg); affected parts may be contralateral or ipsilateral. Simple hemihyperplasia is involvement of a single limb. Hemifacial hyperplasia (133900) is involvement of one side of the face.

In the course of reviewing their experience with 18 patients with a referring diagnosis of Proteus syndrome (176920), Biesecker et al. (1998) found that Proteus syndrome is frequently confused with hemihyperplasia. They defined a distinct subtype of hemihyperplasia with static or mildly progressive hemihyperplasia and multiple lipomata. They referred to this as the hemihyperplasia-multiple lipomatosis syndrome and suggested that this provisional designation may prove to be a useful description for a group of patients that show moderate abnormalities of asymmetry and overgrowth with subcutaneous lipomata. The manifestations in all but 1 of these patients were relatively stable from infancy through preadolescence. One subject showed evidence of severe progressive lipomatosis in his first 2 years of life.

See also Meadows et al. (1974)

Biesecker, L. G.; Peters, K. F.; Darling, T. N.; Choyke, P.; Hill, S.; Schimke, N.; Cunningham, M.; Meltzer, P.; Cohen, M. M., Jr.: Clinical differentiation between Proteus syndrome and hemihyperplasia: description of a distinct form of hemihyperplasia. Am. J. Med. Genet. 79: 311-318, 1998.

Fraumeni, J. F., Jr.; Geiser, C. F.; Manning, M. D.: Wilms' tumor and congenital hemihypertrophy: report of five new cases and review of literature. Pediatrics 40: 886-899, 1967.

Frota-Pessoa, O.: Personal Communication. Sao Paulo, Brazil 12/10/1979.

Hoyme, H. E.; Seaver, L. H.; Jones, K. L.; Procopio, F.; Crooks, W.; Feingold, M.: Isolated hemihyperplasia (hemihypertrophy): report of a prospective multicenter study of the incidence of neoplasia and review. Am. J. Med. Genet. 79: 274-278, 1998.

Mannens, M.; Slater, R. M.; Heyting, C.; Bliek, J.; Hoovers, J.; Bleeker-Wagemakers, E. M.; Voute, P. A.; Coad, N.; Frants, R. R.; Pearson, P. L.: Chromosome 11, Wilms' tumour and associated congenital diseases. (Abstract) Cytogenet. Cell Genet. 46: 655 only, 1987.

Meadows, A. T.; Lichtenfeld, J. L.; Koop, C. E.: Wilms' tumor in three children of a woman with congenital hemihypertrophy. New Eng. J. Med. 291: 23-24, 1974.

Nudleman, K.; Andermann, E.; Andermann, F.; Bertrand, G.; Rogala, E.: The hemi 3 syndrome: hemihypertrophy, hemihypaesthesia, hemiareflexia and scoliosis. Brain 107: 533-546, 1984.

Viljoen, D.; Pearn, J.; Beighton, P.: Manifestations and natural history of idiopathic hemihypertrophy: a review of eleven cases. Clin. Genet. 26: 81-86, 1984.

MULTIPLE TUMOR—ASSOCIATED CHROMOSOME REGION 1 (*194071)

JJM NOTE

Here is further proof that most stories get more complicated before they become simple, and that every cancer family might contribute to the understanding of cancer genes.

Using a range of probes for chromosome 11, Mannens et al. (1988) demonstrated that loss of heterozygosity in Wilms tumors may not necessarily involve the proposed Wilms tumor locus at 11p13 (194070) and may be limited to 11p15.5. Jeanpierre et al. (1990) found loss of maternal alleles from the 11p15 area of the maternal chromosome in Wilms tumor tissue and a constitutional deletion of 11p13 of the maternal chromosome. There have been other instances in which the 11p region involved in loss of heterozygosity (11p15) is different from the region involved in hereditary predisposition (11p13).

In a family with multiple cases of Wiedemann-Beckwith syndrome in multiple sibships and at least 2 generations (Aleck and Hadro, 1989),

Koufos et al. (1989) demonstrated tight linkage of the mutation causing the syndrome to markers located at 11p15.5. Stimulated by this observation, Koufos et al. (1989) studied Wilms tumors of sporadic nature, not associated with WBS, and identified a subset of these tumors which had attained somatic homozygosity through mitotic recombination, with the smallest shared region of overlap being distal to the beta-globin complex at 11p15.5. The data suggested that there is a second locus, distinct from that involved in the WAGR syndrome, which plays a role in the association of Wilms tumor with WBS. See also Henry et al. (1989). Reeve et al. (1989) found loss of allelic heterozygosity at polymorphic 11p15 loci in Wilms tumor cells from 5 patients. In cells of 1 tumor, mitotic recombination had occurred distal to the gamma-globin gene on 11p15.5. Wadey et al. (1990) found that some Wilms tumors had allele loss restricted to 11p15 or to 11p13 and distal sequences. Dowdy et al. (1991) provided in vivo functional evidence for the existence of this second genetic locus involved in suppressing the tumorigenic phenotype of Wilms tumor: they constructed a chromosome 11 that had one or the other region deleted, and this chromosome was introduced into the tumorigenic Wilms tumor cell line G401. When assayed for tumor-forming activity in nude mice, the 11p13-deleted, but not the 11p15.5-p14.1-deleted chromosome, retained its ability to suppress tumor formation.

See also Henry et al. (1989)

Aleck, K. A.; Hadro, T. A.: Dominant inheritance of Wiedemann-Beckwith syndrome: further evidence for transmission of 'unstable premutation' through carrier women. Am. J. Med. Genet. 33: 155-160, 1989.

Dowdy, S. F.; Fasching, C. L.; Araujo, D.; Lai, K.-M.; Livanos, E.; Weissman, B. E.; Stanbridge, E. J.: Suppression of tumorigenicity in Wilms tumor by the p15.5-p14 region of chromosome 11. Science 254: 293-295, 1991.

Henry, I.; Grandjouan, S.; Couillin, P.; Barichard, F.; Huerre-Jeanpierre, C.; Glaser, T.; Philip, T.; Lenoir, G.; Chaussain, J. L.; Junien, C.: Tumor-specific loss of 11p15.5 alleles in del11p13 Wilms tumor and in familial adrenocortical carcinoma. Proc. Nat. Acad. Sci. 86: 3247-3251, 1989.

Jeanpierre, C.; Antignac, C.; Beroud, C.; Lavedan, C.; Henry, I.; Saunders, G.; Williams, B.; Glaser, T.; Junien, C.: Constitutional and somatic deletions of two different regions of maternal chromosome 11 in Wilms tumor. Genomics 7: 434-438, 1990.

Koufos, A.; Grundy, P.; Morgan, K.; Aleck, K. A.; Hadro, T.; Lampkin, B. C.; Kalbakji, A.; Cavenee, W. K.: Familial Wiedemann-Beckwith syndrome and a second Wilms tumor locus both map to 11p15.5. Am. J. Hum. Genet. 44: 711-719, 1989.

Mannens, M.; Slater, R. M.; Heyting, C.; Bliek, J.; de Kraker, J.; Coad, N.; de Pagter-Holthuizen, P.; Pearson, P. L.: Molecular nature of genetic changes resulting in loss of heterozygosity of chromosome 11 in Wilms' tumours. Hum. Genet. 81: 41-48, 1988.

Reeve, A. E.; Sih, S. A.; Raizis, A. M.; Feinberg, A. P.: Loss of allelic heterozygosity at a second locus on chromosome 11 in sporadic Wilms' tumor cells. Molec. Cell. Biol. 9: 1799-1803, 1989.

Wadey, R. B.; Pal, N.; Buckle, B.; Yeomans, E.; Pritchard, J.; Cowell, J. K.: Loss of heterozygosity in Wilms' tumour involves two distinct regions of chromosome 11. Oncogene 5: 901-907, 1990.

RENAL HAMARTOMAS, NEPHROBLASTOMATOSIS, AND FETAL GIGANTISM (PERLMAN SYNDROME) (*267000)

JJM NOTE

This disorder should be considered whenever the diagnosis of Beckwith-Wiedemann syndrome is entertained and, by extension, when Wilms tumor has occurred.

Liban and Kozenitzky (1970) and Perlman et al. (1973) described 5 offspring, of Jewish-Yemenite second-cousin parents, with a disorder manifested by large birth size, bilateral renal hamartomas with or without nephroblastomatosis, hypertrophy of the islets of Langerhans, and unusual facies. The longest survival was 27 days. There are some obvious similarities to the Beckwith-Wiedemann syndrome (130650) but the facies is thought to be characteristic with depressed nasal bridge and anteverted upper lip. Neri et al. (1984) reported affected brother and sister with unaffected, unrelated parents. One developed Wilms tumor; the other died suddenly as an infant. Hyperinsulinism is probably an important feature and may be a preventable cause of death. The cases of 4,5,6:Greenberg et al. (1984, 1985, 1986) were in 2 sibs with polyhydramnios, fetal ascites, abdominal muscular hypoplasia, visceromegaly, and subsequent development of bilateral Wilms tumor in one of them. This disorder should be considered in the differential diagnosis of fetal ascites without hydrops. Perlman (1986) published very instructive photographs of the 2 sibs that he and his colleagues reported in 1973 and 1975. The remarkably similar and distinctive facies consisted of round fullness, hypotonic appearance with open mouth, a long upper lip with inverted V-shape, upsweep of anterior scalp hair, and mild micrognathia. Dao et al. (1987) studied chromosome 11p markers in a patient with this syndrome and found the same loss of 11p DNA sequences that occurs in Wilms tumor (194070). Genetic differences between 2 tumors indicated that they developed independently, the results of different genetic events. Greenberg et al. (1988) reported an infant with manifestations of Perlman syndrome including polyhydramnios, macrosomia, bilateral nephromegaly with nephroblastomatosis, visceromegaly and cryptorchidism. The patient also had diaphragmatic hernia and interrupted aortic arch, suggesting that these may be find ings of Perlman syndrome. Hamel et al. (1989) described additional features of volvulus and distal ileal atresia as well as agenesis of the corpus callosum. From a study of 11p markers, they found no evidence of rearrangements. Because of the overlapping of clinical features with the Beckwith-Wiedemann syndrome, the findings of a cytogenetic abnormality of chromosome 11 is of interest. Chernos et al. (1990) described a newborn infant with macroglossia and bilateral hydronephrosis which had been diagnosed in utero. The infant had prominent forehead, broad flat facies, hypertelorism, bilateral epicanthic folds, deep-set eyes, a short upturned nose with wide nostrils, and prominent upper lip. Cytogenetic studies demonstrated an extra band on the tip of the short arm of one chromosome 11 which was G-dark, Q-bright, and C-negative.

See also Perlman et al. (1975)

Chernos, J. E.; Fowlow, S. B.; Cox, D. M.: A case of Perlman syndrome associated with a cytogenetic abnormality of chromosome 11. (Abstract) Am. J. Hum. Genet. 47 (suppl.): A28 only, 1990.

Dao, D. D.; Schroeder, W. T.; Chao, L.-Y.; Kikuchi, H.; Strong, L. C.; Riccardi, V. M.; Pathak, S.; Nichols, W. W.; Lewis, W. H.; Saunders, G. F.: Genetic mechanisms of tumor-specific loss of 11p DNA sequences in Wilms tumor. Am. J. Hum. Genet. 41: 202-217, 1987.

Greenberg, F.; Copeland, K.; Gresik, M. V.: Expanding the spectrum of the Perlman syndrome. Am. J. Med. Genet. 29: 773-776, 1988.

Greenberg, F.; Stein, F.; Gresik, M. V.; Finegold, M. J.; Carpenter, R. J.; Riccardi, V. M.; Beaudet, A. L.: Fetal ascites, 'prune belly' sequence, hepatomegaly, and nephromegaly associated with Wilms tumor. (Abstract) Proc. Greenwood Genet. Center 3: 133 only, 1984.

Greenberg, F.; Stein, F.; Gresik, M. V.; Finegold, M. J.; Carpenter, R. J.; Riccardi, V. M.; Beaudet, A. L.: Perlman syndrome: familial nephroblastomatosis, fetal ascites, polyhydramnios, macrosomia, and Wilms' tumor—follow-up. (Abstract) Proc. Greenwood Genet. Center 4: 150 only, 1985.

Greenberg, F.; Stein, F.; Gresik, M. V.; Finegold, M. J.; Carpenter, R. J.; Riccardi, V. M.; Beaudet, A. L.: The Perlman familial nephroblastomatosis syndrome. Am. J. Med. Genet. 24: 101-110, 1986.

Hamel, B. C. J.; Mannens, M.; Bokkerink, J. P. M.: Perlman syndrome: report of a case and results of molecular studies. (Abstract) Am. J. Hum. Genet. 45 (suppl.): A48 only, 1989.

Liban, E.; Kozenitzky, I. L.: Metanephric hamartomas and nephroblastomatosis in siblings. Cancer 25: 885-888, 1970.

Neri, G.; Martini-Neri, M. E.; Katz, B. E.; Opitz, B. E.: The Perlman syndrome: familial renal dysplasia with Wilms tumor, fetal gigantism and multiple congenital anomalies. Am. J. Med. Genet. 19: 195-207, 1984.

Perlman, M.: Perlman syndrome: familial renal dysplasia with Wilms tumor, fetal gigantism, and multiple congenital anomalies. (Letter) Am. J. Med. Genet. 25: 793-795, 1986.

Perlman, M.; Goldberg, G. M.; Bar-Ziv, J.; Danovitch, G.: Renal hamartomas and nephroblastomatosis with fetal gigantism: a familial syndrome. J. Pediat. 83: 414-418, 1973.

Perlman, M.; Levin, M.; Wittels, B.: Syndrome of fetal gigantism, renal hamartomas and nephroblastomatosis with Wilms' tumour. Cancer 35: 1212-1217, 1975.

WILMS TUMOR AND PSEUDOHERMAPHRODITISM (#194080)

JJM NOTE

In my opinion, Drash syndrome has three major elements: pseudohermaphroditism, Wilms tumor, and congenital nephropathy, which can be glomerular or tubular. Case reports illustrate all combinations of these three elements; hence, like associations of birth defects, when one feature is seen in an infant, the other features should be sought. Allan L. Drash is an endocrinologist and, in fact, leaves the continuing management of patients to the nephrologists.

A number sign (#) is used with this entry because this syndrome is the result of mutation in the Wilms tumor suppressor gene (WT1; 194070).

Drash et al. (1970) reported 2 unrelated children with Wilms tumor, pseudohermaphroditism, and nephropathy. They were initially evaluated for sexual ambiguity. Barakat et al. (1974) reported 3 cases and made reference to 2 additional unreported cases. Nothing is known of possible point mutation or chromosomal basis. Habib et al. (1985) reported a well-documented series of 10 patients. Eight of the 10 were 46,XY. The usual renal lesion is diffuse mesangial sclerosis of early onset. Turleau et al. (1987) reported partial androgen receptor deficiency and mixed gonadal dysgenesis in Drash syndrome. They urged that androgen receptor deficiency be sought in cases of the WAGR complex (194070). Friedman and Finlay (1987) indicated that the sexual abnormality is male pseudohermaphroditism (XY gonadal dysgenesis) and noted that renal failure or Wilms tumor may develop late. They presented a patient without ambiguous genitalia and suggested that all girls with Wilms tumor should be considered at risk for the Drash syndrome. The ovaries should be evaluated at surgery. Moorthy et al. (1987) suggested that some of the patients reported as cases of Drash syndrome in fact had a different disorder for which they suggested the designation Frasier syndrome (Frasier et al., 1964). Moorthy et al. (1987) discussed 6 previously reported patients with streak gonads, pseudohermaphroditism, and renal failure. In several of the patients the diagnosis was established only after a successful kidney transplantation during evaluation for primary amenorrhea. Gonadoblastoma arising from the streak gonad was noted in 5 of the 6 patients. See 136680 for molecular evidence that the WT1 gene is not the site of the mutation in Frasier syndrome.

In 1 patient with Drash syndrome, including bilateral Wilms tumor and gonadoblastoma, Dao et al. (1987) found tumor-specific loss of heterozygosity for markers on chromosome 11p as in other types of Wilms tumor. From molecular analysis of chromosome region 11p13, Jadresic et al. (1991) found no evidence of deletion or major rearrangement in that area. The nephropathy of the Drash triad is characterized by the early onset of proteinuria, which, in most cases, is sufficiently severe to cause nephrotic syndrome. Impairment of renal function commonly progresses to end-stage renal failure before the age of 3 years. Histologically the kidneys show varying degrees of focal or diffuse mesangial sclerosis. Deposition of fibrillary material in the cytoplasm leads to mesangial cell expansion. Drash patients who develop Wilms tumor usually have bilateral tumors and present at a mean age of about 18 months. The genital abnormality is most often male pseudohermaphroditism with children having either ambiguous external genitalia or a normal female phenotype with an XY karyotype. 11,12:Jadresic et al. (1990, 1991) studied 10 affected children, 5 males and 5 females. All the males had a 46,XY karyotype as did 3 of the 5 phenotypic females; the other 2 females had a 46,XX karyotype. Pelletier et al. (1991) demonstrated that germline mutations in the Wilms tumor suppressor gene are associated with abnormal urogenital development in Denys-Drash syndrome. (Pelletier et al. (1991) preferred the eponym Denys-Drash syndrome because the constellation of anomalies was first described by Denys et al. (1967) in the French literature.) Baird et al. (1992) demonstrated heterozygous constitutional mutations in 6 of 8 patients with the Denys-Drash syndrome. The clinical features of 6 of the patients had previously been described by Jadresic et al. (1990). Several of the mutations had been reported by others, presumably independently, so that they represented recurrences (see 194070.0003, 1 94070.0004, and 194070.0006). Four of the mutations were in exon 9, all resulting in missense substitutions. Three were at nucleotide position 1180 resulting in an arg-to-trp amino acid change. The other was at position 1186 resulting in an asp-to-asn change in the predicted resultant protein. Hastie (1992) reviewed the evidence showing that dominant-negative mutations in the WT1 gene cause Denys-Drash syndrome; this is proof that the WT1 tumor suppressor gene plays a crucial role in normal genitourinary development.

Mueller (1994) reviewed the Denys-Drash syndrome in detail on the basis of 150 reported cases. A tabulation was provided, furthermore, of 25 reported mutations in the WT1 gene, including 6 arg394-to-trp mutations (194070.0003), 1 arg394-to-pro mutation (194070.0007), and 2 different mutations at codon 396, namely, (in 2 cases) asp396-to-asn (194070.0006) and (in 1 case) asp396-to-gly (194070.0005).

Devriendt et al. (1995) described a newborn infant with male pseudohermaphroditism and glomerular lesions but without Wilms tumor. A constitutional heterozygous arg366-to-his mutation (194070.0004) was identified. In addition, the child had a large diaphragmatic hernia, a previously undescribed feature of Denys-Drash syndrome. The expression of the WT1 gene in pleural and abdominal mesothelium and the occurrence of diaphragmatic hernia in transgenic mice with a homozygous WT1 deletion strongly suggested that the diaphragmatic hernia in this patient was part of the malformation pattern caused by the WT1 mutation.

Baird, P. N.; Santos, A.; Groves, N.; Jadresic, L.; Cowell, J. K.: Constitutional mutations in the WT1 gene in patients with Denys-Drash syndrome. Hum. Molec. Genet. 1: 301-305, 1992.

Barakat, A. Y.; Papadopoulou, Z. L.; Chandra, R. S.; Hollerman, C. E.; Calcagno, P. L.: Pseudohermaphroditism, nephron disorder and Wilms' tumor: a unifying concept. Pediatrics 54: 366-369, 1974.

Dao, D. D.; Schroeder, W. T.; Chao, L.-Y.; Kikuchi, H.; Strong, L. C.; Riccardi, V. M.; Pathak, S.; Nichols, W. W.; Lewis, W. H.; Saunders, G. F.: Genetic mechanisms of tumor-specific loss of 11p DNA sequences in Wilms tumor. Am. J. Hum. Genet. 41: 202-217, 1987.

Denys, P.; Malvaux, P.; van den Berghe, H.; Tanghe, W.; Proesmans, W.: Association d'un syndrome anatomo-pathologique de pseudohermaphrodisme masculin, d'une tumeur de Wilms, d'une nephropathie parenchymateuse et d'un mosaicisme XX/XY. Arch. Franc. Pediat. 24: 729-739, 1967.

Devriendt, K.; Deloof, E.; Moerman, P.; Legius, E.; Vanhole, C.; de Zegher, F.; Proesmans, W.; Devlieger, H.: Diaphragmatic hernia in Denys-Drash syndrome. Am. J. Med. Genet. 57: 97-101, 1995.

Drash, A.; Sherman, F.; Hartmann, W. H.; Blizzard, R. M.: A syndrome of pseudohermaphroditism, Wilms' tumor, hypertension, and degenerative renal disease. J. Pediat. 76: 585-593, 1970.

Frasier, S. D.; Bashore, R. A.; Mosier, H. D.: Gonadoblastoma associated with pure gonadal dysgenesis in monozygotic twins. J. Pediat. 64: 740-745, 1964.

Friedman, A. L.; Finlay, J. L.: The Drash syndrome revisited: diagnosis and follow-up. Am. J. Med. Genet. Suppl. 3: 293-296, 1987.

Habib, R.; Loirat, C.; Gubler, M. C.; Niaudet, P.; Bensman, A.; Levy, M.; Broyer, M.: The nephropathy associated with male pseudohermaphroditism and Wilms' tumor (Drash syndrome): a distinctive glomerular lesion—report of 10 cases. Clin. Nephrol. 24: 269-278, 1985.

Hastie, N. D.: Dominant negative mutations in the Wilms tumour (WT1) gene cause Denys-Drash syndrome—proof that a tumour-suppressor gene plays a crucial role in normal genitourinary development. Hum. Molec. Genet. 1: 293-295, 1992.

Jadresic, L.; Leake, J.; Gordon, I.; Dillon, M. J.; Grant, D. B.; Pritchard, J.; Risdon, R. A.; Barratt, T. M.: Clinicopathologic review of twelve children with nephropathy, Wilms tumor, and genital abnormalities (Drash syndrome). J. Pediat. 117: 717-725, 1990.

Jadresic, L.; Wadey, R. B.; Buckle, B.; Barratt, T. M.; Mitchell, C. D.; Cowell, J. K.: Molecular analysis of chromosome region 11p13 in patients with Drash syndrome. Hum. Genet. 86: 497-501, 1991.

Moorthy, A. V.; Chesney, R. W.; Lubinsky, M.: Chronic renal failure and XY gonadal dysgenesis: 'Frasier' syndrome—a commentary on reported cases. Am. J. Med. Genet. Suppl. 3: 297-302, 1987.

Mueller, R. F.: The Denys-Drash syndrome. J. Med. Genet. 31: 471-477, 1994.

Pelletier, J.; Bruening, W.; Kashtan, C. E.; Mauer, S. M.; Manivel, J. C.; Striegel, J. E.; Houghton, D. C.; Junien, C.; Habib, R.; Fouser, L.; Fine, R. N.; Silverman, B. L.; Haber, D. A.; Housman, D.: Germline mutations in the Wilms' tumor suppressor gene are associated with abnormal urogenital development in Denys-Drash syndrome. Cell 67: 437-447, 1991.

Turleau, C.; Niaudet, P.; Sultan, C.; Rault, G.; Mahfoud, A.; Nihoul-Fekete, C.; Iris, L.; de Grouchy, J.: Partial androgen receptor deficiency and mixed gonadal dysgenesis in Drash syndrome. Hum. Genet. 75: 81-83, 1987.

WILMS TUMOR 3 (*194090)

JJM NOTE

One would expect this third major Wilms tumor gene to be mapped by family studies to confirm the loss of heterozygosity results.

A form of autosomal dominant Wilms tumor that does not map to 11p is indicated by the work of Grundy et al. (1988) and Huff et al. (1988). Both groups used polymorphic DNA markers mapped to chromosome 11p. In a multipoint linkage analysis of a family segregating for Wilms tumor, Grundy et al. (1988) excluded the mutation from both 11p13 and 11p15, which have been implicated in Wilms tumor (see 194070 and 194071). In a second family, the 11p15 alleles lost in the tumor were derived from the affected parent, thus excluding this region as the location of the inherited mutation. Huff et al. (1988) likewise studied a Wilms tumor family with 7 DNA markers that spanned the 11p13 region and found no linkage. Thus there may be at least 3 types of Wilms tumor: one related to 11p13, another related to 11p15, and the last related to neither. Schwartz et al. (1991) used a highly informative CA repeat polymorphism within the WT1 gene to study a family with 6 persons with Wilms tumor in 3 generations. They demonstrated that a predisposition to WT did not segregate with the polymorphism; furthermore, linkage analysis excluded WT predisposition from 11p15 also. With loss of heterozygosity studies, Maw et al. (1992) concluded that a third Wilms tumor locus is on 16q. In addition to loss on chromosome 11p (11 of 25 informative Wilms tumors), there was significant loss on 16q (9 of 45 informative tumors), while the total frequency of allele loss excluding these loci was low (9 of 426 total informative loci). They screened loci on 33 autosomal arms. The parental origin of the lost chromosome 16q allele was paternal in 4 and maternal in 4 sporadic tumors tested. Thus, unlike chromosome 11p, alleles of either parental origin are lost on 16q.

See also Koufos et al. (1989)

Grundy, P.; Koufos, A.; Morgan, K.; Li, F. P.; Meadows, A. T.; Cavenee, W. K.: Familial predisposition to Wilms' tumour does not map to the short arm of chromosome 11. Nature 336: 374-376, 1988.

Huff, V.; Compton, D. A.; Chao, L.-Y.; Strong, L. C.; Geiser, C. F.; Saunders, G. F.: Lack of linkage of familial Wilms' tumour to chromosomal band 11p13. Nature 336: 377-378, 1988.

Koufos, A.; Grundy, P.; Morgan, K.; Aleck, K. A.; Hadro, T.; Lampkin, B. C.; Kalbakji, A.; Cavenee, W. K.: Familial Wiedemann-Beckwith syndrome and a second Wilms tumor locus both map to 11p15.5. Am. J. Hum. Genet. 44: 711-719, 1989.

Maw, M. A.; Grundy, P. E.; Millow, L. J.; Eccles, M. R.; Dunn, R. S.; Smith, P. J.; Feinberg, A. P.; Law, D. J.; Paterson, M. C.; Telzerow, P. E.; Callen, D. F.; Thompson, A. D.; Richards, R. I.; Reeve, A. E.: A third Wilms' tumor locus on chromosome 16q. Cancer Res. 52: 3094-3098, 1992.

Schwartz, C. E.; Haber, D. A.; Stanton, V. P.; Strong, L. C.; Skolnick, M. H.; Housman, D. E.: Familial predisposition to Wilms tumor does not segregate with the WT1 gene. Genomics 10: 927-930, 1991.

WILMS TUMOR 4 (*601363)

JJM NOTE

The results are from a single family, of which the usual "flags" of a genetic cancer were violated. The tumors were unilateral and appeared later than the average age of onset.

Wilms tumor (WT) is one of the most common solid tumors of childhood, occurring in 1 in 10,000 children and accounting for 8% of childhood cancers. It is believed to result from malignant transformation of abnormally persistent renal stem cells which retain embryonic differentiation potential (Breslow and Beckwith, 1982). Rahman et al. (1996) noted that several genes or chromosomal areas had been associated with the development of WT: WT1 (194070) at 11p13 and loci at 11p15.5 (194071) and 16q (194090). Rahman et al. (1996) noted that these loci had not been implicated in familial WT. The investigators carried out a genome-wide linkage search in a large Canadian family with 7 confirmed cases of WT in 3 generations. The results provided strong evidence for the localization of a familial WT predisposition gene, which they symbolized FWT1, to an 18-cM interval on 17q12-q21. No congenital abnormalities or other cancers had been observed in the family. The average age of presentation was 5 years (age range of 2 to 12 years), which is older than the average age of diagnosis of sporadic WT (3 to 4 years). Typical triphasic histology with stromal, blastemal, and epithelial elements was found in 5 tumors, while the sixth tumor was predominantly myogenic. Five tumors were reported to be unilateral. The myogenic WT was reported to be bilateral at preoperative CT scan. However, at surgery one kidney appeared normal and no biopsy was performed. The predominance of unilateral WTs in this pedigree was consistent with reports from other WT families.

Breslow, N. E.; Beckwith, J. R.: Epidemiological features of Wilms' tumor: results of the national Wilms' tumor study. J. Nat. Cancer Inst. 68: 429-436, 1982.

Rahman, N.; Arbour, L.; Tonin, P.; Renshaw, J.; Pelletier, J.; Baruchel, S.; Pritchard-Jones, K.; Stratton, M. R.; Narod, S. A.: Evidence for a familial Wilms' tumour gene (FWT1) on chromosome 17q12-q21. Nature Genet. 13: 461-463, 1996.

WILMS TUMOR-SUPPRESSOR LOCUS ON CHROMOSOME 7 (*601583)

JJM NOTE

Although the generalizability of this gene to sporadic Wilms tumor is uncertain, the findings show the value of intense study of a single patient.

Wilmore et al. (1994) observed a Wilms tumor (WT) patient who had a constitutional balanced translocation between chromosomes 1 and 7, t(1;7)(q42;p15), and suggested that the breakpoints might represent a Wilms tumor predisposition gene. Cytogenetic analysis of the tumor from this patient revealed an acquired abnormality of the other chromosome 7, resulting in an isochromosome of the long arm. This confirmed the monosomy of 7p and trisomy of 7q in the tumor of the translocation patient, and in addition a loss of 7p alleles was identified in a WT from a bilaterally affected patient. Also, 2 WTs were shown to have an extra copy of chromosome 7 alleles. Multiple copies of chromosome 1q alleles, probably resulting from secondary changes, were observed in 2 WTs, 1 of which was also associated with a trisomy of chromosome 7. These results suggested that 7p may contain a tumor suppressor gene involved in WT development, and that duplication of 7q also may play a role in WT development.

Miozzo et al. (1996) carried out an analysis on 11 sporadic Wilms tumors to test the hypothesis that alterations in chromosome 7 genes play a role in the development of these tumors. They identified a region of loss of heterozygosity (LOH) in the tumors analyzed. This region is flanked by the loci D7S506 and D7S526 and spans approximately 25 cM on chromosome 7p11.2 to 7p15. Reynolds et al. (1996) investigated a patient who presented with 1 Wilms tumor and bilateral nephrogenic rests (suggestive of a predisposition) associated with a de novo t(1;7)(q42;p15) constitutional translocation as the only visible cytogenetic abnormality. He also had bilateral radial aplasia and other skeletal abnormalities, but there was no manifestation of any syndrome previously associated with Wilms tumors. In the tumor, the translocation was retained, and the other 7p region was lost by formation of an isochromosome i(7q). Reynolds et al. (1996) localized the chromosome 7 breakpoint within a YAC contig by using fluorescence in situ hybridization. They suggested that the region disrupted by the translocation is the site of a novel tumor suppressor gene that is involved in Wilms tumor and also in normal renal development.

Grundy et al. (1998) performed LOH in studies of 7p in 40 sporadic Wilms tumors, using a panel of 10 microsatellite polymorphic markers distributed along the length of the chromosome arm. Four tumors (10%) showed allelic loss for 7p markers which was twice the background rate of LOH in WT. The shortest common region of overlap of LOH was located between markers D7S517 and D7S503 in band 7p21-p15. In 1 tumor there was evidence for a homozygous interstitial deletion at a locus within this region.

Grundy, R. G.; Pritchard, J.; Scambler, P.; Cowell, J. K.: Loss of heterozygosity for the short arm of chromosome 7 in sporadic Wilms tumour. Oncogene 17: 395-400, 1998.

Miozzo, M.; Perotti, D.; Minoletti, F.; Mondini, P.; Pilotti, S.; Luksch, R.; Fossati-Bellani, F.; Pierotti, M. A.; Sozzi, G.; Radice, P.: Mapping of a putative tumor suppressor locus to proximal 7p in Wilms tumors. Genomics 37: 310-315, 1996.

Reynolds, P. A.; Powlesland, R. M.; Keen, T. J.; Inglehearn, C. F.; Cunningham, A. F.; Green, E. D.; Brown, K. W.: Localization of a novel t(1;7) translocation associated with Wilms' tumor predisposition and skeletal abnormalities. Genes Chromosomes Cancer 17: 151-155, 1996.

Wilmore, H. P.; White, G. F. J.; Howell, R. T.; Brown, K. W.: Germline and somatic abnormalities of chromosome 7 in Wilms' tumor. Cancer Genet. Cytogenet. 77: 93-98, 1994.

INSULIN-LIKE GROWTH FACTOR 2 (*147470)

JJM NOTE

This locus not only is genetically linked to the *WT1* locus but also seems to be associated with its manifestations in a pathogenetic way.

Insulin-like growth factors I and II, also known as somatomedin C and somatomedin A, respectively (see 147440), are single chain polypeptides which share an amino acid sequence homology of about 47% with insulin (INS; 176730) and about 31% with relaxin (179730) and with them comprise the insulin family of polypeptide growth factors. Their functions include mediation of growth hormone action, stimulation of growth of cultured cells, stimulation of the action of insulin, and involvement in development and growth. They appear to be autocrine regulators of cell proliferation. Structural defects may be involved in some forms of short stature (see 265850). Bell et al. (1984) and Dull et al. (1984) studied the sequence and organization of the gene for IGF-II in comparison with that for insulin. Using cDNA probes in the analysis of somatic cell hybrids, Brissenden et al. (1984) and Tricoli et al. (1984) independently assigned IGF2 to 11p15-p11. By analysis of human-Chinese hamster somatic cell hybrids, de Pagter-Holthuizen et al. (1985) assigned the IGF2 gene to chromosome 11. The location on chromosome 11 of this locus, structurally homologous to the proinsulin locus (also on 11), is noteworthy. The assignment of IGF1 to chromosome 12 adds to the known homology of 11 and 12. By comparing the restriction enzyme cleavage maps of the IGF2 and INS genes, including their flanking regions and hybridization with an IGF2 cDNA probe, Bell et al. (1985) concluded that these 2 genes are adjacent to one another. They have the same polarity and are separated by 12.6 kbp of intergenic DNA that includes a dispersed middle repetitive Alu sequence. The order of the genes is 5-prime—INS—IGF2-3-prime. Comings (1973) proposed that dominantly inherited tumors may arise through inactivation or loss of a pair of regulatory genes that normally suppress the expression of a structural transforming gene. In 4 cases of Wilms tumor, Reeve et al. (1985) found that transcripts of IGF2 were highly elevated as compared with adjacent normal kidney. Furthermore, by in situ hybridization, they mapped the IGF2 gene to 11p14.1, close to the WAGR locus. They proposed that IGF2 is the (or a) transforming gene in Wilms tumor. (Their positioning of the IGF2 locus is inconsistent with that of others who place it in a somewhat more distal band.) By in situ hybridization, Morton et al. (1986) assigned IGF2 to 11p15.

Scott et al. (1985) pointed out that Wilms tumor is histologically indistinguishable from the early stages of kidney development. In 12 sporadic cases of Wilms tumor, Scott et al. (1985) found that expression of the IGF2 gene was markedly increased relative to adult tissues, but was comparable to the level of expression in several fetal tissues including kidney, liver, adrenal, and striated muscle. Although this may merely reflect the stage of tumor differentiation, the possibility that IGF2 is involved in the transformation process was raised. Shen et al. (1988) demonstrated alternative forms of IGF-II transcripts which may be products of alternative splicing. Daughaday et al. (1988) demonstrated that recurrent hypoglycemia in a woman with leiomyosarcoma was the result of secretion of IGF-II by the tumor. Burgisser et al. (1991) studied the function of 3 experimentally produced mutants of IGF-II expressed in NIH 3T3 cells.

Since the IGF2 gene maps to the terminal region of 11p which is also involved in the Beckwith-Wiedemann syndrome (BWS; 130650) and since BWS shows an imprinting phenomenon, it is attractive to think that IGF2 is a candidate gene for BWS. However, see 130650 for evidence for and against involvement of IGF2 in that disorder (VAM).

DeChiara et al. (1991) produced targeted disruption of the Igf-2 gene in mice by homologous recombination in ES (embryonic stem) cells. Transmission of this mutation through the male germline resulted in heterozygous progeny that were growth deficient. In contrast, when the disrupted gene was transmitted maternally, the heterozygous offspring were phenotypically normal. Homozygous mutants were indistinguishable in appearance from growth-deficient heterozygous sibs. Nuclease protection and in situ hybridization analyses of the transcripts from the wildtype and mutated alleles indicated that only the paternal allele was expressed in embryos, while the maternal allele was silent. An exception was found in the choroid plexus and leptomeninges, where both alleles were transcriptionally active. These results demonstrated that IGF-II is indispensable for normal embryonic growth and that the gene is subject to tissue-specific parental imprinting. Zhang and Tycko (1992) referred to this as monoallelic expression. The IGF-II protein is overexpressed in Wilms tumors (194070). The major fetal IGF-II promoter, as defined in transient transfection assays, is a region spanning from nucleotides -295 to +135, relative to the transcription start site. Drummond et al. (1992) demonstrated that the WT1 gene product binds to multiple sites in this region and functions as a potent repressor of IGF-II transcription in vivo. Maximal repression was dependent on the presence of WT1 binding sites on each side of the transcriptional initiation site. These findings provided a molecular basis for overexpression of IGF-II in Wilms tumors and suggested that WT1 negatively regulates blastemal cell proliferation by limiting the production of a fetal growth factor in the developing vertebrate kidney. Rainier et al. (1993) reported relaxation (or loss) of imprinting of IGF2 in sporadic Wilms tumor; the same was reported in a group of patients with Beckwith-Wiedemann syndrome by Weksberg et al. (1993). The loss of imprinting was demonstrated by finding biallelic expression of the locus. Ogawa et al. (1993) examined the imprinting of the IGF2 gene in 10 normal kidney samples from children with renal embryonal neoplasms. In kidney samples from 9 children with normal growth profiles, IGF2 mRNA was transcribed monoallelically, consistent with normal imprinting of the gene. On the other hand, in 1 child who had generalized somatic overgrowth, IGF2 was transcribed from both alleles in her kidney, peripheral blood leukocytes, and Wilms tumor. These findings suggested that a defect in genomic imprinting can occur constitutionally, leading to growth abnormalities and predisposition to Wilms tumor. The patient was a 9-year-old Pacific Islander female (the report was from New Zealand) who at 2.5 years had treatment of Wilms tumor. The tumor showed epithelial and rhabdomyoblastic differentiation and there was an adjacent large intralobar nephrogenic rest. Karyotype was normal. Both height and weight had remained 3 to 4 standard deviations above the mean for age. Her growth excess was proportionate and there were no features typical of Beckwith-Wiedemann syndrome or other syndromal causes of excessive growth. The parents had normal height. The patient's plasma growth hormone and serum IGF1 and IGF2 were normal.

IGF2 is a mitogen for many cell types and an important modulator of muscle growth and differentiation. The gene is widely expressed during prenatal development and its activity is regulated by genomic imprinting, the gene being inactive on the chromosome inherited from the mother in most normal tissues. Pedone et al. (1994) showed that monoallelic expression of IGF2 is conserved in normal adult muscle tissue, whereas 2 or more copies of active IGF2 alleles, arising by either relaxation of imprinting or duplication of the active allele, were found in 9 out of 11 (82%) rhabdomyosarcomas retaining heterozygosity at 11p15, regardless of the histologic subtype. This observation, together with the finding that IGF2 acts as an autocrine growth factor for rhabdomyosarcoma cells, indicates that acquisition of a double dosage of active IGF2 gene is an important step for the initiation or progression of rhabdomyosarcoma tumorigenesis. Among different types of muscle tumors, relaxation of imprinting seems to arise predominantly in rhabdomyosarcomas, since Pedone et al. (1994) found only 1 case of partial reactivation of the maternal IGF2 allele out of 7 leiomyosarcomas tested.

Morison et al. (1996) reported constitutional relaxation of imprinting of IGF2 in 4 children with somatic overgrowth who did not show signs of Beckwith-Wiedemann syndrome. All 4 children showed nephromegaly and 2 developed Wilms tumors. Imprinting of IGF2 was examined using a nested RNA PCR assay in which a portion of the gene encompassing the exon 9 ApaI/AvaII restriction site polymorphism was amplified. In all 4 patients, IGF2 was found to be biallelically expressed. In 28 informative children and 22 informative normal adults, this assay showed monoallelic expression of IGF2. Studies carried out in 13 children from informative families confirmed that the paternally derived allele was expressed. Disruption of H19 (103280) methylation was observed in 3 of the 4 patients. On the basis of their findings, Morison et al. (1996) proposed that IGF2 overgrowth disorder represents an entity and Beckwith-Wiedemann syndrome represents an extreme manifestation of this disorder.

Brown et al. (1996) described evidence of alteration of imprinting in a BWS family with an inversion, inv(11)(p11.2;p15.5). They demonstrated that the inversion led to biallelic expression of IGF2 and altered DNA replication patterns in the IGF2 region. The H19 imprinting in affected individuals was normal, suggesting an H19-independent pathway to biallelic IGF2 transcription. DNA methylation in IGF2 remained monoallelic, suggesting to Brown et al. (1996) that the mutation caused by the translocation had uncoupled allele-specific methylation from gene expression.

Further evidence of the role of IGF2 in BWS was presented by Sun et al. (1997). They introduced Igf2 transgenes into the mouse genome by using ES cells and thereby caused transactivation of the endogenous Igf2 gene. The consequent overexpression of Igf2 resulted in most of the symptoms of Beckwith-Wiedemann syndrome, including prenatal overgrowth, polyhydramnios, fetal and neonatal lethality, disproportionate organ overgrowth including tongue enlargement, and skeletal abnormalities. This was presented as evidence that IGF2 overexpression is a key determinant of BWS.

Lighten et al. (1997) demonstrated that IGF2 is parentally imprinted during human preimplantation development. For this purpose they used an ApaI polymorphism at the 3-prime untranslated region of the IGF2 gene to distinguish between mRNA derived from each copy of the gene in fertilized embryos at the 4- and 8-cell stage. They found that at the 8-cell stage and beyond only the paternal IGF2 gene was expressed.

See also Brissenden et al. (1984); Haselbacher et al. (1987); Jansen et al. (1985); Roth (1988); Tricoli et al. (1984)

Bell, G. I.; Gerhard, D. S.; Fong, N. M.; Sanchez-Pescador, R.; Rall, L. B.: Isolation of the human insulin-like growth factor genes: insulin-like growth factor II and insulin genes are contiguous. Proc. Nat. Acad. Sci. 82: 6450-6454, 1985.

Bell, G. I.; Merryweather, J. P.; Sanchez-Pescador, R.; Stempien, M. M.; Priestley, L.; Scott, J.; Rall, L. B.: Sequence of a cDNA clone encoding human preproinsulin-like growth factor II. Nature 310: 775-777, 1984.

Brissenden, J. E.; Ullrich, A.; Francke, U.: Chromosomal mapping of loci for insulin-like growth factors I and II and for epidermal growth factor in man. (Abstract) Am. J. Hum. Genet. 36: 133S, 1984.

Brissenden, J. E.; Ullrich, A.; Francke, U.: Human chromosomal mapping of genes for insulin-like growth factors I and II and epidermal growth factor. Nature 310: 781-784, 1984.

Brown, K. W.; Villar, A. J.; Bickmore, W.; Clayton-Smith, J.; Catchpoole, D.; Maher, E. R.; Reik, W.: Imprinting mutation in the Beckwith-Wiedemann syndrome leads to biallelic IGF2 expression through an H19-independent pathway. Hum. Molec. Genet. 5: 2027-2032, 1996.

Burgisser, D. M.; Roth, B. V.; Giger, R.; Luthi, C.; Weigl, S.; Zarn, J.; Humbel, R. E.: Mutants of human insulin-like growth factor II with altered affinities for the type 1 and type 2 insulin-like growth factor receptor. J. Biol. Chem. 266: 1029-1033, 1991.

Comings, D. E.: A general theory of carcinogenesis. Proc. Nat. Acad. Sci. 70: 3324-3328, 1973.

Daughaday, W. H.; Emanuele, M. A.; Brooks, M. H.; Barbato, A. L.; Kapadia, M.; Rotwein, P.: Synthesis and secretion of insulin-like growth factor II by a leiomyosarcoma with associated hypoglycemia. New Eng. J. Med. 319: 1434-1440, 1988.

DeChiara, T. M.; Robertson, E. J.; Efstratiadis, A.: Parental imprinting of the mouse insulin-like growth factor II gene. Cell 64: 849-859, 1991.

de Pagter-Holthuizen, P.; Hoppener, J. W. M.; Jansen, M.; Geurts van Kessel, A. H. M.; van Ommen, G. J. B.; Sussenbach, J. S.: Chromosomal localization and preliminary characterization of the human gene encoding insulin-like growth factor II. Hum. Genet. 69: 170-173, 1985.

Drummond, I. A.; Madden, S. L.; Rohwer-Nutter, P.; Bell, G. I.; Sukhatme, V. P.; Rauscher, F. J., III: Repression of the insulin-like growth factor II gene by the Wilms tumor suppressor WT1. Science 257: 674-678, 1992.

Dull, T. J.; Gray, A.; Hayflick, J. S.; Ullrich, A.: Insulin-like growth factor II precursor gene organization in relation to insulin gene family. Nature 310: 777-781, 1984.

Haselbacher, G. K.; Irminger, J.-C.; Zapf, J.; Ziegler, W. H.; Humbel, R. E.: Insulin-like growth factor II in human adrenal pheochromocytoma and Wilms

tumors: expression at the mRNA and protein level. Proc. Nat. Acad. Sci. 84: 1104-1106, 1987.
Jansen, M.; van Schaik, F. M. A.; van Tol, H.; Van den Brande, J. L.; Sussenbach, J. S.: Nucleotide sequences of cDNAs encoding precursors of human insulin-like growth factor II (IGF-II) and an IGF-II variant. FEBS Lett. 179: 243-246, 1985.
Lighten, A. D.; Hardy, K.; Winston, R. M. L.; Moore, G. E.: IGF2 is parentally imprinted in human preimplantation embryos. (Letter) Nature Genet. 122-123, 1997.
Morison, I. M.; Becroft, D. M.; Taniguchi, T.; Woods, C. G.; Reeve, A. E.: Somatic overgrowth associated with overexpression of insulin-like growth factor II. Nature Med. 2: 311-316, 1996.
Morton, C. C.; Byers, M. G.; Nakai, H.; Bell, G. I.; Shows, T. B.: Human genes for insulin-like growth factors I and II and epidermal growth factor are located on 12q22-q24.1, 11p15, and 4q25-q27, respectively. Cytogenet. Cell Genet. 41: 245-249, 1986.
Ogawa, O.; Becroft, D. M.; Morison, I. M.; Eccles, M. R.; Skeen, J. E.; Mauger, D. C.; Reeve, A. E.: Constitutional relaxation of insulin-like growth factor II gene imprinting associated with Wilms' tumour and gigantism. Nature Genet. 5: 408-412, 1993.
Pedone, P. V.; Tirabosco, R.; Cavazzana, A. O.; Ungaro, P.; Basso, G.; Luksch, R.; Carli, M.; Bruni, C. B.; Frunzio, R.; Riccio, A.: Mono- and bi-allelic expression of insulin-like growth factor II gene in human muscle tumors. Hum. Molec. Genet. 3: 1117-1121, 1994.
Rainier, S.; Johnson, L. A.; Dobry, C. J.; Ping, A. J.; Grundy, P. E.; Feinberg, A. P.: Relaxation of imprinted genes in human cancer. Nature 362: 747-749, 1993.
Reeve, A. E.; Eccles, M. R.; Wilkins, R. J.; Bell, G. I.; Millow, L. J.: Expression of insulin-like growth factor-II transcripts in Wilms' tumour. Nature 317: 258-260, 1985.
Roth, R. A.: Structure of the receptor for insulin-like growth factor II: the puzzle amplified. Science 239: 1269-1271, 1988.
Scott, J.; Cowell, J.; Robertson, M. E.; Priestley, L. M.; Wadey, R.; Hopkins, B.; Pritchard, J.; Bell, G. I.; Rall, L. B.; Graham, C. F.; Knott, T. J.: Insulin-like growth factor-II gene expression in Wilms' tumour and embryonic tissues. Nature 317: 260-262, 1985.
Shen, S.-J.; Daimon, M.; Wang, C.-Y.; Jansen, M.; Ilan, J.: Isolation of an insulin-like growth factor II cDNA with a unique 5-prime untranslated region from human placenta. Proc. Nat. Acad. Sci. 85: 1947-1951, 1988.
Sun, F.-L.; Dean, W. L.; Kelsey, G.; Allen, N. D.; Reik, W.: Transactivation of IGF2 in a mouse model of Beckwith-Wiedemann syndrome. Nature 389: 809-815, 1997.
Tricoli, J. V.; Rall, L. B.; Scott, J.; Bell, G. I.; Shows, T. B.: Insulin-like growth factor genes: chromosome organization and association with disease. (Abstract) Am. J. Hum. Genet. 36: 121S, 1984.
Tricoli, J. V.; Rall, L. B.; Scott, J.; Bell, G. I.; Shows, T. B.: Localization of insulin-like growth factor genes to human chromosomes 11 and 12. Nature 310: 784-786, 1984.
Weksberg, R.; Shen, D. R.; Fei, Y.-L.; Song, Q. L.; Squire, J.: Disruption of insulin-like growth factor 2 imprinting in Beckwith-Wiedemann syndrome. Nature Genet. 5: 143-150, 1993.
Zhang, Y.; Tycko, B.: Monoallelic expression of the human H19 gene. Nature Genet. 1: 40-44, 1992.

QM GENE (312173)

JJM NOTE

This gene on chromosome X may interact with the product of the *WT1* gene in tumorigenesis.

Weissman et al. (1987) provided a functional assay for the identification of the suppressor gene(s) involved in Wilms tumor, by showing that introduction of a normal chromosome 11 into a Wilms tumor cell line completely abolished its tumorigenic action in nude mice. Dowdy et al. (1991) showed that this suppressor activity resided in the 11p15 and not in the 11p13 region; see Wilms tumor, type II (WT2; 194071). In an attempt to isolate a suppressor gene, Dowdy et al. (1991) performed a subtractive hybridization assay with the tumorigenic Wilms tumor cell line and its nontumorigenic derivative which contained an extra t(X;11) translocation chromosome. A single novel cDNA clone, designated QM, was identified and, as the QM mRNA was modulated between tumorigenic and nontumorigenic cell lines, the QM gene was an attractive tumor-suppressor candidate gene. In the course of investigating disease genes in the Xqter region, van den Ouweland et al. (1992) found a cosmid that appeared to harbor the QM gene as it demonstrated 100% identity with the cDNA sequence published by Dowdy et al. (1991). With an exon-specific polymerase chain reaction, van den Ouweland et al. (1992) demonstrated that the genomic homolog of the QM cDNA is located in Xq28 in the region of the G6PD and color vision genes. No homologous sequences could be detected on 11p. Thus, the QM gene is not, per se, involved in the suppression of Wilms tumor.

As an essential step toward an exhaustive analysis of the coding potential of a large region of the genome, Korn et al. (1992) developed a protocol allowing the rapid isolation of transcripts defined by overlapping clone libraries. The technique was applied to the analysis of transcripts from 2 cosmid contigs, which together encompassed a region of 900 kb in Xq28. They identified 81 cDNA clones, of which 54 clones were mapped back to the cosmid contigs. Of the 54 clones placed on the contig maps, 12 cDNA clones were shown to belong to 2 genes that had previously been reported, L1CAM (308840) and QM.

In the process of characterizing cosmid clones from the Xq28 region, Kaneko et al. (1992) identified a sequence conserved across species in a particular clone. Using the conserved sequence as a probe, they isolated cDNA clones and demonstrated that the nucleotide sequence was the same as that of the QM cDNA. Kaneko et al. (1992) went on to demonstrate that the QM gene consists of at least 7 exons. Dowdy et al. (1991) had demonstrated that the QM cDNA encodes a predicted hydrophilic 25 kD basic protein, primarily consisting of alpha-helices. The homology search did not reveal any functional domains related to known proteins. Southern blot analysis of a somatic cell hybrid panel suggested that the QM gene belongs to a multigene family with other representatives on chromosomes 2, 3, 6, 14, 16, and possibly other chromosomes. Northern blot analysis demonstrated that the QM gene is expressed in all adult tissues examined as well as in cell lines including HeLa cells, fibroblasts, and somatic cell hybrids, with increased expression in liver, spleen, testis, and adrenal gland. Its location in Xq28 makes the QM gene a candidate gene for one of the many X-linked disorders of unknown basic defect that map to that region.

QM is a 214-amino acid polypeptide encoded by a gene previously designated DXS648. It contains a high percentage of charged amino acids and binds to the JUN oncogene (165160) and to DNA. Although they found no matches between QM and any other known transcription factors in searches of DNA databases, Farmer et al. (1994) found a high degree of conservation throughout the first 175 residues of the protein when studies were performed on a diverse array of eukaryotes. Most notable was the considerable conservation of charged amino acids within specific regions. The rate of sequence divergence of the various homologs was found to be slow, of the order of 1% change every 22 million years, consistent with a critical role of QM in eukaryotic cells. Farmer et al. (1994) suggested that QM belongs to a novel class of transcription regulatory proteins.

See also Dowdy et al. (1991)

Dowdy, S. F.; Fasching, C. L.; Araujo, D.; Lai, K.-M.; Livanos, E.; Weissman, B. E.; Stanbridge, E. J.: Suppression of tumorigenicity in Wilms' tumor by the p15.5-p14 region of chromosome 11. Science 254: 293-295, 1991.
Dowdy, S. F.; Lai, K.-M.; Weissman, B. E.; Matsui, Y.; Hogan, B. L. M.; Stanbridge, E. J.: The isolation and characterization of a novel cDNA demonstrating an altered mRNA level in nontumorigenic Wilms' microcell hybrid cells. Nucleic Acids Res. 19: 5763-5769, 1991.
Farmer, A. A.; Loftus, T. M.; Mills, A. A.; Sato, K. Y.; Neill, J. D.; Tron, T.; Yang, M.; Trumpower, B. L.; Stanbridge, E. J.: Extreme evolutionary conservation of QM, a novel c-Jun associated transcription factor. Hum. Molec. Genet. 3: 723-728, 1994.
Kaneko, K.; Kobayashi, H.; Onodera, O.; Miyatake, T.; Tsuji, S.: Genomic organization of a cDNA (QM) demonstrating an altered mRNA level in nontumorigenic Wilms' microcell hybrid cells and its localization to Xq28. Hum. Molec. Genet. 1: 529-533, 1992.
Korn, B.; Sedlacek, Z.; Manca, A.; Kioschis, P.; Konecki, D.; Lehrach, H.; Poustka, A.: A strategy for the selection of transcribed sequences in the Xq28 region. Hum. Molec. Genet. 1: 235-242, 1992.
van den Ouweland, A. M. W.; Verdijk, M.; Mannens, M. M. A. M.; van Oost, B. A.: The QM gene is X-linked and therefore not involved in suppression of tumorigenesis in Wilms' tumor. Hum. Molec. Genet. 90: 144-146, 1992.
Weissman, B. E.; Saxon, P. J.; Pasquale, S. R.; Jones, G. R.; Geiser, A. G.; Stanbridge, E. J.: Introduction of a normal human chromosome 11 into a Wilms' tumor cell line controls its tumorigenic expression. Science 236: 175-180, 1987.

PREMATURE CENTROMERE DIVISION (*176430)

JJM NOTE

One patient with the cytogenetic phenomenon had Wilms tumor. My colleagues and I had one patient with the chromosome finding and SC phocomelia syndrome complicated by cutaneous malignant melanoma (Parry DM et al.: SC phocomelia syndrome, premature centromere separation, and congenital cranial nerve paralysis in two sisters, one with malignant melanoma. Am. J. Med. Genet. 24:653–72, 1986).

In short-term lymphocyte cultures from 3 unrelated persons, Rudd et al. (1983) found an increased frequency of mitoses with separated centromeres and splayed chromatids in the presence of colcemid. They referred to the phenomenon as premature centromere division (PCD). In 2 of the 3 patients, the frequency of PCD was reduced when colcemid was omitted but was still higher than in controls. Cultured fibroblasts from the third patient, whose cells showed no reduction with colcemid, exhibited increased tetraploidy and multinucleated cells. Transmission in each of the families was consistent with autosomal dominant inheritance; male-to-male transmission was shown in 3 instances in 2 families. Fitzgerald et al. (1986) reported findings in a clinically normal 28-year-old woman who had 3 conceptuses with trisomy 21 (190685) and 1 normal child. She showed minimal evidence of mosaicism: 4% of blood cells and 6% of skin fibroblasts had trisomy 21. Also, 7% of her blood cells showed aneuploidy of the X chromosome which was associated with PCD of the X; 6% of fibroblasts showed trisomy 18; 10% of fibroblasts showed PCD of 21; 1% of fibroblasts showed PCD of 18. Since it was unlikely that this woman was a constitutional mosaic for trisomies X, 18, and 21, all at low levels, the authors suggested she was liable to PCD especially of those chromosomes. Gabarron et al. (1986) found PCD in lymphocyte cultures from 4 members of a subfertile family in Spain. PCD was observed both in the presence and in the absence of colchicine. Cultured fibroblasts from the proband, on the other hand, showed only normal diploid metaphases. PCD cells seem to have a shorter cell cycle. PCD was observed in the grandmother, the mother (who was the proband) and a son and daughter, the only living children. Both the grandmother and the mother had had multiple spontaneous abortions. The mother was healthy, as were her children, but was only 144 cm tall. Madan et al. (1987) described a family with 4 phenotypically normal persons in 3 successive generations who had an increased frequency of cells with premature centromere division of all chromosomes. This phenomenon is apparently different from the PCD of the X chromosome, which may be recessive (Fitzgerald, 1975; see 212790), and from the centromere splitting in cells of patients with Roberts syndrome (268300). Miller et al. (1990) described a boy with mental retardation, dysmorphic features and combined immunodeficiencies. Trisomies and monosomies of most autosomes and gonosomal aberrations were found separately or in combination in most of the proband's lymphocytes and fibroblasts. The chromosome number varied from 44 to 50. A high proportion of metaphases showed PCD or had the appearance of C-anaphases (C = colchicine; i.e., colchicine arrested anaphases).

Mehes (1997) stated that he was the first to suggest that nonrandom centromere division may be a mechanism of nondisjunction in autosomal disorders (Mehes, 1978). See also 257300.

Kajii et al. (1998) referred to this condition as total premature chromatid separation (total PCS). The phenomenon was observed in 67 to 87.5% of repeated cultures of peripheral blood lymphocytes from 2 unrelated infants. Also noted was a variety of mosaic aneuploidies, especially trisomies, double trisomies, and monosomies (mosaic variegated aneuploidy). Both infants showed pre- and postnatal growth retardation, profound developmental retardation, uncontrolled seizures, severe microcephaly, hypoplasia of the brain, Dandy-Walker anomaly, abnormal facial appearance, and bilateral cataract. Patient 1, a girl, also had a cleft palate, multiple renal cysts, and Wilms tumor of the left kidney. Patient 2, a boy, had ambiguous external genitalia. Both infants died under 2 years of age. In the 2 families, their parents and 3 other members showed 2.5 to 47% lymphocytes with total PCS but without mosaic variegated aneuploidy or phenotypic abnormalities. Another 10 relatives showed 0 to 1% cells with total PCS and so were judged negative for the total PCS trait. It was deduced that the total PCS trait in the 2 families was transmitted in an autosomal dominant fashion and that the 2 affected infants were homozygous for the trait.

Kajii et al. (1998) preferred the designation total premature chromatid separation to the term premature centromere division because the phenomenon involves not only the centromeres but also the entire chromatids.

Fitzgerald, P. H.: A mechanism of X chromosome aneuploidy in lymphocytes of aging women. Humangenetik 28: 153-158, 1975.

Fitzgerald, P. H.; Archer, S. A.; Morris, C. M.: Evidence for the repeated primary non-disjunction of chromosome 21 as a result of premature centromere division (PCD). Hum. Genet. 72: 58-62, 1986.

Gabarron, J.; Jimenez, A.; Glover, G.: Premature centromere division dominantly inherited in a subfertile family. Cytogenet. Cell Genet. 43: 69-71, 1986.

Kajii, T.; Kawai, T.; Takumi, T.; Misu, H.; Mabuchi, O.; Takahashi, Y.; Tachino, M.; Nihei, F.; Ikeuchi, T.: Mosaic variegated aneuploidy with multiple congenital abnormalities: homozygosity for total premature chromatid separation trait. Am. J. Med. Genet. 78: 245-249, 1998.

Madan, K.; Lindhout, D.; Palan, A.: Premature centromere division (PCD): a dominantly inherited cytogenetic anomaly. Hum. Genet. 77: 193-196, 1987.

Mehes, K.: Non-random centromere division: a mechanism of non-disjunction causing aneuploidy? Hum. Hered. 28: 255-260, 1978.

Mehes, K.: Personal Communication. Pecs, Hungary 5/28/1997.

Miller, K.; Muller, W.; Winkler, L.; Hadam, M. R.; Ehrich, J. H. H.; Flatz, S. D.: Mitotic disturbance associated with mosaic aneuploidies. Hum. Genet. 84: 361-364, 1990.

Rudd, N. L.; Teshima, I. E.; Martin, R. H.; Sisken, J. E.; Weksberg, R.: A dominantly inherited cytogenetic anomaly: a possible cell division mutant. Hum. Genet. 65: 117-121, 1983.

NEPHROSIS, CONGENITAL (FINNISH) (#256300)

JJM NOTE

This disorder is included because of the case reports of two siblings with congenital nephrosis and Wilms tumor (nephroblastoma) (Zunin and Soave 1964). Could they represent Drash or Perlman syndromes? Could a prenatal nephrosis of any type predispose to Wilms tumor? Or, could this mutant gene lead separately to neoplasia and nephrosis? Additional cases must be studied to address these and other questions.

A number sign (#) is used with this entry because Finnish congenital nephrosis has been found to be caused by mutation in the nephrin gene (602716).

Whereas the usual idiopathic nephrotic syndrome of childhood almost never has its onset before the age of 18 months, congenital nephrosis shows itself in the first days or weeks of life. Furthermore, the familial occurrence including parental consanguinity is that of an autosomal recessive trait. Otherwise the clinical, chemical, and pathologic features are identical with those of the idiopathic condition. This disorder has a relatively high frequency in Finland (Norio et al., 1964), where the incidence is about 1 in 8000 (Norio, 1980). A large series of cases was collected by Hallman and Hjelt (1959) in Finland and by Vernier et al. (1957) and Worthen et al. (1959) in Minnesota, where many persons of Finnish extraction live. The latter group was impressed with the high frequency of maternal toxemia in these cases. Giles et al. (1957) reported 2 affected sibs from a first-cousin marriage and a third patient who was the child of cousins. Ongre (1961) described sibs with nephrosis starting in the neonatal period and with cystic-like dilation of renal tubules. It is likely that this is not congenital cystic disease but rather congenital nephrosis, as in the other series mentioned. Nephrosis of later onset is occasionally familial but not necessarily mendelian (Roy and Pitcock, 1971).

Seppala et al. (1976) demonstrated that this disorder can be diagnosed antenatally by elevated levels of alpha-fetoprotein in amniotic fluid. Nephrosis with congenital cardiac malformation (Fournier et al., 1963) may be a distinct autosomal recessive disorder. Zunin and Soave (1964) observed nephrosis in association with nephroblastoma in 2 sibs. In one of them, removal of the tumor was accompanied by amelioration of the nephrotic syndrome. Grahame-Smith et al. (1988) described twins with Finnish type congenital nephrosis. One twin was stillborn; the second presented with a diagnosis of pyloric stenosis.

Twelve percent of 41 infants with congenital nephrotic syndrome described by Mahan et al. (1984) presented with pyloric stenosis. Nine of the 41 patients (22%), observed in Minneapolis, Minnesota, were shown to have Finnish ancestry. Steroids or cytotoxic drugs, alone or in combination, were without benefit. Intensive medical therapy to control bacterial infections, combined with renal transplantation, was judged to offer a good opportunity for survival with an acceptable quality of life for infants with congenital nephrotic syndrome.

Morris et al. (1995) described the Finnish type of congenital nephrosis in 2 of 3 successive pregnancies of a nonconsanguineous couple with no known Finnish ancestry. They confirmed the usefulness of amniotic fluid alpha-fetoprotein determination in the prenatal diagnosis. Morris et al. (1995) observed that, because of the involvement of the kidneys in NPHS1, the fetus looses large amounts of AFP in the urine.

Using radioimmunoassay methods, Risteli et al. (1982) found accumulation of type IV collagen in the renal cortex which was out of proportion to another basement membrane protein, laminin. They interpreted this to mean that metabolism of type IV collagen is disturbed in this disorder. The normal barrier to penetration of the renal glomerular basement membrane by anionic plasma proteins depends in part on the existence of negatively charged sites within the membrane (Cotran and Rennke, 1983). Vernier et al. (1983) found that normal subjects had such sites distributed at regular intervals in the lamina rara externa, with a frequency of 23.8 sites per 1,000 nm length of membrane, whereas 5 patients with congenital nephrosis had 8.9 sites. An in vitro histochemical technique was used in these studies. Vernier et al. (1983) concluded that the basic defect in congenital nephrosis is failure of heparan sulfate-rich anionic sites to develop in the lamina rara externa of the glomerular basement membrane.

Dressler and Douglass (1992) showed that, in transgenic mice, deregulation of expression of the gene coding for the Pax-2 DNA-binding protein (167409) resulted in severe kidney abnormalities resembling the clinical and pathologic findings in patients with Finnish nephrosis. Kestila et al. (1994) showed, however, that the disorder in these patients is not linked to the PAX2 gene locus on chromosome 10. In fact, Kestila et al. (1994) assigned the locus for congenital nephrotic syndrome of the Finnish type (symbolized CNF by them) to 19q12-q13.1 on the basis of linkage analyses in 17 Finnish families.

Olsen et al. (1996) assembled a 1-Mb cosmid contig and restriction map spanning the candidate region for NPHS1 in 19q13.1. Mannikko et al. (1996) applied haplotype analysis to several non-Finnish CNF families to determine whether the same genetic locus is involved in these families as in Finnish families. The results indicated linkage to the 19q13.1 region. It was also observed that, in most cases, alleles typically found on CNF chromosomes of Finnish families were also found on CNF chromosomes of non-Finnish families from North America and Europe.

The Finnish type of congenital nephrosis is only one of many disorders, numbering more than 30, that are absent or infrequent elsewhere and exist in the Finnish population, sometimes at high carrier frequencies. Conversely, recessive autosomal diseases common in other European populations, such as cystic fibrosis (219700), phenylketonuria (261600), or galactosemia (230400), are rare or absent in Finland. Sajantila et al. (1996) noted that single mutations embedded in chromosomal regions exhibiting linkage disequilibrium have been demonstrated in the case of several of these 'Finnish' genetic disorders. In contrast, outside Finland, the rare cases of these disorders are usually due to several different mutations. Furthermore, many of the disorders occur in locally restricted areas in Finland. Sajantila et al. (1996) found that Y-chromosomal haplotypes in several European populations revealed an almost monomorphic pattern in the Finns, whereas Y-chromosomal diversity was significantly higher in other populations. Furthermore, analyses of nucleotide positions in the mitochondrial control region that evolves slowly showed a decrease in genetic diversity in Finns. Thus, relatively few men and women contributed to the genetic lineages that today survive in the Finnish population. This is likely to have caused the 'Finnish disease heritage,' i.e., the occurrence of several genetic diseases in the Finnish population that are rare elsewhere. A preliminary analysis of the mitochondrial mutations that had accumulated subsequent to the bottleneck suggested that it occurred about 4,000 years ago, presumably when populations using agriculture and animal husbandry arrived in Finland. The results suggested that genetic founder effects have played a role also in the biologic history of Estonians and the Basques.

By use of positional cloning strategies, Kestila et al. (1998) isolated the gene responsible for NPHS1. Mutations in Finnish patients with NPHS1 were found in this gene, which they termed nephrin. The most common Finnish mutation was a deletion of 2 nucleotides in exon 2 (602716.0001), resulting in a frameshift and a truncated protein. The predicted nephrin protein belongs to the immunoglobulin family of cell adhesion molecules and is specifically expressed in renal glomeruli.

See also Autio-Harmainen and Rapola (1983); Bader et al. (1974); Hallman et al. (1956); Hallman et al. (1967); Huttunen (1976); Inferrera et al. (1980); Morgan et al. (1981); Norio (1966)

Autio-Harmainen, H.; Rapola, J.: The thickness of the glomerular basement membrane in congenital nephrotic syndrome of the Finnish type. Nephron 34: 48-50, 1983.

Bader, P. I.; Grove, J.; Trygstad, C. W.; Nance, W. E.: Familial nephrotic syndrome. Am. J. Med. 56: 34-43, 1974.

Cotran, R. S.; Rennke, H. G.: Anionic sites and the mechanisms of proteinuria. (Editorial) New Eng. J. Med. 309: 1050-1051, 1983.

Dressler, G. R.; Douglass, E. C.: Pax-2 is a DNA-binding protein expressed in embryonic kidney and Wilms tumor. Proc. Nat. Acad. Sci. 89: 1179-1183, 1992.

Fournier, A.; Paget, M.; Pauli, A.; Devin, P.: Syndromes nephrotiques familiaux. Syndrome nephrotique associe a une cardiopathie congenitale chez quatre soeurs. Pediatrie 18: 677-685, 1963.

Giles, H. M.; Pugh, R. C. B.; Darmady, E. M.; Stranack, F.; Woolf, L. I.: The nephrotic syndrome in early infancy: a report of 3 cases. Arch. Dis. Child. 32: 167-180, 1957.

Grahame-Smith, H. N.; Ward, P. S.; Jones, R. D.: Finnish type congenital nephrotic syndrome in twins: presentation with pyloric stenosis. J. Roy. Soc. Med. 81: 358 only, 1988.

Hallman, N.; Hjelt, L.: Congenital nephrotic syndrome. J. Pediat. 55: 152-162, 1959.

Hallman, N.; Hjelt, L.; Ahvenainen, E. K.: Nephrotic syndrome in newborn and young infants. Ann. Paediat. Fenn. 2: 227-241, 1956.

Hallman, N.; Norio, R.; Kouvalainen, K.: Main features of the congenital nephrotic syndrome. Acta Paediat. Scand. 172 (suppl.): 75-78, 1967.

Huttunen, N.-P.: Congenital nephrotic syndrome of Finnish type: study of 75 patients. Arch. Dis. Child. 51: 344-348, 1976.

Inferrera, C.; Barresi, G.; Chimicata, S.; De Luca, F.; Baviera, G.; Gulli, V.; Gemelli, M.: Morphologic considerations on the placenta in congenital nephrotic syndrome of Finnish type. Virchows Arch. A 389: 13-26, 1980.

Kestila, M.; Lenkkeri, U.; Mannikko, M.; Lamerdin, J.; McCready, P.; Putaala, H.; Ruotsalainen, V.; Morita, T.; Nissinen, M.; Herva, R.; Kashtan, C. E.; Peltonen, L.; Holmberg, C.; Olsen, A.; Tryggvason, K.: Positionally cloned gene for a novel glomerular protein—nephrin—is mutated in congenital nephrotic syndrome. Molec. Cell 1: 575-582, 1998.

Kestila, M.; Mannikko, M.; Holmberg, C.; Gyapay, G.; Weissenbach, J.; Savolainen, E.-R.; Peltonen, L.; Tryggvason, K.: Congenital nephrotic syndrome of the Finnish type maps to the long arm of chromosome 19. Am. J. Hum. Genet. 54: 757-764, 1994.

Kestila, M.; Mannikko, M.; Holmberg, C.; Tryggvason, K.; Peltonen, L.: Congenital nephrotic syndrome of the Finnish type is not associated with the Pax-2 gene despite the promising transgenic animal model. Genomics 19: 570-572, 1994.

Mahan, J. D.; Mauer, S. M.; Sibley, R. K.; Vernier, R. L.: Congenital nephrotic syndrome: evolution of medical management and results of renal transplantation. J. Pediat. 105: 549-557, 1984.

Mannikko, M.; Lenkkeri, U.; Kashtan, C. E.; Kestila, M.; Holmberg, C.; Tryggvason, K.: Haplotype analysis of congenital nephrotic syndrome of the Finnish type in non-Finnish families. J. Am. Soc. Nephrol. 7: 2700-2703, 1996.

Morgan, G.; Postlethwaite, R. J.; Lendon, M.; Houston, I. B.; Savage, J. M.: Postural deformities in congenital nephrotic syndrome. Arch. Dis. Child. 56: 959-962, 1981.

Morris, J.; Ellwood, D.; Kennedy, D.; Knight, J.: Amniotic alpha-fetoprotein in the prenatal diagnosis of congenital nephrotic syndrome of the Finnish type. Prenatal Diag. 15: 482-485, 1995.

Norio, R.: Heredity in the congenital nephrotic syndrome. A genetic study of 57 Finnish families with a review of reported cases. Ann. Paediat. Fenn. 12 (suppl. 27): 1-94, 1966.

Norio, R.: Congenital nephrotic syndrome of Finnish type (CNF).In: Eriksson, A. W.; Forsius, H. R.; Nevanlinna, H. R.; Workman, P. L.; Norio, R. K.: Population Structure and Genetic Disorders. New York: Academic Press (pub.) 1980. Pp. 600-604.

Norio, R.; Hjelt, L.; Hallman, N.: Congenital nephrotic syndrome: an inherited disease? A preliminary report. Ann. Paediat. Fenn. 10: 223-227, 1964.

Olsen, A. S.; Georgescu, A.; Johnson, S.; Carrano, A. V.: Assembly of a 1-Mb restriction-mapped cosmid contig spanning the candidate region for Finnish congenital nephrosis (NPHS1) in 19q13.1. Genomics 34: 223-225, 1996.

Ongre, A. A.: Nephrotic syndrome with cyst-like dilations of renal tubules: report of 2 cases in siblings in early infancy. Acta Path. Microbiol. Scand. 51: 1-8, 1961.

Risteli, L.; Autio-Harmainen, H.; Huttunen, N.-P.; Risteli, J.: Slow accumulation of basement membrane collagen in kidney cortex in congenital nephrotic syndrome. Lancet I: 712-714, 1982.

Roy, S.; Pitcock, J. A.: Idiopathic nephrosis in identical twins. Am. J. Dis. Child. 121: 428-430, 1971.

Sajantila, A.; Salem, A.-H.; Savolainen, P.; Bauer, K.; Gierig, C.; Paabo, S.: Paternal and maternal DNA lineages reveal a bottleneck in the founding of the Finnish population. Proc. Nat. Acad. Sci. 93: 12035-12039, 1996.

Seppala, M.; Rapola, J.; Huttunen, N.-P.; Aula, P.; Karjalainen, O.; Ruoslahti, E.: Congenital nephrotic syndrome: prenatal diagnosis and genetic counselling by estimation of amniotic-fluid and maternal serum alpha-fetoprotein. Lancet II: 123-124, 1976.

Vernier, R. L.; Brunson, J.; Good, R. A.: Studies on familial nephrosis. I. Clinical and pathologic study of four cases in a single family. Am. J. Dis. Child. 93: 469-485, 1957.

Vernier, R. L.; Klein, D. J.; Sisson, S. P.; Mahan, J. D.; Oegema, T. R.; Brown, D. M.: Heparan sulfate-rich anionic sites in the human glomerular basement membrane. New Eng. J. Med. 309: 1001-1009, 1983.

Worthen, H. G.; Vernier, R. L.; Good, R. A.: Infantile nephrosis: clinical, biochemical, and morphologic studies of the syndrome. Am. J. Dis. Child. 98: 731-748, 1959.

Zunin, C.; Soave, F.: Association of nephrotic syndrome and nephroblastoma in siblings. Ann. Paediat. (Basel) 203: 29-38, 1964.

HYPERPROLINEMIA 1 (*239500)

JJM NOTE

If not by coincidence, a single patient having this rare inborn error of metabolism and a rare childhood cancer, Wilms tumor, could give a clue to the biochemical pathogenesis of Wilms tumor.

Scriver et al. (1961) and Schafer et al. (1962) described a family in which the male proband and 3 sibs had elevated plasma levels of L-proline. Probably the anomaly bore no relation to 2 other genetic defects in the same family: familial nephropathy and photogenic epilepsy. The aminoaciduria includes hydroxyproline and glycine as well as proline, presumably because these 3 amino acids share a renal tubular active transport mechanism which is overloaded by the high level of proline in the glomerular filtrate. Two types of hyperprolinemia appear to exist. In type I, referred to above, the defect involves the enzyme proline oxidase (Efron, 1965). Renal abnormalities occur in this form. In type II, which is usually characterized by mental retardation and convulsions, the enzyme defect concerns delta-1-pyrroline-5-carboxylate dehydrogenase, and the substance normally acted on by this enzyme is excreted in the urine; see 239510. The family reported by Efron (1965) was Italian; that reported by Scriver et al. (1961) and Schafer et al. (1962) was Scottish-Irish. Perry et al. (1968) reported type I hyperprolinemia in 2 generations of a consanguineous American Indian family. Hereditary renal abnormalities occurred in members of 3 generations. The proband later developed Wilms tumor. Marked elevations of plasma proline occur in homozygotes and normal or moderately elevated levels in heterozygotes, in the view of the authors. Both genotypes are accompanied by renal abnormalities. Mental retardation is not a feature of classic type I hyperprolinemia. Goyer et al. (1968) observed hereditary nephritis, neurosensory hearing loss, prolinuria and ichthyosis in various combinations in 23 members of a kindred. The relation to the hyperprolinurias and to Alport syndrome is not clear. The patient reported by Rokkones and Loken (1968) had uncle-niece parents. Emery et al. (1968) described an affected mentally retarded 18-year-old girl whose retarded sister had died, pre sumably of the same disorder. Scriver and Efron (1972) concluded that renal disease was only a coincidental occurrence in type I hyperprolinemia. Oyanagi et al. (1987) described a severely retarded infant with type I hyperprolinemia and partial duplication of the short arm of chromosome 10. Proline oxidase activity of liver tissue obtained by biopsy was about 9% of that of controls. Restriction of dietary proline resulted in a prompt fall of plasma proline levels with satisfactory growth, but no improvement in mental development. The patient had the facial appearance characteristic of hyperprolinemia and suffered from convulsions beginning at age 10 months. By mixing experiments, it was possible to exclude the deficiency of an activator of proline oxidase. Whether the findings in this patient indicate that a gene for proline oxidase is located on 10p could not be determined; it is possible that the extra genetic material on 10p was derived from somewhere else. The karyotype in both parents was normal.

Campbell et al. (1997) cloned the complete coding region for a human homolog of the Drosophila melanogaster 'sluggish-A' and yeast 'PUT1' genes, previously shown to encode proline oxidase activity in these organisms. The predicted 516-residue human protein shows 51% amino acid sequence identity to the Drosophila protein, indicating that this human gene may encode proline oxidase. Northern analysis showed that the gene is expressed in human lung, skeletal muscle, and brain, to a lesser extent in heart and kidney, and weakly in liver, placenta, and pancreas. By fluorescence in situ hybridization (FISH) and by tritium-labeled in situ hybridization, they mapped the gene (PRODH) to 22q11.2, a region previously implicated in type 1 hyperprolinemia in a case of CATCH22 syndrome (188400), a contiguous gene deletion syndrome involving 22q11. Jaeken et al. (1996) described a patient with type 1 hyperprolinemia in whom a submicroscopic 22q11 deletion was demonstrated by FISH analysis. The patient exhibited symptoms of the CATCH22 syndrome and also appeared to be heterozygous for heparin cofactor II (142360), which maps to 22q11. The parents and sister of the patient had normal levels of proline and heparin cofactor II, and did not have the deletion at 22q11. The results of Campbell et al. (1997), taken together with the patient reported by Jaeken et al. (1996), suggested that the gene mutant in type 1 hyperprolinemia, proline oxidase, is located on 22q11.2.

Nomenclature: The gene symbols for proline oxidase in Drosophila and yeast are slgA and PUT1, respectively. PRODH (for 'proline dehydrogenase') is used here as a symbol for proline oxidase.

See also Blake et al. (1974); Potter and Waickman (1973); Scriver (1978)

Blake, R. L.; Grillo, R. V.; Russell, E. S.: Increased taurine excretion in hereditary hyperprolinemia of the mouse. Life Sci. 14: 1285-1290, 1974.

Campbell, H. D.; Webb, G. C.; Young, I. G.: A human homologue of the Drosophila melanogaster sluggish-A (proline oxidase) gene maps to 22q11.2, and is a candidate gene for type-I hyperprolinaemia. Hum. Genet. 101: 69-74, 1997.

Efron, M. L.: Familial hyperprolinemia: report of second case, associated with congenital renal malformation, hereditary hematuria and mild mental retardation, with demonstration of enzyme defect. New Eng. J. Med. 272: 1243-1254, 1965.

Emery, F. A.; Goldie, L.; Stern, J.: Hyperprolinaemia type 2. J. Ment. Defic. Res. 12: 187-195, 1968.

Goyer, R. A.; Reynolds, J., Jr.; Burke, J.; Burkholder, P.: Hereditary renal disease with neurosensory hearing loss, prolinuria and ichthyosis. Am. J. Med. Sci. 256: 166-179, 1968.

Jaeken, J.; Goemans, N.; Fryns, J.-P.; Francois, I.; de Zegher, F.: Association of hyperprolinaemia type I and heparin cofactor II deficiency with CATCH 22 syndrome: evidence for a contiguous gene syndrome locating the proline oxidase gene. J. Inherit. Metab. Dis. 19: 275-277, 1996.

Oyanagi, K.; Tsuchiyama, A.; Itakura, Y.; Tamura, Y.; Nakao, T.; Fujita, S.; Shiono, H.: Clinical, biochemical and enzymatic studies in type I hyperprolinemia associated with chromosomal abnormality. Tohoku J. Exp. Med. 151: 465-475, 1987.

Perry, T. L.; Hardwick, D. F.; Lowry, R. B.; Hansen, S.: Hyperprolinaemia in two successive generations of a North American Indian family. Ann. Hum. Genet. 31: 401-408, 1968.

Potter, J. L.; Waickman, F. J.: Hyperprolinemia I: study of a large family. J. Pediat. 83: 635-637, 1973.

Rokkones, T.; Loken, A. C.: Congenital renal dysplasia, retinal dysplasia and mental retardation associated with hyperprolinuria and hyper-OH-prolinuria. Acta Paediat. Scand. 57: 225-229, 1968.

Schafer, I. A.; Scriver, C. R.; Efron, M. L.: Familial hyperprolinemia, cerebral dysfunction and renal anomalies occurring in a family with hereditary nephropathy and deafness. New Eng. J. Med. 267: 51-60, 1962.

Scriver, C. R.: Disorders of proline and hydroxyproline and 5-oxoproline metabolism.In: Stanbury, J. B.; Wyngaarden, J. B.; Fredrickson, D. S.: Metabolic Basis of Inherited Disease. New York: McGraw-Hill (pub.) (4th ed.): 1978. Pp. 336-361.

Scriver, C. R.; Efron, M. L.: Disorders of proline and hydroxyproline metabolism.In: Stanbury, J. B.; Wyngaarden, J. B.; Frederickson, D. S.: Metabolic Basis of Inherited Disease. New York: McGraw-Hill (pub.) (3rd ed.): 1972. Pp. 351-370.

Scriver, C. R.; Schafer, I. A.; Efron, M. L.: New renal tubular amino-acid transport system and a new hereditary disorder of amino-acid metabolism. Nature 192: 672-673, 1961.

Other Urinary Tract Tumors

URETER, CANCER OF (191600)

JJM NOTE

For this rare cancer to be considered familial, additional cases are needed. At the National Naval Medical Center, Bethesda, Maryland, I was presented with a 29-year-old male with metastatic colon cancer, whose brother and two paternal cousins were said to have ureteral cancer. The man's father had died of lymphoma at age 56 years (Mulvihill JJ: Reproductive outcomes among men treated for cancer. In Olshan AF, Mattison DR, eds.: *Male Mediated Developmental Toxicity.* New York: Plenum, 197–204, 1994). 1994). Ureteral cancer is part of the Lynch cancer family syndrome of hereditary nonpolypotic colorectal cancer and the related Muir-Torre syndrome. Ureteral polyps occur in Peutz-Jeghers syndrome.

Burkland and Juzek (1966) reported cancer of the right ureter in mother and son.

Burkland, C. E.; Juzek, R. H.: Familial occurrence of carcinoma of the ureter. J. Urol. 96: 697-701, 1966.

BLADDER CANCER (#109800)

JJM NOTE

In order of frequency, chromosomes involved in nonrandom changes in urinary bladder carcinomas are 1, 11, 3, 9, 7, 17, 10, 5, and 13. In theory, familial aggregation could arise from shared environmental factors, such as schistomiasis of the bladder, a common infection found in local rivers (e.g., in Egypt), chronic arsenic ingestion from contaminated well water (e.g., in rural Taiwan), or tobacco use. A gene on chromosome 11 reverses in vitro genetic damage of urinary bladder carcinoma cells manifested as micronucleus induction by promoter-activated neutrophils, similar to what might occur after schistosomiasis infection (Rosin MP et al.: Inflammation, chromosomal instability, and cancer—the schistosomiasis model. Cancer Res. 54:1929–33, 1994). Finally, a number of polymorphic metabolic enzymes could predispose certain individuals to bladder cancer (e.g., P450IIC, P450IID6, P450IIA4, *N*-acetyl transferase, and glutathione *S*-transferase) (Silverman DT et al.: Epidemiology of bladder cancer. Hem./Onc. Clin. North Am. 6:1–31, 1992). A study of 296 patients led to the conclusion that heritable glutathione *S*-reductase deficiency is responsible for 17% of bladder cancers (etiologic fraction, 95%; confidence limits, 2–30%) (Brockmoller J et al.: Glutathione S-transferase M1 and its variants A and B as host factors of bladder cancer susceptibility: a case-control study. Cancer Res. 54:4103–11, 1994). Studies of loss of heterozygosity suggest specific genotypic patterns with progressive stages of bladder cancer, akin to those seen in colorectal carcinoma (Dalbagni G et al.: Genetic alterations in bladder cancer. Lancet 342:469–71, 1993). Aberrant Mdm2 and p53 protein phenotypes were seen excessively in 87 bladder tumors (Lianes P et al.: Altered patterns of *MDM2* and *TP53* expression in human bladder-cancer. JNCI 86:1325-1330, 1994).

A number sign (#) is used with this entry because mutations at several loci, e.g., HRAS (190020) and RB1 (180200), have been implicated as the cause of bladder cancer.

Fraumeni and Thomas (1967) observed affected father and 3 sons. I have encountered 2 instances of affected father and son (P0135 and P7658). McCullough et al. (1975) found transitional cell carcinoma in 6 persons in 3 sibships of 2 generations of a kindred. Goldfarb et al. (1982) studied the DNA from T24, a cell line derived from a human bladder carcinoma, which can induce the morphologic transformation of nonmalignant cells. The gene responsible for this transformation was cloned by techniques of gene rescue. It was shown to be human in origin and less than 5 kb long. Blot analysis showed extensive restriction endonuclease polymorphism near this gene in human DNAs. See Bishop (1982) for a discussion of oncogenes. By Southern blot analysis of human-rodent hybrid cell DNA, de Martinville et al. (1983) found that the cellular homolog of the transforming DNA sequence isolated from the bladder carcinoma line EJ is located on the short arm of chromosome 11. The locus also contains sequences homologous to the Harvey ras oncogene. No evidence of gene amplification was found. These workers also found karyologically 'a complex rearrangement of the short arm in two of the four copies of chromosome 11 present in this heteroploid cell line' (EJ). Region 11p15 was the site of a breakpoint in a t(3;11) translocation found in tumor cells from a patient with hereditary renal cell carcinoma (144700). Shih et al. (1981) found that DNA from mouse and rabbit bladder cancers as well as from a human bladder cancer cell line (EJ) induced foci of transformed cells when applied to monolayer cultures of NIH 3T3 cells. In a study of loss of heterozygosity (LOH), Shipman et al. (1993) found no evidence of deletion at 17p13, the region known to contain the p53 tumor suppressor gene (TP53; 191170). Analysis of LOH at 11p13, a region containing the Wilms tumor suppressor gene (WT1; 194070), showed deletion at the CAT locus (115500) in 13 of 18 bladder cancers (72%), at the WT1 locus in 7 of 14 (50%), and at the FSHB locus (136530) in 6 of 16 (38%). Risch et al. (1995) demonstrated that the slow N-acetylation genotype (NAT2; 243400) is a susceptibility factor in occupational and smoking-related bladder cancer. Employing PCR-based genotyping, they investigated NAT2 type among 189 Caucasian bladder cancer patients attending a clinic in Birmingham, U.K. The results were compared to those from an age-matched nonmalignant Caucasian control population from the same region. Risch et al. (1995) found a significant excess of genotypic slow acetylators in patients exposed to arylamines as a result of their occupation or cigarette use. A higher proportion of slow acetylators was also found in most bladder cancer patients without identified exposure to arylamines when compared to the nonmalignant controls.

Patients with cancer of the urinary bladder often present with multiple tumors appearing at different times and at different sites in the bladder. This observation had been attributed to a 'field defect' in the bladder that allowed the independent transformation of epithelial cells at a number of sites. Sidransky et al. (1992) tested this hypothesis with molecular genetic techniques and concluded that in fact multiple bladder tumors are of clonal origin. A number of bladder tumors can arise from the uncontrolled spread of a single transformed cell. These tumors can then grow independently with variable subsequent genetic alterations.

Hruban et al. (1994) did a retrospective molecular genetic analysis of the bladder carcinoma that was the cause of death in the case of Hubert H. Humphrey (1911-1978), U.S. senator and vice president. In 1967, hematuria led to a diagnosis of chronic proliferative cystitis. Although urine cytology at that time was thought by one prominent cytopathologist to be diagnostic of carcinoma, a diagnosis of infiltrating carcinoma of the bladder was not made until August 1976. Hruban et al. (1994) analyzed both the invasive bladder carcinoma resected in 1976 and the filters prepared from urine in 1967. Both showed a transversion from adenine to thymine in codon 227, creating a cryptic splice site in exon 7 of the p53 gene. The mutation resulted in the loss of several amino acids and in the production of a shortened, mutant p53 protein. This mutation was not present in nonneoplastic tissue of the resected bladder.

Deletions involving chromosome 9 represent the most frequent genetic change identified in bladder tumors. Several independent studies had reported overall deletion frequencies of 50 to 70% in large series of tumors. It was of particular interest that these deletions were present at similar frequency in bladder tumors of all grades and stages (Tsai et al., 1990). This finding of chromosome 9 deletions as the sole genetic change in many low-grade, early-stage tumors suggests that it may represent an early or initiating genetic event. Keen and Knowles (1994) used a panel of 22 highly informative microsatellite markers, evenly distributed along chromosome 9, to analyze LOH in 95 cases of primary transitional cell carcinoma of the bladder. In 49 tumors (53%), LOH was demonstrated at one or more loci. Of these 49, 30 had LOH at all informative loci, indicating probable monosomy 9. Subchromosomal deletions were found in 19 tumors (22%), 5 of 9p only, 9 of 9q only, and 5 of both 9p and 9q with a clear region of retention of heterozygosity between. The patterns of LOH in these tumors indicated a common region of deletion on 9p between D9S126 (9p21) and the interferon-alpha cluster (IFNA; 147660) located also at 9p21. A single tumor showed a second site of deletion on 9p telomeric to IFNA, indicating the possible existence of 2 target genes on 9p. All deletions of 9q were large, with a common region of deletion between D9S15 (9q13-q21.1) and D9S60 (9q33-q34.1). The results provided evidence for the simultaneous involvement of distinct suppressor loci on 9p and 9q in bladder carcinoma.

In a review of case reports and epidemiologic studies in the literature, Kiemeney and Schoenberg (1996) concluded that first-degree relatives have an increased risk for transitional cell carcinoma by a factor of 2. Familial clustering of smoking did not appear to be the cause of this increased risk.

In Iceland, Kiemeney et al. (1997) studied the first- to third-degree relatives of 190 patients with bladder, ureter, or renal pelvis transitional cell carcinoma diagnosed between 1983 and 1992. In 41 of the 190 pedigrees, at least 1 relative had transitional cell carcinoma of the urinary tract. Of the probands, 38 had only 1 and 3 had 2 affected relatives. The prevalence of family history of transitional cell carcinoma was 3% in first-degree and 10% in first- or second-degree relatives. The risk of transitional cell carcinoma among all relatives was slightly elevated, the observed-to-expected ratio being greater among second- and third-degree relatives than among first-degree relatives. Kiemeney et al. (1997) concluded that the greater risk among distant relatives argues against the existence of a hereditary subtype of bladder transitional cell carcinoma, at least in the founder population of Iceland.

See also Krontiris and Cooper (1981); Leklem and Brown (1976); Lynch et al. (1979); Mahboubi et al. (1981)

Bishop, J. M.: Oncogenes. Sci. Am. 246(3): 80-92, 1982.

de Martinville, B.; Giacalone, J.; Shih, C.; Weinberg, R. A.; Francke, U.: Oncogene from human EJ bladder carcinoma is located on the short arm of chromosome 11. Science 219: 498-501, 1983.

Fraumeni, J. F., Jr.; Thomas, L. B.: Malignant bladder tumors in a family. J.A.M.A. 201: 507-509, 1967.

Goldfarb, M.; Shimizu, K.; Perucho, M.; Wigler, M.: Isolation and preliminary characterization of a human transforming gene from T24 bladder carcinoma cells. Nature 296: 404-409, 1982.

Hruban, R. H.; van der Riet, P.; Erozan, Y. S.; Sidransky, D.: Molecular biology and the early detection of carcinoma of the bladder: the case of Hubert H. Humphrey. New Eng. J. Med. 330: 1276-1278, 1994.

Keen, A. J.; Knowles, M. A.: Definition of two regions of deletion on chromosome 9 in carcinoma of the bladder. Oncogene 9: 2083-2088, 1994.

Kiemeney, L. A.; Moret, N. C.; Witjes, J. A.; Schoenberg, M. P.; Tulinius, H.: Familial transitional cell carcinoma among the population of Iceland. J. Urol. 157: 1649-1651, 1997.

Kiemeney, L. A. L. M.; Schoenberg, M.: Familial transitional cell carcinoma. J. Urol. 156: 867-872, 1996.

Krontiris, T. G.; Cooper, G. M.: Transforming activity of human tumor DNAs. Proc. Nat. Acad. Sci. 78: 1181-1184, 1981.

Leklem, J. E.; Brown, R. R.: Abnormal tryptophan metabolism in a family with a history of bladder cancer. J. Nat. Cancer Inst. 56: 1101-1104, 1976.

Lynch, H. T.; Walzak, M. P.; Fried, R.; Domina, A. H.; Lynch, J. F.: Familial factors in bladder carcinoma. J. Urol. 122: 458-461, 1979.

Mahboubi, A. O.; Ahlvin, R. C.; Mahboubi, E. O.: Familial aggregation of urothelial carcinoma. J. Urol. 126: 691-692, 1981.

McCullough, D. L.; Lamm, D. L.; McLaughlin, A. P., III; Gittes, R. F.: Familial transitional cell carcinoma of the bladder. J. Urol. 113: 629-635, 1975.

Risch, A.; Wallace, D. M. A.; Bathers, S.; Sim, E.: Slow N-acetylation genotype is a susceptibility factor in occupational and smoking related bladder cancer. Hum. Molec. Genet. 4: 231-236, 1995.

Shih, C.; Padhy, L. C.; Murray, M.; Weinberg, R. A.: Transforming genes of carcinomas and neuroblastomas introduced into mouse fibroblasts. Nature 290: 261-264, 1981.

Shipman, R.; Schraml, P.; Colombi, M.; Raefle, G.; Ludwig, C. U.: Loss of heterozygosity on chromosome 11p13 in primary bladder carcinoma. Hum. Genet. 91: 455-458, 1993.

Sidransky, D.; Frost, P.; Von Eschenbach, A.; Oyasu, R.; Preisinger, A. C.; Vogelstein, B.: Clonal origin of bladder cancer. New Eng. J. Med. 326: 737-740, 1992.

Tsai, Y. C.; Nichols, P. W.; Hiti, A. L.; Williams, Z.; Skinner, D. G.; Jones, P. A.: Allelic losses of chromosomes 9, 11, and 17 in human bladder cancer. Cancer Res. 50: 44-47, 1990.

Ovarian Tumors

FAMILIAL OVARIAN TUMOR (SUPPRESSION OF TUMORIGENICITY 8) (*167000)

JJM NOTE

The genetics of ovarian tumors is proving as complex as that of breast and colon cancer. The breast cancer gene on chromosome 17q (*BRCA1*) significantly contributes to ovarian carcinoma, because most of the linked families are not pure breast cancer, but a combination of the two types. *BRCA2* likewise predisposes to ovarian adenocarcinoma, but to a lesser degree. Such families underscore the principle that those tumor types that occur to excess as multiple primary tumors in the same person also tend to aggregate in families. Other syndromic associations are with hereditary nonpolypotic colorectal cancer due to mutations of *hMSH2* (120435), tylosis with esophageal cancer (148500), familial testicular tumors (273300), perhaps the estrogen receptor 1 (133430), and the Cowden multiple hamartoma syndrome (158350) due to mutations in the *PTEN* gene (601728). Specific ovarian cell types are peculiar to Peutz-Jeghers syndrome (175200) with granulosa cell tumors and the nevoid basal cell carcinoma syndrome (109400) with ovarian fibromas. Screening women at high risk is problematic in two ways: the imperfect means of screening and the malignant potential of the remaining peritoneal lining, even after the at-risk tissue, the ovary, is removed by a so-called prophylactic oophorectomy. The current state of screening suggests merit in transvaginal ultrasound examinations, perhaps supplemented by assay of tumor marker Ca-125 in serum and color flow Doppler (van Nagell JR, DePriest PD: Ovarian cancer screening. In Hoskins WJ et al., eds. *Principles and Practice of Gynecology Oncology Updates,* Vol. 1, 1:1-5, Philadelphia: Lippincott 1993, pp. 1–5). A 1994 NIH Consensus Development Conference determined that such screening has not proved effective in the general population.

Jackson (1967) reported a Jamaican family in which grandmother, mother and daughter (i.e., members of 3 generations) developed ovarian tumors which in 2 were known to have been dysgerminomas. Lewis and Davison (1969) described a family in which 5 sisters (out of 6) and their mother had ovarian cancer. One of the 5 had a malignant ovarian cyst but subsequently died of colonic cancer. Prophylactic oophorectomy was performed in the sixth sister and in 5 females of the following generation. Liber (1950) described a family with histologically proven papillary adenocarcinoma of the ovary in 5 sisters and their mother. Li et al. (1970) reported a family in which 7 women (4 of them sisters) were proved to have ovarian carcinoma and this form of malignancy was suspected in 3 others. Ovarian tumors also occur in the Peutz-Jeghers syndrome (175200), in cases of gonadal dysgenesis in which XY cells are present, and in the basal cell nevus syndrome (109400). The ovarian tumor in the Peutz-Jeghers syndrome is characteristically of granulosa cell type (Scully, 1970). Nevo (1978) described 2 families with multiple cases of ovarian papillary adenocarcinoma. In 1 family the tumor was detected in 4 females, of whom 2 had had breast cancer before the development of ovarian cancer. Philip (1979) described a family with multiple cases of poorly differentiated cystadenocarcinoma of the ovary. The 4 relatives with ovarian carcinoma were the proband's mother, maternal aunt, that woman's daughter, and the daughter of another maternal aunt. Lynch et al. (1981) found ovarian cancer in 8 close relatives. Transmission occurred through males in 3 instances; 1 was free of cancer. An apparent increase in cancer at other sites was noted. Among 28 women in 16 families at high risk of ovarian carcinoma, in whom prophylactic oophorectomy was performed, 3 subsequently developed disseminated intraabdominal malignancy (Tobacman et al., 1982). The primary site was uncertain despite extensi ve investigations. The tumors were indistinguishable histopathologically from ovarian carcinoma. The authors concluded that in ovarian-cancer-prone families the susceptible tissue is not limited to the ovary, but includes other derivatives of the coelomic epithelium, from which primary peritoneal neoplasms may arise. Lynch et al. (1986) expanded on this hypothesis and referred to the condition as familial peritoneal ovarian carcinomatosis. Fraumeni et al. (1975) reported 6 families with multiple cases of ovarian cancer, mainly serous cystadenocarcinoma. In 3 of the 6, breast cancer was also aggregated. Prophylactic oophorectomy was performed in 14 asymptomatic women from 4 of the families. Review of the microscopic sections from 8 women showed that 3, representing 2 families, had abnormalities of ovarian surface epithelium and mesothelial tissue, which may be of etiologic significance and portend neoplastic changes. Schildkraut et al. (1989) found a significant genetic correlation between ovarian and breast cancer. On the other hand, evidence for a significant genetic overlap between endometrial cancer and either ovarian or breast cancer was not found. In a multicenter population-based case-control study of 493 women aged 20 to 54 who had been newly diagnosed with epithelial ovarian cancer, Schildkraut and Thompson (1988) found that the odds ratios for ovarian cancer in first- and second-degree relatives were 3.6 and 2.9, respectively, compared with women with no family history of ovarian cancer. The null hypothesis of no association was excluded on both the maternal and paternal sides of the families studied. Among 310 Israeli Jewish women with ovarian cancer of epithelial origin, Menczer and Ben-Baruch (1991) found 24 distributed in 8 families with multiple cases. Of first-degree relatives of these probands, 5 underwent prophylactic oophorectomy, and early ovarian carcinoma was found in 1. Evans et al. (1992) described a family with apparent dominant inheritance of ovarian cancer; 12 female family members had undergone bilateral prophylactic oophorectomy and been given hormone replacement therapy.

Whang-Peng et al. (1984) studied the chromosomes in 72 patients with various forms of ovarian carcinoma; banding techniques were successful in 44. They could not demonstrate specificity of a chromosome 6 deletion (6q-) for ovarian cancer. To define a region on 6q containing a putative tumor suppressor gene for ovarian cancer, Saito et al. (1992) examined loss of heterozygosity (LOH) in 70 ovarian tumors of 3 histologic types, using 9 RFLP markers located at 6q24-q27. Among 33 cancers of serous type that were informative at one or more loci, 17 showed allelic loss at a few or all loci examined, whereas only 1 of 15 mucinous-type tumors and 2 of 12 clear-cell tumors showed LOH. A detailed deletion map narrowed the region to that between 2 markers estimated to be 1.9 cM apart on the basis of linkage analysis. Eccles et al. (1990) found LOH on both 17q and 17p in epithelial ovarian carcinoma. See 164870 for implication of the HER2 gene in ovarian cancer. The HER gene, also symbolized ERBB2, maps to 17q12-q21. Godwin et al. (1994) examined normal and tumor DNA samples from 32 patients with sporadic and 8 patients with familial forms of epithelial ovarian tumors. Evaluation of a set of markers positioned telomeric to BRCA1 (113705) resulted in the highest degree of LOH, 73% (29/40), indicating that a candidate locus involved in ovarian cancer may reside distal to the BRCA1 gene.

Liede et al. (1998) raised the question of the existence of hereditary site-specific ovarian cancer as a genetic entity distinct from hereditary breast-ovarian cancer syndrome. They identified a large Ashkenazi Jewish kindred with 8 cases of ovarian carcinoma and no cases of breast cancer. In all but 1 of the ovarian cancer cases, 185delAG mutation of the BRCA1 gene (113705.0003) segregated with the cancer. Liede et al. (1998) concluded that site-specific ovarian cancer families probably represent a variant of the breast-ovarian cancer syndrome, attributable to mutation in either BRCA1 or BRCA2 (600185).

Because of the high percentage of LOH reported in the 6q25-q27 region in ovarian carcinomas, Colitti et al. (1998) undertook to define the smallest region of common loss using 12 tandem repeat markers spanning a region no more than 18 cM, located between 6q25.1 and 6q26 in 54 fresh and paraffin-embedded invasive ovarian epithelial tumor tissues. Detailed mapping of 6q in these tumor samples identified a 4-cM minimal region of LOH between markers D6S473 and D6S448 (6q25.1-q25.2). Loss of heterozygosity at D6S473 correlated significantly both with serous versus nonserous ovarian tumors and with high-grade versus low-grade specimens.

See also Donnai and Warrell (1983)

Colitti, C. V.; Rodabaugh, K. J.; Welch, W. R.; Berkowitz, R. S.; Mok, S. C.: A novel 4 cM minimal deletion unit on chromosome 6q25.1-q25.2 associated with high grade invasive epithelial ovarian carcinomas. Oncogene 16: 555-559, 1998.

Donnai, D.; Warrell, D. W.: Familial ovarian cancer: gynaecological and genetic management. J. Med. Genet. 20: 300 only, 1983.

Eccles, D. M.; Cranston, G.; Steel, C. M.; Nakamura, Y.; Leonard, R. C. F.: Allele losses on chromosome 17 in human epithelial ovarian carcinoma. Oncogene 5: 1599-1601, 1990.

Evans, D. G. R.; Ribiero, G.; Warrell, D.; Donnai, D.: Ovarian cancer family and prophylactic choices. J. Med. Genet. 29: 416-418, 1992.

Fraumeni, J. F., Jr.; Grundy, G. W.; Creagan, E. T.; Everson, R. B.: Six families prone to ovarian cancer. Cancer 36: 364-369, 1975.

Godwin, A. K.; Vanderveer, L.; Schultz, D. C.; Lynch, H. T.; Altomare, D. A.; Buetow, K. H.; Daly, M.; Getts, L. A.; Masny, A.; Rosenblum, N.; Hogan, M.; Ozols, R. F.; Hamilton, T. C.: A common region of deletion on chromosome 17q in both sporadic and familial epithelial ovarian tumors distal to BRCA1. Am. J. Hum. Genet. 55: 666-677, 1994.

Jackson, S. M.: Ovarian dysgerminoma in three generations? J. Med. Genet. 4: 112-113, 1967.

Lewis, A. C. W.; Davison, B. C. C.: Familial ovarian cancer. Lancet II: 235-237, 1969.

Li, F. P.; Rapoport, A. H.; Fraumeni, J. F., Jr.; Jensen, R. D.: Familial ovarian carcinoma. J.A.M.A. 214: 1559-1561, 1970.

Liber, A. F.: Ovarian cancer in mother and five daughters. Arch. Path. 49: 280-290, 1950.

Liede, A.; Tonin, P. N.; Sun, C. C.; Serruya, C.; Daly, M. B.; Narod, S. A.; Foulkes, W. D.: Is hereditary site-specific ovarian cancer a distinct genetic condition? Am. J. Med. Genet. 75: 55-58, 1998.

Lynch, H. T.; Albano, W.; Black, L.; Lynch, J. F.; Recabaren, J.; Pierson, R.: Familial excess of cancer of the ovary and other anatomic sites. J.A.M.A. 245: 261-264, 1981.

Lynch, H. T.; Bewtra, C.; Lynch, J. F.: Familial peritoneal ovarian carcinomatosis: a new clinical entity. Med. Hypotheses 21: 171-177, 1986.
Menczer, J.; Ben-Baruch, G.: Familial ovarian cancer in Israeli Jewish women. Obstet. Gynec. 77: 276-277, 1991.
Nevo, S.: Familial ovarian carcinoma: a problem in genetic counselling. Clin. Genet. 14: 219-222, 1978.
Philip, E. E.: Familial carcinoma of the ovary: case report. Brit. J. Obstet. Gynaec. 86: 152-153, 1979.
Saito, S.; Saito, H.; Koi, S.; Sagae, S.; Kudo, R.; Saito, J.; Noda, K.; Nakamura, Y.: Fine-scale deletion mapping of the distal long arm of chromosome 6 in 70 human ovarian cancers. Cancer Res. 52: 5815-5817, 1992.
Schildkraut, J. M.; Risch, N.; Thompson, W. D.: Evaluating genetic association among ovarian, breast, and endometrial cancer: evidence for a breast/ovarian cancer relationship. Am. J. Hum. Genet. 45: 521-529, 1989.
Schildkraut, J. M.; Thompson, W. D.: Familial ovarian cancer: a population-based case-control study. Am. J. Epidemiol. 128: 456-466, 1988.
Scully, R. E.: Sex cord tumor with annular tubules—a distinctive ovarian tumor of the Peutz-Jeghers syndrome. Cancer 25: 1107-1121, 1970.
Tobacman, J. K.; Greene, M. H.; Tucker, M. A.; Costa, J.; Kase, R.; Fraumeni, J. F., Jr.: Intra-abdominal carcinomatosis after prophylactic oophorectomy in ovarian-cancer-prone families. Lancet II: 795-797, 1982.
Whang-Peng, J.; Knutsen, T.; Douglass, E. C.; Chu, E.; Ozols, R. F.; Hogan, W. M.; Young, R. C.: Cytogenetic studies in ovarian cancer. Cancer Genet. Cytogenet. 11: 91-106, 1984.

ARRHENOBLASTOMA—THYROID ADENOMA (*107950)

JJM NOTE

Additional cases are needed to clarify a rare but generally benign ovarian tumor with a common benign neoplasm of the thyroid gland. Obviously, when a patient presents with one tumor in the association, the possible presence of the other tumor should be considered. The association points out the lead to understanding human malignant neoplasms by studying benign neoplasms, an underdeveloped strategy, in my opinion. A cardinal pathologic feature of arrhenoblastoma is the appearance of immature testicular tubules. What tumor in a male could be considered equivalent to arrhenoblastoma in a female?

Jensen et al. (1974) described ovarian tumors in a mother and 2 daughters. The tumor proved to be arrhenoblastoma in the 2 daughters. Thyroid adenomas occurred in several members of the family and were found to be associated frequently with ovarian arrhenoblastoma in young women surveyed separately. See 166970 and 167000. O'Brien and Wilansky (1981) described a family in which the 16-year-old proband had a nodular thyroid and a functioning ovarian arrhenoblastoma. Males and females to a total of 6 in 4 generations were known to have nodular thyroids. The disorder was apparently transmitted through an unaffected male. The authors raised the question of testicular tumors in males with the gene.

Jensen, R. D.; Norris, H. J.; Fraumeni, J. F., Jr.: Familial arrhenoblastoma and thyroid adenoma. Cancer 33: 218-223, 1974.
O'Brien, P. K.; Wilansky, D. L.: Familial thyroid nodulation and arrhenoblastoma. Am. J. Clin. Path. 75: 578-581, 1981.

OVARIAN FIBROMAS (166970)

JJM NOTE

This family could be revisited with the chromosome 9 markers that are linked to the gene locus for the nevoid basal cell carcinoma syndrome to test the hypothesis that it is an allele of that disease gene.

Dumont-Herskowitz et al. (1978) reported ovarian fibromata in 6 women in 5 sibships of 4 generations of a family. In most, presentation was in childhood. The pedigree was consistent with either autosomal or X-linked dominant inheritance with, of course, sex limitation. Ovarian tumors occur with the Peutz-Jeghers syndrome (175200) and with the basal cell nevus syndrome (109400). In the latter condition the ovarian tumors are fibromata and undergo calcification. No stigmata of basal cell nevus syndrome was found in the family reported by Dumont-Herskowitz et al. (1978).

Dumont-Herskowitz, R. A.; Safaii, H. S.; Senior, B.: Ovarian fibromata in four successive generations. J. Pediat. 93: 621-624, 1978.

OVARIAN TERATOMA (DERMOID CYST) (166950)

JJM NOTE

My colleagues from Pittsburgh, Pennsylvania, applied the strategy of mapping genes using teratomas to DNA markers on chromosome 1 (Surti U et al.: Genetics and biology of human teratomas. I. Cytogenetic analysis and mechanism of origin. Am. J. Hum. Genet. 47:635–43, 1990; Hoffner L et al.: Genetics and biology of human ovarian teratomas. III. Cytogenetics and origins of malignant ovarian germ cell tumors. Cancer Genet. Cytogenet. 62:58–65, 1992).

Dermoid cysts are generally benign cystic tumors comprised predominantly of ectodermal elements. However, endodermal and mesodermal elements also may be included. These cysts are often filled with hair, skin, teeth, bones, neural tissue and sebaceous material. Plattner and Oxorn (1973) reported the presence of bilateral ovarian dermoid cysts in a mother and her only 2 daughters. This was the first report of such bilateral ovarian dermoid cysts occurring in consecutive generations of a family. These bilateral teratomas were surgically removed from the mother at age 22 and from both her daughters at age 23. The authors stated that the literature contained 18 cases in 6 families with a familial occurrence of dermoid cysts. Hecht et al. (1976) suggested that there may be an important genetic factor in ovarian teratomas. They observed affected grandmother and granddaughter and pointed to early onset, bilaterality (in 10 to 25% of patients) and the demonstrable genetic factor in mice (Stevens and Varnum, 1974) as supporting evidence.

Ovarian teratomas originate through failure of extrusion of the second polar body or refusion of it with the ovum, i.e., self-fertilization. This is a conclusion based on study of chromosome and biochemical polymorphism in teratoma cells (Linder et al., 1975). (Testicular teratomas may have a different origin.) Study of biochemical traits in cells of teratomas is a means of 'centromere mapping.' The farther from the centromere a given locus is situated, the higher is the proportion of teratomas from hosts heterozygous at that locus which will be heterozygous only. Loci close to the centromere will be homozygous; loci farther from the centromere will have a chance to be heterozygous, up to a limiting proportion of two-thirds. The last follows from the fact that a given chromatid has (at the four-strand stage) three other chromatids with which it can crossover. In a heterozygous host two of the three have a different allele at the given locus.

See also Simon et al. (1985)

Hecht, F.; McCaw, B. K.; Patil, S.: Ovarian teratomas and genetics of germ-cell formation. (Letter) Lancet II: 1311 only, 1976.
Linder, D.; McCaw, B. F.; Hecht, F.: Parthenogenic origin of benign ovarian teratomas. New Eng. J. Med. 292: 63-66, 1975.
Plattner, G.; Oxorn, H.: Familial incidence of ovarian dermoid cysts. Canad. Med. Assoc. J. 108: 892-893, 1973.
Simon, A.; Ohel, G.; Neri, A.; Schenker, J. G.: Familial occurrence of mature ovarian teratomas. Obstet. Gynec. 66: 278-279, 1985.
Stevens, L. C.; Varnum, D. S.: The development of teratomas from parthenogenetically activated ovarian mouse eggs. Dev. Biol. 37: 369-380, 1974.

SACRAL AGENESIS, HEREDITARY, WITH PRESACRAL MASS, ANTERIOR MENINGOCELE, AND/OR TERATOMA, AND ANORECTAL MALFORMATION (#176450)

JJM NOTE

This disorder is another example of the overlap of teratogenesis and carcinogenesis that is usually sporadic, but occasionally familial. The presacral mass can be a teratoma, which can become malignant.

Description

A number sign (#) is used with this entry because of evidence that, in at least some cases, the sacral agenesis syndrome (Currarino triad) is caused by mutation in a homeo box gene, HLXB9 (142994).

Hereditary sacral agenesis is an autosomal dominant condition with variable expressivity. Associated features include a presacral mass and urogenital and anorectal anomalies. Linkage analysis has shown that the SCRA1 gene maps to 7q36.

Clinical Features

The hallmark of this autosomal dominant condition is the presence of partial sacral agenesis with intact first sacral vertebra. Association with a presacral mass and anorectal malformation is known as the Currarino triad. Of 10 affected members in the family reported by O'Riordain et al. (1991), only 4 were symptomatic, 3 of these had the full Currarino triad (sacral agenesis, presacral mass, and anorectal malformation), and the fourth had an anterior meningocele with no anorectal problems. All 10 had evidence of partial sacral agenesis on x-ray. Constipation and perianal sepsis are common complaints. Ascending infection resulting in E. coli bacterial meningitis has been reported. Associated malformations include rectovaginal fistula, tethering of the cord, duplex ureter, hydronephrosis, vesicoureteral reflux, bicornuate uterus, and neurogenic bladder. Malignant degeneration of the presacral teratoma has also been reported (Ashcraft and Holder, 1974 and Yates et al., 1983).

Inheritance

Ashcraft and Holder (1974) obtained clear evidence of autosomal dominant inheritance in this condition which has variable expressivity. They described 2 families with 23 affected members.

Mapping

Fellous et al. (1982) described a large French family where 5 of 15 affected members also had spina bifida aperta in association with sacral agenesis. They could find no evidence for linkage to the T-locus using HLA phenotyping. Chatkupt et al. (1994) excluded linkage to the HLA-locus in a large North American pedigree with 40 affected members. Lynch et al. (1995) demonstrated linkage at 7q36 in the families reported by O'Riordain et al. (1991) and Nour et al. (1989). The region 7q36 was considered a good candidate region because of several case reports describing sacral agenesis in association with 7q deletion. The gene maps to the same region as HPE3 (142945), a gene for autosomal dominant holoprosencephaly. Deletions involving 7q are also associated with holoprosencephaly suggesting that haploinsufficiency at this locus predisposes to both disorders. Both disorders are midline embryonic defects. The authors also pointed to the fact that the incidence of both disorders are increased in infants born to diabetic mothers. Lynch et al. (1995) suggested that these 2 conditions are possibly contiguous gene syndromes or even allelic variants.

Nagai et al. (1994) described a Japanese male with typical Currarino triad who showed partial trisomy of 13q and 20p inherited from his mother, suggesting that the disorder may be heterogeneous and that a second locus may be present on 13q or 20p.

Lynch et al. (1995) obtained a maximum lod score of 4.38 at theta = 0.0 with marker D7S559.

Ross et al. (1998) demonstrated mutations in the homeo box gene HLXB9 (142994) as the cause of the sacral agenesis syndrome. The families they studied illustrated a wide range of variability of this autosomal dominant disorder. In the families they studied, some individuals were asymptomatic, although x-ray revealed typical scimitar sacrum or sickle-shaped hemisacrum, whereas others had the full Currarino triad.

Ashcraft, K. W.; Holder, T. M.: Hereditary presacral teratoma. J. Pediat. Surg. 9: 691-697, 1974.

Chatkupt, S.; Speer, M. C.; Ding, Y.; Thomas, M.; Stenroos, E. S.; Dermody, J. J.; Koenigsberger, M. R.; Ott, J.; Johnson, W.G.: Linkage analysis of a candidate locus (HLA) in autosomal dominant sacral defect with anterior meningocele. Am. J. Med. Genet. 52: 1-4, 1994.

Fellous, M.; Boue, J.; Malbrunot, C.; Wollman, E.; Sasportes, M.; Van Cong, N.; Marcelli, A.; Rebourcet, R.; Hubert, C.; Demenais, F.; Elston, R. C.; Namboodiri, K. K.; Kaplan, E. B.: A five-generation family with sacral agenesis and spina bifida: possible similarities with the mouse t-locus. Am. J. Med. Genet. 12: 465-487, 1982.

Lynch, S. A.; Bond, P. M.; Copp, A. J.; Kirwan, W. O.; Nour, S.; Balling, R.; Mariman, E.; Burn, J.; Strachan, T.: A gene for autosomal dominant sacral agenesis maps to the holoprosencephaly region at 7q36. Nature Genet. 11: 93-95, 1995.

Nagai, T.; Katoh, R.; Hasegawa, T.; Ohashi, H.; Fukushima, Y.: Currarino triad (anorectal malformation, sacral bony abnormality and presacral mass) with partial trisomy of chromosomes 13q and 20p. (Letter) Clin. Genet. 45: 22-273, 1994.

Nour, S.; Kumar, D.; Dickson, J.: Anorectal malformations with sacral bony abnormalities.. Arch. Dis. Child. 64: 1618-1620, 1989.

O'Riordain, D. S.; O'Connell, P. R.; Kirwan, W. O.: Hereditary sacral agenesis with presacral mass and anorectal stenosis: the Currarino triad. Br. J. Surg. 78: 536-538, 1991.

Ross, A. J.; Ruiz-Perez, V.; Wang, Y.; Hagan, D.-M.; Scherer, S.; Lynch, S. A.; Lindsay, S.; Custard, E.; Belloni, E.; Wilson, D. I.; Wadey, R.; Goodman, F.; Orstavik, K. H.; Monclair, T.; Robson, S.; Reardon, W.; Burn, J.; Scambler, P.; Strachan, T.: A homeobox gene, HLXB9, is the major locus for dominantly inherited sacral agenesis. Nature Genet. 20: 358-361, 1998.

Yates, V. D.; Wilroy, R. S.; Whitington, L.; Simmons, J. C. H.: Anterior sacral defects: an autosomal dominantly inherited condition. J. Pediat. 102: 239-242, 1983.

SPINA BIFIDA (182940)

JJM NOTE

The occasional association with teratoma justifies this entry.

Because spina bifida and anencephaly are generally considered one entity, see also anencephaly (206500). Lorber (1965) suggested recessive inheritance. However, penetrance must be greatly reduced because he estimated the risk of recurrence of spina bifida cystica, anencephaly or hydrocephalus in subsequently born offspring to be about 8%. Record and McKeown (1950) had estimated the risk at 4%. Taking spina bifida and anencephaly, Carter and Roberts (1967) estimated the risk in England of a third child having major central nervous system malformation, two having previously been affected, to be about 1 in 10. Lorber and Levick (1967) found spina bifida occulta in 14.3% of 188 mothers and 26.8% of 179 fathers of cases, and in 5% of 200 controls. Spina bifida occulta was not more common among parents with more than one affected child, and, in a majority of families, neither parent had it. Carter et al. (1976) studied families of cases of spinal dysraphism. (This is a term used by neurosurgeons to include a condition in which the conus has an abnormal attachment to neighboring structures and may be elongated, with one or more of a variety of anomalies of the spinal cord, vertebrae, and skin. These include diastematomyelia, intradural or extradural lipoma, spina bifida occulta, hairy patch or dimple. Neurologic signs in the lower limbs and urinary incontinence may be associated.) Carter et al. (1976) found as high a frequency of anencephaly and spina bifida cystica among the sibs of spinal dysraphism cases as among the sibs of patients who themselves have these neural tube defects.

From studies in a large kindred, Amos et al. (1975) concluded that there may be a gene locus for spina bifida occulta and asymmetry (symbolized S) linked to the HLA complex. Interest in this possibility was stimulated by the T region of the mouse which determines morphogenesis, especially of the tail, and is linked to the H-2 region on mouse chromosome 17. Mendell et al. (1979) studied 2 extensive North Carolina kindreds suggesting that spina bifida occulta and/or asymmetry of the facet joints is determined by a gene at a locus linked to HLA. The total lod score was 2.21 at theta = 0.05. Bobrow et al. (1975) and de Bruyere et al. (1977) found no linkage with HLA; Fellous et al. (1979) found a suggestion of linkage. Vannier et al. (1981) found a high frequency of spina bifida occulta and lateral asymmetry of the lumbosacral spine in sibs and parents. They could, however, show 'no association between these abnormalities and the HLA system.' Sacral agenesis (also called the caudal regression syndrome) in general appears to have a low recurrence risk.

The genetics of spina bifida was studied by Blumel et al. (1959) and Banta and Nichols (1969). Robert et al. (1974) observed 6 cases in 2 families, one of which suggested irregular dominant inheritance and the other recessive inheritance. A pedigree consistent with autosomal dominant inheritance was published by Say and Coldwell (1975). Finer et al. (1978) observed hypoplasia of the caudal end of the spine and associated anomalies in 2 male sibs who also had congenital heart disease. Features overlapped with the VATER association. Sacral agenesis is often attributable to maternal diabetes. Stewart and Stoll (1979) reported a family in which a diabetic woman gave birth to an affected girl and boy. Welch and Aterman (1984) suggested that the syndrome of caudal dysplasia (CDS), which has maternal diabetes as an etiologic factor (Passarge and Lenz, 1966), should be distinguished from familial sacral dysgenesis, of which they tentatively identified 3 types, all autosomal dominant in their view. Two of the forms involve some degree of 'hemisacrum.' Their familial hemisacrum, type I, is the disorder reported by Cohn and Bay-Nielsen (1969), which is considered to be X-linked dominant (312800) (although the pedigree was equally consistent with autosomal dominance). Kenefick (1973) reported the same disorder. Teratoma was not associated in these cases, but anterior sacral meningocele was relatively frequent. Their familial hemisacrum, type II, is sacral dysgenesis with presacral teratoma as reported by Ashcraft et al. (see 176450). Familial partial sacral agenesis was the Welch-Aterman designation for the disorder that shows some degree of partial sacral agenesis but no characteristic hemisacrum as in the first two familial types.

Margulies et al. (1982) used recombinant DNA techniques to study the family of genes encoding H-2-like molecules in the mouse. They concluded that there are 10 to 15 H-2-like genes in the murine genome, most of them on chromosome 17, and that at least 3 of these genes map outside the MHC in the Tla region. This may be the approach necessary for demonstration of a t-locus in man and mutation therein in certain developmental abnormalities. (Opitz (1982) emphasized the distinction between spina bifida/anencephaly and sacral agenesis with or without spina bifida occulta or aperta. The latter has the greatest evidence of autosomal dominant inheritance and is more likely to be the human equivalent of the t-locus of the mouse.) Erickson (1988) reviewed arguments for the assertion that there is no HLA-linked equivalent of the t-complex of the mouse. He emphasized that the essential attributes of the murine t-complex are distortion of prezygotic transmission ratios and suppression of crossover rather than developmental recessive mutations leading to embryonic arrest. Fellous et al. (1982) reported in detail studies of a 5-generation family with sacral agenesis and spina bifida. Abnormalities ranged from complete absence of the sacrum, with or without spina bifida aperta, to spina bifida occulta. The condition appeared in a man with 4 children who were all affected, and thereafter, to varying degrees, in 17 of his 28 descendants. One hypothesis proposed to account for the finding was a dominant major gene with transmission in excess by heterozygotes, suggesting segregation distortion as for alleles at a t-locus. The only hint of linkage with chromosome 6 loci was with PGM3 (lod score = 1.85 at a recombination fraction of 0.087). Fineman et al. (1982) studied 4 families 'selected randomly on the basis of the occurrence of spina bifida cystica and/or spina bifida occulta in 1 or more family members.' In all, 63 relatives were studied clinically and roentgenographically. Excluding 5 probands, the frequency of all types of spinal/vertebral defects was found to be 30/58 (52%). The distribution fitted autosomal dominant inheritance better than autosomal recessive inheritance or sporadic occurrence. The penetrance probability in the dominant model was 0.749. The authors concluded that the 'occulta' and 'cystica' forms of spina bifida are different expressions of the same dominant gene in these kindreds. Jorde (1983) concluded that linkage to 6p markers can be excluded. Adalsteinsson and Basrur (1984) concluded that spina bifida in Icelandic lambs is an autosomal recessive.

Sacral defect with anterior meningocele (SDAM; 600145) is a form of caudal dysgenesis. It is present at birth and becomes symptomatic later in life, usually because of obstructive labor in females, chronic constipation, rectal fistula and abscess, or meningitis. The inheritance is autosomal dominant. In a 5-generation family with SDAM, Chatkupt et al. (1994) excluded the HLA region as the location of the SDAM gene.

Nickel et al. (1994) reported 3 unrelated children with neural tube defects (NTD) and some additional abnormalities who had deletions of 22q11. Nickel and Magenis (1996) subsequently found that of 295 patients with NTD, 22 had features suggestive of 22q11 deletion. Of 16 children tested, only 3 had cytogenetic or molecular evidence of 22q11.

Fryns et al. (1996) reported a translocation t(X;22)(q27;q121) in a female fetus with lumbosacral spina bifida and no other congenital abnormalities.

Helwig et al. (1995) reported that mice who are doubly heterozygous for the mutants 'undulated' and 'Patch' have a phenotype reminiscent of an extreme form of spina bifida occulta in humans. The unexpected phenotype in double-mutant and not single-mutant mice showed that novel congenital anomalies such as spina bifida can result from interaction between products of independently segregating loci. This is an example of digenic inheritance. (The 'undulated' mutation is sited in the mouse Pax1 gene (PAX1; 167411), and the 'Patch' mutation is related to deletion of the mouse Pdgfra gene (PDGFRA; 173490).)

See also Bennett (1977); Goodfellow and Andrews (1983); Sever (1974); Sever (1983)

Adalsteinsson, S.; Basrur, P. K.: Inheritance of spina bifida in Icelandic lambs. J. Hered. 75: 378-382, 1984.
Amos, D. B.; Ruderman, N.; Mendell, N.; Johnson, A. H.: Linkage between HLA and spinal development. Transplant. Proc. 7: 93-95, 1975.
Banta, J. V.; Nichols, O.: Sacral agenesis. J. Bone Joint Surg. 51A: 693-703, 1969.
Bennett, D.: L. C. Dunn and his contribution to T-locus genetics. Ann. Rev. Genet. 11: 1-12, 1977.
Blumel, J.; Evans, E. B.; Eggers, G. W. N.: Partial and complete agenesis or malformation of the sacrum with associated anomalies. J. Bone Joint Surg. 41A: 497-518, 1959.
Bobrow, M.; Bodmer, J. G.; Bodmer, W. F.; McDevitt, H. O.; Lorber, J.; Swift, P. N.: The search for a human equivalent of the mouse T-locus. Negative results from a study of HLA types in spina bifida. Tissue Antigens 5: 234-237, 1975.
Carter, C. O.; Evans, K. A.; Till, K.: Spinal dysraphism: genetic relation to neural tube malformations. J. Med. Genet. 13: 343-350, 1976.
Carter, C. O.; Roberts, J. A. F.: The risk of recurrence after two children with central-nervous-system malformations. Lancet I: 306-308, 1967.
Chatkupt, S.; Speer, M. C.; Ding, Y.; Thomas, M.; Stenroos, E. S.; Dermody, J. J.; Koenigsberger, M. R.; Ott, J.; Johnson, W. G.: Linkage analysis of a candidate locus (HLA) in autosomal dominant sacral defect with anterior meningocele. Am. J. Med. Genet. 52: 1-4, 1994.
Cohn, J.; Bay-Nielsen, E.: Hereditary defects of the sacrum and coccyx with anterior sacral meningocele. Acta Paediat. Scand. 58: 268-274, 1969.
de Bruyere, M.; Kulakowski, S.; Melchaire, J.; Delire, M.; Sokal, G.: HLA gene and haplotype frequencies in spina bifida: population and family studies. Tissue Antigens 10: 399-402, 1977.
Erickson, R. P.: The 6's and 17's of developmental mutants near the major histocompatibility complex: the mouse t-complex does not have a human equivalent. Am. J. Hum. Genet. 43: 115-118, 1988.
Fellous, M.; Boue, J.; Malbrunot, C.; Wollman, E.; Sasportes, M.; Van Cong, N.; Marcelli, A.; Rebourcet, R.; Hubert, C.; Demenais, F.; Elston, R. C.; Namboodiri, K. K.; Kaplan, E. B.: A five-generation family with sacral agenesis and spina bifida: possible similarities with the mouse T-locus. Am. J. Med. Genet. 12: 465-487, 1982.
Fellous, M.; Hors, J.; Bone, J.; Dausset, J.; Jacob, F.: Are there human analogs of the mouse T-locus in central nervous system malformations?. Birth Defects Orig. Art. Ser. XV(3): 93-104, 1979.
Fineman, R. M.; Jorde, L. B.; Martin, R. A.; Hasstedt, S. J.; Wing, S. D.; Walker, M. L.: Spinal dysraphia as an autosomal dominant defect in four families. Am. J. Med. Genet. 12: 457-464, 1982.
Finer, N. N.; Bowen, P.; Dunbar, L. G.: Caudal regression anomalad (sacral agenesis) in siblings. Clin. Genet. 13: 353-358, 1978.
Fryns, J. P.; Devriendt, K.; Moerman, P.: Lumbosacral spina bifida and myeloschizis in a female foetus with de novo X/autosomal translocation (t(X;22)(q27;q121)). (Letter) Genet. Counsel. 7: 159-160, 1996.
Goodfellow, P. N.; Andrews, P. W.: Is there a human T/t locus?. Nature 302: 657-658, 1983.
Helwig, U.; Imai, K.; Schmahl, W.; Thomas, B. E.; Varnum, D. S.; Nadeau, J. H.; Balling, R.: Interaction between undulated and Patch leads to an extreme form of spina bifida in double-mutant mice. Nature Genet. 11: 60-63, 1995.
Jorde, L.: Personal Communication. Salt Lake City, Utah 9/27/1983.
Kenefick, J. S.: Hereditary sacral agenesis associated with presacral tumors. Brit. J. Surg. 60: 271-274, 1973.
Lorber, J.: The family of spina bifida cystica. Pediatrics 35: 589-595, 1965.
Lorber, J.; Levick, K.: Spina bifida cystica: incidence of spina bifida occulta in parents and in controls. Arch. Dis. Child. 42: 171-173, 1967.
Margulies, D. H.; Evans, G. A.; Flaherty, L.; Seidman, J. G.: H-2-like genes in the Tla region of mouse chromosome 17. Nature 295: 168-170, 1982.
Mendell, N. R.; Ruderman, R. J.; Demenais, F.; Ruderman, J. G.; Johnson, A. H.; Amos, D. B.; Elston, R. C.: The genetics of spinal development in man: similarities with the mouse T locus. (Abstract) Cytogenet. Cell Genet. 25: 184 only, 1979.
Nickel, R. E.; Magenis, R. E.: Neural tube defects and deletions of 22q11. Amer. J. Med. Genet. 66: 25-27, 1996.
Nickel, R. E.; Pillers, D. A.; Merkens, M.; Magenis, R. E.; Driscoll, D. A.; Emanuel, B. S.; Zonana, J.: Velo-cardio-facial syndrome and DiGeorge sequence with meningomyelocele and deletions of the 22q11 region. Am. J. Med. Genet. 52: 445-449, 1994.
Opitz, J. M.: Personal Communication. Helena, Mont. 4/1982.
Passarge, E.; Lenz, W.: Syndrome of caudal regression in infants of diabetic mothers: observation of further cases. Pediatrics 37: 672-679, 1966.
Record, R. G.; McKeown, T.: Congenital malformation of the central nervous system. III. Risk of malformations in sibs of malformed individuals. Brit. J. Prev. Soc. Med. 4: 217-220, 1950.
Robert, J. M.; Pernod, J.; Bonnet, R.: L'agenesie sacro-coccygienne familiale. J. Genet. Hum. 22: 45-60, 1974.
Say, B.; Coldwell, J. G.: Hereditary defect of the sacrum. Humangenetik 27: 231-234, 1975.
Sever, L. E.: A case of meningomyelocele in a kindred with multiple cases of spondylolisthesis and spina bifida occulta. J. Med. Genet. 11: 94-96, 1974.
Sever, L. E.: Spinal anomalies and neural tube defects. (Letter) Am. J. Med. Genet. 15: 343-345, 1983.
Stewart, J. M.; Stoll, S.: Familial caudal regression anomalad and maternal diabetes. J. Med. Genet. 16: 17-20, 1979.
Vannier, J. P.; Lefort, J.; Cavelier, B.; Ledosseur, P.; Assailly, C.; Feingold, J.: Spina bifida cystica families x-ray examination and HLA typing. Pediat. Res. 15: 326-329, 1981.
Welch, J. P.; Aterman, K.: The syndrome of caudal dysplasia: a review, including etiologic considerations and evidence of heterogeneity. Pediat. Path. 2: 313-327, 1984.

SCHINZEL-GIEDION MIDFACE-RETRACTION SYNDROME (269150)

JJM NOTE

This multiple congenital anomaly syndrome clearly has excessive malignancies, mostly sacrococcygeal teratomas, but also hepatoblastomas.

Schinzel and Giedion (1978) described a unique syndrome in brother and sister, who lived 24 hours and 16 months, respectively. Severe midface retraction, multiple skull anomalies (short and sclerotic base, multiple wormian bones, wide cranial sutures and fontanels), congenital heart defect, hydronephrosis, clubfeet, and hypertrichosis were features. Long tubular bones also showed increased density. Another syndrome carries Schinzel's name (181450), and Giedion's is combined with Langer's for a disorder resembling the trichorhinophalangeal syndrome (150230). Despite these possibilities for confusion, I have assigned the above designation for want of a better one. Donnai and Harris (1979) reported a case. Multiple telangiectases were observed over the nose and cheeks. By personal communication and observation, Schinzel (1982) was aware of a total of 8 other unreported cases, all sporadic, including at least 2 offspring of consanguineous parents. Al-Gazali et al. (1988) reported 3 cases. Hypoplasia of dermal ridges and genital anomalies were mentioned as abnormalities. The skeletal abnormalities also included broad ribs and hypoplasia of the distal phalanges. Two of the 7 children reported in the literature died in the neonatal period, and the remainder developed intractable epilepsy and spasticity along with profound growth and developmental retardation. Maclennan et al. (1991) described increasing ventriculomegaly, intraventricular bands, and subependymal pseudocysts in an affected infant born prematurely who died at the age of 9 weeks. Al-Gazali et al. (1990) reviewed 5 previously reported cases and 5 new cases. They stated that consanguinity in fact has not been observed, but that 2 sibs of unlike sex are known. Thus, a dominant disorder with gonadal mosaicism in one parent or an unbalanced structural chromosome abnormality that went undetected by present methods cannot be excluded. Robin et al. (1993) described malignant sacrococcygeal teratoma discovered so on after birth. They pointed out that Burck (1982) described hepatoblastoma in this disorder and suggested that embryonal malignancy may be a common complication. Labrune et al. (1994) added 3 new cases to the 13 previously reported.

Rodriguez et al. (1994) reported the third example of a postmortem study. In addition to having the major manifestations of the syndrome, the 11-month-old boy with SGS had a primitive neuroectodermal tumor in the lumbosacral region, bilateral syndactyly of the hands and feet, and brain anomalies. Rodriguez et al. (1994) commented that of 19 SGS patients reported, 4 had syndactyly and 3 had embryonal tumors; the 3 published necropsy studies showed brain anomalies. They noted that the only familial cases were the affected sibs in the initial report by Schinzel and Giedion (1978). Antich et al. (1995) reported the cases of a brother and sister. The sister had a malignant sacrococcygeal teratoma (the third case of malignancy in Schinzel-Giedion syndrome). The brother died shortly after birth. McPherson et al. (1998) reported a fourth patient with Schinzel-Giedion syndrome and a sacrococcygeal tumor (the previous cases being those of Robin et al. (1993), Rodriguez et al. (1994), and Antich et al. (1995)).

Santos et al. (1994) described a girl with characteristic manifestations of SGS (including typical face, skeletal abnormalities, and hydronephrosis) who also had hypothyroidism and diabetes insipidus. The TSH value in her newborn period was normal. MRI of her brain at 9 months of age showed malformation of the posterior fossa and arachnoid cysts in both temporal regions and in the left cerebellar hemisphere. The brain stem seemed to be pushed backwards by an anterior cyst, and the pituitary stalk was stretched. Santos et al. (1994) suggested that hypothyroidism and diabetes insipidus were secondary to these central defects.

Okamoto et al. (1995) reported a patient they claimed to be the twenty-first in the literature and the first reported instance of the syndrome in an Oriental patient.

A male infant described by Elliott et al. (1996) had, in addition to typical manifestations (characteristic face, choanal stenosis, tricuspid regurgitation, hypospadias), seizures, hearing loss, camptodactyly, and dysplastic sternum. Autopsy disclosed small brain with diminution of white matter (but no evidence of any brain storage), hypoplastic corpus callosum, steatosis of liver, and lipid vacuolization of the zona fasciculata of the adrenals.

See also Kelley et al. (1982)

Al-Gazali, L. I.; Farndon, P.; Burn, J.; Flannery, D. B.; Davison, C.; Mueller, R. F.: The Schinzel-Giedion syndrome. J. Med. Genet. 27: 42-47, 1990.

Al-Gazali, L. I.; Farndon, P. A.; Burn, J.; Massey, R.; Mueller, R. F.: The Schinzel-Giedion syndrome: clinical features and natural history. (Abstract) J. Med. Genet. 25: 275 only, 1988.

Antich, J.; Manzanares, R.; Camarasa, F.; Krauel, X.; Vila, J.; Cusi, V.: Schinzel-Giedion syndrome: report of two sibs. Am. J. Med. Genet. 59: 96-99, 1995.

Burck, U.: Mittelgesichtshypoplasie, Skelettanomalien, Apnoen, Retardierung-eine weitere Beobachtung.In: Tolksdorf, M.; Spranger, J.: Klinishce Genetik in der Padiatrie. Symposium in Kiel, Jahrgang 8, Heft 5 (pub.) 1982.

Donnai, D.; Harris, R.: A further case of a new syndrome including midface retraction, hypertrichosis, and skeletal anomalies. J. Med. Genet. 16: 483-486, 1979.

Elliott, A. M.; Meagher-Villemure, K.; Oudjhane, K.; Der Kaloustian, V. M.: Schinzel-Giedion syndrome: further delineation of the phenotype. Clin. Dysmorph. 5: 135-142, 1996.

Kelley, R. I.; Zackai, E. H.; Charney, E. B.: Congenital hydronephrosis, skeletal dysplasia, and severe developmental retardation: the Schinzel-Giedion syndrome. J. Pediat. 100: 943-946, 1982.

Labrune, P.; Lyonnet, S.; Zupan, V.; Imbert, M.-C.; Goutieres, F.; Hubert, P.; Le Merrer, M.: Three new cases of the Schinzel-Giedion syndrome and review of the literature. Am. J. Med. Genet. 50: 90-93, 1994.

Maclennan, A. C.; Doyle, D.; Simpson, R. M.: Neurosonography and pathology in the Schinzel-Giedion syndrome. J. Med. Genet. 28: 547-549, 1991.

McPherson, E.; Clemens, M.; Hoffner, L.; Surti, U.: Sacral tumors in Schinzel-Giedion syndrome. (Letter) Am. J. Med. Genet. 79: 62-63, 1998.

Okamoto, N.; Takeuchi, M.; Kitajima, H.; Hosokawa, S.: A patient with Schinzel-Giedion syndrome and a review of 20 patients. Jpn. J. Hum. Genet. 40: 189-193, 1995.

Robin, N. H.; Grace, K.; DeSouza, T. G.; McDonald-McGinn, D.; Zackai, E. H.: New finding of Schinzel-Giedion syndrome: a case with a malignant sacrococcygeal teratoma. Am. J. Med. Genet. 47: 852-856, 1993.

Rodriguez, J. I.; Jimenez-Heffernan, J. A.; Leal, J.: Schinzel-Giedion syndrome: autopsy report and additional clinical manifestations. Am. J. Med. Genet. 53: 374-377, 1994.

Santos, H.; Cordeiro, I.; Medeira, A.; Mendonca, E.; Lobo Antunes, N.; Coelho Rosa, F.: Schinzel-Giedion syndrome: a patient with hypothyroidism and diabetes insipidus. Genet. Counsel. 5: 187-189, 1994.

Schinzel, A.: A syndrome of midface retraction, multiple radiological anomalies, renal malformations and hypertrichosis. (Letter) Hum. Genet. 62: 382 only, 1982.

Schinzel, A.; Giedion, A.: A syndrome of severe midface retraction, multiple skull anomalies, clubfeet, and cardiac and renal malformations in sibs. Am. J. Med. Genet. 1: 361-375, 1978.

STEIN-LEVENTHAL SYNDROME (POLYCYSTIC OVARIAN DISEASE, FAMILIAL) (#184700)

JJM NOTE

This disorder (and, by implication, polycystic ovarian disease) is included because polycystic ovaries are in the differential diagnosis of ovarian masses and are a risk factor for endometrial carcinoma.

A number sign (#) is used with this entry because of evidence that polymorphism at the CYP11A locus (118485) is involved in the pathogenesis of Stein-leventhal syndrome, at least in some cases.

Stein and Leventhal (1935) demonstrated that obesity, hirsutism, and amenorrhea are clinical correlates of enlarged polycystic ovaries. Thereafter, the criteria were established as bilateral enlarged ovaries, normal 24-hr urinary ketosteroids, and lack of virilization. Response to ovarian wedge resection was included as a criterion by some. The fathers tend to be abnormally hairy, female sibs are hirsute, and mothers and sisters often have oligomenorrhea. Culdoscopy has often shown signs of Stein-Leventhal syndrome, e.g., 8 of 12 sisters of cases showed ovarian changes consistent with that diagnosis (Cooper et al., 1968). Urinary steroid determinations also suggest a genetic basis. Ovarian hyperthecosis was the term used by Givens et al. (1971), who found 41 women (in 2 kindreds) who had hirsutism and/or oligomenorrhea. Ovarian histology performed in 8 showed hyperplasia of theca cells in atretic follicles, a paucity of primordial and developing follicles, and stromal hyperplasia. Elevated levels of androstanedione and/or testosterone and of luteinizing hormone were found. Estradiol and follicle-stimulating hormone levels were low. These levels tended to return to normal after bilateral wedge resection of the ovaries. Some men of the families had low plasma testosterone and had abnormally high LH-FSH ratio as in the women. The pedigrees were consistent with dominant inheritance, probably autosomal because in 1 kindred the disorder was apparently transmitted through a father and son. Mandel et al. (1983) excluded linkage to HLA. They studied 4 families, each with 2 affected sisters; in 1 family, the mother and a maternal aunt were likewise affected. The diagnosis of PCO was confirmed by increased serum testosterone, androstenedione, and LH levels compared to those in normal women. Elevated concentrations of dehydroepiandrosterone sulfate indicated excess adrenal androgen secretion. Zumoff et al. (1983) presented evidence for a chronobiologic abnormality in secretion of lu teinizing hormone. Whether the abnormality resides in the hypothalamus or pituitary is not clear. Kuttenn et al. (1985) concluded that their 24 hirsute women with late-onset adrenal hyperplasia due to 21-hydroxylase deficiency (201910) were indistinguishable from those with idiopathic hirsutism or Stein-Leventhal syndrome. Chrousos et al. (1982) made the same observation. Stein-Leventhal syndrome is probably heterogeneous; see 264300 for information that some patients may have deficiency of estradiol 17-beta-dehydrogenase. It appears that 'idiopathic hirsutism,' which at times is familial, is sometimes due to increased skin sensitivity to androgen and occurs in the absence of elevated plasma androgens (Kuttenn et al., 1977). On the basis of a large family study, Hague et al. (1986) arrived at an hypothesis of dominant inheritance with meiotic drive accounting for the anomalous segregation. They found PCO in 96% of daughters of affected females and in 82% of daughters of carrier males. The pathogenesis of polycystic ovary syndrome was discussed by McKenna (1988) and Givens (1988). Lunde et al. (1989) reported that 19.7% of male first-degree relatives of PCO patients had early baldness or excessive hairiness, as opposed to 6.5% of relatives of controls. For female first-degree relatives, the percentages for PCO-related symptoms were 31.4 and 3.2, respectively. In a subgroup of 52 families of PCO patients in which 1 of the parents was reported to have symptoms, 35% of brothers and 58% of sisters had symptoms. The findings were considered consistent with autosomal dominant inheritance for a 'sizeable fraction of families.' The possibility of X-linked dominant inheritance was eliminated. Stewart et al. (1990) presented evidence suggesting that the fundamental defect in many cases of PCO disease is increased 5-alpha-reductase activity in the liver and skin. As a result, testosterone is converted to the more potent and rogen dihydrotestosterone, leading to hirsutism.

Carey et al. (1993) described an autosomal dominant syndrome of polycystic ovaries and premature male-patterned baldness (109200). Carey et al. (1994) demonstrated an association with a change in the 5-prime promoter region of the CYP17 gene (202110), which modified the expression of the syndrome in some families but did not appear to be the primary genetic defect.

Gharani et al. (1997) presented linkage and association data indicating a strong relationship between the CYP11A (118485) locus and the polycystic ovary syndrome with hyperandrogenism.

To assess the potential contribution of leptin to the pathogenesis of PCO, Mantzoros et al. (1997) measured leptin levels in obese women with PCO and in controls to determine if alterations in hyperinsulinemia produced by administration of the insulin-sensitizing agent troglitazone had an effect on serum leptin levels. Baseline leptin levels were not different between PCO and control subjects and remained unchanged after treatment with troglitazone. They concluded that leptin levels in patients with PCO do not differ from those in controls and that increased circulating insulin due to insulin resistance does not appear to alter circulating leptin levels in women with PCO.

Franks et al. (1997) favored an oligogenic basis (over autosomal dominant inheritance) for polycystic ovary syndrome. They stated that it is the most common endocrinopathy in women of reproductive age.

See also Borghi et al. (1972); Cohen et al. (1975)

Borghi, A.; Maiello, M.; Giusti, G.: Stein-Leventhal syndrome in sisters. The possible role of genetic factors in the 'polycystic ovary syndrome.' Acta Genet. Med. Gemellol. 21: 79-93, 1972.
Carey, A. H.; Chan, K. L.; Short, F.; White, D.; Williamson, R.; Franks, S.: Evidence for a single gene effect causing polycystic ovaries and male pattern baldness. Clin. Endocr. 38: 653-658, 1993.
Carey, A. H.; Waterworth, D.; Patel, K.; White, D.; Little, J.; Novelli, P.; Franks, S.; Williamson, R.: Polycystic ovaries and premature male pattern baldness are associated with one allele of the steroid metabolism gene CYP17. Hum. Molec. Genet. 3: 1873-1876, 1994.
Chrousos, G. P.; Loriaux, D. L.; Mann, D. L.; Cutler, G. B., Jr.: Late onset 21-hydroxylase deficiency mimicking idiopathic hirsutism or polycystic ovarian disease; an allelic variant of congenital virilizing adrenal hyperplasia with a milder enzymatic defect. Ann. Intern. Med. 96: 143-148, 1982.
Cohen, P. N.; Givens, J. R.; Wiser, W. L.; Wilroy, R. S., Jr.; Summitt, R. L.; Coleman, S. A.; Andersen, R. N.: Polycystic ovarian disease, maturation arrest of spermiogenesis, and Klinefelter's syndrome in siblings of a family with familial hirsutism. Fertil. Steril. 26: 1228-1238, 1975.
Cooper, H. E.; Spellacy, W. N.; Prem, K. A.; Cohen, W. D.: Hereditary factors in the Stein-Leventhal syndrome. Am. J. Obstet. Gynec. 100: 371-387, 1968.
Franks, S.; Gharani, N.; Waterworth, D.; Batty, S.; White, D.; Williamson, R.; McCarthy, M.: The genetic basis of polycystic ovary syndrome. Hum. Reprod. 12: 2641-2648, 1997.
Gharani, N.; Waterworth, D. M.; Batty, S.; White, D.; Gilling-Smith, C.; Conway, G. S.; McCarthy, M.; Franks, S.; Williamson, R.: Association of the steroid synthesis gene CYP11a with polycystic ovary syndrome and hyperandrogenism. Hum. Molec. Genet. 6: 397-402, 1997.
Givens, J. R.: Familial polycystic ovarian disease. Endocr. Metab. Clin. N. Am. 17: 771-783, 1988.
Givens, J. R.; Wiser, W. L.; Coleman, S. A.; Wilroy, R. S.; Andersen, R. N.; Fish, S. A.; Watson, B. S.: Familial ovarian hyperthecosis: a study of two families. Am. J. Obstet. Gynec. 110: 959-972, 1971.
Hague, W.; Adams, J.; Reeders, S.; Jacobs, H.: Non-mendelian segregation ratios in familial polycystic ovaries. (Abstract) 7th Int. Cong. Hum. Genet., Berlin 277-278, 1986.
Kuttenn, F.; Couillin, P.; Girard, F.; Billaud, L.; Vincens, M.; Boucekkine, C.; Thalabard, J.-C.; Maudelonde, T.; Spritzer, P.; Mowszowicz, I.; Boue, A.; Mauvais-Jarvis, P.: Late-onset adrenal hyperplasia in hirsutism. New Eng. J. Med. 313: 224-231, 1985.
Kuttenn, F.; Mowszowicz, I.; Schaison, G.; Mauvais-Jarvis, P.: Androgen production and skin metabolism in hirsutism. J. Endocr. 75: 83-91, 1977.
Lunde, O.; Magnus, P.; Sandvik, L.; Hoglo, S.: Familial clustering in the polycystic ovarian syndrome. Gynec. Obstet. Invest. 28: 23-30, 1989.
Mandel, F. P.; Chang, R. J.; Dupont, B.; Pollack, M. S.; Levine, L. S.; New, M. I.; Lu, J. K. H.; Judd, H. L.: HLA genotyping in family members and patients with familial polycystic ovarian disease. J. Clin. Endocr. Metab. 56: 862-864, 1983.
Mantzoros, C. S.; Dunaif, A.; Flier, J. S.: Leptin concentrations in the polycystic ovary syndrome. J. Clin. Endocr. Metab. 82: 1687-1691, 1997.
McKenna, T. J.: Pathogenesis and treatment of polycystic ovary syndrome. New Eng. J. Med. 318: 558-562, 1988.
Stein, I. F.; Leventhal, M. L.: Amenorrhea associated with bilateral polycystic ovaries. Am. J. Obstet. Gynec. 29: 181-191, 1935.
Stewart, P. M.; Shackleton, C. H. L.; Beastall, G. H.; Edwards, C. R. W.: 5-alpha-reductase activity in polycystic ovary syndrome. Lancet 335: 431-433, 1990.
Zumoff, B.; Freeman, R.; Coupey, S.; Saenger, P.; Markowitz, M.; Kream, J.: A chronobiologic abnormality in luteinizing hormone secretion in teenage girls with the polycystic-ovary syndrome. New Eng. J. Med. 309: 1206-1209, 1983.

SYSTEMIC CYSTIC ANGIOMATOSIS AND SEIP SYNDROME (272500)

JJM NOTE

This vascular disorder is included because of polycystic ovarian disease.

In a black family, Brunzell et al. (1968) observed a combination of congenital generalized lipodystrophy and systemic cystic angiomatosis in 5 of 12 sibs. The authors found previous reports of only 14 cases of cystic angiomatosis, none of which was familial. Progressive incapacitating bone involvement occurred. Two had soft tissue (e.g., subcutaneous) angiomas. The lipodystrophy was accompanied by acanthosis nigricans, large hands and feet, acromegaloid facial features, lipemia, and hepatosplenomegaly and was in all ways identical to that of Seip syndrome (269700). Thus, the same gene may be responsible for the syndrome in reported cases of Seip disease and in the affected persons reported by Brunzell et al. (1968). Cystic angiomatosis may have been late in developing or overlooked in reported cases. That Seip syndrome and the syndrome described by Brunzell et al. (1968) are the same is supported by the black family with 3 affected sibs reported by Huseman et al. (1978). The patients had cystic angiomatosis of the long bones and one had polycystic ovarian disease. In the product of an uncle-niece mating, Van Maldergem et al. (1992) described features characteristic of total lipodystrophy (Berardinelli-Seip syndrome), including absent subcutaneous fat, hyperlipidemia, acanthosis nigricans, facial dysmorphia, and mental retardation. In addition, at the age of 13 years, the girl was found to have large, multilocular cysts at the ends of the long bones, particularly the humeri and femori. At the age of 17 years, oligospaniomenorrhea (few and scanty menses) was a complaint, and polycystic ovaries were demonstrated. Insulin resistance was progressive. IQ was 50. Van Maldergem et al. (1992) favored the distinctness of Brunzell syndrome, which they suggested includes polycystic ovaries.

Brunzell, J. D.; Shankle, S. W.; Bethune, J. E.: Congenital generalized lipodystrophy and systemic cystic angiomatosis: the simultaneous occurrence of two unusual syndromes in a single family. Ann. Intern. Med. 69: 501-516, 1968.
Huseman, C.; Johanson, A. J.; Varma, M.; Blizzard, R. M.: Congenital total lipodystrophy: an endocrine study in 3 siblings. I. Disorders of carbohydrate metabolism. J. Pediat. 93: 221-226, 1978.
Van Maldergem, L.; Bacq, C.; Mommen, N.; Fourneau, C.; Hilbert, P.; Gillerot, Y.: Total lipodystrophy, polycystic ovaries and cystic angiomatosis of bones (Brunzell syndrome): confirmation of a separate entity. (Abstract) Am. J. Hum. Genet. 51 (suppl.): A109 only, 1992.

Lower Female Genital Tract Tumors

TOOTH-AND-NAIL SYNDROME (*189500)

JJM NOTE

This entity, clearly in the heuristic category of genodermatosis, is included because of polycystic ovaries.

Changes are limited largely to teeth (some of which are missing) and nails (which are poorly formed early in life, especially toenails). This condition is distinguished from anhidrotic ectodermal dysplasia by autosomal dominant inheritance and little involvement of hair and sweat glands. The teeth are not as severely affected. Witkop (1965) stated that the condition is frequent among Dutch Mennonites in Canada. He presented a pedigree supporting autosomal dominant inheritance. The teeth are not affected in the autosomal dominant hidrotic ectodermal dysplasia. Giansanti et al. (1974) reported a single case. The main features were hypoplastic nails and hypodontia. Eyebrows and eyelashes were normal, but the scalp hair was fine. The patient showed bilateral polycystic ovaries. Redpath and Winter (1969) probably reported cases. Hudson and Witkop (1975) presented clinical details on 23 cases in 6 families, with several instances of male-to-male transmission. Characteristic, centrally hollowed, dysplastic toenails were frequently apparent only in childhood. The condition is usually not detected until the permanent teeth fail to erupt. Mandibular incisors, second molars, and maxillary canines are most often absent. (In simple hypodontia, premolars and maxillary lateral incisors are most often missing.) Somewhat pouting lower lip was described.

Giansanti, J. S.; Long, S. M.; Rankin, J. L.: The 'tooth and nail' type of autosomal dominant ectodermal dysplasia. Oral Surg. 37: 576-582, 1974.
Hudson, C. D.; Witkop, C. J., Jr.: Autosomal dominant hypodontia with nail dysgenesis. Oral Surg. 39: 409-423, 1975.

Redpath, T. H.; Winter, G. B.: Autosomal dominant ectodermal dysplasia with significant dental defects. Brit. Dent. J. 126: 123-128, 1969.
Witkop, C. J., Jr.: Genetic disease of the oral cavity. In: Tiecke, R. W.: Oral Pathology. New York: McGraw-Hill (pub.) 1965. Pp. 810-814.

DELETED IN ENDOMETRIAL CARCINOMA (*602084)

JJM NOTE

Typically, genes discovered by loss of heterozygosity studies are put in the tumor-suppressor section, but this is the best entry to comment on the clinical genetics of endometrial cancer. Pure endometrial cancer families have not been reported. The large U.S. Cancer and Steroid Hormone Study had 455 women with endometrial cancer and showed a relative risk of 2.8 for the same cancer among first-degree relatives and of 1.9 for colorectal cancer (Gruber SB, Thompson WD: A population-based study of endometrial cancer and familial risk in younger women. Cancer Epidemiol. Biomarkers Prev. 5:411–17, 1996). Among 215 women with endometrial cancer, the greatest familial pattern was seen in the 14% who have a familial history of colorectal cancer (Fornasarig M et al.: Risk factors for endometrial cancer according to familial susceptibility. Int. J. Cancer 77:29–32, 1998). In a similar, but smaller, study in Japan, 6 of 142 women with endometrial cancer gave a strong familial history of cancer, in general (Ichikawa Y et al.: Nippon Sanka Fujinka Gakkai Zasshi 48:835–40, 1996). The clinical syndromes that feature endometrial cancer are the Lynch cancer family syndrome, hereditary nonpolypotic colorectal cancer, and Muir-Torre syndrome, which, of course, have the same underlying genes.

In 9 of 19 (47%) cases of endometrial carcinoma, Safara et al. (1997) demonstrated loss of heterozygosity (LOH) for markers in the 10q23-q26 region. The pattern of deletion suggested a role for a tumor suppressor gene located between markers D10S587 and D10S216, a region spanning 2 cM and important for the development or progression of endometrial carcinoma. Safara et al. (1997) named the putative tumor suppressor gene DEC for 'deleted in endometrial carcinoma.'

Loss-of-heterozygosity studies by Peiffer et al. (1995) and Nagase et al. (1996) also pointed to the 10q25-q26 region as the likely site of a tumor suppressor involved in endometrial tumorigenesis. In an attempt to define further the localization of a tumor suppressor gene at 10q25, Peiffer-Schneider et al. (1998) screened a panel of 123 endometrioid adenocarcinomas for LOH of 10q25.3 sequences. LOH at one or more loci was found in 43 (35%) of the cases. Observed patterns of allelic loss defined a minimum consensus region of deletion between D10S221 and D10S610. A sequence-ready bacterial clone contig and a long-range restriction map for a 1-Mb interval spanning the deletion region were developed as the first step toward discovery of the 10q25 tumor suppressor.

Nagase, S.; Sato, S.; Tezuka, F.; Wada, Y.; Yajima, A.; Horii, A.: Deletion mapping on chromosome 10q25-q26 in human endometrial cancer. Brit. J. Cancer 74: 1979-1983, 1996.
Peiffer, S. L.; Herzog, T. J.; Tribune, D. J.; Mutch, D. G.; Gersell, D. J.; Goodfellow, P. J.: Allelic loss of sequences from the long arm of chromosome 10 and replication errors in endometrial cancers. Cancer Res. 55: 1922-1926, 1995.
Peiffer-Schneider, S.; Noonan, F. C.; Mutch, D. G.; Simpkins, S. B.; Herzog, T.; Rader, J.; Elbendary, A.; Gersell, D. J.; Call, K.; Goodfellow, P. J.: Mapping an endometrial cancer tumor suppressor gene at 10q25 and development of a bacterial clone contig for the consensus deletion interval. Genomics 52: 9-16, 1998.
Safara, A.; Maia, A. T.; Martins, A. P.; Roriz, M. L.; Ramos, S.; Leal, F.; Rueff, J.; Monteiro, C.: A putative tumour suppressor gene DEC involved in endometrial cancer located in the q26 region of chromosome 10. (Abstract) Am. J. Hum. Genet. 61 (suppl.): A81 only, 1997.

GESTATIONAL TROPHOBLASTIC DISEASE (HYDATIDIFORM MOLE) (231090)

JJM NOTE

Aaron E. Szulman (personal communication, 1991), a dean of molar pathology, pointed out the possibility that the predisposition of some moles to malignancy may reflect mutations in recessive tumor-suppressor genes on the Y chromosome, since it is the homozygous paternal genome that predisposes to malignancy.

In India, Ambani et al. (1980) observed gestational trophoblastic disease in multiple pregnancies of sisters in 3 unrelated kindreds. In 1 family a first cousin of 2 'affected' sisters also had a mole pregnancy and the 3 husbands of the 'affected' females had a common ancestral couple, i.e., were related as second cousins, although not related to their wives. Kajii and Ohama (1977) presented evidence for solely paternal genome in hydatidiform moles. On the basis of genetic origin, hydatidiform mole can be divided into 3 types. Approximately 25% of hydatidiform moles ascertained clinically are partial moles. They are triploid, the additional set of chromosomes generally being paternally derived. They do not seem to be associated with the development of choriocarcinoma. A second type of mole that can be classified pathologically is the complete mole. These moles are genetically diploid but are unusual in that all chromosomes are paternally derived, although the cytoplasm has been shown to be maternally derived as in normal conceptions. Complete hydatidiform moles may have 1 of 2 different origins. Most, about 90%, are homozygous, arising from duplication of a haploid sperm. More rarely, complete mole arises by dispermy, the fertilization of an anucleate egg by 2 sperm, and are therefore heterozygous. In 1 choriocarcinoma following pregnancy with hydatidiform mole, Fisher et al. (1988) demonstrated homozygosity.

Ambani, L. M.; Vaidya, R. A.; Rao, C. S.; Daftary, S. D.; Motashaw, N. D.: Familial occurrence of trophoblastic disease—report of recurrent molar pregnancies in sisters in three families. Clin. Genet. 18: 27-29, 1980.
Fisher, R. A.; Lawler, S. D.; Povey, S.; Bagshawe, K. D.: Genetically homozygous choriocarcinoma following pregnancy with hydatidiform mole. Brit. J. Cancer 58: 788-792, 1988.
Kajii, T.; Ohama, K.: Androgenetic origin of hydatidiform mole. Nature 268: 633-634, 1977.

PAPILLOMAVIRUS TYPE 18 INTEGRATION SITE (*167960)

JJM NOTE

This entry could be placed in the protein section, but is used to comment on familial cervical cancer. Few "pure" families with cervical cancer have been reported, and strong environmental determinants, especially human papilloma virus 16 and 18, are well recognized. The integration site provides the possibility of inherited variation in susceptibility to cervical cancer (ecogenetics). Patients with dyskeratosis congenita have an excess of cervical cancer, as do carriers of *BRCA2* mutations in Iceland.

The DNA of human papillomavirus (HPV) types 16 and 18 has been found closely associated with human genital cancers, supporting an etiologic role for these viruses. Both HPV16 and HPV18 DNA sequences have also been detected in cell lines derived from cervical cancer. For example, the most famous such line, HeLa, contains HPV18 DNA amplified about 10- to 50-fold together with the flanking cellular sequences. Most, if not all, HPV-positive mRNAs from HeLa are virus-cell fusion transcripts. The host cell sequences are spliced to the 5-prime HPV18 exons. In a primary cervical carcinoma, Durst et al. (1987) found that HPV-16 flanking sequences were localized to chromosome regions 20pter-q13 and 3p25-qter, regions that also contain the protooncogenes SRC (190090) and RAF1 (164760), respectively. In the SiHa cervical carcinoma-derived cell line, Durst et al. (1987) demonstrated that the HPV16 integration site was into chromosome region 13q14-q32. The HPV18 integration site in SW756 cervical carcinoma cells was in chromosome 12 but was not closely linked to the KRAS2 gene (190070). In 2 cervical carcinoma cell lines, HeLa and C4-I, HPV18 DNA was integrated in chromosome 8, 5-prime to the MYC gene (190080). All this suggested that at least in some genital tumors, cis-activation of cellular oncogenes by HPV may be involved in malignant transformation of cervical cells. Popescu et al. (1987) also found that the HPV18 DNA in the cervical carcinoma cell line SW756 was integrated at a single site, namely, 12q13. Multiple copies were present at this site, which was determined by in situ hybridization. The single integration site corresponds to a heritable fragile site which, Popescu et al. (1987) suggested, may have facilitated the integration of the viral DNA.

Durst, M.; Croce, C. M.; Gissmann, L.; Schwarz, E.; Huebner, K.: Papilloma virus sequences integrate near cellular oncogenes in some cervical carcinomas. Proc. Nat. Acad. Sci. 84: 1070-1074, 1987.

Popescu, N. C.; Amsbaugh, S. C.; DiPaolo, J. A.: Human papillomavirus type 18 DNA is integrated at a single chromosome site in cervical carcinoma cell line SW756. J. Virol. 51: 1682-1685, 1987.

LICHEN SCLEROSUS ET ATROPHICUS (151590)

JJM NOTE

The fact that only the vulva is affected raises the hypothesis that an interacting virus and a genetic predisposition are contributing to this condition and its malignant complication.

Lichen sclerosus et atrophicus is a relatively uncommon cutaneous disorder that may affect any area. There is a particular predilection, however, for involvement of the female genitalia. Although typically the patient is a middle-aged female, prepubertal females may also develop this disorder. Shirer and Ray (1987) stated that about 18 cases of familial LSA had been reported. They concluded that 'heredity plays a role in some, if not all, cases.' They and others reported instances of affected mother-daughter pairs. The eruption of LSA consists of ivory-colored, polygonal, flat-topped lichenoid papules. These papules may coalesce into plaques, which then become atrophic. Adults with vulvar LSA may have an increased risk of developing squamous cell carcinoma of the vulva. Carcinoma has not been known to occur in extragenital lesions. Friedrich and MacLaren (1984) found 13 instances of familial lichen sclerosis in the literature and added 2 families.

Clay et al. (1994) found an association between lichen sclerosis and a specific variable number tandem repeat (VNTR) polymorphism in intron 2 of the gene encoding interleukin-1 receptor antagonist (IL1RN; 147679).

Clay, F. E.; Cork, M. J.; Tarlow, J. K.; Blakemore, A. I. F.; Harrington, C. I.; Lewis, F.; Duff, G. W.: Interleukin 1 receptor antagonist gene polymorphism association with lichen sclerosis. Hum. Genet. 94: 407-410, 1994.

Friedrich, E. G., Jr.; MacLaren, N. K.: Genetic aspects of vulvar lichen sclerosis. Am. J. Obstet. Gynec. 150: 161-165, 1984.

Shirer, J. A., Jr.; Ray, M. C.: Familial occurrence of lichen sclerosus et atrophicus: case reports of a mother and daughter. Arch. Derm. 123: 485-488, 1987.

WHIM SYNDROME (193670)

JJM NOTE

Given the immunodeficiency and the cervical dysplasia, I wonder if case reports of frank malignancy will emerge.

Wetzler et al. (1990) described a family in which 2 sisters and their father had a combination of chronic papillomavirus and bacterial sinopulmonary infections, low immunoglobulin levels, and peripheral neutropenia in the face of bone marrow hypercellularity, shift to the right in granulopoiesis, and distinctive neutrophil morphology. The occurrence of neutropenia and retention of bone marrow neutrophils has been called myelokathexis (kathexis = retention) (Zuelzer, 1964; O'Regan et al., 1977). In addition to extensive verrucae vulgaris of the hands, both sisters, aged 23 and 22 years, had cervical papillomatosis with severe dysplasia of the cervix, and 1 had vulval condylomata acuminata. Neutrophils had cytoplasmic vacuoles and hypersegmented nuclei with dense, pycnotic lobes connected by long filaments. There were no morphologic abnormalities observed in lymphoid cells, erythroid series or megakaryocytes and the karyotype was normal in both. The father had severe neutropenia and warts as well as frequent episodes of pneumonia necessitating partial lung resection. He died at the age of 31 years of meningitis and septicemia. A sister of the father died at age 1 month of pneumonia. The same disorder appears to have been described by Mentzer et al. (1977). Wetzler et al. (1990) suggested that this disorder be designated the WHIM syndrome; the acronym stands for warts, hypogammaglobulinemia, infections, and myelokathexis.

Gorlin (1998) stated that he had studied a 3-generation family with 5 affected individuals. He noted in this family and in the reported families that no male-to-male transmission had been observed and affected males transmitted the disorder to all of their daughters. On this basis he suggested X-linked dominant inheritance.

Gorlin, R. J.: Personal Communication. Minneapolis, Minn. 3/12/1998.

Mentzer, W. C.; Johnston, R. B., Jr.; Baehner, R. L.; Nathan, D. G.: An unusual form of chronic neutropenia in a father and a daughter with hypogammaglobulinaemia. Brit. J. Haemat. 36: 313-322, 1977.

O'Regan, S.; Newman, A. J.; Graham, R. C.: Myelokathexis: neutropenia with marrow hyperplasia. Am. J. Dis. Child. 131: 655-658, 1977.

Wetzler, M.; Talpaz, M.; Kleinerman, E. S.; King, A.; Huh, Y. O.; Gutterman, J. U.; Kurzrock, R.: A new familial immunodeficiency disorder characterized by severe neutropenia, a defective marrow release mechanism, and hypogammaglobulinemia. Am. J. Med. 89: 663-672, 1990.

Zuelzer, W. W.: Myelokathexis: a new form of chronic granulocytopenia. New Eng. J. Med. 270: 699-704, 1964.

Gonadoblastomas

GONADOBLASTOMA (424500)

JJM NOTE

Seven MIM entries relate to gonadoblastoma. My colleagues and I produced a Venn diagram that emphasized the apparently exclusive predisposition to gonadoblastoma of Y chromosome–bearing dysgenetic gonad (Mulvihill JJ et al.: Gonadoblastoma in dysgenetic gonads with a Y chromosome. Lancet 1:863, 1975). The biologic challenge is to explain the differential risk for malignancy in histologically identical gonads. The clinical challenge is to rule out a Y chromosome component in individuals with various gonadal dysgenesis syndromes, including the most commonest—Turner syndrome. Gonadectomy is indicated when Y chromosome material is present—another example of prophylactic surgery to prevent cancer in high-risk patients.

An increased risk for gonadoblastoma in phenotypic females with dysgenetic gonads and the presence of Y-chromosomal material is well established (Verp and Simpson, 1987). Page (1987) postulated the existence of a gene (GBY) on the Y chromosome with an undefined physiologic function in normal males but with the ability to predispose dysgenetic gonads to develop malignancy when present in females. On the basis of 2 reported cases of females with dysgenetic gonads, deleted Y chromosomes, and gonadoblastoma, Page (1987) argued that the GBY gene is located either near the centromere or on the long arm of the Y chromosome and is distinct from the testis-determining factor. Petrovic et al. (1992) described a 12-year-old girl who was referred for chromosome analysis because of short stature and was found to have a mosaic karyotype 45,X/46,X,+mar. The marker chromosome was found in 58% of the blood lymphocytes. DNA analysis using Y-specific probes showed absence of the testis-determining region and the presence of some short arm and centromeric Y-chromosomal material. In situ hybridization confirmed that the Y-chromosomal material was associated with the marker chromosome. At laparotomy the patient was found to have streak gonads with a dysgerminoma arising from a gonadoblastoma in the left gonad. The case demonstrated that even very small Y-derived marker chromosomes with pericentric material can predispose the phenotypic female to gonadal neoplasia.

Tsuchiya et al. (1995) used sequence-tagged sites to perform deletion mapping of the Y chromosome in sex-reversed female patients with a Y chromosome and gonadoblastoma. The GBY gene was sublocalized to a small region near the centromere. They estimated the size of the GBY critical region to be approximately 1-2 Mb. Their analysis indicated that copies of the 2 dispersed Y-linked gene families, TSPY (480100) and YRRM (Y-chromosome RNA recognition motif; 415000) were present in all patients. Copies of TSPY but not YRRM fall within the GBY critical region as formally defined by deletion mapping. Two tumor samples showed expression of both genes and in one patient this expression was limited to a unilateral gonadoblastoma but absent in the contralateral streak gonad. Although the results of Tsuchiya et al. (1995) did not directly implicate TSPY or YRRM in the etiology of the tumor, they raised the issue of whether there is one GBY gene in the critical region or possibly multiple GBY loci dispersed on the Y chromosome.

Salo et al. (1995) attempted to map GBY by making use of a dense map of Y-chromosome sequence-tagged sited (STSs). In 2 female patients with gonadoblastoma, small marker chromosomes contained portions of the Y chromosome, and a single region of overlap could be defined extending from probe pDP97 in interval 4B, which contains the centromere, to marker sY182 in interval 5E of the proximal long arm. This interval is contained in a YAC contig that comprises approximately 4 Mb of DNA. Their result confirmed the previous localization of GBY and greatly refined it. The localization overlaps with the region to which GCY (475000), a putative growth determinant, had been assigned.

Page, D. C.: Hypothesis: a Y-chromosomal gene causes gonadoblastoma in dysgenetic gonads. Development 101 (suppl.): 151-155, 1987.

Petrovic, V.; Nasioulas, S.; Chow, C. W.; Voullaire, L.; Schmidt, M.; Dahl, H.: Minute Y chromosome derived marker in a child with gonadoblastoma: cytogenetic and DNA studies. J. Med. Genet. 29: 542-546, 1992.
Salo, P.; Kaariainen, H.; Petrovic, V.; Peltomaki, P.; Page, D. C.; de la Chapelle, A.: Molecular mapping of the putative gonadoblastoma locus on the Y chromosome. Genes Chromosomes Cancer 14: 210-214, 1995.
Tsuchiya, K.; Reijo, R.; Page, D. C.; Disteche, C. M.: Gonadoblastoma: molecular definition of the susceptibility region on the Y chromosome. Am. J. Hum. Genet. 57: 1400-1407, 1995.
Verp, M. S.; Simpson, J. L.: Abnormal sexual differentiation and neoplasia. Cancer Genet. Cytogenet. 25: 191-218, 1987.

SEX-DETERMINING REGION Y; TESTIS-DETERMINING FACTOR (*480000)

JJM NOTE

A search of gonadoblastomas for loss of heterozygosity in this gene would seem to be a logical next step.

Description

SRY encodes a transcription factor that is a member of the high mobility group (HMG)-box family of DNA binding proteins. Mutations in this gene give rise to XY females with gonadal dysgenesis (one type of Swyer syndrome, 306100) and translocation of part of the Y chromosome containing this gene to the X chromosome in XX males (278850).

Genotype

Goodfellow and Lovell-Badge (1993) provided a major review of SRY and sex determination in mammals.

From the study of normal males and females, persons with abnormal numbers of sex chromosomes, and persons carrying variant Y chromosomes, a factor (or factors) that determines the differentiation of the indifferent gonads into testes is known to be located on the Y chromosome and specifically on the short arm; this was designated testis-determining factor (TDF) in the 1960s. (See Mendelian Inheritance in Man, 4th ed., fig. 1, p. lix, 1975.)

Mittwoch (1992) argued that the 'dogma' that all differences distinguishing male and female mammals can be traced to the presence or absence of a single gene encoding a testis-determining factor lacks, as she said, 'biological validity.' She suggested that the genotype of the functional, i.e., fertile, male differs from that of a functional female by the presence of multiple Y-chromosomal genes in association with but a single X chromosome.

Mapping

TDF was ultimately mapped to the human Y chromosome by molecular examination of sex-reversed patients. Analysis of 4 XX males with testes who had minute portions of the Y material translocated to the X chromosome were critical in defining the sex-determining region on the human Y chromosome (Palmer et al., 1989; Sinclair et al., 1990). The sex-determining region on the human Y chromosome was later defined to a 35-kb region of Y-specific DNA adjacent to the pseudoautosomal boundary (Sinclair et al., 1990).

Page et al. (1987) cloned part or all of what they thought to be the TDF gene, found that some sequences were highly conserved in mammals and even birds, and showed that the nucleotide sequence of the conserved DNA codes for zinc finger domains. ZFY (zinc finger protein, Y-linked; 314980) was the designation approved by the HGM workshop committee, with ZFX being the X-linked counterpart. ZFY proved, however, not to be TDF (Palmer et al., 1989).

After detailed analysis of this 35-kb region, Sinclair et al. (1990) identified a gene that they named SRY (sex-determining region Y) gene. The mouse homolog Sry was subsequently cloned and found to be present in Sxr-prime mice, which have the smallest part of the Y chromosome known to be sex-determining (Gubbay et al., 1990). Furthermore, Sry was deleted from a mutant Y chromosome that was no longer sex-determining (Gubbay et al., 1990).

Behlke et al. (1993) found that 2 RNAs hybridized to a 4,741-bp genomic segment of the sex-determining region of the human Y chromosome: one transcript deriving from SRY, and a second transcript cross-hybridizing to a pseudogene located 2.5 kb 5-prime of the SRY open reading frame. Analysis of the SRY transcript suggested that the entire SRY protein is encoded by a single exon.

Molecular Genetics

Lahr et al. (1995) used RT-PCR to investigate the transcription of the Sry gene in mice. The gene was transcribed in the hypothalamus, midbrain, and testis of adult male but not adult female mice. Whereas the transcripts in the adult testis were circular, those in brain were linear and therefore capable of translation. Lahr et al. (1995) hypothesized that some male specific properties of the brain may be generated directly by the SRY gene product.

Hawkins (1993) gave a mutational analysis of the SRY gene in XY females. He noted that 11 mutations had been described at that time, and all were within the DNA-binding HMG-box region of the protein.

Braun et al. (1993) reported a 46,XY true hermaphrodite who had a mutation of SRY in gonadal DNA but not in leukocyte DNA, suggesting that the mutation was postzygotic. Because of this finding Fuqua et al. (1997) attempted to determine whether postzygotic mutations of SRY might explain the numerous cases of gonadal dysgenesis in which no SRY mutation was detected in leukocyte DNA. They evaluated 16 subjects with 46,XY gonadal dysgenesis who had a normal SRY sequence in leukocyte DNA, 5 of them having 46,XY complete gonadal dysgenesis. They did not find mutations in gonadal DNA from any of 16 subjects and concluded that postzygotic mutations of SRY are a rare cause of 46,XY gonadal dysgenesis.

Cameron and Sinclair (1997) stated that 26 different mutations in the SRY gene have been found in individuals with a 46,XY karyotype. They cited reports stating that no polymorphisms had been described in SRY among 50 normal males. De novo mutations in the SRY HMG-box region almost always resulted in 46,XY unambiguous females with no testicular differentiation. They found 5 reports of familial 46,XY complete gonadal dysgenesis associated with mutations in the SRY HMG-box region. In 4 of these reports, the father carried the same SRY mutation as his 46,XY daughter. None of the mutations appeared to be polymorphisms. Explanations for the sex reversal associated with these familial SRY mutations included paternal gonadal mosaicism for the mutation (yet to be proven) and variable penetrance of the mutation. Support for a penetrance effect came from murine studies in which at least 3 autosomal recessive alleles were found to interact with Y-chromosome alleles, resulting in the genesis of XY ovaries and true hermaphrodites (Eicher and Washburn, 1986). Cameron and Sinclair (1997) noted that timing and expression of SRY are exquisitely regulated and probably must reach a threshold. Consequently, a given mutation in SRY against a particular genetic background might produce sufficient SRY expression to reach the threshold required; testis formation can then ensue, accounting for an unaffected father.

Zenteno et al. (1997) described a Mexican family in which 2 brothers, aged 28 and 26, were thought to be instances of 'classic' XX males without genital ambiguity but were found to be negative for several Y-chromosome sequences, including SRY. The data suggested that an inherited loss-of-function mutation in a gene participating in the sex-determining cascade can induce normal male sexual differentiation in the absence of SRY.

Evolution

Using a human SRY probe, Foster et al. (1992) identified and cloned related genes from the Y chromosome of 2 marsupial species. Comparisons of eutherian ('placental') and metatherian (marsupial) Y-located SRY sequences suggested rapid evolution of these genes, especially outside the region encoding the DNA-binding 'high mobility group' domain (HMG box). The SRY homolog and the homolog of the mouse Ube1y were the first genes to be identified on the marsupial Y chromosome. Whitfield et al. (1993) and Tucker and Lundrigan (1993) likewise found that whereas the central 'high mobility group' domain of about 78 amino acids of the SRY protein is highly conserved, evolution in primates and in mice and rats has been rapid in the regions flanking the conserved domain. The high degree of sequence divergence and the frequency of nonsynonymous mutations suggested either that the majority of the coding sequence has no functional significance and therefore is not functionally constrained or that it has been subject to directional selection with species-specific adaptive divergence.

Animal Model

In 6 sterile heifers which were female in appearance and in genital organs, Kawakura et al. (1996) found that blood, skin, spleen, and kidney showed a normal male 60,XY karyotype. Although the SRY gene was detected by PCR in a normal bulls, it was not detected in normal cows or in 3 60,XY female bovine cases studied.

See also Andersson et al. (1986); Cherfas (1991); Clepet et al. (1993); Davis (1981); Disteche et al. (1986); Evans et al. (1979); Ferguson-Smith (1966); Giese et al. (1994); Koopman et al. (1991); Koopman et al. (1990); Schmitt-Ney et al. (1995); Taylor et al. (1966); Werner et al. (1995)

Affara, N. A.; Chalmers, I. J.; Ferguson-Smith, M. A.: Analysis of the SRY gene in 22 sex-reversed XY females identifies four new point mutations in the conserved DNA binding domain. Hum. Molec. Genet. 2: 785-789, 1993.
Andersson, M.; Page, D. C.; de la Chapelle, A.: Chromosome Y-specific DNA is transferred to the short arm of X chromosome in human XX males. Science 233: 786-788, 1986.

Behlke, M. A.; Bogan, J. S.; Beer-Romero, P.; Page, D. C.: Evidence that the SRY protein is encoded by a single exon on the human Y chromosome. Genomics 17: 736-739, 1993.
Berkovitz, G. D.; Fechner, P. Y.; Zacur, H. W.; Rock, J. A.; Snyder, H. M., III; Migeon, C. J.; Perlman, E. J.: Clinical and pathologic spectrum of 46,XY gonadal dysgenesis: its relevance to the understanding of sex differentiation. Medicine 70: 375-383, 1991.
Berta, P.; Hawkins, J. R.; Sinclair, A. H.; Taylor, A.; Griffiths, B. L.; Goodfellow, P. N.; Fellous, M.: Genetic evidence equating SRY and the testis-determining factor. Nature 348: 448-450, 1990.
Braun, A.; Kammerer, S.; Cleve, H.; Lohrs, U.; Schwarz, H.-P.; Kuhnle, U.: True hermaphroditism in a 46,XY individual, caused by a postzygotic somatic point mutation in the male gonadal sex-determining locus (SRY): molecular genetics and histological findings in a sporadic case. Am. J. Hum. Genet. 52: 578-585, 1993.
Brown, S.; Yu, C. C.; Lanzano, P.; Heller, D.; Thomas, L.; Warburton, D.; Kitajewski, J.; Stadtmauer, L.: A de novo mutation (gln2stop) at the 5-prime end of the SRY gene leads to sex reversal with partial ovarian function. (Letter) Am. J. Hum. Genet. 62: 189-192, 1998.
Cameron, F. J.; Sinclair, A. H.: Mutations in SRY and SOX9: testis-determining genes. Hum. Mutat. 9: 388-395, 1997.
Cherfas, J.: Sex and the single gene. Science 252: 782, 1991.
Clepet, C.; Schafer, A. J.; Sinclair, A. H.; Palmer, M. S.; Lovell-Badge, R.; Goodfellow, P. N.: The human SRY transcript. Hum. Molec. Genet. 2: 2007-2012, 1993.
Davis, R. M.: Localisation of male determining factors in man: a thorough review of structural anomalies of the Y chromosome. J. Med. Genet. 18: 161-195, 1981.
Disteche, C. M.; Casanova, M.; Saal, H.; Friedman, C.; Sybert, V.; Graham, J.; Thuline, H.; Page, D.; Fellous, M.: Small deletions of the short arm of the Y chromosome in 46,XY females. Proc. Nat. Acad. Sci. 83: 7841-7844, 1986.
Domenice, S.; Nishi, M. Y.; Billerbeck, A. E. C.; Latronico, A. C.; Medeiros, M. A.; Russell, A. J.; Vass, K.; Carvalho, F. M.; Frade, E. M. C.; Arnhold, I. J. P.; Mendonca, B. B.: A novel missense mutation (S18N) in the 5-prime non-HMG box region of the SRY gene in a patient with partial gonadal dysgenesis and his normal male relatives. Hum. Genet. 102: 213-215, 1998.
Dork, T.; Stuhrmann, M.; Miller, K.; Schmidtke, J.: Independent observation of SRY mutation I90M in a patient with complete gonadal dysgenesis. (Letter) Hum. Mutat. 11: 90-91, 1998.
Eicher, E. M.; Washburn, L. L.: Genetic control of primary sex determination in mice. Annu. Rev. Genet. 20: 327-360, 1986.
Evans, H. J.; Buckton, K. E.; Spowart, G.; Carothers, A. D.: Heteromorphic X chromosomes in 46,XX males: evidence for the involvement of X-Y interchange. Hum. Genet. 49: 11-31, 1979.
Ferguson-Smith, M. A.: X-Y chromosomal interchange in the aetiology of true hermaphroditism and of XX Klinefelter's syndrome. Lancet II: 475-476, 1966.
Foster, J. W.; Brennan, F. E.; Hampikian, G. K.; Goodfellow, P. N.; Sinclair, A. H.; Lovell-Badge, R.; Selwood, L.; Renfree, M. B.; Cooper, D. W.; Graves, J. A. M.: Evolution of sex determination and the Y chromosome: SRY-related sequences in marsupials. Nature 359: 531-533, 1992.
Fuqua, J. S.; McLaughlin, J.; Perlman, E. J.; Berkowitz, G. D.: Analysis of the SRY gene in gonadal tissue of subjects with 46,XY gonadal dysgenesis (Letter) J. Clin. Endocr. Metab. 82: 701-702, 1997.
Giese, K.; Pagel, J.; Grosschedl, R.: Distinct DNA-binding properties of the high mobility group domain of murine and human SRY sex-determining factors. Proc. Nat. Acad. Sci. 91: 3368-3372, 1994.
Goodfellow, P. N.; Lovell-Badge, R.: SRY and sex determination in mammals. Annu. Rev. Genet. 27: 71-92, 1993.
Gubbay, J.; Collignon, J.; Koopman, P.; Capel, B.; Economou, A.; Munsterberg, A.; Vivian, N.; Goodfellow, P. N.; Lovell-Badge, R.: A gene mapping to the sex-determining region of the mouse Y chromosome is a member of a novel family of embryonically expressed genes. Nature 346: 245-250, 1990.
Haqq, C. M.; King, C.-Y.; Ukiyama, E.; Falsafi, S.; Haqq, T. N.; Donahoe, P. K.; Weiss, M. A.: Molecular basis of mammalian sexual determination: activation of Mullerian inhibiting substance gene expression by SRY. Science 266: 1494-1500, 1994.
Hawkins, J. R.: Mutational analysis of SRY in XY females. Hum. Mutat. 2: 347-350, 1993.
Hawkins, J. R.; Taylor, A.; Berta, P.; Levilliers, J.; Van der Auwera, B.; Goodfellow, P. N.: Mutational analysis of SRY: nonsense and missense mutations in XY sex reversal. Hum. Genet. 88: 471-474, 1992.
Hawkins, J. R.; Taylor, A.; Goodfellow, P. N.; Migeon, C. J.; Smith, K. D.; Berkovitz, G. D.: Evidence for increased prevalence of SRY mutations in XY females with complete rather than partial gonadal dysgenesis. Am. J. Hum. Genet. 51: 979-984, 1992.
Iida, T.; Nakahori, Y.; Komaki, R.; Mori, E.; Hayashi, N.; Tsutsumi, O.; Taketani, Y.; Nakagome, Y.: A novel missense mutation in the HMG box of the SRY gene in a patient with XY sex reversal. Hum. Molec. Genet. 3: 1437-1438, 1994.
Jager, R. J.; Anvret, M.; Hall, K.; Scherer, G.: A human XY female with a frame shift mutation in the candidate testis-determining gene SRY. Nature 348: 452-454, 1990.
Jager, R. J.; Harley, V. R.; Pfeiffer, R. A.; Goodfellow, P. N.; Scherer, G.: A familial mutation in the testis-determining gene SRY shared by both sexes. Hum. Genet. 90: 350-355, 1992.
Kawakura, K.; Miyake, Y.-I.; Murakami, R.-K.; Kondoh, S.; Hirata, T.-I.; Kaneda, Y.: Deletion of the SRY region on the Y chromosome detected in bovine gonadal hypoplasia (XY female) by PCR. Cytogenet. Cell Genet. 72: 183-184, 1996.
Koopman, P.; Gubbay, J.; Vivian, N.; Goodfellow, P.; Lovell-Badge, R.: Male development of chromosomally female mice transgenic for Sry. Nature 351: 117-121, 1991.
Koopman, P.; Munsterberg, A.; Capel, B.; Vivian, N.; Lovell-Badge, R.: Expression of a candidate sex-determining gene during mouse testis differentiation. Nature 348: 450-452, 1990.
Lahr, G.; Maxson, S. C.; Mayer, A.; Just, W.; Pilgrim, C.; Reisert, I.: Transcription of the Y chromosomal gene, Sry, in adult mouse brain. Molec. Brain Res. 33: 179-182, 1995.
Lovell-Badge, R.; Robertson, E.: XY female mice resulting from a heritable mutation in the primary testis-determining gene, Tdy. Development 109: 635-646, 1990.
McElreavey, K. D.; Vilain, E.; Boucekkine, C.; Vidaud, M.; Jaubert, F.; Richaud, F.; Fellous, M.: XY sex reversal associated with a nonsense mutation in SRY. Genomics 13: 838-840, 1992.
Mittwoch, U.: Sex determination and sex reversal: genotype, phenotype, dogma and semantics. Hum. Genet. 89: 467-479, 1992.
Muller, J.; Schwartz, M.; Skakkebaek, N. E.: Analysis of the sex-determining region of the Y-chromosome (SRY) in sex reversed patients: point-mutation in SRY causing sex-reversion in a 46,XY female. J. Clin. Endocr. Metab. 75: 331-333, 1992.
Page, D. C.; Mosher, R.; Simpson, E. M.; Fisher, E. M. C.; Mardon, G.; Pollack, J.; McGillivray, B.; de la Chapelle, A.; Brown, L. G.: The sex-determining region of the human Y chromosome encodes a finger protein. Cell 51: 1091-1104, 1987.
Palmer, M. S.; Sinclair, A. H.; Berta, P.; Ellis, N. A.; Goodfellow, P. N.; Abbas, N. E.; Fellous, M.: Genetic evidence that ZFY is not the testis-determining factor. Nature 342: 937-939, 1989.
Peters, R.; King, C.-Y.; Ukiyama, E.; Falsafi, S.; Donahoe, P. K.; Weiss, M. A.: An SRY mutation causing human sex reversal resolves a general mechanism of structure-specific DNA recognition: application to the four-way DNA junction. Biochemistry 34: 4569-4576, 1995.
Schmitt-Ney, M.; Thiele, H.; Kaltwasser, P.; Bardoni, B.; Cisternino, M.; Scherer, G.: Two novel SRY missense mutations reducing DNA binding identified in XY females and their mosaic fathers. Am. J. Hum. Genet. 56: 862-869, 1995.
Sinclair, A. H.; Berta, P.; Palmer, M. S.; Hawkins, R.; Griffiths, B. L.; Smith, M.; Foster, J. W.; Frischauf, A.-M.; Lovell-Badge, R.; Goodfellow, P. N.: A gene from the human sex-determining region encodes a protein with homology to a conserved DNA-binding motif.. Nature 346: 240-245, 1990.
Taylor, H.; Barter, R. H.; Jacobson, C. B.: Neoplasms of dysgenetic gonads. Am. J. Obstet. Gynec. 96: 816-823, 1966.
Tucker, P. K.; Lundrigan, B. L.: Rapid evolution of the sex determining locus in Old World mice and rats. Nature 364: 715-717, 1993.
Veitia, R.; Ion, A.; Barbaux, S.; Jobling, M. A.; Souleyreau, N.; Ennis, K.; Ostrer, H.; Tosi, M.; Meo, T.; Chibani, J.; Fellous, M.; McElreavey, K.: Mutations and sequence variants in the testis-determining region of the Y chromosome in individuals with a 46,XY female phenotype. Hum. Genet. 99: 648-652, 1997.
Vilain, E.; McElreavey, K.; Jaubert, F.; Raymond, J.-P.; Richaud, F.; Fellous, M.: Familial case with sequence variant in the testis-determining region associated with two sex phenotypes. Am. J. Hum. Genet. 50: 1008-1011, 1992.
Werner, M. H.; Huth, J. R.; Gronenborn, A. M.; Clore, G. M.: Molecular basis of human 46X,Y (sic) sex reversal revealed from the three-dimensional solution structure of the human SRY-DNA complex. Cell 81: 705-714, 1995.
Whitfield, L. S.; Lovell-Badge, R.; Goodfellow, P. N.: Rapid sequence evolution of the mammalian sex-determining gene SRY. Nature 364: 713-715, 1993.
Zeng, Y.; Ren, Z.; Zhang, M.; Huang, Y.; Zeng, F.; Huang, S.: A new de novo mutation (A113T) in HMG box of the SRY gene leads to XY gonadal dysgenesis. J. Med. Genet. 30: 655-657, 1993.
Zenteno, J. C.; Lopez, M.; Vera, C.; Mendez, J. P.; Kofman-Alfaro, S.: Two SRY-negative XX male brothers without genital ambiguity. Hum. Genet. 100: 606-610, 1997.

FRASIER SYNDROME (#136680)

JJM NOTE

Whether the risk of gonadoblastoma arises solely from the gonadal dysgenesis or is augmented by other elements of the syndrome awaits further study.

A number sign (#) is used with this entry because of evidence that Frasier syndrome is caused by mutation in the WT1 gene (194070).

Moorthy et al. (1987) suggested that some of the patients reported as cases of Denys-Drash syndrome (194080) in fact had a different disorder for which they suggested the designation Frasier syndrome (Frasier et al., 1964). Moorthy et al. (1987) discussed 6 previously reported patients with streak gonads, pseudohermaphroditism, and renal failure. In several of the patients the diagnosis was established only after successful kidney transplantation during evaluation for primary amenorrhea. Gonadoblastoma arising from the streak gonad was noted in 5 of the 6 patients. Poulat et al. (1993) found no mutations within exons 8 or 9 of the WT1 gene in 3 patients with Frasier syndrome; mutations in these exons had been found in 10 unrelated cases of the syndrome by Pelletier et al. (1991). Although the findings suggested that there was a molecular difference between Frasier syndrome and Denys-Drash syndrome, the possibility of mutations elsewhere in the gene was not completely excluded. Poulat et al. (1993) sought alterations in the coding regions of exons 1-10 of the WT1 gene by use of single-strand conformation polymorphism (SSCP) analysis. Although no abnormality was found, the method detects only an estimated 80% of point mutations. Berta et al. (1992) found no large deletion in the sex-determining region of the Y chromosome or mutations in the SRY gene (480000) in 2 girls with XY gonadal dysgenesis and chronic renal failure (Frasier syndrome).

Barbaux et al. (1997) found mutations in the donor splice site of intron 9 of the WT1 gene (e.g., 194070.0018). Normally, an alternative splice site in intron 9 allows the addition of 3 amino acids (KTS) between the third and fourth zinc fingers of the WT1 protein. All 3 patients had male pseudohermaphroditism, nephrotic syndrome progressing to end-stage renal failure, and 46,XY gonadal dysgenesis. In 1 patient, gonadoblastoma was diagnosed at the age of 19.

Klamt et al. (1998) showed that no mutant protein is produced by the mutations in WT1 causing Frasier syndrome. Instead, the mutation results in an altered ratio of the 2 splice isoforms of the protein, those with and those without the extra 3 amino acids (KTS). In the Denys-Drash syndrome, the tumor risk is much greater than in Frasier syndrome. The dominant-negative mutant allele is defective and loss of the second allele (according to the 2-hit model) may be an important step in tumor formation. In contrast, Frasier patients have 1 normal copy of WT1 and 1 that can only produce a shorter isoform. Allele loss would thus lead to cells that cannot produce the +KTS isoform of WT1, but still have large amounts of the -KTS isoform. In this respect, it is interesting to note that tumorigenicity of the G401 Wilms tumor cell line in nude mice can be suppressed to the same extent by +KTS and -KTS isoforms. Gonadoblastoma is frequent in Frasier patients.

Barbaux, S.; Niaudet, P.; Gubler, M.-C.; Grunfeld, J.-P.; Jaubert, F.; Kuttenn, F.; Fekete, C. N.; Souleyreau-Therville, N.; Thibaud, E.; Fellous, M.; McElreavey, K.: Donor splice-site mutations in WT1 are responsible for Frasier syndrome. Nature Genet. 17: 467-470, 1997.

Berta, P.; Morin, D.; Poulat, F.; Taviaux, S.; Lobaccaro, J. M.; Sultan, C.; Dumas, R.: Molecular analysis of the sex-determining region from the Y chromosome in two patients with Frasier syndrome. Hormone Res. 37: 103-106, 1992.

Frasier, S. D.; Bashore, R. A.; Mosier, H. D.: Gonadoblastoma associated with pure gonadal dysgenesis in monozygotic twins. J. Pediat. 64: 740-745, 1964.

Klamt, B.; Koziell, A.; Poulat, F.; Wieacker, P.; Scambler, P.; Berta, P.; Gessler, M.: Frasier syndrome is caused by defective alternative splicing of WT1 leading to an altered ratio of WT1 +/-KTS splice isoforms. Hum. Molec. Genet. 7: 709-714, 1998.

Moorthy, A. V.; Chesney, R. W.; Lubinsky, M.: Chronic renal failure and XY gonadal dysgenesis: 'Frasier' syndrome—a commentary on reported cases. Am. J. Med. Genet. Suppl. 3: 297-302, 1987.

Pelletier, J.; Bruening, W.; Kashtan, C. E.; Mauer, S. M.; Manivel, J. C.; Striegel, J. E.; Houghton, D. C.; Junien, C.; Habib, R.; Fouser, L.; Fine, R. N.; Silverman, B. L.; Haber, D. A.; Housman, D.: Germline mutations in the Wilms' tumor suppressor gene are associated with abnormal urogenital development in Denys-Drash syndrome. Cell 67: 437-447, 1991.

Poulat, F.; Morin, D.; Konig, A.; Brun, P.; Giltay, J.; Sultan, C.; Dumas, R.; Gessler, M.; Berta, P.: Distinct molecular origins for Denys-Drash and Frasier syndromes. Hum. Genet. 91: 285-286, 1993.

GONADAL DYSGENESIS, XY FEMALE TYPE (306100)

JJM NOTE

This entity predisposes to gonadoblastoma.

Description

Gonadal dysgenesis, XY female type, is associated with point mutations or deletions of the SRY gene (480000), but also in some cases with changes in the X chromosome.

At birth the patients with the XY female type of gonadal dysgenesis (Swyer syndrome) appear to be normal females; however, they do not develop secondary sexual characteristics at puberty, do not menstruate, and have 'streak gonads.' They are chromatin negative and have a 46,XY karyotype.

Clinical Features

Affected sisters were reported by Cohen and Shaw (1965), and twins by Frasier et al. (1964). Sternberg et al. (1968) observed 3 cases, each in a different sibship of a family connected through normal females (proposita, maternal cousin, and maternal aunt). A high incidence of neoplasia (gonadoblastomas and germinomas) in streak gonads of patients with the XY karyotype was claimed by Taylor et al. (1966).

Patients are of normal stature and have no somatic stigmata of Turner syndrome except, of course, the lack of secondary sexual characteristics and streak gonads. In this condition, as in the testicular feminization syndrome (300068), it was at first unclear whether the gene that was responsible was on the X chromosome or on an autosome and expressed only in chromosomal males. Whether the abnormal gene directly suppresses testis-determining loci on the chromosome or blocks some early stage of testicular morphogenesis was also unknown. The sisters reported by Cohen and Shaw (1965) had a marker autosome, which was present also in the mother. They referred to another instance of XY 'sisters' with an abnormal autosome. One of their 2 patients had gonadoblastoma.

Two sisters reported by Fine et al. (1962) were of normal stature but were chromatin negative. One of these cases and 1 of those reported by Baron et al. (1962) had gonadoblastoma. In the last family, 2 'females' and a male were affected, the male showing no testes. All 3 sibs were sex-chromatin negative. Barr et al. (1967) reported on a sibship containing 2 genetic males. The first, who had male pseudohermaphroditism, was reared as a female; he developed signs of masculinization at puberty and had undescended but otherwise normal testes and small fallopian tubes. The second genetic male (180 cm tall) had pure gonadal dysgenesis with small uterus and streak gonads. This patient was at first thought to have the testicular feminization syndrome. An unaffected sister had a son with perineal hypospadias (urethral orifice at the base of the penis). The sibship reported by Chemke et al. (1970) was similar to that of Barr et al. (1967). Espiner et al. (1970) described 5 XY females in 3 sibships of 2 generations. They emphasized that the affected persons were unusually tall for females. The height of patients with XY gonadal dysgenesis (unusually great for females) is probably explained by androgen production in the streak gonad (Rose et al., 1974). Clitoromegaly is present in some cases.

Rushton (1979) pointed out that the streak gonads of this disorder differ from those of the 45,X Turner syndrome in the presence of calcification and the increased hazard of gonadoblastoma. Comparative studies of the frequency of gonadoblastoma in Turner mosaics with normal or rearranged Y chromosomes have suggested that the integrity of the Y chromosome, and in particular the presence of the distal fluorescent band Yqh, is required in these mosaics for the tumor to develop; no cases with distal deletions of the fluorescent band on Yq had been reported (Lukusa et al., 1986).

Moreira-Filho et al. (1979) suggested that there are 3 forms of Swyer syndrome (defined as streak gonads without other somatic features of the Turner syndrome and with a normal 46,XY karyotype). (1) Sporadic testicular agenesis syndrome (STAS) corresponds to H-Y negative Swyer syndrome. (2) Familial testicular agenesis syndrome (FTAS) is H-Y negative Swyer syndrome showing an X-linked recessive pedigree pattern. The mutation is probably homologous to that of the wood lemming. The phenotype of STAS and FTAS is identical even though the mutation is probably on the Y in STAS and on the X in FTAS. (3) In familial testicular dysgenesis syndrome (FTDS), the patients are H-Y positive and have a female phenotype and streak gonads; the streak gonads may contain testis-like tumoral structures. (See report of 3 sisters by Moreira-Filho et al. (1979) and cases of Wolf (1979).) The XY gonadal agenesis syndrome is a separate disorder (see 273250).

Passarge and Wolf (1981) pointed out that there are 2 groups of patients with XY gonadal dysgenesis (Swyer syndrome) and that each of these may be heterogeneous. One group is the H-Y antigen-positive form, which may represent a 'receptor disease.' The second is the H-Y antigen-negative form, which may be due to mutation in the H-Y generating system, either of the structural gene (presumably autosomal) or of a controlling gene (on the sex chromosomes). It may be only the H-Y antigen-positive cases that are at risk for gonadoblastoma or dysgerminoma.

See 233300 for discussion of the XX type of gonadal dysgenesis.

Inheritance

Simpson et al. (1981) reported 3 pedigrees of XY gonadal dysgenesis consistent with X-linked inheritance.

German et al. (1978) suggested that there is a gene on the X chromosome that blocks the testis-determining function of H-Y (which was then a leading candidate for TDF, testis-determining factor). However, it was later shown that TDF and H-Y antigen map to different parts of the Y chromosome with TDF being absent and H-Y antigen being present in XY females with Y short arm deletions (Simpson et al., 1987). See 278850. It appeared that 46,XY women had premature ovarian involution, with resulting 'streak gonads.' Families such as that of Barr et al. (1967) described above may indicate that the mutation is 'leaky.' The pedigree pattern was equally consistent with X-linked recessive or autosomal dominant inheritance. Indeed, Allard et al. (1972) observed transmission through a normal male, arguing for autosomal inheritance.

Nazareth et al. (1979) found H-Y positivity in a sporadic case occurring in an offspring of first-cousin parents. They favored recessive inheritance; see 233420.

Cytogenetics

De Arce et al. (1992) contributed further support of this hypothesis by demonstrating lack of gonadoblastoma in a 14-year-old girl who was a mosaic for 45X/46X-isodicentric Y. The anomalous Y chromosome showed no fluorescent distal Yq. In another patient, an 8-year-old girl with 45X/46XY karyotype, bilateral gonadoblastoma developed in her rudimentary ovaries at the age of 8. Her normal Y chromosome showed the characteristic distal fluorescence seen in her father's Y chromosome. Using Y chromosome probes, De Arce et al. (1992) demonstrated the Y chromosome in the paraffin blocks of the ovarian tissue of both girls.

Wachtel (1979) and Wachtel et al. (1980) suggested the existence of 4 'causes' of XY gonadal dysgenesis: (1) mutational suppression of H-Y structural genes by regulatory elements of the X chromosome or failure of an X-linked structural gene (in association with H-Y negative somatic cell phenotype); (2) failure of H-Y antigen to engage its gonadal receptor (in association with the H-Y positive somatic cell phenotype); (3) loss of the critical moiety of H-Y genes in deleted or translocated Y chromosome (in association with H-Y negative or intermediate somatic cell phenotype); and (4) presence of XY-XO mosaicism.

(Small deletions in the short arm of the Y chromosome can result in 46,XY females (Disteche et al., 1986). The 2 patients reported by Disteche et al. (1986) had some signs of Turner syndrome, including congenital lymphedema and primary amenorrhea with streak gonads, but were of normal height. One of the patients had bilateral gonadoblastoma. Several Y-chromosome-specific DNA probes were found to be deleted in the 2 patients. DNA analysis showed that the 2 deletions were different, but included a common overlapping region likely to contain the testis-determining factor (TDF) gene.)

Bernstein et al. (1980) observed an abnormal band on Xp in a 46,XY female and her 46,XY female fetal sib. Despite the presence of an intact Y chromosome, neither had testicular differentiation and both were H-Y negative. Giemsa banding suggested duplication of p21 and p22. The maternal grandmother, mother and a younger sister, all phenotypically normal, had a karyotype 46,XXp+. The proband had profound psychomotor retardation, and both sibs had multiple congenital malformations. (The second sib was ascertained by amniocentesis for prenatal diagnosis followed by elective abortion.) Multiple congenital anomalies in the proband included ventricular septal defect, cleft palate, asymmetric skull and facies, prognathic jaw, low-set ears, and clinodactyly V. When the girl died at 5 year of age, postmortem studies showed hypoplastic uterus and fallopian tubes. Histologic examination of the uterine adnexa revealed an area of ovarian stroma with scattered degenerative follicles. There was no testicular morphology, and the external genitalia were those of a normal 5-year-old female. The second affected sib, the product of a pregnancy terminated at 20 weeks, showed ovaries containing numerous follicles and germ cells. As in the proband, there was no evidence of testicular morphology. Wachtel (1998) referred to other cases of XY sex reversal in subjects with Xp duplication and chromosomal abnormalities resembling those in the family reported by Bernstein et al. (1980). This suggested occurrence of a gene on Xp, duplication of which can block development of the testis in an XY fetus. The gonads begin to develop as ovaries, but in the absence of the second X chromosome, the germ cells die, the follicles become atretic, and the ovaries degenerate.

Cytogenetic duplication of the X chromosome in males is a rare event usually characterized by a significant degree of phenotypic abnormality, which can include sex reversal despite an apparently normal Y chromosome. Arn et al. (1994) reported 2 half-brothers with maternally inherited cytogenetic duplications of Xp and sex reversal; the absence of dysmorphic features in mother and children was thought to be because of the relatively small extent of the duplication. Comparison with previous reports allowed the putative sex reversing locus (SRVX) to be assigned to a 5- to 10-Mb segment between Xp22.11 and Xp21.2, which includes the DMD locus. The regional assignment may help in the isolation of SRVX mutations that may cause sex reversal in the 90% of sex-reversed women with XY gonadal dysgenesis who do not have detectable mutations of the SRY gene.

Mapping

Mapping studies by hybridization to DNA from somatic cell hybrids containing various fragments of the X chromosome suggested that the sequence on the X chromosome maps to region Xp22.3-p21.

Molecular Genetics

Page et al. (1987) cloned a 230-kb segment of the human Y chromosome thought to contain some or all of the TDF gene. The cloned region spanned the deletion in a female who carried all but 160 kb of the Y. Homologous sequences were found within the sex-determining region of the mouse Y chromosome.

Jager et al. (1990) demonstrated a mutation in SRY in 1 out of 12 sex-reversed XY females with gonadal dysgenesis who had no large deletions of the short arm of the Y chromosome. They found a 4-nucleotide deletion in the part of the SRY gene that encodes a conserved DNA-binding motif. A frameshift presumably led to a nonfunctional protein. Mutation occurred de novo, because the father had a normal SRY sequence. This is strong evidence that SRY is TDF. The de novo G-to-A mutation led to a change from methionine to isoleucine at a residue that lies within the putative DNA-binding motif of SRY and is identical in all SRY and SRY-related genes. (TDF and SRY are written Tdy and Sry in the mouse.)

Heterogeneity

Vilain et al. (1992) described a family in which all 5 XY individuals in 2 generations had a single basepair substitution resulting in an amino acid change in the conserved domain of the SRY open reading frame (480000.0004). A G-to-C change at nucleotide 588 resulted in substitution of leucine for valine. Three of the individuals were XY sex-reversed females and 2 were XY males. One of the males had 8 children; all were phenotypic females, 2 of whom were sex-reversed XY females carrying the mutation mentioned. Several models were proposed to explain association between a sequence variant in SRY and 2 alternative sex phenotypes. These included the existence of alleles at an unlinked locus.

McElreavey et al. (1992) described an XY sex-reversed female with pure gonadal dysgenesis who harbored a de novo nonsense mutation in SRY, which resulted directly in the formation of a stop codon in the putative DNA-binding motif. A C-to-T transition at nucleotide 687 changed a glutamine codon (CAG) to a termination codon (TAG); see 480000.0005. The patient, referred to as the 'propositus,' was a phenotypic female who presented at age 20 years for primary amenorrhea. Treatment with estrogen induced menstruation and slight enlargement of the breasts which were underdeveloped. Laparotomy showed 2 streak gonads without germ cells or remnants of tubes.

Harley et al. (1992) found point mutations in the region of the SRY gene encoding the high mobility group (HMG) box in 5 XY females. (The HMG box is related to that present in the T-cell-specific, DNA binding protein TCF-1 (142410).) In 4 cases, the binding activity of mutant SRY protein for the AACAAAG core sequence was negligible; in the fifth case, DNA binding was reduced. In the SRY gene in a 46,XY female, Muller et al. (1992) demonstrated an A-to-T transversion of nucleotide 684 in the open reading frame, resulting in a change of lysine (AAG) to a stop codon (UAG) The patient had gonadoblastoma.

Pathogenesis

Page et al. (1987) advanced several hypotheses to explain the existence of the X-linked locus. One hypothesis was inconsistent with the prevailing notion of a dominantly acting sex-determining factor unique to the Y chromosome and suggested that the X and Y loci are functionally interchangeable, that both are testis determining, and that the X locus is subject to X-chromosome inactivation. According to this model, sex is determined by the total number of expressed X and Y loci: a single dose is female determining, while a double (or greater) dose is male determining. The addition of an X-derived transgene to the genome of an XX embryo should result in testis differentiation, as long as that transgene is not subject to X inactivation. Increased expression of the X-chromosomal locus could explain the presence of testicular tissue in XX hermaphrodites and the rare Y-negative XX males, who lack the TDF locus of the Y chromosome. Although some XY females lack TDF as judged by Y-DNA analysis, others do not have discernible deletions. These unexplained XY females may have point mutations in TDF or in genes that function in conjunction with or downstream of TDF. The model mentioned above suggests that mutation in the X-chromosomal locus (at Xp22.3-p21) could cause XY embryos to develop as females.

However, Berta et al. (1990) and Jager et al. (1990) presented compelling evidence that the mutation in one type of XY female gonadal dysgenesis is not on the X but on the Y chromosome. In the human sex-determining region in a 35-kb interval near the pseudoautosomal boundary of the Y chromosome, there is a candidate gene for testis-determining factor, termed SRY ('sex-reversed, Y,' from mouse terminology), which is conserved and specific to the Y chromosome in all mammals tested (Sinclair et al., 1990); see 480000. (Cherfas (1991) stated that SRY stands for 'sex-determining region Y.' This is a nice presumption and perhaps in its present usage should be so considered, but it does not indicate the true historical derivation.)

Clinical Management

Moreira-Filho et al. (1979) suggested that the H-Y antigen status in the Swyer syndrome may be a useful indicator of whether removal of the gonads is necessary to avoid malignancy.

See also Boczkowski (1976); Ghosh et al. (1978); Herbst et al. (1978); Judd et al. (1970); Koopman et al. (1991); Koopman et al. (1990); Mann et al. (1983); Wolf et al. (1980)

Allard, S.; Cadotte, M.; Boivin, Y.: Dysgenesie gonadique pure familiare et gonadoblastome. Un. Med. Canada 101: 448-452, 1972.

Arn, P.; Chen, H.; Tuck-Muller, C. M.; Mankinen, C.; Wachtel, G.; Li, S.; Shen, C. C.; Wachtel, S. S.: SRVX, a sex reversing locus in Xp21.2-p22.11. Hum. Genet. 93: 389-393, 1994.
Baron, J.; Rucki, T.; Simm, S.: Familial gonadal malformations. Gynaecologia 153: 298-308, 1962.
Barr, M. L.; Carr, D. H.; Plunkett, E. R.; Soltan, H. C.; Wiens, R. G.: Male pseudohermaphroditism and pure gonadal dysgenesis in sisters. Am. J. Obstet. Gynec. 99: 1047-1055, 1967.
Bernstein, R.; Koo, G. C.; Wachtel, S. S.: Abnormalities of the X-chromosome in 46,XY female sibs with ovaries. Science 207: 768-769, 1980.
Berta, P.; Hawkins, J. R.; Sinclair, A. H.; Taylor, A.; Griffiths, B. L.; Goodfellow, P. N.; Fellous, M.: Genetic evidence equating SRY and the testis-determining factor. Nature 348: 448-450, 1990.
Boczkowski, K.: Familial occurrence of gonadal tumors in XY females with breast development. Hum. Genet. 33: 289-294, 1976.
Chemke, J.; Carmichael, R.; Stewart, J. M.; Geer, R. H.; Robinson, A.: Familial XY gonadal dysgenesis. J. Med. Genet. 7: 105-111, 1970.
Cherfas, J.: Sex and the single gene. Science 252: 782 only, 1991.
Cohen, M. M.; Shaw, M. W.: Two XY siblings with gonadal dysgenesis and a female phenotype. New Eng. J. Med. 272: 1083-1088, 1965.
De Arce, M. A.; Costigan, C.; Gosden, J. R.; Lawler, M.; Humphries, P.: Further evidence consistent with Yqh as an indicator of risk of gonadal blastoma in Y-bearing mosaic Turner syndrome. Clin. Genet. 41: 28-32, 1992.
Disteche, C. M.; Casanova, M.; Saal, H.; Friedman, C.; Sybert, V.; Graham, J.; Thuline, H.; Page, D.; Fellous, M.: Small deletions of the short arm of the Y chromosome in 46,XY females. Proc. Nat. Acad. Sci. 83: 7841-7844, 1986.
Espiner, E. A.; Veale, A. M.; Sands, V. E.; Fitzgerald, P. H.: Familial syndrome of streak gonads and normal male karyotype in five phenotypic females. New Eng. J. Med. 283: 6-11, 1970.
Fine, G.; Mellinger, R. C.; Canton, J. N.: Gonadoblastoma occurring in a patient with familial gonadal dysgenesis. Am. J. Clin. Path. 38: 615-629, 1962.
Frasier, S. D.; Bashore, R. A.; Mosier, H. D.: Gonadoblastoma associated with pure gonadal dysgenesis in monozygous twins. J. Pediat. 64: 740-745, 1964.
German, J.; Simpson, J. L.; Chaganti, R. S. K.: Genetically determined sex-reversal in 46,XY humans. Science 202: 53-56, 1978.
Ghosh, S. N.; Shah, P. M.; Gharpure, H. M.: Absence of H-Y antigen in XY females with dysgenetic gonads. Nature 276: 180-181, 1978.
Harley, V. R.; Jackson, D. I.; Hextall, P. J.; Hawkins, J. R.; Berkovitz, G. D.; Sockanathan, S.; Lovell-Badge, R.; Goodfellow, P. N.: DNA binding activity of recombinant SRY from normal males and XY females. Science 255: 453-456, 1992.
Herbst, E. W.; Fredga, K.; Frank, F.; Winking, H.; Gropp, A.: Cytological identification of two X-chromosome types in the wood lemming (Myopus schisticolor). Chromosoma 69: 185-191, 1978.
Jager, R. J.; Anvret, M.; Hall, K.; Scherer, G.: A human XY female with a frame shift mutation in the candidate testis-determining gene SRY. Nature 348: 452-454, 1990.
Judd, H. L.; Scully, R. E.; Atkins, L.; Neer, R. M.; Kliman, B.: Pure gonadal dysgenesis with progressive hirsutism: demonstration of testosterone production by gonadal streaks. New Eng. J. Med. 282: 881-885, 1970.
Koopman, P.; Gubbay, J.; Vivian, N.; Goodfellow, P.; Lovell-Badge, R.: Male development of chromosomally female mice transgenic for Sry. Nature 351: 117-121, 1991.
Koopman, P.; Munsterberg, A.; Capel, B.; Vivian, N.; Lovell-Badge, R.: Expression of a candidate sex-determining gene during mouse testis differentiation. Nature 348: 450-452, 1990.
Lukusa, T.; Fryns, J. P.; van den Berghe, H.: Gonadoblastoma and Y chromosome fluorescence. Clin. Genet. 29: 311-316, 1986.
Mann, J. R.; Corkery, J. J.; Fisher, H. J. W.; Cameron, A. H.; Mayerova, A.; Wolf, U.; Kennaugh, A. A.; Woolley, V.: The X-linked recessive form of XY gonadal dysgenesis with a high incidence of gonadal germ cell tumours: clinical and genetic studies. J. Med. Genet. 20: 264-270, 1983.
McElreavey, K. D.; Vilain, E.; Boucekkine, C.; Vidaud, M.; Jaubert, F.; Richaud, F.; Fellous, M.: XY sex reversal associated with a nonsense mutation in SRY. Genomics 13: 838-840, 1992.
Moreira-Filho, C. A.; Toledo, S. P. A.; Bagnolli, V. R.; Frota-Pessoa, O.; Bisi, H.; Wajntal, A.: H-Y antigen in Swyer syndrome and the genetics of XY gonadal dysgenesis. Hum. Genet. 53: 51-56, 1979.
Muller, J.; Schwartz, M.; Skakkebaek, N. E.: Analysis of the sex-determining region of the Y chromosome (SRY) in sex reversed patients: point-mutation in SRY causing sex-reversion in a 46,XY female. J. Clin. Endocr. Metab. 75: 331-333, 1992.
Nazareth, H. R. S.; Moreira-Filho, C. A.; Cunha, A. J. B.; Vieira-Filho, J. P. B.; Lengyel, A. M. J.; Lima, M. C.: H-Y antigens in 46,XY pure testicular dysgenesis. Am. J. Med. Genet. 3: 149-154, 1979.
Page, D. C.; Mosher, R.; Simpson, E. M.; Fisher, E. M. C.; Mardon, G.; Pollack, J.; McGillivray, B.; de la Chapelle, A.; Brown, L. G.: The sex-determining region of the human Y chromosome encodes a finger protein. Cell 51: 1091-1104, 1987.
Passarge, E.; Wolf, U.: Genetic heterogeneity of XY gonadal dysgenesis (Swyer syndrome): H-Y antigen-negative XY gonadal dysgenesis associated with inflammatory bowel disease. Am. J. Med. Genet. 8: 437-441, 1981.
Rose, L. I.; Underwood, R. H.; Williams, G. H.; Pinkus, G. S.: Pure gonadal dysgenesis: studies of in vitro androgen metabolism. Am. J. Med. 57: 957-961, 1974.
Rushton, D. I.: XY gonadal dysgenesis. (Letter) Lancet I: 209 only, 1979.
Simpson, E.; Chandler, P.; Goulmy, E.; Disteche, C. M.; Ferguson-Smith, M. A.; Page, D. C.: Separation of the genetic loci for the H-Y antigen and for testis determination on human Y chromosome. Nature 326: 876-878, 1987.
Simpson, J. L.; Blagowidow, N.; Martin, A. O.: XY gonadal dysgenesis: genetic heterogeneity based upon clinical observations, H-Y antigen status, and segregation analysis. Hum. Genet. 58: 91-97, 1981.
Sinclair, A. H.; Berta, P.; Palmer, M. S.; Hawkins, J. R.; Griffiths, B. L.; Smith, M. J.; Foster, J. W.; Frischauf, A.-M.; Lovell-Badge, R.; Goodfellow, P. N.: A gene from the human sex-determining region encodes a protein with homology to a conserved DNA-binding motif. Nature 346: 240-244, 1990.
Sternberg, W. H.; Barclay, D. L.; Kloepfer, H. W.: Familial XY gonadal dysgenesis. New Eng. J. Med. 278: 695-700, 1968.
Taylor, H.; Barter, R. H.; Jacobson, C. B.: Neoplasms of dysgenetic gonads. Am. J. Obstet. Gynec. 96: 816-823, 1966.
Vilain, E.; McElreavey, K.; Jaubert, F.; Raymond, J.-P.; Richaud, F.; Fellous, M.: Familial case with sequence variant in the testis-determining region associated with two sex phenotypes. Am. J. Hum. Genet. 50: 1008-1011, 1992.
Wachtel, S. S.: X-linked sex-reversing genes. Cytogenet. Cell Genet. 80: 222-225, 1998.
Wachtel, S. S.: The genetics of intersexuality: clinical and theoretic perspectives. Obstet. Gynec. 54: 671-685, 1979.
Wachtel, S. S.; Koo, G. C.; de la Chapelle, A.; Kallio, H.; Heyman, J. M.; Miller, O. J.: H-Y antigen in 46,XY gonadal dysgenesis. Hum. Genet. 54: 25-30, 1980.
Wolf, U.: XY gonadal dysgenesis and the H-Y antigen. Hum. Genet. 47: 269-277, 1979.
Wolf, U.; Fraccaro, M.; Mayerova, A.; Hecht, T.; Maraschio, P.; Hameister, H.: A gene controlling H-Y antigen on the X chromosome: tentative assignment by deletion mapping to Xp223. Hum. Genet. 54: 149-154, 1980.

GONADAL DYSGENESIS, XY TYPE (233420)

JJM NOTE

This entity predisposes to gonadal neoplasm.

Wachtel et al. (1980) concluded that the XY female type of gonadal dysgenesis may arise through any of 4 ways: loss of the H-Y structural genes from the Y through mutation or deletion; loss of the function of an X-chromosomal regulator of H-Y (see 306100); loss of the H-Y receptor; or XY-XO mosaicism. The third of these mechanisms is probably involved in an autosomal recessive form of XY gonadal dysgenesis. These individuals are H-Y positive (Wachtel et al. (1980) reported 5 such cases), but the H-Y antigen is ineffectual because of lack of receptor on target cells (see 143150). One model of the H-Y antigen generating system suggests the requirement of at least 3 genes: the H-Y inducer gene (I) on the Y chromosome, one or more repressor genes (R) on the X, and the structural gene (S) on an autosome. The testis-determining function of the ubiquitously distributed H-Y antigen relies on a specific embryonic gonadal somatic cell receptor. The X-Y molecule may have two parts: one responsible for its organogenetic function, and one determining its antigenic activity. Moreira-Filho et al. (1982) reported an H-Y negative case of XY gonadal dysgenesis in an offspring of first-cousin parents. They suggested that the patient may be homozygous for an S locus mutation affecting both parts of the H-Y molecule. An alternative designation they used for this condition was familial testicular agenesis syndrome (FTAS). They pointed to the cases of Haseltine and Ohno (1981) and Ghosh et al. (1978) as likewise representing S mutations. Cases of H-Y antigen-positive XY gonadal dysgenesis differ from the cases just mentioned by the findings in the gonads which show testicular primordia; hence the term familial testicular dysgenesis syndrome (FTDS) suggested by Moreira-Filho et al. (1979). Parental consanguinity (Nazareth et al., 1979) and familial occurrence (Moltz et al., 1981) support an autosomal mutation, which may alter the H-Y antigen receptor or a regulator function involved in normal testicular organogenesis. Alternatively, mutation at the S locus might alter organogenetic function without destroying antigenic activity. Gonadal tumors occur frequently in FTDS but apparently not in FTAS. Thus, the H-Y antigen status can be used as a practical guide as to whether extirpation of the streak gonads is necessary.

Prompted by the observation that mutations of the SOX9 gene cause campomelic dysplasia (114290), which is frequently associated with 46,XY sex reversal, Kwok et al. (1996) examined the entire coding region of the SOX9 gene in 30 46,XY patients with abnormalities of sexual development but without any skeletal abnormalities. Of the 30 patients, gonadal dysgenesis was diagnosed in 12 and partial gonadal dysgenesis in 14. Except for a C-to-T polymorphism at nucleotide 507 in 1 person, no other abnormalities of the SOX9 were found.

Fuqua et al. (1996) used linkage and sequence analyses to test the role of X-linked and various autosomal genes in the etiology of the familial 46,XY partial gonadal dysgenesis. No linkage was detected between the trait and 29 markers on the X chromosome. Linkage analysis using a polymorphism within the 3-prime untranslated region of the WT1 gene excluded the involvement of autosomal loci. Similarly, linkage to 4 microsatellites on the distal short arm of chromosome 9 was excluded. Also, linkage analysis of a locus close to the SOX9 gene as well as analysis of the coding region of the SOX9 gene suggested that this gene was not associated with the trait.

Fuqua, J. S.; Sher, E. S.; Fechner, P. Y.; Ostrer, H.; Oddeux, C.; Schafer, A. J.; Rosales, T. O.; Migeon, C. J.; Berkovitz, G. D.: Linkage analysis of a kindred

with inherited 46,XY partial gonadal dysgenesis. J. Clin. Endocr. Metab. 81: 4479-4483, 1996.
Ghosh, S. N.; Shah, P. M.; Gharpure, H. M.: Absence of H-Y antigen in XY females with dysgenetic gonads. Nature 276: 180 only, 1978.
Haseltine, F. P.; Ohno, S.: Mechanisms of gonadal differentiation. Science 211: 1272-1278, 1981.
Kwok, C.; Goodfellow, P. N.; Hawkins, J. R.: Evidence to exclude SOX9 as a candidate gene for XY sex reversal without skeletal malformation. J. Med. Genet. 33: 800-801, 1996.
Moltz, L.; Schwartz, U.; Pickartz, H.; Hammerstein, J.; Wolf, U.: XY gonadal dysgenesis: aberrant testicular differentiation in the presence of H-Y antigen. Obstet. Gynec. 58: 17-25, 1981.
Moreira-Filho, C. A.; Frota-Pessoa, O.; Vianna-Morgante, A. M.; Chu, T. H.; Bisi, H.; Gollop, T. R.: H-Y antigen generating and receptor systems in abnormal sexual development. Am. J. Med. Genet. 13: 401-411, 1982.
Moreira-Filho, C. A.; Toledo, S. P. A.; Bagnolli, V. R.; Frota-Pessoa, O.; Bisi, H.; Wajntal, A.: H-Y antigen in Swyer syndrome and the genetics of XY gonadal dysgenesis. Hum. Genet. 53: 51-56, 1979.
Nazareth, H. R. S.; Moreira-Filho, C. A.; Cunha, A. J. B.; Vieira-Filho, J. P. B.; Lengyel, A. M. J.; Lima, M. C.: H-Y antigen in 46,XY pure testicular dysgenesis. Am. J. Med. Genet. 3: 149-154, 1979.
Wachtel, S. S.; Koo, G. C.; de la Chapelle, A.; Kallio, H.; Heyman, J. M.; Miller, O. J.: H-Y antigen in 46,XY gonadal dysgenesis. Hum. Genet. 54: 25-30, 1980.

CRYPTOPHTHALMOS WITH OTHER MALFORMATIONS (FRASER SYNDROME) (*219000)

JJM NOTE

This entity is included because of one case with gonadoblastoma. The patient also had gonadal dysgenesis, which is the likely pathway to malignancy. Hence, the Fraser syndrome per se is not now known to predispose to neoplasia.

In each of 2 sibships, Fraser (1962) observed 2 sisters affected at birth by various combinations: cryptophthalmos; absent or malformed lacrimal ducts; middle and outer ear malformations; high palate; cleavage along the midplane of nares and tongue; hypertelorism; laryngeal stenosis; syndactyly; wide separation of symphysis pubis; displacement of umbilicus and nipples; primitive mesentery of small bowel; maldeveloped kidneys; fusion of labia and enlargement of clitoris; and bicornuate uterus and malformed fallopian tubes. In each sibship, 1 sister was stillborn and the other viable. Sex chromatin was positive in both surviving infants. Neither set of parents was consanguineous. See Bowen syndrome (211200) for a comparable but probably distinct syndrome of multiple congenital malformations.

Gupta and Saxena (1962) reported cryptophthalmos in 2 offspring of consanguineous parents. In 1 case it was unilateral and death occurred at 1 month. In the other the cryptophthalmos was bilateral and was accompanied by congenital deafness, undescended testes, small penis with hypospadias, and other deformities. The older literature on cryptophthalmos with associated malformations was reviewed by Duke-Elder (1963).

Francois (1969) described affected brother and sister and gave a comprehensive review, pointing out the rather frequent examples of parental consanguinity (about 15% of cases) and of familial cases. Syndactyly is a feature of many of the cases. Francannet et al. (1990) reported instances of parental consanguinity: in 1 family, 3 children were born of a first-cousin marriage, and in a second, an affected fetus resulted from a double first-cousin marriage. An isolated case was reported by Ide and Wollschlaeger (1969). The parents were not related. The patient had syndactyly, hair extending forward over the temples, deformity of the nares and external auditory meati, etc. Azevedo et al. (1973) reported 4 cases in 2 sibships, each with consanguineous parents. Syndactyly and other malformations were present in some.

Lurie and Cherstvoy (1984) reported 4 families with the cryptophthalmos-syndactyly syndrome. Perinatal death occurred in 9 affected members. Autopsy, performed in 6 cases, showed renal agenesis (bilateral in 3, unilateral in 3). One family had at least 4 affected sibs and 2 others had 2 affected children each. No consanguinity was established. Burn and Marwood (1982) reported 3 sibs with Fraser syndrome presenting as bilateral renal agenesis. Only 1 of the 3 sibs had cryptophthalmos (which was unilateral). Because cryptophthalmos is not an essential part of the syndrome, Fraser syndrome is a preferable designation. Mortimer et al. (1985) reported monozygotic twins concordant for bilateral renal agenesis and other features consistent with Fraser syndrome: syndactyly and laryngeal stenosis.

From their personal experience and a review of the literature, Koenig and Spranger (1986) identified 5 cases of the cryptophthalmos-syndactyly syndrome without cryptophthalmos. Thus, the eye lesions are apparently nonobligatory components of the syndrome. The diagnosis of Fraser syndrome should be entertained in patients with a combination of acrofacial and urogenital malformations with or without cryptophthalmos. Thomas et al. (1986) also emphasized the occurrence of the cryptophthalmos syndrome without cryptophthalmos—an argument for using the eponym Fraser syndrome. In a literature review, they found 124 cases of cryptophthalmos of which 27 were isolated and 97 associated with multiple malformations. They found that isolated cryptophthalmos was sporadic in 16 cases and familial in 11. The familial cases were in 3 families with vertical transmission (123570). Some of the most characteristic malformations of the Fraser syndrome occur in areas that remain temporarily fused in utero: the eyelids, the digits, and the vagina. Since separation of the eyelids and digits involves a process of controlled necrosis, Thomas et al. (1986) speculated about a defect in programmed cell death. They also drew attention to the parallel with experimental teratogenesis by means of hypovitaminosis A in the pig and rat. Malformations reminiscent of the Fraser syndrome are produced. Thomas et al. (1986) asked whether defects in the metabolism of retinoids may play a role in pathogenesis. Meinecke (1986) also contributed to the understanding of the Fraser syndrome without cryptophthalmos. Greenberg et al. (1986) described associated gonadal dysgenesis in an infant girl with Fraser syndrome: gonadoblastoma developed in a dysgenetic ovary.

Gattuso et al. (1987) gave a quantitative estimate of the frequency of the several clinical manifestations on the basis of 3 new cases and 68 published cases. Craniofacial abnormalities were reported in all, cryptophthalmos in 93%, and syndactyly in 54%. They quoted Warkany (1971) as citing the description by Pliny the Elder in the first century A.D. of a family with 3 children born with a membrane over the eye. Bialer and Wilson (1988) reported a child who according to the strict criteria proposed by Thomas et al. (1986) might not be considered to have syndromal cryptophthalmos; however, the child had several malformations considered minor in that classification and was born of parents who were related in a complex way, giving the proband a coefficient of consanguinity of 0.086.

Boyd et al. (1988) presented postmortem findings in 11 cases of probable Fraser syndrome; 8 were neonatal deaths, 1 was a stillbirth, and 2 were midtrimester fetuses. The authors concluded that prenatal diagnosis by ultrasound examination of eyes, digits, and kidneys should detect the severe form of the syndrome. Serville et al. (1989) demonstrated the feasibility of ultrasonographic diagnosis of the Fraser syndrome at 18 weeks' gestation. They suggested that the diagnosis could be made if 2 of the following signs are present: obstructive uropathy, microphthalmia, syndactyly, and oligohydramnios. Schauer et al. (1990) made the diagnosis at 18.5 weeks' gestation on the basis of sonography. Both the female fetus and the phenotypically normal father had a chromosome anomaly: inv(9)(p11q21). An earlier born infant had Fraser syndrome and the same chromosome 9 inversion. Ramsing et al. (1990) presented the clinical and autopsy findings in 2 fetuses and 1 newborn infant with Fraser syndrome.

Stevens et al. (1994) observed pulmonary hyperplasia and laryngeal stenosis in 2 sibs with Fraser cryptothalmos syndrome. Markedly enlarged echogenic lungs were demonstrable at 16 and 17 weeks of gestation. A third unrelated patient with Fraser syndrome had laryngeal stenosis, renal agenesis, and normal lung development. Reports of 3 additional cases of pulmonary hyperplasia in the Fraser syndrome were found. In all of these cases the likely mechanism for pulmonary hyperplasia was retention of fetal lung fluid by laryngeal or tracheal obstruction.

Pankau et al. (1994) described a newborn female with total syndactyly of the 3rd and 4th fingers, partial syndactyly of all other fingers, postaxial polydactyly of feet, bilateral renal agenesis, and bilateral anophthalmia. The child was born of consanguineous Turkish parents. Although the infant had normal eyelids with palpebrae and eyelashes and had histologically confirmed anophthalmia and postaxial polydactyly, Pankau et al. (1994) considered this case to be an unusual manifestation of Fraser syndrome.

On the occasion of the retirement of George Fraser, Reardon (1997) provided a biographical and historical note on Fraser's contributions to clinical genetics.

See also Codere et al. (1981); Fraser (1966)

Azevedo, E. S.; Biondi, J.; Ramalho, M.: Cryptophthalmos in two families from Bahia, Brazil. J. Med. Genet. 10: 389-392, 1973.
Bialer, M. G.; Wilson, W. G.: Syndromic cryptophthalmos. (Letter) Am. J. Med. Genet. 30: 835-837, 1988.
Boyd, P. A.; Keeling, J. W.; Lindenbaum, R. H.: Fraser syndrome (cryptophthalmos-syndactyly syndrome): a review of eleven cases with postmortem findings. Am. J. Med. Genet. 31: 159-168, 1988.
Burn, J.; Marwood, R. P.: Fraser syndrome presenting as bilateral renal agenesis in three sibs. J. Med. Genet. 19: 360-361, 1982.
Codere, F.; Brownstein, S.; Chen, M. F.: Cryptophthalmos syndrome with bilateral renal agenesis. Am. J. Ophthal. 91: 737-742, 1981.
Duke-Elder, S.: System of Ophthalmology, Normal and Abnormal Development. St. Louis: C. V. Mosby (pub.) 3, Part 2: 1963. Pp. 829-834.

Francannet, C.; Lefrancois, P.; Dechelotte, P.; Robert, E.; Malpuech, G.; Robert, J. M.: Fraser syndrome with renal agenesis in two consanguineous Turkish families. Am. J. Med. Genet. 36: 477-479, 1990.
Francois, J.: Syndrome malformatif avec cryptophtalmie. Acta Genet. Med. Gemellol. 18: 18-50, 1969.
Fraser, G. R.: XX chromosomes and renal agenesis. (Letter) Lancet I: 1427 only, 1966.
Fraser, G. R.: Our genetical 'load:' a review of some aspects of genetical variation. Ann. Hum. Genet. 25: 387-415, 1962.
Gattuso, J.; Patton, M. A.; Baraitser, M.: The clinical spectrum of the Fraser syndrome: report of three new cases and review. J. Med. Genet. 24: 549-555, 1987.
Greenberg, F.; Keenan, B.; De Yanis, V.; Finegold, M.: Gonadal dysgenesis and gonadoblastoma in situ in a female with Fraser (cryptophthalmos) syndrome. J. Pediat. 108: 952-954, 1986.
Gupta, S. P.; Saxena, R. C.: Cryptophthalmos. Brit. J. Ophthal. 46: 629-632, 1962.
Ide, C. H.; Wollschlaeger, P. B.: Multiple congenital abnormalities associated with cryptophthalmia. Arch. Ophthal. 81: 638-644, 1969.
Koenig, R.; Spranger, J.: Cryptophthalmos-syndactyly syndrome without cryptophthalmos. Clin. Genet. 29: 413-416, 1986.
Lurie, I. W.; Cherstvoy, E. D.: Renal agenesis as a diagnostic feature of the cryptophthalmos-syndactyly syndrome. Clin. Genet. 25: 528-532, 1984.
Meinecke, P.: Cryptophthalmos-syndactyly syndrome without cryptophthalmos. (Letter) Clin. Genet. 30: 527-528, 1986.
Mortimer, G.; McEwan, H. P.; Yates, J. R. W.: Fraser syndrome presenting as monozygotic twins with bilateral renal agenesis. J. Med. Genet. 22: 76-78, 1985.
Pankau, R.; Partsch, C. J.; Janig, U.; Meinecke, R.: Fraser (cryptophthalmos-syndactyly) syndrome: a case with bilateral anophthalmia but presence of normal eyelids. Genet. Counsel. 5: 191-194, 1994.
Ramsing, M.; Rehder, H.; Holzgreve, W.; Meinecke, P.; Lenz, W.: Fraser syndrome (cryptophthalmos with syndactyly) in the fetus and newborn. Clin. Genet. 37: 84-96, 1990.
Reardon, W.: Historical note: Dr George Fraser. J. Audiol. Med. 6: 185-189, 1997.
Schauer, G. M.; Dunn, L. K.; Godmilow, L.; Eagle, R. C., Jr.; Knisely, A. S.: Prenatal diagnosis of Fraser syndrome at 18.5 weeks gestation, with autopsy findings at 19 weeks. Am. J. Med. Genet. 37: 583-591, 1990.
Serville, F.; Carles, D.; Broussin, B.: Fraser syndrome: prenatal ultrasonic detection. (Letter) Am. J. Med. Genet. 32: 561-563, 1989.
Stevens, C. A.; McClanahan, C.; Steck, A.; Shiel, F. O'M.; Carey, J. C.: Pulmonary hyperplasia in the Fraser cryptophthalmos syndrome. Am. J. Med. Genet. 52: 427-431, 1994.
Thomas, I. T.; Frias, J. L.; Felix, V.; Sanchez de Leon, L.; Hernandez, R. A.; Jones, M. C.: Isolated and syndromic cryptophthalmos. Am. J. Med. Genet. 25: 85-98, 1986.
Warkany, J.: Congenital Malformations. Chicago: Year Book Med. Publ. (pub.) 1971. Pp. 18 only.

Testicular Tumors

TESTICULAR TUMOR (*273300)

JJM NOTE

Testicular tumors, in general, are emerging to have considerable prenatal and familial determinants. Cryptorchidism, seen in about 0.5% of newborns, is a large risk factor of testicular tumors, perhaps via the mechanism of dysgenetic gonads. Regardless, orchiopexy is probably the most common operation done to prevent morbidity from malignancy. Familial aggregation is seen in many case reports and is often the highest relative risk factor identified by case-control studies. Further investigations might pay more attention to histologic subtypes than previous ones have.

Human male germ cell tumors (GCT) result from malignant transformation of premeiotic or early meiotic germ cells and exhibit embryonal-like differentiation of the 3 germinal layers. The genetic basis of malignant transformation and expression of differentiated phenotypes by the transformed cells are poorly understood. Germ cell tumors (GCTs) are the most common malignant neoplasm of males aged 15 to 34 and a major cause of death due to cancer in this age group. Hutter et al. (1967) reviewed the reports of testicular tumors in brothers and in twins and reported affected brothers. Gustavson et al. (1975) reported bilateral testicular teratoma in 2 infant brothers with XXY Klinefelter syndrome. One of them also had hydrocephalus due to stenosis of the aqueduct of Sylvius.

Familial occurrence of the Klinefelter syndrome is rare. The association of the Klinefelter syndrome and testicular teratoma may be more than coincidental because they have been observed together in other cases and many testicular teratoma are both X-chromatin and Y-chromatin positive suggesting that they are XXY. Raghavan et al. (1980) reported a father who had sequential bilateral seminomas and a son who had embryonal cell carcinoma and seminoma. The authors reviewed 5 other reports of testicular tumors in father and son, as well as 7 reports of concordant monozygotic twin pairs and 11 reports of nontwin brothers. The report of Raghavan et al. (1980) illustrates the dominant inheritance of hereditary tumors and their bilaterality (e.g., acoustic neuroma, retinoblastoma, pheochromocytoma, etc.). The sons (and other first-degree relatives) of men with bilateral tumors may be at particular risk.

Shinohara et al. (1980) reported mature testicular teratoma in 2 first cousins. Furthermore, the common grandparents were consanguineous, being related as first cousins. The parent (i.e., the parent involved in the consanguinity) of the teratoma-carrying boys was the mother in one case and the father in the other. In a 10-member sibship in a Spanish-American family, DiBella (1983) described testicular neoplasm in 3 brothers, benign ovarian neoplasms in 2 sisters, suspected benign tumors of the uterus in 2 additional sisters, and a suspected testicular mass in a fourth brother. Lynch et al. (1985) described the infantile form of embryonal carcinoma of the testis in a 5-year-old boy and in a 23-year-old man who was the maternal half-brother of his mother. Copeland et al. (1986) reported testicular embryonal carcinoma in 2 brothers and a first cousin.

Von der Maase et al. (1986) found carcinoma in situ in the contralateral testis in 27 of 500 patients (5.4%) with unilateral testicular germ cell cancer. The estimated risk of developing invasive growth from the contralateral testicular cancer was 40% within 3 years and 50% within 5 years. None of the 473 patients without carcinoma in situ detected by screening biopsy developed contralateral testicular cancer after an observation time ranging from 12 to 96 months. These observations suggested to the authors that there is a subset of this type of testicular cancer that is genetic and has a bilateral predisposition. The authors suggested that all such carcinomas begin as carcinoma in situ. Von der Maase et al. (1986) recommended that all patients with unilateral testicular germ cell cancer should be offered biopsy of the contralateral testis. Of the 27 patients, 16 had a cancer that was labeled seminoma and 11 had a cancer that was considered to be nonseminoma. It would be of great interest to know the median age of the patients with contralateral carcinoma in situ as contrasted with the others. If these represent a subset who had inherited 1 of the 2 mutations according to the Knudson theory, then the patients with contralateral carcinoma in situ should have an earlier average age of development of carcinoma.

Patel et al. (1990) reported 6 cases of familial testicular cancer: 4 father-son pairs, a pair of brothers, and a 23-year-old man who had a maternal uncle with testicular cancer. In the U.K., according to Forman et al. (1992), 42 families with 2 or more cases of testicular cancer were reported to the familial testicular cancer registry. These families included 2 pairs of identical twins, 27 sets of other brothers (25 pairs, 2 triples), 9 father-son pairs, 2 pairs of first cousins, and 2 uncle-nephew pairs. In all, 91 testicular tumors were described in 86 persons. Pure seminoma was present in 46% and other germ cell tumors in 54%. The median age at diagnosis was significantly younger than in a comparable series of nonfamilial patients. The cumulative risk of developing testicular cancer by the age of 50 years for a brother of a patient was estimated to be 2.2%, which results in a relative risk of 9.8 in comparison with the general population. No significant peculiarity of class I HLA type was found in a study of 21 affected sib pairs.

Studying direct preparations and 24-hr cultures, Atkin and Baker (1982) found an isochromosome for the short arm of chromosome 12 in all of 10 seminomas, 1 malignant teratoma, and 1 combined seminoma and teratoma of the testis. (The same workers found a possible isochromosome for 5p in 12 of 18 carcinomas of the cervix.) They also noted a relative excess of normal chromosomes 12 in 4 of 5 of the seminomas analyzed in detail. Castedo et al. (1989) found at least 1 copy of a 12p isochromosome in 8 of 10 seminomas. Thus, the authors concluded that amplification of 1 or more genes on the short arm of chromosome 12 may be important in the development of malignant testicular tumors. Chromosomal changes presumably lead to the malignant phenotype by gene loss, gene modification or gene amplification.

Lothe et al. (1989) found loss of heterozygosity (LOH) for 3p or 11p sequences in 40% of testicular cancers. Samaniego et al. (1990) analyzed the karyotype of 24 male germ cell tumors from both testicular and extragonadal sites and belonging to the histologic categories seminoma, teratoma, embryonal carcinoma, choriocarcinoma, and endodermal sinus tumor. In 90% of tumors, including all histologic subtypes and both gonadal and extragonadal presentation, they found isochromosome 12p. In contrast, they found del(12)(q13-q22) exclusively in nonseminomatous GCTs, and mixed GCTs occurring in 44% of such lesions. They developed a method based on DNA analysis for detecting i(12p) as increased copy number of 12p. Furthermore, they detected cytologic evidence of gene amplification in 12p in the form of homogeneously staining regions (HSRs) and double minute chromosomes in both treated and untreated primary extragonadal and metastatic GCTs.

26,25:Suijkerbuijk et al. (1991, 1992) applied competitive in situ hybridization (CISH) techniques (Kievits et al., 1990) to show that the aberrant chromosome in testicular germ cell tumors is indeed an isochromosome-12p. Other marker chromosomes representing translocation products that involve chromosome 12 were also identified. In the studies, DNAs from 2 rodent-human somatic cell hybrids, containing either a normal chromosome 12 or the p arm of chromosome 12 as their unique human material, were used as probes. (Competitive in situ hybridization, also referred to as chromosome painting, employs large pools of cloned genomic sequences originating from a single human chromosome as probe and involves a preannealing step in the presence of an excess of sonicated total human DNA. It results in complete staining of the particular chromosome in metaphase spreads and in interphase nuclei. Kievits et al., 1990 stated that the approach permits detection of hitherto undetectable chromosomal aberrations.)

In a cytogenetic analysis of 65 consecutively ascertained GCTs with chromosomal abnormalities, Rodriguez et al. (1992) found that an isochromosome for the short arm of chromosome 12 (i(12p)), monosomy 12, and deletions in 12q occurred with frequencies of 86%, 11%, and 20% respectively. Their data strongly suggested that loss of genetic material on 12q characterizes the development of these tumors. To define the region of common deletion in GCTs at the molecular level, Murty et al. (1992) compared germline and tumor genotypes for 8 polymorphic loci in paired normal/tumor DNA samples from 45 GCT patients. Analysis demonstrated 2 regions of loss of constitutional heterozygosity, one at 12q13 and the other at 12q22. One tumor exhibited homozygous deletion of a region of 12q22 which includes the MGF gene (184745). The MGF and KIT (164920) genes have been shown to play key roles in embryonal and postnatal development of germ cells. The MGF gene product constitutes the ligand for the receptor encoded by the KIT protooncogene. They evaluated the expression of these 2 genes by Northern blot analysis in a panel of 3 GCT cell lines and 24 fresh GCT biopsies. Deregulated expression of MGF and KIT, which was discordant between seminomatous and nonseminomatous lesions, was observed. Murty et al. (1994) refined their data on the mapping of male germ cell tumors (MGCTs). Using 5 dinucleotide repeats mapping to 12q22, they found LOH in approximately 41% of tumors; one of the loci, D12S218, showed LOH in 37% of tumors, suggesting the presence of a tumor suppressor gene in its vicinity. In this study, a panel of 66 tumor DNA samples and their corresponding normal cells were investigated.

Because a marker chromosome interpreted as isochromosome 12p is present in most testicular tumors of germ cell origin, Peltomaki et al. (1992) investigated 22 patients with testicular germ-cell tumors by Southern blot hybridization to characterize changes in chromosome 12. In comparison with normal DNA, tumor DNA of 18 patients showed increased dosages of 12p accompanied by a comparable or smaller increase or no change in the dosage of centromeric sequences of chromosome 12. The interpretation offered by the authors was that most testicular tumors had one or several isochromosomes for 12p that were formed by somatic division of the centromere and that the points of breakage and reunion in the centromeric region were different in different tumors. Sex-limited parental imprinting was excluded by the fact that allelic 12p fragments showing increased intensity were paternal in 4 and maternal in 3 of 7 informative cases. Furthermore, the observed patterns of allelic fragments suggested that the marker isochromosome was formed by sister chromatids of 1 homolog number 12 rather than the result of interchange of genetic material between different homologs.

Forman et al. (1992) reported an epidemiologic study that showed an 8- to 10-fold increase in relative risk of testicular cancer to brothers of patients and a 4-fold increase in risk to fathers and sons. Families with multiple cases of testicular cancer are rare and almost all those reported have only 2 affected members. Leahy et al. (1995) performed a sib-pair analysis on 35 families in which there were either 2 or 3 affected brothers. These families were typed for 220 autosomal microsatellite markers spaced 10-20 cM throughout the genome. Six regions that gave a lod score of more than 1.0 on formal linkage analysis or a P value of 0.05 or less using a non-parametric approach were considered as candidate regions for a susceptibility gene. Of particular interest was one region on chromosome 4. A positive lod score of 2.6 on multipoint analysis was obtained with 2 neighboring probes in the region of 4cen-q13.

Although testicular teratomas occur as part of the spectrum of cancers in the Li-Fraumeni syndrome (151623) due to mutations in the P53 gene (191170), the molecular basis of most familial cases of testicular teratoma is unknown. Huddart et al.(1996) raised the question of whether the familial predisposition extends to familial germ cell tumors as well as male germ cell tumors. They identified 3 families suggesting this possibility. In 1, the proband presented with a seminoma at the age of 51, his brother had had a testicular teratoma at the age of 28, and their cousin had an endodermal sinus tumor of the ovary diagnosed at 32 years. In the second family, the index case presented with an undifferentiated malignant teratoma at 28 years of age and his sister was diagnosed with bilateral mature teratomatous cysts at the age of 39. In the third family, the index case presented with a retroperitoneal teratoma at 26 years and his sister was diagnosed with an ovarian dysgerminoma at 45 years. Huddart et al. (1996) noted that none of these families had any features indicative of the Li-Fraumeni syndrome or any other cancer family syndrome. Trentini and Palmieri (1974) and Yule et al. (1994) reported single families with ovarian and testicular germ cell tumors and Jackson (1967) presented a family with multiple cases of dysgerminoma.

Heimdal et al. (1996) found that 51 of 922 (5.5%) Norwegian patients with testicular cancer and 5 of 237 (2.1%) Swedish patients had a relative with confirmed testicular cancer. It was a first-degree relative who was affected in the case of 32 of the probands. Standardized incidence ratios (SIRs) were 10.2 for brothers, 4.3 for fathers, and 5.7 for sons. The estimate for the risk to brothers in the Norwegian part of the sample for development of testicular cancer by the age of 60 was 4.1%. Patients with familial testicular cancer had bilateral tumors more often than sporadic cases (9.8% bilaterality in familial vs 2.8% in sporadic cases; P = 0.02). For patients with seminoma, age of onset was lower in familial than in sporadic cases (32.9 vs 37.6 years; P = 0.06). Heimdal et al. (1996) stated that the prevalence of undescended testis did not seem to be higher in familial than in sporadic testicular cancer.

In a detailed deletion mapping analysis of 67 normal-tumor DNA comparisons using 20 polymorphic markers mapped to 12q22-q24, Murty et al. (1996) identified the limit of the minimal region of deletion at 12q22 between D12S377 (proximal) and D12S296 (distal). They constructed a YAC contig map of a 3-cM region of this band and developed a radiation hybrid (RH) map of the region. The consensus order developed by RH mapping was in good agreement with the YAC STS-content map order. The RH map estimated the distance between the D12S101 and D12S346 to be 246 cR(8000) and the minimal region of deletion to be 141 cR(8000). In addition, 4 genes that were previously mapped to 12q22 were excluded as candidate genes: mast cell growth factor (184745), B-cell translocation gene-1 (109580), thymopoietin (188380), and neural precursor cell expressed, developmentally down-regulated-1 (600372).

Murty and Chaganti (1998) reviewed the genetics of male germ cell tumors. A characteristic of GCTs is high sensitivity to cisplatin-based chemotherapy. Chromosomal and molecular cytogenetic studies identified multiplication of 12p, manifested in i(12p) or tandem duplication of 12p, as a unique change in GCTs which serves as a diagnostic marker. Ectopic overexpression of cyclin D2 (CCND2; 123833), which maps to 12p, as early as in carcinoma in situ, identified CCND2 as a candidate gene in germ cell transformation. Genetic alterations identified in the tumor suppressor genes DCC (120470), RB1 (180200), and nonmetastatic protein-23 (NME1; 156490) in GCTs suggested that their inactivation plays a key role in transformation or differentiation. The exquisite sensitivity of these tumors to chemotherapy is reflected in their overexpression of wildtype p53 protein and lack of TP53 mutations.

See also Zevallos et al. (1983)

Atkin, N. B.; Baker, M. C.: Specific chromosome change, i(12p), in testicular tumours? (Letter) Lancet II: 1349, 1982.

Castedo, S. M. M. J.; de Jong, B.; Oosterhuis, J. W.; Seruca, R.; te Meerman, G. J.; Dam, A.; Koops, H. S.: Cytogenetic analysis of ten human seminomas. Cancer Res. 49: 439-443, 1989.

Copeland, G. P.; Shennan, J. M.; Muranda, C.; Griffiths, D.: Familial occurrence of testicular neoplasia: case report of testicular embryonal carcinoma in 2 brothers and a first cousin. J. Urol. 136: 676-677, 1986.

DiBella, N. J.: Familial gonadal neoplasms. (Letter) New Eng. J. Med. 309: 1389, 1983.

Forman, D.; Oliver, R. T. D.; Brett, A. R.; Marsh, S. G. E.; Moses, J. H.; Bodmer, J. G.; Chilvers, C. E. D.; Pike, M. C.: Familial testicular cancer: a report of the UK family register, estimation of risk and an HLA class 1 sib-pair analysis. Brit. J. Cancer 65: 255-262, 1992.

Gustavson, K.-H.; Gamstorp, I.; Meurling, S.: Bilateral teratoma of testis in two brothers with 47,XXY Klinefelter's syndrome. Clin. Genet. 8: 5-10, 1975.

Heimdal, K.; Olsson, H.; Tretli, S.; Flodgren, P.; Borresen, A.-L.; Fossa, S. D.: Familial testicular cancer in Norway and southern Sweden. Brit. J. Cancer 73: 964-969, 1996.

Huddart, R. A.; Thompson, C.; Houlston, R.; Huddart, R. A.; Nicholls, E. J.; Horwich, A.: Familial predisposition to both male and female germ cell tumours? (Letter) J. Med. Genet. 86, 1996.

Hutter, A. M.; Lynch, J. J.; Shnider, B. I.: Malignant testicular tumors in brothers: a case report. J.A.M.A. 199: 1009-1010, 1967.

Jackson, S. M.: Ovarian dysgerminoma in three generations? J. Med. Genet. 4: 112-113, 1967.

Kievits, T.; Devilee, P.; Wiegant, J.; Wapenaar, M. C.; Cornelisse, C. J.; van Ommen, G. J. B.; Pearson, P. L.: Direct nonradioactive in situ hybridization of somatic cell hybrid DNA to human lymphocyte chromosomes. Cytometry 11: 105-109, 1990.

Leahy, M. G.; Tonks, S.; Moses, J. H.; Brett, A. R.; Huddart, R.; Forman, D.; Oliver, R. T. D.; Bishop, D. T.; Bodmer, J. G.: Candidate regions for a testicular cancer susceptibility gene. Hum. Molec. Genet. 4: 1551-1555, 1995.

Lothe, R. A.; Fossa, S. D.; Stenwig, A. E.; Nakamura, Y.; White, R.; Borresen, A.-L.; Brogger, A.: Loss of 3p or 11p alleles is associated with testicular cancer tumors. Genomics 5: 134-138, 1989.

Lynch, H. T.; Katz, D.; Bogard, P.; Voorhees, G. J.; Lynch, J.; Wagner, C.: Familial embryonal carcinoma in a cancer-prone kindred. Am. J. Med. 78: 891-896, 1985.
Murty, V. V. V. S.; Bosi, G.; Le Blanc-Straceski, J.; Kucherlapati, R.; Chaganti, R. S. K.: Molecular mapping of 12q22 deletions in male germ cell tumors. (Abstract) Cytogenet. Cell Genet. 67: 271-272, 1994.
Murty, V. V. V. S.; Chaganti, R. S. K.: A genetic perspective of male germ cell tumors. Semin. Oncol. 25: 133-144, 1998.
Murty, V. V. V. S.; Houldsworth, J.; Baldwin, S.; Reuter, V.; Hunziker, W.; Besmer, P.; Bosl, G.; Chaganti, R. S. K.: Allelic deletions in the long arm of chromosome 12 identify sites of candidate tumor suppressor genes in male germ cell tumors. Proc. Nat. Acad. Sci. 89: 11006-11010, 1992.
Murty, V. V. V. S.; Renault, B.; Falk, C. T.; Bosl, G. J.; Kucherlapati, R.; Chaganti, R. S. K.: Physical mapping of a commonly deleted region, the site of a candidate tumor suppressor gene, at 12q22 in human male germ cell tumors. Genomics 35: 562-570, 1996.
Patel, S. R.; Kvols, L. K.; Richardson, R. L.: Familial testicular cancer: report of six cases and review of the literature. Mayo Clin. Proc. 65: 804-808, 1990.
Peltomaki, P.; Lothe, R. A.; Borresen, A.-L.; Fossa, S. D.; Brogger, A.; de la Chapelle, A.: Chromosome 12 in human testicular cancer: dosage changes and their parental origin. Cancer Genet. Cytogenet. 64: 21-26, 1992.
Raghavan, D.; Jelihovsky, T.; Fox, R. M.: Father-son testicular malignancy: does genetic anticipation occur? Cancer 45: 1005-1009, 1980.
Rodriguez, E.; Mathew, S.; Mukherjee, A. B.; Reuter, V. E.; Bosl, G. J.; Chaganti, R. S. K.: Analysis of chromosome 12 aneuploidy in interphase cells from human male germ cell tumors by fluorescence in situ hybridization. Genes Chromosomes Cancer 5: 21-29, 1992.
Samaniego, F.; Rodriguez, E.; Houldsworth, J.; Murty, V. V. V. S.; Ladanyi, M.; Lele, K. P.; Chen, Q.; Dmitrovsky, E.; Geller, N. L.; Reuter, V.; Jhanwar, S. C.; Bosl, G. J.; Chaganti, R. S. K.: Cytogenetic and molecular analysis of human male germ cell tumors: chromosome 12 abnormalities and gene amplification. Genes Chromosomes Cancer 1: 289-300, 1990.
Shinohara, M.; Komatsu, H.; Karamura, T.; Yokoyama, M.: Familial testicular teratoma in 2 children: familial report and review of the literature. J. Urol. 123: 552-555, 1980.
Suijkerbuijk, R. F.; Looijenga, L.; de Jong, B.; Oosterhuis, J. W.; Cassiman, J. J.; Geurts van Kessel, A.: Verification of isochromosome 12p and identification of other chromosome 12 aberrations in gonadal and extragonadal human germ cell tumors by bicolor double fluorescence in situ hybridization. Cancer Genet. Cytogenet. 63: 8-16, 1992.
Suijkerbuijk, R. F.; van de Veen, A. Y.; van Echten, J.; Buys, C. H. C. M.; de Jong, B.; Oosterhuis, J. W.; Warburton, D. A.; Cassiman, J. J.; Schonk, D.; Geurts van Kessel, A.: Demonstration of the genuine iso-12p character of the standard marker chromosome of testicular germ cell tumors and identification of further chromosome 12 aberrations by competitive in situ hybridization. Am. J. Hum. Genet. 48: 269-273, 1991.
Trentini, G. P.; Palmieri, B.: An unusual case of gonadic germinal tumor in a brother and sister. Cancer 33: 250-255, 1974.
von der Maase, H.; Rorth, M.; Walbom-Jorgensen, S.; Sorensen, B. L.; Christophersen, I. S.; Hald, T.; Jacobsen, G. K.; Berthelsen, J. G.; Skakkebaek, N. E.: Carcinoma in situ of contralateral testis in patients with testicular germ cell cancer: study of 27 cases in 500 patients. Brit. Med. J. 293: 1398-1401, 1986.
Yule, S. M.; Dawes, P. J.; Malcolm, A. J.; Pearson, A. D.: Occurrence of seminoma and dysgerminoma in father and daughter. Pediat. Hemat. Oncol. 11;-213, 1994.
Zevallos, M.; Snyder, R. N.; Sadoff, L.; Cooper, J. F.: Testicular neoplasm in identical twins: a case report. J.A.M.A. 250: 645-646, 1983.

ICHTHYOSIS, X-LINKED (*308100)

JJM NOTE

This entity, a gendodermatosis, is included because of two cases of testicular tumor: a seminoma and an embryonal cancer.

In addition to the genetic difference between X-linked ichthyosis and autosomal dominant ichthyosis vulgaris (146700), clinical and histologic differences exist (Wells and Jennings, 1967). In the X-linked form, onset is at birth and scalp, ears, neck and one or more flexures are involved, with more striking scaling on the abdomen than on the back, and extension of the scaling down the front of the leg onto the dorsum of the foot. Histologically the epidermis is atrophic in ichthyosis vulgaris and hypertrophic in the X-linked variety. Lake et al. (1991) developed a histochemical method for demonstration of hexanol dehydrogenase activity by use of a simple staining method on cryostat sections of skin biopsies as an adjunct to the biochemical assay in the differentiation of the various forms of ichthyosis.

Csorsz (1928) described 2 presumed homozygous, affected females. See biographic account of Czeizel (1979). Schlammadinger et al. (1987) restudied the Hungarian family first reported by Csorsz (1928). A female member of the family who as a child was described as having ichthyosis was found on restudy to have no sign thereof and by enzyme studies was thought to be heterozygous. A second affected female, presumably homozygous, was apparently not investigated from the point of view of enzyme levels. No pregnancy-related complications, no hypogenitalism or cryptorchidism (Traupe and Happle, 1983), and no deep corneal opacities (Lykkesfeldt et al., 1985) were found in the family.

In 1929 Orel found in the literature 10 families with the X-linked form. Sever et al. (1968) described deep corneal opacities in all of 17 affected males and in 7 of 8 heterozygous females; there were no symptoms. Went et al. (1969) found mild abnormality of the skin in about one-fourth of heterozygotes. Schnyder (1970) gave a useful classification of the inherited ichthyoses. Solomon and Schoen (1971) reported a patient with XO Turner syndrome and ichthyosis which by the pedigree and by its clinical features was X-linked. Passarge et al. (1971) described an X-linked pedigree in which the clinical picture was intermediate between those of the 'classic' X-linked form and the autosomal dominant form. Possibly this is an allelic form.

The human placenta is rich in the enzymes 3-beta-steroid sulfatase and arylsulfatase. France and Liggins (1969) and others described deficiency of 3-beta-steroid sulfatase in the placenta. France et al. (1973) showed that arylsulfatase is also deficient in these instances. France and Downey (1974) showed that the deficiency is limited to the placenta; the mother did not show deficiency in her tissues. Multiple sibs with the defect have been observed. All those that France and Downey (1974) specifically referred to were males. 62,61:Koppe et al. (1977, 1978), Shapiro and Weiss (1977), and Shapiro et al. (1978) showed that the deficiency is X-linked and that it is expressed in postnatal life as X-linked ichthyosis. Placental steroid sulfatase (EC 3.1.6.2.) deficiency is manifested by low estriol levels in urine and plasma, delay in the onset of labor, relative refractoriness to oxytoxic agents, and increased stillbirth frequency. The liveborn infants are, however, clinically normal at birth. Gant et al. (1977) showed that steroid sulfatase resides in the chorion laeve; the amnion is totally devoid of this activity. Hameister et al. (1979) observed partial placental steroid sulfatase deficiency, detected by the finding of low estriol excretion and failure of induction of labor. The fibroblasts and placenta in one case showed 34% normal activity. Enzyme activity in the mother was normal. None of the cases had developed ichthyosis by the age of 6 months. Cocultivation with normal fibroblasts resulted in mixing (intermediate enzyme level), not cross-correction.

X-linked ichthyosis is fundamentally the same entity as placental steroid sulfatase deficiency. Here is an example of affinity ('lumping') of phenotypes thought previously to be separate—the opposite of genetic heterogeneity. Priority for publication of sulfatase deficiency in X-linked ichthyosis rests with Jobsis et al. (1976). In multiple sulfatase deficiency (272200), ichthyosis is observed also (Shapiro, 1977). Shapiro et al. (1978) were not able to find a case of X-linked ichthyosis without steroid sulfatase deficiency. Placental steroid sulfatase deficiency was identified through the practice of measuring estriol in the urine of pregnant women as an indication of an abnormality of gestation. Koppe et al. (1978) described 3 children with placental sulfatase deficiency who developed ichthyosis of the X-linked form at 2 to 8 months of age.

Kerr et al. (1964) presented evidence suggesting that the X-linked ichthyosis locus may be within 'mappable' distance of the Xg locus (314700). Closer situation of the Xg and ichthyosis loci was indicated by studies of Adam et al. (1969) who estimated the recombination fraction as 0.105 and of Went et al. (1969) who found a value of 0.115. Close linkage with the deutan, protan and G6PD loci was excluded (Filippi and Meera Khan, 1968; Adam et al., 1969). In studies of X-chromosome anomalies in man-mouse hybrids, Mohandas et al. (1979) concluded that the steroid sulfatase locus is in the p22-pter segment of X. Because of the linkage of X-linked ichthyosis to Xg, the somatic cell hybridization work indicated that Xg is also on Xp.

It appears that the steroid sulfatase locus does not lyonize; in fibroblasts doubly heterozygous for steroid sulfatase and G6PD, steroid sulfatase was expressed in all clones regardless of whether the X chromosome was active or not as indicated by the G6PD activity of the clone (Shapiro et al., 1979). Conflicting results were obtained by Balazs et al. (1979), who concluded that the STS locus is situated between Xq13 and Xq24 and that it lyonizes regularly. Tracing back from an STS-deficient mouse cell line, they showed that STS is X-linked in the mouse from which the line was derived. Muller et al. (1980) demonstrated that steroid sulfatase activity is higher in normal females than in normal males, a finding consistent with nonlyonization. Furthermore, they demonstrated that heterozygotes are clearly distinguished both from normal females and from hemizygous affected males. Thus, heterozygote detection is not impeded by the usual vagaries of lyonization. It is curious that the STS locus has maintained its X chromosomal localization despite the fact that it is not inactivated. Although not normally inactivated, the STS locus may be inactivated when located on an aberrant X chromosome (Ropers et al., 1981). Similar inactivation patterns have been reported for the Xg locus.

Craig and Tolley (1986) reviewed the relationship between STS and mammalian X-chromosome conservation. They surveyed information available on the female:male ratio for STS levels in fibroblasts and placenta and came up with the average value of 1.6 rather than the expected 2.0 if no lyonization occurs. Ratios observed for peripheral white cells were even lower. Furthermore, individuals with multiple X chromosomes do not exhibit the proportionately high levels of enzymes anticipated. Migeon et al. (1982) pursued the possibility of incomplete inactivation (or incomplete escape from inactivation) by examining STS levels in fibroblast clones from women heterozygous for deficiency of both STS and G6PD. Their study demonstrated that the wildtype STS allele was expressed at about half the level from the inactive chromosome compared with its expression from the active homolog. Partial inactivation is the most satisfactory explanation for this observation and would account for the deviation from a strict dosage relationship observed for the female:male ratio. XO persons and STS-deficiency heterozygotes have enzyme levels below the normal male range.

Epstein et al. (1981) found that low-density lipoproteins from patients with X-linked ichthyosis have abnormally rapid anodic electrophoretic mobility. The finding is explained by increased plasma cholesterol sulfate concentration in these patients. The cholesterol sulfate is found predominantly in the low-density lipoprotein fraction of plasma.

Tiepolo et al. (1980) found steroid sulfatase to be severely deficient in a boy with ichthyosis and nullisomy for the distal portion of Xp; the mother, who was monosomic for this segment, had steroid sulfatase levels in the heterozygous range. Muller et al. (1981) also used deletion mapping to assign the STS locus to Xp22.3; they found 'almost undetectable' levels of the enzyme in 2 brothers with the same defect as in the patient of Tiepolo et al. (1980) and levels like those of heterozygotes in the mother. Wieacker et al. (1983) studied the linkage between the restriction fragment length polymorphism defined by the cloned DNA sequence RC8 and X-linked ichthyosis. At least 2 crossovers were found among 9 meioses in an informative family, suggesting that RC8 and STS may be about 25 cM apart. Since STS is 15 cM proximal to the Xg locus and since the RC8 and Duchenne muscular dystrophy are closely linked, DMD may be 50 cM or more from Xg. Metaxotou et al. (1983) described a 14-year-old boy with nullisomy for the Xp22-pter segment and with hypogonadism, ichthyosis, and mental retardation. There was, furthermore, t(X;Y)(p22;q11). The mother was monosomic for the deleted segment of Xp and had the same X;Y translocation.

In a prospective half-side trial, Lykkesfeldt and Hoyer (1983) found that a topical cream containing 10% cholesterol effected considerable improvement, suggesting that reduction in the cholesterol content of the stratum corneum may be responsible for abnormal cornification in this disorder. Traupe and Happle (1983) found 7 instances of cryptorchidism in a series of 25 patients with STS deficiency and suggested a causal relationship. Lykkesfeldt et al. (1983) reported 2 men with STS deficiency and testicular cancer. One patient had the left testis removed for seminoma at age 21 and the right testis removed for embryonal cancer at age 26. The second had the left testis removed for seminoma at age 31. Both had normally descended testes but a nephew of the second patient had STS deficiency ichthyosis and bilateral inguinal cryptorchidism. The testis has potent STS activity and may have a role in gonadal steroid regulation but this is not known in detail. Lykkesfeldt et al. (1985) studied 76 cases in 50 kindreds; 42 kindreds had multiple cases. Maldescent of the testes was noted in 9 patients. Testicular cancer occurred in 2 males with normally descended testes (as noted earlier). Corneal opacities, not impairing visual acuity, were seen in 14 of 28 males by slit-lamp examination. Jay et al. (1968) found stromal opacities as a late feature of X-linked ichthyosis including the carrier state; in their series, all patients over 25 years (but only 5 of 8 patients under 25 years of age) showed this abnormality. Macsai and Doshi (1994) described a 73-year-old man with X-linked ichthyosis and steroid sulfatase deficiency in whom superficial corneal opacities were found. The opacities were granular in nature, involving the subepithelial and anterior stromal layers.

Traupe et al. (1984) described a typical instance of X-linked ichthyosis with severe hypogenitalism and hypogonadism and deficiency of steroid sulfatase. The patient had an affected maternal first cousin, and the maternal grandfather of the 2 affected males was affected. They also described a presumably isolated case in a Pakistani male; in this instance, STS activity and serum lipoprotein electrophoresis were normal. The disorders in the 2 males were clinically indistinguishable. Munke et al. (1983) also documented heterogeneity in the syndrome of ichthyosis and hypogonadism, with STS deficiency in some cases and not in others. See 308200 for a discussion of Rud syndrome, in which ichthyosis and hypogonadism are combined with neurologic abnormality. Pike et al. (1989) described X-linked ichthyosis, steroid sulfatase deficiency, and hypogonadism, in males in 3 separate sibships connected through females. Endocrinologic testing suggested hypogonadotropic hypogonadism with deficiency of luteinizing hormone. Garcia Perez and Crespo (1981) and Stoll et al. (1983) reported 2 families containing 5 boys with the combination of hypertrophic pyloric stenosis and X-linked recessive ichthyosis. Vogel et al. (1984) confirmed the nonactivation of the STS gene as well as of the ARSC gene (the f form of ARSC; 301780). Elias et al. (1984) concluded that accumulation of undegraded cholesterol sulfate is responsible for scale-formation in steroid sulfatase deficiency. Curry et al. (1984) found that the steroid sulfatase, Xg, and MIC2X loci as well as the locus for X-linked chondrodysplasia punctata were apparently absent in males with deletion of Xp22.32.

Ross et al. (1985) described a family in which 4 brothers had X-linked ichthyosis and nullisomy for Xpter-Xp22.3 resulting from an Xp to Yq translocation with the entire Y short arm and Xpter-Xp22.3 deleted: 46,Y,t(x;y)(Xqter-Xp22.3 Yq11-Yqter). Cultured cells were completely deficient in steroid sulfatase. Xg blood group data were not provided. Ballabio et al. (1986) found apparent close linkage of STS and Kallmann syndrome (308700). Shapiro (1987) suggested that the STS locus must be very close to the exchanging (pseudoautosomal) segment of Xp. He found that it is usually deleted in XX males in whom TDF is presumably translocated to the end of Xp. Anomalous exchange between the X and Y chromosomes may account for a high frequency of X-linked ichthyosis, about 1 in 5,000 in many populations (Shapiro, 1987). Conary et al. (1987) analyzed a cDNA clone specific for steroid sulfatase and showed that DNA from 2 patients with steroid sulfatase deficiency and X-linked ichthyosis did not hybridize with it. DNA from a third patient showed a normal hybridization pattern. Yen et al. (1987) isolated and sequenced cDNA clones containing the entire coding sequence of STS. A pseudogene was found on Yq, suggesting a recent pericentric inversion. Of 10 patients with inherited STS deficiency, 8 were found to have complete gene deletions. Yen et al. (1987) raised the possibility that STS deficiency results from aberrant X-Y interchange. Comparative in situ hybridization in various primate species demonstrated a pseudoautosomal location of the human ANT3 gene (300151) and an X-specific location for the STS gene throughout the higher primate species up to the New World monkeys. However, Toder et al. (1995) found that ANT3 and STS map together on an autosome of 2 prosimian species of the genera Lemur and Eulemur. These results suggested an autosome-to-X/Y translocation after the simians radiated from the prosimians, resulting in a pseudoautosomal location of genes, such as ANT3 and STS. In simian primates, STS then became X-specific by a pericentric inversion in the Y chromosome followed by mutational inactivation of the Y allele.

Gillard et al. (1987) used a RFLP closely linked to STS to demonstrate deletion of the STS locus in 8 of 9 families with X-linked ichthyosis. In an extension of these studies, Gillard et al. (1987) found deletion of a DNA marker tightly linked to STS in affected males in 12 of 15 families with X-linked ichthyosis. From an expression library, Ballabio et al. (1987) isolated several cDNA clones by screening with antibodies prepared against steroid sulfatase. One of these clones had been assigned to Xp22.3 by mapping with a somatic cell hybrid panel and by in situ hybridization. In 12 steroid sulfatase-deficient patients, including 8 cases of classic ichthyosis, they found that the genomic sequences detected by an STS clone had been deleted. Andria et al. (1987) suggested that the 2 patients described by Sunohara et al. (1986) and the 2 families with several affected males that they had reported (Andria et al., 1984; Ballabio et al., 1986) may have had a microdeletion of the short arm of the X chromosome involving the steroid sulfatase gene and a closely located gene responsible for X-linked Kallmann syndrome (308700). Linkage studies in 1 of their families showed no crossing over with a probe that maps to Xpter-p22, suggesting that the disease locus was indeed in that area. They pointed out that the patients described by Sunohara et al. (1986) showed peculiar manifestations that have been reported in patients with Kallmann syndrome, including renal agenesis and neurologic disorders such as mirror movements. Shapiro and Yen (1987) responded to the suggestion that the condition in these patients may represent a microcytogenetic disorder (Schmickel, 1986). They described their progress in characterization of the STS gene. They stated that STS is encoded by about 80 kb of DNA and that homologous but nonfunctional sequences were found on the long arm of the Y chromosome in the patients of Sunohara et al. (1986). Indeed, they found a complete deletion of the STS gene with continued presence of MIC2 sequences, which are located more distally on the X chromosome, in both the X and Y chromosomes. In studies of 9 unrelated patients with simple X-linked ichthyosis, they found 7 with complete deletion of the STS gene and 1 with a partial 5-prime deletion. Only 1 subject had an intact STS gene. Shapiro et al. (1987) reviewed 4 studies with a total of 45 unrelated cases of X-linked ichthyosis; 41 were found to have a deletion. In 16 men from 10 unrelated Italian families affected with STS deficiency, Ballabio et al. (1987) found no hybridization signal when Southern blotting was done with an STS cDNA probe. The sample included 2 families with a total of 7 cases combining X-linked ichthyosis and Kallmann syndrome.

Bonifas et al. (1987) cloned a cDNA for STS from a placental cDNA library. Hybridization of this cDNA with EcoRI-digested genomic DNA indicated that patients from 14 of 15 apparently unrelated families had gross deletions of the STS gene. One patient had genomic DNA fragments that were identical to those from normal individuals. Ballabio et al. (1988) described a family in which an X/Y translocation was transmitted through females in at least 3 generations and was accompanied by X-linked ichthyosis in 2 male cousins connected through carrier sisters. The breakpoints of the translocation were at Xp22 and Yq11. Both boys showed mental retardation and facies consistent with chondrodysplasia punctata. One of the boys was short and was demonstrated to have tracheal and tarsal stippling, consistent with chondrodysplasia punctata, as well as distal phalangeal hypoplasia at the age of 11 months. Ballabio et al. (1988) demonstrated that the translocated region of the Y chromosome includes the steroid sulfatase pseudogene; the X/Y translocation may have thus been derived from a recombination event between homologous regions located on the short arm of the X chromosome and the long arm of the Y chromosome. They pointed out that all cases of X/Y translocation involving the STS gene showed mental retardation, whereas in those patients with X/Y translocation without involvement of the STS gene, mental retardation has not been present. Their review of families with X/Y translocations, including the family reported by them, showed 24 affected persons in 30 informative meioses. This suggests preferential fertilization of the oocyte carrying the aberrant X chromosome. Wirth et al. (1988) found deletion involving the STS locus in 8 of 9 unrelated families with X-linked ichthyosis. Three patients in the ninth family had no evident deletion when studied with 2 probes. Since approximately 90% of STS-deficient persons have large deletions at the STS locus, Shapiro et al. (1989) investigated t he breakpoints to identify potential sequences prone to undergo either intrachromosomal or interchromosomal nonhomologous recombination. Most of the breakpoints occurred at a distance from the STS gene itself and were, therefore, difficult to characterize. They found 1 subject who had an entirely intragenic deletion of 40 kb, permitting cloning and sequencing of the deletion junction. Bernatowicz et al. (1992) stated that only 2 patients with partial deletions of the STS gene had been reported. One was the person studied by Shapiro et al. (1989) who had an intragenic deletion extending from intron 1 to intron 5. The breakpoints contained no detectable secondary structure or repetitive elements except an 8-bp direct repeat located 8 bp 5-prime of the deletion junction. Bernatowicz et al. (1992) characterized the breakpoints in the other patient who had a deletion of the 3-prime end of the STS gene. The deletion started within intron 7 of the gene and extended over 150 kb downstream toward the centromere. Analysis of sequences flanking the deletion breakpoints revealed 3 bp of homology. (The 5-prime end of the STS gene is oriented toward the telomere.)

By several approaches, Mohandas et al. (1990) demonstrated that in XX males in whom the testis-determining factor gene (TDF) is transferred from a Y chromatid to an X chromatid, the breakpoint is distal to the STS locus, which is retained on the TDF-bearing X chromosome.

Robledo et al. (1995) described a Sardinian kindred in which congenital ichthyosis was associated with normal levels of steroid sulfatase and a normal pattern on Southern blot analysis suggesting the presence of an intact STS gene. Although the pedigree pattern was entirely consistent with X-linked recessive inheritance, the ichthyosis was found to segregate independently of genetic polymorphisms detected by probes mapping to Xp22.3, where the STS locus maps. The search for linkage to markers elsewhere on the X chromosome had not been successful. Robledo et al. (1995) concluded that there may be a form of X-linked ichthyosis due to some mechanism other than STS deficiency.

Shapiro (1997) pointed out that ichthyosis with STS deficiency is related to increased water loss in the skin with resulting desiccation. Ichthyosis occurs with other disorders of lipid metabolism in the skin, including Refsum disease (266500) and Sjogren-Larsson syndrome (270200). The detection of STS deficiency, which may have a frequency of 1 in 3,000 males, has increased greatly by routine use of screening of estrogen metabolism in pregnancy.

Most individuals with X-linked ichthyosis have extensive deletions of the STS gene; however, point mutations had been reported in 3 patients. Alperin and Shapiro (1997) added 3 more point mutations. All 6 were unique single basepair substitutions. The mutations were located in a 105-amino acid region of the C-terminal half of the polypeptide. Five of the 6 mutations involved the substitution of 1 amino acid for another (e.g., 308100.0004); the sixth mutation involved a splice junction and resulted in frameshift causing premature termination (308100.0006). The authors noted that all of the affected residues are conserved to some degree within the sulfatase family. The 6 mutations were reproduced in normal STS cDNA and transiently expressed in STS-deficient cells. All 6 mutant sequences directed the expression of STS protein that lacked enzymatic activity.

Animal Model

Eicher (1974) speculated that the 'scurfy' (sf) mutation in the mouse may be homologous to X-linked ichthyosis of man. Buckle et al. (1985) alluded to ichthyosis with male hypogonadism (see 308200) as an entity separate from ichthyosis with steroid sulfatase deficiency and homologous to 'scurfy' in the mouse. From comparative mapping of the X chromosomes of mouse and man, they predicted that this possibly separate human condition may be determined by a mutation on Xp near OTC (311250).

Gartler and Rivest (1983) confirmed X-linkage of STS in the mouse by the study of oocytes of XX and XO mice. Assays of STS in kidney tissue of these mice indicated dosage compensation for the gene, which is different from the situation in man.

Keitges et al. (1985) resolved the question of whether STS is autosomal or X-linked in the mouse. They showed that it is X-linked, or, if you will, pseudoautosomal. Their results indirectly indicated the existence of a functional STS allele on the Y chromosome which undergoes obligatory recombination during meiosis with the X-linked allele. Their experiments consisted of crosses between STS-deficient C3H/An male mice and STS-normal XO animals. STS should map to the same region as Sxr ('sex-reversed') which, from its equal transmission to male and female offspring, appears also to be on a homologous pairing segment of X and Y. Keitges and Gartler (1986) confirmed the existence of functional Y-linked and X-linked alleles for Sts in the mouse by dosage studies in XO, XX, and XY mice and by clonal analysis in fibroblast cell cultures from mice heterozygous for steroid sulfatase deficiency and for the X-linked electrophoretic variant of phosphoglycerate kinase (311800).

As an explanation for the fact that the STS locus is in the pseudoautosomal segment of the X and Y in the mouse but not in man, Yen et al. (1988) suggested that a pericentric inversion of the Y chromosome occurred during primate evolution, disrupting the former pseudoautosomal arrangement of these genes. In man, an STS pseudogene is present on the long arm of the Y chromosome. Obligatory recombination of pairing segments of the human X and Y appears to be excluded by results from study of a polymorphism of red cell antigen 12E7 (coded by MIC2; 313470) which showed a complex sex-limited expression of variation in 12E7 levels (Goodfellow and Tippett, 1981).

Lyon et al. (1990) described hematologic abnormalities in the 'scurfy' mouse and raised a question of homology to Wiskott-Aldrich syndrome (301000) rather than STS deficiency. In comparing gene order in mouse and human, Laval and Boyd (1993) found evidence for a partial inversion of gene order within a homologous segment of the X chromosome between DXS255 and TIMP. In the 2 species, the scurfy/WAS phenotypes and the GATA1 (305371)/Gf-1 loci mapped to the same region of the X chromosome. This supported the possibility that scurfy and WAS are indeed homologous.

Salido et al. (1996) used a degenerate-primer RT-PCR approach to amplify a conserved fragment of a rat STS cDNA that was then used to clone a mouse Sts cDNA. This 2.3-kb cDNA shows 75% similarity with the rat STS, while it is only 63% similar to human STS. Sts was mapped physically by Salido et al. (1996) to the distal end of the mouse sex chromosomes (X and Y). Backcross studies placed Sts distal to the 'obligatory' crossover in male meiosis. Cloning the mouse Sts gene using human reagents (STS cDNA and anti-STS antibodies) had previously failed, suggesting to them a substantial divergence between the human and mouse genes. The fact that the gene is pseudoautosomal in the mouse and not in the human likewise suggested divergence. Kipling et al. (1996) reported physical linkage of 3 mouse pseudoautosomal region (PAR) probes: DXYHgu1, DXYMov15, and the telomeric sequence (TTAGGG)n. They found that in the mouse the Sts locus maps distal to these 3 probes, indicating that there is an internal array of the telomere sequence in the PAR. They found that pseudoautosomal PacI restriction fragments, up to 2 Mb in size, are unstable in C57BL/6 x C57BL/6 crosses. New alleles, often several hundred kilobases different in size, occurred at a sex-averaged rate of approximately 30% per allele. Kipling et al. (1996) noted that such frequent large-scale germline genome rearrangements were without precedent in mammals.

Several attempts to clone the mouse homolog of the human STS gene failed, suggesting a substantial divergence between these genes. However, Kawano et al. (1989) showed that partial N-terminal sequence from purified rat liver STS was very similar to its human counterpart, and sequence comparisons revealed several domains that are conserved among all the sulfatases characterized to date. Li et al. (1996) used a degenerate-primer RT-PCR approach to amplify a 321-bp fragment from rat liver cDNA, which was used as a probe to clone and characterize the complete cDNA. Comparison of the complete coding region between the rat and human genes shows 66% homology both at the DNA and the protein levels. Authenticity of the cloned cDNA was indicated by the fact that STS activity was conferred to STS(-) A9 cells upon transfection with a rat Sts expression construct. While Sts is located in the mouse pseudoautosomal region, both physical and genetic mapping demonstrated that STS is not pseudoautosomal in the rat. The rat gene is only 8.2 kb long, while the human gene spans over 146 kb.

Ropers and Wiberg (1982) demonstrated that STS is also X-linked and noninactivated in the wood lemming, Myopus schisticolor. According to the work of Cooper et al. (1984), steroid sulfatase is not X-linked in Australian marsupials. Correlated with this are the facts that the 'basic' marsupial X is smaller than the 'basic' eutherian X, and the X and Y of Australian marsupials lack a pairing segment.

Freiberg et al. (1997) used X-linked ichthyosis to develop a model of corrective gene therapy to human skin in vivo. They produced a retroviral expression vector and used it for STS gene transfer to primary keratinocytes in patients with this disorder. Transduction was associated with restoration of full-length STS protein expression as well as steroid sulfatase activity in proportion to the number of proviral integrations in XLI cells. Transduced and uncorrected XLI keratinocytes, along with normal controls, were then grafted onto immunodeficient mice to regenerate full thickness human epidermis. Unmodified XLI keratinocytes regenerated a hyperkeratotic epidermis lacking STS expression with defective skin barrier function, effectively recapitulating the human disease in vivo. Transduced XLI keratinocytes from the same patients, however, regenerated epidermis histologically indistinguishable from that formed by keratinocytes from patients with normal skin. Transduced XLI epidermis demonstrated STS expression in vivo by immunostaining, as well as a normalization of histologic appearance at 5 weeks post-grafting. In addition, transduced XLI epidermis demonstrated a return of barrier function parameters to normal.

Cuevas-Covarrubias et al. (1995) investigated 5 apparently non-familial cases of X-linked recessive ichthyosis and their mothers. None of the XLI patients showed STS activity; 4 mothers had an activity level compatible with the carrier state and only 1 showed a normal level, indicating that her son had a de novo mutation.

See also Ballabio et al. (1985); Ballabio et al. (1987); Burns (1983); Chance and Gartler (1983); Chang et al. (1981); Cockayne (1933); DeUnamuno et al. (1977); Epstein and Bonifas (1985); Epstein et al. (1981); Gartler and Andina (1976); Gillard et al. (1987); Gladstein et al. (1979); Happle (1979); Harris (1947); Jobsis et al. (1983); Jobsis et al. (1980); Kerr and Wells (1965); Kubilus et al. (1979); Lykkesfeldt et al. (1985); Lykkesfeldt et al. (1984); Marinkovic-Ilsen et al. (1978); Meyer et al. (1979); Mohandas et al. (1980); Monroe and Chang (1987); Muller et al. (1980); Okano et al. (1985); Orel (1929); Shapiro et al. (1978); Steinmann et al. (1981); Tabei and Heinrichs (1976); Traupe and Ropers (1982); Turpin et al. (1945); Willard and Holmes (1984)

Adam, A.; Ziprkowski, L.; Feinstein, A.; Sanger, R.; Tippett, P.; Gavin, J.; Race, R. R.: Linkage relations of X-borne ichthyosis to the Xg blood groups and to other markers of the X in Israelis. Ann. Hum. Genet. 32: 323-332, 1969.

Alperin, E. S.; Shapiro, L. J.: Characterization of point mutations with X-linked ichthyosis: effects on the structure and function of the steroid sulfatase protein. J. Biol. Chem. 272: 20756-20763, 1997.

Andria, G.; Ballabio, A.; Parenti, G.: X-linked ichthyosis due to steroid sulfatase deficiency associated with hypogonadism and anosmia. (Letter) Ann. Neurol. 22: 98, 1987.

Andria, G.; Ballabio, A.; Parenti, G.; Di Maio, S.; Piccirillo, A.: Steroid sulphatase deficiency is present in patients with the syndrome 'ichthyosis and male hypogonadism' and with 'Rud syndrome.'. J. Inherit. Metab. Dis. 7: 159-160, 1984.

Bairoch, A.: Personal Communication. Geneva, Switzerland 5/13/1994.

Balazs, I.; Filippi, G.; Rinaldi, A.; Grzeschik, K.-H.; Siniscalco, M.: Studies on X-linked ichthyosis and steroid sulfatase in man, mice and their hybrids. (Abstract) Cytogenet. Cell Genet. 25: 133, 1979.

Ballabio, A.: Personal Communication. Houston, Tex. 5/18/1994.

Ballabio, A.; Parenti, G.; Carrozzo, R.; Coppa, G.; Felici, L.; Migliori, V.; Silengo, M.; Franceschini, P.; Andria, G.: X/Y translocation in a family with X-linked ichthyosis, chondrodysplasia punctata, and mental retardation: DNA analysis reveals deletion of the steroid sulphatase gene and translocation of its Y pseudogene. Clin. Genet. 34: 31-37, 1988.

Ballabio, A.; Parenti, G.; Carrozzo, R.; Sebastio, G.; Andria, G.; Buckle, V.; Fraser, N.; Craig, I.; Rocchi, M.; Romeo, G.; Jobsis, A. C.; Persico, M. G.: Isolation and characterization of a steroid sulfatase cDNA clone: genomic deletions in patients with X-chromosome-linked ichthyosis. Proc. Nat. Acad. Sci. 84: 4519-4523, 1987.

Ballabio, A.; Parenti, G.; Napolitano, E.; Di Natale, P.; Andria, G.: Genetic complementation of steroid sulphatase after somatic cell hybridization of X-linked ichthyosis and multiple sulphatase deficiency. Hum. Genet. 70: 315-317, 1985.

Ballabio, A.; Parenti, G.; Tippett, P.; Mondello, C.; Di Maio, S.; Tenore, A.; Andria, G.: X-linked ichthyosis, due to steroid sulphatase deficiency, associated with Kallmann syndrome (hypogonadotropic hypogonadism and anosmia): linkage relationships with Xg and cloned DNA sequences from the distal short arm of the X chromosome. Hum. Genet. 72: 237-240, 1986.

Ballabio, A.; Sebastio, G.; Carrozzo, R.; Parenti, G.; Piccirillo, A.; Persico, M. G.; Andria, G.: Deletions of the steroid sulphatase gene in 'classical' X-linked ichthyosis and in X-linked ichthyosis associated with Kallmann syndrome. Hum. Genet. 77: 338-341, 1987.

Basler, E.; Grompe, M.; Parenti, G.; Caskey, C. T.; Ballabio, A.: Identification of point mutations in three patients with steroid sulfatase deficiency. (Abstract) Am. J. Hum. Genet. 47 (suppl.): A150, 1990.

Basler, E.; Grompe, M.; Parenti, G.; Yates, J.; Ballabio, A.: Identification of point mutations in the steroid sulfatase gene of three patients with X-linked ichthyosis. Am. J. Hum. Genet. 50: 483-491, 1992.

Bernatowicz, L. F.; Li, X.-M.; Carrozzo, R.; Ballabio, A.; Mohandas, T.; Yen, P. H.; Shapiro, L. J.: Sequence analysis of a partial deletion of the human steroid sulfatase gene reveals 3 bp of homology at deletion breakpoints. Genomics 13: 892-893, 1992.

Bonifas, J. M.; Morley, B. J.; Oakey, R. E.; Kan, Y. W.; Epstein, E. H., Jr.: Cloning of a cDNA for steroid sulfatase: frequent occurrence of gene deletions in patients with recessive X chromosome-linked ichthyosis. Proc. Nat. Acad. Sci. 84: 9248-9251, 1987.

Buckle, V. J.; Edwards, J. H.; Evans, E. P.; Jonasson, J. A.; Lyon, M. F.; Peters, J.; Searle, A. G.: Comparative maps of human and mouse X chromosomes. (Abstract) Cytogenet. Cell Genet. 40: 594-595, 1985.

Burns, G.: On the identity of arylsulphatase C and steroid sulphatase. Hum. Genet. 65: 189, 1983.

Chance, P. F.; Gartler, S. M.: Evidence for a dosage effect at the X-linked steroid sulfatase locus in human tissues. Am. J. Hum. Genet. 35: 234-240, 1983.

Chang, P. L.; Lafferty, K. I.; Rosa, N. E.; Davidson, R. G.: Arylsulfatase-C isozymes in human tissues. (Abstract) Am. J. Hum. Genet. 33: 38A, 1981.

Cockayne, E. A.: Inherited Abnormalities of the Skin and Its Appendages. London: Oxford Univ. Press (pub.) 1933. Pp. 213 only.

Conary, J. T.; Lorkowski, G.; Schmidt, B.; Pohlmann, R.; Nagel, G.; Meyer, H. E.; Krentler, C.; Cully, J.; Hasilik, A.; von Figura, K.: Genetic heterogeneity of steroid sulfatase deficiency revealed with cDNA for human steroid sulfatase. Biochem. Biophys. Res. Commun. 144: 1010-1017, 1987.

Cooper, D. W.; McAllan, B. M.; Donald, J. A.; Dawson, G.; Dobrovic, A.; Marshall Graves, J. A.: Steroid sulphatase is not detected on the X chromosome of Australian marsupials. (Abstract) Cytogenet. Cell Genet. 37: 439, 1984.

Craig, I. W.; Tolley, E.: Steroid sulphatase and the conservation of mammalian X chromosomes. Trends Genet. 2: 201-204, 1986.

Csorsz, K.: Ichthyosis (X-linked). Mschr. Unfallheilk. Med. 2: 180, 1928. Note: Alternate: Haut. Geschlechtskr. 26: 463, 1928. . .

Cuevas-Covarrubias, S. A.; Kofman-Alfaro, S.; Orozco Orozco, E.; Diaz-Zagoya, J. C.: The biochemical identification of carrier state in mothers of sporadic cases of X-linked recessive ichthyosis. Genet. Counsel. 6: 103-107, 1995.

Curry, C. J. R.; Magenis, R. E.; Brown, M.; Lanman, J. T., Jr.; Tsai, J.; O'Lague, P.; Goodfellow, P.; Mohandas, T.; Bergner, E. A.; Shapiro, L. J.: Inherited chondrodysplasia punctata due to a deletion of the terminal short arm of an X chromosome. New Eng. J. Med. 311: 1010-1015, 1984.

Czeizel, A.: A historical evaluation of the doctrine of heredodegeneration. Orv. Hetil. 120: 722, and 840, and 963, 1979.

DeUnamuno, P.; Martin-Pascual, A.; Garcia-Perez, A.: X-linked ichthyosis. Brit. J. Derm. 97: 53-58, 1977.

Eicher, E. M.: Personal Communication. Bar Harbor, Me. 1974.

Elias, P. M.; Williams, M. L.; Maloney, M. E.; Bonifas, J. A.; Brown, B. E.; Grayson, S.; Epstein, E. H., Jr.: Stratum corneum lipids in disorders of cornification: steroid sulfatase and cholesterol sulfate in normal desquamation and the pathogenesis of recessive X-linked ichthyosis. J. Clin. Invest. 74: 1414-1421, 1984.

Epstein, E. H., Jr.; Bonifas, J. M.: Recessive X-linked ichthyosis: lack of immunologically detectable steroid sulfatase enzyme protein. Hum. Genet. 71: 201-205, 1985.

Epstein, E. H., Jr.; Krauss, R. M.; Shackleton, C. H. L.: X-linked ichthyosis: increased blood cholesterol sulfate and electrophoretic mobility of low-density lipoprotein. Science 214: 659-660, 1981.

Epstein, E. H., Jr.; Williams, M. L.; Elias, P. M.: Steroid sulfatase, X-linked ichthyosis, and stratum corneum cell cohesion. Arch. Derm. 117: 761-763, 1981.

Filippi, G.; Meera Khan, P.: Linkage studies on X-linked ichthyosis in Sardinia. Am. J. Hum. Genet. 20: 564-569, 1968.

France, J. T.; Downey, J. A.: A study of arylsulfatase activity in children born of pregnancies affected with placental sulfatase deficiency. Biochem. Med. 10: 167-174, 1974.

France, J. T.; Liggins, G. C.: Placental sulfatase deficiency. J. Clin. Endocr. 29: 138-141, 1969.

France, J. T.; Seddons, R. J.; Liggins, G. C.: A study of a pregnancy with low estrogen production due to placental sulfatase deficiency. J. Clin. Endocr. 36: 1-9, 1973.

Freiberg, R. A.; Choate, K. A.; Deng, H.; Alperin, E. S.; Shapiro, L. J.; Khavari, P. A.: A model of corrective gene transfer in X-linked ichthyosis. Hum. Molec. Genet. 6: 927-933, 1997.

Gant, N. F.; Milewich, L.; Calvert, M. E.; MacDonald, P. C.: Steroid sulfatase activity in human fetal membranes. J. Clin. Endocr. 45: 965-972, 1977.

Garcia Perez, A.; Crespo, M.: X-linked ichthyosis associated with hypertrophic pyloric stenosis in three brothers. Clin. Exp. Derm. 6: 159-161, 1981.

Gartler, S. M.; Andina, R. J.: Mammalian X-chromosome inactivation. Adv. Hum. Genet. 7: 99-140, 1976.

Gartler, S. M.; Rivest, M.: Evidence for X-linkage of steroid sulfatase in the mouse: steroid sulfatase levels in oocytes of XX and XO mice. Genetics 103: 137-141, 1983.

Gillard, E. F.; Affara, N. A.; Yates, J. R. W.; Goudie, D. R.; Cooke, A.; Lambert, J.; Aitken, D. A.; Ferguson-Smith, M. A.: Deletion of a DNA probe DXS237 (GMGX9) in twelve of fifteen males with X-linked ichthyosis (steroid sulphatase deficiency). (Abstract) Cytogenet. Cell Genet. 46: 620, 1987.

Gillard, E. F.; Affara, N. A.; Yates, J. R. W.; Goudie, D. R.; Lambert, J.; Aitken, D. A.; Ferguson-Smith, M. A.: Deletion of a DNA sequence in eight of nine families with X-linked ichthyosis (steroid sulphatase deficiency). Nucleic Acids Res. 15: 3977-3985, 1987.

Gladstein, K.; Shapiro, L. J.; Spence, M. A.: Estimating sex ratio biases in X-linked disorders: is there an excess of males in families with X-linked ichthyosis?. Am. J. Hum. Genet. 31: 741-746, 1979.

Goodfellow, P. N.; Tippett, P.: A human quantitative polymorphism related to Xg blood groups. Nature 289: 404-409, 1981.

Hameister, H.; Wolff, G.; Laurietzen, C. H.; Lehmann, W. O.; Hauser, A.; Ropers, H. H.: Clinical and biochemical investigations on patients with partial deficiency of placental steroid sulfatase. Hum. Genet. 46: 199-207, 1979.

Happle, R.: X-linked dominant ichthyosis. Clin. Genet. 15: 239-240, 1979.

Harris, H.: A pedigree of sex-linked ichthyosis vulgaris. Ann. Eugen. 14: 9, 1947.

Jay, B.; Blach, R. K.; Wells, R. S.: Ocular manifestations of ichthyosis. Brit. J. Ophthal. 52: 217-226, 1968.
Jobsis, A. C.; De Groot, W. P.; Meijer, A. E. F. H.; van der Loos, C. M.: A new method for the determination of steroid sulphatase activity in leukocytes in X-linked recessive ichthyosis. Brit. J. Derm. 108: 567-572, 1983.
Jobsis, A. C.; De Groot, W. P.; Tigges, A. J.; De Bruijn, H. W.; Rijken, Y.; Meijer, A. E. F. H.; Marinkovic-Ilsen, A.: X-linked ichthyosis and X-linked placental sulfatase deficiency: a disease entity. Histochemical observations. Am. J. Path. 99: 279-290, 1980.
Jobsis, A. C.; van Duuren, C. Y.; de Vries, G. P.; Koppe, J. G.; Rijken, Y.; van Kempen, G. M. J.; de Groot, W. P.: Trophoblast sulphatase deficiency associated with X-chromosomal ichthyosis. (Abstract) Nederl. T. Geneesk. 120: 1980, 1976.
Kawano, J.; Kotani, T.; Ohtaki, S.; Mimamino, N.; Matsuo, H.; Oinuma, T.; Aikawa, E.: Characterization of rat and human steroid sulfatases. Biochim. Biophys. Acta 997: 199-205, 1989.
Keitges, E.; Gartler, S. M.: Dosage of the Sts gene in the mouse. Am. J. Hum. Genet. 39: 470-476, 1986.
Keitges, E.; Rivest, M.; Siniscalco, M.; Gartler, S. M.: X-linkage of steroid sulphatase in the mouse is evidence for a functional Y-linked allele. Nature 315: 226-227, 1985.
Kerr, C. B.; Wells, R. S.: Sex-linked ichthyosis. Ann. Hum. Genet. 29: 33-50, 1965.
Kerr, C. B.; Wells, R. S.; Sanger, R.: X-linked ichthyosis and the Xg groups. Lancet II: 1369-1370, 1964.
Kipling, D.; Salido, E. C.; Shapiro, L. J.; Cooke, H. J.: High frequency de novo alterations in the long-range genomic structure of the mouse pseudoautosomal region. Nature Genet. 13: 78-82, 1996.
Koppe, J. G.; Marinkovic-Ilsen, A.; Rijken, Y.; De Groot, W. P.; Jobsis, A. C.: X-linked ichthyosis: a sulphatase deficiency. Arch. Dis. Child. 53: 803-806, 1978.
Koppe, J. G.; Rijken, Y.; Jobsis, A. C.; Marinkovic-Ilsen, A.: X-linked ichthyosis, a sulfatase deficiency. (Abstract) Vth Int. Conf. on Birth Defects, Montreal , 8/1977.
Kubilus, J.; Tarascio, A. J.; Baden, H. P.: Steroid-sulfatase deficiency in sex-linked ichthyosis. Am. J. Hum. Genet. 31: 50-53, 1979.
Lake, B. D.; Smith, V. V.; Judge, M. R.; Harper, J. I.; Besley, G. T. N.: Hexanol dehydrogenase activity shown by enzyme histochemistry on skin biopsies allows differentiation of Sjogren-Larsson syndrome from other ichthyoses. J. Inherit. Metab. Dis. 14: 338-340, 1991.
Laval, S. H.; Boyd, Y.: Partial inversion of gene order within a homologous segment on the X chromosome. Mammalian Genome 4: 119-123, 1993.
Li, X. M.; Salido, E. C.; Gong, Y.; Kitada, K.; Serikawa, T.; Yen, P. H.; Shapiro, L. J.: Cloning of the rat steroid sulfatase gene (Sts), a non-pseudoautosomal X-linked gene that undergoes X inactivation. Mammalian Genome 7: 420-424, 1996.
Lykkesfeldt, G.; Bennett, P.; Lykkesfeldt, A. E.; Micic, S.; Moller, S.; Svenstrup, B.: Abnormal androgen and oestrogen metabolism in men with steroid sulphatase deficiency and recessive X-linked ichthyosis. Clin. Endocr. 23: 385-393, 1985.
Lykkesfeldt, G.; Hoyer, H.: Topical cholesterol treatment of recessive X-linked ichthyosis. Lancet II: 1337-1338, 1983.
Lykkesfeldt, G.; Hoyer, H.; Ibsen, H. H.; Brandrup, F.: Steroid sulphatase deficiency disease. Clin. Genet. 28: 231-237, 1985.
Lykkesfeldt, G.; Hoyer, H.; Lykkesfeldt, A. E.; Skakkebaek, N. E.: Steroid sulphatase deficiency associated with testis cancer. Lancet II: 1456, 1983.
Lykkesfeldt, G.; Lykkesfeldt, A. E.; Skakkebaek, N. E.: Steroid sulphatase in man: a non inactivated X-locus with partial gene dosage compensation. Hum. Genet. 65: 355-357, 1984.
Lyon, M. F.; Peters, J.; Glenister, P. H.; Ball, S.; Wright, E.: The scurfy mouse mutant has previously unrecognized hematological abnormalities and resembles Wiskott-Aldrich syndrome. Proc. Nat. Acad. Sci. 87: 2433-2437, 1990.
Macsai, M. S.; Doshi, H.: Clinical pathologic correlation of superficial corneal opacities in X-linked ichthyosis. Am. J. Ophthal. 118: 477-484, 1994.
Marinkovic-Ilsen, A.; Koppe, J. G.; Jobsis, A. C.; De Groot, W. P.: Enzymatic basis of typical X-linked ichthyosis. (Letter) Lancet II: 1097, 1978.
Metaxotou, C.; Ikkos, D.; Panagiotopoulou, P.; Alevizaki, M.; Mavrou, A.; Tsenghi, C.; Matsaniotis, N.: A familial X/Y translocation in a boy with ichthyosis, hypogonadism and mental retardation. Clin. Genet. 24: 380-383, 1983.
Meyer, J. C.; Weiss, H.; Grundemann, H. P.; Wurseh, T. G.; Schnyder, U. W.: Deficiency of arylsulfatase C in cultured skin fibroblasts of X-linked ichthyosis. Hum. Genet. 53: 115-116, 1979.
Migeon, B. R.; Shapiro, L. J.; Norum, R. A.; Mohandas, T.; Axelman, J.; Dabora, R. L.: Differential expression of steroid sulphatase locus on active and inactive human X chromosome. Nature 299: 838-840, 1982.
Mohandas, T.; Shapiro, L. J.; Sparkes, R. S.; Sparkes, M. C.: Regional assignment of the steroid sulfatase—X-linked ichthyosis locus: implications for a non-inactivated region on the short arm of the human X-chromosome. Proc. Nat. Acad. Sci. 76: 5779-5783, 1979.
Mohandas, T.; Sparkes, R. S.; Hellkuhl, B.; Grzeschik, K. H.; Shapiro, L. J.: Expression of an X-linked gene from an inactive human X chromosome in mouse-human hybrid cells: further evidence for the noninactivation of the steroid sulfatase locus in man. Proc. Nat. Acad. Sci. 77: 6759-6763, 1980.
Mohandas, T. K.; Stern, H. J.; Meeker, C. A.; Passage, M. B.; Muller, U.; Page, D. C.; Yen, P. H.; Shapiro, L. J.: Steroid sulfatase gene in XX males. Am. J. Hum. Genet. 46: 369-376, 1990.
Monroe, D. G.; Chang, P. L.: Tissue-specific expression of human arylsulfatase-C isozymes and steroid sulfatase. Am. J. Hum. Genet. 40: 102-114, 1987.
Muller, C. R.; Migl, B.; Ropers, H.-H.: X-linked steroid sulfatase: evidence for different gene-dosage in males and females. Hum. Genet. 54: 197-199, 1980.
Muller, C. R.; Migl, B.; Ropers, H.-H.; Happle, R.: Heterozygote detection in steroid sulphatase deficiency. (Letter) Lancet I: 546-547, 1980.
Muller, C. R.; Wahlstrom, J.; Ropers, H.-H.: Further evidence for the assignment of the steroid sulfatase X-linked ichthyosis locus to the telomer of Xp. (Letter) Hum. Genet. 58: 446, 1981.
Munke, M.; Kruse, K.; Goos, M.; Ropers, H. H.; Tolksdorf, M.: Genetic heterogeneity of the ichthyosis, hypogonadism, mental retardation, and epilepsy syndrome: clinical and biochemical investigations on two patients with Rud syndrome and review of the literature. Europ. J. Pediat. 141: 8-13, 1983.
Okano, M.; Kitano, Y.; Nakamura, T.; Matsuzawa, Y.: Detection of heterozygotes of X-linked ichthyosis by measuring steroid sulphatase activity of lymphocytes: mode of inheritance in three families. Brit. J. Derm. 113: 645-649, 1985.
Orel, H.: Die Vererbung der Ichthyosis congenita und der Ichthyosis vulgaris. Z. Kinderheilk. 47: 312-340, 1929.
Passarge, E.; Post, B.; Schopf, E.: Possible genetic heterogeneity of X-linked ichthyosis. Birth Defects Orig. Art. Ser. VII(8): 46-49, 1971.
Pike, M. G.; Hammerton, M.; Edge, J.; Atherton, D. J.; Grant, D. B.: A family with X-linked ichthyosis and hypogonadism. Europ. J. Pediat. 148: 442-444, 1989.
Robledo, R.; Melis, P.; Schillinger, E.; Casciano, I.; Balazs, I.; Rinaldi, A.; Siniscalco, M.; Filippi, G.: X-linked ichthyosis without STS deficiency: clinical, genetical, and molecular studies. Am. J, Med. Genet. 59: 143-148, 1995.
Ropers, H.-H.; Migl, B.; Zimmer, J.; Fraccaro, M.; Maraschio, P. P.; Westerveld, A.: Activity of steroid sulfatase in fibroblasts with numerical and structural X chromosome aberrations. Hum. Genet. 57: 354-356, 1981.
Ropers, H.-H.; Wiberg, U.: Evidence for X-linkage and non-inactivation of steroid sulphatase locus in wood lemming. Nature 296: 766-767, 1982.
Ross, J. B.; Allderdice, P. W.; Shapiro, L. J.; Aveling, J.; Eales, B. A.; Simms, D., Jr.: Familial X-linked ichthyosis, steroid sulfatase deficiency, mental retardation, and nullisomy for Xp223-pter. Arch. Derm. 121: 1524-1528, 1985.
Salido, E. C.; Li, X. M.; Yen, P. H.; Martin, N.; Mohandas, T. K.; Shapiro, L. J.: Cloning and expression of the mouse pseudoautosomal steroid sulphatase gene (Sts). Nature Genet. 13: 83-86, 1996.
Schlammadinger, J.; Meyer, J. C.; Vajda, I.; Szabo, G.: X-linked recessive ichthyosis: reinvestigation of a family first described in 1928. Dermatologica 175: 217-223, 1987.
Schmickel, R. D.: Chromosomal deletions and enzyme deficiencies. J. Pediat. 108: 244-246, 1986.
Schnyder, U. W.: Inherited ichthyoses. Arch. Derm. 102: 240-252, 1970.
Sever, R. J.; Frost, P.; Weinstein, G.: Eye changes in ichthyosis. J.A.M.A. 206: 2283-2286, 1968.
Shapiro, L. J.: Personal Communication. San Francisco, Calif. 5/20/1997.
Shapiro, L. J.: Personal Communication. Torrance, Calif. 1977.
Shapiro, L. J.: Personal Communication. Los Angeles, Calif. 3/1987.
Shapiro, L. J.; Mohandas, T.; Weiss, R.; Romeo, G.: Non-activation of a X-chromosome locus in man. Science 204: 1224-1226, 1979.
Shapiro, L. J.; Weiss, R.: Diminished cholesterol sulfatase activity in fibroblasts of placental sulfatase deficiency patients. (Abstract) Vth Int. Conf. on Birth Defects, Montreal , 8/1977.
Shapiro, L. J.; Weiss, R.; Buxman, M. M.; Vidgoff, J.; Dimond, R. L.: Enzymatic basis of typical X-linked ichthyosis. Lancet II: 756-757, 1978.
Shapiro, L. J.; Weiss, R.; Webster, D.; France, J. T.: X-linked ichthyosis due to steroid-sulphatase deficiency. Lancet I: 70-72, 1978.
Shapiro, L. J.; Yen, P.; Pomerantz, D.; Martin, E.; Rolewic, L.; Mohandas, T.: Molecular studies of deletions at the human steroid sulfatase locus. Proc. Nat. Acad. Sci. 86: 8477-8481, 1989.
Shapiro, L. J.; Yen, P. H.: Reply to letter of Andria et al., 1987. (Letter) Ann. Neurol. 22: 99, 1987.
Shapiro, L. J.; Yen, P. H.; Marsh, B.; Mohandas, T.: Frequent deletions at the steroid sulfatase (STS) locus. (Abstract) Am. J. Hum. Genet. 41: A238, 1987.
Solomon, I. L.; Schoen, E. J.: Sex-linked ichthyosis in XO gonadal dysgenesis. (Letter) Lancet I: 1304-1305, 1971.
Stein, C.; Hille, A.; Seidel, J.; Rijnbout, S.; Waheed, A.; Schmidt, B.; Geuze, H.; von Figura, K.: Cloning and expression of human steroid-sulfatase: membrane topology, glycosylation, and subcellular distribution in BHK-21 cells. J. Biol. Chem. 264: 13865-13872, 1989.
Steinmann, B.; Mieth, D.; Gitzelmann, R.: A newly recognized cause of low urinary estriol in pregnancy: multiple sulfatase deficiency of the fetus. Gynec. Obstet. Invest. 12: 107-109, 1981.
Stoll, C.; Grosshans, E.; Binder, P.; Roth, M.: Hypertrophic pyloric stenosis associated with X-linked ichthyosis in two brothers. Clin. Exp. Derm. 8: 61-64, 1983.
Sunohara, N.; Sakuragawa, N.; Satoyoshi, E.; Tanae, A.; Shapiro, L. J.: A new syndrome of anosmia, ichthyosis, hypogonadism, and various neurological manifestations with deficiency of steroid sulphatase and arylsulphatase C. Ann. Neurol. 19: 174-181, 1986.
Tabei, F.; Heinrichs, W. L.: Diagnosis of placental sulfatase deficiency. Am. J. Obstet. Gynec. 124: 409-414, 1976.
Tiepolo, L.; Zuffardi, O.; Fraccaro, M.; di Natale, D.; Gargantini, L.; Muller, C. R.; Ropers, H.-H.: Assignment by deletion mapping of the steroid sulfatase X-linked ichthyosis locus to Xp223. Hum. Genet. 54: 205-206, 1980.
Toder, R.; Rappold, G. A.; Schiebel, K.; Schempp, W.: ANT3 and STS are autosomal in prosimian lemurs: implications for the evolution of the pseudoautosomal region. Hum. Genet. 95: 22-28, 1995.
Traupe, H.; Happle, R.: Clinical spectrum of steroid sulfatase deficiency: X-linked recessive ichthyosis, birth complications, and cryptorchidism. Europ. J. Pediat. 140: 19-21, 1983.
Traupe, H.; Muller-Migl, C. R.; Kolde, G.; Happle, R.; Kovary, P. M.; Hameister, H.; Ropers, H. H.: Ichthyosis vulgaris with hypogenitalism and hypogonadism: evidence for different genotypes by lipoprotein electrophoresis and steroid sulfatase testing. Clin. Genet. 25: 42-51, 1984.
Traupe, H.; Ropers, H.-H.: Cryptorchidism and hypogenitalism in X-linked recessive ichthyosis vulgaris. (Letter) Hum. Genet. 60: 206, 1982.
Turpin, R.; Desvignes, P.; Demassieux, J.-L.: Sur une variete d'ichthyose hereditaire avec alteration du fond d'oeil. Nouveau syndrome ectoblastique?. Sem. Hop. Paris 21: 343, 1945.

Vogel, W.; Grompe, M.; Storz, R.; Pentz, S.: A comparative study on steroid sulfatase and arylsulfatase C in fibroblast clones from 45,X/47,XXX and 69,XXY. Hum. Genet. 66: 367-369, 1984.
Wells, R. S.; Jennings, M. C.: X-linked ichthyosis and ichthyosis vulgaris: clinical and genetic distinctions in a second series of families. J.A.M.A. 202: 485-488, 1967.
Went, L. N.; DeGroot, W. P.; Sanger, R.; Tippett, P.; Gavin, J.: X-linked ichthyosis: linkage relationship with the Xg blood groups and other studies in a large Dutch kindred. Ann. Hum. Genet. 32: 333-346, 1969.
Wieacker, P.; Davies, K. E.; Mevorah, B.; Ropers, H. H.: Linkage studies in a family with X-linked recessive ichthyosis employing a cloned DNA sequence from the distal short arm of the X chromosome. Hum. Genet. 63: 113-116, 1983.
Willard, H. F.; Holmes, M. T.: A sensitive and dependable assay for distinguishing hamster and human X-linked steroid sulfatase activity in somatic cell hybrids. Hum. Genet. 66: 272-275, 1984.
Wirth, B.; Herrmann, F. H.; Neugebauer, M.; Gillard, E. F.; Wulff, K.; Stein, C.; von Figura, K.; Ferguson-Smith, M. A.; Gal, A.: Linkage analysis in X-linked ichthyosis (steroid sulfatase deficiency). Hum. Genet. 80: 191-192, 1988.
Yen, P. H.; Allen, E.; Marsh, B.; Mohandas, T.; Wang, N.; Taggart, R. T.; Shapiro, L. J.: Cloning and expression of steroid sulfatase cDNA and the frequent occurrence of deletions in STS deficiency: implications for X-Y interchange. Cell 49: 443-454, 1987.
Yen, P. H.; Marsh, B.; Allen, E.; Tsai, S. P.; Ellison, J.; Connolly, L.; Neiswanger, K.; Shapiro, L. J.: The human X-linked steroid sulfatase gene and a Y-encoded pseudogene: evidence for an inversion of the Y chromosome during primate evolution. Cell 55: 1123-1135, 1988.

ADRENAL HYPERPLASIA 3 (*201910)

JJM NOTE

The 21-hydroxylase deficiency can present as testicular masses that are not clearly malignant nor even neoplastic, for they can regress with corticosteroid replacement therapy. For this phenomenon, Jaresch et al. (1992) used the term *incidentaloma.* Hence, the real neoplastic association with this entity may be the Milwaukee experience with sarcoma and astrocytoma and one case of adrenal myelolipoma (Boudreaux et al. 1979).

Description

Congenital adrenal hyperplasia (CAH) results from a deficiency in one or another of the enzymes of cortisol biosynthesis. In about 95% of cases, 21-hydroxylation is impaired in the zona fasciculata of the adrenal cortex so that 17-hydroxyprogesterone (17-OHP) is not converted to 11-deoxycortisol. Because of defective cortisol synthesis, ACTH levels increase, resulting in overproduction and accumulation of cortisol precursors, particularly 17-OHP, proximal to the block. This causes excessive production of androgens, resulting in virilization.

Slominski et al. (1996) presented evidence that the CYP21A2, CYP11A1 (118485), CYP17 (202110), and ACTHR (202200) genes are expressed in skin (see 202200). The authors suggested that expression of these genes may play a role in skin physiology and pathology and that cutaneous proopiomelanocortin activity may be autoregulated by a feedback mechanism involving glucocorticoids synthesized locally.

Nomenclature

Since the symbol CAH is used for all forms of congenital adrenal hypoplasia (240200, 300200) and since CAH1 contains no intrinsic information indicating the particular type of adrenal hyperplasia, CA21H enjoyed favor as the preferred symbol after about 1983.

Regarding nomenclature of the gene, the functional gene, previously designated CYP21B, is symbolized simply as CYP21, and the nonfunctional gene, previously CYP21A, is symbolized CYP21P (for pseudogene) (McAlpine, 1988).

Clinical Features

There are 4 recognized clinical forms of congenital adrenal hyperplasia, the majority of cases being associated with 21-hydroxylase deficiency: salt-wasting (SW), simple virilizing (SV), nonclassic (NC) late onset (also called attenuated and acquired), and cryptic. All 4 forms are closely linked to HLA and represent the effects of various combinations of alleles.

In female newborns, the external genitalia are masculinized; gonads and internal genitalia are normal. Postnatally, untreated males as well as females may manifest rapid growth, penile or clitoral enlargement, precocious adrenarche, and ultimately early epiphyseal closure and short stature. In about half of cases, an additional defect in aldosterone synthesis (conversion of progesterone to 11-deoxycorticosterone) in the zona glomerulosa is present; untreated, it can result in shock or death in the neonatal period from inability to conserve urinary sodium (severe salt wasting similar to Addisonian crisis).

Hypoglycemia sometimes occurs. Recurrent fever also may occur and may be related to etiocholanolone, although this remains to be clarified. Hypertension occurs in this form in addition to the other features. A mild form of late-onset adrenal hyperplasia due to 21-hydroxylase deficiency can occur in adults and has hirsutism as the only manifestation in the most attenuated form.

All types of adrenal hyperplasia were reviewed exhaustively by Bongiovanni and Root (1963). Prader et al. (1962) reported an enormous interlocking Swiss kindred. (See precocious puberty (176400) for a simulating condition.)

Galal et al. (1968) concluded that the 2 clinical forms of 21-hydroxylase deficiency (with and without salt loss) correlate with the extent of the defect in the cortisol pathway. Some had suggested the existence of 2 different 21-hydroxylating systems, one specific for progesterone and concerned with aldosterone synthesis and the other specific for 17-alpha-hydroxyprogesterone involved in cortisol synthesis. However, Orta-Flores et al. (1976) presented evidence that there is only one 21-hydroxylation system with 2 active sites: one active on progesterone only and a second active on either substrate indiscriminately. The authors suggested that both sites are defective in the salt-losing variety and only the second in the nonsalt-losing form.

Presentation with gynecomastia and bilateral testicular masses was reported by Kadair et al. (1977) in a case of 21-hydroxylase deficiency. Others have reported bilateral testicular tumors. Lewis et al. (1968) found that intelligence is increased in the adrenogenital syndrome, a remarkable and possibly significant feature from the point of view of selection and gene frequency. However, McGuire and Omenn (1975) presented data indicating that patients with congenital adrenal hyperplasia do not have higher IQs than expected from the family background. Wenzel et al. (1978) found similar results.

Blankstein et al. (1980) reported a possible allelic form of 21-hydroxylase deficiency in 2 sisters, aged 28 and 30 years, who had primary infertility and mild hirsutism but normal puberty, regular menses, and normal female sexual characteristics. Two sibs were normal. The affected sibs were HLA-identical; their healthy sibs were of different HLA type.

Levine et al. (1980) studied serum androgen and 17-hydroxyprogesterone levels as well as HLA genotypes in 124 families of patients with classic 21-hydroxylase deficiency. In 8 kindreds, 16 pubertal or postpubertal persons of either sex were found to have biochemical evidence of 21-hydroxylase deficiency without clinical symptoms of excess virilism, amenorrhea, or infertility. They designated the disorder 'cryptic 21-hydroxylase deficiency.' Within each generation, the family members with the cryptic form were HLA identical. They suggested that these persons were compound heterozygotes for the classic gene and a cryptogenic gene. Of 42 pediatric patients with 21-hydroxylase deficiency (from 36 families) treated in Milwaukee between 1965 and 1981, 4 developed a malignant tumor: sarcoma or astrocytoma (Duck, 1981).

Kuttenn et al. (1985) found that 21-hydroxylase deficiency was the basis of hirsutism in 24 of 400 women (6%). The diagnosis was made by a high plasma level of 17-hydroxyprogesterone and its marked increase after ACTH stimulation. From genotyping of the 24 families, a high correlation with HLA-B14 and Aw33 was found. Nine HLA-identical sibs showed similar biologic profiles but had no hirsutism; skin sensitivity to androgens may be important in determining clinical expression of the disorder. (It was previously known that unusual sensitivity to androgens can lead to hirsutism despite normal plasma levels of androgen (Kuttenn et al., 1977).) The patients were not distinguishable from women with idiopathic hirsutism or polycystic ovarian disease (184700), either clinically or in plasma androgen levels.

Knochenhauer et al. (1997) hypothesized that heterozygosity for CYP21 mutations in women increases their risk of developing clinically evident hyperandrogenism, and that this risk is related to the severity of the mutation of CYP21 and/or the 17-hydroxyprogesterone (17-OHP) response to ACTH stimulation. To test these hypotheses, they studied 38 obligate carriers for 21-hydroxylase deficiency (i.e., mothers of children with CAH1 or nonclassic CAH), comparing them to 27 controls. Their data indicate that heterozygosity for CYP21 mutations does not appear to increase the risk of clinically evident hyperandrogenism, although carrying the defect was associated with higher mean and free T levels. Finally, due to the low frequency of androgen excess in their heterozygote population, they were unable to correlate the severity of CYP21 mutations and/or 17-OHP responses to ACTH stimulation with the presence of the phenotype.

Sinnott et al. (1989) presented analyses of families that showed profound discordance between the clinical features of sibs with 21-hydroxylase deficiency who appeared to be HLA identical, both in terms of serologically defined HLA polymorphism and in gene organization at the 21-hydroxylase and C4 loci. For example, in 1 family a boy had the simple virilizing form while his 2 younger sisters, who were both HLA identical to their brother, had additional salt-wasting features. In 1 family they made the unusual observation of HLA-Bw47-bearing haplotypes that appeared to carry a functional 21-hydroxylase gene.

Jaresch et al. (1992) found a high frequency of asymptomatic adrenal tumors in association with homozygosity (82%) and heterozygosity (45%) for 21-hydroxylase deficiency. Jaresch et al. (1992) suggested that CAH should always be ruled out in the case of incidentally detected adrenal masses. Since CAH is a relatively frequent disorder and adrenal carcinoma belongs to the rarest malignant tumors, they concluded that malignant transformation of these tumors is unlikely.

Ravichandran et al. (1996) pointed out that both homozygous and heterozygous patients with congenital adrenal hyperplasia have an increased cross-sectional area of their adrenal glands as well as an increased prevalence of adrenal incidentalomas, i.e., adrenal tumors discovered incidentally in the course of imaging studies performed for unrelated reasons (Jaresch et al., 1992). The prevalence of adrenal tumors may be more than 70% in nonclassic CAH and 'unmasked heterozygotes.' Ravichandran et al. (1996) presented 2 patients, female pseudohermaphrodites with the simple virilizing form of CAH and 21-hydroxylase deficiency, who functioned successfully as married phenotypic males. Both came to medical attention in their sixth decade by virtue of massive adrenal incidentalomas encountered in the evaluation of recurrent urinary tract infections. Each had a 46,XX karyotype, no palpable testes, and markedly elevated baseline levels of 17-hydroxyprogesterone. Both responded appropriately to dexamethasone suppression. Histologic and autopsy examination of the first patient's tumor and computed tomographic characteristics of the second patient revealed benign adenoma and mild lipoma, respectively. Ravichandran et al. (1996) concluded that these observations extended and confirmed previous recommendations that CAH be included in the differential diagnosis of adrenal incidentaloma and that baseline 17-hydroxyprogesterone levels be obtained, with ACTH stimulation if necessary, to diagnose the presence of nonclassic CAH.

In a follow-up study of 52 males with congenital virilizing adrenal hyperplasia seen at Johns Hopkins between 1950 and 1978, 51 had 21-hydroxylase deficiency and 1 had 11-hydroxylase deficiency (Urban et al., 1978).

Because little is known about the relation between endogenous TSH and cortisol secretion under physiologic or slightly disturbed conditions, Ghizzoni et al. (1997) evaluated the pulsatility, circadian rhythmicity, and 24-hour secretory patterns of cortisol and TSH in 8 prepubertal children with nonclassical CAH and 8 age-matched short normal children. In both groups, TSH and cortisol were secreted in a pulsatile and circadian fashion, with a clear nocturnal TSH surge. Although no difference in mean 24-hour TSH levels was observed between the 2 groups, daytime TSH levels were lower in the nonclassical CAH group than in controls (P less than 0.05). Cross-correlation analysis showed that TSH and cortisol were negatively correlated, possibly reflecting a negative glucocorticoid effect on TSH under physiologic conditions. The authors concluded that the hypothalamic-pituitary-adrenal axis has a primarily negative influence on endogenous TSH secretion and that even mild disturbances in cortisol biosynthesis can be associated with slight alterations in TSH secretion.

Other Features

Winqvist et al. (1992) demonstrated that 21-hydroxylase, which is prominent in the zona glomerulosa of the adrenal cortex, is a major autoantigen in idiopathic Addison disease. This is another example of the way in which genetic disease can be mimicked by the development of autoantibodies against the gene product that is genetically deficient in the inherited disorder. Other examples are hemophilia A (306700), dystrophic epidermolysis bullosa (131750), hereditary angioedema (106100), and perhaps congenital myasthenia gravis (254210). All of these hereditary disorders appear to have an acquired mimic which is an autoimmune disorder.

Inheritance

Congenital adrenal hyperplasia-1 is an autosomal recessive disorder.

Mapping

Patients with 21-hydroxylase deficiency also show genetic linkage disequilibrium with complement allotypes. Different forms of 21-hydroxylase deficiency are associated with characteristic HLA haplotypes. Holler et al. (1985) studied HLA types and plasma 17-hydroxyprogesterone levels after ACTH stimulation in 134 German families of patients with salt-wasting (SW), simple virilizing (SV), or nonclassic (NC) late-onset CAH. Hormone evidence for CAH was found in 6 otherwise healthy relatives who, therefore, were thought to be NC cryptic cases. The SW form was strongly associated with HLA Bw47, whereas the SV form was associated with B5(w51). The almost complete association of the NC form with HLA B14 was confirmed. These alleles, especially Bw47 and B14, are components of normally rare haplotypes. Thus, all or almost all persons in the general population with 1 of these haplotypes will be heterozygotes.

Dupont et al. (1977) demonstrated close linkage of 21-hydroxylase deficiency and the HLA complex (lod score = 3.394 at a recombination fraction of 0.00). One patient had inherited a maternal recombinant between HLA-A and HLA-B. Studies in this family indicated that the abnormal gene is close to the HLA-B locus. Both the salt-losing and nonsalt-losing forms of 21-hydroxylase deficiency show linkage to HLA, suggesting allelism. Murtaza et al. (1978) identified possible genetic compounds.

Levine et al. (1978) obtained a lod score of 9.5 with a 0.00 recombination fraction. In a study of 48 patients, 48 sibs and their parents, all patients were HLA-different from their unaffected sibs. When 2 or more children were affected in a sibship they were always HLA-B identical. In 34 unrelated patients no selective increase of a particular haplotype was observed, thus excluding association or linkage disequilibrium. Klouda et al. (1980) found a lod score of almost 9.0 for the linkage of HLA-B and 21-hydroxylase deficiency at a recombination fraction of 0.03. They pointed to the association of an excess of HLA-Bw47 with a deficiency of HLA-B8 persons. The workers concluded that the 21-hydroxylase locus 'lies outside the HLA system and is closely linked to the HLA-DR locus.' Fleischnick et al. (1983) demonstrated that extended MHC haplotypes are markers for different mutations causing 21-hydroxylase deficiency, just as the extended restriction nonalpha-globin haplotypes are markers for different beta-thalassemia mutations. In studying 29 families, more than 20% were found to have a very rare extended haplotype (taking into consideration complement loci and glyoxalase I as well). Furthermore, 3 other haplotypes were each found twice in unrelated patients concordant for their disease phenotype and ethnic background. Previously, striking linkage disequilibrium was noted; e.g., in Sheffield, England, the frequency of Bw47 was 27.3% in the patient population and 0.4% in the general European population. They commented on the fact that Klouda et al. (1980) as well as at least 1 other group placed CAH between D/DR and GLO1, whereas others place it between HLA-A and HLA-D/DR (Pucholt et al., 1980; Bias et al., 1981). Sobel et al. (1980) pointed out that heterozygotes can be detected by the linkage principle. They also reported the first instance of presumed recombination between AH3 and the HLA-B locus.

Presumably because of linkage disequilibrium, the common severe form of 21-hydroxylase deficiency is positively associated with Bw47 and negatively with B8, while the late-onset type is positively associated with B14 (reviewed by Petersen et al., 1982). HLA haplotyping was used to confirm the genetic compound nature of the cryptic form of 21-hydroxylase deficiency (Zachmann and Prader, 1979; 60,61:Levine et al., 1980, 1981).

Patients with the HLA-Bw47 antigen invariably show simultaneous deficiencies of 21-hydroxylase activity and the C4A (Rodgers) form of C4. The HLA-Bw47(w4) antigen is very similar serologically and otherwise to the more common antigen HLA-B13(w4). Therefore, it was proposed that a deletion or rearrangement simultaneously affected the B13 gene and the closely linked 21-hydroxylase locus. White et al. (1984) used a plasmid with bovine adrenal cDNA insert encoding part of the cytochrome P450 polypeptide to examine this hypothesis. The hybridization patterns of normal DNA and that from 21-hydroxylase-deficient persons were compared. One band from both EcoRI and TaqI digests was absent in DNA from a patient homozygous for HLA-Bw47. Of 6 unrelated patients homozygous for CAH and heterozygous for HLA-Bw47, 5 had a relative intensity of this band consistent with heterozygosity and one had complete absence. The deletion segregated with HLA-Bw47 in a large pedigree with 21-hydroxylase deficiency and HLA-Bw47. These authors referred to the structural gene for P450(C21). Close linkage of said gene and that for C4A is indicated by the occurrence of the null allele at that locus. Apparently only one of the two 21-hydroxylase genes is mutant. Several alternative explanations might be considered. The second gene may in fact be mutated also. The second gene may be regulated by the renin-angiotensin system and be involved in aldosterone synthesis in the zona glomerulosa. The second gene may be a pseudogene or may be expressed only at certain times in ontogeny or in other organs (the kidney and liver also contain 21-hydroxylase activity). In an addendum, White et al. (1984) stated that reexamination of the C4 allotypes associated with HLA-Bw47 led to the conclusion that the data are consistent with the P450(C21) gene being near the C4B (Chido) gene and both of those genes being deleted in the case of the HLA-Bw47 haplotype.

Carroll et al. (1985) identified two 21-hydroxylase genes situated in the following relationship to C4A and C4B: 5-prime—C4A—21-OHA—C4B—21-OHB—3-prime. White et al. (1985) presented evidence for the existence of 2 genes encoding steroid 21-hydroxylase in the C4 gene region, i.e., among the MHC class III genes. The order appears to be: centromere—GLO—DP—DQ—DR—C2—BF—C4A—21OHA—C4B—21OHB—B—C—A. The 21-hydroxylase B gene and the adjacent C4B gene appear to be deleted on the chromosome carrying HLA-Bw47 and the allele for salt-wasting 21-hydroxylase deficiency. In contrast, the chromosome carrying the HLA-A1;B8;DR3 haplotype is not associated with 21-hydroxylase deficiency and in the conclusions of White et al. (1985) based on restriction enzyme analysis may have a deletion of the C4A and 21OHA genes. This suggests that the latter is not functional. (In the human, the 21-hydroxylase B gene is functional; the A gene is missing 8 basepairs from exon 2. In the mouse, the 21-hydroxylase A gene is functional; the B gene is missing 215 basepairs from exon 2.)

Speiser et al. (1985) studied the frequency of 21-hydroxylase deficiencies in several ethnic groups and showed that the gene for the nonclassic form is in linkage disequilibrium with HLA-B14. The classic form shows linkage disequilibrium with HLA-Bw47;DR7.

Higashi et al. (1986) confirmed close linkage to C4 by finding that the cloned P450C21 genes hybridized with the 5-prime or 3-prime end regions of human C4 DNA. The P450C21 gene that is nonfunctional is identical to the other except for 3 mutations, each of which is capable of causing premature termination: a 1-base insertion, an 8-base deletion, and a transition mutation. Higashi et al. (1986) suggested that tandem arrangement of the highly homologous pseudo- and genuine genes in close proximity could account for the high incidence of P450C21 gene deficiency or defect through nonhomologous pairing and unequal crossing-over.

By analysis of data collected on 157 families ascertained through a proband with the classic form of 21-hydroxylase deficiency, Aston et al. (1988) could not arrive at a definitive conclusion as to whether the gene is closer to HLA-B or to HLA-DR. They pointed out the limitations of present methods of estimating genetic distance when recombination frequencies are of the order of 0.005.

Molecular Genetics

White et al. (1986) found that cDNA corresponding to 21-hydroxylase is 2 kb long. The encoded protein is predicted to contain 494 amino acids with a molecular weight of 55,000. The enzyme is at most 28% homologous to other cytochrome P450 enzymes that have been studied.

White et al. (1984) demonstrated that the mutations in the several forms of congenital adrenal hyperplasia due to 21-hydroxylase deficiency involve the structural gene for the adrenal microsomal cytochrome P450 specific for steroid 21-hydroxylation (EC 1.14.99.10). Rodrigues et al. (1987) pointed out that 21-hydroxylation was the first enzymatic activity ascribed to any cytochrome P450 (Cooper et al., 1965). Since 1988, CYP21 has come into general use as the preferred symbol.

Rodrigues et al. (1987) determined the nucleotide sequence of the 21-hydroxylase B gene in a patient with congenital adrenal hyperplasia. Eleven nucleotide differences from the normal were found: 2 in the 5-prime flanking region, 4 in introns, 1 in the 3-prime untranslated region, and 4 in exons. Two of the differences in exons caused codon changes: serine-269 to threonine and asparagine-494 to serine. Rodrigues et al. (1987) confirmed that the 21-hydroxylase A gene is a pseudogene due to 3 deleterious mutations in the exons. Comparison of published sequences with those they determined suggested that the 21-hydroxylase B gene is polymorphic. They suggested, as had others, that the 4 distinct clinical forms of 21-hydroxylase deficiency (simple virilizing, salt-wasting, late-onset, and cryptic) may be the consequence of different allelic mutations in the 21-hydroxylase B gene.

The CYP21 pseudogene has 3 main defects: an 8-bp deletion in exon 3, a T insertion in exon 7, and a stop codon in exon 8. Kawaguchi et al. (1992) demonstrated that the 8-bp deletion is present in the chimpanzee also, whereas the other 2 defects are not found in the chimpanzee, gorilla, or orangutan. In the gorilla and orangutan, however, extra CYP21 copies are inactivated by other defects so that the number of functional copies is reduced in each species. Comparison of the sequences revealed evidence for intraspecific homogenization (concerted evolution) of the CYP21 genes, presumably through an expansion-contraction process effected by relatively frequent unequal but homologous crossing-over.

Jospe et al. (1987) performed genomic restriction analysis of 14 unrelated patients with salt-losing congenital adrenal hyperplasia, identifying 3 patterns of mutation in the CA21HB gene: in 16 of the 28 chromosomes (or haplotypes) analyzed, there was no detectable restriction fragment abnormality suggesting that these were point mutations or small deletions or insertions. Complete deletion of CA21HB was found in 9 of 28 haplotypes (32%). In 3 of 28 haplotypes (11%), apparent conversion of CA21HB to the pseudogene CA21HA had occurred. Jospe et al. (1987) described how apparent gene conversion could be detected in the restriction fragment patterns. An alternative explanation to conversion is that unequal crossingover occurred between a haplotype of 2 CA21HA genes and 1 CA21HB gene, and a normal haplotype to produce loss of the CA21HB gene from the first haplotype but retention of 2 CA21HA genes. CA21HB deletion was associated with HLA-Bw47 in 6 haplotypes and with absent C4B expression in 7 haplotypes of the 9.

The gene encoding 21-hydroxylase contains 10 exons; the genes for other P450 enzymes contain 7, 8, or 9 exons. The inactive A gene has an 8-base deletion in codons 110-112, resulting in a frameshift that brings a stop codon into the reading frame at codon 130; a second frameshift and a nonsense mutation occur farther downstream. The two P450C21 genes have 9 introns and are about 3.4 kb long (Higashi et al., 1986). In studies of DNA from 20 patients with 21-hydroxylase deficiency, Rumsby et al. (1986) found one homozygous for a deletion encompassing the C4B and 21-hydroxylase genes. They presented evidence that this originated by recombination between homologous regions of 21-hydroxylase A and B. No alteration in the 21-hydroxylase gene was detected in 12 patients. Seven patients appeared to be heterozygous for the above deletion; i.e., they were genetic compounds.

In molecular studies of the C4/21-hydroxylase genes in patients with the classic salt-wasting form, Schneider et al. (1986) found deletion of C4B and 21-hydroxylase B genes in some. In 2, only the 21-hydroxylase B gene was deleted. Werkmeister et al. (1986) found deletion of the active CA21H gene in almost one-fourth of classic cases of 21-hydroxylase deficiency, whereas mild 'nonclassic' 21-hydroxylase deficiency was associated with a duplicated CA21H gene.

Using multiple restriction enzymes in the analysis of the 21-hydroxylase gene in 10 families, each of which included 2 or more affected persons, Matteson et al. (1987) concluded that the 'deletions' that have been reported as a frequent finding in CAH patients probably represent gene conversions, unequal crossovers, and polymorphisms rather than simple gene deletions. Miller (1987) challenged the interpretation of a high frequency of gene deletion underlying 21-hydroxylase deficiency. White et al. (1987) defended their interpretation. They reiterated their view that probes for the closely linked and highly polymorphic HLA genes should be used for prenatal diagnosis, not CYP21 probes (Mornet et al., 1986).

By Southern blot analysis of genomic DNA using a 21-hydroxylase DNA probe, Harada et al. (1987) found an apparent absence of restriction fragments corresponding to the 21-hydroxylase B gene. They found that this apparent absence was not due to deletion of the gene but rather to a conversion of the functional 21-hydroxylase B gene into the nonfunctional 21-hydroxylase A pseudogene. In 2 patients studied, the affected HLA haplotypes were different, suggesting that conversion had occurred as independent events in the 2 instances. Harada et al. (1987) suggested that gene conversion-like events may be a relatively common cause of 21-hydroxylase deficiency in Japanese. They suggested that this mechanism might also account in part for the predominance of congenital adrenal hyperplasia due to 21-hydroxylase deficiency over that due to deficiency of other steroidogenic P450 enzymes. There may be other examples of gene conversion-like events that are responsible for monogenic disorders when related homologous genes reside in tandem array.

Origin of Mutations

Mornet et al. (1991) estimated that gene conversions involving small DNA segments probably account for 74% of cases of 21-hydroxylase deficiency. Complete deletion of the CYP21B gene accounted for about 20% of cases of the classic form of the disease. Complete deletion of CYP21B was associated with the salt-wasting form, as was an 8-bp deletion in the third exon. A G-to-C transversion in the seventh exon was associated with the late-onset form of the disease. Ghanem et al. (1990) concluded that about 70% of the mutations in the CYP21B gene causing classic and nonclassic CAH are point mutations, because the defective gene was indistinguishable from its structurally intact corresponding gene in Southern blot analysis. Due to the presence of a varying number of C4/21-hydroxylase repeat units, this gene region varies in length among haplotypes. Haplotypes carrying one C4/21-hydroxylase repeat unit with a CYP21P gene transmit the severe form of 21-hydroxylase deficiency. Haglund-Stengler et al. (1991) found association between triplication of the C4/21-hydroxylase repeat unit and the mild form of 21-hydroxylase deficiency.

Gene conversion, a nonreciprocal exchange of homologous genetic information, has been studied extensively in lower eukaryotes in which all the products of a single meiosis can be recovered and analyzed. Because the latter is not possible in mammals, nonreciprocality of the genetic exchange cannot formally be demonstrated. Despite this limitation, the designation 'gene conversion' has been applied to exchanges observed in mammalian genomes that involve an alteration of an allele at a specific locus in such a way as to suggest that an internal portion of its sequence has been replaced by a homologous segment copied from another allele or locus. Gaucher disease (230800) is another example of a disorder in which conversion events occur between the functional gene and a neighboring pseudogene. Gene conversion has been postulated in other clustered gene families, including those for globins (e.g., 142200), immunoglobulins (e.g., 147070), red-green visual pigments (e.g., 303800), and others. With the notable exception of the HLA genes, in which many of the presumed gene conversion events involve allelic exchanges, the postulated gene conversion events in the other systems involve interlocus exchange. The evidence for gene conversion in the human genome had been circumstantial until the description by Collier et al. (1993) of a de novo mutation which permitted the direct comparison of the 'converted' allele with its original form. They observed the de novo introduction of a CYP21A pseudogene-specific mutation into a CYP21B allele. Despite extensive investigations, not a single mutant CYP21B allele has been reported to lack pseudogene-specific mutations that are incompatible with normal gene expression. Consequently, the pathogenesis of 21-hydroxylase deficiency appears to be due almost exclusively to gene-pseudogene exchanges. Tajima et al. (1993) concluded that approximately 90% of the genes in patients with 21-hydroxylase deficiency are accounted for either by a causative mutation from the pseudogene or by a deletion and suggested that the remaining 10% may represent new mutations that do not exist in the pseudogene. Tajima et al. (1993) described a de novo mutation of the CYP21B gene causing CAH. HLA-identical affected and unaffected sibs were observed. Both inherited a missense mutation in exon 4 from the father, but only the affected sib received an intron 2 mutation that caused aberrant RNA splicing from the mother, who was homozygous normal.

White et al. (1994) reviewed mutations in the CYP21 gene which are responsible for more than 90% of cases of the inherited inability to synthesize cortisol. Most of the mutations in CYP21 causing CAH are generated by recombinations between CYP21 and CYP21P which either delete CYP21 or transfer deleterious mutations from CYP21P to CYP21.

Miller (1988) discussed gene conversion in relation to the monogenic form of adrenal hyperplasia. Higashi et al. (1988) presented evidence for either unequal intragenic or intergenic recombination and/or gene conversion events taking place between the pseudogene and the functional gene.

In 4 steroid 21-hydroxylase B mutations from three 21-hydroxylase-deficient patients, Higashi et al. (1988) observed several base changes as compared with the functional B gene. Many of these base changes were identical to those in the CYP21A pseudogene. Two of them were shown to have a point mutation in the second intron, causing aberrant splicing. A third carried 3 clustered missense mutations in the sixth exon, which impaired 21-hydroxylase activity. Since all of these critical mutations could be seen in the corresponding site of the CYP21A pseudogene, the data strongly suggested the involvement of gene conversion in this genetic disease.

Using a genomic probe, Morel et al. (1989) defined 5 haplotypes that identified the mutations in 57 families. Specifically, of 116 CAH-bearing chromosomes, 114 could be sorted into 1 of these 5 haplotypes, based on blots of DNA digested with TaqI and BglII. Haplotype 1, present in 65.6%, was indistinguishable from the normal, and therefore bore very small lesions, presumably point mutations. Haplotype 2, present in 3.4%, and haplotype 3, present in 6.9%, had deletions and duplications of the CYP21 pseudogene but had especially intact functional genes, presumably bearing point mutations. Thus, point mutation was the genetic defect in 75.9% of the chromosomes. Haplotypes 4 and 5, present in 11.2%, appeared to represent a gene that had undergone a gene conversion event. Haplotype 5, present in 11.2%, appeared to have a deletion of about 30 kb of DNA, resulting in a single hybrid CYPA/B gene.

Donohoue et al. (1989) concluded that a single unequal crossingover between the CYP21A and CYP21B genes yields deletion of the latter active gene to result in salt-losing CAH; furthermore, these crossovers do not occur randomly within the complex. In a patient with 21-hydroxylase deficiency, Sinnott et al. (1990) demonstrated a maternally inherited haplotype that carried a de novo deletion of an approximately 30-kb segment including the CYP21B gene and the associated C4B gene. The disease haplotype appeared to have been generated through meiotic unequal crossingover. One of the maternal haplotypes was the frequently occurring HLA-DR3,B8,A1 haplotype that normally carries a deletion of an approximately 30-kb segment including the CYP21A gene and C4A gene. Haplotypes of this type may act as premutations, increasing the susceptibility to development of a 21-hydroxylase deficiency mutation by facilitating unequal chromosome pairing. Mutations in the pseudogene CYP21A include a C-to-T change that leads to a termination codon, TAG, in the eighth exon. Urabe et al. (1990) found that same change in a mutant CYP21B gene isolated from a patient with 21-hydroxylase deficiency. Furthermore, a reciprocal change, i.e., a T-to-C change in the eighth exon of the CYP21A gene, was observed in the Japanese population. This was considered evidence for gene conversion.

Wu and Chung (1991) studied the effects of induced missense mutations at cysteine-428, valine-281, and serine-268 of the 21-hydroxylase gene. A ser268-to-thr mutation (201910.0005) had been found in a patient suffering from CAH and a val281-to-leu mutation (201910.0002) was identified in a patient with nonclassic CAH characterized by partial enzyme deficiency. Cysteine-428 is the invariant cys among all cytochrome P450s and is presumed to be the heme ligand. Wu and Chung (1991) mutated ser268 to thr, cys, and met to see if these changes altered the function of 21-hydroxylase. They changed val281 to leu, ile, and thr, similarly, to study the effects on structure and function of 21-hydroxylase. Val, leu, and ile share properties; therefore, substituting one with another should not drastically disturb the structure of the protein. Wu and Chung (1991) changed cys428 to thr, met, and ser to study the effects of these mutations. They found that the cys428, val281, and ser268 mutations resulted in complete, partial, or no loss of enzymatic activity, respectively. All the cys428 mutants had neither enzymatic activity nor P450 absorption, thus supporting the notion that cys428 is the heme ligand. All the 268-mutants exhibited the same activity as normal 21-hydroxylase, demonstrating that the clinically observed ser268-to-thr change represents a polymorphism rather than the cause of enzyme deficiency.

Tusie-Luna and White (1995) pointed out that steroid 21-hydroxylase deficiency is unusual among genetic diseases in that approximately 95% of the mutant alleles have apparently been generated by recombination between a normally active gene (CYP21) and a closely linked pseudogene (CYP21P). Approximately 20% of mutant alleles carry DNA deletions of 30 kb that have presumably been generated by unequal meiotic crossingover, whereas 75% carry one or more mutations in CYP21 that are normally found in the CYP21P pseudogene. These latter mutations are termed 'gene conversions,' although the mechanism by which they are generated is not well understood. To assess the frequency at which these different recombination events occur, Tusie-Luna and White (1995) used PCR to detect de novo deletions and gene conversions in matched sperm and peripheral blood leukocyte DNA samples from normal persons. Deletions with breakpoints in a 100-bp region in intron 2 and exon 3 were detected in sperm DNA samples with frequencies of approximately 1 in 10(5)-10(6) genomes but were not detected in the matching leukocyte DNA. Gene conversions in the same region occurred in approximately 1 in 10(3)-10(5) genomes in both sperm and leukocyte DNA. These data suggested to the authors that whereas deletions occur exclusively in meiosis, gene conversions occur during both meiosis and mitosis, or perhaps only during mitosis. Thus, the authors concluded that gene conversions must occur by a mechanism distinct from unequal crossingover.

Phenotype/genotype

Speiser et al. (1992) correlated genotype and phenotype in 88 families with congenital adrenal hyperplasia due to 21-hydroxylase deficiency. Mutations were detected on 95% of chromosomes examined. The most common mutations were an A-to-G change in the second intron affecting pre-mRNA splicing in 26% (201910.0006), large deletions in 21%, the ile172-to-asn mutation (201910.0001) in 16%, and the val281-to-leu mutation (201910.0002) in 11%. Patients were classified into 3 mutation groups based on the degree of predicted enzymatic compromise. Mutation group A with no enzymatic activity consisted principally of severely affected salt-wasting patients, group B with 2% activity of simple virilizing patients, and group C with 10 to 20% activity of nonclassic mildly affected patients, but each group contained patients with phenotypes either more or less severe than predicted. The data suggested that most of the phenotypic variability in 21-hydroxylase deficiency results from allelic variation in CYP21. They postulated that phenotypic severity might be influenced by parental imprinting or by negative allelic complementation giving an exaggerated gene dosage effect. However, there was no evidence of either of these phenomena in the group of families studied.

Nikoshkov et al. (1997) studied a rare allele in 2 sibs with late-onset CAH1. This allele contained 3 sequence alterations: a C-to-T transition located 4 bases upstream of translation initiation, a pro105-to-leu substitution, and a pro453-to-ser substitution. The last mutation has been found in other ethnic groups, whereas pro105 to leu seems to be unique to this family. They tested the function of the -4, pro105-to-leu, and pro453-to-ser mutations by in vitro translation after expression of the mutant enzymes in cultured cells. While the -4 substitution had no measurable effect, the pro105-to-leu and pro453-to-ser mutations reduced enzyme activity to 62 and 68% for 17-hydroxyprogesterone and 64 and 46% for progesterone, respectively. When present in combination, these 2 mutations caused a reduction of enzyme activity to 10% for 17-hydroxyprogesterone and 7% for progesterone. These results indicate that pro105-to-leu and pro453-to-ser alleles should only cause very subtle disease when not in combination but may be considered when genotyping patients with the mildest forms of CAH1.

Using allele-specific oligonucleotide hybridization, SSCP, and heteroduplex analyses, Witchel et al. (1996) identified 38 subjects from 21 different families who had 2 deleterious CYP21 mutations. All 38 were homozygous or compound heterozygotes for the intron 2 splicing mutation, which as mentioned earlier, is often identified in 21-hydroxylase deficiency. Comparison of their phenotypic CAH features with their

CYP21 genotypes showed phenotypic heterogeneity extending from classic salt-losing 21-hydroxylase deficiency to asymptomatic phenotypes. Witchel et al. (1996) suggested 3 possibilities for this phenotypic heterogeneity: the presence of other (compensating splice) mutations; the presence of additional functional copies of the CYP21 gene; or leakiness of the splice mutation. Miller (1997) noted a fourth possibility, i.e., the activity of other genes encoding proteins other than P450C21 that have steroid 21-hydroxylase activity. Cytochrome P450 enzymes tend to be 'promiscuous' enzymes that bind many different substrates and catalyze a wide variety of hydroxylations. The author hypothesized that adrenal expression of such an enzyme could account for the cryptic 21-hydroxylase activity seen in patients with known CYP21 deletions who experience apparent recovery of their ability to synthesize mineralocorticoids. According to Miller (1997), the identification of such enzymes may constitute the next major advance in the clinical biology of congenital adrenal hyperplasia due to 21-hydroxylase deficiency.

As indicated earlier, the majority of mutations causing steroid 21-hydroxylase deficiency result from recombinations between the functional gene and the closely related, highly homologous pseudogene. Levo and Partanen (1997) analyzed mutations and recombination breakpoints in the CYP21 gene and determined the associated haplotypes in 51 unrelated Finnish families with CAH. These represented at least half of all CYP21 deficiency patients in Finland. The results indicated multiple founder mutation-haplotype combinations in this population. The 3 most common haplotypes constituted half of all affected chromosomes; only one-sixth of the haplotypes represented single cases. Several of the frequent mutation-haplotype combinations in Finland had been found in other populations of patients of European origin, thus suggesting that these haplotypes were of ancient origin.

Jaaskelainen et al. (1997) reported a population-wide analysis of 120 patients with 21-hydroxylase deficiency found in Finland. Blood samples for CYP21 genotyping were obtained from 78 patients (65%), and their phenotypes were compared with their genotypes. In general, the severity of gene defects correlated well with clinical expression. All patients carrying mutations with severe effects on enzymatic activity had the salt-wasting form of 21-hydroxylase deficiency. Those with the I2 splice mutation (201910.0006), which in some reports has a variable phenotype, had severe mineralocorticoid deficiency. In contrast, patients with the I172N mutation (201910.0001) expressed a wide spectrum of phenotypes that could not be attributed to additional mutations.

Diagnosis

Merkatz et al. (1969) could not diagnose the disorder early in pregnancy by amniocentesis and hormone assay of the amniotic fluid.

Levine et al. (1980) expressed the opinion that experience is still so limited with HLA typing of amniotic cells and with hormonal measurements of amniotic fluid that both approaches to prenatal diagnosis should be used. Gueux et al. (1988) found significant elevations of both 21-deoxycortisol and 17-hydroxyprogesterone in the amniotic fluids of affected pregnancies, as determined by HLA typing and linkage analysis to HLA probes. Hughes et al. (1987) determined the concentration of 17-OH-progesterone in the amniotic fluid collected from 55 pregnant women who had previously had a child with 21-hydroxylase deficiency. In 8 pregnancies the levels were raised. These parents elected to terminate in 4 cases, and examination of the fetus confirmed the diagnosis of CAH. In each case, the affected sib had been a salt-loser. The remaining 4 affected pregnancies proceeded to term, and each infant had salt-losing 21-hydroxylase deficiency. All 47 infants predicted to be unaffected were normal at birth; however, an increased plasma concentration of 17-OH-progesterone was documented in a male nonsalt-loser at 3 months of age. Hughes et al. (1987) concluded that prenatal diagnosis of congenital adrenal hyperplasia by amniotic fluid steroid analysis is reliable only for the salt-losing form. They published a photograph of the external genitalia of a female fetus with 21-hydroxylase deficiency which showed clitoromegaly and fusion of the labia.

New et al. (1983) published nomograms relating baseline and ACTH-stimulated levels of adrenal hormones. These nomograms distinguished the milder symptomatic and asymptomatic nonclassic forms of 21-hydroxylase deficiency (termed late-onset and cryptic forms, respectively), as well as heterozygotes for all of the forms, from normal subjects.

Mornet et al. (1986) demonstrated that one can use linkage of HLA-DNA probes in chorion villus samples in the first trimester diagnosis. They also used determination of 17-hydroxyprogesterone in the first trimester amniotic fluid in the diagnosis.

Reindollar et al. (1988) described the use of a RFLP of the 21-hydroxylase gene for prenatal diagnosis.

Lee et al. (1996) developed primers for differential PCR-amplification of the CYP21 gene and the nonfunctional CYP21P gene. Using the amplification created restriction site (ACRS) approach for direct mutational detection, a secondary PCR was then performed using a panel of primers specific for 11 mutations associated with CAH. Subsequent restriction analysis allowed not only the detection but also the determination of the zygosity of the mutations analyzed. In the analysis of 20 independent chromosomes in 11 families of CAH patients in Taiwan, Lee et al. (1996) detected 4 CYP21 mutation types besides deletion. In 5 different alleles, the CYP21P pseudogene contained some polymorphisms that the authors believed to be associated with the CYP21 gene. This finding suggested that gene conversion events are occurring in both CYP21P and CYP21. The combined differential PCR-ACRS protocol was described as simple, direct, and applicable to prenatal diagnosis of CAH using chorionic villi or amniotic cells.

During the course of genetic analysis of CYP21 mutations in CAH families, Day et al. (1996) noticed a number of relatives genotyped as nucleotide 656G (201910.0006) homozygotes who showed no clinical signs of disease. They proposed that the putative asymptomatic nucleotide 656G/G individuals are incorrectly typed due to a dropout of 1 haplotype during PCR amplification of CYP21. They recommended that for prenatal diagnosis, microsatellite typing be used as a supplement to CYP21 genotyping in order to resolve ambiguities at nucleotide 656.

Clinical Management

Jones (1978) found cases of mild 'adult' adrenal hyperplasia manifest by oligomenorrhea and treated like the usual form with adrenocorticosteroids.

Even after being present for several years, hypertension associated with 21-hydroxylase deficiency is relieved by steroid therapy.

Cutfield et al. (1983) described 2 male cousins with partial 21-hydroxylase deficiency presenting with bilateral testicular masses and infertility. In both cases, the testicular masses, consisting of adrenocorticotropic hormone-dependent pluripotential interstitial cells, were thought to play a major etiologic role in infertility. Nighttime (11 p.m.), low-dose dexamethasone therapy led to disappearance of the masses and restoration of fertility. Hydrocortisone, 10 mg 3 times daily, had failed to accomplish this reversal.

Cutler and Laue (1990) investigated the use of a new form of therapy which would combine hydrocortisone in strictly physiologic dosage with antiandrogen and aromatase-inhibitor therapy. By blocking androgen and the conversion of testosterone to estrogen, they hoped to achieve normal growth. The proposal, which remained to be tested, was suggested by the success of a similar program in the treatment of familial male precocious puberty (176410).

A multicentric study of prenatal treatment of 21-hydroxylase deficiency with dexamethasone administered by mouth to the mother was undertaken in France (Forest et al., 1989). Wudy et al. (1994) reported successful treatment of a single case with dexamethasone (0.5 mg, tid, p.o.) starting from the beginning of the eighth week of gestation.

Cornean et al. (1998) studied 22 prepubertal children with 21-hydroxylase deficiency whose steroid therapy was considered to be optimal in terms of linear growth and skeletal maturity. The authors reported a significant increase in body mass index (BMI) as a consequence of increased body fat. This was consistent with an early 'rebound' of BMI, which is associated with obesity in later childhood and an increased risk of long-term health problems related to adult obesity.

Population Genetics

Congenital adrenal hyperplasia affects about 1 in 5,000 births.

In the canton of Zurich, Switzerland, Prader (1958) estimated the frequency of the congenital adrenogenital syndrome to be 1 in 5,041 live births, giving a frequency of carriers of 1 in 35. Childs et al. (1956) had estimated the frequency in Maryland to be 1 in 67,000 births.

In Toronto, Qazi and Thompson (1972) estimated the minimum frequency of salt-losing C-21 hydroxylase deficiency as 1 per 26,292. Presumably it is a salt-losing variety of 21-hydroxylase deficiency that is present in relatively high frequency in Eskimos of Alaska (Hirschfeld and Fleshman, 1969). Other recessive conditions of high frequency among the Alaskan Eskimos include Kuskokwim disease (208200), methemoglobinemia (250800), and pseudocholinesterase deficiency (see 177400). The forms of adrenal hyperplasia that may present in adulthood are 21- and 11-hydroxylase deficiencies.

Speiser et al. (1985) concluded that nonclassic 21-hydroxylase deficiency is probably the most frequent autosomal recessive genetic disease. It is especially frequent in Ashkenazim (3.7%), Hispanics (1.9%), Yugoslavs (1.6%), and Italians (0.3%). With the exception of the Yugoslavs, the gene for the nonclassic form is in linkage disequilibrium with HLA-B14. The classic form shows linkage disequilibrium with HLA-Bw47;DR7. Sherman et al. (1988) estimated the frequency of the gene for the nonclassic form of 21-hydroxylase D to be as high as 0.223 among Ashkenazi Jews. Segregation analysis of families ascertained through a nonclassic proband and those ascertained through a classic proband showed essentially identical results. The authors concluded that the possibility that the gene is incompletely penetrant in a small number of homozygotes is likely for the nonclassic form and unlikely for the classic form.

Layrisse et al. (1987) studied 19 Venezuelan families of mixed ethnic origin having 20 affected newborns with the salt-wasting form of 21-hydroxylase deficiency. HLA haplotypes and complotypes were determined. The results were markedly different from those reported in the literature which show an association at the population level with HLA-Bw47 and the extended haplotype HLA-Bw47,DR7,FC91,0. Four of the unrelated patients were homozygous for all MHC loci tested while 3 others were homozygous for at least 2 HLA loci. The findings were interpreted as indicating that among Venezuelan patients, salt-wasting 21-hydroxylase deficiency results in the main from founder effect of relatively few independent mutations. The mutation marked by HLA-Bw47 was not observed in this population.

Thilen and Larsson (1990) performed a retrospective study of all Swedish patients with CAH born between 1969 and 1986, to determine possible benefits of neonatal screening. Information was obtained concerning 67 males and 83 females. Of these, 143 were regarded as classic and 7 as nonclassic (symptoms after 5 years of age or cryptic). All but 2 (a girl with 11-hydroxylase deficiency and a boy with beta-hydroxysteroid dehydrogenase deficiency) had 21-hydroxylase deficiency. The prevalence was 1 in 11,500. Salt loss was displayed by 93 patients (48 male, 45 female), all before the age of 3 months. The median age at diagnosis for boys in this group was 21 days. Gender assignment was a major problem in 38 of 57 girls, with ambiguous genitalia noticed at birth. Of these girls, 15 were considered to be male before the diagnosis of CAH was made. In a similar study in Kuwait, Lubani et al. (1990) found 60 children with CAH diagnosed between 1978 and 1988, giving an estimated prevalence of 1 in 9,000 live births. In addition, there was presumptive evidence of CAH resulting in the death of 20 other children, giving a prevalence figure of 1 in 7,000. In 54 patients (90%), 21-hydroxylase deficiency was diagnosed; in 3 patients each, the diagnosis was 3-beta-hydroxysteroid dehydrogenase deficiency and 11-beta-hydroxylase deficiency.

Chrousos et al. (1982) estimated that 6 to 12% of hirsute women have 21-hydroxylase deficiency because of homozygosity for a mild allele of the 21-hydroxylase gene. They calculated that the frequency of the gene for the attenuated form of the disease is 0.015 to 0.057.

From 1991 to 1994, approximately 4.5 million infants had newborn screening for CAH in Japan. In this cohort, Tajima et al. (1997) identified 2 sibs and 2 unrelated newborns who had mild elevations of serum 17-hydroxyprogesterone levels at 5 days of age but no symptoms of CAH. These 4 cases were diagnosed as having probable nonclassic steroid 21-hydroxylase deficiency. The 2 sibs had ile172-to-asn 201910.0001 and arg356-to-trp 201910.0003 mutations in 1 allele and a gene conversion that included the pro30-to-leu 201910.0004 mutation in the other allele. The first unrelated case had a gene conversion encoding the same pro30-to-leu mutation in 1 allele. The second allele had an intron 2 mutation (668-12 A-to-G), which perturbed splicing, and the arg356-to-trp mutation. The second unrelated case was a compound heterozygote for an arg356-to-trp and a 707del8 mutation. Since the the estimated rate of detection of the nonclassic form by mass screening (1 in 1,000,000) seemed low compared to the established detection rate for the classic form (1 in 18,000), the authors concluded that detection by neonatal screening may be particularly difficult for nonclassic cases in which both alleles contain only nonclassic associated mutations.

Witchel et al. (1997) hypothesized that those heterozygous for 21-hydroxylase deficiency have a survival advantage. They found significantly elevated cortisol responses in 28 proven carriers compared to 22 mutation-negative controls (30 min cortisol levels: normal, 24.2 microg/dL; carrier, 28.1 microg/dL; P less than 0.005). The authors proposed that the higher cortisol response observed in carriers may enable a rapid return to homeostasis in response to infectious, inflammatory, or other environmental stresses and may protect from inappropriate immune responses, such as autoimmune diseases.

Animal Model

In the mouse, Chaplin et al. (1986) showed that only one of the two 21-hydroxylase genes is expressed. The authors presented the complete primary structure of both 21-hydroxylase encoding genes. The active gene in the mouse is referred to as A, whereas in man it is referred to as B. In the mouse, Chaplin et al. (1986) found a deletion of 215 nucleotides spanning the second exon in the 21-hydroxylase B gene; other nucleotide changes introduced frameshifts and premature termination codons. A hybrid gene composed of the 21-hydroxylase B promoter placed 5-prime of the 21-hydroxylase A structural sequences was efficiently transcribed following transfection into adrenocortical tumor cells. These findings demonstrated that the lack of expression was due to mutations within the 21-hydroxylase B structural gene and not due to a defect of the promoter. In the human, the CA21HA gene is a pseudogene (184754) and the nature of the gene deletions that prevent expression is different from that in the mouse. Specifically, the 21-hydroxylase A gene has an 8-base deletion within the third exon, introducing a premature termination codon (White et al., 1986; Higashi et al., 1986). See review by White et al. (1987).

Gotoh et al. (1988) described deletion of the 21-hydroxylase gene in mice of a particular H-2 recombinant haplotype. They found, furthermore, that newborn homozygous mice are deficient in 21-hydroxylase activity and that homozygosity results in death at an early postnatal stage. Morphologic changes in the adrenal glands of newborn homozygotes were observed.

Pseudogenes

The 21-hydroxylase pseudogene, symbolized CYP21P or CYP21A, is situated on 6p, close to the functional gene, CYP21. Higashi et al. (1986) suggested that this particular genomic anatomy predisposes the functional gene to mutation through gene conversion or through deletion by homologous recombination and unequal crossingover.

See also Bongiovanni (1972); Boudreaux et al. (1979); Couillin et al. (1980); Dumic et al. (1985); Flori et al. (1980); Gordon et al. (1985); Grosse-Wilde et al. (1979); Gutai et al. (1979); Higashi et al. (1988); Kirkland et al. (1977); Levine et al. (1980); Libber et al. (1985); Lorenzen et al. (1979); Lorenzen et al. (1980); Murtaza et al. (1980); New et al. (1981); New et al. (1981); New and Levine (1984); Pang et al. (1985); Petersen et al. (1980); Pollack et al. (1986); Pollack et al. (1981); Pollack et al. (1979); Pollack et al. (1981); Rosenmann et al. (1980); Speiser and New (1985); Speiser et al. (1992); Tajima et al. (1993); Warsof et al. (1980); Webb et al. (1980); Werder et al. (1980); White et al. (1987); White et al. (1984); Yang et al. (1978)

Amor, M.; Parker, K. L.; Globerman, H.; New, M. I.; White, P. C.: Mutation in the CYP21B gene (ile172-to-asn) causes steroid 21-hydroxylase deficiency. Proc. Nat. Acad. Sci. 85: 1600-1604, 1988.

Aston, C. E.; Sherman, S. L.; Morton, N. E.; Speiser, P. W.; New, M. I.: Genetic mapping of the 21-hydroxylase locus: estimation of small recombination frequencies. Am. J. Hum. Genet. 43: 304-310, 1988.

Bias, W. B.; Urban, M. D.; Migeon, C. J.; Hsu, S. H.; Lee, P. A.: Intra-HLA recombinations localizing the 21-hydroxylase deficiency gene within the HLA complex. Hum. Immun. 2: 139-145, 1981.

Blankstein, J.; Faiman, C.; Reyes, F. I.; Schroeder, M. L.; Winter, J. S. D.: Adult-onset familial adrenal 21-hydroxylase deficiency. Am. J. Med. 68: 441-448, 1980.

Bongiovanni, A. M.: Disorders of adrenocortical steroid biogenesis.In: Stanbury, J. B.; Wyngaarden, J. B.; Fredrickson, D. S.: The Metabolic Basis of Inherited Disease. New York: McGraw-Hill (pub.) (3rd ed.): 1972. Pp. 857-885.

Bongiovanni, A. M.; Root, A. W.: The adrenogenital syndrome. New Eng. J. Med. 268: 1283-1289; 1342-1351; 1391-1399, 1963.

Boudreaux, D.; Waisman, J.; Skinner, D. G.; Low, R.: Giant adrenal myelolipoma and testicular interstitial cell tumor in a man with congenital 21-hydroxylase deficiency. Am. J. Surg. Path. 3: 109-123, 1979.

Carroll, M. C.; Campbell, R. D.; Porter, R. R.: Mapping of steroid 21-hydroxylase genes adjacent to complement component C4 genes in HLA, the major histocompatibility complex in man. Proc. Nat. Acad. Sci. 82: 521-525, 1985.

Chaplin, D. D.; Galbraith, L. J.; Seidman, J. G.; White, P. C.; Parker, K. L.: Nucleotide sequence analysis of murine 21-hydroxylase gene: mutations affecting gene expression. Proc. Nat. Acad. Sci. 83: 9601-9605, 1986.

Childs, B.; Grumbach, M. M.; Van Wyk, J. J.: Virilizing adrenal hyperplasia: genetic and hormonal studies. J. Clin. Invest. 35: 213-222, 1956.

Chiou, S.-H.; Hu, M.-C.; Chung, B.: A missense mutation at ile172-to-asn or arg356-to-trp causes steroid 21-hydroxylase deficiency. J. Biol. Chem. 265: 3549-3552, 1990.

Chrousos, G. P.; Loriaux, D. L.; Mann, D. L.; Cutler, G. B., Jr.: Late-onset 21-hydroxylase deficiency mimicking idiopathic hirsutism or polycystic ovarian disease: an allelic variant of congenital virilizing adrenal hyperplasia with a milder enzymatic defect. Ann. Intern. Med. 96: 143-148, 1982.

Collier, S.; Tassabehji, M.; Strachan, T.: A de novo pathological point mutation at the 21-hydroxylase locus: implications for gene conversion in the human genome. Nature Genet. 3: 260-265, 1993.

Cooper, D. Y.; Levin, S.; Narasimhulu, S.; Rosenthal, O.; Estabrook, R. W.: Photochemical action spectrum of the terminal oxidase of mixed function oxidase systems. Science 147: 400-402, 1965.

Cornean, R. E.; Hindmarsh, P. C.; Brook, C. G. D.: Obesity in 21-hydroxylase deficient patients. Arch. Dis. Child. 78: 261-263, 1998.

Couillin, P.; Kottler-Missonnier, M. L.; Grisard, M. C.; Hors, J.; Feingold, J.; Boue, J.; Boue, A.: HLA-A, B, C, DR alleles in congenital adrenal hyperplasia. Hum. Genet. 53: 389-392, 1980.

Cutfield, R. G.; Bateman, J. M.; Odell, W. D.: Infertility caused by bilateral testicular masses secondary to congenital adrenal hyperplasia (21-hydroxylase deficiency). Fertil. Steril. 40: 809-814, 1983.

Cutler, G. B., Jr.; Laue, L.: Congenital adrenal hyperplasia due to 21-hydroxylase deficiency. New Eng. J. Med. 323: 1806-1813, 1990.

Day, D. J.; Speiser, P. W.; Schulze, E.; Bettendorf, M.; Fitness, J.; Barany, F.; White, P. C.: Identification of non-amplifying CYP21 genes when using PCR-based diagnosis of 21-hydroxylase deficiency in congenital adrenal hyperplasia (CAH) affected pedigrees. Hum. Molec. Genet. 5: 2039-2048, 1996.

Donohoue, P. A.; Jospe, N.; Migeon, C. J.; Van Dop, C.: Two distinct areas of unequal crossingover within the steroid 21-hydroxylase genes produce absence of CYP21B. Genomics 5: 397-406, 1989.

Duck, S. C.: Malignancy associated with congenital adrenal hyperplasia. J. Pediat. 99: 423-424, 1981.

Dumic, M.; Brkljacic, L.; Mardesic, D.; Plavsic, V.; Lukenda, M.; Kastelan, A.: 'Cryptic' form of congenital adrenal hyperplasia due to 21-hydroxylase deficiency in the Yugoslav population. Acta Endocr. 109: 386-392, 1985.
Dupont, B.; Smithwick, E. M.; Oberfield, S. E.; Lee, T. D.; Levine, L. S.: Close genetic linkage between HLA and congenital adrenal hyperplasia (21-hydroxylase deficiency). Lancet II: 1309-1312, 1977.
Ezquieta, B.; Oliver, A.; Gracia, R.; Gancedo, P. G.: Analysis of steroid 21-hydroxylase gene mutations in the Spanish population. Hum. Genet. 96: 198-204, 1995.
Fleischnick, E.; Awdeh, Z. L.; Raum, D.; Granados, J.; Alosco, S. M.; Crigler, J. F., Jr.; Gerald, P. S.; Giles, C. M.; Yunis, E. J.; Alper, C. A.: Extended MHC haplotypes in 21-hydroxylase-deficiency congenital adrenal hyperplasia: shared genotypes in unrelated patients. Lancet I: 152-156, 1983.
Flori, J.; Tongio, M. M.; Kurtz, F.; Mayer, S.; Juif, J. G.: Liaison entre le gene responsible de l'hyperplasie surrenalienne par deficit en 21 hydroxylase et les marqueurs du bras court du chromosome 6. J. Genet. Hum. 28: 185-193, 1980.
Forest, M. G.; Betuel, H.; David, M.: Prenatal treatment in congenital adrenal hyperplasia due to 21-hydroxylase deficiency: up-date 88 of the French multicentric study. Endocr. Res. 15: 277-301, 1989.
Galal, O. M.; Rudd, B. T.; Drayer, N. M.: Evaluation of deficiency of 21-hydroxylation in patients with congenital adrenal hyperplasia. Arch. Dis. Child. 43: 410-414, 1968.
Ghanem, N.; Lobaccaro, J. M.; Buresi, C.; Abbal, M.; Halaby, G.; Sultan, C.; Lefranc, G.: Defective, deleted or converted CYP21B gene and negative association with a rare restriction fragment length polymorphism allele of the factor B gene in congenital adrenal hyperplasia. Hum. Genet. 86: 117-125, 1990.
Ghizzoni, L.; Mastorakos, G.; Street, M. E.; Vottero, A.; Mazzardo, G.; Vanelli, M.; Chrousos, G. P.; Bernasconi, S.: Spontaneous thyrotropin and cortisol secretion interactions in patients with nonclassical 21-hydroxylase deficiency and control children. J. Clin. Endocr. Metab. 82: 3677-3683, 1997.
Globerman, H.; Amor, H.; Parker, K. L.; New, M. I.; White, P. C.: A nonsense mutation causing steroid 21-hydroxylase deficiency. J. Clin. Invest. 82: 139-144, 1988.
Gordon, M. T.; Conway, D. I.; Anderson, D. C.; Harris, R.: Genetics and biochemical variability of variants of 21 hydroxylase deficiency. J. Med. Genet. 22: 354-360, 1985.
Gotoh, H.; Sagai, T.; Hata, J.-I.; Shiroishi, T.; Moriwaki, K.: Steroid 21-hydroxylase deficiency in mice. Endocrinology 123: 1923-1927, 1988.
Grosse-Wilde, H.; Weil, J.; Albert, E.; Scholz, S.; Bidlingmaier, F.; Sippel, W. G.; Knorr, D.: Genetic linkage studies between congenital adrenal hyperplasia and the HLA blood group system. Immunogenetics 8: 41-49, 1979.
Gueux, B.; Fiet, J.; Couillin, P.; Raux-Demay, M.-C.; Mornet, E.; Galons, H.; Villette, J.-M.; Boue, J.; Dreux, C.: Prenatal diagnosis of 21-hydroxylase deficiency congenital adrenal hyperplasia by simultaneous radioimmunoassay of 21-deoxycortisol and 17-hydroxyprogesterone in amniotic fluid. J. Clin. Endocr. Metab. 66: 534-537, 1988.
Gutai, J. P.; Lee, P. A.; Johnsonbaugh, R. E.; Gareis, F.; Urban, M. D.; Migeon, C. J.: Detection of the heterozygous state in siblings of patients with congenital adrenal hyperplasia due to 21-hydroxylase deficiency. J. Pediat. 94: 770-772, 1979.
Haglund-Stengler, B.; Ritzen, E. M.; Gustafsson, J.; Luthman, H.: Haplotypes of the steroid 21-hydroxylase gene region encoding mild steroid 21-hydroxylase deficiency. Proc. Nat. Acad. Sci. 88: 8352-8356, 1991.
Harada, F.; Kimura, A.; Iwanaga, T.; Shimozawa, K.; Yata, J.; Sasazuki, T.: Gene conversion-like events cause steroid 21-hydroxylase deficiency in congenital adrenal hyperplasia. Proc. Nat. Acad. Sci. 84: 8091-8094, 1987.
Helmberg, A.; Tusie-Luna, M. T.; Tabarelli, M.; Kofler, R.; White, P. C.: R339H and P453S: CYP21 mutations associated with nonclassic steroid 21-hydroxylase deficiency that are not apparent gene conversions. Molec. Endocr. 6: 1318-1322, 1992.
Higashi, Y.; Tanae, A.; Inoue, H.; Fujii-Kuriyama, Y.: Evidence for frequent gene conversion in the steroid 21-hydroxylase P-450(C21) gene: implications for steroid 21-hydroxylase deficiency. Am. J. Hum. Genet. 42: 17-25, 1988.
Higashi, Y.; Tanae, A.; Inoue, H.; Hiromasa, T.; Fujii-Kuriyama, Y.: Aberrant splicing and missense mutations cause steroid 21-hydroxylase (P-450[C21]) deficiency in humans: possible gene conversion products. Proc. Nat. Acad. Sci. 85: 7486-7490, 1988.
Higashi, Y.; Yoshioka, H.; Yamane, M.; Gotoh, O.; Fujii-Kuriyama, Y.: Complete nucleotide sequence of two steroid 21-hydroxylase genes tandemly arranged in human chromosome: a pseudogene and a genuine gene. Proc. Nat. Acad. Sci. 83: 2841-2845, 1986.
Hirschfeld, A. J.; Fleshman, J. K.: An unusually high incidence of salt-losing congenital adrenal hyperplasia in the Alaskan Eskimo. J. Pediat. 75: 492-494, 1969.
Holler, W.; Scholz, S.; Knorr, D.; Bidlingmaier, F.; Keller, E.; Albert, E. D.: Genetic differences between the salt-wasting, simple virilizing, and nonclassical types of congenital adrenal hyperplasia. J. Clin. Endocr. Metab. 60: 757-763, 1985.
Hughes, I. A.; Dyas, J.; Riad-Fahmy, D.; Laurence, K. M.: Prenatal diagnosis of congenital adrenal hyperplasia: reliability of amniotic fluid steroid analysis. J. Med. Genet. 24: 344-347, 1987.
Jaaskelainen, J.; Levo, A.; Voutilainen, R.; Partanen, J.: Population-wide evaluation of disease manifestation in relation to molecular genotype in steroid 21-hydroxylase (CYP21) deficiency: good correlation in a well defined population. J. Clin. Endocr. Metab. 82: 3293-3297, 1997.
Jaresch, S.; Kornely, E.; Kley, H.-K.; Schlaghecke, R.: Adrenal incidentaloma and patients with homozygous or heterozygous congenital adrenal hyperplasia. J. Clin. Endocr. Metab. 74: 685-689, 1992.
Jones, H. W., Jr.: Personal Communication. Baltimore, Md. 2/10/1978.
Jospe, N.; Donohoue, P. A.; Van Dop, C.; McLean, R. H.; Bias, W. B.; Migeon, C. J.: Prevalence of polymorphic 21-hydroxylase gene (CA21HB) mutations in salt-losing congenital adrenal hyperplasia. Biochem. Biophys. Res. Commun. 142: 798-804, 1987.
Kadair, R. G.; Block, M. B.; Katz, F. H.; Hofeldt, F. D.: 'Masked' 21-hydroxylase deficiency of the adrenal presenting with gynecomastia and bilateral testicular masses. Am. J. Med. 62: 278-282, 1977.
Kawaguchi, H.; O'hUigin, C.; Klein, J.: Evolutionary origin of mutations in the primate cytochrome P450c21 gene. Am. J. Hum. Genet. 50: 766-780, 1992.
Kirby-Keyser, L.; Porter, C. C.; Donohoue, P. A.: E380D: a novel point mutation of CYP21 in an HLA-homozygous patient with salt-losing congenital adrenal hyperplasia due to 21-hydroxylase deficiency. Hum. Mutat. 9: 181-182, 1997.
Kirkland, R. T.; Kirkland, J. L.; Keenan, B. S.; Bongiovanni, A. M.; Rosenberg, H. S.; Clayton, G. W.: Bilateral testicular tumors in congenital adrenal hyperplasia. J. Clin. Endocr. Metab. 44: 369-378, 1977.
Klouda, P. T.; Harris, R.; Price, D. A.: Linkage and association between HLA and 21-hydroxylase deficiency. J. Med. Genet. 17: 337-341, 1980.
Knochenhauer, E. S.; Cortet-Rudelli, C.; Cunningham, R. D.; Conway-Myers, B. A.; Dewailly, D.; Azziz, R.: Carriers of 21-hydroxylase deficiency are not at increased risk for hyperandrogenism. J. Clin. Endocr. Metab. 82: 479-485, 1997.
Kuttenn, F.; Couillin, P.; Girard, F.; Billaud, L.; Vincens, M.; Boucekkine, C.; Thalabard, J.-C.; Maudelonde, T.; Spritzer, P.; Mowszowicz, I.; Boue, A.; Mauvais-Jarvis, P.: Late-onset adrenal hyperplasia in hirsutism. New Eng. J. Med. 313: 224-231, 1985.
Kuttenn, F.; Mowszowicz, I.; Schaison, G.; Mauvais-Jarvis, P.: Androgen production and skin metabolism in hirsutism. J. Endocr. 75: 83-91, 1977.
Layrisse, A.; White, C.; Gunczler, P.; Gafaro Valera, L.; Arias, S.; Yunis, E. J.; Alper, C. A.; Awdeh, Z. L.: Sharing of MHC haplotypes among apparently unrelated patients with congenital adrenal hyperplasia due to 21-hydroxylase deficiency. Immunogenetics 25: 99-103, 1987.
Lee, H.-H.; Chao, H.-T.; Ng, H.-T.; Choo, K.-B.: Direct molecular diagnosis of CYP21 mutations in congenital adrenal hyperplasia. J. Med. Genet. 33: 371-375, 1996.
Levine, L. S.; Dupont, B.; Lorenzen, F.; Pang, S.; Pollack, M.; Oberfield, S.; Kohn, B.; Lerner, A.; Cacciari, E.; Mantero, F.; Cassio, A.; Scaroni, C.; Chiumello, G.; Rondanini, G. F.; Gargantini, L.; Giovannelli, G.; Virdis, R.; Bartolotta, E.; Migliori, C.; Pintor, C.; Tato, L.; Barboni, F.; New, M. I.: Cryptic 21-hydroxylase deficiency in families of patients with classical congenital adrenal hyperplasia. J. Clin. Endocr. Metab. 51: 1316-1324, 1980.
Levine, L. S.; Dupont, B.; Lorenzen, F.; Pang, S.; Pollack, M.; Oberfield, S. E.; Kohn, B.; Lerner, A.; Cacciari, E.; Mantero, F.; Cassio, A.; Scaroni, C.; Chiumello, G.; Rondanini, G. F.; Gargantini, L.; Giovannelli, G.; Virdis, R.; Bartolotta, E.; Migliori, C.; Pintor, C.; Tato, L.; Barboni, F.; New, M. I.: Genetic and hormonal characterization of cryptic 21-hydroxylase deficiency. J. Clin. Endocr. Metab. 53: 1193-1198, 1981.
Levine, L. S.; Pang, S.; Dupont, B.; Pollack, M.; Lorenzen, F.; New, M. I.: Detection of heterozygote of 21-hydroxylase deficiency. (Letter) Lancet I: 603-604, 1980.
Levine, L. S.; Zachmann, M.; New, M. I.; Prader, A.; Pollack, M. S.; O'Neill, G. J.; Yang, S. Y.; Oberfield, S. E.; Dupont, B.: Genetic mapping of the 21-hydroxylase-deficiency gene within the HLA linkage group. New Eng. J. Med. 299: 911-915, 1978.
Levo, A.; Partanen, J.: Mutation-haplotype analysis of steroid 21-hydroxylase (CYP21) deficiency in Finland. Implications for the population history of defective alleles. Hum. Genet. 99: 488-497, 1997.
Lewis, V. G.; Money, J.; Epstein, R.: Concordance of verbal and nonverbal ability in the adrenogenital syndrome. Johns Hopkins Med. J. 122: 192-195, 1968.
Libber, S. M.; Migeon, C. J.; Bias, W. B.: Ascertainment of 21-hydroxylase deficiency in individuals with HLA-B14 haplotype. J. Clin. Endocr. Metab. 60: 727-730, 1985.
Lorenzen, F.; Pang, S.; New, M. I.; Dupont, B.; Pollack, M.; Chow, D. M.; Levine, L. S.: Hormonal phenotype and HLA-genotype in families of patients with congenital adrenal hyperplasia (21-hydroxylase deficiency). Pediat. Res. 13: 1356-1360, 1979.
Lorenzen, F.; Pang, S.; New, M. I.; Oberfield, S.; Dupont, B.; Chow, D.; Schneider, B.; Levine, L.: Studies of the C-21 and C-19 steroids and HLA genotyping in siblings and parents of patients with congenital adrenal hyperplasia due to 21-hydroxylase deficiency. J. Clin. Endocr. Metab. 50: 572-577, 1980.
Lubani, M. M.; Issa, A.-R. A.; Bushnaq, R.; Al-Saleh, Q. A.; Dudin, K. I.; Reavey, P. C.; El-Khalifa, M. Y.; Manandhar, D. S.; Abdul Al, Y. K.; Ismail, E. A.; Teebi, A. S.: Prevalence of congenital adrenal hyperplasia in Kuwait. Europ. J. Pediat. 149: 391-392, 1990.
Matteson, K. J.; Phillips, J. A., III; Miller, W. L.; Chung, B.-C.; Orlando, P. J.; Frisch, H.; Ferrandez, A.; Burr, I. M.: P450XXI (steroid 21-hydroxylase) gene deletions are not found in family studies of congenital adrenal hyperplasia. Proc. Nat. Acad. Sci. 84: 5858-5862, 1987.
McAlpine, P. J.: Personal Communication. Winnipeg, Manitoba, Canada 2/26/1988.
McGuire, L. S.; Omenn, G. S.: Congenital adrenal hyperplasia. I. Family studies of IQ. Behav. Genet. 5: 165-173, 1975.
Merkatz, I. R.; New, M. I.; Peterson, R. E.; Seaman, M. P.: Prenatal diagnosis of adrenogenital syndrome by amniocentesis. J. Pediat. 75: 977-982, 1969.
Miller, W. L.: Personal Communication. San Francisco, Calif. 1/3/1996.
Miller, W. L.: Gene conversions, deletions and polymorphisms in congenital adrenal hyperplasia. Am. J. Hum. Genet. 42: 4-7, 1988.
Miller, W. L.: Congenital adrenal hyperplasia. (Letter) New Eng. J. Med. 317: 1413-1414, 1987.
Miller, W. L.: Phenotypic heterogeneity associated with the splicing mutation in congenital adrenal hyperplasia due to 21-hydroxylase deficiency. J. Clin. Endocr. Metab. 82: 1304 only, 1997.
Morel, Y.; Andre, J.; Uring-Lambert, B.; Hauptmann, G.; Betuel, H.; Tossi, M.; Forest, M. G.; David, M.; Bertrand, J.; Miller, W. L.: Rearrangements and point mutations of P450c21 genes are distinguished in five restriction endonuclease haplotypes identified by a new probing strategy in 57 families with congenital adrenal hyperplasia. J. Clin. Invest. 83: 527-536, 1989.

Mornet, E.; Boue, J.; Raux-Demay, M.; Couillin, P.; Oury, J. F.; Dumez, Y.; Dausset, J.; Cohen, D.; Boue, A.: First trimester prenatal diagnosis of 21-hydroxylase deficiency by linkage analysis to HLA-DNA probes and by 17-hydroxyprogesterone determination. Hum. Genet. 73: 358-364, 1986.
Mornet, E.; Crete, P.; Kuttenn, F.; Raux-Demay, M.-C.; Boue, J.; White, P. C.; Boue, A.: Distribution of deletions and seven point mutations on CYP21B genes in three clinical forms of steroid 21-hydroxylase deficiency. Am. J. Hum. Genet. 48: 79-88, 1991.
Murtaza, L.; Sibert, J. R.; Hughes, I.; Balfour, I. C.: Congenital adrenal hyperplasia—a clinical and genetic survey: are we detecting male salt-losers? Arch. Dis. Child. 55: 622-625, 1980.
Murtaza, L. M.; Hughes, I. A.; Sibert, J. R.; Balfour, I. C.: HLA and congenital adrenal hyperplasia. (Letter) Lancet II: 524 only, 1978.
New, M. I.; Dupont, B.; Pang, S.; Pollack, M.; Levine, L. S.: An update of congenital adrenal hyperplasia. Recent Prog. Horm. Res. 37: 105-181, 1981.
New, M. I.; Dupont, B.; Pollack, M. S.; Levine, L. S.: The biochemical basis for genotyping 21-hydroxylase deficiency. Hum. Genet. 58: 123-127, 1981.
New, M. I.; Levine, L. S.: Recent advances in 21-hydroxylase deficiency. Ann. Rev. Med. 35: 649-663, 1984.
New, M. I.; Lorenzen, F.; Lerner, A. J.; Kohn, B.; Oberfield, S. E.; Pollack, M. S.; Dupont, B.; Stoner, E.; Levy, D. J.; Pang, S.; Levine, L. S.: Genotyping steroid 21-hydroxylase deficiency: hormonal reference data. J. Clin. Endocr. Metab. 57: 320-326, 1983.
Nikoshkov, A.; Lajic, S.; Holst, M.; Wedell, A.; Luthman, H.: Synergistic effect of partially inactivating mutations in steroid 21-hydroxylase deficiency. J. Clin. Endocr. Metab. 82: 194-199, 1997.
Orta-Flores, Z.; Cantu, J. M.; Dominguez, O. V.: Reciprocal interactions of progesterone and 17-alpha-OH-progesterone as exogenous substrates for rat adrenal 21-hydroxylase. J. Steroid Biochem. 7: 761-767, 1976.
Owerbach, D.; Crawford, Y. M.; Draznin, M. B.: Direct analysis of CYP21B genes in 21-hydroxylase deficiency using polymerase chain reaction amplification. Molec. Endocr. 4: 125-131, 1990.
Owerbach, D.; Sherman, L.; Ballard, A.-L.; Azziz, R.: Pro453-to-ser mutation in CYP21 is associated with nonclassic steroid 21-hydroxylase deficiency. Molec. Endocr. 6: 1211-1215, 1992.
Pang, S.; Murphey, W.; Levine, L. S.; Spence, D. A.; Leon, A.; LaFranchi, S.; Surve, A. S.; New, M. I.: A pilot newborn screening program for congenital adrenal hyperplasia in Alaska. J. Clin. Endocr. Metab. 55: 413-420, 1982.
Pang, S.; Pollack, M. S.; Loo, M.; Green, O.; Nussbaum, R.; Clayton, G.; Dupont, B.; New, M. I.: Pitfalls of prenatal diagnosis of 21-hydroxylase deficiency congenital adrenal hyperplasia. J. Clin. Endocr. Metab. 61: 89-97, 1985.
Partanen, J.; Campbell, R. D.: Substitution of ile172-to-asn in the steroid 21-hydroxylase B (P450c21B) gene in a Finnish patient with the simple virilizing form of congenital adrenal hyperplasia. Hum. Genet. 87: 716-720, 1991.
Petersen, F.; Knudsen, F. U.; Nielsen, M. D.; Mikkelsen, M.: Congenital adrenal hyperplasia associated with a balanced 13-18 translocation. Europ. J. Pediat. 133: 283-285, 1980.
Petersen, K. E.; Svejgaard, A.; Nielsen, M. D.; Dissing, J.: Heterozygotes and cryptic patients in families of patients with congenital adrenal hyperplasia (21-hydroxylase deficiency): HLA and glyoxalase I typing and hormonal studies. Hormone Res. 16: 151-159, 1982.
Pollack, M. S.; Keenan, B.; Christiansen, F. T.; Cobain, T. J.; Dawkins, R. L.; Clayton, G.: The immunological detection of a 21-OH deficiency mutation HLA supratype. Am. J. Hum. Genet. 38: 688-698, 1986.
Pollack, M. S.; Levine, L. S.; O'Neill, G. J.; Pang, S.; Lorenzen, F.; Kohn, B.; Rondanini, G. F.; Chiumello, G.; New, M. I.; Dupont, B.: HLA linkage and B14,DR1,BfS haplotype association with the genes for late onset and cryptic 21-hydroxylase deficiency. Am. J. Hum. Genet. 33: 540-550, 1981.
Pollack, M. S.; Levine, L. S.; Pang, S.; Owens, R. P.; Nitowsky, H. M.; Maurer, D.; New, M. I.; Duchon, M.; Merkatz, I. R.; Sachs, G.; Dupont, B.: Prenatal diagnosis of congenital adrenal hyperplasia (21-hydroxylase deficiency) by HLA typing. Lancet I: 1107-1108, 1979.
Pollack, M. S.; New, M. I.; O'Neill, G. J.; Levine, L. S.; Callaway, C.; Pang, S.; Cacciari, E.; Mantero, F.; Cassio, A.; Scaroni, C.; Chiumello, G.; Rondanini, G. F.; Gargantini, L.; Giovannelli, G.; Virdis, R.; Bartolotta, E.; Migliori, C.; Pintor, C.; Tato, L.; Barboni, F.; Dupont, B.: HLA genotypes and HLA-linked genetic markers in Italian patients with classical 21-hydroxylase deficiency. Hum. Genet. 58: 331-337, 1981.
Prader, A.: Die Haeufigkeit des kongenitalen adrenogenitalen Syndroms. Helv. Paediat. Acta 13: 426-431, 1958.
Prader, A.; Anders, G. J. P. A.; Habich, H.: Zur Genetik des kongenitalen adrenogenitalen Syndroms (virilisierende Nebennierenhyperplasia). Helv. Paediat. Acta 17: 271-284, 1962.
Pucholt, V.; Fitzsimmons, J. S.; Gelsthorpe, K.; Reynolds, M. A.; Milner, R. D. G.: Location of the gene for 21-hydroxylase deficiency. J. Med. Genet. 17: 447-452, 1980.
Qazi, Q. H.; Thompson, M. W.: Incidence of salt-losing form of congenital virilizing adrenal hyperplasia. Arch. Dis. Child. 47: 302-303, 1972.
Ravichandran, R.; Lafferty, F.; McGinniss, M. J.; Taylor, H. C.: Congenital adrenal hyperplasia presenting as massive adrenal incidentalomas in the sixth decade of life: report of two patients with 21-hydroxylase deficiency. J. Clin. Endocr. Metab. 81: 1776-1779, 1996.
Reindollar, R. H.; Lewis, J. B.; White, P. C.; Fernhoff, P. M.; McDonough, P. G.; Whitney, J. B., III: Prenatal diagnosis of 21-hydroxylase deficiency by the complementary deoxyribonucleic acid probe for cytochrome P-450(C-21OH). Am. J. Obstet. Gynec. 158: 545-547, 1988.
Rodrigues, N. R.; Dunham, I.; Yu, C. Y.; Carroll, M. C.; Porter, R. R.; Campbell, R. D.: Molecular characterization of the HLA-linked steroid 21-hydroxylase B gene from an individual with congenital adrenal hyperplasia. EMBO J. 6: 1653-1661, 1987.
Rosenmann, A.; Schumert, Z.; Theodor, R.; Cohen, T.; Brautbar, C.: Amniotic 17-alpha hydroxyprogesterone and HLA typing for the prenatal diagnosis of 21-alpha hydroxylase deficiency—congenital adrenal hyperplasia. Am. J. Med. Genet. 6: 295-300, 1980.
Rumsby, G.; Carroll, M. C.; Porter, R. R.; Grant, D. B.; Hjelm, M.: Deletion of the steroid 21-hydroxylase and complement C4 genes in congenital adrenal hyperplasia. J. Med. Genet. 23: 204-209, 1986.
Schneider, P. M.; Carroll, M. C.; Alper, C. A.; Rittner, C.; Whitehead, A. S.; Yunis, E. J.; Colten, H. R.: Polymorphism of the human complement C4 and steroid 21-hydroxylase genes: restriction fragment length polymorphisms revealing structural deletions, homoduplications, and size variants. J. Clin. Invest. 78: 650-657, 1986.
Sherman, S. L.; Aston, C. E.; Morton, N. E.; Speiser, P. W.; New, M. I.: A segregation and linkage study of classical and nonclassical 21-hydroxylase deficiency. Am. J. Hum. Genet. 42: 830-838, 1988.
Sinnott, P.; Collier, S.; Costigan, C.; Dyer, P. A.; Harris, R.; Strachan, T.: Genesis by meiotic unequal crossover of a de novo deletion that contributes to a steroid 21-hydroxylase deficiency. Proc. Nat. Acad. Sci. 87: 2107-2111, 1990.
Sinnott, P. J.; Dyer, P. A.; Price, D. A.; Harris, R.; Strachan, T.: 21-hydroxylase deficiency families with HLA identical affected and unaffected sibs. J. Med. Genet. 26: 10-17, 1989.
Slominski, A.; Ermak, G.; Mihm, M.: ACTH receptor, CYP11A1, CYP17 and CYP21A2 genes are expressed in skin. J. Clin. Endocr. Metab. 81: 2746-2749, 1996.
Sobel, D. O.; Gutai, J. P.; Jones, J. C.; Wagener, D. D.; Smith, W.: Detection of heterozygote of 21 hydroxylase deficiency. (Letter) Lancet I: 47 only, 1980.
Speiser, P. W.; Dupont, B.; Rubinstein, P.; Piazza, A.; Kastelan, A.; New, M. I.: High frequency of nonclassical steroid 21-hydroxylase deficiency. Am. J. Hum. Genet. 37: 650-667, 1985.
Speiser, P. W.; Dupont, J.; Zhu, D.; Serrat, J.; Buegeleisen, M.; Tusie-Luna, M.-T.; Lesser, M.; New, M. I.; White, P. C.: Disease expression and molecular genotype in congenital adrenal hyperplasia due to 21-hydroxylase deficiency. J. Clin. Invest. 90: 584-595, 1992.
Speiser, P. W.; New, M. I.: Genetics of steroid 21-hydroxylase deficiency. Trends Genet. 1: 275-278, 1985.
Speiser, P. W.; New, M. I.; Tannin, G. M.; Pickering, D.; Yang, S. Y.; White, P. C.: Genotype of Yupik Eskimos with congenital adrenal hyperplasia due to 21-hydroxylase deficiency. Hum. Genet. 88: 647-648, 1992.
Speiser, P. W.; New, M. I.; White, P. C.: Molecular genetic analysis of nonclassic steroid 21-hydroxylase deficiency associated with HLA-B14,DR1. New Eng. J. Med. 319: 19-23, 1988.
Speiser, P. W.; New, M. I.; White, P. C.: Clinical and genetic characterization of nonclassic 21-hydroxylase deficiency. Endocr. Res. 15: 257-276, 1989.
Tajima, T.; Fujieda, K.; Fujii-Kuriyama, Y.: De novo mutation causes steroid 21-hydroxylase deficiency in one family of HLA-identical affected and unaffected individuals. J. Clin. Endocr. Metab. 77: 86-89, 1993.
Tajima, T.; Fujieda, K.; Nakae, J.; Toyoura, T.; Shimozawa, K.; Kusuda, S.; Goji, K.; Nagashima, T.; Cutler, G. B., Jr.: Molecular basis of nonclassical steroid 21-hydroxylase deficiency detected by neonatal mass screening in Japan. J. Clin. Endocr. Metab. 82: 2350-2356, 1997.
Tajima, T.; Fujieda, K.; Nakayama, K.; Fujii-Kuriyama, Y.: Molecular analysis of patient and carrier genes with congenital steroid 21-hydroxylase deficiency by using polymerase chain reaction and single strand conformation polymorphism. J. Clin. Invest. 92: 2182-2190, 1993.
Thilen, A.; Larsson, A.: Congenital adrenal hyperplasia in Sweden 1969-1986: prevalence, symptoms and age at diagnosis. Acta Paediat. Scand. 79: 168-175, 1990.
Tusie-Luna, M.-T.; Speiser, P. W.; Dumic, M.; New, M. I.; White, P. C.: A mutation (pro30-to-leu) in CYP21 represents a potential nonclassic steroid 21-hydroxylase deficiency allele. Molec. Endocr. 5: 685-692, 1991.
Tusie-Luna, M.-T.; White, P. C.: Gene conversions and unequal crossovers between CYP21 (steroid 21-hydroxylase gene) and CYP21P involve different mechanisms. Proc. Nat. Acad. Sci. 92: 10796-10800, 1995.
Tusie-Luna, M. T.; Traktman, P.; White, P. C.: Determination of functional effects of mutations in the steroid 21-hydroxylase gene (CYP21) using recombinant vaccinia virus. J. Biol. Chem. 265: 20916-20922, 1990.
Urabe, K.; Kimura, A.; Harada, F.; Iwanaga, T.; Sasazuki, T.: Gene conversion in steroid 21-hydroxylase genes. Am. J. Hum. Genet. 46: 1178-1186, 1990.
Urban, M. D.; Lee, P. A.; Migeon, C. J.: Adult height and fertility in men with congenital virilizing adrenal hyperplasia. New Eng. J. Med. 299: 1392-1396, 1978.
Warsof, S. L.; Larsen, J. W.; Kent, S. G.; Rosenbaum, K. N.; August, G. P.; Migeon, C. J.; Schulman, J. D.: Prenatal diagnosis of congenital adrenal hyperplasia. Obstet. Gynec. 55: 751-754, 1980.
Webb, T.; Mackintosh, P.; Wells, L. J.: Cytogenetic evidence for the localisation of the gene for congenital adrenal hyperplasia. Clin. Genet. 17: 349-354, 1980.
Wedell, A.; Luthman, H.: Steroid 21-hydroxylase deficiency: two additional mutations in salt-wasting disease and rapid screening of disease-causing mutations. Hum. Molec. Genet. 2: 499-504, 1993.
Wedell, A.; Ritzen, E. M.; Haglund-Stengler, B.; Luthman, H.: Steroid 21-hydroxylase deficiency: three additional mutated alleles and establishment of phenotype-genotype relationships of common mutations. Proc. Nat. Acad. Sci. 89: 7232-7236, 1992.
Wedell, A.; Thilen, A.; Ritzen, E. M.; Stengler, B.; Luthman, H.: Mutational spectrum of the steroid 21-hydroxylase gene in Sweden: implications for genetic diagnosis and association with disease malformation. J. Clin. Endocr. Metab. 78: 1145-1152, 1994.
Wenzel, U.; Schneider, M.; Zachmann, M.; Knorr-Murset, G.; Weber, A.; Prader, A.: Intelligence of patients with congenital adrenal hyperplasia due to 21-hydroxylase deficiency, their parents and unaffected siblings. Helv. Paediat. Acta 33: 11-16, 1978.
Werder, E. A.; Siebenmann, R. E.; Knorr-Murset, G.; Zimmermann, A.; Sizonenko, P. C.; Theintz, P.; Girard, J.; Zachmann, M.; Prader, A.: The incidence of congenital adrenal hyperplasia in Switzerland—a survey of patients born in 1960 to 1974. Helv. Paediat. Acta 35: 5-11, 1980.
Werkmeister, J. W.; New, M. I.; Dupont, B.; White, P. C.: Frequent deletion and duplication of the steroid 21-hydroxylase genes. Am. J. Hum. Genet. 39: 461-469, 1986.

White, P. C.; Chaplin, D. D.; Weis, J. H.; Dupont, B.; New, M. I.; Seidman, J. G.: Two steroid 21-hydroxylase genes are located in the murine S region. Nature 312: 465-467, 1984.
White, P. C.; Grossberger, D.; Onufer, B. J.; Chaplin, D. D.; New, M. I.; Dupont, B.; Strominger, J. L.: Two genes encoding steroid 21-hydroxylase are located near the genes encoding the fourth component of complement in man. Proc. Nat. Acad. Sci. 82: 1089-1093, 1985.
White, P. C.; New, M.; Dupont, B.: Congenital adrenal hyperplasia. (Letter) New Eng. J. Med. 317: 1414-1415, 1987.
White, P. C.; New, M. I.; Dupont, B.: Congenital adrenal hyperplasia. New Eng. J. Med. 316: 1519-1524, 1987.
White, P. C.; New, M. I.; Dupont, B.: HLA-linked congenital adrenal hyperplasia results from a defective gene encoding a cytochrome P-450 specific for steroid 21-hydroxylation. Proc. Nat. Acad. Sci. 81: 7505-7509, 1984.
White, P. C.; New, M. I.; Dupont, B.: Structure of human steroid 21-hydroxylase genes. Proc. Nat. Acad. Sci. 83: 5111-5115, 1986.
White, P. C.; Tusie-Luna, M.-T.; New, M. I.; Speiser, P. W.: Mutations in steroid 21-hydroxylase (CYP21). Hum. Mutat. 3: 373-378, 1994.
White, P. C.; Vitek, A.; Dupont, B.; New, M. I.: Characterization of frequent deletions causing steroid 21-hydroxylase deficiency. Proc. Nat. Acad. Sci. 85: 4436-4440, 1988.
Winqvist, O.; Karlsson, F. A.; Kampe, O.: 21-Hydroxylase, a major autoantigen in idiopathic Addison's disease. Lancet 339: 1559-1562, 1992.
Witchel, S. F.; Bhamidipati, D. K.; Hoffman, E. P.; Cohen, J. B.: Phenotypic heterogeneity associated with the splicing mutation in congenital adrenal hyperplasia due to 21-hydroxylase deficiency. J. Clin. Endocr. Metab. 81: 4081-4088, 1996.
Witchel, S. F.; Lee, P. A.; Suda-Hartman, M.; Trucco, M.; Hoffman, E. P.: Evidence for a heterozygote advantage in congenital adrenal hyperplasia due to 21-hydroxylase deficiency. J. Clin. Endocr. Metab. 82: 2097-2101, 1997.
Wu, D.-A.; Chung, B.: Mutations of P450c21 (steroid 21-hydroxylase) at cys-428, val-281, and ser-268 result in complete, partial, or no loss of enzymatic activity, respectively. J. Clin. Invest. 88: 519-523, 1991.
Wudy, S. A.; Homoki, J.; Teller, W. M.: Successful prenatal treatment of congenital adrenal hyperplasia due to 21-hydroxylase deficiency. Europ. J. Pediat. 153: 556-559, 1994.
Yang, S. Y.; Levine, L. S.; Zachmann, M.; New, M. I.; Prader, A.; Oberfield, S. E.; O'Neill, G. J.; Pollack, M. S.; Dupont, B.: Mapping of the 21-hydroxylase deficiency gene within the HLA linkage group. Transplant. Proc. 10: 753-755, 1978.
Zachmann, M.; Prader, A.: Unusual heterozygotes of congenital adrenal hyperplasia due to 21-hydroxylase deficiency confirmed by HLA tissue typing. Acta Endocr. 92: 542-546, 1979.

Prostate Tumors

PROSTATE CANCER; PRCA1 (#176807)

JJM NOTE

Since prostate cancer was highest on the list of tumors that shared familial aggregation in the match of the Mormon genealogies with the Utah Cancer Registry, it is surprising that clinical reports of familial aggregation were not more common. Furthermore, the heterogeneity that is being recognized follows the pattern of discovering the genes that predispose to breast cancer. Familial aggregation is significant enough and screening good enough that brothers and sons of affected men should be counseled for screening. But, there is no call for prophylactic prostatectomy. In addition to the genes mentioned in the first paragraph by McKusick, prostate cancer has occurred in individuals with germline mutations of *BRCA1* and the androgen insensitivity syndrome. The finding of genetic associations with specific polymorphism of the androgen receptor gene and ADP-ribosyltransferase raises the possibility that a multifactorial risk calculation could be devised, akin to the Gail model for breast cancer.

A number sign (#) is used with this entry because of evidence that several genes are involved in the origin and/or progression of this neoplasm: prostate cancer, hereditary 1 (HPC1; 601518) on 1q, MAX interacting protein 1 (MXI1; 600020) on 10q, and KAI1 prostate cancer antimetastasis gene (KAI1; 600623) on 11p. Point mutations in the MXI1 gene have been observed in prostate cancer (e.g., 600020.0001). A second predisposing locus for early-onset prostate cancer (HPC2; 602759) has been assigned to 1q42.2-q43, at a distance from the HPC1 locus, which is also on 1q. A prostate cancer susceptibility locus on Xq27-q28 (HPCX; 300147) has also been proposed.

In American men, prostate cancer is the most common malignancy and the second most common cause of cancer deaths. Over 100,000 cases of prostate cancer occurred annually in the United States in recent years, with 20,000 of these cases occurring in men under the age of 65 years (Silverberg, 1986). It is estimated that 9% of white males and 10% of black males in the U.S. will develop clinical prostate cancer in their lifetime (Seidman et al., 1985).

Studies by Woolf (1960), Cannon et al. (1982), Meikle et al. (1985) and others suggested the significance of familial factors in prostate cancer. See HPC1 (601518) for discussion of a chromosome 1-linked prostate cancer susceptibility locus identified by Smith et al. (1996). Whereas the susceptibility locus found by Smith et al. (1996) was located at 1q24-q25, Berthon et al. (1998) found by linkage a second putative predisposing gene for prostate cancer located on 1q42.2-q43 (HPC2; 602759), significantly distant from the previously proposed locus.

Steinberg et al. (1990) investigated the frequency of prostate cancer in the relatives of 691 men with prostate cancer and 640 of their spouses. In 15% of the cases, but only 8% of the controls, was a father or brother affected with prostate cancer (P less than 0.001). Men with a father or brother affected were twice as likely to develop prostate cancer as men with no relatives affected. There was a trend of increasing risk with increasing number of affected family members such that men with 2 or 3 first-degree relatives affected had 5- and 11-fold increased risk of developing prostate cancer, respectively. Carter et al. (1992) found in the 691 prostate cancer families that early onset of disease in the proband was also an important determinant of risk. Complex segregation analysis led to the conclusion that the family clustering was best explained by autosomal dominant inheritance of a rare (q = 0.0030) high-risk allele leading to an early onset of prostate cancer. The estimated cumulative risk of prostate cancer for carriers showed that the allele was highly penetrant; by age 85, 88% of carriers compared to only 5% of noncarriers were projected to be affected with prostate cancer. Twin studies support the existence of a significant genetic factor (Walsh, 1992).

To determine whether family history is associated with an increased prevalence of prostate cancer in an unselected group of men attending a hospital-based screening clinic, Narod et al. (1995) inquired about affected relatives before prostate cancer screening in 6,390 men, aged 50 to 80 years, in the region of Quebec City. Of these 6,390 men, 1,563 (24.5%) had a positive test by either rectal examination or blood test for prostate-specific antigen (APS; 176820). Among the 6,390 men, 264 were found to have prostate cancer (4.13%). The prevalence was increased in those men with any first-degree relative affected (prevalence = 6.7%; relative risk = 1.72 as compared with men with no first-degree relative affected; prevalence = 3.89; relative risk = 1.00). Most of the increase in relative risk was contributed by affected brothers (prevalence = 10.2%; relative risk = 2.62; P = 0.0002).

Ichikawa et al. (1991) and Ichikawa et al. (1992) demonstrated that somatic cell hybridization of nonmetastatic rat prostatic cancer cells with highly metastatic rat prostatic cancer cells resulted in suppression of metastatic potential, although the cell hybrids remained tumorigenic. They also showed that the 11p11.2 region of the human genome contains a gene that suppresses metastasis as indicated by the results of microcell-mediated chromosome transfer. Dong et al. (1995) isolated the 11p11.2 gene, which they designated KAI1 (600623).

Eagle et al. (1995) demonstrated that mutation in the MXI1 gene (600020) is involved in either the pathogenesis or the neoplastic evolution of some prostatic cancer. The MXI1 protein negatively regulated MYC oncoprotein (190080) activity and thus potentially serves a tumor suppressor function. Furthermore, MXI1 maps to chromosome 10q25, a region that is deleted in some cases of prostate cancer. Eagle et al. (1995) detected mutations in the retained MXI1 allele in 4 primary prostate tumors with 10q24-q25 deletions, thus satisfying the Knudson hypothesis.

In Los Angeles County, California, prostate cancer incidence differs considerably among the various racial-ethnic groups: highest in African-Americans (116 per 1,000 person-years), intermediate in non-Hispanic whites (71 per 100,000 person-years), and lowest among Asians (Japanese at 39 per100,000 person-years and Chinese at 28 per 100,000 person-years). One of the critical functions of the product of the androgen receptor gene (313700) is to activate the expression of target genes. This transactivation activity resides in the N-terminal domain of the protein, which is encoded in exon 1 and contains polymorphic CAG and GGC repeats (microsatellites). A smaller size of the CAG repeat is associated with a higher level of receptor transactivation function, thereby possibly resulting in a higher risk of prostate cancer. Schoenberg et al. (1994) demonstrated contraction in this microsatellite from 24 to 18 CAG units in an adenocarcinoma of the prostate, and the affects of the shorter allele were implicated in the development of the tumor. Edwards et al. (1992) and Irvine et al. (1995) showed that the prevalence of short CAG alleles was highest in African-American males with the highest risk for prostate cancer, intermediate in intermediate-risk non-Hispanic whites, and lowest in Asians at very low risk for prostate cancer. Irvine et al. (1995) found that high-risk African Americans also had the lowest frequency of the GGC allele. Consistent with the interracial variation in CAG and GGC distributions, there was an excess of white patients with short CAG repeats relative to white controls. Irvine et al. (1995) found a statistically significant negative association between the number of CAG and GGC repeats among the prostate cancer patients. Overall, the data were interpreted to suggest a possible association between the microsatellites of the androgen receptor gene and the development of prostate cancer.

Schaid et al. (1998) performed a family history cancer survey on 5,486 men who underwent radical prostatectomy at the Mayo Clinic for clinically localized prostate cancer; 4,288 men responded to the survey. The best-fitting model that explained familial aggregation was a rare autosomal dominant susceptibility gene, and this model fit best when probands were diagnosed under 60 years of age. The model predicted that the frequency of a susceptibility gene in the population is 0.006 and that the risk of prostate cancer by age 85 years is 89% among carriers of the gene and 3% among noncarriers. Genetic heterogeneity was suggested by the study. Other evidence suggesting genetic complexity included the significantly elevated age-adjusted risk of prostate cancer among brothers of probands, compared with their fathers.

See also Silverberg and Lubera (1988)

Berthon, P.; Valeri, A.; Cohen-Akenine, A.; Drelon, E.; Paiss, T.; Wohr, G.; Latil, A.; Millaseau, P.; Mellah, I.; Cohen, N.; Blanche, H.; Bellane-Chantelot, C.; and 14 others: Predisposing gene for early-onset prostate cancer, localized on chromosome 1q42.2-43. Am. J. Hum. Genet. 62: 1416-1424, 1998.
Cannon, L.; Bishop, D. T.; Skolnick, M.; Hunt, S.; Lyon, J. L.; Smart, C. R.: Genetic epidemiology of prostate cancer in the Utah Mormon genealogy. Cancer Surv. 1: 47-69, 1982.
Carter, B. S.; Beaty, T. H.; Steinberg, G. D.; Childs, B.; Walsh, P. C.: Mendelian inheritance of familial prostate cancer. Proc. Nat. Acad. Sci. 89: 3367-3371, 1992.
Dong, J.-T.; Lamb, P. W.; Rinker-Schaeffer, C. W.; Vukanovic, J.; Ichikawa, T.; Isaacs, J. T.; Barrett, J. C.: KAI1, a metastasis suppressor gene for prostate cancer on human chromosome 11p11.2. Science 268: 884-886, 1995.
Eagle, L. R.; Yin, X.; Brothman, A. R.; Williams, B. J.; Atkin, N. B.; Prochownik, E. V.: Mutation of the MXI1 gene in prostate cancer. Nature Genet. 9: 249-255, 1995.
Edwards, A.; Hammond, H. A.; Jin, L.; Caskey, C. T.; Chakraborty, R.: Genetic variation at five trimeric and tetrameric tandem repeat loci in four human population groups. Genomics 12: 241-253, 1992.
Ichikawa, T.; Ichikawa, Y.; Dong, J.; Hawkins, A. L.; Griffin, C. A.; Isaacs, W. B.; Oshimura, M.; Barrett, J. C.; Isaacs, J. T.: Localization of metastasis suppressor gene(s) for prostatic cancer to the short arm of human chromosome 11. Cancer Res. 52: 3486-3490, 1992.
Ichikawa, T.; Ichikawa, Y.; Isaacs, J. T.: Genetic factors and suppression of metastatic ability of prostatic cancer. Cancer Res. 51: 3788-3792, 1991.
Irvine, R. A.; Yu, M. C.; Ross, R. K.; Coetzee, G. A.: The CAG and GGC microsatellites of the androgen receptor gene are in linkage disequilibrium in men with prostate cancer. Cancer. Res. 55: 1937-1940, 1995.
Meikle, A. W.; Smith, J. A.; West, D. W.: Familial factors affecting prostatic cancer risk and plasma sex steroid levels. Prostate 6: 121-128, 1985.
Narod, S. A.; Dupont, A.; Cusan, L.; Diamond, P.; Gomez, J.-L.; Suburu, R.; Labrie, F.: The impact of family history on early detection of prostate cancer. (Letter) Nature Med. 1: 99-101, 1995.
Schaid, D. J.; McDonnell, S. K.; Blute, M. L.; Thibodeau, S. N.: Evidence for autosomal dominant inheritance of prostate cancer. Am. J. Hum. Genet. 62: 1425-1438, 1998.
Schoenberg, M. P.; Hakimi, J. M.; Wang, S.; Bova, G. S.; Epstein, J. I.; Fischbeck, K. H.; Isaacs, W. B.; Walsh, P. C.; Barrack, E. R.: Microsatellite mutation (CAG (24-to-18)) in the androgen receptor gene in human prostate cancer Biochem. Biophys. Res. Commun. 198: 74-80, 1994.
Seidman, H.; Mushinski, M. H.; Gelb, S. K.; Silverberg, E.: Probabilities of eventually developing or dying of cancer: United States. Chem. Abstr. 35: 36-56, 1985.
Silverberg, E.: Statistical and epidemiologic data on urologic cancer. Cancer 60: 692-717, 1986.
Silverberg, E.; Lubera, J. A.: Cancer statistics. Cancer 38: 5-22, 1988.
Smith, J. R.; Freije, D.; Carpten, J. D.; Gronberg, H.; Xu, J.; Isaacs, S. D.; Brownstein, M. J.; Bova, G. S.; Guo, H.; Bujnovszky, P.; Nusskern, D. R.; Damber, J.-E.; Bergh, A.; Emanuelsson, M.; Kallioniemi, O. P.; Walker-Daniels, J.; Bailey-Wilson, J. E.; Beaty, T. H.; Meyers, D. A.; Walsh, P. C.; Collins, F. S.; Trent, J. M.; Isaacs, W. B.: Major susceptibility locus for prostate cancer on chromosome 1 suggested by a genome-wide search. Science 274: 1371-1374, 1996.
Steinberg, G. D.; Carter, B. S.; Beaty, T. H.; Childs, B.; Walsh, P. C.: Family history and the risk of prostate cancer. Prostate 17: 337-347, 1990.
Walsh, P. C.: Personal Communication. Baltimore, Md. 1/9/1992.
Woolf, C. M.: An investigation of the familial aspects of carcinoma of the prostate. Cancer 13: 739-744, 1960.

PROSTATE CANCER, HEREDITARY 1 (*601518)

JJM NOTE

The first mapping was aided by an avalanche of potential study subjects who volunteered after a story in the national Sunday magazine *Parade.*

Smith et al. (1996) noted that segregation analysis of familial prostate cancer suggests the existence of at least 1 dominant susceptibility locus and predicts that high-risk alleles at such loci account in the aggregate for 9% of all prostate cancers and for more than 40% of early-onset disease. They stated further that prostate cancer presents a number of serious obstacles to linkage analysis, including occurrence of phenocopies, late age of onset, and absence of clinical features that might allow subgrouping to reflect potential genetic heterogeneity. Smith et al. (1996) undertook linkage analysis in 79 North American and 12 Swedish pedigrees, each having at least 3 first-degree relatives with prostate cancer. In these families there was also evidence of bilineal inheritance. Affected individuals had an average age at diagnosis of 65 years, with a total of 34 men diagnosed before the age of 55. The investigators analyzed a total of 341 dinucleotide repeat markers in a subgroup of 66 North American families. For the parametric analysis of the data they used a model of dominant inheritance that included a fixed phenocopy rate of 15% and the assumption that unaffected men over 75 were not likely to be gene carriers. The highest lod score observed was 2.75 with the marker D1S218, which maps to the distal long arm of chromosome 1 at 1q24-q25. The investigators then typed additional markers in this region in all 79 North American families and the 12 Swedish families. This analysis provided additional evidence for linkage with a maximum 2-point lod score of 3.65 at theta = 0.18 with marker D1S2883. Significant evidence for locus heterogeneity was obtained by an admixture test, which estimated that 34% of families were linked to the 1q24-q25 region. The maximum multipoint lod score with markers D1S2883, D1S158, and D1S422 under the assumption of heterogeneity was 5.43, with the postulated susceptibility locus mapping close to D1S422. Smith et al. (1996) found no clinical features that ap peared to distinguish families showing linkage to chromosome 1 from unlinked families. They noted that Cher et al. (1996) reported that a large portion of chromosome 1, including 1q24-q25, is frequently increased in copy number in advanced prostate cancer specimens examined by comparative genomic hybridization. Smith et al. (1996) proposed the designation of HPC1 (hereditary prostate cancer 1) for the locus. They emphasized that early diagnosis can be life saving in prostate cancer and that the potential ability to identify individuals at high genetic risk, when combined with physical exam, transrectal ultrasound, and prostate-specific antigen (176820) analysis, could ultimately be of significant medical benefit.

McIndoe et al. (1997) reported results of linkage analysis in the region of the putative HPC1 locus on 1q24-q25. The study was part of the Prostate Cancer Genetic Research Study (PROGRESS) at the Fred Hutchinson Cancer Research Center in Seattle. Analysis by 2 parametric methods and 1 nonparametric method failed to confirm linkage to this region. Additionally, they were unable to demonstrate heterogeneity within the data set.

Cooney et al. (1997) confirmed 1q24-q25 as a likely location of a prostate cancer susceptibility gene. Nonparametric multipoint linkage analysis of data on 6 polymorphic marker sequences covering the candidate chromosomal region were performed in 59 unrelated families selected for analysis on the sole criterion that more than 1 living family member was affected by prostate cancer. In the entire set of 59 families, tight linkage to D1S466 was found; maximum lod = 1.58. In 20 families fulfilling more rigorous criteria, i.e., 3 or more affected individuals within one nuclear family, affected individuals in 3 successive generations, and/or clustering of 2 or more individuals affected before the age of 55 years a maximum lod score of 1.72 was observed with the same marker. The 6 African-American families in this study contributed disproportionately to the observation of linkage, with a maximum nonparametric linkage (NPL) Z score at marker D1S158 of 1.39.

Using both parametric and nonparametric methods, Eeles et al. (1998) attempted to confirm the assignment of a locus (HPC1) to 1q, in 60 affected related pairs and in 76 families with 3 or more cases of prostate cancer, but could find no significant evidence of linkage. The estimated proportion of linked families, under a standard autosomal dominant model, was 4%, with an upper 95% confidence limit of 31%. They concluded that the HPC1 locus is responsible for only a minority of familial prostate cancer cases and that it is likely to be most important in families with at least 4 cases of the disease.

See entries for prostate cancer (176807) and MXI1 (600020) for discussion of a locus on chromosome 10q25 which may be associated with some prostate cancer.

Cher, M. L.; Bova, G. S.; Moore, D. H.; Small, E. J.; Carroll, P. R.; Pin, S. S.; Epstein, J. I.; Isaacs, W. B.; Jensen, R. H.: Genetic alterations in untreated metastases and androgen-independent prostate cancer detected by comparative genomic hybridization and allelotyping. Cancer Res. 56: 3091-3102, 1996.
Cooney, K. A.; McCarthy, J. D.; Lange, E.; Huang, L.; Miesfeldt, S.; Montie, J. E.; Oesterling, J. E.; Sandler, H. M.; Lange, K.: Prostate cancer susceptibility locus on chromosome 1q: a confirmatory study. J. Nat. Cancer Inst. 89: 955-999, 1997.
Eeles, R. A.; Durocher, F.; Edwards, S.; Teare, D.; Badzioch, M.; Hamoudi, R.; Gill, S.; Biggs, P.; Dearnaley, D.; Ardern-Jones, A.; Dowe, A.; Shearer, R.;

and 13 others: Linkage analysis of chromosome 1q markers in 136 prostate cancer families. Am. J. Hum. Genet. 62: 653-658, 1998.

McIndoe, R. A.; Stanford, J. L.; Gibbs, M.; Jarvik, G. P.; Brandzel, S.; Neal, C. L.; Li, S.; Gammack, J. T.; Gay, A. A.; Goode, E. L.; Hood, L.; Ostrander, E. A.: Linkage analysis of 49 high-risk families does not support a common familial prostate cancer-susceptibility gene at 1q24-25. Am. J. Hum. Genet. 61: 347-353, 1997.

Smith, J. R.; Freije, D.; Carpten, J. D.; Gronberg, H.; Xu, J.; Isaacs, S. D.; Brownstein, M. J.; Bova, G. S.; Guo, H.; Bujnovszky, P.; Nusskern, D. R.; Damber, J.-E.; Bergh, A.; Emanuelsson, M.; Kallioniemi, O. P.; Walker-Daniels, J.; Bailey-Wilson, J. E.; Beaty, T. H.; Meyers, D. A.; Walsh, P. C.; Collins, F. S.; Trent, J. M.; Isaacs, W. B.: Major susceptibility locus for prostate cancer on chromosome 1 suggested by a genome-wide search. Science 274: 1371-1374, 1996.

PROSTATE CANCER, HEREDITARY 2 (*602759)

JJM NOTE

As with breast cancer gene mapping on chromosome 17, once the investigative spotlight falls on one chromosomal region, several associated genes are discovered. I conclude that many genes influence the occurrence of every tumor type.

Several studies have indicated that approximately 10% of all prostate cancer is hereditary with an autosomal dominant inheritance (Gronberg et al., 1997). Smith et al. (1996) proposed a susceptibility locus for prostate cancer on 1q24-q25, designated HPC1 (601518). The locus has been confirmed by 2 other groups (Cooney et al., 1997; Hsieh et al., 1997) and through re-analysis and extension of the original study (Gronberg et al., 1997). Conversely, HPC1 was not confirmed by other investigations (Eeles et al., 1997; McIndoe et al., 1997).

Berthon et al. (1998) performed linkage study using a set of European families (French and German) with 3 or more prostate cancer patients per family. They concluded that there is a prostate cancer susceptibility locus on 1q42.2-q43, significantly distant from the proposed locus on 1q24-q25 (Smith et al., 1996). The primary localization was confirmed with several markers by use of 3 different genetic models. They obtained a maximum 2-point lod score of 2.7 with marker D1S2785. Multipoint parametric and nonparametric linkage (NPL) analysis yielded maximum HLOD and NPL scores of 2.2 and 3.1, respectively, with an associated P value of 0.001. Homogeneity analysis with multipoint lod scores gave an estimate of the proportion of families with linkage to the 1q42.2-q43 locus of 50%, with a likelihood ratio of 157/1 in favor of heterogeneity. Furthermore, 9 of the 47 families with early-onset prostate cancer (at an age less than 60 years) gave multipoint lod and NPL scores of 3.31 and 3.32, respectively, with $P = 0.001$.

Berthon, P.; Valeri, A.; Cohen-Akenine, A.; Drelon, E.; Paiss, T.; Wohr, G.; Latil, A.; Millasseau, P.; Mellah, I.; Cohen, N.; Blanche, H.; Bellane-Chantelot, C.; and 14 others: Predisposing gene for early-onset prostate cancer, localized on chromosome 1q42.2-43. Am. J. Hum. Genet. 62: 1416-1424, 1998.

Cooney, K. A.; McCarthy, J. D.; Lange, E.; Huang, L.; Miesfeldt, S.; Montie, J. E.; Oesterling, J. E.; Sandler, H. M.; Lange, K.: Prostate cancer susceptibility locus on chromosome 1q: a confirmatory study. J. Nat. Cancer Inst. 89: 955-999, 1997.

Eeles, R. A.; Durocher, F.; Edwards, S.; Teare, D.; Badzioch, M.; Hamoudi, R.; Gill, S.; Biggs, P.; Dearnaley, D.; Ardern-Jones, A.; Dowe, A.; Shearer, R.; and 12 others: Linkage analysis of chromosome 1q markers in 136 prostate cancer families. Am. J. Hum. Genet. 62: 653-658, 1997.

Gronberg, H.; Damber, L.; Damber, J.-E.; Iselius, L.: Segregation analysis of prostate cancer in Sweden: support for dominant inheritance. Am. J. Epidemiol. 146: 552-557, 1997.

Gronberg, H.; Xu, J.; Smith, J. R.; Carpten, J. D.; Isaacs, S. D.; Freije, D.; Bova, G. S.; Walsh, P. C.; Collins, F. S.; Trent, J. M.; Meyers, D. A.; Isaacs, W. B.: Early age at diagnosis in families providing evidence of linkage to the hereditary prostate cancer locus (HPC1) on chromosome 1. Cancer Res. 57: 4707-4709, 1997.

Hsieh, C.-L.; Oakley-Girvan, I.; Gallagher, R. P.; Wu, A. H.; Kolonel, L. N.; Teh, C.-Z.; Halpern, J.; West, D. W.; Paffenbarger, R. S., Jr.; Whittemore, A. S.: Re: prostate cancer susceptibility locus on chromosome 1q: a confirmatory study (Letter) J. Nat. Cancer Inst. 89: 1893-1894, 1997.

McIndoe, R. A.; Stanford, J. L.; Gibbs, M.; Jarvik, G. P.; Brandzel, S.; Neal, C. L.; Li, S.; Gammack, J. T.; Gay, A. A.; Goode, E. L.; Hood, L.; Ostrander, E. A.: Linkage analysis of 49 high-risk families does not support a common familial prostate cancer-susceptibility gene at 1q24-25. Am. J. Hum. Genet. 61: 347-353, 1997.

Smith, J. R.; Freije, D.; Carpten, J. D.; Gronberg, H.; Xu, J.; Isaacs, S. D.; Brownstein, M. J.; Bova, G. S.; Guo, H.; Bujnovszky, P.; Nusskern, D. R.; Damber, J.-E.; and 11 others: Major susceptibility locus for prostate cancer on chromosome 1 suggested by a genome-wide search. Science 274: 1371-1374, 1996.

ANDROGEN INSENSITIVITY SYNDROME (#300068)

JJM NOTE

Constitutional mutations in this gene lead to various syndromes of ambiguous genitalia in rare individuals, but to a slightly increased risk for prostate cancer in a wider population of men.

A number sign (#) is used with this entry because the disorder is caused by mutations in the gene for the androgen receptor (AR; 313700).

Description

The androgen insensitivity syndrome is an X-linked recessive disorder in which affected males have female external genitalia, female breast development, blind vagina, absent uterus and female adnexa, and abdominal or inguinal testes, despite a normal male (2A + XY) karyotype. Partial androgen insensitivity results in hypospadias and micropenis with gynecomastia (Reifenstein syndrome; 312300).

Nomenclature

The androgen insensitivity syndrome was referred to earlier, in both the human and the mouse, as testicular feminization (TFM). The designation androgen insensitivity more accurately reflects the basic pathophysiology and is clearly more satisfactory to patients and their families.

Clinical Features

Patients with androgen insensitivity syndrome often come to medical attention because of a presumed inguinal hernia. Many have absent pubic and axillary hair ('hairless pseudofemale'). The hair of the head is luxuriant, without temporal balding. The phenotype is often voluptuously feminine: Netter et al. (1958) reported this disorder in a famous photographic model, Marshall and Harder (1958) reported affected monozygotic twins who worked as airline stewardesses, and Polaillon (1891) described prostitution in an affected person.

In a patient studied by Wilkins (1957), the hair follicles of the axillary and pubic areas, although anatomically normal, were unresponsive to local or parenteral administration of androgens and the beard, voice, and clitoris were similarly unresponsive. This was the first demonstration that the basic defect in cases of the hairless pseudofemale type is end-organ unresponsiveness to androgen, a situation comparable to nephrogenic diabetes insipidus and pseudohypoparathyroidism. (These conditions are analogous to the situation in the Sebright Bantam cock which has a female comb structure despite obvious demonstrations of virility. Albright et al. (1942) misspelled 'Sebright' in their classic article.) It is likely that more than one distinct entity is included in the testicular feminization syndrome. Wilkins stated: 'in about one-third of the cases of male pseudohermaphroditism 'of feminine type' sexual hair has been entirely lacking.'

Morris (1962) called attention to the following case of Gayral et al. (1960): a woman, who was sister, mother, and grandmother of affected males, showed asymmetry in the development of the breasts, body hair, and vulva. The right breast was smaller than the left and there was no pubic hair to the right of the mid-line. She had always had menstrual irregularity but had 3 children, an affected male, a carrier daughter, and a daughter who was the mother of 3 unaffected sons. The findings may be best explained by an X-linked recessive (or incompletely recessive) gene whose effects are to render tissues resistant to male hormone, the patchy changes in the heterozygous female representing the Lyon phenomenon.

According to Wilson (1976), Morris (1953) first described incomplete testicular feminization and concluded that the complete and incomplete forms never occur in the same family. The incomplete syndrome resembles the complete form in respect to female phenotype, bilateral testes and 46,XY karyotype, but differs by clitoral enlargement from birth and virilization at puberty. The abnormality of the external genitalia is characteristic; fusion of the labioscrotal folds occurs for about half of the dorsal portion. Although the degree of masculinization of the external genitalia is variable, most patients are raised as females. In the family described by Lubs et al. (1959), some spermatogenesis was found. There is partial responsiveness to androgen (Winterborn et al., 1970) in this form of the disorder.

It can be difficult to distinguish clinically the incomplete testicular feminization syndrome from pseudovaginal perineoscrotal hypospadias (264600), which is autosomal recessive. Opitz et al. (1972) concluded that the consanguineous family reported by Philip and Trolle (1965) had pseudovaginal perineoscrotal hypoplasia. Boczkowski and Teter (1965) described 3 cases of incomplete testicular feminization among the children of 2 sisters. Wilson (1981) studied 35 families with 1 of the 4 forms of androgen insensitivity: testicular feminization, incomplete testicular feminization, Reifenstein syndrome, or infertile male syndrome.

In 31 of the families, he found an abnormality of the androgen receptor: abnormal binding, qualitatively abnormal receptor or decreased amount of receptor. In the other 4, no abnormality of receptor could be demonstrated. Bals-Pratsch et al. (1990) found qualitative and quantitative abnormalities of the androgen receptor in 3 brothers with prepenile scrotum (congenital transposition of the penis), bifid scrotum, scrotal hypospadias, and bilateral undescended testes. In 2 brothers with perineal hypospadias, Batch et al. (1993) found a qualitative androgen binding defect and a point mutation in the AR gene (313700.0020); they suggested that familial hypospadias is part of the phenotypic spectrum of partial androgen sensitivity.

Kaufman et al. (1984) studied an XY patient, with ambiguous genitalia at birth and breast development at puberty, whose cultured fibroblasts showed normal initial formation of low-affinity androgen-receptor complexes but defective transformation of these complexes to a higher affinity state. They presumed that the defect was in the X-linked structural gene for androgen receptor. A qualitative defect of the androgen receptor was demonstrated (Kovacs et al., 1984); although its binding properties were normal, it was unstable on sucrose density gradient centrifugation.

Hughes and Evans (1986) described 2 sibs with classic complete androgen insensitivity syndrome but increased androgen receptor concentrations in genital skin fibroblasts. The steroid-receptor complex appeared to be translocated normally into the nucleus. They concluded that 'the gene coding for the androgen receptor is intact and does not account for the androgen insensitivity.' But is it not possible that the mutation is in the part of the receptor that is concerned with its effects on DNA? Pinsky et al. (1987) described a family in which the proposita and her aunt had partial androgen resistance of a type different from those previously described. Although there was normal maximum binding capacity, there was an increased apparent equilibrium dissociation constant with dihydrotestosterone and 2 synthetic androgens.

Grino et al. (1988) described a family in which gynecomastia and undervirilization occurred in 5 men, 4 of whom had fathered children, in a pedigree pattern consistent with X-linked recessive inheritance. In fibroblasts cultured from genital skin from 2 of the men, the levels of androgen receptor and the affinity of receptor for dihydrotestosterone were normal. However, androgen binding in fibroblast monolayers was thermolabile, up-regulation of receptor levels did not occur after prolonged incubation with dihydrotestosterone or methyltrienolone, and dissociation rates at 37 degrees centigrade were increased with the synthetic androgen mibolerone. In addition, in cytosol preparations the androgen receptor protein was unstable. Grino et al. (1988) suggested that this disorder represents the most subtle functional abnormality of androgen receptor characterized to date, since it was compatible with normal male phenotypic development and in some affected men with fertility.

Davies et al. (1997) described 2 patients with complete androgen insensitivity syndrome (CAIS) and mental retardation associated with submicroscopic deletion of the AR gene. They pointed to the report of another patient with associated CAIS and MR. They postulated that the deletion involves, in addition to the AR gene, 1 or more neighboring genes that are implicated in nonspecific MR.

Additional abridged information regarding clinical features is available in the clinical synopsis.

Biochemical Features

Amrhein et al. (1976) presented evidence for 2 types of testicular feminization: in one, the receptor for dihydrotestosterone (DHT) was deficient; in the other, the receptor (androgen receptor, AR; 313700) was apparently present but the receptor-DHT complex was for some reason ineffective. The second type, 'receptor-positive' cases, included the 3 sibs pictured by McKusick (1964). They displayed some pubic hair. The first type included a patient with the 'hairless female' phenotype, also pictured by McKusick (1964). All were longtime patients of Dr. Lawson Wilkins, and it was in the last patient that he demonstrated unresponsiveness to locally administered androgens.

Griffin (1979) found a qualitative abnormality of androgen receptor, manifested by thermolability, in some cases of testicular feminization. Binding overlapped the normal range at 26 degrees C. It was half-normal at 37 degrees and less than 20% of normal at 42 degrees. Gerli et al. (1979) described a case of complete testicular feminization syndrome in a person with the 47,XXY karyotype. Obviously, nondisjunction occurred in the carrier mother, who was 40 years old. Two sibs and a daughter of each of 2 sisters of the patient also had testicular feminization. Unlike the usual cases, the patient had low plasma testosterone and high gonadotropins. German and Vesell (1966) reported this situation in monozygotic twins. Kaufman et al. (1979) reported 2 'receptor-positive' cases of complete androgen insensitivity. One of these had maternally related affected relatives in 3 successive generations.

Kaufman et al. (1981) suggested that whereas one class of mutation that affects the structural domain of the androgen receptor confers increased dissociability and defective up-regulation (a term they coined), a second impairs up-regulation only.

Inheritance

The means for establishing X-linked inheritance include demonstration of linkage with an X chromosome marker, demonstration of lyonization in heterozygous females, and demonstration that the proportion of new mutation cases is one-third rather than one-half (expected of an autosomal dominant). Meyer et al. (1975) found 2 clones of fibroblasts in heterozygous females, one with androgen-binding and one without, thus clinching the X-linkage of this disorder.

Cytogenetics

Muller et al. (1990) described an almost 12-year-old black female with testicular feminization and 47,XXY Klinefelter syndrome. Using DNA markers, they demonstrated that the supernumerary X chromosome resulted from maternal nondisjunction during meiosis II. The error at this stage provided the basis for homozygosity of the mutation at the androgen receptor locus.

Mapping

In the most extensively affected kindred known with complete androgen insensitivity, one living in the Dominican Republic, Imperato-McGinley et al. (1990) found linkage to DXS1 and PGK1, localizing the AR gene to an area between Xq11 and Xq13. Linkage between DXS1 and AR showed a peak lod score of 3.2 at theta = 0.06. No recombination was found between PGK1 and AR; peak lod score was 2.9 at theta = 0.0. Although both AR and PGK1 are distal to DXS1, it was not possible to determine the sequence of the 2. Using 3 cDNA probes spanning various parts of the AR gene, they could demonstrate no abnormality in restriction fragment patterns, suggesting that the gene defect is not a deletion but rather a point mutation or a small insertion/deletion.

Also see the mapping section under androgen receptor (AR; 313700).

Molecular Genetics

See androgen receptor (AR; 313700).

Pathogenesis

French et al. (1966) found that testosterone failed to affect the urinary excretion of nitrogen, phosphorus and citric acid when given in a dosage much greater than that which in controls decreased excretion of all three. Plasma estrogen levels were the same as those observed in the normal female. Leydig cell stimulation to estrogen production occurs probably because of failure of the feedback repression of the pituitary which shares the unresponsiveness to testosterone. Southern and Saito (1961) showed normal testosterone levels in this disorder.

Population Genetics

Mainly using data on the frequency of inguinal hernia in females, Jagiello and Atwell (1962) estimated the frequency of testicular feminization as being about 1 in 65,000 males.

Edwards et al. (1992) demonstrated that the distribution of the number of CAG repeats in exon 1 of the AR gene was lowest in African-Americans, intermediate in non-Hispanic whites, and highest in Asians. The distribution of allele size was bimodal in African-Americans, and only in African-Americans was there a deviation from Hardy-Weinberg equilibrium. Irvine et al. (1995) studied the distribution of the CAG and GC repeats (microsatellites) in exon 1 of the AR gene in African-Americans, non-Hispanic whites, and Asians (Japanese and Chinese) and confirmed the findings of Edwards et al. (1992). The frequency of prostatic cancer (176807) in the 3 racial groups is inversely proportional to the length of the repeats. One of the critical functions of the product of the AR gene is to activate the expression of target genes. This transactivation activity resides in the N-terminal domain of the protein which is encoded in exon 1 which contains the polymorphic repeats. The smaller size of the CAG repeat is associated with a higher level of receptor transactivation function, thereby possibly resulting in a higher risk of prostate cancer. Irvine et al. (1995) noted that Schoenberg et al. (1994) had observed a somatic mutation resulting in a contraction of the CAG repeat from 24 to 18 in an adenocarcinoma prostate and the effects of the shorter allele were implicated in the development of the tumor.

Animal Model

Lyon and Hawkes (1970) described a homologous phenotype in the mouse and showed that it is genetic, the Tfm locus being situated in the middle of the X chromosome. Ohno and Lyon (1970) showed that in these mice certain enzymes of the mouse kidney, e.g., alcohol dehydrogenase, are not inducible by testosterone as is usually possible. They postulated that the Tfm locus is a repressive regulatory locus controlling many testosterone inducible enzymes. In affected hemizygotes all these enzymes become noninducible. According to their suggestion, this is a regulator mutation like the noninducible mutation in the lac-repressor locus of E. coli as elucidated by Jacob and Monod (1963). Bardin et al. (1970) described studies of the pseudohermaphroditic rat which seems to have a disorder analogous to testicular feminization. Androgen-dependent differentiation is absent. Defective formation of dihydrotestosterone was apparently not the explanation. Goldstein and Wilson (1972) studied the Tfm mouse and showed, by giving dihydrotestosterone to pregnant mothers, that there is resistance to androgen-mediated sexual differentiation in embryos. Low serum testosterone and low production of testosterone in adult Tfm testis of the mouse were features different from those in man, but were considered by them as secondary to the defect in differentiation. They showed deficient binding of testosterone in the nuclei of the submaxillary gland of these adult Tfm animals, but again this may be the result of incomplete differentiation of an androgen-sensitive cell line.

Bullock and Bardin (1972) concluded that androgen-binding proteins are absent from the cytosol of preputial gland of Tfm rats and from the kidney of Tfm mice. Testicular feminization rats, despite female external sexual development, show masculine sexual behavior and little feminine sexual behavior. In the Tfm mouse, Charest et al. (1991) demonstrated a single base deletion in the N-terminal domain of the androgen receptor, resulting in a frameshift mutation. Gaspar et al. (1991) independently demonstrated the same abnormality. They found no structural aberration in the coding region of the messenger by a series of RNase-protection assays. However, cell-free translation of RNAs transcribed in vitro from enzymatically amplified overlapping segments of exon 1 demonstrated a truncated receptor protein. Sequence analysis showed deletion of a single nucleotide in the hexacytidine stretch at position 1107-1112 altering the reading frame of the messenger and introducing 41 missense amino acids before the premature termination codon at position 1235-1237.

In female mice heterozygous for the Tfm gene, Takeda et al. (1987) demonstrated mosaicism in 2 androgen target tissues by steroid autoradiographic techniques, thus documenting X-linked inheritance. See also Takeda et al. (1987).

History

The variety of sex anomaly described in this entry has been of relatively long interest to geneticists, largely through the publication of Pettersson and Bonnier (1937), who concluded that the affected persons are genetic males. Dieffenbach (1912), an American geneticist, had pointed out the hereditary pattern. Morris (1953), in a classic paper, first used the term testicular feminization.

Miller (1961) considered 'feminizing labial testes' of the type described by Lubs et al. (1959) to be a separate form of male pseudohermaphroditism. However, Wilson et al. (1984) described well-studied cases that indicated that the Lubs syndrome (Lubs et al., 1959), like classic testicular feminization, is due to mutation in the androgen receptor. The patients were first cousins; their mothers were sisters.

See also Adachi and Kano (1970); Burgermeister (1953); Eil (1983); French et al. (1965); Griffin and Wilson (1980); Grumbach and Barr (1958); Hauser (1961); Jukier et al. (1984); Kaufman et al. (1976); Keenan et al. (1974); Liao and Witte (1985); Lin and Ohno (1981); Long and David (1981); Madden et al. (1975); Mauvais-Jarvis et al. (1970); Morris and Mahesh (1963); Northcutt et al. (1969); Ohno (1977); Ohno (1971); Perez-Palacios et al. (1975); Pinsky et al. (1984); Pinsky et al. (1981); Puck et al. (1960); Schreiner (1959); Shapiro et al. (1980); Southern (1965); Stenchever et al. (1969); Strickland and French (1969)

Adachi, K.; Kano, M.: Adenyl cyclase in human hair follicles: its inhibition by dihydrotestosterone. Biochem. Biophys. Res. Commun. 41: 884-890, 1970.

Albright, F.; Burnett, C. H.; Smith, P. H.; Parson, W.: Pseudo-hypoparathyroidism, an example of 'Seabright-Bantam syndrome': report of 3 cases. Endocrinology 30: 922-932, 1942.

Amrhein, J. A.; Meyer, W. J., III; Jones, H. W., Jr.; Migeon, C. J.: Androgen insensitivity in man: evidence for genetic heterogeneity. Proc. Nat. Acad. Sci. 73: 891-894, 1976.

Bals-Pratsch, M.; Schweikert, H.-U.; Nieschlag, E.: Androgen receptor disorder in three brothers with bifid prepenile scrotum and hypospadias. Acta Endocr. 123: 271-276, 1990.

Bardin, C. W.; Bullock, L.; Schneider, G.; Allison, J. E.; Stanley, A. J.: Pseudohermaphrodite rat: end organ insensitivity to testosterone. Science 167: 1136-1137, 1970.

Batch, J. A.; Evans, B. A. J.; Hughes, I. A.; Patterson, M. N.: Mutations of the androgen receptor gene identified in perineal hypospadias. J. Med. Genet. 30: 198-201, 1993.

Boczkowski, K.; Teter, J.: Familial male pseudohermaphroditism. Acta Endocr. 49: 497-509, 1965.

Bullock, L. P.; Bardin, C. W.: Androgen receptors in testicular feminization. J. Clin. Endocr. 35: 935-937, 1972.

Burgermeister, J. J.: Contribution a l'etude d'un type familial d'intersexualite. J. Genet. Hum. 2: 51-82, 1953.

Charest, N. J.; Zhou, Z.; Lubahn, D. B.; Olsen, K. L.; Wilson, E. M.; French, F. S.: A frameshift mutation destabilizes androgen receptor messenger RNA in the Tfm mouse. Molec. Endocr. 5: 573-581, 1991.

Davies, H. R.; Hughes, I. A.; Savage, M. O.; Quigley, C. A.; Trifiro, M.; Pinsky, L.; Brown, T. R.; Patterson, M. N.: Androgen insensitivity with mental retardation: a contiguous gene syndrome? J. Med. Genet. 34: 158-160, 1997.

Dieffenbach, H.: Familiaerer Hermaphroditismus. Inaug. Dissert.: Stuttgart (pub.) 1912.

Edwards, A.; Hammond, H. A.; Jin, L.; Caskey, C. T.; Chakraborty, R.: Genetic variation at five trimeric and tetrameric tandem repeat loci in four human population groups. Genomics 12: 241-253, 1992.

Eil, C.: Familial incomplete male pseudohermaphroditism associated with impaired nuclear androgen retention: studies in cultured skin fibroblasts. J. Clin. Invest. 71: 850-858, 1983.

French, F. S.; Baggett, B.; Van Wyk, J. J.; Talbert, L. M.; Hubbard, W. R.; Johnston, F. R.; Weaver, R. P.; Forchielli, E.; Rao, G. S.; Sarda, I. R.: Testicular feminization: clinical, morphological and biochemical studies. J. Clin. Endocr. 25: 661-677, 1965.

French, F. S.; Van Wyk, J. J.; Baggett, B.; Easterling, W. E.; Talbert, L. M.; Johnston, F. R.; Forchielli, E.: Further evidence of a target organ defect in the syndrome of testicular feminization. J. Clin. Endocr. 26: 493-503, 1966.

Gaspar, M.-L.; Meo, T.; Bourgarel, P.; Guenet, J.-L.; Tosi, M.: A single base deletion in the Tfm androgen receptor gene creates a short-lived messenger RNA that directs internal translation initiation. Proc. Nat. Acad. Sci. 88: 8606-8610, 1991.

Gayral, L.; Barraud, M.; Carrie, J.; Candebat, L.: Pseudo-hermaphrodisme a type de 'testicule feminisant': 11 cas. Etude hormonale et etude psychologique. Toulouse Med. 61: 637-647, 1960.

Gerli, M.; Migliorini, G.; Bocchini, V.; Venti, G.; Ferrarese, R.; Donti, E.; Rosi, G.: A case report of complete testicular feminization and 47,XXY karyotype. J. Med. Genet. 16: 480-483, 1979.

German, J.; Vesell, M.: Testicular feminization in monozygotic twins with 47 chromosomes (XXY). Ann. Genet. 9: 5-8, 1966.

Goldstein, J. L.; Wilson, J. D.: Studies on the pathogenesis of the pseudohermaphroditism in the mouse with testicular feminization. J. Clin. Invest. 51: 1647-1658, 1972.

Griffin, J. E.: Testicular feminization associated with a thermolabile androgen receptor in cultured human fibroblasts. J. Clin. Invest. 64: 1624-1631, 1979.

Griffin, J. E.; Wilson, J. D.: The syndromes of androgen resistance. New Eng. J. Med. 302: 198-209, 1980.

Grino, P. B.; Griffin, J. E.; Cushard, W. G., Jr.; Wilson, J. D.: A mutation of the androgen receptor associated with partial androgen resistance, familial gynecomastia, and fertility. J. Clin. Endocr. Metab. 66: 754-761, 1988.

Grumbach, M. M.; Barr, M. L.: Cytologic tests of chromosomal sex in relation to sexual anomalies in man. Recent Prog. Horm. Res. 14: 255-334, 1958.

Hauser, G. A.: Testiki laere Feminisierung.In: Overzier, C.: Die Intersexualitaet. Stuttgart: Georg Thieme Verlag (pub.) 1961. Pp. 261-282.

Hughes, I. A.; Evans, B. A. J.: Complete androgen insensitivity syndrome characterized by increased concentration of a normal androgen receptor in genital skin fibroblasts. J. Clin. Endocr. Metab. 63: 309-315, 1986.

Imperato-McGinley, J.; Ip, N. Y.; Gautier, T.; Neuweiler, J.; Gruenspan, H.; Liao, S.; Chang, C.; Balazs, I.: DNA linkage analysis and studies of the androgen receptor gene in a large kindred with complete androgen insensitivity. Am. J. Med. Genet. 36: 104-108, 1990.

Irvine, R. A.; Yu, M. C.; Ross, R. K.; Coetzee, G. A.: The CAG and GGC microsatellites of the androgen receptor gene are in linkage disequilibrium in men with prostate cancer. Cancer. Res. 55: 1937-1940, 1995.

Jacob, F.; Monod, J.: General repression, allosteric inhibition, and cellular differentiation.In: Locke, M.: Cytodifferentiation and Macromolecular Synthesis. London: Academic Press (pub.) 1963. Pp. 30-64.

Jagiello, G.; Atwell, J. D.: Prevalence of testicular feminisation. Lancet I: 329, 1962.

Jukier, L.; Kaufman, M.; Pinsky, L.; Peterson, R. E.: Partial androgen resistance associated with secondary 5-alpha-reductase deficiency: identification of a novel qualitative androgen receptor defect and clinical implications. J. Clin. Endocr. Metab. 59: 679-688, 1984.

Kaufman, M.; Pinsky, L.; Baird, P. A.; McGillivray, B. C.: Complete androgen insensitivity with a normal amount of 5-alpha-dihydrotestosterone-binding activity in labium majus skin fibroblasts. Am. J. Med. Genet. 4: 401-411, 1979.

Kaufman, M.; Pinsky, L.; Bowin, A.; Au, M. W. S.: Familial external genital ambiguity due to a transformation defect of androgen-receptor complexes that is expressed with 5-alpha-dihydrotestosterone and the synthetic androgen methyltrienolone. Am. J. Med. Genet. 18: 493-507, 1984.

Kaufman, M.; Pinsky, L.; Feder-Hollander, R.: Defective up-regulation of the androgen receptor in human androgen insensitivity. Nature 293: 735-737, 1981.

Kaufman, M.; Straisfeld, C.; Pinsky, L.: Male pseudohermaphroditism presumably due to target organ unresponsiveness to androgens: deficient 5-alpha-dihydrotestosterone binding in cultured skin fibroblasts. J. Clin. Invest. 58: 345-350, 1976.

Keenan, B. S.; Meyer, W. J., III; Hadjian, A. J.; Jones, H. W.; Migeon, C. J.: Syndrome of androgen insensitivity in man: absence of 5-alpha-dihydrotestosterone binding protein in skin fibroblasts. J. Clin. Endocr. 38: 1143-1146, 1974.

Kovacs, W. J.; Griffin, J. E.; Weaver, D. D.; Carlson, B. R.; Wilson, J. D.: A mutation that causes lability of the androgen receptor under conditions that normally promote transformation to the DNA-binding state. J. Clin. Invest. 73: 1095-1104, 1984.

Liao, S.; Witte, D.: Autoimmune anti-androgen-receptor antibodies in human serum. Proc. Nat. Acad. Sci. 82: 8345-8348, 1985.

Lin, S.-Y.; Ohno, S.: The binding of androgen receptor to DNA and RNA. Biochim. Biophys. Acta 654: 181-186, 1981.

Long, S. E.; David, J. S. E.: Testicular feminization in an Ayrshire cow. Vet. Rec. 109: 116-118, 1981.

Lubs, H. A., Jr.; Vilar, O.; Bergenstal, D. M.: Familial male pseudohermaphroditism with labial testes and partial feminization: endocrine studies and genetic aspects. J. Clin. Endocr. 19: 1110-1120, 1959.
Lyon, M. F.; Hawkes, S. G.: X-linked gene for testicular feminization in the mouse. Nature 227: 1217-1219, 1970.
Madden, J. D.; Walsh, P. C.; MacDonald, P. C.; Wilson, J. D.: Clinical and endocrinologic characterization of a patient with the syndrome of incomplete testicular feminization. J. Clin. Endocr. 41: 751-760, 1975.
Marshall, H. K.; Harder, H. I.: Testicular feminizing syndrome in male pseudohermaphrodite: report of two cases in identical twins. Obstet. Gynec. 12: 284-293, 1958.
Mauvais-Jarvis, P.; Bercovici, J. P.; Crepy, O.; Gauthier, F.: Studies on testosterone metabolism in subjects with testicular feminization syndrome. J. Clin. Invest. 49: 31-40, 1970.
McKusick, V. A.: On the X Chromosome of Man. Washington: Am. Inst. Biol. Sci. (pub.) 1964.
Meyer, W. J., III; Migeon, B. R.; Migeon, C. J.: Locus on human X chromosome for dihydrotestosterone receptor and androgen insensitivity. Proc. Nat. Acad. Sci. 72: 1469-1472, 1975.
Miller, O. J.: Developmental sex abnormalities.In: Penrose, L. S.: Recent Advances in Human Genetics. London: J. and A. Churchill Ltd. (pub.) 1961. Pp. 39-55.
Morris, J. M.: Personal Communication. New Haven, Conn. 1962.
Morris, J. M.: The syndrome of testicular feminization in male pseudohermaphrodites. Am. J. Obstet. Gynec. 65: 1192-1211, 1953.
Morris, J. M.; Mahesh, V. B.: Further observations on the syndrome, 'testicular feminization.'. Am. J. Obstet. Gynec. 87: 731-748, 1963.
Muller, U.; Schneider, N. R.; Marks, J. F.; Kupke, K. G.; Wilson, G. N.: Maternal meiosis II nondisjunction in a case of 47,XXY testicular feminization. Hum. Genet. 84: 289-292, 1990.
Netter, A.; Lumbrosa, P.; Yaneva, H.; Belaisch, J.: Le testicule feminisant. Ann. Endocr. 19: 994-1014, 1958.
Northcutt, R. C.; Island, D. P.; Liddle, G. W.: An explanation for the target organ unresponsiveness to testosterone in the testicular feminization syndrome. J. Clin. Endocr. 29: 422-425, 1969.
Ohno, S.: The Y-linked antigen locus and the X-linked Tfm locus as major regulatory genes of the mammalian sex determining mechanism. J. Steroid Biochem. 8: 585-592, 1977.
Ohno, S.: Simplicity of mammalian regulatory systems inferred by single gene determination of sex phenotypes. Nature 234: 134-137, 1971.
Ohno, S.; Lyon, M. F.: X-linked testicular feminization in the mouse as a non-inducible regulatory mutation of the Jacob-Monod type. Clin. Genet. 1: 121-127, 1970.
Opitz, J. M.; Simpson, J. L.; Sarto, G. E.; Summitt, R. L.; New, M.; German, J.: Pseudovaginal perineoscrotal hypospadias. Clin. Genet. 3: 1-26, 1972.
Perez-Palacios, G.; Ortiz, S.; Lopez-Amor, E.; Morato, T.; Febres, F.; Lisker, R.; Saglia, H.: Familial incomplete virilization due to partial end organ insensitivity to androgens. J. Clin. Endocr. 41: 946-952, 1975.
Pettersson, G.; Bonnier, G.: Inherited sex-mosaic in man. Hereditas 23: 49-69, 1937.
Philip, J.; Trolle, D.: Familial male hermaphroditism with delayed and partial masculinization. Am. J. Obstet. Gynec. 93: 1076-1083, 1965.
Pinsky, L.; Kaufman, M.; Killinger, D. W.; Burko, B.; Shatz, D.; Volpe, R.: Human minimal androgen insensitivity with normal dihydrotestosterone-binding capacity in cultured genital skin fibroblasts: evidence for an androgen-selective qualitative abnormality of the receptor. Am. J. Hum. Genet. 36: 965-978, 1984.
Pinsky, L.; Kaufman, M.; Levitsky, L. L.: Partial androgen resistance due to a distinctive qualitative defect of the androgen receptor. Am. J. Med. Genet. 27: 459-466, 1987.
Pinsky, L.; Kaufman, M.; Summitt, R. L.: Congenital androgen insensitivity due to a qualitatively abnormal androgen receptor. Am. J. Med. Genet. 10: 91-99, 1981.
Polaillon, J. F. B.: Observation d'hermaphrodisme. Bull. Mem. Soc. Obstet. Gynec. Paris 123-130, 1891. Note: Cited by: Morris, J. M.: Am. J. Obstet. Gynec. 65: 1192-1211, 1953.
Puck, T. T.; Robinson, A.; Tjio, J. H.: Familial primary amenorrhea due to testicular feminization: a human gene affecting sex differentiation. Proc. Soc. Exp. Biol. Med. 103: 192-196, 1960.
Schoenberg, M. P.; Hakimi, J. M.; Wang, S.; Bova, G. S.; Epstein, J. I.; Fischbeck, K. H.; Isaacs, W. B.; Walsh, P. C.; Barrack, E. R.: Microsatellite mutation (CAG (24-to-18)) in the androgen receptor gene in human prostate cancer Biochem. Biophys. Res. Commun. 198: 74-80, 1994.
Schreiner, W. E.: Ueber eine hereditaere Form von Pseudohermaphrodismus masculinus (testiculaere Feminisierung). Gynaecologia 148: 355-357, 1959.
Shapiro, B. H.; Levine, D. C.; Adler, N. T.: The testicular feminized rat: a naturally occurring model of androgen independent brain masculinization. Science 209: 418-420, 1980.
Southern, A. L.: The syndrome of testicular feminization.In: Levine, R.; Luft, R.: Advances in Metabolic Diseases. New York: Academic Press (pub.) 1965. Pp. 227-256.
Southern, A. L.; Saito, A.: The syndrome of testicular feminization: a report of three cases with chromatographic analysis of the urinary neutral 17-ketosteroids. Ann. Intern. Med. 55: 925-931, 1961.
Stenchever, M. A.; Ng, A. B. P.; Jones, G. K.; Jarvis, J. A.: Testicular feminization syndrome: chromosomal, histologic, and genetic studies in a large kindred. Obstet. Gynec. 33: 649-657, 1969.
Strickland, A. L.; French, F. S.: Absence of response to dihydrotestosterone in the syndrome of testicular feminization. J. Clin. Endocr. 29: 1284-1286, 1969.
Takeda, H.; Lasnitzki, I.; Mizuno, T.: Change of mosaic pattern by androgens during prostatic bud formation in X(Tfm)/X(+) heterozygous female mice. J. Endocr. 114: 131-167, 1987.
Takeda, H.; Suzuki, M.; Lasnitzki, I.; Mizuno, T.: Visualization of X-chromosome inactivation mosaicism of Tfm gene in X(Tfm)/X(+) heterozygous female mice. J. Endocr. 114: 125-129, 1987.
Wilkins, L.: The Diagnosis and Treatment of Endocrine Disorders in Childhood and Adolescence. Springfield, Ill.: Charles C Thomas (pub.) (2nd ed.): 1957.
Wilson, J. D.: Personal Communication. Dallas, Tex. 10/30/1981.
Wilson, J. D.: Personal Communication. Dallas, Tex. 1976.
Wilson, J. D.; Carlson, B. R.; Weaver, D. D.; Kovacs, W. J.; Griffin, J. E.: Endocrine and genetic characterization of cousins with male pseudohermaphroditism: evidence that the Lubs phenotype can result from a mutation that alters the structure of the androgen receptor. Clin. Genet. 26: 363-370, 1984.
Winterborn, M. H.; France, N. E.; Raiti, S.: Incomplete testicular feminization. Arch. Dis. Child. 45: 811-812, 1970.

ADP-RIBOSYLTRANSFERASE (*173870)

JJM NOTE

This entity is another gene that has a specific allele that is statistically a risk factor for prostate cancer.

The chromatin-associated enzyme poly(ADP-ribose) polymerase (ADPRT; EC 2.4.2.30) is a 116-kD protein that uses NAD as substrate to catalyze both the covalent transfer of ADP-ribose to a variety of nuclear protein acceptors and subsequently catalyzes the transfer of additional 60 to 80 ADP-ribose units to the initial moiety. Nuclear proteins that become predominantly poly(ADP-ribosyl)ated include nucleosomal core histones, histone H1, HMG proteins, and topoisomerases I and II. ADP ribosyltransferase is required for cellular repair. Inhibitors of this enzyme potentiate the lethal effects of noxious agents. During repair, NAD(+) is consumed and the NAD(+) content of the cell decreases. Concomitantly, nuclear proteins are ADP-ribosylated. The enzyme is induced by single-strand breaks in DNA which serve as cosubstrate for the reaction.

Loetscher et al. (1987) proposed that poly(ADP-ribose) may signal altered metabolic conditions to the chromatin. They were led to this proposal from the finding that the constitutive level of posttranslational poly(ADP-ribose) modification of chromatin proteins in mammalian cells is related to the availability of NAD, which varies in different physiological and pathological states.

ADP-ribosylation is a eukaryotic posttranslational modification of proteins that is strongly induced by the presence of DNA strand breaks and plays a role in DNA repair and the recovery of cells from DNA damage. Grube and Burkle (1992) found a strong positive correlation (r = 0.84; P much less than 0.001) between PARP activity and life span, with human cells displaying, for example, about 5 times the activity of rat cells. The cells studied were mononuclear leukocytes. Grube and Burkle (1992) suggested that higher poly(ADP-ribosyl)ation capacity may contribute to the efficient maintenance of genome integrity

Alkhatib et al. (1987) isolated cDNA clones for this enzyme from a human hepatoma lambda library and studied its expression. McBride et al. (1987) concluded that a large functional PARP gene of more than 15 to 20 kb is located on 1q and that sequences on chromosomes 13 and 14 most likely represent processed pseudogenes. These localizations were achieved by probing of the DNA from panels of somatic cell hybrids. Auer et al. (1989) demonstrated that the gene is 43 kb long and split into 23 exons.

Herzog et al. (1988) cloned a cDNA for ADPRT and localized the gene to 1q21-q22 by in situ hybridization. Herzog et al. (1989) mapped the ADPRT gene to 1q41-q42 by in situ hybridization. Using high resolution in situ hybridization techniques, Zabel et al. (1989) localized PPOL to 1q41-q42. With the conditions used, only 1 additional site of hybridization, 14q22, could be detected; this probably represented a pseudogene which had previously been identified and called ADPRTP2. By nonisotopic in situ hybridization, Baumgartner et al. (1992) confirmed localization of the functional gene to 1q42. Two other hybridization peaks, one at 13q34 and one at 14q24, suggested the location of pseudogenes.

Schweiger et al. (1987) suggested that the defect in Fanconi anemia (227650) is one of impaired ADP-ribosylation. Several independent observations have suggested that ADPRT might be the site of the mutation in Fanconi anemia; however, Flick et al. (1992) could find no abnormality in cells from an FA patient of complementation group A (cell line GM6914). It is also noteworthy that a gene involved in complementation of the defect in xeroderma pigmentosum cells of group A (278700) can be localized to 1q42-qter (Keijzer et al., 1987).

A processed pseudogene or a gene with extensive identity to the ADPRT gene was studied by Bhatia et al. (1990), who mapped it to 13q33-qter. The gene was deleted in a polymorphism that was 3 times higher in frequency among blacks than Caucasians. Bhatia et al. (1990)

suggested that this deletion might be a predisposing factor in several forms of malignancy.

Lyn et al. (1993) studied the 2-allele (A/B) polymorphism of the gene on 13q34. An elevated B-allele frequency was found in germ-line DNA in blacks with multiple myeloma, prostate cancer, and colon cancer. They found that the A allele has a close sequence similarity (91.8%) to the PPOL cDNA coded by 1q42 and is intronless, suggesting that the gene on 13q is a processed pseudogene. They presented data indicating that the polymorphism reflects a 193-bp duplication within the processed-pseudogene sequence, with absence of this duplicated region being characteristic of the B genotype. Doll et al. (1996) confirmed the association between prostate cancer in black Americans and an allele of the pseudogene locus (which they symbolized PADPRP) on chromosome 13. Two genes in the region 13q33-q34 were viewed as potential candidates for prostate cancer: ERCC5 (133530) and RAP2A (179540).

See also Cherney et al. (1987)

Alkhatib, H. M.; Chen, D.; Cherney, B.; Bhatia, K.; Notario, V.; Giri, C.; Stein, G.; Slattery, E.; Roeder, R. G.; Smulson, M. E.: Cloning and expression of cDNA for human poly(ADP-ribose) polymerase. Proc. Nat. Acad. Sci. 84: 1224-1228, 1987.

Auer, B.; Nagl, U.; Herzog, H.; Schneider, R.; Schweiger, M.: Human nuclear NAD+ ADP-ribosyltransferase(polymerizing): organization of the gene. DNA 8: 575-580, 1989.

Baumgartner, M.; Schneider, R.; Auer, B.; Herzog, H.; Schweiger, M.; Hirsch-Kauffmann, M.: Fluorescence in situ mapping of the human nuclear NAD+ ADP-ribosyltransferase gene (ADPRT) and two secondary sites to human chromosomal bands 1q42, 13q34, and 14q24. Cytogenet. Cell Genet. 61: 172-174, 1992.

Bhatia, K. G.; Cherney, B. W.; Huppi, K.; Magrath, I. T.; Cossman, J.; Sausville, E.; Barriga, F.; Johnson, B.; Gause, B.; Bonney, G.; Neequayi, J.; DeBernardi, M.; Smulson, M.: A deletion linked to a poly(ADP-ribose) polymerase gene on chromosome 3q33-qter occurs frequently in the normal black population as well as in multiple tumor DNA. Cancer Res. 50: 5406-5413, 1990.

Cherney, B. W.; McBride, O. W.; Chen, D.; Alkhatib, H.; Bhatia, K.; Hensley, P.; Smulson, M. E.: cDNA sequence, protein structure, and chromosomal location of the human gene for poly(ADP-ribose) polymerase. Proc. Nat. Acad. Sci. 84: 8370-8374, 1987.

Doll, J. A.; Suarez, B. K.; Donis-Keller, H.: Association between prostate cancer in black Americans and an allele of the PADPRP pseudogene locus on chromosome 13. (Letter) Am. J. Hum. Genet. 58: 425-428, 1996.

Flick, K.; Schneider, R.; Auer, B.; Hirsch-Kauffmann, M.; Schweiger, M.: No abnormalities in the NAD(+) ADP-ribosyltransferase (polymerizing) gene of transformed cells from a Fanconi's anemia patient. (Letter) Hum. Genet. 89: 690-691, 1992.

Grube, K.; Burkle, A.: Poly(ADP-ribose) polymerase activity in mononuclear leukocytes of 13 mammalian species correlates with species-specific life span. Proc. Nat. Acad. Sci. 89: 11759-11763, 1992.

Herzog, H.; Zabel, B.; Schneider, R.; Auer, B.; Hirsch-Kauffmann, M.; Schweiger, M.: Human NAD(+):protein ADP ribosyltransferase (ADPRT): synthesis of active enzyme in E. coli, sequence of its cDNA from Hela cells and chromosomal localization. Biol. Chem. Hoppe-Seyler 369: 836-837, 1988.

Herzog, H.; Zabel, B. U.; Schneider, R.; Auer, B.; Hirsch-Kauffmann, M.; Schweiger, M.: Human nuclear NAD(+) ADP-ribosyltransferase: localization of the gene on chromosome 1q41-q42 and expression of an active human enzyme in Escherichia coli. Proc. Nat. Acad. Sci. 86: 3514-3518, 1989.

Keijzer, W.; Stefanini, M.; Bootsma, D.; Verkerk, A.; Geurts van Kessel, A. H. M.; Jongkind, J. F.; Westerveld, A.: Localization of a gene involved in complementation of the defect in xeroderma pigmentosum group A cells on human chromosome 1. Exp. Cell Res. 169: 490-501, 1987.

Loetscher, P.; Alvarez-Gonzalez, R.; Althaus, F. R.: Poly(ADP-ribose) may signal changing metabolic conditions to the chromatin of mammalian cells. Proc. Nat. Acad. Sci. 84: 1286-1289, 1987.

Lyn, D.; Cherney, B. W.; Lalande, M.; Berenson, J. R.; Lichtenstein, A.; Lupold, S.; Bhatia, K. G.; Smulson, M.: A duplicated region is responsible for the poly(ADP-ribose) polymerase polymorphism, on chromosome 13, associated with a predisposition to cancer. Am. J. Hum. Genet. 52: 124-134, 1993.

McBride, O. W.; Cherney, B.; Takourdin, C.; Smulson, M. E.: Human poly(ADP-ribose) polymerase sequences are located on chromosomes 1, 13, and 14. (Abstract) Cytogenet. Cell Genet. 46: 659-660, 1987.

Schweiger, M.; Auer, B.; Burtscher, H. J.; Hirsch-Kauffmann, M.; Klocker, H.; Schneider, R.: DNA repair in human cells: biochemistry of the hereditary diseases Fanconi's anaemia and Cockayne syndrome. Europ. J. Biochem. 165: 235-242, 1987.

Zabel, B. U.; Herzog, H.; Schneider, R.; Auer, B.; Hirsch-Kauffmann, M.; Schweiger, M.: Chromosomal sublocalization of the gene for human poly (ADP-ribose) polymerase (NAD+ ADP-ribosyltransferase) at 1q41-42. (Abstract) Cytogenet. Cell Genet. 51: 1115, 1989.

7. CARDIORESPIRATORY SYSTEM

Lung Tumors

CANCER OF THE LUNG (211980)

JJM NOTE

Elsewhere, I review the cytogenetic, molecular genetic, and single-gene traits that predispose to lung cancer, as well as the literature on familial aggregation (Mulvihill JJ: Lung cancer. In King RA, Rotter JI, Motulsky AG, eds. *The Genetic Basis of Common Disease.* New York, Oxford University Press, 1992, pp. 691–700). In general, when a person gets lung cancer, first-degree relatives should be considered to have a threefold risk for lung cancer. Major susceptibility genes are some cytochrome P450 genes (especially *CYP1A1, CYP2D*), glutathione *S*-transferase, *N*-acetyltransferase, ABO blood group, and O[6]-methylguanine repair. Segregation analysis shows codominant inheritance of a rare major dominant gene (Sellers TA et al.: Evidence for Mendelian inheritance in the pathogenesis of lung cancer. J. Natl. Cancer Inst. 82:1272–79, 1990). Lung cancer occurs to slight excess in carriers of mutations of *TP53* (Li-Fraumeni cancer family syndrome) and *BRCA1.* Other familial disorders complicated by lung cancer are the lung diseases listed below.

Joishy et al. (1977) described identical twins who developed symptoms of alveolar cell carcinoma almost simultaneously. In the DNA from 1 colon and 2 lung carcinoma cell lines, Perucho et al. (1981) demonstrated the same or closely related transforming elements. By DNA-mediated gene transfer, mouse fibroblasts could be morphologically transformed and rendered tumorigenic in nude mice. In 3 varieties of non-small-cell cancer of the lung, Weston et al. (1989) found evidence of loss of heterozygosity in 17p and chromosome 11. Only a minority had loss of heterozygosity involving a chromosomal locus previously shown to be lost consistently in small-cell cancer of the lung, a locus on 3p (182280).

Braun et al. (1994) conducted a genetic analysis of lung cancer mortality in the National Academy of Sciences/National Research Council Twin Registry. The registry is composed of 15,924 male twin pairs who were born in the U.S. between 1917 and 1927 and who served in the armed forces during World War II. As evidence for a genetic effect on lung cancer, they required concordance for lung cancer death to be greater among monozygotic than among dizygotic twin pairs. No genetic effect on lung cancer mortality was observed. The ratio of observed to expected concordance among monozygotic twins did not exceed that among dizygotic twins, even though monozygotic twin pairs were more likely to be concordant for smoking than dizygotic twin pairs in this population. A cohort analysis (accounting for age, sex, race, and smoking intensity) of lung cancer mortality found no lung cancer deaths during the 300 person-years of follow-up among 47 monozygotic twins smokers whose smoking twin died of lung cancer, even though smoking histories were very similar within twin pairs.

In a multicenter study of lung cancer in lifetime nonsmokers in the United States, 646 female lung cancer patients and 1,252 population controls were interviewed regarding history of cancer in their first-degree relatives. Wu et al. (1996) summarized the findings. A 30% increased risk was associated with a history of respiratory tract cancer in parents or sibs after adjustment for exposure to environmental tobacco smoke in adult life. Lung cancer, which represented approximately two-thirds of the respiratory tract cancers, occurred more frequently in first-degree relatives of lung cancer patients than in comparable relatives of population controls. A significant 3-fold increased risk for lung cancer was associated with lung cancer diagnosed in mothers and sisters. Wu et al. (1996) also observed the increased risk in relation to family history of lung cancer among parents and sibs who were smokers as well as in those who were nonsmokers. The association with family history of lung cancer was strengthened when the analysis was restricted to adenocarcinoma of the lung. However, the authors pointed out that there was no association between family history of other cancers and risk of lung cancer in nonsmokers.

Sjalander et al. (1996) found an increased frequency of the p21 codon 31ARG allele (116899.0001) in lung cancer patients, especially in comparison with patients with chronic obstructive pulmonary disease (COPD); $p = 0.004$. Thus allelic variants of both p53 and its effector protein p21 may have an influence on lung cancer.

See also Goffman et al. (1982)

Braun, M. M.; Caporaso, N. E.; Page, W. F.; Hoover, R. N.: Genetic component of lung cancer: cohort study of twins. Lancet 344: 440-443, 1994.

Goffman, T. E.; Hassinger, D. D.; Mulvihill, J. J.: Familial respiratory tract cancer: opportunities for research and prevention. J.A.M.A. 247: 1020-1023, 1982.

Joishy, S. K.; Cooper, R. A.; Rowley, P. T.: Alveolar cell carcinoma in identical twins: similarity in time of onset, histochemistry, and site of metastasis. Ann. Intern. Med. 87: 447-450, 1977.

Perucho, M.; Goldfarb, M.; Shimizu, K.; Lama, C.; Fogh, J.; Wigler, M.: Human-tumor-derived cell lines contain common and different transforming genes. Cell 27: 467-476, 1981.

Sjalander, A.; Birgander, R.; Rannug, A.; Alexandrie, A.-K.; Tornling, G.; Beckman, G.: Association between the p21 codon 31A1 (arg) allele and lung cancer. Hum. Hered. 46: 221-225, 1996.

Weston, A.; Willey, J. C.; Modali, R.; Sugimura, H.; McDowell, E. M.; Resau, J.; Light, B.; Haugen, A.; Mann, D. L.; Trump, B. F.; Harris, C. C.: Differential DNA sequence deletions from chromosomes 3, 11, 13, and 17 in squamous-cell carcinoma, large-cell carcinoma, and adenocarcinoma of the human lung. Proc. Nat. Acad. Sci. 86: 5099-5103, 1989.

Wu, A. H.; Fontham, E. T. H.; Reynolds, P.; Greenberg, R. S.; Buffler, P.; Liff, J.; Boyd, P.; Correa, P.: Family history of cancer and risk of lung cancer among lifetime nonsmoking women in the United States. Am. J. Epidemiol. 143: 535-542, 1996.

SMALL CELL CANCER OF THE LUNG (*182280)

JJM NOTE

Probably the most common cytogenetic abnormality in any cancer is the deletion of the short arm of chromosome 3 in small-cell carcinoma of the lung, given the prevalence of lung cancer. Several genes and proteins have been identified from the region and hence incriminated in small-cell cancer of the lung (see Chapter 11).

Small-cell cancer of the lung accounts for about a fourth of the 110,000 new cases of lung cancer that occur annually in the United States. It is clinically distinctive: usually metastases are already present at the time of discovery so that surgery is not used. In contrast to adeno- and squamous carcinoma, SCCL is sensitive to chemotherapy and radiotherapy. Whang-Peng et al. (1982) found a specific, acquired chromosomal abnormality (deletion 3p) in at least one chromosome 3 in all metaphases in all 12 cell lines cultured from human SCCL tissue in 2-day tumor culture specimens from 3 patients. The shortest region of overlap showed the deletion to involve 3p23-p14. No other type of lung cancer showed this deletion, nor did lymphoblastoid lines cultured from SCCL patients whose tumors had the 3p deletion. SCCL is 'caused' by cigarette smoking as are other types of lung cancer. Thus, like chronic myeloid leukemia, this is an example of an exogenously induced malignancy with a specific chromosomal change. Cytogenetic effects of cigarette smoke are relevant in this connection (Madle et al., 1981).

Several biochemical markers were found to be associated with small cell cancer of the lung (Gazdar et al., 1981; Tapia et al., 1981). Perhaps genes in the 3p14-23 region have something to do with these markers as well as with the genesis of SCCL. (The cell of origin of SCCL is thought to be the Kulchitsky cell, an argentaffine cell situated in the bronchial epithelium, although this is not proved.) Erisman et al. (1982) showed that SCCL contains bombesin, a tetradecapeptide from anuran skin. It had been identified in human fetal and neonatal lung but not in adult lung. Some symptoms of SCCL may be attributable to bombesin. The syndrome of inappropriate secretion of antidiuretic hormone and Cushing syndrome, occurring with SCCL, are due to ectopic production of antidiuretic hormone and ACTH, respectively. The relation between ectopic hormone production and the aberration involving chromosome 3 is unknown. Baylin et al. (1982) found 12 distinguishing surface proteins on SCCL that were not shared by any of the 3 other carcinogen-induced forms of lung cancer (squamous, adeno-, and large cell undifferentiated carcinoma) or by human lymphoblastoid cells and fibroblasts. The neuroendocrine nature of SCCL was supported by the fact that 6 of the 12 were shared by human neuroblastoma cells. On human SCCL cells and tumors, Ruff and Pert (1984) demonstrated 4 surface antigens previously recognized only in macrophages. They suggested that cancerous cells may arise from macrophage precursors in bone marrow, and these precursors migrate to lung to participate in the repair of tissue damage produced by continuous heavy smoking. About 5% of SCCL patients have no apparent pulmonary involvement and the early, rapid and widespread dissemination of tumor to extrathoracic sites requires explanation. Naylor et al. (1984) used an anonymous, polymorphic DNA probe, D3S3, to confirm the presence of deletion of 3p in SCCL. This probe had been assigned to 3p21-cen. Studying 7 SCCL tumors and nor mal tissue from the same persons, they found that 6 of the 'normal' DNA samples were heterozygous for the D3S3 MspI polymorphism, whereas in all cases the tumor tissues were homozygous. De Leij et al. (1985) isolated 3 new, well-growing cell lines from SCCL. Deletions in 3p, with 3p23-p21 as the smallest region of overlap, were found. Mooibroek et al. (1987) used a recombinant DNA fragment detecting a RFLP presumably at 3p21 to probe DNA isolated from leukocytes of 12 patients with small-cell lung cancer. Four of these patients were heterozygous. Analysis of tumor material from the 4 patients showed homozygosity for either one or the other restriction fragment in every case. Gerber and Scoggin (1987) and Naylor et al. (1987) demonstrated loss of constitutional heterozygosity in SCCL. Comparing tumor and constitutional genotypes of 9 patients with small-cell lung cancer, Naylor et al. (1987) found a loss of alleles of chromosome 3p markers in tumor DNA of all 9 patients. Brauch et al. (1987) concluded that loss of alleles on 3p is a consistent change in small-cell lung cancer but occasionally occurs in non-small-cell lung cancer as well. Using a molecular genetic approach, Kok et al. (1987) found evidence for a consistent deletion at the 3p21 region not only in SCCL but in all major types of lung cancer. Yokota et al. (1987) found loss of heterozygosity for RFLPs on chromosome 3p in 7 of 7 patients, on 13q in 10 of 11 patients, and on 17p in 5 of 5 patients. Deletions at these loci in small-cell carcinomas were observed even in tumors without any clinical evidence of metastasis. Furthermore, loss of heterozygosity on 3p and 13q occurred before NMYC amplification and before chromosome 11p deletion. (Loss of heterozygosity on 3p was also detected in the adenocarcinomas from 5 of 6 patients. Heterozygosity of chromosomes 13q and 17p was lost in 10 of 31 patients and in 3 of 12 patients, respectively, of lung cancers other t h an small-cell carcinomas.) Johnson et al. (1988) found unequivocal loss of heterozygosity in the DNA from tumor tissue of 23 of 25 patients who were constitutionally heterozygous for at least 1 marker in the region 3p14-p21.

Birrer and Minna (1988) pointed to at least 3 molecular mechanisms involved in the development of lung cancer: deletion of 3p, deregulated expression of the MYC family of genes, and growth factors such as gastrin-releasing hormone (137260). Drabkin et al. (1988) demonstrated that the SCLC 'locus' is proximal to ERBA2 (190160) and also to the constitutive 3p14.2 fragile site. Using 15 chromosome 3 probes that identified 19 different RFLPs, Daly et al. (1991) identified a single 3p deletion extending proximal to the D3S2 locus at 3p21-p14.2 and including at least 3p14-p13. The locus D3F15S2 was excluded from the deleted region, an uncharacteristic feature of SCLC deletions. Moreover, D3S30 and D3S4 were included within this deletion, and thus map within the proximal half of chromosome 3p. Leduc et al. (1989) found that virtually all cases of SCLC had lost heterozygosity at the ERBA2 locus. A smaller but substantial portion of non-small-cell carcinomas of the lung had lost heterozygosity at this locus. Among all of the non-small-cell tumors, some had lost heterozygosity at a proximal locus but not at ERBA2, whereas none were found where the reverse was true. Thus, the locus that plays a role in non-small-cell tumorigenesis probably lies proximal to ERBA2 and is almost certainly not the ERBA2 gene. Sellers et al. (1989) presented epidemiologic data supporting the role of mendelian factors in the susceptibility to human lung cancer. The specific pathology of the tumors studied was not stated in the report.

From studies of allele loss by use of 13 RFLP probes on 3p, Hibi et al. (1992) concluded that 3 distinct regions on 3p are frequently deleted in lung cancer: 3p25, 3p21.3, and 3p14-cen.

Killary et al. (1992) described a rapid genetic assay system that allowed functional analysis of defined areas of 3p in the suppression of tumorigenicity in vivo. Human/mouse microcell hybrids containing fragments of chromosome 3p were constructed and screened for tumorigenicity in athymic nude mice. Hybrid clones were obtained that showed a dramatic tumor suppression and contained a 2-megabase fragment of human chromosomal material encompassing the region 3p21 near the interface with 3p22. This should be the first step toward isolating the tumor suppressor gene.

Chen et al. (1996) analyzed microsatellite repeat markers in plasma, tumor samples, and normal cells from 21 patients with a confirmed diagnosis of small cell lung carcinoma (SCLC). They reported that microsatellite alteration (including loss of heterozygosity or the appearance of new size forms) occurred in 16 out of 21 SCLC tumors and in 15 out of 21 plasma samples. In 52% of cases the marker UT762 on chromosome 21 was altered, and in 38% of cases the marker AR (313700) on the X chromosome was altered. Chen et al. (1996) concluded that analysis of plasma DNA may constitute a new tool for tumor staging, management, or possibly detection. See also 601400.

See also Falor et al. (1985); Graziano et al. (1987); Naylor et al. (1987); Whang-Peng et al. (1982)

Baylin, S. B.; Gazdar, A. F.; Minna, J. D.; Bernal, S. D.; Shaper, J. H.: A unique cell-surface protein phenotype distinguishes human small-cell from non-small-cell lung cancer. Proc. Nat. Acad. Sci. 79: 4650-4654, 1982.

Birrer, M. J.; Minna, J. D.: Molecular genetics of lung cancer. Semin. Oncol. 15: 226-235, 1988.

Brauch, H.; Johnson, B.; Hovis, J.; Yano, T.; Gazdar, A.; Pettengill, O. S.; Graziano, S.; Sorenson, G. D.; Poiesz, B. J.; Minna, J.; Linehan, M.; Zbar, B.: Molecular analysis of the short arm of chromosome 3 in small-cell and non-small-cell carcinoma of the lung. New Eng. J. Med. 317: 1109-1113, 1987.

Chen, X. Q.; Stroun, M.; Magnenat, J.-L.; Nicod, L. P.; Kurt, A.-M.; Lyautey, J.; Lederrey, C.; Anker, P.: Microsatellite alterations in plasma DNA of small cell lung cancer patients. Nature Med. 2: 1033-1034, 1996.

Daly, M. C.; Douglas, J. B.; Bleehen, N. M.; Hastleton, P.; Twentyman, P. R.; Sundaresan, V.; Carritt, B.; Bergh, J.; Rabbitts, P. H.: An unusually proximal deletion on the short arm of chromosome 3 in a patient with small cell lung cancer. Genomics 9: 113-119, 1991.

de Leij, L.; Postmus, P. E.; Buys, C. H. C. M.; Elema, J. D.; Ramaekers, F.; Poppema, S.; Brouwer, M.; van der Veen, A. Y.; Mesander, G.; The, T. H.: Characterization of three new variant type cell lines derived from small cell carcinoma of the lung. Cancer Res. 45: 6024-6033, 1985.

Drabkin, H.; Kao, F.-T.; Hartz, J.; Hart, I.; Gazdar, A.; Weinberger, C.; Evans, R.; Gerber, M.: Localization of human ERBA2 to the 3p22-3p24.1 region of chromosome 3 and variable deletion in small cell lung cancer. Proc. Nat. Acad. Sci. 85: 9258-9262, 1988.

Erisman, M. D.; Linnoila, R. I.; Hernandez, O.; DiAugustine, R. P.; Lazarus, L. H.: Human lung small-cell carcinoma contains bombesin. Proc. Nat. Acad. Sci. 79: 2379-2383, 1982.

Falor, W. H.; Ward-Skinner, R.; Wegryn, S.: A 3p deletion in small cell lung carcinoma. Cytogenet. Cell Genet. 16: 175-177, 1985.

Gazdar, A. F.; Zweig, M. H.; Carney, D. N.; Van Steirteghen, A. C.; Baylin, S. B.; Minna, J. D.: Levels of creatine kinase and its BB isoenzyme in lung cancer specimens and cultures. Cancer Res. 41: 2773-2777, 1981.

Gerber, M. J.; Scoggin, C. H.: Loss of constitutional heterozygosity in small cell lung cancer. (Abstract) Am. J. Hum. Genet. 41: A27 only, 1987.

Graziano, S. L.; Cowan, B. Y.; Carney, D. N.; Bryke, C. R.; Mitter, N. S.; Johnson, B. E.; Mark, G. E.; Planas, A. T.; Catino, J. J.; Comis, R. L.; Pioesz, B. J.: Small cell lung cancer cell line derived from a primary tumor with a characteristic deletion of 3p. Cancer Res. 47: 2148-2155, 1987.

Hibi, K.; Takahashi, T.; Yamakawa, K.; Ueda, R.; Sekido, Y.; Ariyoshi, Y.; Suyama, M.; Takagi, H.; Nakamura, Y.; Takahashi, T.: Three distinct regions involved in 3p deletion in human lung cancer. Oncogene 7: 445-449, 1992.

Johnson, B. E.; Sakaguchi, A. Y.; Gazdar, A. F.; Minna, J. D.; Burch, D.; Marshall, A.; Naylor, S. L.: Restriction fragment length polymorphism studies show consistent loss of chromosome 3p alleles in small cell lung cancer patients' tumors. J. Clin. Invest. 82: 502-507, 1988.

Killary, A. M.; Wolf, M. E.; Giambernardi, T. A.; Naylor, S. L.: Definition of a tumor suppressor locus within human chromosome 3p21-p22. Proc. Nat. Acad. Sci. 89: 10877-10881, 1992.

Kok, K.; Osinga, J.; Carritt, B.; Davis, M. B.; van der Hout, A. H.; van der Veen, A. Y.; Landsvater, R. M.; de Leij, L. F. M. H.; Berendsen, H. H.; Postmus, P. E.; Poppema, S.; Buys, C. H. C. M.: Deletion of a DNA sequence at the chromosomal region 3p21 in all major types of lung cancer. Nature 330: 578-581, 1987.

Leduc, F.; Brauch, H.; Hajj, C.; Dobrovic, A.; Kaye, F.; Gazdar, A.; Harbour, J. W.; Pettengill, O. S.; Sorenson, G. D.; van den Berg, A.; Kok, K.; Campling, B.; Paquin, F.; Bradley, W. E. C.; Zbar, B.; Minna, J.; Buys, C.; Ayoub, J.: Loss of heterozygosity in a gene coding for a thyroid hormone receptor in lung cancers. Am. J. Hum. Genet. 44: 282-287, 1989.

Madle, S.; Korte, A.; Obe, G.: Cytogenetic effects of cigarette smoke condensates in vitro and in vivo. Hum. Genet. 59: 349-352, 1981.

Mooibroek, H.; Osinga, J.; Postmus, P. E.; Carritt, B.; Buys, C. H. C. M.: Loss of heterozygosity for a chromosome 3 sequence presumably at 3p21 in small cell lung cancer. Cancer Genet. Cytogenet. 27: 361-365, 1987.

Naylor, S. L.; Johnson, B. E.; Minna, J. D.; Sakaguchi, A. Y.: High frequency loss of heterozygosity of chromosome 3p markers in small cell lung cancer. (Abstract) Am. J. Hum. Genet. 41: A33 only, 1987.

Naylor, S. L.; Johnson, B. E.; Minna, J. D.; Sakaguchi, A. Y.: Loss of heterozygosity of chromosome 3p markers in small-cell lung cancer. Nature 329: 451-454, 1987.

Naylor, S. L.; Minna, J.; Johnson, B.; Sakaguchi, A. Y.: DNA polymorphisms confirm the deletion in the short arm of chromosome 3 in small cell lung cancer. (Abstract) Am. J. Hum. Genet. 36: 35S only, 1984.

Ruff, M. R.; Pert, C. B.: Small cell carcinoma of the lung: macrophage-specific antigens suggest hemopoietic stem cell origin. Science 225: 1034-1036, 1984.

Sellers, T. A.; Bailey-Wilson, J. E.; Elston, R. C.; Wilson, A. F.; Rothschild, H.: Evidence for mendelian inheritance of susceptibility to lung cancer in humans. (Abstract) Am. J. Hum. Genet. 45 (suppl.): A33 only, 1989.

Tapia, F. J.; Polak, J. M.; Barbosa, A. J. A.; Bloom, S. R.; Marangos, P. J.; Dermody, C.; Pearse, A. G. E.: Neuron-specific enolase is produced by neuroendocrine tumours. Lancet I: 808-811, 1981.

Whang-Peng, J.; Bunn, P. A., Jr.; Kao-Shan, C. S.; Lee, E. C.; Carney, D. N.; Gazdar, A.; Minna, J. D.: A nonrandom chromosomal abnormality, del 3p(14-23), in human small cell lung cancer (SCLC). Cancer Genet. Cytogenet. 6: 119-134, 1982.

Whang-Peng, J.; Kao-Shan, C. S.; Lee, E. C.; Bunn, P. A.; Carney, D. N.; Gazdar, A. F.; Minna, J. D.: Specific chromosome defect associated with human small-cell lung cancer: deletion 3p(14-23). Science 215: 181-182, 1982.

Yokota, J.; Wada, M.; Shimosato, Y.; Terada, M.; Sugimura, T.: Loss of heterozygosity on chromosomes 3, 13, and 17 in small-cell carcinoma and on chromosome 3 in adenocarcinoma of the lung. Proc. Nat. Acad. Sci. 84: 9252-9256, 1987.

GASTRIN-RELEASING POLYPEPTIDE (*137260)

JJM NOTE

This entry is selected because of its relevance to small-cell carcinoma of the lung (see previous entry) and lung carcinoid tumors.

Gastrin-releasing polypeptide is the mammalian equivalent of the amphibian tetradecapeptide bombesin. In nanogram amounts, both GRP and bombesin infused into humans increase plasma gastrin (137250), pancreatic polypeptide (167780), glucagon (138030), gastric inhibitory peptide, and insulin (176730). Bombesin is also a potent neuroregulator. Bombesin-like immunoreactivity has a high level in neuroendocrine cells of the lung. High levels are found in the human lung just after birth and levels decrease thereafter in parallel with the observed disease in number of pulmonary neuroendocrine cells. Bombesin immunoreactivity is high in pulmonary carcinoid tumors and in some small cell carcinomas of the lung (182280). Spindel et al. (1984) used RNA from a lung carcinoid tumor to prepare cDNA encoding GRP. They showed that the GRP gene codes for a precursor of 148 amino acids. GRP is a 27-amino acid neuropeptide. In human-mouse somatic cell hybrids, 4,3:Naylor et al. (1985, 1987) found that a GRP probe segregated with chromosome 18 and the marker peptidase A. Spindel et al. (1986) found 2 alternative mRNAs read from 1 gene by a frameshift. The sequence of the 2 product peptides differed by 27 amino acids. By comparison of the sequence of cDNA with that of a genomic prepro-GRP clone, Sausville et al. (1986) found 3 forms of mRNA arising from a single primary transcript which undergoes alternative processing from 2 splice donor sites to 2 splice acceptor sites. By in situ hybridization, the localization was narrowed to 18q21 (Lebacq-Verheyden et al., 1987). Using molecular probes in the analysis of an interspecific backcross between C57BL/6J and Mus spretus, Justice et al. (1992) mapped the corresponding gene to mouse chromosome 18.

Justice, M. J.; Gilbert, D. J.; Kinzler, K. W.; Vogelstein, B.; Buchberg, A. M.; Ceci, J. D.; Matsuda, Y.; Chapman, V. M.; Patriotis, C.; Makris, A.; Tsichlis, P. N.; Jenkins, N. A.; Copeland, N. G.: A molecular genetic linkage map of mouse chromosome 18 reveals extensive linkage conservation with human chromosomes 5 and 18. Genomics 13: 1281-1288, 1992.

Lebacq-Verheyden, A.-M.; Bertness, V.; Kirsch, I.; Hollis, G. F.; McBride, O. W.; Battey, J.: Human gastrin-releasing peptide gene maps to chromosome band 18q21. Somat. Cell Molec. Genet. 13: 81-86, 1987.

Naylor, S. L.; Sakaguchi, A. Y.; Spindel, E.; Chin, W. W.: Human gastrin-releasing peptide gene is located on chromosome 18. Somat. Cell Molec. Genet. 13: 87-91, 1987.

Naylor, S. L.; Spindel, E.; Chin, W. W.; Sakaguchi, A. Y.: Gastrin releasing peptide gene is located on human chromosome 18. (Abstract) Cytogenet. Cell Genet. 40: 711 only, 1985.

Sausville, E. A.; Lebacq-Verheyden, A.-M.; Spindel, E. R.; Cuttitta, F.; Gazdar, A. F.; Battey, J. F.: Expression of the gastrin-releasing peptide gene in human small cell lung cancer: evidence for alternative processing resulting in three distinct mRNAs. J. Biol. Chem. 261: 2451-2457, 1986.

Spindel, E. R.; Chin, W. W.; Price, J.; Rees, L. H.; Besser, G. M.; Habener, J. F.: Cloning and characterization of cDNAs encoding human gastrin-releasing peptide. Proc. Nat. Acad. Sci. 81: 5699-5703, 1984.

Spindel, E. R.; Zilberberg, M. D.; Habener, J. F.; Chin, W. W.: Two prohormones for gastrin-releasing peptide are encoded by two mRNAs differing by 19 nucleotides. Proc. Nat. Acad. Sci. 83: 19-23, 1986.

MESOTHELIOMA, MALIGNANT (156240)

JJM NOTE

A tumor that has such conspicuous determinants as asbestos-induced mesothelioma underscores the distinction between *familial* (aggregating in relatives, usually in the same household) and *genetic* (determined by DNA sequences). Most cases of familial mesothelioma arise from asbestos workers who carry fibers on their work clothes. Still, the question can be raised, Is there a genetic susceptibility to carcinogenesis by asbestos?

In connection with the etiology of mesothelioma, primary attention has appropriately been focused on environmental factors, particularly asbestos exposure. Risberg et al. (1980) described a family in Sweden in which the father, 3 brothers and a sister died of malignant mesothelioma. Four of the 5 probably had had asbestos exposure in the building industry. All were smokers. The area showed low incidence of malignant mesothelioma. There were 8 other sibs who were unaffected at the time of report (2 had died of other causes). In addition to smoking and asbestos, genetic factors may be involved in pathogenesis. Li et al. (1978) reported pleural mesothelioma in the wife and daughter of a man who worked for about 25 years as a pipe insulator at a shipyard and who also developed pulmonary asbestosis and lung cancer. The wife and daughter had no asbestos exposure other than that from the man's clothing. Although common household or occupational exposure may be responsible for familial aggregation, Lynch et al. (1985) raised the question of a host factor in the occurrence and/or the histologic characteristics of mesothelioma. They reported brothers who died of malignant pleural mesothelioma. Martensson et al. (1984) observed malignant mesothelioma in 2 pairs of sibs and also raised a question of a hereditary predisposing factor. Hammar et al. (1989) reported 3 brothers who worked in the asbestos insulation business and developed mesothelioma. In a second family, a father, who was occupationally exposed to asbestos, died from a tubulopapillary peritoneal mesothelioma 11 years before his son died from a peritoneal mesothelioma of identical histologic type. Although it is possible that the son was secondarily exposed to asbestos from the father's work clothes, quantitative asbestos analysis of the son's lung tissue showed numbers of asbestos bodies well within the lower limits seen in the general urban population with no occupational exposure to asbestos. The simulation of mendelian do minant inheritance was indicated by the occurrence of familial mesothelioma contracted as an infant by a woman who died of this disorder at age 32.

See also Anderson et al. (1976); Li et al. (1989)

Anderson, H. A.; Lilis, R.; Daum, S. M.; Fischbein, A. S.; Selikoff, I. J.: Household-contact asbestos neoplastic risk. Ann. N.Y. Acad. Sci. 271: 311-323, 1976.

Hammar, S. P.; Bockus, D.; Remington, F.; Freidman, S.; LaZerte, G.: Familial mesothelioma: a report of two families. Hum. Path. 20: 107-112, 1989.

Li, F. P.; Dreyfus, M. G.; Antman, K. H.: Asbestos-contaminated nappies and familial mesothelioma. (Letter) Lancet I: 909-910, 1989.

Li, F. P.; Lokich, J.; Lapey, J.; Neptune, W. B.; Wilkins, E. W., Jr.: Familial mesothelioma after intense asbestos exposure at home. J.A.M.A. 240: 467 only, 1978.

Lynch, H. T.; Katz, D.; Markvicka, S. E.: Familial mesothelioma: review and family study. Cancer Genet. Cytogenet. 15: 25-35, 1985.

Martensson, G.; Larsson, S.; Zettergre, L.: Malignant mesothelioma in two pairs of siblings: Is there a hereditary predisposing factor?. Europ. J. Resp. Dis. 65: 179-184, 1984.

Risberg, B.; Nickels, J.; Wagermark, J.: Familial clustering of malignant mesothelioma. Cancer 45: 2422-2427, 1980.

PULMONARY FIBROSIS, IDIOPATHIC (HAMMAN-RICH DISEASE) (178500)

JJM NOTE

This disorder may be present in an occasional patient with alveolar cell carcinoma.

Jacox et al. (1964) described a family in which idiopathic pulmonary fibrosis had been observed in 8 definite and 3 probable instances in a dominant pedigree pattern. Apparent male-to-male transmission had occurred in 1 instance. Increase of a gamma globulin fraction was thought to be a possible integral part of the syndrome. Hughes (1964) described the disorder in a mother and 2 daughters. Danies and Potts (1964) observed affected brothers in whom clubbing of the fingers was present for many years before the development of respiratory symptoms. It is by no means certain that the condition in these reports is an entity separate from that referred to elsewhere as fibrocystic pulmonary dysplasia (135000). Wagley (1972) described a family in which 3 brothers and a sister had well-documented pulmonary fibrosis and their mother, 2 of their sibs, and the son of 1 brother probably had pulmonary fibrosis. Gadek et al. (1979) found high concentrations of collagenase in the lower respiratory tract of patients with idiopathic pulmonary fibrosis. Beaumont et al. (1981) reported the occurrence of interstitial pulmonary fibrosis and alveolar cell carcinoma (ACC) in a family. At the time of report, 5 members had pulmonary fibrosis, of whom 3 had ACC. A sixth member had ACC without proven pulmonary fibrosis. Ten Kate (1981) wrote me that 1 of the 2 with pulmonary fibrosis without ACC had developed ACC. ACC with familial pulmonary fibrosis was reported by McKusick and Fisher (1958). In familial idiopathic pulmonary fibrosis, there is a characteristic pattern of alveolar inflammation that features neutrophil accumulation and macrophage activation. Bitterman et al. (1986) found that, despite being clinically normal in every respect, about half the first-degree relatives of patients with familial idiopathic fibrosis (they studied 3 families) had a similar pattern of alveolar inflammation. From studies of 5 patients with idiopathic pulmonary fibrosis, Antoniades et al. (1 990) suggested that overproduction of platelet-derived growth factor-beta (190040) may play an important role in pathogenesis.

After reviewing histologic material from 3 of the 4 original cases described by Hamman and Rich (1944) as well as 29 cases of their own, Olson et al. (1990) concluded that the disorder is an acute rather than a chronic interstitial pneumonia. The patients present with the adult respiratory distress syndrome (ARDS). Of their 29 patients, 12 survived, some after long and complicated hospitalization.

See also Bonanni et al. (1965); Javaheri et al. (1980); Swaye et al. (1969)

Antoniades, H. N.; Bravo, M. A.; Avila, R. E.; Galanopoulos, T.; Neville-Golden, J.; Maxwell, M.; Selman, M.: Platelet-derived growth factor in idiopathic pulmonary fibrosis. J. Clin. Invest. 86: 1055-1064, 1990.
Beaumont, F.; Jansen, H. M.; Elema, J. D.; Ten Kate, L. P.; Sluiter, H. J.: Simultaneous occurrence of pulmonary interstitial fibrosis and alveolar cell carcinoma in one family. Thorax 36: 252-258, 1981.
Bitterman, P. B.; Rennard, S. I.; Keogh, B. A.; Wewers, M. D.; Adelberg, S.; Crystal, R. G.: Familial idiopathic pulmonary fibrosis: evidence of lung inflammation in unaffected family members. New Eng. J. Med. 314: 1343-1347, 1986.
Bonanni, P. P.; Frymoyer, J. W.; Jacox, R. F.: A family study of idiopathic pulmonary fibrosis, a possible dysproteinemic and genetically determined disease. Am. J. Med. 39: 411-421, 1965.
Danies, G. M.; Potts, M. W.: Chronic diffuse interstitial pulmonary fibrosis in brothers. Guy's Hosp. Rep. 113: 36-44, 1964.
Gadek, J. E.; Kelman, J. A.; Fells, G.; Weinberger, S. E.; Horwitz, A. L.; Reynolds, H. V.; Fulmer, J. D.; Crystal, R. G.: Collagenase in the lower respiratory tract of patients with idiopathic pulmonary fibrosis. New Eng. J. Med. 301: 737-742, 1979.
Hamman, L.; Rich, A. R.: Acute diffuse interstitial fibrosis of the lungs. Bull. Johns Hopkins Hosp. 74: 177-212, 1944.
Hughes, E. W.: Familial interstitial pulmonary fibrosis. Thorax 19: 515-525, 1964.
Jacox, R. F.; Frymoyer, J. W.; Bonanni, P. P.: A family study of idiopathic pulmonary fibrosis: a possible dysproteinemic and genetically determined disease. Trans. Assoc. Am. Phys. 77: 232-238, 1964.
Javaheri, S.; Lederer, D. H.; Pella, J. A.; Mark, G. J.; Levine, B. W.: Idiopathic pulmonary fibrosis in monozygotic twins: the importance of genetic predisposition. Chest 78: 591-594, 1980.
McKusick, V. A.; Fisher, A. M.: Congenital cystic disease of the lung with progressive pulmonary fibrosis and carcinomatosis. Ann. Intern. Med. 48: 774-790, 1958.
Olson, J.; Colby, T. V.; Elliott, C. G.: Hamman-Rich syndrome revisited. Mayo Clin. Proc. 65: 1538-1548, 1990.
Swaye, P.; Van Ordstrand, H. S.; McCormick, L. J.; Wolpaw, S. E.: Familial Hamman-Rich syndrome: report of eight cases. Dis. Chest 55: 7-12, 1969.
Ten Kate, L. P.: Personal Communication. Groningen, The Netherlands 5/22/1981.
Wagley, P. F.: A new look at the Hamman-Rich syndrome. Johns Hopkins Med. J. 131: 412-424, 1972.

FIBROCYSTIC PULMONARY DYSPLASIA (*135000)

JJM NOTE

This disorder may be the precedent in an occasional patient with lung cancer.

The features of this disorder, which may be the same as idiopathic pulmonary fibrosis (178500), are progressive dyspnea and cyanosis, digital clubbing, pulmonary hypertension, polycythemia, and diffuse pulmonary fibrosis by x-ray. McKusick and Fisher (1958) described 3 cases. Donohue et al. (1959) reported a Canadian family with 8 cases in 4 generations, and Rezek and Talbert (1962) reported father and daughter. Koch (1965) observed a family with 3 definite and 5 probable cases. The definite cases included an instance of father-to-son transmission; one patient developed bronchial carcinoma. Swaye et al. (1969) described 8 cases in 3 generations. In 1 case, the diagnosis was made at age 3.5 years by lung biopsy. Two brothers had coexistent pulmonary fibrosis and bronchogenic cancer. Solliday et al. (1973) described father-to-son transmission. Idiopathic parenchymal respiratory disorders of childhood, such as desquamative interstitial pneumonitis (DIP), often lead to progressive pulmonary fibrosis (see 263000). The Hamman-Rich syndrome (178500) is the end stage of a heterogeneous group of interstitial lung diseases. Study of these cases is impeded by the fact, among others, that most patients who have undergone lung biopsy were in an advanced stage of the disease when studied.

See also Adelman et al. (1966); Young (1966)

Adelman, A. G.; Chertkow, G.; Hayton, R. C.: Familial fibrocystic pulmonary dysplasia: a detailed family study. Canad. Med. Assoc. J. 95: 603-610, 1966.
Donohue, W. L.; Laski, B.; Uchida, I. A.; Munn, J. D.: Familial fibrocystic pulmonary dysplasia and its relation to the Hamman-Rich syndrome. Pediatrics 24: 786-813, 1959.
Koch, B.: Familial fibrocystic pulmonary dysplasia: observations in one family. Canad. Med. Assoc. J. 92: 801-808, 1965.
McKusick, V. A.; Fisher, A. M.: Congenital cystic disease of the lung with progressive pulmonary fibrosis and carcinomatosis. Ann. Intern. Med. 48: 774-790, 1958.
Rezek, P. R.; Talbert, W. R., Jr.: Kongenitale (familiaere) zystische Fibrose der Lunge. Wien. Klin. Wschr. 74: 869-873, 1962.
Solliday, N. H.; Williams, J. A.; Gaensler, E. A.; Coutu, R. E.; Carrington, C. B.: Familial chronic interstitial pneumonia. Am. Rev. Resp. Dis. 108: 193-204, 1973.
Swaye, P.; Van Ordstrand, H. S.; McCormick, L. J.; Wolpaw, S. E.: Familial Hamman-Rich syndrome. Dis. Chest 55: 7-12, 1969.
Young, W. A.: Familial fibrocystic pulmonary dysplasia: a new case in a known affected family. Canad. Med. Assoc. J. 94: 1059-1061, 1966.

SARCOIDOSIS (BOECK SARCOID) (181000)

JJM NOTE

This disorder may account for an occasional patient with lung cancer, although chance association is always a possibility. In 2,544 patients with sarcoidosis seen in 10 years, 9 had lung tumors, compared with an expected number of 2.8 (Brincker H, Wilbek E: The incidence of malignant tumours in patients with respiratory sarcoidosis. Br. J. Cancer 29:247–51, 1974).

Familial aggregation was studied by Buck and McKusick (1961) and by Allison (1964), among others. The familial aggregation in this disease of unknown etiology may have a nongenetic basis. The much greater frequency in U.S. blacks than in U.S. whites suggests a genetic contribution to etiology. The family pattern does not conform to a simple mendelian mode of inheritance. In Allison's family affected persons were 2 brothers out of 4 sibs and 2 of the 4 children of one of these. Willoughby et al. (1971) described 3 sibs of whom 2 had sarcoidosis and 1 had Crohn disease. In a family of 9 sibs, Sharma et al. (1971) described sarcoidosis in 4. The mode of onset and clinical representation were acute in 3. The fourth was asymptomatic. Headings et al. (1976) favored multifactorial inheritance of susceptibility. Gronhagen-Riska et al. (1983) commented on the occurrence of sarcoidosis and regional enteritis in the same family. Nowack and Goebel (1987) found an unusually high frequency of HLA-DR5 (relative risk, 6.6). Their study was done in Marburg, West Germany, where, during the last 18 years, 440 patients with sarcoidosis 'regularly attended the University Hospital of Marburg or neighboring clinics.' Kronauer et al. (1988) described sarcoidosis in a father and son of 1 family and in 2 sisters of a second family.

See also Sharma et al. (1976)

Allison, J. R., Jr.: Sarcoidosis. I. Familial occurrence. II. Pseudotumor cerebri and unusual skin lesions. Sth. Med. J. 57: 27-32, 1964.
Buck, A. A.; McKusick, V. A.: Epidemiologic investigations of sarcoidosis. III. Serum proteins, syphilis, association with tuberculosis: familial aggregation. Am. J. Hyg. 74: 174-188, 1961.
Gronhagen-Riska, C.; Fyhrquist, F.; Hortling, L.; Koskimies, S.: Familial occurrence of sarcoidosis and Crohn's disease. (Letter) Lancet I: 1287-1288, 1983.
Headings, V. E.; Weston, D.; Young, R. C., Jr.; Hackney, R. L., Jr.: Familial sarcoidosis with multiple occurrences in eleven families: a possible mechanism of inheritance. Ann. N.Y. Acad. Sci. 278: 377-385, 1976.
Kronauer, C. M.; Medici, T. C.; Russi, E. W.: Familiare Sarkoidose: 4 Falle aus 2 Familien. Schweiz. Med. Wschr. 118: 1482-1486, 1988.
Nowack, C.; Goebel, K. M.: Genetic aspects of sarcoidosis: class II histocompatibility antigens and a family study. Arch. Intern. Med. 147: 481-483, 1987.
Sharma, O. P.; Johnson, C. S.; Balchum, O. J.: Familial sarcoidosis: report of four siblings with acute sarcoidosis. Am. Rev. Resp. Dis. 104: 255-257, 1971.
Sharma, O. P.; Neville, E.; Walker, A. N.; James, D. G.: Familial sarcoidosis: a possible genetic influence. Ann. N.Y. Acad. Sci. 278: 386-400, 1976.
Willoughby, J. M. T.; Mitchell, D. N.; Wilson, J. D.: Sarcoidosis and Crohn's disease in siblings. Am. Rev. Resp. Dis. 104: 249-254, 1971.

SCLERODERMA, FAMILIAL PROGRESSIVE (*181750)

JJM NOTE

This disorder may present in an occasional patient with alveolar cell carcinoma (Petitjean R et al.: Une nouvelle observation d'association sclerodermie et cancer alveolaire du poumon. J. Fr. Med. Chir. Thorac. 22:43–52, 1968; Driessen APPM et al.: Familiair voorkomende diffuse interstitiele longfibrose gecompliceerd door alveolaire-cellencarcinoom. Ned. Tijdschr. Geneeskd. 114:2041–45, 1970; Beaumont F et al.: Simultaneous occurrence of pulmonary interstitial fibrosis and alveolar cell carcinoma in one family. Thorax 36:252–258, 1981; Beaumont F: Interstitial fibrosis and alveolar cell carcinoma in a family. Thorax 36:560, 1981; McDonnell L et al.: Familial pulmonary fibrosis and lung cancer. Ir. J. Med. Sci. 151:315–17, 1982). Scleroderma may predispose to alveolar cell carcinoma via immune dysfunction, but it is also a curious entry on the list of chromosome breakage syndromes, epitomized by ataxia-telangiectasia, Fanconi pancytopenia, and Bloom syndrome. Excessive chromosomal breaks are seen in peripheral lymphocytes and bone marrow cells from scleroderma patients, but also in cells from persons without scleroderma on incubation with serum from scleroderma patients (Emerit I et al.: Sclérodermie généralisée et cassures chromosomiques: Mise en évidence d'un "facteur cassant" dans le sérum des malades. Ann. Genet. 16:135–18, 1973; Tolchin SF et al.: Chromosome abnormalities from cyclophosphamide therapy in rheumatoid arthritis and progressive systemic sclerosis (scleroderma). Arthritis Rheum. 17:375–82, 1974; Pan SF et al.: Chromosomal abnormalities in progressive systemic sclerosis (scleroderma) with consideration of radiation effects. J. Lab. Clin. Med. 86:300–308, 1975). See also Rittner et al. (1988).

Greger (1975) described 3 cases of progressive systemic scleroderma (systemic sclerosis) in 3 males, including father and son, in an inbred triracial isolate of southern Maryland, known as the Brandywine group. A sister of the aforementioned son had rheumatoid arthritis. Because of the inbreeding, dominant, as opposed to recessive, inheritance cannot be defended. Indeed, the pedigree gives no clear proof of mendelian nature of the process that seemed to be in every way typical of severe systemic sclerosis.

Winterbauer (1964) defined and named the CRST syndrome (calcinosis, Raynaud syndrome, sclerodactyly, telangiectasia), a variant of scleroderma, when he was a Johns Hopkins medical student. (The CRST syndrome simulates hereditary telangiectasia; see 187300.) The initialism CRST was expanded to the acronym CREST by Shulman's group (Frayha et al., 1973; Velayos et al., 1979), also at Johns Hopkins, who added esophageal involvement to the cardinal manifestations. (The same group (Stevens et al., 1964) had pointed out a correlation between Raynaud phenomenon and aperistalsis of the esophagus in connective tissue disorders. They proposed that the connection may indicate that the esophageal abnormality is due to autonomic dysfunction rather than sclerosis. Stevens (1984) suggested that if this is the case, E of CREST may stand for 'epiphenomenon.') The disorder had been called Thibierge-Weissenbach syndrome. Thibierge and Weissenbach (1911) described 'subcutaneous calcareous concretions and scleroderma' in a single patient and found 8 other similar cases in the literature. Although telangiectases were not noted in the summaries of the earlier reported cases or in their discussion and certainly were not conceived by them as an integral part of the syndrome, the authors commented that 'there exists also in our patient a third type of skin lesion, namely telangiectases, remarkably prominent on the face, neck and thorax.' Frayha et al. (1977) described the CRST syndrome in mother and daughter. McColl and Buchanan (1994) described the CREST syndrome in 2 sisters of 1 family and a grandmother and grandson in a second family.

Sheldon et al. (1981) described a kindred in which 3 sibs had scleroderma and 2 others had Raynaud phenomenon. The father died at age 43, having had, over the previous 20 years, 'flexion deformity of the fingers, cold intolerance, cyanosis, and pallor of the fingers with sores described as bone felons requiring lancing several times.' Sheldon et al. (1981) found reports of 19 instances of familial occurrence of scleroderma. Black et al. (1983) found in scleroderma due to vinyl chloride the same relationship to HLA types B8 and DR3 and the same anticentromere and other antibodies as in idiopathic scleroderma. Rittner et al. (1988) found increased chromosomal breakage rate (ICBR) in 27 of 28 patients with systemic sclerosis; 5 patients with CREST syndrome, 4 with incomplete CREST, 1 with overlapping syndrome, and 18 with progressive systemic sclerosis were studied. In addition to the patients, about half of their first-degree relatives showed increased chromosomal breakage rate, segregating as a dominant marker in 9 families. In the 6 informative of the 9 families, the ICBR trait showed close linkage with HLA (maximum lod = 5.5 at theta = 0). ICBR was predominantly observed in linkage disequilibrium with HLA haplotype A1, Cw7, B8, C4AQ0B1, DR3, which is frequently observed in autoimmune diseases. In a 67-year-old male patient with the CREST variant of systemic sclerosis and in his 34-year-old daughter, Schmid et al. (1989) found a supernumerary chromosome in over half of metaphases. The microchromosome consisted of constitutive heterochromatin and contained nuclear antigens reacting with specific antikinetichore antibodies. Furthermore, it showed a close association with the centromere of normal chromosomes. Kinetichore-specific autoantibodies are demonstrable in the sera of over 90% of CREST patients (Fritzler and Kinsella, 1980; Moroi et al., 1980; Tan et al., 1980). The daughter, who inherited the microchromosome from her father, was clinically healthy, and n one of 15 other CREST patients was found to have a supernumerary microchromosome.

See also Soppi et al. (1982)

Black, C. M.; Welsh, K. I.; Walker, A. E.; Bernstein, R. M.; Catoggio, L. J.; McGregor, A. R.; Lloyd Jones, J. K.: Genetic susceptibility to scleroderma-like syndrome induced by vinyl chloride. Lancet I: 53-56, 1983.

Frayha, R. A.; Scarola, J. A.; Shulman, L. E.: Calcinosis in scleroderma: a reevaluation of the CREST syndrome. (Abstract) Arthritis Rheum. 16: 542 only, 1973.

Frayha, R. A.; Tabbara, K. F.; Geha, R. S.: Familial CRST syndrome with sicca complex. J. Rheum. 4: 53-58, 1977.

Fritzler, M. J.; Kinsella, T. D.: The CREST syndrome: a distinct serologic entity with anticentromere antibodies. Am. J. Med. 69: 520-526, 1980.

Greger, R. E.: Familial progressive systemic scleroderma. Arch. Derm. 111: 81-85, 1975.

McColl, G. J.; Buchanan, R. R. C.: Familial CREST syndrome. J. Rheumat. 21: 754-756, 1994.

Moroi, Y.; Peebles, C.; Fritzler, M.; Steigerwald, J.; Tan, E. M.: Autoantibody to centromere (kinetochore) in scleroderma sera. Proc. Nat. Acad. Sci. 77: 1627-1631, 1980.

Rittner, G.; Schwanitz, G.; Baur, M. P.; Black, C. M.; Welsh, K. I.; Kuhnl, P.; Rittner, C.: Family studies in scleroderma (systemic sclerosis) demonstrating an HLA-linked increased chromosomal breakage rate in cultured lymphocytes. Hum. Genet. 81: 64-70, 1988.

Schmid, M.; Haaf, T.; Schindler, D.; Meurer, M.: Centromeric association of a microchromosome: a new category of non-random arrangement of metaphase chromosomes. Hum. Genet. 81: 127-136, 1989.

Sheldon, W. B.; Lurie, D. P.; Maricq, H. R.; Kahaleh, M. B.; DeLustro, F. A.; Gibofsky, A.; LeRoy, E. C.: Three siblings with scleroderma (systemic sclerosis) and two with Raynaud's phenomenon from a single kindred. Arthritis Rheum. 24: 668-676, 1981.

Soppi, E.; Lehtonen, A.; Toivanen, A.: Familial progressive systemic sclerosis (scleroderma): immunological analysis of two patients and six siblings from a single kindred. Clin. Exp. Immun. 50: 275-282, 1982.

Stevens, M. B.: Personal Communication. Baltimore, Md. 3/29/1984.

Stevens, M. B.; Hookman, P.; Siegal, C.; Esterly, J.; Shulman, L. E.; Hendrix, T. R.: Aperistalsis of the esophagus in connective tissue disorders and Raynaud's phenomenon. New Eng. J. Med. 270: 1218-1222, 1964.

Tan, E. M.; Rodnan, G. P.; Garcia, I.; Moroi, Y.; Fritzler, M.; Peebles, C.: Diversity of antinuclear antibodies in progressive systemic sclerosis: anti-centromere antibody and its relationship to CREST syndrome. Arthritis Rheum. 23: 617-625, 1980.

Thibierge, G.; Weissenbach, R.-J.: Concretions calcaires sous-cutanees et sclerodermie. Ann. Derm. Syph. 2: 129-155, 1911.

Velayos, E. E.; Masi, A. T.; Stevens, M. B.; Shulman, L. E.: The 'CREST' syndrome: comparison with systemic sclerosis (scleroderma). Arch. Intern. Med. 139: 1240-1244, 1979.

Winterbauer, R. H.: Multiple telangiectasia, Raynaud's phenomenon, sclerodactyly and subcutaneous calcinosis: a syndrome mimicking hereditary hemorrhagic telangiectasia. Bull. Johns Hopkins Hosp. 114: 361-383, 1964.

KARTAGENER SYNDROME (*244400)

JJM NOTE

A 52-year-old male with this disorder and lung cancer has been reported (Baruah BD, Chari MV: Kartagener's syndrome with bronchogenic carcinoma. J. Indian Med. Assoc. 21:438–39, 1952). Obviously, chance association is the best explanation.

Description

The Kartagener (pronounced KART-agayner) syndrome is an autosomal recessive disorder characterized by bronchiectasis, sinusitis, dextrocardia, and infertility.

Clinical Features

Moreno et al. (1965) found the full syndrome in 2 of 5 offspring of first-cousin parents. Another sib had bronchiectasis as did the father, and the other 2 children were 'chronic coughers.' Holmes et al. (1968) found low serum levels of gamma A globulin in some cases. It is likely that the Kartagener syndrome is a mendelian subgroup of situs inversus viscerum (270100), which for the most part has no simple mendelian basis (Torgersen, 1950). There is no association with left-handedness. Dextrocardia may be absent and the patient may present with bronchiectasis, sinusitis, and infertility (Guerrant et al., 1978). See immotile cilia syndrome (242650). Patients with the Kartagener syndrome may have anosmia (Goldstein, 1979).

Afzelius (1976) found that the sperm of males with Kartagener syndrome are immobile and that this is true also of the cilia in the respiratory tract. By electron microscopy he demonstrated that the sperm and the respiratory epithelium show an absence of dynein arms, structures that form temporary cross bridges between adjacent ciliary filaments and are believed to be responsible for generating movement in cilia and sperm tails. He postulated that normal visceral asymmetry is determined by movement of cilia in certain embryonic epithelial tissues. This is an example of a congenital malformation with functional abnormality later in life. Sterility of males and semisterility of females have been observed (Arge, 1960; Afzelius, 1976).

Afzelius (1980) pointed out that electron microscopy shows different kinds of ultrastructural defects of cilia and sperm tails in the immotile cilia syndrome. Since the cilium is constructed of some 100 different polypeptides, there is ample room for genetic heterogeneity. He speculated that embryonic cilia bring the heart to the left side in the mid-fourth week in the mouse. Afzelius (1979) gave an extensive review of cilia and their disorders. He pointed out that the ependyma of the brain is ciliated epithelium: '. . . .two-thirds of the investigated persons complained about rather severe, chronic headaches, and. . .many had sought medical advice for this complaint. Some regarded the headaches as the symptom that caused the greatest suffering.'

Eliasson et al. (1977) concluded that situs inversus may be absent. What they called the immobile cilia syndrome may not be a separate entity when situs inversus is missing. A lower proportion of affected sibs, at least with situs inversus, than would be expected on the basis of autosomal recessive inheritance may have the following explanation: The gene involved may control normal situs of the viscera. In the homozygote for Kartagener mutation, the control may be lacking so that situs is a random matter. Thus, some homozygotes would be expected to have the normal levocardia, for example. Moreno and Murphy (1981) reexamined the surmised autosomal recessive inheritance of situs inversus in Kartagener syndrome on the hypothesis that the homozygous state replaces lateralization of the viscera during ontogeny by an indifferent lateralization. The segregation ratio in the progeny of 2 carriers would then be not one-fourth but one-eighth. Within the limitations of published data, they concluded that there is agreement with the hypothesis.

Neustein et al. (1980) reported a case with absence of only the inner dynein arm in respiratory cilia rather than total absence of the dynein arms. The patient had repair of duodenal atresia at birth. Corneal abnormalities, headaches, and a poor sense of smell have been attributed to the primary defect in the immotile cilia syndrome. The inside of the cornea is monociliated; each cell carries a single cilium. Embryonic cilia are thought to cause the normal embryo to bend into a right-handed helical twist, shifting the heart to the left. The first studies showed that the axoneme lacks dynein arms completely or almost completely (Afzelius, 1981). Later studies showed that some patients have defective ciliary spokes; others have short spokes and no central sheath; still others have no inner dynein arms but prominent outer dynein arms. Studies in Sweden, France, and Canada indicate that lack of dynein arms is the most common cause of the immotile cilia syndrome. In rare cases, no morphologic abnormality of cilia is detectable, even though ciliary function is abnormal and the clinical syndrome is typical (Herzon and Murphy, 1980). In one clinically typical case, no cilia could be found in biopsies taken from 3 areas that normally are ciliated (Jahrsdoerfer et al., 1979). Thus, there must be much heterogeneity in the immotile cilia syndrome.

Jonsson et al. (1982) described a 21-year-old man with recurrent sinusitis, bronchitis and otitis media, situs inversus viscerum including left-sided appendix with appendicitis at age 12, and a normal 4-year-old son. Electron microscopy of nasal and bronchial mucosa showed abnormal orientation of the basal processes of the cilia and absent dynein arms but completely normal sperm.

Samuel (1987) presented the case of a 26-year-old man with typical Kartagener syndrome except for the presence of normal spermatozoa. The patient had chronic sinopulmonary symptoms, situs inversus, and absent frontal sinuses, while his spermatozoa had normal motility, ultrastructure, and fertilizing capacity. Repeated brush biopsies of the bronchial epithelium showed only keratinized squamous epithelium and no ciliated cells. The patient had a twin brother, probably monozygotic, who died in early infancy following a respiratory infection. Schidlow and Katz (1983) pointed out that structurally normal respiratory cilia may be found in otherwise typical Kartagener syndrome. Schidlow et al. (1982) reported a family in which 1 sib had Kartagener syndrome and another had the polysplenia syndrome. Only a small percentage (less than 20%) of the respiratory cilia of these 2 children were abnormal. Two female third cousins had Kartagener syndrome.

Rott (1983) listed 7 types of primary ciliary dyskinesia based on different axonemal defects found in 6 of them and 'no visible change' in a seventh. The most frequent form (type I), accounting for 74% of cases, is characterized by lack of both dynein arms. Affected sibs, supporting recessive inheritance, have been observed in types I, IV (lack of spoke structures), and V (lack of central structures). Rott (1983) suggested that among the others there may be instances of new dominant mutations.

In vitro studies have shown that various patterns of abnormal ciliary beating (Rossman et al., 1980; Rutland and Cole, 1980) are the most frequently observed abnormalities in the Kartagener syndrome and the immotile cilia syndrome. Hence, the term primary ciliary dyskinesia (PCD) has merit for this class of disorder. Valerius et al. (1983) found depressed motility of polymorphonuclear leukocytes in 10 patients with PCD of whom 6 had typical Kartagener syndrome. Bactericidal capacity was normal. They suggested that a generalized defect of microtubules extends to the polymorphs. 22,23:Gagnon et al. (1980, 1982) found reduced protein-carboxyl methylation in infertile males with immotile sperm. They studied protein-carboxyl methylase (EC 2.1.1.24) because it is an enzyme involved in the regulation of cellular locomotion in both bacteria and leukocytes. Low enzyme activity in 9 infertile patients with nonmotile spermatozoa was, they thought, not due to a primary genetic defect, since the enzyme activity was normal in red cells of these patients and spontaneous recovery of motility was associated with return of enzyme activity. Of the 9 patients, 2 had bronchiectasis, 1 with sinusitis and dextrocardia and the electron microscopic changes of the Kartagener syndrome in sperm tails.

Eavey et al. (1986) found significantly fewer ciliary outer dynein arms in all 4 patients with full-blown Kartagener syndrome and in 2 of 5 patients with sinusitis and bronchiectasis but no dextrocardia. No changes were found in carriers or in any other persons studied. The count of outer dynein arms was consistent in a given individual. Sturgess et al. (1986) analyzed 46 cases from 38 families: 20 males and 26 females. Situs inversus was present in 26. The ultrastructural change in respiratory tract cilia was deficiency in outer dynein arms (in 19), inner dynein arms (in 3), both inner and outer dynein arms (in 15), and radial spokes (in 5) and involved a microtubular transposition anomaly in 4. Segregation analysis was consistent with autosomal recessive inheritance. The finding of various structural defects suggests that there are several genetic determinants. Examination of paternal age and birth order gave no evidence of new dominant mutation.

Women with Kartagener syndrome are fertile, which speaks against the notion that ciliary currents waft eggs down the fallopian tube. The spermatozoa from Kartagener patients are perfectly normal; if micromanipulated so that they come to lie adjacent to the plasma membrane of the oocyte, they will fuse with the membrane and fertilize the egg (Aitken et al., 1983). The possibility of using in vitro fertilization in the management of such patients is a possibility.

Inheritance

Gorham and Merselis (1959) concluded that the disorder is inherited as a recessive with incomplete penetrance. Family studies were done by Knox et al. (1960) and by Cook et al. (1962). Knox et al. (1960) suggested linkage with the Rhesus locus. Probably all familial cases have been confined to sibs, although Torgersen (1947) suggested dominant inheritance.

Narayan et al. (1994) described a family consisting of a mother and her 5 male children, the offspring of 3 different fathers, all of whom had primary ciliary dyskinesia. None of the mother's marriages was consanguineous. The findings suggested either X-linked or autosomal dominant inheritance. Cytogenetic and fluorescence in situ hybridization analyses on the mother and 1 son showed no abnormality, specifically, on human chromosomes 12 and 14, which show homology of synteny with chromosomes carrying the hpy and iv mutations in the mouse. One of the 6 affected individuals had dextrocardia; all had sinopulmonary symptoms with abnormalities of the cilia by electron microscopy. Narayan et al. (1994) reviewed reports of other families with inheritance patterns other than autosomal recessive.

Pathogenesis

The human cytoplasmic dynein heavy chain (DNECL; 600112) is encoded by a gene which Narayan et al. (1994) mapped to 14qter by fluorescence in situ hybridization. The homologous gene is located on chromosome 14 in the mouse. For reasons of homology of synteny, the mutation responsible for Kartagener syndrome has been suggested to be on chromosome 14. (See 270100 for a discussion of the possibility that situs inversus viscerum is determined by mutation at 14q32.)

Clinical Management

Miralles et al. (1992) presented a method of overcoming some of the technical difficulties of heart-lung transplantation in situs inversus. Macchiarini et al. (1994) reported en bloc double lung transplantation with bilateral bronchial anastomoses in 3 patients with complete situs inversus and end-stage Kartagener syndrome. One patient died of obliterative bronchiolitis 36 months after the operation. The remaining 2 were alive and doing well after 48 months. The patients were a 41-year-old woman, a 38-year-old man, and a 49-year-old man.

Animal Model

Bryan et al. (1977) found that mice homozygous for the hydrocephaly-polydactyly (hpy) mutation have a similar dynein defect in cilia and flagella. Bryan (1977) wrote that 'the A-tubules of the outer doublets appear to lack dynein arms.' The hpy mutation is on mouse chromosome 6 (which carries triosephosphate isomerase and glyceraldehyde-3-phosphate dehydrogenase, which are syntenic, on 12p, in man). Thus, 12p is a candidate map site for the Kartagener gene in man.

Fifty percent of mice homozygous for the mutation situs inversus, iv, show situs inversus (Layton, 1976; Brueckner et al., 1989). Handel and Kennedy (1984) found no abnormality in the ultrastructure and motility of tracheal cilia or sperm tails in these mice. Layton (1986) speculated that the centriole may be the 'compass of the cell' and that the defect in these mice may be in the ability to read the compass. Afzelius (1979) referred to Kartagener syndrome in the dog.

Merlino et al. (1991) created a transgenic mouse line exhibiting an immotile sperm syndrome that duplicated the features of Kartagener syndrome. The mice were generated by inserting the cDNA for a human epidermal growth factor receptor (EGFR; 131550) into the mouse genome, driven by the chicken beta-actin promoter. One of the resulting transgenic lines expressed extremely high levels of EGFR RNA in the testes but not in any other tissue examined. In situ hybridization studies indicated that the transgene was transcribed during meiosis, as early as the primary spermatocyte stage, whereas the appearance of human EGF receptors was delayed until the post-meiotic stages of sperm differentiation. About half the males homozygous for the transgene were infertile. Moreover, their infertility was associated with a lack of sperm motility due to disorganization of the axoneme.

Genetic heterogeneity has been well documented in Chlamydomonas, a unicellular green alga with 2 flagellae that have the same axonemal structure as human bronchial cilia and sperm tails. Immotile strains of Chlamydomonas appear to have 'the same disease' as patients with primary ciliary dyskinesia. There are 3 immotile strains with different mutations as indicated by breeding experiments. All show identical defects by electron microscopy, namely defects in the central microtubular structures.

Pan et al. (1998) described a child with complete situs inversus who developed respiratory distress at 10 days of age. At 6 months of age, the child was growth retarded with signs of significant respiratory disease. A tracheal aspirate grew Pseudomonas aerugenosa and Staphylococcus aureus, and a sweat chloride test confirmed the clinical diagnosis of cystic fibrosis. Bronchial cilia obtained by respiratory mucosal brushings during bronchoscopy revealed no normal ciliary motion despite the lack of an inflammatory infiltrate. Electron microscopy, however, demonstrated structurally normal cilia. A peripheral lymphocyte karyotype was normal. DNA testing showed that the patient was homozygous for the delta-F508 mutation, despite the fact that his mother lacked the mutation. A sample from the father was not available for study. Investigation of a total of 21 loci distributed throughout chromosome 7 were used to demonstrate paternal isodisomy for the chromosome. The patient was homozygous at all 21 loci, with 10 of the markers being fully informative for paternal inheritance (i.e., the mother did not carry the marker). Maternity was established by testing 10 highly polymorphic markers for autosomal loci other than those on chromosome 7. To address the possibility of a second independent etiology for situs abnormalities, the patient's DNA was examined for deletions within the critical region of the X-linked heterotaxy locus on Xq26.2 (HTX1; 306955). Furthermore, the coding region of the HTX1 gene, also symbolized ZIC3 for 'ZIC family member 3,' was sequenced. No abnormality was found. Males with X-linked heterotaxy typically had complex heart anomalies in addition to visceral situs ambiguous.

Pan et al. (1998) suggested that the most likely explanation for the uniparental disomy in that case was (1) monosomy duplication associated with a nullisomic maternal gamete, or (2) a paternal meiosis II nondisjunction resulting in a trisomic conceptus, with subsequent reduction to disomy through loss of the maternal chromosome 7, so-called trisomic rescue. It is possible that the child inherited 2 recessive disorders from his father: CF and immotile cilia associated with complete situs inversus. It is also possible that the 2 disorders are unrelated in this patient; for example, he may coincidentally have autosomal dominant or recessive heterotaxy (the father was not available for detailed clinical evaluation). However, laterality defects have been associated with complex chromosome rearrangements involving chromosome 7. Pan et al. (1998) suggested that chromosome 7 should be examined in linkage studies of a large collection of families with Kartagener syndrome.

History

Kartagener, an internist in Zurich, and Horlacher described a familial form of bronchiectasis with dextrocardia and nasal polyps (Kartagener and Horlacher, 1936). Kartagener and Stucki (1962) found 334 cases in the literature and added 2 more. Kartagener syndrome is sometimes known as the Siewert syndrome (Siewert, 1904).

(Arvid Afzelius, the grandfather of B. Afzelius, who first observed the EM changes in Kartagener syndrome, probably was the first to report the disorder we now know as Lyme disease (Garfield, 1989; Afzelius, 1989). In the early 1900s, the grandfather reported a characteristic skin rash, which he termed erythema chronicum migrans, following bites of the tick Ixodes ricinus. Lyme disease in the United States and the less severe erythema migrans in Europe are caused by infection of related spirochetes injected by ticks. Berglund et al. (1995) reported that Lyme disease is very common in southern Sweden, with a relatively high frequency of neurologic complications and arthritis, but a lower incidence of carditis than is found in the United States.)

See also Aarons and Powell (1969); Adams et al. (1981); Afzelius (1976); Afzelius and Eliasson (1979); Aitken (1991); Antonelli et al. (1981); Fox (1980); Guggenheim (1971); Hartline and Zelkowitz (1971); Logan et al. (1965); Resler and Walker (1969); Rossman et al. (1980); Rott (1979); Stowater (1976); Torgersen (1946); Torgersen (1949); Whitelaw et al. (1981); Woodring et al. (1982)

Aarons, G. H.; Powell, I. J.: The familial occurrence of the Kartagener triad in Jamaica. Postgrad. Med. J. 45: 736-744, 1969.

Adams, G. M. W.; Huang, B.; Pipirno, G.; Luck, D. J. L.: Central-pair microtubular complex of Chlamydomonas flagella: polypeptide composition as revealed by analysis of mutants. J. Cell Biol. 91: 69-76, 1981.

Afzelius, B.: Afzelius's disease—or is it Lyme?. Curr. Contents 32(49): 9, 1989.

Afzelius, B. A.: The immotile-cilia syndrome and other ciliary diseases. Int. Rev. Exp. Path. 19: 1-43, 1979.

Afzelius, B. A.: 'Immotile-cilia' syndrome and ciliary abnormalities induced by infection and injury. (Editorial) Am. Rev. Resp. Dis. 124: 107-109, 1981.

Afzelius, B. A.: A human syndrome caused by immobile cilia. Science 193: 317-319, 1976.

Afzelius, B. A.: Personal Communication. Stockholm, Sweden 1976.

Afzelius, B. A.: Genetic aspects of the immotile-cilia syndrome. (Abstract) Clin. Genet. 17: 52, 1980.

Afzelius, B. A.; Eliasson, R.: Flagellar mutants in man: on the heterogeneity of the immotile-cilia syndrome. J. Ultrastruct. Res. 69: 43-52, 1979.

Aitken, J.: A clue to Kartagener's. (Abstract) Nature 353: 306, 1991.

Aitken, R. J.; Ross, A.; Lees, M. M.: Analysis of sperm function in Kartagener's syndrome. Fertil. Steril. 40: 696-698, 1983.

Antonelli, M.; Modesti, A.; De Angelis, M.; Marcolini, P.; Lucarelli, N.; Crifo, S.: Immotile cilia syndrome: radial spokes deficiency in a patient with Kartagener's triad. Acta Paediat. Scand. 70: 571-573, 1981.

Arge, E.: Transposition of the viscera and sterility in men. Lancet I: 412-414, 1960.

Berglund, J.; Eitrem, R.; Ornstein, K.; Lindberg, A.; Ringner, A.; Elmrud, H.; Carlsson, M.; Runehagen, A.; Svanborg, C.; Norrby, R.: An epidemiologic study of Lyme disease in southern Sweden. New Eng. J. Med. 333: 1319-1324, 1995.

Brueckner, M.; D'Eustachio, P.; Horwich, A. L.: Linkage mapping of a mouse gene, iv, that controls left-right asymmetry of the heart and viscera. Proc. Nat. Acad. Sci. 86: 5035-5038, 1989.

Bryan, J.: Spermatogenesis revisited. IV. Abnormal spermiogenesis in mice homozygous for another male-sterility-inducing mutation, hpy (hydrocephalic-polydactyly). Cell Tissue Res. 180: 187-201, 1977.

Bryan, J. H. D.; Bates, T. J.; Chandler, D. B.: Cilia and flagella dysgenesis in hydrocephalic mutant mice. (Abstract) Vth Int. Conf. on Birth Defects, Montreal , 1977.

Cook, C. D.; Geller, F.; Hutchison, G. B.; Gerald, P. S.; Allen, F. H., Jr.: Blood grouping in three families with Kartagener's syndrome. Am. J. Hum. Genet. 14: 290-294, 1962.

Eavey, R. D.; Nadol, J. B., Jr.; Holmes, L. B.; Laird, N. M.; Lapey, A.; Joseph, M. P.; Strome, M.: Kartagener's syndrome: a blinded, controlled study of cilia ultrastructure. Arch. Otolaryng. Head Neck Surg. 112: 646-650, 1986.

Eliasson, R.; Mossberg, B.; Camner, P.; Afzelius, B. A.: The immobile-cilia syndrome: a congenital ciliary abnormality as an etiologic factor in chronic airway infections and male sterility. New Eng. J. Med. 297: 1-6, 1977.

Fox, B.: Ciliary dyskinesia. (Letter) Lancet II: 859, 1980.

Gagnon, C.; Sherins, R. J.; Mann, T.; Bouchard, P.; Phillips, D. M.; Bardin, C. W.: Defective protein carboxyl methylation in patients with immotile spermatozoa. (Abstract) Clin. Res. 28: 479A, 1980.

Gagnon, C.; Sherins, R. J.; Phillips, D. M.; Bardin, C. W.: Deficiency of protein-carboxyl methylase in immotile spermatozoa of infertile men. New Eng. J. Med. 306: 821-825, 1982.

Garfield, E.: Lyme disease research uncovers a case of delayed recognition: Arvid Afzelius and his successors. Curr. Contents 32(49): 3-8, 1989.

Goldstein, J. L.: Personal Communication. Dallas, Texas 1/25/1979.

Gorham, G. W.; Merselis, J. G., Jr.: Kartagener's triad: a family study. Bull. Johns Hopkins Hosp. 104: 11-16, 1959.

Guerrant, J. L.; Douty, T.; Tegtmeyer, C.; Jahrsdoerfer, R. A.: Bronchiectasis in the immotile-cilia syndrome. (Letter) New Eng. J. Med. 298: 282, 1978.

Guggenheim, F.: Kartagener's syndrome in an Arab family. Israel J. Med. Sci. 7: 1079-1081, 1971.

Handel, M. A.; Kennedy, J. R.: Situs inversus in homozygous mice without immotile cilia. (Abstract) J. Hered. 75: 498, 1984.

Hartline, J. V.; Zelkowitz, P. S.: Kartagener's syndrome in childhood. Am. J. Dis. Child. 121: 349-352, 1971.

Herzon, F. S.; Murphy, S.: Normal ciliary ultrastructure in children with Kartagener's syndrome. Ann. Otol. Rhinol. Laryng. 89: 81-83, 1980.

Holmes, L. B.; Blennerhassett, J. B.; Austen, K. F.: A reappraisal of Kartagener's syndrome. Am. J. Med. Sci. 255: 13-28, 1968.

Jahrsdoerfer, R.; Feldman, P. S.; Rubel, E. W.; Guerrant, J. L.; Eggleston, P. A.; Selden, R. F.: Otitis media and the immotile cilia syndrome. Laryngoscope 89: 769-778, 1979.
Jonsson, M. S.; McCormick, J. R.; Gillies, C. G.; Gondos, B.: Kartagener's syndrome with motile spermatozoa. New Eng. J. Med. 307: 1131-1133, 1982.
Kartagener, M.; Horlacher, A.: Situs viscerum inversus und Polyposis nasi in einem Falle familiaerer Bronchiektasien. Beitr. Klin. Tuberk. 87: 331-333, 1936.
Kartagener, M.; Stucki, P.: Bronchiectasis with situs inversus. Arch. Pediat. 79: 193-207, 1962.
Knox, G.; Murray, S.; Strang, L. B.: A family with Kartagener's syndrome: linkage data. Ann. Hum. Genet. 24: 137-140, 1960.
Layton, W. M., Jr.: Random determination of a developmental process: reversal of normal visceral asymmetry in the mouse. J. Hered. 67: 336-338, 1976.
Layton, W. M., Jr.: Personal Communication. Hanover, N. H. 2/10/1986.
Logan, W. D., Jr.; Abbott, O. A.; Hatcher, C. R., Jr.: Kartagener's triad. Dis. Chest 48: 613-616, 1965.
Macchiarini, P.; Chapelier, A.; Vouhe, P.; Cerrina, J.; Ladurie, F. L. R.; Parquin, F.; Brenot, F.; Simonneau, G.; Dartevelle, P.: Double lung transplantation in situ inversus with Kartagener's syndrome. J. Thorac. Cardiovasc. Surg. 108: 86-91, 1994.
Merlino, G. T.; Stahle, C.; Jhappan, C.; Linton, R.; Mahon, K. A.; Willingham, M. C.: Inactivation of a sperm motility gene by insertion of an epidermal growth factor receptor transgene whose product is overexpressed and compartmentalized during spermatogenesis. Genes Dev. 5: 1395-1406, 1991.
Miralles, A.; Muneretto, C.; Gandjbakhch, I.; Lecompte, Y.; Pavie, A.; Rabago, G.; Bracamonte, L.; Desruennes, M.; Cabrol, A.; Cabrol, C.: Heart-lung transplantation in situs inversus: a case report in a patient with Kartagener's syndrome. J. Thorac. Cardiovasc. Surg. 103: 307-313, 1992.
Moreno, A.; Murphy, E. A.: Inheritance of Kartagener syndrome. Am. J. Med. Genet. 8: 305-313, 1981.
Moreno, J.; Ortega, L.; Montero, E.: Sindrome de Kartagener: referencia de dos casos familiares con analysis citogenetico. An. Desarrollo 13: 207-213, 1965.
Narayan, D.; Desai, T.; Banks, A.; Patanjali, S. R.; Ravikumar, T. S.; Ward, D. C.: Localization of the human cytoplasmic dynein heavy chain (DNECL) to 14qter by fluorescence in situ hybridization. Genomics 22: 660-661, 1994.
Narayan, D.; Krishnan, S. N.; Upender, M.; Ravikumar, T. S.; Mahoney, M. J.; Dolan, T. F., Jr.; Teebi, A. S.; Haddad, G. G.: Unusual inheritance of primary ciliary dyskinesia (Kartagener's syndrome). J. Med. Genet. 31: 493-496, 1994.
Neustein, H. B.; Nickerson, B.; O'Neal, M.: Kartagener's syndrome with absence of inner dynein arms of respiratory cilia. Am. Rev. Resp. Dis. 122: 979-981, 1980.
Pan, Y.; McCaskill, C. D.; Thompson, K. H.; Hicks, J.; Casey, B.; Shaffer, L. G.; Craigen, W. J.: Paternal isodisomy of chromosome 7 associated with complete situs inversus and immotile cilia. (Letter) Am. J. Hum. Genet. 62: 1551-1555, 1998.
Resler, D. R.; Walker, E. A., Jr.: Kartagener's syndrome in sisters. Okla. State Med. Assoc. J. 62: 559-563, 1969.
Rossman, C. M.; Forrest, J. B.; Lee, R. M. K. W.; Newhouse, M. T.: The dyskinetic cilia syndrome: ciliary motility in immotile cilia syndrome. Chest 78: 580-582, 1980.
Rossman, C. M.; Forrest, J. B.; Ruffin, R. E.; Newhouse, M. T.: Immotile cilia syndrome in persons with and without Kartagener's syndrome. Am. Rev. Resp. Dis. 121: 1011-1016, 1980.
Rott, H.-D.: Kartagener's syndrome and the syndrome of immotile cilia. Hum. Genet. 46: 249-261, 1979.
Rott, H.-D.: Genetics of Kartagener's syndrome. Europ. J. Respir. Dis. 64 (suppl. 127): 1-4, 1983.
Rutland, J.; Cole, P.: Ciliary dyskinesia. (Letter) Lancet II: 859, 1980.
Samuel, I.: Kartagener's syndrome with normal spermatozoa. (Letter) J.A.M.A. 258: 1329-1330, 1987.
Schidlow, D. V.; Katz, S. M.: Immotile cilia syndrome. (Letter) New Eng. J. Med. 308: 595, 1983.
Schidlow, D. V.; Katz, S. M.; Turtz, M. G.; Donner, R. M.; Capasso, S.: Polysplenia and Kartagener syndromes in a sibship: association with abnormal respiratory cilia. J. Pediat. 100: 401-403, 1982.
Siewert, A. K.: Ueber einen Fall von Bronchiectasie bei einem Patienten mit Situs inversus viscerum. Berl. Klin. Wochr. 41: 139-141, 1904.
Stowater, J. L.: Kartagener's syndrome in a dog. Am. J. Vet. Radiol. Soc. 17: 174-177, 1976.
Sturgess, J. M.; Thompson, M. W.; Czegledy-Nagy, E.; Turner, J. A. P.: Genetic aspects of immotile cilia syndrome. Am. J. Med. Genet. 25: 149-160, 1986.
Torgersen, J.: Familial transposition of viscera. Acta Med. Scand. 126: 319-322, 1946.
Torgersen, J.: Transposition of viscera, bronchiectasis, and nasal polyps. A genetical analysis and contributions to the problem of constitution. Acta Radiol. 28: 17-24, 1947.
Torgersen, J.: Genetic factors in visceral asymmetry and in the development and pathologic changes of the lungs, heart and abdominal organs. Arch. Path. 47: 566-593, 1949.
Torgersen, J.: Situs inversus, asymmetry, and twinning. Am. J. Hum. Genet. 2: 361-370, 1950.
Valerius, N. H.; Knudsen, B. B.; Pedersen, M.: Defective neutrophil motility in patients with primary ciliary dyskinesia. Europ. J. Clin. Invest. 13: 489-494, 1983.
Whitelaw, A.; Evans, A.; Corrin, B.: Immotile cilia syndrome: a new cause of neonatal respiratory distress. Arch. Dis. Child. 56: 432-435, 1981.
Woodring, J. H.; Royer, J. M.; McDonagh, D.: Kartagener's syndrome. J.A.M.A. 247: 2814-2816, 1982.

ARYL HYDROCARDON HYDROXYLASE INDUCIBILITY (#108340)

JJM NOTE

This trait is of historic interest as the first ecogenetic trait that seemed frequent in the population and a major determinant of an epidemic cancer. Since the end point is a cumbersome bioassay, it is understandable that the focus of current research is the molecular genetic basis of the trait, chiefly *CYP1A1* and *CYP2D* (P450 genes).

A number sign (#) is used with this entry because the phenotype is due to characteristics of the P450 genes CYP1A1 (108330) and CYP1A2 (124060). See Price Evans (1993).

AHH is one of the mixed function oxidases in the microsomal fraction. Busbee et al. (1972) found three distinct groups—low, intermediate, and high—in regard to inducibility of AHH measured in cultured lymphocytes 24 hours after introduction of 3-methylcholanthrene. Family studies indicated diallelic determination at a single locus. Using the same inducer, Kellermann et al. (1973) found polymorphic inducibility of lymphocyte AHH. Since AHH is an enzyme involved in metabolism of carcinogens, the genetic difference might be relevant to the occurrence of cancer. In a normal white U.S. population, Kellermann et al. (1973) found low, intermediate and high inducibility in the following proportions: 44.7%, 45.9%, 9.4%, respectively. Among 50 patients with bronchogenic cancer, they found the following proportions: 4.0%, 66.0%, and 30.0%, respectively. Genetically determined high inducibility of AHH may be associated with enhanced risk of cancer in cigarette smokers (Kouri et al., 1982). In the mouse it was shown by Shichi et al. (1978) that homozygotes and heterozygotes for the Ah(b) allele (which renders the mouse susceptible to AHH induction by 3-methylcholanthrene) developed an irreversible opacity of the anterior portion of the lens, resembling a senile cataract, within 6 hours after a large intraperitoneal dose of acetaminophen. Whether the same occurs in man is not known. Fletcher et al. (1978) emphasized the poor reproducibility of AHH inducibility in lymphocytes. From studies of AHH in twins, Paigen et al. (1978) concluded that AHH inducibility may be determined by a single or a few polymorphic genes. From a twin study, Borresen et al. (1981) concluded that inducibility (but not basal level) is heritable (heritability = 0.7). Major control of inducibility by one locus was considered possible.

As discussed by Price Evans (1993), the aryl hydrocarbon receptor, which is encoded by a gene (AHR; 600253) located on 7p15, combines with the aryl hydrocarbon which is inducer. The resulting inducer-receptor complex is transported to the nucleus where it binds to the regulatory region of the CYP1A1 gene (108330), located on 15q. Induction of synthesis of enzyme in the cytoplasm leads to oxidation of the hydrocarbon.

See also Bickers and Kappas (1978); Emery et al. (1978); Kellermann et al. (1973); Paigen et al. (1977); Trell et al. (1976)

Bickers, D. R.; Kappas, A.: Human skin aryl hydrocarbon hydroxylase: induction by coal tar. J. Clin. Invest. 62: 1061-1068, 1978.
Borresen, A.-L.; Berg, K.; Magnus, P.: A twin study of aryl hydrocarbon hydroxylase (AHH) inducibility in cultured lymphocytes. Clin. Genet. 19: 281-289, 1981.
Busbee, D. L.; Shaw, C. R.; Cautrell, E. T.: Aryl hydrocarbon hydroxylase induction in human leucocytes. Science 178: 315-316, 1972.
Emery, A. E. H.; Danford, N.; Anand, R.; Duncan, W.; Paton, L.: Aryl-hydrocarbon-hydroxylase inducibility in patients with cancer. Lancet I: 470-472, 1978.
Fletcher, K. A.; Evans, D. A. P.; Canning, M. V.: Inducibility of aryl hydrocarbon hydroxylase in cultured human lymphocytes: a study of repeatability. J. Med. Genet. 15: 182-188, 1978.
Kellermann, G.; Luyter-Kellermann, M.; Shaw, C. R.: Genetic variation of aryl hydrocarbon hydroxylase in human lymphocytes. Am. J. Hum. Genet. 25: 327-331, 1973.
Kellermann, G.; Shaw, C. R.; Luyter-Kellermann, M.: Aryl hydrocarbon hydroxylase inducibility and bronchogenic carcinoma. New Eng. J. Med. 289: 934-937, 1973.
Kouri, R. E.; McKinney, C. E.; Slomiany, D. J.; Snodgrass, D. R.; Wray, N. P.; McLemore, T. L.: Positive correlation between high aryl hydrocarbon hydroxylase activity and primary lung cancer as analyzed in cryopreserved lymphocytes. Cancer Res. 42: 5030-5037, 1982.
Paigen, B.; Gurtoo, H. L.; Minowada, J.; Houten, L.; Vincent, R. A., Jr.; Paigen, K.; Parker, N. B.; Ward, E.; Hayner, N. T.: Questionable relation of aryl hydrocarbon hydroxylase to lung-cancer risk. New Eng. J. Med. 297: 346-350, 1977.
Paigen, B.; Ward, E.; Steenland, K.; Houten, L.; Gurtoo, H. L.; Minowada, J.: Aryl-hydrocarbon hydroxylase in cultured lymphocytes of twins. Am. J. Hum. Genet. 30: 561-571, 1978.
Price Evans, D. A.: Genetic Factors in Drug Therapy: Clinical and Molecular Pharmacogenetics. Cambridge: Cambridge Univ. Press , 1993. Pp. 19-34.

Shichi, H.; Gaasterland, D. E.; Jensen, N. M.; Nebert, D. W.: Ah locus: genetic differences in susceptibility to cataracts induced by acetaminophen. Science 200: 539-541, 1978.

Trell, E.; Korsgaard, R.; Hood, B.; Kitzing, P.; Norden, G.; Simonsson, B. G.: Aryl hydrocarbon hydroxylase inducibility and laryngeal carcinomas. (Letter) Lancet II: 140 only, 1976.

CYTOCHROME P450, SUBFAMILY I, POLYPEPTIDE 1; CYP1A1 (*108330)

JJM NOTE

Polymorphisms of this xenobiotic metabolizing protein have been associated with an increased risk of lung cancer, but complex interactions with the environment (degree of smoking), race, and other genes, such as *TP53,* are emerging.

From study of mouse-human hybrid cells, Brown et al. (1976) concluded that a structural gene for AHH is on chromosome 2 and that possibly a regulatory gene is there also. Ocraft et al. (1985) localized the gene to 2q31-2pter. According to McBride (1985), the gene mapped to chromosome 2 by expression assays is almost certainly not the structural locus; the structural locus is that assigned to chromosome 15: dioxin-inducible P1-450. (Possibly the locus on chromosome 2 is concerned with AHH inducibility; see 108340.) Nebert (1988) recommended that the AHH locus held to be on chromosome 2 be removed from that listing. The assignment was based on measurements of AHH inducibility in tissue culture, and effects of dibutyryl cAMP or other factor on the enzyme activity might have been observed. Neither the Ah receptor nor the P(1)450 or P(3)450 genes that it regulates map to chromosome 2 or to its mouse or hamster homolog.

Cytochrome P1-450 is the form of P-450 most closely associated with polycyclic-hydrocarbon-induced AHH activity. Chen et al. (1983) cloned a portion of the genomic gene. The compound 2,3,7,8-tetrachlorodibenzo-p-dioxin (TCDD) is a potent inducer of many proteins including drug-metabolizing enzymes such as the cytochrome P-450 proteins. The P1-450 that is induced by TCDD is the same as AHH. Jaiswal et al. (1985) used a human cell line in which TCDD resulted in high levels of AHH (P1-450) activity and of human P1-450. Jaiswal et al. (1985) estimated that the TCDD-inducible P-450 gene family diverged from the phenobarbital-inducible P-450 gene family (123960) more than 200 million years ago. Nebert and Gonzalez (1987) estimated that this divergence occurred more than 750 million years ago. Jaiswal et al. (1985) expressed hope that finding RFLPs representing high and low inducibility may make it possible to predict risk for persons exposed to various environmental pollutants. Kouri et al. (1982) reported that individuals with the high-inducibility phenotype (present in approximately 10% of the human population) might be at greater risk than low-inducibility individuals for cigarette smoke-induced bronchogenic carcinoma. In a 3-generation family of 15 individuals, Petersen et al. (1991) showed that the high-CYP1A1-inducibility phenotype segregated concordantly with an infrequent polymorphic site located 450 bases downstream from the CYP1A1 gene. These findings were consistent with those of 16,15:Kawajiri et al. (1986, 1990), who demonstrated an association between this polymorphism and an increased incidence of squamous-cell lung cancer. Quattrochi et al. (1985) cloned human P450DX genes and concluded that there are at least 2 in humans. Hildebrand et al. (1985) used a full-length cDNA for human 2,3,7,8-tetrachlorodibenzo-p-dioxin (TCDD)-inducible cytochrome P1-450 to study DNA from somatic hybrid cells. They assigned the gene to chromosome 15. Jaiswal and Nebert (1986) indicated that this locus is in the 15q22-qter segment, near MPI (154550). The P3-450 gene (CYP1A2) has also been located on chromosome 15; see 124060. See also CYP1B1 (601771).

The nomenclature and symbolization of the P450 enzymes and their genes have gone through many changes. The currently preferred system (Nebert, 1988) uses the symbol CYP followed by a number for family and a letter for subfamily. CYP1 is the designation of the family of P450 genes located on human chromosome 15 and mouse chromosome 9. (CYP1 was previously used for a P450 gene on chromosome 19 (123960), which is now called CYP2.) The number assigned to the family is sometimes arbitrary or selected for reasons of historical priority; in other cases it has specific significance, e.g., in the case of CYP21 on 6p and CYP17 on 10, which are genes for the enzymes of classes designated P450XXI (steroid 21-hydroxylase) and P450XVII (steroid 17-alpha-hydroxylase), respectively. Hildebrand et al. (1985) showed that in the mouse, which has 2 dioxin-inducible P-450 genes, P1-450 and P3-450, the 2 genes are situated in the middle portion of chromosome 9 near the Mpi-1 locus, between Thy-1 and Pk-3. Treatment of mice with polycyclic aromatic hydrocarbons results in induction of P1-450 and P3-450. Their genes have been cloned and shown to be coordinately regulated by the cytosolic receptor which is coded by the Ah locus and specifically binds the inducing chemicals. By Southern blot analysis of DNA from hamster-mouse somatic cell hybrids, Tukey et al. (1984) demonstrated that the genes for P1-450 and P3-450 map to chromosome 9 in the mouse. The major regulatory gene controlling P1-450 induction in the mouse is located in the centromeric region of chromosome 12. Mouse chromosome 9 shows other homology of synteny with human 15. Jaiswal et al. (1985) and Kawajiri et al. (1986) isolated and analyzed the complete nucleotide sequence of a human genomic clone highly homologous to the rat cytochrome P-450 that is induced by methylcholanthrene and TCDD. A fusion gene, which was constructed by ligating the 5-prime flanking region of the gene to the structural gene for prokaryotic chloramphenicol acetyltransferase (CAT), expressed the CAT activity in mouse cells in response to administered methylcholanthrene. Thus, the isolated human gene was indeed one for methylcholanthrene inducibility. Jones et al. (1991) coupled a DNA fragment containing the murine Cyp1a-1 enhancer elements and promoter region to the chloramphenicol acetyltransferase (CAT) reporter gene and used it to create transgenic mice. Treatment with 3-methylcholanthrene increased hepatic expression levels by as much as 10,000-fold. Differences in the response to induction between male and female mice suggested that Cyp1a-1 expression may be governed in a gender-related manner.

(The chloramphenicol acetyltransferase (CAT) assay system for monitoring gene expression was reported by Gorman et al. (1983). Gorman (1993) described the circumstances surrounding the development of the method. The initial report was turned down by the journal Nature, whose editorial staff charged that the work was not of wide enough interest for publication there.)

Polycyclic aromatic hydrocarbons (PAHs) generated from the combustion of fossil fuels, and aromatic amines, which are present in cigarette smoke and other environmental media, present 2 classic environmental carcinogens. Perera (1997) reviewed evidence on variation and susceptibility to the effects of these carcinogens. CYP1A1 encodes a phase I cytochrome P450 enzyme that metabolizes PAHs such as benzo[a]pyrene (BP). About 10% of Caucasians have a highly inducible form of the enzyme that is associated with an increased risk of lung cancer in smokers. Although not all studies have been positive, in Japanese and certain Caucasian populations, increased lung cancer risk was correlated with 1 or both CYP1A1 polymorphisms: the so called MSPI polymorphism and the closely-linked exon 7 (isoleucine-valine) polymorphism (Kawajiri et al., 1996; Nakachi et al., 1991; Xu et al., 1996). The greatest incremental lung cancer risk from the 'susceptible' CYP1A1 genotype was seen in light smokers (7 times the risk of light smokers without the genotype), whereas heavy smokers with this genotype had less than twice the risk of heavy smokers without the genotype. The proposed mechanism for the increased risk is higher CYP1A1 inducibility or enhanced catalytic activity of the valine-type CYP1A1 enzyme. Consistent with these mechanisms, Mooney et al. (1997) found that U. S. smoking volunteers with the exon 7 mutation had more PAH-DNA adducts in their white blood cells than did smokers without the variant. Perera (1997) stated that PAH-DNA adducts were also elevated in cord blood and placenta of newborns with the CYP1A1 MSP1 polymorphism, which suggested that the genetic polymorphism may increase risk from transplacental PAH exposure. In lung tissue of adults, adduct concentration correlated with CYP1A1 expression or enzyme activity. Perera (1997) noted that lung tumors of Japanese smokers were found to be significantly more likely to have p53 (191170) mutations if they had the sus ceptible CYP1A1 genotype. A failure to demonstrate genetic susceptibility through CYP1A1 polymorphism when exposure to the environmental carcinogen is heavy is observed with some other polymorphisms and carcinogenic exposures. It is possible that at higher exposures, the effects of the genetic traits are overwhelmed by the environmental insults.

See also Hildebrand et al. (1985); Jaiswal et al. (1987); Jaiswal et al. (1985); Jaiswal et al. (1986); Jaiswal et al. (1987); Wiebel et al. (1981)

Brown, S.; Wiebel, F. J.; Gelboin, H. V.; Minna, J. D.: Assignment of a locus required for flavoprotein-linked monooxygenase expression to human chromosome 2. Proc. Nat. Acad. Sci. 73: 4628-4632, 1976.

Chen, Y. T.; Tukey, R. H.; Swan, D. C.; Negishi, N.; Nebert, D. W.: Characterization of the human P1-450 genomic gene. (Abstract) Clin. Res. 31: 456A, 1983.

Gorman, C.; Padmanabhan, R.; Howard, B. H.: High efficiency DNA-mediated transformation of primate cells. Science 221: 551-553, 1983.

Gorman, C. M.: CAT: an easy assay for gene expression (citation classic). Current Contents (Life Sciences) 36(22): 8, 1993.

Hildebrand, C. E.; Gonzalez, F. J.; Kozak, C. A.; Nebert, D. W.: Regional linkage analysis of the dioxin-inducible P-450 gene family on mouse chromosome 9. Biochem. Biophys. Res. Commun. 130: 396-406, 1985.

Hildebrand, C. E.; Gonzalez, F. J.; McBride, O. W.; Nebert, D. W.: Assignment of the human 2,3,7,8-tetrachlorodibenzo-p-dioxin-inducible cytochrome P1-450 gene to chromosome 15. Nucleic Acids Res. 13: 2009-2016, 1985.

Jaiswal, A. K.; Gonzalez, F. J.; Nebert, D. W.: Comparison of human mouse P(1)450 upstream regulatory sequences in liver- and nonliver-derived cell lines. Molec. Endocr. 1: 312-320, 1987.
Jaiswal, A. K.; Gonzalez, F. J.; Nebert, D. W.: Human P(1)-450 gene sequence and correlation of mRNA with genetic differences in benzo(a)pyrene metabolism. Nucleic Acids Res. 13: 4503-4520, 1985.
Jaiswal, A. K.; Gonzalez, F. J.; Nebert, D. W.: Human dioxin-inducible cytochrome P1-450: complementary DNA and amino acid sequence. Science 228: 80-83, 1985.
Jaiswal, A. K.; Nebert, D. W.: Two RFLPs associated with the human P(1)450 gene linked to the MPI locus on chromosome 15 (HGM8 D15S8). Nucleic Acids Res. 14: 4376, 1986.
Jaiswal, A. K.; Nebert, D. W.; Gonzalez, F. J.: Human P(3)450: cDNA and complete amino acid sequence. Nucleic Acids Res. 14: 6773-6774, 1986.
Jaiswal, A. K.; Nebert, D. W.; McBride, O. W.; Gonzalez, F. J.: Human P(3)450: cDNA and complete protein sequence, repetitive Alu sequences in the 3-prime nontranslated region, and localization of gene to chromosome 15. J. Exp. Path. 3: 1-17, 1987.
Jones, S. N.; Jones, P. G.; Ibarguen, H.; Caskey, C. T.; Craigen, W. J.: Induction of the Cyp1a-1 dioxin-responsive enhancer in transgenic mice. Nucleic Acids Res. 19: 6547-6551, 1991.
Kawajiri, K.; Eguchi, H.; Nakachi, K.; Sekiya, T.; Yamamoto, M.: Association of CYP1A1 germ line polymorphisms with mutations of the p53 gene in lung cancer. Cancer Res. 56: 72-76, 1996.
Kawajiri, K.; Nakachi, K.; Imai, K.; Yoshii, A.; Shinoda, N.; Watanabe, J.: Identification of genetically high risk individuals to lung cancer by DNA polymorphisms of the cytochrome P450IA1 gene. FEBS Lett. 263: 131-133, 1990.
Kawajiri, K.; Watanabe, J.; Gotoh, O.; Tagashira, Y.; Sogawa, K.; Fujii-Kuriyama, Y.: Structure and drug inducibility of the human cytochrome P-450c gene. Europ. J. Biochem. 159: 219-225, 1986.
Kouri, R. E.; McKinney, C. E.; Slomiany, D. J.; Snodgrass, D. R.; Wray, N. P.; McLemore, T. L.: Positive correlation between high aryl hydrocarbon hydroxylase activity and primary lung cancer as analyzed in cryopreserved lymphocytes. Cancer Res. 42: 5030-5037, 1982.
McBride, O. W.: Personal Communication. Bethesda, Md. 9/16/1985.
Mooney, L. A.; Bell, D. A.; Santella, R. M.; Van Bennekum, A. M.; Ottman, R.; Paik, M.; Blaner, W. S.; Lucier, G. W.; Covey, L.; Young, T. L.; Cooper, T. B.; Glassman, A. H.; Perera, F. P.: Contribution of genetic and nutritional factors to DNA damage in heavy smokers. Carcinogenesis 18: 503-509, 1997.
Nakachi, K.; Imai, K.; Hayashi, S.; Watanabe, J.; Kawajiri, K.: Genetic susceptibility to squamous cell carcinoma of the lung in relation to cigarette smoking dose. Cancer Res. 51: 5177-5180, 1991.
Nebert, D. W.: Personal Communication. Bethesda, Md. 2/1/1988.
Nebert, D. W.; Gonzalez, F. J.: P450 genes: structure, evolution, and regulation. Annu. Rev. Biochem. 56: 945-993, 1987.
Ocraft, K. P.; Muskett, J. M.; Brown, S.: Localization of the human arylhydrocarbon hydroxylase gene to the 2q31-2pter region of chromosome 2. Ann. Hum. Genet. 49: 237-239, 1985.
Perera, F. P.: Environment and cancer: who are susceptible? Science 278: 1068-1073, 1997.
Petersen, D. D.; McKinney, C. E.; Ikeya, K.; Smith, H. H.; Bale, A. E.; McBride, O. W.; Nebert, D. W.: Human CYP1A1 gene: cosegregation of the enzyme inducibility phenotype and an RFLP. Am. J. Hum. Genet. 48: 720-725, 1991.
Quattrochi, L. C.; Okino, S. T.; Pendurthi, U. R.; Tukey, R. H.: Cloning and isolation of human cytochrome P-450 cDNAs homologous to dioxin-inducible rabbit mRNAs encoding P-450 4 and P-450 6. DNA 4: 395-400, 1985.
Tukey, R. H.; Lalley, P. A.; Nebert, D. W.: Localization of cytochrome P1-450 and P3-450 genes to mouse chromosome 9. Proc. Nat. Acad. Sci. 81: 3163-3166, 1984.
Wiebel, F. J.; Hlavica, P.; Grzeschik, K. H.: Expression of aromatic polycyclic hydrocarbon-induced monooxygenase (aryl hydrocarbon hydroxylase) in man-mouse hybrids is associated with human chromosome 2. Hum. Genet. 59: 277-280, 1981.
Xu, X.; Kelsey, K. T.; Wiencke, J. K.; Wain, J. C.; Christiani, D. C.: Cytochrome P450 CYP1A1 MspI polymorphism and lung cancer susceptibility. Cancer Epidemiol. Biomarkers Prev. 5: 687-692, 1996.

CYTOCHROME P450, FAMILY II, SUBFAMILY D; CYP2D (*124030)

JJM NOTE

Oral legend has it that this polymorphism was discovered when the head of the pharmacology unit that was testing debrisoquine collapsed with vascular hypotension on taking a trial dose of the new drug. He was found to be a poor metabolizer. The literature that confirms the association of this gene with lung cancer is growing (Law MR et al.: Debrisoquine metabolism and genetic predispositon to lung cancer. Br. J. Cancer 59:686–87, 1989; Caporaso N et al.: The distribution of debrisoquine metabolic phenotypes and implications for the suggested association with lung cancer risk. Genet. Epidemiol. 6:517–24, 1989; Caporaso NE et al.: Lung cancer and the debrisoquine metabolic phenotype. J. Natl. Cancer Inst. 82:1264–72, 1990; Roots I et al.: Debrisoquine hydroxylation phenotype, acetylation phenotype, and ABO blood groups as genetic host factors of lung cancer risk. Klin. Wochenschr. 66 (suppl. 11): 87–97, 1988).

Description

The P-450II family comprises at least 5 subfamilies, designated A through E by the system of nomenclature recommended by an international committee (Nebert et al., 1987). The P450IID subfamily comprises at least 2 genes in the rat, one of which is highly specific for debrisoquine 4-hydroxylase activity. The literature is growing that confirms the association of this gene with lung cancer (Law et al., 1989; 7,8:Caporaso et al., 1989, 1990; Roots et al., 1988). Enhanced CYP2D6 activity has been related to malignancies of the bladder, liver, pharynx, and stomach and, especially, to cigarette-smoking-induced lung cancer. The data suggest that enhanced CYP2D6-mediated metabolism of one or more dietary and other environmental agents, to form a reactive intermediate, plays a role in cancer initiation and/or promotion in various tissues (Nebert, 1997). Reduced CYP2D6 activity resulting in the 'poor metabolizer' phenotype has been thought to be related to an increased risk of Parkinson disease (168600).

CYP2D6 polymorphism, which is responsible for the variation in metabolism of debrisoquine 4-hydroxylase, is important in the metabolism of more than 30 drugs and environmental chemicals, including as much as 20% of all commonly prescribed drugs. Although humans probably have approximately 60 unique P450 genes (Nelson et al., 1996), only approximately a half dozen appear to be responsible for metabolism of the vast majority of prescribed and over-the-counter drugs: CYP1A2 (124060), CYP2C17, CYP2D6, CYP2E1 (124040), CYP3A4 (124010), and CYP4A11 (601310).

Clinical Features

Oral legend has it that this polymorphism was discovered when the head of the pharmacology unit that was testing debrisoquine as an antihypertensive agent collapsed with vascular hypotension on taking a trial dose of the new drug. He was found to be a poor metabolizer.

Debrisoquine ('Declinax' Roche) is an adrenergic-blocking drug used in England for the treatment of hypertension. Wide variation in hypotensive response had been noted. A clinical consequence of low metabolism is greater sensitivity to the antihypertensive effects of debrisoquine (Idle et al., 1978). Mahgoub et al. (1977) showed that a polymorphism in hydroxylation of the drug is responsible. They used the ratio of urinary debrisoquine and 4-hydroxydebrisoquine after a single oral dose of 10 mg of debrisoquine. The distribution ratio showed a sharp bimodality. About 3% of the study population were so-called nonmetabolizers. These individuals, they suggested, are homozygous for a recessive gene. Family data supported the suggestion although an extensive and systemic family study was not reported. Waring et al. (1981) suggested that sulfoxidation of mucodyne may be regulated by the same gene that controls the oxidation of debrisoquine.

Poor hydroxylators of debrisoquine also have impaired metabolism of some other drugs such as phenacetin, nortriptyline, phenformin, sparteine, encainide, propranolol, bufuralol, guanoxan, perhexilene, and amitriptyline (Mellstrom et al., 1981; Oates et al., 1982; Inaba et al., 1980; Harmer et al., 1986; Nebert, 1987).

Lennard et al. (1983) identified at least 14 other drugs that are metabolized by the same mechanism as debrisoquine. The antiarrhythmic agent propafenone, a sodium-channel blocker with some structural similarity to propranolol, is effective in the management of atrial and ventricular arrhythmias. It has a variable amount of beta-blockade effect. Propafenone is biotransformed to a 5-hydroxy derivative which is at least as effective a channel-blocking agent as the parent drug but is less potent as a beta-blocker. The metabolism to the 5-hydroxy derivative is catalyzed by the enzyme coded by CYP2D. Lee et al. (1990) presented evidence that genetically determined variations in the conversion of propafenone to its 5-hydroxy metabolite accounts for variations in the drug's beta-blocking action. Slow converters showed greater beta-blockade.

Lennard et al. (1982) demonstrated that metabolism of metoprolol, a beta-1-selective adrenoceptor antagonist, exhibits genetic polymorphism of the debrisoquine type. They provided an explanation for the 17-fold difference in plasma concentrations of metoprolol in patients taking the same dose. In poor hydroxylators, a single daily dose of metoprolol may control angina, whereas in 'extensive hydroxylators,' 2 or 3 doses a day may be necessary. As compared with propranolol, metoprolol has relatively little effect on beta-2-receptors in the bronchi and peripheral vessels. It has, therefore, been preferred to propranolol in patients with bronchial asthma. The beta-1-selectivity is, however, relative; asthmatics who are poor hydroxylators may be at the same risk if treated with standard doses of metoprolol.

Barbeau et al. (1985) postulated that Parkinson disease is the result of environmental factors acting on genetically susceptible persons against a background of 'normal' aging. Many potential neurotoxic xenobiotics are detoxified by hepatic cytochrome P-450. Barbeau et al. (1985) studied one such system in 40 parkinsonians and 40 controls. Significantly more patients than controls had partially or totally defective 4-hydroxylation of debrisoquine. Poor metabolizers had earlier onset of disease. In a study of 229 patients with Parkinson disease and 720 controls, Smith et al. (1992) found that those with the poor metabolizer phenotype had a 2.54-fold (95% confidence interval 1.51-4.28) increased risk of Parkinson disease.

Biochemical Features

Guengerich et al. (1986) stated that 9 forms of cytochrome P-450 had been purified to electrophoretic homogeneity from human liver microsomes. These included the enzymes involved in debrisoquine 4-hydroxylation, phenacetin O-deethylation (124060), and mephenytoin 4-hydroxylation (124020), 3 reactions characterized by genetic polymorphism in humans. Different forms of P-450 are involved in the 3 reactions. Guengerich et al. (1986) presented evidence that the P-450 nifedipine oxidase is separate from these others (see 124010). Nifedipine is a calcium-channel blocker which about 17% of the population cannot metabolize rapidly as judged by in vivo studies (Kleinbloesem et al., 1984). Although the extensive metabolizers have a reduced risk of overdosage effects of debrisoquine and related drugs, they have a far greater risk (20- to 40-fold) of developing cancers of the liver, gastrointestinal tract, and lung than have poor metabolizers (Nebert, 1987).

Harmer et al. (1986) found no correlation between the acetylator and the sparteine hydroxylation phenotypes. This is perhaps not surprising inasmuch as the N-acetyltransferase enzyme is cytosolic in liver and jejunal mucosa, whereas the polymorphic enzyme that governs hydroxylation of sparteine, debrisoquine and other drugs is a liver P-450.

Other Features

Distlerath and Guengerich (1984) found that antibodies prepared to a cytochrome P-450 shown to be responsible for debrisoquine 4-hydroxylation in rats inhibited the oxidation of debrisoquine and sparteine, encainide, and propranolol, 3 other drugs suggested to be associated with the poor- metabolizer phenotype, in human liver microsomes. The antibodies did not inhibit the oxidation of 7 other cytochrome P-450 substrates.

Inheritance

Deficient hydroxylation of debrisoquine and some other drugs previously noted is inherited as an autosomal recessive trait.

Mapping

This gene subfamily maps to human chromosome 22 by Southern analysis of DNA from somatic cell hybrids (Gonzalez et al., 1988). CYP2D is said to be linked to blood group P1 (111410; HGM9). Eichelbaum et al. (1987) mapped the CYP2D gene to chromosome 22. Gough et al. (1993) used PCR for analysis of somatic cell hybrids, in situ hybridization, and linkage analysis to map the CYP2D gene to 22q13.1.

Molecular Genetics

Gonzalez et al. (1988) showed that poor metabolizers have negligible amounts of the cytochrome P450 enzyme product of the CYP2D locus (called P450db1 by them). They cloned cDNA for the gene. By directly cloning and sequencing cDNAs from several poor-metabolizer livers, they identified 3 variant mRNAs that are products of mutant genes producing incorrectly spliced pre-mRNA, thus providing a molecular explanation for one of man's most commonly defective genes (frequency of mutant alleles, 35-43%). Skoda et al. (1988) demonstrated restriction fragment length polymorphisms associated with the P450db1 locus and observed more frequently in individuals with the poor-metabolizer phenotype. A different, 29-kb XbaI fragment was present in all individuals with the extensive-metabolizer phenotype. The authors proposed that these polymorphisms identified 2 independent mutant alleles of the P450db1 gene. Eight additional RFLPs associated with this gene were also reported.

Kimura et al. (1989) identified 3 genes that compose the CYP2D gene cluster located distal to IGLC (147220) on chromosome 22. One of these, CYP2D8P, was found to be a pseudogene while the CYP2D7 gene contained a single reading frame-disrupting insertion in its first exon. The CYP2D6 gene is that associated with polymorphic debrisoquine metabolism.

Heim and Meyer (1990) used allele-specific PCR amplification of parts of the mutant genes for what they termed hepatic cytochrome P450IID6 to genotype poor metabolizers of debrisoquine. Over 95% of poor metabolizers were identified. Mura et al. (1993) offered an explanation for the fact that extensive metabolizers exhibit highly variable metabolic activity. In a large set of French Caucasian families, they found, as have other authors, that among extensive metabolizers part of the heterogeneity is explained by a subgroup of individuals heterozygous for a mutant allele. However, a second level of heterogeneity was detected among individuals not carrying mutations, and this was related to a polymorphic BamHI-defined DNA haplotype. Different combinations of haplotypes were associated with differences in CYP2D6 metabolic activity. They suggested that this finding may help to clarify the conflicting data on the relation between CYP2D6 activity and susceptibility to lung cancer.

Saxena et al. (1994) stated that 3 well-defined variant alleles, designated CYP2D6(A), CYP2D6(B), and CYP2D6(D), represent 90 to 95% of the CYP2D6 variant alleles associated with the 'poor metabolizer' (PM) phenotype as documented by mutation-specific PCR amplification and Southern analysis in persons of known metabolic phenotype. Additional variant alleles had been described, CYP2D6(C) and CYP2D6(J), but their frequency of occurrence and effects on metabolic phenotypes were not yet entirely clear. In a search for additional CYP2D6 gene alterations that result in the PM phenotype, Saxena et al. (1994) sequenced all 9 exons and their intron/exon boundaries from 15 unrelated PM individuals who lacked one of the known PM-associated alleles. They reported the discovery of a 1-bp deletion in exon 3, which resulted in reduced CYP2D6 activity (see 124030.0003).

See Sachse et al. (1997) and McLellan et al. (1997) (in the POPULATION GENETICS section) for discussion of CYP2D6 alleles showing gene duplication in different populations.

Nomenclature: Based on recommendations for human genome nomenclature, Daly et al. (1996) proposed that alleles at this locus be designated by CYP2D6 followed by an asterisk and a combination of Roman letters and Arabic numerals distinct for each allele, with the number specifying the key mutation and, where appropriate, a letter specifying additional mutations. Criteria for classification as a separate allele and protein nomenclature were also presented.

Heterogeneity

In addition to deficient hydroxylation of debrisoquine, an autosomal recessive trait that affects approximately 7% of Caucasians, the opposite phenomenon exists; certain individuals metabolize debrisoquine and several other commonly used drugs very rapidly, resulting in subtherapeutic plasma concentrations at normal doses.

Johansson et al. (1993) found that the molecular genetic basis for ultrarapid metabolism was the presence of an amplified variant CYP2D6 gene, called CYP2D6L. In the father and his 2 children in one family, the CYP2D6L gene was amplified at 12-fold; 2 copies of the CYP2D6L gene were present in 'affected' members of the other family. The CYP2D6L gene had an open reading frame and carried 2 mutations causing amino acid substitutions: one in exon 6, yielding an arg296-to-cys exchange, and one in exon 9, causing a ser486-to-thr exchange. The metabolic rate in subjects carrying one copy of the CYP2D6L gene did not differ significantly from that of those with the wildtype gene, indicating that the structural alterations were not of importance for the catalytic properties of the gene product. Examination of the metabolic rate among subjects carrying wildtype CYP2D6, CYP2D6L, or deficient alleles revealed a relationship between the number of active genes and metabolic rate. Johansson et al. (1993) commented that the finding of a specific haplotype with 2 or more active CYP2D6 genes would allow genotyping for ultrarapid drug metabolizers. This genotyping could be of predictive value for individualized and more efficient drug therapy.

Population Genetics

Evans et al. (1980) estimated the frequency of the poor-hydroxylator phenotype to be about 9% in the United Kingdom, but it varies widely among ethnic groups, being about 1% in Arabs and 30% in Hong Kong Chinese (Kalow, 1982). Zhou et al. (1989) found that Chinese subjects had at least a 2-fold greater sensitivity to the beta-blocking effects of propranolol than white subjects, as measured by the mean plasma concentrations producing a 20% reduction in heart rate in both the supine and the upright positions and after exercise testing. In addition, Chinese subjects had much greater sensitivity to the hypotensive effects of propranolol, as shown by the concentrations that reduced blood pressure by 10% in the supine or in the upright position. No difference in beta-receptor density or affinity of lymphocytes was found between the groups. The Chinese group metabolized propranolol more rapidly than the white group, which resulted in a 76% higher clearance of an oral dose because of increased metabolism through multiple metabolic pathways. The Chinese group had a 45% higher free fraction of propranolol in plasma, which may have contributed to the increased drug effect but cannot explain it entirely. Although poor metabolizers of sparteine are relatively rare in Japan, Chiba et al. (1988) found 1 kindred in which autosomal recessive inheritance seemed quite clear.

CYP2D6 activity ranges from complete deficiency to ultrafast metabolism, depending on at least 16 different known alleles. Sachse et al. (1997) determined their frequencies in 589 unrelated German volunteers and correlated enzyme activity measure by phenotyping with dextromethorphan or debrisoquine. For genotyping, they developed nested PCR-RFLP tests from a PCR amplification of the entire CYP2D6 gene. The frequency of the CYP2D6*1 allele coding for extensive metabolizer (EM) phenotype was 0.364. The alleles coding for slightly (CYP2D6*2) or moderately (*9 and *10) reduced activity (intermediate metabolizer phenotype, IM) showed frequencies of 0.324, 0.018, and 0.015, respectively. By use of novel PCR tests for discrimination, CYP2D6 gene duplication alleles were found with frequencies of 0.005 (*1x2), 0.013 (*2x2), and 0.001 (*4x2). Frequencies of alleles with complete deficiency (poor metabolizer phenotype, PM) were 0.207 (*4), 0.020 (*3 and *5), 0.009 (*6), and 0.001 (*7, *15, and *16). The defective CYP2D6 alleles *8, *11, *12, *13, and *14 were not found. All 41 PMs (7.0%) in this sample were explained by 5 mutations detected by 4 PCR-RFLPs, which may suffice, together with the gene duplication test, for clinical prediction of CYP2D6 capacity. Sachse et al. (1997) found 3 novel variants of known CYP2D6 alleles. Analysis of variants showed significant differences in enzymatic activity measured by the dextromethorphan metabolic ratio (MR) between carries of EM/PM (mean = 0.006) and IM/PM (mean = 0.014) alleles and between carriers of 1 (mean MR = 0.009) and 2 (mean MR = 0.003) functional alleles. They commented that their results provided a solid basis for prediction of CYP2D6 capacity, as required in drug research and routine drug treatment.

In contrast with poor metabolizers, individuals with a capacity to metabolize CYP2D6 substrate drugs very rapidly also exist. 4,5:Bertilsson et al. (1985, 1993) reported cases where higher than normal doses of drugs were required to attain therapeutic concentrations. Bertilsson et al. (1993) and Johansson et al. (1993) showed that the genetic basis of this phenomenon is gene duplication or amplification of functionally active CYP2D6 genes, resulting in higher levels of enzyme being expressed. In addition, subjects with multiduplicated CYP2D6 genes present in 3, 4, or 5 copies on 1 allele have been seen (Dahl et al., 1995; Akullu et al., 1996). The occurrence of gene duplication has been found to vary between populations, as individuals carrying extra genes are seen in Ethiopia at a frequency of 29% (Akullu et al., 1996), whereas a frequency of 1 to 2% was observed in Swedish, German, Chinese, and black Zimbabwean populations. McLellan et al. (1997) found duplication of the CYP2D6 gene in 21 of 101 Saudi Arabians studied. In contrast, only 2 individuals were heterozygous for a deletion of the whole gene. The allele frequency of CYP2D6*4, the most common defective allele among Caucasians, was only 3.5% in the Saudi population. These findings were in agreement with earlier Saudi Arabian phenotyping studies that reported a low frequency (1 to 2%) of poor metabolizers for CYP2D6-probe drugs.

Marez et al. (1997) undertook to define the mutations and alleles of the CYP2D6 gene and their frequencies in Europeans. Using PCR-SSCP analysis, 672 unrelated subjects were screened for mutations in the 9 exons of the gene and their exon-intron boundaries. A total of 48 point mutations were identified, of which 29 were novel.

Bathum et al. (1998) examined the frequency of CYP2D6 gene duplications in a Danish population and validated a long PCR method for identification of ultrarapid metabolizers. Based on EcoRI restriction fragment length polymorphism analysis, 8 individuals were found with a duplicated CYP2D6 gene, whereas using a long PCR method, 9 individuals with a 3.6-kb fragment indicative of 2 CYP2D6 genes in tandem were found among 60 individuals with a low metabolic ratio. They estimated that the frequency of individuals with CYP2D6 duplication in the Danish population would be 0.8%, which is comparable to the frequency in the Swedish and the German populations, but considerably lower than that in Spanish or African populations. They concluded that the long PCR assay is simple and reliable for detection of duplications of the CYP2D6 gene.

Evolution

Heim and Meyer (1992) indicated that the CYP2D cluster on chromosome 22 contains the functional gene CYP2D6 and 2 or 3 highly homologous pseudogenes. To understand the evolutionary sequence of mutational events as well as the interracial differences, they analyzed the arrangement of the CYP2D haplotype containing a common mutant allele of CYP2D6 associated with an XbaI 44-kb restriction fragment. This haplotype was found to contain 4 CYP2D genes instead of 3. Comparison of the sequences of these genes with those of previously characterized haplotypes suggested that an early point mutation was followed by a crossover and a gene conversion event, the latter found preferentially in Caucasians. The data were considered consistent with the rapid evolution of the CYP2D 'locus' during 'plant-animal warfare' with practical consequences for present-day defense against environmental adversity. Kimura et al. (1989) suggested that these enzymes evolved to metabolize plant toxins; since current human diets rely almost totally on cultivation, while early man was a hunter-gatherer, the genes may now be in the process of being lost. Gene conversion events involving the closely linked and similar CYP2D7 and CYP2D8P genes may hasten this process.

Nebert (1997) reviewed current thinking in the field of drug-metabolizing enzymes (DMEs) and DME-receptor research and thinking concerning how DMEs have evolved through animal-plant interactions.

Animal Model

Matsunaga et al. (1989) studied the DA rat, which has impaired ability to metabolize debrisoquine and therefore may be useful as a model of the human genetic defect in debrisoquine metabolism.

See also Eichelbaum et al. (1987); Eichelbaum et al. (1986); Eichelbaum and Woolhouse (1985); Gonzalez et al. (1988); Silas et al. (1977); Tucker et al. (1977)

Akullu, E.; Persson, I.; Bertilsson, L.; Johansson, I.; Rodrigues, F.; Ingelman-Sundberg, M.: Frequent distribution of ultrarapid metabolizers of debrisoquine in an Ethiopian population carrying duplicated and multiduplicated functional CYP2D6 alleles. J. Pharm. Exp. Therapy. 278: 441-446, 1996.

Barbeau, A.; Cloutier, T.; Roy, M.; Plasse, L.; Paris, S.; Poirier, J.: Ecogenetics of Parkinson's disease: 4-hydroxylation of debrisoquine. Lancet II: 1213-1216, 1985.

Bathum, L.; Johansson, I.; Ingelman-Sundberg, M.; Horder, M.; Brosen, K.: Ultrarapid metabolism of sparteine: frequency of alleles with duplicated CYP2D6 genes in a Danish population as determined by restriction fragment length polymorphism and long polymerase chain reaction. Pharmacogenetics 8: 119-123, 1998.

Bertilsson, L.; Aberg-Wistedt, A.; Gustafsson, L. L.; Nordin, C.: Extremely rapid hydroxylation of debrisoquine: a case report with implication for treatment with nortriptyline and other tricyclic antidepressants. Ther. Drug Monit. 7: 478-480, 1985.

Bertilsson, L.; Dahl, M.-L.; Sjoqvist, F.; Aberg-Wistedt, A.; Humble, M.; Johansson, I.; Lundqvist, E.; Ingelman-Sundberg, M.: Molecular basis for rational megaprescribing in ultrarapid hydroxylators of debrisoquine. (Letter) Lancet 341: 63 only, 1993.

Broly, F.; Marez, D.; Lo Guidice, J.-M.; Sabbagh, N.; Legrand, M.; Boone, P.; Meyer, U. A.: A nonsense mutation in the cytochrome P450 CYP2D6 gene identified in a Caucasian with an enzyme deficiency. Hum. Genet. 96: 601-603, 1995.

Caporaso, N.; Pickle, L. W.; Bale, S.; Ayesh, R.; Hetzel, M.; Idle, J.: The distribution of debrisoquine metabolic phenotypes and implications for the suggested association with lung cancer risk. Genet. Epidemiol. 6: 517-524, 1989.

Caporaso, N. E.; Tucker, M. A.; Hoover, R. N.; Hayes, R. B.; Pickle, L. W.; Issaw, H. J.; Muschik, G. M.; Green-Gallo, L.; Buivys, D.; Aisner, S.; Resau, J. H.; Trump, B. F.; Tollerud, D.; Weston, A.; Harris, C. C.: Lung cancer and the debrisoquine metabolic phenotype. J. Natl. Cancer Inst 82: 1264-1272, 1990.

Chen, X.; Xia, Y.; Alford, M.; DeTeresa, R.; Hansen, L.; Klauber, M. R.; Katzman, R.; Thal, L.; Masliah, E.; Saitoh, T.: The CYP2D6B allele is associated with a milder synaptic pathology in Alzheimer's disease. Ann. Neurol. 38: 653-658, 1995.

Chiba, K.; Kato, J.; Hashimoto, K.; Ishizaki, T.: Apparent mendelian recessive inheritance of sparteine metabolism in an extended Japanese family. Europ. J. Clin. Pharm. 34: 661-662, 1988.

Dahl, M.-L.; Johansson, I.; Bertilsson, L.; Ingelman-Sundberg, M.; Sjoqvist, F.: Ultrarapid hydroxylation of debrisoquine in a Swedish population: analysis of the molecular genetic basis. J. Pharm. Exp. Therapy. 274: 516-520, 1995.

Daly, A. K.; Brockmoller, J.; Broly, F.; Eichelbaum, M.; Evans, W. E.; Gonzalez, E.-J.; Huang, J.-D.; Idle, J. R.; Ingelman-Sundberg, M.; Ishizaki, T.; Jacqz-Aigrain, E.; Meyer, U. A.; Nebert, D. W.; Steen, V. M.; Wolf, C. R.; Zanger, U. M.: Nomenclature for human CYP2D6 alleles. Pharamacogenetics 6: 193-201, 1996.

Distlerath, L. M.; Guengerich, F. P.: Characterization of a human liver cytochrome P-450 involved in the oxidation of debrisoquine and other drugs by using antibodies raised to the analogous rat enzyme. Proc. Nat. Acad. Sci. 81: 7348-7352, 1984.

Eichelbaum, M.; Baur, M. P.; Dengler, H. J.; Osikowska-Evers, B. O.; Tieves, G.; Zekorn, C.; Rittner, C.: Chromosomal assignment of human cytochrome P-450 (debrisoquine/sparteine type) to chromosome 22. (Abstract) Cytogenet. Cell Genet. 46: 610, 1987.

Eichelbaum, M.; Baur, M. P.; Dengler, H. J.; Osikowska-Evers, B. O.; Tieves, G.; Zekorn, C.; Rittner, C.: Chromosomal assignment of human cytochrome P-450 (debrisoquine/sparteine type) to chromosome 22. Brit. J. Clin. Pharm. 23: 455-458, 1987.

Eichelbaum, M.; Reetz, K. P.; Schmidt, E. K.; Zekorn, C.: The genetic polymorphism of sparteine metabolism. Xenobiotica 16: 465-481, 1986.

Eichelbaum, M.; Woolhouse, N. M.: Inter-ethnic difference in sparteine oxidation among Ghanaians and Germans. Europ. J. Clin. Pharm. 28: 79-83, 1985.

Evans, D. A. P.; Mahgoub, A.; Sloan, T. P.; Idle, J. R.; Smith, R. L.: A family and population study of the genetic polymorphism of debrisoquine oxidation in a white British population. J. Med. Genet. 17: 102-105, 1980.

Gaedigk, A.; Blum, M.; Gaedigk, R.; Eichelbaum, M.; Meyer, U. A.: Deletion of the entire cytochrome P450 CYP2D6 gene as a cause of impaired drug metabolism in poor metabolizers of the debrisoquine/sparteine polymorphism. Am. J. Hum. Genet. 48: 943-950, 1991.

Gonzalez, F. J.; Skoda, R. C.; Kimura, S.; Umeno, M.; Zanger, U. M.; Nebert, D. W.; Gelboin, H. V.; Hardwick, J. P.; Meyer, U. A.: Characterization of the common genetic defect in humans deficient in debrisoquine metabolism. Nature 331: 442-446, 1988.

Gonzalez, F. J.; Vilbois, F.; Hardwick, J. P.; McBride, O. W.; Nebert, D. W.; Gelboin, H. V.; Meyer, U. A.: Human debrisoquine 4-hydroxylase (P450IID1)

cDNA and deduced amino acid sequence and assignment of the CYP2D locus to chromosome 22. Genomics 2: 174-179, 1988.
Gough, A. C.; Miles, J. S.; Spurr, N. K.; Moss, J. E.; Gaedigk, A.; Eichelbaum, M.; Wolf, C. R.: Identification of the primary gene defect at the cytochrome P(450) CYP2D locus. Nature 347: 773-776, 1990.
Gough, A. C.; Smith, C. A. D.; Howell, S. M.; Wolf, C. R.; Bryant, S. P.; Spurr, N. K.: Localization of the CYP2D gene locus to human chromosome 22q13.1 by polymerase chain reaction, in situ hybridization, and linkage analysis. Genomics 15: 430-432, 1993.
Guengerich, F. P.; Distlerath, L. M.; Reilly, P. E. B.; Wolff, T.; Shimada, T.; Umbenhauer, D. R.; Martin, M. V.: Human-liver cytochromes P-450 involved in polymorphisms of drug oxidation. Xenobiotica 16: 367-378, 1986.
Hanioka, N.; Kimura, S.; Meyer, U. A.; Gonzalez, F. J.: The human CYP2D locus associated with a common genetic defect in drug oxidation: a G(1934)-to-A base change in intron 3 of a mutant CYP2D6 allele results in an aberrant 3-prime splice recognition site. Am. J. Hum. Genet. 47: 994-1001, 1990.
Harmer, D.; Evans, D. A. P.; Eze, L. C.; Jolly, M.; Whibley, E. J.: The relationship between the acetylator and the sparteine hydroxylation polymorphisms. J. Med. Genet. 23: 155-156, 1986.
Heim, M.; Meyer, U. A.: Genotyping of poor metabolisers of debrisoquine by allele-specific PCR amplification. Lancet 336: 529-532, 1990.
Heim, M. H.; Meyer, U. A.: Evolution of a highly polymorphic human cytochrome P450 gene cluster: CYP2D6. Genomics 14: 49-58, 1992.
Idle, J. R.; Mahgoub, A.; Lancaster, R.; Smith, R. L.: Hypotensive response to debrisoquine and hydroxylation phenotype. Life Sci. 22: 979-984, 1978.
Inaba, T.; Otton, S. V.; Kalow, W.: Deficient metabolism of debrisoquine and sparteine. Clin. Pharm. Therap. 27: 547-549, 1980.
Johansson, I.; Lundqvist, E.; Bertilsson, L.; Dahl, M.-L.; Sjoqvist, F.; Ingelman-Sundberg, M.: Inherited amplification of an active gene in the cytochrome P450 CYP2D locus as a cause of ultrarapid metabolism of debrisoquine. Proc. Nat. Acad. Sci. 90: 11825-11829, 1993.
Kagimoto, M.; Heim, M.; Kagimoto, K.; Zeugin, T.; Meyer, U. A.: Multiple mutations of the human cytochrome P450IID6 gene (CYP2D6) in poor metabolizers of debrisoquine: study of the functional significance of individual mutations by expression of chimeric genes. J. Biol. Chem. 265: 17209-17214, 1990.
Kalow, W.: The metabolism of xenobiotics in different populations. Canad. J. Physiol. Pharm. 60: 1-12, 1982.
Kimura, S.; Umeno, M.; Skoda, R. C.; Meyer, U. A.; Gonzalez, F. J.: The human debrisoquine 4-hydroxylase (CYP2D) locus: sequence and identification of the polymorphic CYP2D6 gene, a related gene, and a pseudogene. Am. J. Hum. Genet. 45: 889-904, 1989.
Kleinbloesem, C. H.; van Brummelen, P.; Faber, H.; Danhof, M.; Vermeulen, N. P. E.; Breimer, D. D.: Variability in nifedipine pharmacokinetics and dynamics: a new oxidation polymorphism in man. Biochem. Pharm. 33: 3721-3724, 1984.
Law, M. R.; Hetzel, M. R.; Idle, J. R.: Debrisoquine metabolism and genetic predisposition to lung cancer. Br. J. Cancer 59: 686-687, 1989.
Lee, J. T.; Kroemer, H. K.; Silberstein, D. J.; Funck-Brentano, C.; Lineberry, M. D.; Wood, A. J. J.; Roden, D. M.; Woosley, R. L.: The role of genetically determined polymorphic drug metabolism in the beta-blockade produced by propafenone. New Eng. J. Med. 322: 1764-1768, 1990.
Lennard, M. S.; Ramsey, L. E.; Silas, J. H.; Tucker, G. T.; Woods, H. F.: Protecting the poor metaboliser: clinical consequences of genetic polymorphism of drug oxidation. Pharm. Int. 4: 61-65, 1983.
Lennard, M. S.; Silas, J. H.; Freestone, S.; Ramsay, L. E.; Tucker, G. T.; Woods, H. F.: Oxidation phenotype—a major determinant of metoprolol metabolism and response. New Eng. J. Med. 307: 1558-1560, 1982.
Mahgoub, A.; Idle, J. R.; Dring, L. G.; Lancaster, R.; Smith, R. L.: Polymorphic hydroxylation of debrisoquine in man. Lancet II: 584-586, 1977.
Marez, D.; Legrand, M.; Sabbagh, N.; Lo Guidice, J.-M.; Spire, C.; Lafitte, J.-J.; Meyer, U. A.; Broly, F.: Polymorphism of the cytochrome P450 CYP2D6 gene in a European population: characterization of 48 mutations and 53 alleles, their frequencies and evolution. Pharmacogenetics 7: 193-202, 1997.
Matsunaga, E.; Zanger, U. M.; Hardwick, J. P.; Gelboin, H. V.; Meyer, U. A.; Gonzalez, F. J.: The CYP2D gene subfamily: analysis of the molecular basis of the debrisoquine 4-hydroxylase deficiency in DA rats. Biochemistry 28: 7349-7355, 1989.
McLellan, R. A.; Oscarson, M.; Seidegard, J.; Price Evans, D. A.; Ingelman-Sundberg, M.: Frequent occurrence of CYP2D6 gene duplication in Saudi Arabians. Pharmacogenetics 7: 187-191, 1997.
Mellstrom, B.; Bertilsson, L.; Sawe, J.; Schulz, H.-U.; Sjoqvist, F.: E- and Z-10-hydroxylation of nortriptyline: relationship to polymorphic debrisoquine hydroxylation. Clin. Pharm. Therap. 30: 189-193, 1981.
Mura, C.; Panserat, S.; Vincent-Viry, M.; Galteau, M. M.; Jacqz-Aigrain, E.; Krishnamoorthy, R.: DNA haplotype dependency of debrisoquine-4-hydroxylase (CYP2D6) expression among extensive metabolisers. Hum. Genet. 92: 367-372, 1993.
Nebert, D. W.: Polymorphisms in drug-metabolizing enzymes: what is their clinical relevance and why do they exist? (Editorial) Am. J. Hum. Genet. 60: 265-271, 1997.
Nebert, D. W.: Personal Communication. Bethesda, Md. 2/1987.
Nebert, D. W.; Adesnik, M.; Coon, M. J.; Estabrook, R. W.; Gonzalez, F. J.; Guengerich, F. P.; Gunsalus, I. C.; Johnson, E. F.; Kemper, B.; Levin, W.; Phillips, I. R.; Sato, R.; Waterman, M. R.: The P450 gene superfamily: recommended nomenclature. DNA 6: 1-11, 1987.
Nelson, D. R.; Koymans, L.; Kamataki, T.; Stegeman, J. J.; Feyereisen, R.; Waxman, D. J.; Waterman, M. R.; Gotoh, O.; Coon, M. J.; Estabrook, R. W.; Gunsalas, I. C.; Nebert, D. W.: Cytochrome P450 superfamily: update on new sequences, gene mapping, accession numbers, and nomenclature. Pharmacogenetics 6: 1-42, 1996.
Oates, N. S.; Shah, R. R.; Idle, J. R.; Smith, R. L.: Genetic polymorphism of phenformin 4-hydroxylation. Clin. Pharm. Therap. 32: 81-89, 1982.
Roots, I.; Drakoulis, N.; Ploch, M.; Heinmeyer, G.; Loddenkemper, R.; Minks, T.; Nitz, M.; Otte, F.; Koch, M.: Debrisoquine hydroxylation phenotype, acetylation phenotype, and ABO blood groups as genetic host factors of lung cancer risk. Lin. Wochenschr. 66: 87-97, 1988.
Sachse, C.; Brockmoller, J.; Bauer, S.; Roots, I.: Cytochrome P450 2D6 variants in a Caucasian population: allele frequencies and phenotypic consequences. Am. J. Hum. Genet. 60: 284-295, 1997.
Saxena, R.; Shaw, G. L.; Relling, M. V.; Frame, J. N.; Moir, D. T.; Evans, W. E.; Caporaso, N.; Weiffenbach, B.: Identification of a new variant CYP2D6 allele with a single base deletion in exon 3 and its association with the poor metabolizer phenotype. Hum. Molec. Genet. 3: 923-926, 1994.
Silas, J. H.; Lennard, M. S.; Tucker, G. T.; Smith, A. J.; Malcolm, S. L.; Marten, T. R.: Why hypertensive patients vary in their response to oral debrisoquine. Brit. Med. J. 1: 422-425, 1977.
Skoda, R. C.; Gonzalez, F. J.; Demierre, A.; Meyer, U. A.: Two mutant alleles of the human cytochrome P-450db1 gene (P450C2D1) associated with genetically deficient metabolism of debrisoquine and other drugs. Proc. Nat. Acad. Sci. 85: 5240-5243, 1988.
Smith, C. A. D.; Gough, A. C.; Leigh, P. N.; Summers, B. A.; Harding, A. E.; Maranganore, D. M.; Sturman, S. G.; Schapira, A. H. V.; Williams, A. C.; Spurr, N. K.; Wolf, C. R.: Debrisoquine hydroxylase gene polymorphism and susceptibility to Parkinson's disease. Lancet 339: 1375-1377, 1992.
Tucker, G. T.; Silas, J. H.; Iyun, A. O.; Lennard, M. S.; Smith, A. J.: Polymorphic hydroxylation of debrisoquine. (Letter) Lancet II: 718, 1977.
Waring, R. H.; Mitchell, S. C.; Idle, J. R.; Smith, R. L.: Genetically determined impaired drug sulphoxidation. (Letter) Lancet I: 778, 1981.
Zhou, H.-H.; Koshakji, R. P.; Silberstein, D. J.; Wilkinson, G. R.; Wood, A. J. J.: Racial differences in drug response: altered sensitivity to and clearance of propranolol in men of Chinese descent as compared with American whites. New Eng. J. Med. 320: 565-570, 1989.

Nasopharyngeal Tumors

NASOPHARYNGEAL CANCER (161550)

JJM NOTE

This is another example of an ecogenetic cancer: many environmental determinants are recognized, but evidence grows for a genetic predisposition as well. One case of nasopharyngeal cancer in a patient with the Saethre-Chotzen syndrome of craniosynostosis and facial assymetry is likely a chance occurrence (McKeen EA et al.: The concurrence of Saethre-Chotzen syndrome and malignancy in a family with in vitro immune dysfunction. Cancer 54:2946–51, 1984).

Although perhaps not mendelian, a strong genetic factor in nasopharyngeal cancer is suggested by the high frequency in Cantonese Chinese living in China or elsewhere. Familial aggregation has been observed. The frequency of nasopharyngeal carcinoma (NPC) is said to be nearly 100-fold higher in southern Chinese than in most European populations. Males are more frequently affected than females. The age group affected is younger than most adult cancer patients. Japanese and Koreans, although related to the Chinese, have a low frequency. The Epstein-Barr virus may be implicated. Gajwani et al. (1980) described nasopharyngeal carcinoma in 4 sibs, 2 of them dizygotic twins. They came from a sibship of 11, all born in the U.S. of Poland-born nonconsanguineous parents. Fischer et al. (1984) described 2 affected brothers with an affected male first cousin, all born in Australia of Greek extraction. The brothers were nonsmokers and used little alcohol. In 3 generations of an American family of Scandinavian descent, 3,4:Coffin et al. (1989, 1991) found 5 cases of nasopharyngeal carcinoma and 6 other malignancies. Moreover, there was a history of autoimmune disorders, including pernicious anemia, thyroid disease, and psoriasis, in this family. The authors observed an increased frequency of the HLA haplotype A1-B37-DR6 in affected family members. Chan et al. (1983) found relative risks slightly over 2-fold for the haplotype A2,Bw46 and the antigen B17. Lu et al. (1990) reported a linkage study based on affected sib pairs which suggested that a gene closely linked to the HLA region confers a greatly increased risk of nasopharyngeal carcinoma. The maximum likelihood estimate placed the relative risk at approximately 21.

See 191170.0024 for description of a point mutation in the p53 gene in nasopharyngeal carcinoma. Chakrani et al. (1995) studied 41 primary tumors of the undifferentiated nasopharyngeal cancer type from Hong Kong and the Guangxi province of southeastern China. Four point mutations were found clustered in exon 5 within the DNA stretch from codon 175 to 177.

See also Kirk et al. (1978); Wen (1975)

Chakrani, F.; Armand, J.-P.; Lenoir, G.; Ju, L.; Liang, J.-P.; May, E.; May, P.: Mutations clustered in exon 5 of the p53 gene in primary nasopharyngeal carcinomas from southeastern Asia. Int. J. Cancer 61: 316-320, 1995.

Chan, S. H.; Day, N. E.; Kunaratnam, N.; Chia, K. B.; Simons, M. J.: HLA and nasopharyngeal carcinoma in Chinese—a further study. Int. J. Cancer 32: 171-176, 1983.
Coffin, C. M.; Dehner, L. P.; Rich, S. S.: Familial aggregation of nasopharyngeal carcinoma and other malignancies in a single Caucasian kindred. (Abstract) Am. J. Hum. Genet. 45 (suppl.): A18 only, 1989.
Coffin, C. M.; Rich, S. S.; Dehner, L. P.: Familial aggregation of nasopharyngeal carcinoma and other malignancies: a clinicopathologic description. Cancer 68: 1323-1328, 1991.
Fischer, A.; Fischer, G. O.; Cooper, E.: Familial nasopharyngeal carcinoma. Pathology 16: 23-24, 1984.
Gajwani, B. W.; Devereaux, J. M.; Beg, J. A.: Familial clustering of nasopharyngeal carcinoma. Cancer 46: 2325-2327, 1980.
Kirk, R. L.; Blake, N. M.; Serjeantson, S.; Simons, M. J.; Chan, S. H.: Genetic components in susceptibility to nasopharyngeal carcinoma.In: de The, G.; Ito, Y.: Nasopharyngeal Carcinoma: Etiology and Control. Lyon, France: 1 ARC (pub.) 1978. Pp. 283-297.
Lu, S.; Day, N. E.; Degos, L.; Lepage, V.; Wang, P.-C.; Chan, S.-H.; Simons, M.; McKnight, B.; Easton, D.; Zeng, Y.; de-The, G.: Linkage of a nasopharyngeal carcinoma susceptibility locus to the HLA region. Nature 346: 470-471, 1990.
Wen, C.-P.: Nasopharyngeal cancer.In: Quinn, J. R.: China Medicine as We Saw It. Bethesda, Md.: DHEW (pub.) (NIH) 75-684: 1975. Pp. 289-344.

ACROCEPHALOSYNDACTYLY 3 (SAETHRE-CHOTZEN SYNDROME) (#101400)

JJM NOTE

This entity is included because of one affected patient with nasopharyngeal carcinoma. Although likely a chance coincidence, the craniofacial dysmorphology might alter air passage enough to contribute to carcinogenesis by an airborne agent.

A number sign (#) is used with this entry because of evidence that the Saethre-Chotzen syndrome is caused by mutations in the TWIST transcription factor gene (601622). Although most patients with the Saethre-Chotzen syndrome have mutations in the TWIST gene, some patients with an overlapping phenotype have a mutation in the FGFR3 gene (134934.0014) and at least 1 individual has been described with a mutation in the FGFR2 gene (176943.0023).

In the family described by Saethre (1931), a mother, 2 daughters, and probably other maternal relatives showed mild acrocephaly, asymmetry of the skull, and partial soft tissue syndactyly of fingers 2 and 3 and toes 3 and 4. Chotzen (1932) found identical malformations in a father and 2 sons. Bartsocas et al. (1970) described a Lithuanian kindred living in the United States in which 10 persons in 3 generations were affected, with several instances of male-to-male transmission. Waardenburg et al. (1961) reported asymmetry of the skull and orbits (plagiocephaly), strabismus, and a thin, long, pointed nose in 6 generations of a kindred. Some affected persons had bifid terminal phalanges of digits 2 and 3 and absence of the first metatarsal. Cleft palate, hydrophthalmos, cardiac malformation, and contractures of elbows and knees were present in some. Aase and Smith (1970) described a syndrome comprising asymmetry of the face (hypoplasia of the left side), unusually shaped ear with prominent crus (see their Fig. 2), and Simian crease in 5 members of 3 generations (with 1 instance of male-to-male transmission). They pointed out similarities to and differences from the asymmetry of the face and skull with abnormalities of the digits described by Waardenburg et al. (1961). Gorlin (1971) thought the syndrome described by Aase and Smith (1970) was Chotzen syndrome. Carter et al. (1982) recognized 9 patients, including familial cases. Like Aase and Smith (1970), they recognized a long and prominent ear crus as a valuable sign. Kurczynski and Casperson (1988) described mother and daughter with craniosynostosis and symmetrical syndactyly involving the fourth and fifth toes. In addition, both had a short columella and small pinnae. Kurczynski and Casperson (1988) concluded that this represented a new form of acrocephalosyndactyly and suggested the designation auralcephalosyndactyly (109050). Legius et al. (1989) described mother and son with bilate ral symmetrical syndactyly of the third, fourth and fifth toes, mild craniosynostosis, and small pinnae. In addition, the mother had fusion of 2 cervical vertebrae and partial duplication of the first metatarsal. Furthermore, the distal phalanges of both great toes were bifid. These skeletal changes in combination with cutaneous syndactyly of the toes, abnormal auricles, and acrocephaly have been described in the Saethre-Chotzen syndrome (Kopysc et al., 1980) and also in the Robinow-Sorauf syndrome (Carter et al., 1982). Legius et al. (1989) concluded that the Saethre-Chotzen, auralcephalosyndactyly, and Robinow-Sorauf (180750) syndromes may be somewhat different expressions of the same dominant gene. Marini et al. (1991) presented a family illustrating the mild and easily missed expression of the gene in a parent. Niemann-Seyde et al. (1991) observed ACS III in 9 members of 4 generations of a family; 5 of them were severely affected. Russo et al. (1991) described a case of renotubular dysgenesis (267430) in an infant who had widely patent cranial fontanels and whose father and sister showed acrocephalosyndactyly of the Saethre-Chotzen type. This was probably a coincidental association between a recessive disorder and a dominant disorder.

See craniosynostosis (123100) for well-established mapping to 7p21.3-p21.2 on the basis of structural alterations in that region. The gene for Greig cephalopolysyndactyly syndrome (GCPS; 175700) appears to be located at 7p13. Brueton et al. (1992) presented molecular genetic linkage studies suggesting localization of the gene for the Saethre-Chotzen syndrome on distal 7p. Sixteen families with involvement in 2 or more generations were available for study. One of their families (number 16) had characteristics suggesting the Jackson-Weiss syndrome (123150). Excluding this family and pedigree number 15 which had a Pfeiffer-like syndrome (101600), Brueton et al. (1992) found tight linkage to D7S370 (maximum lod = 3.00 at theta = 0.00) and with D7S10 (maximum lod = 2.39 at theta = 0.00). The relationship to other forms of craniosynostosis with hand anomalies that map to 7p remains to be determined. In linkage analysis on 6 ACS III families using 5 CA repeat polymorphisms from 7p, Malcolm et al. (1993) found evidence suggesting location between D7S493 and D7S516. Two patients, a father and daughter, were found with ACS III and a balanced translocation t(7;10)(p21;q21.2). Reid et al. (1993) reported 2 additional patients, a male infant and his mother, with an apparently balanced translocation t(2;7)(p23;p22). Lewanda et al. (1994) confirmed linkage of the Saethre-Chotzen syndrome to 7p. The tightest linkage was to D7S493; linkage and haplotype analyses refined the location of the gene to the region between D7S513 and D7S516. On the basis of 4 patients with apparently balanced translocations at 7p21.2, Rose et al. (1994) narrowed the localization of the ACS3 gene to a 6-cM region. By fluorescence in situ hybridization, they showed that the breakpoints were situated within the region flanked by genetic markers D7S488 and D7S493 in distal 7p. Lewanda et al. (1994) used linkage and haplotype analyses to narrow the disease locus to an 8-cM region between D7S664 and D7S507. The tightest linkage was to D7S664; maximum lod = 7.16 at theta = 0.00. Studying the t(2;7)(p23;p22) in a patient with Saethre-Chotzen syndrome, Lewanda et al. (1994) found that the D7S664 locus lay distal to the 7p22 breakpoint, whereas the D7S507 locus was deleted from the translocation chromosome. Wilkie et al. (1995) reported 3 further families, each segregating a different reciprocal chromosomal translocation involving 7p21. A total of 7 apparently balanced carriers were identified and all manifest features of the Saethre-Chotzen syndrome, although only 2 had overt craniosynostosis. In 1 family, the carriers were immediately recognized by their unusual ears, and clefts of the hard or soft palate were present in all 3 families. The abnormally configured ear was pictured in 1 member from each of 3 generations.

Ma et al. (1996) studied 3 further families to provide additional support to the localization of a disease gene between D7S493 and D7S664. There was a suspicion that at least 2 disease-causing genes may map to 7p, 1 distal and 1 proximal to D7S488. The MEOX2 gene (600535) maps to the same region of 7p (as does SCS), and is a major candidate gene in SCS, as it is expressed in the mesenchyma of craniofacial and limb structures during early mouse embryogenesis.

Reardon and Winter (1994) wrote as follows: 'Clinical geneticists are inured to anecdotes recounting odd presentations of dysmorphic syndromes. Saethre-Chotzen syndrome is a case in point. A consultation for schizophrenia led to the first report from the Norwegian psychiatrist, Haakon Saethre. . .' (Saethre, 1931). Chotzen (1932) reported a father and 2 sons with the syndrome that came to carry his name.

Howard et al. (1997) and El Ghouzzi et al. (1997) demonstrated that the Saethre-Chotzen syndrome results from mutations in the TWIST gene (601622). They were prompted to evaluate the TWIST gene, which encodes a basic helix-loop-helix transcription factor, because its expression pattern and mutant phenotypes in Drosophila and mouse are consistent with the SCS phenotype in humans. Howard et al. (1997) mapped the human TWIST gene by PCR analysis of somatic cell hybrids to 7p22-p21 in a region homologous to the region of mouse chromosome 12 where the murine TWIST gene had been mapped. They assigned it to a specific YAC which was known to contain the breakpoint of a chromosome translocation in 1 Saethre-Chotzen syndrome case. Bourgeois et al. (1996) had previously cloned human TWIST and mapped it to 7p21. Howard et al. (1997) identified nonsense, missense, insertion, and deletion mutations in TWIST in patients with Saethre-Chotzen syndrome. El Ghouzzi et al. (1997) reported 21-bp insertions and nonsense mutations in the TWIST gene in 7 probands with SCS.

Paznekas et al. (1998) screened 32 unrelated patients with features of Saethre-Chotzen syndrome for mutations in the TWIST, FGFR2 (176943), and FGFR3 (134934) genes. Nine novel and 3 recurrent TWIST mutations were found in 12 families. Seven families were found to have the FGFR3 P250R mutation (134934.0014), and 1 individual was found to have a VV269-270 deletion of the FGFR2 gene

(176943.0023). The detection rate for TWIST or FGFR mutations, including 5 previously reported patients with TWIST mutations, was 68% in Saethre-Chotzen syndrome patients. More than 35 different TWIST mutations had been reported. The most common phenotypic features, present in more than one-third of Saethre-Chotzen patients with TWIST mutations, were coronal synostosis, brachycephaly, low frontal hairline, facial asymmetry, ptosis, hypertelorism, broad great toes, and clinodactyly. Significant intra- and interfamilial phenotypic variability was present for either TWIST or FGFR mutations.

See also Bianchi et al. (1985); Escobar et al. (1977); Kreiborg et al. (1972); Lewanda et al. (1994); McKeon-Kern and Mamunes (1977); Pantke et al. (1975)

Aase, J. M.; Smith, D. W.: Facial asymmetry and abnormalities of palms and ears: a dominantly inherited developmental syndrome. J. Pediat. 76: 928-930, 1970.

Bartsocas, C. S.; Weber, A. L.; Crawford, J. D.: Acrocephalosyndactyly type 3: Chotzen's syndrome. J. Pediat. 77: 267-272, 1970.

Bianchi, E.; Arico, M.; Podesta, A. F.; Grana, M.; Fiori, P.; Beluffi, G.: A family with the Saethre-Chotzen syndrome. Am. J. Med. Genet. 22: 649-658, 1985.

Bourgeois, P.; Stoetzel, C.; Bolcato-Bellemin, A. L.; Mattei, M. G.; Perrin-Schmitt, F.: The human H-twist gene is located at 7p21 and encodes a B-HLH protein that is 96% similar to its murine M-twist counterpart. Mammalian Genome 7: 915-917, 1996.

Brueton, L. A.; van Herwerden, L.; Chotai, K. A.; Winter, R. M.: The mapping of a gene for craniosynostosis: evidence for linkage of the Saethre-Chotzen syndrome to distal chromosome 7p. J. Med. Genet. 29: 681-685, 1992.

Carter, C. O.; Till, K.; Fraser, V.; Coffey, R.: A family study of craniosynostosis, with probable recognition of a distinct syndrome. J. Med. Genet. 19: 280-285, 1982.

Chotzen, F.: Eine eigenartige familiaere Entwicklungsstoerung (Akrocephalosyndaktylie, Dysostosis craniofacialis und Hypertelorismus). Mschr. Kinderheilk. 55: 97-122, 1932.

El Ghouzzi, V.; Le Merrer, M.; Perrin-Schmitt, F.; Lajeunie, E.; Benit, P.; Renier, D.; Bourgeois, P.; Bolcato-Bellemin, A.-L.; Munnich, A.; Bonaventure, J.: Mutations of the TWIST gene in the Saethre-Chotzen syndrome. Nature Genet. 15: 42-46, 1997.

Escobar, V.; Brandt, I. K.; Bixler, D.: Unusual association of Saethre-Chotzen syndrome and congenital adrenal hyperplasia. Clin. Genet. 11: 365-371, 1977.

Gorlin, R. J.: Personal Communication. Minneapolis, Minn. 1971.

Howard, T. D.; Paznekas, W. A.; Green, E. D.; Chiang, L. C.; Ma, N.; Ortiz De Luna, R. I.; Delgado, C. G.; Gonzalez-Ramos, M.; Kline, A. D.; Jabs, E. W.: Mutations in TWIST, a basic helix-loop-helix transcription factor, in Saethre-Chotzen syndrome. Nature Genet. 15: 36-41, 1997.

Kopysc, Z.; Stanska, M.; Ryzko, J.; Kulczyk, B.: The Saethre-Chotzen syndrome with partial bifid of the distal phalanges of the great toes: observations of three cases in one family. Hum. Genet. 56: 195-204, 1980.

Kreiborg, S.; Pruzansky, S.; Pashayan, H.: The Saethre-Chotzen syndrome. Teratology 6: 287-294, 1972.

Kurczynski, T. W.; Casperson, S. M.: Auralcephalosyndactyly: a new hereditary craniosynostosis syndrome. J. Med. Genet. 25: 491-493, 1988.

Legius, E.; Fryns, J. P.; Van den Berghe, H.: Auralcephalosyndactyly: a new craniosynostosis syndrome or a variant of the Saethre-Chotzen syndrome?. J. Med. Genet. 26: 522-524, 1989.

Lewanda, A. F.; Cohen, M. M., Jr.; Jackson, C. E.; Taylor, E. W.; Li, X.; Beloff, M.; Day, D.; Clarren, S. K.; Ortiz, R.; Garcia, C.; Hauselman, E.; Figueroa, A.; Wulfsberg, E.; Wilson, M.; Warman, M. L.; Padwa, B. L.; Whiteman, D. A. H.; Mulliken, J. B.; Jabs, E. W.: Genetic heterogeneity among craniosynostosis syndromes: mapping the Saethre-Chotzen syndrome locus between D7S513 and D7S516 and exclusion of Jackson-Weiss and Crouzon syndrome loci from 7p. Genomics 19: 115-119, 1994.

Lewanda, A. F.; Green, E. D.; Weissenbach, J.; Jerald, H.; Taylor, E.; Summar, M. L.; Phillips, J. A., III; Cohen, M.; Feingold, M.; Mouradian, W.; Clarren, S. K.; Jabs, E. W.: Evidence that the Saethre-Chotzen syndrome locus lies between D7S664 and D7S507, by genetic analysis and detection of a microdeletion in a patient. Am. J. Hum. Genet. 55: 1195-1201, 1994.

Ma, H. W.; Lajeunie, E.; de Parseval, N.; Munnich, A.; Renier, D.; Le Merrer, M.: Possible genetic heterogeneity in the Saethre-Chotzen syndrome. Hum. Genet. 98: 228-232, 1996.

Malcolm, S.; Rose, C. P. S.; van Herwerden, L.; Reardon, W.; Brueton, L.; Weissenbach, J.; Winter, R. M.: Mapping of Saethre-Chotzen syndrome (ACS III) to 7p21. (Abstract) Am. J. Hum. Genet. 53 (suppl.): A136 only, 1993.

Marini, R.; Temple, K.; Chitty, L.; Genet, S.; Baraitser, M.: Pitfalls in counselling: the craniosynostoses. J. Med. Genet. 28: 117-121, 1991.

McKeon-Kern, C.; Mamunes, P.: A case of Saethre-Chotzen syndrome. Med. Coll. Va. Quart. 13(4): 186-188, 1977.

Niemann-Seyde, S. C.; Eber, S. W.; Zoll, B.: Saethre-Chotzen syndrome (ACS III) in four generations. Clin. Genet. 40: 271-276, 1991.

Pantke, O. A.; Cohen, M. M., Jr.; Witkop, C. J., Jr.; Feingold, M.; Schaumann, B.; Pantke, H. C.; Gorlin, R. J.: The Saethre-Chotzen syndrome. Birth Defects Orig. Art. Ser. XI(2): 190-225, 1975.

Paznekas, W. A.; Cunningham, M. L.; Howard, T. D.; Korf, B. R.; Lipson, M. H.; Grix, A. W.; Feingold, M.; Goldberg, R.; Borochowitz, Z.; Aleck, K.; Mulliken, J.; Yin, M.; Jabs, E. W.: Genetic heterogeneity of Saethre-Chotzen syndrome, due to TWIST and FGFR mutations. Am. J. Hum. Genet. 62: 1370-1380, 1998.

Reardon, W.; Winter, R. M.: Saethre-Chotzen syndrome. J. Med. Genet. 31: 393-396, 1994.

Reid, C. S.; McMorrow, L. E.; McDonald-McGinn, D. M.; Grace, K. J.; Ramos, F. J.; Zackai, E. H.; Cohen, M. M., Jr.; Jabs, E. W.: Saethre-Chotzen syndrome with familial translocation at chromosome 7p22. Am. J. Med. Genet. 47: 637-639, 1993.

Rose, C. S. P.; King, A. A. J.; Summers, D.; Palmer, R.; Yang, S.; Wilkie, A. O. M.; Reardon, W.; Malcolm, S.; Winter, R. M.: Localization of the genetic locus for Saethre-Chotzen syndrome to a 6 cM region of chromosome 7 using four cases with apparently balanced translocations at 7p21.2. Hum. Molec. Genet. 3: 1405-1408, 1994.

Russo, R.; D'Armiento, M.; Vecchione, R.: Renal tubular dysgenesis and very large cranial fontanels in a family with acrocephalosyndactyly S.C. type. Am. J. Med. Genet. 39: 482-485, 1991.

Saethre, M.: Ein Beitrag zum Turmschaedelproblem (Pathogenese, Erblichkeit und Symptomatologie). Dtsch. Z. Nervenheilk. 119: 533-555, 1931.

Waardenburg, P. J.; Franceschetti, A.; Klein, D.: Genetics and Ophthalmology. Springfield, Ill.: Charles C Thomas (pub.) 1: 1961. Pp. 301-354.

Wilkie, A. O. M.; Yang, S. P.; Summers, D.; Poole, M. D.; Reardon, W.; Winter, R. M.: Saethre-Chotzen syndrome associated with balanced translocations involving 7p21: three further families. J. Med. Genet. 32: 174-180, 1995.

Cardiac Tumors

CARDIOMYOPATHY, FAMILIAL IDIOPATHIC (*115200)

JJM NOTE

This entity is included because of the hypertrophy that occurs.

See also CDCD2 (601154), which maps to chromosome 3p25-p22. Whitfield (1961) described a family in which 10 members were suffering or had died from cardiomyopathy and 6 others were probably affected. Although both males and females were affected, transmission seemingly occurred only through the female. Schrader et al. (1961) described 2 sisters with familial idiopathic cardiomegaly. Almost certainly the mother, who died at age 34, and probably 1 brother, who died at age 16, had the same condition. In the family reported by Battersby and Glenner (1961), affected persons were limited to 1 sibship and deposits of a nonmetachromatic, diastase-resistant, PAS-positive polysaccharide were described in the myocardium. Undoubtedly heterogeneity exists in the group of cardiomyopathies. Boyd et al. (1965) suggested that there may be 3 types: (1) form with predominant fibrosis, (2) form with predominant hypertrophy (see ventricular hypertrophy, hereditary; 192600), and (3) form with deposits described above. See amyloidosis III (176300.0007) for another familial myocardopathy. Kariv et al. (1966) observed 6 affected persons in 3 generations. In 2 of these persons, Adams-Stokes attacks required an artificial pacemaker. The affected males showed significant increase in the serum levels of multiple muscle-derived enzymes. Heterogeneity was suggested by the finding of normal serum enzyme levels in affected members of a second family. Rywlin et al. (1969) favored the view that obstructive and nonobstructive forms of familial cardiopathy are different expressions of a single entity. Classification into 'hypertrophic' and 'congestive' clinical types by Goodwin (1970) implies the same. Sommer et al. (1972) took an opposite view, i.e., that there is a separate nonobstructive familial cardiomyopathy. They described an Amish family with affected persons in 3 generations. Severity varied widely. The most severely affected pursued a rapidly fatal course whereas others manifested mainly conduction defects compatible with long survival. Machida et al. (1971) described a Japanese family with affected persons in 2 and perhaps 3 generations with male-to-male transmission. Emanuel et al. (1971) suggested that both dominant and recessive forms may exist. The possibility of an autosomal recessive form of congestive cardiomyopathy was raised by Yamaguchi et al. (1977), who found an astoundingly high rate of parental consanguinity (about 64%) and a segregation ratio of 0.196 consistent with autosomal recessive inheritance.

Fragola et al. (1988) studied 44 first-degree relatives of 12 probands with idiopathic dilated cardiomyopathy. Affected relatives were identified in 4 of 12 families. In each case, the affected relatives were sibs. This may be due to a late age of onset for expression of genetic factors involved in the etiology of this condition. An asterisk seems justified with this entry since there appears to be at least one autosomal dominant form of cardiomyopathy separate from other entries. We studied a kindred in which bizarre ventricular arrhythmia dominated the clinical picture in some, congestive cardiomyopathy in others, in 3 generations (P15207). Buchner et al. (1978) reviewed studies of the hereditary cardiomyopathy, a recessive in the golden hamster, discovered by Hamburger (1962). The genetic defect is thought to concern actomyosin. Moller et al. (1979) described an autosomal dominant form of congestive cardiomyopathy. The earliest sign of the disease was arrhythmia and/or conduction defects. Symptoms of pump failure had their onset in adulthood. Three members of the extensively affected kindred had died suddenly. Septal hypertrophy was found in 2 affected persons. O'Connell et al. (1984) used endomyocardial biopsy and gallium-67 scans in patients with dilated cardiomyopathy to demonstrate a subset of patients with myocardial inflammation. Histologic confirmation was found at autopsy. A defect in suppressor lymphocyte function was found in 1 patient, who showed improvement with immunosuppressive therapy. In 1 family, 5 persons in 3 generations were affected; in another, a father and 2 brothers were affected. A puzzling feature of a family with cardiomyopathy I once saw was striking pericardial effusion (Battersby and Glenner, 1961). Other early reports (e.g., Evans, 1949) have commented on inflammatory changes found at necropsy. Pericardial effusion occurs episodically with the iron-overload cardiomyopathy of multitransfused thalassemia and occurs also in the cardio myopathy of Friedreich ataxia (229300).

Ozick et al. (1984) reported identical twin sisters with congestive cardiomyopathy and autoimmune thyroid disease. Both had antithyroid microsomal antibodies and cytolytic antiheart myolemmal antibodies. The postpartum state may have been a factor in one of the twins; both cardiomyopathy and autoimmune thyroid disease may become clinically apparent in the postpartum period. Gardner et al. (1985) evaluated a kindred in which 12 persons had cardiomegaly with poor ventricular function and/or dysrhythmia. The disorder was evident by echocardiogram in a 6-month-old infant. Skeletal muscle biopsies showed subtle myopathic alterations. The pedigree, spanning 5 generations, was consistent with autosomal dominant inheritance. Gardner et al. (1987) described a family in which multiple members in 3 and probably 4 generations had dilated cardiomyopathy with overt clinical onset between the fourth and seventh decades. Dysrhythmia was frequent. They concluded that there might be an associated skeletal myopathy manifested by very mild proximal weakness or detectable only on biopsy. MacLennan et al. (1987) described 8 affected individuals, 4 of whom were males in 3 generations. Average age at presentation was 39.5 years. Average time to death from onset of symptoms suggestive of cardiomyopathy in 6 affected members was 16 months. One member died suddenly after being asymptomatic. The myocardium showed variation in muscle fiber size and interstitial fibrosis.

Graber et al. (1986) described a large kindred with an autosomal dominant form of disease of the cardiac conduction system and of the myocardium. Stage I occurred in the second and third decades and was characterized by absence of symptoms, normal heart size, sinus bradycardia, and premature atrial contractions. Stage II was marked by first-degree AV block in the third and fourth decades. Stage III occurred in the fourth and fifth decades and was accompanied by chest pain, fatigue, lightheadedness, and advanced AV block, followed by the development of atrial fibrillation or flutter. Stage IV, in the fifth and sixth decades of life, was characterized by congestive heart failure and recurrent ventricular arrhythmias. Right ventricular endomyocardial biopsy specimens showed progressive changes. At autopsy in the proband, the atrial changes were more severe than the ventricular ones. This suggested that the disorder discussed in entry 108770 is the same as this condition. While there was a range in the phenotypic expression of the inherited gene defect in this kindred, the dilated cardiomyopathy was less impressive than the dysrhythmia. Arrhythmias were the earliest manifestation of the disease (in the second to third decade). By linkage studies, Kass et al. (1994) demonstrated linkage of the disease locus to polymorphic loci near the centromere of chromosome 1; maximum multipoint lod score = 13.2 in the interval between D1S305 and D1S176. Based on the disease phenotype and the map location, Kass et al. (1994) speculated that the gap junction protein connexin 40 (121013) is a candidate for the site of mutations that result in conduction system disease and dilated cardiomyopathy.

Koike et al. (1987) described 2 families with dilated cardiomyopathy. In 1 of these families, the mode of inheritance was autosomal dominant; in the other, it appeared to be autosomal recessive. In both families, the pattern of inheritance was consistent with linkage to the HLA locus; however, because the families were small, the lod scores were low.

Schmidt et al. (1988) studied familial dilated cardiomyopathy in 6 families. The familial nature of the disorder was not readily apparent in 3 of these families until thorough family investigations were performed. The authors suggested that the family history should be reviewed in all patients with dilated cardiomyopathy and that further investigation of relatives should be performed if there are cases of unexplained heart disease, sudden unexpected death, or syncopal episodes. Echocardiography is a convenient noninvasive tool for these investigations. Early diagnosis is indicated for 2 reasons: treatment of significant arrhythmias may prevent sudden unexpected death, and genetic counseling can be provided. In studies of the first-degree relatives of 59 index cases with idiopathic dilated cardiomyopathy, Michels et al. (1992) found that 18 relatives from 12 families had dilated cardiomyopathy. Thus, 12 of the 59 index patients (20.3%) had familial disease. No differences in age, sex, severity of disease, exposure to selected environmental factors, or electrocardiographic or echocardiographic features were detected between the index patients with familial disease and those with nonfamilial disease. A noteworthy finding was that 22 of 240 healthy relatives (9.2%) with normal ejection fractions had increased left ventricular diameters during systole or diastole (or both), as compared with 2 of 112 healthy control subjects (1.8%) who were studied separately. In a case-control study of idiopathic dilated cardiomyopathy in Baltimore, a roughly 3-fold increase in risk was observed among blacks after adjustment for potential confounding variables (Coughlin et al., 1990). The increased frequency of dilated cardiomyopathy in black males was the basis in the past of the designation 'Osler-2 myocarditis'; Osler-2 was the black male ward at The Johns Hopkins Hospital.

Michels et al. (1993) performed PCR-based assays and Southern blot analysis of the dystrophin gene (DMD; 310200) in 27 males with idiopathic dilated cardiomyopathy. Five families had familial disease, without male-to-male transmission in 4 families. In the fifth family, there was no evidence of male-to-male transmission when the family was entered into the study, but on follow-up the index patient's son was found to have developed the disease. None of the patients had clinical evidence of skeletal muscle disease or any systemic illness that could cause heart disease. The mean age of the patients was 50.2 years; the range of age was 5 to 72 years. No dystrophin gene defects were found.

From linkage studies in 12 families, Olson et al. (1995) excluded genetic linkage between the disease phenotype and a 21-cM region spanning the HLA cluster in at least 60% of the families.

Csanady et al. (1995) compared 31 familial and 209 nonfamilial cases of dilated cardiomyopathy. They concluded that the familial form is more malignant: it occurs at an earlier age and progresses more rapidly than the nonfamilial form.

Among 100 patients with dilated cardiomyopathy, McKenna et al. (1997) found that familial prevalence was definite in 14 of 56 (25%) and possible in 25 of 56 (45%). The HLA-DR4 frequency in the 100 patients with dilated cardiomyopathy was similar to that in 9,000 controls; however, the DR4 subtype was significantly more common in the 25 probands with a familial tendency to dilated cardiomyopathy than in the 31 probands with nonfamilial dilated cardiomyopathy. McKenna et al. (1997) concluded that there is an HLA-linked predisposition to familial dilated cardiomyopathy.

See also Barry and Hall (1962); Biorck and Orinius (1964); Bishop et al. (1962); Michels et al. (1989)

Barry, M.; Hall, M.: Familial cardiomyopathy. Brit. Heart J. 24: 613-624, 1962.

Battersby, E. J.; Glenner, G. G.: Familial cardiomyopathy. Am. J. Med. 30: 382-391, 1961.

Biorck, G.; Orinius, E.: Familial cardiomyopathies. Acta Med. Scand. 176: 407-424, 1964.

Bishop, J. M.; Campbell, M.; Jones, E. W.: Cardiomyopathy in four members of a family. Brit. Heart J. 24: 715-728, 1962.

Boyd, D. L.; Mishkin, M. E.; Feigenbaum, H.; Genovese, P. D.: Three families with familial cardiomyopathy. Ann. Intern. Med. 63: 386-401, 1965.

Buchner, F.; Onishi, S.; Wada, A.: Cardiomyopathy Associated with Systemic Myopathy: Genetic Defect of Actomyocin Influencing Muscular Structure and Function. Baltimore and Munich: Urban and Schwarzenberg (pub.) 1978.

Coughlin, S. S.; Szklo, M.; Baughman, K.; Pearson, T. A.: The epidemiology of idiopathic dilated cardiomyopathy in a biracial community. Am. J. Epidemiol. 131: 48-56, 1990.

Csanady, M.; Hogye, M.; Kallai, A.; Forster, T.; Szarazajtai, T.: Familial dilated cardiomyopathy: a worse prognosis compared with sporadic forms. Brit. Heart J. 74: 171-173, 1995.

Emanuel, R.; Withers, R.; O'Brien, K.: Dominant and recessive modes of inheritance in idiopathic cardiomyopathy. Lancet II: 1065-1067, 1971.

Evans, W.: Familial cardiomegaly. Brit. Heart J. 11: 68-82, 1949.

Fragola, P. V.; Autore, C.; Picelli, A.; Sommariva, L.; Cannata, D.; Sangiorgi, M.: Familial idiopathic dilated cardiomyopathy. Am. Heart J. 115: 912-914, 1988.

Gardner, R. J. M.; Ardinger, H. H.; Florentine, M. S.; Hanson, J. W.; Hart, M. N.; Hinrichs, R. L.; Ionasescu, V. V.; Mahoney, L. T.; Rose, E. E.; Skorton, D. J.: Dominantly inherited dilated cardiomyopathy with skeletal myopathy. (Abstract) Am. J. Hum. Genet. 37: A54, 1985.

Gardner, R. J. M.; Hanson, J. W.; Ionasescu, V. V.; Ardinger, H. H.; Skorton, D. J.; Mahoney, L. T.; Hart, M. N.; Rose, E. F.; Smith, W. L.; Florentine, M. S.; Hinrichs, R. L.: Dominantly inherited dilated cardiomyopathy. Am. J. Med. Genet. 27: 61-73, 1987.

Goodwin, J. F.: Congestive and hypertrophic cardiomyopathies. Lancet I: 731-739, 1970.

Graber, H. L.; Unverferth, D. V.; Baker, P. B.; Ryan, J. M.; Baba, N.; Wooley, C. F.: Evolution of a hereditary cardiac conduction and muscle disorder: a study involving a family with six generations affected. Circulation 74: 21-35, 1986.

Kariv, I.; Szeinberg, A.; Fabian, I.; Sherf, L.; Kreisler, B.; Zeltzer, M.: A family with cardiomyopathy. Am. J. Med. 40: 140-148, 1966.

Kass, S.; MacRae, C.; Graber, H. L.; Sparks, E. A.; McNamara, D.; Boudoulas, H.; Basson, C. T.; Baker, P. B., III; Cody, R. J.; Fishman, M. C.; Cox, N.; Kong, A.; Wooley, C. F.; Seidman, J. G.; Seidman, C. E.: A gene defect that causes conduction system disease and dilated cardiomyopathy maps to chromosome 1p1-1q1. Nature Genet. 7: 546-551, 1994.

Koike, S.; Kawa, S.; Yabu, K.; Endo, R.; Sasaki, Y.; Furuta, S.; Ota, M.: Familial dilated cardiomyopathy and human leucocyte antigen: a report of two family cases. Jpn. Heart J. 28: 941-945, 1987.

Machida, K.; Iguchi, K.; Yoshimi, S.; Saito, Y.; Sugishita, Y.; Murayama, M.; Mori, M.; Yamaguchi, H.; Ito, I.; Uede, H.: Familial cardiomyopathy: immunological studies and review of literatures on autopsied cases in Japan. Jpn. Heart J. 12: 40-49, 1971.

MacLennan, B. A.; Tsoi, E. Y.; Maguire, C.; Adgey, A. A. J.: Familial idiopathic congestive cardiomyopathy in three generations: a family study with eight affected members. Quart. J. Med. 63: 335-347, 1987.
McKenna, C. J.; Codd, M. B.; McCann, H. A.; Sugrue, D. D.: Idiopathic dilated cardiomyopathy: familial prevalence and HLA distribution. Heart 77: 549-552, 1997.
Michels, V. V.; Moll, P. P.; Miller, F. A.; Tajik, A. J.; Chu, J. S.; Driscoll, D. J.; Burnett, J. C.; Rodeheffer, R. J.; Chesebro, J. H.; Tazelaar, H. D.: The frequency of familial dilated cardiomyopathy in a series of patients with idiopathic dilated cardiomyopathy. New Eng. J. Med. 326: 77-82, 1992.
Michels, V. V.; Moll, P. P.; Miller, F. A.; Tajik, A. J.; Driscoll, D. J.; Chu, J. S.; Burnett, J. C.; Chesebro, J. H.; Rodeheffer, R. J.: Frequency of familial dilated cardiomyopathy in an unselected series of patients with idiopathic dilated cardiomyopathy. (Abstract) Am. J. Hum. Genet. 45 (suppl.): A55, 1989.
Michels, V. V.; Pastores, G. M.; Moll, P. P.; Driscoll, D. J.; Miller, F. A.; Burnett, J. C.; Rodeheffer, R. J.; Tajik, J. A.; Beggs, A. H.; Kunkel, L. M.; Thibodeau, S. N.: Dystrophin analysis in idiopathic dilated cardiomyopathy. J. Med. Genet. 30: 955-957, 1993.
Moller, P.; Lunde, P.; Hovig, T.; Nitter-Hauge, S.: Familial cardiomyopathy: autosomally, dominantly inherited congestive cardiomyopathy with two cases of septal hypertrophy in one family. Clin. Genet. 16: 233-243, 1979.
O'Connell, J. B.; Fowles, R. E.; Robinson, J. A.; Subramanian, R.; Henkin, R. E.; Gunnar, R. M.: Clinical and pathologic findings of myocarditis in two families with dilated cardiomyopathy. Am. Heart J. 107: 127-135, 1984.
Olson, T. M.; Thibodeau, S. N.; Lundquist, P. A.; Schaid, D. J.; Michels, V. V.: Exclusion of a primary defect at the HLA locus in familial idiopathic dilated cardiomyopathy. J. Med. Genet. 32: 876-880, 1995.
Ozick, H.; Hollander, G.; Greengart, A.; Shani, J.; Lichstein, E.: Dilated cardiomyopathy in identical twins. Chest 86: 878-880, 1984.
Rywlin, A. M.; Barold, S. S.; Linhart, J. W.; Kramer, H. C.; Meitus, M. L.; Samet, P.: Idiopathic familial cardiopathy: a study of two families. J. Genet. Hum. 17: 453-470, 1969.
Schmidt, M. A.; Michels, V. V.; Edwards, W. D.; Miller, F. A.: Familial dilated cardiomyopathy. Am. J. Med. Genet. 31: 135-143, 1988.
Schrader, W. H.; Pankey, G. A.; Davis, R. B.; Theologides, A.: Familial idiopathic cardiomegaly. Circulation 24: 599-606, 1961.
Sommer, A.; Sanz, G.; Craenen, J. M.; Newton, W. A., Jr.: Familial cardiomyopathy. Birth Defects Orig. Art. Ser. VIII(5): 178-181, 1972.
Whitfield, A. G. W.: Familial cardiomyopathy. Quart. J. Med. 30: 119-134, 1961.
Yamaguchi, M.; Toshima, H.; Yanase, T.; Ikeda, H.; Koga, Y.; Yoshioka, H.: Ito, M.; Fujino, T.; Yasuda, H.: A family study of idiopathic cardiomyopathy. Proc. Jpn. Acad. 53 (ser. B): 209-214, 1977.

VENTRICULAR HYPERTROPHY; HEREDITARY HYPERTROPHIC CARDIOMYOPATHY (#192600)

JJM NOTE

This entity is included because, in a sense, it represents hypertrophy, perhaps even hyperplasia of the interventricular septum. A study of cellular clonality of the pathologic tissue is needed (e.g., using DNA probes from regions of chromosome X that are inactivated in females with the disorder).

Description

A number sign (#) is used with this entry because at least one form of the disorder has been shown to be due to mutation in a cardiac myosin heavy chain gene (160760); by linkage studies, other genes causing this disorder have been identified, e.g., CMH2 (115195) and CMH3 (115196).

Hereditary ventricular hypertrophy (CMH1, ASH, or IHSS) in early stages produces a presystolic gallop due to an atrial heart sound, and EKG changes of ventricular hypertrophy. Progressive ventricular outflow obstruction may cause palpitation associated with arrhythmia, congestive heart failure, and sudden death.

Maron et al. (1996) collected information on 158 sudden deaths that had occurred in trained athletes throughout the United States from 1985 through 1995. In 24 athletes (15%), noncardiovascular causes were found. Among the 134 athletes who had cardiovascular causes of sudden death, the median age was 17 years. The most common competitive sports involved were basketball (47 cases) and football (45 cases), together accounting for 68% of sudden deaths. The most common structural cardiovascular diseases identified at autopsy as the primary cause of death were hypertrophic cardiomyopathy (48 athletes, 36%), which was disproportionately prevalent in black athletes compared with white athletes (48% vs 26% of deaths; P = 0.01), and malformations involving anomalous coronary artery origin (17 athletes, 13%). Of 115 athletes who had a standard preparticipation medical evaluation, only 4 (3%) were suspected of having cardiovascular disease, and the cardiovascular anomaly responsible for sudden death was correctly identified in only 1 athlete (0.9%).

Clinical Features

In the first demonstration of asymmetric hypertrophy of the heart in young adults, Teare (1958) reported the autopsy findings in 9 cases of sudden death in young subjects distributed in 6 families. This condition has been called muscular subaortic stenosis but more generalized ventricular hypertrophy is often an earlier and more impressive feature, and obstruction to outflow from the right ventricle can also occur. Study of the families of probands with the full-blown condition shows that an atrial heart sound ('presystolic gallop') and EKG changes of ventricular hypertrophy are the earliest signs. Sudden death occurs in some cases. Braunwald et al. (1964) reported in detail on 64 patients; multiple cases were observed in 11 families, which contained in all at least 41 definite or probable cases. As pointed out by Nasser et al. (1967), outflow obstruction may be absent in some affected members of families in which others do have outflow obstruction. Maron et al. (1974) studied 4 infants that died with ASH in the first 5 months of life, including 1 stillborn. ASH was demonstrated in one first-degree relative of each infant. Maron et al. (1976) analyzed the clinical picture of 46 children with ASH. On the basis of a study of an outpatient population, Spirito et al. (1989) suggested that the prognosis in hypertrophic cardiomyopathy may be less grave than has usually been considered on the basis of hospital-study patients.

On morphologic grounds, four types of hypertrophic cardiomyopathy have been described: type 1 with hypertrophy confined to the anterior segment of the ventricular septum; type 2 with hypertrophy of both the anterior and the posterior segments of the ventricular septum; type 3 with involvement of both the ventricular septum and the free wall of the left ventricle and type 4 with involvement of the posterior segment of the septum, the anterolateral free wall, or the apical half of the septum (Maron et al., 1982; Ciro et al., 1983). Apical hypertrophic cardiomyopathy is, therefore, one form of type IV. It was first described by Yamaguchi et al. (1979) in Japan (where it appears to be more frequent than elsewhere) and later by Maron et al. (1982). The cases of apical hypertrophic cardiomyopathy described by Maron et al. (1982) belonged to families with different forms of hypertrophic cardiomyopathy. Malouf et al. (1985) reported apical hypertrophic cardiomyopathy in father and daughter of a Lebanese Christian family. The parents were not related; an only sib was normal on examination and echocardiogram as were 2 sisters of the father and their 6 children.

In a meta-analysis of sudden death from cardiac causes in children and young adults, Liberthson (1996) found that hypertrophic cardiomyopathy was the most frequent cause of sudden death in young persons in association with strenuous physical exertion or sports.

Inheritance

In the family reported by Horlick et al. (1966), 10 persons in 4 generations were thought to have been affected. Pare et al. (1961) described this disorder in 30 out of 87 members of a French-Canadian kindred. The genealogic survey was carried back to the original emigrant from France in the 1600s. The pattern of occurrence over 5 generations and 160 years since the death of the man believed to be the first instance of the heart disease indicated autosomal dominant inheritance. Elevated paternal age of sporadic (possible fresh mutation) cases was observed by Jorgensen (1968). The family study of Clark et al. (1973), using echocardiography, indicated that 28 of 30 probands (93%) had an affected parent. This agrees well with estimates of the extent to which this disorder, on the average, reduces reproductive fitness.

Greaves et al. (1987) performed echocardiographic studies of 193 first-degree relatives of 50 patients with hypertrophic cardiomyopathy. More males than females were affected. In 28 of 50 families, familial occurrence was observed. In 15 families the pattern of inheritance was consistent with autosomal dominant inheritance; in the other 13 the affected members were in a single generation and the pattern of inheritance could not be determined.

The family reported by Yamaguchi et al. (1979) suggested X-linked recessive inheritance. Burn (1985) felt that the existence of a recessive form of hypertrophic cardiomyopathy (Emanuel et al., 1971; Branzi et al., 1985) could neither be established nor disproved at the time of his writing. Branzi et al. (1985) claimed the existence of an autosomal recessive form because of a family they found with 2 affected sisters and both parents normal by careful study. Formal segregation analysis supported the existence of 2 classes: one with a segregation ratio close to 50% and one with a value close to 25%.

Mapping

Darsee et al. (1979) found a lod score of 7.7 for linkage between ASH and HLA. They concluded that, in addition to the hereditary form linked to HLA, a sporadic unlinked form is associated with severe systemic hypertension. White patients with ASH were B12; black patients were B5. This presumably strong evidence placing a gene for hypertrophic subaortic stenosis on 6p by linkage to HLA was invalidated when the infamous John R. Darsee confessed fabrication of the data.

Nutter also published a retraction. Motulsky (1979) wrote a laudatory editorial to accompany the original article.

In his retraction letter, Darsee stated: 'The lod scores were calculated, in part, by one of the journal referees who felt they should be included, and partly by my own calculations. The biometrist I consulted at Emory regarding these calculations was not familiar with lod scores and unable to provide assistance.' Before Darsee confessed, Darsee and Heymsfield (1981) wrote: 'It is the pinhole through which we are forced to view this disease or these diseases that has helped confer a degree of homogeneity. The pinhole is the limited collection of tools we have to study hypertrophic cardiomyopathy—the angiogram, the echocardiogram, and the autopsy table. It is a common practice of even the most perspicacious and critical investigators to conclude that diseases that look the same on canvas were painted with the same brush.' Although these words are true in general terms and are a fine statement of the principle of genetic heterogeneity, the falsified data do not support them, of course.

Jarcho et al. (1989) did studies with DNA markers in the Canadian family originally reported by Pare et al. (1961). At the time of the study, hypertrophic cardiomyopathy had occurred in 20 surviving and 24 deceased family members. With a polymorphic DNA probe with the trivial name CRI-L436, which identified a DNA segment designated D14S26, they found no recombination (lod score = 9.37 at theta = 0). This probe had been assigned to chromosome 14 on the basis of somatic cell hybrid analysis (Donis-Keller et al., 1987). The gene encoding the alpha chain of the T-cell receptor (186880) was located approximately 20 cM from D14S26 (Mitchell et al., 1989). Solomon et al. (1990) mapped the probe CRI-L436 to 14q11-q12 by in situ hybridization. Because the cardiac myosin heavy chain genes (MYH6, 160710; MYH7) map to the same chromosomal band, they determined the genetic distance between the gene for the beta heavy chain of cardiac myosin, D14S26, and the CMH1 locus. They presented data indicating that these 3 loci are linked within 5 cM of each other. The data were consistent with the possibility that the CMH1 mutation is in either the alpha or the beta gene.

Hejtmancik et al. (1991) found that the gene for familial hypertrophic cardiomyopathy was located at 14q1 in 8 unrelated families of varied ethnic origins. Of 5 families with hypertrophic cardiomyopathy, Epstein et al. (1992) found linkage to chromosome 14 markers in one and suggestive linkage in a second. However, linkage to chromosome 14 markers was excluded in the other 3 kindreds. Ko et al. (1992) excluded linkage to D14S26 in a Chinese family, likewise indicating genetic heterogeneity. See 160760.0001 for a description of the molecular defect in chromosome 14-linked hypertrophic cardiomyopathy (symbolized CMH1). The symbols CMH2 and CMH3 are used for the forms of this disorder that map to 1q3 (115195) and 15q2 (115196), respectively. Since some families are not linked to any one of these 3 loci, there must be at least one other genetic form of familial hypertrophic cardiomyopathy.

Molecular Genetics

To screen for mutations that cause familial hypertrophic cardiomyopathy, Rosenzweig et al. (1991) capitalized on the fact that 'ectopic' or 'illegitimate' transcription of beta cardiac myosin heavy chain gene can be detected in blood lymphocytes. Preclinical or prenatal screening will make it possible to study the disorder longitudinally and to develop preventive interventions. The findings again illustrate the important application of PCR. Clarke and Harper (1992) suggested that 'the parallels between this cardiomyopathy and Huntington's disease are sufficiently striking that we would be very cautious about testing for it in childhood. The emotional consequences of being brought up under a cloud of doom may be damaging, and the lack of any uncertainty in identifying gene carriers by mutation analysis might paradoxically make this worse.' Watkins et al. (1992) countered this view, saying that children with the condition face a 4 to 6% risk of sudden death each year. Genetic diagnosis will allow evaluation of prophylactic use of antiarrhythmic agents or implantable defibrillator devices. It will also provide parents and physicians an appropriate basis on which to make decisions regarding the participation of children in competitive sports. They suggested that in their experience '. . .any perception of a cloud of doom comes as much from a lack of knowledge of and research into this inherited cardiomyopathy as from anything else.'

Phenotype/genotype

Fananapazir et al. (1993) demonstrated by biopsy of the soleus muscle the presence of central core disease of skeletal muscle (117000) in association with hypertrophic cardiomyopathy due to any of 4 different mutations in the MYH7 gene. Soleus muscle samples from patients in 4 kindreds in which hypertrophic cardiomyopathy was not linked to the MYH7 locus showed no myopathy or central core disease. In 1 family with the leu908-to-val mutation of the MYH7 gene (160760.0010), central core disease was demonstrated on soleus muscle biopsy, although cardiac hypertrophy was absent on echocardiogram in 2 adults and 3 children. Almost all patients had no significant muscle weakness, despite the histologic changes.

Heterogeneity

In affected members of an Italian family, Ferraro et al. (1990) found that 7 affected members and none of 3 unaffected members showed a fragile site on 16q (FRA16B). Solomon et al. (1990) demonstrated heterogeneity in familial hypertrophic cardiomyopathy: 1 of 3 families showed mapping of the gene to the 14q11-q12 region, whereas the other 2 were not linked to that region.

Pathogenesis

Wagner et al. (1989) investigated a possible role of adrenergic innervation or of cellular calcium regulation in pathogenesis, as suggested by the presence of hyperdynamic left ventricular function and by the clinical and symptomatic improvement seen in patients treated with beta-receptor antagonists or calcium antagonists. They found that calcium-antagonist binding sites, measured as the amount of dihydropyridine bound to atrial tissue, were increased by 33% in patients with hypertrophic cardiomyopathy. The densities of saxitoxin-binding sites on voltage-sensitive sodium channels and beta-adrenoceptors did not differ from controls. Wagner et al. (1989) interpreted the findings as suggesting that abnormal calcium fluxes through voltage-sensitive calcium channels may play a pathophysiologic role in the disease.

Clinical Management

Wilson et al. (1983) observed marked improvement in the manifestations of familial hypertrophic cardiomyopathy when affected persons with hyperthyroidism were treated for the latter condition. This prompted them to suggest that antithyroid therapy 'should be considered in this form of cardiomyopathy.'

In discussing the management of hypertrophic cardiomyopathy, Spirito et al. (1997) reviewed heterogeneity of clinical and genetic features and stated that 'the diverse clinical and genetic features of hypertrophic cardiomyopathy make it impossible to define precise guidelines for management.' The treatment of symptoms to improve quality of life and the identification of patients who are at high risk for sudden death and require aggressive therapy are 2 distinct issues that must be addressed by largely independent strategies. The stratification of risk and the prevention of sudden death were discussed.

Population Genetics

In a discussion of hypertrophic cardiomyopathy, Maron et al. (1987) stated that approximately 45% of cases are sporadic. New mutation cannot be the explanation for all of the sporadic cases; hence, there may be other etiologically distinct disorders represented in the group of hypertrophic cardiomyopathies. Systematic echocardiographic surveys of families of patients with hypertrophic cardiomyopathy have identified relatives older than 50 years of age with mild and localized left ventricular hypertrophy. Thus, the true proportion of sporadic cases may not be as high as 45%.

See also Bingle et al. (1975); Bulkley et al. (1977); Criley et al. (1965); Gardin et al. (1982); Goodwin and Krikler (1976); Hardarson et al. (1973); Haugland et al. (1986); Henry et al. (1973); Manchester (1963); Masuya et al. (1982); Powell et al. (1973); Smith et al. (1976); Solomon et al. (1990); Wei et al. (1980); Wood et al. (1962)

Bingle, G. J.; Dillon, J.; Hurwitz, R.: Asymmetric septal hypertrophy in a large Amish kindred. Clin. Genet. 7: 255-261, 1975.

Branzi, A.; Romeo, G.; Specchia, S.; Lolli, C.; Binetti, G.; Devoto, M.; Bacchi, M.; Magnani, B.: Genetic heterogeneity of hypertrophic cardiomyopathy. Int. J. Cardiol. 7: 129-133, 1985.

Braunwald, E.; Lambrew, C. T.; Rockoff, S. D.; Ross, J., Jr.; Morrow, A. G.: Idiopathic hypertrophic subaortic stenosis. I. A description of the disease based on an analysis of 64 patients. Circulation 30 (suppl. 4): 3-119, 1964.

Bulkley, B. H.; Wiesfeldt, M. L.; Hutchins, G. M.: Isometric cardiac contraction: a possible cause of the disorganized myocardial pattern of idiopathic hypertrophic subaortic stenosis. New Eng. J. Med. 296: 135-139, 1977.

Burn, J.: The genetics of hypertrophic cardiomyopathy. (Editorial) Int. J. Cardiol. 7: 135-138, 1985.

Ciro, E.; Nichols, P. F.; Maron, B. J.: Heterogeneous morphologic expression of genetically transmitted hypertrophic cardiomyopathy: two-dimensional echocardiographic analysis. Circulation 67: 1227-1233, 1983.

Clark, C. E.; Henry, W. L.; Epstein, S. E.: Familial prevalence and genetic transmission of idiopathic hypertrophic subaortic stenosis. New Eng. J. Med. 289: 709-714, 1973.

Clarke, A.; Harper, P.: Genetic testing for hypertrophic cardiomyopathy. (Letter) New Eng. J. Med. 327: 1175-1176, 1992.

Criley, J.; Lewis, K. B.; White, R. I., Jr.; Ross, R. S.: Pressure gradients without obstruction: a new concept of 'hypertrophic subaortic stenosis.'. Circulation 32: 881-887, 1965.

Darsee, J. R.; Heymsfield, S. B.: Decreased myocardial taurine levels and hypertaurinuria in a kindred with mitral-valve prolapse and congestive cardiomyopa-

thy. New Eng. J. Med. 304: 129-135, 1981. Note: Retraction: Heymsfield and Glenn, New Eng. J. Med. 308: 1400 only, 1983.
Darsee, J. R.; Heymsfield, S. B.; Nutter, D. O.: Hypertrophic cardiomyopathy and human leukocyte antigen linkage: differentiation of two forms of hypertrophic cardiomyopathy. New Eng. J. Med. 300: 877-882, 1979. Note: Retraction: Nutter, Heymsfield, and Glenn, New Eng. J. Med. 308: 1400 only, 1983.
Donis-Keller, H.; Green, P.; Helms, C.; Cartinhour, S.; Weiffenbach, B.; Stephens, K.; Keith, T. P.; Bowden, D. W.; Smith, D. R.; Lander, E. S.; et al.: A genetic linkage map of the human genome. Cell 51: 319-337, 1987.
Emanuel, R.; Withers, R.; O'Brien, K.: Dominant and recessive modes of inheritance in idiopathic cardiomyopathy. Lancet II: 1065-1067, 1971.
Epstein, N. D.; Fananapazir, L.; Lin, H. J.; Mulvihill, J.; White, R.; Lalouel, J.-M.; Lifton, R. P.; Nienhuis, A. W.; Leppert, M.: Evidence of genetic heterogeneity in five kindreds with familial hypertrophic cardiomyopathy. Circulation 85: 635-647, 1992.
Fananapazir, L.; Dalakas, M. C.; Cyran, F.; Cohn, G.; Epstein, N. D.: Missense mutations in the beta-myosin heavy-chain gene cause central core disease in hypertrophic cardiomyopathy. Proc. Nat. Acad. Sci. 90: 3993-3997, 1993.
Ferraro, M.; Scarton, G.; Ambrosini, M.: Cosegregation of hypertrophic cardiomyopathy and a fragile site on chromosome 16 in a large Italian family. J. Med. Genet. 27: 363-366, 1990.
Gardin, J. M.; Gottdiener, J. S.; Radvany, R.; Maron, B. J.; Lesch, M.: HLA linkage vs association in hypertrophic cardiomyopathy: evidence for the absence of an association in a heterogeneous Caucasian population. Chest 81: 466-472, 1982.
Goodwin, J. F.; Krikler, D. M.: Arrhythmia as a cause of sudden death in hypertrophic cardiomyopathy. Lancet II: 937-940, 1976.
Greaves, S. C.; Roche, A. H. G.; Neutze, J. M.; Whitlock, R. M. L.; Veale, A. M. O.: Inheritance of hypertrophic cardiomyopathy: a cross sectional and M mode echocardiographic study of 50 families. Brit. Heart J. 58: 259-266, 1987.
Hardarson, T.; Curiel, R.; de la Calzada, C. S.; Goodwin, J. F.: Prognosis and mortality of hypertrophic obstructive cardiomyopathy. Lancet II: 1462-1467, 1973.
Haugland, H.; Ohm, O.-J.; Boman, H.; Thorsby, E.: Hypertrophic cardiomyopathy in three generations of a large Norwegian family: a clinical, echocardiographic, and genetic study. Brit. Heart J. 55: 168-175, 1986.
Hejtmancik, J. F.; Brink, P. A.; Towbin, J.; Hill, R.; Brink, L.; Tapscott, T.; Trakhtenbroit, A.; Roberts, R.: Localization of gene for familial hypertrophic cardiomyopathy to chromosome 14q1 in a diverse U.S. population. Circulation 83: 1592-1597, 1991.
Henry, W. L.; Clark, C. E.; Epstein, S. E.: Asymmetric septal hypertrophy (ASH): the unifying link in the IHSS disease spectrum—observations regarding its pathogenesis, pathophysiology and course. Circulation 47: 827-832, 1973.
Horlick, L.; Petkovich, N. J.; Bolton, C. F.: Idiopathic hypertrophic subvalvular stenosis. A study of a family involving four generations. Clinical, hemodynamic and pathologic observations. Am. J. Cardiol. 17: 411-418, 1966.
Jarcho, J. A.; McKenna, W.; Pare, J. A. P.; Solomon, S. D.; Holcombe, R. F.; Dickie, S.; Levi, T.; Donis-Keller, H.; Seidman, J. G.; Seidman, C. E.: Mapping a gene for familial hypertrophic cardiomyopathy to chromosome 14q1. New Eng. J. Med. 321: 1372-1378, 1989.
Jorgensen, G.: Genetische Untersuchungen bei funktionell-obstruktiver subvalvulaerer Aortenstenose (irregulaer hypertrophische Kardiomyopathie). Humangenetik 6: 13-28, 1968.
Ko, Y.-L.; Lien, W.-P.; Chen, J.-J.; Wu, C.-W.; Tang, T.-K.; Liew, C.-C.: No evidence for linkage of familial hypertrophic cardiomyopathy and chromosome 14q1 locus D14S26 in a Chinese family: evidence for genetic heterogeneity. Hum. Genet. 89: 597-601, 1992.
Liberthson, R. R.: Sudden death from cardiac causes in children and young adults. New Eng. J. Med. 334: 1039-1044, 1996.
Malouf, J.; Ratl, H.; Der Kaloustian, V. M.: Apical hypertrophic cardiomyopathy in a father and daughter. Am. J. Med. Genet. 22: 75-80, 1985.
Manchester, G. H.: Muscular subaortic stenosis. New Eng. J. Med. 269: 300-306, 1963.
Maron, B. J.; Bonow, R. O.; Cannon, R. O., III; Leon, M. B.; Epstein, S. E.: Hypertrophic cardiomyopathy: interrelations of clinical manifestations, pathophysiology, and therapy. New Eng. J. Med. 316: 780-789, 1987.
Maron, B. J.; Bonow, R. O.; Seshagiri, T. N. R.; Roberts, W. C.; Epstein, S. E.: Hypertrophic cardiomyopathy with ventricular septal hypertrophy localized to the apical region of the left ventricle (apical hypertrophic cardiomyopathy). Am. J. Cardiol. 49: 1838-1848, 1982.
Maron, B. J.; Edwards, J. E.; Henry, W. L.; Clark, C. E.; Bingle, G. J.; Epstein, S. E.: Asymmetric septal hypertrophy (ASH) in infancy. Circulation 50: 809-820, 1974.
Maron, B. J.; Henry, W. L.; Clark, C. E.; Redwood, D. R.; Roberts, W. C.; Epstein, S. E.: Asymmetric septal hypertrophy in childhood. Circulation 53: 9-19, 1976.
Maron, B. J.; Shirani, J.; Poliac, L. C.; Mathenge, R.; Roberts, W. C.; Mueller, F. O.: Sudden death in young competitive athletes: clinical, demographic, and pathological profiles. J.A.M.A. 276: 199-204, 1996.
Masuya, K.; Murakami, E.; Takekoshi, N.; Matsui, S.; Murakami, H.; Nomura, M.; Fujita, S.; Tsuji, S.; Chadani, T.; Emoto, J.; Tsugawa, H.; Hashimoto, A.; Noumi, I.: Hypertrophic cardiomyopathy in two elderly siblings. Jpn. Heart J. 23: 253-262, 1982.
Mitchell, A. L.; Bale, A. E.; Wang, M.; Pirtle, R.; McBride, O. W.: Localization of TCRA gene and LPT tRNA gene cluster on chromosome 14. (Abstract) Cytogenet. Cell Genet. 51: 1045-1046, 1989.
Motulsky, A. G.: The HLA complex and disease: some interpretations and new data in cardiomyopathy. (Editorial) New Eng. J. Med. 300: 918-919, 1979.
Nasser, W. K.; Williams, J. F.; Mishkin, M. E.; Childress, R. H.; Helmen, C.; Merritt, A. D.; Genovese, P. D.: Familial myocardial disease with and without obstruction to left ventricular outflow: clinical, hemodynamic and angiographic findings. Circulation 35: 638-652, 1967.
Pare, J. A. P.; Fraser, R. G.; Pirozynski, W. J.; Shanks, J. A.; Stubington, D.: Hereditary cardiovascular dysplasia: a form of familial cardiomyopathy. Am. J. Med. 31: 37-62, 1961.
Powell, W. J.; Whiting, R. B.; Dinsmore, R. E.; Sanders, C. A.: Symptomatic prognosis in patients with idiopathic hypertrophic subaortic stenosis (IHSS). Am. J. Med. 55: 15-24, 1973.
Rosenzweig, A.; Watkins, H.; Hwang, D.-S.; Miri, M.; McKenna, W.; Traill, T. A.; Seidman, J. G.; Seidman, C. E.: Preclinical diagnosis of familial hypertrophic cardiomyopathy by genetic analysis of blood lymphocytes. New Eng. J. Med. 325: 1753-1760, 1991.
Smith, E. R.; Heffernan, L. P.; Sangalang, V. E.; Vaughan, L. M.; Flemington, C. S.: Voluntary muscle involvement in hypertrophic cardiomyopathy: a study of eleven patients. Ann. Intern. Med. 85: 566-572, 1976.
Solomon, S. D.; Geisterfer-Lowrance, A. A. T.; Vosberg, H.-P.; Hiller, G.; Jarcho, J. A.; Morton, C. C.; McBride, W. O.; Mitchell, A. L.; Bale, A. E.; McKenna, W. J.; Seidman, J. G.; Seidman, C. E.: A locus for familial hypertrophic cardiomyopathy is closely linked to the cardiac myosin heavy chain genes, CRI-L436, and CRI-L329 on chromosome 14 at q11-q12. Am. J. Hum. Genet. 47: 389-394, 1990.
Solomon, S. D.; Jarcho, J. A.; McKenna, W.; Geisterfer-Lowrance, A.; Germain, R.; Salerni, R.; Seidman, J. G.; Seidman, C. E.: Familial hypertrophic cardiomyopathy is a genetically heterogeneous disease. J. Clin. Invest. 86: 993-999, 1990.
Spirito, P.; Chiarella, F.; Carratino, L.; Berisso, M. Z.; Bellotti, P.; Vecchio, C.: Clinical course and prognosis of hypertrophic cardiomyopathy in an outpatient population. New Eng. J. Med. 320: 749-755, 1989.
Spirito, P.; Seidman, C. E.; McKenna, W. J.; Maron, B. J.: The management of hypertrophic cardiomyopathy. New Eng. J. Med. 336: 775-782, 1997.
Teare, D.: Asymmetrical hypertrophy of the heart in young adults. Brit. Heart J. 20: 1-8, 1958.
Wagner, J. A.; Sax, F. L.; Weisman, H. F.; Porterfield, J.; McIntosh, C.; Weisfeldt, M. L.; Snyder, S. H.; Epstein, S. E.: Calcium-antagonist receptors in the atrial tissue of patients with hypertrophic cardiomyopathy. New Eng. J. Med. 320: 755-761, 1989.
Watkins, H.; Seidman, J. G.; Seidman, C. E.: Genetic testing for hypertrophic cardiomyopathy. (Letter) New Eng. J. Med. 327: 1176, 1992.
Wei, J. Y.; Weiss, J. L.; Bulkley, B. H.: The heterogeneity of hypertrophic cardiomyopathy: an autopsy and one dimensional echocardiographic study. Am. J. Cardiol. 45: 24-32, 1980.
Wilson, R.; Gibson, T. C.; Terrien, C. M., Jr.; Levy, A. M.: Hyperthyroidism and familial hypertrophic cardiomyopathy. Arch. Intern. Med. 143: 378-380, 1983.
Wood, R. S.; Taylor, W. J.; Wheat, M. W.; Schiebler, G. L.: Muscular subaortic stenosis in childhood: report of occurrence in three siblings. Pediatrics 30: 749-758, 1962.
Yamaguchi, H.; Ishimura, T.; Nishiyama, S.; Nagasaki, F.; Nakanishi, S.; Takatsu, F.; Nishijo, T.; Umeda, T.; Machii, K.: Hypertrophic nonobstructive cardiomyopathy with giant negative T waves (apical hypertrophy): ventriculographic and echocardiographic features in 30 patients. Am. J. Cardiol. 44: 401-412, 1979.

MYXOMA, INTRACARDIAC (*255960)

JJM NOTE

Myxoma is a rare cardiac neoplasm, but has several associated syndromes suggesting that it has major genetic determinants. When diagnosing a patient with myxoma, the clinician should initiate a search for related disorders in the patient and relatives. See Carney syndrome (160980) for various myoxomas and polyostotic fibrous dysplasia (174800) for cardiac ones.

Krause et al. (1971) treated a 34-year-old patient with a pulmonic valve myxoma complicated by bacterial endocarditis. A brother of the patient had died at age 25 years of left atrial myxoma. Two other sibs had had rheumatic heart disease. Heydorn et al. (1973) reported the occurrence of atrial myxoma in 2 teenage brothers. Kleid et al. (1973) described left atrial myxoma in a 14-year-old boy and a right atrial myxoma in his 16-year-old brother. Farah (1975) reported affected brother and sister. Siltanen et al. (1976) reported myxoma in a mother and all 3 of her sons. Powers et al. (1979) described myxoma in father and daughter. The daughter had an infected right ventricular myxoma that was mistaken for valvular bacterial endocarditis. The father had a right atrial myxoma associated with atrial septal defect and mitral valve prolapse, and findings suggestive of paradoxical emboli. Atrial myxoma is a component of the complex of lentigines, myxoma and endocrine hyperfunction (Carney syndrome; 160980). Grauer and Grauer (1983) reported affected father and daughter. Farah (1994) used 2-dimensional echocardiography to study relatives of 14 patients with cardiac myxoma. Four family members from 2 different families were found to have cardiac myxoma. The first family included a brother and a sister, both with acromegaly, suggesting a syndromal form of myxoma. Farah (1994) stated that 'none of the family members had skin tumors, abnormal pigmentation, other tumors or evidence of endocrine disease.' In the family in which a brother and sister had atrial myxoma and acromegaly, he also stated that 'a complete study of all members of this family was not possible because some members refused the echocardiographic studies provided free.' It should be noted that the association of acromegaly with familial myxoma was reported by Carney et al. (1985). Thus, it seems likely that the sibs in fact had Carney syndrome. Indeed, the sister would appear to have been case 14 of Carney et al. (1985). In that report, the brother was in fact said to have pigmented spots of the face and lips and had had nodular and pedunculated myxomas of the skin removed. The mother and 3 out of 7 children had cardiac myxoma. Farah (1994) concluded that cardiac myxoma was more frequently found in family members when the proband had right-sided or bilateral myxoma; that patients with familial cardiac myxoma are younger than the nonfamilial cases; and that long-term recurrence is more frequent.

In a review of the literature, van Gelder et al. (1992) noted 15 families with myxoma of the heart, to which they added 2 more. The patients were young (mean age 27 years) with multicenteric lesions in 22% of the cases. In 61%, the tumors occurred in the left atrium, with a recurrence rate of 10% after removal. The mode of inheritance appeared to be autosomal dominant.

Dandolu et al. (1995) described a family in which the mother had biatrial myxoma with stalks growing from either side of the interatrial septum. A son and a daughter of hers had atrial myxoma and a 12-year-old son had died suddenly after having cardiac symptoms consistent with myxoma. No signs indicative of Carney syndrome were found. Dandolu et al. (1995) suggested autosomal dominant inheritance.

See also Dewald et al. (1987); Liebler et al. (1976)

Carney, J. A.; Gordon, H.; Carpenter, P. C.; Shenoy, B. V.; Go, V. L. W.: The complex of myxomas, spotty pigmentation, and endocrine overactivity. Medicine 64: 270-283, 1985.
Dandolu, B. R.; Iyer, K. S.; Das, B.; Venugopal, P.: Nonsyndrome familial atrial myxoma in two generations. J. Thorac. Cardiovasc. Surg. 110: 872-874, 1995.
Dewald, G. W.; Dahl, R. J.; Spurbeck, J. L.; Carney, J. A.; Gordon, H.: Chromosomally abnormal clones and nonrandom telomeric translocations in cardiac myxomas. Mayo Clin. Proc. 62: 558-567, 1987.
Farah, M. G.: Familial atrial myxoma. Ann. Intern. Med. 83: 358-360, 1975.
Farah, M. G.: Familial cardiac myxoma: a study of relatives of patients with myxoma. Chest 105: 65-68, 1994.
Grauer, K.; Grauer, M. C.: Familial atrial myxoma with bilateral recurrence. Heart Lung 12: 600-602, 1983.
Heydorn, W. H.; Gomez, A. C.; Kleid, J. J.; Haas, J. J.: Atrial myxoma in siblings. J. Thorac. Cardiovasc. Surg. 65: 484-486, 1973.
Kleid, J. J.; Klugman, J.; Haas, J. M.; Battock, D.: Familial atrial myxoma. Am. J. Cardiol. 32: 361-364, 1973.
Krause, S.; Adler, L. N.; Reddy, P. S.; Magovern, G. J.: Intracardiac myxoma in siblings. Chest 60: 404-406, 1971.
Liebler, G. A.; Magovern, G. J.; Park, S. B.; Cushing, W. J.; Begg, F. R.; Joyner, C. R.: Familial myxomas in four siblings. J. Thorac. Cardiovasc. Surg. 71: 605-608, 1976.
Powers, J. C.; Falkoff, M.; Heinle, R. A.; Nanda, N. C.; Ong, L. S.; Weiner, R. S.; Barold, S. S.: Familial cardiac myxoma: emphasis on unusual clinical manifestations. J. Thorac. Cardiovasc. Surg. 77: 782-788, 1979.
Siltanen, P.; Tuuteri, L.; Norio, R.; Tala, P.; Ahrenberg, P.; Halonen, P. I.: Atrial myxoma in a family. Am. J. Cardiol. 38: 252-256, 1976.
van Gelder, H. M.; O'Brien, D. J.; Staples, E. D.; Alexander, J. A.: Familial cardiac myxoma. Ann. Thorac. Surg. 53: 419-424, 1992.

8. SKELETAL SYSTEM

Bone Tumors

OSTEOSARCOMA (#259500)

JJM NOTE

This entity is the most common form of primary malignant neoplasia of bone. Of course, when a family history of "bone cancer" is elicited, one must be skeptical. Most likely, the tumor is metastatic from another site, such as breast, prostate, or colon. Many single genes surely contribute to the phenotype known as osteosarcoma: *TP53* (Li-Fraumeni cancer family syndrome), retinoblastoma, Gardner variant of familial polyposis coli, Werner and Okamoto premature aging syndromes, and even sickle cell anemia (sarcoma arising in bone infarctions). Childhood brain and bone cancer occur to excess in siblings, as detected by vital statistics and cancer registries.

A number sign (#) is used with this entry because of the evidence that the RB1 gene (180200) is the site of the causative mutation in many instances of osteogenic sarcoma. An osteosarcoma tumor-suppressor gene that maps to 18q in the region of a gene involved with Paget disease of the bone (PDB2; 602080) is thought to be involved in the high frequency of osteosarcoma in that disorder; see osteosarcoma tumor suppressor (OSTS; 603045).

Harmon and Morton (1966) reported osteogenic sarcoma in 4 sibs, with onset at 11, 15, 20, and 22 years. On the other hand, Epstein et al. (1970) observed osteogenic sarcoma in a father and daughter. See chondrosarcoma (215300). The demonstration of evidence of immune response (lysis of radiolabeled tumor cells by donor lymphocytes) in household contacts of patients with osteosarcoma (Levin et al., 1974) suggests that the familial aggregation may be due to a transmitted agent. Osteosarcoma is a component of the acronymically designated OSLAM syndrome (165660). Survivors of the bilateral form of retinoblastoma (RB1) have an increased risk of osteosarcoma. Survivors of unilateral retinoblastoma show the same likelihood of developing osteosarcoma as the general population. It is plausible to think that much of sporadic osteosarcoma is due to homozygosity (or hemizygosity) for a mutation at the RB1 locus on chromosome 13. Gilman et al. (1985) described osteosarcoma developing in 2 prepubertal American Indian sisters at age 8 and 12 years. Rearrangements involving chromosomes 13 and 14 were found in both the surviving sister and the mother. The mother had a typical Robertsonian translocation of 13 and 14. The daughter had a rearrangement of ambiguous nature. Goorin et al. (1985) stated that 16 sets of sibs with osteosarcoma have been identified. Dryja et al. (1986) provided evidence that some human osteosarcomas arise subsequent to the development of homozygosity at loci on the long arm of chromosome 13. They proposed that this is the same locus as the retinoblastoma locus. In all 11 cases of osteosarcoma, Scheffer et al. (1991) found loss of heterozygosity of both chromosomes 13 and 17.

Paget disease of bone, or 'osteitis deformans,' is a bone disorder characterized by rapid bone remodeling resulting in abnormal bone formation. It is the second most common metabolic bone disease after osteoporosis, affecting 3 to 5% of persons aged 40 years or more. Linkage or association with HLA was proposed in the past; see 167250. More recently, genetic linkage analysis of families with multigenerational Paget disease showed linkage to a region of 18q near the polymorphic locus D18S42; see PDB2 (602080). Approximately 1% of Paget patients develop osteosarcoma, which represents an increase in risk that is several thousand-fold over that of the general population. Osteosarcoma in Paget patients is the underlying basis for a substantial fraction of osteosarcomas occurring after age 60 years. Nellissery et al. (1998) identified, by analysis of tumor-specific loss of constitutional heterozygosity (LOH) in 96 sporadic osteosarcomas, a putative osteosarcoma tumor-suppressor locus that mapped to 18q. They localized this tumor-suppressor locus between D18S60 and D18S42, a region tightly linked to familial Paget disease. Analysis of osteosarcomas from patients with Paget disease showed that these tumors also undergo LOH in this region. The findings suggested that the association between Paget disease and osteosarcoma is the result of a single gene or 2 tightly linked genes on chromosome 18.

See also Colyer (1979); Hansen et al. (1985)

Colyer, R. A.: Osteogenic sarcoma in siblings. Johns Hopkins Med. J. 145: 131-135, 1979.

Dryja, T. P.; Rapaport, J. M.; Epstein, J.; Goorin, A. M.; Weichselbaum, R.; Koufos, A.; Cavenee, W. K.: Chromosome 13 homozygosity in osteosarcoma without retinoblastoma. Am. J. Hum. Genet. 38: 59-66, 1986.

Epstein, L. I.; Bixler, D.; Bennett, J. E.: An incident of familial cancer: including 3 cases of osteogenic sarcoma. Cancer 25: 889-891, 1970.

Gilman, P. A.; Wang, N.; Fan, S.-F.; Reede, J.; Khan, A.; Leventhal, B. G.: Familial osteosarcoma associated with 13;14 chromosomal rearrangement. Cancer Genet. Cytogenet. 17: 123-132, 1985.

Goorin, A. M.; Abelson, H. T.; Frei, E., III: Osteosarcoma: fifteen years later. New Eng. J. Med. 313: 1637-1643, 1985.

Hansen, M. F.; Koufos, A.; Gallie, B. L.; Phillips, R. A.; Fodstad, O.; Brogger, A.; Gedde-Dahl, T.; Cavenee, W. K.: Osteosarcoma and retinoblastoma: a shared chromosomal mechanism revealing recessive predisposition. Proc. Nat. Acad. Sci. 82: 6216-6220, 1985.

Harmon, T. P.; Morton, K. S.: Osteogenic sarcoma in four siblings. J. Bone Joint Surg. 48B: 493-498, 1966.

Levin, A. S.; Byers, V. X.; Fudenberg, H. H.; Wybran, J.: Immunologic parameters for monitoring immunotherapy with tumor specific transfer factor. (Abstract) Clin. Res. 22: 570A only, 1974.

Nellissery, M. J.; Padalecki, S. S.; Brkanac, Z.; Singer, F. R.; Roodman, G. D.; Unni, K. K.; Leach, R. J.; Hansen, M. F.: Evidence for a novel osteosarcoma tumor-suppressor gene in the chromosome 18 region genetically linked with Paget disease of bone. Am. J. Hum. Genet. 63: 817-824, 1998.

Scheffer, H.; Kruize, Y. C. M.; Osinga, J.; Kuiken, G.; Oosterhuis, J. W.; Leeuw, J. A.; Schraffordt Koops, H.; Buys, C. H. C. M.: Complete association of loss of heterozygosity of chromosomes 13 and 17 in osteosarcoma. Cancer Genet. Cytogenet. 53: 45-55, 1991.

PAGET DISEASE OF BONE 1 (167250)

JJM NOTE

It is important to consider separately Paget disease of the breast. The risk of malignancy and giant-cell tumor is raised; quantification is lacking, but it likely accounts for a large fraction of the otherwise rare osteosarcoma of the aged. The clear familiality and the tantalizing evidence for environmental determinants make the disorder ripe for study as an ecogenetic trait. An additional summary is available (Beighton P: Other heritable and generalized disorders of connective tissue. In Beighton P, ed. *McKusick's Heritable Disorders of Connective Tissue,* 5th ed. St. Louis: Mosby, 1993, pp. 525–30).

Reports of familial aggregation are rather numerous, including occurrence in successive generations. Montagu (1949) reviewed the reported families with multiple instances of Paget disease of bone (PDB) and concluded that 'when inherited, it is transmitted as an incompletely dominant gene carried on an X chromosome.' McKusick (1960) reviewed 35 pedigrees reported to 1956 and added 2 others. In only 1 family, that of van Bogaert (1933), was there male-to-male transmission. All the persons affected by the bone disease and some members of the family not so affected had retinitis pigmentosa, which may have been an independent, i.e., coincidental, genetic disorder. Jones and Reed (1967) observed 6 cases in 3 generations of a family. Evens and Bartter (1968) described 7 definite and 2 probable cases in 1 kindred. Linkage of HLA on chromosome 6p and Paget disease of bone was found by Fotino et al. (1977). A lod score of 2.44 at a recombination fraction of 0.108 was obtained. Moore and Hoffman (1988) could not demonstrate linkage with MHC in a 2-generation family.

Electron microscopic demonstration of virus-like bodies in bone cells in Paget disease suggests viral etiology (Mills and Singer, 1976). Winfield and Sutherland (1981) could find no serologic support for the suggestion of Rebel et al. (1976) that the inclusions represent measles virus. In all 25 biopsies showing histologic evidence of Paget disease, derived from 22 patients, electron microscopy showed characteristic nuclear and cytoplasmic inclusions in the osteoclasts (Harvey et al., 1982). Osteocytes, osteoblasts, hematopoietic cells and connective tissue cells lacked inclusions. The intranuclear inclusions consisted of stacked rows or complex whorls of tubular filaments, each of 12 to 15 mm diameter, often arranged in a paracrystalline array. The frequency of occurrence of inclusions in osteoclasts and in their individual nuclei correlated with the histologic severity of the disease process. The similarity of the inclusions to paramyxovirus inclusion bodies (particularly those of measles) supported, in the view of the authors, the hypothesis that Paget disease is a slow virus infection.

Concentration of cases in Lancashire, England (Barker et al., 1980) and in Australia (Gardner et al., 1978) suggests an important environmental factor in etiology.

In 5 patients with ancestry in southern Italy, Jacobs et al. (1979) described giant cell tumor as a complication of Paget disease of bone. Three of the patients were related. High doses of dexamethasone resulted in dramatic shrinking of tumors. Wu et al. (1991) reported the cases of 2 sisters and a brother with longstanding (35 years in 1) polyostotic Paget disease. Two of them developed osteogenic sarcoma at sites unrelated to surgical procedures, one in the sacrum and one in the calvaria. Both died shortly after diagnosis of malignancy because of aggressive spread of the tumor. The 3 sibs came from Castallamarre, Italy, which is approximately 10 km from Avellino where the ancestors of the patients of Jacobs et al. (1979) lived. See 147620 for evidence that interleukin-6 is involved in the hyperplasia and hypertrophy of osteoclasts in Paget disease.

Because of some clinical similarities of Paget disease of bone to familial expansile osteolysis (174810), which had been mapped to 18q21-q22, Van Hul et al. (1997) performed linkage analysis in 8 large 'pagetic' families from varied ethnic backgrounds using a panel of polymorphic microsatellite markers spanning the region of interest on 18q. In all families, the pattern of inheritance was consistent with autosomal dominant inheritance with high penetrance by the age of 65. Summated 2-point lod scores were suggestive for linkage with several markers with a maximum of 3.48 for D18S60 at theta = 0.0. HOMOG analysis of the multipoint lod scores obtained for the 8 families suggested genetic heterogeneity, with 4 families indicating a high probability of linkage to 18q21-q22, while 1 family was clearly excluded, and the results on the 3 remaining families were inconclusive. The Paget disease of bone locus possibly linked to 6p is referred to here as PDB1 and the locus tentatively identified on chromosome 18 is symbolized PDB2 (602080).

Singer et al. (1998) treated 13 patients with severe Paget disease of bone (mean serum alkaline phosphatase level 17 times the upper limit of normal) with 30 mg oral risedronate daily for 8 weeks. All patients who completed the study had a decrease in serum alkaline phosphatase levels of at least 77% from the baseline. Oral risedronate was well tolerated by the patients. Only 1 patient discontinued treatment because of an adverse event (diarrhea) thought to be related to risedronate therapy.

See also Sofaer et al. (1983)

Barker, D. J. P.; Chamberlain, A. T.; Guyer, P. B.; Gardner, M. J.: Paget's disease of bone: the Lancashire focus. Brit. Med. J. 280: 1105-1107, 1980.

Evens, R. G.; Bartter, F. C.: The hereditary aspects of Paget's disease (osteitis deformans). J.A.M.A. 205: 900-902, 1968.

Fotino, M.; Haymovits, A.; Falk, C. T.: Evidence for linkage between HLA and Paget's disease. Transplant. Proc. 9: 1867-1868, 1977.

Gardner, M. J.; Guyer, P. B.; Barker, D. J. P.: Radiological prevalence of Paget's disease of bone in British migrants to Australia. Brit. Med. J. I: 1655-1657, 1978.

Harvey, L.; Gray, T.; Beneton, M. N. C.; Douglas, D. L.; Kanis, J. A.; Russell, R. G. G.: Ultrastructural features of the osteoclasts from Paget's disease of bone in relation to a viral aetiology. J. Clin. Path. 35: 771-779, 1982.

Jacobs, T. P.; Michelsen, J.; Polay, J. S.; D'Adamo, A. C.; Canfield, R. E.: Giant cell tumor in Paget's disease of bone: familial and geographic clustering. Cancer 44: 742-747, 1979.

Jones, J. V.; Reed, M. F.: Paget's disease: a family with six cases. Brit. Med. J. 2: 90-91, 1967.

McKusick, V. A.: Paget's disease of the bone. Heritable Disorders of Connective Tissue. St. Louis: C. V. Mosby (pub.) (3rd ed.) : 1960. Pp. 718-723.

Mills, B. G.; Singer, F. R.: Nuclear inclusions in Paget's disease of bone. Science 194: 201-202, 1976.

Montagu, M. F. A.: Paget's disease (osteitis deformans) and heredity. Am. J. Hum. Genet. 1: 94-95, 1949.

Moore, S. B.; Hoffman, D. L.: Absence of HLA linkage in a family with osteitis deformans (Paget's disease of bone). Tissue Antigens 31: 69-70, 1988.

Rebel, A.; Malkani, K.; Basle, M.; Bregeon, C.: Osteoclast ultrastructure in Paget's disease. Calcif. Tissue Res. 20: 187-199, 1976.

Singer, F. R.; Clemens, T. L.; Eusebio, R. A.; Bekker, P. J.: Risedronate, a highly effective oral agent in the treatment of patients with severe Paget's disease. J. Clin. Endocr. Metab. 83: 1906-1910, 1998.

Sofaer, J. A.; Holloway, S. M.; Emery, A. E. H.: A family study of Paget's disease of bone. J. Epidemiol. Comm. Health 37: 226-231, 1983.

Van Bogaert, L.: Ueber eine hereditaere und familiaere Form der Pagetschen ostitis deformans mit Chorioretinitis pigmentosa. Z. Ges. Neurol. Psychiat. 147: 327-345, 1933.

Van Hul, W.; Haslam, S. I; Morales Piga, A.; Balemans, W.; Haites, N. E.; San Millan, J. L.; Navio, T.; Nakatsuka, K.; Willems, P. J.; Ralston, S. H.: Suggestion for a susceptibility locus for Paget's disease of bone on chromosome 18 and for genetic heterogeneity. (Abstract) Am. J. Hum. Genet. 61 (suppl.): A298 only, 1997.

Winfield, J.; Sutherland, S.: Measles antibody in Paget's disease. (Letter) Lancet I: 891 only, 1981.

Wu, R. K.; Trumble, T. E.; Ruwe, P. A.: Familial incidence of Paget's disease and secondary osteogenic sarcoma: a report of three cases from a single family. Clin. Orthop. Rel. Res. 265: 306-309, 1991.

PAGET DISEASE OF BONE 2 (602080)

JJM NOTE

Whether the different genes for Paget disease of bone have differing risks for malignancy awaits further study.

Because of some clinical similarities of Paget disease of bone (PDB) to familial expansile osteolysis (FEO; 174810), which had been mapped to 18q21-q22, Van Hul et al. (1997) performed linkage analysis in 8 large 'pagetic' families from varied ethnic backgrounds using a panel of polymorphic microsatellite markers spanning the region of interest on 18q. In all families, the pattern of inheritance was consistent with autosomal dominant inheritance with high penetrance by the age of 65. Summated 2-point lod scores were suggestive for linkage with several markers with a maximum of 3.48 for D18S60 at theta = 0.0. HOMOG analysis of the multipoint lod scores obtained for the 8 families suggested genetic heterogeneity, with 4 families indicating a high probability of linkage to 18q21-q22, while 1 family was clearly excluded, and the results on the 3 remaining families were inconclusive.

Cody et al. (1997) were likewise impressed with the similarity to FEO. The histologic bone lesions of FEO appeared to be similar to pagetic lesions, although they occurred at a much earlier age and were more severe. Furthermore, as in Paget disease, the osteoclasts in patients with FEO contain paramyxoviral-like nuclear inclusions. By study of a large kindred with a high incidence of Paget disease, Cody et al. (1997) determined that Paget disease was linked to genetic markers in the same region of chromosome 18 as that for FEO. Their analysis yielded a 2-point lod score of 3.40 with the genetic marker D18S42, a marker tightly linked to the FEO locus. Cody et al. (1997) reviewed the evidence suggesting a viral etiology of Paget disease.

See 167250 for possible linkage of PDB to the HLA region of 6p. Although the lod score of 2.44 found by Fotino et al. (1977) did not reach the usually accepted level of proof, the possible locus on chromosome 6 is designated here PDB1; PDB2 is used for the gene locus demonstrated on chromosome 18.

Cody, J. D.; Singer, F. R.; Roodman, G. D.; Otterund, B.; Lewis, T. B.; Leppert, M.; Leach, R. J.: Genetic linkage of Paget disease of the bone to chromosome 18q. Am. J. Hum. Genet. 61: 1117-1122, 1997.

Fotino, M.; Haymovits, A.; Falk, C. T.: Evidence for linkage between HLA and Paget's disease. Transplant. Proc. 9: 1867-1868, 1977.

Van Hul, W.; Haslam, S. I; Morales Piga, A.; Balemans, W.; Haites, N. E.; San Millan, J. L.; Navio, T.; Nakatsuka, K.; Willems, P. J.; Ralston, S. H.: Suggestion for a susceptibility locus for Paget's disease of bone on chromosome 18 and for genetic heterogeneity. (Abstract) Am. J. Hum. Genet. 61 (suppl.): A298 only, 1997.

OSLAM SYNDROME (165660)

JJM NOTE

In the two decades since recognizing this syndrome, I have been unable to find a compelling second family. A computer search of records at the NIH Clinical Center was done to find all patients with osteosarcoma and increased mean corpuscular volumes; review of the charts failed to find a person with limb anomalies as seen in the original family. Perhaps the same exercise at another center would reveal a case. Because the family was Oneida Indian, maybe the pattern of inheritance is really pseudodominant or maybe the traits were coincidental. Fred Li (Boston) told me that Steven Friend (Boston) sought constitutional *TP53* mutations in DNA from the family but did not find any.

Mulvihill et al. (1977) described a family in which 3 of 9 children developed typical osteosarcoma. Limb anomalies (clinodactyly, absence of digital ray in foot, bilateral radioulnar synostosis) and macrocytosis without anemia were present in the surviving child with osteosarcoma, several of her sibs and her father. The authors suggested that this is an autosomal dominant syndrome with impaired regulation of bone and bone marrow development. OSLAM is an acronym for osteosarcoma, limb anomalies and macrocytosis.

Mulvihill, J. J.; Gralnick, H. R.; Whang-Peng, J.; Leventhal, B. G.: Multiple childhood osteosarcomas in an American Indian family with erythroid macrocytosis and skeletal anomalies. Cancer 40: 3115-3122, 1977.

PREMATURE AGING SYNDROME, OKAMOTO TYPE (601811)

JJM NOTE

The true association of osteosarcoma with the syndrome awaits further case reports, but is credible, although, to date, the related Mulvihill-Smith premature aging syndrome (176690) has not had malignancy reported.

Okamoto et al. (1997) described an apparently new premature aging syndrome in a 15-year-old girl with severe growth and developmental abnormalities. Features also included osteosarcoma, cataracts, diabetes mellitus, osteoporosis, and erythroid macrocytosis. A photograph showed a round face, microcephaly, poor hair growth, flat nasal bridge, and low-set somewhat deformed ears. The parents were healthy and unrelated.

Okamoto, N.; Satomura, K.; Hatsukawa, Y.; Hayashida, M.; Saijo, K.; Ohno, T.; Goto, M.: Premature aging syndrome with osteosarcoma, cataracts, diabetes mellitus, osteoporosis, erythroid macrocytosis, severe growth and developmental deficiency. Am. J. Med. Genet. 69: 169-170, 1997.

CHONDROSARCOMA (#215300)

JJM NOTE

Some patients with chondrosarcoma may have the Maffucci syndrome (166000) of multiple enchondromas and hemangiomas. The frequency of chondrosarcoma in the syndrome may be as high as 18% (Sun F-C et al.: Chondrosarcoma in Maffucci's syndrome. J. Bone Joint Surg. 67A:1214–19, 1985). I believe this tumor type has also been seen in the Li-Fraumeni cancer family syndrome.

A number sign (#) is used with this entry because of evidence that mutation, either constitutional or somatic, in 1 of the genes for hereditary multiple exostoses, EXT1 (133700) or EXT2 (133701), may be responsible.

Schajowicz and Bessone (1967) described 3 brothers who, respectively, developed chondrosarcoma of the pelvic bone at 18 years, of the fibula and femur at 16 years, and of the femur at 17 years. Two brothers and a sister were living and well. Karyotypes were normal. See osteogenic sarcoma (259500).

Hecht et al. (1995) studied a large multigenerational family with multiple exostosis and in 1 member a chondrosarcoma. The family demonstrated linkage of the disease to chromosome 11 markers. Loss of marker D11S903 was observed in constitutional DNA from all affected individuals and in the tumor sample. Further studies of constitutional and chondrosarcoma DNA from 6 unrelated individuals, 2 of whom had multiple exostoses, showed 1 tumor from an individual with multiple exostoses that showed LOH for chromosome 8 markers in the region where the EXT1 gene (133700) maps (8q24.1), and a person with a sporadic chondrosarcoma was found to have a tumor-specific LOH and a homozygous deletion of chromosome 11 markers from the pericentric region, where the EXT2 gene (133701) is mapped. These findings suggested to Hecht et al. (1995) that EXT genes may be tumor suppressor genes and that the initiation of tumor development may follow a multistep model.

Hecht, J. T.; Hogue, D.; Strong, L. C.; Hansen, M. F.; Blanton, S. H.; Wagner, M.: Hereditary multiple exostosis and chondrosarcoma: linkage to chromosome 11 and loss of heterozygosity for EXT-linked markers on chromosomes 11 and 8. Am. J. Hum. Genet. 56: 1125-1131, 1995.
Schajowicz, F.; Bessone, J. E.: Chondrosarcoma in 3 brothers: a pathological and genetic study. J. Bone Joint Surg. 49A: 129-141, 1967.

Exostoses

EXOSTOSES, MULTIPLE 1 (*133700)

JJM NOTE

In general, exostosis is included because it represents bony tumors, regardless of the underlying pathogenic process, whether neoplasia, hyperplasia, dysplasia, or something else. Like malignant tumors, the list of Mendelian traits that have exostoses as a feature suggests that many genes can contribute to the tumors. Indeed, the next 20 entries feature exostoses, suggesting that exostosis is a clue for syndrome recognition and possibly for predisposition to malignancy. Malignant evolution probably occurs to excess in some (but not all?) of these disorders, and patients have to be alerted to this possibility. To add precision to the wide ranges of estimates, it would seem best to do a retrospective cohort study on some group defined years ago. Perhaps the Guam population should be revisited (also to help linkage efforts), or the early papers from Denmark (of which follow-up is especially easy).

This disorder is characterized by cartilaginous excrescences near the ends of the diaphyses of the bones of the extremities. These protuberances undergo ossification. The ribs and scapulae may also be affected, but the skull is not involved. Deformity of the legs, forearms (resembling Madelung deformity), and hands (e.g., short metacarpal) is frequent. The patients are often of reduced stature. Del-Rio et al. (1992) described a 14-year-old boy in whom multiple exostoses had an unusual feature: exostoses on the distal phalanges were associated with malalignment and longitudinal dystrophy of the fingernails. Wicklund et al. (1995) conducted a retrospective review of 43 affected probands and 137 of their affected relatives. Only 2.8% of the total affected population had experienced exostosis-related malignancy. Penetrance appeared to be 100%. There was an excess of males within the entire affected population (104:76) and within identified probands (28:15). However, the male-to-female ratio was unskewed in nuclear families (probands, affected sibs, and parents). The excess of males appeared to be related to the fact that males have more severe and more frequent complications of EXT which do not have a primary genetic origin.

Krooth et al. (1961) reported on a study of the families of 6 persons with diaphyseal aclasis (multiple exostoses). The families were Chamorros, a Micronesian people who live in the Mariana Islands. The frequency of diaphyseal aclasis in the Chamorros of Guam was estimated at 1 in 1,000. In published series the disease is more frequent and more severe in males than in females. In the 21 Guam cases, the tumors were evident on inspection in all males but in only half the females. In a study of 56 patients, Solomon (1963) found a sex ratio of 1 and reported that two-thirds of the patients had an affected parent. Solomon (1964) observed 1 family in which all 8 affected persons in 4 sibships of 3 generations showed exostoses on the bones of the hands and fingers with very few elsewhere. In no other patients of his study did the abnormality take this particular form. Other workers have found no correlation between members of the same family as to form and distribution of disease. For these reasons, Solomon suggested that the particular family may suffer from a rare disorder due to a gene distinct from that causing most cases. Two sibs who probably are cases of homozygosity (both parents were affected and came from exostosis families and the children showed unusually severe changes early) were observed by Giedion (1972).

Two conditions in which multiple exostoses occur are metachondromatosis (156250) and the Langer-Giedion syndrome (LGS; 150230). Furthermore, exostosis-like lesions occur with FOP (135100), type IX Ehlers-Danlos syndrome (304150), and the adult stage of hereditary hypophosphatemia (307800); these exostoses are located at sites of tendon and muscle attachment. The supracondylar process, a relatively rare variant, on the anteromedial surface of the distal humerus (Silverman, 1985) can be confused with an exostosis. I wrongly suspected that a young member of a possible exostosis family (NM, JHH 1838624) was affected when I felt this normal variant, which is said to be present in about 1% of persons of European stock (Silverman, 1985). Brooks and Wynne-Davies (1980) reported 2 sisters with diaphyseal aclasis inherited from the father and peripheral dysostosis inherited from the mother.

Voutsinas and Wynne-Davies (1983) suggested that the risk of malignant change in multiple exostoses has been exaggerated and that the true figure is 0.5% (or 1.3% of those over 21 years). Matsuno et al. (1988) described a 19-year-old woman and a 29-year-old man with spindle-cell sarcoma as a complication of multiple exostoses. Both of the patients came from families with many affected members. Matsuno et al. (1988) stated that 20 to 50% of patients with multiple exostoses develop malignant change, as compared with 1 to 4% of persons with a solitary osteochondroma. The estimate of the frequency of malignant change may be abnormally high; the patients who get into trouble are those who come to medical attention. In a review, Hennekam (1991) agreed that malignancies probably occur in about 0.5 to 2% of cases. He quoted Ochsner (1978) who collected data on 59 patients with osteosarcoma in multiple exostoses. The mean age of onset of malignant degeneration was 31 years; it seldom occurred before the tenth or after the fiftieth year. The upper end of the femur and the pelvis were the main locations of exostoses, but they were also found in the shoulder girdle and the ribs. It may be useful to obtain radiographs of the pelvic and shoulder girdles in young affected adults for later comparison. Peterson (1989) discussed the orthopedic management.

Scholz and Murken (1963) did linkage studies with negative results. In a linkage study of multiple exostoses, Beals et al. (1984) found a maximum lod score of 1.96 for theta = 0.19 with ABO. Buhler and Malik (1984) suggested that the mutation of multiple exostoses may be sited on 8q in the region of 8q24.1. This was based on the fact that the multiple exostoses of the Langer-Giedion syndrome are indistinguishable from those of the isolated disorder and that at least 1 case of trichorhinophalangeal syndrome without multiple exostoses has been found to have deletion in this region (Hamers et al., 1983). Thus, 2 closely linked loci may be situated at 8q24 (1 for type I trichorhinophalangeal syndrome (TRPS1; 190350) and 1 for multiple exostoses) and both may be deleted in LGS. Mental retardation present in some cases of LGS may be the result of deletion of additional neighboring genetic material. Hall et al. (1985) could find no abnormality of 8q in cells cultured from an exostosis. Ogle et al. (1991) observed multiple exostoses in association with a balanced reciprocal translocation between 8q and 11p. The breakpoint on chromosome 8 was at proximal 8q24.1 within the critical region reported for Langer-Giedion syndrome. Using probe L48 at locus D8S51, Le Merrer et al. (1992) excluded the EXT gene from the 8q24.1 region. The cumulative lod score was -8.96 at theta = 0. In an addendum they stated that the testing of 3 additional families brought the lod score with probe L48 to -18.32 at theta = 0. Cook et al. (1993) found evidence for linkage of EXT to highly informative, short tandem repeat (STR) markers in the 8q24.11-q24.13 region, with genetic heterogeneity. A model of heterogeneity with linkage of the disease gene to the STR markers in 70% of the families (with a 95% confidence interval of 26 to 96%) produced a maximum lod score of 8.11, with the most likely position of EXT1 between D8S85 and D8S199. In 2 Dutch pedigrees with a total of 22 affected persons, Bakker et al. (1993) excluded linkage to the 8q22-q24 region. See also the chromosome 11-linked locus (EXT2; 133701) and the chromosome 19-linked locus (EXT3; 600209).

Mertens et al. (1994) found structural chromosomal rearrangements leading to loss of the distal part of 8q in 3 of 8 sporadic osteocartilaginous exostoses. Band 8q24.1 was the specific band affected in all. They postulated the existence of a gene which carries a mutant allele that is inherited in the familial form of multiple exostoses and leads to the development of exostosis, when a second mutation occurs as a somatic event in the other chromosome. Studying material from a patient with multiple exostoses and a translocation t(8;13) with a breakpoint at 8q23, Yoshiura et al. (1994) determined the order of 7 loci defined by cosmid clones by means of 2-color fluorescence in situ hybridization on elongated prophase chromosomes. They identified 2 flanking loci and by pulsed field gel electrophoresis with one of the flanking probes concluded that the probe lies less than 600 kb from the chromosomal breakpoint.

Ludecke et al. (1995) and Hou et al. (1995) presented evidence that the Langer-Giedion syndrome is a true contiguous gene syndrome due to loss of functional copies of both the TRPS1 and the EXT1 gene and that the EXT1 gene is more than 1 Mb distal to the TRPS1 gene.

Reviewing a large cohort of 175 multiple exostoses patients referred to them over a 40-year period (1955 to 1995), Legeai-Mallet et al. (1997) found that 109 (62%) cases were familial and 66 (38%) were isolated. The disorder was consistently diagnosed before the age of 12 years and the risk of malignancy, although increased, was modest in this series (0.57%). The observation of 7 unaffected individuals (6 females, 1 male) with a family history and affected offspring supported the incomplete penetrance of the disorder. Moreover, the observation of an unequal sex-ratio with a preponderance of males among the probands in this series (103 to 72, p less than 0.02) and in all reported series to date (198 to 133, p less than 0.001) gave support to the variable penetrance of EXT genes according to sex. Whether this incomplete penetrance associated predominantly with 1 of the disease genes was under investigation.

Blanton et al. (1996) studied 12 large multigenerational EXT families and found that the disorder mapped to 8q24 in 6 and to 11p (EXT2) in 6. None of the families mapped to the chromosome 19 locus (EXT3).

Hecht et al. (1995) and Raskind et al. (1995) presented evidence suggesting that the EXT1 gene on chromosome 8 and the EXT2 gene on chromosome 11 have tumor-suppressor function. They found loss of heterozygosity for markers linked to these 2 genes in chondrosarcomas originating in individuals with multiple exostoses and in sporadic chondrosarcomas.

Ahn et al. (1995) cloned and characterized a cDNA that spans chromosomal breakpoints previously identified in 2 multiple exostoses patients. Furthermore, the gene harbored frameshift mutations (133700.0001) in affected members of 2 EXT1 families. The cDNA had a coding region of 2,238 bp with no apparent homology to other known gene sequences; thus, its precise function in the role of tumor suppressor is unknown.

Ludecke et al. (1997) reported that the EXT1 gene is composed of 11 exons and spans approximately 350 kb.

Lin and Wells (1997) cloned and sequenced a mouse cDNA that is homologous to the human EXT1 gene. No mapping information was provided. Lohmann et al. (1997) showed that the murine Ext1 gene has a high level of sequence similarity with its human homolog and is part of a conserved linkage group on mouse chromosome 15.

Hecht et al. (1997) evaluated 6 families for mutations in the EXT1 gene and tested a series of chondrosarcomas and chondroblastomas for loss of heterozygosity (LOH) and mutations in the EXT1 gene. They noted that, though the EXT1 protein is ubiquitously expressed in many tissues, the only known effect of mutated or inactivated EXT1 appears to be specific to actively growing bone. Hecht et al. (1997) noted further that mutations in this gene apparently allow inappropriate bone growth to be juxtaposed to the growth plate; as the bone continues to grow, the exostoses appear to migrate toward the diaphysis. At puberty, with growth plate fusion, linear growth ceases and no new exostoses develop. Hecht et al. (1997) identified 5 mutations in the EXT1 gene, 4 germline mutations, including a mutation which occurred in 2 unrelated families (133700.0002), and a somatic mutation in a patient with chondrosarcoma (133700.0006). LOH analysis of chondrosarcomas and chondroblastomas revealed multiple LOH events at loci on chromosomes 3q, 8q, 10q, and 19q. The authors reported that a sporadic chondrosarcoma demonstrated LOH for EXT1 and EXT3, while a second chondrosarcoma underwent LOH for EXT2 and chromosome 10.

Legeai-Mallet et al. (1997) tested a series of 29 EXT families for possible linkage to the 3 known disease loci and estimated the probability of linkage of the disease to each locus by using an extension of the admixture test, which made modeling of heterogeneous monogenic disease feasible. A maximum likelihood was obtained for proportions of 44, 28, and 28% of families being linked to chromosomes 8, 11, and 19, respectively. No evidence of a fourth locus for the disease was found.

Wells et al. (1997) identified 6 mutations in the EXT1 gene in 6 unrelated EXT families showing linkage to chromosome 8. One of the mutations was the same 1-bp deletion in exon 6 that was previously reported in 2 independent EXT families (133700.0001). The other 5 mutations were novel. In each case, the mutation was predicted to result in a truncated or nonfunctional EXT1 protein.

Philippe et al. (1997) screened 17 probands with hereditary multiple exostoses for alterations in all translated exons and flanking intronic sequences in the EXT1 and EXT2 genes, using conformation-sensitive gel electrophoresis. They found the disease-causing mutation in 12 families (70%), 7 (41%) of which had EXT1 mutations and 5 (29%) EXT2 mutations. Together with a previously described 1-bp deletion (133700.0001) in exon 6, which was present in 2 of their families, Philippe et al. (1997) reported 5 new mutations in EXT1. Two were missense mutations in exon 2, G339D (133700.0007) and R340C, and the other 3 alterations (a nonsense mutation, a frameshift, and a splicing mutation) were predicted to result in truncated nonfunctional proteins.

Wuyts et al. (1998) analyzed the EXT1 and EXT2 genes in 26 EXT families originating from 9 countries. Of the 26 families, 10 had an EXT1 mutation and 10 had an EXT2 mutation. Twelve of these mutations had not previously been described. From a review of these and previously reported mutations, Wuyts et al. (1998) concluded that mutations in either the EXT1 or the EXT2 gene are responsible for most cases of multiple exostoses. Most of the mutations in these 2 genes cause premature termination of the EXT proteins, whereas missense mutations are rare. The development of exostoses is, therefore, mainly due to loss of function of EXT genes, consistent with the hypothesis that the EXT genes have a tumor-suppressor function.

In a study of 34 families with EXT, Raskind et al. (1998) found evidence of linkage to chromosomes 8 and 11, but no evidence of linkage to chromosome 19. They PCR-amplified and sequenced the 11 EXT1 exons in all 11 isolated cases and in 20 of the 23 familial cases. Twelve different novel EXT1 mutations were detected, including 5 frameshift deletions or insertions, 1 codon deletion, and 6 single basepair substitutions, distributed across 8 of the exons. Only 2 of the mutations were identified in more than 1 family. Three mutations affected sites in which alterations were previously reported. Nonchainterminating missense mutations were identified in codons 280 and 340, both coding for conserved arginine residues. These residues may be crucial to the function of this protein. One of the mutations was identified in a Chamorro native on Guam, where EXT is unusually frequent. They concluded that 45% of the isolated cases and 77% of the familial cases could be attributed to abnormalities in EXT1.

The protein gene product of the EXT1 gene might be referred to as exostosin-1.

McCormick et al. (1998) showed that EXT1 is an endoplasmic reticulum (ER)-resident type II transmembrane glycoprotein whose expression in cells results in the alteration of the synthesis and display of cell surface heparan sulfate glycosaminoglycans (GAGs). Two EXT1 variants containing missense mutations related to multiple exostoses failed to alter cell surface glycosaminoglycans, despite retaining their ER localization.

History

Beighton et al. (1993) analyzed the skeletons of adults in the Museum of Pathological Anatomy in Vienna. The museum was established in 1796 by Emperor Franz I, and is now housed in the Narrenturm, which is situated on the grounds of the Altes Allgemeines Krankenhaus, formerly a facility for the custodial care of persons with insanity. The museum contains 44,000 specimens. Beighton et al. (1993) pictured the skeleton of a man with multiple exostoses who died in 1842 of ruptured aortic aneurysm (probably syphilitic).

See also Buur and Morch (1983); Gardner et al. (1975); Ho and Lipton (1979); Morgan et al. (1962); Shapiro et al. (1979); Shupe et al. (1979); Sugiura et al. (1976); Vinstein and Franken (1971)

Ahn, J.; Ludecke, H.-J.; Lindow, S.; Horton, W. A.; Lee, B.; Wagner, M. J.; Horsthemke, B.; Wells, D. E.: Cloning of the putative tumour suppressor gene for hereditary multiple exostoses (EXT1). Nature Genet. 11: 137-143, 1995.

Bakker, E.; Smoor, M. A.; Maat, A.; van Haeringen, A.; van de Kamp, J. J. P.; van Ommen, G. J. B.: Hereditary multiple exostoses in two Dutch families, LGS region excluded. (Abstract) Human Genome Mapping Workshop 93 15, 1993.

Beals, R.; Weber, C.; Wright, A.; Rowe, S.; Lovrien, E. W.: Multiple exostosis may be loosely linked to ABO. (Abstract) Cytogenet. Cell Genet. 37: 416-417, 1984.

Beighton, P.; Sujansky, E.; Patzak, B.; Portele, K. A.: Genetic skeletal dysplasias in the Museum of Pathological Anatomy, Vienna. Am. J. Med. Genet. 47: 843-847, 1993.

Blanton, S. H.; Hogue, D.; Wagner, M.; Wells, D.; Young, I. D.; Hecht, J. T.: Hereditary multiple exostoses: confirmation of linkage to chromosomes 8 and 11. Am. J. Med. Genet. 62: 150-159, 1996.

Brooks, A. P.; Wynne-Davies, R.: A family with diaphyseal aclasis and peripheral dysostosis. J. Med. Genet. 17: 277-280, 1980.

Buhler, E. M.; Malik, N. J.: The tricho-rhino-phalangeal syndrome(s): chromosome 8 long arm deletion: is there a shortest region of overlap between reported cases? TRP I and TRP II syndromes: are they separate entities?. (Editorial) Am. J. Med. Genet. 19: 113-119, 1984.

Buur, T.; Morch, M. M.: Hereditary multiple exostoses with spinal cord compression. (Letter) J. Neurol. Neurosurg. Psychiat. 46: 96-98, 1983.

Cook, A.; Raskind, W.; Blanton, S. H.; Pauli, R. M.; Gregg, R. G.; Francomano, C. A.; Puffenberger, E.; Conrad, E. U.; Schmale, G.; Schellenberg, G.; Wijsman, E.; Hecht, J. T.; Wells, D.; Wagner, M. J.: Genetic heterogeneity in families with hereditary multiple exostoses. Am. J. Hum. Genet. 53: 71-79, 1993.

Del-Rio, R.; Navarra, E.; Ferrando, J.; Mascaro, J. M.: Multiple exostoses syndrome presenting as nail malalignment and longitudinal dystrophy of fingers. (Letter) Arch. Derm. 128: 1655-1656, 1992.

Gardner, E. J.; Shupe, J. L.; Leone, N. C.; Olson, A. E.: Hereditary multiple exostosis: a comparative genetic evaluation in man and horses. J. Hered. 66: 318-322, 1975.

Giedion, A.: Personal Communication. Zurich, Switzerland 1972.

Hall, J. G.; Wilson, R. D.; Kalousek, D.; Beauchamp, R.: Familial multiple exostoses—no chromosome 8 deletion observed. Am. J. Med. Genet. 22: 639-640, 1985.

Hamers, A.; Peeters, G.; Jongbloet, P.; Geraedts, J.: Microcytogenetics of chromosome 8q. (Abstract) 8th Int. Chrom. Conf., Lubeck Abstract-8, 9/1983.

Hecht, J. T.; Hogue, D.; Strong, L. C.; Hansen, M. F.; Blanton, S. H.; Wagner, M.: Hereditary multiple exostosis and chondrosarcoma: linkage to chromosome 11 and loss of heterozygosity for EXT-linked markers on chromosomes 11 and 8. Am. J. Hum. Genet. 56: 1125-1131, 1995.

Hecht, J. T.; Hogue, D.; Wang, Y.; Blanton, S. H.; Wagner, M.; Strong, L. C.; Raskind, W.; Hansen, M. F.; Wells, D.: Hereditary multiple exostoses (EXT): mutational studies of familial EXT1 cases and EXT-associated malignancies. Am. J. Hum. Genet. 60: 80-86, 1997.

Hennekam, R. C. M.: Hereditary multiple exostoses. J. Med. Genet. 28: 262-266, 1991.

Ho, S. U.; Lipton, H. L.: Hereditary multiple exostoses with myelopathy. Arch. Neurol. 36: 714, 1979.

Hogue, D: Personal Communication. Houston, Tex 2/6/1998.

Hou, J.; Parrish, J.; Ludecke, H.-J.; Sapru, M.; Wang, Y.; Chen, W.; Hill, A.; Siegel-Bartelt, J.; Northrup, H.; Elder, F. F. B.; Chinault, C.; Horsthemke, B.; Wagner, M. J.; Wells, D. E.: A 4-megabase YAC contig that spans the Langer-Giedion syndrome region on human chromosome 8q24.1: use in refining the location of the trichorhinophalangeal syndrome and multiple exostoses genes (TRPS1 and EXT1). Genomics 29: 87-97, 1995.

Krooth, R. S.; Macklin, M. T.; Hilbish, T. F.: Diaphysial aclasis (multiple exostoses) on Guam. Am. J. Hum. Genet. 13: 340-347, 1961.

Legeai-Mallet, L.; Margaritte-Jeannin, P.; Lemdani, M.; Le Merrer, M.; Plauchu, H.; Maroteaux, P.; Munnich, A.; Clerget-Darpoux, F.: An extension of the admixture test for the study of genetic heterogeneity in hereditary multiple exostoses. Hum. Genet. 99: 298-302, 1997.

Legeai-Mallet, L.; Munnich, A.; Maroteaux, P.; Le Merrer, M.: Incomplete penetrance and expressivity skewing in hereditary multiple exostoses. Clin. Genet. 52: 12-16, 1997.

Le Merrer, M.; Ben Othmane, K.; Stanescu, V.; Lyonnet, S.; Van Maldergem, L.; Royer, G.; Munnich, A.; Maroteaux, P.: The gene for hereditary multiple exostoses does not map to the Langer-Giedion region (8q23-q24). J. Med. Genet. 29: 713-715, 1992.

Lin, X.; Wells, D.: Isolation of the mouse cDNA homologous to the human EXT1 gene responsible for hereditary multiple exostoses. DNA Seq. 7: 199-202, 1997.

Lohmann, D. R.; Brandt, B.; Hopping, W.; Passarge, E.; Horsthemke, B.: Spectrum of small length germline mutations in the RB1 gene. Hum. Molec. Genet. 3: 2187-2193, 1994.

Lohmann, D. R.; Buiting, K.; Ludecke, H.-J.; Horsthemke, B.: The murine Ext1 gene shows a high level of sequence similarity with its human homologue and is part of a conserved linkage group on chromosome 15. Cytogenet. Cell Genet. 76: 164-166, 1997.

Ludecke, H.-J.; Ahn, J.; Lin, X.; Hill, A.; Wagner, M. J.; Schomburg, L.; Horsthemke, B.; Wells, D. E.: Genomic organization and promoter structure of the human EXT1 gene. Genomics 40: 351-354, 1997.

Ludecke, H.-J.; Wagner, M. J.; Nardmann, J.; La Pillo, B.; Parrish, J. E.; Willems, P. J.; Haan, E. A.; Frydman, M.; Hamers, G. J. H.; Wells, D. E.; Horsthemke, B.: Molecular dissection of a contiguous gene syndrome: localization of the genes involved in the Langer-Giedion syndrome. Hum. Molec. Genet. 4: 31-36, 1995.

Matsuno, T.; Ichioka, Y.; Yagi, T.; Ishii, S.: Spindle-cell sarcoma in patients who have osteochondromatosis: a report of two cases. J. Bone Joint Surg. 70A: 137-141, 1988.

McCormick, C.; Leduc, Y.; Martindale, D.; Mattison, K.; Esford, L. E.; Dyer, A. P.; Tufaro, F.: The putative tumour suppressor EXT1 alters the expression of cell-surface heparan sulfate. Nature Genet. 19: 158-161, 1998.

Mertens, F.; Rydholm, A.; Kreicbergs, A.; Willen, H.; Jonsson, K.; Heim, S.; Mitelman, F.; Mandahl, N.: Loss of chromosome band 8q24 in sporadic osteocartilaginous exostoses. Genes Chromosomes Cancer 9: 8-12, 1994.

Morgan, J. P.; Carlson, W. D.; Adams, O. R.: Hereditary multiple exostosis in the horse. J. Am. Vet. Med. Assoc. 140: 1320-1322, 1962.

Ochsner, P. E.: Zum Problem der neoplastischen Entartung bei multipen kartilaginaren Exostosen. Z. Orthop. 116: 369-378, 1978.

Ogle, R. F.; Dalzell, P.; Turner, G.; Wass, D.; Yip, M.-Y.: Multiple exostoses in a patient with t(8;11)(q24.11;p15.5). J. Med. Genet. 28: 881-883, 1991.

Peterson, H. A.: Multiple hereditary osteochondromata. Clin. Orthop. 239: 222-230, 1989.

Philippe, C.; Porter, D. E.; Emerton, M. E.; Wells, D. E.; Simpson, A. H. R. W.; Monaco, A. P.: Mutation screening of the EXT1 and EXT2 genes in patients with hereditary multiple exostoses. Am. J. Hum. Genet. 61: 520-528, 1997.

Raskind, W. H.; Conrad, E. U.; Chansky, H.; Matsushita, M.: Loss of heterozygosity in chondrosarcomas for markers linked to hereditary multiple exostoses loci on chromosomes 8 and 11. Am. J. Hum. Genet. 56: 1132-1139, 1995.

Raskind, W. H.; Conrad, E. U., III; Matsushita, M.; Wijsman, E. M.; Wells, D. E.; Chapman, N.; Sandell, L. J.; Wagner, M.; Houck, J.: Evaluation of locus heterogeneity and EXT1 mutations in 34 families with hereditary multiple exostoses. Hum. Mutat. 11: 231-239, 1998.

Scholz, W.; Murken, J. D.: Koppelungsuntersuchungen bei Familien mit multiplen cartilaginaeren Exostosen. Z. Menschl. Vererb. Konstitutionsl. 37: 178-192, 1963.

Shapiro, F.; Simon, S.; Glimcher, M. J.: Hereditary multiple exostoses: anthropometric, roentgenographic, and clinical aspects. J. Bone Joint Surg. 61A: 815-824, 1979.

Shupe, J. L.; Leone, N. C.; Olson, A. E.; Gardner, E. J.: Hereditary multiple exostoses: clinicopathologic features of a comparative study in horses and man. Am. J. Vet. Res. 40: 751-757, 1979.

Silverman, F. N.: Caffey's Pediatric X-ray Diagnosis: An Integrated Imaging Approach. Chicago: Year Book Med. Publ. (pub.) 1: 1985. Pp. 433-434.

Solomon, L.: Hereditary multiple exostosis. J. Bone Joint Surg. 45B: 292-304, 1963.

Solomon, L.: Hereditary multiple exostosis. Am. J. Hum. Genet. 16: 351-363, 1964.

Sugiura, Y.; Sugiura, I.; Iwata, H.: Hereditary multiple exostosis: diaphyseal aclasis. Jpn. J. Hum. Genet. 21: 149-167, 1976.

Vinstein, A. L.; Franken, E. A., Jr.: Hereditary multiple exostoses: report of a case with spinal cord compression. Am. J. Roentgen. 112: 405-407, 1971.

Voutsinas, S.; Wynne-Davies, R.: The infrequency of malignant disease in diaphyseal aclasis and neurofibromatosis. J. Med. Genet. 20: 345-349, 1983.

Wells, D. E.; Hill, A.; Lin, X.; Ahn, J.; Brown, N.; Wagner, M. J.: Identification of novel mutations in the human EXT1 tumor suppressor gene. Hum. Genet. 99: 612-615, 1997.

Wicklund, C. L.; Pauli, R. M.; Johnston, D.; Hecht, J. T.: Natural history study of hereditary multiple exostoses. Am. J. Med. Genet. 55: 43-46, 1995.

Wuyts, W.; Van Hul, W.; De Boulle, K.; Hendrickx, J.; Bakker, E.; Vanhoenacker, F.; Mollica, F.; Ludecke, H.-J.; Sayli, B. S.; Pazzaglia, U. E.; Mortier, G.; Hamel, B.; Conrad, E. U.; Matsushita, M.; Raskind, W. H.; Willems, P. J.: Mutations in the EXT1 and EXT2 genes in hereditary multiple exostoses. Am. J. Hum. Genet. 62: 346-354, 1998.

Yoshiura, K.; Inazawa, J.; Koyama, K.; Nakamura, Y.; Niikawa, N.: Mapping of the 8q23 translocation breakpoint of t(8;13) observed in a patient with multiple exostoses. Genes Chromosomes Cancer 9: 57-61, 1994.

EXOSTOSES, MULTIPLE 2 (*133701)

JJM NOTE

To date, there is no difference in the risk of malignancy among the three types.

Hereditary multiple exostoses is an autosomal dominant disorder characterized by multiple exostoses most commonly arising from the juxtaepiphyseal region of the long bones. Other bones that can be involved include the pectoral and pelvic girdles, rib, and, less frequently, vertebrae, sternum, skull, and carpal and tarsal bones. In addition, abnormal bone modeling, particularly of the long bones, is a feature. This causes bowing, shortness, cortical irregularities, and metaphyseal widening of the involved bones, leading to the deformities of the forearms and disproportionate short stature in severe cases. The exostoses can give rise to complications such as compression or irritation of adjacent nerves, vessels, and tendons, and urinary or intestinal obstruction. Quirini et al. (1996) described depression of the thoracic spinal cord in a 24-year-old male with a calcified osteochondroma contiguous with the upper endplate of T8. The most serious complication is sarcomatous degeneration, which occurs in 0.5 to 2% of affected individuals (Hennekam, 1991). Multiple exostoses are part of the Langer-Giedion syndrome (150230), which appears to be a contiguous gene syndrome, due to deletion in the 8q24 region.

Cook et al. (1993) concluded that about 70% of multiple exostoses families show linkage to markers in the 8q24.11-q24.13 region (see EXT1; 133700). Multiple exostoses in the other families appears to be caused by a mutation at another locus, unlinked to markers in that region. Investigating 2 large exostoses pedigrees in which linkage to markers from 8q24 was excluded, Wu et al. (1994) found evidence of linkage to microsatellite markers from the proximal short and long arms of chromosome 11. The highest lod score by 2-point analysis was found with D11S554; maximum lod = 7.148 at theta = 0.03.

As part of a larger study to determine the frequency of the 3 EXT types in the United States, Hecht et al. (1995) ascertained a large multigenerational family with multiple exostosis and in 1 member a chondrosarcoma. This family demonstrated linkage of the disease to chromosome 11 markers. Constitutional and tumor DNAs from the affected family member were compared using short-tandem-repeat (STR) markers from chromosomes 8, 11, and 19. Loss of heterozygosity (LOH) in the tumor was observed for chromosome 8 and 11 markers, but chromosome 19 markers were intact. Hecht et al. (1995) observed an apparent deletion of D11S903 in constitutional DNA from all affected individuals and in the tumor sample. These results indicated that EXT2 gene maps to the region containing D11S903, which is flanked by D11S1355 and D11S1361. The authors similarly analyzed additional constitutional and chondrosarcoma DNA from 6 unrelated individuals, 2 of whom had EXT. One tumor from an individual with EXT demonstrated LOH for chromosome 8 markers, and a person with a sporadic chondrosarcoma was found to have a tumor-specific LOH and a homozygous deletion of chromosome 11 markers. These findings suggested to Hecht et al. (1995) that EXT genes may be tumor-suppressor genes and that the initiation of tumor development may follow a multistep model. Hecht et al. (1995) cited a prevalence of chondrosarcoma in multiple exostosis of 2 to 5%, compared with the age-adjusted incidence rate of 1 in 100,000 for all bone cancers. Raskind et al. (1995) likewise found loss of constitutional heterozygosity at polymorphic loci linked to genes involved in the multiple exostosis. They detected LOH for markers linked to EXT1 on chromosome 8 in a chondrosarcoma that arose in a man with multiple exostoses. They also found LOH for markers linked to EXT1 in 4 of 17 sporadic chondrosarcomas, and 7 showed LOH for markers linked to EXT2. In all, Raskind et al. (1995) observed LOH for markers linked to EXT1 or EXT2 in 44% of the 18 tumors, whereas heterozygosity was retained for markers on 19p linked to EXT3 (600209). These findings also suggested a tumor suppressor role for EXT genes.

Studying 7 extended multiple exostoses families, all linked to the EXT2 locus, Wuyts et al. (1995) refined the localization of the EXT2 gene to a 3-cM region flanked by D11S1355 and D11S1361/D11S554. The findings indicated that the EXT2 gene is located on 11p12-p11. The refined localization excluded a number of putative candidate genes located in the pericentromeric region of chromosome 11.

Support for the localization of the EXT2 gene was provided also by the report of McGaughran et al. (1995), who described a patient with the combination of multiple exostoses with the WAGR syndrome (Wilms tumor, aniridia, genital anomalies, and mental retardation; 194070), a well documented contiguous gene syndrome resulting from deletion of 11p13. Their patient showed a del(11)(p14.2p11.2). Blanton et al. (1996) studied 12 large multigenerational EXT families and found that the disorder mapped to 8q24 in 6 and to 11p in 6. None of the families mapped to the chromosome 19 locus. As pointed out by Potocki et al. (1995), the description of the contiguous gene syndrome resulting from interstitial deletion of 11p, del(11)(p12p11.2), including multiple exostoses as a feature, provided confirmation of the mapping of EXT2. Other features of this contiguous gene syndrome are mental retardation and parietal foramina, known as Catlin marks (168500). Potocki and Shaffer (1996) reported the clinical and molecular findings in a further patient with an 11(p12p11.2) deletion. Cytogenetic and molecular analysis demonstrated a de novo, paternally-derived deletion for markers known to be tightly linked to EXT2. The patient had an unusual facies (bilateral epicanthal folds, ptosis, short philtrum, and downturned upper lip), mental retardation, multiple exostoses, brachycephaly, and bilateral parietal foramina.

Stickens et al. (1996) reported the identification and characterization of the EXT2 gene on chromosome 11. The EXT2 gene contains 7 exons and encodes a 715-amino acid polypeptide. By Northern analysis, they detected a 3.5-kb transcript in almost all tissues. Alternative splicing generates a minor 3.7-kb transcript in some tissues. The gene shows striking sequence similarity to the EXT1 gene on chromosome 8. Both EXT1 and EXT2 share significant homology with 1 additional expressed sequence tag (EST), defining a new multigene family of proteins with potential tumor suppressor activity. (Preliminary studies suggested that the third gene is expressed but does not map to chromosome 19 and thus was excluded as a candidate for the EXT3 gene product.) Particularly useful in isolating the EXT2 gene was the large multigenerational family with chromosome 11-linked multiple exostoses described by Hecht et al. (1995) in which 1 family member developed a malignant chondrosarcoma. In this family, an apparent deletion of the region including polymorphic marker D11S903 was observed in all affected members. In tumor tissue derived from the patient with chondrosarcoma, Stickens et al. (1996) observed that polymorphic marker D11S903 was deleted and that there was LOH for other chromosome 11 markers. In the initial study of the EXT2 candidate gene for mutations, they performed SSCP analysis using DNA from affected members of 3 unrelated families. In 1 of 2 alleles of 1 patient, deletion of nucleotides 784 to 787 was identified (133701.0001), resulting in a frameshift and a premature termination of the EXT2 gene product. Stickens et al. (1996) speculated that a second mutation event was necessary for the development of exostoses, thus accounting for the asymmetry of exostoses observed in the long bones.

Wuyts et al. (1996) constructed a contig of YAC and P1 clones covering the EXT2 candidate region on chromosome 11p12-p11. The EXT2 region was defined in large families with multiple exostoses by linkage analysis. They also defined the candidate region through deletion analysis in families with the DEFECT 11 microdeletion syndrome (601224). From the EXT2 candidate region, Wuyts et al. (1996) isolated a transcribed sequence with homology to EXT1. Alignment of the EXT1 and EXT2 genes at the amino acid level revealed an overall sequence identity of 30.9%. In the 5-prime gene region, the amino acid homology is 18%, while in the 3-prime region there is 48% homology. Wuyts et al. (1996) used SSCP analysis to identify a mutation in the EXT2 gene in a large family: a nonsense 514C-T mutation leading to a premature stop codon (133701.0002). In a large Belgian family, they described a splice site mutation resulting in exon skipping, loss of 94 nucleotides of the transcript, a frameshift, and truncation of the polypeptide (133701.0003).

Clines et al. (1997) found that the EXT2 gene consists of 14 exons (plus 2 alternative exons) covering an estimated 108 kb of 11p13-p11. They mapped the mouse homolog to chromosome 2. At least 2 homologs of the human and mouse EXT2 genes were found in Caenorhabditis elegans, suggesting to the authors that the genes do not function exclusively as regulators of bone growth. (The protein gene product of the EXT2 gene might be referred to as exostosin-2.)

Stickens and Evans (1997) isolated and characterized the mouse EXT2 gene. They found that the mouse EXT2 cDNA contains an open reading frame of 2,154 bp encoding a predicted protein of 718 amino acids. The nucleic acid sequence is 87% identical to the human EXT2 transcript, resulting in an amino acid sequence that is 95% identical to the human protein. The mouse EXT2 gene also shows significant sequence similarity to the mouse and human EXT1 gene. Northern blot analysis showed that the gene is expressed in early stages of embryonic development, and in situ hybridization suggested that EXT2 plays a role in limb development.

Philippe et al. (1997) studied 17 persons from separate families with hereditary multiple exostoses. Disease-causing mutations were found in 12 families (70%), 7 (41%) of which had EXT1 mutations and 5 (29%) EXT2 mutations. Four new mutations in EXT2 were discovered. A missense mutation, asp227 to asn (D227N; 133701.0004), was found in 2 different families; the other 3 alterations (2 nonsense mutations and 1 frameshift mutation) led directly or indirectly to premature stop codons.

Wuyts et al. (1998) analyzed the EXT1 and EXT2 genes in 26 EXT families originating from 9 countries. Of the 26 families, 10 had an EXT1 mutation and 10 had an EXT2 mutation. Twelve of these mutations had not previously been described. From a review of these and previously reported mutations, Wuyts et al. (1998) concluded that mutations in either the EXT1 or the EXT2 gene are responsible for most cases of multiple exostoses. Most of the mutations in these 2 genes cause premature termination of the EXT proteins, whereas missense mutations are rare. The development of exostoses is, therefore, mainly due to loss of function of EXT genes, consistent with the hypothesis that the EXT genes have a tumor-suppressor function.

By fluorescence in situ hybridization, Ligon et al. (1998) showed that the EXT2 gene is deleted in patients with the del(11)(p11.2p12) syndrome. Since patients with mutations in EXT2 have only multiple exostoses, deletions affecting only the EXT2 locus cannot explain the craniofacial anomalies, biparietal foramina, mental retardation, or genitourinary anomalies observed in patients with proximal 11p deletions. Therefore, deletion of other neighboring genes must be responsible for the additional anomalies, and this chromosomal aberration represents a true contiguous gene deletion syndrome.

Blanton, S. H.; Hogue, D.; Wagner, M.; Wells, D.; Young, I. D.; Hecht, J. T.: Hereditary multiple exostoses: confirmation of linkage to chromosomes 8 and 11. Am. J. Med. Genet. 62: 150-159, 1996.

Clines, G. A.; Ashley, J. A.; Shah, S.; Lovett, M.: The structure of the human multiple exostoses 2 gene and characterization of homologs in mouse and Caenorhabditis elegans. Genome Res. 7: 359-367, 1997.

Cook, A.; Raskind, W.; Blanton, S. H.; Pauli, R. M.; Gregg, R. G.; Francomano, C. A.; Puffenberger, E.; Conrad, E. U.; Schmale, G.; Schellenberg, G.; Wijsman, E.; Hecht, J. T.; Wells, D.; Wagner, M. J.: Genetic heterogeneity in families with hereditary multiple exostoses. Am. J. Hum. Genet. 53: 71-79, 1993.

Hecht, J. T.; Hogue, D.; Strong, L. C.; Hansen, M. F.; Blanton, S. H.; Wagner, M.: Hereditary multiple exostosis and chondrosarcoma: linkage to chromosome 11 and loss of heterozygosity for EXT-linked markers on chromosomes 11 and 8. Am. J. Hum. Genet. 56: 1125-1131, 1995.

Hennekam, R. C. M.: Hereditary multiple exostoses. J. Med. Genet. 28: 262-266, 1991.

Ligon, A. H.; Potocki, L.; Shaffer, L. G.; Stickens, D.; Evans, G. A.: Gene for multiple exostoses (EXT2) maps to 11(p11.2p12) and is deleted in patients with a contiguous gene syndrome. (Letter) Am. J. Med. Genet. 75: 538-540, 1998.

McGaughran, J. M.; Ward, H. B.; Evans, D. G. R.: WAGR syndrome and multiple exostoses in a patient with del(11)(p11.2p14.2). J. Med. Genet. 32: 823-824, 1995.

Philippe, C.; Porter, D. E.; Emerton, M. E.; Wells, D. E.; Simpson, A. H. R. W.; Monaco, A. P.: Mutation screening of the EXT1 and EXT2 genes in patients with hereditary multiple exostoses. Am. J. Hum. Genet. 61: 520-528, 1997.

Potocki, L.; Greenberg, F.; Shaffer, L. G.: Interstitial deletion of 11(p12p11.2): a rare chromosomal syndrome with mental retardation, parietal foramina, and multiple exostoses. (Abstract) Am. J. Hum. Genet. 57: A123, 1995.

Potocki, L.; Shaffer, L. G.: Interstitial deletion of 11(p11.2p12): a newly described contiguous gene deletion syndrome involving the gene for hereditary multiple exostoses (EXT2). Am. J. Med. Genet. 62: 319-325, 1996.

Quirini, G. E.; Meyer, J. R.; Herman, M.; Russell, E. J.: Osteochondroma of the thoracic spine: an unusual cause of spinal cord compression. Am. J. Neuroradiol. 17: 961-964, 1996.

Raskind, W. H.; Conrad, E. U.; Chansky, H.; Matsushita, M.: Loss of heterozygosity in chondrosarcomas for markers linked to hereditary multiple exostoses loci on chromosomes 8 and 11. Am. J. Hum. Genet. 56: 1132-1139, 1995.

Stickens, D.; Clines, G.; Burbee, D.; Ramos, P.; Thomas, S.; Hogue, D.; Hecht, J. T.; Lovett, M.; Evans, G. A.: The EXT2 multiple exostoses gene defines a family of putative tumour suppressor genes. Nature Genet. 14: 25-32, 1996.

Stickens, D.; Evans, G. A.: Isolation and characterization of the murine homolog of the human EXT2 multiple exostoses gene. Biochem. Molec. Med. 61: 16-21, 1997.

Wu, Y.-Q.; Heutink, P.; de Vries, B. B. A.; Sandkuijl, L. A.; van den Ouweland, A. M. W.; Niermeijer, M. F.; Galjaard, H.; Reyniers, E.; Willems, P. J.; Halley, D. J. J.: Assignment of a second locus for multiple exostoses to the pericentromeric region of chromosome 11. Hum. Molec. Genet. 3: 167-171, 1994.

Wuyts, W.; Ramlakhan, S.; Van Hul, W.; Hecht, J. T.; van den Ouweland, A. M. W.; Raskind, W. H.; Hofstede, F. C.; Reyniers, E.; Wells, D. E.; de Vries, B.; Conrad, E. U.; Hill, A.; Zalatayev, D.; Weissenbach, J.; Wagner, M. J.; Bakker, E.; Halley, D. J. J.; Willems, P. J.: Refinement of the multiple exostoses locus (EXT2) to a 3-cM interval on chromosome 11. Am. J. Hum. Genet. 57: 382-387, 1995.

Wuyts, W.; Van Hul, W.; De Boulle, K.; Hendrickx, J.; Bakker, E.; Vanhoenacker, F.; Mollica, F.; Ludecke, H.-J.; Sayli, B. S.; Pazzaglia, U. E.; Mortier, G.; Hamel, B.; Conrad, E. U.; Matsushita, M.; Raskind, W. H.; Willems, P. J.: Mutations in the EXT1 and EXT2 genes in hereditary multiple exostoses. Am. J. Hum. Genet. 62: 346-354, 1998.

Wuyts, W.; Van Hul, W.; Wauters, J.; Nemtsova, M.; Reyniers, E.; Van Hul, E.; De Boulle, K.; de Vries, B. B. A.; Hendrickx, J.; Herrygers, I.; Bossuyt, P.; Balemans, W.; Fransen, E.; Vits, L.; Coucke, P.; Nowak, N. J.; Shows, T. B.; Mallet, L.; van den Ouweland, A. M. W.; McGaughran, J.; Halley, D. J. J.; Willems, P. J.: Positional cloning of a gene involved in hereditary multiple exostoses. Hum. Molec. Genet. 5: 1547-1557, 1996.

EXOSTOSES, MULTIPLE 3 (*600209)

JJM NOTE

Even this single family adds to the understanding of the multiple origins of multiple exostosis.

Le Merrer et al. (1994) tested for linkage to markers in the region 11p12-p11 because of the finding of multiple exostoses in a child with an interstitial deletion of that region; linkage was excluded. However, the locus they termed EXT2 was mapped to 19p by linkage to a microsatellite DNA marker at the D19S221 locus. In studies of 21 families, they found a maximum location score (LS) of 7.22 with a proportion of families linked to chromosome 19 of 35% and a proportion of families linked to chromosome 8 of 65%. Their data supported the location of EXT1 (133700) distal to D8S198 and the location of the other gene, designated EXT3 by the HGM nomenclature committee, between D19S413 and D19S221.

Le Merrer et al. (1994) suggested that the close linkage of EXT3 to the JUNB (165161) and JUND (165162) protooncogenes on 19p makes these genes possible candidate genes, because malignancy is a complication of multiple exostoses and because transgenic mice defective for a functionally related gene (FOS; 164810) have been shown to develop multiple exostoses of long bones (Wang et al., 1991).

Le Merrer, M.; Legeai-Mallet, L.; Jeannin, P. M.; Horsthemke, B.; Schinzel, A.; Plauchu, H.; Toutain, A.; Achard, F.; Munnich, A.; Maroteaux, P.: A gene for hereditary multiple exostoses maps to chromosome 19p. Hum. Molec. Genet. 3: 717-722, 1994.

Wang, Z.-Q.; Grigoriadis, A. E.; Mohle-Steinlein, U.; Wagner, E. F.: A novel target cell for c-fos-induced oncogenesis: development of chondrogenic tumours in embryonic stem cell chimeras. EMBO J. 10: 2437-2450, 1991.

EXOSTOSES, MULTIPLE-LIKE 1 (*601738)

JJM NOTE

It is not yet known whether this particular gene predisposes to malignancy, but it seems likely.

Hereditary multiple exostoses is an autosomal dominant disorder characterized by multiple bony outgrowth from the juxtaepiphyseal region of long bones. Wise et al. (1997) noted that, in a small proportion of cases, these exostoses progress to malignant chondrosarcomas. Genetic linkage studies have described 3 independent loci: EXT1 (123700), on 8q24.1; EXT2 (133701), on 11p13-p11; and EXT3 (600209), on 19p. The EXT1 and EXT2 genes were cloned and shown to have extensive sequence homology. These genes are deleted in exostoses-derived tumors, supporting the hypothesis that they encode tumor suppressors. Wise et al. (1997) identified another gene, termed EXTL by them, that showed striking sequence similarity to both EXT1 and EXT2 at the nucleotide and amino acid sequence levels. Although the mRNA transcribed from this gene is similar in size to that of EXT1 and EXT2, its pattern of expression was quite different. By radiation hybrid analysis and by fluorescence in situ hybridization, Wise et al. (1997) mapped EXTL to 1p36.1 between D1S458 and D1S511, a region that frequently shows loss of heterozygosity (LOH) in a variety of tumor types.

Wise, C. A.; Clines, G. A.; Massa, H.; Trask, B. J.; Lovett, M.: Identification and localization of the gene for EXTL, a third member of the multiple exostoses gene family. Genome Res. 7: 10-16, 1997.

EXOSTOSES, MULTIPLE-LIKE 2 (*602411)

JJM NOTE

This is a gene without a disease (as yet) and is included for convenience.

In patients with multiple exostoses, mutations in 2 different genes have been found: EXT1 (133700) on 8q and EXT2 (133701) on 11p. In addition, linkage has demonstrated a third locus, EXT3 (600209) on 19p as the site of mutations causing multiple exostoses. The family of EXT genes was extended by the identification of an EXT-like gene (EXTL1; 601738) showing a high degree of homology with the EXT genes. Wuyts et al. (1997) described a second EXT-like gene, EXTL2, that is homologous to the EXT and EXTL genes. EXTL2 was found to consist of 5 exons encoding a ubiquitously expressed 330-amino acid protein. In addition, a putative pseudogene, EXTL2P, was identified. The EXTL2 gene was assigned to 1p12-p11 by fluorescence in situ hybridization, whereas EXTL2P was mapped to 2q24-q31.

Wuyts, W.; Van Hul, W.; Hendrickx, J.; Speleman, F.; Wauters, J.; De Boulle, K.; Van Roy, N.; Van Agtmael, T.; Bossuyt, P.; Willems, P. J.: Identification and characterization of a novel member of the EXT gene family, EXTL2. Europ. J. Hum. Genet. 5: 382-389, 1997.

DEFECT 11 CONTIGUOUS GENE SYNDROME (#601224)

JJM NOTE

Like the next entry, the spectrum of this disorder may reflect many missing genes, including that for exostosis 1.

A number sign (#) is used with this entry because of its nature as a contiguous gene syndrome that includes foramina parietalia permagna (168500), multiple exostoses (133701), and in some cases, craniofacial dysostosis and mental retardation. It is caused by deletion in the 11p13-p11 region.

Bartsch et al. (1996) used the designation DEFECT11 for a contiguous gene syndrome due to deletion in the proximal short arm of chromosome 11. They described 8 patients belonging to 4 families. One patient had been reported by Lorenz et al. (1990) as an unusual case of acrocephalosyndactyly. Three of the patients had been reported by Shaffer et al. (1993). Bartsch et al. (1996) used microsatellite markers to characterize the extent of the deletion in each case. In addition to multiple exostoses and parietal foramina, 5 of the 8 patients showed brachycephaly with a somewhat turricephalic skull shape. The patient reported by Lorenz et al. (1990) and 1 of the patients reported by Shaffer et al. (1993) were described in the original publications to have a Saethre-Chotzen-like phenotype (101400). Cutaneous syndactyly between fingers 2 and 5 was present in the first of these 2 patients. The absence of craniofacial dysostosis in 1 family with the contiguous gene syndrome that showed FPP and multiple exostoses and in families with autosomal inheritance of isolated FPP suggested that there is a specific 11p gene involved in craniofacial dysostosis. Five of the patients were severely retarded but 3 patients from 1 family were mentally normal; thus, a specific mental retardation gene may be involved in the deletion. No evidence of imprinting was found; deletions of paternal origin were found in 2 patients and of maternal origin in 5. Bartsch et al. (1996) studied the deletions by cytogenetic and/or molecular methods and found them to be located between the centromere and D11S914 in a region of approximately 20 cM. Their study confirmed the presence of a multiple exostoses gene on 11p and suggested that the gene for isolated foramina parietalia permagna and genes associated with craniofacial dysostosis and mental retardation reside in the same chromosomal region.

Bartsch, O.; Wuyts, W.; Van Hul, W.; Hecht, J. T.; Meinecke, P.; Hogue, D.; Werner, W.; Zabel, B.; Hinkel, G. K.; Powell, C. M.; Shaffer, L. G.; Willems, P. J.: Delineation of a contiguous gene syndrome with multiple exostoses, enlarged parietal foramina, craniofacial dysostosis, and mental retardation, caused by deletions on the short arm of chromosome 11. Am. J. Hum. Genet. 58: 734-742, 1996.

Lorenz, P.; Rupprecht, E.; Tellkamp, H.: An unusual type of acrocephalosyndactyly with bilateral parietooccipital 'encephalocele,' micropenis and severe mental retardation. Am. J. Med. Genet. 35: 453-458, 1990.

Shaffer, L. G.; Hecht, J. T.; Ledbetter, D. H.; Greenberg, F.: Familial interstitial deletion 11(p11.12p12) associated with parietal foramina, brachymicrocephaly, and mental retardation. Am. J. Med. Genet. 45: 581-583, 1993.

LANGER-GIEDION SYNDROME (#150230)

JJM NOTE

This entity is included because of the exostoses. I am not aware of malignancy occurring in a patient with this microdeletion syndrome.

A number sign (#) is used with this entry because of compelling data indicating that it is a contiguous gene syndrome involving loss of functional copies of the TRPS1 (190350) and EXT1 (133700) genes.

This disorder has similarities to the trichorhinophalangeal syndrome type I (190350, 275500), particularly with regard to facies, bulbous nose, sparse hair, and cone-shaped epiphyses. Distinguishing features are mental retardation, microcephaly, multiple exostoses and redundant skin. Less consistent features include hyperextensible joints, recurrent upper respiratory tract infections, and delayed speech development. All cases have been sporadic and a majority have been males. The children may be 'floppy infants.' The exostoses can be striking. The first appearance of exostoses on the back of the scapula can suggest the incorrect diagnosis of FOP (135100). Kozlowski et al. (1977), who reported 2 unrelated patients, a girl and a boy, pointed out that the condition had been described by Ale and Calo (1961). Buhler et al. (1980) reported the case of a teenaged girl with features suggestive of Langer-Giedion syndrome associated with terminal deletion of 8q (the band q24 was missing from one chromosome 8). Pfeiffer (1980) described deletion of a segment (q13-22) of the long arm of chromosome 8 in a mentally retarded boy with Langer-Giedion syndrome. Additional features included colobomata of the iris and defect of the fourth and fifth fingers. Wilson et al. (1981) found interstitial deletion of 8q22.8-q24.1 in a 17-year-old patient with multiple exostoses and developmental delay. Exostoses were first apparent at age 4 years. The patient lacked the typical nose and coned epiphyses of the Langer-Giedion syndrome. Gorlin et al. (1982) found normal chromosomes on prophase banding in 2 patients. Murachi et al. (1981) described affected father and daughter. The father was mildly mentally retarded. They found reports of 9 cases, all sporadic. Chromosome studies were apparently not done. Turleau et al. (1982) concluded that 8q23 is the 'critical segment,' not 8q22. Zaletajev and Marincheva (1983) attributed LGS in their patient to interstitial deletion of 8q22. Langer et al. (1984) described 4 cases without mental retardation and reviewed 32 previously reported cases. Delayed speech development and hearing loss were noted as features. They pointed out that there has been no familial occurrence with the exception of concordant monozygotic twins and no parental consanguinity or ethnic predilection. However, Brenholz et al. (1989) reported LGS in 2 brothers whose mother appears to have been affected. The maternal grandmother and a maternal first cousin may have been affected. Buhler and Malik (1984) suggested that the shortest region of overlap of the 8q deletion is in band q24.1. They raised the question of whether type I trichorhinophalangeal syndrome may be caused by mutation at the same locus or region. Supporting this suggestion is the description of TRPS-I with probable deletion in the same region of 8q (Hamers et al., 1983) and appreciation that the presence or absence of exostoses may be the other 'symptom' that distinguishes types I and II. The fact that the multiple exostoses of LGS are indistinguishable in radiographic features and natural history from those of the long recognized autosomal dominant disorder (133700) may indicate that the latter mutation is sited on 8q. In the context of this interpretation, LGS should be considered a 'contiguous gene syndrome.'

Bowen et al. (1985) described an 18-year-old intellectually normal male with LGS and a small deletion of bands 8q24.11-8q24.12. In addition, he had an apparently balanced de novo translocation (2;9)(q21;q13). Neither abnormality was found in the parents. The risk of LGS in any child of the proband is presumably 50%. Brocas et al. (1986) showed that the thyroglobulin locus, located at 8q24, is intact in LGS. This confirms the distal location previously defined for LGS and assigns the critical region for LGS in the proximal part of band 8q24, i.e., 8q24.11-q24.13. Buhler et al. (1987) concluded that the Langer-Giedion syndrome is due to a deletion extending from 8q24.11 to 8q24.13, whereas TRP I is caused by an even smaller deleted segment, namely, 8q24.12. Buhler et al. (1987) described a case of TRP I with a mosaic deletion of that band. Maroteaux (1987) is of the view that the multiple exostoses of LGS are quite different from those of ordinary exostosis. Okuno et al. (1987) described a typical case with interstitial deletion of 8q24.13-q24.22. They concluded that a part of band 8q24.1 is responsible for the syndrome. Zaletaev et al. (1987) found deletion in 8q in 3 unrelated patients with LGS. The 'critical' region was identified as 8q24.11-q24.13. The findings of Fennell et al. (1989) likewise support the view that the critical segment for LGS is proximal to or involves a proximal part of 8q24.1.

Ludecke et al. (1989) described the microdissection of the Langer-Giedion syndrome region on chromosome 8 from GTG-banded metaphase chromosomes (G-banding with trypsin-Giemsa) and the universal enzymatic amplification of the dissected DNA. Eighty percent of clones from this library (total yield 20,000) identified single-copy DNA sequences. Half of the clones detected deletions in 2 patients with LGS. Ten of the clones have been assigned to the deleted region in Langer-Giedion syndrome (8q23.2-q24.11) based on Southern blot analysis of DNA from 2 patients. The results of Ludecke et al. (1989) demonstrated that thousands of region-specific probes can be isolated in a short period of time. Microdissection and microcloning have been applied successfully to various chromosome regions in Drosophila and mouse, but conventional microtechniques are too coarse and inefficient, especially on unbanded chromosomes, for analysis of the human genome. Ludecke et al. (1989) found 2 RFLPs in an anonymous DNA probe that defines the D8S48 locus within the Langer-Giedion syndrome chromosome region. Both polymorphisms were informative in the family of a Langer-Giedion patient carrying a de novo interstitial deletion 8q23-q24.1. Lack of transmission of a maternal haplotype indicated that the deletion occurred during maternal gametogenesis. Ludecke et al. (1991) used 13 anonymous DNA markers from an 8q24.1-specific microdissection library, as well as MYC (190080) and TG (188450) gene probes, to map the deletion breakpoints in 16 patients with LGS. Twelve patients had a cytogenetically visible deletion, 2 had an apparently balanced translocation, and 2 had an apparently normal karyotype. In all cases except 1 translocation patient, loss of genetic material was detected. The DNA markers fell into 10 deletion intervals. Clone L48 (D8S51) defined the shortest region of deletion overlap, which was estimated to be less than

2 Mbp. The clones that flanked the shortest region of deletion overlap recogn ized evolutionarily conserved sequences. Parrish et al. (1991) isolated 8 DNA clones that were found to lie within the deletion of at least 1 of 3 patients with LGS. One clone identified sequences that were missing from 1 copy of chromosome 8 in all 3 patients. Fryns et al. (1983) and Partington et al. (1991) described hydrometrocolpos and hematometra as complications of this syndrome. Kozlowski et al. (1977) and Partington et al. (1991) described ureteral reflux requiring reimplantation of the ureters in the bladder. In a patient with Langer-Giedion syndrome and interstitial 8q deletion, Ramos et al. (1992) found persistent cloaca and the prune belly sequence (100100).

Using YAC cloning, Southern blotting, PCR analysis, and fluorescence in situ hybridization in the study of chromosome 8 deletions, translocations, an inversion, and an insertion in patients with tricho-rhinophalangeal syndrome type I (TRPS1; 190350), Langer-Giedion syndrome, or multiple exostoses (EXT1; 133700), Ludecke et al. (1995) obtained information indicating that the TRPS1 gene maps more than 1,000-kb proximal to the EXT1 gene and that both genes are affected in Langer-Giedion syndrome. They concluded that the Langer-Giedion syndrome is not due to pleiotropic effects of mutations in a single gene, but that it is a true contiguous gene syndrome.

Hou et al. (1995) constructed a physical map covering 4 Mb of 8q24.1 and used this map to refine the location of the genes responsible for LGS. The map was composed of overlapping YAC clones that were identified and ordered in relation to sequence tagged sites mapped to the Langer-Giedion chromosomal region on somatic cell hybrids. The minimal region of overlap of LGS deletions, previously identified by analysis of 15 patients, was placed on the map by analysis of 2 patients whose deletions define the end points. The chromosome 8 breakpoint of a balanced t(8;9)(q24.1;q33.3) translocation from a patient with TRPS1 was found to be located just within the proximal end of the minimal deletion region. A deletion of 8q24.11-q24.3 in a patient with multiple exostoses was found to overlap the distal end of the LGS deletion region, indicating that the EXT1 gene is distal to the TRPS1 gene and providing further support for the hypothesis that LGS is due to loss of functional copies of both the TRPS1 and the EXT1 genes.

See also Beighle et al. (1977); Buhler (1982); Buhler et al. (1983); Fryns et al. (1980); Fukushima et al. (1983); Giedion (1969); Hall et al. (1974); Langer (1969); Ludecke et al. (1989); Ludecke et al. (1989); Murachi et al. (1979); Oorthuys and Beemer (1979); Shabtai et al. (1985); Stoltzfus et al. (1977); Wilson et al. (1979); Zabel and Baumann (1982)

Ale, G.; Calo, S.: On a case of peripheral dysostosis associated with multiple osteogenic exostoses, dysuniform and dysharmonic hyposomia. Ann. Radiol. Diag. (Bologna) 34: 376-385, 1961.

Beighle, C.; Karp, L. E.; Hanson, J. W.; Hall, J. G.; Hoehn, H.: Small structural changes of chromosome 8. Hum. Genet. 38: 113-121, 1977.

Bowen, P.; Biederman, B.; Hoo, J. J.: The critical segment for the Langer-Giedion syndrome: 8q24.11-q24.12. Ann. Genet. 28: 224-227, 1985.

Brenholz, P.; Swayne, L.; Twersky, S.; Arbeitel, B.; Singer, N.: Dominant inheritance of the Langer-Giedion syndrome. (Abstract) Am. J. Hum. Genet. 45 (suppl.): A41 only, 1989.

Brocas, H.; Buhler, E. M.; Simon, P.; Malik, N. J.; Vassart, G.: Integrity of the thyroglobulin locus in tricho-rhino-phalangeal syndrome II. Hum. Genet. 74: 178-180, 1986.

Buhler, E. M.: Langer-Giedion syndrome and 8q- deletion. (Editorial) Am. J. Med. Genet. 11: 359 only, 1982.

Buhler, E. M.; Buhler, U. K.; Beutler, C.; Fessler, R.: A final word on the tricho-rhino-phalangeal syndromes. Clin. Genet. 31: 273-275, 1987.

Buhler, E. M.; Buhler, U. K.; Christen, R.: Terminal or interstitial deletion in chromosome 8 long arm in Langer-Giedion syndrome (TRP II syndrome)?. Hum. Genet. 64: 163-166, 1983.

Buhler, E. M.; Buhler, U. K.; Stalder, G. R.; Jani, L.; Jurik, L. P.: Chromosome deletion and multiple cartilaginous exostoses. Europ. J. Pediat. 133: 163-166, 1980.

Buhler, E. M.; Malik, N. J.: The tricho-rhino-phalangeal syndrome(s): chromosome 8 long arm deletion: is there a shortest region of overlap between reported cases? TRP I and TRP II syndromes: are they separate entities?. (Editorial) Am. J. Med. Genet. 19: 113-119, 1984.

Fennell, S. J.; Benson, J. W. T.; Kindley, A. D.; Schwarz, M. J.; Czepulkowski, B.: Partial deletion 8q without Langer-Giedion syndrome: a recognisable syndrome. J. Med. Genet. 26: 167-171, 1989.

Fryns, J. P.; Emmery, L.; Timmermans, J.; Pedersen, J. C.; van den Berghe, H.: Tricho-rhino-phalangeal syndrome type II: Langer-Giedion syndrome in a 2.5-year-old boy. J. Genet. Hum. 28: 53-56, 1980.

Fryns, J. P.; Heremans, G.; Marien, J.; Van den Berghe, H.: Langer-Giedion syndrome and deletion of the long arm of chromosome 8: confirmation of the critical segment to 8q23. Hum. Genet. 64: 194-195, 1983.

Fukushima, Y.; Kuroki, Y.; Izawa, T.: Two cases of the Langer-Giedion syndrome with the same interstitial deletion of the long arm of chromosome 8: 46,XY or XX,del(8)(q23.3q24.13). Hum. Genet. 64: 90-93, 1983.

Giedion, A.: Die periphere Dysostose (pD)—ein Sammelbegriff. Fortschr. Roentgenstr. 110: 507-534, 1969.

Gorlin, R. J.; Cervenka, J.; Bloom, B. A.; Langer, L. O., Jr.: No chromosome deletion found on prometaphase banding in two cases of Langer-Giedion syndrome. (Letter) Am. J. Med. Genet. 13: 345-347, 1982.

Hall, B. D.; Langer, L. O., Jr.; Giedion, A.; Smith, D. W.; Cohen, M. M.; Beals, R. K.; Brandner, M.: Langer-Giedion syndrome. Birth Defects Orig. Art. Ser. X(12): 147-164, 1974.

Hamers, A.; Jongbloet, P.; Peeters, G.; Geraedts, J.: Microcytogenetics of chromosome 8q. (Abstract) 8th Int. Chrom. Conf., Lubeck Abstract-8, 9/1983.

Hou, J.; Parrish, J.; Ludecke, H.-J.; Sapru, M.; Wang, Y.; Chen, W.; Hill, A.; Siegel-Bartelt, J.; Northrup, H.; Elder, F. F. B.; Chinault, C.; Horsthemke, B.; Wagner, M. J.; Wells, D. E.: A 4-megabase YAC contig that spans the Langer-Giedion syndrome region on human chromosome 8q24.1: use in refining the location of the trichorhinophalangeal syndrome and multiple exostoses genes (TRPS1 and EXT1). Genomics 29: 87-97, 1995.

Kozlowski, K.; Harrington, G.; Barylak, A.; Bartoszewica, B.: Multiple exostoses-mental retardation syndrome (Ale-Calo or M.E.M.R. syndrome): description of two childhood cases. Clin. Pediat. 16: 219-224, 1977.

Langer, L. O., Jr.: The thoracic-pelvic-phalangeal dystrophy. Birth Defects Orig. Art. Ser. V(4): 55-64, 1969.

Langer, L. O., Jr.; Krassikoff, N.; Laxova, R.; Scheer-Williams, M.; Lutter, L. D.; Gorlin, R. J.; Jennings, C. G.; Day, D. W.: The tricho-rhino-phalangeal syndrome with exostoses (or Langer-Giedion syndrome): four additional patients without mental retardation and review of the literature. Am. J. Med. Genet. 19: 81-111, 1984.

Ludecke, H.-J.; Burdiek, R.; Senger, G.; Claussen, U.; Passarge, E.; Horsthemke, B.: Maternal origin of a de novo chromosome 8 deletion in a patient with Langer-Giedion syndrome. Hum. Genet. 82: 327-329, 1989.

Ludecke, H.-J.; Johnson, C.; Wagner, M. J.; Wells, D. E.; Turleau, C.; Tommerup, N.; Latos-Bielenska, A.; Sandig, K.-R.; Meinecke, P.; Zabel, B.; Horsthemke, B.: Molecular definition of the shortest region of deletion overlap in the Langer-Giedion syndrome. Am. J. Hum. Genet. 49: 1197-1206, 1991.

Ludecke, H.-J.; Senger, G.; Claussen, U.; Horsthemke, B.: Cloning defined regions of the human genome by microdissection of banded chromosomes and enzymatic amplification. Nature 338: 348-350, 1989.

Ludecke, H.-J.; Senger, G.; Claussen, U.; Horsthemke, B.: Molecular approach to the Langer-Giedion syndrome chromosome region. (Abstract) Cytogenet. Cell Genet. 51: 1035 only, 1989.

Ludecke, H.-J.; Wagner, M. J.; Nardmann, J.; La Pillo, B.; Parrish, J. E.; Willems, P. J.; Haan, E. A.; Frydman, M.; Hamers, G. J. H.; Wells, D. E.; Horsthemke, B.: Molecular dissection of a contiguous gene syndrome: localization of the genes involved in the Langer-Giedion syndrome. Hum. Molec. Genet. 4: 31-36, 1995.

Maroteaux, P.: Personal Communication. Paris, France 1987.

Murachi, S.; Itoh, H.; Sugiura, Y.: Tricho-rhino-phalangeal syndrome type II: the Langer-Giedion syndrome. Jpn. J. Hum. Genet. 24: 27-36, 1979.

Murachi, S.; Nogami, H.; Oki, T.; Ogino, T.: Familial tricho-rhino-phalangeal syndrome type II. Clin. Genet. 19: 149-155, 1981.

Okuno, T.; Inoue, A.; Asakura, T.; Nakao, S.: Langer-Giedion syndrome with del8(q24.13-q24.22). Clin. Genet. 32: 40-45, 1987.

Oorthuys, J. W. E.; Beemer, F. A.: The Langer-Giedion-syndrome (tricho-rhino-phalangeal syndrome, type II). Europ. J. Pediat. 132: 55-59, 1979.

Parrish, J. E.; Wagner, M. J.; Hecht, J. T.; Scott, C. I., Jr.; Wells, D. E.: Molecular analysis of overlapping chromosomal deletions in patients with Langer-Giedion syndrome. Genomics 11: 54-61, 1991.

Partington, M. W.; Rae, J.; Payne, M. J.: Haematometra in the Langer-Giedion syndrome. (Letter) J. Med. Genet. 28: 644-645, 1991.

Pfeiffer, R. A.: Langer-Giedion syndrome and additional congenital malformations with interstitial deletion of the long arm of chromosome 8: 46,XY,del8(q13-22). Clin. Genet. 18: 142-146, 1980.

Ramos, F. J.; McDonald-McGinn, D. M.; Emanuel, B. S.; Zackai, E. H.: Tricho-rhino-phalangeal syndrome type II (Langer-Giedion) with persistent cloaca and prune belly sequence in a girl with 8q interstitial deletion. Am. J. Med. Genet. 44: 790-794, 1992.

Shabtai, F.; Sandowski, U.; Nissimov, R.; Klar, D.; Halbrecht, I.: Familial syndrome with some features of the Langer-Giedion syndrome, and paracentric inversion of chromosome 8, inv 8 (q11.23-q21.1). Clin. Genet. 27: 600-605, 1985.

Stoltzfus, E.; Ladda, R. L.; Lloyd-Still, J. D.: Langer-Giedion syndrome: type II tricho-rhino-phalangeal dysplasia. J. Pediat. 91: 277-280, 1977.

Turleau, C.; Chavin-Colin, F.; de Grouchy, J.; Maroteaux, P.; Rivera, H.: Langer-Giedion syndrome with and without del 8q: assignment of critical segment to 8q23. Hum. Genet. 62: 183-187, 1982.

Wilson, W. G.; Herrington, R. T.; Aylsworth, A. S.: The Langer-Giedion syndrome: report of a 22-year-old woman. Pediatrics 64: 542-545, 1979.

Wilson, W. G.; Shah, H.; Wyandt, H. E.: Interstitial deletion of 8q in a patient with multiple exostoses and developmental delay. (Abstract) Am. J. Hum. Genet. 33: 96A only, 1981.

Zabel, B. U.; Baumann, W. A.: Langer-Giedion syndrome with interstitial 8q-deletion. Am. J. Med. Genet. 11: 353-358, 1982.

Zaletaev, D. V.; Kuleshov, N. P.; Lur'e, I. V.; Marincheva, G. S.: Langer-Giedion syndrome and chromosome 8 long arm deletion. Sov. Genet. 23: 629-633, 1987.

Zaletajev, D. V.; Marincheva, G. S.: Langer-Giedion syndrome, in a child with complex structural aberration of chromosome 8. Hum. Genet. 63: 178-182, 1983.

TRICHORHINOPHALANGEAL SYNDROME 1 (*190350)

JJM NOTE

In general, this syndrome does *not* have exostoses, but the proximity of the exostosis gene can result in the previously cited microdeletion syndrome, which has the bony tumors.

This syndrome is entered also in the recessive catalog (275500). Autosomal dominant inheritance seems unequivocal in light of a family in which we have now observed affected grandfather, son and grandson (Murdoch, 1969; McKusick, 1972). The earliest affected male died at age 43 years of a cerebrovascular accident. Beals (1973) described a family in which the father and 2 of 4 children, a male and a female, were affected. Three Japanese families with 19 affected persons in a clear autosomal dominant pedigree pattern were reported by Sugiura et al. (1976). Booth and Maurer (1981) described a sporadic case of this disorder in a girl with de novo 9;11 translocation (p22;q21). The Langer-Giedion syndrome (150230), a similar disorder that includes also multiple exostoses and mental retardation and often has abnormality of chromosome 8, is sometimes called trichorhinophalangeal syndrome II. Sanchez et al. (1985) described a complex chromosome rearrangement in a boy with TRPS I. Skeletal x-rays were normal. The breakpoints in chromosome 8 were at p22 and q13. Goldblatt and Smart (1986) reported a case of TRPS I without exostoses and with a partial microdeletion of 8q23. This is further support for the notion that the Langer-Giedion syndrome is the consequence of deletion or other mutation of 2 or more independent loci, one of which 'causes' exostoses and the other TRPS. Fryns and Van den Berghe (1986) reported a patient with TRPS I and a small interstitial deletion of 8q24.12. Buhler et al. (1987) concluded that the Langer-Giedion syndrome is due to a deletion extending from 8q24.11 to 8q24.13, whereas TRPS I is caused by an even smaller deleted segment, namely, 8q24.12. Buhler et al. (1987) described a case of TRPS I with a mosaic deletion of that band.

Haan et al. (1989) studied a family with an inherited rearrangement of 8q in which persons with the chromosome abnormality had manifestations of both TRPS and the branchiootic syndrome (113650). One breakpoint in the abnormality in this family involved 8q24.11, which is consistent with the previously deduced location of the TRPS gene. Haan et al. (1989) suggested that the gene for the branchiootic syndrome may map to either 8q13.3 or 8q21.13, the 2 other breakpoint sites in this family's chromosome aberration. It was thought that there was no deleted material in this family. Yamamoto et al. (1989) found no evidence of deletion in a study of the chromosomes at the 850-bands stage in a patient with typical familial TRPS I. Naritomi and Hirayama (1989) described a mother and daughter with mild features of TRPS I. Neither had exostosis, microcephaly or mental retardation which are features of the Langer-Giedion syndrome. The mother was a mosaic for an 8q23.3-q24.13 deletion. The daughter had a more complex rearrangement of chromosome 8 resulting in partial trisomy of distal 8q. The daughter, however, showed no features of the trisomy 8 syndrome and was a little more affected than the mother. Yamamoto et al. (1989) and Hamers et al. (1990) independently reported cases of type I TRP syndrome with severe mental retardation but without multiple exostoses; in both cases an interstitial deletion of 8q was found. The findings were interpreted to support the suggestion that TRPS II is due to deletion of 8q24.11-q24.13; that TRPS I patients have a deletion of band 8q24.12; and band 8q24.13 is involved in the development of exostoses. Mental retardation seems to be correlated with the size of the interstitial 8q deletion. Marchau et al. (1993) reported the cases of a brother and sister with TRPS I and an apparently balanced translocation involving 1 breakpoint at 8q24.11. The photograph of the father, who had died of an accident, showed very sparse scalp hair and a bulbous pear-shaped nose. He had short stature.

Ludecke et al. (1995) and Hou et al. (1995) presented evidence that the Langer-Giedion syndrome is a true contiguous gene syndrome due to loss of functional copies of both the TRPS1 and the EXT1 gene (133700) and that the EXT1 gene is distal to the TRPS1 gene.

In contrast to TRPS I patients, most TRPS II patients have cytogenetically visible deletions and are often mentally retarded. Using Southern blot and fluorescence in situ hybridization analysis, Nardmann et al. (1997) searched for submicroscopic deletions in 12 patients with TRPS I and an apparently normal karyotype. One patient of normal intelligence was found to have a deletion of approximately 5 Mb. This suggested that mental retardation in TRPS is caused by genes outside the 5-Mb region. Using 3 Langer-Giedion critical region microsatellite markers, they determined the paternal origin of this TRPS I deletion and of 8 TRPS II deletions. In 6 patients, the deletion was of paternal origin and in 3 patients it was of maternal origin.

See also Felman and Frias (1977); Ferrandez et al. (1980); Giedion et al. (1973); Goodman et al. (1981); Howell and Wynne-Davies (1986); McCloud and Solomon (1977); Peltola and Kuokkanen (1978); Sugiura (1978); Weaver et al. (1974); Yamamoto et al. (1989)

Beals, R. K.: Tricho-rhino-phalangeal dysplasia: report of a kindred. J. Bone Joint Surg. 55: 821-826, 1973.

Booth, C. W.; Maurer, W. F.: De novo 9:11 translocation in a sporadic case of trichorhinophalangeal (I) syndrome. (Abstract) Pediat. Res. 15: 559 only, 1981.

Buhler, E. M.; Buhler, U. K.; Beutler, C.; Fessler, R.: A final word on the tricho-rhino-phalangeal syndromes. Clin. Genet. 31: 273-275, 1987.

Felman, A. H.; Frias, J. L.: Trichorhinophalangeal syndrome—study of 16 patients in one family. Am. J. Roentgen. 129: 631-638, 1977.

Ferrandez, A.; Remirez, J.; Saenz, P.; Calvo, M.: The trichorhinophalangeal syndrome: report of 4 familial cases belonging to 4 generations. Helv. Paediat. Acta 35: 559-567, 1980.

Fryns, J. P.; Van den Berghe, H.: 8q24.12 interstitial deletion in trichorhinophalangeal syndrome type I. Hum. Genet. 74: 188-189, 1986.

Giedion, A.; Burdea, M.; Fruchter, Z.; Meloni, T.; Trosc, V.: Autosomal dominant transmission of the tricho-rhino-phalangeal syndrome: report of 4 unrelated families, review of 60 cases. Helv. Paediat. Acta 28: 249-259, 1973.

Goldblatt, J.; Smart, R. D.: Tricho-rhino-phalangeal syndrome without exostoses, with an interstitial deletion of 8q23. Clin. Genet. 29: 434-438, 1986.

Goodman, R. M.; Trilling, R.; Hertz, M.; Horoszowski, H.; Merlob, P.; Reisner, S.: New clinical observations in the trichorhinophalangeal syndrome. J. Craniofac. Genet. Dev. Biol. 1: 15-29, 1981.

Haan, E. A.; Hull, Y. J.; White, S.; Cockington, R.; Charlton, P.; Callen, D. F.: Tricho-rhino-phalangeal and branchio-oto syndromes in a family with an inherited rearrangement of chromosome 8q. Am. J. Med. Genet. 32: 490-494, 1989.

Hamers, A.; Jongbloet, P.; Peeters, G.; Fryns, J. P.; Geraedts, J.: Severe mental retardation in a patient with tricho-rhino-phalangeal syndrome type I and 8q deletion. Europ. J. Pediat. 149: 618-620, 1990.

Hou, J.; Parrish, J.; Ludecke, H.-J.; Sapru, M.; Wang, Y.; Chen, W.; Hill, A.; Siegel-Bartelt, J.; Northrup, H.; Elder, F. F. B.; Chinault, C.; Horsthemke, B.; Wagner, M. J.; Wells, D. E.: A 4-megabase YAC contig that spans the Langer-Giedion syndrome region on human chromosome 8q24.1: use in refining the location of the trichorhinophalangeal syndrome and multiple exostoses genes (TRPS1 and EXT1). Genomics 29: 87-97, 1995.

Howell, C. J.; Wynne-Davies, R.: The tricho-rhino-phalangeal syndrome: a report of 14 cases in 7 kindreds. J. Bone Joint Surg. 68B: 311-314, 1986.

Ludecke, H.-J.; Wagner, M. J.; Nardmann, J.; La Pillo, B.; Parrish, J. E.; Willems, P. J.; Haan, E. A.; Frydman, M.; Hamers, G. J. H.; Wells, D. E.; Horsthemke, B.: Molecular dissection of a contiguous gene syndrome: localization of the genes involved in the Langer-Giedion syndrome. Hum. Molec. Genet. 4: 31-36, 1995.

Marchau, F. E.; Van Roy, B. C.; Parizel, P. M.; Lambert, J. R.; De Canck, I.; Leroy, J. G.; Gevaert, C. M.; Willems, P. J.; Dumon, J. E.: Tricho-rhino-phalangeal syndrome type I (TRP I) due to an apparently balanced translocation involving 8q24. Am. J. Med. Genet. 45: 450-455, 1993.

McCloud, D. J.; Solomon, L. M.: The tricho-rhino-phalangeal syndrome. Brit. J. Derm. 96: 403-407, 1977.

McKusick, V. A.: Heritable Disorders of Connective Tissue. St. Louis: C. V. Mosby (pub.) (4th ed.): 1972. Note: Fig. 13-38.

Murdoch, J. L.: Tricho-rhino-phalangeal dysplasia with possible autosomal dominant transmission. Birth Defects Orig. Art. Ser. V(2): 218-220, 1969.

Nardmann, J.; Tranebjaerg, L.; Horsthemke, B.; Ludecke, H.-J.: The tricho-rhino-phalangeal syndromes: frequency and parental origin of 8q deletions. Hum. Genet. 99: 638-643, 1997.

Naritomi, K.; Hirayama, K.: Partial trisomy of distal 8q derived from mother with mosaic 8q23.3-24.13 deletion, and relatively mild expression of trichorhinophalangeal syndrome I. Hum. Genet. 82: 199-201, 1989.

Peltola, J.; Kuokkanen, K.: Tricho-rhino-phalangeal syndrome in five successive generations: report on a family in Finland. Acta Derm. Venerol. 58: 65-68, 1978.

Sanchez, J. M.; Labarta, J. D.; De Negrotti, T. C.; Migliorini, A. M.: Complex translocation in a boy with trichorhinophalangeal syndrome. J. Med. Genet. 22: 314-318, 1985.

Sugiura, Y.: Tricho-rhino-phalangeal syndrome associated with Perthes-disease-like bone change and spondylolisthesis. Jpn. J. Hum. Genet. 23: 23-30, 1978.

Sugiura, Y.; Shionoya, M.; Inoue, T.; Tsuruta, T.: Tricho-rhino-phalangeal syndrome: report on three unrelated families. Jpn. J. Hum. Genet. 21: 13-22, 1976.

Weaver, D. D.; Cohen, M. M.; Smith, D. W.: The tricho-rhino-phalangeal syndrome. J. Med. Genet. 11: 312-314, 1974.

Yamamoto, Y.; Oguro, N.; Miyao, M.; Yanagisawa, M.: Tricho-rhino-phalangeal syndrome type I with severe mental retardation due to interstitial deletion of 8q23.3-24.13. Am. J. Med. Genet. 32: 133-135, 1989.

Yamamoto, Y.; Oguro, N.; Miyao, M.; Yanagisawa, M.; Ohsawa, T.: Prometaphase chromosomes in the tricho-rhino-phalangeal syndrome type I. Am. J. Med. Genet. 32: 524-527, 1989.

TRICHORHINOPHALANGEAL SYNDROME, RECESSIVE FORM (275500)

JJM NOTE

This entity is distinct from the previous two disorders only by the apparently different mode of inheritance.

Giedion (1966) delineated a new syndrome consisting of thin and slowly growing hair, pear-shaped nose with high philtrum, brachyphalangy with deformation of the fingers and wedge-shaped epiphyses. Giedion's patient, a girl, had two supernumerary incisors. He found 2 previous reports, each describing 2 affected sibs. Furthermore, the parents were consanguineous in 1 case. One of the pairs of affected sibs was reported as pseudo-pseudohypoparathyroidism (van der Werff Ten Bosch, 1959). We have observed affected brother and sister whose parents are not related and allegedly are unaffected, but the father was not available for examination (Hussels, 1971). While showing that in most instances inheritance is autosomal dominant (190350), Giedion et al. (1973) concluded that a recessive form probably exists. Gonadal mosaicism is a frequent phenomenon, however, and it is probably noteworthy that there is so little evidence for a recessive form.

See also Giedion (1968)

Giedion, A.: Das Tricho-rhino-phalangeal Syndrom. Helv. Paediat. Acta 21: 475-482, 1966.
Giedion, A.: Zapfenepiphysen. Naturgeschichte und diagnostische Bedeutung einer Stoerung des enchondralen Wachstums. Ergeb. Med. Radiol. 8: 59-124, 1968.
Giedion, A.; Burdea, M.; Fruchter, Z.; Meloni, T.; Trosc, V.: Autosomal dominant transmission of the tricho-rhino-phalangeal syndrome. Report of 4 unrelated families, review of 60 cases. Helv. Paediat. Acta 28: 249-259, 1973.
Hussels, I. E.: Trichorhinophalangeal syndrome in two sibs. Birth Defects Orig. Art. Ser. VII(7): 301-303, 1971.
Van der Werff Ten Bosch, J. J.: The syndrome of brachymetacarpal dwarfism ('pseudo-pseudohypoparathyroidism') with and without gonadal dysgenesis. Lancet I: 69-71, 1959.

EAR EXOSTOSES (128300)

JJM NOTE

I am not aware of malignant degeneration of this site of exostosis, but the original article is quite old.

The trait is age and sex dependent, being rare in young children and more frequent in males. Instances of involvement in multiple generations were reviewed by Hrdlicka (1935).

Hrdlicka, A.: Ear exostoses. Smithsonian Miscellaneous Collections 93: 1-100, 1935.

MULTIPLE EXOSTOSES WITH SPASTIC TETRAPARESIS (158345)

JJM NOTE

This entity is included as another syndrome with exostosis.

In 4 generations of a family in Germany, Hamann et al. (1992) observed the combination of multiple exostoses of typical nature (133700) in association with spastic tetraparesis. There were no exostoses in the spine or cranium to account for the tetraspastic disorder. The pedigree pattern was consistent with autosomal dominant inheritance. In the second generation, all of 5 sibs were affected. Hamann et al. (1992) concluded that this represented a new syndrome, although the possibility of a contiguous gene syndrome was considered.

Hamann, G.; Zankl, M.; Schimrigk, K.; Kloss, R.: Spastic disorder in patients with hereditary multiple exostoses, but without spinal cord compression: a new syndrome?. J. Med. Genet. 29: 494-496, 1992.

EXOSTOSES WITH ANETODERMIA AND BRACHYDACTYLY, TYPE E (133690)

JJM NOTE

The family deserves follow-up to clarify the clinical evolution of the disorder. Malignancy was apparently not a feature.

Mollica et al. (1984) reported a kindred in which 6 persons had anetodermia (macular atrophy of the skin), 8 had multiple exostoses, and 2 had type E brachydactyly seemingly unrelated to exostoses. In all, 10 persons had 1 or more of these features. The authors suggested that this is an autosomal dominant syndrome. See 250450 for another skeletal disorder with anetodermia (or anetoderma).

Mollica, F.; Li Volti, S.; Guarneri, B.: New syndrome: exostoses, anetodermia, brachydactyly. Am. J. Med. Genet. 19: 665-667, 1984.

EXOSTOSES OF HEEL (133600)

JJM NOTE

Again, malignancy is not a known complication, but follow-up would be needed.

Gould (1942) described the condition in grandfather, father and son, i.e., males of 3 generations. X-rays were not described. Exostosis of the heel, possibly of the same type, is a manifestation of the Reiter syndrome, a rheumatic disorder that shows a high order of association with a specific HLA type (142800), namely B27 (Brewerton et al., 1973; McClusky et al., 1974; Woodrow et al., 1974). Thus might familial aggregation be accounted for.

Brewerton, D. A.; Nicholls, A.; Oates, J. K.; Caffrey, M.; Walters, D.; James, D. C. O.: Reiter's disease and HL-A 27. Lancet II: 996-998, 1973.
Gould, E. A.: Three generations of exostoses of the heel. Inherited from father to son. J. Hered. 33: 228 only, 1942.
McClusky, O. E.; Lordon, R. E.; Arnett, F. C.: HL-A 27 in Reiter's syndrome and psoriatic arthritis: a genetic factor in disease susceptibility and expression. J. Rheum. 1: 263-268, 1974.
Woodrow, J. C.; Treanor, B.; Usher, N.: The HL-A system in Reiter's syndrome. Tissue Antigens 4: 533-540, 1974.

GASTRIN-RELEASING PEPTIDE RECEPTOR (*305670)

JJM NOTE

Ishikawa-Brush et al. (1997) identified this gene in a young female with exostosis and multiple system defects, including autism, reinforcing the dictum to do chromosomal analysis on patients with disorders that go beyond the clinical spectrum of manifestations for known single-gene traits.

Expression of gastrin-releasing polypeptide (GRP; 137260) and its receptor in tumors suggests that these molecules are part of an autocrine loop for growth. Spindel et al. (1990) cloned the human GRP receptor cDNA from a library made from a small cell lung carcinoma cell line using the mouse GRP receptor. The cDNA was in turn used to screen a human genomic library. Schantz et al. (1991) designed PCR primers that span the exon encoding amino acids 139 to 256 of the GRP receptor and used these for the analysis of somatic cell hybrids. In this way they found that the GRPR gene is located on the X chromosome. A panel of hybrids with translocations of the X chromosome permitted regionalization of the gene to Xp11-q11. Maslen and Boyd (1993) found that GRPR maps to Xp22.3-p21.2 rather than to the Xp11-q11 interval as previously reported. The assignment of GRPR to distal Xp was supported by the comparative map position in the mouse. The mapping in the human was done by means of PCR amplification from a panel of somatic cell hybrids that retained reduced portions of the X chromosome; the mapping in the mouse was done by linkage studies in interspecific backcross matings.

Shiraishi et al. (1996) noted that within the haploid genome there are approximately 1,000 copies of the human endogenous retrovirus-like sequence, HERV-H. Although these sequences are scattered throughout the genome, in situ hybridization experiments showed discrete clusters positioned on 1p and 7q. Shiraishi et al. (1996) located 3 HERV-H sequences that were unexpectedly clustered within a 300-kb region close to the GRPR locus on the X chromosome.

Ishikawa-Brush et al. (1997) reported molecular studies of a 27-year-old female patient with a balanced translocation 46,X,t(X;8)(p22.13;q22.1) and associated with multiple exostoses and autism (209850; Bolton et al., 1995). Multiple exostoses were present around the ankles, knees, wrists, and left clavicle, and the patient had short stature, short span, and small hands with short fourth and fifth metacarpals. She had mild brachycephaly and was diagnosed as autistic according to the International Classification of Diseases Criteria. She was mentally retarded with an IQ of 35 and also suffered from grand mal epilepsy. The translocated X chromosome was demonstrated to be active by bromodeoxyuridine analysis. Although the father was deceased, the mother and a brother were known to have normal karyotypes and no clinical signs of multiple exostoses or autism. Through molecular analysis using YAC and cosmid clones, Ishikawa-Brush et al. (1997) isolated the translocation breakpoint and confirmed that it was reciprocal between a 5-prime-GGCA-3-prime sequence found on both chromosomes X and 8, without gain or loss of a single nucleotide. The translocation breakpoint on the X chromosome occurred in the first intron of GRPR and that on chromosome 8 occurred approximately 30 kb distal to the 3-prime end of the syndecan-2 gene (SDC2; 142460).

The orientation of these genes with respect to the translocation was incompatible with the formation of a fusion gene. The authors speculated that a dosage effect of the GRPR gene and a position effect of the SDC2 gene may have contributed to the phenotype in this patient. They demonstrated that the GRPR gene escapes X inactivation; nevertheless, the gene from the inactive X chromosome may be expressed at only a fraction of the active allele. Such a low dosage might have been sufficient to cause autism in this patient. It is unknown, furthermore, whether inactivation patterns for a given gene are consistent throughout every tissue and cell within the bo dy. This does not appear to be true for imprinted genes, so it is possible that GRPR could escape inactivation in test tissues, such as fibroblasts and lymphoblasts (and other tissues), but could be subject to inactivation in the brain, resulting in no GRPR expression and the observed phenotype. Therefore, Ishikawa-Brush et al. (1997) suggested that examination for rearrangements and point mutations in the GRPR gene and association studies using polymorphisms within and flanking the gene are clearly warranted in patients with autism. The SDC2 gene product is a member of a family of cell surface heparan sulfate proteoglycans that interact with adhesion molecules, growth factors, and a variety of other effectors that support the shaping, maintenance, and repair of an organism. One of its biologic functions is related to aggregating cells in the formation of bone and therefore is consistent with a possible role in multiple exostoses. However, the gene was not disrupted by the translocation breakpoint in this patient and alteration in its expression was not demonstrated. Since translocations occurring as much as 100 kb from the 3-prime end of the PAX6 gene (106210) can be disruptive to the gene through position effect, the translocation reported by Ishikawa-Brush et al. (1997) might be another example of this phenomenon in relation to SDC2 or other genes in 8q22.1.

Bolton, P.; Powell, J.; Rutter, M.; Buckle, V.; Yates, J. R. W.; Ishikawa-Brush, Y.; Monaco, A. P.: Autism, mental retardation, multiple exostoses and short stature in a female with 46,X,t(X;8)(p22.13;q22.1). Psychiat. Genet. 5: 51-55, 1995.

Ishikawa-Brush, Y.; Powell, J. F.; Bolton, P.; Miller, A. P.; Francis, F.; Willard, H. F.; Lehrach, H.; Monaco, A. P.: Autism and multiple exostoses associated with an X;8 translocation occurring within the GRPR gene and 3-prime to the SDC2 gene. Hum. Molec. Genet. 6: 1241-1250, 1997.

Maslen, G. L.; Boyd, Y.: Comparative mapping of the Grpr locus on the X chromosomes of man and mouse. Genomics 17: 106-109, 1993.

Schantz, L. J.; Naylor, S. L.; Giladi, E.; Spindel, E. R.: Assignment of the GRP receptor gene to the human X chromosome. (Abstract) Cytogenet. Cell Genet. 58: 2085-2086, 1991.

Shiraishi, M.; Alitalo, T.; Sekiya, T.: The chromosomal organization of the human endogenous retrovirus-like sequence HERV-H: clustering of the HERV-H sequences in a 300-kb region close to the GRPR locus on the X chromosome. DNA Res. 3: 425-429, 1996.

Spindel, E. R.; Giladi, E.; Brehm, P.; Goodman, R. H.; Segerson, T. P.: Cloning and functional characterization of a complementary DNA encoding the murine fibroblast bombesin/gastrin-releasing peptide receptor. Molec. Endocr. 4: 1956-1963, 1990.

SYNDECAN 2 (*142460)

JJM NOTE

The same patient with exostosis revealed this second gene after study by Ishikawa-Brush et al. (1997).

By polyacrylamide gel electrophoresis of cell surface-associated heparan sulfate proteoglycans after heparitinase treatment, Lories et al. (1987) demonstrated the presence of core proteins with an apparent molecular weight of 125,000, 90,000, 64,000, 48,000 and 35,000. The 90-kD and the 48-kD core proteins share the epitope of the monoclonal antibody 6G12, which Marynen et al. (1989) used in the isolation of the cDNA from a human lung fibroblast expression library. Southern blot analysis indicated the presence of 1 or a limited number of genes corresponding to the 48/90-kD cDNA. By Southern hybridization to a panel of human-mouse somatic cell hybrid DNA and by in situ hybridization, Marynen et al. (1989) showed that the heparan sulfate proteoglycan core protein maps to 8q22-q24. (Heparan sulfate proteoglycan of basement membrane appears to be encoded by a gene on human chromosome 1; see 142461.)

The transmembrane heparan sulfate proteoglycan for which cDNAs were cloned from human lung fibroblasts by Marynen et al. (1989) is also called fibroglycan or syndecan-2 (David et al., 1992). It is located in the same region as the MYC gene. As pointed out by Spring et al. (1994), 4 members of the syndecan family show a remarkable physical relationship with 4 members of the MYC gene family.

Ishikawa-Brush et al. (1997) found that the 3-prime end of the SDC2 gene was located approximately 30 kb proximal to the translocation breakpoint in a 27-year-old female patient with a balanced translocation 46,X,t(X;8)(p22.13;q22.1) associated with multiple exostoses and autism (Bolton et al., 1995). The breakpoint of the X chromosome was located in the first intron of the gastrin-releasing peptide receptor gene (GRPR; 305670) and a possible causal relationship to autism (209850) was posited. Since one of the biologic functions of the SDC2 gene product is related to aggregating cells in the formation of bone, the authors suggested interference with its function as a mechanism of the multiple exostoses. There is precedence for a position effect; translocation occurring in the 3-prime end of the PAX6 gene (106210) as far away as approximately 100 kb appeared to be disruptive to the gene and was associated with aniridia.

See also Marynen et al. (1989)

Bolton, P.; Powell, J.; Rutter, M.; Buckle, V.; Yates, J. R. W.; Ishikawa-Brush, Y.; Monaco, A. P.: Autism, mental retardation, multiple exostoses and short stature in a female with 46,X,t(X;8)(p22.13;q22.1). Psychiat. Genet. 5: 51-55, 1995.

David, G.; van der Schueren, B.; Marynen, P.; Cassiman, J.-J.; van den Berghe, H.: Molecular cloning of amphiglycan, a novel integral membrane heparan sulfate proteoglycan expressed by epithelial and fibroblastic cells. J. Cell Biol. 118: 961-969, 1992.

Ishikawa-Brush, Y.; Powell, J. F.; Bolton, P.; Miller, A. P.; Francis, F.; Willard, H. F.; Lehrach, H.; Monaco, A. P.: Autism and multiple exostoses associated with an X;8 translocation occurring within the GRPR gene and 3-prime to the SDC2 gene. Hum. Molec. Genet. 6: 1241-1250, 1997.

Lories, V.; De Broeck, H.; David, G.; Cassiman, J. J.; Van den Berghe, H.: Heparan sulfate proteoglycans of human lung fibroblasts: structural heterogeneity of the core proteins of the hydrophobic cell-associated forms. J. Biol. Chem. 262: 854-859, 1987.

Marynen, P.; Zhang, J.; Cassiman, J.-J.; David, G.: The gene for the 48kDa and the 90kDa core proteins of cell surface-associated heparan sulphate proteoglycan of human lung fibroblast maps to chromosome 8q2. (Abstract) Cytogenet. Cell Genet. 51: 1040, 1989.

Marynen, P.; Zhang, J.; Cassiman, J.-J.; Van den Berghe, H.; David, G.: Partial primary structure of the 48- and 90-kilodalton core proteins of cell surface-associated heparan sulfate proteoglycans of lung fibroblasts: prediction of an integral membrane domain and evidence for multiple distinct core proteins at the cell surface of human lung fibroblasts. J. Biol. Chem. 264: 7017-7024, 1989.

Spring, J.; Goldberger, O. A.; Jenkins, N. A.; Gilbert, D. J.; Copeland, N. G.; Bernfield, M.: Mapping of the syndecan genes in the mouse: linkage with members of the Myc gene family. Genomics 21: 597-601, 1994.

PSORIASIS SUSCEPTIBILITY 1 (*177900)

JJM NOTE

A single family had both psoriasis and exostosis in several generations (Rosbotham et al. 1994). Although the genes are probably linked and not identical, further study could reveal that they are pathogenetically associated in some way.

Psoriasis is a chronic inflammatory dermatosis that affects approximately 2% of the population. It is characterized by hyperproliferation of epidermal cells and inflammation resulting from infiltration of activated T helper cells and mononuclear cells and release of pro-inflammatory cytokines. It may also be associated with arthritis. Swanbeck et al. (1997) presented empiric data useful in genetic counseling, based on information concerning first-degree relatives of 3095 psoriatic probands. Altogether, 3,717 families with one or both parents who had psoriasis had been analyzed. The lifetime risk of getting psoriasis if no parent, one parent or both parents have psoriasis was found to be 0.04, 0.28, and 0.65, respectively. If there was already one affected child in the family, the corresponding risks were 0.24, 0.51, and 0.83, respectively. The risk of getting psoriasis before the age of 32 years was dependent on the age of onset of psoriasis in one affected parent.

A very large family tree was assembled in North Carolina by Abele et al. (1963). The authors concluded that penetrance was reduced to about 60%. The prevalence of arthritis was not increased in the psoriatic members of the kindred. Lomholt (1965) did a comprehensive study in the Faroe Islands. He found that 91% of patients had affected relatives. Transmission through many generations of many lines of the large kindred reported by Abele et al. (1963) supports dominant inheritance, the mode of inheritance espoused by Romanus (1945). Steinberg et al. (1951) suggested that homozygosity at two separate loci best explains their family data. Russell et al. (1972) found that HLA-A (142800) 13 was present in 12 of 44 unrelated persons with psoriasis and in 3 of 89 controls (a difference significant at a probability less than 0.0001). W17 was present in 10 of 44 unrelated patients and in 17 family members with psoriasis in 4 generations. Two sibs did not have either psoriasis or W17. The study was undertaken because psoriasis is aggravated by streptococcal infection and a protein of group A beta-hemolytic streptococcus cross-reacts with certain HLA-A antigens. The finding of an

HLA-A and disease association is an indication of polygenic inheritance. Even if there is a single major gene, the HLA-A locus must also be a factor. White et al. (1972) likewise found an excess of W17 and HLA-A 13 with a decrease in HLA-A 12 in psoriatic patients. Burch and Rowell (1965) suggested the existence of several distinct genotypes in psoriasis, i.e., genetic heterogeneity. A twin study showed increased concordance in monozygotic twins (Farber and Nall, 1971). Watson et al. (1972) concluded the genetics is multifactorial. Psoriasis is rare in Eskimos, American Indians and Japanese, all of whom have a very low frequency of HLA-A 13 and HLA-A 17. Beckman et al. (1974) confirmed the high frequency of histocompatibility types W17 and HLA-A 13. Familial psoriasis shows an as sociation with HLA-BW17; psoriasis related to the streptococcus shows association with HLA-B13; and spondylitis occurring in psoriasis shows association with HLA-B2 (Arnett, 1977). HLA-Cw6 appears to be a psoriasis gene (Bodmer, 1978), judging by demonstration of close association. Using a sib-pair analysis, Suarez-Almazor and Russell (1990) found that all sib pairs shared at least one HLA haplotype and that 13 of the 15 were HLA identical, compared with an expected frequency of 4. From these results they concluded 'that at least one, and probably more, HLA-linked genes are implicated in the development of psoriasis.' From studies in a 'skin equivalent model,' Saiag et al. (1985) concluded that the primary defect in psoriasis may reside in the dermal fibroblasts. Psoriatic fibroblasts could induce hyperproliferative activity in normal keratocytes. The high rate of proliferation of psoriatic epidermis could not be suppressed by normal fibroblasts. Tiilikainen et al. (1980) found a 20-fold increased risk of developing psoriasis in HLA-Cw6 carriers. The association between psoriasis and certain HLA alleles supports the hypothesis that psoriasis is a T-cell-mediated, autoimmune disorder.

Happle (1991) invoked somatic recombination to explain linear psoriasis. He suggested that through somatic crossingover in early development one of the daughter cells may become homozygous for a psoriasis gene and that this would be the stem cell of a clone proliferating in a linear pattern during development of the skin. For the ultimate manifestation of linear psoriasis, the presence of other predisposing genes as well as environmental factors would presumably be necessary. This would explain why linear psoriasis is usually absent at birth but develops later in life.

Rosbotham et al. (1994) described a British family in which psoriasis and multiple exostoses (133700) was associated in 4 members of 3 successive generations. A fifth member of the family, aged 27 years, in the third generation had only multiple exostoses. Rosbotham et al. (1994) raised the question of genetic linkage accounting for the cosegregation of the 2 disorders.

Studying 8 multiply affected psoriasis kindreds and using polymorphic microsatellite markers, Tomfohrde et al. (1994) localized a gene involved in psoriasis susceptibility to distal 17q. In the family that showed the strongest evidence for linkage, the recombination fraction between a psoriasis susceptibility locus and D17S784 was 0.04 with a maximum 2-point lod score of 5.33. There was also evidence for genetic heterogeneity and although none of the linked families showed any association with HLA-Cw6, 2 unlinked families showed weak association. Tomfohrde et al. (1994) estimated that half the families were unlinked to 17q.

In a large and extensively affected family from the northeast of England, Matthews et al. (1995) excluded linkage between the telomeric region of 17q and psoriasis. Using 24 multiplex psoriasis kindreds and 12 microsatellite markers on chromosome 17, Nair et al. (1995) could find no evidence of linkage to the previously implicated marker D17S784 under any of 8 models varying in mode of inheritance, penetrance, and sporadic cases. All other chromosome 17 markers tested also failed to detect evidence for linkage in any of the kindreds. Although further analysis using affected sib pair methods provided no statistically significant evidence for linkage to any chromosome 17 marker, a cluster of 3 distal 17q loci displayed a trend towards greater than expected allele-sharing values. While not formally confirming the existence of a psoriasis susceptibility locus on distal long arm of chromosome 17, the results suggested possible involvement under a polygenic model.

Caution is necessary in the assessment of linkage to psoriasis susceptibility loci as a number of factors complicate the analyses (Matthews et al., 1996). These include incomplete penetrance, phenocopies, misdiagnosis, and the lack of a robust genetic model that accurately accounts for the observed familial aggregation. Trembath et al. (1997) used a 2-stage approach to search the human genome for genes conferring susceptibility to psoriasis, using a total of 106 affected sib pairs identified from 68 independent families. As over one-third of the extended kindreds included affected relatives besides sibs, in addition to an analysis of allele sharing between affected sibs, a novel linkage strategy was applied that extracted full nonparametric information. Four principal regions of possible linkage were identified on chromosomes 2, 8, and 20, and markers from the major histocompatibility complex (MHC) region at 6p21 showed highly significant evidence of linkage disequilibrium. Data from limited case-control associations had previously implicated the MHC; this study demonstrated that a gene or genes located within the MHC and close to class I HLA loci represent the major determinant of the genetic basis of psoriasis.

In a 12.5-cM genomewide scan for psoriasis susceptibility loci by recombination-based tests, Nair et al. (1997) found linkage to the HLA region (maximum lod = 3.52), as well as suggestive linkage to 2 novel regions: 16q (maximum lod = 2.50) and 20p (maximum lod = 2.62). All 3 regions yielded P values equal to or less than 0.01 by nonparametric analysis. Recombination-based and allele sharing methods also confirmed a previous report of a dominant susceptibility locus on distal 17q. Nair et al. (1997) could not confirm a previously reported locus on distal 4q. Taken together with the demonstrated linkage to HLA-B (142830) and

See also Burch and Rowell (1981); Farber et al. (1974); Kimberling and Dobson (1973); Lomholt (1963); Moll and Wright (1973); Pietrzyk et al. (1982); Propping et al. (1985); Steinberg et al. (1952); Ward and Stephens (1961)

Abele, D. C.; Dobson, R. L.; Graham, J. B.: Heredity and psoriasis: study of a large family. Arch. Derm. 88: 38-47, 1963.

Arnett, F. C., Jr.: Personal Communication. Baltimore, Md. 1977.

Beckman, L.; Bronnestam, R.; Cedergren, B.; Liden, S.: HL-A antigens, blood groups, serum groups and red cell enzyme types in psoriasis. Hum. Hered. 24: 496-506, 1974.

Bodmer, W.: Personal Communication. Oxford, England 1978.

Burch, P. R. J.; Rowell, N. R.: Mode of inheritance in psoriasis. (Letter) Arch. Derm. 117: 251-252, 1981.

Burch, P. R. J.; Rowell, N. R.: Psoriasis: aetiological aspects. Acta Derm. Venerol. 45: 366-380, 1965.

Burden, A. D.; Javed, S.; Bailey, M.; Hodgins, M.; Connor, M.; Tillman, D.: Genetics of psoriasis: paternal inheritance and a locus on chromosome 6p. J. Invest. Derm. 110: 958-960, 1998.

Farber, E. M.; Nall, M. L.: Genetics of psoriasis: twin study.In: Farber, E. M.; Cox, A. J.: Psoriasis (International Symposium). Stanford: Stanford Univ. Press (pub.) 1971. Pp. 7-13.

Farber, E. M.; Nall, M. L.; Watson, W.: Natural history of psoriasis in 61 twin pairs. Arch. Derm. 109: 207-211, 1974.

Happle, R.: Somatic recombination may explain linear psoriasis. J. Med. Genet. 28: 337 only, 1991.

Jenisch, S.; Henseler, T.; Nair, R. P.; Guo, S.-W.; Westphal, E.; Stuart, P.; Kronke, M.; Voorhees, J. J.; Christophers, E.; Elder, J. T.: Linkage analysis of human leukocyte antigen (HLA) markers in familial psoriasis: strong disequilibrium effects provide evidence for a major determinant in the HLA-B/-C region. Am. J. Hum. Genet. 63: 191-199, 1998.

Kimberling, W. J.; Dobson, R. L.: The inheritance of psoriasis. J. Invest. Derm. 60: 538-540, 1973.

Lomholt, G.: Psoriasis: Prevalence, Spontaneous Course, and Genetics. A Census Study on the Prevalence of Skin Disease on the Faroe Islands. Copenhagen: G. E. C. Gad (pub.) 1963.

Lomholt, G.: Psoriasis-Praevalenz, spontaner Verlauf und Vererbung. Eine Zensusuntersuchung von den Farinseln. Z. Haut Geschlechtskr. 38: 223-238, 1965.

Matthews, D.; Fry, L.; Powles, A.; Weber, J.; McCarthy, M.; Fisher, E.; Davies, K.; Williamson, R.: Evidence that a locus for familial psoriasis maps to chromosome 4q. Nature Genet. 14: 231-233, 1996.

Matthews, D.; Fry, L.; Powles, A.; Weissenbach, J.; Williamson, R.: Confirmation of genetic heterogeneity in familial psoriasis. J. Med. Genet. 32: 546-548, 1995.

Moll, J. M. H.; Wright, V.: Familial occurrence of psoriatic arthritis. Ann. Rheum. Dis. 32: 181-201, 1973.

Nair, R. P.; Guo, S.-W.; Jenisch, S.; Henseler, T.; Lange, E. M.; Terhune, M.; Westphal, E.; Christophers, E.; Voorhees, J. J.; Elder, J. T.: Scanning chromosome 17 for psoriasis susceptibility: lack of evidence for a distal 17q locus. Hum. Hered. 45: 219-230, 1995.

Nair, R. P.; Henseler, T.; Jenisch, S.; Stuart, P.; Bichakjian, C. K.; Lenk, W.; Westphal, E.; Guo, S.-W.; Christophers, E.; Voorhees, J. J.; Elder, J. T.: Evidence for two psoriasis susceptibility loci (HLA and 17q) and two novel candidate regions (16q and 20p) by genome-wide scan. Hum. Molec. Genet. 6: 1349-1356, 1997.

Pietrzyk, J. J.; Turowski, G.; Kapinska-Mrowka, M.; Rozanski, B.: Family studies in psoriasis. I. Complex segregation analysis. Arch. Derm. Res. 273: 287-294, 1982.

Propping, P.; Hohenschutz, C.; Voigtlander, V.: Increased birth weight in psoriasis—another expression of a 'thrifty genotype?' (Letter) Hum. Genet. 71: 92 only, 1985.

Romanus, T.: Psoriasis from a Prognostic and Hereditary Point of View. Dissertation: Uppsala (pub.) 1945.

Rosbotham, J. L.; Trembath, R. C.; Glover, M.; Leigh, I.; Barker, J. N. W. N.: An association between psoriasis and hereditary multiple exostoses: a clue for the mapping of a psoriasis susceptibility gene? Brit. J. Derm. 130: 671-674, 1994.

Russell, T. J.; Schultes, L. M.; Kuban, D. J.: Histocompatibility (HL-A) antigens associated with psoriasis. New Eng. J. Med. 287: 738-740, 1972.

Saiag, P.; Coulomb, B.; Lebreton, C.; Bell, E.; Dubertret, L.: Psoriatic fibroblasts induce hyperproliferation of normal keratinocytes in a skin equivalent model in vitro. Science 230: 669-672, 1985.

Steinberg, A. G.; Becker, S. W.; Fitzpatrick, T. B.; Kierland, R. R.: A genetic and statistical study of psoriasis. Am. J. Hum. Genet. 3: 267-281, 1951.

Steinberg, A. G.; Becker, S. W.; Fitzpatrick, T. B.; Kierland, R. R.: A further note on the genetics of psoriasis. Am. J. Hum. Genet. 4: 373-375, 1952.

Suarez-Almazor, M. E.; Russell, A. S.: The genetics of psoriasis: haplotype sharing in siblings with the disease. Arch. Derm. 126: 1040-1042, 1990.
Swanbeck, G.; Inerot, A.; Martinsson, T.; Enerback, C.; Enlund, F.; Samuelsson, L.; Yhr, M.; Wahlstrom, J.: Genetic counselling in psoriasis: empirical data on psoriasis among first-degree relatives of 3095 psoriatic probands. Brit. J. Derm. 137: 939-942, 1997.
Tiilikainen, A.; Lassus, A.; Karvonen, J.; Vartiainen, P.; Julin, M.: Psoriasis and HLA-Cw6. Brit. J. Derm. 102: 179-184, 1980.
Tomfohrde, J.; Silverman, A.; Barnes, R.; Fernandez-Vina, M. A.; Young, M.; Lory, D.; Morris, L.; Wuepper, K. D.; Stastny, P.; Menter, A.; Bowcock, A.: Gene for familial psoriasis susceptibility mapped to the distal end of human chromosome 17q. Science 264: 1141-1145, 1994.
Trembath, R. C.; Clough, R. L.; Rosbotham, J. L.; Jones, A. B.; Camp, R. D. R.; Frodsham, A.; Browne, J.; Barber, R.; Terwilliger, J.; Lathrop, G. M.; Barker, J. N. W. N.: Identification of a major susceptibility locus on chromosome 6p and evidence for further disease loci revealed by a two stage genome-wide search in psoriasis. Hum. Molec. Genet. 6: 813-820, 1997.
Ward, J. H.; Stephens, F. E.: Inheritance of psoriasis in a Utah kindred. Arch. Derm. 84: 589-592, 1961.
Watson, W.; Cann, H. W.; Farber, E. M.; Nall, M. L.: The genetics of psoriasis. Arch. Derm. 105: 197-207, 1972.
White, S. H.; Newcomer, V. D.; Mickey, M. R.; Terasaki, P. I.: Disturbance of HL-A antigen frequency in psoriasis. New Eng. J. Med. 287: 740-743, 1972.

CUTIS LAXA, X-LINKED (#304150)

JJM NOTE

Affected brothers had occipital exostoses (Kuivaniemi et al. 1982).

A number sign (#) is used with this entry because of evidence that the disorder is caused by mutation in the Cu(2+)-transporting ATPase, alpha polypeptide (300011), as is also Menkes syndrome (309400).

Lazoff et al. (1975) described an unusual syndrome in an 11-year-old male and 2 maternal uncles. Bony 'horns,' symmetrically situated on each side of the foramen magnum and pointing caudad, were demonstrable radiographically. A lifelong history of frequent loose stools, obstructive uropathy requiring in 1 uncle ileal loop diversion, and mild mental retardation were other features. Some suspicion that a relative through the maternal grandfather had the same condition (which could not be confirmed because of lack of cooperation) meant that autosomal dominant inheritance with reduced penetrance could not be excluded. This syndrome is the same as X-linked cutis laxa. Byers et al. (1976) found deficiency of lysyl oxidase in affected males in a family with apparent X-linked cutis laxa. Three affected males were observed, each in a different sibship connected through females who as children showed joint laxity but outgrew it. Hooked nose and long philtrum typical of cutis laxa were described. In 1 case, pectus excavatum and carinatum were sufficiently severe to require surgical repair shortly after birth. Two cousins were brought to medical attention because of recurrent urinary tract infection due to multiple large diverticula of the bladder. MacFarlane et al. (1980) described 2 kindreds with an X-linked disorder that in general appeared to fall into the Ehlers-Danlos category but had some unusual features such as bladder diverticula, bladder neck obstruction, marked varicosities, and, by x-ray, occipital horns, short broad clavicles, and fused carpal bones. Hall (1980) found that the children studied with Byers et al. (1976) also had occipital horns, and diarrhea, a feature found in MacFarlane's families, was also present. Thus, these are probably the same disorder. The age at which the affected persons were studied may have been a factor in determining whether the disorder was labeled cutis laxa or EDS. Low levels of ceruloplasmin and serum copper were found in these cases. Po ssibly it is, like Menkes syndrome, a disorder of copper metabolism rather than a primary defect of lysyl oxidase. MacFarlane et al. (1980) suggested the designation Ehlers-Danlos syndrome, type IX. Hollister (1981) pointed out that the patients show hypermobility of the finger joints but limitation of extension of the elbows.

Kuivaniemi et al. (1982) studied 2 brothers with bladder diverticula, inguinal hernias, slight skin laxity and hyperextensibility, and skeletal abnormalities, including occipital exostoses. Lysyl oxidase activity was low in the medium of cultured skin fibroblasts, and conversion of newly synthesized collagen into the insoluble form was reduced. Copper concentrations were markedly elevated in cultured skin fibroblasts but decreased in serum and hair. Serum ceruloplasmin levels were low. Conceivably this disorder is an allelic variant of Menkes disease (Kaitila et al., 1982). Complementation studies might answer the question. Peltonen et al. (1983) found many similar abnormalities of copper and collagen metabolism in the cultured fibroblasts of 13 patients with Menkes syndrome and 2 patients with EDS IX. In both disorders, fibroblasts had markedly increased copper content and rate of incorporation of (64)Cu, and accumulation was in metallothionein or a metallothionein-like protein as previously established for Menkes cells. Histochemical staining showed that copper was distributed uniformly throughout the cytoplasm in both cell types, this location being consistent with accumulation in metallothionein. Both fibroblast types showed very low lysyl oxidase activity and increased extractability of newly synthesized collagen, but no abnormality in cell viability, duplication rate, prolyl 4-hydroxylase activity, or collagen synthesis rate. Skin biopsy specimens from one EDS IX patient showed the same abnormalities in lysyl oxidase activity and collagen extractability. Fibroblasts of the mother of EDS IX patients showed increased (64)Cu incorporation. The possibility that the occipital horn syndrome is allelic to Menkes syndrome might be tested by study of location of the locus on the X chromosome; does it map to the same region as does Menkes syndrome (Xp11-q11)?

Allelism was demonstrated in a definitive way by Das et al. (1995) who identified hemizygosity for a mutation in the copper-transporting ATPase that is mutant in Menkes syndrome. One of the mild mottled mutants in the mouse, 'blotchy,' symbolized Mo-blo, exhibits connective tissue abnormalities reminiscent of those seen in OHS patients, including weak skin and bone abnormalities. In blotchy males, hindlegs are occasionally deformed, vibrissae are kinked at birth, crosslinking of skin collagen and aortic elastin is defective, and death frequently results from aortic rupture. Das et al. (1995) identified similar splicing mutations in both the blotchy mouse and cases of the occipital horn syndrome. Das et al. (1995) reported 2 OHS patients. In each case, there was deletion of 1 exon. In one case, the deletion was caused by an A-to-G transition at the

See also Byers et al. (1980)

Beighton, P.: Personal Communication. Cape Town, South Africa 9/26/1986.
Blackston, R. D.; Hirschhorn, K.; Elsas, L. J.: Ehlers-Danlos syndrome (EDS), type IX: biochemical evidence for X-linkage. (Abstract) Am. J. Hum. Genet. 41: A49, 1987.
Byers, P. H.; Narayanan, A. S.; Bornstein, P.; Hall, J. G.: An X-linked form of cutis laxa due to deficiency of lysyloxidase. Birth Defects Orig. Art. Ser. 12(5): 293-298, 1976.
Byers, P. H.; Siegel, R. C.; Holbrook, K. A.; Narayanan, A. S.; Bornstein, P.; Hall, J. G.: X-linked cutis laxa: defective collagen crosslink formation due to decreased lysyl oxidase activity. New Eng. J. Med. 303: 61-65, 1980.
Das, S.; Levinson, B.; Vulpe, C.; Whitney, S.; Gitschier, J.; Packman, S.: Similar splicing mutations of the Menkes/mottled copper-transporting ATPase gene in occipital horn syndrome and the blotchy mouse. Am. J. Hum. Genet. 56: 570-576, 1995.
Hall, J. G.: Personal Communication. Seattle, Wash. 1980.
Herman, T. E.; McAlister, W. H.; Boniface, A.; Whyte, M. P.: Occipital horn syndrome: additional radiographic findings in two new cases. Pediat. Radiol. 22: 363-365, 1992.
Hollister, D. W.: Clinical features of Ehlers-Danlos syndrome types VIII and IX.In: Akeson, W.; Glimcher, M. J.; Bornstein, P.: Proc. Workshop on Inherited Connective Tissue Disorders. New York: Elsevier/North Holland Press (pub.) 1981.
Kaitila, I. I.; Peltonen, L.; Kuivaniemi, H.; Palotie, A.; Elo, J.; Kivirikko, K. I.: A skeletal and connective tissue disorder associated with lysyl oxidase deficiency and abnormal copper metabolism.In: Papadatos, C. J.; Bartsocas, C. S.: Skeletal Dysplasias. New York: Alan R. Liss (pub.) 1982. Pp. 307-316.
Kaler, S. G.; Gallo, L. K.; Proud, V. K.; Percy, A. K.; Mark, Y.; Segal, N. A.; Goldstein, D. S.; Holmes C. S.; Gahl, W. A.: Occipital horn syndrome and a mild Menkes phenotype associated with splice site mutations at the MNK locus. Nature Genet. 8: 195-202, 1994.
Kuivaniemi, H.; Peltonen, L.; Kivirikko, K. I.: Type IX Ehlers-Danlos syndrome and Menkes syndrome: the decrease in lysyl oxidase activity is associated with a corresponding deficiency in the enzyme protein. Am. J. Hum. Genet. 37: 798-808, 1985.
Kuivaniemi, H.; Peltonen, L.; Palotie, A.; Kaitila, I.; Kivirikko, K. I.: Abnormal copper metabolism and deficient lysyl oxidase activity in a heritable connective tissue disorder. J. Clin. Invest. 69: 730-733, 1982.
Lazoff, S. G.; Rybak, J. J.; Parker, B. R.; Luzzatti, L.: Skeletal dysplasia, occipital horns, intestinal malabsorption, and obstructive uropathy—a new hereditary syndrome. Birth Defects Orig. Art. Ser. XI(5): 71-74, 1975.
Levinson, B.; Conant, R.; Schnur, R.; Das, S.; Packman, S.; Gitschier, J.: A repeated element in the regulatory region of the MNK gene and its deletion in a patient with occipital horn syndrome. Hum. Molec. Genet. 5: 1737-1742, 1996.
Levinson, B.; Gitschier, J.; Vulpe, C.; Whitney, S.; Yang, S.; Packman, S.: Are X-linked cutis laxa and Menkes disease allelic?. (Letter) Nature Genet. 3: 6, 1993.
MacFarlane, J. D.; Hollister, D. W.; Weaver, D. D.; Brandt, K. D.; Luzzatti, L.; Biegel, A. A.: A new Ehlers-Danlos syndrome with skeletal dysplasia. (Abstract) Am. J. Hum. Genet. 32: 118A, 1980.
Peltonen, L.; Kuivaniemi, H.; Palotie, A.; Horn, N.; Kaitila, I.; Kivirikko, K. I.: Alterations in copper and collagen metabolism in the Menkes syndrome and a new subtype of the Ehlers-Danlos syndrome. Biochemistry 22: 6156-6163, 1983.
Tsukahara, M.; Imaizumi, K.; Kawai, S.; Kajii, T.: Occipital horn syndrome: report of a patient and review of the literature. Clin. Genet. 45: 32-35, 1994.
Vulpe, C.; Levinson, B.; Whitney, S.; Packman, S.; Gitschier, J.: Isolation of a candidate gene for Menkes disease and evidence that it encodes a copper-transporting ATPase. Nature Genet. 3: 7-13, 1993.
Wakai, S.; Ishikawa, Y.; Nagaoka, M.; Okabe, M.; Minami, R.; Hayakawa, T.: Central nervous system involvement and generalized muscular atrophy in occipital horn syndrome: Ehlers-Danlos type IX—a first Japanese case. J. Neurol. Sci. 116: 1-5, 1993.

MENKES SYNDROME (#309400)

JJM NOTE

One large kindred had an atypical form of Menkes, and a mutation within the gene also had occipital exostoses (Proud et al. 1996).

Description

A number sign (#) is used with this entry because of evidence that it, as well as the occipital horn syndrome (304150), is caused by mutation in the gene encoding Cu(2+)-transporting ATPase, alpha polypeptide (300011).

Clinical Features

In a family of English-Irish descent living in New York, Menkes et al. (1962) described an X-linked recessive disorder characterized by early retardation in growth, peculiar hair, and focal cerebral and cerebellar degeneration. Severe neurologic impairment began within a month or two of birth and progressed rapidly to decerebration. Five males were affected but the gene could by inference be identified in 4 generations. The failure to grow brought the affected infants to medical attention at the age of a few weeks and death occurred in the first or second year of life. The hair was stubby and white. Microscopically it showed twisting, varying diameter along the length of the shaft, and often fractures of the shaft at regular intervals. Rather extensive biochemical investigations showed elevated plasma glutamic acid as the only consistent abnormality. The anatomic change in the central nervous system was described on the basis of 2 autopsies.

Bray (1965) observed 2 brothers who died as infants with spastic dementia, seizures, and defective hair. Blood and urine amino acids were normal. Whether this is the same disorder as that in Menkes' family was unclear. The condition described by Yoshida et al. (1964) have been the same. French and Sherard (1967) presented evidence that they interpreted as indicating that this disorder may represent an abnormality of lipid metabolism. Their 16-month-old patient showed: (1) scant, whitish, lackluster, kinky hair that microscopically showed pili torti, monilethrix and trichorrhexis nodosa; (2) retarded growth; (3) micrognathia and highly arched palate; (4) decline in mental development; (5) onset of focal and generalized seizures; and (6) spastic quadriparesis with clenched fists, opisthotonos, and scissoring. Biochemical studies showed depressed serum tocopherol and normal amino acid content of hair serum and urine. An abnormal autofluorescence is displayed by hair and by Purkinje cells' axons.

'Kinky hair disease' proved a designation useful in detection of new cases, since the hair change is an easily remembered feature by which physicians can be alerted to the condition (O'Brien, 1968). Changes in the metaphyses of the long bones and tortuosity of cerebral arteries have been described. Hypothermia and acute illness with septicemia were modes of presentation. Patchy abnormality of systemic arteries with stenosis or obliteration was observed by Danks et al. (1971). They also observed toluidine-blue-metachromasia of fibroblasts. Wesenberg et al. (1969) pointed out that the fetal hair does not show pili torti.

Goka et al. (1976) found that cultured fibroblasts have a concentration of copper over 5 times that of normal fibroblasts. Williams et al. (1978) described the cellular pathology of Menkes disease.

Osaka et al. (1977) reported 2 Japanese families. They pointed out that the hair may not be abnormal, that serum copper determination is a simple and reliable diagnostic test, and that 'congenital hypocupraemia' may be a preferred designation. An abnormality in egress of copper from Menkes disease fibroblasts was suggested by studies of Chan et al. (1978). Defective metallothionein (see 156530) was suggested.

Haas et al. (1981) reported an X-linked disorder of copper metabolism which, by my interpretation, may be an allelic variant of Menkes syndrome. The disorder affected 4 males in 3 sibships connected through females. Similarities to Menkes disease were X-linked recessive inheritance, marked psychomotor retardation with seizures, low serum copper and ceruloplasmin levels, and a block in gut copper absorption. Differences from Menkes disease included normal birthweight, no hypothermia, grossly and microscopically normal hair, and radiographically normal bones. Survivorship was much longer than in Menkes disease. The neurologic disorder was static and characterized by hypotonia and choreoathetosis.

Tonnesen et al. (1991) reviewed the cases of atypical Menkes reported by Haas et al. (1981). No pili torti were found in one case and very few (2 in 1,000) in a second. In addition, low levels of copper and ceruloplasmin in the serum were puzzling findings. However, (64)Cu uptake and retention was significantly increased in the range seen for classic Menkes patients, and copper uptake in female relatives gave the same uptake pattern as in heterozygotes in other families with the classic disorder.

Possibly X-linked recessive neurologic disorders accompanied by defects in intestinal absorption of copper were described by Willvonseder et al. (1973) and by Godwin-Austen et al. (1978). Willvonseder et al. (1973) described 3 brothers who had dementia, spastic dysarthria, paresis of vertical eye movements, disturbance of gait, and splenomegaly. The onset was prepubertal and progression slow. Godwin-Austen et al. (1978) described a disorder clinically reminiscent of Wilson disease but without Kayser-Fleischer rings. Symptoms began at age 12 years and defective copper absorption from the distal intestine, with high copper levels in rectal mucosa, was demonstrated.

Proud et al. (1996) reported on the clinical features of 4 affected persons in 3 generations of a family that manifested an unusual variant of the Menkes syndrome. These patients had normal head circumference, moderate-to-severe mental retardation, onset at age of 3 to 4 years, dysarthria, laxity of skin, bladder diverticula, tortuous vessels, chronic diarrhea, and occipital exostoses (evident in 3 persons aged 18 to 38 years). Study of the MNK gene showed an A-to-T change at the +3 position of the splice donor site near the 3-prime end of the MNK coding sequence resulting in abnormal splicing (Kaler et al., 1994). The authors proposed that maintenance of 20% of normal splicing could explain unique phenotypic manifestations in affected persons. Proud et al. (1996) suggested that these patients, as well as a patient with occipital horn syndrome reported by Wakai et al. (1993), represent a new variant of Menkes syndrome.

Tumer and Horn (1997) reviewed the clinical and genetic aspects of Menkes syndrome, including phenotypic expression in females, mutation spectrum, diagnosis, and treatment. They also discussed the mottled mouse as a model for Menkes syndrome and new insights into normal and defective copper metabolism provided by biochemical and genetic studies of Menkes syndrome and Wilson disease (277900).

Biochemical Features

Danks et al. (1972) presented evidence of a defect in the intestinal absorption of copper. Copper deficiency in animals leads to connective tissue changes because formation of lysine-derived cross-links in elastin and collagen is interfered with, the amine oxidase responsible for the initial modification of lysine being copper-dependent. This may explain the arterial abnormalities. The striking hair changes are probably the result of defective formation of disulfide bonds in keratin since this process is copper-dependent, and copper deficiency in sheep leads to the formation of wool with defective cross-linking (Collie et al., 1980). Menkes had sent hair from his original patients to the Australian Wool Commission, but at that early date the Commission could not identify the problem (Menkes, 1972).

Peltonen et al. (1983) found many similar abnormalities of copper and collagen metabolism in the cultured fibroblasts of 13 patients with Menkes syndrome and 2 patients with E-D IX. In both disorders, fibroblasts had markedly increased copper content and rate of incorporation of (64)Cu, and accumulation was in metallothionein (see 156350) or a metallothionein-like protein as previously established for Menkes cells. Histochemical staining showed that copper was distributed uniformly throughout the cytoplasm in both cell types, this location being consistent with accumulation in metallothionein. Both fibroblast types showed very low lysyl oxidase activity and increased extractability of newly synthesized collagen, but no abnormality in cell viability, duplication rate, prolyl 4-hydroxylase activity, or collagen synthesis rate. Skin biopsy specimens from one E-D IX patient showed the same abnormalities in lysyl oxidase activity and collagen extractability. Fibroblasts of the mother of E-D IX patients showed increased (64)Cu incorporation. The similarities in biochemical findings between type IX Ehlers-Danlos syndrome and Menkes syndrome may indicate allelism. In studies of cultured cells from both conditions, Kuivaniemi et al. (1985) could not demonstrate that there was secreted into the medium or contained in the cell any significant amounts of copper-deficient, catalytically inactive lysyl oxidase protein. Although the rapid degradation of a mutant protein could not be excluded, the authors favored the idea that synthesis of the lysyl oxidase protein is impaired.

Scheinberg and Collins (1989) suggested that the primary defect resides in zinc, i.e., that Menkes disease is primarily a disorder of a putative zinc-binding protein, which they symbolized ZBP, whose synthesis is controlled by a gene on the X chromosome. When ionic zinc is present in the liver or intestine it induces the synthesis of metallothionein to which the zinc is bound. Since the affinity of metallothionein for copper is 100,000 times greater than that for zinc, copper in either organ displaces zinc and binds to metallothionein. In the liver such bound copper is probably unavailable for incorporation into specific copper 'apo' proteins. In the intestine copper does not enter the circulation; advantage has been taken of this fact to decrease the intestinal absorption of copper by administering zinc in Wilson disease (277900). Deficiency of ZBP in Menkes disease would presumably result in an increased concentration of nonprotein-bound, ionic zinc—the only form of the element that has been shown to induce synthesis of metallothionein.

Other Features

Menkes (1988) gave a useful review in which he listed 6 cuproenzymes, 5 of which may account for features of the disorder: tyrosinase for depigmentation of hair and skin pallor; lysyl oxidase for frayed and split arterial intima (defect in elastin and collagen cross-linking); monoamine oxidase for kinky hair; cytochrome c oxidase for hypothermia; and ascorbate oxidase for skeletal demineralization. Dopamine-beta-hydroxylase is also a cuproenzyme; what role its deficiency may have in the phenotype of kinky hair disease is unclear.

Cytogenetics

Gerdes et al. (1990) described 3 patients with clinically and biochemically typical Menkes syndrome; a chromosome abnormality was found in only 1 (45X/46XX mosaicism). During a systematic chromosomal survey of 167 unrelated boys with Menkes disease, Tumer et al. (1992) found a unique rearrangement of the X chromosome involving an insertion of the long arm segment Xq13.3-q21.2 into the short arm at band Xp11.4, giving the karyotype 46,XY,ins(X)(p11.4q13.3q21.2). The same rearranged X chromosome was present de novo in the boy's phenotypically normal mother, where it was preferentially inactivated. RFLP and methylation patterns at DXS255 indicated that the rearrangement originated from the maternal grandfather. This finding supported localization of the MNK locus to Xq13 and suggested fine mapping to subband Xq13.3. The chromosomal band associated with the X-inactivation center (XIC; 314670) was present, in this patient, on the proximal long arm of the rearranged X chromosome, in line with the location of XIC proximal to MNK.

Mapping

Wieacker et al. (1983) performed linkage studies in a large kindred with Menkes syndrome using a cloned DNA sequence (RFLP), probe 1.28, that maps to the proximal portion of Xp (between Xcen and Xp113). At least 2 crossovers and an estimated genetic distance of 16 cM were found (lod score = 0.82). Horn et al. (1984) demonstrated linkage between Menkes disease and a centromeric C-banding polymorphism. Other studies of linkage with 2 RFLPs, MGU22 (which is close to the centromere) and L1.28 (which is in the Xp110-Xp113 segment), suggested that the Menkes locus is distal to L1.28 (reviewed by Ropers et al., 1983).

Wienker et al. (1983) suggested the following as the most likely gene order: Xpter—MS—L1.28—MGU22. Comparative mapping suggested to Horn et al. (1984) that the Menkes disease locus is on the long arm close to band q13; on the mouse X-chromosome the homologous Mo locus ('mottled') is located between the structural loci for phosphoglycerate kinase (Pgk-1) and alpha-galactosidase (Ags), closely linked to the Pgk-1 locus, the human equivalent of which, PGK, has been assigned to Xq13. Linkage studies in 5 Dutch families suggested close situation of the Menkes locus and the centromere (recombination fraction 0.5, lod score more than 3.0). Centromeric heteromorphism was used as the 'marker trait.' There was probably no detectable linkage with Xg.

From a 3-point analysis, Tonnesen et al. (1986) concluded that the Menkes locus is on the long arm of the X chromosome proximal to DXYS1. In studies of 4 families in which a characteristic X-centromeric marker was segregating with Menkes disease, Friedrich et al. (1983) found only 1 recombinant out of 18 opportunities, indicating that the gene is near the centromere on either the long arm or the short arm. Kapur et al. (1987) suggested that the Menkes syndrome gene may lie in the Xq13 band because of the finding of Menkes syndrome in a female with a de novo balanced translocation t(2;X). The breakpoint in the X chromosome was at Xq13.1.

Verga et al. (1991) refined the localization of the MNK locus. They established a lymphoblastoid cell line from the patient of Kapur et al. (1987) and used it to isolate the der(2) translocation chromosome in human/hamster somatic cell hybrids. Southern blot analyses using a number of probes specific for chromosomes X and 2 showed that the breakpoint in this patient—and, therefore, probably the Menkes gene—mapped to a small subregion of band Xq13.2-q13.3 proximal to the PGK1 (311800) locus and distal to all other Xq13 loci tested. Hershon (1988) concluded from study of an X/A translocation that the locus lies near the boundary between Xq12 and Xq13, probably at Xq13.1.

Tonnesen et al. (1992) reported on linkage analyses in 11 families in which more than one affected patient had been found. They concluded that the most likely location of MNK is Xq12-q13.3. Working with DNA from the cells from the patient with the translocation t(2;X) reported by Kapur et al. (1987), Consalez et al. (1992) developed a cosmid contig extending 150 kb from a nearby CpG island across the breakpoint on the X chromosome (see erratum indicating additional information on the location of the translocation breakpoint).

Molecular Genetics

Three independent groups, in San Francisco (Vulpe et al., 1993), Oxford (Chelly et al., 1993), and Michigan (Mercer et al., 1993), cloned a candidate gene for Menkes disease. Vulpe et al. (1993), who proceeded directly from the translocation breakpoints at Xq13.3 to the gene by cDNA library screening with a YAC fragment and exon trapping experiments, succeeded in obtaining a complete set of clones corresponding to an 8.5-kb transcript that encodes a 1,500 amino acid protein. The 5-prime region of the same locus was also obtained by Chelly et al. (1993) who took the more cumbersome but genetically rigorous step of first identifying genomic fragments that were deleted in cytogenetically normal patients with Menkes disease, and by Mercer et al. (1993) who employed a strategy involving long range restriction mapping and fluorescence in situ hybridization (FISH) analysis. Two lines of evidence strongly implicated the cloned MNK locus in the etiogenesis of Menkes disease: nonoverlapping portions of the gene were deleted in 16 of 100 unrelated patients (Chelly et al., 1993), and the expression of the transcript was reduced or altered in 23 of 32 patients (Vulpe et al., 1993; Mercer et al., 1993). By a database search of the predicted sequence, Vulpe et al. (1993) found strong homology to P-type ATPases, a family of integral membrane proteins that use an aspartyl phosphate intermediate to transport cations across membranes. The protein has the characteristics of a copper binding protein. Northern blot experiments showed that MNK mRNA is present in a variety of cell types and tissues except liver, in which expression is reduced or absent. This is consistent with the clinical observation that the liver is largely unaffected in Menkes disease and fails to accumulate excess copper.

The MNK protein is localized to the trans-Golgi network (TGN) (Petris et al., 1996). Studies with copper-resistant Chinese hamster ovary cells (CHO) by Petris et al. (1996) suggested that the MNK protein cycles between the TGN and the plasma membrane, depending on the concentration of copper within the cell. TGN38 (603062) is another protein that cycles between the TGN and the plasma membrane (Ladinsky and Howell, 1992; Reaves et al., 1993). A number of Golgi-resident proteins contain specific localization signals, and Francis et al. (1998) showed that this is true also of MNK. By immunofluorescence, they showed that the full-length recombinant Menkes protein, the isoform that is not expressed in the occipital horn syndrome, localizes to the Golgi apparatus, whereas the alternatively spliced form, which lacks sequences for transmembrane domains 3 and 4 encoded by exon 10 and is expressed in the occipital horn syndrome, localizes to the endoplasmic reticulum. Using sequences from exon 10 fused to a non-Golgi reporter molecule, Francis et al. (1998) showed that a 38-amino acid sequence containing transmembrane domain 3 of the MNK protein was sufficient for localization to the Golgi complex. Therefore, the protein sequence encoded by exon 10 may be responsible for this differential localization and both isoforms may be required for comprehensive transport of copper within the cell. By immunogold electron microscopic analyses, La Fontaine et al. (1998) mapped the MNK protein to the TGN. When the extracellular copper concentration was increased, MNK in the CHO cells was redistributed to the cytoplasm and plasma membrane, but returned to the TGN under basal, low copper conditions.

Diagnosis

Carrier status for the Menkes disease gene can usually be determined by examination of multiple hairs from scattered scalp sites for pili torti. Carrier status can, of course, never be completely excluded by negative findings of such scrutiny. Moore and Howell (1985) found pili torti in all affected males and in 43% of 28 obligate carriers or females at risk. When present, pili torti can be considered, in their opinion, a reliable indicator of heterozygosity. Changes in the metaphyses of the long bones resemble scurvy. Ascorbic acid oxidase is copper-dependent. Tumer et al. (1994) described first trimester prenatal diagnosis of Menkes disease using a specific DNA probe.

Clinical Management

Williams et al. (1977) discussed studies of metabolism and long-term copper therapy in Menkes disease. Sander et al. (1988) described a patient who survived to the age of 13.5 years. Most patients have died between the ages of 6 months and 3 years. The administration of copper may have helped in the survival. In studies by De Groot et al. (1989), vitamin C therapy was ineffective. Procopis et al. (1981) described a mild, presumably allelic form. They urged that mentally retarded or ataxic boys with pili torti be investigated with this disorder in mind. Westman et al. (1988) described a second child with the atypical form who had survived to age 9 years and was doing well clinically. Danks (1988) reported on the progress of the patient reported by Procopis et al. (1981). By then aged 10 years, the patient had been treated for many years with injections of copper histidinate. Ataxia and dysarthria had been the principal problems. No radiologic abnormalities had developed in the skull or limb bones; in particular, no 'occipital horn' was discerned.

Sherwood et al. (1989) found excellent results from subcutaneous copper histidinate therapy in 2 unrelated boys with classic Menkes disease. Copper histidinate is probably the form in which copper crosses the blood-brain barrier (Hartter and Barnea, 1988). One of the patients developed orthostatic hypotension such that he preferred to crouch rather than stand. The other patient had 2 massive bladder diverticula.

Whereas parentally administered copper in the form of copper sulfate or copper-EDTA probably does not produce a substantial clinical improvement, Tumer et al. (1996) found evidence for the efficacy of copper-histidine, which is naturally present in the serum and is quantitatively important in copper transport. Copper-histidine appeared to be ineffective when it was given after the first few months of life. However, in 2 unrelated patients with this disorder who were born prematurely and received early copper-histidine treatment, the response was favorable (see Sherwood et al. (1989) and Sarkar et al. (1993)). At the time of the Tumer et al. (1996) report, these patients were aged 19 and 9 years and presented with a milder clinical course, mainly characterized by connective tissue abnormalities resembling the occipital horn syndrome (304150). An unresolved question concerned the severity of the disease in each case. One of the patients had a positive family history suggesting that he was liable to the severe form, but the possibility of intrafamilial clinical variation could not be excluded. To clarify these questions, Tumer et al. (1996) characterized the genetic defects in the ATP7A gene. Using a combination of single-strand conformation analysis and direct sequencing of amplified exons, they detected a single bp deletion in exon 4 in one patient and in exon 12 in the other. Both mutations led to a frameshift and created a premature termination codon within the same exon. RT-PCR analysis of total fibroblast RNA of both patients showed no evidence of exon skipping, indicating that the mutation resulted in severely truncated proteins. They concluded that the disorder would be expected to be severe and that the therapy had been effective. A newspaper photograph of the 9-year-old patient indicated his appearance at the time of the report.

Christodoulou et al. (1998) described the long-term clinical course of 4 boys who had Menkes disease treated from early infancy with parenteral copper-histidine, with follow-up over 10 to 20 years. Male relatives of 3 of the 4 had a severe clinical course compatible with classic Menkes disease. As a consequence of early treatment, their patients had normal or near-normal intellectual development, but developed many of the more severe somatic abnormalities of the related disorder occipital horn syndrome, including severe orthostatic hypotension in 2. In addition, 1 boy developed a previously unreported anomaly: massive splenomegaly and hypersplenism as a consequence of a splenic artery aneurysm. The oldest patient was 20 years of age at the time of report. Hypotension had been a problem from the age of 14 years. A syncopal episode on standing was associated with bradycardia. Treatment with atropine resulted in a brisk increase in heart rate and rapid clinical recovery. There was no increase in blood pressure following immersion of the hand in ice-cold water, or following a mental arithmetic challenge, suggesting that his postural hypotension may have an autonomic basis. It was subsequently treated with a peripheral alpha-adrenergic agonist, midodrine, in combination with fludrocortisone. He suffered from persistent chronic diarrhea since early infancy.

Population Genetics

Danks et al. (1971) suggested that the frequency may be 1 in 40,000 live births in Melbourne and higher than previously thought because some patients may die undiagnosed.

Tonnesen et al. (1991) estimated that the combined frequency of live-born Menkes disease patients in Denmark, France, The Netherlands, the United Kingdom, and West Germany was 1 per 298,000 live-born babies in the period 1976 to 1987. They estimated the mutation rate for Menkes disease to be 1.96 x 10(-6), based on the number of isolated Menkes cases born during that period.

Animal Model

The mottled series of mutations in the mouse may be homologous to Menkes syndrome (Hunt, 1974). The 'mottled' mutation in the hamster is also probably homologous (Yoon, 1973). Brophy et al. (1988) studied aortic aneurysm in the 'blotchy' mouse, one of the mottled series of mutations. Affected animals had a progressive increase in the instance of aneurysms with age, reaching 100% within 6 months. Most aneurysms occurred in the ascending aorta, with some also present in the descending thoracic and abdominal segments. Some animals had multiple aneurysms.

George et al. (1994) analyzed mouse Mnk, a murine locus homologous to Menkes disease, in both normal mice and those with the mottled phenotype.

Male mice with the Mottled-Brindled allele accumulate copper in the intestine, fail to export copper to peripheral organs, and die a few weeks after birth. Much of the intestinal copper found in Mottled-Brindled mice is bound by metallothionein (MT); see 156350. To determine the function of MT in the presence of Atp7a deficiency, Kelly and Palmiter (1996) crossed Mottled-Brindled females with males that bear a targeted disruption of the Mt1 and Mt2 genes. On the metallothionein-deficient background most Mottled males as well as heterozygous Mottled females died before embryonic day 11. The authors explained the lethality in females by preferential inactivation of the paternal X chromosome in extra embryonic tissues and resultant copper toxicity in the absence of MT. In support of this hypothesis, Kelly and Palmiter (1996) found that cell lines derived from metallothionein deficient, Mottled embryos were very sensitive to copper toxicity. They concluded that MT is essential to protect against copper toxicity in embryonic placenta, providing a second line of defense when copper effluxers are defective. They also stated that MT probably protects against hepatic copper toxicity in Wilson disease and the LEC rat model in which a similar copper effluxer, ATP7B (277900), is defective, because MT accumulates to high levels in the liver in those diseases.

The nature of the mutation in the brindled mouse is of importance in understanding the normal role of the protein encoded by ATP7A and for devising treatment strategies for Menkes disease. Grimes et al. (1997) showed that the brindled mouse has a deletion of 2 amino acids in a highly conserved, but functionally uncharacterized, region of the Atp7a gene. They also presented Western blot data for the normal gene product in tissues. In the kidney, immunohistochemistry demonstrated the protein in proximal and distal tubules, with a distribution identical in mutant and normal mice. This distribution was considered consistent with the protein being involved in copper resorption from the urine.

Masson et al. (1997) studied copper uptake and retention in fibroblast cultures established from 4 independent mottled alleles associated with postnatal male survival, 5 independent mottled alleles associated with prenatal death of affected males, and 12 controls. Both groups of mutants were separable from controls on both copper uptake and copper retention assays. Values obtained were the same as those previously reported for human fibroblasts established from patients with Menkes disease, but no significant differences were found between the alleles associated with survival and those associated with prenatal death.

See also Barnard et al. (1978); Billings and Degnan (1971); Bucknall et al. (1973); Camakaris et al. (1980); Collie et al. (1978); Daish et al. (1978); Danks and Cartwright (1973); Danks et al. (1972); Garnica et al. (1977); Grover and Scrutton (1975); Hara et al. (1979); Harcke et al. (1977); Horn (1976); Horn (1981); Horn (1980); Horn (1983); Horn et al. (1978); Horn et al. (1980); Iwata et al. (1979); Leone et al. (1985); McKusick (1972); Prohaska and Lukasewycz (1981); Rowe et al. (1974); Royce et al. (1980); Tonnesen et al. (1985); Tonnesen et al. (1991)

Barnard, R. O.; Best, P. V.; Erdohazi, M.: Neuropathology of Menkes' disease. Dev. Med. Child Neurol. 20: 586-597, 1978.

Billings, D. M.; Degnan, M.: Kinky hair syndrome: a new case and a review. Am. J. Dis. Child. 121: 447-449, 1971.

Bray, P. F.: Sex-linked neurodegenerative disease associated with monilethrix. Pediatrics 36: 417-420, 1965.

Brophy, C. M.; Tilson, J. E.; Braverman, I. M.; Tilson, M. D.: Age of onset, pattern of distribution, and histology of aneurysm development in a genetically predisposed mouse model. J. Vasc. Surg. 8: 45-48, 1988.

Bucknall, W. E.; Haslam, R. H. A.; Holtzman, N. A.: Kinky hair syndrome: response to copper therapy. Pediatrics 52: 653-657, 1973.

Camakaris, J.; Danks, D. M.; Ackland, L.; Cartwright, E.; Borger, P.; Cotton, R. G. H.: Altered copper metabolism in cultured cells from human Menkes' syndrome and mottled mouse mutants. Biochem. Genet. 18: 117-131, 1980.

Chan, W.-Y.; Garnica, A. D.; Rennert, O. M.: Cell culture studies of Menkes kinky hair disease. Clin. Chim. Acta 88: 495-507, 1978.

Chelly, J.; Tumer, Z.; Tonnesen, T.; Petterson, A.; Ishikawa-Brush, Y.; Tommerup, N.; Horn, N.; Monaco, A. P.: Isolation of a candidate gene for Menkes disease that encodes a potential heavy metal binding protein. Nature Genet. 3: 14-19, 1993.

Christodoulou, J.; Danks, D. M.; Sarkar, B.; Baerlocher, K. E.; Casey, R.; Horn, N.; Tumer, Z.; Clarke, J. T. R.: Early treatment of Menkes disease with parenteral cooper (sic)-histidine: long-term follow-up of four treated patients. Am. J. Med. Genet. 76: 154-164, 1998.

Collie, W. R.; Goka, T. J.; Moore, C. M.; Howell, R. R.: Hair in Menkes disease: a comprehensive review.In: Brown, A. C.; Crounse, R. G. (eds.): Hair, Trace Elements, and Human Illness. New York: Praeger Publ. (pub.) 1980. Pp. 197-209.

Collie, W. R.; Moore, C. M.; Goka, T. J.; Howell, R. R.: Pili torti as marker for carriers of Menkes disease.(Letter) Lancet I: 607-608, 1978.

Consalez, G. G.; Gecz, J.; Stayton, C. L.; Dabovic, B.; Pasini, B.; Pezzolo, A.; Bicocchi, M. P.; Fontes, M.; Romeo, G.: Fine mapping and cloning of the breakpoint associated with Menkes syndrome in a female patient. Genomics 14: 557-561, 1992. Note: Erratum: Genomics 16: 304 only, 1993.

Daish, P.; Wheeler, E. M.; Roberts, P. F.; Jones, R. D.: Menkes' syndrome: report of a patient treated from 21 days of age with parenteral copper. Arch. Dis. Child. 53: 956-957, 1978.

Danks, D. M.: The mild form of Menkes disease: progress report on the original case. Am. J. Med. Genet. 30: 859-864, 1988.

Danks, D. M.; Campbell, P. E.; Stevens, B. J.; Mayne, V.; Cartwright, E.: Menkes' kinky hair syndrome: an inherited defect in copper absorption with widespread effects. Pediatrics 50: 188-201, 1972.

Danks, D. M.; Cartwright, E.: Menkes' kinky hair disease: further definition of the defect in copper transport. Science 179: 1140-1141, 1973.
Danks, D. M.; Cartwright, E.; Campbell, P. E.; Mayne, V.: Is Menkes' syndrome a heritable disorder of connective tissue?.(Letter) Lancet II: 1089, 1971.
Danks, D. M.; Stevens, B. J.; Campbell, D. E.; Gillespie, J. M.; Walker-Smith, J.; Bloomfield, J.; Turner, B.: Menkes' kinky-hair syndrome. Lancet I: 1100-1102, 1972.
De Groot, C. J.; Wijburg, F. A.; Barth, P. G.; Bolhuis, P. A.; Peelen, W.; Abeling, N. G. G. M.; van den Hamer, C. J. A.: Vitamin C treatment in Menkes' disease: failure to affect biochemical and clinical parameters. J. Inherit. Metab. Dis. 12 (suppl. 2): 389-392, 1989.
Francis, M. J.; Jones, E. E.; Levy, E. R.; Ponnambalam, S.; Chelly, J.; Monaco, A. P.: A Golgi localization signal identified in the Menkes recombinant protein. Hum. Molec. Genet. 7: 1245-1252, 1998.
French, J. H.; Sherard, E. S.: Studies of the biochemical basis of kinky hair disease.(Abstract) Pediat. Res. 1: 206, 1967.
Friedrich, U.; Horn, N.; Stene, J.: Close linkage of the gene for Menkes disease to the centromere region of the human X-chromosome. (Abstract) Clin. Genet. 23: 230, 1983.
Garnica, A. D.; Frias, J. L.; Rennert, O. M.: Menkes kinky hair syndrome: is it a treatable disorder?. Clin. Genet. 11: 154-161, 1977.
George, A. M.; Reed, V.; Glenister, P.; Chelly, J.; Tumer, Z.; Horn, N.; Monaco, A. P.; Boyd, Y.: Analysis of Mnk, the murine homologue of the locus for Menkes disease, in normal and mottled mice . Genomics 22: 27-35, 1994.
Gerdes, A.-M.; Tonnesen, T.; Horn, N.; Grisar, T.; Marg, W.; Muller, A.; Reinsch, R.; Barton, N. W.; Guiraud, P.; Joannard, A.; Richard, M. J.; Guttler, F.: Clinical expression of Menkes syndrome in females. Clin. Genet. 38: 452-459, 1990.
Godwin-Austen, R. B.; Robinson, A.; Evans, K.; Lascelles, P. T.: An unusual neurological disorder of copper metabolism clinically resembling Wilson's disease but biochemically a distinct entity. J. Neurol. Sci. 39: 85-98, 1978.
Goka, T. J.; Stevenson, R. E.; Hefferan, P. M.; Howell, R.: Menkes disease: a biochemical abnormality in cultured human fibroblasts. Proc. Nat. Acad. Sci. 73: 604-606, 1976.
Grimes, A.; Hearn, C. J.; Lockhart, P.; Newgreen, D. F.; Mercer, J. F. B.: Molecular basis of the brindled mouse mutant (Mo-br): a murine model of Menkes disease. Hum. Molec. Genet. 6: 1037-1042, 1997.
Grover, W. D.; Scrutton, M. C.: Copper infusion therapy in trichopoliodystrophy. J. Pediat. 86: 216-220, 1975.
Haas, R. H.; Robinson, A.; Evans, K.; Lascelles, P. T.; Dubowitz, V.: An X-linked disease of the nervous system with disordered copper metabolism and features differing from Menkes disease. Neurology 31: 852-859, 1981.
Hara, K.; Oohira, A.; Nogami, H.; Watanabe, K.; Miyazaki, S.: Kinky hair disease: biochemical, histochemical, and ultrastructural studies. Pediat. Res. 13: 1222-1226, 1979.
Harcke, H. T.; Capitanio, M. A.; Grover, W. D.; Valdes-Dapena, M.: Bladder diverticula and Menkes' syndrome. Radiology 124: 459-461, 1977.
Hartter, D. E.; Barnea, A.: Brain tissue accumulates (64)copper by two ligand-dependent saturable processes. J. Biol. Chem. 263: 799-805, 1988.
Hershon, K.: Personal Communication. New York, N. Y. 10/13/1988.
Horn, N.: Copper incorporation studies on cultured cells for prenatal diagnosis of Menkes' disease. Lancet I: 1156-1158, 1976.
Horn, N.: Menkes X-linked disease: prenatal diagnosis of hemizygous males and heterozygous females. Prenatal Diag. 1: 107-120, 1981.
Horn, N.: Menkes X-linked disease: heterozygous phenotype in uncloned fibroblast cultures. J. Med. Genet. 17: 257-261, 1980.
Horn, N.: Menkes' X-linked disease: prenatal diagnosis and carrier detection. J. Inherit. Metab. Dis. 6 (suppl. 1): 59-62, 1983.
Horn, N.; Heydorn, K.; Damsgaard, E.; Tygstrup, I.; Vestermark, S.: Is Menkes syndrome a copper storage disorder?. Clin. Genet. 14: 186-187, 1978.
Horn, N.; Mooy, P.; McGuire, V. M.: Menkes X linked disease: two clonal cell populations in heterozygotes. J. Med. Genet. 17: 262-266, 1980.
Horn, N.; Stene, J.; Mollekaer, A.-M.; Friedrich, U.: Linkage studies in Menkes disease: the Xg blood group system and C-banding of the X chromosome. Ann. Hum. Genet. 48: 161-172, 1984.
Hunt, D. M.: Primary defect in copper transport underlies mottled mutants in the mouse. Nature 249: 852-854, 1974.
Iwata, M.; Hirano, A.; French, J. H.: Degeneration of the cerebellar system in X-chromosome-linked malabsorption. Ann. Neurol. 5: 542-549, 1979.
Kaler, S. G.; Gallo, L. K.; Proud, V. K.; Percy, A. K.; Mark, Y.; Segal, N. A.; Goldstein, D. S.; Holmes, C. S.; Gahl, W. A.: Occipital horn syndrome and a mild Menkes phenotype associated with splice site mutations at the MNK locus. Nature Genet. 8: 195-202, 1994.
Kapur, S.; Higgins, J. V.; Delp, K.; Rogers, B.: Menkes syndrome in a girl with X-autosome translocation. Am. J. Med. Genet. 26: 503-510, 1987.
Kelly, E. J.; Palmiter, R. D.: A murine model of Menkes disease reveals a physiological function of metallothionein. Nature Genet. 13: 219-222, 1996.
Kuivaniemi, H.; Peltonen, L.; Kivirikko, K. I.: Type IX Ehlers-Danlos syndrome and Menkes syndrome: the decrease in lysyl oxidase activity is associated with a corresponding deficiency in the enzyme protein. Am. J. Hum. Genet. 37: 798-808, 1985.
Ladinsky, M. S.; Howell, K. E.: The trans-Golgi network can be dissected structurally and functionally from the cisternae of the Golgi complex by brefeldin A. Europ. J. Cell Biol. 59: 92-105, 1992.
La Fontaine, S.; Firth, S. D.; Lockhart, P. J.; Brooks, H.; Parton, R. G.; Camakaris, J.; Mercer, J. F. B.: Functional analysis and intracellular localization of the human Menkes protein (MNK) stably expressed from a cDNA construct in Chinese hamster ovary cells (CHO-K1). Hum. Molec. Genet. 7: 1293-1300, 1998.
Leone, A.; Pavlakis, G. N.; Hamer, D. H.: Menkes' disease: abnormal metallothionein gene regulation in response to copper. Cell 40: 301-309, 1985.
Masson, W.; Hughes, H.; Papworth, D.; Boyd, Y.; Horn, N.: Abnormalities of copper accumulation in cell lines established from nine different alleles of mottled are the same as those found in Menkes disease. J. Med. Genet. 34: 729-732, 1997.
McKusick, V. A.: Heritable Disorders of Connective Tissue. St. Louis: C. V. Mosby (pub.) (4th ed.): 1972. Pp. 712-713. Note: Fig. 12-10.
Menkes, J. H.: Kinky hair disease. Pediatrics 50: 181-182, 1972.
Menkes, J. H.: Kinky hair disease: twenty five years later. Brain Dev. 10: 77-79, 1988.
Menkes, J. H.; Alter, M.; Steigleder, G. K.; Weakley, D. R.; Sung, J. H.: A sex-linked recessive disorder with retardation of growth, peculiar hair and focal cerebral and cerebellar degeneration. Pediatrics 29: 764-779, 1962.
Mercer, J. F. B.; Livingston, J.; Hall, B.; Paynter, J. A.; Begy, C.; Chandrasekharappa, S.; Lockhart, P.; Grimes, A.; Bhave, M.; Siemieniak, D.; Glover, T. W.: Isolation of a partial candidate gene for Menkes disease by positional cloning. Nature Genet. 3: 20-25, 1993.
Moore, C. M.; Howell, R. R.: Ectodermal manifestations in Menkes disease. Clin. Genet. 28: 532-540, 1985.
O'Brien, J. S.: Personal Communication. Los Angeles, Calif. 1968.
Osaka, K.; Sato, N.; Matsumoto, S.; Ogino, H.; Kadama, S.; Yokoyama, S.; Sugiyama, T.: Congenital hypocupraemia syndrome with and without steely hair: report of two Japanese infants. Dev. Med. Child Neurol. 19: 62-68, 1977.
Peltonen, L.; Kuivaniemi, H.; Palotie, A.; Horn, N.; Kaitila, I.; Kivirikko, K. I.: Alterations in copper and collagen metabolism in the Menkes syndrome and a new subtype of the Ehlers-Danlos syndrome. Biochemistry 22: 6156-6163, 1983.
Petris, M. J.; Mercer, J. F.; Culvenor, J. G.; Lockhart, P.; Gleeson, P. A.; Camakaris, J.: Ligand-regulated transport of the Menkes copper P-type ATPase efflux pump from the Golgi apparatus to the plasma membrane: a novel mechanism of regulated trafficking. EMBO J. 15: 6084-6095, 1996.
Procopis, P.; Camakaris, J.; Danks, D. M.: A mild form of Menkes steely hair syndrome. J. Pediat. 98: 97-99, 1981.
Prohaska, J. R.; Lukasewycz, O. A.: Copper deficiency suppresses the immune response of mice. Science 213: 559-561, 1981.
Proud, V. K.; Mussell, H. G.; Kaler, S. G.; Young, D. W.; Percy, A. K.: Distinctive Menkes disease variant with occipital horns: delineation of natural history and clinical phenotype. Am. J. Med. Genet. 65: 44-51, 1996.
Reaves, B.; Horn, M.; Banting, G.: TGN38/41 recycles between the cell surface and the TGN: brefeldin A affects its rate of return to the TGN. Mol. Biol. Cell 4: 93-105, 1993.
Ropers, H.-H.; Wieacker, P.; Wienker, T. F.; Davies, K.; Williamson, R.: On the genetic length of the short arm of the human X chromosome. Hum. Genet. 65: 53-55, 1983.
Rowe, D. W.; McGoodwin, E. B.; Martin, G. R.; Sussman, M. D.; Grahn, D.; Faris, B.; Franzblau, C.: A sex-linked defect in the cross-linking of collagen and elastin associated with the mottled locus in mice. J. Exp. Med. 139: 180-192, 1974.
Royce, P. M.; Camakaris, J.; Danks, D. M.: Reduced lysyl oxidase activity in skin fibroblasts from patients with Menkes' syndrome. Biochem. J. 192: 579-586, 1980.
Sander, C.; Niederhoff, H.; Horn, N.: Life-span and Menkes kinky hair syndrome: report of a 13-year course of this disease. Clin. Genet. 33: 228-233, 1988.
Sarkar, B.; Ligertat-Walsh, K.; Clarke, J. T. R.: Copper-histidine therapy for Menkes disease. J. Pediat. 123: 828-830, 1993.
Scheinberg, I. H.; Collins, J. C.: Menkes' disease: a disorder of zinc metabolism? (Letter) Lancet I: 619, 1989.
Sherwood, G.; Sarkar, B.; Sass Kortsak, A.: Copper histidinate therapy in Menkes' disease: prevention of progressive neurodegeneration. J. Inherit. Metab. Dis. 12 (suppl. 2): 393-396, 1989.
Tonnesen, T.; Garrett, C.; Gerdes, A.-M.: High (64)Cu uptake and retention values in two clinically atypical Menkes patients. J. Med. Genet. 28: 615-618, 1991.
Tonnesen, T.; Gerdes, A. M.; Horn, N.; Friedrich, U.; Grisar, T.; Muller, A.: Localization of the Menkes gene to the long arm of the X-chromosome.(Abstract) 7th Int. Cong. Hum. Genet., Berlin 627, 1986.
Tonnesen, T.; Horn, N.; Sondergaard, F.; Mikkelsen, M.; Boue, J.; Damsgaard, E.; Heydorn, K.: Measurement of copper in chorionic villi for first-trimester diagnosis of Menkes disease.(Letter) Lancet I: 1038-1039, 1985.
Tonnesen, T.; Kleijer, W. J.; Horn, N.: Incidence of Menkes disease. Hum. Genet. 86: 408-410, 1991.
Tonnesen, T.; Petterson, A.; Kruse, T. A.; Gerdes, A.-M.; Horn, N.: Multipoint linkage analysis in Menkes disease. Am. J. Hum. Genet. 50: 1012-1017, 1992.
Tumer, Z.; Horn, N.: Menkes disease: recent advances and new aspects. J. Med. Genet. 34: 265-274, 1997.
Tumer, Z.; Horn, N.; Tonnesen, T.; Christodoulou, J.; Clarke, J. T. R.; Sarkar, B.: Early copper-histidine treatment for Menkes disease.(Letter) Nature Genet. 12: 11-13, 1996.
Tumer, Z.; Tommerup, N.; Tonnesen, T.; Kreuder, J.; Craig, I. W.; Horn, N.: Mapping of the Menkes locus to Xq13.3 distal to the X-inactivation center by an intrachromosomal insertion of the segment Xq13.3-q21.2. Hum. Genet. 88: 668-672, 1992.
Tumer, Z.; Tonnesen, T.; Bohmann, J.; Marg, W.; Horn, N.: First trimester prenatal diagnosis of Menkes disease by DNA analysis. J. Med. Genet. 31: 615-617, 1994.
Verga, V.; Hall, B. K.; Wang, S.; Johnson, S.; Higgins, J. V.; Glover, T. W.: Localization of the translocation breakpoint in a female with Menkes syndrome to Xq13.2-q13.3 proximal to PGK-1. Am. J. Hum. Genet. 48: 1133-1138, 1991.
Vulpe, C.; Levinson, B.; Whitney, S.; Packman, S.; Gitschier, J.: Isolation of a candidate gene for Menkes disease and evidence that it encodes a copper-transporting ATPase. Nature Genet. 3: 7-13, 1993.
Wakai, S.; Ishikawa, Y.; Nagaoka, M.; Okabe, M.; Minami, R.; Hayakawa, T.: Central nervous system involvement and generalized muscular atrophy in occipital horn syndrome: Ehlers-Danlos type IX—a first Japanese case. J. Neurol. Sci. 116: 1-5, 1993.

Wesenberg, R. L.; Gwinn, J. L.; Barnes, G. R., Jr.: Radiological findings in the kinky-hair syndrome. Radiology 92: 500-506, 1969.
Westman, J. A.; Richardson, D. C.; Rennert, O. M.; Morrow, G., III: Atypical Menkes steely hair disease. Am. J. Med. Genet. 30: 853-858, 1988.
Wieacker, P.; Horn, N.; Pearson, P.; Wienker, T. F.; McKay, E.; Ropers, H. H.: Menkes kinky hair disease: a search for closely linked restriction fragment length polymorphism. Hum. Genet. 64: 139-142, 1983.
Wienker, T. F.; Wieacker, P.; Cooke, H. J.; Horn, N.; Ropers, H.-H.: Evidence that the Menkes locus maps on proximal Xp. Hum. Genet. 65: 72-73, 1983.
Williams, D. M.; Atkin, C. L.; Frens, D. B.; Bray, P. F.: Menkes kinky hair syndrome—studies of copper metabolism and long term copper therapy. Pediat. Res. 11: 823-826, 1977.
Williams, R. S.; Marshall, P. C.; Lott, I. T.; Caviness, V. S., Jr.: The cellular pathology of Menkes' steely hair syndrome. Neurology 28: 575-583, 1978.
Willvonseder, R.; Goldstein, N. P.; McCall, J. T.; Yoss, R. E.; Tauxe, W. N.: A hereditary disorder with dementia, spastic dysarthria, vertical eye movement paresis, gait disturbance, splenomegaly, and abnormal copper metabolism. Neurology 23: 1039-1049, 1973.
Yoon, C. H.: Recent advances in Syrian hamster genetics. J. Hered. 64: 305-307, 1973.
Yoshida, T.; Tada, K.; Mizuno, T.; Wada, Y.; Akabane, J.; Ogasawara, J.; Minagawa, A.; Morikawa, T.; Okamura, T.: A sex-linked disorder with mental and physical retardation characterized by cerebrocortical atrophy and increase of glutamic acid in the cerebrospinal fluid. Tohoku J. Med. Sci. 83: 261-269, 1964.

COLLAGEN, TYPE XI, ALPHA-2 (#120290)

JJM NOTE

A family with the Stickler syndrome (without eye involvement) has a 27-bp deletion and multiple exostosis, not part of the classic Stickler syndrome (Sirko-Osadsa et al. 1998).

See COL11A1 120280. 7,6:Law et al. (1989, 1990) used a cosmid clone containing the COL11A2 gene as a probe in the Southern blot analysis of DNA from a panel of human/hamster somatic cell hybrids containing different numbers and combinations of human chromosomes. They concluded that the gene is located on chromosome 6, and study of a cell hybrid containing only 6q indicated that the COL11A2 gene is on 6p. By a combination of somatic cell hybrid mapping and in situ hybridization, 2,4:Hanson et al. (1989, 1991) localized the COL11A2 gene to 6p21.3. By physical mapping of the class II HLA region using pulsed field gel electrophoresis, Hanson et al. (1991) demonstrated that the COL11A2 gene is about 45 kb centromeric to HLA-DPA1 (142880) and is transcribed in the opposite (i.e., telomeric) direction. Kimura et al. (1989) assigned the gene to 6p21.2 by in situ hybridization. By Northern blot analysis, they showed that the gene is expressed in cartilage but not in adult liver, skin, and tendon. The nucleotide sequence showed that although type XI collagen belongs to the fibril-forming class of collagens, there are substantial differences in exon sizes at the 3-prime end of the gene when comparing the COL11A2 gene with the genes for types I, II, and III collagens. It is thought that the alpha-3 chain of type XI collagen is a posttranslational variant of the type II, or cartilage, collagen subunit, i.e., is encoded by the COL2A1 gene (120140). Stubbs et al. (1993) showed that the homologous gene in the mouse is also 'embedded' within the major histocompatibility complex on chromosome 17.

Vuoristo et al. (1995) analyzed the COL11A2 gene from 2 overlapping cosmid clones that had previously been isolated in the course of searching the human major histocompatibility region. Nucleotide sequence defined over 28,000 bp of the gene. It was shown to contain 66 exons. As with most genes for fibrillar collagens, the first intron was among the largest, and the introns at the 5-prime end of the gene were in general larger than the introns at the 3-prime end. Analysis of the exons coding for the major triple helical domain indicated that the gene structure had not evolved with the genes for the major fibrillar collagen and that there were marked differences in the number of exons, the exon sizes, and codon usage. The gene was located close to the gene for the retinoid X receptor beta (180246) in a head-to-tail arrangement similar to that previously seen with the 2 mouse genes. Also, there was marked interspecies homology in the intergenic sequences.

Lui et al. (1996) showed that COL11A2 contains at least 62 exons spanning 30.5 kb. The gene differs from other collagens in that the amino propeptide is encoded by 14 exons rather than the usual 5 to 8. The promoter is GC-rich and lacks a TATA box. The authors believed that the gene is likely to undergo alternative splicing. The gene lies within the MHC region and is only 1.1-kb from the retinoid X receptor-beta (180246) and about 40 kb from DPB2 (142880).

Type II collagen is closely associated with both type IX collagen (120210) and type XI collagen in thin collagen fibrils of hyaline cartilage. This prompted Brunner et al. (1994) to study linkage of a Stickler syndrome phenotype (184840) to polymorphic markers within or near genes encoding these cartilage collagens. In a large Dutch kindred, Stickler syndrome was found to be linked to the 6p22-p21.3 region where COL11A2 maps. Unlike the usual cases of Stickler syndrome which show high myopia, vitreoretinal degeneration, and retinal detachment, ocular features were lacking in all affected individuals in the Dutch family. This may be explained by absence of the alpha-2 chain of type XI collagen in the vitreous. The skeletal and otologic manifestations of the Stickler syndrome were present in the family, however. Implication of type XI collagen in human osteochondrodysplasias is supported by the fact that in mice with chondrodysplasia (cho), an autosomal recessive disorder in which there is neonatal lethality, small mandible, cleft palate, small thorax, disproportionate limbs, and fragile cartilage (Seegmiller et al., 1971), were shown. to have abnormality in the alpha-1 chain of type 11 collagen.

Vikkula et al. (1995) established that the affected members of the Dutch kindred with Stickler syndrome reported by Brunner et al. (1994) had a mutation in the COL11A2 gene (120290.0001). Furthermore, they found a mutation in the COL11A2 gene in an autosomal recessive disorder characterized by spondyloepiphyseal dysplasia and sensorineural hearing loss, similar to the otospondylomegaepiphyseal dysplasia syndrome (OSMED; 215150).

Sirko-Osadsa et al. (1998) pointed out that absence of eye involvement is characteristic of the form of Stickler syndrome caused by mutations in COL11A2. They described a family in which there was no eye involvement and demonstrated that the phenotype was due to heterozygosity for a 27-bp deletion in the COL11A2 gene. The affected members of the family had sensorineural hearing loss, cleft palate/uvula, micrognathia, malar flattening, joint pains, and multiple hereditary exostoses.

See also Hanson et al. (1989)

Brunner, H. G.; van Beersum, S. E. C.; Warman, M. L.; Olsen, B. R.; Ropers, H.-H.; Mariman, E. C. M.: A Stickler syndrome gene is linked to chromosome 6 near the COL11A2 gene. Hum. Molec. Genet. 3: 1561-1564, 1994.
Hanson, I. M.; Cheah, K. S. E.; Gorman, P. A.; Solomon, E.; Trowsdale, J.: The pro-alpha2(XI) collagen gene, COL11A2, maps to the centromeric border of the major histocompatibility complex. (Abstract) Cytogenet. Cell Genet. 51: 1010-1011, 1989.
Hanson, I. M.; Gorman, P.; Lui, V. C. H.; Cheah, K. S. E.; Solomon, E.; Trowsdale, J.: The human alpha-2(XI) collagen gene (COL11A2) maps to the centromeric border of the major histocompatibility complex on chromosome 6. Genomics 5: 925-931, 1989.
Hanson, I. M.; Poustka, A.; Trowsdale, J.: New genes in the class II region of the human major histocompatibility complex. Genomics 10: 417-424, 1991.
Kimura, T.; Cheah, K. S. E.; Chan, S. D. H.; Lui, V. C. H.; Mattei, M.-G.; van der Rest, M.; Ono, K.; Solomon, E.; Ninomiya, Y.; Olsen, B. R.: The human alpha-2(XI) collagen (COL11A2) chain: molecular cloning of cDNA and genomic DNA reveals characteristics of a fibrillar collagen with differences in genomic organization. J. Biol. Chem. 264: 13910-13916, 1989.
Law, M. L.; Chan, S. D. H.; Berger, R.; Jones, C.; Kao, F. T.; Solomon, E.; Cheah, K. S. E.: The gene for the alpha-2 chain of the human fibrillar collagen type XI (COL11A2) assigned to the short arm of chromosome 6. Ann. Hum. Genet. 54: 23-29, 1990.
Law, M. L.; Chan, S. D. H.; Berger, R.; Jones, C. A.; Kao, F. T.; Solomon, E.; Cheah, K. S. E.: The gene for the alpha2 chain of the human fibrillar collagen, type XI (COL11A2) is on the short arm of chromosome 6. (Abstract) Cytogenet. Cell Genet. 51: 1029-1030, 1989.
Lui, V. C. H.; Ng, L. J.; Sat, E. W. Y.; Cheah, K. S. E.: The human alpha-2(XI) collagen gene (COL11A2): completion of coding information, identification of the promoter sequence, and precise localization within the major histocompatibility complex reveal overlap with the KE5 gene. Genomics 32: 401-412, 1996.
Seegmiller, R.; Fraser, F. C.; Sheldon, H.: A new chondrodystrophic mutant in mice: electron microscopy of normal and abnormal chondrogenesis. J. Cell Biol. 48: 580-593, 1971.
Sirko-Osadsa, D. A.; Murray, M. A.; Scott, J. A.; Lavery, M. A.; Warman, M. L.; Robin, N. H.: Stickler syndrome without eye involvement is caused by mutations in COL11A2, the gene encoding the alpha-2(XI) chain of type XI collagen. J. Pediat. 132: 368-371, 1998.
Stubbs, L.; Lui, V. C. H.; Ng, L. J.; Cheah, K. S. E.: The alpha-2(XI) collagen gene lies within 8 kb of Pb in the proximal portion of the murine major histocompatibility complex. Mammalian Genome 4: 95-103, 1993.
Vikkula, M.; Mariman, E. C. M.; Lui, V. C. H.; Zhidkova, N. I.; Tiller, G. E.; Goldring, M. B.; van Beersum, S. E. C.; de Waal Malefijt, M. C.; van den Hoogen, F. H. J.; Ropers, H.-H.; Mayne, R.; Cheah, K. S. E.; Olsen, B. R.; Warman, M. L.; Brunner, H. G.: Autosomal dominant and recessive osteochondrodysplasias associated with the COL11A2 locus. Cell 80: 431-437, 1995.
Vuoristo, M. M.; Pihlajamaa, T.; Vandenberg, P.; Prockop, D. J.; Ala-Kokko, L.: The human COL11A2 gene structure indicates that the gene has not evolved with the genes for the major fibrillar collagens. J. Biol. Chem. 270: 22873-22881, 1995.

PULMONARY ALVEOLAR MICROLITHIASIS (*265100)

JJM NOTE

A 5-year-old male with this rare disorder also had multiple exostoses, not a previously recognized association (Richie et al. 1992).

The condition is characterized by multiple minute calcifications located in the alveoli and producing a typical radiographic appearance. Sibs have been affected in a number of cases. Information on consanguinity has apparently not been collected in a systematic manner; however, in several of the reported cases, parental consanguinity was noted. Caffrey and Altman (1965) described the disorder in premature twins who died at age 12 hours. They reviewed 68 cases in the literature (including theirs): 34 were familial, occurring in 13 families. A disproportionately large number of cases may be of Spanish extraction. In Spain, Lopez-Areal et al. (1965) described 2 affected sisters in one family, and a boy and his 2 sisters in a second family. O'Neill et al. (1967) observed 3 affected sibs. Affected brother and sister with first-cousin parents were reported by Burguet and Reginster (1967). In Beirut, Balikian et al. (1968) described the disorder in 2 pairs of brothers and an unrelated girl. Tachibana (1984) found many cases in Japan, including cases in multiple sibs. Richie et al. (1992) found pulmonary alveolar microlithiasis in a 5-year-old boy being investigated for multiple exostoses (133700). Neither disorder was identified in other members of the immediate family. Ucan et al. (1993) stated that only 173 cases of this disorder had been reported worldwide. The condition is unusually common in Turkey; they reported 52 Turkish cases, 49 of which had previously been described only in Turkish publications. A positive family history was noted in 19 of 39 patients. All the familial cases were sibs except for 3 who had a maternal aunt/nephew relationship. Of the 52 cases, 10 were free of symptoms.

See also Cale et al. (1983); Drinkovic et al. (1962); Gomez et al. (1959); Prakash et al. (1983); Sosman et al. (1957); Viswanathan (1962); Yang and Lin (1963)

Balikian, J. P.; Fuleihan, F. J. D.; Nucho, C. N.: Pulmonary alveolar microlithiasis: report of five cases with special reference to roentgen manifestations. Am. J. Roentgen. 103: 509-518, 1968.
Burguet, W.; Reginster, A.: L'heredite de la microlithiase alveolaire pulmonaire. A propos d'une nouvelle observation familiale. Ann. Genet. 10: 75-81, 1967.
Caffrey, P. R.; Altman, R. S.: Pulmonary alveolar microlithiasis in premature twins. J. Pediat. 66: 758-763, 1965.
Cale, W. F.; Petsonk, E. L.; Boyd, C. B.: Transbronchial biopsy of pulmonary alveolar microlithiasis. Arch. Intern. Med. 143: 358-359, 1983.
Drinkovic, I.; Strohal, K.; Sablijica, B.: Mikrolithiasis alveolaris pulmonum. Fortschr. Geb. Roentgenstr. 97: 180-185, 1962.
Gomez, G.; Gomez, G. E.; Lichlemberger, E.; Santamaria, A.; Carvajal, L.; Jimenez-Penulea, B.; Saaibi, E.; Barrera, A. R.; Orduz, E.; Correa-Henao, A.: Familial pulmonary alveolar microlithiasis: four cases from Colombia, S. A.: Is microlithiasis also an environmental disease?. Radiology 72: 550-561, 1959.
Lopez-Areal, L.; Zumarraga, R.; Turner, C. G.; Granizo, I. F. M.; Vara Cuadrado, F.; Duque Fraile, J.: Microlitiasis alveolar pulmonar familiar e infantil. (Descripcion de cinco casos in dos familias.). Rev. Clin. Esp. 97: 389-395, 1965.
O'Neill, R. P.; Cohn, J. E.; Pellegrino, E. D.: Pulmonary alveolar microlithiasis—a family study. Ann. Intern. Med. 67: 957-967, 1967.
Prakash, U. B. S.; Barham, S. S.; Rosenow, E. C., III; Brown, M. L.; Payne, W. S.: Pulmonary alveolar microlithiasis: a review including ultrastructural and pulmonary function studies. Mayo Clin. Proc. 58: 290-300, 1983.
Richie, D. A.; O'Connor, S. A.; McGivern, D.: An unusual presentation of pulmonary alveolar microlithiasis and diaphyseal aclasia. Brit. J. Radiol. 65: 178-181, 1992.
Sosman, M. C.; Dodd, G. D.; Jones, W. D.; Pillmore, G. U.: The familial occurrence of pulmonary alveolar microlithiasis. Am. J. Roentgen. 77: 947-1012, 1957.
Tachibana, T.: Personal Communication. Osaka, Japan 9/17/1984.
Ucan, E. S.; Keyf, A. I.; Aydilek, R.; Yalcin, Z.; Sebit, S.; Kudu, M.; Ok, U.: Pulmonary alveolar microlithiasis: review of Turkish reports. Thorax 48: 171-173, 1993.
Viswanathan, R.: Pulmonary alveolar microlithiasis. Thorax 17: 251-256, 1962.
Yang, S.-P.; Lin, C.-C.: Pulmonary alveolar microlithiasis: a report of two youngest cases in a family. Dis. Chest 44: 163-167, 1963.

METACHONDROMATOSIS (156250)

JJM NOTE

This entity is included as another syndrome with exostosis.

Maroteaux (1971) described 2 families with skeletal radiologic features of both multiple exostoses (133700) and Ollier disease (166000). He called the disorder metachondromatosis and suggested autosomal dominant inheritance on the basis of 1 family with 5 affected persons. Lachman et al. (1974) reported a case. Kennedy (1983) presented the case of a 9.5-year-old boy in whom 'bumps' on the hands, feet and knees had been noted at age 5. In the next few years, some of these enlarged, new ones appeared, and others regressed. Peculiar striations were noted radiologically in the metaphyses of the long bones and iliac crests. The mother had a similar although milder history; x-ray showed a single exostosis in the wrist. The maternal grandfather had had lesions of the hands, feet, and knees, but none was evident at the time of radiologic study. Two cousins were said to have similar lesions. Differentiation from multiple exostoses of the classic type (133700) is important because of the usual regression with little or minimal residual deformity. In metachondromatosis involvement of the hands and feet is the rule but is said by Kennedy (1983) to be unusual in classic multiple exostoses. Furthermore, in metachondromatosis, the exostoses point toward the epiphysis. Dorst (1983) observed metachondromatosis in a brother and sister of Korean extraction. The radiologic findings combined those of multiple exostoses, multiple enchondromatosis (Ollier disease), and dysplasia epiphysealis hemimelica (127800). No other affected relatives were known but the parents were not available for study. In the family studied by Vanek (1982), the mother of 2 of the affected persons had a history of exostoses as a child, which regressed as she grew older. X-ray studies showed no peripheral exostoses but the proximal end of one humerus was wide and a first metatarsal was abnormally thin. A brother was said also to have had nodular growths near joints as a child. Bassett and Cowell (1985) stud ied 4 members of a kindred that had at least 8 cases in all. They pointed out that the enchondromatous lesions involve the iliac crest and metaphyseal region of the long bones of the lower extremities as in Ollier disease. The exostoses of metachondromatosis, unlike those of hereditary multiple exostosis, point toward the nearby joint and do not cause shortening or bowing of the long bone, joint deformity, or subluxation. They affect particularly the digits and may resolve spontaneously.

See also De la Cruz and Garcia-Castro (1976); Giedion et al. (1975)

Bassett, G. S.; Cowell, H. R.: Metachondromatosis: report of four cases. J. Bone Joint Surg. 67A: 811-814, 1985.
De la Cruz, J.; Garcia-Castro, J. M.: Metachondromatosis: a diagnostic dilemma. Apropos of studies in a Puerto Rican family. Bol. Assoc. Med. PR 68: 340-344, 1976.
Dorst, J. P.: Personal Communication. Baltimore, Md. 6/14/1983.
Giedion, A.; Kesztler, R.; Muggiasca, F.: The widened spectrum of multiple cartilaginous exostosis (MCE). Pediat. Radiol. 3: 93-100, 1975.
Kennedy, L. A.: Metachondromatosis. Radiology 148: 117-118, 1983.
Lachman, R. S.; Cohen, A.; Hollister, D. W.; Rimoin, D. L.: Metachondromatosis. Birth Defects Orig. Art. Ser. X(9): 171-178, 1974.
Maroteaux, P.: La metachondromatose. Z. Kinderheilk. 109: 246-261, 1971.
Vanek, V. J.: Metachondromatose: 3 Beobachtungen mit erblichen Vorkommen. Beitr. Orthop. Traumatol. 29: 103-107, 1982.

Other Bone Tumors

OSTEOCHONDROMATOSIS (166000)

JJM NOTE

This entity is included because of the complication of chondrosarcoma. The patients with osteochondromatosis and granulosa cell tumor of the ovary suggest an overlap with Peutz-Jeghers syndrome.

When hemangiomata are associated, the condition is known as Maffucci syndrome. Neither condition seems to be genetically determined in a simple mendelian manner. There are a few instances of familial occurrence of Ollier disease, however. Steudel (1892) described 2 affected brothers, and Rossberg (1959) reported affected brother and sister whose paternal grandfather was also affected. Lamy et al. (1954) observed 3 affected sibs, and Carbonell Juanico and Vineta Teixido (1962) reported affected brother and sister. Dominant inheritance with reduced penetrance is possible. Sun et al. (1985) reported that 9 patients with Maffucci syndrome seen at the Mayo Clinic developed chondrosarcoma. From a review of the English literature since 1973, they concluded that the incidence of chondrosarcoma in this disorder is 17.8%. This conclusion is suspect. The difficulties of stating the frequency of malignancy in von Recklinghausen neurofibromatosis and multiple exostoses, from hospital records or reports, is well known. Vaz and Turner (1986) described a case of Ollier disease associated with ovarian juvenile granulosa cell tumor and precocious pseudopuberty. Tamimi and Bolen (1984) had described the same association.

Kaplan et al. (1993) reported 2 cases, a 42-year-old man and a 44-year-old man. The latter had had multiple enchondromas and subcutaneous hemangiomas since birth on both the arms and the legs. A painful mass that had been growing in the region of the right elbow in recent months was found to be a chondrosarcoma. The patient underwent right shoulder disarticulation. Kaplan et al. (1993) reviewed 63 other cases, concluding that the syndrome is nonhereditary, occurs in all races, and occurs in both sexes equally. They suggested that chondrosarcoma occurs in 30% of patients with Maffucci syndrome. As in the case of multiple exostoses and neurofibromatosis type I, the frequency of malignancy may be exaggerated through bias; the patients in whom malignancy develops are, of course, more likely to come to medical attention.

See also Anderson (1965); Andren et al. (1963); Cauble and Bowman (1968); Kaplan et al. (1994); Loewinger et al. (1977); Slagsvold and Larsen (1977)

Anderson, I. F.: Maffucci's syndrome: report of a case with a review of the literature. S. Afr. Med. J. 39: 1066-1070, 1965.
Andren, L.; Dymling, J. F.; Elner, A.; Hogeman, K. E.: Maffucci's syndrome: report of four cases. Acta Chir. Scand. 126: 397-405, 1963.
Carbonell Juanico, M.; Vineta Teixido, J.: Otro caso de discondroteosis generalizada congenita, tipo Ollier. Rev. Esp. Pediat. 18: 91-99, 1962.
Cauble, W. G.; Bowman, H. S.: Dyschondroplasia and hemangiomas (Maffucci's syndrome): presentation of a case. Arch. Surg. 97: 678-681, 1968.
Kaplan, R. P.; Wang, J. T.; Amron, D. M.; Kaplan, L.: Maffucci's syndrome: two case reports with a literature review. J. Am. Acad. Derm. 29: 894-899, 1994.
Lamy, M.; Aussannaire, M.; Jammet, M. L.; Nezelof, C.: Trois cas de maladie d'Ollier dans une fratrie. Bull. Mem. Soc. Med. Hop. Paris 70: 62-70, 1954.
Loewinger, R. J.; Lichtenstein, J. R.; Dodson, W. E.; Eisen, A. Z.: Maffucci's syndrome: a mesenchymal dysplasia and multiple tumour syndrome. Brit. J. Derm. 96: 317-322, 1977.
Rossberg, A.: Zur Erblichkeit der Knochenchondromatose. Fortschr. Roentgenstr. 90(1): 138-139, 1959.
Slagsvold, J. E.; Larsen, J. L.: Fibromuscular dysplasia of intracranial arteries in a patient with enchondromas (Ollier disease). Neurology 27: 1168-1171, 1977.
Steudel, (NI): Multiple Enchondrome der Knochen in Verbindung mit venoesen Angiomen der Weichteile. Bruns' Beitr. Klin. Chir. 8: 503-521, 1892.
Sun, T.-C.; Swee, R. G.; Shives, T. C.; Unni, K. K.: Chondrosarcoma in Maffucci's syndrome. J. Bone Joint Surg. 67A: 1214-1219, 1985.
Tamimi, H. K.; Bolen, J. W.: Enchondromatosis (Ollier's disease) and ovarian juvenile granulosa cell tumor. Cancer 53: 1605-1608, 1984.
Vaz, R. M.; Turner, C.: Ollier disease (enchondromatosis) associated with ovarian juvenile granulosa cell tumor and precocious pseudopuberty. J. Pediat. 108: 945-947, 1986.

GENOCHONDROMATOSIS (137360)

JJM NOTE

This entity is included as another syndrome with enchondromas.

Le Merrer et al. (1991) described 4 patients from the same family with a characteristic localization of chondromatosis: clavicle, upper end of humerus, and lower end of femur (hence, 'geno-' for knee). The patients were a mother and her daughter and 2 sons. The condition showed a benign course and was clearly different from metachondromatosis (156250), generalized encondromatosis (225795) and spondyloenchondrodysplasia (271550).

Le Merrer, M.; Fressinger, P.; Maroteaux, P.: Genochondromatosis. J. Med. Genet. 28: 485-489, 1991.

SPONDYLOENCHONDRODYSPLASIA (*271550)

JJM NOTE

This entity is included for its possible predisposition to malignancy, by analogy to the other enchondromatoses.

Schorr et al. (1976) described enchondromatosis, similar to that of Ollier disease, in association with platyspondyly in 2 sons of first-cousin parents of Iraqi Jewish background. Both boys were of normal intelligence but of short stature. Spranger et al. (1978) identified 6 types of enchondromatosis: Ollier disease, Maffucci disease, metachondromatosis, the disorder in the patients of Schorr et al. (1976), and 2 other forms represented by patients they reported. There was no evidence of a genetic basis of the latter 2 types. Sauvegrain et al. (1980) described enchondromatosis with spinal involvement. Van Creveld et al. (1971) had a patient (Case 1) with striking symmetric enchondroma-like changes in the metaphyses and an unidentified mucopolysacchariduria; see 215050. They called the condition 'chondrodysplasia calcificans metaphysealis.' Frydman et al. (1986) studied 4 cases. The height of the patients was more than 3 SDs below the mean. Kyphoscoliosis with flat vertebrae and with 'cystic' radiolucencies in the posterior part of the bodies was observed. The metaphyses of long bones showed symmetric radiolucencies that extended into the shafts with sclerotic changes in the diaphyses. The proximal fibula and distal ulna were more severely affected than the corresponding areas of the tibia and radius. Menger et al. (1989) reported 4 patients, 3 of whom were members of 1 family containing complex consanguinity. They pointed to the 2 cases reported by Gustavson et al. (1978) as 'autosomal recessive spondylometaphyseal dysplasia,' as well as to the case of Chagnon et al. (1985). They documented considerable clinical and radiologic variability. One of the patients in the consanguineous family had normal physical appearance and a height at the 10th centile, yet the x-rays showed marked structural irregularities in the dorsal third of the vertebral bodies. Their unrelated patient was the product of an incestuous (father-daughter) union. Frydman et al. (1990) s uggested that there may be 2 forms of spondyloenchondrodysplasia. Their type I was represented by 3 patients whose findings were similar to those reported originally by Schorr et al. (1976). Two of the patients were sibs of Iraqi-Jewish origin who were apparently unrelated to the patients of the same origin reported by Schorr et al. (1976). The third patient was Arab. Parents in both instances were consanguineous. Their type II was represented by 3 patients who had similar radiographic findings and body proportions but who also had CNS calcifications, particularly in the basal ganglia. Progressive spastic quadriparesis was observed in 2 of the type II patients. Hypoplasia of the odontoid process did not appear to be the cause.

Zack and Beighton (1995) described spondyloenchondromatosis in 1 of 6 children of a consanguineous couple of mixed ancestry. When first seen at the age of 2.5 years, a tentative diagnosis of pseudoachondroplasia had been made but the features later evolved to a radiologic appearance diagnostic of spondyloenchondromatosis. With hindsight, the configuration of the pelvis at age 2.5 years was somewhat suggestive of asphyxiating thoracic dysplasia. Subsequently, marked constriction of the chest developed, as indicated by the photographs taken at the age of 13 years.

Chagnon, S.; Lacert, P.; Blery, M.: Spondylo-enchondrodysplasie. J. Radiol. 66: 75-77, 1985.
Frydman, M.; Bar-Ziv, J.; Preminger-Shapiro, R.; Brezner, A.; Brand, N.; Ben-Ami, T.; Lachman, R. S.; Gruber, H. E.; Rimoin, D. L.: Possible heterogeneity in spondyloenchondrodysplasia: quadriparesis, basal ganglia calcifications, and chondrocyte inclusions. Am. J. Med. Genet. 36: 279-284, 1990.
Frydman, M.; Preminger-Shapiro, R.; Bar-Ziv, J.: Spondylometaphyseal dysplasia with 'enchondromatous-like' changes—a distinctive type. (Abstract) 7th Int. Cong. Hum. Genet., Berlin 257-258, 1986.
Gustavson, K.-H.; Holmgren, G.; Probst, F.: Spondylometaphyseal dysplasia in two sibs of normal parents. Pediat. Radiol. 7: 90-96, 1978.
Menger, H.; Kruse, K.; Spranger, J.: Spondyloenchondrodysplasia. J. Med. Genet. 26: 93-99, 1989.
Sauvegrain, J.; Maroteaux, P.; Ribier, J.; Garel, L.; Tato, L.; Rochiccioli, P.; De Magalhaes, J.; Duhamel, B.: Chondromes multiples avec atteinte rachidienne: spondylo-enchondroplasie et autres formes. J. Radiol. 61: 495-501, 1980.
Schorr, S.; Legum, C.; Ochshorn, M.: Spondyloenchondrodysplasia: enchondromatosis with severe platyspondyly in two brothers. Radiology 118: 133-139, 1976.
Spranger, J.; Kemperdieck, H.; Bakowski, H.; Opitz, J. M.: Two peculiar types of enchondromatosis. Pediat. Radiol. 7: 215-219, 1978.
van Creveld, S.; Kozlowski, K.; Pietron, K.; van der Valk, A.: Metaphyseal chondrodysplasia calcificans: a report of two cases. Brit. J. Radiol. 44: 773-779, 1971.
Zack, P.; Beighton, P.: Spondyloenchondromatosis: syndromic identity and evolution of the phenotype. Am. J. Med. Genet. 55: 478-482, 1995.

DYSPLASIA EPIPHYSEALIS HEMIMELICA WITH CHONDROMAS AND OSTEOCHONDROMAS (127820)

JJM NOTE

Follow-up of this family should reveal the natural history and even the diagnosis and mapping of the trait.

Hensinger et al. (1974) described a kindred in which 7 members in 2 generations in a pattern consistent with autosomal dominant inheritance had dysplasia epiphysealis hemimelica, intracapsular or periarticular chondromas of the knee, extraskeletal chondromas, and osteochondromas in various combinations. None of these individually has a known mendelian basis. Under the designation of dominant carpotarsal osteochondromatosis, Maroteaux et al. (1993) described a mother and son with what appeared to be the same condition as that in the family reported by Hensinger et al. (1974). The son was first seen at the age of 16 years for swelling and restricted mobility of the left ankle. About one year later he presented with swelling on the palmar surface of the right wrist and complained of limited flexion of the right knee with one episode of locking. The mother had first noted deformity of her right wrist in adolescence. At the age of 52 years, a swelling could be palpated on the palmar side of the left wrist. These had been demonstrated by x-rays taken at the age of 33 years. Maroteaux et al. (1993) suggested that this, in fact, is a condition distinct from dysplasia epiphysealis hemimelica, which is usually sporadic. Genetic counseling is clearly quite different.

Hensinger, R. N.; Cowell, H. R.; Ramsey, P. L.; Leopold, R. G.: Familial dysplasia epiphysealis hemimelica, associated with chondromas and osteochondromas: report of a kindred with variable presentations. J. Bone Joint Surg. 56A: 1513-1516, 1974.

Maroteaux, P.; Le Merrer, M.; Bensahel, H.; Freisinger, P.: Dominant carpotarsal osteochondromatosis. J. Med. Genet. 30: 704-706, 1993.

ENCHONDROMATOSIS, GENERALIZED (225795)

JJM NOTE

This entity is included as another syndrome with enchondromas.

Halal and Azouz (1991) reported the case of a boy who had platyspondyly and metaphyseal manifestations of enchondromatosis with severe involvement of the hands and feet compatible with generalized enchondromatosis, or Spranger's type VI enchondromatosis (Spranger et al., 1978). The father was short of stature and had only moderate platyspondyly. Both the father and the son had consanguineous parents. Halal and Azouz (1991) suggested that platyspondyly may be: (1) a manifestation of the carrier state for an autosomal recessive trait; (2) a minor expression of the same autosomal recessive trait in an affected individual since the father's parents were also consanguineous and some of his sibs were reported to have prominent joints; or (3) less likely, variable expression of an autosomal dominant trait in the father and son. In their classification of the chondromatoses, Spranger et al. (1978) called Ollier disease and Maffucci syndrome (166000) types I and II enchondromatosis, respectively; metachondromatosis (156250), type III; and spondyloenchondrodysplasia (271550), type IV. Halal and Azouz (1991) added 3 tentative categories to the 6 in the classification of Spranger et al. (1978).

Halal, F.; Azouz, E. M.: Generalized enchondromatosis in a boy with only platyspondyly in the father. Am. J. Med. Genet. 38: 588-592, 1991.

Spranger, J.; Kemperdieck, H.; Bakowski, H.; Opitz, J. M.: Two peculiar types of enchondromatosis. Pediat. Radiol. 7: 215-219, 1978.

UPINGTON DISEASE (*191520)

JJM NOTE

This seems to be another private syndrome that deserves revisit.

In 5 sibships and 3 generations of a family originating from Upington district of the Cape Province, South Africa, Schweitzer et al. (1971) described a new 'dyschondroplasia' in which father-son transmission was noted. The features were Perthes-like hip changes, enchondromata and ecchondromata. Some similarities to Ollier disease and Maffucci disease were noted. The latter two conditions are, however, not mendelian.

Schweitzer, G.; Jones, B.; Timme, H.: Upington disease: a familial dyschondroplasia. S. Afr. Med. J. 45: 994-1000, 1971.

COLLAGEN, TYPE VI, ALPHA-1 CHAIN (*120220)

JJM NOTE

This entity is included because of the presumed chondrodysplasia mentioned by Murata et al. (1987).

Chung et al. (1976) isolated a collagen from the intima of human aorta that differs from types IV and V collagen. Seemingly, the same collagen was isolated from bovine placenta by Jander et al. (1981) and from human placenta by Furuto and Miller (1981). It is cysteine-rich and appears to have 3 peptides: a single relatively acidic peptide plus 2 more basic peptides. Type VI collagen appears to be unusual among collagens in the small size of its collagenous domains and in its supramolecular structure. It has been called 'short-chain collagen.' It is relatively resistant to bacterial collagenase and has a glycine content less than one-third of the protein, suggesting interrupted helical regions. Electron microscopy shows additional unique features. Collagen VI is a component of microfibrillar structures in many tissues (Engel et al., 1985). These microfibrils localize close to cells, nerves, blood vessels, and large collagen fibrils and are considered to have an anchoring function. Consistent with such a function are the biochemical findings that type VI collagen binds cells and that its fusion protein binds type I collagen. The binding activity also implies that, in addition to a structural role, type VI collagen may be involved in cell migration and differentiation and embryonic development.

Trueb and Winterhalter (1986) showed that type VI collagen consists of 2 different 140-kD subunits (alpha-1 and alpha-2) and a 200-kD subunit (alpha-3). The alpha-3(VI) chain is synthesized by cells in culture as a precursor of 260 kD, while no precursor forms of the other 2 chains could be detected. Working with pepsin-solubilized collagen VI from human placenta, Chu et al. (1987) characterized the 3 constituent chains by peptide-sequences and cDNA clones.

By somatic cell hybrid analysis and in situ hybridization, Weil et al. (1988) localized the alpha-1 and alpha-2 (COL6A2) collagen VI genes to chromosome 21q22.3 and the alpha-3(VI) gene (COL6A3; 120250) to chromosome 2q37. By Southern hybridization analysis of DNA from Chinese hamster-human somatic cell hybrids segregating different portions of human chromosome 21, Cutting et al. (1988) showed that COL6A1 and COL6A2 map distal to the locus D21S3. Linkage analysis in 40 CEPH families showed 12.5% recombination between COL6A1 and ETS2 (164740). No recombinants were observed between COL6A1 and COL6A2. Recombinants between the Marfan phenotype (154700) and COL6A1 markers in 3 families excluded mutation of COL6A1 and COL6A2 as the cause of the Marfan syndrome. By pulsed field gel electrophoresis (PFGE), Cutting et al. (1988) showed that COL6A1 and COL6A2 hybridized to the same size restriction fragments, the smallest of which was 205 kb. Petersen et al. (1991) presented a genetic linkage map of chromosome 21 involving 27 markers (10 genes and 17 anonymous sequences). The length of the male map was 132 cM and of the female map 161 cM. In both sexes, approximately one-half of the crossovers occurred distally in terminal band 21q22.3, which also contained 16 of the markers studied, including 8 of the 10 genes. Band 21q22.3 spans about 75 cM in the female map. The Petersen map illustrated the homogeneity in the distribution of genes in the genome, with a concentration of genes in Giemsa-light bands and near the ends of chromosomes. Francomano et al. (1991) used PFGE and somatic cell hybrids to demonstrate that COL6A1 and COL6A2 form a gene cluster on the most distal part of chromosome 21. They detected several DNA polymorphisms (both restriction site and VNTRs) associated with these loci. Using a slot-blot method for the dosage of single copy sequences, Delabar et al. (1992) assessed the copy number of 30 chromosome 21 markers in the blood DNA of 11 patients with partial trisomy or monosomy 21. The physical order of the markers on chromosome 21 was thereby determined. The results showed that COL6A1 and S100B (176990) were in the most terminal region, with CD18 (116920) in the penultimate segment of 21q and PFKL (171860) in the next segment toward the centromere. Using a single interspecific backcross, Justice et al. (1990) mapped the homologous gene to mouse chromosome 10 and determined its location relative to 17 other markers.

Murata et al. (1987) found predominant production of type VI collagen by the tumors in a patient with 'multiple fibromatosis occurring at the sites of multiple cartilaginous dysplasia.' They stated that the patient had 'a hereditary disease, with regions of multiple articular dysplasia surrounded by numerous protuberant tumors. Elastic globe-shaped tumors, weighed (sic) about 100 g., were removed from his cervical regions at operation.' The age of the patient was not given, and the nature of the ailment is unclear.

Davies et al. (1995) studied genetic variation in the COL6A1/COL6A2 gene cluster on chromosome 21 in 113 controls and 58 European families (including control and family subgroups of British/Irish origin) having a child with trisomy 21. They found statistically significant differences among subgroups of trisomy children with and without congenital heart defects in distributions of definitive, 3-RFLP haplotype classes received from their nondisjoining and disjoining parents. The haplotypes received by trisomy children with congenital heart defects from the disjoining parents were not a random sample of controls' haplotypes. Analysis of parental single-RFLP genotypes and linkage disequilibrium patterns confirmed this parent subgroup differed from a random sample of controls. There was no significant difference in parent subgroup genotype distribution at any of 9 control loci distributed along chromosome 21q. The study by Davies et al. (1995) showed an association between genetic variation in the COL6A1 region and congenital heart defects in trisomy 21.

By high-resolution fluorescence in situ hybridization (FISH) techniques, Heiskanen et al. (1995) determined the distance separating the COL6A1 and COL6A2 genes (150 kb), the size of the COL6A1 gene (29 kb), and the 5-prime/3-prime orientation of these genes. By fiber-FISH, they showed that the orientation is 5-prime COL6A1 3-prime/5-

prime COL6A2 3-prime. This appeared to be the first collagen gene pair found to be in a head-to-tail configuration. Other closely located collagen gene pairs are either in a head-to-head configuration (type IV collagen genes) with a common promoter region or in a tail-to-tail configuration (COL3A1 and COL5A2) on chromosome 2q.

In 9 kindreds with the Bethlem form of autosomal dominant myopathy with contractures (158810), Jobsis et al. (1996) demonstrated genetic linkage to the COL6A1-COL6A2 cluster on 21q22.3. By sequence analysis in 4 families Jobsis et al. (1996) identified a mutation in COL6A1 (120220.0001) in 1 and a COL6A2 mutation (120240.0001) in 2 other kindreds. Both mutations disrupted the gly-X-Y motif of the triple helical domain by substitution of gly for either val or ser. Analogous to the putative perturbation of the anchoring function of the dystrophin-associated complex in congenital muscular dystrophy with mutations in the alpha-2-subunit of laminin (156225), the observation suggested to the authors a similar mechanism in Bethlem myopathy.

Lamande et al. (1998) found that functional haploinsufficiency of collagen VI causes Bethlem myopathy. They identified a heterozygous COL6A1 mutation (120220.0002) resulting in a 1-bp deletion from the mRNA and a premature stop codon. Mutant mRNA was unstable, subject to nonsense-mediated mRNA decay and was almost completely absent both from patient fibroblasts and skeletal muscle. They stated that Bethlem myopathy is the first example of a muscular dystrophy caused by haploinsufficiency of a structural protein or member of the dystrophin-glycoprotein complex, and identifies collagen VI as a critical contributor to cell-matrix adhesion in skeletal muscle.

Using in situ hybridization to sectioned embryonic mouse hearts, Klewer et al. (1998) delineated the developmental expression pattern of the COL6A1, COL6A2, and COL6A3 genes. COL6A1 mRNA was first detected at embryonic day 11-11.5 in endothelial cells overlying the atrioventricular endocardial cushions. Expression was seen throughout the atrioventricular endocardial cushions following epithelial-mesenchymal transition. High expression levels were seen after embryonic day 13 in the remodeling atrioventricular endocardial cushions as they contributed to the developing atrioventricular valves and atrioventricular septum. The authors noted that type VI collagen is present in higher amounts than usual in the extracellular matrix of skin from individuals with trisomy 21, but commented that it is uncertain whether overexpression of COL6A1 and COL6A2 is sufficient to disrupt normal atrioventricular valvuloseptal morphogenesis and lead to the congenital heart malformations typical of trisomy 21.

See also Hessle and Engvall (1984)

Chu, M.-L.; Mann, K.; Deutzmann, R.; Pribula-Conway, D.; Hsu-Chen, C.-C.; Bernard, M. P.; Timpl, R.: Characterization of three constituent chains of collagen type VI by peptide sequences and cDNA clones. Europ. J. Biochem. 168: 309-317, 1987.

Chung, E.; Rhodes, R. K.; Miller, E. J.: Isolation of three collagenous components of probable basement membrane origin from several tissues. Biochem. Biophys. Res. Commun. 71: 1167-1174, 1976.

Cutting, G.; Francomano, C. A.; Chu, M. L.; Timpl, R.; McCormick, M. K.; Warren, A. C.; Hong, H. K.; Pyeritz, R. E.; Antonarakis, S. E.: Genetic linkage analysis and macrorestriction mapping of COL6A1 and COL6A2, structural genes of type VI collagen. (Abstract) Am. J. Hum. Genet. 43: A141 only, 1988.

Davies, G. E.; Howard, C. M.; Farrer, M. J.; Coleman, M. M.; Bennett, L. B.; Cullen, L. M.; Wyse, R. K. H.; Burn, J.; Williamson, R.; Kessling, A. M.: Genetic variation in the COL6A1 region is associated with congenital heart defects in trisomy 21 (Down's syndrome). Ann. Hum. Genet. 59: 253-269, 1995.

Delabar, J.-M.; Chettouh, Z.; Rahmani, Z.; Theophile, D.; Blouin, J.-L.; Bono, R.; Kraus, J.; Barton, J.; Patterson, D.; Sinet, P.-M.: Gene-dosage mapping of 30 DNA markers on chromosome 21. Genomics 13: 887-889, 1992.

Engel, J.; Furthmayr, H.; Odermatt, E.; Von der Mark, H.; Aumailley, M.; Fleishmajer, R.; Timpl, R.: Structure and macromolecular organization of type VI collagen. Ann. N.Y. Acad. Sci. 460: 25-37, 1985.

Francomano, C. A.; Cutting, G. R.; McCormick, M. K.; Chu, M. L.; Timpl, R.; Hong, H. K.; Antonarakis, S. E.: The COL6A1 and COL6A2 genes exist as a gene cluster and detect highly informative DNA polymorphisms in the telomeric region of human chromosome 21q. Hum. Genet. 87: 162-166, 1991.

Furuto, D. K.; Miller, E. J.: Characterization of a unique collagenous fraction from limited pepsin digests of human placental tissue: molecular organization of the native aggregate. Biochemistry 20: 1635-1640, 1981.

Heiskanen, M.; Saitta, B.; Palotie, A.; Chu, M.-L.: Head to tail organization of the human COL6A1 and COL6A2 genes by fiber-FISH. Genomics 29: 801-803, 1995.

Hessle, H.; Engvall, E.: Type VI collagen: studies on its localization, structure, and biosynthetic form with monoclonal antibodies. J. Biol. Chem. 259: 3955-3961, 1984.

Jander, R.; Rauterberg, J.; Voss, B.; von Bassewitz, D. B.: A cysteine-rich collagenous protein from bovine placenta: isolation of its constituent polypeptide chains and some properties of the non-denatured protein. Europ. J. Biochem. 114: 17-25, 1981.

Jobsis, G. J.; Bolhuis, P. A.; Boers, J. M.; Baas, F.; Wolterman, R. A.; Hensels, G. W.; de Visser, M.: Genetic localization of Bethlem myopathy. Neurology 46: 779-782, 1996.

Jobsis, G. J.; Keizers, H.; Vreijling, J. P.; de Visser, M.; Speer, M. C.; Wolterman, R. A.; Baas, F.; Bohlhuis, P. A.: Type VI collagen mutations in Bethlem myopathy, an autosomal dominant myopathy with contractures. Nature Genet. 14: 113-115, 1996.

Justice, M. J.; Siracusa, L. D.; Gilbert, D. J.; Heisterkamp, N.; Groffen, J.; Chada, K.; Silan, C. M.; Copeland, N. G.; Jenkins, N. A.: A genetic linkage map of mouse chromosome 10: localization of eighteen molecular markers using a single interspecific backcross. Genetics 125: 855-866, 1990.

Klewer, S. E.; Krob, S. L.; Kolker, S. J.; Kitten, G. T.: Expression of type VI collagen in the developing mouse heart. Dev. Dyn. 211: 248-255, 1998.

Lamande, S. R.; Bateman, J. F.; Hutchison, W.; McKinlay Gardner, R. J.; Bower, S. P.; Byrne, E.; Dahl, H.-H. M.: Reduced collagen VI causes Bethlem myopathy: a heterozygous COL6A1 nonsense mutation results in mRNA decay and functional haploinsufficiency. Hum. Molec. Genet. 7: 981-989, 1998.

Murata, K.; Motoyama, T.; Suka, M.; Ohno, M.; Kuboki, Y.: High production of type VI collagen in multiple fibromatosis with multiple articular dysplasia. Biochem. Biophys. Res. Commun. 147: 275-281, 1987.

Petersen, M. B.; Slaugenhaupt, S. A.; Lewis, J. G.; Warren, A. C.; Chakravarti, A.; Antonarakis, S. E.: A genetic linkage map of 27 markers on human chromosome 21. Genomics 9: 407-419, 1991.

Roest, P. A. M.; Roberts, R. G.; Sugino, S.; van Ommen, G.-J. B.; den Dunnen, J. T.: Protein truncation test (PTT) for rapid detection of translation-terminating mutations. Hum. Molec. Genet. 2: 1719-1721, 1993.

Trueb, B.; Winterhalter, K. H.: Type VI collagen is composed of a 200 kD subunit and two 140 kD subunits. EMBO J. 5: 2815-2819, 1986.

Weil, D.; Mattei, M.-G.; Passage, E.; Van Cong, N.; Pribula-Conway, D.; Mann, K.; Deutzmann, R.; Timpl, R.; Chu, M.-L.: Cloning and chromosomal localization of human genes encoding the three chains of type VI collagen. Am. J. Hum. Genet. 42: 435-445, 1988.

BONE DYSPLASIA WITH MEDULLARY FIBROSARCOMA (*112250)

JJM NOTE

Continued follow-up of the families may clarify the disorder. The stored DNA allowed mapping of the gene locus.

Arnold (1973) described several generations of a Vermont and New York kindred demonstrating multiple areas of necrosis in the diaphyses of the large tubular bones. The radiographic appearance of this skeletal condition resembled radiation osteitis, a highly premalignant condition; however, no source of radiation exposure was found in this family. Medullary fibrosarcoma, an uncommon bone tumor, was noted in 4 of the 12 affected members. Death had occurred from widespread metastases at ages varying from 23 to 48 years. Occurrence of fibrosarcoma in idiopathic bone infarcts (Furey et al., 1960) and in an infarct in a caisson worker (Dorfman et al., 1966) has been reported. Hardcastle et al. (1986) gave follow-up information on the original American family and reported 2 other families, one English and the other Australian. They could find no reports of any hereditary or acquired condition similar to that in these 3 families. They suggested that the malignant change should be labeled 'malignant fibrous histiocytoma' rather than fibrosarcoma because the tumors were markedly aggressive. The malignancy occurred generally in the second to fifth decades of life. They defined the skeletal dysplasia as a diaphyseal medullary stenosis with overlying cortical bone thickening. The occurrence with minimal trauma was emphasized.

Martignetti et al. (1997) conducted a linkage study in 3 multigenerational kindreds with this disorder using microsatellite markers. With 2-point linkage analysis, they obtained a maximum lod score of 5.53 at theta = 0.03 with marker D9S171. Haplotype analysis of recombinants with available polymorphic markers further defined the locus, which they referred to as HBD (for 'hereditary bone dysplasia'), to a 3-cM region on 9p22-p21. This region contains a number of tumorigenic genes and cell-cycle and growth regulating genes.

Arnold, W. H.: Hereditary bone dysplasia with sarcomatous degeneration. Ann. Intern. Med. 78: 902-906, 1973.

Dorfman, H. D.; Norman, A.; Wolff, H.: Fibrosarcoma complicating bone infarction in a caisson worker. J. Bone Joint Surg. 48A: 528-532, 1966.

Furey, J. G.; Ferrer-Torells, M.; Reagan, J. W.: Fibrosarcoma arising at the site of bone infarcts. J. Bone Joint Surg. 42A: 802-810, 1960.

Hardcastle, P.; Nade, S.; Arnold, W.: Hereditary bone dysplasia with malignant change: report of three families. J. Bone Joint Surg. 68A: 1079-1089, 1986.

Martignetti, J. A.; Desnick, R. J.; Norton, K.; Hardcastle, P.; Nade, S.; Gelb, B. D.: Genetic linkage of hereditary bone dysplasia with malignant changes to chromosome 9p21-22. (Abstract) Am. J. Hum. Genet. 61 (suppl.): A284 only, 1997.

CHERUBISM (*118400)

JJM NOTE

This entity is included as a tumor, even though the process may not be neoplastic.

Swelling of the lower face begins around the third or fourth year of life and progresses until the late teens. The enlargement may be exaggerated by enlargement of submandibular lymph nodes. X-ray reveals multilocular cystic changes in the mandible and maxilla and often in the anterior ends of the ribs. Though clinical swelling usually abates by the third decade, radiographic changes commonly persist into the fourth decade. The condition must be differentiated from Caffey disease (114000) in which the x-ray appearance is different and involvement of the skeleton, e.g., the tibia, is more widespread. It is, like Caffey disease, a benign self-limited condition. The disorder has also been called familial benign giant-cell tumor of the jaw, familial multilocular cystic disease of the jaw, etc. Jones (1965) pointed out that lack of signs or history in either parent does not exclude the possibility of one's being affected. In one of his cases (he was the first to describe the entity), the disorder would not have been discovered, or even suspected, were it not that x-rays were made in childhood in a deliberate search for the entity because of its occurrence in other members of the family. Salinas et al. (1983) reported 2 cases of cherubism with multilocular cystic lesions of the ribs in addition to those of the mandible. In 1 of the patients, biopsy of both the jaw and the rib lesions showed numerous multinucleated giant cells in cellular fibrous tissue.

Quan et al. (1995) described cherubism in association with mental retardation due to mosaicism for expansion and deletion of the FMR1 CGG repeat, i.e., the fragile X syndrome (309550). Although these were probably independent mutations, Quan et al. (1995) pointed out the peculiarities of the inheritance of cherubism, which has been thought to be an autosomal dominant: twice as many males are affected as females and, while penetrance in males is 100%, penetrance in females is only 50-70%.

See also Anderson and McClendon (1962); Burland (1962); Khosla and Korobkin (1970); Peters (1979); Salzano and Ebling (1966); Thompson (1959)

Anderson, D. E.; McClendon, J. L.: Cherubism-hereditary fibrous dysplasia of the jaws. I. Genetic considerations. Oral Surg. 15 (suppl. 2): 5-16, 1962.

Burland, J. G.: Cherubism: familial bilateral osseous dysplasia of the jaws. Oral Surg. 15 (suppl. 2): 43-68, 1962.

Jones, W. A.: Cherubism: a thumbnail sketch of its diagnosis and a conservative method of treatment. Oral Surg. 20: 648-653, 1965.

Khosla, V. M.; Korobkin, M.: Cherubism. Am. J. Dis. Child. 120: 458-461, 1970.

Peters, W. J. N.: Cherubism: a study of twenty cases from one family. Oral Surg. 47: 307-311, 1979.

Quan, F.; Grompe, M.; Jakobs, P.; Popovich, B. W.: Spontaneous deletion in the FMR1 gene in a patient with fragile X syndrome and cherubism. Hum. Molec. Genet. 4: 1681-1684, 1995.

Salinas, C. F.; Bradford, B. F.; Laden, S. A.; Neville, B. W.: Cherubism associated with rib anomalies. (Abstract) Proc. Greenwood Genet. Center 2: 129-130, 1983.

Salzano, F. M.; Ebling, H.: Cherubism in a Brazilian kindred. Acta Genet. Med. Gemellol. 15: 296-301, 1966.

Thompson, N.: Cherubism: familial fibrous dysplasia of the jaws. Brit. J. Plast. Surg. 12: 89-103, 1959.

OSTEOID OSTEOMA (259550)

JJM NOTE

Surely there are other instances that have not been reported.

Kaye and Arnold (1977) observed almost simultaneous onset of symptoms of osteoid osteoma in 2 brothers, aged 17 and 12 years. A viral or other etiology was suggested.

Kaye, J. J.; Arnold, W. D.: Osteoid osteoma in siblings: case reports. Clin. Orthop. 126: 273-275, 1977.

OSTEOMA OF MIDDLE EAR (259650)

JJM NOTE

What has happened to the family in the intervening three decades?

Thomas (1964) reported bilateral osteoma of the middle ear in a 10-year-old boy and unilateral osteoma in his 6-year-old sister. There was no history of deafness in other relatives.

Thomas, R.: Familial osteoma of the middle ear. J. Laryng. 78: 805-807, 1964.

OSTEOMA OF MANDIBLE (166400)

JJM NOTE

This entity is another lead that deserves research follow-up. Could the original family be located?

Multiple smoothly outlined globoid osteomas occur on the jaw in Gardner syndrome (175100). Whether this tumor ever occurs as an inherited trait independent of intestinal polyps and other bony and soft-tissue tumors is not clear. Frangenheim (1914) described this type of tumor in a father and 3 of his children but intestinal polyps were not excluded.

Frangenheim, P.: Familiaere Hyperostosen der Kiefer. Beitr. Klin. Chir. 90: 139-152, 1914.

9. INTEGUMENTARY SYSTEM

Breast Tumors

CANCER OF THE BREAST, FAMILIAL (#114480)

JJM NOTE

In October 1998 querying OMIM on "breast cancer" produced 111 hits. By inspection, some 28 represent clinical entities of which breast cancer is a part. Some are falsely positive hits. The remainder are oncogenes, cellular phenotypes, proteins, and so forth, seen to date as tumor markers or from in vitro studies of breast cancer tissue or cell lines; any of the other 50 or so genes might eventually account for hereditary predisposition to breast cancer. The two dozen seen as germline predispositions could explain an unknown portion of breast cancer, even sporadic cases. Ten of the entries are clustered in this section of the *Catalog*; others can be located via the index.

Description

A number sign (#) is used with this entry because of evidence that mutation at more than one locus can be involved in different families or even in the same case. These loci include BRCA1 (113705) on 17q and BRCA2 (600185) on 13q.

Breast cancer (referring to mammary carcinoma, not mammary sarcoma) is histopathologically and almost certainly etiologically and genetically heterogeneous. Important genetic factors have been indicated by familial occurrence and bilateral involvement. At least one causative mutant gene (BRCA1) has been identified on chromosome 17q by linkage analysis.

Clinical Features

Cady (1970) described a family in which 3 sisters had bilateral breast cancer. Together with reports in the literature, this suggested to him the existence of families with a particular tendency to early-onset, bilateral breast cancer. The genetic basis might, of course, be multifactorial.

Anderson (1974) concluded that the sisters of women with breast cancer whose mothers also had breast cancer have a risk 47 to 51 times that in control women; a revised estimate was 39 times (Anderson, 1976). The disease in these women usually developed before menopause, was often bilateral, and seemed to be associated with ovarian function. About 30% of daughters with early-onset, bilateral breast cancer inherited the susceptibility. The risk of breast cancer to women with affected relatives is higher when the diagnosis is made at an early age and when the disease is bilateral. Ottman et al. (1983) provided tables that give the cumulative risk of breast cancer to mothers and sisters at various ages. The highest risk group is sisters of premenstrual probands with bilateral disease. Among the sisters of women with breast cancer, Anderson and Badzioch (1985) found the highest lifetime risks when the proband had bilateral disease, an affected mother (25 +/- 7.2%), or an affected sister (28 +/- 11%). The risks were reduced to 18 +/- 3.3% and 14 +/- 2.6%, respectively, with unilateral disease. An early example of familial breast cancer was provided by Broca (1866). According to the pedigree drawn by Lynch (1976), 10 women in 4 generations of the family of Broca's wife died of breast cancer. Eisinger et al. (1998) called attention to an even earlier report of hereditary breast cancer by Le Dran (1757), who related the experience of a colleague in Avignon who had diagnosed a 19-year-old nun with cancer of the right breast. The patient refused a mastectomy not only because of the pain of surgery, but also because of a belief that the operation would be futile. Her grandmother and a grandmaternal uncle died with breast cancer, and she was convinced that this malady was hereditary and that 'her blood was corrupted by a cancerous ferment natural to her family.'

Two families with an extraordinary incidence of male breast cancer and father-to-son transmission of same was reported by Everson et al. (1976). They found a suggestion of elevated urinary estrogen in 3 of the affected males. Teasdale et al. (1976) described breast cancer in 2 brothers and in a daughter of 1 brother. Kozak et al. (1986) reported breast cancer in 2 related males, an uncle and nephew. In this family and in several reported families with male breast cancer, Kozak et al. (1986) found women in the same family with breast cancer.

Soft tissue sarcomas are associated with breast cancer in the Li-Fraumeni syndrome (151623). Mulvihill (1982) used the term cancer family syndrome of Lynch (114400) for the association of colon and endometrial carcinoma and other neoplasms including breast cancer.

Seltzer et al. (1990) concluded that dermatoglyphics can help in the identification of women either with or at risk for breast cancer. They found that the presence of 6 or more whorls is associated in a statistically significant manner with breast cancer.

Marger et al. (1975) presented the cases of 2 brothers with breast cancer and reviewed the courses of 28 other previously unreported male patients. In one of the brothers, breast cancer was preceded by prostatic cancer and estrogen administration, raising the possibility that the breast cancer was a metastatic deposit. The possibility of prostatic metastases was raised in 2 other patients. Demeter et al. (1990) reported breast cancer in a 64-year-old man who had had bilateral gynecomastia since childhood. His maternal grandfather had been found to have adenocarcinoma of the breast at the age of 65. His maternal grandmother had radical mastectomy for breast cancer at the age of 66 and 2 years later underwent radiation therapy for rib metastases. The proband's sister developed breast cancer at the age of 31 years and despite aggressive therapy died 1 year later with extensive metastases.

Hauser et al. (1992) reported a family in which 2 females and 2 males in 2 generations had breast cancer. Two females in the family had prophylactic bilateral mastectomy at a young age. One male developed a left breast mass and axillary node at age 59 and died of metastatic disease at age 62. His paternal uncle presented at age 57 years with bleeding from his right breast. Biopsy suggested Paget disease of the breast and he underwent mastectomy. He subsequently died at age 75 years of prostatic carcinoma. He had a daughter who developed breast cancer at age 27 years and died at age 30 with disseminated disease, and a son who developed infiltrating grade 4 adenocarcinoma of the breast at age 54.

Other Features

Chang et al. (1987) showed that the noncancerous skin fibroblasts of members of a family with Li-Fraumeni syndrome (which show resistance to the killing effect of ionizing radiation) have a 3- to 8-fold elevation in expression of the MYC oncogene (190080) and an apparent activation of the RAF1 gene (164760). Normal fetal and adult skin fibroblasts show distinctive migratory behavior when plated on 3-dimensional collagen gels.

Haggie et al. (1987) found that skin fibroblasts from 13 of 15 patients with hereditary breast cancer showed fetal-like behavior compared with only 1 of 12 age-matched healthy controls. In addition, 10 of 15 first-degree relatives of patients with hereditary breast cancer showed a fetal-like fibroblast phenotype, compared with none of 7 surgical controls.

Inheritance

Petrakis (1977) listed the evidence for a genetic role in breast cancer as follows: 1) family history of breast cancer, especially bilateral breast cancer; 2) marked differences in rates between certain racial groups (lower in Orientals); 3) lack of major change in incidence over many years despite dramatic decline in other cancers; 4) concordance in monozygotic twins; and 5) concordance of laterality in closely related persons. Lynch et al. (1984) found evidence consistent with a hereditary breast cancer syndrome in 5% of 225 consecutively ascertained patients with verified breast cancer. From a maximum-likelihood mendelian model, the frequency of the susceptibility allele was 0.0006 in the general population, and the lifetime risk of breast cancer was 0.82 among susceptible women and 0.08 among women without the susceptibility allele. They concluded that inherited susceptibility affected only 4% of the families in the sample; multiple cases of this relatively common disease occurred in other families by chance. They pictured an extended pedigree with 14 cases of breast cancer, 3 of them in men.

The Danish twin registry (Holm et al., 1980) had 5 out of 45 MZ twins and 4 out of 77 DZ twins concordant for breast cancer; heritability was calculated at 0.3-0.4.

From complex segregation analysis of 200 Danish breast cancer pedigrees, Williams and Anderson (1984) concluded that the distribution of cases was compatible with transmission of an autosomal dominant gene. Newman et al. (1988) used complex segregation analysis to investigate patterns of breast cancer occurrence in 1,579 nuclear families. They concluded that an autosomal dominant model with a highly penetrant susceptibility allele fully explains disease clustering.

Iselius et al. (1992) reanalyzed the Danish breast cancer data collected by Jacobsen (1946), using morbid risks that incorporate mortality due to breast cancer. They interpreted the results to favor a dominant gene for familial breast cancer. No evidence of heterogeneity was found. Cases with bilateral breast cancer and males with breast cancer all belonged to families favoring a major gene. Of the cancer sites frequently reported to be associated with familial breast cancer, only ovarian cancer was significant in this study.

Houlston et al. (1992) showed that the risk of breast cancer increased progressively in inverse relationship to the age of the index patient. First-degree relatives of patients with bilateral breast cancer had a 6.43-fold increase in risk. Houlston et al. (1992) estimated that the genetic contribution to overall lifetime liability to breast cancer in relatives declined with increasing age of onset of breast cancer in the index case from 37% at 20 years to 8% by 45 years. In Iceland, Tulinius et al. (1992) likewise found that early onset and bilaterality of breast cancer increased the risk to relatives. In an analysis of a prospective cohort study, Sellers et al. (1992) found that the increase in the risk of breast cancer associated with a high waist-to-hip ratio (the circumference of the waist divided by that of the hips), low parity, or greater age at first pregnancy was more pronounced among women with a family history of breast cancer. They concluded that there are etiologic differences between familial breast cancer and the sporadic form.

Mapping

For linkage between glutamate-pyruvate transaminase (138200) and breast cancer, King et al. (1980) found a lod score of 1.84 for 6 families showing linkage and 1.43 for all 11 breast cancer families studied. In Mormon breast cancer pedigrees, however, McLellan et al. (1984) obtained a cumulative lod score of -3.86 for breast cancer and GPT, thus eliminating this possible linkage.

Pathak and Goodacre (1986) found reciprocal translocations involving 1q21 and chromosomes 3, 5, 10, 11, and 12. Chen et al. (1989) demonstrated loss of heterozygosity (LOH) in the region 1q23-q32.

Because of the possibility that inherited rare alleles at the HRAS locus (190020) might be associated with susceptibility to breast cancer, Hall et al. (1990) studied linkage to markers on 11p in 12 high-risk families. Linkage could be excluded within 17 cM of HRAS; the lod score for close linkage to HRAS was -19.9.

Goldstein et al. (1989) found a suggestion of linkage to acid phosphatase (ACP1; 171500); maximum lod score = 1.01 at theta = 0.001. Very close linkage was excluded between breast cancer and ABO, GC, GPT, MNS, and PGM1. By linkage studies, Bowcock et al. (1990) excluded the retinoblastoma gene (180200) and 13q in general as the site of the primary lesion. Abnormality there was sought because of observation of LOH of alleles on 13q in some ductal breast tumors and because 2 breast cancer lines had been found to have an alteration in the retinoblastoma gene. Narod and Amos (1990) analyzed the effects of phenocopies and genetic heterogeneity on the demonstration of linkage between a putative cancer susceptibility gene and polymorphic DNA markers.

See 113705 for a review of the extensive linkage studies that demonstrate the existence of a gene for susceptibility to early-onset breast cancer (BRCA1) on 17q.

Molecular Genetics

A previously reported loss of alleles at the HRAS locus, located at 11p14, in about 20% of tumors was confirmed by Mackay et al. (1988). Comparing tumor and blood leukocyte DNA from a consecutive series of patients with primary breast cancer, Mackay et al. (1988) found that 61% of the tumors had allele loss demonstrated with a probe located at 17p13.3.

Coles et al. (1990) mapped regions of LOH on chromosome 17 by comparing DNA of paired tumor and blood leukocyte samples. They confirmed a high frequency of LOH on 17p, where 2 distinct regions of LOH were identified in bands p13.3 and p13.1. The latter probably involves the structural gene TP53 (191170). The frequency of LOH was higher, however, at 17p13.3, and there was no correlation between allele loss at the 2 sites. Since LOH at 17p13.3 was associated with overexpression of p53 mRNA, Coles et al. (1990) suggested the existence of a gene some 20 megabases telomeric of TP53 that regulates its expression; see 113721. They concluded that lesions of this regulatory gene are involved in the majority of breast cancers. Devilee et al. (1991) reported LOH data.

Davidoff et al. (1991) found that in 11 (22%) of 49 primary invasive human breast cancers, widespread overexpression of p53 was indicated by immunohistochemical staining. The p53 gene was directly sequenced in 7 of the tumors with elevated levels of protein, and in each case a mutation that altered the coding sequence for p53 was found in a highly conserved region of the gene. Whereas 4 of these tumors contained only a mutant p53 allele, the other 3 exhibited coding sequences from both a mutant and a wildtype allele. Six tumors that were deleted at or near the p53 locus but did not express high levels of the protein were sequenced and all retained a wildtype p53 allele. This was interpreted as indicating that overexpression of the p53 protein, not allelic loss, was associated with mutation of the p53 gene.

Animal Model

Parallels may exist with breast cancer in mice, which has long been studied from the viewpoint of genetic-viral etiology and pathogenesis. This story begins with Bittner's 'milk agent,' originally discovered by Bittner (1936); using reciprocal matings between high tumor and low tumor strains, the Jackson Laboratory staff showed in 1933 that the tumor incidence in F1 females was a function of the strain of the mother. Virologists demonstrated that the mouse mammary tumor virus (MMTV, also called MuMTV) is indeed transmitted through the milk and is an RNA virus seen in its mature form as the B particle. This was the first virus universally accepted in this country as a cancer-causing virus. Some mouse strains have been shown to carry a potent MMTV transmitted in milk and also in the egg and sperm (see review by Heston and Parks, 1977). Strains of mice purged of the MMTV by foster-nursing the young on a clean strain still show a low incidence of breast cancer developing at a late age. By introducing the cancer-enhancing gene A(vy), the incidence could be raised to 90%; however, the agent was not transmitted through the milk but by both eggs and sperm.

In one strain developed by Muhlbock (1965), Bentvelzen (1972) demonstrated that the high incidence of mammary tumors was caused by an MMTV transmitted in milk, eggs, and sperm. Particles resembling B-type retroviruses have been identified in human milk (Moore et al., 1971); MMTV-related RNA has been found in some breast cancers (Axel et al., 1972) and a breast cancer cell line that releases retrovirus-like particles has been established (McGrath et al., 1974). Callahan et al. (1982) and Westley and May (1984) demonstrated sequences in normal human DNA that appear to be homologous to endogenous retroviral sequences. By transfection of NIH 3T3 mouse cells, Lane et al. (1981) demonstrated a transforming gene in a human mammary tumor cell line (MCF-7). See 164820 for information on the human homolog of the putative mammary tumor oncogene.

See also Anderson (1972); Armstrong and Davies (1978); Lynch (1981); Lynch et al. (1985); Miyagi et al. (1992)

Anderson, D. E.: A genetic study of human breast cancer. J. Nat. Cancer Inst. 48: 1029-1034, 1972.

Anderson, D. E.: Genetic study of breast cancer: identification of a high risk group. Cancer 34: 1090-1097, 1974.

Anderson, D. E.: Genetic predisposition to breast cancer. Recent Results Cancer Res. 57: 10-20, 1976.

Anderson, D. E.; Badzioch, M. D.: Risk of familial breast cancer. Cancer 56: 383-387, 1985.

Armstrong, A. E.; Davies, J. M.: Familial breast cancer: report of a family pedigree. Brit. J. Cancer 37: 294-307, 1978.

Axel, R.; Schlom, J.; Spiegelman, S.: Presence in human breast cancer of RNA homologous to mouse mammary tumour virus RNA. Nature 235: 32-36, 1972.

Bentvelzen, P.: Hereditary infection with mammary tumor viruses in mice.In: Emmelot, P.; Bentvelzen, P.: RNA Viruses and Host Genome in Oncogenesis. Amsterdam: North Holland (pub.) 1972.

Bittner, J. J.: Some possible effects of nursing on the mammary gland tumor incidence in mice. Science 84: 162 only, 1936.

Bowcock, A. M.; Hall, J. M.; Hebert, J. M.; King, M.-C.: Exclusion of the retinoblastoma gene and chromosome 13q as the site of a primary lesion for human breast cancer. Am. J. Hum. Genet. 46: 12-17, 1990.

Broca, P. P.: Traite des Tumeurs. Paris: P. Asselin (pub.) 1: 1866. Pp. 80 only.

Cady, B.: Familial bilateral cancer of the breast. Ann. Surg. 172: 264-272, 1970.

Callahan, R.; Drohan, W.; Tronick, S.; Schlom, J.: Detection and cloning of human DNA sequences related to the mouse mammary tumor virus genome. Proc. Nat. Acad. Sci. 79: 5503-5507, 1982.

Chang, E. H.; Pirollo, K. F.; Zou, Z. Q.; Cheung, H.-Y.; Lawler, E. L.; Garner, R.; White, E.; Bernstein, W. B.; Fraumeni, J. W., Jr.; Blattner, W. A.: Oncogenes in radioresistant, noncancerous skin fibroblasts from a cancer-prone family. Science 237: 1036-1039, 1987.

Chen, L.-C.; Dollbaum, C.; Smith, H. S.: Loss of heterozygosity on chromosome 1q in human breast cancer. Proc. Nat. Acad. Sci. 86: 7204-7207, 1989.

Coles, C.; Thompson, A. M.; Elder, P. A.; Cohen, B. B.; Mackenzie, I. M.; Cranston, G.; Chetty, U.; Mackay, J.; Macdonald, M.; Nakamura, Y.; Hoyheim, B.; Steel, C. M.: Evidence implicating at least two genes on chromosome 17p in breast carcinogenesis. Lancet 336: 761-763, 1990.

Davidoff, A. M.; Humphrey, P. A.; Iglehart, J. D.; Marks, J. R.: Genetic basis for p53 overexpression in human breast cancer. Proc. Nat. Acad. Sci. 88: 5006-5010, 1991.

Demeter, J. G.; Waterman, N. G.; Verdi, G. D.: Familial male breast carcinoma. Cancer 65: 2342-2343, 1990.

Devilee, P.; van Vliet, M.; van Sloun, P.; Kuipers Dijkshoorn, N.; Hermans, J.; Pearson, P. L.; Cornelisse, C. J.: Allelotype of human breast carcinoma: a second major site for loss of heterozygosity is on chromosome 6q. Oncogene 6: 1705-1711, 1991.

Eisinger, F.; Sobol, H.; Serin, D.; Whorton, J. C.: Hereditary breast cancer, circa 1750. (Letter) Lancet 351: 1366 only, 1998.

Everson, R. B.; Li, F. P.; Fraumeni, J. F., Jr.; Fishman, J.; Wilson, R. E.; Stout, D.; Norris, H. J.: Familial male breast cancer. Lancet I: 9-12, 1976.

Goldstein, A. M.; Haile, R. W.; Spence, M. A.; Sparkes, R. S.; Paganini-Hill, A.: A genetic epidemiologic investigation of breast cancer in families with bilateral breast cancer. II. Linkage analysis. Clin. Genet. 36: 100-106, 1989.

Haggie, J. A.; Sellwood, R. A.; Howell, A.; Birch, J. M.; Schor, S. L.: Fibroblasts from relatives of patients with hereditary breast cancer show fetal-like behaviour in vitro. Lancet I: 1455-1457, 1987.

Hall, J. M.; Huey, B.; Morrow, J.; Newman, B.; Lee, M.; Jones, E.; Carter, C.; Buehring, G. C.; King, M.-C.: Rare HRAS alleles and susceptibility to human breast cancer. Genomics 6: 188-191, 1990.

Hauser, A. R.; Lerner, I. J.; King, R. A.: Familial breast cancer. (Letter) Am. J. Med. Genet. 44: 839-840, 1992.
Heston, W. E.; Parks, W. P.: Mammary tumors and mammary tumor virus expression in hybrid mice of strains C57BL and GR. J. Exp. Med. 146: 1206-1220, 1977.
Holm, N. V.; Hauge, M.; Harvald, B.: Etiologic factors of breast cancer elucidated in a study of unselected twins. J. Nat. Cancer Inst. 65: 285-298, 1980.
Houlston, R. S.; McCarter, E.; Parbhoo, S.; Scurr, J. H.; Slack, J.: Family history and risk of breast cancer. J. Med. Genet. 29: 154-157, 1992.
Iselius, L.; Littler, M.; Morton, N.: Transmission of breast cancer—a controversy resolved. Clin. Genet. 41: 211-217, 1992.
Jacobsen, O.: Heredity and Breast Cancer. London: H. K. Lewis (pub.) 1946.
King, M.-C.; Go, R. C. P.; Elston, R. C.; Lynch, H. T.; Petrakis, N. L.: Allele increasing susceptibility to human breast cancer may be linked to the glutamate-pyruvate transaminase locus. Science 208: 406-408, 1980.
Kozak, F. K.; Hall, J. G.; Baird, P. A.: Familial breast cancer in males: a case report and review of the literature. Cancer 58: 2736-2739, 1986.
Lane, M.-A.; Sainten, A.; Cooper, G. M.: Activation of related transforming genes in mouse and human mammary carcinomas. Proc. Nat. Acad. Sci. 78: 5185-5189, 1981.
Le Dran, H.: Memoire avec un precis de plusieurs observations sur le cancer. Mem. Acad. R. Chir. 3: 1-54, 1757.
Lynch, H. T.: Introduction to cancer genetics.In: Lynch, H. T.: Cancer Genetics. Springfield, Ill.: Charles C Thomas (pub.) 1976. Pp. 3-31.
Lynch, H. T.: Genetics and Breast Cancer. New York: Van Nostrand-Reinhold (pub.) 1981.
Lynch, H. T.; Albano, W. A.; Danes, B. S.; Layton, M. A.; Kimberling, W. J.; Lynch, J. F.; Cheng, S. C.; Costello, K. A.; Mulcahy, G. M.; Wagner, C. A.; Tindall, S. L.: Genetic predisposition to breast cancer. Cancer 53: 612-622, 1984.
Lynch, H. T.; Katz, D. A.; Bogard, P. J.; Lynch, J. F.: The sarcoma, breast cancer, lung cancer, and adrenocortical carcinoma syndrome revisited. Am. J. Dis. Child. 139: 134-136, 1985.
Mackay, J.; Steel, C. M.; Elder, P. A.; Forrest, A. P. M.; Evans, H. J.: Allele loss on short arm of chromosome 17 in breast cancers. Lancet II: 1384-1385, 1988.
Marger, D.; Urdaneta, N.; Fischer, J. J.: Breast cancer in brothers: case reports and a review of 30 cases of male breast cancer. Cancer 36: 458-461, 1975.
McGrath, C. M.; Grant, P. M.; Soule, H. D.; Glancy, T.; Rich, M. A.: Replication of oncornavirus-like particle in human breast carcinoma cell line, MCF-7. Nature 252: 247-250, 1974.
McLellan, T.; Cannon, L. A.; Bishop, D. T.; Skolnick, M. H.: The cumulative lod score between a breast cancer susceptibility locus and GPT is -3.86. (Abstract) Cytogenet. Cell Genet. 37: 536-537, 1984.
Miyagi, M.; Inazawa, J.; Takita, K.; Nakamura, Y.: Cloning and characterization of an interstitial deletion at chromosome 11p15 in a sporadic breast cancer. Hum. Molec. Genet. 1: 705-708, 1992.
Moore, D. H.; Charney, J.; Kramarsky, B.; Lasfargues, E. Y.; Sarkar, N. H.; Brennan, M. J.; Burrows, J. H.; Sirsat, S. M.; Paymaster, J. C.; Vaidya, A. B.: Search for a human breast cancer virus. Nature 229: 611-615, 1971.
Muhlbock, O.: Note of a new inbred mouse strain GR/A. Europ. J. Cancer 1: 123-124, 1965.
Mulvihill, J. J.: Personal Communication. Bethesda, Md. 6/11/1982.
Narod, S. A.; Amos, C.: Estimating the power of linkage analysis in hereditary breast cancer. Am. J. Hum. Genet. 46: 266-272, 1990.
Newman, B.; Austin, M. A.; Lee, M.; King, M.-C.: Inheritance of human breast cancer: evidence for autosomal dominant transmission in high-risk families. Proc. Nat. Acad. Sci. 85: 3044-3048, 1988.
Ottman, R.; Pike, M. C.; King, M.-C.; Henderson, B. E.: Practical guide for estimating risk for familial breast cancer. Lancet II: 556-558, 1983.
Pathak, S.; Goodacre, A.: Specific chromosome anomalies and predisposition to human breast, renal cell, and colorectal carcinoma. Cancer Genet. Cytogenet. 19: 29-36, 1986.
Petrakis, N. L.: Genetic factors in the etiology of breast cancer. Cancer 39: 2709-2715, 1977.
Sellers, T. A.; Kushi, L. H.; Potter, J. D.; Kaye, S. A.; Nelson, C. L.; McGovern, P. G.; Folsom, A. R.: Effect of family history, body-fat distribution, and reproductive factors on the risk of postmenopausal breast cancer. New Eng. J. Med. 326: 1323-1329, 1992.
Seltzer, M. H.; Plato, C. C.; Fox, K. M.: Dermatoglyphics in the identification of women either with or at risk for breast cancer. Am. J. Med. Genet. 37: 482-488, 1990.
Teasdale, C.; Forbes, J. F.; Baum, M.: Familial male breast cancer. (Letter) Lancet I: 360-361, 1976.
Tulinius, H.; Sigvaldason, H.; Olafsdottir, G.; Tryggvadottir, L.: Epidemiology of breast cancer in families in Iceland. J. Med. Genet. 29: 158-164, 1992.
Westley, B.; May, F. E. B.: The human genome contains multiple sequences of varying homology to mouse mammary tumour virus DNA. Gene 28: 221-227, 1984.
Williams, W. R.; Anderson, D. E.: Genetic epidemiology of breast cancer: segregation analysis of 200 Danish pedigrees. Genet. Epidemiol. 1: 7-20, 1984.

BREAST CANCER 1; BRCA1 (*113705)

JJM NOTE

It will be quite interesting to reread the early generalizations about *BRCA1,* including the current dogma that 50% of hereditary breast cancer families are due to mutations in the gene. Its mapping and isolation are milestones in understanding and, one hopes, controlling this killer cancer, but, in truth, only a new starting line has been drawn for the racers, who include investigators, therapists, and women (and men) at risk and with breast cancer and their families. The lessons learned in applying *BRCA1* testing in the high- and average-risk populations will set the pattern for the individual and public health applications of other genes that predispose to adult-onset common disorders. For now, clinical use should be rendered by experts in the area, chiefly geneticists and genetic counselors.

Hall et al. (1990) studied 23 extended families with 146 cases of breast cancer. All were Caucasian and came from a variety of ancestries. The 329 participating relatives lived in 40 states of the United States, Puerto Rico, Canada, the United Kingdom, and Columbia. The families shared the epidemiologic features characteristic of familial, versus sporadic, breast cancer: younger age at diagnosis, frequent bilateral disease, and frequent occurrence of disease among men. Hall et al. (1990) found a lod score of 5.98 for linkage of breast cancer susceptibility in early-onset families to D17S74, which is located in band 17q21. Negative lod scores were found in families with late-onset disease. Likelihood ratios in favor of linkage heterogeneity among families ranged from 2000:1 to greater than 10(6):1 on the basis of multipoint analysis of 4 loci in the region of 17q21. Candidate genes in the region include HER2 (164870), which is thought to be identical to ERBB2; estradiol-17-beta-dehydrogenase (109684); a cluster of homeo box 2 genes (e.g., 142960); retinoic acid receptor alpha (180240); and INT4 (165330). In studies of 103 women from 20 kindreds that were selected for the presence of 2 first-degree relatives with breast cancer and of 31 control women, Skolnick et al. (1990) found, by 4-quadrant fine-needle breast aspirates, evidence of proliferative breast disease (PBD) in 35% of clinically normal female first-degree relatives of breast cancer cases and in 13% of controls. Genetic analysis suggested that genetic susceptibility caused both PBD, a precursor lesion, and breast cancer in these kindreds. The study supported the hypothesis that this susceptibility is responsible for a considerable proportion of breast cancer, including unilateral and postmenopausal breast cancer. Linkage analyses failed to show linkage with D17S74 in either early- or late-age onset.

The gene mapped by Hall et al. (1990) may be the same as that deduced by Claus et al. (1991). In a data-set based on 4,730 histologically confirmed breast cancer patients aged 20 to 54 years and on 4,688 controls, the latter group of workers presented evidence for the existence of a rare autosomal dominant allele ($q = 0.0033$) leading to increased susceptibility to breast cancer. The cumulative lifetime risk of breast cancer for women who carry the susceptibility allele was predicted to be approximately 92%, while the cumulative lifetime risk for noncarriers was estimated to be approximately 10%. Narod et al. (1991) investigated 5 large families with a hereditary predisposition to cancer of the breast and ovary. Three families showed linkage with the D17S74 marker used by Hall et al. (1990). For the largest family the lod score was 2.72 at a recombination fraction of 0.07. Narod et al. (1991) suggested that about 60% of breast cancer families have linkage of the susceptibility to the chromosome 17q locus. Lynch and Watson (1992) reported extension of the linkage work to 19 families, most of which showed the HBOC (hereditary breast-ovarian cancer) syndrome. In 70% of families, linkage to 17q was demonstrated. Lynch and Watson (1992) reported the first experience with genetic counseling and targeted management of patients demonstrated to be at risk for HBOC by use of multipoint linkage analysis in the largest and most informative of the kindreds studied to date. The single family provided a lod score of 3.03. In those persons shown by linkage to be at risk, they recommended completing their families before the age of 35 so that prophylactic oophorectomy could be performed at an early age. Margaritte et al. (1992) found that when account is made for the higher relative probability of sporadic rather than inherited disease for late-onset cases of breast cancer, later-onset families are much less informative and linkage heterogeneity based on age at onset is no longer significant. Furthermore, for the sample of families as a whole, linkage is significant at a recombination fraction in the 17q21 region. Although there is probably more than one gene for inherited breast cancer, age at onset may not be a reflection of this heterogeneity.

Hall et al. (1992) indicated that the proportion of older-onset breast cancer attributable to BRCA1 was not yet determinable, because both inherited and sporadic cases occur in older-onset families. Hall et al. (1992) found that the most closely linked marker in their repertoire was D17S579, a highly informative CA repeat polymorphism located at 17q21. There were no recombinants with inherited breast or ovarian cancer in 79 informative meioses in the 7 families with early-onset disease; lod score = 9.12 at 0 recombination. Sobol et al. (1992) also pointed to genetic heterogeneity of early-onset familial breast cancer; in an extensively affected family they found no evidence of linkage to markers on 17q. Goldgar et al. (1992) identified a Utah kindred in which the BRCA1 locus was linked to 17q markers with odds in excess of a million to one. The kindred included 170 descendents of 2 Utah pioneers of 1847, containing a total of 24 cancer cases (16 breast, 8 ovarian). The median age of onset was 48 for breast cancer and 53 for ovarian cancer. The penetrance of the BRCA1 gene was estimated to be 0.92 by age 70. Easton et al. (1993) reported the results of genetic linkage analysis in 214 families. In 15 accompanying papers, confirmatory evidence on the linkage was reported from Icelandic, Scottish, Dutch, Swedish, and other families including one African-American family. In Icelandic studies, Arason et al. (1993) suggested that male carriers of the BRCA1 gene may have an increased risk of prostatic cancer.

If the gene predisposing to breast cancer (and ovarian cancer) mapped to 17q12-q21 is a tumor suppressor gene, one would expect, based on the Knudson hypothesis, that tumors from affected family members would show loss of heterozygosity (LOH) affecting the wild-type chromosome. In 4 multiple case breast-ovarian cancer families, Smith et al. (1992) indeed found that in each of 9 tumors that showed allele loss, the losses were from the wildtype chromosome. Kelsell et al. (1993) found the same for each of 7 breast tumors from a single multi-affected breast/ovarian cancer pedigree. In the same family, they generated linkage data which, in combination with previously published information, suggested that the BRCA1 gene is contained in a region estimated to be 1 to 1.5 Mb long.

Because of the finding of genetic recombination between the BRCA1 locus and the gene for retinoic acid receptor alpha (180240), Simard et al. (1993) was able to exclude that candidate gene. BRCA1 and the gene for estradiol 17-beta-hydroxysteroid dehydrogenase II (109685) map to a 6-cM interval (between THRA1 and D17S579) and no recombination was observed between the 2 genes; however, direct sequencing of overlapping PCR products containing the entire EDH17B2 gene in 4 unrelated affected woman did not uncover any sequence variation other than previously described polymorphisms. They concluded, therefore, that mutations in the EDH17B2 gene are probably not responsible for hereditary breast-ovarian cancer syndrome. Kelsell et al. (1993) sequenced the two 17-beta-estradiol dehydrogenase genes (EDH17B1, EDH17B2), which had been suggested as candidate genes for BRCA1, in 4 members of the same family; no germline mutations were detected. (Actually, the EDH17B1 and EDH17B2 genes appear to map proximal to the BRCA1 locus.)

Piver et al. (1993) presented data suggesting that mucinous carcinomas of the ovary may be underrepresented in familial ovarian cancer. Mutations in the BRCA1 gene account for most families with the hereditary breast-ovarian cancer syndrome. To address whether or not there is an association between the presence of a BRCA1 mutation and the subtype of epithelial ovarian carcinoma, Narod et al. (1994) reviewed the histology of 49 ovarian cancers seen in 16 hereditary breast-ovarian cancer families shown to be linked to BRCA1 markers. Of the 49 cancers, 5 (10.2%) were mucinous. By haplotype analysis with 17q markers, they determined the BRCA1 carrier status of 40 of the cases; 36 occurred in women who were BRCA1 mutation carriers and 4 were sporadic in that they occurred in noncarriers. Only 2 of the 36 ovarian cancers found in BRCA1 carriers were mucinous, compared with 3 or 4 mucinous carcinomas observed in BRCA1 noncarriers. Albertsen et al. (1994) used simple sequence repeat (SSR) markers to construct a high-resolution genetic map of a 40-cM region around 17q21. For 5 of the markers, genotypes were 'captured' by using an ABI sequencing instrument and stored in a locally developed database as a step toward automated genotyping. In a second report, Albertsen et al. (1994) described construction of a physical map of a 4-cM region containing the BRCA1 gene. The map comprised a contig of 137 overlapping YACs and P1 clones, onto which they had placed 112 PCR markers. They localized more than 20 genes on the map, 10 of which had not been mapped to the region previously, and isolated 30 cDNA clones representing partial sequences of as yet unidentified genes. They failed to find any deleterious mutations on sequencing of 2 genes that lie within a narrow region defined by meiotic breakpoints in BRCA1 patients. O'Connell et al. (1994) developed a radiation hybrid map of the BRCA1 region as the basis of YAC cloning and pulsed field gel electrophoretic mapping of the candidate region for the BRCA1 gene.

Miki et al. (1994) identified a strong candidate for the BRCA1 gene by positional cloning methods. Probable predisposing mutations were detected in 5 of 8 kindreds thought to segregate BRCA1 susceptibility alleles. The mutations included an 11-bp deletion, a 1-bp insertion, a stop codon, a missense substitution, and an inferred regulatory mutation. The BRCA1 gene is expressed in numerous tissues, including breast and ovary, and encodes a predicted protein of 1,863 amino acids. The protein contains a zinc finger domain in its amino-terminal region, but is otherwise unrelated to previously described proteins. Futreal et al. (1994) extended the observations to studies of primary breast and ovarian tumors that show allele loss at the BRCA1 locus. Mutations were detected in 3 of 32 breasts and 1 of 12 ovarian carcinomas; all 4 mutations were germline alterations and occurred in cancers of early-onset type. These results were interpreted as indicating that mutation in the BRCA1 gene may not be critical to the development of most breast and ovarian cancers that arise in the absence of a mutant germline allele. This situation is unlike that in the APC gene (175100), which is involved in both hereditary polyposis coli and sporadic colorectal cancer, and that of some other genes involved in both familial and sporadic cancer.

Using single-strand conformation polymorphism (SSCP) analysis on PCR-amplified genomic DNA in an analysis of 50 probands with a family history of breast and/or ovarian cancer, Castilla et al. (1994) found 8 putative disease-causing alterations: 4 frameshift mutations, 2 nonsense mutations, and 2 missense mutations. The data were considered consistent with a tumor suppressor model. The heterogeneity of mutations, coupled with the large size of the gene, indicated that clinical application of BRCA1 mutation testing would be technically challenging. Simard et al. (1994) identified mutations in the BRCA1 gene in 12 of 30 Canadian families. Six frameshift mutations accounted for all 12 mutant alleles, including nucleotide insertions (2 mutations) and deletions (4 mutations). The same 1-bp insertion mutation in codon 1,755 was found in 4 independent families, whereas 4 other families shared a 2-bp deletion mutation in codons 22 to 23. These families were not known to be related, but haplotype analysis suggested that the carriers of each of these mutations had common ancestors. Friedman et al. (1994) likewise used SSCP analysis and direct sequencing to identify 9 different mutations in 10 families. The mutations in 7 instances led to protein truncation at sites throughout the gene. A missense mutation, which occurred independently in 2 families, led to loss of a cysteine in the zinc-binding domain. An intronic single basepair substitution destroyed an acceptor site and activated a cryptic splice site, leading to a 59-bp insertion and chain termination. In 4 families with both breast and ovarian cancer, chain termination mutations were found in the N-terminal half of the protein.

In 4 of 47 sporadic ovarian cancers, Merajver et al. (1995) examined tumor DNAs by SSCP and found 4 somatic mutations in the BRCA1 gene; all 4 had loss of heterozygosity at a BRCA1 intragenic marker. The findings supported a tumor-suppressor mechanism for BRCA1; somatic mutation on one chromosome and LOH on the other may result in inactivation of BRCA1 in some sporadic ovarian cancers. In sporadic breast cancer, Thompson et al. (1995) found that BRCA1 mRNA levels are markedly decreased during the transition from carcinoma in situ to invasive cancer. Thompson et al. (1995) found that experimental inhibition of BRCA1 expression with antisense oligonucleotides produced accelerated growth of normal and malignant mammary cells but had no effect on nonmammary epithelial cells. They interpreted these results as indicating that BRCA1 may normally serve as a negative regulator of mammary epithelial cell growth and that this function is compromised in breast cancer either by direct mutation or by alterations in gene expression.

Bennett et al. (1995) found that the mouse Brca1 gene shares 75% identity of the coding region with the human sequence at the nucleotide level, whereas the predicted amino acid identity was only 58%. By an intersubspecific backcross using a DNA sequence variant in the Brca1 locus, they mapped the gene to distal mouse chromosome 11 in a region of extensive homology of synteny to human chromosome 17. Schrock et al. (1996) likewise mapped the Brca1 gene to mouse chromosome 11, specifically 11D. De Gregorio et al. (1996) also mapped the gene to mouse chromosome 11.

Cornelis et al. (1995) sought criteria that would identify breast cancer families with a high prior probability that the tumors were caused by a BRCA1 mutation. They performed a linkage study in 59 consecutively collected Dutch breast cancer families, including 16 families with at least 1 case of ovarian cancer. They used a family intake cutoff of at least 3 first-degree relatives with breast and/or ovarian cancer at any age. Significant evidence for linkage was found only among the 13 breast cancer families with a mean age at diagnosis of less than 45 years. An unexpectedly low proportion of breast-ovarian cancer families were estimated to be linked to BRCA1, which could be due to a founder effect in the Dutch population. Cornelis et al. (1995) proposed that, during an interim period, BRCA1 mutation testing be offered only to families with a strong positive family history for early onset breast and/or ovarian cancer. Tonin et al. (1995) studied 26 Canadian families with hereditary breast or ovarian cancer for linkage to markers flanking BRCA1. Of the 15 families that contained cases of ovarian cancer, 94% were estimated to be linked to BRCA1. In contrast, there was no overall evidence of linkage in the group of 10 families with breast cancer without ovarian cancer.

Since more than 75% of the reported mutations in the BRCA1 gene result in truncated proteins, Hogervorst et al. (1995) used the protein truncation test (PTT) to screen for mutations in exon 11 which encodes 61% of the BRCA1 protein. In 45 patients from breast and/or ovarian cancer families, they found 6 novel mutations: 2 single nucleotide insertions, 3 small deletions (of 1-5 bp), and a nonsense mutation identified in 2 unrelated families. Furthermore, they were able to amplify the remaining coding region by RT-PCR using lymphocyte RNA. Combined with the protein truncation test, they detected aberrantly spliced products affecting exons 5 and 6 in 1 of 2 BRCA1-linked families examined.

Struewing et al. (1995) stated that more than 50 unique mutations had been detected in the BRCA1 gene in the germline of individuals with breast and ovarian cancer. In high-risk pedigrees, female carriers of a BRCA1 mutation had an 80%-90% lifetime risk of breast cancer and a 40%-50% risk of ovarian cancer. Not known, however, was the mutation status of women unselected for breast or ovarian cancer, and it was not known whether mutations in such women confer the same risk of cancer as in women from the high-risk families. Following the finding of a 185delAG frameshift mutation (113705.0003) in several Ashkenazi Jewish breast/ovarian families, Struewing et al. (1995) determined the frequency of this mutation in 858 Ashkenazim seeking genetic testing for conditions unrelated to cancer, and in 815 reference persons not selected for ethnic origin. They found the 185delAG mutation in 0.9% of Ashkenazim (95% confidence limit, 0.4%-1.8%) and in none of the reference samples. The results suggested that 1 in 100 women of Ashkenazi descent may be at especially high risk of developing breast and/or ovarian cancer. The possibility was raised in an accompanying editorial by Goldgar and Reilly (1995) that a high frequency of mortality from breast cancer in Nassau County, New York over the last 2 decades might be related to the high proportion of Ashkenazim (roughly 16%) in that population; the pathogenetic collaboration of exposure to an environmental pollutant was also raised. Ethical, legal, and social issues raised by these findings were also discussed by Goldgar and Reilly (1995).

In a study of 37 families with 4 or more cases of breast cancer or breast and ovarian cancer, Friedman et al. (1995) found that 5 families of Ashkenazi Jewish descent carried the 185delAG mutation and shared the same haplotype at 8 polymorphic markers spanning approximately 850 kb at BRCA1. Expressivity of 185delAG in these families varied from early-onset bilateral breast cancer and ovarian cancer to late-onset breast cancer without ovarian cancer. Overall, BRCA1 mutations were detected in 26 of the families: 16 with positive BRCA1 linkage lod scores, 7 with negative lod scores (reflecting multiple sporadic breast cancers), and 3 not tested for linkage.

Among 5,318 Jewish subjects, Struewing et al. (1997) found 120 carriers of a BRCA1 or BRCA2 (600185) mutation. The BRCA1 mutations studied were 185delAG (113705.0003) and 5382insC (113705.0018); the BRCA2 mutation studied was 6174delT (600185.0009). By the age of 70, the estimated risk of breast cancer among carriers was 56%; of ovarian cancer, 16%; and of prostate cancer, 16%. There were no significant differences in the risk of breast cancer between carriers of BRCA1 mutations and carriers of BRCA2 mutations, and the incidence of colon cancer among the relatives of carriers was not elevated. They concluded that over 2% of Ashkenazi Jews carried mutations in BRCA1 or BRCA2 that conferred increased risks of breast, ovarian, and prostate cancer. Krainer et al. (1997) found definite BRCA2 mutations in 2 of 73 women with early onset (by age 32) breast cancer, suggesting that BRCA2 is associated with fewer cases than BRCA1 ($P = 0.03$).

In a series of 268 Ashkenazi Jewish women with breast cancer, regardless of family history or age at onset, Fodor et al. (1998) determined the frequency of the common BRCA1 and BRCA2 mutations: 185delAG (113705.0003), 5382insC (113705.0018), and 6174delT (600185.0009). DNA was analyzed for the 3 mutations by allele-specific oligonucleotide (ASO) hybridization. Eight patients (3%) were heterozygous for the 185delAG mutation, 2 (0.75%) for the 5382insC mutation, and 8 (3%) for the 6174delT mutation. The lifetime risk for breast cancer in Ashkenazi Jewish carriers of the BRCA1 185delAG or BRCA2 6174delT mutations was estimated to be 36%, approximately 3 times the overall risk for the general population (relative risk 2.9). The results differed markedly from previous estimates based on high-risk breast cancer families.

Couch et al. (1997) identified BRCA1 mutations in 16% of women with a family history of breast cancer. Only 76% of women from families with a history of breast cancer but not ovarian cancer had BRCA1 mutations. They concluded that even in a referral clinic specializing in screening women from high risk families, most tests for BRCA1 mutations will be negative and, therefore, uninformative.

Stratton et al. (1994) examined 22 families with at least 1 case of male breast cancer for linkage to the BRCA1 locus. They found strong evidence against linkage to BRCA1 (lod score, -16.63) and the best estimate of the proportion of linked families was 0% (95% confidence interval, 0-18%).

Narod et al. (1995) reported the results of linkage analysis of 145 breast-ovarian families, each of which had 3 or more cases of early-onset breast cancer (age less than 60) or of ovarian cancer. All families had at least 1 case of ovarian cancer (there were 9 site-specific ovarian cancer families). Overall, they estimated that 76% of families were linked to the BRCA1 locus. At that time, the group stated that none of the 13 families with cases of male breast cancer appeared to be linked to BRCA1. In their letter, Narod et al. (1995) summarized their updated findings and reported a family with male breast cancer that showed a mutation (113705.0003) in BRCA1; Struewing et al. (1995) had also reported such a family. Their final results indicated that BRCA1 and BRCA2 account for the most breast-ovarian cancer families. Although a third breast cancer locus may be found, Narod et al. (1995) felt it unlikely that it would account for a significant proportion of breast-ovarian cancer families.

Chen et al. (1995) identified the BRCA1 gene product as a 220-kD nuclear phosphoprotein in normal cells, including breast ductal epithelial cells, and in 18 of 20 tumor cell lines derived from tissues other than breast and ovary. However, in 16 of 17 breast and ovarian cancer lines and in 17 of 17 samples of cells obtained from malignant effusions, BRCA1 localized mainly in the cytoplasm. Absence of BRCA1 or aberrant subcellular location was also observed to a variable extent in histologic sections of many breast cancer biopsies. The findings suggested to the authors that BRCA1 abnormalities may be involved in the pathogenesis of many breast cancers, sporadic as well as familial. Scully et al. (1996), however, reported results that did not support the hypothesis that wildtype BRCA1 is specifically excluded from the nucleus in sporadic breast and ovarian cancer.

Coene et al. (1997) reported a well-defined localization of BRCA1 in the perinuclear compartment of the endoplasmic reticulum-Golgi complex and in tubes invaginating the nucleus. The nuclear detection was fixation dependent, which helped to explain the controversial findings previously reported. The nuclear tubes were not seen in every cell, and therefore the authors suggested that an involvement in the cell cycle was possible. These tubes probably enhance nuclear-cytoplasmic interactions by increasing the surface area.

Chen et al. (1996) raised mouse polyclonal antibodies to 3 regions of the human BRCA1 protein and confirmed their earlier finding of a 220-kD nuclear phosphoprotein. They reported that expression and phosphorylation of the BRCA1 gene and protein are cell cycle dependent in a synchronized population of bladder carcinoma cells. The greatest levels of both expression and phosphorylation occurred in S and M phases.

Chen et al. (1998) used mammalian expression vectors to transfect cells with BRCA1 and BRCA2 as well as with several antibodies to recognize these proteins in order to study their subcellular localizations. They showed that BRCA1 and BRCA2 coexist in a biochemical complex and colocalize in subnuclear foci in somatic cells and on the axial elements of developing synaptonemal complexes. Like BRCA1 and RAD51, BRCA2 relocates to replication sites following exposure of S phase cells to hydroxyurea or UV irradiation. Thus, BRCA1 and BRCA2 participate together in a pathway (or pathways) associated with the activation of double-strand break repair and/or homologous recombination. Dysfunction of this pathway may be a general phenomenon in the majority of cases of hereditary breast and/or ovarian cancer.

Gayther et al. (1995) analyzed 60 families with a history of breast and/or ovarian cancer for germline mutations in BRCA1. In 32 families (53%), a total of 22 different mutations were detected, of which 14 were previously unreported. They observed a significant correlation between the location of the mutation in the gene and the ratio of breast to ovarian cancer incidence within the family. The data suggested to the authors a transition in risk such that mutations in the 3-prime third of the gene are associated with a lower proportion of ovarian cancer. Haplotype analysis supported previous data suggesting that some BRCA1 mutation carriers have common ancestors; however, Gayther et al. (1995) found at least 2 examples where recurrent mutations appeared to have arisen independently, judging from the different haplotype background.

Serova et al. (1996) analyzed 20 breast-ovarian cancer families, most of which showed evidence of linkage to 17q12, for germline mutations in BRCA1. Mutations in this gene cosegregating with breast and ovarian cancer susceptibility were identified in 16 of the 20 families, including 1 family with a case of male breast cancer. Nine of these mutations had not been reported previously. Most of them generated a premature stop codon leading to the formation of a truncated BRCA1 protein of 2 to 88% of the expected normal length. The RING-finger domain was altered by 2 of the mutations. A reduced quantity of BRCA1 transcript was associated with 8 of the mutations. Of the 4 families with no detectable BRCA1 mutation, only 1 was clearly linked to the BRCA1 locus.

Holt et al. (1996) demonstrated that retroviral transfer of the wildtype BRCA1 gene inhibits growth in vitro of all breast cancer and ovarian cancer cell lines tested, but not colon or lung cancer cells or fibroblasts. Mutant BRCA1, however, had no effect on growth of breast cancer cells; ovarian cancer cell growth was not affected by BRCA1 mutations in the 5-prime portion of the gene but was inhibited by 3-prime BRCA1 mutations. Development of MCF-7 tumors in nude mice was inhibited when MCF-7 cells were transfected with wildtype, but not mutant, BRCA1. Among mice with established MCF-7 tumors, peritoneal treatment with a retroviral vector expressing wildtype BRCA1 significantly inhibited tumor growth and increased survival. The results of Holt et al. (1996) were consistent with the previous observation that the site of BRCA1 mutation is associated with relative susceptibility to ovarian versus breast cancer.

Jensen et al. (1996) demonstrated that BRCA1 encodes a 190-kD protein with sequence homology and biochemical analogy to the granin protein family. BRCA2 also includes a motif similar to the granin consensus at the C terminus of the protein. Both BRCA1 and the granins localized to secretory vesicles, are secreted by a regulated pathway, are posttranslationally glycosylated, and are responsive to hormones. The authors stated that, as a regulated secretory protein, BRCA1 appears to function by a mechanism not previously described for tumor suppressor products. The granins with which BRCA1 and BRCA2 were compared included chromogranin A (118910), chromogranin B (118920), and secretogranin II, also known as chromogranin C (118930). As reviewed by Steeg (1996), granins are a family of acidic proteins that bind calcium and aggregate in its presence. Known members of the granin family have been solely neuroendocrine or endocrine in origin; if BRCA1 is a granin it will necessarily expand the families boundaries.

Studies of a number of diseases have indicated that fine-structure haplotype analysis can provide insight into the 'genetic history' of a particular mutation (or presumed mutation for rare diseases where the disease gene is not yet identified). To address both the question of mutation origin and the relationship between mutation and phenotype, Neuhausen et al. (1996) constructed a haplotype of 9 polymorphic STR markers within or immediately flanking the BRCA1 locus in a set of 61 families (selected to contain 1 of 6 BRCA1 mutations that had been identified a minimum of 4 times). The mutation appeared to have an affect on the relative proportion of cases of breast and ovarian cancer: 57% of women presumed affected because of the 1294 del 40 mutation had ovarian cancer, compared with 14% of affected women with the splice site mutation in intron 5 of BRCA1. A high degree of haplotype conservation across the region was observed. Any haplotype differences found were most often due to mutations in the short-tandem-repeat markers, although some likely instances of recombination also were observed. One mutation, 4184 del 4, had the same ancestral haplotype in two-thirds of the families studied. Neuhausen et al. (1996) estimated that this mutation had arisen 170 generations ago.

Langston et al. (1996) found germline BRCA1 mutations in 6 of 80 women in whom breast cancer was diagnosed before the age of 35 and who were not selected on the basis of family history. Four additional rare sequence variants of unknown functional significance were also identified. Two of the mutations and 3 of the rare sequence variants were found among the 39 women who reported no family history of breast or ovarian cancer. None of the mutations and only 1 of the rare variants was identified in a reference population of 73 unrelated subjects. Similar results were found by FitzGerald et al., 1996 in a study of 30 women with breast cancer before the age of 30: 4 (13%) had chain-terminating mutations and 1 had a missense mutation. The 185delAG mutation (113705.0003) was found in 2 of the 4 Jewish women in this cohort. Among the 39 Jewish women with breast cancer before the age of 40, FitzGerald et al. (1996) found that 8 (21%) carried the 185delAG mutation (95% confidence interval, 9-36%). FitzGerald et al. (1996) concluded that germline BRCA1 mutations can be present in young women with breast cancer who do not belong to families with multiple affected members.

Women who carry a mutation in the BRCA1 gene have an 80% risk of breast cancer and a 40% risk of ovarian cancer by the age of 70 years. Phelan et al. (1996) demonstrated that a modifier of this risk is the HRAS1 (190020) variable number of tandem repeats (VNTR) polymorphism, located 1 kb downstream of the HRAS1 oncogene. Individuals who have rare alleles of this VNTR had been found to have an increased risk of certain types of cancer, including breast cancer. Phelan et al. (1996) claimed that this was the first study to show the effect of a modifying gene on the penetrance of an inherited cancer syndrome.

Johannsson et al. (1996) identified 9 different germline mutations in the BRCA1 gene in 15 of 47 kindreds from southern Sweden, by use of SSCP and heteroduplex analysis of all exons and flanking intron region and by a protein-truncation test for exon 11, followed by direct sequencing. All but one of the mutations were predicted to give rise to premature translation termination and included 7 frameshift insertions or deletions, a nonsense mutation, and a splice acceptor site mutation. The remaining mutation was a missense mutation (cys61-to-gly) in the zinc-binding motif. They also identified 4 novel Swedish founding mutations: deletion of 2595A in 5 families, the C-to-T nonsense mutation of nt1806 in 3 families, the insertion of TGAGA after nt3166 in 3 families, and the deletion of 11 nucleotides after nt1201 in 2 families. Analysis of the intragenic polymorphism D17S855 supported common origins of the mutations. Eleven of the 15 kindreds manifesting BRCA1 mutations were breast-ovarian cancer families, several of which had a predominant ovarian cancer phenotype. Among the 32 families in which no BRCA1 alteration was detected, there was 1 breast-ovarian cancer kindred showing clear linkage to the BRCA1 region and loss of the wildtype chromosome in associated tumors. Other tumor types found in BRCA1 mutation or haplotype carriers included prostatic, pancreas, skin, and lung cancer, a malignant melanoma, an oligodendroglioma, and a carcinosarcoma. In all, 12 of the 16 kindreds manifesting BRCA1 mutation or linkage contained ovarian cancer, as compared with only 6 of the remaining 31 families. Gayther et al. (1996) stated that more than 65 distinct mutations scattered throughout the coding region of BRCA1 had been detected.

Langston et al. (1996) studied the BRCA1 gene in 61 men who met one or more of these criteria: (1) under 53 years of age at diagnosis of prostate cancer; (2) a family history of breast cancer in a first-degree female relative diagnosed under 51 years of age; or (3) a family history of prostate cancer in 2 or more male relatives, with at least 1 relative diagnosed at less than 56 years of age. They found 1 germline mutation, 185delAG (113705.0003), in 1 subject and 5 different rare sequence variants (1 of which was detected in 2 unrelated men). None of the rare variants were found in population-based controls. Isaacs et al. (1995) failed to identify a significantly increased risk of breast cancer among relatives of prostate cancer probands. The findings of Langston et al. (1996) are not necessarily in conflict, since the contribution of germline BRCA1 mutations to the overall incidence of prostate cancer appears to be small, at most, and may be limited to specific subgroups of patients.

Couch et al. (1996) reported a total of 254 BRCA1 mutations, 132 (52%) of which were unique. These represented mutations entered into a database established by the Breast Cancer Information Core (BIC). A total of 221 (87%) of all mutations or 107 (81%) of the unique mutations are small deletions, insertions, nonsense point mutations, splice variants, and regulatory mutations that result in truncation or absence of the BRCA1 protein. A total of 11 disease-associated missense mutations (5 unique) and 21 variants (19 unique) as yet unclassified as missense mutations or polymorphisms had been detected. Thirty-five independent benign polymorphisms had been described. The most common mutations were 185delAG (113705.0003) and 5382insC (113705.0018), which accounted for 30 (11.7%) and 26 (10.1%), respectively, of all the mutations.

Stoppa-Lyonnet et al. (1996) described 2 independent BRCA1 mutations in a single family. A woman with breast cancer diagnosed at age 25 inherited a deleterious allele from her father. Her mother had ovarian and breast cancer caused by a separate mutation, which was the basis of breast cancer in 5 or more of her relatives. The authors pointed out that the segregation of 2 BRCA1 mutations resulted in the failure to demonstrate linkage to either chromosome 17 or chromosome 13 and could leads to the erroneous hypothesis of the involvement of a third locus in familial breast cancer. Narod et al. (1995) suggested that the fraction of familial breast cancer that is not accounted for by BRCA1 or BRCA2 may be small.

Rebbeck et al. (1996) performed specific studies of 23 families identified through 2 high-risk breast cancer research programs. In 14 (61%) it was possible to attribute the pattern of hereditary cancer to BRCA1 by a combination of linkage and mutation analyses. No families were attributed to BRCA2. In 5 families (22%), evidence against linkage to both BRCA1 and BRCA2 was found; no BRCA1 or BRCA2 mutations were detected in these 5 families. The BRCA1 or BRCA2 status of the 4 remaining families (17%) could not be determined.

Brown et al. (1996) determined the detailed structure of the BRCA1 genomic region. They showed that this region of chromosome 17 contains a tandem duplication of approximately 30 kb which results in 2 copies of BRCA1 exons 1 and 2, of exons 1 and 3 of the adjacent gene that Brown et al. (1994) designated 1A1-3B (M17S2; 166945), and of a previously reported 295-bp intergenic region. Sequence analysis of the duplicated exons of BRCA1, 1A1-3B, and flanking genomic DNA revealed to Brown et al. (1996) that there was maintenance of exon/intron structure and a high degree of nucleotide sequence identity, which suggested that these duplicated exons are non-processed pseudogenes. They noted that these findings could not only confound BRCA1 mutation analysis but could have implications for the normal and abnormal regulation of BRCA1 transcription, translation and function.

Although many distinct mutations were identified in the breast-ovarian cancer susceptibility gene BRCA1 with loss of the wildtype allele in more than 90% of tumors from patients with inherited BRCA1 mutation, a very low incidence of somatic mutations were found in sporadic tumors, suggesting the BRCA1 inactivation occurs by alternative mechanisms, such as interstitial chromosomal deletion or reduced transcription. To identify possible features of the BRCA1 genomic region that may contribute to chromosomal instability, as well as potential transcriptional regulatory elements, Smith et al. (1996) sequenced 117,143 bp from human chromosome 17 encompassing BRCA1. The 24 exons of BRCA1 spanned an 81-kb region that had an unusually high density of Alu repetitive DNA (41.5%), but a relatively low density (4.8%) of other repetitive sequences. BRCA1 intron lengths ranged in size from 403 bp to 9.2 kb and contained 3 intragenic microsatellite markers located in introns 12, 19, and 20. In addition to BRCA1, the contig contained 2 complete genes which they called RHO7 (601555) and VAT1. RHO7 is a member of the RHO family of GTP binding proteins and VAT1 is an abundant membrane protein of cholinergic synaptic vesicles. The order of genes on the chromosome was found to be as follows: centromere-IFP 35 (600735)-VAT1-RHO7-BRCA1-1A1-3B-telomere.

In an effort to understand the function of BRCA1, Wu et al. (1996) used a yeast 2-hybrid system to identify proteins that associate with BRCA1 in vivo. This analysis led to the identification of a novel protein that interacts with the N-terminal region of BRCA1. Wu et al. (1996) designated this protein BARD1 (601593) and determined that it maps to chromosome 2q.

In a population-based series of 54 breast cancer cases from southern California, Friedman et al. (1997) found no instance of germline mutation in the BRCA1 gene but found 2 male breast cancer patients who carried novel truncating mutations in the BRCA2 gene. Only 1 of the 2 had a family history of cancer, namely, ovarian cancer in a first-degree relative.

Hamann et al. (1997) studied 45 German breast/ovarian cancer families for germline mutations in the BRCA1 gene. They identified 4 germline mutations in 3 breast cancer families and in 1 breast/ovarian cancer family. One of these, a missense mutation, was also found in 2.8% of the general population, suggesting that this was not disease associated. Hamann et al. (1997) concluded that the low incidence of BRCA1 germline mutations in these families suggests the involvement of other susceptibility genes.

In addition to the deletion mutations in BRCA1 and BRCA2 genes that appear to be highly penetrant in the causation of young onset breast cancer and ovarian cancer, there are several common polymorphisms in the BRCA1 gene that generate amino acid substitutions. Dunning et al. (1997) raised the question whether these common variants may confer more modest individual risks which might, however, be of significance. They examined the frequency of 4 of these polymorphisms in a large series of breast and ovarian cancer cases and matched controls. Due to strong linkage disequilibrium, the 4 sites generated only 3 haplotypes with a frequency more than 1.3%. The 2 most common haplotypes had frequencies of 0.57 and 0.32, respectively, and these frequencies did not differ significantly between patient and control groups. Thus, Dunning et al. (1997) concluded that the most common polymorphisms of BRAC1 gene do not make a significant contribution to breast or ovarian cancer risk. However, the data suggested that the arg356 allele may have a different genotype distribution in breast cancer patients than that in controls; arg356 homozygotes were more frequent in the control group ($P = 0.01$), indicating that it may be protective against breast cancer.

Hacia et al. (1996) noted that all of the current methods used to detect BRCA1 mutations begin with PCR amplification and require gel electrophoresis, which seriously complicates the challenge of scale-up, automation, and cost reduction. They demonstrated the feasibility of using oligonucleotide arrays in a DNA chip-based assay to screen for a wide range of heterozygous mutations in the 3.45-kb exon 11 of the BRCA1 gene. They concluded that DNA chip-based assays provided a valuable new technology for high throughput, cost-efficient detection of genetic alterations.

The detection of inactivating mutations in tumor suppressor genes is critical to their characterization, as well as to the development of diagnostic testing. Most approaches for mutational screening of germline specimens are complicated by the fact that mutations are heterozygous and that missense mutations are difficult to interpret in the absence of information about protein function. Ishioka et al. (1997) described a novel method using Saccharomyces cerevisiae for detecting protein-truncating mutations in any gene of interest. In their procedure, the PCR-amplified coding sequence of the gene is inserted by homologous recombination into a yeast URA3 fusion protein, and transformants are assayed for growth in the absence of uracil. The high efficiency of homologous recombination in yeast ensures that both alleles are represented among transformants and achieves separation of alleles, which facilitates subsequent nucleotide sequencing of the mutated transcript. The specificity of translational initiation of the URA3 gene lead to minimal enzymatic activity in transformants harboring an inserted stop codon, and hence to reliable distinction between specimens with wild-type alleles and those with a heterozygous truncating mutation. This yeast-based codon assay accurately detected heterozygous truncating mutations in the BRCA1 gene in patients with early onset of breast cancer and in the APC gene (175100) in patients with familial adenomatous polyposis.

Szabo and King (1997) collated information on the population genetics of BRCA1 and BRCA2 in populations from many countries of Europe as well as the U.S., Canada, and Japan.

Scully et al. (1997) found that the BRCA1 gene product is a component of the RNA polymerase II holoenzyme (polII) by several criteria. BRCA1 was found to copurify with the holoenzyme over multiple chromatographic steps. Other tested transcription activators that could potentially contact the holoenzyme were not stably associated with the holoenzyme as determined by copurification. Antibody specific for the holoenzyme component SRB7 specifically purified BRCA1. (SRB proteins are a key component of the holoenzyme and were discovered in a yeast genetic screen as suppressors of RNA polymerase B mutations; hence, the designation SRB. A benign difference is that SRB proteins bind to the C-terminal domain of yeast polII and are found only in the polII holoenzyme.) Immunopurification of BRCA1 complexes also specifically purified transcriptionally active RNA polII and transcription factors TFIIF (see 189968), TFIIE (see 189962), and TFIIH (see 189972), which are known components of the holoenzyme. Moreover, a BRCA1 domain, which is deleted in about 90% of clinically relevant mutations, participated in binding to the holoenzyme complex in cells. These data were considered consistent with other data identifying transcription activation domains in the BRCA1 protein, and link the BRCA1 tumor suppressor protein with the transcription process as a holoenzyme-bound protein.

RNA helicase A, or RHA (140 kD) was identified by Lee and Hurwitz (1993) and Zhang and Grosse (1997) as a helicase of unknown function with homology to the Drosophila 'maleless' gene, which functions to increase expression of genes from the male X chromosome. Anderson et al. (1998) showed that RHA protein links BRCA1 to the holoenzyme complex. These results were the first to identify specific protein interaction with the BRCA1 C-terminal domain and were consistent with the model that BRCA1 functions as a transcriptional coactivator.

According to the conclusions of the Breast Cancer Linkage Consortium (1997), the histology of breast cancers in women predisposed by reason of carrying BRCA1 and BRCA2 mutations differs from that in sporadic cases, and there are differences between breast cancers in carriers of BRCA1 and BRCA2 mutations. The findings were interpreted as suggesting that breast cancer due to BRCA1 has a different natural history to BRCA2 or apparently sporadic disease, which may have implications for screening and management.

Petrij-Bosch et al. (1997) reported that the mutation spectrum of BRCA1 available at that time had been biased by PCR-based mutation-screening methods, such as SSCP, the protein truncation test (PPT), and direct sequencing, using genomic DNA as template. Three large genomic deletions that were not detectable by those approaches comprised 36% of all BRCA1 mutations found in Dutch breast cancer families up to that time. A 510-bp Alu-mediated deletion comprising exon 22 was found in 8 of 170 breast cancer families recruited for research purposes and in 6 of 49 probands referred to the Amsterdam Family Cancer Clinic for genetic counseling. In addition, a 3,835-bp Alu-mediated deletion encompassing exon 13 was detected in 6 of the 170 research families, while a deletion of approximately 14 kb was detected in a single family. Haplotype analyses indicated that each recurrent mutation had a single common ancestor.

In a screening of Hungarian breast/ovarian cancer families for germline mutations in BRCA1 and BRCA2, Ramus et al. (1997) found 1 individual who carried the 185delAG mutation (113705.0003) in BRCA1, as well as the 6174delT mutation (600185.0009) in BRCA2. Each mutation had been shown to have a frequency of approximately 1% in the Ashkenazi Jewish population. Although the patient was not recorded as having a Jewish origin, haplotype analysis suggested that both mutations were of the Ashkenazi type. There was a maternal family history of breast cancer and the paternal family history was unknown. The patient was found to have breast cancer at age 48 and ovarian cancer at age 50 years. The ages at diagnosis and the tumor types were not different from those of patients with either BRCA1 or BRCA2 mutations. Both mutations were present in 3 different samples from the patient: breast tumor, ovarian tumor, and lymphocyte DNA. There was no evidence of loss of heterozygosity on either chromosome 13 or chromosome 17. Liede et al. (1998) found mutations of both BRCA1 and BRCA2 in a breast cancer patient of Scottish descent. Grade II adenocarcinoma of the breast was diagnosed at the age of 35 years. Simultaneous screening by protein truncation tests of both BRCA genes detected a 2508G-T mutation of the BRCA1 gene (113705.0023) and a 3295insA mutation of BRCA2 (600185.0011). The patient had both a maternal and a paternal history of breast cancer. The maternal side contained cases of postmenopausal breast cancer; the paternal side contained cases of premenopausal breast cancer. The mother, however, did not have either mutation, suggesting that both BRCA1 and BRCA2 germline mutations originated from the father of the proband.

Ford et al. (1998) assessed the contribution of BRCA1 and BRCA2 to inherited breast cancer by linkage and mutation analysis in 237 families, each with at least 4 cases of breast cancer, collected by the Breast Cancer Linkage Consortium. Families were included without regard to the occurrence of ovarian or other cancers. Overall, disease was linked to BRCA1 in an estimated 52% of families, to BRCA2 in 32% of families, and to neither gene in 16%, suggesting other predisposition genes. The majority (81%) of the breast-ovarian cancer families were due to BRCA1, with most others (14%) due to BRCA2. Conversely, the majority (76%) of families with both male and female breast cancer were due to BRCA2. The largest proportion (67%) of families due to other genes were families with 4 or 5 cases of female breast cancer only. Among those families with disease due to BRCA1 that were tested by one of the standard screening methods, mutations were detected in the coding sequence or splice sites in an estimated 63%. The estimated sensitivity was identical for direct sequencing and other techniques.

The risk of ovarian cancer is reduced by 50% or more in unselected women with long-term use of oral contraceptives (Franceschi et al., 1991; Whittemore et al., 1992). To evaluate the potential benefit of oral contraceptive use in women at high risk for ovarian cancer, Narod et al. (1998) studied 207 patients with BRCA1 or BRCA2 mutations and ovarian cancer and 161 of their sisters, who served as controls. Their findings suggested that oral contraceptive use protects against ovarian cancer in carriers of either the BRCA1 or BRCA2 mutation.

Association of the BRCA1 protein with the DNA repair gene RAD51 (179617) and changes in the phosphorylation and cellular localization of the protein after exposure to DNA-damaging agents are consistent with a role for BRCA1 in DNA repair. Gowen et al. (1998) showed that mouse embryonic stem cells deficient in BRCA1 are defective in the ability to carry out transcription-coupled repair of oxidative DNA damage, and are hypersensitive to ionizing radiation and hydrogen peroxide. These results suggested that BRCA1 participates, directly or indirectly, in transcription-coupled repair of oxidative DNA damage.

Patients with germline BRCA1 mutations may develop papillary serous carcinoma of the peritoneum (PSCP), a malignancy that diffusely involves peritoneal surfaces, sparing or only superficially involving the ovaries. PSCP is histologically indistinguishable from serous epithelial ovarian carcinoma, and it may develop years after oophorectomy. Schorge et al. (1998) used the androgen receptor (AR; 313700) gene locus to test the hypothesis that some cases of PSCP have a multifocal origin and to determine if patients with germline BRCA1 mutations develop multifocal PSCP. Specimens were studied from 22 women with PSCP. The AR gene locus was evaluated for patterns of loss of heterozygosity and X-chromosome inactivation. The methylation-sensitive HpaII restriction enzyme was used to differentiate the active and inactive X chromosomes. They found patterns of selective LOH at the AR locus in 5 (23%) of the 22 subjects, consistent with multifocal, polyclonal disease origin. Two patients with selective LOH also had alternating X-chromosome inactivation patterns. Patients with germline BRCA1 mutations were more likely to have evidence of multifocal disease.

Three Jewish founder mutations, 185delAG (113705.0003) and 5382insC (113705.0018) in BRCA1 and 6174delT (600185.0009) in BRCA2, have been identified in breast cancer and ovarian cancer Ashkenazi patients. Friedman et al. (1998) pooled results from 4 cancer/genetic centers in Israel to analyze approximately 1,500 breast-ovarian cancer Ashkenazi patients for the presence of double heterozygosity as well as homozygosity for any of these mutations. Although the small number of cases precluded definite conclusions, the results suggested that the phenotypic effects of double heterozygosity for BRCA1 and BRCA2 germline mutations were not cumulative. This was in agreement with the observation that the phenotype of mice that are homozygous knockouts for the BRCA1 and BRCA2 genes is similar to that of mice that were BRCA1 knockouts. This suggests that the BRCA1 mutation is epistatic over the BRCA2 mutation. Two of the double heterozygotes described had had reproductive problems: one with primary sterility and irregular menses and another with premature menopause at the age of 37 years. Friedman et al. (1998) suggested that identification of additional carriers of more than one mutation will increase our understanding between various mutations and will improve genetic counseling.

Animal Model

Gowen et al. (1996) described homozygous mice lacking the mouse Brca1 gene. The mice, possessing a deletion of the large exon 11, died between days 10 and 13 of embryonic development, suffering from a variety of neuroepithelial defects. Hakem et al. (1996) described another strain of homozygous mice for a putative Brca1 null mutation produced by targeted deletion of exons 5 and 6. These mutant mice were more severely affected, dying at about embryonic day 7.5 with no signs of mesoderm formation and exhibiting reduced cell proliferation. There were also strong signs of disruptive cell cycle regulation via altered expression levels of cyclin E (123837), mdm2 (164785) and p21 (116899). Hakem et al. (1996) speculated that the death of mutant embryos was due to failure of the proliferative burst required for germ layer development. Hakem et al. (1996) reported that after about 1 year of age, Brca1 heterozygous female mice showed no evidence of cancer. Gowen et al. (1996) also had been unable to detect tumors in 1-year-old heterozygotes.

See also Albertsen et al. (1994); Narod et al. (1991); Narod et al. (1995); Struewing et al. (1995)

Albertsen, H.; Plaetke, R.; Ballard, L.; Fujimoto, E.; Connolly, J.; Lawrence, E.; Rodriguez, P.; Robertson, M.; Bradley, P.; Milner, B.; Fuhrman, D.; Marks, A.; Sargent, R.; Cartwright, P.; Matsunami, N.; White, R.: Genetic mapping of the BRCA1 region on chromosome 17q21. Am. J. Hum. Genet. 54: 516-525, 1994.

Albertsen, H. M.; Smith, S. A.; Mazoyer, S.; Fujimoto, E.; Stevens, J.; Williams, B.; Rodriguez, P.; Cropp, C. S.; Slijepcevic, P.; Carlson, M.; Robertson, M.; Bradley, P.; Lawrence, E.; Harrington, T.; Mei Sheng, Z.; Hoopes, R.; Sternberg, N.; Brothman, A.; Callahan, R.; Ponder, B. A. J.; White, R.: A physical map and candidate genes in the BRCA1 region on chromosome 17q12-21. Nature Genet. 7: 472-479, 1994.

Anderson, S. F.; Schlegel, B. P.; Nakajima, T.; Wolpin, E. S.; Parvin, J. D.: BRCA1 protein is linked to the RNA polymerase II holoenzyme complex via RNA helicase A. Nature Genet. 19: 254-256, 1998.

Arason, A.; Barkardottir, R. B.; Egilsson, V.: Linkage analysis of chromosome 17q markers and breast-ovarian cancer in Icelandic families, and possible relationship to prostatic cancer. Am. J. Hum. Genet. 52: 711-717, 1993.

Bandera, C. A.; Muto, M. G.; Schorge, J. O.; Berkowitz, R. S.; Rubin, S. C.; Mok, S. C.: BRCA1 gene mutations in women with papillary serous carcinoma of the peritoneum. Obstet. Gynec. 92: 596-600, 1998.

Bar-Sade, R. B.; Kruglikova, A.; Modan, B.; Gak, E.; Hirsh-Yechezkel, G.; Theodor, L.; Novikov, I.; Gershoni-Baruch, R.; Risel, S.; Papa, M. Z.; Ben-Baruch, G.; Friedman, E.: The 185delAG BRCA1 mutation originated before the dispersion of Jews in the Diaspora and is not limited to Ashkenazim. Hum. Molec. Genet. 7: 801-805, 1998.

Bar-Sade, R. B.; Theodor, L.; Gak, E.; Kruglikova, A.; Hirsch-Yechezkel, G.; Modan, B.; Kuperstein, G.; Seligsohn, U.; Rechavi, G.; Friedman, E.: Could the 185delAG BRCA1 mutation be an ancient Jewish mutation? Europ. J. Hum. Genet. 5: 413-416, 1997.

Barker, D. F.; Almeida, E. R. A.; Casey, G.; Fain, P. R.; Liao, S. Y.; Masunaka, I.; Noble, B.; Kurosaki, T.; Anton-Culver, H.: BRCA1 R841W: a strong candidate for a common mutation with moderate phenotype. Genet. Epidemiol. 13: 595-604, 1996.

Bennett, L. M.; Haugen-Strano, A.; Cochran, C.; Brownlee, H. A.; Fiedorek, F. T., Jr.; Wiseman, R. W.: Isolation of the mouse homologue of BRCA1 and genetic mapping to mouse chromosome 11. Genomics 29: 576-581, 1995.

Breast Cancer Linkage Consortium: Pathology of familial breast cancer: differences between breast cancers in carriers of BRCA1 and BRCA2 mutations and sporadic cases. Lancet 349: 1505-1510, 1997.

Brown, M. A.; Nicolai, H.; Xu, C.-F.; Griffiths, B. L.; Jones, K. A.; Solomon, E.; Hosking, L.; Trowsdale, J.; Black, D. M.; McFarlane, R.: Regulation of BRCA1. (Letter) Nature 372: 733 only, 1994.

Brown, M. A.; Xu, C.-F.; Nicolai, H.; Griffiths, B.; Chambers, J. A.; Black, D.; Solomon, E.: The 5-prime end of the BRCA1 gene lies within a duplicated region of human chromosome 17q21. Oncogene 12: 2507-2513, 1996.

Castilla, L. H.; Couch, F. J.; Erdos, M. R.; Hoskins, K. F.; Calzone, K.; Garber, J. E.; Boyd, J.; Lubin, M. B.; Deshano, M. L.; Brody, L. C.; Collins, F. S.; Weber, B. L.: Mutations in the BRCA1 gene in families with early-onset breast and ovarian cancer. Nature Genet. 8: 387-391, 1994.

Chen, J.; Silver, D. P.; Walpita, D.; Cantor, S. B.; Gazdar, A. F.; Tomlinson, G.; Couch, F. J.; Weber, B. L.; Ashley, T.; Livingston, D. M.; Scully, R.: Stable interaction between the products of the BRCA1 and BRCA2 tumor suppressor genes in mitotic and meiotic cells. Molec. Cell 2: 317-328, 1998.

Chen, Y.; Chen, C.-F.; Riley, D. J.; Allred, D. C.; Chen, P.-L.; Von Hoff, D.; Osborne, C. K.; Lee, W.-H.: Aberrant subcellular localization of BRCA1 in breast cancer. Science 270: 789-791, 1995.

Chen, Y.; Farmer, A. A.; Chen, C.-F.; Jones, D. C.; Chen, P.-L.; Lee, W.-H.: BRCA1 is a 220-kDa nuclear phosphoprotein that is expressed and phosphorylated in a cell cycle-dependent manner. Cancer Res. 56: 3168-3172, 1996.

Claus, E. B.; Risch, N.; Thompson, W. D.: Genetic analysis of breast cancer in the cancer and steroid hormone study. Am. J. Hum. Genet. 48: 232-242, 1991.

Coene, E.; Van Oostveldt, P.; Willems, K.; van Emmelo, J.; De Potter, C. R.: BRCA1 is localized in cytoplasmic tube-like invaginations in the nucleus. (Letter) Nature Genet. 116: 122-124, 1997.

Cornelis, R. S.; Vasen, H. F. A.; Meijers-Heijboer, H.; Ford, D.; van Vliet, M.; van Tilborg, A. A. G.; Cleton, F. J.; Klijn, J. G. M.; Menko, F. H.; Khan, P. M.; Cornelisse, C. J.; Devilee, P.: Age at diagnosis as an indicator of eligibility for BRCA1 DNA testing in familial breast cancer. Hum. Genet. 95: 539-544, 1995.

Couch, F. J.; DeShano, M. L.; Blackwood, M. A.; Calzone, K.; Stopfer, J.; Campeau, L.; Ganguly, A.; Rebbeck, T.; Weber, B. L.: BRCA1 mutations in women attending clinics that evaluate the risk of breast cancer. New Eng. J. Med. 336: 1409-1415, 1997.

Couch, F. J.; Weber, B. L.; Breast Cancer Information Core: Mutations and polymorphisms in the familial early-onset breast cancer (BRCA1) gene. Hum. Mutat. 8: 8-18, 1996.

De Gregorio, L.; Harshman, K.; Rosenthal, J.; Dragani, T. A.; Pierotti, M. A.: Genetic mapping of the Brca1 gene on mouse chromosome 11. Mammalian Genome 7: 242, 1996.

Dunning, A. M.; Chiano, M.; Smith, N. R.; Dearden, J.; Gore, M.; Oakes, S.; Wilson, C.; Stratton, M.; Peto, J.; Easton, D.; Clayton, D.; Ponder, B. A. J.: Common BRCA1 variants and susceptibility to breast and ovarian cancer in the general population. Hum. Molec. Genet. 6: 285-289, 1997.

Easton, D. F.; Bishop, D. T.; Ford, D.; Crockford, G. P.; Breast Cancer Linkage Consortium: Genetic linkage analysis in familial breast and ovarian cancer: results from 214 families. Am. J. Hum. Genet. 52: 678-701, 1993.

FitzGerald, M. G.; MacDonald, D. J.; Krainer, M.; Hoover, I.; O'Neil, E.; Unsal, H.; Silva-Arrieto, S.; Finkelstein, D. M.; Beer-Romero, P.; Englert, C.; Sgroi, D. C.; Smith, B. L.; Younger, J. W.; Garber, J. E.; Duda, R. B.; Mayzel, K. A.; Isselbacher, K. J.; Friend, S. H.; Haber, D. A.: Germ-line BRCA1 mutations in Jewish and non-Jewish women with early-onset breast cancer. New Eng. J. Med. 334: 143-149, 1996.

Fodor, F. H.; Weston, A.; Bleiweiss, I. J.; McCurdy, L. D.; Walsh, M. M.; Tartter, P. I.; Brower, S. T.; Eng, C. M.: Frequency and carrier risk associated with common BRCA1 and BRCA2 mutations in Ashkenazi Jewish breast cancer patients. Am. J. Hum. Genet. 63: 45-51, 1998.

Ford, D.; Easton, D. F.; Stratton, M.; Narod, S.; Goldgar, D.; Devilee, P.; Bishop, D. T.; Weber, B.; Lenoir, G.; Chang-Claude, J.; Sobol, H.; Teare, M. D.; and 27 others: Genetic heterogeneity and penetrance analysis of the BRCA1 and BRCA2 genes in breast cancer families. Am. J. Hum. Genet. 62: 676-689, 1998.

Franceschi, S.; Parazzini, F.; Negri, E.; Booth, M.; La Vecchia, C.; Beral, V.; Tzonov, A.; Trichopoulos, D.: Pooled analysis of 3 European case-control studies of epithelial ovarian cancer. III. Oral contraceptive use. Int. J. Cancer 49: 61-65, 1991.

Friedman, E.; Bar-Sade, R. B.; Kruglikova, A.; Risel, S.; Levy-Lahad, E.; Halle, D.; Bar-On, E.; Gershoni-Baruch, R.; Dagan, E.; Kepten, I.; Peretz, T.; Lerer, I.; Wienberg, N.; Shushan, A.; Abeliovich, D.: Double heterozygotes for the Ashkenazi founder mutations in BRCA1 and BRCA2 genes. (Letter) Am. J. Hum. Genet. 63: 1224-1227, 1998.

Friedman, L. S.; Gayther, S. A.; Kurosaki, T.; Gordon, D.; Noble, B.; Casey, G.; Ponder, B. A. J.; Anton-Culver, H.: Mutation analysis of BRCA1 and BRCA2 in a male breast cancer population. Am. J. Hum. Genet. 60: 313-319, 1997.

Friedman, L. S.; Ostermeyer, E. A.; Szabo, C. I.; Dowd, P.; Lynch, E. D.; Rowell, S. E.; King, M.-C.: Confirmation of BRCA1 by analysis of germline mutations linked to breast and ovarian cancer in ten families. Nature Genet. 8: 399-404, 1994.

Friedman, L. S.; Szabo, C. I.; Ostermeyer, E. A.; Dowd, P.; Butler, L.; Park, T.; Lee, M. K.; Goode, E. L.; Rowell, S. E.; King, M.-C.: Novel inherited mutations and variable expressivity of BRCA1 alleles, including the founder mutation 185delAG in Ashkenazi Jewish families. Am. J. Hum. Genet. 57: 1284-1297, 1995.

Futreal, P. A.; Liu, Q.; Shattuck-Eidens, D.; Cochran, C.; Harshman, K.; Tavtigian, S.; Bennett, L. M.; Haugen-Strano, A.; Swensen, J.; Miki, Y.; Eddington, K.; McClure, M.; Frye, C.; Weaver-Feldhaus, J.; Ding, W.; Gholami, Z.; Soderkvist, P.; Terry, L.; Jhanwar, S.; Berchuck, A.; Iglehart, J. D.; Marks, J.; Ballinger, D. G.; Barrett, J. C.; Skolnick, M. H.; Kamb, A.; Wiseman, R.: BRCA1 mutation in primary and ovarian carcinomas. Science 266: 120-122, 1994.

Gayther, S. A.; Harrington, P.; Russell, P.; Kharkevich, G.; Garkavtseva, R. F.; Ponder, B. A. J.; UKCCCR Familial Ovarian Cancer Study Group : Rapid detection of regionally clustered germ-line BRCA1 mutations by multiplex heteroduplex analysis. Am. J. Hum. Genet. 58: 451-456, 1996.

Gayther, S. A.; Warren, W.; Mazoyer, S.; Russell, P. A.; Harrington, P. A.; Chiano, M.; Seal, S.; Hamoudi, R.; van Rensburg, E. J.; Dunning, A. M.; Love, R.; Evans, G.; Easton, D.; Clayton, D.; Stratton, M. R.; Ponder, B. A. J.: Germline mutations of the BRCA1 gene in breast and ovarian cancer families provide evidence for a genotype-phenotype correlation. Nature Genet. 11: 428-433, 1995.

Goldgar, D. E.; Fields, P.; Lewis, C. M.; Cannon-Albright, L. A.; Linker, G.; Tran, T.; Skolnick, M.: A large kindred with 17q-linked susceptibility to breast and ovarian cancer: relationship between genotype and phenotype. (Abstract) Am. J. Hum. Genet. 51 (suppl.): A27, 1992.

Goldgar, D. E.; Reilly, P. R.: A common BRCA1 mutation in the Ashkenazim. Nature Genet. 11: 113-114, 1995.

Gowen, L. C.; Avrutskaya, A. V.; Latour, A. M.; Koller, B. H.; Leadon, S. A.: BRCA1 required for transcription-coupled repair of oxidative DNA damage. Science 281: 1009-1012, 1998.

Gowen, L. C.; Johnson, B. L.; Latour, A. M.; Sulik, K. K.; Koller, B. H.: Brca1 deficiency results in early embryonic lethality characterized by neuroepithelial abnormalities. Nature Genet. 12: 191-194, 1996.

Hacia, J. G.; Brody, L. C.; Chee, M. S.; Fodor, S. P. A.; Collins, F. S.: Detection of heterozygous mutations in BRCA1 using high density oligonucleotide arrays and two-colour fluorescence analysis. Nature Genet. 14: 441-447, 1996.

Hakem, R.; de la Pompa, J. L.; Sirard, C.; Mo, R.; Woo, M.; Hakem, A.; Wakeham, A.; Potter, J.; Reitmair, A.; Billia, F.; Firpo, E.; Hui, C. C.; Roberts, J.; Rossant, J.; Mak, T. W.: The tumor suppressor gene Brca1 is required for embryonic cellular proliferation in the mouse. Cell 85: 1009-1023, 1996.

Hall, J. M.; Friedman, L.; Guenther, C.; Lee, M. K.; Weber, J. L.; Black, D. M.; King, M.-C.: Closing in on a breast cancer gene on chromosome 17q. Am. J. Hum. Genet. 50: 1235-1242, 1992.

Hall, J. M.; Lee, M. K.; Newman, B.; Morrow, J. E.; Anderson, L. A.; Huey, B.; King, M.-C.: Linkage of early-onset familial breast cancer to chromosome 17q21. Science 250: 1684-1689, 1990.

Hamann, U.; Haner, M.; Stosiek, U.; Bastert, G.; Scott, R. J.: Low frequency of BRCA1 germline mutations in 45 German breast/ovarian cancer families. J. Med. Genet. 34: 884-888, 1997.

Hogervorst, F. B. L.; Cornelis, R. S.; Bout, M.; van Vliet, M.; Oosterwijk, J. C.; Olmer, R.; Bakker, B.; Klijn, J. G. M.; Vasen, H. F. A.; Meijers-Heijboer, H.; Menko, F. H.; Cornelisse, C. J.; den Dunnen, J. T.; Devilee, P.; van Ommen, G.-J. B.: Rapid detection of BRCA1 mutations by the protein truncation test. Nature Genet. 10: 208-212, 1995.

Holt, J. T.; Thompson, M. E.; Szabo, C.; Robinson-Benion, C.; Arteaga, C. L.; King, M.-C.; Jensen, R. A.: Growth retardation and tumour inhibition by BRCA1. Nature Genet. 12: 298-302, 1996.

Isaacs, S. D.; Kiemeney, L. A. L. M.; Baffoe-Bonnie, A.; Beaty, T. H.; Walsh, P. C.: Risk of cancer in relatives of prostate cancer probands. J. Nat. Cancer Inst. 87: 991-996, 1995.

Ishioka, C.; Suzuki, T.; Fitzgerald, M.; Krainer, M.; Shimodaira, H.; Shimada, A.; Nomizu, T.; Isselbacher, K. J.; Haber, D.; Kanamaru, R.: Detection of heterozygous truncating mutations in the BRCA1 and APC genes by using a rapid screening assay in yeast. Proc. Nat. Acad. Sci. 94: 2449-2453, 1997.

Jensen, R. A.; Thompson, M. E.; Jetton, T. L.; Szabo, C. I.; van der Meer, R.; Helou, B.; Tronick, S. R.; Page, D. L.; King, M.-C.; Holt, J. T.: BRCA1 is secreted and exhibits properties of a granin. Nature Genet. 12: 303-308, 1996.

Johannsson, O.; Ostermeyer, E. A.; Hakansson, S.; Friedman, L. S.; Johansson, U.; Sellberg, G.; Brondum-Nielsen, K.; Sele, V.; Olsson, H.; King, M.-C.; Borg, A.: Founding BRCA1 mutations in hereditary breast and ovarian cancer in Southern Sweden. Am. J. Hum. Genet. 58: 441-450, 1996.

Kelsell, D. P.; Black, D. M.; Bishop, D. T.; Spurr, N. K.: Genetic analysis of the BRCA1 region in a large breast/ovarian family: refinement of the minimal region containing BRCA1. Hum. Molec. Genet. 2: 1823-1828, 1993.

Krainer, M.; Silva-Arrieta, S.; Fitzgerald, M. G.; Shimada, A.; Ishioka, C.; Kanamaru, R.; MacDonald, D. J.; Unsal, H.; Finkelstein, D. M.; Bowcock, A.; Isselbacher, K. J.; Haber, D. A.: Differential contributions of BRCA1 and BRCA2 to early-onset breast cancer. New Eng. J. Med. 336: 1416-1421, 1997.

Langston, A. A.; Malone, K. E.; Thompson, J. D.; Daling, J. R.; Ostrander, E. A.: BRCA1 mutations in a population-based sample of young women with breast cancer. New Eng. J. Med. 334: 137-142, 1996.

Langston, A. A.; Stanford, J. L.; Wicklund, K. G.; Thompson, J. D.; Blazej, R. G.; Ostrander, E. A.: Germ-line BRCA1 mutations in selected men with prostate cancer. (Letter) Am. J. Hum. Genet. 58: 881-885, 1996.

Lee, C. G.; Hurwitz, J.: Human RNA helicase A is homologous to the maleless protein of Drosophila. J. Biol. Chem. 268: 16822-16830, 1993.

Liede, A.; Rehal, P.; Vesprini, D.; Jack, E.; Abrahamson, J.; Narod, S. A.: A breast cancer patient of Scottish descent with germ-line mutations in BRCA1 and BRCA2. (Letter) Am. J. Hum. Genet. 62: 1543-1544, 1998.

Lynch, H. T.; Watson, P.: Genetic counselling and hereditary breast/ovarian cancer. (Letter) Lancet 339: 1181 only, 1992.

Margaritte, P.; Bonaiti-Pellie, C.; King, M.-C.; Clerget-Darpoux, F.: Linkage of familial breast cancer to chromosome 17q21 may not be restricted to early-onset disease. Am. J. Hum. Genet. 50: 1231-1234, 1992.

Merajver, S. D.; Pham, T. M.; Caduff, R. F.; Chen, M.; Poy, E. L.; Cooney, K. A.; Weber, B. L.; Collins, F. S.; Johnston, C.; Frank, T. S.: Somatic mutations in the BRCA1 gene in sporadic ovarian tumours. Nature Genet. 9: 439-443, 1995.

Miki, Y.; Swensen, J.; Shattuck-Eidens, D.; Futreal, P. A.; Harshman, K.; Tavtigian, S.; Liu, Q.; Cochran, C.; Bennett, L. M.; Ding, W.; Bell, R.; Rosenthal, J.; and 33 others: A strong candidate for the breast and ovarian cancer susceptibility gene BRCA1. Science 266: 66-71, 1994.

Narod, S.; Feunteun, J.; Lynch, H.; Watson, P.; Conway, T.; Lynch, J.; Lenoir, G.: A breast-ovarian cancer locus on chromosome 17. (Abstract) Am. J. Hum. Genet. 49 (suppl.): 352, 1991.

Narod, S.; Ford, D.; Devilee, P.; Barkardottir, R. B.; Eyfjord, J.; Lenoir, G.; Serova, O.; Easton, D.; Goldgar, D.; Breast Cancer Linkage Consortium: Genetic heterogeneity of breast-ovarian cancer revisited. (Letter) Am. J. Hum. Genet. 57: 957-958, 1995.

Narod, S.; Tonin, P.; Lynch, H.; Watson, P.; Feunteun, J.; Lenoir, G.: Histology of BRCA1-associated ovarian tumours. (Letter) Lancet 343: 236 only, 1994.

Narod, S. A.; Feunteun, J.; Lynch, H. T.; Watson, P.; Conway, T.; Lynch, J.; Lenoir, G. M.: Familial breast-ovarian cancer locus on chromosome 17q12-q23. Lancet 338: 82-83, 1991.

Narod, S. A.; Ford, D.; Devilee, P.; Barkardottir, R. B.; Lynch, H. T.; Smith, S. A.; Ponder, B. A. J.; Weber, B. L.; Garber, J. E.; Birch, J. M.; Cornelis, R. S.; Kelsell, D. P.; and 17 others: An evaluation of genetic heterogeneity in 145 breast-ovarian cancer families. Am. J. Hum. Genet. 56: 254-264, 1995.

Narod, S. A.; Risch, H.; Moslehi, R.; Dorum, A.; Neuhausen, S.; Olsson, H.; Provencher, D.; Radice, P.; Evans, G.; Bishop, S.; Brunet, J.-S.; Ponder, B. A. J.; Hereditary Ovarian Cancer Clinical Study Group: Oral contraceptives and the risk of hereditary ovarian cancer. New Eng. J. Med. 339: 424-428, 1998.

Neuhausen, S. L.; Mazoyer, S.; Friedman, L.; Stratton, M.; Offit, K.; Caligo, A.; Tomlinson, G.; Cannon-Albright, L.; Bishop, T.; Kelsell, D.; Solomon, E.; Weber, B.; Couch, F.; Struewing, J.; Tonin, P.; Durocher, F.; Narod, S.; Skolnick, M. H.; Lenoir, G.; Serova, O.; Ponder, B.; Stoppa-Lyonnet, D.; Easton, D.; King, M.-C.; Goldgar, D. E.: Haplotype and phenotype analysis of six recurrent BRCA1 mutations in 61 families: results of an international study. Am. J. Hum. Genet. 58: 271-280, 1996.

O'Connell, P.; Albertsen, H.; Matsunami, N.; Taylor, T.; Hundley, J. E.; Johnson-Pais, T. L.; Reus, B.; Lawrence, E.; Ballard, L.; White, R.; Leach, R. J.: A radiation hybrid map of the BRCA1 region. Am. J. Hum. Genet. 54: 526-534, 1994.

Petrij-Bosch, A.; Peelen, T.; van Vliet, M.; van Eijk, R.; Olmer, R.; Drusedau, M.; Hogervorst, F. B. L.; Hageman, S.; Arts, P. J. W.; Ligtenberg, M. J. L.; Meijers-Heijboer, H.; Klijn, J. G. M.; Vasen, H. F. A.; Cornelisse, C. J.; van't Veer, L. J.; Bakker, E.; van Ommen, G.-J. B.; Devilee, P.: BRCA1 genomic deletions are major founder mutations in Dutch breast cancer patients. Nature Genet. 17: 341-345, 1997.

Phelan, C. M.; Rebbeck, T. R.; Weber, B. L.; Devilee, P.; Ruttledge, M. H.; Lynch, H. T.; Lenoir, G. M.; Stratton, M. R.; Easton, D. F.; Ponder, B. A. J.; Cannon-Albright, L.; Larsson, C.; Goldgar, D. E.; Narod, S. A.: Ovarian cancer risk in BRCA1 carriers is modified by the HRAS1 variable number of tandem repeat (VNTR) locus. Nature Genet. 12: 309-311, 1996.

Piver, M. S.; Baker, T. R.; Jishi, M. F.; Sandecki, A. M.; Tsukada, Y.; Natarajan, N.; Mettlin, C. J.; Blake, C. A.: Familial ovarian cancer: a report of 658 families from the Gilda Radner Familial Ovarian Cancer Registry 1981-1991. Cancer 71: 582-588, 1993.

Ramus, S. J.; Friedman, L. S.; Gayther, S. A.; Ponder, B. A. J.; Bobrow, L. G.; van der Looji, M.; Papp, J.; Olah, E.: A breast/ovarian cancer patient with germline mutations in both BRCA1 and BRCA2. (Letter) Nature Genet. 15: 14-15, 1997.

Rebbeck, T. R.; Couch, F. J.; Kant, J.; Calzone, K.; DeShano, M.; Peng, Y.; Chen, K.; Garber, J. E.; Weber, B. L.: Genetic heterogeneity in hereditary breast cancer: role of BRCA1 and BRCA2. Am. J. Hum. Genet. 59: 547-553, 1996.

Roa, B. B.; Boyd, A. A.; Volcik, K.; Richards, C. S.: Ashkenazi Jewish population frequencies for common mutations in BRCA1 and BRCA2. Nature Genet. 14: 185-187, 1996.

Schorge, J. O.; Muto, M. G.; Welch, W. R.; Bandera, C. A.; Rubin, S. C.; Bell, D. A.; Berkowitz, R. S.; Mok, S. C.: Molecular evidence for multifocal papillary serous carcinoma of the peritoneum in patients with germline BRCA1 mutations. J. Nat. Cancer Inst. 90: 841-845, 1998.

Schrock, E.; Badger, P.; Larson, D.; Erdos, M.; Wynshaw-Boris, A.; Ried, T.; Brody, L.: The murine homolog of the human breast and ovarian cancer susceptibility gene Brca1 maps to mouse chromosome 11D. Hum. Genet. 97: 256-259, 1996.

Scully, R.; Anderson, S. F.; Chao, D. M.; Wei, W.; Ye, L.; Young, R. A.; Livingston, D. M.; Parvin, J. D.: BRCA1 is a component of the RNA polymerase II holoenzyme. Proc. Nat. Acad. Sci. 94: 5605-5610, 1997.

Scully, R.; Ganesan, S.; Brown, M.; De Caprio, J. A.; Cannistra, S. A.; Feunteun, J.; Schnitt, S.; Livingston, D. M.: Location of BRCA1 in human breast and ovarian cancer cells. Science 272: 123-125, 1996.

Serova, O.; Montagna, M.; Torchard, D.; Narod, S. A.; Tonin, P.; Sylla, B.; Lynch, H. T.; Feunteun, J.; Lenoir, G. M.: A high incidence of BRCA1 mutations in 20 breast-ovarian cancer families. Am. J. Hum. Genet. 58: 42-51, 1996.

Simard, J.; Feunteun, J.; Lenoir, G.; Tonin, P.; Normand, T.; The, V. L.; Vivier, A.; Lasko, D.; Morgan, K.; Rouleau, G. A.; Lynch, H.; Labrie, F.; Narod, S. A.: Genetic mapping of the breast-ovarian cancer syndrome to a small interval on chromosome 17q12-21: exclusion of candidate genes EDH17B2 and RARA. Hum. Molec. Genet. 2: 1193-1199, 1993.

Simard, J.; Tonin, P.; Durocher, F.; Morgan, K.; Rommens, J.; Gingras, S.; Samson, C.; Leblanc, J.-F.; Belanger, C.; Dion, F.; Liu, Q.; Skolnick, M.; Goldgar, D.; Shattuck-Eidens, D.; Labrie, F.; Narod, S. A.: Common origins of BRCA1 mutations in Canadian breast and ovarian cancer families. Nature Genet. 8: 392-398, 1994.

Skolnick, M. H.; Cannon-Albright, L. A.; Goldgar, D. E.; Ward, J. H.; Marshall, C. J.; Schumann, G. B.; Hogle, H.; McWhorter, W. P.; Wright, E. C.; Tran, T. D.; Bishop, D. T.; Kushner, J. P.; Eyre, H. J.: Inheritance of proliferative breast disease in breast cancer kindreds. Science 250: 1715-1720, 1990.

Smith, S. A.; Easton, D. F.; Evans, D. G. R.; Ponder, B. A. J.: Allele losses in the region 17q12-21 in familial breast and ovarian cancer involve the wild-type chromosome. Nature Genet. 2: 128-131, 1992.

Smith, T. M.; Lee, M. K.; Szabo, C. I.; Jerome, N.; McEuen, M.; Taylor, M.; Hood, L.; King, M.-C.: Complete genomic sequence and analysis of 117 kb of human DNA containing the gene BRCA1. Genome Res. 6: 1029-1049, 1996.

Sobol, H.; Mazoyer, S.; Narod, S. A.; Smith, S. A.; Black, D. M.; Kerbrat, P.; Jamot, B.; Solomon, E.; Ponder, B. A. J.; Guerin, D.: Genetic heterogeneity of early-onset familial breast cancer. Hum. Genet. 89: 381-383, 1992.

Steeg, P. S.: Granin expectations in breast cancer? Nature Genet. 12: 223-225, 1996.

Stoppa-Lyonnet, D.; Fricker, J. P.; Essioux, L.; Pages, S.; Limacher, J. M.; Sobol, H.; Laurent-Puig, P.; Thomas, G.: Segregation of two BRCA1 mutations in a single family. (Letter) Am. J. Hum. Genet. 59: 479-481, 1996.

Stratton, M. R.; Ford, D.; Neuhasen, S.; Seal, S.; Wooster, R.; Friedman, L. S.; King, M.-C.; Egilsson, V.; Devilee, P.; McManus, R.; Daly, P. A.; Smyth, E.; Ponder, B. A. J.; Peto, J.; Cannon-Albright, L.; Easton, D. F.; Goldgar, D. E.: Familial male breast cancer is not linked to the BRCA1 locus on chromosome 17q. Nature Genet. 7: 103-107, 1994.

Struewing, J. P.; Abeliovich, D.; Peretz, T.; Avishai, N.; Kaback, M. M.; Collins, F. S.; Brody, L. C.: The carrier frequency of the BRCA1 185delAG mutation is approximately 1 percent in Ashkenazi Jewish individuals. Nature Genet. 11: 198-200, 1995.

Struewing, J. P.; Brody, L. C.; Erdos, M. R.; Kase, R. G.; Giambarresi, T. R.; Smith, S. A.; Collins, F. S.; Tucker, M.A.: Detection of eight BRCA1 mutations in 10 breast/ovarian cancer families, including 1 family with male breast cancer. Am. J. Hum. Genet. 57: 1-7, 1995.

Struewing, J. P.; Hartge, P.; Wacholder, S.; Baker, S. M.; Berlin, M.; McAdams, M.; Timmerman, M. M.; Brody, L. C.; Tucker, M. A.: The risk of cancer associated with specific mutations of BRCA1 and BRCA2 among Ashkenazi Jews. New Eng. J. Med. 336: 1401-1408, 1997.

Szabo, C. I.; King, M.-C.: Population genetics of BRCA1 and BRCA2. (Editorial) Am. J. Hum. Genet. 60: 1013-1020, 1997.

Thompson, M. E.; Jensen, R. A.; Obermiller, P. S.; Page, D. L.; Holt, J. T.: Decreased expression of BRCA1 accelerates growth and is often present during sporadic breast cancer progression. Nature Genet. 9: 444-450, 1995.

Tonin, P.; Moslehi, R.; Green, R.; Rosen, B.; Cole, D.; Boyd, N.; Cutler, C.; Margolese, R.; Carter, R.; McGillivray, B.; Ives, E.; Labrie, F.; Gilchrist, D.; Morgan, K.; Simard, J.; Narod, S. A.: Linkage analysis of 26 Canadian breast and breast-ovarian cancer families. Hum. Genet. 95: 545-550, 1995.

Whittemore, A. S.; Harris, R.; Itnyre, J.; Collaborative Ovarian Cancer Group: Characteristics relating to ovarian cancer risk: collaborative analysis of 12 U.S. case-control studies. II. Invasive epithelial ovarian cancers in white women. Am. J. Epidemiol. 136: 1184-1203, 1992.

Wu, L. C.; Wang, Z. W.; Tsan, J. T.; Spillman, M. A.; Phung, A.; Xu, X. L.; Yang, M.-C. W.; Hwang, L.-Y.; Bowcock, A. M.; Baer, R.: Identification of a RING protein that can interact in vivo with the BRCA1 gene product. Nature Genet. 14: 430-440, 1996.

Zhang, S.; Grosse, F.: Domain structure of human nuclear DNA helicase II (RNA helicase A). J. Biol. Chem. 272: 11487-11494, 1997.

BREAST CANCER 2; BRCA2 (*600185)

JJM NOTE

BRCA2 may bear even greater relevance to cancer in general than *BRCA1,* if early observations on its role in male breast cancer, chronic lymphatic leukemia, and cancers of the pancreas, prostate, colon, and thyroid hold up (cf. Thorlacius et al. 1996). In 245 consecutive patients operated on for pancreatic cancer, 3 had germline mutations in *BRCA2;* in 41 tumors and cell lines, 15 had loss of heterozygosity at *BRCA2* (Goggins M et al.: Germline BRCA2 gene mutations in patients with apparently sporadic pancreatic carcinomas. Cancer Res. 56:5360–364, 1996).

A small proportion of breast cancer, in particular those cases arising at a young age, is due to the inheritance of dominant susceptibility genes conferring a high risk of the disease. The first of these genes to be confirmed was BRCA1 (113705), located on chromosome 17. Wooster et al. (1994) performed a genomic linkage search in 15 high-risk breast cancer families that were unlinked to the BRCA1 locus on 17q21. This analysis uncovered a second breast cancer susceptibility locus, BRCA2, located in a 6-cM interval on chromosome 13q12-q13. It appeared that BRCA2, while conferring a high risk of breast cancer, does not confer a substantially elevated risk of ovarian cancer as is the case with BRCA1.

Schutte et al. (1995) demonstrated a homozygous deletion in a 1-cM region of 13q12.3 lying within the 6-cM region identified as the BRCA2 'locus.' Thus it is possible that the BRCA2 gene is involved in multiple tumor types and that its function is that of a tumor suppressor rather than that of a dominant oncogene.

Weber et al. (1996) analyzed 3 large exons of BRCA2 (exons 10, 11, and 27) in 69 unselected samples of frozen breast tumor sections using the protein truncation test (PTT). They identified a truncating somatic mutation of BRCA2 in a primary ductal breast carcinoma: a 1-bp deletion of nucleotide 2882 in exon 11 that results in a frameshift, addition of 9 novel amino acids, and translation termination at codon 894. Loss of heterozygosity (LOH) was also demonstrated in 2 microsatellite markers, D13S260 and D13S171, which flank the BRCA2 locus.

Thorlacius et al. (1995) described a family with multiple cases of male breast cancer but no increase in female breast cancer. Seven males in 3 generations had breast cancer. Other members of the family had other forms of cancer. Linkage of breast cancer to the BRCA2 region was demonstrated and all the affected men shared the same haplotype for BRCA2 markers and loss of the other alleles in their tumors.

Wooster et al. (1995) reported identification of a gene from the 13q12-q13 region in which they detected 6 different germline mutations in breast cancer families. Each mutation caused serious disruption to the open reading frame of the transcriptional unit. By fine mapping in Icelandic breast cancer families using microsatellite markers from the region of chromosome 13, they had established that BRCA2 is likely to be located in a 600-kb interval centered around D13S171. An unexpected contribution to the fine localization of BRCA2 was provided by the detection of a homozygous somatic deletion in a single pancreatic cancer (Schutte et al., 1995). Northern analysis demonstrated that BRCA2 is encoded by a transcript of 10-12 kb which is present in normal breast epithelial cells, placenta, and a breast cancer cell line (MCF7). Loss of heterozygosity on 13q in sporadic breast and other cancers suggested to the authors that BRCA2 is a tumor suppressor gene.

Tavtigian et al. (1996) determined the complete coding sequence and exonic structure of BRCA2 and examined its pattern of expression. They reported the sequences for a set of PCR primers sufficient to screen the entire coding sequence of BRCA2 using genomic DNA. They also reported mutation analysis of BRCA2 in families selected on the basis of linkage analysis and/or the presence of one or more cases of male breast cancer. The composite BRCA2 cDNA sequence assembled by Tavtigian et al. (1996) consisted of 11,385 bp but did not include the polyadenylation signal or poly(A) tail. Conceptual translation of the cDNA revealed an ORF beginning at nucleotide 229 and encoding a protein of 3,418 amino acids. There was no similarity to other proteins. There was no signal sequence at the end of terminus, and no obvious membrane-spanning regions. Tavtigian et al. (1996) noted that, like BRCA1, the BRCA2 protein is highly charged; roughly one-quarter of the residues are acidic or basic. However, there were few clues as to the biochemical function of BRCA2. BRCA2 is composed of 27 exons distributed over roughly 70 kb of genomic DNA. The highest levels of expression were observed in breast and thymus, with slightly lower levels in lung, ovary, and spleen. Potentially deleterious sequence alterations were detected in 9 of 18 kindreds. All except 1, a deletion of 3 nucleotides, involved nucleotide deletions that altered the reading frame, leading to truncation of the BRCA2 protein. No missense or nonsense mutations were found. The authors noted that BRCA2 is remarkably similar to BRCA1 in other respects. Both have a large exon 11; both have translational start sites in exon 2; both have coding sequences that are AT-rich; both span approximately 70 kb of genomic DNA; and both are expressed at high levels in testis. The fact that BRCA2 participates in male breast cancer suggests that the 2 genes may not function in the same genetic pathway. The mutational profile of BRCA2 differs from that of BRCA1 , however. Microinsertions and point mutations are about as common in BRCA1 as microdeletions (which predominate in BRCA2).

Friedman et al. (1997) analyzed a population-based series of 54 male breast cancer cases from southern California for germline mutations in the BRCA1 and BRCA2 genes. A family history of breast and/or ovarian cancer in at least one first-degree relative was found in 9 patients (17%). A further 7 (13%) reported breast/ovarian cancer in at least one second-degree relative and in no first-degree relatives. The 54 patients showed no germline BRCA1 mutations. On the other hand, 2 of the male breast cancer patients (4% of the total) were found to carry novel truncating mutations in the BRCA2 gene. Only 1 of the 2 had a family history of cancer, with 1 case of ovarian cancer in a first-degree relative.

To define the role of BRCA2 in sporadic breast and ovarian cancer, Lancaster et al. (1996) screened the entire BRCA2 gene for mutations using a combination of techniques in 70 primary breast carcinomas and in 55 primary epithelial ovarian carcinomas. They found alterations in 2 of 70 breast tumors and none of the ovarian carcinomas. One alteration found in the breast cancers was a 2-bp deletion (4710delAG) which was subsequently shown to be a germline mutation; the other was a somatic missense mutation (asp3095-to-glu) of unknown significance. The results suggested to Lancaster et al. (1996) that BRCA2 is a very infrequent target for somatic inactivation in breast and ovarian carcinomas. Teng et al. (1996) had similar results; mutations in BRCA2 appeared to be infrequent in all cancers including breast carcinoma. However, a probable germline mutation in a pancreatic tumor cell line suggested a role for BRCA2 in that neoplasm. Miki et al. (1996) screened 100 primary breast cancers from Japanese patients using PCR-SSCP. They found 2 germline mutations and 1 somatic mutation. One of the germline mutations was an insertion of an Alu element into exon 22, which resulted in alternative splicing that skipped exon 22.

Thorlacius et al. (1996) studied 21 Icelandic families selected on the basis of a high frequency of breast cancer in females or the occurrence of one or more cases of male breast cancer. Strong evidence for linkage to the BRCA2 region was found in 16 families; these families shared a common haplotype in the BRCA2 region, suggesting a founder effect. In all 16 families there was evidence for a 5-bp deletion in exon 9, starting at nucleotide 999 in codon 257. In the Icelandic population, the 999del5 mutation has been found in individuals with different tumor types including cancer of the prostate, pancreas, ovary, colon, stomach, thyroid, cervix, and endometrium.

Among 7 large Icelandic breast cancer families, Gudmundsson et al. (1996) found that 5 showed strong evidence of linkage to the BRCA2 region. The maximum 2-point lod scores in the 5 families ranged from 1.06 to 3.19. Furthermore, haplotype analyses revealed a region with identical allele sizes in the families, suggesting to the authors that they inherited the mutation from a common ancestor. Cancer types other than breast cancer occurred in individuals segregating the affected haplotype in these families in both males and females.

Jensen et al. (1996) demonstrated that BRCA1 encodes a 190-kD protein with sequence homology and biochemical analogy to the granin protein family. They noted that BRCA2 also includes a motif similar to the granin consensus at the C terminus of the protein. Studying the expression of Brca2 in murine mammary epithelial cells as a function of proliferation and differentiation, Rajan et al. (1996) demonstrated that Brca2 mRNA expression is tightly regulated during mammary epithelial proliferation and differentiation, and that this regulation occurs coordinately with Brca1. Both genes showed mRNA expression that was upregulated in rapidly proliferating cells; was down-regulated in response to serum deprivation; was expressed in a cell cycle-dependent manner, peaking at the G1/S boundary; and was upregulated in the differentiating mammary epithelial cells in response to glucocorticoids. The results suggested that these genes are induced by, and may function in, overlapping regulatory pathways involved in the control of cell proliferation and differentiation.

Couch et al. (1996) generated a detailed transcription map of the 1.0-Mb region containing the BRCA2 gene. Evidence for 7 genes, 2 putative pseudogenes, and 9 additional putative transcription units was obtained. BRCA2 was one of the genes but all others were novel genes of unknown function.

Garcia-Marco et al. (1996) used fluorescence in situ hybridization to analyze chromosome 13 deletions in chronic lymphocytic leukemia (CLL). They demonstrated deletion of the 1-Mb 13q12.3 locus, which encompasses the BRCA2 gene, in 80% of 35 cases with CLL. Homozygous deletion of BRCA2 was detected in 60% of cases. Deletion of DBM (109543), a previously described 13q14 locus detected with the probe D13S25, was seen in 63% of cases. Garcia-Marco et al. (1996) concluded that their data provided evidence for the existence of a new tumor suppressor locus in B-cell CLL located at 13q12.3. They postulated that BRCA2, which is located within the minimal deletion region, is a candidate for this new B-cell CLL tumor suppressor locus.

Connor et al. (1997) described the mouse Brca2 gene. The gene maps to mouse chromosome 5, consistent with its localization on human 13q12. They sequenced cDNA for the entire 3,329-amino acid Brca2 protein and found that, like Brca1, Brca2 is relatively poorly conserved between humans and mice. Brca2 is transcribed in a diverse range of mouse tissues, and the pattern of expression is strikingly similar to that of Brca1.

The mouse Brca2 gene is expressed in adult thymus, testis and ovary, and in the midgestation embryo. It is also expressed in mammary epithelial cells and appears to be coordinately regulated with Brca1 expression. Both the Brca1 and Brca2 genes are highly expressed in rapidly proliferating cells, with the expression peaking at the G1/S boundary of the cell cycle. This expression profile and the common breast cancer phenotype seen in patients with mutations in these genes suggests that Brca1 and Brca2 may be acting in the same pathway. Sharan et al. (1997) identified an interaction of the Brca2 protein with the DNA-repair protein Rad51 (179617). Developmental arrest in Brca2-deficient embryos, their radiation sensitivity, and the association of Brca2 with Rad51 indicated that Brca2 may be an essential cofactor in the Rad51-dependent DNA repair of double-strand breaks, thereby explaining the tumor-suppressor function of Brca2.

Milner et al. (1997) showed that the portion of human BRCA2 encoded by its third exon shares homology with a known transcription factor and is capable of activating transcription, thus indicating a potential function of BRCA2. The exon 3 sequence at the N-terminus of BRCA2 (within a region highly conserved between human and mouse) showed sequence similarity to the activation domain of JUN (165160). They found that the activation potential within exon 3 is under negative control of inhibitory regions (IR1 and IR2) present immediately on either side of exon 3. The finding that BRCA2, like BRCA1, has transcriptional activation potential provides functional evidence of a relationship between the 2 proteins. Indeed, the fact that mutations found naturally in breast cancers disrupt the activation potential of both BRCA1 and BRCA2, indicates that compromising this activity may be an important step in the generation of a subset of familial breast cancers. Mutations found outside these activation domains may affect other, as yet unidentified, functions.

Among 5,318 Jewish subjects, Struewing et al. (1997) found 120 carriers of a BRCA1 or BRCA2 mutation. The BRCA1 mutations studied were 185delAG (113705.0003) and 5382insC (113705.0018); the BRCA2 mutation studied was 6174delT (600185.0009). By the age of 70, the estimated risk of breast cancer among carriers was 56%; of ovarian cancer, 16%; and of prostate cancer, 16%. There were no significant differences in the risk of breast cancer between carriers of BRCA1 mutations and carriers of BRCA2 mutations, and the incidence of colon cancer among the relatives of carriers was not elevated. They concluded that over 2% of Ashkenazi Jews carried mutations in BRCA1 or BRCA2 that conferred increased risks of breast, ovarian, and prostate cancer. Krainer et al. (1997) found definite BRCA2 mutations in 2 of 73 women with early onset (by age 32) breast cancer, suggesting that BRCA2 is associated with fewer cases than BRCA1 ($P = 0.03$).

Kinzler and Vogelstein (1997) made a distinction between gatekeeper genes and caretaker genes in the determination of cancer. Gatekeepers are genes that directly regulate the growth of tumors by inhibiting growth or promoting death. Each cell type has only one (or a few) gatekeepers, and inactivation of the given gatekeeper leads to a very specific tissue distribution of cancer; for example, inherited mutations of the retinoblastoma (RB1; 180200), von Hippel-Lindau (VHL; 193300), neurofibromatosis type I (NF1; 162200), and adenomatous polyposis coli (APC; 175100) genes lead to tumors of the retina, kidney, Schwann cells, and colon, respectively. Both the maternal and the paternal copies of the gene must be altered for tumor development. It is in connection with these gatekeeper, or tumor suppressor, genes that the Knudson 2-hit hypothesis was advanced. In contrast, inactivation of a caretaker gene does not promote tumor initiation directly. Rather, neoplasia occurs indirectly; inactivation leads to genetic instability that results in increased mutation of all genes, including gatekeepers. Once a tumor is initiated by inactivation of a gatekeeper gene, it may progress rapidly due to an accelerated rate of mutation in other genes that directly control cell birth or death. Known caretaker genes include the nucleotide-excision-repair genes that are responsible for xeroderma pigmentosum, mismatch-repair genes that cause hereditary nonpolyposis colorectal cancer, and probably the ATM gene, which is responsible for ataxia-telangiectasia. Kinzler and Vogelstein (1997) proposed that BRCA1 and BRCA2 should be added to the list of caretaker genes. Consistent with this hypothesis, mutations in BRCA1 and BRCA2 are rarely found in sporadic cancers, and the risk of cancer arising in people with BRCA mutations is relatively low. The distinction between gatekeepers and caretakers has important practical, as well as theoretical, ramifications. Tumors that have defective caretaker genes present an additional therapeutic target. Such tumors would be expected to respond favorably to therapeutic agents that induce the type of genomic damage that is normally detected or repaired by the particular caretaker gene involved. The discovery by Sharan et al. (1997) that most cells with defective Brca2 genes are sensitive to gamma-irradiation suggests that tumors from breast cancer patients with inherited BRCA mutations should be more sensitive to such radiation than other breast cancers.

Patel et al. (1998) showed that in culture, mouse cells harboring truncated Brca2 exhibited a proliferative impediment that worsened with successive passages. Arrest in the G1 and G2/M phases was accompanied by elevated p53 and p21 expression. Increased sensitivity to genotoxic agents, particularly ultraviolet light and methylmethanesulfonate, showed that Brca2 function was essential for the ability to survive DNA damage. Checkpoint activation and apoptotic mechanisms were largely unaffected, thereby implicating Brca2 in repair. This was substantiated by the spontaneous accumulation of chromosomal abnormalities, including breaks and aberrant chromatid exchanges. These findings defined a function of Brca2 in DNA repair, whose loss precipitates replicative failure, mutagen sensitivity, and genetic instability reminiscent of Bloom syndrome (210900) and Fanconi anemia (see 227650) (Patel et al., 1998).

Chen et al. (1998) used mammalian expression vectors to transfect cells with BRCA1 and BRCA2 as well as with several antibodies to recognize these proteins in order to study their subcellular localizations. They showed that BRCA1 and BRCA2 coexist in a biochemical complex and colocalize in subnuclear foci in somatic cells and on the axial elements of developing synaptonemal complexes. Like BRCA1 and RAD51, BRCA2 relocates to replication sites following exposure of S phase cells to hydroxyurea or UV irradiation. Thus, BRCA1 and BRCA2 participate together in a pathway (or pathways) associated with the activation of double-strand break repair and/or homologous recombination. Dysfunction of this pathway may be a general phenomenon in the majority of cases of hereditary breast and/or ovarian cancer.

In 2 large, systematically ascertained families in which breast cancer showed linkage to 13q12-q13, Easton et al. (1997) determined the penetrance of the BRCA2 gene by use of a maximum-likelihood method to incorporate both cancer-incidence data and 13q marker typings. The cumulative risk of breast cancer in female gene carriers was estimated to be 59.8% by age 50 years (95% confidence interval (CI) 25.9 to 78.5%) and 79.5% by age 70 years (95% CI 28.9 to 97.5%). Cumulative risk of breast cancer in male carriers was estimated to be 6.3% (95% CI 1.4 to 25.6%) by age 70 years. There was no evidence of any risk difference between the 2 families. The results indicated that the lifetime breast cancer risk in BRCA2 carriers, for at least a subset of mutations, is comparable to that for BRCA1, in which Easton et al. (1995) estimated the risk of breast cancer to be 51% by age 50 years. The corresponding estimates by age 60 years were 54% for BRCA1-mutation carriers, compared with 71% for BRCA2. A significant excess of ovarian cancer in BRCA2 gene carriers was observed (relative risk 17.69, based on 3 cases), but the absolute risk of ovarian cancer was less than that reported for BRCA1. Significant excesses of laryngeal cancer (relative risk 7.67, based on 2 possible carriers) and prostate cancer (relative risk 2.89, based on 5 possible carriers) were also observed. One case of ocular melanoma, as well as a second eye cancer of unspecificed histology, occurred in obligate gene carriers.

Szabo and King (1997) collated information on the population genetics of BRCA1 and BRCA2 in populations from many countries of Europe as well as the U.S., Canada, and Japan.

According to the conclusions of the Breast Cancer Linkage Consortium (1997), the histology of breast cancers in women predisposed by reason of carrying BRCA1 and BRCA2 mutations differs from that in sporadic cases, and there are differences between breast cancers in carriers of BRCA1 and BRCA2 mutations. The findings were interpreted as suggesting that breast cancer due to BRCA1 has a different natural history to BRCA2 or apparently sporadic disease, which may have implications for screening and management.

The most frequent type of complete hydatidiform mole is a 46,XX homozygote formed by the fertilization of an empty ovum by a single haploid sperm that later duplicates its chromosomes to give a diploid tumor. The homozygous nature of these complete hydatidiform moles makes them a unique resource for human genome studies. Taillon-Miller et al. (1997) pointed out that they can serve as homozygous controls in the development of single nucleotide polymorphism (SNP) markers and provide a way to obtain long-range haplotypes that are useful in population studies. The use of a homozygous control makes it possible to estimate the allele frequencies of the SNP markers in any population by sequencing pooled DNA samples. Taillon-Miller et al. (1997) presented evidence of homozygosity of a complete hydatidiform mole using 20 diallelic markers distributed across the genome. Its usefulness as a homozygous control in SNP development and as a resource for long-range haplotype determination was demonstrated using 11 newly discovered loci in the BRCA2 region of 13q12-q13.

To investigate both mutation origin and mutation-specific phenotypes due to BRCA2 mutations, Neuhausen et al. (1998) constructed a haplotype of 10 polymorphic short tandem repeat (STR) markers flanking the BRCA2 locus, in a set of 111 breast or breast/ovarian cancer families selected for having 1 of 9 recurrent BRCA2 mutations. Six of the individual mutations were estimated to have arisen 400 to 2,000 years ago. In particular, the 6174delT mutation (600185.0009), found in approximately 1% of individuals of Ashkenazi Jewish ancestry, was estimated to have arisen 29 generations ago (1-lod support interval = 22-38). This is substantially more recent than the estimated age of the BRCA1 185delAG (113705.0003) mutation, 46 generations, derived from an analogous study of BRCA1 mutations. In general, Neuhausen et al. (1998) found no evidence of multiple origins of identical BRCA2 mutations. The data were consistent with the previous report of a higher incidence of ovarian cancer in families with mutations in a 3.3-kb region of exon 11, known as the ovarian cancer cluster region (OCCR), but the higher incidence was not statistically significant. There was significant evidence that age at diagnosis of breast cancer varied by mutation, although only 8% of the variance in age at diagnosis could be explained by the specific mutation, and there was no evidence of family-specific effects. Cases associated with mutations in the OCCR had a significantly older mean age at diagnosis than were seen in those outside this region (48 vs 42 years; P = 0.0005).

Three Jewish founder mutations, 185delAG (113705.0003) and 5382insC (113705.0018) in BRCA1 and 6174delT (600185.0009) in BRCA2, have been identified in breast cancer and ovarian cancer Ashkenazi patients. Friedman et al. (1998) pooled results from 4 cancer/genetic centers in Israel to analyze approximately 1,500 breast/ovarian cancer Ashkenazi patients for the presence of double heterozygosity as well as homozygosity for any of these mutations. Although the small number of cases precluded definite conclusions, the results suggested that the phenotypic effects of double heterozygosity for BRCA1 and BRCA2 germline mutations were not cumulative. This was in agreement with the observation that the phenotype of mice that are homozygous knockouts for the BRCA1 and BRCA2 genes is similar to that of mice that are BRCA1 knockouts. This suggests that the BRCA1 mutation is epistatic over the BRCA2 mutation. Two of the double heterozygotes described had had reproductive problems: one with primary sterility and irregular menses and another with premature menopause at the age of 37 years. Friedman et al. (1998) suggested that identification of additional carriers of more than one mutation will increase our understanding between various mutations and will improve genetic counseling.

Breast Cancer Linkage Consortium: Pathology of familial breast cancer: differences between breast cancers in carriers of BRCA1 and BRCA2 mutations and sporadic cases. Lancet 349: 1505-1510, 1997.

Chen, J.; Silver, D. P.; Walpita, D.; Cantor, S. B.; Gazdar, A. F.; Tomlinson, G.; Couch, F. J.; Weber, B. L.; Ashley, T.; Livingston, D. M.; Scully, R.: Stable interaction between the products of the BRCA1 and BRCA2 tumor suppressor genes in mitotic and meiotic cells. Molec. Cell 2: 317-328, 1998.

Connor, F.; Smith, A.; Wooster, R.; Stratton, M.; Dixon, A.; Campbell, E.; Tait, T.-M.; Freeman, T.; Ashworth, A.: Cloning, chromosomal mapping and expression pattern of the mouse Brca2 gene. Hum. Molec. Genet. 6: 291-300, 1997.

Couch, F. J.; Rommens, J. M.; Neuhausen, S. L.; Belanger, C.; Dumont, M.; Abel, K.; Bell, R.; Berry, S.; Bogden, R.; Cannon-Albright, L.; Farid, L.; Frye, C.; and 30 others: Generation of an integrated transcription map of the BRCA2 region on chromosome 13q12-q13. Genomics 36: 86-99, 1996.

Easton, D. F.; Ford, D.; Bishop, D. T.; Breast Cancer Linkage Consortium: Breast and ovarian cancer incidence in BRCA1-mutation carriers. Am. J. Hum. Genet. 56: 265-271, 1995.

Easton, D. F.; Steele, L.; Fields, P.; Ormiston, W.; Averill, D.; Daly, P. A.; McManus, R.; Neuhausen, S. L.; Ford, D.; Wooster, R.; Cannon-Albright, L. A.; Stratton, M. R.; Goldgar, D. E.: Cancer risks in two large breast cancer families linked to BRCA2 on chromosome 13q12-13. Am. J. Hum. Genet. 61: 120-128, 1997.

Friedman, E.; Bar-Sade, R. B.; Kruglikova, A.; Risel, S.; Levy-Lahad, E.; Halle, D.; Bar-On, E.; Gershoni-Baruch, R.; Dagan, E.; Kepten, I.; Peretz, T.; Lerer, I.; Wienberg, N.; Shushan, A.; Abeliovich, D.: Double heterozygotes for the Ashkenazi founder mutations in BRCA1 and BRCA2 genes. (Letter) Am. J. Hum. Genet. 63: 1224-1227, 1998.

Friedman, L. S.; Gayther, S. A.; Kurosaki, T.; Gordon, D.; Noble, B.; Casey, G.; Ponder, B. A. J.; Anton-Culver, H.: Mutation analysis of BRCA1 and BRCA2 in a male breast cancer population. Am. J. Hum. Genet. 60: 313-319, 1997.

Garcia-Marco, J. A.; Caldas, C.; Price, C. M.; Wiedemann, L. M.; Ashworth, A.; Catovsky, D.: Frequent somatic deletion of the 13q12.3 locus encompassing BRCA2 in chronic lymphocytic leukemia. Blood 88: 1568-1575, 1996.

Gudmundsson, J.; Johannesdottir, G.; Arason, A.; Bergthorsson, J. T.; Ingvarsson, S.; Egilsson, V.; Barkardottir, R. B.: Frequent occurrence of BRCA2 linkage in Icelandic breast cancer families and segregation of a common BRCA2 haplotype. Am. J. Hum. Genet. 58: 749-756, 1996.

Jensen, R. A.; Thompson, M. E.; Jetton, T. L.; Szabo, C. I.; van der Meer, R.; Helou, B.; Tronick, S. R.; Page, D. L.; King, M.-C.; Holt, J. T.: BRCA1 is secreted and exhibits properties of a granin. Nature Genet. 12: 303-308, 1996.
Kinzler, K. W.; Vogelstein, B.: Gatekeepers and caretakers. Nature 386: 761-763, 1997.
Krainer, M.; Silva-Arrieta, S.; Fitzgerald, M. G.; Shimada, A.; Ishioka, C.; Kanamaru, R.; MacDonald, D. J.; Unsal, H.; Finkelstein, D. M.; Bowcock, A.; Isselbacher, K. J.; Haber, D. A.: Differential contributions of BRCA1 and BRCA2 to early-onset breast cancer. New Eng. J. Med. 336: 1416-1421, 1997.
Lancaster, J. M.; Wooster, R.; Mangion, J.; Phelan, C. M.; Cochran, C.; Gumbs, C.; Seal, S.; Barfoot, R.; Collins, N.; Bignell, G.; Patel, S.; Hamoudi, R.; Larsson, C.; Wiseman, R. W.; Berchuck, A.; Iglehart, J. D.; Marks, J. R.; Ashworth, A.; Stratton, M. R.; Futreal, P. A.: BRCA2 mutations in primary breast and ovarian cancers. Nature Genet. 13: 238-240, 1996.
Lerer, I.; Wang, T.; Peretz, T.; Sagi, M.; Kaduri, L.; Orr-Urtreger, A.; Stadler, J.; Gutman, H.; Abeliovich, D.: The 8765delAG mutation in BRCA2 is common among Jews of Yemenite extraction. (Letter) Am. J. Hum. Genet. 63: 272-274, 1998.
Liede, A.; Rehal, P.; Vesprini, D.; Jack, E.; Abrahamson, J.; Narod, S. A.: A breast cancer patient of Scottish descent with germ-line mutations in BRCA1 and BRCA2. (Letter) Am. J. Hum. Genet. 62: 1543-1544, 1998.
Miki, Y.; Katagiri, T.; Kasumi, F.; Yoshimoto, T.; Nakamura, Y.: Mutation analysis in the BRCA2 gene in primary breast cancers. Nature Genet. 13: 245-247, 1996.
Milner, J.; Ponder, B.; Hughes-Davies, L.; Seltmann, M.; Kouzarides, T.: Transcriptional activation functions in BRCA2. (Letter) Nature 386: 772-773, 1997.
Neuhausen, S.; Gilewski, T.; Norton, L.; Tran, T.; McGuire, P.; Swensen, J.; Hampel, H.; Borgen, P.; Brown, K.; Skolnick, M.; Shattuck-Eidens, D.; Jhanwar, S.; Goldgar, D.; Offit, K.: Recurrent BRCA2 6174delT mutations in Ashkenazi Jewish women affected by breast cancer. Nature Genet. 13: 126-128, 1996.
Neuhausen, S. L.; Godwin, A. K.; Gershoni-Baruch, R.; Shubert, E.; Garber, J.; Stoppa-Lyonnet, D.; Olah, E.; Csokay, B.; Serova, O.; Lalloo, F.; Osorio, A.; Stratton, M.; and 18 others: Haplotype and phenotype analysis of nine recurrent BRCA2 mutations in 111 families: results of an international study. Am. J. Hum. Genet. 62: 1381-1388, 1998.
Oddoux, C.; Struewing, J. P.; Clayton, C. M.; Neuhausen, S.; Brody, L. C.; Kaback, M.; Haas, B.; Norton, L.; Borgen, P.; Jhanwar, S.; Goldgar, D.; Ostrer, H.; Offit, K.: The carrier frequency of the BRCA2 6174delT mutation among Ashkenazi Jewish individuals is approximately 1%. Nature Genet. 14: 188-190, 1996.
Ozcelik, H.; Schmocker, B.; Di Nicola, N.; Shi, X.-H.; Langer, B.; Moore, M.; Taylor, B. R.; Narod, S. A.; Darlington, G.; Andrulis, I. L.; Gallinger, S.; Redston, M.: Germline BRCA2 6174delT mutations in Ashkenazi Jewish pancreatic cancer patients. (Letter) Nature Genet. 16: 17-18, 1997.
Patel, K. J.; Yu, V. P. C. C.; Lee, H.; Corcoran, A.; Thistlethwaite, F. C.; Evans, M. J.; Colledge, W. H.; Friedman, L. S.; Ponder, B. A. J.; Venkitaraman, A. R.: Involvement of Brca2 in DNA repair. Molec. Cell 1: 347-357, 1998.
Phelan, C. M.; Lancaster, J. M.; Tonin, P.; Gumbs, C.; Cochran, C.; Carter, R.; Ghadirian, P.; Perret, C.; Moslehi, R.; Dion, F.; Faucher, M. C.; Dole, K.; Karimi, S.; Foulkes, W.; Lounis, H.; Warner, E.; Goss, P.; Anderson, D.; Larsson, C.; Narod, S. A.; Futreal, P. A.: Mutation analysis of the BRCA2 gene in 49 site-specific breast cancer families. Nature Genet. 13: 120-122, 1996.
Rajan, J. V.; Wang, M.; Marquis, S. T.; Chodosh, L. A.: Brca2 is coordinately regulated with Brca1 during proliferation and differentiation in mammary epithelial cells. Proc. Nat. Acad. Sci. 93: 13078-13083, 1996.
Roa, B. B.; Boyd, A. A.; Volcik, K.; Richards, C. S.: Ashkenazi Jewish population frequencies for common mutations in BRCA1 and BRCA2. Nature Genet. 14: 185-187, 1996.
Schutte, M.; da Costa, L. T.; Hahn, S. A.; Moskaluk, C.; Hoque, A. T. M. S.; Rozenblum, E.; Weinstein, C. L.; Bittner, M.; Meltzer, P. S.; Trent, J. M.; Yeo, C. J.; Hruban, R. H.; Kern, S. E.: Identification by representational difference analysis of a homozygous deletion in pancreatic carcinoma that lies with the BRCA2 region. Proc. Nat. Acad. Sci. 92: 5950-5954, 1995.
Sharan, S. K.; Morimatsu, M.; Albrecht, U.; Lim, D.-S.; Regel, E.; Dinh, C.; Sands, A.; Eichele, G.; Hasty, P.; Bradley, A.: Embryonic lethality and radiation hypersensitivity mediated by Rad51 in mice lacking Brca2. Nature 386: 804-810, 1997.
Stratton, M. R.; Ford, D.; Neuhasen, S.; Seal, S.; Wooster, R.; Friedman, L. S.; King, M.-C.; Egilsson, V.; Devilee, P.; McManus, R.; Daly, P. A.; Smyth, E.; Ponder, B. A. J.; Peto, J.; Cannon-Albright, L.; Easton, D. F.; Goldgar, D. E.: Familial male breast cancer is not linked to the BRCA1 locus on chromosome 17q. Nature Genet. 7: 103-107, 1994.
Struewing, J. P.; Hartge, P.; Wacholder, S.; Baker, S. M.; Berlin, M.; McAdams, M.; Timmerman, M. M.; Brody, L. C.; Tucker, M. A.: The risk of cancer associated with specific mutations of BRCA1 and BRCA2 among Ashkenazi Jews. New Eng. J. Med. 336: 1401-1408, 1997.
Szabo, C. I.; King, M.-C.: Population genetics of BRCA1 and BRCA2. (Editorial) Am. J. Hum. Genet. 60: 1013-1020, 1997.
Taillon-Miller, P.; Bauer-Sardina, I.; Zakeri, H.; Hillier, L.; Mutch, D. G.; Kwok, P.-Y.: The homozygous complete hydatidiform mole: a unique resource for genome studies. Genomics 46: 307-310, 1997.
Tavtigian, S. V.; Simard, J.; Rommens, J.; Couch, F.; Shattuck-Eidens, D.; Neuhausen, S.; Merajver, S.; Thorlacius, S.; Offit, K.; Stoppa-Lyonnet, D.; Belanger, C.; Bell, R.; and 37 others: The complete BRCA2 gene and mutations in chromosome 13q-linked kindreds. Nature Genet. 12: 333-337, 1996.
Teng, D. H.-F.; Bogden, R.; Mitchell, J.; Baumgard, M.; Bell, R.; Berry, S.; David, T.; Ha, P. C.; Kehrer, R.; Jammulapati, S.; Chen, Q.; Offit, K.; Skolnick, M. H.; Tavtigian, S. V.; Jhanwar, S.; Swedlund, B.; Wong, A. K. C.; Kamb, A.: Low incidence of BRCA2 mutations in breast carcinoma and other cancers. Nature Genet. 13: 241-244, 1996.
Thorlacius, S.; Olafsdottir, G.; Tryggvadottir, L.; Neuhausen, S.; Jonasson, J. G.; Tavtigian, S. V.; Tulinius, H.; Ogmundsdottir, H. M.; Eyfjord, J. E.: A single BRCA2 mutation in male and female breast cancer families from Iceland with varied cancer phenotypes. Nature Genet. 13: 117-122, 1996.
Thorlacius, S.; Sigurdsson, S.; Bjarnadottir, H.; Olafsdottir, G.; Jonasson, J. G.; Tryggvadottir, L.; Tulinius, H.; Eyfjord, J. E.: Study of a single BRCA2 mutation with high carrier frequency in a small population. Am. J. Hum. Genet. 60: 1079-1084, 1997.
Thorlacius, S.; Tryggvadottir, L.; Olafsdottir, G. H.; Jonasson, J. G.; Ogmundsdottir, H. M.; Tulinius, H.; Eyfjord, J. E.: Linkage to BRCA2 region in hereditary male breast cancer. Lancet 346: 544-545, 1995.
Weber, B. H. F.; Brohm, M.; Stec, I.; Backe, J.; Caffier, H.: A somatic truncating mutation in BRCA2 in a sporadic breast tumor. Am. J. Hum. Genet. 59: 962-964, 1996.
Wooster, R.; Bignell, G.; Lancaster, J.; Swift, S.; Seal, S.; Mangion, J.; Collins, N.; Gregory, S.; Gumbs, C.; Micklem, G.; Barfoot, R.; Hamoudi, R.; Patel, S.; Rice, C.; Biggs, P.; Hashim, Y.; Smith, A.; Connor, F.; Arason, A.; Gudmundsson, J.; Ficenec, D.; Keisell, D.; Ford, D.; Tonin, P.; Biship D. T.; Spurr, N. K.; Ponder, B. A. J.; Eeles, R.; Peto, J.; Devilee, P.; Cornelisse, C.; Lynch, H.; Narod, S.; Lenoir, G.; Egilsson, V.; Barkadottir, R. B.; Easton, D. F.; Bentley, D. R.; Futreal, P. A.; Ashworth, A.; Stratton, M. R.: Identification of the breast cancer susceptibility gene BRCA2. Nature 378: 789-792, 1995.
Wooster, R.; Neuhausen, S. L.; Mangion, J.; Quirk, Y.; Ford, D.; Collins, N.; Nguyen, K.; Seal, S.; Tran, T.; Averill, D.; Fields, P.; Marshall, G.; Narod, S.; Lenoir, G. M.; Lynch, H.; Feunteun, J.; Devilee, P.; Cornelisse, C. J.; Menko, F. H.; Daly, P. A.; Ormiston, W.; McManus, R.; Pye, C.; Lewis, C. M.; Cannon-Albright, L. A.; Peto, J.; Ponder, B. A. J.; Skolnick, M. H.; Easton, D. F.; Goldgar, D. E.; Stratton, M. R.: Localization of a breast cancer susceptibility gene, BRCA2, to chromosome 13q12-13. Science 265: 2088-2090, 1994.

BREAST CANCER, DUCTAL 1 (*211410)

JJM NOTE

Pursuit of this gene seems to be suspended, in view of more compelling prospects.

Familial breast cancer shares several features with hereditary tumors that satisfy the conditions predicted by the 2-hit hypothesis of Knudson (1971); tumors are frequently bilateral and multifocal. They tend to occur in premenopausal women, while the overall incidence of breast cancer shows a peak at postmenopausal age; and male relatives in high-risk families are more often affected than are males in the general population. Lundberg et al. (1987) tested their hypothesis that the pathogenesis of breast cancer in males and young females involves a chromosomal rearrangement that serves to unmask a recessive cancer gene. Lundberg et al. (1987) studied 10 cases of ductal breast cancer: 8 premenopausal females and 2 males. In 3 females and 1 male, somatic loss of constitutional heterozygosity was observed at loci on chromosome 13 in primary tumor tissue. In 2 cases, specific loss of heterozygosity at 3 distinct genetic loci along the length of chromosome 13 was observed. In a third case, concurrent loss of alleles at loci on chromosomes 2, 13, 14, and 20 was detected, whereas a fourth case showed loss of heterozygosity for chromosomes 5 and 13. In each instance, the data were consistent with loss of one of the homologous chromosomes by mitotic nondisjunction. The relative specificity of the events was suggested by the fact that analysis of loci on several other chromosomes showed retention of constitutional heterozygosity. On the other hand, analyses of other breast cancers, including comedocarcinoma, medullary carcinoma, and juvenile secretory carcinoma, showed no loss of alleles at loci on chromosome 13. Lundberg et al. (1987) interpreted these data as suggesting that in a substantial proportion of cases, the pathogenesis of ductal breast cancer involves the unmasking of a recessive locus on chromosome 13 and involvement of the same locus in heritable forms of this disease. Lundberg et al. (1987) raised the possibility of using molecular cytogenetics as an adjunct to histopath ology in the diagnosis of breast tumors. Friend et al. (1987) studied the same 10 ductal carcinomas of the breast. They found no deletion of the RB gene (180200), even though karyotypic analysis of tumor cells indicated specific deletion of the chromosomal segment mapping to 13q14.1.

Cytogenetic studies indicated that structural rearrangements, including deletions and duplications, of chromosome 1 are the most common aberrations in breast carcinoma (Kovacs, 1978; Rodgers et al., 1984). Furthermore, Anderson et al. (1985) found weak but suggestive evidence of linkage between a hereditary form of breast cancer and the Rh locus (maximum lod score = 2.28 at theta = 0.0). On the basis of these findings, Genuardi et al. (1989) studied loss of heterozygosity for DNA markers in 37 ductal breast carcinomas and in constitutional DNA from the same persons. A high frequency (41%) of reduction to homozygosity was detected with a probe that recognizes the highly polymorphic locus D1Z2, localized at 1p36. Somatic loss of heterozygosity at D1Z2 was most frequent in patients with a strong family history of breast cancer (60%), in patients with diagnosis before 45 years of age (70%), and in those with multiple tumors or tumor foci (50%) than in patients with none of the characteristics of hereditary tumors (21%). Genuardi et al. (1989) suggested that inactivation of a tumor suppressor gene (BCDS2; 211420) located on the distal portion of 1p, alone or in combination with other genetic changes, may represent a fundamental step in the pathogenesis of ductal carcinoma of the breast. In 9 primary breast cancers, Rodgers et al. (1984) found changes in 1q21-qter as the only cytogenetic change shared by all. In studies of 48 breast cancers, Chen et al. (1989) found allelic loss at SPTA1 (182860) in 6 of 26 (23%) informative tumors, at AT3 (107300) in 3 of 13 (23%), and at D1S53 in 5 of 19 (26%). They suggested that the region 1q23-q32 may contain gene(s) that, when inactivated, contribute to breast cancer.

Anderson, D. E.; Ferrell, R. E.; Williams, W. R.: A linkage study of human breast cancer. Cytogenet. Cell Genet. 40: 568-569, 1985.
Chen, L.-C.; Dollbaum, C.; Smith, H. S.: Loss of heterozygosity on chromosome 1q in human breast cancer. Proc. Nat. Acad. Sci. 86: 7204-7207, 1989.
Friend, S. H.; Horowitz, J. M.; Gerber, M. R.; Wang, X.-F.; Bogenmann, E.; Li, F. P.; Weinberg, R. A.: Deletions of a DNA sequence in retinoblastomas and mesenchymal tumors: organization of the sequence and its encoded protein. Proc. Nat. Acad. Sci. 84: 9059-9063, 1987.
Genuardi, M.; Tsihira, H.; Anderson, D. E.; Saunders, G. F.: Distal deletion of chromosome 1p in ductal carcinoma of the breast. Am. J. Hum. Genet. 45: 73-82, 1989.
Knudson, A. G., Jr.: Mutation and cancer: a statistical study of retinoblastoma. Proc. Nat. Acad. Sci. 68: 820-823, 1971.
Kovacs, G.: Abnormalities of chromosome no. 1 in human solid malignant tumors. Int. J. Cancer 21: 688-694, 1978.
Lundberg, C.; Skoog, L.; Cavenee, W. K.; Nordenskjold, M.: Loss of heterozygosity in human ductal breast tumors indicates a recessive mutation on chromosome 13. Proc. Nat. Acad. Sci. 84: 2372-2376, 1987.
Rodgers, C. S.; Hill, S. M.; Hulten, M. A.: Cytogenetic analysis in human breast carcinoma. I. Nine cases in the diploid range investigated using direct preparations. Cancer Genet. Cytogenet. 13: 95-119, 1984.

BREAST CANCER, DUCTAL 2 (*211420)

JJM NOTE

Pursuit of this gene seems to be suspended, in view of more compelling prospects.

See 211410.

BREAST CANCER, 11;22 TRANSLOCATION ASSOCIATED (*600048)

JJM NOTE

This represents another lead that should be picked up again.

The most frequently occurring constitutional reciprocal translocation in man is t(11;22)(q23;q11), which has been described in more than 100 unrelated families (Iselius et al., 1983). Lindblom et al. (1994) observed a patient with this translocation and breast cancer, prompting a study of the relationship between the 2 conditions. Among 8 families with a total of 22 balanced carriers, 1 case of breast cancer was found in each of 5 families. In another family, an unknown malignancy was reported in 1 member. No other malignancies were found among these patients. The number of breast cancer cases was significantly higher than expected among the translocation carriers (P less than 0.001). In the 7 families studied, the breakpoints showed the same localization with the markers used. The information should be useful in connection with counseling and screening and suggests the involvement of a gene on 11q and/or 22q in the pathogenesis of breast cancer. (At least 4 loci involved in hereditary breast cancer had previously been identified: the BRCA1 locus (113705) on 17q, the BRCA2 (600185) locus on 13q12-q13, the TP53 gene (191170) on 17p, and the androgen receptor gene (AR; 313700) on the X chromosome in which mutations have been found in cases of male breast cancer (313700.0016).)

Iselius, L.; Lindsten, J.; Aurias, A.; Fraccaro, M.; Bastard, C.; Bottelli, A. M.; Bui, T.-H.; Caufin, D.; Dalpra, L.; Delendi, N.; Dutrillaux, B.; Fukushima, Y.; Geraedts, J. P. M.; de Grouchy, J.; Gyftodimou, J.; Hanley, A. L.; Hansmann, I.; Ishii, T.; Jalbert, P.; Jingeleski, S.; Kajii, T.; von Koskull, H.; Niikawa, N.; Noel, B.; Pasquali, F.; Probeck, H. D.; Robinson, A.; Roncarati, E.; Sachs, E.; Scappaticci, S.; Schwinger, E.; Simoni, G.; Veenema, H.; Vigi, V.; Volpato, S.; Wegner, R.-D.; Welch, J. P.; Winsor, E. J. T.; Zhang, S.; Zuffardi, O.: The 11q;22q translocation: a collaborative study of 20 new cases and analysis of 110 families. Hum. Genet. 64: 343-355, 1983.
Lindblom, A.; Sandelin, K.; Iselius, L.; Dumanski, J.; White, I.; Nordenskjold, M.; Larsson, C.: Predisposition for breast cancer in carriers of constitutional translocation 11q;22q. Am. J. Hum. Genet. 54: 871-876, 1994.

CERUMEN, VARIATION IN (*117800)

JJM NOTE

This phenotype would seem to be a mappable gene that could have public health implications as an easily screenable risk factor.

In Japanese, Matsunaga (1962) described a dimorphism of ear wax, the 2 types being wet and dry. This variation has been studied extensively in Japan since at least 1934. Less attention has been given to this variation elsewhere, probably because Caucasians and blacks have only the wet type of cerumen. In 80 to 85% of Japanese, the cerumen is gray, dry and brittle. It is referred to as 'rice-bran ear wax' in Japanese. In the other Japanese, the cerumen is brown, sticky and wet. This is referred to as 'honey ear wax,' 'oily ear wax' or 'cat ear wax.' In all except about 0.5% of Japanese, classification is simple. Family studies indicate monofactorial inheritance, with the rarer phenotype, wet wax, being dominant. Wet cerumen is often associated with axillary odor, which because of its rarity in Japan, is considered in the lay mind a pathologic state requiring medical attention. Petrakis et al. (1967) found a high frequency of dry cerumen in pure-blooded American Indians. Ibraimov (1991) presented data on the high frequency of dry cerumen in Mongoloid populations and low frequency among Europoids. Intermediate frequencies were found among peoples of subequatorial Africa. No qualitative differences in chemical composition have been identified (Kataura and Kataura, 1967). Petrakis (1971) noted the positive correlation between wet ear wax and breast cancer in several countries and suggested an association. This hypothesis seems reasonable because the ceruminous gland and breast are both apocrine and share biochemical characteristics. Ing et al. (1973), in a study of Chinese women in Hong Kong, could not confirm the association.

See also Hyslop (1971); Kataura and Kataura (1967); Martin and Jackson (1969); Petrakis (1977)

Hyslop, N. E., Jr.: Ear wax and host defense. (Editorial) New Eng. J. Med. 284: 1099-1100, 1971.
Ibraimov, A. I.: Brief communication: cerumen phenotypes in certain populations of Eurasia and Africa. Am. J. Phys. Anthrop. 84: 209-211, 1991.
Ing, R.; Petrakis, N. L.; Ho, H. C.: Evidence against association between wet cerumen and breast cancer. Lancet I: 41 only, 1973.
Kataura, A.; Kataura, K.: The comparison of free and bound amino acids between dry and wet types of cerumen. Tohoku J. Exp. Med. 91: 215-225, 1967.
Kataura, A.; Kataura, K.: The comparison of lipids between dry and wet types of cerumen. Tohoku J. Exp. Med. 91: 227-237, 1967.
Martin, L. M.; Jackson, J. F.: Cerumen types in Choctaw Indians. Science 163: 677-678, 1969.
Matsunaga, E.: The dimorphism in human normal cerumen. Ann. Hum. Genet. 25: 273-286, 1962.
Petrakis, N. L.: Cerumen genetics and human breast cancer. Science 173: 347-349, 1971.
Petrakis, N. L.: Genetic cerumen type, breast secretory activity, and breast cancer epidemiology. In: Mulvihill, J. J.; Miller, R. W.; Fraumeni, J. F., Jr.: Genetics of Human Cancer. New York: Raven Press (pub.) 1977. Pp. 297-300.
Petrakis, N. L.; Molohan, K. T.; Tepper, D. J.: Cerumen in American Indians: genetic implications of sticky and dry types. Science 158: 1192-1193, 1967.

PAGET DISEASE, EXTRAMAMMARY (167300)

JJM NOTE

This entity is supported by just one incompletely documented family.

Extramammary Paget disease is a cancerous disease seen at various sites, most often in the anogenital region. The clinical features are usually those of eczematous eruptions with weeping and crust formation. This disease has been shown to be a skin manifestation of internal malignancy. In a study of 40 patients with Paget disease of the anogenital region, Helwig and Graham (1963) did not find a family history of the disease in any of the cases. Kuehn et al. (1973) described a case occurring in a father and son. The father, aged 66, presented with extramammary Paget disease in the right scrotal area. No mention was made of other family history.

Helwig, E. B.; Graham, F. N.: Anogenital (extramammary Paget's disease). Cancer 16: 387-403, 1963.
Kuehn, P. G.; Tennant, R.; Brennerman, A. R.: Familial occurrence of extramammary Paget's disease. Cancer 31: 145-148, 1973.

USHER SYNDROME 3 (*276902)

JJM NOTE

This entity is included because of an affected patient who also had breast cancer, cited in the following entry (Shinkawa and Nadol 1986).

Karjalainen et al. (1985) described an apparently distinct type of Usher syndrome characterized by progressive hearing loss and vestibular hypoactivity. Smith et al. (1992) described a family in the French-Acadian population of southwestern Louisiana in which 2 males had a clinical phenotype like that in the family reported by Karjalainen et al. (1985). Both USH1 and USH2 are frequent among the Louisiana Acadians but this family appeared to have a separate form. USH3 has been estimated to comprise 2% of all Usher syndrome cases; however, based on clinical criteria, 42% of patients with Usher syndrome in Finland are thought to have USH3, suggesting gene enrichment by a founder effect. 4,3:Sankila et al. (1994, 1995) excluded 4 other chromosomal sites at which Usher syndrome had been mapped: 14q32 (USH1A; 276900), 11q13.5 (USH1B; 276903), 11p (USH1C; 176904), and 1q32 (USH2; 276901). Using highly polymorphic microsatellite markers for a systematic search for the USH3 locus by genetic linkage analyses of 11 multiply affected families, 4,3:Sankila et al. (1994, 1995) assigned the locus to 3q21-q25. Of 20 parental disease chromosomes, 15 had identical alleles at 3 marker loci covering a 3-cM genetic distance, including the putative USH3 locus. This was the fifth distinctive form of Usher syndrome to be identified. Although others had estimated USH3 comprises 2% of all Usher syndrome cases, based on clinical criteria, in Finland 42% of Usher syndrome patients showed progressive hearing loss.

Joensuu et al. (1996) typed a total of 32 pedigrees from a geographically isolated Finnish founder population for polymorphisms in the USH3 region of chromosome 3. By analysis of linkage disequilibrium and historical recombinations in 77 USH3 chromosomes, the location of the Finnish USH3 mutation was narrowed to an interval of approximately 1 cM between markers D3S1299 and D3S3625. The profilin-2 gene (PFN2; 176590) was mapped to the vicinity but was excluded as a candidate for USH3 by sequencing. The putative mouse homolog of PFN2 was mapped to mouse chromosome 3, thus suggesting a localization for the mouse homolog of USH3.

Joensuu, T.; Blanco, G.; Pakarinen, L.; Sistonen, P.; Kaariainen, H.; Brown, S.; de la Chapelle, A.; Sankila, E.-M.: Refined mapping of the Usher syndrome type III locus on chromosome 3, exclusion of candidate genes, and identification of the putative mouse homologous region. Genomics 38: 255-263, 1996.

Karjalainen, S.; Vartiainen, E.; Terasvirta, M.; Karja, J.; Kaariainen, H.: An unusual otological manifestation of Usher's syndrome in 4 siblings. Adv. Audiol. 3: 32-40, 1985.

Sankila, E.-M.; Pakarinen, L.; Kaariainen, H.; Aittomaki, K.; Karjalainen, S.; Sistonen, P.; de la Chapelle, A.: Assignment of an Usher syndrome type III (USH3) gene to chromosome 3q. Hum. Molec. Genet. 4: 93-98, 1995.

Sankila, E.-M.; Pakarinen, L.; Sistonen, P.; Aittomaki, K.; Kaariainen, H.; Karjalainen, S.; de la Chapelle, A.: The existence of Usher syndrome type III proven by assignment of its locus to chromosome 3q by linkage. (Abstract) Am. J. Hum. Genet. 55 (suppl.): A15 only, 1994.

Smith, R. J. H.; Pelias, M. Z.; Daiger, S. P.; Keats, B.; Kimberling, W.; Hejtmancik, J. F.: Clinical variability and genetic heterogeneity within the Acadian Usher population. Am. J. Med. Genet. 43: 964-969, 1992.

USHER SYNDROME 1 (*276900)

JJM NOTE

This entry is included because of one case of breast cancer, probably coincidental (Shinkawa and Nadol 1986); in fact, Usher syndrome 3 was the diagnosis.

Usher syndrome, or more appropriately the Usher syndromes, are named for Charles Usher (1914), a British ophthalmologist who emphasized their hereditary nature. The earliest descriptions were given by Von Graefe (1858), Liebreich (1861), who commented on a relatively high frequency in Jews in Berlin, and Hammerschlag (1907). Lindenov (1945) wrote on deaf-mutism associated with retinitis pigmentosa and feeblemindedness. Lang (1959) observed 5 affected children out of 10 from a first-cousin marriage. Hallgren (1959) found 177 affected persons in 102 families. In addition to the features noted in the title of his paper, cataract developed by age 40 in most. Mental deficiency and psychosis occurred in about one-quarter of cases. A large majority had a disturbance of gait attributed to a lesion of the labyrinth. Kloepfer et al. (1966) identified 537 persons with hearing loss in a French-Acadian ('Cajun') group in Louisiana. Of the 468 living persons with hearing loss, at least 158 or about 30% were known to have retinitis pigmentosa and cataract. (As the heterogeneity in Usher syndrome has become defined by clinical and linkage studies, the distinctive Acadian form is termed type IC (276904).) In Finland, Nuutila (1970) found 133 persons with retinitis pigmentosa and congenital sensory deafness, 4 with RP and progressive sensory deafness. Numerous studies suggest genetic heterogeneity of this phenotype. On the basis of 133 patients in Finland, Forsius et al. (1971) concluded that there are two distinct forms of the Usher syndrome: one characterized by congenital deafness and severe retinitis pigmentosa, and a second less frequent form in which the inner ear and retina are less severely affected. Whether these were allelic forms or not was unknown. Holland et al. (1972) found gyrate atrophy in a few heterozygotes. Davenport et al. (1978) found that about 90% of reported cases had profound congenital deafness with onset of RP before puberty, w hereas the rest had moderate to severe hearing loss from birth and RP beginning after puberty. Ataxia, probably labyrinthine in origin, occurred in a great majority of the first type and in a few of the second. The possibility of an X-linked form was suggested by 2 pairs of affected brothers whose mothers were sisters.

Gorlin et al. (1979) summarized the classification of Davenport and Omenn (1977) as follows: type I—profound congenital deafness with onset of RP by age 10; type II—moderate-to-severe congenital deafness with onset of RP in late teens (276901); type III—RP first noted at puberty with progressive hearing loss; type IV—possible X-linked form. The fourth type was based on the observation of 4 affected brothers reported by Davenport et al. (1978). In fact, autosomal recessive inheritance was considered most likely; the heterozygous parents showed unilateral high-frequency hearing loss with normal retinal and vestibular function. Maat-Kievit (1991) found no evidence of X-linked pedigrees in the literature. Jay (1982) found 16 Usher syndrome families out of 571 RP families in the experience of the Moorfields Eye Hospital in London. Other numbers were: autosomal dominant, 130 families; X-linked, 27; autosomal recessive, 5; male multiplex, 24; mixed multiplex, 76; simplex, 292 and adopted, 1. In 4 of 10 sibs, Karjalainen et al. (1983) described an unusual form of Usher syndrome. In 2, hearing loss developed in school age; in the other 2, it developed in the thirties. In 1, retinitis pigmentosa was diagnosed before hearing impairment was evident. In a study of 70 patients, Fishman et al. (1983) also suggested the existence of 2 distinct types of Usher syndrome. In their experience, the deafness is congenital and nonprogressive, whereas the retinitis pigmentosa is progressive. In their type I, onset of night blindness was earlier, visual field loss occurred earlier and in greater severity, hearing impairment was more severe, speech was more likely to be unintelligible, vestibular reflexes and clinically evident ataxia were more frequently found—all as contrasted with type II. Of the 70 patients, 46 were type II. Boughman et al. (1983) reviewed information on 600 cases of deaf-blindness in the registry of the Helen Keller National Center for Deaf Blind Youths and Adults. Of these, 54% satisfied criteria for the diagnosis of Usher syndrome, although only 23.8% had been so diagnosed. From the Louisiana School for the Deaf, they ascertained 30 males and 18 females in 26 nuclear families, reflecting the recognized high frequency in the Louisiana Acadian population. They considered a prevalence estimate for the U.S. of 4.4 per 100,000 as conservative. Grondahl and Mjoen (1986) found 18 cases of Usher syndrome among 89 probands selected for tapetoretinal degeneration. Among the relatives, another 10 cases of Usher syndrome were found. These fell into the 3 types as follows: type I, 14 cases; type II, 10 cases; type III (according to Davenport and Omenn (1977)), 4 cases. In 12 families the pattern of inheritance was autosomal recessive; the remaining 6 probands were solitary cases without parental consanguinity. There was a high intrafamilial correlation with respect to hearing function. Vestibular response was abolished in 3 patients with type I and was normal in 3 patients with type II and in 1 patient with type III. Shinkawa and Nadol (1986) studied the inner ear of a patient with type III Usher syndrome who died at age 52 of metastatic cancer. They found hair cell loss in the basal turn, severe loss of spiral ganglion cells, widespread neural degeneration in the cochlea, and discrete collections of degenerating supporting cells in the organ of Corti. The pattern of neural degeneration bore some similarity to abnormalities in the retina in retinitis pigmentosa. In Norway, Grondahl (1987) found 28 patients from 18 families with Usher syndrome. The prevalence of Usher syndrome was calculated to be 3.6/100,000. Both retinitis pigmentosa and Usher syndrome were more prevalent in Lapps than in other Norwegians. Davenport et al. (1988) recognized 2 main types and a third rare type. Type I not only has congenital profound deafness and early onset of RP, but also congenitally absent vestibular function. Their type II has hearing loss which is congenital and of high frequency type, with little deterioration and with later onset of RP and normal vestibular function. In type III both hearing and vision start out normal or near-normal and progressively deteriorate over several decades. Type I children, because of the vestibular defect, have delayed motor milestones and clumsiness.

Type II children are usually 'mainstreamed' with no problems until teen age. In Colombia, Tamayo et al. (1991) found that about 70% of the Usher syndrome cases were type I, about 26% type II, and 4% type III.

In the French-Acadian ('Cajun') population of southwestern Louisiana where Usher syndrome is frequent (Kloepfer et al., 1966), Daiger et al. (1987) and Pelias et al. (1988) found a suggestion of linkage to GC (139200) on 4q (maximum lod score = 1.41 at theta = 0.17). In the same large Louisiana kindred, 43,42:Smith et al. (1988, 1989) excluded the Usher syndrome locus from a large part of chromosome 4 by multipoint linkage analysis. Notably, a 17-cM region around the GC locus was excluded by reason of a lod score less than -2.0. Bonneau et al. (1990) also excluded linkage to GC. Usher syndrome in the 'Cajuns' is predominantly type I, although some type II cases have been identified (Kimberling, 1990; Smith et al., 1992). Linkage studies demonstrated that the common Acadian type is genetically distinct; see 276904.

Kaplan et al. (1990) excluded linkage of USH1 and the D1S81 marker, which is known to be linked to USH2. Kaplan et al. (1991) demonstrated linkage between the USH1 gene and a probe at the D14S13 locus, suggesting location of the gene on 14q (maximum lod = 2.70 at theta = 0.0). In a study of 10 families of western France ancestry, Kaplan et al. (1992) found a lod score of 4.13 at theta = 0.0 for linkage with D14S13. Of the 10 families, 8 originated from a small area of the Poitou-Charentes region, suggesting that a founder effect might be involved. The D14S13 marker is located between PI (107400) at 14q32.1 and D14S1 (107750) at 14q32.3 (Donis-Keller et al., 1992). Larget-Piet et al. (1994) combined linkage and geographic mapping to demonstrate heterogeneity of Usher syndrome type I in French families. In their sample of 20 families split into 2 groups according to the geographic origin of the probands' grandparents, they confirmed the presence of a gene on 14q32 (USH1A) in 9 families originating from the Poitou region in western France. Moreover, they refined the genetic mapping of USH1A by showing that the disease gene maps to the D14S13 locus, within the genetic interval defined by loci D14S78 and D14S250.

The symbol USH1A and the additional trivial designation of 'French variety' of Usher syndrome has been assigned to the form that was mapped to chromosome 14; the symbol USH1C has been assigned to the 'Acadian variety' of the disease on 11p (276904), and USH1B for the 'non-Acadian variety' on 11q (276903).

Smith et al. (1994) described criteria for the clinical diagnosis of Usher syndrome, adopted by the Usher Syndrome Consortium. They pointed out that there was evidence for at least 3 distinct USH1 loci (USH1A, USH1B, USH1C) and 2 distinct USH2 loci. They pointed to the need to exclude congenital infections, such as rubella, syphilis, and cytomegalovirus, and problems associated with gestation, delivery, or the perinatal period that also can cause profound hearing loss and retinal damage.

Photoreceptors, auditory hair cells, and vestibular hair cells develop from ciliated progenitors. Several lines of evidence suggest that a generalized abnormality of axoneme structure is present in patients with Usher syndrome. Hunter et al. (1986) found a high proportion of abnormal axonemes in retinal photoreceptor cells of a patient with Usher syndrome. Shinkawa and Nadol (1986) found a decrease in outer ciliary cells in the lower part of the cochlea in this syndrome. Structural and functional evidence for abnormal nasal cilia has been found in this disorder as in other patients with retinitis pigmentosa (Arden and Fox, 1979). Finally, sperm motility, velocity, and structure have been found abnormal in Usher syndrome, a feature probably related to the markedly decreased fertility of these patients (Hunter et al., 1986; Nuutila, 1970). Kaplan et al. (1992) noted that there is reason to think that a gene for normal ciliary function is located in the 14q32 region. Brueckner et al. (1989) found that the iv (situs inversus) mutation in the mouse maps to a corresponding region; this mouse mutation may be homologous to Kartagener syndrome (244400). Lake and Sharma (1973) reported the association of Kartagener syndrome with retinitis pigmentosa and congenital deafness. Bonneau et al. (1993) reported the association of type I Usher syndrome with bronchiectasis, chronic sinusitis and reduced nasal mucociliary clearance in 2 brothers and suggested that USH1 could be a primary ciliary disorder.

Schaefer et al. (1998) performed quantitative analysis of magnetic resonance imaging studies of 19 patients with Usher syndrome (8 with type I, 11 with type II). They found a significant decrease in intracranial volume and in size of the brain and cerebellum with a trend toward an increase in the size of the subarachnoid spaces. These data suggested that the disease process in Usher syndrome involves the entire brain and is not limited to the posterior fossa or auditory and visual systems.

See also Bascom et al. (1992); Beatty et al. (1979); De Haas et al. (1970); Nuutila (1980); Smith et al. (1989); Usher (1935); Vernon (1969)

Arden, G. B.; Fox, B.: Increased incidence of abnormal nasal cilia in patients with retinitis pigmentosa. Nature 279: 534-536, 1979.

Bascom, R. A.; Liu, L.; Chen, J.; Duncan, A.; Kimberling, W. J.; Moller, C. G.; Humphries, P.; Nathans, J.; McInnes, R. R.: ROM1: a candidate gene for autosomal dominant retinitis pigmentosa (ADRP), Usher syndrome type 1, and Best vitelliform macular dystrophy. (Abstract) Am. J. Hum. Genet. 51 (suppl.): A6 only, 1992.

Beatty, C. W.; McDonald, T. J.; Colvard, D. M.: Usher's syndrome with unusual otologic manifestations. Mayo Clin. Proc. 54: 543-546, 1979.

Bonneau, D.; Kaplan, J.; Berthier, M.; Seger, J.; Dufier, J. L.; Frezal, J.: Evidence against linkage of the gene for Usher's syndrome type 1 with group specific component (GC) on chromosome 4. Ann. Genet. 33: 103-104, 1990.

Bonneau, D.; Raymond, F.; Kremer, C.; Klossek, J.-M.; Kaplan, J.; Patte, F.: Usher syndrome type I associated with bronchiectasis and immotile nasal cilia in two brothers. J. Med. Genet. 30: 253-254, 1993.

Boughman, J. A.; Vernon, M.; Shaver, K. A.: Usher syndrome: definition and estimate of prevalence from two high-risk populations. J. Chronic Dis. 36: 595-603, 1983.

Brueckner, M.; D'Eustachio, P.; Horwich, A. L.: Linkage mapping of a mouse gene, 'iv,' that controls left-right asymmetry of the heart and viscera. Proc. Nat. Acad. Sci. 86: 5035-5038, 1989.

Daiger, S. P.; Bertin, T. K.; Smith, R. J. H.; Pelias, M. Z.: Tentative linkage of Usher's syndrome to Gc on human chromosome 4. (Abstract) Cytogenet. Cell Genet. 46: 602 only, 1987.

Davenport, S. L. H.; Moller, C.; Biscone-Halterman, K.; Kimberling, W. J.: Early diagnosis of Usher syndrome. (Abstract) Am. J. Hum. Genet. 43: A45 only, 1988.

Davenport, S. L. H.; O'Nuallain, S.; Omenn, G. S.; Wilkus, R. J.: Usher syndrome in four hard-of-hearing siblings. Pediatrics 62: 578-583, 1978.

Davenport, S. L. H.; Omenn, G. S.: The heterogeneity of Usher syndrome. (Abstract) Vth Int. Conf. on Birth Defects, Montreal , 8/1977.

De Haas, E. B. H.; Van Lith, G. H. M.; Rijnders, J.; Rumke, A. M. L.; Volmer, C. H.: Usher's syndrome, with special reference to heterozygous manifestations. Docum. Ophthal. 28: 166-190, 1970.

Donis-Keller, H.; Weaver, R.; Ramachandra, S.; Warlick, C.; Burgess, A. K.; Weber, J. L.; Litt, M.; Nickerson, D. A.; Boysen, C. A.; Hood, L.: A comprehensive genetic linkage map of the human genome 'chromosome 14.'. Science 258: 67-86, 1992.

Fishman, G. A.; Kumar, A.; Joseph, M. E.; Torok, N.; Anderson, R. J.: Usher's syndrome: ophthalmic and neuro-otologic findings suggesting genetic heterogeneity. Arch. Ophthal. 101: 1367-1374, 1983.

Forsius, H.; Erikkson, A.; Nuutila, A.; Vainio-Mattila, B.; Krause, U.: A genetic study of three rare retinal disorders: dystrophia retinae dysacusis syndrome, X-chromosomal retinoschisis and grouped pigments of the retina. Birth Defects Orig. Art. Ser. 7(3): 83-98, 1971.

Gorlin, R. J.; Tilsner, T. J.; Feinstein, S.; Duvall, A. J.: Usher's syndrome type III. Arch. Otolaryng. 105: 353-354, 1979.

Grondahl, J.: Estimation of prognosis and prevalence of retinitis pigmentosa and Usher syndrome in Norway. Clin. Genet. 31: 255-264, 1987.

Grondahl, J.; Mjoen, S.: Usher syndrome in four Norwegian counties. Clin. Genet. 30: 14-28, 1986.

Hallgren, B.: Retinitis pigmentosa combined with congenital deafness; with vestibulo-cerebellar ataxia and mental abnormality in a proportion of cases: a clinical and genetico-statistical study. Acta Psychiat. Neurol. Scand. 34 (suppl. 138): 9-101, 1959.

Hammerschlag, V.: Zur Kenntnis der hereditaer-degenerativen Taubstummen und ihre differential-diagnostische Bedeutung. Z. Ohrenheilk. 54: 18-36, 1907.

Holland, M. G.; Cambie, E.; Kloepfer, W.: An evaluation of genetic carriers of Usher's syndrome. Am. J. Ophthal. 74: 940-947, 1972.

Hunter, D. G.; Fishman, G. A.; Mehta, R. S.; Kretzer, F. L.: Abnormal sperm and photoreceptor axonemes in Usher's syndrome. Arch. Ophthal. 104: 385-389, 1986.

Jay, M.: Figures and fantasies: the frequencies of the different genetic forms of retinitis pigmentosa. Birth Defects Orig. Art. Ser. 18(6): 167-173, 1982.

Kaplan, J.; Gerber, S.; Bonneau, D.; Rozet, J.; Briard, M.; Dufier, J.; Munnich, A.; Frezal, J.: Probable location of Usher type I gene on chromosome 14q by linkage with D14S13 (MLJ14 probe). (Abstract) Cytogenet. Cell Genet. 58: 1988 only, 1991.

Kaplan, J.; Gerber, S.; Bonneau, D.; Rozet, J. M.; Delrieu, O.; Briard, M. L.; Dollfus, H.; Ghazi, I.; Dufier, J. L.; Frezal, J.; Munnich, A.: A gene for Usher syndrome type I (USH1A) maps to chromosome 14q. Genomics 14: 979-987, 1992.

Kaplan, J.; Guasconi, G.; Bonneau, D.; Melki, J.; Briard, M.-L.; Munnich, A.; Dufier, J. L.; Frezal, J.: Usher syndrome type I is not linked to D1S81 (pTHH 33): evidence for genetic heterogeneity. Ann. Genet. 33: 105-108, 1990.

Karjalainen, S.; Terasvirta, M.; Karja, J.; Kaariainen, H.: An unusual otological manifestation of Usher's syndrome in four siblings. Clin. Genet. 24: 273-279, 1983.

Kimberling, W. J.: Personal Communication. Omaha, Nebraska 3/5/1990.

Kloepfer, H. W.; Laguaite, J. K.; McLaurin, J. W.: The hereditary syndrome of congenital deafness and retinitis pigmentosa: (Usher's syndrome). Laryngoscope 76: 850-862, 1966.

Lake, K.; Sharma, O. P.: Kartagener's syndrome and deaf-mutism: an unusual association. Chest 64: 661-663, 1973.

Lang, H. A.: Retinal degeneration and nerve deafness. Brit. Med. J. 2: 1096 only, 1959.

Larget-Piet, D.; Gerber, S.; Bonneau, D.; Rozet, J.-M.; Marc, S.; Ghazi, I.; Dufier, J.-L.; David, A.; Bitoun, P.; Weissenbach, J.; Munnich, A.; Kaplan, J.: Genetic heterogeneity of Usher syndrome type 1 in French families. Genomics 21: 138-143, 1994.

Liebreich, R.: Abkunft aus Ehen unter Blutsverwandten als Grund von Retinitis pigmentosa. Dtsch. Klin. 13: 53 only, 1861.

Lindenov, H.: The Etiology of Deaf-mutism with Special Reference to Heredity. Copenhagen: E. Munksgaard (pub.) 1945.

Maat-Kievit, J. A.: Personal Communication. Leiden, The Netherlands 3/25/1991.

Nuutila, A.: Dystrophia retinae pigmentosa-dysacusis syndrome (DRD): a study of the Usher or Hallgren syndrome. J. Genet. Hum. 18: 57-88, 1970.

Nuutila, A.: Dystrophia retinae pigmentosa-dysacusis syndrome (DRD).In: Eriksson, A. W.; Forsius, H. R.; Nevanlinna, H. R.; Workman, P. L.; Norio, R. K.: Population Structure and Genetic Disorders. New York: Academic Press (pub.) 1980. Pp. 614-616.
Pelias, M. Z.; Lemoine, D. R.; Kossar, A. L.; Ward, L. J.; Wilson, A. F.; Elston, R. C.: Linkage studies of Usher syndrome: analysis of an Acadian kindred in Louisiana. Cytogenet. Cell Genet. 47: 111-112, 1988.
Schaefer, G. B.; Bodensteiner, J. B.; Thompson, J. N., Jr.; Kimberling, W. J.; Craft, J. M.: Volumetric neuroimaging in Usher syndrome: evidence of global involvement. Am. J. Med. Genet. 79: 1-4, 1998.
Shinkawa, H.; Nadol, J. B., Jr.: Histopathology of the inner ear in Usher's syndrome as observed by light and electron microscopy. Ann. Otol. Rhinol. Laryng. 95: 313-318, 1986.
Smith, R. J. H.; Berlin, C. I.; Hejtmancik, J. F.; Keats, B. J. B.; Kimberling, W. J.; Lewis, R. A.; Moller, C. G.; Pelias, M. Z.; Tranebjaerg, L.: Clinical diagnosis of the Usher syndromes. Am. J. Med. Genet. 50: 32-38, 1994.
Smith, R. J. H.; Daiger, S. P.; Pelias, M. Z.; Giesenschlag, N.; Hejtmancik, J. F.: Multipoint linkage analysis of Usher syndrome: exclusion of chromosome 4. (Abstract) Cytogenet. Cell Genet. 51: 1082 only, 1989.
Smith, R. J. H.; Holcomb, J. D.; Caskey, C. T.; Pelias, M. Z.; Daiger, S. P.; Heitmancik, J. F.: Multipoint linkage analysis of Usher syndrome: exclusion of chromosome 4. (Abstract) Am. J. Hum. Genet. 43: A159 only, 1988.
Smith, R. J. H.; Holcomb, J. D.; Daiger, S. P.; Caskey, C. T.; Pelias, M. Z.; Alford, B. R.; Fontenot, D. D.; Hejtmancik, J. F.: Exclusion of Usher syndrome gene from much of chromosome 4. Cytogenet. Cell Genet. 50: 102-106, 1989.
Smith, R. J. H.; Pelias, M. Z.; Daiger, S. P.; Keats, B.; Kimberling, W.; Hejtmancik, J. F.: Clinical variability and genetic heterogeneity within the Acadian Usher population. Am. J. Med. Genet. 43: 964-969, 1992.
Tamayo, M. L.; Bernal, J. E.; Tamayo, G. E.; Frias, J. L.; Alvira, G.; Vergara, O.; Rodriguez, V.; Uribe, J. I.; Silva, J. C.: Usher syndrome: results of a screening program in Colombia. Clin. Genet. 40: 304-311, 1991.
Usher, C. H.: Bowman's lecture: on a few hereditary eye affections. Trans. Ophthal. Soc. U.K. 55: 164-245, 1935.
Vernon, M.: Usher's syndrome-deafness and progressive blindness. Clinical cases, prevention, theory and literature survey. J. Chronic Dis. 22: 133-151, 1969.
Von Graefe, A.: Exceptionelles Verhalten des Gesichtsfeldes bei Pigmententartung der Netzhaut. Graefes Arch. Ophthal. 4: 250-253, 1858.

Mastocytosis

MAST CELL DISEASE (MASTOCYTOSIS) (154800)

JJM NOTE

Given the pervasive role of mast cells in disease, including neurofibromatosis 1, the genetics of mast cells and their disorders should be interesting. Perhaps cytogenetics is the place to start.

The cutaneous manifestation is termed urticaria pigmentosa. Generalized involvement, which may be fatal, is sometimes observed. Burgoon et al. (1968) observed the disorder in father and daughter. Selmanowitz et al. (1970) described cutaneous mastocytosis in 8 females in 3 generations with a male carrier representing a 'skipped generation.' They also reviewed experience in twins. Selmanowitz and Orentreich (1970) stated that about 40 familial cases and 6 concordant pairs of monozygotic twins are known. Both dominant and recessive inheritance has been postulated (Shaw, 1968). Fowler et al. (1986) reported urticaria pigmentosa in 4 persons in 2 generations (father and son and 2 sisters of the father). No consistent HLA haplotypes were found; the affected boy and his unaffected brother, for instance, were of identical haplotype. Bazex et al. (1971) reviewed 35 families reported between 1891 and 1971. Rockoff (1978) compiled 11 sets of twins and 1 set of triplets with urticaria pigmentosa to which he added his own report. Fowler et al. (1986) added 14 further families. Boyano et al. (1990) described monozygotic male twins who simultaneously developed urticaria pigmentosa. Boyano et al. (1990) suggested that this was the tenth such pair to be reported. Noto et al. (1995) gave a 5-year follow-up on identical twins strikingly concordant for specific clinical features of urticaria pigmentosa.

Of 50 reported families with familial urticaria pigmentosa and cutaneous mastocytosis, Anstey et al. (1991) found that 22 had 2 or more generations affected; in 20 families, only females were affected, and in 17 families, members of both sexes were affected. Oku et al. (1990) described bullous mastocytosis in a mother and 2 daughters and a son. The appearance of the mast cells differed in the adult from that in the children and improvement in the clinical features occurred spontaneously over time. Clark et al. (1990) described urticaria pigmentosa in a sister and brother and in their maternal grandfather who had a daughter (aunt of the sibs) who had histologically diagnosed urticaria pigmentosa. An unusual feature was the marked diversity of the skin disease: classic lesions, telangiectasia macularis eruptiva perstans, and clinically normal skin with positive dermal infiltrates of excess mast cells. Anstey et al. (1991) described a woman and her daughter who both developed diffuse cutaneous mastocytosis in early infancy. Mast-cell granules in both cases were found on electron microscopy to have an abnormal ultrastructure. Similar giant granules had been noted in familial cases by James and Eady (1981).

Anstey, A.; Lowe, D. G.; Kirby, J. D.; Horton, M. A.: Familial mastocytosis: a clinical, immunophenotypic, light and electron microscopic study. Brit. J. Derm. 125: 583-587, 1991.
Bazex, A.; Dupre, A.; Christol, B.; Andrieu, H.: Les mastocytoses familiales. Presentation de deux observations. Revue general. Interet nosologique. Ann. Derm. Syph. 98: 241-260, 1971.
Boyano, T.; Carrascosa, T.; Porta, N.; Agud, J. L.; Garcia, M. J.: Urticaria pigmentosa in monozygotic twins.(Letter) Arch. Derm. 126: 1375-1376, 1990.
Burgoon, C. F., Jr.; Graham, J. H.; McCaffree, D. L.: Mast cell disease: a cutaneous variant with multisystem involvement. Arch. Derm. 98: 590-605, 1968.
Clark, D. P.; Buescher, L.; Havey, A.: Familial urticaria pigmentosa. Arch. Intern. Med. 150: 1742-1744, 1990.
Fowler, J. F.; Parsley, W. M.; Cotter, P. G.: Familial urticaria pigmentosa. Arch. Derm. 122: 80-81, 1986.
James, M. P.; Eady, R. A. J.: Familial urticaria pigmentosa with giant mast cell granules: a clinical, light, and electron microscopic study. Arch. Derm. 117: 713-718, 1981.
Noto, G.; Pravata, G.; Arico, M.: Concordant urticaria pigmentosa in a couple of identical twins: a five-year follow-up.(Letter) Acta Derm. Venerol. 75: 499, 1995.
Oku, T.; Hashizume, H.; Yokote, R.; Sano, T.; Yamada, M.: The familial occurrence of bullous mastocytosis (diffuse cutaneous mastocytosis). Arch. Derm. 126: 1478-1484, 1990.
Rockoff, A. S.: Urticaria pigmentosa in identical twins. Arch. Derm. 114: 1227-1228, 1978.
Selmanowitz, V. J.; Orentreich, N.: Mastocytosis: a clinical genetic evaluation. J. Hered. 61: 91-94, 1970.
Selmanowitz, V. J.; Orentreich, N.; Tiangco, C. C.; Demis, D. J.: Uniovular twins discordant for cutaneous mastocytosis. Arch. Derm. 102: 34-41, 1970.
Shaw, J. M.: Genetic aspects of urticaria pigmentosa. Arch. Derm. 97: 137-138, 1968.

Melanoma

MELANOMA, CUTANEOUS MALIGNANT 1 (*155600)

JJM NOTE

In a memorandum dated March 24, 1975, from my old NIH files, Robert W. Miller recounts etiologically interesting anecdotes he heard in the course of a lecture tour. The first paragraph begins, "At the Jeanes Hospital in Fox Chase (Philadelphia) I met Dr. Michael Mastrangelo who has under observation a family in which four members have melanoma . . . a brother, sister, father, and his brother. The boy and the uncle each have three primary melanomas." About a year later, the National Cancer Institute team of Mark Green, Nancy Ebby, and I joined the Philadelphia team at a Trenton, New Jersey, hospital emergency room. While I was doing skin biopsies for fibroblast lines (which might still be used for DNA for the controversial linkage studies, for all I know), Wallace Clark was doing total skin examinations. At one point, he almost yelled out for us all to assemble around one member of the K family (as in B-K). Then he pointed out in astonishment that the common thread was not the melanomas but the precursor lesion, the dysplastic nevus. I understand that, in the original NCI cohort of families, followed for some 20 years now with education, self-examination, and aggressive biopsying, no deaths from new melanoma have occurred in families that had anticipated inevitable melanoma deaths in the past—a triumph of the clinical effectiveness of genetic counseling and education!

Description

Malignant melanoma is a malignant neoplasm of melanocytes, arising de novo or from a pre-existing benign nevus, which occurs most often in the skin but also may involve other sites.

Clinical Features

Several writers (e.g., Moschella, 1961; Schoch, 1963; Salamon et al., 1963) commented on the usual fair complexion, blue eyes and multiple ephelides in patients with familial melanoma.

In a questionnaire study, Kopf et al. (1986) found that a positive family history for melanoma was correlated with a younger age at first diagnosis in the proband, a smaller diameter of the lesion, lower Clark level, decreased frequency of nonmelanoma skin cancer, and reduced prevalence of noncutaneous cancer. (The Clark index refers to the level of invasion.) A comparison of monozygotic and dizygotic twins for melanoma might be important because of cases of melanoma in non-blood-related members of the same household (Robinson and Manheimer, 1972).

Lynch et al. (1978) suggested that a cutaneous marker indicative of susceptibility to malignant melanoma is characterized by large moles, variable in number, reddish brown to pink in color, and with an irregular border. Histologically, they show a bizarre intraepidermal pattern. The authors also described a melanoma family with distinctive freckling and dryness of the skin, suggesting xeroderma pigmentosum (278700) but with normal unscheduled DNA repair and a dominant pedigree pattern. Other malignancies such as colon cancer had an increased frequency in these families.

Clark et al. (1978), Greene et al. (1978), and Reimer et al. (1978) pointed out distinctive clinical and histologic features of the moles that are precursors of familial malignant melanomas. They termed these features the 'B-K mole syndrome' after the family names of 2 patients; later, Greene et al. (1980) and Elder et al. (1980) expressed a preference for the designation 'hereditary dysplastic nevus syndrome.' The same lesion underlies some cases of nonfamilial malignant melanoma. Greene et al. (1980) referred to this as 'dysplastic nevus syndrome, sporadic type.' The clinical features include between 10 and 100 moles on the upper trunk and limbs, and variability of mole size (from 5 to 15 mm), outline, and color. Histologically, B-K moles show atypical melanocytic hyperplasia, lymphocytic infiltration, delicate fibroplasia, and new blood vessel formation. Lynch et al. (1980) referred to this as FAMMM (familial atypical mole—malignant melanoma syndrome). Arndt (1984) and Greene et al. (1985) provided photographic illustration of the familial dysplastic nevus syndrome.

Lynch et al. (1980) studied 3 kindreds of the FAMMM syndrome. Father-to-son transmission was observed. One patient had 9 separate primary melanomas in 18 years. Expressivity was highly variable. Management is difficult because one cannot be certain which moles require biopsy and then, following histologic study, which require wide excision. The possibility of increased risk of cancer at other sites was raised. Hartley et al. (1987) described several cases of malignant melanoma in close relatives of children with osteosarcoma (259500) and chondrosarcoma (215300). They proposed that in certain families malignant melanoma may be a manifestation of the same gene defect that results in susceptibility to tumors characteristic of the SBLA syndrome (151623).

Other Features

Tumor-specific antigens have been found in malignant melanoma (Hawkins et al., 1981; Pellegris et al., 1982).

Inheritance

Multiple authors have documented familial inheritance in malignant melanoma: see Cawley (1952); Smith et al. (1966); Andrews (1968). Katzenellenbogen and Sandbank (1967) described dizygotic twins with malignant melanoma.

Anderson et al. (1967) described malignant melanoma in at least 15 members of 3 generations of 1 kindred. Early age of onset and a tendency for multiple primary lesions were features. Lynch and Krush (1968) described 2 families with malignant melanoma in 2 generations in 1 family and 3 generations in the other. Anderson (1971) reported 36 pedigrees in which a total of 106 members had cutaneous melanoma. He noted that in addition to earlier age at onset and increased frequency of multiple primary lesions, familial cases have a higher survival rate than nonfamilial cases.

Rhodes et al. (1985) found that the prevalence rate of congenital nevomelanocytic nevi was 11 times greater in sibs of probands than in the general population. They had some families with 2 generations affected.

In the families with CMM studied by Greene et al. (1983), further studies (7,8:Bale et al., 1985, 1986) showed that dysplastic nevus (DN), a lesion known to be a precursor of melanoma, also segregates in an autosomal dominant manner. Pascoe (1987) challenged the concept of a single dominant gene as proposed by Bale et al. (1986). Bale and Chakravarti (1987) defended their conclusion.

Traupe et al. (1989) also challenged the autosomal dominant hypothesis for dysplastic nevus syndrome on the basis of the lack of a genetic equilibrium between eliminated and newly arising mutations. Happle et al. (1982) had advanced arguments in favor of polygenic inheritance of dysplastic nevi: 1) lack of a consistent family pattern; 2) frequent sporadic occurrence of the trait; 3) continuous transition between ordinary and dysplastic nevi; and 4) analogy with an animal model.

Kraemer et al. (1983) found 4 persons affected with the dysplastic nevus phenotype. The risk of developing melanoma is not constant but increases with the number of melanoma patients in the family. This is a feature typical of polygenic inheritance.

Bergman et al. (1986) studied extensively affected kindreds in 'an ancient fishing village in the neighborhood of Leiden,' The Netherlands. Autosomal dominant inheritance of DNS was confirmed. In 6 pedigrees, 33 patients with melanoma occurred. Fifteen unaffected persons were identified as gene carriers by their position in the pedigrees.

Cytogenetics

In 4 of 5 cases of malignant melanoma, Trent et al. (1983) found chromosome alterations, including deletion and translocation in the long arm of chromosome 6, specifically in the 6q15-q23 region. They pointed out that the MYB oncogene maps to this region. Becher et al. (1983), reviewing cytologic findings in malignant melanoma in their own and reported cases, likewise pointed to a high incidence of structural aberration of 6q (segment q11-q31), whereas the short arm remains structurally unchanged, though its genetic material is often duplicated, as in the case of isochromosome-6p in one of their cases. These findings accentuate the interest, they pointed out, in the relationships found between specific HLA haplotypes and familial malignant melanoma (Hawkins et al., 1981; Pellegris et al., 1982).

Pathak et al. (1983), Rao et al. (1984), Balaban et al. (1984), and Rey et al. (1985) also reported preferential abnormalities of chromosome 6. Hecht et al. (1989) found a marked increase in chromosomal rearrangements in dysplastic nevi from patients with CMM and in their normal-looking skin but not in their lymphocytes.

Mapping

Linkage studies of DN and CMM showed lod = 3.857 at theta = 0.08. All families giving evidence on linkage were in coupling and the maximum likelihood estimate of recombination was not significantly different from 0 (7,8:Bale et al., 1985, 1986) Bale et al. (1985) excluded linkage of CMM to HLA.

In the Dutch family described by Bergman et al. (1986), Frants et al. (1989) and van Haeringen et al. (1989) excluded linkage with Rh; indeed, data from additional markers excluded the DNS gene from the entirety of 1p. Bergman et al. (1994) found in this family that a melanoma-associated gene was linked to several markers on 9p21. In a linkage analysis in which only melanoma patients were considered as affected, marker D9S171 showed a maximum lod score of 3.11 at theta = 0.0. After introducing family members with 10 or more, or 5 or more, atypical nevi as affected in addition to the melanoma patients, the maximum lod score rose to 4.88 at theta = 0.05 and 3.79 at theta = 0.07, respectively. Interestingly, the sharing of a unique 9p21 haplotype among most melanoma patients in the families from 2 different villages suggested that an old common mutation is present in the Leiden region.

Multipoint linkage analysis appeared to support the assignment of CMM to 1p (Bale et al., 1987). In 3 Utah kindreds ascertained through multiple cases of melanoma, Cannon-Albright et al. (1990) could find no evidence of linkage with the 2 markers most closely linked in the Bale study. Both melanoma alone and a combined melanoma/dysplastic nevus syndrome phenotype were analyzed. Furthermore, multipoint linkage analysis excluded the CM/DNS locus from an area of 55 cM. Bale et al. (1989) presented further evidence supporting assignment of the CMM locus to chromosome 1p36, 7.6 cM distal to PND and flanked by D1S47.

Dracopoli et al. (1989) found loss of heterozygosity at loci on 1p in 43% of melanomas and 52% of melanoma cell lines. Analysis of multiple metastases derived from the same patient and of melanoma and lymphoblastoid samples from a family with hereditary melanoma showed that loss of heterozygosity at loci on distal 1p is a late event in tumor progression rather than the second mutation that would occur if melanoma were due to a cellular recessive mechanism. In neuroblastoma and in type II endocrine neoplasia also, 1p loss of heterozygosity is frequent, suggesting that this loss is a common late event of neuroectodermal tumor progression. By multipoint linkage analysis of 6 families, Dracopoli et al. (1989) found evidence that the familial melanoma gene maps to 1p36 about 8 cm distal to PND (108780). The lod score was 5.42. Goldstein et al. (1993) extended the linkage studies to updated versions of these 6 families plus 7 new families. They concluded that there was 'significant evidence of heterogeneity,' and considered that this was responsible for the failure of some previous studies to confirm linkage to 1p in some families.

Millikin et al. (1991) used RFLPs to look for loss of constitutional heterozygosity (LOH) for markers on 6q. LOH on chromosome 6q was identified in 21 of 53 informative loci (40%). The chromosomal region bearing the highest frequency of 6q allelic loss was defined by the marker loci MYB (189990) and ESR (133430) located at 6q22-q23 and 6q24-q27, respectively. Possibly contradictory to chromosome 6 information is the report of Greene et al. (1983) of possible linkage to Rh (which is on 1p). A maximum lod score of 2.0 at theta 0.30 was observed.

Nancarrow et al. (1992) reviewed the contradictory findings of linkage in this disorder and presented studies of 7 Australian kindreds. Both Cannon-Albright et al. (1990) and Kefford et al. (1991) had questioned the validity of dysplastic nevi as a marker for familial melanoma and excluded linkage to markers on 1p when familial melanoma alone (symbolized MLM) was used as the phenotype. Several of the Australian families studied by Kefford et al. (1991) showed little or no history of dysplastic nevus syndrome or surgical removal of histologically characterized dysplastic nevi. Of the 7 other Australian kindreds studied by Nancarrow et al. (1992), 3 had the largest number of affected individuals reported worldwide. Because they also had families without dysplastic nevi and because the data used to calculate the parameters of the model used by Kefford et al. (1991) were estimated from a population-based survey, Nancarrow et al. (1992) used the latter model but also analyzed the data with the model of Bale et al. (1989). The Kefford model was applied to MLM alone and took into account variable penetrance with age and variable frequency of sporadic cases with age. With this approach, they excluded MLM from a 40-cM region that spanned the interval between D1S47 and PND and extended approximately 15 cM on either side of these markers to a total of 70 cM. In addition, they excluded a region of about 20 cM around the D1S57/MYCL1 (164850) loci at 1p32. Nancarrow et al. (1992) carried out linkage analysis in 3 large Australian melanoma pedigrees, using 172 microsatellite markers spread across all autosomes. Three additional smaller families were typed for 70 of the same markers. In 5 of the 6 families, they found lod scores between 1.0 and 2.3, which suggested localization of melanoma genes in proximity to some of the markers. This may indicate genetic heterogeneity since there was no marker for which all families gave significantly high lods. Their data prov ided the basis of an exclusion map; regions of chromosome 6, 9cen, and 10qter could not be excluded in these studies.

The cutaneous malignant melanoma locus on chromosome 1 has been designated CMM1. Several lines of evidence point to the existence of a second locus on chromosome 9 designated CMM2 (see 155601). There may be a third locus (CMM3) on chromosome 6 (Dracopoli, 1992).

Following up on previous linkage analyses of 19 cutaneous malignant melanoma/dysplastic nevi (CMM/DN) kindreds which showed significant evidence of linkage and heterogeneity to both chromosomes 1p and 9p, Goldstein et al. (1996) examined 2-locus hypotheses. The lod scores for CMM alone were highest using the single locus-heterogeneity model. They found much stronger evidence of linkage to 9p than to 1p for CMM alone; the lod scores were approximately 2 times greater on 9p than on 1p. A change in lod scores from an evaluation of CMM alone to CMM/DN suggested to the authors that a chromosome 1p locus contributed to both CMM and CMM/DN, whereas a 9p locus contributed more to CMM alone. For 2-locus models, the lod scores from 1p were greater for CMM/DN than for CMM alone. After conditioning on linkage to the other locus, only the 9p locus consistently showed significant evidence for linkage to CMM alone.

Clinical Management

The familial dysplastic nevus syndrome is a good example of a genetic disorder which lends itself to the practice of preventive genetics, i.e., preventive medicine, at the family level (Greene et al., 1985). Since 1960, mortality from cutaneous melanoma in the U.S. has risen more than mortality from any other cancer except carcinoma of the lung.

History

The earliest report may be that of Norris (1820). In describing a case of malignant melanoma, Norris wrote: 'It is remarkable that this gentleman's father, about thirty years ago, died of a similar disease. A surgeon of this town attended him, and he informed me that a number of small tumours appeared between the shoulders. . .This tumour, I have remarked, originated in a mole, and it is worth mentioning, that not only my patient and his children had many moles on various parts of their bodies, but also his own father and brothers had many of them. The youngest son had one of these marks exactly in the same place where the disease in his father first manifested itself. These facts, together with a case that has come under my notice, rather similar, would incline me to believe that this disease is hereditary.' See commentary by Hecht (1989).

Nomenclature: Knudson (1997) referred to the form of melanoma caused by mutation in the p16 gene (600160) and mapping to 9p21 (155601) as melanoma-1, since it was the first form of melanoma to be defined at the DNA level. Furthermore, he referred to the form of melanoma second to be defined molecularly, that mapping to chromosome 12 and caused by mutations in the CDK4 gene (123829), as melanoma-2. Presumably, the form of melanoma that maps to 1p should, according to this scheme, be designated melanoma-3 when the gene is identified.

See also Bale et al. (1989); Bale et al. (1985); Bale et al. (1985); Becher et al. (1983); Dracopoli et al. (1987); Dracopoli et al. (1989); Elder et al. (1982); Fountain et al. (1992); Howell et al. (1984); Lynch et al. (1983); Ochi et al. (1984); Turkington (1965); Vasen et al. (1989); Wallace et al. (1973); Wallace et al. (1971)

Anderson, D. E.: Clinical characteristics of the genetic variety of cutaneous melanoma in man. Cancer 28: 721-725, 1971.

Anderson, D. E.; Smith, J. L., Jr.; McBride, C. M.: Hereditary aspects of malignant melanoma. J.A.M.A. 200: 741-746, 1967.

Andrews, J. C.: Malignant melanoma in siblings. Arch. Derm. 98: 282-283, 1968.

Arndt, K. A.: Precursors to malignant melanoma: congenital and dysplastic nevi. J.A.M.A. 251: 1882-1883, 1984.

Balaban, G.; Herlyn, M.; Guerry, D.; Bartolo, R.; Koprowski, H.; Clark, W. H.; Nowell, P. C.: Cytogenetics of human malignant melanoma and premalignant lesions. Cancer Genet. Cytogenet. 11: 175-183, 1984.

Bale, S. J.; Chakravarti, A.: Evidence for autosomal dominance and pleiotropy of the cutaneous malignant melanoma (CMM)/dysplastic nevus (DN) gene. (Letter) Am. J. Hum. Genet. 40: 466-467, 1987.

Bale, S. J.; Chakravarti, A.; Greene, M. H.: Dysplastic nevi (DN) and cutaneous malignant melanoma (CMM) represent pleiotropic effects of a single gene. (Abstract) Cytogenet. Cell Genet. 40: 575, 1985.

Bale, S. J.; Chakravarti, A.; Greene, M. H.: Cutaneous malignant melanoma and familial dysplastic nevi: evidence for autosomal dominance and pleiotropy. Am. J. Hum. Genet. 38: 188-196, 1986.

Bale, S. J.; Dracopoli, N. C.; Greene, M. H.; Gerhard, D. S.; Housman, D. E.: Linkage analysis of melanoma and dysplastic nevus syndrome with polymorphic loci on human chromosome 1p. (Abstract) Cytogenet. Cell Genet. 46: 575, 1987.

Bale, S. J.; Dracopoli, N. C.; Tucker, M. A.; Clark, W. H., Jr.; Fraser, M. C.; Stanger, B. Z.; Green, P.; Donis-Keller, H.; Housman, D. E.; Greene, M. H.: Mapping the gene for hereditary cutaneous malignant melanoma-dysplastic nevus to chromosome 1p. New Eng. J. Med. 320: 1367-1372, 1989.

Bale, S. J.; Dracopoli, N. C.; Tucker, M. A.; Clark, W. H., Jr.; Greene, M. H.; Housman, D. E.: CMM maps to distal 1p. (Abstract) Cytogenet. Cell Genet. 51: 956, 1989.

Bale, S. J.; Greene, M. H.; Chakravarti, A.; Mann, D. L.; Murray, C.; Johnson, A. H.; Gerhard, D.; Housman, D. E.; Payne, C.; Lovrien, E.: Segregation and linkage studies in the familial malignant melanoma (FMM)-dysplastic nevus (DN) syndrome. (Abstract) Am. J. Hum. Genet. 37: A23, 1985.

Bale, S. J.; Greene, M. H.; Murray, C.; Goldin, L. R.; Johnson, A. H.; Mann, D.: Hereditary malignant melanoma is not linked to the HLA complex on chromosome 6. Int. J. Cancer 36: 439-443, 1985.

Becher, R.; Gibas, Z.; Karakousis, C.; Sandberg, A. A.: Non-random chromosome changes in malignant melanoma. Cancer Res. 43: 5010-5016, 1983.

Becher, R.; Gibas, Z.; Sandberg, A. A.: Chromosome 6 in malignant melanoma. Cancer Genet. Cytogenet. 9: 173-175, 1983.

Bergman, W.; Gruis, N. A.; Sandkuijl, L. A.; Frants, R. R.: Genetics of seven Dutch familial atypical multiple mole-melanoma syndrome families: a review of linkage results including chromosomes 1 and 9. J. Invest. Derm. 103: 122S-125S, 1994.

Bergman, W.; Palan, A.; Went, L. N.: Clinical and genetic studies in six Dutch kindreds with the dysplastic naevus syndrome. Ann. Hum. Genet. 50: 249-258, 1986.

Cannon-Albright, L. A.; Goldgar, D. E.; Wright, E. C.; Turco, A.; Jost, M.; Meyer, L. J.; Piepkorn, M.; Zone, J. J.; Skolnick, M. H.: Evidence against the reported linkage of the cutaneous melanoma-dysplastic nevus syndrome locus to chromosome 1p36. Am. J. Hum. Genet. 46: 912-918, 1990.

Cawley, E. P.: Genetic aspects of malignant melanoma. Arch. Derm. Syph. 65: 440-450, 1952.

Clark, W. H., Jr.; Reimer, R. R.; Greene, M.; Ainsworth, A. M.; Mastrangelo, M. J.: Origin of familial malignant melanomas from heritable melanocytic lesions: 'the B-K mole syndrome.'. Arch. Derm. 114: 732-738, 1978.

Dracopoli, N. C.: Personal Communication. Cambridge, Mass. 11/16/1992.

Dracopoli, N. C.; Alhadeff, B.; Houghton, A. N.; Old, L. J.: Loss of heterozygosity at autosomal and X-linked loci during tumor progression in a patient with melanoma. Cancer Res. 47: 3995-4000, 1987.

Dracopoli, N. C.; Bale, S. J.; Housman, D. E.: Assignment of the familial melanoma gene to chromosome 1p36: frequent loss of heterozygosity of this region occurs late in tumor progression. (Abstract) Am. J. Hum. Genet. 45 (suppl.): A19, 1989.

Dracopoli, N. C.; Harnett, P.; Bale, S. J.; Stanger, B. Z.; Tucker, M. A.; Housman, D. E.; Kefford, R. F.: Loss of alleles from the distal short arm of chromosome 1 occurs late in melanoma tumor progression. Proc. Nat. Acad. Sci. 86: 4614-4618, 1989.

Elder, D. E.; Goldman, L. I.; Goldman, S. C.; Greene, M. H.; Clark, W. H., Jr.: Dysplastic nevus syndrome: a phenotypic association of sporadic cutaneous melanoma. Cancer 46: 1787-1794, 1980.

Elder, D. E.; Greene, M. H.; Guerry, D.; Kraemer, K. H.; Clark, W. H., Jr.: The dysplastic nevus syndrome: our definition. Am. J. Dermatopath. 4: 445-460, 1982.

Fountain, J. W.; Karayiorgou, M.; Ernstoff, M. S.; Kirkwood, J. M.; Vlock, D. R.; Titus-Ernstoff, L.; Bouchard, B.; Vijayasaradhi, S.; Houghton, A. N.; Lahti, J.; Kidd, V. J.; Housman, D. E.; Dracopoli, N. C.: Homozygous deletions within human chromosome band 9p21 in melanoma. Proc. Nat. Acad. Sci. 89: 10557-10561, 1992.

Frants, R. R.; Gruis, N. A.; Bergman, W.; Van der Zee, A.; van Haeringen, A.: Linkage studies in Dutch dysplastic nevus syndrome (DNS) families. (Abstract) Cytogenet. Cell Genet. 51: 956, 1989.

Goldstein, A. M.; Dracopoli, N. C.; Ho, E. C.; Fraser, M. C.; Kearns, K. S.; Bale, S. J.; McBride, O. W.; Clark, W. H., Jr.; Tucker, M. A.: Further evidence for a locus for cutaneous malignant melanoma-dysplastic nevus (CMM/DN) on

chromosome 1p, and evidence for genetic heterogeneity. Am. J. Hum. Genet. 52: 537-550, 1993.
Goldstein, A. M.; Goldin, L. R.; Dracopoli, N. C.; Clark, W. H., Jr.; Tucker, M. A.: Two-locus linkage analysis of cutaneous malignant melanoma/dysplastic nevi. Am. J. Hum. Genet. 58: 1050-1056, 1996.
Greene, M. H.; Clark, W. H., Jr.; Tucker, M. A.; Elder, D. E.; Kraemer, K. H.; Fraser, M. C.; Bondi, E. E.; Guerry, D.; Tuthill, R.; Hamilton, R.; LaRossa, D.: Precursor naevi in cutaneous malignant melanoma: a proposed nomenclature. (Letter) Lancet II: 1024, 1980.
Greene, M. H.; Clark, W. H., Jr.; Tucker, M. A.; Elder, D. E.; Kraemer, K. H.; Guerry, D., IV.; Witmer, W. K.; Thompson, J.; Matozzo, I.; Fraser, M. C.: Acquired precursors of cutaneous malignant melanoma: the familial dysplastic nevus syndrome. New Eng. J. Med. 312: 91-97, 1985.
Greene, M. H.; Goldin, L. R.; Clark, W. H., Jr.; Lovrien, E.; Kraemer, K. H.; Tucker, M. A.; Elder, D. E.; Fraser, M. C.; Rowe, S.: Familial cutaneous malignant melanoma: autosomal dominant trait possibly linked to the Rh locus. Proc. Nat. Acad. Sci. 80: 6071-6075, 1983.
Greene, M. H.; Reimer, R. R.; Clark, W. H., Jr.; Mastrangelo, M. J.: Precursor lesions in familial melanoma. Semin. Onco. 5: 85-87, 1978.
Happle, R.; Traupe, H.; Vakilzadeh, F.; Macher, E.: Arguments in favor of a polygenic inheritance of precursor nevi. J. Am. Acad. Derm. 6: 540-543, 1982.
Hartley, A. L.; Birch, J. M.; Marsden, H. B.; Harris, M.: Malignant melanoma in families of children with osteosarcoma, chondrosarcoma, and adrenal cortical carcinoma. J. Med. Genet. 24: 664-668, 1987.
Hawkins, B. R.; Dawkins, R. L.; Hockey, A.; Houliston, J. B.; Kirk, R. L.: Evidence for linkage between HLA and malignant melanoma. Tissue Antigens 17: 540-541, 1981.
Hecht, F.: The annals of cancer genetics: the description by Norris of hereditary malignant melanoma of the skin. Cancer Genet. Cytogenet. 42: 153-156, 1989.
Hecht, F.; Bixenman, H. A.; Cannizzaro, L. A.; Hecht, B. K.; Ramesh, H. R.; Lynch, H. T.; Fusaro, R. M.: Chromosome instability is found in hereditary malignant melanoma which is linked to Rh in 1p36.2-p34. (Abstract) Cytogenet. Cell Genet. 51: 1012, 1989.
Howell, J. N.; Greene, M. H.; Corner, R. C.; Maher, V. M.; McCormick, J. J.: Fibroblasts from patients with hereditary cutaneous malignant melanoma are abnormally sensitive to the mutagenic effect of simulated sunlight and 4-nitroquinoline 1-oxide. Proc. Nat. Acad. Sci. 81: 1179-1183, 1984.
Katzenellenbogen, I.; Sandbank, M.: Malignant melanoma in twins. Arch. Derm. 94: 331-332, 1967.
Kefford, R. F.; Salmon, J.; Shaw, H. M.; Donald, J. A.; McCarthy, W. H.: Hereditary melanoma in Australia: variable association with dysplastic nevi and absence of genetic linkage to chromosome 1p. Cancer Genet. Cytogenet. 51: 45-55, 1991.
Knudson,: Personal Communication. Philadelphia, Pa. 2/21/1997.
Kopf, A. W.; Hellman, L. J.; Rogers, G. S.; Gross, D. F.; Rigel, D. S.; Friedman, R. J.; Levenstein, M.; Brown, J.; Golomb, F. M.; Roses, D. F.; Gumport, S. L.; Mintzis, M. M.: Familial malignant melanoma. J.A.M.A. 256: 1915-1919, 1986.
Kraemer, K. H.; Greene, M. H.; Tarone, R.; Elder, D. E.; Clark, W. H.; Guerry, D.: Dysplastic nevi and cutaneous melanoma risk. Lancet II: 1076-1077, 1983.
Lynch, H. T.; Frichot, B. C., III.; Lynch, J. F.: Familial atypical multiple mole-melanoma syndrome. J. Med. Genet. 15: 352-356, 1978.
Lynch, H. T.; Fusaro, R. M.; Albano, W. A.; Pester, J.; Kimberling, W. J.; Lynch, J. F.: Phenotypic variation in the familial atypical multiple mole-melanoma syndrome (FAMMM). J. Med. Genet. 20: 25-29, 1983.
Lynch, H. T.; Fusaro, R. M.; Pester, J.; Lynch, J. F.: Familial atypical multiple mole melanoma (FAMMM) syndrome: genetic heterogeneity and malignant melanoma. Brit. J. Cancer 42: 58-70, 1980.
Lynch, H. T.; Krush, A. J.: Heredity and malignant melanoma: implications for early cancer detection. Canad. Med. Assoc. J. 99: 17-21, 1968.
Millikin, D.; Meese, E.; Vogelstein, B.; Witkowski, C.; Trent, J.: Loss of heterozygosity for loci on the long arm of chromosome 6 in human malignant melanoma. Cancer Res. 51: 5449-5453, 1991.
Moschella, S. L.: A report of malignant melanoma of the skin in sisters. Arch. Derm. 84: 1024-1025, 1961.
Nancarrow, D. J.; Palmer, J. M.; Walters, M. K.; Kerr, B. M.; Hafner, G. J.; Garske, L.; McLeod, G. R.; Hayward, N. K.: Exclusion of the familial melanoma locus (MLM) from the PND/D1S47 and MYCL1 regions of chromosome arm 1p in 7 Australian pedigrees. Genomics 12: 18-25, 1992.
Norris, W.: A case of fungoid disease. Edinburgh Med. Surg. J. 16: 562-565, 1820.
Ochi, H.; Wake, N.; Rao, U.; Takeuchi, J.; Slocum, H. K.; Rustum, Y. C.; Karakousis, C.; Sandberg, A. A.: Serial cytogenetic analysis of a recurrent malignant melanoma. Cancer Genet. Cytogenet. 11: 175-183, 1984.
Pascoe, L.: The inheritance of cutaneous malignant melanoma (CMM) and dysplastic nevus syndrome (DNS). (Letter) Am. J. Hum. Genet. 40: 464-465, 1987.
Pathak, S.; Drwinga, H. L.; Hsu, T. C.: Involvement of chromosome 6 in rearrangements in human malignant melanoma cell lines. Cytogenet. Cell Genet. 36: 573-579, 1983.
Pellegris, G.; Illeni, M. T.; Rovini, D.; Vaglini, M.; Cascinelli, N.; Ghidoni, A.: HLA complex and familial malignant melanoma. Int. J. Cancer 29: 621-623, 1982.
Reimer, R. R.; Clark, W. H., Jr.; Greene, M. H.; Ainsworth, A. M.; Fraumeni, J. F., Jr.: Precursor lesions in familial melanoma: a new genetic preneoplastic syndrome. J.A.M.A. 239: 744-746, 1978.
Rey, J. A.; Bello, M. J.; de Campos, J. M.; Ramos, M. C.; Benitez, J.: Cytogenetic findings in a human malignant melanoma metastatic to the brain. Cancer Genet. Cytogenet. 16: 179-183, 1985.
Rhodes, A. R.; Slifman, N. R.; Korf, B. R.: Familial aggregation of small congenital nevomelanocytic nevi. Am. J. Med. Genet. 22: 315-326, 1985.
Robinson, M. J.; Manheimer, L.: Familial melanomas. (Letter) J.A.M.A. 220: 277, 1972.
Salamon, T.; Schnyder, U. W.; Storck, H.: A contribution to the question of heredity of malignant melanomas. Dermatologica 126: 65-75, 1963.
Schoch, E. P., Jr.: Familial malignant melanoma: a pedigree and cytogenetic study. Arch. Derm. 88: 445-455, 1963.
Smith, F. E.; Henly, W. S.; Knox, J. M.; Lane, M.: Familial melanoma. Arch. Intern. Med. 117: 820-823, 1966.
Traupe, H.; Macher, E.; Hamm, H.; Happle, R.: Mutation rate estimates are not compatible with autosomal dominant inheritance of the dysplastic nevus 'syndrome'. Am. J. Med. Genet. 32: 155-157, 1989.
Trent, J. M.; Rosenfeld, S. B.; Meyskens, F. L.: Chromosome 6q involvement in human malignant melanoma. Cancer Genet. Cytogenet. 9: 177-180, 1983.
Turkington, R. W.: Familial factor in malignant melanoma. J.A.M.A. 192: 77-82, 1965.
van Haeringen, A.; Bergman, W.; Nelen, M. R.; van der Kooij-Meijs, E.; Hendrikse, I.; Wijnen, J. T.; Meera Khan, P.; Klasen, E. C.; Frants, R. R.: Exclusion of the dysplastic nevus syndrome (DNS) locus from the short arm of chromosome 1 by linkage studies in Dutch families. Genomics 5: 61-64, 1989.
Vasen, H. F. A.; Bergman, W.; van Haeringen, A.; Scheffer, E.; Van Slooten, E. A.: The familial dysplastic nevus syndrome: natural history and the impact of screening on prognosis—a study of nine families in the Netherlands. Europ. J. Cancer Clin. Oncol. 25: 337-341, 1989.
Wallace, D. C.; Beardmore, G. L.; Exton, L. A.: Familial malignant melanoma. Ann. Surg. 177: 15-20, 1973.
Wallace, D. C.; Exton, L. A.; McLeod, G. R.: Genetic factor in malignant melanoma. Cancer 27: 1262-1266, 1971.

MELANOMA, CUTANEOUS MALIGNANT 2 (*155601)

JJM NOTE

To date, no clinical heterogeneity has been suggested, despite the fact that at least two genetic loci are involved, perhaps even in the same family with dysplastic nevus syndrome.

A gene (CMM) involved in the early stages of cutaneous malignant melanoma has been mapped to chromosome 1; see 155600. A locus on 9p21 has been identified as the site of mutations in malignant melanoma (see later). Further evidence indicates that mutation in the p16 gene (CDKN2; 600160) is the basis of 9p21-related melanoma. For this reason, a number sign (#) is used with this entry. Indeed, Puig et al. (1995) presented evidence suggesting the existence of several tumor suppressor genes on 9p that are involved in the predisposition to and/or progression of CMM and excluded p16 from involvement in the early development of some melanoma tumors.

Four independent reports supported the existence of a tumor-suppressor locus on 9p that is mutated during the early stages of melanoma development (Dracopoli et al., 1987; Cowan et al., 1988; Pedersen and Wang, 1989; Kacker et al., 1990). Rearrangements of this chromosome occur in nearly half of all melanomas, with the most frequently targeted region extending from 9pter to 9q13 (Fountain et al., 1990). The alpha-interferon (147660) and beta-1 interferon (147640) genes are located within this region and are viewed as candidates for involvement in cutaneous melanoma because they have growth-inhibitory properties and, as therapeutic agents, have caused the regression of metastatic melanoma. In addition, interferon-like molecules have been detected in the basal layer of the epidermis, suggesting that they may contribute to the control of normal epidermal growth or regeneration. Also residing in this region of 9p are the tyrosinase-related protein (115501) and the glycoprotein 4-beta-galactosyltransferase (137060) genes which might play a role. Using markers on 9p in studies of loss of heterozygosity, Fountain et al. (1992) identified a small critical region of 2 to 3 Mb on 9p21 in which a putative melanoma tumor-suppressor gene resides. In the course of these studies they excluded IFNB1, the IFNA gene cluster, GGTB2, and TYRP as the site of mutations in melanoma because they are located outside the homozygously deleted regions. In 10 Utah kindreds and 1 Texas kindred with multiple cases of cutaneous malignant melanoma, Cannon-Albright et al. (1992) provided evidence for a locus determining susceptibility in the region 9p13-p22. The markers that determined this location by linkage studies resided in a candidate region on 9p21, previously implicated by the presence of homozygous deletions in melanoma tumors and by the presence of a germline deletion in a person with 8 independent melanomas. Linkage studies placed the susceptibility locus (which Cannon-Albrigh t et al. (1992) symbolized MLM) near IFNA and D9S126 with a maximum location score of 12.71. The location of a gene predisposing to CMM on 9p was supported by the description of de novo constitutional unbalanced reciprocal translocation involving the short arms of chromosomes 5 and 9 with a cytogenetically visible deletion of one or the other, in a woman with atypical moles and multiple primary melanomas, who lacked a family history of hereditary cutaneous malignant melanoma (Petty et al., 1993). In a second article, Petty et al. (1993) demonstrated deletion of the IFNA locus and several markers located at 9p21. By multipoint linkage analysis using the 9p markers IFNA and D9S126 in 26 Australian melanoma kindreds, Nancarrow et al. (1993) obtained a peak lod score of 4.43, 15 cM centromeric to D9S126; a lod score of 4.13 was also found 15 cM telomeric of IFNA. These data suggest that the CMM2 locus lies outside of the IFNA-D9S126 interval.

The identification of the CDKN2 gene (p16; 600160) in melanoma cell lines by study of homozygous deletions for the 9p21 region and the identification of nonsense, missense, and frameshift mutations in those cell lines that had at least one CDKN2 allele present suggest that mutations in this gene may be the basis of familial malignant melanoma in many instances. The mutations in this tumor suppressor gene are involved in many other malignancies, rivaling the p53 gene (191170) in its catholicity.

Using 2 highly informative (CA)n repeats, D9S126 and IFNA (147660), Goldstein et al. (1994) performed linkage studies in 13 families previously investigated for linkage to chromosome 1p (Bale et al., 1989; Goldstein et al., 1993). Significant evidence for linkage to IFNA was found. The analyses were performed using CMM alone (all individuals without either confirmed melanoma or borderline lesions were considered unaffected); CMM/DN with both variable age at onset and sporadics; or CMM affecteds only (all individuals either without confirmed melanoma or with borderline lesions were designated 'unknown'). No significant evidence of linkage was found for D9S126 with either CMM alone or CMM/DN. Significant evidence for heterogeneity was found when a homogeneity test allowing for linkage to chromosome 9p or chromosome 1p or neither region was used. Since some families showed linkage to both 1p and 9p, a 2-locus model was being investigated also.

Weaver-Feldhaus et al. (1994) isolated genomic clones that span a large region in 9p21 surrounding the presumptive tumor suppressor gene(s) thought to be involved in susceptibility to melanoma and to influence progression of certain other tumors. A set of sequence-tagged sites in this region were developed. By using these markers and others previously reported, the 9p21 region was studied by physical mapping in 84 melanoma cell lines. Homozygous deletions of the 9p21 region were found in 56% of melanoma tumors tested. A putative tumor suppressor gene was localized to a region of less than 40 kb that lies proximal (centromeric) to the IFNA gene cluster. The results were consistent with previous genetic studies of MLM that mapped the gene to the region between IFNA1 and D9S126.

Cannon-Albright et al. (1994) narrowed the assignment of the MLM gene to a 2-cM region that is proximal to D9S736 and distal to D9S171. They commented that 2 of the linkage studies that confirmed the chromosome 9 assignment for malignant melanoma provided no evidence for the existence of genetic heterogeneity (Gruis et al., 1993; Nancarrow et al., 1993). There is evidence both for and against linkage of melanoma to the 1p region as indicated in 155600. Cannon-Albright et al. (1994) suggested that the question of genetic heterogeneity for melanoma susceptibility will be best answered when the 9p susceptibility locus has been cloned and tested in multiple kindreds and cases.

Following up on previous linkage analyses of 19 cutaneous malignant melanoma/dysplastic nevi (CMM/DN) kindreds which showed significant evidence of linkage and heterogeneity to both chromosomes 1p and 9p, Goldstein et al. (1996) examined 2-locus hypotheses. The lod scores for CMM alone were highest using the single locus-heterogeneity model. They found much stronger evidence of linkage to 9p than to 1p for CMM alone; the lod scores were approximately 2 times greater on 9p than on 1p. A change in lod scores from an evaluation of CMM alone to CMM/DN suggested to the authors that a chromosome 1p locus contributed to both CMM and CMM/DN, whereas a 9p locus contributed more to CMM alone. For 2-locus models, the lod scores from 1p were greater for CMM/DN than for CMM alone. After conditioning on linkage to the other locus, only the 9p locus consistently showed significant evidence for linkage to CMM alone.

Nomenclature: Knudson (1997) referred to the form of melanoma caused by mutation in the p16 gene (CDKN2) and mapping to 9p21 as melanoma-1, since it was the first form of melanoma to be defined at the DNA level. Furthermore, he referred to the form of melanoma second to be defined molecularly, that mapping to chromosome 12 and caused by mutations in the CDK4 gene (123829), as melanoma-2. Presumably, the form of melanoma that maps to 1p (155600) should, according to this scheme, be designated melanoma-3 when the gene is identified.

See also Petty et al. (1993)

Bale, S. J.; Dracopoli, N. C.; Tucker, M. A.; Clark, W. H., Jr.; Fraser, M. C.; Stanger, B. Z.; Green, P.; Donis-Keller, H.; Housman, D. E.; Greene, M. H.: Mapping the gene for hereditary cutaneous malignant melanoma-dysplastic nevus to chromosome 1p. New Eng. J. Med. 320: 1367-1372, 1989.

Cannon-Albright, L. A.; Goldgar, D. E.; Meyer, L. J.; Lewis, C. M.; Anderson, D. E.; Fountain, J. W.; Hegi, M. E.; Wiseman, R. W.; Petty, E. M.; Bale, A. E.; Olopade, O. I.; Diaz, M. O.; Kwiatkowski, D. J.; Piepkorn, M. W.; Zone, J. J.; Skolnick, M. H.: Assignment of a locus for familial melanoma, MLM, to chromosome 9p13-p22. Science 258: 1148-1152, 1992.

Cannon-Albright, L. A.; Goldgar, D. E.; Neuhausen, S.; Gruis, N. A.; Anderson, D. E.; Lewis, C. M.; Jost, M.; Tran, T. D.; Nyguen, K.; Kamb, A.; Weaver-Feldhaus, J.; Meyer, L. J.; Zone, J. J.; Skolnick, M. H.: Localization of the 9p melanoma susceptibility locus (MLM) to a 2-cM region between D9S736 and D9S171. Genomics 23: 265-268, 1994.

Cowan, J. M.; Halaban, R.; Francke, U.: Cytogenetic analysis of melanocytes from premalignant nevi and melanomas. J. Nat. Cancer Inst. 80: 1159-1164, 1988.

Dracopoli, N. C.; Alhadeff, B.; Houghton, A. N.; Old, L. J.: Loss of heterozygosity at autosomal and X-linked loci during tumor progression in a patient with melanoma. Cancer Res. 47: 3995-4000, 1987.

Fountain, J. W.; Bale, S. J.; Housman, D. E.; Dracopoli, N. C.: Genetics of melanoma. Cancer Surv. 9: 645-671, 1990.

Fountain, J. W.; Karayiorgou, M.; Ernstoff, M. S.; Kirkwood, J. M.; Vlock, D. R.; Titus-Ernstoff, L.; Bouchard, B.; Vijayasaradhi, S.; Houghton, A. N.; Lahti, J.; Kidd, V. J.; Housman, D. E.; Dracopoli, N. C.: Homozygous deletions within human chromosome band 9p21 in melanoma. Proc. Nat. Acad. Sci. 89: 10557-10561, 1992.

Goldstein, A. M.; Dracopoli, N. C.; Engelstein, M.; Fraser, M. C.; Clark, W. H., Jr.; Tucker, M. A.: Linkage of cutaneous malignant melanoma/dysplastic nevi to chromosome 9p, and evidence for genetic heterogeneity. Am. J. Hum. Genet. 54: 489-496, 1994.

Goldstein, A. M.; Dracopoli, N. C.; Ho, E. C.; Fraser, M. C.; Kearns, K. S.; Bale, S. J.; McBride, O. W.; Clark, W. H., Jr.; Tucker, M. A.: Further evidence for a locus for cutaneous malignant melanoma-dysplastic nevus (CMM/DN) on chromosome 1p, and evidence for genetic heterogeneity. Am. J. Hum. Genet. 52: 537-550, 1993.

Goldstein, A. M.; Goldin, L. R.; Dracopoli, N. C.; Clark, W. H., Jr.; Tucker, M. A.: Two-locus linkage analysis of cutaneous malignant melanoma/dysplastic nevi. Am. J. Hum. Genet. 58: 1050-1056, 1996.

Gruis, N. A.; Sandkuijl, L. A.; Weber, J. L.; van der Zee, A.; Borgstein, A.-M.; Bergman, W.; Frants, R. R.: Linkage analysis in Dutch familial atypical multiple mole-melanoma (FAMMM) syndrome families: effect of naevus count. Melanoma Res. 3: 271-277, 1993.

Kacker, R. K.; Giovanella, B. C.; Pathak, S.: Consistent karyotypic abnormalities in human malignant melanomas. Anticancer Res. 10: 859-872, 1990.

Knudson,: Personal Communication. Philadelphia, Pa. 2/21/1997.

Nancarrow, D. J.; Mann, G. J.; Holland, E. A.; Walker, G. J.; Beaton, S. C.; Walters, M. K.; Luxford, C.; Palmer, J. M.; Donald, J. A.; Weber, J. L.; Fountain, J. W.; Kefford, R. F.; Hayward, N. K.: Confirmation of chromosome 9p linkage in familial melanoma. Am. J. Hum. Genet. 53: 936-942, 1993.

Pedersen, M. I.; Wang, N.: Consistent karyotypic abnormalities in human malignant melanomas. Cancer Genet. Cytogenet. 41: 185-201, 1989.

Petty, E. M.; Bolognia, J. L.; Bale, A. E.; Yang-Feng, T.: Cutaneous malignant melanoma and atypical moles associated with a constitutional rearrangement of chromosomes 5 and 9. Am. J. Med. Genet. 45: 77-80, 1993.

Petty, E. M.; Gibson, L. H.; Fountain, J. W.; Bolognia, J. L.; Yang-Feng, T. L.; Housman, D. E.; Bale, A. E.: Molecular definition of a chromosome 9p21 germ-line deletion in a woman with multiple melanomas and a plexiform neurofibroma: implications for 9p tumor-suppressor gene(s). Am. J. Hum. Genet. 53: 96-104, 1993.

Puig, S.; Ruiz, A.; Lazaro, C.; Castel, T.; Lynch, M.; Palou, J.; Vilalta, A.; Weissenbach, J.; Mascaro, J.-M.; Estivill, X.: Chromosome 9p deletions in cutaneous malignant melanoma tumors: the minimal deleted region involves markers outside the p16 (CDKN2) gene. Am. J. Hum. Genet. 57: 395-402, 1995.

Weaver-Feldhaus, J.; Gruis, N. A.; Neuhausen, S.; Le Paslier, D.; Stockert, E.; Skolnick, M. H.; Kamb, A.: Localization of a putative tumor suppressor gene by using homozygous deletions in melanomas. Proc. Nat. Acad. Sci. 91: 7563-7567, 1994.

MELANOSIS, NEUROCUTANEOUS (249400)

JJM NOTE

Fine histopathologic dissection and loss of heterozygosity studies could be done to test the hypothesis that this condition could be mosaicism for some melanoma-predisposing gene.

This rare condition, associated skin and meningeal pigmentation, is potentially highly malignant. Death usually occurs in early childhood. No certain evidence of a mendelian basis has been found. The condition is thought to be a congenital dysplasia of the neural crest. Ferris et al. (1987) described a case and suggested that the segmental distribution, sporadic nature, and invariable discordance in identical twins for the occurrence of giant, hairy nevi is consistent with the hypothesis that the disease is a consequence of one or more somatic mutations which would result in prenatal lethality if they occurred in the germ line cells. Demirci et al. (1995) described the MRI appearance of the brain in 2 cases of neurocutaneous melanosis. This demonstrated intraparenchymal melanin deposits but no detectable leptomeningeal abnormality.

See also Fox et al. (1964); Kaplan et al. (1975); Reed et al. (1965); Tveten (1965)

Demirci, A.; Kawamura, Y.; Sze, G.; Duncan, C.: MR of parenchymal neurocutaneous melanosis. AJNR 16: 603-606, 1995.
Ferris, M. K.; Proud, V. K.; Narva, S. F.; Nance, W. E.: Neurocutaneous melanosis syndrome. (Abstract) Am. J. Hum. Genet. 41: A57 only, 1987.
Fox, H.; Emery, J. L.; Goodbody, R. A.; Yates, P. O.: Neuro-cutaneous melanosis. Arch. Dis. Child. 39: 508-516, 1964.
Kaplan, A. M.; Itabashi, H. H.; Hanelin, L. G.; Lu, A. T.: Neurocutaneous melanosis with malignant leptomeningeal melanoma. Arch. Neurol. 32: 669-671, 1975.
Reed, W. B.; Becker, S. W.; Becker, S. W., Jr.; Nickel, W. R.: Giant pigmented nevi melanoma, and leptomeningeal melanocytosis: a clinical and histopathological study. Arch. Derm. 91: 100-119, 1965.
Tveten, L.: Primary meningeal melanosis: a clinico-pathological report of two cases. Acta Path. Microbiol. Scand. 63: 1-10, 1965.

MELANOMA, MALIGNANT INTRAOCULAR (155700)

JJM NOTE

Perhaps a series of patients with pigmentary lesions or nevi of the fundus could be assembled to determine whether the analogy of the cutaneous dysplastic nevus applies. A complementary study would involve updating the published pedigrees. Occasionally, I see an ocular or uveal melanoma reported in a breast cancer family.

Bowen et al. (1964) reported malignant intraocular melanoma in a 45-year-old white female and her 26-year-old daughter. Davenport (1927) reported this malignancy in 3 successive generations. The occurrence of cutaneous melanoma and intraocular melanoma as double primary cancers in the same patient and in different members of the same family has suggested that these 2 forms of melanoma may be etiologically related. From their family studies, Greene et al. (1983) concluded that these associations may be coincidental.

See also Cawley (1952); Simons et al. (1983)

Bowen, S. F.; Brady, H.; Jones, V. L.: Malignant melanoma of eye occurring in two successive generations. Arch. Ophthal. 71: 805-806, 1964.
Cawley, E. P.: Genetic aspects of malignant melanoma. Arch. Derm. Syph. 65: 440-450, 1952.
Davenport, R. C.: Familial history of choroidal sarcoma. Brit. J. Ophthal. 11: 443-445, 1927.
Greene, M. H.; Sanders, R. J.; Chu, F. C.; Clark, W. H., Jr.; Elder, D. E.; Cogan, D. G.: The familial occurrence of cutaneous melanoma, intraocular melanoma, and the dysplastic nevus syndrome. Am. J. Ophthal. 96: 238-245, 1983.
Simons, K. B.; Hale, L. M.; Morrison, H. M., Jr.; Eifrig, D. E.; Peiffer, R. L., Jr.: Choroidal malignant melanoma in siblings. Am. J. Ophthal. 96: 675-680, 1983.

MELANOMA, UVEAL (155720)

JJM NOTE

Is there a precursor lesion to uveal melanoma? Do relatives of the rare familial cases have an excess of other cancers? One hopes that the results of genetic linkage studies will be clear. Clearly, progress on this rare familial tumor must involve the interinstitutional collection of families.

Uveal melanoma is the most common primary intraocular malignancy. Canning and Hungerford (1988) found reports of 12 families with adequate documentation of the occurrence of uveal melanomas in 2 or more members. They described 2 more families, each with 2 affected members: a brother and sister who presented at ages 69 and 63 years, respectively, and a mother and son who presented at ages 33 and 20, respectively. Young et al. (1994) reported 11 families in which 2 or more members had uveal melanoma. The median age at diagnosis was 56.5 years, similar to the median age at diagnosis of the other patients treated at their institution. Of the 24 affected persons, 13 were male and 11 were female.

In 19 patients with uveal melanoma, 4,3:Mukai and Dryja (1985, 1986) sought evidence of homozygosity or hemizygosity for DNA polymorphisms that were heterozygous in the host. The rationale follows that which has been successful in the study of retinoblastoma and Wilms tumor. They found that 2 of the 15 informative patients had, in their tumors, lost alleles at loci on chromosome 2. Using DNA markers, Prescher et al. (1992) found loss of chromosome 3 alleles and multiplication of 8q alleles in uveal melanoma. Cytogenetic analysis had previously shown monosomy of chromosome 3 and multiplication of chromosome 8q material as nonrandom chromosomal aberrations. Horsthemke et al. (1992) interpreted the findings as indicating that a tumor suppressor gene on chromosome 3 and an oncogene on chromosome 8 are involved in the formation or progression of uveal melanoma. In 4 of 10 posterior uveal melanomas, all derived from the ciliary body, Sisley et al. (1992) demonstrated monosomy 3 and i(8q). See also Prescher et al. (1994).

In a review of medical charts of 4,500 patients with uveal melanoma, Singh et al. (1996) found 56 patients in 27 families with a family history. The uveal melanoma was unilateral in all 56 familial cases. In 17 cases (63%), the second affected relative was a first-degree relative; second-degree relatives accounted for 22% and third-degree relatives for 15%. In 25 families (93%), only 2 members were affected and in 2 families (7%), 3 members had uveal melanoma. Patients with familial uveal melanoma were 4 times as likely to have a second primary malignant neoplasm than persons in the general population; however, unaffected relatives of these patients were apparently not at higher risk of having another primary malignant neoplasm.

Canning, C. R.; Hungerford, J.: Familial uveal melanoma. Brit. J. Ophthal. 72: 241-243, 1988.
Horsthemke, B.; Prescher, G.; Bornfeld, N.; Becher, R.: Loss of chromosome 3 alleles and multiplication of chromosome 8 alleles in uveal melanoma. Genes Chromosomes Cancer 4: 217-221, 1992.
Mukai, S.; Dryja, T. P.: Loss of alleles at polymorphic loci on chromosome 2 in uveal melanoma. Cancer Genet. Cytogenet. 22: 45-53, 1986.
Mukai, S.; Dryja, T. P.: Loss of alleles at polymorphic loci on chromosome 2 in uveal melanoma.(Abstract) Am. J. Hum. Genet. 37: A33, 1985.
Prescher, G.; Bornfeld, N.; Becher, R.: Two subclones in a case of uveal melanoma: relevance of monosomy 3 and multiplication of chromosome 8q. Cancer Genet. Cytogenet. 77: 144-146, 1994.
Prescher, G.; Bornfeld, N.; Horsthemke, B.; Becher, R.: Chromosomal aberrations defining uveal melanoma of poor prognosis.(Letter) Lancet 339: 691-692, 1992.
Singh, A. D.; Shields, C. L.; De Potter, P.; Shields, J. A.; Trock, B.; Cater, J.; Pastore, D.: Familial uveal melanoma: clinical observations on 56 patients. Arch. Ophthal. 114: 392-399, 1996.
Sisley, K.; Cottam, D. W.; Rennie, I. G.; Parsons, M. A.; Potter, A. M.; Potter, C. W.; Rees, R. C.: Non-random abnormalities of chromosomes 3, 6, and 8 associated with posterior uveal melanoma. Genes Chromosomes Cancer 5: 197-200, 1992.
Young, L. H. Y.; Egan, K. M.; Walsh, S. M.; Gragoudas, E. S.: Familial uveal melanoma. Am. J. Ophthal. 117: 516-520, 1994.

NEVI; PIGMENTED MOLES (*162900)

JJM NOTE

The same loci linked to the hereditary dysplastic nevus syndrome could be tested in families with "ordinary" pigmented nevi.

Although it is a common observation that nevi occur in families, probably with dominant transmission, the study by Denaro (1944) is one of the few which has examined the matter specifically. Multiple pigmented moles are a feature of one chromosomal aberration, the Turner syndrome, as pointed out by Sharpey-Schafer (1941) in the pre-chromosomal era. Estabrook (1928) reported affected persons in 5 generations of a family. (His term 'nevus spilus' comes from the Greek 'spilos' for 'spot.') Goodman et al. (1971) described 3 unrelated cases of giant pigmented hairy nevi in which other members of pedigrees exhibited multiple small pigmented nevi. They considered the presence of 6 such small nevi as indication of the presence of a mutant gene. See giant pigmented hairy nevus (137550) and malignant melanoma (155600).

See also Aschinberg et al. (1977); Meirowski (1942)

Aschinberg, L. C.; Solomon, L. M.; Zeis, P. M.; Justice, P.; Rosenthal, I. M.: Vitamin D-resistant rickets associated with epidermal nevus syndrome: demonstration of a phosphaturic substance in the dermal lesions. J. Pediat. 91: 56-60, 1977.
Denaro, S. J.: The inheritance of nevi. J. Hered. 35: 215-218, 1944.
Estabrook, A. H.: A family with birthmarks (nevus spilus) for five generations. (Abstract) Eugen. News 13: 90-92, 1928.
Goodman, R. M.; Caren, J.; Ziprkowski, M.; Padeh, B.; Ziprkowski, L.; Cohen, B. E.: Genetic considerations in giant pigmented hairy naevus. Brit. J. Derm. 85: 150-157, 1971.
Meirowski, E.: Moles and malformations of the skin and their relationship to inheritance and phylogenesis (new and old investigations). Brit. J. Derm. 54: 99-121, 1942.
Sharpey-Schafer, E. P.: Case of pterygo-nuchal infantilism (Turner's syndrome), with post-mortem findings. Lancet II: 559-560, 1941.

SC PHOCOMELIA SYNDROME (#269000)

JJM NOTE

This entity is included because of a case report of a patient with cutaneous malignant melanoma (Parry et al. 1986). The occurrence may well be coincidental (especially since ascertainment was made by the Surgery Branch of the NCI); however, she was one of the oldest reported patients, so the tumor could be a late (and infrequent) manifestation.

A number sign (#) is used with this entry because of the near-certainty that it is the same as Roberts syndrome (268300).

In a family with surname beginning with S and another with surname beginning with C, Herrmann et al. (1969) described a syndrome consisting of the following features: (1) nearly symmetrical reductive malformations of the limbs, resembling phocomelia; (2) flexion contractures of various joints; (3) multiple minor anomalies, including capillary hemangioma of the face, forehead and ears, hypoplastic cartilages of the ears and nose, micrognathia, scanty, silvery-blond hair, and cloudy corneas; (4) intrauterine and extrauterine growth retardation; (5) possibly mental retardation, and (6) autosomal recessive inheritance. Feingold (1992) provided follow-up on the 'C-patient' reported by Herrmann et al. (1969). At age 14 the patient developed Moyamoya disease (252350) causing left hemiparesis. Chromosome studies showed premature centromere separation characteristic of this syndrome, i.e., splaying of the heterochromatin, repulsion of the short arms of the acrocentric chromosomes, and splaying of Y chromosome heterochromatin. At the age of 23 the patient developed chest pain and electrocardiographic changes indicative of anterior myocardial infarction. Subsequently he was found to have an apical aneurysm of the left ventricle with apical thrombosis and died of congestive heart failure. He remained short throughout his life; his final height was 139 cm. Although his facial characteristics remained constant, his scanty silvery blond hair became brown and less sparse. Mild valvular aortic stenosis was diagnosed at one time. The same syndrome was probably described by O'Brien and Mustard (1921) in 3 of 8 children of normal parents who were related as double first cousins. Hall and Greenberg (1972) described an affected 8-year-old girl who was mentally normal. They were impressed with hypotrichosis and midfacial hemangioma, for which reason they proposed the designation hypomelia-hypotrichosis-facial hemangioma syndrome. They emphasized a characteristic appearance of the face. Although the SC syndrome is sometimes referred to as the pseudothalidomide syndrome, some cases of the Holt-Oram syndrome (142900) may be more deserving of that name (Lenz et al., 1974). Humeroradial synostosis was also a feature; see 143050 and 236400.

Grosse et al. (1975) described a case of what they termed the tetraphocomelia-cleft palate syndrome. Capillary hemangioma of the glabella and lids and fine silvery-blond hair were described. The authors questioned the separateness of Roberts syndrome (268300) and this syndrome. By an analysis of phenotype, Herrmann et al. (1977) could not tell whether the SC phocomelia syndrome and the Roberts syndrome were 'due to different recessive genes, different alleles, or the same recessive gene.' Tomkins et al. (1979) noted the uncertainty as to whether Roberts syndrome and the SC syndrome are distinct entities. They found a consistent centromeric abnormality of the chromosomes, namely, puffing and splitting, in 4 patients who had certain clinical features in common: bilateral corneal opacities, microcephaly, absence of radii, limitation of extension at the elbow and knees, enlargement of the phallus, and survival beyond the neonatal period. The usual absence of cleft palate in the SC syndrome may be a difference. Qazi et al. (1979) described chromosomal abnormalities like those of Roberts syndrome. Because of increasing evidence (reviewed in 268300) that the Roberts and SC syndromes are one entity, an asterisk is not used with this entry. Opitz and Lowry (1987) stated that early on they had the impression that the Roberts syndrome and the SC-pseudothalidomide syndrome may occur in different members of the same sibship and that the identification of identical cytological markers complements the conclusion of nosologic identity (as well as effective prenatal diagnosis).

In a Christian Lebanese family, BenEzra et al. (1982) observed 2 brothers with radial aplasia and abnormality of anterior chamber cleavage in the eye. Both showed a normal XY karyotype except for unusual sister chromatid disjunction at the centromeric region during prophase and metaphase. In the first brother, the radius and thumbs were absent and the tibia and fibula were relatively short; bilateral microphthalmia with central corneal leukoma was described, as well as moderate micrognathia and aplasia of the alae nasi. The second brother had extensive facial hemangioma, total cleft palate, total aplasia of the radius and ulna, marked shortening of the humerus, absent thumbs, absent fibula, and deformed tibia. Comment concerning parental consanguinity was not made. Presumably, this disorder falls into the same group as Roberts syndrome and SC phocomelia syndrome. Parry et al. (1986) reported the cases of 2 middle-aged sisters with most of the manifestations of the SC phocomelia syndrome including postnatal growth retardation, symmetric limb deficiencies with radial aplasia and absent thumbs, facial anomalies with microcephaly, microphthalmia, hypoplastic nasal alae, and borderline-to-mild mental retardation. Unusual findings included congenital paralysis of some cranial nerves in both and malignant melanoma in 1. Tumor cells and cultured lymphocytes and fibroblasts of both patients showed premature separation of centromeric heterochromatin (PCS). Holmes-Siedle et al. (1990) described what they called the Roberts/SC phocomelia syndrome in 3 sibs of second-cousin parents. One child survived to age 9. A somewhat unusual feature was craniosynostosis; overlap with the Baller-Gerold syndrome (218600), which has craniosynostosis as a feature, was discussed.

See also Bokesoy et al. (1982); Klein et al. (1978); Leonard et al. (1982)

BenEzra, D.; Abulafia, H.; Maftzir, G.; Yatziv, S.; Paez, J.; Zelikovitch, A.: Radial aplasia, chromosomal aberration, and anterior chamber cleavage manifestations in two siblings. Birth Defects Orig. Art. Ser. 18(6): 571-575, 1982.

Bokesoy, I.; Balci, S.; Bilgic, S.: A case of SC-phocomelia syndrome with non-random centromere separation.In: Papadatos, C. J.; Bartsocas, C. S.: Skeletal Dysplasias. New York: Alan R. Liss (pub.) 1982. Pp. 351-356.

Feingold, M.: History of C-patient with SC-Roberts/pseudothalidamide (sic) syndrome. (Letter) Am. J. Med. Genet. 43: 898-899, 1992.

Grosse, F. R.; Pandel, C.; Wiedemann, H. R.: The tetraphocomelia—cleft palate syndrome: description of a new case. Humangenetik 28: 353-356, 1975.

Hall, B. D.; Greenberg, M. H.: Hypomelia-hypotrichosis-facial hemangioma syndrome (pseudothalidomide, SC syndrome, SC phocomelia syndrome). Am. J. Dis. Child. 123: 602-604, 1972.

Herrmann, J.; Feingold, M.; Tuffli, G. A.; Opitz, J. M.: A familial dysmorphogenetic syndrome of limb deformities, characteristic facial appearance and associated anomalies: the 'pseudothalidomide' or 'SC-syndrome.'. Birth Defects Orig. Art. Ser. V(3): 81-89, 1969.

Holmes-Siedle, M.; Seres-Santamaria, A.; Crocker, M.; Hall, J. G.; Crouchman, M.: A sibship with Roberts/SC phocomelia syndrome. Am. J. Med. Genet. 37: 18-22, 1990.

Klein, D.; Tobler, R.; Konig, H.: Hypomelia-hypotrichosis-facial-hemangioma syndrome associated with congenital glaucoma and other anomalies. Docum. Ophthal. Proc. Ser. 17: 383-391, 1978.

Lenz, W. D.; Marquardt, E.; Weicker, H.: Pseudothalidomide syndrome. Birth Defects Orig. Art. Ser. X(5): 97-107, 1974.

Leonard, P.; Rendle-Short, J.; Skardoon, L.: Roberts'—SC phocomelia syndrome with cytogenetic findings. Hum. Genet. 60: 379-380, 1982.

O'Brien, H. R.; Mustard, H. S.: An adult living case of total phocomelia. J.A.M.A. 77: 1964-1967, 1921.

Opitz, J. M.; Lowry, R. B.: Lincoln vs. Douglas again; comments on the papers by Curry et al, Greenberg et al, and Belmont et al. (Editorial) Am. J. Med. Genet. 26: 69-71, 1987.

Parry, D. M.; Mulvihill, J. J.; Tsai, S.; Kaiser-Kupfer, M. I.; Cowan, J. M.: SC phocomelia syndrome, premature centromere separation, and congenital cranial nerve paralysis in two sisters, one with malignant melanoma. Am. J. Med. Genet. 24: 653-672, 1986.

Qazi, Q. H.; Kassner, E. G.; Masakawa, A.; Madahar, C.; Choi, S. J.: The SC phocomelia syndrome: report of two cases with cytogenetic abnormality. Am. J. Med. Genet. 4: 231-238, 1979.

Tomkins, D.; Hunter, A.; Roberts, M.: Cytogenetic findings in Roberts-SC phocomelia syndrome(s). Am. J. Med. Genet. 4: 17-26, 1979.

ROBERTS SYNDROME (*268300)

JJM NOTE

This entity is included because it is widely considered to be the same as the previous entry. In fact, the case of melanoma studied by Krassikoff et al. (1988) is the same case as in the previous entry.

Roberts (1919) described 3 affected sibs of first-cousin Italian parents. Pictures were included. The bones of the legs were almost absent and those of the arms hypoplastic. Bilateral cleft lip and cleft palate were present. The skull looked oxycephalic with prominent eyes (as in Crouzon disease). The patient of Stroer (1939), also of first-cousin parents, may have had the same disorder. Appelt et al. (1966) described cases and pointed out that clitoral or penile enlargement is a feature. Corneal opacities occur in this disorder. In the view of Opitz (1982), 'deficiency' is a better term than 'absence deformity' for the abnormality of the limbs. Freeman et al. (1974) presented a good survey. Maserati et al. (1991) described 5 cases in 4 nuclear families; in 3 of the families, the parents were consanguineous. They pictured affected sisters at ages 23 and 15 and emphasized the wide range of variability in the phenotype. The affected sisters had bilateral radial aplasia, hypoplastic ulnas and malformed hands; in the lower limbs, they had aplasia of the fibula with a bent tibia and bilateral clubfoot. At the other extreme was severe tetraphocomelia with death at or soon after birth. Satar et al. (1994) described a male infant who, in addition to typical manifestations of Roberts syndrome, had atrial septal defect, rudimentary gallbladder, and accessory spleen.

Temtamy (1974) concluded that Roberts syndrome and the SC phocomelia syndrome (269000) are the same. Furthermore, we observed a case of apparent Roberts syndrome in which thrombocytopenia occurred and an aunt had well-confirmed TAR syndrome (274000). Thus, the relationship of the Roberts and TAR syndromes awaits clarification. Cleft palate has not been observed in the TAR syndrome, to my knowledge. By an analysis of phenotype, Herrmann and Opitz (1977) concluded that they could not tell whether the SC phocomelia syndrome and the Roberts syndrome are 'due to different recessive genes, different alleles, or the same recessive gene.' Because of overlapping features in their patient, Waldenmaier et al. (1978) suggested that the SC phocomelia syndrome and the TAR syndrome are not separate from the Roberts syndrome. Tomkins et al. (1979) noted the uncertainty as to whether Roberts syndrome and the SC syndrome are separate entities. They found a consistent centromeric abnormality of the chromosomes, namely, puffing and splitting, in 4 patients who had certain clinical features in common: bilateral corneal opacities, microcephaly, absence of radii, limitation of extension at the elbows and knees, enlargement of the phallus, and survival beyond the neonatal period. Fryns et al. (1980) reported identical twins concordant for the tetraphocomelia-cleft palate syndrome. Since the twins showed the severe tetraphocomelia of Roberts syndrome and the less prominent craniofacial abnormalities of the pseudothalidomide syndrome, the authors favored the view that these two entities are one. Stoll et al. (1979) raised the question of phenocopy resulting from maternal ingestion of clonidine, an antihypertensive medication. Da Silva and Bezerra (1982) reported 4 affected sibs of first-cousin parents.

Zergollern and Hitrec (1982) also concluded that the Roberts and SC syndromes are one entity. In a sibship with 4 affected, they found silver-blond hair, typical of SC, in 1; 2 had cloudy corneas typical of SC. They described chromosomal changes like those described by others and proposed their use in prenatal diagnosis. The chromosomal abnormality involves the heterochromatic, C-banding regions of most chromosomes. In addition to the above noted puffing of heterochromatic regions around the centromeres and nucleolar organizers, the heterochromatin of the long arms of the Y chromosome is often widely separated in metaphase spreads. German (1979) suggested that these configurations result from a repulsion or lack of attraction between the chromatids in these regions leading to premature separation during prophase and metaphase. Maserati et al. (1991) used indirect immunofluorescence with serum antibodies from patients with CREST (181750) to demonstrate that the centromeric structures are normal in Roberts syndrome, thus confirming the assumption of German (1979) that chromatid repulsion is confined to the heterochromatin. Studies suggested that heterozygotes can be screened by the phenomenon of centromeric heterochromatin separation. Tomkins and Sisken (1984) suggested that impediment to cellular growth is responsible for reduced pre- and postnatal growth rates and also for the developmental abnormalities. Premature centromere separation (PCS) has been reported in lymphocytes and/or fibroblasts from at least 17 patients whose clinical phenotypes cover the range of the Roberts syndrome at the severe end and the SC phocomelia syndrome at the milder end (Parry et al., 1986). This argues that the 2 syndromes may represent the same clinical entity. Romke et al. (1987) reported a family in which 3 sibs had various manifestations of Roberts syndrome or SC phocomelia, leading them to conclude that the 2 syndromes are the same genetic entity. Krassikoff et al. (1986) found that aneuploid cells from a metastatic melanoma in a patient with the Roberts/SC phocomelia syndrome, aged 32 years, showed a reduced frequency of PCS. Furthermore, when the patient's fibroblasts, which showed a high frequency of PCS, were cocultivated with either an immortal hamster cell line or with a human male fibroblast strain carrying a t(4;6) translocation, the phenomenon was neither corrected in the patient's cells nor induced in the other cells. In each experiment, only the patient's metaphase spreads showed PCS. In fusion hybrids between the patient's fibroblasts and an established Chinese cell line, however, the human chromosomes behaved normally. No chromatid repulsion (PCS) was observed, suggesting that the missing or mutant gene product in Roberts/SC phocomelia syndrome is supplied by the Chinese hamster genome.

Fryns et al. (1987) described 2 sibs with tetraphocomelia typical of Roberts syndrome: there was almost complete reduction of the midparts of the upper and lower limbs, and characteristic oligodactyly with absent nails. Neither cleft lip/cleft palate nor eye anomalies were present. Furthermore, premature centromere separation was not observed. The facies was unusual, consisting of a beaked nose, short philtrum, and triangular mouth. Keppen et al. (1991) described an infant with the clinical diagnosis of Roberts syndrome, but without the premature separation of centromeric heterochromatin and typical abnormalities of the cell division cycle reported in this condition.

Jabs et al. (1991) presented evidence that Roberts syndrome is a 'mitotic mutant.' They emphasized that, in addition to previously described changes, aneuploidy with random chromosome loss and micronuclei and/or nuclear lobulation in the interphase cells are characteristic. They considered it unlikely that the defect in this disorder is in one of the structural proteins of the kinetochore. They suggested, however, that the defect might lie in one of the proteins transiently associated with the kinetochore and involved in its function. Stioui et al. (1992) detected premature centromere separation on chorionic villus sampling at 8 weeks' gestation in a woman at risk of recurrence of Roberts syndrome. Hirschhorn and Kaffe (1992) pointed out that they had made a prenatal diagnosis of Roberts syndrome in a family at risk by detection of skeletal and renal abnormalities (Kaffe et al., 1977).

Van Den Berg and Francke (1993) provided a review of 100 cases of Roberts syndrome and defined a new rating system for quantitating severity.

Allingham-Hawkins and Tomkins (1995) pointed out that some Roberts syndrome patients (referred to as RS+), but not others (RS-), have an abnormality of their constitutive heterochromatin and show cellular hypersensitivity to DNA damaging agents such as mitomycin C. Lymphoblastoid cell lines from 2 unrelated RS+ patients were fused and hybrid cells examined for correction of these 2 defects. Neither cellular defect was corrected in the 2 hybrid cell lines, suggesting that the patients represent a single complementation group. On the other hand, fusions between 1 RS+ cell line and 2 different RS- cell lines produced in each case hybrids demonstrating correction of both defects. This suggested that RS+ and RS- patients belong to different complementation groups and do not arise from the same single gene mutation.

Sinha et al. (1994) reviewed clinical heterogeneity of the skeletal dysplasia in Roberts syndrome. Sabry (1995) suggested that, in the light of contemporary molecular and developmental genetics, such heterogeneity would not be surprising with different mutations in the same gene or with mutations in closely related genes of the same family. Sabry (1995) raised the possibility that the mutations may lie in centromere-related proteins, which may also have a role in body patterning.

Lopez-Allen et al. (1996) pictured an afflicted newborn infant and the chromosome changes.

Urban et al. (1997) described a specimen of tetraphocomelia and bilaterally cleft lip recovered from what had been the Virchow Museum of Humboldt University in Berlin and showed convincingly that this was the specimen reported by Virchow (1898). This was clearly an instance of Roberts syndrome.

See also Ekong and Rozidlsky (1978); Graham et al. (1983); Herrmann et al. (1969); Louie and German (1981); Petrinelli et al. (1984); Pfeiffer and Zwerner (1982); Zergollern and Hitrec (1976)

Allingham-Hawkins, D. J.; Tomkins, D. J.: Heterogeneity in Roberts syndrome. Am. J. Med. Genet. 55: 188-194, 1995.

Appelt, J.; Gerken, H.; Lenz, W.: Tetraphokomelie mit Lippen-Kiefer-Gaumenspalte und Klitorishypertrophie—ein Syndrom. Paediat. Paedol. 2: 119-124, 1966.

da Silva, E. O.; Bezerra, L. H. G. E.: The Roberts syndrome. Hum. Genet. 61: 372-374, 1982.

Ekong, C. E. U.; Rozidlsky, B.: Hydranencephaly in association with Roberts syndrome. J. Canad. Sci. Neurol. 5: 253-255, 1978.

Freeman, M. V. R.; Williams, D. W.; Schimke, R. N.; Temtamy, S. A.: The Roberts syndrome. Clin. Genet. 5: 1-16, 1974.

Fryns, J.; Goddeeris, P.; Moerman, F.; Herman, F.; van den Berghe, H.: The tetraphocomelia-cleft palate syndrome in identical twins. Hum. Genet. 53: 279-281, 1980.

Fryns, J. P.; Kleczkowska, A.; Moerman, P.; Van Den Berghe, K.; Van Den Berghe, H.: The Roberts tetraphocomelia syndrome: identical limb defects in two siblings. Ann. Genet. 30: 243-245, 1987.

German, J.: Robert's syndrome. I. Cytological evidence for a disturbance in chromatid pairing. Clin. Genet. 16: 441-447, 1979.

Graham, J. M., Jr.; Stephens, T. D.; Shepard, T. H.: Nuchal cystic hygroma in a fetus with presumed Roberts syndrome. (Letter) Am. J. Med. Genet. 15: 163-167, 1983.

Herrmann, J.; Feingold, M.; Tuffli, G. A.; Opitz, J. M.: A familial dysmorphogenetic syndrome of limb deformities, characteristic facial appearance and associated anomalies: the 'pseudothalidomide' or 'SC-syndrome.'. Birth Defects Orig. Art. Ser. V(3): 81-89, 1969.

Herrmann, J.; Opitz, J. M.: The SC phocomelia and the Roberts syndrome: nosologic aspects. Europ. J. Pediat. 125: 117-134, 1977.

Hirschhorn, K.; Kaffe, S.: Prenatal diagnosis of Roberts syndrome. (Letter) Prenatal Diag. 12: 976 only, 1992.

Jabs, E. W.; Tuck-Muller, C. M.; Cusano, R.; Rattner, J. B.: Studies of mitotic and centromeric abnormalities in Roberts syndrome: implications for a defect in the mitotic mechanism. Chromosoma 100: 251-261, 1991.

Kaffe, S.; Rose, J. S.; Godmilow, L.; Walker, B. A.; Kerenyi, T.; Beratis, N.; Reyes, P.; Hirschhorn, K.: Prenatal diagnosis of renal anomalies. Am. J. Med. Genet. 1: 241-251, 1977.

Keppen, L. D.; Gollin, S. M.; Seibert, J. J.; Sisken, J. E.: Roberts syndrome with normal cell division. Am. J. Med. Genet. 38: 21-24, 1991.

Krassikoff, N. E.; Cowan, J. M.; Parry, D. M.; Francke, U.: Chromatid repulsion associated with Roberts/SC phocomelia syndrome is reduced in malignant cells and not expressed in interspecies somatic-cell hybrids. Am. J. Hum. Genet. 39: 618-630, 1986.
Lopez-Allen, G.; Hutcheon, R. G.; Shaham, M.; Tunnessen, W. W., Jr.: Picture of the month. Arch. Pediat. Adolesc. Med. 150: 645-646, 1996.
Louie, E.; German, J.: Roberts's syndrome. II. Aberrant Y-chromosome behavior. Clin. Genet. 19: 71-74, 1981.
Maserati, E.; Pasquali, F.; Zuffardi, O.; Buttitta, P.; Cuoco, C.; Defant, G.; Gimelli, G.; Fraccaro, M.: Roberts syndrome: phenotypic variation, cytogenetic definition and heterozygotic detection. Ann. Genet. 34: 239-246, 1991.
Opitz, J. M.: Personal Communication. Helena, Mont. 4/1982.
Parry, D. M.; Mulvihill, J. J.; Tsai, S.; Kaiser-Kupfer, M. I.; Cowan, J. M.: SC phocomelia syndrome, premature centromere separation and congenital cranial nerve paralysis in two sisters, one with malignant melanoma. Am. J. Med. Genet. 24: 653-672, 1986.
Petrinelli, P.; Antonelli, A.; Marcucci, L.; Dallapiccola, B.: Premature centromere splitting in a presumptive mild form of Roberts syndrome. Hum. Genet. 66: 96-99, 1984.
Pfeiffer, R. A.; Zwerner, H.: The Roberts syndrome: report of a case without anomaly of the centromeric region. Mschr. Kinderheilk. 130: 296-298, 1982.
Roberts, J. B.: A child with double cleft of lip and palate, protrusion of the intermaxillary portion of the upper jaw and imperfect development of the bones of the four extremities. Ann. Surg. 70: 252-254, 1919.
Romke, C.; Froster-Iskenius, U.; Heyne, K.; Hohn, W.; Hof, M.; Grzejszczyk, G.; Rauskolb, R.; Rehder, H.; Schwinger, E.: Roberts syndrome and SC phocomelia: a single genetic entity. Clin. Genet. 31: 170-177, 1987.
Sabry, M. A.: Possible genetic heterogeneity in the Roberts-SC phocomelia syndrome. (Letter) Hum. Hered. 45: 241-242, 1995.
Satar, M.; Atici, A.; Bisak, U.; Tunali, N.: Roberts-SC phocomelia syndrome: a case with additional anomalies. Clin. Genet. 45: 107-108, 1994.
Sinha, A. K.; Verma, R. S.; Mani, V. J.: Clinical heterogeneity of skeletal dysplasia in Roberts syndrome: a review. Hum. Hered. 44: 121-126, 1994.
Stioui, S.; Privitera, O.; Brambati, B.; Zuliani, G.; Lalatta, F.; Simoni, G.: First-trimester prenatal diagnosis of Roberts syndrome. Prenatal Diag. 12: 145-149, 1992.
Stoll, C.; Levy, J.-M.; Beshara, D.: Roberts's syndrome and clonidine. J. Med. Genet. 16: 486-487, 1979.
Stroer, W. F. H.: Ueber das Zusammentreffen von Hasenscharte mit ernsten Extremitaeten-Missbildungen. Erbarzt 7: 101-104, 1939.
Temtamy, S. A.: Personal Communication. Baltimore, Md. and Cairo, Egypt 1974.
Tomkins, D.; Hunter, A.; Roberts, M.: Cytogenetic findings in Roberts-SC phocomelia syndrome(s). Am. J. Med. Genet. 4: 17-26, 1979.
Tomkins, D. J.; Sisken, J. E.: Abnormalities in the cell-division cycle in Roberts syndrome fibroblasts: a cellular basis for the phenotypic characteristics?. Am. J. Hum. Genet. 36: 1332-1340, 1984.
Urban, M.; Rogalla, P.; Tinschert, S.; Krietsch, P.: Tetraphocomelia and bilateral cleft lip in a historical case of Roberts syndrome (Virchow, 1988). Am. J. Med. Genet. 72: 307-314, 1997.
Van Den Berg, D. J.; Francke, U.: Roberts syndrome: a review of 100 cases and a new rating system for severity. Am. J. Med. Genet. 47: 1104-1123, 1993.
Virchow, R.: Die Phokomelen und das Barenweib. Verh. Berl. Gesell. Anthrop. 30: 55-61, 1898.
Waldenmaier, C.; Aldenhoff, P.; Klemm, T.: Roberts' syndrome. Hum. Genet. 40: 345-349, 1978.
Zergollern, L.; Hitrec, V.: Three siblings with Robert's syndrome. Clin. Genet. 9: 433-436, 1976.
Zergollern, L.; Hitrec, V.: Four siblings with Robert's syndrome. Clin. Genet. 21: 1-6, 1982.

Adnexal Skin Tumors

EPITHELIOMA, SELF-HEALING SQUAMOUS (*132800)

JJM NOTE

Keratoacanthoma may be a skin tumor with conspicuous genetic determinants: it is a feature of this syndrome, Muir-Torre syndrome (158320), and xeroderma pigmentosum. To the extent that an "epithelioma" is a benign neoplasm, this gene is one that gives rise to a precursor lesion to squamous cell carcinoma. A brief review of the ecogenetics of skin cancer (Kraemer KH: Sunlight and skin cancer: Another link revealed. Proc. Natl. Acad. Sci. (USA) 94:11–14, 1997) predicted an understanding of squamous cell carcinoma through the gene for this disorder (*ESS1*), just as sporadic basal cell carcinoma is being understood through the *PTCH* (patch gene) that is mutated in the nevoid basal cell carcinoma syndrome.

This is considered to be a variety of multiple keratoacanthoma. It goes under many different names. Ereaux and Schopflocher (1965) observed affected brother and sister. Sommerville and Milne (1950) reported 2 cases in each of 2 successive generations. Affected father and son were referred to by Epstein et al. (1957). Degos et al. (1964) described the condition in a woman and 2 daughters. Ferguson Smith (1934) first described this disorder in a single case, that of a 23-year-old miner, who first developed spots on the legs at age 16 years. These healed spontaneously but were replaced by others at neighboring sites and later on the face and ears. He continued to work except for a few months when lesions on the knees prevented him from kneeling. A depressed scar remained after healing. Both clinically and histologically, the lesions had the appearance of squamous carcinoma. Ferguson Smith (1948) provided a follow-up. Treatment of large lesions on his right leg with radium was 'followed by necrosis of the tibia, and ultimately, at the patient's own request, Mr. G. T. Mowat amputated below the knee.' He was wearing a prosthesis to cover an extensive destruction of the nose. Although other cases were described, the familial nature of the disorder was not noted. Ferguson-Smith, geneticist son of the dermatologist who originally described this condition, and his colleagues (1971) assembled reliable information on 62 cases in the West of Scotland. Ferguson-Smith et al. (1971) considered it possible that all the Scottish cases derived from a single mutation which occurred before 1790. The lesions were found more frequently on exposed areas of the skin. Some of the U.S. and Canada cases may have the same origin. The Scottish cases were in 11 independently ascertained families but the genealogic connections of some could be demonstrated. Two girls had their first lesion during their 13th year; the oldest onset in a male was at age 56 and in a female 55. The mean age of onset in women was 25.5 and in men 26.9. Many examples of male-to-male transmission, equal involvement of the sexes, and precise agreement with the 50:50 segregation ratio proved autosomal dominant inheritance. A single instance of 'skipped generation' was found: the daughter of an affected male and mother of 2 affected daughters had unblemished skin when fully examined at age 57. Ferguson-Smith (1974) provided later follow-up of the original case. The patient died of 'suppurative meningitis' in December 1948. Autopsy findings were reported by Currie and Ferguson Smith (1952). In addition to the face, ears, arms and legs, the skin of the anus, scrotum and anterior abdominal wall were affected. All tumors were well-differentiated squamous epitheliomata with lymphatic infiltration of the anal and aural lesions. The anal tumor infiltrated the sphincter and muscle coats of the anal canal.

In studies of 13 affected families, Goudie et al. (1991) demonstrated linkage between the MSSE locus and several markers on 9q. The most closely linked marker was D9S29, which had been localized to 9q31 by in situ hybridization; maximum lod score = 8.05 at theta = 0.02. Goudie et al. (1993) found tight linkage of ESS1 to D9S53 (at 9q31); maximum lod score = 9.02 at theta = 0.03. Multipoint linkage analysis demonstrated that the disease locus probably lies between D9S58 (9q22.3-q31) and ASSP3 (9q11-q22). Comparison of markers associated with ESS1 in independently ascertained families suggested a common origin of the disease and defined the location of ESS1 further.

Richards et al. (1997) studied the question of whether XPA (278700) or PTCH (601309), both genes that map to the same region as MSSE, is the site of the mutation in this disorder. By analysis of a polymorphism in the 5-prime untranslated region of XPA, they excluded XPA as the site of the mutation. The Patched (PTCH) gene is mutated in Gorlin syndrome (109400); MSSE families were shown to share a common haplotype at 3 novel intragenic PTCH polymorphisms. Although no mutations were detected in MSSE families, PTCH had not been excluded as the MSSE gene.

See also Haydey et al. (1980)

Currie, A. R.; Ferguson Smith, J.: Multiple primary spontaneous-healing squamous-cell carcinomata of the skin. J. Path. Bact. 64: 827-839, 1952.
Degos, R.; Civatte, J.; Touraine, B.; Guilaine, J.: Spontan-heilende Epitheliome Ferguson-Smith und multiple familiaere Keratoacanthome. Hautarzt 15: 7-11, 1964.
Epstein, N. N.; Biskind, G. R.; Pollack, R. S.: Multiple primary self-healing squamous-cell 'epitheliomas' of the skin: generalized keratoacanthoma. Arch. Derm. 75: 210-223, 1957.
Ereaux, L. P.; Schopflocher, P.: Familial primary self-healing squamous epithelioma of skin. Arch. Derm. 91: 589-594, 1965.
Ferguson-Smith, M. A.: Personal Communication. Glasgow, Scotland 1974.
Ferguson-Smith, M. A.; Wallace, D. C.; James, Z. H.; Renwick, J. H.: Multiple self-healing squamous epithelioma. Birth Defects Orig. Art. Ser. VII(8): 157-163, 1971.
Ferguson Smith, J.: A case of multiple primary squamous-celled carcinomata of the skin in a young man, with spontaneous healing. Brit. J. Derm. 46: 267-272, 1934.
Ferguson Smith, J.: Multiple primary, self-healing squamous epithelioma of the skin. Brit. J. Derm. 60: 315-319, 1948.
Goudie, D. R.; Yuille, M. A. R.; Affara, N. A.; Ferguson-Smith, M. A.: Localisation of the gene for multiple self-healing squamous epithelioma (Ferguson-Smith type) to the long arm of chromosome 9. (Abstract) Cytogenet. Cell Genet. 58: 1939 only, 1991.
Goudie, D. R.; Yuille, M. A. R.; Leversha, M. A.; Furlong, R. A.; Carter, N. P.; Lush, M. J.; Affara, N. A.; Ferguson-Smith, M. A.: Multiple self-healing squamous epitheliomata (ESS1) mapped to chromosome 9q22-q31 in families with common ancestry. Nature Genet. 3: 165-169, 1993.

Haydey, R. P.; Reed, M. L.; Dzubow, L. M.; Shupack, J. L.: Treatment of keratoacanthomas with oral 13-cis-retinoic acid. New Eng. J. Med. 303: 560-562, 1980.
Richards, F. M.; Goudie, D. R.; Cooper, W. N.; Jene, Q.; Barroso, I.; Wicking, C.; Wainwright, B. J.; Ferguson-Smith, M. A.: Mapping the multiple self-healing squamous epithelioma (MSSE) gene and investigation of xeroderma pigmentosum group A (XPA) and patched (PTCH) as candidate genes. Hum. Genet. 101: 317-322, 1997.
Sommerville, J.; Milne, J. A.: Familial primary self-healing squamous epithelioma of the skin. Brit. J. Derm. 62: 485-490, 1950.

EPITHELIOMA, HEREDITARY BENIGN CYSTIC (CYLINDROMATOSIS) (*132700)

JJM NOTE

The various tumors of skin appendages have long histories of definition and nomenclature. Embryologically they are closely related. A simultaneous and contemporary comparison taking all clinical, syndromic, pathologic, and eventually molecular genetic aspects into consideration would probably yield a much simplified list of distinct entities. Mapping and isolating a gene for any one of the large kindreds with these overlapping tumor types would produce candidate genes that could be examined in others. Despite the dramatic picture of this disorder, the predisposition to frank malignancy may not be high. The exception might be an enormous risk for a very rare tumor—malignant lymphoepithelial lesion (MLEL) of the parotid gland. The apparent excess of MLEL in Mongoloid Chinese and Eskimo peoples surely cannot be explained by the gene for cylindromatosis. But perhaps this entity is another example of ecogenetics in action. Trichoepithelioma is another skin tumor that may have major genetic determinants. It occurs in Cowden disease (158350), familial polyposis coli (175100), and probably Muir-Torre syndrome (158320), as well as in this and the following two entries.

Fliegelman and Kruse (1948) described 10 cases in 3 generations. They indicated that despite some clinical similarities the disorder could be distinguished from syringocystadenoma, adenoma sebaceum, and cylindroma. Gartler et al. (1966), studying members of a family with affected members in 4 generations, found that females heterozygous for G6PD-deficiency had shown both G6PD-deficient and G6PD-normal cells in the same tumor, thus indicating multicellular origin. Ziprkowski and Schewach-Millet (1966) reported the dermatologic features in the same family. The skin tumors show differentiation in the direction of hair structures, hence the synonym trichoepithelioma. One affected person developed basosquamous cell carcinoma. Some think that this and cylindroma are the same entity. Welch et al. (1968) presented family data supporting the view that Ancell-Spiegler cylindromas (313100) and Brooke-Fordyce trichoepitheliomas are manifestations of a single entity. (Trichoepithelioma is one characteristic skin lesion of Cowden multiple hamartoma syndrome (158350).) The term cylindroma was also applied by Billroth (1859) to a type of adenocarcinoma arising in salivary gland tissue (Evans et al., 1966). Anderson and Howell (1976) employed the useful abbreviation EAC. Gaul (1953) described the case of a woman who was apparently homozygous: both parents were affected and 8 children by 2 husbands were affected. The homozygous woman was particularly severely affected. Autio-Harmainen et al. (1988) described a Finnish kindred in which many members had dominantly inherited trichoepithelioma and 2 members had, in addition, a rare, malignant lymphoepithelial lesion (MLEL) of the parotid gland. At least 4 generations and 7 sibships were affected, particularly with facial trichoepithelioma, and at least 4 members of the family were reported to have scalp cylindromas. One patient had eccrine spiradenoma, including painful spiradenomas on the upper chest. MLEL of the parotid gland was present in mother and daughter. Salivary gland MLEL has been reported predominantly in the Mongoloid race, with a preponderance of Canadian and Greenland Eskimos and Chinese. Familial occurrence has been noted among the Eskimos (Merrick et al., 1986).

Hart (1973) suggested that one can discern in ancient Parthian coins the lesions of hereditary multiple benign cystic epithelioma.

Multiple cylindromas, or cylindromatosis, appears to be inherited as an autosomal dominant, although a female preponderance raised questions of X-linked dominant inheritance in the past (see 313100). Furthermore, this is fundamentally the same disorder as multiple trichoepithelioma. Schuermann and Weber (1937) presented a pedigree with 9 persons affected in 4 generations. Six of the 9 were female. Although no male-to-male transmission was noted, 2 daughters of an affected male were unaffected. See Harper (1971) for a dramatic example of cylindromatosis. Vernon et al. (1988) described a solitary, apparently benign lung cylindroma in a 42-year-old woman with multiple cutaneous cylindromas who died from coronary occlusion which occurred precociously in other members of her family. Vernon et al. (1988) thought, however, that there was no connection between the coronary artery disease and the cylindromas. Van Balkom and Hennekam (1994), who preferred the designation dermal eccrine cylindromatosis for this disorder, provided a review.

Gerretsen et al. (1995) described a large family in which cutaneous cylindromas, trichoepitheliomas, and milia occurred in members of 5 generations with 30 affected persons (11 male, 19 female). Female-to-female, female-to-male, male-to-female, and male-to-male inheritance was observed. Penetrance reached to 100% in adult life. Harada et al. (1996) appeared to have demonstrated by linkage studies that multiple familial trichoepithelioma (MFT; 601606) is an entity distinct from cylindromatosis. They performed linkage analysis with microsatellite markers in 3 families with multiple trichoepithelioma and found linkage to 9p21 in a 4-cM region between IFNA (147660) and D9S126.

Studying 2 families with cylindromatosis, Biggs et al. (1995) found strong evidence for linkage to 16q12-q13. Using markers close to the cylindromatosis gene, consistent loss of the wildtype allele was observed in 19 tumors from 4 individuals in the 2 families, indicating that the gene is likely to be a tumor suppressor gene. In the linkage studies, individuals were classified as affected if they had multiple tumors with classical clinical characteristics and distribution; unaffected if they were over 40 years of age and did not show multiple tumors; and classified if they were below 40 with no tumors or an ambiguous clinical presentation.

Biggs et al. (1996) examined polymorphic markers on each chromosome, some of which are close to known tumor suppressor genes, in 25 tumors from 4 individuals with familial cylindromatosis. No loss of heterozygosity (LOH) was detected other than that at loci on chromosome 16q. Observations suggested to them that the gene for cylindromatosis (symbolized CYLD1 by them) may be the only tumor suppressor gene implicated in the development of cylindromas. They also demonstrated LOH using markers on chromosome 16q in 8 of 14 (57%) sporadic cylindromas, indicating that the CYLD1 gene is probably involved in the genesis of both familial and sporadic cylindromas.

Schramm et al. (1996) reported the cases of a 49-year-old woman and her 20-year-old daughter with multiple papular and nodular lesions on the face and scalp, showing the typical clinical and histopathologic features of trichoepithelioma. In addition, the mother developed multiple, blue-colored and painful lesions on her breast and back, showing the histopathologic characteristics of eccrine spiradenomas. In a tissue specimen, both the trichoepitheliomas and spiradenomas were found in close proximity.

See also Baden (1962); Brooke (1892)

Anderson, D. E.; Howell, J. B.: Epithelioma adenoides cysticum: genetic update. Brit. J. Derm. 95: 225-232, 1976.
Autio-Harmainen, H.; Paakko, P.; Alavaikko, M.; Karvonen, J.; Leisti, J.: Familial occurrence of malignant lymphoepithelial lesion of the parotid gland in a Finnish family with dominantly inherited trichoepithelioma. Cancer 61: 161-166, 1988.
Baden, H. P.: Cylindromatosis simulating neurofibromatosis. New Eng. J. Med. 267: 296-297, 1962.
Biggs, P. J.; Chapman, P.; Lakhani, S. R.; Burn, J.; Stratton, M. R.: The cylindromatosis gene (cyld1) on chromosome 16q may be the only tumour suppressor gene involved in the development of cylindromas. Oncogene 12: 1375-1377, 1996.
Biggs, P. J.; Wooster, R.; Ford, D.; Chapman, P.; Mangion, J.; Quirk, Y.; Easton, D. F.; Burn, J.; Stratton, M. R.: Familial cylindromatosis (turban tumour syndrome) gene localised to chromosome 16q12-q13: evidence for its role as a tumour suppressor gene. Nature Genet. 11: 441-443, 1995.
Billroth, T.: Beobachtungen ueber Geschwuelste der Speicheldruesen. Virchows Arch. Path. Anat. 17: 357-375, 1859.
Brooke, H. G.: Epithelioma adenoides cysticum. Brit. J. Derm. 4: 269-287, 1892.
Evans, J. C.; Efskind, J.; Roberts, T. W.: Cylindroma. Am. J. Roentgen. 96: 191-196, 1966.
Fliegelman, M. T.; Kruse, W. T.: Hereditary multiple benign cystic epithelioma. J. Invest. Derm. 11: 189-196, 1948.
Gartler, S. M.; Ziprkowski, L.; Krakowski, A.; Ezra, R.; Szeinberg, A.; Adam, A.: Glucose-6-phosphate dehydrogenase mosaicism as a tracer in the study of hereditary multiple trichoepithelioma. Am. J. Hum. Genet. 18: 282-287, 1966.
Gaul, L. E.: Heredity of multiple benign cystic epithelioma. Arch. Derm. Syph. 68: 517-524, 1953.
Gerretsen, A. L.; Beemer, F. A.; Deenstra, W.; Hennekam, F. A. M.; van Vloten, W. A.: Familial cutaneous cylindromas: investigations in five generations of a family. J. Am. Acad. Derm. 33: 199-206, 1995.
Harada, H.; Hashimoto, K.; Ko, M. S. H.: The gene for multiple familial trichoepithelioma maps to chromosome 9p21. J. Invest. Derm. 107: 41-43, 1996.
Harper, P. S.: Turban tumors (cylindromatosis). Birth Defects Orig. Art. Ser. VII(8): 338-341, 1971.
Hart, G. D.: The diagnosis of disease from ancient coins. Archaeology 26: 123-127, 1973.

Merrick, Y.; Albeck, H.; Nielsen, N. H.; Hansen, H. S.: Familial clustering of salivary gland carcinoma in Greenland. Cancer 57: 2097-2102, 1986.
Schramm, M.; Blume-Peytavi, U.; Kruger, K.; Gollnick, H.: A rare association of multiple hereditary trichoepitheliomas with multiple spiradenomas. Europ. J. Derm. 6: 259-261, 1996.
Schuermann, H.; Weber, K.: Beitrag zur Kenntnis der Spieglerschen Tumoren (Cylindrome) nebst einigen Bemerkungen zum Epithelioma adenoides cysticum. Arch. Derm. Syph. 175: 682-695, 1937.
van Balkom, I. D. C.; Hennekam, R. C. M.: Dermal eccrine cylindromatosis. J. Med. Genet. 31: 321-324, 1994.
Vernon, H. J.; Olsen, E. A.; Vollmer, R. T.: Autosomal dominant multiple cylindromas associated with solitary lung cylindroma. J. Am. Acad. Derm. 19: 397-400, 1988.
Welch, J. P.; Wells, R. S.; Kerr, C. B.: Ancell-Spiegler cylindromas (turban tumours) and Brooke-Fordyce trichoepitheliomas: evidence for a single genetic entity. J. Med. Genet. 5: 29-35, 1968.
Ziprkowski, L.; Schewach-Millet, M.: Multiple trichoepithelioma in a mother and two children. Dermatologica 132: 248-256, 1966.

TRICHOEPITHELIOMA, MULTIPLE FAMILIAL (*601606)

JJM NOTE

This skin tumor is also seen in Cowden multiple hamartoma syndrome and Muir-Torre syndrome.

Trichoepithelioma is a benign skin tumor originating from hair follicles. Multiple familial trichoepithelioma (MFT) is a relatively common genetic dermatosis characterized by the presence of many small tumors predominantly on the face, inherited in an autosomal dominant pattern. Because histologic features are dermal aggregates of basaloid cells with connection to or differentiation toward hair follicles, this disorder has been thought to represent a benign hamartoma of the pilosebaceous apparatus. Trichoepithelioma can usually be differentiated from basal cell carcinoma because of its characteristic clinical presentation. However, MFT can degenerate into basal cell carcinoma (Johnson and Bennett, 1993). In 3 families with multiple familial trichoepithelioma, 2 African-American and 1 Caucasian, Harada et al. (1996) found linkage of the disorder to a 4-cM region between IFNA (147660) and D9S126 on 9p21; maximum combined lod = 3.31 at D9S171 at theta = 0.0. Multiple familial trichoepithelioma is occasionally associated with familial cylindromatosis (132700). Gerretsen et al. (1995) argued that the 2 types of dermatoses may be caused by dysfunction of the same gene, because both tumors can occur in the same patient or in different patients within a single family. A gene for familial cylindromatosis was assigned to 16q12-q13 by Biggs et al. (1995). It may be that there are 2 forms of multiple trichoepithelioma, one that occurs in isolation and is determined by a gene on 9p, and another that is associated with cylindromatosis and is determined by a gene on chromosome 16.

Correa-Cerro et al. (1997) described hereditary multiple trichoepithelioma in a mother and daughter. Manifestations began in both at age 7 years, the daughter being more severely affected than the mother. The mother was the product of the second pregnancy of a sibship of 11. At 7 years of age the daughter had small skin lesions like comedones located on the surface of the nose and on the lower palpebral fissure. The mother had lesions involving the entire face by age 11 years. Small tumors of 2 to 10 mm in diameter most striking on the dorsal region of the nose were distributed over the entire face and ears. Correa-Cerro et al. (1997) commented on the excess of affected women. It is, of course, unclear whether this family suffered from the disorder linked to chromosome 9 or to that linked to chromosome 16 (132700).

Biggs, P. J.; Wooster, R.; Ford, D.; Chapman, P.; Mangion, J.; Quirk, Y.; Easton, D. F.; Burn, J.; Stratton, M. R.: Familial cylindromatosis (turban tumour syndrome) gene localised to chromosome 16q12-q13: evidence for its role as a tumour suppressor gene. Nature Genet. 11: 441-443, 1995.
Correa-Cerro, L. S.; Garcia-Cruz, D.; Sarralde, A.; Morales-Peralta, E.; Gonzalez-Mendoza, A.; Sanchez-Corona, J.: Hereditary multiple benign cystic epithelioma (multiple trichoepithelioma) with onset at early age. Genet. Counsel. 8: 83-86, 1997.
Gerretsen, A. L.; Beemer, F. A.; Deenstra, W.; Hennekam, F. A. M.; van Vloten, W. A.: Familial cutaneous cylindromas: investigations in five generations of a family. J. Am. Acad. Derm. 33: 199-206, 1995.
Harada, H.; Hashimoto, K.; Ko, M. S. H.: The gene for multiple familial trichoepithelioma maps to chromosome 9p21. J. Invest. Derm. 107: 41-43, 1996.
Johnson, S. C.; Bennett, R. G.: Occurrence of basal cell carcinoma among multiple trichoepitheliomas. J. Am. Acad. Derm. 28: 322-326, 1993.

PAPILLOMATOSIS, FAMILIAL CUTANEOUS (167900)

JJM NOTE

Never having seen patients with this disorder, I still venture to guess that there are many undiagnosed such clusters and that there is more to the phenotype than the papillomas.

Baden (1965) described a confluent, reticular type of papillomatosis in 2 sisters and the daughter of one. Henning and de Wit (1981) described the disorder in a 44-year-old woman and her 15-year-old daughter and 18-year-old son. The lesions became more pigmented during the premenstrual phase of the women's cycle.

Baden, H. P.: Familial cutaneous papillomatosis. Arch. Derm. 92: 394-395, 1965.
Henning, J. P. H.; de Wit, R. F. E.: Familial occurrence of confluent and reticulated papillomatosis. Arch. Derm. 117: 809-810, 1981.

SPIEGLER-BROOKE TUMORS (313100)

JJM NOTE

This somewhat duplicative entry makes the point that X-linked inheritance of these rare skin tumors remains a theoretical possibility.

The demonstration, in the same patients and same families, of cylindroma as described by Spiegler (1899) and of epithelioma adenoides cysticum as described by Brooke (1892) supports their genetic identity (Guggenheim and Schnyder, 1961). (See hereditary benign cystic epithelioma, 132700, for further discussion of nosology.) Schmidt-Baumler (1931) raised the question of X-linked dominant inheritance. The pedigree of Blandy et al. (1961) showed an affected male who had all daughters affected and all sons unaffected. Up to the end of 1954, Evans found 47 reported cases, of which 30 were female. Guggenheim and Schnyder found that 132 of 212 reported cases were in females. However, against X-linkage is the fact that male-to-male transmission has been frequently observed. We have 1 pedigree with at least 9 affected persons and at least 1 instance of male-to-male transmission. The disorder is probably an autosomal dominant with stronger expression in the female. Stronger expression in presumed heterozygous females than in hemizygous males is contrary to what is expected in X-linked dominant inheritance. The possibility of X-linkage in a small proportion of families is not completely excluded, however.

See also Anderson and Howell (1976); Chalstrey (1955); Evans (1954); Regan (1956); Wiedmann (1929)

Anderson, D. E.; Howell, J. B.: Epithelioma adenoides cysticum: genetic update. Brit. J. Derm. 95: 225-232, 1976.
Blandy, J. P.; Gammie, W. F. P.; Stovin, P. G. I.; Swettenham, K.: Turban tumours in brother and sister. Brit. J. Surg. 49: 136-140, 1961.
Brooke, H. G.: Epithelioma adenoides cysticum. Brit. J. Derm. 4: 269-287, 1892.
Chalstrey, L. J.: Turban tumors. St. Bart's Hosp. J. 59: 378-383, 1955.
Evans, C. D.: Turban tumour. Brit. J. Derm. 66: 434-443, 1954.
Guggenheim, W.; Schnyder, U. W.: Zur Nosologie der Spiegler-Brookeschen Tumoren. Dermatologica 122: 274-278, 1961.
Regan, W. J.: Turban tumours. Proc. Roy. Soc. Med. 49: 337-339, 1956.
Schmidt-Baumler, H.: Familiares Cylindrome. Ein Beitrag zur Frage der geschlechtsbegrenzten Vererbung. Arch. Derm. Syph. 163: 114-125, 1931.
Spiegler, E.: Ueber Endotheliome der Haut. Arch. Derm. Syph. 50: 163-176, 1899.
Wiedmann, A.: Weitere Beitraege zur Kenntnis der sogenannten Zylindrome der Kopfhaut. Arch. Derm. Syph. 159: 180-187, 1929.

SYRINGOMAS, MULTIPLE (186600)

JJM NOTE

A clinical search for syndromic associations would seem worthwhile; I am not aware of any.

Multiple syringomas or sweat gland tumors occur particularly on the face and around the eyes. They are not to be confused with milia, which are intraepithelial cysts. Familial occurrence is, it seems, a commonplace observation of dermatologists and autosomal dominant inheritance is likely (Reed, 1967). Reed (1970) described a family in which 7 females and 1 male in 4 generations were affected. Yesudian and Thambiah (1975) described identically affected brothers. Familial occurrence was reported also by Headington et al. (1972) and by Woringer and Eichler (1951).

Headington, J. T.; Koski, J.; Murphy, P. J.: Clear cell glycogenosis in multiple syringomas. Arch. Derm. 106: 353-356, 1972.
Reed, W. B.: Personal Communication. Burbank, Calif. 1967.
Reed, W. B.: Genetische Aspekte in der Dermatologie. Hautarzt 21: 8-16, 1970.
Woringer, F.; Eichler, A.: Constatation et reflexion au sujet d'un cas d'hidradenome eruptive. Ann. Derm. Syph. 78: 152-164, 1951.
Yesudian, P.; Thambiah, A.: Familial syringoma. Dermatologica 150: 32-35, 1975.

TRICHODISCOMAS, FAMILIAL MULTIPLE (190340)

JJM NOTE

A genodermatologic clinical team might address heterogeneity and syndromic associations of this unusual tumor. Given its complex origins, the insights should prove interesting and widely relevant.

Trichodiscoma, as first described by Pinkus et al. (1974), is a small benign fibrovascular tumor of the dermal part of the hair disk. The hair disk is a richly vascularized dermal pad in close association with a hair. It is supplied by a thick myelinated nerve and is considered to be a slow-adapting mechanoreceptor. Trichodiscomas are small, flat or dome-shaped, skin-colored, firm papules with a telangiectatic surface. Many of the lesions show a hair at the periphery or just outside it. Starink et al. (1985) described a family in which 7 persons in 3 generations had multiple trichodiscomas. The lesions had onset early in life. In the opinion of Starink et al. (1985), this disorder is distinct from autosomal dominant fibrofolliculomas with trichodiscomas and acrochordons (135150).

Pinkus, H.; Coskey, R.; Burgess, G. H.: Trichodiscoma: a benign tumor related to haarscheibe (hair disk). J. Invest. Derm. 63: 212-218, 1974.
Starink, T. M.; Kisch, L. S.; Meijer, C. J. L. M.: Familial multiple trichodiscomas: a clinicopathologic study. Arch. Derm. 121: 888-891, 1985.

STEATOCYSTOMA MULTIPLEX (#184500)

JJM NOTE

Although the number of benign subcutaneous cysts can be large, malignancy does not seem to be a complication. An additional family had hair cysts as well (Van der Kley AM et al.: Eruptive vellus hair cysts and steatocytoma multiplex in a family. Ann. Dermatol. Venereol. 117:790–92, 1990).

A number sign (#) is used with this entry because of evidence that steatocystoma multiplex is caused by mutations in the keratin-17 gene (KRT17; 148069) which also has been found to be mutated in cases of pachyonychia congenita of the Jackson-Lawler type.

In typical cases the patient may exhibit 100 to 2,000 round or oval cystic tumors widely distributed on the back, anterior trunk, arms, scrotum, and thighs. Noojin and Reynolds (1948) observed 12 cases in 3 generations. Sebaceous cysts presenting mainly as wens of the scalp were reported by Stephens (1959) in a large number of individuals in 5 generations in a dominant pedigree pattern. Sebaceous and other soft tissue tumors occur as part of Gardner syndrome (175100); actually, the so-called sebaceous cysts of Gardner syndrome are usually epidermoid cysts. Bushkell and Gorlin (1975) found leukonychia totalis, multiple sebaceous cysts, and renal calculi in grandfather, father and son, and some of these features in 2 other relatives. Koilonychia (149300) was also found in 3 of the affected persons. The association of steatocystoma multiplex with pachyonychia congenita (167200) may represent a distinct mutation distinct (Hodes and Norins, 1977); this appears to be the condition which Feinstein et al. (1988) referred to as type II pachyonychia congenita. Malignant degeneration of a cyst in a steatocystoma case was reported by Harper and Davis (1971). Cuccia-Belvedere et al. (1989) studied steatocystoma multiplex in 13 persons in 2 unrelated Italian families.

Pamoukian and Westreich (1997) described their experiences in the treatment of 7 members of a family of Jewish Turkish origin over a period of 17 years. Steatocystoma multiplex congenita, especially the cysts in exposed areas of the body, created serious self-image problems. Because of the number of cysts and their fragility, standard cyst excision techniques were impractical and difficult. They described a modified excision method that yielded good results. Their method consisted of making a stab incision into the cyst and manually evacuating its contents. Thereafter, a fine mosquito hemostat was inserted to grasp the cyst wall from within and strip out its lining. At one sitting, 50 to 150 cysts could be removed. Excisions were performed once or twice a year.

Smith et al. (1997) described mutations in the KRT17 gene in 2 families diagnosed as steatocystoma multiplex. On reevaluation, mild changes in the nails were found in some but not all members of these 2 families, compatible with pachyonychia congenita, and some members had mild focal nonepidermolytic palmoplantar keratoderma. Just as there are 2 recognized forms of pachyonychia congenita distinguished by the presence of natal teeth (PC-1; 167200) or the absence of natal teeth (PC-2; 167210), steatocystoma multiplex with (184510) and without natal teeth occurs. Indeed, steatocystoma multiplex and pachyonychia congenita must be considered to be at opposite ends of a phenotypic expression of the same basic disorder.

Bushkell, L. L.; Gorlin, R. J.: Leukonychia totalis, multiple sebaceous cysts, and renal calculi: a syndrome. Arch. Derm. 111: 899-901, 1975.
Cuccia-Belvedere, M.; Brazzelli, V.; Martinetti, M.; Berardesca, E.; Dugoujon, J. M.; De Paoli, F.; Borroni, G.; Rabbiosi, G.: Familial steatocystoma multiplex: HLA, Gm, Km genotyping and chromosomal analysis in two unrelated families. Clin. Genet. 36: 136-140, 1989.
Feinstein, A.; Friedman, J.; Schewach-Millet, M.: Pachyonychia congenita. J. Am. Acad. Derm. 19: 705-711, 1988.
Harper, P. S.; Davis, J. K.: Steatocystoma multiplex (multiple sebaceous cysts) with familial incidence in the first case. Birth Defects Orig. Art. Ser. XII(8): 342 only, 1971.
Hodes, M. E.; Norins, A. L.: Pachyonychia congenita and steatocystoma multiplex. Clin. Genet. 11: 359-364, 1977.
Noojin, R. O.; Reynolds, J. P.: Familial steatocystoma multiplex. Twelve cases in three generations. Arch. Derm. Syph. 57: 1013-1018, 1948.
Pamoukian, V. N.; Westreich, M.: Five generations with steatocystoma multiplex congenita: a treatment regimen. Plast. Reconst. Surg. 99: 1142-1146, 1997.
Smith, F. J. D.; Corden, L. D.; Rugg, E. L.; Ratnavel, R.; Leigh, I. M.; Moss, C.; Tidman, M. J.; Hohl, D.; Huber, M.; Kunkeler, L.; Munro, C. S.; Lane, E. B.; McLean, W. H. I.: Missense mutations in keratin 17 cause either pachyonychia congenita type 2 or a phenotype resembling steatocystoma multiplex. J. Invest. Derm. 108: 220-223, 1997.
Stephens, F. E.: Hereditary multiple sebaceous cysts. J. Hered. 50: 299-301, 1959.

STEATOCYSTOMA MULTIPLEX WITH NATAL TEETH (184510)

JJM NOTE

This entity is an example of syndrome delineation by starting with a rare skin tumor.

King and Lee (1987) described a 5-generation Chinese family in which at least 21 individuals had natal teeth and generalized multiple steatocystomas. There were no manifestations to suggest pachyonychia congenita (167200, 167210), which is also accompanied by natal teeth. McDonald and Reed (1976) reported the same combination in a single case.

Smith et al. (1997) observed that demonstration of mutations in keratin genes in pachyonychia congenita without natal teeth (the Jackson-Lawler type; 167210), the demonstration that patients with pachyonychia congenita of both types have mutations in type I keratin, and the finding that families with steatocystoma multiplex without natal teeth (184500) have specific mutations in the same gene (KRT17; 148067) as that affected in the Jackson-Lawler type of pachyonychia congenita, all suggest that steatocystoma multiplex with natal teeth is, in fact, fundamentally the same disorder as type 1 pachyonychia congenita (167200).

King, N. M.; Lee, A. M. P.: Natal teeth and steatocystoma multiplex: a newly recognized syndrome. J. Craniofac. Genet. Dev. Biol. 7: 311-317, 1987.
McDonald, R. M.; Reed, W. B.: Natal teeth and steatocystoma multiplex complicated by hidradenitis suppurativa. Arch. Derm. 112: 1132-1134, 1976.
Smith, F. J. D.; Corden, L. D.; Rugg, E. L.; Ratnavel, R.; Leigh, I. M.; Moss, C.; Tidman, M. J.; Hohl, D.; Huber, M.; Kunkeler, L.; Munro, C. S.; Lane, E. B.; McLean, W. H. I.: Missense mutations in keratin 17 cause either pachyonychia congenita type 2 or a phenotype resembling steatocystoma multiplex. J. Invest. Derm. 108: 220-223, 1997.

FIBROFOLLICULOMAS WITH TRICHODISCOMAS AND ACROCHORDONS (135150)

JJM NOTE

Recall of the Birt family could address linkage in general and the Cowden and the *RET* loci in particular.

This dermatologic condition, first described by Birt et al. (1977), has fibrofolliculomas as its hallmark and trichodiscomas (190340) and acrochordons as associated features. Onset is invariably in adulthood. A central hair is often visible in the lesions. Birt et al. (1977) observed the disorder in 15 persons in 3 generations, with male-to-male transmission. Hereditary medullary carcinoma of the thyroid was also segregating, apparently independently, in the kindred. In a sibship of 9, 6 had medullary carcinoma of the thyroid. Two of those with thyroid cancer and 2 without had numerous small papular skin lesions that Birt et al. (1977) labeled fibrofolliculoma. The lesion was characterized by abnormal hair follicles with epithelial strands extending out from the infundibulum of the hair follicle into a hyperplastic mantle of specialized fibrous tissue. Associated skin lesions were trichodiscomas (tumor of the hair disc) and acrochordons ('wart with a thin neck'; skin tag). It is possible that the disorder in this kindred was the Cowden multiple hamartoma syndrome (158350) (Mulvihill, 1990).

Birt, A. R.; Hogg, G. R.; Dube, W. J.: Hereditary multiple fibrofolliculomas with trichodiscomas and acrochordons. Arch. Derm. 113: 1674-1677, 1977.
Mulvihill, J. J.: Personal Communication. Bethesda, Md. 1/31/1990.

OSTEOMA CUTIS (*166350)

JJM NOTE

The genes that are being identified with Albright hereditary osteodystrophy and fibrodysplasia ossificans progressiva might be assayed in individuals and families with osteoma cutis.

Fawcett and Marsden (1983) reported osteoma cutis in 3 generations. The 3-year-old proposita developed hard nodules in the skin at age 6 months. Celiac disease was diagnosed at age 3 years. Skin biopsies showed multiple spicules of bone in the skin. Normal membranous bone structures (osteocytes, cement lines, Haversian canals, occasional osteoblasts, and osteoclasts) were seen. The surrounding dermis showed no scarring or inflammatory changes. Her 6-year-old sister had skin lesions of the legs and trunk starting about the same age but at age 2 years she had a painful, swollen right ankle and soft tissue calcification was found by x-ray. At age 6, such calcifications were visualized in both the hands and feet. The father of the girls had several cutaneous osteomas on the arms. His father, deceased, was known to have had similar lesions on his shoulders for many years. Osteoma cutis has been described in many ostensibly normal persons, including a mother and son (Peterson and Mandel, 1963). In the latter case, the mother also had multiple pigmented nevi; the son died at 15 months of alveolar sarcoma of the cerebellum. Some of these cases, however, may have been instances of Albright hereditary osteodystrophy (AHO; 103580, 203330, 300800) with which ectopic ossification is known to occur (Brook and Valman, 1971). Fawcett and Marsden (1983) felt that AHO was unlikely but could not be excluded completely. Gardner et al. (1988) described a family with affected members in 2 generations and 3 sibships. In most members of the family the disorder was characterized by childhood onset of multifocal subcutaneous ossifications (primary osteoma cutis), which were of trivial clinical significance. One family member had extensive ectopic ossification involving the right leg, first noted at the age of 3 weeks and severely interfering with growth of the limb by age 8 years. Gardner et al. (1988) speculated that somatic mutation might have caused conversion to homozygosity. McKusic k (1989) observed osteoma cutis in a 38-year-old man who also had unerupted teeth, severe bilateral carpal tunnel syndrome, and unusual changes in the bones of the forearms and legs that were quite different from those of pseudohypoparathyroidism (J.T., JHH 2367189, P20,519). Maclean et al. (1966) described a mother and daughter with multifocal connective tissue ossification. The daughter was a 7-year-old mentally retarded girl of normal stature. Pseudopseudohypoparathyroidism (103580) seems likely in these cases. Izraeli et al. (1992) described congenital osteoma cutis in members of 3 generations with convincing clinical and laboratory evidence of Albright hereditary osteodystrophy and with one instance of male-to-male transmission. In the propositus small, flat, and hard subcutaneous plaques were noted on the back at birth and biopsy at the age of 3 days confirmed the diagnosis of osteoma cutis. In a sister similar nodules were noted in the first month of life and the mother had had a hard subcutaneous nodule on the hand from birth. No deficiency of Gs activity was detected by the assay used.

Kaplan et al. (1994) reported 2 new cases and follow-up on 3 previously reported cases, 1 of which was that reported by Gardner et al. (1988). At follow-up examination of this family, the proband was 31 years old, was married, and had a daughter who was clinically normal. This patient's family was the only one in which relatives were affected; individuals in 4 sibships in 2 generations showed minor cutaneous ossification only. A maculopapular eruption was observed in the early stages of the disorder. Kaplan et al. (1994) referred to the condition as progressive osseous heteroplasia and concluded that it is distinct from fibrodysplasia ossificans progressiva (FOP; 135100) and from McCune-Albright syndrome (174800), both of which are accompanied by ectopic ossification. It is also distinct from tumoral calcinosis (211900).

Rosenfeld and Kaplan (1995) found reports of 8 classic cases, all in female patients, and documented typical features of progressive osseous heteroplasia in 2 boys, 1 of which (a 3-year-old) showed left-sided hemimelic heterotopic ossification. Athanasou et al. (1994) reported a typical case in an 18-year-old girl. Schmidt et al. (1994) reported the case of a patient who had progressive ossification of the soft tissues on the right side of the face and body. Histologic analysis of biopsy material showed intramembranous, subcutaneous ossification. The patient was a 9-year-old Hispanic girl.

Athanasou, N. A.; Benson, M. K. D'A.; Brenton, D. P.; Smith, R.: Progressive osseous heteroplasia: a case report. Bone 15: 471-475, 1994.
Brook, C. G. D.; Valman, H. B.: Osteoma cutis and Albright's hereditary osteodystrophy. Brit. J. Derm. 85: 471-475, 1971.
Fawcett, H. A.; Marsden, R. A.: Hereditary osteoma cutis. J. Roy. Soc. Med. 76: 697-699, 1983.
Gardner, R. J. M.; Yun, K.; Craw, S. M.: Familial ectopic ossification. J. Med. Genet. 25: 113-117, 1988.
Izraeli, S.; Metzker, A.; Horev, G.; Karmi, D.; Merlob, P.; Farfel, Z.: Albright hereditary osteodystrophy with hypothyroidism, normocalcemia, and normal Gs protein activity: a family presenting with congenital osteoma cutis. Am. J. Med. Genet. 43: 764-767, 1992.
Kaplan, F. S.; Craver, R.; MacEwen, G. D.; Gannon, F. H.; Finkel, G.; Hahn, G.; Tabas, J.; Gardner, R. J. M.; Zasloff, M. A.: Progressive osseous heteroplasia: a distinct developmental disorder of heterotopic ossification: two new case reports and follow-up of three previously reported cases. J. Bone Joint Surg. 76A: 425-436, 1994.
Maclean, G. D.; Main, R. A.; Anderson, T. E.; Best, P. V.: Connective tissue ossification presenting in the skin. Arch. Derm. 94: 168-174, 1966.
Peterson, W. C., Jr.; Mandel, S. L.: Primary osteomas of skin. Arch. Derm. 87: 626-632, 1963.
Rosenfeld, S. R.; Kaplan, F. S.: Progressive osseous heteroplasia in male patients: two new case reports. Clin. Orthop. Rel. Res. 317: 243-245, 1995.
Schmidt, A. H.; Vincent, K. A.; Aiona, M. D.: Hemimelic progressive osseous heteroplasia: a case report. J. Bone Joint Surg. 76-A: 907-912, 1994.

LEIOMYOMAS, HEREDITARY MULTIPLE, OF SKIN (*150800)

JJM NOTE

Additional families are needed for gene-mapping efforts. My Pittsburgh colleagues see two distinct regions of deletion in 7q in sporadic uterine leiomyomas (Ishwad CS et al.: Two discrete regions of deletion at 7q in uterine leiomyomas. Genes Chromo. Cancer 19:156–60, 1997).

Multiple small tumors composed of smooth muscle fibers develop in the skin. Malignant transformation is rare. The tumors are thought to arise from the erector pilorum muscles. Kloepfer et al. (1958) described 3 Italian half first cousins with multiple leiomyomata of the skin. The parents and common grandparent were not known to be affected, but all critical individuals were not examined. As the authors pointed out, in this type of anomaly one anticipates dominant inheritance, probably with reduced penetrance. Dominant inheritance with incomplete penetrance is supported by the pedigree of Mezzadra (1965), who described cutaneous leiomyomata in 3 generations of an Italian family. Uterine myomata were associated. Guillet et al. (1987) described a nonfamilial case of associated multiple cutaneous leiomyomas and uterine fibromas. Weilbaecher (1967) observed a Swedish family with 5 affected in 3 generations and male-to-male transmission. Rudner et al. (1972) described identical twins with multiple cutaneous leiomyomata and a history of hysterectomy for uterine leiomyomata. Reed et al. (1973) also emphasized the association of uterine myomata. Engelke and Christophers (1979) commented on the unusually early age of onset of uterine myofibromas. Fryns et al. (1985) described a severely mentally retarded woman with 9p trisomy/18pter monosomy. The patient was judged to have phenotypic features typical of 9p trisomy (Rethore et al., 1970) but in addition had multiple cutaneous leiomyomata, of which some were nodular, some linear, and all looked rather like keloids. The authors raised the interesting question of whether this was another example of a specific chromosomal deletion (18pter) in a dominantly inherited multiple tumor, like retinoblastoma and nephroblastoma.

See also Berendes et al. (1971)

Berendes, U.; Kuhner, A.; Schnyder, U. W.: Segmentary and disseminated lesions in multiple hereditary cutaneous leiomyoma. Humangenetik 13: 81-82, 1971.
Engelke, H.; Christophers, E.: Leiomyomatosis cutis et uteri. Acta Derm. Venerol. 59 (suppl. 85): 51-54, 1979.
Fryns, J. P.; Haspeslagh, M.; de Muelenaere, A.; van den Berghe, H.: 9p trisomy/18p distal monosomy and multiple cutaneous leiomyomata: another specific chromosomal site (18pter) in dominantly inherited multiple tumors?. Hum. Genet. 70: 284-286, 1985.
Guillet, G.; Grau, P.; Sassolas, B.; Zagnoli, A.; Leroy, J. P.; Labouche, F.: Leiomyomes cutanes multiples et fibromes uterins: a propos d'une observation d'un cas non familial. Semin. Hop. Paris 63: 65-67, 1987.
Kloepfer, H. W.; Krafchuk, J.; Derbes, V.; Burks, J.: Hereditary multiple leiomyoma of the skin. Am. J. Hum. Genet. 10: 48-52, 1958.
Mezzadra, G.: Leiomioma cutaneo multiplo ereditario. Studio di un caso sistematizzato in soggetto maschile appartenente a famiglia portatrice di leiomiomatosi cutanea e fibromiomatosi uterina. Minerva Derm. 40: 388-393, 1965.
Reed, W. B.; Walker, R.; Horowitz, R.: Cutaneous leiomyomata with uterine leiomyomata. Acta Derm. Venerol. 53: 409-416, 1973.
Rethore, M. O.; Larget-Piet, L.; Abonyi, D.; Boeswillwald, M.; Berger, P.; Carpentier, S.; Cruveiller, J.; Dutrillaux, B.; Lafourcade, J.; Penneau, M.; Lejeune, J.: Sur quatre cas de trisomie pour le bras court du chromosome 9: individualisation d'une nouvelle entite morbide. Ann. Genet. 13: 217-232, 1970.
Rudner, E. J.; Schwartz, O. D.; Greekin, J. N.: Multiple cutaneous leiomyoma in identical twins. Arch. Derm. 104: 81-82, 1972.
Weilbaecher, R. G.: Personal Communication. New Orleans, La. 1967.

HEMANGIOMAS, CAVERNOUS, OF FACE AND SUPRAUMBILICAL MIDLINE RAPHE (140850)

JJM NOTE

This entity is an example of a specific syndrome that needs further documentation.

This combination of unknown etiology was described by Leiber (1982) in 2 unrelated German infants and by Igarashi et al. (1985) in a Japanese infant. In the last case, supraumbilical raphe appeared as a whitish groove between the umbilicus and xiphoid with interruptions. In the German cases, it resembled a healing wound.

Gorlin et al. (1994) found no evidence of sex predilection among 42 reported examples of a combination of sternal nonunion and supraumbilical raphe. However, among an additional 31 cases in which the triad included facial hemangioma, there was almost exclusive female occurrence.

Gorlin, R. J.; Kantaputra, P.; Aughton, D. J.; Mulliken, J. B.: Marked female predilection in some syndromes associated with facial hemangiomas. Am. J. Med. Genet. 52: 130-135, 1994.
Igarashi, M.; Uchida, H.; Kajii, T.: Supraumbilical midabdominal raphe and facial cavernous hemangiomas. Clin. Genet. 27: 196-198, 1985.
Leiber, B.: Angeborene supraumbilikale Mittelbauchrhaphe (SMBR) und kavernose Gesichtshamangiomatose—ein neues Syndrom?. Mschr. Kinderheilk. 130: 84-90, 1982.

HEMANGIOMA-THROMBOCYTOPENIA SYNDROME (KASABACH-MERRITT SYNDROME) (141000)

JJM NOTE

In Pittsburgh, I saw an infant with neurofibromatosis 1 and nearly lethal Kasabach-Merritt syndrome.

With giant hemangiomas in small children, thrombocytopenia and red cell changes compatible with trauma ('microangiopathic hemolytic anemia') have been observed. The mechanism of the hematologic changes is obscure. No evidence of a simple genetic basis has been discovered. Mulvihill (1982) pointed out to me that hemangioma of the placenta can cause symptomatic thrombocytopenia of the newborn. Experience with the therapy of 6 children with capillary hemangiomas in association with consumptive coagulopathy (Kasabach-Merritt syndrome) was reviewed by Larsen et al. (1987). Sencer et al. (1987) reported the case of a newborn infant with splenic hemangioendothelioma with thrombocytopenia, anemia, and disseminated intravascular coagulation. A splenectomy was indicated despite the known risks of subsequent overwhelming sepsis by encapsulated organisms. Vellodi and Bini (1988) described a severe hyperkalemia resulting in 'malignant ventricular arrhythmias.' They attributed the hyperkalemia to breakdown of erythrocytes. Breakdown of platelets is another possible source.

Enjoiras et al. (1997) concluded that the vascular lesion underlying KMS is not a 'true,' classic, involuting type of hemangioma of infancy. It is a different vascular tumor with a resemblance pathologically to either tufted angioma or kaposiform hemangioendothelioma in association with lymphatic-like vessels.

See also Brizel and Raccuglia (1965); David et al. (1983); Propp and Scharfman (1966); Rodriguez-Erdmann et al. (1970)

Brizel, H. E.; Raccuglia, G.: Giant hemangioma with thrombocytopenia. Radioisotopic demonstration of platelet sequestration. Blood 26: 751-756, 1965.
David, T. J.; Evans, D. I. K.; Stevens, R. F.: Haemangioma with thrombocytopenia (Kasabach-Merritt syndrome). Arch. Dis. Child. 58: 1022-1023, 1983.
Enjoiras, O.; Wassef, M.; Mazoyer, E.; Frieden, I. J.; Rieu, P. N.; Drouet, L.; Taieb, A.; Stalder, J.-F.; Escande, J.-P.: Infants with Kasabach-Merritt syndrome do not have 'true' hemangiomas. J. Pediat. 130: 631-640, 1997.
Larsen, E. C.; Zinkham, W. H.; Eggleston, J. C.; Zitelli, B. J.: Kasabach-Merritt syndrome: therapeutic considerations. Pediatrics 79: 971-980, 1987.
Mulvihill, J. J.: Personal Communication. Bethesda, Md. 4/1982.
Propp, R. P.; Scharfman, W. B.: Hemangioma-thrombocytopenia syndrome associated with microangiopathic hemolytic anemia. Blood 28: 623-633, 1966.
Rodriguez-Erdmann, F.; Murray, J. E.; Moloney, W. C.: Consumption-coagulopathy in Kasabach-Merritt syndrome. Trans. Assoc. Am. Phys. 83: 168-175, 1970.
Sencer, S.; Coulter-Knoff, A.; Day, D.; Foker, J.; Thompson, T.; Burke, B.: Splenic hemangioma with thrombocytopenia in a newborn. Pediatrics 79: 960-966, 1987.
Vellodi, A.; Bini, R. M.: Malignant ventricular arrhythmias caused by hyperkalaemia complicating the Kasabach-Merritt syndrome. J. Roy. Soc. Med. 81: 167-168, 1988.

ANGIOMA SERPIGINOSUM (106050)

JJM NOTE

Perhaps the published kindreds can be recontacted for further study.

This uncommon dermatosis was first described by Jonathan Hutchinson (1889) in Plate IX of Vol. 1 of his Archives of Surgery. More common in females, the condition begins before puberty as pin-sized capillary puncta affecting any part of the body surface except the palms and soles and sparing also mucous membranes. Marriott et al. (1975) reported 2 kindreds with several affected individuals, consistent with dominant inheritance and reduced penetrance; no male-to-male transmission was observed.

Hutchinson, J.: . Arch. Surg. 1: 1889. Note: Plate IX.
Marriott, P. J.; Munro, D. D.; Ryan, T.: Angioma serpiginosum—familial incidence. Brit. J. Derm. 93: 701-706, 1975.

ANGIOMA, HEREDITARY NEUROCUTANEOUS (*106070)

JJM NOTE

This disorder seems ripe for mapping studies, including recontact of published families.

Zaremba et al. (1979) reported 4 affected persons in 3 generations, including a father and his 2 sons. One patient died at age 28 of multiple dilated thin-walled vessels in the cerebral substance; an extensive, irregularly shaped, pink hemangioma planum, which faded on pressure, was present on the skin of the left shoulder, arm and forearm. His brother developed left hemiparesis at age 13 and died at age 19 after an unsuccessful attempt was made to resect a spinal angioma in the C6-T1 region (producing the Horner syndrome and the Brown-Sequard syndrome). He had an angioma in the left frontotemporal area and a second over the right mastoid process. Their father developed left hemiparesis at age 58 and had episodes of urinary and gastrointestinal bleeding. Angiomas were present on the chest and left thigh. A daughter of the oldest of his sons (who died at age 28) had 3 angiomas in the lumbosacral area and 1 on the left palm. None of the patients had retinal angiomas or telangiectases typical of Osler-Rendu-Weber syndrome. The involvement of the central nervous system resembled that in the Icelandic family reported by Kidd and Cumings (1947) but that family had no skin angiomas. Burke et al. (1964) described 2 unrelated infants with a large number of small hemangiomata in many areas of the skin and also in the brain. Kaplan et al. (1976) described a 16-month-old girl with cutaneomeningospinal angiomatosis leading to paraplegia because of intraspinal AV malformation. Skin hemangiomas occurred in 3 generations of the family (with no instance of male-to-male transmission). Hemangioma of the skin in the same dermatome as the symptoms of a space-occupying spinal lesion can be a clue to early diagnosis of the nature of the latter. Foo et al. (1980) reported the case of a 33-year-old man who developed cervical anterior cord syndrome from spontaneous bleeding of an arteriovenous malformation in the cervical epidural space. Follow-up (Foo et al., 1980) revealed cutaneous vascular ma lformations in 3 generations. The proband's mother had 4 hemangiomas removed (from the neck, back, right thigh and face). A maternal aunt had a left ankle hemangioma removed at age 20. One of his younger sisters had a hemangioma resected from the right shoulder at age 15 and another from the pelvis at 31. This sister passed the gene to her 2 sons; one son had a hemangioma removed from the forehead at age 2, and the other had a hemangioma removed from the left side of the head at age 3. The proband's brother had a hemangioma removed from above the right ear at age 10. Hurst and Baraitser (1988) reported 2 families with this disorder. In 1 family there was father-to-son transmission; the father had cutaneous hemangiomata of the nose, arm, and trunk, and the son had a temporal lobe arteriovenous malformation. Four generations and 5 individuals were affected in the other family. Although the evidence is not ironclad, this syndrome of hereditary neurocutaneous angioma is probably distinct from familial cavernous malformations of the CNS and retina (116860).

See also Foo et al. (1980)

Burke, E. C.; Winkelmann, R. K.; Strickland, M. K.: Disseminated hemangiomatosis: the newborn with central nervous system involvement. Am. J. Dis. Child. 108: 418-424, 1964.

Foo, D.; Chang, Y. C.; Rossier, A. B.: Spontaneous cervical epidural hemorrhage, anterior cord syndrome, and familial vascular malformation: case report. Neurology 30: 308-311, 1980.

Foo, D.; Chang, Y. C.; Rossier, A. B.: Spontaneous cervical epidural hemorrhage, anterior cord syndrome, and familial vascular malformation. (Letter) Neurology 30: 1253-1254, 1980.

Hurst, J.; Baraitser, M.: Hereditary neurocutaneous angiomatous malformations: autosomal dominant inheritance in two families. Clin. Genet. 33: 44-48, 1988.

Kaplan, P.; Hollenberg, R. D.; Fraser, F. C.: A spinal arteriovenous malformation with hereditary cutaneous hemangiomas. Am. J. Dis. Child. 130: 1329-1331, 1976.

Kidd, H. A.; Cumings, J. N.: Cerebral angiomata in an Icelandic family. Lancet I: 747-748, 1947.

Zaremba, J.; Stepien, M.; Jelowicka, M.; Ostrowska, D.: Hereditary neurocutaneous angioma: a new genetic entity?. J. Med. Genet. 16: 443-447, 1979.

CEREBRAL CAVERNOUS ANGIOMA, FAMILIAL (*116860)

JJM NOTE

Recognizing the familiality and now gene mapping of this disorder can be lifesaving.

Cavernous angiomas are relatively rare vascular malformations that may involve any part of the central nervous system. Some are clinically silent, whereas others cause seizures, hemorrhage, or focal neurologic deficit. Identification of these lesions is important because surgical removal of many is relatively easy. Magnetic resonance imaging (MRI) is replacing computerized axial tomography as the diagnostic modality of choice. Bicknell et al. (1978) found 3 reports of familial incidence and added 2 from their own experience. In 1 family a woman, 2 of her sons, and 1 of her son's sons were affected; in the second family a woman and her daughter were affected. Successive generations were affected in families reported by Michael and Levin (1936), Kidd and Cumings (1947), and Clark (1970). Michael and Levin (1936) described a Swedish family in which a mother, her 2 brothers, and 3 daughters had multiple telangiectases of the brain. Convulsions and migraine attacks were observed. Autopsy in one case demonstrated calcification in the vascular lesions of the brain. (The lesions in the Michael-Levin cases seem more appropriately called cavernous angiomata.) Clark (1970) described cavernous angioma of the brain in a man who died in 1945 at age 27 and in his daughter who died in 1969 at age 28. Hayman et al. (1982) examined 43 relatives in 1 kindred by cranial computed tomography (CCT) and found 15 affected with cerebral vascular angiomas. Angiography failed to detect lesions in 5 patients who had positive CCT. Expression was variable and in 2 individuals, each the parent of an affected offspring, the CCT was normal. Familial cavernous angioma should be included in the differential diagnosis of any young person with cerebrovascular impairment, seizures, intracranial calcifications or hemorrhage. Gorlin (1985) told me of an extensively affected 3-generation family.

Mason et al. (1988) described cavernous angiomas in 10 of 22 members of a large Hispanic family. The authors commented that 2 families previously reported by them (Bicknell et al., 1978), the family reported by Hayman et al. (1982), and 5 of the 6 families reported in abstract by Rigamonti et al. (1987) were Hispanic as well. Dobyns et al. (1987) described a family in which 4 persons from 3 generations had multiple cavernous malformations ('angiomas') of the CNS and/or retina. They found reports of 16 other families containing a total of 50 cases. Excluding the probands, 68% of the patients were symptomatic. Cutaneous vascular lesions were an inconsistent manifestation. They recommended that any patient with a vascular malformation, especially a cavernous one of the brain, spinal cord, or retina, be evaluated for the possibility of this syndrome which they referred to as FCMCR. All first-degree relatives should undergo a full evaluation if multiple vascular malformations are detected in the index patient or if the family history is suggestive because of seizures, cutaneous vascular lesions, recognized intracranial hemorrhage, or sudden unexplained death. Presymptomatic diagnosis in affected relatives will permit genetic counseling and close monitoring to allow prompt treatment if symptoms occur. Dobyns et al. (1987) concluded that there is a second group of patients with multiple cutaneous lesions and inconsistent CNS lesions referred to as hereditary neurocutaneous angioma (106070). The vascular lesions in this group were always arteriovenous malformations and were often located in the spinal cord.

Rigamonti et al. (1988) reviewed familial occurrence, presenting signs and symptoms, and radiographic features of the disorder in 24 patients with histologically verified cerebral cavernous malformations. Eleven patients had no evidence of a heritable trait and had negative family histories. The other 13 patients were members of 6 unrelated Mexican-American families. Among 64 first-degree and second-degree relatives, 11% had seizures. MRI was performed in 16 relatives (5 of whom were asymptomatic); 14 studies showed cavernous malformations and 11 studies identified multiple lesions. MRI was far more accurate in detecting these lesions than computerized tomography or angiography. Rigamonti et al. (1988) concluded that a familial form of this disorder is particularly frequent among Mexican-Americans. Bicknell (1989) described cavernous angioma of the brain stem in a 23-year-old Hispanic woman whose mother had died of brain hemorrhage. After moving to Baltimore from the southwestern part of the United States, Rigamonti (1993) concluded that there is not an unusual frequency of the disorder among Mexican-Americans. He emphasized that cavernous angiomas are not arteriovenous malformations; they represent a honeycomb of veins. They are not demonstrated by arteriography and therefore have been referred to as angiographically silent. Epilepsy is the most frequent symptom; bleeds occur in some cases. Dellemijn and Vanneste (1993) investigated 20 relatives of a 23-year-old woman with cavernous angiomatosis of the central nervous system. Studies revealed 4 additional patients with symptomatic cavernous angioma and 1 with asymptomatic cavernous angioma. The basis of the neurologic symptoms had not previously been identified in the symptomatic patients. The pedigree pattern was consistent with autosomal dominant inheritance.

Steichen-Gersdorf et al. (1992) reported a family in which cavernous angiomas of the brain were documented in 6 individuals in 5 sibships of 4 generations of a family. Two brothers in the third generation were asymptomatic but showed changes on MRI. 12,11:Filling-Katz et al. (1989, 1992) described a family with cavernous angiomatosis in which 2 members had terminal transverse defects at the midforearm. Multiple family members had had episodic bleeding from cavernous angiomas of the central nervous system. Two had had retinal cavernous angiomas, one hepatic angioma, and 2 cavernous angiomas of soft tissue; skin angiomas were frequent. Studies of the forearm in 1 of the affected individuals showed abrupt termination distal to the normal radius and ulnar heads and apparently normal blood vessels. 12,11:Filling-Katz et al. (1989, 1992) suggested that acute vascular disruption is the cause and that this is related to the fundamental defect in familial cavernous angiomatosis. Corboy and Galetta (1989) described a family in which the proband had suffered for 9 years from recurrent 'acute chiasmal syndrome,' diagnosed at first as retrobulbar neuritis.

Angiomatous malformations, arteriovenous malformations, hemangiomata, nevus flammeus (port-wine stain), etc., occur in many syndromes, some of them mendelian, and may occur as isolated mendelian traits. In some, intracranial vascular malformations are associated with hemangiomas of the skin. Pasyk et al. (1984) described a remarkable family in which multiple vascular malformations, including cavernous hemangiomas, arteriovenous malformations, and capillary hemangiomas, occurred in 25 persons in 5 generations. Slightly reduced penetrance was suggested by the fact that a clinically unaffected woman had a child with a hemangioma on the foot and that in the part of the pedigree with the most complete documentation, the ratio of affected to unaffected was 15:20. Norwood and Everett (1964) reported the case of a 21-year-old black female who during pregnancy developed large hemangiomas at many sites, such as earlobe and axilla, and heart failure as a result. After delivery, the hemangiomas rapidly subsided. The patient's mother and 6-year-old son had macular hemangiomas of the face and trunk and her brother had classical Klippel-Trenaunay-Weber syndrome of the right lower extremity. Beers and Clark (1942) described a family with cutaneous hemangiomas ranging in size from a millimeter to many centimeters in diameter, in 12 persons in 3 generations. Metatarsus atavicus (second toe longer than the first toe) was an independent dominant trait in this family. (See toes, relative length of 1st and 2nd; 189200.) Michels et al. (1985) stated that 19 families with 77 persons with cavernous angiomas of the central nervous system and retina have been described. They described a 3-generation family ascertained through an 8-year-old boy with seizures and 2 unexplained lesions on CT and MRI. His mother presented a year later with a seizure and similar brain lesions. Angiography and eye examination were normal. The asymptomatic grandfather had 5 intracranial lesions on MRI scan. Keret et al. (1990) described an 18-year-old male with left scrotal cavernous hemangioma. Cutaneous hemangiomata were found in 34 relatives (21 males and 13 females). Only the proband had a genital lesion. The differentiation of scrotal hemangioma from varicocele was discussed.

Computed tomography and MRI led to reassessment of the incidence of cavernous angioma of the brain including its familial occurrence. Drigo et al. (1994) described an Italian family with multiple cavernous angiomas of the brain, sometimes in association with liver angiomas, in 10 members of 4 generations. No neurologic symptoms were detected in subjects from the first 2 generations but symptoms were found in adult age in members of the third generation; 2 fourth-generation members came under medical observation at 2.5 years of age. Symptoms included partial epileptic fits which sometimes became generalized later and were generally controlled adequately by therapy. None of the patients was mentally retarded or restricted in daily life. Because of symptomatic hepatomegaly and postmortem finding of multiple liver and brain angiomas in a member of the first generation, liver ultrasonography was performed in all members of the family with detection of liver angiomas in members of the second and third generation. Retinal angioma was detected in 1 patient.

Using linkage analysis and a set of short tandem repeat polymorphisms, Dubovsky et al. (1995) mapped the gene responsible for cavernous malformations in a large Hispanic kindred to 7q11-q22. The maximum pairwise lod score of 4.2 was obtained at zero recombination with a marker at locus D7S804. Lod scores in excess of 3.0 were obtained with 4 additional markers closely linked to D7S804. A chromosome 7q haplotype of 33 cM on the sex-averaged map was shared by all affected individuals, indicating that the gene lies between D7S502 and D7S479. Using a linkage approach in 2 extended cavernous malformation kindreds, Gunel et al. (1995) also linked CAM to 7q, specifically 7q11.2-q21. Multipoint linkage analysis yielded a maximum lod score of 6.88 with zero recombination with D7S669 and localized the gene to a 7-cM region in the interval between ELN (130160) and D7S802. The preferred symbol for this gene is CCM1 for cerebral cavernous malformations-1.

Marchuk et al. (1995) likewise mapped the gene in this disorder to proximal 7q by linkage methods. In 2 families, 1 of Italian-American origin and 1 of Mexican-American origin, they found a combined maximum lod score of 3.92 at theta = 0.0 for marker D7S479. Haplotype analysis placed the locus between D7S502 proximally and D7S515 distally, an interval of approximately 41 cM. The chromosomal location distinguishes this disorder from the autosomal dominant vascular malformation syndrome (VMCM; 600195) in which lesions are primarily cutaneous; VMCM is due to a gene that maps to 9p21.

Gunel et al. (1996) found that 47 affected members of 14 Hispanic American kindreds shared identical alleles for up to 15 markers linked to the cavernous-malformation gene in a short segment of proximal 7q. Ten patients with sporadic cases also shared these same alleles, indicating that they too had inherited the same mutation. Thirty-three asymptomatic carriers of the disease gene were identified, demonstrating the variability and age dependence of the development of symptoms and explaining the appearance of apparently sporadic cases. Gunel et al. (1996) concluded that virtually all cases of familial and sporadic cavernous malformation among Hispanic Americans of Mexican descent are due to the inheritance of the same mutation from a common ancestor.

Craig et al. (1998) reported analysis of linkage in 20 non-Hispanic Caucasian kindreds with familial CCM. Analyses showed linkage to 2 new loci, CCM2 (603284) at 7p15-p13 and CCM3 (603285) at 3q25.2-q27. Multilocus analysis yielded a maximum lod score of 14.11, with 40% of kindreds linked to CCM1, 20% linked to CCM2, and 40% linked to CCM3, with highly significant evidence for linkage to 3 loci. Linkage to these 3 loci could account for inheritance of CCM in all kindreds studied. The penetrance of symptomatic disease among apparent gene carriers for kindreds linked to CCM1, CCM2, and CCM3 was 88, 100, and 63%, respectively. These differences were not explained by differences in age or gender of gene carriers among families and none of the asymptomatic gene carriers in this analysis was under age 20.

Beers, C. V.; Clark, L. A.: Tumors and short-toe—a dihybrid pedigree: a family history showing the inheritance of hemangioma and metatarsus atavicus. J. Hered. 33: 366-368, 1942.

Bicknell, J. M.: Familial cavernous angioma of the brain stem dominantly inherited in Hispanics. Neurosurgery 24: 102-105, 1989.

Bicknell, J. M.; Carlow, T. J.; Kornfeld, M.; Stovring, J.; Turner, P.: Familial cavernous angiomas. Arch. Neurol. 35: 746-749, 1978.

Clark, J. V.: Familial occurrence of cavernous angiomata of the brain. J. Neurol. Neurosurg. Psychiat. 33: 871-876, 1970.

Corboy, J. R.; Galetta, S. L.: Familial cavernous angiomas manifesting with an acute chiasmal syndrome. Am. J. Ophthal. 108: 245-250, 1989.

Craig, H. D.; Gunel, M.; Cepeda, O.; Johnson, E. W.; Ptacek, L.; Steinberg, G. K.; Ogilvy, C. S.; Berg, M. J.; Crawford, S. C.; Scott, R. M.; Steichen-Gersdorf, E.; Sabroe, R.; Kennedy, C. T. C.; Mettler, G.; Beis, M. J.; Fryer, A.; Awad, I. A.; Lifton, R. P.: Multilocus linkage identifies two new loci for a Mendelian form of stroke, cerebral cavernous malformation, at 7p15-13 and 3q25.2-27. Hum. Molec. Genet. 7: 1851-1858, 1998.

Dellemijn, P. L. I.; Vanneste, J. A. L.: Cavernous angiomatosis of the central nervous system: usefulness of screening the family. Acta Neurol. Scand. 88: 259-263, 1993.

Dobyns, W. B.; Michels, V. V.; Groover, R. V.; Mokri, B.; Trautmann, J. C.; Forbes, G. S.; Laws, E. R., Jr.: Familial cavernous malformations of the central nervous system and retina. Ann. Neurol. 21: 578-583, 1987.

Drigo, P.; Mammi, I.; Battistella, P. A.; Riccheri, G.; Carollo, C.: Familial cerebral, hepatic, and retinal cavernous angiomas: a new syndrome. Child's Nerv. Syst. 10: 205-209, 1994.

Dubovsky, J.; Zabramski, J. M.; Kurth, J.; Spetzler, R. F.; Rich, S. S.; Orr, H. T.; Weber, J. L.: A gene responsible for cavernous malformations of the brain maps to chromosome 7q. Hum. Molec. Genet. 4: 453-458, 1995.

Filling-Katz, M. R.; Levin, S. W.; Patronas, N. J.; Katz, N. N. K.: Terminal transverse limb defects associated with familial cavernous angiomatosis. Am. J. Med. Genet. 42: 346-351, 1992.

Filling-Katz, M. R.; Levin, S. W.; Patronas, N. J.; Katz, N. N. K.: Terminal transverse defects are associated with familial cavernous angiomatosis. (Abstract) Am. J. Hum. Genet. 45 (suppl.): A45, 1989.

Gorlin, R. J.: Personal Communication. Minneapolis, Minn. 3/12/1985.

Gunel, M.; Awad, I. A.; Anson, J.; Lifton, R. P.: Mapping a gene causing cerebral cavernous malformation to 7q11.2-q21. Proc. Nat. Acad. Sci. 92: 6620-6624, 1995.

Gunel, M.; Awad, I. A.; Finberg, K.; Anson, J. A.; Steinberg, G. K.; Batjer, H. H.; Kopitnik, T. A.; Morrison, L.; Giannotta, S. L.; Nelson-Williams, C.; Lifton, R. P.: A founder mutation as a cause of cerebral cavernous malformation in Hispanic Americans. New Eng. J. Med. 334: 946-951, 1996.

Hayman, L. A.; Evans, R. A.; Ferrell, R. E.; Fahr, L. M.; Ostrow, P.; Riccardi, V. M.: Familial cavernous angiomas: natural history and genetic study over a 5-year period. Am. J. Med. Genet. 11: 147-160, 1982.

Keret, D.; Kam, I.; Ben-Arieh, Y.; Hashmonai, M.: Scrotal cavernous haemangioma with a family history of cutaneous angiomata. J. Roy. Soc. Med. 83: 402-403, 1990.

Kidd, H. A.; Cumings, J. N.: Cerebral angiomata in an Icelandic family. Lancet I: 747-748, 1947.

Marchuk, D. A.; Gallione, C. J.; Morrison, L. A.; Clericuzio, C. L.; Hart, B. L.; Kosofsky, B. E.; Louis, D. N.; Gusella, J. F.; Davis, L. E.; Prenger, V. L.: A locus for cerebral cavernous malformations maps to chromosome 7q in two families. Genomics 28: 311-314, 1995.

Mason, I.; Aase, J. M.; Orrison, W. W.; Wicks, J. D.; Seigel, R. S.; Bicknell, J. M.: Familial cavernous angiomas of the brain in an Hispanic family. Neurology 38: 324-326, 1988.

Michael, J. C.; Levin, P. M.: Multiple telangiectases of brain: a discussion of hereditary factors in their development. Arch. Neurol. Psychiat. 36: 514-536, 1936.

Michels, V. V.; Dobyns, W. B.; Groover, R. V.; Mokri, B.; Forbes, G. S.; Laws, E. R.: Familial cavernous angiomas of the central nervous system and retina. (Abstract) Am. J. Hum. Genet. 37: A69, 1985.

Norwood, O. T.; Everett, M. A.: Cardiac failure due to endocrine dependent hemangiomas. Arch. Derm. 89: 759-760, 1964.

Pasyk, K. A.; Argenta, L. C.; Erickson, R. P.: Familial vascular malformations: report of 25 members of one family. Clin. Genet. 26: 221-227, 1984.

Rigamonti, D.: Personal Communication. Baltimore, Md. 9/15/1993.

Rigamonti, D.; Drayer, B.; Johnsen, S.; Johnson, P.; Sidell, A.; Tarby, T.; Spetzler, R.: Cavernous malformations, MRI, and epilepsy. (Abstract) Neurology 37: 322, 1987.

Rigamonti, D.; Hadley, M. N.; Drayer, B. P.; Johnson, P. C.; Hoenig-Rigamonti, K.; Knight, J. T.; Spetzler, R. F.: Cerebral cavernous malformations: incidence and familial occurrence. New Eng. J. Med. 319: 343-347, 1988.

Steichen-Gersdorf, E.; Felber, S.; Fuchs, W.; Russeger, L.; Twerdy, K.: Familial cavernous angiomas of the brain: observations in a four generation family. Europ. J. Pediat. 151: 861-863, 1992.

VENOUS MALFORMATION, MULTIPLE CUTANEOUS AND MUCOSAL (#600195)

JJM NOTE

The power of positional gene cloning identified the underlying mutation, which never would have been predicted from biologic reasoning, as in the gene for TEK, an endothelial tyrosine kinase.

A number sign (#) is used with this entry because of evidence that the phenotype is caused by an activating mutation in the receptor tyrosine kinase TIE2 (TEK; 600221).

In 15 members of 3 generations of a kindred and by implication in a 16th member in an earlier generation, in a pedigree pattern consistent with autosomal dominant inheritance, Boon et al. (1994) described cutaneous and mucosal venous malformations. They pictured an 18-year-old male with a 'slow-flow' venous malformation in the tongue which bled intermittently and caused deformation of the maxilla and mandible. They also pictured characteristic small cutaneous venous malformations located on the posterior aspect of the ear in 1 member. The lesions varied in size from 0.5 to 2 cm, except for the lesion on the tongue which measured 5 cm. Lesions were located on the arms and legs, face, oral mucosa, or genitalia. Some were present at birth; most lesions appeared by puberty. None of the family members had a history of gastrointestinal bleeding and occult blood was not demonstrated by guaiac testing of the stool.

Boon et al. (1994) demonstrated that the disorder did not map to 9q, the site of the mutation in hereditary hemorrhagic telangiectasia (HHT; 187300). On the other hand, they were able to demonstrate that the locus for multiple cutaneous and mucosal venous malformations in this family lies within a 24-cM interval on 9p, defined by the markers D9S157 and D9S163. Gallione et al. (1995) confirmed linkage to 9p in a second family. The location of the venous malformations was almost identical for several family members. In most cases, the lesions were small, enlarged with time, and were asymptomatic. A few of the females observed an increase in size of the vascular lesions approximately 1 week before menstruation and a subsequent decrease in size during menstruation. The lesions were present on the face and mucous membranes, including the lips, tongue, cheeks, tonsils, and larynx. Some members had venous malformations within internal organs. One developed 2 vascular tumors in the large intestine and required blood transfusions because of excessive bleeding. Two members of the family died from complications of these vascular abnormalities. Pathologic examination in one of these patients showed vascular tumors within the stomach, liver, pancreas, and spleen.

The endothelial cell-specific receptor tyrosine kinase TIE2 (TEK; 600221) had been previously mapped to 9p21. To test whether TIE2 is localized to the interval where the VMCM1 locus is situated, Vikkula et al. (1996) used 2 human melanoma cell lines containing previously defined homozygous deletions of 9p and a YAC contig covering the 8-cM region from the IFN gene cluster to the marker D9S161. They were able to show that indeed the TIE2 gene lies within the linked interval. Then analyzing TIE2 as a candidate gene in the 2 VMCM families, they identified a C-to-T transition at position 2545 in the TIE2 gene, predicted to result in an arg489-to-trp amino acid change (600221.0001).

See blue rubber bleb nevus (112200) for another form of autosomal dominant venous malformation of uncertain relationship to the disorder reported in the family of Boon et al. (1994). Gallione et al. (1995) were of the opinion that the disorder in the family they reported was indeed identical to the blue rubber bleb nevus syndrome. They pointed out that the family originally described by Bean (1958), like their family, had gastrointestinal bleeding from vascular lesions.

Multiple cerebrovenous anomalies without cutaneous lesions are also familial; see familial cavernous angioma (116860).

DNA pooling is a powerful and efficient method for rapid genome-wide linkage scans in autosomal recessive diseases in inbred populations, where affected individuals are likely to be homozygous for alleles at loci surrounding the disease gene. This strategy exploits the fact that affected individuals in inbred families share a chromosomal region inherited from a common ancestor surrounding the disease locus. Traditionally, genotype results are generated for individual DNA samples using a given marker or markers. In the DNA pooling strategy, DNA from multiple affected and unaffected individuals from each family, as well as population controls, are pooled in separate lanes and simultaneously amplified by PCR with a given marker. Markers linked to and in association with the disease gene will show loss of heterozygosity and hence a different pattern of alleles from appropriate control groups. Unlinked markers should demonstrate an allele distribution that is similar to control lanes. Damji et al. (1998) investigated whether this approach would detect linkage in autosomal dominant disorders, as suggested by Kanis et al. (1995), where affected individuals within a given family may share 1 allele identical by descent at loci tightly linked to the disease. They used 2 outbred pedigrees in which familial venous malformations and hereditary hemorrhagic telangiectasia (HHT1; 187300) are linked to sites at opposite ends of chromosome 9. These were the kindreds reported by Gallione et al. (1995) and McDonald et al. (1994), respectively. Separate pools of DNA from 21 family members affected with VMCM and 17 with HHT1, 9 unaffected family members, and 25 unrelated population controls were established. In both autosomal dominant diseases investigated, quantitative DNA pooling detected shifts in allele frequency, thus identifying areas of known linkage. Although the false-positive rate appeared to be high, the approach was thought still to serve the purpose of reducing the amount of genotyping required.

Bean, W. B.: Vascular Spiders and Related Lesions of the Skin. Springfield, Ill.: Charles C. Thomas (pub.) 1958. Pp. 178-185.

Boon, L. M.; Mulliken, J. B.; Vikkula, M.; Watkins, H.; Seidman, J.; Olsen, B. R.; Warman, M. L.: Assignment of a locus for dominantly inherited venous malformations to chromosome 9p. Hum. Molec. Genet. 3: 1583-1587, 1994.

Damji, K. F.; Gallione, C. J.; Allingham, R. R.; Slotterbeck, B.; Guttmacher, A. E.; Pasyk, K. A.; Vance, J. M.; Pericak-Vance, M. A.; Speer, M. C.; Marchuk, D. A.: Quantitative DNA pooling to increase the efficiency of linkage analysis in autosomal dominant disease. Hum. Genet. 102: 207-212, 1998.

Gallione, C. J.; Pasyk, K. A.; Boon, L. M.; Lennon, F.; Johnson, D. W.; Helmbold, E. A.; Markel, D. S.; Vikkula, M.; Mulliken, J. B.; Warman, M. L.; Pericak-Vance, M. A.; Marchuk, D. A.: A gene for familial venous malformations maps to chromosome 9p in a second large kindred. J. Med. Genet. 32: 197-199, 1995.

Kanis, A. B.; Beck, J. S.; Rokhlina, T.; Tchida, C.; Stone, E. M.; Sheffield, V. C.: Use of DNA pooling strategies for mapping dominant disorders and the potential application to mapping complex diseases.(Abstract) Am. J. Hum. Genet. 57 (suppl.): A195 only, 1995.

McDonald, M. T.; Papenberg, K. A.; Ghosh, S.; Glatfelter, A. A.; Biesecker, B. B.; Helmbold, E. A.; Markel, D. S.; Zolotor, A.; McKinnon, W. C.; Vanderstoep, J. L.; Jackson, C. E.; Iannuzzi, M.; Collins, F. S.; Boehnke, M.; Porteous, M. E.; Guttmacher, A. E.; Marchuk, D. A.: A disease locus for hereditary haemorrhagic telangiectasia maps to chromosome 9q33-34. Nature Genet. 6: 197-204, 1994.

Vikkula, M.; Boon, L. M.; Carraway, K. L., III; Calvert, J. T.; Diamonti, A. J.; Goumnerov, B.; Pasyk, K. A.; Marchuk, D. A.; Warman, M. L.; Cantley, L. C.; Mulliken, J. B.; Olsen, B. R.: Vascular dysmorphogenesis caused by an activating mutation in the receptor tyrosine kinase TIE2. Cell 87: 1181-1190, 1996.

NODULI CUTANEI, MULTIPLE, WITH URINARY TRACT ABNORMALITIES (163850)

JJM NOTE

The skin tumors seem to be fibromas.

Selmanowitz et al. (1970) suggested that fibrosis is associated with hydronephrosis and urinary tract abnormalities on a familial basis. They reported a single case, a 44-year-old woman with multiple nodules on the legs and two elsewhere and a double collecting system of the right kidney. Gelfarb and Hyman (1962) described mother and daughter with multiple nodules and associated hydronephrosis.

Gelfarb, M.; Hyman, A. B.: Multiple noduli cutanei. Arch. Derm. 85: 89-94, 1962.

Selmanowitz, V. J.; Lerer, W. N.; Orentreich, N.: Multiple noduli cutanei and urinary tract abnormalities: a possible significant association. Cancer 26: 1256-1260, 1970.

NUCHAL BLEB, FAMILIAL; FETAL CYSTIC HYGROMA (257350)

JJM NOTE

This disorder seems conspicuously genetic in origin, although rarely familial. Clinically, the neck mass in an older infant can be life threatening by putting pressure on the trachea.

Bieber et al. (197) reported a sibship in which the first and third pregnancies ended in stillborn female infants with a 'bag of water' at the nape of the neck. The mother sought genetic counseling in relation to the fourth pregnancy. Ultrasonography suggested occipital encephalocele, and alpha-1-fetoprotein (AFP) greatly elevated. Elective abortion was performed at 21 weeks of gestation. The fetus had a normal 46,XX karyotype and a large nuchal bleb with no evidence of encephalocele. Nuchal bleb in the XO Turner syndrome is also accompanied by elevation of AFP in the amniotic fluid. Chervenak et al. (1983) defined fetal cystic hygroma as a congenital malformation of the lymphatic system appearing as single or multiloculated fluid-filled cavities, most often in the neck. They are thought to arise from failure of the lymphatic system to communicate with the venous system in the neck. They often progress to hydrops and cause fetal death. Chervenak et al. (1983) studied 15 consecutive cases of nuchal hygroma detected prenatally. None survived. In 11, the karyotype was consistent with the Turner syndrome and another probably had XY gonadal dysgenesis. Three had a 46,XX karyotype and 2 of these had multiple malformations. Although cystic hygroma is a conspicuous feature of the Turner syndrome and may occur in some mendelian disorders, its occurrence as an isolated autosomal recessive is suggested by the report of Watson et al. (1990): a 29-year-old white female had consecutive pregnancies in which the fetus was found to have cystic hygroma with hydrops. In 2 of the fetuses, a normal karyotype was documented. Dallapiccola et al. (1984) described a similar patient who had 2 consecutive fetuses with cystic hygroma, both with a normal karyotype.

Williams and Josephson (1997) described an Amish family in which 2 brothers had isolated cystic hygroma. A second Amish family had 2 children, a boy and a girl, with congenital chylothorax; both died as a consequence of this condition, one prenatally and one neonatally. The authors stated that familial cystic hygroma not associated with hydrops fetalis and neonatal death had not previously been reported.

See also Bieber et al. (1979)

Bieber, F. R.; Petres, R. E.; Bieber, J. M.; Nance, W. E.: Prenatal detection of a familial nuchal bleb simulating encephalocele. Birth Defects Orig. Art. Ser. XV(5A): 51-61, 1979.

Chervenak, F. A.; Isaacson, G.; Blakemore, K. J.; Breg, W. R.; Hobbins, J. C.; Berkowitz, R. L.; Tortora, M.; Mayden, K.; Mahoney, M. J.: Fetal cystic hygroma: cause and natural history. New Eng. J. Med. 309: 822-825, 1983.

Dallapiccola, B.; Zelante, L.; Perla, G.; Villani, G.: Prenatal diagnosis of recurrent cystic hygroma with normal chromosomes. Prenatal Diag. 4: 383-386, 1984.

Watson, W. J.; Thorp, J. M., Jr.; Seeds, J. W.: Familial cystic hygroma with normal karyotype. Prenatal Diag. 10: 37-40, 1990.

Williams, M. S.; Josephson, K. D.: Unusual autosomal recessive lymphatic anomalies in two unrelated Amish families. Am. J. Med. Genet. 73: 286-289, 1997.

WHITE SPONGE NEVUS OF CANNON (#193900)

JJM NOTE

This is an intraoral benign neoplastic lesion, but is included in skin for convenience.

A number sign (#) is used with this entry because of evidence (Rugg et al., 1995) that the disorder can be caused by mutation in mucosal keratin K4 (123940). Rugg et al. (1995) predicted that in cases of white sponge nevus, mutations will be found also in keratin-13 (148065), which is coexpressed and forms obligate heteropolymers with keratin-4. Richard et al. (1995) indeed demonstrated a keratin-13 point mutation in this disorder.

This disorder is manifested by thickened spongy-fold mucosa in the mouth with a white opalescent tint. The vagina, rectum and nasal cavity may be similarly involved. It is differentiated from benign intraepithelial dyskeratosis (127600) by the presence of vaginal and anal lesions and the absence of conjunctival involvement and the characteristic cell-within-cell histologic change. Scott (1966) found 3 generations affected. Haye and Whitehead (1968) reported on 7 cases in 3 generations with male-to-male transmission. Linkage studies were uninformative (Browne et al., 1969). Verma (1967) described the disorder in Asiatic Indians. Ultrastructural studies were carried out by Whitten (1970) and McGinnis and Turner (1975). Jorgenson and Levin (1981) pointed out that there are no extramucosal lesions associated with WSN and that the plaques of WSN undergo alternate periods of remission and exacerbation. In 2 members of a family penicillin was thought to have led to remission.

In affected members of 2 Scottish kindreds with WSN, Rugg et al. (1995) found heterozygosity for a 3-bp deletion in the 1A domain; see 123940.0001.

See also Rugg et al. (1995); Witkop and Gorlin (1961)

Browne, W. G.; Izatt, M. M.; Renwick, J. H.: White sponge naevus of the mucosa: clinical and linkage data. Ann. Hum. Genet. 32: 271-282, 1969.

Haye, K. R.; Whitehead, F. I. H.: Hereditary leukokeratosis of the mucous membranes. Brit. J. Derm. 80: 529-533, 1968.

Jorgenson, R. J.; Levin, L. S.: White sponge nevus. Arch. Derm. 117: 73-76, 1981.

McGinnis, J. P., Jr.; Turner, J. E.: Ultrastructure of the white sponge nevus. Oral Surg. 40: 644-651, 1975.

Richard, G.; De Laurenzi, V.; Didona, B.; Bale, S. J.; Compton, J. G.: Keratin 13 point mutation underlies the hereditary mucosal epithelia disorder white sponge nevus. Nature Genet. 11: 453-455, 1995.

Rugg, E. L.; McLean, W. H. I.; Allison, W. E.; Lunny, D. P.; Lane, E. B.; Macleod, R. I.; Felix, D. H.; Munro, C. S.: White sponge nevus (WSN) is caused by mutations in mucosal keratin K4. (Abstract) Am. J. Hum. Genet. 57: A47 only, 1995.

Rugg, E. L.; McLean, W. H. I.; Allison, W. E.; Lunny, D. P.; Macleod, R. I.; Felix, D. H.; Lane, E. B.; Munro, C. S.: A mutation in the mucosal keratin K4 is associated with oral white sponge nevus. Nature Genet. 11: 450-452, 1995.

Scott, C. R.: Hereditary leukokeratosis; white mouth. J. Pediat. 68: 768-772, 1966.

Verma, B. S.: Hereditary mucosal keratosis. Indian J. Med. Sci. 21: 310-313, 1967.

Whitten, J. B.: The electron microscopic examination of congenital keratosis of the oral mucous membranes. Oral Surg. 29: 69-84, 1970.

Witkop, C. J., Jr.; Gorlin, R. J.: Four hereditary mucosal syndromes. Arch. Derm. 84: 762-771, 1961.

NEVUS SEBACEUS OF JADASSOHN (163200)

JJM NOTE

Many nevi have no excess risk of malignancy, but basal cell carcinoma has arisen in the lesion of this syndrome.

Feuerstein and Mims (1962) described 2 unrelated patients with linear nevus sebaceus of the midline of the face associated with epilepsy, focal EEG abnormalities and mental retardation. Nothing is known about a possible genetic basis. Multiple developmental abnormalities are often associated. Mehregan and Pinkus (1965) pointed out characteristics of the natural history. In the first stage there is alopecia with absent or primitive hair follicles and numerous small hypoplastic sebaceous glands. At puberty the lesions become verrucous with hyperplastic sebaceous glands. In late stages benign or malignant tumors develop. Monk and Vollum (1982) reported mother and daughter with nevus sebaceus of Jadassohn. The lesion was in the scalp in each case. They suggested that this was the first familial occurrence reported. Baker et al. (1987) discussed neurologic complications in the epidermal nevus syndrome. A similar condition, inflammatory linear verrucous epidermal nevus (ILVEN), was described by Hamm and Happle (1986) in mother and daughter. In a black man, his daughter and a granddaughter, Sahl (1990) described nevus sebaceus of the scalp. A basal cell carcinoma had arisen in the nevus in the grandfather. Benedetto et al. (1990) described the disorder in a boy and his maternal half-brother. Kousseff (1992) posited that Jadassohn nevus phakomatosis (JNP), as he called the disorder, is, like other phakomatoses, a paracrine growth regulation disorder or paracrinopathy, i.e., disregulation of paracrine growth and of transforming growth factors at cellular and extracellular matrix levels, leading to localized over- and under-growth anomalies. For the deficiencies, he adopted the term used by Gomez (1988), i.e., hamartias, parallel to the term hamartomas. He reviewed 13 cases, including one of extraordinarily severe and widespread distribution.

See also Lantis et al. (1968)

Baker, R. S.; Ross, P. A.; Baumann, R. J.: Neurologic complications of the epidermal nevus syndrome. Arch. Neurol. 44: 227-232, 1987.

Benedetto, L.; Sood, U.; Blumenthal, N.; Madjar, D.; Sturman, S.; Hashimoto, K.: Familial nevus sebaceus. J. Am. Acad. Derm. 23: 130-132, 1990.

Feuerstein, R. C.; Mims, L. C.: Linear nevus sebaceus with convulsions and mental retardation. Am. J. Dis. Child. 104: 675-679, 1962.

Gomez, M. R.: Varieties of expression of tuberous sclerosis. Neurofibromatosis 1: 330-338, 1988.

Hamm, H.; Happle, R.: Inflammatory linear verrucous epidermal nevus (ILVEN) in a mother and her daughter. Am. J. Med. Genet. 24: 685-690, 1986.

Kousseff, B. G.: Hypothesis: Jadassohn nevus phakomatosis: a paracrinopathy with variable phenotype. Am. J. Med. Genet. 43: 651-661, 1992.

Lantis, S.; Leyden, J.; Thew, M.; Heaton, C.: Nevus sebaceus of Jadassohn. Arch. Derm. 98: 117-123, 1968.

Mehregan, A. H.; Pinkus, H.: Life history of organoid nevi. Arch. Derm. 91: 574-588, 1965.

Monk, B. E.; Vollum, D. I.: Familial naevus sebaceus. J. Roy. Soc. Med. 75: 660-661, 1982.

Sahl, W. J., Jr.: Familial nevus sebaceus of Jadassohn: occurrence in three generations. J. Am. Acad. Derm. 22: 853-854, 1990.

SEBACEOUS NEVUS SYNDROME AND HEMIMEGALENCEPHALY (601359)

JJM NOTE

Although I am not aware of frank malignancy, this entity is included because of the similarity to nevus sebaceus of Jadassohn, the local mass effect, and the hemihypertrophy.

Sebaceous nevus syndrome consists of sebaceous nevus of the face or scalp associated with ipsilateral defects of the brain, eye, and connective tissue (Happle, 1991). Dodge and Dobyns (1995) described a girl, born of unrelated parents, with sebaceous nevus syndrome, hemihypertrophy, coloboma of the left iris, talipes equinovarus, genu recurvatum, syndactyly 3-4 on the left foot, left hemimegalencephaly, and pachygyria. In addition, Dandy-Walker malformation and agenesis of the corpus callosum were found on cranial MRI. Consistent with the proposal by Happle (1991), Dodge and Dobyns (1995) suggested that sebaceous nevus syndrome may be caused by mosaic mutation of a gene which would be lethal if expressed in all cells.

Dodge, N. N.; Dobyns, W. B.: Agenesis of the corpus callosum and Dandy-Walker malformation associated with hemimegalencephaly in the sebaceous nevus syndrome. Am. J. Med. Genet. 56: 147-150, 1995.
Happle, R.: How many epidermal nevus syndromes exist? A clinicogenetic classification. J. Am. Acad. Derm. 25: 550-556, 1991.

SEBACEOUS GLAND HYPERPLASIA, FAMILIAL PRESENILE (601700)

JJM NOTE

This entity is included because of hyperplasia, not malignancy.

Sebaceous gland hyperplasia occurs frequently, particularly in men past middle age (Nomland, 1930). A few cases of sebaceous gland hyperplasia have been described in younger age groups (De Villez and Roberts, 1982; Bhawan and Calhoun, 1983). Dupre et al. (1980) described 2 brothers with premature sebaceous hyperplasia. Boonchai and Leenutaphong (1997) described a Thai family with premature sebaceous hyperplasia in 5 consecutive generations. There were 6 instances of male-to-male transmission. The disorder presented as solitary or several elevated, soft, yellow papules with central umbilication on the face, particularly the forehead. Lesions may spread to the neck and upper part of the thorax. A premature form has its appearance during puberty or just afterwards, male predominance, and excessive sebaceous secretion. Most cases are sporadic.

Bhawan, J.; Calhoun, J.: Premature sebaceous gland hyperplasia. (Letter) J. Am. Acad. Derm. 8: 136 only, 1983.
Boonchai, W.; Leenutaphong, V.: Familial presenile sebaceous gland hyperplasia. J. Amer. Acad. Derm. 36: 120-122, 1997.
De Villez, R. L.; Roberts, L. C.: Premature sebaceous gland hyperplasia. J. Am. Acad. Derm. 6: 933-935, 1982.
Dupre, A.; Bonafe, J. L.; Lamon, R.: Functional familial sebaceous hyperplasia of the face. Reverse of the Cunliffe acne-free nevus? Its inclusion among naevoid sebaceous receptor diseases. Clin. Exp. Derm. 5: 203-207, 1980.
Nomland, R.: Senile sebaceous adenoma. Arch. Derm. Syphil. 22: 1004-1009, 1930.

Actinic Skin Tumors

PIEBALD TRAIT (#172800)

JJM NOTE

Of the 46 OMIM entries that have albinism as a feature, only a few have actually been reported to have skin or other cancers. One syndrome with albinism, Chediak-Higashi syndrome, predisposes to leukemia and lymphoma. Clinical suspicion for possible malignancy and sun protection should be exercised in all hypopigmented persons, if the mechanism of carcinogenesis is skin cell mutation by ultraviolet light. This entity has epitheliomas.

A number sign (#) is used with this entry because of evidence that piebaldism can be caused by mutation in the KIT protooncogene. See 164920 for discussion of the findings of Giebel and Spritz (1991) demonstrating mutation in the KIT protooncogene as the cause of autosomal dominant piebaldism. The previous mapping of the KIT locus to 4q11-q12 is consistent with the location of piebald trait to 4q12 on the basis of chromosomal aberrations and homology with the mouse.

The features are white forelock and absence of pigmentation of the medial portion of the forehead, eyebrows and chin and of the ventral chest, abdomen, and extremities. The borders of unpigmented areas are hyperpigmented. Heterochromia iridis occurs in some. Keeler (1934) described a Louisiana black family in which the disorder could be traced back to a woman born in 1853. Sundfor (1939) described a family in which many persons had a white forelock, often with unpigmented patches on the forehead, limbs, body, etc. Comings and Odland (1966) found the trait in 6 generations. A genetic defect in melanoblast differentiation was postulated. Loewenthal (1959) assigned the name albinoidism to a dominantly inherited condition characterized by a white 'blaze' in the scalp hair, usually the forelock, and/or patches of leukoderma. Epitheliomas occurred with increased frequency. The designation albinoidism is better reserved for the recessive condition simulating true albinism. George Catlin (1796-1872), painter of the American Indians, painted an affected Mandan Indian. Multiple members of the group were said to have been affected.

The statement that deafness does not occur in persons with the piebald trait as a pleiotropic effect of the gene may not be true. Reed et al. (1967) noted profound deafness with piebaldism in 2 patients. Some of the patients of Comings and Odland (1966) were deaf. Selmanowitz et al. (1977) published a pedigree with at least 10 affected persons in 4 generations. Winship et al. (1991) described 7 affected persons in 3 generations. Two other affected individuals were deceased. The disorder seemed quite distinct from Waardenburg syndrome (193500). White forelock and patches of leukoderma occur also in Waardenburg syndrome and in Fanconi anemia (227650).

Spritz and Beighton (1998) described a South African girl of Xhosa stock with severe piebaldism and profound congenital sensorineural deafness. In this patient they identified a novel missense substitution, arg796 to gly, at a highly conserved residue in the intracellular kinase domain of the KIT protooncogene (164920.0016).

Farag et al. (1992) described a Bedouin kindred with 19 affected persons in 5 generations.

In mice, aganglionic megacolon is associated with the piebald trait (Bielschowsky and Schofield, 1962), inherited probably as an autosomal recessive. Dennis (1978) observed a black newborn with Hirschsprung disease (142623) (diagnosed by rectal biopsy), white forelock, V-shaped extension of the anterior fontanel into the metopic suture, and canthal distances of 2.5 cm (inner) and 6 cm (outer). From India, Mahakrishnan and Srinivasan (1980) reported Hirschsprung disease in 2 brothers who had piebaldness (white forelock, patches of depigmentation over the upper third of the forearms and the lower part of the arms, diffuse hypopigmentation of the abdomen and chest, and heterochromia iridis); their father had a white forelock also. The piebald character, which is widely distributed in the animal kingdom, is an intriguing problem in developmental genetics. Does the skin have pigmented and unpigmented clones of melanocytes? Is there a deficient number of melanocytes in unpigmented areas? Is there a patchy peculiarity of tissues that is inimical to normal melanocyte development or function? There are probably other possibilities. As in spotting mutations in all species, the details of the gene-determined abnormality in development is not clear. Presumably it is a peculiarity of neural crest migration.

Funderburk and Crandall (1974) reported a 3-year-old boy with moderate mental retardation, short stature, and integumentary pigment changes typical of the autosomal dominant piebald syndrome. The patient's chromosomes showed a reciprocal translocation and an intercalary deletion of one chromosome 4. Lacassie et al. (1977) found a similar case that illustrated the association of piebald trait with interstitial deletion of the long arm of chromosome 4 (4q13). The deleted segment was adjacent to centromeric heterochromatin, raising the question of position effect. Hoo et al. (1986) described a case of de novo deletion in 4q and pointed out that several of the patients with comparable deletions have had abnormal skin pigmentation compatible with the piebald trait. Further analysis suggested that the piebald trait locus may be situated in band 4q12. Yamamoto et al. (1989) reported piebald trait in a child with de novo interstitial deletion of 4q, specifically 4q12-q21.1. Other features included mental and motor retardation despite normal somatic growth, aplasia cutis of the scalp, flat nasal root and tip, micrognathia, widely spaced nipples, and agenesis of the right kidney. Hulten et al. (1987) reported a presumed homozygote; the severely affected child was born to heterozygous parents. He had complete absence of hair and pigmentation and had blue irides. Lyon (1988) pointed out that the location of the W locus on mouse chromosome 5 supports the location of a piebald trait gene on chromosome 4 of man since there is a large, conserved synteny group on those chromosomes of the 2 species. Geissler et al. (1988) identified cloned DNA markers near the W locus and determined the genetic distance from a number of other loci.

See also Comings and Odland (1965); Cooke (1952); Cromwell (1940); Fitch (1937); Froggatt (1951); Jahr and McIntyre (1954)

Bielschowsky, M.; Schofield, G. C.: Studies on megacolon in piebald mice. Aust. J. Exp. Biol. Med. Sci. 40: 395-403, 1962.
Comings, D. E.; Odland, G. F.: Electron microscope study of partial albinism. (Abstract) Clin. Res. 13: 265, 1965.
Comings, D. E.; Odland, G. F.: Partial albinism. J.A.M.A. 195: 510-523, 1966.
Cooke, J. V.: Familial white skin spotting (piebaldness) ('partial albinism') with white forelock. J. Pediat. 4: 1-12, 1952.
Cromwell, A. M.: Inheritance of white forelock in a mulatto family. J. Hered. 31: 94-96, 1940.
Dennis, N. R.: Personal Communication. Buffalo, N. Y. 2/22/1978.
Farag, T. I.; El-Ramly, M. A.; Sakr, M. F.: A Bedouin kindred with 19 piebalds in 5 generations. (Letter) Clin. Genet. 42: 326-328, 1992.
Fitch, L.: Inheritance of a white forelock: through five successive generations in the Logsdon family. J. Hered. 28: 413-414, 1937.
Froggatt, P.: An outline with bibliography of human pie-baldism and white forelock. Irish J. Med. Sci. 398: 86-94, 1951.

Funderburk, S. J.; Crandall, B. F.: Dominant piebald trait in a retarded child with a reciprocal translocation and small intercalary deletion. Am. J. Hum. Genet. 26: 715-722, 1974.
Geissler, E. N.; Cheng, S. V.; Gusella, J. F.; Housman, D. E.: Genetic analysis of the dominant white-spotting (W) region on mouse chromosome 5: identification of cloned DNA markers near W. Proc. Nat. Acad. Sci. 85: 9635-9639, 1988.
Giebel, L. B.; Spritz, R. A.: Mutation of the KIT (mast/stem cell growth factor receptor) protooncogene in human piebaldism. Proc. Nat. Acad. Sci. 88: 8696-8699, 1991.
Hoo, J. J.; Haslam, R. H. A.; van Orman, C.: Tentative assignment of piebald trait gene to chromosome band 4q12. Hum. Genet. 73: 230-231, 1986.
Hulten, M. A.; Honeyman, M. M.; Mayne, A. J.; Tarlow, M. J.: Homozygosity in piebald trait. J. Med. Genet. 24: 568-571, 1987.
Jahr, H. M.; McIntyre, M. S.: Piebaldness, or familial white skin spotting (partial albinism). Am. J. Dis. Child. 88: 481-484, 1954.
Keeler, C. E.: The heredity of a congenital white spotting in Negroes. J.A.M.A. 103: 179-180, 1934.
Lacassie, Y.; Thurmon, T. F.; Tracy, M. C.; Pelias, M. Z.: Piebald trait in a retarded child with interstitial deletion of chromosome 4. (Letter) Am. J. Hum. Genet. 29: 641-642, 1977.
Loewenthal, L. J. A.: Albinoidism with epitheliomatosis. Brit. J. Derm. 71: 37-38, 1959.
Lyon, M. F.: Personal Communication. Harwell, England 6/9/1988.
Mahakrishnan, A.; Srinivasan, M. S.: Piebaldness with Hirschsprung's disease. (Letter) Arch. Derm. 116: 1102, 1980.
Reed, W. B.; Stone, V. M.; Boder, E.; Ziprkowski, L.: Pigmentary disorders in association with congenital deafness. Arch. Derm. 95: 176-186, 1967.
Selmanowitz, V. J.; Rabinowitz, A. D.; Orentreich, N.; Went, E.: Pigmentary correction of piebaldism by autografts. I. Procedures and clinical findings. J. Derm. Surg. Oncol. 3: 615-622, 1977. Note: Fig. 1..
Spritz, R. A.; Beighton, P.: Piebaldism with deafness: molecular evidence for an expanded syndrome. Am. J. Med. Genet. 75: 101-103, 1998.
Sundfor, H.: A pedigree of skin-spotting in man: 42 piebalds in a Norwegian family. J. Hered. 30: 67-77, 1939.
Winship, I.; Young, K.; Martell, R.; Ramesar, R.; Curtis, D.; Beighton, P.: Piebaldism: an autonomous autosomal dominant entity. Clin. Genet. 39: 330-337, 1991.
Yamamoto, Y.; Nishimoto, H.; Ikemoto, S.: Interstitial deletion of the proximal long arm of chromosome 4 associated with father-child incompatibility within the Gc-system: probable reduced gene dosage effect and partial piebald trait. Am. J. Med. Genet. 32: 520-523, 1989.

ALBINISM 1 (*203100)

JJM NOTE

Although cancer is not specifically mentioned in the text, there is at least one report of squamous cell carcinoma in tyrosinase-negative albinism—but of the genitalia, not an area usually exposed to ultraviolet light, even in albino individuals! (Claudy AL, Ortonne JP: Tyrosinase-negative albinism with congenital malformations and squamous cell carcinoma of the genitalia. Acta Derm. Venereol. 62:260–62, 1982.)

Description

Oculocutaneous albinism type I is an autosomal recessive disorder characterized by absence of pigment in hair, skin, and eyes, and does not vary with race or age. Severe nystagmus, photophobia, and reduced visual acuity are common features.

Clinical Features

In describing albinism in the Caribe Cuna Indians, Keeler (1953) commented on the abundant straight white down consisting of hairs up to 2.5 cm long that develops on the body and extremities. It is not clear that this indicates genetic distinctness; it may somehow be related to the exposure of the subjects—partial hirsutism.

Witkop (1971) suggested that the 'yellow mutant' form of albinism first observed among the Amish (Nance et al., 1970) and later observed in non-Amish families (Witkop et al., 1971) may be distinct from both tyrosinase-positive and tyrosinase-negative albinism. As in type II albinism (203200), the homozygote is 'dead white' at birth, with serious ocular abnormalities, but rather rapidly develops normal skin pigmentation and yellow hair. The condition differs from albinism II in the yellow hair and the fact that incubation with L-tyrosine or L-DOPA yields equivocal results. That this is a temperature-sensitive mutant seems to have been excluded. See 203100.0006 and 203100.0007 for evidence of mutations in the tyrosinase gene as a cause of the Amish form of albinism.

Taylor (1978) pointed out that in the albino the ganglion cell layer does not thin out in the foveolar pit but shows a layer 6 to 8 cells thick where there should be none. He commented that 'this must degrade the retinal image. . . .There is therefore ample reason for the uncorrectable defective central fixation, and. . .the ocular nystagmus, in this case of the optical variety.' In 60% of his patients he noted an abnormal head posture, which minimized the nystagmus with slight improvement in visual acuity, at least for reading. All types of conditions with oculocutaneous or ocular hypopigmentation in man and animals with nystagmus tested to date have shown either electrophysiologic or anatomic evidence of a decussation defect in the optic tracts. Patients without nystagmus do not (Witkop et al., 1982).

Evidence that anomalous decussation exists also in the auditory system was presented by Creel et al. (1980). The amount of pigment in the inner ear correlates directly with the amount in the iris; otic pigment is lacking in albinos. In homozygotes an abnormal proportion of fibers from the ganglion cells of the temporal retina decussate to the contralateral cerebral hemisphere; this can be demonstrated by monocular visual evoked potential asymmetry (Apkarian et al., 1983).

Van Dorp (1987) suggested that patients with autosomal recessive albinism may have normal pigmentation. In a family with several albinos, they found a cousin, the offspring of a consanguineous mating, who was normally pigmented but had absence of macromelanosomes on skin biopsy as well as ocular and electrophysiologic signs of albinism. Van Dorp (1987) also concluded that patients with X-linked ocular albinism (300500, 300600) may be generally underpigmented and that patients with the Hermansky-Pudlak syndrome (203300) may have a dark complexion.

Witkop et al. (1989) concluded, mainly on the basis of a family reported by Hu et al. (1980), that yellow albinism is probably allelic to tyrosinase-negative albinism. Hu et al. (1980) described 3 sisters, including monozygotic twins, with clinical, ultrastructural, and histochemical features typical of yellow mutant albinism. Clinically, as the authors pointed out, this form of albinism is similar to the tyrosinase-positive albinism, but the hair bulbs show organelles similar to the pheomelanosomes of red hair and absence of tyrosinase activity. Classic tyrosinase-negative albinism was found in a maternal cousin of the probands. The authors suggested that the probands are genetic compounds for the yellow mutant allele and the classic albinism allele. They supported the view that the locus is homologous to the c-locus in the mouse and should be so designated in man. Hu et al. (1981) gave a revised (and corrected) table of proposed genotypes of oculocutaneous albinism involving the c-locus. King (1987) suggested that 'mutant' be dropped from the designation and that the condition be called simply 'yellow albinism' because patients find it offensive to be referred to as 'yellow mutant.'

Summers et al. (1991) observed striking discordance in ocular expression of albinism in 2 brothers. Both had foveal hypoplasia and misrouting of the optic fibers at the chiasm, as judged in the usual manner by visual evoked potentials (VEP): one had strabismus and nystagmus with maximum correction of vision to 20/100, whereas the other brother had vision with myopic correction of 20/20 in both eyes and no nystagmus or strabismus.

Summers and King (1994) described minimal pigment oculocutaneous albinism, a tyrosinase-related form of the disorder. The findings in 9 patients followed for up to 11 years were described. Patients were born with white scalp hair and skin, and nystagmus developed. Visual acuity was reduced, but in 1 patient vision improved with maturity. The irides were blue. In 7 of the 9 patients, including the 1 patient with improved visual acuity, iris pigment developed as demonstrated by transillumination with slit-lamp biomicroscopy.

Biochemical Features

Amelanic melanocytes are present in the skin of albinos. These contain granules similar to the premelanosomes of normal melanocytes. Tyrosinase is a monomer in mammals. It is avidly membrane bound and thus it has been difficult to separate. In the test developed by King and Witkop (1977), which determines free (unbound) tyrosine, heterozygotes have shown little or no tyrosinase activity. Witkop et al. (1989) postulated that what tyrosinase is synthesized in the heterozygote is immediately bound to the melanosome matrix.

Kwon et al. (1987) screened a lambda-gt11 human melanocyte cDNA library with antibodies against hamster tyrosinase to isolate a clone for human tyrosinase. The amino acid sequence deduced from the nucleotide sequence showed that the human enzyme is composed of 548 amino acids with a molecular weight of 62,610. The deduced protein contained glycosylation sites and histidine-rich sites that could be used for copper binding. Southern blot analysis of DNA derived from newborn mice carrying lethal albino deletion mutations showed that the clone maps near or at the c-albino locus on mouse chromosome 7, which is known to be the structural gene for tyrosinase. The enzyme tyrosinase (EC 1.14.18.1) catalyzes the first 2 steps, and at least 1 subsequent step, in the conversion of tyrosine to melanin (Spritz, 1994).

Inheritance

Pipkin and Pipkin (1942) claimed dominant inheritance for total albinism without other features in one family, but a quasidominant pedigree pattern of the usual recessive forms seems likely.

McLeod and Lowry (1976) observed seemingly dominant inheritance in 2 generations of 1 family but concluded that partial penetrance of the albinism II gene in heterozygotes was responsible.

Mapping

Nance et al. (1970) found no conclusive evidence for linkage between the locus for the yellow mutant and any of 15 polymorphic loci tested. Abundant phaeomelanin but little eumelanin was found naturally in the Amish patients who were referred to as representing the yellow mutant (ym).

The only GPI locus that has been assigned in man is known to be on chromosome 19. Of interest in connection with the possible linkage of beta-globin and albinism (suggested by homology to the mouse) are the reports of a family with both albinism and sicklemia (Massie and Hartmann, 1957) and of a Sicilian boy with albinism and an unusual combination of hemoglobinopathies (Schiliro et al., 1983).

Barton et al. (1988) used a human tyrosinase cDNA to map the human TYR locus to 11q14-q21 by Southern blot analysis of somatic cell hybrid DNA and by in situ chromosomal hybridization. A second site of tyrosinase-related sequences was detected on the short arm of chromosome 11 near the centromere (p11.2-cen).

Spritz et al. (1988) identified a TaqI RFLP at the TYR locus. The human tyrosinase gene contains 4 introns, with exon-intron boundaries identical to those in the mouse gene. Using RFLPs identified within the human tyrosinase gene, Spritz et al. (1989) did linkage analysis in albinism families demonstrating absolute linkage. They found no cases of tyrosinase gene deletions or other rearrangements, even in DNAs from patients with both tyrosinase-deficient oculocutaneous albinism and mental retardation. The families studied exhibited several different pigmentation phenotypes suggesting that tyrosinase-deficient OCA results from heteroallelism for different small defects of the tyrosinase gene.

16,14:Giebel et al. (1990, 1991) demonstrated genetic linkage between classic tyrosinase-negative oculocutaneous albinism (type IA) and the tyrosinase gene RFLP (lod = 6.17 at theta = 0). Giebel and Spritz (1990) reported an MboI RFLP in the TYR gene. The frequencies of alleles A1 and A2 were estimated to be 0.48 and 0.52, respectively.

Molecular Genetics

In a patient with classic tyrosinase-negative OCA, Spritz et al. (1989) found a thr-to-lys substitution that abolished 1 of 6 putative N-linked glycosylation sites that are completely conserved between humans and mice; see 203100.0003. In a patient with 'yellow' OCA, Spritz et al. (1989) identified a pro-to-leu substitution that may interfere with the normal folding of the tyrosinase polypeptide; see 203100.0002.

Giebel et al. (1991) demonstrated that the tyrosinase gene contains 5 exons and spans more than 50 kb of DNA. A tyrosinase-related 'gene,' which contains only exons 4 and 5, is located on 11p; see 191270. Ponnazhagan et al. (1994) concluded from their restriction mapping studies that the gene spans over 65 kb. Furthermore, they characterized the promoter region.

On the basis of an analysis of 16 missense mutations, King et al. (1991) pointed out that most of the mutations cluster in 4 areas of the gene. Two clusters involve the copper A and copper B binding sites and mutations in these areas could disrupt the metal ion-protein interaction necessary for enzyme function. The other 2 clusters are in exons 1 and 4 and could indicate important functional domains of the enzyme.

Tripathi et al. (1992) stated that more than 60 independent albinism-producing alleles had been described at the TYR locus. They reviewed 29 of these and commented on 2 additional novel missense substitutions in a 'note added in proof.' They commented that type I OCA in Caucasians clearly results from a great variety of different uncommon alleles. About 90% of OCA in Caucasians was accounted for by the 29 mutations they described. More than 80% of the then-known missense substitutions clustered within 2 relatively small regions of the tyrosinase polypeptide, suggesting that these may represent functionally critical sites within the enzyme. Oetting and King (1992) reviewed 17 reported missense mutations and 10 nonsense and frameshift mutations causing tyrosinase-negative OCA and added 2 and 3 new mutations of the respective types.

Although a separate locus on chromosome 15 has been identified as the site of mutations responsible for tyrosinase-positive OCA, it turns out, on the basis of the mutation analyses of Tripathi et al. (1992), that some patients clinically defined as 'tyrosinase-positive' OCA in fact have mutations in the tyrosinase gene. Oetting and King (1993) tabulated 36 mutations identified in type I OCA: 24 missense, 4 nonsense, and 8 frameshift mutations. The affected individuals in these cases were compound heterozygotes. They also listed 6 polymorphic sites useful in haplotype analysis: 2 in the promoter region, 2 in the coding region associated with alternative amino acids in the tyrosinase protein, and 2 RFLPs in the first intron.

Diagnosis

Commenting on the availability of prenatal diagnosis in albinism, Taylor (1987) argued that elective abortion of albino fetuses is difficult to defend because of the satisfactory adjustment and even success in some areas of activity of albino individuals. Persistent ocular albinism and nystagmus permit accurate diagnosis in the adult.

Shimizu et al. (1994) made the prenatal diagnosis of tyrosinase-negative OCA by an electron-microscopic DOPA reaction test of fetal skin at 20 weeks' gestation. A previous child born with albinism was 9 years old at the time; the pregnancy in which the diagnosis had been made was terminated at 21 weeks.

Population Genetics

Froggatt (1960) estimated a phenotype frequency of albinism I to be 1 in 10,000 in Northern Ireland. First-cousin marriages occurred in 4.5% of the parents. An excess of males was almost exclusively in the probands and the sex ratio of secondary cases was about 1; therefore, bias of ascertainment probably accounted for the excess of males. The mutation rate was estimated to be between 3.3 and 7 x 10(-5) per gene per generation. Abnormal iris translucency, occurring in 70% of the parents and children of albinos, was interpreted as a heterozygous manifestation.

In British Columbia, McLeod and Lowry (1976) found the incidence of type I albinism to be 1 in 67,800 live births and of type II albinism (203200) to be 1 in 35,700 live births.

Jay et al. (1982) tabulated the frequency of different types of albinism in England: tyrosine-negative OCA, 54; tyrosinase-positive OCA, 50; yellow mutant OCA, 7; Hermansky-Pudlak syndrome, 2; X-linked ocular albinism hemizygotes, 21, and heterozygotes, 15; autosomal recessive ocular albinism, 16. In surveying congenital anomaly syndromes in a Spanish gypsy population, Martinez-Frias and Bermejo (1992) found an impressive frequency of albinism. In one of the pedigrees, there were 2 examples of pseudodominant inheritance, i.e., apparent parent-to-child transmission.

Animal Model

Working with albino melanomas and tyrosinase inhibitor in animals, Chian and Wilgram (1967) found that the inhibitor is effective against soluble tyrosinase but not against tyrosinase aggregated into melanosomes. In one type of albino mutation, tyrosinase apparently could not aggregate because of genetic alteration in its protein carrier and therefore was vulnerable to the effects of the inhibitor. These workers suggested that a similar situation may obtain in some type of albinism of man.

In a wide variety of animals, the albinism gene is known to have a pleiotropic effect on the visual pathways (Guillery, 1974). Some of the optic nerve fibers do not decussate as in the normal. This structural abnormality, the mechanism of which is unknown, can be associated with crossed eyes in albino animals. Carroll et al. (1980) presented evidence that the human albino has the same anatomic peculiarity of the visual pathways, resulting in misrouting of the retinogeniculate projections, that has been found in albinos of other species.

Leventhal et al. (1985) studied cats who were obligatory heterozygotes for a c-locus tyrosinase-negative allele (Cc) and had no relationship to 'deaf white cats' (W). In these normally pigmented animals abnormalities of the retinogeniculocortical pathways were found to be similar to those in homozygous albinos. By unilateral injection of horseradish peroxidase (HRP) into the dorsal lateral geniculate nucleus, they could map the retrogradely labeled retinal ganglion cells. Compared to homozygous normal controls, heterozygotes showed labeling of an abnormally large number of cells, especially large alpha cells, in the contralateral temporal retina. The authors pointed out that 1 to 2% of the human population may be heterozygous for albinism and that the above described abnormality may have an adverse effect on binocular depth perception. The locus ceruleus and substantia nigra are normally pigmented in albinos. They owe their pigmentation to neuromelanin, which is synthesized by tyrosine hydroxylase rather than tyrosinase (Witkop et al., 1989).

The human albinism locus is presumably homologous to the c-locus in the mouse and, as in the mouse, there are presumably multiple alleles. One of these is the allele for 'yellow;' another is the allele for 'minimal pigment.' The genetic compound of the tyrosinase-negative albinism allele and the 'minimal pigment' allele might represent autosomal recessive ocular albinism (203310) (King, 1987).

Snyder (1980) pointed out that in Mus musculus, Rattus norvegicus, and Peromyscus maniculatus, glucosephosphate isomerase (Gpi-1), albinism (c), and beta-type globin (Hbb) are linked. In the first 2 species, pink-eyed dilution (p) is also known to be in this same cluster. (Indeed, the pink-eye—albinism linkage in the mouse was the first to be demonstrated in any mammal, by Haldane et al., 1915.) See Lyon et al. (1992) and Gardner et al. (1992) for genetic and molecular analyses of the pink-eyed dilution gene in the mouse. Gardner et al. (1992) demonstrated that the human homolog of the mouse p mutation is a gene located on 15q11.2-q12, a region associated with Prader-Willi syndrome (176270) and Angelman syndrome (105830); see 203200. It is not linked to the human homolog of any of the 3 above-mentioned genes that are located on mouse chromosome 7 in linkage with 'pink-eye.'

O'Brien et al. (1986) found that the albino locus in the domestic cat is linked to the beta-globin locus at a distance of approximately 8 cM. Evolutionary conservation of syntenic homology of feline chromosome D1 and human chromosome 11 is extensive. High resolution G-trypsin-banded preparations of the 2 chromosomes showed similarities. (A tyrosinase-related gene, which contains only exons 4 and 5, is located on 11p; see 191270.)

'Dark-eyed albino' is a recessive mutation at the mouse albino 'c' locus, which encodes tyrosinase. Similar to type IB OCA in humans, overall production of pigment is greatly reduced in dark-eyed albino mice and obvious only in the eyes.

Using a human tyrosinase cDNA clone, Barton et al. (1988) and Kwon et al. (1989) isolated mouse tyrosinase genomic clones and used them to map the mouse tyrosinase locus to a site at or near the albino locus on mouse chromosome 7. Mouse tyrosinase mRNA was found to measure approximately 2.4 kb. A mutation in tyrosinase responsible for the albino mouse appears to be a change of cysteine-85 to serine (Kwon et al., 1988), resulting from a change of guanine 390 to cytosine. Jackson and Bennett (1990) studied revertant cells and found that loss of the mutant allele was responsible.

History

Famous albinos include Noah of flood fame and the Rev. Dr. Spooner. Evidence that Noah was an albino was presented by Sorsby (1958). Spooner was a brilliant classicist at Oxford whose amusing tendency to errors of speech came to be known as spoonerisms. Although probably elaborated on by students, the aberration appears to have been marked. As a classicist, Spooner must have read extensively. The aberration of speech was probably related to his nystagmus which caused a jumbling of information from the printed page. His intelligence was such that his mind comprehended despite the jumbling, but a jumbling of sorts occurred with oral output (Edwards, 1980). (Spooner was warden of New College, Oxford University, where his brightly colored portrait hangs (Gibson, 1980).) It would be of interest to know whether spoonerism (as a process and phenomenon) is more frequent in albinos or others with nystagmus. An 'albino society' has been formed in England.

See also Collewijn et al. (1978); Fitzpatrick and Quevedo (1972); Giebel et al. (1990); Giebel et al. (1991); Giebel et al. (1991); Giebel et al. (1991); Hanhart (1952); King et al. (1991); Kwon et al. (1989); Schmidt and Beermann (1994); Sears (1974); Spritz et al. (1990); Tripathi et al. (1992); van Dorp et al. (1984); Witkop et al. (1978)

Apkarian, P.; Reits, D.; Spekreijse, H.; van Dorp, D.: A decisive electrophysiological test for human albinism. Electroenceph. Clin. Neurophysiol. 55: 513-531, 1983.
Bard, L. A.: Heterogeneity in Waardenburg's syndrome: report of a family with ocular albinism. Arch. Ophthal. 96: 1193-1198, 1978.
Barton, D. E.; Kwon, B. S.; Francke, U.: Human tyrosinase gene, mapped to chromosome 11 (q14-q21), defines second region of homology with mouse chromosome 7. Genomics 3: 17-24, 1988.
Carroll, W. M.; Jay, B. S.; McDonald, W. I.; Halliday, A. M.: Two distinct patterns of visual evoked response asymmetry in human albinism. Nature 286: 604-606, 1980.
Chian, L. T. Y.; Wilgram, G. F.: Tyrosinase inhibition: its role in suntanning and in albinism. Science 155: 198-200, 1967.
Collewijn, H.; Winterson, B. J.; Dubois, M. F. W.: Optokinetic eye movements in albino rabbits: inversion in anterior visual field. Science 199: 1351-1353, 1978.
Creel, D.; Garber, S. R.; King, R. A.; Witkop, C. J., Jr.: Auditory brainstem anomalies in human albinos. Science 209: 1253-1255, 1980.
Edwards, J. H.: Personal Communication. Oxford, England 8/1980.
Fitzpatrick, T. B.; Quevedo, W. C., Jr.: Albinism.In: Stanbury, J. B.; Wyngaarden, J. B.; Fredrickson, D. S.: The Metabolic Basis of Inherited Disease. New York: McGraw-Hill (pub.) (3rd ed.) : 1972. Pp. 326-337.
Froggatt, P.: Albinism in Northern Ireland. Ann. Hum. Genet. 24: 213-238, 1960.
Fukai, K.; Holmes, S. A.; Lucchese, N. J.; Siu, V. M.; Weleber, R. G.; Schnur, R. E.; Spritz, R. A.: Autosomal recessive ocular albinism associated with a functionally significant tyrosinase gene polymorphism. Nature Genet. 9: 92-95, 1995.
Gardner, J. M.; Nakatsu, Y.; Gondo, Y.; Lee, S.; Lyon, M. F.; King, R. A.; Brilliant, M. H.: The mouse pink-eyed dilution gene: association with human Prader-Willi and Angelman syndromes. Science 257: 1121-1124, 1992.
Gibson, W. C.: Oxford revisited. J.A.M.A. 244: 577-579, 1980.
Giebel, L. B.; Musarella, M. A.; Spritz, R. A.: A nonsense mutation in the tyrosinase gene of Afghan patients with tyrosinase negative (type IA) oculocutaneous albinism. J. Med. Genet. 28: 464-467, 1991.
Giebel, L. B.; Spritz, R. A.: RFLP for MboI in the human tyrosinase (TYR) gene detected by PCR. Nucleic Acids Res. 18: 3103, 1990.
Giebel, L. B.; Strunk, K. M.; Jackson, C. E.; Hanifin, J. M.; King, R. A.; Spritz, R. A.: Tyrosinase gene mutations in patients with type IB ('yellow') oculocutaneous albinism. (Abstract) Am. J. Hum. Genet. 47 (suppl.): A156, 1990.
Giebel, L. B.; Strunk, K. M.; King, R. A.; Hanifin, J. M.; Spritz, R. A.: A frequent tyrosinase gene mutation in classic, tyrosinase-negative (type IA) oculocutaneous albinism. Proc. Nat. Acad. Sci. 87: 3255-3258, 1990.
Giebel, L. B.; Strunk, K. M.; Spritz, R. A.: Organization and nucleotide sequences of the human tyrosinase gene and a truncated tyrosinase-related segment. Genomics 9: 435-445, 1991.
Giebel, L. B.; Tripathi, R. K.; King, R. A.; Spritz, R. A.: A tyrosinase gene missense mutation in temperature-sensitive type I oculocutaneous albinism: a human homologue to the Siamese cat and the Himalayan mouse. J. Clin. Invest. 87: 1119-1122, 1991.
Giebel, L. B.; Tripathi, R. K.; Strunk, K. M.; Hanifin, J. M.; Jackson, C. E.; King, R. A.; Spritz, R. A.: Tyrosinase gene mutations associated with type IB ('yellow') oculocutaneous albinism. Am. J. Hum. Genet. 48: 1159-1167, 1991.
Guillery, R. W.: Visual pathways in albinos. Sci. Am. 230(5): 44-54, 1974.
Haldane, J. B. S.; Sprunt, A. D.; Haldane, N. M.: Reduplication in mice. J. Genet. 5: 133-135, 1915.
Hanhart, E.: Ueber 18 lebende und 13 verstorbene Albinos in einem Dorf des Piemont nebst weiteren Beitraegen zur Populationsgenetik des Albinismus universalis. Arch. Klaus Stift. Vererbungsforsch. 27: 178-188, 1952.
Hu, F.; Hanifin, J. M.; Prescott, G. H.; Tongue, A. C.: Yellow mutant albinism: cytochemical, ultrastructural, and genetic characterization suggesting multiple allelism. Am. J. Hum. Genet. 32: 387-395, 1980.
Hu, F.; Hanifin, J. M.; Prescott, G. H.; Tongue, A. C.: Response to Warren's letter. (Letter) Am. J. Hum. Genet. 33: 479-480, 1981.
Jackson, I. J.; Bennett, D. C.: Identification of the albino mutation of mouse tyrosinase by analysis of an in vitro revertant. Proc. Nat. Acad. Sci. 87: 7010-7014, 1990.
Jay, B.; Witkop, C. J.; King, R. A.: Albinism in England. Birth Defects Orig. Art. Ser. 18(6): 319-325, 1982.
Keeler, C. E.: The Caribe Cuna moon-child and its heredity. J. Hered. 44: 163-171, 1953.
Kikuchi, H.; Hara, S.; Ishiguro, S.; Tamai, M.; Watanabe, M.: Detection of point mutation in the tyrosinase gene of a Japanese albino patient by a direct sequencing of amplified DNA. Hum. Genet. 85: 123-124, 1990.
King, R. A.: Personal Communication. Minneapolis, Minn. 6/15/1987.
King, R. A.; Mentink, M. M.; Oetting, W. S.: Non-random distribution of missense mutations within the human tyrosinase gene in type I (tyrosinase-related) oculocutaneous albinism. Molec. Biol. Med. 8: 19-29, 1991.
King, R. A.; Townsend, D.; Oetting, W.; Summers, C. G.; Olds, D. P.; White, J. G.; Spritz, R. A.: Temperature-sensitive tyrosinase associated with peripheral pigmentation in oculocutaneous albinism. J. Clin. Invest. 87: 1046-1053, 1991.
King, R. A.; Townsend, D.; Oetting, W. S.; Spritz, R. A.: An unusual pigment pattern in type I oculocutaneous albinism (OCA) resulting from a temperature-sensitive enzyme. (Abstract) Am. J. Hum. Genet. 45 (suppl.): A8, 1989.
King, R. A.; Witkop, C. J., Jr.: Detection of heterozygotes for tyrosinase-negative oculocutaneous albinism by hairbulb tyrosinase assay. Am. J. Hum. Genet. 29: 164-168, 1977.
Kwon, B. S.; Halaban, R.; Chintamaneni, C.: Molecular basis of mouse Himalayan mutation. Biochem. Biophys. Res. Commun. 161: 252-260, 1989.
Kwon, B. S.; Haq, A. K.; Pomerantz, S. H.; Halaban, R.: Isolation and sequence of a cDNA clone for human tyrosinase that maps at the mouse c-albino locus. Proc. Nat. Acad. Sci. 84: 7473-7477, 1987.
Kwon, B. S.; Haq, A. K.; Wakulchik, M.; Kestler, D.; Barton, D. E.; Francke, U.; Lamoreux, M. L.; Whitney, J. B., III; Halaban, R.: Isolation, chromosomal mapping, and expression of the mouse tyrosinase gene. J. Invest. Derm. 93: 589-594, 1989.
Kwon, B. S.; Wakulchik, M.; Haq, A. K.; Halaban, R.; Kestler, D.: Sequence analysis of mouse tyrosinase cDNA and the effect of melanotropin on its gene expression. Biochem. Biophys. Res. Commun. 153: 1301-1309, 1988.
Leventhal, A. G.; Vitek, D. J.; Creel, D. J.: Abnormal visual pathways in normally pigmented cats that are heterozygous for albinism. Science 229: 1395-1397, 1985.
Lyon, M. F.; King, T. R.; Gondo, Y.; Gardner, J. M.; Nakatsu, Y.; Eicher, E. M.; Brilliant, M. H.: Genetic and molecular analysis of recessive alleles at the pink-eyed dilution (p) locus of the mouse. Proc. Nat. Acad. Sci. 89: 6968-6972, 1992.
Martinez-Frias, M. L.; Bermejo, E.: Prevalence of congenital anomaly syndromes in a Spanish gypsy population. J. Med. Genet. 29: 483-486, 1992.
Massie, R. W.; Hartmann, R. C.: Albinism and sicklemia in a Negro family. Am. J. Hum. Genet. 9: 127-132, 1957.
McLeod, R.; Lowry, R. B.: Incidence of albinism in British Columbia: separation by hairbulb test. Clin. Genet. 9: 77-80, 1976.
Morell, R.; Spritz, R. A.; Ho, L.; Pierpont, J.; Guo, W.; Friedman, T. B.; Asher, J. H., Jr.: Apparent digenic inheritance of Waardenburg syndrome type 2 (WS2) and autosomal recessive ocular albinism (AROA). Hum. Molec. Genet. 6: 659-664, 1997.
Nance, W. E.; Jackson, C. E.; Witkop, C. J., Jr.: Amish albinism: a distinctive autosomal recessive phenotype. Am. J. Hum. Genet. 22: 579-586, 1970.
O'Brien, S. J.; Haskins, M. E.; Winkler, C. A.; Nash, W. G.; Patterson, D. F.: Chromosomal mapping of beta-globin and albino loci in the domestic cat: a conserved mammalian chromosome group. J. Hered. 77: 374-378, 1986.
Oetting, W. S.; King, R. A.: Molecular basis of type I (tyrosinase-related) oculocutaneous albinism: mutations and polymorphisms of the human tyrosinase gene. Hum. Mutat. 2: 1-6, 1993.
Oetting, W. S.; King, R. A.: Molecular analysis of type I-A (tyrosine negative) oculocutaneous albinism. Hum. Genet. 90: 258-262, 1992.
Oetting, W. S.; King, R. A.: Mutations within the promoter region of the tyrosinase gene in type I (tyrosinase-related) oculocutaneous albinism. (Abstract) Clin. Res. 39: 267A, 1991.

Oetting, W. S.; Mentink, M. M.; Summers, C. G.; Lewis, R. A.; White, J. G.; King, R. A.: Three different frameshift mutations of the tyrosinase gene in type IA oculocutaneous albinism. Am. J. Hum. Genet. 49: 199-206, 1991.
Oetting, W. S.; Witkop, C. J., Jr.; Brown, S. A.; Colomer, R.; Fryer, J. P.; Bloom, K. E.; King, R. A.: A frequent tyrosinase gene mutation associated with type I-A (tyrosinase-negative) oculocutaneous albinism in Puerto Rico. Am. J. Hum. Genet. 52: 17-23, 1993.
Pipkin, A. C.; Pipkin, S. B.: Albinism in Negroes. J. Hered. 33: 419-427, 1942.
Ponnazhagan, S.; Hou, L.; Kwon, B. S.: Structural organization of the human tyrosinase gene and sequence analysis and characterization of its promoter region. J. Invest. Derm. 102: 744-748, 1994.
Schiliro, G.; Pavone, L.; Romeo, M. A.; Russo, A.; Musumeci, S.; Russo, G.: Unusual combination of genetic defects in a Sicilian boy: G-gamma/A-gamma delta-beta thalassemia, G-gamma/A-gamma heterocellular HPFH, beta-zero-thalassemia, and albinism. Am. J. Med. Genet. 15: 225-231, 1983.
Schmidt, A.; Beermann, F.: Molecular basis of dark-eyed albinism in the mouse. Proc. Nat. Acad. Sci. 91: 4756-4760, 1994.
Sears, M. L.: Browning of the lens in generalized albinism. Am. J. Ophthal. 77: 819-823, 1974.
Shimizu, H.; Ishiko, A.; Kikuchi, A.; Akiyama, M.; Suzumori, K.; Nishikawa, T.: Prenatal diagnosis of tyrosinase-negative oculocutaneous albinism by an electron microscopic DOPA reaction test of fetal skin. Prenatal Diag. 14: 443-450, 1994.
Snyder, L. R. G.: Evolutionary conservation of linkage groups: additional evidence from Murid and Cricetid rodents. Biochem. Genet. 18: 209-220, 1980.
Sorsby, A.: Noah—an albino. Brit. Med. J. 2: 1587-1589, 1958.
Spritz, R.; Strunk, K.; Oetting, W.; King, R.: RFLP for TaqI at the human tyrosinase locus. Nucleic Acids Res. 16: 9890, 1988.
Spritz, R. A.: Personal Communication. Madison, Wis. 5/1997.
Spritz, R. A.: Molecular genetics of oculocutaneous albinism. Hum. Molec. Genet. 3: 1469-1475, 1994.
Spritz, R. A.; Giebel, L. B.; Tripathi, R. K.; Strunk, K. M.: Structure and polymorphisms of the human tyrosinase gene and a truncated tyrosinase pseudogene. (Abstract) Am. J. Hum. Genet. 47 (suppl.): A117, 1990.
Spritz, R. A.; Strunk, K.; King, R. A.: Molecular analyses of the tyrosinase gene in patients with tyrosinase-deficient oculocutaneous albinism. (Abstract) Am. J. Hum. Genet. 45 (suppl.): A221, 1989.
Spritz, R. A.; Strunk, K. M.; Giebel, L. B.; King, R. A.: Detection of mutations in the tyrosinase gene in a patient with type IA oculocutaneous albinism. New Eng. J. Med. 322: 1724-1728, 1990.
Spritz, R. A.; Strunk, K. M.; Hsieh, C.-L.; Sekhon, G. S.; Francke, U.: Homozygous tyrosinase gene mutation in an American black with tyrosinase-negative (type IA) oculocutaneous albinism. Am. J. Hum. Genet. 48: 318-324, 1991.
Summers, C. G.; Creel, D.; Townsend, D.; King, R. A.: Variable expression of vision in sibs with albinism. Am. J. Med. Genet. 40: 327-331, 1991.
Summers, C. G.; King, R. A.: Ophthalmic features of minimal pigment oculocutaneous albinism. Ophthalmology 101: 906-914, 1994.
Takeda, A.; Tomita, Y.; Matsunaga, J.; Tagami, H.; Shibahara, S.: Molecular basis of tyrosinase-negative oculocutaneous albinism: a single base mutation in the tyrosinase gene causing arginine to glutamine substitution at position 59. J. Biol. Chem. 265: 17792-17797, 1990.
Taylor, W. O. G.: Prenatal diagnosis in albinism. Lancet I: 1307-1308, 1987.
Taylor, W. O. G.: Visual disabilities of oculocutaneous albinism and their alleviation. Trans. Ophthal. Soc. U.K. 98: 423-445, 1978.
Tomita, Y.; Takeda, A.; Okinaga, S.; Tagami, H.; Shibahara, S.: Human oculocutaneous albinism caused by single base insertion in the tyrosinase gene. Biochem. Biophys. Res. Commun. 164: 990-996, 1989.
Tripathi, R. K.; Droetto, S.; Spritz, R. A.: Many patients with 'tyrosinase-positive' oculocutaneous albinism have tyrosinase gene mutations. (Abstract) Am. J. Hum. Genet. 51 (suppl.): A179, 1992.
Tripathi, R. K.; Strunk, K. M.; Giebel, L. B.; Weleber, R. G.; Spritz, R. A.: Tyrosinase gene mutations in type I (tyrosinase-deficient) oculocutaneous albinism define two clusters of missense substitutions. Am. J. Med. Genet. 43: 865-871, 1992.
van Dorp, D. B.: Albinism, or the NOACH syndrome (the book of Enoch c.v. 1-20). Clin. Genet. 31: 228-242, 1987.
van Dorp, D. B.; Delleman, J. W.; Loewer-Sieger, D. H.: Oculocutaneous albinism and anterior chambre cleavage malformations: not a coincidence. Clin. Genet. 26: 440-444, 1984.
Witkop, C. J., Jr.: Albinism.In: Harris, H.; Hirschhorn, K.: Advances in Human Genetics. New York: Plenum Press (pub.) 2: 1971. Pp. 61-142.
Witkop, C. J., Jr.; Jay, B.; Creel, D.; Guillery, R. W.: Optic and otic neurologic abnormalities in oculocutaneous and ocular albinism. Birth Defects Orig. Art. Ser. 18(6): 299-318, 1982.
Witkop, C. J., Jr.; Quevedo, W. C., Jr.; Fitzpatrick, T. B.: Albinism.In: Stanbury, J. B.; Wyngaarden, J. B.; Fredrickson, D. S.: Metabolic Basis of Inherited Disease. New York: McGraw-Hill (pub.) (4th ed.): 1978. Pp. 283-316.
Witkop, C. J., Jr.; Quevedo, W. C., Jr.; Fitzpatrick, T. B.; King, R. A.: Albinism.In: Scriver, C. R.; Beaudet, A. L.; Sly, W. S.; Valle, D.: The Metabolic Basis of Inherited Disease. New York: McGraw-Hill (pub.) (6th ed.) II: 1989. Pp. 2905-2947.
Witkop, C. J., Jr.; White, J. G.; Nance, W. E.; Jackson, C. E.; Desnick, S.: Classification of albinism in man. Clinical Delineation of Birth Defects. Skin, Hair and Nails. New York: National Foundation—March of Dimes (pub.) XII: 1971. Pp. 13-25.

ALBINISM 2 (*203200)

JJM NOTE

The first trial for chemoprevention of cancer, I understand, took place among albinos in Africa, to assess the benefit of 5-fluorouracil to reduce the frequency of skin cancer (Dogliotti M: Actinic keratoses in Bantu albinos: Clinical experiences with the topical use of 5-fluorouracil. South Afr. Med. J. 47:2169–172, 1973). The types of cancers found in albinos have been squamous cell and basal cell carcinomas, and occasionally cutaneous and choroidal melanomas (Kheterpal S et al.: Choroidal melanoma in an African-American albino. Am. J. Ophthalmol. 122:901–03, 1966; Lookingbill DP et al.: Actinic damage and skin cancer in albinos in northern Tanzania: findings in 164 patients enrolled in an outreach skin care program. Am. Acad. Dermatol. 32:653–58, 1995; Ihn H et al.: Amelanotic metastatic melanoma in a patient with oculocutaneous albinism. J. Am. Acad. Dermatol. 28:895–900, 1993; Levine EA et al.: Malignant melanoma in a child with oculocutaneous albinism. Surg. Oncol. 51:138–42, 1992; Aquaron R: Oculocutaneous albinism in Cameroon: A 15-year follow-up study. Ophthalmic Paediatr. Genet. 11:255–63, 1990). One report of atrial myxoma in a Japanese albino could be by chance or another of the myxoma syndromes (Mochizuki N et al.: A case of cardiac myxoma associated with congenital oculocutaneous albinism. Kokyu To Junkan 37:1147–49, 1989).

The evidence for a second nonallelic form of recessive albinism was provided by a family reported by 35,36:Trevor-Roper (1952, 1963). Two albino parents had 4 normally pigmented children. Assuming paternity (and blood groups provided no reason to question it), it is possible that the father had X-linked ocular albinism and a light complexion. However, if such were the case, his daughters, necessarily heterozygous for the X-linked gene, should have had the mosaic pigmentary pattern characteristic of the fundus oculi in the heterozygote. Such was not found. The existence of more than one locus may also be supported by the fact that the rate of parental consanguinity is higher than would be expected if only one locus were involved. Finally, applying the chemical method of Kugelman and Van Scott (1961), Witkop (1962) found suggestive evidence of separate forms of albinism. Witkop (1966) examined Trevor-Roper's family and found that, whereas the mother did not show pigmentation in the Kugelman-Van Scott test, the father did show pigment. Clinically, it is difficult to distinguish tyrosinase-positive from tyrosinase-negative albinism, especially in Caucasoids. Pigmented nevi in tyr-positive cases may be the only clue. In blacks with this form of albinism, the hair is yellow and many pigmented spots develop in the skin. There appears to be a block in the formation of eumelanin with a continuing formation of pheomelanin. In heterozygotes, Castle et al. (1988) could demonstrate no asymmetry on monocular testing of visual evoked potentials to suggest an abnormality of decussation. Witkop et al. (1989) stated that tyrosinase-positive albinism was the form present in the Hopi and Zuni Indians studied by Woolf (1965) and Woolf and Dukepoo (1969). This form was also found in the Brandywine triracial isolate (Witkop et al., 1972). In a study of 20 informative families, Heim et al. (1988) excluded linkage between the tyrosinase-positive oculocutaneous albinism locu s and the beta-globin locus (141900); the maximum lod score was -9.85 at theta = 0.05. The albino c locus, the p (pink-eyed dilution) locus, and the hemoglobin beta-chain locus constitute a conserved linkage group in the cat, rat, mouse, deermouse, and rabbit. The c and p loci in lower mammals have been considered equivalent to the tyrosinase-negative and tyrosinase-positive albinism loci in man, respectively. In the mouse, these loci are situated on chromosome 7 in the sequence centromere—p—c—Hbb. About 14 cM separate p and c, and about 8 cM separate c and Hbb. In a family with a balanced translocation, 46,XY,t(2;4)(q31.2;q31.22), Walpole and Mulcahy (1991) found a child with tyrosinase-positive albinism in addition to the familial translocation. They postulated that the OCA2 gene resides at either 2q31.1 or 4q31.22 and was disrupted by the translocation, but did not lead to albinism until it was paired up with a mutant gene at that locus through mating of a translocation carrier with a carrier of the mutant gene.

Tyrosinase-positive oculocutaneous albinism (OCA, type II) is the most prevalent type of albinism throughout the world. Lee et al. (1994) gave the overall frequency of OCA2 in the United States as approximately 1 per 36,000; however, the incidence is about 1 per 10,000 among African-Americans and is said to have a prevalence of 1/1,100 in the Ibo of Nigeria (Okoro, 1975) and a rate of about 1/3,900 in negroids of South Africa (17,18:Kromberg and Jenkins, 1982, 1984) where it is the most common recessive genetic disorder of this group. Throughout sub-Saharan Africa, it is responsible for a great deal of morbidity, with skin cancer and gross visual impairment being important sequelae. Whereas OCA2 is much less frequent among Caucasians, OCA1 (203100) is uncommon in African-Americans and Africans. In OCA2 in both Africans and Caucasians, some pigment is present at birth and lost later. On the contrary, most persons with tyrosinase-related albinism show completely white hair at birth, even though in some forms of tyrosinase deficiency a certain amount of pigment may develop later (King, 1992).

(It has been assumed, on the basis of a small number of observed matings, that tyrosinase-negative oculocutaneous albinism (203100) is distinct from the tyrosinase-positive form discussed in this entry. However, to their surprise, Tripathi et al. (1992) identified novel tyrosinase gene mutations in OCA type II patients from several different ethnic groups; specifically, 6 of 17 patients with 'tyrosinase-positive' OCA had tyrosinase gene mutations, none of which were common polymorphisms. Some of these mutations had previously been identified in association with type IB OCA; the remainder were entirely novel and would be expected to have relatively mild effects on tyrosinase enzymatic activity.)

Studying Bantu subjects in South Africa, Ramsay et al. (1992) demonstrated linkage of the OCA2 locus to 2 DNA markers, D15S10 and D15S13, in the region 15q11.2-q12. Since the pink-eyed dilution locus, p, on mouse chromosome 7, maps to a region of homology of synteny with human 15q, Ramsay et al. (1992) postulated that the OCA2 and p loci are homologous.

The pink-eye (or pink-eyed dilution) locus in the mouse is of historic significance because the first mammalian linkage to be discovered, by Haldane et al. (1915), was the linkage of this locus with the albinism locus. Subsequently, the gene was demonstrated to be on mouse chromosome 7 and the molecular nature of the mutation identified (Lyon et al., 1992; Gardner et al., 1992). Gardner et al. (1992) isolated cDNA clones from the p locus from murine melanoma and melanocyte libraries. The transcript from this gene was missing or altered in 6 independent mutant alleles of the p locus, suggesting that disruption of this gene is responsible for the hypopigmentation that defines mutant p alleles. Gardner et al. (1992) obtained the human counterpart of the murine p cDNA by screening a human melanoma cDNA library with a fragment of mouse genomic DNA. The predicted amino acid sequence of the human gene product showed 84% identity from amino acids 283 to 414 of the predicted mouse protein. During the course of experiments to establish the long-range restriction maps of the p locus in man and mouse by pulsed field gel electrophoresis, they noticed that probes derived from the p gene gave identical patterns of hybridization to those observed with the human genomic probe IR-10, designated as the D15S12 locus on human chromosome 15. Further studies indicated that DN10 is the human homolog of the mouse p gene. This locus had been mapped to 15q11.2-q12 (Donlon et al., 1986; Knoll et al., 1990; Robinson et al., 1991), the region of abnormality associated with Prader-Willi syndrome (PWS; 176270) and Angelman syndrome (AS; 105830). Hypopigmentation characterizes both of these disorders, but other features of PWS and AS are not found in the pink-eyed dilution in the mouse. Although some of the other features are apparent in mice with certain radiation-induced mutations of the p locus, none of the murine mutations appears to be affected by imprinting as is the case in the hum an. Mice heterozygous for mutant p alleles are fully pigmented regardless of the parental origin of the mutant allele. In a review, Brilliant (1992) suggested that mutation in the human homolog of the pink-eyed dilution gene may be responsible also for a form of hypomelanosis of Ito (146150). Gondo et al. (1993) studied the mouse pink-eyed unstable mutation, p(un), affecting coat color, which exhibits one of the highest reported reversion frequencies of any mammalian mutation and is associated with a duplication of genomic DNA at the p locus. They showed that DNA from p(un) was distinguished from wildtype and revertant DNA by a head-to-tail tandem duplication of approximately 70 kb. No differences were detected between revertant and wildtype DNAs. Thus, the reversion in phenotype was coupled with a loss of one copy of the 70-kb duplicated segment. They commented that the mouse p locus has a human homolog in D15S12 and that duplications of this gene occur in about 6% of Prader-Willi syndrome (176270) patients (Hamabe et al., 1991).

To characterize the protein encoded by the p gene, Rosemblat et al. (1994) performed immunoblot analysis of extracts of melanocytes cultured from wildtype mice with an antiserum from rabbits immunized with a peptide corresponding to amino acids 285 to 298 of the predicted protein product of the murine p gene. This antiserum recognized a 110-kD protein. The protein was absent from extracts of melanocytes cultured from mice with 2 allelic p mutations in which transcripts of the p gene are absent or greatly reduced. Introduction of the cDNA for the p gene into melanocytes carrying one of these mutations, by electroporation, resulted in expression of the 3.3-kb mRNA and the 110-kD protein. Upon subcellular fractionation of cultured melanocytes, the 110-kD protein was found to be present in melanosomes but absent from the vesicular fraction. Studies confirmed the hydrophobic nature of the protein. The results demonstrated that the p gene encodes an integral melanosomal membrane protein and established a framework by which mutations at this locus, which diminish pigmentation, can be analyzed at the cellular and biochemical levels.

Rinchik et al. (1993) demonstrated that the human cDNA DN10, linked to the p locus in mice, identifies the human homolog (P) of the mouse p gene, and appears to encode an integral membrane transporter protein. The human P protein is an 838-amino acid polypeptide that contains 12 putative transmembrane domains and exhibits structural homology to transporters of small organic molecules. In addition, Rinchik et al. (1993) reported a mutation in both copies of the human P gene in a case of OCA type II.

In linkage studies in negroid peoples of South Africa, Colman et al. (1993) excluded the tyrosinase locus (203100) on chromosome 11 and the CAS2/TRP1 locus (115501) on chromosome 9 as the site of the mutation in tyrosinase-positive oculocutaneous albinism.

In southern Africa, negroids with tyrosinase-positive ocular cutaneous albinism present with 2 distinctly different phenotypes, with or without darkly pigmented patches (ephelides, or dendritic freckles) on exposed areas of the skin. These phenotypes are concordant within families. Kromberg et al. (1989) found a correlation between the presence of ephelides and a lower risk of developing skin cancer, possibly as the result of the presence of some melanin pigment offering photoprotection. Kedda et al. (1994) analyzed for linkage with markers on 15q11-q13 in 41 southern African negroid families, each containing at least 1 affected tyrosinase-positive albino. The affected persons in 13 of the families had ephelides; those in 23 families did not have ephelides; and those in 5 families were of unknown ephelus status, with 100% concordance with respect to ephelus status in all 36 families. Analysis showed no obligatory crossovers between the alleles at the D15S12 locus and a tyrosinase-positive OCA, suggesting that the D15S12 locus is very close to or part of the disease locus, P. Unlike caucasoid 'tyrosinase-positive OCA' individuals, negroid tyrosinase-positive OCA individuals did not show evidence of locus heterogeneity. No haplotype showed statistically significant association with tyrosinase-positive OCA, and thus none could be used to predict the origins of the tyrosinase-positive OCA mutations.

Durham-Pierre et al. (1994) described a small deletion encompassing 1 exon of the P gene which they concluded is a frequent OCA2 mutation in both African-American and African patients. However, Lee et al. (1994) studied 7 unrelated African-American patients with OCA2 and identified different abnormalities of the P gene in all 7. In addition to the single exon deletion found by Durham-Pierre et al. (1994), they observed 2 large deletions, 2 small in-frame deletions, and 6 different point mutations.

Stevens et al. (1995) also found that the intragenic deletion first identified by Durham-Pierre et al. (1994) is a frequent cause of OCA2 in South African negroids, being found in 114 of 146 (78%) of OCA2 chromosomes. A common haplotype was found in 43 of 55 (78%) OCA2 chromosomes studied. Thus, the African origin of this allele was confirmed. On the basis of haplotype data, it appeared that at least 7 additional, less-frequent OCA2 mutations had occurred in this population. Spritz et al. (1995) found that the same 2.7-kb deletion allele accounted for most of the mutant P alleles in Tanzania. The 2.7-kb deletion includes exons 7 and results in a frameshift and premature truncation of the predicted polypeptide product. In a screening of filter blood spots from 470 newborn African-Americans in Michigan, Durham-Pierre et al. (1996) found that 2 were heterozygous for the 2.7-kb deletion.

Kagore and Lund (1995) found that the prevalence of OCA2 in school children in Harare, the capital city of Zimbabwe, was 1 in 2,833. On the basis of this prevalence, the gene frequency for OCA2 was estimated to be 0.0188, with a carrier frequency of 1 in 27. Most of the school children with albinism belonged to the majority Shoney ethnic group. As consanguineous marriages were discouraged in the Shoney culture, this high rate was probably the result of genetic drift in a relatively small population showing limited mobility.

Stevens et al. (1997) characterized OCA2 as the most common autosomal recessive disorder in the South African Negroid population, occurring with a prevalence of 1/3,900 persons. The common 2.7 kb P gene deletion was found in 0.013 (10/780) of the normally pigmented southern African Negroid population and at a lower frequency in individuals from central Africa (0.002; 2/834). The deletion was found in 131/170 (0.77) of OCA2 chromosomes in southern Africa, 11/14 (0.79) in Zambia, and 4/12 (0.33) in central African Republic. The study confirmed the African origin of the deletion allele. Haplotype analysis suggested that the deletion mutation occurred only once and that it arose before the divergence of these African populations, which was estimated to be about 2,000 to 3,000 years ago. The usually high frequency of OCA2 mutations, in particular the 2.7-kb deletion, suggested selection or genetic drift.

Lee et al. (1995) reported that the human P gene (OCA2) consists of 25 exons spanning between 250 and 600 kb of chromosome 15q11-q13. The 838 amino acid protein contains 12 transmembrane domains arranged similarly to various transporters and appears to be an integral membrane protein of melanosomes. Sequence comparisons suggested to Lee et al. (1995) that the P protein is one member of a family of transporters that includes an E. coli Na+/H+ antiporter and that it may be a tyrosine transporter. The authors also reported several nonpathologic polymorphisms.

See also Nance et al. (1970); Nance et al. (1971); Waardenburg (1961); Witkop et al. (1963)

Brilliant, M. H.: The mouse pink-eyed dilution locus: a model for aspects of Prader-Willi syndrome, Angelman syndrome, and a form of hypomelanosis of Ito. Mammalian Genome 3: 187-191, 1992.
Castle, D.; Kromberg, J.; Kowalsky, R.; Moosa, R.; Gillman, N.; Zwane, E.; Fritz, V.: Visual evoked potentials in Negro carriers of the gene for tyrosinase positive oculocutaneous albinism. J. Med. Genet. 25: 835-837, 1988.
Colman, M. A.; Stevens, G.; Ramsay, M.; Kwon, B.; Jenkins, T.: Exclusion of two candidate pigment loci, c and b, part of chromosome 11p, and 33 random polymorphic markers as the locus for tyrosinase-positive oculocutaneous albinism. Hum. Genet. 90: 556-560, 1993.
Donlon, T. A.; Lalande, M.; Wyman, A.; Bruns, G.; Latt, S. A.: Isolation of molecular probes associated with the chromosome 15 instability in the Prader-Willi syndrome. Proc. Nat. Acad. Sci. 83: 4408-4412, 1986.
Durham-Pierre, D.; Gardner, J. M.; Nakatsu, Y.; King, R. A.; Francke, U.; Ching, A.; Aquaron, R.; del Marmol, V.; Brilliant, M. H.: African origin of an intragenic deletion of the human P gene in tyrosinase positive oculocutaneous albinism. Nature Genet. 7: 176-179, 1994.
Durham-Pierre, D.; King, R. A.; Naber, J. M.; Laken, S.; Brilliant, M. H.: Estimation of carrier frequency of a 2.7 kb deletion allele of the P gene associated with OCA2 in African-Americans. Hum. Mutat. 7: 370-373, 1996.
Gardner, J. M.; Nakatsu, Y.; Gondo, Y.; Lee, S.; Lyon, M. F.; King, R. A.; Brilliant, M. H.: The mouse pink-eyed dilution gene: association with human Prader-Willi and Angelman syndromes. Science 257: 1121-1124, 1992.
Gondo, Y.; Gardner, J. M.; Nakatsu, Y.; Durham-Pierre, D.; Deveau, S. A.; Kuper, C.; Brilliant, M. H.: High-frequency genetic reversion mediated by a DNA duplication: the mouse pink-eyed unstable mutation. Proc. Nat. Acad. Sci. 90: 297-301, 1993.
Haldane, J. B. S.; Sprunt, A. D.; Haldane, N. M.: Reduplication in mice. J. Genet. 5: 133-135, 1915.
Hamabe, J.; Fukushima, Y.; Harada, N.; Abe, K.; Matsuo, N.; Nagai, T.; Yoshioka, A.; Tonoki, H.; Tsukino, R.; Niikawa, N.: Molecular study of the Prader-Willi syndrome: deletion, RFLP, and phenotype analyses of 50 patients. Am. J. Med. Genet. 41: 54-63, 1991.
Heim, R. A.; Dunn, D. S.; Candy, S. E.; Zwane, E.; Kromberg, J. G. R.; Jenkins, T.: The tyrosine-positive oculocutaneous albinism locus is not linked to the beta-globin locus in man. Hum. Genet. 79: 89, 1988.
Kagore, F.; Lund, P. M.: Oculocutaneous albinism among schoolchildren in Harare, Zimbabwe. J. Med. Genet. 32: 859-861, 1995.
Kedda, M. A.; Stevens, G.; Manga, P.; Viljoen, C.; Jenkins, T.; Ramsay, M.: The tyrosinase-positive oculocutaneous albinism gene shows locus homogeneity on chromosome 15q11-q13 and evidence of multiple mutations in Southern African negroids. Am. J. Hum. Genet. 54: 1078-1084, 1994.
King, R. A.: Personal Communication. Minneapolis, Minn. 12/31/1992.
Knoll, J. H. M.; Nicholls, R. D.; Magenis, R. E.; Glatt, K.; Graham, J. M., Jr.; Kaplan, L.; Lalande, M.: Angelman syndrome: three molecular classes identified with chromosome 15q11q13-specific DNA markers. Am. J. Hum. Genet. 47: 149-155, 1990.
Kromberg, J. G. R.; Castle, D.; Zwane, E. M.; Jenkins, T.: Albinism and skin cancer in southern Africa. Clin. Genet. 36: 43-52, 1989.
Kromberg, J. G. R.; Jenkins, T.: Prevalence of albinism in the South African Negro. S. Afr. Med. J. 61: 383-386, 1982.
Kromberg, J. G. R.; Jenkins, T.: Albinism in the South African Negro. III. Genetic counselling issues. J. Biosoc. Sci. 16: 99-108, 1984.
Kugelman, T. P.; Van Scott, E. J.: Tyrosinase activity in melanocytes of human albinos. J. Invest. Derm. 37: 73-76, 1961.
Lee, S.-T.; Nicholls, R. D.; Bundey, S.; Laxova, R.; Musarella, M.; Spritz, R. A.: Mutations of the P gene in oculocutaneous albinism, ocular albinism, and Prader-Willi syndrome plus albinism. New Eng. J. Med. 330: 529-534, 1994.
Lee, S.-T.; Nicholls, R. D.; Jong, M. T. C.; Fukai, K.; Spritz, R. A.: Organization and sequence of the human P gene and identification of a new family of transport proteins. Genomics 26: 354-363, 1995.
Lee, S.-T.; Nicholls, R. D.; Schnur, R. E.; Guida, L. C.; Lu-Kuo, J.; Spinner, N. B.; Zackai, E. H.; Spritz, R. A.: Diverse mutations of the P gene among African-Americans with type II (tyrosinase-positive) oculocutaneous albinism (OCA2). Hum. Molec. Genet. 3: 2047-2051, 1994.
Ludowese, C. J.; Thompson, K. J.; Sekhon, G. S.; Pauli, R. M.: Absence of predictable phenotypic expression in proximal 15q duplications. Clin. Genet. 40: 194-201, 1991.
Lyon, M. F.; King, T. R.; Gondo, Y.; Gardner, J. M.; Nakatsu, Y.; Eicher, E. M.; Brilliant, M. H.: Genetic and molecular analysis of recessive alleles at the pink-eyed dilution (p) locus of the mouse. Proc. Nat. Acad. Sci. 89: 6968-6972, 1992.
Nance, W. E.; Jackson, C. E.; Witkop, C. J., Jr.: Amish albinism: a distinctive autosomal recessive phenotype. Am. J. Hum. Genet. 22: 579-586, 1970.
Nance, W. E.; Witkop, C. J., Jr.; Rawls, R. F.: Genetic and biochemical evidence for two forms of oculocutaneous albinism in man.In: The Clinical Delineation of Birth Defects. Eye. Baltimore: Williams and Wilkins (pub.) VIII: 1971. Pp. 125-128.
Okoro, A. N.: Albinism in Nigeria: a clinical and social study. Brit. J. Derm. 92: 485-492, 1975.
Ramsay, M.; Colman, M.-A.; Stevens, G.; Zwane, E.; Kromberg, J.; Farrall, M.; Jenkins, T.: The tyrosinase-positive oculocutaneous albinism locus maps to chromosome 15q11.2-q12. Am. J. Hum. Genet. 51: 879-884, 1992.
Rinchik, E. M.; Bultman, S. J.; Horsthemke, B.; Lee, S.-T.; Strunk, K. M.; Spritz, R. A.; Avidano, K. M.; Jong, M. T. C.; Nicholls, R. D.: A gene for the mouse pink-eyed dilution locus and for human type II oculocutaneous albinism. Nature 361: 72-76, 1993.
Robinson, W. P.; Bottani, A.; Yagang, X.; Balakrishman, J.; Binkert, F.; Machler, M.; Prader, A.; Schinzel, A.: Molecular, cytogenetic, and clinical investigations of Prader-Willi syndrome patients. Am. J. Hum. Genet. 49: 1219-1234, 1991.
Rosemblat, S.; Durham-Pierre, D.; Gardner, J. M.; Nakatsu, Y.; Brilliant, M. H.; Orlow, S. J.: Identification of a melanosomal membrane protein encoded by the pink-eyed dilution (type II oculocutaneous albinism) gene. Proc. Nat. Acad. Sci. 91: 12071-12075, 1994.
Spritz, R. A.; Fukai, K.; Holmes, S. A.; Luande, J.: Frequent intragenic deletion of the P gene in Tanzanian patients with type II oculocutaneous albinism (OCA2). Am. J. Hum. Genet. 56: 1320-1323, 1995.
Stevens, G.; Ramsay, M.; Jenkins, T.: Oculocutaneous albinism (OCA2) in sub-Saharan Africa: distribution of the common 2.7-kb P gene deletion mutation. Hum. Genet. 99: 523-527, 1997.
Stevens, G.; van Beukering, J.; Jenkins, T.; Ramsay, M.: An intragenic deletion of the P gene is the common mutation causing tyrosinase-positive oculocutaneous albinism in Southern African Negroids. Am. J. Hum. Genet. 56: 586-591, 1995.
Trevor-Roper, P. D.: Marriage of two complete albinos with normally pigmented offspring. Brit. J. Ophthal. 36: 107-108, 1952.
Trevor-Roper, P. D.: Albinism. Proc. Roy. Soc. Med. 56: 21-24, 1963.
Tripathi, R. K.; Droetto, S.; Spritz, R. A.: Many patients with 'tyrosinase-positive' oculocutaneous albinism have tyrosinase gene mutations. (Abstract) Am. J. Hum. Genet. 51 (suppl.): A179, 1992.
Waardenburg, P. J.: Genetics and Ophthalmology. Springfield, Ill.: Charles C Thomas (pub.) 1: 1961. Pp. 732 only.
Walpole, I. R.; Mulcahy, M. T.: Tyrosinase positive albinism with familial 46,XY,t(2;4)(q31.2;q31.22) balanced translocation. J. Med. Genet. 28: 482-484, 1991.
Witkop, C. J., Jr.: Dental problems of an hereditary nature.In: Witkop, C. J.: Genetics and Dental Health. New York: McGraw-Hill (pub.) 1962.
Witkop, C. J., Jr.: Personal Communication. Minneapolis, Minnesota 1966.
Witkop, C. J., Jr.; Niswander, J. D.; Bergsma, D. R.; Workman, P. L.; White, J. G.: Tyrosinase-positive oculocutaneous albinism among the Zuni and the Brandywine triracial isolate: biochemical and clinical characteristics and fertility. Am. J. Phys. Anthrop. 36: 397-405, 1972.
Witkop, C. J., Jr.; Quevedo, W. C., Jr.; Fitzpatrick, T. B.; King, R. A.: Albinism.In: Scriver, C. R.; Beaudet, A. L.; Sly, W. S.; Valle, D.: The Metabolic Basis of Inherited Disease. New York: McGraw-Hill (pub.) (6th ed.) II: 1989. Pp. 2905-2947.
Witkop, C. J., Jr.; Van Scott, E. J., Jr.; Jacoby, G. A.: Evidence for two forms of autosomal recessive albinism in man.(Abstract) Proc. Second Int. Cong. Hum. Genet., Rome, Sept. 6-12, 1961 2: 1064-1065, 1963.
Woolf, C. M.: Albinism among Indians in Arizona and New Mexico. Am. J. Hum. Genet. 17: 23-35, 1965.
Woolf, C. M.; Dukepoo, F. C.: Hopi Indians, inbreeding and albinism. Science 164: 30-37, 1969.

HERMANSKY-PUDLAK SYNDROME (*203300)

JJM NOTE

One patient with this syndrome had cutaneous malignant melanoma (Kinnear and Tuddenham 1985).

Hermansky and Pudlak (1959) described 2 unrelated albinos with lifelong bleeding tendency and peculiar pigmented reticular cells in the bone marrow as well as in lymph node and liver biopsies. One was male and one female; both were 33 years old. The female has since died and was found to have large amounts of the pigment in reticuloendothelial cells everywhere and in the walls of small blood vessels (Hermansky, 1963). Two families, each with 2 sibs affected with this syndrome, had come to Hermansky's attention (1963). This syndrome is clearly different from the Chediak-Higashi syndrome (214500) because no qualitative changes of leukocytes are found in Hermansky syndrome and no pigmented macrophages are found in the Chediak-Higashi syndrome. Report of a family by Verloop et al. (1964) supports this conclusion. Logan et al. (1971) described a patient with albinism and bleeding diathesis in whom a defect in platelet ADP-release was demonstrated. Prolonged bleeding time and defective platelet aggregation were found. Two other albino patients with a defect in ADP-release had been reported, as well as 12 albino patients with prolonged bleeding. In 6 of 7 in whom the bone marrow was studied, histiocytes were found to contain abnormal granules. Weiss et al. (1979) studied the platelet defect, which they called delta-storage pool disease, in 7 patients; 6, including 3 sisters, were of Puerto Rican ancestry. Four other unrelated patients had the same platelet defect but did not have albinism. Depinho and Kaplan (1985) reported 3 affected sibs from a consanguineous Puerto Rican kindred. The proband, a 31-year-old woman with 2 children, had fatal restrictive lung disease, a complication pointed out by others (Davies and Tuddenham, 1976; Garay et al., 1979). Indeed, the first patient (Hermansky and Pudlak, 1959) was a 33-year-old farmer who developed chronic interstitial pulmonary fibrosis. Davies and Tuddenham (1976) described 4 sibs with HPS-associated pulmonary fibrosis. The onset of pulmonary fibrosis is most often in the third or fourth decade.

Inflammatory bowel disease, with onset of symptoms between 12 and 30 years of age, has been reported by several (e.g., Schinella et al., 1980). Most of the patients with HPS-related bowel disease have been Puerto Rican. Response to medical therapy is said to be poor. Epistaxis is the most frequent hemorrhagic manifestation. All 3 sibs studied by Depinho and Kaplan (1985) had recurrent infections and were anergic. Vitamin E was thought to be of some benefit for the hemorrhagic problem (Depinho and Kaplan, 1985). Wijermans and van Dorp (1989) reported that 1-desamino-8D-arginine vasopressin (DDAVP) is useful in the treatment of some cases of HPS. The authors suggested that it be combined with drugs that inhibit fibrinolysis because DDAVP also increases plasminogen activator activity. Kinnear and Tuddenham (1985) reported 4 cases. Cutaneous malignant melanoma developed in 1. The ocular features were similar to those of the more common form of tyrosinase-positive oculocutaneous albinism (203200). The triad of the syndrome is albinism, platelets lacking dense bodies, and storage of ceroid-like material in tissues. The manifestations of the storage disease include ulcerative colitis, restrictive lung disease, kidney failure, and cardiomyopathy. The autofluorescent material stored in HPS is histochemically similar to that stored in neuronal ceroid lipofuscinosis (204200). The Hermansky-Pudlak syndrome appears to be a disorder of membranes of lysosomes, melanosomes, and platelet dense bodies. It is said to have a frequency of about 1 in 2,000 among Puerto Ricans (Witkop, 1986). The origin of HPS in Puerto Rico has been traced to a region of southern Spain, and a connection to cases in Holland is likely (King, 1987). Mahadeo et al. (1991) reported 2 children, aged 7 years and 3 years, with granulomatous colitis in association with HPS. One of the children was Puerto Rican.

White et al. (1992) showed that the 'pallid' mutation of the mouse is a platelet storage pool disease associated with the protein 4.2 gene (177070), referred to as pallidin. It may be an authentic model for the Hermansky-Pudlak syndrome. White et al. (1992) mapped the murine gene for protein 4.2 (Epb-4.2) to chromosome 2. The homologous gene maps to human chromosome 15 which has an extensive segment of homology to mouse chromosome 2.

Hearing (1993) reviewed the complexities of melanogenesis, including the parallels between mouse and the human. The complex nature of the process was anticipated since, in the mouse, more than 150 distinct mutations affect visible pigmentation, and those occur at more than 50 genetic loci.

Nishibori et al. (1993) found evidence to suggest that the primary defect may reside in CD63, also known as melanoma-associated antigen (155740). However, whereas CD63 maps to chromosome 12, Fukai et al. (1995) found that HPS maps to chromosome 10. They used the linkage disequilibrium mapping approach to localize the HPS gene in 2 groups in whom the disorder is particularly frequent: a group in Puerto Rico and a group in an isolated village in the Swiss Alps. They localized the HPS gene in both groups to a 0.6 cM interval in chromosome segment 10q23.1-q23.2. These results indicated that the Puerto Rican and Swiss forms of HPS are either allelic or the result of mutations in very closely linked genes. This region of distal 10q is syntenic to the region of mouse chromosome 19 that includes 'pale ear' (ep) and 'ruby-eye' (ru), which must be considered potential murine homologs of human HPS. Wildenberg et al. (1995) likewise mapped the HPS gene to 10q. They collected blood samples from a relatively homogeneous population in Puerto Rico where HPS has a frequency of about 1/1,800, giving a carrier frequence estimated to be 1/21. Analysis of pooled DNA samples allowed them to screen the genome rapidly for candidate loci, and identify linkage with a marker on 10q. The result was verified with additional markers, and a maximum lod score of 5.07 at theta = 0.001 was calculated for marker D10S198. Haplotype analysis placed the HPS gene in a region of approximately 14 cM that contains the markers D10S198 and D10S1239.

The Hermansky-Pudlak syndrome observed in the isolated Swiss Alps village (Lattion et al., 1983 and Schallreuter et al., 1993) usually shows a relatively mild clinical course with normal life expectancy and the patients lack manifestations of ceroid storage. The 2 most common complications associated with ceroid deposition in the usual form of HPS are pulmonary fibrosis and granulomatous colitis, and the major cause of death in HPS after 1 year of age is pulmonary fibrosis. There is striking variability in pigmentation, which can vary from an almost total absence of pigment to an amount that is nearly normal. Obligate heterozygotes are normally pigmented. The third feature of HPS is the absence of dense bodies in the platelets.

HPS may be the most frequent single-gene disorder in Puerto Rico and is frequent in an isolated mountain village high in the Swiss Alps (Schallreuter et al., 1993). Oh et al. (1996) identified the HPS gene by positional cloning and found homozygous frameshifts in this gene in Puerto Rican, Swiss, Irish, and Japanese HPS patients. The HPS polypeptide is a novel transmembrane protein that is likely to be a component of multiple cytoplasmic organelles and is apparently crucial for their normal development and function. The different clinical phenotypes associated with different HPS frameshifts suggested that differentially truncated HPS polypeptides may have somewhat different consequences for subcellular function.

As indicated earlier, 2 genetically distinct mouse loci, 'pale ear' (ep) and 'ruby-eye' (ru), both with mutant phenotypes similar to human HPS, map close together in the homologous region of murine chromosome 19, suggesting that 1 of these loci might be homologous to human HPS. Feng et al. (1997) characterized the mouse Hps cDNA and genomic locus, and identified pathologic Hps gene mutations in ep but not in ru mice, establishing mouse 'pale ear' as an animal model for human HPS. The phenotype of homozygous ep mutant mice encompasses those of both HPS and Chediak-Higashi syndrome, suggesting that these disorders may be closely related. In addition, the mouse and human HPS genes both contain a rare 'AT-AC' intron (i.e., an intron with nonconsensus splice sites), and comparison of the sequences of this intron in the mouse and human genes identified conserved sequences that suggest a possible role for pre-mRNA secondary structure in excision of this rare class of introns. (See 601428 and 601429 for a discussion of small nuclear RNAs necessary for AT-AC intron splicing and see Hall and Padgett (1994) for a discussion of introns with nonconsensus splice sites.)

Hazelwood et al. (1997) ascertained 2 individuals with Hermansky-Pudlak syndrome from central Puerto Rico who lacked the 16-bp duplication (203300.0001), exhibited significant amounts of normal-size HPS gene mRNA by Northern blot analysis, and had haplotypes in the HPS region of chromosome 10 that were different from the haplotype of every 16-bp duplication patient. Moreover, these 2 individuals displayed no mutations in their HPS cDNA sequences. Both patients exhibited pigment dilution, impaired visual acuity, nystagmus, bleeding diathesis, and absent platelet dense bodies, confirming the diagnosis of HPS. The findings appear to indicate locus heterogeneity of this disorder in Puerto Rico, consistent with the existence of several mouse strains manifesting both pigment dilution and a platelet storage-pool deficiency.

Oh et al. (1998) performed mutation analysis on 44 unrelated Puerto Rican and 24 unrelated non-Puerto Rican HPS patients. A 16-bp frameshift duplication (203300.0001), the result of an apparent founder effect, was nearly ubiquitous among Puerto Rican patients. A frameshift at codon 322 may be the most frequent HPS mutation in Europeans. The mutation in these cases was a 1-bp insertion (or duplication) in a poly(C) tract at codons 322 to 324.

Oh et al. (1998) also described 6 novel HPS mutations: a 5-prime splice-junction mutation of IVS5, 3 frameshifts, a nonsense mutation, and a 1-codon in-frame deletion. These mutations defined an apparent frameshift hotspot at codons 321-322. Overall, however, they detected mutations in the HPS gene in only about half of non-Puerto Rican patients, and presented evidence suggesting locus heterogeneity for HPS.

One evidence of locus heterogeneity for HPS was provided by findings of homozygosity analysis of 4 inbred HPS patients in whom Oh et al. (1998) detected no mutations. If these patients were homozygous (by descent) for occult HPS mutations, they should also be homozygous for the polymorphic markers that Oh et al. (1998) had found immediately flanking the gene. However, 3 of these patients were heterozygous for these markers, thus apparently excluding the HPS locus in these cases. Furthermore, genetic linkage analysis of the extended family of 1 of these 4 patients, as well as in another non-inbred family, showed no evidence for linkage.

Wildenberg et al. (1998) demonstrated an apparent alternate transcript of the HPS gene. By RT-PCR and Northern blot analysis, 2 polyadenylated transcripts were found in normal human melanocytes, human bone marrow cells, human melanoma cells, lymphoblastoid cell lines, and megakaryocytic leukemia cells. Thus, the HPS gene is complex and may have more than 1 biologically active transcript. The existence of alternative transcripts has been observed also in the gene that is mutant in Chediak-Higashi syndrome (214500) by Barbosa et al. (1997). Wildenberg et al. (1998) noted that the same type of alternative splicing occurs in the IGHM gene (147020), resulting in membrane-bound and secreted protein products.

See also Rendu et al. (1978); Takahashi and Yokoyama (1984); White et al. (1971)

Barbosa, M. D. F. S.; Barrat, F. J.; Tchernev, V. T.; Nguyen, Q. A.; Mishra, V. S.; Colman, S. D.; Pastural, E.; Dufourcq-Lagelouse, R.; Fischer, A.; Holcombe, R. F.; Wallace, M. R.; Brandt, S. J.; de Saint Basile, G.; Kingsmore S. F.: Identification of mutations in two major mRNA isoforms of the Chediak-Higashi syndrome gene in human and mouse. Hum. Molec. Genet. 6: 1091-1098, 1997.

Davies, B. H.; Tuddenham, E. G. D.: Familial pulmonary fibrosis associated with oculocutaneous albinism and platelet function defect: a new syndrome. Quart. J. Med. 45: 219-232, 1976.

Depinho, R. A.; Kaplan, K. L.: The Hermansky-Pudlak syndrome: report of three cases and review of pathophysiology and management considerations. Medicine 64: 192-202, 1985.

Feng, G. H.; Bailin, T.; Oh, J.; Spritz, R. A.: Mouse pale ear (ep) is homologous to human Hermansky-Pudlak syndrome and contains a rare 'AT-AC' intron. Hum. Molec. Genet. 6: 793-797, 1997.

Fukai, K.; Oh, J.; Frenk, E.; Almodovar, C.; Spritz, R. A.: Linkage disequilibrium mapping of the gene for Hermansky-Pudlak syndrome to chromosome 10q23.1-q23.3. Hum. Molec. Genet. 4: 1665-1669, 1995.

Gahl, W. A.; Brantly, M.; Kaiser-Kupfer, M. I.; Iwata, F.; Hazelwood, S.; Shotelersuk, V.; Duffy, L. F.; Kuehl, E. M.; Troendle, J.; Bernardini, I.: Genetic defects and clinical characteristics of patients with a form of oculocutaneous albinism (Hermansky-Pudlak syndrome). New Eng. J. Med. 338: 1258-1264, 1998.

Garay, S. M.; Gardella, J. E.; Fazzini, E. P.; Goldring, R. M.: Hermansky-Pudlak syndrome: pulmonary manifestations of a ceroid storage disorder. Am. J. Med. 66: 737-747, 1979.

Hall, S. L.; Padgett, R. A.: Conserved sequences in a class of rare eukaryotic nuclear introns with non-consensus splice sites. J. Molec. Biol. 239: 357-365, 1994.

Hazelwood, S.; Shotelersuk, V.; Wildenberg, S. C.; Chen, D.; Iwata, F.; Kaiser-Kupfer, M. I.; White, J. G.; King, R. A.; Gahl, W. A.: Evidence for locus heterogeneity in Puerto Ricans with Hermansky-Pudlak syndrome. Am. J. Hum. Genet. 61: 1088-1094, 1997.

Hearing, V. J.: Unraveling the melanocyte. (Editorial) Am. J. Hum. Genet. 52: 1-7, 1993.

Hermansky, F.: Personal Communication. Prague, Czechoslovakia 1963.

Hermansky, F.; Pudlak, P.: Albinism associated with hemorrhagic diathesis and unusual pigmented reticular cells in the bone marrow: report of two cases with histochemical studies. Blood 14: 162-169, 1959.

King, R. A.: Personal Communication. Minneapolis, Minn. 6/15/1987.

Kinnear, P. E.; Tuddenham, E. G. D.: Albinism with haemorrhagic diathesis: Hermansky-Pudlak syndrome. Brit. J. Ophthal. 69: 904-908, 1985.

Lattion, F.; Schneider, P.; Da Prada, M.; Lorez, H. P.; Richards, J. G.; Picotti, G. B.; Frenck, E.: Syndrome d'Hermansky-Pudlak dans un village valaisan. Helv. Paediat. Acta 38: 495-512, 1983.

Logan, L. J.; Rapaport, S. I.; Maher, I.: Albinism and abnormal platelet function. New Eng. J. Med. 284: 1340-1345, 1971.

Mahadeo, R.; Markowitz, J.; Fisher, S.; Daum, F.: Hermansky-Pudlak syndrome with granulomatous colitis in children. J. Pediat. 118: 904-906, 1991.

Nishibori, M.; Cham, B.; McNicol, A.; Shalev, A.; Jain, N.; Gerrard, J. M.: The protein CD63 is in platelet dense granules, is deficient in a patient with Hermansky-Pudlak syndrome, and appears identical to granulophysin. J. Clin. Invest. 91: 1775-1782, 1993.

Oh, J.; Bailin, T.; Fukai, K.; Feng, G. H.; Ho, L.; Mao, J.; Frenk, E.; Tamura, N.; Spritz, R. A.: Positional cloning of a gene for Hermansky-Pudlak syndrome, a disorder of cytoplasmic organelles. Nature Genet. 14: 300-306, 1996.

Oh, J.; Ho, L.; Ala-Mello, S.; Amato, D.; Armstrong, L.; Bellucci, S.; Carakushansky, G.; Ellis, J. P.; Fong, C.-T.; Green, J. S.; Heon, E.; Legius, E.; Levin, A. V.; Nieuwenhuis, H. K.; Pinckers, A.; Tamura, N.; Whiteford, M. L.; Yamasaki, H.; Spritz, R. A.: Mutation analysis of patients with Hermansky-Pudlak syndrome: a frameshift hot spot in the HPS gene and apparent locus heterogeneity. Am. J. Hum. Genet. 62: 593-598, 1998.

Rendu, F.; Breton-Gorius, J.; Trugman, G.; Castro-Melaspina, H.; Andrieu, J.-M.; Bereziat, G.; Libret, M.; Caen, J. P.: Studies on a new variant of the Hermansky-Pudlak syndrome: quantitative, ultrastructural, and functional abnormalities of the platelet-dense bodies associated with a phospholipase A defect. Am. J. Hemat. 4: 387-399, 1978.

Schallreuter, K. U.; Frenk, E.; Wolfe, L. S.; Witkop, C. J.; Wood, J. M.: Hermansky-Pudlak syndrome in a Swiss population. Dermatology 187: 248-256, 1993.

Schinella, R. A.; Greco, M. A.; Cobert, B. L.; Denmark, L. W.; Cox, R. P.: Hermansky-Pudlak syndrome with granulomatous colitis. Ann. Intern. Med. 92: 20-23, 1980.

Shotelersuk, V.; Hazelwood, S.; Larson, D.; Iwata, F.; Kaiser-Kupfer, M. I.; Kuehl, E.; Bernardini, I.; Gahl, W. A.: Three new mutations in a gene causing Hermansky-Pudlak syndrome: clinical correlations. Molec. Genet. Metab. 64: 99-107, 1998.

Takahashi, A.; Yokoyama, T.: Hermansky-Pudlak syndrome with special reference to lysosomal dysfunction: a case report and review of the literature. Virchows Arch. Path. Anat. 402: 247-258, 1984.

Verloop, M. C.; Von Wieringen, A.; Vuylsteke, J.; Hart, H. C.; Huizinga, J.: Albinismus, hemorrhagische Diathese und anormale Pigmentzellen im Knockenmark. Med. Klin. 59: 408-412, 1964.

Weiss, H. J.; Witte, L. D.; Kaplan, K. L.; Lages, B. A.; Chernoff, A.; Nossel, H. L.; Goodman, D. S.; Baumgartner, H. R.: Heterogeneity in storage pool deficiency: studies in granule-bound substances in 18 patients including variants deficient in alpha-granules, platelet factor 4, beta-thromboglobin, and platelet-derived growth factor. Blood 54: 1296-1319, 1979.

White, J. G.; Edson, J. R.; Desnick, S. J.; Witkop, C. J., Jr.: Studies of platelets in a variant of the Hermansky-Pudlak syndrome. Am. J. Path. 63: 319-332, 1971.

White, R. A.; Peters, L. L.; Adkison, L. R.; Korsgren, C.; Cohen, C. M.; Lux, S. E.: The murine pallid mutation is a platelet storage pool disease associated with the protein 4.2 (pallidin) gene. Nature Genet. 2: 80-83, 1992.

Wijermans, P. W.; van Dorp, D. B.: Hermansky-Pudlak syndrome: correction of bleeding time by 1-desamino-8D-arginine vasopressin. Am. J. Hemat. 30: 154-157, 1989.

Wildenberg, S. C.; Fryer, J. P.; Gardner, J. M.; Oetting, W. S.; Brilliant, M. H.; King, R. A.: Identification of a novel transcript produced by the gene responsible for the Hermansky-Pudlak syndrome in Puerto Rico. J. Invest. Derm. 110: 777-781, 1998.

Wildenberg, S. C.; Oetting, W. S.; Almodovar, C.; Krumwiede, M.; White, J. G.; King, R. A.: A gene causing Hermansky-Pudlak syndrome in a Puerto Rican population maps to chromosome 10q2. Am. J. Hum. Genet. 57: 755-765, 1995.

Witkop, C. J., Jr.: Personal Communication. Minneapolis, Minn. 2/1986.

XERODERMA PIGMENTOSUM 1 (*278700)

JJM NOTE

Of the 18 OMIM entries that mention *xeroderma* or *excision-repair,* all might be suspected of having an excess of skin cancer; in fact, only 7 entries do. All types of skin cancers have been seen to excess: squamous and basal cell carcinomas, as well as cutaneous and ocular malignant melanomas. Another rare tumor is carcinoma of the tip (not the base) of the tongue, also likely explained by ultraviolet exposure. Lip cancer is common also.

Sensitivity to sunlight with the development of carcinomata at an early age is a leading feature of xeroderma pigmentosum. Onset, with freckle-like lesions in exposed areas, usually occurs in the first years of life. The possibility of partial sex-linked recessive inheritance was suggested by Haldane (1936) but is now considered unlikely. Parental consanguinity is frequent. The sex ratio is about 1. Ruder (cited by Cockayne, 1933) observed the condition in 7 out of 13 sibs. It is not clear whether heterozygotes show changes; increased freckling has been claimed to be such a manifestation. Cleaver (1968) showed that whereas normal skin fibroblasts can repair ultraviolet radiation damage to DNA by inserting new bases into DNA, cells from patients with xeroderma pigmentosum lack this capacity or have a much reduced capacity for repair. Goldstein and Lin (1972) showed that XP-hamster hybrid cells had normal repair and survived ultraviolet irradiation.

Cell-fusion studies indicate that complementation can take place between the fibroblasts from certain pairs of patients. Merely mixing cells does not effect repair of DNA, but this can be done by fusion of cells. By 1974, 4 complementation groups had been found, indicating that mutation at any one of at least 4 loci can cause defective DNA repair leading to the clinical state of xeroderma pigmentosum (Robbins et al., 1974). In addition, patients with xeroderma pigmentosum and a normal in vitro DNA repair rate (see 278750) were observed (Robbins et al., 1974; Cleaver, 1972). The 4 complementation groups, called A-D, showed no consistent clinical correlations (as to neurologic signs, for example) but did show correlations with DNA repair rate: group A—less than 2% of normal; group B—3 to 7%; group C—10 to 25%; group D—25 to 55%. By 1975, 5 forms of XP with deficiency in excision repair had been identified by cell fusion studies of complementation (Kraemer et al., 1975; Cleaver, 1975). By 1979, 7 complementation groups (A-G) were identified in the class of XP patients with defective excision of pyrimidine dimers (excision-deficient XP). In addition, at least one form has a defect in postreplicative repair. Bootsma and Hoeijmakers (1991) compared the clinical and cell-biological characteristics of 7 excision repair-deficient XP complementation groups (XPD included the patient originally classified as group H). XPA and XPC have high relative frequency; XPD and XPF (278760) have intermediate frequency, and the other 3 are rare. For each of the 7 types, the occurrence of clinical manifestations in 4 areas (skin cancer, neurologic abnormality, symptoms of Cockayne syndrome, and trichothiodystrophy) and the levels of unscheduled DNA synthesis after UV exposure and of UV sensitivity were tabulated.

Andrews et al. (1978) demonstrated a correlation between the degree of sensitivity of cultured fibroblasts to UV (as measured by colony forming ability) and the severity of neurologic manifestations. From this they concluded that DNA repair is required for maintenance of the functional integrity of the nervous system by preventing premature death of neurons. Kanda et al. (1990) indicated that most patients with group A XP, the most common form in Japan, show involvement of the central and peripheral nervous systems in addition to cutaneous lesions (De Sanctis-Cacchione syndrome; 278800). Kanda et al. (1990) detailed the pathologic changes in the peripheral nervous system in this disorder. Stefanini et al. (1980) observed a discrepancy in clinical severity and residual enzyme activity in sibs and suggested somatic back-mutation in the milder sib.

Some aspect of DNA repair mechanisms is deficient in 4 other inherited diseases: ataxia-telangiectasia (208900), Fanconi anemia (227650),

Bloom syndrome (210900), and Cockayne syndrome (216400). Welshimer and Swift (1982) studied families of homozygotes for ataxia-telangiectasia (AT), Fanconi anemia (FA), and xeroderma pigmentosum (XP) to test the hypothesis that heterozygotes may be predisposed to some of the same congenital malformations and developmental disabilities that are common among homozygotes. Among XP relatives, 11 of 1100 had unexplained mental retardation, whereas only 3 of 1439 relatives of FA and AT homozygotes showed mental retardation. Four XP relatives and no FA or AT relatives had microcephaly. Idiopathic scoliosis and vertebral anomalies occurred in excess in AT relatives, while genitourinary and distal limb malformations were found in FA families. Salob et al. (1992) reported a Pakistani girl, the offspring of second cousins, who developed photosensitivity at the age of 2 years; at the age of 9 years, she was found to have typical xeroderma pigmentosum together with bone marrow hypoplasia of a type seen in Fanconi anemia. Salob et al. (1992) suggested that this was not a chance association but that the bone marrow hypoplasia reflected an increased susceptibility of the patient's hematopoietic stem cells to non-UV mutagens.

El-Hefnawi et al. (1965) presented useful Egyptian pedigrees and suggested linkage with the ABO blood group locus. Bootsma and Keijzer (1979) studied 8 patients from 6 Egyptian families. Three were assigned to complementation group A and 5 to group C. Thus, El-Hefnawi's 34 families may have comprised at least 2 different complementation groups. Groups A and C are, the world over, the most common XP mutants. Thus, for at least 2 reasons, 'XP Egyptian' is not a useful, meaningful, or reliable designation.

Keijzer et al. (1982) observed that cytoplasts from normal human fibroblasts can complement the defect of xeroderma pigmentosum A cells, with fast kinetics, after fusion with UV-irradiated XPA cells. The genetic determination of XPA complementation was studied by Keijzer et al. (1984) using cytoplasts from hamster-human hybrid cells with various chromosomal constitution. The X chromosome, earlier thought to be involved (Stefanini et al., 1982), was excluded and strong evidence was found for localization of the genetic determinant on the long arm of chromosome 1. Is this to be taken to mean that the gene mutant in XPA is on chromosome 1? Alternatively, chromosome 13 may be the site of the mutation in XPA. A lesion in repair of UV damage in the mouse that seems to be homologous to the human disorder was corrected by normal human chromosome 13 (Hori et al., 1983; see 133530). Keijzer et al. (1987) fused human, Chinese hamster or Chinese hamster/human hybrid cytoplasts with UV-irradiated xeroderma pigmentosum group A cells. Unscheduled DNA synthesis in the XP nucleus showed correction of the defect immediately after fusion of human cytoplasts, whereas the Chinese hamster cytoplasts did not show this rapid increase in excision repair. The results obtained after fusion of cytoplasts isolated from a panel of 26 Chinese hamster-human hybrids showed that chromosome 1 bears genetic information that is necessary for the rapid correction of the XP defect. This genetic information was regionally assigned to 1q42-qter by analyzing hybrid cell lines retaining various segments of chromosome 1. Correction was also observed with cytoplasts from a hybrid cell made between a Chinese hamster cell and an XP cell containing human chromosome 1. This result suggested that the correcting factor consists of both human and Chinese hamster components and that the gene mapped to chromosome 1 may not be the one that is mutated in XP of the A group. In fact, the mapping to chromosome 1 was never con firmed (Cleaver, 1993) and subsequently (see later) the gene was mapped to chromosome 9.

In studies performed in mouse-human hybrid cell lines, Kaur and Athwal (1989) achieved complementation of the repair defect in xeroderma pigmentosum cells of complementation group A by the transfer of human chromosome 9. Obviously, this result is in conflict with the suggestion that the XPA gene is located on 1q. Henning et al. (1990) showed that microcell-mediated transfer of a single rearranged human chromosome from a human-mouse somatic cell hybrid resulted in specific complementation of defective repair and UV sensitivity in XPA cells. Preliminary studies had shown that this complementing chromosome contained a translocation involving human chromosome 9 and suggested that the XPA gene was located on 9q. Cytogenetic analysis of the complementing human chromosome suggested location of the gene on segment 9q22.2-q34.3. Southern blot analysis using probes for genes known to map in this area confirmed this location and indicated that the XPA locus is distal to ASSP3, a pseudogene for argininosuccinate synthetase gene (215700), located at 9q11-q22. 74,73:Tanaka et al. (1989, 1990) cloned a mouse gene that on transfection into either of 2 group-A XP cell lines restored resistance to ultraviolet light. No correction was observed in relation to groups C, D, F, or G. Tanaka et al. (1989) concluded that they had succeeded in cloning the mouse homolog of the gene that is mutant in XPA. By transferring single human chromosomes derived from normal human fibroblasts into XPA cells by means of microcell fusion, Ishizaki et al. (1990) demonstrated that chromosomes 1 and 12 had no effect on UV sensitivity of microcell hybrid cells, whereas after transferring chromosome 9, 7 of 11 recipient clones were as UV-resistant as normal human cells. Southern hybridization analysis using the polymorphic probe homologous to a sequence at the D9S17 locus on chromosome 9 confirmed that at least part of a normal human chromosome 9 had been transferred into recipient clones. Tanaka et al. (1990) cloned cDNAs for human and mouse XPAC. They found a 1.0-1.1 kb mRNA in normal mouse cells and 2 mRNAs, 1.3-1.4 kb and 1.0-1.1 kb, in normal human cells. Almost all group A cell lines tested showed abnormality or absence of XPAC mRNA. Tanaka et al. (1990) localized the human and mouse XPAC genes to 9q34.1 and 4C2, respectively, by in situ hybridization. The significance of the observation of Kaur et al. (1992) of partial complementation in XP group A cells by a gene that mapped to human chromosome 8 is unclear.

Farndon et al. (1994) stated that the XPAC gene, like the basal cell nevus syndrome gene (NBCCS; 109400), the gene for self-healing squamous epithelioma (ESS1; 132800), and the group C Fanconi anemia gene (FACC; 227645), all map to 9q22.3-q31.

The XPA-correcting gene encodes a hydrophilic protein of relative molecular mass 31,000. It displays a distinct zinc finger motif, indicating that it interacts directly with DNA, presumably as part of an enzyme complex that makes an incision near damaged sites. Most Japanese patients with type A XP were found by Tanaka et al. (1990) to have a splice site mutation (see 278700.0001). In these patients, reduced amount of mRNA of reduced size was found on Northern blot analysis. In 2 non-Japanese patients, they found absent mRNA by Northern blot analysis. In 2 other group A XP patients, 1 Japanese and 1 non-Japanese, who showed mild skin symptoms with few, if any, skin tumors or neurologic abnormalities, normal XPAC mRNAs were detected. De Jonge et al. (1983) found that microinjection of cell extracts from either human placenta or HeLa cells into XP cells of complementation group A restores unscheduled DNA synthesis (UDS) in these cells after UV irradiation. These cells show, furthermore, normal resistance to UV irradiation. Yamaizumi et al. (1986) found that the half-life of the injected factor was 14 hrs in XP cells of group A with maximal effects exerted 2 hrs after microinjection. The factors were sensitive to protease but not to RNase and had a molecular weight of about 160 and 90 kD by gel filtration. The 2 fractions of the factor(s) did not work in groups B, C, D, F and G and probably not in E and H.

Kraemer et al. (1987) reviewed 850 published cases of XP in articles dating from 1874 to 1982. They stated that xeroderma pigmentosum was first described in Vienna by the Hungarian-born professor of dermatology Moriz Kaposi (born Moritz Kohn), in the textbook of dermatology he published in 1870 with his father-in-law, Ferdinand Hebra (Hebra and Kaposi, 1874). They believed that xeroderma pigmentosum with neurologic abnormalities (De Sanctis-Cacchione syndrome) was first reported in 1883 by Albert Neisser of Breslau, Germany (who also discovered the bacterial cause of gonorrhea, the agent named Neisseria).

Kraemer et al. (1988) found that 5 patients with XP had a reduction in the number of skin cancers during treatment with isotretinoin: they had a total of 121 tumors in a 2-year interval before treatment and a total of 25 tumors during 2 years of treatment. After the drug was discontinued, the tumor frequency increased at a mean of 8.5-fold over the frequency during treatment. All patients experienced mucocutaneous toxic effects, however, and abnormalities of triglycerides, liver function, or the skeleton developed in some. Parshad et al. (1990) identified XP heterozygotes by the finding of elevated frequencies of chromatid breaks and gaps after G2 phase x-irradiation of peripheral blood lymphocytes or skin fibroblasts.

Rinaldy et al. (1988) designed a competition hybridization strategy for the cloning of genes whose mRNAs are either regulated transcriptionally or vary in abundance as a function of cell line or cell cycle. They used this strategy to construct cDNA libraries from a particular size fraction of mRNA from 3 members of an XPA family: father, mother, and homozygous child. At least 14 gamma-gt11 clones were found to represent the same gene that was present in greater abundance in the father and mother than in the homozygous affected child.

In the yeast Saccharomyces cerevisiae, any one of 5 genes, RAD1, RAD2, RAD3, RAD4, and RAD10, causes a total defect in the incision step of excision repair of DNA damaged by ultraviolet light and an extreme sensitivity to ultraviolet light. Bankmann et al. (1992) reported the characterization of the yeast RAD14 gene. Studies with a RAD14 deletion mutation indicated the absolute requirement of the gene in the incision process. Bankmann et al. (1992) found, furthermore, that RAD14 encodes a highly hydrophilic protein of 247 amino acids containing zinc-finger motifs and that it is similar to the protein encoded by the human XPAC gene that complements XP group A cell lines.

Shimamoto et al. (1991) compared the homologous genes in chicken, Xenopus laevis, and Drosophila melanogaster with the human gene. A high level of conservation was found in the COOH-terminal domain where the frequency of identical amino acids was about 50% with preservation of the zinc-finger motif, suggesting that the gene plays an important role in the function of the protein. Miyamoto et al. (1992)

showed that the nuclear localization signal is located in the region encoded by exon 1, whereas exons 2 through 6 are essential for the DNA repair function. All 4 cysteines forming a zinc-finger structure and also the glutamic acid cluster in the region encoded by exon 2 were found to be important for DNA repair function.

By way of summary of the mapping information, the mutation in XPA is clearly located at 9q34.1; the assignment of a gene resulting in complementation of XPA on 1q has not been substantiated and the suggestion of a complementation group A-related gene on chromosome 8 (Mori et al., 1993) is likewise unclear because complementation was only partial (Cleaver, 1993).

Cleaver (1994) outlined recent information on the human diseases due to defects in DNA repair, including the mismatch repair system in familial nonpolyposis colorectal cancer (MSH2; 120435). He provided a highly useful block diagram of the steps involved in major branches of nucleotide excision repair. The initial step involves the DNA-binding proteins XPA and XPE; the latter has 2 subunits. Thereafter, 2 branches, transcription dependent and transcription independent, operate in different regions of the genome to change DNA structure by a helicase action for excision. Some gene products appear unique to these branches (XPC for transcription independent; CSA and CSB for transcription dependent), whereas others are involved in both (XPB and XPD). The damaged region is cut out in a 29 to 30 nucleotide fragment by the action of 2 excision nucleases: one is a complex of ERCC1 (126380) and ERCC4 (133520) that can act on single-stranded and duplex-supercoiled DNA; the other, the XPG product, acts on single-stranded DNA. The remaining single-strand template is protected by single-strand binding protein RPA, and a patch is then synthesized by proliferating cell nuclear antigen (PCNA), DNA polymerase delta/epsilon, and ligase. Cleaver (1994) noted that the ERCC4, ERCC11, and XPF proteins may be identical, a statement proven by Sijbers et al. (1996).

Li et al. (1994) demonstrated that the XPA protein associates in a specific manner with ERCC1. They interpreted their findings as suggesting that one possible function of XPA is the loading and perhaps the orientation of an incision complex, containing ERCC1 and other factors, to the site of DNA damage. Park and Sancar (1994) concluded from their studies that XPA, ERCC1, and ERCC4 proteins form a ternary complex that participates in both damage recognition and incision activities. Li et al. (1995) used site-specific mutagenesis of XPA to demonstrate the functional relevance of the interaction between XPA and ERCC1. The results suggested that the association between XPA and ERCC1 is a required step in the nucleotide excision repair pathway and that the probable role of the interaction is to recruit the ERCC1 incision complex to the damaged site.

Excision repair of bulky DNA adducts, such as those formed by the cancer chemotherapy agent cisplatin, appears to be mediated by an aggregate of genes, with the XPAC and ERCC1 proteins being involved in DNA damage recognition and excision. Dabholkar et al. (1994) assessed mRNA levels of ERCC1 and XPAC in malignant ovarian cancer tissues from 28 patients that were harvested before the administration of platinum-based chemotherapy. They found that cancer tissues from patients whose tumors were clinically resistant to therapy showed greater levels of total ERCC1 mRNA, full-length transcript of ERCC1 mRNA, and XPAC mRNA, as compared with tumor tissues from those individuals clinically sensitive to therapy. The number of patients in the 2 groups were 13 and 15, respectively.

Cleaver et al. (1995) reported novel deletion and insertion mutations in a family with 2 daughters previously classified as homozygous for defects in the XPA gene (Davis et al., 1994). One mutation was a deletion of 20 nucleotides in exon IV; the other was an insertion of 1 adenine in exon V in a region normally consisting of 6 adenines. Owing to the presence of multiple As in affected regions of both alleles, the precise bases that were lost or inserted to generate the observed alterations were ambiguous. The 20-nucleotide deletion was found to be from the maternal allele. Neither of these mutations had been defined in studies in Japanese and Tunisian populations where a high degree of consanguinity is present and most patients are homozygous for 1 of a small number of mutations.

Nomenclature of human DNA repair genes: Lehmann et al. (1994) made the following proposals: 1) For xeroderma pigmentosum genes, the final C from the XPAC, XPBC, etc., genes should be omitted. The final C should be retained in excision-repair cross-complementing (ERCC) genes; 2) where an ERCC gene has been found to be unequivocally identical to a xeroderma pigmentosum, Cockayne syndrome, or trichothiodystrophy gene, the name should eventually be replaced by the corresponding XP, CS, or TTD gene. Thus, ERCC2, ERCC3, and ERCC6 would become the XPD, XPB, and CSB genes, respectively. (The identification of inactivating mutations in the ERCC5 (133530) gene in an XPG patient justifies changing the ERCC5 gene to XPG.)

Nakane et al. (1995) established XPA-deficient mice by gene targeting of mouse embryonic stem cells. The XPA-deficient mice showed neither obvious physical abnormalities nor pathologic alterations, but were defective in nucleotide-excision repair (NER) and highly susceptible to ultraviolet-B- and 9,10-dimethyl-1,2-benz[a]anthracene-induced skin carcinogenesis. Findings provided in vivo evidence that the XPA protein protects mice from carcinogenesis initiated by ultraviolet or chemical carcinogen. Comparable findings were independently reported by de Vries et al. (1995).

States et al. (1998) performed a mutation analysis on XPA cell lines from 19 American and European patients. Most mutations were deletions and splice site mutations, observed previously in other XPA patients in exon III, intron III, or exon IV, that resulted in frameshifts within the DNA-binding region. One new mutation was a point mutation within intron III causing a new splice acceptor site that may compete with the original splice acceptor site. Mutations in the DNA-binding region of XPA were from patients with the more severe disease often associated with neurologic complications, whereas mutations in the C-terminus of the protein, which interacts with the TFIIH transcription factor (see 189972), were from patients with milder skin disease only. The rarity of naturally occurring missense mutations in the DNA binding region of XPA suggests that amino acid changes may be sufficiently tolerated such that patients would have mild symptoms and escape detection.

See also Afifi et al. (1972); Bootsma (1978); Cleaver (1978); Cleaver (1990); Day (1975); de Grouchy et al. (1967); De Weerd-Kastelein et al. (1974); Freidberg (1978); German et al. (1984); Hodge et al. (1980); Kraemer et al. (1975); Lynch et al. (1978); Macklin (1944); Maher et al. (1976); Maher et al. (1982); Pawsey et al. (1979); Regan et al. (1971); Satokata et al. (1992); Schultz et al. (1985); Yang et al. (1984)

Afifi, A. K.; Der Kaloustian, V. M.; Mire, J. J.: Muscular abnormality in xeroderma pigmentosum: high resolution light-microscopy and electron-microscopic observations. J. Neurol. Sci. 17: 435-442, 1972.

Andrews, A. D.; Barrett, S. F.; Robbins, J. H.: Xeroderma pigmentosum neurological abnormalities correlated with colony-forming ability after ultraviolet radiation. Proc. Nat. Acad. Sci. 75: 1984-1988, 1978.

Bankmann, M.; Prakash, L.; Prakash, S.: Yeast RAD14 and human xeroderma pigmentosum group A DNA-repair genes encode homologous proteins. Nature 355: 555-558, 1992.

Bootsma, D.: Xeroderma pigmentosum.In: Hanawalt, P. C.; Friedberg, E. C.; Fox, C. F.: DNA Repair Mechanisms. ICN-UCLA Symposium on Molecular and Cellular Biology. New York: Academic Press (pub.) IX: 1978. Pp. 589-601.

Bootsma, D.; Hoeijmakers, J. H. J.: The genetic basis of xeroderma pigmentosum. Ann. Genet. 34: 143-150, 1991.

Bootsma, D.; Keijzer, W.: Genetic analysis of xeroderma pigmentosum including Egyptian families. (Abstract) Cytogenet. Cell Genet. 25: 139, 1979.

Cleaver, J. E.: Xeroderma pigmentosum.In: Stanbury, J. B.; Wyngaarden, J. B.; Fredrickson, D. S.: Metabolic Basis of Inherited Disease. New York: McGraw-Hill (pub.) (4th ed.): 1978. Pp. 1072-1095.

Cleaver, J. E.: Do we know the cause of xeroderma pigmentosum? Carcinogenesis 11: 875-882, 1990.

Cleaver, J. E.: Defective repair replication of DNA in xeroderma pigmentosum. Nature 218: 652-656, 1968.

Cleaver, J. E.: Xeroderma pigmentosum: variants with normal DNA repair and normal sensitivity to ultraviolet light. J. Invest. Derm. 58: 124-128, 1972.

Cleaver, J. E.: Xeroderma pigmentosum: biochemical and genetic characteristics. Ann. Rev. Genet. 9: 19-38, 1975.

Cleaver, J. E.: It was a very good year for DNA repair. Cell 76: 1-4, 1994.

Cleaver, J. E.: Personal Communication. San Francisco, Calif. 1/15/1993.

Cleaver, J. E.; Charles, W. C.; Thomas, G. H.; McDowell, M. L.: A deletion and an insertion in the alleles for the xeroderma pigmentosum (XPA) DNA-binding protein in mildly affected patients. Hum. Molec. Genet. 4: 1685-1687, 1995.

Cockayne, E. A.: Inherited Abnormalities of the Skin and Its Appendages. London: Oxford Univ. Press (pub.) 1933.

Dabholkar, M.; Vionnet, J.; Bostick-Bruton, F.; Yu, J. J.; Reed, E.: Messenger RNA levels of XPAC and ERCC1 in ovarian cancer tissue correlate with response to platinum-based chemotherapy. J. Clin. Invest. 94: 703-708, 1994.

Davis, B. E.; Koh, H. K.; Rohrer, T. E.; Gonzalez, E.; Cleaver, J. E.: Sunlight avoidance and cancer prevention in xeroderma pigmentosum. Arch. Derm. 130: 806-808, 1994.

Day, R. S.: Xeroderma pigmentosum variants have decreased repair of ultraviolet-damaged DNA. Nature 253: 748-749, 1975.

de Grouchy, J.; De Nava, C.; Feingold, J.; Frezal, J.; Lamy, M.: Asynchronie chromosomique dans un cas de xeroderma pigmentosum. Ann. Genet. 10: 224-225, 1967.

de Jonge, A. J.; Vermeulen, W.; Klein, B.; Hoeijmakers, J. H.: Microinjection of human cell extracts corrects xeroderma pigmentosum defect. EMBO J. 2: 637-641, 1983.

de Vries, A.; van Oostrom, C. Th. M.; Hofhuis, F. M. A.; Dortant, P. M.; Berg, R. J. W.; de Gruijl, F. R.; Wester, P. W.; van Kreijl, C. F.; Capel, P. J. A.; van Steeg, H.; Verbeek, S. J.: Increased susceptibility to ultraviolet-B and carcinogens of mice lacking the DNA excision repair gene XPA. Nature 377: 169-173, 1995.

De Weerd-Kastelein, E. A.; Keijzer, W.; Bootsma, D.: A third complementation group in xeroderma pigmentosum. Mutat. Res. 22: 87-91, 1974.

El-Hefnawi, H.; Smith, S. M.; Penrose, L. S.: Xeroderma pigmentosum—its inheritance and relationship to the ABO blood-group system. Ann. Hum. Genet. 28: 273-290, 1965.

Farndon, P. A.; Morris, D. J.; Hardy, C.; McConville, C. M.; Weissenbach, J.; Kilpatrick, M. W.; Reis, A.: Analysis of 133 meioses places the genes for

nevoid basal cell carcinoma (Gorlin) syndrome and Fanconi anemia group C in a 2.6-cM interval and contributes to the fine map of 9q22.3. Genomics 23: 486-489, 1994.
Freidberg, E. C.: Xeroderma pigmentosum: recent studies on the DNA repair defects. Arch. Path. Lab. Med. 102: 3-7, 1978.
German, J.; Hashem, N.; El-Hefnawi, M.; Cleaver, J. E.: Xeroderma pigmentosum in Egypt. III. ABO blood grouping in 22 affected families. Ann. Hum. Genet. 48: 61-64, 1984.
Goldstein, S.; Lin, C. C.: Survival and DNA repair of somatic cell hybrids after ultraviolet irradiation. Nature N.B. 239: 142-145, 1972.
Haldane, J. B. S.: A search for incomplete sex-linkage in man. Ann. Eugen. 17: 28-57, 1936.
Hebra, F.; Kaposi, M.: On Diseases of the Skin, Including Exanthemata. London: New Sydenham Society (pub.) 3: 1874. Pp. 252-258.
Henning, K. A.; Schultz, R. A.; Sekhon, G. S.; Friedberg, E. C.: Gene complementing xeroderma pigmentosum group A cells maps to distal human chromosome 9q. Somat. Cell Molec. Genet. 16: 395-400, 1990.
Hodge, S. E.; Berkel, A. I.; Gatti, R. A.; Boder, E.; Spence, M. A.: Ataxia-telangiectasia and xeroderma pigmentosum: no evidence of linkage to HLA. Tissue Antigens 15: 313-317, 1980.
Hori, T.; Shiomi, T.; Sato, K.: Human chromosome 13 compensates a DNA repair defect in UV-sensitive mouse cells by mouse-human cell hybridization. Proc. Nat. Acad. Sci. 80: 5655-5659, 1983.
Ishizaki, K.; Oshimura, M.; Sasaki, M. S.; Nakamura, Y.; Ikenaga, M.: Human chromosome 9 can complement UV sensitivity of xeroderma pigmentosum group A cells. Mutat. Res. 235: 209-215, 1990.
Kanda, T.; Oda, M.; Yonezawa, M.; Tamagawa, K.; Isa, F.; Hanakago, R.; Tsukagoshi, H.: Peripheral neuropathy in xeroderma pigmentosum. Brain 113: 1025-1044, 1990.
Kaur, G. P.; Athwal, R. S.: Complementation of a DNA repair defect in xeroderma pigmentosum cells by transfer of human chromosome 9. Proc. Nat. Acad. Sci. 86: 8872-8876, 1989.
Kaur, G. P.; Rinaldy, A.; Lloyd, R. S.; Athwal, R. S.: A gene that partially complements xeroderma pigmentosum group A cells maps to human chromosome 8. Somat. Cell Molec. Genet. 18: 371-379, 1992.
Keijzer, W.; Stefanini, M.; Bootsma, D.; Verkerk, A.; Geurts van Kessel, A. H. M.; Jongkind, J. F.; Westerveld, A.: Localization of a gene involved in complementation of the defect in xeroderma pigmentosum group A cells on human chromosome 1. Exp. Cell Res. 169: 490-501, 1987.
Keijzer, W.; Stefanini, M.; Westerveld, A.; Bootsma, D.: Mapping of the XPAC gene involved in complementation of the defect in xeroderma pigmentosum group A cells. (Abstract) Cytogenet. Cell Genet. 37: 508, 1984.
Keijzer, W.; Verkerk, A.; Bootsma, D.: Phenotypic correction of the defect in xeroderma pigmentosum cells after fusion with isolated cytoplasts. Exp. Cell Res. 140: 119-125, 1982.
Kore-eda, S.; Tanaka, T.; Moriwaki, S.; Nishigori, C.; Imamura, S.: A case of xeroderma pigmentosum group A diagnosed with a polymerase chain reaction (PCR) technique: usefulness of PCR in the detection of point mutation in a patient with a hereditary disease. Arch. Derm. 128: 971-974, 1992.
Kraemer, K. H.; Coon, H. G.; Petinga, R. A.; Barrett, S. F.; Rahe, A. E.; Robbins, J. H.: Genetic heterogeneity in xeroderma pigmentosum: complementation groups and their relationship to DNA repair rates. Proc. Nat. Acad. Sci. 72: 59-63, 1975.
Kraemer, K. H.; De Weerd-Kastelein, E. A.; Robbins, J. H.; Keijzer, W.; Barrett, S. F.; Petinga, R. A.; Bootsma, D.: Five complementation groups in xeroderma pigmentosum. Mutat. Res. 33: 327-340, 1975.
Kraemer, K. H.; DiGiovanna, J. J.; Moshell, A. N.; Tarone, R. E.; Peck, G. L.: Prevention of skin cancer in xeroderma pigmentosum with the use of oral isotretinoin. New Eng. J. Med. 318: 1633-1637, 1988.
Kraemer, K. H.; Lee, M. M.; Scotto, J.: Xeroderma pigmentosum: cutaneous, ocular, and neurologic abnormalities in 830 published cases. Arch. Derm. 123: 241-250, 1987.
Lehmann, A. R.; Bootsma, D.; Clarkson, S. G.; Cleaver, J. E.; McAlpine, P. J.; Tanaka, K.; Thompson, L. H.; Wood, R. D.: Nomenclature of human DNA repair genes. Mutat. Res. 315: 41-42, 1994.
Li, L.; Elledge, S. J.; Peterson, C. A.; Bales, E. S.; Legerski, R. J.: Specific association between the human DNA repair proteins XPA and ERCC1. Proc. Nat. Acad. Sci. 91: 5012-5016, 1994.
Li, L.; Peterson, C. A.; Lu, X.; Legerski, R. J.: Mutations in XPA that prevent association with ERCC1 are defective in nucleotide excision repair. Molec. Cell. Biol. 15: 1993-1998, 1995.
Lynch, H. T.; Frichot, B. C., III; Fisher, J.; Smith, J. L., Jr.; Lynch, J. F.: Spontaneous regression of metastatic malignant melanoma in 2 sibs with xeroderma pigmentosum. J. Med. Genet. 15: 357-362, 1978.
Macklin, M. T.: Xeroderma pigmentosum: report of a case and consideration of incomplete sex linkage in inheritance of the disease. Arch. Derm. Syph. 49: 157-171, 1944.
Maeda, T.; Sato, K.; Minami, H.; Taguchi, H.; Yoshikawa, K.: Chronological difference in walking impairment among Japanese group A xeroderma pigmentosum (XP-A) patients with various combinations of mutation sites. Clin. Genet. 48: 225-231, 1995.
Maher, V. M.; Ouellette, L. M.; Curren, R. D.; McCormick, J. J.: Frequency of ultraviolet light-induced mutation is higher in xeroderma pigmentosum variant cells than in normal human cells. Nature 261: 593-595, 1976.
Maher, V. M.; Rowan, L. A.; Silinskas, K. C.; Kateley, S. A.; McCormick, J. J.: Frequency of UV-induced neoplastic transformation of diploid human fibroblasts is higher in xeroderma pigmentosum cells than in normal cells. Proc. Nat. Acad. Sci. 79: 2613-2617, 1982.
Miyamoto, I.; Miura, N.; Niwa, H.; Miyazaki, J.; Tanaka, K.: Mutational analysis of the structure and function of the xeroderma pigmentosum group A complementing protein: identification of essential domains for nuclear localization and DNA excision repair. J. Biol. Chem. 267: 12182-12187, 1992.
Mori, T.; Rinaldy, T. L.; Athwal, R. S.; Kaur, G. P.; Nikaido, O.; Lloyd, R. S.; Rinaldy, A.: A xeroderma pigmentosum complementation group A related gene: confirmation using monoclonal antibodies against the cyclobutane dimer and (6-4) photoproduct. Mutat. Res. 293: 143-150, 1993.
Nakane, H.; Takeuchi, S.; Yuba, S.; Saijo, M.; Nakatsu, Y.; Murai, H.; Nakatsuru, Y.; Ishikawa, T.; Hirota, S.; Kitamura, Y.; Kato, Y.; Tsunoda, Y.; Miyauchi, H.; Horio, T.; Tokunaga, T.; Matsunaga, T.; Nikaido, O.; Nishimune, Y.; Okada, Y.; Tanaka, K.: High incidence of ultraviolet-B- or chemical-carcinogen-induced skin tumours in mice lacking the xeroderma pigmentosum group A gene. Nature 377: 165-168, 1995.
Nishigori, C.; Zghal, M.; Yagi, T.; Imamura, S.; Komoun, M. R.; Takebe, H.: High prevalence of the point mutation in exon 6 of the xeroderma pigmentosum group A-complementing (XPAC) gene in xeroderma pigmentosum group A patients in Tunisia. Am. J. Hum. Genet. 53: 1001-1006, 1993.
Park, C.-H.; Sancar, A.: Formation of a ternary complex by human XPA, ERCC1, and ERCC4 (XPF) excision repair proteins. Proc. Nat. Acad. Sci. 91: 5017-5021, 1994.
Parshad, R.; Sanford, K. K.; Kraemer, K. H.; Jones, G. M.; Tarone, R. E.: Carrier detection in xeroderma pigmentosum. J. Clin. Invest. 85: 135-138, 1990.
Pawsey, S. A.; Magnus, I. A.; Ramsay, C. A.; Benson, P. F.; Giannelli, F.: Clinical, genetic and DNA repair on a consecutive series of patients with xeroderma pigmentosum. Quart. J. Med. 48: 179-210, 1979.
Regan, J. D.; Setlow, R. B.; Kaback, M. M.; Howell, R. R.; Klein, E.; Burgess, G.: Xeroderma pigmentosum: a rapid sensitive method for prenatal diagnosis. Science 174: 147-150, 1971.
Rinaldy, A.; Dodson, M. L.; Darling, T. L.; Lloyd, R. S.: Gene cloning using cDNA libraries in a differential competition hybridization strategy: application to cloning XP-A related genes. DNA 7: 563-570, 1988.
Robbins, J. H.; Kraemer, K. H.; Lutzner, M. A.; Festoff, B. W.; Coon, H. G.: Xeroderma pigmentosum: an inherited disease with sun sensitivity, multiple cutaneous neoplasms, and abnormal DNA repair. Ann. Intern. Med. 80: 221-248, 1974.
Salob, S. P.; Webb, D. K. H.; Atherton, D. J.: A child with xeroderma pigmentosum and bone marrow failure. Brit. J. Derm. 126: 372-374, 1992.
Satokata, I.; Tanaka, K.; Miura, N.; Miyamoto, I.; Satoh, Y.; Kondo, S.; Okada, Y.: Characterization of a splicing mutation in group A xeroderma pigmentosum. Proc. Nat. Acad. Sci. 87: 9908-9912, 1990.
Satokata, I.; Tanaka, K.; Miura, N.; Narita, M.; Mimaki, T.; Satoh, Y.; Kondo, S.; Okada, Y.: Three nonsense mutations responsible for group A xeroderma pigmentosum. Mutat. Res. 273: 193-202, 1992.
Satokata, I.; Tanaka, K.; Okada, Y.: Molecular basis of group A xeroderma pigmentosum: a missense mutation and two deletions located in a zinc finger consensus sequence of the XPAC gene. Hum. Genet. 88: 603-607, 1992.
Schultz, R. A.; Barbis, D. P.; Friedberg, E. C.: Studies on gene transfer and reversion to UV resistance in xeroderma pigmentosum cells. Somat. Cell Molec. Genet. 11: 617-624, 1985.
Shimamoto, T.; Kohno, K.; Tanaka, K.; Okada, Y.: Molecular cloning of human XPAC gene homologs from chicken, Xenopus laevis and Drosophila melanogaster. Biochem. Biophys. Res. Commun. 181: 1231-1237, 1991.
Sijbers, A. M.; de Laat, W. L.; Ariza, R. R.; Biggerstaff, M.; Wei, Y.-F.; Moggs, J. G.; Carter, K. C.; Shell, B. K.; Evans, E.; de Jong, M. C.; Rademakers, S.; de Rooij, J.; Jaspers, N. G. J.; Hoeijmakers, J. H. J.; Wood, R. D.: Xeroderma pigmentosum group F caused by a defect in a structure-specific DNA repair endonuclease. Cell 86: 811-822, 1996.
States, J. C.; McDuffie, E. R.; Myrand, S. P.; McDowell, M.; Cleaver, J. E.: Distribution of mutations in the human xeroderma pigmentosum group A gene and their relationships to the functional regions of the DNA damage recognition protein. Hum. Mutat. 12: 103-113, 1998.
Stefanini, M.; Keijzer, W.; Dalpra, L.; Elli, R.; Porro, M. N.; Nicoletti, B.; Nuzzo, F.: Differences in the levels of UV repair and in clinical symptoms in two sibs affected by xeroderma pigmentosum. Hum. Genet. 54: 177-182, 1980.
Stefanini, M.; Keijzer, W.; Westerveld, A.; Geurts van Kessel, A.; Jongkind, J. F.; Bootsma, D.: Complementation and mapping of genes involved in deficient DNA repair in xeroderma pigmentosum cells. (Abstract) Cytogenet. Cell Genet. 32: 321-322, 1982.
Tanaka, K.; Miura, N.; Satokata, I.; Miyamoto, I.; Yoshida, M. C.; Satoh, Y.; Kondo, S.; Yasui, A.; Okayama, H.; Okada, Y.: Analysis of a human DNA excision repair gene involved in group A xeroderma pigmentosum and containing a zinc-finger domain. Nature 348: 73-76, 1990.
Tanaka, K.; Satokata, I.; Ogita, Z.; Uchida, T.; Okada, Y.: Molecular cloning of a mouse DNA repair gene that complements the defect of group-A xeroderma pigmentosum. Proc. Nat. Acad. Sci. 86: 5512-5516, 1989.
Welshimer, K.; Swift, M.: Congenital malformations and developmental disabilities in ataxia-telangiectasia, Fanconi anemia, and xeroderma pigmentosum families. Am. J. Hum. Genet. 34: 781-793, 1982.
Yamaizumi, M.; Sugano, T.; Asahina, H.; Okada, Y.; Uchida, T.: Microinjection of partially purified protein factor restores DNA damage specifically in group A of xeroderma pigmentosum cells. Proc. Nat. Acad. Sci. 83: 1476-1479, 1986.
Yang, L. L.; Kouri, R. E.; Curren, R. D.: Xeroderma pigmentosum fibroblasts are more sensitive to asbestos fibers than are normal human fibroblasts. Carcinogenesis 5: 291-294, 1984.

XERODERMA PIGMENTOSUM, COMPLEMENTATION GROUP C (*278720)

JJM NOTE

The suggestion of Lynch et al. (1984) that melanoma occurs to particular excess in this complementation group needs to be addressed as experience has been gained. Indeed, the cancer spectrum of each group deserves analysis.

See 278700. Hananian and Cleaver (1980) described systemic lupus erythematosus (see 152700) in a patient with type C xeroderma pigmentosum. Lynch et al. (1984) suggested that complementation group C patients may be particularly prone to malignant melanoma. Teitz et al. (1987) were able to correct UV-sensitivity in a group C cell line by transfection with DNA. UV resistance appeared to be closely linked to G418 resistance. G418 is an antibiotic (GIBCO). They used the gene symbol XPCC (for xeroderma pigmentosum group C-complementing). Seguin et al. (1988) studied the frequency of ultraviolet light-induced chromosomal breaks in lymphoblastoid cell lines from patients with Cockayne syndrome and from patients with xeroderma pigmentosum. Cells established from patients with either disorder had the same abnormal increase in the number of induced aberrations. The authors concluded that the frequency of UV-induced breakage in the chromosomes of xeroderma pigmentosa group C cells does not explain the susceptibility of these patients to sunlight-induced skin cancer.

Peterson and Legerski (1991) devised a simple, highly efficient cDNA expression system for use in human cells. Legerski and Peterson (1992) used this system to isolate a cDNA clone that restored the ultraviolet sensitivity and unscheduled DNA synthesis of XPC cells to normal levels. The cloned XPC gene was found to encode a highly hydrophilic protein that is composed of a predicted 823 amino acids and shares limited homology with the product of the yeast DNA repair gene RAD4. The XPC transcript was undetectable by Northern blotting in most XPC cell lines examined. They concluded tentatively that the gene is located on chromosome 3 (Cleaver, 1993). However, by transferring single human chromosomes into XPC cells by microcell fusion, Kaur and Athwal (1992) found that human chromosome 5 conferred enhanced levels of resistance to UV irradiation and capacity to repair UV-induced DNA damage. Presence of the transferred chromosome 5 in complemented clones was confirmed by polymorphism for dinucleotide repeats at 2 loci on chromosome 5. Revertant clones isolated for the loss of the introduced chromosome 5 regained UV sensitivity.

Following up on the isolation of the XPC gene by Peterson and Legerski (1991), Li et al. (1993) identified changes in the gene in 5 XPC cell lines. In 4 of them, Northern blot analysis of RNAs demonstrated subnormal levels of the XPC transcript, whereas the fifth exhibited a near normal level. Each cell line exhibited a unique mutation that correlated well with the cellular DNA repair deficiency and the clinical severity of the disease. In later studies by analysis of a panel of chromosome 3 deletions in somatic cell hybrids and by in situ hybridization, the XPC gene was reliably mapped to 3p25 (Legerski et al., 1994). Although it is closely linked to the RAD23B gene (600062) in the human, the mouse homolog of RAD23B maps to chromosome 4 and the mouse homolog of XPC maps to chromosome 6 (van der Spek et al., 1996). The physical disconnection of the the genes in the mouse argues against a functional significance of the colocalization of these genes in the human.

Masutani et al. (1994) reported the purification to homogeneity and subsequent cDNA cloning of a repair complex by in vitro complementation of the XPC defect in a cell-free repair system containing UV-damaged SV40 minichromosomes. The complex had a high affinity for single-stranded DNA and consisted of 2 tightly associated proteins of 125 and 58 kD. The 125-kD subunit was an N-terminally extended version of the XPCC gene product which is thought to represent the human homolog of the RAD4 nucleotide excision repair (NER) gene of Saccharomyces cerevisiae. The 58-kD species turned out to be a human homolog of yeast RAD23. Unexpectedly, a second human counterpart of RAD23 was identified. Masutani et al. (1994) referred to the 2 as HHR23A (600061) and HHR23B. The 2 RAD23 homologs were expressed in the same cells; in the XPCC purification scheme, however, only the HHR23B protein was found in a complex with p125/XPCC. Masutani et al. (1994) pointed out that no human mutant defective in HHR23A had been identified. By fluorescence in situ hybridization, van der Spek et al. (1994) showed that the XPC and HHR23B genes, the products of which form a tight complex, colocalize to 3p25.1. Pulsed-field gel electrophoresis revealed that the 2 genes may share a MluI restriction fragment of about 625 kb.

Sands et al. (1995) generated XPC-deficient mice by 'knockout' of the mouse homolog of the human XPC gene using embryonic stem cell technology. Mice homozygous for mutant alleles were viable and did not exhibit an increased susceptibility to spontaneous tumor generation at 1 year of age. However, they were found to be highly susceptible to ultraviolet-induced carcinogenesis compared to mice heterozygous for the mutant allele and to wildtype controls. Homozygous mutant mice also displayed a spectrum of ultraviolet-exposure-related pathologic skin and eye changes consistent with those found in the human disease xeroderma pigmentosum group C. The deficient mice showed marked hyperplasia of the epidermis with focal areas of hyperkeratosis in varying degrees of dysplasia, acantholysis, and/or dyskeratosis, similar to the human lesions known as actinic or solar keratosis. Changes in the eye included severe keratitis and corneal ulceration.

See also Cleaver (1986); Halley et al. (1979)

Cleaver, J. E.: Personal Communication. San Francisco, Calif. 1/15/1993.

Cleaver, J. E.: DNA repair in human xeroderma pigmentosum group C cells involves a different distribution of damaged sites in confluent and growing cells. Nucleic Acids Res. 14: 8155-8165, 1986.

Halley, D. J. J.; Keijzer, W.; Jaspers, N. G. J.; Niermeijer, M. F.; Kleijer, W. J.; Boue, J.; Boue, A.; Bootsma, D.: Prenatal diagnosis of xeroderma pigmentosum (group C) using assays of unscheduled DNA synthesis and postreplication repair. Clin. Genet. 16: 137-146, 1979.

Hananian, J.; Cleaver, J. E.: Xeroderma pigmentosum exhibiting neurological disorders and systemic lupus erythematosus. Clin. Genet. 17: 39-45, 1980.

Kaur, G. P.; Athwal, R. S.: Identification of a DNA repair gene on human chromosome 5 that complements xeroderma pigmentosum cells of group C. (Abstract) Am. J. Hum. Genet. 51 (suppl.): A51, 1992.

Legerski, R.; Peterson, C.: Expression cloning of a human DNA repair gene involved in xeroderma pigmentosum group C. Nature 359: 70-73, 1992.

Legerski, R. J.; Liu, P.; Li, L.; Peterson, C. A.; Zhao, Y.; Leach, R. J.; Naylor, S. L.; Siciliano, M. J.: Assignment of xeroderma pigmentosum group C (XPC) gene to chromosome 3p25. Genomics 21: 266-269, 1994.

Li, L.; Bales, E. S.; Peterson, C. A.; Legerski, R. J.: Characterization of molecular defects in xeroderma pigmentosum group C. Nature Genet. 5: 413-417, 1993.

Lynch, H. T.; Fusaro, R. M.; Johnson, J. A.: Xeroderma pigmentosum: complementation group C and malignant melanoma. Arch. Derm. 120: 175-179, 1984.

Masutani, C.; Sugasawa, K.; Yanagisawa, J.; Sonoyama, T.; Ui, M.; Enomoto, T.; Takio, K.; Tanaka, K.; van der Spek, P.; Bootsma, D.; Hoeijmakers, J. H. J.; Hanaoka, F.: Purification and cloning of a nucleotide excision repair complex involving the xeroderma pigmentosum group C protein and a human homologue of yeast RAD23. EMBO J. 13: 1831-1843, 1994.

Peterson, C.; Legerski, R.: High-frequency transformation of human repair-deficient cell lines by an Epstein-Barr virus-based cDNA expression vector. Gene 107: 279-284, 1991.

Sands, A. T.; Abuin, A.; Sanchez, A.; Conti, C. J.; Bradley, A.: High susceptibility to ultraviolet-induced carcinogenesis in mice lacking XPC. Nature 377: 162-165, 1995.

Seguin, L. R.; Tarone, R. E.; Liao, K.; Robbins, J. H.: Ultraviolet light-induced chromosomal aberrations in cultured cells from Cockayne syndrome and complementation group C xeroderma pigmentosum patients: lack of correlation with cancer susceptibility. Am. J. Hum. Genet. 42: 468-475, 1988.

Teitz, T.; Naiman, T.; Avissar, S. S.; Bar, S.; Okayama, H.; Canaani, D.: Complementation of the UV-sensitive phenotype of a xeroderma pigmentosum human cell line by transfection with a cDNA clone library. Proc. Nat. Acad. Sci. 84: 8801-8804, 1987.

van der Spek, P. J.; Smit, E. M. E.; Beverloo, H. B.; Sugasawa, K.; Masutani, C.; Hanaoka, F.; Hoeijmakers, J. H. J.; Hagemeijer, A.: Chromosomal localization of three repair genes: the xeroderma pigmentosum group C gene and two human homologs of yeast RAD23. Genomics 23: 651-658, 1994.

van der Spek, P. J.; Visser, C. E.; Hanaoka, F.; Smit, B.; Hagemeijer, A.; Bootsma, D.; Hoeijmakers, J. H. J.: Cloning, comparative mapping, and RNA expression of the mouse homologues of the Saccharomyces cerevisiae nucleotide excision repair gene RAD23. Genomics 31: 20-27, 1996.

XERODERMA PIGMENTOSUM 4 (#278730)

JJM NOTE

This clinical phenotype has, in a sense, been replaced by the specific underlying genotype.

A number sign (#) is used with this entry because of the work reported by Flejter et al. (1992) indicating that the ERCC2 DNA repair gene (126340) corrects the defect in XPD cells.

See 278700. Linkage of trichothiodystrophy (TTD) and xeroderma pigmentosum of complementation group D was suggested by Nuzzo et al. (1986) on the basis of 3 Italian families in which 4 persons had the 2 disorders. Study of surnames and genealogies suggested that the 3 sibships probably had a common ancestral couple. The characteristics of the TTD were hair shaft abnormalities, ichthyosis, immature sexual development, short stature, and peculiar facies. It may be the disorder called Tay syndrome (see 601675). The form in the same complementation group as XPD might be referred to as TTD1; the second type (TTD2) is also described in 601675.

Johnson and Squires (1992) stated that more than 30 unrelated persons with XP of group D were known and that less than half of this number showed major abnormalities of the CNS, once considered to be the hallmark of XPD. They also commented on the fact that some patients with trichothiodystrophy or with Cockayne syndrome have mutations in the XPD gene. Stefanini et al. (1986) pointed out that the only known patient with complementation group B xeroderma pigmentosum (278720) had clinical features of Cockayne syndrome (216400) and, further, that the only known patient thought to have complementation group H xeroderma pigmentosum also had Cockayne syndrome.

Stefanini et al. (1986) reported studies on the defect in DNA repair with creation of heterokaryons. Seetharam et al. (1987) used a shuttle vector plasmid to assess the types of mutations that cells from a patient with xeroderma pigmentosum of complementation group D introduce into UV-damaged, replicating DNA. In comparison to UV-treated plasmids replicated in normal cells, there were fewer surviving plasmids, a higher frequency of plasmids with mutations, fewer plasmids with 2 or more mutations in the marker gene (coding for a tyrosine suppressor transfer RNA), and a new mutagenic hotspot. The major type of base substitution was the G:C to A:T transition. Similar findings were reported with cells from a xeroderma pigmentosum patient in complementation group A (278700). Ichihashi et al. (1988) described mild skin lesions and no apparent neurologic abnormalities at the age of 31 years despite the usual group D level of DNA repair deficiency and cultured fibroblasts in a case classified as complementation group D by cell-fusion complementation methods. Arrand et al. (1989) cloned and characterized a hamster DNA repair gene that is able to confer an increase in resistance to ultraviolet irradiation on 2 XPD cell lines but not on an XPA line. They found no obvious similarities to 2 human excision repair genes, ERCC1 (126380) and ERCC2, which correct repair-defective hamster cells but have no effect on XP cells. (Flejter et al. (1992) found that ERCC2 did correct the defect in XPD cells.) Homologs of the hamster gene were identified in normal human genomic DNA and mRNA. The transcription pattern was not altered in XPD or XPA cells. The gene complementing group D xeroderma pigmentosum has been cloned (Takebe, 1989).

Moshell et al. (1983) defined complementation group H xeroderma pigmentosum on the basis of a single patient who had both xeroderma pigmentosum and Cockayne syndrome. Johnson et al. (1989) found that hybrids between XPD cells and cells from groups A, B, C, E, F, G, and I showed cross-correction. However, no correction was observed when hybrids were created with XPH cells; thus, it is possible that XPD and XPH are allelic. Nonetheless, Robbins (1989) defended the separateness of XPH, quoting studies from the laboratory of Fujiwara indicating complete restoration of unscheduled DNA synthesis (UDS) in heterokaryons found between the group H strain and each of the 3 group D strains tested. Johnson (1989) continued to defend his negative result. Later, Robbins (1991) concluded that the so-called group H is indeed group D.

Kleijer et al. (1994) described a child with what Crovato et al. (1983) and Rebora and Crovato (1988) referred to as the PIBI(D)S syndrome with trichothiodystrophy. The girl had photosensitivity (P), ichthyosis (I), brittle hair (B), impaired intelligence (I), possibly decreased fertility (D), and short stature (S). A remarkable feature was the intermittent character of the scalp hair loss during infectious periods, such as with pneumonia. The child died unexpectedly at home during sleep at the age of 2 years and 8 months. Easy suntanning suggested photosensitivity and prompted DNA repair studies which demonstrated reduced UV-induced DNA repair synthesis. Other studies assigned this patient to xeroderma pigmentosum group D.

Arrand, J. E.; Bone, N. M.; Johnson, R. T.: Molecular cloning and characterization of a mammalian excision repair gene that partially restores UV resistance to xeroderma pigmentosum complementation group D cells. Proc. Nat. Acad. Sci. 86: 6997-7001, 1989.

Crovato, F.; Borrone, F.; Rebora, A.: Trichothiodystrophy—BIDS, IBIDS and PIBIDS?. (Letter) Brit. J. Derm. 108: 247 only, 1983.

Flejter, W. L.; McDaniel, L. D.; Johns, D.; Friedberg, E. C.; Schultz, R. A.: Correction of xeroderma pigmentosum complementation group D mutant cell phenotypes by chromosome and gene transfer: involvement of the human ERCC2 DNA repair gene. Proc. Nat. Acad. Sci. 89: 261-265, 1992.

Ichihashi, M.; Yamamura, K.; Hiramoto, T.; Fujiwara, Y.: No apparent neurologic defect in a patient with xeroderma pigmentosum complementation group D. Arch. Derm. 124: 256-260, 1988.

Johnson, R. T.: Reply to letter by J. H. Robbins. (Letter) Hum. Genet. 84: 101 only, 1989.

Johnson, R. T.; Elliott, G. C.; Squires, S.; Joysey, V. C.: Lack of complementation between xeroderma pigmentosum complementation groups D and H. Hum. Genet. 81: 203-210, 1989.

Johnson, R. T.; Squires, S.: The XPD complementation group: insights into xeroderma pigmentosum, Cockayne's syndrome and trichothiodystrophy. Mutat. Res. 273: 97-118, 1992.

Kleijer, W. J.; Beemer, F. A.; Boom, B. W.: Intermittent hair loss in a child with PIBI(D)S syndrome and trichothiodystrophy with defective DNA repair-xeroderma pigmentosum group D. Am. J. Med. Genet. 52: 227-230, 1994.

Moshell, A. N.; Ganges, M. B.; Lutzner, M. A.; Coon, H. G.; Barrett, S. F.; Dupuy, J. M.; Robbins, J. H.: A new patient with both xeroderma pigmentosum and Cockayne syndrome establishes the new xeroderma pigmentosum complementation group H.In: Friedberg, E. C.; Bridges, B. A.: Cellular Responses to DNA Damage. New York: Alan R. Liss (pub.) 1983. Pp. 209-213.

Nuzzo, F.; Stefanini, M.; Colognola, R.; Zei, G.; Santachiara, A. S.; Lagomarsini, P.; Casati, S.; Marinoni, S.: Association of two rare hereditary disorders, xeroderma pigmentosum and trichothiodystrophy, in three families from northeast Italy. (Abstract) 7th Int. Cong. Hum. Genet., Berlin 249 only, 1986.

Rebora, A.; Crovato, F.: Trichothiodystrophy, xeroderma pigmentosum and PIBI(D)S syndrome. (Letter) Hum. Genet. 78: 106-108, 1988.

Robbins, J. H.: No lack of complementation for unscheduled DNA synthesis between xeroderma pigmentosum complementation groups D and H. (Letter) Hum. Genet. 84: 99-100, 1989.

Robbins, J. H.: Xeroderma pigmentosum complementation group H is withdrawn and reassigned to group D. Hum. Genet. 88: 242 only, 1991.

Seetharam, S.; Protic-Sabljic, M.; Seidman, M. M.; Kraemer, K. H.: Abnormal ultraviolet mutagenic spectrum in plasmid DNA replicated in cultured fibroblasts from a patient with the skin cancer-prone disease, xeroderma pigmentosum. J. Clin. Invest. 80: 1613-1617, 1987.

Stefanini, M.; Lagomarsini, P.; Arlett, C. F.; Marinoni, S.; Borrone, C.; Crovato, F.; Trevisan, G.; Cordone, G.; Nuzzo, F.: Xeroderma pigmentosum (complementation group D) mutation is present in patients affected by trichothiodystrophy with photosensitivity. Hum. Genet. 74: 107-112, 1986.

Takebe, H.: Personal Communication. Kyoto, Japan 1989.

XERODERMA PIGMENTOSUM 5 (*278740)

JJM NOTE

The interaction with differing DNA damage-binding proteins (DDB1 and DDB2) may result in differing phenotypes and illustrates the generalization that there is no such thing as a single-gene disease.

See 278700. Kondo et al. (1988) studied 3 Japanese patients, aged 50, 42, and 41 years. All had mild XP symptoms and no neurologic abnormalities. Two had developed basal cell carcinomas at ages 46 and 41 years. Chu and Chang (1988) used an extension of the gel electrophoresis binding assay to identify at least 1 nuclear factor that binds to DNA damaged by ultraviolet radiation or the antitumor drug cisplatin, but is notably absent in cells from complementation group E. The defect was not the result of a failure of protein transport to the nucleus since binding activity was absent in both the nuclear and cytoplasmic extracts. The nuclear factor was normally present in cells from groups A, B, C, D, F, G, and H. The absence of specific binding activity for damaged DNA in group E cells could be corrected by the addition of extracts from any 1 of these other groups, thus ruling out the possibility that group E cells contain a factor that inhibits binding activity.

The 2 patients with group E xeroderma pigmentosum studied by Chu and Chang (1988) were related as second cousins and suffered from a skin disease less severe than that seen in patients of several other complementation groups. Keeney et al. (1993) purified to near homogeneity from HeLa cells a DNA damage-binding protein (DDB1; 600045) implicated in xeroderma pigmentosum E. The protein was abundant and had a native molecular weight of about 160,000 as estimated by gel filtration and glycerol gradient sedimentation. DNA damage binding activity copurified with polypeptides of 124 and 41 kD. It appeared that the 2 polypeptides are subunits of a heterodimeric protein. Keeney et al. (1994) injected purified human DDB protein into XPE cells and showed that DNA repair was stimulated to normal levels in those strains that lacked the DDB activity but not in cells from other xeroderma pigmentosum groups or in the subset of XPE cells that contain the activity. These results provided direct evidence that defective DDB activity causes the repair defect in a subset of XPE patients, which in turn establishes a role for this activity in nucleotide-excision repair in vivo. Thus, there appear to be 2 types of xeroderma pigmentosum, group E: a DDB-positive form and a DDB-negative form. Mutations in the smaller component of the heterodimeric protein, encoded by the DDB2 gene (600811) were demonstrated by Nichols et al. (1996) in cases of the DDB-negative form. Nichols et al. (1996) demonstrated mutations in the DDB2 gene, which encodes the b48 subunit of the DDB heterodimer (600811.0001).

Chu, G.; Chang, E.: Xeroderma pigmentosum group E cells lack a nuclear factor that binds to damaged DNA. Science 242: 564-567, 1988.

Keeney, S.; Chang, G. J.; Linn, S.: Characterization of a human DNA damage binding protein implicated in xeroderma pigmentosum E. J. Biol. Chem. 268: 21293-21300, 1993.

Keeney, S.; Eker, A. P. M.; Brody, T.; Vermeulen, W.; Bootsma, D.; Hoeijmakers, J. H. J.; Linn, S.: Correction of the DNA repair defect in xeroderma pigmentosum group E by injection of a DNA damage-binding protein. Proc. Nat. Acad. Sci. 91: 4053-4056, 1994.

Kondo, S.; Fukuro, S.; Mamada, A.; Kawada, A.; Satoh, Y.; Fujiwara, Y.: Assignment of three patients with xeroderma pigmentosum to complementation group E and their characteristics. J. Invest. Derm. 90: 152-157, 1988.

Nichols, A. F.; Ong, P.; Linn, S.: Mutations specific to the xeroderma pigmentosum group E Ddb- phenotype. J. Biol. Chem. 271: 24317-24320, 1996.

XERODERMA PIGMENTOSUM, COMPLEMENTATION GROUP F (#278760)

JJM NOTE

Of additional note is the occurrence of bile duct cancer in an affected 50-year-old male (Moriwaki et al. 1993).

A number sign (#) is used with this entry because of evidence that the disorder is caused by mutations in the ERCC4 gene (133520).

See 278700 for a general discussion of xeroderma pigmentosum. Group F xeroderma pigmentosum had probably been observed only in Japan (Fujiwara et al., 1985) until the report by Norris et al. (1988) of a case in an English woman. Yamamura et al. (1989) described a 61-year-old female patient with XP assigned to complementation group F by the cell fusion-complementation method. Her fibroblasts in culture exhibited a defective DNA repair capacity of 10 to 15% unscheduled DNA synthesis and a 3-fold increased sensitivity to the lethal effects of ultraviolet light. The patient had mild clinical symptoms consisting of numerous pigmented freckles and a small number of seborrheic keratosis-like papules. She had no skin cancers in the sun-exposed areas and no neurologic abnormalities. Yamamura et al. (1989) reviewed 11 Japanese group F patients demonstrating very mild skin symptoms and no ocular or neuropsychiatric abnormalities. Single skin cancers occurred in only 3 of the 11 patients, with an average age of 52 years for their first skin malignancy. Moriwaki et al. (1993) reported the case of a 48-year-old Japanese man suffering from xeroderma pigmentosum associated with mental retardation, cerebral atrophy, and cerebellar ataxia. Genetic complementation tests by cell fusion with polyethylene glycol revealed that the patient belonged to group F. He died of bile duct cancer at the age of 50. Moriwaki et al. (1993) stated that this was the first report of an XPF patient with neurologic abnormalities.

Saxon et al. (1989) found that microcell-mediated transfer of a single human chromosome from repair-proficient human cells to XP cells of complementation group F resulted in partial complementation of repair-defective phenotypes. They identified the complementing chromosome as human chromosome 15 by cytogenetic and molecular analysis. About 20% of the resistance of wildtype cells to killing by UV radiation or by the UV-mimetic chemical 4-nitroquinoline-1-oxide, as well as partial repair synthesis of DNA measured as unscheduled DNA synthesis, resulted from the transfer of the chromosome. Although the work seemed to indicate that the gene on chromosome 15 carries a mutation determining XPF, the reason for the incomplete correction was unclear. Because of incomplete correction, the assignment of XPF to chromosome 15 was considered in limbo (Cleaver, 1993). By means of fluorescence in situ hybridization, Sijbers et al. (1996) mapped the XPF gene to 16p13.2-p13.1. This corresponds to the ERCC4 locus, a human repair gene complementing rodent nucleotide excision repair mutants of group 4, that was originally identified using cell hybrids (Liu et al., 1993) and a genomic clone (Thompson et al., 1994).

See also Nishigori et al. (1986)

Cleaver, J. E.: Personal Communication. San Francisco, Calif. 1/15/1993.
Fujiwara, Y.; Ichihashi, M.; Uehara, Y.; Matsumoto, A.; Yamamoto, Y.; Kano, Y.; Tanakura, Y.: Xeroderma pigmentosum groups C and F: additional assignments and a review of the subjects in Japan. J. Radiat. Res. 26: 443-449, 1985.
Liu, P.; Siciliano, J.; White, B.; Legerski, R.; Callen, D.; Reeders, S.; Siciliano, M. J.; Thompson, L. H.: Regional mapping of human DNA excision-repair gene ERCC4 to chromosome 16p13.13-p13.2. Mutagenesis 8: 199-205, 1993.
Moriwaki, S.; Nishigori, C.; Imamura, S.; Yagi, T.; Takahashi, C.; Fujimoto, N.; Takebe, H.: A case of xeroderma pigmentosum complementation group F with neurological abnormalities. Brit. J. Derm. 128: 91-94, 1993.
Nishigori, C.; Ishizaki, K.; Takebe, H.; Imamura, S.; Hayakawa, M.: A case of xeroderma pigmentosum group F with late onset of clinical symptoms. (Letter) Arch. Derm. 122: 510-511, 1986.
Norris, P. G.; Hawk, J. L. M.; Avery, J. A.; Giannelli, F.: Xeroderma pigmentosum complementation group F in a non-Japanese patient. J. Am. Acad. Derm. 18: 1185-1188, 1988.
Saxon, P. J.; Schultz, R. A.; Stanbridge, E. J.; Friedberg, E. C.: Human chromosome 15 confers partial complementation of phenotypes to xeroderma pigmentosum group F cells. Am. J. Hum. Genet. 44: 474-485, 1989.
Sijbers, A. M.; de Laat, W. L.; Ariza, R. R.; Biggerstaff, M.; Wei, Y.-F.; Moggs, J. G.; Carter, K. C.; Shell, B. K.; Evans, E.; de Jong, M. C.; Rademakers, S.; de Rooij, J.; Jaspers, N. G. J.; Hoeijmakers, J. H. J.; Wood, R. D.: Xeroderma pigmentosum group F caused by a defect in a structure-specific DNA repair endonuclease. Cell 86: 811-822, 1996.
Thompson, L. H.; Brookman, K. W.; Weber, C.-A.; Salazar, E. P.; Reardon, J. T.; Sancar, A.; Deng, Z. M.; Siciliano, M. J.: Molecular cloning of the human nucleotide excision-repair gene ERCC4. Proc. Nat. Acad. Sci. 91: 6855-6859, 1994.
Yamamura, K.; Ichihashi, M.; Hiramoto, T.; Ogoshi, M.; Nishioka, K.; Fujiwara, Y.: Clinical and photobiological characteristics of xeroderma pigmentosum complementation group F: a review of cases from Japan. Brit. J. Derm. 121: 471-480, 1989.

XERODERMA PIGMENTOSUM WITH NORMAL DNA REPAIR RATES (278750)

JJM NOTE

There are many routes to the same phenotype and cancer predisposition.

As indicated in 278700, some patients with xeroderma pigmentosum have been found to have normal DNA repair rates. Presumably this disorder is autosomal recessive, but the proof is not yet available. Although excision repair is normal, postreplication repair is defective (Lehman et al., 1975). The XP variant class is characterized by a defect in conversion of newly synthesized DNA from low to high molecular weight after UV irradiation. 'Pigmentary xerodermoid' is apparently identical to the XP variant (Cleaver et al., 1980). Fujiwara et al. (1981) reported studies of cultured cells from an 8-year-old boy, the son of first-cousin parents, who, they suggested, had a 'new' form of photodermatosis with a defect in the recovery of post-UV DNA synthesis. Except for sun sensitivity, he had no growth retardation, microcephaly, congenital malformations, or other abnormalities. The authors identified 3 categories of disorders of DNA repair and replication. Category 1 manifests deficiency in excision repair of DNA lesions induced by UV. In this category fall the seven xeroderma pigmentosum (XP) complementation groups and the XP variant group. In category 2, represented by the Cockayne syndrome (216400), cells show defective recovery of DNA synthesis after UV irradiation despite apparently normal nucleotide excision repair. The patient whose cells were studied by Fujiwara et al. (1981) falls in category 2. Category 3, represented by the syndromes of Bloom (210900) and Werner (277700), is characterized by a retarded rate of DNA replication without detectable defects in known DNA repair mechanisms.

Despite having the clinical characteristics of XP, including increased frequency of skin cancer, these patients have cells that are only slightly more sensitive than normal cells to the cytotoxic effect of UV radiation but that are significantly more sensitive to its mutagenic effect. To examine the mechanism for the hypermutability, Wang et al. (1991) transfected an XP variant cell line with a UV-irradiated shuttle vector carrying a gene as a target for mutation, allowed replication of the plasmid, determined the frequency and spectrum of mutations induced, and compared the results with those obtained previously when irradiated plasmids carrying the same target gene replicated in a normal cell line. The frequency of mutants increased linearly with dose, but with a slope 5 times steeper than that seen with normal cells. Most of the mutants contained base substitutions and substitutions involving a dipyrimidine. Twenty eight percent of the mutations involved AT basepairs as compared to 11% in normal cells. If the rate of excision of lesions from the marker gene and the 2 cell lines is equal, then the hypermutability of XP variant cells may be explained.

Cleaver, J. E.; Arutyunyan, R. M.; Sarkisian, T.; Kaufmann, W. K.; Greene, A. E.; Coriell, L.: Similar defects in DNA repair and replication in the pigmented xerodermoid and the xeroderma pigmentosum variants. Carcinogenesis 1: 647-655, 1980.
Fujiwara, Y.; Ichihashi, M.; Kano, Y.; Goto, K.; Shimizu, K.: A new human photosensitive subject with a defect in the recovery of DNA synthesis after ultraviolet-light irradiation. J. Invest. Derm. 77: 256-263, 1981.
Lehman, A. R.; Kirk-Bell, S.; Arlett, C. F.; Paterson, M. C.; Lohman, P. H. M.; De Weerd-Kastelein, E. A.; Bootsma, D.: Xeroderma pigmentosum cells with normal levels of excision repair have a defect in DNA synthesis after UV-irradiation. Proc. Nat. Acad. Sci. 72: 219-223, 1975.
Wang, Y.-C.; Maher, V. M.; McCormick, J. J.: Xeroderma pigmentosum variant cells are less likely than normal cells to incorporate dAMP opposite photoproducts during replication of UV-irradiated plasmids. Proc. Nat. Acad. Sci. 88: 7810-7814, 1991.

XERODERMA PIGMENTOSUM 9, COMPLEMENTATION GROUP I (278810)

JJM NOTE

This entity may be a remnant of a laboratory accident!

See 278730. A ninth complementation group described by Fischer et al. (1985) was withdrawn by Bootsma et al. (1989). The initial classification had been done on the basis of studies of 1 cell line only. Study with a second biopsy from both of the 2 sibs showed that the cells behaved exactly like those of the XPC complementation group. Bootsma et al. (1989) suspected that the original cell strain was contaminated with another XP cell strain, at least at the stage when complementation experiments with XPC fibroblasts were performed. If these contaminating cells belong to an XP complementation group other than C, one might expect to find complementation in all instances, even with itself. Unfortunately, the latter control experiment was not performed in the original study.

Bootsma, D.; Keijzer, W.; Jung, E. G.; Bohnert, E.: Xeroderma pigmentosum complementation group XP-1 withdrawn. Mutat. Res. 218: 149-151, 1989.
Fischer, E.; Keijzer, W.; Thielmann, H. W.; Popanda, O.; Bohnert, E.; Edler, L.; Jung, E. G.; Bootsma, D.: A ninth complementation group in xeroderma pigmentosum, XP I. Mutat. Res. 145: 217-225, 1985.

XERODERMIC IDIOCY OF DESANCTIS AND CACCHIONE (#278800)

JJM NOTE

Skin cancer is the major concern, but there is one affected patient with acute lymphatic leukemia, perhaps a coincidence (Reed et al. 1969).

A number sign (#) is used with this entry because the particular phenotype may be displayed by patients with any of several different forms of xeroderma pigmentosum, although it is said to occur most often in those of complementation group D (XPD; 278730).

The condition entitled xerodermic idiocy by de Sanctis and Cacchione (1932) shows, in addition to xeroderma pigmentosum, mental deficiency, dwarfism, and gonadal hypoplasia. Reed et al. (1965) described the syndrome in a Caucasian brother and sister and 2 'Japanese' brothers. Choreoathetoid neurologic signs occurred in their cases. Cerebral and olivopontocerebellar atrophy was found at autopsy of a Japanese case (Yano, 1950) and a Japanese-American case (Reed et al., 1969). Of the 5 patients described by Reed et al. (1969), 4 had associated neurologic features (de Sanctis-Cacchione syndrome). One of the patients with the latter syndrome developed acute lymphatic leukemia at the age of 3 years. De Weerd-Kastelein et al. (1972) found complementation when cells from classic xeroderma pigmentosum were hybridized with cells from a case of the de Sanctis-Cacchione syndrome. Death of neurons may be occurring more rapidly in persons with a defect in DNA-repair than in normals (Robbins et al., 1974). Areflexia is a common finding. Cerebellar ataxia often develops. The de Sanctis-Cacchione phenotype may be displayed by patients in any one of the 7 complementation groups, but occurs most often in those of complementation group D (278730) (Kraemer et al., 1987). Kraemer et al. (1987) pointed out that xeroderma pigmentosum with neurologic abnormalities was first reported in 1883 by Albert Neisser of Breslau, Germany (who also discovered the bacterial cause of gonorrhea, the agent Neisseria). The patients Neisser described were 2 sibs who had XP with progressive neurologic degeneration beginning in the second decade. In a brother and 2 sisters in a consanguineous Hispanic family, 5,Greenhaw et al. (1989, 1992) described features consistent with the de Sanctis-Cacchione syndrome except that CNS changes were much more striking and the cutaneous manifestations less striking than is usually the case and postultraviolet exposure DNA repair rate, as measured by percent of unscheduled DNA synthesis, was 70% of normal control, demonstrating that the cells had no significant DNA repair defect. In patients with the de Sanctis-Cacchione syndrome (278700), this value is usually less than 10%. The de Sanctis-Cacchione syndrome is usually present in the group A form of xeroderma pigmentosum (Kanda et al., 1990) but may occur in other forms of XP.

See also Elsasser et al. (1950); Greenhaw et al. (1992); Roytta and Anttinen (1986)

de Sanctis, C.; Cacchione, A.: L'idiozia xerodermica. Rivista Sperimentale di Freniatria e Medicina Legale delle Alienazioni Mentali 56: 269-292, 1932.
De Weerd-Kastelein, E. A.; Keijzer, W.; Bootsma, D.: Genetic heterogeneity of xeroderma pigmentosum demonstrated by somatic cell hybridization. Nature N.B. 238: 80-83, 1972.
Elsasser, G.; Freusberg, O.; Theml, F.: Das Xeroderma pigmentosum und die 'xerodermische Idiotie.'. Arch. Derm. 188: 651-655, 1950.
Greenhaw, G. A.; Hebert, A.; Duke-Woodside, M. E.; Butler, I. J.; Hecht, J. T.; Cleaver, J. E.; Thomas, G. H.; Horton, W. A.: Xeroderma pigmentosum and Cockayne syndrome: overlapping clinical and biochemical phenotypes. Am. J. Hum. Genet. 50: 677-689, 1992.
Greenhaw, G. A.; Hecht, J. T.; Hebert, A. A.; Butler, I. J.; Cleaver, J. E.; Horton, W. A.: Xeroderma pigmentosa with severe neurological involvement without significant DNA repair defect. (Abstract) Am. J. Hum. Genet. 45 (suppl.): A47, 1989.
Kanda, T.; Oda, M.; Yonezawa, M.; Tamagawa, K.; Isa, F.; Hanakago, R.; Tsukagoshi, H.: Peripheral neuropathy in xeroderma pigmentosum. Brain 113: 1025-1044, 1990.
Kraemer, K. H.; Lee, M. M.; Scotto, J.: Xeroderma pigmentosum: cutaneous, ocular, and neurologic abnormalities in 830 published cases. Arch. Derm. 123: 241-250, 1987.
Reed, W. B.; Landing, B. H.; Sugarman, G. I.; Cleaver, J. E.; Melnyk, J.: Xeroderma pigmentosum: clinical and laboratory investigation of its basic defect. J.A.M.A. 207: 2073-2079, 1969.
Reed, W. B.; May, S. B.; Nickel, W. R.: Xeroderma pigmentosum with neurological complications. Arch. Derm. 91: 224-226, 1965.
Robbins, J. H.; Kraemer, K. H.; Lutzner, M. A.; Festoff, B. W.; Coon, H. G.: Xeroderma pigmentosum: an inherited disease with sun sensitivity, multiple cutaneous neoplasms and abnormal DNA repair. Ann. Intern. Med. 80: 221-248, 1974.
Roytta, M.; Anttinen, A.: Xeroderma pigmentosum with neurological abnormalities: a clinical and neuropathological study. Acta Neurol. Scand. 73: 191-199, 1986.
Yano, K.: Xeroderma pigmentosum mit Stoerungen des Zentralnervensystems: eine histopathologische Untersuchung. Folia Psychiat. Neurol. Jpn. 4: 143-151, 1950.

XERODERMA PIGMENTOSUM (194400)

JJM NOTE

This entity follows a dominant mode of inheritance, but may not be xeroderma pigmentosum as we now understand it.

In addition to the usual severe, recessively inherited xeroderma pigmentosum (278700-278780), the existence of a milder form behaving as a dominant was claimed by Anderson and Begg (1950) who described 11 affected persons in 5 sibships of 4 generations of a Scottish family by the name of MacPherson. The patients showed freckling and multiple skin cancers as in the recessive form but did not get trouble as early in life and survived longer. Indeed, Anderson and Begg examined 1 affected member of the family who was 74 years of age. No affected member was said to have died of the disease. Sedano (1984) raised doubts about this family. In the Scottish family, Anderson personally examined 7 members: 'Mr. George MacPherson, sen., his wife, George, jun., Harold, Lena, Douglas, and the son of George, jun. (Ian).' All these except the wife were said to be affected; the authors even raised the question of whether the wife might be 'a mild case;' was she in fact a carrier and is this an example of pseudodominance? Actually the father of 'George, sen.,' the authors stated, 'was known to have been affected.' Anderson and Begg (1950) were attracted to the possibility that this man's wife and the wife of 'George, sen.' were both heterozygotes. Skin tumors, mostly squamous epitheliomas, had onset as early as age 14. All the affected persons were 'rufous' (red-haired). Tests for increased sensitivity to ultraviolet light were equivocal or negative. Although the disorder in this kindred was milder than ordinary xeroderma pigmentosum ('George was in a tropical climate for many years, being invalided home on two occasions on account of the disease' and 'Douglas was invalided home from North Africa in 1943'), 'Douglas was considered to be a case of xeroderma pigmentosum by all the members present when he was shown at a meeting of the North British Dermatological Society in 1946.' The means are now available, of course, to answer the question of whether this is true xeroderma pigmentosum. I am inclin ed to think that this is an example of unusually fair skin, prone to develop changes including malignancy, but not xeroderma pigmentosum as we now know it. The inheritance may be dominant, or more likely multifactorial.

Imray et al. (1986) described an Australian kindred with persons in 5 generations affected by a mild form of XP. The features were marked xeroderma, generalized dark pigmentation, excessive freckling, and severe erythema in response to sunlight. Skin cancer was not common in this family; the proband had a melanoma and her mother had a basal cell carcinoma removed at age 70. Cleaver (1981) described autosomal dominant predisposition to multiple skin cancers in exposed facial skin. Fibroblasts from affected persons showed no defects in excision repair. On the other hand, in the Australian kindred, Imray et al. (1986) found, in studying lymphoblastoid cells and fibroblasts, cellular sensitivity to ultraviolet light as judged by diminished clonogenicity and increased frequency of UV-induced chromosome aberrations. After UV exposure, cells showed abnormally great depression of replicative DNA synthesis and the level of UV-induced DNA repair synthesis was lower than normal. The number of sister-chromatid exchanges and 6-thioguanine-resistant mutants was not increased.

Anderson, T. E.; Begg, M.: Xeroderma pigmentosum of mild type. Brit. J. Derm. 62: 402-407, 1950.
Cleaver, J. E.: An autosomal dominant inheritance for multiple sunlight-induced malignancy in a patient without abnormalities in DNA repair or replication. Cytogenet. Cell Genet. 29: 122-124, 1981.
Imray, F. P.; Hockey, A.; Relf, W.; Ramsay, R. G.; Kidson, C.: Sensitivity to ultraviolet radiation in a dominantly inherited form of xeroderma pigmentosum. J. Med. Genet. 23: 72-78, 1986.
Sedano, H. O.: Personal Communication. Minneapolis, Minn. 9/18/1984.

Genodermatoses with Skin Cancers

EPIDERMOLYSIS BULLOSA DYSTROPHICA, PASINI TYPE (#131750)

JJM NOTE

Skin malignancies seem to complicate only the scarring, dystrophic type of epidermolysis, so the mechanism of carcinogenesis, as in bony infarctions in sickle cell anemia, is likely to be chronic inflammation and tissue repair.

A number sign (#) is used with this entry because of the evidence that the causative mutation resides in the gene for type VII collagen (COL7A1; 120120).

Scarring and malignancy occur with healing of lesions in the dystrophic form of epidermolysis bullosa but not in the 'simplex' forms. In the simplex type the blisters occur within the epidermis and are subcorneal, whereas the blisters are subepidermal in the dystrophic form. Davison (1965) had 6 families with the dystrophic type, of which 4 were dominant and 2 recessive. Whether changes in the nails occur or not is a simple way to distinguish the dystrophic and simplex forms, according to the suggestion of Gedde-Dahl (1977). The dominant forms of EBD are those of Cockayne-Touraine (131800), Pasini, and Bart (132000). Aside from the peculiar Pasini papules, the feature distinguishing the Cockayne-Touraine and Pasini types of EBD is the greater severity of the second. Both are distinguished from the several forms of EB simplex (131800, 131900, 131950, 131960) by the predilection for blisters on the dorsal aspects of the extremities, by milia formation as a sequel of recent bullae, and by red atrophic scarring after recurrent blisters. Blisters are present at birth or from the first weeks of life. The limbs are predominantly involved, but blisters may occur on the ears and rarely the face during infancy, and later also on the buttocks. Blisters in the mouth are common. The albopapuloid Pasini papule appears on the trunk during adolescence. In the Cockayne-Touraine type the nails may be hypertrophic, leading to Touraine's designation (1942) of EB hyperplastica, but this is not a consistent finding. In Faroe Islanders, Joensen et al. (1979) found that dystrophic epidermolysis bullosa was unlinked to GPT and thus distinct from the Ogna type of EBD simplex (131950).

A mutation in the structural gene for anchoring fibril protein was postulated by Anton-Lamprecht and Hashimoto (1976). Because type VII collagen is a major collagenous component of the anchoring fibrils, the COL7A1 gene (120120) was, in the opinion of Ryynanen et al. (1991), a candidate gene for the mutation in some families with the dystrophic forms of epidermolysis bullosa. Indeed, they found linkage between a PvuII RFLP and the disorder; maximum lod = 5.37 at theta = 0.0. Furthermore, they demonstrated that the COL7A1 gene is located on chromosome 3 by hybridization to a panel of human-rodent somatic cell hybrids. In 3 British families, Al-Imara et al. (1992) found close linkage of EBDD to D3S2, a marker known to be close to the COL7A1 locus; combined lod score = 6.75 at theta = 0. Ryynanen et al. (1992) presented further evidence of the EBDD and COL7A1 linkage; maximum lod = 8.77 at theta = 0.0. Later, Uitto et al. (1992) extended the linkage studies to a higher lod score value with no recombinants. In 2 Dutch kindreds with intrafamilial characteristics of both the Cockayne-Touraine type and the Bart type of autosomal dominant dystrophic epidermolysis bullosa, Gruis et al. (1992) found linkage to COL7A1 with no recombination; maximum lod = 6.08 at theta = 0.00. This contributes to the accumulating evidence that the 3 types of autosomal dominant epidermolysis bullosa dystrophica, namely, the Cockayne-Touraine type, the Pasini type, and Bart syndrome, are due to mutations in the same gene, that for type VII collagen. Interestingly, generalized mutilating recessive dystrophic epidermolysis bullosa (226600) likewise shows absolute linkage to COL7A1. The difference in the nature of the clinical picture may be related to differences in the mutation. Heterozygous expression may be due to 'protein suicide' resulting from some mutations, whereas other mutations may require homozygosity for expression.

Anton-Lamprecht (1978) pointed out that electron microscopy is particularly revealing in dominant disorders in which structural abnormality of a protein is likely to be found whereas biochemistry is more likely to be revealing in recessive disorders. The examples he used from dermatology to illustrate electron microscopic abnormalities in dominant disorders were: structural defects of tonofibrils in hystrix-like ichthyoses, of the anchoring fibrils in dominant dystrophic epidermolysis bullosa of Pasini, and of keratohyalin in autosomal dominant ichthyosis vulgaris. Bauer et al. (1979) found that cultured fibroblasts from patients with this disorder display deranged glycosaminoglycan metabolism. The cells accumulated increased amounts of sulfated glycosaminoglycans, due, it seems, to increased synthesis. Secretion of glycosaminoglycans by the cells was also increased.

Pretibial epidermolysis bullosa is a rare form of dominant dystrophic epidermolysis bullosa. Naeyaert et al. (1995) described a large Belgian family in which this disorder was remarkable for its late age of onset and its misleading clinical presentation in the proband in whom the disorder strongly resembled keratosis lichenoides chronica. Since both recessively and dominantly inherited forms of dystrophic epidermolysis bullosa had been shown to be linked to COL7A1, they performed a 2-point linkage analysis using 2 intragenic COL7A1 polymorphisms. Strong genetic linkage was found between the disease in this family and COL7A1; lod score = 4.45 at theta = 0.0 for 1 of the 2 polymorphisms. The observed intrafamilial variability of clinical phenotypes contradicted the existing classification of dominantly inherited dystrophic EB. Pretibial EB is characterized by recurrent blistering and scarring, mainly in the pretibial area. The findings from this family indicate that pretibial EB is yet another form of COL7A1 related EB. In this family, the onset was after 10 years of age and the lesions were strikingly lichenoid.

See also Anton-Lamprecht and Schnyder (1973); Gedde-Dahl (1978); Touraine (1942)

Al-Imara, L.; Richards, A. J.; Eady, R. A. J.; Leigh, I. M.; Farrall, M.; Pope, F. M.: Linkage of autosomal dominant dystrophic epidermolysis bullosa in three British families to the marker D3S2 close to the COL7A1 locus. J. Med. Genet. 29: 381-382, 1992.

Anton-Lamprecht, I.: Electron microscopy in the early diagnosis of genetic disorders of the skin. Dermatologica 157: 65-85, 1978.

Anton-Lamprecht, I.; Hashimoto, I.: Epidermolysis bullosa dystrophica dominans (Pasini)—a primary structural defect of the anchoring fibrils. Hum. Genet. 32: 69-76, 1976.

Anton-Lamprecht, I.; Schnyder, U. W.: Epidermolysis bullosa dystrophica dominans—ein Defekt der anchoring fibrils?. Dermatologica 147: 289-298, 1973.

Bauer, E. A.; Fiehler, W. K.; Esterly, N. B.: Increased glycosaminoglycan accumulation as a genetic characteristic in cell cultures of one variety of dominant dystrophic epidermolysis bullosa. J. Clin. Invest. 64: 32-39, 1979.

Davison, B. C. C.: Epidermolysis bullosa. J. Med. Genet. 2: 233-242, 1965.

Gedde-Dahl, T., Jr.: Personal Communication. Oslo, Norway 1977.

Gedde-Dahl, T., Jr.: Classification of epidermolysis bullosa. In: Herzberg, J. J.; Korting, G. W.: Padiatrische Dermatologie. Stuttgart: F. K. Schattauer (pub.) 1978. Pp. 65-91.

Gruis, N. A.; Bouwes Bavinck, J. N.; Steijlen, P. M.; van der Schroeff, J. G.; van Haeringen, A.; Happle, R.; Mariman, E.; van Beersum, S. E. C.; Uitto, J.; Vermeer, B. J.; Frants, R. R.: Genetic linkage between the collagen VII (COL7A1) gene and the autosomal dominant form of dystrophic epidermolysis bullosa in two Dutch kindreds. Invest. Derm. 99: 528-530, 1992.

Joensen, H. D.; Hansen, H. E.; Henningsen, K.; Svejgaard, A.; Andersen, I.: A study of the linkage relations of epidermolysis bullosa dystrophica. Hum. Hered. 29: 221-225, 1979.

Naeyaert, J. M.; Nuytinck, L.; De Bie, S.; Beele, H.; Kint, A.; De Paepe, A.: Genetic linkage between the collagen type VII gene COLA7A and pretibial epidermolysis bullosa with lichenoid features. J. Invest. Derm. 104: 803-805, 1995.

Ryynanen, M.; Knowlton, R. G.; Parente, M. G.; Chung, L. C.; Chu, M.-L.; Uitto, J.: Human type VII collagen: genetic linkage of the gene (COL7A1) on chromosome 3 to dominant dystrophic epidermolysis bullosa. Am. J. Hum. Genet. 49: 797-803, 1991.

Ryynanen, M.; Ryynanen, J.; Sollberg, S.; Iozzo, R. V.; Knowlton, R. G.; Uitto, J.: Genetic linkage of type VII collagen (COL7A1) to dominant dystrophic epidermolysis bullosa in families with abnormal anchoring fibrils. J. Clin. Invest. 89: 974-980, 1992.

Touraine, M. A.: Classification des epidermolyses bulleuses. Ann. Derm. Syph. 8: 138-144, 1942.

Uitto, J.; Ryynanen, M.; Christiano, A. M.; Hovnanian, A.; Frantz, R.; Bauer, E. A.; Knowlton, R. G.: Genetic linkage of the type VII collagen gene (COL7A1) to dominant dystrophic epidermolysis bullosa (DDEB) in families with abnormal anchoring fibrils. (Abstract) Clin. Res. 40: 188A only, 1992.

EPIDERMOLYSIS BULLOSA DYSTROPHICA, HALLOPEAU-SIEMENS TYPE (#226600)

JJM NOTE

The multiplicity of cancers may be related to the amount and duration of scarring.

A number sign (#) is used with this entry because of the evidence that the causative mutation resides in the gene for type VII collagen (COL7A1; 120120).

This severe and destructive form of epidermolysis bullosa may be present at birth or appear in infancy. Hands, feet, elbows, and knees are sites of predilection. Bullae also develop on the mucosal surfaces and even the conjunctiva and cornea may be involved. The impressive kindred reported by Hofman is diagrammed on page 279 of von Verschuer (1959). Destro et al. (1987) described an affected 40-year-old woman who had corneal involvement. She had fusion of all fingers and toes into mittenlike deformities and severe contractures of all 4 limbs. She had survived a spontaneous esophageal perforation and had had 15 squamous cell carcinomas removed from the limbs. A similarly affected sister died at the age of 26 years from metastatic squamous cell carcinoma. Gedde-Dahl (1977) thought that some clinical variants of epidermolysis bullosa are the result of the presence of two nonidentical recessive alleles. Thus, the genes responsible for some or all of the several forms of recessive epidermolysis bullosa may be at the same locus. The Hallopeau-Siemens form of recessive EBD and EBD inversa show an ultrastructural similarity, viz., dermolytic separation. They are distinguished on clinical grounds. The H-S type shows wide variability in severity but within families tends to 'breed true,' suggesting to Gedde-Dahl (1977) the existence of multiple allelic forms and genetic compounds. Some of the cases are of the severe 'mutilans' type leading to an acquired syndactyly and esophageal stricture. The dental enamel is usually visibly defective in severe cases. On the basis of an analysis of 246 patients with epidermolysis bullosa of various types, Travis et al. (1992) reported that dysphagia developed in 76% of those with recessive dystrophic EB, in 20% of those with dominant dystrophic EB (131750), in 15% of those with junctional EB (226700), and in 2% of those with simplex forms. Lingual adhesions or microstomia occurred in dystrophic epidermolysis bullosa only, but were 8 times more common in the recessive form than in the dominant form. These lesions are provoked by the trauma of eating and reduce food intake which exacerbates constipation caused by anal blisters and results in malnutrition. Stricture of the esophagus was frequent, with single or multiple esophageal webs.

Bourke et al. (1995) observed fatal systemic amyloidosis of the AA type. One sister was 22 years old when the diagnosis of amyloidosis was made. Despite rapidly deteriorating renal function, dialysis was deemed impossible because of her extensive cutaneous infection. The older sister had negative findings of a search for amyloidosis at the age of 26 years. Although her skin disease was equally as severe as her sister's, she did not develop amyloid nephropathy until the age of 35 years.

Bauer (1977) studied the properties of procollagenase purified from fibroblasts of 2 patients and found peculiarities that he concluded were consistent with either a structural gene mutation, defective posttranslational modification of the enzyme, or a mutation in a gene regulating normal degradation of collagenase. Collagenolytic changes are demonstrated by electron microscopy. Fibroblasts cultured from patients with autosomal recessive dystrophic epidermolysis bullosa demonstrate an increased capacity to synthesize and secrete collagenase (Bauer and Eisen, 1978). Increased levels of immunoreactive collagenase are found in unaffected and affected areas of the skin. A structural abnormality of the collagenase produced by cultured fibroblasts is suspected. Bauer et al. (1986) found that enhanced expression of collagenase by fetal recessive dystrophic epidermolysis bullosa (RDEB) skin fibroblasts can serve as a biochemical adjunct and possibly an alternative to morphologic examination of tissue for antenatal diagnosis. Phenytoin was thought to be effective in the systemic treatment of RDEB (Bauer et al., 1980); however, in a controlled study, Caldwell-Brown et al. (1992) showed that it is without effect.

Anton-Lamprecht et al. (1981) achieved prenatal diagnosis by inspection of the skin through the fetoscope, confirmed by electron microscopic examination of a skin biopsy. Bass et al. (1993) described the case of a prematurely born female with this disorder whose mother had strikingly elevated mid-trimester serum and amniotic fluid concentrations of alpha-fetoprotein (AFP; 104150), a positive amniotic fluid acetylcholinesterase band, and negative serial ultrasound studies.

In a family with consanguineous parents and 3 affected children, Hovnanian et al. (1991) excluded linkage between recessive dystrophic epidermolysis bullosa and the collagenase gene (CLG; 120353). In 8 other families, in 2 of which 2 affected sibs were affected, they could not exclude the possibility of linkage. RDEB may be heterogeneous. Indeed, Bauer and Eisen (1978) found enhanced collagenase production by cultured skin fibroblasts in only 8 of 10 patients, and Winberg et al. (1989) found collagenase overexpression in only 4 of 18 RDEB patients. In the family with 3 affected sibs in which linkage studies excluded the involvement of the collagenase locus, Hovnanian et al. (1991) found high levels of collagenase mRNA in only 2 of the 3 affected sibs, suggesting that it was not the primary defect in that family. By linkage studies, Colombi et al. (1992) excluded the interstitial collagenase gene as the one responsible for severe generalized recessive epidermolysis bullosa dystrophica. They also excluded stromelysin I (STMY1; 185250) and stromelysin II (STMY2; 185260). The same family had other members affected with a form of cerebellar ataxia of postpubertal onset similar to a form that may be due to a gene in the same area as CLG, STMY1, and STMY2 on 11q; in their family cerebellar ataxia was not linked to any of these. They also excluded linkage of both pathologic phenotypes with the fibronectin gene (FN1; 135600) on chromosome 2.

In a dominant form of EBD (131750), Ryynanen et al. (1991) demonstrated linkage to the type VII collagen gene (COL7A1; 120120), suggesting that this is the site of the mutation in that form of the disease. Type VII collagen, the major component of anchoring fibrils, is susceptible to degradation by collagenase (Seltzer et al., 1989). Hovnanian et al. (1992) demonstrated linkage between a PvuII polymorphic site in the type VII collagen gene and recessive dystrophic epidermolysis bullosa, the generalized mutilating disorder. In 19 informative families, they found a maximum lod score of 3.95 at theta = 0.0, with no obligatory recombinants. It is of interest that the dominant dystrophic epidermolysis bullosa is probably also caused by mutation in the COL7A1 gene because it shows absolute linkage. This is likely to be an example of different mutations in the same gene causing either recessive inheritance or dominant inheritance depending on the type of change produced in the protein gene product. Heterozygous expression is probably due to 'protein suicide' resulting from disruptive effects of the mutation in the formation of the trimeric collagen molecule, whereas the mutations in the case of the recessive form are the type that must be homozygous to be expressed.

Melville et al. (1996) reported 2 unrelated children with autosomal recessive dystrophic EB who developed fatal dilated cardiomyopathy. Both were malnourished and showed severely retarded growth. The authors suggested that the likely cause for the cardiomyopathy was a micronutrient deficiency, most probably selenium deficiency, since the serum selenium level was reduced in the case in which they measured it, and also in 14 of 25 other children with dystrophic epidermolysis bullosa. Echocardiographic screening of 18 other patients with recessive dystrophic epidermolysis bullosa showed no evidence of cardiomyopathy.

See also Anton-Lamprecht (1981); Bauer et al. (1977); Book (1952); Church et al. (1983); Davison (1965); Didolkar et al. (1974); Gerhard et al. (1987); Goldberg et al. (1986); Heinrichsbauer (1928); Kanan et al. (1977); Reed et al. (1974); Robinson (1946); Schnyder and Eichhoff (1963); Sorsby (1953); Sorsby et al. (1951); Spurr et al. (1988); Stricklin et al. (1982); Thompson et al. (1980); Tidman and Eady (1984)

Anton-Lamprecht, I.: Prenatal diagnosis of genetic disorders of the skin by means of electron microscopy. Hum. Genet. 59: 392-405, 1981.

Anton-Lamprecht, I.; Rauskolb, R.; Jovanovic, V.; Kern, B.; Arnold, M.-L.; Schenck, W.: Prenatal diagnosis of epidermolysis bullosa dystrophica Hallopeau-Siemens with electron microscopy of fetal skin. Lancet II: 1077-1079, 1981.

Bass, H. N.; Miranda, C.; Oei, R.; Crandall, B. F.: Association of generalized dystrophic epidermolysis bullosa with positive acetylcholinesterase and markedly elevated maternal serum and amniotic fluid alpha-fetoprotein. Prenatal Diag. 13: 55-59, 1993.

Bauer, E. A.: Recessive dystrophic epidermolysis bullosa: evidence for an altered collagenase in fibroblast cultures. Proc. Nat. Acad. Sci. 74: 4646-4650, 1977.

Bauer, E. A.; Cooper, T. W.; Tucker, D. R.; Esterly, N. B.: Phenytoin therapy of recessive dystrophic epidermolysis bullosa: clinical trial and proposed mechanism of action on collagenase. New Eng. J. Med. 303: 776-781, 1980.

Bauer, E. A.; Eisen, A. Z.: Recessive dystrophic epidermolysis bullosa: evidence for increased collagenase as a genetic characteristic in cell culture. J. Exp. Med. 148: 1378-1387, 1978.

Bauer, E. A.; Gedde-Dahl, T., Jr.; Eisen, A. Z.: The role of human skin collagenase in epidermolysis bullosa. J. Invest. Derm. 68: 119-124, 1977.

Bauer, E. A.; Ludman, M. D.; Goldberg, J. D.; Berkowitz, R. L.; Holbrook, K. A.: Antenatal diagnosis of recessive dystrophic epidermolysis bullosa: collagenase expression in cultured fibroblasts as a biochemical marker. J. Invest. Derm. 87: 597-601, 1986.

Book, J. A.: Frequence de mutation de la chondrodystrophie et de l'epidermolyse bulleuse dans une population du sud de la Suede. J. Genet. Hum. 1: 24-26, 1952.

Bourke, J. F.; Browne, G.; Gaffney, E. F.; Young, M.: Fatal systemic amyloidosis (AA type) in two sisters with dystrophic epidermolysis bullosa. J. Am. Acad. Derm. 33: 370-372, 1995.

Caldwell-Brown, D.; Stern, R. S.; Lin, A. N.; Carter, D. M.; the Epidermolysis Bullosa Study Group: Lack of efficacy of phenytoin in recessive dystrophic epidermolysis bullosa. New Eng. J. Med. 327: 163-167, 1992.

Church, R. L.; Bauer, E. A.; Eisen, A. Z.: Human skin collagenase: assignment of the structural gene to chromosome 11 in both normal and recessive dystrophic epidermolysis bullosa cells using human-mouse somatic cell hybrids. Collagen Rel. Res. 3: 115-124, 1983.

Colombi, M.; Gardella, R.; Zoppi, N.; Moro, L.; Marini, D.; Spurr, N. K.; Barlati, S.: Exclusion of stromelysin-1, stromelysin-2, interstitial collagenase and fibronectin genes as the mutant loci in a family with recessive epidermolysis bullosa dystrophica and a form of cerebellar ataxia. Hum. Genet. 89: 503-507, 1992.

Davison, B. C. C.: Epidermolysis bullosa. J. Med. Genet. 2: 233-242, 1965.

Destro, M.; Wallow, I. H. L.; Brightbill, F. S.: Recessive dystrophic epidermolysis bullosa. Arch. Ophthal. 105: 1248-1252, 1987.

Didolkar, M. S.; Gerner, R. E.; Moore, G. E.: Epidermolysis bullosa dystrophica and epithelioma of the skin: review of published cases and report of an additional patient. Cancer 33: 198-202, 1974.

Gedde-Dahl, T., Jr.: Personal Communication. Oslo, Norway 1977.

Gerhard, D. S.; Jones, C.; Bauer, E. A.; Eisen, A. Z.; Goldberg, G. I.: Human collagenase gene is localized to 11q. (Abstract) Cytogenet. Cell Genet. 46: 619 only, 1987.
Goldberg, G. I.; Wilhelm, S. M.; Kronberger, A.; Bauer, E. A.; Grant, G. A.; Eisen, A. Z.: Human fibroblast collagenase: complete primary structure and homology to an oncogene transformation-induced rat protein. J. Biol. Chem. 261: 6600-6605, 1986.
Heinrichsbauer, F.: Ein weiterer Beitrag zur Frage angeborener Hautdefekte. (Ueber ein familiaeres letales Krankheitsbild mit Blasenbildung und angeborenen Defekten der Haut.). Arch. Gynaek. 134: 673-692, 1928.
Hovnanian, A.; Duquesnoy, P.; Amselem, S.; Blanchet-Bardon, C.; Lathrop, M.; Dubertret, L.; Goossens, M.: Exclusion linkage between the collagenase gene and generalized recessive dystrophic epidermolysis bullosa phenotype. J. Clin. Invest. 88: 1716-1721, 1991.
Hovnanian, A.; Duquesnoy, P.; Blanchet-Bardon, C.; Knowlton, R. G.; Amselem, S.; Lathrop, M., II; Dubertret, L.; Uitto, J.; Goossens, M.: Genetic linkage of recessive dystrophic epidermolysis bullosa to the type VII collagen gene. (Abstract) Clin. Res. 40: 188A only, 1992.
Kanan, M. W.; Francis, M. J. O.; Sykes, B.; Reed, W. B.; Ryan, T. J.; van Diest, P.; Marsden, A.: Preponderance of lysosomal bodies in cultured fibroblasts from patients with recessive epidermolysis bullosa dystrophica: an electron microscopic study. Brit. J. Derm. 96: 521-532, 1977.
Melville, C.; Atherton, D.; Burch, M.; Cohn, A.; Sullivan, I.: Fatal cardiomyopathy in dystrophic epidermolysis bullosa. Brit. J. Derm. 135: 603-606, 1996.
Reed, W. B.; College, J., Jr.; Francis, M. J. O.; Zachariae, H.; Mohs, F.; Sher, M. A.; Sneddon, I. B.: Epidermolysis bullosa dystrophica with epidermal neoplasm. Arch. Derm. 110: 894-902, 1974.
Robinson, M. M.: Epidermolysis bullosa hereditaria. Urol. Cutan. Rev. 50: 545-561, 1946.
Ryynanen, M.; Knowlton, R. G.; Parente, M. G.; Chung, L. C.; Chu, M.-L.; Uitto, J.: Human type VII collagen: genetic linkage of the gene (COL7A1) on chromosome 3 to dominant dystrophic epidermolysis bullosa. Am. J. Hum. Genet. 49: 797-803, 1991.
Schnyder, U. W.; Eichhoff, D.: Zur Klinik und Genetik der dominant-dystrophischen Epidermolysis bullosa hereditaria. Arch. Derm. 218: 62-90, 1963.
Seltzer, J. L.; Eisen, A. Z.; Bauer, E. A.; Morris, N. P.; Glanville, R. W.; Burgeson, R. E.: Cleavage of type VII collagen by interstitial collagenase and type IV collagenase (gelatinase) derived from human skin. J. Biol. Chem. 264: 3822-3826, 1989.
Sorsby, A.: Clinical Genetics. St. Louis: C. V. Mosby (pub.) 1953. Pp. 136 only.
Sorsby, A.; Roberts, J. A. F.; Brain, R. T.: Essential shrinking of conjunctiva in hereditary affection allied to epidermolysis bullosa. Docum. Ophthal. 5-6: 118-150, 1951.
Spurr, N. K.; Gough, A. C.; Gosden, J.; Rout, D.; Porteous, D. J.; van Heyningen, V.; Docherty, A. J. P.: Restriction fragment length polymorphism analysis and assignment of the metalloproteinases stromelysin and collagenase to the long arm of chromosome 11. Genomics 2: 119-127, 1988.
Stricklin, G. P.; Welgus, H. G.; Bauer, E. A.: Human skin collagenase in recessive dystrophic epidermolysis bullosa: purification of a mutant enzyme from fibroblast cultures. J. Clin. Invest. 69: 1373-1383, 1982.
Thompson, J. W.; Ahmed, A. R.; Dudley, J. P.: Epidermolysis bullosa dystrophica of the larynx and trachea: acute airway obstruction. Ann. Otol. 89: 428-429, 1980.
Tidman, M. J.; Eady, R. A. J.: Evidence for a functional defect of the lamina lucida in recessive dystrophic epidermolysis bullosa demonstrated by suction blisters. Brit. J. Derm. 111: 379-387, 1984.
Travis, S. P. L.; McGrath, J. A.; Turnbull, A. J.; Schofield, O. M.; Chan, O.; O'Connor, A. F.; Mayou, B.; Eady, R. A. J.; Thompson, R. P. H.: Oral and gastrointestinal manifestations of epidermolysis bullosa. Lancet 340: 1505-1506, 1992.
von Verschuer, O. F.: Genetik des Menschen. Lehrbuch der Humangenetik. Berlin: Urban und Schwarzenberg (pub.) 1959.
Winberg, J.-O.; Gedde-Dahl, T., Jr.; Bauer, E. A.: Collagenase expression in skin fibroblasts from families with recessive dystrophic epidermolysis bullosa. J. Invest. Derm. 92: 82-85, 1989.

COLLAGEN, TYPE VII, ALPHA-1; COL7A1 (*120120)

JJM NOTE

Since this is the underlying gene defect in the previous clinical entity, it is included here as the cancer-predisposing gene.

From human chorioamniotic membranes, Bentz et al. (1983) isolated a distinctive type of collagen which from its amino acid composition and other characteristics must be the product of a previously unrecognized gene. It consists of 3 identical alpha chains, each with a molecular weight of about 170,000. The authors gave this collagen the trivial name long-chain (LC) collagen and suggested that it be referred to as type VII collagen. Collagen VII has a triple-helical domain almost half again longer than the type I collagen triple helix. Collagen VII is the main constituent of anchoring fibrils, which in the skin are located below the basal lamina at the dermal-epidermal basement membrane zone. The collagen VII molecules form disulfide bond stabilized dimeric aggregates by lateral accretion in a nonstaggered array (Burgeson et al., 1985). By Northern hybridization studies using a cDNA, Ryynanen et al. (1992) demonstrated a 9-kb mRNA transcript with a high level of expression in epidermal keratinocytes and in an oral epidermoid carcinoma cell line and with considerably lower expression in skin fibroblasts. Indirect immunofluorescence of skin from a 19-week fetus showed type VII collagen gene expression at the dermal-epidermal basement membrane zone. They interpreted these results as indicating that epidermal keratinocytes may be the primary cell source of type VII collagen in developing human skin. Tanaka et al. (1992) isolated a cDNA of approximately 1 kb from a human keratinocyte library. The deduced primary structure of the clone reflected the noncollagenous domain of type VII collagen that may be involved in cell attachment. The region showed weak homology (approximately 23%) to the cell attachment domain of fibronectin. Greenspan (1993) described the carboxyl-terminal half of type VII collagen and the intron/exon organization of the corresponding region of the COL7A1 gene. Christiano et al. (1994) found that the COL7A1 gene has 118 exons, more than any previously des cribed gene. Despite this complexity, COL7A1 is compact. Consisting of 31,132 bp from transcription start site to polyadenylation site, it is only about 3 times the size of type VII collagen mRNA. Thus, COL7A1 introns are small. A 71-nucleotide COL7A1 intron is the smallest intron reported in a collagen gene, and only 1 COL7A1 intron is greater than 1 kb long. Christiano et al. (1994) stated that the type VII collagen mRNA is approximately 9.2 kb long, with an open reading frame of 8,833 nucleotides encoding 2,944 amino acids. They reported the complete cDNA and primary amino acid sequences and also described polymorphisms in the gene.

Parente et al. (1991) cloned a cDNA for type VII collagen and used it for chromosomal in situ hybridization to localize the COL7A1 gene to 3p21. By fluorescence in situ hybridization, Greenspan et al. (1993) narrowed the assignment to 3p21.3. Knowlton et al. (1991) and Ryynanen et al. (1991) mapped the COL7A1 gene to chromosome 3 by analysis of human-rodent somatic cell hybrids. By use of interspecific backcross mapping, Li et al. (1993) showed that the corresponding gene is located on chromosome 9 in the mouse.

Knowlton et al. (1991) and Ryynanen et al. (1991) demonstrated close linkage (with no recombination) of COL7A1 to autosomal dominant dystrophic epidermolysis bullosa (DEB; see 131750). In a study of 19 informative families with recessive DEB (226600), Hovnanian et al. (1992) found a maximum lod score of 3.95 at theta = 0.0 with no recombinants. Also, Uitto et al. (1992) demonstrated absolute linkage between a PvuII RFLP of the COL7A1 gene and dominant dystrophic epidermolysis bullosa; in 4 informative families a combined lod score of 14.6 at theta = 0 was found, with no recombinants. Thus, 2 forms of epidermolysis bullosa, one dominant and one recessive, are due to mutations in the same gene. It is likely that the mutations with heterozygous expression produce a change in the protein gene product causing 'protein suicide,' i.e., disruption in the formation of the trimeric collagen molecule. On the other hand, the recessive mutations may cause a change in the gene product of such a nature that all the gene product must be of the mutant type for the phenotype to be expressed. Christiano et al. (1995) described nonsense mutations resulting in a premature termination codon in the amino-terminal portion of the COL7A1 gene in 4 COL7A1 alleles from 3 unrelated patients with severe, mutilating recessive DEB (see 120120.0005). One of the patients was a compound heterozygote. Heterozygous carriers of the nonsense mutations were clinically unaffected although they showed a 50% reduction in anchoring fibrils.

Type VII collagen appears to be restricted to the basement membrane zone beneath stratified squamous epithelia. Within the cutaneous basement membrane zone, type VII collagen localizes to the lamina densa and sub-lamina densa areas in the upper papillary dermis. More precisely, immunolocalization demonstrated that type VII collagen is a major collagenous component of anchoring fibrils. The acquired form of epidermolysis bullosa, EB acquisata (EBA), is an autoimmune disorder resulting from autoantibodies to type VII collagen. By Western immunoblotting analysis with sera from 19 patients with EBA, using bacterial collagenase- or pepsin-resistant portions of type VII collagen and a panel of 12 recombinant fusion proteins corresponding to approximately 80% of the primary sequence of the COL7A1 polypeptide, Lapiere et al. (1993) identified 4 major immunodominant epitopes within the noncollagenous (NC1) domain. The pattern of epitopes recognized by the sera from 3 patients with bullous systemic lupus erythematosus (BSLE) was similar to that found with EBA, suggesting that the same epitopes could serve as autoantigens in both blistering conditions. In contrast, sera from healthy controls or from patients with unrelated blistering skin diseases did not react with type VII collagen epitopes. Lapiere et al. (1993) postulated that such antibodies could disrupt the assembly of type VII collagen into anchoring fibrils and/or interfere with their interactions with other extracellular matrix molecules within the cutaneous basement membrane zone.

Hovnanian et al. (1995) used COL7A1 gene analysis for successful first-trimester prenatal diagnosis in 6 families at risk for recurrence of generalized recessive DEB. The disorder was of the severe Hallopeau-

Siemens form in 5 families and the generalized nonmutilating form in 1. In all cases analysis of fetal DNA from amniotic fluid cells showed that the fetus had inherited at least one normal COL7A1 allele.

Zelickson et al. (1995) studied the descendants of the original family with Bart syndrome (132000), a skin disorder involving congenital localized absence of skin and deformity of nails, and found linkage of the disorder to 3p, at or near the site of the COL7A1 gene. Christiano et al. (1996) performed mutation analysis in this family using electrophoretic heteroduplex analysis followed by direct nucleotide sequencing (see 120120.0008).

Christiano et al. (1996) demonstrated the wide variability in the phenotype of DEB and the differences in patterns of inheritance with various glycine substitutions in the triple-helical region of type VII collagen. They reported 6 new families and compared the results with other data. Among the 6 families, 2 demonstrated a mild phenotype, and the inheritance was consistent with the dominantly inherited form of DEB. In the 4 other families, the mutation was silent in the heterozygous state but, when present in the homozygous state or combined with a second mutation, resulted in a recessively inherited DEB phenotype. Christiano et al. (1996) stated that the COL7A1 gene is, therefore, unique among the collagen genes in that different glycine substitutions can be either silent in heterozygotes or can result in a dominantly inherited DEB. Inspection of the location of the glycine substitutions along the COL7A1 polypeptide suggested to the authors that the consequences of the mutations, in terms of phenotype and pattern of inheritance, are position-independent. In an accompanying article Christiano et al. (1996) described compound heterozygosity in twins with severe DEB. The twins had inherited from the father a recessive deletion/insertion mutation and from the mother a dominant-negative maternal glycine substitution. Careful questioning of the mother revealed that she and her father had a history of shedding of toenails and an occasional poor healing of erosions, consistent with a mild form of dominantly inherited DEB. The mutation from the father was a 2-bp deletion/1-bp insertion in exon 56 (designated 5103CC-to-G by the authors) which resulted in a frameshift and a downstream premature termination codon. The mutation from the mother was a glycine-to-arginine substitution in exon 91 (G2351R).

Hovnanian et al. (1997) characterized 21 mutations in the COL7A1 gene, 18 of which had not been previously reported, in patients from 15 unrelated families with recessive dystrophic epidermolysis bullosa (RDEB). Mutations in both alleles were identified by screening the 118 exons of COL7A1 and flanking intronic regions. Fourteen mutations created premature termination codons and consisted of nonsense mutations, small insertions, deletions, and splice site mutations. A further 7 mutations predicted glycine or arginine substitutions in the collagenous domain of the molecule. Two mutations were found in more than 1 family, and 6 of the 7 missense mutations showed clustering within exons 72 to 74 next to the hinge region of the protein. Patients who were homozygous or compound heterozygotes displayed both absence or drastic reduction of COL7A1 transcripts and undetectable type VII collagen in skin. In contrast, missense mutations were associated with clearly detectable COL7A1 transcripts and with normal or reduced expression of type VII collagen at the dermo/epidermal junction. The results provided evidence for at least 2 distinct molecular mechanisms underlying defective anchoring fibril formation in RDEB: one involving premature termination codons leading to mRNA instability and absence of protein synthesis, the other implicating missense mutations resulting in the synthesis of type VII collagen with decreased stability and/or altered function. Genotype/phenotype correlations suggested to Hovnanian et al. (1997) that the nature and location of these mutations are important determinants of the disease phenotype and showed evidence for interfamilial phenotypic variability.

Jarvikallio et al. (1997) reviewed mutations in the COL7A1 gene as maintained in a database in Philadelphia. Hammami-Hauasli et al. (1998) investigated naturally occurring COL7A1 mutations and showed that some, but not all, glycine substitutions in collagen VII interfere with biosynthesis of the protein in a dominant-negative manner. Three point mutations in exon 73 caused glycine substitutions, G2006D, G2034R, and G2015E, in the triple helical domain of collagen VII and interfered with its folding and secretion. Confocal laser scanning studies and semiquantitative immunoblotting determined that dystrophic epidermolysis bullosa keratinocytes retained up to 2.5-fold more procollagen VII within the rough endoplasmic reticulum than controls. In contrast, the glycine substitution G1519D in another segment of the triple helix affected neither procollagen VII secretion nor anchoring fibril function and remained phenotypically silent. Thus, collagen VII is a remarkable exception among collagens in that not all glycine substitutions within the triple helix exert dominant-negative interference. The biologic consequences of the substitutions probably depend on their position within the triple helix.

Bart, B. J.; Gorlin, R. J.; Anderson, V. E.; Lynch, F. W.: Congenital localized absence of skin and associated abnormalities resembling epidermolysis bullosa: a new syndrome. Arch. Derm. 93: 296-304, 1966.

Bentz, H.; Morris, N. P.; Murray, L. W.; Sakai, L. Y.; Hollister, D. W.; Burgeson, R. E.: Isolation and partial characterization of a new human collagen with an extended triple-helical structural domain. Proc. Nat. Acad. Sci. 80: 3168-3172, 1983.

Burgeson, R. E.; Morris, N. P.; Murray, L. W.; Duncan, K. G.; Keene, D. R.; Sakai, L. Y.: The structure of type VII collagen. Ann. N.Y. Acad. Sci. 460: 47-57, 1985.

Christiano, A. M.; Anton-Lamprecht, I.; Amano, S.; Ebschner, U.; Burgeson, R. E.; Uitto, J.: Compound heterozygosity for COL7A1 mutations in twins with dystrophic epidermolysis bullosa: a recessive paternal deletion/insertion mutation and a dominant negative maternal glycine substitution result in a severe phenotype. Am. J. Hum. Genet. 58: 682-693, 1996.

Christiano, A. M.; Bart, B. J.; Epstein, E. H., Jr.; Uitto, J.: Genetic basis of Bart's syndrome: a glycine substitution mutation in the type VII collagen gene. J. Invest. Derm. 106: 778-780, 1996.

Christiano, A. M.; Greenspan, D. S.; Hoffman, G. G.; Zhang, X.; Tamai, Y.; Lin, A. N.; Dietz, H. C.; Hovnanian, A.; Uitto, J.: A missense mutation in type VII collagen in two affected siblings with recessive dystrophic epidermolysis bullosa. Nature Genet. 4: 62-66, 1993.

Christiano, A. M.; Greenspan, D. S.; Lee, S.; Uitto, J.: Cloning of human type VII collagen: complete primary sequence of the alpha-1(VII) chain and identification of intragenic polymorphisms. J. Biol. Chem. 269: 20256-20262, 1994.

Christiano, A. M.; Hoffman, G. G.; Chung-Honet, L. C.; Lee, S.; Cheng, W.; Uitto, J.; Greenspan, D. S.: Structural organization of the human type VII collagen gene (COL7A1), composed of more exons than any previously characterized gene. Genomics 21: 169-179, 1994.

Christiano, A. M.; Lee, J. Y.-Y.; Chen, W. J.; LaForgia, S.; Uitto, J.: Pretibial epidermolysis bullosa: genetic linkage to COL7A1 and identification of a glycine-to-cysteine substitution in the triple-helical domain of type VII collagen. Hum. Molec. Genet. 4: 1579-1583, 1995.

Christiano, A. M.; McGrath, J. A.; Tan, K. C.; Uitto, J.: Glycine substitutions in the triple-helical region of type VII collagen result in a spectrum of dystrophic epidermolysis bullosa phenotypes and patterns of inheritance. Am. J. Hum. Genet. 58: 671-681, 1996.

Christiano, A. M.; Ryynanen, M.; Uitto, J.: Dominant dystrophic epidermolysis bullosa: identification of a gly-to-ser substitution in the triple-helical domain of type VII collagen. Proc. Nat. Acad. Sci. 91: 3549-3553, 1994.

Christiano, A. M.; Suga, Y.; Greenspan, D. S.; Ogawa, H.; Uitto, J.: Premature termination codons on both alleles of the type VII collagen gene (COL7A1) in three brothers with recessive dystrophic epidermolysis bullosa. J. Clin. Invest. 95: 1328-1334, 1995.

Gardella, R.; Belletti, L.; Zoppi, N.; Marini, D.; Barlati, S.; Colombi, M.: Identification of two splicing mutations in the collagen type VII gene (COL7A1) of a patient affected by the localisata variant of recessive dystrophic epidermolysis bullosa. Am. J. Hum. Genet. 59: 292-300, 1996.

Greenspan, D. S.: The carboxyl-terminal half of type VII collagen, including the non-collagenous NC-2 domain and intron/exon organization of the corresponding region of the COL7A1 gene. Hum. Molec. Genet. 2: 273-278, 1993.

Greenspan, D. S.; Byers, M. G.; Eddy, R. L.; Hoffman, G. G.; Shows, T. B.: Localization of the human collagen gene COL7A1 to 3p21.3 by fluorescence in situ hybridization. Cytogenet. Cell Genet. 62: 35-36, 1993.

Hammami-Hauasli, N.; Schumann, H.; Raghunath, M.; Kilgus, O.; Luthi, U.; Luger, T.; Bruckner-Tuderman, L.: Some, but not all, glycine substitution mutations in COL7A1 result in intracellular accumulation of collagen VII, loss of anchoring fibrils, and skin blistering. J. Biol. Chem. 273: 19228-19234, 1998.

Hovnanian, A.; Duquesnoy, P.; Blanchet-Bardon, C.; Knowlton, R. G.; Amselem, S.; Lathrop, M., II; Dubertret, L.; Uitto, J.; Goossens, M.: Genetic linkage of recessive dystrophic epidermolysis bullosa to the type VII collagen gene.(Abstract) Clin. Res. 40: 188A, 1992.

Hovnanian, A.; Hilal, L.; Blanchet-Bardon, C.; Bodemer, C.; de Prost, Y.; Stark, C. A.; Christiano, A. M.; Dommergues, M.; Terwilliger, J. D.; Izquierdo, L.; Conteville, P.; Dumez, Y.; Uitto, J.; Goossens, M.: DNA-based prenatal diagnosis of generalized recessive dystrophic epidermolysis bullosa in six pregnancies at risk for recurrence. J. Invest. Derm. 104: 456-461, 1995.

Hovnanian, A.; Rochat, A.; Bodemer, C.; Petit, E.; Rivers, C. A.; Prost, C.; Fraitag, S.; Christiano, A. M.; Uitto, J.; Lathrop, M.; Barrandon, Y.; de Prost, Y.: Characterization of 18 new mutations in COL7A1 in recessive dystrophic epidermolysis bullosa provides evidence for distinct molecular mechanisms underlying defective anchoring fibril formation. Am. J. Hum. Genet. 61: 599-610, 1997.

Jarvikallio, A.; Pulkkinen, L.; Uitto, J.: Molecular basis of dystrophic epidermolysis bullosa: mutations in the type VII collagen gene (COL7A1). Hum. Mutat. 10: 338-347, 1997.

Knowlton, R. G.; Ryynanen, M.; Parente, M. G.; Chung, L. C.; Chu, M.-L.; Uitto, J.: Genetic linkage of dominant dystrophic epidermolysis bullosa to the type VII collagen gene on chromosome 3.(Abstract) Am. J. Hum. Genet. 49 (suppl.): 16, 1991.

Lapiere, J.-C.; Woodley, D. T.; Parente, M. G.; Iwasaki, T.; Wynn, K. C.; Christiano, A. M.; Uitto, J.: Epitope mapping of type VII collagen: identification of discrete peptide sequences recognized by sera from patients with acquired epidermolysis bullosa. J. Clin. Invest. 92: 1831-1839, 1993.

Li, K.; Christiano, A. M.; Copeland, N. G.; Gilbert, D. J.; Chu, M.-L.; Jenkins, N. A.; Uitto, J.: cDNA cloning and chromosomal mapping of the mouse type VII collagen gene (Col7a1): evidence for rapid evolutionary divergence of the gene. Genomics 16: 733-739, 1993.

Parente, M. G.; Chung, L. C.; Ryynanen, J.; Woodley, D. T.; Wynn, K. C.; Bauer, E. A.; Mattei, M.-G.; Chu, M.-L.; Uitto, J.: Human type VII collagen: cDNA cloning and chromosomal mapping of the gene. Proc. Nat. Acad. Sci. 88: 6931-6935, 1991.

Ryynanen, J.; Sollberg, S.; Parente, M. G.; Chung, L. C.; Christiano, A. M.; Uitto, J.: Type VII collagen gene expression by cultured human cells and in fetal skin: abundant mRNA and protein levels in epidermal keratinocytes. J. Clin. Invest. 89: 163-168, 1992.

Ryynanen, M.; Knowlton, R. G.; Parente, M. G.; Chung, L. C.; Chu, M.-L.; Uitto, J.: Human type VII collagen: genetic linkage of the gene (COL7A1) on chromosome 3 to dominant dystrophic epidermolysis bullosa. Am. J. Hum. Genet. 49: 797-803, 1991.

Tanaka, T.; Takahashi, K.; Furukawa, F.; Imamura, S.: Molecular cloning and characterization of type VII collagen cDNA. Biochem. Biophys. Res. Commun. 183: 958-963, 1992.

Uitto, J.; Ryynanen, M.; Christiano, A. M.; Hovnanian, A.; Frantz, R.; Bauer, E. A.; Knowlton, R. G.: Genetic linkage of the type VII collagen gene (COL7A1) to dominant dystrophic epidermolysis bullosa (DDEB) in families with abnormal anchoring fibrils.(Abstract) Clin. Res. 40: 188A, 1992.

Zelickson, B.; Matsumura, K.; Kist, D.; Epstein, E. H., Jr.; Bart, B. J.: Bart's syndrome: ultrastructure and genetic linkage. Arch. Derm. 131: 663-668, 1995.

EPIDERMODYSPLASIA VERRUCIFORMIS (226400)

JJM NOTE

This condition may illustrate the concept of genetic susceptibility to a specific virus or even a sequence of specific viruses to lead to an excess of basal cell carcinomas of the skin.

Sullivan and Ellis (1939) found that of the 16 previously reported families, 4 had consanguineous parents. The lesions often resemble verrucae planae. The mucous membranes, hair, and nails are not affected. Malignant degeneration, usually of the superficial basal cell type, is frequent. Characteristic changes in the epidermal cells with peculiar vacuolization are observed. Ellis (1953) stated that this disorder occurs most frequently in Orientals. It is by no means proved that this is a mendelizing disorder. The view that epidermodysplasia verruciformis is an extensive form of viral verrucae planae is supported by successful autoinoculation and heteroinoculation experiments. Lutz (1957) was one of the first to describe the condition; he accepted that it is not an entity but suggested that genetic predisposition may account for the extensiveness of the eruption of warts. Familial aggregation was described by Midana (1949) and by Jablonska et al. (1966). Hermann (1955) found parental consanguinity. By electron microscopy, Baker (1968), as well as others, demonstrated particles suggesting papovavirus. This is an autosomal recessive disorder in which children develop multiple warts (Lutzner, 1977). The common wart virus can be demonstrated in the warts by both electron and fluorescent microscopy (Yabe and Sadakane, 1975). Warts appear to progress to squamous cell carcinoma in about 10% of cases (Lutzner, 1977). As in Shope papilloma, virus is no longer demonstrable in the cancers. The family reported by Jablonska et al. (1979) suggested autosomal dominant inheritance. The fact that spouses and some family members stayed free of the disease speaks against intrafamilial infection as the cause. The authors observed papillomaviruses (HPVs), either HPV3 or HPV4 and sometimes both, in cases and found that the clinical picture differed depending on which virus was involved. Malignancies developed only in family members infected with HPV4. The same group (Orth et al., 1979) pointed to papilloma virus type 5 as the determinant of malignant evolution of the warts. Feuerman et al. (1979) reported the cases of 2 Arab brothers. The parents and 7 sibs were unaffected.

See also Jablonska and Formas (1959)

Baker, H.: Epidermodysplasia verruciformis with electron microscopic demonstration of virus. Proc. Roy. Soc. Med. 61: 589-590, 1968.

Ellis, F.: In discussion of Barker and Sachs. Arch. Derm. 67: 443-455, 1953.

Feuerman, E. J.; Sandbank, M.; David, M.: Two siblings with epidermodysplasia verruciformis with large clear cells in the epidermis: electron microscope and immunological findings. Acta Derm. Venerol. 59: 513-520, 1979.

Hermann, H.: Epidermodysplasia verruciformis: Erb-und Erscheinungsbild. Z. Menschl. Vererb. Konstitutionsl. 32: 409-417, 1955.

Jablonska, S.; Fabjanska, L.; Formas, I.: On the viral etiology of epidermodysplasia verruciformis. Dermatologica 132: 369-385, 1966.

Jablonska, S.; Formas, I.: Weitere positive Ergebnisse mit Auto-und Heteroinokulation bei Epidermodysplasia verruciformis Lewandowsky-Lutz. Dermatologica 118: 86-93, 1959.

Jablonska, S.; Orth, G.; Jarzabek-Chorzelska, M.; Glinski, W.; Obalek, S.; Rzesa, G.; Croissant, O.; Favre, M.: Twenty-one years of follow-up studies of familial epidermodysplasia verruciformis. Dermatologica 158: 309-327, 1979.

Lutz, W.: Zur Epidermodysplasia verruciformis. Dermatologica 115: 309-314, 1957.

Lutzner, M. A.: Nosology among the neoplastic genodermatoses. In: Mulvihill, J. J.; Miller, R. W.; Fraumeni, J. F., Jr.: Genetics of Human Cancer. New York: Raven Press (pub.) 1977. Pp. 145-167.

Midana, A.: Sulla questione dei rapporti tra epidermodysplasia verruciformis e verrucosi generalizzata. Dermatologica 99: 1-23, 1949.

Orth, G.; Jablonska, S.; Jarzabek-Chorzelska, M.; Obalek, S.; Rzesa, G.; Favre, M.; Croissant, O.: Characteristics of the lesions and risk of malignant conversion associated with the type of human papillomavirus involved in epidermodysplasia verruciformis. Cancer Res. 39: 1074-1082, 1979.

Sullivan, M.; Ellis, F. A.: Epidermodysplasia verruciformis (Lewandowsky and Lutz). Arch. Derm. Syph. 40: 422-432, 1939.

Yabe, Y.; Sadakane, H.: Virus of epidermodysplasia verruciformis: electron microscopic and fluorescent antibody studies. J. Invest. Derm. 65: 324-330, 1975.

EPIDERMODYSPLASIA VERRUCIFORMIS, X-LINKED (305350)

JJM NOTE

Follow-up of the family might clarify whether the squamous cell carcinomas are incidental or, as I believe, related to the interaction of genetic predisposition, specific infections, and ultraviolet light.

Androphy et al. (1985) described a kindred in which a 56-year-old man had EDV, none of his 5 sons or 5 daughters had EDV, and 4 of his grandsons (through 2 daughters) had EDV. All were infected with human papillomavirus 3 (HPV 3) and with HPV 8. The proband, who had onset of warts at age 5 years with no regression over the next 50 years and with extension to cover about 10% of his skin surface, had squamous carcinoma arising on sun-exposed areas of the face, ears, neck, back, arms, and hands over the previous 25 years. Other pedigrees have suggested autosomal inheritance although whether dominant as suggested by some families or recessive as suggested by parental consanguinity (see 226400) is not certain.

Androphy, E. J.; Dvoretzky, I.; Lowy, D. R.: X-linked inheritance of epidermodysplasia verruciformis: genetic and virologic studies of a kindred. Arch. Derm. 121: 864-868, 1985.

SCLEROTYLOSIS (*181600)

JJM NOTE

The peculiarities of this cancer predisposition are the special virulence, that is, aggressive behavior of the early-age squamous cell carcinomas of teh skin and the possible excess of bowel cancer.

Huriez et al. (1968) described a 'new' genodermatosis in 44 members of 3 French kindreds. The characteristics were atrophic fibrosis of the skin of the limbs, hypoplasia of nails, and keratodermia of the palms and soles. Skin cancer and bowel cancer were frequent. Linkage with the MN blood group (111300) locus has been established (Mennecier, 1967). Thus, the TYS locus is located in the segment 4q28-q31. Lambert et al. (1977) studied 3 cases in a family reported by Huriez et al. (1968); the family lived in Lille, France. Fischer (1978) reported a fourth family, living in Bourges, France, at a considerable distance from Lille. Mother and 2 daughters were affected. MNSs blood groups were consistent with linkage. The deceased maternal grandfather was also affected. The triad of manifestations was scleroatrophic lesions predominantly of the hands, hypoplastic nail changes, and palmoplantar keratodermia. Changes were present at birth.

Delaporte et al. (1995) stated that, since the initial description of the disease, 3 other families, and possibly a fourth, had been reported. They reassessed the clinical, pathologic, and genetic data in 114 members of one of the families originally described by Huriez et al. (1968), of whom 27 who were affected by the disorder were still alive. Squamous cell carcinomas (SSCs) had occurred in 3. One patient had 7 successive SCCs on different fingers, the first at the age of 35; he subsequently died from metastatic SCC at the age of 54. Delaporte et al. (1995) challenged the linkage to MNSs; a maximum lod score of 1.68 was obtained at recombination fraction theta = 0.18. It would appear that, in fact, the authors' findings support loose linkage and that further studies using highly polymorphic DNA markers should be conducted before linkage is rejected.

Lucker et al. (1997) described a mother and son with this form of palmoplantar keratoderma characterized by scleroatrophy, sclerodactyly, and nail anomalies. In the affected skin of the mother, 3 squamous cell carcinomas developed at a young age. Retinoid treatment was started prophylactically. Malignant degeneration of affected skin is a distinctive feature of the syndrome. A high mortality rate for this type of skin cancer has also been reported.

Hamm et al. (1996) reported the first family from Germany affected with this syndrome. Immunohistochemical and ultrastructural studies of their index case revealed an absence of Langerans cells in involved skin. The authors suggested that this might account for the tendency of scleroatrophic skin to undergo malignant change.

Kavanagh et al. (1997) reported the first family from the U.K. with the scleroatrophic syndrome of Huriez. As in the study of Delaporte et al. (1995), they were unable to confirm the linkage to 4q28-q31. The proband of Kavanagh et al. (1997) was a 29-year-old nurse who, in addition to a worsening hand dermatitis, had hands small from birth with abnormal finger- and toenails that did not require cutting. There was notable absence of sweating on the extremities. A sister had died at the age of 22 years of metastatic squamous cell carcinoma that developed on scleroatrophic skin on the finger. An aunt also had an SCC excised at 45 years of age and later developed bowel carcinoma. The proband and other members of the family showed poikiloderma-like changes of the nose and telangiectases of the lips.

See also Huriez et al. (1968)

Delaporte, E.; N'guyen-Mailfer, C.; Janin, A.; Savary, J. B.; Vasseur, F.; Feingold, N.; Piette, F.; Bergoend, H.: Keratoderma with scleroatrophy of the extremities or sclerotylosis (Huriez syndrome): a reappraisal. Brit. J. Derm. 133: 409-416, 1995.

Fischer, S.: La genodermatose scleroatrophiante et keratodermique des extremites (au sujet de trois nouveaux cas familiaux). Ann. Derm. Venerol. 105: 1079-1082, 1978.

Hamm, H.; Traupe, H.; Brocker, E. B.; Schubert, H.; Kolde, G.: The scleroatrophic syndrome of Huriez: a cancer-prone genodermatosis. Brit. J. Derm. 134: 512-518, 1996.

Huriez, C.; Deminatti, M.; Agache, P.; Mennecier, M.: A propos de 28 cas d'epidermolyse bulleuse dans 11 familles dont une famille etudiee du point de une genetique, sans mise en evidence de linkage. Bull. Soc. Franc. Derm. Syph. 75: 750-755, 1968.

Huriez, C.; Deminatti, M.; Agache, P.; Mennecier, M.: Une genodysplasie non encore individualisee: la genodermatose sclero-atrophiante et keratodermique des extremites frequemment degenerative. Sem. Hop. Paris 44: 481-488, 1968.

Kavanagh, G. M.; Jardine, P. E.; Peachey, R. D.; Murray, J. C.; De Berker, D.: The scleroatrophic syndrome of Huriez. Brit. J. Derm 137: 114-118, 1997.

Lambert, D.; Planche, H.; Chapuis, J.-L.: La genodermatose sclero-atrophiante et keratodermique des extremites. Ann. Derm. Venerol. 104: 654-657, 1977.

Lucker, G. P. H.; Zeedijk, N.; Steijlen, P. M.: The Huriez syndrome: scleroatrophic palmoplantar keratoderma. Europ. J. Derm. 7: 155-157, 1997.

Mennecier, M.: Individualisation d'une nouvelle entite: la genodermatose scleroatrophiante et keratodermique des extremites frequemment degenerative: etude clinique et genetique (possibilite de linkage avec le systeme MNSs). M.D. Thesis: Univ. de Lille (pub.) 1967.

POROKERATOSIS OF MIBELLI (*175800)

JJM NOTE

An interaction with sunlight among the mostly Italian family allows this entity to be on the list of ecogenetic cancer traits. The presence of adenocarcinoma of the colon in an effected adult male who also had Werner Syndrome (not strongly associated with a risk for bowel cancer) suggests a possible association that awaits confirmation with additional cases.

This is a rare hereditary keratoatrophoderma characterized by centrifugally spreading patches surrounded by narrow horny ridges and with central atrophy. The lesions are craterlike. More cases have been described in Italians than in other nationalities, according to some, although Bloom and Abramowitz (1943) were impressed with the wide ethnic distribution of cases including blacks. They described the disorder in an Italian man and his 2 sons. The grandfather was said to be affected. Autosomal dominant inheritance, probably with some reduction in penetrance in females, seems quite certain. The prefix 'poro' comes from the Greek for callus. Mibelli (1860-1910), who described this condition, was an Italian dermatologist. His cases came from the province of Parma. He called the disorder porokeratosis because he believed it started from sweat glands. Taylor et al. (1973) found a high proportion of cultured fibroblasts exhibiting a variety of chromosomal aberrations with no specific aberration consistently present. The existence of a nonfamilial form has been proposed (Bhutani et al., 1977). The male/female ratio of affected cases is said to be 3 to 1 (Goerttler and Jung, 1975). Skin eruption usually begins by middle age and about 7% of patients eventually develop skin cancer, usually on the extremities. Quiescent lesions become active and extensive after immunosuppression for renal transplantation (MacMillan and Roberts, 1974). Machino et al. (1984) described a Japanese patient who had features also of Werner syndrome (277700), a relatively frequent disorder in Japan. He died of metastatic squamous carcinoma which developed in a lesion on the foot. Several carcinomas-in-situ were found in other lesions and at autopsy adenocarcinoma of the descending colon was also found. Two brothers and a sister had porokeratosis as did their father. Scappaticci et al. (1989) found clonal chromosomal abnormalities with preferential involvement of chromosome 3 in 3 sibs and 1 sporadic case with this disorder. Region 3p14-p12, which includes the most common fragile site in humans, was particularly involved. Scappaticci et al. (1989) concluded that porokeratosis of Mibelli is associated with chromosomal instability and that this in turn may predispose to malignancy.

Schamroth et al. (1997) reviewed the wide range of clinical presentations and the resulting confusion with numerous classifications of the several forms of this disorder. Its association with immunosuppression and occasional malignant transformation were discussed. At least 6 different types were tabulated and autosomal dominant inheritance was suggested for at least 5 of them.

See also Cort (1972); Reed and Leone (1970); Saunders (1961); Taylor et al. (1973)

Bhutani, L. K.; Kanwar, A. J.; Singh, O. P.: Porokeratosis of Mibelli with unusual features. Dermatologica 155: 296-300, 1977.

Bloom, D.; Abramowitz, E. W.: Porokeratosis Mibelli: report of three cases in one family: histologic studies. Arch. Derm. Syph. 47: 1-15, 1943.

Cort, D. F.: Epithelioma arising in porokeratosis of Mibelli. Brit. J. Plast. Surg. 25: 318-328, 1972.

Goerttler, E. A.; Jung, E. G.: Porokeratosis of Mibelli and skin carcinoma: a critical review. Humangenetik 26: 291-296, 1975.

Machino, H.; Miki, Y.; Teramoto, T.; Shiraishi, S.; Sasaki, M. S.: Cytogenetic studies in a patient with porokeratosis of Mibelli, multiple cancers and a forme fruste of Werner's syndrome. Brit. J. Derm. 111: 579-586, 1984.

MacMillan, A. L.; Roberts, S. O.: Porokeratosis of Mibelli after renal transplantation. Brit. J. Derm. 90: 45-51, 1974.

Reed, R. J.; Leone, P.: Porokeratosis—a mutant clonal keratosis of the epidermis: histogenesis. Arch. Derm. 101: 340-347, 1970.

Saunders, T. S.: Porokeratosis. A disease of epidermal eccrine-sweat-duct units. Arch. Derm. 84: 980-988, 1961.

Scappaticci, S.; Lambiase, S.; Orecchia, G.; Fraccaro, M.: Clonal chromosome abnormalities with preferential involvement of chromosome 3 in patients with porokeratosis of Mibelli. Cancer Genet. Cytogenet. 43: 89-94, 1989.

Schamroth, J. M.; Zlotogorski, A.; Gilead, L.: Porokeratosis of Mibelli: overview and review of the literature. Acta. Derm. Venerol. 77: 207-213, 1997.

Taylor, A. M. R.; Harnden, D. G.; Fairburn, E. A.: Chromosomal instability associated with susceptibility to malignant disease in patients with porokeratosis of Mibelli. J. Nat. Cancer Inst. 51: 371-378, 1973.

Taylor, A. M. R.; Harnden, D. G.; Fairburn, E. A.: Chromosomal instability and malignant disease in patients with porokeratosis of Mibelli. Brit. J. Cancer 28: 88 only, 1973.

BAZEX SYNDROME (*301845)

JJM NOTE

Besides the risk of skin cancer, recent notes have emphasized the possibility of metastatic cancer arising from the upper aerodigestive tract (Wareing MJ et al. Acrokeratosis paraneoplastica: Bazex Syndrome. J. Laryngol. Otol. 110:899–900, 1996, Sarker B et al. Bazex syndrome (acrokeratosis paraneoplastic). Eur. Arch. Otorhinolaryngol. 255:205–10, 1998).

Viksnins and Berlin (1977) described a kindred in which 8 persons in 3 generations had the condition described by 2,1:Bazex et al. (1964, 1966) in 6 members of a family. Affected persons had lesions suggesting 'multiple ice-pick marks' on the dorsum of the hands and elbows dating from early infancy and basal cell carcinomas that developed on the face between ages 15 and 26 years. 'Follicular atrophoderma,' although a well-established term, is not appropriate because histologic studies do not show atrophy. Hypotrichosis was also present in the cases of Bazex et al. (1964). Identical skin lesions are seen in Conradi disease (215100) and in basal cell nevus syndrome (109400). A relation, possibly allelic, of the latter condition to Bazex syndrome remains to be proved. Viksnins and Berlin (1977) raised the question of X-linked dominant inheritance because no male-to-male transmission was observed in either reported family and in the second family all 3 daughters of an affected male (but none of his sons) were affected. Gould and Barker (1978) described affected individuals in 4 generations of a British family. Prominent pitting of the skin on the face and back of the hands was noted, as well as hypotrichosis in 2 patients studied. Hairs had a twisted and flattened appearance on scanning electron microscopy. In their family, likewise, there was no instance of male-to-male transmission and the only affected male in the family had a single child, an unaffected son. Vabres and de Prost (1993) suggested that the family reported by Oley et al. (1992) had Bazex syndrome. They pointed out that Pierard et al. (1971) and Gould and Barker (1978) favored X-linked dominant inheritance. They also pointed out that the pedigree pattern was consistent with X-linked dominant inheritance as proposed for Bazex syndrome.

From the report by Rapelanoro et al. (1994), with follow-up by Lacombe and Taieb (1995), it appears that Bazex syndrome is indeed an X-linked dominant, and that congenital hypotrichosis with milia (146530) and basal cell carcinomas with milia and coarse, sparse hair (109390) are probably the same entity. Rombo syndrome (180730) has similar features but male-to-male transmission has apparently been observed in that disorder.

Kidd et al. (1996) described a Scottish family with typical manifestations of Bazex syndrome and reviewed 15 previously reported families with approximately 120 affected persons. The authors noted that 'pinched' nose with hypoplastic alae and prominent columella may be another characteristic manifestation of the disorder.

Using microsatellite markers of the X chromosome in 3 families with Bazex syndrome, Vabres et al. (1995) found evidence for X-linkage and regional assignment to Xq24-q27. The maximum lod score was 5.26 at theta = 0 with the DXS1192 locus.

Bazex, A.; Dupre, A.; Christol, B.: Atrophodermic folliculaire, proliferations baso-cellulaires et hypotrichose. Ann. Derm. Syph. 93: 241-254, 1966.

Bazex, A.; Dupre, A.; Christol, B.: Genodermatose complexe de type indetermine associant une hypotrichose, un etat atrophodermique generalise et des degenerescences cutanees multiples (epitheliomas-basocellulaires). Bull. Soc. Franc. Derm. Syph. 71: 206 only, 1964.

Gould, D. J.; Barker, D. J.: Follicular atrophoderma with multiple basal cell carcinomas (Bazex). Brit. J. Derm. 99: 431-435, 1978.

Kidd, A.; Carson, L.; Gregory, D. W.; de Silva, D.; Holmes, J.; Dean, J. C. S.; Haites, N. S.: A Scottish family with Bazex-Dupre-Christol syndrome: follicular atrophoderma, congenital hypotrichosis, and basal cell carcinoma. J. Med. Genet. 33: 493-497, 1996.

Lacombe, D.; Taieb, A.: Overlap between the Bazex syndrome and congenital hypotrichosis and milia. (Letter) Am. J. Med. Genet. 56: 423-424, 1995.

Oley, C. A.; Sharpe, H.; Chenevix-Trench, G.: Basal cell carcinomas, coarse sparse hair, and milia. Am. J. Med. Genet. 43: 799-804, 1992.

Pierard, J.; Dhondt, F.; Geerts, M. L.; Kriekemans, J.: Atrophodermie folliculaire, proliferations baso-cellulaires et hypotrichose. Arch. Belg. Derm. 27: 55-68, 1971.

Rapelanoro, R.; Taieb, A.; Lacombe, D.: Congenital hypotrichosis and milia: report of a large family suggesting X-linked dominant inheritance. Am. J. Med. Genet. 52: 487-490, 1994.

Vabres, P.; de Prost, Y.: Bazex-Dupre-Christol syndrome: a possible diagnosis for basal cell carcinomas, coarse sparse hair, and milia. (Letter) Am. J. Med. Genet. 45: 786 only, 1993.

Vabres, P.; Lacombe, D.; Rabinowitz, L. G.; Aubert, G.; Anderson, C. E.; Taieb, A.; Bonafe, J.-L.; Hors-Cayla, M.-C.: The gene for Bazex-Dupre-Christol syndrome maps to chromosome Xq. J. Invest. Derm. 105: 87-91, 1995.

Viksnins, P.; Berlin, A.: Follicular atrophoderma and basal cell carcinomas: the Bazex syndrome. Arch. Derm. 113: 948-951, 1977.

HYPOTRICHOSIS WITH LIGHT-COLORED HAIR AND FACIAL MILIA (*146530)

JJM NOTE

Basal cell carcinoma of the skin is the complicating cancer.

Parrish et al. (1972) described this syndrome in 11 members of 3 generations with no male-to-male transmission but 1 normal daughter of an affected male making X-linked dominance unlikely. A markedly reduced density of scalp hair was the only abnormality identified. Sparcity of hair was less marked in adults than in children. A defect in the induction phase of hair development during fetal life was postulated. 'Melanization' of the hair shaft was reduced. Milia was confined to the face. The family was of Italian origin. Oley et al. (1992) reported a similar condition in an Australian family of Italian origin. Hypertrichosis was present at birth and improved in adult life. Multiple milia involved the face, limbs, and buttocks, and disappeared spontaneously by adolescence. Furthermore, basal cell carcinomas or nevi developed in adult life. The pedigree was consistent with either autosomal dominant or X-linked dominant inheritance. The disorder in the family of Oley et al. (1992) is discussed as a separate entity in 109390. Some similarities were demonstrated to the so-called Rombo syndrome (180730) and the Bazex syndrome (301845).

Rapelanoro et al. (1994) observed a 30-month-old boy with coarse, sparse hair and multiple milia on the face, chest, axillae, and pubic region. A sister, aged 16 years, had apparently normal scalp, axillary, and pubic hair although she had had coarse and very sparse scalp hair and multiple milia on the face and limbs in childhood. Most of the milia disappeared at puberty; a few persisted on the forehead. The same manifestations were present in the mother from birth and disappeared at 40 years of age. There were no abnormalities of teeth and nails. Polarizing light microscopy showed an increased diameter of the hair shaft. Using a questionnaire concerning hair defects, skin, nails and teeth, Rapelanoro et al. (1994) investigated the family. The condition had been present for 4 generations with 20 individuals reported as affected. Both the males and females were affected. There was no transmission from an unaffected parent, and no male-to-male transmission was observed. Intelligence and behavior of affected individuals were normal. The pattern of inheritance was considered compatible with either autosomal or X-linked dominant inheritance. Lacombe and Taieb (1995) examined 17 members of the family reported by Rapelanoro et al. (1994) and affirmed striking intrafamilial phenotypic variability similar to that of Bazex syndrome. In 3 of 8 patients, follicular atrophoderma on the back of the hand was found in addition to hypertrichosis and milia, and 2 patients had been treated for basal cell carcinoma. They concluded that the same family had been reported by Le Coulant et al. (1967). Doubts were raised at that time concerning the X-linked mode of inheritance of Bazex syndrome because of an instance of male-to-male transmission in the pedigree. Lacombe and Taieb (1995) found that the suspected affected male was in fact unaffected. Thus there is no remaining data in the literature against the X-linked inheritance of Bazex syndrome.

See also Oley et al. (1992)

Goldsmith, L. A.; Baden, H. P.: The analysis of genetically determined hair defects. Birth Defects Orig. Art. Ser. VII(2): 86-90, 1971.

Lacombe, D.; Taieb, A.: Overlap between the Bazex syndrome and congenital hypotrichosis and milia. (Letter) Am. J. Med. Genet. 56: 423-424, 1995.

Le Coulant, P.; Texier, L.; Tamisier, J. M.; Caubet, P.: Atrophodermie folliculaire, kystes epidermiques et hypotrichose: a propos d'une observation familiale de genodermatose. Bull. Soc. Franc. Derm. Syph. 74: 502-504, 1967.

Oley, C. A.; Sharpe, H.; Chenevix-Trench, G.: Basal cell carcinomas, coarse sparse hair, and milia. Am. J. Med. Genet. 43: 799-804, 1992.

Parrish, J. A.; Baden, H. P.; Goldsmith, L. A.; Matz, M. H.: Studies of the density and the properties of the hair in a new inherited syndrome of hypotrichosis. Ann. Hum. Genet. 35: 349-356, 1972.

Rapelanoro, R.; Taieb, A.; Lacombe, D.: Congenital hypotrichosis and milia: report of a large family suggesting X-linked dominant inheritance. Am. J. Med. Genet. 52: 487-490, 1994.

BASAL CELL CARCINOMAS WITH MILIA AND COARSE, SPARSE HAIR (109390)

JJM NOTE

Recognition of this disorder lengthens the list of conditions that should be considered when a patient has multiple basal cell carcinomas of the skin.

Oley et al. (1992) described a family in which 9 individuals in 4 generations and 6 sibships had basal cell carcinomas, coarse and sparse scalp hair, sparse body hair, and multiple milia on face and limbs spontaneously disappearing by adolescence. Since there was no instance of male-to-male transmission, the pedigree was consistent with either autosomal dominant or X-linked dominant transmission. All 3 daughters of 2 males who had children were affected. The family was originally ascertained as part of a search for examples of Gorlin-Goltz syndrome (109400). The hair abnormalities, excessive sweating, and enormous number of large milia in childhood were features considered as distinguishing it from that disorder as well as the lack of keratocysts of the jaw, plantar or palmar pits, rib or vertebral abnormalities, or abnormal calcification on skull x-ray. Some features resembled those of the Bazex syndrome (301845) and the Rombo syndrome (180730), both of which have multiple basal cell carcinomas as features. The hair changes were similar to those in the Marie Unna form of hypotrichosis except that further loss of hair did not occur with aging. One 32-year-old patient with multiple milia over her face had markedly increased pigmentation of the face particularly around the eyes; she had had about 20 basal cell carcinomas removed, starting at the age of 22. Vabres and de Prost (1993) were of the opinion that the family reported by Oley et al. (1992) in fact had Bazex syndrome. They noted that the presence or absence of follicular atrophoderma was not mentioned by Oley et al. (1992) but pointed out that 'this manifestation may be undiagnosed, even by experienced dermatologists, if not carefully sought.'

Oley, C. A.; Sharpe, H.; Chenevix-Trench, G.: Basal cell carcinomas, coarse sparse hair, and milia. Am. J. Med. Genet. 43: 799-804, 1992.

Vabres, P.; de Prost, Y.: Bazex-Dupre-Christol syndrome: a possible diagnosis for basal cell carcinomas, coarse sparse hair, and milia. (Letter) Am. J. Med. Genet. 45: 786 only, 1993.

ROMBO SYNDROME (180730)

JJM NOTE

In addition to basal cell carcinomas, trichoepithelioma has been seen in affected patients.

Michaelsson et al. (1981) designated an apparently hitherto undescribed disorder Rombo syndrome after the oldest affected family member. Skin changes began at ages 7 to 10 years. At that time a pronounced, somewhat cyanotic redness of the lips and hands was evident as well as moderate follicular atrophy of the skin on the cheeks. In adulthood, whitish-yellow, milia-like papules and telangiectatic vessels developed. The papules were present particularly on the cheeks and forehead, gradually becoming very conspicuous and dominating the clinical picture. Trichoepitheliomas were found in 1 case. In adults, the eyelashes and eyebrows were either missing or irregularly distributed with defective and maldirected growth. Basal cell carcinomas were a frequent complication. The skin atrophy was referred to as vermiculate atrophoderma. Basal cell carcinomas developed around the age of 35. Peripheral vasodilation with cyanosis was commented on. Histologic sections showed the dermis to be almost devoid of elastin in most areas with clumping of elastic material in other areas. The disorder had been transmitted through at least 4 generations with instances of male-to-male transmission. Some similarity to the Bazex syndrome (301845) was noted.

Michaelsson, G.; Olsson, E.; Westermark, P.: The Rombo syndrome: a familial disorder with vermiculate atrophoderma, milia, hypotrichosis, trichoepitheliomas, basal cell carcinomas and peripheral vasodilation with cyanosis. Acta Derm. Venerol. 61: 497-503, 1981.

PACHYONYCHIA CONGENITA, JACKSON-LAWLER TYPE (#167210)

JJM NOTE

The complicating benign neoplasms are steatocystomas and epidermoid cysts.

A number sign (#) is used with this entry because of evidence that the disorder is caused by mutation in the gene for keratin 17 (KRT17; 148069). Mutation of keratin K6B (KRT6B; 148042) has also been identified in type 2 pachyonychia congenita.

Gorlin et al. (1976) suggested that 2 distinct syndromes are subsumed under the designation pachyonychia congenita. PC1, the Jadassohn-Lewandowski syndrome (167200), shows oral leukokeratosis. PC2, the form of Jackson and Lawler (1951-52), has natal teeth and epidermoid cysts (cylindromas), but no oral leukoplakia. Corneal dystrophy may be a feature exclusively of the Jackson-Lawler type. Both disorders are clearly autosomal dominant. In contrast to PC1, PC2 has minimal oral involvement and milder keratoderma, and multiple steatocystomas is a major clinical feature (Smith et al., 1998). Steatocystoma, also known as eruptive vellus cyst, is a cystic hamartoma lined by sebaceous ductal epithelium.

Clementi et al. (1986) described a family in which 3 members in 2 generations had pachyonychia congenita, hyperkeratosis, and hyperhidrosis of the palms and soles, follicular keratosis, neonatal teeth, and epidermoid cysts. They pointed out that this is the rarer form of pachyonychia congenita. Armpit folliculitis was described in the 21-year-old proposita and in her 61-year-old mother. Cysts in the metacarpal and elbow regions were observed at puberty in the daughter. They were noted to adhere to the overlying skin on the elbows, dorsal surface of the proximal phalanges, and on the knees and pretibial regions.

In a large kindred extensively affected with pachyonychia congenita of the Jackson-Lawler type, Munro et al. (1994) found tight linkage of the disease locus to markers mapping within, or very close to, the keratin type I cluster (e.g., KRT14; 148066) at 17q12-q21. Maximum lod scores for linkage of the disease to a KRT10 (148080) polymorphism and to D17S800, a marker known to be very tightly linked to KRT10, were, respectively, 4.51 and 7.73, both at theta = 0.00. The findings suggested a role for keratins in the phenomenon of natal dentition. In the kindred studied by Munro et al. (1994), McLean et al. (1995) showed that a heterozygous missense asn92-to-asp mutation in the helix initiation motif of keratin 17 cosegregated with the disease.

The Jadassohn-Lewandowski type of pachyonychia congenita is caused by mutations in the keratin 16 gene (KRT16; 148067), located on chromosome 17, or in the keratin-6A gene (KRT6A; 148041), located on chromosome 12, whereas the Jackson-Lawler type is caused by mutations in the gene for keratin 17 on chromosome 17; thus, the 2 types of pachyonychia congenita are not allelic.

Clementi, M.; Cardin de Stefani, E.; Dei Rossi, C.; Avventi, V.; Tenconi, R.: Pachyonychia congenita Jackson-Lawler type: a distinct malformation syndrome. Brit. J. Derm. 114: 367-370, 1986.
Gorlin, R. J.; Pindborg, J. J.; Cohen, M. M., Jr.: Syndromes of the Head and Neck. New York: McGraw-Hill (pub.) (2nd ed.): 1976. Pp. 600-603.
Jackson, A. D. M.; Lawler, S. D.: Pachyonychia congenita: a report of six cases in one family. Ann. Eugen. 16: 142-146, 1951.
McLean, W. H. I.; Rugg, E. L.; Lunny, D. P.; Morley, S. M.; Lane, E. B.; Swensson, O.; Dopping-Hepenstal, P. J. C.; Griffiths, W. A. D.; Eady, R. A. J.; Higgins, C.; Navsaria, H. A.; Leigh, I. M.; Strachan, T.; Kunkeler, L.; Munro, C. S.: Keratin 16 and keratin 17 mutations cause pachyonychia congenita. Nature Genet. 9: 273-278, 1995.
Munro, C. S.; Carter, S.; Bryce, S.; Hall, M.; Rees, J. L.; Kunkeler, L.; Stephenson, A.; Strachan, T.: A gene for pachyonychia congenita is closely linked to the keratin gene cluster on 17q12-q21. J. Med. Genet. 31: 675-678, 1994.
Smith, F. J. D.; Jonkman, M. F.; van Goor, H.; Coleman, C. M.; Covello, S. P.; Uitto, J.; McLean, W. H. I.: A mutation in human keratin K6b produces a phenocopy of the K17 disorder pachyonychia congenita type 2. Am. J. Hum. Genet. 7: 1143-1148, 1998.

HYPERKERATOSIS LENTICULARIS PERSTANS (*144150)

JJM NOTE

Further families are needed to clarify whether or not squamous cell and basal cell carcinomas are really associated with this condition.

Flegel (1958) first described the condition. Beveridge and Langlands (1973) described 2 generations of a kindred manifesting HLP. The mother and 4 children showed the typical hyperkeratotic lesions on the lower leg and dorsum of the foot; however, involvement of the trunk, thigh, arms and dorsum of the hand may occur. The pink or reddish-brown scaly papules from 1 to 5 mm in size developed in the third or fourth decade of life. Interestingly, this family had a high incidence of skin tumors, including squamous and basal cell carcinomas, in areas other than those affected by the hyperkeratotic lesions. The authors suggested that there may be an inherited tendency to epithelial neoplasia in this family. Bean (1969) first reported the familial occurrence of HLP without skin tumors. In 1970 he reported finding HLP in 3 generations of his previously documented family. Frenk and Tapernoux (1974) described formation of a compact hyperkeratotic stratum corneum which appeared to be related to absence of Odland bodies in the underlying epidermis. Odland bodies are round or oval lamellar cytoplasmic organelles found in the upper layers of keratinizing epithelia. Their function is not known.

See also Bean (1972)

Bean, S. F.: Hyperkeratosis lenticularis perstans. A clinical, histopathological and genetic study. Arch. Derm. 99: 705-709, 1969.
Bean, S. F.: The genetics of hyperkeratosis lenticularis perstans. Arch. Derm. 106: 72 only, 1972.
Beveridge, G. W.; Langlands, A. O.: Familial hyperkeratosis lenticularis perstans associated with tumors of the skin. Brit. J. Derm. 88: 453-458, 1973.
Flegel, H.: Hyperkeratosis lenticularis perstans. Hautarzt 9: 362-364, 1958.
Frenk, E.; Tapernoux, B.: Hyperkeratosis lenticularis perstans (Flegel): an autosomal dominant skin disease due to lack of a keratinocyte organelle. Humangenetik 24: 151-153, 1974.

KERATOSIS PALMOPLANTARIS WITH PERIODONTOPATHIA (*245000)

JJM NOTE

This entity is included because of the cysts or hidrocystomas of the eyelids. Side-by-side comparison of the histopathology may clarify the syndrome, an apparently preventable cause of dental loss, if recognized and treated early.

Both the milk teeth and the permanent teeth are lost prematurely. The skin lesions are very similar or identical to those of mal de Meleda (248300). Gorlin et al. (1964) suggested that calcification of the dura mater is a third component of the syndrome.

Nazzaro et al. (1988) reported 4 sibs with Papillon-Lefevre syndrome, ranging in age from 2 to 11 years. The parents were double first cousins. Nazzaro et al. (1988) reported that histologic abnormalities improved markedly during treatment with acitretin, the free acid of etretinate. They suggested that if treatment is started at an early age, patients with PLS should be able to have normal adult dentition. Hattab et al. (1995) reported on 4 cases of PLS affecting 2 Jordanian families with a total of 8 children. The patients were between 4.5 and 12 years of age and their parents, who were first cousins, were not affected. In all patients, there was a relationship between increased severity of skin lesions and seasonal variations and intensified periodontal destruction. There was an early eruption of the permanent teeth. The teeth were caries-free with no sign of root resorption.

In 2 consanguineous families of Turkish origin and 3 multiplex families, 1 Ethiopian and 2 German, with altogether 10 affected and 5 unaffected sibs, Laass et al. (1997) demonstrated linkage of PLS with D11S937 on 11q13-q14 (maximum lod = 5.1 at theta = 0.0). The affected members of the Turkish and Ethiopian families were homozygous by descent for markers from an 18-cM interval to the gene for ultra-high-sulfur keratin (KRN1; 148021) between marker loci D11S937 and D11S4120. Laass (1997) indicated that the frequency of PLS is approximately 1 to 4 per million; that the primary teeth are usually lost by age 4 and permanent teeth by age 17; that periodontitis is gone after the teeth are lost; and that retinoid therapy is valuable for both the keratitis and the periodontitis. Study of further families allowed refinement of the PLS gene location to a 9.2-cM interval on 11q22.

Fischer et al. (1997) likewise conducted a primary genomewide search by homozygosity mapping in a large consanguineous family with 4 affected sibs. Homozygosity and linkage were demonstrated in the region 11q14. Linkage was confirmed in 4 additional families with diverse ethnic and geographic backgrounds, 2 of which were consanguineous. A maximum 2-point lod score of 8.19 was obtained for marker D11S901 at theta = 0.0. The analysis of recombination events placed the gene within a 7-cM interval between D11S901 nad D11S4175. No shared haplotype was found in the 5 families analyzed.

Laass et al. (1997) did homozygosity mapping of PLS on the basis of 3 consanguineous families, 2 of Turkish and 1 of German origin. A traditional linkage analysis was also performed in these and 3 multiplex families. Linkage was obtained with marker D11S937 with a maximum 2-point lod score of 6.1 at recombination fraction theta = 0.00 on 11q14-q21 near the metalloproteinase gene cluster (e.g. MMP7; 178990), which has been thought to map to 11q21-q22, or 11q22-q23, or 11q22.3, or 11q23. Multipoint likelihood calculations gave a maximum lod score of 7.35 between D11S901 and D11S1358. A 9.2-cM region homozygous by descent in the affected members of the 3 consanguineous families was positioned between markers D11S1989 and D11S4176. Haplotype analyses in all the families studied supported this localization.

See also Cheung et al. (1982); Greither (1959); Haneke (1979); Jansen and Dekker (1956); Ziprkowski et al. (1963)

Cheung, H. S.; Landow, R. K.; Bauer, M.: Increased collagen synthesis by gingival fibroblasts derived from a Papillon-Lefevre patient. J. Dent. Res. 61: 378-381, 1982.

Fischer, J.; Blanchet-Bardon, C.; Prud'homme, J.-F.; Pavek, S.; Steijlen, P. M.; Dubertret, L.; Weissenbach, J.: Mapping of Papillon-Lefevre syndrome to the chromosome 11q14 region. Europ. J. Hum. Genet. 5: 156-160, 1997.

Gorlin, R. J.; Sedano, H.; Anderson, V. E.: The syndrome of palmar-plantar hyperkeratosis and premature periodontal destruction of the teeth: a clinical and genetic analysis of the Papillon-Lefevre syndrome. J. Pediat. 65: 895-908, 1964.

Greither, A.: Keratosis palmo-plantaris mit Periodontopathie (Papillon-Lefevre). Dermatologica 119: 248-263, 1959.

Haneke, E.: The Papillon-Lefevre syndrome: keratosis palmoplantaris with periodontopathy: report of a case and review of the cases in the literature. Hum. Genet. 51: 1-35, 1979.

Hattab, F. N.; Rawashdeh, M. A.; Yassin, O. M.; Al-Momani, A. S.; Al-Ubosi, M. M.: Papillon-Lefevre syndrome: a review of the literature and report of 4 cases. J. Periodont. 66: 413-420, 1995.

Jansen, L. H.; Dekker, G.: Hyperkeratosis palmo-plantaris with periodontosis (Papillon-Lefevre). Dermatologica 113: 207-219, 1956.

Laass, M. W.: Personal Communication. Berlin, Germany 5/20/1997.

Laass, M. W.; Hennies, H. C.; Jung, M.; Preis, S.; Stevens, H. P.; Wienker, T. F.; Reis, A.: Papillon-Lefevre syndrome: localisation of a gene to chromosome 11q14-21 by homozygosity mapping. (Abstract) Medizinische Genetik 9: 16 only, 1997.

Laass, M. W.; Hennies, H. C.; Preis, S.; Stevens, H. P.; Jung, M.; Leigh, I. M.; Wienker, T. F.; Reis, A.: Localisation of a gene for Papillon-Lefevre syndrome to chromosome 11q14-q21 by homozygosity mapping. Hum. Genet. 101: 376-382, 1997.

Nazzaro, V.; Blanchet-Bardon, C.; Mimoz, C.; Revuz, J.; Puissant, A.: Papillon-Lefevre syndrome: ultrastructural study and successful treatment with acitretin. Arch. Derm. 124: 533-539, 1988.

Ziprkowski, L.; Raymon, Y.; Brish, M.: Hyperkeratosis palmoplantaris with periodontosis (Papillon-Lefevre). Arch. Derm. 88: 207-209, 1963.

TORTICOLLIS, KELOIDS, CRYPTORCHIDISM, AND RENAL DYSPLASIA (*314300)

JJM NOTE

This is a complex overlap of carcinogenesis, teratogenesis, and dysplasia, a theme epitomized by the nevoid basal cell carcinoma syndrome. The entity includes basal cell carcinoma, benign neoplasia/dysplasia (pigmented nevi and, in a sense, keloids), and developmental defects (torticollis and cryptorchidism) that are not early frank congenital malformations.

Goeminne (1968) described a syndrome that is probably inherited as an X-linked trait with incomplete dominance. None of the affected males reproduced. Affected persons included: (1) a male with congenital muscular torticollis, (2) a male with torticollis, cryptorchidism and varicose veins, (3) a male with torticollis, many spontaneous keloids, unilateral cryptorchidism, oligospermia, chronic pyelonephritis with unilateral renal atrophy, multiple cutaneous nevi, a basal cell epithelioma and varicose veins, (4) a male with torticollis, keloids and cryptorchidism, (5) a female with torticollis and pigmented nevi, and (6) a female with facial asymmetry, chronic pyelonephritis and nevi. Zuffardi and Fraccaro (1982) mapped the gene for this syndrome to Xq28, distal to G6PD, based on reports they discovered in the literature of occurrence in 2 unrelated females with a balanced X-autosome translocation. In mouse-human cell hybrids containing the active der(X) chromosome from 1 of these patients, G6PD was expressed (Hellkuhl et al., 1982). Edwards (1982) pointed out that because of the disruption of genes that are located at the breakpoint on the X chromosome and because of the inactivation of the normal X chromosome in females with balanced X-autosome translocations, useful mapping information is provided by females with X/A translocations. Duchenne muscular dystrophy (310200) and Aicardi syndrome (304050) are disorders mapped by this method, and the Aarskog-Scott syndrome (305400) may be an example.

Edwards, J. H.: Chromosomal abnormalities in mendelian disorders. Lancet II: 322-323, 1982.

Goeminne, L.: A new probably X-linked inherited syndrome: congenital torticollis, multiple keloids, cryptorchidism and renal dysplasia. Acta Genet. Med. Gemellol. 17: 439-467, 1968.

Hellkuhl, B.; de la Chapelle, A.; Grzeschik, K.-H.: Different patterns of X chromosome inactivity in lymphocytes and fibroblasts of a human balanced X;autosome translocation. Hum. Genet. 60: 126-129, 1982.

Zuffardi, O.; Fraccaro, M.: Gene mapping and serendipity: the locus for torticollis, keloids, cryptorchidism and renal dysplasia (31430, McKusick) is at Xq28, distal to the G6PD locus. Hum. Genet. 62: 280-281, 1982.

FANCONI-LIKE SYNDROME (227850)

JJM NOTE

This entity is included because of the skin cancers.

In 2 brothers, Abels and Reed (1973) described a disorder with both similarities to and differences from the Fanconi syndrome. One brother died in his mid-twenties after a prolonged course characterized by pancytopenia, recurrent infections, low IgA, chronic lung infections complicated by multiple bilateral pneumothoraces, osteomyelitis, and multiple cutaneous malignancies with lymph node metastases. The other brother had severe pancytopenia responsive to methyltestosterone therapy.

Abels, D.; Reed, W. B.: Fanconi-like syndrome. Immunologic deficiency, pancytopenia, and cutaneous malignancies. Arch. Derm. 107: 419-423, 1973.

DYSKERATOSIS CONGENITA (ZINSSER-COLE-ENGMAN SYNDROME) (#305000)

JJM NOTE

Pancytopenia is the major cause of premature death—in a sense, a hematologic malignancy. Major cancers are squamous cell carcinomas of the skin and mucosae, including the mouth, esophagus, and anus. A few odd ones included pancreatic adenocarcinoma, Hodgkin disease, and spinal cord tumors (Davidson and Connor 1988). A recent addition is adenocarcinoma of the stomach in an Indian teenager (Chatura KR et al.: Case report: Gastric carcinoma as a complication of dyskeratosis congenita in an adolescent boy. Dig. Dis. Sci. 41:2340–42, 1996).

A number sign (#) is used with this entry following the identification of a novel gene by positional cloning, mutations in which have been shown to cause dyskeratosis congenita. DKC1 is the symbol of the gene and dyskerin is the designation of the gene product (602574).

The features of dyskeratosis congenita (DKC, or DC) are cutaneous pigmentation, dystrophy of the nails, leukoplakia of the oral mucosa, continuous lacrimation due to atresia of the lacrimal ducts, often thrombocytopenia, anemia, and in most cases testicular atrophy. Only males are affected in a pattern consistent with X-linked recessive inheritance. Milgrom et al. (1964) described the condition in a black male. They pointed out that the 2 serious complications are anemia and cancer, which develops in the leukoplakia of the anus or mouth, or in the skin. Bryan and Nixon (1965) reported a pedigree with 4 and possibly 5 affected males in a relationship consistent with X-linked recessive inheritance. The patients had pancytopenia which led the authors (incorrectly, I think) to the conclusion that Fanconi panmyelopathy (227650) and dyskeratosis congenita are the same entity. Selmanowitz and van Voolen (1971) also pointed out the phenotypic overlap with Fanconi anemia and raised the question whether Fanconi anemia and dyskeratosis congenita might be causally related. Because of the difference in inheritance (X-linked vs autosomal) this possibility can be rejected out of hand. Addison and Rice (1965) described a male with seemingly typical skin and mucosal changes, as well as pancytopenia. His sister had poikiloderma and oral leukoplakia, progressing to squamous carcinoma fatal at age 24 years. Sorrow and Hitch (1963) described a female patient who had fatal cervical and vaginal squamous carcinoma. The relation of the disorder in these female patients to the condition which is clearly X-linked is not known. Sirinavin and Trowbridge (1975) reported a particularly instructive kindred in which 9 males in 4 sibships and 3 generations were affected. They excluded close linkage with the Xg locus. Pancytopenia and malignancy were features, and opportunistic infections were also a major problem. Nail dystrophy, reticulated atrophic telangiectatic hyper- and hypopigmented skin lesions, oral leukoplakia and mental retardation were described. An extensive review of the literature was provided. Gutman et al. (1978) observed 2 maternal male cousins. Linkage analysis indicated that dyskeratosis, Xg, and G6PD are far apart. In an extensively affected kindred, Connor and Teague (1981) excluded close linkage to Xg. They noted 3 previously unreported complications: Hodgkin disease, pancreatic adenocarcinoma, and deafness. Normal chromosomal stability was found in the 3 patients studied. Studies uncovered no early generalized defect of cell-mediated immunity.

Womer et al. (1983) reported 2 brothers who showed reticular hyperpigmentation, dystrophic nails, leukoplakia oris, and aplastic anemia. Less common features included prenatal and postnatal growth retardation, mental retardation, elevated immunoglobulin levels, and gastrointestinal hemorrhage from mucosal ulcerations. 'New' features were intracranial calcifications and nutmeg-like cirrhosis of the liver. No increased chromosomal breakage was noted. Death occurred at ages 18 and 14 years. Kalb et al. (1986) described a 33-year-old man with typical features of this disorder and with avascular necrosis of the femoral head. Such had previously been reported in cases of this disorder but only in patients who had received systemic adrenocorticosteroids for pancytopenia or thrombocytopenia. Kalb et al. (1986) emphasized the dermatologic triad of reticulated hyperpigmentation, nail dystrophy, and mucosal leukoplakia. Reichel et al. (1992) found reports of 15 cases of elevated fetal hemoglobin in association with DKC and added another case, an 11-year-old boy who, in addition to DKC and elevated fetal hemoglobin, had X-linked ocular albinism and juvenile-onset diabetes mellitus.

Connor et al. (1986) assigned DKC to Xq28 by linkage of DNA markers. Its relationships to other loci in this segment were not determined. Arngrimsson et al. (1993) confirmed the linkage with study of 3 further families which brought the maximum lod score for DKC and DXS52 to 5.33 at zero recombination. Knight et al. (1996) studied 5 families with additional Xq28 polymorphic markers and concluded that the DKC locus is between GABRA3 (305660) and DXS1108, an interval of approximately 4 Mb.

Davidson and Connor (1988) provided an extensive review of 104 published cases of which 51 had previously been reviewed by Sirinavin and Trowbridge (1975). DeBauche et al. (1990) found increased frequency of chromatid breaks and chromatid gaps after x-radiation during the G-2 phase of the cell cycle. Ning et al. (1992) found that the mean number of chromosome breaks per cell in bleomycin-treated lymphocytes was higher in patients with dyskeratosis congenita and in obligatory heterozygotes than in normals. Unequivocal heterozygote detection was not possible owing to overlap of values. Phillips et al. (1992) described a 5-year-old boy who was treated with bone marrow transplantation for aplastic anemia at the age of 2 years. The diagnosis of dyskeratosis congenita was not made until 18 months after the bone marrow transplant. He had diffuse nonscarring alopecia and problems related to bilateral lacrimal duct blockage. At the age of 5 years, all nails were hypoplastic and irregular, and he had reticulate hyperpigmentation on his neck. He required a hearing aid and had poor vision in one eye. Lacrimal duct stenosis is said to be present in about 80% of cases. In 70%, pancytopenia is the cause of death. Because of a previously widely held view that the outcome of bone marrow transplantation in this disorder is poor, this treatment option was sometimes not considered when pancytopenia developed. Phillips et al. (1992) suggested that the results may be satisfactory if radiation is avoided.

In a review, Dokal (1996) noted that late fatal vascular complications had been reported in some cases following bone marrow transplantation (Berthou et al., 1991). This may be related to preexisting endothelial damage in DKC patients, as evidenced by raised von Willebrand factor levels in the plasma. Dokal (1996) suggested that DKC may be a good candidate for gene therapy for several reasons: first, it is a single gene disorder; second, the main cause of mortality relates to bone marrow failure and hemopoietic cells are accessible for targeting; and third, hemopoietic stem cells transfected with the normal DKC gene would be expected to have a selective growth advantage in the hypoplastic marrow.

In 4 patients with DKC, 3 from 1 family and 1 from another, Dokal et al. (1992) found that primary skin fibroblast cultures were abnormal both in morphology (polygonal cell shape, ballooning, and dendritic-like projections) and in growth rate (doubling time about twice normal). Fibroblast survival studies using 4 clastogens and gamma radiation showed no significant difference between DKC and normal fibroblasts. Furthermore, cytogenetic studies performed on peripheral blood lymphocytes showed no difference between DKC and normal lymphocytes with or without prior incubation with clastogens. However, bone marrow metaphases from 1 of 3 patients and fibroblasts from 2 of 4 patients showed numerous unbalanced chromosomal rearrangements (dicentrics, tricentrics, and translocations) in the absence of any clastogenic agents. A higher rate of chromosomal rearrangements was found in the older patients and this, together with the cell-specific differences, appeared to correlate with the clinical evolution of the disease.

Caux et al. (1996) presented a patient and reviewed the pathogenesis of the disorder.

Devriendt et al. (1997) reported that dyskeratosis congenita was quite clearly X-linked because linkage analysis with markers in the factor VIII gene at Xq28 yielded a lod score of 2 at a recombination of 0.0, and clinical manifestations of DKC were present in 2 obligate carrier females, e.g., skin lesions following the Blaschko lines. Highly skewed X inactivation was observed in white blood cells, cultured skin fibroblasts, and buccal mucosa from female carriers of DKC in this family. The skewing suggested a critical role for the DKC gene in proliferation of fibroblasts and bone marrow cells. Devriendt et al. (1997) presented photographs of the linear skin lesions of the palmar aspects of the hands.

Bone marrow failure has been reported in approximately 50% of cases of dyskeratosis congenita (Dokal, 1996), and in some patients symptoms related to aplastic anemia may precede the diagnosis of DC (Forni et al., 1993). In vitro clonogenic assays, as well as long-term bone marrow culture studies (Marsh et al., 1992), suggested that symptoms of aplastic anemia in DC may be due to a defect at the level of the hematopoietic stem cell. Ferraris et al. (1997) hypothesized that, at least in some DC families, the selective pressure in the heterozygote might be strong enough to determine negative selection of progenitors bearing the mutant allele, resulting in extreme skewing of X-chromosome inactivation in cells of hemopoietic descent. The pattern of methylation of HpaII and HhaI sites with a highly polymorphic CAG repeat in the coding region of the first exon of the androgen receptor gene (Allen et al., 1992) was used in these studies. Ferraris et al. (1997) studied 2 families and found that indeed carrier females showed nonrandom X inactivation in whole blood leukocytes, granulocytes, and mononuclear cells.

Using the methylation-sensitive HpaII site in the androgen receptor gene (313700), Vulliamy et al. (1997) found that in 5 different families in which the inheritance of DC appeared to be X-linked, all 16 carriers showed skewed X-inactivation patterns. The results indicated that in the hematopoiesis of heterozygous females cells expressing the normal DC allele had a growth advantage over cells that express the mutant allele. In 7 other families with sporadic cases of DC or with an uncertain pattern of inheritance, both skewed and normal patterns of X-inactivation were observed. In these families, or where crucial family members are unavailable, the study of X-inactivation patterns will add to linkage analysis in providing information about carrier status.

As a result of a large-scale positional cloning and sequencing of the region of Xq28 containing the DKC1 gene, virtually all DKC positional candidates had been identified. By hybridization screening with 28 candidate cDNAs, Heiss et al. (1998) detected a 3-prime deletion in 1

DKC patient with a cDNA probe derived from XAP101. They subsequently identified 5 different missense mutations in 5 unrelated patients in the XAP101 gene, indicating that it is the gene responsible for X-linked DKC (DKC1). DKC1 is highly conserved across species barriers and is the ortholog of rat NAP57 and S. cerevisiae CBF5. The peptide, referred to as dyskerin, was found to contain 2 TruB pseudouridine synthase motifs, multiple phosphorylation sites, and a carboxy-terminal lysine-rich repeat domain. By analogy to the function of the known dyskerin orthologs, involvement in the cell cycle and nucleolar function was predicted for the protein.

See also Carter et al. (1978); Civatte et al. (1978); Friedland et al. (1985); Garb (1958); Giannetti and Seidenari (1980); Inoue et al. (1973); Koszewski and Hubbard (1956); Steier et al. (1972); Wiedemann et al. (1984)

Addison, M.; Rice, M. S.: The association of dyskeratosis congenita and Fanconi's anaemia. Med. J. Aust. 1: 797-799, 1965.

Allen, R. C.; Zoghbi, H. Y.; Moseley, A. B.; Rosenblatt, H. M.; Belmont, J. W.: Methylation of HpaII and HhaI sites near the polymorphic CAG repeat in the human androgen-receptor gene correlates with X chromosome inactivation. Am. J. Hum. Genet. 51: 1229-1239, 1992.

Arngrimsson, R.; Dokal, I.; Luzzatto, L.; Connor, J. M.: Dyskeratosis congenita: three additional families show linkage to a locus in Xq28. J. Med. Genet. 30: 618-619, 1993.

Berthou, C.; Devergie, A.; D'Agay, M. F.; Sonsino, E.; Scrobohaci, M. L.; Loirat, C.; Gluckman, E.: Late vascular complications after bone marrow transplantation for dyskeratosis congenita. Brit. J. Haemat. 79: 335-336, 1991.

Bryan, H. G.; Nixon, R. K.: Dyskeratosis congenita and familial pancytopenia. J.A.M.A. 192: 203-208, 1965.

Carter, D. M.; Pan, M.; Gaynor, A.; Sibrack, L.; McGuire, J. S.: Delayed excision of psoralen-DNA cross-linking photoadducts in dyskeratosis congenita. (Abstract) Clin. Res. 26: 568A only, 1978.

Caux, F.; Aractingi, S.; Sawaf, M. H.; Ouhayoun, J.-P.; Dubertret, L.; Gluckman, E.: Dyskeratosis congenita. Europ. J. Derm. 6: 332-334, 1996.

Civatte, J.; Belaich, S.; Pluquet, C.: Syndrome de Zinsser-Cole-Engman, incomplet chez deux freres. Ann. Derm. Venerol. 105: 1071-1072, 1978.

Connor, J. M.; Gatherer, D.; Gray, F. C.; Pirrit, L. A.; Affara, N. A.: Assignment of the gene for dyskeratosis congenita to Xq28. Hum. Genet. 72: 348-351, 1986.

Connor, J. M.; Teague, R. H.: Dyskeratosis congenita: report of a large kindred. Brit. J. Derm. 105: 321-325, 1981.

Davidson, H. R.; Connor, J. M.: Dyskeratosis congenita. J. Med. Genet. 25: 843-846, 1988.

DeBauche, D. M.; Shashidhar Pai, G.; Stanley, W. S.: Enhanced G2 chromatid radiosensitivity in dyskeratosis congenita fibroblasts. Am. J. Hum. Genet. 46: 350-357, 1990.

Devriendt, K.; Matthijs, G.; Legius, E.; Schollen, E.; Blockmans, D.; van Geet, C.; Degreef, H.; Cassiman, J.-J.; Fryns, J.-P.: Skewed X-chromosome inactivation in female carriers of dyskeratosis congenita. Am. J. Hum. Genet. 60: 581-587, 1997.

Dokal, I.: Dyskeratosis congenita: an inherited bone marrow failure syndrome. Brit. J. Haemat. 92: 775-779, 1996.

Dokal, I.; Bungey, J.; Williamson, P.; Oscier, D.; Hows, J.; Luzzatto, L.: Dyskeratosis congenita fibroblasts are abnormal and have unbalanced chromosomal rearrangements. Blood 80: 3090-3096, 1992.

Ferraris, A. M.; Forni, G. L.; Mangerini, R.; Gaetani, G. F.: Nonrandom X-chromosome inactivation in hemopoietic cells from carriers of dyskeratosis congenita. (Letter) Am. J. Hum. Genet. 61: 458-461, 1997.

Forni, G. L.; Melevendi, C.; Jappelli, S.; Rasore-Quartino, A.: Dyskeratosis congenita: unusual presenting features within a kindred. Pediat. Hemat. Oncol. 10: 145-149, 1993.

Friedland, M.; Lutton, J. D.; Spitzer, R.; Levere, R. D.: Dyskeratosis congenita with hyoplastic anemia: a stem cell defect. Am. J. Hemat. 20: 85-87, 1985.

Garb, J.: Dyskeratosis congenita with pigmentation, dystrophia unguium and leukoplakia oris. Arch. Derm. 77: 704-712, 1958.

Giannetti, A.; Seidenari, S.: Deficit of cell-mediated immunity, chromosomal alterations and defective DNA repair in a case of dyskeratosis congenita. Dermatologica 160: 113-117, 1980.

Gutman, A.; Frumkin, A.; Adam, A.; Bloch-Shatacher, N.; Rosenszajn, L. A.: X-linked dyskeratosis congenita with pancytopenia. Arch. Derm. 114: 1667-1671, 1978.

Heiss, N. S.; Knight, S. W.; Vulliamy, T. J.; Klauck, S. M.; Wiemann, S.; Mason, P. J.; Poustka, A.; Dokal, I.: X-linked dyskeratosis congenita is caused by mutations in a highly conserved gene with putative nucleolar functions. Nature Genet. 19: 32-38, 1998.

Inoue, S.; Mekanik, G.; Mahallati, M.; Zuelzer, W. W.: Dyskeratosis congenita with pancytopenia: another constitutional anemia. Am. J. Dis. Child. 126: 389-393, 1973.

Kalb, R. E.; Grossman, M. E.; Hutt, C.: Avascular necrosis of bone in dyskeratosis congenita. Am. J. Med. 80: 511-513, 1986.

Knight, S. W.; Vulliamy, T.; Forni, G. L.; Oscier, D.; Mason, P. J.; Dokal, I.: Fine mapping of the dyskeratosis congenita locus in Xq28. J. Med. Genet. 33: 993-995, 1996.

Koszewski, B. J.; Hubbard, T. F.: Congenital anemia in hereditary ectodermal dysplasia. Arch. Derm. 74: 159-166, 1956.

Marsh, J. C. W.; Will, A. J.; Hows, J. M.; Sartori, P.; Darbyshire, P. J.; Williamson, P. J.; Oscier, D. G.; Dexter, T. M.; Testa, N. G.: 'Stem cell' origin of the hematopoietic defect in dyskeratosis congenita. Blood 79: 3138-3144, 1992.

Milgrom, H.; Stoll, H. L., Jr.; Crissey, J. T.: Dyskeratosis congenita: a case with new features. Arch. Derm. 89: 345-349, 1964.

Ning, Y.; Yongshan, Y.; Pai, G. S.; Gross, A. J.: Heterozygote detection through bleomycin-induced G(2) chromatid breakage in dyskeratosis congenita families. Cancer Genet. Cytogenet. 60: 31-34, 1992.

Phillips, R. J.; Judge, M.; Webb, D.; Harper, J. I.: Dyskeratosis congenita: delay in diagnosis and successful treatment of pancytopenia by bone marrow transplantation. Brit. J. Derm. 127: 278-280, 1992.

Reichel, M.; Grix, A. C.; Isseroff, R. R.: Dyskeratosis congenita associated with elevated fetal hemoglobin, X-linked ocular albinism, and juvenile-onset diabetes mellitus. Pediat. Derm. 9: 103-106, 1992.

Selmanowitz, V. J.; van Voolen, G. A.: Fanconi's anemia and dyskeratosis congenita. (Letter) J.A.M.A. 216: 2015 only, 1971.

Sirinavin, C.; Trowbridge, A. A.: Dyskeratosis congenita: clinical features and genetic aspects: report of a family and review of the literature. J. Med. Genet. 12: 339-354, 1975.

Sorrow, J. M., Jr.; Hitch, J. M.: Dyskeratosis congenita: first report of its occurrence in a female and a review of the literature. Arch. Derm. 88: 340-347, 1963.

Steier, W.; van Voolen, G. A.; Selmanowitz, V. J.: Dyskeratosis congenita: relationship to Fanconi's anemia. Blood 34: 510-521, 1972.

Vulliamy, T. J.; Knight, S. W.; Dokal, I.; Mason, P. J.: Skewed X-inactivation in carriers of X-linked dyskeratosis congenita. Blood 90: 2213-2216, 1997.

Wiedemann, H. P.; McGuire, J.; Dwyer, J. M.; Sabetta, J.; Gee, J. B. L.; Smith, G. J. W.; Loke, J.: Progressive immune failure in dyskeratosis congenita: report of an adult in whom Pneumocystis carinii and fatal disseminated candidiasis developed. Arch. Intern. Med. 144: 397-399, 1984.

Womer, R.; Clark, J. E.; Wood, P.; Sabio, H.; Kelly, T. E.: Dyskeratosis congenita: two examples of this multisystem disorder. Pediatrics 71: 603-607, 1983.

DYSKERATOSIS CONGENITA, SCOGGINS TYPE (127550)

JJM NOTE

This entity is included because of the overlap with classic dyskeratosis congenita, although frank cancer has not been cited (yet?).

Scoggins et al. (1971) described a black family with a form of dyskeratosis congenita (DKC) obviously inherited as an autosomal dominant. Features included reticular hyperpigmentation of the skin (due to dermal pigmentation, melanin having been released by melanocytes and taken up by dermal phagocytes), dystrophic nails, osteoporosis, premalignant leukokeratosis of the mouth mucosa, absent fingerprints, scant hair, poor dentition, absent lacrimal puncta, palmar hyperkeratosis, anemia, endoreduplication on chromosome studies, and a defect of the immune mechanism (probably in the afferent limb). The hematologic, immunologic and chromosomal changes were similar to those of Fanconi panmyelopathy (227650). Three generations were affected with male-to-male transmission. Dyskeratosis congenita is usually X-linked recessive (305000). Tchou and Kohn (1982) also reported a presumed autosomal dominant form of DKC. Of the 104 patients with DKC reviewed by Davidson and Connor (1988), 92 were males. Five of the 12 affected females were from the family of Tchou and Kohn (1982). In this family the age of onset was later and the features milder and less typical than usual in the X-linked form, particularly in the 3 affected males.

In an isolated case in a female patient, the daughter of young, unrelated parents, Kehrer et al. (1992) found a markedly increased frequency of chromosomal breaks, hypodiploidy, and premature centromere disjunction in skin fibroblast cultures. The frequency of mitotic disturbances was almost as great as that in cultures from 2 severely affected patients with Fanconi anemia.

Reticulated hyperpigmentation is usually the first cutaneous manifestation of dyskeratosis congenita. The pigmentary change may be limited to neck, upper chest, and proximal parts of the limbs initially, but within affected areas the involvement is always diffuse. Baselga et al. (1998) described a patient with typical diffuse cutaneous signs of DKC superimposed with hyperpigmentation that was more pronounced along Blaschko lines. To explain this phenomenon, they assumed that the patient had the autosomal dominant type and that loss of heterozygosity occurred in a somatic cell giving rise to a population of cells that migrated along these lines during embryogenesis. The cells that migrated along Blaschko lines and expressed an intensified clinical picture would be homozygous or hemizygous for the defect. Happle (1996) suggested this mechanism for the occurrence of severe segmental lesions superimposed on a milder, diffuse manifestation of autosomal skin disorders such as neurofibromatosis, epidermolytic hyperkeratosis, or porokeratosis. This theory is particularly attractive in the case of DKC, which appears to be a disorder of chromosomal instability.

Baselga, E.; Drolet, B. A.; van Tuinen, P.; Esterly, N. B.; Happle, R.: Dyskeratosis congenita with linear areas of severe cutaneous involvement. Am. J. Med. Genet. 75: 492-496, 1998.

Davidson, H. R.; Connor, J. M.: Dyskeratosis congenita. J. Med. Genet. 25: 843-846, 1988.

Happle, R.: Segmental forms of autosomal dominant skin disorders: different types of severity reflect different states of zygosity. (Letter) Am. J. Med. Genet. 66: 241-242, 1996.

Kehrer, H.; Krone, W.; Schindler, D.; Kaufmann, R.; Schrezenmeier, H.: Cytogenetic studies of skin fibroblast cultures from a karyotypically normal female with dyskeratosis congenita. Clin. Genet. 41: 129-134, 1992.

Scoggins, R. B.; Prescott, K. J.; Asher, G. H.; Blaylock, W. K.; Bright, R. W.: Dyskeratosis congenita with Fanconi-type anemia: investigations of immunologic and other defects. (Abstract) Clin. Res. 19: 409 only, 1971.

Tchou, P. K.; Kohn, T.: Dyskeratosis congenita: an autosomal dominant disorder. J. Am. Acad. Derm. 6: 1034-1039, 1982.

FUCOSIDOSIS (*230000)

JJM NOTE

This entity is included because of the angiokeratoma corporis diffusum.

Description

Fucosidosis is an autosomal recessive lysosomal storage disease caused by defective alpha-L-fucosidase with accumulation of fucose in the tissues. Different phenotypes include clinical features such as neurologic deterioration, growth retardation, visceromegaly, and seizures in a severe early form; coarse facial features, angiokeratoma corporis diffusum, spasticity and delayed psychomotor development in a longer surviving form; and an unusual spondylometaphyseoepiphyseal dysplasia in yet another form.

Clinical Features

Consistent with the classification used by Kousseff et al. (1976) and by others, type 1, the severe disorder, is characterized by neurologic deterioration beginning at about 6 months of age and by death within the first decade of life. Type 2, which represents a majority of patients, shows a neurologic deterioration that is slower in course and less severe and is accompanied by development of angiokeratoma corporis diffusum and survival to the second, third, or fourth decades.

Schoonderwaldt et al. (1980) subscribed to classification of fucosidosis into types 1, 2, and 3 on the basis of age of onset. Types 1 and 2 have their onset at 10 and 18 months, respectively, and death occurs before age 6 years. Durand et al. (1969) described type 1 and Loeb et al. (1969) type 2. Type 3 is a juvenile form of the disease, with less rapid psychomotor and neurologic deterioration than in types 1 and 2, with survival into the twenties in some cases, and in all cases the typical rash of Fabry disease (angiokeratomata). Gordon et al. (1995) described dystonia of lower extremities in an affected 7-year-old boy. There was no alpha-fucosidase activity in his cultured lymphoblasts. He was found to be homozygous for a Q422X mutation, which resulted in loss of an EcoRI restriction site. The authors stated that no previously reported fucosidosis families with this same homozygous mutation showed dystonia.

Van Hoof and Hers (1968) found deficiency of alpha-fucosidase activity in the liver of patients with a Hurler-like disorder described by 17,16:Durand et al. (1967, 1968). Fucose accumulated in all tissues (Durand et al., 1969). The patient of Bernard et al. (1966) was female. Durand et al. (1968) called the condition fucosidosis. The Belgian patient studied by Loeb et al. (1969) was probably related to the 2 Italian patients reported by Durand et al. (1969).

Yet another phenotype with deficiency of alpha-L-fucosidase was reported by Schafer et al. (1971), who found deficiency of this enzyme in a 9-year-old child with an unusual spondylometaphyseoepiphyseal dysplasia.

Patel et al. (1972) described a different phenotype resulting from deficiency of alpha-L-fucosidase. The patient, who was not Hurler-like in appearance, showed unusual survival (to at least 20 years) and from age 4 had angiokeratoma of the skin as in Fabry disease. Differing from Fabry disease were severe mental and physical retardation and normal renal function. He also showed anhidrosis and inability to control body temperature. Urinary and leukocyte alpha-L-fucosidase was 10% of normal; obligate heterozygotes had intermediate values. This is, presumably, an example of allelism with production of quite a different clinical picture. (Compare the different phenotypes of the Hurler and Scheie syndromes (252800), allelic mutations in the alpha-L-iduronidase gene.)

Schoonderwaldt et al. (1980) reported 2 Dutch brothers with some characteristics of type 2, but more like type 3 from the point of view of rate of progression and length of survival; however, the brothers did not show angiokeratoma. The skin was distinctively dry and thin, a feature not described in other patients. Hurler-like dysmorphic features occur in types 1 and 2 and sometimes in type 3. Ikeda et al. (1984) studied 3 sisters with the adult form (type 2 in the classification of Kousseff et al., 1976). They had prominent psychomotor retardation, gargoyle features, and angiokeratoma. Abnormalities were found in macrophages, endothelial cells, fibroblasts, and Schwann cells on rectal biopsy.

Willems et al. (1988) described 2 families, 1 with 3 patients and a second with 2 patients, in each of which both type 1 and type 2 fucosidosis were represented. In one family described by Willems et al. (1988), 1 patient survived to almost 20 years of age and showed angiokeratoma, whereas 2 other patients in a cousin sibship died under age 5 years. All 4 parents were traced to common ancestors; thus, the children were presumably homozygous for the same gene. In the second family, one patient was alive at almost 25 years of age, whereas a sib had died at age 4 years. Both had angiokeratoma. Willems et al. (1988) suggested that environmental factors or 'modifying genes' separate from the fucosidase structural gene may contribute to the phenotype. Willems et al. (1991) reviewed the literature on 77 patients with fucosidosis. They presented 4 lines of evidence suggesting that the distinction between a severe type 1 form and a less severe type 2 form with survival into adulthood may not reflect true genetic heterogeneity. First, family pedigrees have shown both types within a single family. Second, clinical studies failed to show a clear distinction between types 1 and 2, but rather a seemingly continuous clinical spectrum. Third, no biochemical heterogeneity has been observed. Very low to negligible residual fucosidase enzyme activity has been found in patients of both types. Fourth, the identical DNA mutation has been found in the homozygous state in patients with the rapidly or slowly progressive form of the disease. The authors speculated that types 1 and 2 fucosidosis, rather than representing 2 distinct clinical entities, represent the extremes of a continuous clinical spectrum.

Mapping

Turner et al. (1975) demonstrated a common polymorphism of alpha-fucosidase by the technique of isoelectric focusing on thin layer acrylamide gel. Alpha-fucosidase is one of the enzymes for which both polymorphism and deficiency are known (cf. G6PD, 305900). By family studies, Corney et al. (1977) showed that Rh and alpha-fucosidase are closely linked. The Goss-Harris method of mapping combines features of recombinational study in families and synteny tests in hybrid cells. As applied to chromosome 1, the method shows that AK2 (103020) and UMPK (191710) are distal to PGM1 (171900), and that the order of the loci is PGM1: UMPK: (AK2, alpha-FUC): ENO1 (172430) (Goss and Harris, 1977).

On the basis of a family segregating for elliptocytosis and PGD (172200) as well as the common polymorphisms Rh, PGM1, and alpha-fucosidase, Cook et al. (1977) concluded that the map of 1p is, in the male, 1pter—PGD—18%—El—2%—Rh—2%—alpha-FUC—25%—PGM1—centromere. In the female the above intervals were estimated to be 22%, 4%, 2%, and 37%, respectively. In a linkage map of chromosome 1 prepared by Rouleau et al. (1990), FUCA1 was placed 8 cM distal to RH.

Carritt et al. (1982) presented evidence that FUCA is on 1p34. (See 136820 for FUCA2, which was thought to map to chromosome 4, but was later shown to be on 6q.) On the basis of somatic cell hybridization studies, Fowler et al. (1986) stated that the most likely location of the FUCA gene is the distal region of 1p34. Southern analysis of somatic cell hybrid DNA detected a second site of homology on chromosome 2. They designated the second site as FUCA1L. Using a cDNA in somatic cell hybrid studies, Carritt and Welch (1987) demonstrated an apparent processed FUCA pseudogene on chromosome 2. Coucke et al. (1991) assigned a pseudogene, FUCA1P, to 2q31-q32 by fluorescence in situ hybridization and digital imaging microscopy. The localization was confirmed by linkage analysis between FUCA1P and COL3A1 (120180) in 2q24-q32, which gave a maximal lod score of 4.03 at 3% recombination.

Molecular Genetics

Kretz et al. (1992) reported the structure of the FUCA gene and the fucosidase pseudogene. FUCA is composed of 8 exons spanning 23 kb. The sequence 5-prime to the open reading frame shows multiple transcription factor binding sites but no TATA box. Northern blot analysis demonstrated an mRNA of 2.3 kb in lymphoblasts, testis, and epithelial cells. The pseudogene was found to be 80% identical to FUCA cDNA but did not contain an open reading frame.

Phenotype/Genotype

Gatti et al. (1973) emphasized heterogeneity in fucosidosis. Johnson and Dawson (1985) presented molecular evidence for heterogeneity among patients with fucosidosis. They studied 11 patients with less than 5% of normal alpha-L-fucosidase activity: in 8 patients, fibroblasts synthesized no detectable enzyme protein; in 2, they synthesized normal amounts of the 53,000-dalton precursor, but none of the mature 50,000-dalton form was detectable; and in 1, they contained small amounts of crossreacting material. Patients with this disorder have difficulty in degrading fucose-containing blood group H and Lewis substances.

Wood (1977) argued that plasma and leukocyte alpha-L-fucosidases are under the same genetic control and that the electrophoretic polymorphism of the enzyme in leukocytes and the enzyme in fucosidosis is determined by mutations at the same locus. In fucosidosis, deficiency of alpha-L-fucosidase is found in both plasma and leukocytes. Normal individuals with low plasma enzyme but normal leukocyte enzyme were described by Ng et al. (1976); however, plasma enzyme is different electrophoretically from leukocyte enzyme. The level of plasma fucosidase could be explained on the basis of 2 alleles leading to 3 phenotypes of high, intermediate, and low enzyme levels.

A majority of earlier-reported cases of fucosidosis were Italian; most of these patients originated from 2 neighboring villages, Grotteria and Mammola, in the southern Italian province of Reggio Calabria (Sangiorgi et al., 1982). The family with the 2 sibs they reported was of Italian extraction. Lewis A and B antigens were very high in both red cells and saliva. However, H specificity showed no increase. Because Lewis-specific alpha-fucose is bound to beta-N-acetyl-D-glucosamine by an alpha-1 to -4 linkage, whereas H-specific alpha-fucose is linked to beta-galactose by an alpha-1 to -2 linkage, the fucosidase that is deficient in fucosidosis may be the one for the 1-to-4 linkage, or, alternatively, the mutation may be such that only that specificity is lost. The last possibility is based on the notion that different mutations of the same enzyme molecule are responsible for the 2 types of fucosidosis (and perhaps for the skeletal dysplasia described by Schafer et al., 1971). This idea follows from the one-gene/one-enzyme/many-substrates idea of O'Brien (1975) and the notion that mutation can interfere with one or another but not all substrate specificities.

Darby et al. (1988) found 2 RFLPs in the FUCA gene in Caucasians; the polymorphism information content (PIC) of the combined DNA markers was 0.38. No recombinants were observed between fucosidosis phenotype and these RFLP markers, suggesting that the lesion responsible for fucosidosis is located in the FUCA1 gene. In the course of Southern blot analysis of the FUCA1 gene in 23 patients with fucosidosis, Willems et al. (1988) found in 5 patients obliteration of an EcoRI restriction site in the open reading frame encoding mature alpha-L-fucosidase. This abnormality was not observed in 80 controls and was thought to be the basic defect responsible for fucosidosis in these patients. Both patients with the severe type 1 form and patients with the less severe type 2 were shown to be homozygous for this presumed mutation. In the remaining 18 patients, no EcoRI site obliteration, major gene deletions, or insertions were detected. The heterogeneity found at the DNA level was not present at the protein level, as all fucosidosis patients investigated had low fucosidase protein (less than 6% of normal) and negligible fucosidase activity in fibroblasts and lymphoblastoid cell lines. In 6 unrelated Italian patients with fucosidosis, Guazzi et al. (1989) found 3 patterns of mRNA. One patient showed a variant pattern of DNA hybridization with a partial length cDNA; the change represented loss of an EcoRI site. The same patient showed a markedly decreased amount of mRNA on Northern blot hybridization. Two patients apparently lacked mRNA. The other 3 patients showed a transcript similar in size and amount to that observed in controls.

Cragg et al. (1997) investigated the molecular basis of the deficiency of alpha-L-fucosidase in 8 patients in whom the diagnosis of fucosidosis had been made on clinical and enzymatic grounds. None of the patients had a deletion or gross alteration of the FUCA1 gene. SSCP analysis followed by direct sequencing of amplified exons and flanking regions identified putative disease-causing mutations in 6 of the 8 patients, who had severe forms of the disease and very low residual fucosidase activity and protein. Five of the 6 mutations had not previously been described.

Data on gene frequencies of allelic variants were tabulated by Roychoudhury and Nei (1988).

Animal Model

In the rat, alpha-L-fucosidase, like hemoglobin, is a tetramer, with 2 structurally different subunits, alpha and beta (Carlsen and Pierce, 1972). If the same is true in man, there is opportunity for at least 2 nonallelic varieties of fucosidosis.

In a dog with alpha-L-fucosidase deficiency, Taylor et al. (1986) found that bone marrow transplantation after total lymphoid irradiation raised the level of enzyme activity in both visceral and neural tissues with consequent reduction in the severity of the storage lesions. These results offered hope that early bone marrow transplantation may prevent the development of disease in neurovisceral storage disorders.

Studying the canine model of fucosidosis in a colony of English Springer Spaniels, Occhiodoro and Anson (1996) determined the nucleotide sequence and predicted amino acid sequence of canine fucosidase, which showed a high level of identity with the human and rat sequences. The study of the gene in the affected animals demonstrated a 14-bp deletion in mRNA. The deletion created a frameshift and introduced a premature translation termination codon at amino acid 152 and was shown to correspond to a deletion of the last 14 bp of exon 1. Rapid PCR-based screening for the mutation was performed on genomic DNA from dogs within the colony, enabling detection of both carrier and homozygous dogs.

See also Beratis et al. (1977); Beyer and Wiederschain (1982); Blitzer et al. (1985); Brill et al. (1975); Di Matteo et al. (1976); Epinette et al. (1973); Fisher et al. (1978); Fukushima et al. (1985); Koch et al. (1978); Kousseff et al. (1973); Poenaru et al. (1976); Romeo et al. (1977); Turner et al. (1978); Willems et al. (1988); Zielke et al. (1972); Zielke et al. (1972)

Beratis, N. G.; Turner, B. M.; Labadie, G.; Hirschhorn, K.: Alpha-L-fucosidase in cultured skin fibroblasts from normal subjects and fucosidosis patients. Pediat. Res. 11: 862-866, 1977.

Beyer, E. M.; Wiederschain, G. Y.: Further evidence of human alpha-L-fucosidase polymorphism. Clin. Chim. Acta 123: 251-259, 1982.

Blitzer, M. G.; Sutton, M.; Miller, J. B.; Shapira, E.: A thermolabile variant of alpha-L-fucosidase—clinical and laboratory findings. Am. J. Med. Genet. 20: 535-539, 1985.

Brill, P. W.; Beratis, N. G.; Kousseff, B. G.; Hirschhorn, K.: Roentgenographic findings in fucosidosis type 2. Am. J. Roentgen. 124: 75-82, 1975.

Carlsen, R. B.; Pierce, J. G.: Purification and properties of an alpha-L-fucosidase from rat epididymis. J. Biol. Chem. 247: 23-32, 1972.

Carritt, B.; King, J.; Welch, H. M.: Gene order and localization of enzyme loci on the short arm of chromosome 1. Ann. Hum. Genet. 46: 329-335, 1982.

Carritt, B.; Welch, H. M.: An alpha-fucosidase pseudogene on human chromosome 2. Hum. Genet. 75: 248-250, 1987.

Cook, P. J. L.; Noades, J. E.; Newton, M. S.; de Mey, R.: On the orientation of the Rh:E1-1 linkage group. Ann. Hum. Genet. 41: 157-162, 1977.

Corney, G.; Fisher, R. A. F.; Cook, P. J. L.; Noades, J.; Robson, E. B.: Linkage between alpha-fucosidase and the Rhesus blood group. Ann. Hum. Genet. 40: 403-405, 1977.

Coucke, P.; Mangelschots, K.; Speleman, F.; Bossuyt, P.; Van Oostveldt, P.; Van der Auwera, B.; Carritt, B.; Willems, P. J.: Assignment of the fucosidase pseudogene FUCA1P to chromosome region 2q31-q32. Cytogenet. Cell Genet. 57: 120-122, 1991.

Cragg, H.; Williamson, M.; Young, E.; O'Brien, J.; Alhadeff, J.; Fang-Kircher, S.; Paschke, E.; Winchester, B.: Fucosidosis: genetic and biochemical analysis of eight cases. J. Med. Genet. 34: 105-110, 1997.

Cragg, H.; Winchester, B.; Seo, H.-C.; O'Brien, J.; Swallow, D.: Molecular basis of the common electrophoretic polymorphism (Fu1/Fu2) in human alpha-L-fucosidase. (Letter) J. Med. Genet. 31: 659-660, 1994.

Darby, J. K.; Willems, P. J.; Nakashima, P.; Johnsen, J.; Ferrell, R. E.; Wijsman, E. M.; Gerhard, D. S.; Dracopoli, N. C.; Housman, D.; Henke, J.; Fowler, M. L.; Shows, T. B.; O'Brien, J. S.; Cavalli-Sforza, L. L.: Restriction analysis of the structural alpha-L-fucosidase gene and its linkage to fucosidosis. Am. J. Hum. Genet. 43: 749-755, 1988.

Di Matteo, G.; Durand, P.; Gatti, R.; Maresca, A.; Orfeo, M.; Urbano, F.; Romeo, G.: Human alpha-fucosidase: single residual enzymatic form in fucosidosis. Biochim. Biophys. Acta 427: 538-545, 1976.

Durand, P.; Borrone, C.; Della Cella, G.: Fucosidosis. J. Pediat. 75: 665-674, 1969.

Durand, P.; Borrone, C.; Della Cella, G.; Philippart, M.: Fucosidosis. (Letter) Lancet I: 1198, 1968.

Durand, P.; Philippart, M.; Borrone, C.; Della Cella, G.: A new glycolipid storage disease. (Abstract) Pediat. Res. 1: 416 only, 1967.

Epinette, W. W.; Norins, A. L.; Zeman, W.; Patel, V.: Angiokeratoma corporis diffusum with alpha-L-fucosidase deficiency. Arch. Derm. 107: 754-757, 1973.

Fisher, R. A.; Noades, J. E.; Robson, E. B.: Studies on the Rh:alpha-fucosidase linkage group. Cytogenet. Cell Genet. 22: 381-383, 1978.

Fleming, C. J.; Sinclair, D. U.; White, E. J.; Winchester, B.; Whiteford, M. L.; Connor, J. M.: A fucosidosis patient with relative longevity and a missense mutation in exon 7 of the alpha-fucosidase gene. J. Inherit. Metab. Dis. 21: 688-689, 1998.

Fowler, M. L.; Nakai, H.; Byers, M. G.; Fukushima, H.; Eddy, R. L.; Henry, W. M.; Haley, L. L.; O'Brien, J. S.; Shows, T. B.: Chromosome 1 localization of the human alpha-L-fucosidase structural gene with a homologous site on chromosome 2. Cytogenet. Cell Genet. 43: 103-108, 1986.

Fukushima, H.; de Wet, J. R.; O'Brien, J. S.: Molecular cloning of a cDNA for human alpha-L-fucosidase. Proc. Nat. Acad. Sci. 82: 1262-1265, 1985.

Gatti, R.; Borrone, C.; Trias, X.; Durand, P.: Genetic heterogeneity in fucosidosis. (Letter) Lancet II: 1024 only, 1973.

Gordon, B. A.; Gordon, K. E.; Seo, H. C.; Yang, M.; DiCioccio, R. A.; O'Brien, J. S.: Fucosidosis with dystonia. Neuropediatrics 26: 325-327, 1995.

Goss, S. J.; Harris, H.: Gene transfer by means of cell fusion. II. The mapping of 8 loci on human chromosome 1 by statistical analysis of gene assortment in somatic cell hybrids. J. Cell Sci. 25: 39-57, 1977.

Guazzi, S.; Persici, P.; Gatti, R.; Durand, P.; O'Brien, J. S.; Romeo, G.: Heterogeneity of mRNA expression in Italian fucosidosis patients. Hum. Genet. 82: 63-66, 1989.

Ikeda, S.; Kondo, K.; Oguchi, K.; Yanagisawa, N.; Horigome, R.; Murata, F.: Adult fucosidosis: histochemical and ultrastructural studies of rectal mucosa biopsy. Neurology 34: 451-456, 1984.

Johnson, K.; Dawson, G.: Molecular defect in processing alpha-fucosidase in fucosidosis. Biochem. Biophys. Res. Commun. 133: 90-97, 1985.

Koch, G. A.; Brown, J. A.; Shows, T. B.: Alpha-L-fucosidase (alpha-FUC) localized to the pter-p21 region of chromosome 1 in man. Cytogenet. Cell Genet. 22: 389-391, 1978.

Kousseff, B. G.; Beratis, N. G.; Danesino, C.; Hirschhorn, K.: Genetic heterogeneity in fucosidosis. Lancet II: 1387-1388, 1973.

Kousseff, B. G.; Beratis, N. G.; Strauss, L.; Brill, P. W.; Rosenfield, R. E.; Kaplan, B.; Hirschhorn, K.: Fucosidosis type 2. Pediatrics 57: 205-213, 1976.
Kretz, K. A.; Cripe, D.; Carson, G. S.; Fukushima, H.; O'Brien, J. S.: Structure and sequence of the human alpha-L-fucosidase gene and pseudogene. Genomics 12: 276-280, 1992.
Kretz, K. A.; Darby, J. K.; Willems, P. J.; O'Brien, J. S.: Characterization of EcoRI mutation in fucosidosis patients: a stop codon in the open reading frame. J. Molec. Neurosci. 1: 177-180, 1989.
Loeb, H.; Tondeur, M.; Jonniaux, G.; Mockel-Pohl, S.; Vamos-Hurwitz, E.: Biochemical and ultrastructural studies in a case of mucopolysaccharidosis F (fucosidosis). Helv. Paediat. Acta 24: 519-537, 1969.
Ng, W. G.; Donnell, G. N.; Koch, R.; Bergren, W. R.: Biochemical and genetic studies of plasma and leukocyte alpha-L-fucosidase. Am. J. Hum. Genet. 28: 42-50, 1976.
O'Brien, J. S.: Molecular genetics of GM1 beta-galactosidase. Clin. Genet. 8: 303-313, 1975.
Occhiodoro, T.; Anson, D. S.: Isolation of the canine alpha-L-fucosidase cDNA and definition of the fucosidosis mutation in English Springer Spaniels. Mammalian Genome 7: 271-274, 1996.
Patel, V.; Watanabe, I.; Zeman, W.: Deficiency of alpha-L-fucosidase. Science 176: 426-428, 1972.
Poenaru, L.; Dreyfus, J.-C.; Boue, J.; Nicolesco, H.; Ravise, N.; Bamberger, J.: Prenatal diagnosis of fucosidosis. Clin. Genet. 10: 260-264, 1976.
Primrose, D. A.: A note on fucosidosis in a mentally subnormal female. J. Mental Defic. Res. 19: 267 only, 1975.
Romeo, G.; Borrone, C.; Gatti, R.; Durand, P : Fucosidosis in Calabria: founder effect or high gene frequency? (Letter) Lancet I: 368-369, 1977.
Rouleau, G. A.; Bazanowski, A.; Gusella, J. F.; Haines, J. L.: A genetic map of chromosome 1: comparison of different data sets and linkage programs. Genomics 7: 313-318, 1990.
Roychoudhury, A. K.; Nei, M.: Human Polymorphic Genes: World Distribution. New York: Oxford Univ. Press (pub.) 1988.
Sangiorgi, S.; Mochi, M.; Beretta, M.; Prosperi, L.; Costantino, G.; Romeo, G.: Genetic and demographic characterization of a population with high incidence of fucosidosis. Hum. Hered. 32: 100-105, 1982.
Schafer, I. A.; Powell, D. W.; Sullivan, J. C.: Lysosomal bone disease. (Abstract) Pediat. Res. 5: 391-392, 1971.
Schoonderwaldt, H. C.; Lamers, K. J. B.; Kleijnen, F. M.; van den Berg, C. J. M. G.; de Bruyn, C. H. M. M.: Two patients with an unusual form of type II fucosidosis. Clin. Genet. 18: 348-354, 1980.
Seo, H.-C.; Willems, P. J.; O'Brien, J. S.: Six additional mutations in fucosidosis: three nonsense mutations and three frameshift mutations. Hum. Molec. Genet. 2: 1205-1208, 1993.
Taylor, R. M.; Farrow, B. R. H.; Stewart, G. J.; Healy, P. J.: Enzyme replacement in nervous tissue after allogeneic bone-marrow transplantation for fucosidosis in dogs. Lancet II: 772-774, 1986.
Turner, B. M.; Smith, M.; Turner, V. S.; Kucherlapati, R. S.; Ruddle, F. H.; Hirschhorn, K.: Assignment of the gene locus for human alpha-L-fucosidase to chromosome 1 by analysis of somatic cell hybrids. Somat. Cell Genet. 4: 45-54, 1978.
Turner, B. M.; Turner, V. S.; Beratis, N. G.; Hirschhorn, K.: Polymorphism of alpha-fucosidase. Am. J. Hum. Genet. 27: 651-661, 1975.
Van Hoof, F.; Hers, H. G.: Mucopolysaccharidosis by absence of alpha-fucosidase. (Letter) Lancet I: 1198 only, 1968.
Willems, P. J.; Darby, J. K.; DiCioccio, R. A.; Nakashima, P.; Eng, C.; Kretz, K. A.; Cavalli-Sforza, L. L.; Shooter, E. M.; O'Brien, J. S.: Identification of a mutation in the structural alpha-L-fucosidase gene in fucosidosis. Am. J. Hum. Genet. 43: 756-763, 1988.
Willems, P. J.; Garcia, C. A.; De Smedt, M. C. H.; Martin-Jimenez, R.; Darby, J. K.; Duenas, D. A.; Granado-Villar, D.; O'Brien, J. S.: Intrafamilial variability in fucosidosis. Clin. Genet. 34: 7-14, 1988.
Willems, P. J.; Gatti, R.; Darby, J. K.; Romeo, G.; Durand, P.; Dumon, J. E.; O'Brien, J. S.: Fucosidosis revisited: a review of 77 patients. Am. J. Med. Genet. 38: 111-131, 1991.
Williamson, M.; Cragg, H.; Grant, J.; Kretz, K.; O'Brien, J.; Willems, P. J.; Young, E.; Winchester, B.: A 5-prime splice site mutation in fucosidosis. J. Med. Genet. 30: 218-223, 1993.
Wood, S.: Genetic control of alpha-L-fucosidase. (Letter) Lancet I: 368 only, 1977.
Yang, M.; Allen, H.; DiCioccio, R. A.: A mutation generating a stop codon in the alpha-L-fucosidase gene of a fucosidosis patient. Biochem. Biophys. Res. Commun. 189: 1063-1068, 1992.
Zielke, K.; Okada, S.; O'Brien, J. S.: Fucosidosis: diagnosis by serum assay of alpha-L-fucosidase. J. Lab. Clin. Med. 79: 164-169, 1972.
Zielke, K.; Veath, M. L.; O'Brien, J. S.: Fucosidosis: deficiency of alpha-L-fucosidase in cultured skin fibroblasts. J. Exp. Med. 136: 197-199, 1972.

ALPHA-GALACTOSIDASE B (*104170)

JJM NOTE

This entity is included because of the angiokeratomas.

In a study of man-rodent somatic cell hybrids, de Groot et al. (1978) assayed human N-acetyl-alpha-D-galactosaminidase activity and concluded that alpha-galactosidase B and mitochondrial aconitase (ACO2; 100850), known to be on chromosome 22, are syntenic. They also obtained evidence for direct assignment of alpha-galactosidase B to chromosome 22. Alpha-NAGA was thought to be a more appropriate designation for this enzyme than alpha-galactosidase B by de Groot et al. (1978), who claimed that there was no structural relationship between alpha-gal A (on the X chromosome; GLA; 301500) and so-called alpha-gal B. However, DNA studies (Wang et al., 1990; Wang and Desnick, 1991), described later, led to a different conclusion. In man-rodent cell hybrids, 5,6:Geurts van Kessel et al. (1979, 1980) studied chronic myeloid leukemia cells to determine the site of the break on 22q relative to markers assigned to chromosomes 22 and 9. Alpha-NAGA remained with the Ph-1 chromosome, whereas ACO2 went with chromosome 9. Thus, the former is probably in band 22q11, whereas the latter is between it and 22qter.

Wang et al. (1990) isolated a full-length 2.2-kb cDNA and a genomic cosmid clone containing the entire NAGA gene. Sequence analysis revealed striking similarities between the NAGA locus and exons 1-6 of alpha-galactosidase A, suggesting that the 2 genes evolved by duplication and divergence from a common ancestral locus. Wang and Desnick (1991) also pointed to remarkable amino acid identity between the NAGA and GLA genes.

In 2 sons of a German couple with remote consanguinity, 14,15:van Diggelen et al. (1987, 1988) described the clinical and biochemical features of lysosomal alpha-N-acetylgalactosaminidase deficiency. The boys showed neurologic abnormalities starting at age 9 months, followed by progressive psychomotor deterioration. By the age of 2.5 and 4 years, they had 'largely lost their previously acquired motor and language skills.' Growth had been normal. Computerized tomographic scans were normal, and there was no organomegaly, obvious coarsening of the facies, or skeletal dysplasia. A uniquely abnormal pattern of urinary oligosaccharides was demonstrated by thin-layer chromatography. Among the carbohydrate-hydrolyzing lysosomal enzymes, only alpha-NAGA had not previously been associated with a disorder. The levels of this enzyme were very low in cultured fibroblasts, leukocytes and plasma, whereas these levels were normal in a healthy brother. Both parents had low normal or reduced activity. A major neutral oligosaccharide from the urine of 1 patient was identified as the blood group A determinant, a trisaccharide with terminal alpha-N-acetylgalactosamine. The concentration of this product in the urine of the older boy, who was a secretor and had blood group A, was 5 times normal. The younger boy, who had blood group O, did not excrete this trisaccharide. Schindler et al. (1988) described the clinical findings as consisting of severe psychomotor retardation with myoclonic seizures, decorticate posture, optic atrophy, blindness, marked long tract signs, and total loss of contact with the environment. No features of other lysosomal storage diseases were present. Ultrastructural examination of peripheral nerves was unremarkable, whereas the rectal mucosa contained dystrophic autonomic axons with 'tubulovesicular' material. A unique pattern of abnormal urinary oligosaccharides/glycopeptides was found by thin layer chromatography. Wang et al. (1988) pointed out that the brothers reported by van Diggelen et al. (1987) had a clinical course and neuropathologic findings similar to those in Seitelberger disease, the infantile form of neuroaxonal dystrophy (256600). The characteristic 'spheroids' were observed histologically and ultrastructurally in terminal axons in gray matter. This disorder, which they referred to as Schindler disease, must represent, therefore, a form of infantile axonal dystrophy, the first in which a specific enzymatic defect has been identified. The disorder is autosomal recessive. Schindler et al. (1989) also characterized the disorder as a neuroaxonal dystrophy. They pointed out that although the disorder is caused by deficiency of a lysosomal enzyme, no lysosomal storage could be identified. It has been proposed that the dystrophic axons in infantile neuroaxonal dystrophy result from defective retrograde axonal transport. How deficiency of alpha-N-acetylgalactosaminidase might lead to a similar problem is not clear. Using PCR amplification and sequence analysis of PCR product from type I and type II offspring of consanguineous matings, Wang et al. (1990) demonstrated single basepair mutations in the homozygous state in both type I and type II. (Type I is classic Schindler disease; type II is an adult disorder with angiokeratoma as a prominent feature (104170.0002). Type II might appropriately be called Kanzaki disease (Kanzaki et al., 1989).)

Keulemans et al. (1996) reported the genotypes of 5 more patients with NAGA deficiency. One of them, related to the first reported German family (van Diggelen et al., 1987), had classic Schindler disease and the same homozygous mutation, i.e., glu325to-lys (104170.0001). The only manifestations in another patient, a 5-year-old Dutch girl whose family was clinically described by de Jong et al. (1994), were convulsions during fever and psychomotor retardation starting after the age of 1 year. She had 2 different mutations: glu325-to-lys inherited from her father and ser160-to-cys (104170.0004) inherited from her mother. The same genotype was found in a clinically unaffected 3.5-year-old brother of the proband. Keulemans et al. (1996) suggested that the brother might be a preclinical case of NAGA deficiency detected through screening. A homozygous nonsense mutation, glu193-to-ter, was found in 2 adult Spanish sibs who had angiokeratoma, lymphedema, and vacuolization in dermal cells, but no neurologic signs. These sibs, previously reported by Chabas et al. (1994), were clinically similar to the original patient described by Kanzaki et al. (1989). Although at the metabolic level the patients with NAGA deficiency are similar, extreme differences between the infantile form(s) and the adult form (Kanzaki disease) suggested to Keulemans et al. (1996) that other factors or genes contribute to the clinical heterogeneity.

See also Wang et al. (1990)

Chabas, A.; Coll, M. J.; Aparicio, M.; Rodriguez Diaz, E.: Mild phenotypic expression of alpha-N-acetylgalactosaminidase deficiency in two adult siblings. J. Inherit. Metab. Dis. 17: 724-731, 1994.

de Groot, P. G.; Westerveld, A.; Meera Khan, P.; Tager, J. M.: Localization of a gene for human alpha-galactosidase B (=N-acetyl-alpha-D-galactosaminidase) on chromosome 22. Hum. Genet. 44: 305-312, 1978.

de Jong, J; van den Berg, C; Wijburg, H.; Willemsen, R.; van Diggelen, O.; Schindler, D.; Hoevenaars, F.; Wevers, R.: Alpha-N-acetylgalactosaminidase deficiency with mild clinical manifestations and difficult biochemical diagnosis. J. Pediat. 125: 385-391, 1994.

Desnick, R. J.: Personal Communication. New York, N. Y. 1/15/1991.

Geurts van Kessel, A. H. M.; ten Brinke, H.; de Groot, P. G.; Hagemeijer, A.; Westerveld, A.; Meera Khan, P.; Pearson, P. L.: Regional localization of NAGA and ACO2 on human chromosome 22. (Abstract) Cytogenet. Cell Genet. 25: 161 only, 1979.

Geurts van Kessel, A. H. M.; Westerveld, A.; de Groot, P. G.; Meera Khan, P.; Hagemeijer, A.: Regional localization of the genes coding for human ACO2, ARSA, and NAGA on chromosome 22. Cytogenet. Cell Genet. 28: 169-172, 1980.

Kanzaki, T.; Wang, A. M.; Desnick, R. J.: Lysosomal alpha-N-acetylgalactosaminidase deficiency, the enzymatic defect in angiokeratoma corporis diffusum with glycopeptiduria. J. Clin. Invest. 88: 707-711, 1991.

Kanzaki, T.; Yokota, M.; Irie, F.; Hirabayashi, Y.; Wang, A. M.; Desnick, R. J.: Angiokeratoma corporis diffusum with glycopeptiduria due to deficient lysosomal alpha-N-acetylgalactosaminidase activity: clinical, morphologic, and biochemical studies. Arch. Derm. 129: 460-465, 1993.

Kanzaki, T.; Yokota, M.; Mizuno, N.: Clinical and ultrastructural studies of novel angiokeratoma corporis diffusum. (Abstract) Clin. Res. 36: 377A only, 1988.

Kanzaki, T.; Yokota, M.; Mizuno, N.; Matsumoto, Y.; Hirabayashi, Y.: Novel lysosomal glycoaminoacid storage disease with angiokeratoma corporis diffusum. Lancet I: 875-876, 1989.

Keulemans, J. L. M.; Reuser, A. J. J.; Kroos, M. A.; Willemsen, R.; Hermans, M. M. P.; van den Ouweland, A. M. W.; de Jong, J. G. N.; Wevers, R. A.; Renier, W. O.; Schindler, D.; Coll, M. J.; Chabas, A.; Sakuraba, H.; Suzuki, Y.; van Diggelen, O. P.: Human alpha-N-acetylgalactosaminidase (alpha-NAGA) deficiency: new mutations and the paradox between genotype and phenotype. J. Med. Genet. 33: 458-464, 1996.

Schindler, D.; Bishop, D. F.; Wallace, S.; Wolfe, D. E.; Desnick, R. J.: Characterization of alpha-N-acetylgalactosaminidase deficiency: a new neurodegenerative lysosomal disease. (Abstract) Pediat. Res. 23: 333A only, 1988.

Schindler, D.; Bishop, D. F.; Wolfe, D. E.; Wang, A. M.; Egge, H.; Lemieux, R. U.; Desnick, R. J.: Neuroaxonal dystrophy due to lysosomal alpha-N-acetylgalactosaminidase deficiency. New Eng. J. Med. 320: 1735-1740, 1989.

van Diggelen, O. P.; Schindler, D.; Kleijer, W. J.; Huijmans, J. G. M.; Galjaard, H.; Linden, H. U.; Peter-Katalinic, J.; Egge, H.; Dabrowski, U.; Cantz, M.: Lysosomal alpha-N-acetylgalactosaminidase deficiency: a new inherited metabolic disease. (Letter) Lancet II: 804 only, 1987.

van Diggelen, O. P.; Schindler, D.; Willemsen, R.; Boer, M.; Kleijer, W. J.; Huijmans, J. G. M.; Blom, W.; Galjaard, H.: Alpha-N-acetylgalactosaminidase deficiency, a new lysosomal storage disorder. J. Inherit. Metab. Dis. 11: 349-357, 1988.

Wang, A. M.; Bishop, D. F.; Desnick, R. J.: Human alpha-N-acetylgalactosaminidase-molecular cloning, nucleotide sequence, and expression of a full-length cDNA: homology with human alpha-galactosidase A suggests evolution from a common ancestral gene. J. Biol. Chem. 265: 21859-21866, 1990.

Wang, A. M.; Desnick, R. J.: Structural organization and complete sequence of the human alpha-N-acetylgalactosaminidase gene: homology with the alpha-galactosidase A gene provides evidence for evolution from a common ancestral gene. Genomics 10: 133-142, 1991.

Wang, A. M.; Kanzaki, T.; Desnick, R. J.: The molecular lesion in the alpha-N-acetylgalactosaminidase gene that causes angiokeratoma corporis diffusum with glycopeptiduria. J. Clin. Invest. 94: 839-845, 1994.

Wang, A. M.; Schindler, D.; Bishop, D. F.; Lemieux, R. U.; Desnick, R. J.: Schindler disease: biochemical and molecular characterization of a new neuroaxonal dystrophy due to alpha-N-acetylgalactosaminidase deficiency. (Abstract) Am. J. Hum. Genet. 43: A99 only, 1988.

Wang, A. M.; Schindler, D.; Desnick, R. J.: Schindler disease: the molecular lesion in the alpha-N-acetylgalactosaminidase gene that causes an infantile neuroaxonal dystrophy. J. Clin. Invest. 86: 1752-1756, 1990.

Wang, A. M.; Schindler, D.; Kanzaki, T.; Desnick, R. J.: Alpha-N-acetylgalactosaminidase gene: homology with human alpha-galactosidase A, and identification and confirmation of the mutations causing type I and II Schindler disease. (Abstract) Am. J. Hum. Genet. 47 (suppl.): A169 only, 1990.

Yamauchi, T.; Hiraiwa, M.; Kobayashi, H.; Uda, Y.; Miyatake, T.; Tsuji, S.: Molecular cloning of two species of cDNAs for human alpha-N-acetylgalactosaminidase and expression in mammalian cells. Biochem. Biophys. Res. Commun. 170: 231-237, 1990.

MANNOSIDOSIS, BETA (*248510)

JJM NOTE

This is another metabolic condition with angiokeratomas.

The innermost mannosyl residue in the N-linked oligosaccharide side-chain of glycoproteins is in a beta-linkage. Wenger et al. (1986) described a 46-month-old boy with a severe deficiency in beta-mannosidase activity, whose parents had a level of enzyme consistent with autosomal recessive inheritance. Beta-mannosidase activity levels near zero were found in all available samples from the patient, including leukocytes, plasma, and cultured skin fibroblasts. In addition to the deficiency of beta-mannosidase, this patient also had a low level of heparin sulfamidase activity in cultured skin fibroblasts. The patient's urine contained excess disaccharide that was identified as mannosyl(1-4)-N-acetylglucosamine and heparan sulfate. The patient showed coarsening facial features, mild bone disease, delayed speech development, hyperactivity, and mental retardation. Many of the findings resembled those in patients with Sanfilippo syndrome as well as those found to have defects in the catabolism of the oligosaccharide chains of glycoproteins such as fucosidosis (230000), sialidosis (256550), and aspartylglucosaminuria (208400). The deficiency of heparin sulfamidase and the excess of heparan sulfate in the urine, both findings consistent with Sanfilippo syndrome, type A (252900), may indicate the presence of 2 disorders. Indeed, complementation studies (Hu et al., 1990) indicated this to be the case.

Cooper et al. (1986) reported a case of beta-mannosidosis in an Indian-Hindu man with mental retardation whose parents were not related. Intellectual impairment was first observed when the patient began to attend school. Clinical assessment at the age of 44 years showed mental retardation but no other neurologic signs. The patient had angiokeratomas on the scrotum and the shaft of the penis and marked tortuosity of conjunctival vessels. He did not have facial dysmorphism, hepatosplenomegaly, or radiologic changes of the bones. The urine was free of albumin, red cells, and casts. CT scan of the brain was normal. The patient's 19-year-old mentally retarded brother had similar skin lesions. Both brothers had deficiency of beta-mannosidase and the parents had intermediate levels. Difference in the storage material in the caprine disease may account for the clinical differences; a trisaccharide accumulates in the goat.

Rodriguez-Serna et al. (1996) described a 22-year-old woman who since the age of 12 years had progressive angiokeratoma affecting the lower limbs and buttocks. Enzyme studies revealed beta-mannosidase deficiency in cultured fibroblasts and in serum and leukocytes. The patient's parents exhibited intermediate enzyme levels, thus confirming autosomal recessive inheritance. The patient demonstrated no other anomalies. Rodriguez-Serna et al. (1996) stated that, to that time, only 11 cases of beta-mannosidase deficiency, occurring in 8 families, had been reported. Mental retardation and neurologic disorders had been present in most reported patients.

Cooper et al. (1988) described the biochemical findings in plasma, white cells, fibroblasts, and urine obtained from the 2 brothers and their parents (the younger brother was stated to be 29 years old). Biopsy of a skin lesion from the 'bathing trunk' area showed marked cytoplasmic vacuolization. Beta-mannosidase activity was markedly deficient in plasma, leukocytes, fibroblasts, and urine; all other lysosomal enzymes measured, including sulfamidase, showed normal activity. Both parents showed reduced plasma and leukocyte activity. Urinary mucopolysaccharide excretion was normal, but thin layer chromatography of urinary oligosaccharides showed an abnormal band shown to be a disaccharide.

By analyzing urinary oligosaccharides and by measuring enzyme activity in leukocytes and plasma, Dorland et al. (1988) diagnosed beta-mannosidosis in 2 sons of a first-cousin Turkish couple. Behavioral abnormalities and deafness with speech retardation were leading clinical features. Poenaru et al. (1992) described a further case in which speech impairment was the first symptom of infantile onset. Beta-mannosidase activity was completely lacking in the patient and a heterozygous level was found in the parents. In addition, mannosyl-N-acetylglucosamine was identified in the patient's urine. In a gypsy family, Kleijer et al. (1990) described the first female patient with beta-mannosidosis. The girl was severely affected; an older brother had milder manifestations. Heterozygous levels of beta-mannosidase were found in the fibroblasts and/or plasma of the parents and a sister. In a black African 14-year-old boy, Levade et al. (1994) described severe deficiency of beta-mannosidase activity manifested by bilateral thenar and hypothenar amyotrophy, electrophysiologically demonstrable demyelinating peripheral neuropathy, and cytoplasmic vacuolation of skin fibroblasts and lymphoid cells. They could find reports of only 10 patients in 7 families.

In goats, deficiency of beta-mannosidase causes a severe disorder that affects peripheral and central nervous system myelin resulting in tremor, nystagmus, ataxia and early death (Jones and Dawson, 1981; Malachowski and Jones, 1983; Lovell and Jones, 1983; Jones et al., 1983). Indeed, the caprine disorder was thoroughly described several years before the human disorder as a severe neurologic disease with CNS demyelination and early death. Although several laboratories sought the human equivalent, it was not found until 1986. Bovine beta-mannosidosis was reported in 1991 in neonatal Salers calves in North America and New Zealand (Jones and Abbitt, 1993).

Although the beta-mannosidase locus had not been mapped in man, Lundin (1987) succeeded in mapping it to mouse chromosome 3 by demonstration of linkage to a gene for cadmium resistance (cdm) as well as to the Adh-3 locus (see 103730). The ADH3 locus in man is located on chromosome 4, which carries at least 2 other genes that are on mouse chromosome 3: EGF (131530) and FABP2 (134640). Thus, arguing from homology, the gene for beta-mannosidase may be located on human chromosome 4.

Leipprandt et al. (1996) determined the complete sequence of the caprine beta-mannosidase cDNA coding region and reported a mutation associated with caprine beta-mannosidosis. The cDNA codes for an 879-amino acid peptide that has 4 potential N-glycosylation sites. Comparison of the caprine and bovine cDNAs revealed that 96.3% of the nucleotides and 95% of the deduced amino acids were identical. A single-base deletion at position 1398 of coding sequence was identified in the DNA isolated from a goat with beta-mannosidosis. This deletion resulted in a shift in the reading frame and premature termination of translation, yielding a deduced peptide of 481 amino acids. Affected animals were homozygous for the mutation and obligate carriers were heterozygous. The assay for the presence or absence of the mutation was used for prenatal diagnosis using DNA collected from fetal fluids. The assay also confirmed chimerism in a goat with atypically mild beta-mannosidosis phenotype. Leipprandt et al. (1996) stated that with this assay it will be possible to assess the efficacy of engraftment of hematopoietic stem cells after prenatal transfer from donor sources.

As indicated earlier, unlike the severe clinical manifestations of beta-mannosidosis in ruminants, in which it leads to neonatal death, the human disease phenotype is generally milder. In addition, the phenotypic manifestations vary, even among members of the same family. To understand the molecular basis of the human disease and the mechanisms for such clinical variability, Alkhayat et al. (1998) sequenced the entire coding region of the human MANB1 gene using a combination of cDNA library screening, RT-PCR, and 5-prime rapid amplification of cDNA ends (RACE). The composite cDNA was 3,293 nucleotides, consisting of an 87-nucleotide 5-prime untranslated region, 2,640-nucleotide coding region, and 566-nucleotide 3-prime untranslated region. They mapped the gene to 4q22-q25 by study of a panel of somatic cell hybrid lines, including somatic cell hybrid lines harboring various deletions in overlapping regions of human chromosome 4. Analysis of a multiple tissue Northern blot demonstrated a single 3.7-kb transcript. Mutation analysis in a Czech Gypsy family with 2 affected sibs, reported by Kleijer et al. (1990), demonstrated homozygosity for an acceptor splice site mutation (248510.0001).

Alkhayat, A. H.; Kraemer, S. A.; Leipprandt, J. R.; Macek, M.; Kleijer, W. J.; Friderici, K. H.: Human beta-mannosidase cDNA characterization and first identification of a mutation associated with human beta-mannosidosis. Hum. Molec. Genet. 7: 75-83, 1998.

Cooper, A.; Hatton, C.; Thornley, M.; Sardharwalla, I. B.: Human beta-mannosidase deficiency: biochemical findings in plasma, fibroblasts, white cells and urine. J. Inherit. Metab. Dis. 11: 17-29, 1988.

Cooper, A.; Sardharwalla, I. B.; Roberts, M. M.: Human beta-mannosidase deficiency. (Letter) New Eng. J. Med. 315: 1231, 1986.

Dorland, L.; Duran, M.; Hoefnagels, F. E. T.; Breg, J. N.; Fabery de Jonge, H.; Cransberg, K.; van Sprang, F. J.; van Diggelen, O. P.: Beta-mannosidosis in two brothers with hearing loss. J. Inherit. Metab. Dis. 11 (suppl. 2): 255-258, 1988.

Hu, P.; Wenger, D. A.; van Diggelen, O. P.; Klejer, W. J.: Complementation studies in human and caprine beta-mannosidosis. J. Inherit. Metab. Dis. 14: 13, 1990.

Jones, M. Z.; Abbitt, B.: Animal model of human disease: bovine beta-mannosidosis. Am. J. Path. 142: 957-960, 1993.

Jones, M. Z.; Cunningham, J. G.; Dade, A. W.; Alessi, D. M.; Mostosky, U. V.; Vorro, J. R.; Benitez, J. T.; Lovell, K. L.: Caprine beta-mannosidosis: clinical and pathological features. J. Neuropath. Exp. Neurol. 42: 268-285, 1983.

Jones, M. Z.; Dawson, G.: Caprine beta-mannosidosis: inherited deficiency of beta-D-mannosidase. J. Biol. Chem. 256: 5185-5188, 1981.

Kleijer, W. J.; Hu, P.; Thoomes, R.; Boer, M.; Huijmans, J. G. M.; Blom, W.; Van Diggelen, O. P.; Seemanova, E.; Macek, M.: Beta-mannosidase deficiency: heterogeneous manifestation in the first female patient and her brother. J. Inherit. Metab. Dis. 13: 867-872, 1990.

Leipprandt, J. R.; Kraemer, S. A.; Haithcock, B. E.; Chen, H.; Dyme, J. L.; Cavanagh, K. T.; Friderici, K. H.; Jones, M. Z.: Caprine beta-mannosidase: sequencing and characterization of the cDNA and identification of the molecular defect of caprine beta-mannosidosis. Genomics 37: 51-56, 1996.

Levade, T.; Graber, D.; Flurin, V.; Delisle, M.-B.; Pieraggi, M.-T.; Testut, M.-F.; Carriere, J.-P.; Salvayre, R.: Human beta-mannosidase deficiency associated with peripheral neuropathy. Ann. Neurol. 35: 116-119, 1994.

Lovell, K. L.; Jones, M. Z.: Distribution of central nervous system lesions in beta-mannosidosis. Acta Neuropath. 62: 121-126, 1983.

Lundin, L.-G.: A gene (Bmn) controlling beta-mannosidase activity in the mouse is located in the distal part of chromosome 3. Biochem. Genet. 25: 603-610, 1987.

Malachowski, J. A.; Jones, M. Z.: Beta-mannosidosis: lesions of the distal peripheral nervous system. Acta Neuropath. 61: 95-100, 1983.

Poenaru, L.; Akli, S.; Rocchiccioli, F.; Eydoux, P.; Zamet, P.: Human beta-mannosidosis: a 3-year-old boy with speech impairment and emotional instability. Clin. Genet. 41: 331-334, 1992.

Rodriguez-Serna, M.; Botella-Estrada, R.; Chabas, A.; Coll, M.-J.; Oliver, V.; Febrer, M.-I.; Aliaga, A.: Angiokeratoma corporis diffusum associated with beta-mannosidase deficiency. Arch. Derm. 132: 1219-1222, 1996.

Wenger, D. A.; Sujansky, E.; Fennessey, P. V.; Thompson, J. N.: Human beta-mannosidase deficiency. New Eng. J. Med. 315: 1201-1205, 1986.

ANGIOKERATOMA, DIFFUSE (*301500)

JJM NOTE

Also known as Fabry disease, this entity has multiple angiokeratomas, especially in the umbilicus.

Skin lesions of vascular nature are the basis of the name 'angiokeratoma.' Since angiokeratoma occurs with some other lysosomal diseases and may be missing in the disorder discussed here (see later), the use of the eponym 'Fabry disease' has much to recommend it, even though it ignores the important contribution of Anderson (1898). Fabry of Dortmund, Germany, wrote about the disorder first in 1898 and again in 1930. In his first paper, Fabry (1898) called the skin lesions 'purpura papulosa haemorrhagica Hebrae,' suggesting that they had previously been described by Hebra. Attacks of pain in the abdomen are often misdiagnosed as appendicitis or other 'surgical abdomen.' Such pains and those elsewhere, such as in the extremities, probably have their basis in lipid changes in ganglion cells of the autonomic nervous system. Vascular lesions of lipid nature occur at other sites such as the ocular fundi and kidney. Renal failure is the usual cause of death. Heterozygous females almost never have skin lesions and survive longer despite renal involvement.

Hasholt et al. (1990) described a girl in whom Fabry disease was suspected at the age of 18 years because of corneal opacities and tortuous retinal vessels. She had had attacks of high fever, arthralgia, and abdominal pain. Studies indicated that she suffered from a new mutation for Fabry disease; she was heterozygous. Hamburger et al. (1964) described a familial nephropathy, manifested clinically by proteinuria and renal insufficiency. Renal biopsy showed that the epithelial cells of the glomerular tufts and to a lesser extent the tubular epithelial cells, glomerular endocapillary cells and arteriolar muscular cells were severely deformed with a large amount of cytoplasmic inclusion material. The inclusion material was thought to be lipoid in nature. The findings resemble those of angiokeratoma corporis diffusum but the absence of other signs of this disease suggested that a new entity may be involved. The mother's father died of uremia. Skin lesions are easily overlooked. It is clear, however, that they may be lacking even in patients with severe visceral manifestations (Johnston, 1967); thus, the eponym Fabry disease is a better designation than angiokeratoma. Furthermore, identical angiokeratoma skin lesions occur in other lysosomal disorders: alpha-L-fucosidase deficiency (230000) (Patel et al., 1972); type II fucosidosis and a form of beta-galactosidase deficiency compatible with survival to adulthood (Loonen et al., 1974); occasionally aspartylglycosaminuria (208400); and an adult form of Schindler syndrome (104170). Flynn et al. (1972) described a family without skin lesions. One affected male had severe enteropathy. Franceschetti et al. (1969) reexamined the family with 'cornea verticillata' reported by Gruber (1946) and showed that Fabry disease was the 'cause' of the corneal change. The extent of involvement of the cornea is about the same in males and females. Thus, carrier females can be identified. The corneal condition was formerly ca lled also Fleischer vortex dystrophy, or whorl-like corneal dystrophy. The antimalarial atabrine can produce a phenocopy.

Mutoh et al. (1988) described an unusual degree of autonomic dysfunction, manifested by severe orthostatic hypotension, in a 21-year-old carrier. Rosenberg et al. (1980) pointed out that deposition of sphingolipid in epithelial cells of the respiratory tract leads to chronic obstruction to airflow. The effects are greatest in smokers.

The observations of Ogawa et al. (1990), Elleder et al. (1990), and von Scheidt et al. (1991) indicate that manifestations of Fabry disease can be limited to the heart. Fabry disease should be considered in patients who have cardiac symptoms (angina, exercise intolerance, and electrocardiographic changes), but normal coronary arteries, heart size, and hemodynamic findings. Cases of Fabry disease limited to the cardiac manifestation may be identified by ultrastructural examination of endomyocardial biopsy specimens or, less invasively, by determining the plasma alpha-galactosidase A activity in male patients with unexplained cardiac symptoms. The oldest of the patients reported with this form of the disease was a Japanese patient who died at the age of 71 years (Ogawa et al., 1990). Nagao et al. (1991) reported 2 unrelated male hemizygotes who developed hypertrophic cardiomyopathy after the age of 50 as their first manifestation. The molecular defect in this late-onset variety is discussed in 301500.0003 and 301500.0005. Hillsley et al. (1995) described the case of a 74-year-old woman with restrictive cardiomyopathy complicating conventional coronary artery disease that had been treated with angioplasty and with coronary artery bypass grafting.

Elleder et al. (1994) demonstrated storage material in the external arachnoidal epithelium of the leptomeninges in a case of cardiocyte-restricted Fabry disease. The patient, a 65 year-old male with Fabry disease who died from bile duct carcinoma, showed only cardiac symptoms and signs premortem. This pattern contrasted with the generalized leptomeningeal storage seen in typical Fabry disease.

Sawada et al. (1996) found a G-to-A transition in exon 6 of the GLA gene that resulted in an arg301-to-gln substitution (301500.0003) in a 45-year-old man who developed moderate proteinuria and was found to have the renal histologic findings of Fabry disease, together with a decrease in activity of alpha-galactosidase A in his plasma, urine, leukocytes, and skin fibroblasts. The mutation was inherited from his mother. The patient was unique in that he demonstrated only renal manifestations, whereas all other patients with atypical Fabry disease, including a case with the identical point mutation (Sakuraba et al., 1990), presented with cardiomyopathy.

Nakao et al. (1995) found 7 unrelated patients with atypical variants of Fabry disease among 230 men with left ventricular hypertrophy. These 7 unrelated men, ranging in age from 55 to 72 years, did not have angiokeratoma, acroparesthesias, hypohidrosis, or corneal opacities. Endomyocardial biopsy, performed in 5 patients, revealed marked sarcoplasmic vacuolization in all 5. Electron microscopy, performed on tissues from 4 of these patients, revealed typical lysosomal inclusions with a concentric lamellar configuration. Two patients had novel missense mutations in exon 1 (301500.0051) and exon 6 (301500.0052). The remaining 5 had no mutations in the coding region of the GLA gene; however, in addition to low plasma alpha-galactosidase activity, GLA mRNA was markedly lower than normal.

Kint (1970) showed that the activity of alpha-galactosidase is deficient in leukocytes of male patients with Fabry disease and that carrier females can be identified by this method. Moser (1983) considered the urinary trihexoside assay, described by Cable et al. (1982), to be the most reliable way to identify carriers of Fabry disease. Kirkilionis et al. (1991) described a Nova Scotia kindred known to contain 30 affected males and 50 possible female carriers. They found that there was no major alteration of the gene but that it was linked to the rarer allele (frequency = 0.20) of the polymorphic NcoI site located 3-prime to the gene. This helped carrier identification; all of 17 obligate carriers examined were identified, including 6 who were not identified as carriers by enzyme assay.

In a period of a few months at a Montreal hospital, Clarke et al. (1971) saw 2 men with Fabry disease with clear corneas and without skin lesions, suggesting that it may be a more frequent cause of proteinuria or renal failure than realized. Romeo et al. (1972) studied one of Clarke's cases and concluded that enzymatically there were differences from the classic cases. A difference from the usual form of Fabry disease is suggested by the fact that leukocyte alpha-galactosidase deficiency was only partial rather than complete (Kint, 1970).

Ko et al. (1996) described a so-called oligosymptomatic variant of Fabry disease. A renal biopsy performed for evaluation of trace proteinuria revealed histologic and ultrastructural findings compatible with Fabry disease in a 34-year-old man with no history of renal disease in other family members. Biopsy of a few initially unrecognized, scattered, dark-pinkish scrotal papules showed typical angiokeratoma. Alpha-galactosidase was markedly decreased in urine and plasma.

Brown et al. (1997) pursued the suggested association between Fabry disease and airway obstruction by study of 25 unselected, consecutive, enzymatically diagnosed men. Dyspnea was a complaint in 36%, and 24% had cough and/or wheezing. Symptoms were similar in smokers and nonsmokers. Nine (36%) had airway obstruction on spirometry; this finding was associated with age of more than 26 years and with dyspnea or wheezing, but only weakly associated with smoking. Five of 8 patients responded to bronchodilators, but all 10 methylcholine challenges were negative. Chest radiographs showed normal lung fields in 24 patients and streaky bibasilar densities in 1. Specific alpha-galactosidase A mutations were identified in 17 patients; all 3 patients with frameshift mutations and both subjects with the asp264-to-val mutation (310500.0021) had obstructive impairment. Brown et al. (1997) concluded that airway obstruction occurs commonly in patients with Fabry disease regardless of smoking history, and that it increases with age. The presence of obstruction may be associated with certain mutations and most likely results from fixed narrowing of the airways by accumulated glycosphingolipid.

Romeo and Migeon (1970) presented evidence for a structural change in the mutant enzyme (slower heat inactivation than in the normal and different K(m) values). Halsted and Rowe (1975) described a 59-year-old man with Fabry disease. In addition to his unusually advanced age, he had celiac sprue. The last may have been related to the presence of the HLA-8 antigen, which is found in about 80% of persons with gluten-sensitive enteropathy. Peltier et al. (1977) reported male twins with Fabry disease but normal alpha-galactosidase and normality of several other lysosomal enzymes including alpha-fucosidase.

In 2 patients, Mapes et al. (1970) demonstrated a decline in the plasma level of galactosylgalactosylglucosylceramide when normal plasma was infused to provide active enzyme (ceramide trihexosidase). Clement et al. (1982) reported that successful renal transplantation not only corrects the anemia but also produces marked improvement in other clinical manifestations of the disease. Friedlaender et al. (1987) presented renal biopsy data on a patient with Fabry disease before and 8 years after successful renal transplantation. The graft maintained normal function, and graft histology showed no abnormalities. The patient had been able to return to work and his severe symptoms of acroparesthesia had been cured. Bannwart (1982) described a postmortem examination 12 years after transplantation; no histologic recurrence of Fabry disease in the kidney graft was found by light or electron microscopy. Findings in other reports have not been as satisfactory. For example, Faraggiana et al. (1981) examined the kidney of a patient who died 6 months following transplantation and found widespread recurrence of the disease in the graft. In that case, there was evidence of a circulating inhibitor of galactosidase A activity. It is intriguing to speculate that in the successful cases there was not a complete absence of enzyme so that antibodies did not develop as they may have in the case of Faraggiana et al. (1981). Hasholt and Sorensen (1986) found that GLA enzyme activity is particularly high in endothelial cells cultured from umbilical cord and absent in hemizygotes, thus accounting for the characteristic pathology of Fabry disease.

Johnston et al. (1969) estimated the recombination fraction of angiokeratoma versus Xg to be 0.24 (95% probability limits, 8-49.8%) and of angiokeratoma versus deutan to be 0.17 (95% probability limits, 1-50%). The Fabry locus does 'lyonize' (Romeo and Migeon, 1970). Localization of the alpha-galactosidase (alpha-GAL; EC 3.2.1.22) locus to the X chromosome had been achieved also by cell hybridization (Grzeschik, 1972). From study of radiation-induced segregants (irradiated human cells 'rescued' by fusion with hamster cells), Goss and Harris (1977) showed that the order of the following 4 loci is PGK—alpha-GAL—HPRT—G6PD and that the 3 intervals between these 4 loci are, in relative terms, 0.33, 0.30, and 0.23. Johnston and Sanger (1981) reanalyzed all data on Xg and Fabry linkage and obtained negative lod scores at all recombination rates. Alpha-GAL, HPRT, PGK and G6PD are X-linked in the rabbit, according to mouse-rabbit hybrid cell studies (Cianfriglia et al., 1979; Echard and Gillois, 1979). By comparable methods, Hors-Cayla et al. (1979) found them to be X-linked also in cattle. Francke and Taggart (1979) assigned HPRT and alpha-GAL to the X-chromosome in the Chinese hamster by study of mouse-Chinese hamster hybrid cells. MacDermot et al. (1987) found no recombination between Fabry disease and 3 DNA markers, DXS87, DXS88, and DXS17, which gave maximum lod scores of 6.4, 6.4, and 5.8, respectively, at theta = 0.00 (upper confidence limit 0.10). DXYS1 was not linked.

By Southern analysis of the GLA gene in affected males from 130 unrelated families, Bernstein et al. (1989) found 6 with different gene rearrangements, and one with an exonic point mutation resulting in the obliteration of an MspI restriction site. Five partial gene deletions were detected, ranging in size from 0.4 to more than 5.5 kb. Four of the 5 deletions had breakpoints in exon 2, a region containing multiple Alu repeat sequences. A sixth genomic rearrangement was identified in which a region of about 8 kb, containing exons 2-6, was duplicated by a homologous, but unequal crossover event. The MspI site obliteration, which mapped to exon 7, was found to result from a C-to-T transition at nucleotide 1066 in the coding sequence. The resulting change, the first point mutation identified in Fabry disease, resulted in an arg356-to-trp substitution, which altered the enzyme's kinetic properties and stability. Kornreich et al. (1990) assessed the 6 gene arrangements identified by Bernstein et al. (1989) to determine the possible role of Alu repetitive elements and short direct and/or inverted repeats in the generation of these germinal mutations. Although the alpha-galactosidase A gene contains 12 Alu repetitive elements (representing about 30% of the 12-kb gene) only 1 deletion resulted from an Alu-Alu recombination. The remaining 5 rearrangements involved illegitimate recombinational events between short direct repeats of 2 to 6 bp at the deletion or duplication breakpoints.

The GLA gene is about 12 kb long and contains 7 exons encoding a precursor protein of 429 amino acids. The processed message for the entire alpha-galactosidase A subunit is about 1.45 kb long (Bishop et al., 1986). Bishop et al. (1988) cloned the entire GLA gene, characterized its intron-exon organization, and described the 5-prime regulatory elements and the 3-prime flanking sequence. The unusual lack of a 3-prime untranslated sequence in the GLA cDNA was confirmed. Kornreich et al. (1989) presented the complete nucleotide sequence of the GLA gene, including an extensive segment of 5-prime and 3-prime flanking sequences. Sakuraba et al. (1992) gave a useful listing of exon skipping and use of cryptic splice sites resulting from various mammalian (mainly human) 5-prime consensus splice site mutations. In the first exon of the GLA gene, which contains 60 bp of 5-prime untranslated sequence before the methionine initiation codon, Davies et al. (1993) demonstrated 3 polymorphic variants by SSCP screening. Such a high level of polymorphism had not previously been reported in the 5-prime untranslated region of the human gene and was unusual in such a short stretch of DNA. Davies et al. (1993) suggested that the polymorphisms should be of great assistance in family studies of angiokeratoma.

Davies et al. (1993) found 7 putative disease-causing single base substitutions and 1 small insertion mutation in 9 unrelated families with the classic form of Fabry disease. Eng and Desnick (1994) reviewed and tabulated 15 GLA gene rearrangements, 3 GLA mRNA processing defects, and 31 GLA point mutations causing Fabry disease.

During a study of the GLA gene, Novo et al. (1995) found that RT-PCR amplification of alpha-galactosidase mRNAs obtained from 3 different tissues isolated from males revealed a substantial number of clones with a U2A conversion at nucleotide position 1187 of their sequence. Such a modification of the coding sequence would result in an amino acid substitution in the C-terminal region, phe396-to-tyr, of the enzyme. Neither PCR analysis of the genomic sequence nor the RT-PCR amplification of RNA obtained by in vitro transcription of the wildtype cDNA showed this change in the sequence. Multiple genes, pseudogenes or allelic variants were excluded. Hence, Novo et al. (1995) proposed RNA editing as a mechanism responsible for this base change in the GLA RNA, similar to that which has been demonstrated for the nuclear encoded RNA for intestinal apoB (107730) and several subunits of brain L-glutamate receptors, such as GLUR2 (138247), GLUR5 (138245), and GLUR6 (138244).

Blanch et al. (1996) amplified the 7 exons of the alpha-galactosidase A gene and the adjacent intron boundaries from a member of each of 9 kindreds and analyzed them for sequence alterations by SSCP analysis followed by sequencing. By this method they detected the causative mutations in 9 patients with classic severe Fabry disease. Counting a previously reported mutation, this strategy successfully detected all the mutations present in 10 kindreds analyzed.

Miyamura et al. (1996) identified a novel mutation of the GLA gene, which converted a tyrosine at codon 365 to a stop (tyr365ter) and resulted in a truncation of the C-terminus by 65 amino acid residues in a family with Fabry disease. In a heterozygote of this family, although the mutant and normal alleles were equally transcribed in cultured fibroblasts, lymphocyte alpha-galactosidase A activity was approximately 30% of the normal control and severe clinical symptoms were apparent. COS-1 cells transfected with this mutant cDNA showed a complete loss of its enzymatic activity. Furthermore, cells cotransfected with mutant and wildtype cDNAs showed a lower enzyme activity than those with wildtype alone (approximately 30% wildtype alone), which suggested a dominant-negative effect of this mutation and implied the importance of the C terminus for its activity. Generating mutant cDNAs with various deletions of the C terminus, they found that enzyme activity was enhanced up to 6-fold compared with wildtype when from 2 to 10 amino acid residues were deleted. In contrast, deletion of 12 or more amino acid residues resulted in a complete loss of enzyme activity. These data suggested that the C-terminal region of the GLA protein plays an important role in the regulation its enzyme activity.

Ohshima et al. (1997) generated GLA-deficient mice by gene targeting. The knockout mice displayed a complete lack of alpha-galactosidase A activity. They appeared clinically normal at 10 weeks of age, however. Ultrastructual analysis revealed concentric lamellar inclusions in the kidneys, and confocal microscopy using a fluorescent-labeled lectin specific for alpha-D-galactosyl residues showed accumulation of substrate in the kidneys as well as in cultured fibroblasts. Lipid analysis revealed a marked accumulation of ceramide trihexaside in the liver and the kidneys. The deficiency of enzyme activity and the accumulation of material containing terminal alpha-galactosyl residues in cultured embryonic fibroblasts derived from the knockout mice were corrected by transducing these cells with human GLA cDNA.

See also Bach et al. (1982); Beaudet and Caskey (1978); Bird and Lagunoff (1978); Brady et al. (1967); Broadbent et al. (1981); Cable et al. (1982); Cable et al. (1982); Calhoun et al. (1985); Colucci et al. (1982); Davies et al. (1993); Eng et al. (1993); Friedman et al. (1984); Frost et al. (1966); Gruber (1946); Hamers et al. (1979); Kornreich et al. (1989); Lusis and West (1976); MacDermot et al. (1987); Maisey and Cosh (1980); O'Brien (1980); Opitz et al. (1965); Philippart et al. (1969); Pierides et al. (1976); Pyeritz et al. (1980); Rahman et al. (1961); Rodriguez et al. (1985); Ropers et al. (1977); Sakuraba et al. (1986); Sheth et al. (1981); Shows et al. (1978); Sorensen and Hasholt (1980); Spence et al. (1978); Spence et al. (1976); Sweeley and Klionsky (1963); Tagliavini et al. (1982); Wise et al. (1962)

Anderson, W.: A case of 'angeio-keratoma.'. Brit. J. Derm. 10: 113-117, 1898.

Bach, G.; Rosenmann, E.; Karni, A.; Cohen, T.: Pseudodeficiency of alpha-galactosidase A. Clin. Genet. 21: 59-64, 1982.

Bannwart, F.: Morbus Fabry: Licht- und Elektronenmikroskopischer Herzbefund 12 Jahre nach erfolgreicher Nierentransplantation. Schweiz. Med. Wschr. 112: 1742-1747, 1982.

Beaudet, A. L.; Caskey, C. T.: Detection of Fabry's disease heterozygotes by hair root analysis. Clin. Genet. 13: 251-258, 1978.

Bernstein, H. S.; Bishop, D. F.; Astrin, K. H.; Kornreich, R.; Eng, C. M.; Sakuraba, H.; Desnick, R. J.: Fabry disease: six gene rearrangements and an exonic point mutation in the alpha-galactosidase gene. J. Clin. Invest. 83: 1390-1399, 1989.

Bird, T. D.; Lagunoff, D.: Neurological manifestations of Fabry disease in female carriers. Ann. Neurol. 4: 537-540, 1978.

Bishop, D. F.; Calhoun, D. H.; Bernstein, H. S.; Hantzopoulos, P.; Quinn, M.; Desnick, R. J.: Human alpha-galactosidase A: nucleotide sequence of a cDNA clone encoding the mature enzyme. Proc. Nat. Acad. Sci. 83: 4859-4863, 1986.

Bishop, D. F.; Kornreich, R.; Desnick, R. J.: Structural organization of the human alpha-galactosidase A gene: further evidence for the absence of a 3-prime untranslated region. Proc. Nat. Acad. Sci. 85: 3903-3907, 1988.

Blanch, L. C.; Meaney, C.; Morris, C. P.: A sensitive mutation screening strategy for Fabry disease: detection of nine mutations in the alpha-galactosidase A gene. Hum. Mutat. 8: 38-43, 1996.

Brady, R. O.; Gal, A. E.; Bradley, R. M.; Martensson, E.; Warshaw, A. L.; Laster, L.: Enzymatic defect in Fabry's disease: ceramidetrihexosidase deficiency. New Eng. J. Med. 276: 1163-1167, 1967.

Broadbent, J. C.; Edwards, W. D.; Gordon, H.; Hartzler, G. O.; Krawisz, J. E.: Fabry cardiomyopathy in the female confirmed by endomyocardial biopsy. Mayo Clin. Proc. 56: 623-628, 1981.

Brown, L. K.; Miller, A.; Bhuptani, A.; Sloane, M. F.; Zimmerman, M. I.; Schilero, G.; Eng, C. M.; Desnick, R. J.: Pulmonary involvement in Fabry disease. Am. J. Resp. Crit. Care Med. 155: 1004-1010, 1997.

Cable, W. J. L.; Dvorak, A. M.; Osage, J. E.; Kolodny, E. H.: Fabry disease: significance of ultrastructural localization of lipid inclusions in dermal nerves. Neurology 32: 347-353, 1982.

Cable, W. J. L.; Kolodny, E. H.; Adams, R. D.: Fabry disease: impaired autonomic function. Neurology 32: 498-502, 1982.

Cable, W. J. L.; McCluer, R. H.; Kolodny, E. H.; Ullman, M. D.: Fabry disease: detection of heterozygotes by examination of glycolipids in urinary sediment. Neurology 32: 1139-1145, 1982.

Calhoun, D. H.; Bishop, D. F.; Bernstein, H. S.; Quinn, M.; Hantzopoulos, P.; Desnick, R. J.: Fabry disease: isolation of a cDNA clone encoding human alpha-galactosidase A. Proc. Nat. Acad. Sci. 82: 7364-7368, 1985.

Cariolou, M. A.; Christodoulides, M.; Manoli, P.; Kokkofitou, A.; Tsambaos, D.: Novel trinucleotide deletion in Fabry's disease. Hum. Genet. 97: 468-470, 1996.

Chen, C.-H.; Shyu, P.-W.; Wu, S.-J.; Sheu, S.-S.; Desnick, R. J.; Hsiao, K.-J.: Identification of a novel point mutation (S65T) in alpha-galactosidase A gene in Chinese patients with Fabry disease. Hum. Mutat. 11: 328-330, 1998.

Chen, E. Y.; Liao, Y.-C.; Smith, D. H.; Barrera-Saldana, H. A.; Gelinas, R. E.; Seeburg, P. H.: The human growth hormone locus: nucleotide sequence, biology, and evolution. Genomics 4: 479-497, 1989.

Cianfriglia, M.; Miggiano, V. C.; Meo, T.; Muller, H. J.; Muller, E.; Battistuzzi, G.: Evidence for synteny between the rabbit gene loci coding for HPRT, PGK and G6PD in mouse-rabbit somatic cell hybrids. (Abstract) Cytogenet. Cell Genet. 25: 142, 1979.

Clarke, J. T.; Knaack, J.; Crawhall, J. C.; Wolfe, L. S.: Ceramide trihexosidosis (Fabry's disease) without skin lesions. New Eng. J. Med. 284: 233-235, 1971.

Clement, M.; McGonigle, R. J. S.; Monkhouse, P. M.; Keogh, A. M.; Marten, R. H.; Bewick, M.; Parsons, V.: Renal transplantation in Anderson-Fabry disease. J. Roy. Soc. Med. 75: 557-560, 1982.

Colucci, W. S.; Lorell, B. H.; Schoen, F. J.; Warhol, M. J.; Grossman, W.: Hypertrophic obstructive cardiomyopathy due to Fabry's disease. New Eng. J. Med. 307: 926-931, 1982.

Davies, J. P.; Winchester, B. G.; Malcolm, S.: Sequence variations in the first exon of alpha-galactosidase A. J. Med. Genet. 30: 658-663, 1993.

Davies, J. P.; Winchester, B. G.; Malcolm, S.: Mutation analysis in patients with the typical form of Anderson-Fabry disease. Hum. Molec. Genet. 2: 1051-1053, 1993.

deJong, J. G. N.; Jansen, P. P. M.; van den Berg, C. J. M. G.; Hamel, B. C. J.; Wevers, R. A.; Ploos van Amstel, J. K.: Genetic heterogeneity in Fabry's disease: mutations in the alpha-galactosidase A gene..In: Proc. 2nd Int. Duodecim Symposium: Molecular Biology of Lysosomal Disease, Majik, Finland. 1993.

Echard, G.; Gillois, M.: Rabbit gene mapping: G6PD—alpha-GAL-PGK—HPRT synteny. (Abstract) Cytogenet. Cell Genet. 25: 148, 1979.

Elleder, M.; Bradova, V.; Smid, F.; Budesinsky, M.; Harzer, K.; Kustermann-Kuhn, B.; Ledvinova, J.; Belohlavek, (NI); Kral, V.; Dorazilova, V.: Cardiocyte storage and hypertrophy as a sole manifestation of Fabry's disease: report on a case simulating hypertrophic non-obstructive cardiomyopathy. Virchows Arch. A 417: 449-455, 1990.

Elleder, M.; Christomanou, H.; Kustermann-Kuhn, B.; Harzer, K.: Leptomeningeal lipid storage patterns in Fabry disease. Acta Neuropath. 88: 579-582, 1994.

Eng, C. M.; Desnick, R. J.: Molecular basis of Fabry disease: mutations and polymorphisms in the human alpha-galactosidase A gene. Hum. Mutat. 3: 103-111, 1994.

Eng, C. M.; Niehaus, D. J.; Desnick, R. J.: Molecular analysis of classical and variant phenotypes. (Abstract) Pediat. Res. 33: 128A, 1993.

Eng, C. M.; Resnick-Silverman, L. A.; Niehaus, D. J.; Astrin, K. H.; Desnick, R. J.: Nature and frequency of mutations in the alpha-galactosidase A gene that cause Fabry disease. Am. J. Hum. Genet. 53: 1186-1197, 1993.

Fabry, J.: Ein Beitrag Zur Kenntnis der Purpura haemorrhagica nodularis (Purpura papulosa hemorrhagica Hebrae). Arch. Derm. Syph. 43: 187-200, 1898.

Faraggiana, T.; Churg, J.; Grishman, E.; Strauss, L.; Prado, A.; Bishop, D. F.; Schuchman, E.; Desnick, R. J.: Light and electron-microscopic histochemistry of Fabry's disease. Am. J. Path. 103: 247-262, 1981.

Flynn, D. M.; Lake, B. D.; Boothby, C. B.; Young, E. P.: Gut lesions in Fabry's disease without a rash. Arch. Dis. Child. 47: 26-33, 1972.

Franceschetti, A. T.; Philippart, M.; Franceschetti, A.: A study of Fabry's disease. I. Clinical examination of a family with cornea verticillata. Dermatologica 138: 209-221, 1969.

Francke, U.; Taggart, R. T.: Regional mapping of SOD-1 on mouse chromosome 16, and of HPRT and alpha-GAL (Ags) on mouse X, using Chinese hamster-mouse T(X;16)16H somatic cell hybrids. (Abstract) Cytogenet. Cell Genet. 25: 155-156, 1979.

Friedlaender, M. M.; Kopolovic, J.; Rubinger, D.; Silver, J.; Drukker, A.; Ben-Gershon, Z.; Durst, A. L.; Popovtzer, M. M.: Renal biopsy in Fabry's disease eight years after successful renal transplantation. Clin. Nephrol. 27: 206-211, 1987.

Friedman, L. S.; Kirkham, S. E.; Thistlethwaite, J. R.; Platika, D.; Kolodny, E. H.; Schuffler, M. D.: Jejunal diverticulosis with perforation as a complication of Fabry's disease. Gastroenterology 86: 558-563, 1984.

Frost, P.; Tanaka, Y.; Spaeth, G. L.: Fabry's disease—glycolipid lipidoses. Histochemical and electron microscopic studies of two cases. Am. J. Med. 40: 618-627, 1966.

Fukuhara, Y.; Sakuraba, H.; Oshima, A.; Shimmoto, M.; Nagao, Y.; Nadaoka, Y.; Suzuki, T.; Suzuki, Y.: Partial deletion of human alpha-galactosidase A gene in Fabry disease: direct repeat sequences as a possible cause of slipped mispairing. Biochem. Biophys. Res. Commun. 170: 296-300, 1990.

Goss, S. J.; Harris, H.: Gene transfer by means of cell fusion. I. Statistical mapping of the human X-chromosome by analysis of radiation-induced gene segregation. J. Cell Sci. 25: 17-37, 1977.

Gruber, M.: Cornea verticillata. II. Mitteilung. Ophthalmologica 112: 88-91, 1946.

Gruber, M.: Cornea verticillata. (Eine einfach-dominante Variante der Hornhaut des menschlichen Auges). Ophthalmologica 111: 120-129, 1946.

Grzeschik, K.-H.: Personal Communication. Leiden, The Netherlands 1972.

Halsted, C. H.; Rowe, J. W.: Occurrence of celiac sprue in a patient with Fabry's disease. Ann. Intern. Med. 83: 524-525, 1975.

Hamburger, J.; Dormont, J.; De Montera, H.; Hinglais, N.: Sur une singuliere malformation familiale de l'epithelium renal. Schweiz. Med. Wschr. 94: 871-876, 1964.

Hamers, M. N.; Wise, D.; Ejiofor, A.; Strijland, A.; Robinson, D.; Tager, J. M.: Relationship between biochemical and clinical features in an English Anderson-Fabry family. Acta Med. Scand. 206: 5-10, 1979.

Hasholt, L.; Sorensen, S. A.: Lysosomal alpha-galactosidase in endothelial cell cultures established from a Fabry hemizygous and normal umbilical veins. Hum. Genet. 72: 72-76, 1986.

Hasholt, L.; Sorensen, S. A.; Wandall, A.; Andersen, E. B.; Arlien-Soborg, P.: A Fabry's disease heterozygote with a new mutation: biochemical, ultrastructural, and clinical investigations. J. Med. Genet. 27: 303-306, 1990.

Hillsley, R. E.; Hernandez, E.; Steenbergen, C.; Bashore, T. M.; Harrison, J. K.: Inherited restrictive cardiomyopathy in a 74-year-old woman: a case of Fabry's disease. Am. Heart J. 129: 199-202, 1995.

Hors-Cayla, M. C.; Heuertz, S.; Van Cong, N.; Benne, F.: Cattle gene mapping by somatic cell hybridization. (Abstract) Cytogenet. Cell Genet. 25: 165-166, 1979.

Ishii, S.; Sakuraba, H.; Shimmoto, M.; Minamikawa-Tachino, R.; Suzuki, T.; Suzuki, Y.: Fabry disease: detection of a 13-bp deletion in alpha-galactosidase A gene and its application to gene diagnosis of heterozygotes.. Ann. Neurol. 29: 560-564, 1991.

Ishii, S.; Sakuraba, H.; Suzuki, Y.: Point mutations in the upstream region of the alpha-galactosidase A gene exon 6 in an atypical variant of Fabry disease. Hum. Genet. 89: 29-32, 1992.

Johnston, A. W.: Fabry's disease without skin lesions. (Letter) Lancet I: 1277, 1967.

Johnston, A. W.; Frost, P.; Spaeth, G. L.; Renwick, J. H.: Linkage relationships of the angiokeratoma (Fabry) locus. Ann. Hum. Genet. 32: 369-374, 1969.

Johnston, A. W.; Sanger, R.: Linkage relationship of the loci for Anderson-Fabry disease and the Xg blood groups. Ann. Hum. Genet. 45: 155-157, 1981.

Kint, J. A.: Fabry's disease: alpha-galactosidase deficiency. Science 167: 1268-1269, 1970.

Kirkilionis, A. J.; Riddell, D. C.; Spence, M. W.; Fenwick, R. G.: Fabry disease in a large Nova Scotia kindred: carrier detection using leucocyte alpha-galactosidase activity and an NcoI polymorphism detected by an alpha-galactosidase cDNA clone. J. Med. Genet. 28: 232-240, 1991.

Ko, Y.-H.; Kim, H.-J.; Roh, Y.-S.; Park, C.-K.; Kwon, C.-K.; Park, M.-H.: Atypical Fabry's disease: an oligosymptomatic variant. Arch. Path. Lab. Med. 120: 86-89, 1996.

Koide, T.; Ishiura, M.; Iwai, K.; Inoue, M.; Kaneda, Y.; Okada, Y.; Uchida, T.: A case of Fabry's disease in a patient with no alpha-galactosidase A activity caused by a single amino acid substitution of pro40 by ser. FEBS Lett. 259: 353-356, 1990.

Kornreich, R.; Bishop, D. F.; Desnick, R. J.: The gene encoding alpha-galactosidase A and gene rearrangements causing Fabry disease. Trans. Assoc. Am. Phys. 102: 30-43, 1989.

Kornreich, R.; Bishop, D. F.; Desnick, R. J.: Alpha-galactosidase A gene rearrangements causing Fabry disease: identification of short direct repeats at breakpoints in an Alu-rich gene. J. Biol. Chem. 265: 9319-9326, 1990.

Kornreich, R.; Desnick, R. J.; Bishop, D. F.: Nucleotide sequence of the human alpha-galactosidase A gene. Nucleic Acids Res. 17: 3301-3302, 1989.

Loonen, M. C. B.; van de Lugt, L.; Franke, C. L.: Angiokeratoma corporis diffusum and lysosomal enzyme deficiency. (Letter) Lancet II: 785, 1974.

Lusis, A. J.; West, J. D.: X-linked inheritance of a structural gene for alpha-galactosidase in Mus musculus. Biochem. Genet. 14: 849-855, 1976.

MacDermot, K. D.; Morgan, S. H.; Cheshire, J. K.; Wilson, T. M.: Anderson-Fabry disease: close linkage with DXS17, DXS87 and DXS88. (Abstract) Cytogenet. Cell Genet. 46: 654, 1987.

MacDermot, K. D.; Morgan, S. H.; Cheshire, J. K.; Wilson, T. M.: Anderson Fabry disease, a close linkage with highly polymorphic DNA markers DXS17, DXS87, and DXS88. Hum. Genet. 77: 263-266, 1987.

Madsen, K. M.; Hasholt, L.; Fermer, M. L.; Dahl, N.: Identification of mutations in Danish families with Fabry's disease..In: Proc. 2nd Int. Duodecim Symposium: Molecular Biology of Lysosomal Disease, Majik, Finland. 1993.

Maisey, D. N.; Cosh, J. A.: Basilar artery aneurysm and Anderson-Fabry disease. J. Neurol. Neurosurg. Psychiat. 43: 85-87, 1980.

Mapes, C. A.; Anderson, R. L.; Sweeley, C. C.; Desnick, R. J.; Krivit, W.: Enzyme replacement in Fabry's disease, an inborn error of metabolism. Science 169: 987-989, 1970.

Miyamura, N.; Araki, E.; Matsuda, K.; Yoshimura, R.; Furukawa, N.; Tsuruzoe, K.; Shirotani, T.; Kishikawa, H.; Yamaguchi, K.; Shichiri, M.: A carboxy-terminal truncation of human alpha-galactosidase A in a heterozygous female with Fabry disease and modification of the enzymatic activity by the carboxy-terminal domain: increased, reduced, or absent enzyme activity depending on number of amino acid residues deleted. J. Clin. Invest. 98: 1809-1817, 1996.

Moser, H. W.: Personal Communication. Baltimore, Md. 12/14/1983.

Mutoh, T.; Senda, Y.; Sugimura, K.; Koike, Y.; Matsuoka, Y.; Sobue, I.; Takahashi, A.; Naoi, M.: Severe orthostatic hypotension in a female carrier of Fabry's disease. Arch. Neurol. 45: 468-472, 1988.

Nagao, Y.; Nakashima, H.; Fukuhara, Y.; Shimmoto, M.; Oshima, A.; Ikari, Y.; Mori, Y.; Sakuraba, H.; Suzuki, Y.: Hypertrophic cardiomyopathy in late-onset variant of Fabry disease with high residual activity of alpha-galactosidase A. Clin. Genet. 39: 233-237, 1991.

Nakao, S.; Takenaka, T.; Maeda, M.; Kodama, C.; Tanaka, A.; Tahara, M.; Yoshida, A.; Kuriyama, M.; Hayashibe, H.; Sakuraba, H.; Tanaka, H.: An atypical variant of Fabry's disease in men with left ventricular hypertrophy. New Eng. J. Med. 333: 288-293, 1995.

Novo, F. J.; Kruseweski, A.; MacDermot, K. D.; Goldspink, G.; Gorecki, D. C.: Editing of human alpha-galactosidase RNA resulting in a pyrimidine to purine conversion. Nucleic Acids Res. 23: 2636-2640, 1995.

O'Brien, S. J.: The extent and character of biochemical genetic variation in the domestic cat. J. Hered. 71: 2-8, 1980.

Ogawa, K.; Sugamata, K.; Funamoto, N.; Abe, T.; Sato, T.; Nagashima, K.; Ohkawa, S.: Restricted accumulation of globotriaosylceramide in the hearts of atypical cases of Fabry's disease. Hum. Path. 21: 1067-1073, 1990.

Ohshima, T.; Murray, G. J.; Swaim, W. D.; Longenecker, G.; Quirk, J. M.; Cardarelli, C. O.; Sugimoto, Y.; Pastan, I.; Gottesman, M. M.; Brady, R. O.; Kulkarni, A. B.: Alpha-galactosidase A deficient mice: a model of Fabry disease. Proc. Nat. Acad. Sci. 94: 2540-2544, 1997.

Opitz, J. M.; Stiles, F. C.; Wise, D.; Race, R. R.; Sanger, R.; Von Gemmingen, G. R.; Kierland, R. R.; Cross, E. G.; DeGroot, W. P.: The genetics of angiokeratoma corporis diffusum (Fabry's disease) and its linkage relations with the Xg locus. Am. J. Hum. Genet. 17: 325-342, 1965.

Patel, V.; Watanabe, I.; Zeman, W.: Deficiency of alpha-L-fucosidase. Science 176: 426-428, 1972.

Peltier, A.; Herebeuval, E.; Baondeau, M. T.; Belleville, F.; Nabet, P.: Pseudo-clinical Fabry's disease without alpha-galactosidase deficiency. Biomedicine 26: 194-201, 1977.

Philippart, M.; Sarlieve, L.; Manacorda, A.: Urinary glycolipids in Fabry's disease. Their examination in the detection of atypical variants and the pre-symptomatic state. Pediatrics 43: 201-206, 1969.

Pierides, A. M.; Holti, G.; Crombie, A. L.; Roberts, D. F.; Gardiner, S. E.; Colling, A.; Anderson, J.: Study on a family with Anderson-Fabry's disease and associated familial spastic paraplegia. J. Med. Genet. 13: 455-461, 1976.

Pyeritz, R. E.; Ullman, M. D.; Moser, A. B.; Braine, H. G.; Moser, H. W.: Plasma exchange removes glycosphingolipid in Fabry disease. Am. J. Med. Genet. 7: 301-308, 1980.

Rahman, A. N.; Simeone, F. A.; Hackel, D. B.; Hall, P. W., III; Hirsch, E. Z.; Harris, J. W.: Angiokeratoma corporis diffusum universale (hereditary dystopic lipidosis). Trans. Assoc. Am. Phys. 74: 366-377, 1961.

Rodriguez, F. H., Jr.; Hoffmann, E. O.; Ordinario, A. T.; Baliga, M.: Fabry's disease in a heterozygous woman. Arch. Path. Lab. Med. 109: 89-91, 1985.

Romeo, G.; Childs, B.; Migeon, B. R.: Genetic heterogeneity of alpha-galactosidase in Fabry's disease. FEBS Lett. 27: 161-166, 1972.

Romeo, G.; Migeon, B. R.: Genetic inactivation of the alpha-galactosidase locus in carriers of Fabry's disease. Science 170: 180-181, 1970.

Ropers, H.-H.; Wienker, T.; Grimm, T.; Schroetter, K.; Bender, K.: Evidence for preferential X-chromosome inactivation in a family with Fabry's disease. Am. J. Hum. Genet. 29: 361-370, 1977.

Rosenberg, D. M.; Ferrans, V. J.; Fulmer, J. D.; Line, B. R.; Barranger, J. A.; Brady, R. O.; Crystal, R. G.: Chronic airflow obstruction in Fabry's disease. Am. J. Med. 68: 898-905, 1980.

Sakuraba, H.; Eng, C. M.; Desnick, R. J.; Bishop, D. F.: Invariant exon skipping in the human alpha-galactosidase A pre-mRNA: a g(+1) to t substitution in a 5-prime-splice site causing Fabry disease. Genomics 12: 643-650, 1992.

Sakuraba, H.; Oshima, A.; Fukuhara, Y.; Shimmoto, M.; Nagao, Y.; Bishop, D. F.; Desnick, R. J.; Suzuki, Y.: Identification of point mutations in the alpha-

galactosidase A gene in classical and atypical hemizygotes with Fabry disease. Am. J. Hum. Genet. 47: 784-789, 1990.
Sakuraba, H.; Yanagawa, Y.; Igarashi, T.; Suzuki, Y.; Suzuki, T.; Watanabe, K.; Ieki, K.; Shimoda, K.; Yamanaka, T.: Cardiovascular manifestations in Fabry's disease: a high incidence of mitral valve prolapse in hemizygotes and heterozygotes. Clin. Genet. 29: 276-283, 1986.
Sawada, K.; Mizoguchi, K.; Hishida, A.; Kaneko, E.; Koide, Y.; Nishimura, K.; Kimura, M.: Point mutation in the alpha-galactosidase A gene of atypical Fabry disease with only nephropathy. Clin. Nephrol. 45: 289-294, 1996.
Sheth, K. J.; Good, T. A.; Murphy, J. V.: Heterozygote detection in Fabry disease utilizing multiple enzyme activities. Am. J. Med. Genet. 10: 141-146, 1981.
Shows, T. B.; Brown, J. A.; Haley, L. L.; Goggin, A. P.; Eddy, R. L.; Byers, M. G.: Assignment of alpha-galactosidase (alpha-GAL) to the q22-qter region of the X chromosome in man. Cytogenet. Cell Genet. 22: 541-544, 1978.
Sorensen, S. A.; Hasholt, L.: Alpha-galactosidase isozymes in normal individuals, and in Fabry homozygotes and heterozygotes. Ann. Hum. Genet. 43: 313-321, 1980.
Spence, M. W.; Clarke, J. T. R.; D'Entremont, D. M.; Sapp, G. A.; Smith, E. R.; Goldbloom, A. L.; Davar, G.: Angiokeratoma corporis diffusum (Anderson-Fabry disease) in a single family in Nova Scotia. J. Med. Genet. 15: 428-434, 1978.
Spence, M. W.; MacKinnon, K. E.; Burgess, J. K.; d'Entremont, D. M.; Belitsky, P.; Lannon, S. G.; MacDonald, A. S.: Failure to correct the metabolic defect by renal allotransplantation in Fabry's disease. Ann. Intern. Med. 84: 13-16, 1976.
Sweeley, C. C.; Klionsky, B.: Fabry's disease: classification as a sphingolipidosis and partial characterization of a novel glycolipid. J. Biol. Chem. 238: 3148-3150, 1963.
Tagliavini, F.; Pietrini, V.; Gemignani, F.; Lechi, A.; Pallini, R.; Federico, A.: Anderson-Fabry's disease: neuropathological and neurochemical investigation. Acta Neuropath. 56: 93-98, 1982.
von Scheidt, W.; Eng, C. M.; Fitzmaurice, T. F.; Erdmann, E.; Hubner, G.; Olsen, E. G. J.; Christomanou, H.; Kandolf, R.; Bishop, D. F.; Desnick, R. J.: An atypical variant of Fabry's disease with manifestations confined to the myocardium. New Eng. J. Med. 324: 395-399, 1991.
Wise, D.; Wallace, H. J.; Jellinek, E. H.: Angiokeratoma corporis diffusum: a clinical study of eight affected families. Quart. J. Med. 31: 177-206, 1962.
Yokoi, T.; Shinoda, K.; Ohno, I.; Kato, K.; Miyawaki, T.; Taniguchi, N.: A 3-prime splice site consensus sequence mutation in the intron 3 of the alpha-galactosidase A gene in a patient with Fabry disease. Jpn. J. Hum. Genet. 36: 245-250, 1991.

GANGLIOSIDOSIS, GENERALIZED GM1, TYPE 3, OR ADULT TYPE (#230650)

JJM NOTE

This entity is included because of the occasional manifestation of angiokeratoma, one of the few tumors that should make a clinician think of a classic inborn error of metabolism.

A number sign (#) is used with this entry because the phenotype is due to mutation in the same gene as that which causes type I, or infantile-type, generalized gangliosidosis (GLB1; 230500).

Adults with beta-galactosidase deficiency but less severe abnormality than in types I and II Gm(1)-gangliosidosis were described by Loonen et al. (1974), Wenger et al. (1974), and O'Brien et al. (1976). The phenotype was not consistent. The patient of Loonen et al. (1974) had angiokeratoma. The patient of O'Brien et al. (1976) had spondyloepiphyseal dysplasia. All 3 of the above patients were found by O'Brien and Norden (1977) to be carrying CRM-positive mutations. Galjaard et al. (1975) reported a patient who had a biochemical defect that was complemented when his cells were hybridized with those from patients with either type I or type II Gm(1)-gangliosidosis. Before the CRM studies of O'Brien and Norden (1977), it was thought that the case of Loonen et al. (1974) might represent a regulatory mutation. Since beta-galactosidase appears to have a single polypeptide chain, the basis of the complementation is not clear; O'Brien and Norden (1977) discussed several possible explanations. Beta-galactosidase deficiency has been demonstrated in young adults (Wenger et al., 1974). Some of these have had diffuse angiokeratoma, which is more characteristic of alpha-galactosidase deficiency (Fabry disease, 301500), but has also been seen with fucosidase deficiency (230000). The diversity of phenotype in beta-galactosidase deficiency is further indicated by the 3 adults reported by Stevenson et al. (1978). None had visceromegaly or macular red spots. Koster et al. (1976) described somatic cell complementation studies in an adult with deficiency of beta-galactosidase. Loonen et al. (1974) had previously described the clinical features, which included angiokeratoma beginning at age 8 years, cerebellar dysfunction and diminishing vision beginning at age 16, and myoclonic fits, intellectual deterioration and coarsening of the face beginning at age 22. When cells of the patient were fused with those from types I and II beta-galactosidase deficiency, mutual correction occurred. No complementation occurred with cells from a patient labeled type III by the laboratory and by Pinsky et al. (1974). Reuser et al. (1979) described a case in which fibroblasts complemented both type I and type II but did not complement fibroblasts from clinical variants called type III and adult type IV. Taylor et al. (1980) reported further studies of 2 of the patients reported by Stevenson et al. (1978). Beta-galactosidase differed in pH optima, heat denaturation, NaCl kinetics, and electrophoretic mobility from each other and from the normal. No complementation of the two was found in cell fusion studies. The authors concluded that the 2 patients had different primary mutations at the beta-galactosidase locus which are probably structural in nature. Wenger et al. (1980) described a brother and sister, aged 19 and 25, respectively, with ataxia, mild intellectual deterioration, slurred speech, mild vertebral changes and 'little, if any, visceromegaly.' A primary defect in beta-galactosidase was indicated by the half-normal values in many relatives, including both parents, and by the normal levels of sialidase. Furthermore, complementation with infantile type I GM1-gangliosidosis did not occur. Uyama et al. (1992) described 3 Japanese brothers with type III GM1-gangliosidosis presenting as dystonia. The patients were examined at ages 28, 31, and 33 years. They had developed dysarthria with facial grimacing in early childhood. Generalized dystonia was the predominant feature. Both dystonic postures and dystonic movements did not disappear when the patients were lying or sitting relaxed. Magnetic resonance imaging (MRI) showed bilaterally symmetric high density lesions only in the putamen. Changes there were thought to be responsible for dystonia in this disorder.

See also Galjaard et al. (1975)

Galjaard, H.; Hoogeveen, A.; Keijzer, W.; DeWit-Verbeek, H. A.; Reuser, A. J. J.: Different gene mutations in variants of Gm(1)- and Gm(2)-gangliosidosis demonstrated by enzyme analysis of (single) somatic hybrid cells. Birth Defects Orig. Art. Ser. XI(3): 150-156, 1975. Note: Alternate: Cytogenet. Cell Genet. 14: 320-326, 1975.
Galjaard, H.; Hoogeveen, A.; Keijzer, W.; deWit-Verbeek, H. A.; Reuser, A. J. J.; Ho, M. W.; Robinson, D.: Genetic heterogeneity in GM1-gangliosidosis. Nature 257: 60-62, 1975.
Koster, J. F.; Niermeijer, M. F.; Loonen, M. C. B.; Galjaard, H.: Beta-galactosidase deficiency in an adult: a biochemical and somatic cell genetic study on a variant of GM-1-gangliosidosis. Clin. Genet. 9: 427-432, 1976.
Loonen, M. C. B.; van de Lugt, L.; Franke, C. L.: Angiokeratoma corporis diffusum and lysosomal enzyme deficiency. (Letter) Lancet II: 785 only, 1974.
O'Brien, J. S.; Gugler, E.; Giedion, A.; Weissmann, U.; Herschkowitz, N.; Neier, C.; Leroy, J. G.: Spondyloepiphyseal dysplasia, corneal clouding, normal intelligence and beta-galactosidase deficiency. Clin. Genet. 9: 495-504, 1976.
O'Brien, J. S.; Norden, A. G. W.: Nature of the mutation in adult beta-galactosidase deficient patients. Am. J. Hum. Genet. 29: 184-190, 1977.
Reuser, A. J. J.; Andria, G.; de Wit-Verbeek, E.; Hoogeveen, A.; del Giudice, E.; Halley, D.: A two-year-old patient with an atypical expression of Gm(1)-beta-galactosidase deficiency: biochemical, immunological, and cell genetic studies. Hum. Genet. 46: 11-19, 1979.
Stevenson, R. E.; Taylor, H. A., Jr.; Parks, S. E.: Beta-galactosidase deficiency: prolonged survival in three patients following early central nervous system deterioration. Clin. Genet. 13: 305-313, 1978.
Taylor, H. A.; Stevenson, R. E.; Parks, S. E.: Beta-galactosidase deficiency: studies of two patients with prolonged survival. Am. J. Med. Genet. 5: 235-245, 1980.
Uyama, E.; Terasaki, T.; Watanabe, S.; Naito, M.; Owada, M.; Araki, S.; Ando, M.: Type 3 GM1 gangliosidosis: characteristic MRI findings correlated with dystonia. Acta Neurol. Scand. 86: 609-615, 1992.
Wenger, D. A.; Goodman, S. I.; Myers, G. B.: Beta-galactosidase deficiency in young adults. (Letter) Lancet II: 1319-1320, 1974.
Wenger, D. A.; Sattler, M.; Mueller, O. T.; Myers, G. G.; Schneiman, R. S.; Nixon, G. W.: Adult GM1 gangliosidosis: clinical and biochemical studies on two patients and comparison to other patients called variant or adult GM1 gangliosidosis. Clin. Genet. 17: 323-334, 1980.

10. HEMOLYMPHORETICULAR SYSTEM

Red Cell Proliferations

ERYTHROCYTOSIS, FAMILIAL (*133100)

JJM NOTE

Given the computer-generated and -archived hematology records that exist worldwide, it would seem feasible to do a record linkage study of the familiality of erythrocytosis and even to sort out some etiologic heterogeneity. Then, some large families might be assembled for mapping additional genes that control the erythron.

This disorder is characterized by an increase in red blood cell mass with no increase in platelets and leukocytes, and by a relatively benign course and familial incidence. (It is best to avoid 'benign' in the title of this condition. Although it is distinct from polycythemia vera, a malignant disorder, patients are subject to the complications of hyperviscosity and the improved sense of well-being after phlebotomy indicates some ill effects. Polycythemia vera is more frequent in Jews than in non-Jews in the United States (Modan, 1965) but shows no simple mendelian pattern.) A patient reported by Auerbach et al. (1958) was again reported by Cassileth and Hyman (1966) with family study. Engelking (1920) and Wieland (1932) separately reported a family in which 11 members of 3 generations were polycythemic. In some the abnormality was noted in childhood. Such families should be studied for a hemoglobinopathy. Polycythemia is a feature of several variant hemoglobins: Chesapeake, J (Capetown), Yakima, Kempsey, Rainier (Weatherall, 1969), Ypsilanti, and Hiroshima. The heterozygotes show polycythemia; hence the phenotype is dominant. Alperin et al. (1967) reported finding elevated levels of erythropoietin in affected members of 1 family. Alperin (1982) stated 15 years later that he had not yet published this report. Geary et al. (1967) observed polycythemia in 5 persons in 3 generations of a family and showed that the basis was thalassemia minor. The red cell count was elevated but total hemoglobin was normal, thus giving hypochromia. Hemoglobin A2 was elevated. Thus, thalassemia minor leads to 'microcytic polycythemia,' which is quite different from the other conditions discussed here. More often thalassemia minor presents as refractory hypochromic anemia. Davey et al. (1968) found erythrocytosis in a brother and sister, offspring of a second-cousin marriage, and raised the question of recessive inheritance. The father had slight but persistent erythrocyto sis. Apparently hemoglobin electrophoresis was not performed. Hemoglobin Olympia (q.v.) is an electrophoretically silent mutant form of hemoglobin which leads to erythrocytosis. Some instances of familial polycythemia with normal hemoglobin may represent a defect in the mechanisms for maintaining intra-erythrocytic pH (Charache, 1974). Stamatoyannopoulos (1972) suggested that one form of dominant erythrocytosis may be due to a mutation in the regulation of 2-3 diphosphoglycerate. Distelhorst et al. (1981) described erythrocytosis due to autonomous erythropoietin production in 4 members of 3 successive generations. Dainiak et al. (1979) and Yonemitsu et al. (1973) also attributed familial erythrocytosis to increased erythropoietin production. A leading difficulty in defining the mechanism in familial erythrocytosis is lack of a suitably sensitive assay for erythropoietin. It is likely that eventually multiple mechanisms for familial erythrocytosis will be recognized, each related to one of the steps in the regulation of erythropoiesis (Spivak, 1981). Increased erythropoietin production and increased target cell sensitivity to the hormone cannot at present be distinguished with assurance. Ly et al. (1983) reported 5 persons in 3 generations with familial erythrocytosis caused by some factor other than increased erythropoietin production. Hemoglobin function and 2,3-DPG were normal. Erythropoietin assays showed low values. Quiesser et al. (1988) described erythrocytosis in 7 members of a family in 4 generations with male-to-male transmission. He reviewed the reported cases. The propositus was first diagnosed at age 26 years. He had headaches and marked plethora. Erythropoietin levels were not elevated. The disorder was characterized in middle age in other members of the family by hypertension, cardiovascular and thromboembolic phenomena, and abnormal bleeding. He was treated successfully with repeated venous blood lettings. E manuel et al. (1992) reported studies of 2 extensively affected families with familial and congenital polycythemia. Although 13 individuals in 4 generations were demonstrated in 1 family, there was no instance of male-to-male transmission; in the second family, however, there were 2 such instances. Another patient was an isolated case. In all 3 families, secondary causes of polycythemia were ruled out. Erythropoietin levels were normal or low, even after phlebotomy. No evidence of rearrangement or amplification of the erythropoietin receptor gene (EPOR; 133171) could be demonstrated. Functional studies examining the number and binding affinity of the EPO receptor on erythroid progenitors from 3 of their patients demonstrated no abnormalities. Emanuel et al. (1992) concluded that the erythrocytosis in these families that did not involve the EPO receptor may result from alterations in the postreceptor responses. In a large and extensively affected Finnish kindred with affected members in 5 generations and many instances of male-to-male transmission (Juvonen et al., 1991), de la Chapelle et al. (1993) demonstrated linkage of the erythrocytosis phenotype to a highly informative, simple sequence repeat (SSR) polymorphism in the 5-prime region of the EPOR gene. There was no recombination between the disease phenotype and the polymorphism, the lod score being 6.37 at theta = 0.0.

See also Adamson (1975); Charache (1970); Spodaro and Forkner (1933); Stamatoyannopoulos et al. (1973)

Adamson, J. W.: Familial polycythemia. Seminars Hemat. 12: 383-396, 1975.
Alperin, J. B.: Personal Communication. Galveston, Texas 1982.
Alperin, J. B.; Levin, W. C.; Alexanian, R.; Houston, E. W.: Familial erythrocytosis: a disorder due to increased erythropoietin production. (Unpublished) 1967.
Auerbach, M. L.; Wolff, J. A.; Mettier, S. R.: Benign familial polycythemia in childhood. Pediatrics 21: 54-58, 1958.
Cassileth, P. A.; Hyman, G. A.: Benign familial erythrocytosis: report of three cases and a review of the literature. Am. J. Med. Sci. 251: 692-697, 1966.
Charache, S.: Familial polycythemia. Mt. Sinai J. Med. 37: 418-425, 1970.
Charache, S.: Personal Communication. Baltimore, Md. 1/14/1974.
Dainiak, N.; Hoffman, R.; Lebowitz, A. I.; Solomon, L.; Maffei, L.; Ritchey, K.: Erythropoietin-dependent primary pure erythrocytosis. Blood 53: 1076-1084, 1979.
Davey, M. G.; Lawrence, J. R.; Lander, H.; Robson, H. N.: Familial erythrocytosis: a report of two cases, and a review. Acta Haemat. 39: 65-74, 1968.
de la Chapelle, A.; Sistonen, P.; Lehvaslaiho, H.; Ikkala, E.; Juvonen, E.: Familial erythrocytosis genetically linked to erythropoietin receptor gene. Lancet 341: 82-84, 1993.
Distelhorst, C. W.; Wagner, D. S.; Goldwasser, E.; Adamson, J. W.: Autosomal dominant familial erythrocytosis due to autonomous erythropoietin production. Blood 58: 1155-1158, 1981.
Emanuel, P. D.; Eaves, C. J.; Broudy, V. C.; Papayannopoulou, T.; Moore, M. R.; D'Andrea, A. D.; Prchal, J. F.; Eaves, A. C.; Prchal, J. T.: Familial and congenital polycythemia in three unrelated families. Blood 79: 3019-3030, 1992.
Engelking, E.: Ueber familiaere Polyzythaemie und die dabei beobachteten Augenveraenderungen. Klin. Mbl. Augenheilk. 64: 645-664, 1920.
Geary, C. G.; Amos, H. E.; MacIver, J. E.: Benign familial polycythemia. J. Clin. Path. 20: 158-160, 1967.
Juvonen, E.; Ikkala, E.; Fyhrquist, F.; Ruutu, T.: Autosomal dominant erythrocytosis caused by increased sensitivity to erythropoietin. Blood 78: 3066-3069, 1991.
Ly, B.; Meberg, A.; Kannelonning, K.; Refsum, H. E.; Berg, K.: Dominant familial erythrocytosis with low plasma erythropoietin activity: studies on four cases. Scand. J. Haemat. 30 (suppl. 39): 11-17, 1983.
Modan, B.: An epidemiological study of polycythemia vera. Blood 26: 657-667, 1965.
Quiesser, W.; Heim, M. E.; Schmitz, J. M.; Worst, P.: Idiopathische familiaere Polyglobulie: Beobachtung einer Familie mit autosomal dominantem Erbgang. Dtsch. Med. Wschr. 113: 851-856, 1988.
Spivak, J. L.: Personal Communication. Baltimore, Md. 6/26/1981.
Spodaro, A.; Forkner, C. E.: Benign familial polycythemia. Arch. Intern. Med. 52: 593-602, 1933.
Stamatoyannopoulos, G.: Familial erythrocytosis. Birth Defects Orig. Art. Ser. 8(3): 39-45, 1972.
Stamatoyannopoulos, G.; Nute, P. E.; Adamson, J. W.; Bellingham, A. J.; Funk, D.; Hornung, S.: Hemoglobin Olympia (beta 20 valine to methionine): an electrophoretically silent variant associated with high oxygen affinity and erythrocytosis. J. Clin. Invest. 52: 342-349, 1973.
Weatherall, D. J.: Polycythemia resulting from abnormal hemoglobins. New Eng. J. Med. 280: 604-606, 1969.
Wieland, W.: Weitere Untersuchungen ueber Polycythaemia vera im Kindesalter. Z. Kinderheilk. 53: 703-715, 1932.
Yonemitsu, H.; Yamaguchi, K.; Shigeta, H.; Okuda, K.; Takaku, F.: Two cases of familial erythrocytosis with increased erythropoietin activity in plasma and urine. Blood 42: 793-797, 1973.

POLYCYTHEMIA, BENIGN FAMILIAL (*263400)

JJM NOTE

This disorder is distinguished from the prior one by the recessive mode of inheritance, although the nuances are arguable.

In addition to families with dominant transmission of erythrocytosis and a demonstrable abnormality of hemoglobin (such as Hb Chesapeake, the first such to be found) in heterozygous state, families with apparent recessive inheritance have been observed (Stamatoyannopoulos, 1972): the patients are offspring of hematologically normal parents, the segregation ratio is probably appropriate for recessive inheritance, and parental consanguinity has been documented (Davey et al., 1968). Auerbach et al. (1958) reported 3 families. In 1 family, 2 brothers and a sister were affected and in a second the proband and an aunt. The parents were normal. Unlike polycythemia vera, the subjects demonstrated no increase in white count, platelets or uric acid and the process was benign. Indeed, benign familial erythrocytosis is a designation much to be preferred because only the erythroid series is affected. Nadler and Cohn (1939) described a family in which 4 of 11 children showed polycythemia. The mother stated that these 4 children had red faces from the time of birth. Yonemitsu et al. (1973) described 2 affected sons of parents related as half first cousins. Both had a marked increase in erythroprotein concentration in plasma and urine. Adamson et al. (1973) studied 2 families with recessive erythrocytosis and found increased erythropoietin production uninfluenced by alterations in the oxygen-carrying capacity of the blood when the hematocrit was lowered by phlebotomy. Hemoglobin and red cell function and renal vasculature were normal. A genetic defect in regulation of erythropoietin production was postulated. Whitcomb et al. (1980) studied 3 cases of congenital erythrocytosis and found an absolute or relative elevation of erythropoietin. Urinary excretion of erythropoietin was more than doubled by phlebotomy. An inherited defect, 'likely residing in the renal sensor responsible for the production of erythropoietin,' was postulated.

Vasserman et al. (1998) indicated that autosomal recessive benign erythrocytosis is 'rather frequent in a small area of Chuvashiya,' Russia. They studied 12 unrelated families and excluded the erythropoietin gene (EPO; 133170) and the erythropoietin receptor gene (EPOR; 133171) as candidates by linkage analysis. Using genomewide searching, they demonstrated linkage of the disorder with markers on 11q (maximum lod = 5.04).

Adamson, J. W.; Stamatoyannopoulos, G.; Kontras, S.; Lascari, A.; Detter, J. C.: Recessive familial erythrocytosis: aspects of marrow regulation in two families. Blood 41: 641-652, 1973.
Auerbach, M. L.; Wolff, J. A.; Mettier, S. R.: Benign familial polycythemia in childhood: report of two cases. Pediatrics 21: 54-58, 1958.
Davey, M. G.; Lawrence, J. R.; Lander, H.; Robson, H. N.: Familial erythrocytosis: a report of two cases and a review. Acta Haemat. 39: 65-74, 1968.
Nadler, S. B.; Cohn, I.: Familial polycythemia. Am. J. Med. Sci. 198: 41-48, 1939.
Stamatoyannopoulos, G.: Familial erythrocytosis. Birth Defects Orig. Art. Ser. 8(3): 39-45, 1972.
Vasserman, N.; Tverskaya, S.; Karzakova, L.; Evrafov, O.: Locus for autosomal recessive benign erythrocytosis maps to chromosome 11q. (Abstract) Europ. J. Hum. Genet. 6 (suppl. 1): 35 only, 1998.
Whitcomb, W. H.; Peschle, C.; Moore, M.; Nitschke, R.; Adamson, J. W.: Congenital erythrocytosis: a new form associated with an erythropoietin-dependent mechanism. Brit. J. Haemat. 44: 17-24, 1980.
Yonemitsu, H.; Yamaguchi, K.; Shigeta, H.; Okuda, K.; Takaku, F.: Two cases of familial erythrocytosis with increased erythroprotein activity in plasma and urine. Blood 42: 793-797, 1973.

POLYCYTHEMIA RUBRA VERA (263300)

JJM NOTE

For such a well-known entity, familiality seems remarkably infrequent. Perhaps the issue could be addressed as part of a larger study for other purposes, such as therapy or environmental epidemiology.

Modan (1965) suggested that in only 2 reports of familial PV was the diagnosis completely documented (Lawrence and Goetsch, 1950; Erf, 1956). Lawrence and Goetsch (1950) described 3 affected sibs. Two patients in the series of Erf (1956) were brothers and 3 others had 'a definite family history.' Owen (1924) emphasized the familial nature of polycythemia vera and presented a possible example. Levin et al. (1967) reported a curious case of 2 brothers with polycythemia vera and the Philadelphia chromosome. Subsequently, this was shown to be an instance of familial small Y chromosome (Levin, 1974). The precise mode of inheritance is unknown. Greenberg and Golde (1977) studied 2 brothers, aged 26 and 28, whose erythrocytosis had been discovered incidentally. The parents were hematologically normal. Studies suggested that the erythropoietin system was functioning normally and that the primary defect was an expansion of the erythroid precursor pool. Polycythemia vera has been identified as an acquired clonal stem cell disorder (Adamson et al., 1976). Some instances of familial erythrocytosis may have a similar erythropoietic defect of a genetic basis. Clonality is the prime criterion of polycythemia vera, but unfortunately can at present be investigated only in relatively rare persons heterozygous for the G6PD A-B polymorphism. As more X-chromosome markers, including restriction enzyme polymorphisms, become available, clonality can be checked more easily. Ratnoff and Gress (1980) described polycythemia vera in a father and son, both of whom had had intermittent exposure to organic solvents, including tetrachloroethylene and Stoddard solvent, which may have been of etiologic significance. Although they found 3 other well-substantiated familial occurrences of the disorder, none of them encompassed successive generations.

Chen et al. (1998) presented results indicating that amplification of a gene or genes on 9p plays a crucial role in the pathogenesis of PV. They observed 2 cases of PV with an extra i(9)(p10) isochromosome as the sole anomaly.

See also Friedland et al. (1981); Manoharan and Garson (1976)

Adamson, J. W.; Fialkow, P. J.; Murphy, S.; Prchal, J. F.; Steinmann, L.: Polycythemia vera: stem-cell and probably clonal origin of the disease. New Eng. J. Med. 295: 913-916, 1976.
Chen, Z.; Notohamiprodjo, M.; Guan, X.-Y.; Paietta, E.; Blackwell, S.; Stout, K.; Turner, A.; Richkind, K.; Trent, J. M.; Lamb, A.; Sandberg, A. A. Gain of 9p in the pathogenesis of polycythemia vera. Genes Chromosomes Cancer 22: 321-324, 1998.
Erf, L. A.: Radioactive phosphorus in the treatment of primary polycythemia. Prog. Hemat. 1: 153-165, 1956.
Friedland, M. L.; Wittels, E. G.; Robinson, R. J.: Polycythemia vera in identical twins. Am. J. Hemat. 10: 101-103, 1981.
Greenberg, B. R.; Golde, D. W.: Erythropoiesis in familial erythrocytosis. New Eng. J. Med. 296: 1080-1084, 1977.
Lawrence, J. H.; Goetsch, A. T.: Familial occurrence of polycythemia and leukemia. Calif. Med. 73: 361-364, 1950.
Levin, W. C.: Personal Communication. Galveston, Texas 1/9/1974.
Levin, W. C.; Houston, E. W.; Ritzmann, S. E.: Polycythemia vera with Ph-1 chromosomes in two brothers. Blood 30: 503-512, 1967.
Manoharan, A.; Garson, O. M.: Familial polycythemia vera: a study of three sisters. Scand. J. Haemat. 17: 10-16, 1976.
Modan, B.: Polycythemia: a review of epidemiological and clinical aspects. J. Chronic Dis. 18: 605-645, 1965.
Owen, T.: A case of polycythaemia vera with special reference to the familial features and treatment with phenylhydrazine. Johns Hopkins Hosp. Bull. 35: 258-262, 1924.
Ratnoff, W. D.; Gress, R. E.: The familial occurrence of polycythemia vera: report of a father and son, with consideration of the possible etiologic role of exposure to organic solvents, including tetrachloroethylene. Blood 56: 233-236, 1980.

ERYTHROLEUKEMIA, FAMILIAL (DIGUGLIELMO DISEASE) (133180)

JJM NOTE

These three tantalizing reports of familial aggregation of such a rare leukemia suggest the need for a consortium effort to document and study all available living families (e.g., by a few international cooperative clinical trials groups).

This is a leukemic or preleukemic state in which red cell proliferation is the predominant feature. Hematologic characteristics include particularly ineffective and hyperplastic erythropoiesis with megaloblastic components accompanied by myeloblastic proliferation of varying degree. Davidson et al. (1978) gave clinical and hematologic details on affected brother and sister. Their father presumably had died of a similar disease. Both sibs showed chromosomal changes, a finding previously reported in a majority of sporadic cases. The authors could find no 'documented evidence of the familial occurrence' of erythroleukemia. Nissenblatt et al. (1982) diagnosed erythroleukemia in 3 brothers in a 6-month period in 1976. A son of 1 brother had died with erythroleukemia 5 years earlier. Among close relatives of these men, 14 of 16 tested had elevated IgM (mean, 352.8 mg per dl; normal, less than 145 mg per dl). This was neither a monoclonal protein nor rheumatoid factor. Erythrocyte hexokinase was elevated in 23 of 24 persons (mean, 35 units per 100 ml RBC; normal, less than 18 units). Karyotypes in 18 relatives were normal; one leukemic showed hypoploidy and a marker chromosome. Peterson et al. (1984) described a family in which 4 of 11 sibs were affected. A fifth sib developed unexplained marrow hypoplasia. Cytogenetic and environmental studies revealed no explanation for the familial clustering. The authors suggested that the natural history of the disorder in this sibship was initial reactive changes in the marrow with subsequent progression to myelodysplasia with sideroblastosis, and finally to DiGuglielmo syndrome. The proband, aged 55, was investigated because 'most of my family have leukemia' and was found to have marrow hypocellularity. The deceased members of the family, all with well-documented erythroleukemia, were a brother who died at age 26, 14 months after clinical presentation; a 28-year-old man, who died 6 months after presentation, and his MZ twin who died at age 39, 5 mon ths after presentation and a 41-year-old woman who died 7.5 years after mild anemia and leukopenia were discovered incidentally. The only consanguinity in the family was probably irrelevant; the paternal grandparents were distantly related.

See also Di Guglielmo (1928)

Davidson, R. J. L.; Walker, W.; Watt, J. L.; Page, B. M.: Familial erythroleukemia: a cytogenetic and haematological study. Scand. J. Haemat. 20: 351-359, 1978.
Di Guglielmo, G.: Le Eritremie. Haematologica 9: 301-347, 1928.
Nissenblatt, M. J.; Bias, W.; Borgaonkar, D.; Dixon, S.; Cody, R. P.: Familial erythroleukemia: four cases of the Diguglielmo syndrome in close relatives. Johns Hopkins Med. J. 150: 1-9, 1982.
Peterson, H. R., Jr.; Bowlds, C. F.; Yam, L. T.: Familial DiGuglielmo syndrome. Cancer 54: 932-938, 1984.

Platelet Disorders

ESSENTIAL THROMBOCYTHEMIA (#187950)

JJM NOTE

The initial disorder and rare familial aggregations could not be sorted out by bioassays. Cytogenetic findings gave a false lead at first, with the chromosome 21 abnormality, but continued work pointed to chromosome 3 as the location of the underlying genetic defect. The association of acute nonlymphocytic leukemia with thrombocythemia allows two speculations: that acquired cytogenetic abnormalities could produce the control defect in platelet production and that leukemia might complicate familial thrombocythemia.

A number sign is used with this entry because of evidence that essential thrombocythemia can be caused by mutation of the thrombopoietin gene (THPO; 600044), which is located on chromosome 3.

Kaywin et al. (1978) studied 4 male patients with essential thrombocythemia. Two of them failed to aggregate or release serotonin in response to concentrations of epinephrine that aggregated platelets of normal controls. Studies with 3H-DHE (tritiated dihydroergocryptine), an alpha-adrenergic antagonist, showed that the platelets from these men contained only about half as many binding sites as were found on control platelets. The other 2 patients showed normal epinephrine responses and receptor site numbers. In 5 cases of primary thrombocytosis, Zaccaria and Tura (1978) found partial deletion of the long arm of chromosome 21. Petit and Van den Berghe (1979) observed a sixth such patient. A site for primary thrombocytosis was assigned to 21q11-21qter on the basis of this chromosomal aberration (Petit and Van den Berghe, 1979; Rajendra et al., 1981). Verhest and Monsieur (1983) found the Philadelphia chromosome in a case of essential thrombocytopenia. In a study of 33 patients with essential thrombocythemia, Case (1984) could not confirm the reported association of 21q-. All patients met rigid criteria for diagnosis of this myeloproliferative disorder. Constitutional symptoms were present in 90%, with 25% having thrombohemorrhagic manifestations. All patients had hypercellular marrows with megakaryocytic hyperplasia and dysplasia. The median platelet count at diagnosis was 17 million with a range as high as 70 million per cu mm. Emilia et al. (1985) concluded that no chromosomal abnormality is consistently associated with essential thrombocytosis. Specifically, they suggested that the evidence does not indicate a clear relationship to deletion of 21q. Essential thrombocythemia was observed in a brother-sister pair from a family in which 6 other persons had platelet counts in excess of 400,000 per cubic mm (Fickers and Speck, 1974; Slee et al., 1981). A second familial report (Eyster et al., 1986) concerned 5 members of 3 successive generations with ages ranging from 2 to 62 years. The proposita, a 26-year-old woman, had persistent elevation of platelet count, mild splenomegaly, normal hemoglobin level, normal white blood cell count, and abnormal platelet aggregation. Megakaryocytes were increased in number and size. Platelet arachidonic acid metabolites and serum thrombopoietin levels were normal. Her 2 sons, aged 4 and 2 years, had platelet counts of 951,000 and 1,070,000 per cu mm. Her mother had platelet counts of about 900,000 per cu mm. Bone marrow karyotypes were normal. The proband's mother and her sister had thrombocythemia; both had had myocardial infarction at ages 52 and 57, respectively. The authors concluded that young women and children with essential thrombocythemia have long survival without treatment.

Sweet et al. (1979), Norrby et al. (1982), Carbonell et al. (1982), and Bernstein et al. (1982) described cases of leukemia or preleukemia associated with thrombocytosis and an insertion in chromosome 3, ins(3;3)(q26;q21q26). Furthermore, Bernstein et al. (1982), who referred to this as homologous translocation (transposition is another possible designation), described 3 patients with acute nonlymphocytic leukemia (ANLL) and thrombocytosis associated with inv(3)(q21q26).

Carroll et al. (1986) described sideroblastic anemia and thrombocytosis causing recurrent and ultimately fatal thromboembolism in a patient with 46,XY,ins(3;3)(q26;q21q26) in bone marrow cells. All of these observations suggested that 'a gene on the long arm of chromosome 3 may encode a protein that, at least in part, regulates megakaryopoiesis.' This inference was supported by the demonstration that the lineage-restricted thrombopoietin gene (THPO; 600044) maps to 3q26.3-q27 and that a mutation in the THPO gene segregated with hereditary thrombocythemia in a 4-generation Dutch kindred (Wiestner et al., 1998).

Schlemper et al. (1994) identified 11 persons with essential thrombocythemia in 4 generations of a family. Of these, 5 were asymptomatic, 3 had both vasoocclusive and hemorrhagic manifestations, and 3 patients had only vasoocclusive features. The platelet count ranged from 500 to 1700 x 10(9)/l. Symptoms correlated with age but not with platelet count. ADP-induced platelet aggregation distinguished best between patients and unaffected members of the family. No chromosome abnormalities were found. The propositus presented at the age of 32 years with recurrent transient symptoms of 'cold tip feeling,' erythromelalgia, and acrocyanosis of the toes, which eventually led to gangrene and amputation of a toe in 1968. He was treated for hypertension from the age of 40 years. At the age of 51, he had a recurrence of erythromelalgia and acrocyanosis. At the age of 55 years, he developed symptoms due to stenosis of the left carotid artery which required surgical 'desobstruction.' The family contained 1 instance of male-to-male transmission. See 133020 for a discussion of familial primary erythromelalgia.

Familial thrombocytosis was also reported by Fernandez-Robles et al. (1990). Kikuchi et al. (1995) observed 4 cases of thrombocytosis in 3 successive generations of a family with no male-to-male transmission. A high peripheral platelet count was found incidentally in the proband with cutaneous malignant lymphoma which was thought to be coincidental. Bone marrow examinations showed megakaryocytic hyperplasia. Neither Philadelphia chromosome nor chimeric BCR/ABL junction was detected in marrow cells.

In a family with multiple cases of thrombocytosis in 4 generations, van Dijken et al. (1996) observed chronic thrombocytosis in an infant. The mother was known to have familial thrombocytosis for which she was treated with low-dose aspirin. Her infant son was seen at the age of 6 weeks to check his platelet count. Unexpectedly, he was found to be malnourished with low weight and enlarged liver and spleen and marked leukocytosis with an abnormal differential white cell count. Although leukemia was suspected, the hematologic picture began to normalize about the age of 1 year and hepatosplenomegaly decreased. At the age of 5 years he was apparently well with only a just palpable spleen, like most of his family members with thrombocytosis.

In addition to the description of familial cases, primary thrombocytosis has been described in monozygotic twins (Dodsworth, 1980). There was 1 instance of male-to-male transmission in the family reported by van Dijken et al. (1996).

See also Fialkow et al. (1981); Gaetani et al. (1982); Singal et al. (1983)

Bernstein, R.; Pinto, M. R.; Behr, A.; Mendelow, B.: Chromosome 3 abnormalities in acute nonlymphocytic leukemia (ANLL) with abnormal thrombopoiesis: report of three patients with a 'new' inversion anomaly and a further case of homologous translocation. Blood 60: 613-617, 1982.
Carbonell, F.; Hoelzer, D.; Thiel, E.; Bartl, R.: Ph(1)-positive CML associated with megakaryocytic hyperplasia and thrombocythemia and an abnormality of chromosome no. 3. Cancer Genet. Cytogenet. 6: 153-161, 1982.
Carroll, A. J.; Poon, M.-C.; Robinson, N. C.; Crist, W. M.: Sideroblastic anemia associated with thrombocytosis and a chromosome 3 abnormality. Cancer Genet. Cytogenet. 22: 183-187, 1986.
Case, D. C., Jr.: Absence of a specific chromosomal marker in essential thrombocythemia. Cancer Genet. Cytogenet. 12: 163-165, 1984.
Dodsworth, H.: Primary thrombocythaemia in monozygotic twins. Brit. Med. J. 280: 1506 only, 1980.
Emilia, G.; Torelli, G.; Sacchi, S.; Donelli, A.: Chromosomal abnormalities in essential thrombocythemia. (Letter) Cancer Genet. Cytogenet. 18: 91-93, 1985.
Eyster, M. E.; Saletan, S. L.; Rabellino, E. M.; Karanas, A.; McDonald, T. P.; Locke, L. A.; Luderer, J. R.: Familial essential thrombocythemia. Am. J. Med. 80: 497-502, 1986.
Fernandez-Robles, E.; Vermylen, C.; Martiat, P.; Ninane, J.; Cornu, G.: Familial essential thrombocythemia. Pediat. Hemat. Oncol. 7: 373-376, 1990.
Fialkow, P. J.; Faguet, G. B.; Jacobson, R. J.; Vaidya, K.; Murphy, S.: Evidence that essential thrombocythemia is a clonal disorder with origin in a multipotent stem cell. Blood 58: 916-919, 1981.
Fickers, M.; Speck, B.: Thrombocythaemia: familial occurrence and transition into blastic crisis. Acta Haemat. 51: 257-265, 1974.
Gaetani, G. F.; Ferraris, A. M.; Galiano, S.; Giuntini, P.; Canepa, L.; d'Urso, M.: Primary thrombocythemia: clonal origin of platelets, erythrocytes, and granulocytes in a Gd(B)-Gd(Mediterranean) subject. Blood 59: 76-79, 1982.

Kaywin, P.; McDonough, M.; Insel, P. A.; Shattil, S. J.: Platelet function in essential thrombocythemia: decreased epinephrine responsiveness associated with a deficiency of platelet alpha-adrenergic receptors. New Eng. J. Med. 299: 505-509, 1978.

Kikuchi, M.; Tayama, T.; Hayakawa, H.; Takahashi, I.; Hoshino, H.; Ohsaka, A.: Familial thrombocytosis. Brit. J. Haemat. 89: 900-902, 1995.

Norrby, A.; Ridell, B.; Swolin, B.; Westin, J.: Rearrangement of chromosome no. 3 in a case of preleukemia with thrombocytosis. Cancer Genet. Cytogenet. 5: 257-263, 1982.

Petit, P.; Van den Berghe, H.: A chromosomal abnormality (21q-) in primary thrombocytosis. Hum. Genet. 50: 105-106, 1979.

Rajendra, B. R.; Lee, M.; Nissenblatt, M. J.; Gartenberg, G.; Rose, D. V.; Sciorra, L. J.: The occurrence of the Philadelphia chromosome in essential thrombocytosis. Hum. Genet. 56: 287-291, 1981.

Schlemper, R. J.; van der Maas, A. P. C.; Eikenboom, J. C. J.: Familial essential thrombocythemia: clinical characteristics of 11 cases in one family. Ann. Hemat. 68: 153-158, 1994.

Singal, U.; Prasad, A. S.; Halton, D. M.; Bishop, C.: Essential thrombocythemia: a clonal disorder of hematopoietic stem cell. Am. J. Hemat. 14: 193-196, 1983.

Slee, P. H. T. J.; van Everdingen, J. J. E.; Geraedts, J. P. M.; te Velde, J.; den Ouolander, G. J.: Familial myeloproliferative disease: hematological and cytogenetic studies. Acta Med. Scand. 210: 321-327, 1981.

Sweet, D. L.; Golomb, H. M.; Rowley, J. D.; Vardiman, J. W.: Acute myelogenous leukemia and thrombocythemia associated with an abnormality of chromosome no. 3. Cancer Genet. Cytogenet. 1: 33-37, 1979.

van Dijken, P. J.; Woldendorp, K. H.; van Wouwe, J. P.: Familial thrombocytosis in infancy presenting with a leukaemoid reaction. Acta. Paediat. 85: 1132-1134, 1996.

Verhest, A.; Monsieur, R.: Philadelphia chromosome-positive thrombocythemia with leukemic transformation. (Letter) New Eng. J. Med. 308: 1603 only, 1983.

Wiestner, A.; Schlemper, R. J.; van der Maas, A. P. C.; Skoda, R. C.: An activating splice donor mutation in the thrombopoietin gene causes hereditary thrombocythaemia. Nature Genet. 18: 49-52, 1998.

Zaccaria, A.; Tura, S.: A chromosomal abnormality in primary thrombocythemia. New Eng. J. Med. 298: 1422-1423, 1978.

THROMBOPOIETIN (*600044)

JJM NOTE

Different mutations in this gene underlie two kindreds with familial essential thrombocythemia.

As dramatically indicated by Metcalf (1994), the isolation of a postulated thrombopoietin eluded investigators for more than 30 years. Finally, cDNAs for both human and murine thrombopoietin were cloned and shown to encode a glycoprotein with selective action on megakaryocyte proliferation in vitro. As often happens, the success story had an unlikely beginning in the findings in a different field. Vigon et al. (1992) cloned the human homolog of the myeloproliferative leukemia virus (MPL; 159530) and found that it showed striking homologies with members of the hematopoietin receptor superfamily. Methia et al. (1993) demonstrated that MPL may have special relevance as a receptor for cells of the megakaryocyte lineage. They did this by using MPL antisense oligodeoxynucleotides to show that these inhibited megakaryocyte colony formation in vitro, but had no inhibitory effects on erythroid or granulocyte-macrophage colonies. Thus, the power of the 'receptor first' approach for detecting growth factors was demonstrated. Four groups used cell lines engineered to express the MPL receptor as essential tools for monitoring and cloning the MPL ligand. De Sauvage et al. (1994) purified the ligand from the plasma of irradiated pigs by using MPL affinity columns and, with amino acid sequence data, isolated a porcine genomic clone. A human genomic MPL ligand clone was then isolated by PCR, using porcine-based oligonucleotides, and a cDNA clone isolated from a human fetal liver library. De Sauvage et al. (1994) then went on to show that recombinant MPL ligand increases levels of platelets when injected into animals. Lok et al. (1994) and Kaushansky et al. (1994) adopted a quite different strategy. They based their approach on the observations, made with a number of hemopoietic cell lines, that when autonomous transformation occurs, it is often due to the acquired autocrine production by the cells of the relevant growth factor for a receptor being expressed on the cells. Therefore, they analyzed clones of autonomous mutants of cells expressing the inserted candidate receptor, Mpl, in the hope of detecting one where transformation was based on Mpl ligand production. One such cell line was found and cDNA libraries were prepared from it. Expression screening was undertaken using Mpl-bearing target cells. In this way, they isolated a cDNA for murine MPL ligand. The recombinant protein proved capable of increasing platelet levels in mice to far higher levels than those attainable with known hemopoietic regulators. Analysis in vitro showed that it could also stimulate megakaryocyte formation, particularly in combination with other factors. Wendling et al. (1994) completed the evidence that MPL ligand is indeed thrombopoietin. They showed that all the megakaryocyte colony-stimulating activity and platelet-elevating activity in thrombocytopenic plasma can be removed through binding to recombinant MPL (the receptor), indicating that both biologic properties can be ascribed to the MPL ligand. Metcalf (1994), a pioneer researcher in the field of hematopoietic factors, commented that 'a busy time now lies ahead for biologists and clinicians. . . .it will also be a busy time for the lawyers arguing who has the patent rights to what should prove to be a lucrative and valuable new therapeutic agent.'

Bartley et al. (1994) reported the identification and cloning of a megakaryocyte growth and development factor (MGDF) from canine, murine, and human sources. Human, dog, and mouse cDNAs for MGDF are highly conserved and encode open reading frames for proteins of 353, 352, and 356 amino acids, respectively. Both canine and recombinant human MGDF support the development of megakaryocytes from human CD34(+) progenitor cell populations. MGDF binds to Mpl affinity columns, and its biological effects are inhibited by the soluble extracellular domain of Mpl, suggesting that MGDF is a cytokine that regulates megakaryocyte development and is a ligand for the MPL receptor.

Farese et al. (1996) reported that administration of a pegylated form of recombinant MGDF (PEG-MGDF) significantly induced bone marrow regeneration compared with rMGDF untreated with PEG. When combined with recombinant G-CSF, PEG-MGDF significantly enhanced multilineage hematopoietic recovery with no evidence of lineage competition.

Ratajczak et al. (1997) found that in the presence of erythropoietin (133170) and interleukin-3 (147740), THPO was able to stimulate a small increase in erythroid colony formation in culture. Other studies suggested that THPO has little direct stimulatory effect on erythroid progenitor cells but may indirectly enhance erythropoiesis by preventing very early erythroid progenitor cells from undergoing apoptotic cell death.

Essential thrombocythemia (187950) is a chronic myeloproliferative syndrome due to sustained proliferation of megakaryocytes, which results in elevated numbers of circulating platelets, thrombotic or hemorrhagic episodes, and occasional leukemic transformation. Hereditary thrombocythemia with autosomal dominant transmission has been described with manifestations similar to those of sporadic essential thrombocythemia. As the THPO gene encodes a lineage-restricted growth factor with profound stimulatory effects on megakaryopoiesis and platelet production, Wiestner et al. (1998) tested the hypothesis that mutation in this gene might be responsible for the disorder in a Dutch family with 11 affected individuals. TPO protein concentrations in serum were consistently elevated in affected members of this family. Using an intragenic CA marker for the THPO gene, they demonstrated linkage to the disorder; lod score = 3.5 at theta = 0.0. Affected members were found to carry a splice donor mutation (600044.0001) that led to THPO mRNAs with shortened 5-prime untranslated regions that were more efficiently translated than the normal THPO transcripts.

Sohma et al. (1994) cloned the complete THPO gene by screening a human genomic library using human THPO cDNA as a probe. They found that the gene is 6.2 kb long and contains 6 exons and 5 introns. According to Southern blot analysis, the human genome contains a single copy. By fluorescence in situ hybridization, Sohma et al. (1994) mapped the gene to 3q27.

Using a murine cDNA for the mpl ligand as a probe, Foster et al. (1994) isolated a gene encoding human THPO from a human genomic library. They found that the THPO locus spans over 6 kb and has a structure similar to that of the erythropoietin gene (EPO; 133170). Southern blot analysis of human genomic DNA showed a hybridization pattern consistent with a single gene locus. By fluorescence in situ hybridization, Foster et al. (1994) mapped the THPO gene to 3q26-q27, a site where a number of chromosomal abnormalities associated with thrombocythemia in cases of acute myeloid leukemia have been mapped. A human THPO cDNA isolated by PCR from kidney mRNA was found to encode a protein with 80% identity to murine THPO.

Chang et al. (1995) showed that the MGDF gene consists of 7 exons and 6 introns spanning 8 kilobases. The protein is encoded by exons 3 through 7. By fluorescence in situ hybridization, they mapped the MGDF gene to 3q26.3. In addition to the previously described full-length cDNA, 2 cDNA variants were isolated from human fetal liver. Comparison of these with the genomic sequence indicated that they arise by differential splicing. Chang et al. (1995) mapped the murine Thpo gene to chromosome 16 by linkage analysis in an interspecific backcross.

Bartley, T. D.; Bogenberger, J.; Hunt, P.; Li, Y.-S.; Lu, H. S.; Martin, F.; Chang, M.-S.; Samal, B.; Nichol, J. L.; Swift, S.; Johnson, M. J.; Hsu, R.-Y.; and 41 others: Identification and cloning of a megakaryocyte growth and development factor that is a ligand for the cytokine receptor Mpl. Cell 77: 1117-1124, 1994.

Chang, M.; McNinch, J.; Basu, R.; Shutter, J.; Hsu, R.; Perkins, C.; Mar, V.; Suggs, S.; Welcher, A.; Li, L.; Lu, H.; Bartley, T.; Hunt, P.; Martin, F.; Samal, B.; Bogenberger, J.: Cloning and characterization of the human megakaryocyte

growth and development factor (MGDF) gene. J. Biol. Chem. 270: 511-514, 1995.
Chang, M.-S.; Hsu, R.-Y.; McNinch, J.; Copeland, N. G.; Jenkins, N. A.: The gene for murine megakaryocyte growth and development factor (thrombopoietin, Thpo) is located on mouse chromosome 16. Genomics 26: 636-637, 1995.
de Sauvage, F. J.; Hass, P. E.; Spencer, S. D.; Malloy, B. E.; Gurney, A. L.; Spencer, S. A.; Darbonne, W. C.; Henzel, W. J.; Wong, S. C.; Kuang, W.-J.; Oles, K. J.; Hultgren, B.; Solberg, L. A., Jr.; Goeddel, D. V.; Eaton, D. L.: Stimulation of megakaryocytopoiesis and thrombopoiesis by the c-Mpl ligand. Nature 369: 533-538, 1994.
Farese, A. M.; Hunt, P.; Grab, L. B.; MacVittie, T. J.: Combined administration of recombinant human megakaryocyte growth and development factor and granulocyte colony-stimulating factor enhances multilineage hematopoietic reconstitution in nonhuman primates after radiation-induced marrow aplasia. J. Clin. Invest. 97: 2145-2151, 1996.
Foster, D. C.; Sprecher, C. A.; Grant, F. J.; Kramer, J. M.; Kuijper, J. L.; Holly, R. D.; Whitmore, T. E.; Heipel, M. D.; Bell, L. A.; Ching, A. F. T.; McGrane, V.; Hart, C.; O'Hara, P. J.; Lok, S.: Human thrombopoietin: gene structure, cDNA sequence, expression, and chromosomal localization. Proc. Nat. Acad. Sci. 91: 13023-13027, 1994.
Kaushansky, K.; Lok, S.; Holly, R. D.; Broudy, V. C.; Lin, N.; Bailey, M. C.; Forstrom, J. W.; Buddle, M. M.; Oort, P. J.; Hagen, F. S.; Roth, G. J.; Papayannopoulou, T.; Foster, D. C.: Promotion of megakaryocyte progenitor expansion and differentiation by the c-Mpl ligand thrombopoietin. Nature 369: 568-571, 1994.
Kondo, T.; Okabe, M.; Sanada, M.; Kurosawa, M.; Suzuki, S.; Kobayashi, M.; Hosokawa, M.; Asaka, M.: Familial essential thrombocythemia associated with one-base deletion in the 5-prime-untranslated region of the thrombopoietin gene. Blood 92: 1091-1096, 1998.
Lok, S.; Kaushansky, K.; Holly, R. D.; Kuijper, J. L.; Lofton-Day, C. E.; Oort, P. J.; Grant, F. J.; Heipel, M. D.; Burkhead, S. K.; Kramer, J. M.; Bell, L. A.; Sprecher, C. A.; Blumberg, H.; Johnson, R.; Prunkard, D.; Ching, A. F. T.; Mathewes, S. L.; Bailey, M. C.; Forstrom, J. W.; Buddle, M. M.; Osborn, S. G.; Evans, S. J.; Sheppard, P. O.; Presnell, S. R.; O'Hara, P. J.; Hagen, F. S.; Roth, G. J.; Foster, D. C.: Cloning and expression of murine thrombopoietin cDNA and stimulation of platelet production in vivo. Nature 369: 565-568, 1994.
Metcalf, D.: Thrombopoietin: at last. Nature 369: 519-520, 1994.
Methia, N.; Louache, F.; Vainchenker, W.; Wendling, F.: Oligodeoxynucleotides antisense to the proto-oncogene c-mpl specifically inhibit in vitro megakaryocytopoiesis. Blood 82: 1395-1401, 1993.
Ratajczak, M. Z.; Ratajczak, J.; Marlicz, W.; Pletcher, C. H., Jr.; Machalinski, B.; Moore, J.; Hung, H.; Gewirtz, A. M.: Recombinant human thrombopoietin (TPO) stimulates erythropoiesis by inhibiting erythroid progenitor cell apoptosis. Brit. J. Haemat. 98: 8-17, 1997.
Sohma, Y.; Akahori, H.; Seki, N.; Hori, T.; Ogami, K.; Kato, T.; Shimada, Y.; Kawamura, K.; Miyazaki, H.: Molecular cloning and chromosomal localization of the human thrombopoietin gene. FEBS Lett. 353: 57-61, 1994.
Vigon, I.; Mornon, J.-P.; Cocault, L.; Mitjavila, M.-T.; Tambourin, P.; Gisselbrecht, S.; Souyri, M.: Molecular cloning and characterization of MPL, the human homolog of the v-mpl oncogene: identification of a member of the hematopoietic growth factor receptor superfamily. Proc. Nat. Acad. Sci. 89: 5640-5644, 1992.
Wendling, F.; Maraskovsky, E.; Debili, N.; Florindo, C.; Teepe, M.; Titeux, M.; Methia, N.; Breton-Gorius, J.; Cosman, D.; Vainchenker, W.: c-Mpl ligand is a humoral regulator of megakaryocytopoiesis. Nature 369: 571-574, 1994.
Wiestner, A.; Schlemper, R. J.; van der Maas, A. P. C.; Skoda, R. C.: An activating splice donor mutation in the thrombopoietin gene causes hereditary thrombocythaemia. Nature Genet. 18: 49-52, 1998.

PLATELET DISORDER, UNDEFINED (173420)

JJM NOTE

This family awaits confirmation and further delineation by current methods, including genetic mapping. The family is also mentioned in the entry on neuroblastoma, which rarely runs in families but was present in two members of this kindred.

Dowton et al. (1985) found that at least 22 members of a large kindred had a bleeding tendency resulting from an autosomal dominant disorder of platelet production and function. Phenotypic features included mild-to-moderate thrombocytopenia, prolonged bleeding time, and abnormal platelet aggregation. Platelet survival time was normal. The disorder seemed to be separate from any known thrombocytopenic or thrombocytopathic syndrome. Hematologic neoplasms had developed in 6 family members; 2 had neuroblastoma (256700). The authors listed a considerable number of reported families with an autosomal dominant platelet disorder of ill-defined or undefined nature.

Dowton, S. B.; Beardsley, D.; Jamison, D.; Blattner, S.; Li, F. P.: Studies of a familial platelet disorder. Blood 65: 557-563, 1985.

Granulocyte Disorders

NEUTROPENIA, CYCLIC (162800)

JJM NOTE

This disorder is included because of the unusual control of cell proliferation, not because a girl from an affected family had leukemia (because she probably did not have cyclic neutropenia from birth).

Cyclic neutropenia is a rare blood disease characterized by regular 21-day cyclic fluctuations in the numbers of blood neutrophils, monocytes, eosinophils, lymphocytes, platelets, and reticulocytes. Patients with the disease typically have regularly recurring symptoms of fever, malaise, mucosal ulcers, and, occasionally, life-threatening infections during periods of neutropenia. Most patients with cyclic neutropenia are given the diagnosis in childhood, but its onset later in life in some patients suggests that it may also be an acquired disorder. Hahneman and Alt (1958) described a 29-year-old man who from an early age had neutropenia that recurred every 21 days and was accompanied by infection. Complete remission occurred at age 18 years. The man's daughter was seen at the age of 2 years with similar periodic disease recurring every 14 days. Torrioli-Riggio (1958) also reported cases. Morley et al. (1967) described 20 cases in 5 families. Clinical manifestations usually began in childhood and improved thereafter. The commonest were fever, oral ulcerations and skin infections. Neutropenia occurred at intervals of 15 to 35 days. It was often accompanied by monocytosis and sometimes by anemia, eosinophilia or thrombocytopenia. Male-to-male transmission occurred. Cyclic neutropenia in the collie dog is accompanied by gray fur, leads to early death from pyogenic infections, and is an autosomal recessive (Dale et al., 1970). Weiden et al. (1974) showed by transplantation of grey collie bone marrow into normal dogs which had been irradiated that the basic defect is in the stem cell. There are sufficient similarities between the canine and human diseases (Guerry et al., 1972) to suggest that the same may be true in man. Krance et al. (1982) confirmed this when a patient, in the process of bone marrow transplantation as treatment for acute lymphoblastic leukemia in relapse, acquired cyclic neutropenia from her histocompatible donor sib. Seven persons in 4 sibships of her family had cyclic neutropenia. In dogs, the disease can be transferred and cured through bone marrow transplantation (Jones et al., 1975). The disease in collie dogs differs from the human disease principally in the length of the cycle (12 rather than 21 days). Hammond et al. (1989) found that granulocyte-colony stimulating factor (138970) is effective treatment. Inoue et al. (1992) identified cyclic neutropenia in a 34-year-old woman, her 3 sons, and her mother. Oscillations in the blood neutrophil counts were rather regular, with a periodicity of 21 days. The granulocyte colony-stimulating factor level in her serum was persistently high, with the peak occurring during the neutropenic phase. The patient's serum appeared to contain an inhibitory factor. To control infections, Inoue et al. (1992) administered recombinant hG-CSF for 7 days around the early neutropenic phase.

On the basis of studies in 9 families, no clinical evidence of penetrance or heterogeneity was found by Palmer et al. (1996). However, the pattern of expression suggested anticipation: no families appeared to display milder phenotypes in successive generations, and the most severe cases occurred in children in the youngest generations. The spectrum of severity included death from necrotizing enterocolitis in 4 subjects ranging in age from 6 to 17 years.

See also Dale et al. (1972); Dale et al. (1972); Meuret and Fliedner (1970); Page and Good (1957); Wright et al. (1981)

Dale, D. C.; Alling, D. W.; Wolff, S. M.: Cyclic hematopoiesis: the mechanism of cyclic neutropenia in grey collie dogs. J. Clin. Invest. 51: 2197-2204, 1972.
Dale, D. C.; Kimball, H. R.; Wolff, S. M.: Studies of cyclic neutropenia in gray collie dogs. (Abstract) Clin. Res. 18: 402 only, 1970.
Dale, D. C.; Ward, S. B.; Kimball, H. R.; Wolff, S. M.: Studies of neutrophil production and turnover in grey collie dogs with cyclic neutropenia. J. Clin. Invest. 51: 2190-2196, 1972.
Guerry, D. D.; Dale, D. C.; Omine, M.; Perry, S.; Wolff, S. M.: Studies on the mechanism of human cyclic neutropenia. (Abstract) Brit. J. Haemat. 40: 951 only, 1972.
Hahneman, B. M.; Alt, H. L.: Cyclic neutropenia in a father and daughter. J.A.M.A. 168: 270-272, 1958.
Hammond, W. P., IV; Price, T. H.; Souza, L. M.; Dale, D. C.: Treatment of cyclic neutropenia with granulocyte colony-stimulating factor. New Eng. J. Med. 320: 1306-1311, 1989.
Inoue, T.; Tani, K.; Tajiri, M.; Ishida, Y.; Seguchi, M.; Tanaka, H.; Asano, S.; Kaneko, T.; Matsumoto, N.: A case report of familial cyclic neutropenia. Tohoku J. Exp. Med. 167: 107-113, 1992.

Jones, J. B.; Yang, T. J.; Dale, J. B.; Lange, R. D.: Canine cyclic haematopoiesis: marrow transplantation between littermates. Brit. J. Haemat. 30: 215-223, 1975.

Krance, R. A.; Spruce, W. E.; Forman, S. J.; Rosen, R. B.; Hecht, T.; Hammond, W. P.; Blume, K. G.: Human cyclic neutropenia transferred by allogeneic bone marrow grafting. Blood 60: 1263-1266, 1982.

Meuret, G.; Fliedner, T. M.: Zellkinetik der Granulopoiese und des Neutrophilensystems bei einem Fall von zyklischer Neutropenie. Acta Haemat. 43: 48-63, 1970.

Morley, A. A.; Carew, J. P.; Baikie, A. G.: Familial cyclical neutropenia. Brit. J. Haemat. 13: 719-738, 1967.

Page, A. R.; Good, R. A.: Studies on cyclic neutropenia. A clinical and experimental investigation. Am. J. Dis. Child. 94: 623-661, 1957.

Palmer, S. E.; Stephens, K.; Dale, D. C.: Anticipation in autosomal dominant cyclic hematopoiesis. (Abstract) Am. J. Hum. Genet. 59 (suppl.): A37 only, 1996.

Torrioli-Riggio, G.: Considerazioni su una famiglia di granulopenici. Acta Genet. Med. Gemellol. 7: 237-248, 1958.

Weiden, P. L.; Robinett, B.; Graham, T. C.; Adamson, J.; Storb, R.: Canine cyclic neutropenia. A stem cell defect. J. Clin. Invest. 53: 950-953, 1974.

Wright, D. G.; Dale, D. C.; Fauci, A. S.; Wolff, S. M.: Human cyclic neutropenia: clinical review and long-term follow-up of patients. Medicine 60: 1-13, 1981.

LEUKEMIA, ACUTE, X-LINKED (308960)

JJM NOTE

This single family raises a hypothesis that could be tested with another family or even follow-up of the original family. Or, is this disorder a manifestation of the next entry, since both are X-linked?

Li et al. (1979) described a kindred in which 8 males died of acute leukemia or a potentially preleukemic blood disease. They suggested X-linked inheritance.

Li, F. P.; Marchetto, D. J.; Vawter, G. F.: Acute leukemia and preleukemia in eight males in a family: an X-linked disorder?. Am. J. Hemat. 6: 61-69, 1979.

HEMOPOIETIC PROLIFERATION (306930)

JJM NOTE

A single family awaits replication or follow-up, but the original observation represents a clever hypothesis.

Luzzatto et al. (1979) concluded that an X-chromosomal gene affects growth of hemopoietic cells. The conclusion was based on study of a Nigerian family segregating for a G6PD variant called Ilesha. In heterozygous females one or the other allele was almost exclusively expressed. The data were consistent with random inactivation of one X chromosome followed by selection for one of the two resulting cell types on the basis of an unlinked X-borne gene that affects the rate of proliferation of hemopoietic cells. Deviation from 1:1 ratio of cell types in females heterozygous for X-linked mutations has been observed for HGPRT (where almost all erythroid cells are wildtype). X-chromosome structural aberrations also deviate from the 1:1 ratio; no chromosome abnormality was found in the family of Luzzatto et al. (1979).

Luzzatto, L.; Usanga, E. A.; Bienzle, U.; Esan, G. F. J.; Fusuan, F. A.: Imbalance in X-chromosome expression: evidence for a human X-linked gene affecting growth of hemopoietic cells. Science 205: 1418-1420, 1979.

LEUKEMIA, CHRONIC MYELOID (*151410)

JJM NOTE

I wonder why hereditary (familial) translocation cases of the Philadelphia chromosome are not known or are so rare. The fusion gene has such a specific function in adults, yet the consequences to the embryo must be drastic, to have no congenital discoveries of the chromosome.

Chronic myeloid leukemia is fundamentally a genetic disorder, specifically a somatic cell genetic disorder, which has environmental causes such as irradiation and chemical (e.g., benzene) exposure (Jacobs, 1989). Fitzgerald (1976) described a family in which a man and 2 of his 3 children had a Philadelphia-like chromosome t(11;22)(q25;q13). The fact that they did not have leukemia or other hematologic disorder was thought by Fitzgerald to relate to the finding that the break in chromosome 22 was distal to that in CML. In these cases it was at the q12-q13 interface, whereas in the Philadelphia chromosome it is at the q11-q12 band interface. Thus the 22q12 band may contain critical genetic material concerned with normal (and abnormal) myeloid proliferation. The Philadelphia chromosome of chronic myeloid leukemia (a G group chromosome with part of its long arm missing) was found by Nowell and Hungerford in 1960. It was presumed to be a deleted chromosome 21, the same chromosome as that which is trisomic in Down syndrome (190685). The elevated alkaline phosphatase activity in Down syndrome and depressed activity in CML was viewed as consistent with this interpretation. (It was called the Philadelphia chromosome because it was thought to be useful to follow the practice of hemoglobinologists and name anomalous chromosomes after the city of discovery.) Using the improved definition provided by 'banding' methods, Rowley (1973) showed that in fact there is a translocation of the distal part of chromosome 22 (not 21) onto another chromosome, usually 9q. Court Brown and Doll (1965) followed up more than 14,000 patients irradiated in the treatment of ankylosing spondylitis in British clinics between 1935 and 1954. A high frequency of Philadelphia-chromosome-positive CML was found. This is an example of a chromosomal change (specific translocation) being the common oncogenetic mechanism for various 'causes' which may include viral infection and chemicals in addition to ionizing radiation.

De Klein et al. (1982) demonstrated that the Abelson oncogene (189980) is translocated from chromosome 9 to chromosome 22 in the formation of the Philadelphia chromosome. This indicated that the translocation is reciprocal and suggested a role for the ABL gene in the generation of CML. Not only is the ABL oncogene translocated from 9q to 22 but the SIS oncogene (190040) is presumably translocated from 22 to 9 since it is situated distal to the breakpoint that creates the Philadelphia chromosome (Swan et al., 1982). What is involved in the variant 22q- Philadelphia chromosomes with translocations to other chromosomes? Some 18 different ones were found by Mittelman and Levan (1978). Do all these other chromosomes contribute oncogenes to the 22q- chromosome? Does the translocation of SIS to the recipient chromosome play some role in the usual CML and the variant forms? These were questions raised by Klein (1983), who also asked, Will the microevolutionary process that leads tumor cells towards increased independence from host control, usually referred to as tumor progression, turn out to depend on the sequential activation of multiple oncogenes by genetic rearrangement?

Prakash and Yunis (1984) located the breakpoints in CML to subbands 22q11.21 and 9q34.1. Although the position of the breakpoint in chromosome 9 is quite variable, the breakpoint in chromosome 22 is clustered in an area called bcr for 'breakpoint cluster region.' Shtivelman et al. (1985) referred to bcr as a gene and stated that the ABL oncogene is transferred 'into the bcr gene of chromosome 22.' They found that an 8-kb RNA specific to CML is a fused transcript of the 2 genes. The fused protein is presumably involved in the malignant process. The protein has bcr information at its amino terminus and retains most but not all of the normal abl protein sequences. Since the breakpoint in 9 may be as much as 30 or 40 kb 5-prime to ABL, a large amount must be 'looped out' in the fusion process. The fusion protein has tyrosine kinase activity.

Lillicrap and Sterndale (1984) reported 3 cases of CML in 3 successive generations with a myeloproliferative disorder in a 4th member of the kindred.

About 10% of patients with acute lymphocytic leukemia (ALL; see 159555) have the translocation t(9;22)(q34;q11) indistinguishable from that of CML. Erikson et al. (1986), however, found in 3 of 5 such cases of ALL that the bcr region was not involved and that the 22q11 chromosome breakpoint was proximal (5-prime) to the bcr region. Furthermore, the bcr and c-abl transcripts were of normal size in an ALL line carrying the t(9;22) translocation. The breakpoints of the t(9;22) CML, the t(9;22) of acute lymphocytic leukemia, and the t(8;22) of Burkitt lymphoma fall into 22q11 and are cytologically indistinguishable. By chromosomal in situ hybridization, however, they can be distinguished (Emanuel et al., 1984; Erikson et al., 1986). To this experience, Griffin et al. (1986) added observations on t(11;22), both constitutional and tumor-related (see 133450). In the hybrid gene of CML, the bcr contribution is 5-prime to the abl contribution. In the creation of the gene, splicing can occur across an interval as great as 100 kb. The product of the bcr/abl hybrid gene is a 210-kD protein. It was suggested that the hybrid gene and its product protein be designated PHL. Verhest and Monsieur (1983) found the Philadelphia chromosome in a case of essential thrombocytopenia. Ganesan et al. (1986) found that the BCR gene was rearranged in 7 cases of Philadelphia chromosome-negative CML. In 5 cases hematologic findings were indistinguishable from those of patients with the Philadelphia chromosome. Mes-Masson et al. (1986) isolated overlapping cDNA clones defining the complete coding region for the hybrid protein generated from the ABL and BCR genes. Hariharan and Adams (1987) isolated cDNA clones that collectively spanned the entire BCR coding region. The nucleotide sequence indicated that the BCR polypeptide comprises 1,271 amino acid residues. Stam et al. (1987) showed that the normal gene on chromosome 22 involved at the breakpoint cluster region in the transloc ation between chromosomes 9 and 22 encodes a 160,000-Da phosphoprotein with serine or threonine kinase activity. Designated PHL, this gene is oriented with its 5-prime end toward the centromere of chromosome 22. As a result of the CML translocation, the 3-prime PHL exons are removed and remain on chromosome 22. The PHL and ABL genes are fused in a head-to-tail fashion.

Schaefer-Rego et al. (1987) found that the breakpoints in 8 of 9 patients in blast crisis were in the 3-prime portion of BCR, whereas the breakpoints in 17 patients in the chronic phase were clustered in the 5-prime portion. Croce et al. (1987) demonstrated that there are in fact 4 BCR genes, all located in the 22q11.2 band. By studying mouse-human hybrid cells with breakpoints at various sites in that region, they concluded that the order of loci is centromere—BCR2, BCR4, IGL—BCR1—BCR3—SIS. This linear order was confirmed by Budarf et al. (1988), who also confirmed that all of the BCR-like genes map proximal to the 22q11-q12 breakpoint of a t(11;22) in a Ewing sarcoma. Hermans et al. (1987) showed that a distinct and different fusion of BCR and ABL occurs in creation of the Philadelphia chromosome associated with acute lymphoblastic leukemia. Rubin et al. (1988) similarly reported a distinctive fusion in acute lymphoblastic leukemia. Mills et al. (1988) found a striking correlation between the site of the breakpoint within BCR and the length of time between presentation and onset of acute phase: on average, patients with a 5-prime breakpoint had a 4-fold longer chronic phase than those with a 3-prime breakpoint. The median times were 203 weeks versus 52 weeks. Grossman et al. (1989) also presented evidence consistent with but not proving a relationship between the site of the breakpoint in BCR and the length of the clinical course before onset of blast crisis. Patients with 3-prime breakpoints progressed to acute disease after a shorter period (average 36.6 months) than did patients with 5-prime breakpoints (average 56.1 months), although the difference was not statistically significant.

To determine whether the P210(bcr/abl) hybrid protein can induce leukemia, Daley et al. (1990) infected murine bone marrow with a retrovirus encoding this protein and transplanted the bone marrow into irradiated syngeneic recipients. Transplant recipients developed several hematologic malignancies, prominent among which was a myeloproliferative syndrome closely resembling the chronic phase of human chronic myelogenous leukemia. Tumor tissue from diseased mice harbored the provirus encoding P210(bcr/abl). Heisterkamp et al. (1990) satisfied Koch postulates in relation to leukemia by demonstrating leukemia in mice transgenic for a bcr/abl p190 construct of the type found in acute lymphoblastic leukemia. Tkachuk et al. (1990) used 2-color fluorescence in situ hybridization (FISH) with probes from portions of the BCR and ABL genes to detect the BCR-ABL fusion in individual blood and bone marrow cells from 6 patients. The fusion event was detected in all samples analyzed, of which 3 were cytogenetically Ph(1)-negative. One of the Ph(1)-negative samples was also PCR-negative.

The BCR gene is the site of breakpoints used in the generation of the 2 alternative forms of the Philadelphia chromosome translocation found in chronic myeloid leukemia and acute lymphocytic leukemia (Groffen et al., 1984; Shtivelman et al., 1985; Hermans et al., 1987). These alternative breakpoints join different exon sets of BCR to a common subset of the exons of the ABL gene located on chromosome 9. This fusion results in 2 alternative chimeric oncogene products called p210(BCR-ABL) and p185(BCR-ABL). The activation of ABL tyrosine kinase activity is necessary for the oncogenic potential of the chimeric oncogene. Sequences within the first exon of BCR appear to be essential for this activation and probably work through direct physical binding to the kinase regulatory domain of ABL. The normal cellular BCR gene encodes a 160-kD phosphoprotein associated with a serine/threonine kinase activity. Maru and Witte (1991) demonstrated that the first exon of BCR encodes a novel serine/threonine kinase activity.

Haas et al. (1992) demonstrated a 'parent of origin' bias in cases of Philadelphia-chromosome-positive leukemia by use of unique specific chromosome band polymorphisms. They showed that the translocated chromosome 9 was always of paternal origin, whereas the translocated chromosome 22 was derived exclusively from the maternal copy. Preferential retention of paternal alleles has been found in sporadic tumors such as Wilms tumor (194070), rhabdomyosarcoma (268210), osteosarcoma, and retinoblastoma (180200). The finding of preferential participation of the paternally derived chromosome 9 and the maternally derived chromosome 22 in the Ph-translocation strongly indicates that the chromosomal regions involved are imprinted. Feinberg (1993) reviewed 11 examples of preferential parental allelic alterations in cancer. Fioretos et al. (1994) were unable to find evidence for genomic imprinting of the human BCR gene. They identified a BamHI polymorphism in the coding region of BCR exon 1 and by RT-PCR assay showed that both BCR alleles are expressed in the peripheral blood cells of normal persons. Furthermore, Litz and Copenhaver (1994) used PvuII and MaeII restriction site polymorphisms in the BCR gene to study 3 cases of Philadelphia chromosome-positive CML. In all 3 cases, the rearranged allele was paternal in origin. Melo et al. (1994) could find no evidence of parental imprinting of the ABL gene and cited evidence which appeared to exclude imprinting of the BCR gene and to suggest that there is, in fact, no preferential involvement of the maternal BCR or paternal ABL alleles in the formation of the BCR-ABL fusion gene. Melo et al. (1995) further reviewed the evidence they interpreted as indicating that there is no parental bias in the origin of the translocated ABL gene and no evidence for genomic imprinting of ABL in CML. On the other hand, Haas (1995) argued that it still remains likely that ABL and BCR are imprinted.

Using an interspecific backcross, Justice et al. (1990) mapped the Bcr gene to chromosome 10 of the mouse.

When the Philadelphia-chromosome-positive chronic myeloid leukemia-blast crisis cell line BV173 is injected into SCID mice, a disease process closely resembling that seen in leukemia patients results. For example, BCR-ABL transcripts are detectable in bone marrow, spleen, peripheral blood, liver, and lungs. Skorski et al. (1994) found that systemic treatment of the leukemic mice with a 26-mer BCR-ABL antisense oligodeoxynucleotide induced disappearance of leukemic cells and a marked decrease in BCR-ABL mRNA in mouse tissues. Untreated mice or mice treated with a BCR-ABL sense oligodeoxynucleotide or a 6-base-mismatched antisense oligodeoxynucleotide were dead 8 to 13 weeks after leukemia cell injection; in marked contrast, mice treated with BCR-ABL antisense oligodeoxynucleotide died of leukemia 18 to 23 weeks after injection of leukemic cells. Findings were interpreted as indicating the in vivo effectiveness of an anticancer therapy based on antisense oligodeoxynucleotides targeting a tumor-specific gene.

In summary, the normal BCR gene occupies a region of about 135 kb on chromosome 22. It is expressed as mRNAs of 4.5- and 6.7-kb, which apparently encode for the same cytoplasmic 160-kD protein, and contains 23 exons as well as an unusual inverted repeat flanking the first exon. The BCR protein reportedly contains a unique serine/threonine kinase activity and at least two SH2 binding sites encoded in its first exon and a C-terminal domain that functions as a GTPase activating protein for p21(rac) (Diekmann et al., 1991); see rac serine/threonine protein kinase (179090). Chissoe et al. (1995) sequenced the complete BCR gene and greater than 80% of the human ABL gene, which are both involved in the t(9;22) translocation (Philadelphia chromosome) associated with more than 90% of chronic myelogenous leukemia, 25 to 30% of adult and 2 to 10% of childhood acute lymphoblastic leukemia, and rare cases of acute myelogenous leukemia. Comparison of the gene with its cDNA sequence revealed the positions of 23 BCR exons and putative alternative BCR first and second exons. From the sequence of 4 newly studied Philadelphia chromosome translocations and a review of several other previously sequenced breakpoints, Chissoe et al. (1995) could discern no consistent breakpoint features. No clear-cut mechanism for Philadelphia chromosome translocation was evident.

See also Brunning (1980); Grosveld et al. (1986); Heisterkamp et al. (1985); Koeffler and Golde (1981); Kohno and Sandberg (1980); Nowell and Hungerford (1960); Nowell and Hungerford (1960); Pegoraro et al. (1983); Priest et al. (1980); Shtivelman et al. (1987); Stam et al. (1985); Teyssier et al. (1985); Verma and Dosik (1980)

Brunning, R. D.: Philadelphia chromosome positive leukemia. Hum. Path. 11: 307-309, 1980.

Budarf, M.; Canaani, E.; Emanuel, B. S.: Linear order of the four BCR-related loci in 22q11. Genomics 3: 168-171, 1988.

Chissoe, S. L.; Bodenteich, A.; Wang, Y.-F.; Wang, Y.-P.; Burian, D.; Clifton, S. W.; Crabtree, J.; Freeman, A.; Iyer, K.; Jian, L.; Ma, Y.; McLaury, H.-J.; Pan, H.-Q.; Sarhan, O. H.; Toth, S.; Wang, Z.; Zhang, G.; Heisterkamp, N.; Groffen, J.; Roe, B. A.: Sequence and analysis of the human ABL gene, the BCR gene, and regions involved in the Philadelphia chromosomal translocation. Genomics 27: 67-82, 1995.

Court Brown, W. M.; Doll, R.: Mortality from cancer and other causes after radiotherapy for ankylosing spondylitis. Brit. Med. J. 2: 1327-1332, 1965.

Croce, C. M.; Huebner, K.; Isobe, M.; Fainstain, E.; Lifshitz, B.; Shtivelman, E.; Canaani, E.: Mapping of four distinct BCR-related loci to chromosome region 22q11: order of BCR loci relative to chronic myelogenous leukemia and acute lymphoblastic leukemia breakpoints. Proc. Nat. Acad. Sci. 84: 7174-7178, 1987.

Daley, G. Q.; Van Etten, R. A.; Baltimore, D.: Induction of chronic myelogenous leukemia in mice by the P210(bcr/abl) gene of the Philadelphia chromosome. Science 247: 824-830, 1990.

de Klein, A.; Geurts van Kessel, A.; Grosveld, G.; Bartram, C. R.; Hagemeijer, A.; Bootsma, D.; Spurr, N. K.; Heisterkamp, N.; Groffen, J.; Stephenson, J. R.: A cellular oncogene is translocated to the Philadelphia chromosome in chronic myelocytic leukaemia. Nature 300: 765-767, 1982.

Diekmann, D.; Brill, S.; Garrett, M. D.; Totty, N.; Hsuan, J.; Monfries, C.; Hall, C.; Lim, L.; Hall, A.: Bcr encodes a GTPase-activating protein for p21(rac). Nature 351: 400-402, 1991.
Emanuel, B. S.; Selden, J. R.; Wang, E.; Nowell, P. C.; Croce, C. M.: In situ hybridization and translocation breakpoint mapping. I. Non-identical 22q11 breakpoints for the t(9;22) of CML and the t(8;22) of Burkitt lymphoma. Cytogenet. Cell Genet. 38: 127-131, 1984.
Erikson, J.; Griffin, C.; ar-Rushdi, A.; Valtieri, M.; Hoxie, J.; Finan, J.; Emanuel, B. S.; Rovera, G.; Nowell, P. C.; Croce, C. M.: Heterogeneity of chromosome 22 breakpoint in Philadelphia-positive (Ph+) acute lymphocytic leukemia. Proc. Nat. Acad. Sci. 83: 1807-1811, 1986.
Feinberg, A. P.: Genomic imprinting and gene activation in cancer. Nature Genet. 4: 110-113, 1993.
Fioretos, T.; Heisterkamp, N.; Groffen, J.: No evidence for genomic imprinting of the human BCR gene. Blood 83: 3441-3444, 1994.
Fitzgerald, P. H.: Evidence that chromosome band 22q12 is concerned with cell proliferation in chronic myeloid leukemia. Hum. Genet. 33: 269-274, 1976.
Ganesan, T. S.; Rassool, F.; Guo, A.-P.; Th'ng, K. H.; Dowding, C.; Hibbin, J. A.; Young, B. D.; White, H.; Kumaran, T. O.; Galton, D. A. G.; Goldman, J. M.: Rearrangement of the bcr gene in Philadelphia chromosome-negative chronic myeloid leukemia. Blood 68: 957-960, 1986.
Griffin, C. A.; McKeon, C.; Israel, M. A.; Gegonne, A.; Ghysdael, J.; Stehelin, D.; Douglass, E. C.; Green, A. A.; Emanuel, B. S.: Comparison of constitutional and tumor-associated 11;22 translocations: nonidentical breakpoints on chromosomes 11 and 22. Proc. Nat. Acad. Sci. 83: 6122-6126, 1986.
Groffen, J.; Stephenson, J. R.; Heisterkamp, N.; de Klein, A.; Bartram, C. R.; Grosveld, G.: Philadelphia chromosomal breakpoints are clustered within a limited region, bcr, on chromosome 22. Cell 36: 93-99, 1984.
Grossman, A.; Silver, R. T.; Arlin, Z.; Coleman, M.; Camposano, E.; Gascon, P.; Benn, P. A.: Fine mapping of chromosome 22 breakpoints within the breakpoint cluster region (bcr) implies a role for bcr exon 3 in determining disease duration in chronic myeloid leukemia. Am. J. Hum. Genet. 45: 729-738, 1989.
Grosveld, G.; Verwoerd, T.; van Agthoven, T.; de Klein, A.; Ramachandran, K. L.; Heisterkamp, N.; Stam, K.; Groffen, J.: The chronic myelocytic cell line K562 contains a breakpoint in bcr and produces a chimeric bcr/c-abl transcript. Molec. Cell. Biol. 6: 607-616, 1986.
Haas, O. A.: Are ABL and BCR imprinted? No definitive answers, but more questions. Leukemia 9: 740-745, 1995.
Haas, O. A.; Argyriou-Tirita, A.; Lion, T.: Parental origin of chromosomes involved in the translocation t(9;22). Nature 359: 414-416, 1992.
Hariharan, I. K.; Adams, J. M.: cDNA sequence for human bcr, the gene that translocates to the abl oncogene in chronic myeloid leukaemia. EMBO J. 6: 115-119, 1987.
Heisterkamp, N.; Jenster, G.; ten Hoeve, J.; Zovich, D.; Pattengale, P. K.; Groffen, J.: Acute leukaemia in bcr/abl transgenic mice. Nature 344: 251-253, 1990.
Heisterkamp, N.; Stam, K.; Groffen, J.; de Klein, A.; Grosveld, G.: Structural organization of the bcr gene and its role in the Ph-1 translocation. Nature 315: 758-761, 1985.
Hermans, A.; Heisterkamp, N.; von Lindern, M.; van Baal, S.; Meijer, D.; van der Plas, D.; Wiedemann, L. M.; Groffen, J.; Bootsma, D.; Grosveld, G.: Unique fusion of bcr and c-abl genes in Philadelphia chromosome positive acute lymphoblastic leukemia. Cell 51: 33-40, 1987.
Jacobs, A.: Benzene and leukemia. Brit. J. Haemat. 72: 119-121, 1989.
Justice, M. J.; Siracusa, L. D.; Gilbert, D. J.; Heisterkamp, N.; Groffen, J.; Chada, K.; Silan, C. M.; Copeland, N. G.; Jenkins, N. A.: A genetic linkage map of mouse chromosome 10: localization of eighteen molecular markers using a single interspecific backcross. Genetics 125: 855-866, 1990.
Klein, G.: Specific chromosomal translocations and the genesis of the B-cell-derived tumors in mice and men. Cell 32: 311-315, 1983.
Koeffler, H. P.; Golde, D. W.: Chronic myelogenous leukemia—new concepts. New Eng. J. Med. 304: 1201-1209 and 1269-1274, 1981.
Kohno, S.-I.; Sandberg, A. A.: Chromosomes and causation of human cancer and leukemia. XXXIX. Usual and unusual findings in Ph(1)-positive CML. Cancer 46: 2227-2237, 1980.
Lillicrap, D. A.; Sterndale, H.: Familial chronic myeloid leukaemia. (Letter) Lancet II: 699, 1984.
Litz, C. E.; Copenhaver, C. M.: Paternal origin of the rearranged major breakpoint cluster region in chronic myeloid leukemia. Blood 83: 3445-3448, 1994.
Maru, Y.; Witte, O. N.: The BCR gene encodes a novel serine/threonine kinase activity within a single exon. Cell 67: 459-468, 1991.
Melo, J. V.; Yan, X.-H.; Diamond, J.; Goldman, J. M.: Lack of imprinting of the ABL gene. (Letter) Nature Genet. 8: 318-319, 1994.
Melo, J. V.; Yan, X.-H.; Diamond, J.; Goldman, J. M.: Balanced parental contribution to the ABL component of the BCR-ABL gene in chronic myeloid leukemia. Leukemia 9: 734-745, 1995.
Mes-Masson, A.-M.; McLaughlin, J.; Daley, G. Q.; Paskind, M.; Witte, O. N.: Overlapping cDNA clones define the complete coding region for the P210(c-abl) gene product associated with chronic myelogeneous leukemia cells containing the Philadelphia chromosome. Proc. Nat. Acad. Sci. 83: 9768-9772, 1986.
Mills, K. I.; MacKenzie, E. D.; Birnie, G. D.: The site of the breakpoint within the bcr is a prognostic factor in Philadelphia-positive CML patients. Blood 72: 1237-1241, 1988.
Mittelman, F.; Levan, G.: Clustering of aberrations to specific chromosomes in human neoplasms. III. Incidence and geographic distribution of chromosome aberrations in 856 cases. Hereditas 89: 207-232, 1978.
Nowell, P. C.; Hungerford, D. A.: A minute chromosome in human chronic granulocytic leukemia. Science 132: 1497, 1960.
Nowell, P. C.; Hungerford, D. A.: Chromosome studies on normal and leukemic human leukocytes. J. Nat. Cancer Inst. 25: 85-109, 1960.
Pegoraro, L.; Matera, L.; Ritz, J.; Levis, A.; Palumbo, A.; Biagini, G.: Establishment of a Ph(1)-positive human cell line (BV173). J. Nat. Cancer Inst. 70: 447-451, 1983.
Prakash, O.; Yunis, J. J.: High resolution chromosomes of the t(9;22) positive leukemias. Cancer Genet. Cytogenet. 11: 361-367, 1984.
Priest, J. R.; Robison, L. L.; McKenna, R. W.; Lindquist, L. L.; Warkentin, P. I.; LeBien, T. W.; Woods, W. G.; Kersey, J. H.; Coccia, P. F.; Nesbit, M. E., Jr.: Philadelphia chromosome positive childhood acute lymphoblastic leukemia. Blood 56: 15-22, 1980.
Rowley, J. D.: A new consistent chromosomal abnormality in chronic myelogenous leukemia identified by quinacrine fluorescence and Giemsa staining. Nature 243: 290-293, 1973.
Rubin, C. M.; Carrino, J. J.; Dickler, M. N.; Leibowitz, D.; Smith, S. D.; Westbrook, C. A.: Heterogeneity of genomic fusion of BCR and ABL in Philadelphia chromosome-positive acute lymphoblastic leukemia. Proc. Nat. Acad. Sci. 85: 2795-2799, 1988.
Schaefer-Rego, K.; Dudek, H.; Popenoe, D.; Arlin, Z.; Mears, J. G.; Bank, A.; Leibowitz, D.: CML patients in blast crisis have breakpoints localized to a specific region of the BCR. Blood 70: 448-455, 1987.
Shtivelman, E.; Gale, R. P.; Dreazen, O.; Berrebi, A.; Zaizov, R.; Kubonishi, I.; Miyoshi, I.; Canaani, E.: bcr-abl RNA in patients with chronic myelogenous leukemia. Blood 69: 971-973, 1987.
Shtivelman, E.; Lifshitz, B.; Gale, R. P.; Canaani, E.: Fused transcript of abl and bcr genes in chronic myelogenous leukaemia. Nature 315: 550-554, 1985.
Skorski, T.; Nieborowska-Skorska, M.; Nicolaides, N. C.; Szczylik, C.; Iversen, P.; Iozzo, R. V.; Zon, G.; Calabretta, B.: Suppression of Philadelphia-1 leukemia cell growth in mice by BCR-ABL antisense oligodeoxynucleotide. Proc. Nat. Acad. Sci. 91: 4504-4508, 1994.
Stam, K.; Heisterkamp, N.; Grosveld, G.; de Klein, A.; Verma, R. S.; Coleman, M.; Dosik, H.; Groffen, J.: Evidence of a new chimeric bcr/c-abl mRNA in patients with chronic myelocytic leukemia and the Philadelphia chromosome. New Eng. J. Med. 313: 1429-1433, 1985.
Stam, K.; Heisterkamp, N.; Reynolds, F. H., Jr.; Groffen, J.: Evidence that the phl gene encodes a 160,000-dalton phosphoprotein with associated kinase activity. Molec. Cell. Biol. 7: 1955-1960, 1987.
Swan, D. C.; McBride, O. W.; Robbins, K. C.; Keithley, D. A.; Reddy, E. P.; Aaronson, S. A.: Chromosomal mapping of the simian sarcoma virus onc gene analogue in human cells. Proc. Nat. Acad. Sci. 79: 4691-4695, 1982.
Teyssier, J. R.; Bartram, C. R.; Deville, J.; Potron, G.; Pigeon, F.: C-abl oncogene and chromosome 22 'bcr' juxtaposition in chronic myelogenous leukemia. New Eng. J. Med. 312: 1393-1394, 1985.
Tkachuk, D. C.; Westbrook, C. A.; Andreeff, M.; Donlon, T. A.; Cleary, M. L.; Suryanarayan, K.; Homge, M.; Redner, A.; Gray, J.; Pinkel, D.: Detection of bcr-abl fusion in chronic myelogeneous leukemia by in situ hybridization. Science 250: 559-562, 1990.
Verhest, A.; Monsieur, R.: Philadelphia chromosome-positive thrombocythemia with leukemic transformation. (Letter) New Eng. J. Med. 308: 1603, 1983.
Verma, R. S.; Dosik, H.: Heteromorphisms of the Philadelphia (Ph-1) chromosome in patients with chronic myelogenous leukaemia (CML). I. Classification and clinical significance. Brit. J. Haemat. 45: 215-222, 1980.

PLATELET DISORDER, FAMILIAL, WITH ASSOCIATED MYELOID MALIGNANCY (*601399)

JJM NOTE

Could this gene relate to the risk of leukemia in Down syndrome?

Using a large pedigree originally described by Dowton et al. (1985) with an autosomal dominant familial 'aspirin-like' platelet disorder associated with development of acute myelogenous leukemia, Ho et al. (1996) found linkage of the platelet disorder to markers D21S263, D21S216, IFNAR (107450), and D21S65, with lod scores of 6.19, 4.89, 9.40, and 10.48, respectively (theta = 0.001). They proposed a familial platelet disorder critical region of 15.2 cM in addition to approximate placement for markers D21S211 and UT7582/D21S1413. This critical region contains 5 candidate genes, including AML1 (151385), IFNAR, CRF2-4 (123889), GART (138440), and SON (182465). See also 176630.

In a family of mixed Czechoslovakian and Hungarian background, Arepally et al. (1998) reported a new kindred with a similar phenotype which the authors mapped to the same region of chromosome 21. The proband presented at the age of 4 years for hypospadias repair. He had easy bruising since infancy and was found preoperatively to have a platelet count of 106,000/microliter. The mother, maternal aunt, and maternal grandmother had low platelet counts and a bleeding diathesis. In addition to thrombocytopenia, both the index patient and his brother had hypospadias and umbilical hernia. At 41 years of age, the proband's mother developed progressive anemia and neutropenia and was shown to have refractory anemia. Within 6 months of diagnosis of myelodysplasia, increased marrow blast counts above 30% were noted. Cytogenetics yielded normal results, as did fluorescence in situ hybridization studies for monosomy 5 and 7. Immunohistochemical typing of her blasts confirmed acute myelogenous leukemia. Genotyping and 2-point lod scores at 10 previously mapped chromosome 21 polymorphic marker loci showed evidence for linkage to chromosome 21; maximum lod score at D21S65 = 1.682. Multipoint analyses placed the putative disease gene locus between D21S263 and D21S211. Maximum lod scores greater than 3 were obtained at each map interval. No evidence for allele sharing between this kindred and the French-Canadian pedigree reported by Ho et al. (1996) was found.

Arepally, G.; Rebbeck, T. R.; Song, W.; Gilliland, G.; Maris, J. M.; Poncz, M.: Evidence for genetic homogeneity in a familial platelet disorder with predisposition to acute myelogenous leukemia (FPD/AML). (Letter) Blood 92: 2600-2602, 1998.
Dowton, S. B.; Beardsley, D.; Jamison, D.; Blattner, S.; Li, F. P.: Studies of a familial platelet disorder. Blood 65: 557-563, 1985.
Ho, C. Y.; Otterud, B.; Legare, R. D.; Varvil, T.; Saxena, R.; DeHart, D. B.; Kohler, S. E.; Aster, J. C.; Dowton, S. B.; Li, F. P.; Leppert, M.; Gilliland, D. G.: Linkage of a familial platelet disorder with a propensity to develop myeloid malignancies to human chromosome 21q22.1-22.2. Blood 87: 5218-5224, 1996.

MYELOCYTIC LEUKEMIA-LIKE SYNDROME (600080)

JJM NOTE

Could further molecular analysis on the family reported by Lardi (1994) be informative?

Lardi et al. (1994) described 3 Saudi sibs, 2 girls and a boy, with clinical features of chronic myelocytic leukemia in early infancy. Their parents were first cousins. Peripheral blood counts of the parents and available relatives were normal and there was no family history of leukemia. A familial form of CML that closely resembles the juvenile type, which lacks the Philadelphia chromosome, had previously been reported in at least 3 family sets (Anderson, 1951; Holton and Johnson, 1968). The Philadelphia chromosome was absent in all 3 children reported by Lardi et al. (1994). The short survival of the first 2 children (18 and 2 months, respectively) favored a diagnosis of leukemia rather than leukemoid reaction. The possibility of viral or other infection being responsible for familial clustering was mentioned.

Anderson, R. C.: Familial leukemia. A report of leukemia in five siblings, with a brief review of the genetic aspects of this disease. Am. J. Dis. Child 81: 313-322, 1951.
Holton, C. P.; Johnson, W. W.: Chronic myelocytic leukemia in infant siblings. J. Pediat. 72: 377-383, 1968.
Lardi, A. A.; Taha, O. M. A.; Al-Jefri, A.; Ahmed, M. A. M.: Familial chronic myelocytic leukaemia-like syndrome probably of congenital origin. Acta. Paediat. 83: 558-560, 1994.

FAMILIAL MIXOPLOIDY MITOTIC MUTANTS, INCLUDED (158250)

JJM NOTE

This entity gives details of the only familial aggregation of Philadelphia-positive chronic myelogenous leukemia. It is worthy of follow-up with current techniques.

Chromosomal mosaicism based on a mendelizing tendency to nondisjunction (257300) is discussed elsewhere. A curious example of familial mosaicism is that for the Philadelphia chromosome, as reported by Hirschhorn (1968) and Weiner (1965). The proband, a 65-year-old man, had classic chronic myeloid leukemia (CML) with Ph(1) chromosome in all bone marrow and peripheral blood cells examined without phytohemagglutinin (PHA). His father died of leukemia of undetermined type and a brother and sister died of well-documented CML without chromosome studies. The sister's son died of thymoma at age 32. The proband's daughter, aged 29, and 2 of his sons, aged 7 and 5, showed Ph(1) chromosome in some cells of the peripheral blood cultured without PHA. Lymphocytes stimulated with PHA showed normal karyotypes. De Bault and Halmi (1975) described mosaicism for trisomy 7 in mother and daughter. Zellweger and Abbo (1965) described mosaicism in 3 successive generations. Clones with normal karyotype, D-G and D-D translocations, partial trisomy 21 and monosomy X were found. Because of an unusual frequency of satellite association, they postulated the existence of a 'dominant gene' that leads to mosaicism of somatic origin. The family was described in full by Abbo et al. (1966), who concluded that inheritance of a postulated gene for increased satellite association from both parents might lead to mosaicism. Holliday et al. (1988) described a family in which females in 3 generations showed sex-chromosomal mosaicism. Phenotypic effect, if any, appeared to be limited to reduced fertility. Juberg et al. (1990) reported sex chromosome mosaicism in females of 3 generations. The phenotypic effect, if any, appeared to be limited to reduced fertility.

See also Hecht (1981); Zellweger et al. (1966)

Abbo, G. N.; Zellweger, H.; Cuany, R.: Satellite association (SA) in familial mosaicism. Helv. Paediat. Acta 21: 293-299, 1966.
De Bault, L. E.; Halmi, K. A.: Familial trisomy 7 mosaicism. J. Med. Genet. 12: 200-203, 1975.
Hecht, F.: Familial sex chromosome mosaicism (yes) and interchromosomal effects (no). (Letter) Clin. Genet. 19: 77-78, 1981.
Hirschhorn, K.: Cytogenetic alterations in leukemia.In: Dameshek, W.; Dutcher, R. M.: Perspectives in Leukemia. New York: Grune and Stratton (pub.) 1968. Pp. 113-122.
Holliday, D. J.; Juberg, R. C.; Hennessy, V. S.: Familial sex chromosomal mosaicism. (Abstract) Am. J. Hum. Genet. 43: A235 only, 1988.
Juberg, R. C.; Holliday, D. J.; Hennessy, V. S.: Familial sex chromosomal mosaicism. Am. J. Med. Genet. 37: 15-17, 1990.
Weiner, L.: A family with high incidence of leukemia and unique Ph(1) chromosome findings. (Abstract) 8th Annual Meeting, Am. Soc. Hemat., Philadelphia , 1965.
Zellweger, H.; Abbo, G. N.: About a new gene as a cause of increased satellite association. (Abstract) J. Pediat. 67: 935 only, 1965.
Zellweger, H.; Abbo, G. N.; Cuany, R.: Satellite association and translocation mongolism. J. Med. Genet. 3: 186-189, 1966.

LEUKEMIA, ACUTE MONOCYTIC (151380)

JJM NOTE

At least this trait is represented by two families, but think how much more could be learned with follow-up and current tools!

Kjellstrom et al. (1979) reported the cases of acute monocytic leukemia in 2 brothers in 1 family and in 3 persons in 3 generations of another family (grandfather, grandson, and paternal aunt of the latter).

Kjellstrom, T.; Barkenius, G.; Malmquist, J.; Rausing, A.: Familial leukaemia: a report of two families. Scand. J. Haemat. 23: 272-276, 1979.

LEUKEMIA, FAMILIAL (*601626)

JJM NOTE

This reference is, at least, a great literature review. Of all the disorders which ascertainment could lead to an artifactual anticipation, acute leukemia would be a likely one, since it has been so rapidly lethal until recent years. Perhaps this phenomenon is best sought by a consortium of population-based cancer registries.

Horwitz et al. (1996) reported evidence of anticipation in familial acute myelogenous leukemia (AML). Horwitz et al. (1996) further studied those pedigrees and others from the literature. In 49 affected individuals from 9 families transmitting autosomal dominant AML, the mean age of onset was 57 years in the grandparental generation, 32 years in the parental generation, and 13 years in the youngest generation (P less than 0.001). Horwitz et al. (1996) also reported evidence of anticipation in autosomal dominant chronic lymphocytic leukemia (CLL; 151400) (P = 0.008). In 18 affected individuals from 7 pedigrees with autosomal dominant CLL, the mean age of onset in the parental generation was 66 years, versus 51 years in the younger generation. Based on this evidence of anticipation, Horwitz et al. (1996) suggested that dynamic mutations of unstable DNA sequence repeats could be a common mechanism of inherited hematopoietic malignancy. They proposed 3 possible candidate chromosomal regions for familial leukemia with anticipation: 21q22.1-22.2, 11q23.3 in the vicinity of the CBL2 gene (165360), and 16q22 in the vicinity of the CBFB gene (121360).

Horwitz, M.; Goode, E. L.; Jarvik, G. P.: Anticipation in familial leukemia. Am. J. Hum. Genet. 59: 990-998, 1996.
Horwitz, M.; Sabath, D. E.; Smithson, W. A.; Raddich, J.: A family inheriting different subtypes of acute myelogenous leukemia. Am. J. Hemat. 52: 295-304, 1996.

MYELOPROLIFERATIVE DISEASE (254700)

JJM NOTE

A single family can define a syndrome, but additional information is needed, either from another family or from long-term follow-up of the original family.

Randall et al. (1965) observed a severe myeloproliferative disorder with features resembling chronic or subacute myeloid leukemia in 9 children related as first or second cousins. Two children recovered completely after a chronic illness of 10 to 12 years. No consistent chromosomal aberration was found. Low leukocyte alkaline phosphatase was found in all affected children and in 18 of 20 asymptomatic relatives.

Randall, D. L.; Reiquam, C. W.; Githens, J. H.; Robinson, A.: Familial myeloproliferative disease. A new syndrome closely simulating myelogenous leukemia in childhood. Am. J. Dis. Child. 110: 479-500, 1965.

EOSINOPHILIA, FAMILIAL (*131400)

JJM NOTE

My colleagues and I reported a 67-year-old farmer who presented with large-cell carcinoma of the lung and died within two months of a fulminant hypereosinophilia. In retrospect, the eosinophilia was called a leukemia (sometimes a difficult diagnosis) because of a clone of abnormal cells with a deletion of the long arm of chromosome 15. At autopsy a cerebral meningioma, prostatic carcinoma, and thyroid adenoma were discovered. Various relatives had lymphomas and two had prostatic cancer (Goffman TE et al.: Fatal hypereosinophilia with chromosome 15q– in a patient with multiple primary and familial neoplasms. Cancer Genet. Cytogenet. 8:197–202, 1983).

Naiman et al. (1964) observed eosinophilia in 3 generations of a family. No allergies were recorded. Zeni et al. (1964) observed eosinophilia in 21 members of 3 generations of a kindred. Sparrevohn (1967) described an 18-month-old girl with recurrent asthmatic bronchitis, recurrent pulmonary infiltrates, leukocytosis, persistent marked eosinophilia with 'shift to the left,' intermittent thrombocytopenia, eosinophilia of liver and bone marrow, cellular infiltration including mast cells and eosinophils in skin and muscle, no signs of allergy by usual skin tests or of parasitism, and a chronic but benign course. The mother and a brother had transient eosinophilia and similar changes in skin and muscle biopsy. Zuelzer and Apt (1949) described the same syndrome in young children. One of their patients had a sister with marked eosinophilia.

In a 5-generation kindred with familial eosinophilia, Lin et al. (1997) identified 19 affected persons. Four subjects with sustained eosinophilia had documented cardiac involvement and 2 of them also had neurologic symptoms. Affected individuals had significantly higher white cell and absolute eosinophil counts, as well as lower red cell counts. Serum cytokine levels of interleukin-5 (IL5; 147850), interleukin-3 (IL3; 147740), and granulocyte-macrophage colony-stimulating factor (CSF2; 138960) and serology for parasitic helminth infection revealed no differences between the affected and unaffected individuals. No individuals studied had positive serology for parasitic infection. Using polymorphic markers for a genomewide scan, Lin et al. (1997) and Rioux et al. (1998) demonstrated linkage to 5q. Multipoint analysis yielded a maximum lod score of 7.8 in the 5q31-q33 region (Lin et al., 1997). In the same family, Lin et al. (1998) found a pericentric inversion of chromosome 10, inv(10)(p11.2q21.2), in 2 of 8 affected persons who had peripheral blood or bone marrow karyotype analysis.

The region to which Rioux et al. (1998) mapped the EOS gene contains a cytokine gene cluster, which includes 3 genes whose products play important roles in the development and proliferation of eosinophils: IL3, IL5, and CSF2. Rioux et al. (1998) screened these 3 cytokine genes for potential disease-specific mutations by resequencing a subgroup of affected individuals from the large kindred. No functional sequence polymorphisms were found within the promoter, the exons, or the introns of any of these genes or within the IL3/CSF2 enhancer, suggesting that the primary defect in familial eosinophilia is not mutation in any one of these genes but, rather, in another gene in the area.

See also Stewart (1933)

Lin, A. Y.; Nutman, T. B.; Kaslow, D.; Mulvihill, J. J.; Fontaine, L.; White, B. J.; Knutsen, T.; Theil, K. S.; Raghuprasad, P. K.; Goldstein, A. M.; Tucker, M. A.: Familial eosinophilia: clinical and laboratory results on a U.S. kindred. Am. J. Med. Genet. 76: 229-237, 1998.
Lin, A. Y.; Rioux, J. D.; Nutman, T.; Daly, M.; Stone, V.; Nguyen, H.; Green, T.; Hudson, T.; Zimmerman, P. A.; Fontaine, L.; Tucker, M. A.; Lander, E.; Goldstein, A. M.: A gene for familial eosinophilia maps to chromosome 5q31-q33. (Abstract) Am. J. Hum. Genet. 61 (suppl.): A283 only, 1997.
Naiman, J. L.; Oski, F. A.; Allen, F. H.; Diamond, L. K.: Hereditary eosinophilia. Report of a family and review of the literature. Am. J. Hum. Genet. 16: 195-203, 1964.
Rioux, J. D.; Stone, V. A.; Daly, M. J.; Cargill, M.; Green, T.; Nguyen, H.; Nutman, T.; Zimmerman, P. A.; Tucker, M. A.; Hudson, T.; Goldstein, A. M.; Lander, E.; Lin, A. Y.: Familial eosinophilia maps to the cytokine gene cluster on human chromosomal region 5q31-q33. Am. J. Hum. Genet. 63: 1086-1094, 1998.
Sparrevohn, S.: Disseminated eosinophilic collagenosis and familial eosinophilia. Acta Paediat. Scand. 56: 307-312, 1967.
Stewart, S. G.: Familial eosinophilia. Am. J. Med. Sci. 185: 21-29, 1933.
Zeni, G.; Nardi, F.; Frezza, M.: In tema di ipereosinofilia constituzionale familiare idiopatica. Acta Med. Patav. 24: 589-602, 1964.
Zuelzer, W. W.; Apt, L.: Disseminated visceral lesions associated with extreme eosinophilia: pathological and clinical observations on syndrome of young children. Am. J. Dis. Child. 78: 153-181, 1949.

Lymphocyte Proliferations

LEUKEMIA, ACUTE T-CELL (*151390)

JJM NOTE

The family reported by Sarin et al. (1983) emphasizes the point that familial neoplasia of any type could be due to an environmental determinant.

TCL2 is the designation for the gene located in 11p13 that is activated by translocation of part of the T-cell receptor alpha-chain locus (TCRA; 186880) to its vicinity in cases of acute T-cell leukemia (Showe and Croce, 1986). Direct involvement of the TCRA locus was demonstrated in cases of T-cell acute lymphocytic leukemia (T-ALL) with t(11;14)(p13q11) by Erikson et al. (1985) and Lewis et al. (1985). Finger et al. (1986) found that in T-cell leukemias carrying a t(8;14)(q24;q11) translocation, the TCRA gene cluster was implicated with translocation of a constant gene locus to a region 3-prime to the MYC oncogene (190080). Thus, in this T-cell neoplasia, a mechanism operates comparable to that in B-cell neoplasms such as Burkitt lymphoma. Le Beau et al. (1986) did studies by in situ chromosomal hybridization and Southern blot analysis in leukemic cells in a child with acute T-cell lymphoblastic leukemia and a translocation involving chromosome 14 (band q11) and chromosome 11 (band p15). They found that the break at 14q11 occurred within the variable region sequences of the TCRA gene whereas the break at 11p15 occurred between the HRAS1 gene (190020) and the genes for insulin and IGF2. (Although to our knowledge not the determinant of an inherited autosomal dominant phenotype, TCL2 is a specific DNA coding segment that is implicated in a specific form of neoplasia through somatic cell mutation.)

(A type-C retrovirus known as human T-cell leukemia (lymphoma) virus (HTLV) was first isolated in the United States in cases of adult T-cell lymphoma and leukemia (6,7:Poiesz et al., 1980, 1981; Gallo et al., 1982). Subsequently it was found associated with T-cell leukemia in patients in Japan and the West Indies by detection of HTLV-specific antibodies in the serum. Sarin et al. (1983) found a Japanese family in which one member, a 21-year-old college student, had ATL and his mother had morphologically abnormal lymphocytes with convoluted nuclei typically found in T-cell leukemia or lymphoma patients. Other family members, with the exception of the patient's sister, either had HTLV-related serum antibodies or expressed HTLV-related antigens (or both) in cultured T cells and expressed type-C virus particles. ATL is relatively frequent in natives of Kyushu and Shikoku in southwestern Japan. The family described by Sarin et al. (1983) was from Honshu in northwestern Japan and had no family ties to the 2 endemic areas. No consistent cytogenetic abnormality was found.)

Erikson, J.; Williams, D. L.; Finan, J.; Nowell, P. C.; Croce, C. M.: Locus of the alpha-chain of the T-cell receptor is split by chromosome translocation in T-cell leukemias. Science 229: 784-786, 1985.
Finger, L. R.; Harvey, R. C.; Moore, R. C. A.; Showe, L. C.; Croce, C. M.: A common mechanism of chromosomal translocation in T- and B-cell neoplasia. Science 234: 982-985, 1986.
Gallo, R. C.; Mann, D.; Broder, S.; Ruscetti, F. W.; Maeda, M.; Kalyanaraman, V. S.; Robert-Guroff, M.; Reitz, M. S.: Human T-cell leukemia-lymphoma virus (HTLV) is in T but not B lymphocytes from a patient with cutaneous T-cell lymphoma. Proc. Nat. Acad. Sci. 79: 5680-5683, 1982.
Le Beau, M. M.; McKeithan, T. W.; Shima, E. A.; Goldman-Leikin, R. E.; Chan, S. J.; Bell, G. I.; Rowley, J. D.; Diaz, M. O.: T-cell receptor alpha-chain gene is split in a human T-cell leukemia cell line with a t(11;14)(p15;q11). Proc. Nat. Acad. Sci. 83: 9744-9748, 1986.
Lewis, W. H.; Michalopoulos, E. E.; Williams, D. L.; Minden, M. D.; Mak, T. W.: Breakpoints in the human T-cell antigen receptor alpha-chain locus in two T-cell leukaemia patients with chromosomal translocations. Nature 317: 544-546, 1985.
Poiesz, B. J.; Ruscetti, F. W.; Gazdar, A. F.; Bunn, P. A.; Minna, J. D.; Gallo, R. C.: Detection and isolation of type C retrovirus particles from fresh and cultured lymphocytes of a patient with cutaneous T-cell lymphoma. Proc. Nat. Acad. Sci. 77: 7415-7419, 1980.
Poiesz, B. J.; Ruscetti, F. W.; Reitz, M. S.; Kalyanaraman, V. S.; Gallo, R. C.: Isolation of a new type C retrovirus (HTLV) in primary uncultured cells of a patient with Sezary T-cell leukemia. Nature 294: 268-271, 1981.

Sarin, P. S.; Aoki, T.; Shibata, A.; Ohnishi, Y.; Aoyagi, Y.; Miyakoshi, H.; Emura, I.; Kalyanaraman, V. S.; Robert-Guroff, M.; Popovic, M.; Sarngadharan, M.; Nowell, P. C.; Gallo, R. C.: High incidence of human type-C retrovirus (HTLV) in family members of a HTLV-positive Japanese T-cell leukemia patient. Proc. Nat. Acad. Sci. 80: 2370-2374, 1983.
Showe, L. C.; Croce, C. M.: Chromosome translocations in B and T cell neoplasias. Seminars Hemat. 23: 237-244, 1986.

LYMPHOBLASTIC LEUKEMIA, ACUTE, WITH LYMPHOMATOUS FEATURES (247640)

JJM NOTE

One possible pair of affected second-degree relatives is in the report of Chilcote et al. (1985).

Patients with acute lymphoblastic leukemia (ALL) who present with bulky disease of the lymph nodes, spleen, and mediastinum, so-called lymphomatous ALL (LALL), appear clinically to represent a distinct category of ALL of T-cell lineage. The biologic basis of this distinction was pointed out by Chilcote et al. (1985) who found that 6 of 8 patients with clinical features of LALL had karyotypic abnormalities leading to loss of bands 9p22-p21. The mechanisms varied and included deletions, unbalanced translocations, and loss of the entire chromosome. Only 1 of 57 patients without LALL had an abnormality of chromosome 9 at diagnosis. Loss of a 'suppressor' gene comparable to that of retinoblastoma (RB1; 180200) was postulated. A relationship to methylthioadenosine phosphorylase (156540) was postulated because the structural gene for this enzyme maps to the same region and patients with LALL may lack this enzyme in malignant cells during relapse. Lymphoblasts lacking this enzyme are unable to salvage adenine and methionine and are therefore especially sensitive to inhibitors of de novo purine synthesis (Kamatani et al., 1981). Kowalczyk and Sandberg (1981) had earlier found changes in 9p in a subgroup of ALL cases. Chilcote et al. (1985) pointed out that there is a fragile site at 9p21 and raised the question of familial predisposition on this basis. (This fragile site is the breakpoint in the translocation t(9;11)(p21-22;q23), which is associated with acute nonlymphocytic leukemia with monocytic features, ANLL-AMoL-M5a.) The aunt of one of the patients of Chilcote et al. (1985) had died as a child from ALL with lymphomatous features. If the analogy to RB1 holds, there is the same dilemma as to whether this should be called dominant or recessive; by the Chilcote hypothesis, it is presumably recessive. In a large series, Murphy et al. (1985) confirmed an abnormality of 9p in 10 to 11% of cases (33 out of more than 300) of acute lymphoblastic leukemia. The breakpoints in 9p c lustered in the p22-p21 region. They could not, however, corroborate the specific association with T-cell origin or so-called lymphomatous clinical features.

(Although to our knowledge not the determinant of an inherited phenotype, dominant or recessive, LALL appears to be a specific DNA coding segment that is involved in causation of a specific neoplasm through somatic cell mutation.)

Chilcote, R. R.; Brown, E.; Rowley, J. D.: Lymphoblastic leukemia with lymphomatous features associated with abnormalities of the short arm of chromosome 9. New Eng. J. Med. 313: 286-291, 1985.
Kamatani, N.; Nelson-Rees, W. A.; Carson, D. A.: Selective killing of human malignant cell lines deficient in methylthioadenosine phosphorylase, a purine metabolic enzyme. Proc. Nat. Acad. Sci. 78: 1219-1223, 1981.
Kowalczyk, J.; Sandberg, A. A.: A possible subgroup of ALL with 9p-. Cancer Genet. Cytogenet. 9: 383-385, 1981.
Murphy, S. B.; Raimondi, S.; Williams, D. L.; Carroll, A. J.; Castleberry, R. P.; Crist, W. M.: Lymphomatous ALL with abnormalities of the short arm of chromosome 9. (Letter) New Eng. J. Med. 313: 1611 only, 1985.

ZUNICH NEUROECTODERMAL SYNDROME (280000)

JJM NOTE

One case report of pre-B acute lymphatic leukemia in a rare syndrome is enough to raise the suggestion of an association. At relapse the bone marrow had a Philadelphia chromosome.

5,6:Zunich and Kaye (1983, 1984) and Zunich et al. (1985) described 2 unrelated children with a seemingly new neuroectodermal syndrome, the major features of which included early-onset migratory ichthyosiform dermatosis, bilateral ocular coloboma, conductive hearing loss, seizures, mental retardation, and remarkably similar facial features. The male patient had cleft palate, and the female patient had tetralogy of Fallot. Later, as reported by Zunich et al. (1988), the male had a brother born with similar features. Bifid uvula and submucous cleft were present. Transposition of the great vessels was found on cardiac catheterization. Retinal colobomas were present bilaterally on ophthalmologic evaluation, and hearing loss was detected by auditory evoked responses. The skin changes were less severe than in the other 2 patients. Zunich et al. (1988) referred to a fourth patient, an Amish male infant, with similar features.

Shashi et al. (1995) stated that ichthyosis and neurologic abnormalities, including seizures and mental retardation, have been described in several genetic syndromes, including Ruds syndrome (308200), Refsum disease (266500), Sjogren-Larsson syndrome (270200), Netherton syndrome (256500), the keratitis, ichthyosis, and deafness (KID) syndrome (148210), and the ichthyosis, brittle hair, impaired intelligence, decreased fertility, and short stature (IBIDS) syndrome (601675). Although there is some overlap in the clinical features, certain characteristics help to distinguish one disorder from another. Shashi et al. (1995) reported a fifth case of the Zunich neuroectodermal syndrome. They suggested the disorder be called CHIME syndrome (for oculo colobomas, congenital heart disease, early onset ichthyosiform dermatosis, mental retardation and ear anomalies (conductive hearing loss)). Alternatively, the 'E' could stand for seizures (epilepsy).

Schnur et al. (1997) reported a sixth child with this syndrome who developed acute lymphoblastic leukemia at age 4.5 years. Her major problems included a migratory ichthyosiform dermatosis, multiple skin infections and infestations, bilateral retinal coloboma, developmental delay, seizures, infantile macrosomia, facial anomalies, a duplicated renal collecting system, and conductive hearing loss.

Schnur, R. E.; Greenbaum, B. H.; Heymann, W. R.; Christensen, K.; Buck, A. S.; Reid, C. S.: Acute lymphoblastic leukemia in a child with the CHIME neuroectodermal dysplasia syndrome. Am. J. Med. Genet. 72: 24-29, 1997.
Shashi, V.; Zunich, J.; Kelly, T. E.; Fryburg, J. S.: Neuroectodermal (CHIME) syndrome: an additional case with long term follow up of all reported cases. J. Med. Genet. 32: 465-469, 1995.
Zunich, J.; Esterly, N. B.; Holbrook, K. A.; Kaye, C. I.: Congenital migratory ichthyosiform dermatosis with neurologic and ophthalmologic abnormalities. Arch. Derm. 121: 1149-1156, 1985.
Zunich, J.; Esterly, N. B.; Kaye, C. I.: Autosomal recessive transmission of neuroectodermal syndrome. (Letter) Arch. Derm. 124: 1188-1189, 1988.
Zunich, J.; Kaye, C. I.: New syndrome of congenital ichthyosis with neurologic abnormalities. Am. J. Med. Genet. 15: 331-333, 1983.
Zunich, J.; Kaye, C. I.: Additional case report of new neuroectodermal syndrome. (Letter) Am. J. Med. Genet. 17: 707-710, 1984.

T-CELL LEUKEMIA TRANSLOCATION-ASSOCIATED GENE (*600690)

JJM NOTE

A candidate gene whose role in leukemias must be clarified.

Aplan et al. (1995) cloned a gene at the site of a t(1;3)(p34;p21) translocation breakpoint in T-cell acute lymphoblastic leukemia. A cDNA for the gene, designated TCTA (T-cell leukemia translocation-associated gene), was identified from a human bone marrow library. The cDNA has a short open reading frame of 103 amino acids. Northern blots showed the gene was expressed ubiquitously with highest levels in the kidney. TCTA, located at 3p21, is deleted in a subset of small cell lung cancer (SCLC) cell lines. By comparative Southern blotting the authors detected related sequences in the DNA of other mammals, chicken and Drosophila.

Aplan, P. D.; Johnson, B. E.; Russell, E.; Chervinsky, D. S.; Kirsch, I. R.: Cloning and characterization of TCTA, a gene located at the site of a t(1;3) translocation. Cancer Res. 55: 1917-1921, 1995.

T-CELL LYMPHOMA INVASION AND METASTASIS 1 (*600687)

JJM NOTE

This is a candidate gene whose role in lymphomas must be clarified.

In an effort to identify genes involved in metastasis by insertional mutagenesis, Habets et al. (1994) infected mouse BW5147 T lymphoma cells with Moloney murine leukemia virus resulting in 5-20 proviral insertions per cell. By this 'proviral tagging' method, invasive variants were selected on monolayers of fibroblasts. Disrupting proviral insertions were found in these cells within the coding exons of a gene Designated Tiam-1 (T lymphoma invasion and metastasis-1). Selected clones were also metastatic in nude mice, and transfection of a truncated Tiam-1 cDNA into noninvasive cells transmitted the invasive phenotype. Northern blots demonstrated highest expression in the brain and testis. The mouse cDNA was isolated from a brain library and shown to encode a 1,591-amino acid protein which is serine rich and has regions of similarity to the Dbl-homologous (DH) domain found in GDP-GTP exchangers. The protein sequence also contains a pleckstrin-homologous (PH) domain thought to be involved in protein-protein interactions. Habets et al. (1995) cloned the human TIAM1 cDNA from a brain library using the mouse cDNA as a probe. The human protein was 95% identical to the mouse homolog. The authors speculated that TIAM1 may function in cellular signaling by activation of a Rho-like GTPase that regulates the cytoskeletal organization. Habets et al. (1995) mapped the mouse gene to the distal end of chromosome 16, 3.8 cM centromeric of Ets2. The human homolog was mapped to the syntenic human region 21q22 between the centromere and AML1 (151385).

Chen and Antonarakis (1995) cloned human TIAM1 by exon-trapping and localized the gene to 21q22.1 between the markers D21S298 and D21S404.

Chen, H.; Antonarakis, S. E.: Localization of a human homolog of the mouse Tiam-1 gene to chromosome 21q22.1. Genomics 30: 123-127, 1995.
Habets, G. G. M.; Scholtes, E. H. M.; Zuydgeest, D.; van der Kammen, R. A.; Stam, J. C.; Berns, A.; Collard, J. G.: Identification of an invasion-inducing gene, Tiam-1, that encodes a protein with homology to GDP-GTP exchangers for Rho-like proteins. Cell 77: 537-549, 1994.
Habets, G. G. M.; van der Kammen, R. A.; Jenkins, N. A.; Gilbert, D. J.; Copeland, N. G.; Hagemeijer, A.; Collard, J. G.: The invasion-inducing TIAM1 gene maps to human chromosome band 21q22 and mouse chromosome 16. Cytogenet. Cell. Genet. 70: 48-51, 1995.
Habets, G. G. M.; van der Kammen, R. A.; Stam, J. C.; Michiels, F.; Collard, J. G.: Sequence of the human invasion-inducing TIAM1 gene, its conservation in evolution and its expression in tumor cell lines of different tissue origin. Oncogene 10: 1371-1376, 1995.

T-CELL IMMUNODEFICIENCY, CONGENITAL ALOPECIA, AND NAIL DYSTROPHY (601705)

JJM NOTE

This entity is included not because of specific patients with malignancy, but because of high predisposition.

Pignata et al. (1996) reported 2 sisters with an association of congenital alopecia, severe T-cell immunodeficiency, and ridging and pitting of all nails. A decrease of mature T lymphocytes was mainly due to a low number of helper T cells, whereas the number of suppressor/cytotoxic T cells was relatively normal. Mitogen stimulation did not produce an adequate proliferative response. Since a normal proliferative response was seen following phorbol myristate acetate and ionomycin stimulation of T cells, the authors suggested that in these girls there was a block in the signal transmission upstream to protein kinase C. One sister died at the age of 1 year. Another girl underwent bone marrow transplantation (BMT) at 5 months and was alive 1 year after BMT. Pignata et al. (1996) suggested this association is a new syndrome presumably with autosomal recessive inheritance. Although alopecia has been reported in several patients with Omenn syndrome (267700), the sisters reported by Pignata et al. (1996) had alopecia at birth, before clinical evidence of erythrodermia, and the alopecia persisted after BMT.

Pignata, C.; Fiore, M.; Guzzetta, V.; Castaldo, A.; Sebastio, G.; Porta, F.; Guarino, A.: Congenital alopecia and nail dystropy associated with severe functional T-cell immunodeficiency in two sibs. Am. J. Med. Genet. 65: 167-170, 1996.

DOHLE BODIES AND LEUKEMIA (223350)

JJM NOTE

Follow-up would be instructive.

Goudsmit et al. (1971) reported a family in which 2 sisters and 3 brothers had Dohle bodies. Two of these 5 died of acute myeloblastic leukemia and 2 others had iron-resistant anemia. The parents and another sib did not have Dohle bodies. No statement concerning parental consanguinity was made. Dohle bodies of polymorphonuclear leukocytes are also seen in the May-Hegglin anomaly (155100).

Goudsmit, R.; v. Leeuwen, A. M.; James, J.: Dohle bodies and acute myeloblastic leukemia in one family: a new familial disorder. Brit. J. Haemat. 20: 557-562, 1971.

PELGER-HUET ANOMALY (*169400)

JJM NOTE

This trait is included because of one case of acute lymphoblastic leukemia. I wonder if the rabbit model deserves another look with modern morphologic, physiologic, and genetic tools.

The nucleus of the granulocytes is hyposegmented, being rodlike, dumbbell- or peanut-shaped, or spectaclelike. In Spokane, Washington, Ludden and Harvey (1962) found 4 cases among 43,000 persons. Affected persons were of German or Dutch descent. In Cleveland, Skendzel and Hoffman (1962) found a frequency of 1 in 4,785 routine smears. All figures in this country and also that of Davidson in England (1 in 6,000) are lower than that of Nachtsheim (1 in 1,020). This anomaly is also found in the rabbit. The homozygote in the rabbit has chondrodystrophy (Nachtsheim, 1950). Skeletal abnormality apparently does not occur in the human homozygote (Stobbe and Jorke, 1965). However, Aznar and Vaya (1981) described a presumed homozygote in whom postaxial polydactyly of all 4 limbs was also present. See also Haverkamp Begemann and van Lookeren Campagne (1952). Rioux et al. (1968) reported an extensively affected French-Canadian kindred. The nuclei of leukocytes had a pince-nez appearance. Oneson et al. (1987) described a child with familial Pelger-Huet anomaly who developed acute lymphoblastic leukemia. The disorder is effectively detected by the average lobe index (ALI) of the neutrophils. The ALI is the total number of nuclear lobes in 100 neutrophils divided by 100. The normal range for ALI is 2.5 to 3.1 (mean, 2.8). ALI in the affected nonleukemic members of this family varied from 1.12 to 1.60, with the lowest values in children and the highest values in adults. That folate deficiency increases segmentation was indicated by the fact that the ALI of the proband increased during 6-mercaptopurine and methotrexate therapy. Elevation of temperature to 42 degrees C resulted in an increase in the ALI of both normal cells and cells with the Pelger-Huet anomaly. Fishbein and Falletta (1991) described a newborn with Pelger-Huet anomaly associated with multiple congenital anomalies (diaphragmatic hernia, coarse facies, and distal limb anomalies) suggestive of Fryns syndrome. The parents did not have the blood anomaly. See proximal muscular dystrophy (159000) for information on linkage.

See also Jensson et al. (1977); Latimer et al. (1985); Rosse and Gurney (1959); Ware et al. (1988)

Aznar, J.; Vaya, A.: Homozygous form of the Pelger-Huet leukocyte anomaly in man. Acta Haemat. 66: 59-62, 1981.
Fishbein, J. D.; Falletta, J. M.: Pelger-Huet anomaly in an infant with multiple congenital anomalies. Am. J. Hemat. 38: 240-242, 1991.
Haverkamp Begemann, N.; van Lookeren Campagne, A.: Homozygous form of Pelger-Huet's nuclear anomaly in man. Acta Haemat. 7: 295-302, 1952.
Jensson, O.; Arnason, K.; Johannesson, G. M.; Ulfarsson, J.: Studies on the Pelger anomaly in Iceland. Acta Med. Scand. 201: 183-185, 1977.
Latimer, K. S.; Rakich, P. M.; Thompson, D. F.: Pelger-Huet anomaly in cats. Vet. Path. 22: 370-374, 1985.
Ludden, T. E.; Harvey, M.: Pelger-Huet anomaly of leukocytes: report of a case and survey of incidence. Am. J. Clin. Path. 37: 302-304, 1962.
Nachtsheim, H.: The Pelger-anomaly in man and rabbit: mendelian character of the nuclei of the leucocytes. J. Hered. 41: 131-137, 1950.
Oneson, R.; Sabio, H.; Innes, D. J., Jr.: Acute lymphoblastic leukaemia in a child with familial Pelger-Huet anomaly. Brit. J. Haemat. 66: 193-197, 1987.
Rioux, E.; St. Arneault, G.; Brosseau, C.: The Pelger-Huet anomaly of leukocytes: description of a Quebec kindred. Canad. Med. Assoc. J. 99: 621-624, 1968.
Rosse, W. F.; Gurney, C. W.: The Pelger-Huet anomaly in three families and its uses in determining the disappearance of transfused neutrophils from the peripheral blood. Blood 14: 170-186, 1959.
Skendzel, L. P.; Hoffman, G. C.: The Pelger anomaly of leukocytes: forty-one cases in seven families. Am. J. Clin. Path. 37: 294-301, 1962.
Stobbe, V. H.; Jorke, D.: Befunde an homozygoten Pelger-merkmalstragern. Schweiz. Med. Wschr. 95: 1524-1529, 1965.
Ware, R.; Kurtzberg, J.; Brazy, J.; Falletta, J. M.: Congenital Pelger-Huet anomaly in triplets. Am. J. Hemat. 27: 226-227, 1988.

LEUKEMIA, CHRONIC LYMPHATIC 1 (*151400)

JJM NOTE

The ever-improving tools of immunogenetics have to be applied to this relatively common familial type of neoplasia. The tenacity and recurrent investigations of Blattner et al. (1978, 1979) on the same families deserve admiration and emulation wherever such families have been encountered.

Chronic lymphatic leukemia seems especially prone to familial occurrence. Furbetta and Solinas (1963) reported affected grandfather, son, and grandson. Branda et al. (1978) studied lymphocytes in a mother and son with chronic lymphatic leukemia. Morphologic, functional and surface marker characteristics were very similar, as was the impairment of cellular and humoral immunity. Fraumeni et al. (1969) reported familial aggregation of chronic lymphocytic leukemia associated with immune defects in 3 sibs. Blattner et al. (1979) reported additional information on the family, including HLA studies and description of nodular, poorly differentiated lymphocytic lymphoma in the daughter of the proband. Conley et al. (1980) found an increased frequency of chronic lymphatic leukemia and also of autoimmune disease (hyperthyroidism, pernicious anemia, rheumatoid arthritis, systemic lupus erythematosus) in families of patients with CLL. They concluded that genetic factors in these families disturb the regulation of the immune system. Blattner et al. (1976) reported CLL in 4 of 5 sibs and their father. Follow-up by Neuland et al. (1983) showed spontaneous regression of the disease in 1 sib and shifts in the clinical pattern in the others. The unaffected sib developed lung cancer. Two CLL patients had abnormality of chromosome 12: trisomy in 1 and a mixture of dicentrics and translocations involving no. 12 in the other. In this same family, Shen et al. (1987) found that 2 brothers had rearrangement of the same heavy-chain variable region gene. Shen et al. (1987) suggested that the particular V(H) gene, which had an unusual DNA sequence, defined a previously unrecognized V(H) gene family. It was thought to be representative of the genome by, at most, 4 homologous sequences.

Tsujimoto et al. (1984) cloned the chromosomal breakpoint of chronic lymphocytic leukemia cells of the B-cell type carrying t(11;14)(q13;q32). The breakpoint was in the joining segment of the heavy chain locus on chromosome 14. A probe that is specific for chromosome 11 and maps immediately 5-prime to the breakpoint on 14q+ was isolated. The probe detected a rearrangement of the homologous genomic DNA segment in CLL cells and in DNA from a diffuse large cell lymphoma with the t(11;14) translocation. This rearranged DNA segment was not present in Burkitt lymphoma cells with the t(8;14) translocation or in nonneoplastic human lymphoblastoid cells. The probe thus can be used to identify and characterize a gene located on 11q13 involved in the malignant transformation of B cells in the t(11;14) translocation. Tsujimoto et al. (1984) referred to this gene as bcl-1; it appears to be related to none of the known retrovirus oncogenes. In 2 different cases of B-cell chronic lymphatic leukemia, Tsujimoto et al. (1985) found that the breakpoints on chromosome 11 were within 8 nucleotides of each other and on chromosome 14 involved the J4 DNA segment of the Ig heavy chain segment. Because they detected a 7mer-9mer signallike sequence with a 12-base-long spacer on the normal chromosome 11, close to the breakpoint, they speculated that the t(11;14) chromosome translocation in CLL may be sequence specific and may involve the recombination system for immunoglobulin V-D-J gene segment joining.

Translocation t(14;19)(q32;q13.1) is also a recurring translocation in the neoplastic cells of patients with chronic lymphocytic leukemia. In 1 such patient, McKeithan et al. (1987) analyzed the leukemic cells with probes from the immunoglobulin heavy-chain locus. Using a probe for the IGHA1 gene (146900), they detected a rearranged band by Southern blot analysis. The rearranged band was cloned and mapped. A subclone was shown to be from chromosome 19 by analysis of human-mouse somatic cell hybrids, confirming that the rearranged band contains the translocation breakpoint junction. See 109560.

Rosenberg et al. (1991) presented evidence that the PRAD1 gene (168461) on 11q13 is the 'BCL1 oncogene.'

Szepetowski et al. (1992) concluded that the BCL1 breakpoint at 11q13 is a boundary between 2 areas of human/mouse homology present in conserved syntenic regions on mouse chromosomes 7 (on the telomeric side) and 19 (on the centromeric side).

DNA amplification in human tumors involves a stretch of DNA much larger than the selected gene that is amplified. Amplicons are heterogeneous and dynamic structures that demonstrate discontinuities, i.e., nonamplified regions separating amplified ones. Szepetowski et al. (1993) hypothesized that discontinuities can be mapping tools. They tested the hypothesis with respect to the part of 11q13 that is centromeric from BCL1, the most frequently amplified 11q13 locus in breast cancer. They found that 14 genes, including BCL1, fell into 4 clusters in terms of coamplification. Consistent results were obtained when physical mapping was used and each probe was hybridized on Southern blots of human DNA fragments generated by rare-cutting restriction endonucleases separated by pulsed field gel electrophoresis. For example, one cluster in the same group as BCL1 had an order of GST3—PP1A—BARK—ADORA2; the last 3 were in a 380-kb fragment.

Horwitz et al. (1996) reported evidence of anticipation in autosomal dominant chronic lymphocytic leukemia (P = 0.008). In 18 affected individuals from 7 pedigrees with autosomal dominant CLL, the mean age of onset in the parental generation was 66 years, versus 51 years in the younger generation. Based on this evidence of anticipation, Horwitz et al. (1996) suggested that dynamic mutations of unstable DNA sequence repeats could be a common mechanism of inherited hematopoietic malignancy. They proposed 3 possible candidate chromosomal regions for familial leukemia with anticipation: 21q22.1-22.2, 11q23.3 in the vicinity of the CBL2 gene (165360), and 16q22 in the vicinity of the CBFB gene (121360).

See also Blattner et al. (1978); Erikson et al. (1984); Gunz and Dameshek (1957); McPhedran et al. (1969); Wisniewski and Weinreich (1966)

Blattner, W. A.; Dean, J. H.; Fraumeni, J. F., Jr.: Familial lymphoproliferative malignancy: clinical and laboratory follow-up. Ann. Intern. Med. 90: 943-944, 1979.

Blattner, W. A.; Naiman, J. L.; Mann, D. L.; Wimer, R. S.; Dean, J. H.; Fraumeni, J. F., Jr.: Immunogenetic determinants of familial acute lymphocytic leukemia. Ann. Intern. Med. 89: 173-176, 1978.

Blattner, W. A.; Strober, W.; Muchmore, A. V.; Blaese, R. M.; Broder, S.; Fraumeni, J. F., Jr.: Familial chronic lymphocytic leukemia: immunologic and cellular characterization. Ann. Intern. Med. 84: 554-557, 1976.

Branda, R. F.; Ackerman, S. K.; Handwerger, B. S.; Howe, R. B.; Douglas, S. D.: Lymphocyte studies in familial chronic lymphatic leukemia. Am. J. Med. 64: 508-514, 1978.

Conley, C. L.; Misiti, J.; Laster, A. J.: Genetic factors predisposing to chronic lymphocytic leukemia and to autoimmune disease. Medicine 59: 323-334, 1980.

Erikson, J.; Finan, J.; Tsujimoto, Y.; Nowell, P. C.; Croce, C. M.: The chromosome 14 breakpoint in neoplastic B cells with the t(11;14) translocation involves the immunoglobulin heavy chain locus. Proc. Nat. Acad. Sci. 81: 4144-4148, 1984.

Fraumeni, J. F.; Vogel, C. L.; DeVita, V. T.: Familial chronic lymphocytic leukemia. Ann. Intern. Med. 71: 279-284, 1969.

Furbetta, D.; Solinas, P.: Hereditary chronic lymphatic leukemia.In: Proc. 2nd Int. Cong. Hum. Genet., Rome, Sept. 6-12, 1961. (pub.) 2: 1963. Pp. 1078-1079.

Gunz, F.; Dameshek, W.: Chronic lymphatic leukemia in a family, including twin brothers and a son. J.A.M.A. 164: 1323-1325, 1957.

Horwitz, M.; Goode, E. L.; Jarvik, G. P.: Anticipation in familial leukemia. Am. J. Hum. Genet. 59: 990-998, 1996.

McKeithan, T. W.; Rowley, J. D.; Shows, T. B.; Diaz, M. O.: Cloning of the chromosome translocation breakpoint junction of the t(14;19) in chronic lymphocytic leukemia. Proc. Nat. Acad. Sci. 84: 9257-9260, 1987.

McPhedran, P.; Heath, C. W., Jr.; Lee, J.: Patterns of familial leukemia. Ten cases of leukemia in two interrelated families. Cancer 24: 403-407, 1969.

Neuland, C. Y.; Blattner, W. A.; Mann, D. L.; Fraser, M. C.; Tsai, S.; Strong, D. M.: Familial chronic lymphocytic leukemia. J. Nat. Cancer Inst. 71: 1143-1150, 1983.

Rosenberg, C. L.; Wong, E.; Petty, E. M.; Bale, A. E.; Tsujimoto, Y.; Harris, N. L.; Arnold, A.: PRAD1, a candidate BCL1 oncogene: mapping and expression in centrocytic lymphoma. Proc. Nat. Acad. Sci. 88: 9638-9642, 1991.

Shen, A.; Humphries, C.; Tucker, P.; Blattner, F.: Human heavy-chain variable region gene family nonrandomly rearranged in familial chronic lymphocytic leukemia. Proc. Nat. Acad. Sci. 84: 8563-8567, 1987.

Szepetowski, P.; Perucca-Lostanlen, D.; Gaudray, P.: Mapping genes according to their amplification status in tumor cells: contribution to the map of 11q13. Genomics 16: 745-750, 1993.

Szepetowski, P.; Simon, M.-P.; Grosgeorge, J.; Huebner, K.; Bastard, C.; Evans, G. A.; Tsujimoto, Y.; Birnbaum, D.; Theillet, C.; Gaudray, P.: Localization of 11q13 loci with respect to regional chromosomal breakpoints. Genomics 12: 738-744, 1992.

Tsujimoto, Y.; Jaffe, E.; Cossman, J.; Gorham, J.; Nowell, P. C.; Croce, C. M.: Clustering of breakpoints on chromosome 11 in human B-cell neoplasms with the t(11;14) chromosome translocation. Nature 315: 340-343, 1985.

Tsujimoto, Y.; Yunis, J.; Onorato-Showe, L.; Erikson, J.; Nowell, P. C.; Croce, C. M.: Molecular cloning of the chromosomal breakpoint of B-cell lymphomas and leukemias with the t(11;14) chromosome translocation. Science 224: 1403-1406, 1984.

Wisniewski, D.; Weinreich, J.: Lymphatische Leukaemie bei Vater und Sohn. Blut 12: 241-244, 1966.

LEUKEMIA, CHRONIC LYMPHATIC 2 (*151430)

JJM NOTE

This translocation break point is especially interesting because it may define a new class of oncogenes (regulators of cell death) and because it arises spontaneously with aging. However, it has not been associated with familial chronic lymphatic leukemia. It is perhaps only a matter of investigating further?

Description

One of the most frequent hematologic malignancies, follicular lymphoma is associated with a reciprocal translocation t(14;18)(q32;q21). This results in deregulated expression of a chromosome 18 gene, BCL2, which interferes with normal apoptosis of B-lymphocytes.

Biochemical Features

Ngan et al. (1988) found immunoreactive BCL-2 protein in the neoplastic cells of almost all follicular lymphomas whereas no BCL-2 protein was detected in follicles affected by nonneoplastic processes or in normal lymphoid tissue. Every tumor with molecular-genetic evidence of t(14;18) translocation expressed detectable levels of BCL-2 protein, regardless of whether the breakpoint was located in or at a distance from the BCL2 gene. Demonstration of the protein with anti-BCL-2 antibodies should be useful in the diagnosis of many non-Hodgkin lymphomas.

By means of immunolocalization studies, Hockenbery et al. (1990) demonstrated that BCL2 is an integral inner mitochondrial membrane protein of relative molecular mass 25,000. Overexpression of BCL2 blocks the apoptotic death of a pro-B-lymphocyte cell line. Thus, BCL2 is unique among protooncogenes, being localized in mitochondria and interfering with programmed cell death independent of promoting cell division.

Wang et al. (1996) showed that BCL2 can target the protein kinase RAF1 (164760) to the mitochondria. Active RAF1 improved BCL2-mediated resistance to apoptosis.

Vaux et al. (1988) undertook to determine the biologic effects of the BCL2 gene by introducing BCL2 cDNA into bone marrow cells. They found that BCL2 cooperated with MYC to promote proliferation of B-cell precursors, some of which became tumorigenic. Reed et al. (1988) also demonstrated the oncogenic potential of BCL2 by gene transfer.

Tsujimoto (1989) used Epstein-Barr virus-infected human lymphoblastoid B-cell lines transfected with BCL-2 sequences and driven by the simian virus 40 promoter and enhancer to demonstrate that overproduction of the BCL-2 protein results in a distinct cellular growth advantage. Nunez et al. (1989) demonstrated the protooncogene role of BCL2 in B-cell growth and B-cell neoplasm formation. ERV1 is separate from BCL2 (Croce, 1989).

Williams (1991) reviewed the evidence that BCL2 acts by inhibiting cell loss by apoptosis (programmed cell death) rather than by stimulating cell production. (The term apoptosis is derived from ancient Greek for 'falling off of tree leaves.') Fesus et al. (1991) reviewed molecular mechanisms of apoptosis.

Jacobson et al. (1993) showed that the action of BCL2 in protecting cells from apoptosis is not by altering mitochondrial function; they found that human mutant cell lines that lack mitochondrial DNA can still be induced to die by apoptosis and that they can be protected from apoptosis by the overexpression of BCL2. Migheli et al. (1994) performed a light-microscopic immunohistochemical analysis of BCL2 protein expression in autopsy specimens of human brain and spinal cord in normal, aged individuals and those who had suffered from neurodegenerative diseases. BCL2 was strongly enriched within lipofuscin and autophagic vacuoles of neurons, and glial and vascular cells. Deng and Podack (1993) showed that transcription of the BCL2 gene is downregulated by interleukin-2 (IL2; 147680) deprivation and upregulated by IL2 addition. Deregulated expression of BCL2 was found to prolong the survival of cells of the cytotoxic T-cell line CTLL2 in the absence of IL-2. Hengartner and Horvitz (1994) presented evidence that the cell survival gene in Caenorhabditis elegans called ced-9 is a homolog of BCL2; thus, molecular mechanisms of programmed cell death have been conserved from nematodes to mammals.

Martinou et al. (1994) generated transgenic mice in which neurons overexpress the human BCL2 protein under control of neuron-specific enolase or phosphoglycerate kinase promoters. These transgenic mice had reduced neuronal loss during the period of naturally occurring cell death with a resulted hypertrophy of the nervous system. The facial nucleus and the ganglion cell layer of the retina had 40 to 50% more neurons than in control animals. In addition, these transgenic mice were more resistant to ischemic damage induced by middle cerebral artery occlusion than were control mice.

Farlie et al. (1995) reported observations in transgenic mice expressing BCL2 under the control of the neuron-specific enolase promoter, suggesting that the role of BCL2 is wider than merely its role in lymphocytes. Sensory neurons isolated from dorsal ganglia of newborn mice normally require nerve growth factor for their survival in culture, but those from the BCL2 transgenic mice showed enhanced survival in its absence. Furthermore, apoptotic death of motor neurons after axotomy of the sciatic nerve was inhibited in these mice. The number of neurons in the 2 neuronal populations from the central and peripheral nervous system was increased by 30%, indicating that BCL2 expression can protect neurons from cell death during development. Thus, BCL2 may play an important role in survival of neurons both during development and throughout adult life.

Nunez et al. (1991) reported that this protooncogene maintains immune responsiveness. Transgenic mice overproducing Bcl-2 showed long-term persistence of immunoglobulin-secreting cells and an extended lifetime for memory B cells. Through studies in transgenic mice, Sentman et al. (1991) and Strasser et al. (1991) concluded that modulation of BCL2 expression is 'a determinant of life and death in normal lymphocytes.'

Ji et al. (1996) examined the promoter activities of the normal and translocated BCL2 alleles in the DHL-4 cell line with the t(14;18) translocation. They demonstrated that cAMP response-binding protein (CREB; 123810) binds to a cAMP response element (CRE) in the BCL2 5-prime flanking region of the translocated allele. Access to this CRE site is blocked in the normal BCL2 allele. They concluded that the CRE site of the translocated BCL2 allele functions as a positive regulatory site in t(14;18) lymphomas.

Cytogenetics

Yunis et al. (1982) found a translocation between chromosomes 18 and 14 in 16 of 19 patients with follicular lymphomas, one of the most common forms of B-cell neoplasia. They found 8;14 translocation in 5 of 6 patients with small noncleaved-cell (non-Burkitt) lymphoma or large-cell immunoblastic lymphoma and trisomy 12 in 4 of 11 patients with small-cell lymphocytic lymphoma. They stated: 'It is conceivable that when DNA at the breakpoint site of the donor chromosome 8 or 18 becomes contiguous with genes involved in immunoglobulin synthesis, lymphomatous transformation is initiated.' The assignment of an oncogene to chromosome 18 (ERV1; 131150) suggested to (O'Brien et al., 1983) that it may serve the same role in neoplastic transformation in follicular lymphoma as the MYC gene (190080) appears to fill in the case of Burkitt lymphoma.

From a young male with acute lymphoblastic leukemia, Pegoraro et al. (1984) established a cell line that showed an 8;14 and a 14;18 translocation, which are characteristic of Burkitt lymphoma and of follicular lymphoma, respectively. The cell line was Epstein-Barr virus antigen-negative, reacted with monoclonal antibodies specific for B cells and contained rearranged heavy and light chain genes, but did not express immunoglobulins. One of the J(H) segments of one of the 14q+ chromosomes was rearranged with a segment of chromosome 8, where the MYC gene is situated; the other 14q+ chromosome was rearranged with a segment of chromosome 18 where a putative oncogene they called BCL2 was thought to reside. The breakpoint in chromosome 18 was at q21. The translocated MYC gene was in its germline configuration and was located more than 14 kb from the chromosomal breakpoint.

In a study of 71 patients with follicular lymphoma, Yunis et al. (1987) observed 14;18 translocation in 85% of the patients and concluded that this is the main determinant of a follicular pattern. Different changes were found in other patients; the changes seemed to correlate with the evolution of the patients' malignant disease.

Haluska et al. (1987) discussed a general hypothesis for the mechanism of chromosome translocation in B- and T-cell neoplasia. They suggested that it is a perversion of normal translocation processes; specifically in B-cell chronic lymphocytic leukemia, there appears to be involvement of the immunoglobulin V-D-J recombinase. There is, at 14q32 in IgH, a consensus that is mimicked by heptamer-nonamer sequences at 11q13, 18q21, and 8q24. In these areas are located cell growth factors BCL1, BCL2, and MYC, respectively.

Molecular Genetics

Pegoraro et al. (1984) postulated 2 steps in the malignant process. First, the 14;18 translocation, occurring in an activated B cell and involving the excluded heavy chain allele on 14q32 and the BCL2 gene on chromosome 18, brought a heavy chain enhancer close to the BCL2 gene. Constitutive expression of BCL2 led to clonal expansion of t(14;18) cells and a relatively low-grade malignancy. Second, within the malignant clone of B cells, the t(8;14) translocation occurred, leading to high-grade malignancy through activation of MYC. Mufti et al. (1983) reported a double translocation of the same type in a case of acute leukemia.

From the cell line studied by Pegoraro et al. (1984), Tsujimoto et al. (1985) derived a DNA clone that was specific for chromosome 18 and was flanked by the heavy chain joining (J) region of the immunoglobulin heavy chain locus (147100) on chromosome 14—thus, it was derived from the breakpoint on chromosome 18 involved in the creation of the t(14;18)(q32;q21). This probe detected rearrangement of the homologous DNA segment in the leukemic cells and in follicular lymphoma cells with the t(14;18) chromosome translocation but not in other neoplastic or normal B or T cells. These workers concluded that the probe identifies BCL2, a gene locus on 18q21 that is unrelated to known oncogenes and may be important in the pathogenesis of B-cell neoplasms with this translocation.

By studying cloned recombinant DNA probes from the area flanking the breakpoints in chromosome 18 in cells from patients with acute lymphocytic leukemia of the B-cell type carrying t(14;18), Tsujimoto et al. (1985) found 2 that detected DNA rearrangements in about 60% of cases of follicular lymphoma screened. Most of the breakpoints in 18q21 in follicular lymphoma were clustered in a stretch of DNA about

2.1 kb long. The BCL2 gene seemed to be interrupted in most cases of follicular lymphoma carrying the t(14;18) translocation. This study involved chromosome walking to identify the putative BCL2 gene.

Bakhshi et al. (1985) also cloned the breakpoints of t(14;18) in 4 cases. The breakpoints clustered within a 4.3-kb region on chromosome 18. The breakpoint on chromosome 14 brought the Ig enhancer region close to a newly identified transcriptional unit on 18q21. Since none of the oncogenes are known to map to 18q21, cloning this element may provide an opportunity to characterize a new transforming gene.

Tsujimoto and Croce (1986) showed that the BCL2 gene consists of at least 2 exons separated by an intron of more than 50 kb. The first exon is transcribed into a 3.5-kb mRNA. This exon contains a splicing donor signal so that BCL2 mRNA is spliced to the second exon to produce a 5.5- and an 8.5-kb mRNA. The 5.5- and 3.5-kb mRNAs code for the BCL2 alpha and beta proteins, respectively, which are identical except for the carboxyl terminal portion.

Tsujimoto et al. (1985) showed that about 60% of the breakpoints on chromosome 18 in cases of t(14;18) translocation are tightly clustered in the 3-prime noncoding region of the BCL2 gene and about 10% are clustered at a region 3-prime to the BCL2 gene. From analysis of a panel of follicular lymphoma DNAs with probes for the first exon of the BCL2 gene, Tsujimoto et al. (1987) showed that DNA rearrangements may also occur 5-prime to the involved BCL2 gene.

By nucleotide sequence analyses, Cleary et al. (1986) showed that the 6-kb BCL2 mRNA potentially encodes a 26-kD protein that is homologous to a predicted Epstein-Barr virus protein. Most 14;18 translocation breakpoints cluster within a narrow region of a 5.4-kb exon that contains a long 3-prime untranslated region of the BCL2 mRNA. As a result of the translocation, hybrid BCL2/immunoglobulin heavy chain transcripts are produced that consist of the 5-prime half of the BCL2 mRNA fused to a 'decapitated' immunoglobulin heavy chain mRNA. Nucleotide sequence analyses confirmed that the hybrid transcripts continue to encode a normal BCL-2 protein. Bakhshi et al. (1987) concluded that immunoglobulin recombinase plays no role in the chromosome 18 breakage. Instead, a direct repeat duplication of chromosome 18 sequences was discovered at both chromosomal junctures, typical of the repair of a naturally occurring staggered double-stranded DNA break. The translocation in t(14;18) occurs in a B cell as an illegitimate recombination at the first step in the rearrangement of the heavy chain gene cluster, the step in which D(H) is fused to J(H)—see 146910 and 147010.

Lee et al. (1987) used a method of DNA amplification to detect t(14;18) hybrid DNA sequences in a patient with follicular lymphoma in whom remission marrow and blood samples showed no abnormality by morphologic examination and conventional Southern blot analysis. By Southern blotting in cases of non-Hodgkin lymphoma, Aisenberg et al. (1988) found frequent rearrangement of the BCL2 gene.

Pulsed field gel electrophoresis suggested that the 3-exon BCL2 gene is several hundred kb long. Silverman et al. (1990) isolated 2 YACs, each containing part of the BCL2 gene and overlapping by 60 kb. They sought to take advantage of the high recombination frequency in yeast to induce physical recombination between the 2 clones. Analysis of resulting tetrads revealed a spore containing a single recombinant YAC of 800 kb. Further analysis showed that this recombined YAC contained the entire BCL2 gene of about 230 kb, without overt rearrangements or deletions.

In a review, Korsmeyer (1992) pointed out that the effects of deregulated BCL2 suggest that it does not qualify as either a category I (promoter of cell growth and proliferation) or category II (tumor suppressor) oncogene. Whereas the product of the fusion of the PML gene (102578) and the retinoic acid receptor-alpha gene (180240) and the product of the fusion of BCR (151410) and ABL (189980) represent perturbation of both genes contributing to pathogenesis, disruption of the normal expression of the BCL2 locus alone appears to contribute to the development of lymphoma (Frankel, 1993).

Using a nested PCR assay for quantitation of rare t(14;18) translocations, Liu et al. (1994) demonstrated that this oncogenic mutation occurs in persons without lymphoid neoplasia and in an age-dependent fashion. The translocations copurified with B lymphocytes. The frequency of the translocations increased with age in both peripheral blood lymphocytes and spleens, as does human risk for lymphoma. Average translocation frequency was more than 40 times greater in the spleen and 13 times greater in the peripheral blood in the oldest individuals (61 years and older) compared with the youngest individuals (20 years or younger). Particular t(14;18)-bearing clones persisted over a period of 5 months in 2 persons. On the basis of these findings, Liu et al. (1994) proposed a multihit model of lymphomagenesis involving t(14;18) translocation followed by antigen stimulation. They suggested that the t(14;18)-bearing B-cell clones accumulate other mutations in growth-dysregulatory or lymphomagenic loci.

DiCroce and Krontiris (1995) observed that most translocations involving BCL2 are very narrowly targeted to 3 breakpoint clusters evenly spaced over a 100-bp region of the gene's terminal exon. The immediate upstream boundary of this major breakpoint region (MBR) is a specific recognition site for single-strand DNA (ssDNA) binding proteins on the sense and antisense strands. The downstream flank of the MBR is a helicase binding site. DiCroce and Krontiris (1995) demonstrated that the helicase and ssDNA binding proteins show reciprocal changes in binding activity over the cell cycle. The helicase is maximally active in G1 and early S phases; the ssDNA binding proteins are maximally active in late S and G2/M phases. One component of the helicase binding complex is the Ku antigen (152690). Thus, the authors showed that a protein with helicase activity implicated in repair of double-strand breaks, V(D)J recombination, and, potentially, cell-cycle regulation is targeted to the BCL2 MBR.

Brag1: A Spurious 'bcl2-related' Gene

Das et al. (1996) reported the isolation and molecular characterization of a novel Bcl2-related cDNA from a human glioma. They designated the gene BRAG1 for 'brain related apoptosis gene.' They gene showed extensive homology to the BCL2 family of genes. The gene is expressed in normal brain as a 4.5-kb transcript and as a 1.8-kb transcript in human gliomas, which suggested to the authors that it may be rearranged in these tumors. The open reading frame of the BRAG1 gene encodes for a protein of 31 kD. Using a bacterial expression vector, Das et al. (1996) produced BRAG1 protein that was found to crossreact with a BCL2 monoclonal antibody, further suggesting structural and immunologic similarity to BCL2. In an erratum, the authors concluded that the gene in question was in fact from the genome of Escherichia coli and occurred as a contaminant in cDNA libraries.

Clinical Management

Yunis et al. (1989) found that response to therapy was poorer in those patients with B-cell nonimmunoblastic lymphoma or follicular lymphoma who had a BCL2 oncogene rearrangement than it was in patients without BCL2 rearrangement.

Animal Model

Tsujimoto et al. (1985) found that their most specific probe hybridized to cellular DNA from hamster and mouse under stringent conditions, indicating that at least part of the BCL2 gene is conserved across mammalian species as are many cellular oncogenes.

Negrini et al. (1987) cloned the murine gene homologous to BCL2 in man. The gene is composed of 2 exons separated by more than 15 kb and maps to mouse chromosome 1. Mock et al. (1988) demonstrated linkage to markers on mouse chromosome 1 and concluded that bcl-2 is centromeric to the renin locus in that species.

Nakayama et al. (1994) reported results from the study of BCL2-deficient mice created through the injection of clones containing 1 mutated bcl-2 allele into C57BL/6 blastocysts to generate chimeric mice. Animals homozygous for the mutation were smaller but viable, although about half of them died by 6 weeks of age. As shown earlier with somatic bcl-2 gene-targeted mice, the number of lymphocytes markedly decreased within a few weeks after birth while other hematopoietic lineages remained unaffected. Among lymphocytes, CD8(+) T cells disappeared most quickly followed by CD4(+) T cells, whereas B cells were least affected. The homozygously defective lymphocytes could, however, respond normally to various stimuli including anti-CD3, Con A, interleukin-2, lipopolysaccharide, and anti-IgM antibody. Abnormalities in nonlymphoid organs included smaller auricles, hair color turning gray at 4 to 5 weeks of age, and polycystic kidney disease-like change of renal tubules. These results suggested that bcl-2 may be involved during morphogenesis where inductive interactions between epithelium and mesenchyme are important such as in the kidneys, hair follicles, and perichondrium of auricles. Surprisingly, the nervous system, intestines, and skin appeared normal despite the fact that these organs show high levels of endogenous bcl-2 expression in normal mice.

McDonnell et al. (1989) found that transgenic mice bearing a bcl-2-immunoglobulin minigene that mimicked the t(14;18) translocation displayed a polyclonal follicular hyperplasia with a 4-fold increase in resting B cells. B cells accumulated because of extended cell survival rather than increased proliferation.

By crossing mice with motoneuron disease (pmn) with mice that overexpressed Bcl2, Sagot et al. (1995) demonstrated rescue of facial motoneurons with restoration of normal soma size and expression of choline acetyltransferase. However, Bcl2 overexpression did not prevent degeneration of myelinated axons and did not increase the life span of the animals.

Apoptosis of photoreceptors occurs infrequently in adult retina and can be triggered in inherited and environmentally induced retinal degenerations. BCL2 is known to be a potent regulator of cell survival in neurons. Chen et al. (1996) created lines of transgenic mice overexpressing Bcl2 to test for its ability to increase photoreceptor survival.

They found that Bcl2 increased photoreceptor survival in 3 mouse models: a line of transgenic mice expressing a C-terminal truncated form of rhodopsin (180380) associated with rapid degeneration of photoreceptors; homozygous rd mice with nonfunctional rod-cGMP-phosphodiesterase (see 180071); and albino mice exposed to sustained illumination. Bcl2 increased photoreceptor survival in the first 2 mouse models and decreased the damaging effects of constant light exposure in the albino mice. Apoptosis was induced in normal photoreceptors by very high levels of Bcl2. Chen et al. (1996) concluded that Bcl2 is an important regulator of photoreceptor cell death in retinal degenerations.

See also Cleary and Sklar (1985); Graninger et al. (1987); Tsujimoto et al. (1985)

Aisenberg, A. C.; Wilkes, B. M.; Jacobson, J. O.: The bcl-2 gene is rearranged in many diffuse B-cell lymphomas. Blood 71: 969-972, 1988.

Bakhshi, A.; Jensen, J. P.; Goldman, P.; Wright, J. J.; McBride, O. W.; Epstein, A. L.; Korsmeyer, S. J.: Cloning the chromosomal breakpoint of t(14;18) human lymphomas: clustering around J(H) on chromosome 14 and near a transcriptional unit on 18. Cell 41: 899-906, 1985.

Bakhshi, A.; Wright, J. J.; Graninger, W.; Seto, M.; Owens, J.; Cossman, J.; Jensen, J. P.; Goldman, P.; Korsmeyer, S. J.: Mechanism of the t(14;18) chromosomal translocation: structural analysis of both derivative 14 and 18 reciprocal partners. Proc. Nat. Acad. Sci. 84: 2396-2400, 1987.

Chen, J.; Flannery, J. G.; LaVail, M. M.; Steinberg, R. H.; Xu, J.; Simon, M. I.: Bcl-2 overexpression reduces apoptotic photoreceptor cell death in three different retinal degenerations. Proc. Nat. Acad. Sci. 93: 7042-7047, 1996.

Cleary, M. L.; Sklar, J.: Nucleotide sequence of a t(14;18) chromosomal breakpoint in follicular lymphoma and demonstration of a breakpoint-cluster region near a transcriptionally active locus on chromosome 18. Proc. Nat. Acad. Sci. 82: 7439-7443, 1985.

Cleary, M. L.; Smith, S. D.; Sklar, J.: Cloning and structural analysis of cDNAs for bcl-2 and a hybrid bcl-2/immunoglobulin transcript resulting from the t(14;18) translocation. Cell 47: 19-28, 1986.

Croce, C. M.: Personal Communication. Philadelphia, Pa. 1989.

Das, R.; Reddy, E. P.; Chatterjee, D.; Andrews, D. W.: Identification of a novel Bcl-2 related gene, BRAG-1, in human glioma. Oncogene 12: 947-951, 1996. Note: Erratum: Oncogene 16: 429 only, 1998.

Deng, G.; Podack, E. R.: Suppression of apoptosis in a cytotoxic T-cell line by interleukin 2-mediated gene transcription and deregulated expression of the protooncogene bcl-2. Proc. Nat. Acad. Sci. 90: 2189-2193, 1993.

DiCroce, P. A.; Krontiris, T. G.: The BCL2 major breakpoint region is a sequence- and cell-cycle-specific binding site of the Ku antigen. Proc. Nat. Acad. Sci. 92: 10137-10141, 1995.

Farlie, P. G.; Dringen, R.; Rees, S. M.; Kannourakis, G.; Bernard, O.: bcl-2 transgene expression can protect neurons against developmental and induced cell death. Proc. Nat. Acad. Sci. 92: 4397-4401, 1995.

Fesus, L.; Davies, P. J. A.; Piacentini, M.: Apoptosis: molecular mechanisms in programmed cell death. Europ. J. Cell Biol. 56: 170-177, 1991.

Frankel, S. R.: Correction: location of bcl-2 oncogene. (Letter) New Eng. J. Med. 328: 141, 1993.

Graninger, W. B.; Seto, M.; Boutain, B.; Goldman, P.; Korsmeyer, S. J.: Expression of Bcl-2 and Bcl-2-Ig fusion transcripts in normal and neoplastic cells. J. Clin. Invest. 80: 1512-1515, 1987.

Haluska, F. G.; Tsujimoto, Y.; Croce, C. M.: Mechanisms of chromosome translocation in B- and T-cell neoplasia. Trends Genet. 3: 11-15, 1987.

Hengartner, M. O.; Horvitz, H. R.: C. elegans cell survival gene ced-9 encodes a functional homolog of the mammalian proto-oncogene bcl-2. Cell 76: 665-676, 1994.

Hockenbery, D.; Nunez, G.; Milliman, C.; Schreiber, R. D.; Korsmeyer, S. J.: Bcl-2 is an inner mitochondrial membrane protein that blocks programmed cell death. Nature 348: 334-336, 1990.

Jacobson, M. D.; Burne, J. F.; King, M. P.; Miyashita, T.; Reed, J. C.; Raff, M. C.: Bcl-2 blocks apoptosis in cells lacking mitochondrial DNA. Nature 361: 365-369, 1993.

Ji, L.; Mochon, E.; Arcinas, M.; Boxer, L. M.: CREB proteins function as positive regulators of the translocated bcl-2 allele in t(14;18) lymphomas. J. Biol. Chem. 271: 22687-22691, 1996.

Korsmeyer, S. J.: Bcl-2 initiates a new category of oncogenes: regulators of cell death. Blood 80: 879-886, 1992.

Lee, M.-S.; Chang, K.-S.; Cabanillas, F.; Freireich, E. J.; Trujillo, J. M.; Stass, S. A.: Detection of minimal residual cells carrying the t(14;18) by DNA sequence amplification. Science 237: 175-178, 1987.

Liu, Y.; Hernandez, A. M.; Shibata, D.; Cortopassi, G. A.: BCL2 translocation frequency rises with age in humans. Proc. Nat. Acad. Sci. 91: 8910-8914, 1994.

Martinou, J.-C.; Dubois-Dauphin, M.; Staple, J. K.; Rodriguez, I.; Frankowski, H.; Missotten, M.; Albertini, P.; Talabot, D.; Catsicas, S.; Pietra, C.; Huarte, J.: Overexpression of BCL-2 in transgenic mice protects neurons from naturally occurring cell death and experimental ischemia. Neuron 13: 1017-1030, 1994.

McDonnell, T. J.; Deane, N.; Platt, F. M.; Nunez, G.; Jaeger, U.; McKearn, J. P.; Korsmeyer, S. J.: Bcl-2-immunoglobulin transgenic mice demonstrate extended B cell survival and follicular lymphoproliferation. Cell 57: 79-88, 1989.

Migheli, A.; Cavalla, P.; Piva, R.; Giordana, M. T.; Schiffer, D.: bcl-2 protein expression in aged brain and neurodegenerative diseases. NeuroReport 5: 1906-1908, 1994.

Mock, B. A.; Givol, D.; D'Hoostelaere, L. A.; Huppi, K.; Seldin, M. F.; Gurfinkel, N.; Unger, T.; Potter, M.; Mushinski, J. F.: Mapping of the bcl-2 oncogene on mouse chromosome 1. Cytogenet. Cell Genet. 47: 11-15, 1988.

Mufti, G. J.; Hamblin, T. J.; Oscier, D. G.; Johnson, S.: Common ALL with pre-B-cell features showing (8;14) and (14;18) chromosome translocations. Blood 62: 1141-1146, 1983.

Nakayama, K.; Nakayama, K.; Negishi, I.; Kuida, K.; Sawa, H.; Loh, D. Y.: Targeted disruption of Bcl-2-alpha-beta in mice: occurrence of gray hair, polycystic kidney disease, and lymphocytopenia. Proc. Nat. Acad. Sci. 91: 3700-3704, 1994.

Negrini, M.; Silini, E.; Kozak, C.; Tsujimoto, Y.; Croce, C. M.: Molecular analysis of mbcl-2: structure and expression of the murine gene homologous to the human gene involved in follicular lymphoma. Cell 49: 455-463, 1987.

Ngan, B.-Y.; Chen-Levy, Z.; Weiss, L. M.; Warnke, R. A.; Cleary, M. L.: Expression in non-Hodgkin's lymphoma of the BCL-2 protein associated with the t(14;18) chromosomal translocation. New Eng. J. Med. 318: 1638-1644, 1988.

Nunez, G.; Hockenbery, D.; McDonnell, T. J.; Sorensen, C. M.; Korsmeyer, S. J.: Bcl-2 maintains B cell memory. Nature 353: 71-73, 1991.

Nunez, G.; Seto, M.; Seremetis, S.; Ferrero, D.; Grignani, F.; Korsmeyer, S. J.; Dalla-Favera, R.: Growth- and tumor-promoting effects of deregulated BCL2 in human B-lymphoblastoid cells. Proc. Nat. Acad. Sci. 86: 4589-4593, 1989.

O'Brien, S. J.; Bonner, T. I.; Cohen, M.; O'Connell, C.; Nash, W. G.: Mapping of an endogenous retroviral sequence to human chromosome 18. Nature 303: 74-77, 1983.

Pegoraro, L.; Palumbo, A.; Erikson, J.; Falda, M.; Giovanazzo, B.; Emanuel, B. S.; Rovera, G.; Nowell, P. C.; Croce, C. M.: A 14;18 and an 8;14 chromosome translocation in a cell line derived from an acute B-cell leukemia. Proc. Nat. Acad. Sci. 81: 7166-7170, 1984.

Reed, J. C.; Cuddy, M.; Slabiak, T.; Croce, C. M.; Nowell, P. C.: Oncogenic potential of bcl-2 demonstrated by gene transfer. Nature 336: 259-261, 1988.

Sagot, Y.; Dubois-Dauphin, M.; Tan, S. A.; de Bilbao, F.; Aebischer, P.; Martinou, J.-C.; Kato, A. C.: Bcl-2 overexpression prevents motoneuron cell body loss but not axonal degeneration in a mouse model of a neurodegenerative disease. J. Neurosci. 15: 7727-7733, 1995.

Sentman, C. L.; Shutter, J. R.; Hockenbery, D.; Kanagawa, O.; Korsmeyer, S. J.: bcl-2 inhibits multiple forms of apoptosis but not negative selection in thymocytes. Cell 67: 879-888, 1991.

Silverman, G. A.; Green, E. D.; Young, R. L.; Jockel, J. I.; Domer, P. H.; Korsmeyer, S. J.: Meiotic recombination between yeast artificial chromosomes yields a single clone containing the entire BCL2 protooncogene. Proc. Nat. Acad. Sci. 87: 9913-9917, 1990.

Strasser, A.; Harris, A. W.; Cory, S.: bcl-2 transgene inhibits T cell death and perturbs thymic self-censorship. Cell 67: 889-899, 1991.

Tsujimoto, Y.: Overexpression of the human BCL-2 gene product results in growth enhancement of Epstein-Barr virus-immortalized B cells. Proc. Nat. Acad. Sci. 86: 1958-1962, 1989.

Tsujimoto, Y.; Bashir, M. M.; Givol, I.; Cossman, J.; Jaffe, E.; Croce, C. M.: DNA rearrangements in human follicular lymphoma can involve the 5-prime or the 3-prime region of the bcl-2 gene. Proc. Nat. Acad. Sci. 84: 1329-1331, 1987.

Tsujimoto, Y.; Cossman, J.; Jaffe, E.; Croce, C. M.: Involvement of the bcl-2 gene in human follicular lymphoma. Science 228: 1440-1443, 1985.

Tsujimoto, Y.; Croce, C. M.: Analysis of the structure, transcripts, and protein products of bcl-2, the gene involved in human follicular lymphoma. Proc. Nat. Acad. Sci. 83: 5214-5218, 1986.

Tsujimoto, Y.; Finger, L. R.; Yunis, J.; Nowell, P. C.; Croce, C. M.: Cloning of the chromosome breakpoint of neoplastic B cells with the t(14;18) chromosome translocation. Science 226: 1097-1099, 1985.

Tsujimoto, Y.; Gorham, J.; Cossman, J.; Jaffe, E.; Croce, C. M.: The t(14;18) chromosome translocations involved in B-cell neoplasms result from mistakes in VDJ joining. Science 229: 1390-1393, 1985.

Vaux, D. L.; Cory, S.; Adams, J. M.: Bcl-2 gene promotes haemopoietic cell survival and cooperates with c-myc to immortalize pre-B cells. Nature 335: 440-442, 1988.

Wang, H.-G.; Rapp, U. R.; Reed, J. C.: Bcl-2 targets the protein kinase Raf-1 to mitochondria. Cell 87: 629-638, 1996.

Williams, G. T.: Programmed cell death: apoptosis and oncogenesis. Cell 65: 1097-1098, 1991.

Yunis, J. J.; Frizzera, G.; Oken, M. M.; McKenna, J.; Theologides, A.; Arnesen, M.: Multiple recurrent genomic defects in follicular lymphoma: a possible model for cancer. New Eng. J. Med. 316: 79-84, 1987.

Yunis, J. J.; Mayer, M. G.; Arnesen, M. A.; Aeppli, D. P.; Oken, M. M.; Frizzera, G.: Bcl-2 and other genomic alterations in the prognosis of large-cell lymphoma. New Eng. J. Med. 320: 1047-1054, 1989.

Yunis, J. J.; Oken, M. M.; Kaplan, M. E.; Ensrud, K. M.; Howe, R. R.; Theologides, A.: Distinctive chromosomal abnormalities in histologic subtypes of non-Hodgkin's lymphoma. New Eng. J. Med. 307: 1231-1236, 1982.

MYCOSIS FUNGOIDES (254400)

JJM NOTE

Maybe the next research plan should be to compare haplotypes in affected sib-pairs to search for common predisposing genes.

Mycosis fungoides is a malignant T-cell lymphoma of the skin, first reported (and named) by Alibert (1835). Sandbank and Katzenellenbogen (1968) observed mycosis fungoides in brother and sister. Cameron (1933) reported the condition in mother and daughter. Shelley (1980) observed mycosis fungoides in father and daughter and reviewed the sparse reports of other familial occurrences. Hodgkin disease (236000) and various lymphomas also show familial aggregation. Greene et al. (1982) found that of 526 consecutive patients with cutaneous T-cell lymphomas, 21 had first-degree relatives with lymphoproliferative or hematopoietic malignancies—29 such malignancies in 21 kindreds. Hodgkin disease accounted for a third, with various leukemias (11 cases), non-Hodgkin lymphoma (5 cases), and multiple myeloma (3 cases) accounting for the rest.

Alibert, J. L. M.: Monographie des dermatoses. Paris: Germer Balliere (pub.) (2nd ed.): 1835.
Cameron, O. J.: Mycosis fungoides in mother and in daughter. Arch. Derm. 27: 232-236, 1933.
Greene, M. H.; Pinto, H. A.; Kant, J. A.; Siler, K.; Vonderheid, E. C.; Lamberg, S. I.; Dalager, N. A.: Lymphomas and leukemias in the relatives of patients with mycosis fungoides. Cancer 49: 737-741, 1982.
Sandbank, M.; Katzenellenbogen, I.: Mycosis fungoides of prolonged duration in siblings. Arch. Derm. 98: 620-627, 1968.
Shelley, W. B.: Familial mycosis fungoides revisited. Arch. Derm. 116: 1177-1178, 1980.

BURKITT LYMPHOMA (#113970)

JJM NOTE

In addition to the sibship mentioned by McKusick, Fred Li reported another family with members having Burkitt lymphoma and nasopharyngeal carcinoma (Li FP: Familial Burkitt lymphoma and nasopharyngeal carcinoma. Lancet 1:687–88, 1976). Given the high rate of Burkitt lymphoma in parts of Africa, the report of three sib-pairs in Nigeria might not be an excess (Salawu L et al.: Familial Burkitt's lymphoma in Nigerians. Ann. Trop. Paediatr. 17:375–79, 1997). On the other hand, perhaps four instances from a non-African population are more than expected (Winnett A et al.: Familial Burkitt's lymphoma in Papua New Guinea. Br. J. Cancer 75:757–61, 1997).

A number sign (#) is used with this entry because the phenotype does not reflect a single gene locus but rather the interaction of 2 separately discussed gene loci.

Burkitt lymphoma results from chromosomal translocations that involve the MYC gene (190080) and either the lambda or the kappa light chain immunoglobulin genes (147200, 147220). Burkitt lymphoma is causally related to the Epstein-Barr virus although the pathogenetic mechanisms are not clear. Most BL cell lines show a specific translocation involving chromosome 8 (breakpoint at 8q24) and either 2, 14 or 22. The type of immunoglobulins produced by this B-cell tumor correlates with the type of translocation (Lenoir et al., 1982): those with the 8;2 translocation produce predominantly kappa light chains; those with the 8;22 translocation produce lambda light chains; those with the 8;14 translocation produce immunoglobulins with both types of light chains. Furthermore, the kappa and lambda light chains map to the regions of 2p and 22q, respectively, that are involved in the breakpoint creating the translocations; in the 8;14 translocations, the breakpoint is the 14q32 band where the genes for immunoglobulin heavy chains map (Kirsch et al., 1982). Klein (1981) suggested that the consistent involvement of 8q24 may indicate that activation of an onc gene underlies this tumor. In this connection, it is noteworthy that the mos onc gene (190060) has been assigned to chromosome 8; the regional localization will be of interest, as well as information on mos DNA sequences in BL. In Burkitt lymphoma of the t(8;22) type, the breakpoint in chromosome 22 is proximal to the lambda immunoglobulin constant gene cluster (147220), whereas in the CML t(9;22) it is distal (Emanuel et al., 1984). Burkitt lymphoma and related neoplasms have their analog in murine plasmacytomas (also referred to as myelomas) in which a specific translocation occurs between mouse chromosome 15 and either mouse chromosome 12 (which in the mouse carries the heavy chain genes) or mouse chromosome 6 (which carries the kappa light chain genes). Calame et al. (1982) identified a region of DNA on mouse chromosome 1 5 that is commonly rearranged in transformed mouse lymphocytes.

Anderson et al. (1986) described 2 sisters in an American family who died of Burkitt lymphoma at ages 11 and 22 years. The mother and 2 healthy brothers had abnormality of lymphocyte subsets. An inherited disturbance of lymphocytes was thought to underlie the familial aggregation for Burkitt lymphoma. Haluska et al. (1987) suggested the following scenario for African Burkitt lymphoma: EBV is a polyclonal activator of B lymphocytes, and infection of normal B cells in vitro by EBV is associated with immortalization. In regions of equatorial Africa where Burkitt lymphoma is endemic, 80% of children demonstrate evidence of EBV infection. Malaria is also hyperendemic in the area and causes immunosuppression. Polyclonal B-lymphocyte proliferation therefore proceeds unchecked in the absence of T-cell suppression, probably enlarging the population of cells susceptible to translocation. Translocation involving the IgH locus leads to deregulation of the MYC oncogene. In Europe and North America, childhood EBV infection is less frequent, as is malaria. Burkitt lymphoma appears to occur in mature B cells following antigenic stimulation and during isotype switching. Haluska et al. (1987) presented evidence that the t(8;14) chromosome translocation of the Burkitt lymphoma cell line Daudi occurred during immunoglobulin gene rearrangement and involved the heavy chain diversity region (146910). They suggested that the translocation resulted from a recombinase error. Neri et al. (1988) showed that the endemic, sporadic, and AIDS-associated forms of Burkitt lymphoma carrying t(8;14) chromosomal translocations display different breakpoints within the immunoglobulin heavy-chain locus. Cloning and sequencing of the t(8;14) chromosomal junctions from 2 endemic BL cell lines and 1 endemic BL biopsy sample showed that the recombinations did not involve IGH-specific recombination signals on chromosome 14 or homologous sequences on chromosome 8. Thus, these events probably were not mediated by the same mechanisms or enzymes as in IGH rearrangement.

Denis Parsons Burkitt, who died in 1993 at the age of 82, was famed for the distinctive lymphoma he described and for the dietary fiber hypothesis he developed and espoused (Heaton, 1993).

See also Burkitt (1958); Burkitt (1983); Haluska et al. (1987); Pelicci et al. (1986); Zech et al. (1976)

Anderson, K. C.; Jamison, D. S.; Peters, W. P.; Li, F. P.: Familial Burkitt's lymphoma: association with altered lymphocyte subsets in family members. Am. J. Med. 81: 158-162, 1986.
Burkitt, D.: A sarcoma involving the jaws in African children. Brit. J. Surg. 46: 218-223, 1958.
Burkitt, D. P.: The discovery of Burkitt's lymphoma. Cancer 51: 1777-1786, 1983.
Calame, K.; Kim, S.; Lalley, P.; Hill, R.; Davis, M.; Hood, L.: Molecular cloning of translocations involving chromosome 15 and the immunoglobulin C-alpha gene from chromosome 12 in two murine plasmacytomas. Proc. Nat. Acad. Sci. 79: 6994-6998, 1982.
Emanuel, B. S.; Selden, J. R.; Wang, E.; Nowell, P. C.; Croce, C. M.: In situ hybridization and translocation breakpoint mapping. I. Nonidentical 22q11 breakpoints for the t(9;22) of Burkitt lymphoma. Cytogenet. Cell Genet. 38: 127-131, 1984.
Haluska, F. G.; Tsujimoto, Y.; Croce, C. M.: Mechanisms of chromosome translocation in B- and T-cell neoplasia. Trends Genet. 3: 11-15, 1987.
Haluska, F. G.; Tsujimoto, Y.; Croce, C. M.: The t(8;14) chromosome translocation of the Burkitt lymphoma cell line Daudi occurred during immunoglobulin gene rearrangement and involved the heavy chain diversity region. Proc. Nat. Acad. Sci. 84: 6835-6839, 1987.
Heaton, K.: Denis Burkitt. Lancet 341: 951-952, 1993.
Kirsch, I. R.; Morton, C. C.; Nakahara, K.; Leder, P.: Human immunoglobulin heavy chain genes map to a region of translocations in malignant B lymphocytes. Science 216: 301-303, 1982.
Klein, G.: The role of gene dosage and genetic transpositions in carcinogenesis. Nature 294: 313-318, 1981.
Lenoir, G. M.; Preud'homme, J. L.; Bernheim, A.; Berger, R.: Correlation between immunoglobulin light chain expression and variant translocation in Burkitt's lymphoma. Nature 298: 474-476, 1982.
Neri, A.; Barriga, F.; Knowles, D. M.; Magrath, I. T.; Dalla-Favera, R.: Different regions of the immunoglobulin heavy-chain locus are involved in chromosomal translocations in distinct pathogenetic forms of Burkitt lymphoma. Proc. Nat. Acad. Sci. 85: 2748-2752, 1988.
Pelicci, P.-G.; Knowles, D. M., II; Magrath, I.; Dalla-Favera, R.: Chromosomal breakpoints and structural alterations of the c-myc locus differ in endemic and sporadic forms of Burkitt lymphoma. Proc. Nat. Acad. Sci. 83: 2984-2988, 1986.
Zech, L.; Haglund, U.; Nilsson, K.; Klein, G.: Characteristic chromosomal abnormalities in biopsies and lymphoid-cell lines from patients with Burkitt and non-Burkitt lymphomas. Int. J. Cancer 17: 47-56, 1976.

MACROGLOBULINEMIA, WALDENSTROM (153600)

JJM NOTE

With a handful of tantalizing family reports, this disorder seems ripe for a concentrated effort to collect additional families, follow up the original ones, and apply current investigational tools.

Vannotti (1963) observed this in mother and son. Seligmann et al. (1963) had an instance of mother and 2 sons affected. Brown et al. (1967) found an abnormal chromosome in some lymphocytes of 5 members of 1 family; 3 of the 5 had protein abnormalities. See also Elves and Brown (1968). Fraumeni et al. (1975) described a kindred in which, in 1 sibship of 9 adults, 4 died of lymphocytic or histiocytic lymphomas and 1, a male, of Waldenstrom macroglobulinemia complicated by adenocarcinoma of the lung. In the next generation, 1 person died of Hodgkin disease; 4 of 9 healthy persons had impaired lymphocyte transformation with phytohemagglutinin, and 3 of these had polyclonal elevation of IgM. Subsequent to the studies, adenocarcinoma of the lung developed in one of those with an immune defect, a woman, and her 3-year-old grandson developed lymphocytic leukemia. This is the first suggestion of a genetic or immunologic basis of lung adenocarcinoma. In an Icelandic kindred, Bjornsson et al. (1978) observed a woman with macroglobulinemia who had 2 brothers with monoclonal macroglobulinemia—one asymptomatic and one with polyneuropathy and deposits of IgM in peripheral nerves. A third brother died of lymphoreticular disease, which presented as polyneuropathy. Protein abnormalities were found in 3 other sibs and 7 descendants. Blattner et al. (1980) found WM in a father and 3 children. Clinical and subclinical autoimmune disorders were also frequent in the family. All persons with WM and all but 1 with autoimmune manifestations had HLA haplotype A2/B8/DRw3. A lod score of 4.86 favored linkage of HLA and a gene predisposing to lymphoproliferative and autoimmune disorders. Fine et al. (1986) reported the occurrence of Waldenstrom macroglobulinemia in monozygotic twins. The monoclonal IgM in these 2 cases differed in their light chain type and their idiotypic determinants. Renier et al. (1989) reported 4 brothers with WM. Two of them had kappa type light chain IgM monoclo nal components, and 2 had lambda type light chain IgM monoclonal components. Anti-idiotypic rabbit antisera, prepared for each monoclonal component, showed no cross-reactivity. The 4 brothers did not share a common HLA haplotype, and a genetic linkage to the major histocompatibility complex could not be demonstrated. Five of 12 relatives had high serum immunoglobulin concentration without monoclonal components.

See also Getaz and Staples (1977); Massari et al. (1962); San Roman et al. (1985)

Bjornsson, O. G.; Arnason, A.; Gudmundsson, S.; Jensson, O.; Olafsson, S.; Valdimarsson, H.: Macroglobulinaemia in an Icelandic family. Acta Med. Scand. 203: 283-288, 1978.

Blattner, W. A.; Garber, J. E.; Mann, D. L.; McKeen, E. A.; Henson, R.; McGuire, D. B.; Fisher, W. B.; Bauman, A. W.; Goldin, L. R.; Fraumeni, J. F.: Waldenstrom's macroglobulinemia and autoimmune disease in a family. Ann. Intern. Med. 93: 830-832, 1980.

Brown, A. K.; Elves, M. W.; Gunson, H. H.; Pell-Ilderton, R.: Waldenstrom's macroglobulinaemia. A family study. Acta Haemat. 38: 184-192, 1967.

Elves, M. W.; Brown, A. K.: Cytogenetic studies in a family with Waldenstrom's macroglobulinaemia. J. Med. Genet. 5: 118-122, 1968.

Fine, J. M.; Muller, J. Y.; Rochu, D.; Marneux, M.; Gorin, N. C.; Fine, A.; Lambin, P.: Waldenstrom's macroglobulinemia in monozygotic twins. Acta Med. Scand. 220: 369-373, 1986.

Fraumeni, J. F., Jr.; Wertelecki, W.; Blattner, W. A.; Jensen, R. D.; Leventhal, B. G.: Varied manifestations of a familial lymphoproliferative disorder. Am. J. Med. 59: 145-151, 1975.

Getaz, E. P. S.; Staples, W. G.: Familial Waldenstrom's macroglobulinaemia: a case report. S. Afr. Med. J. 51: 891-892, 1977.

Massari, R.; Fine, J. M.; Metais, R.: Waldenstrom's macroglobulinaemia observed in two brothers. Nature 196: 176-178, 1962.

Renier, G.; Ifrah, N.; Chevailler, A.; Saint-Andre, J. P.; Boasson, M.; Hurez, D.: Four brothers with Waldenstrom's macroglobulinemia. Cancer 64: 1554-1559, 1989.

San Roman, C.; Ferro, T.; Guzman, M.; Odriozola, J.: Clonal abnormalities in patients with Waldenstrom's macroglobulinemia with special reference to a Burkitt-type t(8;14). Cancer Genet. Cytogenet. 18: 155-158, 1985.

Seligmann, M.; Danon, F.; Fine, J. M.: Immunological studies in familial beta 2-macroglobulinaemias. Proc. Soc. Exp. Biol. Med. 114: 482-486, 1963.

Vannotti, A.: Etude clinique d'un cas de macroglobulinemie de Waldenstrom a caractere familial, associe a des troubles endocriniens. Schweiz. Med. Wschr. 93: 1744-1746, 1963.

MYELOMA, MULTIPLE (#254500)

JJM NOTE

Again, the case reports have accumulated to the point where a renewed interdisciplinary effort would likely be productive.

A number sign (#) is used with this entry because a number of different molecular changes have been found underlying this disorder. Dysregulation of oncogenes by translocations to the IgH locus (147100) on 14q32 is a seminal event in the pathogenesis of B-cell tumors, including multiple myeloma. In the latter condition, translocations to the IgH locus have been reported at an incidence of 20 to 60%. Cyclin D1 (168461), located on 11q13, is frequently involved; the fibroblast growth factor receptor 3 gene (134934) has also been implicated. See also multiple myeloma oncogene-1 (IRF4; 601900).

Leoncini and Korngold (1964) described multiple myeloma in 2 sisters and reviewed the literature on familial cases. Manson (1961) reported affected sisters, one of whom also had pernicious anemia. Myeloma has also been observed in father and son (Nadeau et al., 1956). Thomas (1964) observed myeloma in a brother and sister. Alexander and Benninghoff (1965) described 3 affected black sibs. Whitehouse (1971) observed affected brother and sister. In a large population survey in Sweden, Axelsson and Hallen (1965) found 2 families, one with 2 and one with 3 sibs, showing high M-component. In a third family, 2 persons with high M-component were more remotely related. These 7 were from a total group of 59 (out of 7,918) found to have M-component. Their condition was considered to be a variety of essential benign monoclonal hypergammaglobulinemia. Berlin et al. (1968) described familial occurrence of M-components. One possible explanation for familial paraproteinemia is that plasma cell clones with similar structural genes for the paraprotein synthesized by these cells proliferate in related individuals. This hypothesis predicts that paraproteins from 2 members of the same family would be identical. The paraproteins of a mother with multiple myeloma and a son with probably benign monoclonal gammopathy were isolated by Grant et al. (1971). Light chains were of the lambda type, but had differences on peptide map in both the common and variable regions of the proteins. These data show that the structural genes operative in paraprotein light chain production in these first-degree relatives are different. The presence of a genetic basis is suggested by the occurrence of 2 different monoclonal gammopathies in 1 patient. Humphrey (1973), for example, told me of a patient who had an intracranial plasmacytoma which was surgically removed. Six years later she developed a plasmacytoma of one kidney. The second tumor produced a different gamma globulin from that release d into the cerebrospinal fluid by the brain plasmacytoma.

Zawadzki et al. (1977) described 19 cases of familial immunopathy, distributed in 9 families. Ten members of 5 families had multiple myeloma, 5 members of 2 families had lanthanic paraproteinemia, and 4 members of 2 families had one or the other of these. 'Lanthanic' is from a Greek word meaning 'to escape.' It is used in place of 'benign' because malignant immunocytic dyscrasia has been known to emerge. The term is intended to convey that the condition was asymptomatic and came to attention only by serendipity. (Actually, in the course of a specific study of relatives of clinically affected probands, this is not serendipity; Walpole's Prince of Serendip set out to find one thing and instead found something else (Cannon, 1945).) Blattner (1980) gave an excellent review, with a classification of monoclonal gammopathies. Multiple myeloma and Waldenstrom macroglobulinemia (153600) are presumably closely related; both are malignant monoclonal gammopathies. Multiple myeloma is about 2 times more frequent in U.S. blacks than in U.S. whites; it is the eleventh and twentieth most frequent malignancy in the 2 races, respectively.

Horwitz et al. (1985) reported 3 affected sibs and stated that a review of the literature revealed reports of 38 affected pairs of sibs, 8 families with 3 affected sibs, and 4 families with another affected relative (in addition to the pair of affected sibs). Comotti et al. (1987) and Judson et al. (1985) reported identical twins concordant for multiple myeloma. Gertz et al. (1986) reported primary immunoglobulin-related amyloidosis (AL) in 2 members of each of 3 families: 2 brothers, a brother and a sister, and 2 first cousins. Primary amyloidosis of this type may be closely akin to multiple myeloma and to Waldenstrom macroglobulinemia. Closely akin also may be monoclonal gammopathy of undetermined significance (MGUS), that is, without current signs of malignant lymphocytic or plasmocytic disease. Jensen et al. (1988) described a brother and sister with progressive mixed axonal and demyelinating polyneuropathy in association with a monoclonal IgM gammopathy of kappa and lambda type, respectively. Sera from both patients contained antibodies directed to bovine peripheral nerve myelin as determined by ELISA technique and to normal human peripheral nerve myelin as demonstrated by indirect immunofluorescence histochemistry. These sibs may have had a genetic predisposition to the formation of autoantibodies with peripheral nerve myelin as the target.

Miliani et al. (1996) described 3 Italian sibs (2 brothers and a sister) with immunoglobulin-related amyloidosis. Systemic amyloidosis was associated with monoclonal gammopathy in all 3. One of the sibs had Waldenstrom macroglobulinemia, whereas the other 2 had no evidence of multiple myeloma or related diseases. All 3 sibs showed a common pattern of polyneuropathy to different degrees; 2 presented a sicca syndrome and 1 also suffered from nephropathy.

Plasma cell tumor induction in mice by pristane is under multigenic control. Backcross and congenic strain analyses indicated that at least 4 genes determine the susceptibility to mouse plasmacytomagenesis. One of these genes, Pctr1, resides in the mid-portion of mouse chromosome 4 near the alpha-interferon locus. Zhang et al. (1998) presented evidence that Cdkn2a (600160) is a strong candidate for the Pctr1 locus.

See also Goldstone et al. (1973); Herrell et al. (1958); Maldonado and Kyle (1974); Rostoker et al. (1986)

Alexander, L. L.; Benninghoff, D. L.: Familial multiple myeloma. J. Nat. Med. Assoc. 57: 471-475, 1965.

Axelsson, U.; Hallen, J.: Familial occurrence of pathological serum-proteins of different gamma-globulin groups. Lancet II: 369-370, 1965.

Berlin, S. O.; Odeberg, H.; Weingart, L.: Familial occurrence of M-components. Acta Med. Scand. 183: 347-350, 1968.

Blattner, W. A.: Epidemiology of multiple myeloma and related plasma cell disorders: an analytic review.In: Potter, M.: Progress in Myeloma. Amsterdam: Elsevier/North Holland (pub.) 1980. Pp. 1-65.
Cannon, W. B.: The Way of the Investigator. New York: W. W. Norton (pub.) 1945.
Comotti, B.; Bassan, R.; Buzzetti, M.; Finazzi, G.; Barbui, T.: Multiple myeloma in a pair of twins. Brit. J. Haemat. 65: 123-124, 1987.
Gertz, M. A.; Garton, J. P.; Kyle, R. A.: Primary amyloidosis (AL) in families. Am. J. Hemat. 22: 193-198, 1986.
Goldstone, A. H.; Wood, J. K.; Cook, M. K.: Myeloma in mother and daughter. Acta Haemat. 49: 176-181, 1973.
Grant, J. A.; Blumenschein, G. R.; Buckley, C. E., III: Familial paraproteinemia. Arch. Intern. Med. 128: 427-431, 1971.
Herrell, W. E.; Ruff, J. D.; Bayrd, E. D.: Multiple myeloma in siblings. J.A.M.A. 167: 1485-1487, 1958.
Horwitz, L. J.; Levy, R. N.; Rosner, F.: Multiple myeloma in three siblings. Arch. Intern. Med. 145: 1449-1450, 1985.
Humphrey, R. L.: Personal Communication. Baltimore, Md. 1973.
Jensen, T. S.; Schroder, H. D.; Jonsson, V.; Ernerudh, J.; Stigsby, B.; Kamieniecka, Z.; Hippe, E.; Trojaborg, W.: IgM monoclonal gammopathy and neuropathy in two siblings. J. Neurol. Neurosurg. Psychiat. 51: 1308-1315, 1988.
Judson, I. R.; Wiltshaw, E.; Newland, A. C.: Multiple myeloma in a pair of monozygotic twins: the first reported case. Brit. J. Haemat. 60: 551-554, 1985.
Leoncini, D. L.; Korngold, L.: Multiple myeloma in 2 sisters: an immunochemical study. Cancer 17: 733-737, 1964.
Maldonado, J. E.; Kyle, R. A.: Familial myeloma: report of eight families and a study of serum proteins in their relatives. Am. J. Med. 57: 875-884, 1974.
Manson, D. I.: Multiple myeloma in sisters. Scot. Med. J. 6: 188, 1961.
Miliani, A.; Bergesio, F.; Salvadori, M.; Amantini, A.; Macucci, M.; Arbustini, E.; Becucci, A.; Sodi, A.; Zuccarini, S.; Menicucci, A.; Torricelli, F.; Capobianco, T.; Di Lollo, S.; Piazza, E.; Gemmi, F.; Cozzolino, F.; Merlini, G.: Familial AL-amyloidosis in three Italian siblings. Haematologica 81: 105-109, 1996.
Nadeau, L. A.; Magalini, S. I.; Stefanini, M.: Familial multiple myeloma. Arch. Path. 61: 101-106, 1956.
Rostoker, G.; Uzzan, B.; Baumelou, E.; Chapman, A.: Myelome familial IgG kappa chez une mere et sa fille: revue de la litterature. Nouv. Rev. Franc. Hemat. 28: 27-32, 1986.
Thomas, T. F.: Multiple myeloma in siblings. New York J. Med. 64: 2096-2099, 1964.
Whitehouse, S.: Personal Communication. Baltimore, Md. 1971.
Zawadzki, Z. A.; Aizawa, Y.; Kraj, M. A.; Haradin, A. R.; Fisher, B.: Familial immunopathies: report of nine families and survey of literature. Cancer 40: 2094-2101, 1977.
Zhang, S.; Ramsay, E. S.; Mock, B. A.: Cdkn2a, the cyclin-dependent kinase inhibitor encoding p16-INK4-alpha and p19-ARF, is a candidate for the plasmacytoma susceptibility locus, Pctr1. Proc. Nat. Acad. Sci. 95: 2429-2434, 1998.

PLASMACYTOMA-ASSOCIATED NEURONAL GLYCOPROTEIN (*601325)

JJM NOTE

This is a candidate gene that may prove to be involved in human plasmacytoma.

Pang, a mouse gene encoding a neuronal adhesion molecule, was isolated as a plasmacytoma-specific transcript using an RT-PCR-based strategy in an attempt to isolate Myc-like genes in murine plasmacytomas (Connelly et al., 1994). Pang is a member of the immunoglobulin/fibronectin superfamily of adhesion molecules; its closest relatives, TAG1 (190197) and contactin 1 (600016), promote axon growth and migration. The normal site of Pang expression is the brain, where it is detected as 4.0- and 6.1-kb RNAs on Northern blots; Pang is not detected in other normal tissues. Abnormally sized Pang transcripts were uniquely found in murine plasmacytomas, where it is ectopically activated by intracisternal A-type particle long terminal repeats. Mock et al. (1996) mapped the Pang gene to mouse chromosome 6 by somatic cell hybrid analysis and further positioned it on the chromosome between Wnt7a and Pcp1. Southern blot analysis of human-rodent somatic cell hybrids together with predictions from the mouse map location indicated that human PANG is located at 3p26.

Connelly, M. A.; Grady, R. C.; Mushinski, J. F.; Marcu, K. B.: PANG, a gene encoding a neuronal glycoprotein, is ectopically activated by intracisternal A-type particle long terminal repeats in murine plasmacytomas. Proc. Nat. Acad. Sci. 91: 1337-1341, 1994.
Mock, B. A.; Connelly, M. A.; McBride, O. W.; Kozak, C. A.; Marcu, K. B.: Plasmacytoma-associated neuronal glycoprotein, Pang, maps to mouse chromosome 6 and human chromosome 3. Genomics 34: 226-228, 1996.

RENAL DYSPLASIA, DIFFUSE CYSTIC (601331)

JJM NOTE

This new entity is included not because of known cases of cancer, but because of the likely predisposition represented by this type of dysplasia.

Sase et al. (1996) described a family in which 3 sibs, a male and 2 females, were affected with nonsyndromal diffuse cystic dysplasia of the kidneys. The parents were nonconsanguineous and had no renal abnormality. A possibly similar condition was reported by Cole et al. (1976) in 3 sibs (2 females and 1 male) who were born to nonconsanguineous parents with apparently normal kidneys. The first affected sib died soon after birth; the other 2 were found to have the disorder by ultrasonography and the pregnancies were terminated. Autopsies were performed in all 3 cases. There were no other external or internal malformations except for hypoplastic lungs related to the renal abnormality. Sase et al. (1996) suggested that the disorder is an autosomal recessive. Diffuse cystic renal dysplasia is often seen in malformation syndromes, of which the investigators listed a number.

Cole, B. R.; Kaufman, R. L.; McAlister, W. H.; Kissane, J. M.: Bilateral renal dysplasia in three siblings: report of a survivor. Clin. Nephrol. 5: 83-87, 1976.
Sase, M.; Tsukahara, M.; Oga, A.; Kaneko, N.; Nakata, M.; Saito, T.; Kato, H.: Diffuse cystic renal dysplasia: nonsyndromal familial case. Am. J. Med. Genet. 63: 332-334, 1996.

CRYOGLOBULINEMIA, FAMILIAL MIXED (MELTZER SYNDROME) (*123550)

JJM NOTE

With all the automatic analyses of serum proteins, additional families should be identifiable for further studies.

Nightingale et al. (1977) described affected mother and 4 children. The proband developed hematuria and anasarca at age 12, and showed progressively deteriorating renal function thereafter. At age 39 the mother was found to have mild hypertension, proteinuria, an 'active urine sediment' and elevated serum creatinine. The proband, mother, and 3 sibs had a mixed IgG-IgA-IgM cryoglobulinemia. In later studies, Nightingale and Pelley (1981) found 10 members of 3 generations of the family to have IgM-IgG cryoglobulins and rheumatoid factor; another member had rheumatoid factor but no cryoglobulins. Nightingale and Pelley (1981) described an antigen, first identified on the IgM cryoglobulin of the proband, which was present in the serum of all 11 members of the family with rheumatoid factor. This antigen had the serologic properties of an IgM rheumatoid factor idiotype. Nightingale et al. (1981) concluded that inheritance of 'essential' mixed cryoglobulinemia was autosomal dominant; there were 2 instances of father-to-son transmission. No underlying disease that could account for cryoglobulinemia was found in any patient. No linkage to HLA, blood groups, or immunoglobulin Gm allotypes was found. Although the rheumatoid factors in this kindred reacted with some human IgG, they were not antibodies to any known Gm or Km allotype. Cryoglobulinemia with systemic manifestations has acquired the name of Meltzer syndrome.

See also Meltzer and Franklin (1966); Sitomer et al. (1963)

Meltzer, M.; Franklin, E. C.: Cryoglobulinemia—a clinical and laboratory study. Am. J. Med. 40: 837-856, 1966.
Nightingale, S. D.; Pelley, R. P.: A shared cryoglobulin antigen in familial cryoglobulinemia. Am. J. Hum. Genet. 33: 722-734, 1981.
Nightingale, S. D.; Pelley, R. P.; Delaney, N. L.; Bias, W. B.; Hamburger, M. I.; Fries, L. F.; Steinberg, A. G.: Inheritance of mixed cryoglobulinemia. Am. J. Hum. Genet. 33: 735-744, 1981.
Nightingale, S. D.; Solez, K.; Humphrey, R. L.: Familial cryoglobulinemia. Johns Hopkins Med. J. 140: 267-274, 1977.
Sitomer, G.; Blum, J. J.; Slavin, R. E.: Cryoglobulinemia: an inherited molecular disease?. Am. J. Med. 34: 565-571, 1963.

THYMIC NEOPLASIA (274260)

JJM NOTE

Thymic carcinoid has occurred in multiple endocrine neoplasia 2, Bruton agammaglobulinemia, and myasthenia gravis, which itself has significant familial aggregation.

Wick et al. (1982) described 2 brothers who developed malignant epithelial tumors of the thymus. One had thymic carcinoma with death in his late 50s; the other had an invasive spindle cell thymoma with associated hypogammaglobulinemia. The authors thought the occurrence particularly significant because thymic tumors are rare and malignant forms especially so.

Wick, M. R.; Scheithauer, B. W.; Dines, D. E.: Thymic neoplasia in two male siblings. Mayo Clin. Proc. 57: 653-656, 1982.

THYMOMA, FAMILIAL (274230)

JJM NOTE

This is another family that deserves follow-up investigation and report.

Matani and Dristsas (1973) reported a Greek sibship of 3, demonstrating familial occurrence of thymoma. One of the 3 sibs, a 2-year-old girl, died of respiratory insufficiency resulting from a lymphocytic thymoma. Her 9-month-old brother died 2 years earlier of the same cause. The eldest sib, a male, and the parents were healthy. No reference was made concerning parental consanguinity. Thymomas are notably rare in persons of this young age.

See also Lattes (1962); Legg and Brady (1965)

Lattes, R.: Thymoma and other tumors of the thymus: an analysis of 107 cases. Cancer 15: 1224-1260, 1962.
Legg, M. A.; Brady, W. J.: Pathology and clinical behavior of thymomas: a survey of 51 cases. Cancer 18: 1131-1144, 1965.
Matani, A.; Dristsas, C.: Familial occurrence of thymoma. Arch. Path. 95: 90-91, 1973.

HODGKIN DISEASE (236000)

JJM NOTE

Hodgkin disease has occurred in patients or families with ataxia-telangiectasia, mycosis fungoides, cartilage-hair hypoplasia, Swiss-type agammaglobulinemia, dyskeratosis congenita, Bloom syndrome, Li-Fraumeni syndrome, and Muir-Torre syndrome. The familial aggregation of young age Hodgkin disease is conspicuous, but tends to be ascribed to environmental (perhaps viral) determinants, not to genetic susceptibility. The argument of Bodmer (1986) deserves emphasis: especially for recessive susceptibilty genes, familial aggregation can be infrequent, even in the presence of a single major gene determinant.

Manigand et al. (1964) described a brother and sister with Hodgkin disease and reviewed the literature on familial occurrence. In familial cases Vianna et al. (1974) found that the time-intervals between diagnoses were shorter than the age differences. Also the time-intervals were shorter for relatives living together than for those living apart. These findings suggested an environmental basis for familial occurrence. On the other hand, the similarity of Rye histologic type in relatives, regardless of proximity, suggested a genetic factor in host reactivity. Fraumeni et al. (1975) described a kindred in which, in 1 sibship of 9 adults, 4 died of lymphocytic or histiocytic lymphomas and 1, a male, died of Waldenstrom macroglobulinemia complicated by adenocarcinoma of the lung. In the next generation, 1 person died of Hodgkin disease; 4 of 9 healthy persons had impaired lymphocyte transformation with phytohemagglutinin, and 3 of these had polyclonal elevation of IgM. Subsequent to the studies, adenocarcinoma of the lung developed in 1 of those with an immune defect, a woman, and her 3-year-old grandson developed lymphocytic leukemia. This was the first suggestion of a genetic or immunologic basis of lung adenocarcinoma.

Grufferman et al. (1977) found a 7-fold increased risk of Hodgkin disease in sibs under 45 years of age. Twelve of 13 sib pairs were sex-concordant. The series showed an excess of nodular sclerosis type and pairs showed an excess of concordance for this type. This type may be the form most likely to have an infectious basis (Cole et al., 1968). Apparently no excess sib risk exists for Hodgkin disease diagnosed after age 45. MacMahon (1966) suggested that the cause may differ in the young and old. In general, a primarily nongenetic basis is suggested by these findings. Conte et al. (1983) studied 4 families, each with 2 cases of Hodgkin disease of the nodular sclerosis type. All 8 patients showed B18 antigen. The affected persons included father-son, brother-sister, mother-son, and father-daughter pairs.

Bodmer (1986) emphasized that the sib pair method is an effective way to demonstrate linkage of a disease with HLA. It is necessary to look at only affected persons, if reduced penetrance is likely. The expected distribution of both marker phenotype 1, one marker-1 with the other marker-2, and both marker phenotype 2 is 1 in 4, 1 in 2, and 1 in 4. The distribution for Hodgkin disease in 32 sib pairs (both affected) was 16, 11 and 5—a difference from 8, 16, and 8 significant at the 0.005 level. Bodmer (1986) concluded from this that there must be a gene for Hodgkin disease in or near the HLA complex. He pointed out, furthermore, that rarity of familial clustering does not exclude an important genetic component. His argument was as follows: Assume that there is a gene d with a frequency of 1 in 1000, and that 100% of dd persons develop Hodgkin disease whereas only 5% of dD persons and 0% of DD persons develop Hodgkin disease. Under these circumstances, only about 1 of every 100 cases of Hodgkin disease will be a homozygote; the overwhelming majority will be heterozygotes. Under the terms of this model, the frequency of 2-child families with both affected can be calculated to be about 1 in 400,000. This value is calculated from the product of the proportion of 2-sib families that have dD x DD parents (about 1 in 250) and the proportion of 2-sib families in which both sibs inherit the gene (1 in 4) and express the disease phenotype (1 in 400). This model permits the possibility that in some populations the recessive gene may achieve considerably higher frequency through drift/founder effect, resulting in more impressive familial clustering.

A link between Hodgkin disease and Epstein Barr virus (EBV) was first hinted at through epidemiologic studies (Munoz et al., 1978; Mueller et al., 1989). A possible role for EBV in the pathogenesis of Hodgkin disease was suggested by direct detection of EBV DNA in tumor biopsies in up to 50% of Hodgkin disease cases (25,24:Weiss et al., 1987, 1989; Wu et al., 1990) and the finding of a consistent pattern of EBV latent gene transcription in EBV-positive Reed-Sternberg cells (Pallesen et al., 1991; Deacon et al., 1993), as demonstrated by in situ hybridization. Performing molecular typing of HLA class II loci, Klitz et al. (1994) found an overall association of the nodular sclerosing (NSHD) group with the HLA class II region. The results of the study suggested that susceptibility to NSHD is influenced by more than 1 locus within the class II region. Haplotypes and alleles defined for the 4 loci typed (DRB1, DQA1, DQB1, and DPB1; see 142857 and 142858) were present in both excess and deficit in the sample of NSHD (N = 155; age range 9-62 years). Klitz et al. (1994) suggested that the data explain why earlier studies showed HLA linkage but not association and that they substantiate the specific involvement of the immune system in certain neoplastic diseases.

Relatives of young adults with Hodgkin disease are at increased risk of Hodgkin disease, and, as indicated earlier, lines of evidence implicate both inheritance and environment. Mack et al. (1995) identified and followed 432 sets of twins affected by Hodgkin disease. The twins were identified through weekly advertisements between 1980 and 1992 requesting that 'twins with cancer' contact the group. The number of cases of Hodgkin disease observed before the age of 50 years in the healthy monozygotic and dizygotic cotwins of the patients with Hodgkin disease was compared with the number expected from national age-specific incidence rates. None of the 187 pairs of dizygotic twins became concordant for Hodgkin disease, whereas 10 of the 179 pairs of monozygotic twins did; in 5 of the 10 pairs, the second case appeared after the original ascertainment. During the observation period, 0.1 (monozygotic) and 0.1 (dizygotic) cases in the unaffected twins were expected. Mack et al. (1995) concluded that genetic susceptibility underlies Hodgkin disease in young adulthood.

Ferraris et al. (1997) quoted an estimate of 4.5% as the proportion of all cases of Hodgkin disease represented by familial Hodgkin disease. They reviewed 28 articles on familial Hodgkin disease, published between 1972 and 1995, with detailed analysis of data from 18 papers, reporting on 328 patients. The male-to-female ratio was 1.5, similar to that reported for sporadic Hodgkin disease. A significant difference was found between sporadic and familial Hodgkin disease in the age at diagnosis: only 1 major peak between 15 and 34 years was present in the group of patients with familial Hodgkin disease. For the sporadic disease, the age of onset is characteristically bimodal, with a peak in the late twenties and another after the age of 50. A notable exception is Japan, where the young-adult incidence peak of Hodgkin disease is barely detectable.

See also Bowers et al. (1977); Carter et al. (1978); Creagan and Fraumeni (1972); Fraumeni (1974); Lynch et al. (1976); Marshall et al. (1977); Marshall and Thompson (1981); Torres et al. (1980)

Bodmer, W. F.: Personal Communication. London, England 9/26/1986.
Bowers, T. K.; Moldow, C. F.; Bloomfield, C. D.; Yunis, E. J.: Familial Hodgkin's disease and the major histocompatibility complex. Vox Sang. 33: 273-277, 1977.

Carter, N. D.; Van Loghem, E.; Marshall, W. H.; Newton, R. M.; West, C. M.: Serum genetic markers in a Newfoundland isolate with a familial aggregate of Hodgkin's disease. Hum. Hered. 28: 372-379, 1978.

Cole, P.; MacMahon, B.; Aisenberg, A.: Mortality from Hodgkin's disease in the United States: evidence for the multiple aetiology hypothesis. Lancet II: 1371-1376, 1968.

Conte, R.; Lauria, F.; Zucchelli, P.: HLA in familial Hodgkin's disease. J. Immunogenet. 10: 251-255, 1983.

Creagan, E. T.; Fraumeni, J. F., Jr.: Familial Hodgkin's disease. (Letter) Lancet II: 547 only, 1972.

Deacon, E. M.; Pallesen, G.; Niedobitek, G.; Crocker, J.; Brooks, L.; Rickinson, A. B.; Young, L. S.: Epstein-Barr virus and Hodgkin's disease: transcriptional analysis of virus latency in the malignant cells. J. Exp. Med. 177: 339-349, 1993.

Ferraris, A. M.; Racchi, O.; Rapezzi, D.; Gaetani, G. F.; Boffetta, P.: Familial Hodgkin's disease: a disease of young adulthood? Ann. Hemat. 74: 131-134, 1997.

Fraumeni, J. F., Jr.: Family studies in Hodgkin's disease. Cancer Res. 34: 1164-1165, 1974.

Fraumeni, J. F., Jr.; Wertelecki, W.; Blattner, W. A.; Jensen, R. D.; Leventhal, B. G.: Varied manifestations of a familial lymphoproliferative disorder. Am. J. Med. 59: 145-151, 1975.

Grufferman, S.; Cole, P.; Smith, P. G.; Lukes, R. J.: Hodgkin's disease in siblings. New Eng. J. Med. 296: 248-250, 1977.

Klitz, W.; Aldrich, C. L.; Fildes, N.; Horning, S. J.; Begovich, A. B.: Localization of predisposition to Hodgkin disease in the HLA class II region. Am. J. Hum. Genet. 54: 497-505, 1994.

Lynch, H. T.; Salvidar, V. A.; Guirgis, H. A.; Terasaki, P. I.; Bardawil, W. A.; Harris, R. E.; Lynch, J. F.; Thomas, R.: Familial Hodgkin's disease and associated cancer: a clinical-pathologic study. Cancer 38: 2033-2041, 1976.

Mack, T. M.; Cozen, W.; Shibata, D. K.; Weiss, L. M.; Nathwani, B. N.; Hernandez, A. M.; Taylor, C. R.; Hamilton, A. S.; Deapen, D. M.; Rappaport, E. B.: Concordance for Hodgkin's disease in identical twins suggesting genetic susceptibility to the young-adult form of the disease. New Eng. J. Med. 332: 413-418, 1995.

MacMahon, B.: Epidemiology of Hodgkin's disease. Cancer Res. 26: 1189-1200, 1966.

Manigand, G.; Macrez, C.; Chome, J.; Bosson, C. H.; Delamare, J.; Deparis, M.: Maladie de Hodgkin familiale. Presse Med. 72: 1871-1874, 1964.

Marshall, W. H.; Barnard, J. M.; Buehler, S. K.; Crumley, J.; Larsen, B.: HLA in familial Hodgkin's disease. Results and a new hypothesis. Int. J. Cancer 19: 450-455, 1977.

Marshall, W. H.; Thompson, E. A.: Inferring the existence of a Hodgkin disease gene in North West Newfoundland. (Abstract) Sixth Int. Cong. Hum. Genet., Jerusalem 308 only, 1981.

Mueller, N.; Evans, A.; Harris, N. L.; Comstock, G. W.; Jellum, E.; Magnus, K.; Orentreich, N.; Polk, B. F.; Vogelman, J.: Hodgkin's disease and Epstein-Barr virus: altered antibody pattern before diagnosis. New Eng. J. Med. 320: 689-695, 1989.

Munoz, N.; Davidson, R. J. L.; Witthoff, B.; Ericsson, J. E.; de-The, G.: Infectious mononucleosis and Hodgkin's disease. Int. J. Cancer 22: 10-13, 1978.

Pallesen, G.; Hamilton-Dutoit, S. J.; Rowe, M.; Young, L. S.: Expression of Epstein-Barr virus latent gene products in tumour cells of Hodgkin's disease. Lancet 337: 320-322, 1991.

Torres, A.; Martinez, F.; Gomez, P.; Gomez, C.; Garcia, J. M.; Nunez-Roldan, A.: Simultaneous Hodgkin's disease in three siblings with identical HLA-genotype. Cancer 46: 838-843, 1980.

Vianna, N. J.; Davies, J. N. P.; Polan, A. K.; Wolfgang, P.: Familial Hodgkin's disease: an environmental and genetic disorder. Lancet II: 854-857, 1974.

Weiss, L. M.; Movahed, L. A.; Warnke, R. A.; Sklar, J.: Detection of Epstein-Barr viral genomes in Reed-Sternberg cells of Hodgkin's disease. New Eng. J. Med. 320: 502-506, 1989.

Weiss, L. M.; Strickler, J. G.; Warnke, R. A.; Purtilo, D. T.; Sklar, J.: Epstein-Barr viral DNA in tissues of Hodgkin's disease. Am. J. Path. 129: 86-91, 1987.

Wu, T.-C.; Mann, R. B.; Charache, P.; Hayward, S. D.; Staal, S.; Lambe, B. C.; Ambinder, R. F.: Detection of EBV gene expression in Reed-Sternberg cells of Hodgkin's disease. Int. J. Cancer 46: 801-804, 1990.

HYPERKERATOSIS, LOCALIZED EPIDERMOLYTIC (*144200)

JJM NOTE

This entity is included next to Hodgkin disease because of one affected family, now reported fully, in which 23% of 43 adults had the skin disease as well as Hodgkin disease and adenocarcinomas of the kidney, breast, pancreas, and colon (Stevens HP et al.: Punctate palmoplantar keratoderma and malignancy in a four-generation family. Br. J. Dermatol. 134:720–26, 1996).

Localized epidermolytic hyperkeratosis was first described by Vorner (1901). Klaus et al. (1970) demonstrated dominant inheritance and male-to-male transmission. Blasik et al. (1981), Fritsch et al. (1978), and Camisa and Williams (1985) reported affected families. In an affected father and daughter, Moriwaki et al. (1988) noted a decrease in 67-kD keratin in the involved epidermis and the appearance of 48-kD keratin. Southern blot analysis using 67-kD keratin cDNA showed no abnormality in the gene for 67-kD keratin. Alsaleh and Teebi (1990) described 2 affected sons of phenotypically normal, consanguineous Arab parents. They presented with patchy eczematous skin lesions followed by palmoplantar keratoderma and raised serum levels of IgE. Alsaleh and Teebi (1990) suggested the existence of an autosomal recessive form, but because of the high background frequency of consanguinity in Kuwait where the patients were observed, and because of the possibility of gonadal mosaicism, the evidence for recessive inheritance is not strong in this small family. Nogita et al. (1991) described the disorder in a 32-year-old Japanese man who had knuckle-pad-like lesions over the dorsal aspect of the proximal interphalangeal joints since the age of 3 or 4 months and whose father had had the same skin disease.

Keratosis palmaris et plantaris, first described by Thost (1880) and Unna (1883) and later delineated by Greither (1952), is characterized by diffuse hyperkeratosis of the palms and soles which usually becomes evident between the ages of 3 and 12 months. Low serum vitamin A has been found in some cases. Goette (1974) described successful use of topical vitamin A. The family of Anderson and Klintworth (1961) had clinodactyly also, possibly as an independent trait (see 148520). In addition to the diffuse type referred to here, a punctate type (see keratosis palmoplantaris papulosa, 148600) and a linear or striate form (see keratosis palmoplantaris striata, 148700) are recognized on morphologic grounds. It is possible that these are genetically distinct from the diffuse type but such cannot be considered proved. Gamborg Nielsen (1985) did a follow-up study on hereditary palmoplantar keratoderma originally surveyed in the northernmost county of Sweden (Norrbotten) by Bergstrom (1967). Two clinical types were found: a common form with the usual autosomal dominant inheritance and a severe form thought to have autosomal recessive inheritance (see 244850). Mal de Meleda (248300) is a recessive form of keratosis palmoplantaris.

The Unna-Thost form of palmoplantar keratoderma is clinically identical to localized epidermolytic hyperkeratosis of the palms and soles as described by Vorner (1901). Although they were earlier thought to be histopathologically distinguishable, reappraisal of the kindred originally analyzed by Thost (1880) and study of a descendant supported identity of the 2 conditions (Kuster and Becker, 1992). In a large family with epidermolytic palmoplantar keratoderma of the Vorner-Unna-Thost type, Reis et al. (1992) mapped the gene to 17q11-q23 by linkage analysis using microsatellite DNA polymorphisms; maximum lod score = 6.66 with D17S579 at theta = 0.00. Thus, the disease mutation maps to the same region as the type I (acidic) keratin gene cluster (e.g., 148080). One acidic keratin, keratin 9 (KRT9), is expressed only in the terminally differentiated epidermis of palms and soles; thus, the KRT9 gene may be the site of the mutation in epidermolytic palmoplantar keratoderma. After the KRT9 gene was cloned, 15,14:Hennies et al. (1993, 1994) confirmed its localization on chromosome 17 by PCR amplification from a chromosome 17-only hybrid. Mutation search by solid phase sequencing demonstrated an arg162-to-trp mutation in exon 1 of the KRT9 gene (144200.0005) in the family studied by Reis et al. (1992). The R162W mutation was found in 5 other apparently unrelated EPPK families. A second mutation, gln171-to-pro, was detected in an additional large German kindred by restriction analysis of PCR products. Both mutations were located in the coil 1A domain, a highly conserved region in all keratins which is thought to be relevant for dimer formation in intermediate filaments. The R162W mutation is in the residue corresponding to that which is altered in KRT14 (148066) and KRT10 in epidermolysis bullosa simplex and epidermolytic hyperkeratosis, respectively. Reis et al. (1994) refined the assignment of the KRT9 gene to 17q21.1-q21.2 by fluorescence in situ hybridization.

Stevens et al. (1994) reported a large family in which 38 persons in 4 generations had keratosis punctata. Ten of 34 affected adults developed different malignancies (Hodgkin disease and renal, breast, pancreatic, and colonic adenocarcinomas). In 5 persons malignancies developed before the age of 50. Stevens et al. (1994) proposed that in this family mutation in the type I acidic keratin gene cluster at 17q12-q21 may be related to pathology of tumor suppressor gene(s) in the region of 17q21.

In connection with the question of whether the nonepidermolytic form of palmoplantar keratoderma (Unna-Thost type) is the same as the epidermolytic form, it should be noted that both epidermolytic and nonepidermolytic forms of palmoplantar keratoderma have been observed with various mutations in the keratin-1 gene (139350). Indeed, mutations of the KRT1 and KRT9 genes that are associated with the epidermolytic form of PPK affect the central regions of the protein that are important for filament assembly and stability, and for that reason lead to cellular degeneration or disruption. On the other hand, the mutation of the KRT1 gene that Kimonis et al. (1994) found in association with nonepidermolytic PPK involved the amino-terminal variable end region, which may be involved in supramolecular interactions of keratin filaments rather than stability.

Kobayashi et al. (1996) obtained experimental evidence that the arg162-to-gln point mutation (144200.0005) observed in patients with epidermolytic hereditary palmoplantar keratoderma (Reis et al., 1994) has a dominant-negative effect on keratin network formation. Transient transfection of KRT9 cDNA into both epithelial- and epidermal-derived cell lines showed normal keratin network formation in cells transfected with normal KRT9 cDNA and disrupted keratin filaments with droplet formation in cells transfected with the mutant KRT9 cDNA. Kobayashi et al. (1996) also summarized published reports of mutations in KRT9 associated with EPPK (their Figure 3) and commented that amino acid residues 156-171, which correspond to the head of the keratin 9 alpha-helical domain, may be crucial for keratin filament assembly and intermediate filament network formation. (Allelic variants 144200.0001 through 144200.0008 are within the same region of the KRT9 polypeptide cited by the authors. PMC.)

Rogaev et al. (1993) found tight linkage of keratosis palmaris et plantaris without epidermolysis to a rare, high molecular weight allele of an insertion-deletion polymorphism in the C-terminal coding region of the keratin 10 gene (maximum lod = 8.36 at theta = 0.00). As noted earlier, keratin 10, like keratin 9, maps to chromosome 17.

See also Chung (1937); Klintworth and Anderson (1961)

Alsaleh, Q. A.; Teebi, A. S.: Autosomal recessive epidermolytic palmoplantar keratoderma. J. Med. Genet. 27: 519-522, 1990.
Anderson, I. F.; Klintworth, G. K.: Hypovitaminosis-A in a family with tylosis and clinodactyly. Brit. Med. J. 1: 1293-1297, 1961.
Bergstrom, C.: Keratodermia palmaris et plantaris. Nord. Med. 78: 155-156, 1967.
Berth-Jones, J.; Hutchinson, P. E.: A family with palmoplantar epidermolytic hyperkeratosis. Clin. Exp. Derm. 14: 313-316, 1989.
Blasik, L. G.; Dimond, R. L.; Baughman, R. D.: Hereditary epidermolytic palmoplantar keratoderma. Arch. Derm. 117: 229-231, 1981.
Bonifas, J. M.; Matsumura, K.; Chen, M. A.; Berth-Jones, J.; Hutchinson, P. E.; Zloczower, M.; Fritsch, P. O.; Epstein, E. H., Jr.: Mutations of keratin 9 in two families with palmoplantar epidermolytic hyperkeratosis. J. Invest. Derm. 103: 474-477, 1994.
Camisa, C.; Williams, H.: Epidermolytic variant of hereditary palmoplantar keratoderma. Brit. J. Derm. 112: 221-225, 1985.
Chung, H.-L.: Keratoma palmare et plantare hereditarium, with special reference to its mode of inheritance as traced in six and seven generations, respectively, in two Chinese families. Arch. Derm. Syph. 36: 303-313, 1937.
Endo, H.; Hatamochi, A.; Shinkai, H.: A novel mutation of a leucine residue in coil 1A of keratin 9 in epidermolytic palmoplantar keratoderma. J. Invest. Derm. 109: 113-115, 1997.
Fritsch, P.; Honigsmann, H.; Jaschke, E.: Epidermolytic hereditary palmoplantar keratoderma. Brit. J. Derm. 99: 561-568, 1978.
Gamborg Nielsen, P.: Two different clinical and genetic forms of hereditary palmoplantar keratoderma in the northernmost county of Sweden. Clin. Genet. 28: 361-366, 1985.
Goette, D. K.: Familial congenital epidermolytic hyperkeratosis confined to the palms and soles. Sth. Med. J. 67: 1126-1128, 1974.
Greither, A.: Keratosis extremitatum hereditaria progrediens mit dominantem Erbgang. Hautarzt 3: 198-203, 1952.
Hennies, H.-C.; Zehender, D.; Kunze, J.; Kuster, W.; Reis, A.: Keratin 9 gene mutational heterogeneity in patients with epidermolytic palmoplantar keratoderma. Hum. Genet. 93: 649-654, 1994.
Hennies, H. C.; Kuster, W.; Langbein, L.; Digweed, M.; Mischke, D.; Franke, W. W.; Sperling, K.; Reis, A.: Keratin 9 gene mutations in epidermolytic palmoplantar keratoderma (EPPK). (Abstract) Am. J. Hum. Genet. 53 (suppl.): A211, 1993.
Kimonis, V.; DiGiovanna, J. J.; Yang, J.-M.; Doyle, S. Z.; Bale, S. J.; Compton, J. G.: A mutation in the V1 end domain of keratin 1 in non-epidermolytic palmar-plantar keratoderma. J. Invest. Derm. 103: 764-769, 1994.
Klaus, S.; Weinstein, G. D.; Frost, P.: Localized epidermolytic hyperkeratosis: a form of keratoderma of the palms and soles. Arch. Derm. 101: 272-275, 1970.
Klintworth, G. K.; Anderson, I. F.: Tylosis palmaris et plantaris familiaris associated with clinodactyly. S. Afr. Med. J. 35: 170-175, 1961.
Kobayashi, S.; Tanaka, T.; Matsuyoshi, N.; Imamura, S.: Keratin 9 point mutation in the pedigree of epidermolytic hereditary palmoplantar keratoderma perturbs keratin intermediate filament network formation. FEBS Lett. 386: 149-155, 1996.
Kuster, W.; Becker, A.: Indication for the identity of palmoplantar keratoderma type Unna-Thost with type Vorner: Thost's family revisited 110 years later. Acta Derm. Venerol. 72: 120-122, 1992.
Moriwaki, S.; Tanaka, T.; Horiguchi, Y.; Danno, K.; Imamura, S.: Epidermolytic hereditary palmoplantar keratoderma: histologic, ultrastructural, protein-chemical, and DNA analysis in two patients. Arch. Derm. 124: 555-559, 1988.
Nogita, T.; Nakagawa, H.; Ishibashi, Y.: Hereditary epidermolytic palmoplantar keratoderma with knuckle pad-like lesions over the finger joints. (Letter) Brit. J. Derm. 125: 496, 1991.
Reis, A.; Hennies, H.-C.; Langbein, L.; Digweed, M.; Mischke, D.; Drechsler, M.; Schrock, E.; Royer-Pokora, B.; Franke, W. W.; Sperling, K.; Kuster, W.: Keratin 9 gene mutations in epidermolytic palmoplantar keratoderma (EPPK). Nature Genet. 6: 174-179, 1994.
Reis, A.; Kuster, W.; Eckardt, R.; Sperling, K.: Mapping of a gene for epidermolytic palmoplantar keratoderma to the region of the acidic keratin gene cluster at 17q12-q21. Hum. Genet. 90: 113-116, 1992.
Rogaev, E. I.; Rogaeva, E. A.; Ginter, E. K.; Korovaitseva, G. I.; Farrer, L. A.; Shlensky, A. B.; Pritkov, A. N.; Mordovtsev, V. N.; St. George-Hyslop, P. H.: Identification of the genetic locus for keratosis palmaris et plantaris on chromosome 17 near the RARA and keratin type I genes. Nature Genet. 5: 158-162, 1993.
Rothnagel, J. A.; Wojcik, S.; Liefer, K. M.; Dominey, A. M.; Huber, M.; Hohl, D.; Roop, D. R.: Mutations in the 1A domain of keratin 9 in patients with epidermolytic palmoplantar keratoderma. J. Invest. Derm. 104: 430-433, 1995.
Stevens, H. P.; Rustin, M. H. A.; MacDermot, K.: Keratosis punctata palmaris et plantaris: an autosomal dominantly inherited keratoderma associated with malignancy. (Abstract) J. Med. Genet. 31: 168 only, 1994.
Thost, A.: Ueber erbliche Ichthyosis palmaris et plantaris cornea. Dissertation: Heidelberg (pub.) 1880.
Torchard, D.; Blanchet-Bardon, C.; Serova, O.; Langbein, L.; Narod, S.; Janin, N.; Goguel, A. F.; Bernheim, A.; Franke, W. W.; Lenoir, G. M.; Feunteun, J.: Epidermolytic palmoplantar keratoderma cosegregates with a keratin 9 mutation in a pedigree with breast and ovarian cancer. Nature Genet. 6: 106-110, 1994.
Unna, P. G.: Ueber das Keratoma palmare et plantare hereditarium. Arch. Derm. Syph. 15: 231-270, 1883.
Vorner, H.: Zur Kenntniss des Keratoma hereditarium palmare et plantare. Arch. Derm. Syph. 56: 3-31, 1901.

RETICULUM CELL SARCOMA (267730)

JJM NOTE

This family needs to be updated, including review of histopathology.

Escobar and Bixler (1975) described a family with reticulum cell sarcoma in at least 4 successive generations. All 5 in the sibship of the first generations were affected.

Escobar, V.; Bixler, D.: Reticulum cell sarcoma—a hereditary disease?. (Abstract) Birth Defects Conference, Kansas City, Mo. June only, 1975.

FAMILIAL HEMOPHAGOCYTIC LYMPHOHISTIOCYTOSIS (#267700)

JJM NOTE

Establishing the genetic determinants of this condition is a major and clinically important challenge, in part because one therapy is bone marrow transplantation, including that from relatives. One effort failed to distinguish the familial and nonfamilial cases based on immunophenotyping (Imashuku S et al.: Heterogeneity of immune markers in hemophagocytic lymphohistiocytosis: Comparative study of 9 familial and 14 familial inheritance-unproved cases. J. Pediatr. Hematol. Oncol. 20:207–14, 1998). A new entity, deserving a separate OMIM entry, is one inbred kindred, already clinically characterized and mapped (Moynihan LM et al.: Autozygosity mapping, to chromosome 11q25, of a rare autosomal recessive syndrome causing histiocytosis, joint contractures, and sensorineural deafness. Am. J. Hum. Genet. 62:1123–28, 1998). An international registry should speed the clarification of nosology and genetics, as well as therapy (Arico M et al.: Hemophagocytic lymphohistiocytosis: Report of 122 children from the International Registry. FHL Study Group of the Histiocyte Society. Leukemia 10:197–203, 1996).

A number sign (#) is used with this entry because the disorder has been found to be caused by mutation in the RAG1 (179615) and the RAG2 (179616) genes.

Anemia, granulocytopenia and thrombocytopenia are produced in part by phagocytosis of blood cells, and in part by replacement of the marrow by histiocytic infiltration. Families have been reported by Marrian and Sanerkin (1963) and by Farquhar and Claireaux (1952) and Farquhar et al. (1958). In the latter family 4 sibs were affected. The father showed autoantibody and shortened red cell life span. Farquhar et al. (1958) concluded that the minor changes observed in the father and one sib represented the heterozygous state. They were not concerned about the lack of changes in the mother since expression in the heterozygote is often variable. The disorder discussed by Omenn (1965) and by Miller (1966) is considered the same by some. Omenn (1965) described an inbred American family of Irish extraction with a large number of affected persons in many related sibships. Miller (1966) described 5 sisters—a complete sibship, including a pair of twins—with clinical features of failure to thrive, recurrent infections, lymphadenopathy, hepatosplenomegaly, pulmonary infiltration, and terminal pancytopenia and hypergammaglobulinemia. Death occurred between ages 20 months and 57 months. Autopsy showed diffuse reticulum cell infiltration of most organs including the central nervous system, obliteration of architecture of lymph glands and marked plasmacytosis. Barth et al. (1972) concluded that the familial reticuloendotheliosis with eosinophilia described by Omenn is a distinct entity. Pneumocystis carinii, responsible for eosinophilia in other immune deficiency disorders, was not detected in any of Omenn's cases. The family reported by Farquhar and Claireaux (1952) and Farquhar et al. (1958) was Scottish. Another Scottish family, with 3 affected sibs, was reported by Goodall et al. (1965). Bell et al. (1968) described affected brothers born 11 years apart. Meningoencephalitis during infancy was a feature in each. Hemophagocytosis in bone marrow p reparations made the diagnosis.

Donohue (1968) has autopsy information on 6 cases that occurred in an inbred Mennonite group in Ontario. De Veber (1974) provided further information on these cases, which he considered to be in the general group of histiocytoses. (These cases are also discussed in entry 246400.) A major difference from other reported families was raised platelet counts. Cutbush et al. (1974) identified 22 cases in an inbred Mennonite kindred. Six of them were still living. The disease presented at about 3 months of age with hepatomegaly and variable lymphadenopathy, and untreated cases died rapidly with high fever, hemolytic anemia, and a picture resembling acute leukemia. Some cases responded to prednisone. Others died despite prednisone and cytotoxic agents. Healthy relatives may have high platelet counts.

The familial histiocytoses are nosologically confused. Even the terminology, based on histopathology, is confusing: lymphohistiocytic, erythrophagocytic, lymphoreticular, etc. The confused group of histiocytoses includes monocytic leukemia, histiocytic lymphoma, Letterer-Siwe disease, malignant histiocytosis, Hand-Schuller-Christian disease, eosinophilic granuloma, histiocytosis X, reticuloendotheliosis, histiocytic reticulosis, disseminated lipogranulomatosis, and familial hemophagocytic reticulosis. Newton and Hamoudi (1973) gave a useful review, but a convincing classification has not been achieved. Confusion is compounded by failure to distinguish infectious diseases such as histoplasmosis and graft-versus-host reaction such as occurs in infants with severe combined immunodeficiency transfused with fresh whole blood or receiving maternal blood transplacentally. Some would suggest that familial erythrophagocytic lymphohistiocytosis, familial reticuloendotheliosis with eosinophilia and Letterer-Siwe disease (246400) can be lumped together. It seems more likely to me that they are separate entities. Nelson et al. (1961) claimed that the disorder they described was quite different from Letterer-Siwe disease and presumably also from the disorder described here. Chediak-Higashi syndrome (214500) has only partial similarity.

Mozziconacci et al. (1965) described 2 brothers, aged 6 and 8, with this fatal disease characterized by a high and irregular fever, hepatosplenomegaly, purpura, and, later, jaundice, polyneuritis, meningeal reaction, choked disks, moderate anemia, and severe granulocytopenia. A possible relationship to ceroid storage disease (214200) is only speculative. Price et al. (1971) described 4 out of 12 sibs with a progressive neurologic disease characterized by diffuse lymphohistiocytic infiltrations of the central nervous system in association with multiple foci of parenchymal destruction. The range of age at death was 15 months to 12 years. The spinal fluid showed pleocytosis and increased protein. Histologically the disorder resembled familial hemophagocytic reticulosis or familial erythrophagocytic lymphohistiocytosis but unlike these conditions the process was largely confined to CNS. Since lymphocytic and histiocytic infiltration of viscera was present in some of the patients and leukopenia with hypocellular bone marrow was described, most would consider this the same as familial histiocytic reticulosis. Ladisch et al. (1978) demonstrated abnormal lipid metabolism and defects in both humoral and cellular immunity, together with a plasma inhibitor of in vitro lymphocyte blastogenesis. Because a component of the immunodeficiency is plasma-mediated inhibition of lymphocyte proliferation, Ladisch et al. (1982) tested whether repeated plasma or blood exchange would help in FEL. Clinical improvement was complete in 2 and partial in 1 of 3 patients studied. Laboratory values also improved. Relapse, which was ultimately fatal, was accompanied by recurrence of the immune defects. Bergholz et al. (1978) reviewed the evidence that 'congenital allogeneity' with graft-versus-host reaction might be involved. Jouan et al. (1987) suggested that the lesions, particularly the skin and bone marrow changes, are reminiscent of those observed in acute graft-versus-host reaction. Althoug h a blood chimerism has never been demonstrated, Jouan et al. (1987) suggested that these pathologic observations supported the hypothesis of graft-versus-host disease in a primary cellular immunodeficiency and the persistence of proliferating maternal cells in peripheral target organs. Nemoto and Ohnishi (1987) reported histologic studies of the spleen, liver, and mesenteric lymph nodes from a patient in whom splenectomy and biopsy of the other tissues were performed. They described lymphoid cells of various sizes, with atypical features including mitotic figures and convoluted nuclei; these were suspected to be T cells on the basis of immunohistochemical findings. The authors suggested that this may be a special type of lymphoproliferative disease accompanied by severe proliferation of erythrophagocytic histiocytes rather than being a true histiocytic disease.

Stark et al. (1984) studied 11 patients in 4 Jewish families of Iranian and Iraqi origin. Parental consanguinity was found in 3. The age of onset varied from 6 weeks to 36 months. All had fever, wasting, and hepatosplenomegaly. Lymph node enlargement and neurologic abnormalities were common. Pancytopenia, atypical lymphomonocytoid cells in the peripheral blood, abnormal liver function tests, and increased CSF protein were the most consistent laboratory findings. In 9 patients death occurred in 2 weeks to 3 months after presentation. The longest survival was 2 years after presentation. Gelfand et al. (1984) found absent or greatly diminished activity of ecto-5-prime-nucleotidase in Omenn syndrome. Cohen et al. (1980) concluded, however, that this finding is not specific for Omenn syndrome. Schofer et al. (1991) described an infant who had features compatible with Omenn syndrome but who also had short-limbed dwarfism caused by metaphyseal chondrodysplasia. Gatti et al. (1969) described identical bone changes in 2 sibs who likewise had the features of Omenn syndrome. (See 200900 for discussion of metaphyseal chondrodysplasia associated with immune defect.) De Saint-Basile et al. (1991) studied 5 unrelated patients; 2 were born of consanguineous parents. One of the patients had a brother and a sister who had died with the same syndrome and the brother of another patient was said to have died with typical alymphocytosis-type severe combined immunodeficiency (SCID; 202500). They described the syndrome as being characterized by T-cell infiltration of skin, gut, liver, and spleen, leading to diffuse erythroderma, protracted diarrhea, failure to thrive, and hepatosplenomegaly. Although the lesions resembled those in graft-versus-host disease, the blood T cells were shown by DNA haplotype analysis to belong to the patients; the same was true of the T cells infiltrating the gut and skin in 1 patient. A restricted heterogeneity of the T-cell repertoire was indicated by oligoclonality of the T-cell receptor gene rearrangements. The occurrence of an alymphocytosis type of SCID in the brother of 1 of the patients suggested that the restricted heterogeneity of T-cell receptor gene usage in Omenn syndrome may rise from leakiness, within the context of a genetically determined faulty T-cell differentiation. Cavazzana-Calvo et al. (1993) presented further evidence that the Omenn syndrome is a leaky T-negative/B-negative SCID phenotype; see 202500.

Janka (1983) reviewed 121 cases, and Henter and Elinder (1991) provided a clinical review based on the findings in 7 children. The latter review indicated successful results with chemotherapy. Nespoli et al. (1991) treated 3 patients, aged 6 to 20 months, with allogeneic bone marrow transplantation (BMT). They concluded that when an HLA-identical family donor exists, the BMT followed by intensive chemotherapy can produce a stable remission. Gomez et al. (1995) reported the outcome of allogenic BMT in 9 consecutive patients with Omenn syndrome treated between 1980 and 1989. They concluded that both HLA-identical and haploidentical BMT can cure Omenn syndrome, provided that parenteral nutrition and immunosuppressive therapy are given before transplantation.

In Sweden, Henter et al. (1991) studied the incidence of this disorder in children during a 16-year period, 1971-1986; the incidence was 1.2/1 million children per year. One child per 50,000 liveborns developed the disorder during this period. The sex ratio was approximately 1:1. Prominent early clinical signs were fever, splenomegaly, hepatomegaly, rash, and lymph node enlargement. Neurologic symptoms, which developed in 47%, could totally dominate the clinical picture and develop before other symptoms and signs. In only 11 of 32 children was the diagnosis made during their lifetime. Henter et al. (1991) demonstrated elevated levels of circulating interferon-gamma (147570), tumor necrosis factor (191160), and interleukin-6 (147620) in children during active phases of FHL. Soluble CD8 was also increased in all of 7 children tested. Henter et al. (1991) suggested that a genetic defect in cytokine regulation underlies this disorder.

Mache et al. (1994) reported 2 sibs who were thought to have disseminated T-cell lymphoma at the time of diagnosis of familial hemophagocytic lymphohistiocytosis. DNA hybridization studies provided evidence of a monoclonal T-cell receptor beta-chain gene rearrangement. One sib received an allogeneic bone marrow transplant from a related healthy donor. A Moroccan male infant born to consanguineous parents (Rybojad et al., 1996) manifested nephrotic syndrome, in addition to typical findings. Renal biopsy confirmed minimal change glomerular disease.

The entity described here should not be confused with Langerhans cell histiocytosis, which has little or no tendency to familial aggregation. The Langerhans cell is a dendritic cell that was in the epidermis and was described by medical student Paul Langerhans who thought that it was part of the nervous system Langerhans, 1868. Birbeck et al. (1961) found that the Langerhans cell displays a unique electron-microscopic morphology. The discoveries that these cells are not confined to skin and that they make up a sizable portion of the cellular infiltrate in histiocytosis X, along with other evidence, suggest that they play an immunologic role in protecting against environmental antigens. Egeler and D'Angio (1995) presented a classification of histiocytosis syndromes in children: class I, Langerhans cell histiocytosis; class II, histiocytosis of mononuclear macrophages other than Langerhans cells, including familial hemophagocytic lymphohistiocytosis; and class III, malignant histiocytic disorders, including histiocytic lymphoma.

Henter and Elinder (1995) pointed out that hemophagocytic lymphohistiocytosis can be divided into 2 categories, a primary and a secondary form. The primary form is hereditary, whereas the secondary form is a reactive condition, commonly associated with immunosuppressive therapy, malignancies and/or infections, which often are of viral origin. Henter et al. (1991) published diagnostic guidelines for hemophagocytic lymphohistiocytoses. Henter and Elinder (1995) pointed out that a viral infection may elicit a bout of familial hemophagocytic lymphohistiocytosis in a genetically predisposed child.

Allogeneic bone marrow transplantation (BMT) from an HLA-identical related donor is the treatment of choice in patients with familial hemophagocytic lymphohistiocytosis. However, fewer than 20% of patients have a disease-free HLA-identical sib. Jabado et al. (1997) described BMTs from 13 HLA-nonidentical related donors and from a matched unrelated donor that were performed in 2 centers and 14 consecutive cases of this disorder. Remission of disease was achieved before BMT in 10 patients. Marrow was T-cell-depleted to minimize graft-versus-host disease (GVHD). In addition to a conditioning regiment of cytotoxic drugs or antithymocyte globulin (ATG), antiadhesion antibodies specific for the alpha chain of the leukocyte function-associated antigen-1 (ITGAL; 153370) and CD2 (186990) were infused pre-BMT and post-BMT to help prevent graft rejection. Acute GVHD greater than stage 1 was not observed, and 1 patient had mild cutaneous chronic GVHD that resolved. Sustained engraftment was obtained in 11 of 17 transplants (3 patients had 2 transplants), and disease-free survival in 9 patients, with a follow-up period of 8 to 69 months (mean, 33), was observed.

Villa et al. (1998) reported that patients with Omenn syndrome have missense mutations in either the RAG1 (179615) or the RAG2 (179616) gene that result in partial activity of the 2 proteins. Two of the amino acid substitutions map within the RAG1 homeodomain and decrease DNA binding activity, while 3 others lower the efficiency of RAG1/RAG2 interaction. These findings indicated that the immunodeficiency manifested in patients with Omenn syndrome arises from mutations that decrease the efficiency of V(D)J recombination.

See also Botha et al. (1975); Buist et al. (1971); Delaney et al. (1984); Friedman and Steigbigel (1965); Gencik et al. (1984); Gross-Kieselstein et al. (1981); Henter et al. (1991); Herbert (1983); Hong et al. (1985); Kletzel et al. (1986); Landing et al. (1961); MacMahon et al. (1963); Martin and Cras (1985); Rybojad et al. (1996); Soffer et al. (1984); Velders et al. (1983)

Barth, R. F.; Khurana, S. K.; Vergara, G. G.; Lowman, J. T.; Beckwith, J. B.: Rapidly fatal familial histiocytosis associated with eosinophilia and primary immunological deficiency. Lancet II: 503-506, 1972.

Bell, R. J. M.; Brafield, A. J. E.; Barnes, N. D.; France, N. E.: Familial haemophagocytic reticulosis. Arch. Dis. Child. 43: 601-606, 1968.

Bergholz, M.; Rahlf, G.; Doering, K.-M.: Familial hemophagocytic reticulosis (Farquhar). Path. Res. Pract. 163: 267-280, 1978.

Birbeck, M. S.; Breathnach, A. S.; Everall, J. D.: An electron microscope study of basal melanocytes and high-level clear cells (Langerhans cells) in vitiligo. J. Invest. Derm. 37: 51-64, 1961.

Botha, J. B. C.; Kahn, L. B.; Kaschula, R. O. C.: Familial haemophagocytic reticulosis: report of 2 cases in sibs. S. Afr. Med. J. 49: 1305-1308, 1975.

Buist, N. R. M.; Jones, R. N.; Cavens, T. R.: Familial haemophagocytic reticulosis in first cousins. Arch. Dis. Child. 46: 728-729, 1971.

Cavazzana-Calvo, M.; Le Deist, F.; De Saint Basile, G.; Papadopoulo, D.; De Villartay, J. P.; Fischer, A.: Increased radiosensitivity of granulocyte macrophage colony-forming units and skin fibroblasts in human autosomal recessive severe combined immunodeficiency. J. Clin. Invest. 91: 1214-1218, 1993.

Cohen, A.; Mansour, A.; Dosch, H.-M.; Gelfand, E. W.: Association of a lymphocyte purine enzyme deficiency (5-prime-nucleotidase) with combined immunodeficiency. Clin. Immun. Immunopath. 15: 245-250, 1980.

Cutbush, W.; De Veber, L. L.; Rathbun, J. C.: Familial histiocytosis. (Abstract) Canad. Res. Soc. Meeting, St. John's Newfoundland , 1974.

Delaney, M. M.; Shafford, E. A.; Al-Attar, A.; Pritchard, J.: Familial erythrophagocytic reticulosis: complete response to combination chemotherapy. Arch. Dis. Child. 59: 173-175, 1984.

de Saint-Basile, G.; Le Deist, F.; de Villartay, J.-P.; Cerf-Bensussan, N.; Journet, O.; Brousse, N.; Griscelli, C.; Fischer, A.: Restricted heterogeneity of T lymphocytes in combined immunodeficiency with hypereosinophilia (Omenn's syndrome). J. Clin. Invest. 87: 1352-1359, 1991.

De Veber, L. L.: Personal Communication. London, Ontario, Canada 1974.

Donohue, W. L.: Personal Communication. Toronto, Ontario, Canada 1968.

Egeler, R. M.; D'Angio, G. J.: Langerhans cell histiocytosis. J. Pediat. 127: 1-11, 1995.

Farquhar, J. W.; Claireaux, A. E.: Familial haemophagocytic reticulosis. Arch. Dis. Child. 27: 519-525, 1952.

Farquhar, J. W.; MacGregor, A. R.; Richmond, J.: Familial haemophagocytic reticulosis. Brit. Med. J. 2: 1561-1564, 1958.

Friedman, R. M.; Steigbigel, N. H.: Histiocytic medullary reticulosis. Am. J. Med. 38: 130-133, 1965.

Gatti, R. A.; Platt, N.; Pomerance, H. H.; Hong, R.; Langer, L. O.; Kay, H. E. M.; Good, R. A.: Hereditary lymphopenic agammaglobulinemia associated with a distinctive form of short-limbed dwarfism and ectodermal dysplasia. J. Pediat. 75: 675-684, 1969.

Gelfand, E. W.; McCurdy, D.; Rao, C. P.; Cohen, A.: Absence of lymphocyte ecto-5-prime-nucleotidase in infants with reticuloendotheliosis and eosinophilia (Omenn's syndrome). Blood 63: 1475-1480, 1984.

Gencik, A.; Signer, E.; Muller, H.: Genetic analysis of familial erythrophagocytic lymphohistiocytosis. Europ. J. Pediat. 142: 248-252, 1984.

Gomez, L.; Le Deist, F.; Blanche, S.; Cavazzana-Calvo, M.; Griscelli, C.; Fischer, A.: Treatment of Omenn syndrome by bone marrow transplantation. J. Pediat. 127: 76-81, 1995.

Goodall, H. B.; Guthrie, W.; Buist, N. R. M.: Familial haemophagocytic reticulosis. Scot. Med. J. 10: 425-438, 1965.

Gross-Kieselstein, E.; Navon, P.; Branski, D.; Abrahamov, A.; Dollberg, L.: Familial erythrophagocytic lymphohistiocytosis in infancy. Europ. J. Pediat. 136: 223-225, 1981.

Henter, J.-I.; Elinder, G.: Familial hemophagocytic lymphohistiocytosis: clinical review based on the findings in seven children. Acta Paediat. Scand. 80: 269-277, 1991.

Henter, J.-I.; Elinder, G.: Haemophagocytic lymphohistiocytosis: an inherited primary form and a reactive secondary form. Brit. J. Haemat. 91: 774-775, 1995.

Henter, J.-I.; Elinder, G.; Ost, A.; FHL Study Group of the Histiocyte Society: Diagnostic guidelines for hemophagocytic lymphohistiocytosis. Semin. Oncol. 18: 29-33, 1991.

Henter, J.-I.; Elinder, G.; Soder, O.; Hansson, M.; Andersson, B.; Andersson, U.: Hypercytokinemia in familial hemophagocytic lymphohistiocytosis. Blood 78: 2918-2922, 1991.

Henter, J.-I.; Elinder, G.; Soder, O.; Ost, A.: Incidence in Sweden and clinical features of familial hemophagocytic lymphohistiocytosis. Acta Paediat. Scand. 80: 428-435, 1991.

Herbert, P. N.: Familial erythrophagocytic lymphohistiocytosis: an association with serum lipid abnormalities. J. Pediat. 102: 270-273, 1983.

Hong, R.; Gilbert, E. F.; Opitz, J. M.: Omenn disease: termination in lymphoma. Pediat. Path. 3: 143-154, 1985.

Jabado, N.; de Graeff-Meeder, E. R.; Cavazzana-Calvo, M.; Haddad, E.; Le Deist, F.; Benkerrou, M.; Dufourcq, R.; Caillat, S.; Blanche, S.; Fischer, A.: Treatment of familial hemophagocytic lymphohistiocytosis with bone marrow transplantation from HLA genetically nonidentical donors. Blood 90: 4743-4748, 1997.

Janka, G. E.: Familial hemophagocytic lymphohistiocytosis. Europ. J. Pediat. 140: 221-230, 1983.

Jouan, H.; Le Deist, F.; Nezelof, C.: Omenn's syndrome—pathologic arguments in favor of a graft versus host pathogenesis: a report of nine cases. Hum. Path. 18: 1101-1108, 1987.

Kletzel, M.; Gollin, S. M.; Gloster, E. S.; Jimenez, J. F.; Golladay, E. S.; Berry, D. H.: Chromosome abnormalities in familial hemophagocytic lymphohistiocytosis. Cancer 57: 2153-2157, 1986.

Ladisch, S.; Ho, W.; Matheson, D.; Pilkington, R.; Hartman, G.: Immunologic and clinical effects of repeated blood exchange in familial erythrophagocytic lymphohistiocytosis. Blood 60: 814-821, 1982.

Ladisch, S.; Holiman, B.; Poplack, D. G.; Blaese, R. M.: Immunodeficiency in familial erythrophagocytic lymphohistiocytosis. Lancet I: 581-583, 1978.

Landing, B. H.; Strauss, L.; Crocker, A. C.; Braunstein, H.; Henley, W. L.; Will, J. R.; Sanders, M. D.: Thrombocytopenic purpura with histiocytosis of the spleen. New Eng. J. Med. 265: 572-576, 1961.

Langerhans, P.: Ueber die Nervender menschlicher. Haut. Virchows Arch. (Pathol. Anat.) 44: 325, 1868.

Mache, C. J.; Slavc, I.; Schmid, C.; Hoefler, G.; Urban, C. E.; Schwinger, W.; Winter, E.; Hulla, W.; Zenz, W.; Holter, W.: Familial hemophagocytic lymphohistiocytosis associated with disseminated T-cell lymphoma: a report of two siblings. Ann. Hemat. 69: 85-91, 1994.

MacMahon, H. E.; Bedizel, M.; Ellis, C. A.: Familial erythrophagocytic lymphohistiocytosis. Pediatrics 32: 868-879, 1963.

Marrian, V. J.; Sanerkin, N. G.: Familial histiocytic reticulosis (familial haemophagocytic reticulosis). J. Clin. Path. 16: 65-69, 1963.

Martin, J. J.; Cras, P.: Familial erythrophagocytic lymphohistiocytosis: a neuropathologic study. Acta Neuropath. 66: 140-144, 1985.

Miller, D. R.: Familial reticuloendotheliosis: concurrence of disease in five siblings. Pediatrics 38: 986-995, 1966.

Mozziconacci, P.; Nezelof, C.; Attal, C.; Girard, F.; Pham-Huu-Trung; Weil, J.; Desbuquois, B.; Gadot, M.: La lympho-histiocytose familiale. Arch. Franc. Pediat. 22: 385-408, 1965.

Nelson, P.; Santamaria, A.; Olson, R. L.; Nayak, N. C.: Generalized lymphohistiocytic infiltration. A familial disease not previously described and different from Letterer-Siwe disease and Chediak-Higashi syndrome. Pediatrics 27: 931-950, 1961.

Nemoto, K.; Ohnishi, Y.: Familial hemophagocytic reticulosis: clinicopathologic findings, and cytochemical, immunohistochemical and electron microscopic studies. Acta Path. Jpn. 37: 1811-1822, 1987.

Nespoli, L.; Locatelli, F.; Bonetti, F.; Porta, F.; Zecca, M.; Gibardi, A.; Gambarana, D.; Vitale, V.; Burgio, G. R.; Severi, F.: Familial haemophagocytic lymphohistiocytosis treated with allogeneic bone marrow transplantation. Bone Marrow Transplant. 7: 139-142, 1991.

Newton, W. A., Jr.; Hamoudi, A. B.: Histiocytosis: a histological classification with clinical correlation. Perspect. Pediat. Path. 1: 251-253, 1973.

Omenn, G. S.: Familial reticuloendotheliosis with eosinophilia. New Eng. J. Med. 273: 427-432, 1965.

Price, D. L.; Woolsey, J. E.; Rosman, N. P.; Richman, E. P., Jr.: Familial lymphohistiocytosis of the nervous system. Arch. Neurol. 24: 270-283, 1971.

Rybojad, M.; Cambiaghi, S.; Moraillon, I.; Vignon-Pennamen, M. D.; Morel, P.; Baudoin, V.; Loirat, C.: Omenn's reticulosis associated with the nephrotic syndrome. Brit. J. Derm. 135: 124-127, 1996.

Schofer, O.; Blaha, I.; Mannhardt, W.; Zepp, F.; Stallmach, T.; Spranger, J.: Omenn phenotype with short-limbed dwarfism. J. Pediat. 118: 86-89, 1991.

Soffer, D.; Okon, E.; Rosen, N.; Stark, B.; Hershko, C.: Familial hemophagocytic lymphohistiocytosis in Israel. II. Pathologic findings. Cancer 54: 2423-2431, 1984.

Stark, B.; Hershko, C.; Rosen, N.; Cividalli, G.; Karsai, H.; Soffer, D.: Familial hemophagocytic lymphohistiocytosis (FHLH) in Israel. I. Description of 11 patients of Iranian-Iraqi origin and review of the literature. Cancer 54: 2109-2121, 1984.

Velders, A. J.; Kuis, W.; van Dijk, H. A.; Poppema, S.; Elema, J. D.; Klokke, A. H.; van Voorst Vader, P. C.: Omenn syndrome: familial reticulo-endotheliosis with eosinophilia and combined immunodeficiency. Brit. J. Derm. 108: 118-120, 1983.

Villa, A.; Santagata, S.; Bozzi, F.; Giliani, S.; Frattini, A.; Imberti, L.; Gatta, L. B.; Ochs, H. D.; Schwarz, K.; Notarangelo, L. D.; Vezzoni, P.; Spanopoulou, E.: Partial V(D)J recombination activity leads to Omenn syndrome. Cell 93: 885-896, 1998.

LETTERER-SIWE DISEASE (*246400)

JJM NOTE

This entity seems ripe for an interinstitutional, interdisciplinary investigation to clarify pathophysiology and to map and isolate the responsible gene(s).

Christie et al. (1954) described the disease in sibs who were never in contact, thus tending to discredit an infectious hypothesis. Rogers and Benson (1962) reported affected sibs and reviewed the literature. Falk and Gellei (1963) also observed a family. Schoeck et al. (1963) described 2 sibs with L-S disease. Ten other families with multiple affected sibs were reviewed, including Farquhar 'familial hemophagocytic reticulosis' (267700) and Nelson 'generalized lymphohistiocytic infiltration' (also see 267700), which Schoeck et al. suggested are all the same entity. In a survey of deaths from Letterer-Siwe disease in a 5-year period in the U.S., Glass and Miller (1968) found 5 sib pairs among 270 deaths, a pair of concordant like-sex twins, and a peak of mortality under 1 year of age. Freundlich et al. (1972) reported 2 families with multiple cases of consanguineous parents. Hirsch and Kong (1973) reported a father and son with histiocytosis X of the lung. The father presented with cough and exertional dyspnea, whereas the son was asymptomatic. Biopsies confirmed histiocytosis X (eosinophilic granuloma) in both. Neither father nor son had any evidence of disseminated histiocytosis. I have information from Cook (1967) concerning 8 cases of histiocytosis X occurring in 2 sibships in an inbred Mennonite group in Waterloo County, Ontario. In each case the parents were related as second cousins and all 4 parents shared in common a grandparental couple. In 2 cases, treatment with adrenocorticosteroids was begun early and both patients survived. The other cases pursued an identical course. They were well until ages 8 to 18 weeks, following which they developed general irritability, especially on being touched or moved. There was pallor, dyspnea, distended abdomen, fever, and, in the terminal stages, usually jaundice. Medical investigations showed hepatosplenomegaly, anemia, neutropenia, and thrombocytopenia. Autopsy showed histiocytic infiltration of the liver, s pleen and lymph nodes. There was no persistent skin rash and no bone lesions were identified. The course of the illness in all 6 fatal cases was rapid, ranging from 2 to 5 weeks. None of the children had contact with each other. These cases are also discussed under familial histiocytic reticulosis (267700). The nosology of this category is confused, as is the terminology. For example, Donohue and Thompson (1972) concluded that the Mennonite cases represent the disorder reported as 'familial Letterer-Siwe disease' and not histiocytosis X. The occurrence of 'encephalopathy' in both the Mennonite cases and reported familial Letterer-Siwe disease, but not in either nonfamilial L-S disease or histiocytosis X, was one reason for their conclusion. Kloepfer et al. (1972) observed the same or a similar disorder in an inbred triracial group in Louisiana, known locally as 'Redbones.'

The entity described here should not be confused with Langerhans cell histiocytosis, which has little or no tendency to familial aggregation. The Langerhans cell is a dendritic cell that was in the epidermis and was described by medical student Paul Langerhans who thought that it was part of the nervous system Langerhans, 1868. Birbeck et al. (1961) found that the Langerhans cell displays a unique electron-microscopic morphology. The discoveries that these cells are not confined to skin and that they make up a sizable portion of the cellular infiltrate in histiocytosis X, along with other evidence, suggest that they play an immunologic role in protecting against environmental antigens. Egeler and D'Angio (1995) presented a classification of histiocytosis syndromes in children: class I, Langerhans cell histiocytosis; class II, histiocytosis of mononuclear macrophages other than Langerhans cells, including familial hemophagocytic lymphohistiocytosis; and class III, malignant histiocytic disorders, including histiocytic lymphoma.

See also Birbeck et al. (1961); Frisell et al. (1977); Juberg et al. (1970)

Birbeck, M. S.; Breathnach, A. S.; Everall, J. D.: An electron microscope study of basal melanocytes and high-level clear cells (Langerhans cells) in vitiligo. J. Invest. Derm. 37: 51-64, 1961.

Birbeck, M. S.; Breathnach, A. S.; Everall, J. D.: An electron microscope study of basal melanocytes and high-level clear cells (Langerhans cells) in vitiligo. J. Invest. Derm. 37: 51-63, 1961.

Christie, A.; Batson, R.; Shapiro, J.; Riley, H. D., Jr.; Laughmiller, R.; Stahlman, M.: Acute disseminated (non-lipid) reticuloendotheliosis. Acta Paediat. 43 (suppl. 100): 65-76, 1954.

Cook, M. S.: Personal Communication. Toronto, Ontario, Canada 1967.

Donohue, W. L.; Thompson, M. W.: Familial lymphohistiocytosis. Birth Defects Orig. Art. Ser. VIII(3): 105-111, 1972.

Egeler, R. M.; D'Angio, G. J.: Langerhans cell histiocytosis. J. Pediat. 127: 1-11, 1995.

Falk, W.; Gellei, B.: Letterer-Siwe disease (non-lipoid reticuloendotheliosis).In: Goldschmidt, E.: Genetics of Migrant and Isolate Populations. Baltimore: Williams and Wilkins (pub.) 1963. Pp. 312 only. Note: See Also: The familial occurrence of Letterer-Siwe disease. Acta Paediat. 46: 471-480, 1957. . . . . .

Freundlich, E.; Amit, S.; Montag, Y.; Suprun, H.; Nevo, S.: Familial occurrence of Letterer-Siwe disease. Arch. Dis. Child. 47: 122-125, 1972.

Frisell, E.; Bjorksten, B.; Holmgren, G.; Angstrom, T.: Familial occurrence of histiocytosis. Clin. Genet. 11: 163-170, 1977.

Glass, A. G.; Miller, R. W.: U.S. mortality from Letterer-Siwe disease, 1960-1964. Pediatrics 42: 364-367, 1968.

Hirsch, M. S.; Kong, C. H.: Familial pulmonary histiocytosis X. Am. Rev. Resp. Dis. 107: 831-835, 1973.

Juberg, R. C.; Kloepfer, H. W.; Oberman, H. A.: Genetic determination of acute disseminated histiocytosis X (Letterer-Siwe syndrome). Pediatrics 45: 753-765, 1970.

Kloepfer, H. W.; Ichinose, H.; White, T. R.: Fulminating disseminated histiocytosis simulating Letterer-Siwe disease. Birth Defects Orig. Art. Ser. 8(3): 112-114, 1972.

Langerhans, P.: Ueber die Nervender menschlicher. Haut. Virchows Arch. (Pathol. Anat.) 44: 325, 1868.

Rogers, D. L.; Benson, T. E.: Familial Letterer-Siwe disease. Report of a case. J. Pediat. 60: 550-554, 1962.

Schoeck, V. W.; Peterson, R. D. A.; Good, R. A.: Familial occurrence of Letterer-Siwe disease. Pediatrics 32: 1055-1063, 1963.

RETICULOENDOTHELIOSIS, X-LINKED (*312500)

JJM NOTE

Maybe this entity can be mapped, simply from archival specimens. Maybe there are transmitting females at risk in the next generation.

Falletta et al. (1973) described a Latin American family in which 17 males in 2 generations died under the age of 6 years, following an illness characterized by fever, pallor, jaundice, hepatosplenomegaly, and lymphadenopathy. Median age of onset was 14 months (4-62 months) and median duration of illness was 22 days (1-50 days). Histologic changes were consistent with malignant reticuloendotheliosis. All affected males were related through their mothers.

Falletta, J. M.; Fernbach, D. J.; Singer, D. B.; Smith, M. A.; Landing, B. H.; Heath, C. W., Jr.; Shore, N. A.; Barrett, F. F.: A fatal X-linked recessive reticuloendothelial syndrome with hyperglobulinemia: X-linked recessive reticuloendotheliosis. J. Pediat. 83: 549-556, 1973.

MYELOFIBROSIS, FAMILIAL (254450)

JJM NOTE

A single family awaits follow-up for confirmation.

Sieff and Malleson (1980) described a brother and sister who developed fulminant fatal myeloproliferative disease at 7 and 8 weeks of age. The bone marrow showed reduced hemopoiesis with generalized fibrosis. Although clinically resembling familial hemophagocytic reticulosis, the disorder did not show the characteristic hemophagocytosis as a prominent feature. The parents were not related.

Sieff, C. A.; Malleson, P.: Familial myelofibrosis. Arch. Dis. Child. 55: 888-893, 1980.

Immunodeficiencies with Neoplasia

IMMUNODEFICIENCY, X-LINKED PROGRESSIVE COMBINED VARIABLE (*308240)

JJM NOTE

Any young male presenting with a manifestation of this multifaceted syndrome, but especially those with ileal or brain lymphoma, should be suspected of having this susceptibility gene. Family history may be highly suggestive, and DNA probes are available for confirming a diagnosis. For ease of reference, this disorder has been labeled *Purtilo syndrome.*

In a kindred by the name of Duncan, 17,18:Purtilo et al. (1974, 1975) observed 6 males who died between the ages of 2 and 19 years from a lymphoproliferative disease. The subtle, progressive combined variable immunodeficiency disease was characterized by benign or malignant proliferation of lymphocytes and histiocytosis, and alterations in concentrations of serum immunoglobulins. In at least 3 of 6 boys, infectious mononucleosis occurred during or preceding the terminal events. Fever, pharyngitis, lymphadenopathy, hepatosplenomegaly, atypical lymphocytosis, and a spectrum ranging from agammaglobulinemia to polyclonal hypergammaglobulinemia occurred. At necropsy, the thymus glands and thymic-dependent areas in the lymph nodes and spleen were depleted of lymphocytes. Hematopoietic organs, viscera, and central nervous system were diffusely infiltrated by lymphocytes, plasma cells, and histiocytes, some containing erythrocytes. Two of the 6 males, half-sibs, had lymphomas of the ileum and central nervous system. The authors raised the possibility that 'the Epstein-Barr virus or other viruses triggered the fatal proliferation of lymphocytes and that progressive attrition of T-cell function allowed uncontrolled lymphoproliferation.' In addition to the kindred described by Purtilo and his colleagues, the kindred in which 4 young male cousins died of infectious mononucleosis, as reported by Bar et al. (1974), and the kindred with agammaglobulinemia developing after infectious mononucleosis in 3 maternal male cousins, as reported by Provisor et al. (1975), may be examples of Duncan disease. Exposure to Epstein-Barr virus results in fatal infectious mononucleosis, agammaglobulinemia or B-cell lymphoma (Purtilo et al., 1977). Hamilton et al. (1980) abbreviated the designation of this disease as XLP (X-linked lymphoproliferative) syndrome. They reported on studies of 59 affected males in 7 unrelated kindreds ascertained through an XLP registry. Thirty-four patients died of infectious mo nonucleosis, 8 had fatal infectious mononucleosis with immunoblastic sarcoma, 9 had depressed immunity following Epstein-Barr virus infection, and 8 developed lymphoma. Sullivan et al. (1980) found deficient activity of natural killer cells. Purtilo et al. (1982) reviewed 100 cases of XLP in 25 kindreds, demonstrating 4 major interrelated phenotypes: infectious mononucleosis (IM), malignant B-cell lymphoma (ML), aplastic anemia (AA), and hypogammaglobulinemia (HGG). Eighty-one of the patients died; 2 were asymptomatic but showed immunodeficiency to EBV; 75 had IM and, concurrently, 17 of this group had AA; all with AA died within a week. On the other hand, AA did not accompany HGG or ML. In 9, IM appeared to evolve into ML; however, most patients with ML showed no obvious antecedent IM. In 1, IM occurred after recurrent ML. Twenty-six of 35 lymphomas were in the terminal ileum. Heterozygous women (mothers of boys with XLP) showed abnormally elevated titers of antibodies to EBV.

Lyon and Loutit (1983) suggested the existence of an X-linked gene that determines susceptibility to the acquired immunodeficiency syndrome (AIDS). She pointed to the rather similar X-linked lymphoproliferative disease which, like AIDS, has a propensity for special malignancies and has 70% early mortality. (Various X-linked immunodeficiencies were reviewed by Scher (1982).) Males represent 95% of AIDS cases. If the 5% affected females are homozygous for an X-linked susceptibility gene and if males and females are equally exposed to a causative agent, then the frequency of the relevant gene in males is about 5% (female:male = p(2)/p). If females are less exposed than males, then the gene frequency in males would be higher. Sullivan et al. (1983) studied 2 males before and during acute fatal Epstein-Barr virus infection. Before EBV infection both showed normal cellular and humoral immunity. Death in both cases was caused by liver failure: one developed extensive hepatic necrosis; the other developed massive infiltration of the liver with EBV-infected immunoblasts after aggressive immunosuppressive therapy. Sullivan et al. (1983) proposed that an aberrant immune response triggered by acute EBV infection results in unregulated anomalous killer and natural killer cell activity against EBV infected and uninfected cells. They further suggested that the global cellular immune defects in males with XLPD who survive EBV infection represent an epiphenomenon. An autosomal form of susceptibility to EBV has also been suggested (see 226990).

Skare et al. (1987) demonstrated linkage of XLPD with DXS42, which maps to Xq24-q27. The most likely recombination frequency was 4% (lod = 5.26). Haplotype analysis showed that XLPD is distal to DXS42 but proximal to DXS99, which maps to Xq26-q27. With this information it became possible to identify carrier females and to diagnose the disorder prenatally. Skare et al. (1987) concluded that IMD5 is probably in the Xq26 band. Skare et al. (1989) extended the linkage information on the family they reported in 1987. Skare et al. (1989) studied 6 additional families, all of which corroborated the close linkage to DXS42 (1% recombination; lod = 17.5) and DXS37 (maximum lod = 13.3 at theta = 0.0.) Hayoz et al. (1988) identified a large Swiss family with XLPD. In this family a new mutation for hemophilia A (306700) had occurred, resulting in a woman who was a carrier of both genes. She had a son who had hemophilia, a daughter who was a carrier of both, and a son who was free of both disorders. The doubly-heterozygous daughter had 1 son with XLPD and 1 son who was doubly affected. These observations suggest that the hemophilia A locus and the XLPD locus are far apart, a conclusion that is supported by other mapping data. Hayoz et al. (1988) reported an extensively affected Swiss family ascertained through a patient with acquired hypogammaglobulinemia associated with a mononucleosis syndrome at the age of 18 years; the patient died at 19. Harris et al. (1988) excluded linkage to 28 X-linked probes. Harris and Docherty (1988) found no particular chromosomal abnormalities in this disorder. One of their patients was found to have the Klinefelter syndrome, having inherited 2 copies of his maternal XLPD-carrying X chromosome. Wyandt et al. (1989) found deletion of part of band Xq25 in a male with this disorder; others found the same deletion in the mother and a sister. Skare et al. (1989) reported family linkage studies of this gene with several Xq DNA markers. They also reported on 1 male out of 14 unrelated affected persons who had a deletion in Xq which was thought to involve about one-half of Xq25. This male retained sequences at all 5 loci that had been found to be closely linked to XLP, but lacked DXS6 which by somatic hybrid mapping had been assigned to Xq26-qter. Sanger et al. (1990) demonstrated an interstitial deletion involving a portion of Xq25 in an affected male as well as in one sister and their mother. In a large Swiss family, Sylla et al. (1989) demonstrated linkage of XLP to marker DXS37, which is located in Xq25-q26. Multipoint linkage analysis showed that the XLP locus is distal to DXS11 but proximal to HPRT (308000). Skare et al. (1993) used blot hybridization to find DNA sequences missing in a male who lacked two-thirds of Xq25. Using the probes, they discovered 2 additional males with deletions resulting in XLP. All 3 deletions had a region in common, and DXS739 was within this region. The new deletions were also detectable using chromosome banding; although the smallest removed only one-third of Xq25, XLP was the only consequence of the deletion in all 3 cases.

Purtilo and Grierson (1991) reported that during the previous decade 240 males with XLP within 59 unrelated kindreds had been identified worldwide. One-half of the patients had developed fatal infectious mononucleosis at an average age of about 2.5 years and median survival of only 33 days following onset of illness. About one-third had acquired hypogammaglobulinemia and another one-fourth had developed malignant lymphoma. The latter 2 phenotypes were associated with longer survivals, but no patient had been documented as living into the fifth decade of life. Purtilo and Grierson (1991) concluded that the diagnosis of XLP in affected males and female carriers was 99% accurate based on results of linkage to DXS42 RFLPs (lod = 19.4) and 95% accurate with RFLP probes to DXS37 (lod = 11.8). By linkage they detected males with the XLP gene before EBV infection occurred. The malignant lymphomas occurring in patients with XLP were often of the Burkitt type and most involved the ileocecal region. In the American Burkitt Lymphoma Registry, 39 of 41 patients were males (Levine et al., 1982). Finding a Burkitt lymphoma in the ileocecal region should prompt investigation for XLP in the patient and his family. Skare et al. (1992) made the diagnosis of XLP on the basis of analysis of closely linked RFLP markers applied to cells obtained at amniocentesis at 15 weeks' gestation. By use of DNA markers applied to chorionic villus sampling (CVS) material, Mulley et al. (1992) identified with high reliability an unaffected male fetus, brother of an affected male. By HLA DR typing of the CVS, they also showed that the fetus was DR-identical to the affected sib.

Grierson et al. (1993) reported studies of 10 families in which a single male had died of infectious mononucleosis. In such families, Epstein-Barr virus seronegative males must be considered at risk for XLP until they seroconvert normally to the virus. Given that 50% of males inheriting the defective XLP gene die following primary EBV infection, it is vital that they be identified pre-EBV infection. In 3 families, XLP was ruled out (except as a result of a new mutation). One family was determined to have XLP; a male identified as having the genotype of XLP through the use of DNA markers subsequently became symptomatic. In 3 families, the possibility of XLP was ruled out in 6 males, 4 potential carriers were identified in 3 other families, and carrier status was ruled out in 5 women from 4 families.

Williams et al. (1993) reported successful bone marrow transplantation in an 11-year-old boy with Duncan syndrome, with restoration of an apparently normal host immune response to EBV. They presented this as evidence that the primary abnormality in this disorder resides in bone marrow-derived cells. Vowels et al. (1993) described a boy with this disorder in whom transplantation of cord-blood stem cells from an HLA-identical sib resulted in correction of the genetic defect and the hypogammaglobulinemia. Cord blood collected at birth contains 5 to 10 times more marrow progenitor cells than the peripheral blood of older infants or children, and the volume of blood and nucleated cells that can be collected is substantial. Cord blood had successfully been transplanted into patients with aplastic anemia and leukemia and had resulted in repopulation of the bone marrow and immune systems.

Seemayer et al. (1995) reviewed this disorder 25 years after Purtilo's first observations in 1969. Purtilo established a registry in 1980 to serve as a worldwide resource for the diagnosis, treatment, and research of this condition. After Purtilo's death in late 1992, the registry and research unit continued to function as a worldwide consultative service and contributed the following information: First, the number of affected boys continued to grow; some 272 affected members of 80 kindreds had been identified. Second, approximately 10% of the boys who inherited the mutated XLP gene were immunologically abnormal, even before evidence of EBV exposure. Third, the search for the XLP gene had been narrowed to a small region on Xq25, and attempts to clone the gene were progressing.

In summary, XLP is characterized by extreme sensitivity to EBV. The average age of onset is 2.5 years, with 100% mortality by the age of 40 years. Following infection with EBV, patients mount a vigorous, uncontrolled polyclonal expansion of T and B cells. The phenotype is that of severe or fatal infectious mononucleosis in approximately half of cases, acquired hypogammaglobulinemia in about 30%, or malignant lymphoma in about 25%, alone or in combination. Most of the malignant lymphomas are extranodal non-Hodgkin lymphomas, usually of the Burkitt type, and most involve the ileocaecal region of the intestine. The primary cause of death is hepatic necrosis and bone marrow failure. The extensive tissue destruction of the liver and bone marrow appears to stem from the uncontrolled cytotoxic T-cell response.

Coffey et al. (1998) used a positional cloning approach to identify the gene mutated in XLP. In 9 unrelated patients, mutations were found in a novel gene designated 'SH2 domain protein 1A' (SH2D1A). The SH2D1A cDNA was 2,530 bp long and contained an open reading frame (ORF) of 462 bp, with an initiation codon 79 bp from the start of the ORF. Northern blot analysis with probes generated from the cDNA showed expression of an approximately 2.5-kb mRNA at high level in thymus and lung, with a lower level of expression in spleen and liver. SH2D1A expression was also detected by RT-PCR in all lymphocyte populations assayed, but was not detected in a range of EBV-transformed lymphoblastoid cell lines. SH2D1A encodes a protein of 128 amino acids consisting of a 5-amino acid N-terminal sequence, an SH2 domain, and a 25-amino acid C-terminal tail. No hydrophobic signal sequence was present, suggesting that SH2D1A is localized in the cytoplasm. The SH2D1A gene comprises 4 exons.

Independently, Sayos et al. (1998) identified the SH2D1A gene, which they referred to as SAP for 'SLAM-associated protein.' Sayos et al. (1998) reached the same goal as did Coffey et al. (1998) but by a different route. They studied SLAM (signaling lymphocyte activation molecule), a glycosylated transmembrane protein also known as CDw150. SLAM has a high affinity for itself and is centrally involved in the bidirectional stimulation of T and B cells. When activated, it mediates expansion of activated T cells during immune responses, induces production of interferon-gamma, and changes the functional profile of subsets of T cells. Signaling through SLAM-SLAM binding during mutual interaction between B cells, and between B cells and T cells, increases the expansion and differentiation of activated B cells. Sayos et al. (1998) used the human SLAM cytoplasmic domain as bait in the yeast 2-hybrid system. From a human T-cell library, they isolated 8 cDNA clones that encoded a 128-amino acid polypeptide, designated SAP. Both in humans and in mice, SAP is expressed in T cells, but not in B cells. Using a clone that contained all 4 exons of mouse Sap, Sayos et al. (1998) localized the gene to the part of the mouse X chromosome corresponding to human Xq25. To test the possibility that SAP was encoded by the XLP gene, SAP cDNAs were isolated from the blood cells of 3 XLP patients. All 3 cDNAs were mutated or deleted, in contrast to those from a healthy brother and other controls. The mutant SAPs did not bind SLAM. Sayos et al. (1998) suggested that SAP is a natural inhibitor of SLAP. Consistent with this idea, they found that SAP binding blocks the recruitment of the tyrosine phosphatase SHP2 (176876) to the phosphorylated cytoplasmic domain of SLAM. Upon T-cell activation, SLAM may switch from a signaling cascade that is dependent on SAP, and probably on FYN (137025), a member of the src tyrosine kinase family, to one that depends on SHP2. Put more generally, Sayos et a l. (1998) proposed that SAP controls the signal-transduction pathways initiated by interactions between SLAM molecules at the interface between T and B cells. Klein and Klein (1998) commented on the unusual characteristics of the Epstein-Barr virus as the most potent transforming and potentially tumorigenic virus known and the remarkable equilibrium that is usually maintained by a finely poised balance between the regulation of viral gene expression and the immune system, particularly T cells. The work defining the defect in XLP goes a long way toward illuminating that equilibrium.

See also Loeffel et al. (1985); Purtilo (1976); Purtilo (1981); Purtilo et al. (1978); Purtilo et al. (1977); Skare et al. (1989); Skare et al. (1987); Skare et al. (1989); Steinherz et al. (1985)

Bar, R. S.; DeLor, C. J.; Clausen, K. P.; Hurtubise, P.; Henle, W.; Hewetson, J. F.: Fatal infectious mononucleosis in a family. New Eng. J. Med. 290: 363-367, 1974.

Coffey, A. J.; Brooksbank, R. A.; Brandau, O.; Oohashi, T.; Howell, G. R.; Bye, J. M.; Cahn, A. P.; Durham, J.; Heath, P.; Wray, P.; Pavitt, R.; Wilkinson, J.; and 31 others: Host response to EBV infection in X-linked lymphoproliferative disease results from mutations in an SH2-domain encoding gene. Nature Genet. 20: 129-135, 1998.

Grierson, H. L.; Skare, J.; Church, J.; Silberman, T.; Davis, J. R.; Kobrinsky, N.; McGregor, R.; Israels, S.; McCarty, J.; Andrews, L. G.; Blecha, T.; Erdman, S.; Obringer, A.; Scharnhorst, D.; Purtilo, D. T.: Evaluation of families wherein a single male manifests a phenotype of X-linked lymphoproliferative disease (XLP). Am. J. Med. Genet. 47: 458-463, 1993.

Hamilton, J. K.; Paquin, L. A.; Sullivan, J. L.; Maurer, H. S.; Cruzi, F. G.; Provisor, A. J.; Steuber, C. P.; Hawkins, E.; Yawn, D.; Cornet, J.; Clausen, K.; Finkelstein, G. Z.; Landing, B.; Grunnet, M.; Purtilo, D. T.: X-linked lymphoproliferative syndrome registry report. J. Pediat. 96: 669-673, 1980.

Harris, A.; Docherty, Z.: X-linked lymphoproliferative disease: a karyotype analysis. Cytogenet. Cell Genet. 47: 92-94, 1988.

Harris, A.; Lenoir, G. M.; Lankester, S. A.: X-linked lymphoproliferative disease: linkage studies using DNA probes. Clin. Genet. 33: 162-168, 1988.

Hayoz, D.; Lenoir, G. M.; Nicole, A.; Pugin, P.; Regamey, C.: X-linked lymphoproliferative syndrome: identification of a large family in Switzerland. Am. J. Med. 84: 529-534, 1988.

Klein, G.; Klein, E.: Sinking surveillance's flagship. Nature 395: 441-445, 1998.

Levine, P. H.; Kamaraju, L. S.; Connelly, R. R.; Berard, C. W.; Dorfman, R. F.; Magrath, I.; Easton, J. M.: The American Burkitt's Lymphoma Registry: eight years' experience. Cancer 49: 1016-1022, 1982.

Loeffel, S.; Chang, C.-H.; Heyn, R.; Harada, S.; Lipscomb, H.; Sinangil, F.; Volsky, D. J.; McClain, K.; Ochs, H.; Purtilo, D. T.: Necrotizing lymphoid vasculitis in X-linked lymphoproliferative syndrome. Arch. Path. Lab. Med. 109: 546-550, 1985.

Lyon, M. F.; Loutit, J. F.: X-linked factor in acquired immunodeficiency syndrome?. (Letter) Lancet I: 768 only, 1983.

Mulley, J. C.; Turner, A. M.; Gedeon, A. K.; Berdoukas, V. A.; Huang, T. H. M.; Ledbetter, D. H.; Grierson, H.; Purtilo, D. T.: X-linked lymphoproliferative disease: prenatal detection of an unaffected histocompatible male. Clin. Genet. 42: 76-79, 1992.

Provisor, A. J.; Iacuone, J. J.; Chilcote, R. R.; Neiburger, R. G.; Crussi, F. G.; Baehner, R. L.: Acquired agammaglobulinemia after a life-threatening illness with clinical and laboratory features of infectious mononucleosis in three related male children. New Eng. J. Med. 293: 62-65, 1975.

Purtilo, D. T.: Pathogenesis and phenotypes of an X-linked recessive lymphoproliferative syndrome. Lancet II: 882-885, 1976.

Purtilo, D. T.: X-linked lymphoproliferative syndrome: an immunodeficiency disorder with acquired agammaglobulinemia, fatal infectious mononucleosis, or malignant lymphoma. Arch. Path. Lab. Med. 105: 119-121, 1981.

Purtilo, D. T.; Bhawan, J.; Hutt, L. M.; De Nicola, L.; Szymanski, I.; Yang, J. P. S.; Boto, W.; Naier, R.; Thorley-Lawson, D.: Epstein-Barr virus in the X-linked recessive lymphoproliferative syndrome. Lancet I: 798-801, 1978.

Purtilo, D. T.; Cassel, C. K.; Yang, J. P. S.: Fatal infectious mononucleosis in familial lymphohistiocytosis. (Letter) New Eng. J. Med. 201: 736 only, 1974.

Purtilo, D. T.; Cassel, C. K.; Yang, J. P. S.; Harper, R.; Stephenson, S. R.; Landing, B. H.; Vewter, G. F.: X-linked recessive progressive combined variable immunodeficiency (Duncan's disease). Lancet I: 935-941, 1975.

Purtilo, D. T.; DeFlorio, D., Jr.; Hutt, L. M.; Bhawan, J.; Yang, J. P. S.; Otto, R. L.; Edwards, W.: Variable phenotypic expression of an X-linked recessive lymphoproliferative syndrome. New Eng. J. Med. 297: 1077-1081, 1977.

Purtilo, D. T.; Grierson, H. L.: Methods of detection of new families with X-linked lymphoproliferative disease. Cancer Genet. Cytogenet. 51: 143-153, 1991.

Purtilo, D. T.; Sakamoto, K.; Barnabei, V.; Seeley, J.; Bechtold, T.; Rogers, G.; Yetz, J.; Harada, S.; the XLP collaborators: Epstein-Barr virus-induced diseases in boys with the X-linked lymphoproliferative syndrome (XLP): update on studies of the registry. Am. J. Med. 73: 49-56, 1982.

Purtilo, D. T.; Yang, J. P. S.; Allegra, S.; DeFlorio, D.; Hutt, L. M.; Soltani, M.; Vawter, G. F.: Hematopathology and pathogenesis of the X-linked recessive lymphoproliferative syndrome. Am. J. Med. 62: 225-233, 1977.

Sanger, W. G.; Grierson, H. L.; Skare, J.; Wyandt, H.; Pirruccello, S.; Fordyce, R.; Purtilo, D. T.: Partial Xq25 deletion in a family with the X-linked lymphoproliferative disease (XLP). Cancer Genet. Cytogenet. 47: 163-169, 1990.

Sayos, J.; Wu, C.; Morra, M.; Wang, N.; Zhang, X.; Allen, D.; van Schaik, S.; Notarangelo, L.; Gehat, R.; Roncarolo, M. G.; Oettgen, H.; De Vries, J. E.; Aversall, G.; Terhorst, C.: The X-linked lymphoproliferative-disease gene product SAP regulates signals induced through the co-receptor SLAM. Nature 395: 462-469, 1998.

Scher, I.: The CBA/N mouse strain: an experimental model illustrating the influence of the X-chromosome on immunity. Adv. Immun. 33: 1-71, 1982.
Seemayer, T. A.; Gross, T. G.; Egeler, R. M.; Pirruccello, S. J.; Davis, D. J.; Kelly, C. M.; Okano, M.; Lanyi, A.; Sumegi, J.: X-linked lymphoproliferative disease: twenty-five years after the discovery. Pediat. Res. 38: 471-478, 1995.
Shapiro, M. B.; Senapathy, P.: RNA splice junctions of different classes of eukaryotes: sequence statistics and functional implications in gene expression. Nucleic Acids Res. 15: 7155-7174, 1987.
Skare, J.; Grierson, H.; Wyandt, H.; Sanger, W.; Milunsky, J.; Purtilo, D.; Sullivan, J.; Milunsky, A.: Genetics of the X-linked lymphoproliferative syndrome. (Abstract) Am. J. Hum. Genet. 45 (suppl.): A161 only, 1989.
Skare, J.; Madan, S.; Glaser, J.; Purtilo, D.; Nitowsky, H.; Pulijaal, V.; Milunsky, A.: First prenatal diagnosis of X-linked lymphoproliferative disease. Am. J. Med. Genet. 44: 79-81, 1992.
Skare, J.; Milunsky, A.; Byron, K.; Sullivan, J.: The mutation causing X-linked lymphoproliferative syndrome lies in Xq26. (Abstract) Am. J. Hum. Genet. 41: A185 only, 1987.
Skare, J.; Wu, B.-L.; Madan, S.; Pulijaal, V.; Purtilo, D.; Haber, D.; Nelson, D.; Sylla, B.; Grierson, H.; Nitowsky, H.; Glaser, J.; Wissink, J.; White, B.; Holden, J.; Housman, D.; Lenoir, G.; Wyandt, H.; Milunsky, A.: Characterization of three overlapping deletions causing X-linked lymphoproliferative disease. Genomics 16: 254-255, 1993.
Skare, J. C.; Grierson, H. L.; Sullivan, J. L.; Nussbaum, R. L.; Purtilo, D. T.; Sylla, B. S.; Lenoir, G. M.; Reilly, D. S.; White, B. N.; Milunsky, A.: Linkage analysis of seven kindreds with the X-linked lymphoproliferative syndrome (XLP) confirms that the XLP locus is near DXS42 and DXS37. Hum. Genet. 82: 354-358, 1989.
Skare, J. C.; Milunsky, A.; Byron, K. S.; Sullivan, J. L.: Mapping the X-linked lymphoproliferative syndrome. Proc. Nat. Acad. Sci. 84: 2015-2018, 1987.
Skare, J. C.; Sullivan, J. L.; Milunsky, A.: Mapping the mutation causing the X-linked lymphoproliferative syndrome in relation to restriction fragment length polymorphisms on Xq. Hum. Genet. 82: 349-353, 1989.
Steinherz, R.; Levy, Y.; Litwin, A.; Nitzan, M.; Friedman, E.; Levin, S.: X-linked lymphoproliferative syndrome: a new kindred with variable phenotypic expression. Am. J. Dis. Child. 139: 191-193, 1985.
Sullivan, J. L.; Byron, K. S.; Brewster, F. E.; Baker, S. M.; Ochs, H. D.: X-linked lymphoproliferative syndrome: natural history of the immunodeficiency. J. Clin. Invest. 71: 1765-1778, 1983.
Sullivan, J. L.; Byron, K. S.; Brewster, F. E.; Purtilo, D. T.: Deficient natural killer cell activity in X-linked lymphoproliferative syndrome. Science 210: 543-545, 1980.
Sylla, B. S.; Wang, Q.; Hayoz, D.: Lathrop, G. M. and Lenoir, G. M.: Multipoint linkage mapping of the Xq25-q26 region in a family affected by the X-linked lymphoproliferative syndrome. Clin. Genet. 36: 359-462, 1989.
Vowels, M. R.; Lam-Po-Tang, R.; Berdoukas, V.; Ford, D.; Thierry, D.; Purtilo, D.; Gluckman, E.: Correction of X-linked lymphoproliferative disease by transplantation of cord-blood stem cells. New Eng. J. Med. 329: 1623-1625, 1993.
Williams, L. L.; Rooney, C. M.; Conley, M. E.; Brenner, M. K.; Krance, R. A.; Heslop, H. E.: Correction of Duncan's syndrome by allogeneic bone marrow transplantation. Lancet 342: 587-588, 1993.
Wyandt, H. E.; Skare, J. C.; Grierson, H. L.; Purtilo, D. T.; Milunsky, A.: Detection of a chromosomal deletion of Xq25 in an affected male with X-linked lymphoproliferative disease. (Abstract) Am. J. Hum. Genet. 45 (suppl.): A108 only, 1989.

SEVERE COMBINED IMMUNODEFICIENCY DISEASE (SCID), AUTOSOMAL RECESSIVE, T-NEGATIVE/B-POSITIVE TYPE (#600802)

JJM NOTE

Because cancer has not been described with specific genotypes of SCID, this entity is provisionally included.

A number sign (#) is used with this entry because of the likelihood that the phenotype has several possible molecular bases, one of which is known to be deficiency of Janus kinase-3 (600173). Deficiency of protein tyrosine kinase ZAP-70 (176947) also causes SCID with a selective defect of T and natural killer (NK) cells (Elder, 1996). A T cell-negative, B cell/NK cell-positive form of SCID is caused by mutations in the interleukin 7 receptor (IL7R; 146661.0001). Other pathogenetically distinct forms of autosomal recessive SCID include HLA class II-negative SCID or the bare lymphocyte syndrome (209920), which has 2 fundamental forms, that due to a defect in C2TA (600005) and that due to a defect in RFX (600006). The classic type of autosomal recessive SCID is that due to deficiency of adenine deaminase (102700). Omenn syndrome (267700) is a leaky T(-)B(-) phenotype.

Elder, M. E.: Severe combined immunodeficiency due to a defect in the tyrosine kinase ZAP-70. Pediat. Res. 39: 743-748, 1996.

SEVERE COMBINED IMMUNODEFICIENCY, B-CELL NEGATIVE (#601457)

JJM NOTE

It is not clear whether the risk for malignancy varies between B-cell-negative- and -positive cases.

A number sign (#) is used with this entry because of evidence that the disorder is due to mutation in the RAG1 (179615) and RAG2 (179616) genes.

Patients affected with severe combined immunodeficiency (SCID) can be divided into 2 main classes: those with B lymphocytes (B+ SCID) and those without (B- SCID). Schwarz et al. (1996) stated that approximately 70% of all SCID cases are of the B+ type and are caused by various defects, including mutations in receptors for interleukins (e.g., 308380), adenosine deaminase deficiency (102700), and improper regulation of HLA class II expression (see 209920). The etiology of B-SCID, however, had been undefined. Six of 14 B- SCID patients studied by Schwarz et al. (1996) were found to carry mutations of the recombinase activating genes RAG1 (179615) and RAG2 (179616). The mutations resulted in a functional inability to form antigen receptors through genetic recombination. In 4 families, 4 B- SCID patients exhibited an altered migration pattern for RAG1 amplimers as assayed by SSCP analysis. In 2 families, 3 patients exhibited an altered migration pattern for RAG2 amplimers. Upon sequence analysis of the amplimers, Schwarz et al. (1996) detected 5 missense, 3 nonsense, and 1 deletion mutation in RAG1 and RAG2. None of these mutations were detected in B+ SCID patients or in 35 healthy subjects. Several of the SCID patients had unaffected sibs who were heterozygous with 1 wildtype allele for RAG1 or RAG2. Schwarz et al. (1996) therefore concluded that 1 wildtype allele is sufficient for normal lymphocyte development. Transient transfection assays revealed that the SCID-associated RAG1 and RAG2 mutations exhibit either a complete loss or a marked reduction of V(D)J recombination activity. Schwarz et al. (1996) concluded that structural mutations of the RAG genes account for a substantial proportion of human SCID cases.

Schwarz, K.; Gauss, G. H.; Ludwig, L.; Pannicke, U.; Li, Z.; Linder, D.; Friedrich, W.; Seger, R. A.; Hansen-Hagge, T. E.; Desiderio, S.; Lieber, M. R.; Bartram, C. R.: RAG mutations in human B cell-negative SCID. Science 274: 97-99, 1996.

RECOMBINATION ACTIVATING GENE 1 (*179615)

JJM NOTE

This gene underlies familial hemophagocytic lymphohistiocytosis and combined immunodeficiency.

Schatz et al. (1989) isolated the RAG1 gene which activates the V(D)J recombination when introduced into NIH 3T3 fibroblasts. Nucleotide sequencing of human and mouse RAG1 cDNA clones encode predicted 119-kD proteins of 1,043 and 1,040 amino acids, respectively, with 90% sequence identity. The RAG1 gene has been conserved between species that carry out the V(D)J recombination, and its pattern of expression correlates exactly with the pattern of expression of V(D)J recombinase activity. Schatz et al. (1989) raised the question as to whether RAG1 activates the V(D)J recombination indirectly or whether it may encode the V(D)J recombinase itself. Mapping the gene in mouse and man may help determine whether the RAG1 gene is involved in any form of severe combined immunodeficiency in mouse or man. It is found, for example, that the scid mouse, caused by a recessive mutation on chromosome 16 of that species, has disruption of normal V(D)J recombination. Schatz et al. (1989) reported that Southern blots of DNA from a variety of somatic cell hybrids demonstrated that the murine equivalent of RAG1 does not map to chromosome 16, thus indicating that it is not the site of the mutation in the scid mouse. There is some reason to think that the scid gene does not encode the V(D)J recombinase because myeloid and fibroblastoid cells show an increased sensitivity to radiation-induced damage in that mutant mouse, suggesting that the defect is in a ubiquitously expressed factor involving both V(D)J recombination and the repair of chromosomal damage (Fulop and Phillips, 1990). Oettinger et al. (1990) gave a corrected length of 6.6 kb for the RAG1 cDNA. Mombaerts et al. (1992) introduced a mutation in the V(D)J recombination activating gene RAG1 into the germline of mice via gene targeting in embryonic stem cells. They found that such mice had small lymphoid organs that did not contain mature B and T lymphocytes. The phenotype was that of 'nonleaky' scid mice. Although RAG1 expression had been reported in the central nervous system of the mouse, no obvious neuroanatomical or behavioral abnormalities were found in the RAG1-deficient mice.

Blanquet et al. (1992) determined by in situ hybridization that the RAG1 gene is located in the 14q21.3-q22.2 region. This assignment must have been in error; possibly the probe used was not in fact from that gene (Oettinger, 1993). Oettinger et al. (1992) mapped the RAG1 and RAG2 loci to 11p by Southern analysis of hybrid cell lines derived from patients with the WAGR syndrome (194070) and from mutagenized cell hybrids selected for deletions in chromosome 11. The RAG locus defined a new interval of human 11p which was not known to contain any genetically mapped human disease. Guided by the localization of the human genes, they mapped the homologous loci to mouse chromosome 2. Sherrington et al. (1992) confirmed the assignment of the RAG1 and RAG2 genes to human 11p13. The results of study of a somatic cell hybrid panel placed the RAG1 gene near CD44 and proximal to CAT. They pointed out that these recombinase-activating genes are thus not linked to ataxia-telangiectasia complementation groups A, C, or D, which have been mapped to the region 11q22-q23, and are presumably not directly responsible for the phenotype of that disorder (see 208900). Ichihara et al. (1992) mapped both RAG1 and RAG2 to 11p13-p12 by fluorescence in situ hybridization.

Formation of double-strand breaks at recombination signal sequences is an early step in V(D)J recombination. McBlane et al. (1995) showed that purified RAG1 and RAG2 proteins are sufficient to carry out this reaction. The cleavage reaction can be divided into 2 distinct steps, nicking and hairpin formation, each of which requires the presence of a signal sequence and both RAG proteins.

Roman and Baltimore (1996) presented genetic evidence that RAG1 is directly involved in the recognition of the DNA substrate. The RAG1 genomic locus was originally isolated by its ability to activate recombination in the fibroblast line 3TGR. 3TGR harbors an integrated retroviral recombination substrate that contains a neomycin (neo) resistance gene, which is dependent on inversion via V(D)J recombination for its transcriptional activation. Two murine RAG1 cDNAs, called M2 and M6 by them, were originally isolated by their group (Schatz et al., 1989), but only 1 of the clones (M2) encoded a protein that complemented recombination in 3TGR cells; M6 was inactive. Roman and Baltimore (1996) showed that the M6 cDNA contained a single amino acid substitution (H109L) in the RAG1 gene that rendered its activity sensitive to the sequence of the V(D)J coding region abutting the heptamer site in the recombination signal sequence. These results indicated to Roman and Baltimore (1996) that RAG1 interacts directly with DNA.

Schwarz et al. (1996) reported that patients with severe combined immunodeficiency can be divided into those with B lymphocytes (B-positive SCID) and those without (B-negative SCID; 601457). They searched for RAG1 and RAG2 (179616) mutations in B-negative SCID patients through the use of SSCP analysis with primer cassettes overlapping the entire RAG1 and RAG2 coding regions. Six of 14 B-negative SCID patients were found to carry mutations of the recombinase activating genes. Mutations resulted in a functional inability to form antigen receptors through genetic recombination. In 4 families, 4 B-negative SCID patients exhibited an altered migration pattern for RAG1 amplimers. They identified 2 missense mutations (179615.0001 and 179615.0004) and 2 nonsense mutations (179615.0002 and 179615.0003) in RAG1. In one case there was a paternal deletion which encompassed the RAG1 and RAG2 loci on chromosome 11p13. Transient transfection assays revealed that the SCID-associated RAG1 and RAG2 mutations exhibited either a complete loss or a marked reduction of V(D)J recombination activity.

Hikida et al. (1996) reported that RAG1 and RAG2 are expressed in mature mouse B cells after culture with interleukin-4 (147780) in association with costimuli (lipopolysaccharide and other cytokines). Reexpression was also detected in draining lymph nodes from immunized mice. Hikida et al. (1996) noted that previously reported studies had indicated that RAG1 and RAG2 were expressed only in immature B cells.

Villa et al. (1998) reported that patients with Omenn syndrome (familial histiocytic reticulosis; 267700), a severe immunodeficiency characterized by the presence of activated, anergic, oligoclonal T cells, hypereosinophilia, and high IgE levels, have missense mutations in either the RAG1 or RAG2 (179616) genes that result is partial activity of the 2 proteins. Two of the amino acid substitutions map within the RAG1 homeodomain and decrease DNA binding activity, while 3 others lower the efficiency of RAG1/RAG2 interaction. These findings provided evidence indicating that the immunodeficiency manifested in patients with Omenn syndrome arises from mutations that decrease the efficiency of V(D)J recombination.

Immunoglobulin and T-cell receptor genes are assembled from component gene segments in developing lymphocytes by a site-specific recombination reaction which mediates V(D)J joining. Agrawal et al. (1998) showed that RAG1 and RAG2 are essential to this reaction. Together they form a transposase capable of excising a piece of DNA-containing recombination signals from a donor site and inserting it into a target DNA molecule. The products formed contain a short duplication of target DNA immediately flanking the transposed reactions. The results supported the theory that RAG1 and RAG2 were once components of a transposable element, and that the split nature of immunoglobulin and T-cell receptor genes derived from germline insertion of this element into an ancestral receptor gene soon after the evolutionary divergence of jawed and jawless vertebrates. Thus the repertoire of the human immune system may owe to 1 transposon insertion, which occurred 450 million years ago in an ancestor of the jawed vertebrates. Vertebrates seemed have tamed this ancient transposon for generation of the immune repertoire. It was surprising when RAG1 and RAG2 were discovered (by Schatz and Oettinger working as graduate students in the laboratory of David Baltimore (Schatz et al., 1989)) to be located within such a small segment of the genome. This was a lucky circumstance since the selection system they used required that both be present in the fragment. The work of Agrawal et al. (1998) explained the reason for this close situation: they once had to fit in a small transposable element.

Like Agrawal et al. (1998), Hiom et al. (1998) concluded that the RAG-mediated V(D)J recombination system evolved from an ancient mobile DNA element. They suggested that repeated transposition may have promoted the expansion of the antigen receptor loci. They stated further that the inappropriate diversion of V(D)J rearrangement to a transpositional pathway may help explain certain types of DNA translocation associated with lymphatic tumors.

Agrawal, A.; Eastman, Q. M.; Schatz, D. G.: Transposition mediated by RAG1 and RAG2 and its implications for the evolution of the immune system. Nature 394: 744-751, 1998.

Blanquet, V.; Turleau, C.; Goossens, M.; Besmond, C.: Assignment of the human recombination activating gene 1 (RAG1) to the 14q21.3-q22.2 region. Genomics 13: 488-489, 1992.

Fulop, G. M.; Phillips, R. A.: The scid mutation in mice causes a general defect in DNA repair. Nature 347: 479-482, 1990.

Hikida, M.; Mori, M.; Takai, T.; Tomochika, K.; Hamatani, K.; Ohmori, H.: Reexpression of RAG-1 and RAG-2 genes in activated mature mouse B cells. Science 274: 2091-2093, 1996.

Hiom, K.; Melek, M.; Gellert, M.: DNA transposition by the RAG1 and RAG2 proteins: a possible source of oncogenic translocations. Cell 94: 463-470, 1998.

Ichihara, Y.; Hirai, M.; Kurosawa, Y.: Sequence and chromosome assignment to 11p13-p12 of human RAG genes. Immun. Lett. 33: 277-284, 1992.

McBlane, J. F.; van Gent, D. C.; Ramsden, D. A.; Romeo, C.; Cuomo, C. A.; Gellert, M.; Oettinger, M. A.: Cleavage at a V(D)J recombination signal requires only RAG1 and RAG2 proteins and occurs in two steps. Cell 83: 387-395, 1995.

Mombaerts, P.; Iacomini, J.; Johnson, R. S.; Herrup, K.; Tonegawa, S.; Papaioannou, V. E.: RAG-1-deficient mice have no mature B and T lymphocytes. Cell 68: 869-877, 1992.

Oettinger, M. A.: Personal Communication. Boston, Mass. 1/15/1993.

Oettinger, M. A.; Schatz, D. G.; Gorka, C.; Baltimore, D.: RAG-1 and RAG-2, adjacent genes that synergistically activate V(D)J recombination. Science 248: 1517-1523, 1990.

Oettinger, M. A.; Stanger, B.; Schatz, D. G.; Glaser, T.; Call, K.; Housman, D.; Baltimore, D.: The recombination activating genes, RAG1 and RAG2, are on chromosome 11p in humans and chromosome 2p in mice. Immunogenetics 35: 97-101, 1992.

Roman, C. A. J.; Baltimore, D.: Genetic evidence that the RAG1 protein directly participates in V(D)J recombination through substrate recognition. Proc. Nat. Acad. Sci. 93: 2333-2338, 1996.

Schatz, D. G.; Oettinger, M. A.; Baltimore, D.: The V(D)J recombination activating gene, RAG-1. Cell 59: 1035-1048, 1989.

Schwarz, K.; Gauss, G. H.; Ludwig, L.; Pannicke, U.; Li, Z.; Linder, D.; Friedrich, W.; Seger, R. A.; Hansen-Hagge, T. E.; Desiderio, S.; Lieber, M. R.; Bartram, C. R.: RAG mutations in human B cell-negative SCID. Science 274: 97-99, 1996.

Sherrington, P. D.; Forster, A.; Seawright, A.; van Heyningen, V.; Rabbitts, T. H.: Human RAG2, like RAG1, is on chromosome 11 band p13 and therefore not linked to ataxia telangiectasia complementation groups. Genes Chromosomes Cancer 5: 404-406, 1992.

Villa, A.; Santagata, S.; Bozzi, F.; Giliani, S.; Frattini, A.; Imberti, L.; Gatta, L. B.; Ochs, H. D.; Schwarz, K.; Notarangelo, L. D.; Vezzoni, P.; Spanopoulou, E.: Partial V(D)J recombination activity leads to Omenn syndrome. Cell 93: 885-896, 1998.

RECOMBINATION ACTIVATING GENE 2 (*179616)

JJM NOTE

This is another gene that underlies familial hemophagocytic lymphohistiocytosis and combined immunodeficiency.

Oettinger et al. (1990) demonstrated that cotransfection of the RAG1 gene (179615) with an adjacent gene, RAG2, results in at least a 1,000-

fold increase in the frequency of V(D)J recombination, whereas RAG1 alone inefficiently induces recombinase activity. Oettinger et al. (1990) reported that the 2.1-kb RAG2 cDNA encodes a putative protein of 527 amino acids whose sequence is unrelated to that of RAG1. The genomic size of RAG2 is approximately 18 kb. Like RAG1, RAG2 is conserved between species that carry out V(D)J recombination, and its pattern of expression correlates precisely with that of V(D)J recombinase activity. In addition to being located only 8 kb apart, these convergently transcribed genes are unusual in that most, if not all, of their coding and 3-prime untranslated sequences are contained in a single exon. It is likely that the RAG1 and RAG2 gene products participate directly in the recombination reaction. In transgenic mice carrying a germline mutation in which a large portion in the RAG2 coding region was deleted, Shinkai et al. (1992) found that although homozygotes were viable, they failed to produce mature B or T lymphocytes. Immature lymphoid cells were present in primary lymphoid organs; however, these cells did not rearrange their immunoglobulin or T-cell receptor loci. Thus, loss of RAG2 function results in total inability to initiate V(D)J rearrangement, leading to a severe combined immunodeficiency (SCID) phenotype. Since the SCID phenotype was the only obvious abnormality detected in these mice, RAG2 function and V(D)J recombinase activity, per se, must not be required for development of cells other than lymphocytes.

Lin and Desiderio (1994) found that in an immature B-cell line and in normal thymocytes, RAG2 protein accumulated preferentially in the G0/G1 phases of the cell cycle and declined by at least 20-fold before cells entered S phase. The amount of RAG2 protein remained low throughout the S, G2, and M phases. The amount of RAG1 protein showed considerably less fluctuation. The variation in RAG2 protein is likely to be established, at least in part, by a posttranscriptional mechanism. These observations suggested to Lin and Desiderio (1994) that V(D)J rearrangement occurs entirely or preferentially within G0/G1.

Formation of double-strand breaks at recombination signal sequences is an early step in V(D)J recombination. McBlane et al. (1995) showed that purified RAG1 and RAG2 proteins are sufficient to carry out this reaction. The cleavage reaction can be divided into 2 distinct steps, nicking and hairpin formation, each of which requires the presence of a signal sequence and both RAG proteins.

Schwarz et al. (1996) reported that patients with severe combined immunodeficiency can be divided into those with B lymphocytes (B-positive SCID) and those without (B-negative SCID; 601457). They searched for RAG1 and RAG2 mutations in B-negative SCID patients through the use of SSCP analysis with primer cassettes overlapping the entire RAG1 and RAG2 coding regions. Six of 14 B-negative SCID patients were found to carry mutations of the recombinase activating genes. Mutations resulted in a functional inability to form antigen receptors through genetic recombination. In 2 families, 3 patients exhibited an altered migration pattern for RAG2 amplimers. The PCR products were then sequenced. Two related patients were found to be homozygous for a missense mutation leading to cys476-to-tyr mutation in RAG2 (179616.0001). One patient was found to have inherited a RAG2 missense mutation (179616.0002) from the mother and a deletion involving RAG1 and RAG2 from the father. Transient transfection assays revealed that the SCID-associated RAG2 mutations exhibited either a complete loss or a marked reduction of V(D)J recombination activity.

Hikida et al. (1996) reported that RAG1 and RAG2 are expressed in mature mouse B cells after culture with interleukin-4 (147780) in association with costimuli (lipopolysaccharide and other cytokines). Reexpression was also detected in draining lymph nodes from immunized mice. Hikida et al. (1996) noted that previously reported studies had indicated that RAG1 and RAG2 were expressed only in immature B cells.

Villa et al. (1998) reported that patients with Omenn syndrome (familial histiocytic reticulosis; 267700), a severe immunodeficiency characterized by the presence of activated, anergic, oligoclonal T cells, hypereosinophilia, and high IgE levels, have missense mutations in either the RAG1 (179615) or RAG2 genes that result in partial activity of the 2 proteins. Two of the amino acid substitutions map within the RAG1 homeodomain and decrease DNA binding activity, while 3 others lower the efficiency of RAG1/RAG2 interaction. These findings provided evidence indicating that the immunodeficiency manifested in patients with Omenn syndrome arises from mutations that decrease the efficiency of V(D)J recombination.

Hikida, M.; Mori, M.; Takai, T.; Tomochika, K.; Hamatani, K.; Ohmori, H.: Reexpression of RAG-1 and RAG-2 genes in activated mature mouse B cells. Science 274: 2091-2093, 1996.

Lin, W.-C.; Desiderio, S.: Cell cycle regulation of V(D)J recombination-activating protein RAG-2. Proc. Nat. Acad. Sci. 91: 2733-2737, 1994.

McBlane, J. F.; van Gent, D. C.; Ramsden, D. A.; Romeo, C.; Cuomo, C. A.; Gellert, M.; Oettinger, M. A.: Cleavage at a V(D)J recombination signal requires only RAG1 and RAG2 proteins and occurs in two steps. Cell 83: 387-395, 1995.

Oettinger, M. A.; Schatz, D. G.; Gorka, C.; Baltimore, D.: RAG-1 and RAG-2, adjacent genes that synergistically activate V(D)J recombination. Science 248: 1517-1523, 1990.

Schwarz, K.; Gauss, G. H.; Ludwig, L.; Pannicke, U.; Li, Z.; Linder, D.; Friedrich, W.; Seger, R. A.; Hansen-Hagge, T. E.; Desiderio, S.; Lieber, M. R.; Bartram, C. R.: RAG mutations in human B cell-negative SCID. Science 274: 97-99, 1996.

Shinkai, Y.; Rathbun, G.; Lam, K.-P.; Oltz, E. M.; Stewart, V.; Mendelsohn, M.; Charron, J.; Datta, M.; Young, F.; Stall, A. M.; Alt, F. W.: RAG-2-deficient mice lack mature lymphocytes owing to inability to initiate V(D)J rearrangement. Cell 68: 855-867, 1992.

Villa, A.; Santagata, S.; Bozzi, F.; Giliani, S.; Frattini, A.; Imberti, L.; Gatta, L. B.; Ochs, H. D.; Schwarz, K.; Notarangelo, L. D.; Vezzoni, P.; Spanopoulou, E.: Partial V(D)J recombination activity leads to Omenn syndrome. Cell 93: 885-896, 1998.

SYK-RELATED TYROSINE KINASE (*176947)

JJM NOTE

Because cancer has not been described with specific genotypes of SCID, this entity is provisionally included.

Protein-tyrosine kinases (PTKs) play an integral role in T-cell activation. Stimulation of the T-cell antigen receptor results in tyrosine phosphorylation of a number of cellular substrates. One of these is the TCR-zeta chain (186780), which can mediate the transduction of extracellular stimuli into cellular effector functions. Chan et al. (1992) isolated a cDNA encoding a 70-kD tyrosine phosphoprotein, ZAP70, that associates with the zeta chain and undergoes tyrosine phosphorylation following TCR stimulation. The ZAP70 gene is expressed in T- and natural killer cells.

Watts et al. (1996) identified a ZAP70 phosphorylation site on the erythrocyte band-3 protein (see 109270) cytoplasmic fragment. They found that this phosphorylation site is distinct from the phosphorylation sites for other PTKs.

Chan et al. (1994) mapped the ZAP70 gene to chromosome 2q12 by hybridization of a genomic fragment to human/rodent somatic hybrid cell DNA and by fluorescence in situ hybridization (FISH). Ku et al. (1994) likewise assigned the ZAP70 gene to 2q12 by isotopic in situ hybridization. They mapped the murine homolog to mouse chromosome 1 by the same method. Saito et al. (1997) also mapped the gene to mouse chromosome 1 and rat chromosome 9q22.1 by FISH; by genetic linkage analysis, they further refined the location of the gene on mouse chromosome 1.

Chan et al. (1994) described 2 mutations in the ZAP70 gene occurring in compound heterozygous state in 3 sibs, 2 boys and a girl, with severe combined immunodeficiency (SCID). Inherited from the father was a C-to-A transition at nucleotide 1763, resulting in substitution of arginine for serine at residue 518. Inherited from the mother was a 9-bp insertion, CTTGAGCAG, resulting in the addition of leu-glu-gln within the catalytic domain. The insertion was due to a G-to-A transition within an intron 9 bp upstream of the normal splice acceptor site. The mutation resulted in a more favored splice acceptor site, AGC, rather than AGA.

Roifman et al. (1989) described a novel type of human immunodeficiency characterized by a selective T-cell defect (STD). Peripheral circulating T cells from these patients exclusively expressed CD4, CD3, and T-cell receptor-alpha/beta, but not CD8 molecules on their surface. The inability to produce peripheral CD8 single-positive cells was traced to an intrathymic developmental defect. While CD4(+)/CD8(+)-positive cells were present in the thymic cortex of these patients, only CD4, not CD8, single-positive cells could be detected in the thymic medulla, suggesting a selective block of positive selection of CD8(+) cells. Peripheral CD4(+) T cells from STD patients failed to proliferate in response to mitogens or to treatment with anti-CD3 antibody. Since the phenotype of STD T cells is consistent with a deficiency of a T cell-specific protein-tyrosine kinase, Arpaia et al. (1994) screened the cells of the patients for a possible mutation in these genes and showed that STD patients carry a mutation of ZAP70 resulting in loss of the activity of this kinase (see 176947.0001). They studied 3 patients in 2 sibships. The disorder is autosomal recessive as indicated by the fact that both parents and 3 sibs were heterozygous.

Elder et al. (1994) reported studies of a 1-year-old girl, the daughter of first cousins, in whom SCID was associated with a highly unusual T-cell subset distribution in the blood. T cells are normally divided into 2 major subsets that express either CD4 or CD8 protein on their surfaces. In this patient, the overall T-cell count was moderately elevated, but CD8(+) cells were virtually absent and nearly all circulating T cells were of the CD4(+) type. They found that the child was homozygous for a 13-bp deletion involving nucleotides 1719 to 1731. Both parents and 2 unaffected sibs were heterozygous. The patient was subsequently cured of her disease by bone marrow transplantation. The deletion would be expected to cause a translational frameshift after amino acid 503, resulting in premature termination 35 codons downstream and yielding a mutant protein 82 residues shorter than wildtype ZAP70.

Arpaia, E.; Shahar, M.; Dadi, H.; Cohen, A.; Roifman, C. M.: Defective T cell receptor signaling and CD8(+) thymic selection in humans lacking Zap-70 kinase. Cell 76: 947-958, 1994.

Chan, A. C.; Iwashima, M.; Turck, C. W.; Weiss, A.: ZAP-70: a 70 kd protein-tyrosine kinase that associates with the TCR zeta chain. Cell 71: 649-662, 1992.

Chan, A. C.; Kadlecek, T. A.; Elder, M. E.; Filipovich, A. H.; Kuo, W.-L.; Iwashima, M.; Parslow, T. G.; Weiss, A.: ZAP-70 deficiency in an autosomal recessive form of severe combined immunodeficiency. Science 264: 1559-1601, 1994.

Elder, M. E.; Lin, D.; Clever, J.; Chan, A. C.; Hope, T. J.; Weiss, A.; Parslow, T. G.: Human severe combined immunodeficiency due to a defect in ZAP-70, a T cell tyrosine kinase. Science 264: 1596-1599, 1994.

Ku, G.; Malissen, B.; Mattei, M.-G.: Chromosomal location of the Syk and ZAP-70 tyrosine kinase genes in mice and humans. Immunogenetics 40: 300-302, 1994.

Roifman, C. M.; Hummel, D.; Martinez-Valdez, H.; Thorner, P.; Doherty, P. J.; Pan, S.; Cohen, F.; Cohen, A.: Depletion of CD8(+) cells in human thymic medulla results in selective immune deficiency. J. Exp. Med. 170: 2177-2182, 1989.

Saito, T.; Matsuda, Y.; Ito, H.; Fusaki, N.; Hori, T.; Yamamoto, T.: Localization of Zap70, the gene for a T cell-specific protein tyrosine kinase, to mouse and rat chromosomes by fluorescence in situ hybridization and molecular genetic linkage analyses. Mammalian Genome 8: 45-46, 1997.

Watts, J. D.; Brabb, T.; Bures, E. J.; Wange, R. L.; Samelson, L. E.; Aebersold, R.: Identification and characterization of a substrate specific for the T cell protein tyrosine kinase ZAP-70. FEBS Lett. 398: 217-222, 1996.

BARE LYMPHOCYTE SYNDROME (#209920)

JJM NOTE

Because cancer has not been described with specific genotypes of SCID, this entity is provisionally included.

A number sign (#) is used with this entry because of evidence that this severe primary immunodeficiency, characterized by the absence of major histocompatibility complex (MHC) class II gene expression, results from defects in at least 4 different transacting regulatory genes required for transcription of MHC class II genes. There are at least 4 complementation groups: group A with a defect in the MHC2TA gene (600005); group C with a defect in the RFX5 gene (601863); group D with a defect in the RFXAP gene (601861); and group B with a defect in the RFXANK gene (603200). The disorder is inherited as an autosomal recessive.

The bare lymphocyte syndrome is a member of the relatively heterogeneous class of SCID, or severe combined immunodeficiency. It is associated with, and probably results from, the lack of expression of HLA antigens on some cells of hematopoietic origin (Touraine et al., 1978). In addition to being of interest in its own right as a 'cause' of SCID, BLS provides insight into the role of MHC determinants in lymphocyte differentiation. In France, Touraine et al. (1980) reported 2 affected brothers, and a family in the Netherlands was reported by 17,16:Schuurman et al. (1979, 1980). In the French family, 4 other sibs had died in infancy. BLS lymphocytes lacked HLA-A, -B, and -C antigens; they also lacked beta-2-microglobulin (109700) to a degree approaching that of the Daudi cell line which has a deletion of chromosome 15. Schuurman et al. (1980) reported studies of 2 unrelated families, 1 Turkish, the other Algerian. Consanguinity was likely in the former and certain in the latter. The affected children were of both sexes. First symptoms presented after the age of 3 or 4 months. All children had severe and persistent diarrhea, mucocutaneous candidiasis, interstitial pneumonia and various bacterial infections but no proved systemic viral infections. The findings of special studies supported the important role of class I HLA antigens in antigen recognition by T lymphocytes. It seemed that the abnormality in this syndrome might lie in the HLA genes or the B2M gene.

Marcadet et al. (1985) reported BLS in a 5-year-old girl born of first-cousin parents who had had repeated infections of the upper and lower respiratory tract, protracted diarrhea, and malabsorption for several years. The organisms included Hemophilus influenzae, Candida albicans, herpes simplex, cytomegalovirus, and adenovirus. Despite normal numbers of T and B lymphocytes, there were no delayed hypersensitivity reactions in vivo; panhypogammaglobulinemia was present, and antigen-induced lymphocyte proliferation and cell-mediated lymphocytotoxicity were absent in vitro. They also reported the disorder in a 4-year-old girl born of presumably unrelated parents of Algerian descent; 2 sisters and 1 brother had died of severe infections before age 4 years. In these 2 patients, Marcadet et al. (1985) showed by molecular genetic techniques that the HLA genes, although poorly expressed, were in fact present. Furthermore, therapeutic decisions concerning bone marrow transplantation were possible. In a family with 2 affected sibs, Sullivan et al. (1985) investigated the molecular basis of this syndrome by means of cDNA probes for both beta-2-microglobulin and class I MHC genes. Southern blots showed no gross internal defect in either. Northern blot analysis also showed no qualitative difference between affected and unaffected family members. In contrast, quantitation of transcripts demonstrated that both B2M and class I MHC were decreased and decreased in a coordinate fashion. The authors interpreted this to mean that BLS represents a pretranslational regulatory defect of expression of 2 genes. Baxter-Lowe et al. (1989) used in vitro amplification of the HLA genes and sequence-specific oligonucleotide probe hybridization (SSOPH) to investigate the HLA status of an infant with HLA-deficient severe combined immunodeficiency. The technology used permitted detection of HLA polymorphism at the level of single amino acid differences and eliminated the requirement for HLA expression. After d etermining the patient's haplotypes by this method, they selected a donor for bone marrow transplantation with a successful result.

Two types of BLS have been identified: in type I, the defect concerns only HLA class I molecules (Touraine et al., 1978; Schuurman et al., 1979), whereas in type II, both HLA class I and class II molecules are affected (Griscelli et al., 1980). In the latter disease, also called HLA class II-negative SCID, the abnormal expression of HLA molecules has been shown to be secondary to defective synthesis (Lisowska-Grospierre et al., 1985) due in turn to an abnormal transacting regulatory gene located outside the major histocompatibility complex (Marcadet et al., 1985; de Preval et al., 1985). The transacting regulatory factor, known as RF-X, binds to class II promoters and is defective in hereditary HLA deficiency type II, otherwise known as the 'bare lymphocyte syndrome.' The failure of HLA expression leads to immunodeficiency affecting both cellular and humoral responses to antigens. Death due to chronic diarrhea and repeated bacterial and viral infections frequently occurs in childhood. Lisowska-Grospierre et al. (1985) studied patients with an autosomal recessive combined immunodeficiency and an HLA-negative phenotype of activated T and B lymphocytes. The synthesis of the HLA A, B, and C heavy chain was markedly decreased, whereas beta-2-microglobulin was made in normal amounts. No mRNA for either alpha or beta chains of HLA-DR was found. The Ii-chain, the invariant polypeptide associated intracellularly with HLA-DR, and its mRNA were made in normal amounts. Since the structural genes coding for class II polypeptides seemed to be unaffected, the genetic defect in these patients must have concerned the regulation of the expression of the HLA-DR genes. There appeared to be a transactive pleiotropic MHC-regulating gene located perhaps outside the MHC region. Will et al. (1990) described an offspring of healthy consanguineous Tunisian parents who had type II bare lymphocyte syndrome with the additional feature of neutropenia and neutrophil dysfunction (defective spontaneous migration and chemotaxis). Prenatal diagnosis was performed by Durandy et al. (1987) in 6 fetuses at risk for the disease, using membrane immunofluorescence on blood lymphocytes and monocytes. Two pregnancies were found to be affected.

Not only do the BLS trait and the MHC locus segregate independently (de Preval et al., 1985), but class II gene expression can be restored by fusion to a class II MHC-positive cell (Accolla et al., 1985). Several complementation groups have been defined by fusion studies (Hume and Lee, 1989). By studies of a panel of class II MHC-negative regulatory mutants, Kara and Glimcher (1991) demonstrated at least 2 distinct defects that affected gene expression by mechanisms other than direct promoter mutations. Some BLS cells had protein-promoter interactions, both in vitro and in vivo, suggesting a defect in gene activation. A mutation in an activation domain of a transcription factor or in a coactivator protein that does not directly contact DNA could be responsible. In other cases of BLS, Kara and Glimcher (1991) found no promoter interactions in vivo, despite the observation that such interactions occur in vitro. This finding suggested that the defect in these instances may involve promoter accessibility and, thus, chromatin structure. A mutation affecting a factor that interacts with a common locus activation region could prevent the organization of closed chromatin into an open domain. Alternatively, the defect might reside in a common transcription factor that serves not only to activate transcription but also to displace or reorganize nucleosomes locally. By the fusion complementation assay, using a relatively large number of class II-defective patient cell lines and mutant B-lymphoblastoid cell lines, Benichou and Strominger (1991) showed that genetic defects in at least 3, and probably 4, transactivating factors can lead to this phenotype, a failure to transcribe and express class II genes. Several specific DNA binding factors that bind to various class II gene promoters had been cloned. Lisowska-Grospierre et al. (1994) performed fusion experiments with B-cell and fibroblast cell lines from 22 HLA class II-deficient patients, representing two-thirds of all know n cases. They found that 2 complementation groups accounted for 20 of the 22 cases. These 2 complementation groups corresponded closely to 2 ethnic groups: most patients of north African origin were classified into one group, while all patients originating from Spain were classified into a second main group.

In a comprehensive review, Mach et al. (1994) pointed to 2 general types of defects in BLS, which is a disease of gene regulation: in complementation group A, the defect resides in the C2TA gene, RFX binding is normal in vitro, and HLA class II promoter occupancy is normal in vivo. All patients in complementation groups B and C are defective in RFX binding in vitro and exhibit unoccupied class II promoters in vivo. Mach et al. (1994) noted that only a small number of patients had been assigned to complementation group A.

Wolf et al. (1995) reported the cases of identical twin brothers with a defect in constitutive and inducible surface expression of MHC class II molecules on B cells, monocytes, and activated T cells. Unlike previously reported patients, the 2 infants did not have a regulatory defect in the coordinated control of MHC class II gene transcription. Low levels of HLA-DR alpha-chain and beta-chain protein (142860) were detected and, most surprisingly, cellular and humoral immune responses were induced in both infants after tetanus vaccination. For 18 months after birth up until the time of report, the 2 boys had had a benign clinical course.

Kovats et al. (1995) demonstrated that the function of antigen-presenting cells is deficient in multiple genetic complementation groups of BLS type II. This they demonstrated by studying antigen-presenting cell (APC) function in DR-transfected BLS cells derived from multiple complementation groups. Each BLS cell line displayed the same defective APC phenotype: an inability to mediate class II-restricted presentation of exogenous protein antigens and structurally altered class II alpha/beta dimers. Expression of the HLA class II-like genes DMA (142855) and DMB (142856), previously implicated in antigen presentation, was reduced or absent in the BLS cells. Fusion of BLS cells with a cell line that has a genomic deletion of HLA class II genes coordinately restored class II structural gene and DM gene expression and a wildtype APC phenotype. Thus, each of the molecular defects that silences class II structural gene transcription also results in a defective APC phenotype, providing strong evidence for coregulation of these 2 functionally linked pathways.

See also Clement et al. (1988); Griscelli et al. (1989)

Accolla, R. S.; Carra, G.; Guardiola, J.: Reactivation by a trans-acting factor of human major histocompatibility complex Ia gene expression in interspecies hybrids between an Ia-negative human B-cell variant and an Ia-positive mouse B-cell lymphoma. Proc. Nat. Acad. Sci. 82: 5145-5149, 1985.

Baxter-Lowe, L. A.; Hunter, J. B.; Casper, J. T.; Gorski, J.: HLA gene amplification and hybridization analysis of polymorphism: HLA matching for bone marrow transplantation of a patient with HLA-deficient severe combined immunodeficiency syndrome. J. Clin. Invest. 84: 613-618, 1989.

Benichou, B.; Strominger, J. L.: Class II-antigen-negative patient and mutant B-cell lines represent at least three, and probably four, distinct genetic defects defined by complementation analysis. Proc. Nat. Acad. Sci. 88: 4285-4288, 1991.

Clement, L. T.; Plaeger-Marshall, S.; Haas, A.; Saxon, A.; Martin, A. M.: Bare lymphocyte syndrome: consequences of absent class II major histocompatibility antigen expression for B lymphocyte differentiation and function. J. Clin. Invest. 81: 669-675, 1988.

de Preval, C.; Lisowska-Grospierre, B.; Loche, M.; Griscelli, C.; Mach, B.: A trans-acting class II regulatory gene unlinked to the MHC controls expression of HLA class II genes. Nature 318: 291-293, 1985.

Durandy, A.; Cerf-Bensussan, N.; Dumez, Y.; Griscelli, C.: Prenatal diagnosis of severe combined immunodeficiency with defective synthesis of HLA molecules. Prenatal Diag. 7: 27-34, 1987.

Griscelli, C.; Durandy, A.; Virelizier, J. L.; Hors, J.; Lepage, V.; Colombani, J.: Impaired cell-to-cell interaction in partial combined immunodeficiency with variable expression of HLA antigens.In: Seligmann, M.; Hitzig, W. H.: Primary Immunodeficiencies, INSERM Symposium No. 16. Amsterdam: Elsevier/North Holland (pub.) 1980. Pp. 499-503.

Griscelli, C.; Lisowska-Grospierre, B.; Mach, B.: Combined immunodeficiency with defective expression in MHC class II genes. Immunodef. Rev. 1: 135-153, 1989.

Hume, C. R.; Lee, J. S.: Congenital immunodeficiencies associated with absence of HLA class II antigens on lymphocytes result from distinct mutations in trans-acting factors. Hum. Immun. 26: 288-309, 1989.

Kara, C. J.; Glimcher, L. H.: In vivo footprinting of MHC class II genes: bare promoters in the bare lymphocyte syndrome. Science 252: 709-712, 1991.

Kovats, S.; Nepom, G. T.; Coleman, M.; Nepom, B.; Kwok, W. W.; Blum, J. S.: Deficient antigen-presenting cell function in multiple genetic complementation groups of type II bare lymphocyte syndrome. J. Clin. Invest. 96: 217-223, 1995.

Lisowska-Grospierre, B.; Charron, D. J.; de Preval, C.; Durandy, A.; Griscelli, C.; Mach, B.: A defect in the regulation of major histocompatibility complex class II gene expression in HLA-DR negative lymphocytes from patients with combined immunodeficiency syndrome. J. Clin. Invest. 76: 381-385, 1985.

Lisowska-Grospierre, B.; Fondaneche, M.-C.; Rols, M.-P.; Griscelli, C.; Fischer, A.: Two complementation groups account for most cases of inherited MHC class II deficiency. Hum. Molec. Genet. 3: 953-958, 1994.

Mach, B.; Steimle, V.; Reith, W.: MHC class II-deficient combined immunodeficiency: a disease of gene regulation. Immun. Rev. 138: 207-221, 1994.

Marcadet, A.; Cohen, D.; Dausset, J.; Fischer, A.; Durandy, A.; Griscelli, C.: Genotyping with DNA probes in combined immunodeficiency syndrome with defective expression of HLA. New Eng. J. Med. 312: 1287-1292, 1985.

Schuurman, R. K. B.; Gelfand, E. W.; Touraine, J. L.; Van Rood, J. J.: Lymphocyte-membrane abnormalities associated with primary immunodeficiency disease.In: Seligmann, M.; Hitzig, W. H.: Primary Immunodeficiencies, INSERM Symposium No. 16. Amsterdam: Elsevier/North Holland (pub.) 1980. Pp. 87-99.

Schuurman, R. K. B.; Van Rood, J. J.; Vossen, J. M.; Schellekens, P. T. A.; Feltkamp-Vroom, T. M.; Doyer, E.; Gmelig-Meyling, F.; Visser, H. K. A.: Failure of lymphocyte-membrane HLA-A and B expression in two siblings with combined immunodeficiency. Clin. Immun. Immunopath. 14: 418-434, 1979.

Sullivan, K. E.; Stobo, J. D.; Peterlin, B. M.: Molecular analysis of the bare lymphocyte syndrome. J. Clin. Invest. 76: 75-79, 1985.

Touraine, J. L.; Betuel, H.; Souillet, G.; Jeune, M.: Combined immunodeficiency disease associated with absence of cell-surface HLA-A and -B antigens. J. Pediat. 93: 47-51, 1978.

Touraine, J. L.; Betuel, H.; Touraine, F.; Philippe, N.; Betend, B.; Francois, R.: Role of MHC determinants in immunodeficiency diseases as shown by the 'bare lymphocyte syndrome' and by chimeric patients.In: Seligmann, M.; Hitzig, W. H.: Primary Immunodeficiencies, INSERM Symposium No. 16. Amsterdam: Elsevier/North Holland (pub.) 1980. Pp. 79-86.

Will, N.; Seger, R. A.; Betzler, C.; Dockter, G.; Graf, N.; Buttner, M.; Irle, C.; Tiercy, J. M.; Sitzmann, F. C.: Bare lymphocyte syndrome: combined immunodeficiency and neutrophil dysfunction. Europ. J. Pediat. 149: 700-704, 1990.

Wolf, H. M.; Hauber, I.; Gulle, H.; Thon, V.; Eggenbauer, H.; Fischer, M. B.; Fiala, S.; Eibl, M. M.: Brief report: Twin boys with major histocompatibility complex class II deficiency but inducible immune responses. New Eng. J. Med. 332: 86-90, 1995.

MHC CLASS II TRANSACTIVATOR (*600005)

JJM NOTE

Because cancer has not been described with specific genotypes of SCID, this entity is provisionally included.

Hereditary MHC class II deficiency, or bare lymphocyte syndrome (209920), occurs in several complementation groups. Steimle et al. (1993) isolated a transactivator of MHC class II gene expression using an MHC class II-negative mutant cell line for complementation cloning. The gene, which they called CIITA for 'class II transactivator,' restored expression of all MHC class II isotypes in mutant cells and fully corrected the MHC class II regulatory defect of cells from patients with the bare lymphocyte syndrome. In the latter disease, Steimle et al. (1993) identified a splicing mutation that resulted in a 24-amino acid deletion in C2TA, resulting in loss of function of the transactivator. In the RJ2.25 cells, which originated from a Burkitt lymphoma, one C2TA allele was completely deleted, while in the other allele an internal deletion removed over half of the coding region. The splice donor site mutation was identified as a G-to-A transition in the splice donor site 3-prime of a 72-bp exon (600005.0001). Exon skipping due to such a splice site mutation would be expected. Only the deleted C2TA mRNA was expressed and only the mutated allele was detected in the patient, suggesting that the mutation was homozygous. The full complementation of MHC class II deficiency in cell lines from these patients with SCID by transfection of C2TA cDNA raised hopes for gene therapy in this disorder; bone marrow transplantation had had poor success.

Mach et al. (1994) reviewed the topic of MHC class II-deficient combined immunodeficiency. In addition to the splice donor site mutation with exon skipping, homozygosity for deletion of the C2TA gene was demonstrated in another cell line of complementation group A. The C2TA cDNA codes for a protein of 1,130 amino acids which contains no obvious structural motifs that shed light on its mode of action. No evidence for direct binding to the MHC class II promoter was found, suggesting that CIITA is not a DNA-binding protein. C2TA mRNA is differentially expressed and can be detected only in HLA class II-positive cell lines and tissues. This tight correlation suggests that expression of HLA class II genes is to a large extent, if not entirely, under the control of C2TA. Mach et al. (1994) referred to experimental results indicating that the C2TA gene is induced by interferon-gamma (147570) and that CIITA is an essential mediator of inducible MHC class II gene expression.

CIITA is a positive regulator of class II major histocompatibility complex gene transcription that is defective in 1 of the 5 complementation groups of class II MHC-negative cell lines. Its N-terminal region is capable of activating transcription from a reporter gene when fused to a DNA binding domain. Mahanta et al. (1997) found that transactivation by CIITA was extremely sensitive to a mutation in TATA box binding protein (TBP; 600075) that in yeast is known to abolish VP16-mediated transcription but leaves basal transcription unaffected. Overall the mechanism of transactivation by the human B-cell-specific CIITA was very similar to that mediated by the herpesvirus transactivator VP16 in the ways that were tested.

According to Reith (1997), the C2TA gene has been assigned to chromosome 16.

Two of the genes defective in the 5 complementation groups identified in class II-negative bare lymphocyte syndrome or in corresponding laboratory mutants have been cloned (Mach et al., 1996). One gene encodes RFX5 (601863), a member of the RFX family of DNA binding proteins; the other, CIITA, encodes a large protein with a defined acidic transcriptional activation domain. The latter protein does not interact with DNA. Scholl et al. (1997) demonstrated that RFX5 can activate transcription only in cooperation with CIITA. RFX5 and CIITA associate to form a complex capable of activating transcription from class II major histocompatibility complex promoters. In this complex, promoter specificity is determined by the DNA binding domain of RFX5 and the general transcription apparatus is recruited by the acidic activation domain of CIITA.

Bontron, S.; Steimle, V.; Ucla, C.; Eibl, M. M.; Mach, B.: Two novel mutations in the MHC class II transactivator CIITA in a second patient from MHC class II deficiency complementation group A. Hum. Genet. 99: 541-546, 1997.
Mach, B.; Steimle, V.; Martinez-Soria, E.; Reith, W.: Regulation of MHC class II genes: lessons from a disease. Annu. Rev. Immun. 14: 301-331, 1996.
Mach, B.; Steimle, V.; Reith, W.: MHC class II-deficient combined immunodeficiency: a disease of gene regulation. Immun. Rev. 138: 207-221, 1994.
Mahanta, S. K.; Scholl, T.; Yang, F.-C.; Strominger, J. L.: Transactivation by CIITA, the type II bare lymphocyte syndrome-associated factor, requires participation of multiple regions of the TATA box binding protein. Proc. Nat. Acad. Sci. 94: 6324-6329, 1997.
Reith, W.: Personal Communication. Geneva, Switzerland 5/30/1997.
Scholl, T.; Mahanta, S. K.; Strominger, J. L.: Specific complex formation between the type II bare lymphocyte syndrome-associated transactivators CIITA and RFX5. Proc. Nat. Acad. Sci. 94: 6330-6334, 1997.
Steimle, V.; Otten, L. A.; Zufferey, M.; Mach, B.: Complementation cloning of an MHC class II transactivator mutated in hereditary MHC class II deficiency (or bare lymphocyte syndrome). Cell 75: 135-146, 1993.

REGULATORY FACTOR 1 (*600006)

JJM NOTE

Because cancer has not been described with specific genotypes of SCID, this entity is provisionally included.

The bare lymphocyte syndrome (BLS; 209920) is a form of severe combined immunodeficiency (SCID) that is associated with—and appears to result from—the lack of expression of HLA antigens on some cells of hematopoietic origin. As reviewed by Mach et al. (1994), several complementation groups have been defined by cell fusion studies (Hume and Lee, 1989). Complementation group A, which represents a minority of cases, shows a defect in MHC class II transactivator (600005). In complementation groups B and C, the defect resides in the MHC class II regulatory factor RF-X.

The RFX1 gene product is a transactivator of the human hepatitis B viral enhancer I. Reith et al. (1994) commented that the RFX family members, particularly RFX1 and RFX3 (601337), constitute the nuclear complexes referred to previously as enhancer factor C (EF-C), EP, and methylation-dependent DNA-binding protein (MDBP), or rpL30-alpha. Reith et al. (1994) identified and cloned 3 members of this gene family from both human and mouse using lambda gt11 cDNA libraries. Homology between the 3 RFX proteins is restricted largely to 5 conserved regions, including the 2 domains required for DNA binding and dimerization. Reith et al. (1994) found that RFX1, RFX2 (142765), and RFX3 have similar DNA-binding specificities. The RFX monomers can heterodimerize both in vivo and in vitro, but all 3 are capable of binding DNA as monomers. They showed that the RFX1 transcript is expressed in many mouse tissues.

Pugliati et al. (1989) used an RFX cDNA for in situ hybridization to show that the gene maps to 19p13. (Pugliatti's surname was misspelled in the 1989 abstract.) By in situ hybridization using tritium-labeled RFX1 cDNA, Pugliatti et al. (1992) identified 2 distinct targets on the short arm of chromosome 19, 19p13.1 and 19p13.3-p13.2. By the use of in situ hybridization with a biotinylated genomic cosmid clone they localized RFX1 to 19p13.1. RFX1 is close to another DNA-binding factor, LYL1 (151440). Doyle et al. (1996) physically mapped RFX1 to cosmid contigs for chromosome 19. The gene was localized between MEL (165040) and LYL1. Doyle et al. (1996) used interspecific backcross mapping to place Rfx1 in the central region of mouse chromosome 8 near the homolog of JUNB (165161).

RFX1 is not mutant in the bare lymphocyte syndrome (Reith, 1997). Previous information suggesting that it is involved in BLS was in error.

Emery et al. (1996) reviewed RFX1, RFX5, and other members of the RFX family of DNA binding proteins.

Doyle, J.; Hoffman, S.; Ucla, C.; Reith, W.; Mach, B.; Stubbs, L.: Locations of human and mouse genes encoding the RFX1 and RFX2 transcription factor proteins. Genomics 35: 227-230, 1996.
Emery, P.; Durand, B.; Mach, B.; Reith, W.: RFX proteins, a novel family of DNA binding proteins conserved in the eukaryotic kingdom. Nucleic Acids Res. 24: 803-807, 1996.
Hume, C. R.; Lee, J. S.: Congenital immunodeficiencies associated with absence of HLA class II antigens on lymphocytes result from distinct mutations in trans-acting factors. Hum. Immun. 26: 288-309, 1989.
Mach, B.; Steimle, V.; Reith, W.: MHC class II-deficient combined immunodeficiency: a disease of gene regulation. Immun. Rev. 138: 207-221, 1994.
Pugliati, L.; Reith, W.; Fey, S.; Mach, B.: Mapping the RF-X gene, encoding a DNA-binding protein controlling HLA class II gene expression, to 19p13.(Abstract) Cytogenet. Cell Genet. 51: 1061 only, 1989.
Pugliatti, L.; Derre, J.; Berger, R.; Ucla, C.; Reith, W.; Mach, B.: The genes for MHC class II regulatory factors RFX1 and RFX2 are located on the short arm of chromosome 19. Genomics 13: 1307-1310, 1992.
Reith, W.: Personal Communication. Geneva, Switzerland 5/30/1997.
Reith, W.; Ucla, C.; Barras, E.; Gaud, A.; Durand, B.; Herrero-Sanchez, C.; Kobr, M.; Mach, B.: RFX1, a transactivator of hepatitis B virus enhancer I, belongs to a novel family of homodimeric and heterodimeric DNA-binding proteins. Molec. Cell. Biol. 14: 1230-1244, 1994.

WISKOTT-ALDRICH SYNDROME (*301000)

JJM NOTE

In a multi-institutional survey of 154 patients, the classic triad of thrombocytopenia with small platelets, recurrent otitis media, and eczema was seen in only 27%; patients with autoimmune disorders were at increased risk of having a malignancy (Sullivan KE et al.: A multi-institutional survey of the Wiskott-Aldrich syndrome. J. Pediatr. 125:876-885, 1994). By early 1978, the Immunodeficiency-Cancer Registry of the University of Minnesota (Spector BD et al.: Genetically determined immunodeficiency diseases (GDID) and malignancy: report from the immunodeficiency cancer registry. Clin. Immunol. Immunopathol. 1:12–29, 1978) had 45 cases of Wiskott-Aldrich syndrome with cancer: 12 histiocytic lymphoma; 9 malignant reticuloendotheliosis; 2 lymphocytic lymphoma; 11 other lymphoreticular malignancies; 3 Hodgkin disease; 3 myeloid leukemia; 2 nervous system tumors; and 3 others. Minor signs of Wiskott-Aldrich syndrome should be sought in young males with such malignancies, in part because of the implications of X-linked inheritance of even distant relatives on the maternal side.

The manifestations are eczema, thrombocytopenia, proneness to infection, and bloody diarrhea. Death usually occurs before age 10 years. The original American kindred reported by Aldrich et al. (1954) was of Dutch extraction; the 2 patients of Wiskott (1937) were German. Wiskott, who worked in Munich, referred to the disorder in his patients as 'Werlhof's disease,' the eponymic designation for thrombocytopenic purpura. Van den Bosch and Drukker (1964) described several families in the Netherlands. In 3 of 5 female carriers, the platelet count was below the lower limit of normal. Several groups (Blaese et al., 1968; Cooper et al., 1968) presented evidence that the immune defect is in the afferent limb, i.e., is one of antigen processing or recognition. In an obligate heterozygote who was heterozygous for the AB polymorphism of G6PD (305900), Gealy et al. (1980) found that only the B isoenzyme was present in platelets and T lymphocytes, although both were present in erythrocytes and neutrophils. The findings suggested selection against the WAS gene in these tissues, which are also the ones that express the defect in the hemizygous affected male. Prchal et al. (1980) pursued the implications of this finding for genetic counseling. Although G6PD is likely to be useful in only a limited number of potential carriers, the large number of X-chromosome markers, DNA polymorphisms and other markers now available make it likely that carrier detection will be possible. See later discussion of the work of Fearon et al. (1988), Puck et al. (1990), de Saint-Basile et al. (1991), and Notarangelo et al. (1991). Shapiro et al. (1978) concluded that carriers can be identified by study of platelets, which show a defect in oxidative phosphorylation.

Perry et al. (1980) found an incidence of 4.0 per million live male births in the United States. Median survival increased from 8 months for patients born before 1935 to 6.5 years for those born after 1964. One patient had survived to age 36 years at the time of the survey. Causes of death were mainly infections or bleeding, but 36 of the 301 patients

(12%) developed malignancies: lymphoreticular tumors in 23 and leukemia in 7. Ten Bensel et al. (1966) called attention to the occurrence of malignancy of the reticuloendothelial system, which they saw in 2 of 4 sibs and found in 5 reported cases.

Parkman et al. (1981) studied the surface proteins of lymphocytes and platelets by radioiodination followed by SDS-polyacrylamide gel electrophoresis and autoradiography. All 3 patients studied showed, in lymphocytes, absence of a protein, molecular weight 115,000, found in normals. Platelets also showed an abnormality of surface glycoproteins. CD43 (182160), or sialophorin, is a cell-surface sialoglycoprotein that is deficient in quantity and/or is defective in lymphocytes of patients with this disorder (Parkman et al., 1981; Remold-O'Donnell et al., 1984). Mentzer et al. (1987) suggested that sialophorin functions in T-cell activation. Hara et al. (1993) found that the gene for UDP-galactose translocator (314375) maps to the same region of the X chromosome as WAS. As a consequence, they concluded that mutation in that gene may be responsible for this disorder.

Corash et al. (1985) studied the mechanism of the usual improvement in thrombocytopenia in WAS after splenectomy. The thrombocytopenia is accompanied by elevated platelet-associated IgG and low mean platelet size. Both return to normal after splenectomy. Patients who relapse redevelop elevated IgG but maintain normal platelet size.

Holmberg et al. (1983) found that normal midtrimester fetuses have platelets of the same size as normal newborns and adults. They used these data 'to exclude Wiskott-Aldrich syndrome in an 18-week fetus at 50% risk of being affected.' Unfortunately, we do not know that the platelets of the WAS fetus are abnormally small. Capsoni et al. (1986) described a 19-year-old man with WAS. Only 7 affected persons over age 18 had been described previously. Standen et al. (1986) reported a kindred with 13 males in 6 sibships, related through females, with inherited thrombocytopenia thought to be a variant of WAS because it was associated with elevated serum IgA and mild nephropathy. Five suffered from severe eczema since infancy but had no unusual susceptibility to infections. Platelet volume was reduced. Gutenberger et al. (1970) reported a similar family. Renal biopsy was performed in 3 patients. In the first, advanced membranoproliferative glomerulonephritis was found with deposition of complement and IgG on the basement membrane. In the second, mesangial glomerulonephritis with focal glomerulosclerosis and deposition of complement and IgA were found. The third showed minimal glomerulonephritis. Standen et al. (1986) concluded that despite the clinical similarities and the elevated IgA in both conditions, the disorder is distinct from Berger disease (161950). Spitler et al. (1980) found nephropathy in 5 of 32 patients with WAS who participated in a study of treatment with transfer factor, a dialyzable extract of leukocytes that enhances cellular immunity. Although nephropathy occurred without such treatment, the temporal relationships suggested that transfer factor aggravated the problem. Webb et al. (1993) described their experience with renal transplantation in a 46-year-old man with the syndrome of thrombocytopenia with raised IgA levels and impaired renal function. The man had a strong family history of hereditary thrombocytopenia and had presented in early childhood with allergic eczema, asthma, thrombocytopenic purpura, and recurrent middle ear infections. He had a normal platelet count after splenectomy was performed at the age of about 30. In his mid-thirties, he had subtotal colectomy and ileostomy for severe ulcerative colitis. This disorder later recurred, associated with keratitis and arthritis of large joints. He was later admitted to the hospital with a febrile illness, biopsy-proven cutaneous vasculitis, raised IgA levels, and impaired renal function. Renal biopsy demonstrated mesangioproliferative glomerulonephritis, old crescents, and mesangial IgA deposition. After renal transplant, a 'reduced immunosuppressive protocol' was instituted because of his underlying immunologic disorder. Despite this, no rejection episodes occurred. The occurrence of a milder variant raises questions about the distinctness of X-linked thrombocytopenia (313900) from WAS (Weiden and Blaese, 1972). Knox-Macaulay et al. (1993) described a Saudi Arabian family in which 3 brothers presented in early childhood with thrombocytopenia but without immunologic abnormalities. Results of DNA analysis with the probe M27-beta were consistent with X-linkage and indicated also that the locus of the relevant gene lies close to or is identical to the WAS locus. However, there were distinguishing features which included the presence of large and normal sized platelets (rather than small platelets) and, as stated, freedom from immune deficiency. Mean platelet volume was increased and the bone marrow showed increased numbers of megakaryocytes.

The first reports of successful bone marrow transplantation for severe combined immunodeficiency (XSCID, 300400) and for WAS were provided by Gatti et al. (1968) and Bach et al. (1968). Fischer et al. (1986) gave a retrospective analysis of results in 162 patients who had undergone transplantation in 14 European centers between 1969 and 1985. Brochstein et al. (1991) reported on the bone marrow transplantation in 17 patients with WAS.

McEnery and Nash (1973) described 2 unrelated males with the association of WAS and infantile cortical hyperostosis (Caffey disease; 114000), and Abinun et al. (1988) also described a case. Thus, an immunologic defect may play a role in the pathogenesis of infantile cortical hyperostosis. Meropol et al. (1992) reported the case of a 24-year-old man with WAS complicated by T-cell large cell lymphoma and Kaposi sarcoma (148000). Kaposi sarcoma is well known in connection with the immunosuppression used with allograft transplantation and in patients with HIV infection but this was the first incidence of its occurrence in this form of immunodeficiency.

Peacocke and Siminovitch (1987) studied 10 kindreds for linkage with RFLPs. Significant linkage was found between WAS and 2 loci, DXS14 and DXS7, that mapped to the proximal short arm of the X chromosome. Maximal lod scores were 4.29 (at theta = 0.03) and 4.12 (at theta = 0.00), respectively. Arveiler et al. (1987) found a strong suggestion of linkage between IMD2 and DXS1, which is located in Xq11-q12. Kwan et al. (1988) concluded from linkage studies that the WAS gene lies between DXS7 (Xp11.3) and DXS14 (Xp11); the likelihood of this position was at least 128 times higher than that of any other interval studied. In a study of 12 WAS families, Kwan et al. (1989) demonstrated linkage to another DNA marker, DXS255, located at Xp11.22; peak lod score = 4.65 at theta = 0.05. Greer et al. (1989) showed linkage between WAS and DXZ1 (lod score = 7.08 at theta = 0.03) and between WAS and the TIMP locus (305370) (lod score = 5.09 at theta = 0.0). Greer et al. (1990) extended the linkage studies, demonstrating strongest linkage (maximum lod score = 10.19 at theta = 0.0) between WAS and the hypervariable DXS255 locus, a marker already mapped between DXS7 and DXS14. De Saint Basile et al. (1989) found close linkage of WAS to DXS255 (maximum lod = 5.42 at theta = 0.00). Kwan et al. (1991) likewise concluded that DXS255 is the closest marker identified; WAS showed a multipoint maximum lod score of 8.59 at 1.2 cM distal to DXS255. Furthermore, they concluded that the TIMP gene must lie distal to WAS; thus, WAS was thought to lie between DXS255 (Xp11.22) and TIMP (Xp11.3). Greer et al. (1992) demonstrated close linkage between the WAS and OATL1 (311240) loci; maximum lod = 6.08 at theta = 0.00. The finding localized the TIMP, OATL1, and WAS loci distal to DXS146 and the OATL1 and WAS loci proximal to TIMP.

Schwartz et al. (1989) described the first-trimester diagnosis and exclusion of WAS by means of closely linked DNA markers. Arveiler et al. (1990) showed that failure to demonstrate linkage of WAS to markers known from other families to be closely situated was attributable to germ cell mosaicism in the grandfather of affected males. (Hendriks et al. (1989) described the same phenomenon in X-linked agammaglobulinemia; see 300300.) Fearon et al. (1988) studied the pattern of X-chromosome inactivation in various cell populations from female relatives of patients with WAS, through analysis of the methylation patterns of X-linked genes that display RFLPs. They found that carriers could be accurately identified by the fact that peripheral blood T cells, granulocytes, and B cells of obligate heterozygotes display specific patterns of X-chromosome inactivation that are clearly different from those of normal controls. Puck et al. (1990) pointed out that the diagnosis of WAS may be difficult in infancy when sporadic thrombocytopenia with no, or only questionable, immunologic abnormalities are present. In the case of 2 unrelated males with this problem, X-chromosome inactivation in the T cells of the mothers showed each of them to have a highly skewed X-chromosome inactivation pattern typical of WAS carriers. In one of the patients, a T-cell defect was subsequently demonstrated directly by studies of the lymphocytes, which failed to proliferate in periodate and anti-CD43. Notarangelo et al. (1991) reported a similar case of a boy presenting as idiopathic thrombocytopenia. De Saint-Basile et al. (1991) studied a family in which 4 members had X-linked thrombocytopenia. Linkage studies showed mapping to the same region of the X chromosome as that affected in WAS. Although polymorphonuclear leukocytes showed a normal pattern of X-inactivation, a skewed pattern was demonstrated in lymphocytes. De Saint-Basile et al. (1991) concluded that this was consistent with allelic mutations at the same locus, with the severity of disease varying according to the distinct patterns of hematopoietic cell involvement in obligate carriers. Notarangelo et al. (1991) studied a presumably heterozygous, thrombocytopenic female from a WAS pedigree. Her carrier status was confirmed by linkage studies. Both small-sized and normal-sized platelets were present, suggesting that, unlike the vast majority of WAS carriers, she did not manifest nonrandom X-chromosome inactivation in the thrombopoietic cell lineage. Studies of X-chromosome inactivation by means of RFLP and methylation analysis showed that the pattern of X-chromosome inactivation was nonrandom in T lymphocytes but random in granulocytes. Notarangelo et al. (1993) reviewed the use of the biased inactivation of the X chromosome in hematopoietic cells as a tool for carrier detection in connection with genetic counseling. A closely linked hypervariable marker, M27-beta (DXS255), was used.

Wengler et al. (1995) stated that obligate female carriers of the gene for X-linked agammaglobulinemia (300300) show nonrandom X-chromosome inactivation only in B lymphocytes, and obligate female carriers of the gene for X-linked severe combined immunodeficiency (SCIDX; 300400) show nonrandom X-chromosome inactivation in both T and B lymphocytes, as well as natural killer cells. However, all formed elements of the blood appear to be affected, as a rule, in obligate carriers of WAS, as judged by the criteria of nonrandom X-chromosome inactivation and segregation of G6PD alleles in informative females. Wengler et al. (1995) demonstrated that CD34(+) hematopoietic progenitor cells collected from obligate carriers of WAS by apheresis showed nonrandom inactivation. They used PCR analysis of a polymorphic VNTR within the X-linked androgen receptor gene (313700) to demonstrate nonrandom inactivation which clearly must occur early during hematopoietic differentiation. Parolini et al. (1998) reported X-linked WAS in an 8-year-old girl. She had a sporadic mutation, glu133 to lys, on the paternally derived X chromosome, but had nonrandom X inactivation of the maternal X chromosome in both blood and buccal mucosa. Her mother and maternal grandmother also had nonrandom X inactivation, which suggested to the authors the possibility of a defect in XIST (314670) or some other gene involved in the X-inactivation process. Puck and Willard (1998) commented on the subject of X inactivation in females with X-linked disease in reference to the paper by Parolini et al. (1998).

Symons et al. (1996) reported that the Wiskott-Aldrich protein has a GTPase binding site and that it interacts specifically with activated CDC42Hs, a member of the Rho-like GTPase family. They noted that WASP localizes in the cytoplasm in clusters that are enriched in polymerized actin. They proposed that WASP provides a link between CDC42Hs and the actin cytoskeleton.

Simon et al. (1992) presented experimental results indicating the association of WAS with a defect in the coupling of surface immunoglobulin (sIg) on B cells to signal transduction pathways considered prerequisite for B-cell activation, probably at the level of tyrosine phosphorylation.

To isolate the WAS gene, Derry et al. (1994) used a positional cloning strategy that involved the construction of a clone contig in the region Xp11.23-p11.22, bounded by the markers DXS255 and TIMP (305370), demonstrated to contain the locus. Evaluation of several candidate cDNAs led to the identification of a sequence whose expression was limited to lymphocytic and megakaryocytic cell lineages and which was altered in affected individuals. The gene was not expressed in 2 unrelated patients, 1 of whom had a single base deletion that produced a frameshift and premature termination of translation. Two additional patients were identified with point mutations that changed the same arginine residue to either a histidine or a leucine. Derry et al. (1994) referred to the gene as WASP and showed that it encodes a 501-amino acid proline-rich protein. In a subsequent erratum, Derry et al. (1994) stated that WASP contained 502 amino acids. The gene spans about 9 kb of genomic DNA and has 11 exons.

Villa et al. (1995) presented proof that mutations in the WAS gene can result in X-linked thrombocytopenia characterized by thrombocytopenia with small-sized platelets as an isolated finding (313900). Why some mutations impair only the megakaryocytic lineage and have no apparent effect on the lymphoid lineage was unclear.

Kwan et al. (1995) updated the coding and genomic sequences of the WAS gene. They found 11 additional mutations that involved single base changes, small deletions, and an insertion. They also isolated and characterized a polymorphic CA dinucleotide repeat, DXS6940, that lies within 30 kb of the gene. The tightly linked polymorphism should prove a powerful tool for linkage analysis, especially in families where the WAS mutation proved difficult to identify. They tabulated 12 mutations in all. These were located in 6 different exons. Kolluri et al. (1995) used PCR-SSCP analysis to screen for WAS gene mutations in 19 unrelated boys with the diagnosis of classic or attenuated WAS or isolated thrombocytopenia. All 19 patients had WAS mutations, each of which localized to the initial 3 or terminal 3 exons of the gene, and most of which were unique in each case. However, the arg86-to-his mutation (301000.0003) was found in 1 boy with severe WAS and an arg86-to-cys mutation was found in 2 boys with severe WAS and 1 boy presenting with thrombocytopenia alone. While the 3 mutations found in isolated thrombocytopenia patients left the reading frame intact, about one-half of the gene alterations detected in both severe and attenuated WAS patients resulted in frameshifted transcript and premature translation termination. Derry et al. (1995) described the spectrum of novel mutations in 12 additional unrelated families: missense, nonsense, and frameshift mutations involving 8 of the 12 exons of the gene. Two mutations creating premature termination codons were associated with lack of detectable mRNA on Northern blots. Four amino acid substitutions were found in patients with congenital thrombocytopenia and no clinically evident immune defect. A T-cell line from a WAS patient contained 2 independent DNA alterations, a constitutional frameshift mutation also present in peripheral blood lymphocytes from the patient, and a compensatory splice site mutation unique to the cell line. Although RNA fr om untransformed cells was not available for analysis, Derry et al. (1995) proposed that the splice site mutation was introduced into the T-cell during transformation with HTLV-1 and was then selected for during expansion of the culture. The selection hypothesis implies that a mutant protein lacking the 14 amino acids of exon 8 was produced and conveyed a growth advantage to the cells that carried it. In support of this idea was the observation of nonrandom X-inactivation of circulating lymphocytes in female carriers for WAS, due to selection for cells with the normal X chromosome active. Sites in the first 2 exons of the WAS gene appear to be hotspots for mutation; 20 of 31 unrelated families studied by Kolluri et al. (1995) and Derry et al. (1995) had mutations in exon 1 or exon 2.

Sullivan et al. (1994) reported on a multiinstitutional survey of WAS in the U.S. In this survey, laboratory and clinical data were collected on 154 affected individuals. There was a family history of the disorder in the case of 74 of the patients. Thrombocytopenia was a prerequisite for entry into the study; however, only 27% of patients had the typical set of three symptoms described originally by Aldrich et al. (1954). The immunologic findings in particular varied considerably with the most distinctive finding: that 61% of the patients had a low CD8+ count. Eczema developed in 81% but was not always present at diagnosis. In those patients where platelet size was measured, Sullivan et al. (1994) found them to be small, although they did increase in size following splenectomy. The average age at diagnosis was 21 months; the average age at death was 8 years. There were 16 patients who lived beyond 18 years, and the prognosis for the disorder had improved considerably in recent years. Bone marrow transplantation had been carried out in 47 cases and a good outcome was reported in two-thirds of them. Autoimmune disorders occurred in 40% of patients; this group had a poor prognosis as they were more likely to develop a malignancy. Malignancies were seen in 13% of patients and were mainly of the lymphoreticular system.

Derry et al. (1995) isolated and sequenced the mouse WAS gene which they called Wasp. The predicted amino acid sequence was found to be 86% identical to the human WAS sequence. A distinct feature of the mouse gene was an expanded polymorphic GGA trinucleotide repeat that codes for polyglycine and varies from 15 to 17 triplets in different mouse strains. The genomic structure of the mouse gene closely resembles that in the human with respect to exon/intron positions and intron lengths. The mouse WAS gene is expressed as a mRNA of approximately 2.4 kb in thymus and spleen. Chromosomal mapping and interspecific backcross placed the Wasp locus near the centromere of the mouse X chromosome, inseparable from Gata1 (305371), Tcfe3 (314310), and 'scurfy' (sf). Derry et al. (1995) stated that this localization made Wasp a candidate for involvement in 'scurfy,' a T cell-mediated fatal lymphoreticular disease of mice that had previously been proposed as a mouse homolog of Wiskott-Aldrich syndrome (Lyon et al., 1990). Northern analysis of sf tissue samples indicated the presence of Wasp mRNA in liver and skin, presumably as a consequence of lymphocyte infiltration, but no abnormalities in the amount or size of mRNA were identified.

Wengler et al. (1995) identified 15 novel mutations in the WAS gene in patients with full-blown Wiskott-Aldrich syndrome. These mutations involve single basepair changes, or small insertions or deletions, all of which resulted in premature stop codon, frameshift with secondary premature stop codon, or splice site defect. Zhu et al. (1995) likewise identified 12 unique mutations distributed throughout the WAS gene. Patients with classic WAS had 'more complex' mutations, resulting in termination codons, frameshifts, and early termination. By contrast, 4 unrelated patients with the X-linked thrombocytopenia phenotype had missense mutations affecting exon 2 (in 3) and a splice site mutation affecting exon 9 (in 1). De Saint Basile et al. (1996) also found single point mutations in exon 2 of the WAS gene in 2 unrelated families with a history of isolated X-linked thrombocytopenia.

Schindelhauer et al. (1996) found 7 novel and 10 known mutations in the course of mutation analysis in 19 families of German, Swiss, and Turkish descent by using SSCP analysis and sequencing. They noted a striking clustering of missense mutations in the first 4 exons that contrasted with a random distribution of nonsense mutations. More than 85% of all known missense mutations were located in the amino-terminal stretch of the WAS gene product; this region contained a mutational hotspot at codon 86 (see 301000.0002 and 301000.0003); R86C, R86H, and R86P were observed in this study and R86H was found in 2 unrelated families. Schindelhauer et al. (1996) found no genotype/phenotype correlation emerge after a comparison of the identified mutations with the resulting clinical picture for a classical WAS phenotype. A mild course, reminiscent of X-linked thrombocytopenia, or an attenuated phenotype was more often associated with missense than with the other types of mutations. Greer et al. (1996) examined the genotypes and phenotypes of 24 patients with WAS and compared them with other known mutations of the WASP gene. They demonstrated clustering of WASP mutations within the 4 most N-terminal exons of the gene and identified arg86 as the most prominent hotspot for WASP mutations. They noted the prominence of missense mutations among patients with milder forms of WAS, while noting that missense mutations also comprise a substantial portion of mutations in patients with severe forms of the disease. Greer et al. (1996) concluded that phenotypes and genotypes of WAS are not well correlated; phenotypic outcome cannot be reliably predicted on the basis of WASP genotype.

T lymphocytes of affected males with WAS exhibit a severe disturbance of the actin cytoskeleton, suggesting that the WAS protein may regulate its organization. Kolluri et al. (1996) showed that WAS protein interacts with Cdc42 (116952), a member of the RHO family of GTPases. This interaction, which is GTP-dependent, was detected in cell lysates, in transient transfections, and with purified recombinant proteins. Different mutant WAS proteins from 3 unrelated affected males retained their ability to interact with Cdc42 but the level of expression of the WAS protein in these mutants was only 2-5% of normal. Taken together, these data suggested to Kolluri et al. (1996) that the WAS protein may function as a signal transduction adaptor downstream of Cdc42, and that, in affected males, the cytoskeletal abnormalities may result from a defect in Cdc42 signaling.

Stewart et al. (1996) used monoclonal anti-WASP antibodies in Western immunoblots to show that WASP is present in the cytoplasmic but not the nuclear fraction of normal human peripheral blood mononuclear cells, platelets, T lymphocytes, non-T lymphocytes, and monocytes. WASP was present in 2 of 4 EBV-transformed cell lines from WAS patients. Sequence studies in a WAS patient showed a C-to-G transversion at nucleotide position 155 which caused an arginine-to-glycine substitution at codon 41; in a second patient, a C insertion after nucleotide 1016 produced a frameshift resulting in amino acid substitutions at codons 328, 329, 331, and 332. Deletion of a G just after nucleotide 1029 returned the reading frame to normal. The failure to find WASP in the nucleus suggested to Stewart et al. (1996) that it is not a transcription factor.

See also Blaese et al. (1971); Derry et al. (1995); Diaz-Buxo et al. (1974); Filipovich et al. (1979); Gelzer and Gasser (1961); Hutter and Jones (1981); Kapoor et al. (1981); Krivit and Good (1959); Levin et al. (1970); Lum et al. (1980); Nathan (1980); Notarangelo et al. (1991); Ochs et al. (1980); Parkman et al. (1978); Steinberg (1959); Wolff (1967)

Abinun, M.; Mikuska, M.; Filipovic, B.: Infantile cortical hyperostosis associated with the Wiskott-Aldrich syndrome. Europ. J. Pediat. 147: 518-519, 1988.

Aldrich, R. A.; Steinberg, A. G.; Campbell, D. C.: Pedigree demonstrating a sex-linked recessive condition characterized by draining ears, eczematoid dermatitis and bloody diarrhea. Pediatrics 13: 133-139, 1954.

Ariga, T.; Yamada, M.; Sakiyama, Y.; Tatsuzawa, O.: A case of Wiskott-Aldrich syndrome with dual mutations in exon 10 of the WASP gene: an additional de novo one-base insertion, which restores frame shift due to an inherent one-base deletion, detected in the major population of the patient's peripheral blood lymphocytes. (Letter) Blood 92: 699-701, 1998.

Arveiler, B.; de Saint-Basile, G.; Fischer, A.; Griscelli, C.; Mandel, J. L.: Germline mosaicism simulates genetic heterogeneity in Wiskott-Aldrich syndrome. Am. J. Hum. Genet. 46: 906-911, 1990.

Arveiler, B.; de Saint Basile, G.; Debre, M.; Fischer, A.; Griscelli, C.; Mandel, J. L.: Linkage analysis of the Wiskott-Aldrich syndrome (IMD2) using X-linked DNA polymorphisms. (Abstract) Cytogenet. Cell Genet. 46: 573 only, 1987.

Bach, F. H.; Albertini, R. J.; Anderson, J. L.; Joo, P.; Bortin, M.: Bone marrow transplantation in a patient with the Wiskott Aldrich syndrome. Lancet II: 1364-1366, 1968.

Blaese, R. M.; Strober, W.; Brown, R. S.; Waldmann, T. A.: The Wiskott-Aldrich syndrome: a disorder with a possible defect in antigen processing or recognition. Lancet I: 1056-1060, 1968.

Blaese, R. M.; Strober, W.; Levy, A. L.; Waldmann, T. A.: Hypercatabolism of IgG, IgA, IgM, and albumin in the Wiskott-Aldrich syndrome: a unique disorder of serum protein metabolism. J. Clin. Invest. 50: 2331-2338, 1971.

Brochstein, J. A.; Gillio, A. P.; Ruggiero, M.; Kernan, N. A.; Emanuel, D.; Laver, J.; Small, T.; O'Reilly, R. J.: Marrow transplantation from human leukocyte antigen-identical or haploidentical donors for correction of Wiskott-Aldrich syndrome. J. Pediat. 119: 907-912, 1991.

Capsoni, F.; Acerbi, L.; Bonora, G.; Perletti, L.; Ongari, A. M.; Vanoli, M.; Zanussi, C.: Phagocyte function and immunological findings in a Wiskott-Aldrich syndrome long-term survivor. J. Lab. Clin. Immun. 19: 91-97, 1986.

Cooper, M. D.; Chae, H. P.; Lowman, J. T.; Krivit, W.; Good, R. A.: Wiskott-Aldrich syndrome: an immunologic deficiency disease involving the afferent limb of immunity. Am. J. Med. 44: 499-513, 1968.

Corash, L.; Shafer, B.; Blaese, R. M.: Platelet-associated immunoglobulin, platelet size, and the effect of splenectomy in the Wiskott-Aldrich syndrome. Blood 65: 1439-1443, 1985.

Derry, J. M. J.; Kerns, J. A.; Weinberg, K. I.; Ochs, H. D.; Volpini, V.; Estivill, X.; Walker, A. P.; Francke, U.: WASP gene mutations in Wiskott-Aldrich syndrome and X-linked thrombocytopenia. Hum. Molec. Genet. 4: 1127-1135, 1995.

Derry, J. M. J.; Ochs, H. D.; Francke, U.: Isolation of a novel gene mutated in Wiskott-Aldrich syndrome. Cell 78: 635-644, 1994.

Derry, J. M. J.; Wiedemann, P.; Blair, P.; Wang, Y.; Kerns, J. A.; Lemahieu, V.; Godfrey, V. L.; Wilkinson, J. E.; Francke, U.: The mouse homolog of the Wiskott-Aldrich syndrome protein (WASP) gene is highly conserved and maps near the scurfy (sf) mutation on the X chromosome. Genomics 29: 471-477, 1995.

de Saint-Basile, G.; Schlegel, N.; Caniglia, M.; Le Deist, F.; Kaplan, C.; Lecompte, T.; Piller, F.; Fischer, A.; Griscelli, C.: X-linked thrombocytopenia and Wiskott-Aldrich syndrome: similar regional assignment but distinct X-inactivation pattern in carriers. Ann. Hemat. 63: 107-110, 1991.

de Saint Basile, G.; Arveiler, B.; Fraser, N. F.; Boyd, Y.; Graig, I. W.; Griscelli, G.; Fischer, A.: Close linkage of hypervariable marker DXS255 to disease locus of Wiskott-Aldrich syndrome. Lancet II: 1319-1321, 1989.

de Saint Basile, G.; Lagelouse, R. D.; Lambert, N.; Schwarz, K.; Le Mareck, B.; Odent, S.; Schlegel, N.; Fischer, A.: Isolated X-linked thrombocytopenia in two unrelated families is associated with point mutations in the Wiskott-Aldrich syndrome protein gene. J. Pediat. 129: 56-62, 1996.

Diaz-Buxo, J. A.; Hermans, P. E.; Ritts, R. E., Jr.: Wiskott-Aldrich syndrome in an adult. Mayo Clin. Proc. 49: 455-459, 1974.

Fearon, E. R.; Kohn, D. B.; Winkelstein, J. A.; Vogelstein, B.; Blaese, R. M.: Carrier detection in the Wiskott Aldrich syndrome. Blood 72: 1735-1739, 1988.

Filipovich, A. H.; Krivit, W.; Kersey, J. H.; Burke, B. A.: Fatal arteritis as a complication of Wiskott-Aldrich syndrome. J. Pediat. 95: 742-744, 1979.

Fischer, A.; Friedrich, W.; Levinsky, R.; Vossen, J.; Griscelli, C.; Kubanek, B.; Morgan, G.; Wagemaker, G.; Landais, P.: Bone-marrow transplantation for immunodeficiencies and osteopetrosis: European survey, 1968-1985. Lancet II: 1080-1084, 1986.

Gatti, R. A.; Meuwissen, J. J.; Allen, H. D.; Hong, R.; Good, R. A.: Immunological reconstitution of sex-linked lymphopenic immunological deficiency. Lancet I: 1366-1369, 1968.

Gealy, W. J.; Dwyer, J. M.; Harley, J. B.: Allelic exclusion of glucose-6-phosphate dehydrogenase in platelets and T lymphocytes from a Wiskott-Aldrich syndrome carrier. Lancet I: 63-65, 1980.

Gelzer, J.; Gasser, C.: Wiskott-Aldrich-Syndrom. Helv. Paediat. Acta 16: 17-39, 1961.

Greer, W. L.; Mahtani, M. M.; Kwong, P. C.; Rubin, L. A.; Peacocke, M.; Willard, H. F.; Siminovitch, K. A.: Linkage studies of the Wiskott-Aldrich syndrome: polymorphisms at TIMP and the X chromosome centromere are informative markers for genetic prediction. Hum. Genet. 83: 227-230, 1989.

Greer, W. L.; Peacocke, M.; Siminovitch, K. A.: The Wiskott-Aldrich syndrome: refinement of the localization on Xp and identification of another closely linked marker locus, OATL1. Hum. Genet. 88: 453-456, 1992.

Greer, W. L.; Shehabeldin, A.; Schulman, J.; Junker, A.; Siminovitch, K. A.: Identification of WASP mutations, mutation hotspots and genotype-phenotype disparities in 24 patients with the Wiskott-Aldrich syndrome. Hum. Genet. 98: 685-690, 1996.

Greer, W. L.; Somani, A.-K.; Kwong, P. C.; Peacocke, M.; Rubin, L. A.; Siminovitch, K. A.: Linkage relationships of the Wiskott-Aldrich syndrome to 10 loci in the pericentromeric region of the human X chromosome. Genomics 6: 568-571, 1990.

Gutenberger, J.; Trygstad, C. W.; Stiehm, E. R.; Opitz, J. M.; Thatcher, L. G.; Bloodworth, J. M. B.: Familial thrombocytopenia, elevated serum IgA and renal disease. Am. J. Med. 49: 729-741, 1970.

Hara, T.; Yamauchi, M.; Takahashi, E.; Hoshino, M.; Aoki, K.; Ayusawa, D.; Kawakita, M.: The UDP-galactose translocator gene is mapped to band Xp11.23-p11.22 containing the Wiskott-Aldrich syndrome locus. Somat. Cell Molec. Genet. 19: 571-575, 1993.

Hendriks, R. W.; Mensink, E. J. B. M.; Kraakman, M. E. M.; Thompson, A.; Schuurman, R. K. B.: Evidence for male X chromosomal mosaicism in X-linked agammaglobulinemia. Hum. Genet. 83: 267-270, 1989.

Hirschhorn, R.; Yang, D. R.; Puck, J. M.; Huie, M. L.; Jiang, C.-K.; Kurlandsky, L. E.: Spontaneous in vivo reversion to normal of an inherited mutation in a patient with adenosine deaminase deficiency. Nature Genet. 13: 290-295, 1996.

Holmberg, L.; Gustavii, B.; Jonsson, A.: A prenatal study of fetal platelet count and size with application to fetus at risk for Wiskott-Aldrich syndrome. J. Pediat. 102: 773-776, 1983.

Hutter, J. J., Jr.; Jones, J. F.: Results of a thymic epithelial transplant in a child with Wiskott-Aldrich syndrome and central nervous system lymphoma. Clin. Immun. Immunopath. 18: 121-125, 1981.

Kapoor, N.; Kirkpatrick, D.; Blaese, R. M.; Oleske, J.; Hilgartner, M. H.; Chaganti, R. S. K.; Good, R. A.; O'Reilly, R. J.: Reconstitution of normal megakaryocytopoiesis and immunologic functions in Wiskott-Aldrich syndrome by marrow transplantation following myelo-ablation and immunosuppression with busulphan and cyclophosphamide. Blood 57: 692-696, 1981.

Knox-Macaulay, H. H. M.; Bashawri, L.; Davies, K. E.: X linked recessive thrombocytopenia. J. Med. Genet. 30: 968-969, 1993.

Kolluri, R.; Shehabeldin, A.; Peacocke, M.; Lamhonwah, A.-M.; Teichert-Kuliszewska, K.; Weissman, S. M.; Siminovitch, K. A.: Identification of WASP mutations in patients with Wiskott-Aldrich syndrome and isolated thrombocytopenia reveals allelic heterogeneity at the WAS locus. Hum. Molec. Genet. 4: 1119-1126, 1995.

Kolluri, R.; Tolias, K. F.; Carpenter, C. L.; Rosen, F. S.; Kirchhausen, T.: Direct interaction of the Wiskott-Aldrich syndrome protein with the GTPase Cdc42. Proc. Nat. Acad. Sci. 93: 5615-5618, 1996.

Krivit, W.; Good, R. A.: Aldrich's syndrome (thrombocytopenia, eczema and infection in infants). Studies of the defense mechanisms. Am. J. Dis. Child. 97: 137-153, 1959.

Kwan, S.-P.; Hagemann, T. L.; Radtke, B. E.; Blaese, R. M.; Rosen, F. S.: Identification of mutations in the Wiskott-Aldrich syndrome gene and characterization of a polymorphic dinucleotide repeat at DXS6940, adjacent to the disease gene. Proc. Nat. Acad. Sci. 92: 4706-4710, 1995.

Kwan, S.-P.; Lehner, T.; Hagemann, T.; Lu, B.; Blaese, M.; Ochs, H.; Wedgwood, R.; Ott, J.; Craig, I. W.; Rosen, F. S.: Localization of the gene for the Wiskott-Aldrich syndrome between two flanking markers, TIMP and DXS255, on Xp11.22-Xp11.3. Genomics 10: 29-33, 1991.

Kwan, S.-P.; Lehner, T.; Lu, B.; Raghu, G.; Blaese, M.; Sandkuyl, L. A.; Ott, J.; Fraser, N.; Boyd, Y.; Craig, I. W.; Fischer, S.; Rosen, F.: Linkage of DXS255 to the Wiskott Aldrich syndrome gene. (Abstract) Cytogenet. Cell Genet. 51: 1027 only, 1989.

Kwan, S.-P.; Sandkuyl, L. A.; Blaese, M.; Kunkel, L. M.; Bruns, G.; Parmley, R.; Skarshaug, S.; Page, D. C.; Ott, J.; Rosen, F. S.: Genetic mapping of the Wiskott-Aldrich syndrome with two highly-linked polymorphic DNA markers. Genomics 3: 39-43, 1988.

Levin, A. S.; Spitler, L. E.; Stiles, D. P.; Fudenberg, H. H.: Wiskott-Aldrich syndrome, a genetically determined cellular immunologic deficiency: clinical and laboratory responses to therapy with transfer factor. Proc. Nat. Acad. Sci. 67: 821-828, 1970.

Lum, L. G.; Tubergen, D. G.; Corash, L.; Blaese, R. M.: Splenectomy in the management of the thrombocytopenia of the Wiskott-Aldrich syndrome. New Eng. J. Med. 302: 892-896, 1980.

Lyon, M. F.; Peters, J.; Glenister, P. H.; Ball, S.; Wright, E.: The scurfy mouse mutant has previously unrecognized hematological abnormalities and resembles Wiskott-Aldrich syndrome. Proc. Nat. Acad. Sci. 87: 2433-2437, 1990.

McEnery, G.; Nash, F. W.: Wiskott-Aldrich syndrome associated with idiopathic infantile cortical hyperostosis (Caffey's disease). Arch. Dis. Child. 48: 818-821, 1973.

Mentzer, S. J.; Remold-O'Donnell, E.; Crimmins, M. A. V.; Bierer, B. E.; Rosen, F. S.; Burakoff, S. J.: Sialophorin, a surface sialoglycoprotein defective in the Wiskott-Aldrich syndrome, is involved in human T lymphocyte proliferation. J. Exp. Med. 165: 1383-1392, 1987.

Meropol, N. J.; Hicks, D.; Brooks, J. J.; Siminovitch, K. A.; Fishman, N. O.; Kant, J. A.; Bennett, J. S.: Coincident Kaposi sarcoma and T-cell lymphoma in a patient with the Wiskott-Aldrich syndrome. Am. J. Hemat. 40: 126-134, 1992.

Nathan, D. G.: Splenectomy in the Wiskott-Aldrich syndrome. (Editorial) New Eng. J. Med. 302: 916-917, 1980.

Notarangelo, L. D.; Candotti, F.; Parolini, O.; Mantuano, E.; Giliani, S.; Lanfranchi, A.; Albertini, A.: Application of molecular analysis to genetic counseling in the Wiskott-Aldrich syndrome (WAS). DNA Cell Biol. 12: 645-649, 1993.

Notarangelo, L. D.; Parolini, O.; Faustini, R.; Porteri, V.; Albertini, A.; Ugazio, A. G.: Presentation of Wiskott-Aldrich syndrome as isolated thrombocytopenia. (Letter) Blood 77: 1124-1128, 1991.

Notarangelo, L. D.; Parolini, O.; Porta, F.; Locatelli, F.; Lanfranchi, A.; Marconi, M.; Nespoli, L.; Albertini, A.; Craig, I. W.; Ugazio, A. G.: Analysis of X-chromosome inactivation and presumptive expression of the Wiskott-Aldrich syndrome (WAS) gene in hematopoietic cell lineages of a thrombocytopenic carrier female of WAS. Hum. Genet. 88: 237-241, 1991.

Ochs, H. D.; Slichter, S. J.; Harker, L. A.; Von Behrens, W. E.; Clark, R. A.; Wedgwood, R. J.: The Wiskott-Aldrich syndrome: studies of lymphocytes, granulocytes, and platelets. Blood 55: 243-252, 1980.

Parkman, R.; Kenney, D. M.; Remold-O'Donnell, E.; Perrine, S.; Rosen, F. S.: Surface protein abnormalities in lymphocytes and platelets from patients with Wiskott-Aldrich syndrome. Lancet II: 1387-1389, 1981.

Parkman, R.; Rappeport, J.; Geha, R.; Belli, J.; Cassady, R.; Levey, R.; Nathan, D. G.; Rosen, F. S.: Complete correction of the Wiskott-Aldrich syndrome by allogeneic bone marrow transplantation. New Eng. J. Med. 298: 921-927, 1978.

Parolini, O.; Ressmann, G.; Haas, O. A.; Pawlowsky, J.; Gadner, H.; Knapp, W.; Holter, W.: X-linked Wiskott-Aldrich syndrome in a girl. New Eng. J. Med. 338: 291-295, 1998.

Peacocke, M.; Siminovitch, K. A.: Linkage of the Wiskott-Aldrich syndrome with polymorphic DNA sequences from the human X chromosome. Proc. Nat. Acad. Sci. 84: 3430-3433, 1987.

Perry, G. S., III; Spector, B. D.; Schuman, L. M.; Mandel, J. S.; Anderson, V. E.; McHugh, R. B.; Hanson, M. R.; Fahlstrom, S. M.; Krivit, W.; Kersey, J. H.: The Wiskott-Aldrich syndrome in the United States and Canada (1892-1979). J. Pediat. 97: 72-78, 1980.

Prchal, J. T.; Carroll, A. J.; Prchal, J. F.; Crist, W. M.; Skalka, H. W.; Gealy, W. J.; Harley, J.; Malluh, A.: Wiskott-Aldrich syndrome: cellular impairments and their implication for carrier detection. Blood 56: 1048-1054, 1980.

Puck, J. M.; Siminovitch, K. A.; Poncz, M.; Greenberg, C. R.; Rottem, M.; Conley, M. E.: Atypical presentation of Wiskott-Aldrich syndrome: diagnosis in two unrelated males based on studies of maternal T cell X chromosome inactivation. Blood 75: 2269-2374, 1990.

Puck, J. M.; Willard, H. F.: X inactivation in females with X-linked disease. New Eng. J. Med. 338: 325-327, 1998.

Remold-O'Donnell, E.; Kenney, D. M.; Parkman, R.; Cairns, L.; Savage, B.; Rosen, F. S.: Characterization of a human lymphocyte surface sialoglycoprotein that is defective in Wiskott-Aldrich syndrome. J. Exp. Med. 159: 1705-1723, 1984.

Schindelhauer, D.; Weiss, M.; Hellebrand, H.; Golla, A.; Hergersberg, M.; Seger, R.; Belohradsky, B. H.; Meindl, A.: Wiskott-Aldrich syndrome: no strict genotype-phenotype correlations but clustering of missense mutations in the amino-terminal part of the WASP gene product. Hum. Genet. 98: 68-76, 1996.

Schwartz, M.; Mibashan, R. S.; Nicolaides, K. H.; Millar, D. S.; Jenkins, E.; Rodeck, C. H.; Orstavik, K. H.; Stormorken, H.: First-trimester diagnosis of Wiskott-Aldrich syndrome by DNA markers. (Letter) Lancet II: 1405 only, 1989.

Shapiro, R. S.; Perry, G. S., III; Krivit, W.; Gerrard, J. M.; White, J. G.; Kersey, J. H.: Wiskott-Aldrich syndrome: detection of carrier state by metabolic stress of platelets. Lancet I: 121-123, 1978.

Siminovitch, K. A.; Greer, W. L.; Novogrodsky, A.; Axelsson, B.; Somani, A.-K.; Peacocke, M.: A diagnostic assay for the Wiskott-Aldrich syndrome and its variant forms. J. Invest. Med. 43: 159-169, 1995.

Simon, H.-U.; Mills, G. B.; Hashimoto, S.; Siminovitch, K. A.: Evidence for defective transmembrane signaling in B cells from patients with Wiskott-Aldrich syndrome. J. Clin. Invest. 90: 1396-1405, 1992.

Spitler, L. E.; Wray, B. B.; Mogerman, S.; Miller, J. J., III; O'Reilly, R. J.; Lagios, M.: Nephropathy in the Wiskott-Aldrich syndrome. Pediatrics 66: 391-398, 1980.

Standen, G. R.; Lillicrap, D. P.; Matthews, N.; Bloom, A. L.: Inherited thrombocytopenia, elevated serum IgA and renal disease: identification as a variant of the Wiskott-Aldrich syndrome. Quart. J. Med. 59: 401-408, 1986.

Steinberg, A. G.: Methodology in human genetics. J. Med. Educ. 34: 315-334, 1959.

Stewart, D. M.; Treiber-Held, S.; Kurman, C. C.; Facchetti, F.; Notarangelo, L. D.; Nelson, D. L.: Studies of the expression of the Wiskott-Aldrich syndrome protein. J. Clin. Invest. 97: 2627-2634, 1996.

Sullivan, K. E.; Mullen, C. A.; Blaese, R. M.; Winkelstein, J. A.: A multiinstitutional survey of Wiskott-Aldrich syndrome. J. Pediat. 125: 876-885, 1994.

Symons, M.; Derry, J. M. J.; Karlak, B.; Jiang, S.; Lemahieu, V.; McCormick, F.; Francke, U.; Abo, A.: Wiskott-Aldrich syndrome protein, a novel effector for the GTPase CDC42Hs, is implicated in actin polymerization. Cell 84: 723-734, 1996.

ten Bensel, R. W.; Stadlan, E. M.; Krivit, W.: The development of malignancy in the course of the Aldrich syndrome. J. Pediat. 68: 761-767, 1966.

Van den Bosch, J.; Drukker, J.: Het Syndroom van Aldrich: een klinisch en genetisch Onderzoek van enige nederlandse Families. Maandschr. Kindergeneesk. 32: 359-373, 1964.

Villa, A.; Notarangelo, L.; Macchi, P.; Mantuano, E.; Cavagni, G.; Brugnoni, D.; Strina, D.; Patrosso, M. C.; Ramenghi, U.; Sacco, M. G.; Ugazio, A.; Vezzoni, P.: X-linked thrombocytopenia and Wiskott-Aldrich syndrome are allelic diseases with mutations in the WASP gene. Nature Genet. 9: 414-417, 1995.

Webb, M. C.; Andrews, P. A.; Koffman, C. G.; Cameron, J. S.: Renal transplantation in Wiskott-Aldrich syndrome. Transplantation 56: 747-748, 1993.

Weiden, P. L.; Blaese, R.: Hereditary thrombocytopenia: relation to Wiskott-Aldrich syndrome with special reference to splenectomy. J. Pediat. 80: 226-234, 1972.

Wengler, G.; Gorlin, J. B.; Williamson, J. M.; Rosen, F. S.; Bing, D. H.: Nonrandom inactivation of the X chromosome in early lineage hematopoietic cells in carriers of Wiskott-Aldrich syndrome. Blood 85: 2471-2477, 1995.

Wengler, G. S.; Notarangelo, L. D.; Berardelli, S.; Pollonni, G.; Mella, P.; Fasth, A.; Ugazio, A. G.; Parolini, O.: High prevalence of nonsense, frame shift, and splice-site mutations in 16 patients with full-blown Wiskott-Aldrich syndrome. Blood 86: 3648-3654, 1995.

Wiskott, A.: Familiarer, angeborener Morbus Werlhofii? Mschr. Kinderheilk. 68: 212-216, 1937.

Wolff, J. A.: Wiskott-Aldrich syndrome: clinical, immunologic, and pathologic observations. J. Pediat. 70: 221-232, 1967.

Zhu, Q.; Zhang, M.; Blaese, R. M.; Derry, J. M. J.; Junker, A.; Francke, U.; Chen, S.-H.; Ochs, H. D.: The Wiskott-Aldrich syndrome and X-linked congenital thrombocytopenia are caused by mutations of the same gene. Blood 86: 3797-3804, 1995.

WISKOTT-ALDRICH SYNDROME, AUTOSOMAL DOMINANT FORM (600903)

JJM NOTE

This entity, represented by only one family, is not associated (yet) with neoplasia, but it is included for completeness.

Neri et al. (1995) raised the possibility of an autosomal dominant form of Wiskott-Aldrich syndrome on the basis of a 3-generation family in which several members presented clinical and laboratory findings of WAS (301000), including decreased CD43 expression on T lymphocytes. The gene for CD43, or sialophorin (SPN; 182160), is located on 16p11.2. However, no alteration of CD43 was found: Southern blot analysis failed to detect gross abnormalities of the CD43 gene and genotype analysis showed that the affected family members did not share a common CD43 allele.

Neri, G.; Bellacosa, A.; Rocca, B.; De Cristofaro, R.; Della Ventura, M.; Maggiano, N.; Landolfi, R.: Wiskott-Aldrich syndrome: an autosomal dominant variant. (Abstract) Am. J. Hum. Genet. 57: A98 only, 1995.

AGAMMAGLOBULINEMIA (*300300)

JJM NOTE

The Immunodeficiency-Cancer Registry of the University of Minnesota (Clin. Immunol. Immunopathol. 11:12–29, 1978) tallied 14 cases: 3 lymphatic leukemia; 1 each of lymphocytic lymphoma, histiocytic lymphoma, and Hodgkin disease; 2 other lymphoreticular malignancies; 2 malignant reticuloendotheliosis; 2 other mesenchymal; and 2 other epithelial cancers. A short report was made of three cases with colorectal cancer at ages 25, 30, and 36 years (van der Keer et al. 1993).

Description

Agammaglobulinemia is an X-linked immunodeficiency characterized by failure to produce mature B lymphocyte cells and associated with a failure of Ig heavy chain rearrangement. The defect in this disorder resides in Bruton tyrosine kinase (BTK, also known as BPK or ATK), a key regulator in B-cell development (Rawlings and Witte, 1994).

Clinical Features

Patients are unusually prone to bacterial infection but not to viral infection. A clinical picture resembling rheumatoid arthritis develops in many. Before antibiotics, death occurred in the first decade. In the more usual X-linked form of the disease, plasma cells are lacking. A rarer form of agammaglobulinemia (Hitzig and Willi, 1961), which is inherited as an autosomal recessive (202500), shows marked depression of the circulating lymphocytes, and lymphocytes are absent from the lymphoid tissue.

The alymphocytotic type (also see 300400) is even more virulent than the Bruton form, leading to death in the first 18 months after birth from severe thrush, chronic diarrhea, and recurrent pulmonary infections.

Seligmann et al. (1968) proposed a classification of immunologic deficiencies. Ament et al. (1973) pointed out that gastrointestinal infestation with Giardia lamblia is frequent in this and other forms of immunodeficiency. Infection with Campylobacter jejuni and Salmonella spp is also frequent (Melamed et al., 1983). Giardiasis may lead to malabsorption, while C. jejuni infection may result in recurrent fever (van der Meer et al., 1986; Kerstens et al., 1992).

Geha et al. (1973) showed that males with proven X-linked agammaglobulinemias lacked bone marrow-derived (B) lymphocytes from the circulating blood whereas progenitor and thymus (T) cells were normal. See 301000 and 308230 for other X-linked deficiencies of immunoglobulins.

Although patients have recurrent bacterial infections, they generally have a normal response to viral infection, presumably because cell-mediated immunity is intact. A notable exception is the usually fatal echovirus-induced meningoencephalitis, which is often associated with the 'dermatomyositis-like' syndrome first described by Janeway et al. (1956). Mease et al. (1981) successfully treated a 32-year-old man who developed signs of myopathy and encephalopathy over a period of 3 months. Echo 11 virus was recovered from muscle and spinal fluid. In vitro lymphocyte transformation was temporarily markedly depressed by the infection. High doses of immune globulin given intravenously cured the man of this usually fatal complication.

Rosen et al. (1984) reviewed primary immunodeficiencies, giving a classification according to whether the immunodeficiency was predominantly one of antibody formation, was predominantly one of cell-mediated immunity, or was associated with other defects as in ataxia-telangiectasia.

Lederman and Winkelstein (1985) collected data from 96 patients cared for in 26 North American medical centers and representing a total experience of almost 1,200 patient-years.

Boys with agammaglobulinemia lack circulating B cells. Landreth et al. (1985) described 4 boys with agammaglobulinemia who lacked pre-B lymphocytes. In classic agammaglobulinemia, pre-B cells are present in normal numbers in the bone marrow but appear to be either blocked or aborted in their ability to mature, express surface immunoglobulins, or produce antibody. In the boys who lacked pre-B cells, clinical presentation with recurrent infections was delayed until the second or third year. None of the 4 boys had a history of recurrent infection or similar disease in maternal first cousins or uncles. Two of the patients were brothers. The mode of inheritance is unclear. The immune defect resembled that of the thymoma-agammaglobulinemia syndrome, but thymoma was not present in any of the 4.

Thorsteinsson et al. (1990) described studies in 3 brothers with IMD1, the first of whom was diagnosed in 1963 at the age of 9 years and died at the age of 23.

Van der Meer et al. (1993) reported the cases of 3 unrelated men with XLA who developed colorectal cancer at the ages of 26, 29, and 36 years. Van der Meer et al. (1993) suggested that there is an increased risk of colorectal cancer in these individuals and that it may be related to intestinal infections.

Biochemical Features

Edwards et al. (1978) showed reduced ecto-5-prime-nucleotidase (129190) activity in peripheral blood lymphocytes. This is an ectoenzyme that regulates the uptake of AMP into lymphocytes by converting the nontransportable nucleotide to its readily transported nucleoside, adenosine.

Other Features

Pearl et al. (1978) showed that precursor B lymphocytes containing IgM heavy chains can be demonstrated in the bone marrow in XLA. This suggested that an arrest in the differentiation of precursor B lymphocytes into B lymphocytes may be involved. Schwaber et al. (1983) found that about 5% of normal pre-B cells and 100% of XLA pre-B cells produce incomplete mu chains (147020), i.e., C(mu) polypeptide without associated V(H). Thus, XLA represents a block in differentiation secondary to failure to express V(H) genes. (Cytoplasmic mu-chain protein has served as a marker for pre-B cells. Mu-chain gene expression precedes rearrangement and expression of light-chain genes.) Presumably the X chromosome codes for enzyme(s) specific for translocation of V(H) genes or a regulatory mechanism necessary for pre-B cells to differentiate to a stage using these enzymes.

In 2 sisters heterozygous for both XLA and G6PD A-/B polymorphism, Conley et al. (1986) found that B cells showed activity of only the A- form of G6PD, whereas T cells and neutrophils had about equal amounts of A- and B enzyme activity. This indicates that the basic defect in XLA is intrinsic to the B cell.

Vorechovsky et al. (1993) pointed out that common variable immunodeficiency (CVID) is sometimes clinically and immunologically indistinguishable from XLA if it starts early in childhood and occurs sporadically in males with a decreased number of B cells. Using a cDNA clone that represented the full-length ATK cDNA, Vorechovsky et al. (1993) did Southern blot analysis of 39 Swedish male patients diagnosed with CVID or possible CVID. One man in his late 40s, who had had recurrent respiratory infections from infancy, lacked immunoglobulins of all isotypes, and had less than 1% B cells among peripheral blood mononuclear cells, had an abnormality of the ATK gene. The abnormality was missing in his mother but had been inherited by both of his daughters.

Vorechovsky et al. (1993) failed to find the arg28-to-cys mutation, which is found in xid in mice (Rawlings et al., 1993), in 13 unrelated patients with XLA and 2 patients with the syndrome of XLA and growth hormone deficiency (307200). They pointed to the milder phenotype of the xid mouse compared to XLA cases and suggested that if this particular mutation occurs in the human BTK gene, it might result in a milder phenotype with normal or only moderately reduced B cells and more selective immunoglobulin deficiency in boys, which may or may not increase susceptibility to infections.

Inheritance

Lau et al. (1988) discussed the calculation of genetic risks in XLA, including allowance for nonallelic genetic heterogeneity.

Hendriks et al. (1989) described a family in which each of 2 sisters had a son with XLA. The 2 sisters with affected sons and another sister all showed exclusive inactivation of the paternal X chromosome in B lymphocytes, indicating that the gene for XLA came from their father, who, however, had no agammaglobulinemia. He was presumed to be an X chromosomal mosaic. RFLP segregation analyses in other XLA pedigrees suggested that this may be a frequent situation.

Mapping

Race and Sanger (1975) thought that the agammaglobulinemia locus was possibly linked to Xg; the lods were positive but low at a recombination fraction of 30%.

In 12 families, including an extensively affected Dutch kindred of 8 generations, Mensink et al. (1984) studied linkage with Xg (314700) and the 12E7 polymorphism that is closely linked to Xg. They concluded that XLA and Xg are at least 20 cM apart. Cohen et al. (1985) isolated a cDNA probe recognizing a family of genes, called XLR (see 300113), on the mouse X chromosome, at least some members of which are closely linked to the xid trait. In accompanying studies, Cohen et al. (1985) presented data which, combined with the RFLP analysis closely linking the XLR gene family to the xid mutation, suggest that the xid defect resides in a member of this family. From a study of the comparative mapping of the human and mouse X chromosomes, Buckle et al. (1985) predicted that the XLA locus of man may be on Xq between PGK1 (311800) and GLA (301500), i.e., in the segment Xq13-Xq22. This remarkable prediction was borne out by the findings of Kwan et al. (1986).

By RFLP studies in 11 families, they showed that XLA is linked to 2 markers, DXS3 and DXS17, both localized in region Xq21.3-q22 (lod = 3.65 at theta = 0.04 and lod = 2.17 at theta = 0.0, respectively). In a single 8-generation Dutch kindred, Mensink et al. (1986) found a maximum lod score of 3.30 at a recombination fraction of 0.06 for linkage of XLA and marker p19-2 (DXS3).

In another pedigree, similar linkage to DXS3 was excluded (lod = -3.14 at theta 0.06). This suggests the existence of a second form of X-linked agammaglobulinemia; data obtained by Mensink et al. (1986) from all pedigrees suggested localization of a second XLA gene in the Xp22 band as defined by marker p99-6 (DXS41); see 300310. This is a possible parallel to the historic demonstration of heterogeneity in elliptocytosis (130500) by the linkage principle. Mensink et al. (1986) predicted that more detailed molecular studies 'will ultimately reveal phenotypic differences, reflecting different XLA gene loci, one of them probably coding for a recombinase involved in immunoglobulin heavy-chain rearrangements (Schwaber et al., 1983) and the other(s) being involved in later stages of precursor B cell differentiation (Levitt et al., 1984). 'With a multipoint linkage analysis in 9 families with XLA, Ott et al. (1986) concluded that there was 'clear evidence for heterogeneity of XLA.' The finding of possible linkage to Xg by Race and Sanger (1975) may have been related to their having a mixture of 'Xp' and 'Xq' families. Malcolm et al. (1989) presented further evidence, based on linkage data, for the existence of 2 loci.

Mensink et al. (1987) mapped XLA to Xq21.3-q22. No recombination was found between XLA and DXS17 (lod = greater than 6 at theta = 0); no recombinants were found between XLA and DXS17 in this study or in the study by Kwan et al. (1986)—with the exception of the remarkable Z pedigree which may have carried a different form of agammaglobulinemia. Malcolm et al. (1987) demonstrated close linkage to DNA markers in the Xq21.3-q22 region in studies of 15 families. Guioli et al. (1989) found close linkage of IMD1 and DXS178. No recombinants were observed, giving a maximum lod score of 5.92 at theta = 0.

Kwan et al. (1990) demonstrated another marker closely linked to XLA, DXS178.

Molecular Genetics

Fearon et al. (1987) used a strategy similar to that of Conley et al. (1986) to show that the defect in XLA is intrinsic to B cells as well as to detect the carrier state. According to their strategy, recombinant DNA probes simultaneously detect RFLPs and patterns of methylation of X-chromosome genes. (Different DNA methylation patterns reflect whether the X chromosome is active or inactive and these differences in methylation can be monitored by restriction endonucleases that have the capacity to recognize methylated cytosine residues.) Random X-inactivation patterns were observed in isolated peripheral blood granulocytes, T lymphocytes, and B lymphocytes of women who were not carriers. In contrast, 1 of the 2 X chromosomes was preferentially active in the B cells, but not the T cells or granulocytes, of 3 carriers of the disorder. Fearon et al. (1987) used X-chromosome inactivation analysis to demonstrate that the X chromosome with the wildtype allele at the agammaglobulinemia locus was the active one in all the B cells. Allen et al. (1994) tested carrier status by study of B lymphocytes and T lymphocytes separated by means of antibodies to the B-cell specific antigen CD19 (107265). B lymphocytes were isolated from the mononuclear cell fraction of 20 cc of blood by using anti-CD19 immunomagnetic beads. Quantitative PCR at the androgen-receptor locus was then used to examine patterns of X inactivation in CD19-positive B cells. The trinucleotide repeat at the androgen receptor locus (AR; 313700) is within approximately 100 bp of 2 HpaII restriction-enzyme sites that are methylated on the inactive X chromosome but unmethylated on the active X chromosome. Obligate carriers of XLA demonstrated more than 95% skewing of X inactivation in CD19-positive cells but not in CD19-negative cells. Allen et al. (1994) suggested that refinements in techniques for primary carrier testing and genetic mapping of XLA make possible an ordered approach to prenatal diagnosis and genetic cou nseling.

Schwaber et al. (1988) found an unusual phenotype of B cells in a patient with XLA, and cellular evidence for lyonization of B cells from the mother and sister. The patient had a failure of B-cell maturation at the stage of early B lymphocytes, associated with production of truncated mu and delta heavy chains composed of D-J(H)-C resulting from abortive rearrangement of variable region genes. There was also delayed expression of L chains. Peripheral blood and B-cell lines from the patient's mother and sister included 50% cells that expressed H chain without L chain. B-cell lines from the mother and sister produced both full-length mu and gamma H chains and truncated mu and delta chains corresponding to the H chains produced by the patient's B cells. Schwaber and Chen (1988) concluded that failure of variable region gene rearrangement may underlie the failure of B lymphoid development in XLA. They observed that immature B cells from a patient produced truncated mu and delta immunoglobulin H chains. In cases of XLA there is variability in the stage at which the arrest of development occurs; the major phenotype is arrested at the stage of pre-B cells, while a minor phenotype is arrested at the stage of immature B lymphocytes. The failure of B lymphoid development is associated in both phenotypes with a failure of Ig heavy chain variable region rearrangement. The immature B cells of a patient with the minor phenotype of XLA produce truncated gamma and delta heavy chains composed of a D-J-constant complex resulting from failure to rearrange a V segment. Schwaber et al. (1988) demonstrated that the fusion of these cells with mouse myeloma complemented the failure of V(H) gene rearrangement. H chains produced by such hybrid cells are composed of V(H)-D-J(H)-C. The genes encoding each of these elements were of human parental origin, indicating that the mouse myeloma provided a trans-acting regulatory element necessary for V(H) rearrangement which the XLA B cells lack. Complementation occurred in all hybrid cells examined, regardless of whether the human X chromosome was retained.

Schwaber (1992) presented direct evidence that there is a failure of V(D)J recombination which causes arrest in the transition from pre-B cell to B lymphocyte. He pointed out that the arrest in B-cell development is not absolute: rare B lymphocytes have been identified in peripheral blood of some patients, and B-cell lines have been established from these cells by Epstein-Barr virus transformation. Leakiness of the mutation would not be inconsistent with the proposed mechanism.

Vetrie et al. (1993) isolated a novel gene that mapped to the XLA locus, was expressed in B cells, and showed mutations in families with the disorder. They identified the gene as a member of the SRC family of protooncogenes that encode protein-tyrosine kinases. This was, to their knowledge, the first evidence that mutations in an SRC-related gene are involved in human genetic disease. The cloning strategy was based on map location—positional cloning. The XLA locus had been localized to a 2-cM interval. Marker DXS178 was shown to lie 70-140 kb distal to GLA (301500) and about 500 kb proximal to DXS101. A number of putative CpG islands were found in the region around DXS178, indicative of the presence of genes. A 640-kb YAC clone containing the GLA and DXS178 loci was chosen from a 6.5-megabase contig. The YAC was used as a hybridization probe to identify 48 cosmid clones in a library constructed using DNA from cells of karyotype 49,XXXXX. Transcribed sequences within the 640-kb YAC were identified by screening a cDNA library derived from a Burkitt lymphoma cell line. Several cDNAs were isolated that localized to a 20-kb region about 50-70 kb centromeric to GLA. Indeed, 25 subcloned cDNAs were shown to be derived from a single transcript which by in situ hybridization was shown to map to a unique site at the Xq21.3-q22 boundary. Using probes so derived for the Southern analysis of DNA from 33 unrelated families and 150 normal X chromosomes,

Using probes so derived for the Southern analysis of DNA from 33 unrelated families and 150 normal X chromosomes, Vetrie et al. (1993) detected restriction pattern abnormalities in 8 families. Five of them had deletions which were shown to be entirely intragenic, confirming the involvement of this gene in XLA. Determination of the complete nucleotide sequence demonstrated that the cDNA compromises 2,560 nucleotides and has an open reading frame encoding a 659-amino acid polypeptide beginning with the initiation codon at nucleotide 133. The reading frame is closed 15 codons upstream of this start site. Two alternative initiation codons were located at nucleotides 397 and 619 within the same open reading frame as the first; initiation of translation from these 2 start signals would result in peptide chains of 571 and 497 amino acids, respectively. Across a stretch of about 1,200 bp of its open reading frame, the gene showed a high degree of similarity with 2 members of the SRC gene family. On this basis, Vetrie et al. (1993) symbolized the gene ATK (for agammaglobulinemia tyrosine kinase).

Northern analysis of RNAs derived from lymphoid lineages demonstrated that the 2.6-kb ATK mRNA was expressed in a B-cell line and in B cells of 2 patients with chronic lymphocytic leukemia but not in T cells or a T-cell line. Two single-base missense mutations were identified in XLA patients. The failure of pre-B cells in the bone marrow of XLA males to develop into mature, circulating B cells could be the result of the product of the mutant ATK gene failing to fulfill its role in B-cell signaling. Vetrie et al. (1993) noted that inactivation of the LCK gene (153390) by homologous recombination in mice results in a thymocyte differentiation defect; LCK is also a member of the SRC family of tyrosine kinases. Desiderio (1993) compared the structure of ATK and LTK with SRC.

Tsukada et al. (1993) independently described a cytoplasmic tyrosine kinase which they termed BPK (for B-cell progenitor kinase) and which is expressed in all cells of the B lineage and in myeloid cells. They found that BPK mRNA, protein expression, and kinase activity were all reduced or absent in XLA pre-B and B cell lines. They found that a BPK probe hybridized with a 200-kb YAC which had been previously isolated using the probe (p212) most closely linked to XLA. Further restriction mapping demonstrated that the BPK gene was within 60 kb of the end of the YAC and within 100 kb of the XLA-linked 212 probe. Tsukada et al. (1993) concluded that BPK is not a member of the SRC family of protein tyrosine kinases based on the following differences: (1) the kinase catalytic domain contains the sequence DLAARN which is similar to ABL, FPS, and CSK but different from the SRC family (DLRAAN); (2) BPK lacks the consensus myristoylation signal (glycine at position 2 and lysine (or arginine) at position 7); (3) BPK lacks the equivalent of tyrosine 527 in SRC in the C-terminal domain following the kinase sequences (found to be important in regulation of kinase activity); and (4) the N-terminal region of BPK is unusually long. BPK is a member of a new subfamily of tyrosine kinases. Rohrer et al. (1994) determined the genomic organization of the gene. BTK is divided into 19 exons spanning 37 kb of genomic DNA. The region 5-prime to the first untranslated exon revealed no consensus TATAA or CAAT boxes; however, 3 retinoic acid-binding sites were identified. Oeltjen et al.

(1995) concluded that the 3-prime end of the GLA gene is 9 kb from the 5-prime end of the BTK gene and found 2 additional genes in the region immediately 5-prime to BTK.

Duriez et al. (1994) found an exon-skipping mutation in the BTK gene which appeared to account for the syndrome of X-linked agammaglobulinemia and isolated growth hormone deficiency in a sporadic case (see 300300.0004).

Ohta et al. (1994) reported the DNA sequence of the 18 coding exons of BTK and their flanking regions. Correlations were made between the nature of mutations and the organization of the BTK gene. They found several examples of the same mutation occurring in unrelated patients, and one of these mutations occurred at the same codon that is substituted in the xid mouse. However, in xid, the mutation occurs at the first position in the conserved arginine codon, C214-to-T, and results in an arg28-to-cys amino acid change, whereas in human cases it occurs in the second nucleotide, G215-to-A, and results in an arg28-to-his amino acid change (300300.0005). The observations suggest that a limited number of deleterious changes in BTK produce clinically recognizable XLA. XLA patients have been classified in 2 general groups: those presenting at an early age with particularly severe infections and those with less severe disease in which production of immunoglobulin is sustained at low-to-normal levels well into the first decade of life. In the latter cases, an oncogenetic change may occur in which the defective tyrosine kinase no longer can sustain the B-cell population, and a progressive reduction in immunoglobulin production occurs.

Ohta et al. (1994) described the arg525-to-gln mutation (300300.0001) in patients whose disorder might have been classified as common variable immunodeficiency disease. These patients had low levels of circulating B cells at an early age with mildly decreased IgM and variable IgG levels, although all were IgA deficient. Kornfeld et al. (1995) described the case of a 16-year-old boy who had recurrent upper respiratory tract infections at 13 months of age and was diagnosed as having transient hypogammaglobulinemia of infancy on the basis of low immunoglobulin levels, normal diphtheria and tetanus antibody responses, normal anterior and posterior cervical nodes, normal tonsillar tissue, and normal numbers of B cells in the blood. IgA levels returned to normal at 15 months of age and remained within normal limits over the next 12 months, and IgG and IgM levels remained relatively unchanged. At age 10, he began receiving intravenous gammaglobulin, which resulted in cessation of infections. The clinical picture was thought to be that of common variable immunodeficiency disease. However, gene studies revealed the deletion of exon 16 of the BTK gene resulting from a splice junction defect. The patient represents an example of the extreme variation that can occur in the XLA phenotype.

Vihinen et al. (1994) used a 3-dimensional model for the BTK kinase domain, based on the core structure of cAMP-dependent protein kinase, to interpret the structural basis for disease in 8 independent point mutations in patients with XLA. Because arg525 of BTK had been thought to substitute functionally for a critical lysine residue in protein-serine kinases, they studied the arg525-to-gln mutation and found that it abrogated the tyrosine kinase activity of BTK. All of the 8 mutations, including lys430-to-glu (300300.0002), were located on one face of the BTK kinase domain, indicating structural clustering of functionally important residues.

Hagemann et al. (1995) described 6 mutations in the BTK gene as the cause of XLA; 5 were novel. The mutations included 2 nonsense and 2 missense mutations, a single base deletion at an intron acceptor splice site, and a 16-bp insertion.

Kobayashi et al. (1996) reported abnormalities in the BTK gene in 12 unrelated Japanese families with X-linked agammaglobulinemia. Gene rearrangement in the kinase domain was found in 2 patients by Southern blotting. Seven point mutations, 2 small deletions, and 1 small insertion were detected by SSCP analysis and sequencing. Phenotypic heterogeneity was observed in affected family members with the same mutation. The authors concluded that analyzing BTK gene alterations with SSCP is valuable for the diagnosis of XLA patients and for carrier detection; however, the correlation between gene abnormalities and clinical features remains unclear.

Among 26 unrelated patients with XLA, Vorechovsky et al. (1997) found 24 different mutations of the BTK gene. Most resulted in the premature termination of translation. Mutations were detected in most BTK exons with a predominance of frameshift and nonsense mutations in the 5-prime end of the gene and missense mutations in its 3-prime part, corresponding to the catalytic domain of the enzyme.

Conley et al. (1998) analyzed 101 families in which affected males were diagnosed as having XLA. Mutations in the BTK gene were identified in 38 of 40 families with more than 1 affected family member and in 56 of 61 families with sporadic disease. Excluding the patients in whom the marked decrease in B cell numbers characteristic of XLA could not be confirmed by immunofluorescence studies, mutations in BTK were identified in 43 of 46 patients with presumed sporadic XLA. Two of the 3 remaining patients had defects in other genes required for normal B cell development, namely the mu heavy chain gene (IGHM1; 147020), as reported by Yel et al. (1996) or the lambda-5/14.1 surrogate light chain gene (IGLL1; 146770), as reported by Minegishi et al. (1998). Both of these patients were compound heterozygotes and there were no clinical features that would distinguish them from patients with typical XLA. An Epstein-Barr virus-transformed cell line from a third patient had normal BTK cDNA by SSCP, normal BTK message by Northern blot, and normal BTK protein by Western blot. Therefore, it is unlikely that this patient had XLA. Ten mutations were found in more than one family; 1 of these occurred in 3 families. Of the 83 unique mutations included in the study of Conley et al. (1998), 43 had been described previously by their laboratory, 5 had been reported by other groups, and 35 had not been previously described.

Parolini et al. (1993) identified a family in which a healthy father transmitted the XLA defect to 2 of his daughters, indicating gonadal or somatic mosaicism. To assess the frequency of this phenomenon, Conley et al. (1998) evaluated 11 sisters of 7 women who were carriers of XLA and whose mutation occurred on the paternal haplotype. None of the 11 sisters were carriers of the mutations seen in their nephews.

Pathogenesis

Although evidence from the study of X-linked agammaglobulinemia indicated that BTK plays a crucial role in B-lymphocyte differentiation and activation, its precise mechanism of action remained unknown, primarily because the proteins that it interacts with had not been identified until the work of Cheng et al. (1994). They showed that BTK interacts with SRC homology 3 domains of FYN (137025), LYN (165120), and HCK (142370). All of these are protein-tyrosine kinases that are activated upon stimulation of B- and T-cell receptors. These interactions are mediated by two 10-amino acid motifs in BTK. An analogous site with the same specificity is also present in ITK (186973), the T-cell-specific homolog of BTK. The findings of Cheng et al. (1994) extended the range of interactions mediated by SRC homology 3 domains and provided an indication of a link between BTK and previously established signaling pathways in B lymphocytes.

Uckun et al. (1996) noted that a number of human diseases including immune deficiencies apparently stem from inherited or acquired deficiencies of checkpoints that regulate the rate of apoptosis in lymphoid cells. Uckun et al. (1996) reported that DT-40 lymphoma B cells rendered BTK deficient through targeted disruption of the BTK gene did not undergo radiation-induced apoptosis. They further demonstrated that the tyrosine kinase domain of BTK was necessary for triggering radiation-induced apoptosis.

Diagnosis

Schuurman et al. (1988) demonstrated the usefulness of linked RFLP markers in identifying the carrier state and in the early diagnosis of XLA in a newborn son.

Journet et al. (1992) demonstrated that the pregnant mother of a boy with XLA but no family history of immune disease was a carrier by demonstrating with a methylation-sensitive probe that the X-inactivation pattern was skewed in the woman's B cells but random in her polymorphonuclear cells. Using RFLP probes flanking the XLA locus on each side, they excluded the diagnosis of XLA in the fetus on the basis of a chorionic villus sample (risk of error less than 0.003). Subsequent studies of the baby confirmed normality.

Animal Model

Presumably the X-linked B lymphocyte defect of mice, studied by Marshall-Clarke et al. (1979), is homologous. This defect is characteristic of the CBA-N strain of mice (Scher et al., 1975). Defective mice lack the subpopulation of B lymphocytes responsive to certain T-independent antigens of which trinitrophenylated (TNP)-Ficoll is the prototype. Their responses to T-dependent antigens may also be impaired and they are unable to respond to the hapten phosphorylcholine (PC). They lack those B cells that form colonies when cultured in vitro.

Cohen et al. (1985) isolated a cDNA probe recognizing a family of genes, called XLR, on the mouse X chromosome, at least some members of which are closely linked to the xid trait.

Linkage studies involving 1,114 progeny backcross revealed colocalization of X-linked immunodeficiency (xid) mutation in mice with the gene for Bruton agammaglobulinemia tyrosine kinase (btk) (Thomas et al., 1993). The xid mutation was associated in mice with a missense mutation that altered the highly conserved arginine near the amino-terminus of the btk protein. Because this region of the protein lies outside any obvious kinase domain, the xid mutation may define another aspect of tyrosine kinase. Rawlings et al. (1993) likewise mapped the xid and the btk gene to the same region and demonstrated the same missense mutation, an arg28-to-cys change.

Drabek et al. (1997) generated transgenic mice in which expression of the human BTK gene was driven by the murine class II major histocompatibility complex Ea gene locus control region, which provides gene expression from the pre-B cell stage onwards. When these transgenic mice were mated onto a Btk(-) background, correction of the xid B cell defects was observed: B cells differentiated to mature low IgM/high IgD stages, peritoneal CD5(+) B cells were present, and serum immunoglobulin levels and in vivo responses to antigens were in the normal ranges. A comparable rescue by transgenic Btk expression was also observed in heterozygous Btk(+/-) female mice in those B-lineage cells that were Btk-deficient as a result of X-chromosome inactivation.

History

X-linked agammaglobulinemia, the first genetic immunodeficiency to be specifically identified, was described by Bruton (1952).

Vihinen et al. (1996) described a database of BTK mutations (BTKbase) listing entries from 189 unrelated families showing 148 unique molecular events. Information was included regarding the phenotype. Mutations in all 5 domains of the BTK had been observed to cause XLA, the most common class of changes being missense mutations. The mutations appeared almost uniformly throughout the molecule and frequently affected CpG sites forming arginine residues.

Ochs and Smith (1996) provided a comprehensive review of the clinical and molecular aspects of X-linked agammaglobulinemia.

See also Berning et al. (1980); Cohen et al. (1985); Erlendsson et al. (1985); Garvie and Kendall (1961); Gitlin and Craig (1963); Janeway et al. (1953); Perryman et al. (1983); Saulsbury et al. (1979); Schwaber et al. (1988); Thompson et al. (1980); Vorechovsky et al. (1993)

Allen, R. C.; Nachtman, R. G.; Rosenblatt, H. M.; Belmont, J. W.: Application of carrier testing to genetic counseling for X-linked agammaglobulinemia. Am. J. Hum. Genet. 54: 25-35, 1994.

Ament, M. E.; Ochs, H. D.; Davis, S. D.: Structure and function of the gastrointestinal tract in primary immunodeficiency syndromes: a study of 39 patients. Medicine 52: 227-248, 1973.

Berning, A. K.; Eicher, E. M.; Paul, W. E.; Scher, I.: Mapping of the X-linked immune deficiency mutation (xid) of CBA/N mice. J. Immun. 124: 1875-1877, 1980.

Bradley, L. A. D.; Sweatman, A. K.; Lovering, R. C.; Jones, A. M.; Morgan, G.; Levinsky, R. J.; Kinnon, C.: Mutation detection in the X-linked agammaglobulinemia gene, BTK, using single strand conformation polymorphism analysis. Hum. Molec. Genet. 3: 79-83, 1994.

Bruton, O. C.: Agammaglobulinemia. Pediatrics 9: 722-727, 1952.

Buckle, V. J.; Edwards, J. H.; Evans, E. P.; Jonasson, J. A.; Lyon, M. F.; Peters, J.; Searle, A. G.: Comparative maps of human and mouse X chromosomes. (Abstract) Cytogenet. Cell Genet. 40: 594-595, 1985.

Buckley, R. H.: Assessing inheritance of agammaglobulinemia. (Editorial) New Eng. J. Med. 330: 1526-1528, 1994.

Buckley, R. H.; Rowlands, D. T., Jr.: Agammaglobulinemia, neutropenia, fever, and abdominal pain. J. Allergy Clin. Immun. 51: 308-318, 1973.

Bykowsky, M. J.; Haire, R. N.; Ohta, Y.; Tang, H.; Sung, S.-S. J.; Veksler, E. S.; Greene, J. M.; Fu, S. M.; Litman, G. W.; Sullivan, K. E.: Discordant phenotype in siblings with X-linked agammaglobulinemia. Am. J. Hum. Genet. 58: 477-483, 1996.

Cheng, G.; Ye, Z.-S.; Baltimore, D.: Binding of Bruton's tyrosine kinase to Fyn, Lyn, or Hck through a Src homology 3 domain-mediated interaction. Proc. Nat. Acad. Sci. 91: 8152-8155, 1994.

Cohen, D. I.; Hedrick, S. M.; Nielsen, E. A.; D'Eustachio, P.; Ruddle, F.; Steinberg, A. D.; Paul, W. E.; Davis, M. M.: Isolation of a cDNA clone corresponding to an X-linked gene family (XLR) closely linked to the murine immunodeficiency disorder xid. Nature 314: 372-374, 1985.

Cohen, D. I.; Steinberg, A. D.; Paul, W. E.; Davis, M. M.: Expression of an X-linked gene family (XLR) in late-stage B cells and its alteration by the xid mutation.. Nature 314: 372-374, 1985.

Conley, M. E.; Brown, P.; Pickard, A. R.; Buckley, R. H.; Miller, D. S.; Raskind, W. H.; Singer, J. W.; Fialkow, P. J.: Expression of the gene defect in X-linked agammaglobulinemia. New Eng. J. Med. 315: 564-567, 1986.

Conley, M. E.; Fitch-Hilgenberg, M. E.; Cleveland, J. L.; Parolini, O.; Rohrer, J.: Screening of genomic DNA to identify mutations in the gene for Bruton's tyrosine kinase. Hum. Molec. Genet. 3: 1751-1756, 1994.

Conley, M. E.; Mathias, D.; Treadaway, J.; Minegishi, Y.; Rohrer, J.: Mutations in Btk in patients with presumed X-linked agammaglobulinemia. Am. J. Hum. Genet. 62: 1034-1043, 1998.

Conley, M. E.; Puck, J. M.: Carrier detection in typical and atypical X-linked agammaglobulinemia. J. Pediat. 112: 688-694, 1988.

Desiderio, S.: Becoming B cells. (Abstract) Nature 361: 202, 1993.

de Weers, M.; Mensink, R. G. J.; Kraakman, M. E. M.; Schuurman, R. K. B.; Hendriks, R. W.: Mutation analysis of the Bruton's tyrosine kinase gene in X-linked agammaglobulinemia: identification of a mutation which affects the same codon as is altered in immunodeficient xid mice. Hum. Molec. Genet. 3: 161-166, 1994.

Drabek, D.; Raguz, S.; De Wit, T. P. M.; Dingjan, G. M.; Savelkoul, H. F. J.; Grosveld, F.; Hendriks, R. W.: Correction of the X-linked immunodeficiency phenotype by transgenic expression of human Bruton tyrosine kinase under the control of the class II major histocompatibility complex Ea locus control region. Proc. Nat. Acad. Sci. 94: 610-615, 1997.

Duriez, B.; Duquesnoy, P.; Dastot, F.; Bougneres, P.; Amselem, S.; Goossens, M.: An exon-skipping mutation in the btk gene of a patient with X-linked agammaglobulinemia and isolated growth hormone deficiency. FEBS Lett. 346: 165-170, 1994.

Edwards, N. L.; Magilavy, D. B.; Cassidy, J. T.; Fox, I. H.: Lymphocyte ecto-5-prime-nucleotidase deficiency in agammaglobulinemia. Science 201: 628-630, 1978.

Erlendsson, K.; Swartz, T.; Dwyer, J. M.: Successful reversal of echovirus encephalitis in X-linked hypogammaglobulinemia by intraventricular administration of immunoglobulin. New Eng. J. Med. 312: 351-353, 1985.

Fearon, E. R.; Winkelstein, J. A.; Civin, C. I.; Pardoll, D. M.; Vogelstein, B.: Carrier detection in X-linked agammaglobulinemia by analysis of X-chromosome inactivation. New Eng. J. Med. 316: 427-431, 1987.

Garvie, J. M.; Kendall, A. C.: Congenital agammaglobulinaemia: report of two further cases. Brit. Med. J. 1: 548-550, 1961.

Geha, R. S.; Rosen, F. S.; Merler, E.: Identification and characterization of subpopulations of lymphocytes in human peripheral blood after fractionation on discontinuous gradients of albumin: the cellular defect in X-linked agammaglobulinemia. J. Clin. Invest. 52: 1726-1734, 1973.

Gitlin, D.; Craig, J. M.: The thymus and other lymphoid tissues in congenital agammaglobulinemia. I. Thymic alymphoplasia and lymphocytic hypoplasia and their relation to infection. Pediatrics 32: 517-530, 1963.

Guioli, S.; Arveiler, B.; Bardoni, B.; Notarangelo, L. D.; Panina, P.; Duse, M.; Ugazio, A.; Oberle, I.; de Saint Basile, G.; Mandel, J. L.; Camerino, G.: Close linkage of probe p212 (DXS178) to X-linked agammaglobulinemia. Hum. Genet. 84: 19-21, 1989.

Hagemann, T. L.; Chen, Y.; Rosen, F. S.; Kwan, S.-P.: Genomic organization of the Btk gene and exon scanning for mutations in patients with X-linked agammaglobulinemia. Hum. Molec. Genet. 3: 1743-1749, 1994.

Hagemann, T. L.; Rosen, F. S.; Kwan, S.-P.: Characterization of germline mutations of the gene encoding Bruton's tyrosine kinase in families with X-linked agammaglobulinemia. Hum. Mutat. 5: 296-302, 1995.

Hendriks, R. W.; Mensink, E. J. B. M.; Kraakman, M. E. M.; Thompson, A.; Schuurman, R. K. B.: Evidence for male X chromosomal mosaicism in X-linked agammaglobulinemia. Hum. Genet. 83: 267-270, 1989.

Hitzig, W. H.; Willi, H.: Hereditary lymphoplasmocytic dysgenesis ('alymphocytose mit agammaglobulinamia'). Schweiz. Med. Wschr. 91: 1625-1633, 1961.

Janeway, C. A.; Apt, L.; Gitlin, D.: Agammaglobulinemia. Trans. Assoc. Am. Phys. 66: 200-202, 1953.

Janeway, C. A.; Gitlin, D.; Craig, J. M.; Grice, D. C.: 'Collagen disease' in patients with congenital agammaglobulinemia. Trans. Assoc. Am. Phys. 69: 93-97, 1956.

Jones, A.; Bradley, L.; Alterman, L.; Tarlow, M.; Thompson, R.; Kinnon, C.; Morgan, G.: X linked agammaglobulinaemia with a 'leaky' phenotype. Arch. Dis. Child. 74: 548-549, 1996.

Journet, O.; Durandy, A.; Doussau, M.; Le Deist, F.; Couvreur, J.; Griscelli, C.; Fischer, A.; de Saint-Basile, G.: Carrier detection and prenatal diagnosis of X-linked agammaglobulinemia. Am. J. Med. Genet. 43: 885-887, 1992.

Kerstens, P. J. S. M.; Endtz, H. P.; Meis, J. F. G. M.; Oyen, W. J. G.; Koopman, R. J. I.; van den Broek, P. J.; van der Meer, J. W. M.: Erysipelas-like skin lesions associated with Campylobacter jejuni septicemia in patients with hypogammaglobulinemia. Europ. J. Clin. Microbiol. Infect. Dis. 11: 842-847, 1992.

Kobayashi, S.; Iwata, T.; Saito, M.; Iwasaki, R.; Matsumoto, H.; Naritaka, S.; Kono, Y.; Hayashi, Y.: Mutations of the Btk gene in 12 unrelated families with X-linked agammaglobulinemia in Japan. Hum. Genet. 97: 424-430, 1996.

Kornfeld, S. J.; Kratz, J.; Haire, R. N.; Litman, G. W.; Good, R. A.: X-linked agammaglobulinemia presenting as transient hypogammaglobulinemia of infancy. J. Allergy Clin. Immun. 95: 915-917, 1995.

Kwan, S.-P.; Kunkel, L.; Bruns, G.; Wedgwood, R. J.; Latt, S.; Rosen, F. S.: Mapping of the X-linked agammaglobulinemia locus by use of restriction fragment-length polymorphism. J. Clin. Invest. 77: 649-652, 1986.

Kwan, S.-P.; Terwilliger, J.; Parmley, R.; Raghu, G.; Sandkuyl, L. A.; Ott, J.; Ochs, H.; Wedgwood, R.; Rosen, F.: Identification of a closely linked DNA marker, DXS178, to further refine the X-linked agammaglobulinemia locus. Genomics 6: 238-242, 1990.

Landreth, K. S.; Engelhard, D.; Anasetti, C.; Kapoor, N.; Kincade, P. W.; Good, R. A.: Pre-B cells in agammaglobulinemia: evidence for disease heterogeneity among affected boys. J. Clin. Immun. 5: 84-89, 1985.

Lau, Y. L.; Levinsky, R. J.; Malcolm, S.; Goodship, J.; Winter, R.; Pembrey, M.: Genetic prediction in X-linked agammaglobulinaemia. Am. J. Med. Genet. 31: 437-448, 1988.

Lederman, H. M.; Winkelstein, J. A.: X-linked agammaglobulinemia: an analysis of 96 patients. Medicine 64: 145-156, 1985.

Levitt, D.; Ochs, H.; Wedgwood, R. J.: Epstein-Barr virus-induced lymphoblastoid cell lines derived from the peripheral blood of patients with X-linked agammaglobulinemia can secrete IgM. J. Clin. Immun. 4: 143-150, 1984.

Malcolm, S.; de Saint Basile, G.; Arveiler, B.; Lau, Y. L.; Szabo, P.; Fischer, A.; Griscelli, C.; Debre, M.; Mandel, J. L.; Callard, R. E.; Robertson, M. E.; Goodship, J. A.; Pembrey, M. E.; Levinsky, R. J.: Close linkage of random DNA fragments from Xq21.3-22 to X-linked agammaglobulinaemia (XLA). Hum. Genet. 77: 172-174, 1987.

Malcolm, S.; Goodship, J.; Read, A. P.; Donnai, D.; Levinsky, R. J.: Linkage of Bruton's agammaglobulinemia (IMD1) to probes in Xq21.3-22: evidence for a second gene coding for X linked agammaglobulinemia. (Abstract) Cytogenet. Cell Genet. 51: 1038, 1989.

Marshall-Clarke, S.; Cooke, A.; Hutchings, P. R.: Deficient production of anti-red cell autoantibodies by mice with an X-linked B-lymphocyte defect. Europ. J. Immun. 9: 820-823, 1979.

Mease, P. J.; Ochs, H. D.; Wedgwood, R. J.: Successful treatment of echovirus meningoencephalitis and myositis-fasciitis with intravenous immune globulin therapy in a patient with X-linked agammaglobulinemia. New Eng. J. Med. 304: 1278-1281, 1981.

Melamed, I.; Bujanover, Y.; Igra, Y. S.; Schwartz, D.; Zakuth, V.; Spirer, Z.: Campylobacter enteritis in normal and immunodeficient children. Am. J. Dis. Child. 137: 752-753, 1983.

Mensink, E. J. B. M.; Schot, J. D. L.; Tippett, P.; Ott, J.; Schuurman, R. K. B.: X-linked agammaglobulinemia and the red blood cell determinants Xg and 12E7 are not closely linked. Hum. Genet. 68: 303-309, 1984.

Mensink, E. J. B. M.; Thompson, A.; Schot, J. D. L.; Kraakman, M. E. M.; Sandkuyl, L. A.; Schuurman, R. K. B.: Genetic heterogeneity in X-linked agammaglobulinemia complicates carrier detection and prenatal diagnosis. Clin. Genet. 31: 91-96, 1987.
Mensink, E. J. B. M.; Thompson, A.; Schot, J. D. L.; van de Greef, W. M. M.; Sandkuyl, L. A.; Schuurman, R. K. B.: Mapping of a gene for X-linked agammaglobulinemia and evidence for genetic heterogeneity. Hum. Genet. 73: 327-332, 1986.
Minegishi, Y.; Coustan-Smith, E.; Wang, Y.-H.; Cooper, M. D.; Campana, D.; Conley, M. E.: Mutations in the human lambda-5/14.1 gene result in B cell deficiency and agammaglobulinemia. J. Exp. Med. 187: 71-77, 1998.
Ochs, H. D.; Smith, C. I. E.: X-linked agammaglobulinemia: a clinical and molecular analysis. Medicine 75: 287-299, 1996.
Oeltjen, J. C.; Liu, X.; Lu, J.; Allen, R. C.; Muzny, D.; Belmont, J. W.; Gibbs, R. A.: Sixty-nine kilobases of contiguous human genomic sequence containing the alpha-galactosidase A and Bruton's tyrosine kinase loci. Mammalian Genome 6: 334-338, 1995.
Ohta, Y.; Haire, R. N.; Litman, R. T.; Fu, S. M.; Nelson, R. P.; Kratz, J.; Kornfeld, S. J.; de la Morena, M.; Good, R. A.; Litman, G. W.: Genomic organization and structure of Bruton agammaglobulinemia tyrosine kinase: localization of mutations associated with varied clinical presentations and course in X chromosome-linked agammaglobulinemia. Proc. Nat. Acad. Sci. 91: 9062-9066, 1994.
Ott, J.; Mensink, E. J. B. M.; Thompson, A.; Schot, J. D. L.; Schuurman, R. K. B.: Heterogeneity in the map distance between X-linked agammaglobulinemia and a map of nine RFLP loci. Hum. Genet. 74: 280-283, 1986.
Parolini, O.; Hejtmancik, J. F.; Allen, R. C.; Belmont, J. W.; Lassiter, G. L.; Henry, M. J.; Barker, D. F.; Conley, M. E.: Linkage analysis and physical mapping near the gene for X-linked agammaglobulinemia at Xq22. Genomics 15: 342-349, 1993.
Pearl, E. R.; Vogler, L. B.; Okos, A. J.; Crist, W. M.; Lawton, A. R., III; Cooper, M. D.: B-lymphocyte precursors in human bone marrow: an analysis of normal individuals and patients with antibody-deficient states. J. Immun. 120: 1169-1175, 1978.
Perryman, L. E.; McGuire, T. C.; Banks, K. L.: Infantile X-linked agammaglobulinemia: agammaglobulinemia in horses. Am. J. Path. 111: 125-127, 1983.
Race, R.; Sanger, R.: Blood Groups in Man. Oxford: Blackwell (pub.) (6th ed.): 1975. Pp. 601 only.
Rawlings, D. J.; Saffran, D. C.; Tsukada, S.; Largaespada, D. A.; Grimaldi, J. C.; Cohen, L.; Mohr, R. N.; Bazan, J. F.; Howard, M.; Copeland, N. G.; Jenkins, N. A.; Witte, O. N.: Mutation of unique region of Bruton's tyrosine kinase in immunodeficient xid mice. Science 261: 358-361, 1993.
Rawlings, D. J.; Witte, O. N.: Bruton's tyrosine kinase is a key regulator in B-cell development. Immun. Rev. 138: 105-119, 1994.
Rohrer, J.; Parolino, O.; Belmont, J. W.; Conley, M. E.: The genomic structure of human BTK, the defective gene in X-linked agammaglobulinemia. Immunogenetics 40: 319-324, 1994.
Rosen, F. S.; Cooper, M. D.; Wedgwood, R. J. P.: The primary immunodeficiencies. New Eng. J. Med. 311: 235-242 and 300-310, 1984.
Saffran, D. C.; Parolini, O.; Fitch-Hilgenberg, M. E.; Rawlings, D. J.; Afar, D. E. H.; Witte, O. N.; Conley, M. E.: A point mutation in the SH2 domain of Bruton's tyrosine kinase in atypical X-linked agammaglobulinemia. New Eng. J. Med. 330: 1488-1491, 1994.
Saulsbury, F. T.; Bernstein, M. T.; Winkelstein, J. A.: Pneumocystis carinii pneumonia as the presenting infection in congenital hypogammaglobulinemia. J. Pediat. 95: 559-561, 1979.
Scher, I.; Steinberg, A. D.; Berning, A. K.; Paul, W. E.: X-linked B-lymphocyte immune defect in CBA-N mice. II. Studies of the mechanisms underlying the immune defect. J. Exp. Med. 142: 637-650, 1975.
Schuurman, R. K. B.; Mensink, E. J. B. M.; Sandkuyl, L. A.; Post, E. D. M.; van Velzen-Blad, H.: Early diagnosis in X-linked agammaglobulinaemia. Europ. J. Pediat. 147: 93-95, 1988.
Schwaber, J.: Evidence for failure of V(D)J recombination in bone marrow pre-B cells from X-linked agammaglobulinemia. J. Clin. Invest. 89: 2053-2059, 1992.
Schwaber, J.; Chen, R. H.: Premature termination of variable gene rearrangement in B lymphocytes from X-linked agammaglobulinemia. J. Clin. Invest. 81: 2004-2009, 1988.
Schwaber, J.; Koenig, N.; Girard, J.: Correction of the molecular defect in B lymphocytes from X-linked agammaglobulinemia by cell fusion. J. Clin. Invest. 82: 1471-1476, 1988.
Schwaber, J.; Molgaard, H.; Orkin, S. H.; Gould, H. J.; Rosen, F. S.: Early pre-B cells from normal and X-linked agammaglobulinaemia produce C(mu) without an attached V(H) region. Nature 304: 355-358, 1983.
Schwaber, J.; Payne, J.; Chen, R.: B lymphocytes from X-linked agammaglobulinemia: delayed expression of light chain and demonstration of lyonization in carriers. J. Clin. Invest. 81: 514-522, 1988.
Seligmann, M.; Fudenberg, H. H.; Good, R. A.: A proposed classification of primary immunologic deficiencies. (Editorial) Am. J. Med. 45: 817-825, 1968.
Thomas, J. D.; Sideras, P.; Smith, C. I. E.; Vorechovsky, I.; Chapman, V.; Paul, W. E.: Colocalization of X-linked agammaglobulinemia and X-linked immunodeficiency genes. Science 261: 355-358, 1993.
Thompson, L. F.; Boss, G. R.; Spiegelberg, H. L.; Bianchino, A.; Seegmiller, J. E.: Ecto-5'-nucleotidase activity in lymphoblastoid cell lines derived from heterozygotes for congenital X-linked agammaglobulinemia. J. Immun. 125: 190-193, 1980.
Thorsteinsson, L.; Ogmundsdottir, H. M.; Sigfusson, A.; Arnason, A.; Eyjolfsson, G.; Jensson, O.: The first Icelandic family with X-linked agammaglobulinaemia: studies of genetic markers and immune function. Scand. J. Immun. 32: 273-280, 1990.
Tsukada, S.; Saffran, D. C.; Rawlings, D. J.; Parolini, O.; Allen, R. C.; Klisak, I.; Sparkes, R. S.; Kubagawa, H.; Mohandas, T.; Quan, S.; Belmont, J. W.; Cooper, M. D.; Conley, M. E.; Witte, O. N.: Deficient expression of a B cell cytoplasmic tyrosine kinase in human X-linked agammaglobulinemia. Cell 72: 279-290, 1993.
Uckun, F. M.; Waddick, K. G.; Mahajan, S.; Jun, X.; Takata, M.; Bolen, J.; Kurosaki, T.: BTK as a mediator of radiation-induced apoptosis in DT-40 lymphoma B cells. Science 273: 1096-1099, 1996.
van der Meer, J. W. M.; Mouton, R. P.; Daha, M. R.; Schuurman, R. K. B.: Campylobacter jejuni bacteraemia as a cause of recurrent fever in a patient with hypogammaglobulinaemia. J. Infect. 12: 235-239, 1986.
van der Meer, J. W. M.; Weening, R. S.; Schellekens, P. T. A.; van Munster, I. P.; Nagengast, F. M.: Colorectal cancer in patients with X-linked agammaglobulinaemia. Lancet 341: 1439-1440, 1993.
Vetrie, D.; Vorechovsky, I.; Sideras, P.; Holland, J.; Davies, A.; Flinter, F.; Hammarstrom, L.; Kinnon, C.; Levinsky, R.; Bobrow, M.; Smith, C. I. E.; Bentley, D. R.: The gene involved in X-linked agammaglobulinaemia is a member of the src family of protein-tyrosine kinases. Nature 361: 226-233, 1993.
Vihinen, M.; Iwata, T.; Kinnon, C.; Kwan, S.-P.; Ochs, H. D.; Vorechovsky, I.; Smith, C. I. E.: BTKbase, mutation database for X-linked agammaglobulinemia (XLA). Nucleic Acids Res. 24: 160-165, 1996.
Vihinen, M.; Vetrie, D.; Maniar, H. S.; Ochs, H. D.; Zhu, Q.; Vorechovsky, I.; Webster, A. D. B.; Notarangelo, L. D.; Nilsson, L.; Sowadski, J. M.; Smith, C. I. E.: Structural basis for chromosome X-linked agammaglobulinemia: a tyrosine kinase disease. Proc. Nat. Acad. Sci. 91: 12803-12807, 1994.
Vorechovsky, I.; Luo, L.; Hertz, J. M.; Froland, S. S.; Klemola, T.; Fiorini, M.; Quinti, I.; Paganelli, R.; Ozsahin, H.; Hammarstrom, L.; Webster, A. D. B.; Smith, C. I. E.: Mutation pattern in the Bruton's tyrosine kinase gene in 26 unrelated patients with X-linked agammaglobulinemia. Hum. Mutat. 9: 418-425, 1997.
Vorechovsky, I.; Zhou, J.-N.; Hammarstrom, L.; Smith, C. I. E.; Thomas, J. D.; Paul, W. E.; Notarangelo, L. D.; Bernatowska-Matuszkiewicz, E.: Absence of xid mutation in X-linked agammaglobulinaemia. (Letter) Lancet 342: 552, 1993.
Vorechovsky, I.; Zhou, J.-N.; Vetrie, D.; Bentley, D.; Bjorkander, J.; Hammarstrom, L.; Smith, C. I. E.: Molecular diagnosis of X-linked agammaglobulinaemia. (Letter) Lancet 341: 1153, 1993.
Yel, L.; Minegishi, Y.; Coustan-Smith, E.; Buckley, R. H.; Trubel, H.; Pachman, L. M.; Kitchingman, G. R.; Campana, D.; Rohrer, J.; Conley, M. E.: Mutations in the mu heavy-chain gene in patients with agammaglobulinemia. New Eng. J. Med. 335: 1486-1493, 1996.
Zhu, Q.; Zhang, M.; Winkelstein, J.; Chen, S.-H.; Ochs, H. D.: Unique mutations of Bruton's tyrosine kinase in fourteen unrelated X-linked agammaglobulinemia families. Hum. Molec. Genet. 3: 1899-1900, 1994.

AGAMMAGLOBULINEMIA, X-LINKED, TYPE 2 (*300310)

JJM NOTE

Although a different gene, there is no reason to think the risk for malignancy is any different from that of Bruton agammaglobulinemia.

In linkage studies, Mensink et al. (1986) found evidence suggesting at least 2 types of X-linked agammaglobulinemia: one located at Xq21.3-q22 and a second not linked to the marker in this area and possibly located in the Xp22 band; see 300300.

Mensink, E. J. B. M.; Thompson, A.; Schot, J. D. L.; van de Greef, W. M. M.; Sandkuyl, L. A.; Schuurman, R. K. B.: Mapping of a gene for X-linked agammaglobulinemia and evidence for genetic heterogeneity. Hum. Genet. 73: 327-332, 1986.

AGAMMAGLOBULINEMIA, SWISS TYPE (#300400)

JJM NOTE

The Immunodeficiency-Cancer Registry of the University of Minnesota (Spector BD et al.: Genetically determined immunodeficiency diseases (GDID) and malignancy: report from the immunodeficiency cancer registry. Clin. Immunol. Immunopathol. 11:12–29, 1978) tallied 12 cases: 3 malignant reticuloendotheliosis; 2 histiocytic lymphoma; 1 each of lymphatic leukemia, lymphocytic lymphoma, and Hodgkin disease; and 4 other lymphoreticular malignancies. Swiss type was not specific, but, rather, "severe combined immunodeficiency"; hence, the number is uncertain.

A number sign (#) is used with this entry because of the demonstration that the defect resides in the gene for interleukin-2 receptor, gamma chain (IL2RG; 308380); see review by Leonard et al. (1994). This disorder, now designated X-linked SCID, was earlier referred to as

Swiss-type agammaglobulinemia, thymic epithelial hypoplasia, or immunodeficiency 4.

This type of immunodeficiency differs from the Bruton type (300300) of agammaglobulinemia by the presence of lymphocytopenia ('alymphocytosis'), earlier age at death, vulnerability to viral and fungal as well as bacterial infections, lack of delayed hypersensitivity, atrophy of the thymus, and lack of benefit by gamma globulin administration. Severe combined immunodeficiency, as it came to be known, was inherited as an autosomal recessive in the cases in which it was first described in Switzerland by Hitzig and Willi (1961) (see 202500), but all cases in 3 families studied by Rosen et al. (1966) were male and 1 kindred (family T) had 9 affected males in 5 sibships in 3 generations connected through females in a typical X-linked recessive pedigree pattern. Indeed, Gitlin and Craig (1963) reported 15 boys with hypogammaglobulinemia and noted that they could be divided into 2 groups of almost equal size based on their clinical course. The first group had onset of infections early in life, often before 3 months of age. Their course was characterized by lymphopenia and persistent pneumonitis, moniliasis, and frequent rashes. This disorder was uniformly fatal in infancy even in children treated with gammaglobulin. Autopsy showed an abnormally small thymus with thymic alymphoplasia. The second group of patients had onset of infections somewhat later, usually between 6 and 18 months of age. Infection was intermittent rather than persistent, and gamma globulin was clinically useful. These patients did not have lymphopenia, and in those who died, the thymus was not found to be small although lymph nodes lacked germinal follicles and plasma cells. About half the patients in each group had a family history of severe infections in male relatives. The first group would be known now to have X-linked severe combined immunodeficiency and the second group X-linked agammaglobulinemia of Bruton. Miller and Schieken (1967) suggested that one form of thymic dysplasia is X-linked. An impressive pedigree with 6 affected males in 3 generations was published by Dooren et al. (1968), who, following the recommendations of a workshop on immunological deficiency diseases in man (Sanibel Island, Fort Myers, Fla., Feb. 1-5, 1967), called the condition thymic epithelial hypoplasia. In the same workshop Rosen et al. (1968) pointed out that a difference from the autosomal recessive is less profound lymphocytopenia.

Yount et al. (1978) studied a child with the X-linked form. Adenosine deaminase and nucleoside phosphorylase levels were normal. The patient showed (1) virtual absence of lymphocytes capable of rosetting with sheep red blood cells, absent skin tests and lack of in vitro responses to mitogens, antigens or allogeneic cells and (2) profound humoral immunodeficiency despite a plethora of B lymphocytes. Apparently, the latter cells suffered from an inability to undergo terminal differentiation into plasma cells capable of synthesizing and secreting immunoglobulins. A brother of the patient they studied died at 10 months of Pneumocystis carinii pneumonia complicated by disseminated influenza infection (Hong Kong strain). Autopsy showed hypoplastic thymus without epithelial corpuscles and absence of germinal centers in lymph nodes and bowel lamina propria. In 2 unrelated males with SCID, Conley et al. (1984) found that T cells demonstrated a typical XX female karyotype and were probably of maternal origin, whereas the B cells had an XY male karyotype. They suggested that this may represent maternal lymphoid engraftment and that the SCID in these patients was the result of graft-versus-host disease. Since this would presumably affect only males, repetition in the family would simulate X-linked recessive inheritance.

19,20:Puck et al. (1986, 1987) showed that carriers for X-linked SCID can be detected; when a carrier's T cells are fused with rodent cells, the resultant hybrids have RFLP markers of only one X chromosome, i.e., the X chromosome not carrying the mutant gene. Thus, those cells in the T-cell lineage (the cell line affected in XSCID) that have the mutant-bearing X chromosome as the active one do not survive just as they do not survive in the hemizygous affected male. Puck et al. (1992) found that X-inactivation ratios determined in a control group of noncarrier women exhibited a wide range; 20 to 86% of T cells had the paternal X active. The maximum-likelihood analysis of these data suggested that, in humans, mature T cells are derived from a pool of only about 10 randomly inactivated stem cells. Despite the wide variability in normal X-inactivation ratios, X-inactivation in XSCID carriers appeared to be far more markedly skewed. Therefore, a maximum-likelihood odds-ratio test was developed and proved to be successful in predicting the carrier status of women in such pedigrees.

Wengler et al. (1993) demonstrated that all 4 lymphoid cell populations studied (NK cells, B cells, CD4+ cells, CD8+ cells) from 3 heterozygous women exhibited exclusive use of a single X as the active X, whereas both X chromosomes were used as the active X in neutrophils and monocytes. The study was done by means of a PCR technique based on 2 observations: that the active and inactive X chromosomes differ in methylation and that throughout the genome there are highly polymorphic sites consisting of sequences of 2-to-5 nucleotides that are repeated a variable number of times. The first exon of the androgen receptor gene (AR; 313700) on Xq contains a site that is both methylated on the inactive X and unmethylated on the active X and has within 100 nucleotides a highly polymorphic trinucleotide repeat. After digesting DNA with the methylation-sensitive enzyme HpaII, primers that flanked both the polymorphic site and the methylation site were used to amplify the intervening DNA. If there was random X-chromosome inactivation, then a portion of both the maternally and the paternally derived X chromosomes would be intact after the HpaII digestion, and both alleles would be amplified. However, if there was nonrandom X-chromosome inactivation, the allele on the active X chromosome would be completely digested and only the other allele would be amplified.

The importance of XSCID is indicated by the fact that about 50% of SCID in the United States is of this variety. About 15% is due to ADA deficiency (102700) or nucleoside phosphorylase deficiency (164050). About 35% of SCID is autosomal recessive, cause unknown; see 202500. The manifestations of XSCID are persistent infection from the age of about 6 months, absent mitogenic response, hypogammaglobulinemia, absent or severely depressed T cells, and normal or somewhat increased numbers of B cells. In inherited cases it is the paternal X that is present in all T cells.

Conley et al. (1990) pointed out that over 80% of infants with SCID of unknown genetic etiology are males, yet less than a third of these affected males have a family history of X-linked disease. To help identify new mutations for SCIDX, they studied the X-chromosome inactivation patterns in T cells from 16 women who had sons with sporadic SCID. By analysis of human/hamster hybrids that selectively retained the active human X chromosome and use of an X-linked RFLP for which the woman in question was heterozygous, they showed exclusive use of a single X as the active X in T-cell hybrids from 7 of the 16 women, identifying these as carriers of the disorder. Studies on additional family members confirmed the mutant nature of the inactive X and showed the source of the new mutation in 3 of the families. The clinical records of the males with proved X-linked SCID were compared with the others. The most consistent finding in 21 patients with X-linked SCID was an elevated proportion of B cells.

De Saint Basile et al. (1987) mapped the X-linked SCID locus to Xq11-q13 by linkage analysis with RFLPs. No recombination was observed with marker DXS159. According to Mensink and Schuurman (1987), J. L. Mandel found close linkage with the DXS159 marker at Xq12-q13 in 6 pedigrees. They also indicated that there may be more than one X-linked SCID locus because there is immunologic heterogeneity. Puck et al. (1988) found linkage with loci in Xq12-q21.3, but concluded that the exact localization remained uncertain and that heterogeneity might exist. Puck et al. (1989) performed linkage analysis in 6 kindreds using a random pattern of T-cell X-inactivation to rule out the carrier state in at-risk women. Their findings, combined with analysis of Xq interstitial deletions, permitted assignment of the locus to Xq13.1-q21.1 and defined flanking markers for prenatal diagnosis and carrier testing. Smead et al. (1989) found no recombination among SCID, PGK1, and DXS72; DXS72 is known to be distal to SCID because males with normal immunity have been described with Xq21 interstitial deletions involving DXS72. DXS159 and DXS3 appeared to be flanking markers for SCID. Goodship et al. (1989) demonstrated no recombination between IMD4 and DXS159, PGK1 (311800), or DXS72; the maximum lod score for linkage to PGK1 was 5.03.

Goodship et al. (1988) showed a nonrandom pattern of X-chromosome inactivation in T-lymphocyte DNA of 2 obligate carriers of XSCID. This was done by examining a differential pattern of methylation by the method of Vogelstein et al. (1987). The method was used further to distinguish the autosomal recessive and X-linked forms of the disease and to demonstrate carrier status in the mother of a sporadic case. To determine if the failure of B cells to produce immunoglobulin might be due to expression of the XSCID gene defect in B-lineage cells as well as T cells, Conley et al. (1988) analyzed patterns of X-chromosome inactivation in B cells from 9 obligate carriers of this disorder. A series of somatic cell hybrids that selectively retained the active X chromosome because of deficiency of HPRT in the Chinese hamster fibroblast line was produced from Epstein-Barr virus-stimulated B cells from each of 9 obligate carriers. To distinguish between the 2 X chromosomes, the hybrids from each woman were analyzed using an X-linked RFLP for which the woman in question was heterozygous. All obligate carriers showed preferential use of a single X chromosome in the B-cell hybrids; namely, all had the nonmutant X chromosome as the active X. The authors showed further that the small number of B-cell hybrids that contained the mutant X were derived from an immature subset of B cells; lymphocytes from 3 carriers were separated into surface IgM positive and surface IgM negative B cells before exposure to EBV and production of B-cell hybrids. The results indicate that the XSCID gene product has a role in B cells as well as T cells and is required during B-cell maturation. By the study of X-chromosome inactivation patterns, Goodship et al. (1991) demonstrated that the mutation is expressed in B lymphocytes and in granulocytes as well as in T lymphocytes. They concluded that this disorder is not in a T-lymphocyte differentiation gene but rather in a metabolic pathway as in ADA deficiency and PNP deficiency. In 6 individuals in 2 sibships of a French family, de Saint-Basile et al. (1992) observed severe infections. The propositus, a 5-year-old boy, had severe and progressive T- and B-cell functional immunodeficiency. The mother and 1 sister showed nonrandom X-chromosome inactivation of T cells and, partially, of B cells but not of polymorphonuclear leukocytes, a pattern similar to that observed in X-linked SCID carriers. RFLP studies identified a haplotype segregating with the abnormal locus that may be localized in the proximal part of the long arm of the X chromosome. They thought that this might represent a new X-linked immunodeficiency, but an alternative explanation is that it represents an 'attenuated phenotype' of X-linked SCID. Hendriks et al. (1992) likewise raised the possibility of 2 distinct XSCID defects. They determined the pattern of X-chromosome inactivation in 14 females from 3 unrelated pedigrees with XSCID. Since all the females were heterozygous for the hypervariable DXS255 locus, differential methylation of this locus in peripheral blood cells was studied. In 2 pedigrees the granulocyte population of all carriers showed a random distribution of X-chromosome inactivation; in the third pedigree, the granulocytes of the 3 carriers who were analyzed manifested complete inactivation of the X chromosome that carried the XSCID mutation, exposing a selective disadvantage of granulocytes that express the XSCID defect. This pedigree-dependent difference in the involvement of granulocytes suggested the existence of 2 distinct XSCID defects.

By somatic cell hybrid analysis and methylation differences, Deschenes et al. (1994) demonstrated that female dogs carrying X-linked SCID have the same lymphocyte-limited skewed X-chromosome inactivation patterns as human carriers. Henthorn et al. (1994) demonstrated a 4-bp deletion in the first exon of the IL2RG gene, which precluded the production of a functional protein, as the basis of canine XSCID.

Flake et al. (1996) reported the successful treatment of a fetus with X-linked SCID by the in utero transplantation of paternal bone marrow that was enriched with hematopoietic cell progenitors. The mother had lost a previous son at 7 months of age to this disease. Studies of that child's DNA identified a splice-donor-site mutation at the +5 position of intron 6 of the IL2RG gene.

See also Rosen and Janeway (1966); Simar et al. (1972)

Conley, M. E.; Buckley, R. H.; Hong, R.; Guerra-Hanson, C.; Roifman, C. M.; Brochstein, J. A.; Pahwa, S.; Puck, J. M.: X-linked severe combined immunodeficiency: diagnosis in males with sporadic severe combined immunodeficiency and clarification of clinical findings. J. Clin. Invest. 85: 1548-1554, 1990.

Conley, M. E.; Lavoie, A.; Briggs, C.; Brown, P.; Guerra, C.; Puck, J. M.: Nonrandom X chromosome inactivation in B cells from carriers of X chromosome-linked severe combined immunodeficiency. Proc. Nat. Acad. Sci. 85: 3090-3094, 1988.

Conley, M. E.; Nowell, P. C.; Henle, G.; Douglas, S. D.: XX T cells and XY B cells in two patients with severe combined immune deficiency. Clin. Immun. Immunopath. 31: 87-95, 1984.

de Saint-Basile, G.; Le Deist, F.; Caniglia, M.; Lebranchu, Y.; Griscelli, C.; Fischer, A.: Genetic study of a new X-linked recessive immunodeficiency syndrome. J. Clin. Invest. 89: 861-866, 1992.

de Saint Basile, G.; Arveiler, B.; Oberle, I.; Malcolm, S.; Levinsky, R. J.; Lau, Y. L.; Hofker, M.; Debre, M.; Fischer, A.; Griscelli, C.; Mandel, J. L.: Close linkage of the locus for X chromosome-linked severe combined immunodeficiency to polymorphic DNA markers in Xq11-q13. Proc. Nat. Acad. Sci. 84: 7576-7579, 1987.

Deschenes, S. M.; Puck, J. M.; Dutra, A. S.; Somberg, R. L.; Felsburg, P. J.; Henthorn, P. S.: Comparative mapping of canine and human proximal Xq and genetic analysis of canine X-linked severe combined immunodeficiency. Genomics 23: 62-68, 1994.

Dooren, L. J.; de Vries, M. J.; van Bekkum, D. W.; Cleton, F. J.; de Koning, J.: Sex-linked thymic epithelial hypoplasia in two siblings: attempt at treatment by transplantation with fetal thymus and adult bone marrow. J. Pediat. 72: 51-62, 1968.

Flake, A. W.; Roncarolo, M.-G.; Puck, J. M.; Almeida-Porada, G.; Evans, M. I.; Johnson, M. P.; Abella, E. M.; Harrison, D. D.; Zanjani, E. D.: Treatment of X-linked severe combined immunodeficiency by in utero transplantation of paternal bone marrow. New Eng. J. Med. 335: 1806-1810, 1996.

Gitlin, D.; Craig, J. M.: The thymus and other lymphoid tissues in congenital agammaglobulinemia. I. Thymic alymphoplasia and lymphocytic hypoplasia and their relation to infection. Pediatrics 32: 517-530, 1963.

Goodship, J.; Levinsky, R.; Malcolm, S.: Linkage of PGK1 to X-linked severe combined immunodeficiency (IMD4) allows predictive testing in families with no surviving male. Hum. Genet. 84: 11-14, 1989.

Goodship, J.; Malcolm, S.; Lau, Y. L.; Pembrey, M. E.; Levinsky, R. J.: Use of X chromosome inactivation analysis to establish carrier status for X-linked severe combined immunodeficiency. Lancet I: 729-732, 1988.

Goodship, J.; Malcolm, S.; Levinsky, R. J.: Evidence that X-linked severe combined immunodeficiency is not a differentiation defect of T lymphocytes. Clin. Exp. Immun. 83: 4-9, 1991.

Hendriks, R. W.; Kraakman, M. E. M.; Schuurman, R. K. B.: X chromosome inactivation patterns in haematopoietic cells of female carriers of X-linked severe combined immunodeficiency determined by methylation analysis at the hypervariable DXS255 locus. Clin. Genet. 42: 114-121, 1992.

Henthorn, P. S.; Somberg, R. L.; Fimiani, V. M.; Puck, J. M.; Patterson, D. F.; Felsburg, P. J.: IL-2R-gamma gene microdeletion demonstrates that canine X-linked severe combined immunodeficiency is a homologue of the human disease. Genomics 23: 69-74, 1994.

Hitzig, W. H.; Willi, H.: Hereditary lymphoplasmocytic dysgenesis ('alymphocytose mit agammaglobulinamia'). Schweiz. Med. Wschr. 91: 1625-1633, 1961.

Leonard, W. J.; Noguchi, M.; Russell, S. M.; McBride, O. W.: The molecular basis of X-linked severe combined immunodeficiency: the role of the interleukin-2 receptor gamma-chain as a common gamma-chain, gamma-c. Immun. Rev. 138: 61-86, 1994.

Mensink, E. J. B. M.; Schuurman, R. K. B.: Immunodeficiency disease genes on the X chromosome. Dis. Markers 5: 129-140, 1987.

Miller, M. E.; Schieken, R. M.: Thymic dysplasia: a separable entity from 'Swiss agammaglobulinemia.'. Am. J. Med. Sci. 253: 741-750, 1967.

Puck, J. M.; Nussbaum, R. L.; Briggs, C.; Brown, P.; Conley, M. E.: Carrier detection in X-linked forms of agammaglobulinemia and severe combined immunodeficiency. (Abstract) Am. J. Hum. Genet. 39: A99 only, 1986.

Puck, J. M.; Nussbaum, R. L.; Conley, M. E.: Carrier detection in X-linked severe combined immunodeficiency based on patterns of X chromosome inactivation. J. Clin. Invest. 79: 1395-1400, 1987.

Puck, J. M.; Nussbaum, R. L.; Smead, D. L.; Conley, M. E.: Regional mapping of X-linked severe combined immunodeficiency to Xq13-21.3 using nonrandom X inactivation as a carrier test. (Abstract) Am. J. Hum. Genet. 43: A156 only, 1988.

Puck, J. M.; Nussbaum, R. L.; Smead, D. L.; Conley, M. E.: X-linked severe combined immunodeficiency: localization within the region Xq13.1-q21.1 by linkage and deletion analysis. Am. J. Hum. Genet. 44: 724-730, 1989.

Puck, J. M.; Stewart, C. C.; Nussbaum, R. L.: Maximum-likelihood analysis of human T-cell X chromosome inactivation patterns: normal women versus carriers of X-linked severe combined immunodeficiency. Am. J. Hum. Genet. 50: 742-748, 1992.

Rosen, F. S.; Craig, J. M.; Vawter, G. F.; Janeway, C. A.: The dysgammaglobulinemias and X-linked thymic hypoplasia.In: Bergsma, D.; Good, R. A.: Immunologic Deficiency Diseases in Man. New York: National Foundation (pub.) 1968. Pp. 67-70.

Rosen, F. S.; Gotoff, S. P.; Craig, J. M.; Ritchie, J.; Janeway, C. A.: Further observations on the Swiss type of agammaglobulinemia (alymphocytosis): the effect of syngeneic bone-marrow cells. New Eng. J. Med. 274: 18-21, 1966.

Rosen, F. S.; Janeway, C. A.: The gamma globulins. III. The antibody deficiency syndromes. New Eng. J. Med. 275: 709-715 and 769-775, 1966.

Simar, J.; Farriaux, J.-P.; Pauli, A.; Fournier, A.; Fontaine, G.: Le syndrome de Giltin ou deficit immunitaire primitif, mixte et severe de transmission recessive liee au sexe. Ann. Paediat. 19: 889-896, 1972.

Smead, D. L.; Conley, M. E.; Puck, J. M.: Improved regional localization of X-linked severe combined immunodeficiency of Xq13.1-q21.1. (Abstract) Am. J. Hum. Genet. 45 (suppl.): A162 only, 1989.

Vogelstein, B.; Fearon, E. R.; Hamilton, S. R.; Preisinger, A. C.; Willard, H. F.; Michelson, A. M.; Riggs, A. D.; Orkin, S. H.: Clonal analysis using recombinant DNA probes from the X-chromosome. Cancer Res. 47: 4806-4813, 1987.

Wengler, G. S.; Allen, R. C.; Parolini, O.; Smith, H.; Conley, M. E.: Nonrandom X chromosome inactivation in natural killer cells from obligate carriers of X-linked severe combined immunodeficiency. J. Immun. 150: 700-704, 1993.

Yount, W. J.; Utsinger, P. D.; Whisnant, J.; Folds, J. D.: Lymphocyte subpopulations in X-linked severe combined immunodeficiency (SCID): evidence against a stem cell defect; transformation response to calcium ionophore A23187. Am. J. Med. 65: 847-854, 1978.

Aplastic Disorders

FANCONI PANCYTOPENIA 1 (*227650)

JJM NOTE

Malignancies seen to excess are nonlymphocytic leukemia, especially acute myelomonocytic leukemia, squamous cell carcinoma of the esophagus, and hepatic adenomas and carcinomas. The latter may represent ecogenetic tumors because long-term anabolic steroids may interact with the mutant gene to cause liver tumors. On the other hand, most long-time survivors of Fanconi anemia must be on such steroids to prolong life.

Description

Type 1 Fanconi pancytopenia is an autosomal recessive disorder affecting all bone marrow elements and associated with cardiac, renal, and limb malformations as well as dermal pigmentary changes.

Clinical Features

All marrow elements are usually affected, resulting in anemia, leukopenia, and thrombopenia. Pigmentary changes in the skin and malformations of the heart, kidney, and limbs (aplasia of the radius, thumb deformity) are associated features. Leukemia is a fatal complication (Garriga and Crosby, 1959) and may occur in family members lacking full-blown features. Hagerman and Williams (1993) illustrated the characteristically short thumb and a cafe-au-lait spot in a patient with Fanconi anemia, together with cytogenetic studies showing chromatid fragments and a dicentric chromosome. Giampietro et al. (1993) pointed to the 'extreme clinical heterogeneity' among patients with Fanconi anemia. Giampietro et al. (1993) analyzed clinical data from 370 patients enrolled in the International Fanconi Anemia Registry. Of these, 220 (60%) represented probands with congenital malformations. In addition to short stature, cafe-au-lait spots, and radial-ray and renal malformations, affected patients presented with cardiac, gastrointestinal, central nervous system, and various skeletal abnormalities. Genital anomalies were common in male patients. Approximately 50% of the patients had radial-ray abnormalities, which ranged from bilateral absent thumbs and radii to a unilateral hypoplastic thumb or bifid thumb. Among the patients with congenital malformations, the diagnosis of Fanconi anemia was made in only 28% before the onset of hematologic manifestations. About one-third of all patients enrolled in the registry did not have congenital malformations; of these patients, 85% had at least one of the following: skin pigmentation abnormalities, microphthalmos, or height, weight, or head circumference in the lowest 5% for their age. Minor congenital anomalies were noted in approximately 20% of these patients.

Macdougall et al. (1990) described FA in 25 black African children seen in Johannesburg over an 11-year period. Seventeen (68%) of the children died during the period of observation. Leukemia was the terminal event in 2. Response to androgens was poor and most patients required regular transfusion. Mean age of death was 9.8 years and the mean time between diagnosis and death 2.3 years. This particular mutation predisposes to multiple chromosomal breaks (Schroeder et al., 1964; Bloom et al., 1966).

Thrombocytopenia-absent radius (274000) is a separate entity. Zaizov et al. (1969) described 2 sisters and a brother with pancytopenia similar to that of Fanconi anemia but without congenital malformations. Chromosomal changes similar to those of Fanconi anemia were present and patchy areas of hyperpigmentation were noted in 2 of the sibs.

Swift et al. (1974) concluded that male heterozygotes for Fanconi anemia have a risk of malignant neoplasm 3.4 times that of the general population.

Bloom syndrome (210900) is another single gene disorder accompanied by chromosomal breakage and predisposition to leukemia. Schroeder and German (1974) showed that aberrations were more numerous in Fanconi cells than in Bloom cells. In Bloom syndrome most interchanges were between homologous chromosomes, whereas in Fanconi anemia they were usually between nonhomologous chromosomes.

Observations by Li and Potter (1978) can be interpreted as indicating that Blackfan-Diamond hypoplastic anemia (205900) is allelic with Fanconi anemia: they reported typical Fanconi anemia in a close relative of the 5 sibs with hypoplastic anemia reported by Estren and Dameshek (1947). The parents of the Fanconi patient were second cousins and both were first cousins of the 5 sibs reported by Estren and Dameshek (1947). Li and Potter (1978) suggested that these 5 sibs may have been genetic compounds.

Welshimer and Swift (1982) studied families of homozygotes for ataxia-telangiectasia (AT; 208900), Fanconi anemia, and xeroderma pigmentosum (XP) to test the hypothesis that heterozygotes may be predisposed to some of the same congenital malformations and developmental disabilities that are common among homozygotes. Among XP relatives, 11 of 1,100 had unexplained mental retardation, whereas only 3 of 1,439 relatives of FA and AT homozygotes showed mental retardation. Four XP relatives and no FA or AT relatives had microcephaly. Idiopathic scoliosis and vertebral anomalies occurred in excess in AT relatives, while genitourinary and distal limb malformations were found in FA families. Considerable intergenic heterogeneity has been found in xeroderma pigmentosum and some in ataxia-telangiectasia.

Berkovitz et al. (1984) concluded that abnormal sexual development in Fanconi anemia represents hypergonadotropic hypogonadism.

De Vroede et al. (1982) observed simultaneous onset of pancytopenia in a brother and sister, 5 years apart in age, suggesting possible exposure to a common external agent. One of the patients showed ropalocytosis, i.e., club-shaped cell processes, affecting the erythropoietic series from basophilic erythroblasts to reticulocytes.

Auerbach et al. (1989) reported results from analysis of data in the International Fanconi Anemia Registry. They also concluded that hypersensitivity to the clastogenic effect of diepoxybutane (DEB) is a useful discriminator for FA. Their data indicated wider clinical diversity in this disorder than previously recognized.

According to Auerbach (1992), a review of all cases of FA reported to the International Fanconi Anemia Registry indicated that at least 15% manifested acute myelogenous leukemia (AML) or preleukemia. These patients usually have karyotypically abnormal bone marrow clones but do not exhibit chromosomal translocations involving breakpoints associated with specific oncogenes.

Alter (1993) gave a review in which she pointed out that Fanconi anemia should really be called Fanconi syndrome because the primary defect or defects are not hematopoietic, dermatologic, or orthopedic, but presumably related in some as yet undefined manner to DNA repair. However, the designation Fanconi syndrome is already used to describe a specific constellation of renal tubular dysfunction.

On the basis of complementation analysis (see later) of 47 FA patients from Europe and U.S./Canada, the following frequencies of the various subtypes were identified by Buchwald (1995): 31 were group A (66%), 2 were group B (4.3%), 6 were group C (12.7%), 2 were group D (4.3%), and 6 were group E (12.7%). The above data were compiled from several reports. Reporting for the European Fanconi Anaemia Research Group, Joenje (1996) found that among ethnically and clinically unselected FA patients from Germany and the Netherlands, FA-A was most prevalent in Germany (13/22, 59%), whereas in the Netherlands a majority of patients were FA-C (4/6, 67%).

Jakobs et al. (1997) determined the complementation group represented by each of 16 unrelated FA patients from North America. The majority of the patients belonged to FA complementation group A (69%), followed by FA-C (18%), FA-D (4%) and FA-B or FA-E (9%).

Biochemical Features

In 2 brothers and a third unrelated patient, Lohr et al. (1965) found marked reduction of red cell, leukocyte, and platelet hexokinase activity. This is apparently a different defect from the hexokinase deficiency (235700) that is limited to red cells and results in hemolytic anemia alone. A consistent defect in hexokinase cannot be considered as proved (Brunetti et al., 1966).

Joenje et al. (1979) found deficiency of red cell superoxide dismutase in Fanconi anemia by 2 independent methods of assay. The activity per antigenic unit and the electrophoretic mobility of the enzyme were normal, suggesting that the deficiency is due to a regulatory disturbance, not a mutation in the structural gene for the enzyme.

Other Features

The G2 phase of the cell cycle is very long in Fanconi anemia (Dutrillaux et al., 1982), a feature that might be used for diagnosis when no other manifestations are present (Schindler et al., 1985). Schindler et al. (1985) performed tests with BrdU-Hoechst flow cytometry, with the ratio of G2 to G1 as the measure. The results showed arrest at the G2 phase in lymphocytes. Similar findings have been found in ataxia-telangiectasia (Schindler et al., 1987). Heterozygotes for both conditions have intermediate values (Schinzel, 1991).

Inheritance

Using family data on Fanconi anemia, Rogatko and Auerbach (1988) tested a new method of segregation analysis when no information about mode of ascertainment is available. The results confirmed a monogenic autosomal recessive mode of inheritance.

Cytogenetics

Spontaneous chromosome breakage is a feature of FA. In a review of cytogenetic studies of FA-associated leukemias, Auerbach and Allen (1991) found a high frequency of monosomy 7 and duplications involving 1q. There were no occurrences of t(8;21), t(15;17), or abnormalities of 11q, which are associated with M2, M3, and M5 leukemias, respectively. The mean age of death in FA patients developing leukemia was 15 years.

Mapping

By linkage studies in families from the International Fanconi Anemia Registry which contained 2 or more affected offspring, one or more offspring from consanguineous marriages, or multiple affected children in collateral sibships, Auerbach et al. (1989) excluded the long arm of chromosome 19 as the location for the Fanconi anemia mutation. Chromosome 19q had been considered a candidate location for the gene because 3 DNA repair genes (126340, 126380, 194360) are located there. In a linkage study involving 34 families enrolled in the International Fanconi Anemia Registry, Mann et al. (1991) found a lod score of 3.04 at theta = 0.12 for linkage of FA to anonymous marker D20S20 on 20q. Tests with another marker, D20S19, which is highly polymorphic and tightly linked to D20S20, indicated linkage heterogeneity. In 2 of the largest kindreds studied, lod scores of 2.6 and 2.1 at theta = 0.001 were obtained, thus supporting a provisional assignment of the Fanconi anemia gene to chromosome 20q.

As a continuation of the complementation analysis of Duckworth-Rysiecki et al. (1985), Strathdee et al. (1992) constructed a panel of somatic cell hybrids using polyethylene glycol as a fusigen. Three new complementation groups were identified, designated FA(B), FA(C), and FA(D). The gene defective in FA(C) (FACC; 227645) was mapped to 9q22.3. Strathdee et al. (1992) suggested that there are at least 4 different FA genes, mutations at any one of which can lead to the FA phenotype.

Steinlein et al. (1992) pointed out that FA is in the same region as EEGV1 (130180) and BFNC (121200), i.e., 20q13.2-q13.3, because all are linked to the VNTR marker CMM6 (D20S19).

Pronk et al. (1995) concluded that the FA1 gene (which they called FAA) is, in fact, not on 20q which was excluded in their linkage studies, but rather linked to microsatellite markers on 16q24.3. Pronk et al. (1995) established a panel of families classified as Fanconi anemia, type A, by complementation analysis, and used them to search for the FAA gene by linkage analysis. Pronk et al. (1995) found strong evidence of allelic association between the disease and the marker D16S303 in the Afrikaner population of South Africa, indicating founder effect, which had been previously suggested by Rosendorff et al. (1987).

Gschwend et al. (1996) used homozygosity mapping in the study of 23 inbred families with Fanconi anemia and identified linkage to a locus near marker D16S520 at 16q24.3. Although approximately 65% of the families displayed clear linkage to D16S250, the authors found strong evidence of genetic heterogeneity. Family ascertainment was biased against the previously identified FAC gene on chromosome 9, and no linkage was observed to that locus. Simultaneous search analysis suggested to Gschwend et al. (1996) several additional chromosomal regions that could account for a small fraction of Fanconi anemia in the families studied, but the sample size was insufficient to provide statistical significance. They also demonstrated the strong effect of marker allele frequencies on lod scores obtained in homozygosity mapping and discussed ways to avoid false positives arising from this effect.

Based on linkage analysis and allelic association in Afrikaner FA families the Fanconi Anaemia/Breast Cancer Consortium (1996) refined the map interval for FAA to the region D16S3026-D16S303 (maximum lod = 3.15 at theta = 0.00 for D16S303) They used a positional cloning strategy to isolate a candidate cDNA for the FAA gene (see Molecular Genetics section).

Savoia et al. (1997) performed linkage analysis in 11 FA-A and 16 unclassified FA Italian families using microsatellite markers. All the families were consistent with linkage at 16q24, the largest lod score being observed with D16S1320. In 2 genetic isolates common haplotypes were observed. Autozygosity mapping and haplotype analysis suggested that FAA is located distal to D16S305.

Molecular Genetics

Poon et al. (1974) showed that cells from patients with Fanconi anemia are deficient in their ability to excise UV-induced pyrimidine dimers from their DNA. They are capable, however, of single strand break production and unscheduled DNA synthesis. From this the authors inferred deficiency in an exonuclease which specifically recognizes and excises distortions in the tertiary structure of DNA. Hirsch-Kauffmann et al. (1978), like some other workers, could find no defect in exonuclease but found reduction in DNA ligase activity in both a patient and the heterozygous mother.

Fujiwara et al. (1977) presented evidence that Fanconi anemia fibroblasts have an impaired capacity of removing DNA interstrand crosslinks induced by mitomycin C. They favored the view that a DNA crosslink repair deficiency is responsible for chromosomal damage in this disorder. Wunder et al. (1981) suggested that the defect in Fanconi anemia is in the passage of DNA-repair-related enzymes from the site of synthesis in the cytoplasm to the site of action in the nucleus. Studying the placenta of an affected infant, an unusual distribution of DNA topoisomerase was noted: high in the cytoplasm, very low in the nuclear sap. Whether the defect resides in the nuclear membrane or in the enzyme molecule is not clear. Wunder (1984) extended the studies suggesting that relatively high cytoplasmic DNA topoisomerase I in Fanconi placenta and fibroblasts may be due to an impediment to entry into the nucleus or perhaps binding to chromatin.

In somatic cell hybrid studies, Duckworth-Rysiecki et al. (1985) presented evidence for the existence of at least 2 FA complementation groups. They correspond to phenotypically distinct classes of cells exhibiting different rates of recovery of semiconservative DNA synthesis after treatment with DNA crosslinking agents in culture (Moustacchi et al., 1987) and different rates of removal of DNA crosslinks as shown by electron microscopy (Rousset et al., 1990). However, these studies do not provide a reliable method for determining the complementation group of a given patient, nor is there any apparent correlation between clinical phenotype and genetic class.

Cultured FA cells are unusually sensitive to DNA crosslinking agents such as mitomycin C whereas their sensitivity to radiation is close to normal.

In the hands of Zakrzewski and Sperling (1982), complementation studies based on mitomycin C sensitivity showed no evidence of heterogeneity when fusion was done between cells from different ethnic groups. Complementation studies with hybrids of cell lines derived from 4 patients in whom different biochemical lesions had been postulated led Zakrzewski et al. (1983) to conclude that the mutations are allelic.

Heterogeneous responses of various cell lines to DNA crosslinking treatments suggest genetic heterogeneity (Moustacchi and Diatloff-Zito, 1985), as do complementation studies (see 227660). Diatloff-Zito et al. (1986) found that normal DNA transfected into FA cells rendered the cells resistant to the effects of mitomycin C. Transfection of DNA of their own cells or DNA of yeast or salmon sperm did not give resistance.

Shaham et al. (1987) likewise found by transfection experiments that DNA sequences present in both the human and the Chinese hamster will correct the 2 cellular defects that are hallmarks of FA: spontaneous chromosome breakage and hypersensitivity to the cell-killing and clastogenic effects of the difunctional alkylating agent diepoxybutane. These observations open the way for cloning 'the FA gene,' mapping it, and determining its gene product and precise function. Chaganti and Houldsworth (1991) gave a review.

While granting that the molecular defect is unknown, Schweiger et al. (1987) suggested that the defect in DNA repair is located beyond incision, exonuclease reaction, and DNA synthesis, and that it most likely represents impaired metabolism of ADP-ribose.

Auerbach (1992) suggested that the cellular defect in FA results in chromosomal instability, hypersensitivity to DNA damage, and hypermutability for allele-loss mutations, thus predisposing to leukemia as a multistep process. Auerbach (1992) pointed to topoisomerase I (TOP1; 126420) and proliferating cell nuclear antigen (PCNA; 176740) as candidate genes for FA of complementation group A because of their location on chromosome 20 as well as their known function. Saito et al. (1994) performed a mutation analysis on topoisomerase I cDNA from FA cells by using chemical cleavage mismatch scanning and nucleotide sequencing. No mutation was detected from GM1309, an FA cell line of group A.

Lo Ten Foe et al. (1996) used a cDNA expression library to complement the mitomycin C-sensitive phenotype of FA lymphoblastoid cells. By this method they identified a cDNA clone that corrected the crosslinker hypersensitivity of FAA cells but not of FAC cells. Lo Ten Foe et al. (1996) stated (data not shown) that this cDNA mapped to the telomere of chromosome 16q by fluorescence in situ hybridization (FISH); they noted that the FAA locus was previously mapped to 16q24.3 on the basis of linkage studies. The cDNA clone for the gene (called FAA by them) has an open reading frame of 4,368 bp and encodes a 1,455-amino acid polypeptide predicted to contain 2 overlapping bipartite nuclear localization signals and a partial leucine zipper consensus sequence, suggesting that the protein is localized in the nucleus. Northern blot analysis detected a 5.5-kb transcript in controls and in FA patients from complementation groups other than FAA. In an FAA cell line, the 5.5-kb band was virtually undetectable, suggesting reduced transcription or reduced stability of the FAA mRNA transcript.

Through exon trapping and cDNA selection the Fanconi Anaemia/Breast Cancer Consortium (1996) identified a candidate cDNA clone for FAA and through searches in the dbEST database they were able to identify additional overlapping cDNA clones. Sequencing of these clones revealed a full-length FAA candidate gene that encodes a polypeptide of 1,455 amino acids. Sequence analysis revealed that the full-length cDNA isolated by positional cloning was identical to the cDNA clone isolated by complementation and reported by Lo Ten Foe et al. (1996), except that the clone isolated by positional cloning had an additional 13 bp of 5-prime untranslated sequence and 2 amino acid substitutions (S501G and I717M). RT-PCR and direct sequencing of PCR products from FAA patients led to the identification of 4 different mutations (227650.0001 through 227650.0004). The Fanconi Anaemia/Breast Cancer Consortium (1996) postulated that the introns of the FAA gene may contain repetitive unstable sequences susceptible to deletion.

Joenje et al. (1997) defined 3 new complementation groups among Fanconi anemia patients, adding FA-F, FA-G, and FA-H, to the 5 previously identified groups and bringing the total to a minimum of 8 distinct FA genes.

The fact that there are 5 FA complementation groups displaying similar phenotypes suggested that FA genes are functionally related. The cloned FA genes, FAA and FAC (227645), encode proteins that are unrelated to each other or to other proteins in the databases. Kupfer et al. (1997) demonstrated that FAA and FAC bind each other and form a complex. Protein binding correlated with the functional activity of FAA and FAC, as patient-derived mutant FAC, leu554 to pro (227645.0001), failed to bind FAA. Although unbound FAA and FAC localized predominantly to the cytoplasm, the FAA-FAC complex was found in similar abundance in both cytoplasm and nucleus. The results confirmed the interrelatedness of the FA genes in a pathway and suggested the cooperation of FAA and FAC in a nuclear function.

Levran et al. (1997) used SSCP analysis to screen genomic DNA from a panel of 97 racially and ethnically diverse FA patients from the International Fanconi Anemia Registry for mutations in the FAA gene. A total of 85 variant bands were detected. Forty-five of the variants were probably benign polymorphisms, of which 9 are common and can be used for various applications, including mapping studies for other genes in this region of 16q. Amplification refractory mutation system assays were developed to simplify their detection. Forty variants were considered probable pathogenic mutations. Seventeen of these were microdeletions/microinsertions associated with short direct repeats or homonucleotide tracts, a type of mutation thought to be generated by a mechanism of slipped-strand mispairing during DNA replication. A screening of 350 FA probands from the registry for 2 of these deletions (1115-1118del and 3788-3790del) revealed that they are carried on about 2% and 5% of the FA alleles, respectively. The second of these was found in a variety of ethnic groups and existed on at least 2 different haplotypes. Levran et al. (1997) suggested that FAA is hypermutable, and that slipped-strand mispairing, a mutational mechanism recognized as important for the generation of germline and somatic mutations in a variety of cancer-related genes, including p53 (191170), APC (175100), RB1 (180200), WT1 (194070), and BRCA1 (113705), may be a major mechanism for FAA mutagenesis. They speculated that FAA is a member of the 'caretaker' gene family according to the caretaker-gatekeeper model of Kinzler and Vogelstein (1997): inactivation of a caretaker gene results in a higher mutation rate in all genes, including gatekeeper genes that directly regulate tumor growth.

Levran et al. (1998) reported the occurrence of Alu-mediated genomic deletions in the FACA gene. Two different deletions of 1.2 kb and 1.9 kb were found in patients with Fanconi anemia. Both deletions included exons 16 and 17 and removed a 156-bp segment from the transcript, causing a shorter in-frame message. Introns 15 and 17 are rich in partial and complete Alu repeats. Sequence analysis of the deletion showed that the 5-prime breakpoints occurred at different sites in the same Alu element in intron 15, while the 3-prime breakpoints were located in different Alu repeats in intron 17. Numerous Alu repeats are present in the FACA gene, suggesting that Alu-mediated recombination may be an important mechanism for the generation of Fanconi anemia-producing mutations.

Diagnosis

Rosendorff and Bernstein (1988) concluded that in vitro enhancement of chromosome breakage by DEB and mitomycin C is usually a reliable technique to identify FA homozygotes but could not be depended on to identify individual FA heterozygotes.

Prenatal Diagnosis

Auerbach et al. (1985) attempted prenatal diagnosis in 30 fetuses at risk, using increased baseline and DEB-induced chromosomal breakage in amniotic fluid cells (and in 4 cases, chorion villus cells) as the measure of affection. Seven of the fetuses were diagnosed as affected; 2 were carried to term and 5 were terminated. The 2 who went to term were clinically affected; 2 of the abortuses showed congenital malformations, including abnormalities of the thumb and radius. No clinical suggestion of FA was found in the other 23 cases with diagnosis of no FA type abnormality.

Auerbach et al. (1986) extended the studies to the first trimester of pregnancy by the study of chromosomal breakage induced by DEB. Baseline chromosomal breakage and breakage after the agent were analyzed in 10 pregnancies: in 2, Fanconi anemia was diagnosed; in 8, FA was excluded even though the fetus was at risk. The results were unambiguous.

Poole et al. (1992) described monozygotic twin girls in whom the diagnosis of Fanconi anemia had been made at birth on the basis of limb anomalies and an apparently increased baseline chromosomal breakage frequency in one twin. Over a 13-year follow-up, they had not developed aplastic anemia or other hematologic manifestations of FA. Furthermore, repeat studies in 2 laboratories showed no evidence for increased baseline or DEB-induced chromosomal breakage in either twin. Using the scoring system for FA diagnosis developed by Auerbach et al. (1989), an FA probability coefficient of 0.98 was obtained. Through the International FA Registry, 15 additional patients were identified who had an FA probability score of 0.75 or greater but who had not developed aplastic anemia and were DEB negative. Poole et al. (1992) suggested that these patients should not be considered as instances of FA and that they probably represent a heterogeneous group of disorders with genetic as well as nongenetic causes such as Holt-Oram syndrome (142900), VATER and VACTERL association (192350), and IVIC syndrome (147750).

Clinical Management

Deeg et al. (1983) performed allogeneic marrow transplantation in 8 patients with Fanconi anemia. Seven were pretreated with cyclophosphamide alone and one with that agent plus procarbazine and antithymocyte globulin. All had engraftment. Three died of graft-versus-host disease (GVHD) and one of cerebral hemorrhage. Four were surviving 647 to 3,435 days after grafting. Two were well; 2 had chronic GVHD that was improving.

Porfirio et al. (1989) found that the iron chelator desferrioxamine (DFO) partially corrected the chromosome instability in Fanconi anemia. The study was undertaken on the assumption that one of the mechanisms involved in the pathogenesis of Fanconi anemia may be impaired capacity of the cells to remove active oxygen species. There appears to be a relationship between intraleukocyte chelatable iron pool and free radical formation. Porfirio et al. (1989) concluded that it is tempting to envisage a therapeutic trial with DFO.

Gluckman et al. (1989) achieved hematopoietic reconstitution in a 5-year-old boy with severe Fanconi anemia by administration of cryopreserved umbilical cord blood from a sister shown by prenatal testing to be unaffected by the disorder, to have a normal karyotype, and to be HLA-identical to the patient. They used a pretransplantation conditioning procedure developed specifically for the treatment of such patients. This technique makes use of the hypersensitivity of the abnormal cells to alkylating agents that crosslink DNA and to irradiation. Broxmeyer et al. (1989) had proposed that cord blood might be useful for such hematopoietic reconstitution.

Gluckman et al. (1992) gave a preliminary report on results of allogeneic bone marrow transplants (BMT) based on the records of the International Bone Marrow Transplant Registry. They suggested that because of the good results of BMT, the adverse effect of previous blood transfusions, the possible toxicity of long-term, high-dose androgen therapy, and the risk of leukemic transformation, it seemed advisable to transplant all patients with an HLA-identical sib as soon as pancytopenia requiring androgen therapy developed. Kohli-Kumar et al. (1994) reported results with bone marrow transplantation from matched sibs in 18 patients.

The findings of clastogenic factor in plasma from patients with Fanconi anemia by Emerit et al. (1995) has possible therapeutic implications. As had previously been reported for ataxia-telangiectasia and Bloom syndrome, transferable clastogenic material could be demonstrated in the plasma from patients with Fanconi anemia. While all plasma ultrafiltrates from homozygotes had chromosome damaging properties, the clastogenic material had to be concentrated in most heterozygotes to reach detectable levels. The clastogenic effect was exerted via the intermediacy of superoxide radicals, since it was regularly inhibited by superoxide dismutase. This added further evidence for a prooxidant state in Fanconi anemia. The clastogenic activity possibly plays a role in the progressive impairment of blood cell-producing bone marrow and may predispose patients to develop cancer and leukemia. Prophylactic use of antioxidants may be recommended, using clastogenic plasma activity as a guide.

Rackoff et al. (1996) reported that prolonged administration of granulocyte colony stimulating factor (GCSF; 138970) exerts a stimulatory effect on the bone marrow of Fanconi anemia patients and may be used to maintain a clinically adequate absolute neutrophil count in these patients. In some patients GCSF had beneficial effects on multiple hematopoietic lineages and may be used in combination with cytokine protocols for patients with progressive aplastic anemia. Rackoff et al. (1996) also reported that GCSF increases the number of circulating CD34 cells in Fanconi anemia patients.

Population Genetics

Rosendorff et al. (1987) estimated that the birth incidence of FA in white, Afrikaans-speaking South Africans is at least 1 in 22,000, the calculated heterozygote prevalence being approximately 1 in 77. They attributed this unusually high gene frequency to founder effect. Founder effect was strongly supported by the demonstration of allelic association between the disease and marker D16S303 in the Afrikaner population (Pronk et al., 1995). Alter (1992) concluded that Fanconi anemia in the Afrikaners represents the most clearly differentiated form of this heterogeneous disorder. She concluded that Fanconi anemia in blacks is clinically indistinguishable from that in other groups with the exception of the Afrikaners.

Savoia et al. (1996) found that 11 of 12 Fanconi anemia patients analyzed by complementation belonged to complementation group A. Four and 7 families came from 2 geographic clusters in the Veneto and Campania regions, respectively, which are thought to consist of aggregates of related families in reproductive isolation. The clinical characteristics of the patients showed both intra- and interfamilial heterogeneity, although overall the disease had a relatively mild course. Since the populations of both regions are likely to represent genetic isolates, the findings of Savoia et al. (1996) predicted linkage disequilibrium for markers flanking the FAA gene on chromosome 16. Thus, they concluded that DNAs from these FA families may be useful for positional cloning of the gene through haplotype disequilibrium mapping.

History

Guido Fanconi first described this syndrome in 1927 (Fanconi, 1927). The disorder described by Estren and Dameshek (1947) has been referred to as the Estren-Dameshek variant of Fanconi anemia (Nowell et al., 1984). Its characteristics are in all ways identical to full-blown Fanconi anemia except for the absence of the congenital malformations typical of Fanconi anemia.

See also Auerbach et al. (1989); Auerbach et al. (1979); Barrett et al. (1977); Berger et al. (1982); Bernstein et al. (1971); Cohen et al. (1975); Eldar et al. (1979); Fanconi (1966); Gebhart et al. (1985); Giampietro et al. (1994); Glanz and Fraser (1982); Gleadhill et al. (1975); Joenje et al. (1981); McDonald and Goldschmidt (1960); Nordenson et al. (1980); Obeid et al. (1980); Poll et al. (1985); Sasaki (1975); Schmid (1967); Schmid et al. (1965); Schroeder (1975); Schroeder et al. (1979); Schroeder et al. (1976); Shoyab et al. (1981); Strathdee et al. (1992); Swift and Hirschhorn (1966); Swift and Sholman (1972); Voss et al. (1981); Wunder and Fleischer-Reischmann (1983); Yoshida (1980); Yoshida (1982); Zachmann et al. (1972)

Alter, B. P.: Fanconi's anaemia and its variability. Brit. J. Haemat. 85: 9-14, 1993.

Alter, B. P.: Fanconi anemia in blacks. (Letter) Am. J. Med. Genet. 42: 393, 1992.

Auerbach, A. D.: Fanconi anemia and leukemia: tracking the genes. Leukemia 6 (suppl. 1): 1-4, 1992.

Auerbach, A. D.; Allen, R. G.: Leukemia and preleukemia in Fanconi anemia patients: a review of the literature and report of the International Fanconi Anemia Registry. Cancer Genet. Cytogenet. 51: 1-12, 1991.

Auerbach, A. D.; Mann, W. R.; Venkatraj, V. S.; Ghosh, R.; Liu, W.; Olsen, D.; Sherman, S. S.: Genetic linkage studies in Fanconi anemia. (Abstract) Cytogenet. Cell Genet. 51: 954, 1989.

Auerbach, A. D.; Min, Z.; Ghosh, R.; Pergament, E.; Verlinsky, Y.; Nicolas, H.; Boue, J.: Clastogen-induced chromosomal breakage as a marker for first trimester prenatal diagnosis of Fanconi anemia. Hum. Genet. 73: 86-88, 1986.

Auerbach, A. D.; Rogatko, A.; Schroeder-Kurth, T. M.: International Fanconi Anemia Registry: relation of clinical symptoms to diepoxybutane sensitivity. Blood 73: 391-396, 1989.

Auerbach, A. D.; Sagi, M.; Adler, B.: Fanconi anemia: prenatal diagnosis in 30 fetuses at risk. Pediatrics 76: 794-800, 1985.

Auerbach, A. D.; Warburton, D.; Bloom, A. D.; Chaganti, R. S. K.: Prenatal detection of the Fanconi anemia gene by cytogenic methods. Am. J. Hum. Genet. 31: 77-81, 1979.

Barrett, A. J.; Brigden, W. D.; Hobbs, J. R.; Hugh-Jones, K.; Humble, J. G.; James, D. C. O.; Retsas, S.; Roberts, T. R. F.; Selwyn, S.; Sneath, P.; Watson, J. G.: Successful bone marrow transplant for Fanconi's anaemia. Brit. Med. J. 1: 420-421, 1977.

Berger, N. A.; Berger, S. J.; Catino, D. M.: Abnormal NAD+ levels in cells from patients with Fanconi's anaemia. Nature 299: 271-273, 1982.

Berkovitz, G. D.; Zinkham, W. H.; Migeon, C. J.: Gonadal function in two siblings with Fanconi's anemia. Hormone Res. 19: 137-141, 1984.

Bernstein, M. S.; Hunter, R. L.; Yachnin, S.: Hepatoma and peliosis hepatis developing in a patient with Fanconi's anemia. New Eng. J. Med. 284: 1135-1136, 1971.

Bloom, G. E.; Warner, S.; Gerald, P. S.; Diamond, L. K.: Chromosome abnormalities in constitutional aplastic anemia. New Eng. J. Med. 274: 8-14, 1966.

Broxmeyer, H. E.; Douglas, G. W.; Hangoc, G.; Cooper, S.; Bard, J.; English, D.; Arny, M.; Thomas, L.; Boyse, E. A.: Human umbilical cord blood as a potential source of transplantable hematopoietic stem/progenitor cells. Proc. Nat. Acad. Sci. 86: 3828-3832, 1989.

Brunetti, P.; Neuci, G. G.; Vaccaro, R.; Puxeddu, A.; Migliorini, E.: Fanconi's anaemia. (Letter) Lancet II: 1194-1195, 1966.

Buchwald, M.: Complementation groups: one or more per gene?. Nature Genet. 11: 228-230, 1995.

Chaganti, R. S. K.; Houldsworth, J.: Fanconi anemia: a pleotropic (sic) mutation with multiple cellular and developmental abnormalities. Ann. Genet. 34: 206-211, 1991.

Cohen, M. M.; Shaham, M.; Dagan, J.; Shmueli, E.; Kohn, G.: Cytogenetic investigations in families with ataxia-telangiectasia. Cytogenet. Cell Genet. 15: 338-356, 1975.

Deeg, H. J.; Storb, R.; Thomas, E. D.; Appelbaum, F.; Buckner, C. D.; Clift, R. A.; Doney, K.; Johnson, L.; Sanders, J. E.; Stewart, P.; Sullivan, K. M.; Witherspoon, R. P.: Fanconi's anemia treated by allogeneic marrow transplantation. Blood 61: 954-959, 1983.

de Vroede, M.; Feremans, W.; de Maertelaere-Laurent, E.; Mandelbaum, I.; Toppett, M.; Vamos, E.; Fondu, P.: Fanconi's anaemia: simultaneous onset in 2 siblings and unusual cytological findings. Scand. J. Haemat. 28: 431-440, 1982.

Diatloff-Zito, C.; Papadopoulo, D.; Averbeck, D.; Moustacchi, E.: Abnormal response to DNA crosslinking agents of Fanconi anemia fibroblasts can be corrected by transfection with normal human DNA. Proc. Nat. Acad. Sci. 83: 7034-7038, 1986.

Duckworth-Rysiecki, G.; Cornish, K.; Clarke, C. A.; Buchwald, M.: Identification of two complementation groups in Fanconi anemia. Somat. Cell Molec. Genet. 11: 35-41, 1985.

Dutrillaux, B.; Aurias, A.; Dutrillaux, A.-M.; Buriot, D.; Prieur, M.: The cell cycle of lymphocytes in Fanconi anemia. Hum. Genet. 62: 327-332, 1982.

Eldar, M.; Shoenfeld, Y.; Zaizov, R.; Fogel, R.; Asherov, J.; Liban, E.; Pinkhas, J.: Pulmonary alveolar proteinosis associated with Fanconi's anemia. Respiration 38: 177-178, 1979.

Emerit, I.; Levy, A.; Pagano, G.; Pinto, L.; Calzone, R.; Zatterale, A.: Transferable clastogenic activity in plasma from patients with Fanconi anemia. Hum. Genet. 96: 14-20, 1995.

Estren, S.; Dameshek, W.: Familial hypoplastic anemia of childhood: report of eight cases in two families with beneficial effect of splenectomy in one case. Am. J. Dis. Child. 73: 671-687, 1947.

Fanconi, G.: Familiaere infantile perniziosaartige Anaemie (pernizioeses Blutbild und Konstitution). Jahrb. Kinderheilk. 117: 257-280, 1927.

Fanconi, G.: Familial constitutional panmyelocytopathy, Fanconi's anemia (F.A.). I. Clinical aspects. Seminars Hemat. 4: 233-240, 1966.

Fanconi Anaemia/Breast Cancer Consortium: Positional cloning of the Fanconi anaemia group A gene. Nature Genet. 14: 324-328, 1996.

Fujiwara, Y.; Tatsumi, M.; Sasaki, M.: Cross-link repair in human cells and its possible defect in Fanconi's anemia cells. J. Molec. Biol. 113: 635-649, 1977.

Garriga, S.; Crosby, W. H.: The incidence of leukemia in families of patients with hypoplasia of the marrow. Blood 14: 1008-1014, 1959.

Gebhart, E.; Kysela, D.; Matthee, H.; Nikol, M.: Cytogenetic analyses utilizing various clastogens in two sibs with Fanconi anemia, their relatives, and control individuals. Hum. Genet. 69: 309-315, 1985.

Giampietro, P. F.; Adler-Brecher, B.; Verlander, P. C.; Pavlakis, S. G.; Davis, J. G.; Auerbach, A. D.: The need for more accurate and timely diagnosis in Fanconi anemia: a report from the International Fanconi Anemia Registry. Pediatrics 91: 1116-1120, 1993.

Giampietro, P. F.; Davis, J. G.; Auerbach, A. D.: Fanconi's anemia. (Letter) New Eng. J. Med. 330: 720-721, 1994.

Glanz, A.; Fraser, F. C.: Spectrum of anomalies in Fanconi anemia. J. Med. Genet. 19: 412-416, 1982.

Gleadhill, V.; Bridges, J. M.; Hadden, D. R.: Fanconi's aplastic anaemia with short stature: absence of response to human growth hormone. Arch. Dis. Child. 50: 318-320, 1975.

Gluckman, E.; Auerbach, A.; Ash, R. C.; Biggs, J. C.; Bortin, M. M.; Camitta, B. M.; Champlin, R. E.; Friedrich, W.; Gale, R. P.; Good, R. A.; Gordon-Smith, E. C.; Horwitz, M. M.; Massaoka, T.; Ortega, J. J.; Pasquini, R.; Ramsay, N. K. C.; Rozman, C.; Speck, B.; Vowels, M. R.: Allogeneic bone marrow transplants for Fanconi anemia: a preliminary report from the International Bone Marrow Transplant Registry. Bone Marrow Transplant. 10 (suppl. 1): 53-57, 1992.

Gluckman, E.; Broxmeyer, H. E.; Auerbach, A. D.; Friedman, H. S.; Douglas, G. W.; Devergie, A.; Esperou, H.; Thierry, D.; Socie, G.; Lehn, P.; Cooper, S.; English, D.; Kurtzberg, J.; Bard, J.; Boyse, E. A.: Hematopoietic reconstitution in a patient with Fanconi's anemia by means of umbilical-cord blood from an HLA-identical sibling. New Eng. J. Med. 321: 1174-1178, 1989.

Gschwend, M.; Levran, O.; Kruglyak, L.; Ranade, K.; Verlander, P. C.; Shen, S.; Faure, S.; Weissenbach, J.; Altay, C.; Lander, E. S.; Auerbach, A. D.; Botstein, D.: A locus for Fanconi anemia on 16q determined by homozygosity mapping. Am. J. Hum. Genet. 59: 377-384, 1996.

Hagerman, D. A.; Williams, G. P.: Some features of Fanconi's anemia. New Eng. J. Med. 329: 1168, 1993.

Hirsch-Kauffmann, M.; Schweiger, M.; Wagner, E. F.; Sperling, K.: Deficiency of DNA ligase activity in Fanconi's anemia. Hum. Genet. 45: 25-32, 1978.

Jakobs, P. M.; Fiddler-Odell, E.; Reifsteck, C.; Olson, S.; Moses, R. E.; Grompe, M.: Complementation group assignments in Fanconi anemia fibroblast cell lines from North America. Somat. Cell Molec. Genet. 23: 1-7, 1997.

Joenje, H.: Fanconi anaemia complementation groups in Germany and The Netherlands. Hum. Genet. 97: 280-282, 1996.

Joenje, H.; Arwert, F.; Eriksson, A. W.; de Koning, H.; Oostra, A. B.: Oxygen-dependence of chromosomal aberrations in Fanconi's anaemia. Nature 290: 142-143, 1981.

Joenje, H.; Frants, R. R.; Arwert, F.; De Bruin, G. J. M.; Kostense, P. J.; van de Kamp, J. J. P.; de Koning, J.; Eriksson, A. W.: Erythrocyte superoxide dismutase deficiency in Fanconi's anaemia established by two independent methods of assay. Scand. J. Clin. Lab. Invest. 39: 759-764, 1979.

Joenje, H.; Oostra, A. B.; Wijker, M.; di Summa, F. M.; van Berkel, C. G. M.; Rooimans, M. A.; Ebell, W.; van Weel, M.; Pronk, J. C.; Buchwald, M.; Arwert, F.: Evidence for at least eight Fanconi anemia genes. Am. J. Hum. Genet. 61: 940-944, 1997.

Kinzler, K. W.; Vogelstein, B.: Gatekeepers and caretakers. Nature 386: 761-763, 1997.

Kohli-Kumar, M.; Morris, C.; DeLaat, C.; Sambrano, J.; Masterson, M.; Mueller, R.; Shahidi, N. T.; Yanik, G.; Desantes, K.; Friedman, D. J.; Auerbach, A. D.; Harris, R. E.: Bone marrow transplantation in Fanconi anemia using matched sibling donors. Blood 84: 2050-2054, 1994.

Kupfer, G. M.; Naf, D.; Suliman, A.; Pulsipher, M.; D'Andrea, A. D.: The Fanconi anaemia proteins, FAA and FAC, interact to form a nuclear complex. Nature Genet. 17: 487-490, 1997.

Levran, O.; Doggett, N. A.; Auerbach, A. D.: Identification of Alu-mediated deletions in the Fanconi anemia gene FAA. Hum. Mutat. 12: 145-152, 1998.

Levran, O.; Erlich, T.; Magdalena, N.; Gregory, J. J.; Batish, S. D.; Verlander, P. C.; Auerbach, A. D.: Sequence variation in the Fanconi anemia gene FAA. Proc. Nat. Acad. Sci. 94: 13051-13056, 1997.

Li, F. P.; Potter, N. U.: Classical Fanconi anemia in a family with hypoplastic anemia. J. Pediat. 92: 943-944, 1978.

Lohr, G. W.; Waller, H. D.; Anschutz, F.; Knopp, A.: Biochemische Defekte in den Blutzellen bei familiaerer Panmyelopathie (Typ Fanconi). Humangenetik 1: 383-387, 1965.

Lo Ten Foe, J. R.; Rooimans, M. A.; Bosnoyan-Collins, L.; Alon, N.; Wijker, M.; Parker, L.; Lightfoot, J.; Carreau, M.; Callen, D. F.; Savoia, A.; Cheng, N. C.; van Berkel, C. G. M.; Strunk. M. H. P.; Gille. J. J. P.; Pals, G.; Kruyt, F. A. E.; Pronk, J. C.; Arwert, F.; Buchwald, M.; Joenje, H.: Expression cloning of a cDNA for the major Fanconi anaemia gene, FAA. Nature Genet. 14: 320-323, 1996.

Macdougall, L. G.; Greeff, M. C.; Rosendorff, J.; Bernstein, R.: Fanconi anemia in black African children. Am. J. Med. Genet. 36: 408-413, 1990.

Mann, W. R.; Venkatraj, V. S.; Allen, R. G.; Liu, Q.; Olsen, D. A.; Adler-Brecher, B.; Mao, J.-I.; Weiffenbach, B.; Sherman, S. L.; Auerbach, A. D.: Fanconi anemia: evidence for linkage heterogeneity on chromosome 20q. Genomics 9: 329-337, 1991.
McDonald, R.; Goldschmidt, B.: Pancytopenia with congenital defects (Fanconi's anaemia). Arch. Dis. Child. 35: 367-372, 1960.
Moustacchi, E.; Diatloff-Zito, C.: DNA semi-conservative synthesis in normal and Fanconi anemia fibroblasts following treatment with 8-methoxypsoralen and near ultraviolet light or with x-rays. Hum. Genet. 70: 236-242, 1985.
Moustacchi, E.; Papadopoulo, D.; Diatloff-Zito, C.; Buchwald, M.: Two complementation groups of Fanconi's anemia differ in their phenotypic response to a DNA-crosslinking treatment. Hum. Genet. 75: 45-47, 1987.
Nordenson, I.; Bjorksten, B.; Lundh, B.: Prevention of chromosomal breakage in Fanconi's anemia by cocultivation with normal cells. Hum. Genet. 56: 169-171, 1980.
Nowell, P.; Bergman, G.; Besa, E.; Wilmoth, D.; Emanuel, B.: Progressive preleukemia with a chromosomally abnormal clone in a kindred with the Estren-Dameshek variant of Fanconi's anemia. Blood 64: 1135-1138, 1984.
Obeid, D. A.; Hill, F. G. H.; Harnden, D.; Mann, J. R.; Wood, B. S. B.: Fanconi anemia: oxymetholone hepatic tumors, and chromosome aberrations associated with leukemic transition. Cancer 46: 1401-1404, 1980.
Poll, E. H. A.; Arwert, F.; Joenje, H.; Wanamarta, A. H.: Differential sensitivity of Fanconi anaemia lymphocytes to the clastogenic action of cis-diamminedichloroplatinum (II) and trans-diamminedichloroplatinum (II). Hum. Genet. 71: 206-210, 1985.
Poole, S. R.; Smith, A. C. M.; Hays, T.; McGavran, L.; Auerbach, A. D.: Monozygotic twin girls with congenital malformations resembling Fanconi anemia. Am. J. Med. Genet. 42: 780-784, 1992.
Poon, P. K.; O'Brien, R. L.; Parker, J. W.: Defective DNA repair in Fanconi anaemia. Nature 250: 223-225, 1974.
Porfirio, B.; Ambroso, G.; Giannella, G.; Isacchi, G.; Dallapiccola, B.: Partial correction of chromosome instability in Fanconi anemia by desferrioxamine. Hum. Genet. 83: 49-51, 1989.
Pronk, J. C.; Gibson, R. A.; Savoia, A.; Wijker, M.; Morgan, N. V.; Melchionda, S.; Ford, D.; Temtamy, S.; Ortega, J. J.; Jansen, S.; Havenga, C.; Cohn, R. J.; de Ravel, T. J.; Roberts, I.; Westerveld, A.; Easton, D. F.; Joenje, H.; Mathew, C. G.; Arwert, F.: Localisation of the Fanconi anaemia complementation group A gene to chromosome 16q24.3. Nature Genet. 11: 338-340, 1995.
Rackoff, W. R.; Orazi, A.; Robinson, C. A.; Cooper, R. J.; Alter, B. P.; Freedman, M. H.; Harris, R. E.; Williams, D. A.: Prolonged administration of granulocyte colony-stimulating factor (Filgrastim) to patients with Fanconi anemia: a pilot study. Blood 88: 1588-1593, 1996.
Rogatko, A.; Auerbach, A. D.: Segregation analysis with uncertain ascertainment: application to Fanconi anemia. Am. J. Hum. Genet. 42: 889-897, 1988.
Rosendorff, J.; Bernstein, R.: Fanconi's anemia—chromosome breakage studies in homozygotes and heterozygotes. Cancer Genet. Cytogenet. 33: 175-183, 1988.
Rosendorff, J.; Bernstein, R.; Macdougall, L.; Jenkins, T.: Fanconi anemia: another disease of unusually high prevalence in the Afrikaans population of South Africa. Am. J. Med. Genet. 27: 793-797, 1987.
Rousset, S.; Nocentini, S.; Revet, B.; Moustacchi, E.: Molecular analysis by electron microscopy of the removal of psoralen-photoinduced DNA cross-links in normal and Fanconi's anemia fibroblasts. Cancer Res. 50: 2443-2448, 1990.
Saito, H.; Grompe, M.; Neeley, T. L.; Jakobs, P. M.; Moses, R. E.: Fanconi anemia cells have a normal gene structure for topoisomerase I. Hum. Genet. 93: 583-586, 1994.
Sasaki, M. S.: Is Fanconi's anaemia defective in a process essential to the repair of DNA cross links?. Nature 257: 501-503, 1975.
Savoia, A.; Piemontese, M. R.; Savino, M.; Zatterale, A.; Pronk, J.; Arwert, F.; Joenje, H.; Ramenghi, U.; Dagna-Bricarelli, F.; Dallapiccola, B.; Zelante, L.: Linkage analysis of Fanconi anaemia in Italy and mapping of the complementation group A gene. Hum. Genet. 99: 93-97, 1997.
Savoia, A.; Zatterale, A.; Del Principe, D.; Joenje, H.: Fanconi anaemia in Italy: high prevalence of complementation group A in two geographic clusters. Hum. Genet. 97: 599-603, 1996.
Schindler, D.; Kubbies, M.; Hoehn, H.; Schinzel, A.; Rabinovitch, P. S.: Presymptomatic diagnosis of Fanconi's anaemia. (Letter) Lancet I: 937, 1985.
Schindler, D.; Seyschab, H.; Poot, M.; Hoehn, H.; Schinzel, A.; Fryns, J. P.; Tommerup, N.; Rabinovitch, P. S.: Screening test for ataxia telangiectasia. (Letter) Lancet II: 1398-1399, 1987.
Schinzel, A.: Personal Communication. Zurich, Switzerland 10/3/1991.
Schmid, W.: Familial constitutional panmyelocytopathy, Fanconi's anemia (F. A.). II. A discussion of the cytogenetic findings in Fanconi's anemia. Seminars Hemat. 4: 241-249, 1967.
Schmid, W.; Scharer, K.; Baumann, T.; Fanconi, G.: Chromosomenbruechigkeit bei der familiaeren Panmyelopathie (typus Fanconi). Schweiz. Med. Wschr. 95: 1461-1464, 1965.
Schroeder, T. M.: Sister chromatid exchanges and chromatid interchanges in Bloom's syndrome. Humangenetik 30: 317-323, 1975.
Schroeder, T. M.; Anschutz, F.; Knopp, A.: Spontane Chromosomenaberrationen bei familiaerer Panmyelopathie. Humangenetik 1: 194-196, 1964.
Schroeder, T. M.; German, J.: Bloom's syndrome and Fanconi's anemia: demonstration of two distinctive patterns of chromosome disruption and rearrangement. Humangenetik 25: 299-306, 1974.
Schroeder, T. M.; Pohler, E.; Hufnagel, H. D.; Stahl-Mauge, C.: Fanconi's anemia: terminal leukemia and 'forme fruste' in one family. Clin. Genet. 16: 260-268, 1979.
Schroeder, T. M.; Tilgen, D.; Kruger, J.; Vogel, F.: Formal genetics of Fanconi's anemia. Hum. Genet. 32: 257-288, 1976.
Schweiger, M.; Auer, B.; Burtscher, H. J.; Hirsch-Kauffmann, M.; Klocker, H.; Schneider, R.: DNA repair in human cells: biochemistry of the hereditary diseases Fanconi's anaemia and Cockayne syndrome. Europ. J. Biochem. 165: 235-242, 1987.
Shaham, M.; Adler, B.; Ganguly, S.; Chaganti, R. S. K.: Transfection of normal human and Chinese hamster DNA corrects diepoxybutane-induced chromosomal hypersensitivity of Fanconi anemia fibroblasts. Proc. Nat. Acad. Sci. 84: 5853-5857, 1987.
Shoyab, M.; Gunnell, M.; Lubiniecki, A. S.: Reduced uptake and incorporation of 3-H-thymidine in Fanconi anemia fibroblasts. Hum. Genet. 57: 296-299, 1981.
Steinlein, O.; Fischer, C.; Keil, R.; Smigrodzki, R.; Vogel, F.: D20S19, linked to low voltage EEG, benign neonatal convulsions, and Fanconi anaemia, maps to a region of enhanced recombination and is localized between CpG islands. Hum. Molec. Genet. 1: 325-329, 1992.
Strathdee, C. A.; Duncan, A. M. V.; Buchwald, M.: Evidence for at least four Fanconi anaemia genes including FACC on chromosome 9. Nature Genet. 1: 196-198, 1992.
Strathdee, C. A.; Gavish, H.; Shannon, W. R.; Buchwald, M.: Cloning of cDNAs for Fanconi's anaemia by functional complementation. Nature 356: 763-767, 1992.
Swift, M. R.; Cohen, J.; Pinlchham, R.: Maximum-likelihood method for estimating the disease predisposition of heterozygotes. Am. J. Hum. Genet. 26: 304-317, 1974.
Swift, M. R.; Hirschhorn, K.: Fanconi's anemia: inherited susceptibility to chromosome breakage in various tissues. Ann. Intern. Med. 65: 496-503, 1966.
Swift, M. R.; Sholman, L.: Diabetes mellitus and the gene for Fanconi's anemia. Science 178: 308-310, 1972.
Voss, R.; Kohn, G.; Shaham, M.; Benzur, Z.; Arnon, J.; Ornoy, A.; Yaffe, H.; Golbus, M.; Auerbach, A. D.: Prenatal diagnosis of Fanconi anemia. Clin. Genet. 20: 185-190, 1981.
Welshimer, K.; Swift, M.: Congenital malformations and developmental disabilities in ataxia-telangiectasia, Fanconi anemia, and xeroderma pigmentosum families. Am. J. Hum. Genet. 34: 781-793, 1982.
Wunder, E.: Further studies on compartmentalisation of DNA-topoisomerase I in Fanconi anemia tissue. Hum. Genet. 68: 276-281, 1984.
Wunder, E.; Burghardt, U.; Lang, B.; Hamilton, L.: Fanconi's anemia: anomaly of enzyme passage through the nuclear membrane? Anomalous intracellular distribution of topoisomerase activity in placental extracts in a case of Fanconi's anemia. Hum. Genet. 58: 149-155, 1981.
Wunder, E.; Fleischer-Reischmann, B.: Response of lymphocytes from Fanconi's anemia patients and their heterozygous relatives to 8-methoxy-psoralene in a cloning survival test system. Hum. Genet. 64: 167-172, 1983.
Yoshida, M. C.: Suppression of spontaneous and mitomycin C-induced chromosome aberrations in Fanconi's anemia by cell fusion with normal human fibroblasts. Hum. Genet. 55: 223-226, 1980.
Yoshida, M. C.: The absence of genetic heterogeneity in Fanconi's anemia in five patients demonstrated by somatic cell hybridization. (Abstract) Cytogenet. Cell Genet. 32: 329-330, 1982.
Zachmann, M.; Illig, R.; Prader, A.: Fanconi's anemia with isolated growth hormone deficiency. (Letter) J. Pediat. 80: 159, 1972.
Zaizov, R.; Matoth, Y.; Mamon, Z.: Familial aplastic anaemia without congenital malformations. Acta Paediat. Scand. 58: 151-156, 1969.
Zakrzewski, S.; Koch, M.; Sperling, K.: Complementation studies between Fanconi's anemia cells with different DNA repair characteristics. Hum. Genet. 64: 55-57, 1983.
Zakrzewski, S.; Sperling, K.: Analysis of heterogeneity in Fanconi's anemia patients of different ethnic origin. Hum. Genet. 62: 321-323, 1982.

FANCONI PANCYTOPENIA 2 (*227660)

JJM NOTE

To my knowledge, predisposition to malignancy has not varied by complementation group, but definitive studies have not been conducted.

The existence of at least two separate loci, homozygosity at either of which can result in the Fanconi syndrome, was indicated by the complementation observed by Zakrzewski and Sperling (1980) in cell hybrid studies. They found no complementation between cells of classic Fanconi syndrome and those from patients lacking skeletal malformation. However, cells from a late-onset case complemented cells from an early-onset case. Sensitivity to the cytogenetic effects of mitomycin C was the phenotype of which complementation was studied. From comparable complementation studies, Duckworth-Rysiecki et al. (1985) likewise concluded that there are at least 2 complementation groups. Digweed et al. (1988) showed that the complementation grouping established by fibroblast fusion studies correlates with the rate of semiconservative DNA synthesis after 8-methoxypsoralen/UVA-irradiation treatment (Moustacchi et al., 1987). Because of the general nature of the disorder, FA appears to be a prime candidate for somatic cell gene therapy. Diatloff-Zito et al. (1990) reported studies in which transfectants were obtained by mouse DNA-mediated gene transfer into FA primary fibroblasts. These studies again demonstrated differences between complementation groups A and B. Cells from group A, which are the most sensitive to the effects of crosslinking agents, were partially corrected for both the chromosomal aberrations and the cytotoxicity to 8-methoxypsoralen photoaddition. In contrast, after treatment with mitomycin C, only the chromosomal sensitivity was reestablished to a near-normal level. The opposite was true for group B cells, i.e., cell survival to MMC was partially corrected, whereas the frequency of MMC-induced chromosomal aberrations remained close to that of the untransfected cells.

Strathdee et al. (1992) indicated that there are at least 4 FA complementation groups, designated complementation groups A through D.

Diatloff-Zito, C.; Rosselli, F.; Heddle, J.; Moustacchi, E.: Partial complementation of the Fanconi anemia defect upon transfection by heterologous DNA: phenotypic dissociation of chromosomal and cellular hypersensitivity to DNA cross-linking agents. Hum. Genet. 86: 151-161, 1990.

Digweed, M.; Zakrzewski-Ludcke, S.; Sperling, K.: Fanconi's anaemia: correlation of genetic complementation group with psoralen/UVA response. Hum. Genet. 78: 51-54, 1988.

Duckworth-Rysiecki, G.; Cornish, K.; Clarke, C. A.; Buchwald, M.: Identification of two complementation groups in Fanconi anemia. Somat. Cell Molec. Genet. 11: 35-41, 1985.

Moustacchi, E.; Papadopoulo, D.; Diatloff-Zito, C.; Buchwald, M.: Two complementation groups of Fanconi's anemia differ in their phenotypic response to a DNA-crosslinking treatment. Hum. Genet. 75: 45-47, 1987.

Strathdee, C. A.; Gavish, H.; Shannon, W. R.; Buchwald, M.: Cloning of cDNAs for Fanconi's anaemia by functional complementation. Nature 356: 763-767, 1992.

Zakrzewski, S.; Sperling, K.: Genetic heterogeneity of Fanconi's anemia demonstrated by somatic cell hybrids. Hum. Genet. 56: 81-84, 1980.

FANCONI ANEMIA, COMPLEMENTATION GROUP E (*600901)

JJM NOTE

This entity is not (yet) associated with neoplasia, but there is no reason to think otherwise.

Joenje et al. (1995) presented evidence for a fifth subtype of Fanconi anemia, designated group E. Buchwald (1995) stated that 6 of 31 patients (12.7%) could be classified as group E. The FACE group is defined as being different from groups A (227650), B (227660), C (227645), and D (227646) and may itself be heterogeneous.

Wegner et al. (1996) also described the fifth Fanconi anemia complementation group. Their first patient, a Turkish boy, presented with psychomotor retardation, growth retardation, retarded bone age, brachycephaly, hypotelorism, epicanthus, syndactyly, brachydactyly, renal dystopia, and cryptorchidism. In addition, an asymmetric skeletal anomaly was seen with a double distal phalanx of the left thumb and hypoplasia of the right thumb. Typical hematologic features of the disorder developed at the age of 2.5 years, about 1 year after initial diagnosis. Cytogenetic studies confirmed the clinical diagnosis and revealed a spontaneous chromosomal instability and hypersensitivity to the crosslinking agents diepoxybutane and Trenimon. The findings in the patient were compared with data reported for other patients with Fanconi anemia.

Following the identification of the first FACE patient (Joenje, 1996), additional patients were identified whose cell lines complemented groups A-D. To assess their possible assignment to group E, Joenje et al. (1997) extended the complementation analysis by introducing suitable selection markers into the original FACE cell line and 3 cell lines classified as non-ABCD. The results indicated that each of these 3 cell lines represented a new complementation group, implying existence of at least 8 complementation groups in Fanconi anemia, including the 3 new groups FACF, FACG, and FACH.

Buchwald, M.: Complementation groups: one or more per gene? Nature Genet. 11: 228-230, 1995.

Joenje, H.: Fanconi anaemia complementation groups in Germany and The Netherlands. Hum. Genet. 97: 280-282, 1996.

Joenje, H.; Lo Ten Foe, J. R.; Oostra, A. B.; van Berkel, C. G. M.; Rooimans, M. A.; Schroeder-Kurth, T.; Wegner, R.-D.; Gille, J. J. P.; Buchwald, M.; Arwert, F.: Classification of Fanconi anemia patients by complementation analysis: evidence for a fifth genetic subtype. Blood 86: 2156-2160, 1995.

Joenje, H.; Oostra, A. B.; Wijker, M.; di Summa, F. M.; van Berkel, C. G. M.; Rooimans, M. A.; Ebell, W.; van Weel, M.; Pronk, J. C.; Buchwald, M.; Arwert, F.: Evidence for at least eight Fanconi anemia genes. Am. J. Hum. Genet. 61: 940-944, 1997.

Wegner, R.-D.; Henrichs, I.; Joenje, H.; Schroeder-Kurth, T.: Fanconi anemia complementation group E: clinical and cytogenetic data of the first patient. Clin. Genet. 50: 479-482, 1996.

AGRANULOCYTOSIS, INFANTILE GENETIC (*202700)

JJM NOTE

Unusual cell type is a flag of a malignancy of possible genetic origins. In Kostmann agranulocytosis, the complicating cancer is acute monocytic leukemia.

A number sign (#) is used with this entry because of evidence that the infantile agranulocytosis that goes by the name of Kostmann (1956), who first clearly delineated it, is caused by a defect in granulocyte colony-stimulating factor receptor (CSF3R; 138971).

Not all patients with congenital neutropenia have mutations in the GCSFR gene. Dong et al. (1995) found mutations in GCSFR in 4 of 14 patients with severe congenital neutropenia; in all 4, the mutation truncated the C-terminal region. Dong et al. (1995) stated that no such mutations were found in 3 patients with severe congenital neutropenia who were members of the Swedish families in which the disease was originally described (Kostmann, 1956).

In addition to Kostmann agranulocytosis, recessively inherited neutropenic syndromes include congenital neutropenia with eosinophilia, Chediak-Higashi syndrome, and Fanconi pancytopenic syndrome. Hedenberg (1959) found that addition of sulfur-containing amino acids to tissue cultures led to maturation of white cells. L'Esperance et al. (1973) showed that the disease could be reproduced in tissue culture. Experiments such as Hedenburg's should be repeated using this technique. Barak et al. (1971) also cultured marrow cells from a patient with this disease. Gilman et al. (1970) described prolonged survival and death from acute monocytic leukemia at age 14 years and 10 months. About three-fourths of patients die before age 3 years. Fungal and viral infections have not been a problem. L'Esperance et al. (1975) proposed heterogeneity of this disorder because in soft agar cultures of bone marrow one patient showed 'loose' colonies developing only to promyelocytes, whereas a second produced normal neutrophil colonies. Maturation arrest occurs at the promyelocyte stage. Hansen et al. (1977) found association with HLA-B12 and postulated linkage disequilibrium. A gene controlling neutrophil differentiation is, presumably, closely linked to the HLA complex. The relationship may reflect a basic function of the histocompatibility system, namely, coding for cell-surface determinants fundamental to cell-cell recognition and to control of cellular differentiation.

Starting with 19 sibships collected by Kostmann (1975), Iselius and Gustavson (1984) assembled evidence that a single founder was responsible for the cases observed in Sweden. The likely origin of the gene was thought to be the parish of Overkalix in the county of Norrbotten in the most northern part of Sweden. In a retrospective survey of 162 patients in whom bone marrow transplantation was performed in 14 European centers between 1969 and 1985, Fischer et al. (1986) found 1 patient with Kostmann syndrome. Olofsson et al. (1976) reported 2 patients, a boy and his niece. In each instance the parents were consanguineous. On the first day of life, 1 of the 2 patients had granulocytopenia; complete agranulocytosis developed in 3 weeks after an interval during which the patient had normal or even increased neutrophil levels, due possibly to meningitis. In addition to persistent severe absolute neutropenia (500 neutrophils per microliter or fewer) and bone marrow morphology that suggests maturational arrest of neutrophil precursors at the promyelocyte stage, variable degrees of monocytosis, eosinophilia, hypergammaglobulinemia, and thrombocytosis may be found. Bonilla et al. (1989) administered recombinant human granulocyte colony-stimulating factor (138970) to 5 patients. All 5 patients showed a response and had sustained neutrophil counts of 1,000 cells per microliter or more for 9 to 13 months while receiving subcutaneous maintenance therapy. Preexisting chronic infections resolved and the number of new infectious episodes decreased. Bonilla et al. (1989) stated that further studies would be necessary to determine whether the GCSF gene is deleted or mutated. Another possibility is that the receptors are defective and do not respond to GCSF unless it is administered in pharmacologic doses. The latter possibility appears to be confirmed by the findings of Dong et al. (1994) of a mutation in the GCSFR gene, as discussed later.

Corcione et al. (1993) characterized a spontaneous lymphoblastoid cell line raised from the peripheral blood of a patient with Kostmann congenital neutropenia. Neutropenia had first been discovered at the age of 1 month. When examined at the age of 2 years and 9 months, the girl had a history of recurrent bacterial infections (pneumonitis, otitis media, cutaneous furuncles, and tonsillitis) since the age of 3 months. Her general condition was poor, and mild hepatosplenomegaly and diffuse lymph node enlargement were found. Corcione et al. (1993) found that the lymphoblastoid cell line was composed of EBV-infected polyclonal B cells. The supernatant contained a colony-inhibiting activity that was demonstrated to be transforming growth factor beta-1 (190180). Corcione et al. (1993) hypothesized that the B cells latently infected by EBV in vivo, and possibly expanded as a consequence of the infection, contributed to the inhibition of the patient's granulopoiesis by releasing TGF-beta-1.

Dong et al. (1994) used RT-PCR to amplify cDNA for granulocyte colony-stimulating factor receptor (138971) in patients with Kostmann syndrome and screened for mutations by single-strand conformation polymorphism (SSCP) analysis. In 1 patient, they identified a somatic point mutation that resulted in the cytoplasmic truncation of the GCSF receptor protein. Further functional characterization demonstrated that the truncated receptor was unable to transduce a maturation signal. The patient was heterozygous for the mutation. Dong et al. (1994) suggested that the mutant receptor chain may act in a dominant-negative manner to block granulocyte maturation. Dong et al. (1994) commented that Kostmann disease may be a heterogeneous group of disorders with different basic etiologies. They also commented that cases of this disorder that terminated in acute leukemia have been reported (Gilman et al., 1970; Lui et al., 1978; Rosen and Kang, 1979) and that some patients with the disorder have developed leukemia or myelodysplastic syndrome following treatment with GCSF.

Koren et al. (1989) reported a family in which a male and female sib with first-cousin parents had what was called congenital dysgranulopoietic neutropenia. The female suffered from omphalitis due to enterobacter at age 2 weeks and subsequently died from sepsis at age 1 month. The male was admitted at age 2 months with an abscess in the right inguinal region due to Pseudomonas aeruginosa. He suffered from frequent severe pyogenic infections. At the time of the report by Koren et al. (1994) he was 11 years old and had been treated successfully with GCFS from the age of 8. In 1992, the mother became pregnant and sonographically guided fetal blood sampling was performed by cordocentesis. The results of the blood studies indicated that the fetus was not at risk. The newborn baby was indeed healthy with normal neutrophil counts at 2 and 4 months of age.

Dong et al. (1995) described mutations in the GCSFR gene in 2 patients with acute myeloid leukemia and histories of severe congenital neutropenia. Like the mutation in the patient reported by Dong et al. (1994), the mutations truncated the C-terminal cytoplasmic region of the GCSF receptor. The mutation in one of the patients was already present in the neutropenic phase that preceded the development of acute myeloid leukemia.

Myeloid precursor cells from patients with severe congenital neutropenia (SCN) require pharmacologic dosages of recombinant human granulocyte colony-stimulating factor to differentiate normal neutrophils. Because JAK2 (147796), a nonreceptor tyrosine kinase, is involved in the signaling pathway of GCSF, Rauprich et al. (1995) studied the expression and activity of JAK2 in neutrophils from SCN patients during therapy. The immunoprecipitated JAK2 protein showed increased tyrosine phosphorylation in neutrophils from SCN patients as compared with that in neutrophils from healthy controls. Rauprich et al. (1995) pointed out that only a few patients, who subsequently develop acute myeloid leukemia, have a point mutation in the cytoplasmic region of the GCSF receptor, resulting in a truncation from the C terminus of the receptor and an inability of the receptor to transduce the signal on GCSF stimulation. Thus they suspected that various defects are responsible for SCN. That pharmacologic doses of GCSF are required to overcome the neutropenia suggested a defect of other specific molecules in the GCSF signal transduction pathway. The primary defect does not appear to be in JAK2; it may be that the phosphotyrosines on the receptor create binding sites for STAT proteins (signal transducers and activators of transcription; see 600555).

Yakisan et al. (1997) assessed bone mineral content and density in 30 patients with SCN who were treated with recombinant methionyl-human granulocyte colony stimulating factor (r-metHuGCSF). The authors noted that the availability of r-metHuGCSF for treatment of children with severe congenital neutropenia has substantially improved the patients' quality of life and life expectancy. However, bone pain and unusual fractures have been reported in treated patients. Yakisan et al. (1997) reviewed roentgenograms in 29 of 30 patients to evaluate bone loss before and during treatment and assessed bone mineral status in 17 of the 30 patients. Their data indicated a high incidence of bone mineral loss in children with severe congenital neutropenia. The investigators concluded that it is more likely that the bone loss was caused by the pathophysiologic features of the underlying disease; however, they could not rule out the possibility that r-metHuGCSF accelerates bone mineral loss.

In SCN, absolute neutrophil counts are usually less than 200 cells per cubic millimeter, with a remainder of the blood counts relatively normal. Treatment with granulocyte colony-stimulating factor leads to an increase in neutrophil counts to more than 1,000 cells per cubic millimeter in 90% of patients and results in significant improvements in survival and quality of life (Dale et al., 1993; Bonilla et al., 1994).

SCN patients are at increased risk of developing acute myelogenous leukemia (AML) or myelodysplasia (MDS). In the series of Welte and Dale (1996), 10% of the patients with SCN followed for 5 or more years developed AML or MDS. Patients with GCSFR mutations appeared to be at the greatest risk; Welte and Touw (1997) found that 8 of 16 patients with SCN and GCSFR mutations developed AML or MDS. Conversely, no patients with SCN and without a mutation of the CSF3R gene had been reported who developed AML or MDS. This striking association led to speculation that CSF3R mutations may contribute to leukemogenesis in these patients. To test this hypothesis directly, McLemore et al. (1998) generated mice carrying a targeted, 'knock-in' mutation of their Csf3r gene that reproduced the mutation found in a patient with SCN and AML. A point mutation (C to T at nucleotide 2403) was introduced into exon 17 of the Csf3r gene, using homologous recombination in embryonic stem cells. The mutation generated a premature stop codon that led to truncation of the C-terminal 96 amino acids and reproduced the mutation found in a patient with SCN by Dong et al. (1995). The mutant allele was expressed in a myeloid-specific fashion at levels comparable to the wildtype allele. Mice heterozygous or homozygous for this mutation had normal levels of circulating neutrophils and no evidence for a block in myeloid maturation, indicating that resting granulopoiesis was normal. However, in response to GCSF treatment, these mice demonstrated a significantly greater increase in the level of circulating neutrophils. This effect appeared to be due to increased neutrophil production as the absolute number of GCSF-responsive progenitors in the bone marrow and their proliferation in response to GCSF was increased. Furthermore, the in vitro survival and GCSF-dependent suppression of apoptosis of mutant neutrophils were normal. Despite this evidence for a hyperproliferative response to GCSF, no cases of AML were detected. These data demonstrated that the GCSFR mutation found in patients with SCN is not sufficient to induce either an SCN phenotype or AML in mice. McLemore et al. (1998) suggested that the results represent strong evidence that these mutations are not responsible for the impaired granulopoiesis present in patients with SCN. In fact, the results of the study suggested that expression of the mutant GCSFR on myeloid progenitors may render them hyperresponsive to GCSF. Whether this altered GCSF-responsiveness contributes to the development of AML and/or MDS in patients with SCN will require further study.

See also Andrews et al. (1960); Bjure et al. (1962)

Andrews, J. P.; McClellan, J. T.; Scott, C. H.: Lethal congenital neutropenia with eosinophilia occurring in two siblings. Am. J. Med. 29: 358-362, 1960.

Barak, Y.; Paran, M.; Levin, S.; Sachs, L.: In vitro induction of myeloid proliferation and maturation in infantile genetic agranulocytosis. Blood 38: 74-80, 1971.

Bjure, J.; Nilsson, L. R.; Plum, C. M.: Familial neutropenia possibly caused by deficiency of a plasma factor. Acta Paediat. 51: 497-508, 1962.

Bonilla, M. A.; Dale, D.; Zeidler, C.; Last, L.; Reiter, A.; Ruggeiro, M.; Davis, M.; Koci, B.; Hammond, W.; Gillio, A.; Welte, K.: Long-term safety of treatment with recombinant human granulocyte colony-stimulating factor (r-metHuG-CSF) in patients with severe congenital neutropenias. Brit. J. Haemat. 88: 723-730, 1994.

Bonilla, M. A.; Gillio, A. P.; Ruggeiro, M.; Kernan, N. A.; Brochstein, J. A.; Abboud, M.; Fumagalli, L.; Vincent, M.; Gabrilove, J. L.; Welte, K.; Souza, L. M.; O'Reilly, R. J.: Effects of recombinant human granulocyte colony-stimulating factor on neutropenia in patients with congenital agranulocytosis. New Eng. J. Med. 320: 1574-1580, 1989.

Corcione, A.; Roncella, S.; Cutrona, G.; Mori, P. G.; Ferrarini, M.; Pistoia, V.: Transforming growth factor beta-1 (TGF-beta-1) released by an Epstein-Barr virus (EBV) positive spontaneous lymphoblastoid cell line from a patient with Kostmann's congenital neutropenia inhibits the growth of normal committed haemopoietic progenitors in vitro. Brit. J. Haemat. 85: 684-691, 1993.

Dale, D. C.; Bonilla, M. A.; Davis, M. W.; Nakanishi, A. M.; Hammond, W. P.; Kurtzberg, J.; Wang, W.; Jakubowski, A.; Winton, E.; Lalezari, P.; Robinson, W.; Glaspy, J. A.; Emerson, S.; Gabrilove, J.; Vincent, M.; Boxer, L. A.: A randomized controlled phase III trial of recombinant human granulocyte colony-stimulating factor (filgrastim) for treatment of severe chronic neutropenia. Blood 81: 2496-2502, 1993.

Dong, F.; Brynes, R. K.; Tidow, N.; Welte, K.; Lowenberg, B.; Touw, I. P.: Mutations in the gene for the granulocyte colony-stimulating-factor receptor in patients with acute myeloid leukemia preceded by severe congenital neutropenia. New Eng. J. Med. 333: 487-493, 1995.

Dong, F.; Hoefsloot, L. H.; Schelen, A. M.; Broeders, L. C. A. M.; Meijer, Y.; Veerman, A. J. P.; Touw, I. P.; Lowenberg, B.: Identification of a nonsense mutation in the granulocyte-colony-stimulating factor receptor in severe congenital neutropenia. Proc. Nat. Acad. Sci. 91: 4480-4484, 1994.

Fischer, A.; Friedrich, W.; Levinsky, R.; Vossen, J.; Griscelli, C.; Kubanek, B.; Morgan, G.; Wagemaker, G.; Landais, P.: Bone-marrow transplantation for immunodeficiencies and osteopetrosis: European survey, 1968-1985. Lancet II: 1080-1084, 1986.

Gilman, P. A.; Jackson, D. P.; Guild, H. G.: Congenital agranulocytosis: prolonged survival and terminal acute leukemia. Blood 36: 576-585, 1970.

Hansen, J. A.; Dupont, B.; L'Esperance, P. L.; Good, R. A.: Congenital neutropenia: abnormal neutrophil differentiation associated with HLA. Immunogenetics 4: 327-332, 1977.

Hedenberg, F.: Infantile agranulocytosis of probably congenital origin. Acta Paediat. 48: 77-84, 1959.

Iselius, L.; Gustavson, K. H.: Spatial distribution of the gene for infantile genetic agranulocytosis. Hum. Hered. 34: 358-363, 1984.
Koren, A.; Bar Sela, P.; Barak, Y.; Jammalieh, J.; Katzuni, E.: Congenital dysgranulopoietic neutropenia in two siblings: clinical, ultrastructural and in vitro bone marrow culture studies. Pediat. Hemat. Oncol. 6: 293-305, 1989.
Koren, A.; Cohen, H.; Shneyour, I.; Shalev, E.: Prenatal diagnosis of congenital dysgranulopoietic neutropenia. Brit. J. Haemat. 88: 207-208, 1994.
Kostmann, R.: Infantile Genetic Agranulocytosis (agranulocytosis infantilis hereditaria): a new recessive lethal disease in man. Uppsala: Almqvist and Wiksells Boktryckeri (pub.) 1956.
Kostmann, R.: Infantile genetic agranulocytosis: a review with presentation of ten new cases. Acta Paediat. Scand. 64: 362-368, 1975.
L'Esperance, P. L.; Brunning, R.; Deinard, A. S.; Park, B. H.; Biggar, W. D.; Good, R. A.: Congenital neutropenia: impaired maturation with diminished stem-cell input.In: Bergsma, D.: Immunodeficiency in Man and Animals. New York: National Foundation-March of Dimes (pub.) 1975. Pp. 59-65.
L'Esperance, P. L.; Brunning, R.; Good, R. A.: Congenital neutropenia: in vitro growth of colonies mimicking the disease. Proc. Nat. Acad. Sci. 70: 669-672, 1973.
Lui, V.; Ragab, A. H.; Findley, H.; Frauen, B.: Infantile genetic agranulocytosis and acute lymphocytic leukemia in two sibs. J. Pediat. 92: 1028, 1978.
McLemore, M. L.; Poursine-Laurent, J.; Link, D. C.: Increased granulocyte colony-stimulating factor responsiveness but normal resting granulopoiesis in mice carrying a targeted granulocyte colony-stimulating factor receptor mutation derived from a patient with severe congenital neutropenia. J. Clin. Invest. 102: 483-492, 1998.
Olofsson, T.; Olsson, I.; Kostmann, R.; Malmstrom, S.; Thilen, A.: Granulopoiesis in infantile genetic agranulocytosis: in vitro cloning of marrow cells in agar culture. Scand. J. Haemat. 16: 18-24, 1976.
Rauprich, P.; Kasper, B.; Tidow, N.; Welte, K.: The protein tyrosine kinase JAK2 is activated in neutrophils from patients with severe congenital neutropenia. Blood 86: 4500-4505, 1995.
Rosen, R. B.; Kang, S.-J.: Congenital agranulocytosis terminating in acute myelomonocytic leukemia. J. Pediat. 94: 406-408, 1979.
Welte, K.; Dale, D.: Pathophysiology and treatment of severe chronic neutropenia. Ann. Hemat. 72: 158-165, 1996.
Welte, K.; Touw, I. P.: G-CSF receptor mutations in patients with severe chronic neutropenia: a step in leukemogenesis? (Abstract) Blood 90: 1921A only, 1997.
Yakisan, E.; Schirg, E.; Zeidler, C.; Bishop, N. J.; Reiter, A.; Hirt, A.; Riehm, H.; Welte, K.: High incidence of significant bone loss in patients with severe congenital neutropenia (Kostmann's syndrome). J. Pediat. 131: 592-597, 1997.

ANEMIA, CONGENITAL HYPOPLASTIC, OF BLACKFAN AND DIAMOND (205900)

JJM NOTE

The complicating neoplasm is acute nonlymphatic leukemia.

The Diamond-Blackfan syndrome (DBS), or Diamond-Blackfan anemia (DBA), characteristically presents as hypoplastic macrocytic anemia in the first year of life. A defect in the erythroid stem cell is suspected (Glader et al., 1983). Patients show a propensity for leukemia (Krishnan et al., 1978; Wasser et al., 1978). Familial cases have been reported by Burgert et al. (1954) and by Diamond et al. (1961). This disorder is sometimes encountered in the newborn and is progressive nonregenerative. There is no erythroblastosis, hemolysis or hepatosplenomegaly (until many transfusions have been given). Leukocytes and platelets are usually normal. Occasionally cortisone is effective. An abnormality of tryptophane metabolism, manifested by urinary excretion of anthranilic acid, has been found in some patients. Hirschman et al. (1969) reported 2 brothers with aplastic anemia similar to Fanconi anemia but without associated congenital anomalies. Both responded to androgen therapy and showed increased chromosomal breakage as in Fanconi anemia. One had a stable translocation chromosome in bone marrow cells. The other's skin fibroblasts showed increased susceptibility to 'malignant' transformation by SV40 virus, as in Fanconi anemia. Skin fibroblasts of the mother and a sister, both normal, also showed increased susceptibility to 'malignant' transformation. (Alter (1981) considered the cases of Hirshman et al. to be instances of Fanconi anemia.) The father was, it seems, not studied from this point of view. See 205600 for the syndrome of triphalangeal thumbs with congenital hypoplastic anemia. Dominant inheritance is suggested by at least 5 reported families; see 105650. Lawton et al. (1974) described father and son with documented anemia from infancy. The father's anemia remitted at age 6 years but he continued to have macrocytosis, reticulocytosis, and raised fetal hemoglobin. Forare (1963) observed affected brother and sister with the same father but different mothers. Falter and Robinson (1972) described affected mother and daughter. Hamilton et al. (1974) described affected mother and daughter. In addition, Wallman (1956) described a father and daughter with erythroid hypoplasia but the ages of onset (34 and 6 years, respectively) were beyond the usual limits of the Diamond-Blackfan syndrome. Sensenbrenner (1972) described affected brother and sister. Pallor was first noted in the male at age 4 months and heart failure from anemia occurred at 10 months. Prednisone controlled the anemia. His height was below the 3rd percentile. At age 16 he was 147 cm tall and showed good sexual maturation. He had aseptic necrosis of the left hip. The sister was below the 3rd percentile for height from at least 2 years. At age 11 she showed good sexual maturation but her height was only 127 cm. Cathie (1950) described a similar facial appearance in 4 unrelated affected children: snub noses, thick upper lips and widely separated eyes.

Nathan et al. (1978) suggested that Diamond-Blackfan anemia may be a 'congenital abnormality of erythropoietin (133170) responsiveness that causes a functional, if not absolute, deficiency of erythroid precursors.' Li and Potter (1978) found Fanconi anemia (227650) in a close relative of the 5 sibs with hypoplastic anemia reported by Estren and Dameshek (1947). (The parents of the Fanconi patient were second cousins and both were first cousins of the sibs.) The authors raised the question of possible allelism of hypoplastic and Fanconi anemia, and specifically whether the patients of Estren and Dameshek might have been genetic compounds. Familial cases, which are the exception, were observed also by Wang et al. (1978) and Gray (1982).

Glader et al. (1983) found increased adenosine deaminase (102700) activity in red cells. Whitehouse et al. (1984) extended the findings of Glader et al. (1983). They found heterogeneity in DBS in respect to erythrocyte ADA activity and concluded that increased ADA activity is not limited to erythroid cells. Two sibs in 1 family showed increased red cell ADA activity over 4 months of multiple blood sampling. That the high activity was not due to mutation at the ADA locus was indicated by the fact that both patients had the ADA 2-1 electrophoretic pattern and that both allelozymes showed hyperactivity. Thus, a trans-acting regulatory mutation is likely. Schmid and Fanconi (1978) referred to the disorder in patients with the hematologic and cytogenetic abnormalities similar to those of Fanconi anemia as 'Fanconi anemia, type Estren-Dameshek' and Bloom et al. (1966) called it 'constitutional aplastic anemia type II.' In studies of a family with 2 affected sisters, Nowell et al. (1984) extended observations indicating the similarity of the Estren-Dameshek hypoplastic anemia and Fanconi anemia: increased chromosomal fragility, progression to preleukemia (with a clone of chromosomally abnormal cells in the bone marrow), and then to acute nonlymphocytic leukemia. In 2 out of 6 patients, Dunbar et al. (1991) observed sustained remission following treatment with interleukin-3 (147740).

Abkowitz et al. (1991) cultured marrow and blood mononuclear cells from 10 Diamond-Blackfan patients with various hematopoietic growth factors in the presence or absence of stem cell factor (SCF; mast cell growth factor; Steel factor; SF; 184745). Because of erythroid bursts observed in cultures containing SCF, they speculated that the c-kit receptor/ligand axis may be involved in the pathogenesis of Diamond-Blackfan anemia and suggested that a therapeutic trial of SCF in patients would be worthwhile. Similar results were obtained by Bagnara et al. (1991) and by Carow et al. (1991). Olivieri et al. (1991) did in vitro trials in 10 Diamond-Blackfan patients and observed 3 patterns of marrow response: those that responded quantitatively and qualitatively almost as well as cells from normal marrow, those that responded at an intermediate level, and those that did not respond at all. The results supported a fundamental heterogeneity in the pathogenesis of this disorder which is suggested also by the fact that both recessive and dominant (105650) forms of the disorder may exist. Olivieri et al. (1991) found no gross abnormality in the structure of either the KIT (164920) or Steel locus in these patients. Spritz and Freedman (1993) found no mutations in either the mast/stem cell growth factor or in its tyrosine kinase receptor, KIT.

Sieff et al. (1992) investigated whether DBA is due to hyporesponsiveness to or hypoproduction of Steel factor. By studying long-term bone marrow cultures, they found that the DBA patients studied responded to SF and produced SF mRNA normally, indicating that SF itself is not involved in DBA pathophysiology. The effects suggested, however, that, despite lack of evidence for a causative role, SF may prove to be effective treatment for such patients.

Dianzani et al. (1996) stated that there is neither molecular nor clinical evidence for the involvement of stem cell factor, or interleukin-3, in the pathogenesis of DBA. DBA patients display an impaired in vivo and in vitro response to pharmacologic doses of erythropoietin (Bagnara et al., 1991; Nathan et al., 1978). Bagnara et al. (1991) also found that serum erythropoietin concentration is high in proportion to the severity of anemia. However, Halperin and Freedman (1989) stated that the structure of EPO is normal and anti-EPO antibodies are absent, indicating that abnormalities in EPO receptor expression, EPO binding, or EPO signal transduction may be present. Dianzani et al. (1996) could find no abnormality of the coding sequence of the EPOR gene (133171) in 23 DBA patients using SSCP. A Southern blot hybridization with an EPOR probe was also negative in 7 patients. Furthermore, in 2 informative families linkage studies showed that the disorder did not segregate with the EPOR gene.

Ball et al. (1996) analyzed retrospective data from 80 cases of DBA (33 male, 47 female) born in the U.K. in a 20-year period (1975-1994), representing an annual incidence of 5 per million live births. Ten children from 7 families had an apparently familial disorder. Thirteen percent were anemic at birth, and 72.5% had presented by the age of 3 months. Sixty-seven percent had macrocytosis at presentation, 72% responded initially to steroids, and at the time of study 61% were transfusion-dependent (45% steroid-dependent) and 39% required regular transfusions. Unequivocal physical anomalies, predominantly craniofacial, were present in 37%, and were more likely in boys (52%) than girls (25%). Eighteen percent had thumb anomalies. Height was below the third centile for age in 28%, and 31% had neither short stature nor physical anomalies. In 4 children without physical abnormalities, red cell indices were normal and steroid-independent remission was achieved, suggesting transient erythroblastopenia of childhood rather than DBA. The birth month distribution of children with sporadic DBA and craniofacial dysmorphism suggested a possible seasonality, consistent with a viral etiology. In the familial cases affected males had unequivocal anomalies, whereas females had only short stature or equivocal anomalies. Two generations were involved in 3 families and in 1 family a boy in the cohort studied had both an affected father and affected paternal grandfather.

Gustavsson et al. (1997) were motivated to undertake a linkage analysis with chromosome 19 markers in multiplex Diamond-Blackfan anemia families because of observations of a patient with DBA and a chromosomal translocation involving 19q. They studied 26 families with DBA. Of the 6 pedigrees presented, autosomal recessive inheritance was thought likely in 2 families, whereas dominant inheritance with variable expressivity appeared to be the case in 4 families. In 1 family the disease was evident in 3 generations with 2 instances of male-to-male transmission. In 2 of the families the mother showed a mild anemia. In a fourth family no phenotype was detected in the parents but the segregation of haplotypes indicated dominant inheritance from the mother. Gustavsson et al. (1997) found linkage for both the dominant and recessive inherited DBA to 19q13 with a peak lod score at D19S197 (maximum lod = 7.08, theta = 0.00). Within this region, a submicroscopic de novo deletion of 3.3 Mb was identified in a patient with DBA. The deletion coincided with the translocation breakpoint observed in the patient mentioned earlier and, together with key recombinations, restricted the DBA gene to a 1.8-Mb region. Their results suggested that despite the heterogeneity in clinical presentation and inheritance pattern, DBA is genetically homogeneous with the causative gene located at 19q13.

Using polymorphic 19q13 markers, including a newly identified short-tandem repeat in the critical gene region, Gustavsson et al. (1998) studied 29 multiplex DBA families and 50 families with sporadic DBA cases. In 26 of the 29 multiplex families, DNA analysis yielded results consistent with a DBA gene on 19q within a 4.1-cM interval restricted by D19S200 and D19S178; however, in 3 multiplex families, the DBA candidate region on 19q13 was excluded from the segregation of marker alleles. This result suggested genetic heterogeneity for DBA, but indicated that the gene region on 19q segregates with the majority of familial cases. Among the 50 families comprising sporadic DBA cases, Gustavsson et al. (1998) identified 2 de novo and overlapping microdeletions on 19q13. In combination, the 3 known microdeletions associated with DBA restricted the critical gene region to approximately 1 Mb.

'Congenital hypoplastic anemia' was the term used by Diamond et al. (1961), subsequently called Diamond-Blackfan anemia (DBA). Confusingly, Estren and Dameshek (1947) had used the designation for the disorder in 2 families that were subsequently shown to have Fanconi anemia (Li and Potter, 1978). Young and Alter (1994) concluded that the proportion of FA homozygotes without external anomalies is underestimated by literature review. Such patients represent one end of the spectrum of FA. At least 25% of homozygotes identified because they had affected sibs lacked anomalies.

See also Altman and Miller (1953); Hoffman et al. (1976); Hunter and Hakami (1972); Kass and Sundal (1953); Mott et al. (1969); Pearson and Cone (1957)

Abkowitz, J. L.; Sabo, K. M.; Nakamoto, B.; Blau, C. A.; Martin, F. H.; Zsebo, K. M.; Papayannopoulou, T.: Diamond-Blackfan anemia: in vitro response of erythroid progenitors to the ligand for c-kit. Blood 78: 2198-2202, 1991.

Alter, B. P.: Personal Communication. New York, N. Y. 10/12/1981.

Altman, K. I.; Miller, G.: A disturbance of tryptophan metabolism in congenital hypoplastic anemia. Nature 172: 868, 1953.

Bagnara, G. P.; Zauli, G.; Vitale, L.; Rosito, P.; Vecchi, V.; Paolucci, G.; Avanzi, G. C.; Ramenghi, U.; Timeus, F.; Gabutti, V.: In vitro growth and regulation of bone marrow enriched CD34(+) hematopoietic progenitors in Diamond-Blackfan anemia. Blood 78: 2203-2210, 1991.

Ball, S. E.; McGuckin, C. P.; Jenkins, G.; Gordon-Smith, E. C.: Diamond-Blackfan anaemia in the U.K.: analysis of 80 cases from a 20-year birth cohort. Brit. J. Haemat. 94: 645-653, 1996.

Bloom, G. E.; Warner, S.; Gerald, P. S.; Diamond, L. K.: Chromosome abnormalities in constitutional aplastic anemia. New Eng. J. Med. 274: 8-14, 1966.

Burgert, E. O., Jr.; Kennedy, R. L. J.; Pease, G. L.: Congenital hypoplastic anemia. Pediatrics 13: 218-226, 1954.

Carow, C. E.; Hangoc, G.; Cooper, S. H.; Williams, D. E.; Broxmeyer, H. E.: Mast cell growth factor (c-kit ligand) supports the growth of human multipotential progenitor cells with a high replating potential. Blood 78: 2216-2221, 1991.

Cathie, I. A. B.: Erythrogenesis imperfecta. Arch. Dis. Child. 25: 313-324, 1950.

Diamond, L. K.; Allen, D. W.; Magill, F. B.: Congenital (erythroid) hypoplastic anemia: a 25 year study. Am. J. Dis. Child. 102: 403-415, 1961.

Dianzani, I.; Garelli, E.; Dompe, C.; Crescenzio, N.; Locatelli, F.; Schiliro, G.; Castaman, G.; Bagnara, G. P.; Olivieri, N. F.; Gabutti, V.; Ramenghi, U.: Mutations in the erythropoietin receptor gene are not a common cause of Diamond-Blackfan anemia. Blood 87: 2568-2572, 1996.

Dunbar, C. E.; Smith, D. A.; Kimball, J.; Garrison, L.; Nienhuis, A. W.; Young, N. S.: Treatment of Diamond-Blackfan anaemia with haematopoietic growth factors, granulocyte-macrophage colony stimulating factor and interleukin 3: sustained remissions following IL-3. Brit. J. Haemat. 79: 316-321, 1991.

Estren, S.; Dameshek, W.: Familial hypoplastic anemia of childhood: report of eight cases in two families with beneficial effect of splenectomy in one case. Am. J. Dis. Child. 73: 671-687, 1947.

Falter, M. L.; Robinson, M. G.: Autosomal dominant inheritance and aminoaciduria in Blackfan-Diamond anemia. J. Med. Genet. 9: 64-66, 1972.

Forare, S. A.: Pure red cell anemia in step siblings. Acta Paediat. 52: 159-160, 1963.

Glader, B. E.; Backer, K.; Diamond, L. K.: Elevated erythrocyte adenosine deaminase activity in congenital hypoplastic anemia. New Eng. J. Med. 309: 1486-1490, 1983.

Gray, P. H.: Pure red-cell aplasia: occurrence in three generations. Med. J. Aust. 1: 519-521, 1982.

Gustavsson, P.; Garelli, E.; Draptchinskaia, N.; Ball, S.; Willig, T.-N.; Tentler, D.; Dianzani, I.; Punnett, H. H.; Shafer, F. E.; Cario, H.; Ramenghi, U.; Glomstein, A.; Pfeiffer, R. A.; Goringe, A.; Olivieri, N. F.; Smibert, E.; Tchernia, G.; Elinder, G.; Dahl, N.: Identification of microdeletions spanning the Diamond-Blackfan anemia locus on 19q13 and evidence for genetic heterogeneity. Am. J. Hum. Genet. 63: 1388-1395, 1998.

Gustavsson, P.; Willig, T.-N.; van Haeringen, A.; Tchernia, G.; Dianzani, I.; Donner, M.; Elinder, G.; Henter, J.-I.; Nilsson, P.-G.; Gordon, L.; Skeppner, G.; van't Veer-Korthof, L.; Kreuger, A.; Dahl, N.: Diamond-Blackfan anaemia: genetic heterogeneity for a gene on chromosome 19q13 restricted to 1.8 Mb. Nature Genet. 16: 368-371, 1997.

Halperin, D. S.; Freedman, M. H.: Diamond-Blackfan anemia: etiology, pathophysiology, and treatment. Am. J. Pediat. Hemat. Oncol. 11: 380-394, 1989.

Hamilton, P. J.; Dawson, A. A.; Galloway, W. H.: Congenital erythroid hyperplastic anaemia in mother and daughter. Arch. Dis. Child. 49: 71-73, 1974.

Hirschman, R. J.; Shulman, N. R.; Abuelo, J. G.; Whang-Peng, J.: Chromosomal aberrations in two cases of inherited aplastic anemia with unusual clinical features. Ann. Intern. Med. 71: 107-117, 1969.

Hoffman, R.; Zanjani, E. D.; Vila, J.; Zalusky, R.; Lutton, J. D.; Wasserman, L. R.: Diamond-Blackfan syndrome: lymphocyte-mediated suppression of erythropoiesis. Science 193: 899-900, 1976.

Hunter, R. E.; Hakami, N.: The occurrence of congenital hypoplastic anemia in half-brothers. J. Pediat. 81: 346-348, 1972.

Kass, A.; Sundal, A.: Anaemia hypoplastica congenita (anaemia typus Josephs-Diamond-Blackfan). Report of a case treated with adrenocorticotropin with effect. Acta Paediat. 42: 265-274, 1953.

Krishnan, E. U.; Wegner, K.; Garg, S. K.: Congenital hypoplastic anemia terminating in acute promyelocytic leukemia. Pediatrics 61: 898-901, 1978.

Lawton, J. W. M.; Aldrich, J. E.; Turner, T. L.: Congenital erythroid hypoplastic anemia: autosomal dominant transmission. Scand. J. Haemat. 13: 276-280, 1974.

Li, F. P.; Potter, N. U.: Classical Fanconi anemia in a family with hypoplastic anemia. J. Pediat. 92: 943-944, 1978.

Mott, M. G.; Apley, J.; Raper, A. B.: Congenital (erythroid) hypoplastic anaemia: modified expression in males. Arch. Dis. Child. 44: 757-760, 1969.

Nathan, D. G.; Clarke, B. J.; Hillman, D. G.; Alter, B. P.; Housman, D. E.: Erythroid precursors in congenital hypoplastic (Diamond-Blackfan) anemia. J. Clin. Invest. 61: 489-498, 1978.

Nowell, P.; Bergman, G.; Besa, E.; Wilmoth, D.; Emanuel, B.: Progressive preleukemia with a chromosomally abnormal clone in a kindred with the Estren-Dameshek variant of Fanconi's anemia. Blood 64: 1135-1138, 1984.

Olivieri, N. F.; Grunberger, T.; Ben-David, Y.; Ng, J.; Williams, D. E.; Lyman, S.; Anderson, D. M.; Axelrad, A. A.; Correa, P.; Bernstein, A.; Freedman, M. H.: Diamond-Blackfan anemia: heterogenous response of hematopoietic progenitor cells in vitro to the protein product of the Steel locus. Blood 78: 2211-2215, 1991.

Pearson, H. A.; Cone, T. E., Jr.: Congenital hypoplastic anemia. Pediatrics 19: 192-200, 1957.

Schmid, W.; Fanconi, G.: Fragility and spiralization anomalies of the chromosome in three cases, including fraternal twins, with Fanconi's anaemia, type Estren-Dameshek. Cytogenet. Cell Genet. 20: 141-149, 1978.

Sensenbrenner, J. A.: Congenital hypoplastic anemia of Blackfan and Diamond in sibs. Birth Defects Orig. Art. Ser. 8(3): 166-170, 1972.

Sieff, C. A.; Yokoyama, C. T.; Zsebo, K. M.; Trammell, J.; Andersen, J. W.; Nathan, D. G.; Williams, D. A.: The production of Steel factor mRNA in Diamond-Blackfan anaemia long-term cultures and interactions of Steel factor with erythropoietin and interleukin-3. Brit. J. Haemat. 82: 640-647, 1992.

Spritz, R. A.; Freedman, M. H.: Lack of mutations of the MGF and KIT genes in Diamond-Blackfan anemia. (Letter) Blood 81: 3165, 1993.

Wallman, I. S.: Hereditary red cell aplasia. Med. J. Aust. 2: 488-490, 1956.

Wang, W. C.; Mentzer, W.; Alter, B.: Congenital hypoplastic anemia: Diamond-Blackfan syndrome: comments and additional data on clinical aspects of Diamond-Blackfan syndrome. Blood Cells 4: 215-218, 1978.

Wasser, J. S.; Yolken, R.; Miller, D. R.; Diamond, L.: Congenital hypoplastic anemia (Diamond-Blackfan syndrome) terminating in acute myelogenous leukemia. Blood 51: 991-995, 1978.
Whitehouse, D. B.; Hopkinson, D. A.; Evans, D. I. K.: Adenosine deaminase activity in Diamond-Blackfan syndrome. (Letter) Lancet II: 1398-1399, 1984.
Young, N. S.; Alter, B. P.: Clinical features of Fanconi's Anemia.In: Aplastic Anemia Acquired and Inherited. Philadelphia, Pa.: W. B. Saunders Co. 1994. P. 281.

PANCREATIC INSUFFICIENCY AND BONE MARROW DYSFUNCTION (*260400)

JJM NOTE

The similarity of risk for acute leukemia in Shwachman syndrome and Fanconi pancytopenia suggests a final common pathway from a primary bone marrow failure or insufficiency to frank malignancy. Again, an odd cell type has been seen, acute monoblastic leukemia.

Shwachman et al. (1964) described a syndrome of pancreatic insufficiency (suggesting cystic fibrosis of the pancreas but with normal sweat electrolytes and no respiratory difficulties) and pancytopenia. One sibship contained 2 affected brothers and an affected female. The same syndrome was described by Nezelof and Watchi (1961) and later by other authors such as Pringle et al. (1968). Goldstein (1968) and others before him called this condition congenital lipomatosis of the pancreas. He described one affected fraternal twin girl. Affected sibs were referred to by Burke et al. (1967) and Pringle et al. (1968) observed associated skeletal changes of the metaphyseal dysostosis type. These are of interest because of the digestive abnormalities (not yet well characterized) and hematologic changes in cartilage-hair hypoplasia (250250), a form of metaphyseal chondrodysplasia. The exocrine pancreas is replaced by fat, whereas the islets of Langerhans are normal. Although dwarfing is usually moderate and becomes apparent only after 1 or 2 years of life, Danks et al. (1976) described 2 pairs of brothers who showed neonatal respiratory distress resembling that of Jeune syndrome (208500). The true nature of the osseous disorder became clear in the second or third year of life. Susceptibility to infection was marked in 1 family and led to death of 1 of the brothers. Patients with Shwachman syndrome may have exocrine pancreatic insufficiency without hematologic abnormalities. They are predisposed to hematologic malignancies similar to those that occur with Fanconi anemia (Woods et al., 1981). Smith et al. (1995) described a 5-year-old boy with this disorder in whom acute monoblastic leukemia developed following a period of myelodysplasia associated with a clonal cytogenetic abnormality involving chromosome 7.

Rothbaum et al. (1982) postulated that abnormal polymorphonuclear chemotaxis reflects defective cytoskeletal integrity in the Shwachman syndrome. In support of this idea, they demonstrated abnormal distribution of concanavalin-A receptors on polymorphonuclear leukocytes. Genieser et al. (1982) demonstrated the usefulness of computed tomography (CT scan) in the diagnosis. From the early paper of Bartholomew et al. (1959) it appears that so-called primary atrophy of the pancreas may be, in some instances, the same disorder and that manifestations may develop first after the fifth decade of life. Tada et al. (1987) found increased frequencies of spontaneous chromosome aberrations in a patient's PHA-stimulated circulating lymphocytes; however, the lymphocytes did not show increased sensitivity to mitomycin C. In 2 affected sisters, Fraccaro et al. (1988) were unable to confirm the observation of Tada et al. (1987) of increased chromosome aberrations. Wulfeck et al. (1991), who referred to this disorder as Shwachman-Diamond syndrome, evaluated 2 affected sisters, aged 8 and 13 years, in whom the most prominent neurologic abnormality was global apraxia, which affected their motor skills. Generalized weakness and hypotonia were also observed.

Masuno et al. (1995) observed a de novo and apparently balanced reciprocal translocation t(6;12)(q16.2;q21.2) in an 18-month-old girl with Shwachman syndrome, characterized by exocrine pancreatic insufficiency in bone marrow dysfunction. They suggested that the translocation breakpoints in this patient are candidate regions for a gene responsible for Shwachman syndrome.

Scott Hamilton, 1984 Olympic Gold Medallist figure skater, was ill as a child with Shwachman syndrome.

Smith et al. (1996) reported hematologic abnormalities in 21 children diagnosed with Shwachman-Diamond syndrome at their institution over 25 years. Anemia was found in 14 patients, thrombocytopenia in 5, and pancytopenia in 2 patients. Bone marrow cellularity was decreased in 5 and increased in 3 out of 13 patients studied. Cytogenetic examination of the bone marrow showed clonal abnormalities in 4 out of 12 children at the time of diagnosis, and 1 boy developed a clone later in the course of his illness. Chromosome 7 was involved in rearrangements in 4 children. Myelodysplastic syndrome developed in 7 patients (including all 5 with clonal bone marrow abnormalities); 5 of these persons developed acute myeloid leukemia and died. Smith et al. (1996) showed that the actual risk of leukemic transformation in the patients with Shwachman-Diamond syndrome is much higher than 5% (as it was previously considered), and clonal cytogenetic abnormalities in the bone marrow predispose to such transformation.

Mack et al. (1996) reviewed findings in 25 patients. Mean birth weight was at the twenty-fifth percentile; however, by 6 months of age, mean heights and weights were less than the fifth percentile. After 6 months of age, growth velocity was normal. Neutropenia was the most common hematologic abnormality (88%), but leukopenia, thrombocytopenia, and anemia were also frequently encountered. Patients with hypoplasia of all 3 bone marrow cellular lines (n = 11) had the worst prognosis; 5 patients died, 2 of sepsis and 3 of acute myelogeneous leukemia.

Dokal et al. (1997) described 3 men (2 of whom were brothers) with Shwachman-Diamond syndrome who presented with acute myeloid leukemia in adulthood. The brothers were 37 and 43 at time of presentation. The third patient was 25 years old. Dokal et al. (1997) pointed out that of the cases of acute myeloid leukemia in SDS, approximately one-quarter (5 in 18) have M6 morphology. They suggested that the only therapy likely to be successful is allogeneic bone marrow transplantation, success with which had been reported in several cases.

See also Bodian et al. (1964); McLennan and Steinbach (1974); Saint-Martin et al. (1969); Saunders et al. (1979); Shmerling et al. (1969); Shwachman and Holsclaw (1972); Smith et al. (1996); Taybi et al. (1969)

Bartholomew, L. G.; Baggenstoss, A. H.; Morlock, C. G.; Comfort, M. W.: Primary atrophy and lipomatosis of the pancreas. Gastroenterology 36: 563-572, 1959.
Bodian, M.; Sheldon, W.; Lightwood, R.: Congenital hypoplasia of the exocrine pancreas. Acta Paediat. 53: 282-293, 1964.
Burke, V.; Colebatch, J. H.; Anderson, C. M.; Simons, M. J.: Association of pancreatic insufficiency and chronic neutropenia in childhood. Arch. Dis. Child. 42: 147-157, 1967.
Danks, D. M.; Haslam, R. H. A.; Mayne, V.; Kaufmann, H. J.; Holtzapple, P. G.: Metaphyseal chondrodysplasia, neutropenia, and pancreatic insufficiency presenting with respiratory distress in the neonatal period. Arch. Dis. Child. 51: 697-701, 1976.
Dokal, I.; Rule, S.; Chen, F.; Potter, M.; Goldman, J.: Adult onset of acute myeloid leukaemia (M6) in patients with Shwachman-Diamond syndrome. Brit. J. Haemat. 99: 171-173, 1997.
Fraccaro, M.; Scappaticci, S.; Arico, M.: Shwachman syndrome and chromosome breakage. (Letter) Hum. Genet. 79: 194, 1988.
Genieser, N. B.; Halac, E. R.; Greco, M. A.; Selvarani Richards, H. M.: Shwachman-Bodian syndrome. J. Comput. Assist. Tomogr. 6: 1191-1192, 1982.
Goldstein, R.: Congenital lipomatosis of the pancreas. Malabsorption, dwarfism, leukopenia with relative granulocytopenia and thrombocytopenia. Clin. Pediat. 7: 419-422, 1968.
Mack, D. R.; Forstner, G. G.; Wilschanski, M.; Freedman, M. H.; Durie, P. R.: Shwachman syndrome: exocrine pancreatic dysfunction and variable phenotypic expression. Gastroenterology 111: 1593-1602, 1996.
Masuno, M.; Imaizumi, K.; Nishimura, G.; Nakamura, M.; Saito, I.; Akagi, K.; Kuroki, Y.: Shwachman syndrome associated with de novo reciprocal translocation t(6;12)(q16.2;q21.2). J. Med. Genet. 32: 894-895, 1995.
McLennan, T. W.; Steinbach, H. L.: Shwachman's syndrome: the broad spectrum of bony abnormalities. Radiology 112: 167-173, 1974.
Nezelof, C.; Watchi, M.: L'hypoplasie congenitale lipomateuse du pancreas exocrine chez l'enfant (Deux observations et revue de la litterature). Arch. Franc. Pediat. 18: 1135-1172, 1961.
Pringle, E. M.; Young, W. F.; Haworth, E. M.: Syndrome of pancreatic insufficiency, blood dyscrasia and metaphyseal dysplasia. Proc. Roy. Soc. Med. 61: 776-777, 1968.
Rothbaum, R. J.; Williams, D. A.; Daugherty, C. C.: Unusual surface distribution of concanavalin A reflects a cytoskeletal defect in neutrophils in Shwachman's syndrome. Lancet II: 800-801, 1982.
Saint-Martin, J.; Fournet, J. P.; Charlas, J.; Schaison, G.; Nodot, A.; Meyer, B.; Vialatte, J.: Insuffisance pancreatique externe avec granulopenie chronique. Arch. Franc. Pediat. 26: 861-871, 1969.
Saunders, E. F.; Gall, G.; Freedman, M. H.: Granulopoiesis in Shwachman's syndrome (pancreatic insufficiency and bone marrow dysfunction). Pediatrics 64: 515-519, 1979.
Shmerling, D. H.; Prader, A.; Hitzig, W. H.; Giedion, A.; Hadorn, B.; Kuhni, M.: The syndrome of exocrine pancreatic insufficiency, neutropenia, metaphyseal dysostosis and dwarfism. Helv. Paediat. Acta 24: 547-575, 1969.
Shwachman, H.; Diamond, L. K.; Oski, F. A.; Khaw, K. T.: The syndrome of pancreatic insufficiency and bone marrow dysfunction. J. Pediat. 65: 645-663, 1964.
Shwachman, H.; Holsclaw, D.: Some clinical observations on the Shwachman syndrome (pancreatic insufficiency and bone marrow hypoplasia). Birth Defects Orig. Art. Ser. 8(3): 46-49, 1972.
Smith, O. P.; Chan, M. Y.; Evans, J.; Veys, P.: Shwachman-Diamond syndrome and matched unrelated donor BMT. Bone Marrow Trans. 16: 717-718, 1995.

Smith, O. P.; Hann, I. M.; Chessells, J. M.; Reeves, B. R.; Milla, P.: Haematological abnormalities in Shwachman-Diamond syndrome. Brit. J. Haemat. 94: 279-284, 1996.

Smith, O. P.; Hann, I. M.; Chessells, J. M.; Reeves, B. R.; Milla, P.: Haematological abnormalities in Shwachman-Diamond syndrome. Brit. J. Haemat. 94: 279-284, 1996.

Tada, H.; Ri, T.; Yoshida, H.; Ishimoto, K.; Kaneko, M.; Yamashiro, Y.; Shinohara, T.: A case of Shwachman syndrome with increased spontaneous chromosome breakage. Hum. Genet. 77: 289-291, 1987.
Taybi, H.; Mitchell, A. D.; Friedman, G. D.: Metaphyseal dysostosis and associated syndrome of pancreatic insufficiency and blood disorders. Radiology 93: 563-571, 1969.
Woods, W. G.; Roloff, J. S.; Lukens, J. N.; Krivit, W.: The occurrence of leukemia in patients with the Shwachman syndrome. J. Pediat. 99: 425-428, 1981.
Wulfeck, B.; Trauner, D.; Marsden, D.; Nyhan, W.; Tallal, P.: Neurobehavioral profiles in two sisters with Shwachman-Diamond syndrome. Dysmorph. Clin. Genet. 5: 15-22, 1991.

MYELOCEREBELLAR DISORDER (ATAXIA-PANCYTOPENIA SYNDROME) (159550)

JJM NOTE

Another rare pattern was signaled by the presence of an unusual cell type: acute myelomonocytic leukemia.

Li et al. (1978, 1981) described a family with ataxia and pancytopenia. The propositus, the youngest of a sibship of 5, had cerebellar ataxia, developed hypoplastic anemia at age 3 years, and died of acute myelomonocytic leukemia at age 7. Serial cytogenetic studies of his hypoplastic bone marrow over a 25-month period showed progressive expansion of a clone of cells with monosomy 7. Cerebellar ataxia was found in the proband's father and all 4 sibs. Two brothers, including 1 with monosomy 7, died with hypoplastic anemia at ages 5 and 9 years, and a third brother died with acute myelocytic leukemia at age 10 years. The only surviving sib at the time of the report of Li et al. (1981) was a 19-year-old sister who had unexplained anemia, decreased mitotic activity in the bone marrow, and a slowly progressive cerebellar ataxia. In addition to progressive cerebellar ataxia, the father had marked progressive cerebellar ataxia, brisk deep tendon reflexes, bilateral Babinski reflexes, and diminished vibratory sensation in the legs. Unsteady gait first brought the father to medical attention at age 28 years and he was not able to work thereafter. Computed tomographic scans of the father, the sister, and 1 brother showed cerebellar atrophy. Autopsy in 1 brother showed marked cerebellar atrophy and loss of 90% of the Purkinje cells; the spinal column was normal. By 1989, the sister had given up her career in nursing because of increasingly severe cerebellar ataxia and the father was largely confined at home (Li, 1989). In a family reported by Daghistani et al. (1989), the proband, a 5-year-old boy, presented with left facial abscess and pancytopenia, following a 2-year history of mild incoordination. He had nystagmus and dysmetria. Monosomy 7 was found. He died at the age of about 6 years from pancytopenia. The sister of the proband, 12 years old at the time of report, had the onset of mild progressive unsteadiness at the age of 7 years. She had nystagmus and dysarthria. Nerve conduction studies showed reduced velocities in both sibs. The mother of the 2 children had had onset of incoordination before age 5. She showed severe gait ataxia and pronounced dysarthria with upper extremity dysmetria and horizontal and vertical nystagmus. Nerve conduction studies showed reduced velocities. The father was normal neurologically and hematologically. None of the 3 patients had oculocutaneous telangiectasia, scoliosis, or cardiac dysfunction.

Daghistani, D.; Curless, R.; Toledano, S. R.; Ayyar, D. R.: Ataxia-pancytopenia and monosomy 7 syndrome. J. Pediat. 115: 108-110, 1989.
Li, F. P.: Personal Communication. Boston, Mass. 10/3/1989.
Li, F. P.; Hecht, F.; Kaiser-McCaw, B.; Baranko, P. V.; Potter, N. U.: Ataxia-pancytopenia: syndrome of cerebellar ataxia, hypoplastic anemia, monosomy 7, and acute myelogenous leukemia. Cancer Genet. Cytogenet. 4: 189-196, 1981.
Li, F. P.; Potter, N. U.; Buchanan, G. R.; Vawter, G.; Whang-Peng, J.; Rosen, R. B.: A family with acute leukemia, hypoplastic anemia and cerebellar ataxia: association with bone marrow C-monosomy. Am. J. Med. 65: 933-940, 1978.

WT LIMB-BLOOD SYNDROME (*194350)

JJM NOTE

This syndrome, which deserves revisiting, represents perhaps the only dominant pattern of frank leukemia.

Gonzalez et al. (1977) reported 2 families, with surnames beginning with W and T, in which multiple members had severe hypoplastic anemia, congenital anomalies, and leukemia. Important characteristics differentiating this condition from Fanconi anemia (227650) were autosomal dominant inheritance and absence of chromosome breakage. In addition, the congenital anomalies were somewhat different. The W family was ascertained through a 13-year-old girl without congenital anomalies who developed acute lymphocytic leukemia at the age of 11.5 years. Her father had clinodactyly V bilaterally and at 26 years was found to have pancytopenia refractory to treatment and leading to death 1 year later. His oldest brother was born with congenital malformation of the elbow and hands and was healthy until age 38 when he was found to have idiopathic anemia evolving into refractory pancytopenia. He died at 42 years of age. Their mother had died at age 71 of leukemia. The propositus of the T family had malformations of both hands at birth and at 21 months was found to have anemia for which he received transfusions. Several paternal relatives had severe hand anomalies and a history of childhood anemia. The paternal grandfather died at the age of 51 of acute monocytic leukemia. Gonzalez et al. (1977) concluded that some earlier reports of presumed Fanconi anemia actually are instances of the WT syndrome. For example, McDonald and Goldschmidt (1960) reported 3 children and their mother who had severe anemia in childhood. All the children had abnormal thumbs. McDonald and Mibashan (1968) reported that 2 of the children recovered spontaneously from serious hematologic abnormalities. Smith et al. (1987) described a third family. The 13-year-old male proband, referred for evaluation of possible Fanconi anemia, had bilateral absent thumbs with flexion contractures of the fingers, irregular hyperpigmentation, radioulnar synostosis, and bilateral sensorineural hearing loss. The karyotype was 47,XXY; cyto genetic studies using clastogenic agents were negative. Nine other members of the family spanning 3 generations were affected with findings ranging from transient to chronic anemia and/or radial defects of digitalized, hypoplastic or absent thumbs. Of 10 affected individuals, 4 had both hematologic and radiologic abnormalities and 5 had only radial defects. Shahidi (1987) gave comparisons with Fanconi anemia and dyskeratosis congenita (305000).

Gonzalez, C. H.; Durkin-Stamm, M. V.; Geimer, N. K.; Shahidi, N. T.; Schilling, R. F.; Rubira, F.; Opitz, J. M.: The WT syndrome—a 'new' autosomal dominant pleiotropic trait of radial-ulnar hypoplasia with high risk of bone marrow failure and/or leukemia. Birth Defects Orig. Art. Ser. XIII(3B): 31-38, 1977.
McDonald, R.; Goldschmidt, B.: Pancytopenia with congenital defects (Fanconi's anaemia). Arch. Dis. Child. 35: 367-372, 1960.
McDonald, R.; Mibashan, R. S.: Prolonged remission in Fanconi-type anemia. Helv. Paediat. Acta 6: 566-576, 1968.
Shahidi, N. T.: Fanconi anemia, dyskeratosis congenita, and WT syndrome. Am. J. Med. Genet. Suppl. 3: 263-278, 1987.
Smith, A. C. M.; Hays, T.; Harvey, L. A.; Dowman, C.: WT syndrome: a third family. (Abstract) Am. J. Hum. Genet. 41: A84 only, 1987.

Other Myeloproliferative Predispositions

TRISOMY 21 (#190685)

JJM NOTE

The excess of malignancy, up to 20-fold, is primarily of acute lymphoblastic leukemia, with an average age of onset that is a year earlier than that of leukemia in children without Down syndrome. The leukemoid reaction of newborns with Down syndrome may be a harbinger of subsequent leukemia; it must involve the same genetic mechanisms that give rise to frank leukemia. The responsible genes on chromosome 21 are being pursued (e.g., Ives JH et al.: Increased levels of a chromosome 21-encoded tumour invasion and metastasis factor (TIAMI) mRNA in bone marrow of Down syndrome children during the acute phase of AML(M7). Genes Chromosomes Cancer 23:61-66, 1998). With regard to solid tumors, a recent review concludes, "Malignant solid tumors seem to be globally underrepresented as compared with those in the general population. However, among these tumors, some tumors are in excess: lymphomas, gonadal and extragonadal germ cell tumors, and possibly retinoblastomas and pancreatic and bone tumors. Neoplasms in excess are seen earlier, sometimes in fetal life (leukemias and testicular germ cell tumors) or neontally (leukemias and lymphoma) and affect mainly male subjects . . . . Other tumors are underrepresented, particularly neuroblastomas and nephroblastomas, in young children, and perhaps common epithelial tumors in adults" (Satge D et al.:

A tumor profile in Down syndrome. Am. J. Med. Genet. 78:207-16, 1998).

Description

A number sign (#) is used with this entry because it concerns a chromosomal aberration and many genes are involved in the phenotype.

Down syndrome, the most frequent form of mental retardation caused by a microscopically demonstrable chromosomal aberration, is characterized by well-defined and distinctive phenotypic features and natural history. It is caused by triplicate state (trisomy) of all or a critical portion of chromosome 21.

Clinical Features

Trisomy 21 (Lejeune et al., 1959), one of the most common chromosomal abnormalities in liveborn children, causes Down syndrome (Down, 1866), a particular combination of phenotypic features that includes mental retardation and characteristic facies.

It has long been recognized that the risk of having a child with trisomy 21 increases with maternal age (Penrose, 1933). For example, the risk of having a liveborn with Down syndrome at maternal age 30 is 1 in 1,000 and at maternal age 40 is 9 in 1,000 (Hook, 1982; Hook et al., 1983).

Individuals with Down syndrome often have specific major congenital malformations such as those of the heart (30-40% in some studies), particularly the atrioventricular canal, and of the gastrointestinal tract, such as duodenal stenosis or atresia, imperforate anus, and Hirschsprung disease (142623). Some of these clinical features have been incorporated into the preliminary phenotypic maps of chromosome 21 developed by Korenberg (1993), Korenberg et al. (1992), and Delabar et al. (1993).

Leukemia (both ALL and AML) and leukemoid reactions show increased incidence in Down syndrome (Fong and Brodeur, 1987; Robinson, 1992). Estimates of the relative risk have ranged from 10 to 20 times higher than the normal population; in particular, acute megakaryocytic leukemia occurs 200 to 400 times more frequently in the Down syndrome than in the chromosomally normal population (Zipursky et al., 1987). Transient leukemoid reactions have been reported repeatedly in the neonatal period, and this phenotype has been tentatively mapped to the proximal long arm of chromosome 21 (Niikawa et al., 1991).

Ninety percent of all Down syndrome patients have a significant hearing loss, usually of the conductive type (Mazzoni et al., 1994).

For additional defects and phenotypic characteristics, see Epstein (1989).

Patients with Down syndrome develop the neuropathologic hallmarks of Alzheimer disease at a much earlier age than individuals with Alzheimer disease without trisomy 21 (Wisniewski et al., 1985). Characteristic senile plaques and neurofibrillary tangles are present in the brain of all individuals with Down syndrome over the age of 40 years (Wisniewski et al., 1985). It is not known if the triplication of the amyloid precursor protein gene (APP; 104760) is the cause of this phenomenon. Several mutations in the APP gene have been described in families with early-onset Alzheimer disease without trisomy 21 (e.g., Goate et al., 1991).

Cytogenetics

Most individuals (95%) with trisomy 21 have 3 free copies of chromosome 21; in about 5% of patients, 1 copy is translocated to another acrocentric chromosome, most often chromosome 14 or 21 (Thuline and Pueschel, 1982; Hook, 1982). In 2 to 4% of cases with free trisomy 21 there is recognizable mosaicism for a trisomic and a normal cell line (Mikkelsen, 1977).

Origin of Free Trisomy 21

The availability of highly informative DNA markers has allowed the parental origin of the extra chromosome 21 and the meiotic/mitotic origin to be determined. More than 400 families have been studied (3,4:Antonarakis et al., 1991, 1992; Antonarakis, 1993; 30,29:Sherman et al., 1991, 1992) and the results are as follows: 1) Errors in meiosis that lead to trisomy 21 are overwhelmingly of maternal origin; only about 5% occur during spermatogenesis. 2) Most errors in maternal meiosis occur in meiosis I and the mean maternal age associated with these is 32 years (the mean maternal age of the general population is approximately 27 years). Thus, meiosis I errors account for 76 to 80% of maternal meiotic errors and 67 to 73% of all instances of free trisomy 21. 3) Maternal meiosis II errors constitute 20 to 24% of maternal errors and 18 to 20% of all cases of free trisomy 21. The mean maternal age is also advanced and is 31.4 in one study and 34.1 in another. 4) In rare families in which there is paternal nondisjunction, most of the errors occur in meiosis II. The mean maternal and paternal ages are similar to the mean reproductive age in western societies. 5) In 5% of trisomic individuals the supernumerary chromosome 21 appears to result from an error in mitosis. In these cases there is no advanced maternal age and there is no preference for which chromosome 21 is duplicated in the mitotic error.

Warren et al. (1987) and Sherman et al. (1991) have described an association between trisomy 21 and reduced recombination in meioses. A significant proportion (at least 30%) of maternal meiosis I nondisjunctions of chromosome 21 is associated with failure to recombine. It is not known whether the paucity of recombination is related to maternal age; moreover, the mechanism of recombination failure (asynapsis vs abnormalities during or after synapsis) is as yet unclear.

Origin of Translocation Trisomy 21

All de novo t(14;21) trisomies studied have originated in maternal germ cells (Petersen et al., 1991; Shaffer et al., 1992). The mean maternal age was 29.2 years. In de novo t(21;21) Down syndrome the situation is different (Grasso et al., 1989; Antonarakis et al., 1990; Shaffer et al., 1992). In most cases (14 out of 17) the t(21;21) is an isochromosome (dup21q) rather than the result of a Robertsonian translocation caused by a fusion between 2 heterologous chromatids. About half were of paternal and half of maternal origin. In the 3 de novo t(21;21) true Robertsonian trisomy 21 cases, the extra chromosome 21 was maternal.

Down Syndrome Critical Region

Mapping of the chromosomal region that, if triplicated, results in the phenotypic characteristics of Down syndrome has been facilitated by the use of DNA samples from individuals who have partial trisomy 21 with or without features of the Down syndrome phenotype (Rahmani et al., 1989; McCormick et al., 1989; Korenberg et al., 1990; Delabar et al., 1993; Korenberg, 1993). Although detailed analysis of these DNAs is still under way, an area of approximately 5 Mb between loci D21S58 and D21S42 has been identified that is associated with mental retardation and most of the facial features of the syndrome. In particular, a subregion that includes D21S55 and MX1 (interferon-induced protein p78; 147150), the latter being located in band 21q22.3, has been associated with mental retardation and several morphologic features, including oblique eye fissure, epicanthus, flat nasal bridge, protruding tongue, short broad hands, clinodactyly of the fifth finger, gap between first and second toes, hypotonia, short stature, Brushfield spots, and characteristic dermatoglyphics (Delabar et al., 1993). Additional phenotypic characteristics may map outside the minimum critical region (symbolized DCR). Material from other rare patients who have features of Down syndrome but no visible chromosomal abnormality may help to narrow down the critical region. In several such studies, however, no triplicated region has been identified (McCormick et al., 1989; Delabar et al., 1993). It is possible that these patients do not have any chromosome 21 abnormality and their phenotype is a phenocopy of Down syndrome.

Fuentes et al. (1995) cloned a gene (DSCR1; 602917) from the Down syndrome critical region that is highly expressed in brain and heart, and suggested it as a candidate for involvement in the pathogenesis of DS, in particular mental retardation and/or cardiac defects.

Pathogenesis

As a first step in identifying the genes responsible for individual features of Down syndrome and their pathophysiology, Korenberg et al. (1994) established a panel of cell lines derived from 16 individuals with Down syndrome caused by duplication of small regions of chromosome 21. The molecular breakpoints were determined using fluorescence in situ hybridization and Southern blot dosage analysis of 32 markers unique to chromosome 21. Combining this information with detailed clinical evaluations of the subjects, Korenberg et al. (1994) constructed a 'phenotypic map' that included 25 features and assigned regions of 2 to 20 Mb as likely to contain the genes responsible. This study provided evidence for a significant contribution of genes outside the D21S55 region to the DS phenotypes, including the facies, microcephaly, short stature, hypotonia, abnormal dermatoglyphics, and mental retardation. The results strongly suggest that DS is a contiguous gene syndrome and make it unlikely that a single DS chromosomal region is responsible for most of the DS phenotypic features.

Population Genetics

The frequency of trisomy 21 in the population is 1 in 650 to 1,000 live births (Hook, 1982).

See also Rex and Preus (1982)

Antonarakis, S. E.: Human chromosome 21: genome mapping and exploration circa 1993. Trends Genet. 9: 142-148, 1993.

Antonarakis, S. E.; Adelsberger, P. A.; Petersen, M. B.; Binkert, F.; Schinzel, A. A.: Analysis of DNA polymorphism suggests that most de novo dup(21q) chromosomes in patients with Down syndrome are isochromosomes and not translocations. Am. J. Hum. Genet. 47: 968-972, 1990.

Antonarakis, S. E.; Lewis, J. G.; Adelsberger, P. A.; Petersen, M. B.; Schinzel, A. A.; Binkert, F.; Schmid, W.; Pangalos, C.; Raoul, O.; Chakravarti, A.; Hafez, M.; Cohen, M. M.; Roulston, D.; Schwartz, S.; Mikkelsen, M.; Tranebjaerg,

L.; Greenberg, F.; Hoar, D. I.; Rudd, N. L.; Warren, A. C.; Metaxotou, C.; Bartsocas, C.; Down Syndrome Collaborative Group: Parental origin of the extra chromosome in trisomy 21 using DNA polymorphism analysis. New Eng. J. Med. 324: 872-876, 1991.

Antonarakis, S. E.; Petersen, M. B.; McInnis, M. G.; Adelsberger, P. A.; Schinzel, A. A.; Binkert, F.; Pangalos, C.; Raoul, O.; Slaugenhaupt, S. A.; Hafez, M.; Cohen, M. M.; Roulson, D.; Schwartz, S.; Mikkelsen, M.; Tranebjaerg, L.; Greenberg, F.; Hoar, D. I.; Rudd, N. L.; Warren, A. C.; Metaxotou, C.; Bartsocas, C.; Chakravarti, A.: The meiotic stage of nondisjunction in trisomy 21: determination using DNA polymorphisms. Am. J. Hum. Genet. 50: 544-550, 1992.

Delabar, J. M.; Theophile, D.; Rahmani, Z.; Chettouh, Z.; Blouin, J. L.; Prieur, M.; Noel, B.; Sinet, P. M.: Molecular mapping of twenty-four features of Down syndrome on chromosome 21. Europ. J. Hum. Genet. 1: 114-124, 1993.

Down, J. L. H.: Observations on an ethnic classification of idiots. London Hosp. Clin. Lect. Rep. 3: 259 only, 1866.

Epstein, C. J.: Down syndrome, trisomy 21.In: Scriver, C. R.; Beaudet, A. L.; Sly, W. S.; Valle, D.: Metabolic Basis of Inherited Disease. New York: McGraw-Hill (pub.) 1989. Pp. 291-326.

Fong, C.-T.; Brodeur, G. M.: Down's syndrome and leukemia: epidemiology, genetics, cytogenetics and mechanisms of leukemogenesis. Cancer Genet. Cytogenet. 28: 55-76, 1987.

Fuentes, J.-J.; Pritchard, M. A.; Planas, A. M.; Bosch, A.; Ferrer, I.; Estivill, X.: A new human gene from the Down syndrome critical region encodes a proline-rich protein highly expressed in fetal brain and heart. Hum. Molec. Genet. 4: 1935-1944, 1995.

Goate, A.; Chartier-Harlin, M. C.; Mullan, M.; Brown, J.; Crawford, F.; Fidayi, L.; Giuffra, L.; Haynes, A.; Irvine, N.; James, L.; Mant, R.; Newton, P.; Rooke, K.; Roques, P.; Talbot, C.; Pericak-Vance, M.; Roses, A.; Williamson, R.; Rorsov, M.; Owen, M.; Hardy, J.: Segregation of a missense mutation in the amyloid precursor protein gene with familial Alzheimer disease. Nature 349: 704-706, 1991.

Grasso, M.; Giovannucci, M. L.; Pierluigi, M.; Tavellini, F.; Perroni, L.; Dagna-Bricarelli, F.: Isochromosome, not translocation in trisomy 21q21q. Hum. Genet. 84: 63-65, 1989.

Hook, E. B.; Cross, P. K.; Schreinemachers, D. M.: Chromosomal abnormality rates at amniocentesis and in live-born infants. J.A.M.A. 249: 2034-2038, 1983.

Hook, E. G.: Epidemiology of Down syndrome.In: Pueschel, S. M.; Rynders, J. E.: Down Syndrome. Advances in Biomedicine and the Behavioral Sciences. Cambridge: Ware Press (pub.) 1982. Pp. 11 only.

Korenberg, J.; Bradley, C.; Disteche, C.: Down syndrome: molecular mapping of congenital heart disease and duodenal stenosis. Am. J. Hum. Genet. 50: 294-302, 1992.

Korenberg, J.; Kawashima, H.; Pulst, S.; Ikeuchi, T.; Ogasawara, N.; Yamamoto, K.; Schonberg, S.; Kojis, T.; Allen, L.; Magenis, E.; Ikawa, H.; Taniguchi, N.; Epstein, C.: Molecular definition of the region of chromosome 21 that causes features of the Down syndrome phenotype. Am. J. Hum. Genet. 47: 236-246, 1990.

Korenberg, J. R.: Toward a molecular understanding of Down syndrome.In: Epstein, C. J.: The Phenotypic Man. Prog. Clin. Biol. Res.. 384: 1993. Pp. 87-115.

Korenberg, J. R.; Chen, X.-N.; Schipper, R.; Sun, Z.; Gonsky, R.; Gerwehr, S.; Carpenter, N.; Daumer, C.; Dignan, P.; Disteche, C.; Graham, J. M., Jr.; Hugdins, L.; McGillivray, B.; Miyazaki, K.; Ogasawara, N.; Park, J. P.; Pagon, R.; Pueschel, S.; Sack, G.; Say, B.; Schuffenhauer, S.; Soukup, S.; Yamanaka, T.: Down syndrome phenotypes: the consequences of chromosomal imbalance. Proc. Nat. Acad. Sci. 91: 4997-5001, 1994.

Lejeune, J.; Gautier, M.; Turpin, R.: Etude des chromosomes somatiques de neuf enfants mongoliens. C. R. Acad. Sci. 248: 1721-1722, 1959.

Mazzoni, D. S.; Ackley, R. S.; Nash, D. J.: Abnormal pinna type and hearing loss correlations in Down's syndrome. J. Intellect. Disabil. Res. 38: 549-560, 1994.

McCormick, M.; Schinzel, A.; Petersen, M.; Stetten, G.; Driscoll, D.; Cantu, E.; Tranebjaerg, L.; Mikkelsen, M.; Watkins, P.; Antonarakis, S.: Molecular genetic approach to the characterization of the Down syndrome region of chromosome 21. Genomics 5: 325-331, 1989.

Mikkelsen, M.: Down's syndrome cytogenetic epidemiology. Hereditas 86: 45-59, 1977.

Niikawa, N.; Deng, H. X.; Abe, K.; Harada, N.; Okada, T.; Tsuchiya, H.; Akaboshi, I.; Matsuda, I.; Fukushima, Y.; Kaneko, Y.: Possible mapping of the gene for transient myeloproliferative syndrome at 21q11.2. Hum. Genet. 87: 561-566, 1991.

Penrose, L. S.: The relative effects of paternal and maternal age in mongolism. J. Genet. 27: 219 only, 1933.

Petersen, M. B.; Adelsberger, P. A.; Schinzel, A. A.; Binkert, F.; Hinkel, G. K.; Antonarakis, S. E.: Down syndrome due to de novo Robertsonian translocation t14;21: DNA polymorphism analysis suggests that the origin of the extra 21q is maternal. Am. J. Hum. Genet. 49: 529-536, 1991.

Rahmani, Z.; Blouin, J.; Creau-Goldberg, N.; Watkins, P.; Mattei, J.; Poissonnier, M.; Prieur, M.; Chettouh, Z.; Nicole, A.; Aurias, A.; Sinet, P.; Delabar, J.: Critical role of D21S55 region on chromosome 21 in the pathogenesis of Down syndrome. Proc. Nat. Acad. Sci. 86: 5958-5962, 1989.

Rex, A. P.; Preus, M.: A diagnostic index for Down syndrome. J. Pediatr. 100: 903-906, 1982.

Robinson, L. L.: Down syndrome and leukemia. Leukemia 6: 5-7, 1992.

Shaffer, L. G.; Jackson-Cook, C. K.; Stasiowski, B. A.; Spence, J. E.; Brown, J. A.: Parental origin determination in 30 de novo Robertsonian translocations. Am. J. Med. Genet. 43: 957-963, 1992.

Sherman, S. L.; Freeman, S. B.; Grantham, M.; Peters, J.; Jacobs, P. A.; Kurnit, D. M.; Uchida, I.; Petersen, M. B.; Mikkelsen, M.; Hassold, T. J.: Non-disjunction of trisomy 21: comparison of centromere maps resulting from maternal meiosis I and II non-disjunction. (Abstract) Am. J. Hum. Genet. 51 (suppl.): A24 only, 1992.

Sherman, S. L.; Takaesu, N.; Freeman, S. B.; Grantham, M.; Phillips, C.; Blackston, R. D.; Jacobs, P. A.; Cockwell, A. E.: Association between reduced recombination and nondisjunction. Am. J. Hum. Genet. 49: 608-620, 1991.

Thuline, H. C.; Pueschel, S. M.: Cytogenetics in Down syndrome.In: Pueschel, S. M.; Rynders, J. E.: Down Syndrome. Advances in Biomedicine and the Behavioral Sciences. Cambridge: Ware Press (pub.) 1982. Pp. 133 only.

Warren, A. C.; Chakravarti, A.; Wong, C.; Slaugenhaupt, S. A.; Halloran, S. L.; Watkins, P. C.; Metaxotou, C.; Antonarakis, S. E.: Evidence for reduced recombination on the nondisjoined chromosomes 21 in Down syndrome. Science 237: 652-654, 1987.

Wisniewski, K. E.; Wisniewski, H. M.; Wen, G. Y.: Occurrence of neuropathological changes and dementia of Alzheimer's disease in Down's syndrome. Ann. Neurol. 17: 278-282, 1985.

Zipursky, A.; Peeters, M.; Poon, A.: Megakaryoblastic leukemia and Down syndrome—A review.In: McCoy, E. E.; Epstein, C. J.: Oncology and Immunology of Down Syndrome. New York: Alan R. Liss (pub.) 1987. Pp. 33-56.

5Q– SYNDROME (#153550)

JJM NOTE

Acute nonlymphocytic or myeloid leukemia is the complicating neoplasm for what is a clonal disorder; that is, the deleted chromosome is present in only a fraction of somatic cells as a mosaic cell lineage and not as a constitutional deletion present in all cells from the zygote. At least 10 genes are credibly involved in the refractory anemia syndrome (Boultwood J et al.: Novel genes mapping to the critical region of the 5q- syndrome. Genomics 45:88-96, 1997; Jaju RJ et al.: Molecular cytogenetic delineation of the critical deleted region in the 5q- syndrome. Genes Chromosomes Cancer 22:251-56, 1998).

A number sign (#) is used with this entry because it appears that it does not represent a single-gene disorder, but rather a contiguous gene syndrome resulting from defects in several genes that are discussed elsewhere. IRF1 (147575), a gene whose product manifests antioncogenic activity, is probably one of these genes, and neighboring genes that may be involved include IL5 (147850), CDC25C (157680), IL3 (147740), and CSF2 (138960).

Van den Berghe et al. (1974) first described refractory macrocytic anemia associated with deletion of the long arm of chromosome 5, which was known as Belgian disease or 'anemie refractaire de type belge.' It was, of course, found not to be limited to Belgium and the 5q-change was found in other hematologic malignancies (see review of Van den Berghe et al., 1985). Tinegate et al. (1983) counted 34 recorded cases to the time of their report; 25 were female. A characteristic bone marrow finding is the presence of numerous unilobular nucleated megakaryocytes; the nucleus is often eccentric with copious granular cytoplasm exhibiting plentiful production of large platelets. The clinical course is relatively mild. Transformation into acute nonlymphocytic leukemia is rare when there is no other chromosome abnormality than 5q-. Desferrioxamine administration is recommended to lessen the complications of hemosiderosis. Treatment is difficult. Stuart and Mangan (1986) described successful treatment with syngeneic bone marrow transplantation. When first seen at age 29 years, the patient had the picture of pure red cell aplasia and normal marrow karyotype but with hypolobulated megakaryocytic nuclei. The anemia did not respond to any medical therapy. Two years later he showed the 5q- abnormality in 70% and later 100% of marrow metaphases. Because of transfusion-induced hemosiderosis and the availability of a cytogenetically normal monozygotic twin, bone marrow transplantation was performed. The patient's marrow was ablated with a busulfan plus cyclophosphamide regimen used for patients with nonlymphocytic leukemia. Sustained engraftment of cytogenetically normal marrow ensued. The 5q- myelodysplastic syndrome typically occurs in older persons, particularly females. The deletion is usually interstitial; the distal breakpoint is usually in band q32 and the proximal breakpoint in q12 or q14.

Oncogene FMS (164770) is situated near the distal breakpoint of the 5q- deletion. Nienhuis et al. (1985) found that the FMS oncogene was deleted in a case of the 5q- syndrome. Huebner et al. (1985) showed that the gene for granulocyte-macrophage colony-stimulating factor (138960) is located in the 5q21-5q32 segment commonly deleted in the 5q- syndrome. By in situ hybridization, Le Beau et al. (1986) assigned FMS to 5q33 and GMCSF to 5q23-q31. Both genes were deleted in the 5q- chromosome from bone marrow cells of 2 patients with refractory anemia and del(5)(q15q33.3). From study of other cases they concluded that FMS is located in band 5q33.2 or 5q33.3 rather than 5q34-q35 as reported earlier. See review by Bunn (1986). Mathew et al. (1993) reported the experience with 43 consecutive patients seen at the Mayo Clinic in whom the diagnosis was defined by strict morphologic criteria and the finding of a solitary 5q- cytogenetic defect. The median age at diagnosis was 68 years, with a female predominance (7:3). Transfusion-dependence was present in 80% of patients at the time of the diagnosis, and all untransfused patients had macrocytic indices. In contrast, significant neutropenia or thrombocytopenia was rare.

By deletion mapping, Boultwood et al. (1994) established the critical region of gene loss in the 5q- chromosome, giving the location for a putative tumor-suppressor gene in the 5.6-Mb region between FGFA (131220) and IL12B (161561). (They referred to the IL12B gene as NKSF1; NKSF2 is the symbol used here for the subunit of interleukin-12 encoded by chromosome 5, and NKSF1, the symbol for the subunit encoded by chromosome 3 (IL12A; 161560).)

See also Van den Berghe (1986)

Boultwood, J.; Fidler, C.; Lewis, S.; Kelly, S.; Sheridan, H.; Littlewood, T. J.; Buckle, V. J.; Wainscoat, J. S.: Molecular mapping of uncharacteristically small 5q deletions in two patients with the 5q- syndrome: delineation of the critical region on 5q and identification of a 5q- breakpoint. Genomics 19: 425-432, 1994.
Bunn, H. F.: 5q- and disordered haematopoiesis. Clin. Haemat. 15: 1023-1035, 1986.
Huebner, K.; Isobe, M.; Croce, C. M.; Golde, D. W.; Kaufman, S. E.; Gasson, J. C.: The human gene encoding GM-CSF is at 5q21-q32, the chromosome region deleted in the 5q- anomaly. Science 230: 1282-1285, 1985.
Le Beau, M. M.; Westbrook, C. A.; Diaz, M. O.; Larson, R. A.; Rowley, J. D.; Gasson, J. C.; Golde, D. W.; Sherr, C. J.: Evidence for the involvement of GM-CSF and FMS in the deletion (5q) in myeloid disorders. Science 231: 984-987, 1986.
Mathew, P.; Tefferi, A.; Dewald, G. W.; Goldberg, S. L.; Su, J.; Hoagland, H. C.; Noel, P.: The 5q- syndrome: a single-institution study of 43 consecutive patients. Blood 81: 1040-1045, 1993.
Nienhuis, A. W.; Bunn, H. F.; Turner, P. H.; Gopal, T. V.; Nash, W. G.; O'Brien, S. J.; Sherr, C. J.: Expression of the human c-fms proto-oncogene in hematopoietic cells and its deletion in the 5q syndrome. Cell 42: 421-428, 1985.
Stuart, R. K.; Mangan, K. F.: Hematologic and cytogenetic remission of 5q-refractory anemia after syngeneic bone marrow transplantation. Am. J. Med. 80: 503-507, 1986.
Tinegate, H.; Gaunt, L.; Hamilton, P. J.: The 5q- syndrome: an underdiagnosed form of macrocytic anaemia. Brit. J. Haemat. 53: 103-110, 1983.
Van den Berghe, H.: The 5q- syndrome. Scand. J. Haemat. 36 (suppl. 45): 78-81, 1986.
Van den Berghe, H.; Cassiman, J.; David, G.; Fryns, J.-P.; Michaux, J.; Sokal, G.: Distinct haematological disorder with deletion of long arm of no. 5 chromosome. Nature 251: 437-438, 1974.
Van den Berghe, H.; Vermaelen, K.; Mecucci, C.; Barbieri, D.; Tricot, G.: The 5q- anomaly. Cancer Genet. Cytogenet. 17: 189-255, 1985.

DK–PHOCOMELIA SYNDROME (223340)

JJM NOTE

This entity is included because of the bone marrow dysfunction.

Cherstvoy et al. (1980) referred to a case reported by workers in Kiel and described an apparently identical case of a syndrome of phocomelia, thrombocytopenia, encephalocele and urogenital abnormalities. They called it the 'DK—phocomelia syndrome' from the surname of the 2 patients.

Bird et al. (1994) described a female fetus with manifestations of DK—phocomelia syndrome (occipital encephalocele, cleft palate, absence of the left radius and digits 1 and 2 of the left hand, single horseshoe kidney, anal atresia, abnormal branching of coronary arteries, abnormal lobation of the lungs) and diaphragmatic agenesis. Her brother had diaphragmatic agenesis, omphalocele, accessory spleen, abnormal lobation of the lungs, and absence of papillary muscles in one leaflet of the tricuspid valve. Diaphragmatic agenesis was the only primary defect common to both sibs. Bird et al. (1994) suggested that this family provides evidence for an autosomal gene controlling the morphogenesis of midline structures.

The MURCS association consists of Mullerian duct and renal agenesis and cervical thoracic somite dysplasia. Lin et al. (1996) described the combination of MURCS association and occipital encephalocele in a stillborn girl of 41 weeks' gestation. The malformations in this disorder are compatible with a defect in the organization of the paraxial mesoderm that gives rise to occipital, cervical, and thoracic somites and adjoining intermediate mesoderm. These structures contribute to the occipital bone, cervical spine, upper limbs, and urogenital system.

All previously reported cases of DK phocomelia had been reported as having a normal chromosomal constitution. Bamforth and Lin (1997) reported a case in which the patient had normal lymphocyte chromosomes but was mosaic in fibroblasts for deletion of 13q12. The proposita was an 8-year-old girl who had a small parietooccipital encephalocele in the midline at birth as well as asymmetry of the face with flattening of the right side and elevation of the right ear. There was absence of the right fifth finger, hypoplasia of the thumb, and a single, palmar crease. On the other side there was hypoplasia of the fifth finger with dislocation and normal sized thumb. The hips were dislocated. By 15 months of age, the facial asymmetry had resolved. Renal pelviureteric junction obstruction required surgical relief. When her hair started to grow, she had bald patches. At the age of 9 years, the child developed seizures. There was a wide gap between the central incisors.

Bamforth, J. S.; Lin, C. C.: DK phocomelia phenotype (von Voss-Cherstvoy syndrome) caused by somatic mosaicism for del(13q). Am. J. Med. Genet. 73: 408-411, 1997.
Bird, L. M.; Newbury, R. O.; Ruiz-Velasco, R.; Jones, M. C.: Recurrence of diaphragmatic agenesis associated with multiple midline defects: evidence for an autosomal gene regulating the midline. Am. J. Med. Genet. 53: 33-38, 1994.
Cherstvoy, E.; Lazjuk, G.; Lurie, I.; Ostrovskaya, T.; Shved, I.: Syndrome of multiple congenital malformations including phocomelia, thrombocytopenia, encephalocele, and urogenital abnormalities. (Letter) Lancet II: 485 only, 1980.
Lin, H. J.; Cornford, M. E.; Hu, B.; Rutgers, J. K. L.; Beall, M. H.; Lachman, R. S.: Occipital encephalocele and MURCS association: case report and review of central nervous system anomalies in MURCS patients. Am. J. Med. Genet. 61: 59-62, 1996.

POLAND SYNDROME (173800)

JJM NOTE

This entity is included because of convincing numbers of children with the syndrome complicated by acute leukemia. The case report literature now includes other ages, other types of leukemia, and a sprinkling of other tumor types, including Wilms tumor, ipsilateral thoracic neuroblastoma, and leiomyosarcoma (Parikh PM et al.: Poland's syndrome with acute lymphoblastic leukemia in an adult. Med. Pediatr. Oncol. 16:290-92, 1988; Costa R et al.: Poland's syndrome associated with chronic granulocytic leukemia. Sangre (Barc) 36:417-18, 1991; Athale UH, Warrier R: Poland's syndrome and Wilms tumor: an unusual association. Med. Pediatr. Oncol. 30:67-68, 1998; Caksen H et al.: Neuroblastoma and Poland syndrome in a 15-year-old boy. Acta Paediatr. Jpn. 39:701-4, 1997; Shaham D et al.: Leiomyosarcoma in Poland's syndrome: A case report. Acta Radiol. 33:444-46, 1992.)

This condition consists of unilateral symbrachydactyly and ipsilateral aplasia of the sternal head of the pectoralis major muscle. Sometimes called Poland sequence, this disorder was first described by Poland (1841). Trier (1965) found 2 instances of parent and child with Poland syndrome. Fuhrmann et al. (1971) reported a family with father-to-son transmission and referred to other cases. Later information suggested that the grandfather may also have been affected (Fuhrmann, 1972). Females show aplasia of the breast and either sex may have patchy absence of axillary hair. David (1972) was impressed with a high frequency of 'potentially noxious social and physical ante-natal influence.' Of 10 patients, 5 were adopted and the mothers of the other 5 had probably attempted abortion in early pregnancy. In British Columbia, McGillivray and Lowry (1977) found an incidence of 1 per 32,000 live births. No familial cases were found. Paternal age was elevated, but all 24 children of 8 affected adults were unaffected. The experience of others is that this syndrome is overwhelmingly sporadic, indicating that it is a causally nonspecific developmental field defect (Opitz, 1982).

David (1982) described 2 second cousins with typical left-sided Poland anomaly. Discounting for one or another reason the previously reported familial cases, he claimed that this was the first recorded instance of familial occurrence. The conclusion, 'For purposes of genetic counseling, the Poland anomaly can be regarded as a sporadic condition with an extremely low recurrence risk,' cannot be gainsaid. David and Winter (1985) reported a family in which males in 3 successive generations had unilateral absence of the pectoralis major, serratus anterior, and latissimus dorsi muscles. The authors pointed out that association of absence of other muscles around the shoulder girdle is frequent with pectoralis absence, that the associated abnormality may limit the use of the latissimus muscle in reconstructive surgery (Hester and Bostwick, 1982), and that computerized tomography may be useful in determining the presence of other abnormalities (Suzuki et al., 1983).

Bouwes Bavinck and Weaver (1986) hypothesized that Poland syndrome, Klippel-Feil anomaly (148900), Moebius syndrome (157900), isolated absence of pectoralis major with breast hypoplasia, isolated transverse limb defects, and Sprengel anomaly (184400) are each the result of interruption of the early embryonic blood supply in the subclavian arteries, the vertebral arteries and/or their branches. The term subclavian artery supply disruption sequence (SASDS) was suggested for this group of birth defects. They discussed causative mechanisms such as pressure on the vessel by edema. Soltan and Holmes (1986) and Bartoshesky et al. (1986) observed Poland anomaly in one or more family members and unilateral terminal transverse hemimelia in others. They suggested that some 'inherited liability to a common predisposing event' is likely.

Fraser et al. (1989) reported a family in which a boy had unilateral pectus absence, and his male second cousin had full Poland sequence on the left. Fraser et al. (1989) suggested that isolated pectoralis major muscle defects should be included in the spectrum of anomalies postulated to result from disruption of blood supply in the embryonic subclavian and vertebral arteries, the site and degree of obstruction determining the sites and severity of the resulting anomalies. 15,16:Der Kaloustian et al. (1989, 1991) described 2 families in which the Poland anomalad and the Adams-Oliver syndrome are coexistent. They hypothesized that the Poland anomalad and the Adams-Oliver syndrome result from the interruption of early embryonic blood supply in the subclavian arteries, and that the gene predisposing to this interruption follows an autosomal dominant pattern of inheritance. Gross-Kieselstein and Shalev (1987) reported a Poland-like abnormality in 2 brothers. The features were congenital absence of the trapezius, pectoralis, supraspinatus, and serratus anterior muscles bilaterally.

Riccardi (1978) described unilateral gluteal hypoplasia and brachysyndactyly and proposed that it is the lower limb equivalent of the Poland anomaly. In support of Riccardi's view, Parano et al. (1995) reported on an extended family in whom 3 individuals, all women, had unilateral gluteal hypoplasia and the propositus in addition had unilateral hypoplasia of a pectoral muscle. Another relative in Parano's family had unilateral symbrachydactyly of the distal phalanges of one foot. Corona-Rivera et al. (1997) reported another instance of a possible lower extremity counterpart of Poland sequence. Gluteal and lower extremity hypoplasia with ipsilateral toe brachysyndactyly was noted in a 23-year-old woman. Since no neurologic deficit was found, and electromyographic and nerve conduction studies in the affected limb were normal, they proposed a vascular origin which would involve the external iliac artery supply analogous to the disruption of subclavian artery supply in the upper extremity Poland sequence.

Cobben et al. (1989) described affected mother and daughter. The mother had the defect of the pectoralis major muscle; the daughter had symbrachydactyly as well. Separate platelets may be identified on blood smear because they do not clump normally (Seligsohn, 1990).

Czeizel et al. (1990) reported that in Hungary for the period 1975 to 1984 there was 1 case of full-blown Poland sequence in 87,550 births including 12 cases with Poland sequence-type symbrachydactyly but without pectoralis major defect. The frequency came to 1 in 52,530. Absence of the pectoralis major muscle without upper limb defect could not be ascertained by their approach. Of 18 typical Poland-sequence cases, 1 was familial: the father had absence of the left pectoralis major muscle and severe hypoplasia of the left thumb with partial syndactyly between fingers 2 and 3, while the son had mild hypoplasia of the left pectoralis major muscle and severe hypoplasia of the left thumb (and mild hypoplasia of the right thumb). Czeizel et al. (1990) suggested that the disorder in this family may represent a separate pectoralis-radial defect with dominant inheritance such as was described by Konig and Lenz (1983).

Hennekam and Hofstee (1990) reported a family in which the proband had Poland syndrome. An earlier pregnancy of the mother had produced a baby girl with a large lumbar neural tube defect and congenital hydrocephaly leading to death shortly after birth, and a first cousin of the mother had hemifacial microsomia. Hennekam and Hofstee (1990) proposed that this constellation supports the hypothesis of Stevenson et al. (1987) that neural tube defects may be produced by inadequate nutrient supply to the growing neural folds due to primary vascular abnormalities. Cobben et al. (1992) described a 3-year-old boy with features of both Poland anomaly and Goldenhar syndrome (164210).

Karnak et al. (1998) described what they believed to be the first known case of bilateral Poland anomaly. A 6-year-old girl had bilateral absence of pectoralis major muscles with hand involvement and symmetric chest wall deformity.

See also Bouvet et al. (1978); Brown and McDowell (1940); Castilla et al. (1979); Clarkson (1962); David and Saad (1974); Hegde and Shokeir (1982); Ireland et al. (1976); McDowell (1977); Ravitch (1977); Sujansky et al. (1977)

Bartoshesky, L. E.; Gans, B.; Goldberg, M.: Familial occurrence of malformations possibly attributable to vascular abnormalities. (Letter) J. Pediat. 109: 396, 1986.

Bouvet, J.-P.; Leveque, D.; Bernetieres, F.; Gross, J.-J.: Vascular origin of Poland syndrome? A comparative rheographic study of vascularisation of the arms of eight patients. Europ. J. Pediat. 128: 17-26, 1978.

Bouwes Bavinck, J. N.; Weaver, D. D.: Subclavian artery supply disruption sequence: hypothesis of a vascular etiology for Poland, Klippel-Feil, and Moebius anomalies. Am. J. Med. Genet. 23: 903-918, 1986.

Brown, J. B.; McDowell, F.: Syndactylism with absence of the pectoralis major. Surgery 7: 599-601, 1940.

Castilla, E. E.; Paz, J. E.; Orioli, I. M.: Pectoralis major muscle defect and Poland complex. Am. J. Med. Genet. 4: 263-269, 1979.

Clarkson, P.: Poland's syndactyly. Guy's Hosp. Rep. 111: 335-346, 1962.

Cobben, J. M.; Robinson, P. H.; van Essen, A. J.; van der Wiel, H. L.; ten Kate, L. P.: Poland anomaly in mother and daughter. Am. J. Med. Genet. 33: 519-521, 1989.

Cobben, J. M.; Van Essen, A. J.; McParland, P. C.; Polman, H. A.; ten Kate, L. P.: A boy with Poland anomaly and facio-auriculo-vertebral dysplasia. Clin. Genet. 41: 105-107, 1992.

Corona-Rivera, J. R.; Corona-Rivera, A.; Totsuka-Sutto, S. E.; Corona-Rivera, E.: Corroboration of the lower extremity counterpart of the Poland sequence. Clin. Genet. 51: 257-259, 1997.

Czeizel, A.; Vitez, M.; Lenz, W.: Birth prevalence of Poland sequence and proportion of its familial cases. (Letter) Am. J. Med. Genet. 36: 524, 1990.

David, T. J.: Familial Poland anomaly. J. Med. Genet. 19: 293-296, 1982.

David, T. J.: Nature and etiology of the Poland anomaly. New Eng. J. Med. 287: 487-489, 1972.

David, T. J.; Saad, M. N.: Dermatoglyphic diagnosis of the Poland anomaly in the absence of syndactyly. Hum. Hered. 24: 373-378, 1974.

David, T. J.; Winter, R. M.: Familial absence of the pectoralis major, serratus anterior, and latissimus dorsi muscles. J. Med. Genet. 22: 390-392, 1985.

Der Kaloustian, V. M.; Hoyme, E.; Hogg, H.; Entin, M.; Guttmacher, A.: Possible common pathogenetic mechanisms for the Poland anomalad and Adams-Oliver syndrome. (Abstract) Am. J. Hum. Genet. 45 (suppl.): A43, 1989.

Der Kaloustian, V. M.; Hoyme, H. E.; Hogg, H.; Entin, M. A.; Guttmacher, A. E.: Possible common pathogenetic mechanisms for Poland sequence and Adams-Oliver syndrome. Am. J. Med. Genet. 38: 69-73, 1991.

Fraser, F. C.; Ronen, G. M.; O'Leary, E.: Pectoralis major defect and Poland sequence in second cousins: extension of the Poland sequence spectrum. Am. J. Med. Genet. 33: 468-470, 1989.

Fuhrmann, W.: Personal Communication. Giessen, Germany 1972.

Fuhrmann, W.; Mosseler, U.; Neuss, H.: Zur Klinik und Genetik des Poland-Syndroms. Dtsch. Med. Wschr. 96: 1076-1078, 1971.

Gross-Kieselstein, E.; Shalev, R. S.: Familial absence of the trapezius muscle with associated shoulder girdle abnormalities. Clin. Genet. 32: 145-147, 1987.

Hegde, H. R.; Shokeir, M. H. K.: Posterior shoulder girdle abnormalities with absence of pectoralis major muscle. Am. J. Med. Genet. 13: 285-293, 1982.

Hennekam, R. C. M.; Hofstee, N.: Familial liability to intrauterine vascular impairments. (Letter) Pediatrics 86: 326-327, 1990.

Hester, T. R.; Bostwick, J.: Poland's syndrome: correction with latissimus muscle transposition. Plast. Reconst. Surg. 69: 226-233, 1982.

Ireland, D. C. R.; Takayama, N.; Flatt, A. E.: Poland's syndrome: a review of forty-three cases. J. Bone Joint Surg. 58A: 52-58, 1976.

Karnak, I.; Tanyel, F. C.; Tuncbilek, E.; Unsal, M.; Buyukpamukcu, N.: Bilateral Poland anomaly. Am. J. Med. Genet. 75: 505-507, 1998.

Konig, R.; Lenz, W.: Pektoralis-Handdefekte (Poland-Syndaktylia). Z. Orthop. 121: 244-253, 1983.

McDowell, F.: On the propagation, perpetuation, and parroting of erroneous eponyms such as 'Poland's syndrome'. Plast. Reconst. Surg. 59: 561-562, 1977.

McGillivray, B. C.; Lowry, R. B.: Poland syndrome in British Columbia: incidence and reproductive experience of affected persons. Am. J. Med. Genet. 1: 65-74, 1977.

Opitz, J. M.: Personal Communication. Helena, Mont. 1982.

Parano, E.; Falsaperla, R.; Pavone, V.; Toscano, A.; Bolan, E. A.; Trifiletti, R. R.: Intrafamilial phenotypic heterogeneity of the Poland complex: a case report. Neuropediatrics 26: 217-219, 1995.

Ravitch, M. M.: Poland's syndrome—a study of an eponym. Plast. Reconst. Surg. 59: 508-512, 1977.

Riccardi, V. M.: Unilateral gluteal hypoplasia and brachysyndactyly: lower extremity counterpart of the Poland anomaly. Pediatrics 61: 653-654, 1978.

Seligsohn, U.: Personal Communication. Tel-Aviv, Israel 6/13/1990.

Soltan, H. C.; Holmes, L. B.: Familial occurrence of malformations possibly attributable to vascular abnormalities. J. Pediat. 108: 112-114, 1986.

Stevenson, R. E.; Kelly, J. C.; Aylsworth, A. S.; Phelan, M. C.: Vascular basis for neural tube defects: a hypothesis. Pediatrics 80: 102-106, 1987.

Sujansky, E.; Riccardi, V. M.; Matthew, A. L.: The familial occurrence of Poland syndrome. Birth Defects Orig. Art. Ser. XIII(3A): 117-121, 1977.

Suzuki, T.; Takazawa, H.; Koshino, T.: Computed tomography of the pectoralis muscles in Poland's syndrome. Hand 15: 35-41, 1983.

Trier, W. C.: Complete breast absence: case report and review of the literature. Plast. Reconst. Surg. 36: 431-439, 1965.

INCONTINENTIA PIGMENTI (#308300)

JJM NOTE

A diverse range of very unusual types of childhood malignancies has been associated with incontinentia pigmenti by case reports: Wilms tumor, malignant rhabdoid tumor of the kidney, acute myelogenous leukemia, acute myelomonocytic leukemia, retinoblastoma, and paratesticular rhabdomyosarcoma (Cohen PR, Kurzrock R: Miscellaneous genodermatoses: Beckwith-Wiedemann syndrome, Birt-Hogg-Dube syndrome, familial atypical multiple mole melanoma syndrome, hereditary tylosis, incontinentia pigmenti, and supernumerary nipples. Dermatol. Clin. 13:211-

229, 1995). If there is a common thread, maybe it is the unstable mutation phenomenon that, in part, explains the chronobiology of the disorder (Hatchwell E: Unstable mutation in incontinentia pigmenti? J. Med. Genet. 33:349-350, 1996).

A number sign (#) is used with this entry because of evidence that the sporadic type of incontinentia pigmenti (IP1) is in actuality hypomelanosis of Ito (146150; Happle, 1998). A familial form of incontinentia pigmenti (IP2; 308310) has been mapped to Xq28.

Incontinentia pigmenti is a disturbance of skin pigmentation sometimes associated with a variety of malformations of the eye, teeth, skeleton, heart, etc. The pigmentary disturbance, an autochthonous tattooing, is evident at or soon after birth and may be preceded by a phase suggesting inflammation in the skin. In the fully developed disease, the skin shows swirling patterns of melanin pigmentation, especially on the trunk, suggesting the appearance of 'marble cake.' Histologically, deposits of melanin pigment are seen in the corium: the designation was based on the idea that the basal layer of the epidermis is 'incontinent' of melanin. Garrod (1906) may have described the first case, a girl with typical pigmentary changes together with mental deficiency and tetraplegia. The cutaneous phenotype has other interesting features, namely, that in the first months of life it has some characteristics of an inflammatory process and that the pigmentary changes have usually disappeared completely by the age of 20 years. Caffey disease (infantile hyperostosis; 114000) displays a similar behavior, with pronounced signs suggesting an inflammatory process in many bones with subsequent quiescence and in many cases disappearance of all evidence of previous disease. Kuster and Olbing (1964) reported a mentally retarded woman with incomplete dentition and a history of skin lesions at birth. She had 1 son and 11 daughters. Six of the girls showed incomplete dentition and incontinentia pigmenti. Spallone (1987) examined 7 affected members in a family with a total of 14 affected members in 3 generations. There were many abortions in the family, several of which were identified as male. Spallone (1987) showed that vascular abnormalities of the retina and disorders of the retinal pigment epithelium are the most important ocular lesions. Landy and Donnai (1993) reviewed the disorder in full. They pointed out that classically the dermatologic features occur in 4 stages, although all stages ma y not occur and several stages may overlap. Stage 1 is characterized by erythema, vesicles, and pustules; stage 2 by papules, verrucous lesions, and hyperkeratosis; stage 3 by hyperpigmentation; and stage 4 by pallor, atrophy, and scarring. Dystrophy of the nails is frequent but usually mild. Unilateral breast aplasia is a well-recognized but uncommon feature.

Carney (1976) found 653 cases in the literature (593 females, 16 males, and 44 of unspecified sex). Pfeiffer (1960) proposed female-limited autosomal dominant inheritance. Lenz (1961) suggested X-linked dominant inheritance with lethality in the male. IP in a male with XXY Klinefelter syndrome (Kunze et al., 1977) is consistent with this hypothesis. Ormerod et al. (1987) described incontinentia pigmenti in a boy with XXY Klinefelter syndrome. Pedigree patterns suggest X-linked dominance with lethality in the male. The phenotype in the affected females might be consistent with random X chromosome inactivation as in the Lyon hypothesis. Cytoplasmic (or other nonchromosomal) inheritance with lethality in the male could also account for the pedigree pattern. Features of the histologic and clinical picture have suggested viral etiology to several workers (e.g., Haber, 1952). Cytoplasmic inclusions similar to those of molluscum contagiosum have been identified (Murrell, 1962). No chromosomal abnormality was found in 2 cases of incontinentia pigmenti studied by Benirschke (1962). In the family studied, the mother and 2 daughters were affected; there had been 1 male abortion. Gartler and Francke (1975) suggested that half-chromatid mutations occurring during gametogenesis is a possible mechanism for mosaicism and a possible explanation for the occurrence of fewer than the theoretically expected one-third of cases of X-linked lethal disorders as new mutations. Lenz (1975) suggested that some male cases of incontinentia pigmenti may be mosaics originating in this way. He stated that 355 cases have been reported in females and 6 in males. The pattern of the skin changes is like that of the heterozygous state of some X-linked genes in animals. Mosaicism would account for a similar finding in XY males. Traupe and Vehring (1994) suggested that a more plausible explanation for mosaic skin lesions following the lines of Blaschko in boys with incontinentia pi gmenti would be an unstable premutation that normally remains silent in males during early embryogenesis. Occasionally 'silencing' might be incomplete and give rise to clinically manifest IP reflecting a mosaic state of alleles with the full and the premutation in the same patient. Traupe and Vehring (1994) suggested that this model would account for mother-to-son transmission of IP as in the cases of Kurczynski et al. (1982) and Hecht et al. (1982) and for disparate phenotypes in monozygotic female twins. Traupe and Vehring (1994) pictured a 4-month-old boy with mosaic cutaneous involvement of IP predominantly on the right side; he had subtotal retinal detachment of the right eye and right abducens paralysis, as well as central motor dysfunction predominantly affecting the right arm and leg. An unstable premutation that became partially expressed during early embryogenesis was proposed as the explanation.

It is likely that the 'striated' mutation in mice and 'streaked hairlessness' in cattle (Eldridge and Atkeson, 1953) are homologous to either focal dermal hypoplasia (FDH; 305600) or incontinentia pigmenti in man. In the second of the animal disorders, approximately perpendicular, irregularly narrow streaks of hide on various parts of the cow are affected. No males are affected. A deficiency of sons and an increased length of calving-interval in affected females support X-linked dominant inheritance with lethality in the male at an early embryonic stage. Francke (1986) suggested that 'Tattered' (Td), which produces streaked scars in the skin, is a likely mouse homolog for IP. Its localization near the centromere of the mouse X supports this.

The evolution of lesions can be interpreted as representing death of cells that have the mutant-bearing X chromosome as the active one and replacement of same by cells with the normal X active. The progression is from an erythematous eruption with linear vesiculation in the newborn period (the vesicobullous stage), followed by a verrucous stage. After a few months the verrucous growth drops off and leaves hyperpigmented areas. The third stage persists for several years and usually disappears at about age 20 years. This sequence would be expected to be accompanied by a marked reduction in cells with the mutant X active. Wieacker et al. (1985) tested this prediction. Fibroblasts from normal and hyperpigmented areas were fused with HPRT-deficient mouse RAG cells. From normal skin they isolated 13 hybrid clones and from hyperpigmented skin, 16 hybrid clones. Restriction patterns were consistent with the non-IP X chromosome being the active one in all clones. By way of contrast, in the Aicardi syndrome (304050), X inactivation was apparently at random. A problem in these cases is why there is a mosaic phenotype when the normal X is inactivated in most cells. Why is this not a lethal as in the hemizygous male?

Incontinentia pigmenti was mapped to Xp11 by observations in 2 cases of X/autosome translocation (Hodgson et al., 1985). In both, a de novo X-autosome translocation involved Xp11. Gilgenkrantz et al. (1985) found t(X;9)(p11;q34) in a girl with IP and suggested that the IP gene may be at Xp11. Happle (1987) suggested that the second case of Hodgson et al. (1985) and the case of Gilgenkrantz et al. (1985) in fact had hypomelanosis of Ito (146150) rather than IP. In a girl with IP, mental retardation, and short stature, de Grouchy et al. (1985) found 45,X/46,X,r(X) mosaicism. Since the ring was very small, the finding suggests that the IP locus is juxtacentromeric, e.g., at Xp11. Mules (1985) observed a balanced translocation, t(X;4)(q21;q28), in a girl with probable incontinentia pigmenti. Kajii et al. (1985) described incontinentia pigmenti in a 19-month-old girl who also had bilateral retinoblastoma and carried a de novo translocation t(X;13)(p11.21;q12.3). Q-banding analysis showed that the translocated chromosomes were of paternal origin. The derivative X chromosome was late-replicating in 9% of cultured peripheral blood lymphocytes and in 1% of skin fibroblasts. The red cell esterase D (133280) activity was normal. Kajii et al. (1985) raised the possibility that functional monosomy of 13q14 had occurred in a minority of retinoblasts due to spreading of inactivation from the translocated X chromosome segment. Cannizzaro and Hecht (1987) observed a female with typical IP who displayed an X;10 translocation with breakpoints in Xp11.2 and 10q22. Thus, the breakpoint in the X chromosome in all 4 reported cases of X/autosome translocation in IP has been in band Xp11. Cannizzaro and Hecht (1987) suggested that the IP gene is most likely in subband Xp11.2. Palotta and Dalpra (1988) found no increased chromatid and chromosome gaps and breaks in 4 patients with incontinentia pigmenti. Sefiani et al. (1988) studied linkage in 5 IP families containing 29 potentially informative meioses. Using 10 probes of the Xp arm (including 6 that were precisely localized by somatic cell hybridization using a broken X chromosome derived from an IP patient carrying an X;9 translocation), Sefiani et al. (1988) found negative lod scores, which excluded the previously held view that the IP gene might be located at Xp11. A major part of Xp was also excluded. Sefiani et al. (1989) studied 8 families in which 2 or more females were affected with IP. Using DNA markers, they excluded Xp11 and most of Xq as the site of the IP gene. They concluded that IP is linked to DXS52, which is located in Xq28 (maximum lod = 3.5 at theta = 0.05). Harris et al. (1988) used RFLPs mapping between Xp21 and Xq22.3 in linkage studies of IP. Although 6 independent sporadic cases with findings suggesting IP have been found to have X/autosome translocations involving an Xp11 breakpoint, Harris et al. (1988) could not confirm this localization in familial cases with RFLP markers. The incontinentia pigmenti phenotype associated with translocations may be an indirect, nonspecific consequence of the translocation and not a specific indication of the location of the gene. Another interpretation is that there are at least 2 forms of IP: a sporadic form (IP1) caused by change near the centromere and a familial male-lethal form (IP2; 308310) caused by a mutation in Xq28. Using nonisotopic in situ hybridization technique and an X centromere-specific probe, Crolla et al. (1989) confirmed the Xp11 breakpoint in 3 cases of X/autosome translocation and a ring-X case. However, results from a fifth case, t(X;17), previously reported by Hodgson et al. (1985), showed that the X breakpoint was within the centromeric alphoid repetitive sequences. The clinical presentation in all 5 patients was the same. Gorski et al. (1991) stated that these 5 unrelated patients with nonfamilial IP and a constitutional de novo reciproca l X/autosome translocation all showed a breakpoint involving Xp11.21. They concluded that the breakpoints in 3 of the 5 patients whom they studied was separated by a distance of at least 1,250 kb. In a case of ring X chromosome in a female with IP, Sefiani et al. (1989) found breakpoints in or distal to p11.22 and between q12.2 and q13.1. Bitoun et al. (1992) described incontinentia pigmenti in association with an apparently balanced translocation t(X;5)(p11.2;q35.2). They pointed out that the 7 previously reported cases of type I IP with translocations had the same Xp11 breakpoint. They noted that there was no history of bullous lesions in these cases. They expressed the view that lesions of this type occur only in the 'classical' type II, or familial form of IP, which maps to Xq28.

Nomenclature: Sybert (1994) suggested that the IP1/IP2 nomenclature is 'premature and misleading' and argued for 'the quick and early consignment of the terms. . .to the anomalad graveyard.' She contended that what has been called IP2 is the classic disorder described by Bloch (1926) and Sulzberger (1927), i.e., the male-lethal, X-linked dominant disorder due to a gene located in the Xq28 region. On the other hand, 'IP1' is not incontinentia pigmenti at all. She tabulated 40 cases of X/autosome translocations with a phenotype designated as 'incontinentia pigmenti' and argued that none of them satisfied the diagnostic criteria. She suggested the alternative descriptive phrase 'X-autosome translocation associated with pigmentary abnormality.'

A historical example of a similar nomenclature mistake was the use of the term 'incontinentia pigmenti achromians' interchangeably with the term 'hypomelanosis of Ito.' It suggested an unwarranted association of hypomelanosis of Ito with classic incontinentia pigmenti. Hypomelanosis of Ito is a phenotype well recognized to be associated with chromosomal mosaicism in many cases. None of the patients have peg-shaped teeth or typical retinal vascular changes of incontinentia pigmenti.

Happle (1998) concluded that there is convincing evidence that the 'sporadic type of incontinentia pigmenti' (IP1) does not exist. He stated that the delineation of IP1 represented a historical misunderstanding. A locus proposed for this disorder was Xp11 (Gilgenkrantz et al., 1985; Kajii et al., 1985; Hodgson et al., 1985; Gorski et al., 1991). Because all cases of so-called incontinentia pigmenti showing an association with this locus were sporadic, the name 'sporadic type of incontinentia pigmenti' was given to the phenotype to distinguish it from familial incontinentia pigmenti. Happle (1987) suggested that investigators reporting an association between incontinentia pigmenti and Xp11 had, in fact, not mapped incontinentia pigmenti, but rather hypomelanosis of Ito (146150). Inflammation or blistering of the skin was absent, which would be unusual in incontinentia pigmenti. Thereafter it became clear that hypomelanosis of Ito may not be a discrete entity but a skin disorder that is merely a symptom of many different states of mosaicism. Besides various autosomal regions, Xp11 is rather frequently involved in phenotypes that should be categorized as hypomelanosis of Ito. In cases of pigmentary mosaicism involving Xp11, the skin lesions are never preceded by an inflammatory stage as observed in incontinentia pigmenti. In those cases of hypomelanosis of Ito where no mosaicism can be demonstrated by cytogenetic examination, it is possible that either mosaicism is present at the molecular level or that a minor cytogenetic aberration had gone unnoticed. Jewett et al. (1997) reevaluated a case of de novo unbalanced X;autosome translocation reported by Pettenati et al. (1993) and concluded that it was the case representing confirmation of the association with incontinentia pigmenti with Xp11. Happle (1998) pointed out that the patient lacked typical neonatal inflammatory signs of incontinentia pigmenti and was more likely to represent hypomelanosis of Ito.

See also Bargman and Wyse (1975); Cannizzaro and Hecht (1987); Carney and Carney (1970); Francois (1984); Hecht and Hecht (1983); Lenz (1961); Reed et al. (1967); Sefiani et al. (1989); Sommer and Liu (1984); Wettke-Schafer and Kantner (1983); Wiklund and Weston (1980)

Bargman, H. B.; Wyse, C.: Incontinentia pigmenti in a 21-year-old man. Arch. Derm. 111: 1606-1608, 1975.

Benirschke, K.: Personal Communication. Hanover, N. H. 1962.

Bitoun, P.; Philippe, C.; Cherif, M.; Mulcahy, M.-T.; Gilgenkrantz, S.: Incontinentia pigmenti (type 1) and X;5 translocation. Ann. Genet. 35: 51-54, 1992.

Bloch, B.: Eigentumliche bisher nicht beschriebene Pigmentaffektion (Incontinentia pigmenti). Schweiz. Med. Wschr. 7: 404 only, 1926.

Cannizzaro, L. A.; Hecht, F.: Gene for incontinentia pigmenti maps to band Xp11 with an (X;10)(p11;q22) translocation. Clin. Genet. 32: 66-69, 1987.

Cannizzaro, L. A.; Hecht, F.: The gene for incontinentia pigmenti maps to band Xp11 with an (X;10)(p11;q22) translocation. (Abstract) Cytogenet. Cell Genet. 46: 590 only, 1987.

Carney, R. G.; Carney, R. G., Jr.: Incontinentia pigmenti. Arch. Derm. 102: 157-162, 1970.

Carney, R. G., Jr.: Incontinentia pigmenti: a world statistical analysis. Arch. Derm. 112: 535-542, 1976.

Crolla, J. A.; Gilgenkrantz, S.; de Grouchy, J.; Kajii, T.; Bobrow, M.: Incontinentia pigmenti and X-autosome translocations: non-isotopic in situ hybridization with an X-centromere-specific probe (pSV2X5) reveals a possible X-centromeric breakpoint in one of five published cases. Hum. Genet. 81: 269-272, 1989.

de Grouchy, J.; Turleau, C.; Doussau de Bazignan, M.; Maroteaux, P.; Thibaud, D.: Incontinentia pigmenti (IP) and r(X): tentative mapping of the IP locus to the X juxtacentromeric region. Ann. Genet. 28: 86-89, 1985.

Eldridge, F. E.; Atkeson, F. W.: Streaked hairlessness in Holstein-Friesian cattle: a sex-linked lethal character. J. Hered. 44: 265-271, 1953.

Francke, U.: Personal Communication. New Haven, Conn. 5/27/1986.

Francois, J.: Incontinentia pigmenti (Bloch-Sulzberger syndrome) and retinal changes. Brit. J. Ophthal. 68: 19-25, 1984.

Garrod, A. E.: Peculiar pigmentation of the skin in an infant. Trans. Clin. Soc. London 39: 216 only, 1906.

Gartler, S. M.; Francke, U.: Half chromatid mutations: transmission in humans?. Am. J. Hum. Genet. 27: 218-223, 1975.

Gilgenkrantz, S.; Tridon, P.; Pinel-Briquel, N.; Beurey, J.; Weber, M.: Translocation (X;9)(p11;q34) in a girl with incontinentia pigmenti (IP): implications for the regional assignment of the IP locus to Xp11?. Ann. Genet. 28: 90-92, 1985.

Gorski, J. L.; Burright, E. N.; Harnden, C. E.; Stein, C. K.; Glover, T. W.; Reyner, E. L.: Localization of DNA sequences to a region within Xp11.21 between incontinentia pigmenti (IP1) X-chromosomal translocation breakpoints. Am. J. Hum. Genet. 48: 53-64, 1991.

Haber, H.: The Bloch-Sulzberger syndrome (incontinentia pigmenti). Brit. J. Derm. 64: 129-140, 1952.

Happle, R.: Incontinentia pigmenti versus hypomelanosis of Ito: the whys and wherefores of a confusing issue. (Letter) Am. J. Med. Genet. 79: 64-65, 1998.

Happle, R.: Tentative assignment of hypomelanosis of Ito to 9q33-qter. (Letter) Hum. Genet. 75: 98-99, 1987.

Harris, A.; Lankester, S.; Haan, E.; Beres, J.; Hulten, M.; Szollar, J.; Soutter, L.; Bobrow, M.: The gene for incontinentia pigmenti: failure of linkage studies using DNA probes to confirm cytogenetic localization. Clin. Genet. 34: 1-6, 1988.

Hecht, F.; Hecht, B. K.: The half chromatid mutation model and bidirectional mutation in incontinentia pigmenti. Clin. Genet. 24: 177-179, 1983.

Hecht, F.; Hecht, B. K.; Austin, W. J.: Incontinentia pigmenti in Arizona Indians including transmission from mother to son inconsistent with the half chromatid mutation model. Clin. Genet. 21: 293-296, 1982.

Hodgson, S. V.; Neville, B.; Jones, R. W. A.; Fear, C.; Bobrow, M.: Two cases of X/autosome translocation in females with incontinentia pigmenti. Hum. Genet. 71: 231-234, 1985.

Jewett, T.; Hart, P. S.; Rao, P. N.; Pettenati, M. J.: A case revisited: recent presentation of incontinentia pigmenti in association with a previously reported X;autosome translocation. Am. J. Med. Genet. 69: 96-97, 1997.

Kajii, T.; Tsukahara, M.; Fukushima, Y.; Hata, A.; Matsuo, K.; Kuroki, Y.: Translocation (X;13)(p11.21;q12.3) in a girl with incontinentia pigmenti and bilateral retinoblastoma. Ann. Genet. 28: 219-223, 1985.

Kunze, S.; Frenzel, U. H.; Huttig, E.; Grosse, F.-R.; Wiedemann, H.-R.: Klinefelter's syndrome and incontinentia pigmenti Bloch-Sulzberger. Hum. Genet. 35: 237-240, 1977.

Kurczynski, T. W.; Berns, J. S.; Johnson, W. E.: Studies of a family with incontinentia pigmenti variably expressed in both sexes. J. Med. Genet. 19: 447-451, 1982.

Kuster, F.; Olbing, H.: Incontinentia pigmenti. Bericht ueber neun Erkrankungen in einer Familie und einem Obduktionsbefund. Ann. Paediat. 202: 92-100, 1964.

Landy, S. J.; Donnai, D.: Incontinentia pigmenti (Bloch-Sulzberger syndrome). J. Med. Genet. 30: 53-59, 1993.

Lenz, W.: Medizinische Genetik. Eine Einfuehrung in ihre Grundlagen und Probleme. Stuttgart: Georg Thieme Verlag (pub.) 1961. Pp. 89 only.

Lenz, W.: Zur Genetik der Incontinentia pigmenti. Ann. Paediat. 196: 149-165, 1961.

Lenz, W.: Half chromatid mutations may explain incontinentia pigmenti in males. (Letter) Am. J. Hum. Genet. 27: 690-691, 1975.

Mules, E. H.: Personal Communication. Baltimore, Md. 11/20/1985.

Murrell, T. W., Jr.: Personal Communication. Richmond, Va. 1962.

Ormerod, A. D.; White, M. I.; McKay, E.; Johnston, A. W.: Incontinentia pigmenti in a boy with Klinefelter's syndrome. J. Med. Genet. 24: 439-441, 1987.

Palotta, R.; Dalpra, L.: Chromosomal instability in incontinentia pigmenti: study of four families. Ann. Genet. 31: 27-31, 1988.

Pettenati, M. J.; Teot, L. A.; Smith, C.; Hayworth, R.; Thomas, I. T.; Veille, J. C.; Rao, P. N.: Unbalanced mosaic karyotypes with different structural abnormalities involving a common chromosome region: report of two cases. Am. J. Med. Genet. 45: 365-369, 1993.

Pfeiffer, R. A.: Zur Frage der Vererbung der Incontinentia pigmenti Bloch-Siemens. Z. Menschl. Vererb. Konstitutionsl. 35: 469-493, 1960.

Reed, W. B.; Carter, C.; Cohen, T. M.: Incontinentia pigmenti. Dermatologica 134: 243-250, 1967.

Sefiani, A.; Abel, L.; Heuertz, S.; Sinnett, D.; Lavergne, L.; Labuda, D.; Hors-Cayla, M. C.: The gene for incontinentia pigmenti is assigned to Xq28. Genomics 4: 427-429, 1989.

Sefiani, A.; Heuertz, S.; Turleau, C.; Thibaud, D.; de Grouchy, J.; Hors-Cayla, M. C.: Incontinentia pigmenti: Xp breakpoint is not the same in a case of r(X) and in X/autosome translocations. Ann. Genet. 32: 149-151, 1989.

Sefiani, A.; Sinnett, D.; Abel, L.; Szpiro-Tapia, S.; Heuertz, S.; Craig, I.; Fraser, N.; Kruse, T. A.; Frydman, M.; Peter, M. O.; Schmutz, J. L.; Gilgenkrantz, S.;

Mitchell, G.; Frezal, J.; Melancon, S.; Lavergne, L.; Labuda, D.; Hors-Cayla, M. C.: Linkage studies do not confirm the cytogenetic location of incontinentia pigmenti on Xp11. Hum. Genet. 80: 282-286, 1988.
Sommer, A.; Liu, P. H.: Incontinentia pigmenti in a father and his daughter. Am. J. Med. Genet. 17: 655-659, 1984.
Spallone, A.: Incontinentia pigmenti (Bloch-Sulzberger syndrome): seven case reports from one family. Brit. J. Ophthal. 71: 629-634, 1987.
Sulzberger, M. B.: Ueber eine bisher nicht beschriebene congenitale Pigmentanomalie (Incontinentia pigmenti). Arch. Derm. Syph. 154: 19-32, 1927.
Sybert, V. P.: Incontinentia pigmenti nomenclature. (Letter) Am. J. Hum. Genet. 55: 209-211, 1994.
Traupe, H.; Vehring, K.-H.: Unstable pre-mutation may explain mosaic disease expression of incontinentia pigmenti in males. Am. J. Med. Genet. 49: 397-398, 1994.
Wettke-Schafer, R.; Kantner, G.: X-linked dominant inherited diseases with lethality in hemizygous males. Hum. Genet. 64: 1-23, 1983.
Wieacker, P.; Zimmer, J.; Ropers, H.-H.: X inactivation patterns in two syndromes with probable X-linked dominant, male lethal inheritance. Clin. Genet. 28: 238-242, 1985.
Wiklund, D. A.; Weston, W. L.: Incontinentia pigmenti: a four-generation study. Arch. Derm. 116: 701-703, 1980.

CHEDIAK-HIGASHI SYNDROME (*214500)

JJM NOTE

The neoplasms seen to excess include lymphoma, leukemia, and perhaps the sclerosing stromal tumor of the ovary. The menagerie of animal models should make gene mapping easier than it has been to date.

The features are decreased pigmentation of hair and eyes (partial albinism), photophobia, nystagmus, large eosinophilic, peroxidase-positive inclusion bodies in the myeloblasts and promyelocytes of the bone marrow, neutropenia, abnormal susceptibility to infection, and peculiar malignant lymphoma. Death often occurs before the age of 7 years. (See Hermansky-Pudlak syndrome (203300), a similar but distinct entity.) Kritzler et al. (1964) found the karyotype normal in a 16-year-old patient. Glycolipid inclusions were described in histiocytes, renal tubular epithelium, and neurons. Heterozygotes were identifiable by the presence of a granular anomaly of the lymphocytes. The patient died of massive gastrointestinal hemorrhage. Leukemia and lymphoma have been observed (Efrati and Jonas, 1958).

Windhorst et al. (1966) found large lysosomal granules in leukocytes and giant melanosomes in melanocytes. For this reason, Leader et al. (1966) referred to the condition as 'hereditary leukomelanopathy.' Hargis and Prieur (1985), who studied CHS in cats, quoted White (1966) as providing evidence that many of the enlarged granules in CHS cells are derived from lysosomes. Sheramata et al. (1971) described 3 brothers, aged 31, 34 and 38, who had this disorder and a neurologic picture resembling spinocerebellar degeneration. Neutrophils are deficient in chemotactic and bactericidal activities. Microtubular abnormalities have been demonstrated (Oliver and Zurier, 1976) and ascorbic acid corrects certain functional abnormalities of the cells (Boxer et al., 1976).

Siccardi et al. (1978) described a 4-year-old Italian boy with recurrent infections. Both he and his healthy father had a severe isolated defect in bactericidal activity of circulating neutrophils. The parents of the proband were first cousins once removed. The proband had silvery-blond hair, individual hairs showing silver and blond banding, as well as a slate-gray generalized hyperpigmentation of the skin. Generalized lymph node enlargement and hepatosplenomegaly were present. The boy died at age 4 years and 9 months, following cerebral hemorrhage (probably secondary to thrombocytopenia caused by hypersplenism). No autopsy was performed. Obviously there were similarities to and differences from the Chediak-Higashi syndrome. Inoue et al. (1991) reported the occurrence of sclerosing stromal tumor of the ovary in a 13-year-old girl with CHS.

Ganz et al. (1988) demonstrated deficiency of cathepsin G (116830) and elastase (130130) in all 3 patients with CHS whom they studied. Cathepsin G is a constituent of the azurophil granule; defensins, which are also constituents, were normal or only mildly decreased in the CHS patients. Elastase has an ancillary microbicidal/cytotoxic action. In another disorder with frequent and severe bacterial infections, namely, specific granule deficiency (SGD; 245480), Ganz et al. (1988) found almost complete deficiency of defensins.

Nomenclature: the preferred symbol for the syndrome is CHS1, in order to avoid confusion with the Cohen syndrome (COH1; 216550).

Karim et al. (1997) reported 2 homozygous CHS1 gene mutations. One of these, a frameshift at codon 3197 (214500.0005), supported their assertion that the functional CHS protein is a predicted 3801-amino acid polypeptide encoded by a 13.5 kb mRNA.

Barbosa et al. (1997) reported the sequences of 2 major mRNA isoforms of the CHS gene in human and mouse. These isoforms differ both in size and in sequence at the 3-prime end of their coding domains, with a small isoform (approximately 5.8 kb) arising from incomplete splicing and reading through an intron. These mRNAs also differ in tissue distribution of transcription and in predicted biologic properties. Barbosa et al. (1997) identified novel mutations within the region of the coding domain common to both isoforms in 3 CHS patients: C-to-T transitions that generated stop codons (R50X; 214500.0006 and Q1029X; 214500.0007) were found in 2 patients, and a novel frameshift mutation (deletion of nucleotides 3073 and 3074 of the coding domain) was found in a third. Northern blots of lymphoblastoid mRNA from CHS patients revealed loss of the largest transcript (approximately 13.5 kb) in 2 of 7 CHS patients, while the small mRNA was undiminished in abundance. These results suggested that the small isoform alone cannot complement Chediak-Higashi syndrome. All beige and CHS mutations that had been identified were predicted to result in either truncated or absent proteins.

Mapping

In the mouse, the analog of CHS (beige) is linked to the TCRG gene (186970) on mouse chromosome 13 (Holcombe et al., 1987). The 2 loci show a frequency of recombination of 0.025. Holcombe et al. (1987) found nonlinkage in man between TCRG and the Chediak-Higashi syndrome, however; lod scores were negative through a full range of recombination values and were less than -2.0 at theta = 0.20 and lower. Jenkins et al. (1991) predicted that the CHS gene may reside in distal 1q because in the mouse the homologous condition shows linkage to the nidogen gene (131390) which is located on human 1q.

Fukai et al. (1996) noted that the murine beige (bg) locus is on chromosome 13 in close proximity to the genes Tcrg (186970), Hist1 (142709), Prf (170280), Fim1 (136750), and Nid. This region has, however, undergone considerable genomic rearrangement during mammalian evolution; for example, the human TCRG gene maps to chromosome 7p15 while the human NID gene maps to chromosome 1q43. Fukai et al. (1996) carried out homozygosity mapping in 4 inbred probands with classical childhood CHS using markers derived from the human chromosome segment 1q42-q44 . The lod score between markers in this region (e.g., D1S235, D1S1594, and D1S204) and CHS in the inbred kindreds was 4.82. Fukai et al. (1996) also studied several inbred patients with the atypical adult form of CHS. None of these individuals were homozygous for markers in distal 1q. This finding suggested to the authors that at least some cases of CHS may represent a genetic entity with a different map location.

Barrat et al. (1996) mapped the CHS locus by linkage analysis to a 5-cM interval on chromosome 1q42.1-q42.2. The highest lod score (5.38 at theta = 0) was obtained with the marker D1S235. They used haplotype analysis to define D1S2680 as the telomeric flanking marker and D1S163 as the centromeric flanking marker. The 9 families used in this study were from 7 different countries. There was consanguinity within 5 of the families. Barrat et al. (1996) identified 3 YAC clones which covered the entire region in a contig. The gene is also symbolized LYST for lysosomal trafficking regulator.

Clinical Management

In a girl with Chediak-Higashi syndrome, Aslan et al. (1996) reported on the temporary success (11 months) of high-dose methylprednisolone during the 'accelerated phase' of her condition after unsuccessful treatment with vincristine, prednisolone, ascorbic acid, and antibiotics (ceftriaxone, netilmicin, and co-trimoxazole). After a second trial of high-dose methylprednisolone was unsuccessful, splenectomy continued the child's survival for an additional 29 months. The patient died of neutropenic septicemia. Atypically, this child had pulmonary involvement and no evidence of lymphohistiocytic infiltration in the rectum and sigmoid colon with biopsy proven intestinal polyposis.

Animal Model

Man, mouse, cattle, mink, and killer whale are known to be affected. Kahraman and Prieur (1990) stated that this disorder has been identified in 10 species, including humans. Kahraman and Prieur (1990) succeeded in prenatal diagnosis of the disorder in cats by demonstrating abnormally large lysosomes (stained for acid phosphatase) in cultured amniotic fluid cells. In mink and cattle, the disorder is autosomal recessive (Padgett et al., 1964).

Penner and Prieur (1987) found close morphologic similarities of the CHS fibroblasts from humans, cats, mink, and cattle. Mice homozygous for the 'beige' (bg) gene have a selective deficiency of NK (natural killer) lymphocytes and an increased susceptibility to transplanted tumors. Patients with the homologous Chediak-Higashi syndrome appear to have the same defect of NK cells (Roder et al., 1980). Perou et al. (1997) showed that the mutation in the bg allele is the result of a LINE1 (151626) retrotransposition. Penner and Prieur (1987) found a lack of complementation when human CHS fibroblasts were fused with cat CHS fibroblasts, and also when human CHS fibroblasts were fused with mink CHS fibroblasts. This suggested that the disease has the same cause in these 3 species. NK cells are thought to have an important role in surveillance against tumor development. Virelizier and Griscelli (1980) simultaneously demonstrated the defect in NK cells. They could not modify the NK activity of CHS leukocytes by prolonged in vitro incubation with interferon, or by in vivo administration of interferon. Bone marrow transplantation, however, restored NK activity. Both spontaneous levels of NK activity and its in vitro activation by interferon were restored. Neutrophils kill their targets by means of 2 distinct classes of effector substances: reactive oxygen intermediates (ROI) and microbicidal/cytotoxic proteins. Myeloperoxidase deficiency (254600) and chronic granulomatous disease (306400) are examples of deficient ROI production by polymorphonuclear leukocytes. Perou et al. (1996) succeeded in identifying the murine 'beige' gene by in vitro complementation and positional cloning, and confirmed its identification by defining mutations in 2 independent mutant alleles. The sequence of the 'beige' gene transcript shows strong nucleotide homology to multiple human ESTs, one or more of which may be associated with the Chediak-Higashi syndrome gene. The predicted 2,186-amino ac id polypeptide shows significant amino acid homology to orphan proteins identified in Saccharomyces cerevisiae (e.g., CDC4), Caenorhabditis elegans, and humans (e.g., CDC4L).

Although Perou et al. (1996) and Barbosa et al. (1996) reported identification of the 'beige' gene, the 2 cDNAs were quite different. Nagle et al. (1996) described the sequence of a human cDNA homologous to mouse 'beige,' identified pathologic mutations in patients with Chediak-Higashi syndrome, and clarified the discrepancies of the previous reports of sequence. Analysis of the CHS polypeptide demonstrated that its modular architecture is similar to that of the yeast vacuolar sorting protein VPS15.

Jackson (1997) reviewed homologous pigmentation mutations in human, mouse, and other model organisms. He tabulated the available data on all pigmentation genes cloned from mouse or human, and focused on 3 particular systems. One family of genes, including LYST and HPS (203300), shows the relationship between melanosomes and lysosomes. The G-protein coupled receptor, endothelin receptor-B (EDNRB; 131244) and its ligand, endothelin-3 (EDN3; 131242) are required for the development of both melanocytes and enteric neurons; mutations in either result in the combination of megacolon and pigmentary anomaly (spotting). The melanocortin-1 receptor (MC1R; 155555) is expressed only in melanocytes, but mutations that cause overexpression of agouti protein (ASIP; 600201), an antagonist of the receptor, result in obesity in transgenic mice, thus highlighting a role of melanocortins in weight homoeostasis.

In cells from patients with the Chediak-Higashi syndrome, Faigle et al. (1998) found that peptide loading onto major histocompatibility complex class II molecules and antigen presentation were strongly delayed. Results of other studies suggested that the product of the LYST gene is required for sorting endosomal resident proteins into late multivesicular endosomes by a mechanism involving microtubules.

See also Abo et al. (1982); Blume and Wolff (1972); Chediak (1952); De Beer et al. (1981); Donohue and Bain (1957); Gilloon et al. (1960); Hargis and Prieur (1987); Kanfer et al. (1968); Page et al. (1962); Penner and Prieur (1987); Rausch et al. (1978); Sadan et al. (1965); Spencer and Hogan (1962); Stegmaier and Schneider (1965); Tanaka (1980); Tay et al. (1970); Virelizier et al. (1982); White and Clawson (1979); White and Clawson (1980); Windhorst et al. (1968)

Abo, T.; Roder, J. C.; Abo, W.; Cooper, M. D.; Balch, C. M.: Natural killer (HNK-1+) cells in Chediak-Higashi patients are present in normal numbers but are abnormal in function and morphology. J. Clin. Invest. 70: 193-197, 1982.

Aslan, Y., Erduran, E.; Gedik, Y.; Mocan, H.; Yildiran, A.: The role of high dose methylprednisolone and splenectomy in the accelerated phase of Chediak-Higashi syndrome. Acta Haemat. 96: 105-107, 1996.

Barbosa, M. D. F. S.; Barrat, F. J.; Tchernev, V. T.; Nguyen, Q. A.; Mishra, V. S.; Colman, S. D.; Pastural, E.; Dufourcq-Lagelouse, R.; Fischer, A.; Holcombe, R. F.; Wallace, M. R.; Brandt, S. J.; de Saint Basile, G.; Kingsmore S. F.: Identification of mutations in two major mRNA isoforms of the Chediak-Higashi syndrome gene in human and mouse. Hum. Molec. Genet. 6: 1091-1098, 1997.

Barbosa, M. D. F. S.; Nguyen, Q. A.; Tchernev, V. T.; Ashley, J. A.; Detter, J. C.; Blaydes, S. M.; Brandt, S. J.; Chotai, D.; Hodgman, C.; Solari, R. C. E.; Lovett, M.; Kingsmore, S. F.: Identification of the homologous beige and Chediak-Higashi syndrome genes. Nature 382: 262-265, 1996.

Barrat, F. J.; Auloge, L.; Pastural, E.; Dufourcq Lagelouse, R.; Vilmer, E.; Cant, A. J.; Weissenbach, J.; Le Paslier, D.; Fischer, A.; de Saint Basile, G.: Genetic and physical mapping of the Chediak-Higashi syndrome on chromosome 1q42-43. Am. J. Hum. Genet. 59: 625-632, 1996.

Blume, R. S.; Wolff, S. M.: The Chediak-Higashi syndrome: studies in four patients and a review of the literature. Medicine 51: 247-280, 1972.

Boxer, L. A.; Watanabe, A. M.; Rister, M.; Besch, H. R., Jr.; Allen, J.; Baehner, R. L.: Correction of leukocyte function in Chediak-Higashi syndrome by ascorbate. New Eng. J. Med. 295: 1041-1045, 1976.

Chediak, M.: Nouvelle anomalie leucocytaire de caractere constitutionnel et familial. Rev. Hemat. 7: 362-367, 1952.

De Beer, H. A.; Anderson, R.; Findlay, G. H.: Chediak-Higashi syndrome in a 'black' child: clinical features, immunological studies, and optics of the hair and skin. S. Afr. Med. J. 60: 108-112, 1981.

Donohue, W. L.; Bain, H. W.: Chediak-Higashi syndrome, a lethal familial disease with anomalous inclusions in the leukocytes and constitutional stigmata: report of a case with necropsy. Pediatrics 20: 416-430, 1957.

Efrati, P.; Jonas, W.: Chediak's anomaly of leukocytes in malignant lymphoma associated with leukemic manifestations: case report with necropsy. Blood 13: 1063-1073, 1958.

Faigle, W.; Raposo, G.; Tenza, D.; Pinet, V.; Vogt, A. B.; Kropshofer, H.; Fischer, A.; de Saint-Basile, G.; Amigorena, S.: Deficient peptide loading and MHC class II endosomal sorting in a human genetic immunodeficiency disease: the Chediak-Higashi syndrome. J. Cell Biol. 141: 1121-1134, 1998.

Fukai, K.; Oh, J.; Karim, M. A.; Moore, K. J.; Kandil, H. H.; Ito, H.; Burger, J.; Spritz, R. A.: Homozygosity mapping of the gene for Chediak-Higashi syndrome to chromosome 1q42-q44 in a segment of conserved synteny that includes the mouse beige locus (bg). Am. J. Hum. Genet. 59: 620-624, 1996.

Ganz, T.; Metcalf, J. A.; Gallin, J. I.; Boxer, L. A.; Lehrer, R. I.: Microbicidal/cytotoxic proteins of neutrophils are deficient in two disorders: Chediak-Higashi syndrome and 'specific' granule deficiency. J. Clin. Invest. 82: 552-556, 1988.

Gilloon, J. R.; Pease, G. L.; Mills, S. D.: Chediak-Higashi anomaly of the leukocytes: report of a case. Mayo Clin. Proc. 35: 635-640, 1960.

Hargis, A. M.; Prieur, D. J.: Light and electron microscopy of hepatocytes of cats with Chediak-Higashi syndrome. Am. J. Med. Genet. 22: 659-668, 1985.

Hargis, A. M.; Prieur, D. J.: Renal lesions in cats with Chediak-Higashi-Steinbrinck syndrome. Am. J. Med. Genet. 26: 167-179, 1987.

Holcombe, R. F.; Strauss, W.; Owen, F. L.; Boxer, L. A.; Warren, R. W.; Conley, M. E.; Ferrara, J.; Leavitt, R. Y.; Fauci, A. S.; Taylor, B. A.; Seidman, J. G.: Relationship of the gene for Chediak-Higashi syndrome (beige) and the T-cell receptor gamma chain in mouse and man. Genomics 1: 287-291, 1987.

Inoue, R.; Kondo, N.; Motoyoshi, F.; Hori, Y.; Orii, T.: Chediak-Higashi syndrome: report of a case with an ovarian tumor.(Letter) Clin. Genet. 39: 316-318, 1991.

Jackson, I. J.: Homologous pigmentation mutations in human, mouse and other model organisms. Hum. Molec. Genet. 6: 1613-1624, 1997.

Jenkins, N. A.; Justice, M. J.; Gilbert, D. J.; Chu, M.-L.; Copeland, N. G.: Nidogen/entactin (Nid) maps to the proximal end of mouse chromosome 13 linked to beige (bg) and identifies a new region of homology between mouse and human chromosomes. Genomics 9: 401-403, 1991.

Kahraman, M. M.; Prieur, D. J.: Chediak-Higashi syndrome in the cat: prenatal diagnosis by evaluation of amniotic fluid cells. Am. J. Med. Genet. 36: 321-327, 1990.

Kanfer, J. N.; Blume, R. S.; Yankee, R. A.; Wolff, S. M.: Sphingolipid metabolism in leukocytes in Chediak-Higashi syndrome. New Eng. J. Med. 279: 410-413, 1968.

Karim, M. A.; Nagle, D. L.; Kandil, H. H.; Burger, J.; Moore, K. J.; Spritz, R. A.: Mutations in the Chediak-Higashi syndrome gene (CHS1) indicate requirement for the complete 3801 amino acid CHS protein. Hum. Molec. Genet. 6: 1087-1089, 1997.

Kritzler, R. A.; Terner, J. Y.; Lindenbaum, J.; Magidson, J.; Williams, R.; Preisig, R.; Phillips, G. B.: Chediak-Higashi syndrome: cytologic and serum lipid observations in a case and family. Am. J. Med. 36: 583-594, 1964.

Leader, R. W.; Padgett, G. A.; Gorham, J. R.: Hereditary leukomelanopathy (Chediak-Higashi syndrome of man, mink and cattle).In: Gajdusek, D. C.; Gibbs, C. J., Jr.; Alpers, M.: Slow, Latent and Temperate Virus Infections. Monograph 2. Washington: National Institute of Neurological Diseases and Blindnes (pub.) 1966. Pp. 393-399.

Nagle, D. L.; Karim, M. A.; Woolf, E. A.; Holmgren, L.; Bork, P.; Misumi, D. J.; McGrail, S. H.; Dussault, B. J., Jr.; Perou, C. M.; Boissy, R. E.; Duyk, G. M.; Spritz, R. A.; Moore, K. J.: Identification and mutation analysis of the complete gene for Chediak-Higashi syndrome. Nature Genet. 14: 307-311, 1996.

Oliver, J. M.; Zurier, R. B.: Correction of characteristic abnormalities of microtubule function and granule morphology in Chediak-Higashi syndrome with cholinergic agonists: studies in vitro in man and in vivo in the beige mouse. J. Clin. Invest. 57: 1239-1247, 1976.

Padgett, G. A.; Leader, R. W.; Gorham, J. R.; O'Mary, C. C.: The familial occurrence of the Chediak-Higashi syndrome in mink and cattle. Genetics 49: 505-512, 1964.

Page, A. R.; Berendes, H.; Warner, J.; Good, R. A.: The Chediak-Higashi syndrome. Blood 20: 330-343, 1962.

Penner, J. D.; Prieur, D. J.: Interspecific genetic complementation analysis with fibroblasts from humans and four species of animals with Chediak-Higashi syndrome. Am. J. Med. Genet. 28: 455-470, 1987.

Penner, J. D.; Prieur, D. J.: A comparative study of the lesions in cultured fibroblasts of humans and four species of animals with Chediak-Higashi syndrome. Am. J. Med. Genet. 28: 445-454, 1987.

Perou, C. M.; Moore, K. J.; Nagle, D. L.; Misumi, D. J.; Woolf, E. A.; McGrail, S. H.; Holmgren, L.; Brody, T. H.; Dussault, B. J., Jr.; Monroe, C. A.; Duyk, G. M.; Pryor, R. J.; Li, L.; Justice, M. J.; Kaplan, J.: Identification of the murine beige gene by YAC complementation and positional cloning. Nature Genet. 13: 303-308, 1996.

Perou, C. M.; Pryor, R. J.; Naas, T. P.; Kaplan, J.: The bg allele mutation is due to a LINE1 element retrotransposition. Genomics 42: 366-368, 1997.

Rausch, P. G.; Prygwansky, K. B.; Spitznagel, J. K.: Immunocytochemical identification of azurophilic and specific granule markers in the giant granules of Chediak-Higashi neutrophils. New Eng. J. Med. 298: 693-698, 1978.

Roder, J. C.; Haliotis, T.; Klein, M.; Korec, S.; Jett, J. T.; Ortaldo, J.; Heberman, R. B.; Katz, P.; Fauci, A. S.: A new immunodeficiency disorder in humans involving NK cells. Nature 284: 553-555, 1980.

Sadan, N.; Yaffe, D.; Rozenszajn, L.; Efrati, P.: Chediak's disease: clinical, cytological and hereditary aspects. (Abstract) Israel J. Med. Sci. 1: 850 only, 1965.
Sheramata, W.; Kott, H. S.; Cyr, D. P.: The Chediak-Higashi-Steinbrinck syndrome. Arch. Neurol. 25: 289-294, 1971.
Siccardi, A. G.; Bianchi, E.; Calligari, A.; Clivio, A.; Fortunato, A.; Magrini, U.; Sacchi, F.: A new familial defect in neutrophil bactericidal activity. Helv. Paediat. Acta 33: 401-412, 1978.
Spencer, W. H.; Hogan, M. J.: Ocular manifestations of Chediak-Higashi syndrome: report of a case with histopathologic examination of ocular tissues. Am. J. Ophthal. 50: 1197-1203, 1962.
Stegmaier, O. C.; Schneider, L. A.: Chediak-Higashi syndrome: dermatologic manifestations. Arch. Derm. 91: 1-8, 1965.
Tanaka, T.: Chediak-Higashi syndrome: abnormal lysosomal enzyme levels in granulocytes of patients and family members. Pediat. Res. 14: 901-904, 1980.
Tay, C. H.; Lopez, C. G.; Lazarus, A. R.: The Chediak-Higashi syndrome. Med. J. Aust. 2: 1024-1028, 1970.
Virelizier, J. L.; Griscelli, C.: Interferon administration as an immunoregulatory and antimicrobial treatment in children with defective interferon secretion.In: Seligmann, M.; Hitzig, W. H.: Primary Immunodeficiencies. Amsterdam: Elsevier/North Holland Biomedical Press (pub.) 1980. Pp. 473-484.
Virelizier, J. L.; Lagrue, A.; Durandy, A.; Arenzana, F.; Oury, C.; Griscelli, C.; Reinert, P.: Reversal of natural killer defect in a patient with Chediak-Higashi syndrome after bone-marrow transplantation. (Letter) New Eng. J. Med. 306: 1055-1056, 1982.
White, J. G.: The Chediak-Higashi syndrome: a possible lysosomal disease. Blood 28: 143-156, 1966.
White, J. G.; Clawson, C. C.: The Chediak-Higashi syndrome: microtubules in monocytes and lymphocytes. Am. J. Hemat. 7: 349-356, 1979.
White, J. G.; Clawson, C. C.: The Chediak-Higashi syndrome, the nature of the giant neutrophil granules and their interactions with cytoplasm and foreign particulates. I. Progressive enlargement of the massive inclusions in mature neutrophils. II. Manifestations of cytoplasmic injury and sequestration. III. Interactions between giant organelles and foreign particulates. Am. J. Path. 98: 151-196, 1980.
Windhorst, D. B.; White, J. G.; Zelickson, A. S.; Clawson, C. C.; Dent, P. B.; Pollara, B.; Good, R. A.: The Chediak-Higashi anomaly and the Aleutian trait in mink: homologous defects of lysosomal structure. Ann. N.Y. Acad. Sci. 155: 818-846, 1968.
Windhorst, D. B.; Zelickson, A. S.; Good, R. A.: Chediak-Higashi syndrome: hereditary gigantism of cytoplasmic organelles. Science 151: 81-83, 1966.

CHEDIAK-HIGASHI-LIKE SYNDROME (GRISCELLI SYNDROME) (#214450)

JJM NOTE

Frank malignancy has not been reported, but the entity is included because of its similarity to Chediak-Higashi syndrome and because of the bone marrow dysfunction.

A number sign (#) is used with this entry because of evidence that it is caused by mutations in the gene encoding myosin VA (MYO5A; 160777).

Griscelli et al. (1978) described 2 unrelated patients with a disorder resembling, but probably distinct from, the Chediak-Higashi syndrome (214500). Features were partial albinism, frequent pyogenic infections, and acute episodes of fever, neutropenia and thrombocytopenia. The pigmentary dilution was characterized by large clumps of pigment in the hair shafts and an accumulation of melanosomes in melanocytes. It resembles the 'dilute' mutation of mice. Despite an adequate number of T and B lymphocytes, the patients were hypogammaglobulinemic, deficient in antibody production, and incapable of delayed skin hypersensitivity and skin graft rejection. Their leukocytes did not stimulate normal lymphocytes. A defect of helper T-cells was postulated. One patient was an 11-year-old North African girl with unrelated parents with a brother and sister with silvery hair who had died at 30 and 18 months of age, respectively. Differences from Chediak-Higashi syndrome are morphologic normality of polymorphonuclear leukocytes; the giant granules of the CH syndrome are not found. The CH syndrome has 'beige' as the homologous mutation in mice. The morphologic characteristics of the hypopigmentation also distinguished the disorder from CHS, as well as from all other pigmentary anomalies of man. Another difference was normal leukocyte specific protease activity (which is very low in CHS). Durandy et al. (1993) successfully undertook the prenatal diagnosis of 2 hereditary syndromes associating albinism and immune defects: Chediak-Higashi syndrome and Griscelli syndrome. Because the genes responsible for these diseases had not yet been mapped and the immune abnormalities were too subtle to be diagnosed in utero, prenatal diagnosis was made on morphologic grounds. In the case of CH syndrome, it was based on light-microscopic examination of the hair shaft and on light- and electron-microscopic study of polymorphonuclear cells. In the Griscelli syndrome, only examination of the hair was feasible. The diagnos is was negative in 12 fetuses at risk and positive in 4. Hurvitz et al. (1993) reported studies of 4 members of a highly consanguineous kindred with pigmentation and histologic findings consistent with Griscelli syndrome but without symptoms of immunodeficiency. These patients exhibited a range of neurologic involvement from mild cognitive delay and convulsive delay in one patient to a fatal degenerative course in the other 3. In a review, Klein et al. (1994) stated that Griscelli syndrome can be distinguished from Chediak-Higashi syndrome by pathognomonic histologic features. The prognosis is poor unless early bone marrow transplantation is carried out. They stated that since the original description, only 19 patients had been described. They reported the clinical and biologic features, therapy, and outcome in 7 cases.

Gogus et al. (1995) described 3 Turkish patients with Griscelli syndrome: a girl of consanguineous parents and 2 brothers who were born to another consanguineous couple. The girl and one boy had serious neurological problems with spasticity, rigidity, and convulsions. Despite therapy, both patients died. Hemorrhagic areas and lymphohistiocytic infiltration were found in all sections of the brain. Focal spongy degeneration was found in the cortex. Erythrophagocytosis was present in almost every organ.

Pastural et al. (1997) used linkage analysis in 3 consanguineous families and 1 nonsanguineous family to map Griscelli disease to 15q21. Because of the similarities to the phenotype in the 'dilute' mouse, which is due to mutation in an unconventional myosin gene, they looked at regions of chromosomes 19, 10, and 18 known to contain genes of this type and excluded linkage to those areas. On the other hand, a peak lod score of 4.40 was obtained for a cluster of markers on chromosome 15. This region contains the myosin VA gene (160777). Pastural et al. (1997) demonstrated nucleotide substitutions, 1 nonsense (160777.0002) and 1 missense (160777.0001), in 2 unrelated Turkish patients.

Durandy, A.; Breton-Gorius, J.; Guy-Grand, D.; Dumez, C.; Griscelli, C.: Prenatal diagnosis of syndromes associating albinism and immune deficiencies (Chediak-Higashi syndrome and variant). Prenatal Diag. 13: 13-20, 1993.
Gogus, S.; Topcu, M.; Kucukali, T.; Akcoren, Z.; Berkel, I.; Ersoy, F.; Gunay, M.; Saatci, I.: Griscelli syndrome: report of three cases. Pediat. Path. Lab. Med. 15: 309-319, 1995.
Griscelli, C.; Durandy, A.; Guy-Grand, D.; Daguillard, F.; Herzog, C.; Prunieras, M.: A syndrome associating partial albinism and immunodeficiency. Am. J. Med. 65: 691-702, 1978.
Hurvitz, H.; Gillis, R.; Klaus, S.; Klar, A.; Gross-Kieselstein, F.; Okon, E.: A kindred with Griscelli disease: spectrum of neurological involvement. Europ. J. Pediat. 152: 402-405, 1993.
Klein, C.; Phillipe, N.; Le Deist, F.; Fraitag, S.; Prost, C.; Durandy, A.; Fischer, A.; Griscelli, C.: Partial albinism with immunodeficiency (Griscelli syndrome). J. Pediat. 125: 886-895, 1994.
Pastural, E.; Barrat, F. J.; Dufourcq-Lagelouse, R.; Certain, S.; Sanal, O.; Jabado, N.; Seger, R.; Griscelli, C.; Fischer, A.; de Saint Basile, G.: Griscelli disease maps to chromosome 15q21 and is associated with mutations in the myosin-Va gene. Nature Genet. 16: 289-292, 1997.

MICROCEPHALY WITH NORMAL INTELLIGENCE, IMMUNODEFICIENCY, AND LYMPHORETICULAR MALIGNANCIES (#251260)

JJM NOTE

A wide spectrum of lymphoreticular childhood malignancies has been associated, including acute lymphocytic leukemia and B-cell lymphoma. Given this spectrum, it is unlikely that neuroblastoma occurred, unless by chance, in one patient reported by Seemanova et al. (1985). The caution against radiotherapy cannot be absolute, but the concern is for acute toxicity, not just late second malignancy.

A number sign (#) is used with this entry because the disorder is caused by mutation in the NBS1 gene (602667).

Weemaes et al. (1981) described 2 sons of second-cousin parents who had microcephaly, stunted growth, mental retardation, cafe-au-lait spots, and immunodeficiency. Cytogenetic studies showed a typical form of chromosome instability with multiple rearrangements of chromosomes 7 and 14. A lower frequency of the same chromosome abnormalities was found in the father and 3 of the phenotypically normal sibs. Seemanova et al. (1985) described 9 patients in 6 families with a 'new' disorder characterized by low birthweight for dates, microcephaly with normal intelligence, receding mandibula, cellular and humoral immune defects, and increased risk of lymphoreticular malignancies. No evidence of chromosomal instability was found, but chromosome analysis was difficult because the rate of blastic transformation with phytohemagglutinin was low. Even sex ratio, consanguinity in 1 family and grandparental isonymy in a second, and the occurrence of 2 affected sibs in 3 families supported autosomal recessive inheritance. Bronchiectasis, pneumonia, otitis media, mastoiditis, and sinusitis occurred. Immunoglobulin levels were reduced. In 2 sibs, acute lymphoblastic leukemia developed at ages 9 years and 12 months, respectively. Generalized malignancies, apparently originating in the mediastinum and variously identified as malignant lymphogranuloma, acute undifferentiated hemoblastoma and mediastinal blastoma (probably neuroblastoma) was the cause of death in several. The oldest surviving patient (of 4) was 12.5 years old.

Maraschio et al. (1986) described the case of a 31-year-old woman with primary amenorrhea, microcephaly and immunodeficiency. Her healthy parents were related as first cousins once removed. A younger sister, who also had primary amenorrhea, had died at age 20 years with a malignant lymphoma. Chromosome studies revealed a high proportion of metaphases with multiple chromosome aberrations. The same unbalanced translocation, t(8q;21q), was present in about 59% of metaphases. A few rearrangements involving chromosomes 7 and 14, similar to those described in patients with ataxia-telangiectasia (AT; 208900), were found. Sister-chromatid exchanges were not increased.

Teebi et al. (1987) reported a large inbred Arab kindred in which 8 individuals in 5 sibships had microcephaly and normal intelligence. Two died of acute lymphoreticular malignancy or bronchial pneumonia. Immunologic and chromosomal studies carried out in 3 affected living sibs yielded normal results.

It is now quite clear that at least 1 of the patients of Weemaes et al. (1981) and at least 2 of the patients of Seemanova et al. (1985) had the same basic defect; their cells failed to complement each other in cell fusion studies, using correction of radiosensitivity as the measure (Jaspers et al., 1988). There is at least one other syndrome of immunodeficiency, microcephaly, and chromosomal instability, which did show complementation with the other cases (Jaspers et al., 1988); the case of Conley et al. (1986) and 1 case of Sperling (1983) fell into this category. The patient of Sperling (1983) was further studied by Wegner et al. (1988). Jaspers et al. (1988) found that the disorder in all of these patients, which resembles ataxia-telangiectasia without the neurocutaneous manifestations, is indeed different from AT because fibroblasts from all cases show complementation with the 5 complementation groups of AT. Jaspers et al. (1988) performed complementation studies on fibroblast strains from 50 patients with either AT or NBS. Using the radioresistant DNA replication characteristic as a marker, they demonstrated 6 different genetic complementation groups, 2 of which, groups V1 and V2, involved patients with NBS. An individual with clinical symptoms of both AT and NBS was found in group V2, indicating that the 2 disorders are closely related.

Taalman et al. (1989) reported the findings in 5 families, 2 from the Netherlands and 3 from Czechoslovakia, containing a total of 8 patients. The basic karyotype in these patients was normal, but in a fifth or more of metaphases, rearrangements were found, preferentially involving chromosomes 7 and/or 14 at the sites 7p13, 7q34, and 14q11. The chromosomes of all 5 living patients were very sensitive to ionizing radiation. With ataxia-telangiectasia, NBS shares the occurrence of typical rearrangements of chromosomes 7 and 14, cellular and chromosomal hypersensitivity to x-irradiation, radioresistance of DNA replication, and immunodeficiency. However, NBS patients have microcephaly but neither ataxia nor telangiectasia, and they do not show elevation of serum alpha-fetoprotein, which is characteristic of AT.

Ataxia-telangiectasia variant-1 is the designation applied to the Nijmegen breakage syndrome and AT variant-2, the designation for the Berlin breakage syndrome (600885). Stumm et al. (1995) noted that patients with these disorders share cytogenetic features with ataxia-telangiectasia (208900), such as spontaneous chromosomal instability, clonal occurrence of rearrangements involving, in particular, chromosomes 7 and 14, chromosomal and cellular hypersensitivity to irradiation, and 'radioresistant DNA synthesis.' However, despite the fact that these patients have neither ataxia nor telangiectasia, they historically have been categorized as AT variants. Clinically, they are characterized by pronounced microcephaly, microgenia, 'bird-like' facies, immunodeficiency, and normal serum levels of alpha-fetoprotein. The 2 syndromes can be distinguished from one another only by complementation analysis.

Chrzanowska et al. (1995) reported on 11 patients with Nijmegen breakage syndrome from 8 independent Polish families, with a total of 3 pairs of affected sibs. The clinical pattern included microcephaly, particular 'bird-like' face, growth retardation, and, in some cases, mild-to-moderate mental deficiency. Most of the patients had recurrent respiratory tract infections. One girl developed B-cell lymphoma. Chromosome studies showed structural aberrations with multiple rearrangements, preferentially involving chromosomes 7 and 14, in a proportion of metaphase in all individuals. Profound humoral and cellular immune defects were observed. Serum AFP levels were within normal range. Radioresistant DNA synthesis was strongly increased in all 8 patients who were studied from this point of view.

In 6 of the Polish families reported by Chrzanowska et al. (1995), Stumm et al. (1995) performed haplotype studies and sib-pair analysis and demonstrated lack of linkage to the 11q22-q23 region where the ataxia-telangiectasia locus maps. One of these families had been assigned to complementation group AT-V1 and a second to complementation group AT-V2 by cell-fusion studies. No complementation studies had been done in the other 4 families.

The clinical, immunologic, chromosomal, and cell-biologic findings in 42 patients in the NBS (Nijmegen breakage syndrome) Registry in Nijmegen were reviewed by van der Burgt et al. (1996). Although the immunologic, chromosomal, and cell-biologic findings resembled those in ataxia-telangiectasia, the clinical findings are quite different. The authors stated that NBS appears to be a separate entity not allelic with AT, as indicated by the fact that linkage studies exclude 11q22-q23, where the gene for ataxia-telangiectasia is located, as the site of the NBS gene. None of the patients had signs of cerebellar ataxia, apraxic eye movements, or other neurologic abnormalities except for twin girls described by Curry et al. (1989) who had clinical symptoms of both NBS and AT (see 208900.0014). Complementation studies assigned these cases to NBS complementation group V1. Subtle scleral telangiectasia was noted in 10 of 25 patients. The patients did not have raised serum AFP levels, as in ataxia-telangiectasia. Twelve patients varying in age from 1 to 22 years had developed lymphoma. One patient developed a glioma at the age of 12 years, 1 patient a medulloblastoma at 15 years, and 1 patient a rhabdomyosarcoma at 4 years.

Der Kaloustian et al. (1996) described a boy who in addition to typical manifestations had penoscrotal hypospadias. He had lymphopenia with low percentage of B and T cells, absence of IgE, and low response to mitogen stimulation. At the age of 4 years he developed rhabdomyosarcoma. Cytogenetic study showed multiple chromatid and chromosome breaks, structural rearrangements involving mainly chromosomes 7 and 14, and different monosomies in 57 to 58% of cells. Nijmegen breakage syndrome was diagnosed, although hypospadias and a high percentage of monosomic cells led the authors to suggest he represented a specific variant of this syndrome. Der Kaloustian et al. (1996) suggested that the boy described by Woods et al. (1995) as a patient with Seckel syndrome might have the same variant of Nijmegen breakage syndrome.

In the treatment of malignancies in patients with NBS, van der Burgt et al. (1996) stated that cytostatics are the first choice; however, radiomimetics (for example, bleomycin) should be avoided, and the chemotherapy doses should be reduced. Radiation therapy should be avoided, since x-irradiation can induce malignancies in NBS patients.

NBS and the Berlin breakage syndrome (BBS; 600885) were considered separate entities on the basis of complementation studies which appeared to place them in separate complementation groups. These 2 so-called ataxia-telangiectasia variants are clinically indistinguishable, however. Saar et al. (1997) performed a whole-genome screen in 14 NBS/BBS families and localized the gene to a 1-cM interval on 8q21, between markers D8S271 and D8S270, with a peak lod score of 6.86 at D8S1811. This marker also showed strong allelic association to both Slavic NBS and German BBS patients, suggesting the existence of one major mutation of Slavic origin. The authors stated that since the same allele is seen in both complementation groups, genetic homogeneity of NBS/BBS can be considered as proved. Matsuura et al. (1997) used microcell-mediated chromosome transfer followed by complementation assays based on radiosensitivity to demonstrate that only chromosome 8 complements the sensitivity to ionizing radiation in NBS cell lines. In complementation assays performed after the transfer of a reduced chromosome, merely the long arm of chromosome 8 was sufficient for restoring the defect. The results supported the suggestion that NBS is a homogeneous disorder and that the gene for NBS is located at 8q21-q24.

(Some of the patients studied by Saar et al. (1997) were Germans in whom the Berlin breakage syndrome had been described and others were Slavic patients in whom the Seemanova syndrome (a synonym for NBS) had been described. Saar et al. (1997) noted that it would be interesting to investigate whether Dutch patients also showed an allelic association at D8S1811, similar to what they had found in Slavic and German patients. In the first half of the 17th century, after the battle of Weissenberg in the Thirty Years War, a considerable number of Bohemian Protestants emigrated to the Netherlands from an area presently part of Poland and the Czech Republic. A major NBS mutation may have found its way to the Netherlands by migration.)

Tupler et al. (1997) provided the first report of an Italian case of Nijmegen breakage syndrome. The proband was an immunodeficient, microcephalic, 11-year-old boy with a 'bird-like' face. He developed a T-cell-rich B-cell lymphoma. Spontaneous chromosomal instability was detected in T- and B-lymphocytes and in fibroblasts; chromosomes 7 and 14 were only sporadically involved in the rearrangements and no clonal abnormality was present. The patient appeared to be sensitive both to ionizing radiation and to bleomycin, although his sensitivity did not reach the level of ataxia-telangiectasia reference cells. Although the clinical evaluation suggested to Tupler et al. (1997) a diagnosis of NBS, differences in the cytogenetic and cell-biologic data suggested that the patient might have an allelic form of the disorder.

Stumm et al. (1997) found noncomplementation of radiation-induced chromosome aberrations in heterodikaryons between ataxia-telangiectasia and ataxia-telangiectasia variant cells. They suggested that the results of noncomplementation in AT/AT-V cell hybrids could be explained best by genes whose products contribute to a multisubunit protein involved in the damage response of radiation-induced chromosome aberrations. The data supported the assumption that the AT-V disorders represent a homogeneous genetic trait.

In a geographically diverse group of NBS patients, Cerosaletti et al. (1998) reported linkage to 8q21 in 6 of 7 families, with a maximum lod score of 3.58. Significant linkage disequilibrium was detected for 8 of 13 markers tested in the 8q21 region, including D8S1811. To localize the gene for NBS further, they generated a radiation hybrid map of markers at 8q21 and constructed haplotypes based on this map. Examination of disease haplotypes segregating in 11 NBS pedigrees revealed recombination events that placed the NBS gene between D8S1757 and D8S270. A common founder haplotype was present in 15 of 18 disease chromosomes from 9 of 11 NBS families. Inferred (ancestral) recombination events involving this common haplotype suggested that NBS can be localized further, to an interval flanked by markers D8S273 and D8S88.

Varon et al. (1998) and Carney et al. (1998) isolated the gene responsible for the Nijmegen breakage syndrome (NBS1; 602667). Varon et al. (1998) isolated the gene by positional cloning and named it nibrin, while Carney et al. (1998) isolated the gene by biochemical analysis of the protein components of the MRE11/RAD50 complex, and named it p95. Varon et al. (1998) identified mutations in the nibrin/p95 gene in patients with NBS.

See also Jaspers et al. (1988)

Carney, J. P.; Maser, R. S.; Olivares, H.; Davis, E. M.; Le Beau, M.; Yates, J. R., III; Hays, L.; Morgan, W. F.; Petrini, J. H. J.: The hMre11/hRad50 protein complex and Nijmegen breakage syndrome: linkage of double-strand break repair to the cellular DNA damage response. Cell 93: 477-486, 1998.

Cerosaletti, K. M.; Lange, E.; Stringham, H. M.; Weemaes, C. M. R.; Smeets, D.; Solder, B.; Belohradsky, B. H.; Taylor, A. M. R.; Karnes, P.; Elliott, A.; Komatsu, K.; Gatti, R. A.; Boehnke, M.; Concannon, P.: Fine localization of the Nijmegen breakage syndrome gene to 8q21: evidence for a common founder haplotype. Am. J. Hum. Genet. 63: 125-134, 1998.

Chrzanowska, K. H.; Kleijer, W. J.; Krajewska-Walasek, M.; Bialecka, M.; Gutkowska, A.; Goryluk-Kozakiewicz, B.; Michalkiewicz, J.; Stachowski, J.; Gregorek, H.; Lyson-Wojciechowska, G.; Janowicz, W.; Jozwiak, S.: Eleven Polish patients with microcephaly, immunodeficiency and chromosomal instability: the Nijmegen breakage syndrome. Am. J. Med. Genet. 57: 462-471, 1995.

Conley, M. E.; Spinner, M. B.; Emanuel, B. S.; Nowell, P. C.; Nichols, W. W.: A chromosome breakage syndrome with profound immunodeficiency. Blood 67: 1251-1256, 1986.

Curry, C. J. R.; Tsai, J.; Hutchinson, H. T.; Jaspers, N. G. J.; Wara, D.; Gatti, R. A.: AT(Fresno): a phenotype linking ataxia-telangiectasia with the Nijmegen breakage syndrome. Am. J. Hum. Genet. 45: 270-275, 1989.

Der Kaloustian, V. M.; Kleijer, W.; Booth, A.; Auerbach, A. D.; Mazer, B.; Elliott, A. M.; Abish, S.; Usher, R.; Watters, G.; Vekemans, M.; Eydoux, P.: Possible new variant of Nijmegen breakage syndrome. Am. J. Med. Genet. 65: 21-26, 1996.

Jaspers, N. G. J.; Gatti, R. A.; Baan, C.; Linssen, P. C. M. L.; Bootsma, D.: Genetic complementation analysis of ataxia telangiectasia and Nijmegen breakage syndrome: a survey of 50 patients. Cytogenet. Cell Genet. 49: 259-263, 1988.

Jaspers, N. G. J.; Taalman, R. D. F. M.; Baan, C.: Patients with an inherited syndrome characterized by immunodeficiency, microcephaly, and chromosomal instability: genetic relationship to ataxia telangiectasia. Am. J. Hum. Genet. 42: 66-73, 1988.

Maraschio, P.; Peretti, D.; Lambiase, S.; Lo Curto, F.; Caufin, D.; Gargantini, L.; Minoli, L.; Zuffardi, O.: A new chromosome instability disorder. Clin. Genet. 30: 353-365, 1986.

Matsuura, S.; Weemaes, C.; Smeets, D.; Takami, H.; Kondo, N.; Sakamoto, S.; Yano, N.; Nakamura, A.; Tauchi, H.; Endo, S.; Oshimura, M.; Komatsu, K.: Genetic mapping using microcell-mediated chromosome transfer suggests a locus for Nijmegen breakage syndrome at chromosome 8q21-24. Am. J. Hum. Genet. 60: 1487-1494, 1997.

Saar, K.; Chrzanowska, K. H.; Stumm, M.; Jung, M.; Nurnberg, G.; Wienker, T. F.; Seemanova, E.; Wegner, R.-D.; Reis, A.; Sperling, K.: The gene for the ataxia-telangiectasia variant, Nijmegen breakage syndrome, maps to a 1-cM interval on chromosome 8q21. Am. J. Hum. Genet. 60: 605-610, 1997.

Seemanova, E.; Passarge, E.; Beneskova, D.; Houstek, J.; Kasal, P.; Sevcikova, M.: Familial microcephaly with normal intelligence, immunodeficiency, and risk of lymphoreticular malignancies: a new autosomal recessive disorder. Am. J. Med. Genet. 20: 639-648, 1985.

Sperling, K.: Analisi dell' eterogeneita nell'uomo. Prospettive Pediat. 49: 53-66, 1983.

Stumm, M.; Gatti, R. A.; Reis, A.; Udar, N.; Chrzanowska, K.; Seemanova, E.; Sperling, K.; Wegner, R.-D.: The ataxia-telangiectasia-variant genes 1 and 2 are distinct from the ataxia-telangiectasia gene on chromosome 11q23.1. (Letter) Am. J. Hum. Genet. 57: 960-962, 1995.

Stumm, M.; Sperling, K.; Wegner, R.-D.: Noncomplementation of radiation-induced chromosome aberrations in ataxia-telangiectasia/ataxia-telangiectasia-variant heterodikaryons. (Letter) Am. J. Hum. Genet. 60: 1246-1251, 1997.

Taalman, R. D. F. M.; Hustinx, T. W. J.; Weemaes, C. M. R.; Seemanova, E.; Schmidt, A.; Passarge, E.; Scheres, J. M. J. C.: Further delineation of the Nijmegen breakage syndrome. Am. J. Med. Genet. 32: 425-431, 1989.

Teebi, A. S.; Al-Awadi, S. A.; White, A. G.: Autosomal recessive nonsyndromal microcephaly with normal intelligence. Am. J. Med. Genet. 26: 355-359, 1987.

Tupler, R.; Marseglia, G. L.; Stefanini, M.; Prosperi, E.; Chessa, L.; Nardo, T.; Marchi, A.; Maraschio, P.: A variant of the Nijmegen breakage syndrome with unusual cytogenetic features and intermediate cellular radiosensitivity. J. Med. Genet. 34: 196-202, 1997.

van der Burgt, I.; Chrzanowska, K. H.; Smeets, D.; Weemaes, C.: Nijmegen breakage syndrome. J. Med. Genet. 33: 153-156, 1996.

Varon, R.; Vissinga, C.; Platzer, M.; Cerosaletti, K. M.; Chrzanowska, K. H.; Saar, K.; Beckmann, G.; Seemanova, E.; Cooper, P. R.; Nowak, N. J.; Stumm, M.; Weemaes, C. M. R.; Gatti, R. A.; Wilson, R. K.; Digweed, M.; Rosenthal, A.; Sperling, K.; Concannon, P.; Reis, A.: Nibrin, a novel DNA double-strand break repair protein, is mutated in Nijmegen breakage syndrome. Cell 93: 467-476, 1998.

Weemaes, C. M. R.; Hustinx, T. W. J.; Scheres, J. M. J. C.; van Munster, P. J. J.; Bakkeren, J. A. J. M.; Taalman, R. D. F. M.: A new chromosomal instability disorder: the Nijmegen breakage syndrome. Acta Paediat. Scand. 70: 557-564, 1981.

Wegner, R.-D.; Metzger, M.; Hanefeld, F.; Jaspers, N. G. J.; Baan, C.; Magdorf, K.; Kunze, J.; Sperling, K.: A new chromosomal instability disorder confirmed by complementation studies. Clin. Genet. 33: 20-32, 1988.

Woods, C. G.; Leversha, M.; Rogers, J. G.: Severe intrauterine growth retardation with increased mitomycin C sensitivity: a further chromosome breakage syndrome. J. Med. Genet. 32: 301-305, 1995.

PULMONARY ALVEOLAR PROTEINOSIS (#265120)

JJM NOTE

This entity is included because of the association with various lymphoreticular malignancies, although the etiologic relationship is unclear. I am not aware of malignancy in a patient with congenital pulmonary alveolar proteinosis.

A number sign (#) is used with this entry because of evidence that it is caused, at least in some cases, by mutation in pulmonary surfactant-associated protein-3 (178640). In other cases, this disorder has been shown to be due to a defect in the expression of the beta chain that is common to CSF2RB (138981), the receptor for CSF2, IL3, and IL5.

Pulmonary alveolar proteinosis is a distinctive disorder in which alveoli are filled with periodic acid-Schiff-positive proteinaceous material. It is usually idiopathic, but has been reported in association with hematologic malignancy, lymphoma, and infection (nocardiosis, histoplasmosis, cytomegalovirus infection, etc.). Pulmonary alveolar proteinosis is usually encountered in adults, but has been observed in infants and children and has even been found at birth. Although its histopathologic appearance is similar to that of alveolar proteinosis observed in older children and adults, the congenital form of the illness follows a different clinical course. All reported infants with congenital alveolar proteinosis have died within the first year of life despite maximal medical therapy.

Webster et al. (1980) observed a brother and sister with classic, biopsy-proved pulmonary alveolar proteinosis. Both had low serum and low normal secretory immunoglobulin A (IgA) levels. Although immunologic defects have frequently been associated with pulmonary alveolar proteinosis, 'the literature. . .contains only 2 poorly documented incidental references to the possibility of familial occurrence' (Seard et al., 1970; Tsubura et al., 1974). The early-onset cases appear in many instances to represent a specific autosomal recessive entity. Teja et al. (1981) reported a sibship with second-cousin parents and 4 sibs (2 male) with pulmonary alveolar proteinosis. The 2 girls and 1 of the boys pursued a similar course with onset of respiratory distress at the age of 4 to 7 months and death in 1 to 3 months thereafter. One of the boys was admitted to hospital at the age of 16 months because of failure to thrive. He was cachectic, had clubbing of the fingers, and, by chest radiographs, showed bilateral pulmonary infiltrates with hilar prominence. In the next 17 years, the radiographic findings progressed, with fibrosis and honeycombing. He died at age 19 years. Open lung biopsies and/or autopsy demonstrated pulmonary alveolar proteinosis in all cases. Several families with 2 or more affected sibs were found in the literature. All except the familial cases reported by Webster et al. (1980) had onset in infancy or at birth. Parental consanguinity was present in 1 Mennonite family reported by Haworth et al. (1967), but this family had what the authors labeled 'thymic alymphoplasia' (Swiss type severe combined immunodeficiency; 202500).

Moulton et al. (1992) speculated that the cause of congenital pulmonary alveolar proteinosis is an inborn error of surfactant metabolism.

Nogee et al. (1993) described the cases of an infant brother and sister with this disorder, who died at 5 months and 1 month of age, respectively, after a typical course and with characteristic histopathologic features. Analysis of their lung tissue by immunologic and molecular biologic methods revealed absence of one of the surfactant specific proteins, surfactant protein B (SFTP3; 178640), and its mRNA. In an addendum, Nogee et al. (1993) noted that a third affected sib had been born into the family. There was no detectable protein B in the amniotic and lung-lavage fluid, but markedly increased amounts of surfactant protein C were present. The parents were unrelated. Nogee et al. (1994) subsequently demonstrated a frameshift mutation consisting of a substitution of GAA for C in codon 121 of the surfactant protein B cDNA. The 3 affected infants were homozygous for the mutation, which was present in heterozygous state in both parents and in 2 of 3 healthy sibs. The same mutation was found in 2 other unrelated infants who died from alveolar proteinosis, 1 of whom was also homozygous, and in the parents of an additional unrelated, affected infant, but was not observed in 50 control subjects.

In an infant in whom the diagnosis of PAP was made in the neonatal period and confirmed by bronchoalveolar lavage and lung biopsy, Dirksen et al. (1997) demonstrated no deficiency of surfactant protein-B but a marked deficiency in the synthesis of the common beta chain of the CSF2/IL3/IL5 receptor. Furthermore, at the age of 20 months (at which time the patient was oxygen-dependent), a mutation in the CSF2RB gene was found: a transversion from C to A at nucleotide 1835, resulting in a change of codon 602 from proline to threonine (138981.0001).

Dirksen, U.; Nishinakamura, R.; Groneck, P.; Hattenhorst, U.; Nogee, L.; Murray, R.; Burdach, S.: Human pulmonary alveolar proteinosis associated with a defect in GM-CSF/IL-3/IL-5 receptor common beta chain expression. J. Clin. Invest. 100: 2211-2217, 1997.
Haworth, J. C.; Hoogstraten, J.; Taylor, H.: Thymic alymphoplasia. Arch. Dis. Child. 42: 40-54, 1967.
Moulton, S. L.; Krous, H. F.; Merritt, T. A.; Odell, R. M.; Gangitano, E.; Cornish, J. D.: Congenital pulmonary alveolar proteinosis: failure of treatment with extracorporeal life support. J. Pediat. 120: 297-302, 1992.
Nogee, L. M.; deMello, D. E.; Dehner, L. P.; Colten, H. R.: Deficiency of pulmonary surfactant protein B in congenital alveolar proteinosis. New Eng. J. Med. 328: 406-410, 1993.
Nogee, L. M.; Garnier, G.; Dietz, H. C.; Singer, L.; Murphy, A. M.; deMello, D. E.; Colten, H. R.: A mutation in the surfactant protein B gene responsible for fatal neonatal respiratory disease in multiple kindreds. J. Clin. Invest. 93: 1860-1863, 1994.
Seard, C.; Wasserman, K.; Benfield, J. R.; Cleveland, R. J.; Costley, D. O.; Heimlich, E. M.: Simultaneous bilateral lung lavage (alveolar washing) using partial cardiopulmonary bypass. Report of two cases in siblings. Am. Rev. Resp. Dis. 101: 877-884, 1970.
Teja, K.; Cooper, P. H.; Squires, J. E.; Schnatterly, P. T.: Pulmonary alveolar proteinosis in four siblings. New Eng. J. Med. 305: 1390-1392, 1981.
Tsubura, E.; Kawase, I.; Yamamura, Y.: Hereditary metabolic diseases of the lung. Rec. Med. 26: 1727 only, 1974. Note: Cited by: Webster et al.: Am. J. Med. 69: 786-789, 1980.
Webster, J. R., Jr.; Battifora, H.; Furey, C.; Harrison, R. A.; Shapiro, B.: Pulmonary alveolar proteinosis in two siblings with decreased immunoglobulin A. Am. J. Med. 69: 786-789, 1980.

PRADER-WILLI SYNDROME (*176270)

JJM NOTE

Because other constitutional cytogenetic syndromes and overgrowth syndrome have been associated with childhood neoplasms, there is good reason to suspect that Prader-Willi syndrome could be as well. Despite an early warning of leukemia by Hall (1982), an odd collection of tumor types has accumulated in case reports (Bettio D et al.: Isochromosome 15q of maternal origin in a Prader-Willi patient with pituitary adenoma. Acta Genet. Med. Gemellol. (Roma) 1996;45:213-16; Coppes MJ et al.: Wilms tumor in a patient with Prader-Willi syndrome. J. Pediatr. 1993; 122:730-33; Hashizume K et al.: Prader-Willi syndrome with del(15)(q11,q13) associated with hepatoblastoma. Acta Paediatr. Jpn. 1991;33:718-22; Donhauser G et al.: Pseudo-Kaposi sarcoma in Prader-Labhart-Willi syndrome. Hautarzt. 1991;42:467-70; Robinson AC, Jones WG: Prader Willi syndrome and testicular tumour. Clin. Oncol. (R Coll Radiol) 1990; 2:117)

Description

A number sign (#) is used with this entry because of evidence that Prader-Willi syndrome is in effect a contiguous gene syndrome resulting from deletion of the paternal copies of the imprinted SNRPN gene (182279), the necdin gene (602117), and possibly other genes.

The Prader-Willi syndrome (PWS) is characterized by diminished fetal activity, obesity, muscular hypotonia, mental retardation, short stature, hypogonadotropic hypogonadism, and small hands and feet. It can be considered to be an autosomal dominant disorder and is caused by deletion or disruption of a gene or several genes on the proximal long arm of the paternal chromosome 15 or maternal uniparental disomy 15, because the gene(s) on the maternal chromosome(s) 15 are virtually inactive through imprinting.

Clinical Features

The original paper by Prader et al. (1956) described the full clinical picture.

Prenatal

Mothers with prior experience of normal pregnancies almost without exception report distinctly delayed onset and reduced fetal activity during the pregnancies involving Prader-Willi children. Obstetricians often fail to detect diminished fetal activity with ultrasound investigation. When reduced fetal activity is observed, prenatal cytogenetic examination produces normal results because cytogeneticists were not instructed to look for the characteristic chromosomal changes of PWS (Schinzel, 1986). Alert clinicians should refer CVS material from pregnancies with fetuses that demonstrate poor activity for molecular diagnosis of the syndrome (see below). Other candidates for prenatal diagnosis of PWS are fetuses of pregnancies in which trisomy 15 or mosaic trisomy 15 was determined from CVS, and in which subsequent amniocyte or fetal blood examinations disclosed a normal diploid karyotype. Theoretically, one-third of trisomy 15 fetuses initially with 2 maternal chromosomes 15 and 1 paternal chromosome 15 should give rise to Prader-Willi syndrome patients exhibiting maternal uniparental disomy(Cassidy et al., 1992; Purvis-Smith et al., 1992; Hall, 1992).

Perinatal

Neonates are profoundly hypotonic, which often causes asphyxia. In addition, there is mild prenatal growth retardation with a mean birthweight of about 6 lbs (2.8 kg) at term, hyporeflexia, poor feeding due to diminished swallowing and sucking reflexes, which in many cases necessitates gavage feeding for about 3 to 4 months. Cryptorchidism occurs with hypoplastic penis and scrotum in boys and hypoplastic labiae in girls (Stephenson, 1980). Chitayat et al. (1989) commented on the normal size of hands and feet at birth and in the first year of life.

Infancy and Childhood

Feeding difficulties generally improve by the age of 6 months. From 12 to 18 months onward, uncontrollable hyperphagia causes major somatic as well as psychologic problems. Diminished growth is observed in the majority of infants (Butler and Meaney, 1987). Small hands with delicate and tapering fingers and small feet (acromicria) are seen in most infants and adolescents; hand and foot sizes correlate well with length, but not with age, and foot size tends to be lower than hand size. However, patients of normal height tend to have normally sized hands (Hudgins and Cassidy, 1991). The face is characterized by a narrow bifrontal diameter, almond-shaped eyes (often in mild upslanted position), strabismus, full cheeks, and diminished mimic activity due to muscular hypotonia. Plethoric obesity becomes the most striking feature. From the age of about 6 years onward, many children present scars from scratching due to itching, and later, almost all show abdominal striae.

Depigmentation relative to the familial background is a feature in about three-quarters of the patients. Butler (1989), Hittner et al. (1982), and several authors remarked that this sign is confined to cases with deletions and absent in those with maternal disomy 15. Phelan et al. (1988) presented a black female child with oculocutaneous albinism, PWS, and an interstitial deletion of 15q11.2. Patients with classic albinism (203100) have misrouting of optic fibers, with fibers from 20 degrees or more of the temporal retina crossing at the chiasm instead of projecting to the ipsilateral hemisphere. Misrouting can result in strabismus and nystagmus. Because patients with PWS have hypopigmentation and strabismus, Creel et al. (1986) studied 6 patients, selected for a history of strabismus, with pattern-onset visually evoked potentials on binocular and monocular stimulation. Of the 4 with hypopigmentation, 3 had abnormal evoked potentials indistinguishable from those recorded in albinos. The 2 with normal pigmentation had normal responses. Wiesner et al. (1987) found that 14 of 29 patients with PWS had ocular hypopigmentation. There was possible correlation between hypopigmentation and a deletion of 15q.

MacMillan et al. (1972) described 2 unrelated girls with the features of PWS who additionally showed precocious puberty. They suggested that this is a variant and that a hypothalamic disturbance is responsible for this disorder. Hall and Smith (1972) pointed out narrow bifrontal cranial diameter as a feature. Hall (1985) pointed to a possibly increased risk of leukemia in PWS.

A frequent feature generally overlooked is thick saliva at the edges of the mouth. Patients tend to be relatively insensitive to pain (including that caused by obtaining blood samples)(Prader, 1991).

Adolescence and Adulthood

Greenswag (1987) reported on a survey of 232 adults with PWS, ranging in age from 16 to 64 years. Of 106 patients whose chromosomes were analyzed, 54 had an abnormality of chromosome 15, primarily a deletion. Physical characteristics, health problems, intelligence, psychosocial adjustment, and impact on the family were reviewed. Emotional lability, poor gross motor skills, cognitive impairment, and insatiable hunger were especially remarkable features.

In contrast to infants, adults invariably are small compared to their family members (Butler and Meaney, 1987). Due to high caloric intake, alimentary diabetes frequently sets in during or soon after the period of puberty. Puberty itself is diminished in PWS patients of both sexes. Adolescents and young adults often require digitalization because of cardiac insufficiency; however, it has been shown that substantial weight reduction relieves the need of cardiac therapy. Any attempt to reduce food intake in these adolescents often leads to serious psychologic and behavioral problems, and in some children, the situation in their home environment becomes intolerable (Curfs et al., 1991). Patients rarely survive beyond 25 to 30 years of age, the cause of death being diabetes and cardiac failure. However, if strict weight control is achieved, both diabetes and cardiac failure are greatly reduced and survival is either not or only mildly reduced. Johnsen et al. (1967) studied 7 mentally retarded patients, aged 4 to 19 years. Studies showed that fat synthesis from acetate during fasting was 10 times greater in patients than in unaffected sibs, and that hormone-stimulated lipolysis was depressed. These workers suggested that the condition is comparable to the genetic obese-hyperglycemic mouse. Since during fasting substrate continues to be used for new fat and lipolysis is deficient, survival depends on a continuous supply of exogenous calories. The abundant fat, muscle hypotonia, and small feet and hands are exactly the opposite of the sparse fat, muscle hypertrophy, and large hands and feet in Seip syndrome (269700).

Curfs et al. (1991) concluded that PWS patients score better on visual motor discrimination skills than on auditory verbal processing skills.

Wise et al. (1991) described 5 patients with PWS who experienced recurrent hyperthermia in infancy. On the basis of these patients and other reports of abnormal temperature regulation in PWS patients, particularly hypothermia with exposure to cold, they concluded that defects in temperature regulation may be a manifestation of hypothalamic dysfunction in PWS. On the other hand, Cassidy and McKillop (1991) concluded on the basis of a survey that clinically significant abnormal temperature control is not a common finding in this disorder. Similarly, Williams et al. (1994) concluded on the basis of a survey that the prevalence of febrile convulsions, fever-associated symptoms, and temperature less than 94 degrees F were not unique to PWS but can occur in any neurodevelopmentally handicapped individual and do not necessarily reflect syndrome-specific hypothalamic abnormalities.

Individuals with Prader-Willi syndrome manifest severe skin picking behavior. Bhargava et al. (1996) described 3 adolescent patients in whom an extension of this behavior to rectal picking resulted in significant lower gastrointestinal bleeding and anal rectal disease. Recognition of this behavior is important to avoid misdiagnosing inflammatory bowel disease in PWS patients.

Wharton et al. (1997) presented 6 patients with PWS with dramatic acute gastric distention. In 3 young adult woman with vomiting and apparent gastroenteritis, clinical course progressed rapidly to massive gastric dilatation and gastric necrosis. One patient died of overwhelming sepsis and disseminated intravascular coagulation. In 2 children, gastric dilatation resolved spontaneously. Gastrectomy was performed in 2 cases; in 1, gastrectomy was subtotal and distal, whereas in the other, gastrectomy was combined with partial duodenectomy and pancreatectomy. All specimens showed ischemic gastroenteritis. There was diffuse mucosal infarction with multifocal transmural necrosis.

Inheritance

Familial inheritance of PWS has been described frequently. Gabilan (1962) reported a family with affected brother and sister, as well as a second in which the parents of the proband were first cousins, but his patients were not entirely typical.

Jancar (1971) reported familial incidence. Hall and Smith (1972) reported 2 affected male maternal first cousins. One was of normal stature and intelligence. DeFraites et al. (1975) observed 5 cases in 3 sibships of an inbred Louisiana Acadian kindred. Clarren and Smith (1977) reported affected sibs and affected first cousins. They found a recurrence risk of 1.6% in sibs of probands.

It is clear that chromosomal mechanisms are principally responsible for PWS and that the syndrome is caused by lack of the paternal segment 15q11.2-q12. Basically, there are 2 mechanisms by which such a loss can occur: either through deletion of just the paternal 'critical' segment or through loss of the entire paternal chromosome 15 with presence of 2 maternal homologs (uniparental maternal disomy). The opposite, i.e., maternal deletion or paternal uniparental disomy, causes another characteristic phenotype, the Angelman syndrome (AS; 105830). This indicates that both parental chromosomes are differentially imprinted, and that both are necessary for normal embryonic development.

Occurrence of the Prader-willi Syndrome

The vast majority of PWS cases occur sporadically. These instances include virtually all interstitial deletions, the large majority of de novo unbalanced translocations, all instances of maternal uniparental disomy with normal karyotype or with a de novo rearrangement involving chromosome 15, and almost all cases of maternal uniparental disomy with a familial rearrangement involving chromosome 15. There is no parental age effect whatsoever in the deletion cases.

For full discussion on the mode of inheritance, see Cytogenetics, below.

Recurrence Risk

Monozygotic twins are concordantly affected. However, affected siblings and cousins have repeatedly been reported, and even if a publication bias is considered, their incidence is obviously higher than the estimated incidence in the population of about 1 in 25,000 would suggest. Clarren and Smith (1977) reported affected sibs and first cousins. They found a recurrence risk of 1.6% in sibs of probands. Cassidy (1987) stated that the Prader-Willi Syndrome Association maintained a registry of PWS individuals which, as of December 1986, contained 1,595 names of affected persons in the United States and Canada. While in some of these cases the diagnosis had not been fully confirmed, in only 1 family, that reported by Lubinsky et al. (1987), was there a well-documented recurrence. Thus, it is reasonable to assume that the recurrence risk for PWS is less than 1 in 1,000 and that such recurrence is not likely to occur when a 15q interstitial deletion is identified in the proband. (As pointed out by Kennerknecht (1992), the membership of the PWS association is not limited to affected persons; 'two thirds are families and one third professionals'.)

Ledbetter et al. (1987) summarized a scientific conference on PWS. Of 195 cases studied by high resolution cytogenetic methods, deletion of chromosome 15 was detected in 116 (59.5%); other chromosome 15 abnormalities were found in 7 additional cases (3.6%). It was suggested that the recurrence risk may be as low as 1 in 1,000.

Kennerknecht (1992) used the diagnostic criteria given by Cassidy (1987) to evaluate reported cases of PWS with a view to estimating recurrence risk. Since a deletion at 15q has not been found in familial cases of PWS, except in those where del(15q) is due to familial structural chromosome rearrangement, the recurrence risk with de novo deletion should be nearly zero. In cases with familial translocation, risk estimates depend on the nature of the translocations concerned. If only 1 child is affected and the karyotype is apparently normal, Kennerknecht (1992) estimated an overall recurrence risk of 0.4%. However, if 2 or more sibs are affected, he estimated that the risk to the next sib would be 50%. If every proband were investigated cytogenetically (to ascertain unbalanced chromosome rearrangements), molecularly (with probes to detect invisible deletions and to determine the methylation pattern), and if in each instance of a paternal deletion an examination of the father was carried out, then the few instances with a high recurrence risk could be ascertained before a second child was born.

Mutagenic Factors

Strakowski and Butler (1987) found an increased incidence of paternal periconceptional employment in hydrocarbon-exposing occupations. Among 81 patients with PWS, Cassidy et al. (1989) compared the frequency of possible periconceptional occupational hydrocarbon exposure in those fathers who demonstrated a 15q deletion with the frequency in those who did not. There was no statistically significant difference between the cytogenetically different groups. In both groups, approximately half the fathers had been employed in hydrocarbon-exposing jobs. The data provided additional support for the possibility that hydrocarbon exposure is causally related to the disorder and further suggested lack of etiologic heterogeneity between the cytogenetically different groups.

Cytogenetics

Deletions account for 70 to 80% of cases; the majority are interstitial deletions, many of which can be visualized by prometaphase banding examination. A minority consist of unbalanced translocations, mostly de novo, which are easily detected by routine chromosome examination. The remainder of cases are the result of maternal uniparental disomy. In most of these latter cases, cytogenetic examinations yield normal results. However, in a few cases, either balanced translocations, familial or de novo, or supernumerary small marker chromosomes, are observed.

Deletions

Butler et al. (1986) found an interstitial deletion of chromosome 15 (breakpoints q11 and q13) in 21 of 39 cases and an apparently normal karyotype in the remainder. By studying chromosome 15 heteromorphisms, the del(15q) was demonstrably paternal in origin in all cases, although both parents were normal and all deletions were de novo events. Paternal age was not increased. The exclusively paternal origin of deletions was subsequently confirmed cytogenetically and by molecular marker analysis (Magenis et al., 1990; Zori et al., 1990; Robinson et al., 1991). Examination of other series of patients by different groups resulted in the figures that two-thirds to three-fourths of PWS patients have a deletion of 15q11-q13. In less than 10%, this is due to an unbalanced translocation while the remainder have interstitial deletions.

To analyze the mechanism underlying the interstitial de novo deletions at 15q11-q13 that underlie approximately 70% of PWS cases, Carrozzo et al. (1997) genotyped 10 3-generation families of PWS-deletion patients using microsatellite markers flanking the common deletion region. By FISH and/or other molecular techniques, each patient was known to be deleted for the interval from D15S11 to GABRB3 (137192). In 5 of 7 cases, a different grandparental origin was identified for the alleles flanking the deletion, a finding significantly different from the expected frequency in light of the close position of the markers. This finding was considered highly suggestive of an unequal crossover occurring in the paternal meiosis at the breakpoint as the mechanism leading to deletion. The authors noted that asymmetric exchanges between nonsister chromatids in meiosis I have previously been demonstrated and are the basis of a number of genetic diseases. When the related sequences are part of tandemly arrayed homologous genes, nonhomologous recombination may lead to the formation of chimeric genes, such as those of Lapore hemoglobin and of the red-green pigment genes involved in abnormalities of color vision. In other instances, the deletion/duplication event may arise from the unequal recombination between repetitive elements interspersed throughout a genomic region. A misalignment between Alu-repetitive sequences has been demonstrated in duplications of the LDL-receptor gene (143890; Lehrman et al., 1987) and the HPRT gene (308000; Marcus et al., 1993). Duplications of 15q11-q13 have been reported in only a few instances (Clayton-Smith et al., 1993), and it is unclear whether any of these represent the reciprocal event of deletion by unequal crossingover. The paucity of duplication cases compared with deletions of this region may mean that the duplications occur much less frequently or that a milder phenotype causes them to be ascertained much less often. In 2 PWS families studied b y Carrozzo et al. (1997), the data were consistent with an intrachromosomal mechanism being responsible for the deletion. One of the few precedents for intrachromosomal recombination leading to human disease is provided by the recombination that occurs between the small intronless gene within intron 22 of the factor VIII gene (306700), and a copy of gene A (FSA; 305423) located 500 kb telomeric to the F8 gene, a recombination that causes severe hemophilia (Lakich et al., 1993). This rearrangement arises almost exclusively in male meioses, indicating that it is intrachromosomal. Carrozzo et al. (1997) suggested that the in-cis mechanism leading to the deletions in PWS patients may be related either to an exchange of chromosomal material between sister chromatids or to the formation of an intrachromosomal loop, either during meiosis or as a somatic event, followed by an excision of the chromosomal material lying between the recombining regions.

Maternal Uniparental Disomy

Nicholls et al. (1989), studying cases of PWS in which no deletion was cytologically evident using RFLP analysis, were the first to demonstrate maternal uniparental disomy (UPD) in 2 families. Two different, apparently intact, maternal chromosomes were present ('heterodisomy'), and, as with deletion cases of PWS, there was an absence of paternal genes from the 15q11-q13 segment. Robinson et al. (1991) used cytogenetic and molecular techniques to examine 37 patients with features of PWS. Clinical features in 28 of the patients were thought to fulfill diagnostic criteria for typical PWS. In 21 of these, a deletion of the 15q11.2-q12 region could be identified molecularly, including several cases in which the cytogenetic results were inconclusive. Five cases of maternal heterodisomy and 2 of isodisomy for 15q11-q13 were observed. All 9 patients who did not fulfill clinical criteria for typical PWS showed normal maternal and paternal inheritance of chromosome 15 markers; however, one of these carried a ring-15 chromosome. Thus, all typical PWS cases showed either a deletion or maternal uniparental disomy of 15q11.2-q12. As the disomy patients did not show any additional or more severe features than did the typical deletion patients, it is likely that there is only one imprinted region on chromosome 15. A significantly increased mean maternal age was found in the disomy cases, suggesting an association between increased maternal age and nondisjunction.

Mascari et al. (1992) demonstrated maternal uniparental disomy for chromosome 15 in 18 of 30 patients (60%) without a cytogenetic deletion. Furthermore, they confirmed the observation of Robinson et al. (1991) that the phenomenon was associated with advanced maternal age. In another 8 patients (27%), they identified large molecular deletions. The remaining 4 patients (13%) had evidence of normal biparental inheritance for chromosome 15; 3 of these patients were the only ones in the study which had some atypical clinical features. All told, they estimated that about 20% of cases of PWS result from maternal uniparental disomy and that, by the combined use of cytogenetic and molecular techniques, the genetic basis of PWS can be identified in at least 95% of patients.

Mitchell et al. (1996) compared 79 cases of PWS with UPD and 43 cases with deletions. Although there were no major clinical differences between the 2 classes of patients analyzed as a whole, mean maternal and paternal age were significantly higher in the UPD patients. The UPD group had a predominance of males, yet a gender bias was not seen in the deletion group. Hypopigmentation was found in 77% of the deletion group compared to only 39% of the UPD children. When the groups were analyzed by gender, females with UPD tended to be less severely affected than female deletion patients.

Mutirangura et al. (1993) demonstrated maternal heterodisomy in 10 PWS patients. Since the markers used were 13 cM from the centromere, heterodisomy indicated that maternal meiosis I nondisjunction was primarily involved in the origin of UPD. In contrast, 2 paternal disomy cases of Angelman syndrome (AS) showed isodisomy for all markers tested along the length of chromosome 15. This suggested a paternal meiosis II nondisjunction event (without crossing over) or, more likely, monosomic conception (due to maternal nondisjunction) followed by chromosome duplication. The latter mechanism would indicate that at least some instances of uniparental disomy in PWS and AS initiate as reciprocal products of maternal nondisjunction events.

Robinson et al. (1993) reported data indicating that the majority (82%) of maternal nondisjunction events leading to UPD and causing PWS involve a meiosis I error, whereas most paternal UPD Angelman syndrome cases are meiosis II or, more likely, mitotic errors. Robinson et al. (1993) made the interesting statement that the proportion of UPD cases among all PWS patients in Switzerland is higher than in the United States, which could reflect the higher mean maternal age at birth in Switzerland versus the United States.

Rescuing of Trisomy 15

Maternal nondisjunction does not itself directly lead to uniparental disomy but must also involve a further nondisjunction event to produce a euploid embryo. Purvis-Smith et al. (1992) have confirmed such an origin of uniparental disomy 15 resulting from 'correction' of an initial trisomy 15. Routine chorionic villus sampling performed for advanced maternal age led to detection of placental mosaicism for trisomy 15. Follow-up studies on amniotic fluid indicated a normal 46,XY karyotype with no evidence of trisomy 15, and the pregnancy continued to term. At birth, the baby was found to have PWS. Molecular analysis indicated that the mother was the sole contributor of the chromosome 15 pair in the child. Centromere/short-arm heteromorphisms were different in the 2 chromosome 15 homologs, consistent with meiosis I error. Cassidy et al. (1992) reported a similar case that supported the idea that maternal disomy can result from a 'corrected' trisomy 15 and that maternal age was a predisposing factor to nondisjunction. Thus, in any case in which trisomy or mosaic trisomy 15 has been prenatally determined through CVS examination, a molecular study should follow to exclude uniparental (paternal or) maternal disomy. This type of examination should also be considered in case of pregnancies of translocation carrier parents involving chromosome 15.

Devriendt et al. (1997) proposed partial zygotic trisomy rescue as a mechanism for mosaicism for a de novo jumping translocation of distal chromosome 15q, resulting in partial trisomy for 15q24-qter in a patient with PWS. A maternal uniparental heterodisomy for chromosome 15 was present in all cells and was responsible for the PWS phenotype. The translocated 15q segment was of paternal origin and was present as a jumping translocation, involving chromosomes 14q, 4q, and 16p. The recipient chromosomes were cytogenetically intact. Devriendt et al. (1997) reported that mental retardation was more marked in their patient than is usually observed in PWS, and proposed that this was due to partial trisomy for distal 15q.

Multiple Affected Relatives

There are several mechanisms that explain the simultaneous occurrence of affected first- and second-degree relatives in PWS families. These include translocations that give rise to maternal nondisjunction and hence effective maternal uniparental disomy for the PWS region and translocations which give rise to paternally derived deletions.

The first report of involvement of a D group translocation in PWS (later identified as a 15-15 translocation) dates back to 1963 (Buehler et al., 1963). Additional translocations were found subsequently, and after the introduction of chromosome banding it became obvious that at least one chromosome 15 was involved in all instances (Zuffardi et al., 1978; 74,Kucerova et al. 1979; Guanti, 1980). However, the situation was further complicated by cases in which not only the proband had a translocation involving chromosome 15, but the mother and 2 normal sibs showed the seemingly identical translocation as well (Smith and Noel, 1980). In addition, there were a few cases that did not show a translocation involving chromosome 15, but had a small supernumerary chromosome, presumably an isochromosome for the short arm of an acrocentric (Fleischer-Michaelsen et al., 1979; Fujita et al., 1980; Wisniewski et al., 1980).

Smith and Noel (1980) described a family in which a Prader-Willi girl had the same balanced 4;15 translocation as her mother and other phenotypically normal family members. A second such family was observed by Smith et al. (1983). Nicholls et al. (1989) reported a similar family and demonstrated that the Prader-Willi proband had inherited the maternal translocation chromosome plus the normal maternal homolog, but no paternal 15. Therefore, having a balanced translocation involving chromosome 15 predisposes to PWS offspring via nondisjunction, and this is a much more frequent cause than spontaneous nondisjunction, which may arise from chromosomally normal individuals. The opposite, i.e., Angelman syndrome, could also occur with paternal translocation carriers.

The simplest instance is that of a balanced rearrangement with a breakpoint in 15q13 in related male carriers. Fernandez et al. (1987) reported a family with a 15;22 translocation carrier father who had 2 children with PWS because of an unbalanced segregation. Hulten et al. (1991) described a family in which a balanced translocation involving 15q13 was segregating. Females with the translocation appeared to have an increased risk of having children with AS, whereas male carriers of the translocation had an increased risk of having children with PWS.

Ledbetter et al. (1980) pointed out that apparent balanced translocations involving chromosome 15 have been found. The defect may be an alteration in gene expression, i.e., a regulatory defect. Ledbetter et al. (1981), assuming a small deletion of proximal 15q as the cause of the clinical features in the translocated cases, studied 45 persons with the clinical diagnosis of PWS. Of the 45, 25 had an abnormality of chromosome 15 (which in 23 was an interstitial deletion affecting the q11-q12 region). No relatives of probands showed chromosomal changes.

Orstavik et al. (1992) described 3 sibs thought to have the Prader-Willi syndrome but with no abnormality in the 15q11-q13 region detectable by cytogenetic or molecular genetic methods. One of the sibs, a boy, was born at 32 weeks by cesarean section. He was extremely hypotonic and died at 7 days of age from respiratory distress. The other sibs, a 12-year-old brother and a 7-year-old sister, had an accessory nipple and seemingly typical PWS. A paternally inherited submicroscopic deletion was suggested as one possibility. A very small deletion was later molecularly detected in affected members of this family (Tommerup, 1993).

Ishikawa et al. (1987) described 2 sisters with PWS. No interstitial deletion of 15q was detected in either; 1 sister had a possibly unrelated partial deletion of one X chromosome. No molecular investigations were performed in this family.

Lubinsky et al. (1987) reported the cases of 2 brothers and 2 sisters in a single sibship with PWS but apparently normal chromosomes. Results of chromosome studies in the parents and surviving sibs were normal. The diagnosis was made clinically on the basis of history, behavior, and physical findings in 3 of the sibs. The fourth child had died at the age of 10 months with a history and clinical findings typical of the first phase of PWS. Again, no molecular or fluorescence in situ hybridization (FISH) studies were performed. It seems likely that an undetected structural chromosome rearrangement is the cause for this multiple occurrence of PWS.

Co-occurrence of Prader-willi and Angelman Syndromes

Hasegawa et al. (1984) studied a family in which 2 cousins were claimed to have the Prader-Willi syndrome and found a reciprocal translocation t(14;15)(q11.2;q13) in a single parent of each cousin and in their common grandmother. The affected cousins had the same unbalanced translocation including monosomy of the 15pter-q13 segment. Schinzel et al. (1992) pointed out that the unbalanced karyotype with deletion of 15q11-q13 came from the mother in the case of the proband who had been described to have classic Prader-Willi syndrome and from the father in the case of the cousin; the mother of the proband and the father of the cousin were sister and brother. However, the proband was not hypotonic and had seizures. Schinzel et al. (1992) suggested that the diagnosis in the proband actually may have been Angelman syndrome, consistent with the finding that there has been no reported instance of a patient in which absence of the paternal segment 15q11-q13 does not cause PWS, while the absence of the maternal segment leads to AS.

Another mechanism by which the Prader-Willi syndrome and Angelman syndrome can occur in cousins was reported by Smeets et al. (1992). Two female first cousins were offspring of brothers, both of whom had a familial translocation between chromosome 6 and 15, t(6;15)(p25.3;q11.1). The cousin with the Prader-Willi syndrome had the karyotype 45,XX,-6,-15+t(6;15)(p25.3;q13); DNA studies indicated that there was a large paternally derived deletion of all loci from the Prader-Willi chromosomal region tested. The cousin with Angelman syndrome had the karyotype 45,XX,-6,-15,+t(6;15)(p25.3;q11.1) and DNA studies indicated that she had uniparental heterodisomy, having inherited both the (6;15) translocation and the normal chromosome 15 from her father, but no chromosome 15 from her mother. In an editorial, Hall (1992) suggested that the cousin with Angelman syndrome had started out life as a trisomy and survived only through the loss of extra chromosomal material.

Greenstein (1990) presented a kindred in which both the Prader-Willi and the Angelman syndromes were found; the inheritance pattern was consistent with genetic imprinting.

Marker Chromosomes

Finally, additional small marker chromosomes representing isochromosomes or isodicentric chromosomes from the short arms of acrocentrics have repeatedly been observed (Fleischer-Michaelsen et al., 1979; Fujita et al., 1980; Wisniewski et al., 1980) before Robinson et al. (1993) demonstrated maternal uniparental disomy 15 in a Prader-Willi child mosaic for such a marker and paternal UPD 15 in an Angelman patient also mosaic for a small metacentric marker chromosome.

Investigation of PWS and AS patients with a small inv dup(15) chromosome attributes the abnormal phenotype to uniparental disomy rather than to the extra chromosome (Robinson et al., 1993). The small chromosome may represent either the remnant of the missing parental chromosome 15 or could be associated with nondisjunction.

Park et al. (1998) described an example of maternal disomy and Prader-Willi syndrome consistent with gamete complementation. They considered that the probable event was adjacent-1 segregation of a paternal t(3;15)(p25;q11.2) with simultaneous maternal meiotic nondisjunction for chromosome 15. The patient, a 17-year-old white male with PWS, had 47 chromosomes with a supernumerary, paternal der(15) consisting of the short arm and the proximal long arm of chromosome 15 fused to distal 3p. The t(3;15) was present in the balanced state in the patient's father and a sister. Fluorescence in situ hybridization analysis demonstrated that the PWS critical region resided on the derivative chromosome 3 and that there was no deletion in the PWS region on the normal pair of 15s present in the patient. Maternal disomy was confirmed by 2 methods.

Mapping

Kirkilionis et al. (1991) constructed a long-range restriction map of the PWS region, 15q11.1-q12, using a combination of pulsed-field gel techniques and rare cutting restriction enzymes.

A preliminary YAC contig map was reported by Kuwano et al. (1992), which also localized many common proximal and distal deletion breakpoints to two YACs. Ozcelik et al. (1992) refined the localization of the small nuclear ribonucleoprotein N gene (SNRPN; 182279) within the minimum deletion region. FISH ordering of reference markers in this region was also reported by Knoll et al. (1993) who placed D15S63 in the minimum PWS deletion region between D15S13 and D15S10. Mutirangura et al. (1993) published a complete YAC contig of the PWS/AS critical region and discussed the potential role of uniparental disomy (UPD) in PWS and AS. Buiting et al. (1993) constructed a YAC restriction map of the entire minimum PWS critical region defined by the shortest region of overlap between two key PWS deletion patients. This region is 320 kb and includes D15S63 and SNRPN.

Molecular Genetics

Latt et al. (1987) isolated probes from the proximal region of the long arm of chromosome 15 that are useful in the study of PWS.

Buiting et al. (1992) isolated a putative gene family and candidate genes by microdissection and microcloning from the 15q11-q13 region. One microclone, designated MN7, detected multiple loci in 15q11-q13 and 16p11.2. There were 4 or 5 different MN7 copies spread over a large distance within 15q11-q13. The presence of multiple copies of the MN7 gene family in proximal 15q may be related to the instability of this region and thus to the etiology of PWS and Angelman syndrome.

Using restriction digests with the methyl-sensitive enzymes HpaII and HhaI and probing Southern blots with several genomic and cDNA probes, Driscoll et al. (1992) systematically scanned segments of 15q11-q13 for DNA methylation differences between patients with PWS (20 deletion cases and 20 cases of uniparental disomy) and those with AS (26 deletion cases and 1 case of uniparental disomy). They found that the sequences identified by the cDNA DN34, which is highly conserved in evolution, demonstrate distinct differences in DNA methylation of the parental alleles at the D15S9 locus. Clayton-Smith et al. (1993) used DN34 to perform methylation analysis of 2 first-cousin males, one with AS and the other with PWS. The methylation pattern varied according to the parent of origin, providing further evidence for the association of methylation with genomic imprinting. Thus, DNA methylation can be used as a reliable postnatal diagnostic tool. Dittrich et al. (1992) found that an MspI/HpaII restriction site at the D15S63 locus in 15q11-q13 is methylated on the maternally derived chromosome, but unmethylated on the paternally derived chromosome. Based on this difference, they devised a rapid diagnostic test for patients suspected of having PWS or AS.

The human homolog for the mouse pink-eyed dilution locus (p locus) was found to be equivalent to the D15S12 locus which maps within the PWS/AS deletion region (Rinchik et al., 1993). Mutations in both copies of the P gene were found in a patient with type II oculocutaneous albinism, and it is suggested that deletion of 1 copy of this gene is the cause of hypopigmentation in PWS and AS.

The SNRPN gene was shown by RT-PCR to be expressed in normal and AS individuals, but not in fibroblasts from either deletion or maternal UPD PWS patients who lack a paternal copy of this gene (Glenn et al., 1993). Parent-specific DNA methylation was also identified for the SNRPN gene. Reed and Leff (1994) showed that in the human, as in the mouse, there is maternal imprinting of SNRPN, thus supporting the hypothesis that paternal absence of SNRPN is responsible for the PWS phenotype. See SNRPN (182279) for discussion of evidence indicating that this is a candidate gene in PWS and suggesting that PWS may be caused, in part, by defects in mRNA processing. In 2 sibs with the typical phenotype of PWS but without a cytogenetically detectable deletion in 15q, Ishikawa et al. (1996) demonstrated deletion of SNRPN by FISH.

A DNA transcript, OP2, was identified just centromeric to D15S10 by Woodage et al. (1994). Multiple expressed genes were identified by Sutcliffe (1994) in the region between SNRPN and D15S10. They showed that at least 4 genes are expressed only on the paternal chromosome including SNRPN, PAR1 (600161), PAR5 (600162), and PAR7. A PWS patient with a small paternal deletion showed no expression of these genes, even though the deletion occurs proximal to but not including these maternally imprinted genes, implying a common element involved in regulation of these genes. Wevrick et al. (1994) identified another expressed gene in the region, designated IPW (601491) for 'imprinted gene in the Prader-Willi syndrome region,' that is expressed only from the paternal chromosome 15.

DNA replication was shown by FISH to be asynchronous between maternal and paternal alleles within 15q11-q13 (Knoll et al., 1993). Loci in the PWS-critical region were shown to be early replicating on the paternal chromosome, and alleles within the AS critical region were early replicating on the maternal chromosome. A mosaic replication pattern with maternal and paternal alleles alternatively expressed was noted at the P locus, and is consistent with the presence of hypopigmentation in both PWS and AS due to decreased product.

Cassidy (1997) provided a comprehensive review of the clinical and molecular aspects of Prader-Willi syndrome. Cassidy and Schwartz (1998) provided a similar review of both Prader-Willi syndrome and Angelman syndrome.

PWS and AS are caused by the loss of function of imprinted genes in proximal 15q. In approximately 2 to 4% of patients, this loss of function is the result of an imprinting defect. In some cases, the imprinting defect is the result of a parental imprint-switch failure caused by a microdeletion of the imprinting center (IC). Buiting et al. (1998) described the molecular analysis of 13 PWS patients and 17 AS patients who had an imprinting defect but no IC deletion. Furthermore, heteroduplex and partial sequence analyses did not reveal any point mutations in the known IC elements. All of these patients represented sporadic cases, and some shared the paternal PWS or maternal AS 15q11-q13 haplotype with an unaffected sib. In each of the 5 PWS patients informative for the grandparental origin of the incorrectly imprinted chromosome region and 4 cases described elsewhere, the maternally imprinted paternal chromosome region was inherited from the paternal grandmother. This suggested that the grandmaternal imprint was not erased in the father's germline. In 7 informative AS patients reported by Buiting et al. (1998) and in 3 previously reported patients, the paternally imprinted maternal chromosome region was inherited from either the maternal grandfather or the maternal grandmother. The latter finding was not compatible with an imprint-switch failure, but it suggested that a paternal imprint developed either in the maternal germline or postzygotically. Buiting et al. (1998) concluded that (1) the incorrect imprint in non-IC-deletion cases is the result of a spontaneous prezygotic or postzygotic error; (2) these cases have a low recurrence risk; and (3) the paternal imprint may be the default imprint.

Diagnosis

Seven clinicians experienced with PWS, in consultation with national and international experts, proposed 2 scoring systems as diagnostic criteria: one for children aged 0-36 months and another for children aged 3 years to adults (Holm et al., 1993).

The American Society of Human Genetics/American College of Medical Genetics Test and Technology Transfer Committee (1996) outlined approaches to the laboratory diagnosis of PWS and Angelman syndrome.

White et al. (1996) exploited the allele-specific replication differences that had been observed in imprinted chromosomal regions to obtain a diagnostic test for detecting uniparental disomy. They used FISH of D15S9 and SNRPN (182279) on interphase nuclei to distinguish between Angelman and Prader-Willi syndrome patient samples with uniparental disomy of 15q11-q13 and those with biparental inheritance. They found that the familial recurrence risks are low when the child has de novo uniparental disomy and may be as high as 50% when the child has biparental inheritance. The frequency of interphase cells with asynchronous replication was significantly lower in patients with uniparental disomy than in patients with biparental inheritance. Within the sample population of patients with biparental inheritance, those with altered methylation and presumably imprinting center mutations could not be distinguished from those with no currently detectable mutation. White et al. (1996) considered the test cost-effective because it could be performed on interphase cells from the same hybridized cytologic preparation in which a deletion was included, and additional specimens were not required to determine the parental origin of chromosome 15.

Kubota et al. (1996) noted that neither FISH nor uniparental disomy (UPD) analysis with microsatellite markers will detect rare PWS patients with imprinting mutations, including small deletions or point mutations in the imprinting center region. They reported that as an initial screening test, methylation analysis has the advantage of detecting all of the major classes of molecular defects involved in PWS (deletions, uniparental disomy, and imprinting mutations) without the need for parental blood. Kubota et al. (1996) reported that in 67 patients examined clinically, the methylation results for PW71 were consistent with the clinical diagnosis. They concluded that SNRPN methylation analysis, similar to PW71 methylation analysis, constitutes a reliable diagnostic test for PWS. They emphasized the importance of conventional cytogenetic analysis in parallel with DNA methylation analysis. They noted that a few patients with signs of PWS have balanced translocations within or distal to SNRPN and normal methylation patterns. They noted also that conventional cytogenetic analysis is important to rule out other cytogenetic anomalies in patients who may have similar clinical manifestations but who do not have PWS.

Since the SNRPN gene is not expressed in any patient with PWS regardless of the underlying cytogenetic or molecular cause, Wevrick and Francke (1996) tested for expression of the SNRPN gene and a control gene in 9 patients with PWS and 40 control individuals by PCR analysis of reverse transcribed mRNA from blood leukocytes. SNRPN expression could readily be detected in blood leukocytes by PCR analysis in all control samples but not in samples from known PWS patients. Four suspected PWS cases were negative for SNRPN expression and were found to have chromosome 15 rearrangements, while the diagnosis of PWS was excluded in 7 other patients with normal SNRPN expression based on clinical, molecular, and cytogenetic findings. Thus, Wevrick and Francke (1996) concluded that the SNRPN-expression test is rapid and reliable in the molecular diagnosis of PWS.

Clinical Management

The suggestion of a hypothalamic defect located in the ventromedial or ventrolateral nucleus is plausible, but no such lesion has been reported, nor was such found on careful search in a typical case (Warkany, 1970). Hamilton et al. (1972) showed that the hypogonadism is the hypogonadotropic type and the result of hypothalamic dysfunction. Treatment with clomiphene citrate raised plasma luteinizing hormone, testosterone, and urinary gonadotropin levels to normal and resulted in normal spermatogenesis and physical signs of puberty.

Vagotomy has been successful in correcting obesity in experimental obesity produced by hypothalamic lesions (Hirsch, 1984). Fonkalsrud and Bray (1981) performed truncal vagotomy without pyloroplasty in a 17-year-old boy who had maintained a weight of approximately 264 lb (120 kg) for several years. Initially, he lost weight satisfactorily but by 11 months postoperative he had regained most of the weight. Prader (1991) reported a 17-year-old boy weighing 264 lb (120 kg) who had developed diabetes, required digitalization for cardiac failure, and presented with intolerable behavior problems. Strict dietary control in combination with psychotherapy in a foster environment resulted in a weight reduction to 143 lb (65 kg), cessation of hyperglycemia and glucosuria, and cardiac normalization.

With regard to genetic counseling, the type of cytogenetic aberration and molecular results determine the recurrence risk. Prenatal molecular investigation from chorionic villi should be recommended in every case despite very low recurrence risk. Prenatal ultrasonographic studies of fetal activity may be useful for a first screening since Prader-Willi fetuses will show diminished fetal movement during the second trimester (Schinzel, 1986). Furthermore, a molecular examination for uniparental disomy is indicated in any pregnancy in which a CVS examination disclosed (mosaic) trisomy 15 and a subsequent cytogenetic examination from amniocytes or fetal blood revealed a normal diploid karyotype.

Population Genetics

In a review, Butler (1990) estimated the frequency of PWS at about 1 in 25,000 and suggested that it is the most common syndromal cause of human obesity. In a comprehensive survey of PWS in North Dakota, Burd et al. (1990) identified 17 affected persons, from which they derived a prevalence rate of 1 per 16,062.

Animal Model

Nakatsu et al. (1992) found that the mouse homolog of a human gene within the PWCR is tightly linked to the p locus, which is the site of mutations affecting pigmentation and is often associated with neurologic abnormalities as well. The p locus is located on mouse chromosome 7 near a chromosomal region associated with imprinting effects. Nakatsu et al. (1992) suggested that the hypopigmentation in both PWS and Angelman syndrome may result from an imprinting effect on the human cognate of the mouse p locus.

Although representing only indirectly an animal model in the usual sense, studies focusing on the effects of imprinted genes on brain development by examining the fate of androgenetic (Ag; duplicated paternal genome) and parthenogenetic/gynogenetic (Pg/Gg; duplicated maternal genome) cells in chimeric mouse embryos (Keverne et al., 1996) sheds interesting light on the pathogenesis of the distinctive neuropsychologic features of PWS and Angelman syndrome. Keverne et al. (1996) observed striking cell-autonomous differences in the role of the 2 types of uniparental cells in brain development. Ag cells with a duplicated paternal genome contributed substantially to the hypothalamic structures and not the cerebral cortex. By contrast, Pg/Gg cells with a duplicated maternal genome contributed substantially to the cortex, striatum, and hippocampus but not to the hypothalamic structures. Furthermore, growth of the brain was enhanced by Pg/Gg and retarded by Ag cells. Keverne et al. (1996) proposed that genomic imprinting may represent a change in strategy controlling brain development in mammals. In particular, genomic imprinting may have facilitated a rapid nonlinear expansion of the brain, especially the cortex, during development over evolutionary time. It is noteworthy that Ag cells were seen predominantly in the medial preoptic area and hypothalamus, regions of the brain concerned with neuroendocrine function and primary motivated behavior, including feeding and sexual behavior, which are disturbed in PWS. Contrariwise, MRI shows that the sylvian fissures are anomalous in Angelman patients, who are severely mentally retarded with speech and movement disorders, findings not inconsistent with the distribution of Pg cells.

Yang et al. (1998) created 2 deletion mutations in mice to understand PWS and the mechanism of the 'imprinting center,' or IC, which maps in part to the promoter and first exon of the SNRPN gene (182279). Mice harboring an intragenic deletion of Snrpn were phenotypically normal, suggesting that mutations of SNRPN are not sufficient to induce PWS. Mice with a larger deletion involving both Snrpn and the putative PWS-IC lacked expression of the imprinted genes Zfp127 (mouse homolog of ZNF127), Ndn (602117), and Ipw, and manifested several phenotypes common to PWS infants. Mice heterozygous for the paternally inherited IC-deletion died as neonates, 72% within 48 hours. At birth, the heterozygous mutant mice were present in the expected mendelian ratio. On the day of birth, the affected mice appeared normal but underweight. There was little hypotonia, but one consistently observed difference was that mutant mice were unable to support themselves on their hind feet, resting on their knees instead. No genital or gonadal hypoplasia was observed at the time of birth.

History

Langdon-Down (1828-1896), who described 'mongolism' (Down syndrome), also described PWS (Down, 1887) about 70 years before Prader et al. (1956), and called it polysarcia (see account by Brain, 1967). The patient was a mentally subnormal girl who, when 13 years old, was 4 feet 4 inches tall (1.32 m) and weighed 196 lbs (84 kg). At 25 years of age she weighed 210 lbs (95.4 kg). 'Her feet and hands remained small, and contrasted remarkably with the appendages they terminated. She had no hair in the axillae, and scarcely any on the pubis. She had never menstruated, nor did she exhibit the slightest sexual instinct.'

The diagnostic criteria arrived at by a consensus group were presented in a table by Schulze et al. (1996). In a point system, 1 point each was allowed for each of 5 major criteria, such as feeding problems in infancy and failure to thrive, and one-half point each for 7 minor criteria, such as hypopigmentation. A minimum of 8.5 points was considered necessary for the diagnosis of PWS.

See also Bray et al. (1983); Burke et al. (1987); Butler et al. (1982); Butler et al. (1982); Butler and Palmer (1983); Butler and Palmer (1983); Carpenter (1994); Cassidy (1987); Cassidy et al. (1984); Charrow et al. (1983); Donlon et al. (1986); Duckett et al. (1984); Dunn (1968); Fraccaro et al. (1983); Fryns (1988); Fuhrmann-Rieger et al. (1984); Futterweit et al. (1986); Gabilan and Royer (1968); Gregory et al. (1990); Gregory et al. (1991); Hawkley and Smithies (1976); Hoefnagel et al. (1967); Holm et al. (1981); Katcher et al. (1977); Kousseff (1982); Kucerova et al. (1979); Labidi and Cassidy (1986); Laurance (1967); Laurance et al. (1981); Ledbetter et al. (1982); Mattei et al. (1983); Mattei et al. (1984); Michaelsen et al. (1979); Nicholls et al. (1989); Niikawa and Ishikiriyama (1985); Orenstein et al. (1980); Qumsiyeh et al. (1992); Reed and Butler (1984); Ridler et al. (1971); Rivera et al. (1990); Robinson et al. (1993); Robinson et al. (1993); Robinson et al. (1993); Seyler et al. (1979); Smith et al. (1991); Trent et al. (1991); Veenema et al. (1984); Zellweger and Schneider (1968)

American Society of Human Genetics/American College of Medical Genetics Test and Technology Transfer Committee: Diagnostic testing for Prader-Willi and Angelman syndromes. Am. J. Hum. Genet. 58: 1085-1088, 1996.

Bhargava, S. A.; Putnam, P. E.; Kocoshis, S. A.; Rowe, M.; Hanchett, J. M.: Rectal bleeding in Prader-Willi syndrome. Pediatrics 97: 265-267, 1996.

Brain, R. T.: Historical introduction.In: Wolstenholme, G. E. W.; Porter, R. (eds.): Mongolism. Boston: Little, Brown and Co. (pub.) 1967. Pp. 1-5.

Bray, G. A.; Dahms, W. T.; Swerdloff, R. S.; Fiser, R. H.; Atkinson, R. L.; Carrel, R. E.: The Prader-Willi syndrome: a study of 40 patients and a review of the literature. Medicine 62: 59-80, 1983.

Buehler, E. M.; Rossier, R.; Bodis, I.; Vulliet, V.; Buehler, U. K.; Stalder, G.: Chromosomal translocation in a mentally deficient child with cryptorchidism. Acta Paediat. Scand. 52: 177-182, 1963.

Buiting, K.; Dittrich, B.; Gross, S.; Greger, V.; Lalande, M.; Robinson, W.; Mutirangura, A.; Ledbetter, D.; Horsthemke, B.: Molecular definition of the Prader-Willi syndrome chromosome region and orientation of the SNRPN gene. Hum. Molec. Genet. 2: 1991-1994, 1993.

Buiting, K.; Dittrich, B.; Gross, S.; Lich, C.; Farber, C.; Buchholz, T.; Smith, E.; Reis, A.; Burger, J.; Nothen, M. M.; Barth-Witte, U.; Janssen, B.; and 22 others: Sporadic imprinting defects in Prader-Willi syndrome and Angelman syndrome: implications for imprint-switch models, genetic counseling, and prenatal diagnosis. Am. J. Hum. Genet. 63: 170-180, 1998.

Buiting, K.; Greger, V.; Brownstein, B. H.; Mohr, R. M.; Voiculescu, I.; Winterpacht, A.; Zabel, B.; Horsthemke, B.: A putative gene family in 15q11-13 and 16p11.2: possible implications for Prader-Willi and Angelman syndromes. Proc. Nat. Acad. Sci. 89: 5457-5461, 1992.

Burd, L.; Vesely, B.; Martsolf, J.; Kerbeshian, J.: Prevalence study of Prader-Willi syndrome in North Dakota. Am. J. Med. Genet. 37: 97-99, 1990.

Burke, C. M.; Kousseff, B. G.; Gleeson, M.; O'Connell, B. M.; Devlin, J. G.: Familial Prader-Willi syndrome. Arch. Intern. Med. 147: 673-675, 1987.

Butler, M. G.: Hypopigmentation: a common feature of Prader-Labhart-Willi syndrome. Am. J. Hum. Genet. 45: 140-146, 1989.

Butler, M. G.: Prader-Willi syndrome: current understanding of cause and diagnosis. Am. J. Med. Genet. 35: 319-332, 1990.

Butler, M. G.; Kaler, S. G.; Yu, P. L.; Meaney, F. J.: Metacarpophalangeal pattern profile analysis in Prader-Willi syndrome. Clin. Genet. 22: 315-320, 1982.

Butler, M. G.; Meaney, F. J.: An anthropometric study of 38 individuals with Prader-Labhart-Willi syndrome. Am. J. Med. Genet. 26: 445-455, 1987.

Butler, M. G.; Meaney, F. J.; Kaler, S. G.; Yu, P. L.; Palmer, C. G.: Clinical differences between chromosome 15q deletion and nondeletion Prader-Willi individuals. (Abstract) Am. J. Hum. Genet. 34: 119A only, 1982.

Butler, M. G.; Meaney, F. J.; Palmer, C. G.: Clinical and cytogenetic survey of 39 individuals with Prader-Labhart-Willi syndrome. Am. J. Med. Genet. 23: 793-809, 1986.

Butler, M. G.; Palmer, C. G.: Paternal origin of chromosome 15 deletion in Prader-Willi syndrome. (Abstract) Am. J. Hum. Genet. 35: 128A only, 1983.

Butler, M. G.; Palmer, C. G.: Parental origin of chromosome 15 deletion in Prader-Willi syndrome. (Letter) Lancet I: 1285-1286, 1983.

Carpenter, P. K.: Prader-Willi syndrome in old age. J. Intellect. Disabil. Res. 38: 529-531, 1994.

Carrozzo, R.; Rossi, E.; Christian, S. L.; Kittikamron, K.; Livieri, C.; Corrias, A.; Pucci, L.; Fois, A.; Simi, P.; Bosio, L.; Beccaria, L.; Zuffardi, O.; Ledbetter, D. H.: Inter- and intrachromosomal rearrangements are both involved in the origin of 15q11-q13 deletions in Prader-Willi syndrome. (Letter) Am. J. Hum. Genet. 61: 228-231, 1997.

Cassidy, S. B.: Prader-Willi syndrome. J. Med. Genet. 34: 917-923, 1997.

Cassidy, S. B.: Prader-Willi syndrome, characteristics, management, and etiology. Ala. J. Med. Sci. 24: 169-175, 1987.

Cassidy, S. B.: Recurrence risk in Prader-Willi syndrome. (Letter) Am. J. Med. Genet. 28: 59-60, 1987.

Cassidy, S. B.; Gainey, A. J.; Butler, M. G.: Occupational hydrocarbon exposure among fathers of Prader-Willi syndrome patients with and without deletions of 15q. Am. J. Hum. Genet. 44: 806-810, 1989.

Cassidy, S. B.; Lai, L.-W.; Erickson, R. P.; Magnuson, L.; Thomas, E.; Gendron, R.; Herrmann, J.: Trisomy 15 with loss of the paternal 15 as a cause of Prader-Willi syndrome due to maternal disomy. Am. J. Hum. Genet. 51: 701-708, 1992.

Cassidy, S. B.; McKillop, J. A.: Temperature regulation in Prader-Willi syndrome. (Abstract) Am. J. Med. Genet. 41: 528 only, 1991.

Cassidy, S. B.; Schwartz, S.: Prader-Willi and Angelman syndromes: disorders of genomic imprinting. Medicine 77: 140-151, 1998.
Cassidy, S. B.; Thuline, H. C.; Holm, V. A.: Deletion of chromosome 15(q11-13) in a Prader-Labhart-Willi syndrome clinic population. Am. J. Med. Genet. 17: 485-495, 1984.
Charrow, J.; Balkin, N.; Cohen, M. M.: Translocations in Prader-Willi syndrome. Clin. Genet. 23: 304-307, 1983.
Chitayat, D.; Davis, E. B.; McGillivray, B. C.; Hayden, M. R.; Hall, J. G.: Perinatal and first year follow-up of patients with Prader-Willi syndrome: normal size of hands and feet. Clin. Genet. 35: 161-166, 1989.
Clarren, S. K.; Smith, D. W.: Prader-Willi syndrome: variable severity and recurrence risk. Am. J. Dis. Child. 131: 798-800, 1977.
Clayton-Smith, J.; Driscoll, D. J.; Waters, M. F.; Webb, T.; Andrews, T.; Malcolm, S.; Pembrey, M. E.; Nicholls, R. D.: Difference in methylation patterns within the D15S9 region of chromosome 15q11-13 in first cousins with Angelman syndrome and Prader-Willi syndrome. Am. J. Med. Genet. 47: 683-686, 1993.
Clayton-Smith, J.; Webb, T.; Cheng, X. J.; Pembrey, M. E.; Malcolm, S.: Duplication of chromosome 15 in the region 15q11-13 in a patient with developmental delay and ataxia with similarities to Angelman syndrome. J. Med. Genet. 30: 529-531, 1993.
Creel, D. J.; Bendel, C. M.; Wiesner, G. L.; Wirtschafter, J. D.; Arthur, D. C.; King, R. A.: Abnormalities of the central visual pathways in Prader-Willi syndrome associated with hypopigmentation. New Eng. J. Med. 314: 1606-1609, 1986.
Curfs, L. M. G.; Verhulst, F. C.; Fryns, J. P.: Behavioral and emotional problems in youngsters with Prader-Willi syndrome. Genetic Counseling 2: 33-41, 1991.
Curfs, L. M. G.; Wiegers, A. M.; Sommers, J. R. M.; Borghgraef, M.; Fryns, J. P.: Strengths and weaknesses in the cognitive profile of youngsters with Prader-Willi syndrome. Clin. Genet. 40: 430-434, 1991.
DeFraites, E. B.; Thurmon, T. F.; Farhadian, H.: Familial Prader-Willi syndrome.In: Bergsma, D.: Genetic Forms of Hypogonadism. New York: National Foundation-March of Dimes (pub.) 1975. Pp. 123-126.
Devriendt, K.; Petit, P.; Matthijs, G.; Vermeesch, J. R.; Holvoet, M.; De Muelenaere, A.; Marynen, P.; Cassiman, J.-J.; Fryns, J.-P.: Trisomy 15 rescue with jumping translocation of distal 15q in Prader-Willi syndrome. J. Med. Genet. 34: 395-399, 1997.
Dittrich, B.; Robinson, W. P.; Knoblauch, H.; Buiting, K.; Schmidt, K.; Gillessen-Kaesbach, G.; Horsthemke, B.: Molecular diagnosis of the Prader-Willi and Angelman syndromes by detection of parent-of-origin specific DNA methylation in 15q11-13. Hum. Genet. 90: 313-315, 1992.
Donlon, T. A.; Lalande, M.; Wyman, A.; Bruns, G.; Latt, S. A.: Isolation of molecular probes associated with the chromosome 15 instability in the Prader-Willi syndrome. Proc. Nat. Acad. Sci. 83: 4408-4412, 1986. Note: Correction: Proc. Nat. Acad. Sci. 83: 6964 only, 1986.
Down, J. L.: Mental Affections of Childhood and Youth. London: Churchill (pub.) 1887. Pp. 172 only.
Driscoll, D. J.; Waters, M. F.; Williams, C. A.; Zori, R. T.; Glenn, C. C.; Avidano, K. M.; Nicholls, R. D.: A DNA methylation imprint, determined by the sex of the parent, distinguishes the Angelman and Prader-Willi syndromes. Genomics 13: 917-924, 1992.
Duckett, D. P.; Roberts, S. H.; Davies, P.: Unbalanced reciprocal translocations in cases of Prader-Willi syndrome. Hum. Genet. 67: 156-161, 1984.
Dunn, H. G.: The Prader-Labhart-Willi syndrome: review of the literature and report of nine cases. Acta Paediat. Scand. 186 (suppl.): 1-38, 1968.
Fernandez, F.; Berry, C.; Mutton, D.: Prader-Willi syndrome in siblings, due to unbalanced translocation between chromosomes 15 and 22. Arch. Dis. Child. 62: 841-843, 1987.
Fleischer-Michaelsen, K.; Lundsteen, C.; Hansen, F. J.: Prader-Willi syndrome and chromosomal mosaicism 46,XY/47,XY,+mar in two cases. Clin. Genet. 16: 147-150, 1979.
Fonkalsrud, E. W.; Bray, G.: Vagotomy for treatment of obesity in childhood due to Prader-Willi syndrome. J. Pediat. Surg. 16: 888-889, 1981.
Fraccaro, M.; Zuffardi, O.; Buhler, E.; Schinzel, A.; Simoni, G.; Witkowski, R.; Bonifaci, E.; Caufin, D.; Cignacco, G.; Delendi, N.; Gargantini, L.; Losanowa, T.; Marca, L.; Ullrich, E.; Vigi, V.: Deficiency, transposition, and duplication of one 15q region may be alternatively associated with Prader-Willi (or a similar) syndrome: analysis of seven cases after varying ascertainment. Hum. Genet. 64: 388-394, 1983.
Fryns, J. P.: The Prader-Willi syndrome and the Sotos syndrome: syndromes or sequences? (Letter) Clin. Genet. 33: 457-458, 1988.
Fuhrmann-Rieger, A.; Kohler, A.; Fuhrmann, W.: Duplication or insertion in 15q11-13 associated with mental retardation—short stature and obesity—Prader-Willi or Cohen syndrome? Clin. Genet. 25: 347-352, 1984.
Fujita, H.; Sakamoto, Y.; Hamamoto, Y.: An extra idic(15p)(q11) chromosome in Prader-Willi syndrome. Hum. Genet. 55: 409-411, 1980.
Futterweit, W.; Ritch, R.; Teekhasaenee, C.; Nelson, E. S.: Coexistence of Prader-Willi syndrome, congenital ectropion uveae with glaucoma, and factor XI deficiency. J.A.M.A. 255: 3280-3282, 1986.
Gabilan, J. C.: Syndrome de Prader, Labhardt et Willi. J. Pediat. (Paris) 1: 179, 1962.
Gabilan, J. C.; Royer, P.: Le syndrome de Prader, Labhardt et Willi (etude de onze observations). Arch. Franc. Pediat. 25: 121-149, 1968.
Glenn, C. C.; Nicholls, R. D.; Robinson, W. P.; Saitoh, S.; Niikawa, N.; Schinzel, A.; Horsthemke, B.; Driscoll, D. J.: Modification of the DNA methylation imprint in unique Angelman and Prader-Willi patients. Hum. Molec. Genet. 2: 1377-1382, 1993.
Greenstein, M. A.: Prader-Willi and Angelman syndromes in one kindred with expression consistent with genetic imprinting. (Abstract) Am. J. Hum. Genet. 47 (suppl.): A59 only, 1990.
Greenswag, L. R.: Adults with Prader-Willi syndrome: a survey of 232 cases. Dev. Med. Child Neurol. 29: 145-152, 1987.
Gregory, C. A.; Kirkilionis, A. J.; Greenberg, C. R.; Chudley, A. E.; Hamerton, J. L.: Detection of molecular rearrangements in Prader-Willi syndrome patients by using genomic probes recognizing four loci within the PWCR. Am. J. Med. Genet. 35: 536-545, 1990.
Gregory, C. A.; Schwartz, J.; Kirkilionis, A. J.; Rudd, N.; Hamerton, J. L.: Somatic recombination rather than uniparental disomy suggested as another mechanism by which genetic imprinting may play a role in the etiology of Prader-Willi syndrome. Hum. Genet. 88: 42-48, 1991.
Guanti, G.: A new case of rearrangement of chromosome 15 associated with Prader-Willi syndrome. Clin. Genet. 17: 423-427, 1980.
Hall, B. D.: Leukaemia and the Prader-Willi syndrome. (Letter) Lancet I: 46 only, 1985.
Hall, B. D.; Smith, D. W.: Prader-Willi syndrome: a resume of 32 cases including an instance of affected first cousins, one of whom is of normal stature and intelligence. J. Pediat. 81: 286-293, 1972.
Hall, J. G.: Genomic imprinting and its clinical implications. (Editorial) New Eng. J. Med. 326: 827-829, 1992.
Hamilton, C. R., Jr.; Scully, R. E.; Kliman, B.: Hypogonadotropism in Prader-Willi syndrome: induction of puberty and spermatogenesis by clomiphene citrate. Am. J. Med. 52: 322-329, 1972.
Hasegawa, T.; Hara, M.; Ando, M.; Osawa, M.; Fukuyama, Y.; Takahashi, M.; Yamada, K.: Cytogenetic studies of familial Prader-Willi syndrome. Hum. Genet. 65: 325-330, 1984.
Hawkley, C. J.; Smithies, A.: The Prader-Willi syndrome with a 15-15 translocation: case report and review of the literature. J. Med. Genet. 13: 152-157, 1976.
Hirsch, J.: Personal Communication. New York, N.Y. 4/6/1984.
Hittner, H. M.; King, R. A.; Riccardi, V. M.; Ledbetter, D. H.; Borda, R. P.; Ferrell, R. E.; Kretzer, F. L.: Oculocutaneous albinoidism as a manifestation of reduced neural crest derivatives in the Prader-Willi syndrome. Am. J. Ophthal. 94: 328-337, 1982.
Hoefnagel, D.; Costello, P. J.; Hatoum, K.: Prader-Willi syndrome. J. Ment. Defic. Res. 11: 1-11, 1967.
Holm, V. A.; Cassidy, S. B.; Butler, M. G.; Hanchett, J. M.; Greenswag, L. R.; Whitman, B. Y.; Greenberg, F.: Prader-Willi syndrome: consensus diagnostic criteria. Pediatrics 91: 398-402, 1993.
Holm, V. A.; Sulzbacher, S. J.; Pipes, P. L.: Prader-Willi Syndrome. Baltimore: University Park Press (pub.) 1981.
Hudgins, L.; Cassidy, S. B.: Hand and foot length in Prader-Willi syndrome. Am. J. Med. Genet. 41: 5-9, 1991.
Hulten, M.; Armstrong, S.; Challinor, P.; Gould, C.; Hardy, G.; Leedham, P.; Lee, T.; McKeown, C.: Genomic imprinting in an Angelman and Prader-Willi translocation family. (Letter) Lancet 338: 638-639, 1991.
Ishikawa, T.; Kanayama, M.; Wada, Y.: Prader-Willi syndrome in two siblings: one with normal karyotype, one with a terminal deletion of distal Xq. Clin. Genet. 32: 295-299, 1987.
Ishikawa, T.; Kibe, T.; Wada, Y.: Deletion of small nuclear ribonucleoprotein polypeptide N (SNRPN) in Prader-Willi syndrome detected by fluorescence in situ hybridization: two sibs with the typical phenotype without a cytogenetic deletion in chromosome 15q. Am. J. Med. Genet. 62: 350-352, 1996.
Jancar, J.: Prader-Willi syndrome (hypotonia, obesity, hypogonadism, growth and mental retardation). J. Ment. Defic. Res. 15: 20-29, 1971.
Johnsen, S.; Crawford, J. D.; Haessler, H. A.: Fasting hyperlipogenesis: an inborn error of energy metabolism in Prader-Willi syndrome.. Proc. Am. Pediat. Soc. , 1967.
Katcher, M. L.; Bargman, G. J.; Gilbert, E. F.; Opitz, J. M.: Absence of spermatogonia in the Prader-Willi syndrome. Europ. J. Pediat. 124: 257-260, 1977.
Kennerknecht, I.: Differentiated recurrence risk estimations in the Prader-Willi syndrome. Clin. Genet. 41: 303-308, 1992.
Keverne, E. B.; Fundele, R.; Narasimha, M.; Barton, S. C.; Surani, M. A.: Genomic imprinting and the differential roles of parental genomes in brain development. Develop. Brain Res. 92: 91-100, 1996.
Kirkilionis, A. J.; Gregory, C. A.; Hamerton, J. L.: Long-range restriction mapping and linkage analysis of the Prader-Willi chromosome region (PWCR). Genomics 9: 524-535, 1991.
Knoll, J. H. M.; Sinnett, D.; Wagstaff, J.; Glatt, K.; Schantz-Wilcox, A.; Whiting, P. M.; Wingrove, P.; Sikela, J. M.; Lalande, M.: FISH ordering of reference markers and of the gene for the a5 subunit of the g-aminobutyric acid receptor (GABRA5) with the Angelman and Prader-Willi syndrome chromosomal regions. Hum. Molec. Genet. 2: 183-189, 1993.
Kousseff, B. G.: The cytogenetic controversy in the Prader-Labhart-Willi syndrome. Am. J. Med. Genet. 13: 431-439, 1982.
Kubota, T.; Sutcliffe, J. S.; Aradhya, S.; Gillessen-Kaesbach, G.; Christian, S. L.; Horsthemke, B.; Beaudet, A. L.; Ledbetter, D. H.: Validation studies of SNRPN methylation as a diagnostic test for Prader-Willi syndrome. Am. J. Med. Genet. 66: 77-80, 1996.
Kucerova, M.; Strakova, M.; Polivkova, Z.: The Prader-Willi syndrome with a 15-3 translocation. J. Med. Genet. 16: 234-235, 1979.
Kuwano, A.; Mutirangura, A.; Dittrich, B.; Buiting, K.; Horsthemke, B.; Saitoh, S.; Niikawa, N.; Ledbetter, S. A.; Greenberg, F.; Chinault, A. C.; Ledbetter, D. H.: Molecular dissection of the Prader-Willi/Angelman syndrome region (15q11-13) by YAC cloning and FISH analysis. Hum. Molec. Genet. 1: 417-425, 1992.
Labidi, F.; Cassidy, S. B.: A blind prometaphase study of Prader-Willi syndrome: frequency and consistency in interpretation of del 15q. Am. J. Hum. Genet. 39: 452-460, 1986.
Lakich, D.; Kazazian, H. H., Jr.; Antonarakis, S. E.; Gitschier, J.: Inversions disrupting the factor VIII gene are a common cause of severe haemophilia A. Nature Genet. 5: 236-241, 1993.
Latt, S. A.; Tantravahi, U.; Nicholls, R.; Ringer, S.; Stroh, H.; Fuller, R.; Neve, R.; Kaplan, L.; Wharton, R.: Mapping, characterization, and diagnostic utilization of 10 DNA segments from the proximal long arm of human chromosome 15. (Abstract) Cytogenet. Cell Genet. 46: 644 only, 1987.
Laurance, B. M.: Hypotonia, mental retardation, obesity, and cryptorchidism associated with dwarfism and diabetes in children. Arch. Dis. Child. 42: 126-139, 1967.

Laurance, B. M.; Brito, A.; Wilkinson, J.: Prader-Willi syndrome after age 15 years. Arch. Dis. Child. 56: 181-186, 1981.
Ledbetter, D. H.; Greenberg, F.; Holm, V. A.; Cassidy, S. B.: Conference report: second annual Prader-Willi scientific conference. Am. J. Med. Genet. 28: 779-790, 1987.
Ledbetter, D. H.; Mascarello, J. T.; Riccardi, V. M.; Harper, V. D.; Airhart, S. D.; Strobel, R. J.: Chromosome 15 abnormalities and the Prader-Willi syndrome: a follow-up report of 40 cases. Am. J. Hum. Genet. 34: 278-285, 1982.
Ledbetter, D. H.; Riccardi, V. M.; Airhart, S. D.; Strobel, R. J.; Keenan, B. S.; Crawford, J. D.: Deletions of chromosome 15 as a cause of the Prader-Willi syndrome. New Eng. J. Med. 304: 325-329, 1981.
Ledbetter, D. H.; Riccardi, V. M.; Youngbloom, S. A.; Strobel, R. J.; Keenan, B. S.; Crawford, J. D.; Louro, J. M.: Deletion (15q) as a cause of the Prader-Willi syndrome (PWS). (Abstract) Am. J. Hum. Genet. 32: 77A only, 1980.
Lehrman, M. A.; Goldstein, J. L.; Russell, D. W.; Brown, M. S.: Duplication of seven exons in LDL receptor gene caused by Alu-Alu recombination in a subject with familial hypercholesterolemia. Cell 48: 827-835, 1987.
Lubinsky, M.; Zellweger, H.; Greenswag, L.; Larson, G.; Hansmann, I.; Ledbetter, D.: Familial Prader-Willi syndrome with apparently normal chromosomes. Am. J. Med. Genet. 28: 37-43, 1987.
MacMillan, D. R.; Kim, C. B.; Weisskopf, B.: Syndrome of growth resistance, obesity, and intellectual impairment with precocious puberty. Arch. Dis. Child. 47: 119-121, 1972.
Magenis, R. E.; Toth-Fejel, S.; Allen, L. J.; Black, M.; Brown, M. G.; Budden, S.; Cohen, R.; Friedman, J. M.; Kalousek, D.; Zonana, J.; Lacy, D.; LaFranchi, S.; Lahr, M.; Macfarlane, J.; Williams, C. P. S.: Comparison of the 15q deletions in Prader-Willi and Angelman syndromes: specific regions, extent of deletions, parental origin, and clinical consequences.. Am. J. Med. Genet. 35: 333-349, 1990.
Marcus, S.; Hellgren, D.; Lambert, B.; Fallstrom, S. P.; Wahlstrom, J.: Duplication in the hypoxanthine phosphoribosyl-transferase gene caused by Alu-Alu recombination in a patient with Lesch Nyhan syndrome. Hum. Genet. 90: 477-482, 1993.
Mascari, M. J.; Gottlieb, W.; Rogan, P. K.; Butler, M. G.; Waller, D. A.; Armour, J. A. L.; Jeffreys, A. J.; Ladda, R. L.; Nicholls, R. D.: The frequency of uniparental disomy in Prader-Willi syndrome: implications for molecular diagnosis. New Eng. J. Med. 326: 1599-1607, 1992.
Mattei, J. F.; Mattei, M. G.; Giraud, F.: Prader-Willi syndrome and chromosome 15: a clinical discussion of 20 cases. Hum. Genet. 64: 356-362, 1983.
Mattei, M. G.; Souiah, N.; Mattei, J. F.: Chromosome 15 anomalies and the Prader-Willi syndrome: cytogenetic analysis. Hum. Genet. 66: 313-334, 1984.
Michaelsen, K. F.; Lundsteen, C.; Hansen, F. J.: Prader-Willi syndrome and chromosomal mosaicism 46, XY-47, XY, +mar in two cases. Clin. Genet. 16: 147-150, 1979.
Mitchell, M.; Schinzel, A.; Langlois, S.; Gillessen-Kaesbach, G.; Schuffenhauer, S.; Michaelis, R.; Abeliovich, D.; Lerer, I.; Christian, S.; Guitart, M.; McFadden, D. E.; Robinson, W. P.: Comparison of phenotype in uniparental disomy and deletion Prader-Willi syndrome. Am. J. Med. Genet 65: 133-136, 1996.
Mutirangura, A.; Greenberg, F.; Butler, M. G.; Malcolm, S.; Nicholls, R. D.; Chakravarti, A.; Ledbetter, D. H.: Multiplex PCR of three dinucleotide repeats in the Prader-Willi/Angelman critical region (15q11-q13): molecular diagnosis and mechanism of uniparental disomy. Hum. Molec. Genet. 2: 143-151, 1993.
Nakatsu, Y.; Gondo, Y.; Brilliant, M. H.: The p locus is closely linked to the mouse homolog of a gene from the Prader-Willi chromosomal region. Mammalian Genome 2: 69-71, 1992.
Nicholls, R. D.; Knoll, J. H. M.; Butler, M. G.; Karam, S.; Lalande, M.: Genetic imprinting suggested by maternal heterodisomy in non-deletion Prader-Willi syndrome. Nature 342: 281-285, 1989.
Nicholls, R. D.; Knoll, J. H. M.; Butler, M. G.; Korum, S.; Shook, D.; Glatt, K.; Lalande, M.: Uniparental disomy for chromosome 15 in the Prader-Willi syndrome. (Abstract) Am. J. Hum. Genet. 45 (suppl.): A209 only, 1989.
Niikawa, N.; Ishikiriyama, S.: Clinical and cytogenetic studies of the Prader-Willi syndrome: evidence of phenotype-karyotype correlation. Hum. Genet. 69: 22-27, 1985.
Orenstein, D. M.; Boat, T. F.; Owens, R. P.; Horowitz, J. G.; Primiano, F. P.; Germann, K.; Doershuk, C. F.: The obesity hypoventilation syndrome in children with the Prader-Willi syndrome: a possible role for familial decreased response to carbon dioxide. J. Pediat. 97: 765-767, 1980.
Orstavik, K. H.; Tangsrud, S. E.; Kiil, R.; Hansteen, I.-L.; Steen-Johnsen, J.; Cassidy, S. B.; Martony, A.; Anvret, M.; Tommerup, N.; Brondum-Nielsen, K.: Prader-Willi syndrome in a brother and sister without cytogenetic or detectable molecular genetic abnormality at chromosome 15q11q13. Am. J. Med. Genet. 44: 534-538, 1992.
Ozcelik, T.; Leff, S.; Robinson, W.; Donlon, T.; Lalande, M.; Sanjines, E.; Schinzel, A.; Francke, U.: Small nuclear ribonucleoprotein polypeptide N (SNRPN), an expressed gene in the Prader-Willi syndrome critical region. Nature Genet. 2: 265-269, 1992.
Park, J. P.; Moeschler, J. B.; Hani, V. H.; Hawk, A. B.; Belloni, D. R.; Noll, W. W.; Mohandas, T. K.: Maternal disomy and Prader-Willi syndrome consistent with gamete complementation in a case of familial translocation (3;15)(p25;q11.2). Am. J. Med. Genet. 78: 134-139, 1998.
Phelan, M. C.; Albiez, K. L.; Flannery, D. B.; Stevenson, R. E.: The Prader-Willi syndrome and albinism in a black infant. Proc. Greenwood Genet. Center 7: 27-29, 1988.
Prader, A.: Personal Communication. Zurich, Switzerland 1991.
Prader, A.; Labhart, A.; Willi, H.: Ein Syndrom von Adipositas, Kleinwuchs, Kryptorchismus und Oligophrenie nach Myatonieartigem Zustand im Neugeborenenalter. Schweiz. Med. Wschr. 86: 1260-1261, 1956.
Purvis-Smith, S. G.; Saville, T.; Manass, S.; Yip, M.-Y.; Lam-Po-Tang, P. R. L.; Duffy, B.; Johnston, H.; Leigh, D.; McDonald, B.: Uniparental disomy 15 resulting from 'correction' of an initial trisomy 15. (Letter) Am. J. Hum. Genet. 50: 1348-1350, 1992.
Qumsiyeh, M. B.; Dalton, J. D.; Gordon, P. L.; Wilroy, R. S.; Tharapel, A. T.: Deletion of chromosome 15pter-q11.2 due to t(Y;15) in a boy with Prader-Willi syndrome. Am. J. Med. Genet. 42: 109-111, 1992.
Reed, M. L.; Leff, S. E.: Maternal imprinting of human SNRPN, a gene deleted in Prader-Willi syndrome. Nature Genet. 6: 163-167, 1994.
Reed, T.; Butler, M. G.: Dermatoglyphic features in Prader-Willi syndrome with respect to chromosomal findings. Clin. Genet. 25: 341-346, 1984.
Ridler, M. A. C.; Gerrod, O.; Berg, J. M.: A case of Prader-Willi in a girl with a small extra chromosome. Acta Paediat. Scand. 60: 222-226, 1971.
Rinchik, E. M.; Bultman, S. J.; Horsthemke, B.; Lee, S. T.; Strunk, K. M.; Spritz, R. A.; Avidano, K. M.; Jong, M. T. C.; Nicholls, R. M.: A gene for the mouse pink-eyed dilution locus and for human type II oculocutaneous albinism. Nature 361: 72-76, 1993.
Rivera, H.; Zuffardi, O.; Gargantini, L.: Nonreciprocal and jumping translocations of 15q1-qter in Prader-Willi syndrome. Am. J. Med. Genet. 37: 311-317, 1990.
Robinson, W. P.; Bernasconi, F.; Mutirangura, A.; Ledbetter, D. H.; Langlois, S.; Malcolm, S.; Morris, M. A.; Schinzel, A. A.: Nondisjunction of chromosome 15: origin and recombination. Am. J. Hum. Genet. 53: 740-751, 1993.
Robinson, W. P.; Binkert, F.; Gine, R.; Vazquez, C.; Mueller, W.; Rosenkranz, W.; Schinzel, A.: Clinical and molecular analysis of five inv dup(15) patients. Europ. J. Hum. Genet. 1: 37-50, 1993.
Robinson, W. P.; Bottani, A.; Yagang, X.; Balakrishman, J.; Binkert, F.; Machler, M.; Prader, A.; Schinzel, A.: Molecular, cytogenetic, and clinical investigations of Prader-Willi syndrome patients. Am. J. Hum. Genet. 49: 1219-1234, 1991.
Robinson, W. P.; Lorda-Sanchez, I.; Malcolm, S.; Langlois, S.; Schuffenhauer, S.; Knoblauch, H.; Horsthemke, B.; Schinzel, A. A.: Increased parental ages and uniparental disomy 15: a paternal age effect? Europ. J. Hum. Genet. 1: 280-286, 1993.
Robinson, W. P.; Wagstaff, J.; Bernasconi, F.; Baccichetti, C.; Artifoni, L.; Franzoni, E.; Suslak, L.; Shih, L. Y.; Aviv, H.; Schinzel, A. A.: Uniparental disomy explains the occurrence of the Angelman or Prader-Willi syndrome in patients with an additional small inv dup(15) chromosome. J. Med. Genet. 30: 756-760, 1993.
Schinzel, A.: Approaches to the prenatal diagnosis of the Prader-Willi syndrome. (Letter) Hum. Genet. 74: 327 only, 1986.
Schinzel, A.; Robinson, W. P.; Bottani, A.; Yagang, X.; Prader, A.: Prader-Willi or Angelman syndrome in familial 15q11-q13 deletion of maternal origin? Hum. Genet. 88: 361-362, 1992.
Schulze, A.; Hansen, C.; Skakkebaek, N. E.; Brondum-Nielsen, K.; Ledbetter, D. H.; Tommerup, N.: Exclusion of SNRPN as a major determinant of Prader-Willi syndrome by a translocation breakpoint. Nature Genet. 12: 452-454, 1996.
Seyler, L. E., Jr.; Arulanantham, K.; O'Connor, C. F.: Hypergonadotropic-hypogonadism in the Prader-Labhart-Willi syndrome. J. Pediat. 94: 435-437, 1979.
Smeets, D. F. C. M.; Hamel, B. C. J.; Nelen, M. R.; Smeets, H. J. M.; Bollen, J. H. M.; Smits, A. P. T.; Ropers, H.-H.; van Oost, B. A.: Prader-Willi syndrome and Angelman syndrome in cousins from a family with a translocation between chromosomes 6 and 15. New Eng. J. Med. 326: 807-811, 1992.
Smith, A.; Lindeman, R.; Volpato, F.; Kearney, A.; White, S.; Haan, E.; Trent, R. J.: A de novo unbalanced reciprocal translocation identified as paternal in origin in the Prader-Willi syndrome. Hum. Genet. 86: 534-536, 1991.
Smith, A.; Murray, R.; den Dulk, G.: Familial deletion. Ann. Genet. (Paris) 26: 91-93, 1983.
Smith, A.; Noel, M.: A girl with the Prader-Willi syndrome and Robertsonian translocation 45,XX,t(14;15)(p11;q11) which was present in three normal family members. Hum. Genet. 55: 271-273, 1980.
Stephenson, J. B. P.: Prader-Willi syndrome: neonatal presentation and later development.. Dev. Med. Child Neurol. 22: 792-795, 1980.
Strakowski, S. M.; Butler, M. G.: Paternal hydrocarbon exposure in Prader-Willi syndrome. (Letter) Lancet II: 1458 only, 1987.
Sutcliffe, G. S.: Personal Communication. Oxford, England 1994.
Tommerup, N.: Personal Communication. Zurich, Switzerland 1993.
Trent, R. J.; Volpato, F.; Smith, A.; Lindeman, R.; Wong, M.-K.; Warne, G.; Haan, E.: Molecular and cytogenetic studies of the Prader-Willi syndrome. J. Med. Genet. 28: 649-654, 1991.
Veenema, H.; Beverstock, G. C.; Zvelebil-Tarasevitch, N.; Doorn, J. L.; van Parys, J. A. P.; van de Kamp, J. J. P.: Duplication in the proximal portion of the long arm of chromosome 15, in a girl without phenotypic features of the Prader-Willi syndrome. Clin. Genet. 26: 65-68, 1984.
Warkany, J.: Personal Communication. Cincinnati, Oh. 1970.
Wevrick, R.; Francke, U.: Diagnostic test for the Prader-Willi syndrome by SNRPN expression in blood. Lancet 348: 1068-1069, 1996.
Wevrick, R.; Kerns, J. A.; Francke, U.: Identification of a novel paternally expressed gene in the Prader-Willi syndrome region. Hum. Molec. Genet. 3: 1877-1882, 1994.
Wharton, R. H.; Wang, T.; Graeme-Cook, F.; Briggs, S.; Cole, R. E.: Acute idiopathic gastric dilatation with gastric necrosis in individuals with Prader-Willi syndrome. Am. J. Med. Genet. 73: 437-441, 1997.
White, L. M.; Rogan, P. K.; Nicholls, R. D.; Wu, B.-L.; Korf, B.; Knoll, J. H. M.: Allele-specific replication of 15q11-q13 loci: a diagnostic test for detection of uniparental disomy. Am. J. Hum. Genet. 59: 423-430, 1996.
Wiesner, G. L.; Bendel, C. M.; Olds, D. P.; White, J. G.; Arthur, D. C.; Ball, D. W.; King, R. A.: Hypopigmentation in the Prader-Willi syndrome. Am. J. Hum. Genet. 40: 431-442, 1987.
Williams, M. S.; Rooney, B. L.; Williams, J.; Josephson, K.; Pauli, R.: Investigation of thermoregulatory characteristics in patients with Prader-Willi syndrome. Am. J. Med. Genet. 49: 302-307, 1994.
Wise, M. S.; Zoghbi, H.; Edwards, M.; Byrd, L. K.; Guttmacher, A. E.; Greenberg, F.: Hyperthermia in infants with Prader-Willi syndrome. (Abstract) Am. J. Med. Genet. 41: 528 only, 1991.

Wisniewski, L. P.; Witt, M. E.; Ginsberg-Fellner, F.; Wilner, J.; Desnick, R. J.: Prader-Willi syndrome and a bisatellited derivative of chromosome 15. Clin. Genet. 18: 42-47, 1980.
Woodage, T.; Lindeman, R.; Deng, Z. M.; Fimmel, A.; Smith, A.; Trent, R. J.: Physical mapping studies at D15S10: implications for candidate gene identification in the Angelman syndrome/Prader-Willi syndrome chromosome region of 15q11-q13.. Genomics 19: 170-172, 1994.
Yang, T.; Adamson, T. E.; Resnick, J. L.; Leff, S.; Wevrick, R.; Francke, U.; Jenkins, N. A.; Copeland, N. G.; Brannan, C. I.: A mouse model for Prader-Willi syndrome imprinting-centre mutations. Nature Genet. 19: 25-31, 1998.
Zellweger, H.; Schneider, H. J.: Syndrome of hypotonia-hypomentia-hypogonadism-obesity (HHHO) or Prader-Willi syndrome. Am. J. Dis. Child. 115: 588-598, 1968.
Zori, R.; Williams, C.; Mattei, J. F.; Moncla, A.: Parental origin of del(15)(q11-q13) in Angelman and Prader-Willi syndromes.. Am. J. Med. Genet. 37: 294-295, 1990.
Zuffardi, O.; Buhler, E. M.; Fraccaro, M.: Chromosome 15 and Prader-Willi syndrome. (Abstract) Clin. Genet. 14: 315-316, 1978.

MUTS, E. COLI, HOMOLOG OF, 3; MSH3 (*600887)

JJM NOTE

The association of a mismatch repair gene with myeloid neoplasia alerts us to avoid assuming that such genes influence colorectal cancer only.

The MSH2 gene (120435), encoding a human homolog of the bacterial mismatch repair protein MutS, has been shown to be the gene responsible for type 1 hereditary nonpolyposis colon cancer (120435). Fujii and Shimada (1989) identified a gene encoding another MutS homolog upstream of the human dihydrofolate reductase (DHFR; 126060) gene located at 5q11.2-q13.2. The amino acid sequence encoded by this gene, referred to by them as hMSH3, has 31% identity with that of MutS. In the most conserved region of 156 amino acids in the C-terminal portion, there is more than 50% homology. A similar gene was identified upstream of the mouse DHFR gene (Linton et al., 1989). These 2 genes were the first 2 mammalian genes with apparent sequence homology to the gene for bacterial mismatch repair protein MutS and were named the human mismatch repair protein 1 (MRP1) and the mouse repair gene 3 (Rep3), respectively, by investigators. (Watanabe et al. (1996) noted that the gene had also been called DUG, for Divergently transcribed Upstream Gene.) Thereafter, a number of MutS-related proteins were isolated in Saccharomyces cerevisiae. Human MutS homolog 2 (MSH2) was named on the basis of sequence similarity to S. cerevisiae MSH2. Human MRP1 and mouse Rep3 are most closely related to yeast MSH3 (Smith et al., 1990) and was thus designated MSH3.

Inokuchi et al. (1995) examined the expression of MSH3 genes in bone marrow cells from 40 patients with various hematologic malignancies. MSH3 mRNA was not detectable in 7 cases, including 3 of chronic myelogenous leukemia, 2 of acute myelogeneous leukemia, 1 of acute lymphocytic leukemia, and 1 of myelodysplastic syndrome. In addition, 17 cases showed significantly reduced expression of the MSH3 gene. Southern blot analysis of genomic DNA demonstrated no remarkable change in the structure and copy number of the MSH3 gene in any case. These results suggested to the authors that inactivation of the MSH3 gene may be involved in the development of hematologic malignancies.

Using PCR, Nakajima et al. (1995) identified a polymorphic 9-bp repeat sequence in exon 1 of the MSH3 gene. Five alleles were observed in unrelated Japanese individuals, with heterozygosity of 0.57.

By characterization of cosmid clones, Watanabe et al. (1996) showed that the MSH3 gene consists of 24 exons spanning at least 160 kb. All exon/intron junction sequences matched the classic GT/AG rule, except that intron 6 has AT and AA at the ends. Two major transcripts of 5.0 and 3.8 kb were shown to be derived from the differential use of 2 polyadenylation sites. Expression of this gene and the DHFR gene appear to be regulated by a bidirectional promoter composed of multiple GC boxes and 2 initiator elements.

Many human tumors show length alterations in repetitive sequence elements. Although this microsatellite instability has been attributed to mutations in 4 DNA mismatch repair genes in kindreds with hereditary nonpolyposis colorectal cancer, many sporadic tumors exhibit instability but no detectable mutations in these genes. In yeast, mutations in several genes, including RTH and MSH3, cause microsatellite instability. Risinger et al. (1996) screened 16 endometrial carcinomas with microsatellite instability for alterations in FEN1 (600393), the human homolog of RTH, and in MSH3. Although they found no FEN mutations, an MSH3 frameshift mutation (600887.0001) was observed in an endometrial carcinoma and in an endometrial carcinoma cell line. Extracts of the cell line were deficient in repair of DNA substrates containing mismatches or extra nucleotides. Risinger et al. (1996) introduced chromosome 5 encoding the MSH3 gene by microcell-mediated chromosome transfer into the mutant cell line and observed an increase in the stability of some but not all microsatellites.

Akiyama et al. (1997) screened for somatic mutations of the MSH3 (A)8 repeat in 29 tumors from 23 hereditary nonpolyposis colorectal cancer patients. One or 2 A deletions in the (A)8 repeat were found in 11 (57.9%) of the 19 tumors that showed microsatellite instability (MI) but not in 10 MI-negative ones, indicating secondary mutations after germline mutations of other mismatch repair genes. Moreover, the MI frequency of 3 or more nucleotide repeats was higher in MSH3 (A)8-mutated tumor cells than in nonmutated ones. Their data suggested that a mutation of a mismatch repair gene enhances the frequency of another mismatch repair gene mutation, such as of MSH3, resulting in severe microsatellite instability. Yin et al. (1997) came to a similar conclusion: that DNA mismatch repair genes, such as MSH3 and MSH6 (600678), are targets for the mutagenic activity of upstream mismatch repair gene mutations and that this enhanced genomic instability may accelerate the accumulation of mutations in replication/repair error positive tumors.

Marra et al. (1998) presented data that supported and extended the findings of Drummond et al. (1997) and demonstrated that mismatch repair deficiency can arise not only through mutation or transcriptional silencing of a mismatch repair gene, but also as a result of imbalance in the relative amounts of the MSH3 and MSH6 proteins.

Akiyama, Y.; Tsubouchi, N.; Yuasa, Y.: Frequent somatic mutations of hMSH3 with reference to microsatellite instability in hereditary nonpolyposis colorectal cancer. Biochem. Biophys. Res. Commun. 236: 248-252, 1997.
Drummond, J. T.; Genschel, J.; Wolf, E.; Modrich, P.: DHFR/MSH3 amplification in methotrexate-resistant cells alters the hMutS-alpha/hMutS-beta ratio and reduces the efficiency of base-base mismatch repair. Proc. Nat. Acad. Sci. 94: 10144-10149, 1997.
Fujii, H.; Shimada, T.: Isolation and characterization of cDNA clones derived from the divergently transcribed gene in the region upstream from the human dihydrofolate reductase gene. J. Biol. Chem. 264: 10057-10064, 1989.
Inokuchi, K.; Ikejima, M.; Watanabe, A.; Nakajima, E.; Orimo, H.; Nomura, T.; Shimada, T.: Loss of expression of the human MSH3 gene in hematological malignancies. Biochem. Biophys. Res. Commun. 214: 171-179, 1995.
Linton, J. P.; Yen, J. Y.-J.; Selby, E.; Chen, Z.; Chinsky, J. M.; Liu, K.; Kellems, R. E.; Crouse, G. F.: Dual bidirectional promoters at the mouse dhfr locus: cloning and characterization of two mRNA classes of the divergently transcribed Rep-1 gene. Molec. Cell. Biol. 9: 3058-3072, 1989.
Marra, G.; Iaccarino, I.; Lettieri, T.; Roscilli, G.; Delmastro, P.; Jiricny, J.: Mismatch repair deficiency associated with overexpression of the MSH3 gene. Proc. Nat. Acad. Sci. 95: 8568-8573, 1998.
Nakajima, E.; Orimo, H.; Ikejima, M.; Shimada, T.: Nine-bp repeat polymorphism in exon 1 of the hMSH3 gene. Jpn. J. Hum. Genet. 40: 343-345, 1995.
Risinger, J. I.; Umar, A.; Boyd, J.; Berchuck, A.; Kunkel, T. A.; Barrett, J. C.: Mutation of MSH3 in endometrial cancer and evidence for its functional role in heteroduplex repair. Nature Genet. 14: 102-109, 1996.
Smith, M. L.; Mitchell, P. J.; Crouse, G. F.: Analysis of the mouse Dhfr/Rep-3 major promoter region by using linker-scanning and internal deletion mutations and DNase I footprinting. Molec. Cell. Biol. 10: 6003-6012, 1990.
Watanabe, A.; Ikejima, M.; Suzuki, N.; Shimada, T.: Genomic organization and expression of the human MSH3 gene. Genomics 31: 311-318, 1996.
Yin, J.; Kong, D.; Wang, S.; Zou, T.-T. Souza, R. F.; Smolinski, K. N.; Lynch, P. M.; Hamilton, S. R.; Sugimura, H.; Powell, S. M.; Young, J.; Abraham, J. M.; Meltzer, S. J.: Mutation of hMSH3 and hMSH6 mismatch repair genes in genetically unstable human colorectal and gastric carcinomas. Hum. Mutat. 10: 474-478, 1997.

11. OTHER CANCER GENES

The preceding chapters include genes associated with a clinical phenotype that predisposes to cancer, usually one that is familial in nature, and hence the gene mutation is hereditary, that is, in the germline. Since cancer is a genetic disease at the level of the cell, many other genes have been identified by somatic cell methods, but they have not — as yet — been associated with clinical distinct disease or syndrome, nor have they been documented to exist as germline mutations. Rather, their mutants are somatic cell mutations, usually detected in sporadic cancers. With some unavoidable arbitrariness, therefore, other human genes, not already mentioned in this *Catalog,* are grouped in three categories: (1) oncogenes and proto-oncogenes; (2) translocation breakpoints and fusion proteins; (3) tumor-suppressor genes; and (4) other markers, proteins, and genes.

Oncogenes and Proto-oncogenes

JJM NOTE

The definition of *proto-oncogene* begins with the fact that there are human DNA sequences that are homologous to nucleic acid sequences in retroviruses that cause malignancy in nonhuman species, most often in the laboratory mouse. The retrovirus sequence has the generic name *v-onc,* that is, a viral oncogene. The homologous human sequence is called *c-onc,* a constitutional proto-oncogene. The *proto-* modifier emphasizes that the sequence is not carcinogenic until it is activated by additional molecular events. The wild-type sequences function normally in various roles, including the production of growth factors, growth factor receptors, other transmembrane signaling molecules, or nuclear transcription factors, perhaps only in embryonic or fetal development. The proto-oncogene may be activated to give rise to cancer by various mechanisms. One is point mutation. Another is by a chromosomal translocation that interrupts and juxtaposes two normal genes, sometimes creating a new fusion chimeric protein. Another is by amplification of the sequence documented by molecular genetic techniques or, when large, by cytogenetic techniques showing double minute chromosomes or homogeneously staining regions. One standard definition of an activated proto-oncogene is the ability of DNA extracted from a tumor or tumor cell line to be transfected into and malignantly transform a NIH/3T3 fibroblast cell line. Sometimes a proto-oncogene is defined by its molecular or sequence similarities to other proto-oncogenes. In the end, not all human DNA sequences that are homologous to murine oncogenic retroviruses are associated with human cancer; conversely, not all human transforming sequences or translocation breakpoints have been recognized in oncogenic viruses.

The list below was created by searching the *OMIM* field of title and clinical synopsis for the term *oncogen**. (The asterisk allows for *oncogene, oncogenic,* etc.) This search identified 140 entries. These and their cross-references were read for relevance and possible inclusion. This section includes 53 human oncogenes, specifically proto-oncogenes. Others appear elsewhere in the *Catalog,* especially when germline mutation of the proto-oncogene is associated with a clinical syndrome or familial aggregation of cancer. Hence, all these entries have been seen as somatic cell mutation in fresh human cancer or human cancer cell lines. The order of entry is based on the short common name for the oncogene, such as *ABL, FES,* or *MYC.* The title of the entry has been modified from the name in *OMIM* for uniformity and quick recognition by oncologists.

ONCOGENE ABL1: ABELSON MURINE LEUKEMIA VIRAL ONCOGENE HOMOLOG 1 (*189980)

Heisterkamp et al. (1982) assigned the human cellular homolog of the Abelson strain of murine leukemia virus to chromosome 9. Goff et al. (1982) mapped the mouse homolog on chromosome 2, which bears other homology of synteny to human chromosome 9. Both the FES (190030) and the ABL genes code for proteins with tyrosine-specific protein kinase activity. (Strictly speaking, 'murine' refers to the rodent family Muridae, which includes both rats and mice. By common practice, however, the term is used almost exclusively for mice.)

De Klein et al. (1982) demonstrated that the ABL gene is translocated from chromosome 9 to chromosome 22 in the formation of the Philadelphia (Ph) chromosome. This indicates that the translocation is reciprocal and suggests a role for the ABL gene in the generation of CML. Collins and Groudine (1983) showed amplification of ABL sequences some 4- to 8-fold in K-562, a Philadelphia chromosome-positive cell derived from a patient with CML in blast crises. Furthermore, the lambda light chain immunoglobulin genes were amplified in K-562, but the kappa genes showed no amplification. Whereas in Burkitt lymphoma of the t(8;22) type the lambda light chain genes are translocated to chromosome 8, they remain on chromosome 22 (i.e., on the Philadelphia chromosome) in CML (Selden et al., 1983). On germline chromosomes by in situ hybridization, Jhanwar et al. (1984) assigned ABL to 9q34.1 which is at the breakpoint in no. 9 that gives rise to the Philadelphia chromosome. Heisterkamp et al. (1983) found that the breakpoint in 9q in CML is only 14 kilobases upstream from the beginning of the ABL oncogene. See 151410. Konopka et al. (1985) presented evidence that the translocation of the ABL oncogene in Ph1-positive CML (151410) results in the creation of a chimeric gene leading to production of an abnormal abl protein with tyrosine kinase activity. The said protein probably plays a key role in the malignant transformation. Acute nonlymphocytic leukemia associated with t(6;9)(p23;q34) is not accompanied by alteration in the ABL oncogene (Westbrook et al., 1985), despite the same location of the breakpoint at 9q34 and the frequent association with basophilia as in CML. In about 10% of patients with acute lymphocytic leukemia, patients carry a 9;22 translocation cytogenetically indistinguishable from the Philadelphia chromosome of CML. Clark et al. (1986) demonstrated, however, that Philadelphia chromosome-positive ALL cells express unique abl-derived tyrosine kinases of 1 85 and 180 kD that are distinct from the bcr, abl-derived P210 protein of CML. Kurzrock et al. (1987) found a novel abl protein product in Philadelphia chromosome-positive acute lymphoblastic leukemia. They suggested that alternative mechanisms of activation of abl exist and that a different mechanism may apply in human acute lymphoid leukemia as opposed to myeloid malignancies. Scott et al. (1991) showed in mice that the abl gene can cause a hematologic malignancy similar to but with differences from that caused by the bcr-abl fusion gene. By RT/PCR, Melo et al. (1993) detected ABL-BCR mRNA in cells from 31 of 44 BCR-ABL positive CML patients and in 3 of 5 CML cell lines. Thus, in many patients with CML, 2 reciprocal fusion genes are expressed. Of the 34 positive samples, 31 had classic t(9;22)(q34;q11) translocations; in 3 samples, there were no Philadelphia and/or 9q+ chromosomes.

Bernards et al. (1987) showed by pulsed field gel electrophoresis (PFGE) that a 5-prime exon of the ABL gene lies at least 300 kb upstream of the remaining ABL exons. The very long intron is a target for translocations in chronic myeloid leukemia. Whereas in CML the ABL gene is translocated from chromosome 9 to the center of the BCR gene on chromosome 22 to produce a chimeric BCR-ABL RNA translated into a protein of molecular weight 210 K, Fainstein et al. (1987) found that in acute lymphocytic leukemia, ABL is translocated into the 5-prime region of the BCR gene. The consequence of this is the expression of a fused transcript in which the first exon of BCR is linked to the second ABL exon. This transcript encodes a 190-K protein kinase. The t(9;22)(q34;q11) translocation is found in about 10% of acute lymphocytic leukemia patients. By in situ hybridization, Trakhtenbrot et al. (1990) showed that the murine c-abl oncogene on chromosome 2 lies closer to the centromere than the region commonly deleted in radiation-induced murine myeloid leukemias. Tybulewicz et al. (1991) and Schwartzberg et al. (1991) found that in transgenic mice a disrupted c-abl gene, when homozygous, resulted in runting and death 1 to 2 weeks after birth. Many showed thymic and splenic atrophy and a T- and B-cell lymphopenia. Many showed premature opening of the eyes, which were scarred and progressively opacified, and the head was foreshortened (Schwartzberg et al., 1991).

The ubiquitously expressed ABL tyrosine kinase is localized to the nucleus and binds to DNA. The DNA binding activity is regulated by CDC2-mediated phosphorylation (116940), suggesting a cell cycle function for ABL. Welch and Wang (1993) showed that the tyrosine kinase activity of nuclear ABL is regulated in the cell cycle through a specific interaction with the retinoblastoma protein (RB1; 180200). A domain in the C-terminus of RB binds to the ATP-binding lobe of the ABL tyrosine kinase, resulting in kinase inhibition. The RB-ABL interaction is not affected by the viral oncoproteins that bind to RB. Hyperphosphorylation of RB correlates with release of ABL and activation of the tyrosine kinase in S phase cells. The nuclear ABL tyrosine kinase can enhance transcription, and this activity is inhibited by RB. Thus, nuclear ABL is an S phase-activated tyrosine kinase that might participate directly in the regulation of transcription.

Feller et al. (1994) described the SRC homology domains SH2 and SH3 as molecular adhesives on many proteins involved in signal transduction. They reviewed the interactions of ABL and CRK (164762) as a model of SH2 and SH3 interaction.

Melo et al. (1994) concluded that the normal ABL gene is not imprinted and cited evidence resolving the question of imprinting of the BCR gene in favor of nonimprinting. They felt that their ABL data and the published BCR data excluded the possibility of a parental bias in the origin of the Philadelphia chromosome and cited observations suggesting that there is, in fact, no preferential involvement of maternal BCR or paternal ABL alleles in the formation of the BCR-ABL fusion gene of Ph-positive CML patients, in apparent contradiction to the reported cytogenetic evidence (Haas et al., 1992).

In summary, the t(9;22) translocation occurs in greater than 90% of chronic myelogeneous leukemia, 25 to 30% of adult and 2 to 10% of childhood acute lymphoblastic leukemia, and rare cases of acute myelogenous leukemia. The translocation results in the head-to-tail fusion of the BRC and ABL genes. The ABL gene is about 225 kb in size and is expressed as either a 6- or 7-kb mRNA transcript, with alternatively spliced first exons, exons 1b and 1a, respectively, spliced to the common exons 2-11. Exon 1b is approximately 200 kb 5-prime of exon 1a. The 145-kD ABL protein is classified as a nonreceptor tyrosine kinase. When the N-terminal region of the ABL protein is encoded by exon 1a, the protein is believed to be localized in the nucleus, while when encoded by exon 1b, the resulting N-terminal glycine would be myristylated and thus postulated to direct that protein to the plasma membrane. Chissoe et al. (1995) sequenced the complete BCR gene and greater than 80% of the ABL gene. Included were the common ABL exons 2-11 and the alternative ABL first exons 1a and 1b, as well as a new gene 5-prime to the first ABL exon and an open reading frame with homology to an EST within the BCR fourth intron. One of the most striking observations from analyzing the BCR and ABL gene regions was their extremely high density, 38.83 and 39.35%, respectively, of Alu homologous regions (found by BLASTN homology searches against the pAlu database at the National Center for Biotechnology Information (NCBI)). This contrasts with the previously calculated 4% level that was based on estimates of approximately 500,000 Alu elements in the human genome. Chissoe et al. (1995) analyzed 4 new Philadelphia chromosome translocation breakpoints from chronic myelogenous leukemia patients and several other previously sequenced breakpoints but could find no consistent breakpoint features. No clear-cut mechanism for Philadelphia chromosomal translocation was evident.

The ABL1 protooncogene encodes a cytoplasmic and nuclear protein tyrosine kinase that has been implicated in processes of cell differentiation, cell division, cell adhesion, and stress response. Alterations of ABL1 by chromosomal rearrangement or viral transduction lead to malignant transformation, as in chronic myeloid leukemia. Activity of c-Abl protein is negatively regulated by its SH3 domain through an unknown mechanism, and deletion of the SH3 domain turns ABL1 into an oncogene. Barila and Superti-Furga (1998) presented evidence for an intramolecular inhibitory interaction of the SH3 domain with the catalytic domain and with the linker between the SH2 and catalytic domain. Site-directed mutations in each of these 3 elements activated c-Abl. Mutations in the linker caused a conformational change of the molecule and increased binding of the SH3 domain to peptide ligands. Individual mutation of 2 charged residues in the SH3 and catalytic domain activated c-Abl, while inhibition was restored in the double reciprocal mutant. Barila and Superti-Furga (1998) proposed that regulators of c-Abl will have opposite effects on its activity depending on their ability to favor or disrupt these intramolecular interactions.

See also Bartram et al. (1983); Hoffman-Falk et al. (1983); Lane et al. (1982); Shtivelman et al. (1985)

Barila, D.; Superti-Furga, G.: An intramolecular SH3-domain interaction regulates c-Abl activity. Nature Genet. 18: 280-282, 1998.

Bartram, C. R.; de Klein, A.; Hagemeijer, A.; van Agthoven, T.; Geurts van Kessel, A.; Bootsma, D.; Grosveld, G.; Ferguson-Smith, M. A.; Davies, T.; Stone, M.; Heisterkamp, N.; Stephenson, J. R.; Groffen, J.: Translocation of c-abl oncogene correlates with the presence of a Philadelphia chromosome in chronic myelocytic leukaemia. Nature 306: 277-280, 1983.

Bernards, A.; Rubin, C. M.; Westbrook, C. A.; Paskind, M.; Baltimore, D.: The first intron in the human c-abl gene is at least 200 kilobases long and is a target for translocations in chronic myelogenous leukemia. Molec. Cell. Biol. 7: 3231-3236, 1987.

Chissoe, S. L.; Bodenteich, A.; Wang, Y.-F.; Wang, Y.-P.; Burian, D.; Clifton, S. W.; Crabtree, J.; Freeman, A.; Iyer, K.; Jian, L.; Ma, Y.; McLaury, H.-J.; Pan, H.-Q.; Sarhan, O. H.; Toth, S.; Wang, Z.; Zhang, G.; Heisterkamp, N.; Groffen, J.; Roe, B. A.: Sequence and analysis of the human ABL gene, the BCR gene, and regions involved in the Philadelphia chromosomal translocation. Genomics 27: 67-82, 1995.

Clark, S. S.; McLaughlin, J.; Crist, W. M.; Champlin, R.; Witte, O. N.: Unique forms of the abl tyrosine kinase distinguish Ph-1-positive CML from Ph-1-positive ALL. Science 235: 85-88, 1986.

Collins, S. J.; Groudine, M. T.: Rearrangement and amplification of c-abl sequences in the human chronic myelogenous leukemia cell line K-562. Proc. Nat. Acad. Sci. 80: 4813-4817, 1983.

de Klein, A.; Geurts van Kessel, A.; Grosveld, G.; Bartram, C. R.; Hagemeijer, A.; Bootsma, D.; Spurr, N. K.; Heisterkamp, N.; Groffen, J.; Stephenson, J. R.: A cellular oncogene is translocated to the Philadelphia chromosome in chronic myelocytic leukaemia. Nature 300: 765-767, 1982.

Fainstein, E.; Marcelle, C.; Rosner, A.; Canaani, E.; Gale, R. P.; Dreazen, O.; Smith, S. D.; Croce, C. M.: A new fused transcript in Philadelphia chromosome positive acute lymphocytic leukaemia. Nature 330: 386-388, 1987.

Feller, S. M.; Ren, R. B.; Hanafusa, H.; Baltimore, D.: SH2 and SH3 domains as molecular adhesives: the interactions of crk and abl. Trends Biochem. Sci. 19: 453-458, 1994.

Goff, S. P.; D'Eustachio, P.; Ruddle, F. H.; Baltimore, D.: Chromosomal assignment of endogenous protooncogene c-abl. Science 218: 1317-1319, 1982.

Haas, O. A.; Argyriou-Tirita, A.; Lion, T.: Parental origin of chromosomes involved in the translocation t(9;22). Nature 359: 414-416, 1992.

Heisterkamp, N.; Groffen, J.; Stephenson, J. R.; Spurr, N. K.; Goodfellow, P. N.; Solomon, E.; Carritt, B.; Bodmer, W. F.: Chromosomal localization of human cellular homologues of two viral oncogenes. Nature 299: 747-749, 1982.

Heisterkamp, N.; Stephenson, J. R.; Groffen, J.; Hansen, P. F.; de Klein, A.; Bartram, C. R.; Grosveld, G.: Localization of the c-abl oncogene adjacent to a translocation breakpoint in chronic myelocytic leukaemia. Nature 306: 239-242, 1983.

Hoffman-Falk, H.; Einat, P.; Shilo, B.-Z.; Hoffmann, F. M.: Drosophila melanogaster DNA clones homologous to vertebrate oncogenes: evidence for a common ancestor to the src and abl cellular genes. Cell 32: 589-598, 1983.

Jhanwar, S. C.; Neel, B. G.; Hayward, W. S.; Chaganti, R. S. K.: Localization of the cellular oncogenes ABL, SIS, and FES on human germ-line chromosomes. Cytogenet. Cell Genet. 38: 73-75, 1984.

Konopka, J. B.; Watanabe, S. M.; Singer, J. W.; Collins, S. J.; Witte, O. N.: Cell lines and clinical isolates derived from Ph1-positive chronic myelogenous leukemia patients express c-abl proteins with a common structural alteration. Proc. Nat. Acad. Sci. 82: 1810-1814, 1985.

Kurzrock, R.; Shtalrid, M.; Romero, P.; Kloetzer, W. S.; Talpas, M.; Trujillo, J. M.; Blick, M.; Beran, M.; Gutterman, J. U.: A novel c-abl protein product in Philadelphia-positive acute lymphoblastic leukaemia. Nature 325: 631-635, 1987.

Lane, M.-A.; Neary, D.; Cooper, G. M.: Activation of a cellular transforming gene in tumours induced by Abelson murine leukaemia virus. Nature 300: 659-661, 1982.

Melo, J. V.; Gordon, D. E.; Cross, N. C. P.; Goldman, J. M.: The ABL-BCR fusion gene is expressed in chronic myeloid leukemia. Blood 81: 158-165, 1993.

Melo, J. V.; Yan, X.-H.; Diamond, J.; Goldman, J. M.: Lack of imprinting of the ABL gene. (Letter) Nature Genet. 8: 318-319, 1994.

Schwartzberg, P. L.; Stall, A. M.; Hardin, J. D.; Bowdish, K. S.; Humaran, T.; Boast, S.; Harbison, M. L.; Robertson, E. J.; Goff, S. P.: Mice homozygous for the abl(m1) mutation show poor viability and depletion of selected B and T cell populations. Cell 65: 1165-1175, 1991.

Scott, M. L.; Van Etten, R. A.; Daley, G. Q.; Baltimore, D.: v-abl causes hematopoietic disease distinct from that caused by bcr-abl. Proc. Nat. Acad. Sci. 88: 6506-6510, 1991.

Selden, J. R.; Emanuel, B. S.; Wang, E.; Cannizzaro, L.; Palumbo, A.; Erikson, J.; Nowell, P. C.; Rovera, G.; Croce, C. M.: Amplified C-lambda and c-abl genes are on the same marker chromosome in K562 leukemia cells. Proc. Nat. Acad. Sci. 80: 7289-7292, 1983.

Shtivelman, E.; Lifshitz, B.; Gale, R. P.; Canaani, E.: Fused transcript of abl and bcr genes in chronic myelogenous leukaemia. Nature 315: 550-554, 1985.

Trakhtenbrot, L.; Kelman, Z.; Rotter, V.; Haran-Ghera, N.: Chromosomal mapping of the murine c-abl proto-oncogene by in situ hybridization. Leukemia 4: 136-137, 1990.

Tybulewicz, V. L. J.; Crawford, C. E.; Jackson, P. K.; Bronson, R. T.; Mulligan, R. C.: Neonatal lethality and lymphopenia in mice with a homozygous disruption of the c-abl proto-oncogene. Cell 65: 1153-1163, 1991.

Welch, P. J.; Wang, J. Y. J.: A C-terminal protein-binding domain in the retinoblastoma protein regulates nuclear c-abl tyrosine kinase in the cell cycle. Cell 75: 779-790, 1993.

Westbrook, C. A.; Le Beau, M. M.; Diaz, M. O.; Groffen, J.; Rowley, J. D.: Chromosomal localization and characterization of c-abl in the t(6;9) of acute nonlymphocytic leukemia. Proc. Nat. Acad. Sci. 82: 8742-8746, 1985.

ONCOGENE ABL2: V-ABL ABELSON MURINE LEUKEMIA VIRAL ONCOGENE HOMOLOG 2 (*164690)

Kruh et al. (1986) identified a novel oncogene related to ABL (189980) in DNA sequences from human plasma. The new sequence, called ARG, was localized by in situ hybridization and somatic cell analysis to human chromosome 1q24-q25. The detection of a novel 12-kb transcript from this gene in human normal and tumor cells establishes it as a new member of the tyrosine kinase family that is closely related to but distinct from ABL. A constitutional fragile site is located at 1q24-q25. Arg in the mouse is closely linked to At-3 (Morse, 1988). Seldin and Kruh (1989) mapped the Abll gene to mouse chromosome 1 by analysis of segregation with other distal chromosome 1 genetic polymorphisms in a panel of DNAs from interspecific backcross mice. This defined a region of distal mouse chromosome 1 homologous with human chromosome 1q21-q32.

Kruh, G. D.; King, C. R.; Kraus, M. H.; Popescu, N. C.; Amsbaugh, S. C.; McBride, W. O.; Aaronson, S. A.: A novel human gene closely related to the abl proto-oncogene. Science 234: 1545-1548, 1986.

Morse, H. G.: Personal Communication. Bethesda, Maryland 3/25/1988.

Seldin, M. F.; Kruh, G. D.: Mapping of Abll within a conserved linkage group on distal mouse chromosome 1 syntenic with human chromosome 1 using an interspecific cross. Genomics 4: 221-223, 1989.

ONCOGENE AKT1: V-AKT MURINE THYMOMA VIRAL ONCOGENE HOMOLOG 1 (*164730)

The AKT oncogene was isolated from the directly transforming murine retrovirus AKT8, which was isolated from an AKR mouse thymoma cell line. The human cellular homolog was cloned from a human DNA phage library. Testa et al. (1985) assigned the AKT1 gene to 14q32.3 by in situ hybridization. Somatic hybrid cell data corroborated the assignment to chromosome 14 and indicated that the locus is proximal to the immunoglobulin heavy chain loci. Is AKT1 identical to the oncogene that Croce et al. (1985) termed TCL1 and proposed is located at 14q32.3? They deduced the existence of such an oncogene from observation of chromosomal rearrangements in T-cell leukemias with breakpoint at this site; see 186960 and 186880. Staal (1987) found a 20-fold amplification of the AKT1 gene in 1 of 5 gastric adenocarcinomas tested. Staal (1987) mapped the AKT1 locus to 14q32 by analysis of human-hamster somatic cell hybrids and by in situ hybridization of normal human metaphase chromosome spreads with a radioactive AKT1 probe. By fluorescence in situ hybridization, Bellacosa et al. (1993) mapped akt to mouse chromosome 12; furthermore, they showed that the gene is on rat chromosome 6 and that both are in close proximity to the Igh locus.

The serine-threonine protein kinase encoded by the AKT1 gene is catalytically inactive in serum-starved primary and immortalized fibroblasts. Franke et al. (1995) showed that AKT1 and the related AKT2 (164731) are activated by platelet-derived growth factor (PDGFR; 173410). The activation is rapid and specific, and it is abrogated by mutations in the pleckstrin homology domain of AKT1. Other experiments showed that the activation occurs through phosphatidylinositol 3-kinase.

Dudek et al. (1997) demonstrated that AKT is important for the survival of cerebellar neurons. Thus, the 'orphan' kinase moved center stage as a crucial regulator of life and death decisions emanating from the cell membrane (Hemmings, 1997). The work of Dudek et al. (1997) delineated a signaling pathway by which insulin-like growth factor 1 (IGF1; 147440) promotes the survival of cerebellar neurons. IGF1 activation of phosphoinositide 3-kinase (PI3-K; see 171834) triggered the activation of 2 protein kinases, AKT and the p70 ribosomal protein S6 kinase (180460). Experiments with pharmacologic inhibitors, as well as expression of wildtype and dominant-inhibitory forms of AKT, demonstrated that AKT but not RPS6 mediates PI3-K-dependent survival. The findings suggested that in the developing nervous system AKT is a critical mediator of growth factor-induced neuronal survival. Franke et al. (1997) defined the specific mechanisms by which lipid products of PI3-K regulate AKT.

See also Staal et al. (1988)

Bellacosa, A.; Franke, T. F.; Gonzalez-Portal, M. E.; Datta, K.; Taguchi, T.; Gardner, J.; Cheng, J. Q.; Testa, J. R.; Tsichlis, P. N.: Structure, expression and chromosomal mapping of c-akt: relationship to v-akt and its implications. Oncogene 8: 745-754, 1993.

Croce, C. M.; Isobe, M.; Palumbo, A.; Puck, J.; Ming, J.; Tweardy, D.; Erikson, J.; Davis, M.; Rovera, G.: Gene for alpha-chain of human T-cell receptor: location on chromosome 14 region involved in T-cell neoplasms. Science 227: 1044-1047, 1985.

Dudek, H.; Datta, S. R.; Franke, T. F.; Birnbaum, M. J.; Yao, R.; Cooper, G. M.; Segal, R. A.; Kaplan, D. R.; Greenberg, M. E.: Regulation of neuronal survival by the serine-threonine protein kinase Akt. Science 275: 661-663, 1997.

Franke, T. F.; Kaplan, D. R.; Cantley, L. C.; Toker, A.: Direct regulation of the Akt proto-oncogene product by phosphatidylinositol-3,4-biphosphate. Science 275: 665-667, 1997.

Franke, T. F.; Yang, S.-I.; Chan, T. O.; Datta, K.; Kaziauskas, A.; Morrison, D. K.; Kaplan, D. R.; Tsichlis, P. N.: The protein kinase encoded by the Akt proto-oncogene is a target of the PDGF-activated phosphatidylinositol 3-kinase. Cell 81: 727-736, 1995.

Hemmings, B. A.: Akt signaling: linking membrane events to life and death decisions. Science 275: 628-630, 1997.

Staal, S. P.: Molecular cloning of the akt oncogene and its human homologues AKT1 and AKT2: amplification of AKT1 in a primary human gastric adenocarcinoma. Proc. Nat. Acad. Sci. 84: 5034-5037, 1987.

Staal, S. P.; Huebner, K.; Croce, C. M.; Parsa, N. Z.; Testa, J. R.: The AKT1 proto-oncogene maps to human chromosome 14, band q32. Genomics 2: 96-98, 1988.

Testa, J. R.; Huebner, K.; Croce, C. M.; Staal, S.: The AKT1 gene, the human homologue of a retroviral oncogene, is located on chromosome 14 at band q32. (Abstract) Cytogenet. Cell Genet. 40: 761, 1985.

ONCOGENE AKT2: V-AKT MURINE THYMOMA VIRAL ONCOGENE HOMOLOG 2 (*164731)

Staal (1987) cloned AKT1 (164730) and AKT2, 2 putative cellular homologs of the v-akt oncogene of the retrovirus AKT8. Cheng et al. (1992) reported the cDNA cloning, sequence analysis, and chromosome mapping of AKT2 and demonstrated that this putative oncogene encodes a protein belonging to a subfamily of serine/threonine kinases containing SH2-like (Src homology 2-like) domains. Furthermore, AKT2 was shown to be amplified and overexpressed in 2 of 8 ovarian carcinoma cell lines and 2 of 15 primary ovarian tumors. By fluorescence in situ hybridization (FISH), Cheng et al. (1992) mapped the AKT2 gene to 19q13.1-q13.2. In the 2 ovarian carcinoma cell lines exhibiting amplification of AKT2, the amplified sequences were located within homogeneously staining regions (HSR). In the FISH experiments, all signals on chromosome 19 were located at subband 19q13.1 or at the interface between subbands 19q13.1 and 19q13.2.

Cheng et al. (1996) demonstrated that AKT2 is amplified in approximately 10% of pancreatic carcinomas. They found, furthermore, that expression of AKT2 protein was greatly decreased when pancreatic cancer cells were transfected with antisense AKT2. Furthermore, tumorigenicity in nude mice was markedly reduced in the pancreatic cancer cells expressing antisense AKT2 RNA. Cells expressing antisense AKT2 RNA remained confined in a xenotransplant assay, whereas the untransfected cells showed invasion of surrounding tissues. The data were interpreted as suggesting that overexpression of AKT2 contributes to the malignant phenotype of a subset of human ductal pancreatic cancers.

Altomare et al. (1996) mapped the murine homolog to mouse chromosome 7 in band B1. A pseudogene was mapped to proximal mouse chromosome 11.

Altomare, D. A.; Kozak, C. A.; Sonoda, G.; Testa, J. R.: Chromosome mapping of the mouse Akt2 gene and Akt2 pseudogene. Cytogenet. Cell Genet. 74: 248-251, 1996.

Cheng, J. Q.; Godwin, A. K.; Bellacosa, A.; Taguchi, T.; Franke, T. F.; Hamilton, T. C.; Tsichlis, P. N.; Testa, J. R.: AKT2, a putative oncogene encoding a member of a subfamily of protein-serine/threonine kinases, is amplified in human ovarian carcinomas. Proc. Nat. Acad. Sci. 89: 9267-9271, 1992.

Cheng, J. Q.; Ruggeri, B.; Klein, W. M.; Sonoda, G.; Altomare, D. A.; Watson, D. K.; Testa, J. R.: Amplification of AKT2 in human pancreatic cancer cells and inhibition of AKT2 expression and tumorigenicity by antisense RNA. Proc. Nat. Acad. Sci. 93: 3636-3641, 1996.

Staal, S. P.: Molecular cloning of the akt oncogene and its human homologues AKT1 and AKT2: amplification of AKT1 in a primary human gastric adenocarcinoma. Proc. Nat. Acad. Sci. 84: 5034-5037, 1987.

ONCOGENE AXL: AXL RECEPTOR TYROSINE KINASE (*109135)

In an effort to determine genes involved in the progression of chronic myelogenous leukemia (CML) to acute-phase leukemia, Liu et al. (1988) identified a transforming gene in the DNAs of 2 patients with this disorder. O'Bryan et al. (1991) found by molecular cloning and characterization of this gene, which they termed AXL (from the Greek word 'anexelekto,' or uncontrolled), that it is a receptor tyrosine kinase with a structure novel among tyrosine kinases. They showed that the AXL protein is capable of transforming NIH 3T3 cells. Furthermore, its transforming capacity results from overexpression of AXL mRNA rather than from structural mutation. By fluorescence in situ hybridization, O'Bryan et al. (1991) localized the gene to 19q13.2. Janssen et al. (1991) independently found transforming activity by a tumorigenicity assay using NIH 3T3 cells transfected with DNA from a patient with a chronic myeloproliferative disorder. They reported the cDNA cloning of the corresponding oncogene, which they designated UFO, in allusion to the unidentified function of the protein. Nucleotide sequence analysis revealed a 2,682-bp open reading frame capable of directing the synthesis of an 894-amino acid polypeptide. It was evolutionarily conserved among vertebrate species. The predicted protein showed features characteristic of a transmembrane receptor with tyrosine kinase activity. By nonisotopic in situ hybridization, they mapped the gene to 19q13.1. The gene was transcribed into two 5.0-kb and 3.2-kb mRNAs in human bone marrow and human tumor cell lines.

The transforming activity of AXL demonstrates that the receptor can drive cellular proliferation. Although the function of AXL in nontransformed cells and tissues was unknown, Varnum et al. (1995) suspected that it may involve the stimulation of cell proliferation in response to an appropriate signal, i.e., a ligand that activates the receptor. Varnum et al. (1995) purified an AXL stimulatory factor and identified it as the product of the growth arrest-specific gene-6 (600441) (Manfioletti et al., 1993).

Janssen, J. W. G.; Schulz, A. S.; Steenvoorden, A. C. M.; Schmidberger, M.; Strehl, S.; Ambros, P. F.; Bartram, C. R.: A novel putative tyrosine kinase receptor with oncogenic potential. Oncogene 6: 2113-2120, 1991.

Liu, E.; Hjelle, B.; Bishop, J. M.: Transforming genes in chronic myelogenous leukemia. Proc. Nat. Acad. Sci. 85: 1952-1956, 1988.

Manfioletti, G.; Brancolini, C.; Avanzi, G.; Schneider, C.: The protein encoded by a growth arrest-specific gene (gas6) is a new member of the vitamin K-dependent proteins related to protein S, a negative coregulator in the blood coagulation cascade. Molec. Cell. Biol. 13: 4976-4985, 1993.

O'Bryan, J. P.; Frye, R. A.; Cogswell, P. C.; Neubauer, A.; Kitch, B.; Prokop, C.; Espinosa, R., III; Le Beau, M. M.; Earp, H. S.; Liu, E. T.: Axl, a transforming gene isolated from primary human myeloid leukemia cells, encodes a novel receptor tyrosine kinase. Molec. Cell. Biol. 11: 5016-5031, 1991.

Varnum, B. C.; Young, C.; Elliott, G.; Garcia, A.; Bartley, T. D.; Fridell, Y.-W.; Hunt, R. W.; Trail, G.; Clogston, C.; Toso, R. J.; Yanagihara, D.; Bennett, L.; Sylber, M.; Merewether, L. A.; Tseng, A.; Escobar, E.; Liu, E. T.; Yamane, H. K.: Axl receptor tyrosine kinase stimulated by the vitamin K-dependent protein encoded by growth-arrest-specific gene 6. Nature 373: 623-626, 1995.

ONCOGENE B-LYM: CHICKEN BURSAL LYMPHOMA (*164830)

This oncogene was assigned to chromosome 1; possible regional assignment to 1p32 was suggested (Land et al., 1983). Blym-1 is a transforming gene detected by transfection of NIH 3T3 cells with DNA from Burkitt lymphomas. The gene is homologous by molecular hybridization with the Blym-1 transforming gene isolated from chicken B-cell lymphoma DNA and is not homologous with 12 retroviral transforming genes including MYC and RAS. The involvement of 2 distinct genes in B-cell lymphomas of both chickens and humans supports the hypothesis that they may be involved in different stages of progression to neoplasia. Morton et al. (1984) assigned the Blym-1 gene to 1p32 by in situ hybridization. Diamond et al. (1984) found that BLYM has sequence homology to transferrin and suggested that its transforming gene products may function via a pathway related to transferrin.

See also Diamond et al. (1983)

Diamond, A.; Cooper, G. M.; Ritz, J.; Lane, M.-A.: Identification and molecular cloning of the human Blym transforming gene activated in Burkitt's lymphomas. Nature 305: 112-116, 1983.

Diamond, A.; Devine, J. M.; Cooper, G. M.: Nucleotide sequence of a human Blym transforming gene activated in a Burkitt's lymphoma. Science 225: 516-519, 1984.

Land, H.; Parada, L. F.; Weinberg, R. A.: Cellular oncogenes and multistep carcinogenesis. Science 222: 771-778, 1983.

Morton, C. C.; Taub, R.; Diamond, A.; Lane, M. A.; Cooper, G. M.; Leder, P.: Mapping of the human Blym-1 transforming gene activated in Burkitt lymphomas to chromosome 1. Science 223: 173-175, 1984.

ONCOGENE CBL2 (*165360)

Cas NS-1 is an acutely transforming murine retrovirus that induces pre-B and pro-B cell lymphomas. Molecular cloning showed that it was generated from the ecotropic Cas-Br-M virus by sequential recombinations with endogenous retroviral sequences and a cellular oncogene. Langdon et al. (1989) found that the oncogene sequence shows no homology with known oncogenes, but some similarity to the yeast transcriptional activator GCN4. A 100-kD gag-cbl fusion protein, with no detectable kinase activity, is responsible for the cellular transformation. The cellular homolog of v-cbl, present in mouse and human DNA, is expressed in a range of hemopoietic lineages. Wei et al. (1990) mapped CBL2 to 11q23.3-qter by molecular characterization of the breakpoints in 2 somatic cell hybrids. Savage et al. (1991) studied the relation of CBL2 to the breakpoints in malignancies with translocations involving chromosome 11. CBL2 was translocated from chromosome 11 to 4 in an acute leukemia cell line possessing a t(4;11)(q21;q23) and from chromosome 11 to 14 in a B-cell lymphoma with a t(11;14)(q23;q32). CBL2 remained on chromosome 11 in a Ewing sarcoma cell line (133450) with a t(11;22)(q23;q12). Other studies indicated that NCAM (116930) and the genes for the 3 subunits of CD3 were all proximal to the tightly clustered THY1 (188230), ETS1 (164720), and CBL2 in 11q23.

Jones et al. (1994) localized the rare folate-sensitive fragile site FRA11B (previously mapped to 11q23.3) to an interval of approximately 100 kb containing the 5-prime end of the CBL2 gene, which included a CCG trinucleotide repeat. Since all 4 folate-sensitive fragile sites that had been cloned at that time (i.e., FRAXA, FRAXE, FRAXF, and FRA16A) were known to consist of expanded CCG repeats, Jones et al. (1995) considered the CBL2 repeat as a likely candidate for the location of FRA11B. Furthermore, they presented evidence that expansion of the CCG(n) repeat is involved in the pathogenesis of Jacobsen syndrome (147791), which is the clinical presentation of the loss of part of the long arm of chromosome 11, typically from band 11q23 to 11qter.

Vulval induction during development of Caenorhabditis elegans is mediated by LET-23, a homolog of the mammalian epidermal growth factor receptor tyrosine kinase (131550). The sli-1 gene is a negative regulator of LET-23 and was shown by Yoon et al. (1995) to encode a protein similar to the mammalian proto-oncoprotein CBL2. The sli-1 and CBL2 proteins share approximately 55% amino acid identity over a stretch of 390 residues, which includes a C(3)HC(4) zinc-binding motif known as the RING finger, and multiple consensus binding sites for SRC homology 3 (SH3) domains. The authors stated that sli-1 and CBL2 may define a new class of proteins that modify receptor tyrosine kinase-mediated signal transduction.

Jones, C.; Penny, L.; Mattina, T.; Yu, S.; Baker, E.; Voullaire, L.; Langdon, W. Y.; Sutherland, G. R.; Richards, R. I.; Tunnacliffe, A.: Association of a chromosome deletion syndrome with a fragile site within the proto-oncogene CBL2. Nature 376: 145-149, 1995.

Jones, C.; Slijepcevic, P.; Marsh, S.; Baker, E.; Langdon, W. Y.; Richards, R. I.; Tunnacliffe, A.: Physical linkage of the fragile site FRA11B and a Jacobsen syndrome chromosome deletion breakpoint in 11q23.3. Hum. Molec. Genet. 3: 2123-2130, 1994.

Langdon, W. Y.; Hartley, J. W.; Klinken, S. P.; Ruscetti, S. K.; Morse, H. C., III: v-cbl, an oncogene from a dual-recombinant murine retrovirus that induces early B-lineage lymphomas. Proc. Nat. Acad. Sci. 86: 1168-1172, 1989.

Savage, P. D.; Shapiro, M.; Langdon, W. Y.; Geurts van Kessel, A. D.; Seuanez, H. N.; Akao, Y.; Croce, C.; Morse, H. C., III; Kersey, J. H.: Relationship of the human protooncogene CBL2 on 11q23 to the t(4;11), t(11;22), and t(11;14) breakpoints. Cytogenet. Cell Genet. 56: 112-115, 1991.

Wei, S.; Rocchi, M.; Archidiacono, N.; Sacchi, N.; Romeo, G.; Gatti, R. A.: Physical mapping of the human chromosome 11q23 region containing the ataxia-telangiectasia locus. Cancer Genet. Cytogenet. 46: 1-8, 1990.

Yoon, C. H.; Lee, J.; Jongeward, G. D.; Sternberg, P. W.: Similarity of sli-1, a regulator of vulval development in C. elegans, to the mammalian proto-oncogene c-cbl. Science 269: 1102-1105, 1995.

ONCOGENE COT: CANCER OSAKA THYROID ONCOGENE (*603259)

By transfecting the hamster embryonic cell line SHOK with DNA extracted from a human thyroid carcinoma cell line, Miyoshi et al. (1991) identified the transforming oncogene 'cancer Osaka thyroid' (COT). Sequence analysis revealed that COT is a serine-threonine protein kinase. The authors compared genomic clones of COT from transformed SHOK cells and from human placenta cells and found that the COT oncogene had undergone a rearrangement within the last coding exon, an event which probably occurred during the initial transfection experiment. The COT protooncogene contains 8 exons. Aoki et al. (1993) reported that the predicted normal COT protein has 467 amino acids. In the COT oncoprotein, the C-terminal 70 amino acids of normal COT are replaced by 18 novel residues. Cell fractionation and immunoprecipitation studies demonstrated that the COT protooncogene encodes 58- and 52-kD proteins that are located in the cytosol. Both proteins have serine/threonine kinase activity. The 2 COT isoforms appear to result from the use of alternative translation initiation sites. The 58-kD isoform had stronger transforming activity than the 52-kD protein, although this activity was much weaker than that of the oncoprotein. Aoki et al. (1993) suggested that the N-terminal domain of COT may be necessary for cellular transformation, whereas the C-terminal domain may negatively regulate the transforming activity.

Chan et al. (1993) isolated a Ewing sarcoma cell line cDNA that transformed NIH3T3 cells. They designated the gene EST for 'Ewing sarcoma transformant' and identified it as COT. Since the EST cDNA encodes the normal form of the COT protein, the authors concluded that the COT gene can be activated as an oncogene by overexpression as well as by gene rearrangement. Northern blot analysis revealed that COT is expressed as a 3.2-kb mRNA in human fibroblasts and epithelial cells. Treatment of a lung fibroblast cell line with the tumor promoter okadaic acid induced COT expression.

By analysis of somatic cell hybrids and by fluorescence in situ hybridization, Chan et al. (1993) mapped the COT gene to 10p11.2.

Ohara et al. (1993) reported the sequence of the mouse Cot protooncogene. Like human COT, the mouse Cot gene contains 8 exons, and the intron-exon borders are well-conserved between the 2 species. The predicted mouse and human COT proteins are 94% identical. Northern blot analysis revealed that Cot is expressed in many mouse tissues at various developmental stages.

Aoki, M.; Hamada, F.; Sugimoto, T.; Sumida, S.; Akiyama, T.; Toyoshima, K.: The human cot proto-oncogene encodes two protein serine/threonine kinases with different transforming activities by alternative initiation of translation. J. Biol. Chem. 268: 22723-22732, 1993.

Chan, A. M.-L.; Chedid, M.; McGovern, E. S.; Popescu, N. C.; Miki, T.; Aaronson, S. A.: Expression cDNA cloning of a serine kinase transforming gene. Oncogene 8: 1329-1333, 1993.

Miyoshi, J.; Higashi, T.; Mukai, H.; Ohuchi, T.; Kakunaga, T.: Structure and transforming potential of the human cot oncogene encoding a putative protein kinase. Molec. Cell. Biol. 11: 4088-4096, 1991.

Ohara, R.; Miyoshi, J.; Aoki, M.; Toyoshima, K.: The murine cot proto-oncogene: genome structure and tissue-specific expression. Jpn. J. Cancer Res. 84: 518-525, 1993.

ONCOGENE DJ1 (*602533)

Nagakubo et al. (1997) used yeast 2-hybrid screening of a HeLa cell library to clone a cDNA that encodes a novel protein, termed DJ1. The DJ1 protein has 189-amino acids. Northern blotting revealed that DJ1 is ubiquitously expressed as a 1.0-kb transcript. Western blotting and immunofluorescence revealed that the DJ1 protein is present in both nuclei and cytoplasm of HeLa cells. After addition of serum to cells, DJ1 expression increased and the protein translocated from the cytoplasm to nuclei. Nagakubo et al. (1997) found that the DJ1 gene has weak transforming ability in NIH 3T3 cells, but transformation by DJ1 is synergistically enhanced by cotransfection with HRAS (190020) or MYC (190080).

A search of the GenBank protein database revealed that DJ1 has approximately 40% identity to the 198-amino acid protein product of the E. coli thiazole monophosphate biosynthesis (ThiJ) gene. A homolog also exists in the nematode C. elegans.

Nagakubo, D.; Taira, T.; Kitaura, H.; Ikeda, M.; Tamai, K.; Iguchi-Ariga, S. M. M.; Ariga, H.: DJ-1, a novel oncogene which transforms mouse NIH3T3 cells in cooperation with ras. Biochem. Biophys. Res. Commun. 231: 509-513, 1997.

ONCOGENE EMS1 (*164765)

Amplification of the 11q13 region is frequently found in breast cancer and in squamous cell carcinomas of the head and neck. The known oncogenes within the amplified 11q13 region, INT2 (164950) and FGF4 (164980), are rarely expressed in these tumors, indicating that another, hitherto unidentified gene or genes are involved in the unfavorable clinical course of disease associated with such amplification. To identify the gene or genes, Schuuring et al. (1992) constructed a cDNA library from a cell line with an 11q13 amplification and performed a differential cDNA cloning using labeled cDNAs from human squamous cell carcinoma cell lines with and without an 11q13 amplification. They isolated 2 cDNA clones, U21B31 and U21C8, which recognized genes amplified and overexpressed in cell lines harboring an 11q13 amplification. Sequence analysis of the U21C8 cDNA clone revealed no homology to known genes; they called this gene EMS1. The U21B31 cDNA clone corresponded to the 3-prime end of the PRAD1 protooncogene (168461).

Van Damme et al. (1997) stated that EMS1 is the human homolog of cortactin, an actin-binding protein involved in the restructuring of the cortical actin cytoskeleton. Cortactin is a substrate for the pp60v-src tyrosine kinase (see 190090). Cortactin is overexpressed in carcinoma cells with an amplification of 11q13 and is found in 2 forms, designated p80 and p85. Van Damme et al. (1997) found that in carcinoma cells with the 11q13 amplification, p85 was produced from p80 by posttranslational modification. Also, treatment of these cells with epidermal growth factor (131530) or vanadate caused conversion of p80 to p85 and enhanced phosphorylation of the p85 form. Both overexpression and posttranslational modification of cortactin coincided with its redistribution from the cytoplasm to cell-matrix contact sites, implying a role for cortactin in the modulation of cellular adhesive properties.

Schuuring, E.; Verhoeven, E.; Mooi, W. J.; Michalides, R. J. A. M.: Identification and cloning of two overexpressed genes, U21B31/PRAD1 and EMS1, within the amplified chromosome 11q13 region in human carcinomas. Oncogene 7: 355-361, 1992.

van Damme, H.; Brok, H.; Schuuring-Scholtes, E.; Schuuring, E.: The redistribution of cortactin into cell-matrix contact sites in human carcinoma cells with 11q13 amplification is associated with both overexpression and post-translational modification. J. Biol. Chem. 272: 7374-7380, 1997.

EPHRIN RECEPTOR EPHA1 (*179610)

The EPH and EPH-related receptors comprise the largest subfamily of receptor protein-tyrosine kinases. They have been implicated in mediating developmental events, particularly in the nervous system. Receptors in the Eph subfamily typically have a single kinase domain and an extracellular region containing a Cys-rich domain and 2 fibronectin type III repeats. The ligands for Eph receptors have been named ephrins by the Eph Nomenclature Committee (1997). Based on their structures and sequence relationships, ephrins are divided into the ephrin-A (EFNA) class, which are anchored to the membrane by a glycosylphosphatidylinositol linkage, and the ephrin-B (EFNB) class, which are transmembrane proteins. The Eph family of receptors are divided into 2 groups based on the similarity of their extracellular domain sequences and their affinities for binding ephrin-A and ephrin-B ligands. The Eph Nomenclature Committee (1997) proposed that Eph receptors interacting preferentially with ephrin-A proteins be called EphA and Eph receptors interacting preferentially with ephrin-B proteins be called EphB.

Maru et al. (1988) reported characterization of the novel receptor tyrosine kinase gene, EPH. The splicing points of kinase domain-encoding exons were completely distinct from those of other protein tyrosine kinase genes, suggesting that this is the earliest evolutionary split within this family. In Northern blot analysis, EPH gene mRNA was detected in liver, lung, kidney, and testes of rat; screening of 25 human cancers of various cell types showed preferential expression in cells of epithelial origin. Overexpression of EPH mRNA was found in a hepatoma and a lung cancer without gene amplification. Southern blot analysis of DNAs from human-mouse hybrid clones with an EPH probe showed that this gene is present on human chromosome 7. Two other receptor tyrosine kinase genes, MET (164860) and EGFR (131550), are on the same chromosome. By in situ hybridization, Yoshida et al. (1989) assigned the EPH locus to 7q32-q36.

Eph Nomenclature Committee: Unified nomenclature for the Eph family receptors and their ligands, the ephrins. Cell 90: 403-404, 1997.

Maru, Y.; Hirai, H.; Yoshida, M. C.; Takaku, F.: Evolution, expression, and chromosomal location of a novel receptor tyrosine kinase gene, eph. Molec. Cell. Biol. 8: 3770-3776, 1988.

Yoshida, M. C.; Maru, H.; Hirai, H.; Takau, F.: Chromosomal location of a novel receptor tyrosine kinase gene, EPH, on chromosome 7. (Abstract) Cytogenet. Cell Genet. 51: 1113 only, 1989.

ONCOGENE ERB-B: EPIDERMAL GROWTH FACTOR RECEPTOR (*131550)

Using as one parental cell the mouse A9 line, which is incapable of binding labeled epidermal growth factor (EGF), Shimizu et al. (1980) studied human-mouse cell hybrids and concluded that a receptor for EGF is located on human chromosome 7 in the p22-qter region. Since the EGF receptor is a glycoprotein, EGF may be either a structural gene for receptor protein or a gene for glycosylation of the receptor protein. EGF enhances phosphorylation of several endogenous membrane proteins, including EGF receptor. The EGF receptor is a tyrosine protein kinase. It has 2 components of different molecular weight; both contain phosphotyrosine and phosphothreonine but only the higher molecular weight form contains phosphoserine (Carlin and Knowles, 1982). Carlin et al. (1982) showed that the specific cell surface antigen previously called SA7 (Aden and Knowles, 1976) is identical to the receptor for epidermal growth factor. They indicated 7p22-p12 as the localization. Kondo and Shimizu (1983) concluded that EGFR is in the 7p13-q22 region. The EGFR molecule has 3 regions: one projects outside the cell and contains the site for binding EGF; the second is embedded in the membrane; the third projects into the cytoplasm of the cell's interior. EGFR is a kinase that attaches phosphate groups to tyrosine residues in proteins. Haley et al. (1987) presented detailed structural information about the 110-kb locus encoding the human EGF receptor and the regulation of EGF receptor expression encoded therein.

Yang et al. (1996) demonstrated that treatment with genistein, an inhibitor of tyrosine kinase activity, inhibited EGF-induced tyrosine phosphorylation and degradation of EGFR in HepG2 cells, suggesting to the authors that tyrosine kinase activity is required for either the internalization or the degradation of EGF-EGFR receptor complexes.

Oncogene ERBB may be derived from the gene coding for EGFR (Downward et al., 1984). Spurr et al. (1984) assigned the oncogene ERBB to chromosome 7 by study of mouse-human somatic cell hybrids. The amino acid sequence of the protein encoded by v-erb B (deduced from the nucleotide sequence of the gene) displays strong homologies to tyrosine-specific protein kinases (Privalsky et al., 1984). Both ERBA and ERBB are on mouse chromosome 11, which carries alpha-globin genes and genes for colony-stimulating factor and interleukin-3 (Silver et al., 1985). Neither of the oncogenes is on chromosome 16, which carries alpha-globin genes in man. The most striking and consistent chromosomal finding in a series of human glioblastoma cell lines was an increase in copy number of chromosome 7. Henn et al. (1986) found that in all of them ERBB-specific mRNA was increased to levels even higher than expected from the number of chromosomes 7 present. These changes were not found in benign astrocytomas.

Maternal uniparental disomy (UPD) of chromosome 7 has been reported in approximately 10% of cases of Silver-Russell syndrome (SRS; 180860). This suggests that at least 1 gene on chromosome 7 is imprinted and involved in the pathogenesis of SRS. Wakeling et al. (1998) investigated the EGFR gene as a candidate for imprinting because the gene maps to 7p12, a region homologous to an imprinted region on mouse chromosome 11. Using a restriction fragment length polymorphism, they found, however, biallelic expression of EGFR in a range of normal human fetal tissues. Expression was also demonstrated in fibroblasts and lymphoblasts from SRS patients with maternal UPD7. Thus, no evidence that EGFR is imprinted was found, making its involvement in SRS unlikely. However, EGFR was shown to be widely expressed in the human fetus, providing evidence that it plays an important role in early development. The only gene known to be imprinted on chromosome 7 at that time was MEST, also called paternally expressed gene-1 (601029), which maps to 7q32.

See also Carpenter (1984); Davies et al. (1980); Maciag (1982); Reynolds et al. (1981); Spurr et al. (1984); Ullrich et al. (1984)

Aden, D. P.; Knowles, B. B.: Cell surface antigens coded for by the human chromosome 7. Immunogenetics 3: 209-211, 1976.

Carlin, C. R.; Aden, D. P.; Knowles, B. B.: S6 is the human receptor for epidermal growth factor (EGF). (Abstract) Cytogenet. Cell Genet. 32: 256 only, 1982.

Carlin, C. R.; Knowles, B. B.: Identity of human epidermal growth factor (EGF) receptor with glycoprotein SA-7: evidence for differential phosphorylation of the two components of the EGF receptor from A431 cells. Proc. Nat. Acad. Sci. 79: 5026-5030, 1982.

Carpenter, G.: Properties of the receptor for epidermal growth factor. Cell 37: 357-358, 1984.

Davies, R. L.; Grosse, V. A.; Kucherlapati, R.; Bothwell, M.: Genetic analysis of epidermal growth factor action: assignment of human epidermal growth factor receptor gene to chromosome 7. Proc. Nat. Acad. Sci. 77: 4188-4192, 1980.

Downward, J.; Yarden, Y.; Mayes, E.; Scrace, G.; Totty, N.; Stockwell, P.; Ullrich, A.; Schlessinger, J.; Waterfield, M. D.: Close similarity of epidermal growth factor receptor and v-erb-B oncogene protein sequences. Nature 307: 521-527, 1984.

Haley, J.; Whittle, N.; Bennett, P.; Kinchington, D.; Ullrich, A.; Waterfield, M.: The human EGF receptor gene: structure of the 110 kb locus and identification of sequences regulating its transcription. Oncogene Res. 1: 375-396, 1987.

Henn, W.; Blin, N.; Zang, K. D.: Polysomy of chromosome 7 is correlated with overexpression of the erbB oncogene in human glioblastoma cell lines. Hum. Genet. 74: 104-106, 1986.

Kondo, I.; Shimizu, N.: Mapping of the human gene for epidermal growth factor receptor (EGFR) on the p13-q22 region of chromosome 7. Cytogenet. Cell Genet. 35: 9-14, 1983.

Maciag, T.: The human epidermal growth factor receptor-kinase complex. Trends Biochem. Sci. 7: 1-2, 1982.

Privalsky, M. L.; Ralston, R.; Bishop, J. M.: The membrane glycoprotein encoded by the retroviral oncogene v-erb-B is structurally related to tyrosine-specific protein kinases. Proc. Nat. Acad. Sci. 81: 704-707, 1984.

Reynolds, F. H., Jr.; Todaro, G. J.; Fryling, C.; Stephenson, J. R.: Human transforming growth factors induce tyrosine phosphorylation of EGF receptors. Nature 292: 259-262, 1981.

Shimizu, N.; Behzadian, M. A.; Shimizu, Y.: Genetics of cell surface receptors for bioactive polypeptides: binding of epidermal growth factor is associated with the presence of human chromosome 7 in human-mouse cell hybrids. Proc. Nat. Acad. Sci. 77: 3600-3604, 1980.

Silver, J.; Whitney, J. B., III; Kozak, C.; Hollis, G.; Kirsch, I.: Erbb is linked to the alpha-globin locus on mouse chromosome 11. Molec. Cell. Biol. 5: 1784-1786, 1985.

Spurr, N. K.; Goodfellow, P. N.; Solomon, E.; Parkar, M.; Vennstrom, B.; Bodmer, W. F.: Mapping of cellular oncogenes; erb B on chromosome 7. (Abstract) Cytogenet. Cell Genet. 37: 590 only, 1984.

Spurr, N. K.; Solomon, E.; Jansson, M.; Sheer, D.; Goodfellow, P. N.; Bodmer, W. F.; Vennstrom, B.: Chromosomal localisation of the human homologues to the oncogenes erbA and B. EMBO J. 3: 159-163, 1984.

Ullrich, A.; Coussens, L.; Hayflick, J. S.; Dull, T. J.; Gray, A.; Tam, A. W.; Lee, J.; Yarden, Y.; Libermann, T. A.; Schlessinger, J.; Downward, J.; Mayes, E. L. V.; Whittle, N.; Waterfield, M. D.; Seeburg, P. H.: Human epidermal growth factor receptor cDNA sequence and aberrant expression of the amplified gene in A431 epidermoid carcinoma cells. Nature 309: 418-425, 1984.

Wakeling, E. L.; Abu-Amero, S. N.; Stanier, P.; Preece, M. A.; Moore, G. E.: Human EGFR, a candidate gene for the Silver-Russell syndrome, is biallelically expressed in a wide range of fetal tissues. Europ. J. Hum. Genet. 6: 158-164, 1998.

Yang, E.-B.; Wang, D.-F.; Mack, P.; Cheng, L.-Y.: Genistein, a tyrosine kinase inhibitor, reduces EGF-induced EGF receptor internalization and degradation in human hepatoma HepG2 cells. Biochem. Biophys. Res. Commun. 224: 309-317, 1996.

ONCOGENE ERB-B2 (NEU; HER2): V-ERB-B2 AVIAN ERYTHROBLASTIC LEUKEMIA VIRAL ONCOGENE HOMOLOG 2 (*164870)

The oncogene originally called NEU was derived from rat neuro/glioblastoma cell lines. It encodes a tumor antigen, p185, which is serologically related to EGFR, the epidermal growth factor receptor (131550). EGFR maps to chromosome 7. Yang-Feng et al. (1985) found, however, that the human homolog, which they designated NGL (to avoid confusion with neuraminidase, which is also symbolized NEU), maps to 17q12-q22 by in situ hybridization and to 17q21-qter in somatic cell hybrids. Thus, the SRO is 17q21-q22. Coussens et al. (1985) identified a potential cell surface receptor of the tyrosine kinase gene family and characterized it by cloning the gene. Its primary sequence is very similar to that of the human epidermal growth factor receptor. Because of the seemingly close relationship to the human EGF receptor, the authors called the gene HER2. By Southern blot analysis of somatic cell hybrid DNA and by in situ hybridization, the gene was assigned to 17q21-q22. This chromosomal location of the gene is coincident with the NEU oncogene, which suggests that the 2 genes may in fact be the same; indeed, sequencing indicates that they are identical (Francke, 1988). Van de Vijver et al. (1988) found a correlation between overexpression of NEU protein and the large-cell, comedo growth type of ductal carcinoma. They could find no correlation, however, with lymph-node status or tumor recurrence. Slamon et al. (1989) described the role of HER2/NEU in breast (114480) and ovarian cancer (167000), which together account for one-third of all cancers in women and approximately one-quarter of cancer-related deaths in females.

Semba et al. (1985) identified an ERBB-related gene that is distinct from the ERBB gene (131550), called ERBB1 by these authors. ERBB2 was not amplified in vulva carcinoma cells with EGFR amplification and did not react with EGF receptor mRNA. About 30-fold amplification of ERBB2 was observed in a human adenocarcinoma of the salivary gland. By chromosome sorting combined with velocity sedimentation and Southern hybridization, Fukushige et al. (1986) assigned the ERBB2 gene to chromosome 17. By hybridization to sorted chromosomes and to metaphase spreads with a genomic probe, Fukushige et al. (1986) mapped the ERBB2 locus to 17q21. This is the chromosome 17 breakpoint in acute promyelocytic leukemia (APL). Furthermore, they observed amplification and elevated expression of the ERBB2 gene in a gastric cancer cell line. Akiyama et al. (1986) raised antibodies against a synthetic peptide corresponding to 14 amino acid residues at the COOH-terminus of a protein deduced from the ERBB2 nucleotide sequence. With these antibodies, they precipitated the ERBB2 gene product from adenocarcinoma cells and demonstrated it to be a 185-kD glycoprotein with tyrosine kinase activity. Kaneko et al. (1987) used a cDNA probe for ERBB2 and by in situ hybridization to APL cells with a 15;17 chromosome translocation located the gene to the proximal side of the breakpoint. They suggested that both the gene and the breakpoint are located in band 17q21.1 and, further, that the ERBB2 gene is involved in the development of leukemia. Di Fiore et al. (1987) indicated that NEU and HER2 (see 164870) are both the same as ERBB2. Di Fiore et al. (1987) demonstrated that overexpression alone can convert the gene for a normal growth factor receptor, namely, ERBB2, into an oncogene. Popescu et al. (1989) localized ERBB2 to 17q11-q21 by in situ hybridization. By in situ hybridization to chromosomes derived from fibroblasts carrying a constitutional translocation between 15 and 17, Popescu et al. (1989) showed that the ERBB2 gene was relocated to the derivative chromosome 15; the gene can thus be localized to 17q12-q21.32. By family linkage studies using multiple DNA markers in the 17q12-q21 region, Anderson et al. (1993) placed the ERBB2 gene on the genetic map of the region. By fluorescence in situ hybridization, Muleris et al. (1997) mapped the ERBB2 gene to 17q21.1.

Interleukin-6 (IL6; 147620) is a cytokine that was initially recognized as a regulator of immune and inflammatory responses, but also regulates the growth of many tumor cells, including prostate cancer. Overexpression of ERBB2 and ERBB3 (190151) has been implicated in the neoplastic transformation of prostate cancer. Qiu et al. (1998) showed that treatment of a prostate cancer cell line with IL6 induced tyrosine phosphorylation of ERBB2 and ERBB3, but not ERBB1/EGFR (131550). They also showed the ERBB2 forms a complex with the gp130 subunit of the IL6 receptor (IL6R; 147880) in an IL6-dependent manner. This association was important because the inhibition of ERBB2 activity resulted in abrogation of IL6-induced MAPK activation. Thus, ERBB2 is a critical component of IL6 signaling through the MAP kinase pathway. These findings showed how a cytokine receptor can diversify its signaling pathways by engaging with a growth factor receptor kinase.

See also Bargmann et al. (1986); Fukushige et al. (1986)

Akiyama, T.; Sudo, C.; Ogawara, H.; Toyoshima, K.; Yamamoto, T.: The product of the human c-erbB-2 gene: a 185-kilodalton glycoprotein with tyrosine kinase activity. Science 232: 1644-1646, 1986.

Anderson, L. A.; Friedman, L.; Osborne-Lawrence, S.; Lynch, E.; Weissenbach, J.; Bowcock, A.; King, M.-C.: High-density genetic map of the BRCA1 region of chromosome 17q12-q21. Genomics 17: 618-623, 1993.

Bargmann, C. I.; Hung, M.-C.; Weinberg, R. A.: The NEU oncogene encodes an epidermal growth factor receptor-related protein. Nature 319: 226-230, 1986.

Coussens, L.; Yang-Feng, T. L.; Liao, Y.-C.; Chen, E.; Gray, A.; McGrath, J.; Seeburg, P. H.; Libermann, T. A.; Schlessinger, J.; Francke, U.; Levinson, A.; Ullrich, A.: Tyrosine kinase receptor with extensive homology to EGF receptor shares chromosomal location with NEU oncogene. Science 230: 1132-1139, 1985.

Di Fiore, P. P.; Pierce, J. H.; Kraus, M. H.; Segatto, O.; King, C. R.; Aaronson, S. A.: erbB-2 is a potent oncogene when overexpressed in NIH/3T3 cells. Science 237: 178-182, 1987.

Ehsani, A.; Low, J.; Wallace, R. B.; Wu, A. M.: Characterization of a new allele of the human ERBB2 gene by allele-specific competition hybridization. Genomics 15: 426-429, 1993.

Francke, U.: Personal Communication. New Haven, Connecticut 4/1988.

Fukushige, S.; Murotsu, T.; Matsubara, K.: Chromosomal assignment of human genes for gastrin, thyrotropin (TSH)-beta subunit and c-erb-2 by chromosome sorting combined with velocity sedimentation and southern hybridization. Biochem. Biophys. Res. Commun. 134: 477-483, 1986.

Fukushige, S.-I.; Matsubara, K.-I.; Yoshida, M.; Sasaki, M.; Suzuki, T.; Semba, K.; Toyoshima, K.; Yamamoto, T.: Localization of a novel v-erbB-related gene, c-erbB-2, on human chromosome 17 and its amplification in a gastric cancer cell line. Molec. Cell. Biol. 6: 955-958, 1986.

Kaneko, Y.; Homma, C.; Maseki, N.; Sakurai, M.; Toyoshima, K.; Yamamoto, T.: Human c-erbB-2 remains on chromosome 17 in band q21 in the 15;17 translocation associated with acute promyelocytic leukemia. Jpn. J. Cancer Res. 78: 16-19, 1987.

Muleris, M.; Almeida, A.; Malfoy, B.; Dutrillaux, B.: Assignment of v-erb-b2 avian erythroblastic leukemia viral oncogene homolog 2 (ERBB2) to human chromosome band 17q21.1 by in situ hybridization. Cytogenet. Cell Genet. 76: 34-35, 1997.

Papewalis, J.; Nikitin, A. Y.; Rajewsky, M. F.: G to A polymorphism at amino acid codon 655 of the human erbB-2/HER2 gene. Nucleic Acids Res. 19: 5452 only, 1991.

Popescu, N. C.; King, C. R.; Kraus, M. H.: Localization of the human erbB-2 gene on normal and rearranged chromosomes 17 to bands q12-21.32. Genomics 4: 362-366, 1989.

Qiu, Y.; Ravi, L.; Kung, H.-J.: Requirement of ErbB2 for signalling by interleukin-6 in prostate carcinoma cells. Nature 393: 83-85, 1998.

Semba, K.; Kamata, N.; Toyoshima, K.; Yamamoto, T.: A v-erbB-related protooncogene, c-erbB-2, is distinct from the c-erbB-1/epidermal growth factor-receptor gene and is amplified in a human salivary gland adenocarcinoma. Proc. Nat. Acad. Sci. 82: 6497-6501, 1985.

Slamon, D. J.; Godolphin, W.; Jones, L. A.; Holt, J. A.; Wong, S. G.; Keith, D. E.; Levin, W. J.; Stuart, S. G.; Udove, J.; Ullrich, A.; Press, M. F.: Studies of the HER-2/neu proto-oncogene in human breast and ovarian cancer. Science 244: 707-712, 1989.

van de Vijver, M. J.; Peterse, J. L.; Mooi, W. J.; Wisman, P.; Lomans, J.; Dalesio, O.; Nusse, R.: NEU-protein overexpression in breast cancer: association with comedo-type ductal carcinoma in situ and limited prognostic value in stage II breast cancer. New Eng. J. Med. 319: 1239-1245, 1988.

Yamamoto, T.; Ikawa, S.; Akiyama, T.; Semba, K.; Nomura, N.; Miyajima, N.; Saito, T.; Toyoshima, K.: Similarity of protein encoded by the human c-erb-B-2 gene to epidermal growth factor receptor. Nature 319: 230-234, 1986.

Yang-Feng, T. L.; Schechter, A. L.; Weinberg, R. A.; Francke, U.: Oncogene from rat neuro/glioblastomas (human gene symbol NGL) is located on the proximal long arm of human chromosome 17 and EGFR is confirmed at 7p13-q11.2. (Abstract) Cytogenet. Cell Genet. 40: 784 only, 1985.

#homolog

ONCOGENE ERB-B3 (*190151)

Kraus et al. (1989) detected a DNA fragment related to but distinct from epidermal growth factor receptor (EGFR; 131550) and ERBB2 (164870). cDNA cloning showed a predicted 148-kD transmembrane polypeptide with structural features identifying it as a member of ERBB gene family and prompting the designation ERBB3. Markedly elevated ERBB3 mRNA levels were demonstrated in certain human mammary tumor cell lines, suggesting that it may play a role in some human malignancies just as does EGFR (which is identical to what was previously called ERBB1). Kraus et al. (1989) mapped ERBB3 to 12q13 by in situ hybridization.

Carraway et al. (1994) demonstrated that ERBB3 is a receptor for heregulin (142445) and is capable of mediating HGL-stimulated tyrosine phosphorylation of itself.

Carraway, K. L., III; Sliwkowski, M. X.; Akita, R.; Platko, J. V.; Guy, P. M.; Nuijens, A.; Diamonti, A. J.; Vandlen, R. L.; Cantley, L. C.; Cerione, R. A.: The erbB3 gene product is a receptor for heregulin. J. Biol. Chem. 269: 14303-14306, 1994.

Kraus, M. H.; Issing, W.; Miki, T.; Popescu, N. C.; Aaronson, S. A.: Isolation and characterization of ERBB3, a third member of the ERBB/epidermal growth factor receptor family: evidence for overexpression in a subset of human mammary tumors. Proc. Nat. Acad. Sci. 86: 9193-9197, 1989.

ONCOGENE ERG: V-ETS AVIAN ERYTHROBLASTOSIS VIRUS E26 ONCOGENE RELATED (*165080)

Reddy et al. (1987) isolated a cDNA clone representing the complete coding sequence of oncogene ERG (for 'early response gene'). Nucleotide sequence analysis of this cDNA showed that the gene encodes a 363-residue protein whose predicted amino acid sequence shows homology of about 40% and about 70% to 2 domains corresponding to the 5-prime and 3-prime regions of the ETS oncogene (see 164720), respectively. These results suggested that the ERG oncogene is a member of the ETS oncogene family. Like the ETS oncogene, the ERG oncogene is coded by chromosome 21. Rao et al. (1987) used a cDNA probe to analyze DNA from somatic cell hybrids. By nucleotide sequencing, they found 2 different transcripts, ERG1 and ERG2, the products of alternative splicing and polyadenylation as well as of translation initiation. Murakami et al. (1993) found that ERG2 is a nuclear phosphoprotein and binds to purine-rich sequences. With a half-life of 21 hours, it is considerably more stable than the short-lived ETS1 (164720) or ETS2 (164740) proteins. Like other ETS proteins, ERG2 is a sequence-specific DNA-binding protein and is expressed at higher levels in early myeloid cells than in mature lymphoid cells. From their findings, Murakami et al. (1993) suggested that ERG2 acts as a regulator of genes required for maintenance and/or differentiation of early hematopoietic cells.

By means of family linkage studies to DNA markers that are along with ETS2 translocated to the derivative chromosome 8 in the t(8;21) translocation of acute myelogenous leukemia, Sacchi et al. (1988) concluded that the ETS2 gene is located about 17 cM from the breakpoint. By family linkage studies using RFLPs, Sacchi et al. (1988) demonstrated that ERG is situated just proximal to ETS2. Rao et al. (1988) and Modi et al. (1989) concluded from in situ hybridization and somatic cell hybridization that the ERG gene is located in band 21q22.3. Rao et al. (1988) found that the ERG gene is translocated from chromosome 21 to chromosome 8 in the t(8;21)(q22;q22), a nonrandom translocation found in patients with acute myelogenous leukemia of the subtype M2 (AML-M2). Ichikawa et al. (1994) demonstrated that in human myeloid leukemia with the t(16;21)(p11;q22) translocation the ERG gene is fused with the TLS/FUS gene (137070) on chromosome 16. The TLS/FUS gene is fused with the DDIT3 gene (126337) in the t(12;16) translocation found in myxoid liposarcoma. The chimeric gene product is an RNA-binding protein that is highly homologous to the product of the EWS gene (133450) involved in Ewing sarcoma. Thus, the FUS-ERG gene fusion in t(16;21) leukemia was predicted to produce a protein very similar to the EWS-ERG chimeric protein seen in Ewing sarcoma.

Ewing sarcoma, the second most common malignant bone tumor of children and young adults, is an aggressive osteolytic tumor with a marked propensity for dissemination. It belongs to the heterogeneous group of small-round-cell tumors that are often a diagnostic challenge to the pathologist and clinical oncologist. Ewing sarcoma and related peripheral primitive neuroectodermal tumors (PNETs or 'peanut' tumors) show a t(11;22) translocation or a (21,22) rearrangement that is associated with hybrid transcripts of the EWS gene with the FLI1 (193067) or ERG gene. ERG is a member of the ETS family of transcription factors closely related to FLI1. The chimeric protein resulting from the fusion of EWS with ERG is structurally similar to the typical EWS-FLI1 protein in which the N-terminal portion of EWS is linked to the ETS domain of ERG. Delattre et al. (1994) explored the possibility that these fusion genes may be the defining criterion for the Ewing family of tumors. Samples of RNA from 114 tumors were reverse transcribed and subjected to the polymerase chain reaction with primers designed to amplify the relevant chimeric transcripts. In-frame transcripts were observed in 89 cases. A hybrid transcript was found in 83 of 87 cases (95%) of Ewing sarcoma or peripheral primitive neuroectodermal tumors. Altogether, 78 tumors contained an EWS-FLI1 fusion transcript and 11 tumors had an EWS-ERG fusion transcript. Four different EWS-ERG junctions were found. Kretschmar (1994) discussed the diagnostic and therapeutic potential of these findings. The abnormal fusion transcript might, for example, be targeted and scrambled by specific antisense oligonucleotides, leading to improved eradication of disease in children with these neoplasms based on biology-specific therapy.

Delattre, O.; Zucman, J.; Melot, T.; Garau, X. S.; Zucker, J.-M.; Lenoir, G. M.; Ambros, P. F.; Sheer, D.; Turc-Carel, C.; Triche, T. J.; Aurias, A.; Thomas, G.: The Ewing family of tumors—a subgroup of small-round-cell tumors defined by specific chimeric transcripts. New Eng. J. Med. 331: 294-299, 1994.

Ichikawa, H.; Shimizu, K.; Hayashi, Y.; Ohki, M.: An RNA-binding protein gene, TLS/FUS, is fused to ERG in human myeloid leukemia with t(16;21) chromosomal translocation . Cancer Res. 54: 2865-2868, 1994.

Kretschmar, C. S.: Ewing's sarcoma and the 'peanut' tumors. (Editorial) New Eng. J. Med. 331: 325-327, 1994.

Modi, W. S.; Reddy, S. P.; Rao, V.; Papas, T. S.; O'Brien, S. J.: Chromosomal localization of the ERG cellular proto-oncogene. (Abstract) Cytogenet. Cell Genet. 51: 1046 only, 1989.

Murakami, K.; Mavrothalassitis, G.; Bhat, N. K.; Fisher, R. J.; Papas, T. S.: Human ERG-2 protein is a phosphorylated DNA-binding protein—a distinct member of the ETS family. Oncogene 8: 1559-1566, 1993.

Rao, V. N.; Modi, W. S.; Drabkin, H. D.; Patterson, D.; O'Brien, S. J.; Papas, T. S.; Reddy, E. S.: The human ERG gene maps to chromosome 21, band q22: relationship to the 8;21 translocation of acute myelogenous leukemia. Oncogene 3: 497-500, 1988.

Rao, V. N.; Papas, T. S.; Reddy, E. S. P.: ERG, a human ETS-related gene on chromosome 21: alternative splicing, polyadenylation, and translation. Science 237: 635-639, 1987.

Reddy, E. S. P.; Rao, V. N.; Papas, T. S.: The ERG gene: a human gene related to the ETS oncogene. Proc. Nat. Acad. Sci. 84: 6131-6135, 1987.

Sacchi, N.; Cheng, S. V.; Tanzi, R. E.; Gusella, J. F.; Drabkin, H. A.; Patterson, D.; Haines, J. H.; Papas, T. S.: The ETS genes on chromosome 21 are distal to the breakpoint of the acute myelogenous leukemia translocation (8;21). Genomics 3: 110-116, 1988.

ONCOGENE ETS1: V-ETS AVIAN ERYTHROBLASTOSIS VIRUS E26 ONCOGENE HOMOLOG 1 (*164720)

The oncogene v-ets was originally discovered as a component of a chimeric genome, along with a truncated v-myb (189990) gene, present in the genome of E26, an avian leukosis virus. Members of the ETS gene family have been cloned and sequenced from a variety of species ranging from human to Drosophila; their encoded products are highly conserved, more so than other known nuclear protooncogene products. The several classes of ETS family members could be due to an ancestral gene duplication, followed by genetic recombination(s) or divergence; however, since all of the classes are present in both invertebrates and vertebrates, such an ancestral duplication must predate the emergence of chordates more than 500 million years ago (review by Seth et al., 1992). ETS1 encodes a set of nuclear phosphoproteins ranging in size from 42 to 52 kD. The ETS proteins are transcription factors which activate transcription by interaction with purine-rich sequences present in the promoter/enhancer region of the TCRA gene (186880). The ETS gene family includes, in addition to ETS1 and ETS2 (164740), ERG (165080), ELK1 (311040), and ELK2 (165350). See Fisher et al. (1992).

By in situ hybridization, de Taisne et al. (1984) assigned human oncogene ETS to 11q23-q24. Watson et al. (1985) identified 2 distinct DNA segments homologous to the ETS region of the transforming gene of avian erythroblastosis virus, E26. ETS1, located on chromosome 11, encodes a single mRNA of 6.8 kb; ETS2, on chromosome 21, encodes 3 mRNAs of 4.7, 3.2, and 2.7 kb. Mapping was done by somatic cell hybridization. In 3 patients with acute monocytic leukemia (AMoL) and t(9;11)(p22;q23), Diaz et al. (1986) showed that the breakpoint on 9p split the interferon genes and that the interferon-beta-1 gene was translocated to chromosome 11. The ETS1 gene was translocated from chromosome 11 to the 9p adjacent to interferon genes. They suggested that juxtaposition of interferon and ETS1 genes may be involved in the pathogenesis of AMoL. Diaz et al. (1986) concluded that the fibroblast interferon gene (at least beta-1) is located in 9p22, distal to alpha-interferon. Sacchi et al. (1986) showed that the ETS1 gene is translocated to chromosome 4 in the translocation t(4;11)(q21;q23), which is characteristic of a subtype of leukemia that represents the expansion of a myeloid/lymphoid precursor cell. By in situ hybridization in cells from a person with an 11q23.3-qter deletion, Griffin et al. (1986) refined the assignment of ETS1 to 11q23.3-q24. Rovigatti et al. (1986) found that the ETS1 oncogene was rearranged and amplified 30-fold in a case of acute myelomonocytic leukemia in which a homogeneously stained region (HSR) occurred on 11q23; the oncogene was also rearranged and amplified approximately 10-fold in a case of small lymphocytic cell lymphoma with an inverted insert that also involved band 11q23. In situ hybridization (their figure 3) suggested localization in 11q23.3. By in situ hybridization to pachytene bivalents, Bello et al. (1989) provided sublocalization to 11q23.3. Goyns et al. (1987) found that the ETS1 gene was rearranged in some cases of acute lymph oblastic leukemia. Simmers and Sutherland (1988) refined the assignment of the ETS1 gene to demonstrate that it is separate from fragile site FRA11B which is just distal to the midpoint of band 11q23.3. They specifically assigned ETS1 to 11q24. Gollin et al. (1986) described a patient with Ewing sarcoma (133450) who was heterozygous for FRA11B and had a translocation t(11;22)(q23;q11) in the malignant cells. The localization of ETS1 to 11q24 indicates that the chromosomes of the patient reported by Gollin et al. (1986) did not rearrange at this fragile site to give rise to Ewing sarcoma. Simmers and Sutherland (1988) suggested that this can be added to the 'mounting evidence against individuals with fragile sites being predisposed to developing cancer.' Watson et al. (1988) sequenced human ETS1 cDNA and ETS2 cDNA clones obtained from both human and mouse. The human ETS1 gene is capable of encoding a protein of 441 amino acids which is more than 95% identical to the chicken ETS1 gene product. Human and mouse ETS2 cDNA clones are closely related and contain open reading frames capable of encoding proteins of 469 and 468 residues, respectively. It is of interest that the equivalent of the ETS1 and ETS2 genes are contiguous, i.e., located on the same chromosome, and are coordinately transcribed in birds. Furthermore, whereas the chicken ETS protein, which contains both the ETS-1 and ETS-2 domains, distributes equally between the cytoplasm and nucleus, in the human and other mammals, the ETS-1 protein is cytoplasmic and the ETS2 protein nuclear. This, together with their noncoordinate expression, suggests that ETS-1 and ETS-2 have different biologic functions (Fujiwara et al., 1988). On the basis of multipoint linkage analysis of CEPH families, Concannon et al. (1990) concluded that ETS1 is located approximately 19.2 cM telomeric to THY1. Bhat et al. (1990) found that, on T-cell activation, ETS2 mRNA and proteins are induced , while ETS1 gene expression decreases to very low levels.

Suzuki et al. (1995) found that, when ETS1 is ectopically expressed in 2 highly tumorigenic human colon cancer cell lines devoid of endogenous ETS1 protein, the ETS1 transcription factor reversed the transformed phenotype and tumorigenicity of the cells in a dose-dependent manner. By contrast, expression in 1 cell line of a variant form of ETS1 lacking its native transcriptional activity did not alter the tumorigenic properties of the cells, suggesting that the reduction in tumorigenicity was specific for the wildtype ETS1 gene product. Since these colon cancer cells have multiple genetic alterations, Suzuki et al. (1995) suggested that the system could be a good model for studying suppression of tumorigenicity at a transcriptional level, leading to the design and development of novel drugs for cancer treatment.

Thrombocytopenia or pancytopenia is frequently reported in patients with partial 11q deletion. Gangarossa et al. (1996) described a patient with deletion of the terminal portion of 11q: del(11)(q24.2qter). The classic clinical manifestations including chronic thrombocytopenic purpura were present. They reported for the first time the presence of micromegakaryocytes in the bone marrow aspirate of the patient. Noting that the ETS1 gene maps to the approximate site of the deletion and that nuclear factor related to kappa-B binding protein (164013) maps to 11q24-q25, Gangarossa et al. (1996) raised the possibility that the presence of 1 or both of these DNA-binding proteins that are likely to be involved in hematopoiesis may result, when present in only 1 copy, in thrombocytopenia or pancytopenia.

See also Gold et al. (1987); Watson et al. (1986)

Bello, M. J.; Salagnon, N.; Rey, J. A.; Guichaoua, M. R.; Berge-Lefranc, J. L.; Jordan, B. R.; Luciani, J. M.: Precise in situ localization of NCAM, ETS1, and D11S29 on human meiotic chromosomes. Cytogenet. Cell Genet. 52: 7-10, 1989.

Bhat, N. K.; Thompson, C. B.; Lindsten, T.; June, C. H.; Fujiwara, S.; Koizumi, S.; Fisher, R. J.; Papas, T. S.: Reciprocal expression of human ETS1 and ETS2 genes during T-cell activation: regulatory role for the protooncogene ETS1. Proc. Nat. Acad. Sci. 87: 3723-3727, 1990.

Concannon, P.; Malhotra, U.; Charmley, P.; Reynolds, J.; Lange, K.; Gatti, R. A.: The ataxia-telangiectasia gene (ATA) on chromosome 11 is distinct from the ETS-1 gene. Am. J. Hum. Genet. 46: 789-794, 1990.

de Taisne, C.; Gegonne, A.; Stehelin, D.; Bernheim, A.; Berger, R.: Chromosomal localization of the human proto-oncogene c-ets. Nature 310: 581-583, 1984.

Diaz, M. O.; Le Beau, M. M.; Pitha, P.; Rowley, J. D.: Interferon and c-ets-1 genes in the translocation (9;11)(p22;q23) in human acute monocytic leukemia. Science 231: 265-267, 1986.

Fisher, R. J.; Koizumi, S.; Kondoh, A.; Mariano, J. M.; Mavrothalassitis, G.; Bhat, N. K.; Papas, T. S.: Human ETS1 oncoprotein: purification, isoforms, -SH modification, and DNA sequence-specific binding. J. Biol. Chem. 267: 17957-17965, 1992.

Fujiwara, S.; Fisher, R. J.; Seth, A.; Bhat, N. K.; Showalter, S. D.; Zweig, M.; Papas, T. S.: Characterization and localization of the products of the human homologs of the v-ets oncogene. Oncogene 2: 99-103, 1988.

Gangarossa, S.; Mattina, T.; Romano, V.; Milana, G.; Mollica, F.; Schiliro, G.: Micromegakaryocytes in a patient with partial deletion of the long arm of chromosome 11 (del(11)(q24.2qter)) and chronic thrombocytopenic purpura. Am. J. Med. Genet. 62: 120-123, 1996.

Gold, D. P.; van Dongen, J. J. M.; Morton, C. C.; Bruns, G. A. P.; van den Elsen, P.; Geurts van Kessel, A. H. M.; Terhorst, C.: The gene encoding the epsilon subunit of the T3/T-cell receptor complex maps to chromosome 11 in humans and to chromosome 9 in mice. Proc. Nat. Acad. Sci. 84: 1664-1668, 1987.

Gollin, S. M.; Perrot, L. J.; Gray, B. A.; Kletzel, M.: Spontaneous expression of fra(11)(q23) in a patient with Ewing's sarcoma and t(11;22)(q23;q11). Cancer Genet. Cytogenet. 20: 331-339, 1986.

Goyns, M. H.; Hann, I. M.; Stewart, J.; Gegonne, A.; Birnie, G. D.: The c-ets-1 proto-oncogene is rearranged in some cases of acute lymphoblastic leukaemia. Brit. J. Cancer 56: 611-613, 1987.

Griffin, C. A.; McKeon, C.; Israel, M. A.; Gegonne, A.; Ghysdael, J.; Stehelin, D.; Douglass, E. C.; Green, A. A.; Emanuel, B. S.: Comparison of constitutional and tumor-associated 11;22 translocations: nonidentical breakpoints on chromosomes 11 and 22. Proc. Nat. Acad. Sci. 83: 6122-6126, 1986.

Rovigatti, U.; Watson, D. K.; Yunis, J. J.: Amplification and rearrangement of Hu-ets-1 in leukemia and lymphoma with involvement of 11q23. Science 232: 398-400, 1986.

Sacchi, N.; Watson, D. K.; Geurts van Kessel, A. H. M.; Hagemeijer, A.; Kersey, J.; Drabkin, H. D.; Patterson, D.; Papas, T. S.: Hu-ets-1 and Hu-ets-2 genes are transposed in acute leukemias with (4;11) and (8;21) translocations. Science 231: 379-382, 1986.

Seth, A.; Ascione, R.; Fisher, R. J.; Mavrothalassitis, G. J.; Bhat, N. K.; Papas, T. S.: The ets gene family. Cell Growth Differ. 3: 327-334, 1992.

Simmers, R. N.; Sutherland, G. R.: Further localization of ETS1 indicates that the chromosomal rearrangement in Ewing sarcoma does not occur at fra(11)(q23). Hum. Genet. 78: 144-147, 1988.

Suzuki, H.; Romano-Spica, V.; Papas, T. S.; Bhat, N. K.: ETS1 suppresses tumorigenicity of human colon cancer cells. Proc. Nat. Acad. Sci. 92: 4442-4446, 1995.

Watson, D. K.; McWilliams, M. J.; Lapis, P.; Lautenberger, J. A.; Schweinfest, C. W.; Papas, T. S.: Mammalian ets-1 and ets-2 genes encode highly conserved proteins. Proc. Nat. Acad. Sci. 85: 7862-7866, 1988.

Watson, D. K.; McWilliams-Smith, M. J.; Kozak, C.; Reeves, R.; Gearhart, J.; Nunn, M. F.; Nash, W.; Fowle, J. R., III; Duesberg, P.; Papas, T. S.; O'Brien, S. J.: Conserved chromosomal positions of dual domains of the ets protooncogene in cats, mice, and humans. Proc. Nat. Acad. Sci. 83: 1792-1796, 1986.

Watson, D. K.; McWilliams-Smith, M. J.; Nunn, M. F.; Duesberg, P. H.; O'Brien, S. J.; Papas, T. S.: The ets sequence from the transforming gene of avian erythroblastosis virus, E26, has unique domains on human chromosomes 11 and 21: both loci are transcriptionally active. Proc. Nat. Acad. Sci. 82: 7294-7298, 1985.

ONCOGENE ETS2: V-ETS AVIAN ERYTHROBLASTOSIS VIRUS E26 ONCOGENE HOMOLOG 2 (*164740)

See 164720. By in situ hybridization, Watson et al. (1985) mapped the ETS2 locus to 21q22.1-q22.3. Sacchi et al. (1986) found that the ETS2 gene was translocated to chromosome 8 in the translocation t(8;21)(q22;q22), which is commonly found in patients with acute myeloid leukemia with morphology M2 (AML-M2). In a case of t(8;21)(q22;q22), Le Beau et al. (1986) found that the ETS2 gene was translocated to chromosome 8. The MOS oncogene (190060) was retained at the breakpoint on chromosome 8; thus, one or both of these genes may play a role in the pathogenesis of acute myelogenous leukemia. Using genetic linkage analysis, Sacchi et al. (1988) demonstrated the location of the ETS2 gene relative to other loci and established that the breakpoint in acute myeloid leukemia is about 17 cM from ETS2. Thus, the breakpoint does not affect the ETS2 gene structure. The actual DNA sequence involved in the t(8;21) may reside in a 3-cM genetic region between 2 markers used in these studies. By family linkage studies using RFLPs, Sacchi et al. (1988) demonstrated that ERG is situated just proximal to ETS2. Mavrothalassitis et al. (1990) demonstrated that the ETS2 gene has no TATA box or CAAT box in its promoter. It has an alternative structure that serves a comparable function.

Expression of ETS2, a proto-oncogene and transcription factor, occurs in a variety of cell types. During murine development it is highly expressed in newly formed cartilage, including skull precursor cells and vertebral primordia. Sumarsono et al. (1996) generated transgenic mice to investigate the consequences of overexpression of Ets2. They found that mice with less than 2-fold Ets2 overexpression in particular organs developed neurocranial, visceral cranial, and cervical skeletal abnormalities. These abnormalities had similarities with the skeletal anomalies found in trisomy-16 mice and in humans with Down syndrome, in which the gene dosage of ETS2 is increased. The results were interpreted as indicating that ETS2 has a role in skeletal development and that overexpression is involved in the genesis of some skeletal abnormalities that occur in Down syndrome.

See also Watson et al. (1986)

Le Beau, M. M.; Rowley, J. D.; Sacchi, N.; Watson, D. K.; Papas, T. S.; Diaz, M. O.: Hu-ets-2 is translocated to chromosome 8 in the t(8;21) in acute myelogenous leukemia. Cancer Genet. Cytogenet. 23: 269-274, 1986.

Mavrothalassitis, G. J.; Watson, D. K.; Papas, T. S.: Molecular and functional characterization of the promoter of ETS2, the human c-ets-2 gene. Proc. Nat. Acad. Sci. 87: 1047-1051, 1990.

Sacchi, N.; Cheng, S. V.; Tanzi, R. E.; Gusella, J. F.; Drabkin, H. A.; Patterson, D.; Haines, J. H.; Papas, T. S.: The ETS genes on chromosome 21 are distal to the breakpoint of the acute myelogenous leukemia translocation (8;21). Genomics 3: 110-116, 1988.

Sacchi, N.; Watson, D. K.; Geurts van Kessel, A. H. M.; Hagemeijer, A.; Kersey, J.; Drabkin, H. D.; Patterson, D.; Papas, T. S.: Hu-ets-1 and Hu-ets-2 genes are transposed in acute leukemias with (4;11) and (8;21) translocations. Science 231: 379-382, 1986.

Sumarsono, S. H.; Wilson, T. J.; Tymms, M. J.; Venter, D. J.; Corrick, C. M.; Kola, R.; Lahoud, M. H.; Papas, T. S.; Seth, A.; Kola, I.: Down's syndrome-like skeletal abnormalities in Ets2 transgenic mice. Nature 379: 534-540, 1996.

Watson, D. K.; McWilliams-Smith, M. J.; Kozak, C.; Reeves, R.; Gearhart, J.; Nunn, M. F.; Nash, W.; Fowle, J. R., III; Duesberg, P.; Papas, T. S.; O'Brien, S. J.: Conserved chromosomal positions of dual domains of the ets protooncogene in cats, mice, and humans. Proc. Nat. Acad. Sci. 83: 1792-1796, 1986.

Watson, D. K.; McWilliams-Smith, M. J.; Nunn, M. F.; Duesberg, P. H.; O'Brien, S. J.; Papas, T. S.: The ets sequence from the transforming gene of avian erythroblastosis virus, E26, has unique domains on human chromosomes 11 and 21: both loci are transcriptionally active. Proc. Nat. Acad. Sci. 82: 7294-7298, 1985.

ONCOGENE FES: V-FES FELINE SARCOMA VIRAL/V-FPS FUJINAMI AVIAN SARCOMA VIRAL ONCOGENE (*190030)

Wong-Staal et al. (1981) identified human DNA sequences homologous to cloned DNA fragments containing the oncogenic nucleic acid sequences of a type C mammalian retrovirus, the Snyder-Theilen strain of feline sarcoma virus (FeSV). Non-onc intervening sequences were present in the human counterpart. The onc gene for the feline sarcoma virus was mapped to chromosome 15 (Dalla-Favera et al., 1982). Heisterkamp et al. (1982) also assigned the FES gene to the long arm of chromosome 15. This may have relevance to leukemogenesis in acute promyelocytic leukemia which shows a translocation between chromosomes 15 and 17 (15q+, 17q-). Both the FES and the ABL onc genes are involved in tyrosine phosphorylation. They code for proteins with tyrosine-specific protein kinase activity. A useful general discussion of cellular transforming genes was provided by Cooper (1982). Fourteen transforming genes of acute transforming retroviruses (src, fes, mos, ras, myb, abl, and others) were tabulated. Transforming genes activated in human neoplastic cells had been identified in the following human neoplasms: bladder carcinoma (3 tumors), lung carcinoma (4 tumors), mammary carcinoma (1), colon carcinoma (2), promyelocytic leukemia (1), neuroblastoma (1), pre-B lymphocyte neoplasm (4), B cell lymphoma (6), plasmacytoma/myeloma (2), T cell lymphoma (1), mature T helper cell neoplasm (1), and sarcoma (1). Cooper (1982) concluded 'that oncogenesis can involve dominant genetic alterations resulting in activation of cellular transforming genes.' The same genes are activated in independent neoplasms of the same cell type. The finding that specific transforming genes are activated in neoplasms corresponding to discrete stages of lymphocyte differentiation suggests that the transforming genes activated in neoplasms are closely related to the state of normal differentiation exhibited by the neoplastic cells.

The t(15q+;17q-) of acute promyelocytic leukemia came to light at the First International Workshop on Chromosomes in Leukemia in Helsinki in 1977 and was confirmed by many workers (Rowley et al., 1977; Kondo and Sasaki, 1979). Studying 4 cases of APL, Hagemeijer et al. (1982) defined the breakpoints of the 15;17 translocation as 15q2200 and 17q12. In the 15;17 translocation of acute promyelocytic leukemia, the thymidine kinase locus is on the part of 17q translocated to chromosome 15 (Durnam et al., 1984). Studying somatic cell hybrids constructed between a thymidine kinase-deficient mouse cell line and blood leukocytes from a patient with acute promyelocytic leukemia showing the characteristic 15q+;17q-translocation, Sheer et al. (1983) showed that the FES gene was not present in a hybrid cell showing the 15q+ translocation and little other human chromosomal material. Therefore, FES had been translocated to the 17q- chromosome. B2M (109700) is on 15q+, probably near the breakpoint; this could be relevant to the malignant process. MPI (154550) and PKM2 (179050), like FES, are distal to the breakpoint. The breakpoint on 17 is proximal to the TK and GALK loci; these genes are translocated to 15q+. By in situ hybridization to germ-line chromosomes, Jhanwar et al. (1984) assigned FES to 15q26.1, a site distal to the breakpoint in the translocation commonly seen in acute promyelocytic leukemia, t(15;17)(q24;q22). In all of 27 patients with acute promyelocytic leukemia, including 4 with the microgranular variant, Larson et al. (1984) found a specific translocation: t(15;17)(q22;q21.1). The FES oncogene is located on 15(q25-q26) and the ERBA oncogene on chromosome 17(p11-q21). The relation of the translocation to either of these 2 oncogenes in the causation of neoplasia is problematical (unlike the situation with the ABL oncogene which is within 15 kb of the breakpoint on chromosome 9 that creates the Philadelphia chromosome). By fluorescence in situ hybridization, Mathew et al. (1993) refined the localization of FES to 15q26.1.

See also Hampe et al. (1982); Roebroek et al. (1985); Rowley et al. (1977); Sodroski et al. (1984)

Cooper, G. M.: Cellular transforming genes. Science 217: 801-806, 1982.

Dalla-Favera, R.; Franchini, G.; Martinotti, S.; Wong-Staal, F.; Gallo, R. C.; Croce, C. M.: Chromosomal assignment of the human homologues of feline sarcoma virus and avian myeloblastosis virus onc genes. Proc. Nat. Acad. Sci. 79: 4714-4717, 1982.

Durnam, D. M.; Myerson, D.; McDougall, J. K.; Gelinas, R. E.: Analysis of the t(15;17) specifically associated with acute promyelocytic leukemia.(Abstract) Cytogenet. Cell Genet. 37: 458 only, 1984.

Hagemeijer, A.; Lowenberg, B.; Abels, J.: Analysis of the breakpoints in translocation (15;17) observed in 4 patients with acute promyelocytic leukemia. Hum. Genet. 61: 223-227, 1982.

Hampe, A.; Laprevotte, I.; Galibert, F.; Fedele, L. A.; Sherr, C. J.: Nucleotide sequences of feline retroviral oncogenes (v-fes) provide evidence for a family of tyrosine-specific protein kinase genes. Cell 30: 775-785, 1982.

Heisterkamp, N.; Groffen, J.; Stephenson, J. R.; Spurr, N. K.; Goodfellow, P. N.; Solomon, E.; Carritt, B.; Bodmer, W. F.: Chromosomal localization of human cellular homologues of two viral oncogenes. Nature 299: 747-749, 1982.

Jhanwar, S. C.; Neel, B. G.; Hayward, W. S.; Chaganti, R. S. K.: Localization of the cellular oncogenes ABL, SIS, and FES on human germ-line chromosomes. Cytogenet. Cell Genet. 38: 73-75, 1984.

Kondo, K.; Sasaki, M.: Cytogenetic studies of four cases of acute promyelocytic leukemia (APL). Cancer Genet. Cytogenet. 1: 131-138, 1979.

Larson, R. A.; Kondo, K.; Vardiman, J. W.; Butler, A. E.; Golomb, H. M.; Rowley, J. D.: Evidence for a 15;17 translocation in every patient with acute promyelocytic leukemia. Am. J. Med. 76: 827-841, 1984.

Mathew, S.; Murty, V. V. V. S.; German, J.; Chaganti, R. S. K.: Confirmation of 15q26.1 as the site of the FES protooncogene by fluorescence in situ hybridization. Cytogenet. Cell Genet. 63: 33-34, 1993.

Roebroek, A. J. M.; Schalken, J. A.; Verbeek, J. S.; Van den Ouweland, A. M. W.; Onnekink, C.; Bloemers, H. P. J.; Van de Ven, W. J. M.: The structure of the human c-fes/fps proto-oncogene. EMBO J. 4: 2897-2903, 1985.

Rowley, J. D.; Golomb, H. M.; Dougherty, C.: 15/17 translocation, a consistent chromosomal change in acute promyelocytic leukaemia. (Letter) Lancet I: 549-550, 1977.

Rowley, J. D.; Golomb, H. M.; Vardiman, J.; Fukuhara, S.; Dougherty, C.; Potter, D.: Further evidence for a non-random chromosomal abnormality in acute promyelocytic leukemia. Int. J. Cancer 20: 869-872, 1977.

Sheer, D.; Hiorns, L. R.; Stanley, K. F.; Goodfellow, P. N.; Swallow, D. M.; Povey, S.; Heisterkamp, N.; Groffen, J.; Stephenson, J. R.; Solomon, E.: Genetic analysis of the 15;17 chromosome translocation associated with acute promyelocytic leukemia. Proc. Nat. Acad. Sci. 80: 5007-5011, 1983.

Sodroski, J. G.; Goh, W. C.; Haseltine, W. A.: Transforming potential of a human protooncogene (c-fps/fes) locus. Proc. Nat. Acad. Sci. 81: 3039-3043, 1984.

Wong-Staal, F.; Dalla-Favera, R.; Franchini, G.; Gelmann, E. P.; Gallo, R. C.: Three distinct genes in human DNA related to the transforming genes of mammalian sarcoma retroviruses. Science 213: 226-228, 1981.

ONCOGENE FGR: GARDNER-RASHEED FELINE SARCOMA VIRAL (V-FGR) ONCOGENE (*164940)

The cell-derived domain of Gardner-Rasheed feline sarcoma virus consists of a gamma-actin sequence and a tyrosine-specific protein kinase sequence called v-fgr. By means of a v-fgr probe, Tronick et al. (1985) isolated a human homolog. Analysis showed that the human DNA is distinct from all other retroviral oncogenes. They localized the FGR oncogene to 1p36.2-p36.1 by in situ hybridization. By Southern analysis of somatic cell hybrids, Nishizawa et al. (1986) confirmed the assignment of the FGR locus to chromosome 1. FGR, which is the presently used designation, is the same as the oncogene earlier called SRC2. The latter was mapped to 1p36-p34 by Le Beau et al. (1984) using in situ hybridization and by Lebo et al. (1984) using dual-beam chromosome sorting and spot blot DNA analysis. Dracopoli et al. (1988) assigned SRC2 to 1p by analysis of DNA from a panel of somatic cell hybrids and by linkage analysis in the CEPH families. The latter studies showed that the SRC2 locus is 3.1 cM from the Rh blood group locus. They favored the gene order: 1pter—PND—ALPL—FUCA1—SRC2(FGR)—Rh—D1S57—MYCL.

See also Cheah et al. (1986); Le Beau et al. (1984); Parker et al. (1985)

Cheah, M. S. C.; Ley, T. J.; Tronick, S. R.; Robbins, K. C.: Fgr proto-oncogene mRNA induced in B lymphocytes by Epstein-Barr virus infection. Nature 319: 238-240, 1986.

Dracopoli, N. C.; Stanger, B. Z.; Lager, M.; Housman, D. E.: Localization of the FGR protooncogene on the genetic linkage map of human chromosome 1p. Genomics 3: 124-128, 1988.

Le Beau, M. M.; Westbrook, C. A.; Diaz, M. O.; Rowley, J. D.: In situ hybridization studies of C-SRC: evidence for two distinct loci on human chromosomes no. 1 and no. 20. (Abstract) Am. J. Hum. Genet. 36: 32S, 1984.

Le Beau, M. M.; Westbrook, C. A.; Diaz, M. O.; Rowley, J. D.: Evidence for two distinct c-src loci on human chromosomes 1 and 20. Nature 312: 70-71, 1984.

Lebo, R. V.; Cheung, M.-C.; Bruce, B. D.: Rapid gene mapping by dual laser chromosome sorting and spot blot DNA analysis. (Abstract) Am. J. Hum. Genet. 36: 101S, 1984.

Nishizawa, M.; Semba, K.; Yoshida, M. C.; Yamamoto, T.; Sasaki, M.; Toyoshima, K.: Structure, expression, and chromosomal location of the human c-fgr gene. Molec. Cell. Biol. 6: 511-517, 1986.

Parker, R. C.; Mardon, G.; Lebo, R. V.; Varmus, H. E.; Bishop, J. M.: Isolation of duplicated human c-src genes located on chromosomes 1 and 20. Molec. Cell. Biol. 5: 831-838, 1985.

Tronick, S. R.; Popescu, N. C.; Cheah, M. S. C.; Swan, D. C.; Amsbaugh, S. C.; Lengel, C. R.; DiPaolo, J. A.; Robbins, K. C.: Isolation and chromosomal localization of the human fgr protooncogene, a distinct member of the tyrosine kinase gene family. Proc. Nat. Acad. Sci. 82: 6595-6599, 1985.

ONCOGENE FMS: COLONY-STIMULATING FACTOR 1 RECEPTOR (*164770)

The FMS oncogene was assigned to chromosome 5 by study of mouse-man somatic cell hybrids. The location was narrowed to 5q34 by the study of hamster-human cell hybrids with well-defined deletions of 5q (Groffen et al., 1984). The order on the long arm was found to be centromere—leuS—HEXB—EMTB—CFMS—CHR. By in situ hybridization, Le Beau et al. (1986) assigned FMS to 5q33 and GMCSF (138960) to 5q23-q31. Both genes were deleted in the 5q- chromosome from bone marrow cells of 2 patients with refractory anemia and del(5)(q15q33.3). From study of other cases they concluded that FMS is located in band 5q33.2 or 5q33.3 rather than 5q34-q35 as reported earlier. The FMS oncogene is the same as the receptor for colony-stimulating factor-1, otherwise known as macrophage colony-stimulating factor (120420).

Receptors for the human platelet-derived growth factor-beta (PDGFRB; 173410) and macrophage colony-stimulating factor 1 belong to the same subfamily of receptor tyrosine kinases (Yarden and Ullrich, 1988). The 2 genes are located on 5q and are linked physically in a head-to-tail array, with less than 500 bp between the polyadenylation signal of the PDGFRB gene and the transcription start point of the CSF1R gene (Roberts et al., 1988). Mouse homologs of these genes are also linked in a similar way. How et al. (1996) determined the sequences of the homologs of the human PDGFRB and CSF1R genes in the pufferfish (Fugu rubripes) and demonstrated that in Fugu also the 2 genes are closely linked and are in a head-to-tail array with 2.2 kb of intergenic sequence. Amino acid sequences of the Fugu and human PGFRB and CSF1R genes show an overall homology of 45% and 39%, respectively.

Morgan et al. (1986) pointed out that a break at 5q35 has been found in several cases of malignant histiocytosis, a neoplastic process characterized by fever, progressive wasting, lymphadenopathy, hepatosplenomegaly, and the proliferation of atypical histiocytes at all stages of maturation with frequent phagocytic activity. They suggested that at the molecular level the change in band 5q35 may affect the FMS oncogene responsible for the receptor for mononuclear-phagocyte growth factor and thereby have a role in causing malignant histiocytosis. Benz-Lemoine et al. (1988) also suggested that a breakpoint in 5q35 may be critical to the development of malignant histiocytosis. Roberts et al. (1988) demonstrated that a 5-prime untranslated exon of CSF1R is separated by a 26-kb intron from the 32-kb receptor coding sequences. Furthermore, the 3-prime end of the PDGFR gene is located less than 0.5 kb upstream from this exon. The as-yet-unidentified CSF1R promoter/enhancer sequences may be confined to the nucleotides separating the 2 genes or could potentially lie within the PDGFR gene itself. Similarities in chromosomal localization, organization, and encoded amino acid sequences suggest that the CSF1R and PDGFR genes arose through duplication. The human homolog of the murine Fim-2 proviral integration region corresponds to the 5-prime end of the FMS gene. Roberts et al. (1988) demonstrated that the CSF1R gene and the PDGFR1 gene are physically associated in a head-to-tail array. By pulsed field gel electrophoresis, Eccles (1991) demonstrated that the same is true in the mouse where a 425-kb fragment hybridizes with the 3-prime end of PDGFR1 and the 5-prime end of CSF1R.

Hampe et al. (1989) demonstrated that the FMS gene consists of 21 small exons interrupted by introns ranging in size from 6.3 kb to less than 0.1 kb.

Codons 969 and 301 of the FMS gene are potentially involved in promoting the transforming activity of the gene product (Ridge et al., 1990; see 164770.0001). The tyrosine residue at codon 969 was shown to be involved in a negative regulatory activity, which is disrupted by amino acid substitutions. Mutations at codon 301 lead to neoplastic transformation by ligand independence and constitutive tyrosine kinase activity of the receptor. Boultwood et al. (1991) found loss of both CSF1R alleles in 10 patients with myelodysplasia and a 5q deletion; 6 were hemizygous and 4 were homozygous for CSF1R loss. Boultwood et al. (1991) suggested that loss of this hemopoietic growth factor receptor gene may be important in the pathogenesis of myeloid leukemia also.

During sequence analysis of the first intron of the human FMS gene, Sapi et al. (1994) identified an open reading frame encoding the ribosomal protein L7 (RPL7A; 185640). The sequence was identical to the full-length RPL7 cDNA sequence but lacked any recognizable introns, had a 30-bp poly(A) tail, and was bracketed by 2 perfect direct repeats of 14 bp. Sapi et al. (1994) demonstrated that despite the fact that the 5-prime flanking region of the RPL7 sequence contained a potential TATA box upstream of an intact open reading frame, the pseudogene (RPL7P) was not actively transcribed.

See also De Qi Xu et al. (1985); Gisselbrecht et al. (1987); Hampe et al. (1984); Verbeek et al. (1985); Verbeek et al. (1985)

Benz-Lemoine, E.; Brizard, A.; Huret, J.-L.; Babin, P.; Guilhot, F.; Couet, D.; Tanzer, J.: Malignant histiocytosis: a specific t(2;5)(p23;q35) translocation? Review of the literature. Blood 72: 1045-1047, 1988.

Boultwood, J.; Rack, K.; Kelly, S.; Madden, J.; Sakaguchi, A. Y.; Wang, L.-M.; Oscier, D. G.; Buckle, V. J.; Wainscoat, J. S.: Loss of both CSF1R (FMS) alleles in patients with myelodysplasia and a chromosome 5 deletion. Proc. Nat. Acad. Sci. 88: 6176-6180, 1991.

De Qi Xu; Guilhot, S.; Galibert, F.: Restriction fragment length polymorphism of the human c-fms gene. Proc. Nat. Acad. Sci. 82: 2862-2865, 1985.

Eccles, M. R.: Genes encoding the platelet-derived growth factor (PDGF) receptor and colony-stimulating factor 1 (CSF-1) receptor are physically associated in mice as in humans. Gene 108: 285-288, 1991.

Gisselbrecht, S.; Fichelson, S.; Sola, B.; Bordereaux, D.; Hampe, A.; Andre, C.; Galibert, F.; Tambourin, P. E.: Frequent c-fms activation by proviral insertion in mouse myeloblastic leukaemias. Nature 329: 259-261, 1987.

Groffen, J.; Heisterkamp, N.; Spurr, N. K.; Dana, S. L.; Wasmuth, J. J.; Stephenson, J. R.: Regional assignment of the human c-fms oncogene to band q34 of chromosome 5. (Abstract) Cytogenet. Cell Genet. 37: 484 only, 1984.

Hampe, A.; Gobet, M.; Sherr, C. J.; Galibert, F.: Nucleotide sequence of the feline retroviral oncogene v-fms shows unexpected homology with oncogenes encoding tyrosine-specific protein kinases. Proc. Nat. Acad. Sci. 81: 85-89, 1984.

Hampe, A.; Shamoon, B.-M.; Gobet, M.; Sherr, C. J.; Galibert, F.: Nucleotide sequence and structural organization of the human FMS proto-oncogene. Oncogene Res. 4: 9-17, 1989.

How, G.-F.; Venkatesh, B.; Brenner, S.: Conserved linkage between the puffer fish (Fugu rubripes) and human genes for platelet-derived growth factor receptor and macrophage colony-stimulating factor receptor. Genome Res. 6: 1185-1191, 1996.

Le Beau, M. M.; Westbrook, C. A.; Diaz, M. O.; Larson, R. A.; Rowley, J. D.; Gasson, J. C.; Golde, D. W.; Sherr, C. J.: Evidence for the involvement of GM-

CSF and FMS in the deletion (5q) in myeloid disorders. Science 231: 984-987, 1986.
Morgan, R.; Hecht, B. K.; Sandberg, A. A.; Hecht, F.; Smith, S. D.: Chromosome 5q35 breakpoint in malignant histiocytosis. (Letter) New Eng. J. Med. 314: 1322 only, 1986.
Ridge, S. A.; Worwood, M.; Oscier, D.; Jacobs, A.; Padua, R. A.: FMS mutations in myelodysplastic, leukemic, and normal subjects. Proc. Nat. Acad. Sci. 87: 1377-1380, 1990.
Roberts, W. M.; Look, A. T.; Roussel, M. F.; Sherr, C. J.: Tandem linkage of human CSF-1 receptor (c-fms) and PDGF receptor genes. Cell 55: 655-661, 1988.
Sapi, E.; Flick, M. B.; Kacinski, B. M.: The first intron of human c-fms proto-oncogene contains a processed pseudogene (RPL7P) for ribosomal protein L7. Genomics 22: 641-645, 1994.
Verbeek, J. S.; Roebroek, A. J. M.; van den Ouweland, A. M. W.; Bloemers, H. P. J.; Van de Ven, W. J. M.: Human c-fms proto-oncogene: comparative analysis with an abnormal allele. Molec. Cell. Biol. 5: 422-426, 1985.
Verbeek, J. S.; van Heerikhuizen, H.; de Pauw, B. E.; Haanen, C.; Bloemers, H. P. J.; Van de Ven, W. J. M.: A hereditary abnormal c-fms proto-oncogene in a patient with acute lymphocytic leukaemia and congenital hypothyroidism. Brit. J. Haemat. 61: 135-138, 1985.
Yarden, Y.; Ullrich, A.: Growth factor receptor tyrosine kinases. Ann. Rev. Biochem. 57: 443-478, 1988.

ONCOGENE GLI: GLIOMA-ASSOCIATED ONCOGENE HOMOLOG (*165220)

Kinzler et al. (1987) identified a gene, termed glioma-associated oncogene (GLI) by them, that is amplified more than 50-fold in a malignant glioma. The gene was expressed at high levels in the original tumor and its derived cell line. Kinzler et al. (1988) extended this work by cloning the GLI complementary DNA. Analysis of the nucleotide sequence demonstrated that the gene contains 5 repeats of the DNA-binding consensus sequence (zinc finger). The zinc fingers contain sequence elements that place the GLI gene product among the Kruppel (Kr) family of zinc finger proteins (see 165230). (The German word 'Kruppel' means 'cripple' or 'dwarf.') The Drosophila gene is a member of the gap class of segmentation genes; thoracic and anterior abdominal segments fail to form in Kr mutant embryos. Kr encodes a chromatin-associated phosphoprotein that contains 5 zinc fingers with a consensus sequence connecting the histidine of 1 finger to the cysteine of the next. Conservation of the contiguous stretch of nucleotides encoding this H-C link allowed cloning of Kruppel-related genes from D. melanogaster, mouse, and frog, by hybridization with Kr cDNA at low stringency. Each of these Kr family members were shown to be expressed in embryonic cells, suggesting a role for them in normal development. Northern analysis revealed that GLI is expressed in embryonal carcinoma cells but not in most adult tissue. A fragile site of the folic acid type has been described at 12q13. In addition, translocations involving this region of chromosome 12 have been described in myxoid liposarcoma (151900) and salivary gland tumors. Southern blot analysis with probes for INT1 (164820) and IGF1 (147440), both of which are located on chromosome 12, showed that these genes are not amplified in the malignant glioma.

Kinzler et al. (1987) localized GLI to 12q13-q14.3 by Southern blot analysis of hybrid cell DNA and by in situ hybridization. By in situ hybridization, Arheden et al. (1989) localized the GLI1 gene to 12q13.3-q14.1. By in situ hybridization, Dal Cin et al. (1989) narrowed the assignment of GLI1 to the distal part of band 12q13 (12q13.2 or 12q13.3). The corresponding gene is located on mouse chromosome 10 (Justice et al., 1990).

Chung and Seizinger (1992) reviewed the molecular genetics of neurologic tumors, including the role of the GLI oncogene.

The secreted glycoprotein Sonic hedgehog (SHH; 600725) is thought to act as an endodermal signal that controls hindgut patterning and lung growth. In mice, 3 zinc finger transcription factors, Gli1, Gli2, and Gli3, have been implicated in the transduction of Shh signal. Motoyama et al. (1998) found that mutant mice lacking Gli2 function exhibit foregut defects, including stenosis of the esophagus and trachea, as well as hypoplasia and lobulation defects of the lung. A reduction of 50% in the gene dosage of Gli3 (165240) in a Gli2-/- background resulted in esophageal atresia with tracheoesophageal fistula and a severe disruption of lung development. Mutant mice lacking both Gli2 and Gli3 function did not form esophagus, trachea, or lung. These results indicated that Gli2 and Gli3 possess specific and overlapping functions in Shh signaling during foregut development, and suggested that mutations in GLI genes may be involved in human foregut malformations.

See also Bigner et al. (1988); Ruppert et al. (1988)

Arheden, K.; Ronne, M.; Mandahl, N.; Heim, S.; Kinzler, K. W.; Vogelstein, B.; Mitelman, F.: In situ hybridization localizes the human putative oncogene GLI to chromosome subbands 12q13.3-14.1. Hum. Genet. 82: 1-2, 1989.
Bigner, S. H.; Mark, J.; Burger, P. C.; Mahaley, M. S., Jr.; Bullard, D. E.; Muhlbaier, L. H.; Bigner, D. D.: Specific chromosomal abnormalities in malignant human gliomas. Cancer Res. 88: 405-411, 1988.
Chung, R. Y.; Seizinger, B. R.: Molecular genetics of neurological tumours. J. Med. Genet. 29: 361-367, 1992.
Dal Cin, P.; Turc-Carel, C.; Sandberg, A. A.; Van den Berghe, H.: More precise localization of GLI gene by in situ hybridization. (Abstract) Cytogenet. Cell Genet. 51: 982-983, 1989.
Justice, M. J.; Siracusa, L. D.; Gilbert, D. J.; Heisterkamp, N.; Groffen, J.; Chada, K.; Silan, C. M.; Copeland, N. G.; Jenkins, N. A.: A genetic linkage map of mouse chromosome 10: localization of eighteen molecular markers using a single interspecific backcross. Genetics 125: 855-866, 1990.
Kinzler, K. W.; Bigner, S. H.; Bigner, D. D.; Trent, J. M.; Law, M. L.; O'Brien, S. J.; Wong, A. J.; Vogelstein, B.: Identification of an amplified, highly expressed gene in a human glioma. Science 236: 70-73, 1987.
Kinzler, K. W.; Ruppert, J. M.; Bigner, S. H.; Vogelstein, B.: The GLI gene is a member of the Kruppel family of zinc finger proteins. Nature 332: 371-374, 1988.
Motoyama, J.; Liu, J.; Mo, R.; Ding, Q.; Post, M.; Hui, C.: Essential function of Gli2 and Gli3 in the formation of lung, trachea and oesophagus. Nature Genet. 20: 54-57, 1998.
Ruppert, J. M.; Kinzler, K. W.; Wong, A. J.; Bigner, S. H.; Kao, F.-T.; Law, M. L.; Seuanez, H. N.; O'Brien, S. J.; Vogelstein, B.: The GLI-Kruppel family of human genes. Molec. Cell. Biol. 8: 3104-3113, 1988.

ONCOGENE HST1: FIBROBLAST GROWTH FACTOR 4 (*164980)

Sakamoto et al. (1986) tested the capacity for malignant transformation of DNA from 21 stomach cancers, 16 stomach cancers metastatic to lymph nodes, and 21 specimens of apparently noncancerous stomach mucosa from a total of 26 patients with stomach cancer. The DNA was transferred to NIH 3T3 cells by the calcium precipitation technique. Transforming ability was shown by 3 samples: a primary cancer, a metastatic cancer, and a presumably normal gastric mucosa. The transforming gene from the primary cancer was cloned. It bore no homology with previously reported transforming sequences. Taira et al. (1987) isolated an HST cDNA clone that had an efficient transforming activity in a focus-forming assay when it was inserted into an expression vector. Characterization of this clone allowed Taira et al. (1987) to predict that a 206-amino acid protein product was responsible for this transforming activity. In an addendum, Taira et al. (1987) indicated that 42.3% homology existed between the amino acid residues of 1 ORF (open reading frame) of HST and part of bovine basic fibroblast growth factor. They suggested that further studies would elucidate the role of the HST gene in the development of stomach cancer which, they stated, has the highest incidence of all known cancers. By in situ hybridization, Adelaide et al. (1988) mapped HST to 11q13. This is also the location of the INT2 gene (164950). Furthermore, Adelaide et al. (1988) found the 2 genes to be coamplified in a human melanoma. Both HST and INT2 are in the fibroblast growth factor family of oncogenes. Yoshida et al. (1988) reported expression of HST1 in a human teratoma cell line and in 5 out of 9 surgically resected human testicular germ cell tumors including seminomas and embryonal carcinomas. HST1, for which the designation HSTF1 was proposed for human gene nomenclature, is a heparin-binding growth factor with significant homology to human fibroblast growth factors and the mouse Int-2 protein. Huebner et al. (1988) isolated the same oncogene by transfection of Kaposi sarcoma DNA and demonstrated its significant homology to both basic and acidic FGFs (see 134920 and 131220). Because of its origin from Kaposi sarcoma and because of its similarity to FGFs, they designated the oncogene 'K-FGF.' Huebner et al. (1988) mapped this oncogene to 11q11-q23 by hybridization studies using DNA from rodent-human somatic cell hybrids and then localized it more precisely to 11q13 by in situ hybridization. The 11q13 region is also the site of a potential, although uncharacterized gene, BCL1 (151400), involved in the 11;14 translocation characteristic of some B-cell tumors. The oncogene SEA (165110) has also been mapped to 11q13. Thus, HST, INT2, and SEA are all candidate BCL1 genes. By pulsed-field gel electrophoresis and by analysis of overlapping cosmid clones, Wada et al. (1988) concluded that HST1 is located about 35 kb downstream of INT2 in the same transcriptional orientation. Huebner et al. (1988) found that sequences adjacent to the 3-prime end of the K-FGF coding sequence in transfectants probably derived from the part of the genome lying between 12p12 and 12q13. Whether the juxtaposition of the chromosome 11-linked K-FGF gene to the chromosome 12-linked sequences within the original transfectants was pure accident remained to be determined. (It had earlier been demonstrated that sequences from the FMS oncogene on chromosome 5 were included in the transfectant.)

For a review of the role of this gene in limb development, see Johnson and Tabin (1997).

Adelaide, J.; Mattei, M.-G.; Marics, I.; Raybaud, F.; Planche, J.; De Lapeyriere, O.; Birnbaum, D.: Chromosomal localization of the hst oncogene and its co-amplification with the int.2 oncogene in a human melanoma. Oncogene 2: 413-416, 1988.
Huebner, K.; Ferrari, A. C.; Delli Bovi, P.; Croce, C. M.; Basilico, C.: The FGF-related oncogene, K-FGF, maps to human chromosome region 11q13, possibly near int-2. Oncogene Res. 3: 263-270, 1988.

Johnson, R. L.; Tabin, C. J.: Molecular models for vertebrate limb development. Cell 90: 979-990, 1997.
Sakamoto, H.; Mori, M.; Taira, M.; Yoshida, T.; Matsukawa, S.; Shimizu, K.; Sekiguchi, M.; Terada, M.; Sugimura, T.: Transforming gene from human stomach cancers and a noncancerous portion of stomach mucosa. Proc. Nat. Acad. Sci. 83: 3997-4001, 1986.
Taira, M.; Yoshida, T.; Miyagawa, K.; Sakamoto, H.; Terada, M.; Sugimura, T.: cDNA sequence of human transforming gene hst and identification of the coding sequence required for transforming activity. Proc. Nat. Acad. Sci. 84: 2980-2984, 1987.
Wada, A.; Sakamoto, H.; Katoh, O.; Yoshida, T.; Yokota, J.; Little, P. F. R.; Sugimura, T.; Terada, M.: Two homologous oncogenes, HST1 and INT2, are closely located in human genome. Biochem. Biophys. Res. Commun. 157: 828-835, 1988.
Yoshida, T.; Tsutsumi, M.; Sakamoto, H.; Miyagawa, K.; Teshima, S.; Sugimura, T.; Terada, M.: Expression of the HST1 oncogene in human germ cell tumors. Biochem. Biophys. Res. Commun. 155: 1324-1329, 1988.

ONCOGENE HST2: FIBROBLAST GROWTH FACTOR 6 (*134921)

By screening a mouse cosmid library with a human HST (FGF4; 164980) probe under reduced conditions of stringency, Marics et al. (1989) isolated several positive clones, one of which was identified as a new member of the fibroblast growth factor gene family. They subsequently isolated and sequenced the gene, which they called FGF6. The deduced amino acid sequence showed 70% identity with the HST gene product over the C-terminal two-thirds of the putative protein. Marics et al. (1989) mapped FGF6 to 12p13 by in situ hybridization. The cloned normal gene was shown to transform mouse NIH 3T3 fibroblasts using both focus and tumorigenicity assays. FGF6 is the same as the HST2 oncogene, which was identified by its close homology to the HST1 gene; see Iida et al. (1992). The mapping of HST2 to 12p13 and of HST1 to 11q13 is further evidence of the homeology between these 2 human chromosomes. Studying 2 somatic cell hybrids containing either the der(12) or the der(X) from a mesothelioma with a translocation t(X;12)(q22;p13) as the only chromosomal change, Aerssens et al. (1994) demonstrated that FGF6 lies distal to the 12p13 breakpoint and to VWF (193400).

Aerssens, J.; Chaffanet, M.; Baens, M.; Matthijs, G.; Van Den Berghe, H.; Cassiman, J.-J.; Marynen, P.: Regional assignment of seven loci to 12p13.2-pter by PCR analysis of somatic cell hybrids containing the der(12) or the der(X) chromosome from a mesothelioma showing t(X;12)(q22;p13). Genomics 20: 119-121, 1994.
Iida, S.; Yoshida, T.; Naito, K.; Sakamoto, H.; Katoh, O.; Hirohashi, S.; Sato, T.; Onda, M.; Sugimura, T.; Terada, M.: Human hst-2 (FGF-6) oncogene: cDNA cloning and characterization. Oncogene 7: 303-309, 1992.
Marics, I.; Adelaide, J.; Raybaud, F.; Mattei, M.-G.; Coulier, F.; Planche, J.; de Lapeyriere, O.; Birnbaum, D.: Characterization of the HST-related FGF.6 gene, a new member of the fibroblast growth factor gene family. Oncogene 4: 335-340, 1989.

ONCOGENE JUN: V-JUN AVIAN SARCOMA VIRUS 17 ONCOGENE HOMOLOG (*165160)

The oncogene JUN is the putative transforming gene of avian sarcoma virus 17; it appears to be derived from a gene of the chicken genome and has homologs in several other vertebrate species. (The name JUN comes from the Japanese 'ju-nana,' meaning the number 17.) JUN was thought to be identical to the transcription factor AP1. Bohmann et al. (1987) isolated the human protooncogene JUN and found that the deduced amino acid sequence was more than 80% identical to that of the viral protein. Expression of cloned cDNA in bacteria produced a protein with sequence-specific DNA binding properties identical to the phorbol ester-inducible enhancer-binding protein AP1. Antibodies raised against 2 distinct peptides derived from JUN reacted specifically with human AP1. In addition, partial amino acid sequence of purified AP1 revealed tryptic peptides in common with the JUN protein; thus, AP1 may be encoded by JUN. Nucleotide sequence analysis indicated that the COOH-terminus of JUN is similar to the corresponding part of a yeast transcriptional activator. Haluska et al. (1988) isolated a genomic DNA clone encompassing the JUN gene and used it to determine the chromosomal location. Southern blot analysis of a rodent-human somatic cell hybrid panel indicated that JUN is situated on 1p. In situ hybridization narrowed the assignment to 1p32-p31, a chromosomal region involved in both translocations and deletions in human malignancies. Hattori et al. (1988) isolated a genomic clone of human JUN and determined its primary structure and transcription pattern. A surprising finding was the lack of introns. Transfection experiments showed that the cloned gene is functional, as it encodes a trans-acting factor that stimulates transcription of an AP1-dependent reporter gene. By in situ hybridization, Hattori et al. (1988) mapped JUN to 1p32-p31.

Lamph et al. (1988) investigated regulation of murine c-jun gene transcription and found that both serum and phorbol-ester 12-O-tetradecanoyl phorbol 13-acetate (TPA) induced c-jun gene expression. Marx (1988) reviewed information indicating that the protein encoded by the JUN gene acts directly to activate gene transcription in response to cell stimulation; that the product of the FOS oncogene (164810) cooperates with the JUN product in fostering gene transcription; that there is a structural and functional similarity between JUN and GCN4, which induces the activity of a large set of genes needed for amino acid synthesis in yeast; that specifically both have a DNA-binding domain on the carboxyl end essential for the activation of genes; that the JUN and FOS proteins are held together in a complex by a 'leucine zipper;' and that there are other JUN genes in addition to the original one. Mattei et al. (1990) mapped the mouse homolog to chromosome 4. Bahary et al. (1991) presented molecular genetic linkage maps of mouse chromosome 4 which established the breakpoints in the mouse 4/human 1p region of homology to a 2-cM interval between Ifa and Jun in mouse and to the interval between JUN and ACADM (201450) in the human.

Sphingosylphosphocholine (SPC) is a deacylated derivative of sphingomyelin known to accumulate in Niemann-Pick disease type A (257200). SPC is a potent mitogen that increases intracellular free Ca(2+) and free arachidonate through pathways that are only partly protein kinase C-dependent. Berger et al. (1995) showed that SPC increases specific DNA-binding activity of transcription activator AP1 in electrophoretic mobility-shift assays.

See also Bos et al. (1988)

Bahary, N.; Zorich, G.; Pachter, J. E.; Leibel, R. L.; Friedman, J. M.: Molecular genetic linkage maps of mouse chromosomes 4 and 6. Genomics 11: 33-47, 1991.
Berger, A.; Rosenthal, D.; Spiegel, S.: Sphingosylphosphocholine, a signaling molecule which accumulates in Niemann-Pick disease type A, stimulates DNA-binding activity of the transcription activator protein AP-1. Proc. Nat. Acad. Sci. 92: 5885-5889, 1995.
Bohmann, D.; Bos, T. J.; Admon, A.; Nishimura, T.; Vogt, P. K.; Tjian, R.: Human proto-oncogene c-jun encodes a DNA binding protein with structural and functional properties of transcription factor AP-1. Science 238: 1386-1392, 1987.
Bos, T. J.; Bohmann, D.; Tsuchie, H.; Tjian, R.; Vogt, P. K.: v-jun encodes a nuclear protein with enhancer binding properties of AP-1. Cell 52: 705-712, 1988.
Haluska, F. G.; Huebner, K.; Isobe, M.; Nishimura, T.; Croce, C. M.; Vogt, P. K.: Localization of the human JUN protooncogene to chromosome region 1p31-32. Proc. Nat. Acad. Sci. 85: 2215-2218, 1988.
Hattori, K.; Angel, P.; Le Beau, M. M.; Karin, M.: Structure and chromosomal localization of the functional intronless human JUN protooncogene. Proc. Nat. Acad. Sci. 85: 9148-9152, 1988.
Lamph, W. W.; Wamsley, P.; Sassone-Corsi, P.; Verma, I. M.: Induction of proto-oncogene JUN/AP-1 by serum and TPA. Nature 334: 629-631, 1988.
Marx, J. L.: 'Jun' is bustin' out all over. (Research News). Science 242: 1377-1378, 1988.
Mattei, M. G.; Simon-Chazottes, D.; Hirai, S.-I.; Ryseck, R.-P.; Galcheva-Gargova, Z.; Guenet, J. L.; Mattei, J. F.; Bravo, R.; Yaniv, M.: Chromosomal localization of the three members of the jun proto-oncogene family in mouse and man. Oncogene 5: 151-156, 1990.

ONCOGENE JUN-B (*165161)

Mattei et al. (1990) mapped 3 members of the JUN protooncogene family in the mouse: JUN (165160) to mouse chromosome 4 and JUNB and JUND to the same region of mouse chromosome 8. RFLP analysis of interspecific hybrids confirmed the mapping of JUNB and JUND and showed that they are situated about 7.3 cM apart. Using the same probes for in situ hybridization, Mattei et al. (1990) showed that the human genes are located on 19p13.2, a region that is involved in chromosome translocations in cases of leukemia. By fluorescence in situ hybridization, Trask et al. (1993) assigned the JUNB gene to 19p13.2.

In comparisons of the murine and human junB loci, Phinney et al. (1995) found 9 regions of distal 5-prime and 3-prime flanking DNA that exhibited greater than 72% sequence identity. More than 50% of the JUNB locus is contained in these flanking evolutionarily conserved sequences (FECSs), which may be required for effecting the proper transcriptional regulation of the gene. Comparative sequence analyses involving kilobases of distal flanking DNAs for only a small number of vertebrate genes suggested that FECSs may emerge as common yet important functional components of genes. If they account for a large fraction of the conserved sequence within a disease-related locus, then deleterious mutations might frequently occur in these regions and go undetected with current strategies.

Mattei, M. G.; Simon-Chazottes, D.; Hirai, S.; Ryseck, R. P.; Galcheva-Gargova, Z.; Guenet, J. L.; Mattei, J. F.; Bravo, R.; Yaniv, M.: Chromosomal localization of the three members of the jun proto-oncogene family in mouse and man. Oncogene 5: 151-156, 1990.
Phinney, D. G.; Tseng, S. W.; Ryder, K.: Complex genetic organization of junB: multiple blocks of flanking evolutionarily conserved sequence at the murine and human junB loci. Genomics 28: 228-234, 1995.
Trask, B.; Fertitta, A.; Christensen, M.; Youngblom, J.; Bergmann, A.; Copeland, A.; de Jong, P.; Mohrenweiser, H.; Olsen, A.; Carrano, A.; Tynan, K.: Fluores-

cence in situ hybridization mapping of human chromosome 19: cytogenetic band location of 540 cosmids and 70 genes or DNA markers. Genomics 15: 133-145, 1993.

ONCOGENE JUN-D (*165162)

See 165161. By fluorescence in situ hybridization, Trask et al. (1993) assigned the JUND gene to 19p13.1-p12.

Trask, B.; Fertitta, A.; Christensen, M.; Youngblom, J.; Bergmann, A.; Copeland, A.; de Jong, P.; Mohrenweiser, H.; Olsen, A.; Carrano, A.; Tynan, K.: Fluorescence in situ hybridization mapping of human chromosome 19: cytogenetic band location of 540 cosmids and 70 genes or DNA markers. Genomics 15: 133-145, 1993.

ONCOGENE LIPOSARCOMA (*164953)

Gill et al. (1991) screened a series of adult soft-tissue tumors for the presence of activated oncogenes by transfecting tumor DNA into NIH 3T3 mouse fibroblasts. Following transfection of DNA from a liposarcoma, they identified an activated gene that failed to hybridize to probes prepared from 10 known human oncogenes that had previously been detected in DNA transfection experiments. Using a probe that detects the human Alu family of highly repetitive DNA sequences, they cloned BamHI fragments of this activated gene of 3.5 and 8.5 kb. Repeat-free subclones of these BamHI fragments were used to map the gene to 19p13.2-q13.3. This location does not coincide with the 12q13-q14 location suggested by cytogenetic studies in some forms of lipoma and liposarcoma (151900).

Gill, S.; Stratton, M. R.; Patterson, H.; Spurr, N. K.; Fisher, C.; Gusterson, B. A.; Cooper, C. S.: Detection of transforming genes by transfection of DNA from primary soft-tissue tumours. Oncogene 6: 1651-1656, 1991.

ONCOGENE LIVER CANCER (*165320)

Ochiya et al. (1986) isolated a novel oncogene, designated LCA by them, from a primary human hepatocellular carcinoma by transforming assays using NIH 3T3 cells and a calcium phosphate coprecipitation method. They mapped the gene to human chromosome 2 by Southern blot analysis of DNA from flow-sorted human chromosomes and from human/mouse somatic hybrid cells. Tokino et al. (1988) regionalized the gene to 2q14-q21 by in situ hybridization. The gene symbol proposed by the nomenclature committee of the Human Gene Mapping Workshops was LCO (McAlpine, 1989). (LCA is used as a symbol in 2 other connections: light chain A of clathrin (118960) and leukocyte common antigen (151460).)

McAlpine, P. J.: Personal Communication. Winnipeg, Manitoba, Canada 1989.
Ochiya, T.; Fujiyama, A.; Fukushige, S.; Hatada, I.; Matsubara, K.: Molecular cloning of an oncogene from a human hepatocellular carcinoma. Proc. Nat. Acad. Sci. 83: 4993-4997, 1986.
Tokino, T.; Satoh, H.; Yoshida, M. C.; Ochiya, T.; Matsubara, K.: Regional localization of the LCA oncogene to human chromosome region 2q14-q21. Cytogenet. Cell Genet. 48: 63-64, 1988.

ONCOGENE MAS1 (*165180)

The MAS1 oncogene was isolated from DNA of a human epidermoid carcinoma cell line using the cotransfection and tumorigenicity assay (Young et al., 1984). Based on its deduced amino acid sequence, the MAS1 gene product contains 7 potential transmembrane domains. The mas1 protein is, therefore, probably an integral membrane protein. The mas1 encoded protein may be a receptor that, when activated, modulates a critical component in a growth-regulating pathway to bring about its oncogenic effects.

The structure of the MAS oncogene indicates that it belongs to the class of receptors that are coupled to GTP-binding proteins and share a conserved structural motif, which is described as a '7-transmembrane segment' following the prediction that these hydrophobic segments form membrane-spanning alpha-helices. Examples include the adrenergic receptors (e.g., 104210, 104250, 109690). Jackson et al. (1988) found that the MAS oncogene shows the greatest sequence similarity to the substance-K receptor (162321). On this basis, they predicted that it would encode a peptide receptor with mitogenic activity which would act through the inositol lipid signaling pathways. Expression in Xenopus oocytes and a transfected mammalian cell line demonstrated that the MAS gene product is a functional angiotensin receptor.

Villar and Pedersen (1994), who characterized the MAS gene product as a mitogenic G protein-coupled cell surface receptor, investigated the allele-specific expression pattern of the mouse Mas gene on the basis of its proximity to the known imprinted gene for insulin growth factor type II receptor. Phenotyping of mRNA demonstrated exclusive expression from the paternal allele in all embryonic tissues, including visceral yolk sac, between 11 and 12.5 days of gestation. By 13.5 days of gestation, the paternal allele-specific expression of Mas was restricted to heart, tongue, and visceral yolk sac, whereas all other tissues exhibited relaxation of the parental imprint. The maternally inherited allele appears to be transcriptionally repressed in a developmental and tissue-specific manner.

By in situ hybridization, Rabin et al. (1987) mapped MAS1 to 6q24-q27. The ROS (165020) and MYB (189990) oncogenes are also located on 6q. It is of note that the estrogen receptor gene (133430) maps to the same region as MAS1. Al-Ubaidi et al. (1992) confirmed the assignment of the human MAS1 gene to chromosome 6 by Southern blot analysis of somatic cell hybrids.

Riesewijk et al. (1996) mapped the human MAS gene to 6q25.3-q26 by fluorescence in situ hybridization (FISH) in a region very close to the IGF2R gene (147280), which had been mapped to 6q26. Moreover, FISH signals obtained with YACs carrying the IGF2R gene and the MAS gene were partially overlapping on metaphase chromosomes and in 91 of 100 interphase nuclei. Since for cosmid probes the resolution of interphase FISH is approximately 50 kb, these findings suggested to the authors that in humans the physical distance between MAS and IGF2R is at most several hundred kb (it is less than 300 kb in the mouse). Riesewijk et al. (1996) demonstrated that the MAS gene, like the IGF2R gene, is not imprinted in the human, although both are imprinted in the mouse.

By 2 interspecific backcrosses, Cebra-Thomas et al. (1992) mapped the mouse homolog to the proximal portion of chromosome 17. They further demonstrated that Mas mapped to a 1-Mb region which also contained Igf2r (147280). Al-Ubaidi et al. (1992) also assigned the Mas-1 gene to mouse chromosome 17.

Al-Ubaidi, M. R.; Brock, D.; Siciliano, M. J.: Assignment of the homolog of the MAS1 oncogene to mouse chromosome 17. Genomics 13: 487 only, 1992.
Cebra-Thomas, J. A.; Tsai, J.-Y.; Pilder, S. H.; Copeland, N. G.; Jenkins, N. A.; Silver, L. M.: Localization of the Mas proto-oncogene to a densely marked region of mouse chromosome 17 associated with genomic imprinting. Genomics 13: 444-446, 1992.
Jackson, T. R.; Blair, L. A. C.; Marshall, J.; Goedert, M.; Hanley, M. R.: The mas oncogene encodes an angiotensin receptor. Nature 335: 437-440, 1988.
Rabin, M.; Birnbaum, D.; Young, D.; Birchmeier, C.; Wigler, M.; Ruddle, F. H.: Human ROS1 and MAS1 oncogenes located in regions of chromosome 6 associated with tumor-specific rearrangements. Oncogene Res. 1: 169-178, 1987.
Riesewijk, A. M.; Schepens, M. T.; Mariman, E. M.; Ropers, H.-H.; Kalscheuer, V. M.: The MAS proto-oncogene is not imprinted in humans. Genomics 35: 380-382, 1996.
Villar, A. J.; Pedersen, R. A.: Parental imprinting of the Mas protooncogene in mouse. Nature Genet. 8: 373-379, 1994.
Young, D.; Waitches, G.; Birchmeier, C.; Fasano, O.; Wigler, M.: Isolation and characterization of a new cellular oncogene encoding a protein with multiple transmembrane domains. Cell 45: 711-719, 1984.

MCF.2 CELL LINE DERIVED-TRANSFORMING SEQUENCE (*311030)

MCF2 is the designation of a transforming sequence identified using cotransfection of DNA from a human mammary carcinoma cell line. Noguchi et al. (1987) cloned this sequence and found that it did not cross-hybridize with the known oncogenes tested. By in situ hybridization, Noguchi et al. (1987) localized it to Xq27. This localization was confirmed by hybridization to DNA from a panel of human-rodent somatic cell hybrid lines. By pulsed field gel electrophoresis, Nguyen et al. (1987) found that the MCF2 gene and the F9 gene (306900) are separated by a maximum distance of about 270 kb. Furthermore, they located several HTF islands in this region, i.e., CpG-rich, unmethylated sequences, containing several sites for 'rare cutter' enzymes, which are believed to be associated with expressed 'housekeeping' genes. Anson et al. (1988) further narrowed the interval separating MCF2 and F9, locating MCF2 to a region between 29 and 61 kb 3-prime to F9. In 2 unrelated hemophilia B (306900) patients who raised antibodies to infused factor IX, they found deletions in excess of 273 kb encompassing the F9 and MCF2 genes and a CG-rich island. This appears to have been the first reported nullisomic deletion of a transforming gene. No clinical condition could be attributed to the loss of the MCF2 gene. The CG-rich island may be a marker for an as yet undefined gene lying just 5-prime or just 3-prime off the island. Nguyen et al. (1989) concluded from pulsed field gel electrophoresis studies of the Xq27 region that MCF2 is located telomeric to F9. Grant et al. (1990) demonstrated that Mcf-2 in the mouse lies in the same order as the corresponding gene on the human X chromosome: Hprt—Cf-9—Mcf-2—G6pd. In situ hybridization indicated that the gene lies in the same region as Cf-9 and linkage studies in interspecific mouse backcross populations demonstrated that the Cf-9 and Mcf-2 genes were separated by about 0.5 cM.

A comparison of the restriction maps of the DBL and MCF2 oncogenes, together with their chromosomal localization, indicates that they represent the same genetic locus. The DBL oncogene was initially detected as a transforming gene from a human diffuse B-cell lymphoma and was isolated as a 45-kb transforming human DNA sequence by cosmid cloning (Eva and Aaronson, 1985). By molecular hybridization, DBL lacks detectable homology with a large number of cellular or retroviral oncogenes, including members of the tyrosine kinase family. Srivastava et al. (1986) demonstrated that antiserum from mice bearing tumors induced by this oncogene specifically detected a protein of about 66 kD in DBL transformants. Using DBL cDNA, they isolated mRNA from a transfectant clone and found that it directed the in vitro synthesis of this protein. Subcellular localization studies showed that the protein, also known as p66, is a cytoplasmic protein distributed between cytosol and crude membrane fractions. They showed, furthermore, that p66 is a phosphoprotein, with phosphorylation specific to serine residues. Ron et al. (1988) showed that overexpression of DBL is sufficient to transform NIH 3T3 cells. Ron et al. (1988) cloned and characterized the DBL oncogene. The DBL oncogene was generated by rearrangements involving 3 discontinuous regions of the human genome. By analysis of DNA from human/rodent somatic cell hybrids, Tronick et al. (1989) demonstrated that the DBL gene located on the X chromosome underwent recombination at its 5-prime and 3-prime ends with sequences derived from chromosomes 3 (pter-p21) and 16 (p13-q22), respectively. By in situ hybridization, Tronick et al. (1989) located the DBL gene more precisely to Xq27. Oncogenic activation of the MCF2 gene occurs through substitution of part of its 5-prime coding region by unrelated nonsyntenic sequences. Galland et al. (1992) demonstrated that the upstream replacing sequence, referred to as URS, represents the farthest 5-prime portion of the locus and that it is derived from the D15S93 locus on human chromosome 15q15-q23.

See also Eva et al. (1988)

Anson, D. S.; Blake, D. J.; Winship, P. R.; Birnbaum, D.; Brownlee, G. G.: Nullisomic deletion of the mcf.2 transforming gene in two haemophilia B patients. EMBO J. 7: 2795-2799, 1988.
Eva, A.; Aaronson, S. A.: Isolation of a new human oncogene from a diffuse B-cell lymphoma. Nature 316: 273-275, 1985.
Eva, A.; Vecchio, G.; Rao, C. D.; Tronick, S. R.; Aaronson, S. A.: The predicted DBL oncogene product defines a distinct class of transforming proteins. Proc. Nat. Acad. Sci. 85: 2061-2065, 1988.
Galland, F.; Stefanova, M.; Lafage, M.; Birnbaum, D.: Localization of the 5-prime end of the MCF2 oncogene to human chromosome 15q15-q23. Cytogenet. Cell Genet. 60: 114-116, 1992.
Grant, S. G.; Mattei, M.-G.; Galland, F.; Stephenson, D. A.; Keitz, B. T.; Birnbaum, D.; Chapman, V. M.: Localization of the mouse Mcf-2 (Dbl) protooncogene within a conserved linkage group on the mouse X chromosome. Cytogenet. Cell Genet. 54: 175-181, 1990.
Nguyen, C.; Pontarotti, P.; Birnbaum, D.; Chimini, G.; Rey, J. A.; Mattei, J.-F.; Jordan, B. R.: Large scale physical mapping in the q27 region of the human X chromosome: the coagulation factor IX gene and the mcf.2 transforming sequence are separated by at most 270 kilobase pairs and are surrounded by several 'HTF islands.' EMBO J. 6: 3285-3289, 1987.
Nguyen, C.; Poustka, A.-M.; Djabali, M.; Roux, D.; Mattei, J.-F.; Lehrach, H.; Jordan, B. R.: Large-scale mapping and chromosome jumping in the q27 region of the human X chromosome. Genomics 5: 298-303, 1989.
Noguchi, T.; Mattei, M.-G.; Oberle, I.; Planche, J.; Imbert, J.; Pelassy, C.; Birg, F.; Birnbaum, D.: Localization of the mcf.2 transforming sequence to the X chromosome. EMBO J. 6: 1301-1307, 1987.
Ron, D.; Tronick, S. R.; Aaronson, S. A.; Eva, A.: Molecular cloning and characterization of the human DBL proto-oncogene: evidence that its overexpression is sufficient to transform NIH/3T3 cells. EMBO J. 7: 2465-2473, 1988.
Srivastava, S. K.; Wheelock, R. H. P.; Aaronson, S. A.; Eva, A.: Identification of the protein encoded by the human diffuse B-cell lymphoma (dbl) oncogene. Proc. Nat. Acad. Sci. 83: 8868-8872, 1986.
Tronick, S. R.; McBride, O. W.; Popescu, N. C.; Eva, A.: Chromosomal localization of DBL oncogene sequences. Genomics 5: 546-553, 1989.

ONCOGENE MEL (*165040)

Using DNA transfection into NIH 3T3 cells, Padua et al. (1984) demonstrated that the human malignant melanoma cell line NK14 contains a novel transforming gene. As a result of studies of human-mouse and human-hamster somatic cell hybrids, Spurr et al. (1986) demonstrated that this oncogene, termed MEL, is located in the segment 19p13.2-q13.2. Although MEL was isolated as a transforming gene from a melanoma cell line, no linkage between MEL and malignant melanoma (155600) is demonstrable (Nimmo et al., 1989). By linkage analysis using an NcoI RFLP, 2,3:Nimmo et al. (1989, 1989) mapped the MEL gene to the region of LDLR (143890), i.e., 19p13.2-cen.

Bahler et al. (1997) performed cosmid contig mapping indicating that the MEL locus was 800 kb distal to MY09B (602129) on chromosome 19p13.1.

Bahler, M.; Kehrer, I.; Gordon, L.; Stoffler, H.-E.; Olsen, A. S.: Physical mapping of human myosin-IXB (MYO9B), the human orthologue of the rat myosin myr 5, to chromosome 19p13.1. Genomics 43: 107-109, 1997.
Nimmo, E.; Padua, R.-A.; Hughes, D.; Brook, J. D.; Williamson, R.; Johnson, K. J.: Confirmation and refinement of the localisation of the c-MEL locus on chromosome 19 by physical and genetic mapping. Hum. Genet. 81: 382-384, 1989.
Nimmo, E.; Williamson, R.; Johnson, K.: Localization of the c-MEL gene to 19(cen-p13.2). (Abstract) Cytogenet. Cell Genet. 51: 1053 only, 1989.
Padua, R. A.; Barrass, N.; Currie, G. A.: A novel transforming gene in a human malignant melanoma cell line. Nature 311: 671-673, 1984.
Spurr, N. K.; Hughes, D.; Goodfellow, P. N.; Brook, J. D.; Padua, R. A.: Chromosomal assignment of c-MEL, a human transforming oncogene, to chromosome 19(p13.2-q13.2). Somat. Cell Molec. Genet. 12: 637-640, 1986.

ONCOGENE MYB-L1: V-MYB AVIAN MYELOBLASTOSIS VIRAL ONCOGENE HOMOLOG-LIKE 1 (*159405)

Nomura et al. (1988) isolated and characterized cDNA clones for 2 human MYB-related genes, AMYB and BMYB (601415). Using probes in Southern blot analysis of rodent-human hybrid DNAs, Barletta et al. (1991) localized the MYBL1 locus to 8cen-q22 and refined the localization to 8q22-q23 by in situ hybridization.

Takahashi et al. (1995) found that MYBL1 mRNA is expressed mainly in testis and peripheral blood leukocytes. AMYB could activate transcription from the promoter-containing MYB-binding sites in all cells examined. In addition to the 2 domains (a DNA-binding domain and a transcriptional activation domain), 2 negative regulatory domains were identified in the MYBL1 gene. These results indicated that the gene functions as a transcriptional activator and that the regulatory mechanism of gene activity is similar to that of the MYB (189990) gene.

Barletta, C.; Druck, T.; LaForgia, S.; Calabretta, B.; Drabkin, H.; Patterson, D.; Croce, C. M.; Huebner, K.: Chromosome locations of the MYB related genes, AMYB and BMYB. Cancer Res. 51: 3821-3824, 1991.
Nomura, N.; Takahashi, M.; Matsui, M.; Ishii, S.; Date, T.; Sasamoto, S.; Ishizaki, R.: Isolation of human cDNA clones of MYB-related genes, A-MYB and B-MYB. Nucleic Acids Res. 16: 11075-11089, 1988.
Takahashi, T.; Nakagoshi, H.; Sarai, A.; Nomura, N.; Yamamoto, T.; Ishii, S.: Human A-myb gene encodes a transcriptional activator containing the negative regulatory domains. FEBS Lett. 358: 89-96, 1995.

ONCOGENE MYC: V-MYC AVIAN MYELOCYTOMATOSIS VIRAL ONCOGENE HOMOLOG (*190080)

Sequences of the MYC oncogene have been highly conserved throughout evolution, from Drosophila to vertebrates (Shilo and Weinberg, 1981). Persson and Leder (1984) showed that the product of the MYC gene has a molecular weight of 65,000, is located predominantly in the nucleus, and binds to DNA.

Leder (1982) described in situ hybridization observations suggesting that the MYC locus is on chromosome 8 near 8q24, the breakpoint in Burkitt lymphoma translocations. Collins and Groudine (1982) found that the normal human homolog of the avian myc oncogene was present in multiple copies in the DNA of a malignant promyelocyte cell line derived from the peripheral blood of a patient with acute promyelocytic leukemia. Other human onc genes were not amplified. By the Southern blotting technique applied to somatic cell hybrids, Dalla-Favera et al. (1982) showed that the MYC gene is on chromosome 8. When hybrids between rodent cells and human Burkitt lymphoma cells were analyzed, they could show that the MYC gene is on the part of chromosome 8 (8q24-qter) that is translocated to 2, 14, or 22. Several MYC-related sequences may be pseudogenes. Taub et al. (1982) also mapped MYC to 8q24 and found that in 2 Burkitt cell lines MYC was translocated into a DNA restriction fragment that also encodes the immunoglobulin mu chain gene. In a mouse plasmacytoma, the MYC gene was translocated into the immunoglobulin alpha switch region. Observations such as those of Alitalo et al. (1983) indicate that the same oncogene which in one chromosomal change or rearrangement produces one specific neoplasm results, when altered in a different way, in a different neoplasm. Alitalo et al. (1983) found that the MYC gene, which is involved by translocation in the generation of Burkitt lymphoma, is amplified, resulting in homogeneously staining chromosomal regions (HSRs) in a human neuroendocrine tumor cell line derived from a colon cancer. The HSR resided on a distorted X chromosome; amplification of MYC had been accompanied by translocation of the gene from its normal position on 8q24. Maguire et al. (1983) found that Burkitt and non-Burkitt lymphomas with either an 8;14 or an 8;22 translocation expressed 2- to 5-fold more MYC-specific RNA than B-cell lines without a translocation. Tumor cel l lines of American origin with a translocation of either type expressed similar amounts of MYC-specific RNA. Tumor cell lines of African origin contained slightly higher levels of MYC-specific RNA than American lines, but the level did not correlate with absence or presence of Epstein-Barr virus (EBV). No MOS-related transcripts were found in these tumors. In Burkitt lymphomas bearing the 8;14 translocation, the MYC gene is translocated to a heavy chain switch recombination region (mu or alpha). See Adams et al. (1983). By fluorescence in situ hybridization in combination with R banding, Takahashi et al. (1991) refined the assignment of MYC to 8q24.12-q24.13, distal to fragile site fra(8)(q24.11).

EBV is stably maintained and partially expressed in Burkitt lymphoma and in nasopharyngeal carcinoma. Latently infected cells usually contain multiple episomal copies of nonintegrated viral DNA. In 2 Burkitt cell lines, Henderson et al. (1983) showed that EBV was also integrated into a chromosome, but different chromosomes—nos. 1 and 4. The persistence of EBV in latently infected cells over years of active cell replication may be explained by integration. It is noteworthy that the site of integration is removed from those involved in the translocation. 'The simplest model to explain EBV association with Burkitt tumors is that EBV induces B-cell proliferation and thereby provides enhanced opportunity for chromosomal translocation and malignant degeneration' (Henderson et al., 1983).

The 14q marker in Burkitt lymphoma was first found by Manolov and Manolova (1972). Zech et al. (1976) showed that the extra chromosomal material joined to the end of one chromosome 14 is derived from the distal part of 8q. Bernheim et al. (1981) found either 2;8 or 8;22 translocation in about 10% of cases. The translocations separate the MYC gene from its normal promoter and 5-prime regulatory machinery, and place it under some regulatory element associated with the immunoglobulin gene. By hybrid cell studies of mouse plasmacytoma cells and Burkitt lymphoma cells, Nishikura et al. (1983) showed that cells containing the MYC gene on a translocation chromosome expressed high levels of human specific MYC transcripts whereas hybrid cells containing the untranslocated MYC gene on the normal chromosome did not contain such MYC mRNA. Usually in t(8;14) translocations, the MYC gene is translocated to 14q. When the break occurs between the MYC first and second exons, both segments are transcriptionally active. Croce et al. (1983) studied somatic cell hybrids between mouse myeloma cells and a Burkitt lymphoma human cell line with a t(8;22) chromosome translocation. The MYC gene was found to remain on chromosome 8q+; the normal chromosome 8 remains transcriptionally silent. The lambda constant region is translocated 3-prime to the MYC oncogene. Yokota et al. (1986) concluded that alterations are found in oncogenes MYC, HRAS, or MYB in more than one-third of human solid tumors. Amplification of MYC was found in advanced widespread tumors and in aggressive primary tumors. Apparent allelic deletions of HRAS and MYB could be correlated with progression and metastasis of carcinomas and sarcomas.

Erikson et al. (1986) studied 2 patients with a t(8;14)(q24;q11) chromosome translocation. In 1, rearrangement was detected in a region immediately 3-prime to the MYC locus. In the second, the breakpoint in the chromosome 14 occurred between genes for the variable and constant regions of the T-cell receptor alpha chain (186880). The constant region locus had translocated to a region more than 38 kb 3-prime to the MYC gene, yet as was shown by the study of hybrids between the human cells and mouse cells, only the hybrids carrying the 8q+ chromosome expressed MYC. Thus, deregulation of the MYC locus can occur not only with translocation of the heavy chain locus or one or the other light chain locus to chromosome 8 but also with translocation of the TCRA locus. The involvement of the MYC oncogene in translocations is the prototype in the relationship between chromosomal abnormalities and oncogenes.

Heim and Mitelman (1987) counted a total of 83 bands that have been found to be specifically involved in primary structural chromosome rearrangements in human cancer. They compared the distribution of these breakpoints with the chromosomal sites of 26 cellular oncogenes which had to that time been mapped to individual bands in the human genome. Nineteen of the 26 oncogenes were located in cancer-associated bands. This clustering is statistically significant ($p = 0.0000012$). They pointed out that cancer may be inflated by errors of karyotype interpretation. Furthermore, it appears that only 1 of the 2 breakpoints in cancer-specific translocations is the site of an oncogene, so that the number of cancer-associated breakpoints that are found to contain oncogenes should theoretically approach 50% of the total. Mitelman (1985) provided a useful catalog of chromosome aberrations in cancer. Duesberg (1987) suggested that cellular cancer genes are not activated oncogenes but rather the result of rare truncations and illegitimate recombinations that alter the germ-line configuration of cellular genes. See review by Cole (1986). Since MYC is proximal to the thyroglobulin gene (TG; 188450), which is located in the segment 8q24.2-q24.3, MYC may be located in band 8q24.1.

Morse et al. (1988) found an MYC rearrangement in a breast carcinoma due to insertion of a LINE-1 (L1) element. (Mobile genetic elements are an important source of genetic variability in both prokaryotes and eukaryotes, including Drosophila, yeast, and mouse. At least 10% of the human genome is composed of repetitive retrotransposon-like sequences including short, interspersed Alu sequences, nonviral retrotransposon-like LINE-1 sequences, and the long terminal repeat (LTR)-containing THE-1 elements. Factor VIII mutations due to L1 insertions have been reported (306700) and an Alu transposition event has been found in human lung carcinoma cells (Lin et al., 1988). The RTVL-H family of human endogenous retrovirus-like elements has approximately 1,000 intact members and a similar number of solitary LTRs in the haploid genome. The intact RTVL-H sequences are 5.8 kb long, have 5-prime and 3-prime LTRs and have some segments of sequence similarity to mammalian retroviruses in the putative gag and pol regions. Mager and Goodchild (1989) demonstrated DNA variation in 2 sibs representing a deletion due to homologous recombination between the 5-prime and 3-prime LTRs of a RTVL-H sequence.) Specific types of human papillomavirus (HPV), mostly HPV type 16 (HPV16) and type 18 (HPV18), are associated with genital carcinomas such as those of the cervix and their noninvasive precursors (167959, 167960). In intraepithelial neoplasia, HPV DNA is detected most commonly as episomal molecules, whereas it is found integrated in the cell genome in the majority of invasive carcinomas. By chromosomal in situ hybridization experiments, Couturier et al. (1991) determined the localization of integrated HPV16 or HPV18 genomes in genital cancers. In 3 cancers, HPV sequences were located in band 8q24.1, which contains the MYC gene, and in 1 cancer, HPV sequences were located in band 2p24, which contains the NMYC gene (164840). In 3 of the 4 cases, the protooncogene located near integrated viral sequences was found to be structurally altered and/or overexpressed.

Contrary to the previous belief that MYC is wildtype in both types of tumors, Bhatia et al. (1993) found that 65% of 57 Burkitt lymphomas and 30% of 10 mouse plasmacytomas exhibited at least 1 amino acid substitution. These mutations were apparently homozygous in all Burkitt lymphoma cell lines tested and in 2 tumor biopsies, implying that the mutations often occur before MYC/IG translocation. In the mouse plasmacytomas, only the mutant myc allele was expressed, indicating a functional homozygosity with occurrence of mutations at the translocation. Many of the observed mutations were clustered in regions associated with transcriptional activation and apoptosis, and in the Burkitt lymphomas, they frequently occurred at sites of phosphorylation, suggesting that the mutations had a pathogenetic role. Most of the mutations observed were clearly not polymorphisms, for reasons in addition to the large number of different mutations observed: 1) a high proportion were missense mutations; 2) most tumors contained multiple mutations; and 3) each tumor had a unique pattern of mutations.

Atchley and Fitch (1995) described phylogenetic analyses for 45 MYC protein sequences. A gene duplication early in vertebrate evolution produced the c-MYC lineage and another lineage that later gave rise to the N- and L-MYC lineages by another gene duplication. Evolutionary divergence in the MYC gene family corresponded closely to the known branching order of the major vertebrate groups. The closely related dimerization partner protein MAX (154950) exhibited significantly less variability than MYC. Atchley and Fitch (1995) suggested a reduced variability in MAX stems from natural selection acting to preserve dimerization capability with products of MYC and related genes.

Cell proliferation is regulated by the induction of growth promoting genes and the suppression of growth inhibitory genes. Malignant growth can result from the altered balance of expression of these genes in favor of cell proliferation. Induction of the transcription factor MYC promotes cell proliferation and transformation by activating growth-promoting genes, including the ornithine decarboxylase (ODC1; 165640) and CDC25A (116947) genes. Lee et al. (1997) showed that MYC transcriptionally represses the expression of the growth arrest gene (GAS1; 139185). A conserved MYC structure, MYC box 2, is required for repression of GAS1 and for MYC induction of proliferation and transformation, but not for activation of ODC1.

He et al. (1998) provided a molecular framework for understanding the previously enigmatic overexpression of MYC in colorectal cancers. Inactivating mutations in the adenomatous polyposis coli gene (APC; 175100), found in most colorectal cancers, cause aberrant accumulation of beta-catenin (CTNNB1; 116806), which then binds T-cell factor 4 (TCF4; 602228), causing increased transcriptional activation of unknown genes. He et al. (1998) showed that the MYC oncogene is a target in this signaling pathway. They showed that expression of MYC is repressed by wildtype APC and activated by beta-catenin, and that effects are mediated through TCF4 binding sites in the MYC promoter.

See also Battey et al. (1983); Beimling et al. (1985); Bernard et al. (1983); Colby et al. (1983); Dalla-Favera et al. (1982); Dunnick et al. (1983); Erikson et al. (1983); Hamlyn and Rabbitts (1983); Hayday et al. (1984); Magrath et al. (1983); Marcu et al. (1983); Murphy et al. (1986); Neel et al. (1982); Persson et al. (1984); Peschle et al. (1984); Saito et al. (1983); Sakaguchi et al. (1983); Watt et al. (1983); Watt et al. (1983)

Adams, J. M.; Gerondakis, S.; Webb, E.; Corcoran, L. M.; Cory, S.: Cellular myc oncogene is altered by chromosome translocation to an immunoglobulin locus in murine plasmacytomas and is rearranged similarly in human Burkitt lymphomas. Proc. Nat. Acad. Sci. 80: 1982-1986, 1983.

Alitalo, K.; Schwab, M.; Lin, C. C.; Varmus, H. E.; Bishop, J. M.: Homogeneously staining chromosomal regions contain amplified copies of an abundantly

expressed cellular oncogene (c-myc) in malignant neuroendocrine cells from a human colon carcinoma. Proc. Nat. Acad. Sci. 80: 1707-1711, 1983.
Atchley, W. R.; Fitch, W. M.: Myc and Max: molecular evolution of a family of proto-oncogene products and their dimerization partner. Proc. Nat. Acad. Sci. 92: 10217-10221, 1995.
Battey, J.; Moulding, C.; Taub, R.; Murphy, W.; Stewart, T.; Potter, H.; Lenoir, G.; Leder, P.: The human c-myc oncogene: structural consequences of translocation into the IgH locus in Burkitt lymphoma. Cell 34: 779-787, 1983.
Beimling, P.; Benter, T.; Sander, T.; Moelling, K.: Isolation and characterization of the human cellular myc gene product. Biochemistry 24: 6349-6355, 1985.
Bernard, O.; Cory, S.; Gerondakis, S.; Webb, E.; Adams, J. M.: Sequence of the murine and human cellular myc oncogenes and two modes of myc transcription resulting from chromosome translocation in B lymphoid tumours. EMBO J. 2: 2375-2383, 1983.
Bernheim, A.; Berger, R.; Lenoir, G.: Cytogenetic studies on African Burkitt's lymphoma cell lines: t(8;14), t(2;8) and t(8;22) translocations. Cancer Genet. Cytogenet. 3: 307-315, 1981.
Bhatia, K.; Huppi, K.; Spangler, G.; Siwarski, D.; Iyer, R.; Magrath, I.: Point mutations in the c-Myc transactivation domain are common in Burkitt's lymphoma and mouse plasmacytomas. Nature Genet. 5: 56-61, 1993.
Colby, W. W.; Chen, E. Y.; Smith, D. H.; Levinson, A. D.: Identification and nucleotide sequence of a human locus homologous to the v-myc oncogene of avian myelocytomatosis virus MC29. Nature 301: 722-725, 1983.
Cole, M. D.: The myc oncogene: its role in transformation and differentiation. Annu. Rev. Genet. 20: 361-384, 1986.
Collins, S.; Groudine, M.: Amplification of endogenous myc-related DNA sequences in a human myeloid leukaemia cell line. Nature 298: 679-681, 1982.
Couturier, J.; Sastre-Garau, X.; Schneider-Maunoury, S.; Labib, A.; Orth, G.: Integration of papillomavirus DNA near myc genes in genital carcinomas and its consequences for proto-oncogene expression. J. Virol. 65: 4534-4538, 1991.
Croce, C. M.; Thierfelder, W.; Erikson, J.; Nishikura, K.; Finan, J.; Lenoir, G. M.; Nowell, P. C.: Transcriptional activation of an unrearranged and untranslocated c-myc oncogene by translocation of a C-lambda locus in Burkitt lymphoma cells. Proc. Nat. Acad. Sci. 80: 6922-6926, 1983.
Dalla-Favera, R.; Bregni, M.; Erikson, J.; Patterson, D.; Gallo, R. C.; Croce, C. M.: Human c-myc onc gene is located on the region of chromosome 8 that is translocated in Burkitt lymphoma cells. Proc. Nat. Acad. Sci. 79: 7824-7827, 1982.
Dalla-Favera, R.; Gelmann, E. P.; Martinotti, S.; Franchini, G.; Papas, T. S.; Gallo, R. C.; Wong-Staal, F.: Cloning and characterization of different human sequences related to the onc gene (v-myc) of avian myelocytomatosis virus (MC29). Proc. Nat. Acad. Sci. 79: 6497-6501, 1982.
Duesberg, P. H.: Cancer genes: rare recombinants instead of activated oncogenes (a review). Proc. Nat. Acad. Sci. 84: 2117-2124, 1987.
Dunnick, W.; Shell, B. E.; Dery, C.: DNA sequences near the site of reciprocal recombination between a c-myc oncogene and an immunoglobulin switch region. Proc. Nat. Acad. Sci. 80: 7269-7273, 1983.
Erikson, J.; Finger, L.; Sun, L.; ar-Rushdi, A.; Nishikura, K.; Minowada, J.; Finan, J.; Emanuel, B. S.; Nowell, P. C.; Croce, C. M.: Deregulation of c-myc by translocation of the alpha-locus of the T-cell receptor in T-cell leukemias. Science 232: 884-886, 1986.
Erikson, J.; Nishikura, K.; ar-Rushdi, A.; Finan, J.; Emanuel, B.; Lenoir, G.; Nowell, P. C.; Croce, C. M.: Translocation of an immunoglobulin kappa locus to a region 3-prime of an unrearranged c-myc oncogene enhances c-myc transcription. Proc. Nat. Acad. Sci. 80: 7581-7585, 1983.
Hamlyn, P. H.; Rabbitts, T. H.: Translocation joins c-myc and immunoglobulin gamma-1 genes in a Burkitt lymphoma revealing a third exon in the c-myc oncogene. Nature 304: 135-139, 1983.
Hayday, A. C.; Gillies, S. D.; Saito, H.; Wood, C.; Wiman, K.; Hayward, W. S.; Tonegawa, S.: Activation of a translocated human c-myc gene by an enhancer in the immunoglobulin heavy-chain locus. Nature 307: 334-340, 1984.
He, T.-C.; Sparks, A. B.; Rago, C.; Hermeking, H.; Zawel, L.; da Costa, L. T.; Morin, P. J.; Vogelstein, B.; Kinzler, K. W.: Identification of c-MYC as a target of the APC pathway. Science 281: 1509-1512, 1998.
Heim, S.; Mitelman, F.: Nineteen of 26 cellular oncogenes precisely localized in the human genome map to one of the 83 bands involved in primary cancer-specific rearrangements. Hum. Genet. 75: 70-72, 1987.
Henderson, A.; Ripley, S.; Heller, M.; Kieff, E.: Chromosome site for Epstein-Barr virus DNA in a Burkitt tumor cell line and in lymphocytes growth-transformed in vitro. Proc. Nat. Acad. Sci. 80: 1987-1991, 1983.
Leder, P.: Personal Communication. Boston, Mass. 10/1/1982.
Lee, T. C.; Li, L.; Philipson, L.; Ziff, E. B.: Myc represses transcription of the growth arrest gene gas1. Proc. Nat. Acad. Sci. 94: 12886-12891, 1997.
Lin, C. S.; Goldthwait, D. A.; Samols, D.: Identification of Alu transposition in human lung carcinoma cells. Cell 54: 153-159, 1988.
Mager, D. L.; Goodchild, N. L.: Homologous recombination between the LTRs of a human retrovirus-like element causes a 5-kb deletion in two siblings. Am. J. Hum. Genet. 45: 848-854, 1989.
Magrath, I.; Erikson, J.; Whang-Peng, J.; Sieverts, H.; Armstrong, G.; Benjamin, D.; Triche, T.; Alabaster, O.; Croce, C. M.: Synthesis of kappa light chains by cell lines containing an 8;22 chromosomal translocation derived from a male homosexual with Burkitt's lymphoma. Science 222: 1094-1098, 1983.
Maguire, R. T.; Robins, T. S.; Thorgeirsson, S. S.; Heilman, C. A.: Expression of cellular myc and mos genes in undifferentiated B cell lymphomas of Burkitt and non-Burkitt types. Proc. Nat. Acad. Sci. 80: 1947-1950, 1983.
Manolov, G.; Manolova, Y.: Marker band in one chromosome 14 from Burkitt lymphomas. Nature 237: 33-34, 1972.
Marcu, K. B.; Harris, L. J.; Stanton, L. W.; Erikson, J.; Watt, R.; Croce, C. M.: Transcriptionally active c-myc oncogene is contained within NIARD, a DNA sequence associated with chromosome translocations in B-cell neoplasia. Proc. Nat. Acad. Sci. 80: 519-523, 1983.
Mitelman, F.: Catalog of Chromosome Aberrations in Cancer. New York: Alan R. Liss (pub.) (2nd ed.): 1985.
Morse, B.; Rotherg, P. G.; South, V. J.; Spandorfer, J. M.; Astrin, S. M.: Insertional mutagenesis of the myc locus by a LINE-1 sequence in a human breast carcinoma. Nature 333: 87-90, 1988.
Murphy, W.; Sarid, J.; Taub, R.; Vasicek, T.; Battey, J.; Lenoir, G.; Leder, P.: A translocated human c-myc oncogene is altered in a conserved coding sequence. Proc. Nat. Acad. Sci. 83: 2939-2943, 1986.
Neel, B. G.; Jhanwar, S. C.; Chaganti, R. S. K.; Hayward, W. S.: Two human c-onc genes are located on the long arm of chromosome 8. Proc. Nat. Acad. Sci. 79: 7842-7846, 1982.
Nishikura, K.; ar-Rushdi, A.; Erikson, J.; Watt, R.; Rovera, G.; Croce, C. M.: Differential expression of the normal and of the translocated human c-myc oncogenes in B cells. Proc. Nat. Acad. Sci. 80: 4822-4826, 1983.
Persson, H.; Hennighausen, L.; Taub, R.; DeGrado, W.; Leder, P.: Antibodies to human c-myc oncogene product: evidence of an evolutionarily conserved protein induced during cell proliferation. Science 225: 687-693, 1984.
Persson, H.; Leder, P.: Nuclear localization and DNA binding properties of a protein expressed by human c-myc oncogene. Science 225: 718-721, 1984.
Peschle, C.; Mavilio, F.; Sposi, N. M.; Giampaolo, A.; Care, A.; Bottero, L.; Bruno, M.; Mastroberardino, G.; Gastaldi, R.; Testa, M. G.; Alimena, G.; Amadori, S.; Mandelli, F.: Translocation and rearrangement of c-myc into immunoglobulin alpha heavy chain locus in primary cells from acute lymphocytic leukemia. Proc. Nat. Acad. Sci. 81: 5514-5518, 1984.
Saito, H.; Hayday, A. C.; Wiman, K.; Hayward, W. S.; Tonegawa, S.: Activation of the c-myc gene by translocation: a model for translational control. Proc. Nat. Acad. Sci. 80: 7476-7480, 1983.
Sakaguchi, A. Y.; Lalley, P. A.; Naylor, S. L.: Human and mouse cellular myc protooncogenes reside on chromosomes involved in numerical and structural aberrations in cancer. Somat. Cell Genet. 9: 391-405, 1983.
Shilo, B. Z.; Weinberg, R. A.: DNA sequences homologous to vertebrate oncogenes are conserved in Drosophila melanogaster. Proc. Nat. Acad. Sci. 78: 6789-6792, 1981.
Takahashi, E.; Hori, T.; O'Connell, P.; Leppert, M.; White, R.: Mapping of the MYC gene to band 8q24.12-q24.13 by R-banding and distal to fra(8)(q24.11), FRA8E, by fluorescence in situ hybridization. Cytogenet. Cell Genet. 57: 109-111, 1991.
Taub, R.; Kirsch, I.; Morton, C.; Lenoir, G.; Swan, D.; Tronick, S.; Aaronson, S.; Leder, P.: Translocation of the c-myc gene into the immunoglobulin heavy chain locus in human Burkitt lymphoma and murine plasmacytoma cells. Proc. Nat. Acad. Sci. 79: 7837-7841, 1982.
Watt, R.; Nishikura, K.; Sorrentino, J.; ar-Rushdi, A.; Croce, C. M.; Rovera, G.: The structure and nucleotide sequence of the 5-prime end of the human c-myc oncogene. Proc. Nat. Acad. Sci. 80: 6307-6311, 1983.
Watt, R.; Stanton, L. W.; Marcu, K. B.; Gallo, R. C.; Croce, C. M.; Rovera, G.: Nucleotide sequence of cloned cDNA of human c-myc oncogene. Nature 303: 725-728, 1983.
Yokota, J.; Tsunetsugu-Yokota, Y.; Battifora, H.; Le Fevre, C.; Cline, M. J.: Alterations of myc, myb, and ras(Ha) proto-oncogenes in cancers are frequent and show clinical correlation. Science 231: 261-265, 1986.
Zech, L.; Haglund, U.; Nilsson, K.; Klein, G.: Characteristic chromosomal abnormalities in biopsies and lymphoid-cell lines from patients with Burkitt and non-Burkitt lymphomas. Int. J. Cancer 17: 47-56, 1976.

ONCOGENE MYC-L1: V-MYC AVIAN MYELOCYTOMATOSIS VIRAL ONCOGENE HOMOLOG 1, LUNG CARCINOMA-DERIVED (*164850)

Nau et al. (1985) cloned from DNA of small-cell lung cancer (SCCL) a gene with homology to a small region of both MYC (190080) and NMYC (164840). By somatic cell hybridization and in situ hybridization, they assigned the gene to 1p32. This LMYC sequence was amplified 10- to 20-fold in the DNA of 4 SCCL lines and of 1 SCCL specimen taken directly from a patient. A restriction polymorphism was found. In heterozygotes, only 1 of the 2 alleles was amplified in any 1 genome. In a linkage map of chromosome 1 prepared by Rouleau et al. (1990), it was concluded that MYCL1 is 17 cM proximal to RH. Kaye et al. (1988) found that the LMYC gene is composed of 3 exons and 2 introns spanning 6.6 kilobases. Several distinct mRNAs were produced in all SCCL cell lines that expressed LMYC. These transcripts were generated from a single gene by alternative splicing of introns 1 and 2 and by use of alternative polyadenylation signals. Comparisons with MYC and NMYC demonstrated multiple discrete regions with extensive homology. Kawashima et al. (1988) concluded that a correlation exists between particular RFLP alleles of the MYCL gene and the occurrence of metastasis of lung cancer to lymph nodes and other organs. Among lung cancer patients, those with only the L band (10 kb) had few lymph node metastases, whereas patients with either the S band (6 kb) or the S and L bands almost always had lymph node metastases. A similar correlation was found between the presence of the S band and metastases to other organs. The correlation was particularly marked in cases of adenocarcinoma of the lung. By study of DNA from mouse-hamster somatic cell hybrids, Campbell et al. (1989) mapped 2 L-myc loci provisionally to mouse chromosomes 4 and 12. The locus on chromosome 12 may be a pseudogene.

When studying chromosome 1p breakpoints in neuroblastoma cell lines using fluorescence in situ hybridization (FISH) with region-specific probes, Van Roy et al. (1995) found evidence for a position of MYCL1 more distal than 1p32. To investigate the discrepancy Speleman et al. (1996) used FISH on high-resolution R-banded chromosomes with a YAC clone for MYCL1 and reassigned the gene to 1p34.3.

See also McBride et al. (1985)

Campbell, G. R.; Zimmerman, K.; Blank, R. D.; Alt, F. W.; D'Eustachio, P.: Chromosomal location of N-myc and L-myc genes in the mouse. Oncogene Res. 4: 47-54, 1989.
Kawashima, K.; Shikama, H.; Imoto, K.; Izawa, M.; Naruke, T.; Okabayashi, K.; Nishimura, S.: Close correlation between restriction fragment length polymorphism of the L-MYC gene and metastasis of human lung cancer to the lymph nodes and other organs. Proc. Nat. Acad. Sci. 85: 2353-2356, 1988.
Kaye, F.; Battey, J.; Nau, M.; Brooks, B.; Seifter, E.; De Greve, J.; Birrer, M.; Sausville, E.; Minna, J.: Structure and expression of the human L-myc gene reveal a complex pattern of alternative mRNA processing. Molec. Cell. Biol. 8: 186-195, 1988.
McBride, O. W.; Kirsch, I.; Hollis, G.; Nau, M.; Battey, J.; Minna, J.: Human L-myc (MYCL) proto-oncogene is on chromosome 1p32.(Abstract) Cytogenet. Cell Genet. 40: 694 only, 1985.
Nau, M. M.; Brooks, B. J.; Battey, J.; Sausville, E.; Gazdar, A. F.; Kirsch, I. R.; McBride, O. W.; Bertness, V.; Hollis, G. F.; Minna, J. D.: L-myc, a new myc-related gene amplified and expressed in human small cell lung cancer. Nature 318: 69-73, 1985.
Rouleau, G. A.; Bazanowski, A.; Gusella, J. F.; Haines, J. L.: A genetic map of chromosome 1: comparison of different data sets and linkage programs. Genomics 7: 313-318, 1990.
Speleman, F.; Van Camp, G.; Van Roy, N.: Reassignment of MYCL1 to human chromosome 1p34.3 by fluorescence in situ hybridization. Cytogenet. Cell Genet. 72: 189-190, 1996.
Van Roy, N.; Cheng, N. C.; Laureys, G.; Opdenakker, G.; Versteeg, R.; Speleman, F.: Molecular cytogenetic analysis of 1;17 translocations in neuroblastoma. Europ. J. Cancer 31A: 530-535, 1995.

ONCOGENE N-MYC: V-MYC AVIAN MYELOCYTOMATOSIS VIRAL RELATED ONCOGENE NEUROBLASTOMA-DERIVED (*164840)

Kohl et al. (1983) isolated a genomic DNA segment that is homologous to, but distinct from, the classic MYC oncogene. It is amplified 25- to 700-fold in 8 of 9 human neuroblastoma cell lines which contain either homogeneous staining regions (HSR) or double minutes (DM), the karyologic manifestations of amplified genes. (The ninth cell line showed 30-fold amplification of the MYC oncogene.) Although NMYC is known to be located on the short arm of chromosome 2, none of the 5 HSR-containing cell lines had HSRs on chromosome 2. Amplification of the NMYC gene occurs also in retinoblastoma. Kohl et al. (1984) found amplified expression of NMYC in neuroblastoma cell lines but not in other human tumor cell lines with the exception of a retinoblastoma cell line. Kanda et al. (1983) used human-mouse hybrid cells to map NMYC to chromosome 2. In situ hybridization indicated that NMYC is on 2p. Schwab et al. (1984) assigned NMYC to 2p23 or 2p24. Emanuel et al. (1985) and others have shown that neuroblastoma cell lines show HSRs at various sites, that the NMYC oncogene is amplified at several of these sites, and that there is apparently no preferred site for NMYC integration and amplification. Emanuel et al. (1985) stated that there was no direct evidence of amplification with HSR formation at 2p23-2p24, the site of the NMYC gene. Reviewing gene amplification in neuroblastomas, Brodeur and Seeger (1986) reported that they and other researchers had shown that most DM- and HSR-bearing neuroblastoma cell lines have multiple copies of NMYC. The amplification probably takes place at the level of the extrachromosomal DM, which appear to represent circular molecules, with subsequent linear integration into HSR. Ramsay et al. (1986) identified 2 proteins originating from the same NMYC mRNA. Both were phosphorylated and exceptionally unstable. They were located in the cell nucleus and were bound to both single- and double-stranded DNA.

Garson et al. (1987) used a novel in situ hybridization technique to map NMYC to 2p24. The nonradioactive technique combined the high spatial resolution and rapid signal development of the nonisotope approach with the previously unrivaled sensitivity of autoradiography. The method, which used biotin-labeled DNA probes and a streptavidin-alkaline phosphatase-based detection system, was compatible with G-banding and could be performed on archival material. By study of mouse-hamster somatic cell hybrids, Campbell et al. (1989) mapped the mouse equivalent of NMYC to chromosome 12. By study of RFLPs in recombinant inbred strains, they mapped a second locus, called by them N-myc-2, to mouse chromosome 5. The locus on chromosome 5 may be a pseudogene.

Amplification of the MYCN gene is frequently seen either in extrachromosomal double minutes or in homogeneously staining regions (HSRs) in chromosomes of aggressively growing neuroblastomas. HRSs have never been observed at 2p24-p23, the map location of the MYCN gene. Corvi et al. (1994) used fluorescence in situ hybridization in the study of 5 human neuroblastoma cell lines to demonstrate that, in addition to amplified MYCN in HRSs or double minutes, single-copy MYCN was present at the normal position on chromosome 2. In 1 cell line there was coamplification of MYCN together with DNA of the host chromosome 12 to which MYCN had been transposed. The results suggested that the initial event is transposition of MYCN, with retention of the normal gene in its position on 2p before the occurrence of amplification. Thus, the mechanisms of amplified MYCN are probably different from those leading to amplification of drug-resistance genes.

Reiter and Brodeur (1996) generated a high-resolution restriction map of approximately 500 kb spanning the MYCN locus. They found that deletions and rearrangements of the amplicon occurred less often in primary tumors than in cell lines. They also defined a 130-kb common core region of the MYCN amplicon that was amplified in 32 of 33 neuroblastomas. The authors proposed that, despite the large size of most MYCN amplicons, the core region that is consistently amplified in neuroblastomas probably contains the MYCN gene and little else.

See also Brodeur et al. (1984); Kohl et al. (1986); Lee et al. (1984); Michitsch and Melera (1985); Schwab (1985); Schwab et al. (1983); Schwab et al. (1984); Seeger et al. (1985); Shiloh et al. (1985); Stanton et al. (1986)

Brodeur, G. M.; Seeger, R. C.: Gene amplification in human neuroblastomas: basic mechanisms and clinical implications. Cancer Genet. Cytogenet. 19: 101-111, 1986.
Brodeur, G. M.; Seeger, R. C.; Schwab, M.; Varmus, H. E.; Bishop, J. M.: Amplification of N-myc in untreated human neuroblastomas correlates with advanced disease stage. Science 224: 1121-1124, 1984.
Campbell, G. R.; Zimmerman, K.; Blank, R. D.; Alt, F. W.; D'Eustachio, P.: Chromosomal location of N-myc and L-myc genes in the mouse. Oncogene Res. 4: 47-54, 1989.
Corvi, R.; Amler, L. C.; Savelyeva, L.; Gehring, M.; Schwab, M.: MYCN is retained in single copy at chromosome 2 band p23-24 during amplification in human neuroblastoma cells. Proc. Nat. Acad. Sci. 91: 5523-5527, 1994.
Emanuel, B. S.; Balaban, G.; Boyd, J. P.; Grossman, A.; Negishi, M.; Parmiter, A.; Glick, M. C.: N-myc amplification in multiple homogeneously staining regions in two human neuroblastomas. Proc. Nat. Acad. Sci. 82: 3736-3740, 1985.
Garson, J. A.; van den Berghe, J. A.; Kemshead, J. T.: Novel non-isotopic in situ hybridization technique detects small (1 kb) unique sequences in routinely G-banded human chromosomes: fine mapping of N-myc and beta-NGF genes. Nucleic Acids Res. 15: 4761-4770, 1987.
Kanda, N.; Schreck, R.; Alt, F.; Bruns, G.; Baltimore, D.; Latt, S.: Isolation of amplified DNA sequences from IMR-32 human neuroblastoma cells: facilitation by fluorescence-activated flow sorting of metaphase chromosomes. Proc. Nat. Acad. Sci. 80: 4069-4073, 1983.
Kohl, N. E.; Gee, C. E.; Alt, F. W.: Activated expression of the N-myc gene in human neuroblastomas and related tumors. Science 226: 1335-1337, 1984.
Kohl, N. E.; Kanda, N.; Schreck, R. R.; Bruns, G.; Latt, S. A.; Gilbert, F.; Alt, F. W.: Transposition and amplification of oncogene-related sequences in human neuroblastomas. Cell 35: 359-367, 1983.
Kohl, N. E.; Legouy, E.; DePinho, R. A.; Nisen, P. D.; Smith, R. K.; Gee, C. E.; Alt, F. W.: Human N-myc is closely related in organization and nucleotide sequence to c-myc. Nature 319: 73-77, 1986.
Lee, W.-H.; Murphree, A. L.; Benedict, W. F.: Expression and amplification of the N-myc gene in primary retinoblastoma. Nature 309: 458-460, 1984.
Michitsch, R. W.; Melera, P. W.: Nucleotide sequence of the 3-prime exon of the human N-myc gene. Nucleic Acids Res. 13: 2545-2558, 1985.
Ramsay, G.; Stanton, L.; Schwab, M.; Bishop, J. M.: Human proto-oncogene N-myc encodes nuclear proteins that bind DNA. Molec. Cell. Biol. 6: 4450-4457, 1986.
Reiter, J. L.; Brodeur, G. M.: High-resolution mapping of a 130-kb core region of the MYCN amplicon in neuroblastomas. Genomics 32: 97-103, 1996.
Schwab, M.: Amplification of N-myc in human neuroblastomas. Trends Genet. 1: 271-275, 1985.
Schwab, M.; Alitalo, K.; Klempnauer, K.-H.; Varmus, H. E.; Bishop, J. M.; Gilbert, F.; Brodeur, G.; Goldstein, M.; Trent, J.: Amplified DNA with limited homology to myc cellular oncogene is shared by human neuroblastoma cell lines and a neuroblastoma tumour. Nature 305: 245-248, 1983.
Schwab, M.; Ellison, J.; Busch, M.; Rosenau, W.; Varmus, H. E.; Bishop, J. M.: Enhanced expression of the human gene N-myc consequent to amplification of DNA may contribute to malignant progression of neuroblastoma. Proc. Nat. Acad. Sci. 81: 4940-4944, 1984.
Schwab, M.; Varmus, H. E.; Bishop, J. M.; Grzeschik, K.-H.; Naylor, S. L.; Sakaguchi, A. Y.; Brodeur, G.; Trent, J.: Chromosome localization in normal human cells and neuroblastomas of a gene related to c-myc. Nature 308: 288-291, 1984.
Seeger, R. C.; Brodeur, G. M.; Sather, H.; Dalton, A.; Siegel, S. E.; Wong, K. Y.; Hammond, D.: Association of multiple copies of the N-myc oncogene with rapid progression of neuroblastomas. New Eng. J. Med. 313: 1111-1116, 1985.
Shiloh, Y.; Shipley, J.; Brodeur, G. M.; Bruns, G.; Korf, B.; Donlon, T.; Schreck, R. R.; Seeger, R.; Sakai, K.; Latt, S. A.: Differential amplification, assembly, and relocation of multiple DNA sequences in human neuroblastomas and neuroblastoma cell lines. Proc. Nat. Acad. Sci. 82: 3761-3765, 1985.
Stanton, L. W.; Schwab, M.; Bishop, J. M.: Nucleotide sequence of the human N-myc gene. Proc. Nat. Acad. Sci. 83: 1772-1776, 1986.
#neuroblastoma-derived

ONCOGENE NOV: NEPHROBLASTOMA OVEREXPRESSED GENE (*164958)

The avian nephroblastoma induced by myeloblastosis-associated virus (MAV) is a good model of Wilms tumor because of histologic similarities. Soret et al. (1989) and Joliot et al. (1992) identified in MAV-1-induced avian nephroblastoma a new protooncogene they called nov for nephroblastoma overexpressed gene. Martinerie and Perbal (1991) found that human sequences homologous to nov were expressed in normal hematopoietic cells and in 1 nephroblastoma. By a combination of study of somatic cell hybrids and in situ hybridization, Martinerie et al. (1992) showed that the human homolog maps to 8q24.1, proximal to MYC (190080).

Snaith et al. (1996) stated that the NOV gene encodes a cysteine-rich protein that is overexpressed in avian nephroblastomas. It is a member of the CCN family of proteins that includes connective tissue growth factor (121009). These proteins are encoded by a group of genes known as immediate-early genes, so named because they are expressed after induction by growth factors or certain oncogenes. The proteins share several common structural motifs: a consensus sequence present in IGF (insulin-like growth factor)-binding proteins (the IGFBP motif), an oligomeric complex-forming domain first identified in von Willebrand factor (VWF; 193400), a binding domain to soluble and matrix molecules, and a dimerization (CT) domain (Bork, 1993). All CCN family members are thought to be involved in the control of cell proliferation. Snaith et al. (1996) isolated and characterized genomic and cDNA clones encompassing the mouse nov gene. It is highly conserved with the human and chick nov genes at the level of nucleotide sequence and genomic organization. The exon structure reflected the modular organization of NOV protein in a number of structural domains. These are highly conserved with other members of the CCN family, as is the distribution of 38 of its 40 cysteine residues. Snaith et al. (1996) mapped the nov gene to mouse chromosome 15 in a region of conserved synteny with human chromosome 8.

Kim et al. (1997) noted that NOV and the products of the other immediate-early genes CTGF (121009) and Cyr61 (602369) have significant sequence homology to the IGFBPs (see IGFBP7, 602867). They proposed that, together with IGFBP7, these proteins constitute a family of IGFBPs that bind IGFs with low affinity.

Bork, P.: The modular architecture of a new family of growth regulators related to connective tissue growth factor. FEBS Lett. 327: 125-130, 1993.
Joliot, V.; Martinerie, C.; Dambrine, G.; Plassiart, G.; Brisac, M.; Crochet, J.; Perbal, B.: Proviral rearrangements and overexpression of a new cellular gene (nov) in myeloblastosis-associated virus type 1-induced nephroblastomas. Molec. Cell. Biol. 12: 10-21, 1992.
Kim, H.-S.; Nagalla, S. R.; Oh, Y.; Wilson, E.; Roberts, C. T., Jr.; Rosenfeld, R. G.: Identification of a family of low-affinity insulin-like growth factor binding proteins (IFGBPs): characterization of connective tissue growth factor as a member of the IGFBP superfamily. Proc. Nat. Acad. Sci. 94: 12981-12986, 1997.
Martinerie, C.; Perbal, B.: Expression of a gene encoding a novel potential IGF binding protein in human tissues. C. R. Acad. Sci. (Paris) 313 (ser. 3): 345-351, 1991.
Martinerie, C.; Viegas-Pequignot, E.; Guenard, I.; Dutrillaux, B.; Nguyen, V. C.; Bernheim, A.; Perbal, B.: Physical mapping of human loci homologous to the chicken nov proto-oncogene. Oncogene 7: 2529-2534, 1992.
Snaith, M. R.; Natarajan, D.; Taylor, L. B.; Choi, C.-P.; Martinerie, C.; Perbal, B.; Schofield, P. N.; Boulter, C. A.: Genomic structure and chromosomal mapping of the mouse nov gene. Genomics 38: 425-428, 1996.
Soret, J.; Dambrine, G.; Perbal, B.: Induction of nephroblastoma by myeloblastosis-associated virus type 1: state of proviral DNAs in tumor cells. J. Virol. 63: 1803-1807, 1989.

ONCOGENE OVC (*164759)

By transfection of NIH-3T3 cells with genomic DNA from a human ovarian adenocarcinoma tumor cell line, Halverson et al. (1990) identified OVC, a rearranged human DNA sequence that was generated during transfection and induced both morphologic transformation and tumorigenesis. One fragment mapped to human 9p24 and the other to human chromosome 8. See 167000 for a discussion of familial ovarian cancer. Because rat ovarian surface epithelial cells transformed spontaneously in vitro have been found to have homozygous deletions of the interferon alpha gene (IFNA; 147660), suggesting inactivation of a tumor-suppressor gene in that region may be crucial for the development of ovarian cancer, Chenevix-Trench et al. (1994) used microsatellite markers and Southern analysis to examine the homologous region in humans, 9p, for deletions in sporadic ovarian adenocarcinomas and ovarian tumor cell lines. Loss of heterozygosity occurred in 34 (37%) of 91 informative sporadic tumors, including some benign, low-malignant-potential and early-stage tumors, suggesting that it is an early event in the development of ovarian adenocarcinoma. Furthermore, homozygous deletions on 9p were found in 2 of 10 independent cell lines. Deletion mapping of the tumors and cell lines indicated that the candidate suppressor gene inactivated as a consequence lies between D9S171 and the IFNA locus. This region is deleted in several other tumors and contains the melanoma predisposition gene MLM (155601). In a note added in proof, Chenevix-Trench et al. (1994) suggested that the target of these 9p deletions might be CDKN2 (600160) as described by Kamb et al. (1994).

Chenevix-Trench, G.; Kerr, J.; Friedlander, M.; Hurst, T.; Sanderson, B.; Coglan, M.; Ward, B.; Leary, J.; Khoo, S.-K.: Homozygous deletions on the short arm of chromosome 9 in ovarian adenocarcinoma cell lines and loss of heterozygosity in sporadic tumors. Am. J. Hum. Genet. 55: 143-149, 1994.
Halverson, D.; Modi, W.; Dean, M.; Gelmann, E. P.; Dunn, K. J.; Clanton, D.; Oskarsson, M.; O'Brien, S. J.; Blair, D. G.: An oncogenic chromosome 8-9 gene fusion isolated following transfection of human ovarian carcinoma cell line DNA. Oncogene 5: 1085-1089, 1990.
Kamb, A.; Gruis, N. A.; Weaver-Feldhaus, J.; Liu, Q.; Harshman, K.; Tavtigian, S. V.; Stockert, E.; Day, R. S., III; Johnson, B. E.; Skolnick, M. H.: A cell cycle regulator potentially involved in genesis of many tumor types. Science 264: 436-440, 1994.

ONCOGENE PE1 (*164873)

The ETS oncogene (164720) was first described as part of a fusion gene transduced by the avian retrovirus E26. In E26, v-ets and v-myb (189990) were fused to a portion of GAG to form the transforming gene. The ETS oncogene family shares a conserved peptide motif called the ETS domain that mediates sequence-specific DNA binding. This motif is unique among transcription factor families. Using partially degenerate oligonucleotides from conserved regions of the ETS domain and the polymerase chain reaction, Klemsz et al. (1994) isolated a new member of the ETS family, designated PE-1, from HL60 cells. The PE1 gene was expressed as an approximately 5-kb transcript in most cell lines tested. Using both in situ hybridization and study of human/hamster cell hybrids, Klemsz et al. (1994) demonstrated that the PE1 gene is located on 1q21-q23.

Klemsz, M.; Hromas, R.; Raskind, W.; Bruno, E.; Hoffman, R.: PE-1, a novel ETS oncogene family member, localizes to chromosome 1q21-q23. Genomics 20: 291-294, 1994.

ONCOGENE PIM1 (*164960)

By means of Southern blot analysis of DNA from somatic cell hybrids, Cuypers et al. (1986) assigned the human homolog of the murine PIM1 oncogene to 6pter-q12. By a combination of deletion and linkage mapping, Zoghbi et al. (1990) determined that the PIM1 locus is located between the MUT locus (251000) and the HLA cluster, i.e., between 6p21.3 and 6p12. They identified RFLPs at the PIM1 locus. By fluorescence in situ hybridization, Ragoussis et al. (1992) refined the assignment to 6p21.2. Viral leukemogenesis in mice is often initiated by proviral activation of the highly conserved cellular gene Pim-1. Domen et al. (1987) compared the nucleotide sequence of human and mouse PIM1 cDNAs. In the mouse, this gene is located on chromosome 17. Meeker et al. (1987) also characterized the human PIM1 gene. The deduced amino acid sequence showed significant homology to a number of the protein kinases but did not have a transmembrane region. Genomic clones from a human B cell leukemia line were analyzed. The PIM1 transcript was found to derive from 5 kb of genomic DNA. Six exons and 5 introns were identified. Studies of expression of this gene using Northern blots of human cell lines showed it to be transcribed primarily in B-lymphoid and myeloid cell lines. Amson et al. (1989) showed that the 33-kD product of the PIM gene is highly expressed in the liver and spleen during fetal hematopoiesis. In contrast, it is only slightly expressed in circulating granulocytes in adults. It was overexpressed in hematopoietic malignancies, particularly in myeloid and lymphoid acute leukemias. The results implied a physiologic role of the PIM1 oncogene during hematopoietic development and a deregulation of the gene in various leukemias. Saris et al. (1991) provided evidence that both the murine and the human PIM1 gene products are protein-serine/threonine kinases. In the mouse, at any rate, they showed that the gene encodes both a 44- and a 34-kD protein, the former being an amino-terminal e xtension of the latter which is synthesized by alternative translation initiation at an upstream CUG codon. Ark et al. (1991) provided refined mapping of the Pim-1 locus in the mouse and used the Pim-1 gene as a marker for further genetic analysis of t-haplotypes on mouse chromosome 17.

To understand the function of Pim-1 and its role in hematopoietic development, Laird et al. (1993) generated mice deficient in Pim-1 function. Pim-1-deficient mice were ostensibly normal, healthy, and fertile; however, detailed analysis demonstrated a correlation of Pim-1 deficiency with erythrocyte microcytosis, whereas overexpression of Pim-1 in transgenic mice resulted in erythrocyte macrocytosis.

See also Selten et al. (1986)

Amson, R.; Sigaux, F.; Przedborski, S.; Flandrin, G.; Givol, D.; Telerman, A.: The human protooncogene product p33pim is expressed during fetal hematopoiesis and in diverse leukemias. Proc. Nat. Acad. Sci. 86: 8857-8861, 1989.

Ark, B.; Gummere, G.; Bennett, D.; Artzt, K.: Mapping of the Pim-1 oncogene in mouse t-haplotypes and its use to define the relative map positions of the tcl loci t-0(t-6) and t-w12 and the marker tf (tufted). Genomics 10: 385-389, 1991.

Cuypers, H. T.; Selten, G.; Berns, A.; Geurts van Kessel, A. H. M.: Assignment of the human homologue of Pim-1, a mouse gene implicated in leukemogenesis, to the pter-q12 region of chromosome 6. Hum. Genet. 72: 262-265, 1986.

Domen, J.; von Lindern, M.; Hermans, A.; Breuer, M.; Grosveld, G.; Berns, A.: Comparison of the human and mouse PIM-1 cDNAs: nucleotide sequence and immunological identification of the in vitro synthesized PIM-1 protein. Oncogene Res. 1: 103-112, 1987.

Laird, P. W.; van der Lugt, N. M. T.; Clarke, A.; Domen, J.; Linders, K.; McWhir, J.; Berns, A.; Hooper, M.: In vivo analysis of Pim-1 deficiency. Nucleic Acids Res. 21: 4750-4755, 1993.

Meeker, T. C.; Nagarajan, L.; ar-Rushdi, A.; Rovera, G.; Huebner, K.; Croce, C. M.: Characterization of the human PIM-1 gene: a putative proto-oncogene coding for a tissue specific member of the protein kinase family. Oncogene Res. 1: 87-101, 1987.

Ragoussis, J.; Senger, G.; Mockridge, I.; Sanseau, P.; Ruddy, S.; Dudley, K.; Sheer, D.; Trowsdale, J.: A testis-expressed Zn finger gene (ZNF76) in human 6p21.3 centromeric to the MHC is closely linked to the human homolog of the t-complex gene tcp-11. Genomics 14: 673-679, 1992.

Saris, C. J. M.; Domen, J.; Berns, A.: The pim-1 oncogene encodes two related protein-serine/threonine kinases by alternative initiation at AUG and CUG. EMBO J. 10: 655-664, 1991.

Selten, G.; Cuypers, H. T.; Boelens, W.; Robanus-Maandag, E.; Verbeek, J.; Domen, J.; van Beveren, C.; Berns, A.: The primary structure of the putative oncogene pim-1 shows extensive homology with protein kinases. Cell 46: 603-611, 1986.

Zoghbi, H. Y.; Ballantyne, C. M.; O'Brien, W. E.; McCall, A. E.; Kwiatkowski, T. J., Jr.; Ledbetter, S. A.; Beaudet, A. L.: Deletion and linkage mapping of eight markers from the proximal short arm of chromosome 6. Genomics 6: 352-357, 1990.

ONCOGENE PVT1 (*165140)

In both mouse and man with t(6;15) and t(2;8) translocations, respectively, Graham and Adams (1986) identified a sequence that is involved in some human Burkitt lymphomas and murine plasma cytomas. Termed PVT1, the sequence is coamplified with MYC (190080) and is thought to lie about 100 to 500 kb 3-prime of MYC.

Graham, M.; Adams, J. M.: Chromosome 8 breakpoint far 3-prime of the c-myc oncogene in a Burkitt's lymphoma 2;8 variant translocation is equivalent to the murine pvt-1 locus. EMBO J. 5: 2845-2851, 1986.

ONCOGENE RAF1: V-RAF-1 MURINE LEUKEMIA VIRAL ONCOGENE HOMOLOG 1 (*164760)

Rapp et al. (1983) cloned a unique acutely transforming replication-defective mouse type C virus and characterized its acquired oncogene, called v-raf. The viral genome bears close similarities to the Moloney murine leukemia virus (see MOS, 190060). The cellular homolog, c-raf, is present in 1 or 2 copies per haploid genome in mouse and human DNA. The MIL oncogene, a second oncogene in the avian retrovirus MH2, which contains the MYC oncogene, is the avian equivalent of the murine RAF oncogene, i.e., they are identical. Bonner et al. (1984) showed that RAF2, a processed pseudogene, is on chromosome 4 and that RAF1 is on chromosome 3; RAF1 was assigned to 3p25 by in situ hybridization. This suggests that RAF1 may be involved in mixed parotid gland tumors with the t(3;8)(p25;q21) translocation (9,10:Mark et al., 1980, 1982). Whether deletion of 3p in small cell cancer of the lung (182280) uncovers a deleterious recessive allele of this gene is only speculative. Familial renal cell carcinoma (144700) has been associated with translocations involving 3p. Shimizu et al. (1985) identified the activated RAF gene in the stomach cancer of a Japanese patient. Stomach cancer is the most common cancer in Japan. Fukui et al. (1985) found that transforming DNA in a human glioblastoma line was apparently the RAF gene. Teyssier et al. (1986) presented evidence for a relationship of RAF1 to renal cell carcinoma. Hiroshige et al. (1986) assigned the RAF2 pseudogene to region 4pter-p15 by the study of hybrid cells containing various chromosome 4 regions. Kasid et al. (1987) transfected tumor cell DNA into NIH/3T3 cells to demonstrate that a radiation-resistant laryngeal carcinoma cell line contained altered RAF1 sequences. The karyotype of the tumor cells showed absence of chromosome 3, and transformed cells had double-minute chromosomes. The function of RAF1 was reviewed by Li et al. (1991). Tory et al. (1992) constructed a genetic linkage map of 96 loci on 3p, ext ending from the terminal band to the centromere. Multipoint linkage analysis indicated that the male, female, and sex-averaged maps extend for 102, 147, and 116 cM, respectively. RAF1 and 16 DNA markers were localized by fluorescence in situ hybridization. RAF1 was regionalized to 3p25.

Wang et al. (1996) showed that RAF1 can be targeted to the mitochondria by BCL2 (151430), a regulator of apoptotic cell death. Active RAF1 improved BCL2-mediated resistance to apoptosis. They also showed that RAF1 phosphorylates BAD (603167).

See also Bonner et al. (1985); Bonner et al. (1986); Bonner et al. (1984)

Bonner, T.; O'Brien, S. J.; Nash, W. G.; Rapp, U. R.; Morton, C. C.; Leder, P.: The human homologs of the raf (mil) oncogene are located on human chromosomes 3 and 4. Science 223: 71-74, 1984.

Bonner, T. I.; Kerby, S. B.; Sutrave, P.; Gunnell, M. A.; Mark, G.; Rapp, U. R.: Structure and biological activity of human homologs of the raf/mil oncogene. Molec. Cell. Biol. 5: 1400-1407, 1985.

Bonner, T. I.; Oppermann, H.; Seeburg, P.; Kerby, S. B.; Gunnell, M. A.; Young, A. C.; Rapp, U. R.: The complete coding sequence of the human raf oncogene and the corresponding structure of the c-raf-1 gene. Nucleic Acids Res. 14: 1009-1015, 1986.

Bonner, T. I.; Rapp, U. R.; Nash, W. G.; O'Brien, S. J.: Two human homologues to a new retroviral oncogene raf-1 and raf-2 are assigned to human chromosomes 3 and 4, respectively. (Abstract) Cytogenet. Cell Genet. 37: 424, 1984.

Fukui, M.; Yamamoto, T.; Kawai, S.; Maruo, K.; Toyoshima, K.: Detection of a raf-related and two other transforming DNA sequences in human tumors maintained in nude mice. Proc. Nat. Acad. Sci. 82: 5954-5958, 1985.

Hiroshige, S.; Carlock, L.; Smith, M.: Regional assignment of the RAF2 gene to the region 4pter-4p15. (Abstract) Am. J. Hum. Genet. 39: A157, 1986.

Kasid, U.; Pfeifer, A.; Weichselbaum, R. R.; Dritschilo, A.; Mark, G. E.: The raf oncogene is associated with a radiation-resistant human laryngeal cancer. Science 237: 1039-1041, 1987.

Li, P.; Wood, K.; Mamon, H.; Haser, W.; Roberts, T.: Raf-1: a kinase currently without a cause but not lacking in effects. Cell 64: 479-482, 1991.

Mark, J.; Dahlenfors, R.; Ekedahl, C.; Stenman, G.: The mixed salivary gland tumor—a normally benign human neoplasm frequently showing specific chromosomal abnormalities. Cancer Genet. Cytogenet. 2: 231-234, 1980.

Mark, J.; Dahlenfors, R.; Ekedahl, C.; Stenman, G.: Chromosomal patterns in a benign human neoplasm, the mixed salivary gland tumour. Hereditas 96: 141-148, 1982.

Rapp, U. R.; Goldsborough, M. D.; Mark, G. E.; Bonner, T. I.; Groffen, J.; Reynolds, F. H., Jr.; Stephenson, J. R.: Structure and biological activity of v-raf, a unique oncogene transduced by a retrovirus. Proc. Nat. Acad. Sci. 80: 4218-4222, 1983.

Shimizu, K.; Nakatsu, Y.; Sekiguchi, M.; Hokamura, K.; Tanaka, K.; Terada, M.; Sugimura, T.: Molecular cloning of an activated human oncogene, homologous to v-raf, from primary stomach cancer. Proc. Nat. Acad. Sci. 82: 5641-5645, 1985.

Teyssier, J. R.; Henry, I.; Dozier, C.; Ferre, D.; Adnet, J. J.; Pluot, M.: Recurrent deletion of the short arm of chromosome 3 in human renal cell carcinoma: shift of the c-raf 1 locus. J. Nat. Cancer Inst. 77: 1187-1191, 1986.

Tory, K.; Latif, F.; Modi, W.; Schmidt, L.; Wei, M. H.; Li, H.; Cobler, P.; Orcutt, M. L.; Delisio, J.; Geil, L.; Zbar, B.; Lerman, M. I.: A genetic linkage map of 96 loci on the short arm of human chromosome 3. Genomics 13: 275-286, 1992.

Wang, H.-G.; Rapp, U. R.; Reed, J. C.: Bcl-2 targets the protein kinase Raf-1 to mitochondria. Cell 87: 629-638, 1996.

ONCOGENE ARAF2: V-RAF MURINE SARCOMA 3611 VIRAL ONCOGENE HOMOLOG 2 (*164710)

By screening a fetal liver cDNA library at reduced stringency for v-raf-related sequences, Mark et al. (1986) found a second sequence in addition to the expected RAF1 (164760). The second sequence (called by them PKS, presumably for 'protein kinase sequence') showed 71% nucleotide homology to RAF1. The predicted amino acid sequence of the kinase domain was sufficiently like the sequence of v-raf to suggest that PKS may encode a polypeptide with serine/threonine kinase activity. Mark et al. (1986) found expression of PKS mRNA (2.7 kb long) elevated in peripheral blood mononuclear cells isolated from 2 patients with angioimmunoblastic lymphadenopathy with dysproteinemia, a disease in which autoantibodies are produced following the lymphoproliferative activation of B cells. Further studies by somatic cell and in situ hybridization indicating the existence of two PKS oncogenes, one on chromosome 7 very close to the centromere (7p11-q11.2) and a second on Xp11.4 (311010), were cited.

On screening a mouse cDNA library with a v-raf oncogene probe, Huebner et al. (1986) also isolated a transforming raf-related cDNA, A-raf, that represented a gene distinct from RAF1. As an initial step in the analysis of this RAF1-related cDNA, they isolated a human ARAF cDNA and used it to map the genes in mouse and man. Whereas ARAF1 (311010) was found to map to the X chromosome in both mouse and man, a second ARAF locus, ARAF2, was found in man and mapped to chromosome 7 by studies of mouse-human hybrid cells. Huebner et al. (1986) had not conclusively shown that the ARAF2 locus is transcribed. By in situ hybridization, they localized ARAF2 to 7p11.4-q21. By in situ hybridization, Popescu and Mark (1989) localized the gene to 7p12-q11.21. Lee et al. (1994) referred to ARAF2 as a pseudogene.

Huebner, K.; ar-Rushdi, A.; Griffin, C. A.; Isobe, M.; Kozak, C.; Emanuel, B. S.; Nagarajan, L.; Cleveland, J. L.; Bonner, T. I.; Goldsborough, M. D.; Croce, C. M.; Rapp, U.: Actively transcribed genes in the raf oncogene group, located on the X chromosome in mouse and human. Proc. Nat. Acad. Sci. 83: 3934-3938, 1986.

Lee, J.-E.; Beck, T. W.; Brennscheidt, U.; DeGennaro, L. J.; Rapp, U. R.: The complete sequence and promoter activity of the human A-raf-1 gene (ARAF1). Genomics 20: 43-55, 1994.
Mark, G. E.; Seeley, T. W.; Shows, T. B.; Mountz, J. D.: Pks, a raf-related sequence in humans. Proc. Nat. Acad. Sci. 83: 6312-6316, 1986.
Popescu, N. C.; Mark, G. E.: Localization of the pKs gene, a raf related sequence on human chromosomes X and 7. Oncogene 4: 517-519, 1989.

ONCOGENE H-RAS: V-HA-RAS HARVEY RAT SARCOMA VIRAL ONCOGENE HOMOLOG (*190020)

The 3 RAS oncogenes, HRAS, KRAS (190070), and NRAS (164790), encode 21-kD proteins called p21s. Wong-Staal et al. (1981) identified human DNA sequences homologous to cloned DNA fragments containing the oncogenic nucleic acid sequences of a type C mammalian retrovirus, the Harvey strain of murine sarcoma virus (HaMSV) derived from the rat. Non-onc intervening sequences were present in the human counterpart, which is rather highly conserved in mammalian evolution and probably plays a role in normal cell growth or differentiation. Allelic variation in the human onc HaMSV gene was identified. The transforming genes of retroviruses are derived from a group of cellular genes that are highly conserved evolutionarily. The relationship between viral transforming genes (collectively called v-onc) and their normal cellular counterparts (collectively called c-onc) is obviously of great scientific and medical interest. Chang et al. (1982) studied the Harvey and Kirsten murine sarcoma viruses, 2 closely related rat-derived transforming retroviruses called v-Ha-ras and v-Ki-ras, respectively. They concluded that the human genome contains several copies of the c-ras gene family and that c-Ha-ras-1 (with intervening sequences) has been more highly conserved than has c-Ha-ras-2. Harvey rat sarcoma-1 oncogene (HRAS1) is on chromosome 11; HRAS2 may be on the X chromosome; see 'Pseudogenes' section below. Goyette et al. (1983) found that the number of transcripts of the Harvey ras gene increases during liver regeneration in rats. This appears to indicate regulated change in activity of an 'oncogene' in a physiologic growth process.

By Southern blot analysis of human-rodent hybrid cell DNA, de Martinville et al. (1983) found that the cellular homolog of the transforming DNA sequence isolated from the bladder carcinoma line EJ is located on the short arm of chromosome 11. The locus also contains sequences homologous to the Harvey ras oncogene. No evidence of gene amplification was found. These workers also found karyologically 'a complex rearrangement of the short arm in two of the four copies of chromosome 11 present in this heteroploid cell line' (EJ). Region 11p15 was the site of a breakpoint in a t(3;11) translocation found in tumor cells from a patient with hereditary renal cell carcinoma (144700). By somatic cell hybridization, Junien et al. (1984) found that HRAS1 maps to 11p15.5-p15.1. In 4 cases of deletion of 11p13 with WAGR, they found that the restriction enzyme digestion patterns typical of HRAS1 were present. Thus, HRAS1 is not deleted in WAGR, a finding consistent with the difference in mapping. The HRAS1 and insulin genes appear to be closely situated in the 11pter area. This was found in the in situ hybridization data of Harper (1983) and of Gerhard et al. (1984) and in the genetic linkage data of Gerhard et al. (1984). The latter group found a maximum lod score of 4.1 at theta = 0.0 for the HRAS1 and insulin linkage. Two obligatory recombinants were found. These findings are consistent with the observation that the HRAS gene is not deleted in cases of Wilms tumor with deleted 11p13 (Junien et al., 1984). De Martinville and Francke (1984) likewise mapped HRAS1 and INS (176730), and beta-globin (HBB; 141900) as well, outside the 11p14.1-p11.2 segment. By in situ molecular hybridization studies of meiotic chromosomes (pachytene bivalents), Jhanwar et al. (1983) found that KRAS and HRAS probes mapped to chromomeres corresponding to bands 11p14.1, 12p12.1, and 12q24.2 of somatic chromosomes. HRAS hybridized most avidly at 11p14.1. A weak hybridization at 3p21.3 was no ted. They could not demonstrate hybridization of KRAS to chromosome 6, and in meiotic chromosomes hybridization to the X chromosome could not be studied to investigate the assignment of an HRAS gene to that chromosome.

Fearon et al. (1984) demonstrated that HRAS1 is 8 cM distal to the beta-globin gene and 4 cM proximal to the insulin gene. The beta-globin gene is about 7 cM distal to the parathyroid hormone gene. The length of 11p is estimated to be about 50 cM. By high resolution in situ hybridization to meiotic pachytene chromosomes, Chaganti et al. (1985) concluded that HRAS1 is located at 11p14.1, beta globin at 11p11.22, PTH (not previously assigned regionally) at 11p11.21, and insulin at 11p14.1. Bianchi et al. (1993) mapped the H-ras-1 gene to the beta-globin region of mouse chromosome 7. Russell et al. (1996) constructed a contiguous physical map from the HRAS1 gene to the 11p telomere. The contig spanned approximately 500 kb. Three genes were placed on the contig in the following order: tel—RNH (173320)—HRAS1—HRC (142705).

Der et al. (1982) found that mouse cells transformed by high molecular weight DNAs of a human bladder and a human lung carcinoma cell line contained new sequences homologous, respectively, to the transforming genes of Harvey (ras-H) and Kirsten (ras-K) sarcoma viruses. The HRAS1 oncogene differs from its normal cellular counterpart by the absence of a restriction endonuclease site. This sequence change could be used as the basis of a rapid screening method for this oncogene. Muschel et al. (1983) screened DNA from 34 persons and found that all were homozygous for the normal allele. On the other hand, DNA from a patient's bladder tumor, as well as DNA from his normal bladder and leukocytes, was heterozygous at that restriction endonuclease site. The change was pinpointed to 1 of 2 nucleotides, either of which would change the twelfth amino acid (glycine) in the normal HRAS1 gene product. Thus, the patient appears to be carrying an HRAS1 mutation in his germ line that predisposed him to bladder cancer. The restriction enzyme used in the screen was HpaII or its isoschizomer MspI. However, the authors retracted their data that purported to show an HRAS1 mutation in both tumor tissue and normal tissue; they concluded that the original extractions of DNA from that patient were contaminated by a plasmid DNA containing the HRAS1 oncogene. By restriction analysis, Feinberg et al. (1983) tested 29 human cancers for this mutation and found it in none. Included were 10 primary bladder cancers, 9 colon cancers, and 10 lung cancers. The point mutation altering the twelfth amino acid of the HRAS1 gene product p21, found in a bladder cancer cell line, is the only one known to result in a human transforming gene. Capon et al. (1983) showed that the HRAS1 gene of the T24 human bladder carcinoma line has at least 4 exons and that only a single point mutation in the first exon distinguished the coding region of both alleles of the normal gene from their activated counterpart. Both versions of the gene encode a protein which is predicted to differ from the corresponding viral gene product at 3 amino acid residues, one of which was previously shown to represent the major site of phosphorylation of the viral polypeptide. Pincus et al. (1983) concluded that the bladder oncogene peptide (product of the mutant HRAS1 gene), with valine rather than glycine at position 12 (190020.0001), has a 3-dimensional structure markedly different from the normal.

Sekiya et al. (1984) found a point mutation in the second exon of the HRAS1 gene in a melanoma. Transversion from adenine to thymine resulted in the substitution of leucine for glutamine as amino acid 61 in the predicted p21 protein. Fisher et al. (1984) concluded that HRAS1 is distal to the insulin and beta-globin loci on 11p. In HRAS, KRAS, and NRAS, codons 12 and 61 are 'hotspots' for mutations that activate their malignant transforming properties. Srivastava et al. (1985) showed that mutation at these 3 loci result in changes in electrophoretic mobility of the p21. Changes observed are, for the HRAS gene, gly12val (bladder carcinoma), gly12asp (mammary carcinosarcoma), gln61leu (lung carcinoma), and gln61 to arg (renal pelvic carcinoma) and for the NRAS oncogene, gln61 to arg (lung carcinoma). They proposed that the electrophoretic changes may be a rapid method for identification of activated RAS genes, substituting for the inherently insensitive and time-consuming transfection assay. In 2 of 38 urinary tract tumors, Fujita et al. (1985) detected HRAS oncogenes by transfection, cloned the oncogene in biologically active form, and showed that it contained single base changes at codon 61 leading to substitutions of arginine and leucine, respectively, for glutamine at this position. In 1 tumor, a 40-fold amplification of KRAS was found. In the cell lines isolated from a single colon cancer, Greenhalgh and Kinsella (1985) found a point mutation in codon 12 of HRAS leading to an amino acid change in the gene product. The authors cited experience with KRAS involvement in 3 colon cancers and NRAS involvement in 1, while some 34 other colon cancers failed to demonstrate HRAS activation at codon 12.

Yokota et al. (1986) concluded that alterations are found in oncogenes MYC, HRAS, or MYB in more than one-third of human solid tumors. Amplification of MYC was found in advanced widespread tumors and in aggressive primary tumors. Apparent allelic deletions of HRAS and MYB could be correlated with progression and metastasis of carcinomas and sarcomas. Corell and Zoll (1988) used the restriction enzymes MspI, HpaII, BamHI, and TaqI to analyze 426 alleles of the HRAS locus in DNA samples from 92 healthy individuals, 50 patients with breast cancer, 23 patients with ovarian cancer, and 5 patients with lymphomas. The allelic distribution was comparable among controls and tumor patients, indicating that a genetic predisposition to malignancy is not conferred by unique alleles at the HRAS locus. However, analysis of DNA isolated directly from tumors revealed a discrepancy between the alleles in the white blood cells and those in the tumor tissue. Six patients demonstrated alleles in the tumor tissue which were not observed in DNA from the white blood cells. Tong et al. (1989) determined the structural change in the HRAS gene (called RASH by them) resulting from replacement of glycine 12 by valine.

Ishii et al. (1985) pointed out similarities between the promoter of HRAS and that of EGF receptor (131550). This similarity is intriguing in light of the finding of Hiwasa et al. (1988) that the preferential degradation of EGF receptors by cathepsin L (116880) may be suppressed by HRAS gene products (p21s).

In a study of 118 lung cancer patients and 123 unaffected controls, Ryberg et al. (1990) found striking differences in the distribution of HRAS alleles. Six of 7 rare alleles were unique to the lung cancer group and 1 rare allele for the control group; rare alleles were found in 10 of 236 chromosomes in lung cancer patients as compared to 1 of 246 chromosomes in the controls. The lung cancer group also had a significantly lower frequency of 1 of the common alleles. The authors emphasized the importance of control for ethnic factors and the advantage in studying a relatively homogeneous population such as the Norwegian one.

The HRAS1 gene is tightly linked to a minisatellite located approximately 1 kb downstream from the gene's coding sequences and composed of 30 to 100 units of a 28-bp consensus sequence. Thirty alleles of 1,000 to 3,000 bp have been described. The 4 most common alleles—A1, A2, A3, and A4—represent 94% of all alleles in whites and have apparently served as progenitors for the remaining rare alleles. Rare alleles appear in the genomes of patients with cancer about 3 times as often as in controls without cancer (Krontiris et al., 1985); many such alleles have been observed only in patients with cancer. Krontiris et al. (1993) conducted a case-control study, typing 736 HRAS1 alleles from patients with cancer and 652 from controls by Southern blotting of leukocyte DNA. From analysis of the results and a meta-analysis of 22 other studies, they concluded that there was a significant association of rare HRAS1 alleles with 4 types of cancer: carcinomas of the breast, colorectum, and urinary bladder and acute leukemia. They considered it unlikely that the explanation could be found in linkage disequilibrium between these rare alleles and a pathogenetic lesion in the HRAS1 locus or other neighboring loci. Alternatively, they pointed to observations that new mutations of the HRAS1 minisatellite disrupt the controlled expression of nearby genes, including HRAS1, by interacting directly with transcriptional regulatory mechanisms. Furthermore, the minisatellite is capable of activating and repressing transcription; allele-specific effects have been observed.

Phelan et al. (1996) demonstrated a modifier effect of the HRAS1 locus on the penetrance of the BRCA1 gene (113705) in causing ovarian cancer. The polymorphism in question, a VNTR located 1 kb downstream of the HRAS1 gene, had previously been found to show an association between rare alleles and an increased risk of certain types of cancers, including breast cancer. The risk for ovarian cancer was 2.11 times greater for BRCA1 carriers harboring 1 or 2 rare HRAS1 alleles, compared to carriers with only common alleles (P = 0.015). A magnitude of the risk associated with a rare HRAS1 allele was not altered by adjusting for the other known risk factors for hereditary ovarian cancer. This study was said to have been the first to show the effect of a modifying gene on the penetrance of an inherited cancer syndrome.

Pseudogenes

O'Brien et al. (1983) assigned the HRAS2 gene to the X chromosome. Land et al. (1983) listed HRAS2 as a pseudogene. Hameister and Adolph (1986) concluded there are no oncogenes on the X because the probable trans-acting regulation of oncogenes could not function for them. Furthermore, no specific chromosomal aberration of the X has been related to a specific neoplasm whereas most autosomes have been found to have such changes. No X-linked family cancer syndrome is known. Since the publication of Hameister and Adolph's hypothesis, an expressed oncogene, ARAF1 (311010), has been mapped to the X chromosome.

See also Croce (1983); de Martinville and Francke (1983); Eccles et al. (1984); Fujita et al. (1984); Gibbs et al. (1984); Huerre et al. (1983); Newbold and Overell (1983); Popescu et al. (1985); Seeburg et al. (1984); Stallings et al. (1986)

Bianchi, A. B.; Rinchik, E. M.; Conti, C. J.: Reassignment of the H-ras-1 gene to the Hbb-terminus region of mouse chromosome 7. Mammalian Genome 4: 220-222, 1993.

Capon, D. J.; Chen, E. Y.; Levinson, A. D.; Seeburg, P. H.; Goeddel, D. V.: Complete nucleotide sequences of the T24 human bladder carcinoma oncogene and its normal homologue. Nature 302: 33-37, 1983.

Chaganti, R. S. K.; Jhanwar, S. C.; Antonarakis, S. E.; Hayward, W. S.: Germline chromosomal localization of genes in chromosome 11p linkage: parathyroid hormone, beta-globin, c-Ha-ras-1, and insulin. Somat. Cell Molec. Genet. 11: 197-202, 1985.

Chang, E. H.; Gonda, M. A.; Ellis, R. W.; Scolnick, E. M.; Lowy, D. R.: Human genome contains four genes homologous to transforming genes of Harvey and Kirsten murine sarcoma viruses. Proc. Nat. Acad. Sci. 79: 4848-4852, 1982.

Corell, B.; Zoll, B.: Comparison between the allelic frequency distribution of the Ha-ras-1 locus in normal individuals and patients with lymphoma, breast, and ovarian cancer. Hum. Genet. 79: 225-259, 1988.

Croce, C. M.: Personal Communication. Philadelphia, Pa. 5/9/1983.

de Martinville, B.; Francke, U.: The c-Ha-ras1, insulin and beta-globin loci map outside the deletion associated with aniridia-Wilms' tumour. Nature 305: 641-643, 1983.

de Martinville, B.; Francke, U.: HRAS1, insulin, and beta-globin map outside of 11p11.2-11p14.1. (Abstract) Cytogenet. Cell Genet. 37: 530, 1984.

de Martinville, B.; Giacalone, J.; Shih, C.; Weinberg, R. A.; Francke, U.: Oncogene from human EJ bladder carcinoma is located on the short arm of chromosome 11. Science 219: 498-501, 1983.

Der, C. J.; Krontiris, T. G.; Cooper, G. M.: Transforming genes of human bladder and lung carcinoma cell lines are homologous to the ras genes of Harvey and Kirsten sarcoma viruses. Proc. Nat. Acad. Sci. 79: 3637-3640, 1982.

Eccles, M. R.; Millow, L. J.; Wilkins, R. J.; Reeve, A. E.: Harvey-ras allele deletion detected by in situ hybridization to Wilms' tumor chromosomes. Hum. Genet. 67: 190-192, 1984.

Fearon, E. R.; Antonarakis, S. E.; Meyers, D. A.; Levine, M. A.: c-Ha-ras-1 oncogene lies between beta-globin and insulin loci on human chromosome 11p. Am. J. Hum. Genet. 36: 329-337, 1984.

Fearon, E. R.; Feinberg, A. P.; Hamilton, S. H.; Vogelstein, B.: Loss of genes on the short arm of chromosome 11 in bladder cancer. Nature 318: 377-380, 1985.

Feinberg, A. P.; Vogelstein, B.; Droller, M. J.; Baylin, S. B.; Nelkin, B. D.: Mutation affecting the 12th amino acid of the C-Ha-ras oncogene product occurs infrequently in human cancer. Science 220: 1175-1177, 1983.

Fisher, J. H.; Miller, Y. E.; Sparkes, R. S.; Bateman, J. B.; Kimmel, K. A.; Carey, T. E.; Rodell, T.; Shoemaker, S. A.; Scoggin, C. H.: Wilms' tumor-aniridia association: segregation of affected chromosome in somatic cell hybrids, identification of cell surface antigen associated with deleted area, and regional mapping of c-Ha-ras-1 oncogene, insulin gene, and beta-globin gene. Somat. Cell Molec. Genet. 10: 455-464, 1984.

Fujita, J.; Srivastava, S. K.; Kraus, M. H.; Rhim, J. S.; Tronick, S. R.; Aaronson, S. A.: Frequency of molecular alterations affecting ras protooncogenes in human urinary tract tumors. Proc. Nat. Acad. Sci. 82: 3849-3853, 1985.

Fujita, J.; Yoshida, O.; Yuasa, Y.; Rhim, J. S.; Hatanaka, M.; Aaronson, S. A.: Ha-ras oncogenes are activated by somatic alterations in human urinary tract tumours. Nature 309: 464-466, 1984.

Gerhard, D. S.; Kidd, K. K.; Housman, D.; Gusella, J. F.; Kidd, J. R.: Data on the genetic map of the short arm of chromosome 11 (11p). (Abstract) Cytogenet. Cell Genet. 37: 478, 1984.

Gibbs, J. B.; Ellis, R. W.; Scolnick, E. M.: Autophosphorylation of v-Ha-ras p21 is modulated by amino acid residue 12. Proc. Nat. Acad. Sci. 81: 2674-2678, 1984.

Goyette, M.; Petropoulos, C. J.; Shank, P. R.; Fausto, N.: Expression of a cellular oncogene during liver regeneration. Science 219: 510-512, 1983.

Greenhalgh, D. A.; Kinsella, A. R.: c-Ha-ras not c-Ki-ras activation in three colon tumour cell lines. Carcinogenesis 6: 1533-1535, 1985.

Hameister, H.; Adolph, S.: Oncogenes and the mammalian X chromosome. Hum. Genet. 72: 241-244, 1986.

Harper, M. E.: Personal Communication. La Jolla, Calif. 8/23/1983.

Hiwasa, T.; Sakiyama, S.; Yokoyama, S.; Ha, J.-M.; Fujita, J.; Noguchi, S.; Bando, Y.; Kominami, E.; Katunuma, N.: Inhibition of cathepsin L-induced degradation of epidermal growth factor receptors by c-Ha-ras gene products. Biochem. Biophys. Res. Commun. 151: 78-85, 1988.

Huerre, C.; Despoisse, S.; Gilgenkrantz, S.; Lenoir, G. M.; Junien, C.: c-Ha-ras1 is not deleted in aniridia-Wilms' tumour association. Nature 305: 638-641, 1983.

Ishii, S.; Merlino, G. T.; Pastan, I.: Promoter region of the human Harvey ras proto-oncogene: similarity to the EGF receptor proto-oncogene promoter. Science 230: 1378-1381, 1985.

Jhanwar, S. C.; Neel, B. G.; Hayward, W. S.; Chaganti, R. S. K.: Localization of c-ras oncogene family on human germ-line chromosomes. Proc. Nat. Acad. Sci. 80: 4794-4797, 1983.

Junien, C.; Huerre, C.; Despoisse, S.; Gilgenkrantz, S.; Lenoir, G. M.: c-Ha-ras1 is not deleted in del(11p13) Wilms' tumor (WAGR) and maps to 11p15.1-11p15.5. (Abstract) Cytogenet. Cell Genet. 37: 503, 1984.

Krontiris, T. G.; Devlin, B.; Karp, D. D.; Robert, N. J.; Risch, N.: An association between the risk of cancer and mutations in the HRAS1 minisatellite locus. New Eng. J. Med. 329: 517-523, 1993.

Krontiris, T. G.; DiMartino, N. A.; Colb, M.; Parkinson, D. R.: Unique allelic restriction fragments of the human Ha-ras locus in leukocyte and tumour DNAs of cancer patients. Nature 313: 369-374, 1985.

Land, H.; Parada, L. F.; Weinberg, R. A.: Cellular oncogenes and multistep carcinogenesis. Science 222: 771-778, 1983.

Muschel, R. J.; Khoury, G.; Lebowitz, P.; Koller, R.; Dhar, R.: The human c-ras1(H) oncogene: a mutation in normal and neoplastic tissue from the same patient. Science 219: 853-856, 1983. Note: Retraction: Science 220: 336 only, 1983.

Newbold, R. F.; Overell, R. W.: Fibroblast immortality is a prerequisite for transformation by EJ c-Ha-ras oncogene. Nature 304: 648-651, 1983.

O'Brien, S. J.; Nash, W. G.; Goodwin, J. L.; Lowy, D.; Chang, E. H.: Dispersion of the ras family of transforming genes to four different chromosomes in man. Nature 302: 839-842, 1983.

Phelan, C. M.; Rebbeck, T. R.; Weber, B. L.; Devilee, P.; Ruttledge, M. H.; Lynch, H. T.; Lenoir, G. M.; Stratton, M. R.; Easton, D. F.; Ponder, B. A. J.; Cannon-Albright, L.; Larsson, C.; Goldgar, D. E.; Narod, S. A.: Ovarian cancer risk in BRCA1 carriers is modified by the HRAS1 variable number of tandem repeat (VNTR) locus. Nature Genet. 12: 309-311, 1996.

Pincus, M. R.; van Renswoude, J.; Harford, J. B.; Chang, E. H.; Carty, R. P.; Klausner, R. D.: Prediction of the three-dimensional structure of the transforming region of the EJ/T24 human bladder oncogene product and its normal cellular homologue. Proc. Nat. Acad. Sci. 80: 5253-5257, 1983.

Popescu, N. C.; Amsbaugh, S. C.; DiPaolo, J. A.; Tronick, S. R.; Aaronson, S. A.; Swan, D. C.: Chromosomal localization of three human ras genes by in situ molecular hybridization. Somat. Cell Molec. Genet. 11: 149-155, 1985.

Russell, M. W.; Munroe, D. J.; Bric, E.; Housman, D. E.; Dietz-Band, J.; Riethman, H. C.; Collins, F. S.; Brody, L. C.: A 500-kb physical map and contig from the Harvey ras-1 gene to the 11p telomere. Genomics 35: 353-360, 1996.

Ryberg, D.; Tefre, T.; Ovrebo, S.; Skaug, V.; Stangeland, L.; Naalsund, A.; Baera, R.; Borresen, A.-L.; Haugen, A.: Ha-ras-1 alleles in Norwegian lung cancer patients. Hum. Genet. 86: 40-44, 1990.
Seeburg, P. H.; Colby, W. W.; Capon, D. J.; Goeddel, D. V.; Levinson, A. D.: Biological properties of human c-Ha-ras1 genes mutated at codon 12. Nature 312: 71-75, 1984.
Sekiya, T.; Fushimi, M.; Hori, H.; Hirohashi, S.; Nishimura, S.; Sugimura, T.: Molecular cloning and the total nucleotide sequence of the human c-Ha-ras-1 gene activated in a melanoma from a Japanese patient. Proc. Nat. Acad. Sci. 81: 4771-4775, 1984.
Srivastava, S. K.; Yuasa, Y.; Reynolds, S. H.; Aaronson, S. A.: Effects of two major activating lesions on the structure and conformation of human ras oncogene products. Proc. Nat. Acad. Sci. 82: 38-42, 1985.
Stallings, R. L.; Crawford, B. D.; Black, R. J.; Chang, E. H.: Assignment of RAS proto-oncogenes in Chinese hamsters: implications for mammalian gene linkage conservation and neoplasia. Cytogenet. Cell Genet. 43: 2-5, 1986.
Taparowsky, E.; Suard, Y.; Fasano, O.; Shimizu, K.; Goldfarb, M.; Wigler, M.: Activation of the T24 bladder carcinoma transforming gene is linked to a single amino acid change. Nature 300: 762-765, 1982.
Tong, L.; de Vos, A. M.; Milburn, M. V.; Jancarik, J.; Noguchi, S.; Nishimura, S.; Miura, K.; Ohtsuka, E.; Kim, S.-H.: Structural differences between a RAS oncogene protein and the normal protein. Nature 337: 90-93, 1989.
Wong-Staal, F.; Dalla-Favera, R.; Franchini, G.; Gelmann, E. P.; Gallo, R. C.: Three distinct genes in human DNA related to the transforming genes of mammalian sarcoma retroviruses. Science 213: 226-228, 1981.
Yokota, J.; Tsunetsugu-Yokota, Y.; Battifora, H.; Le Fevre, C.; Cline, M. J.: Alterations of myc, myb, and ras(Ha) proto-oncogenes in cancers are frequent and show clinical correlation. Science 231: 261-265, 1986.

ONCOGENE K-RAS2: V-KI-RAS2 KIRSTEN RAT SARCOMA 2 VIRAL ONCOGENE HOMOLOG 1 (*190070)

See 190020. KRAS2 is on chromosome 12; KRAS1P, a pseudogene, is on chromosome 6. Weinberg (1982) suggested that the then-recognized cellular oncogenes were assignable to a small number of gene families; e.g., the ras family with at least 4 distinct oncogenes, and the src-yes-mos family with another 3. The probable role of at least 2 oncogenes in normal differentiation is indicated by the findings of transcription of KRAS and the McDonough strain of feline sarcoma virus (FMS) during mouse development (Muller et al., 1983). Furthermore, the differences in transcription in different tissues suggested a specific role for each: FMS was expressed in extraembryonic structures or in transport in these tissues, whereas KRAS was expressed ubiquitously. KRAS, with a length of more than 30 kbp, is much larger than HRAS (190020) or NRAS (164790). Although the 3 ras genes, HRAS, KRAS, and NRAS, have different genetic structures, all code for proteins of 189 amino acid residues, generically designated p21. These genes acquire malignant properties by single point mutations that affect the incorporation of the 12th or 61st amino acid residue of their respective p21. KRAS is involved in malignancy much more often than is HRAS. In a study of 96 human tumors or tumor cell lines in the NIH 3T3 transforming system, Pulciani et al. (1982) found a mutated HRAS locus only in T24 bladder cancer cells, whereas transforming KRAS genes were identified in 8 different carcinomas and sarcomas.

In a serous cystadenocarcinoma of the ovary, Feig et al. (1984) showed the presence of an activated KRAS oncogene not activated in normal cells of the same patient. The transforming gene product displayed an electrophoretic mobility in SDS-polyacrylamide gels that differed from the mobility of KRAS transforming proteins in other tumors. Thus, a previously undescribed mutation was responsible for activation of KRAS in this ovarian carcinoma. To study the role of oncogenes in lung cancer, Rodenhuis et al. (1987) used a novel, highly sensitive assay based on oligonucleotide hybridization following an in vitro amplification step. Genomic DNA was examined from 39 tumor specimens obtained at thoracotomy. The KRAS gene was found to be activated by point mutations in codon 12 in 5 of 10 adenocarcinomas. Two of these tumors were less than 2 cm in size and had not metastasized. No HRAS, KRAS, or NRAS mutations were observed in 15 squamous-cell carcinomas, 10 large-cell carcinomas, 1 carcinoid, 2 metastatic adenocarcinomas from primary tumors outside the lung, and 1 small-cell carcinoma. An approximately 20-fold amplification of the unmutated KRAS gene was observed in a tumor that proved to be a solitary lung metastasis of a rectal carcinoma. Yanez et al. (1987) found mutations in codon 12 of the KRAS gene in 4 of 16 colon cancers, 2 of 27 lung cancers, and 1 of 8 breast cancers; no mutations were found at position 61. Of the 6 possible amino acid replacements in codon 12, all but one were represented in the 7 mutations identified. Burmer and Loeb (1989) studied 40 carcinomas and 12 adenomas of the colon for ploidy alterations and mutations in KRAS2. KRAS2 mutations were found in both diploid and aneuploid cells in both adenomas and carcinomas. Twenty-six of 40 colon carcinomas contained mutations at codon 12, and 9 of the 12 adenomas studied contained similar mutations.

By in situ hybridization, Popescu et al. (1985) mapped the KRAS2 gene to 12p12.1-p11.1. By linkage with RFLPs, O'Connell et al. (1985) confirmed the approximate location of KRAS2 on 12p12.1.

Pallister-Killian syndrome is a dysmorphic syndrome characterized by a tissue-limited mosaicism; a majority of fibroblasts have 47 chromosomes with an extra small metacentric chromosome, whereas the karyotype of lymphocytes is normal. Peltomaki et al. (1987) used a KRAS2 gene probe to confirm that the extra chromosome in Pallister-Killian syndrome is composed of 2 short arms of chromosome 12. Hunter et al. (1985) used LDHB (150100) for the same purpose. Peltomaki et al. (1987) showed that the 2 short arms are identical, thus indicating that it represents an isochromosome 12p. Peltomaki et al. (1987) reported findings that may explain the mechanism of origin of the tissue-limited mosaicism: the proportion of abnormal mitoses fell dramatically during long-term culture of fibroblasts. See also Zhang et al. (1989). Wenger et al. (1990) concluded that the tissue-specific mosaicism in the 12p isochromosome syndrome is best explained by in vivo and perhaps also in vitro loss of the isochromosome from 47,XY (or XX), plus i(12p) cells. Soukup and Neidich (1990) made the diagnosis of Pallister-Killian syndrome on routine amniocentesis. In the aborted fetus, various degrees of mosaicism were found in 4 tissues. Schinzel (1991) reviewed the clinical and cytogenetic features of the Pallister-Killian syndrome, which is mosaic tetrasomy 12p. He reported that at birth the infants have temporofrontal balding or sparseness, which, together with the abundant hair over the top of the head, gives a pattern like that of the hair of Iroquois Indians. This pattern disappears later if the child continues to have abnormal hair. Speleman et al. (1991) used fluorescence in situ hybridization with chromosome 12-specific DNA probes for the rapid and reliable detection of the i(12p) chromosome. Detection was possible also in interphase cells. Zakowski et al. (1992) described absence of the pericardium and focal aplasia cutis in the axillary area.

The myelodysplastic syndrome (MDS), or preleukemia, is a hematologic disorder characterized by low blood counts, bone marrow cells of abnormal appearance, and progression to acute leukemia in as many as 30% of patients. In 2 of 4 patients with preleukemia and in 1 who progressed to acute leukemia from MDS, Liu et al. (1987) found a transforming allele of KRAS. In 1 preleukemic patient, they detected a novel mutation in codon 13 of KRAS in bone marrow cells harvested 1.5 years before the acute leukemia developed. The findings provided evidence that RAS mutations may be involved in the early stages of human leukemia.

In 8 of 9 patients with colorectal tumors that contained mutations of KRAS, Sidransky et al. (1992) found that the RAS mutations were detectable in DNA purified from stool. The patients included those with benign and malignant neoplasms from proximal and distal colonic epithelium. Ohashi et al. (1993) used interphase fluorescence in situ hybridization with a chromosome 12-specific alpha satellite probe for diagnosis in buccal mucosal cells. Isochromosome 12p-positive cells showed 3 signals over the nucleus, while diploid cells had 2 signals. Takeda et al. (1993) used a mutant-allele-specific amplification (MASA) method to detect KRAS mutations in cells obtained from the sputum of patients with lung cancer. A mutation was found in 1 of 5 patients studied. They suggested the MASA system could be applied to an examination of metastatic lung carcinomas, particularly from adenocarcinomas of the colon and pancreas in which KRAS mutations are frequently detected, and to mass screening for colorectal tumors, using DNA isolated from feces as template.

Using a sensitive PCR-based method, Lee et al. (1995) investigated the frequency of KRAS codon 12 point mutations in stomach cancer. Such were found in 11 (7.9%) of 140 samples. Seven cases had a 1-bp substitution changing GGT (gly) to AGT (ser)(190070.0007) and 2 had a 1-bp substitution changing GGT to AGT (asp)(190070.0005). Tumors located in the upper third of the stomach had a significantly higher frequency of KRAS codon 12 mutations (3 of 8; 37.5%) compared with tumors located in the middle (4 of 29; 13.8%) or lower (3 of 99; 3%) thirds of the stomach ($P = 0.001$). Among 8 patients with stomach cancer located in the upper part of the stomach, death occurred in 4 of 5 patients with tumors without KRAS gene mutations, but in none of the 3 patients with KRAS gene-mutated tumors.

Although several studies confirmed that approximately 40% of primary colorectal adenocarcinomas in humans contain a mutated form of the KRAS2 gene, the patterns of mutation at codons 12, 13, and 61 are not the same in different populations. Hayashi et al. (1996) used the MASA method to analyze the frequency and type of point mutations in these 3 codons in 319 colorectal cancer tissues collected from patients in Japan, They then compared these results with those from other sources to examine whether different geographic locations and environmental influences might impose distinct patterns on the spectrum of KRAS mutations. Comparing findings in the U.S., France, and Yugoslavia with those in Japan, a number of significant differences were found. A possible explanation put forth by Hayashi et al. (1996) was that an environmental carcinogen prevailing in a geographic region combines with the susceptibility of a particular tissue to dictate which type of DNA lesion will predominate. The predominance of G-to-A mutations among American and Japanese colorectal cancer patients could be attributable to alkylating agents or to the absence of direct interaction with any carcinogens. The prevalence of G-to-T mutations among Yugoslav and French patients might be ascribed to polycyclic aromatic hydrocarbons and heterocyclic amines.

Otori et al. (1997) examined tissue sections from 19 hyperplastic colorectal polyps for mutations in exons 12 and 13 of the KRAS gene. Such polyps are generally thought to be benign with no malignant potential. KRAS mutations were detected in 9 (47%) of 19 polyps, suggesting that some hyperplastic colorectal polyps may be true premalignant lesions.

Andreyev et al. (1997) used PCR amplification and DNA sequencing to investigate KRAS exon 1 mutations (codons 12 and 13) in histologic sections of colorectal adenocarcinomas. They examined samples from 98 patients with Dukes stage A or B fully resected colorectal cancers. Fourteen of these patients had subsequently relapsed. The presence of a KRAS mutation was not associated with tumor stage or histologic grade; neither was there any association with those patients who relapsed. The authors concluded that detection of KRAS mutation in early colorectal adenocarcinomas was of no prognostic value.

Pseudogenes

The KRAS1 gene has been assigned to chromosome 6 (O'Brien et al., 1983; McBride et al., 1983). By in situ hybridization, Popescu et al. (1985) assigned the KRAS1 gene to 6p12-p11. KRAS1 is a pseudogene and is therefore symbolized KRAS1P.

See also Capon et al. (1983); Der and Cooper (1983); McCoy et al. (1983); McGrath et al. (1983); Sakaguchi et al. (1984); Shimizu et al. (1983)

Almoguera, C.; Shibata, D.; Forrester, K.; Martin, J.; Arnheim, N.; Perucho, M.: Most human carcinomas of the exocrine pancreas contain mutant c-K-ras genes. Cell 53: 549-554, 1988.

Andreyev, H. J. N.; Tilsed, J. V. T.; Cunningham, D.; Sampson, S. A.; Norman, A. R.; Schneider, H. J.; Clarke, P. A.: K-ras mutations in patients with early colorectal cancers. Gut 41: 323-329, 1997.

Bollag, G.; Adler, F.; elMasry, N.; McCabe, P. C.; Connor, E., Jr.; Thompson, P.; McCormick, F.; Shannon, K.: Biochemical characterization of a novel KRAS insertion mutation from a human leukemia. J. Biol. Chem. 271: 32491-32494, 1996.

Burmer, G. C.; Loeb, L. A.: Mutations in the KRAS2 oncogene during progressive stages of human colon carcinoma. Proc. Nat. Acad. Sci. 86: 2403-2407, 1989.

Capon, D. J.; Seeburg, P. H.; McGrath, J. P.; Hayflick, J. S.; Edman, U.; Levinson, A. D.; Goeddel, D. V.: Activation of Ki-ras2 gene in human colon and lung carcinomas by two different point mutations. Nature 304: 507-513, 1983.

Der, C. J.; Cooper, G. M.: Altered gene products are associated with activation of cellular ras-k genes in human lung and colon carcinomas. Cell 32: 201-208, 1983.

Feig, L. A.; Bast, R. C., Jr.; Knapp, R. C.; Cooper, G. M.: Somatic activation of ras-K gene in a human ovarian carcinoma. Science 223: 698-701, 1984.

Grimmond, S. M.; Raghavan, D.; Russell, P. J.: Detection of a rare point mutation in Ki-ras of a human bladder cancer xenograft by polymerase chain reaction and direct sequencing. Urol. Res. 20: 121-126, 1992.

Hayashi, N.; Sugai, S.; Ito, I.; Nakamori, S.; Ogawa, M.; Nakamura, Y.: Ethnic difference in the pattern of K-ras oncogene mutations in human colorectal cancers. Hum. Mutat. 8: 258-261, 1996.

Hunter, A. G. W.; Clifford, B.; Cox, D. M.: The characteristic physiognomy and tissue specific karyotype distribution in the Pallister-Killian syndrome. Clin. Genet. 28: 47-53, 1985.

Kozma, S. C.; Bogaard, M. E.; Buser, K.; Saurer, S. M.; Bos, J. L.; Groner, B.; Hynes, N. E.: The human c-Kirsten ras gene is activated by a novel mutation in codon 13 in the breast carcinoma cell line MDA-MB231. Nucleic Acids Res. 15: 5963-5971, 1987.

Lee, K.-H.; Lee, J.-S.; Suh, C.; Kim, S.-W.; Kim, S.-B.; Lee, J.-H.; Lee, M.-S.; Park, M.-Y.; Sun, H.-S.; Kim, S.-H.: Clinicopathologic significance of the K-ras gene codon 12 point mutation in stomach cancer: an analysis of 140 cases. Cancer 75: 2794-2801, 1995.

Liu, E.; Hjelle, B.; Morgan, R.; Hecht, F.; Bishop, J. M.: Mutations of the Kirsten-ras proto-oncogene in human preleukaemia. Nature 330: 186-188, 1987.

McBride, O. W.; Swand, D. C.; Tronick, S. R.; Gol, R.; Klimanis, D.; Moore, D. E.; Aaronson, S. A.: Regional chromosomal localization of N-ras, K-ras-1, K-ras-2 and myb oncogenes in human cells. Nucleic Acids Res. 11: 8221-8236, 1983.

McCoy, M. S.; Toole, J. J.; Cunningham, J. M.; Chang, E. H.; Lowy, D. R.; Weinberg, R. A.: Characterization of a human colon/lung carcinoma oncogene. Nature 302: 79-81, 1983.

McGrath, J. P.; Capon, D. J.; Smith, D. H.; Chen, E. Y.; Seeburg, P. H.; Goeddel, D. V.; Levinson, A. D.: Structure and organization of the human Ki-ras proto-oncogene and a related processed pseudogene. Nature 304: 501-506, 1983.

Motojima, K.; Urano, T.; Nagata, Y.; Shiku, H.; Tsurifune, T.; Kanematsu, T.: Detection of point mutations in the Kirsten-ras oncogene provides evidence for the multicentricity of pancreatic carcinoma. Ann. Surg. 217: 138-143, 1993.

Muller, R.; Slamon, D. J.; Adamson, E. D.; Tremblay, J. M.; Muller, D.; Cline, M. J.; Verma, I. M.: Transcription of c-onc genes c-ras(Ki) and c-fms during mouse development. Molec. Cell. Biol. 3: 1062-1069, 1983.

Nakano, H.; Yamamoto, F.; Neville, C.; Evans, D.; Mizuno, T.; Perucho, M.: Isolation of transforming sequences of two human lung carcinomas: structural and functional analysis of the activated c-K-ras oncogenes. Proc. Nat. Acad. Sci. 81: 71-75, 1984.

O'Brien, S. J.; Nash, W. G.; Goodwin, J. L.; Lowry, D. R.; Chang, E. H.: Dispersion of the ras family of transforming genes to four different chromosomes in man. Nature 302: 839-842, 1983.

O'Connell, P.; Leppert, M.; Hoff, M.; Kumlin, E.; Thomas, W.; Cai, G.; Law, M.; White, R.: A linkage map for human chromosome 12. (Abstract) Am. J. Hum. Genet. 37: A169 only, 1985.

Ohashi, H.; Ishikiriyama, S.; Fukushima, Y.: New diagnostic method for Pallister-Killian syndrome: detection of i(12p) in interphase nuclei of buccal mucosa by fluorescence in situ hybridization. Am. J. Med. Genet. 45: 123-128, 1993.

Otori, K.; Oda, Y.; Sugiyama, K.; Hasebe, T.; Mukai, K.; Fujii, T.; Tajiri, H.; Yoshida, S.; Fukushima, S.; Esumi, H.: High frequency of K-ras mutations in human colorectal hyperplastic polyps. Gut 40: 660-663, 1997.

Peltomaki, P.; Knuutila, S.; Ritvanen, A.; Kaitila, I.; de la Chapelle, A.: Pallister-Killian syndrome: cytogenetic and molecular studies. Clin. Genet. 31: 399-405, 1987.

Popescu, N. C.; Amsbaugh, S. C.; DiPaolo, J. A.; Tronick, S. R.; Aaronson, S. A.; Swan, D. C.: Chromosomal localization of three human ras genes by in situ molecular hybridization. Somat. Cell Molec. Genet. 11: 149-155, 1985.

Pulciani, S.; Santos, E.; Lauver, A. V.; Long, L. K.; Aaronson, S. A.; Barbacid, M.: Oncogene in solid human tumors. Nature 300: 539-542, 1982.

Rodenhuis, S.; van de Wetering, M. L.; Mooi, W. J.; Evers, S. G.; van Zandwijk, N.; Bos, J. L.: Mutational activation of the K-RAS oncogene: a possible pathogenetic factor in adenocarcinoma of the lung. New Eng. J. Med. 317: 929-935, 1987.

Sakaguchi, A. Y.; Zabel, B. U.; Grzeschik, K. H.; Law, M. L.; Ellis, R. W.; Skolnick, E. M.; Naylor, S. L.: Regional localization of two human cellular Kirsten ras genes on chromosomes 6 and 12. Molec. Cell. Biol. 4: 989-993, 1984.

Santos, E.; Martin-Zanca, D.; Reddy, E. P.; Pierotti, M. A.; Della Porta, G.; Barbacid, M.: Malignant activation of a K-ras oncogene in lung carcinoma but not in normal tissue of the same patient. Science 223: 661-664, 1984.

Schinzel, A.: Tetrasomy 12p (Pallister-Killian syndrome). J. Med. Genet. 28: 122-125, 1991.

Shimizu, K.; Birnbaum, D.; Ruley, M. A.; Fasano, O.; Suard, Y.; Edlund, L.; Taparowsky, E.; Goldfarb, M.; Wigler, M.: Structure of the Ki-ras gene of the human lung carcinoma cell line Calu-1. Nature 304: 497-500, 1983.

Sidransky, D.; Tokino, T.; Hamilton, S. R.; Kinzler, K. W.; Levin, B.; Frost, P.; Vogelstein, B.: Identification of RAS oncogene mutations in the stool of patients with curable colorectal tumors. Science 256: 102-105, 1992.

Smit, V. T. H. B. M.; Boot, A. J. M.; Smits, A. M. M.; Fleuren, G. J.; Cornelisse, C. J.; Bos, J. L.: KRAS codon 12 mutations occur very frequently in pancreatic adenocarcinomas. Nucleic Acids Res. 16: 7773-7782, 1988.

Soukup, S.; Neidich, K.: Prenatal diagnosis of Pallister-Killian syndrome. Am. J. Med. Genet. 35: 526-528, 1990.

Speleman, F.; Leroy, J. G.; Van Roy, N.; De Paepe, A.; Suijkerbuijk, R.; Brunner, H.; Looijenga, L.; Verschraegen-Spae, M.-R.; Orye, E.: Pallister-Killian syndrome: characterization of the isochromosome 12p by fluorescent in situ hybridization. Am. J. Med. Genet. 41: 381-387, 1991.

Takeda, S.; Ichii, S.; Nakamura, Y.: Detection of K-ras mutation in sputum by mutant-allele-specific amplification (MASA). Hum. Mutat. 2: 112-117, 1993.

Weinberg, R. A.: Fewer and fewer oncogenes. Cell 30: 3-4, 1982.

Wenger, S. L.; Boone, L. Y.; Steele, M. W.: Mosaicism in Pallister i(12p) syndrome. Am. J. Med. Genet. 35: 523-525, 1990.

Yanez, L.; Groffen, J.; Valenzuela, D. M.: c-K-ras mutations in human carcinomas occur preferentially in codon 12. Oncogene 1: 315-318, 1987.

Zakowski, M. F.; Wright, Y.; Ricci, A., Jr.: Pericardial agenesis and focal aplasia cutis in tetrasomy 12p (Pallister-Killian syndrome). Am. J. Med. Genet. 42: 323-325, 1992.

Zhang, J.; Marynen, P.; Devriendt, K.; Fryns, J.-P.; Van den Berghe, H.; Cassiman, J.-J.: Molecular analysis of the isochromosome 12p in the Pallister-Killian syndrome: construction of a mouse-human hybrid cell line containing an i(12p) as the sole human chromosome. Hum. Genet. 83: 359-363, 1989.

ONCOGENE N-RAS: NEUROBLASTOMA RAS VIRAL ONCOGENE HOMOLOG (*164790)

Hall et al. (1983) cloned an oncogene, which they termed N-ras, from 2 human sarcoma cell lines, HT1080 and RD, and showed that it is a member of the RAS gene family; that it is encoded by a gene on chromosome 1; and that the same gene is activated in HL60, a promyelocytic leukemia line. Hall et al. (1983) settled on the designation NRAS 'after consultation with Wigler and with Weinberg.' De Martinville et al. (1984) assigned NRAS to 1p31-cen. By in situ hybridization, Davis et al. (1983) assigned the NRAS gene to the short arm of chromosome 1. A concentration of grains was observed just above the centromere in band 1p13. They commented on the wide dispersion of the oncogenes in the RAS family; each of the 5 mapped to date was on a separate chromosome. Ryan et al. (1983) confirmed assignment of HRAS to chromosome 11, KRAS to chromosome 12, and NRAS1 to chromosome 1. Addendum in proof indicated that the same laboratory had assigned NRAS1 to 1p21-cen. The NRAS oncogene is distinct from the SK oncogene (164780) in several characteristics (Balazs, 1983) and has a different location on chromosome 1. By somatic cell hybrid studies and by in situ hybridization, Rabin et al. (1984) assigned the NRAS gene to 1p11-p13. By in situ hybridization, Popescu et al. (1985) also assigned the NRAS locus to 1p11-p13. Povey et al. (1985) reviewed the conflicting evidence on the site of NRAS on 1p. They found evidence favoring both 1p22 and 1p12-p11. Dracopoli and Meisler (1990) concluded from linkage analysis and pulsed field gel electrophoresis that TSHB (188540), NGFB (162030), and NRAS form a tightly linked gene cluster located in the same chromosomal band. Their location proximal to the AMY2B gene in 1p21 and close linkage to the alpha-satellite centromeric repeat D1Z5 provided strong evidence that the correct assignment for these 3 loci is 1p13 and not 1p22. Mitchell et al. (1995) localized NRAS to 1p13.2 and CD2 (186990) and NGFB (162030) to 1p13.1. They concluded that the order is as follows: cen—CD2—NGFB—NRAS—tel.

Deka et al. (1989) studied 62 host-teratoma pairs for chromosomal heteromorphisms and polymorphic DNA loci. Their findings contradicted the assumption of Linder et al. (1975) that teratomas originate only by meiosis II error; 27.3% originated by type I error and 15.9% by type IV error. See 166950. Deka et al. (1989) estimated the NRAS-centromere distance (y) to be 0.30. Teratomas originating by meiosis I and II are useful for gene mapping. In this study, at least, 80% of teratomas were found to be useful, a contradiction to the report of Parrington et al. (1984) that most teratomas are not informative for mapping. Nobori et al. (1991) pursued the fact that pCN2, the 3-prime nontranslated region of the NRAS gene, cross-hybridizes with unknown DNA segments on chromosomes 6, 9p, and 22, in addition to the NRAS gene itself on 1p. They mapped the pCN2 hybridizing region more specifically to 9p12-cen and proposed it as a useful marker for that region in studying human malignancies.

See also Bos et al. (1985); Davis et al. (1984); Hall and Brown (1985); Munke et al. (1984); Rabin et al. (1983); Taparowsky et al. (1983); Yuasa et al. (1984)

Balazs, I.: Personal Communication. New York, N. Y. 8/23/1983.
Bos, J. L.; Toksoz, D.; Marshall, C. J.; Verlaan-de Vries, M.; Veeneman, G. H.; van der Eb, A. J.; van Boom, J. H.; Janssen, J. W. G.; Steenvoorden, A. C. M.: Amino-acid substitutions at codon 13 of the N-ras oncogene in human acute myeloid leukaemia. Nature 315: 726-730, 1985.
Davis, M.; Malcolm, S.; Hall, A.: The N-ras oncogene is located on the short arm of chromosome 1. (Abstract) Cytogenet. Cell Genet. 37: 448-449, 1984.
Davis, M.; Malcolm, S.; Hall, A.; Marshall, C. J.: Localisation of the human N-ras oncogene to chromosome 1cen-p21 by in situ hybridisation. EMBO J. 2: 2281-2283, 1983.
Deka, R.; Majumder, P. P.; Warren, A. C.; Surti, U.; Hoffner, L.; Hauselman, E.; Antonarakis, S. E.; Ferrell, R. E.; Chakravarti, A.: Gene-centromere mapping using ovarian teratomas: results from chromosomes 1p, 13q and 21q. (Abstract) Am. J. Hum. Genet. 45 (suppl.): A137 only, 1989.
de Martinville, B.; Cunningham, J. M.; Murray, M. J.; Francke, U.: The N-ras oncogene assigned to chromosome 1 (p31-cen) by somatic cell hybrid analysis. (Abstract) Cytogenet. Cell Genet. 37: 531 only, 1984.
Dracopoli, N. C.; Meisler, M. H.: Mapping the human amylase gene cluster on the proximal short arm of chromosome 1 using a highly informative (CA)n repeat. Genomics 7: 97-102, 1990.
Hall, A.; Brown, R.: Human N-ras: cDNA cloning and gene structure. Nucleic Acids Res. 13: 5255-5268, 1985.
Hall, A.; Marshall, C. J.; Spurr, N. K.; Weiss, R. A.: Identification of transforming gene in two human sarcoma cell lines as a new member of the ras gene family located on chromosome 1. Nature 303: 396-400, 1983.
Linder, D.; McCaw, B. F.; Hecht, F.: Parthenogenic origin of benign ovarian teratomas. New Eng. J. Med. 292: 63-66, 1975.
Mitchell, E. L. D.; Jones, D.; White, G. R. M.; Varley, J. M.; Santibanez Koref, M. F.: Determination of the gene order of the three loci CD2, NGFB, and NRAS at human chromosome band 1p13 and refinement of their localisation at the subband level by fluorescence in situ hybridization. Cytogenet. Cell Genet. 70: 183-185, 1995.
Munke, M.; Lindgren, V.; de Martinville, B.; Francke, U.: Comparative analysis of mouse-human hybrids with rearranged chromosomes 1 by in situ hybridization and Southern blotting: high-resolution mapping of NRAS, NGFB, and AMY on human chromosome 1. Somat. Cell Molec. Genet. 10: 589-599, 1984.
Nitta, N.; Ochiai, M.; Nagao, M.; Sugimura, T.: Amino-acid substitution at codon 13 of the N-ras oncogene in rectal cancer in a Japanese patient. Jpn. J. Cancer Res. 78: 21-26, 1987.
Nobori, T.; Hexdall, L. E.; Carson, D. A.: A polymorphic region defined by pCN2 (the 3-prime nontranslated region of N-ras) maps to chromosome 9cen-p12. Hum. Genet. 87: 433-437, 1991.
Parrington, J. M.; West, L. F.; Povey, S.: The origin of ovarian teratomas. J. Med. Genet. 21: 4-12, 1984.
Popescu, N. C.; Amsbaugh, S. C.; DiPaolo, J. A.; Tronick, S. R.; Aaronson, S. A.; Swan, D. C.: Chromosomal localization of three human ras genes by in situ molecular hybridization. Somat. Cell Molec. Genet. 11: 149-155, 1985.
Povey, S.; Morton, N. E.; Sherman, S. L.: Report of the committee on the genetic constitution of chromosomes 1 and 2 (HGM8). Cytogenet. Cell Genet. 40: 67-106, 1985.
Rabin, M.; Watson, M.; Barker, P.; Ryan, J.; Breg, W. R.; Ruddle, F. H.: Chromosomal assignment of human c-fos and N-ras oncogenes. (Abstract) Am. J. Hum. Genet. 35: 148A only, 1983.
Rabin, M.; Watson, M.; Barker, P. E.; Ryan, J.; Breg, W. R.; Ruddle, F. H.: NRAS transforming gene maps to region p11-p13 on chromosome 1 by in situ hybridization. Cytogenet. Cell Genet. 38: 70-72, 1984.
Ryan, J.; Barker, P. E.; Shimizu, K.; Wigler, M.; Ruddle, F. H.: Chromosomal assignment of a family of human oncogenes. Proc. Nat. Acad. Sci. 80: 4460-4463, 1983.
Taparowsky, E.; Shimizu, K.; Goldfarb, M.; Wigler, M.: Structure and activation of the human N-ras gene. Cell 34: 581-586, 1983.
Yuasa, Y.; Gol, R. A.; Chang, A.; Chiu, I.-M.; Reddy, E. P.; Tronick, S. R.; Aaronson, S. A.: Mechanism of activation of an N-ras oncogene of SW-1271 human lung carcinoma cells. Proc. Nat. Acad. Sci. 81: 3670-3674, 1984.

ONCOGENE ROS (*165020)

ROS is the human homolog of the transforming gene, v-ros, of the avian sarcoma virus, UR2. Using a genomic clone, Nagarajan et al. (1986) mapped the ROS oncogene to 6q16-q22 by Southern analysis of DNA from hybrid cells and by in situ hybridization. They suggested that ROS joins MYB (189990), which is situated more distally on 6q, as a candidate for involvement in 6q deletions and rearrangements seen in various malignancies. Since ROS and MCF3, a transforming gene from a mammary carcinoma, map to the same region, and since the MCF3-deduced amino acid sequence shows homology to ROS, Nagarajan et al. (1986) suggested that the MCF3 and ROS clones 'probably represent the same genomic locus.' Yoshida et al. (1987) assigned the ROS gene to 6q22 by a combination of studies in somatic cell hybrids and in situ hybridization. By in situ hybridization with a fragment of the MCF3 gene from placenta, Rabin et al. (1985) mapped MCF3 to 6q16-q22. Rabin et al. (1987) localized ROS1 to 6q16-q22 by in situ hybridization. Satoh et al. (1987) mapped ROS1 to 6q22 by in situ hybridization. Chromosomal rearrangements in the 6q11-q31 region have been observed in acute lymphoblastic leukemia, malignant melanoma, and ovarian carcinoma. In a survey of 45 different human cell lines, Birchmeier et al. (1987) found ROS1 to be expressed in glioblastoma-derived cell lines at high levels and not to be expressed at all, or expressed at very low levels, in the remaining cell lines. The ROS1 gene was present in normal copy numbers in all cell lines that expressed the gene. However, in 1 particular glioblastoma line, Birchmeier et al. (1987) detected a potentially activating mutation at the ROS1 locus.

See also Birchmeier et al. (1986)

Birchmeier, C.; Birnbaum, D.; Waitches, G.; Fasano, O.; Wigler, M.: Characterization of an activated human ROS gene. Molec. Cell. Biol. 6: 3109-3116, 1986.
Birchmeier, C.; Sharma, S.; Wigler, M.: Expression and rearrangement of the ROS1 gene in human glioblastoma cells. Proc. Nat. Acad. Sci. 84: 9270-9274, 1987.
Nagarajan, L.; Louie, E.; Tsujimoto, Y.; Balduzzi, P. C.; Huebner, K.; Croce, C. M.: The human c-ros gene (ROS) is located at chromosome region 6q16-6q22. Proc. Nat. Acad. Sci. 83: 6568-6572, 1986.
Rabin, M.; Birnbaum, D.; Wigler, M.; Ruddle, F. H.: MCF3 oncogene mapped to region on chromosome 6 associated with malignant transformation. (Abstract) Am. J. Hum. Genet. 37 only, 1985.
Rabin, M.; Birnbaum, D.; Young, D.; Birchmeier, C.; Wigler, M.; Ruddle, F. H.: Human ROS1 and MAS1 oncogenes located in regions of chromosome 6 associated with tumor-specific rearrangements. Oncogene Res. 1: 169-178, 1987.
Satoh, H.; Yoshida, M. C.; Matsushime, H.; Shibuya, M.; Sasaki, M.: Regional localization of the human c-ros-1 on 6q22 and flt on 13q12. Jpn. J. Cancer Res. 78: 772-775, 1987.
Yoshida, M. C.; Satoh, H.; Matsushime, H.; Shibuya, M.; Sasaki, M.: Two ros-related protooncogenes, c-ros-1 and flt, are regionally mapped on human chromosomes 6q22 and 13q12, respectively. (Abstract) Cytogenet. Cell Genet. 46: 724 only, 1987.

ONCOGENE SIS: PLATELET-DERIVED GROWTH FACTOR, BETA POLYPEPTIDE (*190040)

Retroviruses have been identified as the etiologic agents of naturally occurring tumors in several animal species including certain human T-cell leukemias and lymphomas. Some of the viruses rapidly induce tumors when inoculated into animals and can transform cells in vitro. The genomes of these viruses contain sequences, called viral onc genes, which are directly responsible for transformation both in vitro and in vivo. Evidence indicates that these onc genes originated from normal cellular genes by recombination between a parent nontransforming virus and host cellular DNA. Molecular hybridization indicates considerable evolutionary conservation of the cellular genes that give rise to viral transforming genes, suggesting that cellular onc genes may code for important functions in cell growth or tissue differentiation. Dalla-Favera et al. (1981) reported the detection, molecular cloning and genomic organization of the human onc gene (c-sis) related to the transforming gene (v-sis) of simian sarcoma virus (SSV) derived from the wooly monkey. The gene was cloned from the human DNA library of Maniatis. The c-sis gene extends over a region of about 12 kilobases (kb) which includes 1.2 kb of v-sis-related sequences interrupted by four intervening sequences. The protein product of the c-sis gene has not been identified. Many of the known transforming proteins are kinases that have the unusual property of phosphorylating tyrosine residues. Some are structurally and functionally similar to cellular protein kinases involved in the regulation of the Na-K-ATPase pump. In an addendum to another paper (Wong-Staal et al., 1981), the same workers reported finding more than one allelic form of the c-sis locus.

Studies of hybrids between thymidine kinase-deficient mouse cells and human fibroblasts carrying a 22/17 translocation showed that the sis gene is in the region 22q11-22qter (Dalla-Favera et al., 1982). Swan et al. (1982) also mapped c-sis to chromosome 22 by somatic cell hybridization. They pointed out that the simian sarcoma virus was the only known transforming retrovirus of primate origin. Doolittle et al. (1983) concluded that the SIS oncogene is the same as, or very closely related to, the gene for platelet-derived growth factor. The conclusion was based on the demonstration of extensive sequence similarity. Waterfield et al. (1983) likewise pointed out the close structural similarity. PDGF is the major polypeptide mitogen in serum for cells of mesenchymal origin. Josephs et al. (1984) concluded that the SIS gene encodes one chain of human PDGF. PDGF has 2 dissimilar subunits, A and B. It is the B chain that has homology to SIS (Collins et al., 1985). Kelly et al. (1985) showed that the B chain alone is sufficient for mitogenesis. In the mouse SIS maps to chromosome 15 (Kozak et al., 1983), which shows other evidence of homology of synteny to human chromosome 22—diaphorase-1 (250800) and arylsulfatase A (250100) are on MM15 and HSA22. Chromosomal translocations involving 22q12 have been found in most cases of Ewing sarcoma (Aurias et al., 1983; Turc-Carel et al., 1983). Bechet et al. (1984) showed that the SIS oncogene is not activated in Ewing sarcoma. By in situ hybridization, Bartram et al. (1984) concluded that SIS is located in the region 22q12.3-22q13.1, far from the breakpoint 22q11 of CML, and that SIS segregates with the translocated part of chromosome 22 to various chromosomes in Ph1-positive cases of CML but remains on chromosome 22 in Ph1-negative cases. Thus, SIS is probably not involved in the malignant process. On germ-line chromosomes by in situ hybridization, Jhanwar et al. (1984) assigned SIS to 22q13.1, a site distal to the breakpoint involved in formation of the Philadelphia chromosome, 22q11. Julier et al. (1985) concluded from multilocus linkage tests that the oncogene SIS locus is most likely distal to MB and that both are distal to IGL. The following tentative map was derived: cen—IGL—0.10—D22S1–0.20—MB—0.07—(SIS, P1). Hermansson et al. (1988) presented evidence implicating an autocrine growth stimulation by PDGFB in the pathologic proliferation of endothelial cells characteristically found in glioblastomas.

Dermatofibrosarcoma protuberans, an infiltrative skin tumor of intermediate malignancy, presents specific cytogenetic features such as reciprocal translocations t(17;22)(q22;q13) and supernumerary ring chromosomes derived from t(17;22). Simon et al. (1997) characterized the breakpoints from translocations and rings in dermatofibrosarcoma protuberans and its juvenile form, giant cell fibroblastoma, on the genomic and RNA levels. They found that these rearrangements fuse the PDGFB gene and the COL1A1 gene (120150). Simon et al. (1997) commented that PDGFB has transforming activity and is a potent mitogen for a number of cell types, but its role in oncogenic processes was not fully understood. They noted that neither COL1A1 nor PDGFB had hitherto been implicated in tumor translocations. The gene fusions deleted exon 1 of PDGFB and released this growth factor from its normal regulation; see 190040.0002.

Disruption of the Pdgfb gene or the gene for its receptor in mice leads to the development of lethal hemorrhage and edema in late embryogenesis and absence of kidney glomerular mesangial cells. Lindahl et al. (1997) found that mouse embryos deficient in Pdgfb lack microvascular pericytes, which normally form part of the capillary wall, and develop numerous capillary microaneurysms that rupture at late gestation. Endothelial cells of the sprouting capillaries in mutant mice appeared to be unable to attract PDGF-receptor-beta-positive pericyte progenitor cells. Pericytes may contribute to the mechanical stability of the capillary wall. Comparisons made between PDGF-null mouse phenotypes suggested a general role for PDGFs in the development of myofibroblasts.

See also Bishop (1981); Deuel et al. (1983); Devare et al. (1983); Groffen et al. (1983); Josephs et al. (1983); Josephs et al. (1984); Julier et al. (1985); Julier et al. (1985); Leal et al. (1985); Owen et al. (1984); Rao et al. (1986); Robbins et al. (1983); Robbins et al. (1982)

Aurias, A.; Rimbaut, C.; Buffe, D.; Dubousset, J.; Mazabraud, A.: Chromosomal translocations in Ewing's sarcoma.(Letter) New Eng. J. Med. 309: 496-497, 1983.

Bartram, C. R.; de Klein, A.; Hagemeijer, A.; Grosveld, G.; Heisterkamp, N.; Groffen, J.: Localization of the human c-sis oncogene in Ph-1-positive and Ph-1-negative chronic myelocytic leukemia by in situ hybridization. Blood 63: 223-225, 1984.

Bechet, J.-M.; Bornkamm, G.; Freese, U.-K.; Lenoir, G. M.: The c-sis oncogene is not activated in Ewing's sarcoma.(Letter) New Eng. J. Med. 310: 393 only, 1984.

Bishop, J. M.: Enemies within: the genesis of retrovirus oncogenes. Cell 23: 5-6, 1981.

Bolger, G. B.; Stamberg, J.; Kirsch, I. R.; Hollis, G. F.; Schwarz, D. F.; Thomas, G. H.: Chromosomal translocation t(14;22) and oncogene (c-sis) variant in a pedigree with familial meningioma. New Eng. J. Med. 312: 564-567, 1985.

Cohen, J. B.; Levinson, A. D.: A point mutation in the last intron responsible for increased expression and transforming activity of the c-Ha-ras oncogene. Nature 334: 119-124, 1988.

Collins, T.; Ginsburg, D.; Boss, J. M.; Orkin, S. H.; Pober, J. S.: Cultured human endothelial cells express platelet-derived growth factor B chain: cDNA cloning and structural analysis. Nature 316: 748-750, 1985.

Dalla-Favera, R.; Gallo, R. C.; Giallongo, A.; Croce, C.: Chromosomal localization of the human homolog (c-sis) of the simian sarcoma virus onc gene. Science 218: 686-688, 1982.

Dalla-Favera, R.; Gelmann, E. P.; Gallo, R. C.; Wong-Staal, F.: A human onc gene homologous to the transforming gene (v-sis) of simian sarcoma virus. Nature 292: 31-35, 1981.

Deuel, T. F.; Huang, J. S.; Huang, S. S.; Stroobant, P.; Waterfield, M. D.: Expression of a platelet-derived growth factor-like protein in simian sarcoma virus transformed cells. Science 221: 1348-1350, 1983.

Devare, S. G.; Reddy, E. P.; Law, J. D.; Robbins, K. C.; Aaronson, S. A.: Nucleotide sequence of the simian sarcoma virus genome: demonstration that its acquired cellular sequences encode the transforming gene product p28-sis. Proc. Nat. Acad. Sci. 80: 731-735, 1983.

Doolittle, R. F.; Hunkapiller, M. W.; Hood, L. E.; Devare, S. G.; Robbins, K. C.; Aaronson, S. A.; Antoniades, H. N.: Simian sarcoma virus onc gene, v-sis, is derived from the gene (or genes) encoding a platelet-derived growth factor. Science 221: 275-277, 1983.

Groffen, J.; Heisterkamp, N.; Stephenson, J. R.; Geurts van Kessel, A.; de Klein, A.; Grosveld, G.; Bootsma, D.: c-sis is translocated from chromosome 22 to chromosome 9 in chronic myelocytic leukemia. J. Exp. Med. 158: 9-15, 1983.

Hermansson, M.; Nister, M.; Betsholtz, C.; Heldin, C.-H.; Westermark, B.; Funa, K.: Endothelial cell hyperplasia in human glioblastoma: coexpression of mRNA for platelet-derived growth factor (PDGF) B chain and PDGF receptor suggests autocrine growth stimulation. Proc. Nat. Acad. Sci. 85: 7748-7752, 1988.

Jhanwar, S. C.; Neel, B. G.; Hayward, W. S.; Chaganti, R. S. K.: Localization of the cellular oncogenes ABL, SIS, and FES on human germ-line chromosomes. Cytogenet. Cell Genet. 38: 73-75, 1984.

Josephs, S. F.; Dalla-Favera, R.; Gelmann, E. P.; Gallo, R. C.; Wong-Staal, F.: 5-prime viral and human cellular sequences corresponding to the transforming gene of simian sarcoma virus. Science 219: 503-505, 1983.

Josephs, S. F.; Guo, C.; Ratner, L.; Wong-Staal, F.: Human proto-oncogene nucleotide sequences corresponding to the transforming region of simian sarcoma virus. Science 223: 487-491, 1984.

Josephs, S. F.; Ratner, L.; Clarke, M. F.; Westin, E. H.; Reitz, M. S.; Wong-Staal, F.: Transforming potential of human c-sis nucleotide sequences encoding platelet-derived growth factor. Science 225: 636-639, 1984.

Julier, C.; Lathrop, M.; Lalouel, J. M.; Kaplan, J. C.: Use of multilocus tests of gene order: example for chromosome 22.(Abstract) Cytogenet. Cell Genet. 40: 663-664, 1985.

Julier, C.; Lathrop, M.; Lalouel, J. M.; Reghis, A.; Szajnert, M. F.; Kaplan, J. C.: New restriction fragment length polymorphisms on human chromosome 22 at loci SIS, MB and IGLV.(Abstract) Cytogenet. Cell Genet. 40: 664 only, 1985.

Julier, C.; Reghis, A.; Szajnert, M. F.; Kaplan, J. C.; Lathrop, G. M.; Lalouel, J. M.: A preliminary linkage map of human chromosome 22.(Abstract) Cytogenet. Cell Genet. 40: 665 only, 1985.

Kelly, J. D.; Raines, E. W.; Ross, R.; Murray, M. J.: The B chain of PDGF alone is sufficient for mitogenesis. EMBO J. 4: 3399-3405, 1985.

Kozak, C. A.; Sears, J. F.; Hoggan, M. D.: Genetic mapping of the mouse proto-oncogene c-sis to chromosome 15. Science 221: 867-869, 1983.

Leal, F.; Williams, L. T.; Robbins, K. C.; Aaronson, S. A.: Evidence that the v-sis gene product transforms by interaction with the receptor for platelet-derived growth factor. Science 230: 327-330, 1985.

Lindahl, P.; Johansson, B. R.; Leveen, P.; Betsholtz, C.: Pericyte loss and microaneurysm formation in PDGF-B-deficient mice. Science 277: 242-245, 1997.

Owen, A. J.; Pantazis, P.; Antoniades, H. N.: Simian sarcoma virus-transformed cells secrete a mitogen identical to platelet-derived growth factor. Science 225: 54-56, 1984.

Rao, C. D.; Igarashi, H.; Chiu, I.-M.; Robbins, K. C.; Aaronson, S. A.: Structure and sequence of the human c-sis/platelet-derived growth factor 2 (SIS/PDGF2) transcriptional unit. Proc. Nat. Acad. Sci. 83: 2392-2396, 1986.

Robbins, K. C.; Antoniades, H. N.; Devare, S. G.; Hunkapiller, M. W.; Aaronson, S. A.: Structural and immunological similarities between simian sarcoma virus gene product(s) and human platelet-derived growth factor. Nature 305: 605-608, 1983.

Robbins, K. C.; Devare, S. G.; Reddy, E. P.; Aaronson, S. A.: In vivo identification of the transforming gene product of simian sarcoma virus. Science 218: 1131-1133, 1982.

Simon, M.-P.; Pedeutour, F.; Sirvent, N.; Grosgeorge, J.; Minoletti, F.; Coindre, J.-M.; Terrier-Lacombe, M.-J.; Mandahl, N.; Craver, R. D.; Blin, N.; Sozzi, G.; Turc-Carel, C.; O'Brien, K. P.; Kedra, D.; Fransson, I.; Guilbaud, C.; Dumanski, J. P.: Deregulation of the platelet-derived growth factor B-chain gene via fusion with collagen gene COL1A1 in dermatofibrosarcoma protuberans and giant-cell fibroblastoma. Nature Genet. 15: 95-98, 1997.

Smidt, M.; Kirsch, I.; Ratner, L.: Deletion of Alu sequences in the fifth c-sis intron in individuals with meningiomas. J. Clin. Invest. 86: 1151-1157, 1990.

Swan, D. C.; McBride, O. W.; Robbins, K. C.; Keithley, D. A.; Reddy, E. P.; Aaronson, S. A.: Chromosomal mapping of the simian sarcoma virus onc gene analogue in human cells. Proc. Nat. Acad. Sci. 79: 4691-4695, 1982.

Turc-Carel, C.; Philip, I.; Berger, M. P.; Philip, T.; Lenoir, G. M.: Chromosomal translocations in Ewing's sarcoma.(Letter) New Eng. J. Med. 309: 497-498, 1983.

Waterfield, M. D.; Scrace, G. T.; Whittle, N.; Stroobant, P.; Johnsson, A.; Wasteson, A.; Westermark, B.; Heldin, C.-H.; Huang, J. S.; Deuel, T. F.: Platelet-derived growth factor is structurally related to the putative transforming protein p28(sis) of simian sarcoma virus. Nature 304: 35-39, 1983.

Wong-Staal, F.; Dalla-Favera, R.; Franchini, G.; Gelmann, E. P.; Gallo, R. C.: Three distinct genes in human DNA related to the transforming genes of mammalian sarcoma retroviruses. Science 213: 226-228, 1981.

ONCOGENE SRC: V-SRC AVIAN SARCOMA VIRAL ONCOGENE (*190090)

SRC is the symbol for the human gene homologous in sequence to the v-src gene of the Rous sarcoma virus (also called avian sarcoma virus, ASV). The human protooncogene was assigned to chromosome 20 by somatic cell hybrid studies (Sakaguchi et al., 1982). Le Beau et al. (1984) assigned SRC to 20q12-q13 by in situ hybridization. Lebo et al. (1984) and Parker et al. (1985) confirmed the assignment by dual-beam chromosome sorting and spot blot DNA analysis. Le Beau et al. (1985) found that deletions of 20q in myeloid disorders were actually interstitial although they appeared to be terminal; thus, the interstitial deletion had resulted in a shift at the SRC locus from 20q313 to the 20q breakpoint region. (FGR is the designation for an SRC-like oncogene that was called SRC2 at an earlier stage; see 164940.) Azarnia et al. (1988) found that overexpression of the SRC gene in NIH 3T3 cells caused reduction of cell-to-cell transmission of molecules in the 400- to 700-dalton range. Down-regulation was enhanced by point mutation of tyrosine-527, whereas mutation of tyrosine-416 suppressed both the down-regulation of communication by the tyr-527 mutation and that by gene overexpression. The regulation of communication by SRC may be important in the control of embryonic development and cellular growth. By in situ hybridization, Morris et al. (1989) placed the SRC gene at 20q11.2. They observed a secondary peak of grains in the region 20q13.2-qter, the localization of SRC suggested by previous in situ studies. Furthermore, Morris et al. (1989) found that 1 allele of the SRC gene was lost in 2 patients with leukemia and a deletion in 20q. They suggested that the deletions were interstitial. The new assignment, 20q11.2, is consistent with the assignment of HCK (142370) which presumably belongs to the same gene family, having originated from a common ancestral gene.

By homologous recombination in mouse embryonic stem (ES) cells, Soriano et al. (1991) produced mice carrying a null mutation in the SRC gene. Two independently targeted clones were used to generate chimeras that transmitted the mutated allele to their offspring. Intercrossing of heterozygotes gave rise to live-born homozygotes, but most of these mice died within the first few weeks of birth. Histologic and hematologic examination of the homozygous mutants did not show detectable abnormalities in the brain or platelets, where SRC is most highly expressed. Deficiency in bone remodeling, indicating impaired osteoclast function and leading to osteopetrosis, was found, however. These results showed that SRC is not required for general cell viability, possibly because of functional overlap with other related tyrosine kinases such as YES (164880), HCK, FGR, and LYN (165120), and uncovered an essential role for SRC in bone formation. Lowe et al. (1993) used in vitro approaches and fetal liver transplantation into irradiated SRC-deficient mice to demonstrate that the inherent defect resulting in osteopetrosis is in osteoclasts and is autonomous of the bone marrow microenvironment. They identified a cell type in which SRC function is essential and cannot be replaced by other related kinases. Lowe et al. (1993) suggested that it should be possible to isolate the substrate that is specific to SRC.

See also Czernilofsky et al. (1983); Gibbs et al. (1985); Sakaguchi et al. (1985); Sakaguchi et al. (1984)

Azarnia, R.; Reddy, S.; Kmiecik, T. E.; Shalloway, D.; Loewenstein, W. R.: The cellular src gene product regulates junctional cell-to-cell communication. Science 239: 398-401, 1988.

Czernilofsky, A. P.; Levinson, A. D.; Varmus, H. E.; Bishop, J. M.; Tischer, E.; Goodman, H.: Correction to the nucleotide sequence of the src gene of Rous sarcoma virus. Nature 301: 736-738, 1983.

Gibbs, C. P.; Tanaka, A.; Anderson, S. K.; Radul, J.; Baar, J.; Ridgway, A.; Kung, H.-J.; Fujita, D. J.: Isolation and structural mapping of a human c-src gene homologous to the transforming gene (v-src) of Rous sarcoma virus. J. Virol. 53: 19-24, 1985.

Le Beau, M. M.; Westbrook, C. A.; Diaz, M. O.; Rowley, J. D.: Evidence for two distinct c-src loci on human chromosomes 1 and 20. Nature 312: 70-71, 1984.

Le Beau, M. M.; Westbrook, C. A.; Diaz, M. O.; Rowley, J. D.: c-src is consistently conserved in the chromosomal deletion (20q) observed in myeloid disorders. Proc. Nat. Acad. Sci. 82: 6692-6696, 1985.

Lebo, R. V.; Cheung, M.-C.; Bruce, B. D.: Rapid gene mapping by dual laser chromosome sorting and spot blot DNA analysis. (Abstract) Am. J. Hum. Genet. 36: 101S, 1984.

Lowe, C.; Yoneda, T.; Boyce, B. F.; Chen, H.; Mundy, G. R.; Soriano, P.: Osteopetrosis in Src-deficient mice is due to an autonomous defect of osteoclasts. Proc. Nat. Acad. Sci. 90: 4485-4489, 1993.

Morris, C. M.; Honeybone, L. M.; Hollings, P. E.; Fitzgerald, P. H.: Localization of the SRC oncogene to chromosome band 20q11.2 and loss of this gene with deletion (20q) in two leukemic patients. Blood 74: 1768-1773, 1989.

Parker, R. C.; Mardon, G.; Lebo, R. V.; Varmus, H. E.; Bishop, J. M.: Isolation of duplicated human c-src genes located on chromosomes 1 and 20. Molec. Cell. Biol. 5: 831-838, 1985.

Sakaguchi, A. Y.; Mohandas, T.; Naylor, S. L.: A human c-src gene resides on the proximal long arm of chromosome 20 (cen-q13.1). Cancer Genet. Cytogenet. 18: 123-129, 1985.

Sakaguchi, A. Y.; Naylor, S. L.; Weinberg, R. A.; Shows, T. B.: Organization of human proto-oncogenes. (Abstract) Am. J. Hum. Genet. 34: 175A, 1982.

Sakaguchi, A. Y.; Zabel, B. U.; Grzeschik, K. H.; Law, M. L.; Naylor, S. L.: Human proto-oncogene assignments. (Abstract) Cytogenet. Cell Genet. 37: 572-573, 1984.

Soriano, P.; Montgomery, C.; Geske, R.; Bradley, A.: Targeted disruption of the c-src proto-oncogene leads to osteopetrosis in mice. Cell 64: 693-702, 1991.

ONCOGENE TC21 (*600098)

The TC21 oncogene, a member of the RAS superfamily, was initially cloned from a human teratocarcinoma cDNA library by PCR methods using degenerate oligonucleotides corresponding to the conserved region of the RAS genes (Drivas et al., 1990). Chan et al. (1994) found the same oncogene when they generated an expression cDNA library from an ovarian carcinoma line. They found, furthermore, that a single point mutation substituting glutamine for leucine at position 72 was responsible for activation of transforming properties. While the cDNA clone possessed high transforming activity, the ovarian tumor genomic DNA, which contained the mutated TC21 allele, failed to induce transformed foci. Thus, expression cDNA cloning made it possible to identify and isolate a human oncogene that had evaded detection by conventional approaches.

Chan, A. M.-L.; Miki, T.; Meyers, K. A.; Aaronson, S. A.: A human oncogene of the RAS superfamily unmasked by expression cDNA cloning. Proc. Nat. Acad. Sci. 91: 7558-7562, 1994.

Drivas, G. T.; Shih, A.; Coutavas, E.; Rush, M. G.; D'Eustachio, P.: Characterization of four novel ras-like genes expressed in a human teratocarcinoma cell line. Molec. Cell. Biol. 10: 1793-1798, 1990.

ONCOGENE WNT2: WINGLESS-TYPE MMTV INTEGRATION SITE FAMILY, MEMBER 2 (*147870)

Wainwright et al. (1988) cloned from a human lung cDNA library an expressed gene sequence that was identified by isolation of a methylation-free CpG island from human chromosome 7. (Bird et al. (1985) described so-called HTF (HpaII tiny fragments) islands which are relatively rich in G/C and virtually free of methylation at the CpG dinucleotide. Such regions usually occur at discrete units, 1 to 2 kb long, which can be detected by analysis with methylation-sensitive restriction enzymes. HTF islands are particularly associated with the site of initiation of transcription and the first exons of genes. Both tissue-specific and housekeeping genes have been described in association with HTF islands (Bird, 1987). The characteristics of HTF islands make it possible to isolate selectively regions of the genome that are likely to contain structural genes.) As part of the strategy to isolate the gene for cystic fibrosis (219700), Wainwright et al. (1988) found and characterized a coding sequence with marked similarity to the murine protooncogene Int1 and its Drosophila homolog 'wingless,' but which was distinct from the human INT1 gene located on chromosome 12 (164820). Indirect evidence that INT1 is secreted and that the product of 'wingless' is a diffusible gene product suggests that these proteins are secreted growth factors. Wainwright et al. (1988) suggested that the INT1-related protein is an additional member of the INT1 growth factor gene family. Because of its homology to protooncogene INT1, IRP is also symbolized INT1L1. IRP is expressed in a variety of fetal and adult human tissues that do not overlap with the pattern of expression of INT1. Chan et al. (1989) isolated overlapping genomic clones that correspond to the mouse homolog of the IRP gene and demonstrated in mouse-hamster somatic cell hybrids that the gene is located on mouse chromosome 6. In addition, they showed that the mouse Int1l1 and Met genes coamplified in lines of spontaneously transformed mouse NIH 3T3 cells, indicating that these genes are closely linked. IRP is about 500 kb from MET (164860); the CFTR gene (602421) is in the 500-kb interval between IRP and a point 1,000 kb from MET. Anonymous probe J3.11 is about 500 kb farther distally. Nusse et al. (1991) suggested that IRP be referred to as WNT2.

Bird, A.; Taggart, M.; Frommer, M.; Miller, O. J.; Macleod, D.: A fraction of the mouse genome that is derived from islands of nonmethylated, CpG-rich DNA. Cell 40: 91-99, 1985.

Bird, A. P.: CpG islands as gene markers in the vertebrate nucleus. Trends Genet. 3: 342-347, 1987.

Chan, A. M.-L.; Hilkens, J.; Kroezen, V.; Mitchell, P. J.; Scambler, P.; Wainwright, B. J.; Williamson, R.; Cooper, C. S.: Molecular cloning and localization to chromosome 6 of mouse INT1L1 gene. Somat. Cell Molec. Genet. 15: 555-562, 1989.

Nusse, R.; Brown, A.; Papkoff, J.; Scambler, P.; Shackleford, G.; McMahon, A.; Moon, R.; Varmus, H.: A new nomenclature for int-1 and related genes: the Wnt gene family. Cell 64: 231-232, 1991.
Wainwright, B. J.; Scambler, P. J.; Stanier, P.; Watson, E. K.; Bell, G.; Wicking, C.; Estivill, X.; Courtney, M.; Bour, A.; Pedersen, P. S.; Williamson, R.; Farrall, M.: Isolation of a human gene with protein sequence similarity to human and murine int-1 and the Drosophila segment polarity mutant wingless. EMBO J. 7: 1743-1748, 1988.

ONCOGENE WNT7B: WINGLESS-TYPE MMTV INTEGRATION SITE FAMILY, MEMBER 7B (*601967)

For general information about WNTs, see WNT1 (164820).

Gavin et al. (1990) cloned 6 murine Wnt genes, including Wnt7b. Huguet et al. (1994) derived a partial human WNT7B sequence that they used to examine its expression in human breast cell lines; they found that WNT7B was expressed at similar levels in normal and benign tissues but showed an elevation of about 30-fold in 10% of the tumors.

By fluorescence in situ hybridization, van Bokhoven et al. (1997) assigned the WNT7B gene to 22q13.3.

Gavin, B. J.; McMahon, J. A.; McMahon, A. P.: Expression of multiple novel Wnt-1/int-1-related genes during fetal and adult mouse development. Genes Dev. 4: 2319-2332, 1990.
Huguet, E. L.; McMahon, J. A.; McMahon, A. P.; Bicknell, R.; Harris, A. L.: Differential expression of human Wnt genes 2,3,4, and 7B in human breast cell lines and normal and disease states of human breast tissue. Cancer Res. 2615-2621, 1994.
van Bokhoven, H.; Kissing, J.; Schepens, M.; van Beersum, S.; Simons, A.; Riegman, P.; McMahon, J. A.; McMahon, A. P.; Brunner, H. G.: Assignment of WNT7B to human chromosome band 22q13 by in situ hybridization. Cytogenet. Cell Genet. 77: 288-289, 1997.

ONCOGENE WNT10B: WINGLESS-TYPE MMTV INTEGRATION SITE FAMILY, MEMBER 10B (*601906)

Several members of the Wnt gene family have been shown to cause mammary tumors in mice. Using degenerate primer PCR on human genomic DNA and specific PCR of cDNA libraries, Bui et al. (1997) isolated a Wnt gene that had not previously been described in human. It is the human homolog of mouse Wnt10b, which had been shown to be one of the oncogenes cooperating with FGF3 (164950) in the development of mouse mammary tumor virus (MMTV)-induced mammary carcinomas in mice. The human WNT10B sequence is 88 and 95% identical to the murine gene at nucleotide and amino acid levels, respectively. By YAC and fluorescence in situ hybridization (FISH) mapping, Bui et al. (1997) localized the gene to 12q13, a chromosomal region frequently rearranged in human tumors and also containing the WNT1 gene (164820). WNT10B expression was not observed in normal and benign proliferations of human breast tissue but was found to be elevated in 3 of 50 primary breast carcinomas. Southern blot analysis of the carcinoma expressing the highest level of WNT10B showed no amplification or rearrangement of the gene.

Hardiman et al. (1997) demonstrated that the WNT10B gene encodes a 389-amino acid protein with 96.6% sequence identity to mouse Wnt10b. The expression pattern showed that it is synthesized in many adult tissues with the highest levels found in heart and skeletal muscle. By PCR typing of a human/rodent monochromosomal panel and FISH, they mapped WNT10B to 12q13.1.

Bui, T. D.; Rankin, J.; Smith, K.; Huguet, E. L.; Ruben, S.; Strachan, T.; Harris, A. L.; Lindsay, S.: A novel human Wnt gene, WNT10B, maps to 12q13 and is expressed in human breast carcinomas. Oncogene 14: 1249-1253, 1997.
Hardiman, G.; Kastelein, R. A.; Bazan, J. F.: Isolation, characterization and chromosomal localization of human WNT10B. Cytogenet. Cell Genet. 77: 278-282, 1997.

ONCOGENE WNT13: WINGLESS-TYPE MMTV INTEGRATION SITE FAMILY, MEMBER 13 (*601968)

For general information about WNTs, see WNT1 (164820).

Katoh et al. (1996) cloned a novel WNT gene from a human gastric cancer cDNA library. Sequence analysis revealed that WNT13 was the human homolog of Xwnt2, a WNT gene cloned from Xenopus by Wolda and Moon (1992). Katoh et al. (1996) reported that the WNT13 gene encodes a 372-amino acid polypeptide including a putative secretory signal peptide. Northern blotting revealed that WNT13 was expressed as a 2.5-kb mRNA in fetal brain, lung, and kidney. Unlike most other WNTs, which tend to have restricted expression in adult tissues, its transcript was found in a wide variety of adult tissues. Katoh et al. (1996) used fluorescence in situ hybridization to map WNT13 to chromosome 1p13.

Katoh, M.; Hirai, M.; Sugimura, T.; Terada, M.: Cloning, expression and chromosomal localization of Wnt-13, a novel member of the Wnt gene family. Oncogene 13: 873-876, 1996.
Wolda, S. L.; Moon, R. T.: Cloning and developmental expression in Xenopus laevis of seven additional members of the Wnt family. Oncogene 7: 1941-1947, 1992.

ONCOGENE YUASA (164891)

Yuasa et al. (1990) searched for oncogenes associated with familial adenomatous polyposis by a tumorigenicity assay in nude mice. In the course of these studies, a transforming sequence was isolated that did not hybridize with 12 known oncogene probes. It was partially cloned and shown to be located on human chromosome 7. The gene did not hybridize with the MET (164860) and ERBB1 (131550) oncogenes which are located on chromosome 7. Yuasa et al. (1990) interpreted these data to indicate that they had found a new oncogene.

Yuasa, Y.; Kamiyama, T.; Kato, M.; Iwama, T.; Ikeuchi, T.; Tonomura, A.: Transforming genes from familial adenomatous polyposis patient cells detected by a tumorigenicity assay. Oncogene 5: 589-596, 1990.

Translocation Breakpoints and Fusion Proteins

JJM NOTE

This section of the *Catalog* was created by searching the *OMIM* field of title and clinical synopsis for the terms *translocat**, *breakpoint*, and *fusion*. Some entries were assigned to other chapters, especially when a familial phenotype or clinical syndrome was associated with the translocation. Others appear elsewhere in this chapter, usually as oncogenes that have become activated by the mechanism of translocation. Largely these entries represent somatic cell mutations recognized by either karyotypic abnormalities or homology to genes identified by that method. These 73 entries are not germline defects, but rare familial aggregations of four types of cancer that have been seen in association with a familial chromosomal translocation: familial retinoblastoma, familial Wilms tumor, familial meningioma, and familial renal cell carcinoma. Additional instances will emerge, and there may be some justification for karyotyping one member of cancer families, unless another single gene or environmental determinant is very obvious. The predominance of leukemia in the list probably reflects the ease of karyotyping bone marrow aspirate compared to solid tumors, hence the enormous literature on the nonrandom cytogenetic abnormalities associated with leukemia. There are sufficient examples of translocations and breakpoint genes in solid tumors (Ewing sarcoma, lung cancer, and even benign lipomas) to predict that many more such genes will be discovered. Such findings begin, often, with karyotypic analysis of the fresh operative specimen, before it is placed in fixative.

MYELOID/LYMPHOID OR MIXED-LINEAGE LEUKEMIA (ALL1) (*159555)

Recurring chromosomal translocations involving chromosome 11, band q23, have been observed in acute lymphoid leukemias and especially in acute myeloid leukemias. Rowley et al. (1990) demonstrated that the breakpoints in four 11q23 translocations associated with leukemia were contained within a yeast artificial chromosome (YAC) clone bearing the CD3D (186790) and CD3G (186740) gene loci. Within this YAC, Ziemin-van der Poel et al. (1991) identified a transcription unit that spans the breakpoint junctions of 3 of these translocations, 4;11, 9;11, and 11;19. They also described 2 other related transcripts that were upregulated in a translocation cell line. They named the gene MLL for myeloid/lymphoid, or mixed-lineage, leukemia. The types of acute lymphoblastic leukemia and acute myeloid leukemia that are particularly associated with translocations involving 11q23 are acute monoblastic leukemia (AML-M5) and acute myelomonocytic leukemia (AMML-M4). Gu et al. (1992) cloned translocation fragments from leukemic cells from t(4;11) and showed clustering of the breakpoints in an area of 7-8 kb on chromosome 4. Sequencing indicated heptamer and nonamer-like sequences, associated with rearrangements of immunoglobulin and T-cell receptor genes, near the breakpoints. This suggested a direct involvement of the VDJ recombinase in the 11q23 translocations. The gene coding for the RNAs transcribed from genomic DNA adjacent to the breakpoint cluster region was termed ALL1 by Cimino et al. (1991); it is the same as MLL. This large gene spans approximately 100 kb of DNA and contains at least 21 exons. It encodes a protein of more than 3,910 amino acids containing 3 regions with homology to sequences within the Drosophila 'trithorax' gene, including cysteine-rich regions that can be folded into 6 zinc finger-like domains. The breakpoint cluster region within ALL1 spans 8 kb and encompasses several small exons, most of which begin in the same phase of the open reading frame (Gu et al., 1992). Tkachuk et al. (1992) showed that the gene, which they referred to as HRX (for 'homolog of trithorax'), codes for a 431-kD protein. In studies of a t(11;19)-carrying cell line, Tkachuk et al. (1992) identified fusion transcripts expressed from both derivative chromosomes. The more abundant derivative 11 transcript coded for a chimeric protein containing the amino terminal 'AT-hook' motifs of the HRX gene fused to a previously undescribed gene/protein, ENL (MLLT1; 159556), from chromosome 19. (ENL was so named for '11-19 leukemia.') The HRX protein may have effects mediated by DNA binding within the minor groove at AT-rich sites. Tkachuk et al. (1992) referred to this type of leukemia as representing the multilineage leukemias rather than mixed-lineage leukemias. The cell line carrying the t(11;19) was from a patient with T-cell precursor acute lymphocytic leukemia (Smith et al., 1989). McCabe et al. (1992) presented evidence that the breakpoints in all the translocations involving 11q23 in leukemia cells—e.g., t(4;11) t(6;11), t(9;11), and t(11;19)—are clustered within a 9-kb BamHI genomic region of the MLL gene. McCabe et al. (1992) detected rearrangements of DNA in a fragment of the MLL gene by Southern blot hybridization. Djabali et al. (1992) also cloned an 11.5-kb transcript spanning the translocation breakpoint and demonstrated that it encodes a protein highly homologous to the Drosophila 'trithorax' gene, a developmental regulator. They concluded that most of the breakpoints in infant leukemias with t(4;11) and t(9;11) translocations lie within a 5-kb region. Parry et al. (1993) showed that the sequence of a partial DNA contained an open reading frame encoding 1,012 amino acids with extensive homology to the Drosophila trithorax protein, particularly in the zinc finger-like domains. The TRX1 gene appears to be unique in the human genome and has been conserved during evolution. Use of the human cDNA as a probe dem onstrated that the gene is interrupted in both infant and adult acute myeloid (AML) and lymphoid (ALL) leukemia patients with 11q23 translocations. The structure of the TRX1 gene around the breakpoints show that this part of the human gene is interrupted by 9 introns. As a result of the rearrangement, zinc finger domains are translocated in both ALL and AML patients.

Translocations involving 11q23 in leukemia result in the translocation of zinc finger domains with fusion to other genes on chromosome 4, chromosome 9, or chromosome 19. The gene on chromosome 19 with which it is fused is ENL (159556). Nakamura et al. (1993) showed that the genes with which it is fused on chromosome 4 (AF4; 159557) and chromosome 9 (AF9; 159558) show high homology of sequence to ENL. The protein products of the AF4, AF9, and ENL proteins contained nuclear targeting sequences as well as serine-rich and proline-rich regions.

Independently, Domer et al. (1993) characterized the MLL-AF4 fusion product generated by the t(4;11) translocation. The sequence of the complete open reading frame for this fusion transcript revealed that the MLL protein is homologous to DNA methyltransferase. In the fusion gene, the 5-prime portion is derived from the MLL gene and the 3-prime portion from the AF4 gene.

Thirman et al. (1993) demonstrated that MLL gene rearrangements can be detected with a single probe and a single restriction-enzyme digest. The ability to detect an MLL gene rearrangement rapidly and reliably, especially in patients with limited material for cytogenetic analysis, should make it possible to identify patients who have a poor prognosis and therefore require aggressive chemotherapy or marrow transplantation. In 15 out of 26 AML cases in infants, Sorensen et al. (1994) found rearrangement of the MLL gene at the molecular level. These rearrangements were clustered within an 11-kb region containing 9 exons of the gene. In 14 of the 15 cases with rearrangements, the leukemia was associated with myelomonocytic or monocytic phenotypes (M4 or M5 FAB subtypes, respectively), both of which are associated with a poor prognosis in childhood AML. In contrast, only 1 of 11 nonrearranged cases had an M4 or M5 phenotype. Rearrangement also correlated significantly with hyperleukocytosis, another clinical parameter associated with poor outcome. Kobayashi et al. (1993) described a case of acute lymphoblastic leukemia in a 44-year-old woman after adjuvant chemotherapy of breast cancer; they demonstrated rearrangement of the HRX gene.

In a study of patients with acute leukemia but no microscopically visible change at 11q23, Schichman et al. (1994) found molecular evidence of partial duplication of the ALL1 gene. The direct tandem duplication involved a region spanning exons 2 to 6, and a partially duplicated protein gene product was demonstrated. Thus, the ALL1 gene is leukemogenic when it fuses with itself as well as when it fuses with one of the genes on other chromosomes.

In addition to the translocations involving fusion of the ALL1 gene with genes on other chromosomes producing acute lymphoblastic and myelogenous leukemia, the ALL1 gene undergoes self-fusion in acute myeloid leukemias with normal karyotype or trisomy 11. In addition, Baffa et al. (1995) reported rearrangement of the ALL1 gene in a gastric carcinoma cell line. A complex, 3-way translocation involving chromosomes 1 and 11 and resulting in partial duplication of the ALL1 gene was found. Sequencing of RT-PCR products and Northern blot analysis show that only the partially duplicated ALL1 gene was transcribed, producing an mRNA with exon 8 fused to exon 2. Thus, ALL1 gene rearrangement may play a role in the pathogenesis of some solid malignancies. The absence of the normal transcript in this cell line, in association with loss of heterozygosity on 11q23 seen in solid tumors, suggests that ALL1 is involved in tumorigenesis by a loss-of-function mechanism.

Approximately 90% of adult patients with de novo AML and trisomy 11 (+11) as a sole abnormality and 11% of adult patients with de novo AML and normal cytogenetics carry a molecular rearrangement of the ALL1 gene. The rearranged ALL1 gene results from the direct tandem duplication of a portion of ALL1 itself. Caligiuri et al. (1997) showed that in cytogenetically normal cases of AML and cases with +11 as the sole cytogenetic abnormality, only 1 chromosome contains the mutated ALL1 allele. Thus, a single mutated ALL1 allele with the partial tandem duplication is sufficient for ALL1-associated leukemogenesis, irrespective of the number of normal genes present. The frequently occurring specific association of +11 and ALL1 gene mutation in the leukemic clone remained unexplained.

Butler et al. (1997) analyzed the distribution and localization of HRX proteins in cell lines and human tissues, using both polyclonal and monoclonal antibodies. Immunocytochemical analysis showed a punctate distribution of wildtype and chimeric HRX proteins within cell nuclei, suggesting that HRX localizes to nuclear structures in cells with and without 11q23 translocations. Nuclear staining was found in the majority of tissues studied, with the strongest reactivity in cerebral cortex, kidney, thyroid, and lymphoid tissues. Thus, Butler et al. (1997) concluded that HRX is widely expressed in most cell types, including hematopoietic cells, a finding that precludes an immunocytochemical approach for diagnosis of leukemias bearing 11q23 structural abnormalities.

Baffa, R.; Negrini, M.; Schichman, S. A.; Huebner, K.; Croce, C. M.: Involvement of the ALL-1 gene in a solid tumor. Proc. Nat. Acad. Sci. 92: 4922-4926, 1995.

Butler, L. H.; Slany, R.; Cui, X.; Cleary, M. L.; Mason, D. Y.: The HRX proto-oncogene product is widely expressed in human tissues and localizes to nuclear structures. Blood 89: 3361-3370, 1997.

Caligiuri, M. A.; Strout, M. P.; Oberkircher, A. R.; Yu, F.; de la Chapelle, A.; Bloomfield, C. D.: The partial tandem duplication of ALL1 in acute myeloid leukemia with normal cytogenetics or trisomy 11 is restricted to one chromosome. Proc. Nat. Acad. Sci. 94: 3899-3902, 1997.

Cimino, G.; Moir, D. T.; Canaani, O.; Williams, K.; Crist, W. M.; Katzav, S.; Cannizzaro, L.; Lange, B.; Nowell, P. C.; Croce, C. M.; Canaani, E.: Cloning of ALL-1, the locus involved in leukemias with the t(4;11)(q21;q23), t(9;11)(p22;q23), and t(11;19)(q23;p13) chromosome translocations. Cancer Res. 51: 6712-6714, 1991.

Djabali, M.; Selleri, L.; Parry, P.; Bower, M.; Young, B. D.; Evans, G. A.: A trithorax-like gene is interrupted by chromosome 11q23 translocations in acute leukaemias. Nature Genet. 2: 113-118, 1992.

Domer, P. H.; Fakharzadeh, S. S.; Chen, C.-S.; Jockel, J.; Johansen, L.; Silverman, G. A.; Kersey, J. H.; Korsmeyer, S. J.: Acute mixed-lineage leukemia t(4;11)(q21;q23) generates an MLL-AF4 fusion product. Proc. Nat. Acad. Sci. 90: 7884-7888, 1993.

Gu, Y.; Nakamura, T.; Alder, H.; Prasad, R.; Canaani, O.; Cimino, G.; Croce, C. M.; Canaani, E.: The t(4;11) chromosome translocation of human acute leukemias fuses the ALL-1 gene, related to Drosophila trithorax, to the AF-4 gene. Cell 71: 701-708, 1992.

Kobayashi, Y.; Yang, J.; Shindo, E.; Tojo, A.; Tani, K.; Ozawa, K.; Asano, S.: HRX gene rearrangement in acute lymphoblastic leukemia after adjuvant chemotherapy of breast cancer. (Letter) Blood 82: 3220-3223, 1993.

McCabe, N. R.; Burnett, R. C.; Gill, H. J.; Thirman, M. J.; Mbangkollo, D.; Kipiniak, M.; van Melle, E.; Ziemin-van der Poel, S.; Rowley, J. D.; Diaz, M. O.: Cloning of cDNAs of the MLL gene that detect DNA rearrangements and altered RNA transcripts in human leukemic cells with 11q23 translocations. Proc. Nat. Acad. Sci. 89: 11794-11798, 1992.

Nakamura, T.; Alder, H.; Gu, Y.; Prasad, R.; Canaani, O.; Kamada, N.; Gale, R. P.; Lange, B.; Crist, W. M.; Nowell, P. C.; Croce, C. M.; Canaani, E.: Genes on chromosomes 4, 9, and 19 involved in 11q23 abnormalities in acute leukemia share sequence homology and/or common motifs. Proc. Nat. Acad. Sci. 90: 4631-4635, 1993.

Parry, P.; Djabali, M.; Bower, M.; Khristich, J.; Waterman, M.; Gibbons, B.; Young, B. D.; Evans, G.: Structure and expression of the human trithorax-like gene 1 involved in acute leukemias. Proc. Nat. Acad. Sci. 90: 4738-4742, 1993.

Rowley, J. D.; Diaz, M. O.; Espinosa, R., III; Patel, Y. D.; van Melle, E.; Ziemin, S.; Taillon-Miller, P.; Lichter, P.; Evans, G. A.; Kersey, J. H.; Ward, D. C.; Domer, P. H.; Le Beau, M. M.: Mapping chromosome band 11q23 in human acute leukemia with biotinylated probes: identification of 11q23 translocation breakpoints with a yeast artificial chromosome. Proc. Nat. Acad. Sci. 87: 9358-9362, 1990.

Schichman, S. A.; Caligiuri, M. A.; Gu, Y.; Strout, M. P.; Canaani, E.; Bloomfield, C. D.; Croce, C. M.: ALL-1 partial duplication in acute leukemia. Proc. Nat. Acad. Sci. 91: 6236-6239, 1994.

Smith, S. D.; McFall, P.; Morgan, R.; Link, M.; Hecht, F.; Cleary, M.; Sklar, J.: Long-term growth of malignant thymocytes in vitro. Blood 73: 2182-2187, 1989.

Sorensen, P. H. B.; Chen, C.-S.; Smith, F. O.; Arthur, D. C.; Domer, P. H.; Bernstein, I. D.; Korsmeyer, S. J.; Hammond, G. D.; Kersey, J. H.: Molecular rearrangements of the MLL gene are present in most cases of infant acute myeloid leukemia and are strongly correlated with monocytic or myelomonocytic phenotypes. J. Clin. Invest. 93: 429-437, 1994.

Thirman, M. J.; Gill, H. J.; Burnett, R. C.; Mbangkollo, D.; McCabe, N. R.; Kobayashi, H.; Ziemin-van der Poel, S.; Kaneko, Y.; Morgan, R.; Sandberg, A. A.; Chaganti, R. S. K.; Larson, R. A.; Le Beau, M. M.; Diaz, M. O.; Rowley, J. D.: Rearrangement of the MLL gene in acute lymphoblastic and acute myeloid leukemias with 11q23 chromosomal translocations. New Eng. J. Med. 329: 909-914, 1993.

Tkachuk, D. C.; Kohler, S.; Cleary, M. L.: Involvement of a homolog of Drosophila trithorax by 11q23 chromosomal translocations in acute leukemias. Cell 71: 691-700, 1992.

Ziemin-van der Poel, S.; McCabe, N. R.; Gill, H. J.; Espinosa, R., III; Patel, Y.; Harden, A.; Rubinelli, P.; Smith, S. D.; Le Beau, M. M.; Rowley, J. D.; Diaz, M. O.: Identification of a gene, MLL, that spans the breakpoint in 11q23 translocations associated with human leukemias. Proc. Nat. Acad. Sci. 88: 10735-10739, 1991.

MYELOID/LYMPHOID OR MIXED LINEAGE LEUKEMIA, TRANSLOCATED TO 1 (*159556)

Tkachuk et al. (1992) showed that the gene involved in recurring 11q23 leukemogenic translocations codes for an unusually large protein that is a homolog of Drosophila 'trithorax' and is involved in homeotic gene regulation (MLL; 159555). In studies of a t(11;19) translocation, they identified a chimeric protein containing the amino-terminal 'AT-hook' motifs of the MLL gene on chromosome 11 fused to a previously undescribed protein from chromosome 19. The nucleotide sequence determinations demonstrated an open reading frame that coded for a predicted 62-kD protein, which Tkachuk et al. (1992) named ENL for 'eleven-nineteen leukemia.' Nakamura et al. (1993) showed that the gene on chromosome 19 that is fused to the MLL gene (also known as ALL1 gene) in patients with leukemia and translocation t(11;19)(q23;p13) shows high sequence homology to the genes on chromosome 4 (159557) and chromosome 9 (159558) that are fused with the ALL1 gene in patients with translocation t(4;11)(q21;q23) and t(9;11)(p22;q23), respectively. The 3 protein gene products contained nuclear targeting sequences as well as serine-rich and proline-rich regions. The results suggested that the different proteins fused to ALL1 polypeptides. These leukemias provide similar functional domains.

Rubnitz et al. (1994) characterized the transcriptional transactivation properties of ENL. It is a nuclear protein that is capable of activating transcription from synthetic reporter genes in both lymphoid and myeloid cells, as well as in yeast. Deletion mutagenesis demonstrated that the minimal portion of the gene required for activation of transcription was located to its C-terminal 90 amino acids. This region is highly conserved between ENL and the t(9;11) fusion partner AF9 (MLLT3) and is retained in all fusion proteins.

Thirman et al. (1994) pointed out that the breakpoint that results in the fusion of MLL to the ENL gene is located at 19p13.3. Another translocation t(11;19)(q23;p13.1) results in fusion of the MLL gene with the ELL gene (600284) in acute myeloid leukemia. This gene is also symbolized MLLT1.

Nakamura, T.; Alder, H.; Gu, Y.; Prasad, R.; Canaani, O.; Kamada, N.; Gale, R. P.; Lange, B.; Crist, W. M.; Nowell, P. C.; Croce, C. M.; Canaani, E.: Genes on chromosomes 4, 9, and 19 involved in 11q23 abnormalities in acute leukemia share sequence homology and/or common motifs. Proc. Nat. Acad. Sci. 90: 4631-4635, 1993.

Rubnitz, J. E.; Morrissey, J.; Savage, P. A.; Cleary, M. L.: ENL, the gene fused with HRX in t(11;19) leukemias, encodes a nuclear protein with transcriptional activation potential in lymphoid and myeloid cells. Blood 84: 1747-1752, 1994.

Thirman, M. J.; Levitan, D. A.; Kobayashi, H.; Simon, M. C.; Rowley, J. D.: Cloning of ELL, a gene that fuses to MLL in a t(11;19)(q23;p13.1) in acute myeloid leukemia. Proc. Nat. Acad. Sci. 91: 12110-12114, 1994.

Tkachuk, D. C.; Kohler, S.; Cleary, M. L.: Involvement of a homolog of Drosophila trithorax by 11q23 chromosomal translocations in acute leukemias. Cell 71: 691-700, 1992.

ACUTE MYELOGENOUS LEUKEMIA TRANSLOCATION TO 1, CYCLIN-D RELATED (*133435)

In acute myeloid leukemia, especially in the M2 subtype, the t(8;21)(q22;q22) translocation is one of the most frequent karyotypic abnormalities. The translocation produces a chimeric gene made up of the 5-prime region of the AML1 gene (151385) fused to the 3-prime region of the ETO gene (Erickson et al., 1992). Both genes are transcribed from telomere to centromere (Miyoshi et al., 1991). See review by Nucifora and Rowley (1994).

Niwa-Kawakita et al. (1995) isolated and characterized the mouse homolog of the CBFA2T1 gene. Sequence analysis of cDNA clones of the mouse gene revealed an open reading frame encoding a protein of 577 amino acids with 99.3% identity to the human protein. Using a (CA)n dinucleotide repeat polymorphism as a marker, Niwa-Kawakita et al. (1995) mapped the gene to mouse chromosome 4 close to the centromere using recombinant inbred strains and intersubspecific backcross progeny. The chromosomal localization was also confirmed by fluorescence in situ hybridization.

Using a 'knock-in' strategy, Yergeau et al. (1997) generated mice that mimic human t(8;21). Mice heterozygous for an AML1-ETO allele were found to die in midgestation from hemorrhaging in the central nervous system and exhibited a severe block in fetal liver hematopoiesis. This phenotype was very similar to that resulting from homozygous disruption of the AML1 (Cbfa2) or Cbfb genes in the mouse, indicating that AML1-ETO blocks normal AML function. However, yolk sac cells from AML1-ETO knock-in mice differentiated into macrophages in hematopoietic colony forming unit (CFU) assays, unlike homozygous Cbfa2-deficient or Cbfb deficient cells, which formed no colonies in vitro. This indicated to Yergeau et al. (1997) that AML1-ETO may have other functions besides blocking AML1, a property that may be important in leukemogenesis.

When first identified as a partner with AML1 in acute myeloid leukemia (Erickson et al., 1992; Miyoshi et al., 1993), the gene was referred to as MTG8 for 'myeloid translocation gene on 8q22.' Wolford et al. (1998) found that MTG8 mRNAs are expressed at relatively high levels in human adipose tissue. They therefore investigated MTG8 as a candidate gene in obesity, studying the relationship between a highly polymorphic marker in the 3-prime untranslated region of the MTG8 gene and obesity in Pima Indians of Arizona, a population with one of the highest reported rates of obesity. They detected a male-specific association with age-adjusted percentage body fat ($p = 0.0002$), body mass index ($p = 0.01$), waist circumference ($p = 0.008$), and thigh circumference ($p = 0.02$). Comparative analysis of all 13 MTG8 exons in 30 Pimas did not reveal any genetic variants that could explain the association with obesity in males.

Erickson, P.; Gao, J.; Chang, K.-S.; Look, T.; Whisenant, E.; Raimondo, S.; Lasher, R.; Trujillo, J.; Rowley, J. D.; Drabkin, H. A.: Identification of breakpoints in t(8;21) acute myelogenous leukemia and isolation of a fusion transcript, AML1/ETO, with similarity to Drosophila segmentation gene, runt. Blood 80: 1825-1831, 1992.

Miyoshi, H.; Kozu, T.; Shimizu, K.; Enomoto, K.; Maseki, N.; Kaneko, Y.; Kamada, N.; Ohki, M.: The t(8;21) translocation in acute myeloid leukemia results in production of an AML1-MTG8 fusion transcript. EMBO J. 12: 2715-2721, 1993.

Miyoshi, H.; Shimizu, K.; Kozu, T.; Maseki, N.; Kaneko, Y.; Ohki, M.: t(8;21) breakpoints on chromosome 21 in acute myeloid leukemia are clustered within a limited region of a single gene, AML1. Proc. Nat. Acad. Sci. 88: 10431-10434, 1991.

Niwa-Kawakita, M.; Miyoshi, H.; Gotoh, O.; Matsushima, Y.; Nishimura, M.; Shisa, H.; Ohki, M.: Cloning and gene mapping of the mouse homologue of the CBFA2T1 gene associated with human acute myeloid leukemia. Genomics 29: 755-759, 1995.

Nucifora, G.; Rowley, J. D.: The AML1 and ETO genes in acute myeloid leukemia with a t(8;21). Leukemia Lymphoma 14: 353-362, 1994.

Wolford, J. K.; Bogardus, C.; Prochazka, M.: Polymorphism in the 3-prime untranslated region of MTG8 is associated with obesity in Pima Indian males. Biochem. Biophys. Res. Commun. 246: 624-626, 1998.

Yergeau, D. A.; Hetherington, C. J.; Wang, Q.; Zhang, P.; Sharpe, A. H.; Binder, M.; Marin-Padilla, M.; Tenen, D. G.; Speck, N. A.; Zhang, D.-E.: Embryonic lethality and impairment of haematopoiesis in mice heterozygous for an AML1-ETO fusion gene. Nature Genet. 15: 303-306, 1997.

MYELOID/LYMPHOID OR MIXED LINEAGE LEUKEMIA, TRANSLOCATED TO 2 (*159557)

The ALL1 gene (MLL; 159555), located at 11q23, is rearranged in acute leukemias with reciprocal translocations between this region and chromosomes 1, 4, 6 (159559), 9 (159558), 10, or 19 (159556). Gu et al. (1992) studied the t(4;11) translocation which results in 2 reciprocal fusion products coding for chimeric proteins derived from ALL1 and from a gene on chromosome 4. They referred to the latter as AF4 for 'ALL1-fused gene from chromosome 4.' Fused RNAs were found in the three t(4;11) cells lines studied; therefore, Gu et al. (1992) could not establish which of the 2 products is oncogenic.

Nakamura et al. (1993) found that the gene on chromosome 4q21 that is fused with the ALL1 gene (159555) in patients with acute lymphoblastic leukemia and translocation t(4;11)(q21;q23) and the gene on chromosome 9 (159558) that is fused with the ALL1 gene on chromosome 11 in patients with leukemia and the t(9;11)(p22;q23) show high sequence homology with the ENL gene (159556) on chromosome 19 which is fused to the ALL1 gene in patients with leukemia and the translocation t(11;19)(q23;p13). They found further that the protein products of the AF4, AF9 (MLLT3), and ENL (MLLT1) genes contained nuclear targeting sequences as well as serine-rich and proline-rich regions. Stretches abundant in basic amino acids were also present in the 3 proteins. These results indicated that the different proteins fused to ALL1 polypeptides in leukemia provide similar functional domains.

Domer et al. (1993) independently demonstrated that the derivative chromosome 11, carrying a translocated segment of chromosome 4, encodes a fusion RNA which predicts a chimeric protein with features of the MLL gene on 11q23 and the AF4 gene on 4q21. The sequence of the complete open reading frame for this fusion transcript revealed that the MLL protein is homologous with DNA methyltransferase, the Drosophila trithorax gene product, and the 'AT-hook' motif of high-mobility-group proteins. An alternative splice that deletes the AT-hook region of MLL was identified. AF4 is a serine- and proline-rich putative transcription factor with a glutamine-rich carboxyl terminus. The composition of the complete MLL-AF4 fusion product argues that it may act through either a gain-of-function or a dominant-negative mechanism in leukemogenesis. This gene is also symbolized MLLT2.

Uckun et al. (1998) analyzed bone marrow leukemic cells of 17 infants and 127 children with newly diagnosed acute lymphatic leukemia (ALL), as well as fetal liver and bone marrow and normal infant bone marrow samples for the presence of a t(4;11) translocation, using standard cytogenetic techniques and expression of an MLL-AF4 fusion transcript by standard RT-PCR assays as well as nested RT-PCR that is 100-fold more sensitive than the standard RT-PCR. Overall, 9 of the 17 infants and 17 of 127 noninfant pediatric ALL patients were positive for expression of MLL-AF4 fusion transcripts. None of the MLL-AF4(+) cases were positive for E2A-PBX1 (147141; 176310) or BCR-ABL (151410; 189980) fusion transcript expression. Although 8 of 9 MLL-AF4(+) infants had cytogenetically detectable t(4;11) translocation, 15 of the 17 MLL-AF4(+) noninfants were t(4;11) negative. Infants with MLL-AF4(+) ALL had poor outcomes, whereas noninfant fusion-gene-positive, translocation-negative patients has favorable outcomes similar to MLL-AF4(-) patients. Notably, MLL-AF4 transcripts also were detected by nested RT-PCR in 4 of 16 fetal bone marrows, 5 of 13 fetal livers, and 1 of 6 normal infant bone marrows, but not in any of the 44 remission bone marrow specimens from pediatric ALL patients. These results represented unprecedented evidence that MLL-AF4 fusion transcripts can be present in normal hematopoietic cells, indicating that their expression is insufficient for leukemic transformation of normal lymphocyte precursors.

Domer, P. H.; Fakharzadeh, S. S.; Chen, C.-S.; Jockel, J.; Johansen, L.; Silverman, G. A.; Kersey, J. H.; Korsmeyer, S. J.: Acute mixed-lineage leukemia t(4;11)(q21;q23) generates an MLL-AF4 fusion product. Proc. Nat. Acad. Sci. 90: 7884-7888, 1993.

Gu, Y.; Nakamura, T.; Alder, H.; Prasad, R.; Canaani, O.; Cimino, G.; Croce, C. M.; Canaani, E.: The t(4;11) chromosome translocation of human acute leukemias fuses the ALL-1 gene, related to Drosophila trithorax, to the AF-4 gene. Cell 71: 701-708, 1992.

Nakamura, T.; Alder, H.; Gu, Y.; Prasad, R.; Canaani, O.; Kamada, N.; Gale, R. P.; Lange, B.; Crist, W. M.; Nowell, P. C.; Croce, C. M.; Canaani, E.: Genes on chromosomes 4, 9, and 19 involved in 11q23 abnormalities in acute leukemia share sequence homology and/or common motifs. Proc. Nat. Acad. Sci. 90: 4631-4635, 1993.

Uckun, F. M.; Herman-Hatten, K.; Crotty, M.-L.; Sensel, M. G.; Sather, H. N.; Tuel-Ahlgren, L.; Sarquis, M. B.; Bostrom, B.; Nachman, J. B.; Steinherz, P. G.; Gaynon, P. S.; Heerema, N.: Clinical significance of MLL-AF4 fusion transcript expression in the absence of a cytogenetically detectable t(4;11)(q21;q23) chromosomal translocation. Blood 92: 810-821, 1998.

MYELOID/LYMPHOID OR MIXED LINEAGE LEUKEMIA, TRANSLOCATED TO 3 (*159558)

Nakamura et al. (1993) found that the AF4 gene on chromosome 4q21 (159557) that is fused with the ALL1 gene (159555) in patients with acute lymphoblastic leukemia and translocation t(4;11)(q21;q23) and the gene on chromosome 9 that is fused with the ALL1 gene on chromosome 11 in patients with leukemia and the t(9;11)(p22;q23) show high sequence homology with the ENL gene (159556) on chromosome 19 which is fused to the ALL1 gene in patients with leukemia and the translocation t(11;19)(q23;p13). They found further that the protein products of the AF4, AF9, and ENL genes contained nuclear targeting sequences as well as serine-rich and proline-rich regions. Stretches abundant in basic amino acids were also present in the 3 proteins. These results indicated that the different proteins fused to ALL1 polypeptides in leukemia provide similar functional domains. This gene is also symbolized MLLT3.

Nakamura, T.; Alder, H.; Gu, Y.; Prasad, R.; Canaani, O.; Kamada, N.; Gale, R. P.; Lange, B.; Crist, W. M.; Nowell, P. C.; Croce, C. M.; Canaani, E.: Genes on chromosomes 4, 9, and 19 involved in 11q23 abnormalities in acute leukemia share sequence homology and/or common motifs. Proc. Nat. Acad. Sci. 90: 4631-4635, 1993.

MYELOID/LYMPHOID OR MIXED LINEAGE LEUKEMIA, TRANSLOCATED TO 4 (*159559)

Most acute leukemias in infancy and at least 5% of acute lymphoblastic leukemias and acute myeloid leukemias of older children and adults show abnormalities of chromosome band 11q23. In these cases, translocation results in fusion of a gene at 11q23, variously called ALL1, MLL, and the human homolog of Drosophila 'trithorax' (159555), with part of a gene on chromosome 4 (159557), chromosome 9 (159558), or chromosome 19 (159556). Prasad et al. (1993) described the cloning and characterization of the 'partner gene' involved in a fourth common translocation involving 11q23, t(6;11)(q27;q23). The gene, designated AF6 by them, was found to be expressed in a variety of cell types and to encode a protein of 1,612 amino acids. The protein contains short stretches rich in proline, charged amino acids, serines, or glutamines. In addition, the AF6 protein contains the GLGF motif shared with several proteins of vertebrates and invertebrates thought to be involved in signal transduction at special cell-cell junctions.

Using rapid amplification of cDNA ends (RACE) by PCR, Saha et al. (1995) confirmed the breakpoint in AF6 and identified a cDNA clone that was used as a probe to screen a chromosome 6 cosmid library. By fluorescence in situ hybridization, the single clone that was isolated was found to map distal to the critically deleted region associated with ovarian malignancies (167000). AF6 is therefore distinct from and lies telomeric to that region. This gene is also symbolized MLLT4.

Prasad, R.; Gu, Y.; Alder, H.; Nakamura, T.; Canaani, O.; Saito, H.; Huebner, K.; Gale, R. P.; Nowell, P. C.; Kuriyama, K.; Miyazaki, Y.; Croce, C. M.; Canaani, E.: Cloning of the ALL-1 fusion partner, the AF-6 gene, involved in acute myeloid leukemias with the t(6;11) chromosome translocation. Cancer Res. 53: 5624-5628, 1993.

Saha, V.; Lillington, D. M.; Shelling, A. N.; Chaplin, T.; Yaspo, M.-L.; Ganesan, T. S.; Young, B. D.: AF6 gene on chromosome band 6q27 maps distal to the minimal region of deletion in epithelial ovarian cancer. Genes Chromosomes Cancer 14: 220-222, 1995.

MIXED LINEAGE LEUKEMIA, TRANSLOCATED TO 6 (*600328)

Chromosome translocations involving band 11q23 are associated with approximately 10% of patients with acute lymphoblastic leukemia (ALL) and more than 5% of patients with acute myeloid leukemia (AML). The ALL patients are often very young, less than 1 year of age; 11q23 aberrations are the most common cytogenetic change among infants with acute leukemia. These aberrations are also common in secondary myeloid leukemias developing in children with ALL treated with epipodophyllotoxins. Leukemia associated with 11q23 abnormalities tend to show expression of both lymphoid and myeloid markers (hence, 'mixed-lineage leukemia'), massive cell burden, and poor prognosis. The gene at 11q23 involved in the translocations is variously designated ALL1, HRX, MLL, and TRX1 (159555). The most common translocations involving 11q23 and giving rise to mixed-lineage leukemia are t(4;11), t(6;11), t(9;11), and t(11;19). The genes involved in these translocations are referred to as AF4 (159557), AF6 (159559), AF9 (159558), and ENL (159556), respectively. Less frequent translocations involve specific regions on other chromosomes. The partner gene in one of these rarer translocations, t(11;17)(q23;q21), designated AF17, was cloned and characterized by Prasad et al. (1994). The AF17 gene encodes a protein of 1,093 amino acids, containing a leucine-zipper dimerization motif located 3-prime of the fusion point and a cysteine-rich domain at the end terminus. The latter can be arranged in 3 zinc fingers and shows homology to a domain within the protein Br140 (602410); see Thompson et al. (1994). AF17 was found to contain stretches of amino acids previously associated with domains involved in transcriptional repression or activation. Based on features of AF17 and of the proteins encoded by the other partner genes that had been analyzed and in conjunction with other studies, Prasad et al. (1994) proposed a model in which ALL1 rearrangements result in loss of function of that gene. In this model, the partner polypeptide, such as that encoded by AF17, plays an accessory role either by repressing activity of the truncated ALL1 protein or by blocking the function of the normal protein presumably present in leukemic cells.

Prasad, R.; Leshkowitz, D.; Gu, Y.; Alder, H.; Nakamura, T.; Saito, H.; Huebner, K.; Berger, R.; Croce, C. M.; Canaani, E.: Leucine-zipper dimerization motif encoded by the AF17 gene fused to ALL-1 (MLL) in acute leukemia. Proc. Nat. Acad. Sci. 91: 8107-8111, 1994.

Thompson, K. A.; Wang, B.; Argraves, W. S.; Giancotti, F. G.; Schranck, D. P.; Ruoslahti, E.: BR140, a novel zinc-finger protein with homology to the TAF250 subunit of TFIID. Biochem. Biophys. Res. Commun. 198: 1143-1152, 1994.

MYELOID/LYMPHOID OR MIXED-LINEAGE LEUKEMIA, TRANSLOCATED TO 7 (*300033)

A breakpoint in 11q23 is frequently involved in translocations underlying hematologic malignancies, especially acute leukemias. The human homolog of Drosophila 'trithorax,' symbolized MLL (159555), for 'myeloid-lymphoid leukemia' or 'mixed-lineage leukemia,' is located at this breakpoint. Part of the MLL gene is fused with other genes in leukemia: AF4 (159557) in t(4;11)(q21;q23); ENL (159556) in t(11;19)(q23;p13.3), AF9 (159558) in t(9;11)(p22;q23), AF6 (159559) in t(6;11)(q27;q23), and AFX in t(X;11)(q13;q23). Translocations at 11q23 result in the formation of 2 derivative chromosomes that encode chimeric transcripts. The der(11) transcript contains 5-prime MLL sequences fused to 3-prime sequences of the gene located on the partner chromosome, whereas the other derivative chromosome contains the 5-prime sequence of the partner gene potentially fused to the 3-prime sequence of MLL. However, in 25% of patients, translocations are associated with deletions of MLL sequence that is 3-prime to the breakpoint. Thus, in these cases, a fusion transcript from the other derivative chromosome cannot be formed. In addition, analysis of complex 11q23 translocations revealed that the der(11) junction is always conserved. These data indicate that the fusion transcript encoded by the der(11) must be critical to leukemogenesis. Corral et al. (1993) found from a partial sequence of a fusion between MLL and the AFX1 gene from the X chromosome that the latter is rich in ser/pro codons, like the ENL mRNA. Corral et al. (1993) concluded that heterogeneous 11q23 abnormalities may cause attachment of ser/pro-rich segments to the N terminus of MLL, lacking the zinc-finger region, and that translocations occur in early hematopoietic cells, before commitment to distinct lineages.

Parry et al. (1994) cloned and sequenced the t(X;11) breakpoint region from a cell line established from an infant with acute lymphocytic leukemia. The AFX1 gene (also symbolized MLLT7) was expressed in a variety of cell types. Sequence analysis indicated a high degree of homology between AFX1 and the forkhead family of transcription factors. The high degree of identity within the forkhead region and the lack of homology outside that region suggested to the authors that AFX1 represents a novel forkhead family member. It was predicted that a chimeric fusion protein that altered DNA binding activity would result from the translocation.

Borkhardt et al. (1997) cloned and characterized the entire AFX gene, which consists of 2 exons and encodes a protein of 501 amino acids. The single intron of the AFX gene is 3,706-bp long. The AFX protein belongs to the forkhead protein family. It is highly homologous to the FKHR protein (136533), the gene for which it is disrupted by the t(2;13) translocation characteristic of alveolar rhabdomyosarcoma.

The AFX1 gene and the NRB54 gene (300084) map to a yeast artificial chromosome (YAC) contig of Xq13.1 that harbors the X-linked dystonia-parkinsonism locus DYT3 (314250). Peters et al. (1997) found that the AFX1 gene is composed of 3 exons with most of exon 3 being untranslated. The NRB54 gene is made up of 12 exons ranging in size from 40 to 1,227 bp. The start codon is in exon 3 and the stop codon in exon 12. Both genes are expressed in the brain, among other tissues. Peters et al. (1997) excluded both genes as candidates of DYT3 by sequencing of the exons and the flanking intronic sequences in a patient and a control, and by Northern blot analysis.

Borkhardt, A.; Repp, R.; Haas, O. A.; Leis, T.; Harbott, J.; Kreuder, J.; Hammermann, J.; Henn, T.; Lampert, F.: Cloning and characterization of AFX, the gene that fuses to MLL in acute leukemias with a t(X;11)(q13;q23). Oncogene 14: 195-202, 1997.

Corral, J.; Forster, A.; Thompson, S.; Lampert, F.; Kaneko, Y.; Slater, R.; Kroes, W. G.; van der Schoot, C. E.; Ludwig, W.-D.; Karpas, A.; Pocock, C.; Cotter, F.; Rabbitts, T. H.: Acute leukemias of different lineages have similar MLL gene fusions encoding related chimeric proteins resulting from chromosomal translocation. Proc. Nat. Acad. Sci. 90: 8538-8542, 1993.

Parry, P.; Wei, Y.; Evans, G.: Cloning and characterization of the t(X;11) breakpoint from a leukemic cell line identify a new member of the forkhead gene family. Genes Chromosomes Cancer 11: 79-84, 1994.

Peters, U.; Haberhausen, G.; Kostrzewa, M.; Nolte, D.; Muller, U.: AFX1 and p54(nrb): fine mapping, genomic structure, and exclusion as candidate genes of X-linked dystonia parkinsonism. Hum. Genet. 100: 569-572, 1997.

ELEVEN NINETEEN LYSINE-RICH LEUKEMIA GENE (*600284)

Thirman et al. (1994) cloned the gene that fuses to MLL (159555) in patients with acute myeloid leukemia associated with the translocation t(11;19)(q23;p13.1). This translocation is distinct from another type of 11;19 translocation with a 19p13.3 breakpoint that results in the fusion of MLL to the ENL gene (159556). By PCR screening of a cDNA library prepared from a patient's leukemia cells with this translocation, Thirman et al. (1994) obtained a fusion transcript containing exon 7 of MLL and sequence of an unknown gene. The sequence of this gene was amplified and used as a probe to screen a fetal brain cDNA library. On Northern blot analysis, this cDNA detected a 4.4-kb transcript that was abundant in peripheral blood leukocytes, skeletal muscle, placenta, and testis and expressed at lower levels in spleen, thymus, heart, brain, lung, kidney, liver, and ovary. In addition, a 2.8-kb transcript was present in peripheral blood, testis, and placenta. On zoo blots, this gene was shown to be evolutionarily conserved in 10 mammalian species, as well as in chicken, frog, and fish. They named the gene ELL (for 'eleven-nineteen lysine-rich leukemia' gene). A highly basic, lysine-rich motif of the predicted ELL protein was homologous to similar regions of several proteins, including the DNA-binding domain of poly(ADP-ribose) polymerase (173870).

ELL was shown by Shilatifard et al. (1996) to encode an elongation factor that can increase the catalytic rate of RNA polymerase II transcription by suppressing transient pausing by the polymerase at multiple sites along the DNA. The authors noted that ELL is the second elongation factor to be implicated in oncogenesis (elongin (see 600786), which is a transcription factor regulated by the product of the von Hippel-Lindau (VHL) tumor suppressor gene (193300), was the first), and they stated that these findings provide further support for a close connection between the regulation of transcription and cell growth.

Shilatifard, A.; Lane, W. S.; Jackson, K. W.; Conaway, R. C.; Conaway, J. W.: An RNA polymerase II elongation factor encoded by the human ELL gene. Science 271: 1873-1876, 1996.

Thirman, M. J.; Levitan, D. A.; Kobayashi, H.; Simon, M. C.; Rowley, J. D.: Cloning of ELL, a gene that fuses to MLL in a t(11;19)(q23;p13.1) in acute myeloid leukemia. Proc. Nat. Acad. Sci. 91: 12110-12114, 1994.

EXTRA ELEVEN-NINETEEN LEUKEMIA FUSION GENE (*601768)

The MLL gene (159555), the closest human homolog of the Drosophila trithorax gene, undergoes chromosomal translocation with a large number of different partner genes in both acute lymphoid and acute myeloid leukemias. So et al. (1997) identified a new partner gene, designated EEN by them, that was fused to MLL in a case of acute myeloid leukemia. The patient was a 22-month-old female in whom the diagnosis had been made on morphologic and cytochemical criteria. The patient was identified in a general screen of cases of acute leukemia for novel genes fused to MLL based on 2 selection criteria: DNA rearrangement detected with an MLL probe and a negative result in RT-PCR analyses of the 4 more common MLL fusion RNA products: fusion of MLL with AF4 (159557), AF6 (159559), AF9 (159558), and ENL (159556). Since no PCR products were amplified from the patient's DNA when specific primers for these 4 fusion genes were used, So et al. (1997) inferred that MLL in this patient was fused to a novel partner gene sequence. As all characterized MLL fusion genes previously identified had been shown to retain the derivative chromosome 11, the 3-prime RACE (rapid amplification of cDNA ends) method was chosen to isolate the fusion partner of MLL in the patient's leukemic cells. By fluorescence in situ hybridization, the fusion partner was shown to map to 19p13, where 2 other MLL partners genes, ENL and ELL/MEN (600284), had been identified. For that reason, So et al. (1997) designated the new fusion gene EEN for 'extra eleven-nineteen.' The deduced protein of 368 amino acids contains a central alpha-helical region and a C-terminal Src homology 3 (SH3) domain most similar to the C-terminal SH3 domain found in the GRB2 (600180) family of genes. Sequence analysis of the fusion MLL/EEN transcript revealed that exon 6 of MLL is fused to the N-terminal sequence of EEN, a fusion that predicts a chimeric protein that includes the major functional domain of EEN. EEN was expressed in a variety of ti ssue types and encoded a protein of approximately 46 kD. The EEN protein is the human homolog of a member of a recently described murine SH3 domain-containing protein family. It is also highly related to a putative gene identified in C. elegans.

So, C. W.; Caldas, C.; Liu, M.-M.; Chen, S.-J.; Huang, Q.-H.; Gu, L.-J.; Sham, M. H.; Wiedemann, L. M.; Chan, L. C.: EEN encodes for a member of a new family of proteins containing an Src homology 3 domain and is the third gene located on chromosome 19p13 that fuses to MLL in human leukemia. Proc. Nat. Acad. Sci. 94: 2563-2568, 1997.

RETINOIC ACID RECEPTOR, ALPHA (*180240)

Petkovich et al. (1987) cloned a cDNA encoding a protein that binds retinoic acid with high affinity. The protein was found to be homologous to the receptors for steroid hormones, thyroid hormones, and vitamin D3, and appeared to be a retinoic acid-inducible transacting enhancer factor. Thus, the molecular mechanisms of the effect of vitamin A on embryonic development, differentiation and tumor cell growth may be similar to those described for other members of this nuclear receptor family. The genes for all the steroid/thyroid receptors show a common pattern of structure, with 4 regions: A/B, C, D, and E (Robertson, 1987). The function of region A/B is unknown; C encodes the DNA-binding domain; D is believed to be a hinge region; and E encodes the ligand-binding domain. In general, the DNA-binding domain is most highly conserved, both within and between the 2 groups of receptors (steroid and thyroid); the ligand-binding domains show less homology. Arveiler et al. (1988) described a RFLP of the RAR gene. Bale et al. (1988) indicated that the RAR locus had been assigned to chromosome 17 by study of human-rodent somatic cell hybrids. Using a cDNA probe, Mattei et al. (1988) localized the RAR gene to 17q21 by in situ hybridization. Brand et al. (1988) presented evidence for the existence of 2 retinoic acid receptors, RAR-alpha and RAR-beta, mapping to chromosome 17q21.1 and 3p24, respectively. The alpha and beta forms of RAR were found to be more homologous to the 2 closely related thyroid hormone receptors alpha (190120) and beta (190160), located on 17q11.2 and 3p25-p21, respectively, than to any other members of the nuclear receptor family. These observations suggest that the thyroid hormone and retinoic acid receptors evolved by gene, and possibly chromosome, duplications from a common ancestor, which itself diverged rather early in evolution from the common ancestor of the steroid receptor group of the family. They noted that the counterparts of the human RARA an d RARB genes are present in both the mouse and chicken genomes. By genetic linkage studies using multiple DNA markers from the 17q12-q21 region, Anderson et al. (1993) placed the RARA gene on the genetic map of the region.

Acute promyelocytic leukemia (APL), known as acute myeloid leukemia-3, AML3, or M3 in the French-American-British (FAB) classification, is characterized by a predominance of malignant promyelocytes that carry a reciprocal translocation between the long arms of chromosomes 15 and 17: t(15;17)(q22;q11.2-q12). This translocation is diagnostic for APL, as it is present in almost 100% of cases. Borrow et al. (1990) used a NotI linking clone to detect this translocation on pulsed field gel electrophoresis and subsequently with conventional Southern analysis. The breakpoints in 10 APL cases examined were shown to cluster in a 12-kb region of chromosome 17, which contained 2 CpG-rich islands. A comparison of the sequence of cDNA clones from the region of the breakpoint was compared with the EMBL database revealed that the cDNA was that of RARA, which maps to 17q21.1, distal to the APL breakpoint region. They concluded that the cDNAs lay outside the 12-kb breakpoint region and that all of the 15q+ APL breakpoints lie in the first intron of RARA. Since RARA is interrupted in an intron, it is most likely that the product of the translocation is a fusion protein. Borrow et al. (1990) suggested that the chimeric fusion protein encoded by the 15q+ derivative would retain the DNA- and ligand-binding domains of RARA, whereas the transcription-activating function of the 5-prime end of RARA would be replaced with a novel N-terminus, potentially changing the profile of genes activated. The involvement of RARA at the APL breakpoint may explain why the use of retinoic acid as a therapeutic differentiation agent in the treatment of acute myeloid leukemias is limited to APL. Lemons et al. (1990) also cloned the APL breakpoint region. Because RARA maps close to the breakpoint of the t(15;17) translocation specifically associated with acute promyelocytic leukemia, and because retinoic acid has the ability to induce in vivo differentiation of APL cells into mature granulocytes, de The et al. (1990) anal yzed the RARA gene structure and expression in APL cells. In one APL-derived cell line, they found that the RARA gene had been translocated to a locus, MYL (102578), on chromosome 15, resulting in the synthesis of an MYL/RARA fusion mRNA. (PML later became the preferred designation for the chromosome 15 gene that contributed to the chimeric gene product.) Using 2 probes located on either side of the cloned breakpoint, they found genomic rearrangements of one or the other locus in 6 of 8 patients, demonstrating that the RARA and/or MYL genes are frequently rearranged in APL and that the breakpoints are clustered. The findings strongly implicated retinoic acid receptor alpha in leukemogenesis. Alcalay et al. (1991) likewise demonstrated that the chromosome 17 breakpoint in APL lies within the RARA locus. The translocation site occurred in the 3-prime end of the RARA intron 1, 370 bp upstream from the splicing donor site of exon 2, and separated the first coding exon (exon 1) from the rest of the gene. By in situ hybridization, they mapped a probe from the translocation site to normal 15q23-q24, which is more distal than the site determined by others for either the chromosome 17 break in APL or the location of the normal RARA gene.

Mattei et al. (1991) mapped the RARA, RARB (180220), and RARG (180190) genes in man, mouse, and rat, thereby establishing or confirming and extending the following homologies: (1) between human chromosome 17, mouse chromosome 11, and rat chromosome 10, as indicated by the location of RARA; (2) between human chromosome 3, mouse chromosome 14, and rat chromosome 15, as indicated by RARB; and (3) between human chromosome 12, mouse chromosome 15, and rat chromosome 7, as indicated by RARG. The various assignments also indicated the retention of tight linkage between RAR and HOX gene clusters. Nadeau et al. (1992) pointed out that, in the mouse, RARA is located on chromosome 11 near the homeo box-2 complex (see 142960) and the keratin type I complex (148080), whereas RARG is on mouse chromosome 15 near the homeo box-3 complex (see 142970) and the keratin type II complex (see 139350). The close proximity of these genes may be functionally significant but has evolutionary significance, at any rate, indicating the existence of homeologous segments of both in man and mouse.

Hiorns et al. (1994) showed that APL patients with cytogenetically normal chromosomes 15 and 17 may nevertheless have involvement of both PML and RARA genes. Thus there is a subgroup of APL, t(15;17)-negative/PML-RARA-positive, that is analogous to Philadelphia chromosome-negative/BCR-ABL-positive CML. The amount of chromosome 17 material inserted into chromosome 15 in the case studied by Hiorns et al. (1994) was too small to be detected cytogenetically. In cases of Philadelphia chromosome-negative/BCR-ABL-positive CML, the amount of DNA transferred can be substantial (Rassool et al., 1990).

See Wells et al. (1997) and NUMA1 (164009) for discussion of a RAR/NUMA1 fusion protein associated with APL.

See also Giguere et al. (1987)

Alcalay, M.; Zangrilli, D.; Pandolfi, P. P.; Longo, L.; Mencarelli, A.; Giacomucci, A.; Rocchi, M.; Biondi, A.; Rambaldi, A.; Lo Coco, F.; Diverio, D.; Donti, E.; Grignani, F.; Pelicci, P. G.: Translocation breakpoint of acute promyelocytic leukemia lies within the retinoic acid receptor alpha locus. Proc. Nat. Acad. Sci. 88: 1977-1981, 1991.

Anderson, L. A.; Friedman, L.; Osborne-Lawrence, S.; Lynch, E.; Weissenbach, J.; Bowcock, A.; King, M.-C.: High-density genetic map of the BRCA1 region of chromosome 17q12-q21. Genomics 17: 618-623, 1993.

Arveiler, B.; Petkovich, M.; Mandel, J. L.; Chambon, P.: A PstI RFLP for the human retinoic acid receptor in 17q21. Nucleic Acids Res. 16: 6252 only, 1988.

Bale, A. E.; Weinberger, C.; McBride, O. W.: A retinoic acid receptor cDNA probe (RAR2) identifies a moderately frequent RFLP on chromosome 17. Nucleic Acids Res. 16: 7755 only, 1988.

Borrow, J.; Goddard, A. D.; Sheer, D.; Solomon, E.: Molecular analysis of acute promyelocytic leukemia breakpoint cluster region on chromosome 17. Science 249: 1577-1580, 1990.

Brand, N.; Petkovich, M.; Krust, A.; Chambon, P.; de The, H.; Marchio, A.; Tiollais, P.; Dejean, A.: Identification of a second human retinoic acid receptor.(Letter) Nature 332: 850-853, 1988.

de The, H.; Chomienne, C.; Lanotte, M.; Degos, L.; Dejean, A.: The t(15;17) translocation of acute promyelocytic leukaemia fuses the retinoic acid receptor alpha gene to a novel transcribed locus. Nature 347: 558-561, 1990.

Giguere, V.; Ong, E. S.; Segui, P.; Evans, R. M.: Identification of a receptor for the morphogen retinoic acid. Nature 330: 624-629, 1987.

Hiorns, L. R.; Min, T.; Swansbury, G. J.; Zelent, A.; Dyer, M. J. S.; Catovsky, D.: Interstitial insertion of retinoic acid receptor-alpha gene in acute promyelocytic leukemia with normal chromosomes 15 and 17. Blood 83: 2946-2951, 1994.

Lemons, R. S.; Eilender, D.; Waldmann, R. A.; Rebentisch, M.; Frej, A.-K.; Ledbetter, D. H.; Willman, C.; McConnell, T.; O'Connell, P.: Cloning and characterization of the t(15;17) translocation breakpoint region in acute promyelocytic leukemia. Genes Chromosomes Cancer 2: 79-87, 1990.

Mattei, M.-G.; Petkovich, M.; Mattei, J.-F.; Brand, N.; Chambon, P.: Mapping of the human retinoic acid receptor to the q21 band of chromosome 17. Hum. Genet. 80: 186-188, 1988.

Mattei, M.-G.; Riviere, M.; Krust, A.; Ingvarsson, S.; Vennstrom, B.; Islam, M. Q.; Levan, G.; Kautner, P.; Zelent, A.; Chambon, P.; Szpirer, J.; Szpirer, C.: Chromosomal assignment of retinoic acid receptor (RAR) genes in the human, mouse, and rat genomes. Genomics 10: 1061-1069, 1991.

Nadeau, J. H.; Compton, J. G.; Giguere, V.; Rossant, J.; Varmuza, S.: Close linkage of retinoic acid receptor genes with homeobox- and keratin-encoding genes on paralogous segments of mouse chromosomes 11 and 15. Mammalian Genome 3: 202-208, 1992.

Petkovich, M.; Brand, N. J.; Krust, A.; Chambon, P.: A human retinoic acid receptor which belongs to the family of nuclear receptors. Nature 330: 444-450, 1987.

Rassool, F.; Martiat, P.; Klisak, I.; Taj, A.; Goldman, J.: Interstitial insertion of varying amounts of ABL-containing genetic material into chromosome 22 in Ph-negative CML. Leukemia 4: 273-277, 1990.

Robertson, M.: Retinoic acid receptor: towards a biochemistry of morphogenesis. Nature 330: 420-421, 1987.

Wells, R. A.; Catzavelos, C.; Kamel-Reid, S.: Fusion of retinoic acid receptor alpha to NuMA, the nuclear mitotic apparatus protein, by a variant translocation in acute promyelocytic leukaemia. Nature Genet. 17: 109-113, 1997.

ACUTE PROMYELOCYTIC LEUKEMIA, INDUCER OF (*102578)

In the process of analyzing the retinoic acid receptor alpha (RARA; 180240) gene in the t(15;17)(q22;q11.2-q12) translocation specifically associated with acute promyelocytic leukemia (APL), de The et al.

(1990) identified a new gene on chromosome 15 which is involved with the RARA gene in the formation of a fusion product. This gene, which they called MYL for 'myelocytic leukemia,' was transcribed in the same direction as RARA on the translocated chromosome. They identified a 144-bp region, flanked by canonical splice acceptor and donor sequences, that had a high probability of being an exon and showed no significant similarity to any sequence in a protein data bank, thus suggesting that MYL is a previously undescribed gene. In the chimeric gene, the promoter and first exon of the RARA gene were replaced by part of the MYL gene. De The et al. (1990) established that the translocation chromosome generates an MYL/RARA chimeric transcript. The findings strongly implicated retinoic acid receptor alpha in leukemogenesis. The possibility was raised that the altered retinoic acid receptor behaves as a dominant-negative mutant that blocks the expression of retinoic acid target genes involved in granulocytic differentiation. In a later report, de The et al. (1991) changed the name of the gene from MYL to PML. They reported, furthermore, that the gene product contains a novel zinc finger motif common to several DNA-binding proteins. The PML-RARA mRNA encoded a predicted 106-kd chimeric protein containing most of the PML sequences fused to a large part of the RARA gene, including its DNA- and hormone-binding domains. Goddard et al. (1991) demonstrated that PML is a putative zinc finger protein and potential transcription factor that is commonly expressed, with at least 3 major transcription products. The PML breakpoints are clustered in 2 regions on either side of an alternatively spliced exon. Although leukemic cells with translocations characteristically expressed only one fusion prod uct, both PML/RARA (on the 15q+ derivative chromosome) and RARA/PML (on the 17q- derivative) were transcribed. The contribution of PML to the oncogenicity of the fusion products was demonstrated by the following: no mutations affecting RARA alone were observed in 20 APLs analyzed; 2 APLs cytogenetically lacking t(15;17) chromosomes were found to have rearrangements of both PML and RARA; and PML, but not RARA, was molecularly rearranged in a variant APL translocation in which chromosome 15 had been translocated to another chromosome with no visible involvement of chromosome 17. Tong et al. (1992) found that in 20 of 22 patients with a detectable MYL rearrangement the breakpoints were clustered within a 4.4-kb segment, which they designated MYL(bcr). The 2 remaining patients exhibited a more 5-prime rearrangement at about 10-kb upstream of the MYL(bcr) region, indicating the lack of at least one MYL gene exon in the resulting MYL-RARA fusion gene. Cleary (1991) pointed out that detection of the PML-RARA fusion links a specific molecular defect in neoplasia with a characteristic biologic and clinical response to pharmacologic therapy. It is a useful marker for the diagnosis of APL and for the identification of patients who may benefit from retinoid treatment.

PML, the gene involved in the breakpoint on chromosome 15, is a putative transcription factor: it contains a cysteine-rich motif that resembles a zinc finger DNA-binding domain common to several classes of transcriptional factors. Its physiologic role is unknown. Two fusion genes, PML-RARA and RARA-PML, are formed as a result of the characteristic translocation in acute promyelocytic leukemia. Heterogeneity of the chromosome 15 breakpoints accounts for the diverse architecture of the PML-RARA mRNAs isolated from different APL patients, and alternative splicing of PML exons gives rise to multiple isoforms of the PML-RARA mRNAs even within a single patient. Alcalay et al. (1992) investigated the organization and expression pattern of the RARA-PML gene in a series of APL patients. An RARA-PML transcript was present in most, but not all, APL patients. Among 70 patients with APL, Diverio et al. (1992) found an abnormality in intron 2 of the RARA gene in all cases, with clustering of rearrangements within the 20-kb intronic region separating exons 2 and 3. A curious difference was found in the location of breakpoints in males and females: breakpoints at the 5-prime end of intron 2 of the RARA gene occurred in females and 3-prime breakpoints predominated in males.

Grignani et al. (1993) expressed the PML/RARA protein in U937 myeloid precursor cells and showed that they lost the capacity to differentiate under the action of stimuli such as vitamin D3 and transforming growth factor beta-1 (TGFB1; 190180), acquired enhanced sensitivity to retinoic acid, and exhibited a higher growth rate consequent to diminished apoptotic cell death. These results provided evidence of biologic activity of the fusion protein and recapitulated critical features of the promyelocytic leukemia phenotype. While PML does not colocalize with proliferating cell nuclear antigen (PCNA; 176740) or spliceosomes, Dyck et al. (1994) showed that it is part of a macromolecular structure, composed of at least 4 nuclear proteins, that is adhered to the nuclear matrix. This structure shows a labeling pattern resembling spheres that vary in both size and number among individual cells of a given cell line. PML-RAR expression appears to disrupt the integrity of these structures (referred to by Dyck et al. (1994) as PML oncogenic domains, or PODs) and thus appears to be the possible cause of their altered morphology. Retinoid treatment leads to a striking reassembly of the POD, which in turn is linked to differentiation of the leukemic cells. These results identified a novel macromolecular nuclear structure and suggested that it may serve as a target of cellular transformation.

From their analysis of the phosphoamino acids of the PML protein, Chang et al. (1995) concluded that both tyrosine and serine residues are phosphorylated. To investigate whether expression of the PML protein is cell-cycle related, HeLa cells synchronized at various phases of the cell cycle were analyzed by immunofluorescence staining and confocal microscopy. They found that PML was expressed at a lower level in S, G2, and M phases and at a significantly higher level in G1 phase. Other studies showed that PML is a phosphoprotein and is associated with the nuclear matrix. Chang et al. (1995) noted that PML shares many properties with tumor suppressors, such as RB (180200).

Goddard et al. (1995) cloned the murine Pml gene and determined its intron/exon organization. The predicted amino acid sequence of the mouse Pml, a ring-finger protein, shows 80% similarity to that of the human homolog with greater than 90% similarity in the proposed functional domains. Chromosomal localization of the Pml locus by somatic cell hybrids and by linkage analysis indicated that the gene maps to a region of mouse chromosome 9 with known homology of synteny to the region of 15q where PML is located.

Brown et al. (1997) established a transgenic mouse model that documented the ability of the chimeric PMLRAR-alpha gene to initiate leukemogenesis. The mice developed 2 currently unrelated abnormalities. The first was a severe papillomatosis of the skin; the second was a disturbance of hematopoiesis that presented as a partial block of differentiation in the neutrophil lineage of the transgenic mice and then progressed at low frequency to overt APL. The leukemia appeared to be a faithful reproduction of the human disease, including a therapeutic response to retinoic acid that reflected differentiation of the leukemic cells. Both the preleukemic state and the overt leukemia could be transplanted into nontransgenic hosts. Brown et al. (1997) commented that the model should be useful for exploring the pathogenesis and treatment of APL.

From studies in mice with disruption of the Pml gene, Wang et al. (1998) demonstrated that normally, PML regulates hemopoietic differentiation and controls cell growth and tumorigenesis. PML function is essential for the tumor-growth-suppressive activity of retinoic acid (RA) and for its ability to induce terminal myeloid differentiation of precursor cells. PML was needed for the RA-dependent transactivation of the p21(Waf1/Cip1) gene (116899), which regulates cell cycle progression and cellular differentiation. These results provided a framework for understanding the molecular pathogenesis of acute promyelocytic leukemia (APL). Whereas APL might result from the functional interference of PML-RARA with 2 independent pathways, PML and RXR/RAR, Wang et al. (1998) showed that these proteins act, at least in part, in the same pathway. Thus, by simultaneously interacting with RXR and PML, the fusion gene product may inactivate this pathway at multiple levels, leading to the proliferative advantage and the block of hemopoietic differentiation that characterize APL.

Alcalay, M.; Zangrilli, D.; Fagioli, M.; Pandolfi, P. P.; Mencarelli, A.; Lo Coco, F.; Biondi, A.; Grignani, F.; Pelicci, P. G.: Expression pattern of the RAR-alpha-PML fusion gene in acute promyelocytic leukemia. Proc. Nat. Acad. Sci. 89: 4840-4844, 1992.

Brown, D.; Kogan, S.; Lagasse, E.; Weissman, I.; Alcalay, M.; Pelicci, P. G.; Atwater, S.; Bishop, J. M.: A PMLRAR-alpha transgene initiates murine acute promyelocytic leukemia. Proc. Nat. Acad. Sci. 94: 2551-2556, 1997.

Chang, K.-S.; Fan, Y.-H.; Andreeff, M.; Liu, J.; Mu, Z.-M.: The PML gene encodes a phosphoprotein associated with the nuclear matrix. Blood 85: 3646-3653, 1995.

Cleary, M. L.: Oncogenic conversion of transcription factors by chromosomal translocations. Cell 66: 619-622, 1991.

de The, H.; Chomienne, C.; Lanotte, M.; Degos, L.; Dejean, A.: The t(15;17) translocation of acute promyelocytic leukaemia fuses the retinoic acid receptor alpha gene to a novel transcribed locus. Nature 347: 558-561, 1990.

de The, H.; Lavau, C.; Marchio, A.; Chomienne, C.; Degos, L.; Dejean, A.: The PML-RAR-alpha fusion mRNA generated by the t(15;17) translocation in acute promyelocytic leukemia encodes a functionally altered RAR. Cell 66: 675-684, 1991.

Diverio, D.; Lo Coco, F.; D'Adamo, F.; Biondi, A.; Fagioli, M.; Grignani, F.; Rambaldi, A.; Rossi, V.; Avvisati, G.; Petti, M. C.; Testi, A. M.; Liso, V.; Specchia, G.; Fioritoni, G.; Recchia, A.; Frassoni, F.; Ciolli, S.; Pelicci, P. G.: Identification of DNA rearrangements at the retinoic acid receptor-alpha (RAR-alpha) locus in all patients with acute promyelocytic leukemia and mapping of APL breakpoints within the RAR-alpha second intron. Blood 79: 3331-3336, 1992.

Dyck, J. A.; Maul, G. G.; Miller, W. H., Jr.; Chen, J. D.; Kakizuka, A.; Evans, R. M.: A novel macromolecular structure is a target of the promyelocyte-retinoic acid receptor oncoprotein. Cell 76: 333-343, 1994.

Goddard, A. D.; Borrow, J.; Freemont, P. S.; Solomon, E.: Characterization of a zinc finger gene disrupted by the t(15;17) in acute promyelocytic leukemia. Science 254: 1371-1374, 1991.

Goddard, A. D.; Yuan, J. Q.; Fairbairn, L.; Dexter, M.; Borrow, J.; Kozak, C.; Solomon, E.: Cloning of the murine homolog of the leukemia-associated PML gene. Mammalian Genome 6: 732-737, 1995.

Grignani, F.; Ferrucci, P. F.; Testa, U.; Talamo, G.; Fagioli, M.; Alcalay, M.; Mencarelli, A.; Grignani, F.; Peschle, C.; Nicoletti, I.; Pelicci, P. G.: The acute promyelocytic leukemia-specific PML-RAR-alpha fusion protein inhibits differentiation and promotes survival of myeloid precursor cells. Cell 74: 423-431, 1993.

Tong, J.-H.; Dong, S.; Geng, J.-P.; Huang, W.; Wang, Z.-Y.; Sun, G.-L.; Chen, S.-J.; Chen, Z.; Larsen, C.-J.; Berger, R.: Molecular rearrangements of the MYL gene in acute promyelocytic leukemia (APL, M3) define a breakpoint cluster region as well as some molecular variants. Oncogene 7: 311-316, 1992.

Wang, Z. G.; Delva, L.; Gaboli, M.; Rivi, R.; Giorgio, M.; Cordon-Cardo, C.; Grosveld, F.; Pandolfi, P. P.: Role of PML in cell growth and the retinoic acid pathway. Science 279: 1547-1551, 1998.

PROMYELOCYTIC LEUKEMIA ZINC FINGER (*176797)

Almost all patients with acute promyelocytic leukemia (APL) have a chromosomal translocation t(15;17)(q22;q21). Molecular studies reveal that the translocation results in a chimeric gene through fusion between the promyelocytic leukemia gene (PML; 102578) on chromosome 15 and the retinoic acid receptor alpha gene (RARA; 180240) on chromosome 17. Chen et al. (1993) reported studies of a Chinese patient with APL and a variant translocation t(11;17)(q23;21) in which a gene on 11q23.1, designated PLZF (promyelocytic leukemia zinc finger), was fused to the RARA gene on chromosome 17. Fluorescence in situ hybridization using a PLZF-specific probe localized the PLZF gene to 11q23.1. Similar to t(15;17) APL, all-trans retinoic acid treatment produced an early leukocytosis which was followed by a myeloid maturation, but the patient died too early to achieve remission. See Chen et al. (1993) for description of the PLZF gene. Chen et al. (1994) cloned cDNAs encoding PLZF-RARA chimeric proteins and studied their transactivating activities. A 'dominant-negative' effect was observed when PLZF-RARA fusion proteins were cotransfected with vectors expressing RARA and retinoid X receptor alpha (RXRA; 180245). These abnormal transactivation properties observed in retinoic acid-sensitive myeloid cells strongly implicated the fusion proteins in the molecular pathogenesis of APL.

Reid et al. (1995) showed that murine PLZF is expressed at highest levels in undifferentiated, multipotential hematopoietic progenitor cells and its expression declines as cells become more mature and committed to various hematopoietic lineages. In the human there is a lack of PLZF protein expression in mature peripheral blood mononuclear cells and high PLZF levels in the nuclei of CD34+ human bone marrow progenitor cells. Unlike many transcription factors, PLZF protein in these cells shows a distinct punctate distribution, suggesting its compartmentalization in the nucleus.

Ahmad et al. (1998) reported on the crystal structure of the BTB domain of PLCF. The BTB domain (also known as the POZ domain) is an evolutionarily conserved protein-protein interaction motif found at the N terminus of 5 to 10% of C2H2-type zinc finger transcription factors. The BTB domain has transcriptional repression activity and interacts with components of the histone deacetylase complex. The latter association provides a mechanism of linking the transcription factor with enzymatic activities that regulate chromatin conformation.

Ahmad, K. F.; Engel, C. K.; Prive, G. G.: Crystal structure of the BTB domain from PLZF. Proc. Nat. Acad. Sci. 95: 12123-12128, 1998.

Chen, S.-J.; Zelent, A.; Tong, J.-H.; Yu, H.-Q.; Wang, Z.-Y.; Derre, J.; Berger, R.; Waxman, S.; Chen, Z.: Rearrangements of the retinoic acid receptor alpha and promyelocytic leukemia zinc finger genes resulting from t(11;17)(q23;q21) in a patient with acute promyelocytic leukemia. J. Clin. Invest. 91: 2260-2267, 1993.

Chen, Z.; Brand, N. J.; Chen, A.; Chen, S.-J.; Tong, J.-H.; Wang, Z.-Y.; Waxman, S.; Zelent, A.: Fusion between a novel Kruppel-like zinc finger gene and the retinoic acid receptor-alpha locus due to a variant t(11;17) translocation associated with acute promyelocytic leukaemia. EMBO J. 12: 1161-1167, 1993.

Chen, Z.; Guidez, F.; Rousselot, P.; Agadir, A.; Chen, S.-J.; Wang, Z.-Y.; Degos, L.; Zelent, A.; Waxman, S.; Chomienne, C.: PLZF-RAR-alpha fusion proteins generated from the variant t(11;17)(q23;q21) translocation in acute promyelocytic leukemia inhibit ligand-dependent transactivation of wild-type retinoic acid receptors. Proc. Nat. Acad. Sci. 91: 1178-1182, 1994.

Reid, A.; Gould, A.; Brand, N.; Cook, M.; Strutt, P.; Li, J.; Licht, J.; Waxman, S.; Krumlauf, R.; Zelent, A.: Leukemia translocation gene, PLZF, is expressed with a speckled nuclear pattern in early hematopoietic progenitors. Blood 86: 4544-4552, 1995.

CORE-BINDING FACTOR, RUNT DOMAIN, ALPHA SUBUNIT 2; AML1 (*151385)

According to the French-American-British (FAB) classification, the t(8;21)(q22;q22) translocation is one of the most frequent karyotypic abnormalities in acute myeloid leukemia, especially in the M2 subtype. Rowley (1990) estimated that 18% of patients of the M2 subtype have this rearrangement. Johansson et al. (1991) found the t(8;21) in 18% of AML-M2 cases with a remarkable geographic variation. Miyoshi et al. (1991) isolated and sequenced cDNA clones for a gene they named AML1, on chromosome 21, that was rearranged by the t(8;21) translocation. The t(8;21) breakpoints were clustered within a limited region of this gene, probably within the same intron. The receptor for granulocyte-macrophage colony stimulating factor (306250), which probably maps to the pseudoautosomal segment of the X chromosome, may be involved in acute myeloid leukemia of the M2 type. Avramopoulos et al. (1992) detected a polymorphism in the 3-prime untranslated region of the AML1 gene and used it in a genotyping of CEPH families to narrow the assignment to 21q22.3, between markers D21S216 and D21S211. By fluorescence in situ hybridization, Levanon et al. (1994) confirmed the assignment of AML1 to 21q22.

The chromosome 8 contribution to the chimeric gene in the translocation t(8;21) is ETO (133435). AML1 is transcribed from telomere to centromere (Miyoshi et al., 1991). The chimeric gene in the 8;21 translocation contains the 5-prime region of AML1, including the segment homologous to 'runt,' a segmentation gene of Drosophila, fused to the 3-prime region of ETO (Erickson et al., 1992). Two other human genes containing a runt domain, CBFA1 (600210) and CBFA3 (600211), were isolated and characterized by Levanon et al. (1994).

Nucifora et al. (1994) consistently found fusion transcripts between AML1 and EAP (180474) or between AML1 and previously unidentified sequences that they named MDS1 (600049), for 'MDS-associated sequences,' in the leukemic cells of 4 patients with therapy-related myelodysplasia/acute myeloid leukemia and in 1 patient with chronic myelogenous leukemia in blast crisis, all of whom had a t(3;21). In addition, they identified a third chimeric transcript, AML1/EVI1 (165215), in 1 of the therapy-related acute myeloid leukemia patients. Pulsed field gel electrophoresis established the order of the genes as EAP, the most telomeric, and EVI1, the most centromeric, with MDS1 situated between them. The results indicated that translocations can involve multiple genes and affect gene expression over long distances.

The involvement of the AML1 gene in oncogenic transformation is noteworthy since children with trisomy 21 have an increased risk of leukemia. In addition, Down syndrome neonates sometimes have a transient myeloproliferative disorder or transient leukemia that mimics congenital leukemia. In about 50% of leukemic Down syndrome children, the disease is of the acute megakaryoblastic leukemia (AMKL-M7) type (Zipursky et al., 1992). This type of leukemia, which is relatively rare among children, is estimated to be 400 times as common in Down syndrome as in other children.

Okuda et al. (1996) reviewed the literature on AML1 and its role in multiple chromosomal translocations in human leukemia. The AML1-ETO chimeric product resulting from t(8;21) occurs in approximately 15% of cases of acute myeloid leukemia. An inversion of chromosome 16 (inv16) occurs in 15 to 18% of cases of de novo AML and results in the fusion of CDB (121360) to a smooth muscle myosin heavy chain gene, MYH11 (160745), and produces a chimeric product that retains its ability to interact with AML1. In t(3;21) translocations in rare cases of myelodysplasia and blast transformation of chronic myelogenous leukemia, AML1 is fused with either the EVI1 gene (165215), which encodes a known zinc finger-containing transcription factor, or either of 2 alternative genes of unknown function, EAP and MDS1, which are located adjacent to EVI1 on 3q26. Although these findings might suggest that AML1 alterations are limited to leukemia of myeloid lineage, Pui (1995) demonstrated AML1 is frequently mutated in pediatric B progenitor acute lymphoblastic leukemia (ALL), the most common malignancy seen in children. Cloning of the ALL-associated t(12;21) revealed the formation of a chimeric gene that encoded a fusion protein consisting of the N-terminal helix-loop-helix domain of TEL (600618), a member of the Ets-like family of transcription factors, fused with a nearly complete AML1 protein. Analysis of a large number of pediatric ALL cases demonstrated expression of the TEL-AML1 chimeric transcript in approximately 25% of cases with a B progenitor immunophenotype, despite the complete lack of cytogenetic evidence of this translocation in the majority of cases.

Nucifora and Rowley (1995) reviewed the involvement of the AML1 gene in the 8;21 and 3;21 translocations in acute and chronic myeloid leukemia. Three loci closely situated to each other on 3q26 are involved in fusions with AML1 in the 3;21 translocations: EVI1, EAP, and MDS1. They pointed out that the order of the genes on 3q26 is TEL—EAP—MDS1—EVI1 and provided a diagram of the 3q26 region containing these genes and of the various chimeric junctions they had isolated from t(3;21) patients.

Using a large pedigree with an autosomal dominant familial 'aspirin-like' platelet disorder (601399) associated with development of acute myelogenous leukemia, Ho et al. (1996) found linkage to markers D21S263, D21S216, IFNAR, and D21S65, with lod scores of 6.19, 4.89, 9.40, and 10.48, respectively (theta = 0.001). They proposed a familial platelet disorder critical region of 15.2 cM in addition to approximate placement for markers D21S211 and UT7582/D21S1413. This critical region contains 5 candidate genes, including AML1, IF-

NAR (107450), CRF2-4 (123889), GART (138440), and SON (182465).

Animal Model

To investigate the normal biologic function of AML1 in vivo, Okuda et al. (1996) generated mice carrying a disrupted AML1 allele using gene targeting in embryonic stem cells. Mice lacking aml1 died during midembryonic development, secondary to the complete absence of fetal liver-derived hematopoiesis. Moreover, homozygous aml1-deficient cells failed to contribute to hematopoiesis in chimeric animals. These findings indicated that aml1-regulated target genes are essential for definitive hematopoiesis of all lineages. Wang et al. (1996) likewise analyzed the role of CBFA2 in mammalian development by gene disruption in mice. They found that mice lacking a CBF-alpha-2 protein capable of binding DNA died between embryonic days 11.5 and 12.5 due to hemorrhaging in the central nervous system, at the nerve/CNS interfaces of cranial and spinal nerves, and in somitic/intersomitic regions along the presumptive spinal cord. Hemorrhaging was preceded by symmetric, bilateral necrosis in these regions. Definitive erythropoiesis and myelopoiesis did not occur in Cbfa2-deficient embryos, and disruption of 1 copy of the Cbfa2 gene significantly reduced the number of progenitors for erythroid and myeloid cells.

In a review of oncogenic transcription factors in human acute leukemias, Look (1997) diagrammed the distribution of translocation-generated oncogenes among the acute leukemias of children and young adults. The most frequent translocation causing ALL was t(12;21), leading to the TEL-AML1 oncogene and accounting for 20% of ALL cases. The t(8;21)-generated AML1-ETO oncogene (133435) accounted for 12% of AML cases (a myeloblastic endtype).

Look (1997) diagrammed 2 distinct mechanisms by which chromosomal translocations aberrantly activate genes encoding transcription factors. Transcription factor protooncogenes that are silent or expressed at lower levels in the progenitor cells of a particular lineage may be activated when placed under the control of potent enhancer elements within the regulatory region of a gene that is normally highly expressed. Typically, the regulatory region in these cases is contributed by one of the immunoglobulin or T-cell receptor genes present in lymphoid precursors of either the B or T lineage. More commonly, chromosomal breakpoints occur within introns, between the coding sequences of each of 2 transcription factor genes on different chromosomes, producing a fusion gene that encodes a chimeric transcription factor with altered function. The regulatory sequences that drive expression of the hybrid gene generally derive from the gene that contributes the amino-terminal amino acids to the chimeric protein; the carboxy-terminal amino acids often derive from a gene that is not normally expressed in the progenitor cells in which the chimeric oncoprotein arises.

Avramopoulos, D.; Cox, T.; Blaschak, J. E.; Chakravarti, A.; Antonarakis, S. E.: Linkage mapping of the AML1 gene on human chromosome 21 using a DNA polymorphism in the 3-prime untranslated region. Genomics 14: 506-507, 1992.

Erickson, P.; Gao, J.; Chang, K.-S.; Look, T.; Whisenant, E.; Raimondo, S.; Lasher, R.; Trujillo, J.; Rowley, J. D.; Drabkin, H. A.: Identification of breakpoints in t(8;21) acute myelogenous leukemia and isolation of a fusion transcript, AML1/ETO, with similarity to Drosophila segmentation gene, runt. Blood 80: 1825-1831, 1992.

Ho, C. Y.; Otterud, B.; Legare, R. D.; Varvil, T.; Saxena, R.; DeHart, D. B.; Kohler, S. E.; Aster, J. C.; Dowton, S. B.; Li, F. P.; Leppert, M.; Gilliland, D. G.: Linkage of a familial platelet disorder with a propensity to develop myeloid malignancies to human chromosome 21q22.1-22.2. Blood 87: 5218-5224, 1996.

Johansson, B.; Mertens, F.; Mitelman, F.: Geographic heterogeneity of neoplasia-associated chromosome aberrations. Genes Chromosomes Cancer 3: 1-7, 1991.

Levanon, D.; Negreanu, V.; Bernstein, Y.; Bar-Am, I.; Avivi, L.; Groner, Y.: AML1, AML2, and AML3, the human members of the runt domain gene-family: cDNA structure, expression, and chromosomal localization. Genomics 23: 425-432, 1994.

Look, A. T.: Oncogenic transcription factors in the human acute leukemias. Science 278: 1059-1064, 1997.

Miyoshi, H.; Shimizu, K.; Kozu, T.; Maseki, N.; Kaneko, Y.; Ohki, M.: t(8;21) breakpoints on chromosome 21 in acute myeloid leukemia are clustered within a limited region of a single gene, AML1. Proc. Nat. Acad. Sci. 88: 10431-10434, 1991.

Nucifora, G.; Begy, C. R.; Kobayashi, H.; Roulston, D.; Claxton, D.; Pedersen-Bjergaard, J.; Parganas, E.; Ihle, J. N.; Rowley, J. D.: Consistent intergenic splicing and production of multiple transcripts between AML1 at 21q22 and unrelated genes at 3q26 in (3;21)(q26;q22) translocations. Proc. Nat. Acad. Sci. 91: 4004-4008, 1994.

Nucifora, G.; Rowley, J. D.: AML1 and the 8;21 and 3;21 translocations in acute and chronic myeloid leukemia. Blood 86: 1-14, 1995.

Okuda, T.; van Deursen, J.; Hiebert, S. W.; Grosveld, G.; Downing, J. R.: AML1, the target of multiple chromosomal translocations in human leukemia, is essential for normal fetal liver hematopoiesis. Cell 84: 321-330, 1996.

Pui, C.-H.: Childhood leukemias. New Eng. J. Med. 332: 1618-1630, 1995.

Rowley, J. D.: Recurring chromosome abnormalities in leukemia and lymphoma. Seminars Hemat. 27: 122-136, 1990.

Wang, Q.; Stacy, T.; Binder, M.; Marin-Padilla, M.; Sharpe, A. H.; Speck, N. A.: Disruption of the Cbfa2 gene causes necrosis and hemorrhaging in the central nervous system and blocks definitive hematopoiesis. Proc. Nat. Acad. Sci. 93: 3444-3449, 1996.

Zipursky, A.; Poon, A.; Doyle, J.: Leukemia in Down syndrome: a review. Pediat. Hemat. Oncol. 9: 139-149, 1992.

CORE-BINDING FACTOR, BETA SUBUNIT; CBFB (*121360)

Liu et al. (1993) pinpointed the breakpoints in the pericentric inversion of chromosome 16, inv(16)(p13q22), that is characteristic of acute myeloid leukemia, most commonly of the M4Eo subtype. A fusion gene was identified of which the 16p13 contributor was the MYH11 gene (160745), constituting the 3-prime region of that myosin protein. The 16q22 contribution to the fusion gene was provided by the human homolog of a mouse DNA-binding factor CBF-beta (Wang et al., 1993). Core-binding factor (CBF) binds to the core site of murine leukemia virus and also to the enhancers of the T-cell receptor genes. The core site appears to be a major genetic determinant of the tissue specificity of leukemias induced by the murine leukemia virus. Affinity-purified CBF contains at least 2 subunits, alpha and beta. The alpha subunit is identical to AML1 (151385), the gene found to be disrupted in the characteristic t(8;21) translocation in the M2 subtype of acute myeloid leukemia. Another DNA-binding factor, named PEBP2, which binds to similar core sequences in the enhancer of the polyomavirus, was identified in NIH-3T3 cells. PEBP2 also contains 2 subunits, alpha and beta; PEBP2-beta and CBF-beta are identical. Liu et al. (1993) suggested that elucidation of the genes involved as fusion partners in an inversion leading to a common form of adult leukemia should allow the development of a mouse model and a sensitive RT-PCR test for specific diagnosis and assessment of residual disease after treatment. Ogawa et al. (1993) also cloned the mouse Pebp2b gene and cDNAs representing 3 different splice variants.

Liu et al. (1995) provided a review of leukemia pathogenesis related to CBFB. They suggested that it will be interesting to see whether variant fusions between CBFB and another gene exist as a result of translocation between 16q and another chromosome. The study of such variants might shed light on the mechanism of genesis by the inversion 16 fusion gene. Whether the abnormal eosinophils in the circulation in patients with inv(16) are part of the malignant cell population or a result of a secondary response could not be determined. Although the distribution of breakpoints in the introns of the 2 participating genes was heterogeneous, a surprisingly high incidence of breaks was observed in a small (370 bp) intron of the MYH11 gene. CBFB and AML1 encode the 2 subunits of the transcription factor CBF, and alterations of either one are associated with acute myeloid leukemia.

To examine the effects of the inv(16)(p13;q22) on myelopoiesis, Kogan et al. (1998) generated transgenic mice expressing the chimeric fusion protein in myeloid cells. Neutrophil maturation was impaired. Although the transgenic mice had normal numbers of circulating neutrophils, their bone marrow contained increased numbers of immature neutrophilic cells, which exhibited abnormal characteristics. In addition, the fusion protein inhibited neutrophilic differentiation in colonies derived from hematopoietic progenitors. Coexpression of both the fusion protein and activated NRAS (164790) induced a more severe phenotype characterized by abnormal nuclear morphology indicative of granulocytic dysplasia. These results showed that the fusion protein impairs neutrophil development and provided evidence that alterations in Pebp2 can contribute to the genesis of myelodysplasia.

Kogan, S. C.; Lagasse, E.; Atwater, S.; Bae, S.; Weissman, I.; Ito, Y.; Bishop, J. M.: The PEBP2-beta-MYH11 fusion created by inv(16)(p13;q22) in myeloid leukemia impairs neutrophil maturation and contributes to granulocytic dysplasia. Proc. Nat. Acad. Sci. 95: 11863-11868, 1998.

Liu, P.; Tarle, S. A.; Hajra, A.; Claxton, D. F.; Marlton, P.; Freedman, M.; Siciliano, M. J.; Collins, F. S.: Fusion between transcription factor CBF-beta/PEBP2-beta and a myosin heavy chain in acute myeloid leukemia. Science 261: 1041-1044, 1993.

Liu, P. P.; Hajra, A.; Wijmenga, C.; Collins, F. S.: Molecular pathogenesis of the chromosome 16 inversion in the M4Eo subtype of acute myeloid leukemia. Blood 85: 2289-2302, 1995.

Ogawa, E.; Inuzuka, M.; Maruyama, M.; Satake, M.; Naito-Fujimoto, M.; Ito, Y.; Shigesada, K.: Molecular cloning and characterization of PEBP2-beta, the heterodimeric partner of a novel Drosophila runt-related DNA binding protein PEBP2-alpha. Virology 194: 314-331, 1993.

Wang, S.; Wang, Q.; Crute, B. E.; Melnikova, I. N.; Keller, S. R.; Speck, N. A.: Cloning and characterization of subunits of the T-cell receptor and murine leukemia virus enhancer core-binding factor. Molec. Cell. Biol. 13: 3324-3339, 1993.

MYOSIN, SMOOTH MUSCLE, HEAVY CHAIN 11 (*160745)

4,3:Matsuoka et al. (1991, 1993) isolated a smooth muscle myosin heavy-chain gene from a human cDNA library. They confirmed it as a smooth muscle MHC gene by Northern blot hybridization and a partial DNA sequence analysis. By study of human/Chinese hamster and human/mouse hybrid cells and by in situ hybridization, they localized the gene to 16q12.1. This localization was later found to be an error and the gene was shown, in fact, to be located on the short arm of chromosome 16 in a region involved in pericentric inversion inv(16)(p13q22), a characteristic abnormality associated with acute myeloid leukemia, most commonly of the M4Eo subtype. Liu et al. (1993) pinpointed the 16p and 16q breakpoints by yeast artificial chromosome and cosmid cloning and identified the 2 genes involved in the inversion. On 16p, the MYH11 gene was interrupted; on 16q, the inversion occurred near the end of the coding region for CBF-beta (121360), also known as PEBP2-beta, a subunit of a heterodimeric transcription factor regulating genes expressed in T cells. In all of 6 inv(16) patients tested, an in-frame fusion messenger RNA was demonstrated that connected the first 165 amino acids of CBFB with the tail region of MYH11. The repeated coiled coil of MYH11 may result in dimerization of the CBFB fusion protein, which in turn would lead to alterations in transcriptional regulation and contribute to leukemic transformation. Deng et al. (1993) mapped the MYH11 gene to the middle of the short arm of chromosome 16 by fluorescence in situ hybridization. Southern blots of a panel of hybrids containing different portions of human chromosome 16 localized the gene to 16p13.13-p13.12. Studies of DNA from a CHO/mouse hybrid clone mapping panel showed that the gene was located on mouse chromosome 16.

Deng, Z.; Liu, P.; Marlton, P.; Claxton, D. F.; Lane, S.; Callen, D. F.; Collins, F. S.; Siciliano, M. J.: Smooth muscle myosin heavy chain locus (MYH11) maps to 16p13.13-p13.12 and establishes a new region of conserved synteny between human 16p and mouse 16. Genomics 18: 156-159, 1993.
Liu, P.; Tarle, S. A.; Hajra, A.; Claxton, D. F.; Marlton, P.; Freedman, M.; Siciliano, M. J.; Collins, F. S.: Fusion between transcription factor CBF-beta/PEBP2-beta and a myosin heavy chain in acute myeloid leukemia. Science 261: 1041-1044, 1993.
Matsuoka, R.; Yoshida, M. C.; Furutani, Y.; Imamura, S.; Kanda, N.; Yanagisawa, M.; Masaki, T.; Takao, A.: Human smooth muscle myosin heavy chain gene mapped to chromosomal region 16q12. Am. J. Med. Genet. 46: 61-67, 1993.
Matsuoka, R.; Yoshida, M. C.; Kanda, N.; Furutani, Y.; Bruns, G.; Yanagisawa, M.; Masaki, T.; Takao, A.: Human smooth muscle myosin heavy-chain gene mapped to chromosomal region 16q12.1-q12.2. (Abstract) Cytogenet. Cell Genet. 58: 2000-2001, 1991.

ONCOGENE MOS (*190060)

The Moloney murine sarcoma virus (MSV) is a representative of a class of replication-defective retroviruses that transform fibroblasts in culture and induce sarcomas in vivo. It arose by recombination between the Moloney murine leukemia virus and a sequence derived from mouse cells. The mouse cell-derived segment of MSV, termed v-mos, is required for the induction and maintenance of viral transformation. The normal mouse analog of v-mos has been molecularly cloned. Because of the evolutionary conservation of viral onc genes among vertebrate species, Prakash et al. (1982) could use the cloned mouse analog to map the human counterpart (see Watson et al., 1982) to human chromosome 8. Klein (1981) speculated that the consistent occurrence of translocations involving chromosome 8 in Burkitt lymphoma may mean that activation of an onc gene underlies this tumor. The breakpoint in all Burkitt rearrangements is at 8q24. As Prakash et al. (1982) indicated, 'it will be of interest to determine whether c-mos (human) maps at or near the breakpoint in the Burkitt translocations and whether transcripts related to c-mos are detectable in such tumors.' By in situ hybridization, Neel et al. (1982) assigned MOS to 8q22 and MYC to 8q24. They used the method of Kirsch et al. (1982). They pointed out that a familial translocation, t(3;8)(p21;q24), predisposing to renal cell carcinoma (144700) involves the same chromosome region. 10,11:Rowley (1973, 1983) located the MOS oncogene at the breakpoint of the 8;21 translocation associated with acute myeloblastic leukemia. A nonrandom chromosome translocation, t(8;21)(q22;q22), which results in an 8q- and 21q+ chromosome, is seen almost exclusively in the M2 subtype of acute myeloblastic leukemia (AML with maturation). Band 21q22 is critical to the Down syndrome phenotype, which is associated with an increased leukemia risk. In studies of a case of AML-M2, Drabkin et al. (1985) isolated the 21q+ chromosome in a somatic cell hybrid and showe d that the MOS oncogene had not been translocated to chromosome 21. They also could find no rearrangement in a 12.4-kb region surrounding the MOS gene. This does not exclude a key role for MOS, however. Caubet et al. (1985) studied the regional assignment of MOS by in situ hybridization and by hybridization to flow-sorted chromosomes from a cell line with a translocation t(6;8)(q27;q21). Both approaches yielded results indicating 8q11 as the site of MOS, not 8q22 as previously reported. By in situ hybridization, Testa et al. (1988) likewise assigned MOS to 8q11-q12. By in situ hybridization to metaphase chromosomes and by standard genetic backcrosses, Propst et al. (1989) demonstrated that MOS maps near the centromere of mouse chromosome 4. It had previously been assigned to mouse chromosome 4 by use of panels of mouse/hamster somatic cell hybrids (Swan et al., 1982). Dandoy et al. (1989) showed by linkage analysis that the murine Mos gene is located in the region of chromosome 4 compatible with the physical mapping (Threadgill and Womack, 1988) by in situ hybridization.

See also Diaz et al. (1985)

Caubet, J.-F.; Mathieu-Mahul, D.; Bernheim, A.; Larsen, C.-J.; Berger, R.: Human proto-oncogene c-mos maps to 8q11. EMBO J. 4: 2245-2248, 1985.
Dandoy, F.; De Maeyer-Guignard, J.; De Maeyer, E.: Linkage analysis of the murine mos proto-oncogene on chromosome 4. Genomics 4: 546-551, 1989.
Diaz, M. O.; Le Beau, M. M.; Rowley, J. D.; Drabkin, H. A.; Patterson, D.: The role of the c-mos gene in the 8;21 translocation in human acute myeloblastic leukemia. Science 229: 767-769, 1985.
Drabkin, H. A.; Diaz, M.; Bradley, C. M.; Le Beau, M. M.; Rowley, J. D.; Patterson, D.: Isolation and analysis of the 21q+ chromosome in the acute myelogenous leukemia 8;21 translocation: evidence that c-mos is not translocated. Proc. Nat. Acad. Sci. 82: 464-468, 1985.
Kirsch, I. R.; Morton, C.; Nakahara, K.; Leder, P.: Human immunoglobulin heavy chain genes map to a region of translocations in malignant B lymphocytes. Science 216: 301-303, 1982.
Klein, G.: The role of gene dosage and genetic transpositions in carcinogenesis. Nature 294: 313-318, 1981.
Neel, B. G.; Jhanwar, S. C.; Chaganti, R. S. K.; Hayward, W. S.: Two human c-onc genes are located on the long arm of chromosome 8. Proc. Nat. Acad. Sci. 79: 7842-7846, 1982.
Prakash, K.; McBride, O. W.; Swan, D. C.; Devare, S. G.; Tronick, S. R.; Aaronson, S. A.: Molecular cloning and chromosomal mapping of a human locus related to the transforming gene of Moloney murine sarcoma virus. Proc. Nat. Acad. Sci. 79: 5210-5214, 1982.
Propst, F.; Vande Woude, G. F.; Jenkins, N. A.; Copeland, N. G.; Lee, B. K.; Hunt, P. A.; Eicher, E. M.: The MOS proto-oncogene maps near the centromere on mouse chromosome 4. Genomics 5: 118-123, 1989.
Rowley, J. D.: Identification of a translocation with quinacrine fluorescence in a patient with acute leukemia. Ann. Genet. 16: 109-112, 1973.
Rowley, J. D.: Human oncogene location and chromosome aberrations. Nature 301: 290-291, 1983.
Swan, D.; Oskarsson, M.; Keithley, D.; Ruddle, F. H.; D'Eustachio, P.; Vande Woude, G. F.: Chromosomal localization of the Moloney sarcoma virus mouse cellular (c-mos) sequence. J. Virol. 44: 752-754, 1982.
Testa, J. R.; Parsa, N. Z.; Le Beau, M. M.; Vande Woude, G. F.: Localization of the proto-oncogene MOS to 8q11-q12 by in situ chromosomal hybridization. Genomics 3: 44-47, 1988.
Threadgill, D. W.; Womack, J. E.: Regional localization of mouse Abl and Mos proto-oncogenes by in situ hybridization. Genomics 3: 82-86, 1988.
Watson, R.; Oskarsson, M.; Vande Woude, G. F.: Human DNA sequence homologous to the transforming gene (mos) of Moloney murine sarcoma virus. Proc. Nat. Acad. Sci. 79: 4078-4082, 1982.

FIBROBLAST GROWTH FACTOR RECEPTOR 3 (*134934)

The fibroblast growth factors are a family of polypeptide growth factors involved in a variety of activities including mitogenesis, angiogenesis, and wound healing. They contain an extracellular domain with either 2 or 3 immunoglobulin-like domains, a transmembrane domain, and a cytoplasmic tyrosine kinase domain. Keegan et al. (1991) isolated a human cDNA for a 'new' tyrosine kinase receptor that was highly homologous to previously described FGFRs. Expression in COS cells directed formation of a 125-kD glycoprotein. They showed that human acidic and basic fibroblast growth factors activated this receptor as measured by calcium-ion efflux assays. They called the new receptor FGFR3. Thompson et al. (1991) isolated the FGFR3 gene from the Huntington disease (HD; 143100) region of human chromosome 4, p16.3. Histochemical analysis using in situ hybridization showed that the FGFR3 gene is expressed in many areas of the brain, among them the caudate and putamen. The nature of the gene as a growth factor receptor and its map location made it a possible candidate for the HD gene. Of course, this did not prove to be the case. Using an interspecific backcross mapping panel, Avraham et al. (1994) mapped the Fgfr3 gene to mouse chromosome 5 in a region of homology of synteny with human chromosome 4.

Perez-Castro et al. (1997) reported that the FGFR3 gene contains 19 exons spanning 16.5 kb. The overall structure and organization of the human FGFR3 gene are nearly identical to those of the mouse Fgfr3 gene. The human and mouse FGFR3 amino acid sequences show 92% homology.

Shiang et al. (1994) studied the FGFR3 gene as a candidate for the site of the mutation in achondroplasia (ACH; 100800), which had been mapped to the same region. DNA studies revealed point mutations in the FGFR3 gene in both ACH heterozygotes and homozygotes. The mutation on 15 of 16 ACH-affected chromosomes was the same: a G-to-A transition at nucleotide 1138 of the cDNA (134934.0001). The mutation on the other ACH-affected chromosome 4 without the G-to-A transition at nucleotide 1138 had a G-to-C transversion at the same position. Both mutations resulted in the substitution of an arginine residue for a glycine at position 380 of the mature protein, which is in the transmembrane domain of FGFR3. Rousseau et al. (1994) confirmed these mutations by DNA analysis of 17 sporadic cases and 6 unrelated familial forms of achondroplasia. Francomano (1994) reviewed the findings in 154 unrelated achondroplasts. In a later assemblage of data from multiple laboratories (Bellus et al., 1995), 150 were heterozygous for the G-to-A transition at nucleotide 1138 leading to the G380R substitution; 3 were heterozygous for the G-to-C transversion at nucleotide 1138 leading to the same G380R substitution (134934.0002). One achondroplasia patient reported by Superti-Furga et al. (1995) had a G-to-T transversion leading to a G375C (134934.0003) amino acid substitution.

Thanatophoric dysplasia (TD; 187600) resembles homozygous achondroplasia in some respects. Tavormina et al. (1995) found mutations in TD type I (TD1) families which involved the substitution of a cysteine residue for the native amino acid (R248C, 134934.0005; S371C, 134934.0006). In 16 of 16 individuals with type II thanatophoric dysplasia (TD2), they found a sporadic heterozygous mutation causing a lys650-to-glu change in the FGFR3 tyrosine kinase domain (134934.0004). Tavormina et al.(1995) described another TD1-associated cysteine-generating mutation in the extracellular domain of FGFR3 (S249C; 134934.0013). The authors speculated that the unpaired cysteine residue in this region of the protein might result in formation of intermolecular disulfide bonds between 2 mutant FGFR3 monomers and thereby constitutively activate the receptor complex. Rousseau et al. (1996) performed FGFR3 mutation analysis in 26 cases of TD1. Three missense mutations (Y373C, R248C, and S249C) accounted for 73% of the cases. Two stop codon mutations (X807R, 134934.0008; X807C, 134934.0009) and 1 rare G370C mutation were also found. Rousseau et al. (1996) noted that all reported missense mutations created cysteine residues and were located in the extracellular domain of the receptor. The findings provided support for the hypothesis that the newly created cysteine residues may allow disulfide bonds to form between the extracellular domains of mutant monomers, thus inducing constitutive activation of the homodimer receptor complex.

Surprisingly, Meyers et al. (1995) found a mutation in the FGFR3 gene, ala391-to-glu (134934.0011), in 3 unrelated families with a syndrome of Crouzon craniosynostosis (123500) and acanthosis nigricans (100600), a specific skin disorder of hyperkeratosis and hyperpigmentation. The association of non-dwarfing and non-skeletal conditions with an FGFR3 mutation demonstrated the potential for a wide range of FGFR pleiotropic effects, as well as locus heterogeneity in Crouzon syndrome which had previously been found in association only with mutations in the FGFR2 gene (176943).

Naski et al. (1996) studied the effect of the achondroplasia (ACH) and thanatophoric dysplasia (TD) mutations on the activity and regulation of FGFR3 by transient transfection of NIH 3T3 and BaF3 pro-B cells with mutant FGFR3 cDNAs. They showed that each of the mutations studied (R248C, K650E, and G380R) constitutively activates the receptor, as evidenced by ligand-independent receptor tyrosine phosphorylation and cell proliferation. Moreover, the mutations responsible for TD (R248C and K650E) were more strongly activating than the mutation causing ACH (G380R), providing to Naski et al. (1996) a biochemical explanation for the observation that the phenotype of TD is more severe than that of ACH.

Bellus et al. (1996) described a pro250-to-arg mutation (P250R; 134934.0014) in FGFR3. On the basis of 61 individuals from 20 unrelated families where coronal synostosis was due to the P250R mutation in the FGFR3 gene, Muenke et al. (1997) defined a new clinical syndrome distinct from previously defined craniosynostosis syndromes, including the Pfeiffer (101600), Crouzon, Jackson-Weiss (123150), and Apert (101200) syndromes. In addition to the skull findings, some patients had abnormalities on radiographs of hands and feet, including thimble-like middle phalanges, coned epiphyses, and carpal and tarsal fusions. Brachydactyly was seen in some cases; none had clinically significant syndactyly or deviation of the great toe to suggest Apert syndrome or Pfeiffer syndrome, respectively. Sensorineural hearing loss was present in some and developmental delay was seen in a minority. While the radiologic findings of hands and feet can be helpful in the recognition of this syndrome, it was not in all cases clearly distinguishable on a clinical basis from other craniosynostosis syndromes. Therefore, Muenke et al. (1997) suggested that all patients with coronal synostosis should be tested for this mutation. At the urging of Wilkie (1998), we have designated this syndrome caused by the P250R mutation as Muenke syndrome (602849), or Muenke nonsyndromic coronal craniosynostosis. This is in parallel with the usage for Apert syndrome, Pfeiffer syndrome, Crouzon syndrome, Saethre-Chotzen syndrome, etc. The very tight relationship between genotype and phenotype is shared also by achondroplasia, Apert syndrome, and type IIB multiple endocrine neoplasia (MEN2B; 164761.0013).

Dysregulation of oncogenes by translocation to the immunoglobulin heavy chain (IgH) locus (147100) on 14q32 is a seminal event in the pathogenesis of B-cell tumors. In multiple myeloma (254500), translocations to the IgH locus occur in 20 to 60% of cases. For most translocations, the partner chromosome is unknown; for the others, a diverse array of chromosomal partners have been identified, with 11q13 (see cyclin D1; 168461) the only chromosome that is frequently involved. Bergsagel et al. (1996) developed a comprehensive Southern blot assay to identify and distinguish different kinds of IgH switch recombination events. Illegitimate switch recombination fragments (defined as containing sequences from only 1 switch region) are potential markers of translocation events into IgH switch regions and were identified in 15 of 21 myeloma cell lines, including 7 of 8 karyotyped lines that had no detectable 14q32 translocation. These translocation breakpoints involved 6 chromosomal loci: 4p16.3; 6; 8q24.13; 11q13.3; 16q23.1; and 21q22.1. Chesi et al. (1997) found the novel, karyotypically silent translocation t(4;14)(p16.3;q32.3) in 5 myeloma cells lines and in at least 3 of 10 primary tumors. The chromosome-4 breakpoints were clustered in a 70-kb region centromeric to FGFR3, which was thought to be the dysregulated oncogene. Two lines and 1 primary tumor with this translocation selectively expressed an FGFR3 allele containing activating mutations identified previously in thanatophoric dwarfism: tyr373 to cys (134934.0016), lys650 to glu (134934.0004), and lys650 to met (134934.0015). For K650E, the constitutive activation of FGFR3 in the absence of ligand had been proved by transfection experiments. Chesi et al. (1997) proposed that after the t(4;14) translocation, somatic mutation during tumor progression frequently generates an FGFR3 protein that is active in the absence of ligand. Although they could not exclude the possibility that other genes are dysregulated by the translocation t(4;14), several findings pointed to FGFR3. FGFR3 is located no more than 100 kb from the most centromeric breakpoint at 4p16.3, and is on the derivative(14) chromosome that contains the 3-prime IgH enhancer. This is similar to the situation for cyclin D1, which is located 100 to 400 kb from the breakpoint in the translocation t(11;14) that occurs in mantle-cell lymphoma and multiple myeloma tumors. FGFR3 is another example of a gene that can function both as an oncogene and a 'teratogene.'

Lanning and Brown (1997) described an improved method for detecting the common 1138G-A mutation. The mutation had typically been detected by SfcI digestion of amplified genomic DNA. Lanning and Brown (1997) showed that the SfcI digestion protocol does not consistently distinguish between DNA samples heterozygous and homozygous for the G1138A substitution, and illustrated how the misdiagnosis of a homozygous affected fetus for one carrying only 1 copy of the mutation could occur. The simple nonradioactive technique that they described could reliably and consistently detect the presence of the G1138A mutation in both the heterozygous and the homozygous state.

Animal Model

Colvin et al. (1996) reported the findings in mice homozygous for a targeted disruption of Fgfr3. Skeletal defects included kyphosis, scoliosis, crooked tails, and curvature and overgrowth of long bones and vertebrae. Contrasts between the skeletal phenotype of the mice and achondroplasia suggested to the authors that activation of FGFR3 may cause achondroplasia. Furthermore, the mice showed defects of the inner ear, including failure of pillar cell differentiation and tunnel of Corti formation, resulting in profound deafness. The results demonstrated that Fgfr3 is essential for normal endochondral ossification and inner ear development.

Deng et al. (1996) reported studies in mice made FGFR3 deficient by targeted disruption in the Fgfr3 gene by homologous recombination. Fgfr3 +/- mice showed no phenotypic abnormalities. Fgfr -/- mice had phenotypic effects restricted to bones that arise by endochondral ossification, i.e., increased length of the vertebral column and long bones occurred. Histologic studies revealed cellular expansion, involving hypertrophic chondrocytes, in the growth plates of vertebrae and long bones of mutant homozygotes. Deng et al. (1996) proposed that the function of FGFR3 is to limit osteogenesis. They noted that the recessive loss-of-function mutation in Fgfr3 -/- mice produces a phenotype that is the opposite of that observed in achondroplasia and thanatophoric dwarfism. They proposed that the FGFR3 mutations in these disorders lead to constitutive activation (ligand independent activation) of the receptor.

Avraham, K. B.; Givol, D.; Avivi, A.; Yayon, A.; Copeland, N. G.; Jenkins, N. A.: Mapping of murine fibroblast growth factor receptors refines regions of homology between mouse and human chromosomes. Genomics 21: 656-658, 1994.

Bellus, G. A.; Gaudenz, K.; Zackai, E. H.; Clarke, L. A.; Szabo, J.; Francomano, C. A.; Muenke, M.: Identical mutations in three different fibroblast growth factor receptor genes in autosomal dominant craniosynostosis syndromes. Nature Genet. 14: 174-176, 1996.

Bellus, G. A.; Hefferon, T. W.; Ortiz de Luna, R. I.; Hecht, J. T.; Horton, W. A.; Machado, M.; Kaitila, I.; McIntosh, I.; Francomano, C. A.: Achondroplasia is defined by recurrent G380R mutations of FGFR3. Am. J. Hum. Genet. 56: 368-373, 1995.

Bellus, G. A.; McIntosh, I.; Smith, E. A.; Aylsworth, A. S.; Kaitila, I.; Horton, W. A.; Greenhaw, G. A.; Hecht, J. T.; Francomano, C. A.: A recurrent mutation in

the tyrosine kinase domain of fibroblast growth factor receptor 3 causes hypochondroplasia. Nature Genet. 10: 357-359, 1995.
Bergsagel, P. L.; Chesi, M.; Nardini, E.; Brents, L. A.; Kirby, S. L.; Kuehl, W. M.: Promiscuous translocations into immunoglobulin heavy chain switch regions in multiple myeloma. Proc. Nat. Acad. Sci. 93: 13931-13936, 1996.
Breitbart, A. S.; Eaton, C.; McCarthy, J. G.: Crouzon's syndrome associated with acanthosis nigricans: ramifications for the craniofacial surgeon. Ann. Plastic Surg. 22: 310-315, 1989.
Chesi, M.; Nardini, E.; Brents, L. A.; Schrock, E.; Ried, T.; Kuehl, W. M.; Bergsagel, P. L.: Frequent translocation t(4;14)(p16.3;q32.3) in multiple myeloma is associated with increased expression and activating mutations of fibroblast growth factor receptor 3. Nature Genet. 16: 260-264, 1997.
Colvin, J. S.; Bohne, B. A.; Harding, G. W.; McEwen, D. G.; Ornitz, D. M.: Skeletal overgrowth and deafness in mice lacking fibroblast growth factor receptor 3. Nature Genet. 12: 390-397, 1996.
Deng, C.; Wynshaw-Boris, A.; Zhou, F.; Kuo, A.; Leder, P.: Fibroblast growth factor receptor 3 is a negative regulator of bone growth. Cell 84: 911-921, 1996.
Deutz-Terlouw, P. P.; Losekoot, M.; Aalfs, C. M.; Hennekam, R. C. M.; Bakker, E.: Asn540thr substitution in the fibroblast growth factor receptor 3 tyrosine kinase domain causing hypochondroplasia. Hum. Mutat. Suppl. 1: S62-S65, 1998.
Francomano, C. A.: Personal Communication. Baltimore, Md. 12/1/1994.
Francomano, C. A.; Bellus, G. A.; Szabo, J.; McIntosh, I.; Dorst, J.; Lee, R.; Hurko, O.; Fraley, A. E.; Bamshad, M. J.: A new skeletal dysplasia with severe tibial bowing, profound developmental delay and acanthosis nigricans is caused by a Lys 650 Met mutation in fibroblast growth factor receptor 3 (FGFR3). (Abstract) Am. J. Hum. Genet. 59 (suppl.): A25 only, 1996.
Golla, A.; Lichtner, P.; von Gernet, S.; Winterpacht, A.; Fairley, J.; Murken, J.; Schuffenhauer, S.: Phenotypic expression of the fibroblast growth factor receptor 3 (FGFR3) mutation P250R in a large craniosynostosis family. J. Med. Genet. 34: 683-684, 1997.
Grigelioniene, G.; Hagenas, L.; Eklof, O.; Neumeyer, L.; Haereid, P. E.; Anvret, M.: A novel missense mutation ile538-to-val in the fibroblast growth factor receptor 3 in hypochondroplasia. (Abstract) Hum. Mutat. 11: 333 only, 1998.
Gripp, K. W.; McDonald-McGinn, D. M.; Gaudenz, K.; Whitaker, L. A.; Bartlett, S. P.; Glat, P. M.; Cassileth, L. B.; Mayro, R.; Zackai, E. H.; Muenke, M.: Identification of a genetic cause for isolated unilateral coronal synostosis: a unique mutation in the fibroblast growth factor receptor 3. J. Pediat. 132: 714-716, 1998.
Hollway, G. E.; Suthers, G. K.; Battese, K. M.; Turner, A. M.; David, D. J.; Mulley, J. C.: Deafness due to pro250-to-arg mutation of FGFR3. (Letter) Lancet 351: 877-878, 1998.
Ikegawa, S.; Fukushima, Y.; Isomura, M.; Takada, F.; Nakamura, Y.: Mutations of the fibroblast growth factor receptor-3 gene in one familial and six sporadic cases of achondroplasia in Japanese patients. Hum. Genet. 96: 309-311, 1995.
Jabs, E. W.; Li, X.; Scott, A. F.; Meyers, G.; Chen, W.; Eccles, M.; Mao, J. I.; Charnas, L. R.; Jackson, C. E.; Jaye, M.: Jackson-Weiss and Crouzon syndromes are allelic with mutations in fibroblast growth factor receptor 2. Nature Genet. 8: 275-279, 1994.
Keegan, K.; Johnson, D. E.; Williams, L. T.; Hayman, M. J.: Isolation of an additional member of the fibroblast growth factor receptor family, FGFR-3. Proc. Nat. Acad. Sci. 88: 1095-1099, 1991.
Lanning, R. W.; Brown, C. A.: An improved methodology for the detection of the common mutation in the FGFR3 gene responsible for achondroplasia. Hum. Mutat. 10: 496-499, 1997.
McKusick, V. A.; Kelly, T. E.; Dorst, J. P.: Observations suggesting allelism of the achondroplasia and hypochondroplasia genes. J. Med. Genet. 10: 11-16, 1973.
Meyers, G. A.; Orlow, S. J.; Munro, I. R.; Przylepa, K. A.; Jabs, E. W.: Fibroblast growth factor receptor 3 (FGFR3) transmembrane mutation in Crouzon syndrome with acanthosis nigricans. Nature Genet. 11: 462-464, 1995.
Moloney, D. M.; Slaney, S. F.; Oldridge, M.; Wall, S. A.; Sahlin, P.; Stenman, G.; Wilkie, A. O. M.: Exclusive paternal origin of new mutations in Apert syndrome. Nature Genet. 13: 48-53, 1996.
Muenke, M.; Gripp, K. W.; McDonald-McGinn, D. M.; Gaudenz, K.; Whitaker, L. A.; Bartlett, S. P.; Markowitz, R. I.; Robin, N. H.; Nwokoro, N.; Mulvihill, J. J.; Losken, W.; Mulliken, J. B.; Guttmacher, A. E.; Wilroy, R. S.; Clarke, L. A.; Hollway, G.; Ades, L. C.; Haan, E. A.; Mulley, J. C.; Cohen, M. M., Jr.; Bellus, G. A.; Francomano, C. A.; Moloney, D. M.; Wall, S. A.; Wilkie, A. O. M.; Zackai, E. H.: A unique point mutation in the fibroblast growth factor receptor 3 gene (FGFR3) defines a new craniosynostosis syndrome. Am. J. Hum. Genet. 60: 555-564, 1997.
Naski, M. C.; Wang, Q.; Xu, J.; Ornitz, D. M.: Graded activation of fibroblast growth factor receptor 3 by mutations causing achondroplasia and thanatophoric dysplasia. Nature Genet. 13: 233-237, 1996.
Nishimura, G.; Fukushima, Y.; Ohashi, H.; Ikegawa, S.: Atypical radiological findings in achondroplasia with uncommon mutation of the fibroblast growth factor receptor-3 (FGFR-3) gene (gly to cys transition at codon 375). (Letter) Am. J. Med. Genet. 59: 393-395, 1995.
Nishimura, G.; Takada, F.: Another observation of atypical radiologic findings in achondroplasia not due to a common mutation of the FGFR-3 gene: reply to Dr. Gorlin. (Letter) Am. J. Med. Genet. 70: 203-205, 1997.
Orlow, S. J.: Cutaneous findings in craniofacial malformation syndromes. Arch. Derm. 128: 1379-1386, 1992.
Paznekas, W. A.; Cunningham, M. L.; Howard, T. D.; Korf, B. R.; Lipson, M. H.; Grix, A. W.; Feingold, M.; Goldberg, R.; Borochowitz, Z.; Aleck, K.; Mulliken, J.; Yin, M.; Jabs, E. W.: Genetic heterogeneity of Saethre-Chotzen syndrome, due to TWIST and FGFR mutations. Am. J. Hum. Genet. 62: 1370-1380, 1998.
Perez-Castro, A. V.; Wilson, J.; Altherr, M. R.: Genomic organization of the human fibroblast growth factor receptor 3 (FGFR3) gene and comparative sequence analysis with the mouse Fgfr3 gene. Genomics 41: 10-16, 1997.
Pokharel, R. K.; Alimsardjono, H.; Takeshima, Y.; Nakamura, H.; Naritomi, K.; Hirose, S.; Onishi, S.; Matsuo, M.: Japanese cases of type 1 thanatophoric dysplasia exclusively carry a C to T transition at nucleotide 742 of the fibroblast growth factor receptor 3 gene. Biochem. Biophys. Res. Commun. 227: 236-239, 1996.
Prinos, P.; Costa, T.; Sommer, A.; Kilpatrick, M. W.; Tsipouras, P.: A common FGFR3 gene mutation in hypochondroplasia. Hum. Molec. Genet. 4: 2097-2101, 1995.
Prinster, C.; Carrera, P.; Del Maschio, M.; Weber, G.; Maghnie, M.; Vigone, M. C.; Mora, S.; Tonini, G.; Rigon, F.; Beluffi, G.; Severi, F.; Chiumello, G.; Ferrari, M.: Comparison of clinical-radiological and molecular findings in hypochondroplasia. Am. J. Med. Gen. 75: 109-112, 1998.
Ramaswami, U.; Rumsby, G.; Hindmarsh, P. C.; Brook, C. G. D.: Genotype and phenotype in hypochondroplasia. J. Pediat. 133: 99-102, 1998.
Reardon, W.; Wilkes, D.; Rutland, P.; Pulleyn, L. J.; Malcolm, S.; Dean, J. C. S.; Evans, R. D.; Jones, B. M.; Hayward, R.; Hall, C. M.; Nevin, N. C.; Baraitser, M.; Winter, R. M.: Craniosynostosis associated with FGFR3 pro250-to-arg mutation results in a range of clinical presentations including unisutural sporadic craniosynostosis. J. Med. Genet. 34: 632-636, 1997.
Rousseau, F.; Bonaventure, J.; Legeai-Mallet, L.; Pelet, A.; Rozet, J.-M.; Maroteaux, P.; Le Merrer, M.; Munnich, A.: Mutations in the gene encoding fibroblast growth factor receptor-3 in achondroplasia. Nature 371: 252-254, 1994.
Rousseau, F.; El Ghouzzi, V.; Delezoide, A. L.; Legeai-Mallet, L.; Le Merrer, M.; Munnich, A.; Bonaventure, J.: Missense FGFR3 mutations create cysteine residues in thanatophoric dwarfism type I (TD1). Hum. Molec. Genet. 5: 509-512, 1996.
Rousseau, F.; Saugier, P.; Le Merrer, M.; Munnich, A.; Delezoide, A.-L.; Maroteaux, P.; Bonaventure, J.; Narcy, F.; Sanak, M.: Stop codon FGFR3 mutations in thanatophoric dwarfism type 1. Nature Genet. 10: 11-12, 1995.
Shiang, R.; Thompson, L. M.; Zhu, Y.-Z.; Church, D. M.; Fielder, T. J.; Bocian, M.; Winokur, S. T.; Wasmuth, J. J.: Mutations in the transmembrane domain of FGFR3 cause the most common genetic form of dwarfism, achondroplasia. Cell 78: 335-342, 1994.
Superti-Furga, A.; Eich, G.; Bucher, H. U.; Wisser, J.; Giedion, A.; Gitzelmann, R.; Steinmann, B.: A glycine 375-to-cysteine substitution in the transmembrane domain of the fibroblast growth factor receptor-3 in a newborn with achondroplasia. Europ. J. Pediat. 154: 215-219, 1995.
Szabo, J.; Bellus, G. A.; Kaitila, I.; Francomano, C. A.: Fibroblast growth factor receptor 3 (FGFR3) mutations in sporadic cases of achondroplasia occur exclusively on the paternally derived chromosome. Am. J. Hum. Genet.(Abstract) 59 (suppl.): A287 only, 1996.
Tavormina, P. L.; Rimoin, D. L.; Cohn, D. H.; Zhu, Y.-Z.; Shiang, R.; Wasmuth, J. J.: Another mutation that results in the substitution of an unpaired cysteine residue in the extracellular domain of FGFR3 in thanatophoric dysplasia type I. Hum. Molec. Genet. 4: 2175-2177, 1995.
Tavormina, P. L.; Shiang, R.; Thompson, L. M.; Zhu, Y.-Z.; Wilkin, D. J.; Lachman, R. S.; Wilcox, W. R.; Rimoin, D. L.; Cohn, D. H.; Wasmuth, J. J.: Thanatophoric dysplasia (types I and II) caused by distinct mutations in fibroblast growth factor receptor 3. Nature Genet. 9: 321-328, 1995.
Thompson, L. M.; Plummer, S.; Schalling, M.; Altherr, M. R.; Gusella, J. F.; Housman, D. E.; Wasmuth, J. J.: A gene encoding a fibroblast growth factor receptor isolated from the Huntington disease gene region of human chromosome 4. Genomics 11: 1133-1142, 1991.
Wilcox, W. R.; Tavormina, P. L.; Krakow, D.; Kitoh, H.; Lachman, R. S.; Wasmuth, J. J.; Thompson, L. M.; Rimoin, D. L.: Molecular, radiologic, and histopathologic correlations in thanatophoric dysplasia. Am. J. Med. Genet. 78: 274-281, 1998.
Wilkie, A. O. M.: Personal Communication. Oxford, U.K. 6/1998.
Wilkie, A. O. M.: Craniosynostosis: genes and mechanisms. Hum. Molec. Genet. 6: 1647-1656, 1997.

TRANSLOCATED PROMOTER REGION (*189940)

The TPR locus, which has been mapped to chromosome 1, expresses a 10-kb RNA in all human cell lines tested (Dean et al., 1985). In a cell line rendered tumorigenic with the direct-acting carcinogen N-methyl-N-prime-nitronitrosoguanidine (MNNG), Dean et al. (1987) defined an activated MET oncogene that expresses a novel 5.0-kb RNA transcript which is a hybrid RNA derived from a DNA rearrangement involving the TPR locus and the MET locus (on 7q; 164860). Although most of the hybrid RNA is derived from the MET oncogene, the 5-prime portion uses some exons from the TPR gene, presumably the promoter region. Oncogenic activation of MET is reminiscent of the Philadelphia chromosomal translocation in chronic myeloid leukemia that generates the hybrid BCR/ABL tyrosine kinase p210 (see 151410). Gonzatti-Haces et al. (1988) identified the proteins encoded by the human TPR-MET oncogene and the human MET protooncogene.

By fluorescence in situ hybridization, Miranda et al. (1994) assigned the TPR gene to 1q25.

Dean, M.; Park, M.; Le Beau, M. M.; Robins, T. S.; Diaz, M. O.; Rowley, J. D.; Blair, D. G.; Vande Woude, G. F.: The human MET oncogene is related to the tyrosine kinase oncogenes. Nature 318: 385-388, 1985.
Dean, M.; Park, M.; Vande Woude, G. F.: Characterization of the rearranged TPR-MET oncogene breakpoint. Molec. Cell. Biol. 7: 921-924, 1987.
Gonzatti-Haces, M.; Seth, A.; Park, M.; Copeland, T.; Oroszlan, S.; Vande Woude, G. F.: Characterization of the TPR-MET oncogene p65 and the MET protooncogene p140 protein-tyrosine kinases. Proc. Nat. Acad. Sci. 85: 21-25, 1988.

Miranda, C.; Minoletti, F.; Greco, A.; Sozzi, G.; Pierotti, M. A.: Refined localization of the human TPR gene to chromosome 1q25 by in situ hybridization. Genomics 23: 714-715, 1994.

RIBOSOMAL PROTEIN L22; RPL22 (*180474)

A reciprocal translocation between the long arms of chromosomes 3 and 21, at bands 3q26 and 21q22, occurs as an acquired clonal chromosomal abnormality in malignant cells from patients with therapy-related myelodysplastic syndrome or acute myeloid leukemia, as well as in some patients with chronic myeloid leukemia in blast crisis. Nucifora et al. (1993) showed that the gene on chromosome 21 is AML1 (151385), which is fused to the ETO gene (133435) in 8;21 translocations. Nucifora et al. (1993) isolated a fusion cDNA clone from a t(3;21) library derived from a patient with therapy-related myelodysplastic syndrome; this clone contained sequences from AML1 and from EAP, which they localized to 3q26 from the location of the breakpoint on chromosome 3. EAP is a highly expressed small nuclear protein of 128 residues associated with Epstein-Barr virus small RNA. The fusion clone contained the DNA-binding 5-prime part of AML1 that is fused to ETO in the t(8;21) and, in addition, at least 1 other exon. The translocation replaced the last 9 codons of AML1 with the last 96 codons of EAP. The fusion does not maintain the correct reading frame of EAP and may not lead to a functional chimeric protein. EAP has been identified as the ribosomal protein L22.

Nucifora and Rowley (1995) reviewed the involvement of the AML1 gene in the 8;21 and 3;21 translocations in acute and chronic myeloid leukemia. Three loci closely situated to each other on 3q26 are involved in fusions with AML1 in the 3;21 translocations: EVI1 (165215), EAP, and MDS1 (600049). They pointed out that the order of the genes on 3q26 is TEL—EAP—MDS1—EVI1 and provided a diagram (their figure 5) of the 3q26 region containing these genes and of the various chimeric junctions they had isolated from t(3;21) patients.

Nucifora, G.; Begy, C. R.; Erickson, R.; Drabkin, H. A.; Rowley, J. D.: The 3;21 translocation in myelodysplasia results in a fusion transcript between the AML1 gene and the gene for EAP, a highly conserved protein associated with the Epstein-Barr virus small RNA EBER 1. Proc. Nat. Acad. Sci. 90: 7784-7788, 1993.

Nucifora, G.; Rowley, J. D.: AML1 and the 8;21 and 3;21 translocations in acute and chronic myeloid leukemia. Blood 86: 1-14, 1995.

MYELODYSPLASIA SYNDROME 1 (*600049)

In leukemic cells of 4 patients with therapy-related myelodysplasia/acute myeloid leukemia and in 1 patient with chronic myelogenous leukemia in blast crisis, all of whom had a t(3;21)(q26;q22), Nucifora et al. (1994) consistently found fusion transcripts between AML1 (151385) and EAP (180474) or between AML1 and previously unidentified sequences that they named MDS1, for MDS-associated sequences. In addition, they identified a third chimeric transcript, AML1/EVI1 (165215), in 1 of the therapy-related acute myeloid leukemia patients. Pulsed field gel electrophoresis established the order of the genes as EAP, the most telomeric, and EVI1, the most centromeric, with MDS1 situated between them. They used fluorescence in situ hybridization to confirm that EVI1 is centromeric to MDS1.

Nucifora and Rowley (1995) reviewed the involvement of the AML1 gene (151385) in the 8;21 and 3;21 translocations in acute and chronic myeloid leukemia. Three loci closely situated to each other on 3q26 are involved in fusions with AML1 in the 3;21 translocations: EVI1 (165215), EAP (180474), and MDS1. They pointed out that the order of the genes on 3q26 is TEL—EAP—MDS1—EVI1 and provided a diagram (their figure 5) of the 3q26 region containing these genes and of the various chimeric junctions they had isolated from t(3;21) patients.

MDS1 had been identified as a separate gene and also as a previously unreported exon (or exons) of EVI1. Fears et al. (1996) demonstrated that MDS1 exists in normal tissues, both as a unique transcript and as a normal fusion transcript with EVI1, with an additional 188 codons at the 5-prime end of the previously reported EVI1 open reading frame. This additional region has about 40% homology at the amino acid level with the PR domain of the retinoblastoma-interacting zinc-finger protein RIZ (601196). These results are important because EVI1 and MDS1 are involved in leukemia associated with chromosomal translocation breakpoints in the region between the genes.

Fears, S.; Mathieu, C.; Zeleznik-Le, N.; Huang, S.; Rowley, J. D.; Nucifora, G.: Intergenic splicing of MDS1 and EVI1 occurs in normal tissues as well as in myeloid leukemia and produces a new member of the PR domain family. Proc. Nat. Acad. Sci. 93: 1642-1647, 1996.

Nucifora, G.; Begy, C. R.; Kobayashi, H.; Roulston, D.; Claxton, D.; Pedersen-Bjergaard, J.; Parganas, E.; Ihle, J. N.; Rowley, J. D.: Consistent intergenic splicing and production of multiple transcripts between AML1 at 21q22 and unrelated genes at 3q26 in (3;21)(q26;q22) translocations. Proc. Nat. Acad. Sci. 91: 4004-4008, 1994.

Nucifora, G.; Rowley, J. D.: AML1 and the 8;21 and 3;21 translocations in acute and chronic myeloid leukemia. Blood 86: 1-14, 1995.

ECOTROPIC VIRAL INTEGRATION SITE-1 (*165215)

The murine Evi1 gene encodes a protein that has multiple 28-amino acid repeats containing the consensus sequence found in the zinc finger domains of many transcriptional regulatory proteins. Activation of the expression of the Evi1 gene is frequently found in murine myeloid leukemias and leukemia cell lines and is due to retroviral insertions in the 5-prime region of the gene. To examine the role of the EVI1 gene in human leukemias, Morishita et al. (1990) cloned regions of the human locus corresponding to the coding region of the gene. Using these probes they demonstrated that the human EVI1 gene maps to 3q24-q28 in a region that is translocated in acute nonlymphocytic leukemias with a t(3;5)(q25;q34) translocation. By in situ hybridization with metaphase chromosomes from a patient with a 3;5 translocation, they found that the EVI1 gene was translocated to the derivative chromosome 5. However, no rearrangements were detected by Southern blot analysis and no EVI1 transcripts were detected with RNA from leukemic blasts of 1 patient with the translocation. Since retroviral activation of Evi1 gene expression is one of the most common transforming events in murine myeloid leukemias, Morishita et al. (1992) evaluated the role of the EVI1 gene in human acute myelogenous leukemia in 116 patients. In 8 patients, the EVI1 gene was expressed and all but one patient had cytogenetically detectable translocations involving 3q26. To identify breakpoints, a restriction map that spanned 1,700 kb of the EVI1 locus was developed by pulsed field gel electrophoresis. In 2 patients rearrangements were found to be located 5-prime to the gene and, in 1 patient, 150 kb downstream from the 5-prime end of the gene. One of the 5-prime rearrangements, a case of t(3;3)(q21;q26) was shown to have a rearrangement that resulted in fusion of sequences from 3q21-q22 with the EVI1 locus.

Nucifora and Rowley (1995) reviewed the involvement of the AML1 gene (151385) in the 8;21 and 3;21 translocations in acute and chronic myeloid leukemia. Three loci closely situated to each other on 3q26 are involved in fusions with AML1 in the 3;21 translocations: EVI1, EAP (180474), and MDS1 (600049). They pointed out that the order of the genes on 3q26 is TEL—EAP—MDS1—EVI1 and provided a diagram (their figure 5) of the 3q26 region containing these genes and of the various chimeric junctions they had isolated from t(3;21) patients.

Overexpression of the EVI1 gene appears to be a consistent feature of the 3q21q26 syndrome, an association of myeloid leukemia/myelodysplastic syndrome with a specific chromosomal aberration involving both 3q21 and 3q26, such as t(3;3)(q21;q26) or inv(3)(q21q26). The rearrangement in 3q26 occurs near the EVI1 locus, implicating it as the critical gene deregulated in the 3q21q26 syndrome. Ogawa et al. (1996) presented a structural abnormality of the EVI1 protein in a case of the 3q21q26 syndrome associated with the typical chromosome 3 inversion. The 3q26 breakpoint was located within an intron of the EVI1 gene, and resulted in overexpression of an aberrant form of EVI1 protein (normally unexpressed), in which the C-terminal 44 amino acids of the wildtype protein were truncated and replaced by 5 amino acids. The truncated EVI1 protein was shown to increase AP1 (165160) activity when expressed in NIH 3T3 cells as its wildtype counterpart. The results of Ogawa et al. (1996) strongly supported the conclusion that the primary target for the 3q21q26 syndrome is the EVI1 gene which, in the chromosomal aberrations, is juxtaposed to a 3q21 locus containing a putative enhancer element.

Morishita, K.; Parganas, E.; Bartholomew, C.; Sacchi, N.; Valentine, M. B.; Raimondi, S. C.; Le Beau, M. M.; Ihle, J. N.: The human Evi-1 gene is located on chromosome 3q24-q28 but is not rearranged in three cases of acute nonlymphocytic leukemias containing t(3;5)(q25;q34) translocations. Oncogene Res. 5: 221-231, 1990.

Morishita, K.; Parganas, E.; Willman, C. L.; Whittaker, M. H.; Drabkin, H.; Oval, J.; Taetle, R.; Valentine, M. B.; Ihle, J. N.: Activation of EVI1 gene expression in human acute myelogenous leukemias by translocations spanning 300-400 kilobases on chromosome band 3q26. Proc. Nat. Acad. Sci. 89: 3937-3941, 1992.

Nucifora, G.; Rowley, J. D.: AML1 and the 8;21 and 3;21 translocations in acute and chronic myeloid leukemia. Blood 86: 1-14, 1995.

Ogawa, S.; Kurokawa, M.; Tanaka, T.; Mitani, K.; Inazawa, J.; Hangaishi, A.; Tanaka, K.; Matsuo, Y.; Minowada, J.; Tsubota, T.; Yazaki, Y.; Hirai, H.: Structurally altered Evi-1 protein generated in the 3q21q26 syndrome. Oncogene 13: 183-191, 1996.

IMMUNOGLOBULIN LAMBDA CONSTANT REGION 1 (*147220))

The constant region of the lambda light chain of immunoglobulins are of four subtypes, as defined by amino acid substitutions in three monoclonal myeloma L chains (OZ, KERN, Mcg). Subtype 1 is OZ plus KERN, Mcg-. Subtype 2 is OZ-KERN plus Mcg plus. Subtype 3 is OZ-

KERN plus Mcg-. Subtype 4 is OZ-KERN-Mcg-. Although no idiotypic variation of lambda light chains has been found (comparable to Inv and Gm variants of kappa light chains and gamma heavy chains), the existence of at least 1 locus for lambda light chains can be inferred from the amino acid sequence of immunoglobulins. The same type of evidence indicates the existence of mu, delta, and epsilon heavy chain loci which determine the structure of heavy chains in IgM, IgD and IgE, respectively. Each immunoglobulin molecule is composed of 2 heavy chains and 2 light chains. Since there are 5 types of heavy chains and 2 types of light chains, a minimum of 10 classes of immunoglobulins result. Actually, since there are at least four subtypes of gamma heavy chains, there are at least 32 types of immunoglobulins. Six subgroups of lambda L chains are recognized and designated V-lambda I-VI. They are presumably determined by separate but closely linked loci. For assignment of gamma globulin-specific chromosomes, Kucherlapati et al. (1979) created hybridomas between a variant mouse myeloma cell line that produces no immunoglobulin and lymphocytes from patients with chronic lymphocytic leukemia. (Hybrids between mouse myeloma cells and spleen cells from immunized mice had been used for 'rescuing' immunoglobulin producing mouse cells and producing monoclonal antibodies.) They concluded that chromosome 6 and-or 11 is involved in expression of human heavy and-or lambda chain production. The antibody genes provide a unique opportunity for studying the molecular basis of eukaryotic differentiation. Rearrangement of gene segments is correlated with the expression of antibody molecules. The light chains are encoded by 3 gene segments, V(L), J(L) and C(L) , which are separated in the genomes of cells undifferentiated with regard to antibody gene expression. During differentiation of the antibody-producing or B cell, the V(L) and V(J) gene segments are rearranged and joined together while the intervening DNA between the J(L) and C(L) segments remains unmodified. This portion of the transcript is removed by RNA splicing to produce light chain mRNA with contiguous V(L), J(L) and C(L) coding segments. See review in Davis et al. (1980).

Klein (1981) found that B cell-derived tumors (mouse myeloma and human Burkitt lymphoma and B-cell acute lymphoblastic leukemia) have anomalous patterns of immunoglobulin synthesis which correlate with the type of chromosomal aberration. Similar observations were made by Lenoir (1981) who had collected the largest number of variant Burkitt lymphoma translocations. Of 10 tested, all agreed with the hypothesis as to light chain expression: all the 8;22 translocation cells produced lambda as the only light chain; all the 2;8 translocation cells produced only kappa; and 8;14 translocation cells produced either kappa or lambda, with an approximate ratio of 2:1. Erikson et al. (1981) confirmed assignment of the lambda gene cluster to chromosome 22 by the study of derivative clones from somatic cell hybrids between mouse myeloma cells and human B cells.

Hieter et al. (1981) found that the lambda light chain locus of man contains 6 lambda-like genes arranged tandemly on a 50-kb segment of chromosomal DNA. The sequences of 3 of the 6 correspond to 3 known nonallelic lambda chain isotypes: Mcg, Ke(-)Oz(-), and Ke(-)Oz(+). These are situated at the 5-prime end of the cluster of 6. In addition to the 6, three as yet unlinked lambda-like sequences were cloned. The authors suggested that the lambda genes may form an unexpectedly large family within the human genome. At the protein level, at least a fourth nonallelic form of the human lambda constant region has been identified (Solomon, 1977): Kern(+)Oz(-). The amino acids at positions 112, 114, 152, 163, 190, and 216 are, respectively, for Ke(-)Oz(-): ala-ser-ser-thr-arg; for Ke(+)Oz(-): ala-ser-gly-thr-arg; for Ke(-)Oz(+): ala-ser-ser-thr-lys; for Mcg: asn-thr-gly-lys-arg. The 6 genes surround a highly polymorphic and evidently unstable region that was repeatedly deleted when cloned in E. coli. Hereditary restriction fragment length polymorphism was demonstrated in the lambda gene locus. The complete characterization of the lambda locus with regard, for example, to the J regions and the mechanism for achieving diversity remains to be done.

Using a genomic probe and in situ hybridization, Leder (1982) and his colleagues tentatively assigned the lambda gene cluster to 22q11. Using nucleic acid probes prepared from the cloned gene in Southern blots of DNA from somatic cell hybrids, McBride et al. (1982) assigned the kappa constant gene to chromosome 2 and the lambda constant gene to chromosome 22. The human chromosomes carried by each hybrid cell line were identified by isozyme markers.

Wabl and Steinberg (1982) proposed a theory to explain allelic exclusion (only 1 of 2 alleles is functional in any one lymphocyte) and L chain isotypic exclusion (in a given lymphocyte, either kappa or lambda light chain but not both can combine with heavy chain to form a complete Ig molecule). Whereas in Burkitt lymphoma of the t(8;22) type the lambda light chain genes are translocated to chromosome 8, they remain on chromosome 22 (i.e., on the Philadelphia chromosome) in CML (Selden et al., 1983). The rearrangements in Burkitt lymphomas have permitted definition of the normal orientation in the immunoglobulin genes on chromosomes 2p, 14q, and 22q. The 5-prime to 3-prime order is cen—V—J—C—ter for the kappa genes on 2p; ter—V—J—C—cen for the heavy chain genes on 14q; and cen—V—C—ter for the lambda genes on 22q.

According to Croce (1984), the relative frequencies of the 3 forms of Burkitt lymphoma are 75%, 8;14; 16%, 8;22; and 9%, 8;2. One hundred percent of cases show one or another of these 3 types of translocation. The breakpoint in 22q in Burkitt lymphoma is cytogenetically indistinguishable from the breakpoint in CML. Molecular genetic studies indicate that the Burkitt breakpoint is centromeric to the C-lambda locus and the CML breakpoint is distal to C-lambda. Through studies of an 8;22-carrying Burkitt lymphoma cell line by somatic cell genetic and in situ hybridization techniques, Emanuel et al. (1984) concluded that the lambda variable region genes are on the centromeric side of the lambda constant region genes (which lie distal). Six nonallelic immunoglobulin lambda constant region genes have been characterized on a 40-kb stretch of DNA. The nucleotide sequences of the 3 upstream genes of this cluster (Cl1, Cl2, and Cl3) were determined and shown to encode, respectively, Kern(-)Oz(-), and Kern(-)Oz(+) constant regions of the lambda chains.

Dariavach et al. (1987) reported the sequence of the 3 downstream genes of this cluster and showed that 2 of them (Cl4 and Cl5) are pseudogenes. They also showed that Cl6 encodes a Kern(+)Oz(-) chain. By determining the DNA sequence of the complete human C lambda complex, Vasicek and Leder (1990) found a previously undescribed seventh C lambda region that may encode the Ke(+)Oz(-) lambda protein. They demonstrated that the 7 constant regions are organized in a tandem array and that each is preceded by a single J lambda region. Lambda 1, lambda 2, lambda 3, and lambda 7 are apparently active genes, while lambda 4, lambda 5, and lambda 6 are pseudogenes. There are no other J lambda or C lambda regions within a 60-kb region surrounding the C lambda complex; however, there are at least 4 other lambda-like genes and lambda pseudogenes in the human genome. They found a 1,377-bp open reading frame located on the opposite strand in the region containing lambda 7. They had no evidence that it was part of a functional gene, however.

Frippiat et al. (1995) completed a map of the human lambda locus on 22q11.2. They mapped 52 V-lambda genes from 10 V-lambda families and 7 J-lambda and C-lambda genes on an 1140-kb contig constructed from 8 YACs in 129 cosmid clones. Variable genes were contained within 800 kb and the 10 families were arranged in 3 clusters, in contrast to the dispersed organization of the different variable family genes of the heavy-chain and kappa light chain. The VpreB gene (VPREB1; 146770), encoding part of the surrogate light chain, the GGT2 gene (137181), and the BCRL4 (113660) pseudogene were also mapped within the lambda locus.

Pseudogenes

Hollis et al. (1982) postulated the existence of 'processed genes': gene-like sequences that, as opposed to their normal counterparts, bear some evidence of RNA-type processing, e.g., coincident homology to the site of transcriptional initiation, clean loss of intervening sequences, or coincident homology to the site of poly(A) tail. Hollis et al. (1982) described a pseudogene of the human lambda immunoglobulin chain that fulfills these expectations. Lambda-psi-1, as they termed it, is thought not to be on chromosome 22 and the J and C regions, rather than being discontinuous, are cleanly joined in accordance with the rules of RNA splicing. Furthermore, the homology of the pseudogene to the normal gene ends abruptly in a long sequence of adenylic acid residues that resembles a poly(A) tail. The association of gene movement and precise splicing suggested to them that an RNA intermediate may have been involved in the formation of the novel gene as well as in its conveyance to a new location. Pseudogenes of the alpha-globin genes in the mouse show the same phenomenon.

See also de la Chapelle et al. (1983); Frangione et al. (1985); Miranda et al. (1983)

Croce, C. M.: Personal Communication. Philadelphia, Pa. 4/6/1984.

Dariavach, P.; Lefranc, G.; Lefranc, M.-P.: Human immunoglobulin C-lambda-6 gene encodes the Kern(+)Oz(-) lambda chain and C-lambda-4 and C-lambda-5 are pseudogenes. Proc. Nat. Acad. Sci. 84: 9074-9078, 1987.

Davis, M. M.; Calame, K.; Early, P. W.; Livant, D. L.; Joho, R.; Weissman, I. L.; Hood, L.: An immunoglobulin heavy-chain gene is formed by at least two recombinational events. Nature 283: 733-739, 1980.

de la Chapelle, A.; Lenoir, G.; Boue, J.; Boue, A.; Gallano, P.; Huerre, C.; Szajnert, M.-F.; Jeanpierre, M.; Lalouel, J.-M.; Kaplan, J.-C.: Lambda Ig constant region genes are translocated to chromosome 8 in Burkitt's lymphoma with t(8;22). Nucleic Acids Res. 11: 1133-1142, 1983.

Emanuel, B. S.; Cannizzaro, L. A.; Tsujimoto, Y.; Croce, C. M.: Chromosomal orientation of the lambda light chain locus: V-lambda is proximal to C-lambda in 22q11. (Abstract) Am. J. Hum. Genet. 36: 202S only, 1984.

Erikson, J.; Martinis, J.; Croce, C. M.: Assignment of the genes for human lambda immunoglobulin chains to chromosome 22. Nature 294: 173-175, 1981.

Frangione, B.; Moloshok, T.; Prelli, F.; Solomon, A.: Human lambda light-chain constant region gene C-lambda(Mor): the primary structure of lambda-VI Bence Jones protein Mor. Proc. Nat. Acad. Sci. 82: 3415-3419, 1985.
Frippiat, J.-P.; Williams, S. C.; Tomlinson, I. M.; Cook, G. P.; Cherif, D.; Le Paslier, D.; Collins, J. E.; Dunham, I.; Winter, G.; Lefranc, M.-P.: Organization of the human immunoglobulin lambda light-chain locus on chromosome 22q11.2. Hum. Molec. Genet. 4: 983-991, 1995.
Hieter, P. A.; Hollis, G. F.; Korsmeyer, S. J.; Waldmann, T. A.; Leder, P.: Clustered arrangement of immunoglobulin lambda constant region genes in man. Nature 294: 536-540, 1981.
Hollis, G. F.; Hieter, P. A.; McBride, O. W.; Swan, D.; Leder, P.: Processed genes: a dispersed human immunoglobulin gene bearing evidence of RNA-type processing. Nature 296: 321-325, 1982.
Klein, G.: Personal Communication. Stockholm, Sweden 11/4/1981.
Kucherlapati, R.; Dilley, J.; Levy, R.: Mapping human immunoglobin genes by mouse-human hybridomas. (Abstract) Cytogenet. Cell Genet. 25: 176 only, 1979.
Leder, P.: Personal Communication. Boston, Mass. 1/16/1982.
Lenoir, G.: Personal Communication. Lyon, France 1981.
McBride, O. W.; Hieter, P. A.; Hollis, G. F.; Swan, D.; Otey, M. C.; Leder, P.: Chromosomal location of human kappa and lambda immunoglobulin light chain constant region genes. J. Exp. Med. 155: 1480-1490, 1982.
Miranda, J. L. G.; Gomez, A. O.; Ansedes, H. V.; Torres, N. R.; Espinosa, C. G.; Cortabarria, C.; Salgado, G. S.: Monosomy 22 with humoral immunodeficiency: is there an immunoglobulin chain deficit?. J. Med. Genet. 20: 69-72, 1983.
Selden, J. R.; Emanuel, B. S.; Wang, E.; Cannizzaro, L.; Palumbo, A.; Erikson, J.; Nowell, P. C.; Rovera, G.; Croce, C. M.: Amplified C-lambda and c-abl genes are on the same marker chromosome in K562 leukemia cells. Proc. Nat. Acad. Sci. 80: 7289-7292, 1983.
Solomon, A.: Bence Jones proteins and light chains of immunoglobulins. XVI. Immunochemical recognition of the human lambda light-chain constant-region isotype Mcg. Immunogenetics 5: 525-533, 1977.
Vasicek, T. J.; Leder, P.: Structure and expression of the human immunoglobulin lambda genes. J. Exp. Med. 172: 609-620, 1990.
Wabl, M.; Steinberg, C.: A theory of allelic and isotypic exclusion for immunoglobulin genes. Proc. Nat. Acad. Sci. 79: 6976-6978, 1982.

IGG HEAVY CHAIN LOCUS (*147100)

At least 2 separate autosomal loci determining serologic type of gamma globulin were identified in the 1950s and 1960s. One was referred to as the Gm locus and the other as the Inv locus. The genetics of the gamma globulins has been as revealing of general principles as has been that of the hemoglobins. The Gm system is associated with the heavy chains of the IgG molecules encoded, as was later found, by chromosome 14; the Inv system is associated with the kappa light chains (147200). (See Anonymous, 1966 for recommended notation for Gm and Inv types.)

Hood and Ein (1968) presented evidence that antibody light chains are an exception to the rule of 'one gene, one polypeptide chain.' Two separate loci (a specific region locus and a common region locus) appeared to code for a single, continuous polypeptide chain. Three closely linked loci (IgG1, IgG2 and IgG3) were thought to be responsible for the Gm specificities. Van Loghem et al. (1970) presented evidence on the linkage relationship of immunoglobulin markers (gamma 1, 2, 3, Am). That the gamma-G3 and gamma-G1 loci are closely linked was indicated by the findings in a Lepore-type myeloma protein (Kunkel et al., 1969). A fourth IgG locus (gamma-G4) was identifiable in the cluster. A family possibly supporting the sequence (beginning at the N terminus) of alpha-2, gamma-4, gamma-2, gamma-3, and gamma-1 was presented by Lefranc et al. (1977).

20,18:Gedde-Dahl et al. (1972, 1975) presented data on the linkage of Gm-Pi (AAT; 107400). They considered heterogeneity of recombination fraction among males of different Pi type to be very likely. The major difference seemed to be between the Pi(Z) and other alleles. Possible explanations included a chromosomal deletion, inversion or locus regulation recombination in linkage disequilibrium with the Pi locus. Bender et al. (1979) excluded Gm, Pi and C3 from the segment 6q25-qter and Gm and Pi from 6p. See 182900 for evidence of linkage to hereditary spherocytosis. Croce et al. (1979) studied somatic cell hybrids between mouse myeloma cells and either human peripheral lymphocytes or human lymphoblastoid or myeloma cells. They observed that the presence or absence of chromosome 14 correlated with formation of human mu, gamma, and alpha heavy chains. Smith et al. (1981) confirmed assignment of the immunoglobulin heavy chain family of genes to chromosome 14.

Green (1979) reviewed the genetics of the immunoglobulins in mice and proposed a nomenclature. From study of somatic cell hybrids, Hengartner et al. (1978) concluded that the loci for immunoglobulin heavy chains are on chromosome 12 in the mouse. Meo et al. (1980) reported the conclusive mapping of the Igh-1 and the linked prealbumin locus to mouse chromosome 12. In the mouse, the heavy chain variable and constant regions, Igh-V and Igh-C (Green, 1979), occupy a chromosomal segment at least 7-11 units long (Pisetsky and Sachs, 1977), and are linked, probably at the Igh-C end, with the serum prealbumin locus at a distance of about 11 units (Taylor et al., 1975). Steinberg et al. (1975) described polymorphism of both Gm and Inv in baboons of Kenya.

In man and in mouse, fine mapping of the immunoglobulin gene progressed faster than chromosomal and regional assignment. The immunoglobulin loci were thought to be located in three different chromosomal regions carrying heavy chain, kappa light chain and lambda light chain loci. Each region was thought to contain one or more loci specifying the constant region and a larger number of loci specifying the variable region of the particular immunoglobulin chain. Evidence from mice indicated that the codon sequences of each light chain, kappa and lambda, are constructed during differentiation of plasma cell precursors by the joining of DNA segments previously far apart. Davis et al. (1980) showed that the heavy chain genes contain 3 gene segments, V(H), J(H) and C(H), analogous to the 3 segments of the light chain genes and that at least 2 recombinational events take place during differentiation of the antibody-producing or B-cell. The structure of the immunoglobulin genes and their rearrangement during maturation of the lymphocyte were reviewed by Robertson (1981); also see Marx (1981). In man, the immunoglobulin heavy chain family of genes has, beginning from the 5-prime end, 250 or more variable genes, 5 J genes (4 are active), at least 10 D (for diversity) genes, and the genes for the constant part of the mu, delta, gamma, epsilon, and alpha heavy chains of IgM, D, G, E, and A, respectively. In the mouse, the organization of the C(H) gene is 5-prime-J(H)-(6.5 kb)-mu-(4.5 kb)-delta-(unknown kb)-gamma-3-(34 kb)-gamma-1-(21 kb)-gamma-2b-(15 kb)-gamma-2a-(14.5 kb)-epsilon-(12.5 kb)-alpha-3-prime (Shimizu et al., 1981). According to the dogma current by 1981, a complete H chain gene is formed by at least 2 types of combinational events: (1) the recombination between a given V(H), a given J(H), and a given D gene segment to form a V region gene, and (2) a class switch to a particular C(H) gene beginning with mu and later shifting to any one of the others. Klein (1981) found tha t B cell-derived tumors (mouse myeloma and human Burkitt lymphoma and B-cell acute lymphoblastic leukemia) have anomalous patterns of immunoglobulin synthesis which correlate with the type of chromosomal aberration. Similar observations were made by Lenoir et al. (1982) who had collected the largest number of variant Burkitt lymphoma translocations. Of 10 tested, all agreed with the hypothesis as to light chain expression: all the 8;22 translocation cells produced lambda as the only light chain; all the 2;8 translocation cells produced only kappa; and 8;14 translocation cells produced either kappa or lambda, with an approximate ratio of 2:1.

For each of the immunoglobulin chains (heavy, lambda and kappa), I have listed separately the 'genes' for the constant, variable, J, and D regions. One can, with validity, view each of the three as a supergene and the C, V, J, and D coding segments of DNA as exons of that supergene. The immunoglobulin genes are in a chromosomal region noted for its high frequency of breaks associated with chromosome rearrangement, occurring both spontaneously in cultured lymphocytes and in certain malignancies. By means of the same X-14 translocation (known as KOP for Kirby, Opitz and Pallister, the patient, interpreter, and discoverer, respectively) that was used to map G6PD and HGPRT to the long arm of the X chromosome (Ricciuti and Ruddle, 1973), Balazs et al. (1982) concluded that D14S1 is closely linked to the heavy chain immunoglobulin 'locus' and distal to 14q32, i.e., in the subtelomeric region of 14q. A family linkage study showed that the maximum likelihood estimate of recombination between D14S1 and Gm was 3.1% with a 90% upper limit of 11.5%. Cox et al. (1982) reported on the family of a person with a ring chromosome 14 in which one breakpoint was located at 14q32.3. The affected person did not express the C-gamma allotypic heavy chain marker, Gm, of the maternal haplotype, indicating that genetic material necessary for Gm expression is located distal to 14q32.3. From study of 2 families with abnormalities of the long arm of chromosome 14, Cox et al. (1982) localized GM to 14q32.3 and PI to a more proximal position between 14q24.3 and 14q32.1. Cox et al. (1984) refined the assignment to 14q32.33-qter. Linsley et al. (1983) found RFLPs associated with the heavy chain C-gamma genes. The person with the ring-14 had none of the maternal complement of C-gamma gene hybridizing fragments. A C-gamma pseudogene was identified. D14S1 (107750) was not deleted from the ring-14.

Burrows et al. (1983) showed that rearrangement, not differential RNA processing, occurs in heavy chain class switching. They demonstrated loss of DNA sequences between the J(H) and C(G2b) gene segments in a mouse cell line. Deletions of specific constant region genes are prone to occur through nonhomologous pairing and unequal crossing-over. This is a mechanism of evolution and a mechanism of pathogenesis of selective immunoglobulin deficiencies. The study of deletions has been useful for confirmation of the gene order demonstrated by DNA cloning (e.g., Ellison and Hood, 1982; Flanagan and Rabbitts, 1982; Hisajima et al., 1983; Max et al., 1982) and by linkage analysis with both DNA and allotypic markers (e.g., Bech-Hansen et al., 1983; Migone et al., 1983). Lefranc et al. (1983) found, in apparently healthy members of highly inbred communities of Tunisian Berbers, 2 types of multiple IgHC gene deletions. One deletion found by Keyeux et al. (1989) in 6 individuals in 2 different families (family HASS and family TOU) represented a simultaneous absence of the IgG1, IgG2, IgG4, and IgA1 immunoglobulins. This deletion allowed determination of the order of immunoglobulin IgCH genes and localization of a gamma pseudogene between A1 and G2. The TOU family had a second deletion that included only the epsilon pseudogene-1, A1 and the gamma pseudogene. Keyeux et al. (1989) demonstrated that the multigene deletion in the IgCH cluster involves 2 highly homologous regions, called hsg3 and hsg4, which are hotspots of recombination, outside the switch sequences. (Hsg3 is located downstream of G3 and hsg4 is located downstream of G4; hence, their designations.) Migone et al. (1984) identified 2 additional types of multiple heavy chain gene deletions. One included the IgE gene. The deletions were transmitted in a mendelian manner and despite homozygosity seemed to have no ill effects. (Selective absence of single IgG subclasses had been found occasionally by immunologic testing for allotypes and isotypes.) Migone et al. (1984) could confirm the location of the pseudo-gamma gene between the alpha-1 and gamma-2 genes. It is possible that some instances of combined variable hypogammaglobulinemia or selective deficiency of immunoglobulins are caused by deletion or other changes in this region.

Hofker et al. (1989) determined the complete physical map of the heavy chain constant region by means of pulsed field gel electrophoresis. The genes of this region are contained within a 300 kb segment. In both man and mouse the order is 5-prime—IGHM—IGHD—IGHG—IGHE—IGHA—3-prime. The C-gamma gene was duplicated to produce 4 copies in the mouse. In man, after an initial C-gamma duplication, the entire C-gamma/C-gamma/C-epsilon/C-alpha segment was duplicated. Hofker et al. (1989) found that a 60-kb segment separates the C-delta locus and the 5-prime end of the cluster containing IGHG3, IGHG1, IGHCEP1, and IGHCA1, in that order. At a distance 80 kb 3-prime to IGHA1 lies the second cluster of IGHG2, IGHG4, IGHE, and IGHA2. The C-gamma pseudogene lies about 35 kb on the 3-prime side of IGHA1.

Zelaschi et al. (1983) used monoclonal antibodies raised in mice to define 'new' Gm determinants. Jazwinska et al. (1988) demonstrated that RFLP analysis can be substituted for serologic determination of Gm type and pointed out several advantages of the molecular genetic method. Ghanem et al. (1989) found extensive RFLP-type variation in black Africans involving mainly the IGHG3 and IGHG1 genes, the most 5-prime members of the IGHG family. Polymorphism is much more extensive in black Africans than in Caucasoids. The same result was suggested by the study of Gm haplotypes which have been referred to as G1m, G3m, A2m, and Em, corresponding to the IgG and IgA subclasses and the IgE class for which they are markers. The alleles at the respective loci are inherited in fixed combinations, or Gm-Am-Em haplotypes. Although Km gene frequencies showed a random distribution in the populations studied, Zhao and Lee (1989) found that Gm haplotypes were highly useful in mapping the origins of the Chinese nation. A comparison with Gm haplotype frequencies in other populations suggested that during human evolution the Negroid group and the Caucasoid-Mongoloid group diverged first, followed by a divergence between the Caucasoid and Mongoloid. The data appeared to indicate 2 distinct subgroups of the Mongoloid race, northern and southern, corresponding to populations that originated in the Yellow River valley and the Yangtze River valley, respectively. Zhao and Lee (1989) found that the northern and southern Mongoloid populations have Gm(1;21) and Gm(1,3;5) haplotypes as race-associated markers, respectively. They attributed the presence of the Caucasian-associated haplotype Gm(3;5) in several of the minorities living in the northwest part of China to admixture along the Silk Road. The amount of Caucasian admixture was estimated.

The heavy chain constant region 'locus' shows organization in multiple levels of internal homology, suggesting a complex evolutionary history with repeated duplication events. The persisting genetic instability of the region is highlighted by the not uncommon observation of deletions or duplications, which probably originated through unequal crossover events. Bottaro et al. (1989), by pulsed field gel analysis of such deletions, determined the following map (their figure 4): —mu-5 kb-delta—gamma-3–26 kb—gamma-1–19 kb—psi-epsilon—13 kb—alpha-1—?60 kb—psi-gamma—?40 kb—gamma-2–18 kb—gamma-4–23 kb—epsilon—10 kb—alpha-2—.

Data on gene frequencies of allelic variants were tabulated by Roychoudhury and Nei (1988).

See also Bender et al. (1979); Borgaonkar et al. (1973); Ceppellini et al. (1966); Chaabani et al. (1985); Fahey (1965); Gedde-Dahl et al. (1975); Grubb (1970); Hill et al. (1966); Kimberling et al. (1978); Klein (1981); Lennox and Cohn (1967); Migone et al. (1985); Natvig and Kunkel (1973); Oudin (1966); Steinberg (1969)

Anonymous: Notation for genetic factors of human immunoglobulins. Nature 209: 653-655, 1966.

Balazs, I.; Purrello, M.; Rubinstein, P.; Alhadeff, B.; Siniscalco, M.: Highly polymorphic DNA site D14S1 maps to the region of Burkitt lymphoma translocation and is closely linked to the heavy chain gamma 1 immunoglobulin locus. Proc. Nat. Acad. Sci. 79: 7395-7399, 1982.

Bech-Hansen, N. T.; Linsley, P. S.; Cox, D. W.: Restriction fragment length polymorphisms associated with immunoglobulin C-gamma genes reveal linkage disequilibrium and genomic organization. Proc. Nat. Acad. Sci. 80: 6952-6956, 1983.

Bender, K.; Burckhardt, K.; Schroetter, K.: Exclusion of the localization of the Gm, Pi, and C3 genes on 6q25-6qter through blood group analysis of the patients of Schmid, D'Apuzzo and Rossi (Hum. Genet. 46: 279-284, 1979). (Letter) Hum. Genet. 53: 129-130, 1979.

Bender, K.; Muller, C. R.; Schmidt, A.; Strohmier, U.; Wienker, T. F.: Linkage studies on the human Pi, Gm, GLO, and HLA genes. Hum. Genet. 49: 159-166, 1979.

Borgaonkar, D. S.; Bias, W. B.; Chase, G. A.; Sadasivan, G.; Herr, H. M.; Golomb, H. M.; Bahr, G. F.; Kunkel, L. M.: Identification of a C6-G21 translocation chromosome by Q-M and Giemsa banding techniques in a patient with Down's syndrome, with possible assignment of Gm locus. Clin. Genet. 4: 53-57, 1973.

Bottaro, A.; de Marchi, M.; Migone, N.; Carbonara, A. O.: Pulsed-field gel analysis of human immunoglobulin heavy-chain constant region gene deletions reveals the extent of unmapped regions within the locus. Genomics 4: 505-508, 1989.

Burrows, P. D.; Beck-Engeser, G. B.; Wabl, M. R.: Immunoglobulin heavy-chain class switching in a pre-B cell line is accompanied by DNA rearrangement. Nature 306: 243-246, 1983.

Ceppellini, R.; Dray, S.; Fabey, J. L.; Franklin, E. C.; Fudenberg, H.; Gell, P. G. H.; Goodman, H. C.; Grubb, R.; Harboe, M.; Kirk, R. L.; Oudin, J.; Ropartz, C.; Smithies, O.; Steinberg, A. G.; Trnka, Z.: Notation for genetic factors in human immunoglobulins. Genetics 53: 235-241, 1966.

Chaabani, H.; Bech-Hansen, N. T.; Cox, D. W.: A multigene deletion within the immunoglobulin heavy-chain region. Am. J. Hum. Genet. 37: 1164-1171, 1985.

Cox, D. W.; Markovic, V. D.; Teshima, I. E.: Genes for immunoglobulin heavy chains and for alpha-1-antitrypsin are localized to specific regions of chromosome 14q. Nature 297: 428-430, 1982.

Cox, D. W.; Teshima, I.; Linsley, P. S.: Regional localization of the immunoglobulin heavy chain to 14q32.33 to 14qter. (Abstract) Cytogenet. Cell Genet. 37: 441 only, 1984.

Croce, C. M.; Shander, M.; Martinis, J.; Cicurel, L.; D'Ancona, G.; Dolby, T. W.; Koprowski, H.: Chromosomal location of the genes for human immunoglobulin heavy chains. Proc. Nat. Acad. Sci. 76: 3416-3419, 1979.

Davis, M. M.; Calame, K.; Early, P. W.; Livant, D. L.; Joho, R.; Weissman, I. L.; Hood, L.: An immunoglobulin heavy-chain gene is formed by at least two recombinational events. Nature 283: 733-739, 1980.

Ellison, J.; Hood, L.: Linkage and sequence homology of two human immunoglobulin gamma heavy chain constant region genes. Proc. Nat. Acad. Sci. 79: 1984-1988, 1982.

Fahey, J. L.: Antibodies and immunoglobulins. Structure and function. J.A.M.A. 194: 71-74, 1965.

Flanagan, J. G.; Rabbitts, T. H.: Arrangement of human immunoglobulin heavy chain constant region genes implies evolutionary duplication of a segment containing gamma, epsilon, and alpha genes. Nature 300: 709-713, 1982.

Gedde-Dahl, T., Jr.; Cook, P. J. L.; Fagerhol, M. K.; Pierce, J. A.: The Gm-Pi linkage: a summary estimate. Birth Defects Orig. Art. Ser. XI(3): 157-158, 1975. Note: Alternate: Cytogenet. Cell Genet. 14: 327-328, 1975.

Gedde-Dahl, T., Jr.; Cook, P. J. L.; Fagerhol, M. K.; Pierce, J. A.: Improved estimate of the Gm-Pi linkage. Ann. Hum. Genet. 39: 43-50, 1975.

Gedde-Dahl, T., Jr.; Fagerhol, M. K.; Cook, P. J. L.; Noades, J.: Autosomal linkage between the Gm and Pi loci in man. Ann. Hum. Genet. 35: 393-400, 1972.

Ghanem, N.; Bensmana, M.; Dugoujon, J. M.; Constans, J.; Lefranc, M.-P.; Lefranc, G.: BamHI and SacI RFLPs of the human immunoglobulin IGHG genes with reference to the Gm polymorphism in African people: evidence for a major polymorphism. Hum. Genet. 83: 37-44, 1989.

Green, M. C.: Genetic nomenclature for the immunoglobulin loci of the mouse. Immunogenetics 8: 89-97, 1979.

Grubb, R.: The Genetic Markers of Human Immunoglobins. New York: Springer (pub.) 1970.

Hengartner, H.; Meo, T.; Muller, E.: Assignment of genes for immunoglobulin kappa and heavy chains to chromosomes 6 and 12 in mouse. Proc. Nat. Acad. Sci. 75: 4494-4498, 1978.

Hill, R. L.; Delaney, R.; Fellows, R. E., Jr.; Lebovitz, H. E.: The evolutionary origins of the immunoglobulins. Proc. Nat. Acad. Sci. 56: 1762-1769, 1966.

Hisajima, H.; Nishida, Y.; Nakai, S.; Takahashi, N.; Ueda, S.; Honjo, T.: Structure of the human immunoglobulin C-epsilon-2 gene, a truncated pseudogene: implications for its evolutionary origin. Proc. Nat. Acad. Sci. 80: 2995-2999, 1983.

Hofker, M. H.; Walter, M. A.; Cox, D. W.: Complete physical map of the human immunoglobulin heavy chain constant region gene complex. Proc. Nat. Acad. Sci. 86: 5567-5571, 1989.

Hood, L. E.; Ein, D.: Immunoglobulin lambda chain structure: two genes, one polypeptide chain. Nature 220: 764-767, 1968.

Jazwinska, E. C.; Dunckley, H.; Propert, D. N.; Gatenby, P. A.; Serjeantson, S. W.: Gm typing by immunoglobulin heavy-chain gene RFLP analysis. Am. J. Hum. Genet. 43: 175-181, 1988.

Keyeux, G.; Lefranc, G.; Lefranc, M.-P.: A multigene deletion in the human IGH constant region locus involves highly homologous hot spots of recombination. Genomics 5: 431-441, 1989.

Kimberling, W. J.; Taylor, R. A.; Chapman, R. G.; Lubs, H. A.: Linkage gene localization of hereditary spherocytosis (HS). Blood 52: 859-867, 1978.

Klein, G.: Personal Communication. Stockholm, Sweden 11/4/1981.

Klein, G.: The role of gene dosage and genetic transpositions in carcinogenesis. Nature 294: 313-318, 1981.
Kunkel, H. G.; Natvig, J. B.; Joslin, F. G.: A 'Lepore' type of hybrid gamma globulin. Proc. Nat. Acad. Sci. 62: 144-149, 1969.
Lefranc, G.; Chaabani, H.; Van Loghem, E.; Lefranc, M.-P.; De Lange, G.; Helal, A.-N.: Simultaneous absence of the human IgG1, IgG2, IgG4 and IgA1 subclasses: immunological and immunogenetical considerations. Europ. J. Immun. 13: 240-244, 1983.
Lefranc, G.; Rivat, L.; Salier, J. P.; Van Loghem, E.; Aydenian, H.; Zaizal, P.; Chakhachiro, L.; Loiselet, J.; Rozartz, C.: Recombination, mutation, or constitutive expression at a Gm locus and familial hypergammaglobulinemia. Am. J. Hum. Genet. 29: 523-536, 1977.
Lennox, E. S.; Cohn, M.: Immunoglobulins. Annu. Rev. Biochem. 36: 365-402, 1967.
Lenoir, G. M.; Preud'homme, J. L.; Bernheim, A.; Berger, R.: Correlation between immunoglobulin light chain expression and variant translocation in Burkitt's lymphoma. Nature 298: 474-476, 1982.
Linsley, P. S.; Bech-Hansen, N. T.; Siminovitch, L.; Cox, D. W.: Analysis of a break in chromosome 14 mapping to the region of the immunoglobulin heavy chain locus. Proc. Nat. Acad. Sci. 80: 1997-2001, 1983.
Marx, J. L.: Antibodies: getting their genes together. (Annotation). Science 212: 1015-1017, 1981.
Max, E. E.; Battey, J.; Ney, R.; Kirsch, I. R.; Leder, P.: Duplication and deletion in the human immunoglobulin epsilon genes. Cell 29: 691-699, 1982.
Meo, T.; Johnson, J.; Beechey, C. V.; Andrews, S. J.; Peters, J.; Searle, A. G.: Linkage analyses of murine immunoglobulin heavy chain and serum prealbumin genes establish their location on chromosome 12 proximal to the T(5;12)31H breakpoint in band 12F1. Proc. Nat. Acad. Sci. 77: 550-553, 1980.
Migone, N.; de Lange, G.; Piazza, A.; Cavalli-Sforza, L. L.: Genetic analysis of eight linked polymorphisms within the human immunoglobulin heavy-chain region. Am. J. Hum. Genet. 37: 1146-1163, 1985.
Migone, N.; Feder, J.; Cann, H.; van West, B.; Hwang, J.; Takahashi, N.; Honjo, T.; Piazza, A.; Cavalli-Sforza, L. L.: Multiple DNA fragment polymorphisms associated with immunoglobulin mu chain switch-like regions in man. Proc. Nat. Acad. Sci. 80: 467-471, 1983.
Migone, N.; Oliviero, S.; de Lange, G.; Delacroix, D. L.; Boschis, D.; Altruda, F.; Silengo, L.; DeMarchi, M.; Carbonara, A. O.: Multiple gene deletions within the human immunoglobulin heavy-chain cluster. Proc. Nat. Acad. Sci. 81: 5811-5815, 1984.
Natvig, J. B.; Kunkel, H. G.: Human immunoglobulins: classes, subclasses, genetic variants, and idiotypes. Adv. Immun. 16: 1-59, 1973.
Oudin, J.: Genetic regulation of immunoglobulin synthesis. J. Cell. Physiol. 67 (suppl. 1): 77-108, 1966.
Pisetsky, D. S.; Sachs, D. H.: The genetic control of the immune response to staphylococcal nuclease. VI. Recombination between genes determining the A-J anti-nuclease idiotypes and the heavy chain allotype locus. J. Exp. Med. 146: 1603-1612, 1977.
Ricciuti, F. C.; Ruddle, F. H.: Assignment of three gene loci (PGK, HGPRT, G6PD) to the long arm of the human X chromosome by somatic cell genetics. Genetics 74: 661-678, 1973.
Robertson, M.: Genes of lymphocytes I: diverse means to antibody diversity. (Annotation). Nature 290: 625-627, 1981.
Roychoudhury, A. K.; Nei, M.: Human Polymorphic Genes: World Distribution. New York: Oxford Univ. Press (pub.) 1988.
Shimizu, A.; Takahashi, N.; Yamawaki-Kataoka, Y.; Nishida, Y.; Kataoka, T.; Honjo, T.: Ordering of mouse immunoglobulin heavy chain genes by molecular cloning. Nature 289: 149-153, 1981.
Smith, M.; Krinsky, A.; Arredondo-Vega, F.; Wang, A.-L.; Hirschhorn, K.: Confirmation of the assignment of genes for human immunoglobulin heavy chains to chromosome 14 by analyses of Ig synthesis by man-mouse hybridomas. Europ. J. Immun. 11: 852-855, 1981.
Steinberg, A. G.: Gammaglobulin polymorphisms in man. Annu. Rev. Genet. 3: 25-32, 1969.
Steinberg, A. G.; Olivier, T. J.; Buettner-Janusch, J.: Gm and Inv polymorphism among baboons from Kenya. (Abstract) Am. J. Hum. Genet. 27: 86A only, 1975.
Taylor, B. A.; Bailey, D. W.; Cherry, M.; Riblet, R.; Weigert, M.: Genes for immunoglobulin heavy chain and serum prealbumin protein are linked in mouse. Nature 256: 644-646, 1975.
Van Loghem, E.; Natvig, J. B.; Matsumoto, H.: Genetic markers of immunoglobulins in Japanese families. Inheritance of associated markers belonging to one IgA subclass and three IgG subclasses. Ann. Hum. Genet. 33: 351-360, 1970.
Zelaschi, D.; Newby, C.; Parsons, M.; van West, B.; Cavalli-Sforza, L. L.; Herzenberg, L. A.; Herzenberg, L. A.: Human immunoglobulin allotypes: previously unrecognized determinants and alleles defined with monoclonal antibodies. Proc. Nat. Acad. Sci. 80: 3762-3766, 1983.
Zhao, T.; Lee, T. D.: Gm and Km allotypes in 74 Chinese populations: a hypothesis of the origin of the Chinese nation. Hum. Genet. 83: 101-110, 1989.

IMMUNOGLOBULIN KAPPA CONSTANT REGION (*147200)

InV(-1,3) homozygotes have valine at position 191 of the kappa-type light polypeptide chains whereas InV(1,3) heterozygotes have some chains with leucine and some with valine at this position (Terry et al., 1969). By 1977, it was the recommendation that InV be called Km ('kappa marker'). This makes it clear that the kappa chain is involved. When and if allotypes are found on the lambda chains, their generic designation should be Lm. From study of somatic cell hybrids, Hengartner et al. (1978) concluded that the locus (or loci) for kappa light chains is on chromosome 6 in the mouse. The kappa light chain genes were assigned to mouse chromosome 6 by Swan et al. (1979). The genes for at least one variable region subgroup are also on 6. Swan et al. (1979) used nucleic acid probes for nucleic acid hybridization in mouse-hamster hybrids with a variable number of mouse chromosomes. Km and Kidd blood group are linked with a lod score of 3.4 at theta 0.23 (10,11:Keats et al., 1977, 1978). The Lu-Se-Dm linkage group and the Km (Inv)-Jk-Co linkage group were tentatively tied together by a family with myotonic dystrophy reported by Larsen et al. (1979). In the mouse (and there is no reason to think the situation in man is greatly different), the variable regions of kappa chains are coded for by multiple variable (V) gene segments and multiple joining (J) gene segments. In the mouse, the V(kappa) genes code for residues 96 to 108. The V(lambda) regions have a similar determination. Diversity in V(kappa) regions arises from several sources: (1) there are multiple germ-line V(kappa) genes and J(kappa) genes; (2) combination of any one of many V(kappa) genes with any one of many J(kappa) genes may occur; (3) somatic point mutation may occur. (See Weigert et al., 1980.) The kappa light chain cluster of genes was assigned to the chromosome 2 short arm by in situ hybridization by Malcolm et al. (1982), who further localized it to 2cen-p13. A probe for the variable genes of the kappa chain was used. Using nucleic acid probes prepared from the cloned gene in Southern blots of DNA from somatic cell hybrids, McBride et al. (1982) assigned the kappa constant gene to chromosome 2 and the lambda constant gene to chromosome 22. The human chromosomes carried by each hybrid cell line were identified by isozyme markers. Klein (1981) found that B cell-derived tumors (mouse myeloma and human Burkitt lymphoma and B-cell acute lymphoblastic leukemia) have anomalous patterns of immunoglobulin synthesis which correlate with the type of chromosomal aberration. Similar observations were made by Lenoir et al. (1982) who had collected the largest number of variant Burkitt lymphoma translocations. Of 10 tested, all agreed with the hypothesis as to light chain expression: all the 8;22 translocation cells produced lambda as the only light chain; all the 2;8 translocation cells produced only kappa; and 8;14 translocation cells produced either kappa or lambda, with an approximate ratio of 2:1. In the mouse, trisomy 15 is regularly associated with mouse T cell leukemias, even if they are induced by different agents including various leukemia viruses, x-rays, and chemical carcinogens. On the other hand, all mouse plasmacytomas show a consistent translocation of the distal part of chromosome 15 to either chromosome 6 or chromosome 12 (Ohno et al., 1979), both of which are immunoglobulin genes in the mouse. The parallelism with the situation in human Burkitt tumors is evident. Zegers et al. (1976) observed complete absence of immunoglobulin kappa chains in a patient with cystic fibrosis in whom kappa chains were absent in serum immunoglobulins and in blood and bone-marrow lymphocytes. Lymphocytes stimulated by pokeweed mitogen produced no kappa chains. A sister, who also had cystic fibrosis, showed partial deficiency of kappa chains. A few cases with reduced kappa chains have been described (Bernier et al., 1972; Barandun et al., 19 76). In a female offspring of an uncle-niece marriage, Bernier et al. (1972) observed deficient synthesis of kappa-chain-bearing immunoglobulins. The girl had recurrent respiratory infections and diarrhea. Generally the ratio of kappa to lambda light chains in serum immunoglobulin classes is around 2:1. Stavnezer-Nordgren et al. (1985) studied the molecular basis of the absence of kappa chains in the completely deficient person. Their initial hypothesis was that deletion of or mutation in an IGK gene was responsible. They further reasoned that the kappa constant gene was the site of the change because humans have a single constant gene whereas there are 5 J(kappa) genes and at least 25-50 V(kappa) genes. By nucleotide sequencing they found a point mutation in each C(kappa) gene: on one chromosome the mutation had led to the loss of the invariant tryptophan at amino acid position 148; on the other chromosome an invariant cysteine (1 of 2 in the kappa constant chain), that at position 94, was missing. Thus, neither of the proband's C(kappa) domains should be able to form stable intradomain disulfide bonds. Kappa deficiency has little if any effect on health; gamma chains seem able to substitute effectively. The patient's immune responses to a variety of antigens were normal. The patient was male. A sister had a very low amount of kappa chains in the serum; sera from his parents and another sib had approximately normal amounts of kappa chains. Kurth et al. (1991) amplified the kappa constant gene by PCR and used 4 allele-specific oligonucleotides to distinguish the alleles in the amplified PCR products. By physical mapping methods, Weichhold et al. (1993) developed a detailed map of the kappa locus. They concluded that the kappa locus comprises 2 copies showing opposite 5-prime/3-prime polarity. Several immunoglobulin kappa-related sequences were transposed in evolution from the short arm to the long arm of chromosome 2. By in situ hybridization and by pulsed field gel electrophoresis experiments with hybridization probes from both arms of chromosome 2 (i.e., from 2cen-p12 and 2cen-q11), Lautner-Rieske et al. (1993) demonstrated that the common pericentric inversion of chromosome 2 found in the present-day populations results in an apparent reinversion of these sequences to the short arm and the transposition of the kappa and CD8-alpha (CD8A; 186910) loci to the long arm. The fusion between 2 ape chromosomes to form the human chromosome 2 has been well documented (Yunis and Prakash, 1982), and interstitial telomere-like repeat stretches in this region have been discussed in relation to possible breakage, fragility, and recombination (Hastie and Allshire, 1989). The sites of fusion and fragility at 2q13 were studied in molecular detail by 8,9:IJdo et al. (1991, 1992).

Harrington et al. (1997) created de novo human artificial chromosomes (HACs) by combining long synthetic arrays of alpha satellite DNA with telomeric DNA and genomic DNA. The resulting linear microchromosomes were mitotically and cytogenetically stable in the absence of selection for up to 6 months in culture, bound centromere proteins specific for active centromeres, and were estimated to be 6 to 10 Mb in size. They suggested that this first-generation system for the construction of HACs should be suitable for dissecting the sequence requirements of human centromeres, as well as developing constructs useful for therapeutic applications. The next step, the introduction of chromosome fragments into the mouse germline, was achieved by Tomizuka et al. (1997). They used microcell-mediated chromosome transfer (MMCT) to introduce human chromosomes or chromosome fragments derived from normal fibroblasts into mouse embryonic stem (ES) cells and succeeded in producing viable chimeric mice. Transferred chromosomes were stably retained and human genes, including immunoglobulin genes, were expressed in proper tissue-specific manner in adult chimeric tissues. Tomizuka et al. (1997) found that a human chromosome 2-derived fragment was transmitted to the offspring of chimeric mice through the germline. The chimeric transchromosomic mice were found to produce functional human sequences composed of rearranged human V, J, and D segments of human mu, kappa, and lambda immunoglobulin chains. Immunization with human serum albumin (HSA) resulted in production of HSA-specific antibodies, indicating the usefulness of this system for the production of monoclonal antibodies. The investigators suspected that passage through the male germline might be a stumbling block and therefore used both male and female ES cells as recipients in the microcell fusions. They were unable to transmit fragments of human chromosomes 14 or 22 (containing the heavy chain and lambda light chain genes) through the germline of the chimeric mice. In 2 cases, however, they documented transmission of a fragment of human chromosome 2 through the germline, producing offspring that carried copies of this fragment, which appeared to be structurally and functionally unchanged. Transmission of the human fragment from 4 female chimeras and 1 male chimera was achieved. Rastan (1997) reviewed the findings, noting that the fact that the transferred chromosomal fragments were visible cytogenetically means that they were probably of the order of 5 to 20 Mb and would carry hundreds if not thousands of genes.

Data on gene frequencies of allelic variants at the kappa light chain locus were tabulated by Roychoudhury and Nei (1988).

See also Ambrosino et al. (1986)

Ambrosino, D. M.; Barrus, V. A.; DeLange, G. G.; Siber, G. R.: Correlation of the Km(1) immunoglobulin allotype with anti-polysaccharide antibodies in Caucasian adults. J. Clin. Invest. 78: 361-365, 1986.

Barandun, S.; Morell, A.; Skvaril, F.; Oberdorfer, A.: Deficiency of kappa- or lambda-type immunoglobulins. Blood 47: 79-89, 1976.

Bernier, G. M.; Gunderman, J. R.; Ruymann, F. B.: Kappa-chain deficiency. Blood 40: 795-805, 1972.

Harrington, J. J.; Van Bokkelen, G.; Mays, R. W.; Gustashaw, K.; Willard, H. F.: Formation of de novo centromeres and construction of first-generation human artificial microchromosomes. Nature Genet. 15: 345-355, 1997.

Hastie, N. D.; Allshire, R. C.: Human telomeres: fusion and interstitial sites. Trends Genet. 5: 326-331, 1989.

Hengartner, H.; Meo, T.; Muller, E.: Assignment of genes for immunoglobulin kappa and heavy chains to chromosomes 6 and 12 in mouse. Proc. Nat. Acad. Sci. 75: 4494-4498, 1978.

Hieter, P. A.; Max, E. E.; Seidman, J. G.; Maizel, J. V., Jr.; Leder, P.: Cloned human and mouse kappa immunoglobulin constant and J region genes conserve homology in functional segments. Cell 22: 197-207, 1980.

IJdo, J. W.; Baldini, A.; Ward, D. C.; Reeders, S. T.; Wells, R. A.: Origin of human chromosome 2: an ancestral telomere-telomere fusion. Proc. Nat. Acad. Sci. 88: 9051-9055, 1991.

IJdo, J. W.; Baldini, A.; Wells, R. A.; Ward, D. C.; Reeders, S. T.: FRA2B is distinct from inverted telomere repeat arrays at 2q13. Genomics 12: 833-835, 1992.

Keats, B. J. B.; Morton, N. E.; Rao, D. C.: Likely linkage: InV with Jk. Hum. Genet. 39: 157-159, 1977.

Keats, B. J. B.; Morton, N. E.; Rao, D. C.: Possible linkages (lod score over 1.5) and a tentative map of the Jk-Km linkage group. Cytogenet. Cell Genet. 22: 304-308, 1978.

Klein, G.: Personal Communication. Stockholm, Sweden 11/4/1981.

Kurth, J. H.; Bowcock, A. M.; Erlich, H. A.; Nevo, S.; Cavalli-Sforza, L. L.: Km typing with PCR: application to population screening. Am. J. Hum. Genet. 48: 613-620, 1991.

Larsen, B.; Johnson, G.; van Loghem, E.; Newton, R. M.; Pryse-Phillips, W.: Additions to the myotonic dystrophy linkage group. Clin. Genet. 15: 513-517, 1979.

Lautner-Rieske, A.; Hameister, H.; Barbi, G.; Zachau, H. G.: Mapping immunoglobulin gene-related DNA probes to the central region of normal and pericentrically inverted human chromosome 2. Genomics 16: 497-502, 1993.

Lenoir, G. M.; Preud'homme, J. L.; Bernheim, A.; Berger, R.: Correlation between immunoglobulin light chain expression and variant translocation in Burkitt's lymphoma. Nature 298: 474-476, 1982.

Malcolm, S.; Barton, P.; Bentley, D. L.; Ferguson-Smith, M. A.; Murphy, C. S.; Rabbitts, T. H.: Assignment of a IGKV locus for immunoglobulin light chains to the short arm of chromosome 2 (2p13-cen) by in situ hybridisation using a cRNA probe of H(kappa)101(lambda)Ch4A. (Abstract) Cytogenet. Cell Genet. 32: 296 only, 1982.

McBride, O. W.; Hieter, P. A.; Hollis, G. F.; Swan, D.; Otey, M. C.; Leder, P.: Chromosomal location of human kappa and lambda immunoglobulin light chain constant region genes. J. Exp. Med. 155: 1480-1490, 1982.

Milstein, C. P.; Steinberg, A. G.; McLaughlin, C. L.; Solomon, A.: Amino acid sequence change associated with genetic marker Inv(2) of human immunoglobulin. Nature 248: 160-161, 1974.

Ohno, S.; Babonits, M.; Wiener, F.; Spira, J.; Klein, G.; Potter, M.: Nonrandom chromosome changes involving the Ig gene-carrying chromosomes 12 and 6 in pristane-induced mouse plasmacytomas. Cell 18: 1001-1007, 1979.

Rastan, S.: Of men in mice. Nature Genet. 16: 113-114, 1997.

Roychoudhury, A. K.; Nei, M.: Human Polymorphic Genes: World Distribution. New York: Oxford Univ. Press (pub.) 1988.

Stavnezer-Nordgren, J.; Kekish, O.; Zegers, B. J. M.: Molecular defects in a human immunoglobulin kappa chain deficiency. Science 230: 458-461, 1985.

Steinberg, A. G.; Cook, C. E.: The Distribution of the Human Immunoglobulin Allotypes. Oxford: Oxford Univ. Press (pub.) 1981.

Steinberg, A. G.; Milstein, C. P.; McLaughlin, C. L.; Solomon, A.: Structural studies of an Inv(1,-2) kappa light chain. Immunogenetics 1: 108-117, 1974.

Swan, D.; D'Eustachio, P. D.; Leinwand, L.; Seidman, J.; Keithley, D.; Ruddle, F. H.: Chromosomal assignment of the mouse kappa light chain genes. Proc. Nat. Acad. Sci. 76: 2735-2739, 1979.

Terry, W. D.; Fahey, J. L.; Steinberg, A. G.: Gm and Inv factors in subclasses of human IgG. J. Exp. Med. 122: 1087-1102, 1965.

Terry, W. D.; Hood, L. E.; Steinberg, A. G.: Genetics of immunoglobulin kappa chains: chemical analysis of normal human light chains of differing Inv types. Proc. Nat. Acad. Sci. 63: 71-77, 1969.

Tomizuka, K.; Yoshida, H.; Uejima, H.; Kugoh, H.; Sato, K.; Ohguma, A.; Hayasaka, M.; Hanaoka, K.; Oshimura, M.; Ishida, I.: Functional expression and germline transmission of a human chromosome fragment in chimaeric mice. Nature Genet. 16: 133-143, 1997.

Weichhold, G. M.; Ohnheiser, R.; Zachau, H. G.: The human immunoglobulin kappa locus consists of two copies that are organized in opposite polarity. Genomics 16: 503-511, 1993.

Weigert, M.; Perry, R.; Kelley, D.; Hunkapiller, T.; Schilling, J.; Hood, L.: The joining of V and J gene segments creates antibody diversity. Nature 283: 497-499, 1980.

Yunis, J. J.; Prakash, O.: The origin of man: a chromosomal pictorial legacy. Science 215: 1525-1530, 1982.

Zegers, B. J. M.; Maertzdorf, W. J.; van Loghem, E.; Mul, N. A. J.; Stoop, J. W.; van der Laag, J.; Vossen, J. J.; Ballieux, R. E.: Kappa-chain deficiency: an immunoglobulin disorder. New Eng. J. Med. 294: 1026-1030, 1976.

B-CELL TRANSLOCATION GENE 1 (*109580)

Rimokh et al. (1991) cloned the breakpoint of a t(8;12) chromosomal translocation in a case of B-cell chronic lymphocytic leukemia and isolated a coding sequence mapping on 12q22. This sequence detected a 1.8-kb transcript in virtually all tissues tested except in the brain and muscle where the signal was barely detectable. The putative gene corresponding to this sequence, termed BTG1 for B-cell translocation gene 1, was shown to be highly conserved in evolution; a similar 1.8-kb transcript could be detected in murine and chicken tissue by using a human BTG1 DNA probe. Rouault et al. (1992) established the genomic organization of the gene. The full-length cDNA isolated from a lymphoblastoid cell line contained an open reading frame of 171 amino acids. BTG1 expression was maximal in the G0/G1 phases of the cell cycle and downregulated when cells progressed throughout G1. Furthermore, transfection experiments using NIH 3T3 cells indicated that BTG1 negatively regulates cell proliferation. Rouault et al. (1992) postulated that BTG1 is a member of a new family of antiproliferative genes.

Rimokh, R.; Rouault, J. P.; Wahbi, K.; Gadoux, M.; Lafage, M.; Archimbaud, E.; Charrin, C.; Gentilhomme, O.; Germain, D.; Samarut, J.; Magaud, J. P.: A chromosome 12 coding region is juxtaposed to the MYC protooncogene locus in a t(8;12)(q24;q22) translocation in a case of B-cell chronic lymphocytic leukemia. Genes Chromosomes Cancer 3: 24-36, 1991.

Rouault, J.-P.; Rimokh, R.; Tessa, C.; Paranhos, G.; Ffrench, M.; Duret, L.; Garoccio, M.; Germain, D.; Samarut, J.; Magaud, J.-P.: BTG1, a member of a new family of antiproliferative genes. EMBO J. 11: 1663-1670, 1992.

B-CELL TRANSLOCATION GENE 2 (*601597)

Rouault et al. (1996) cloned the human BTG2 gene from a lymphoblastoid cell line cDNA library. The sequence predicted a 158-amino acid protein which shares 93.6% identity with the murine TIS21 protein and 66.4% identity with the BTG1 (109580) protein. Rouault et al. (1996) noted that the only significant difference between the BTG1 and BTG2 protein sequences is a 10-amino acid insertion in the C-terminal part of the BTG1 protein. They mapped the BTG2 protein to human chromosome 1q32 by fluorescent in situ hybridization and by Southern blot analysis of human somatic cell hybrids. The BTG1 protein maps to chromosome 12q22. Rouault et al. (1996) determined that BTG2 was preferentially expressed in quiescent cells and that overexpression of this gene causes a decrease in the growth rate and clonability of NIH 3T3 cells. Btg2 disruption had no detectable effect on the growth of differentiated or undifferentiated embryonic stem (ES) cells. Rouault et al. (1996) reported that Btg2/Tis21 inactivation in ES cells leads to a striking disruption of DNA damage-induced G2/M arrest and to a marked increase in cell death. Rouault et al. (1996) concluded that BTG2 function may be relevant to cell cycle control and to cellular response to DNA damage. They noted that in response to DNA damage, eukaryotic cells delay cell cycle progression from G1 to S and from G2 to M by induction of antiproliferative genes. Arrest in G1 is thought to prevent replication of damaged genetic templates: arrest prior to M allows cells to avoid segregation of defective chromosomes. Rouault et al. (1996) determined that p53 regulates BTG2 gene expression.

As pointed out by Tirone (1997), this gene was isolated in the rat by Bradbury et al. (1991) as an immediate early gene induced by nerve growth factor (NGF; 162030) in pheochromocytoma cells that undergo neural differentiation in the presence of NGF. The gene was called PC3 for pheochromocytoma cell-3. Its induction was independent of new protein synthesis as it could occur in the presence of cycloheximide. PC3 was also induced with similar kinetics, but at lower levels, by membrane depolarization (both in vivo and in vitro) and epidermal growth factor (131530). Furthermore, the antiproliferative activity of PC3 and its involvement in cell cycle regulation was demonstrated by Montagnoli et al. (1996).

Bradbury, A.; Possenti, R.; Shooter, E. M.; Tirone, F.: Molecular cloning of PC3, a putatively secreted protein whose mRNA is induced by nerve growth factor and depolarization. Proc. Nat. Acad. Sci. 88: 3353-3357, 1991.

Montagnoli, A.; Guardavaccaro, D.; Starace, G.; Tirone, F.: Overexpression of the nerve growth factor-inducible PC3 immediate early gene is associated with growth inhibition. Cell Growth Differ. 7: 1327-1336, 1996.

Rouault, J.-P; Falette, N.; Guehenneux, F.; Guillot, C.; Rimokh, R.; Wang, Q.; Berthet, C.; Moyret-Lalle, C.; Savatier, P.; Pain, B.; Shaw, P.; Berger, R.; Samarut, J.; Magaud, J.-P.; Ozturk, M.; Samarut, C.; Puisieux, A.: Identification of BTG2, an antiproliferative p53-dependent component of the DNA damage cellular response pathway. Nature Genet. 14: 482-486, 1996.

Tirone, F.: Personal Communication. Rome, Italy 2/24/1997.

B-CELL LEUKEMIA AND LYMPHOMA 5 (*151441)

Gauwerky et al. (1989) analyzed the oncogene rearrangements involving BCL2 (151430) and MYC (164840) in the leukemia cells of a patient with an aggressive prolymphocytic leukemia that showed an abnormal karyotype including a t(14;18) translocation and a chromosome 17q+. Molecular analysis showed that BCL2 was rearranged in the major breakpoint cluster region and had joined into the immunoglobulin heavy chain gene as in follicular lymphoma. Cloning and sequence analysis of the rearranged MYC gene showed, however, that MYC was truncated at the PvuII site at the end of its first exon and had joined into the regulatory elements of the gene that Gauwerky et al. (1989) called BCL3 but which the HGM10.5 designated as BCL5. By Southern analysis of rodent-human hybrids and by in situ hybridization, they mapped the BCL5 locus to 17q22 and found that it was transcribed as a message of 1.7 kb in many hematopoietic cell lines representing all hematopoietic lineages. In the patient's leukemia cells, the truncated MYC gene was highly expressed under the influence of BCL5 regulatory elements, leading to an aggressive B-cell leukemia that presumably had been derived from an indolent lymphoma carrying a rearranged BCL2 gene. Yano et al. (1993), who referred to the locus as 'BCL-3,' provided a molecular analysis of this locus, called here BCL5, at 17q22. In 264 B-cell malignancies, Southern blot analysis showed rearrangements of the locus in only 4 cases. All were follicular lymphomas and all had BCL2 (151430) rearrangements consistent with their follicular center cell origin. Yano et al. (1993) concluded that although the preferential involvement of BCL5 alterations in a small subset of follicular lymphomas that transform suggests a possible link with progression, further studies are needed to ensure that these alterations are biologically relevant and not simply a manifestation of genomic instability.

(Miki et al. (1994) used the designation BCL5 for the gene they cloned at the 3q27 breakpoint in a patient with Burkitt lymphoma. Here we have used BCL6 for the gene on 3q27 (109565).)

Gauwerky, C. E.; Huebner, K.; Isobe, M.; Nowell, P. C.; Croce, C. M.: Activation of MYC in a masked t(8;17) translocation results in an aggressive B-cell leukemia. Proc. Nat. Acad. Sci. 86: 8867-8871, 1989.

Miki, T.; Kawamata, N.; Arai, A.; Ohashi, K.; Nakamura, Y.; Kato, A.; Hirosawa, S.; Aoki, N.: Molecular cloning of the breakpoint for 3q27 translocation in B-cell lymphomas and leukemias. Blood 83: 217-222, 1994.

Yano, T.; Sander, C. A.; Andrade, R. E.; Gauwerky, C. E.; Croce, C. M.; Longo, D. L.; Jaffe, E. S.; Raffeld, M.: Molecular analysis of the BCL-3 locus at chromosome 17q22 in B-cell neoplasms. Blood 82: 1813-1819, 1993.

B-CELL LYMPHOMA 6 (*109565)

Chromosomal translocations involving chromosome 3q27 and immunoglobulin gene regions are among the most common rearrangements in B-cell non-Hodgkin lymphoma. Using a probe from the immunoglobulin heavy chain joining region locus (147010), Baron et al. (1993) isolated genomic clones from a bacteriophage lambda library prepared from a lymphoma characterized by a translocation t(3;14)(q27;q32). Normal chromosome 3 sequences and the reciprocal breakpoint junction were isolated. DNA probes on each side of the chromosome 3 breakpoint hybridized at high stringency to the DNA of various mammalian species, demonstrating evolutionary conservation. A probe made from partial cDNA clones isolated from a T-cell line hybridized the genomic DNA from both sides of the chromosome 3 breakpoint, indicating that the t(3;14) is associated with a break within the gene on chromosome 3. In situ chromosomal hybridization revealed that the same gene is involved in the t(3;22)(q27;q11). Preliminary nucleotide sequencing showed no identity of the cDNA to gene sequences in available data banks. Baron et al. (1993) proposed the name B-cell lymphoma-6 (BCL6) for this gene, which they presumed plays a role in the pathogenesis of certain B-cell lymphomas. Ye et al. (1993) cloned the BCL6 gene.

Kerckaert et al. (1993) reported the isolation of a gene that was disrupted in 2 patients with non-Hodgkin lymphoma and t(3;14) and t(3;4) translocations. The gene, called LAZ3 (for lymphoma-associated zinc finger gene on chromosome 3) by them, encodes a 79-kD protein containing 6 zinc-finger motifs and sharing amino terminal homology with several transcription factors, including the Drosophila 'tramtrack' and 'Broad-complex' genes, both of which are developmental transcription regulators. LAZ3 is transcribed as a 3.8-kb message predominantly in normal adult skeletal muscle and in several non-Hodgkin lymphomas carrying 3q27 chromosomal defects. Kerckaert et al. (1993) suggested that LAZ3 may act as a transcription regulator and play an important role in lymphoma genesis.

Ye et al. (1993) cloned a gene from chromosomal translocations affecting band 3q27. They showed that the gene, BCL6, encodes a 79-kilodalton protein that is homologous with zinc finger-transcription factors. Chromosomal translocations affecting 3q27 are common in diffuse large cell lymphoma (DLCL). In 13 of 39 DLCL samples, but not in other types of lymphoid malignancies, Ye et al. (1993) found that the BCL6 gene was truncated within its 5-prime noncoding sequences, suggesting that its expression had been deregulated. Thus, BCL6 may be a protooncogene specifically involved in the pathogenesis of DLCL. Miki et al. (1994) cloned the gene located at the 3q27 breakpoint in a patient with Burkitt lymphoma carrying a translocation t(3;22)(q27;q11). The immunoglobulin lambda light chain gene was fused to the gene on 3q27, which Miki et al. (1994) referred to as BCL5. This designation is, however, used for the gene on 17q22 (151441). The characteristics of their BCL5 gene were like those described by Baron et al. (1993) and Ye et al. (1993) for the so-called BCL6 gene.

Because structural alterations of the 5-prime noncoding region of the BCL6 gene are found in 40% of diffuse large cell lymphomas and 5 to 10% of follicular lymphomas, deregulated BCL6 expression may play a role in lymphomagenesis. Nucleotide sequencing of BCL6 cDNA predicts a protein containing 6 zinc-finger domains, suggesting that it may function as a transcription factor. Using antisera raised against N- and C-terminal BCL6 synthetic oligopeptides, Cattoretti et al. (1995) identified the BCL6 gene product as a 95-kD nuclear protein. Western blot analysis of human tumor cell lines representative of various hematopoietic lineages/stages of differentiation showed that the BCL6 protein is predominantly expressed in the B-cell lineage where it was found in mature B cells. Immunohistochemical analysis of normal human lymphoid tissues indicated that BCL6 expression is topographically restricted to germinal centers, including all centroblasts and centrocytes. The results indicated that expression of BCL6 is specifically regulated during B-cell differentiation and suggested a role for BCL6 in germinal center development or function. Because diffuse large cell lymphoma derives from germinal-center B cells, Cattoretti et al. (1995) suggested that deregulated BCL6 expression may contribute to its genesis by preventing postgerminal center differentiation.

Migliazza et al. (1995) reported that in 22/30 (73%) DLCL and 7/15 (47%) follicular lymphoma, but not in other tumor types, the BCL6 gene is also altered by multiple, often biallelic, mutations clustered in its 5-prime noncoding region. These mutations are of somatic origin and are found in cases displaying either normal or rearranged BCL6 alleles indicating their independence from chromosomal rearrangements and association with immunoglobulin genes through translocation. These alterations identify a mechanism of genetic instability and malignant B cells and may have been selected during lymphomagenesis for their role in altering BCL6 expression. A panel of 123 nonhematologic tumors were screened for mutations in the sequences most frequently mutated in non-Hodgkin lymphoma using PCR/SSCP analysis, and no SSCP variant was found accept for previously detected population polymorphisms. Several observations suggested to Migliazza et al. (1995) that BCL6 mutations may be the result of the IgD hypermutation mechanism acting on non-Ig loci. In 10 cases studied in detail, a total of 59 alterations were detected in the BCL6 gene, including 55 single-bp substitutions, 3 small deletions, and 1 insertion.

In 2 B cell-type non-Hodgkin lymphoma patients with the t(3;6)(q27;p21) translocation, Akasaka et al. (1997) found that the translocation fused the H4/m histone gene (602833) to exons 3-9 of BCL6. Since H4 gene expression is tightly coupled to DNA replication, these authors suggested that the translocation causes inappropriate expression of BCL6 during the cell cycle, leading to the development of non-Hodgkins lymphoma.

In summary, approximately 40% of diffuse large cell lymphomas are associated with chromosomal translocations that deregulate the expression of BCL6 by juxtaposing heterologous promoters to BCL6 coding domain. The BCL6 gene encodes a 95-kD protein containing 6 C-terminal zinc finger motifs and an N-terminal POZ domain, suggesting that it may function as a transcription factor. By using a DNA sequence selected for its ability to bind recombinant BCL6 in vitro, Chang et al. (1996) showed that BCL6 is present in DNA-binding complexes in nuclear extracts from various B-cell lines. In transfection experiments, BCL6 can repress transcription from promoters linked to its DNA target sequence and this activity is dependent upon specific DNA-binding and the presence of an intact N-terminal half of the protein. This part of the BCL6 molecule contains an autonomous transrepressor domain, and 2 noncontiguous regions, including the POZ motif, mediate maximum transrepressive activity. Thus, the BCL6 protein can function as a sequence-specific transcriptional repressor and may have a role in normal lymphoid development lymphomagenesis.

Liao et al. (1996) mapped the mouse Bcl6 gene to chromosome 16 by interspecific backcross analysis.

The zinc-finger transcriptional repressor encoded by the BCL6 gene is normally expressed in both B cells and CD4(+) T cells within germinal centers; non-Hodgkin lymphomas are often derived from germinal-center B cells. Ye et al. (1997) showed that mice deficient in BCL6 displayed normal B-cell, T-cell, and lymphoid-organ development but had a selected defect in T-cell-dependent antibody responses. This defect included a complete lack of affinity maturation and was due to the inability of follicular B cells to proliferate and form germinal centers. In addition, BCL6-deficient mice developed an inflammatory response in multiple organs characterized by infiltrations of eosinophils and IgE-bearing B lymphocytes typical of a Th2-mediated hyperimmune response. Thus, Ye et al. (1997) concluded that BCL6 functions as a transcriptional switch that controls germinal center formation and may also modulate specific T-cell-mediated responses. Altered expression of BCL6 in lymphoma represents a deregulation of the pathway normally leading to B-cell proliferation and germinal center formation.

Migliazza et al. (1995) presented evidence that BCL6 may be altered by mechanisms other than chromosomal rearrangements. They found that in approximately 70% of diffuse large cell lymphomas (DLCL) and 50% of follicular lymphomas (FL), BCL6 is altered by multiple, often biallelic, mutations clustering in the 5-prime noncoding regions of the gene. Capello et al. (1997) detected 5-prime mutations of BCL6 in 6 of 21 (28.6%) cases of sporadic Burkitt lymphoma and in 7 of 14 (50%) cases of endemic Burkitt lymphoma.

Somatic hypermutation in B lymphocytes had been assumed to be restricted to the immunoglobulin genes. The somatic point mutations arise between the Ig promoter and 1- to 2-kb downstream, and thus only the variable (V) region, but not the constant (C) region, of an Ig gene is affected. However, if a kappa-chain promoter is artificially inserted upstream of the C region, both C and V are mutated at equal frequencies. This suggested that initiation of transcription of Ig genes is required for somatic hypermutation and that the mutation domain is restricted to the 5-prime end of the gene because a postulated mutation factor acts early in transcript elongation. Hypermutation in B lymphocytes occurs in the immunoglobulin genes of B lymphocytes that are the precursors to memory B cells. Non-Ig promoters are permissible for the mutation process, suggesting that other genes expressed in mutating B cells may be subject to somatic hypermutation. Shen et al. (1998) found that significant mutations were not observed in MYC (190080), S14 (130620), or alpha-fetoprotein (AFP; 104150) genes, but the BCL6 gene was highly mutated in a large proportion of memory B cells of normal individuals. The mutation pattern was similar to that of immunoglobulin genes.

Posttransplantation lymphoproliferative disorders represent a heterogeneous group of Epstein-Barr virus-associated lymphoid proliferations that arise in immunosuppressed transplant recipients. Some of these lesions regress after a reduction in immunosuppressive therapy, whereas some progress despite aggressive therapy. Morphologic, immunophenotypic, and immunogenotypic criteria are not useful in predicting clinical outcome. Cesarman et al. (1998) examined 57 lesions of the posttransplantation lymphoproliferative disorder obtained from 36 solid organ transplant recipients for the presence of mutations in the BCL6 gene. BCL6 mutations were identified in 44% of the specimens and in 44% of the patients; none was identified in the cases classified as plasmacytic hyperplasia. However, mutations were present in 43% of the polymorphic lesions and 90% of the cases diagnosed as non-Hodgkin lymphoma or multiple myeloma. BCL6 mutations predicted shorter survival and refractoriness to reduced immunosuppression and/or surgical excision. Cesarman et al. (1998) suggested that BCL6 is a reliable indicator for the division of this cluster of disorders into the biologic categories of hyperplasia and lymphoma, of which only the former can regress on immune reconstitution.

Akasaka, T.; Miura, I.; Takahashi, N.; Akasaka, H.; Yonetani, N.; Ohno, H.; Fukuhara, S.; Okuma, M.: A recurring translocation, t(3;6)(q27;p21), in non-Hodgkin's lymphoma results in replacement of the 5-prime regulatory region of BCL6 with a novel H4 histone gene. Cancer Res. 57: 7-12, 1997.
Baron, B. W.; Nucifora, G.; McCabe, N.; Espinosa, R., III; Le Beau, M. M.; McKeithan, T. W.: Identification of the gene associated with the recurring chromosomal translocations t(3;14)(q27;q32) and t(3;22)(q27;q11) in B-cell lymphomas. Proc. Nat. Acad. Sci. 90: 5262-5266, 1993.
Capello, D.; Carbone, A.; Pastore, C.; Gloghini, A.; Saglio, G.; Gaidano, G.: Point mutations of the BCL-6 gene in Burkitt's syndrome. Brit. J. Haemat. 99: 168-170, 1997.
Cattoretti, G.; Chang, C.-C.; Cechova, K.; Zhang, J.; Ye, B. H.; Falini, B.; Louie, D. C.; Offit, K.; Chaganti, R. S. K.; Dalla-Favera, R.: BCL-6 protein is expressed in germinal-center B cells. Blood 86: 45-53, 1995.
Cesarman, E.; Chadburn, A.; Liu, Y.-F.; Migliazza, A.; Dalla-Favera, R.; Knowles, D. M.: BCL-6 gene mutations in posttransplantation lymphoproliferative disorders predict response to therapy and clinical outcome. Blood 92: 2294-2302, 1998.
Chang, C.-C.; Ye, B. H.; Chaganti, R. S. K.; Dalla-Favera, R.: BCL-6, a POZ/zinc-finger protein, is a sequence-specific transcriptional repressor. Proc. Nat. Acad. Sci. 93: 6947-6952, 1996.
Kerckaert, J.-P.; Deweindt, C.; Tilly, H.; Quief, S.; Lecocq, G.; Bastard, C.: LAZ3, a novel zinc-finger encoding gene, is disrupted by recurring chromosome 3q27 translocations in human lymphomas. Nature Genet. 5: 66-70, 1993.
Liao, X.; Gilbert, D. J.; Dent, A.; Staudt, L. M.; Jenkins, N. A.; Copeland, N. G.: Mapping of the mouse Bcl6 gene to chromosome 16. Mammalian Genome 7: 621-622, 1996.
Migliazza, A.; Martinotti, S.; Chen, W.; Fusco, C.; Ye, B. H.; Knowles, D. M.; Offit, K.; Changanti, R. S. K.; Dalla-Favera, R.: Frequent somatic hypermutation of the 5-prime noncoding region of the BCL6 gene in B-cell lymphoma. Proc. Nat. Acad. Sci. 92: 12520-12524, 1995.
Miki, T.; Kawamata, N.; Arai, A.; Ohashi, K.; Nakamura, Y.; Kato, A.; Hirosawa, S.; Aoki, N.: Molecular cloning of the breakpoint for 3q27 translocation in B-cell lymphomas and leukemias. Blood 83: 217-222, 1994.
Miki, T.; Kawamata, N.; Hirosawa, S.; Aoki, N.: Gene involved in the 3q27 translocation associated with B-cell lymphoma, BCL5, encodes a Kruppel-like zinc-finger protein. Blood 83: 26-32, 1994.
Shen, H. M.; Peters, A.; Baron, B.; Zhu, X.; Storb, U.: Mutation of BCL-6 gene in normal B cells by the process of somatic hypermutation of Ig genes. Science 280: 1750-1752, 1998.
Ye, B. H.; Cattoretti, G.; Shen, Q.; Zhang, J.; Hawe, N.; de Waard, R.; Leung, C.; Nouri-Shirazi, M.; Orazi, A.; Chaganti, R. S. K.; Rothman, P.; Stall, A. M.; Pandolfi, P.-P.; Dalla-Favera, R.: The BCL-6 proto-oncogene controls germinal-centre formation and Th2-type inflammation. Nature Genet. 16: 161-170, 1997.
Ye, B. H.; Lista, F.; Lo Coco, F.; Knowles, D. M.; Offit, K.; Chaganti, R. S. K.; Dalla-Favera, R.: Alterations of a zinc finger-encoding gene, BCL-6, in diffuse large-cell lymphoma. Science 262: 747-750, 1993.
Ye, B. H.; Rao, P. H.; Chaganti, R. S. K.; Dalla-Favera, R.: Cloning of bcl-6, the locus involved in chromosome translocations affecting band 3q27 in B-cell lymphoma. Cancer Res. 53: 2732-2735, 1993.

B-CELL LEUKEMIA AND LYMPHOMA 7 (*601406)

Chromosome 12q24.1 is the site of a recurrent breakpoint in high-grade B-cell non-Hodgkin lymphoma (BNHL). To identify the gene(s) involved in 12q24.1, molecular cloning of a 3-way translocation t(8;14;12)(q24.1;q32.3;q24.1) in a Burkitt lymphoma cell line was performed by Zani et al. (1996); all 4 translocation breakpoints were cloned and sequenced. Analysis of the der(12)(12;14)(q24.1;q32.3) breakpoint showed a CpG island from 12q24.1 juxtaposed in a tail-to-tail configuration with a productively rearranged V-D-J heavy chain gene. A total of 4.5 kb of genomic DNA was shown to contain a potential 92-bp exon within the centromeric boundary of the CpG island. This was used as a probe to identify an RNA transcript of 3.8 kb expressed at low levels in a wide variety of normal tissues. The longest open reading frame predicted a serine-rich protein of 231 amino acids. This protein, designated BCL7A by the authors, exhibits no recognizable protein motifs but shows homology with the actin-binding protein, caldesmon (114213). In the Burkitt lymphoma cell line Wein 133 containing the 3-way translocation, the BCL7A breakpoint occurred within the first intron and resulted in a free-of-fusion transcript between MYC (190080) and BCL7A, with exon 1 of BCL7A being replaced by MYC exon 1. The normal, untranslocated allele of BCL7A was also expressed without mutation. One of the 11 other BNCLs with 12q24.1 cytogenetic abnormalities showed biallelic rearrangement within the first intron of BCL7A, which was adjacent to the breakpoint observed in Wein 133. Zani et al. (1996) concluded that disruption of the amino-terminus of BCL7A defined a new mechanism in the pathogenesis of a subset of high-grade BNHL.

Zani, V. J.; Asou, N.; Jadayel, D.; Heward, J. M.; Shipley, J.; Nacheva, E.; Takasuki, K.; Catovsky, D.; Dyer, M. J. S.: Molecular cloning of complex chromosomal translocation t(8;14;12)(q24.1;q32.3;q24.1) in a Burkitt lymphoma cell line defines a new gene (BCL7A) with homology to caldesmon. Blood 87: 3124-3134, 1996.

B-CELL LEUKEMIA AND LYMPHOMA 8 (*601889)

Recurrent chromosomal translocations recognized in most lymphomas provide clues to mechanisms of lymphomagenesis. Immunoglobulin genes undergo specific rearrangements during differentiation of lymphoid cells, and errors in recombination lead to chromosome translocation and neoplastic transformation. Dyomin et al. (1997) stated that diffuse large-cell lymphomas (DLCLs) make up approximately 50% of non-Hodgkin lymphomas and show significant variability in terms of their pathologic manifestation, response to therapy, and prognosis. DLCLs are also heterogeneous with respect to karyotypic abnormalities and underlying molecular lesions. The normal pattern of expression of 3 genes is altered by a specific chromosomal translocation in DLCL, namely, myc (190080), BCL2 (151430), and BCL6 (109565), occurring in approximately 10%, 20%, and 25% of DLCL, respectively, as detected by either cytogenetic or molecular methods. One of the remaining subsets of DLCL cases, found in 3 to 4% of cases, shows a rearrangement involving 15q11-q13. A notable feature of this site is the promiscuity of the rearrangement. In addition to translocations involving the immunoglobulin sites 14q32 and 22q11, it exhibits translocations with multiple other sites such as 9p13, 1p32, 7p13, 12q24, and 15q22. Translocations involving 15q11-q13 have been noted also in nonlymphoid tumors. The 15q11-q13 region is implicated in genomic imprinting and contains putative genes for the imprinting-related disorders, Prader-Willi syndrome (176270) and Angelman syndrome (105830). Another unusual feature of the region is the presence of orphan, presumably nonfunctional, copies of V (147070) and D segments of the IgH gene. Dyomin et al. (1997) reported the cloning of the t(14;15)(q32;q11-13) breakpoint from a DLCL patient and the isolation of a new gene which they designated BCL8, located at 15q11-q13, that is potentially involved the pathogenesis of DLCL. A probe derived from the BCL8 locus detected a transcript in human testis and prostate, whereas no expression was found in spleen, thymus, or blood leukocytes. Analysis of the BCL8 cDNA clones isolated from human testis cDNA library showed that the BCL8 gene generates a major transcript of 2.6 kb and a less prominent 4.5-kb species due to differential polyadenylation. By RT-PCR analysis of RNA extracted from frozen DLCL samples and lymphoma cell lines, BCL8 expression was detected in all patients carrying 15q11-q13 abnormalities and in a fraction of randomly selected DLCL patients. These results suggested that the BCL8 gene is not normally expressed in lymphoid tissues, but its expression can be activated by chromosomal translocation or by other mechanisms in DLCL. Ectopic expression of BCL8 in a significant proportion of DLCL suggests an important role for this gene in the molecular pathogenesis of B-cell lymphoma.

Dyomin, V. G.; Rao, P. H.; Dalla-Favera, R.; Chaganti, R. S. K.: BCL8, a novel gene involved in translocations affecting band 15q11-13 in diffuse large-cell lymphoma. Proc. Nat. Acad. Sci. 94: 5728-5732, 1997.

B-CELL LEUKEMIA AND LYMPHOMA 9 (*602597)

Studying a cell line (CEMO-1) from a patient with precursor-B-cell acute lymphoblastic leukemia with a t(1;14)(q21;q32), Willis et al. (1998) identified a fusion partner of the IGHJ gene (147010) on 14q. One allele showed novel sequences upstream of JH5 with no homology to either IGH or any other sequence in databases. Using a single-copy restriction fragment immediately 5-prime of JH5, PAC clones were isolated and mapped to 1q21 on normal metaphases by fluorescence in situ hybridization, confirming that this allele represented the translocation breakpoint. Sequence analysis of the 1q21 restriction fragment showed identity with an expressed sequenced tag, and this probe was therefore used to probe Northern blots.

Willis et al. (1998) designated the gene on 1q21 BCL9. Sequence analysis predicted a protein of 1,394 amino acids, with no recognizable protein motifs or significant homologies to known proteins. The CEMO-1 1q21 breakpoint fell within the 3-prime untranslated region (UTR) of the BCL9 gene. By screening of a panel of 39 B-cell malignancies with 1q abnormalities by Southern blotting, Willis et al. (1998) found 1 additional case with a breakpoint in the 3-prime UTR of BCL9, indicating that this is a recurrent breakpoint. Fluorescence in situ hybridization analysis using an 850-kb YAC spanning BCL9 identified a further case of t(1;22)(q21;q11) causing juxtaposition of BCL9 to the IGL locus (147220).

Willis, T. G.; Zalcberg, I. R.; Coignet, L. J. A.; Wlodarska, M.; Stul, D. M.; Jadayel, D. M.; Bastard, C.; Treleaven, J. G.; Catovsky, D.; Silva, M. L. M.; Dyer, M. J. S.: Molecular cloning of translocation t(1;14)(q21;q32) defines a novel gene (BCL9) at chromosome 1q21. Blood 91: 1873-1881, 1998.

IGA CONSTANT HEAVY CHAIN 1 (146900)

IgA is the predominant immunoglobulin class in body secretions, such as saliva, tears, bronchial secretions, nasal mucosal secretions, prostatic fluid, vaginal secretions and mucous secretions of the small intestines. It may serve both to defend against local infection and to prevent access of foreign antigens to the general immunologic system. Absence of certain immunoglobulins in patients with deleted chromosome 18 suggests the localization of structural and/or controller genes to that chromosome (Finley et al., 1968). Vyas and Fudenberg (1970) described allotype of IgA, using isoantibodies from a patient who displayed an anaphylactoid transfusion reaction. IgA2 has Am(1) and Am(2) genetic markers. The Am(1) allotype of IgA2 lacks the disulfide bond linking the heavy and light chains and instead consists of disulfide-bonded light chains joined noncovalently to a pair of disulfide-bonded heavy chains. Lefranc et al. (1983) first demonstrated the pseudogamma gene, which is located between the alpha-1 and gamma-2 (147110) genes (Migone et al., 1984).

See also Hobart et al. (1981); Smith et al. (1982); Solomon et al. (1982)

Finley, S. C.; Finley, W. H.; Noto, T. A.; Uchida, I. A.; Roddam, R. F.: IgA absence associated with a ring-18 chromosome. (Letter) Lancet I: 1095-1096, 1968.

Hobart, M. J.; Rabbitts, T. H.; Goodfellow, P. N.; Solomon, E.; Chambers, S.; Spurr, N.; Povey, S.: Immunoglobulin heavy chain genes in humans are located on chromosome 14. Ann. Hum. Genet. 45: 331-335, 1981.

Lefranc, M.-P.; Lefranc, G.; de Lange, G.; Out, T. A.; van den Broek, P. J.; van Nieuwkoop, J.; Radl, J.; Helal, A. N.; Chaabani, H.; van Loghem, E.; Rabbitts, T. H.: Instability of the human immunoglobulin heavy chain constant region locus indicated by different inherited chromosomal deletions. Molec. Biol. Med. 1: 207-217, 1983.

Migone, N.; Oliviero, S.; de Lange, G.; Delacroix, D. L.; Boschis, D.; Altruda, F.; Silengo, L.; DeMarchi, M.; Carbonara, A. O.: Multiple gene deletions within the human immunoglobulin heavy-chain cluster. Proc. Nat. Acad. Sci. 81: 5811-5815, 1984.

Smith, M.; Krinsky, A. M.; Arredondo-Vega, F.; Wang, A.-L.; Hirschhorn, K.: Confirmation of the assignment of genes for human immunoglobulin heavy chains to chromosome 14. (Abstract) Cytogenet. Cell Genet. 32: 318 only, 1982.

Solomon, E.; Goodfellow, P.; Chambers, S.; Spurr, N.; Hobart, M. J.; Rabbitts, T. H.; Povey, S.: Confirmation of the assignment of immunoglobulin heavy chain genes to chromosome 14, using cloned DNA as molecular probes. (Abstract) Cytogenet. Cell Genet. 32: 319 only, 1982.

Vyas, G. N.; Fudenberg, H. H.: Immunobiology of human anti-IgA: a serologic and immunogenetic study of immunization to IgA in transfusion and pregnancy. Clin. Genet. 1: 45-64, 1970.

DEAD/H (ASP-GLU-ALA-ASP/HIS) BOX POLYPEPTIDE 6 (*600326)

DEAD box proteins are putative RNA helicases that have a characteristic Asp-Glu-Ala-Asp (DEAD) box as 1 of 8 highly conserved sequence motifs. Akao et al. (1991) cloned the breakpoint of the t(11;14)(q23;q32) in B-cell lymphoma, as represented in the RC-K8 cell line, and named the locus RCK. By pulsed field gel electrophoresis, RCK was shown to be centromeric to the gene for porphobilinogen deaminase (176000), while the breakpoints of t(11;19)(q23;p13) were detected by the CD3D gene probe (186790), which is centromeric to RCK. Akao et al. (1992) did long-range mapping from the CD3 genes to the PBGD gene on 11q23 to determine the relationship between RCK and MLL-ALL1 (159555). They showed that RCK and MLL are on different NotI fragments, indicating that 2 different genes are associated with 11q23 translocations in hematopoietic tumors. Seto et al. (1995) found that the RCK/P54 gene, which had been found to encode a 472 to 483 amino acid-peptide belonging to the RNA helicase/translation initiation factor family, is highly conserved in the mouse. The mouse cDNA showed 93.7% nucleotide identity and 97.7% predicted amino acid identity with human RCK.

Lu and Yunis (1992) cloned a putative human RNA helicase, p54, from a lymphoid cell line with chromosomal breakpoint 11q23.3. The predicted amino acid sequence shared 75% identity with the female germline-specific RNA helicase ME31B gene of Drosophila. Unlike ME31B, however, the new gene expressed an abundant transcript in a large number of adult tissues and its 5-prime noncoding region was found to be split in a t(11;14)(q23.3;q32.3) cell line from a diffuse large B-cell lymphoma. Tunnacliffe et al. (1993) assigned the HLR2 gene more precisely using a panel of sequence tagged sites (STSs) representing 30 markers previously assigned to 11q23. Using fluorescence in situ hybridization, Akao and Matsuda (1996) mapped the Ddx6 gene to mouse chromosome 9.

See also Akao et al. (1990)

Akao, Y.; Matsuda, Y.: Identification and chromosome mapping of the mouse homologue of the human gene (DDX6) that encodes a putative RNA helicase of the DEAD box protein family. Cytogenet. Cell Genet. 75: 38-44, 1996.
Akao, Y.; Seto, M.; Takahashi, T.; Kubonishi, I.; Miyoshi, I.; Nakazawa, S.; Tsujimoti, Y.; Croce, C. M.; Ueda, R.: Molecular cloning of the chromosomal breakpoint of a B-cell lymphoma with the t(11;14)(q23;q32) chromosome translocation. Cancer Res. 51: 1574-1576, 1991.
Akao, Y.; Seto, M.; Yamamoto, K.; Iida, S.; Nakazawa, S.; Inazawa, J.; Abe, T.; Takahashi, T.; Ueda, R.: The RCK gene associated with t(11;14) translocation is distinct from the MLL/ALL-1 gene with t(4;11) and t(11;19) translocations. Cancer Res. 52: 6083-6087, 1992.
Akao, Y.; Tsujimoto, Y.; Finan, J.; Nowell, P. C.; Croce, C. M.: Molecular characterization of a t(11;14)(q23;q32) chromosome translocation in a B-cell lymphoma. Cancer Res. 50: 4856-4859, 1990.
Lu, D.; Yunis, J. J.: Cloning, expression and localization of an RNA helicase gene from a human lymphoid cell line with chromosomal breakpoint 11q23.3. Nucleic Acids Res. 20: 1967-1972, 1992.
Seto, M.; Yamamoto, K.; Takahashi, T.; Ueda, R.: Cloning and expression of a murine cDNA homologous to the human RCK/P54, a lymphoma-linked chromosomal translocation junction gene on 11q23. Gene 166: 293-296, 1995.
Tunnacliffe, A.; Perry, H.; Radice, P.; Budarf, M. L.; Emanuel, B. S.: A panel of sequence tagged sites for chromosome band 11q23. Genomics 17: 744-747, 1993.

INTERFERON REGULATORY FACTOR 4 (*601900)

Grossman et al. (1996) cloned a new human interferon regulatory factor (IRF) that they named LSIRF for 'lymphocyte-specific IRF.' They reported that the gene encodes a 450-amino acid polypeptide with a predicted mass of 51.6 kD. The gene is expressed as a single 5-kb transcript in spleen, lymphocytes, and melanocytes. They found that expression of LSIRF is induced in T cells after cross-linking of the T-cell receptor (see 186880).

In multiple myeloma (254500), chromosomal translocations affecting 14q32 and unidentified partner chromosomes are common, suggesting that they may cause the activation of novel oncogenes. In multiple myeloma cell lines, Iida et al. (1997) identified a t(6;14)(p25;q32) translocation in 2 of 11 cell lines. The translocation juxtaposes the immunoglobulin heavy-chain (IGHG1; 147100) locus to the MUM1 (for 'multiple myeloma oncogene 1') gene, which is also referred to as interferon regulatory factor-4 (IRF4), a member of a gene family known to be active in the control of B-cell proliferation and differentiation. See IRF1 (147575) on chromosome 5 and IRF2 (147576) on chromosome 4. As a result of the translocation, the MUM1/IRF4 gene is overexpressed, an event that may contribute to tumorigenesis, as Iida et al. (1997) showed that MUM1/IRF4 has oncogenic activity in vitro.

Using fluorescence in situ hybridization, Grossman et al. (1996) mapped the LSIRF gene to 6p25-p23.

Grossman, A.; Mittrucker, H.-W.; Nicholl, J.; Suzuki, A.; Chung, S.; Antonio, L.; Suggs, S.; Sutherland, G. R.; Siderovski, D. P.; Mak, T. W.: Cloning of human lymphocyte-specific interferon regulatory factor (hLSIRF/hIRF4) and mapping of the gene to 6p23-p25. Genomics 37: 229-233, 1996.
Iida, S.; Rao, P. H.; Butler, M.; Corradini, P.; Boccadoro, M.; Klein, B.; Chaganti, R. S. K.; Dalla-Favera, R.: Deregulation of MUM1/IRF4 by chromosomal translocation in multiple myeloma. Nature Genet. 17: 226-230, 1997.

NUCLEAR FACTOR KAPPA-B, SUBUNIT 2; ONCOGENE LYT10 (*164012)

See NFKB1 (164011). Liptay et al. (1992) mapped the gene for the p49/p100 subunit of NFKB (NFKB2) to chromosome 10 by Southern blot analysis of panels of human/Chinese hamster cell hybrids. By fluorescence in situ hybridization (FISH), they confirmed the localization and mapped the gene with greater resolution to 10q24. NFKB2 appears to be the same as LYT10.

Chromosomal alterations affecting band 10q24 are recurrently associated with lymphoid malignancies. For example, in about 7% of cases of T-cell acute lymphoblastic leukemia, the t(10;14)(q24;q11) translocation juxtaposes the T-cell receptor alpha/delta chain locus (186880; 186810) at 14q11 with chromosome 10q24. In 7% of cases of low grade B-cell non-Hodgkin lymphoma (B-NHL) and less frequently in intermediate and high grade B-NHL, band 10q24 is involved in heterogeneous aberrations including deletions and translocations with the immunoglobulin heavy chain locus.

Neri et al. (1991) demonstrated that the B-cell lymphoma-associated chromosomal translocation, t(10;14)(q24;q32), juxtaposes the immunoglobulin alpha-1 constant region gene (146900) to a novel gene, which they symbolized LYT10. The normal LYT10 cDNA encodes a 98-kD protein that displays N-terminal homology with the rel (DNA-binding) domain of the NF-kappa(B)-rel family of transcription factors and carboxy-terminal homology with the NF-kappa(B) p50 precursor protein, including the putative proteolytic cleavage domain (poly-G) and the ankyrin-like repeat domains. Chromosomal translocation generated a fusion LYT10-IGHA1 transcriptional unit. Neri et al. (1991) found the LYT10 gene rearrangement in 1 of 2 cases with cytogenetic abnormality at band 10q24, and in 1 case of non-Hodgkin lymphoma and 1 case of chronic lymphocytic leukemia in which cytogenetic data were not available. By FISH, Mathew et al. (1993) confirmed the localization of LYT10 that had been inferred from its isolation from a t(10;14)(q24;q32) translocation.

Liptay, S.; Schmid, R. M.; Perkins, N. D.; Meltzer, P.; Altherr, M. R.; McPherson, J. D.; Wasmuth, J. J.; Nabel, G. J.: Related subunits of NF-kappa-B map to two distinct loci associated with translocations in leukemia, NFKB1 and NFKB2. Genomics 13: 287-292, 1992.
Mathew, S.; Murty, V. V. V. S.; Dalla-Favera, R.; Chaganti, R. S. K.: Chromosomal localization of genes encoding the transcription factors, c-rel, NF-kappa-Bp50, NF-kappa-Bp65, and lyt-10 by fluorescence in situ hybridization. Oncogene 8: 191-193, 1993.
Neri, A.; Chang, C.-C.; Lombardi, L.; Salina, M.; Corradini, P.; Maiolo, A. T.; Chaganti, R. S. K.; Dalla-Favera, R.: B cell lymphoma-associated chromosomal translocation involves candidate oncogene lyt-10, homologous to NF-kappa-B p50. Cell 67: 1075-1087, 1991.

TRANSLOCATION 3;4 (MEMBER H OF THE RAS HOMOLOG GENE FAMILY) (*602037)

Non-Hodgkin lymphoma (NHL) is associated with several translocations between the LAZ3 (109565) oncogene at 3q27 and other unrelated genes. Dallery et al. (1995) stated that translocations involving band 3q27 and the immunoglobulin (Ig) genes are the third most common specific abnormality associated with NHL. LAZ3 rearrangements also occur in several NHLs without the involvement of Ig genes, as in the translocation t(3;4)(q27;p11) observed in the B-cell line VAL. Dallery et al. (1995) demonstrated that the consequence of this translocation is the expression of a chimeric transcript of LAZ3 and a new gene, which they termed TTF for 'translocation three four,' encoding a small G-like protein. Analysis of a full-length TTF cDNA revealed that the gene encodes a 191-amino acid polypeptide with similarity to members of the Ras superfamily (see 139150). The highest degree of identity was found with the Rho-like CDC42 protein (116952). Dallery et al. (1995) reported that the sequence contains residues normally associated with binding and hydrolysis of guanosine triphosphate. Northern blot analysis showed that TTF is expressed in hemopoietic cell lines only. Dallery-Prudhomme et al. (1997) examined the genomic organization of the RhoH/TTF gene and demonstrated that it is encoded on 2 exons spanning 35 kb of the genome. The entire coding region resides on the second exon. Dallery-Prudhomme et al. (1997) used fluorescence in situ hybridization to map the gene to chromosome 4p13. See also ARHA (165390).

Dallery, E.; Galiegue-Zouitina, S.; Collyn-d'Hooghe, M.; Quief, S.; Denis, C.; Hildebrand, M.-P.; Lantoine, D.; Deweindt, C.; Tilly, H.; Bastard, C.; Kerckaert, J.-P.: TTF, a gene encoding a novel small G protein, fuses to the lymphoma-associated LAZ3 gene by t(3;4) chromosomal translocation. Oncogene 10: 2171-2178, 1995.
Dallery-Prudhomme, E.; Roumier, C.; Denis, C.; Preudhomme, C.; Kerckaert, J.-P.; Galiegue-Zouitina, S.: Genomic structure and assignment of the RhoH/TTF small GTPase gene (ARHH) to 4p13 by in situ hybridization. Genomics 43: 89-94, 1997.

FOLLICULAR-VARIANT-TRANSLOCATION GENE (*136440)

The variant t(2;18) and t(18;22) chromosome translocations observed in B-cell chronic lymphocytic leukemias and in follicular lymphomas consistently involve the 5-prime region of the BCL-2 gene (151430) on chromosome 18 and various regions of the Ig light chain loci, either kappa (147200) on chromosome 2 or lambda (147220) on chromosome 22. These are variant translocations in these disorders because the finding in a large proportion of non-Hodgkin B-cell lymphomas, usually of follicular type, is the t(14;18)(q32;q21) translocation. Rimokh et al. (1993) showed that a variant t(2;18) translocation observed in a case of follicular lymphoma led to the juxtaposition of a J-kappa segment to a chromosome 18 transcriptional unit located 10 kb upstream of the BCL2 locus. The cDNA of this evolutionarily conserved gene, termed FVT1 for follicular-variant-translocation gene, coded for a putatively secreted protein of 36 kd that was not homologous with any described protein. It was weakly expressed in all the analyzed normal hematopoietic tissues but showed a very high rate of transcription in some T-cell malignancies and in phytohemagglutinin-stimulated lymphocytes. Rimokh et al. (1993) suggested that in light of the proximity of FVT1 to BCL2, both genes may participate in the tumoral process.

Rimokh, R.; Gadoux, M.; Bertheas, M.-F.; Berger, F.; Garoscio, M.; Deleage, G.; Germain, D.; Magaud, J.-P.: FVT-1, a novel human transcription unit affected by variant translocation t(2;18)(p11;q21) of follicular lymphoma. Blood 81: 136-142, 1993.

T-CELL ACUTE LYMPHOCYTIC LEUKEMIA 1 (*187040)

Finger et al. (1989) analyzed a t(1;14)(p32;q11) chromosomal translocation in a lymphohemopoietic stem cell line derived from a patient with acute T-lymphoblastic leukemia (Kurtzberg et al., 1985). The chromosomal joining of 14 to 1p occurred at the T-cell receptor delta diversity (D-delta-2) segment, and the reciprocal joining on chromosome 14 occurred at the T-cell delta diversity segment D-delta-1. Involvement of delta diversity segments at the translocation junctions suggested that the translocation occurred during an attempt at delta-1/delta-2 joining in a stem cell. Finger et al. (1989) found that the segment of chromosome 1 at band p32, adjacent to the chromosomal breakpoint, encodes a transcriptional unit designated TCL5. Finger et al. (1989) also demonstrated a rearrangement of the TCL5 locus in a human melanoma cell line carrying a deletion at 1p32.

The occurrence of 'biphenotypic' leukemias with lymphoid and myeloid characteristics and evidence of stem cell origin of myeloid, erythroid, megakaryocytic, and lymphoid lineages in chronic myeloid leukemia suggested that leukemias may arise from pluripotent hematopoietic cells. Begley et al. (1989) studied a leukemic stem cell line that was capable of differentiating into either myeloid or lymphoid cells and that carried a translocation between chromosomes 1 and 14, t(1;14)(p33;q11). By means of molecular cloning and sequencing, they showed that as a consequence of the translocation an unusual fusion transcript was generated. The chromosome 1 region involved in the breakpoint was the site of transcriptional activity apparently occurring only in hematopoietic tissues. Begley et al. (1989) concluded that the translocation may identify a gene located on chromosome 1 which is important for hematopoietic development and oncogenesis. They suggested the designation SCL (stem cell leukemia hematopoietic transcription factor).

The discrepancy in location in the findings of Begley et al. (1989) and Finger et al. (1989) resulted from a difference in interpretation of the structural change in the chromosomes. Finger et al. (1989) stated that TCL5 is proximal to LMYC, which supports the location in 1p32 rather than 1p33. Referring to the gene as SLC, Begley et al. (1991) isolated murine SCL cDNA and found that the nucleotide sequence of the coding region shows 88% identity to the human SCL gene; the amino acid sequence is 94% identical. The helix-loop-helix motif and upstream hydrophilic regions are entirely conserved in the murine and human proteins. In the murine SCL cDNA, a long open reading frame begins with the sequence CCCAGGATGA, which is in agreement with the Kozak consensus sequence for translational initiation. (Kozak (1987) compiled 5-prime-noncoding sequences in 699 vertebrate mRNAs and found that (GCC)GCCACCATGG emerged as the consensus sequence for initiation of translation in vertebrates. A most highly conserved position in that motif is the purine in position

Aplan, P. D.; Lombardi, D. P.; Reaman, G. H.; Sather, H. N.; Hammond, G. D.; Kirsch, I. R.: Involvement of the putative hematopoietic transcription factor SCL in T-cell acute lymphoblastic leukemia. Blood 79: 1327-1333, 1992.
Begley, C. G.; Aplan, P. D.; Davey, M. P.; Nakahara, K.; Tchorz, K.; Kurtzberg, J.; Hershfield, M. S.; Haynes, B. F.; Cohen, D. I.; Waldmann, T. A.; Kirsch, I. R.: Chromosomal translocation in a human leukemic stem-cell line disrupts the T-cell antigen receptor delta-chain diversity region and results in a previously unreported fusion transcript. Proc. Nat. Acad. Sci. 86: 2031-2035, 1989.
Begley, C. G.; Visvader, J.; Green, A. R.; Aplan, P. D.; Metcalf, D.; Kirsch, I. R.; Gough, N. M.: Molecular cloning and chromosomal localization of the murine homolog of the human helix-loop-helix gene SCL. Proc. Nat. Acad. Sci. 88: 869-873, 1991.
Finger, L. R.; Kagan, J.; Christopher, G.; Kurtzberg, J.; Hershfield, M. S.; Nowell, P. C.; Croce, C. M.: Involvement of the TCL5 gene on human chromosome 1 in T-cell leukemia and melanoma. Proc. Nat. Acad. Sci. 86: 5039-5043, 1989.
Kozak, M.: An analysis of 5-prime-noncoding sequences from 699 vertebrate messenger RNAs. Nucleic Acids Res. 15: 8125-8148, 1987.
Kurtzberg, J.; Bigner, S. H.; Hershfield, M. S.: Establishment of the DU.528 human lymphohemopoietic stem cell line. J. Exp. Med. 162: 1561-1578, 1985.
Robb, L.; Lyons, I.; Li, R.; Hartley, L.; Kontgen, F.; Harvey, R. P.; Metcalf, D.; Begley, C. G.: Absence of yolk sac hematopoiesis from mice with a targeted disruption of the scl gene. Proc. Nat. Acad. Sci. 92: 7075-7079, 1995.
Shivdasani, R. A.; Mayer, E. L.; Orkin, S. H.: Absence of blood formation in mice lacking the T-cell leukaemia oncoprotein tal-1/SCL. Nature 373: 432-434, 1995.
Xia, Y.; Brown, L.; Yang, C. Y.-C.; Tsou Tsan, J.; Siciliano, M. J.; Espinosa, R., III; Le Beau, M. M.; Baer, R. J.: TAL2, a helix-loop-helix gene activated by the (7;9)(q34;q32) translocation in human T-cell leukemia. Proc. Nat. Acad. Sci. 88: 11416-11420, 1991.

T-CELL ACUTE LYMPHOBLASTIC LEUKEMIA 2 (*186855)

Tumor-specific alteration of the TAL1 gene (187040) occurs in almost 25% of patients with T-cell acute lymphoblastic leukemia (T-ALL). The TAL1 gene product is a helix-loop-helix (HLH) protein which shares 87% amino acid sequence identity with LYL1 (151440), another HLH protein involved in T-cell leukemogenesis. Xia et al. (1991) identified a distinct gene, called TAL2, on the basis of its sequence homology with TAL1. They demonstrated that TAL2 is located 33 kb from the chromosome 9 breakpoint of t(7;9)(q34;q32), a recurring translocation specifically associated with T-ALL. The translocation juxtaposes TAL2 with sequences from the T-cell receptor beta-chain gene (TCRB; 186930) on chromosome 7. The TAL2 gene product includes an HLH protein dimerization domain and a DNA-binding domain that are homologous to those encoded by the TAL1 and LYL1 protooncogenes. The chromosomal localization of TAL2 was performed by using a genomic DNA fragment as a probe in Southern hybridization with DNAs from a panel of human/hamster somatic cell hybrids. The regional localization was determined by fluorescence in situ hybridization. Tycko et al. (1989) had previously characterized t(7;9) at the molecular level in 3 patients with T-ALL and found that the chromosome 9 breakpoints were tightly clustered within a 31-bp sequence. Xia et al. (1991) demonstrated direct linkage between TAL2 and the t(7;9) breakpoint region by chromosome walking.

By interspecific backcross linkage analysis, Pilz et al. (1995) mapped the Tal2 gene to mouse chromosome 4.

Pilz, A.; Woodward, K.; Povey, S.; Abbott, C.: Comparative mapping of 50 human chromosome 9 loci in the laboratory mouse. Genomics 25: 139-149, 1995.
Tycko, B.; Reynolds, T. C.; Smith, S. D.; Sklar, J.: Consistent breakage between consensus recombinase heptamers of chromosome 9 DNA in a recurrent chromosomal translocation of human T cell leukemia. J. Exp. Med. 169: 369-377, 1989.
Xia, Y.; Brown, L.; Yang, C. Y.-C.; Tsou Tsan, J.; Siciliano, M. J.; Espinosa, R., III; Le Beau, M. M.; Baer, R. J.: TAL2, a helix-loop-helix gene activated by the (7;9)(q34;q32) translocation in human T-cell leukemia. Proc. Nat. Acad. Sci. 88: 11416-11420, 1991.

LEUKEMIA, LYMPHOID 1 (*151440)

Cleary et al. (1988) isolated DNA spanning a t(7;19) chromosomal translocation breakpoint from a human T-cell line established from an acute lymphoblastic leukemia. Nucleotide sequence analysis showed that the point of crossover on chromosome 7 occurred immediately adjacent to a joining segment within the TCRB gene (186930), suggesting that this translocation resulted from an error in TCR gene rearrangement. On chromosome 19, the translocation occurred within a previously uncharacterized transcriptional unit for which Cleary et al. (1988) proposed the designation LYL1. An RNA of about 1.5 kb was transcribed from this gene in a wide variety of hematolymphoid cell lines. The t(7;19) resulted in truncation of the LYL1 gene and production of abnormal-sized RNAs, suggesting a role for LYL1 in the pathogenesis of this leukemia. By fluorescence in situ hybridization, Trask et al. (1993) assigned the LYL1 gene to 19p13.2-p13.1. Kuo et al. (1991) mapped the mouse Lyl-1 gene to chromosome 8, thereby defining a new region of homology of synteny with human 19p. The mapping was achieved in an interspecific backcross using a RFLP. The predicted mouse Lyl-1 protein was 78% identical to human LYL1.

Cleary, M. L.; Mellentin, J. D.; Spies, J.; Smith, S. D.: Chromosomal translocation involving the beta T cell receptor gene in acute leukemia. J. Exp. Med. 167: 682-687, 1988.
Kuo, S. S.; Mellentin, J. D.; Copeland, N. G.; Gilbert, D. J.; Jenkins, N. A.; Cleary, M. L.: Structure, chromosome mapping, and expression of the mouse Lyl-1 gene. Oncogene 6: 961-968, 1991.
Trask, B.; Fertitta, A.; Christensen, M.; Youngblom, J.; Bergmann, A.; Copeland, A.; de Jong, P.; Mohrenweiser, H.; Olsen, A.; Carrano, A.; Tynan, K.: Fluorescence in situ hybridization mapping of human chromosome 19: cytogenetic band location of 540 cosmids and 70 genes or DNA markers. Genomics 15: 133-145, 1993.

T-CELL ANTIGEN RECEPTOR, BETA SUBUNIT (*186930)

See 186880. The gene cluster for the beta subunit of T-cell antigen receptor is on chromosome 7 in man and on chromosome 6, near the immunoglobulin kappa light chain, in the mouse—an example of nonhomology of synteny (Caccia et al., 1984; Lee et al., 1984). Barker et al. (1984), using a cDNA probe in the study of mouse-human somatic cell hybrids, assigned the structural gene to chromosome 7. Collins et al. (1984) assigned the TCRB locus to the region 7q22-7qter. Prerequisite to beta-chain gene expression are rearrangements of variable (V), diversity (D), and joining (J) region elements into a transcriptional unit completed by the coding exons of the constant (C) region. Isobe et al. (1985) assigned TCRB to 7q35 by in situ hybridization. They pointed out that this region is unusually prone to develop breaks in vivo and suggested that this may reflect instability generated by somatic rearrangement of T-cell receptor genes during normal differentiation in this cell lineage. Morton et al. (1985), also by in situ hybridization, put TCRB at or near 7q32. Related sequences were found on 7p15-7p21. The authors pointed to the chromosomal rearrangements that have been found at or near these sites in patients with ataxia-telangiectasia (208900) and other disorders. The possibility that the 7p localization found by Caccia et al. (1984) and by Morton et al. (1985) represents the site of TCRG (186970) is worthy of consideration. Ikuta et al. (1985) presented evidence that allelic exclusion occurs in the T-cell receptor gene, i.e., only one of the homologs is active.

Robinson and Kindt (1985) demonstrated RFLPs of the genes encoding the constant region of the T-cell antigen receptor. O'Connor et al. (1985) found rearrangement of the gene coding for the beta-chain of the T-cell receptor in all 6 T-cell leukemias and in 16 of 19 T-cell lymphomas; in all cases the immunoglobulin genes were in the germline arrangement. Contrariwise, all 36 B-cell leukemias and all 16 B-cell lymphomas showed rearranged immunoglobulin genes; the T-cell receptor genes were in germline configuration in all but 3.

Concannon et al. (1986) concluded that the repertoire of expressed human V(beta) genes is in excess of 59, apparently much smaller than the variable regions of the heavy and light chains. To ascertain the extent and organization of the germline human TCRB gene repertoire, Robinson et al. (1993) mapped beta-chain variable region genes by pulsed field gel electrophoresis, cosmid cloning, and in situ hybridization. Probes derived from the 24 known V families mapped to 6 SfiI fragments, 4 of which were linked to the TCRB gene complex on chromosome 7q35. The other 2 fragments, containing about 10% of the known V gene segments, were localized to a cluster of nonfunctional orphon gene segments located on chromosome 9. (Immunoglobulin V genes have also been found to be encoded outside of the immunoglobulin gene complexes (e.g., 147185) and have been designated orphon genes. This term combines the words orphan and exon.)

Rowen et al. (1996) sequenced the TCRB 'locus' on chromosome 7, which comprises a complex family of genes. The locus was found to contain 2 types of coding elements: TCR elements (65 variable gene segments and 2 clusters of diversity, joining, and constant segments) and 8 trypsinogen genes (e.g., 276000). These coding elements constitute 4.6% of the DNA of the locus. The total sequence was 685 kb in length. Rowen et al. (1996) characterized the organization and structures of all of the TCRB elements; 19 pseudogenes and 22 relics could be readily differentiated from the 46 functional genes by computational approaches. Their identification of 8 trypsinogen genes raised a question as to whether their association arose from functional or regulatory constraints, or is inadvertent. They found that a portion of the TCRB locus has been duplicated and translocated from chromosome 7 to chromosome 9, which suggested a possible mechanism for the creation of new multigene families. The extent of sequence variation (polymorphism) within the locus is high. At least 30% of the TCRB locus is composed of genome-wide interspersed repeats, but these do not appear to facilitate the duplication of locus-specific repeats. Eight locus-specific repeats (homology units) have duplicated in a tandem or dispersed manner. Rowen et al. (1996) claimed that the 685-kb TCRB locus was the longest contiguous stretch of DNA analyzed to that time in humans.

Hockertz et al. (1998) studied the association of a gene linked to the TCRB variable region with cases of relapsing-progressive (RP) and relapsing-remitting (RR) multiple sclerosis (MS; 126200). They used a contingency-table test of patient data and affected family-based controls. Control alleles and haplotypes were composed of parental marker alleles and haplotypes not transmitted to the affected child in 90 simplex and 31 multiplex British Columbian families. In HLA class II DRw15(+) MS patients, the authors found that susceptibility to RP MS, but not RR MS, is associated with the recessive inheritance of a gene linked to the TCRB variable region. In DRw15(-) MS patients, they showed that susceptibility to the progressive form of the disease is associated with another gene linked to the TCRB variable region. They commented that the results of this study are supported by findings in 2 animal models of MS related to changes in the TCRB complex.

See also Aisenberg et al. (1985); Bertness et al. (1985); Born et al. (1985); Duby et al. (1985); Hedrick et al. (1985); Jones et al. (1985); Malissen et al. (1986); Patten et al. (1984); Tillinghast et al. (1986); Toyonaga et al. (1985); Trowbridge et al. (1985); Tunnacliffe et al. (1985); Waldmann et al. (1985); Weiss et al. (1985)

Aisenberg, A. C.; Krontiris, T. G.; Mak, T. W.; Wilkes, B. M.: Rearrangement of the gene for the beta chain of the T-cell receptor in T-cell chronic lymphocytic leukemia and related disorders. New Eng. J. Med. 313: 529-533, 1985.

Barker, P. E.; Ruddle, F. H.; Royer, H. D.; Reinherz, E. L.: Chromosomal location of the beta chain gene of the human T-cell receptor complex. (Abstract) Am. J. Hum. Genet. 36: 132S only, 1984.

Bertness, V.; Kirsch, I.; Hollis, G.; Johnson, B.; Bunn, P. A., Jr.: T-cell receptor gene rearrangements as clinical markers of human T-cell lymphomas. New Eng. J. Med. 313: 534-538, 1985.

Born, W.; Yague, J.; Palmer, E.; Kappler, J.; Marrack, P.: Rearrangement of T-cell receptor beta-chain genes during T-cell development. Proc. Nat. Acad. Sci. 82: 2925-2929, 1985.

Caccia, N.; Kronenberg, M.; Saxe, D.; Haars, R.; Bruns, G. A. P.; Goverman, J.; Malissen, M.; Willard, H.; Yoshikai, Y.; Simon, M.; Hood, L.; Mak, T. W.: The T cell receptor beta chain genes are located on chromosome 6 in mice and chromosome 7 in humans. Cell 37: 1091-1099, 1984.

Collins, M. K. L.; Goodfellow, P. N.; Dunne, M. J.; Spurr, N. K.; Solomon, E.; Owen, M. J.: A human T-cell antigen receptor beta chain gene maps to chromosome 7. EMBO J. 3: 2347-2349, 1984.

Concannon, P.; Pickering, L. A.; Kung, P.; Hood, L.: Diversity and structure of human T-cell receptor beta-chain variable region genes. Proc. Nat. Acad. Sci. 83: 6598-6602, 1986.

Duby, A. D.; Klein, K. A.; Murre, C.; Seidman, J. G.: A novel mechanism of somatic rearrangement predicted by a human T-cell antigen receptor beta-chain complementary DNA. Science 228: 1204-1206, 1985.

Hedrick, S. M.; Germain, R. N.; Bevan, M. J.; Dorf, M.; Engel, I.; Fink, P.; Gascoigne, N.; Heber-Katz, E.; Kapp, J.; Kaufmann, Y.; Kaye, J.; Melchers, F.; Pierce, C.; Schwartz, R. H.; Sorensen, C.; Taniguchi, M.; Davis, M. M.: Rearrangement and transcription of a T-cell receptor beta-chain gene in different T-cell subsets. Proc. Nat. Acad. Sci. 82: 531-535, 1985.

Hockertz, M. K.; Paty, D. W.; Beall, S. S.: Susceptibility to relapsing-progressive multiple sclerosis is associated with inheritance of genes linked to the variable region of the TcR beta locus: use of affected family-based controls. Am. J. Hum. Genet. 62: 373-385, 1998.

Ikuta, K.; Ogura, T.; Shimizu, A.; Honjo, T.: Low frequency of somatic mutation in beta-chain variable region genes of human T-cell receptors. Proc. Nat. Acad. Sci. 82: 7701-7705, 1985.

Isobe, M.; Erikson, J.; Emanuel, B. S.; Nowell, P. C.; Croce, C. M.: Location of gene for beta subunit of human T-cell receptor at band 7q35, a region prone to rearrangement in T cells. Science 228: 580-582, 1985.

Jones, N.; Leiden, J.; Dialynas, D.; Fraser, J.; Clabby, M.; Kishimoto, T.; Strominger, J. L.; Andrews, D.; Lane, W.; Woody, J.: Partial primary structure of the alpha and beta chains of human tumor T-cell receptors. Science 227: 311-314, 1985.

Lee, N. E.; D'Eustachio, P.; Pravtcheva, D.; Ruddle, F. H.; Hedrick, S. M.; Davis, M. M.: Murine T cell receptor beta chain is encoded on chromosome 6. J. Exp. Med. 160: 905-913, 1984.

Malissen, M.; McCoy, C.; Blanc, D.; Trucy, J.; Devaux, C.; Schmitt-Verhulst, A.-M.; Fitch, F.; Hood, L.; Malissen, B.: Direct evidence for chromosomal inversion during T-cell receptor beta-gene rearrangements. Nature 319: 28-33, 1986.

Morton, C. C.; Duby, A. D.; Eddy, R. L.; Shows, T. B.; Seidman, J. G.: Genes for beta chain of human T-cell antigen receptor map to regions of chromosomal rearrangements in T cells. Science 228: 582-585, 1985.

O'Connor, N. T. J.; Wainscoat, J. S.; Weatherall, D. J.; Gatter, K. C.; Feller, A. C.; Isaacson, P.; Jones, D.; Lennert, K.; Pallesen, G.; Ramsey, A.; Stein, H.; Wright, D. H.; Mason, D. Y.: Rearrangement of the T-cell-receptor beta-chain gene in the diagnosis of lymphoproliferative disorders. Lancet I: 1295-1297, 1985.

Patten, P.; Yokota, T.; Rothbard, J.; Chien, Y.-H.; Arai, K.-I.; Davis, M. M.: Structure, expression and divergence of T-cell receptor beta-chain variable regions. Nature 312: 40-46, 1984.

Robinson, M. A.; Kindt, T. J.: Segregation of polymorphic T-cell receptor genes in human families. Proc. Nat. Acad. Sci. 82: 3804-3808, 1985.

Robinson, M. A.; Mitchell, M. P.; Wei, S.; Day, C. E.; Zhao, T. M.; Concannon, P.: Organization of human T-cell receptor beta-chain genes: clusters of V-beta genes are present on chromosomes 7 and 9. Proc. Nat. Acad. Sci. 90: 2433-2437, 1993.

Rowen, L.; Koop, B. F.; Hood, L.: The complete 685-kilobase DNA sequence of the human beta T cell receptor locus. Science 272: 1755-1762, 1996.

Tillinghast, J. P.; Behlke, M. A.; Loh, D. Y.: Structure and diversity of the human T-cell receptor beta-chain variable region genes. Science 233: 879-883, 1986.

Toyonaga, B.; Yoshikai, Y.; Vadasz, V.; Chin, B.; Mak, T. W.: Organization and sequences of the diversity, joining, and constant region genes of the human T-cell receptor beta chain. Proc. Nat. Acad. Sci. 82: 8624-8628, 1985.

Trowbridge, I. S.; Lesley, J.; Trotter, J.; Hyman, R.: Thymocyte subpopulation enriched for progenitors with an unrearranged T-cell receptor beta-chain gene. Nature 315: 666-669, 1985.

Tunnacliffe, A.; Kefford, R.; Milstein, C.; Forster, A.; Rabbitts, T. H.: Sequence and evolution of the human T-cell antigen receptor beta-chain genes. Proc. Nat. Acad. Sci. 82: 5068-5072, 1985.

Waldmann, T. A.; Davis, M. M.; Bongiovanni, K. F.; Korsmeyer, S. J.: Rearrangements of genes for the antigen receptor on T cells as markers of lineage and clonality in human lymphoid neoplasms. New Eng. J. Med. 313: 776-783, 1985.

Weiss, L. M.; Hu, E.; Wood, G. S.; Moulds, C.; Cleary, M. L.; Warnke, R.; Sklar, J.: Clonal rearrangements of T-cell receptor genes in mycosis fungoides and dermatopathic lymphadenopathy. New Eng. J. Med. 313: 539-544, 1985.

T-CELL ANTIGEN RECEPTOR, DELTA SUBUNIT (*186810)

Two distinct types of T-cell antigen receptors have been identified: the alpha/beta heterodimer found on functional helper and cytotoxic T cells, and the gamma/delta heterodimer. The latter is first detected approximately 2 days before the appearance of cell-surface alpha/beta heterodimer during T-cell ontogeny. In adult thymus it is found mainly in the least mature cells. Chien et al. (1987) found a new T-cell receptor gene, located just 5-prime to the J-C coding regions of the gene for the alpha chain (TCRA; 186880). The gene shows systematic rearrangement in early thymocytes and appears to use V-gene segments of the TCR alpha-chain family, well before any C(alpha) message of protein is detected. RNA from this locus is expressed at a high level in early thymocytes and adult 'double negative' cells but not in mature T-cell populations. The findings suggest a novel mechanism for the regulation of TCR gene expression during ontogeny, as the new locus was found to be deleted when rearrangement occurred between the V(alpha) and J(alpha) segments. Because the new gene is expressed in cell types that also express gamma-chain messenger RNA (TCRG; 186970) and surface gamma/delta heterodimers, Chien et al. (1987) concluded that the TCR gene encodes the delta-chain molecule. Hata et al. (1987) and Band et al. (1987) identified cDNA clones coding for the delta subunit. The variable region of the T-cell receptor delta-chain can be assembled from as many as 4 distinct gene segments, more than any other antigen-receptor gene (Chien et al., 1987). The TCRD and TCRG loci appear to have coordinated control of rearrangement and expression. See also Loh et al. (1987), Born et al. (1987), and Bonyhadi et al. (1987).

Takihara et al. (1988) described the genomic organization and sequence of the human T-cell receptor delta-chain diversity, joining, and constant genes. The delta-chain constant region is composed of 4 exons, similar in organization to that of the C(alpha) exons; there are 3 joining segments and 2 functional diversity segments. Boehm et al. (1988) described a translocation at the T-cell receptor joining J-delta segment in a human acute T-cell leukemia with a gross chromosomal translocation t(11;14)(p15;q11). The authors hypothesized that the translocation occurred as a result of recombinase error, since the DNA sequence at the chromosome 11 junction is homologous to a recombinase signal sequence. By taking advantage of 'chromosomal walking' techniques, Isobe et al. (1988) obtained clones that encompassed the TCRD gene. They showed that TCRD is located over 80 kb 5-prime of the constant region of TCRA (186880). TCRD was rearranged in a gamma/delta-expressing T-cell line and was deleted in alpha/beta-expressing T-cell lines. Further evidence that the TCRD gene is located in 14q11 came from analysis of the translocation t(8;14)(q24;q11) from a T-cell leukemia: the J(delta) segment was rearranged in cells derived from this tumor and was probably involved in the translocation.

The location of the TCRD gene, which has variable, D, J, and C regions, is curious: it lies in the midst of the TCRA gene. In a 120-kb segment, spanning from V(alpha) at the 5-prime end to C(alpha) at the 3-prime end, these multiple genes or gene units occur in the following sequence: V(alpha)—V(delta)—D(delta)—J(delta)—C(delta)—J(alpha)—C(alpha). Dube et al. (1988) studied the illegitimate recombination that takes place in t(10;14)(q24;q11) in T-cell acute lymphatic leukemia. They identified the first and second D units of the delta gene on the derivative chromosome 10. Hockett et al. (1988) likewise defined the relationship of the TCRD gene to the TCRA gene. Loh et al. (1988) concluded that the TCRD gene may have relatively limited variable-region diversity but a large junctional-region diversity. Lacy et al. (1989) also analyzed this diversity. Begley et al. (1989) studied a translocation between chromosomes 1 and 14, t(1;14)(p33;q11), in a leukemic stem cell line, DU.528. The translocation resulted in a deletion between the delta chain diversity (D) region genes D(delta1) and D(delta2). The authors described an unusual fusion transcript derived from both chromosomes 1 and 14. They hypothesized that the gene on the chromosome 1 side of the translocation is a locus important for hematopoietic development and oncogenesis, for which they proposed the name SCL (stem cell leukemia); see 187040. Redondo et al. (1990) demonstrated that the transcriptional control of the TCRD and TCRA genes is distinct even though the former gene is located within the latter.

See also Chien et al. (1987)

Band, H.; Hochstenbach, F.; McLean, J.; Hata, S.; Krangel, M. S.; Brenner, M. B.: Immunochemical proof that a novel rearranging gene encodes the T cell receptor delta subunit. Science 238: 682-684, 1987.

Begley, C. G.; Aplan, P. D.; Davey, M. P.; Nakahara, K.; Tchorz, K.; Kurtzberg, J.; Hershfield, M. S.; Haynes, B. F.; Cohen, D. I.; Waldmann, T. A.; Kirsch, I. R.: Chromosomal translocation in a human leukemic stem-cell line disrupts the T-cell antigen receptor delta-chain diversity region and results in a previously unreported fusion transcript. Proc. Nat. Acad. Sci. 86: 2031-2035, 1989.

Boehm, T.; Baer, R.; Lavenir, I.; Forster, A.; Waters, J. J.; Nacheva, E.; Rabbitts, T. H.: The mechanism of chromosomal translocation t(11;14) involving the T-cell receptor C-delta locus on human chromosome 14q11 and a transcribed region of chromosome 11p15. EMBO J. 7: 385-394, 1988.

Bonyhadi, M.; Weiss, A.; Tucker, P. W.; Tigelaar, R. E.; Allison, J. P.: Delta is the C(x)-gene product in the gamma/delta antigen receptor of dendritic epidermal cells. Nature 330: 574-576, 1987.

Born, W.; Miles, C.; White, J.; O'Brien, R.; Freed, J. H.; Marrack, P.; Kappler, J.; Kubo, R. T.: Peptide sequences of T-cell receptor delta and gamma chains are identical to predicted X and gamma proteins. Nature 330: 572-574, 1987.

Chien, Y.; Iwashima, M.; Kaplan, K. B.; Elliott, J. F.; Davis, M. M.: A new T-cell receptor gene located within the alpha locus and expressed early in T-cell differentiation. Nature 327: 677-682, 1987.

Chien, Y.; Iwashima, M.; Wettstein, D. A.; Kaplan, K. B.; Elliott, J. F.; Born, W.; Davis, M. M.: T-cell receptor delta gene rearrangements in early thymocytes. Nature 330: 722-727, 1987.

Dube, I. D.; Lu, M.; Minden, M. D.; Raimondi, S. C.; Sutherland, P. M.; Beckett, T. A.: Breakpoints occur within the delta-chain locus of the T-cell receptor gene in the t(10;14)(q24;q11) of T-cell neoplasia. (Abstract) Am. J. Hum. Genet. 43: A23 only, 1988.

Hata, S.; Brenner, M. B.; Krangel, M. S.: Identification of putative human T cell receptor delta complementary DNA clones. Science 238: 678-682, 1987.

Hockett, R. D.; de Villartay, J.-P.; Pollock, K.; Poplack, D. G.; Cohen, D. I.; Korsmeyer, S. J.: Human T-cell antigen receptor (TCR) delta-chain locus and elements responsible for its deletion are within the TCR alpha-chain locus. Proc. Nat. Acad. Sci. 85: 9694-9698, 1988.

Isobe, M.; Russo, G.; Haluska, F. G.; Croce, C. M.: Cloning of the gene encoding the delta-subunit of the human T-cell receptor reveals its physical organization within the alpha-subunit locus and its involvement in chromosome translocations in T-cell malignancy. Proc. Nat. Acad. Sci. 85: 3933-3937, 1988.

Lacy, M. J.; McNeil, L. K.; Roth, M. E.; Kranz, D. M.: T-cell receptor delta-chain diversity in peripheral lymphocytes. Proc. Nat. Acad. Sci. 86: 1023-1026, 1989.

Loh, E. Y.; Cwirla, S.; Serafini, A. T.; Phillips, J. H.; Lanier, L. L.: Human T-cell receptor delta chain: genomic organization, diversity, and expression in populations of cells. Proc. Nat. Acad. Sci. 85: 9714-9718, 1988.

Loh, E. Y.; Lanier, L. L.; Turck, C. W.; Littman, D. R.; Davis, M. M.; Chien, Y.-H.; Weiss, A.: Identification and sequence of a fourth human T cell antigen receptor chain. Nature 330: 569-572, 1987.

Redondo, J. M.; Hata, S.; Brocklehurst, C.; Krangel, M. S.: A T cell-specific transcriptional enhancer within the human T cell receptor delta locus. Science 247: 1225-1229, 1990.

Takihara, Y.; Tkachuk, D.; Michalopoulos, E.; Champagne, E.; Reimann, J.; Minden, M.; Mak, T. W.: Sequence and organization of the diversity, joining, and constant region genes of the human T-cell delta-chain locus. Proc. Nat. Acad. Sci. 85: 6097-6101, 1988.

LIM DOMAIN ONLY 1 (*186921)

Greenberg et al. (1989) studied a previously identified translocation t(11;14)(p15;q11) associated with T cells in acute lymphoblastic leukemia (Boehm et al., 1988). The translocation involves the TCR-delta (186810) gene on 14q11 and also a transcribed sequence on 11p15. They found that the gene on 11p15 is normally expressed in rhombomeres of the developing hindbrain. The gene product, designated rhombosine or rhombotin, has 155 amino acids and no homology to any previously described protein. Henry et al. (1989) determined the position of TAL1 relative to several other genes in that region. Chen et al. (1990) concluded that the TAL1 gene encodes a helix-loop-helix protein with homology to LYL1 (151440), which is also involved in malignant development of lymphocytes. (According to Aplan et al. (1990), the designation tal-1 is also used as a synonym for tcl-5 or scl (187040).) Because human and mouse rhombotin are virtually identical in derived protein sequence, rendering it difficult to identify functionally important regions or motifs, Boehm et al. (1990) isolated and compared rhombotin sequences from disparate species. They showed a sequence that is highly homologous to that of human and mouse rhombotin in Drosophila DNA. This comparison and others suggested that rhombotin, which lacks a DNA-binding homeo domain, may belong to a new class of transcriptional regulators that modulate transcription via intermolecular competitive binding to certain DNA-binding transcription factors. The structure of the predicted rhombotin protein suggests that it consists of 2 tandemly arranged cysteine-rich regions, which were subsequently identified in 3 transcription factors appended to putative DNA binding and activation domains, wherein they were termed LIM domains. The function of LIM domains may be in protein-protein interaction. LIM domains show homology to zinc-binding proteins and to iron-sulfur centers of ferredoxins. Consistent with its close structural homology to the h uman gene, the mouse homolog Ttg1 maps to chromosome 7, distal to the beta-globin locus (Angel et al., 1993).

Boehm et al. (1991) isolated 2 homologs of the rhombotin gene. Rhombotin-2 (RBTN2; 180385) is located close to the T-ALL breakpoint cluster region on 11p13. This was determined by mapping to chromosome 11 by hybridization to a somatic cell that contained only human chromosome 11 and thereafter by hybridizing the human rhombotin exon-2-specific probe to cosmid clones spanning about 100 kb around the T-ALL breakpoint cluster region. Rhom-2 in the mouse is similar to the human homolog. Boehm et al. (1991) identified a third rhombotin-like gene (RBTN3; 180386) and determined that it is not situated on chromosome 11. It shows, however, 93% homology with rhombotin-2 and also encodes a LIM domain protein. It is not known whether the RHOM3 gene is associated with chromosomal abnormalities in leukemia. The rhombotin gene on 11p15 is a consistent but rare feature of a subset of T-ALL.

Angel, J. M.; Moore, J. L.; Pelphrey, A.; Richie, E. R.: The mouse homolog of the rhombotin (Ttg-1) gene maps on chromosome 7 distal to the beta-globin (Hbb) locus. Mammalian Genome 4: 281-282, 1993.
Aplan, P. D.; Lombardi, D. P.; Ginsberg, A. M.; Cossman, J.; Bertness, V. L.; Kirsch, I. L.: Disruption of the human SCL locus by 'illegitimate' V-(D)-J recombinase activity. Science 250: 1426-1429, 1990.
Boehm, T.; Baer, R.; Lavenir, I.; Forster, A.; Waters, J. J.; Nacheva, E.; Rabbitts, T. H.: The mechanism of chromosomal translocation t(11;14) involving the T-cell receptor C-delta locus on human chromosome 14q11 and a transcribed region of chromosome 11p15. EMBO J. 7: 385-394, 1988.
Boehm, T.; Foroni, L.; Kaneko, Y.; Perutz, M. F.; Rabbitts, T. H.: The rhombotin family of cysteine-rich LIM-domain oncogenes: distinct members are involved in T-cell translocations to human chromosomes 11p15 and 11p13. Proc. Nat. Acad. Sci. 88: 4367-4371, 1991.
Boehm, T.; Foroni, L.; Kennedy, M.; Rabbitts, T. H.: The rhombotin gene belongs to a class of transcriptional regulators with a potential novel protein dimerisation motif. Oncogene 5: 1103-1105, 1990.
Chen, Q.; Cheng, J. T.; Tasi, L. H.; Schneider, N.; Buchanan, G.; Carroll, A.; Crist, W.; Ozanne, B.; Siciliano, M. J.; Baer, R.: The TAL gene undergoes chromosome translocation in T cell leukemia and potentially encodes a helix-loop-helix protein. EMBO J. 9: 415-424, 1990.
Greenberg, J.; Boehm, T.; Surane, A.; Keynes, R.; Sofronew, M.; Rabbitts, T. H.: A human T cell translocation involving chromosome 11p15 identifies a gene expressed in developing rhombomeres of the mouse hind brain. (Abstract) Cytogenet. Cell Genet. 51: 1008 only, 1989.
Henry, I.; Puech, A.; Antignac, C.; Couillin, P.; Jeanpierre, M.; Ahnine, L.; Barichard, F.; Boehm, T.; Augereau, P.; Scrable, H.; Rabbitts, T. H.; Rochefort, H.; Cavenee, W.; Junien, C.: Subregional mapping of BWS, CTSD, MYOD1, and a T-ALL breakpoint in 11p15. (Abstract) Cytogenet. Cell Genet. 51: 1013 only, 1989.

HOMOLOG OF DROSOPHILA NOTCH 1 (*190198)

Cytogenetic studies have shown that chromosome 7 band q34-q35, which contains the gene for the beta T-cell receptor (186930), is a common site for translocation in T-cell neoplasms. In a translocation t(7;9)(q34;q34.3) found in a case of acute T-cell lymphoblastic leukemia, Ellisen et al. (1991) found that the locus on chromosome 9 contained a gene highly homologous to the Drosophila gene 'Notch.' Transcripts of the human gene, for which they proposed the name TAN1 (an acronym for translocation-associated 'Notch' homolog), and its murine counterpart were demonstrated in many normal human fetal and adult mouse tissues but were most abundant in lymphoid tissues. In t(7;9)(q34;q34.3) translocations from 3 cases of acute T-cell lymphoblastic leukemia, Ellisen et al. (1991) found breakpoints within 100 bp of an intron in TAN1, resulting in truncation of TAN1 transcripts. They interpreted the observations to indicate that TAN1 is important for normal lymphocyte function and that alterations in TAN1 play a role in the pathogenesis of some T-cell neoplasms. Milner et al. (1994) found that at least 1 'Notch' homolog is expressed in human bone marrow CD34(+) cells, a population enriched for hematopoietic precursors. On the basis of these findings, they suggested that members of the 'Notch' family, including TAN1, may be involved in mediating cell-fate decisions during hematopoiesis.

Franco del Amo et al. (1993) and Pilz et al. (1994) demonstrated that the Notch-1 gene in the mouse is located on chromosome 2.

Larsson et al. (1994) indicated that NOTCH1 is the preferred symbol and provided information on the chromosomal location of the NOTCH2 (600275) and NOTCH3 (600276) genes.

Artavanis-Tsakonas et al. (1995) reviewed the Notch signaling pathway.

Axelrod et al. (1996) reported that the Drosophila Dishevelled gene (601225), which encodes a component of the Wingless (164820) signaling pathway, interacts antagonistically with Notch and one of its ligands, Delta. A direct physical interaction between Dishevelled and the Notch carboxyl terminus suggested to the authors that Dishevelled blocks Notch signaling directly and provides a molecular mechanism for the inhibitory crosstalk observed between the Notch and Wingless signaling pathways.

Artavanis-Tsakonas, S.; Matsuono, K.; Fortini, M.: Notch signaling. Science 268: 225-232, 1995.
Axelrod, J. D.; Matsuno, K.; Artavanis-Tsakonas, S.; Perrimon, N.: Interaction between Wingless and Notch signaling pathways mediated by Dishevelled. Science 271: 1826-1832, 1996.
Ellisen, L. W.; Bird, J.; West, D. C.; Soreng, A. L.; Reynolds, T. C.; Smith, S. D.; Sklar, J.: TAN-1, the human homolog of the Drosophila Notch gene, is broken by chromosomal translocations in T lymphoblastic neoplasms. Cell 66: 649-661, 1991.
Franco del Amo, F.; Gendron-Maguire, M.; Swiatek, P. J.; Jenkins, N. A.; Copeland, N. G.; Gridley, T.: Cloning, analysis, and chromosomal localization of Notch-1, a mouse homolog of Drosophila Notch. Genomics 15: 259-264, 1993.
Larsson, C.; Lardelli, M.; White, I.; Lendahl, U.: The human NOTCH1, 2, and 3 genes are located at chromosome positions 9q34, 1p13-p11, and 19p13.2-p13.1 in regions of neoplasia-associated translocation. Genomics 24: 253-258, 1994.
Milner, L. A.; Kopan, R.; Martin, D. I. K.; Bernstein, I. D.: A human homologue of the Drosophila developmental gene, Notch, is expressed in CD34+ hematopoietic precursors. Blood 83: 2057-2062, 1994.
Pilz, A.; Prohaska, R.; Peters, J.; Abbott, C.: Genetic linkage analysis of the Ak1, Col5a1, Epb7.2, Fpgs, Grp78, Pbx3, and Notch1 genes in the region of mouse chromosome 2 homologous to human chromosome 9q. Genomics 21: 104-109, 1994.

LYMPHOCYTE-SPECIFIC PROTEIN TYROSINE KINASE (*153390)

LCK is the designation for the gene encoding lymphocyte protein tyrosine kinase which has been identified in mouse, where it is coded by chromosome 4, and in man, where molecular analysis by somatic cell hybridization and in situ hybridization indicates that it is coded by the 1p35-p32 segment (Marth et al., 1986). LCK closely resembles other protein tyrosine kinase oncogenes, notably SRC (190090), YES (164880), and FGR (164940), and virus-induced murine lymphomas have been found to have rearrangement of the LCK gene. In man, chromosome abnormalities, deletions, or translocations are seen in 1p34 in neuroblastomas (Gilbert et al., 1982) and in non-Hodgkin lymphoma (Levine et al., 1985). Rearrangement of LCK may play a role in the lymphocyte malignancy, as may FGR (164940), formerly called SRC2, which is also on 1p in the same region but shows only about 70% homology with LCK. The NRAS (164790) oncogene is situated more proximally on 1p. In a patient with T-lymphoblastic leukemia (T-ALL) and a translocation t(1;7)(p34;q34) as the sole karyotypic abnormality, Tycko et al. (1991) found evidence of joining of the LCK gene with the T-cell antigen receptor, beta subunit gene (TCRB; 186930). The chromosomal breakpoint was located 2 kb upstream of the LCK upstream promoter and first nontranslated exon. In a second t(1;7) translocation, the breakpoint was found between 2 alternative LCK promoters, upstream of the second exon. Relative to normal thymus and activated T cells, levels of LCK mRNA were greatly elevated in the first case and moderately elevated in the second. Burnett et al. (1991) also studied the involvement of the LCK gene in a t(1;7) leukemia translocation. Volpi et al. (1994) mapped the LCK gene to 1p35-p34.3 by fluorescence in situ hybridization. By dual-color in situ hybridization with an LCK probe and a probe for D1S57, which was mapped to 1p34.3-p34.2 by Van Roy et al. (1993), they found that the LCK gene was telomeric to the anonymous marker .

During T-cell development, assembly and expression of a complete gene for the beta chain of the T-cell receptor blocks further rearrangement at the TCRB locus (a process referred to as allelic exclusion) and drives the generation and expansion of cells positive for CD4 and CD8. Studies in transgenic animals suggested that the lymphocyte-specific protein tyrosine kinase p56(lck) may impinge on the same pathway used by TCR-beta chains to deliver such signals. This hypothesis predicts that interference with p56(lck) function would mitigate the effects of a simultaneously expressed TCR-beta chain. Anderson et al. (1993) confirmed this prediction through examination of allelic exclusion in mice expressing both a functional TCRB transgene and a catalytically inactive form of p56(lck). Thus, the oncogene LCK is the protein tyrosine kinase that regulates allelic exclusion at the T-cell receptor beta locus.

Burnett et al. (1994) provided a molecular analysis of the t(1;7)(p34;q34) in T-cell acute lymphoblastic leukemia that fuses LCK and TCRB.

The T-cell antigen receptor plays a crucial role in thymocyte differentiation and T-cell activation. After antigen binding to the TCR, and in concert with engagement of other coreceptors and their associated ligands (such as CD4 (186940) and major histocompatibility complex (MHC) class II, CD28 (186760), B7 (112203), CD8 (186910), and MHC I), signal transduction cascades are activated. The earliest measurable biochemical event is the activation of protein tyrosine kinases (PTKs), resulting in the phosphorylation of multiple cellular substrates. At least 3 PTKs are known to be involved in TCR signaling at the level of the receptor, including p59(fyn) (FYN; 137025), p56(lck), and ZAP70 (176947). Individuals with mutations in the gene coding for the ZAP70 protein show severe combined immunodeficiency (SCID) due to selective T-cell defect, e.g., 176947.0001. Patients present with an absence of peripheral CD8(+) T cells and normal to high levels of peripheral CD4(+) T cells which are unable to signal through the TCR. SCID-like phenotypes have also been observed in mice rendered p56lck or p59(fyn) deficient by homologous recombination. Mice lacking p56(lck) as well as transgenic mice expressing dominant-negative mutations of p56(lck) demonstrate severe T-cell developmental defects (Molina et al., 1992; Levin et al., 1993).

SCID is a heterogeneous disorder characterized by profound defects in cellular and humoral immunity. Goldman et al. (1998) described an infant with clinical and laboratory features of SCID and selective CD4 lymphopenia and lack of CD28 expression on CD8(+) T cells. T cells from this patient showed poor blastogenic responses to various mitogens and to interleukin-2 (IL2; 147680). Other T-cell antigen receptor-induced responses, including upregulation of CD69 (107273), were similarly inhibited. However, more proximal T-cell antigen receptor signaling events, such as anti-CD3 induced protein tyrosine phosphorylation, phosphorylation of mitogen-associated protein kinase, and calcium mobilization, were intact. Although p59(fyn) and ZAP70 protein tyrosine kinases were expressed at normal levels, a marked decrease in the level of p56(lck) was noted. Furthermore, this decrease was associated with the presence of an alternatively spliced LCK transcript lacking the exon 7 kinase-encoding domain. These data suggested that deficiency in p56(lck) expression can produce a SCID phenotype in humans. The patient was a male infant from a nonconsanguineous union who was noted to have loose stools and poor weight gain at 1 month of age. He was first hospitalized at 2 months of age for dehydration, failure to thrive, and sepsis. Oral candidiasis and weight below the fifth percentile were found at that time. Rotavirus was isolated from the stool, Enterobacter cloacae from the blood, and cytomegalovirus from the urine and intestinal biopsies. Lymphopenia and hypogammaglobulinemia were noted. Tests for HIV and for other causes of SCID, including deficiencies of adenosine deaminase and purine nucleoside phosphorylase, excluded those diagnoses. A 2-month trial of IL2, begun when the patient was 26 months old, elicited neither a clinical response nor an improvement in lymphocyte numbers (total lymphocyte count ranged between 780 and 1,890). At the age of 32 months, the patient underwent an allogeneic bone marrow transplant from a matched unrelated donor. All sequences normally found in exon 7 were absent in the patient. To test whether the skipping of exon 7 was due to splice site mutations, intron 6, exon 7, and intron 7 were PCR amplified from patient and control genomic DNA and then sequenced. No differences were observed in the splice sequences flanking exon 7 or within 20 bp on either side. The RT-PCR products generated from the patient's cDNA using forward and reverse primers within the 5-prime and 3-prime ends of the molecule yielded 2 bands of approximately equal intensity, whereas 1 band was barely detectable or absent when cDNA from control peripheral blood mononuclear cells, the parents, or Jurkat cDNA was used. While the precise genetic defect responsible for the decreased expression p56(lck) in this patient was not fully understood, Goldman et al. (1998) thought it likely that aberrant alterative splicing of exon 7 was responsible.

See also Marth et al. (1988)

Anderson, S. J.; Levin, S. D.; Perlmutter, R. M.: Protein tyrosine kinase p56(lck) controls allelic exclusion of T-cell receptor beta-chain genes. Nature 365: 552-554, 1993.

Burnett, R. C.; David, J.-C.; Harden, A. M.; Le Beau, M. M.; Rowley, J. D.; Diaz, M. O.: The LCK gene is involved in the t(1;7)(p34;q34) in the T-cell acute lymphoblastic leukemia derived cell line, HSB-2. Genes Chromosomes Cancer 3: 461-467, 1991.

Burnett, R. C.; Thirman, M. J.; Rowley, J. D.; Diaz, M. O.: Molecular analysis of the T-cell acute lymphoblastic leukemia-associated t(1;7)(p34;q34) that fuses LCK and TCRB. Blood 84: 1232-1236, 1994.

Gilbert, F.; Balaban, G.; Moorhead, P.; Bianchi, D.; Schlesinger, H.: Abnormalities of chromosome 1p in human neuroblastoma tumors and cell lines. Cancer Genet. Cytogenet. 7: 33-42, 1982.

Goldman, F. D.; Ballas, Z. K.; Schutte, B. C.; Kemp, J.; Hollenback, C.; Noraz, N.; Taylor, N.: Defective expression of p56lck in an infant with severe combined immunodeficiency. J. Clin. Invest. 102: 421-429, 1998.

Levin, S. D.; Anderson, S. J.; Forbush, K. A.; Perlmutter, R. M.: A dominant-negative transgene defines a role for p56lck in thymopoiesis. EMBO J. 12: 1671-1680, 1993.

Levine, E. G.; Arthur, D. S.; Frizzera, G.; Peterson, B. A.; Hurd, D. D.; Bloomfield, C. D.: There are differences in cytogenetic abnormalities among histologic subtypes of the non-Hodgkin's lymphomas. Blood 66: 1414-1422, 1985.

Marth, J. D.; Disteche, C.; Pravtcheva, D.; Ruddle, F.; Krebs, E. G.; Perlmutter, R. M.: Localization of a lymphocyte-specific protein tyrosine kinase gene (lck) at a site of frequent chromosomal abnormalities in human lymphomas. Proc. Nat. Acad. Sci. 83: 7400-7404, 1986.

Marth, J. D.; Overell, R. W.; Meier, K. E.; Krebs, E. G.; Perlmutter, R. M.: Translational activation of the lck proto-oncogene. Nature 332: 171-173, 1988.

Molina, T. J.; Kishihara, K.; Siderovski, D. P.; van Ewijk, W.; Narendran, A.; Timms, E.; Wakeham, A.; Paige, C. J.; Hartmann, K.-U.; Veillette, A.; Davidson, D.; Mak, T. W.: Profound block in thymocyte development in mice lacking p56'lck'. Nature 357: 161-164, 1992.

Tycko, B.; Smith, S. D.; Sklar, J.: Chromosomal translocations joining LCK and TCRB loci in human T cell leukemia. J. Exp. Med. 174: 867-873, 1991.

Van Roy, N.; Laureys, G.; Versteeg, R.; Opdenakker, G.; Speleman, F.: High-resolution fluorescence mapping of 46 DNA markers to the short arm of human chromosome 1. Genomics 18: 71-78, 1993.

Volpi, E. V.; Romani, M.; Siniscalco, M.: Subregional mapping of the human lymphocyte-specific protein tyrosine kinase gene (LCK) to 1p35-p34.3 and its position relative to the 1p marker D1S57. Cytogenet. Cell Genet. 67: 187-189, 1994.

HOMEO BOX 11 (*186770)

Up to 7% of childhood T-cell acute leukemia is accompanied by a translocation involving 10q24 as a breakpoint, either t(10;14)(q24;q11) or t(7;10)(q35;q24). The gene adjacent to the 10q24 region is transcriptionally activated after translocation to the proximity of either TCRD at 14q11 (186810) or TCRB at 7q35 (186930).

Kagan et al. (1987) fused human leukemic T cells carrying a t(10;14)(q24;q11) translocation with mouse leukemic T cells and examined the hybrids for genetic markers of human chromosomes 10 and 14. Hybrids containing the human 10q+ chromosome had the human genes for terminal deoxynucleotidyltransferase (TDT; 187410), which had been mapped to 10q23-q25, and for the constant region of TCRA (186880). Hybrids containing the human 14q- chromosome retained the variable components of the TCRA gene. Thus the 14q11 breakpoint split the TCRA locus in a region between the variable and constant genes. These results suggested to Kagan et al. (1987) that the leukemic process resulted from translocation of the C(alpha) locus to a putative cellular protooncogene located proximal to the breakpoint at 10q24, with resulting deregulation of said oncogene, for which they proposed the name TCL3. Since the 10q+ chromosome retained the TDT gene, they concluded that the TDT locus is proximal to the TCL3 gene, confirming the location to band 10q23-q24. Zutter et al. (1990) cloned the t(10;14) breakpoint from CD3-negative T-cell acute lymphoblastic leukemia (T-ALL) cells. They identified a locus distinct from TDT at 10q24; this locus was not active in a variety of normal or other neoplastic T cells, but recognized an abundant 2.9-kb RNA in a t(10;14) T-cell leukemia. The authors speculated that this locus is a candidate for the putative TCL3 protooncogene. Kennedy et al. (1991) showed that the 10q24 region encodes a homeo box gene closely related to the developmentally regulated homeotic genes of flies and mammals. The coding capacity of this activated gene, designated HOX11, was undisturbed in a T-cell line carrying the translocation t(7;10)(q35;q24). Hatano et al. (1991) identified the HOX11 gene adjacent to the breakpoint on 10q24 in the t(10;14) translocation of T-ALL. Dube et al. (1991) also implicated the HOX11 gene in the translocation t(10;14)(q24;q11) that occurs in T-ALL. The transloc ation juxtaposes the TCRD gene in chromosome 14q11 with a breakpoint that lies immediately centromeric of the HOX11 gene. The translocation is presumably catalyzed by recombinases normally involved in the generation of immunoglobulin and TCR diversity. Dube et al. (1991) suggested that this was the first example of a human cancer in which deregulated expression of an unaltered homeo box gene is involved in tumorigenesis. Lu et al. (1991) cloned the TCL3 gene and found that its expression was elevated in leukemic cells harboring the t(10;14) translocation. Sequence analysis confirmed that TCL3 is a homeo box-containing gene.

Dear et al. (1993) demonstrated that the HOX11 gene, which is involved in a reciprocal exchange with either TCRD at 14q11 or TCRB at 7q35, encodes a DNA-binding nuclear transcription factor. PCR analysis showed that the HOX11 homeodomain is a member of a distinct class, representatives of which were found in murine and Drosophila genomes. All of these contain a threonine residue in place of the more common isoleucine or valine in helix 3 of the homeodomain. The other two classes of genes encoding transcription factors involved in chromosome translocations in T-cell acute lymphoblastic leukemias are RBTN1 (186921) at 11p15 and RBTN2 (180385) at 11p13, and the genes encoding basic helix-loop-helix proteins, such as LYL1 (151440) at 19p13, TAL1/SCL (187040) at 1p32, and TAL2 (186855) at 9q31.

HOX11 is an orphan homeobox gene since it is located at a site outside the 4 mammalian HOX clusters. Whereas the orthodox HOX genes encode a combinatorial system of positional specification along the anterior-posterior axis of the embryo, the function of orphan homeobox genes is less well understood. As indicated by the findings in the translocation t(10;14), juxtaposition of HOX11 to the T-cell receptor gene redirects the expression of the normal HOX11 product to thymocytes. Transgenic mice in which HOX11 is redirected to the thymus develop T-cell lymphoblastic lymphoma-leukemia. Roberts et al. (1994) isolated a mouse Hox11 genomic clone and found that the homeodomain shared complete amino acid identity with human HOX11. The homologous mouse gene also mapped to the syntenic region within mouse chromosome 19. To assess the functional role of Hox11 in the mouse, Roberts et al. (1994) generated Hox11-deficient mice through gene targeting. Homozygous deficient mice had no spleen, but otherwise appeared normal. Hox11 is normally expressed in the splanchnic anlage arising from the splanchnic mesoderm. The homozygous deficient embryos had no cellular organization at the site of the splanchnic development, but all other splanchnic derivatives developed normally. Roberts et al. (1994) concluded that Hox11 controls the genesis of a single organ; their finding provided new insight into the genetic regulation of morphogenesis. Dear et al. (1995) performed similar studies in 'knockout' mice and found that, in homozygous Hox11-null embryos, spleen formation commences normally up to a specific stage of embryogenesis, at which point the spleen anlage undergoes rapid and complete resorption. Dying spleen cells exhibited characteristics of apoptosis. The authors concluded that Hox11 is not required to initiate spleen development but is essential for the survival of splenic precursors during organogenesis.

Dear, T. N.; Colledge, W. H.; Carlton, M. B. L.; Lavenir, I.; Larson, T.; Smith, A. J. H.; Warren, A. J.; Evans, M. J.; Sofroniew, M. V.; Rabbitts, T. H.: The Hox11 gene is essential for cell survival during spleen development. Development 121: 2909-2915, 1995.

Dear, T. N.; Sanchez-Garcia, I.; Rabbitts, T. H.: The HOX11 gene encodes a DNA-binding nuclear transcription factor belonging to a distinct family of homeobox genes. Proc. Nat. Acad. Sci. 90: 4431-4435, 1993.

Dube, I. D.; Kamel-Reid, S.; Yuan, C. C.; Lu, M.; Wu, X.; Corpus, G.; Raimondi, S. C.; Crist, W. M.; Carroll, A. J.; Minowada, J.; Baker, J. B.: A novel human homeobox gene lies at the chromosome 10 breakpoint in lymphoid neoplasias with chromosomal translocation t(10;14). Blood 78: 2996-3003, 1991.

Hatano, M.; Roberts, C. W. M.; Minden, M.; Crist, W. M.; Korsmeyer, S. J.: Deregulation of a homeobox gene, HOX11, by the t(10;14) in T cell leukemia. Science 253: 79-82, 1991.

Kagan, J.; Finan, J.; Letofsky, J.; Besa, E. C.; Nowell, P. C.; Croce, C. M.: Alpha-chain locus of the T-cell antigen receptor is involved in the t(10;14) chromosome translocation of T-cell acute lymphocytic leukemia. Proc. Nat. Acad. Sci. 84: 4543-4546, 1987.

Kennedy, M. A.; Gonzalez-Sarmiento, R.; Kees, U. R.; Lampert, F.; Dear, N.; Boehm, T.; Rabbitts, T. H.: HOX11, a homeobox-containing T-cell oncogene on human chromosome 10q24. Proc. Nat. Acad. Sci. 88: 8900-8904, 1991.

Lu, M.; Gong, Z. Y.; Shen, W. F.; Ho, A. D.: The TCL-3 proto-oncogene altered by chromosomal translocation in T-cell leukemia codes for a homeobox protein. EMBO J. 10: 2905-2910, 1991.

Roberts, C. W. M.; Shutter, J. R.; Korsmeyer, S. J.: Hox11 controls the genesis of the spleen. Nature 368: 747-750, 1994.

Zutter, M.; Hockett, R. D.; Roberts, C. W. M.; McGuire, E. A.; Bloomstone, J.; Morton, C. C.; Deaven, L. L.; Crist, W. M.; Carroll, A. J.; Korsmeyer, S. J.: The t(10;14)(q24;q11) of T-cell acute lymphoblastic leukemia juxtaposes the delta T-cell receptor with TCL3, a conserved and activated locus at 10q24. Proc. Nat. Acad. Sci. 87: 3161-3165, 1990.

SCL INTERRUPTING LOCUS (*181590)

While studying the transcript of the SCL gene (also called TCL5; 187040), Aplan et al. (1990) identified a new locus which they termed SIL (for SCL-interrupting locus). The locus was identified through a fusion cDNA resulting from an interstitial deletion between SIL and the 5-prime untranslated region of SCL. As with SCL, SIL showed cross-species conservation. Aplan et al. (1990) thought it unlikely that a chimeric SIL/SCL protein is formed, because SIL joins SCL in the SCL 5-prime untranslated region upstream of an in-frame TAA stop codon that precedes the initiation ATG in the SCL message. Similar to 1;14 translocations (see 187040), the deletion brings SIL and SCL close together and disrupts the 5-prime regulatory region of SCL. Essentially identical deletions were found in a total of 3 T-cell lines. Aplan et al. (1990) suspected that the rearrangement was probably mediated by a V-(D)-J recombinase activity, although neither locus is an immunoglobulin or a T-cell receptor. By pulsed field gel analysis of NotI-digested genomic DNA, they concluded that the rearrangement was an interstitial deletion of less than 260 kb of chromosome 1. Brown et al. (1990) reported genomic rearrangement consistent with SIL/SCL fusion in 13 of 50 children with T-cell acute lymphoblastic leukemia. This work illustrates that growth-affecting genes other than immune receptors risk rearrangements. Aplan et al. (1992) found the SIL/SCL rearrangement to be several times more frequent than the t(11;14) translocation and specific for T-cell ALL. Leukemic blasts from 30 patients with B-cell precursor ALL failed to show the change. The fusion occurred in the 5-prime untranslated region of both genes, preserving the SCL coding region. The net result of this rearrangement was that SCL mRNA expression became regulated by the SIL promoter, leading to inappropriate SCL expression. Collazo-Garcia et al. (1995) cloned the mouse cDNA and showed that the predicted 1262-amino acid protein is 75% identica l to the human sequence. RT-PCR was used to show ubiquitous expression which was especially high in rapidly proliferating tissue.

Aplan, P. D.; Lombardi, D. P.; Ginsberg, A. M.; Cossman, J.; Bertness, V. L.; Kirsch, I. L.: Disruption of the human SCL locus by 'illegitimate' V-(D)-J recombinase activity. Science 250: 1426-1429, 1990.

Aplan, P. D.; Lombardi, D. P.; Reaman, G. H.; Sather, H. N.; Hammond, G. D.; Kirsch, I. R.: Involvement of the putative hematopoietic transcription factor SCL in T-cell acute lymphoblastic leukemia. Blood 79: 1327-1333, 1992.

Brown, L.; Cheng, J.-T.; Chen, Q.; Siciliano, M. J.; Crist, W.; Buchanan, G.; Baer, R.: Site-specific recombination of the tal-1 gene is a common occurrence in human T-cell leukemia. EMBO J. 9: 3343-3351, 1990.

Collazo-Garcia, N.; Scherer, P.; Aplan, P. D.: Cloning and characterization of a murine SIL gene. Genomics 30: 506-513, 1995.

BREAKPOINT CLUSTER REGION-LIKE 2; BCRL2 (*113630)

See 151410.

BREAKPOINT CLUSTER REGION-LIKE 3; BCRL3 (*113640)

See 151410.

BREAKPOINT CLUSTER REGION-LIKE 4; BCRL4 (*113660)

See 151410.

T-CELL LYMPHOMA OR LEUKEMIA (*186960)

An inversion of chromosome 14 due to breaks in q11.2 and q32.3 was found in a newly established childhood T-cell lymphoma cell line and confirmed in T-cell chronic lymphocytic leukemia by Hecht et al. (1984). In another T-cell lymphoma cell line, a t(10;14) translocation with a breakpoint at 14q11.2 was found. The authors proposed that a region in or near 14q11.2 is related to T-cell function. It may be significant that nucleotide phosphorylase (164050) and the alpha subunit of the T-cell antigen receptor (186880) map in this region. An alternative interpretation (Croce et al., 1985) is that an oncogene, TCL1, is situated at 14q32.3 and is activated when it comes into juxtaposition with the TCRA locus with inversion; see 186880 and 208900. Erikson et al. (1985) showed that the TCRA locus is split between the proximal V gene and the more distal C gene by the breakpoint at 14q11.2 that creates the t(11;14) of T-cell leukemia. Is the postulated TCL1 oncogene of Croce et al. (1985) the same as AKT1 (164730)? Mathieu-Mahul et al. (1985) cloned a DNA fragment from 14q11 in a case of T-cell malignancy. Mengle-Gaw et al. (1988) and Davey et al. (1988) both studied chromosome rearrangements, inversions, or translocations, involving 14q in patients with ataxia-telangiectasia (208900). These patients are prone to the development of lymphatic neoplasms in association with these abnormalities. Mengle-Gaw et al. (1988) concluded that a single locus at chromosome band 14q32.1, located about 15-20 million basepairs centromeric to the IGH locus, is critical to the development of neoplasia. Davey et al. (1988) reached a similar conclusion, i.e., that a 'growth-effecting' gene in the 14q32 region participates in the leukemogenic process. Presumably this is the locus that has been referred to as TCL1.

Virgilio et al. (1993) took advantage of chromosome-walking techniques and of P1 phage to clone and characterize 450 kb of the germline TCL1 locus, starting from the breakpoints of 2 independent T-cell leukemias. They found that both inversions and translocations occur within a 300-kb region in T-cell leukemias. In an attempt to identify a candidate oncogene responsible for malignant transformation, a CpG island centromeric to the translocations was identified. Two probes near the CpG island detected sequences conserved among species. In further studies, Virgilio et al. (1994) identified a gene within the 14q32.1 region of approximately 350 kb involved in translocations or rearrangements in T-cell leukemias and lymphomas. The gene, encoding a 1.3-kb transcript, was expressed only in restricted subsets of cells within the lymphoid lineage and was expressed in high levels in leukemic cells carrying a t(14;14)(q11;q32) translocation or an inv(14)(q11;q32) inversion. The cognate cDNA sequence revealed an open reading frame of 342 nucleotides encoding a protein of 14 kD. The TCL1 gene sequence appeared to show no homology with other human genes and is preferentially expressed early in T- and B-lymphocyte differentiation.

To show that transcriptional alteration of TCL1 is causally involved in the generation of T-cell neoplasia, Virgilio et al. (1998) generated transgenic mice that carried a TCL1 gene under the transcriptional control of the p56(lck) (153390) promoter element. Their results indicated that transcriptional activation of the TCL1 protooncogene can cause malignant transformation of T lymphocytes.

Croce, C. M.; Isobe, M.; Palumbo, A.; Puck, J.; Ming, J.; Tweardy, D.; Erikson, J.; Davis, M.; Rovera, G.: Gene for alpha-chain of human T-cell receptor: location on chromosome 14 region involved in T-cell neoplasms. Science 227: 1044-1047, 1985.

Davey, M. P.; Bertness, V.; Nakahara, K.; Johnson, J. P.; McBride, O. W.; Waldmann, T. A.; Kirsch, I. R.: Juxtaposition of the T-cell receptor alpha-chain locus (14q11) and a region (14q32) of potential importance in leukemogenesis by a 14;14 translocation in a patient with T-cell chronic lymphocytic leukemia and ataxia-telangiectasia. Proc. Nat. Acad. Sci. 85: 9287-9291, 1988.

Erikson, J.; Williams, D. L.; Finan, J.; Nowell, P. C.; Croce, C. M.: Locus of the alpha-chain of the T-cell receptor is split by chromosome translocation in T-cell leukemias. Science 229: 784-786, 1985.

Hecht, F.; Morgan, R.; Hecht, B. K.-M.; Smith, S. D.: Common region on chromosome 14 in T-cell leukemia and lymphoma. Science 226: 1445-1447, 1984.

Mathieu-Mahul, D.; Caubet, J. F.; Bernheim, A.; Mauchauffe, M.; Palmer, E.; Berger, R.; Larsen, C.-J.: Molecular cloning of a DNA fragment from human chromosome 14(14q11) involved in T-cell malignancies. EMBO J. 4: 3427-3433, 1985.

Mengle-Gaw, L.; Albertson, D. G.; Sherrington, P. D.; Rabbitts, T. H.: Analysis of a T-cell tumor-specific breakpoint cluster at human chromosome 14q32. Proc. Nat. Acad. Sci. 85: 9171-9175, 1988.
Virgilio, L.; Isobe, M.; Narducci, M. G.; Carotenuto, P.; Camerini, B.; Kurosawa, N.; ar-Rushdi, A.; Croce, C. M.; Russo, G.: Chromosome walking on the TCL1 locus involved in T-cell neoplasia. Proc. Nat. Acad. Sci. 90: 9275-9279, 1993.
Virgilio, L.; Lazzeri, C.; Bichi, R.; Nibu, K.; Narducci, M. G.; Russo, G.; Rothstein, J. L.; Croce, C. M.: Deregulated expression of TCL1 causes T cell leukemia in mice. Proc. Nat. Acad. Sci. 95: 3885-3889, 1998.
Virgilio, L.; Narducci, M. G.; Isobe, M.; Billips, L. G.; Cooper, M. D.; Croce, C. M.; Russo, G.: Identification of the TCL1 gene involved in T-cell malignancies. Proc. Nat. Acad. Sci. 91: 12530-12534, 1994.

ETS VARIANT GENE 6; ONCOGENE TEL (*600618)

Cytogenetic abnormalities involving the short arm of chromosome 12 have been documented in a wide variety of hematopoietic malignancies, including acute lymphoblastic leukemia (ALL), acute myeloblastic leukemia, and myelodysplastic syndromes. Kobayashi et al. (1994) showed that most translocations involving band 12p13 are clustered within a small region of the chromosome, leading to the hypothesis that a single gene on 12p13 may be rearranged in these leukemias. In support of this, they showed by fluorescence in situ hybridization that a single YAC spanned most translocation breakpoints involving 12p13. One candidate gene on 12p13 is the TEL gene, encoding an ETS-like putative transcription factor (see ETS1; 164720) which was first identified as a result of its fusion to the platelet-derived growth factor receptor-beta gene (PDGFRB; 173410) in chronic myelomonocytic leukemia with a 5;12 translocation (Golub et al., 1994). Golub et al. (1994) designated the gene TEL for 'translocation, ETS, leukemia.' Golub et al. (1995) documented fusion of TEL to the AML1 gene (151385) on chromosome 21 in 2 pediatric ALL patients with t(12;21) translocations. Thus, TEL can be implicated in the pathogenesis of leukemia through its fusion to either a receptor tyrosine kinase, such as PDGFRB, or a transcription factor, such as AML1.

Papadopoulos et al. (1995) identified a case of ALL in which a previously undescribed fusion between the TEL gene and the ABL gene was identified. The fusion protein showed elevated tyrosine kinase activity. Since both the TEL and the ABL gene (189980) are located near the end of a chromosome, 12 and 9, respectively, it is difficult to identify translocations cytogenetically. Therefore, Janssen et al. (1995) screened a large panel of ALL cases using RT-PCR to determine whether this fusion was present but cytogenetically undetected. They concentrated on adults, because of the high incidence of ABL translocations in that group but, as their index case was a child, they also examined a group of childhood ALL patients. In 186 adult ALL and 30 childhood ALL patients, they were unable to identify any additional cases with a TEL/ABL fusion product.

Since some cases of translocations involving 12p13 in hematologic malignancies have involvement of both TEL alleles resulting in loss of wildtype TEL function in leukemic cells, and since 5% of children with ALL have 12p13-p12 deletions, suggesting the location of a tumor suppressor gene on 12p, Stegmaier et al. (1995) undertook studies that showed that TEL maps between the polymorphic markers D12S89 and D12S98. Using these flanking markers to screen paired diagnosis and remission samples from 81 children with ALL, they found that 15% of informative patients showed TEL loss of heterozygosity that was not evident on cytogenetic analysis. Detailed examination showed that the critically deleted region included 2 candidate suppressor genes: TEL and KIP (600778), the gene encoding the cyclin-dependent kinase inhibitor p27.

Baens et al. (1996) constructed a detailed map of the genomic region containing the ETV6 gene. They developed contigs containing the complete coding sequence and the 5-prime and 3-prime UTRs of ETV6. The helix-loop-helix (HLH) motif is coded by exons 3 and 4, whereas exons 6-8 code for the ETS DNA-binding domain. The ETV6 gene spans 240 kb and is flanked at its 5-prime and 3-prime ends by D12S1697 and D12S98, respectively. Markers D12S1095 and D12S89 are located in the first intron. Baens et al. (1996) identified 2 new DNA polymorphisms in the ETV6 gene that will be useful for the analysis of loss of heterozygosity reported for the ETV6 gene in leukemia.

The ETV6 gene has been shown to be fused to the MN1 gene (156100) at 22q11 in the t(12;22)(p13;q11) translocation observed in myeloid malignancies of different French-American-British (FAB) classification subtypes. Raynaud et al. (1996) reported 5 patients with an identical reciprocal translocation between 3q26 and 12p13. This nonrandom cytogenetic change was observed in 4 patients with myelodysplastic syndrome (MDS) rapidly progressing to acute myeloid leukemia and was found at blast crisis of 1 patient with Philadelphia chromosome-positive CML. The abnormality was associated with a very poor prognosis. Fluorescence in situ hybridization with 3q26 and 12p13 probes was performed on metaphases from these 5 patients. The results were consistent with scattering of the breakpoints previously described in 3q26 rearrangements. Breakpoints at 12p13 involved the ETV6 gene in 3 myelodysplastic syndrome cases.

Berger et al. (1997) described 3 novel translocations involving the TEL/ETV6 gene on chromosome 12. Cave et al. (1997) demonstrated that ETV6 is a target of chromosome 12p deletions in t(12;21) childhood acute lymphocytic leukemia.

Congenital (or infantile) fibrosarcoma (CFS) is a malignant tumor of fibroblasts that occurs in patients aged 2 or younger. CFS is unique among human sarcomas in that it has an excellent prognosis and very low metastatic rate. CFS is histologically identical to adult-type fibrosarcoma (ATFS); however, ATFS is an aggressive malignancy of adults and older children that has a poor prognosis. Knezevich et al. (1998) reported a novel recurrent t(12;15)(p13;q25) rearrangement in CFS that may underlie the distinctive biologic properties of this tumor. By cloning the chromosome breakpoints, they showed that the rearrangement fuses the ETV6 gene from 12p13 with the NTRK3 gene (191316) from 15q25. Analysis of mRNA revealed the expression of ETV6-NTRK3 chimeric transcripts in all 3 CFS tumors analyzed. These were not detected in ATFS or infantile fibromatosis (228550), a histologically similar but benign fibroblastic proliferation occurring in the same age group as CFS. ETV6-NTRK3 fusion transcripts encoded the HLH protein dimerization domain of ETV6 fused to the protein tyrosine kinase (PTK) domain of NTRK3. Presumably, the chimeric protein tyrosine kinase contributed to oncogenesis by dysregulation of NTRK3 signal transduction pathways. Knezevich et al. (1998) commented that the gene fusion should provide a diagnostic marker for CFS.

Ford et al. (1998) reported the extraordinary case of monozygotic twins in whom common acute lymphoblastic leukemia was diagnosed at ages 3.5 years and 4 years. The twins' leukemic DNA shared the same unique (or clonotypic) but nonconstitutive TEL-AML1 fusion sequence. The most plausible explanation for this finding was thought to be a single cell origin of the TEL-AML fusion in 1 fetus in utero, probably as a leukemia-initiating mutation, followed by intraplacental metastasis of clonal progeny to the other twin. Clonal identity was further supported by the finding that the leukemic cells in the 2 twins shared an identical rearranged IGH allele. These data have implications for the etiology and natural history of childhood leukemia.

In a study of a group of 67 cases of chronic myeloid disorders, Nilsson et al. (1998) found no instance of ETV6/ABL fusion.

The TEL gene encodes an Ets family transcription factor and is frequently rearranged in human leukemias of myeloid or lymphoid origins. By gene targeting in mice, Wang et al. (1997) showed that TEL function is required for viability of the developing mouse. The TEL -/- mice suffered a yolk sac angiogenic defect; TEL also appeared essential for the survival of selected neural and mesenchymal populations within the embryo proper. Wang et al. (1998) generated mouse chimeras with TEL -/- embryonic stem cells to examine a possible requirement in adult hematopoiesis. They found that although TEL function is not required for the intrinsic proliferation and/or differentiation of adult-type hematopoietic lineages in the yolk sac and fetal liver, it is essential for the establishment of hematopoiesis of all lineages in the bone marrow. These findings established TEL as the first transcription factor required specifically for hematopoiesis within the bone marrow, as opposed to other sites of hematopoietic activity during development.

Baens, M.; Peeters, P.; Guo, C.; Aerssens, J.; Marynen, P.: Genomic organization of TEL: the human ETS-variant gene 6. Genome Res. 6: 404-413, 1996.
Berger, R.; Le Coniat, M.; Lacronique, V.; Daniel, M.-T.; Lessard, M.; Berthou, C.; Marynen, P.; Bernard, O.: Chromosome abnormalities of the short arm of chromosome 12 in hematopoietic malignancies: a report including three novel translocations involving the TEL/EVT6 gene. Leukemia 11: 1400-1403, 1997.
Cave, H.; Cacheux, V.; Raynaud, S.; Brunie, G.; Bakkus, M.; Cochaux, P.; Preudhomme, C.; Lai, J. L.; Vilmer, E.; Grandchamp, B.: ETV6 is the target of chromosome 12p deletions in t(12;21) childhood acute lymphocytic leukemia. Leukemia 11: 1459-1464, 1997.
Ford, A. M.; Bennett, C. A.; Price, C. M.; Bruin, M. C. A.; Van Wering, E. R.; Greaves, M.: Fetal origins of the TEL-AML1 fusion gene in identical twins with leukemia. Proc. Nat. Acad. Sci. 95: 4584-4588, 1998.
Golub, T. R.; Barker, G. F.; Bohlander, S. K.; Hiebert, S. W.; Ward, D. C.; Bray-Ward, P.; Morgan, E.; Raimondi, S. C.; Rowley, J. D.; Gilliland, D. G.: Fusion of the TEL gene on 12p13 to the AML1 gene on 21q22 in acute lymphoblastic leukemia. Proc. Nat. Acad. Sci. 92: 4917-4921, 1995.
Golub, T. R.; Barker, G. F.; Lovett, M.; Gilliland, D. G.: Fusion of PDGF receptor beta to a novel ets-like gene, tel, in chronic myelomonocytic leukemia with t(5;12) chromosomal translocation. Cell 77: 307-316, 1994.
Janssen, J. W. G.; Ridge, S. A.; Papadopoulos, P.; Cotter, F.; Ludwig, W.-D.; Fonatsch, C.; Rieder, H.; Ostertag, W.; Bartram, C. R.; Wiedemann, L. M.: The fusion of TEL and ABL in human acute lymphoblastic leukaemia is a rare event. Brit. J. Haemat. 90: 222-224, 1995.
Knezevich, S. R.; McFadden, D. E.; Tao, W.; Lim, J. F.; Sorensen, P. H. B.: A novel ETV6-NTRK3 gene fusion in congenital fibrosarcoma. Nature Genet. 18: 184-187, 1998.
Kobayashi, H.; Montgomery, K. T.; Bohlander, S. K.; Adra, C. N.; Lim, B. L.; Kucherlapati, R. S.; Donis-Keller, H.; Holt, M. S.; Le Beau, M. M.; Rowley, J. D.: Fluorescence in situ hybridization mapping of translocations and deletions involving the short arm of human chromosome 12 in malignant hematologic diseases. Blood 84: 3473-3482, 1994.

Nilsson, T.; Andreasson, P.; Hoglund, M.; Fioretos, T.; Billstrom, R.; Garwicz, S.; Mitelman, F.; Johansson, B.: ETV6/ABL fusion is rare in Ph-negative chronic myeloid disorders. Leukemia 12: 1167-1168, 1998.

Papadopoulos, P.; Ridge, S. A.; Boucher, C. A.; Stocking, C.; Wiedemann, L. M.: The novel activation of ABL by fusion to an ets-related gene, TEL. Cancer Res. 55: 34-38, 1995.

Raynaud, S. D.; Baens, M.; Grosgeorge, J.; Rodgers, K.; Reid, C. D. L.; Dainton, M.; Dyer, M.; Fuzibet, J. G.; Gratecos, N.; Taillan, B.; Ayraud, N.; Marynen, P.: Fluorescence in situ hybridization analysis of t(3;12)(q26;p13): a recurring chromosomal abnormality involving the TEL gene (ETV6) in myelodysplastic syndromes. Blood 88: 682-689, 1996.

Stegmaier, K.; Pendse, S.; Barker, G. F.; Bray-Ward, P.; Ward, D. C.; Montgomery, K. T.; Krauter, K. S.; Reynolds, C.; Sklar, J.; Donnelly, M.; Bohlander, S. K.; Rowley, J. D.; Sallan, S. E.; Gilliland, D. G.; Golub, T. R.: Frequent loss of heterozygosity at the TEL gene locus in acute lymphoblastic leukemia of childhood. Blood 86: 38-44, 1995.

Wang, L. C.; Kuo, F.; Fujiwara, Y.; Gilliland, D. G.; Golub, T. R.; Orkin, S. H.: Yolk sac angiogenic defect and intra-embryonic apoptosis in mice lacking the Ets-related factor TEL. EMBO J. 16: 4374-4383, 1997.

Wang, L. C.; Swat, W.; Fujiwara, Y.; Davdison, L.; Visvader, J.; Kuo, F.; Alt, F. W.; Gilliland, D. G.; Golub, T. R.; Orkin, S. H.: The TEL/ETV6 gene is required specifically for hematopoiesis in the bone marrow. Genes Dev. 12: 2392-2401, 1998.

T-CELL ANTIGEN RECEPTOR, ALPHA SUBUNIT (*186880)

T lymphocytes, like B lymphocytes, can recognize a wide range of different antigens. As with B cells, the capability to recognize a given antigen is fixed in any particular clonal line of T cells. However, unlike B cells, T cells recognize antigen in combination with self major histocompatibility complex (MHC) determinants, i.e., the function is 'MHC-restricted.' Many authors commented that 'the chemical nature of the T-cell receptors has been elusive' (e.g., Saito et al., 1984). The development of monoclonal antibodies that recognize and precipitate clone-specific proteins on the surface of T-cells has provided information on these receptor molecules. Hedrick et al. (1984) approached the molecular genetic study of the previously elusive T-cell antigen receptor with 4 assumptions: that they are expressed in T cells but not in B cells; that the mRNAs for the T-cell receptor proteins should be found on membrane-bound polysomes, the nascent receptor polypeptides being attached to the endoplasmic reticulum by a leader peptide (signal sequence); that like immunoglobulin genes, those that encode the T-cell receptor proteins are rearranged in T cells as a mechanism of generating diversity; and that like immunoglobulin genes, they have constant regions that share some functions and variable regions that confer antigen-binding specificity. They found that a cloned T-cell-specific cDNA showed variable, constant and joining regions remarkably similar in size and sequence to those encoding immunoglobulin proteins. Saito et al. (1984) presented the complete deduced primary structure of the T-cell receptor. Siu et al. (1984) stated that the 'T-cell antigen receptor appears to be assembled from 3 gene segments, V, D, and J, and accordingly most closely resembles immunoglobulin heavy chain V genes.' Hannum et al. (1984) presented the sequence of an alpha chain and pointed to its homology to the immunoglobulin polypeptide chains. The antigen-specific receptors of B lymphocytes and T lym phocytes share many similarities. The receptors of the B cells have long been known to be the immunoglobulins. The receptors on T cells consist of immunoglobulin-like integral membrane glycoproteins containing 2 polypeptide subunits, alpha and beta, of similar molecular weight, 40 to 55 kD in the human. Like the immunoglobulins (Ig) of the B cells, each T-cell receptor subunit has, external to the cell membrane, an N-terminal variable (V) domain and a C-terminal constant (C) domain. Like the Ig genes, the genes for the T-cell receptor subunits are assembled from gene segments which are of at least 3 types for alpha, V, joining (J) and C, and at least 4 for beta, V, diversity (D), J and C. Harvey and Showe (1993) pointed out that nearly 60 unique J regions had been identified in TCR-alpha chains, yet fewer than one-third of these had been localized within the TCRA gene. They reported a rapid method for mapping productively rearranged J-alpha regions.

In the rat, Binz et al. (1976) showed linkage between heavy chain immunoglobulin genes and idiotypic T-cell receptors with specificity for MHC antigens but lack of linkage with MHC genes and with kappa light chain genes. If homology exists in man, a likely situation, then a T-cell receptor locus is linked to the Gm loci (147100-147130), which have been mapped to 14q34. In the mouse the alpha subunit is coded by chromosome 14 (Kranz et al., 1985). Barker et al. (1985) assigned the TCRA locus to human chromosome 14, proximal to 14q21. Human chromosome 14 appears to contain 2 regions of syntenic homology to mouse chromosomes: a proximal segment with TCRA and NP (164050) which are on mouse 14 and a distal segment with oncogene FOS (164810) and IGH (147100) which are on mouse 12. By somatic cell hybridization, Croce et al. (1985) assigned the TCRA gene to chromosome 14 and by in situ hybridization further narrowed the assignment to 14q11-14q12. This site is consistently involved in translocations and inversions detectable in human T-cell leukemias and lymphomas. Specifically, an inversion of the segment 14q11.2-q32.2 occurs in T-cell chronic lymphatic leukemia and a t(14;14)(q11;q32) translocation occurs in T-cell malignancies of patients with ataxia-telangiectasia (208900) (McCaw et al., 1975). These observations led Croce et al. (1985) to suggest that the oncogene for which they proposed the designation tcl-1 (186960) is located on band 14q32.3 and becomes activated when it is in proximity to the TCRA gene.

Like the beta chain (186930) of the T-cell antigen receptor, the alpha chain is encoded in separate noncontiguous gene segments, V, J, and C. Using an alpha chain cDNA probe of DNA from somatic cell hybrids, Jones et al. (1985) assigned the gene to chromosome 14. From study of a deletion segregant containing only the distal half of chromosome 14 (14q22-qter), they concluded that the alpha locus is situated proximal to 14q22. They pointed out the high frequency of breaks in the 14q11-q13 segment, possibly involving the alpha locus in T-cell malignancies, and leading Hecht et al. (1984) to suggest the existence of genes relating to T-cell function in this region. Erikson et al. (1985) showed that the TCRA gene was split by chromosome translocation t(11;14)(p13;q11) in 2 cases of T-cell leukemia. The constant segment was translocated to chromosome 11 whereas the variable region remained on chromosome 14. Thus, the V segments are proximal to the C segment within band 14q11.2. Lewis et al. (1985) reported identical findings. In cases of adult T-cell leukemia in Nagasaki Prefecture of Japan, an area of high frequency, Sadamori et al. (1985) found abnormalities at band 14q11. This form of leukemia is associated with HTLV/ATLV viruses. Thus, 14q32 is associated with B-cell lymphoma/leukemia and 14q11 with T-cell lymphoma/leukemia including Sezary syndrome and mycosis fungoides. In an inversion of chromosome 14, inv(14)(q11;q32), in a T cell lymphoma, Baer et al. (1985) showed that on the normal chromosome 14, a V(alpha) segment had rearranged with a J(alpha) segment. In contrast, the inverted chromosome featured an unprecedented rearrangement in which a V-heavy chain segment from 14q32 (147070) had joined with a J(alpha) segment from 14q11. The V(H)-J(alpha)C(alpha) rearrangement was productive at the genomic level and presumably encodes a hybrid immunoglobulin/T cell receptor polypeptide. The MOLT-16 cell line, which was established from the malignant cells of a patient with T-cell acute lymphoblastic leukemia, carries a translocation t(8;14)(q24;q11). By molecular approaches, McKeithan et al. (1986) showed that the breakpoint on 14 occurred close to a joining sequence (J) of the TCRA gene and that the constant region and part of the J region of TCRA are translocated to the 3-prime side of the MYC gene. Le Beau et al. (1986) demonstrated that the TCRA gene was split in a cell line from a child with T-cell acute lymphoblastic leukemia and a t(11;14)(p15;q11). With in situ chromosomal hybridization and with Southern blot analysis, they showed that the break at 14q11 occurred within the variable region of TCRA; the break at 11p15 occurred between the HRAS1 gene (190020) and the genes for insulin and IGF2. By studies of cells from a person with T-cell acute lymphocytic leukemia and a t(10;14) translocation, Kagan et al. (1987) demonstrated that the break in chromosome 14 had occurred in the TCRA locus in a region between the variable and constant genes. The break in chromosome 10 was at 10q24. The derivative 10q+ chromosome had the human gene for terminal deoxynucleotidyltransferase (TDT; 187410), which has been mapped to 10q23-25. These results suggested to Kagan et al. (1987) that the translocation of the TCRA constant locus to a putative cellular protooncogene located proximal to the breakpoint at 10q24, for which they proposed the name TCL3 (186770), had resulted in deregulation of said oncogene, leading to T-cell leukemia. Evidence suggested also that the TDT gene is located proximal to TCL3 at band 10q23-q24. Klein et al. (1987) found considerable variability in the V region and J sequences of the TCRA gene.

Posnett et al. (1986) used 3 different murine monoclonal antibodies to human clonotypic T-cell antigen receptor to demonstrate inherited polymorphism comparable to the allotypic polymorphism of immunoglobulins. Restriction fragment length polymorphisms had previously been identified in human alpha and beta chain genes. These RFLPs mapped to introns; obviously, the polymorphism demonstrated with monoclonal antibodies involved exons. The authors suspected that the polymorphism represented an allotypic system of a variable or joining region. Their results indicated that allelic exclusion governs the expression of the clonotypic receptor by human T-cells and thus is a phenomenon not limited to immunoglobulin-producing cells. Robinson and Kindt (1987) identified 'hotspots' of recombination in the TCRA complex by studying the segregation of 3 RFLPs associated with the C region and 3 RFLPs associated with the V region in 8 families. Oksenberg et al. (1989) found an association between polymorphic markers in the variable and constant regions of the TCR-alpha gene and both multiple sclerosis (126200) and myasthenia gravis (254200).

See review of the TCR genes by Toyonaga and Mak (1987). Marrack and Kappler (1987) reviewed the T-cell receptor from the point of view of structure and particularly of function. Weiss (1990) reviewed the structure and function of the 6-part T-cell antigen receptor complex. Useful diagrams of the structural domains of the receptor proteins (his Figure 1) and of the organization and rearrangement of the T-cell receptor genes (his Figure 2) were presented.

Studies in both mouse and man show that the TCR-delta gene (TCRD; 186810) lies within the TCRA locus, upstream from the estimated 50 to 100 J(alpha) segments and between V(alpha) and J(alpha). Whereas TCRD genes rearrange early in thymic ontogeny, TCRA genes rearrange much later. Further, the utilization of V segments appears to be selective. Satyanarayana et al. (1988) analyzed the germ-line organization of the TCR-alpha/delta locus. Koop et al. (1994) sequenced and analyzed 97.6 kb of DNA containing the TCRA constant gene and the TCRD constant gene as well as the TCRDV3 and 61 different TCRAJ gene segments and compared the organization and structure to the same, previously described region in the mouse. They concluded that this region of the human and mouse genomes is remarkably conserved.

Although the TCR-alpha and TCR-gamma glycoprotein chains are encoded by discrete variable (V), junctional (J), and constant (C) genes, these genes are not represented by separate entries here. (The TCR-beta and TCR-delta chains have additional diversity (D) segments.) The precise number of V-alpha segments in the germline is unknown, but sequence analyses of cDNA clones from a number of individuals have identified 100 different sequences, which can be grouped into 32 subgroups containing sequences sharing greater than, or equal to, 75% homology. During T-cell development the TCR genes rearrange to produce a contiguous V—(D)—J exon. Subsequent splicing of the transcript joins the J and C genes, and the mature mRNA is translated into a complete polypeptide chain. The multiplicity of V, (D), and J segments and the random nature of the V—(D)—J recombination, in addition to junctional variation produced by the enzyme N-terminal transferase (termed N-region diversity), enable the germline repertoire to generate an estimated 10(15) different alpha/beta TCRs (Davis and Bjorkman, 1988).

See also Acuto and Reinherz (1985); Baer et al. (1986); Caccia et al. (1985); Collins et al. (1985); Dembic et al. (1985); Epplen et al. (1987); Hedrick et al. (1984); Ibberson et al. (1995); Mak et al. (1987); Mathieu-Mahul et al. (1986); Minden et al. (1985); Sim et al. (1984); Yanagi et al. (1985); Yoshikai et al. (1985)

Acuto, O.; Reinherz, E. L.: The human T-cell receptor: structure and function. New Eng. J. Med. 312: 1100-1110, 1985.

Baer, R.; Chen, K.-C.; Smith, S. D.; Rabbitts, T. H.: Fusion of an immunoglobulin variable gene and a T cell receptor constant gene in the chromosome 14 inversion associated with T cell tumors. Cell 43: 705-713, 1985.

Baer, R.; Lefranc, M.-P.; Minowada, J.; Forster, A.; Stinson, M. A.; Rabbitts, T. H.: Organization of the T-cell receptor alpha-chain gene and rearrangement in human T-cell leukaemias. Molec. Biol. Med. 3: 265-277, 1986.

Barker, P. E.; Royer, H.-D.; Ruddle, F. H.; Reinherz, E. L.: Human T cell receptor gene TCRA lies in region 14pter-q21. (Abstract) Cytogenet. Cell Genet. 40: 576-577, 1985.

Binz, H.; Wigzell, H.; Bazin, H.: T-cell idiotypes are linked to immunoglobulin heavy chain genes. Nature 264: 639-642, 1976.

Caccia, N.; Bruns, G. A. P.; Kirsch, I. R.; Hollis, G. F.; Bertness, V.; Mak, T. W.: T cell receptor alpha chain genes are located on chromosome 14 at 14q11-14q12 in humans. J. Exp. Med. 161: 1255-1260, 1985.

Collins, M. K. L.; Goodfellow, P. N.; Spurr, N. K.; Solomon, E.; Tanigawa, G.; Tonegawa, S.; Owen, M. J.: The human T-cell receptor alpha-chain gene maps to chromosome 14. Nature 314: 273-274, 1985.

Croce, C. M.; Isobe, M.; Palumbo, A.; Puck, J.; Ming, J.; Tweardy, D.; Erikson, J.; Davis, M.; Rovera, G.: Gene for alpha-chain of human T-cell receptor: location on chromosome 14 region involved in T-cell neoplasms. Science 227: 1044-1047, 1985.

Davis, M. M.; Bjorkman, P. J.: T-cell antigen receptor genes and T-cell recognition. Nature 334: 395-402, 1988.

Dembic, Z.; Bannwarth, W.; Taylor, B. A.; Steinmetz, M.: The gene encoding the T-cell receptor alpha-chain maps close to the Np-2 locus on mouse chromosome 14. Nature 314: 271-273, 1985.

Epplen, J. T.; Chluba, J.; Hardt, C.; Hinkkanen, A.; Steimle, V.; Stockinger, H.: Mammalian T-lymphocyte antigen receptor genes: genetic and nongenetic potential to generate variability. Hum. Genet. 75: 300-310, 1987.

Erikson, J.; Williams, D. L.; Finan, J.; Nowell, P. C.; Croce, C. M.: Locus of the alpha-chain of the T-cell receptor is split by chromosome translocation in T-cell leukemias. Science 229: 784-786, 1985.

Hannum, C. H.; Kappler, J. W.; Trowbridge, I. S.; Marrack, P.; Freed, J. H.: Immunoglobulin-like nature of the alpha-chain of a human T-cell antigen/MHC receptor. Nature 312: 65-67, 1984.

Harvey, R. C.; Showe, L. C.: Localization of 14 functional J regions in the human T-cell receptor alpha-chain locus. Genomics 16: 262-263, 1993.

Hecht, F.; Morgan, R.; Hecht, B. K.-M.; Smith, S. D.: Common region on chromosome 14 in T-cell leukemia and lymphoma. Science 226: 1445-1447, 1984.

Hedrick, S. M.; Cohen, D. I.; Nielsen, E. A.; Davis, M. M.: Isolation of cDNA clones encoding T cell-specific membrane-associated proteins. Nature 308: 149-153, 1984.

Hedrick, S. M.; Nielsen, E. A.; Kavaler, J.; Cohen, D. I.; Davis, M. M.: Sequence relationships between putative T-cell receptor polypeptides and immunoglobulins. Nature 308: 153-158, 1984.

Ibberson, M. R.; Copier, J. P.; So, A. K.: Genomic organization of the human T-cell receptor variable alpha (TCRAV) gene cluster. Genomics 28: 131-139, 1995.

Jones, C.; Morse, H. G.; Kao, F. T.; Carbone, A.; Palmer, E.: Human T-cell receptor alpha-chain genes location on chromosome 14. Science 228: 83-85, 1985.

Kagan, J.; Finan, J.; Letofsky, J.; Besa, E. C.; Nowell, P. C.; Croce, C. M.: Alpha-chain locus of the T-cell antigen receptor is involved in the t(10;14) chromosome translocation of T-cell acute lymphocytic leukemia. Proc. Nat. Acad. Sci. 84: 4543-4546, 1987.

Klein, M. H.; Concannon, P.; Everett, M.; Kim, L. D. H.; Hunkapiller, T.; Hood, L.: Diversity and structure of human T-cell receptor alpha-chain variable region genes. Proc. Nat. Acad. Sci. 84: 6884-6888, 1987.

Koop, B. F.; Rowen, L.; Wang, K.; Kuo, C. L.; Seto, D.; Lenstra, J. A.; Howard, S.; Shan, W.; Deshpande, P.; Hood, L.: The human T-cell receptor TCRAC/TCRDC (C-alpha/C-delta) region: organization, sequence, and evolution of 97.6 kb of DNA. Genomics 19: 478-493, 1994.

Kranz, D. M.; Saito, H.; Disteche, C. M.; Swisshelm, K.; Pravtcheva, D.; Ruddle, F. H.; Eisen, H. N.; Tonegawa, S.: Chromosomal locations of the murine T-cell receptor alpha-chain gene and the T-cell gamma gene. Science 227: 941-945, 1985.

Le Beau, M. M.; McKeithan, T. W.; Shima, E. A.; Goldman-Leikin, R. E.; Chan, S. J.; Bell, G. I.; Rowley, J. D.; Diaz, M. O.: T-cell receptor alpha-chain gene is split in a human T-cell leukemia cell line with a t(11;14)(p15;q11). Proc. Nat. Acad. Sci. 83: 9744-9748, 1986.

Lewis, W. H.; Michalopoulos, E. E.; Williams, D. L.; Minden, M. D.; Mak, T. W.: Breakpoints in the human T-cell antigen receptor alpha-chain locus in two T-cell leukaemia patients with chromosomal translocations. Nature 317: 544-546, 1985.

Mak, T. W.; Caccia, N.; Reis, M.; Ohashi, P.; Sangster, R.; Kimura, N.; Toyonaga, B.: Genes encoding the alpha, beta, and gamma chains of the human T cell antigen receptor. J. Infect. Dis. 155: 418-422, 1987.

Marrack, P.; Kappler, J.: The T cell receptor. Science 238: 1073-1079, 1987.

Mathieu-Mahul, D.; Chen, L. K.; Larsen, C.-J.; Bensussan, A.: Detection of T cell receptor alpha chain locus rearrangements in human antigen-specific T lymphocyte clones using a DNA probe specific for chromosome region 14q11. J. Exp. Med. 163: 1308-1313, 1986.

McCaw, B. K.; Hecht, F.; Harden, D. G.; Teplitz, R. L.: Somatic rearrangement of chromosome 14 in human lymphocytes. Proc. Nat. Acad. Sci. 72: 2071-2075, 1975.

McKeithan, T. W.; Shima, E. A.; Le Beau, M. M.; Minowada, J.; Rowley, J. D.; Diaz, M. O.: Molecular cloning of the breakpoint junction of a human chromosomal 8;14 translocation involving the T-cell receptor alpha-chain gene and sequences on the 3-prime side of MYC. Proc. Nat. Acad. Sci. 83: 6636-6640, 1986.

Minden, M. D.; Toyonaga, B.; Ha, K.; Yanagi, Y.; Chin, B.; Gelfand, E.; Mak, T.: Somatic rearrangement of T-cell antigen receptor gene in human T-cell malignancies. Proc. Nat. Acad. Sci. 82: 1224-1227, 1985.

Oksenberg, J. R.; Sherritt, M.; Begovich, A. B.; Erlich, H. A.; Bernard, C. C.; Cavalli-Sforza, L. L.; Steinman, L.: T-cell receptor V(alpha) and C(alpha) alleles associated with multiple sclerosis and myasthenia gravis. Proc. Nat. Acad. Sci. 86: 988-992, 1989.

Posnett, D. N.; Wang, C. Y.; Friedman, S. M.: Inherited polymorphism of the human T-cell antigen receptor detected by a monoclonal antibody. Proc. Nat. Acad. Sci. 83: 7888-7892, 1986.

Robinson, M. A.; Kindt, T. J.: Genetic recombination within the human T-cell receptor alpha-chain gene complex. Proc. Nat. Acad. Sci. 84: 9089-9093, 1987.

Sadamori, N.; Kusano, M.; Nishino, K.; Tagawa, M.; Yao, E.; Yamada, Y.; Amagasaki, T.; Kinoshita, K.; Ichimaru, M.: Abnormalities of chromosome 14 at band 14q11 in Japanese patients with adult T-cell leukemia. Cancer Genet. Cytogenet. 17: 279-282, 1985.

Saito, H.; Kranz, D. M.; Takagaki, Y.; Hayday, A. C.; Eisen, H. N.; Tonegawa, S.: Complete primary structure of a heterodimeric T-cell receptor deduced from cDNA sequences. Nature 309: 757-762, 1984.

Satyanarayana, K.; Hata, S.; Devlin, P.; Grazia Roncarolo, M.; De Vries, J. E.; Spits, H.; Strominger, J. L.; Krangel, M. S.: Genomic organization of the human T-cell antigen-receptor alpha/delta locus. Proc. Nat. Acad. Sci. 85: 8166-8170, 1988.

Sim, G. K.; Yague, J.; Nelson, J.; Marrack, P.; Palmer, E.; Augustin, A.; Kappler, J.: Primary structure of human T-cell receptor alpha-chain. Nature 312: 771-775, 1984.

Siu, G.; Clark, S. P.; Yoshikai, Y.; Malissen, M.; Yanagi, Y.; Strauss, E.; Mak, T. W.; Hood, L.: The human T cell antigen receptor is encoded by variable, diversity, and joining gene segments that rearrange to generate a complete V gene. Cell 37: 393-401, 1984.

Toyonaga, B.; Mak, T. W.: Genes of the T-cell antigen receptor in normal and malignant T cells. Annu. Rev. Immun. 5: 585-620, 1987.

Weiss, A.: Structure and function of the T cell antigen receptor. J. Clin. Invest. 86: 1015-1022, 1990.

Yanagi, Y.; Chan, A.; Chin, B.; Minden, M.; Mak, T. W.: Analysis of cDNA clones specific for human T cells and the alpha and beta chains of the T-cell receptor heterodimer from a human T-cell line. Proc. Nat. Acad. Sci. 82: 3430-3434, 1985.

Yoshikai, Y.; Clark, S. P.; Taylor, S.; Sohn, U.; Wilson, B. I.; Minden, M. D.; Mak, T. W.: Organization and sequences of the variable, joining and constant region genes of the human T-cell receptor alpha-chain. Nature 316: 837-840, 1985.

T-CELL LEUKEMIA/LYMPHOMA 4 (*186860)

Finger et al. (1988) studied an unusual MYC rearrangement in a human leukemic T-cell line derived from a patient with Sezary syndrome whose serum contained antibodies to human T-cell leukemia virus type 1 (HTLV-I). Translocation between chromosomes 2 and 8 was present in the cell line. They found that a locus derived from chromosome region 2q34 was attached to the 3-prime end of 1 MYC allele. The t(2;8) rearrangement in this cell line resulted in expression of a fused transcript encompassing the MYC gene and a locus designated TCL4 (T-cell leukemia/lymphoma-4). Presumably, the rearrangement was involved in the pathogenesis of the malignancy.

Finger, L. R.; Huebner, K.; Cannizzaro, L. A.; McLeod, K.; Nowell, P. C.; Croce, C. M.: Chromosomal translocation in T-cell leukemia line HUT 78 results in a MYC fusion transcript. Proc. Nat. Acad. Sci. 85: 9158-9162, 1988.

LIM DOMAIN ONLY 2 (*180385)

Rhombotin comprises a multigene family. Rhombotin-1, implicated in an unusual subset of T-cell acute lymphoblastic leukemia, is located in the 11p15 region. A second member of the family, rhombotin-2, maps to 11p13 (see 186921). The known translocation breakpoints in the T-ALL cluster are at the 5-prime end of the RBTN2 gene within 25 kb of the presumed transcriptional start site.

Royer-Pokora et al. (1991) cloned 70 kb of DNA from 11p13 at the site of a recurrent translocation in T-cell leukemia: t(11;14)(p13;q11). The translocation involves the TCR-delta gene (186810) on 14q11 and a new site on 11p13. They found 2 new and 10 previously identified translocations mapping within 25 kb of each other on 11p13 and constituting the so-called 11p13 T-cell translocation cluster (11p13 ttc). A search for expressed sequences surrounding the breakpoint cluster region identified a gene telomeric of all breakpoints that was overexpressed in 3 T-ALL samples with a t(11;14). This new gene, called T-cell translocation gene-2 (TTG2), encodes a small cysteine-rich protein. It showed 48% amino acid identity with T-cell translocation gene-1, or rhombotin (186921), in 11p15. Both proteins contain 2 copies of a cysteine-rich motif. The proteins encoded by these 2 translocation-deregulated genes belong to a class of cysteine-rich proteins with the LIM motif and are important in normal development.

The LIM-finger protein Lmo2, which is activated in T-cell leukemias by chromosomal translocations, is required for yolk sac erythropoiesis. Because Lmo2-null mutant mice died at embryonic day 9-10, assessment of a role in other stages of hematopoiesis was impossible. Yamada et al. (1998) studied the hematopoietic contribution of homozygous mutant Lmo2 -/- mouse embryonic stem cells and found that these cells did not contribute to any hematopoietic lineage in adult chimeric mice, but reintroduction of an Lmo2-expression vector rescued the ability of Lmo2 null embryonic stem cells to contribute to all lineages tested. Yamada et al. (1998) stated that this disruption of hematopoiesis probably occurs because interaction of the Lmo2 protein with factors such as Tal1/Scl (187040) is precluded. Thus, Lmo2 is necessary for early stages of hematopoiesis, and the Lmo2 master gene encodes a protein that has a central and crucial role in hematopoietic development.

See also Boehm et al. (1991)

Boehm, T.; Foroni, L.; Kaneko, Y.; Perutz, M. F.; Rabbitts, T. H.: The rhombotin family of cysteine-rich LIM-domain oncogenes: distinct members are involved in T-cell translocations to human chromosomes 11p15 and 11p13. Proc. Nat. Acad. Sci. 88: 4367-4371, 1991.

Royer-Pokora, B.; Loos, U.; Ludwig, W.-D.: TTG-2, a new gene encoding a cysteine-rich protein with the LIM motif, is overexpressed in acute T-cell leukaemia with the t(11;14)(p13;q11). Oncogene 6: 1887-1893, 1991.

Yamada, Y.; Warren, A. J.; Dobson, C.; Forster, A.; Pannell, R.; Rabbitts, T. H.: The T cell leukemia LIM protein Lmo2 is necessary for adult mouse hematopoiesis. Proc. Nat. Acad. Sci. 95: 3890-3895, 1998.

TRANSLOCATION 12;16 IN MALIGNANT LIPOSARCOMA (*137070)

Rabbitts et al. (1993) showed that the translocation t(12;16)(q13;p11) in malignant myxoid liposarcoma causes the fusion of the CHOP dominant negative transcription factor gene (126337) with a novel gene called FUS. The result is fusion of the FUS glycine-rich protein with the whole CHOP coding region. In independent studies, Crozat et al. (1993) found similar findings of fusion of CHOP with a gene they provisionally named TLS (for 'translocated in liposarcoma'). They showed that TLS (FUS) is a novel nuclear RNA-binding protein with extensive sequence similarity to EWS (133450), the product of a gene commonly translocated in Ewing sarcoma. In TLS-CHOP, the RNA-binding domain of TLS is replaced by the DNA-binding and leucine zipper dimerization domain of CHOP. Targeting of a conserved effector domain of RNA-binding proteins to DNA may play a role in tumor formation.

The location of the FUS gene was identified as 16p11 by the site of the breakpoint in the translocation. The assignment was further narrowed to 16p11.2 by cytogenetic studies of Eneroth et al. (1990) and of Mrozek et al. (1993).

Ichikawa et al. (1994) and Panagopoulos et al. (1994) reported that the FUS gene was fused to ERG in acute myeloid leukemia with t(16:21)(p11;q22).

Aman et al. (1996) defined the genomic structure of FUS and compared it with the EWS structure. The FUS gene consists of 15 exons located within 11 kb of genomic DNA. They analyzed DNA sequence at the intron/exon borders and reported that all acceptor sites match the AG/GT consensus sequence. FUS exon 1 contains a 72-bp untranslated region and the translation initiation codon. Aman et al. (1996) noted that glycine repeats occur in a number of positions in the gene, and that an RNP region is encoded by exons 9, 10 and 11. They reported that the exon/intron structures of FUS and EWS show extensive similarities in the RNP regions, and suggested that the 2 genes may have been derived from a common ancestor. Aman et al. (1996) noted that although the N-terminal ends of FUS and EWS are different, they share extensive homology and they are distinct from the N-terminal regions of other RNP-carrying proteins. They proposed that FUS and EWS may be regarded as the first members of a new family of RNA binding proteins. Sequence analysis of the putative FUS promoter region revealed that the sequence upstream of the transcription start site contains no TATA boxes but did contain stretches of C and G. Several recognition sites for the transcription factors AP2 (107580), GCF (189901), and Sp1 (189906) were found. Aman et al. (1996) stressed that the nature of the promoter regions of FUS and EWS is important for understanding the control of the fusion genes involving FUS and EWS in tumors. All of the tumor-specific translocations lead to the formation of fusion genes with FUS or EWS promoter regions and 5-prime coding regions linked to transcription factor genes. The transcriptional control of the fusion genes is therefore most likely dominated by the FUS and EWS promoters. Aman et al. (1996) reported that exons 1 to 5 of FUS are invariably present in tumor-associated fusion proteins involving FUS. The tumor breakpoints in the EWS and FUS genes are all localized in the same regions of the genes, upstream of the RNP encoding exons. They reported that both FUS and EWS were expressed in all tissues examined.

Aman, P.; Panagopoulos, I.; Lassen, C.; Fioretos, T.; Mencinger, M.; Toresson, H.; Hoglund, M.; Forster, A.; Rabbitts, T. H.; Ron, D.; Mandahl, N.; Mitelman, F.: Expression patterns of the human sarcoma-associated genes FUS and EWS and the genomic structure of FUS. Genomics 37: 1-8, 1996.

Crozat, A.; Aman, P.; Mandahl, N.; Ron, D.: Fusion of CHOP to a novel RNA-binding protein in human myxoid liposarcoma. Nature 363: 640-644, 1993.

Eneroth, M.; Mandahl, N.; Heim, S.; Willen, H.; Rydholm, A.; Alberts, K. A.; Mitelman, F.: Localization of the chromosomal breakpoints of the t(12;16) in liposarcoma to subbands 12q13.3 and 16p11.2. Cancer Genet. Cytogenet. 48: 101-107, 1990.

Ichikawa, H.; Shimizu, K.; Hayashi, Y.; Ohki, M.: An RNA-binding protein gene, TLS/FUS, is fused to ERG in human myeloid leukemia with t(16;21) chromosomal translocation. Cancer Res. 54: 2865-2868, 1994.

Mrozek, K.; Karakousis, C. P.; Bloomfield, C. D.: Chromosome 12 breakpoints are cytogenetically different in benign and malignant lipogenic tumors: localization of breakpoints in lipoma to 12q15 and in myxoid liposarcoma to 12q13.3. Cancer Res. 53: 1670-1675, 1993.

Panagopoulos, I.; Aman, P.; Fioretos, T.; Hoglund, M.; Johansson, B.; Mandahl, N.; Heim, S.; Behrendtz, M.; Mitelman, F.: Fusion of the FUS gene with ERG in acute myeloid leukemia with t(16;21)(p11;q22). Genes Chromosomes Cancer 11: 256-262, 1994.

Rabbitts, T. H.; Forster, A.; Larson, R.; Nathan, P.: Fusion of the dominant negative transcription regulator CHOP with a novel gene FUS by translocation t(12;16) in malignant liposarcoma. Nature Genet. 4: 175-180, 1993.

DNA DAMAGE-INDUCIBLE TRANSCRIPT 3 (*126337)

Ron and Habener (1992) identified a nuclear protein that serves as a dominant-negative inhibitor of the transcription factors C/EBP (116897, 116898) and LAP (189965). The protein was called CHOP for C/EBP-homologous protein but is also referred to as DDIT3 for 'DNA damage-inducible transcript 3' and GADD153 for 'growth arrest and DNA damage-inducible gene.' They found that bacterially expressed CHOP inhibits the DNA-binding activity of C/EBP and LAP by forming heterodimers that cannot bind DNA. Park et al. (1992) isolated the gene from human cells and demonstrated a high level of conservation in both the structural and regulatory regions to the previously described hamster gene. Each is composed of 4 exons with intron/exon junctions maintained at identical positions; they showed 91% identity in amino acid sequence and 78% identity in nucleotide sequence. By study of somatic cell hybrids, they demonstrated that the gene is located on chromosome 12 and by in situ hybridization further localized the gene to 12q13.1-q13.2.

CHOP is consistently rearranged in myxoid liposarcomas (Aman et al., 1992). In the characteristic chromosomal translocation t(12;16)(q13;p11), Crozat et al. (1993) showed that the CHOP gene is fused with a gene on chromosome 16 provisionally designated TLS (for 'translocated in liposarcoma'). The same gene was called FUS by Rabbitts et al. (1993); see 137070. The TLS gene was found to be a novel nuclear RNA-binding protein with extensive sequence similarity to

EWS (133450), the product of a gene commonly translocated in Ewing sarcoma. In TLS-CHOP, the RNA-binding domain of TLS is replaced by the DNA-binding and leucine zipper dimerization domain of CHOP. Targeting of a conserved effector domain of RNA-binding proteins to DNA may play a role in tumor formation.

Mrozek et al. (1993) demonstrated that the chromosome 12 breakpoint in benign lipoma (151900) is at 12q15, whereas the breakpoint in myxoid liposarcoma is at 12q13.3.

In an analysis of peripheral blood samples from 19 patients with myxoid liposarcoma due to t(12;16) and in 1 patient with myxoid liposarcoma due to t(12;22;20), resulting in the fusion of the CHOP and EWS genes, Panagopoulos et al. (1996) found FUS/CHOP hybrid fragments in 3 patients with t(12;16) and a EWS/CHOP hybrid in the patient with the latter translocation.

Aman, P.; Ron, D.; Mandahl, N.; Fioretos, T.; Heim, S.; Arheden, K.; Willen, H.; Rydholm, A.; Mitelman, F.: Rearrangement of the transcription factor gene CHOP in myxoid liposarcomas with t(12;16)(q13;p11). Genes Chromosomes Cancer 5: 278-285, 1992.

Crozat, A.; Aman, P.; Mandahl, N.; Ron, D.: Fusion of CHOP to a novel RNA-binding protein in human myxoid liposarcoma. Nature 363: 640-644, 1993.

Mrozek, K.; Karakousis, C. P.; Bloomfield, C. D.: Chromosome 12 breakpoints are cytogenetically different in benign and malignant lipogenic tumors: localization of breakpoints in lipoma to 12q15 and in myxoid liposarcoma to 12q13.3. Cancer Res. 53: 1670-1675, 1993.

Panagopoulos, I.; Aman, P.; Mertens, F.; Mandahl, N.; Rydholm, A.; Bauer, H. F. C.; Mitelman, F.: Genomic PCR detects tumor cells in peripheral blood from patients with myxoid liposarcoma. Genes Chromosomes Cancer 17: 102-107, 1996.

Park, J. S.; Luethy, J. D.; Wang, M. G.; Fargnoli, J.; Fornace, A. J., Jr.; McBride, O. W.; Holbrook, N. J.: Isolation, characterization and chromosomal localization of the human GADD153 gene. Gene 116: 259-267, 1992.

Rabbitts, T. H.; Forster, A.; Larson, R.; Nathan, P.: Fusion of the dominant negative transcription regulator CHOP with a novel gene FUS by translocation t(12;16) in malignant liposarcoma. Nature Genet. 4: 175-180, 1993.

Ron, D.; Habener, J. F.: CHOP, a novel developmentally regulated nuclear protein that dimerizes with transcription factors C/EBP and LAP and functions as a dominant-negative inhibitor of gene transcription. Genes Dev. 6: 439-453, 1992.

LIM DOMAIN-CONTAINING PREFERRED TRANSLOCATION PARTNER IN LIPOMA (*600700)

In a note added in proof, Schoenmakers et al. (1995) described the isolation and sequencing of cDNA clones of the chromosome 3-derived lipoma-preferred partner gene LPP, which is involved in the translocation t(3;12) in benign lipomas. The 612-amino acid open reading frame of LPP showed approximately 85% similarity to zyxin of chicken. Zyxin is a member of the LIM protein family whose members all possess LIM domains (for example, see 600824). These domains are cysteine-rich, zinc-binding protein sequences found in transcription regulators, proto-oncogene products, and adhesive plaque constituents. Like zyxin, which is present at sites of cell adhesion to the extracellular matrix and other cells, the deduced LPP-encoded protein has 3 LIM domains and lacks classic DNA-binding homeodomains. The t(3;12) translocation in the benign lipomas resulted in the in-phase fusion of the DNA-binding molecules of HMGIC (600698) to the presumptive modular protein-binding interfaces of the LPP-encoded protein (Dal Cin et al., 1988).

By partial cDNA cloning, Petit et al. (1996) established features of the genetic organization of LPP. The gene was found to span a genomic region of over 400 kb. Nucleotide sequence analysis of composite cDNA of LPP revealed an open reading frame of 1,836 nucleotides encoding a proline-rich polypeptide containing a leucine-zipper motif in its N-terminal region and 3 LIM domains in its C-terminal region. Petit et al. (1996) commented that the LPP-encoded protein should be classified as a novel member of the group 3 proteins of the LIM protein gene family. Using RT-PCR analysis of lipoma cell lines and primary lipomas, they concluded that LPP is frequently rearranged in cases without a cytogenetically detectable involvement of 3q27-q28. Two alternative HMGIC/LPP hybrid transcripts were detected; the difference between them was mainly the presence of either 2 or 3 LIM domains in the predicted fusion proteins.

Dal Cin, P.; Turc-Carel, C.; Sandberg, A. A.: Consistent involvement of band 12q14 in two different translocations in three lipomas from the same patient. Cancer Genet. Cytogenet. 31: 237-240, 1988.

Petit, M. M. R.; Mols, R.; Schoenmakers, E. F. P. M.; Mandahl, N.; Van de Ven, W. J. M.: LPP, the preferred fusion partner gene of HMGIC in lipomas, is a novel member of the LIM protein gene family. Genomics 36: 118-129, 1996.

Schoenmakers, E. P. P. M.; Wanschura, S.; Mols, R.; Bullerdiek, J.; Van den Berghe, H.; Van de Ven, W. J. M.: Recurrent rearrangements in the high mobility group protein gene, HMGI-C, in benign mesenchymal tumours. Nature Genet. 10: 436-444, 1995.

HIGH MOBILITY GROUP PROTEIN ISOFORM I-C (*600698)

Cytogenetic studies indicate that the 12q13-q15 region is involved in a variety of malignant and benign solid tumors. The gene for myxoid liposarcoma on 12q was identified as CHOP (126337). The 12q13-q15 gene rearranged in soft tissue clear-cell sarcoma encodes transcription factor ATF1 (123803). Among benign tumors, involvement of 12q13-q15 is frequently observed in benign adipose tissue tumors, pleomorphic adenomas of the salivary glands and uterine leiomyomas (the most common pelvic tumor of women affecting up to 77% of females of reproductive age). Other benign tumors include endometrial polyps, a subgroup of hemangiopericytomas, pulmonary chondroid hamartomas, and giant cell tumors of bone (osteoclatomas). Fluorescence in situ hybridization (FISH), combined with high resolution banding, indicated that it is 12q15 that contains the breakpoints in all these tumors (Rohen et al., 1995). The region has been called MAR for 'multiple aberration region.' Both Ashar et al. (1995) (see below) and Schoenmakers et al. (1995) identified a candidate gene within the 1.7 Mb multiple aberration region on 12q15 that harbors recurrent breakpoints found in these benign tumors. With a 175-kb segment, Schoenmakers et al. (1995) found the gene for high mobility group protein HMGI-C and characterized its genomic organization. The gene contains at least 5 exons distributed over at least 160 kb. By FISH, they showed that most of the breakpoints in benign tumors of 8 different types (lipoma, pleomorphic salivary adenoma, and uterine leiomyoma) fall within the gene. Most of the breaks were within the 140-kb third intron.

In a note added in proof, Schoenmakers et al. (1995) described isolation of cDNA clones of the chromosome 3-derived lipoma-preferred partner LPP (600700), which is fused with the HMGIC gene by the t(3;12) translocation found in benign lipoma.

The viable pygmy (pg) mutation on chromosome 10 of the mouse gives rise to small stature owing to disruptions of growth and development. An insertional mutation facilitated cloning of the locus (Xiang et al., 1990). Subsequently, it was shown that expression of the HMGIC gene is abrogated in 3 pygmy alleles (Zhou et al., 1995). Among the 4 viable spontaneous mouse mutants which disrupt growth, 'pygmy' is unique because its phenotype cannot be explained by aberrations in the growth hormone-insulin-like growth factor endocrine pathway. Zhou et al. (1995) showed that the cause is inactivation of the Hmgi-c gene, a member of the Hmgi family of proteins which function as architectural factors in the nuclear scaffold and are critical in the assembly of stereospecific transcriptional complexes. Hmgi-c and another Hmgi family member, Hmgi(y) (600701), are expressed predominantly during embryogenesis. The HMGI proteins are known to be regulated by cell cycle-dependent phosphorylation which alters their DNA binding affinity. Zhou et al. (1995) suggested that identification of the 'pygmy' gene as Hmgi-c may suggest new avenues of research into the biochemical nature of the African pygmy phenotype (265850) and the numerous growth hormone-resistant human dwarf syndromes.

Ashar et al. (1995) considered the HMGIC gene a good candidate for involvement in lipoma for several reasons, including the facts that it encodes a transcriptional regulating factor, that the pygmy mouse had disproportionately less fat than normal litter mates, and that the mouse gene maps to a region of chromosome 10 with homology of synteny to human 12q14-q15. Therefore, they cloned the human gene and investigated its possible role in lipomas. In FISH studies, Ashar et al. (1995) found apparent deletion of the 3-prime end of the HMGIC gene in translocations associated with lipoma. Chimeric transcripts were isolated from 2 lipomas in which HMGI-C DNA-binding domains (AT hook motifs) were fused to either a LIM or an acidic transactivator domain. The identification of a gene rearranged in a benign neoplastic process suggests a role for HMGI-C in adipogenesis and mesenchymal differentiation.

One of the fusion genes identified by Ashar et al. (1995) involved chromosomes 3 and 15. The situation with fusion of HMGIC with genes on several different chromosomes is comparable to that found with the MLL gene (159555) on 11q23, translocation of which leads to acute leukemia when it is fused with various partners that mostly code for different types of transcriptional regulatory domains.

In 2 pulmonary chondroid hamartomas from different patients and 1 uterine leiomyoma with apparently normal karyotypes, Kazmierczak et al. (1996) found identical RTVL-H 3-prime LTRs (see 190080) fused as ectopic sequences to exon 3 of HMGIC. Screening of genomic cosmid and plasmid clones derived from intron 3 of HMGIC for the presence of these RTVL-H sequences showed a cluster of these retrotransposon-like sequences in this region. Chromosomal abnormalities of the region 12q13-q15 with their breakpoints located in the large intron 3 of HMGIC are frequent in human benign tumors such as uterine leiomyomas, pleomorphic adenomas, lipomas, and pulmonary chondroid hamartomas. Because some RTVL-H LTRs are able to promote the expression of upstream and downstream genes, they may act as transcriptional promoters of the HMGIC. These findings suggested to Kazmierczak et al. (1996) that the reexpression of HMGIC exons 1 to 3, rather than the formation of particular fusion genes, is the critical event in the genesis of the tumors.

Nonrandom involvement of 6p23-p21 in chromosomal aberrations in lipomas has also been observed, and HMGIY has been mapped to 6p21 (Friedmann et al., 1993), raising the possibility of its role also in the pathogenesis of lipoma.

Chau et al. (1995) demonstrated that the HMGIC gene consists of 5 exons, the first and last of which include long untranslated regions. Ashar et al. (1996) also characterized the gene and showed that it spans at least 60 kb, with intron 3 extending at least 25 kb.

Manfioletti et al. (1995) found that the Hmgi-c gene in the mouse also has 5 exons; it is longer than 50 kb. A highly homologous Hmgi-c pseudogene was also identified in the mouse.

Ishwad et al. (1997) identified a highly informative dinucleotide repeat in the 5-prime flanking region of the HMGIC gene. This polymorphism consisted of 18 to 37 copies of a (CT)n repeat with an observed heterozygosity of 82% to 83% in African Americans and Caucasians. Linkage analysis in CEPH pedigrees localized the gene to 12q13-q15 with no recombination observed between HMCIC and markers D12S102 and D12S8.

Ashar, H. R.; Cherath, L.; Przybysz, K. M.; Chada, K.: Genomic characterization of human HMGIC, a member of the accessory transcription factor family found at translocation breakpoints in lipomas. Genomics 31: 207-214, 1996.
Ashar, H. R.; Schoenberg Fejzo, M.; Tkachenko, A.; Zhou, X.; Fletcher, J. A.; Weremowicz, S.; Morton, C. C.; Chada, K.: Disruption of the architectural factor HMGI-C: DNA-binding AT hook motifs fused in lipomas to distinct transcriptional regulatory domains. Cell 82: 57-65, 1995.
Chau, K.-Y.; Patel, U. A.; Lee, K.-L. D.; Lam, H.-Y. P.; Crane-Robinson, C.: The gene for the human architectural transcription factor HMGI-C consists of five exons each coding for a distinct functional element. Nucleic Acids Res. 23: 4262-4266, 1995.
Friedmann, M.; Holth, L. T.; Zoghbi, H. Y.; Reeves, R.: Organization, inducible-expression and chromosome localization of the human HMG-I(Y) nonhistone protein gene. Nucleic Acids Res. 21: 4259-4267, 1993.
Ishwad, C. S.; Shriver, M. D.; Lassige, D. M.; Ferrell, R. E.: The high mobility group I-C gene (HMGI-C): polymorphism and genetic localization. Hum. Genet. 99: 103-105, 1997.
Kazmierczak, B.; Pohnke, Y.; Bullerdiek, J.: Fusion transcripts between the HMGIC gene and RTVL-H-related sequences in mesenchymal tumors without cytogenetic aberrations. Genomics 38: 223-226, 1996.
Manfioletti, G.; Rustighi, A.; Mantovani, F.; Goodwin, G. H.; Giancotti, V.: Isolation and characterization of the gene coding for murine high-mobility-group protein HMGI-C. Gene 167: 249-253, 1995.
Rohen, C.; et al.; et al.: The 12q aberration of a breast hamartoma maps within MAR. Genes Chrom. Cancer in press: . 1995.
Schoenmakers, E. P. P. M.; Wanschura, S.; Mols, R.; Bullerdiek, J.; Van den Berghe, H.; Van de Ven, W. J. M.: Recurrent rearrangements in the high mobility group protein gene, HMGI-C, in benign mesenchymal tumours. Nature Genet. 10: 436-444, 1995.
Xiang, X.; Benson, K. F.; Chada, K.: Mini-mouse: disruption of the pygmy locus in a transgenic insertional mutant. Science 247: 967-969, 1990.
Zhou, X.; Benson, K. F.; Ashar, H. R.; Chada, K.: Mutation responsible for the mouse pygmy phenotype in the developmentally regulated factor HMGI-C. Nature 377: 771-774, 1995.

EWING SARCOMA BREAKPOINT REGION 2 (*193067)

Fli1, previously called Sic1, is a small DNA region identified in the mouse genome as a viral integration site common to 2 retroviruses involved in virus-induced leukemias and lymphomas. The region is near the centromere of mouse chromosome 9 (Bergeron et al., 1991), which is known to have homologous synteny with a portion of the long arm of human chromosome 11. Using an evolutionarily conserved mouse probe and Southern hybridization to rodent/human somatic cell hybrid DNAs, Baud et al. (1991) showed that the human equivalent of Fli1 maps to 11q23-q24. They showed that it lies on a fragment flanked on the centromeric side by the translocation breakpoint in acute lymphoblastic leukemia-associated t(4;11)(q21;q23) and on the telomeric side by the Ewing-associated t(11;22)(q24;q12) breakpoint. Hromas et al. (1993) isolated a gene which they identified as FLI1 from its homology to the mouse gene. By in situ hybridization, they mapped the gene to 11q24, a region of aberrations in Ewing sarcoma and neuroepithelioma. In a neuroepithelioma cell line, they found that the FLI1 transcript was present but had an aberrant structure, indicating that it may be rearranged in neuroepithelioma.

Delattre et al. (1992) demonstrated that the translocation t(11;22)(q24;q12) substitutes a putative RNA-binding domain of the Ewing sarcoma gene on chromosome 22 (EWSR1; 133450) for the DNA-binding domain of the FLI1 gene on chromosome 11. Delattre et al. (1992) found that the breakpoints on chromosome 22 were clustered within a 7-kb region, whereas the breakpoints on chromosome 11 were scattered over a 40-kb region, termed EWSR2. May et al. (1993) clarified the situation further by confirming that the 11;22 translocation of Ewing sarcoma produces a chimeric transcription factor that requires the DNA-binding domain encoded by FLI1 for transformation.

Selleri et al. (1994) cloned in 2 YAC contigs (separated by 200-400 kb) approximately 3.2 Mb (or 70%) of the DNA within human chromosomal band 11q24. One of the contigs encompassed the Ewing sarcoma breakpoint region-2. Selleri et al. (1994) linked ETS1 (164720) and FLI1, 2 members of the ETS family of transcription factors, within 400 kb of intervening DNA on the same contig. They further demonstrated that the 2 genes are most likely transcribed in the same orientation, in a tail-to-head configuration.

Baud, V.; Lipinski, M.; Rassart, E.; Poliquin, L.; Bergeron, D.: The human homolog of the mouse common viral integration region, FLI1, maps to 11q23-q24. Genomics 11: 223-224, 1991.
Bergeron, D.; Poliquin, L.; Kozak, C. A.; Rassart, E.: Identification of a new common viral integration region in Cas- Br-E murine leukemia virus-induced non-T, non-B cell lymphomas. J. Virol. 65: 7-15, 1991.
Delattre, O.; Zucman, J.; Plougastel, B.; Desmaze, C.; Melot, T.; Peter, M.; Kovar, H.; Joubert, I.; de Jong, P.; Rouleau, G.; Aurias, A.; Thomas, G.: Gene fusion with an ETS DNA-binding domain caused by chromosome translocation in human tumours. Nature 359: 162-165, 1992.
Hromas, R.; May, W.; Denny, C.; Raskind, W.; Moore, J.; Maki, R. A.; Beck, E.; Klemsz, M. J.: Human FLI-1 localizes to chromosome 11q24 and has an aberrant transcript in neuroepithelioma. Biochim. Biophys. Acta 1172: 155-158, 1993.
May, W. A.; Gishizky, M. L.; Lessnick, S. L.; Lunsford, L. B.; Lewis, B. C.; Delattre, O.; Zucman, J.; Thomas, G.; Denny, C. T.: Ewing sarcoma 11;22 translocation produces a chimeric transcription factor that requires the DNA-binding domain encoded by FLI1 for transformation. Proc. Nat. Acad. Sci. 90: 5752-5756, 1993.
Selleri, L.; Giovannini, M.; Hermanson, G. G.; Romo, A.; Quackenbush, J.; Penny, L.; Khristich, J. V.; Evans, G. A.: Yeast artificial chromosome cloning of 3.2 megabases within chromosomal band 11q24 closely linking c-ets 1 and fli-1 and encompassing the Ewing sarcoma breakpoint. Genomics 22: 137-147, 1994.

ETS VARIANT GENE 1; ETV1 (*600541)

Jeon et al. (1995) identified a variant Ewing sarcoma translocation (7;22) which fused the EWS gene (133450) to the human homolog of the murine ETS gene ER81. This gene, designated ETV1 (for 'ETS translocation variant'), is located on 7p22 as indicated by the breakpoint in the translocation t(7;22)(p22;q12) and by fluorescence in situ hybridization. Jeon et al. (1995) found that identical EWS nucleotide sequences were found in most of the EWS-FLI1 and EWS-ERG chimeric transcripts and were fused to a portion of the ETV1 gene encoding an ETS domain with sequence-specific DNA-binding activity. These findings confirmed that the fusion of EWS to different ETS family members can result in a similar tumor phenotype.

Jeon, I.-S.; Davis, J. N.; Braun, B. S.; Sublett, J. E.; Roussel, M. F.; Denny, C. T.; Shapiro, D. N.: A variant Ewing's sarcoma translocation (7;22) fuses the EWS gene to the ETS gene ETV1. Oncogene 10: 1229-1234, 1995.

ACTIVATING TRANSCRIPTION FACTOR 1 (*123803)

cAMP-dependent transcription factor-1 (ATF1) was described by Hai et al. (1989) and Yoshimura et al. (1990).

Like Ewing sarcoma and peripheral neuroepithelioma which are associated with a translocation involving chromosomes 11 and 22 that leads to fusion of the EWS gene on chromosome 22 and the FLI1 gene on chromosome 11, malignant melanoma of soft parts (MMSP), or soft tissue clear cell sarcoma, exhibits a recurrent chromosome translocation involving chromosome 22, specifically t(12;22)(q13;q12). MMSP is a rare and aggressive tumor that mainly develops in tendons and aponeuroses of patients between 15 and 35 years of age, and which may also be derived from neuroectoderm (Chung and Enzinger, 1983; Epstein et al., 1984). Zucman et al. (1993) demonstrated that the hybrid transcripts in the t(12;22) cases encoded a chimeric protein consisting of the N-terminal domain of EWS linked to the bZIP domain of ATF-1. Zucman et al. (1993) showed by fluorescence in situ hybridization that the ATF1 gene maps to 12q13. Bicolor FISH on interphase nuclei from an MMSP cell line showed that the EWS locus was split and that the proximal part of EWS was juxtaposed to the ATF1 locus.

Chung, E. B.; Enzinger, F. M.: Malignant melanoma of soft parts: a reassessment of clear cell sarcoma. Am. J. Surg. Path. 7: 405-413, 1983.
Epstein, A. L.; Martin, A. O.; Kempson, R.: Use of a newly established human cell line (SU-CCS-1) to demonstrate the relationship of clear cell sarcoma to malignant melanoma. Cancer Res. 44: 1265-1274, 1984.
Hai, T.; Liu, F.; Coukos, W. J.; Green, M. R.: Transcription factor ATF cDNA clones: an extensive family of leucine zipper proteins able to selectively form DNA-binding heterodimers. Genes Dev. 3: 2083-2090, 1989.
Yoshimura, T.; Fijisawa, J.; Yoshida, M.: Multiple cDNA clones encoding nuclear proteins that bind to the tax-dependent enhancer of HTLV-1: all contain a leucine zipper structure and basic amino acid domain. EMBO J. 9: 2537-2542, 1990.

Zucman, J.; Delattre, O.; Desmaze, C.; Epstein, A. L.; Stenman, G.; Speleman, F.; Fletchers, C. D. M.; Aurias, A.; Thomas, G.: EWS and ATF-1 gene fusion induced by t(12;22) translocation in malignant melanoma of soft parts. Nature Genet. 4: 341-345, 1993.

WAARDENBURG SYNDROME, TYPE 1; PAX3 (*193500)

Description

The features of Waardenburg syndrome are wide bridge of the nose owing to lateral displacement of the inner canthus of each eye, pigmentary disturbance (frontal white blaze of hair, heterochromia iridis, white eye lashes, leukoderma), and cochlear deafness. The severity varies widely and some affected persons escape deafness.

Nomenclature

WS type I (WS1) is distinguished from WS type II (WS2) by the presence in WS1 of dystopia canthorum, a lateral displacement of the inner canthi.

Klein's name is sometimes combined with Waardenburg's in the eponymic designation of this disorder, on the basis of a patient that Klein (1950) described with 'partial albinism,' blue eyes, deaf-mutism, undeveloped muscles and fused joints in the arms, skeletal dysplasia, etc. See 148820 for evidence that the Klein-Waardenburg syndrome, also known as Waardenburg syndrome type III, is due to an allelic mutation in the PAX3 gene or is a contiguous gene syndrome due to deletion of the PAX3 gene and adjacent genes.

Clinical Features

Yoshino et al. (1986) evaluated the incidence of dystopia canthorum in a 3-generation family with Waardenburg syndrome and concluded that it is the most frequently expressed sign of the condition. Laestadius et al. (1969) provided normal standards for the measurement of inner canthal and outer canthal distance. Standards were also presented by Christian et al. (1969). Lateral displacement of the inner canthi is seen also in the oral-facial-digital syndrome type I (311200). Skipped generations and the occurrence of bilateral cleft lip were documented by Giacoia and Klein (1969)

Winship and Beighton (1992) reviewed phenotypic variation on the basis of an analysis of 68 affected children.

In a craniofacial anthropometric study of 51 WS I individuals, da-Silva et al. (1993) concluded that the most discriminating parameters were, from clinical measurements, intercanthal distance (which was increased) and philtrum length (which was decreased) and, from roentgenographic measurements, nasal bone length (which was decreased) and lower facial height (which was increased). In place of the measurement of inner canthal distance, the Waardenburg Consortium (Farrer et al., 1992) recommended the W index: a composite measure including the inner canthal, inner pupillary, and outer canthal distances. Normal and dystopic subjects had W values (mean +/- S.D.) of 1.76 +/- 0.16 and 2.61 +/- 0.19, respectively (Newton, 1989); the Waardenburg Consortium recommended a threshold W value of 2.07. Individual I-2 had a W index of 2.21 but was the only member of the family with a value over 2.07. See 193510 for further discussion of type II Waardenburg syndrome which in some families is not linked to PAX3.

The disorder has been described in American blacks (Hansen et al., 1965) and in Maoris (Houghton, 1964) as well as in Europeans. Cleft palate and/or lip occurs in some cases. Premature graying of the hair is an effect of the gene. The fundus may be completely or partially albinotic and depigmented areas of skin like those of piebald trait (172800) may be present. In the state of South Australia, the Waardenburg syndrome is a leading cause of deafness and 'enjoys' a position comparable to porphyria in South Africa, having been introduced by early settlers who have many descendants (Fraser, 1967). The white forelock may be present at birth and later disappear (Feingold et al., 1967). Arias (1971) observed black forelock in place of white forelock. An affected Chinese family was reported by Chew et al. (1968).

Goodman et al. (1988) observed absence of vagina and of right-sided uterine adnexa in an 18-year-old woman with WS1. They postulated that these are related to Waardenburg syndrome because of altered invasion of neurons in early embryogenesis.

The occurrence of Hirschsprung disease (aganglionic megacolon; 142623) in patients with the Waardenburg syndrome is noteworthy (McKusick, 1973; Lowry, 1975; Omenn and McKusick, 1979). Fraser (1976) described a deaf male with no family history of deafness, complete blue-green heterochromia with hypoplastic stroma in the blue iris, and Hirschsprung disease. Chatkupt et al. (1993) stated that spina bifida had been noted in at least 4 patients with Waardenburg syndrome. Chatkupt et al. (1993) reported the cases of brothers with both Waardenburg syndrome and lumbosacral myelomeningocele. The mother had features of Waardenburg syndrome. Spina bifida occurs with the 'splotch' mutation, which molecular studies indicate is the homologous disorder in the mouse (see later).

Read and Newton (1997) provided a review of the clinical features and molecular basis of Waardenburg syndrome and other auditory pigmentary syndromes.

Inheritance

Jones et al. (1975) found evidence of paternal age effect in new mutations for this autosomal dominant disorder.

Kapur and Karam (1991) described a family in which 3 children with this disorder were born to normal, unrelated parents. Germline mosaicism was postulated.

According to the report of Zlotogora et al. (1995), the homozygous form of Waardenburg syndrome is a very severe disorder that has been called WS type III (148820). In a large kindred, including many individuals affected with Waardenburg syndrome type I, they found a child with a very severe form of type III. The child presented with dystopia canthorum, partial albinism, and very severe upper-limb defects. His parents were first cousins, and both were affected with a mild form of WS1. Molecular analysis showed homozygosity for a point mutation in the PAX3 gene (193500.0009). Since all homozygous PAX3 mutations in mice lead to severe neural tube defects and intrauterine or neonatal death, the survival of the homozygote in this case and the absence of neural tube defects were unexpected. Ayme and Philip (1995) indeed observed exencephaly in a fetus with possible homozygous Waardenburg syndrome. The fetus was the product of a mating between a gypsy brother and sister, both of whom had Waardenburg syndrome.

As noted later, dystopia canthorum, as measured by the W-index, is one of the key diagnostic features of WS1. Reynolds et al. (1996) sought to determine whether the W-index is influenced primarily by allelic variation in the PAX3 disease gene or other major loci, by polygenic background effects, or by all of these potential sources of genetic variation. They studied both WS1-affected individuals and their WS1 unaffected relatives. After adjustment of the W-index for WS1 disease status, segregation analyses by the regression approach indicated major-locus control of this variation, although residual parent-offspring and sib-sib correlations were consistent with additional (possibly polygenic) affects. Separate analyses of WS1-affected and WS1-unaffected individuals suggested that epistatic interactions between disease alleles at the PAX3 WS1 locus and a second major locus influenced variation in dystopia canthorum. Reynolds et al. (1996) suggested that their approach should be applicable for assessing the 'genetic architecture' of variation associated with other genetic diseases.

While mutations in PAX3 seem to be responsible for most, if not all, WS1 cases, it is not clear what accounts for the reduced penetrance of deafness. Stochastic events during development may be the factors that determine whether a person with a PAX3 mutation will be congenitally deaf or not. Alternatively, genetic background, nonrandom environmental factors, or both may be significant. Morell et al. (1997) compared the likelihood for deafness in affected subjects from 24 families with PAX3 mutations and in 7 of the families originally described by Waardenburg. They found evidence that stochastic variation alone does not explain the differences in penetrance of deafness among WS families. Their analyses suggested that genetic background in combination with certain PAX3 alleles may be important factors in the etiology of deafness in WS1.

Mapping

Simpson et al. (1974) and Arias et al. (1975) found a weak suggestion of linkage to the ABO locus, known to be located in 9q34. Read et al. (1989) excluded linkage to ABO. In a study of 2 large kindreds in northeastern Brazil, da-Silva et al. (1990) could not confirm linkage to ABO. Other studies of the same 2 Brazilian kindreds were reported by da-Silva (1991).

Since a plausible mouse model is 'Steel' (Sl), a dominant mutation on mouse chromosome 10 closely linked to Pep-2, Read et al. (1989) studied polymorphic probes for loci on human chromosome 12 close to PEPB, the human homolog, in 7 families. They excluded a sizable region of 12q as the site of this gene.

Ishikiriyama et al. (1989) reported the case of a 20-month-old boy with dystopia canthorum, sensorineural deafness, heterochromia iridis, partially albinotic ocular fundi, and partial leukoderma. Cytogenetic studies showed a paracentric inversion (2)(q35q37.3); his parents had normal chromosomes. Ishikiriyama et al. (1989) suggested that the gene for Waardenburg syndrome type I may be located at 2q35 or 2q37.3. Kirkpatrick et al. (1992) described WS type I in a child with del(2)(q35q36.2). Because of this report and that of Lin et al. (1992) of deletion of 2q37 without features of WS1, Ishikiriyama (1993) concluded that the WS1 gene is located at 2q35.

On the basis of an analysis of mouse and hamster mutants as models for Waardenburg syndrome(s), Asher and Friedman (1990) predicted that the gene(s) would be found to be on chromosome 2q near fibronectin-1, on chromosome 3p near the protooncogene RAF1 or 3q near rhodopsin, or on chromosome 4p near the protooncogene KIT. Foy et al. (1990) demonstrated linkage of the Waardenburg syndrome to placental alkaline phosphatase (ALPP; 171800), which had previously been assigned to 2q37; the peak lod score was 4.76 at a recombination fraction of 0.023. These findings suggest that the distal breakpoint responsible for the paracentric inversion is at the site of the Waardenburg syndrome, namely, 2q37.3. This region of chromosome 2 is homologous to mouse chromosome 1, which contains the 'Splotch' locus (Sp). This patchy pigment mutation is accompanied by a malformation of the inner ear and severe CNS malformation in the homozygote. Whether the heterozygote is deaf is unclear.

By PCR analysis of somatic cell hybrids, Pilz et al. (1993) mapped the PAX3 gene to chromosome 2.

Quoting William Harvey's famous observation that 'Nature is nowhere accustomed more openly to display her secret mysteries than in cases where she shows traces of her workings apart from the beaten path,' Duyk et al. (1992) reviewed the forms of deafness, syndromal and nonsyndromal, for which linkage has been established.

Work in the hamster model for Waardenburg syndrome suggested to Asher and Friedman (1990) that modifier genes may account for the intrafamilial variation in phenotype in Waardenburg syndrome. In a family with 11 affected individuals spanning 4 generations, Asher et al. (1991) confirmed the assignment of WS1 to 2q. No recombination was found with ALPP at 2q37 or with FN1 (135600) at 2q34-q36. In a report from a consortium, Grundfast et al. (1991) concluded that only about 56% of WS families are linked to 2q markers. Furthermore, in the families with apparent linkage, they found no obligatory crossovers between WS1 and ALPP. For the whole collection of families, they obtained a maximum lod of 12.5 at theta = 0.31 for the WS1/ALPP linkage. Farrer et al. (1992) estimated that the WS1 gene on chromosome 2 was responsible for approximately 45% of the 44 families in their sample. Wilcox et al. (1992) used the intron between exons 2 and 3, amplified by PCR, as a probe to determine that the PAX3 gene is located on chromosome 2, from the study of DNA from somatic cell hybrids. They also identified a highly informative CA dinucleotide repeat.

Farrer et al. (1994) found that all families with WS1 showed linkage to the PAX3 region of chromosome 2. Two forms of WS2 (defined as absence of dystopia canthorum, i.e., W index less than 2.07) have been defined on the basis of linkage studies: WS2A (193510) due to mutation in the MITF gene (156845) maps to 3p; WS2B (600193) maps to 1p.

Molecular Genetics

Based on amino acid sequence homology and common genomic exon/intron organization (Goulding et al., 1991), the human homolog of the mouse Pax-3 gene is thought to be the HUP2 gene (Burri et al., 1989). Goulding et al. (1991) found that the product of the Pax-3 gene is a DNA-binding protein expressed during early neurogenesis. The HuP2 sequence contains 3 exons, which Tassabehji et al. (1992) designated exons 2, 3, and 4 because of their correspondence with exons 2, 3, and 4 of the mouse Pax-3 gene which has 5 exons. To identify patient-specific mutations in the human gene, Tassabehji et al. (1992) designed primers to amplify exons of the gene and tested for heteroduplex formation on polyacrylamide gels. Variant bands were seen in 6 out of 17 unrelated WS patients. Three of the variants were seen in exon 2, 2 in exon 3, and 1 in exon 4. No variants were seen in any exon in 50 normal controls. In 3 families that were tested, the variant was found to be familial in 2 and apparently de novo in the third. The variant bands showed perfect linkage to WS in the families studied. The HuP2 sequence showed 92% nucleotide homology and 100% amino acid homology with the mouse Pax-3 sequence over amino acids 2 to 121. In 1 family, an 18-bp deletion was found in the central region of exon 2, resulting in loss of amino acids 29 to 34. The deleted sequence ran between 2 directly repeated GGCCC sequences. Another family had a small change, either a point mutation or a very small deletion, in exon 2 and a third family had a deletion in exon 4. In a large Brazilian family reported by da-Silva (1991), Baldwin et al. (1992) likewise identified a mutation in the HuP2 gene. There were 49 persons in 6 generations in whom the diagnosis of WS1 had been made. More than 78% of the affected members had noticeable hearing loss. Single-strand conformation polymorphism analysis suggested a sequence variation in exon 2 in affected members. Sequencing of exon 2 showed that there were 2 bands (G/ A) 64 bp downstream from the 5-prime end of exon 2. The G band at this position corresponded to the normal sequence in the antisense strand, whereas the unusual A residue substituted the second base of a pro codon (CCG), making it a codon for a leu residue (CTG) in the sense strand. The single base substitution at this position destroyed a HaeIII restriction site in HuP2 exon 2. All 26 affected members of the family were heterozygous for the HaeIII site, whereas all 34 unaffected members and 50 control subjects were homozygous for the presence of the HaeIII site. Baldwin et al. (1992) did not detect the pro-to-leu mutation in 17 unrelated WS1 families.

Tsukamoto et al. (1992) cloned and characterized an inversion breakpoint of the inv(2)(q35q37.3) paracentric inversion reported by Ishikiriyama et al. (1989). Genomic cosmid clones containing the HUP2 gene were isolated from a library constructed from the patient's DNA. One of the clones contained the inversion breakpoint and revealed signals at both 2q35 and 2q37 by fluorescence in situ hybridization, indicating that the HUP2 gene had been disrupted by the inversion. The results suggested that the gene is situated at the more proximal breakpoint 2q35 because one cosmid clone, presumably derived from the normal allele, hybridized only to 2q35.

Baldwin et al. (1995) described 10 additional mutations in the PAX3 gene in families with WS type I. Eight of these mutations were in a region of PAX3 where only 1 mutation had been previously been described. Taken together with previously reported mutations, these mutations covered essentially the entire PAX3 gene. All but 1 of the mutations were 'private;' only 1 mutation had been reported in 2 apparently unrelated families. Preliminary screening for mutations was performed with conformation-sensitive gel electrophoresis (CSGE), as described by Ganguly et al. (1993). Baldwin et al. (1995) cataloged 16 previously reported mutations and 5 chromosomal abnormalities affecting the 2q35 region and associated with WS.

Among 24 unrelated individuals with WS1 mutations, Farrer et al. (1994) found no 2 had the same point mutation in the protein-coding region of PAX3, nor did any of them have a change in the same codon.

In each of 2 families with WS type I, Wildhardt et al. (1996) described the causative PAX3 mutation. One mutation was an insertion in the paired box domain resulting in a protein termination within the paired box. The second mutation was a single-basepair substitution producing an arg271-to-cys amino acid change in the homeobox region.

Phenotype/genotype

In a series of patients with Waardenburg syndrome, Tassabehji et al. (1994) found a number of previously unidentified PAX3 mutations. These included a chromosomal deletion, a splice site mutation, and an amino acid substitution that closely corresponded to the molecular changes seen in the 'splotch-retarded' and 'splotch-delayed' mouse mutants, respectively. These mutations confirmed that Waardenburg syndrome is produced by gene dosage effects and showed that the phenotypic differences between 'splotch' mice and humans with Waardenburg syndrome are caused by differences in genetic background rather than different primary effects of the mutations.

Chalepakis et al. (1994) studied the functional consequence of the mutations described in 193500.0001 and 193500.0006 on DNA binding and compared the results with those in the 'splotch' mouse. Combining the phenotypic features of heterozygous mutants and considering that molecular defects ranging from single point mutations to large deletions cause similar phenotypes, they excluded the possibility that the mutated allele in heterozygotes interferes with the function of the wildtype allele. Contrariwise, they considered both WS and 'splotch' mutants to represent loss-of-function mutations.

Baldwin et al. (1995) stated that their analysis of a total of 30 PAX3 mutations causing WS type I or type III demonstrated little correlation between genotype and phenotype. Deletions of the entire PAX3 gene resulted in phenotypes indistinguishable from those associated with single-base substitutions in the paired domain or homeodomain of the gene. Moreover, 2 similar mutations in close proximity could result in significantly different phenotypes, WS type I in 1 family and WS type III in another.

DeStefano et al. (1998) assessed the relationship between phenotype and gene defect in 48 families containing 271 individuals with WS collected by members of the Waardenburg Consortium. They grouped the 42 unique mutations previously identified in the PAX3 gene in these families into 5 mutation categories: amino acid substitution in the paired domain, amino acid substitution in the homeodomain, deletion of the ser-thr-pro-rich region, deletion of the homeodomain and the ser-thr-pro-rich region, and deletion of the entire gene. This classification of mutations was based on the structure of the PAX3 gene and was chosen to group mutations predicted to have similar defects in the gene product. They found that odds for the presence of eye pigment abnormality, white forelock, and skin hypopigmentation were 2, 8, and 5 times greater, respectively, for individuals with deletions of the homeodomain and the pro-ser-thr-rich region compared to individuals with an amino acid substitution in the homeodomain. Odds ratios that differed significantly from 1.0 for these traits may indicate that the gene products resulting from different classes of mutations act differently in the expression of WS. Although a suggestive association was detected for hearing loss with an odds ratio of 2.6 for amino acid substitution in the paired domain compared with amino acid substitution in the homeodomain, this odds ratio did not differ significantly from 1.0.

Pathogenesis

Mutations in MITF (156845) and PAX3, encoding transcriptions factors, are responsible for Waardenburg syndrome type II (193510) and WS1/WS3, respectively. Tachibana et al. (1996) showed that MITF transactivates the gene for tyrosinase (see 203100), a key enzyme for melanogenesis, and is critically involved in melanocyte differentiation. Absence of melanocytes affects pigmentation in the skin, hair, and eyes, and hearing function in the cochlea. Therefore, hypopigmentation and hearing loss in WS2 are likely to be the results of an anomaly of melanocyte differentiation caused by MITF mutations. However, the molecular mechanism by which PAX3 mutations cause the auditory-pigmentary symptoms in WS1/WS3 had not been explained. Watanabe et al. (1998) showed that PAX3 transactivates the MITF promoter. They further showed that PAX3 proteins associated with WS1 in either the paired domain or the homeodomain failed the recognize and transactivate the MITF promoter. These results provided evidence that PAX3 directly regulates MITF, and suggested that the failure of this regulation due to PAX3 mutations causes the auditory-pigmentary symptoms in at least some individuals with WS1.

Animal Model

The work of Bosher and Hallpike (1966) on an animal analog, deaf white cats, suggested that destruction of the inner ear mechanism occurs in the first days of extrauterine life and was correlated with an inability to regulate properly the constitution of the endolymphatic fluid. The cat, like man, may escape deafness in one or both ears. If more of the factors that lead to retention of hearing were known, deafness might be preventable.

Motohashi et al. (1994) pointed out that melanocytes are a normal component of the inner ear, including the stria vascularis. There are 3 known mutations in the mouse which lead to a deficiency of melanocytes in mast cells: white dominant spotting (W), Steel (Sl), and microphthalmia (mi). All 3 mutants have a thin stria vascularis, without melanocyte-like intermediate cells, and severe impairment of hearing. Thus, the absence of intermediate cells or melanocytes causes severe hearing loss. The absence of melanin has little influence on hearing acuity because albino mice without melanin have no impairment of hearing.

Other genes in the 9q34 band have homologs on mouse chromosome 2. In the mouse, the 'lethal-spotted' (ls) mutation, which results not only in spotting but also in failure of the entire ganglia to colonize the gut, is located on chromosome 2. Chakravarti (1988) suggested that this and other possibly homologous mouse mutations might be used as a clue to the chromosomal location of the WS1 gene in the human. Jacobs-Cohen et al. (1987) found that in lethal spotting piebald (ls/ls) mice, who develop megacolon, the terminal 2 mm of intestine does not become colonized by neural crest cells, resulting in aganglionosis. Neural crest cells transplanted from areas around the neural tube (primary explants) or foregut (secondary explants) did not colonize the terminal portions of the hindgut. These findings suggested that the nonneuronal components are abnormal, preventing migration of normal neural crest derivatives into the bowel wall.

Epstein et al. (1991) studied a deletion of mouse chromosome 1 that involved the 'splotch' locus. The murine equivalent of the ALPP gene was included in the deletion, thus supporting the notion that 'splotch' is the equivalent of WS1. Furthermore, Epstein et al. (1991) mapped the paired box gene Pax-3 to a region near or at the Sp locus on mouse chromosome 1 and found Pax-3 to be deleted in mice heterozygous for a splotch allele. In another allelic variant of splotch, they found deletion of 32 nucleotides in the Pax-3 mRNA transcript and gene. The deletion was located within the paired homeodomain of Pax-3 and was predicted to create a truncated protein as a result of a newly created termination codon at the deletion breakpoint. The splotch mutation is associated in the mouse with spina bifida and exencephaly. The findings were interpreted to indicate that Pax-3 plays a key role in normal neural development. Moase and Trasler (1992) reviewed the subject of the splotch locus in mouse mutants. Steel and Smith (1992) found that, unlike individuals with Waardenburg syndrome, the splotch mouse has normal hearing. They suggested that the difference in expression of the genes in the 2 species may result from different parts of the gene being mutated or from modifying influences as yet undefined.

Epstein et al. (1993) demonstrated that in the original, spontaneously arising Sp allele, a complex mutation in the PAX3 gene had occurred, including an A-to-T transversion at the invariant 3-prime AG splice acceptor of intron 3. This mutation abrogated the normal splicing of intron 3, resulting in the generation of 4 aberrantly spliced mRNA transcripts. Two of the transcripts made use of cryptic 3-prime splice sites within the downstream exon, generating small deletions which disrupted the reading frame of the transcripts. A third aberrant splicing event resulted in the deletion of exon 4, while a fourth retained intron 3. None of these mutations would be expected to result in functional PAX3 proteins.

Asher et al. (1996) stated that over 50 human PAX3 mutations leading to hearing, craniofacial, limb, and pigmentation anomalies had been identified. Variability in penetrance and expressivity is observed in humans with PAX3 mutations and in mice with 'splotch' mutations. In mice with certain 'splotch' mutations, influence of the genetic background and sex of the individual on penetrance and expressivity is demonstrable. Asher et al. (1996) described a murine model for Waardenburg syndrome variation and concluded that a minimum of 2 genes interact with the 'splotch' mutation to influence the craniofacial features of these mice. One of these genes may be either X-linked or sex-influenced, while the other is autosomal. They stated that these studies in mice may lead to the identification of genes that modify the expression of human PAX3 mutations.

See also Arias and Mota (1978); Arias and Mota (1978); David and Warin (1972); Delleman and Hageman (1978); Epstein et al. (1991); Fried and Beer (1980); Goldberg (1966); Hageman (1978); Ishikiriyama et al. (1989); Nutman et al. (1981); Opitz (1980); Preus et al. (1983); Settelmayer and Hogan (1961); Shah et al. (1981)

Arias, S.: Genetic heterogeneity in the Waardenburg syndrome. Birth Defects Orig. Art. Ser. VII(4): 87-101, 1971.

Arias, S.; Mota, M.: Current status of the ABO-Waardenburg syndrome type I linkage. Cytogenet. Cell Genet. 22: 291-294, 1978.

Arias, S.; Mota, M.: Apparent non-penetrance for dystopia in Waardenburg syndrome type I, with some hints on the diagnosis of dystopia canthorum. J. Genet. Hum. 26: 103-131, 1978.

Arias, S.; Mota, M.; de Yanez, A.; Bolivar, M.: Probable loose linkage between the ABO locus and Waardenburg syndrome type I. Humangenetik 27: 145-149, 1975.

Asher, J. H., Jr.; Friedman, T. B.: Mouse and hamster mutants as models for Waardenburg syndromes in humans. J. Med. Genet. 27: 618-626, 1990.

Asher, J. H., Jr.; Harrison, R. W.; Morell, R.; Carey, M. L.; Friedman, T. B.: Effects of Pax3 modifier genes on craniofacial morphology, pigmentation, and viability: a murine model of Waardenburg syndrome variation. Genomics 34: 285-298, 1996.

Asher, J. H., Jr.; Morell, R.; Friedman, T. B.: Waardenburg syndrome (WS): the analysis of a single family with a WS1 mutation showing linkage to RFLP markers on human chromosome 2q. Am. J. Hum. Genet. 48: 43-52, 1991.

Asher, J. H., Jr.; Sommer, A.; Morrell, R.; Friedman, T. B.: Missense mutation in the paired domain of PAX3 causes craniofacial-deafness-hand syndrome. Hum. Mutat. 7: 30-35, 1996.

Ayme, S.; Philip, N.: Possible homozygous Waardenburg syndrome in a fetus with exencephaly. (Letter) Am. J. Med. Genet. 59: 263-265, 1995.

Baldwin, C. T.; Hoth, C. F.; Amos, J. A.; da-Silva, E. O.; Milunsky, A.: An exonic mutation in the HuP2 paired domain gene causes Waardenburg's syndrome. Nature 355: 637-638, 1992.

Baldwin, C. T.; Hoth, C. F.; Macina, R. A.; Milunsky, A.: Mutations in PAX3 that cause Waardenburg syndrome type I: ten new mutations and review of the literature. Am. J. Med. Genet. 58: 115-122, 1995.

Barr, F. G.; Galili, N.; Holick, J.; Biegel, J. A.; Rovera, G.; Emanuel, B. S.: Rearrangement of the PAX3 paired box gene in the paediatric solid tumour alveolar rhabdomyosarcoma. Nature Genet. 3: 113-117, 1993.

Bosher, S. K.; Hallpike, C. S.: Observations on the histogenesis of the inner ear degeneration of the deaf white cat and its possible relationship to the aetiology of certain unexplained varieties of human congenital deafness. J. Laryng. 80: 222-235, 1966.

Burri, M.; Tromvoukis, Y.; Bopp, D.; Frigerio, G.; Noll, M.: Conservation of the paired domain in metazoans and its structure in three isolated human genes. EMBO J. 8: 1183-1190, 1989.

Carezani-Gavin, M.; Clarren, S. K.; Steege, T.: Waardenburg syndrome associated with meningomyelocele. (Letter) Am. J. Med. Genet. 42: 135-136, 1992.

Chakravarti, A.: Personal Communication. Pittsburgh, Pa. 2/9/1988.

Chalepakis, G.; Goulding, M.; Read, A.; Strachan, T.; Gruss, P.: Molecular basis of splotch and Waardenburg Pax-3 mutations. Proc. Nat. Acad. Sci. 91: 3685-3689, 1994.

Chatkupt, S.; Chapkupt, S.; Johnson, W. G.: Waardenburg syndrome and myelomeningocele in a family. J. Med. Genet. 30: 83-84, 1993.

Chew, K. L.; Chen, A. J.; Tan, K. H.: A Chinese family with Waardenburg's syndrome. Am. J. Ophthal. 65: 174-182, 1968.

Christian, J. C.; Bixler, D.; Blythe, S. C.; Merritt, A. D.: Familial telecanthus with associated congenital anomalies. Birth Defects Orig. Art. Ser. V(2): 82-85, 1969.

da-Silva, E. O.: Waardenburg I syndrome: a clinical and genetic study of two large Brazilian kindreds, and literature review. Am. J. Med. Genet. 40: 65-74, 1991.

da-Silva, E. O.; Batista, J. E. M.; Medeiros, M. A. B.; Fonteles, S. M. S.: Craniofacial anthropometric studies in Waardenburg syndrome type I. Clin. Genet. 44: 20-25, 1993.

da-Silva, E. O.; Salzano, F. M.; Weimer, T. A.: Is the Waardenburg I syndrome gene located on chromosome 9? Rev. Brasil. Genet. 13: 551-556, 1990.

David, T. J.; Warin, R. P.: Waardenburg's syndrome in two siblings, both parents and their maternal grandmother. Proc. Roy. Soc. Med. 65: 601-602, 1972.

Delleman, J. W.; Hageman, M. J.: Ophthalmological findings in 34 patients with Waardenburg syndrome. J. Pediat. Ophthal. Strabismus 15: 341-345, 1978.

DeStefano, A. L.; Cupples, L. A.; Arnos, K. S.; Asher, J. H., Jr.; Baldwin, C. T.; Blanton, S.; Carey, M. L.; da Silva, E. O.; Friedman, T. B.; Greenberg, J.; Lalwani, A. K.; Milunsky, A.; Nance, W. E.; Pandya, A.; Ramesar, R. S.; Read, A. P.; Tassabejhi, M.; Wilcox, E. R.; Farrer, L. A.: Correlation between Waardenburg syndrome phenotype and genotype in a population of individuals with identified PAX3 mutations. Hum. Genet. 102: 499-506, 1998.

Duyk, G.; Gastier, J. M.; Mueller, R. F.: Traces of her workings. Nature Genet. 2: 5-8, 1992.
Epstein, D. J.; Malo, D.; Vekemans, M.; Gros, P.: Molecular characterization of a deletion encompassing the splotch mutation on mouse chromosome 1. Genomics 10: 89-93, 1991.
Epstein, D. J.; Vekemans, M.; Gros, P.: Splotch (Sp-2H), a mutation affecting development of the mouse neural tube, shows a deletion within the paired homeodomain of Pax-3. Cell 67: 767-774, 1991.
Epstein, D. J.; Vogan, K. J.; Trasler, D. G.; Gros, P.: A mutation within intron 3 of the Pax-3 gene produces aberrantly spliced mRNA transcripts in the splotch (Sp) mouse mutant. Proc. Nat. Acad. Sci. 90: 532-536, 1993.
Farrer, L. A.; Arnos, K. S.; Asher, J. H., Jr.; Baldwin, C. T.; Diehl, S. R.; Friedman, T. B.; Greenberg, J.; Grundfast, K. M.; Hoth, C.; Lalwani, A. K.; Landa, B.; Leverton, K.; Milunsky, A.; Morell, R.; Nance, W. E.; Newton, V.; Ramesar, R.; Rao, V. S.; Reynolds, J. E.; San Agustin, T. B.; Wilcox, E. R.; Winship, I.; Read, A. P.: Locus heterogeneity for Waardenburg syndrome is predictive of clinical subtypes. Am. J. Hum. Genet. 55: 728-737, 1994.
Farrer, L. A.; Grundfast, K. M.; Amos, J.; Arnos, K. S.; Asher, J. H., Jr.; Beighton, P.; Diehl, S. R.; Fex, J.; Foy, C.; Friedman, T. B.; Greenberg, J.; Hoth, C.; Marazita, M.; Milunsky, A.; Morell, R.; Nance, W.; Newton, V.; Ramesar, R.; San Agustin, T. B.; Skare, J.; Stevens, C. A.; Wagner, R. G., Jr.; Wilcox, E. R.; Winship, I.; Read, A. P.: Waardenburg syndrome (WS) type I is caused by defects at multiple loci, one of which is near ALPP on chromosome 2: first report of the WS Consortium. Am. J. Hum. Genet. 50: 902-913, 1992.
Feingold, M.; Robinson, M. J.; Gellis, S. S.: Waardenburg's syndrome during the first year of life. J. Pediat. 71: 874-876, 1967.
Foy, C.; Newton, V.; Wellesley, D.; Harris, R.; Read, A. P.: Assignment of the locus for Waardenburg syndrome type I to human chromosome 2q37 and possible homology to the splotch mouse. Am. J. Hum. Genet. 46: 1017-1023, 1990.
Fraser, G. R.: The Causes of Profound Deafness in Childhood. Baltimore: Johns Hopkins Univ. Press (pub.) 1976. Pp. 106 only.
Fraser, G. R.: Personal Communication. Adelaide, Australia 1967.
Fried, K.; Beer, S.: Waardenburg's syndrome and Hirschsprung's disease in the same patient. (Letter) Clin. Genet. 18: 91-92, 1980.
Galili, N.; Davis, R. J.; Fredericks, W. J.; Mukhopadhyay, S.; Rauscher, F. J., III; Emanuel, B. S.; Rovera, G.; Barr, F. G.: Fusion of a fork head domain gene to PAX3 in the solid tumour alveolar rhabdomyosarcoma. Nature Genet. 5: 230-235, 1993.
Ganguly, A.; Rock, M. J.; Prockop, D. J.: Conformation-sensitive gel electrophoresis for rapid detection of single-base differences in double-stranded PCR products and DNA fragments: evidence for solvent-induced bends in DNA heteroduplexes. Proc. Nat. Acad. Sci. 90: 10325-10329, 1993.
Giacoia, J. P.; Klein, S. W.: Waardenburg's syndrome with bilateral cleft lip. Am. J. Dis. Child. 117: 344-348, 1969.
Goldberg, M. F.: Waardenburg's syndrome with fundus and other anomalies. Arch. Ophthal. 76: 797-810, 1966.
Goodman, R. M.; Lewithal, I.; Solomon, A.; Klein, D.: Upper limb involvement in the Klein-Waardenburg syndrome. Am. J. Med. Genet. 11: 425-433, 1982.
Goodman, R. M.; Oelsner, G.; Berkenstadt, M.; Admon, D.: Absence of a vagina and right sided adnexa uteri in the Waardenburg syndrome: a possible clue to the embryological defect. J. Med. Genet. 25: 355-357, 1988.
Goulding, M. D.; Chalepakis, G.; Deutsch, U.; Erselius, J. R.; Gruss, P.: Pax-3, a novel murine DNA binding protein expressed during early neurogenesis. EMBO J. 10: 1135-1147, 1991.
Grundfast, K.; Farrer, L.; Amos, J.; Arnos, K. S.; Asher, J.; Beighton, P.; Diehl, S.; Fex, J.; Foy, C.; Friedman, T.; Greenberg, J.; Hoth, C.; Milunsky, A.; Morell, R.; Nance, W.; Newton, V.; Ramesar, R.; Read, A.; Skare, J.; SanAgustin, T.; Stevens, C.; Wagner, R., Jr.; Wilcox, E.; Winship, I.: Waardenburg syndrome is caused by defects at multiple loci, one of which is tightly linked to ALPP on chromosome 2—first report of the WS consortium. (Abstract) Am. J. Hum. Genet. 49 (suppl.): 17, 1991.
Hageman, M. J.: Waardenburg's syndrome in Kenyan Africans. Trop. Geogr. Med. 30: 45-55, 1978.
Hansen, A. C.; Ackaouy, G.; Crump, E. P.: Waardenburg's syndrome: report of a pedigree. J. Nat. Med. Assoc. 57: 8-12, 1965.
Hol, F. A.; Hamel, B. C. J.; Geurds, M. P. A.; Mullaart, R. A.; Barr, F. G.; Macina, R. A.; Mariman, E. C. M.: A frameshift mutation in the gene for PAX3 in a girl with spina bifida and mild signs of Waardenburg syndrome. J. Med. Genet. 32: 52-56, 1995.
Hoth, C. F.; Milunsky, A.; Lipsky, N.; Sheffer, R.; Clarren, S. K.; Baldwin, C. T.: Mutations in the paired domain of the human PAX3 gene cause Klein-Waardenburg syndrome (WS-III) as well as Waardenburg syndrome type I (WS-I). Am. J. Hum. Genet. 52: 455-462, 1993.
Houghton, N. I.: Waardenburg's syndrome with deafness as the presenting symptom: report of two cases. New Zeal. Med. J. 63: 83-89, 1964.
Ishikiriyama, S.: Gene for Waardenburg syndrome type I is located at 2q35, not at 2q37.3. (Letter) Am. J. Med. Genet. 46: 608, 1993.
Ishikiriyama, S.; Tonoki, H.; Shibuya, Y.; Chin, S.; Harada, N.; Abe, K.; Niikawa, N.: Waardenburg syndrome type I in a child with de novo inversion (2)(q35q37.3). Am. J. Med. Genet. 33: 505-507, 1989.
Ishikiriyama, S.; Tonoki, H.; Shibuya, Y.; Chin, S.; Harada, N.; Abe, K.; Niikawa, N.: A sporadic case of Waardenburg syndrome type I associated with de novo inv(2)(q35q37.3). (Abstract) Cytogenet. Cell Genet. 51: 1018, 1989.
Jacobs-Cohen, R. J.; Payette, R. F.; Gershon, M. D.; Rothman, T. P.: Inability of neural crest cells to colonize the presumptive aganglionic bowel of ls/ls mutant mice: requirement for a permissive environment. J. Comp. Neurol. 255: 425-438, 1987.
Jones, K. L.; Smith, D. W.; Harvey, M. A. S.; Hall, B. D.; Quan, L.: Older paternal age and fresh gene mutation: data on additional disorders. J. Pediat. 86: 84-88, 1975.
Kapur, S.; Karam, S.: Germ-line mosaicism in Waardenburg syndrome. Clin. Genet. 39: 194-198, 1991.
Kirkpatrick, S. J.; Kent, C. M.; Laxova, R.; Sekhon, G. S.: Waardenburg syndrome type I in a child with deletion (2)(q35q36.2). (Letter) Am. J. Med. Genet. 44: 699-700, 1992.
Klein, D.: Albinisme partiel (leucisme) avec surdi-mutite, blepharophimosis et dysplasie myo-osteo-articulaire. Helv. Paediat. Acta 5: 38-58, 1950.
Laestadius, N. D.; Aase, J. M.; Smith, D. W.: Normal inner canthal and outer orbital dimensions. J. Pediat. 74: 465-468, 1969.
Lin, S.-P.; Petty, E. M.; Gibson, L. H.; Inserra, J.; Seashore, M. R.; Yang-Feng, T. L.: Smallest terminal deletion of the long arm of chromosome 2 in a mildly affected boy. Am. J. Med. Genet. 44: 500-502, 1992.
Lowry, R. B.: Hirschsprung's disease and congenital deafness. (Letter) J. Med. Genet. 12: 114-115, 1975.
McKusick, V. A.: Congenital deafness and Hirschsprung's disease. (Letter) New Eng. J. Med. 288: 691, 1973.
Milunsky, A.: Personal Communication. Boston, Mass. 3/4/1992.
Moase, C. E.; Trasler, D. G.: Splotch locus mouse mutants: models for neural tube defects and Waardenburg syndrome type I in humans. J. Med. Genet. 29: 145-151, 1992.
Morell, R.; Friedman, T. B.; Asher, J. H., Jr.; Robbins, L. G.: The incidence of deafness is non-randomly distributed among families segregating for Waardenburg syndrome type 1 (WS1). J. Med. Genet. 34: 447-452, 1997.
Morell, R.; Friedman, T. B.; Moeljopawiro, S.; Hartono, (NI); Soewito, (NI); Asher, J. H., Jr.: A frameshift mutation in the HuP2 paired domain of the probable human homolog of murine Pax-3 is responsible for Waardenburg syndrome type 1 in an Indonesian family. Hum. Molec. Genet. 1: 243-247, 1992.
Motohashi, H.; Hozawa, K.; Oshima, T.; Takeuchi, T.; Takasaka, T.: Dysgenesis of melanocytes and cochlear dysfunction in mutant microphthalmia (mi) mice. Hear. Res. 80: 10-20, 1994.
Newton, C. R.; Graham, A.; Heptinstall, L. E.; Powell, S. J.; Summers, C.; Kalsheker, N.; Smith, J. C.; Markham, A. F.: Analysis of any point mutation in DNA: the amplification refractory mutation system (ARMS). Nucleic Acids Res. 17: 2503-2516, 1989.
Newton, V. E.: Waardenburg's syndrome: a comparison of biometric indices used to diagnose lateral displacement of the inner canthi. Scand. Audiol. 18: 221-223, 1989.
Nutman, J.; Nissenkorn, I.; Varsano, I.; Mimouni, M.; Goodman, R. M.: Anal atresia and the Klein-Waardenburg syndrome. J. Med. Genet. 18: 239-241, 1981.
Omenn, G. S.; McKusick, V. A.: The association of Waardenburg syndrome and Hirschsprung megacolon. Am. J. Med. Genet. 3: 217-223, 1979.
Opitz, J. M.: In memoriam: Petrus Johannes Waardenburg, 1886-1979. Am. J. Med. Genet. 7: 35-39, 1980.
Pilz, A. J.; Povey, S.; Gruss, P.; Abbott, C. M.: Mapping of the human homologs of the murine paired-box-containing genes. Mammalian Genome 4: 78-82, 1993.
Preus, M.; Linstrom, C.; Polomeno, R. C.; Milot, J.: Waardenburg syndrome—penetrance of major signs. Am. J. Med. Genet. 15: 383-388, 1983.
Read, A. P.; Foy, C.; Newton, V.: Exclusion mapping in Waardenburg syndrome. (Abstract) Cytogenet. Cell Genet. 51: 1064, 1989.
Read, A. P.; Newton, V. E.: Waardenburg syndrome. J. Med. Genet. 34: 656-665, 1997.
Reynolds, J. E.; Marazita, M. L.; Meyer, J. M.; Stevens, C. A.; Eaves, L. J.; Arnos, K. S.; Ploughman, L. M.; MacLean, C.; Nance, W. E.; Diehl, S. R.: Major-locus contributions to variability of the craniofacial feature dystopia canthorum in Waardenburg syndrome. Am. J. Hum. Genet. 58: 384-392, 1996.
Reynolds, J. E.; Meyer, J. M.; Landa, B.; Stevens, C. A.; Arnos, K. S.; Israel, J.; Marazita, M. L.; Bodurtha, J.; Nance, W. E.; Diehl, S. R.: Analysis of variability of clinical manifestations in Waardenburg syndrome. Am. J. Med. Genet. 57: 540-547, 1995.
Settelmayer, J. R.; Hogan, M.: Waardenburg's syndrome—report of a case in a non-Dutch family. New Eng. J. Med. 264: 500-501, 1961.
Shah, K. N.; Dalal, S. J.; Desai, M. P.; Sheth, P. N.; Joshi, N. C.; Ambani, L. M.: White forelock, pigmentary disorder of irides, and long segment Hirschsprung disease: possible variant of Waardenburg syndrome. J. Pediat. 99: 432-435, 1981.
Sheffer, R.; Zlotogora, J.: Autosomal dominant inheritance of Klein-Waardenburg syndrome. Am. J. Med. Genet. 42: 320-322, 1992.
Simpson, J. L.; Falk, C. T.; Morillo-Cucci, G.; Allen, F. H., Jr.; German, J.: Analysis for possible linkage between the loci for the Waardenburg syndrome and various blood groups and serological traits. Humangenetik 23: 45-50, 1974.
Sommer, A.; Young-Wee, T.; Frye, T.: Previously undescribed syndrome of craniofacial, hand anomalies, and sensorineural deafness. Am. J. Med. Genet. 15: 71-77, 1983.
Steel, K. P.; Smith, R. J. H.: Normal hearing in splotch (Sp/+), the mouse homologue of Waardenburg syndrome type 1. Nature Genet. 2: 75-79, 1992.
Tachibana, M.; Takeda, K.; Nobukuni, Y.; Urabe, K.; Long, J. E.; Meyers, K. A.; Aaronson, S. A.; Miki, T.: Ectopic expression of MITF, a gene for Waardenburg syndrome type 2, converts fibroblasts to cells with melanocytes characteristics. Nature Genet. 14: 50-54, 1996.
Tassabehji, M.; Newton, V. E.; Leverton, K.; Turnbull, K.; Seemanova, E.; Kunze, J.; Sperling, K.; Strachan, T.; Read, A. P.: PAX3 gene structure and mutations: close analogies between Waardenburg syndrome and the 'splotch' mouse. Hum. Molec. Genet. 3: 1069-1074, 1994.
Tassabehji, M.; Newton, V. E.; Liu, X.-Z.; Brady, A.; Donnai, D.; Krajewska-Walasek, M.; Murday, V.; Norman, A.; Obersztyn, E.; Reardon, W.; Rice, J. C.; Trembath, R.; Wieacker, P.; Whiteford, M.; Winter, R.; Read, A. P.: The mutational spectrum in Waardenburg syndrome. Hum. Molec. Genet. 4: 2131-2137, 1995.
Tassabehji, M.; Read, A. P.; Newton, V. E.; Harris, R.; Balling, R.; Gruss, P.; Strachan, T.: Waardenburg's syndrome patients have mutations in the human homologue of the Pax-3 paired box gene. Nature 355: 635-636, 1992.

Tassabehji, M.; Read, A. P.; Newton, V. E.; Patton, M.; Gruss, P.; Harris, R.; Strachan, T.: Mutations in the PAX3 gene causing Waardenburg syndrome type 1 and type 2. Nature Genet. 3: 26-30, 1993.
Tsukamoto, K.; Tohma, T.; Ohta, T.; Yamakawa, K.; Fukushima, Y.; Nakamura, Y.; Niikawa, N.: Cloning and characterization of the inversion breakpoint at chromosome 2q35 in a patient with Waardenburg syndrome type I. Hum. Molec. Genet. 1: 315-317, 1992.
Watanabe, A.; Takeda, K.; Ploplis, B.; Tachibana, M.: Epistatic relationship between Waardenburg syndrome genes MITF and PAX3. Nature Genet. 18: 283-286, 1998.
Wilcox, E. R.; Rivolta, M. N.; Ploplis, B.; Potterf, S. B.; Fex, J.: The PAX3 gene is mapped to human chromosome 2 together with a highly informative CA dinucleotide repeat. Hum. Molec. Genet. 1: 215, 1992.
Wildhardt, G.; Winterpacht, A.; Hibert, K.; Menger, H.; Zabel, B.: Two different PAX3 gene mutations causing Waardenburg syndrome type I. Molec. Cell. Probes 10: 229-231, 1996.
Winship, I.; Beighton, P.: Phenotypic discriminants in the Waardenburg syndrome. Clin. Genet. 41: 181-188, 1992.
Yoshino, M.; Nakao, M.; Shiotsuki, Y.; Nishiyori, A.; Yamashita, F.; Karukaya, S.; Yoshimura, H.; Nishida, H.; Shiotani, N.; Sugita, A.: Incidences of dystopia canthorum and some other signs in a family with Waardenburg syndrome type I. Jpn. J. Hum. Genet. 31: 373-378, 1986.
Zlotogora, J.; Lerer, I.; Bar-David, S.; Ergaz, Z.; Abeliovich, D.: Homozygosity for Waardenburg syndrome. Am. J. Hum. Genet. 56: 1173-1178, 1995.

Tumor-Suppressor Genes

JJM NOTE

Also known as antioncogenes, the human genes that normally act in the cell to suppress the emergence of a neoplasm are called *tumor-suppressor genes*. The classic feature seen in the laboratory is that the tumor, compared to nontumorous tissue, displays the loss of any constitutional heterozygosity of gene loci in the area of the responsible defective gene. That is, compared to DNA from blood lymphocytes, which are shown to have different alleles at the same locus, the tumor DNA shows only one allele. Some well-known tumor-suppressor genes, like retinoblastoma and Wilms tumor genes, are included in the organ system chapters. To comb *OMIM* for other candidate entries, the data fields for titles, text, and clinical synoposis were electronically queried for *loss of heterozygosity*, *LOH*, *delet** (to catch both *deleted* and *deletion*), and *suppress** (to retrieve *suppressor*, *suppressed*, etc.). The 229 unique entries were examined for appropriateness, and those listed here are not listed elsewhere in the *Catalog.*

RLF PROTEIN INVOLVED IN ACTIVATION OF LMYC (*180610)

In 2 independent small cell lung carcinoma (SCLC) cell lines, Makela et al. (1991) found similar novel proteins antigenically related to LMYC (164850). cDNAs corresponding to these proteins were cloned and shown to encode chimeric polypeptides with amino-terminal sequences from a novel gene named RLF (for rearranged LMYC fusion) joined to the LMYC protein. Makela et al. (1991) presented the preliminary characterization of the RLF/LMYC fusion gene which has been found only in cells with an amplified LMYC gene. In addition, they used somatic cell hybrids to assign the normal RLF locus to the same chromosome, chromosome 1, on which LMYC resides. By pulsed field gel electrophoresis, they established a physical linkage between the 2 genes which were found to be separated by less than 800 kb. The presence of independent genetic lesions caused the formation of identical chimeric RLF/LMYC proteins and suggested a role for the fusion protein in the development of tumors.

See also Makela et al. (1991)

Makela, T. P.; Kere, J.; Winqvist, R.; Alitalo, K.: Intrachromosomal rearrangements fusing L-myc and rlf in small-cell lung cancer. Molec. Cell. Biol. 11: 4015-4021, 1991.
Makela, T. P.; Saksela, K.; Evan, G.; Alitalo, K.: A fusion protein formed by L-myc and a novel gene in SCLC. EMBO J. 10: 1331-1335, 1991.

BREAST CANCER-RELATED REGULATOR OF TP53 (113721)

Coles et al. (1990) demonstrated loss of constitutional heterozygosity (LOH) at 17p13.3 in a large proportion of breast cancers. LOH was less frequent in band 17p13.1, which contains the TP53 gene (191170). There was no correlation between allele loss at the 2 sites on 17p. Nevertheless, loss of heterozygosity at 17p13.3 was associated with overexpression of p53 mRNA, suggesting to Coles et al. (1990) the existence of a gene some 20 megabases telomeric of TP53 that regulates its expression. Lesions of this regulatory gene seem to be involved in most breast cancers (114480).

Stack et al. (1995) commented on the high frequency of LOH (50-75%) in the 17p13.3 region distal to TP53 in sporadic breast cancer and also the high frequency in a number of other malignancies including ovarian cancer, astrocytomas, bladder cancer, medulloblastoma, neuroectodermal cancer, and osteosarcoma. LOH was found to be independent of loss at the TP53 gene locus situated some 20 Mb proximal. Stack et al. (1995) investigated loss of heterozygosity in a panel of 40 sporadic breast tumor patients using 8 polymorphic markers which they had ordered within the 17p13.3 region by fluorescence in situ hybridization. Their findings demonstrated a region of high loss (60%) within distal 17p13.3, defined by markers D17S926, D17S695, and D17S849, which map close together. The study presented further evidence for the existence of a breast cancer suppressor locus distal to known genes within 17p13.3.

Coles, C.; Thompson, A. M.; Elder, P. A.; Cohen, B. B.; Mackenzie, I. M.; Cranston, G.; Chetty, U.; Mackay, J.; Macdonald, M.; Nakamura, Y.; Hoyheim, B.; Steel, C. M.: Evidence implicating at least two genes on chromosome 17p in breast carcinogenesis. Lancet 336: 761-763, 1990.
Stack, M.; Jones, D.; White, G.; Liscia, D. S.; Venesio, T.; Casey, G.; Crichton, D.; Varley, J.; Mitchell, E.; Heighway, J.; Santibanez-Koref, M.: Detailed mapping and loss of heterozygosity analysis suggests a suppressor locus involved in sporadic breast cancer within a distal region of chromosome band 17p13.3. Hum. Molec. Genet. 4: 2047-2055, 1995.

CYCLIN-DEPENDENT KINASE INHIBITOR 1A (116899)

The cyclin-dependent kinase CDK2 (116953) associates with cyclins A (123835), D (168461), and E (123837) and has been implicated in the control of the G1 to S phase transition in mammals. To identify potential CDK2 regulators, Harper et al. (1993) used an improved 2-hybrid system to isolate human genes encoding CDK-interacting proteins (CIPs) which they called CIP1. CIP1 was found to encode a novel 21-kd protein that is found in immunoprecipitates of cyclin A, cyclin D1, cyclin E, and CDK2. It is a potent, tight-binding inhibitor of CDKs and can inhibit the phosphorylation of the retinoblastoma protein by several of these complexes. Cotransfection experiments indicated that CIP1 and SV40T antigen function in a mutually antagonistic manner to control cell cycle progression.

The ability of p53 (191170) to activate transcription from specific sequences suggests that genes induced by p53 may mediate its biologic role as a tumor suppressor. Using a subtractive hybridization approach, El-Deiry et al. (1993) identified a gene they called WAF1 (for wildtype p53-activated fragment 1), whose induction was associated with wild-type but not mutant p53 gene expression in a human brain tumor cell line. They mapped the WAF1 gene to 6p21.2 by fluorescence in situ hybridization. They found that the sequence, structure, and activation by p53 was conserved in rodents. Introduction of WAF1 cDNA suppressed the growth of human brain, lung, and colon tumor cells in culture. Using a yeast enhancer trap, they identified a p53-binding site 2.4 kb upstream of WAF1 coding sequences. The WAF1 promoter, including this p53-binding site, conferred p53-dependent inducibility upon a heterologous reporter gene. After acceptance of their paper for publication, El-Deiry et al. (1993) learned that Harper et al. (1993) had identified a gene, called CIP1, whose product binds to cyclin complexes and inhibits the function of cyclin-dependent kinases. They found that the sequence of CIP1, described by Harper et al. (1993) in the same issue of Cell, was identical to that of WAF1. The results provided a dramatic example of the interplay between tumor suppressor genes and the cell cycle.

Chedid et al. (1994) identified a polymorphism at codon 31 where a single point mutation changed AGC (ser) to AGA (arg) (116899.0001). The change resulted in the loss of a restriction site and gain of another, allowing for rapid screening of the polymorphism. Analysis of genomic DNAs from 50 randomly selected individuals revealed that the basepair substitution occurred with an allelic frequency of 0.14. Transfection studies demonstrated that expression of the arg allele was not associated with loss of tumor suppressor activity. Moreover, screening of 22 tumor DNA samples revealed no association between tumor phenotype and the arg allele.

Huppi et al. (1994) cloned and sequenced a mouse p21 cDNA and established that the gene locus, Waf1, lies proximal to H-2 on chromosome 17.

The preferred name and symbol for this gene are cyclin-dependent kinase inhibitor 1A (CDKN1A). Also referred to as p21 and as CDKN1, this protein inhibits cyclin-kinase activity, is tightly regulated at the transcriptional level by p53, and probably serves as the effector of p53 cell cycle control. By fluorescence in situ hybridization, Demetrick et al. (1995) also mapped the gene to 6p21.2.

The WAF1-encoded protein, p21, mediates p53 suppression of tumor cell growth. Overexpression of p21 in a tumor cell line suppresses colony formation similar to that resulting from p53 overexpression. To localize the tumor suppression function within the structure of p21, Zakut and Givol (1995) used vectors constructed with systematic truncations of p21 and tested their efficiency in suppressing tumor cell growth. They demonstrated that the N-terminal half of the p21 molecule shows better tumor cell growth suppression than the entire p21 molecule, whereas the C-terminal half of p21 did not show this effect.

Chedid, M.; Michieli, P.; Lengel, C.; Huppi, K.; Givol, D.: A single nucleotide substitution at codon 31 (ser/arg) defines a polymorphism in a highly conserved region of the p53-inducible gene WAF1/CIP1. Oncogene 9: 3021-3024, 1994.

Demetrick, D. J.; Matsumoto, S.; Hannon, G. J.; Okamoto, K.; Xiong, Y.; Zhang, H.; Beach, D. H.: Chromosomal mapping of the genes for the human cell cycle proteins cyclin C (CCNC), cyclin E (CCNE), p21 (CDKN1) and KAP (CDKN3). Cytogenet. Cell Genet. 69: 190-192, 1995.

El-Deiry, W. S.; Tokino, T.; Velculescu, V. E.; Levy, D. B.; Parsons, R.; Trent, J. M.; Lin, D.; Mercer, E.; Kinzler, K. W.; Vogelstein, B.: WAF1, a potential mediator of p53 tumor suppression. Cell 75: 817-825, 1993.

Harper, J. W.; Adami, G. R.; Wei, N.; Keyomarsi, K.; Elledge, S. J.: The p21 Cdk-interacting protein Cip1 is a potent inhibitor of G1 cyclin-dependent kinases. Cell 75: 805-816, 1993.

Huppi, K.; Siwarski, D.; Dosik, J.; Michieli, P.; Chedid, M.; Reed, S.; Mock, B.; Givol, D.; Mushinski, J. F.: Molecular cloning, sequencing, chromosomal localization and expression of mouse p21 (Waf1). Oncogene 9: 3017-3020, 1994.

Mousses, S.; Ozcelik, H.; Lee, P. D.; Malkin, D.; Bull, S. B.; Andrulis, I. L.: Two variants of the CIP1/WAF1 gene occur together and are associated with human cancer. Hum. Molec. Genet. 4: 1089-1092, 1995.

Sjalander, A.; Birgander, R.; Rannug, A.; Alexandrie, A.-K.; Tornling, G.; Beckman, G.: Association between the p21 codon 31A1 (arg) allele and lung cancer. Hum. Hered. 46: 221-225, 1996.

Zakut, R.; Givol, D.: The tumor suppression function of p21(Waf) is contained in its N-terminal half ('half-WAF'). Oncogene 11: 393-395, 1995.

CYCLIN-DEPENDENT KINASE 4 (123829)

Human cell division is regulated primarily at the G1-to-S or the G2-to-M boundaries. The sequential activation of cyclin-dependent kinases (CDKs) and their subsequent phosphorylation of critical substrates promote orderly progression through the cell cycle. The complexes formed by CDK4 and the D-type cyclins (e.g., D1, 168461; D2, 123833; D3, 123834) are involved in the control of cell proliferation during the G1 phase. CDK4 is inhibited by p16, also known as cyclin-dependent kinase inhibitor 2 (600160). Demetrick et al. (1994) mapped the CDK4 gene to 12q13 by fluorescence in situ hybridization. CDK2 (116953) maps to the same band. By fluorescence in situ hybridization, Mitchell et al. (1995) mapped CDK4 to 12q14 and concluded that it is distal to GLI (165220) and CHOP (126337), which they placed at 12q13.3-q14.1; and proximal to MDM2 (164785), which they placed at 12q14.3-q15.

Wolfel et al. (1995) identified a mutated CDK4 as a tumor-specific antigen recognized by autologous cytolytic T lymphocytes in a human melanoma. The mutated CDK4 allele was present in autologous cultured melanoma cells and metastasis tissue, but not in the patient's lymphocytes. The mutation, an arg24-to-cys (R24C) exchange (123829.0001), was part of the CDK4 peptide recognized by cytolytic T lymphocytes and prevented binding of the CDK4 inhibitor p16(INK4A) (gene symbol = CDKN2A), but not of p21 or of p27 (600778). The same mutation was found in 1 additional melanoma among 28 melanomas analyzed. These results suggested to the authors that mutation of CDK4 can create a tumor-specific antigen and can disrupt the cell-cycle regulation exerted by the tumor suppressor p16. Inactivating mutations of the p16 gene are responsible for genetic predisposition to melanoma. The R24C mutation of CDK4 presumably contributes to malignant transformation in melanoma in addition to creating a tumor-specific antigen. Such antigens are ideally suited as targets of tumor rejection responses. The authors speculated that this may have happened in the first patient in whom it was identified, because the patient had remained free of detectable disease for 7 years.

Nomenclature: Knudson (1997) referred to the form of melanoma caused by mutation in the p16 gene (600160) and mapping to 9p21 (155601) as melanoma-1, since it was the first form of melanoma to be defined at the DNA level. Furthermore, he referred to the form of melanoma second to be defined molecularly, that mapping to chromosome 12 and caused by mutations in the CDK4 gene, as melanoma-2. Presumably, the form of melanoma that maps to 1p (155600) should, according to this scheme, be designated melanoma-3 when the gene is identified.

Demetrick, D. J.; Zhang, H.; Beach, D. H.: Chromosomal mapping of human CDK2, CDK4, and CDK5 cell cycle kinase genes. Cytogenet. Cell Genet. 66: 72-74, 1994.

Knudson,: Personal Communication. Philadelphia, Pa. 2/21/1997.

Mitchell, E. L. D.; White, G. R. M.; Santibanez-Koref, M. F.; Varley, J. M.; Heighway, J.: Mapping of gene loci in the q13-q15 region of chromosome 12. Chromosome Res. 3: 261-262, 1995.

Wolfel, T.; Hauer, M.; Schneider, J.; Serrano, M.; Wolfel, C.; Klehmann-Hieb, E.; De Plaen, E.; Hankeln, T.; Meyer zum Buschenfelde, K.-H.; Beach, D.: A p16(INK4a)-insensitive CDK4 mutant targeted by cytolytic T lymphocytes in a human melanoma. Science 269: 1281-1284, 1995.

Zuo, L.; Weger, J.; Yang, Q.; Goldstein, A. M.; Tucker, M. A.; Walker, G. J.; Hayward, N.; Dracopoli, N. C.: Germline mutations in the p16(INK4a) binding domain of CDK4 in familial melanoma. Nature Genet. 12: 97-99, 1996.

"DELETED IN POLYPOSIS" 1 (125265)

Through study of submicroscopic deletions in 2 unrelated patients with familial adenomatous polyposis coli (APC; 175100), Joslyn et al. (1991) found 3 genes in a 100-kb deleted segment on chromosome 5. MCC (159350), a previous candidate gene, was shown to be located outside the deleted region. One of the new genes contained a sequence identical to SRP19 (182175). The second gene, provisionally designated DP1 ('deleted in polyposis 1'), was transcribed in the same orientation as MCC. Two other cDNAs, DP2 and DP3, were found to overlap, forming a single gene, DP2.5, which was transcribed in the same orientation as SRP19. The DP2.5 gene was shown by Groden et al. (1991) to be the polyposis gene, as indicated in 175100.

Groden, J.; Thliveris, A.; Samowitz, W.; Carlson, M.; Gelbert, L.; Albertsen, H.; Joslyn, G.; Stevens, J.; Spirio, L.; Robertson, M.; Sargeant, L.; Krapcho, K.; Wolff, E.; Burt, R.; Hughes, J. P.; Warrington, J.; McPherson, J.; Wasmuth, J.; Le Paslier, D.; Abderrahim, H.; Cohen, D.; Leppert, M.; White, R.: Identification and characterization of the familial adenomatous polyposis coli gene. Cell 66: 589-600, 1991.

Joslyn, G.; Carlson, M.; Thliveris, A.; Albertsen, H.; Gelbert, L.; Samowitz, W.; Groden, J.; Stevens, J.; Spirio, L.; Robertson, M.; Sargeant, L.; Krapcho, K.; Wolff, E.; Burt, R.; Hughes, J. P.; Warrington, J.; McPherson, J.; Wasmuth, J.; Le Paslier, D.; Abderrahim, H.; Cohen, D.; Leppert, M.; White, R.: Identification of deletion mutations and three new genes at the familial polyposis locus. Cell 66: 601-613, 1991.

FATTY ACID-BINDING PROTEIN 3, MUSCLE AND HEART (134651)

Fatty acid metabolism in mammalian cells depends on a flux of fatty acids, between the plasma membrane and mitochondria or peroxisomes for beta-oxidation, and between other cellular organelles for lipid synthesis. At least 3 different fatty acid-binding proteins (FABPs) play a role as transport vehicles of these hydrophobic compounds throughout the cytoplasm. Different FABPs have been demonstrated in the liver, heart, intestine, skeletal muscle, and brain. Peeters et al. (1991) determined that a cDNA clone for human skeletal muscle FABP had an open reading frame of 399 nucleotides coding for a protein of 133 amino acid residues with a calculated molecular mass of 14,858 Da. Peptide sequence analysis of human skeletal muscle and heart FABP, after carbamoyl methylation and lysyl endopeptidase digestion followed by automatic Edman degradation, showed that the 2 proteins are identical. No evidence was found for isoproteins in muscle. The chromosomal localization of the human muscle FABP gene was determined to be 1pter-q31 by the study of human/rodent somatic cell hybrid lines. By Southern analysis of mouse-hamster somatic cell hybrids, Heuckeroth et al. (1987) mapped the mouse heart FABP gene to 3 chromosomes, 4, 8, and either 10 or 15; 2 of the genes may be pseudogenes. It might be expected that the functional gene is on mouse chromosome 4, which shows homology of synteny with human 1p. For myelin and adipocyte FABP types in man, the chromosomal location remained to be determined. Treuner et al. (1994) placed the gene, which they symbolized Fabph1, on mouse chromosome 4 by interspecies crosses. A pseudogene was assigned to mouse chromosome 10 and another one to mouse chromosome 8.

The deduced and experimentally determined amino acid sequence of human muscle FABP (Peeters et al., 1991) was reported to be identical to that of human heart FABP as determined by Offner et al. (1988) and Borchers et al. (1990). Troxler et al. (1993) found, using either a cDNA or a genomic fragment, that the human heart FABP gene mapped to 1p33-p32 by fluorescence in situ hybridization.

Mammary-derived growth inhibitor (MDGI) was first identified as a growth inhibitor in lactating bovine mammary gland and was shown to play a role in mammary gland differentiation (Bohmer et al. (1987)). Yang et al. (1994) showed that transfection of an MDGI expression construct into breast cancer cell lines reduced proliferation and promoted cellular differentiation; Huynh et al. (1995) showed that it reduced tumor formation in nude mice. Huynh et al. (1995) mapped the

MDGI gene to 1p35-p32, a common region of loss in sporadic breast cancer. Based on the complete cDNA in protein homologies between mouse MDGI and the FABP3 genes, Phelan et al. (1996) assumed that these are the same gene. Independent reports of mapping to the same chromosome region supports their identity. Phelan et al. (1996) demonstrated that the MDGI/FABP3 gene covers approximately 8 kb of genomic DNA and is divided into 4 exons. In an attempt to identify possibly inactivating mutations in the MDGI gene in human breast cancer, they sequenced all 4 exons and their surrounding splice junctions in 30 sporadic breast tumors. Ten of these tumors showed loss of heterozygosity (LOH) in the 1p35-p32 region, with 5 tumors showing LOH in the subregion containing the MDGI gene. They found no mutations in their analysis. Despite experimental evidence that MDGI has tumor suppressor activity, the data suggested to Phelan et al. (1996) that mutations in the coding region are uncommon in human breast tumorigenesis.

See also Peeters et al. (1991)

Bohmer, F. D.; Kraft, R.; Otto, A.; Wernstedt, C.; Hellman, U.; Kurtz, A.; Muller, T.; Rohde, K.; Etzold, G.; Lehmann, W.; Langen, P.; Heldin, C.; Grosse, R.: Identification of a polypeptide growth inhibitor from bovine mammary gland: sequence homology to fatty acid- and retinoid-binding proteins. J. Biol. Chem. 262: 15137-15143, 1987.

Borchers, T.; Hojrup, P.; Nielsen, S. U.; Roepstorff, P.; Spener, F.; Knudsen, J.: Revision of the amino acid sequence of human heart fatty acid binding protein. Molec. Cell. Biochem. 98: 127-133, 1990.

Heuckeroth, R. O.; Birkenmeier, E. H.; Levin, M. S.; Gordon, J. I.: Analysis of the tissue-specific expression, developmental regulation, and linkage relationships of a rodent gene encoding heart fatty acid binding protein. J. Biol. Chem. 262: 9709-9717, 1987.

Huynh, H. T.; Larsson, C.; Narod, S.; Pollak, M.: Tumor suppressor activity of the gene encoding mammary-derived growth inhibitor. Cancer Res. 55: 2225-2231, 1995.

Offner, G. D.; Brecher, P.; Sawlivich, W. B.; Costello, C. E.; Troxler, R. F.: Characterization and amino acid sequence of a fatty acid-binding protein from human heart. Biochem. J. 252: 191-198, 1988.

Peeters, R.; Veerkamp, J.; Kanda, T.; Ono, T.; Geurts van Kessel, A.: The gene encoding skeletal muscle fatty acid binding protein resides on human chromosome 1, region pter-q31. (Abstract) Cytogenet. Cell Genet. 58: 1861, 1991.

Peeters, R. A.; Veerkamp, J. H.; Geurts van Kessel, A.; Kanda, T.; Ono, T.: Cloning of the cDNA encoding human skeletal-muscle fatty-acid-binding protein, its peptide sequence and chromosomal localization. Biochem. J. 276: 203-207, 1991.

Phelan, C. M.; Larsson, C.; Baird, S.; Futreal, P. A.; Ruttledge, M. H.; Morgan, K.; Tonin, P.; Hung, H.; Korneluk, R. G.; Pollak, M. N.; Narod, S. A.: The human mammary-derived growth inhibitor (MDGI) gene: genomic structure and mutation analysis in human breast tumors. Genomics 34: 63-68, 1996.

Treuner, M.; Kozak, C. A.; Gallahan, D.; Grosse, R.; Muller, T.: Cloning and characterization of the mouse gene encoding mammary-derived growth inhibitor/heart-fatty acid-binding protein. Gene 147: 237-242, 1994.

Troxler, R. F.; Offner, G. D.; Jiang, J.-W.; Wu, B.-L.; Skare, J. C.; Milunsky, A.; Wyandt, H. E.: Localization of the gene for human heart fatty acid binding protein to chromosome 1p32-1p33. Hum. Genet. 92: 563-566, 1993.

Yang, Y.; Spitzer, E.; Kenney, N.; Zschiesche, W.; Li, M.; Kromminga, A.; Muller, T.; Spener, F.; Lezius, A.; Veerkamp, J. H.; Smith, G.; Salomon, D.; Grosse, R.: Members of the fatty acid binding protein family are differentiation factors for the mammary gland. J. Cell Biol. 124: 1097-1109, 1994.

HELA TUMOR SUPPRESSION (140750)

The tumorigenicity of HeLa cells in nude mice can be suppressed by the addition of a normal human chromosome 11 in somatic cell hybrids (Stanbridge, 1976; Klinger, 1980); see 191181 for description of a tumor-suppressor gene located on 11q. Lichy et al. (1992) isolated a HeLa cell line that displayed morphologic features of the nontumorigenic hybrids, demonstrated reduced tumorigenicity in nude mice, and showed an 85% reduction in alkaline phosphatase, a consistent marker of the tumorigenic phenotype in these cells. This cell line, designated F2, contained a single exogenous cDNA, which was recovered by polymerase chain reaction (PCR) and designated HTS1 because of its probable association with 'HeLa tumor suppression.' In nontumorigenic hybrids, RNA species of 2.8, 3.1, and 4.6 kb were identified. In 2 tumorigenic hybrid lines, the 2.8-kb species was markedly reduced or absent. Whereas 3 nontumorigenic human keratinocyte lines expressed all 3 RNA species, several tumorigenic cervical carcinoma cell lines lacked the 2.8-kb species. The HTS1 gene was localized to 11p15 by in situ hybridization, confirming the assignment to chromosome 11 by somatic cell hybrid analysis. Lichy et al. (1992) reviewed previous evidence indicating the presence of a tumor suppressor gene in the 11p15 region; see 194071 and 185440.

Klinger, H. P.: Suppression of tumorigenicity in somatic cell hybrids. I. Suppression and reexpression of tumorigenicity in diploid human x D98/AH2 hybrids and independent segregation of tumorigenicity from other cell phenotypes. Cytogenet. Cell Genet. 27: 254-266, 1980.

Lichy, J. H.; Modi, W. S.; Seuanez, H. N.; Howley, P. M.: Identification of a human chromosome 11 gene which is differentially regulated in tumorigenic and nontumorigenic somatic cell hybrids of HeLa cells. Cell Growth Differ. 3: 541-548, 1992.

Stanbridge, E. J.: Suppression of malignancy in human cells. Nature 260: 17-20, 1976.

INSULIN-LIKE GROWTH FACTOR 2 RECEPTOR (147280)

Insulin-like growth factor II (147470) is a polypeptide hormone with structural homologies to insulin and IGF-I. In contrast to these other hormones, the in vivo function of IGF-II is not known. Although IGF-II can stimulate a broad range of biologic responses in isolated cells, these responses appear to be mediated by the insulin and IGF-I receptors (147670, 147370). The receptor for IGF-II was found also to be the receptor for mannose 6-phosphate, which is implicated in targeting of lysosomal enzymes (MacDonald et al., 1988; Roth, 1988; Tong et al., 1988). Purified human and rat IGF2 receptors interact with antibodies to the mannose 6-phosphate receptor and with mannose 6-phosphate. Oshima et al. (1988) cloned and sequenced the full-length cDNA for MPRI; 7,473 nucleotides were found to encode a protein of 2,491 amino acids. The amino acid sequence included a putative signal sequence of 40 amino acids and an extra cytoplasmic domain consisting of 15 homologous repeat sequences of 134-167 amino acids, a transmembrane region of 23 amino acids, and a cytoplasmic domain of 164 amino acids. The predicted molecular size was greater than 270 kD. On comparison to the reported structure of IGF2R (Morgan et al., 1987), MPRI was found to show 99.8% identity at the nucleotide level and 99.4% identity at the amino acid level. By cDNA sequencing, Laureys et al. (1988) showed that the cation-independent mannose 6-phosphate receptor and IGF2 receptor are identical. Laureys et al. (1988) mapped the human gene to 6q25-q27, using cloned cDNAs to probe Southern blots of somatic cell hybrid DNA and for in situ chromosomal hybridization. The corresponding mouse gene mapped to chromosome 17. The cation-dependent mannose 6-phosphate receptor is coded by chromosome 12 (154540). By fluorescence in situ hybridization, Rao et al. (1994) narrowed the assignment of the IGF2R gene to 6q26.

Acquati et al. (1994) described a 2-Mb YAC from the telomeric region of 6q containing the plasminogen-apolipoprotein(a) gene family at its centromeric end and IGF2R at the telomeric end. About 350 kb separated IGF2R from the nearest member of the PLG/LPA cluster of genes.

Mutation in either MPRD or MPRI might result in a clinical disorder resembling a mucolipidosis (252500, 252600). Waheed et al. (1988) showed that the M6P and IGF2 binding sites are located on different segments of the receptor. Kiess et al. (1988) presented biochemical evidence that the IGF2 receptor and the cation-independent mannose 6-phosphate receptor are the same protein but that the binding sites for the 2 ligands are distinct. In its guise as a mannose 6-phosphate receptor, the IGF-2 receptor binds mannose 6-phosphate residues on lysosomal enzymes and transports them into lysosomes (Kornfeld and Mellman, 1989). Mouse embryos produce transcripts of the type II receptor from the maternal chromosome but not from the paternal. This could explain why mice that are heterozygous for deletions of the locus develop normally if their paternal allele is missing but die during early development if the maternal allele is missing. Studies in transgenic mice (Barlow et al., 1991) show parental imprinting, i.e., monoallelic expression, of the Igf2r gene. The Igf2r gene maps to the T-associated maternal effect locus (Tme) on mouse chromosome 17. It is probably no coincidence that the IGF2 gene and the gene for its type 2 receptor are oppositely imprinted (DeChiara et al., 1991). Haig and Graham (1991) proposed that the oppositely imprinted genes function to control the ultimately adverse effects of excessive production of IGF2 from the maternal allele. The suggestion is based on the model of Haig and Westoby (1989) which proposes that the evolution of genomic imprinting is expected in organisms that have both a breeding system in which females carry offspring by more than one male during their life span and a system of parental care in which offspring receive most of their postfertilization nutrients from 1 parent (usually the mother) and thus compete with offspring fathered by other males. Strictly, imprinting is possible whenever an individual's interactions are asymm etric with respect to maternal- and paternal-side relatives.

Szebeny and Rotwein (1994) cloned and characterized the mouse Igf2r gene. They found that it is 93 kb long, with 48 exons, and encodes a predicted protein of 2,482 amino acids.

To determine whether paternal expression of the Igf2r gene is necessary for early development in the mouse, Lau et al. (1994) derived mice in which the gene had been disrupted by targeted mutagenesis in embryonic stem (ES) cells with the subsequent introduction of the mutation into the germline of mice. Lau et al. (1994) found that murine embryos that inherit a nonfunctional Igf2r gene from their father are viable and develop normally into adults; however, most mice inheriting the same mutated allele from their mothers die around the time of birth as a consequence of major cardiac abnormalities. The mice that inherit the mutant allele from their mothers do not express Igf2r in their tissues, are 25 to 30% larger than their normal sibs, have elevated levels of circulating IGF2 and IGF-binding proteins, and exhibit a slight kink in the tail. The findings of overgrowth may support the suggestion that relaxation of maternal imprinting of IGF2 plays a role in the features of Beckwith-Wiedemann syndrome (130650) (Feinberg, 1993).

In summary, the mannose 6-phosphate/insulin-like growth factor II receptor functions in the intracellular trafficking of lysosomal enzymes, the activation of the potent growth inhibitor, transforming growth factor beta, and the degradation of IGF2, a mitogen often overproduced in tumors. De Souza et al. (1995) demonstrated that 70% of human hepatocellular tumors have loss of heterozygosity (LOH) at the M6P/IGF2R locus at 6q26. In a separate report, De Souza et al. (1995) described a mutation screen that identified point mutations in the remaining allele of 25% of human hepatocellular carcinomas with LOH. One mutation created an alternative splice site within an intron (corresponding to intron 40 in mouse) and resulted in a truncated receptor; 2 others (147280.0001, 147280.0002) gave rise to significant amino acid substitutions. These mutations provided evidence to the authors that the M6P/IGF2R gene functions as a tumor suppressor in human liver carcinogenesis.

Although in mice the M6P/IGF2R gene is maternally imprinted (Barlow et al., 1991), in humans imprinting appears to be a polymorphic trait (Xu et al., 1993; Kalscheuer et al., 1993; Ogawa et al., 1993). Thus, mice and any humans who are imprinted may have an increased susceptibility to hepatocellular carcinoma since only 1 mutation would be required to render the gene inactive. De Souza et al. (1995) noted that, since the M6P/IGF2R protein in normally present in the circulation, mutant receptors in the plasma might be helpful in liver tumor detection. Furthermore, they stated that mutated receptors on the plasma membrane of liver tumor cells might provide a surface antigen for the targeting of both therapeutic and diagnostic agents to liver tumors.

Souza et al. (1996) reported that the IGF2R gene contains a number of microsatellite repeats within its coding sequence. They demonstrated microsatellite instability in this gene in 12 of 92 gastrointestinal tumors studied which were replication/repair error-positive. Mutations occurred in 6 of the poorly differentiated tumors. They noted an anticorrespondence of IGF2R and TGFBR2 (190182) mutations. Of 31 gastrointestinal lesions studied with IGF2R or TGFBR2 mutations, 90% (28) contained mutations in one or the other, but not both, of these genes. Souza et al. (1996) demonstrated that all but 1 of the mutations occurred within an 8-polydeoxyguanine tract spanning nucleotides 4089-4096 of the IGF2R coding sequence. In 1 case of gastric adenocarcinoma, mutation occurred in a poly CT tract spanning nucleotides 6169-6180. These mutations all comprised 1- or 2-bp deletions or insertions within the microsatellite region, causing frameshifts and premature stop codons downstream. Souza et al. (1996) noted that the TGFBR2 gene is also subject to microsatellite instability within its coding region. They noted further that IGF2R and TGFBR2 genes comprise serial points in the same tumorigenesis pathway, since mutation of either gene alone Occurred in 90% of the gastrointestinal tumors that they analyzed.

See also De Souza et al. (1995)

Acquati, F.; Malgaretti, N.; Hauptschein, R.; Rao, P.; Gaidano, G.; Taramelli, R.: A 2-Mb YAC contig linking the plasminogen-apoprotein(a) gene family to the insulin-like growth factor 2 receptor (IGF2R) gene on the telomeric region of chromosome 6 (6q26-q27). Genomics 22: 664-666, 1994.

Barlow, D. P.; Stoger, R.; Herrmann, B. G.; Saito, K.; Schweifer, N.: The mouse insulin-like growth factor type-2 receptor is imprinted and closely linked to the Tme locus. Nature 349: 84-87, 1991.

DeChiara, T. M.; Robertson, E. J.; Efstratiadis, A.: Parental imprinting of the mouse insulin-like growth factor II gene. Cell 64: 849-859, 1991.

De Souza, A. T.; Hankins, G. R.; Washington, M. K.; Fine, R. L.; Orton, T. C.; Jirtle, R. L.: Frequent loss of heterozygosity on 6q at the mannose 6-phosphate/insulin-like growth factor II receptor locus in human hepatocellular tumors. Oncogene 10: 1725-1729, 1995.

De Souza, A. T.; Hankins, G. R.; Washington, M. K.; Orton, T. C.; Jirtle, R. L.: M6P/IGF2R gene is mutated in human hepatocellular carcinomas with loss of heterozygosity. Nature Genet. 11: 447-449, 1995.

Feinberg, A. P.: Genomic imprinting and gene activation in cancer. Nature Genet. 4: 110-113, 1993.

Haig, D.; Graham, C.: Genomic imprinting and the strange case of the insulin-like growth factor II receptor. Cell 64: 1045-1046, 1991.

Haig, D.; Westoby, M.: Parent-specific gene expression and the triploid endosperm. Am. Nat. 134: 147-155, 1989.

Kalscheuer, V. M.; Mariman, E. C.; Schepens, M. T.; Rehder, H.; Ropers, H.-H.: The insulin-like-growth factor type-2 receptor gene is imprinted in the mouse but not in humans. Nature Genet. 5: 74-78, 1993.

Kiess, W.; Blickenstaff, G. D.; Sklar, M. M.; Thomas, C. L.; Nissley, S. P.; Sahagian, G. G.: Biochemical evidence that the type II insulin-like growth factor receptor is identical to the cation-independent mannose 6-phosphate receptor. J. Biol. Chem. 263: 9339-9344, 1988.

Kornfeld, S.; Mellman, I.: The biogenesis of lysosomes. Annu. Rev. Cell Biol. 5: 483-525, 1989.

Lau, M. M. H.; Stewart, C. E. H.; Liu, Z.; Bhatt, H.; Rotwein, P.; Stewart, C. L.: Loss of the imprinted IGF2/cation-independent mannose 6-phosphate receptor results in fetal overgrowth and perinatal lethality. Genes Dev. 8: 2953-2963, 1994.

Laureys, G.; Barton, D. E.; Ullrich, A.; Francke, U.: Chromosomal mapping of the gene for the type II insulin-like growth factor receptor/cation-independent mannose 6-phosphate receptor in man and mouse. Genomics 3: 224-229, 1988.

MacDonald, R. G.; Pfeffer, S. R.; Coussens, L.; Tepper, M. A.; Brocklebank, C. M.; Mole, J. E.; Anderson, J. K.: A single receptor binds both insulin-like growth factor II and mannose-6-phosphate. Science 239: 1134-1137, 1988.

Morgan, D. O.; Edman, J. D.; Standring, D. N.; Fried, V. A.; Smith, M. C.; Roth, R. A.; Rutter, W. J.: Insulin-like growth factor II receptor as a multifunctional binding protein. Nature 329: 301-307, 1987.

Ogawa, O.; McNoe, L. A.; Eccles, M. R.; Morison, I. M.; Reeve, A. E.: Human insulin-like growth factor type I and type II receptors are not imprinted. Hum. Molec. Genet. 2: 2163-2165, 1993.

Oshima, A.; Nolan, C. M.; Kyle, J. W.; Grubb, J. H.; Sly, W. S.: The human cation-independent mannose 6-phosphate receptor: cloning and sequence of the full-length cDNA and expression of functional receptor in COS cells. J. Biol. Chem. 263: 2553-2562, 1988.

Rao, P. H.; Murty, V. V. V. S.; Gaidano, G.; Hauptschein, R.; Dalla-Favera, R.; Chaganti, R. S. K.: Subregional mapping of 8 single copy loci to chromosome 6 by fluorescence in situ hybridization. Cytogenet. Cell Genet. 66: 272-273, 1994.

Roth, R. A.: Structure of the receptor for insulin-like growth factor II: the puzzle amplified. Science 239: 1269-1271, 1988.

Souza, R. F.; Appel, R.; Yin, J.; Wang, S.; Smolinski, K. N.; Abraham, J. M.; Zou, T.-T.; Shi, Y.-Q.; Lei, J.; Cottrell, J.; Cymes, K.; Biden, K.; Simms, L.; Leggett, B.; Lynch, P. M.; Frazier, M.; Powell, S. M.; Harpaz, N.; Sugimura, H.; Young, J.; Meltzer, S. J.: Microsatellite instability in the insulin-like growth factor II receptor gene in gastrointestinal tumours. (Letter) Nature Genet. 14: 255-257, 1996.

Szebeny, G.; Rotwein, P.: The mouse insulin-like growth factor II/cation-independent mannose 6-phosphate (IGF-II/MPR) receptor gene: molecular cloning and genomic organization. Genomics 19: 120-129, 1994.

Tong, P. Y.; Tollefsen, S. E.; Kornfeld, S.: The cation-independent mannose 6-phosphate receptor binds insulin-like growth factor II. J. Biol. Chem. 263: 2585-2588, 1988.

Waheed, A.; Braulke, T.; Junghans, U.; von Figura, K.: Mannose 6-phosphate/insulin like growth factor II receptor: the two types of ligands bind simultaneously to one receptor at different sites. Biochem. Biophys. Res. Commun. 152: 1248-1254, 1988.

Xu, Y.; Goodyer, C. G.; Deal, C.; Polychronakos, C.: Functional polymorphism in the parental imprinting of the human IGF2R gene. Biochem. Biophys. Res. Commun. 197: 747-754, 1993.

ADULT SKELETAL MUSCLE GENE (*103280)

Leibovitch et al. (1991) used a rat skeletal muscle probe originating from a rhabdomyosarcoma to isolate a cDNA probe from a human placental cDNA library. In the rat, while the corresponding mRNA and protein were not expressed in fetal muscle, an increasing accumulation of the corresponding mRNA and protein were observed during postnatal development of skeletal muscle, and this accumulation was maximal in adulthood. As no expression was found in any other tissue, the gene was referred to as the adult skeletal muscle (ASM) gene. Leibovitch et al. (1991) mapped the human gene to 11p15 by a combination of somatic hybrid cell analysis and in situ hybridization. (D11S813E was the designation assigned by HGM11 (Nguyen et al., 1991).)

A gene coding for an abundant fetal transcript in mice had been identified by Bartolomei et al. (1991), who designated it H19. The H19 gene is expressed in a number of organs during a restricted period of fetal development, and in embryonal carcinoma cells after induction of differentiation. The gene shows a restricted pattern of expression in adult tissues; expression is confined to skeletal and cardiac muscle. Leibovitch et al. (1991) presented evidence that the human H19 gene has a transcript that gives rise to a 29-kD protein.

The human H19 gene is 2.7 kb long and includes 4 small introns. Zhang and Tycko (1992) found restriction site polymorphisms in the human H19 gene and, by examination of the representation of these polymorphisms in cDNAs from fetal organs, demonstrated that H19 expression was largely or exclusively from a single allele. Expression of the WT1 gene (194070), which, like H19, maps to 11p and shows fetal expression, was found to have biallelic expression. In the context of previous studies of allelic losses in 11p15 in human embryonal tumors, the findings of Zhang and Tycko (1992) supported the possibility of single-step inactivation of monoallelically expressed growth-regulating genes in human oncogenesis. It was not determined in this study whether the expression was uniparental to indicate parental imprinting. The H19 gene and 2 other genes, insulin-like growth factor II (147470) and insulin-like growth factor II receptor (147280), show monoallelic expression in mice. IGF2 is, like H19, located in 11p15. Zhang and Tycko (1992) commented that, if IGF2 also shows monoallelic expression, it may indicate that that region is a 'hotspot' for this phenomenon.

In the mouse, the H19 gene is located on chromosome 7 in a region of conservation of synteny with human 11p (Jones et al., 1992). Like the H19 gene, the Igf2 gene is imprinted in the mouse, although in the opposite parents, one paternally imprinted, the other maternally. Zemel et al. (1992) showed that the Igf2 gene lies about 90 kb 5-prime to H19, in the same transcriptional orientation. Based on similar pulsed field gel analysis, they showed that this physical proximity is conserved in humans. Both genes hybridized to a fragment of about 200 kb. Zemel et al. (1992) proposed a model to account for the imprinting of 2 linked genes in opposite directions, i.e., one (H19) being paternally imprinted and the other (IGF2) maternally imprinted. They pointed out that the IGF2/H19 domain is a candidate for the Beckwith-Wiedemann syndrome (BWS; 130650) since the genes show imprinting and chimeric mouse embryos that are paternally disomic for distal mouse chromosome 7 show an overgrowth phenotype similar to that of BWS (Ferguson-Smith et al., 1991).

From the study of the androgenetic complete hydatidiform mole, Rachmilewitz et al. (1992) presented strong evidence of parental imprinting of the human H19 gene, with the maternally derived allele as the active one. Furthermore, they showed that the paternally derived allele of the IGF2 is expressed. Thus, the situation in the human is the same as that in the mouse. Rainier et al. (1993) found that both H19 and IGF2 show monoallelic expression in human tissues and that, as in mouse, H19 is expressed from the maternal allele and IGF2 from the paternal allele. In contrast, 69% of Wilms tumors not undergoing loss of heterozygosity at 11p showed biallelic expression of one or both genes, suggesting that relaxation or loss of imprinting may represent a new epigenetic mutational mechanism in carcinogenesis.

Mutter et al. (1993) found that normal gestations express H19 only from the maternal allele and express IGF2 from the paternal allele, whereas neither is expressed from the maternal genome of gynogenetic gestations, and both are expressed from the paternal genome of androgenetic gestations. Coexpression of H19 and IGF2 in the androgenetic tissues was in a single population of cells, mononuclear trophoblast—the same cell type expressing these genes in biparental placentas. These results demonstrated that a biparental genome may be required for expression of the reciprocal IGF2/H19 imprint.

In the mouse, the imprinted H19 gene, which encodes an untranslated RNA, lies at the end of a cluster of imprinted genes. Leighton et al. (1995) found that imprinting of the insulin-2 gene and the insulin-like growth factor 2 gene, which lie about 100 kb upstream of H19, can be disrupted by maternal inheritance of a targeted deletion of the H19 gene and its flanking sequence. Animals inheriting the H19 mutation from their mothers were 27% heavier than those inheriting from their fathers. Paternal inheritance of the disruption had no effect, which presumably reflects the normally silent state of the paternal gene. The somatic overgrowth of heterozygotes for the maternal deletion was attributed to a gain-of-function of the Igf2 gene rather than a loss of function of H19.

H19 is abundantly expressed in both extraembryonic and fetal tissues. Jinno et al. (1995) found that H19 is monoallelically (maternally) expressed in the human placenta after 10 weeks of gestation, whereas it is biallelically expressed at earlier stages. Regardless of H19 biallelic or monoallelic expression, IGF2 (147470) is monoallelically (paternally) expressed in the placenta. Furthermore, with in situ mRNA hybridization using placenta showing H19 biallelic and IGF2 monoallelic expression, they demonstrated that defined cell types simultaneously contained both H19 and IGF2 transcripts. Therefore, the reciprocal linkage of H19 and IGF2 expression demonstrated in Wilms tumors is not observed in placentas. Furthermore, Jinno et al. (1995) found that, unlike methylation analyses of the human H19 gene, the promoter region of the human H19 gene is hypomethylated at all stages of placental development. In contrast, allele-dependent methylation of the 3-prime portion of the gene increases with gestational age.

H19 is a developmentally regulated gene with putative tumor suppressor activity; loss of H19 expression may be involved in Wilms tumorigenesis. Han et al. (1996) performed in situ hybridization analysis of H19 expression during normal rabbit development and in human atherosclerotic plaques. They found that H19 expression in developing skeletal and smooth muscles correlated with specific differentiation events in these tissues. Expression of H19 in skeletal muscle correlated with nonproliferative, actin-positive muscle cells. In the prenatal blood vessel, H19 expression was both temporally and spatially regulated with initial loss of expression in the inner smooth muscle layers adjacent to the lumen. Han et al. (1996) also identified H19-positive cells in adult atherosclerotic lesions, suggesting that these cells may recapitulate early developmental events. These results, along with the identification of the insulin family of growth factors as potent regulatory molecules for H19 expression, provided additional clues toward understanding the physiologic regulation and function of H19.

Pfeifer et al. (1996) stated that the product of the H19 gene is an untranslated RNA that is expressed exclusively from the maternal chromosome during mammalian development. The H19 gene and its 5-prime-flanking sequence are required for the genomic imprinting of 2 paternally expressed genes in mice, Ins2 and Igf2, that lie 90 and 115 kb 5-prime to the H19 gene, respectively. Pfeifer et al. (1996) investigated the role of the H19 gene in its own imprinting by introducing a Mus spretus H19 gene into heterologous locations in the mouse genome. They found that multiple copies of the transgene were sufficient for its paternal silencing and DNA methylation. Replacing the H19 structural gene with a luciferase reporter gene resulted in loss of imprinting of the transgene; that is, high expression and low levels of DNA methylation were observed with both paternal and maternal inheritance. Removal of 701 bp at the 5-prime end of the structural H19 gene resulted in a similar loss of paternal-specific DNA methylation, arguing that those sequences are required for both the establishment and maintenance of the sperm-specific gametic mark. The M. spretus H19 transgene could not rescue the loss of IGF-2 imprinting in trans in H19 deletion mice, implying a cis requirement for the H19 gene. In contrast to a previous report (Brunkow and Tilghman, 1991) in which overexpression of a marked H19 gene was a prenatal lethal, Pfeifer et al. (1996) found that expression of the M. spretus transgene had no deleterious effect, leading them to conclude that the 20-bp insertion in the marked gene created a neomorphic mutation.

Bartolomei, M. S.; Zemel, S.; Tilghman, S. M.: Parental imprinting of the mouse H19 gene. Nature 351: 153-155, 1991.

Brunkow, M. E.; Tilghman, S. M.: Ectopic expression of the H19 gene in mice causes prenatal lethality. Genes Dev. 5: 1092-1101, 1991.

Ferguson-Smith, A. C.; Cattanach, B. M.; Barton, S. C.; Beechey, C. V.; Surani, M. A.: Embryological and molecular investigations of parental imprinting on mouse chromosome 7. Nature 351: 667-670, 1991.

Han, D. K. M.; Khaing, Z. Z.; Pollock, R. A.; Haudenschild, C. C.; Liau, G.: H19, a marker of developmental transition, is reexpressed in human atherosclerotic plaques and is regulated by the insulin family of growth factors in cultured rabbit smooth muscle cells. J. Clin. Invest. 97: 1276-1285, 1996.

Jinno, Y.; Ikeda, Y.; Yun, K.; Maw, M.; Masuzaki, H.; Fukuda, H.; Inuzuka, K.; Fujishita, A.; Ohtani, Y.; Okimoto, T.; Ishimaru, T.; Niikawa, N.: Establishment of functional imprinting of the H19 gene in human developing placentae. Nature Genet. 10: 318-324, 1995.

Jones, J. M.; Meisler, M. H.; Seldin, M. F.; Lee, B. K.; Eicher, E. M.: Localization of insulin-2 (Ins-2) and the obesity mutant tubby (tub) to distinct regions of mouse chromosome 7. Genomics 14: 197-199, 1992.

Leibovitch, M. P.; Nguyen, V. C.; Gross, M. S.; Solhonne, B.; Leibovitch, S. A.; Bernheim, A.: The human ASM (adult skeletal muscle) gene: expression and chromosomal assignment to 11p15. Biochem. Biophys. Res. Commun. 180: 1241-1250, 1991.

Leighton, P. A.; Ingram, R. S.; Eggenschwiler, J.; Efstratiadis, A.; Tilghman, S. M.: Disruption of imprinting caused by deletion of the H19 gene region in mice. Nature 375: 34-39, 1995.

Mutter, G. L.; Stewart, C. L.; Chaponot, M. L.; Pomponio, R. J.: Oppositely imprinted genes H19 and insulin-like growth factor 2 are coexpressed in human androgenetic trophoblast. Am. J. Hum. Genet. 53: 1096-1102, 1993.

Nguyen, V. C.; Leibovitch, M.; Gross, M.; Solhonne, B.; Leibovitch, S. A.; Bernheim, A.: Assignment of ASM (adult skeletal muscle) to chromosome 11 (somatic hybrid cell analysis), region 11p15 (in situ hybridization). (Abstract) Cytogenet. Cell Genet. 58: 1968, 1991.

Pfeifer, K.; Leighton, P. A.; Tilghman, S. M.: The structural H19 gene is required for transgene imprinting. Proc. Nat. Acad. Sci. 93: 13876-13883, 1996.

Rachmilewitz, J.; Goshen, R.; Ariel, I.; Schneider, T.; de Groot, N.; Hochberg, A.: Parental imprinting of the human H19 gene. FEBS Lett. 309: 25-28, 1992.

Rainier, S.; Johnson, L. A.; Dobry, C. J.; Ping, A. J.; Grundy, P. E.; Feinberg, A. P.: Relaxation of imprinted genes in human cancer. Nature 362: 747-749, 1993.

Zemel, S.; Bartolomei, M. S.; Tilghman, S. M.: Physical linkage of two mammalian imprinted genes, H19 and insulin-like growth factor 2. Nature Genet. 2: 61-65, 1992.

Zhang, Y.; Tycko, B.: Monoallelic expression of the human H19 gene. Nature Genet. 1: 40-44, 1992.

SUPPRESSOR OF ANCHORAGE INDEPENDENCE 1 (154280)

BHK is an immortal Syrian hamster cell line in which anchorage independence can be shown to arise in a single mutagenic step and to be correlated with the last change leading to tumorigenicity. Stoler and Bouck (1985) fused normal human fibroblasts with carcinogen-transformed BHK cells and found in the hybrid cells suppression of the transformed phenotype. Human chromosome 1 was present in all hybrid cells showing suppression and was lost in all hybrids in which transformation was reexpressed. The presence or absence of chromosome 1 was confirmed by electrophoretic identification of the human isozyme for phosphoglucomutase-1 (171900). Stoler and Bouck (1985) suggested that the rather frequent involvement of chromosome 1 in human neoplasms of various types and at various stages may be related to this locus. They also reasoned from homology of synteny that, as pointed out by Zabel et al. (1985), the locus may be on 1p: the part of 1p distal to 1p22.1 contains an evolutionarily conserved segment that is on mouse chromosome 4. (From 1p22.1 toward the centromere there is a segment homologous to mouse chromosome 3.) Chromosome 4 of the mouse is capable of suppressing the malignancy of several mouse tumor lines (Evans et al., 1982). Benedict et al. (1984) found that loss of 2 copies of human chromosome 1 (and 1 copy of chromosome 4) was associated with regaining ability to form tumors in nude mice by a human fibrosarcoma line of which tumorigenicity was suppressed by fusion to a normal human fibroblast. Thus, the locus identified by Stoler and Bouck (1985) may be on 1p, distal to 1p22.1. Malignancy in the system used by Stoler and Bouck (1985) is a recessive trait comparable to that of retinoblastoma and Wilms tumor; the suppressor, however, behaves as a dominant and the gene is the sort that Knudson (1985) referred to as an anti-oncogene. Szpirer et al. (1990) mapped the SAI1 gene to chromosome 5 of the rat, which carries synteny groups conserved on human chromos omes 1 and 9.

Benedict, W. F.; Weissman, B. E.; Mark, C.; Starbridge, E. J.: Tumorigenicity of human HT1080 fibrosarcoma x normal fibroblast hybrids: chromosome dosage dependency. Cancer Res. 44: 3471-3479, 1984.

Evans, E. P.; Burtenshaw, M. D.; Brown, B. B.; Hennion, R.; Harris, H.: The analysis of malignancy by cell fusion. J. Cell Sci. 56: 113-130, 1982.

Knudson, A. G.: Hereditary cancer, oncogenes and anti-oncogenes. Cancer Res. 45: 1437-1443, 1985.

Stoler, A.; Bouck, N.: Identification of a single chromosome in the normal human genome essential for suppression of hamster cell transformation. Proc. Nat. Acad. Sci. 82: 570-574, 1985.

Szpirer, C.; Riviere, M.; Szpirer, J.; Genet, M.; Dreze, P.; Islam, M. Q.; Levan, G.: Assignment of 12 loci to rat chromosome 5: evidence that this chromosome is homologous to mouse chromosome 4 and to human chromosomes 9 and 1 (1p arm). Genomics 6: 679-684, 1990.

Zabel, B. U.; Eddy, R. L.; Lalley, P. A.; Scott, J.; Bell, G. I.; Shows, T. B.: Chromosomal locations of the human and mouse genes for precursors of epidermal growth factor and the beta subunit of nerve growth factor. Proc. Nat. Acad. Sci. 82: 469-473, 1985.

MUTATED IN COLORECTAL CANCERS (159350)

Hoshino et al. (1991) found that fragments of chromosome 5, including the region containing the APC gene (175100), suppressed tumor activity when transferred into NIH-3T3 cells that had been transformed with Kirsten sarcoma virus. Kinzler et al. (1991), who stated the location of the APC gene as 5q21, detected a somatically rearranged restriction fragment in the DNA from a single tumor as demonstrated by DNA probes from the 5q21 region. Further analysis of this probe revealed 2 regions that were highly conserved in rodent DNA. These were used to identify a gene that they called MCC (for 'mutated in colorectal cancer'), which encodes an 829-amino acid protein with a short region of similarity to a G-protein coupled m3 muscarinic acetylcholine receptor. The rearrangement in the colorectal carcinoma disrupted the coding region of the MCC gene. Moreover, 2 colorectal cancers were found with somatically acquired point mutations in MCC that resulted in amino acid substitutions. Thus, MCC is a candidate for the putative colorectal tumor suppressor gene located at 5q21. By molecular studies in a patient with a cytogenetically evident deletion of 5q, Lindgren et al. (1992) assigned MCC to 5q21. Using molecular probes in the analysis of an interspecific backcross between C57BL/6J and Mus spretus, Justice et al. (1992) mapped the corresponding gene to mouse chromosome 18.

Hoshino, Y.; Horikawa, I.; Oshimura, M.; Yuasa, Y.: Normal human chromosome 5, on which a familial adenomatous polyposis gene is located, has tumor suppressive activity. Biochem. Biophys. Res. Commun. 174: 298-304, 1991.

Justice, M. J.; Gilbert, D. J.; Kinzler, K. W.; Vogelstein, B.; Buchberg, A. M.; Ceci, J. D.; Matsuda, Y.; Chapman, V. M.; Patriotis, C.; Makris, A.; Tsichlis, P. N.; Jenkins, N. A.; Copeland, N. G.: A molecular genetic linkage map of mouse chromosome 18 reveals extensive linkage conservation with human chromosomes 5 and 18. Genomics 13: 1281-1288, 1992.

Kinzler, K. W.; Nilbert, M. C.; Vogelstein, B.; Bryan, T. M.; Levy, D. B.; Smith, K. J.; Preisinger, A. C.; Hamilton, S. R.; Hedge, P.; Markham, A.; Carlson, M.; Joslyn, G.; Groden, J.; White, R.; Miki, Y.; Miyoshi, Y.; Nishisho, I.; Nakamura, Y.: Identification of a gene located at chromosome 5q21 that is mutated in colorectal cancers. Science 251: 1366-1370, 1991.

Lindgren, V.; Bryke, C. R.; Ozcelik, T.; Yang-Feng, T. L.; Francke, U.: Phenotypic, cytogenetic, and molecular studies of three patients with constitutional deletions of chromosome 5 in the region of the gene for familial adenomatous polyposis. Am. J. Hum. Genet. 50: 988-997, 1992.

JUNCTION PLAKOGLOBIN (*173325)

Plakoglobin is a major cytoplasmic protein that occurs in a soluble and a membrane-associated form and is the only known constituent common to the submembranous plaques of both kinds of adhering junctions, the desmosomes and the intermediate junctions. It is a desmoplakin (see 125647) and is referred to as DP III. DP I and DP II are splice variants of the same gene. Using a partial cDNA clone for bovine plakoglobin, Franke et al. (1989) isolated cDNAs encoding human plakoglobin, determined its nucleotide sequence, and deduced the complete amino acid sequence. The polypeptide encoded by the cDNA was synthesized by in vitro transcription and translation and identified by its comigration with authentic plakoglobin in 2-dimensional gel electrophoresis. The protein, which has 744 amino acids and a molecular weight of 81,750 Da, is unrelated to any known protein and is highly conserved between human and bovine tissues. Only one kind of plakoglobin mRNA (3.4 kb) was found in most tissues, but an additional mRNA (3.7 kb) was detected in certain human tumor cell lines. Arnemann et al. (1991) established a PCR assay for the gene coding for plakoglobin and used it to test human/mouse and human/rat somatic cell hybrids with different contents of human chromosomes. In this way, they were able to assign DP3 to chromosome 7. By analysis of progeny from 2 interspecific backcrosses, Guenet et al. (1995) mapped the Jup gene to mouse chromosome 11. Thus, the human JUP gene is probably on 7p because that is the portion of the chromosome showing homology of synteny to mouse 11. However, mouse 11 shows much more extensive homology to human chromosome 17 and, indeed, studies using both Southern blot analysis of human/rodent cell hybrids and linkage studies with an intragenic polymorphism demonstrated that the plakoglobin gene is located on 17q21 (Aberle et al., 1995; see later). Using a monoclonal human-rodent somatic cell hybrid panel, Cowley et al. (1997) likewise showed that the JUP gene is l ocated on chromosome 17.

Plakoglobin associates with the cytoplasmic region of desmoglein I (125670), one of the transmembrane desmosomal proteins (Mathur et al., 1994). It also is a component of the cadherin-catenin complex, which is predominantly localized where actin filaments anchor in adherens junction of epithelial cells (Knudsen and Wheelock, 1992). Aberle et al. (1995) stated that catenins are of central importance for cadherin function in that they mediate the connection of cadherins to actin filaments and are part of a higher order submembranous network by which cadherins are linked to other transmembrane and peripheral cytoplasmic proteins. Interaction of the cadherin-catenin complex with epidermal growth factor receptor (EGFR; 131550) and the finding that beta-catenin (CTNNB1; 116806) and plakoglobin are substrates for tyrosine phosphorylation following EGF stimulation of cells, together with the finding that catenins are associated with the tumor suppressor protein APC (175100), open the possibility that catenins are involved in signaling pathways and tumorigenesis. Aberle et al. (1995) mapped the plakoglobin gene to 17q12-q22 by analysis of DNA from human/hamster or human/mouse hybrid cell lines that contained only human chromosome 17 or parts thereof. Because of evolutionary conservation, human plakoglobin cDNA hybridized also to mouse and hamster DNA; however, the results obtained with DNA from hybrid cells unambiguously demonstrated that the human gene maps to 17q12-q22. The previously reported localization to chromosome 7 was based on PCR analysis using synthetic oligonucleotides and was not confirmed by independent work by Aberle et al. (1995). Furthermore, using the high-density map of polymorphic markers and genes in 17q12-q21 provided by the studies to elucidate the genetic basis of familial breast cancer, Aberle et al. (1995) found linkage between plakoglobin and BRCA1-linked markers. Meiotic recombinants in the 17q region showed that the plakoglobin gene lies betw een KRT10 (148080) and D17S858, in interval that also contains the BRCA1 gene (113705). A single recombination event was found separating the plakoglobin gene from the BRCA1 gene. Three plakoglobin-positive cosmid clones were found to contain both the plasma globin gene and the 17q21 marker locus UM8, in close proximity (less than 40 kb). Aberle et al. (1995) also identified an arginine/histidine polymorphism due to a substitution at nucleotide position 544 in exon 3, leading to replacement of arginine (CGC) by histidine (CAC) at amino acid position 142. This basepair substitution led to a loss of recognition sites for at least 2 restriction enzymes and thereby created an RFLP. The R142H substitution was found to represent a low-frequency polymorphism; from a total of 240 alleles examined, the arg142-to-his allele was found in 10 cases, giving an allele frequency of 0.042 +/- 0.013.

Nomenclature: Since the term plakoglobin is in increasingly accepted usage, it may be desirable to have a gene symbol that is closer to it, e.g., PKGB. Although the development of gene symbols that reflect families of proteins (and genes) to which the entity belongs has some justification, it can be confusing. The symbol JUP is derived from junction plakoglobin.

Aberle, H.; Bierkamp, C.; Torchard, D.; Serova, O.; Wagner, T.; Natt, E.; Wirsching, J.; Heidkamper, C.; Montagna, M.; Lynch, H. T.; Lenoir, G. M.; Scherer, G.; Feunteun, J.; Kemler, R.: The human plakoglobin gene localizes on chromosome 17q21 and is subjected to loss of heterozygosity in breast and ovarian cancers. Proc. Nat. Acad. Sci. 92: 6384-6388, 1995.

Arnemann, J.; Spurr, N. K.; Wheeler, G. N.; Parker, A. E.; Buxton, R. S.: Chromosomal assignment of the human genes coding for the major proteins of the desmosome junction, desmoglein DGI (DSG), desmocollins DGII/III (DSC), desmoplakins DPI/II (DSP), and plakoglobin DPIII (JUP). Genomics 10: 640-645, 1991.

Cowley, C. M. E.; Simrak, D.; Spurr, N. K.; Arnemann, J.; Buxton, R. S.: The plakophilin 1 (PKP1) and plakoglobin (JUP) genes map to human chromosomes 1q and 17, respectively. Hum. Genet. 100: 486-488, 1997.

Franke, W. W.; Goldschmidt, M. D.; Zimbelmann, R.; Mueller, H. M.; Schiller, D. L.; Cowin, P.: Molecular cloning and amino acid sequence of human plakoglobin, the common junctional plaque protein. Proc. Nat. Acad. Sci. 86: 4027-4031, 1989.

Guenet, J.-L.; Simon-Chazottes, D.; Ringwald, M.; Kemler, R.: The genes coding for alpha and beta catenin (Catna1 and Catnb) and plakoglobin (Jup) map to

mouse chromosomes 18, 9, and 11, respectively. Mammalian Genome 6: 363-366, 1995.
Knudsen, K. A.; Wheelock, M. J.: Plakoglobin, or an 83-kD homologue distinct from beta-catenin, interacts with E-cadherin and N-cadherin. J. Cell Biol. 118: 671-679, 1992.
Mathur, M.; Goodwin, L.; Cowin, P.: Interactions of the cytoplasmic domain of the desmosomal cadherin Dsg1 with plakoglobin. J. Biol. Chem. 269: 14075-14080, 1994.

PROHIBITIN (*176705)

Prohibitin is a 30-kD intracellular, antiproliferative protein. White et al. (1991) mapped the gene to chromosome 17 by analysis of human-mouse somatic cell hybrid cell lines using a genomic fragment of human prohibitin DNA isolated from a library using the rat prohibitin cDNA clone. By a study of cell lines containing portions of human chromosome 17, they determined that the PHB gene was located in the 17q11.2-q23 region. By in situ hybridization, they localized the gene to 17q21. Sato et al. (1992) isolated the human homolog of the rat prohibitin gene and mapped it to 17q12-q21 by in situ hybridization. Sato et al. (1993) showed that the human prohibitin gene family consists of 1 functional PHB gene on 17q21 and 4 processed pseudogenes, each on a different chromosome: PHBP1 on 6q25, PHBP2 on 11p11.2, PHBP3 on 1p31.3, and PHBP4 on 2q21.

By DNA sequence analysis of 2 exons in the PHB gene in 23 sporadic breast cancers that showed loss of heterozygosity on 17q or developed in patients 35 years old or younger, they identified 4 cases of somatic mutation: 2 of these were missense mutations, 1 showed a 2-bp deletion resulting in truncation of the gene product due to frameshift, and the fourth had a C-to-T transition in an intron adjacent to an intron-exon boundary. Sato et al. (1993) found no mutations in the PHB gene in other forms of tumors, namely, those of ovary, liver, and lung.

See also White et al. (1991)

Sato, T.; Saito, H.; Swensen, J.; Olifant, A.; Wood, C.; Danner, D.; Sakamoto, T.; Takita, K.; Kasumi, F.; Miki, Y.; Skolnick, M.; Nakamura, Y.: The human prohibitin gene located on chromosome 17q21 is mutated in sporadic breast cancer. Cancer Res. 52: 1643-1646, 1992.
Sato, T.; Sakamoto, T.; Takita, K.-I.; Saito, H.; Okui, K.; Nakamura, Y.: The human prohibitin (PHB) gene family and its somatic mutations in human tumors. Genomics 17: 762-764, 1993.
White, J. J.; Ledbetter, D. H.; Eddy, R. L.; Shows, T. B.; Stewart, D. A.; Nuell, M. J.; Friedman, V.; Wood, C. M.; Owens, G. A.; McClung, J. K.; Danner, D. B.; Morton, C. C.: Assignment of the human prohibitin gene (PHB) to chromosome 17q21. (Abstract) Cytogenet. Cell Genet. 58: 2011 only, 1991.
White, J. J.; Ledbetter, D. H.; Eddy, R. L., Jr.; Shows, T. B.; Stewart, D. A.; Nuell, M. J.; Friedman, V.; Wood, C. M.; Owens, G. A.; McClung, J. K.; Danner, D. B.; Morton, C. C.: Assignment of the human prohibition (sic) gene (PHB) to chromosome 17 and identification of a DNA polymorphism. Genomics 11: 228-230, 1991.

PROTEIN KINASE P58 (*176873)

Bunnell et al. (1990) identified a human cell division control (CDC)-related protein kinase, p58, that is structurally and functionally related to p34(cdc2) (CDC2; 116940). Abnormal expression of the p58 protein kinase in eukaryotic cells had effects suggesting that it is a negative regulator of normal cell cycle progression. The gene is well conserved evolutionarily. Its expression is regulated during murine embryogenesis, and its activity is coordinately regulated with that of p34(cdc2) during the cell cycle. Eipers et al. (1991) assigned the expressed p58 gene to 1p36 by somatic cell hybrid analysis, in situ hybridization, and nested PCR amplification of microdissected chromosomes. The authors stated that this gene, tentatively symbolized PK58, may be implicated in the pathogenesis of tumors that have deletion in the region of 1p36. Eipers et al. (1992) detailed the complete structure of the CDC2L1 gene including its putative promoter region, transcriptional start sites, exonic sequences, and intron/exon boundary sequences. The gene is 10 kb in size and contains 12 exons and 11 introns.

Lahti et al. (1994) characterized the genomic region surrounding CDC2L1 and found that a related gene, CDC2L2 (116951), was located within 100 kb. They referred to CDC2L1 as PITSLRE B, based on the amino acid sequence of the region corresponding to the conserved CDC2 PSTAIRE box. Lahti et al. (1994) demonstrated that 1 allele of the entire gene complex was either deleted or translocated in 18 of 20 neuroblastoma (256700) cell lines investigated. They suggested that the CDC2L gene complex may harbor 1 or more tumor suppressor genes affected by chromosome 1p36 modifications in neuroblastoma. White et al. (1995) showed that CDC2L1 is located outside the consensus region of allelic loss of heterozygosity (LOH) associated with neuroblastomas, namely, 1p36.3-p36.2, and is therefore not the neuroblastoma suppressor gene.

Xiang et al. (1994) isolated multiple, alternatively spliced CDC2L1 and CDC2L2 mRNAs. The CDC2L1 transcripts encode protein isoforms ranging in size from 58 to 110 kD. The isoforms vary in the length and sequence of their N-terminal regions; the protein kinase catalytic and C-terminal domains are identical. All contain a 30-residue acid blob domain comprised primarily of glutamic acid. The larger protein species contained a putative bipartite nuclear localization signal. Western blot analysis revealed that the 58-kD isoform was present only in cells undergoing apoptosis. Xiang et al. (1994) analyzed the expression pattern of the CDC2L1 mRNAs.

Gururajan et al. (1998) determined the structure of the 140-kb CDC2L genomic region. This region consists of 2 identical genomic segments arranged in a tail-to-tail configuration. Each segment contains a CDC2L gene linked to an MMP gene. See MMP21 (603320). In each case, the most widely expressed products of the CDC2L gene are derived from a genomic region that is composed of 20 exons and spans approximately 20 kb. The authors reported that only 15 amino acids of the 773-786 residues that encode the multiple CDC2L isoforms are unique to either CDC2L1 or CDC2L2.

By fluorescence in situ hybridization, Lahti et al. (1994) mapped the CDC2L1 gene to 1p36.3.

Bunnell, B.; Heath, L. S.; Adams, D. E.; Lahti, J. M.; Kidd, V. J.: Elevated expression of a p58 protein kinase leads to changes in the CHO cell cycle. Proc. Nat. Acad. Sci. 87: 7467-7471, 1990.
Eipers, P. G.; Barnoski, B. L.; Han, J.; Carroll, A. J.; Kidd, V. J.: Localization of the expressed human p58 protein kinase chromosomal gene to chromosome 1p36 and a highly related sequence to chromosome 15. Genomics 11: 621-629, 1991.
Eipers, P. G.; Lahti, J. M.; Kidd, V. J.: Structure and expression of the human p58(clk-1) protein kinase chromosomal gene. Genomics 13: 613-621, 1992.
Gururajan, R.; Lahti, J. M.; Grenet, J.; Easton, J.; Gruber, I.; Ambros, P. F.; Kidd, V. J.: Duplication of a genomic region containing the Cdc2L1-2 and MMP21-22 genes on human chromosome 1p36.3 and their linkage to D1Z2. Genome Res. 8: 929-939, 1998.
Lahti, J. M.; Valentine, M.; Xiang, J.; Jones, B.; Amann, J.; Grenet, J.; Richmond, G.; Look, A. T.; Kidd, V. J.: Alterations in the PITSLRE protein kinase gene complex on chromosome 1p36 in childhood neuroblastoma. Nature Genet. 7: 370-375, 1994.
White, P. S.; Maris, J. M.; Beltinger, C.; Sulman, E.; Marshall, H. N.; Fujimori, M.; Kaufman, B. A.; Biegel, J. A.; Allen, C.; Hilliard, C.; Valentine, M. B.; Look, A. T.; Enomoto, H.; Sakiyama, S.; Brodeur, G. M.: A region of consistent deletion in neuroblastoma maps within human chromosome 1p36.2-36.3. Proc. Nat. Acad. Sci. 92: 5520-5524, 1995.
Xiang, J.; Lahti, J. M.; Grenet, J.; Easton, J.; Kidd, V. J.: Molecular cloning and expression of alternatively spliced PITSLRE protein kinase isoforms. J. Biol. Chem. 269: 15786-15794, 1994.

CELL DIVISION CYCLE 2-LIKE 2 (*116951)

Protein kinase p58 (CDC2L1; 176873) is a human cell division control (CDC)-related molecule that is structurally and functionally related to p34 (CDC2; 116940). Whereas the p58 gene is located on 1p36, Eipers et al. (1991) found by mouse-human somatic cell hybrid studies that a highly related p58 sequence maps to chromosome 15. Subsequently, the CDC2L2 gene on chromosome 15 was retired by the HUGO Nomenclature Committee. The designation CDC2L2 represents a gene on 1p36.

Advanced-stage neuroblastoma (256700) and tumor-derived cell lines consistently contain a deletion of 1 allele of the distal short arm of chromosome 1. The deletion correlates with enhanced metastatic potential, suggesting that inactivation of a tumor suppressor gene or genes on 1p may contribute to the pathogenesis of these tumors. While characterizing the structure of the CDC2L1 candidate tumor suppressor gene located on 1p36, Lahti et al. (1994) identified a highly related gene within a contiguous 100-kb genomic region. They referred to CDC2L1 as PITSLRE B, following the convention of naming CDC2-related kinases by the amino acid sequence of the region corresponding to the conserved CDC2 PSTAIRE box. The related gene was designated PITSLRE A, or CDC2L2. Although Lahti et al. (1994) reported that a third gene, PITSLRE C, was also present in this region, Gururajan et al. (1998) found that PITSLRE C does not exist. Lahti et al. (1994) demonstrated that 1 allele of the entire gene complex was either deleted or translocated in 18 of 20 neuroblastoma cell lines investigated. They suggested that the CDC2L gene complex may harbor 1 or more tumor suppressor genes affected by chromosome 1p36 modifications in neuroblastoma.

Xiang et al. (1994) isolated multiple, alternatively spliced CDC2L2 mRNAs. The CDC2L2 transcripts encode 5 protein isoforms ranging in size from 65 to 110 kD. The isoforms vary in the length and sequence of their N-terminal regions; the protein kinase catalytic and C-terminal domains are identical. All contain a 30-residue acid blob domain comprised primarily of glutamic acid. The mRNAs encoding the 3 largest isoforms, CDC2L2 alpha-2-1, -2-2, and -2-3, contain 2 alternative translation initiation sites separated by 58 amino acids. In vitro translation of these mRNAs yields the 90- and 110-kD isoforms observed in extracts of HeLa cells. Subcellular fractionation studies indicated that the 110-kD isoforms are primarily localized in the nucleus. The 65-kD isoform was detected only in cells undergoing apoptosis. Using Northern blot analysis, Xiang et al. (1994) found that the alpha-2 isoforms are encoded by 3.5- and 3.7-kb mRNAs that are expressed in most human cell lines and tissues. The mRNA encoding the 65-kD CDC2L2 alpha-1 isoform had a more restricted expression pattern. The authors also detected an antisense transcript corresponding to the 5-prime untranslated region of the alpha-1 mRNA. The antisense and sense transcripts shared the same expression pattern.

Gururajan et al. (1998) determined the structure of the 140-kb CDC2L genomic region. This region consists of 2 identical genomic segments arranged in a tail-to-tail configuration. Each segment contains a CDC2L gene linked to an MMP gene. See MMP21 (603320). In each case, the most widely expressed products of the CDC2L gene are derived from a genomic region that is composed of 20 exons and spans approximately 20 kb. The authors reported that only 15 amino acids of the 773-786 residues that encode the multiple CDC2L isoforms are unique to either CDC2L1 or CDC2L2.

By fluorescence in situ hybridization, Lahti et al. (1994) mapped the CDC2L2 gene to 1p36.3.

Eipers, P. G.; Barnoski, B. L.; Han, J.; Carroll, A. J.; Kidd, V. J.: Localization of the expressed human p58 protein kinase chromosomal gene to chromosome 1p36 and a highly related sequence to chromosome 15. Genomics 11: 621-629, 1991.

Gururajan, R.; Grenet, J.; Lahti, J. M.; Kidd, V. J.: Isolation and characterization of two novel metalloproteinase genes linked to the Cdc2L locus on human chromosome 1p36.3. Genomics 52: 101-106, 1998.

Gururajan, R.; Lahti, J. M.; Grenet, J.; Easton, J.; Gruber, I.; Ambros, P. F.; Kidd, V. J.: Duplication of a genomic region containing the Cdc2L1-2 and MMP21-22 genes on human chromosome 1p36.3 and their linkage to D1Z2. Genome Res. 8: 929-939, 1998.

Lahti, J. M.; Valentine, M.; Xiang, J.; Jones, B.; Amann, J.; Grenet, J.; Richmond, G.; Look, A. T.; Kidd, V. J.: Alterations in the PITSLRE protein kinase gene complex on chromosome 1p36 in childhood neuroblastoma. Nature Genet. 7: 370-375, 1994.

Xiang, J.; Lahti, J. M.; Grenet, J.; Easton, J.; Kidd, V. J.: Molecular cloning and expression of alternatively spliced PITSLRE protein kinase isoforms. J. Biol. Chem. 269: 15786-15794, 1994.

PROTEIN KINASE C, ALPHA (176960)

Protein kinase C (PKC) is the major phorbol ester receptor. Parker et al. (1986) purified this protein from bovine brain and through the use of oligonucleotide probes based on partial amino acid sequence, derived cDNA clones from bovine cDNA libraries. Thus, the complete amino acid sequence of bovine protein kinase C was determined, revealing a domain structure. Activation of PKC by calcium ions and the second messenger diacylglycerol is thought to play a central role in the induction of cellular responses to a variety of ligand-receptor systems and in the regulation of cellular responsiveness to external stimuli. Coussens et al. (1986) defined a new family of PKC-related genes in bovine, human and rat genomes. Three of these, termed alpha, beta (PRKCB1; 176970), and gamma (PRKCG; 176980), are highly homologous. By Southern analysis of hybrid cell DNA and by in situ hybridization, Coussens et al. (1986) found that the corresponding genes are situated, respectively, in the following segments: 17q22-q24, 16p12-q11.1, and 19q13.2-q13.4. Latos-Bielenska et al. (1991) refined the assignment of PRKCA1 to 17q22-q23.2 by study of a translocation t(2;17) by in situ hybridization. From Southern blot analysis Coussens et al. (1986) concluded that even more PKC genes may exist. In a primary melanoma cell line, Linnenbach et al. (1988) found a tumor-specific deletion within the PKCA gene; the deletion was not detectable cytogenetically. In family linkage studies using the CEPH panel of DNAs and RFLP markers, Summar et al. (1989) estimated the most likely interval between PKCA and GH1 (139250) and PKCA and COL1A1 (120150) to be 0.03 and 0.11, respectively. The theta value calculated for GH1 versus COL1A1 was 0.11. They suggested that the most likely gene order is centromere—COL1A1—PKCA—GH1. On the other hand, Jones et al. (1991) placed PKCA just distal to GH1 by the study of chromosome 17 fragment-containing hybrids. Finkenzeller et al. (1990) reported the nucle otide sequence of cDNA clones of the PKCA gene.

Alvaro, V.; Levy, L.; Dubray, C.; Roche, A.; Peillon, F.; Querat, B.; Joubert, D.: Invasive human pituitary tumors express a point-mutated alpha-protein kinase-C. J. Clin. Endocr. 77: 1125-1129, 1993.

Coussens, L.; Parker, P. J.; Rhee, L.; Yang-Feng, T. L.; Chen, E.; Waterfield, M. D.; Francke, U.; Ullrich, A.: Multiple, distinct forms of bovine and human protein kinase C suggest diversity in cellular signaling pathways. Science 233: 859-866, 1986.

Finkenzeller, G.; Marme, D.; Hug, H.: Sequence of human protein kinase C alpha. Nucleic Acids Res. 18: 2183 only, 1990.

Jones, K. W.; Shapero, M. H.; Chevrette, M.; Fournier, R. E. K.: Subtractive hybridization cloning of a tissue-specific extinguisher: TSE1 encodes a regulatory subunit of protein kinase A. Cell 66: 861-872, 1991.

Latos-Bielenska, A.; Klett, C.; Just, W.; Hameister, H.: Refinement of localization of the human genes for myeloperoxidase (MPO), protein kinase C, alpha polypeptide, PRKCA, and the DNA fragment D17S21 on chromosome 17q. Hereditas 115: 69-72, 1991.

Linnenbach, A. J.; Huebner, K.; Reddy, E. P.; Herlyn, M.; Parmiter, A. H.; Nowell, P. C.; Koprowski, H.: Structural alteration in the MYB protooncogene and deletion within the gene encoding alpha-type protein kinase C in human melanoma cell lines. Proc. Nat. Acad. Sci. 85: 74-78, 1988.

Parker, P. J.; Coussens, L.; Totty, N.; Rhee, L.; Young, S.; Chen, E.; Stabel, S.; Waterfield, M. D.; Ullrich, A.: The complete primary structure of protein kinase C—the major phorbol ester receptor. Science 233: 853-859, 1986.

Summar, M. L.; Phillips, J. A., III; Krishnamani, M. R. S.; Keefer, J.; Trofatter, J.; Schwartz, R. C.; Tsipouras, P.; Willard, H.; Ullrich, A.: Protein kinase C: a new linkage marker for growth hormone and for COL1A1. Genomics 5: 163-165, 1989.

RHABDOMYOSARCOMA, EMBRYONAL, 2 (180295)

Loh et al. (1992) showed that the transfer of a normal human chromosome 11 into an embryonal rhabdomyosarcoma cell line resulted in a dramatic loss of the proliferative capacity of the cells. Cells that possessed only the long arm of chromosome 11 also demonstrated a decreased growth rate. Their functional data supported molecular studies indicating loss of genetic information on 11p15 during the development of embryonal rhabdomyosarcoma (268210). In addition, however, their studies demonstrated the existence of a second gene on the long arm of chromosome 11, previously unrecognized by molecular analyses, which negatively regulates the growth of embryonal rhabdomyosarcoma cell lines.

Loh, W. E., Jr.; Scrable, H. J.; Livanos, E.; Arboleda, M. J.; Cavenee, W. K.; Oshimura, M.; Weissman, B. E.: Human chromosome 11 contains two different growth suppressor genes for embryonal rhabdomyosarcoma. Proc. Nat. Acad. Sci. 89: 1755-1759, 1992.

SUPPRESSOR OF TRANSFORMATION/TUMORIGENICITY 2 (185440)

De Ronde et al. (1988) showed that human fibroblasts derived from persons with deletions in the short arm of one chromosome 11 were susceptible to morphologic transformation by early region BK virus DNA, whereas diploid human fibroblasts were not. They suggested that this difference in susceptibility to transformation might be explained by a putative 'transformation suppressor' locus situated within the deleted region on 11p. The 4 individuals studied had various combinations of aniridia, mental retardation, genitourinary abnormalities, Wilms tumor, or deletion of 1 catalase locus. Two of their 4 subjects were sibs.

de Ronde, A.; Mannens, M.; Slater, R. M.; Hoovers, J.; Heyting, C.; Bleeker-Wagemakers, E. M.; Leschot, N. J.; Van Strien, A.; Sol, C. J. A.; Ter Schegget, J.; van der Noordaa, J.: Morphological transformation by early region human polyomavirus BK DNA of human fibroblasts with deletions in the short arm of one chromosome 11. J. Gen. Virol. 69: 467-471, 1988.

THYROID CARCINOMA, FOLLICULAR (188470)

In all of 6 examples of follicular thyroid carcinoma, Herrmann et al. (1991) found loss of heterozygosity (LOH) for RFLP markers on the short arm of chromosome 3. Such was not found in any of 3 follicular adenomas or 12 papillary thyroid carcinomas (188550). This led Herrmann et al. (1991) to suggest that a tumor suppressor gene on 3p is important for the development or progression of FTC.

Herrmann, M. A.; Hay, I. D.; Bartelt, D. H., Jr.; Ritland, S. R.; Dahl, R. J.; Grant, C. S.; Jenkins, R. B.: Cytogenetic and molecular genetic studies of follicular and papillary thyroid cancers. J. Clin. Invest. 88: 1596-1604, 1991.

THYROID HORMONE RECEPTOR, BETA (190160)

Jansson et al. (1983) cloned the human ERBA2 gene. Middleton et al. (1986) identified an RFLP for the ERBA2 gene. Initially, Spurr et al. (1984) concluded that the gene is located on chromosome 17. By in situ hybridization, Gosden et al. (1986) regionalized the assignment to 17q21.3. By Southern blot analysis of DNA from somatic cell hybrids and in situ hybridization using the same human genomic probe, Rider et al. (1987) concluded that ERBA2 is located in the region 3pter-p21, rather than on chromosome 17 as previously held. Malignant lymphomas and salivary gland tumors consistent with chromosomal changes in that region have been observed. Dobrovic et al. (1987) mapped ERBA2, the thyroid hormone receptor gene, to 3p25-p21 and showed that it was deleted in small cell lung carcinoma (182280) in all 6 cases studied. The assignment was done by study of somatic cell hybrids containing various translocations involving human chromosome 3. Dobrovic et al. (1988) suggested that this sequence is a candidate for a recessive oncogene involved in the genesis of small cell cancer of the lung. Drabkin et al. (1987) mapped an ERBA gene to 3p22-p21.33 by study of somatic cell hybrids and in situ hybridization with a DNA probe. Thompson et al. (1987) presented evidence for the existence of at least 2 thyroid hormone receptors. The receptor that is expressed in the mammalian central nervous system and in most other tissues except liver is coded by a gene on chromosome 17 (THRA; 190120). The receptor that is present in liver and some other tissues is coded by a gene on chromosome 3. Evans (1988) reviewed the molecular biology and physiology of the thyroid hormone receptor superfamily. In a kindred with autosomal dominant generalized thyroid hormone resistance (188570), Bale et al. (1988) found that 1 allele of the ERBA-beta gene cosegregated with the resistance trait; maximum lod score = 3.91 at theta = 0.0. In this kindred a decrease in T3-binding affin ity of salt-extracted fibroblast nuclear receptors had been demonstrated. By both somatic cell hybridization and in situ hybridization, Drabkin et al. (1988) showed that the ERBA2 gene is located on 3p24.1-p22. Drabkin et al. (1988) found, furthermore, that the ERBA2 locus was deleted in some but not in all cases of SCLC (182280). Thus, the putative suppressor gene involved in that neoplasm is probably located centromeric to ERBA2. By nonisotopic in situ hybridization to metaphase chromosomes, Albertson et al. (1989) mapped this gene to 3p24.3. THRB and 95 other loci on 3p were used by Tory et al. (1992) in the construction of a genetic linkage map through studies of 59 CEPH families.

Lazar and Chin (1990) reviewed nuclear thyroid hormone receptors. The review indicates a possible confusion of nomenclature. The so-called placental thyroid hormone receptor encoded by chromosome 3 is the prototype beta-form. Its gene is referred to as ERBA2 (alternatively, THR1 or THRB in the human). So-called beta-1 is found in liver, heart, and brain; beta-2 is specific for pituitary in the mouse and rat. Alpha-1 receptor is encoded by a gene, symbolized THRA or ERBA1, on human chromosome 17. Alternative splicing of the gene transcript yields a species called alpha-2 or variant I, which is identical to alpha-1 for 370 amino acids, including the DNA-binding domain, but then diverges completely. The mRNA for this form is particularly abundant in the brain.

Douglas et al. (1991) discussed the confusion that surrounded the ERBA genes. By screening a genomic library with v-erbA, Jansson et al. (1983) isolated 2 types of clones. One clone, called erbA1, is now known to define the THRA locus on chromosome 17 (190120). The second clone, erbA2, seemed to map both to chromosome 17 and chromosome 3. In response to this inconsistent assignment, the genes for erbA2 and erbA-beta were given different designations (ERBA2 and ERBA1, respectively) by Human Gene Mapping 9. Using pulsed field gel electrophoresis, Douglas et al. (1991) demonstrated that the ERBA2 and ERBA1 loci are both on chromosome 3 separated by 50 to 120 kb. Thus, the erbA-beta probe may recognize a separate but closely linked gene on chromosome 3. Tory et al. (1992) constructed a genetic linkage map of 96 loci on 3p, extending from the terminal band to the centromere. Multipoint linkage analysis indicated that the male, female, and sex-averaged maps extend for 102, 147, and 116 cM, respectively. By fluorescence in situ hybridization, the authors localized RAF1 and 16 DNA markers; they regionalized RAF1 to 3p25.

Takeda et al. (1992) evaluated denaturing gradient gel electrophoresis (DGGE) in the rapid diagnosis of this disorder. In studies of affected members from 21 families, putative mutations were identified in 18 unrelated individuals. Sequencing confirmed the nature of the mutations in 9 [of these] individuals. in 9 by sequencing. All of the mutations were in the hormone-binding domain of the receptor; 13 of the 18 were in its center, exon 7. In 3 families, no mutations in the THRB gene were identified, suggesting the existence of mutations at other loci, possibly the THRA gene or other proteins involved in a thyroid hormone-dependent transactivation system. Of 22 families, the inheritance was recessive in 1, dominant in 15, and unknown in 6. Takeda et al. (1992) stated that they had studied 1 of the 7 families reported by Parrilla et al. (1991), bringing the total of known families close to 30. Nagaya et al. (1992) presented experiments indicating that thyroid hormone receptor mutations that result in generalized growth hormone resistance (GTHR) compete with normal receptors at DNA binding sites in target genes to block normal receptor function; thus, the dominant negative activity of such mutants may be explained by competition for receptor binding to DNA. However, Yen et al. (1992) presented observations of the gly340-to-arg mutation (190160.0001) in the MF family suggesting that the decreased sensitivity to T3 in the cells of the affected persons with this mutation is not due to abnormal interactions with the thyroid hormone receptor. Instead, the dominant-negative activity likely occurs by one or both of the following mechanisms: repression by mutant homodimers that bind to thyroid hormone response elements (TREs) in the presence of T3, and/or diminished ligand-regulated transcription by heterodimers of the mutant receptor or TRAP (thyroid hormone receptor auxiliary protein) heterodimers.

Weiss et al. (1993) indicated that 28 different point mutations in the THRB gene had been associated with GRTH. These mutations are clustered in 2 regions of the T3-binding domain of the gene at codons 310 to 347 and 417 to 453. Weiss et al. (1993) reported 6 examples of mutations occurring in apparently unrelated families; 3 of the mutations occurred in 3 different families each and the other 3 in 2 families each. In 11 of 15 families, the mutation was shown to be distinct either by the fact that de novo mutation was involved or that different polymorphisms were found in the gene. In other instances, ethnic difference, e.g., Japanese and Caucasian, supported distinctness. Weiss et al. (1993) pointed out that 28 of the 38 point mutations so far identified, including all those occurring in more than 1 family, are located in CG-rich areas of the THRB gene. Differences in clinical and laboratory findings in unrelated families harboring the same THRB mutations suggested that genetic variability of other factors modulate the expression of thyroid hormone action.

Beck-Peccoz et al. (1994) presented a revised nomenclature for the THRB mutations causing resistance to thyroid hormone. They presented a tabulation of 27 mutations that had been published using the old notation; more recent publications already using the revised nomenclature were not included. The revision was necessitated by the advances in knowledge of the structure of the THRB gene. Cloning of the cDNA encoding THRB (Weinberger et al., 1986) initially had suggested an open reading frame of 450 amino acids. However, subsequent sequencing of this cDNA, as well as genomic clones, showed a guanine rather than adenine at nucleotide 288, generating a new initiation codon (Sakurai et al., 1990) and leading to a predicted protein sequence that contains 461 amino acids. In addition to differences in the numbering of the amino acids used in reported mutations, the exons had been numbered either from 00 to 8 or from 1 to 10, depending on the designation of noncoding exons. To avoid the confusion arising from a double nomenclature, Beck-Peccoz et al. (1994) recommended that the exons be numbered 1 through 10 with 1 and 2 corresponding to the previously designated 00 and 0, respectively, at the 5-prime end. A deduced sequence for THRB consisting of 461 amino acids and incorporating 5 additional residues (MTPNS) at the amino terminus was to be used. (Codons in the THRB gene are numbered according to the corrected coding sequence, which adds 5 amino acids to the amino terminus (Sakurai et al., 1990).) In the allelic variants listed below, 2 statements of the nature of the mutation are given when both have appeared in the literature.

Hayashi et al. (1994) noted that the growing list of THR1 mutations found in subjects with resistance to thyroid hormone led to the identification of 2 mutational 'hot' areas in the T3-binding domain of the protein. One of these 2 areas spans from codon 310 to 347 and the other from codon 438 to 459, the latter located 2 amino acids upstream of the carboxy terminus. The mutations occur with high frequency in CG-rich sequences. Not a single mutation had been identified in the region between the 2 'hot' areas. Forman and Samuels (1990) described this 'cold region', which encodes 90 amino acids, contains 8 of the 9 heptad repeats, and appears to play an important role in receptor dimerization. Hayashi et al. (1994) produced 10 artificial mutant THR1 genes in this 'cold' region according to the hotspot rule, i.e., C-to-T or G-to-A substitutions in CpGs. The properties of artificial mutants were compared with those of 6 naturally occurring mutants. Among all natural mutants, arg320 to his (190160.0015), manifesting a mild form of thyroid hormone resistance, showed the least impairment of T3-binding affinity. In contrast, T3-binding affinity was normal in 6 artificial mutants and reduced to a lesser extent than that of R320H in 3. One truncated mutant, R410X, did not bind T3. All natural mutants had impaired ability to transactivate T3-responsive elements and exhibited a strong dominant-negative effect on cotransfected wildtype receptor.

Since the serum thyroid hormone levels required to compensate for the reduced binding affinity would be inferior to those found in subjects with R320H, natural mutations occurring in the cold region of the THR1 gene should fail to manifest as thyroid hormone resistance or should escape detection. The artificial mutant R410X, resulting in a truncated protein, manifested thyroid hormone resistance only in the homozygous state. The cold region of the putative T3-binding domain is relatively insensitive to amino acid changes and, thus, may not be involved in a direct interaction with T3.

Pohlenz et al. (1996) stated that 38 different point mutations had been documented in the THRB receptor as the cause of thyroid hormone resistance. Except for 2, all previously reported mutations were located in exons 9 or 10 in the hormone binding domain. In 45% of the mutations, a CpG dinucleotide was involved, indicating 2 hotspots at codons 310 to 347 and 438 to 459. Pohlenz et al. (1996) reported the second mutation in exon 7, R243W. Seto and Weintraub (1996) used PCR-coupled automated direct sequencing of genomic DNA for rapid molecular detection of mutations associated with GTHR to detect 2 novel mutations.

Parma et al. (1997) investigated 33 different, autonomous hot nodules from 31 patients for the presence of somatic mutations in the THRB and Gs-alpha genes (see 139320). Twenty-seven mutations (82%) were found in the THRB gene, affecting a total of 12 different residues or locations. All but 2 of the mutations studied had been identified previously as activating mutations. The authors identified the 2 novel mutations as a point mutation causing a leu629-to-phe substitution (L629F), a deletion of 12 bases removing residues 658-661 (asn-ser-lys-ile) at the C-terminal portion of exoloop 3 (del658-661). Only 2 mutations (6%) were found in Gs-alpha genes. In 4 nodules, no mutation was detected. Five residues (ser281, ile486, ile568, phe631, and asp633) were found to be mutated in 3 or 4 different nodules, making them hotspots for activating mutations. The authors concluded that in a cohort of patients from a moderately iodine-deficient area, somatic mutations increasing the constitutive activity of THRB are the major cause of autonomous thyroid adenomas.

See also Sap et al. (1986)

Adams, M.; Matthews, C.; Collingwood, T. N.; Tone, Y.; Beck-Peccoz, P.; Chatterjee, K. K.: Genetic analysis of 29 kindreds with generalized and pituitary resistance to thyroid hormone: identification of thirteen novel mutations in the thyroid hormone receptor beta gene. J. Clin. Invest. 94: 506-515, 1994.

Adams, M.; Nagaya, T.; Tone, Y.; Jameson, J. L.; Chatterjee, V. K. K.: Functional properties of a novel mutant thyroid hormone receptor in a family with generalized thyroid hormone resistance syndrome. Clin. Endocr. 36: 281-289, 1992.

Albertson, D. G.; Sherrington, P. D.; Rabbitts, P. H.: Localization of polymorphic DNA probes frequently deleted in lung carcinoma. Hum. Genet. 83: 127-132, 1989.

Bale, A. E.; Usala, S. J.; Gesundheit, N.; Weinberger, C.; Lash, R. W.; Wondisford, F. E.; McBride, O. W.; Weintraub, B. D.: The gene for generalized thyroid hormone resistance (GTHR) is tightly linked to and may be identical with the proto-oncogene, c-erbA-beta. (Abstract) Am. J. Hum. Genet. 43: A77, 1988.

Beck-Peccoz, P.; Chatterjee, V. K. K.; Chin, W. W.; DeGroot, L. J.; Jameson, J. L.; Nakamura, H.; Refetoff, S.; Usala, S. J.; Weintraub, B. D.: Nomenclature of thyroid hormone receptor-beta gene mutations in resistance to thyroid hormone: consensus statement from the First Workshop on Thyroid Hormone Resistance, 10-11 July 1993, Cambridge, UK. Clin. Endocr. 40: 697-700, 1994.

Behr, M.; Loos, U.: A point mutation (ala229-to-thr) in the hinge domain of the c-erbA-beta thyroid hormone receptor gene in a family with generalized thyroid hormone resistance. Molec. Endocr. 6: 1119-1126, 1992.

Behr, M.; Ramsden, D. B.; Loos, U.: Deoxyribonucleic acid binding and transcriptional silencing by a truncated c-erbA-beta-1 thyroid hormone receptor identified in a severely retarded patient with resistance to thyroid hormone. J. Clin. Endocr. Metab. 82: 1081-1087, 1997.

Boothroyd, C. V.; Teh, B. T.; Hayward, N. K.; Hickman, P. E.; Ward, G. J.; Cameron, D. P.: Single base mutation in the hormone binding domain of the thyroid hormone receptor beta gene in generalized thyroid hormone resistance demonstrated by single stranded conformation polymorphism analysis. Biochem. Biophys. Res. Commun. 178: 606-612, 1991.

Cooper, D. S.; Ladenson, P. W.; Nisula, B. C.; Dunn, J. F.; Chapman, E. M.; Ridgway, E. C.: Familial thyroid hormone resistance. Metabolism 31: 504-509, 1982.

Cugini, C. D., Jr.; Leidy, J. W., Jr.; Chertow, B. S.; Berard, J.; Bradley, W. E. C.; Menke, J. B.; Hao, E.-H.; Usala, S. J.: An arginine to histidine mutation in codon 315 of the c-erbA-beta thyroid hormone receptor in a kindred with generalized resistance to thyroid hormones results in a receptor with significant 3,5,3-prime-triiodothyronine binding activity. J. Clin. Endocr. Metab. 74: 1164-1170, 1992.

Dobrovic, A.; Houle, B.; Belouchi, A.; Bradley, W. E. C.: The thyroid hormone receptor (THR) maps to the chromosome region 3p21-3p25 and is deleted in small cell lung carcinoma. (Abstract) Cytogenet. Cell Genet. 46: 607, 1987.

Dobrovic, A.; Houle, B.; Belouchi, A.; Bradley, W. E. C.: erbA-related sequence coding for DNA-binding hormone receptor localized to chromosome 3p21-3p25 and deleted in small cell lung carcinoma. Cancer Res. 48: 682-685, 1988.

Douglas, J. B.; Daly, M. C.; Rabbitts, P. H.: A pulsed-field gel electrophoresis map locates the polymorphic probes for ERBA2 and erbA-beta within 120 kb of each other, confirming that THRB (formerly ERBA2) maps to chromosome 3. Genomics 9: 398-400, 1991.

Drabkin, H.; Kao, F.-T.; Hartz, J.; Hart, I.; Gazdar, A.; Weinberger, C.; Evans, R.; Gerber, M.: Localization of human ERBA2 to the 3p22-3p24.1 region of chromosome 3 and variable deletion in small cell lung cancer. Proc. Nat. Acad. Sci. 85: 9258-9262, 1988.

Drabkin, H. A.; Kao, F.-T.; Weinberger, C.; Evans, R.: Human c-erb-A is located at chromosome 3p21.33-p22. (Abstract) Am. J. Hum. Genet. 41: A25, 1987.

Dulgeroff, A. J.; Geffner, M. E.; Koyal, S. N.; Wong, M.; Hershman, J. M.: Bromocriptine and triac therapy for hyperthyroidism due to pituitary resistance to thyroid hormone. J. Clin. Endocr. Metab. 75: 1071-1075, 1992.

Evans, R. M.: The steroid and thyroid hormone receptor superfamily. Science 240: 889-895, 1988.

Forman, B. M.; Samuels, H. H.: Interactions among a subfamily of nuclear hormone receptors: the regulatory zipper model. Molec. Endocr. 4: 1293-1301, 1990.

Geffner, M. E.; Su, F.; Ross, N. S.; Hershman, J. M.; Van Dop, C.; Menke, J. B.; Hao, E.; Stanzak, R. K.; Eaton, T.; Samuels, H. H.; Usala, S. J.: An arginine to histidine mutation in codon 311 of the C-erbA-beta gene results in a mutant thyroid hormone receptor that does not mediate a dominant negative phenotype. J. Clin. Invest. 91: 538-546, 1993.

Gosden, J. R.; Middleton, P. G.; Rout, D.; De Angelis, C.: Chromosomal localization of the human oncogene ERBA2. Cytogenet. Cell Genet. 43: 150-153, 1986.

Hayashi, Y.; Sunthornthepvarakul, T.; Refetoff, S.: Mutations of CpG dinucleotides located in the triiodothyronine (T3)-binding domain of the thyroid hormone receptor (TR)beta gene that appears to be devoid of natural mutations may not be detected because they are unlikely to produce the clinical phenotype of resistance to thyroid hormone.. J. Clin. Invest. 94: 607-615, 1994.

Jansson, M.; Philipson, L.; Vennstrom, B.: Isolation and characterization of multiple human genes homologous to the oncogenes of avian erythroblastosis virus. EMBO J. 2: 561-565, 1983.

Lazar, M. A.; Chin, W. W.: Nuclear thyroid hormone receptors. J. Clin. Invest. 86: 1777-1782, 1990.

Middleton, P. G.; De Angelis, C. L.; Weir-Thompson, E. M.; Steel, C. M.: RFLP for the human erbA2 gene. Nucleic Acids Res. 14: 1925, 1986.

Mixson, A. J.; Parilla, R.; Ransom, S. C.; Wiggs, E. A.; McClaskey, J. H.; Hauser, P.; Weintraub, B. D.: Correlations of language abnormalities with localization of mutations in the beta-thyroid hormone receptor in 13 kindreds with generalized resistance to thyroid hormone: identification of four new mutations. J. Clin. Endocr. Metab. 75: 1039-1045, 1992.

Mixson, A. J.; Renault, J. C.; Ransom, S.; Bodenner, D. L.; Weintraub, B. D.: Identification of a novel mutation in the gene encoding the beta-triiodothyronine receptor in a patient with apparent selective pituitary resistance to thyroid hormone. Clin. Endocr. 38: 227-234, 1993.

Nagaya, T.; Madison, L. D.; Jameson, J. L.: Thyroid hormone receptor mutants that cause resistance to thyroid hormone: evidence for receptor competition for DNA sequences in target genes. J. Biol. Chem. 267: 13014-13019, 1992.

Ono, S.; Schwartz, I. D.; Mueller, O. T.; Root, A. W.; Usala, S. J.; Bercu, B. B.: Homozygosity for a dominant negative thyroid hormone receptor gene responsible for generalized resistance to thyroid hormone. J. Clin. Endocr. Metab. 73: 990-994, 1991.

Parma, J.; Duprez, L.; Van Sande, J.; Hermans, J.; Rocmans, P.; Van Vliet, G.; Costagliola, S.; Rodien, P.; Dumont, J. E.; Vassart, G.: Diversity and prevalence of somatic mutations in the thyrotropin receptor and Gs-alpha genes as a cause of toxic thyroid adenomas. J. Clin. Endocr. Metab. 82: 2695-2701, 1997.

Parrilla, R.; Mixson, A. J.; McPherson, J. A.; McClaskey, J. H.; Weintraub, B. D.: Characterization of seven novel mutations of the c-erbA-beta gene in unrelated kindreds with generalized thyroid hormone resistance: evidence for two 'hot spot' regions of the ligand binding domain. J. Clin. Invest. 88: 2123-2130, 1991.

Pohlenz, J.; Schonberger, W.; Wemme, H.; Winterpacht, A.; Wirth, S.; Zabel, B.: New point mutation (R243W) in the hormone binding domain of the c-erbA beta-1 gene in a family with generalized resistance to thyroid hormone. Hum. Mutat. 7: 79-81, 1996.

Pohlenz, J.; Wirth, S.; Winterpacht, A.; Wemme, H.; Zabel, B.; Schonberger, W.: Phenotypic variability in patients with generalised resistance to thyroid hormone. J. Med. Genet. 32: 393-395, 1995.

Refetoff, S.; DeGroot, L. J.; Benard, B.; DeWind, L. T.: Studies of a sibship with apparent hereditary resistance to the intracellular action of thyroid hormone. Metabolism 21: 723-756, 1972.

Refetoff, S.; DeWind, L. T.; DeGroot, L. J.: Familial syndrome combining deaf-mutism, stippled epiphyses, goiter, and abnormally high PBI: possible target organ refractoriness to thyroid hormone. J. Clin. Endocr. Metab. 27: 279-294, 1967.

Rider, S. H.; Gorman, P. A.; Shipley, J. M.; Moore, G.; Vennstrom, B.; Solomon, E.; Sheer, D.: Localization of the oncogene c-erbA2 to human chromosome 3. Ann. Hum. Genet. 51: 153-160, 1987.

Sakurai, A.; Nakai, A.; DeGroot, L. J.: Structural analysis of human thyroid hormone receptor beta gene. Molec. Cell. Endocr. 71: 83-91, 1990.

Sakurai, A.; Takeda, K.; Ain, K.; Ceccarelli, P.; Nakai, A.; Seino, S.; Bell, G. I.; Refetoff, S.; DeGroot, L. J.: Generalized resistance to thyroid hormone associated with a mutation in the ligand-binding domain of the human thyroid hormone receptor beta. Proc. Nat. Acad. Sci. 86: 8977-8981, 1989.

Sap, J.; Munoz, A.; Damm, K.; Goldberg, Y.; Ghysdael, J.; Leutz, A.; Beug, H.; Vennstrom, B.: The c-erb-A protein is a high-affinity receptor for thyroid hormone. Nature 324: 635-640, 1986.

Sasaki, S.; Nakamura, H.; Tagami, T.; Miyoshi, Y.; Nogimori, T.; Mitsuma, T.; Imura, H.: Pituitary resistance to thyroid hormone associated with a base mutation in the hormone-binding domain of the human 3,5,3-prime-triiodothyronine receptor-beta. J. Clin. Endocr. Metab. 76: 1254-1258, 1993.

Sasaki, S.; Nakamura, H.; Tagami, T.; Miyoshi, Y.; Tanaka, K.; Imura, H.: A point mutation of the T(3) receptor beta-1 gene in a kindred of generalized resistance to thyroid hormone. Molec. Cell. Endocr. 84: 159-166, 1992.

Seto, D.; Weintraub, B. D.: Rapid molecular diagnosis of mutations associated with generalized thyroid hormone resistance by PCR-coupled automated direct sequencing of genomic DNA: detection of two novel mutations. Hum. Mutat. 8: 247-257, 1996.

Shuto, Y.; Wakabayashi, I.; Amuro, N.; Minami, S.; Okazaki, T.: A point mutation of the 3,5,3-prime-triiodothyronine-binding domain of thyroid hormone receptor-beta associated with a family with generalized resistance to thyroid hormone. J. Clin. Endocr. Metab. 75: 213-217, 1992.

Spurr, N. K.; Solomon, E.; Jansson, M.; Sheer, D.; Goodfellow, P. N.; Bodmer, W. F.; Vennstrom, B.: Chromosomal localisation of the human homologues to the oncogenes erbA and B. EMBO J. 3: 159-163, 1984.

Takeda, K.; Balzano, S.; Sakurai, A.; DeGroot, L. J.; Refetoff, S.: Screening of nineteen unrelated families with generalized resistance to thyroid hormone for known point mutations in the thyroid hormone receptor beta gene and the detection of a new mutation. J. Clin. Invest. 87: 496-502, 1991.

Takeda, K.; Weiss, R. E.; Refetoff, S.: Rapid localization of mutations in the thyroid hormone receptor-beta gene by denaturing gradient gel electrophoresis in 18 families with thyroid hormone resistance. J. Clin. Endocr. Metab. 74: 712-719, 1992.

Thompson, C. C.; Weinberger, C.; Lebo, R.; Evans, R. M.: Identification of a novel thyroid hormone receptor expressed in the mammalian central nervous system. Science 237: 1610-1614, 1987.

Tory, K.; Latif, F.; Modi, W.; Schmidt, L.; Wei, M. H.; Li, H.; Cobler, P.; Orcutt, M. L.; Delisio, J.; Geil, L.; Zbar, B.; Lerman, M. I.: A genetic linkage map of 96 loci on the short arm of human chromosome 3. Genomics 13: 275-286, 1992.

Usala, S. J.; Menke, J. B.; Watson, T. L.; Berard, J.; Bradley, W. E. C.; Bale, A. E.; Lash, R. W.; Weintraub, B. D.: A new point mutation in the 3,5,3-prime-triiodothyronine-binding domain of the c-erbA-beta thyroid hormone receptor is tightly linked to generalized thyroid hormone resistance. J. Clin. Endocr. Metab. 72: 32-38, 1991.

Usala, S. J.; Menke, J. B.; Watson, T. L.; Wondisford, F. E.; Weintraub, B. D.; Berard, J.; Bradley, W. E. C.; Ono, S.; Mueller, O. T.; Bercu, B. B.: A homozygous deletion in the c-erbA-beta thyroid hormone receptor gene in a patient with generalized thyroid hormone resistance: isolation and characterization of the mutant receptor. Molec. Endocr. 5: 327-335, 1991.

Usala, S. J.; Tennyson, G. E.; Bale, A. E.; Lash, R. W.; Gesundheit, N.; Wondisford, F. E.; Accili, D.; Hauser, P.; Weintraub, B. D.: A base mutation of the C-erbA-beta thyroid hormone receptor in a kindred with generalized thyroid hormone resistance: molecular heterogeneity in two other kindreds. J. Clin. Invest. 85: 93-100, 1990.

Weinberger, C.; Thompson, C. C.; Ong, E. S.; Lebo, R.; Gruol, D. J.; Evans, R. M.: The c-erb-A gene encodes a thyroid hormone receptor. Nature 324: 641-646, 1986.

Weiss, R. E.; Chyna, B.; Duell, P. B.; Hayashi, Y.; Sunthornthepvarakul, T.; Refetoff, S.: A new point mutation (C446R) in the thyroid hormone receptor-beta gene of a family with resistance to thyroid hormone. J. Clin. Endocr. Metab. 78: 1253-1256, 1994.

Weiss, R. E.; Tunca, H.; Gerstein, H. C.; Refetoff, S.: A new mutation in the thyroid hormone receptor (TR) beta gene (V458A) in a family with resistance to thyroid hormone (RTH). Thyroid 6: 311-312, 1996.

Weiss, R. E.; Weinberg, M.; Refetoff, S.: Identical mutations in unrelated families with generalized resistance to thyroid hormone occur in cytosine-guanine-rich areas of the thyroid hormone receptor beta gene: analysis of 15 families. J. Clin. Invest. 91: 2408-2415, 1993.

Yagi, H.; Pohlenz, J.; Hayashi, Y.; Sakurai, A.; Refetoff, S.: Resistance to thyroid hormone caused by two mutant thyroid hormone receptors beta, R243Q and R243W, with marked impairment of function that cannot be explained by altered in vitro 3,5,3-prime-triiodothyroinine binding affinity. J. Clin. Endocr. Metab. 82: 1608-1614, 1997.

Yen, P. M.; Sugawara, A.; Refetoff, S.; Chin, W. W.: New insights on the mechanism(s) of the dominant negative effect of mutant thyroid hormone receptor in generalized resistance to thyroid hormone. J. Clin. Invest. 90: 1825-1831, 1992.

SUPPRESSOR OF TUMORIGENICITY-3 (191181)

Chromosomally stable intraspecific human cell hybrids derived by the fusion of tumorigenic HeLa cells with normal human cells provided evidence that tumorigenicity is recessive (Stanbridge, 1976); the hybrids were completely nontumorigenic. After a prolonged passage in culture, rare tumorigenic segregants were isolated, possibly owing to the loss of specific chromosomes that contained tumor suppressor sequences. By cytogenetic analysis of these hybrids, Stanbridge et al. (1981) showed a correlation between the loss of a single copy each of chromosome 11 and chromosome 14 and the development of the tumorigenic phenotype. 10,9:Klinger (1980, 1982) confirmed the association between the loss of chromosome 11 and a few other chromosomes and the development of the tumorigenic cells. Use of chromosome-11-specific RFLP probes permitted demonstration that the loss of a single normal chromosome 11 was sufficient for the reexpression of the tumorigenic phenotype (Srivatsan et al., 1986; Kaelbling and Klinger, 1986). Furthermore, Saxon et al. (1986) demonstrated that the introduction of a normal chromosome 11 into a tumorigenic HeLa/fibroblast hybrid cell by the microcell transfer technique suppressed the tumorigenic phenotype. A selection process involving the loss of the introduced chromosome led to the reappearance of the tumorigenic cells. Misra and Srivatsan (1989) demonstrated that tumorigenicity was regained with the loss of region 11q13-q23. By RFLP analysis, Srivatsan et al. (1991) found somatic loss of chromosome 11 heterozygosity in 10 of 33 primary cervical carcinomas. In addition, at least 8-fold amplification of sequences was observed in one of the primary tumors in the q13 region, including sequences coding for the fibroblast growth factor-related gene INT2 (164950).

In the vast majority of cases, invasive carcinoma of the uterine cervix is thought to arise from preinvasive cervical intraepithelial neoplasias, designated CIN I-CIN III, which represent a pathologic continuum from mild to severe epithelial dysplasia. Only some cases of the more dysplasic lesions progress further to invasive carcinoma. Whereas most of the precursor lesions are readily curable, the prognosis for invasive carcinoma is generally poor, making the dissection of the molecular events that cause invasion considerably important. The observations are consistent with models describing a multistep genetic pathway for human tumorigenesis as proposed by Fearon and Vogelstein (1990) for colorectal cancer. Infection of cervical epithelial cells with aggressive subtypes of the human papilloma virus (HPV) is probably one factor in the development of cervical neoplasia (see 167959 and 167960). Pursuing the functional studies showing that human chromosome 11 contains a gene or genes capable of suppressing tumorigenicity in cell lines derived from different histopathologic types of cervical carcinoma, Hampton et al. (1994) carried out a systematic analysis of chromosome 11 in primary tumors of 32 patients with cervical carcinoma. To identify the likely chromosomal position of the relevant gene or genes, they used 16 highly polymorphic markers to compare matched DNA samples from noninvolved tissue and portions of tumor tissue highly enriched for neoplastic cells. Of the 32 patients examined, 14 (44%) demonstrated clonal genetic alterations resulting in loss of heterozygosity for 1 or more markers. From the fact that 7 of the clonal genetic alterations on chromosome 11 were specific to the long arm and by the overlap between these and other allelic deletions, Hampton et al. (1994) concluded that at least one suppressor gene relevant to cervical carcinoma maps to 11q22-q24.

No cell line has been subjected to more extensive study than has the HeLa cell. This cell line was isolated from the cervical carcinoma of a patient named Henrietta Lacks who presented to The Johns Hopkins Hospital in early 1951 at the age of 31. Hers was the only one of many cervical carcinomas from which George O. Gey (1899-1970) succeeded in establishing an immortal cell line. That it was an unusual cervical cancer was indicated by its unusual fungating appearance, suggesting a venereal lesion and prompting a dark-field analysis for spirochetes (which was negative). Although there was no evidence of invasion or metastasis at the time the patient was first seen, and although radium therapy was given, Mrs. Lacks was dead in 8 months. Review of the histology by Jones et al. (1971) indicated that the cancer was an adenosquamous carcinoma rather than the usual squamous cervical carcinoma. The genetic characteristics of the HeLa cell line, including HLA types, were determined by Hsu et al. (1976) and compared with the findings in surviving members of her family. That Mrs. Lacks was a heterozygote for G6PD deficiency, i.e., G6PD A/B, was established by the fact that she had both G6PD-deficient and G6PD-normal sons. The HeLa cell line is G6PD-deficient (i.e., G6PD-A), indicating its monoclonal origin; this fact was established by Fialkow (1977) in studies of the monoclonality of cancers. The vigor of the HeLa cell line is attested to by the extent to which it has contaminated other cell lines in laboratories around the world (Hsu et al., 1976).

To determine whether 11q13 rearrangement is a nonrandom event in cervical carcinomas, Jesudasan et al. (1995) studied 6 different human papilloma virus (HPV)-positive cell lines (including HeLa and Caski) and 2 different HPV-negative cell lines. Long-range restriction mapping using a number of 11q13-specific probes showed molecular rearrangements within 75-kb of an INT2 probe in 3 HPV-positive cell lines and in an HPV-negative cell line. By fluorescence in situ hybridization using an INT2 YAC, Jesudasan et al. (1995) identified a breakpoint within the sequences spanned by this YAC in HeLa and Caski cells.

Chen (1996) published the karyotype of the HeLa cell line and demonstrated that the karyotype reported for a cell line obtained from Dr. George Gey in 1970 in the laboratory of Stella Mamaeva in Russia was identical to the karyotype of the ATCC cell line which had been accessioned from Dr. Gey in 1961.

Chen, T. R.: East HeLa and west HeLa, in memory of Stella Mamaeva, 1939-1995. Cancer Genet. Cytogenet. 91: 91-92, 1996.

Fearon, E. R.; Vogelstein, B.: A genetic model for colorectal tumorigenesis. Cell 61: 759-767, 1990.

Fialkow, P. J.: Clonal origin and stem cell evolution of human tumors.In: Mulvihill, J. J.; Miller, R. W.; Fraumeni, J. F., Jr.: Genetics of Human Cancer. New York: Raven Press (pub.) 1977. Pp. 439-453.

Hampton, G. M.; Penny, L. A.; Baergen, R. N.; Larson, A.; Brewer, C.; Liao, S.; Busby-Earle, R. M. C.; Williams, A. W. R.; Steel, C. M.; Bird, C. C.; Stanbridge, E. J.; Evans, G. A.: Loss of heterozygosity in cervical carcinoma: subchromosomal localization of a putative tumor-suppressor gene to chromosome 11q22-q24. Proc. Nat. Acad. Sci. 91: 6953-6957, 1994.

Hsu, S. H.; Schacter, B. Z.; Delaney, N. L.; Miller, T. B.; McKusick, V. A.; Kennett, R. H.; Bodmer, J. G.; Young, G.; Bodmer, W. F.: Genetic characteristics of the HeLa cell. Science 191: 392-394, 1976.

Jesudasan, R. A.; Rahman, R. A.; Chandrashekharappa, S.; Evans, G. A.; Srivatsan, E. S.: Deletion and translocation of chromosome 11q13 sequences in cervical carcinoma cell lines. Am. J. Hum. Genet. 56: 705-715, 1995.

Jones, H. W., Jr.; McKusick, V. A.; Harper, P. S.; Wuu, K.-D.: George Otto Gey (1899-1970): the HeLa cell and a reappraisal of its origin. Obstet. Gynec. 38: 945-949, 1971.

Kaelbling, M.; Klinger, H. P.: Suppression of tumorigenicity in somatic cell hybrids. III. Cosegregation of human chromosome 11 of a normal cell and suppression of tumorigenicity in intraspecies hybrids of normal diploid times malignant cells. Cytogenet. Cell Genet. 42: 65-70, 1986.

Klinger, H. P.: Suppression of tumorigenicity. Cytogenet. Cell Genet. 32: 68-84, 1982.

Klinger, H. P.: Suppression of tumorigenicity in somatic cell hybrids. I. Suppression and reexpression of tumorigenicity in diploid human times D98AH2 hybrids and independent segregation of tumorigenicity from other cell phenotypes. Cytogenet. Cell Genet. 27: 254-266, 1980.

Misra, B. C.; Srivatsan, E. S.: Localization of HeLa cell tumor-suppressor gene to the long arm of chromosome 11. Am. J. Hum. Genet. 45: 565-577, 1989.

Saxon, P. J.; Srivatsan, E. S.; Stanbridge, E. J.: Introduction of human chromosome 11 via microcell transfer controls tumorigenic expression of HeLa cells. EMBO J. 5: 3461-3466, 1986.

Srivatsan, E. S.; Benedict, W. F.; Stanbridge, E. J.: Implication of chromosome 11 in the suppression of neoplastic expression in human cell hybrids. Cancer Res. 46: 6174-6179, 1986.

Srivatsan, E. S.; Misra, B. C.; Venugopalan, M.; Wilczynski, S. P.: Loss of heterozygosity for alleles on chromosome 11 in cervical carcinoma. Am. J. Hum. Genet. 49: 868-877, 1991.

Stanbridge, E. J.: Suppression of malignancy in human cells. Nature 260: 17-20, 1976.

Stanbridge, E. J.; Flandermeyer, R. R.; Daniels, D. W.; Nelson-Rees, W. A.: Specific chromosome loss associated with the expression of tumorigenicity in human cell hybrids. Somat. Cell Genet. 7: 699-712, 1981.

CADHERIN 1 (192090)

Uvomorulin is a specific calcium ion-dependent cell adhesion molecule. In mice it expresses its adhesive function during preimplantation stage of development and in epithelial cells where it is remarkably concentrated in the intermediate junctions. Mansouri et al. (1987) examined the amino acid sequence of uvomorulin and concluded that the gene is highly conserved. Sequence comparison showed extensive similarity to chicken LCAM (152423). Using the mouse cDNA clone, Mansouri et al. (1987) screened a human liver cDNA library and isolated a 2-kb cDNA clone containing coding sequences for the human uvomorulin. Sequence comparison showed over 80% identity to the mouse in both nucleotide and amino acid sequences. By Southern analysis of DNA from a panel of mouse-human somatic cell hybrids, 13,14:Mansouri et al. (1987, 1988) assigned the UVO gene to 16q (16p11-qter). In the mouse the Um locus is on chromosome 8 (Mansouri et al., 1987; Eistetter et al., 1988), together with a number of other loci that are located on human 16q (Scherer et al., 1989). Using a cDNA probe for the human UVO gene, Natt et al. (1989) did Southern blot analysis of 2 overlapping interstitial deletions on human chromosome 16q and thereby assigned the UVO locus to 16q22.1, distal to LCAT (245900) and proximal to HP (140100) and TAT (276600). Chen et al. (1991) confirmed the assignment to chromosome 16 and concluded that UVO is located near LCAT in band 16q22.1. Starting with a human-hamster cell hybrid carrying a single copy of chromosome 16 as the only human genetic material, Ceccherini et al. (1992) generated radiation hybrids retaining unselected fragments of this human chromosome. The most likely order and location of 38 DNA sequences including the UVO gene and the TAT gene were established by multiple pairwise analysis and scaled to estimate the physical distance in megabases. The experiments illustrated the usefulness of radiation hybrids for mapping.

Berx et al. (1995) cloned the human E-cadherin gene and showed that it has 16 exons spanning approximately 100 kb of genomic DNA. The gene structure is similar to that of other cadherins. The authors confirmed the map position of CDH1 to 16q22.1 by fluorescence in situ hybridization.

Reduced expression of E-cadherin is regarded as one of the main molecular events involved in dysfunction of the cell-cell adhesion system, triggering cancer invasion and metastasis. Becker et al. (1994) suggested that E-cadherin mutations contribute to the histopathologic appearance of stomach cancer because 13 of 26 diffuse-growing gastric carcinomas, which have reduced homophilic cell-to-cell interactions, had abnormal gene transcripts that were not seen in noncancerous tissue from the same patients. Oda et al. (1994) analyzed 10 human cancer cell lines that showed growth characterized morphologically by loose cell-cell adhesion for possible structural abnormalities in the expressed E-cadherin. Strong mRNA and protein expression with no nucleotide sequence abnormalities was found in 4 cell lines, and mRNA was absent in 4 other cell lines. In the remaining 2 gastric carcinoma cell lines, the mRNA sequence was abnormal. One showed a 12-bp in-frame deletion with strong expression of mRNA and protein. In the other, there were 4 mRNA species with insertions of different sizes, among which the major transcripts (with a 7-bp insertion) caused a frameshift, and expression of both mRNA and protein were markedly reduced. In these 2 cell lines, DNA mutations were detected around exon-intron junctions, revealing that aberrant RNA splicing was the cause of the mRNA abnormalities. In addition, the wildtype allele of the E-cadherin locus was lost, suggesting that the E-cadherin gene had been inactivated by 2 hits (mutation and allele loss), similar to the mechanism for inactivation of tumor suppressor genes. To determine if mutations of the E-cadherin gene contribute to epithelial tumorigenesis, Risinger et al. (1994) examined 135 carcinomas of the endometrium and ovary. They found 1 somatic nonsense and 1 somatic missense mutation, both with retention of the wildtype alleles, and 2 missense mutations with somatic loss of heterozygosity in the tumor tissue. Overduin et al. (1995) determin ed the 3-dimensional structure of the amino-terminal repeat of mouse epithelial cadherin. Unexpected structural similarities with the immunoglobulin fold suggested an evolutionary relation between the calcium-dependent cadherin cell adhesion molecules and the calcium-independent immunoglobulin cell adhesion molecules.

Riethmacher et al. (1995) introduced a targeted mutation into the E-cadherin gene by homologous recombination in mouse embryonic stem cells. The mutation removed E-cadherin sequences essential for Ca(2+) binding and for adhesive function. These embryonic stem cells were used to generate mice carrying the mutation. Heterozygous mutant animals appeared normal and were fertile. However, the homozygous mutation was not compatible with life; the homozygous embryos showed severe abnormalities before implantation. Particularly, the adhesive cells of the morula dissociated shortly after compaction had occurred, and their morphologic polarization was then destroyed.

Compelling experimental evidence indicates that E-cadherin serves a potent invasion suppressor role. In addition, a tumor suppressor effect of E-cadherin has been suggested. In human cancers, partial or complete loss of E-cadherin expression correlates with malignancy. To investigate the molecular basis for the altered expression, Berx et al. (1995) performed a PCR/SSCP mutation screen of the CDH1 gene in 49 breast cancer patients. No relevant DNA changes were encountered in any of 42 infiltrative ductal or medullary breast carcinoma samples. In contrast, 4 of 7 infiltrative lobular breast carcinomas harbored protein truncation mutations (3 nonsense and 1 frameshift) in the extracellular part of the E-cadherin protein (e.g., 192090.0004). Each of the 4 lobular carcinomas with E-cadherin mutations showed tumor-specific loss of heterozygosity of chromosomal region 16q22.1 containing the E-cadherin locus, thus supporting the Knudson 2-hit hypothesis. Berx et al. (1995) detected no E-cadherin expression in these 4 tumors by immunohistochemistry.

Ilyas et al. (1997) examined 54 sporadic colorectal cancers and 14 ulcerative colitis-associated colorectal cancers to determine the mechanism underlying reduced E-cadherin protein expression in some bowel tumors. Possible mechanisms were thought to be a CDH1 gene mutation or allele loss, which would affect E-cadherin's function as a suppressor of tumor invasion; an effect secondary to mutations in other regulatory genes; or an epiphenomenon. Exon 16, which at 2,269 bp is the longest exon of the CDH1 gene, codes almost entirely for an unusually long 3-prime untranslated region; poly(T) and poly(A) tracts within this region may be susceptible to replication errors affecting CDH1 mRNA secondary structure and stability. Ilyas et al. (1997) found no association between either allele loss or exon 16 replication errors and low levels of E-cadherin protein expression. Ilyas et al. (1997) suggested that mutations in CDH1 do not account for decreased E-cadherin protein expression in colorectal tumors.

Development of malignant tumors is in part characterized by the ability of a tumor cell to overcome cell-cell adhesion and to invade surrounding tissue. E-cadherin, the main adhesion molecule of epithelia, has been implicated in carcinogenesis because is it frequently lost in human epithelial cancers. Reestablishing the functional cadherin complex in tumor cell lines results in reversion from an invasive to a benign epithelial phenotype. Perl et al. (1998) found that loss of E-cadherin expression coincides with the transition from well-differentiated adenoma to invasive carcinoma in a transgenic mouse model of pancreatic beta-cell carcinogenesis. Intercrossing model mice with transgenic mice that maintained E-cadherin expression in beta-tumor cells resulted in arrest of tumor development at the adenoma stage, whereas expression of a dominant-negative form of E-cadherin induced early invasion and metastasis. The results demonstrated that loss of E-cadherin-mediated cell adhesion is a rate-limiting step in the progression from adenoma to carcinoma and is a cause of that progression, not a consequence.

Berx et al. (1998) found reports of 69 somatic mutations of the CDH1 gene. These comprised, in addition to a few missense mutations, mainly splice site mutations and truncation mutations caused by insertions, deletions, and nonsense mutations. There was a major difference in mutation type between diffuse gastric and infiltrative lobular breast cancers. In diffuse gastric tumors, the predominant defects were exon skippings, which caused inframe deletions. By contrast, most mutations found in infiltrating lobular breast cancers were out-of-frame mutations, which were predicted to yield secreted truncated E-cadherin fragments. In most cases these mutations occurred in combination with loss of heterozygosity.

See also Natt et al. (1989)

Becker, K.-F.; Atkinson, M. J.; Reich, U.; Becker, I.; Nekarda, H.; Siewert, J. R.; Hoefler, H.: E-Cadherin gene mutations provide clues to diffuse type gastric carcinomas. Cancer Res. 54: 3845-3852, 1994.

Berx, G.; Becker, K.-F.; Hofler, H.; van Roy, F.: Mutations of the human E-cadherin (CDH1) gene. Hum. Mutat. 12: 226-237, 1998.

Berx, G.; Cleton-Jansen, A.-M.; Nollet, F.; de Leeuw, W. J. F.; van de Vijver, M. J.; Cornelisse, C.; van Roy, F.: E-cadherin is a tumour/invasion suppressor gene mutated in human lobular breast cancers. EMBO J. 14: 6107-6115, 1995.

Berx, G.; Staes, K.; van Hengel, J.; Molemans, F.; Bussemakers, M. J. G.; van Bokhoven, A.; van Roy, F.: Cloning and characterization of the human invasion suppressor gene E-cadherin (CDH1). Genomics 26: 281-289, 1995.

Ceccherini, I.; Romeo, G.; Lawrence, S.; Breuning, M. H.; Harris, P. C.; Himmelbauer, H.; Frischauf, A. M.; Sutherland, G. R.; Germino, G. G.; Reeders, S. T.; Morton, N. E.: Construction of a map of chromosome 16 by using radiation hybrids. Proc. Nat. Acad. Sci. 89: 104-108, 1992.

Chen, L. Z.; Harris, P. C.; Apostolou, S.; Baker, E.; Holman, K.; Lane, S. A.; Nancarrow, J. K.; Whitmore, S. A.; Stallings, R. L.; Hildebrand, C. E.; Richards, R. I.; Sutherland, G. R.; Callen, D. F.: A refined physical map of the long arm of human chromosome 16. Genomics 10: 308-312, 1991.

Eistetter, H. R.; Adolph, S.; Ringwald, M.; Simon-Chazottes, D.; Schuh, R.; Guenet, J. L.; Kemler, R.: Chromosomal mapping of the structural gene coding for the mouse cell adhesion molecule uvomorulin. Proc. Nat. Acad. Sci. 85: 3489-3493, 1988.

Guilford, P.; Hopkins, J.; Harraway, J.; McLeod, M.; McLeod, N.; Harawira, P.; Taite, H.; Scoular, R.; Miller, A.; Reeve, A. E.: E-cadherin germline mutations in familial gastric cancer. Nature 392: 402-405, 1998.

Ilyas, M.; Tomlinson, I. P. M.; Hanby, A.; Talbot, I. C.; Bodmer, W. F.: Allele loss, replication errors and loss of expression of E-cadherin in colorectal cancers. Gut 40: 654-659, 1997.

Jones, E. G.: Familial gastric cancer. New Zealand Med. J. 63: 287-296, 1964.

Lauren, P.: The two histological main types of gastric carcinoma: diffuse and so-called intestinal-type carcinoma. An attempt at a histo-clinical classification. Acta Path. Microbiol. Scand. 64: 31-49, 1965.

La Vecchia, C.; Negri, E.; Franceschi, S.; Gentile, A.: Family history and the risk of stomach and colorectal cancer. Cancer 70: 50-55, 1992.

Mansouri, A.; Goodfellow, P. N.; Kemler, R.: Molecular cloning and chromosomal localization of the human cell adhesion molecule uvomorulin (UVO). (Abstract) Cytogenet. Cell Genet. 46: 655, 1987.

Mansouri, A.; Spurr, N.; Goodfellow, P. N.; Kemler, R.: Characterization and chromosomal localization of the gene encoding the human cell adhesion molecule uvomorulin. Differentiation 38: 67-71, 1988.

Natt, E.; Magenis, R. E.; Zimmer, J.; Mansouri, A.; Scherer, G.: Regional assignment of the loci for uvomorulin (UVO) and chymotrypsinogen B (CTRB) on human chromosome 16q. (Abstract) Cytogenet. Cell Genet. 51: 1050-1051, 1989.

Natt, E.; Magenis, R. E.; Zimmer, J.; Mansouri, A.; Scherer, G.: Regional assignment of the human loci for uvomorulin (UVO) and chymotrypsinogen B (CTRB) with the help of two overlapping deletions on the long arm of chromosome 16. Cytogenet. Cell Genet. 50: 145-148, 1989.

Oda, T.; Kanai, Y.; Oyama, T.; Yoshiura, K.; Shimoyama, Y.; Birchmeier, W.; Sugimura, T.; Hirohashi, S.: E-cadherin gene mutations in human gastric carcinoma cell lines. Proc. Nat. Acad. Sci. 91: 1858-1862, 1994.

Overduin, M.; Harvey, T. S.; Bagby, S.; Tong, K. I.; Yau, P.; Takeichi, M.; Ikura, M.: Solution structure of the epithelial cadherin domain responsible for selective cell adhesion. Science 267: 386-389, 1995.

Perl, A.-K.; Wilgenbus, P.; Dahl, U.; Semb, H.; Christofori, G.: A causal role for E-cadherin in the transition from adenoma to carcinoma. Nature 392: 190-193, 1998.

Riethmacher, D.; Brinkmann, V.; Birchmeier, C.: A targeted mutation in the mouse E-cadherin gene results in defective preimplantation development. Proc. Nat. Acad. Sci. 92: 855-859, 1995.

Risinger, J. I.; Berchuck, A.; Kohler, M. F.; Boyd, J.: Mutations of the E-cadherin gene in human gynecologic cancers. Nature Genet. 7: 98-102, 1994.

Scherer, G.; Bausch, E.; Gaa, A.; von Deimling, O.: Gene mapping on mouse chromosome 8 by interspecific crosses: new data on a linkage group conserved on human chromosome 16q. Genomics 5: 275-282, 1989.

Zanghieri, G.; Di Gregorio, C.; Sacchetti, C.; Fante, R.; Sassatelli, R.; Cannizzo, G.; Carriero, A.; Ponz de Leon, M.: Familial occurrence of gastric cancer in the 2-year experience of a population-based registry. Cancer 66: 2047-2051, 1990.

ZINC FINGER PROTEIN 35 (194533)

Zinc fingers are protein DNA-binding domains that consist of tandemly repeated units of 28-30 amino acids containing 2 cysteine and 2 histidine residues at invariant positions. The finger motif has been identified in several eukaryotic transcriptional regulatory proteins and members of the finger-gene family are implicated in both proliferation and differentiation of normal cells. By means of hybridization with a finger motif-containing DNA probe, Donti et al. (1990) identified several human finger genes, one of which, HF.10, is expressed at low levels in a variety of human tissues and is downregulated during the in vitro terminal differentiation of human leukemic myeloid cell lines. By in situ hybridization and by analysis of interspecific somatic cell hybrids, Donti et al. (1990) mapped the HF10 gene to 3p22-p21, a chromosome region frequently involved in karyotypic changes associated with lung and renal cancer. Hoovers et al. (1992) also assigned zinc finger protein genes to 3p. Lanfrancone et al. (1992) reported that the ZNF35 gene spans about 13 kb and is interrupted by 3 introns. All 11 zinc finger DNA-binding domains are contiguously encoded within the last 3-prime exon. In situ hybridization experiments showed that both the functional locus and a processed pseudogene mapped to 3p22-p21, a region that is frequently deleted in small cell lung cancer and renal carcinoma. Both the functional gene and the pseudogene were deleted in a case of small cell lung carcinoma with 3p deletion.

Calabro et al. (1995) isolated 3 new genes encoding zinc finger proteins from a YAC contig of 3p22-p21 by using a cDNA hybridization selection. They took this to indicate a clustering of zinc finger protein genes in 3p21.

Calabro, V.; Pengue, G.; Bartoli, P. C.; Pagliuca, A.; Featherstone, T.; Lania, L.: Positional cloning of cDNAs from the human chromosome 3p21-22 region identifies a clustered organization of zinc-finger genes. Hum. Genet. 95: 18-21, 1995.

Donti, E.; Lanfrancone, L.; Huebner, K.; Pascucci, A.; Venti, G.; Pengue, G.; Grignani, F.; Croce, C. M.; Lania, L.; Pelicci, P. G.: Localization of the human HF.10 finger gene on a chromosome region (3p21-p22) frequently deleted in human cancers. Hum. Genet. 84: 391-395, 1990.

Hoovers, J. M. N.; Mannens, M.; John, R.; Bliek, J.; van Heyningen, V.; Porteous, D. J.; Leschot, N. J.; Westerveld, A.; Little, P. F. R.: High-resolution localization of 69 potential human zinc finger protein genes: a number are clustered. Genomics 12: 254-263, 1992.

Lanfrancone, L.; Pengue, G.; Pandolfi, P.-P.; Salcini, A. E.; Giacomucci, A.; Longo, L.; Donti, E.; De Luca, P.; La Mantia, G.; Pelicci, P.-G.; Lania, L.: Structural and functional organization of the HF.10 human zinc finger gene (ZNF35) located on chromosome 3p21-p22. Genomics 12: 720-728, 1992.

ADRENAL UNRESPONSIVENESS TO ACTH (202200)

Migeon et al. (1968) described an entity of adrenal unresponsiveness to ACTH characterized by hypoglycemia, hyperpigmentation, feeding problems in infancy, low urinary 17-OHCS, normal tolerance to salt deprivation, and no elevation of 17-OHCS excretion or plasma cortisol concentration with administration of ACTH. Two of their patients were brothers. Sibs of 2 other patients were probably affected. Affected male and female sibs were reported by Shepard et al. (1959). Franks and Nance (1970) observed the condition in 2 sisters and a brother, offspring of first-cousin parents, and reviewed 8 other familial cases. An excess of males and a deficiency of consanguinity suggested the existence of both autosomal and X-linked recessive forms. Plasma ACTH levels were greatly elevated. Kelch et al. (1972) reported 3 families. They pointed out that variable adrenal pathology from family to family and the possibility of both autosomal and X-linked forms suggest heterogeneity in this condition. Thistlethwaite et al. (1975) described affected brothers who had intermittent hypoglycemia precipitated by the 'stress' of infection. Both were tall and hyperpigmented. Failure of adrenocortical response to ACTH was progressive in the elder boy. Electrolyte balance was normal, even on low sodium diet. Levels of ACTH and deoxycorticosterone in the blood were high. Moshang et al. (1973) studied 5 affected sibs. Because of progression of manifestations, they concluded that a primary unresponsiveness to ACTH is not the lesion, but rather that the disorder is an inherited degenerative process. On the other hand, Spark and Etzkorn (1977) favored a defect at the ACTH receptor or a postreceptor site. Shah et al. (1988) described an Asiatic Indian girl with 2 disorders that they termed familial glucocorticoid deficiency and familial hypophosphatemic rickets. It may have been more appropriate for them to have penned it 'of the familial type' in each case because no family history was provided except that the parents were said to have been nonconsanguineous. Because of the suspicion of an X-linked form, the possibility of a chromosomal abnormality in activating both the X-linked hypophosphatemia locus (307800) and a closely-situated locus for glucocorticoid deficiency is of interest.

ACTH receptors in mononuclear leukocytes have been reported to be similar to those in the adrenal cortex in affinity, immunogenicity, and the coupling to effectors. Smith et al. (1987) reported that a patient with familial ACTH unresponsiveness had a significantly decreased number of high-affinity ACTH receptors in circulating monocytes. Yamaoka et al. (1992) described a form of ACTH unresponsiveness in which the defect appeared to reside distal to the receptor (202355); they described 2 cousins with ACTH unresponsiveness and normal ACTH receptors in monocytes.

Mountjoy et al. (1992) cloned the human ACTH receptor and compared it to the human melanocyte-stimulating hormone receptor (155555). The ACTH receptor is also known as melanocortin-2 receptor or MC2 receptor (MC2R). By fluorescence in situ hybridization (FISH), Gantz et al. (1993) mapped the MC2R locus to 18p11.2. By PCR amplification of somatic cell hybrid DNAs, Vamvakopoulos et al. (1993) assigned the MC2R gene to chromosome 18. They regionalized the assignment to 18p11.21-pter by PCR amplification using genomic DNA from somatic cell hybrids that contained deletions of chromosome 18. By FISH, Magenis et al. (1994) mapped the ACTHR gene to 18p11.2; by study of an intersubspecific backcross mapping panel, they assigned the mouse homolog to chromosome 18.

Mountjoy et al. (1992) showed that the ACTH receptor gene encodes a 297-amino acid protein which is a G protein-coupled receptor. The MC2R gene is intronless. It shows the characteristic 7 transmembrane-spanning domains that form the ligand binding site. Isolated glucocorticoid deficiency was shown by Clark et al. (1993) and Tsigos et al. (1993) to be produced by mutations in the MC2R gene. Previously nephrogenic diabetes insipidus (304800) had been shown to be due to a defect in a G protein-coupled receptor.

In contrast to the report by Mountjoy et al. (1992), Naville et al. (1997) found that the ACTHR gene contains 2 exons. Exon 1 is in the 5-prime untranslated region and is separated from the coding exon by an approximately 18-kb intron. Naville et al. (1997) showed that 1 kb of the 5-prime flanking region has basal promoter activity and several cAMP responsive elements (CREs).

Yamamoto et al. (1995) described ACTH insensitivity in 2 sibs, a 9-year-old boy and his 4-year-old sister. They were referred to a hospital because of cutaneous hyperpigmentation. Peripheral blood mononuclear leukocytes obtained from the patients showed no generation of adenylate cyclase when treated in vitro with ACTH. A lack of high-affinity ACTH binding was observed in the patients. A mutation search in these patients failed to reveal the causative mutation.

Although mutations in the ACTHR gene were demonstrated in some families (e.g., 202200.0001), no mutation was found in the gene in other families. Genetic heterogeneity was confirmed by the demonstration that a highly polymorphic CA repeat marker (D18S40), closely linked to the ACTHR locus, was not linked to the clinical phenotype in these families (Weber and Clark, 1994). Thus, there are at least 2 genetic causes for familial glucocorticoid deficiency. The markers segregated with the syndrome in 4 families and independently of the syndrome in 3 families. In an eighth family, the D18S40 genotype neither excluded nor confirmed linkage to the locus, but no mutations in the ACTHR gene were found. Thus, half of affected families probably have a cause other than mutation in the ACTHR gene.

Clark and Weber (1994) demonstrated that another ACTH resistance syndrome, the Allgrod syndrome (triple-A syndrome; alacrimia-achalasia-addisonianism; 231550), is apparently not due to mutations in the ACTH receptor; they and others had not been able to identify mutations.

Reincke et al. (1997) studied allelic loss of the ACTHR gene in sporadic adrenal tumors using a PstI polymorphism in the promoter region 3 kb upstream of the coding region. The rate of heterozygosity for this polymorphism in 99 unrelated Caucasian individuals was approximately 54%. Using this polymorphism, they analyzed loss of heterozygosity (LOH) of the ACTHR gene in 20 informative cases with benign and malignant adrenocortical tumors. Of 16 patients with benign lesions, LOH was present in 1 oncocytic nonfunctional adenoma, but not in 15 hyperfunctioning adenomas. Of 4 informative patients with adrenocortical carcinomas, LOH was present in 2 cases. Both of these cases had advanced tumor stages and showed a more rapid course than carcinoma patients without LOH. Analysis of the flanking region of the ACTHR using markers D18S37 and D18S40 showed that LOH was confined to the ACTHR gene. Northern blots showed reduced expression of ACTHR mRNA in the tumors with LOH. Reincke et al. (1997) concluded that LOH of the ACTHR gene is possibly involved in adrenal tumorigenesis, contributing to cellular dedifferentiation in adenomas and carcinomas.

Slominski et al. (1996) presented evidence that skin produces glucocorticoids by showing that the ACTHR, CYP11A1 (118485), CYP17 (202110), and CYP21A2 (201910) genes are expressed in skin. Using an RT-PCR assay, they found that the ACTHR gene is expressed in normal corporal skin, scalp tissue, compound nevus tissue, basal cell carcinoma tissue, and cultured melanocytes; the CYP11A1 gene is expressed in squamous cell carcinoma tissue, normal facial skin, normal corporal skin, cultured melanocytes, and basal cell carcinoma tissue; the CYP17 gene is expressed in normal facial skin, corporal skin, cultured melanocytes, squamous cell carcinoma cells, and basal cell carcinoma cells; the CYP21A2 gene is expressed in melanoma tissue, squamous cell carcinoma tissue, basal cell carcinoma tissue, normal facial skin, and cultured melanocytes. Slominski et al. (1996) suggested that cutaneous ACTHR may play a role in skin physiology and pathology and that cutaneous proopiomelanocortin activity may be autoregulated by a feedback mechanism involving glucocorticoids synthesized locally.

See also Stempfel and Engel (1960)

Clark, A. J. L.; McLoughlin, L.; Grossman, A : Familial glucocorticoid deficiency associated with point mutation in the adrenocorticotropin receptor. Lancet 341: 461-462, 1993.

Clark, A. J. L.; Weber, A.: Molecular insights into inherited ACTH resistance syndromes. Trends Endocr. Metab. 5: 209-214, 1994.

Franks, R. C.; Nance, W. E.: Hereditary adrenocortical unresponsiveness to ACTH. Pediatrics 45: 43-48, 1970.

Gantz, I.; Tashiro, T.; Barcroft, C.; Konda, Y.; Shimoto, Y.; Miwa, H.; Glover, T.; Munzert, G.; Yamada, T.: Localization of the genes encoding the melanocortin-2 (adrenocorticotropic hormone) and melanocortin-3 receptors to chromosomes 18p11.2 and 20q13.2-q13.3 by fluorescence in situ hybridization. Genomics 18: 166-167, 1993.

Kelch, R. P.; Kaplan, S. L.; Biglieri, E. G.; Daniels, G. H.; Epstein, C. J.; Grumbach, M. M.: Hereditary adrenocortical unresponsiveness to adrenocorticotropic hormone. J. Pediat. 81: 726-736, 1972.

Magenis, R. E.; Smith, L.; Nadeau, J. H.; Johnson, K. R.; Mountjoy, K. G.; Cone, R. D.: Mapping of the ACTH, MSH, and neural (MC3 and MC4) melanocortin receptors in the mouse and human. Mammalian Genome 5: 503-508, 1994.

Migeon, C. J.; Kenny, F. M.; Kowarski, A. A.; Snipes, C. A.; Spaulding, J. S.; Finkelstein, J. W.; Blizzard, R. M.: The syndrome of congenital adrenocortical unresponsiveness to ACTH. Report of six cases. Pediat. Res. 2: 501-513, 1968.

Moshang, T., Jr.; Rosenfield, R. L.; Bongiovanni, A. M.; Parks, J. S.; Amrhein, J. A.: Familial glucocorticoid insufficiency. J. Pediat. 82: 821-826, 1973.

Mountjoy, K. G.; Robbins, L. S.; Mortrud, M. T.; Cone, R. D.: The cloning of a family of genes that encode the melanocortin receptors. Science 257: 1248-1251, 1992.

Naville, D.; Barjhoux, L.; Jaillard, C.; Faury, D.; Despert, F.; Esteva, B.; Durand, P.; Saez, J. M.; Begeot, M.: Demonstration by transfection studies that mutations in the adrenocorticotropin receptor gene are one cause of the hereditary syndrome of glucocorticoid deficiency. J. Clin. Endocr. Metab. 81: 1442-1448, 1996.

Naville, D.; Jaillard, C.; Barjhoux, L.; Durand, P.; Begeot, M.: Genomic structure and promoter characterization of the human ACTH receptor gene. Biochem. Biophys. Res. Commun. 230: 7-12, 1997.

Reincke, M.; Mora, P.; Beuschlein, F.; Arlt, W.; Chrousos, G. P.; Allolio, B.: Deletion of the adrenocorticotropin receptor gene in human adrenocortical tumors: implications for tumorigenesis. J. Clin. Endocr. Metab. 82: 3054-3058, 1997.

Shah, B. R.; Fiordalisi, I.; Sheinbaum, K.; Finberg, L.: Familial glucocorticoid deficiency in a girl with familial hypophosphatemic rickets. Am. J. Dis. Child. 142: 900-903, 1988.

Shepard, T. H.; Landing, B. H.; Mason, D. G.: Familial Addison's disease. Case reports of two sisters with corticoid deficiency unassociated with hypoaldosteronism. Am. J. Dis. Child. 97: 154-162, 1959.

Slominski, A.; Ermak, G.; Mihm, M.: ACTH receptor, CYP11A1, CYP17 and CYP21A2 genes are expressed in skin. J. Clin. Endocr. Metab. 81: 2746-2749, 1996.

Smith, E. M.; Brosnan, P.; Meyer, W. J.; Blalock, J. E.: An ACTH receptor on human mononuclear leukocytes: relation to adrenal ACTH-receptor activity. New Eng. J. Med. 317: 1266-1269, 1987.

Spark, R. F.; Etzkorn, J. R.: Absent aldosterone response to ACTH in familial glucocorticoid deficiency. New Eng. J. Med. 297: 917-920, 1977.

Stempfel, R. S., Jr.; Engel, F. L.: A congenital, familial syndrome of adrenocortical insufficiency without hypoaldosteronism. J. Pediat. 57: 443-451, 1960.

Thistlethwaite, D.; Darling, J. A. B.; Fraser, R.; Mason, D. A.; Rees, L. H.; Harkness, R. A.: Familial glucocorticoid deficiency: studies of diagnosis and pathogenesis. Arch. Dis. Child. 50: 291-297, 1975.

Tsigos, C.; Arai, K.; Hung, W.; Chrousos, G. P.: Hereditary isolated glucocorticoid deficiency is associated with abnormalities of the adrenocorticotropin receptor gene. J. Clin. Invest. 92: 2458-2461, 1993.

Vamvakopoulos, N. C.; Rojas, K.; Overhauser, J.; Durkin, A. S.; Nierman, W. C.; Chrousos, G. P.: Mapping the human melanocortin 2 receptor (adrenocorticotropic hormone receptor; ACTHR) gene (MC2R) to the small arm of chromosome 18 (18p11.21-pter). Genomics 18: 454-455, 1993.

Weber, A.; Clark, A. J. L.: Mutations of the ACTH receptor gene are only one cause of familial glucocorticoid deficiency. Hum. Molec. Genet. 3: 585-588, 1994.

Yamamoto, Y.; Kawada, Y.; Noda, M.; Yamagishi, M.; Ishida, O.; Fujihira, T.; Shirakawa, F.; Morimoto, I.: Siblings with ACTH insensitivity due to lack of ACTH binding to the receptor. Endocrine J. 42: 171-177, 1995.

Yamaoka, T.; Kudo, T.; Takuwa, Y.; Kawakami, Y.; Itakura, M.; Yamashita, K.: Hereditary adrenocortical unresponsiveness to adrenocorticotropin with a postreceptor defect. J. Clin. Endocr. Metab. 75: 270-274, 1992.

CYCLIN-DEPENDENT KINASE INHIBITOR 2B; CDKN2B (600431)

Transforming growth factor-beta (TGFB; 190180) inhibits cell proliferation by inducing a G1-phase cell cycle arrest. Normal progression through G1 is promoted by the activity of the cyclin-dependent protein kinases CDK4 (123829) and CDK6, which are inhibited by the protein p16(INK4) (CDKN2A; 600160). Hannon and Beach (1994) isolated another member of the p16(INK4) family, which they referred to as p15(INK4B) or p15. The expression of this protein (CDKN2B) was induced approximately 30-fold in human keratinocytes by treatment with TGF-beta, suggesting that p15 may act as an effector of TGFB-mediated cell cycle arrest. Comparison of the sequence of the gene encoding p15(INK4B) with that reported for MTS2 (Kamb et al., 1994) demonstrated that the MTS2 sequence encodes the C-terminal 86 amino acids of p15(INK4B). MTS2 was found by Kamb et al. (1994) to lie adjacent to the p16 gene (CDKN2) at 9p21, and Hannon and Beach (1994) cited unpublished work localizing the p15 gene to that position by fluorescence in situ hybridization. Although a fragment of the p15 gene described by Kamb et al. (1994) had been designated MTS2 (for multiple tumor suppressor-2), Hannon and Beach (1994) favored the name INK4B for 'inhibitor of CDK4-B.'

Cytogenetic abnormalities at 9p21 are common in many types of human tumors. The presence of 2 functional members of the p16 family at 9p21 raises the possibility that loss of tumor suppression may involve inactivation of either or both genes. Deletions of 9p21 that remove both genes (or other mutations that might inactivate both) could simultaneously negate 2 major proliferation control pathways.

The p15(INK4b) gene is located adjacent to p16(INK4a) on 9p21 and is codeleted in a high proportion of established human cancer cell lines (Kamb et al., 1994; Nobori et al., 1994; Hannon and Beach, 1994). Okuda et al. (1995) used interphase fluorescence in situ hybridization, Southern blot analysis, and PCR to analyze the tandemly linked MTS1 (600160) and MTS2 loci in primary leukemic blasts from 43 pediatric patients with newly diagnosed acute lymphoblastic leukemia (ALL). Deletions were identified of 18 of 20 cases with cytogenetically observed abnormalities of 9p and in 5 of 23 cases with apparently normal chromosomes, with the majority containing biallelic deletions (16 homozygous and 7 hemizygous).

Quelle et al. (1995) mapped both the p16 and the p15 gene to mouse chromosome 4 in the C3-C6 region, which is syntenic with human 9p.

Stone et al. (1995) reported the genomic structure of p15 and its pattern of mRNA expression. They showed that ectopic expression of p15 inhibits growth of tumor-derived cell lines. In a search for p15 mutations in tumor cells lines and in 9p21-linked melanoma kindreds, they found, other than the previously described homozygous deletions, no mutations of p15. Collectively, they deduced from these observations that p15 has a role in growth regulation but a limited role in tumor progression.

Hannon, G. J.; Beach, D.: p15(INK4B) is a potential effector of TGF-beta-induced cell cycle arrest. Nature 371: 257-261, 1994.

Kamb, A.; Gruis, N. A.; Weaver-Feldhaus, J.; Liu, Q.; Harshman, K.; Tavtigian, S. V.; Stockert, E.; Day, R. S., III; Johnson, B. E.; Skolnick, M. H.: A cell cycle regulator potentially involved in genesis of many tumor types. Science 264: 436-440, 1994.

Nobori, T.; Miura, K.; Wu, D. J.; Lois, A.; Takabayashi, K.; Carson, D. A.: Deletions of the cyclin-dependent kinase-4 inhibitor gene in multiple human cancers. Nature 368: 753-756, 1994.

Okuda, T.; Shurtleff, S. A.; Valentine, M. B.; Raimondi, S. C.; Head, D. R.; Behm, F.; Curcio-Brint, A. M.; Liu, Q.; Pui, C.-H.; Sherr, C. J.; Beach, D.; Look, A. T.; Downing, J. R.: Frequent deletion of p16(INK4a)/MTS1 and p15(INK4b)/MTS2 in pediatric acute lymphoblastic leukemia. Blood 85: 2321-2330, 1995.

Quelle, D. E.; Ashmun, R. A.; Hannon, G. J.; Rehberger, P. A.; Trono, D.; Richter, K. H.; Walker, C.; Beach, D.; Sherr, C. J.; Serrano, M.: Cloning and characterization of murine p16(INK4a) and p15(INK4b) genes. Oncogene 11: 635-645, 1995.

Stone, S.; Dayananth, P.; Jiang, P.; Weaver-Feldhaus, J. M.; Tavtigian, S. V.; Cannon-Albright, L.; Kamb, A.: Genomic structure, expression and mutational analysis of the P15 (MTS2) gene. Oncogene 11: 987-991, 1995.

FRAGILE X SYNDROME (*309550)

Description

X-linked mental retardation associated with marXq28, or fragile X syndrome, is characterized by moderate to severe mental retardation, macroorchidism, large ears, prominent jaw, and high-pitched jocular speech. Expression is variable, with mental retardation being the most common feature. This phenotype is associated with mutations in the FMR1 gene.

Nomenclature

Hecht et al. (1989) suggested a new system of gene symbols for fragile sites. Their suggestion for the fragile site discussed here was FRAXQ27A, where R means 'rare' and FA indicates that it is of the 'folic acid type.' The gene encoded in this region of the X chromosome which is associated with mental retardation is also called fragile X mental retardation 1 (FMR1).

Clinical Features

The phenotype of the fragile X syndrome (as it came to be most often known) includes moderate to severe mental retardation, macroorchidism, large ears, prominent jaw, and high-pitched, jocular speech. Expression is variable, with mental retardation the most common feature; however, Jacobs (1982) encountered a man and Daker et al. (1981) reported 2 brothers with marXq28 and average intelligence. Fryns and Van Den Berghe (1982) presented a kindred in which the fragile X chromosome was transmitted by at least 3 normal males. These men died at ages 68, 72, and 76 and had a normal phenotype including normal intelligence; one was an administrator and 2 were officers. Johnson et al. (1991) described a large kindred with 10 mentally retarded, fragile-X-positive males, and 2 normal transmitting males. Pellissier et al. (1991) also described a kindred with 2 normal transmitter brothers.

Patients have a prominent symphysis of the mandible rather than true prognathism. Many of the fragile X males have relative macrocephaly (as gauged by the ratio of head circumference to height). Testis volume, calculated as pi/6 x length x width(2), is increased in most postpubertal men and occasionally in boys. Histopathologic examination is unremarkable except for edema. The question of whether ovarian size is increased in females with the fragile X was addressed by Goodman et al. (1987). Stigmata of connective tissue abnormality include finger joint hypermobility, instability of other joints (Opitz et al., 1984; Hagerman et al., 1984), and mitral valve prolapse (Pyeritz et al., 1982). Hagerman and Synhorst (1984) not only confirmed mitral valve prolapse but also demonstrated mild dilatation of the ascending aorta. Lubs et al. (1984) restudied the black family in which Lubs (1969) first described the marker X; they confirmed large testes. Meryash et al. (1984) studied 18 males, aged 18 to 69 years. Of 15 subjects, 13 had macroorchidism. Average height was less than published standards. Of the 18 subjects, 17 had absolute or relative macrocephaly and 12 were dolichocephalic. In a multicenter study in Sweden, Blomquist et al. (1985) found the fragile X in 13 of 83 boys (16%) with infantile autism but in none of 19 girls with infantile autism. Klauck et al. (1997) concluded from molecular genetic studies of 141 patients from 105 simplex and 18 multiplex families that an association of autism with fragile X is nonexistent and that the Xq27.3 region is not a candidate for autism.

Rodewald et al. (1987) described ganglioglioma of the cauda equina in a 17-year-old male with familial Martin-Bell syndrome. Because of the association of neoplasms with autosomal chromosome abnormalities, Rodewald et al. (1987) suggested that this may be more than coincidence. They found, however, no reports of tumors associated with Martin-Bell syndrome. Fryns (1993) pointed out that periorbital hyperpigmentation and scrotal hyperpigmentation about the time of puberty are frequent features.

Hyperkinetic behavior and a problem with concentration are present in most affected males. Longitudinal observations indicate a deterioration of IQ with age; mental retardation may, for example, be moderate at age 12 and severe at age 25.

Loesch and Hay (1988) presented the clinical findings on 113 fragile X female heterozygotes from 44 families. In 85% of a subsample of 92 adult females, the nonverbal IQ score was 85 or less. Verbal ability deficits were much less common. Typical facial characteristics, irregular teeth, and hypermobility of finger joints occurred in approximately 40% of adult females, but facial abnormalities were less common in children. Although the frequency of miscarriages was increased, a moderate increase in the number of children was found in female carriers with borderline intellectual impairment.

Thode et al. (1988) were unable to corroborate the existence of a form of the Martin-Bell syndrome with no detectable fragile X. They identified 32 men with the phenotype who were fragile X negative but concluded that they did not fit the full criteria. Voelckel et al. (1988) reported the findings of 3 brothers with the fragile X, only 2 of whom were mentally retarded. The intellectually normal brother was phenotypically normal in other ways also.

A Prader-Willi-like subphenotype of the fragile X syndrome was described by de Vries et al. (1993). The features were extreme obesity with a full, round face, small, broad hands and feet, and regional skin hyperpigmentation. Unlike the Prader-Willi syndrome (176270), the patients lacked the neonatal hypotonia and feeding problems during infancy followed by hyperphagia during toddlerhood. In a group of 26 patients with suspected Prader-Willi syndrome but without detectable molecular abnormalities of chromosome 15, one fragile X patient was found. Schrander-Stumpel et al. (1994) found the FMR1 mutation in a 3-year-old boy with unexplained extreme obesity and delayed motor and speech development. They compared the clinical features with those in 9 reported patients with the fragile X syndrome and extreme obesity. They suggested that behavioral characteristics such as hyperkinesis, autistic-like behavior, and apparent speech and language deficits may help point toward the diagnosis of the fragile X syndrome.

Davids et al. (1990) found that of 150 male patients with the fragile X syndrome, 75 had flat feet, 85 had excessive laxity of joints, and 10 had scoliosis. In 29 of the patients, flat feet had been evaluated or treated by an orthopedic surgeon before the diagnosis of fragile X syndrome had been made. This condition accounts for about one-half of X-linked mental retardation and is the second most common 'chromosomal' cause of mental impairment after trisomy 21. In Sweden, Blomquist et al. (1982) confirmed the latter finding; in an unselected series of 96 boys with IQ less than 50 born between 1959 and 1970, 6 had fraXq28.

Reiss et al. (1994) showed that the volume of the hippocampus was enlarged in fragile X patients compared to controls. Furthermore, there was an age-related increase in the volume of the hippocampus and an age-related decrease in the volume of the superior temporal gyrus. In another study, Reiss et al. (1991) showed that fragile X males showed a significantly decreased size of the posterior cerebellar vermis and increased size of the fourth ventricle, when compared with age- and sex-matched groups of fragile-X-negative, developmentally disabled subjects and individuals with normal IQ. Reiss et al. (1991) showed that young fragile X females likewise had decreased size of the posterior cerebellar vermis and increased size of the fourth ventricle, when compared with normal age-, sex-, and IQ-matched females. The findings were intermediate between those of the fragile X males and the nonfragile X control groups. Rousseau et al. (1991) observed an age-dependent phenomenon: the 'full' fragile X mutation was found preferentially on the inactive X in leukocytes in adult females but not in younger ones. This phenomenon was not observed in female carriers of a 'premutation,' who have little phenotypic expression. Preliminary data suggested that young females who show preferential presence of a full mutation on the active X in leukocytes may be at increased risk for mental retardation. There is a known decrease with age of the expression of the fragile site.

General overgrowth was described in 4 fragile X patients, all of whom came from families with other affected relatives who showed the classic Martin-Bell phenotype (de Vries et al., 1995).

Inheritance

All mothers of males with the fragile X have been found to be carriers; the mutation must occur either at a low rate or only in males.

In 4 of 27 large fragile X pedigrees, Fryns (1984) found strong evidence of transmission by normal males. Froster-Iskenius et al. (1984) raised the possibility of an autosomal suppressor system to account for the transmission of the marker X syndrome by unaffected males.

Pembrey et al. (1985) advanced a premutation hypothesis to explain unusual characteristics of the genetics of this disorder: transmission occurs through normal males; the heterozygous daughters of such males are never mentally retarded and have few or no fragile sites, and by contrast in the next generation, a third of heterozygous females are mentally subnormal with an average of 29% fragile sites (Sherman et al., 1985). This came to be called the Sherman paradox (Fu et al., 1991). Sherman et al. (1985) suggested that a premutation exists which generates a definitive mutation only when transmitted by a female and that there is a submicroscopic rearrangement at Xq27.3 which per se causes no trouble but generates a significant genetic imbalance when involved in a recombinational event with the other X chromosome. Winter and Pembrey (1986) put the hypothesis of Pembrey et al. (1985) to test by analyzing linkage relationships of flanking genetic markers in daughters of 'normal transmitting males.' A significant reduction in recombination was found in meioses giving rise to affected grandsons of normal transmitting males (NTMs), as compared to families with no apparent normal transmitting males. One interpretation offered was interference related to a recombinational event leading to the full fragile X mutation.

Weaver and Sherman (1987) gave guidelines for counseling families with the Martin-Bell syndrome. Because of the peculiarities of the pedigrees, it is necessary to give different estimates for the risk among the sons and daughters of normal carrier mothers, mentally impaired carrier mothers, and normal transmitting males. Among the sons, the probability for mental impairment is 0.38, 0.5, and 0, respectively, and the chance of a son being a mentally normal carrier is 0.12, 0, and 0, respectively. Among the daughters, the risk of being a mentally impaired carrier is 0.16, 0.28, and 0, respectively, and the chance of being a mentally normal carrier is about 0.34, 0.22, and 1.0, respectively. Given a sporadic case in a male with no fragile X demonstrable in the mother, the estimates for occurrence in a brother of the proband vary from 9 to 27%, depending on the theoretical model used; the estimated risk in first cousins varies from 0.01 to 0.05.

See review by Nussbaum and Ledbetter (1986). Having excluded a mutation rate in male germ cells of the magnitude required by an exclusive mutation hypothesis to explain the high incidence of the fragile X syndrome, Vogel et al. (1990) proceeded to demonstrate an increased fitness of heterozygous females by a comparison with the reproductive performance of 'adequate' controls (mothers and grandparents of Down syndrome patients). Estimates ranged between 1.11 and 1.36. A higher incidence of dizygotic twinning suggested a biologic component for this increased fertility. On the other hand, the fragile X families had a significantly lower social status than the controls. This suggested a sociopsychologic component of their higher fertility.

Imprinting

Laird et al. (1990) concluded from an analysis of data on fragile X that 2 progenitor cells for human oogonia may be present at the time of the initial event that leads to chromosome imprinting. The estimate was based on the fact that one-half of the female's primary oocytes would, on the average, be expected to show imprinting if X-chromosome inactivation is the initial step. Laird (1987) proposed that the fragile X syndrome is caused by abnormal chromosome imprinting. According to this model, the fragile X mutation leads to an imprint, i.e., a stable inactivation of a gene or genes at the fragile X site, because the mutation prevents complete reactivation, before oogenesis, of a mutant fragile X chromosome that had been inactivated in a female for dosage compensation. The basis of this localized block to complete reactivation of a fragile X chromosome was proposed to be late replication of DNA at the fragile site (Laird et al., 1987). The population genetic predictions of the 'X-inactivation imprinting' model indicate that the fraction of carrier males who are nonpenetrant (nonimprinted) would be about 0.5 at equilibrium (Sved and Laird, 1990). Sved and Laird (1988) suggested that this predicted fraction is higher than the reported fraction of 0.2 (Sherman et al., 1985) because of an unusual ascertainment bias. Laird (1991) explained the cytogenetic disappearance of the fragile X site in the few daughters of affected males that have been reported as a consequence of erasure of the imprint when it is passed through males. (Erasure of chromosome imprinting often occurs when the imprinted chromosome is passed through the parental gender opposite from the gender that established the imprint.) Reimprinting apparently can occur, however, in primary oocytes of these daughters (Laird, 1991). According to the X-inactivation imprinting model, proposed as accounting for the unusual pattern of inheritance and expression in the fragile X syndrome, the fragile X m utation causes a local block to the attempted reactivation of the mutant X chromosome prior to oogenesis. This local block results in an imprinted fragile X chromosome that is deleterious in males and in females in whom that chromosome is predominantly the active one.

Follette and Laird (1992) examined the stability of the imprinted state, defining stability as 100% penetrance of the syndrome in sons who receive an imprinted chromosome from the mother. In a preliminary estimate, they concluded that the fragile X imprint was stable in 46 out of 48 female meioses, giving a tentative estimate of about 96% for the stability of the imprint. Kirkilionis et al. (1992) presented the pedigree of a large family that illustrated dramatically the Sherman paradox and was compatible with the predictions of the Laird X-inactivation imprinting model.

Cytogenetics

Lubs (1969) first described a marker X chromosome in mentally retarded males; a secondary constriction on the distal long arm gave the appearance of large satellites. Lubs (1969) suggested that either the anomalous region itself or a closely linked recessive gene might account for X-linked retardation. This observation went unconfirmed for years until cytogeneticists reverted to a folate-deficient medium for tissue culture such as Lubs (1969) employed. Appearance of this secondary constriction (widely referred to as a fragile site) was shown to be dependent on folate deficiency in the culture medium (which leads to deficiency of thymidine monophosphate), localized to the interface between Xq27 and q28, and associated with mental retardation and macroorchidism in males (Giraud et al., 1976; Harvey et al., 1977; Sutherland, 1977). Sutherland was in Melbourne when he made his initial observations on the fragile X. When he went to Adelaide, he upgraded his laboratory, changing from 199 to F10 culture medium to give better chromosomes for banding. The failure to find the fragile X with the new medium led to his discovery of the critical role of folate (Gerald, 1983).

Turner et al. (1978) suggested labeling the marker secondary constriction Xq27; however, convention requires that 'a break suspected at an interface between two bands is identified arbitrarily by the higher of the two band numbers' (ISCN, 1978; section 2.4.4.2). Brookwell and Turner (1983) again concluded that the fragile site is in band Xq27, close to the 27-28 interface. The marker X is not preferentially inactivated in heterozygotes (Lubs, 1969; Martin et al., 1980).

Mapping

Filippi et al. (1983) studied linkage with G6PD and colorblindness in 18 Sardinian pedigrees. In 6 informative pedigrees the fragile X syndrome showed close linkage with G6PD deficiency and deutan colorblindness. The maximum likelihood estimate of recombination was 6% with 90% fiducial limits between 2.5 and 19.5% and odds favoring linkage of 428:1. No hint of linkage of G6PD and the Renpenning form of X-linked mental retardation (309500) or other unspecified type of X-linked mental retardation was found. Patients with the Renpenning form not only lack the facial features and macroorchidism typical of the fragile X syndrome but also have microcephaly.

Szabo et al. (1984) postulated that the fragile X mutation occurs in a region (Xq27.3) that is a 'hotspot' for meiotic recombination; that the microscopically detectable change is probably a minute chromosomal aberration resulting from an inaccurate recombination event; and that recombination is suppressed at the Xq27.3 region in females heterozygous for the fragile X. The hypothesis was based on the finding that the factor IX locus (306900) and the G6PD locus (305900) are closely linked to the fragile X 'locus' but factor IX segregates independently of the hemophilia A, deutan and protan colorblindness loci which are in the G6PD cluster. Warren et al. (1985) reported a family in which 2 brothers with fragile X mental retardation had different factor IX RFLPs, indicating that a recombinational event occurred between the 2 loci and that the 2 loci may not be as tightly linked as previously published data suggested. Camerino et al. (1983) found no recombination out of 17 opportunities for meiotic recombination in 2 families. The combined data gave a peak of 4.02 at a theta of 0.05. Brown et al. (1985) found that pedigrees with nonpenetrant males have tight linkage to factor IX, whereas the linkage is loose in those pedigrees with full penetrance in males.

Giannelli et al. (1987) studied linkage between the Martin-Bell syndrome and the factor IX gene as well as with the anonymous probe 52A. Heterogeneity was found in their own 9 families and in those reported in the literature. One group of families showed tight linkage of fragile X to factor IX while the other showed loose linkage. In the combined collection of families, the 2 groups represented 0.3 and 0.7 of the total, respectively, and showed recombination fractions of 0.0-0.15 and 0.25-0.5, respectively. Families with nonpenetrant carrier males showed tighter linkage to factor IX than did the others. By contrast, no significant difference was found in the recombination between 52A and factor IX in the 2 groups of fragile X families or in these families vs those with Hunter syndrome (309900) examined in the same laboratory. Giannelli et al. (1987) favored the possibility that this indicates the existence of 2 fragile X loci.

Brown et al. (1988) performed a multilocus linkage analysis of the fragile X syndrome in 147 families using 4 flanking markers. As previously observed by Giannelli et al. (1987), significant variation in the recombination distance between F9 and FRAXA was found also in this group of families. Heterogeneity testing showed that 20% of the families had tight F9-FRAXA linkage, whereas 80% demonstrated loose linkage, with an average recombination distance of 0.35. On average, the multipoint distances found were DXS51-F9, 6.9%; F9-FRAXA, 22.4%; FRAXA-DXS52, 12.7%; and DXS52-DXS15, 2.2%. Thibodeau et al. (1988) also reported on linkage data between the fragile X locus and 4 polymorphic markers: DXS51, F9, DXS98, and DXS52. The markers were studied in 14 families with fragile X and 9 normal pedigrees from the CEPH collection. In this set of families, as has been previously observed, there was evidence for genetic heterogeneity between the fragile X locus and the F9 site. The observed recombination frequencies were as follows: DXS51-F9, 0%; F9-DXS52, 45%; DXS51-FRAXA, 15%; F9-FRAXA, 18%; DXS98-FRAXA, 36%; and DXS52-FRAXA, 15%. The authors proposed the following relative order for the 5 loci, based on multipoint linkage analysis: (DXS51, F9, DXS98)—FRAXA—DXS52.

Poustka et al. (1991) described a physical map of the end of Xq encompassing the region from Xq27.2 to the telomere, inclusive of band Xq28. The map covered a total of 12 Mb of DNA and extended from the telomere to a location 3 Mb proximal to the most likely position of the fragile X mutation. The map determined order and position of loci throughout the Xq28 region and localized cell line breakpoints marking the fragile X region to an interval of 300-700 kb between 8 and 8.7 Mb proximal to the Xq telomere.

Molecular Genetics

Note on nomenclature for trinucleotide repeats: The repeat involved in the fragile X syndrome is variously referred to here as (CGG)n or (CCG)n. Furthermore, the identical repeat found in the cloned FRAXE gene (309548) was referred to as (GCC)n by Knight et al. (1993). There are only 10 different trinucleotide repeats; however, each can be written in a number of ways. Sutherland (1993) favored the convention that lists the motif in alphabetical order in the 5-prime to 3-prime direction. Consistent with this, he uses the (CCG)n designation. He preferred, furthermore, the designation (AGC)n for the other clinically significant dinucleotide repeat found in myotonic dystrophy, Huntington disease, Kennedy disease, and SCA1; (CAG)n in the designation most often used. Sutherland (1993) suggested that the same convention can apply to dinucleotides. He wrote: 'It must be very confusing for newcomers to the literature to find (AC)n, (CA)n, (GT)n, and (TG)n repeats, when the cognoscenti know these are synonyms.'

Using the G6PD Mediterranean variant as a marker, Rocchi et al. (1990) investigated the number of somatic cells (fibroblasts or red cells) with an active fragile X chromosome. They found a significant inverse correlation between IQ level in heterozygotes and the percentage of fibroblasts with the fragile X as the active chromosome. In contrast, no significant correlation was found between IQ and the red cell data, suggesting somatic selection against hematopoietic stem cells with an active fragile X.

Using a 275-kb fragment of human DNA isolated in a yeast artificial chromosome (YAC) and thought to span the fragile site, Yu et al. (1991) derived 2 probes that spanned the fragile site as demonstrated by in situ hybridization. Mapping delineated further the sequences that appeared to span the fragile site to about 15 kb. A 5-kb EcoRI fragment was found to contain fragile site breakpoints. When this fragment was used as a probe on the chromosomal DNA of normal and fragile X individuals, alterations in the mobility of the sequences were found only in fragile X DNA. These sequences were of an increased size and varied within families, indicating that the region was unstable. The results were consistent with those of Oberle et al. (1991).

Khalifa et al. (1990) found no evidence that DNA methylation in the vicinity of the fragile X site influences the phenotype of the syndrome. In studies of peripheral lymphocytes, Webb and Jacobs (1990) found that the number of active fra(X) chromosomes was consistently higher in retarded heterozygous females than in mentally normal heterozygous females. A review of the findings in the literature showed the same result. By using microdissected markers close to the fragile site, Bell et al. (1991) demonstrated that the fragile X syndrome is not associated with large structural rearrangements in Xq27.3 but is associated with methylation of DNA sequences distal to the fragile site. Significant differences were observed in the pulsed field gel electrophoresis pattern observed after BssHII digestion of DNA from fragile X-positive, mentally retarded individuals compared with normal male controls. A 600-kb band was either absent or of reduced intensity in affected males. The pattern was normal in normal, fragile X-negative transmitting males, whereas their mentally retarded, fragile X-positive grandsons lacked the fragment. These observations suggested that the absence of the band is the result of methylation of the BssHII site. The findings were considered consistent with Laird's hypothesis of imprinting at this locus and with a 2-step process for the expression of the disease (Laird et al., 1987).

The methylation probably occurs in a CpG island. The lack of expression of a gene associated with this island could result in the disease phenotype. Oberle et al. (1991) found that probes adjacent to a single CpG island that was mapped at or very near the fragile site detected very localized DNA rearrangements that constitute the fragile X mutations. These rearrangements occurred in a 550-bp GC-rich fragment. Phenotypically normal, transmitting males had a 150-bp to 400-bp insertion that was inherited by their daughters either unchanged or with little change in size. Fragile-X-positive persons in the next generation had much larger fragments that differed among sibs and showed a generally heterogeneous pattern indicating somatic mutation. The mutated allele appeared unmethylated in normal transmitting males, methylated only on the inactive X chromosome in their daughters, and totally methylated in most fragile X males. However, some males had a mosaic pattern. Expression of the fragile X syndrome thus appears to result from a 2-step mutation as well as highly localized methylation.

Rousseau et al. (1992) reviewed the 'unstable and methylatable mutations causing the fragile X syndrome.' They pointed out that the CGG repeat is in phase with the following protein coding sequence and, if translated, would code for a stretch of 6 to 54 arginines. The predicted sequence of the FMR1 protein has no strong homology with any other known protein; Verkerk et al. (1991) demonstrated that it contains a nuclear translocation signal. The premutation is meiotically unstable but generally mitotically stable. The full mutation is both meiotically and mitotically unstable; as a result, a heterogeneous pattern of bands of the CGG repeats is seen on Southern blots, e.g., for leukocyte DNA, reflecting mosaicism. Devys et al. (1992) demonstrated that the pattern of mosaicism was strictly identical in 3 pairs of monozygotic twins, indicating that the somatic heterogeneity and abnormal methylation are established early in development. FMR1 mRNA is absent in those with the full mutation, with the exception of mosaics.

Warren et al. (1987) presented evidence that physical breakage occurs at the fragile X site. They demonstrated translocations in somatic cell hybrids that join the fragile site to heterologous DNA; these might be useful in the molecular cloning of the fragile X sequence (193,191:Warren et al., 1988, 1990). Indeed, such proved to be the case. Verkerk et al. (1991) used the symbol FMR1 (for fragile-site mental retardation-1) for the gene they identified in the breakpoint cluster region in the fragile X syndrome. They found the gene, which expresses a 4.8-kb message in human brain, within a 4-cosmid contig of YAC DNA. Within a 7.4-kb EcoRI fragment, containing FMR1 exonic sequences distal to a CpG island previously shown to be hypermethylated in fragile X patients, they demonstrated a fragile X site-induced breakpoint cluster region that exhibits length variation in fragile X chromosomes. This fragment contained a lengthy CGG repeat that is 250 bp distal of the CpG island and maps within an FMR1 exon. Localization of the brain-expressed FMR1 gene to this EcoRI fragment suggested the involvement of this gene in the phenotypic expression of the fragile X syndrome.

As cited earlier, Wohrle et al. (1992) described partial deletion of the FMR1 gene in a patient with the fragile X syndrome. Gedeon et al. (1992) described a patient with typical clinical features of the fragile X syndrome but without cytogenetic expression of the fragile X and without an amplified CCG trinucleotide repeat fragment. The patient had a previously uncharacterized submicroscopic deletion encompassing the CCG repeat, the entire FMR1 gene, and about 2.5 megabases of flanking sequences. Random X inactivation was found in the patient's mother, who was shown to be a carrier of the deletion. Tarleton et al. (1993) described a de novo deletion.

Kirchgessner et al. (1995) studied the expression of the FMR1 gene from inactive X chromosomes by RT-PCR, both in somatic cell hybrids that contained an active or inactive human X chromosome and in a female patient with a large deletion surrounding the FMR gene. In both analyses, the data indicated that FMR1 is normally subject to X inactivation. The authors stated that this finding is consistent with the results of previous studies of DNA methylation of FMR1 on active and inactive X chromosomes and supports the involvement in the variable phenotype of females for with full mutations of the FMR1 gene.

Morton and Macpherson (1992) proposed a model in which the fragile X mutation is postulated to occur as a multistep process. This attractive model provides a framework in which the seemingly contradictory observations of a mutation old enough to establish a founder effect and an apparently high new mutation rate are united. Morton and Macpherson (1992) suggested that 4 types of alleles occur in the fragile X syndrome (see table in the review by Chakravarti, 1992). The 4 types of alleles are: N = normal, with a frequency of 0.9751; S = stable insert with a frequency of 0.0225 and a mean age of about 90 generations; Z = unstable insert with a frequency of 0.0014 and a mean age of 2 generations; and L = mutation with a frequency of 0.0010 and a mean age of 1.4 generations. Thus myotonic dystrophy (160900) and fragile X appear to share both the phenomenon of anticipation and the phenomenon of founder effect. Richards and Sutherland (1992) referred to the amplification mutation involving (CCG)n in the fragile X syndrome and the (AGC)n repeat in myotonic dystrophy and Kennedy disease (313200) as 'dynamic mutations.' In studies using 2 polymorphic CA repeats located close to the 'mutation target' for the fragile X syndrome, Oudet et al. (1993) observed significant differences in allelic and haplotypic distributions between normal and fragile X chromosomes, indicating that a limited number of primary events may have been at the origin of most present-day fragile X chromosomes in Caucasian populations. They proposed a putative scheme with 6 founder chromosomes from which most of the observed fragile X-linked haplotypes can be derived directly or by a single event at one of the marker loci. Such founder chromosomes may have carried a number of CGG repeats in an upper-normal range, from which recurrent multistep expansion mutations have arisen. The diversity of haplotypes at the fragile X locus may reflect genetic heterogeneity but may also be explained by mutations in the markers themselves.

Smits et al. (1993) stated that they had been unable to demonstrate any new mutation in 84 probands referred to them to date. Interestingly, they also demonstrated the same FMR1 gene mutation in 5 fragile X probands with common ancestors married in 1747. Very few mutations in FMR1 other than expanded stretches of CGG repeats in the 5-prime noncoding region have been described as a cause of the fragile X syndrome. The ile367-to-asn mutation (309550.0001) is one of the rare exceptions. This and other such cases involved isolated instances. Meijer et al. (1994), however, reported a family in which 11 individuals had a 1.6-kb deletion proximal to the CGG repeat of the FMR1 gene as the cause of the fragile X syndrome. Although fragile X chromosomes were not cytologically detectable, all 4 affected males and 2 of the carrier females showed the characteristic clinical phenotype. Using RT-PCR, Meijer et al. (1994) demonstrated that FMR1 was not expressed in the affected males, strongly suggesting that the FMR1 promoter sequences 5-prime to the CGG repeat were missing. The deletion patients had approximately 45 CGG repeats in the FMR1 gene; these were not interspersed by AGG triplets that are usually present in both normal and expanded repeats, however. Meijer et al. (1994) hypothesized that prior to the occurrence of the deletion an expansion of the repeat occurred and that the deletion removed the 5-prime part of the CGG repeat containing the AGG triplets. Transmission of the deletion through the family could be traced back to the deceased grandfather of the affected males, which supported the hypothesis that the FMR1 gene product is not required for spermatogenesis.

Eichler et al. (1993) showed by exon-exon PCR and restriction analysis that the FMR1 gene consists of 17 exons spanning 38 kb. From sequencing of intron-exon boundaries, they concluded that the splice donors and acceptors located in the 5-prime portion of the gene demonstrated greater adherence to consensus than did those in the 3-prime end, providing a possible explanation for the finding of alternative splicing in FMR1. Ashley et al. (1993) isolated and characterized cDNA clones encoding the murine homolog, Fmr-1, which exhibited marked sequence identity with the human gene, including the conservation of the CGG repeat. A conserved ATG downstream of the CGG repeat in both human and mouse and an in-frame stop codon in other human 5-prime cDNA sequences demarcate the FMR1 coding region and confine the CGG repeat to the 5-prime untranslated region. Ashley et al. (1993) also presented evidence for alternative splicing of the FMR1 gene in mouse and human brain and showed that one of these splicing events alters the FMR1 reading frame, predicting isoforms with novel carboxy termini. Ashley et al. (1993) identified ribonucleoprotein particle domains within the FMR protein, and RNA was shown to bind in stoichiometric ratios, which suggested that there are 2 RNA binding sites per FMR protein molecule. The protein was able to bind to its own message with high affinity and interacted with approximately 4% of human fetal brain messages. The absence of the normal interaction of the FMR protein with a subset of RNA molecules might result in the pleiotropic phenotype associated with the fragile X syndrome.

The K homology (KH) domain, originally described in the pre-mRNA-binding heterogeneous nuclear ribonucleoprotein (hnRNP) K protein, contains approximately 50 amino acids and is found in a diverse group of proteins of which many, if not all, are RNA-binding proteins. The sequence motif has been evolutionarily conserved. The importance of the KH domain is underscored by the observation that a single point mutation in a conserved residue of a KH domain in the protein product of the FMR1 gene leads to severe fragile X syndrome (309550.0001). To assess the role of KH domains in RNA binding, Siomi et al. (1994) used mutagenesis of KH domains in hnRNP K and FMR1. Conserved residues of all 3 KH domains of hnRNP K are required for its wildtype RNA binding. Furthermore, Siomi et al. (1994) stated that the FMR1 protein containing the ile304-to-asn mutation of De Boulle et al. (1993) showed severe impairment of binding to RNA. (De Boulle et al. (1993) referred to the mutation as ile367-to-asn.) The results demonstrated an essential role for KH domains in RNA binding and strengthened the connection between the fragile X syndrome and loss of RNA-binding activity of FMR1.

Verheij et al. (1993) investigated the nature and function of the protein encoded by the FMR1 gene using polyclonal antibodies raised against the predicted amino acid sequences. Four different protein products, possibly resulting from alternative splicing, were identified by immunoblotting in lymphoblastoid cell lines of healthy persons. All these proteins were missing in cell lines from patients not expressing FMR1 mRNA. The intracellular localization of the FMR1 gene products as demonstrated by transient expression in COS-1 cells was cytoplasmic. In the fragile X syndrome, the amplification of the CGG repeat blocks transcription of the FMR1 gene; this results in the absence of FMR protein with consequent mental retardation. Siomi et al. (1993) demonstrated that the protein product of the FMR1 gene contains 2 types of sequence motifs characteristic of RNA-binding proteins: an RGG box and 2 heterogeneous nuclear RNP K homology domains. They also demonstrated that FMR1 binds RNA in vitro. Using antibodies to FMR1, they detected its expression in cells of unaffected humans but little or no FMR1 in fragile-X-affected patients. It is noteworthy that the point mutation described by de Boulle et al. (1993), ile367-to-asn, is in one of the most highly conserved residues of one of these RNA-binding domains.

In a daughter of a female carrier of the fragile X mutation, van den Ouweland et al. (1994) found a haplotype using flanking markers that predicted she had inherited the fragile X premutation chromosome. However, the CGG repeat sequence and the intragenic polymorphic marker FMRb showed the normal maternal alleles, while 2 other intragenic markers, FMRa and FRAXAC2, and other more distant markers, showed the risk haplotype. These observations were interpreted as indicating gene conversion and might represent back mutation at the FMR1 locus.

It had been suggested that expansion of the CGG repeat in the FMR1 gene is a postzygotic event with the germline protected. From an analysis of intact ovaries of full-mutation fetuses. Malter et al. (1997) showed that only full expansion alleles could be detected in oocyte (but in the unmethylated state). Similarly, the testis of a 13-week full-mutation fetus showed no evidence of premutations, while a 17-week full mutation fetus exhibited some germ cells with attributes of premutations. These data discounted the hypothesis that the germline is protected from full expansion and suggested that full mutation contraction occurs in the immature testis. Thus, full expansion may already exist in the maternal oocyte or postzygotic expansion, if it occurs, arises quite early in development prior to germline segregation.

Kunst et al. (1997) studied the influence of AGG interruptions on CGG repeat stability. Kunst et al. (1997) analyzed the sorted sperm of 2 donors, each with 39 total repeats but distinct AGG interruption patterns. There was approximately 15% variation in repeat length in each case. However, the male with 29 perfect repeats showed 3% expansion changes while the male with 19 perfect repeats had none. Kunst et al. (1997) also noted that all variant sperm showed expansion or contraction of the 3-prime end of the repeat array. Kunst et al. (1997) concluded that these data are consistent with the hypothesis that perfect repeat tracts influence repeat stability and that changes of the FMR1 repeat exhibit polarity.

Phenotype/genotype

In a survey of retarded females who had no obvious physical abnormalities, 7% expressed marXq28 in lymphocytes (Turner et al., 1980). Among obligate heterozygotes, the likelihood of detecting marXq28 correlates with severity of retardation (Howard-Peebles and Stoddard, 1980; Jacobs et al., 1980). In 2 heterozygous sisters who were slow learners, Uchida and Joyce (1982) found that the fragile X was the active one in 100 of 129 cells (77.5%) and 85 of 120 cells (70.8%), whereas 2 heterozygous relatives of normal intelligence had the fragile X active in 40 of 78 cells (51.3%) and 10 of 32 cells (31.3%). An earlier suggestion that the proportion of cells exhibiting marXq28 decreases with increasing heterozygote age (Sutherland, 1979; Jacobs et al., 1980; Turner et al., 1980) is probably an artifact due to ascertaining fewer retarded women in older age groups (Jacobs, 1982). Snyder et al. (1984) showed that culture conditions that promote expression of the fragile X site do not affect expression of lymphocyte HPRT but do cause a marked reduction in G6PD activity. Langenbeck et al. (1984) found that mean corpuscular hemoglobin is increased in this disorder and asked whether this is a reflection of a defect in folate metabolism. Why the marker site is related to retardation is unknown. At least 12 other heritable secondary constrictions ('fragile sites') on other chromosomes were proved by the early 1980s (Sutherland, 1981; Hecht et al., 1982), but none had an association with a particular phenotype. In all pedigrees of marXq28 studied, no crossing-over between the marker and mental retardation had occurred. This suggested that the marker, rather than being closely linked to a gene causing mental retardation, is a direct cytologic indicator of the genetic mutation causing this phenotype (Kaiser-McCaw et al., 1980).

As stated earlier, the designation Sherman paradox came to be applied to the following phenomenon: 20% of males who carry a fragile X chromosome are phenotypically normal; their daughters, to whom they transmit the fragile X chromosome, are likewise normal, but their grandsons are often affected. The brothers of the clinically normal, transmitting males have a low risk, while grandsons and great-grandsons have much higher risks. Fu et al. (1991) demonstrated that the Sherman paradox is related to the particular structure of the (CGG)n repeat found in the coding sequence of the FMR1 gene. (The CGG triplet codes for arginine.) In normal individuals the range of allele sizes varied from a low of 6 to a high of 54 repeats. Premutations showing no phenotypic effect in fragile X families ranged in size from 52 to more than 200 repeats. All alleles with greater than 52 repeats, including those identified in a normal family, were meiotically unstable, with a mutation frequency of 1, while 75 meioses of alleles of 46 repeats and below showed no mutation. Demonstration of mosaicism may indicate that premutation alleles are also mitotically unstable. Fu et al. (1991) showed that the risk of expansion during oogenesis to the full mutation increased with the number of repeats, thus explaining the Sherman paradox, which might be viewed as an example of the phenomenon of anticipation. Remarkably, the premutation expands to a full mutation only when it is transmitted by a female; consequently, daughters of normal transmitting males have only the premutation and never the full mutation, and never show mental retardation or cytogenetic expression of the fragile X syndrome. Furthermore, daughters of affected males do not express the fragile X syndrome at either the clinical or the cytogenetic level. In a study of the sperm in 4 male fragile X patients, Reyniers et al. (1993) found that only the premutation was present, although the full mutation was present in peripheral lymphocytes. This led them to conclude that expansion of the premutation to the full mutation in this disorder does not occur in meiosis but in a postzygotic stage. The same conclusion was supported by the finding of Kruyer et al. (1994) in 2 affected monozygotic brothers who differed in the number of CGG repeats. In another family with 2 monozygotic twin sisters with the same number of repeats, only 1 was mentally retarded. When the methylation status of the FMR1 CpG island was studied, Kruyer et al. (1994) found that the majority of the normal chromosomes had been inactivated in the affected twin.

Using monoclonal antibodies specific for the FMR-1 protein, Devys et al. (1993) found that they could detect 4 or 5 protein bands that appeared identical in cells of normal males and of males carrying a premutation but were absent in affected males with a full mutation. Immunohistochemistry showed a cytoplasmic localization of the FMR1 protein. The highest levels were observed in neurons, while glial cells contained very low levels. In epithelial tissues, levels of FMR1 were higher in dividing layers. In adult testes, FMR1 was detected only in spermatogonia.

To study the involvement of the FMR1 gene in the fragile-X syndrome, Pieretti et al. (1991) studied its expression in lymphoblastoid cell lines and leukocytes derived from patients and normal controls. FMR1 mRNA was absent in most male fragile X patients. Normal persons and carriers all showed expression. The work of Wohrle et al. (1992) supported the suggestion that a gene designated FMR1 is involved in the phenotype of the fragile X syndrome. One exon of this gene is carried on a 5.1-kb EcoRI fragment that exhibits length variation in fragile-X patients because of amplification of or insertion into a CGG-repeat sequence which probably represents the cytogenetically identified fragile site. The EcoRI fragment also includes an HTF island that is hypermethylated in fragile-X patients showing absence of FMR1 mRNA. In a fragile X-negative mentally retarded male who presented with the clinical phenotype of the fragile-X syndrome, Wohrle et al. (1992) found a deletion that included the HTF island and exons of the FMR1 gene. Less than 250 kb of genomic DNA, including DXS548 and at least 5 exons of the FMR1 gene, was deleted. The data supported the hypothesis that loss of function of the FMR1 gene is responsible for the clinical phenotype of the fragile X syndrome. This opens the possibility that pathogenetic mechanisms other than amplification of the CGG repeat can have the same phenotypic consequences. Wohrle et al. (1992) indicated that they had started a systematic screening of sporadic cases of the fragile X syndrome who are cytogenetically negative.

McConkie-Rosell et al. (1993) performed fragile X chromosome studies, physical examinations, and psychological testing on 6 brothers, 3 of whom were affected with fragile X to varying degrees, 2 of whom were nonpenetrant carriers, and 1 of whom was unaffected. Two of the affected brothers and the 2 'nonpenetrant' brothers were found to be methylation mosaics. A correlation was seen between the degree of methylation and the phenotypic expression identified in the 3 affected males. The 2 males initially classified as nonpenetrant were found to have mild phenotypic expression which consisted of minor cognitive deficits and a partial physical phenotype. These two, who were negative on fragile X chromosome studies, were found on DNA analysis to have large, broad smears, with approximately 97% of the DNA unmethylated; they were mosaic for hypermethylation of an expansion of the CGG repeat in the premutation range (100-600 bp). These results indicated that some 'nonpenetrant' carrier males may have varying amounts of methylation of the FMR1 regions, which can result in mild expression of the fragile X syndrome. Expression of the syndrome may not be confined to males with large, hypermethylated expansions (full mutation), but may instead have a gradient effect with a threshold for the full expression of the phenotype.

Abitbol et al. (1993) used in situ hybridization to demonstrate that FMR1 mRNAs are expressed from an early stage in proliferating and migrating cells of the nervous system and retina as well as several nonnervous tissues. In the brain of 25-week old fetuses, it was produced in highest levels in cholinergic neurons of the nucleus basalis magnocellularis and in pyramidal neurons of the hippocampus. The early transcription of the gene and the distribution of mRNAs in human fetuses suggest that alterations of FMR1 gene expression contributes to the pathogenesis of the fragile X syndrome and especially the mental retardation. Bachner et al. (1993) presented evidence suggesting that, in addition to a 'housekeeping' function, which is necessary in every cell, bears no correlation with proliferation or morphogenetic processes, and is reflected by a low level of ubiquitous expression, the FMR1 gene serves a second function in the mature testis, reflected by high expression in spermatogonia and not in Sertoli cells. They suggested further that FMR1 expression in spermatogonia is necessary for germ cell proliferation.

Zhong et al. (1993) described a second mutable sequence within the FMR1 gene. Richards et al. (1991) had described 2 polymorphic markers, designated AC1 and AC2, located within FMR1 and flanking the unstable (CGG)n repeat by approximately 10 kb. Zhong et al. (1993) confirmed linkage disequilibrium of the (CGG)n repeat with AC1 but found linkage equilibrium with AC2, which they also found was highly mutable. A mutation rate of 3.3% was observed but only among fragile X maternally-derived meioses. The finding of a second mutable locus within FMR1 suggested that the target for tandem repeat instability may not be confined to the (CGG)n repeat alone but may also involve microsatellites. Zhong et al. (1993) cited evidence that the AC sequence is located within intron 2 of the FMR1 gene and is adjacent to an Alu element.

A 14-center collaborative study of genotype-phenotype correlations in the fragile X syndrome involved 318 affected families comprising 2,253 individuals, 1,344 of whom carried a fragile X mutation and 693 of whom had a typical full fragile X mutation (Rousseau et al., 1994). There was a significantly higher prevalence of 'mosaic' cases among males who carry a full mutation (12%) than among females who carry a full mutation (6%); the mosaic males had a larger expansion than did the mosaic females. ('Mosaics' are individuals carrying the full mutation who also have some premutations in some of their cells.) The mental status of premutated individuals did not differ from that of those with a normal genotype. Both the abnormal methylation of the FMR1-EagI site and the size of the expansion were highly correlated with cytogenetics, facial dysmorphism, macroorchidism, and mental retardation. Among female carriers of a full mutation, those with mental retardation had significantly larger expansion than did those without MR. Among 164 independent couples, 3 unrelated husbands carried a premutation that suggests that the prevalence of fragile X premutations in the general population is approximately 0.9% of the X chromosomes. The data validated the use of direct DNA testing for fragile X diagnosis and carrier identification.

Quan et al. (1995) described a phenotypically atypical case of fragile X syndrome caused by a deletion that included the entire FMR1 gene and at least 9.0 Mb of flanking DNA. The proband was a 6-year-old mentally retarded male with obesity and anal atresia. The diagnosis of fragile X syndrome was established by the failure of the proband's DNA to hybridize to a restriction fragment specific for the 5-prime end of the FMR1 gene. The analysis of flanking markers in the interval Xq26.3-q28 indicated a deletion extending from between 160- to 500-kb distal, and 9.0-Mb proximal, to the FMR1 gene. High-resolution chromosome banding confirmed a deletion with breakpoints in Xq26.3 and Xq27.3. This deletion was maternally transmitted and arose as a new mutation on the grandpaternal X chromosome. The maternal transmission of the deletion was confirmed by fluorescence in situ hybridization using a 34-kb cosmid containing most of the FMR1 gene. The patient's unusual clinical presentation may indicate the presence of genes located in the deleted interval proximal to the FMR1 locus that are able to modify the fragile X syndrome phenotype.

Quan et al. (1995) described a patient with mosaicism for expansion and deletion of the FMR1 CGG repeat. In addition to mental retardation, the affected male had cherubism (118400). Mosaicism had been described previously by Hirst et al. (1995), among others. The work of Reyniers et al. (1993) and Wohrle et al. (1993) suggested that the expansion of an FMR1 premutation to a full mutation occurs mitotically during a postzygotic stage. In this model of fragile X expansion, carriers of fully expanded FMR1 alleles transmit only premutation-sized alleles. The expansion of the maternally derived allele then occurs mitotically during early embryonic development. The failure to observe the expanded FMR1 alleles in the germline suggest that the expansion event occurs between days 5 and 20 of development, after the separation of the germline but before the divergence of tissue types. The findings of Quan et al. (1995) of mosaicism for expansion and deletion were consistent with this model of fragile X expansion. The detection of premutation-sized alleles in the proband was consistent with the transmission of a maternal premutation-sized FMR1 allele and postzygotic expansion. An unmethylated CpG island upstream of the FMR1 deletion indicated that the expansion and deletion of the CGG repeat occurred before FMR1 methylation. Cherubism was presumably an unrelated phenomenon.

Verheij et al. (1995) demonstrated that the ile367-to-asn substitution in the FMR1 protein did not alter the translation, processing, or localization of FMR1 proteins in lymphoblastoid cells from a patient carrying this mutation. All the high molecular weight FMR1 proteins isolated from normal lymphoblastoid cells and cells from the patient with this substitution were able to bind RNA. However, the FMR1 proteins of the patient had reduced affinity for RNA binding at high salt concentrations.

The FMR protein is cytoplasmic and contains 2 conserved RNA-binding domains, suggesting a possible involvement in RNA metabolism. Khandjian et al. (1996) observed that the FMR protein cosediments with polyribosomes after centrifugation in sucrose density gradients. Specifically, it is associated with the ribosomal 60S subunit and possesses the characteristics of a nonintegral ribosomal protein. Immunofluorescent studies showed a tight colocalization of FMRP with cytoplasmic ribosomes in NIH 3T3 and HeLa cells and in primary cultures of neurons. Thus, the authors concluded that fragile X mental retardation may result from defects in the translational machinery due to absence of FMRP.

By immunofluorescence studies, Sittler et al. (1996) found that splice variants of FMR1 that exclude exon 14 sequences (and have alternate C-terminal regions) are nuclear. Analysis of various deletion mutants suggested the presence of a cytoplasmic retention domain encoded in exon 14 and of a nuclear association domain encoded within the first 8 exons that appear, however, to lack a typical nuclear localization signal.

In 2 clinically normal brothers, Smeets et al. (1995) found expanded CGG repeats in cytogenetically visible fragile sites. The FMR1 promoter was unmethylated and both RNA and protein could be detected. This indicated to the authors that inactivation of the FMR1 gene and not repeat expansion itself resulted in the fragile X phenotype. Smeets et al. (1995) concluded that repeat expansion does not necessarily induce methylation and that methylation is no absolute requirement for the induction of fragile sites. This fragile X family was ascertained through a mentally retarded boy who was the grandson of 1 of the 2 brothers through a daughter.

Subramanian et al. (1996) extended studies of the previously demonstrated replication timing delay of the FMR1 locus and proximal flanking sequences when involved in the full expansion of the repeat element. In this assay, relative times of replication of specific loci were inferred from the ratios of singlet and doublet fluorescence in situ hybridization signals in interphase nuclei. In all individuals with a full expansion of the trinucleotide repeat, a large region of delayed timing was observed; the apparent timing of the earlier-replicating allele in female cells in this region was intermediate between normal and affected alleles in male cells, a finding in accordance with expectations of a mixed population of cells resulting from random X inactivation. Expansion of the nearby FRAXE locus also was found to correlate with replication timing delay, although the extent of the altered region was somewhat less. Trinucleotide repeat expansion thus may be acting in the Xq27.3-q28 region to alter long-range chromatin structure that could influence transcription of gene sequences within the affected domain. Yeshaya et al. (1998) studied replication timing of the fragile X locus relative to the nontranscribed late replicating alpha-satellite region of chromosome X in lymphocytes and amniocytes from normal males, males with fragile X, and males who were premutation carriers. Three distinct populations were identified among the various samples. The first population had a high frequency of cells showing a doublet FMR1; this pattern, indicating early replication of FMR1, characterized cell populations of normal males. The second population had a high frequency of cells showing a singlet FMR1; this pattern, indicating very late replication of FMR1, characterized the population of fragile X patients. The third population had about one-half of the cells showing a singlet FMR1 and the other half with a doublet FMR1, indicating somatic variation in the replication timing of FMR1. This was the pattern seen in the population of premutation carriers.

Nolin et al. (1996) examined transmission of the FMR1 CGGn repeat in 191 families with fragile X and in the general population. They reported that when fathers have (CGG)n expansions in the premutation range (greater than 80 repeats) the daughters frequently inherited smaller repeat expansions. A similar repeat number was inherited more often than expected by chance among a sibship segregating fragile X. They concluded that this familial clustering, observed in the offspring of both males and females with a premutation, implies that there may be an additional factor, independent of parental repeat size, that influences (CGG)n repeat instability. Nolin et al. (1996) found that gray-zone alleles (40 to 60 repeats) in families with no previous history of fragile X syndrome varied in their stability but that there was no repeat expansion to the full fragile X mutation in 1 generation.

Jakala et al. (1997) found that males with the full fragile-X (fM) mutation showed worse cognitive performance than did males with the premutation (pM); deficits in females with the fM were qualitatively similar but less severe than in males with the fM. In a visual memory test, both fM groups were impaired. Hippocampal volumes normalized for intracranial or brain area did not significantly differ between fM and pM groups. Minor abnormalities in temporal lobe structures were found by MRI in fM subjects.

Patients with hereditary nonpolyposis colon cancer resulting from mutation in a mismatch repair gene such as MLH1 (120436) show instability of repeat sequences. Fulchignoni-Lataud et al. (1997) found that peripheral blood leukocytes from HNPCC patients showed roughly twice as great allele variation of the FMR1 CGG repeat as did controls, especially when patients carried mutations in MLH1. The findings suggested that instability within nonneoplastic cells of a subset of HNPCC patients might be a mechanism for transition from normal to the premutation range of the the FMR1 CGG repeat.

In a survey of 222 unrelated mentally retarded individuals attending Spanish special schools, Mila et al. (1997) found 11 boys with full mutations in the FMR1 gene and 1 boy with a CCG repeat expansion in the FMR2 gene.

Diagnosis

Jenkins et al. (1984) described prenatal diagnosis. The testes of 2 fragile-X-positive fetuses appeared large for gestational age. Antibiotics such as Bactrim (Roche) and Septra (Burroughs Wellcome) contain trimethoprim, which can lower folate levels by inhibition of dihydrofolate reductase. Hecht and Glover (1983) urged avoidance of trimethoprim and other folate antagonists in pregnant women who are at risk for having a child with the fragile X syndrome. Lejeune et al. (1982) described severe clinical regression of psychomotor development in a 2-year-old boy with the fragile X syndrome while on trimethoprim.

Jacky and Dill (1980) achieved marXq28 expression in cultured fibroblasts and Jenkins et al. (1982) detected the marker in cultured amniocytes, enabling successful prenatal diagnosis. Thymidine deficiency induced by BUdR or FUdR appeared necessary for expression in cells other than lymphocytes (Glover, 1981; Tommerup et al., 1981). Until reliable expression was possible, however, absence of the marX in amniocytes of male or female fetuses had to be interpreted with caution (see later for prenatal diagnosis using molecular markers). Jacobs et al. (1982) showed that the marX can be demonstrated on lymphoblastoid cell lines and that it is reliably and repeatedly demonstrable after the addition of FUdR to cultures. This simple technique provides a useful in vitro experimental test system.

Sutherland (1989) indicated that there is a fragile site (FRAXD) located at Xq27.2, separate from the classic fragile site at Xq27.3 which is responsible for mental retardation. Obviously, this is a point of diagnostic importance. The FRAXD is inducible by high doses of aphidicolin. Ramos et al. (1992) concluded that the fragile site at Xq27.2 can be demonstrated in normal persons under the conditions of thymidylate stress routinely used for cytogenetic diagnosis of the fragile X syndrome. Furthermore, this fragile site is present at low levels (1-2%) in all persons who express it and, therefore, its expression is unlikely to cause false-positive diagnoses of the syndrome. Lesions at Xq26 are also seen at low levels in lymphocytes of persons without the syndrome.

Oberle et al. (1991) identified carriers of the fragile X mutation regardless of sex or phenotypic expression. Rare apparent false negatives were considered to be the result of genetic heterogeneity or misdiagnosis.

Rousseau et al. (1991) concluded that direct DNA diagnosis of the fragile X syndrome is efficient and reliable. Southern analysis of EcoRI and EagI digests of DNA distinguished clearly in a single test between the normal genotype, the premutation, and the full mutation. Fifteen percent of those with full mutations had some cells carrying only the premutation. All of the mothers of affected children were carriers of either a premutation or a full mutation. Because of the certainty of DNA diagnosis, this method replaces cytogenetic detection of the fragile X chromosome, which carries a rate of misdiagnosis of about 5% for both false positives and the more frequent false negative conclusion, and diagnosis by the linkage principle, which gives a probabilistic result rather than an absolute one. As pointed out by Jacobs (1991), in spite of the advances in the molecular understanding of fragile X, the cytogenetic marker still has an honorable role to play in its diagnosis. As the cytogenetic test is reliable for virtually all males and for the great majority, but not all, of affected females, the most efficient and cost effective methodology for diagnostic specimens still seemed to her to be cytogenetic analysis, followed by molecular studies only when the fra(X) is seen or suspected. Sutherland et al. (1991) illustrated the use of direct DNA diagnosis in establishing the carrier status of a cytogenetically normal woman in a family with the fragile X syndrome and in diagnosing the mutation in her male fetus by detecting the unstable sequence in DNA obtained by chorionic villus sampling. They used a probe referred to as pfxa3. The carrier mother had a normal 1.0-kb band and an abnormal 1.2-kb band. In the villus sample the normal 1.0-kb band was replaced by a 2.3-kb band, a size seen in affected males. Normal carrier males usually have a fragile X band that is between 1.1 and 1.6 kb. Richards et al. (1991) described 2 polymorphic microsatellite AC repeat markers, FRAXAC1 and FRA XAC2, physically located within 10 kb and on either side of the (CCG)n repeat responsible for the fragile site. The two show strong linkage disequilibrium and have heterozygosity of 44 and 71%, respectively. No recombination was observed either between these markers in 40 CEPH pedigrees or with FMR1 in affected pedigrees. These markers provide the means for accurate diagnosis of the fragile X genotype in families by polymerase chain reaction (PCR) analysis. Richards et al. (1991) used them to position FMR1 within the multipoint map of the X chromosome to a position 3.7 cM distal to DXS297 and 1.2 cM proximal to DXS296.

Griffiths and Strachan (1991) described a technique, based on a culture system reported by Wheater and Roberts (1987), that enables the cytogeneticist to do fra(X) screening and prometaphase banding on the same specimen. Yu et al. (1992) found that all persons of the fragile X genotype (where testing was possible) had a parent with amplified p(CCG)n repeat, indicating that few, if any, cases of fragile X syndrome are not familial.

Mandel et al. (1992) reported on the Fifth International Workshop on the Fragile X and X-Linked Mental Retardation held near Strasbourg, France, in August 1991. In addition to their summary, over 50 papers on the fragile X syndrome and 18 papers related to other X-linked mental retardation syndromes presented at the conference were published in the American Journal of Medical Genetics. All aspects of the fragile X syndrome (clinical, cytogenetic, molecular and linkage, and population) were covered in great detail. Mandel et al. (1992) reviewed the hypothesis of Patricia Jacobs which postulates 3 mutations: a change from a normal insert (N) to a small insert that is at low risk of converting to a large insert (S); a change from that type of small insert to a small insert at high risk of converting to a large insert (S*); and a change from the high risk small insert to a large insert (L) which is associated with clinical abnormality. Cytogenetic screening of the mentally handicapped for the fra(X) is equivalent to testing for individuals with a large insert (L) as there is no evidence that a small insert (S) has a deleterious effect on the phenotype. The consensus was that in diagnostic laboratories cytogenetics is still the method of choice, with subsequent molecular investigation of those patients found or suspected of being fra(X) positive; no consensus was reached on the relative merits of cytogenetics and molecular techniques for screening. Mulley et al. (1992) reported a high success rate with the direct molecular diagnosis of fragile X using the pfxa3 probe which detects amplification of an unstable DNA element consisting of variable length CCG repeats. Diagnosis was definitive in all males with only one X chromosome.

Prenatal diagnosis of the fragile X syndrome on the basis of the cytogenetic alteration in amniocytes (Jenkins et al., 1984) was insecure because of technical uncertainties. Yamauchi et al. (1993) used the diagnostic DNA probe pPCRfx1 to confirm that an at-risk fetus was a heterozygous female carrier.

Dreesen et al. (1995) approached preimplantation testing for the fragile-X syndrome by genotyping the polymorphic DXS548 AC-repeat locus, which is closely linked to the FMR1 gene, in unfertilized oocytes and extruded polar bodies. They concluded that a PCR procedure could be performed within 16 hours after blastomere biopsy and that for carrier females heterozygous at the DXS548 locus, preimplantation testing with DXS548 is a possible alternative to prenatal testing.

Willemsen et al. (1995) developed a diagnostic method that allows detection of the fragile X syndrome from a blood smear. The test made use of mouse monoclonal antibodies against the FMR1 protein. This noninvasive test requires only 1 or 2 drops of blood and can be used for screening large groups of mentally retarded persons and neonates for the fragile X syndrome.

Turner et al. (1996) suggested that the clinical definition of this disorder be redefined in the male as mental handicap associated with absolute or relative deficiency of the FMR1 protein. In the absence of a readily available protein test, analysis of the FMR1 trinucleotide repeat size had been used for diagnosis (Snow et al., 1993). They stated that increase in the size of the repeat in the FMR gene over a particular value initiates methylation of the FMR gene promoter site, resulting in the switching off of the FMR1 gene transcription. Turner et al. (1996) noted that this method of testing also defines individuals who lack FMR1 protein as a consequence of deletion of the gene but will not identify those individuals whose FMR1 protein is defective through mutation.

Storm et al. (1998) noted that incomplete EcoRI digestion may lead to false diagnosis of fragile X syndrome and suggested that HindIII digest be used instead of EcoRI to identify premutation vs normal fragment length in genomic DNA.

Population Genetics

Jacobs (1982) indicated that a reasonable estimate of frequency is 0.5 per 1000 males. Many of the cases first ascertained were of northern European descent; subsequently, however, affected males have been found in most ethnic groups. Turner et al. (1996) reported that on the basis of molecular genetic analysis, a prevalence figure of 1:4,000 or 2.4:10,000 was more realistic. The 1:1,000 prevalence figure was based on cytogenetic studies which had proven to be less specific. Filippi et al. (1991) reported findings in a very large Sardinian kindred spanning 6 generations and including 13 patients with Martin-Bell syndrome, several instances of normal transmitting males or females, and the G6PD Mediterranean (305900.0006) mutant segregating in some of its branches. All the fragile X patients and the 15 obligate heterozygous women could be traced through their X-chromosome lineage to a woman in the first generation who must have been heterozygous for a silent premutation at the FRAX locus. Filippi et al. (1991) concluded that this premutation had been converted into a true FRAX mutation at least 9 times during the gametogenesis of this ancestor's X-related descendants, of whom 4 were males. See 309530 for a general discussion of X-linked mental retardation.

Richards et al. (1992) presented haplotype evidence for a founder effect in the fragile X mutation. They found clear evidence of linkage disequilibrium between fragile X and 2 polymorphic microsatellite markers that flank FMR1 and are within 10 kb of the (CCG)n repeat. These markers have 5 to 7 alleles, show no recombination with each other, and define 15 haplotypes. In an analysis of 134 fragile X chromosomes from unrelated affected individuals in Australia and the United States, they found that 58% of the fragile X mutations occurred on the 3 backgrounds that account for 18% of normal chromosomes. Correspondingly, the single most common normal haplotype, which has a frequency of 50%, carries only 18% of fragile X mutations. The data argue for the expected occurrence of multiple, independent mutations but also indicate the unexpectedly long history of some of these fragile X mutations. Using the FRAXAC1 polymorphic marker in the study of a large number of patients, Hirst et al. (1993) found its allele distribution to be strikingly different on fragile X chromosomes, confirming earlier observations and giving further support to the suggestion of a fragile X founder effect (Richards et al., 1992).

In Israel, in a study of 122 families affected with the fragile X syndrome, diagnosed in 7 genetic centers, Dar et al. (1995) found that Tunisian Jews, who comprise only 4% of the general population, accounted for 21% of the fragile X families, suggesting founder effect.

The CGG repeat, which is normally polymorphic in length, is frequently interrupted by AGG triplets, which are believed to stablize the repeat. The absence of AGG triplets, leading to long tracts of perfect CGG repeats, may give rise to predisposed alleles. Kunst et al. (1996) determined the repeat length of 345 chromosomes from 9 populations from various parts of the world and used automated DNA sequencing to assess 14 of them. They found that the FMR1 alleles were very heterogeneous, although the level of variation correlated with the age and/or genetic history of a particular population. Native American alleles, interrupted by 3 AGG repeats, exhibited marked stability over 7,000 years. However, in older African populations, parsimony analysis predicted the occasional loss of an AGG, leading to more perfect CGG repeats.

Animal Model

Faust et al. (1992) and Laval et al. (1992) determined the location of the corresponding gene on the mouse X chromosome by use of human cDNA clones in an interspecific backcross. Both groups found preservation of the order of loci, although no fragile site had been detected microscopically in that region of the mouse X chromosome. The Dutch-Belgian Fragile X Consortium (1994) created a knockout model for the fragile X syndrome in mice. The knockout mice lacked normal FMR1 protein and showed macroorchidism, learning deficits, and hyperactivity.

By mRNA in situ hybridization studies, Hinds et al. (1993) demonstrated that expression of the Fmr-1 gene was localized to several areas of the brain and the tubules of the testes in the adult mouse, whereas universal and very strong expression was observed in early mouse embryos. To identify transcribed sequences rapidly and efficiently, Hanzlik et al. (1993) developed a recombination-based assay to screen bacteriophage lambda-libraries for sequences that share homology with a given probe. This strategy determines whether a given probe is transcribed in a given tissue at a given time of development and may also be used to isolate the transcribed sequence free of the screening probe. Hanzlik et al. (1993) used the technique to demonstrate that the fragile X sequence is transcribed ubiquitously in an 11-week fetus, in a variety of 20-week fetal tissues including brain, spinal cord, eye, liver, kidney, and skeletal muscle, and in adult jejunum.

Eberhart et al. (1996) performed immunolocalization studies to gain insight into the function of the protein FMRP which is encoded by the FMR1 gene. Their studies revealed a nuclear localization signal in the amino terminus and a nuclear export signal encoded by exon 14. Eberhart et al. (1996) further demonstrated that FMRP is found in ribonucleoprotein (RNP) particles. They reported that their data are consistent with nascent FMRP entering the nucleus to assemble into RNP particles prior to export back into the cytoplasm.

History

In retrospect, it was concluded that the X-linked mental retardation reported by Martin and Bell (1943) was the fragile X syndrome. Thus, the Martin-Bell syndrome is a synonym for the fragile X syndrome. According to Opitz and Sutherland (1984), Escalante, a graduate student with Frota-Pessoa in Sao Paulo, Brazil, and Drs. Bryan and Gillian Turner in Sydney, Australia, independently noted macroorchidism in X-linked mental retardation in the late 1960s. Escalante et al. (1971) published their findings (which they had reported in 1969 at the Warsaw Congress of the International Association for the Scientific Study of Mental Deficiency) in the Journal de Genetique Humaine. When I visited the Doctors Turner in Sydney in March 1970, they showed me several mentally retarded patients with macroorchidism. Opitz et al. (1984) referred to this disorder as the Martin-Bell syndrome on the assumption that the family reported from the Queen Square Hospital in London by J. Purdon Martin and Julia Bell (Martin and Bell, 1943) had that disorder. Although macroorchidism was not mentioned by Martin and Bell (1943), one of the patients was described as having a 'big face and jaw'; furthermore, at least 9 of the affected males were maternal grandsons of 2 unaffected brothers. All but 1 of the mothers of affected males were daughters of these 2 brothers, the other being their sister. Martin and Bell (1943) hypothesized that some controlling factor caused suppression of the disease in the 2 grandfathers without affecting their liability to transmit it. For a superb biography of Julia Bell (1879-1979), see Bundey (1996). Richards et al. (1981) followed up on the Martin-Bell kindred, demonstrating that it was indeed the fragile X syndrome. The original index patient was then aged 56. All 4 affected males who had adequate karyotyping showed the fragile-X syndrome in 17 to 50% of their cells. The other major contribution of Julia Bell was in the defining of many hereditary d isorders, such as the forms of brachydactyly, on the basis of massive collections of pedigrees in the famous Treasury of Human Inheritance. She also collaborated with J. B. S. Haldane in the first estimation of linkage in the human, that of colorblindness and hemophilia (Bell and Haldane, 1937).

See also Ashley et al. (1993); Bowen et al. (1978); Bundey (1996); Cantu et al. (1978); Cantu et al. (1976); Carmi et al. (1984); Choo et al. (1984); Cremers et al. (1987); Davies et al. (1985); Gerald (1980); Haataja et al. (1994); Hirst et al. (1993). Hofker et al. (1987); Kinnell (1982); Krawczun et al. (1985); Lubs et al. (1984); Mattei et al. (1981); Mulligan et al. (1985); Park et al. (1994); Parrish et al. (1994); Purrello et al. (1985); Reiss et al. (1991); Rousseau et al. (1991); Ruvalcaba et al. (1977); Soysa et al. (1982); Steinbach (1986); Sutherland and Ashford (1979); Sutherland and Baker (1992); Turner et al. (1980); Turner et al. (1975); Van Roy et al. (1983); Zoll et al. (1985)

Abitbol, M.; Menini, C.; Delezoide, A.-L.; Rhyner, T.; Vekemans, M.; Mallet, J.: Nucleus basalis magnocellularis and hippocampus are the major sites of FMR-1 expression in the human fetal brain. Nature Genet. 4: 147-153, 1993.

Ashley, C. T.; Sutcliffe, J. S.; Kunst, C. B.; Leiner, H. A.; Eichler, E. E.; Nelson, D. L.; Warren, S. T.: Human and murine FMR-1: alternative splicing and translational initiation downstream of the CGG-repeat. Nature Genet. 4: 244-251, 1993.

Ashley, C. T., Jr.; Wilkinson, K. D.; Reines, D.; Warren, S. T.: FMR1 protein: conserved RNP family domains and selective RNA binding. Science 262: 563-566, 1993.

Bachner, D.; Steinbach, P.; Wohrle, D.; Just, W.; Vogel, W.; Hameister, H.; Manca, A.; Poustka, A.: Enhanced Fmr-1 expression in testis. (Letter) Nature Genet. 4: 115-116, 1993.

Bell, J.; Haldane, J. B. S.: The linkage between the genes for colour-blindness and haemophilia in man. Proc. Roy. Soc. London 123B: 119-150, 1937.

Bell, M. V.; Hirst, M. C.; Nakahori, Y.; MacKinnon, R. N.; Roche, A.; Flint, T. J.; Jacobs, P. A.; Tommerup, N.; Tranebjaerg, L.; Froster-Iskenius, U.; Kerr, B.; Turner, G.; Lindenbaum, R. H.; Winter, R.; Pembrey, M.; Thibodeau, S.; Davies, K. E.: Physical mapping across the fragile X: hypermethylation and clinical expression of the fragile X syndrome. Cell 64: 861-866, 1991.

Blomquist, H. K.; Bohman, M.; Edvinsson, S. O.; Gillberg, C.; Gustavson, K.-H.; Holmgren, G.; Wahlstrom, J.: Frequency of the fragile X syndrome in infantile autism: a Swedish multicenter study. Clin. Genet. 27: 113-117, 1985.

Blomquist, H. K.; Gustavson, K.-H.; Holmgren, G.; Nordenson, I.; Sweins, A.: Fragile site X chromosomes and X-linked mental retardation in severely retarded boys in a northern Swedish county: a prevalence study. Clin. Genet. 21: 209-214, 1982.

Bowen, P.; Biederman, B.; Swallow, K. A.: The X-linked syndrome of macroorchidism and mental retardation: further observations. Am. J. Med. Genet. 2: 409-414, 1978.

Brookwell, R.; Turner, G.: High resolution banding and the locus of the Xq fragile site. Hum. Genet. 63: 77, 1983.

Brown, W. T.; Gross, A.; Chan, C.; Jenkins, E. C.; Mandel, J. L.; Oberle, I.; Arveiler, B.; Novelli, G.; Thibodeau, S.; Hagerman, R.; and 18 others: Multilocus analysis of the fragile X syndrome. Hum. Genet. 78: 201-205, 1988.

Brown, W. T.; Gross, A. C.; Chan, C. B.; Jenkins, E. C.: Genetic linkage heterogeneity in the fragile X syndrome. Hum. Genet. 71: 11-18, 1985.

Bundey, S.: Julia Bell, MRCS LRCP FRCP (1879-1979): Steamboat lady, statistician and geneticist. J. Med. Bio. 4: 8-13, 1996.

Camerino, G.; Mattei, M. G.; Mattei, J. F.; Jaye, M.; Mandel, J. L.: Close linkage of fragile X-linked mental retardation syndrome to haemophilia B and transmission through a normal male. Nature 306: 701-707, 1983.

Cantu, J. M.; Scaglia, H. E.; Gonzalez-Diddi, M.; Hernandez-Jauregui, P.; Morato, T.; Moreno, M. E.; Giner, J.; Alcantar, A.; Herrera, D.; Perez-Palacios, G.: Inherited congenital normofunctional testicular hyperplasia and mental deficiency. Hum. Genet. 41: 331-339, 1978.

Cantu, J. M.; Scaglia, H. E.; Medina, M.; Gonzalez-Diddi, M.; Morato, T.; Moreno, M. E.; Perez-Palacios, G.: Inherited congenital normofunctional testicular hyperplasia and mental deficiency. Hum. Genet. 33: 23-33, 1976.

Carmi, R.; Meryash, D. L.; Wood, J.; Gerald, P. S.: Fragile-X syndrome ascertained by the presence of macro-orchidism in a 5-month-old infant. Pediatrics 74: 883-886, 1984.

Chakravarti, A.: Fragile X founder effect?. Nature Genet. 1: 237-238, 1992.

Choo, K. H.; George, D.; Fillby, G.; Halliday, J. L.; Leversha, M.; Webb, G.; Danks, D. M.: Linkage analysis of X-linked mental retardation with and without fragile-X using factor IX gene probe. (Letter) Lancet II: 349, 1984.

Cremers, F. P. M.; Pfeiffer, R. A.; van de Pol, T. J. R.; Hofker, M. H.; Kruse, T. A.; Wieringa, B.; Ropers, H. H.: An interstitial duplication of the X chromosome in a male allows physical fine mapping of probes from the Xq13-q22 region. Hum. Genet. 77: 23-27, 1987.

Daker, M. G.; Chidiac, P.; Fear, C. N.; Berry, A. L.: Fragile X in a normal male: a cautionary tale. Lancet I: 780, 1981.

Dar, H.; Schaap, T.; Bait-Or, H.; Borochowitz, Z.; Gelman-Kohan, Z.; Chemke, T.; Chaki, R.; Cohen, H.; Falik-Borenstein, Z.; Chemke, J.: Ethnic distribution of the fragile X syndrome in Israel: evidence of founder chromosomes(?). Israel J. Med. Sci. 31: 323-325, 1995.

Davids, J. R.; Hagerman, R. J.; Eilert, R. E.: Orthopaedic aspects of fragile-X syndrome. J. Bone Joint Surg. 72A: 889-896, 1990.

Davies, K. E.; Mattei, M. G.; Mattei, J. F.; Veenema, H.; McGlade, S.; Harper, K.; Tommerup, N.; Nielsen, K. B.; Mikkelsen, M.; Beighton, P.; Drayna, D.; White, R.; Pembrey, M. E.: Linkage studies of X-linked mental retardation: high frequency of recombination in the telomeric region of the human X chromosome (fragile site/linkage/recombination/X chromosome). Hum. Genet. 70: 249-255, 1985.

De Boulle, K.; Verkerk, A. J. M. H.; Reyniers, E.; Vits, L.; Hendrickx, J.; Van Roy, B.; Van Den Bos, F.; de Graaff, E.; Oostra, B. A.; Willems, P. J.: A point mutation in the FMR-1 gene associated with fragile X mental retardation. Nature Genet. 3: 31-35, 1993.

de Vries, B. B. A.; Fryns, J.-P.; Butler, M. G.; Canziani, F.; Wesby-van Swaay, E.; van Hemel, J. O.; Oostra, B. A.; Halley, D. J. J.; Niermeijer, M. F.: Clinical and molecular studies in fragile X patients with a Prader-Willi-like phenotype. J. Med. Genet. 30: 761-766, 1993.

de Vries, B. B. A.; Robinson, H.; Stolte-Dijkstra, I.; Tjon Pian Gi, C. V.; Dijkstra, P. F.; van Doorn, J.; Halley, D. J. J.; Oostra, B. A.; Turner, G.; Niermeijer, M. F.: General overgrowth in the fragile X syndrome: variability in the phenotypic expression of the FMR1 gene mutation. J. Med. Genet. 32: 764-769, 1995.

Devys, D.; Biancalana, V.; Rousseau, F.; Boue, J.; Mandel, J. L.; Oberle, I.: Analysis of full fragile X mutations in fetal tissues and monozygotic twins indicate that abnormal methylation and somatic heterogeneity are established early in development. Am. J. Med. Genet. 43: 208-216, 1992.

Devys, D.; Lutz, Y.; Rouyer, N.; Bellocq, J.-P.; Mandel, J.-L.: The FMR-1 protein is cytoplasmic, most abundant in neurons and appears normal in carriers of a fragile X premutation. Nature Genet. 4: 335-340, 1993.

Dreesen, J. C. F. M.; Geraedts, J. P. M.; Dumoulin, J. C. M.; Evers, J. L. H.; Pieters, M. H. E. C.: RS46 (DXS548) genotyping of reproductive cells: approaching preimplantation testing of the fragile-X syndrome. Hum. Genet. 96: 323-329, 1995.

Dutch-Belgian Fragile X Consortium: Fmr1 knockout mice: a model to study fragile X mental retardation. Cell 78: 23-33, 1994.

Eberhart, D. E.; Malter, H. E.; Feng, Y.; Warren, S. T.: The fragile X mental retardation protein is a ribonucleoprotein containing both nuclear localization and nuclear export signals. Hum. Molec. Genet. 5: 1083-1091, 1996.

Eichler, E. E.; Richards, S.; Gibbs, R. A.; Nelson, D. L.: Fine structure of the human FMR1 gene. Hum. Molec. Genet. 2: 1147-1153, 1993.

Escalante, J. A.; Grunspun, H.; Frota-Pessoa, O.: Severe sex-linked mental retardation. J. Genet. Hum. 19: 137-140, 1971.

Faust, C. J.; Verkerk, A. J. M. H.; Wilson, P. J.; Morris, C. P.; Hopwood, J. J.; Oostra, B. A.; Herman, G. E.: Genetic mapping on the mouse X chromosome of human cDNA clones for the fragile X and Hunter syndromes. Genomics 12: 814-817, 1992.

Filippi, G.; Arslanian, A.; Dagna-Bricarelli, F.; Pierluigi, M.; Grasso, M.; Rinaldi, A.; Rocchi, M.; Siniscalco, M.: Premutation for the Martin-Bell syndrome analyzed in a large pedigree segregating also for G6PD-deficiency. I. A working hypothesis on the nature of the FRAX-mutations. Am. J. Med. Genet. 40: 387-394, 1991.

Filippi, G.; Rinaldi, A.; Archidiacono, N.; Rocchi, M.; Balazs, I.; Siniscalco, M.: Linkage between G6PD and fragile-X syndrome. Am. J. Med. Genet. 15: 113-119, 1983.

Follette, P. J.; Laird, C. D.: Estimating the stability of the proposed imprinted state of the fragile-X mutation when transmitted by females. Hum. Genet. 88: 335-343, 1992.

Froster-Iskenius, U.; Schulze, A.; Schwinger, E.: Transmission of the marker X syndrome trait by unaffected males: conclusions from studies of large families. Hum. Genet. 67: 419-427, 1984.

Fryns, J. P.: Personal Communication. Leuven, Belgium 5/28/1993.

Fryns, J. P.: The fragile X syndrome: a study of 83 families. Clin. Genet. 26: 497-528, 1984.

Fryns, J. P.; Van Den Berghe, H.: Transmission of fragile(X)(q27) from normal male(s). Hum. Genet. 61: 262-263, 1982.

Fu, Y.-H.; Kuhl, D. P. A.; Pizzuti, A.; Pieretti, M.; Sutcliffe, J. S.; Richards, S.; Verkerk, A. J. M. H.; Holden, J. J. A.; Fenwick, R. G., Jr.; Warren, S. T.; Oostra, B. A.; Nelson, D. L.; Caskey, C. T.: Variation of the CGG repeat at the fragile X site results in genetic instability: resolution of the Sherman paradox. Cell 67: 1047-1058, 1991.

Fulchignoni-Lataud, M. C.; Olchwang, S.; Serre, J. L.: The fragile X CGG repeat shows a marked level of instability in hereditary non-polyposis colorectal cancer patients. Europ. J. Hum. Genet. 5: 89-93, 1997.

Gedeon, A. K.; Baker, E.; Robinson, H.; Partington, M. W.; Gross, B.; Manca, A.; Korn, B.; Poustka, A.; Yu, S.; Sutherland, G. R.; Mulley, J. C.: Fragile X syndrome without CCG amplification has an FMR1 deletion. Nature Genet. 1: 341-344, 1992.

Gerald, P. S.: X-linked mental retardation and an X-chromosome marker. (Editorial) New Eng. J. Med. 303: 696-697, 1980.

Gerald, P. S.: Personal Communication. Boston, Mass. 8/12/1983.

Giannelli, F.; Morris, A. H.; Garrett, C.; Daker, M.; Thurston, C.; Smith, C. A. B.: Genetic heterogeneity of X-linked mental retardation with fragile X: association of tight linkage to factor IX and incomplete penetrance in males. Ann. Hum. Genet. 51: 107-124, 1987.

Giraud, F.; Ayme, S.; Mattei, J. F.; Mattei, M. G.: Constitutional chromosomal breakage. Hum. Genet. 34: 125-136, 1976.

Glover, T. W.: FUdR induction of the X-chromosome fragile site: evidence for the mechanism of folic acid and thymidine inhibition. Am. J. Hum. Genet. 33: 234-242, 1981.

Goodman, R. M.; Strauss, S.; Friedman, E.; Chaki, R.: Ovarian size in the fragile X mental retardation syndrome. Am. J. Med. Genet. 26: 17-18, 1987.

Griffiths, M. J.; Strachan, M. C.: A single lymphocyte culture for fragile X induction and prometaphase chromosome analysis. J. Med. Genet. 28: 837-839, 1991.

Haataja, R.; Vaisanen, M.-L.; Li, M.; Ryynanen, M.; Leisti, J.: The fragile X syndrome in Finland: demonstration of a founder effect by analysis of microsatellite haplotypes. Hum. Genet. 94: 479-483, 1994.

Hagerman, R. J.; Synhorst, D. P.: Mitral valve prolapse and aortic dilatation in the fragile X syndrome. Am. J. Med. Genet. 17: 123-131, 1984.

Hagerman, R. J.; Van Housen, K.; Smith, A. C. M.; McGavran, L.: Consideration of connective tissue dysfunction in the fragile X syndrome. Am. J. Med. Genet. 17: 111-121, 1984.

Hanzlik, A. J.; Osemlak-Hanzlik, M. M.; Hauser, M. A.; Kurnit, D. M.: A recombination-based assay demonstrates that the fragile X sequence is transcribed widely during development. Nature Genet. 3: 44-48, 1993.

Harvey, J.; Judge, C.; Wiener, S.: Familial X-linked mental retardation with an X chromosome abnormality. J. Med. Genet. 14: 46-50, 1977.

Hecht, F.; Cannizzaro, L. A.; Hecht, B. K.: Gene symbols for fragile sites: a proposal. (Letter) Hum. Genet. 82: 394, 1989.

Hecht, F.; Glover, T. W.: Antibiotics containing trimethoprim and the fragile X chromosome. (Letter) New Eng. J. Med. 308: 285-286, 1983.

Hecht, F.; Jacky, P. B.; Sutherland, G. R.: The fragile X chromosome: current methods. Am. J. Med. Genet. 11: 489-495, 1982.

Hinds, H. L.; Ashley, C. T.; Sutcliffe, J. S.; Nelson, D. L.; Warren, S. T.; Housman, D. E.; Schalling, M.: Tissue specific expression of FMR-1 provides evidence for a functional role in fragile X syndrome. Nature Genet. 3: 36-43, 1993.

Hirst, M.; Grewal, P.; Flannery, A.; Slatter, R.; Maher, E.; Barton, D.; Fryns, J.-P.; Davies, K.: Two new cases of FMR1 deletion associated with mental impairment. Am. J. Hum. Genet. 56: 67-74, 1995.

Hirst, M. C.; Barnicoat, A.; Flynn, G.; Wang, Q.; Daker, M.; Buckle, V. J.; Davies, K. E.; Bobrow, M.: The identification of a third fragile site, FRAXF, in Xq27-q28 distal to both FRAXA and FRAXE. Hum. Molec. Genet. 2: 197-200, 1993.

Hirst, M. C.; Knight, S. J. L.; Christodoulou, Z.; Grewal, P. K.; Fryns, J. P.; Davies, K. E.: Origins of the fragile X syndrome mutation. J. Med. Genet. 30: 647-650, 1993.

Hofker, M. H.; Bergen, A. A. B.; Skraastad, M. I.; Carpenter, N. J.; Veenema, H.; Connor, J. M.; Bakker, E.; van Ommen, G. J. B.; Pearson, P. L.: Efficient isolation of X chromosome-specific single copy probes from a cosmid library of a human X/hamster hybrid-cell line: mapping of new probes close to the locus for X-linked mental retardation. Am. J. Hum. Genet. 40: 312-328, 1987.

Howard-Peebles, P. N.; Stoddard, G. R.: Familial X-linked mental retardation with a marker X chromosome and its relationship to macroorchidism. Clin. Genet. 17: 125-128, 1980.

ISCN: An International System for Human Cytogenetic Nomenclature (1978). Cytogenet. Cell Genet. 21: 309-404, 1978.

Jacky, P. B.; Dill, F. J.: Expression in fibroblast culture of the satellited-X chromosome associated with familial sex-linked mental retardation. Hum. Genet. 53: 267-269, 1980.

Jacobs, P. A.: The fragile X syndrome. (Editorial) J. Med. Genet. 28: 809-810, 1991.

Jacobs, P. A.: Personal Communication. Honolulu, Hawaii 1982.

Jacobs, P. A.; Glover, T. W.; Mayer, M.; Fox, P.; Gerrard, J. W.; Dunn, H. G.; Herbst, D. S.: X-linked mental retardation: a study of 7 families. Am. J. Med. Genet. 7: 471-479, 1980.

Jacobs, P. A.; Hunt, P. A.; Mayer, M.; Wang, J.-C.; Boss, G. R.; Erbe, R. W.: Expression of the marker(X)(q28) in lymphoblastoid cell lines. Am. J. Hum. Genet. 34: 552-557, 1982.

Jakala, P.; Hanninen, T.; Ryynanen, M.; Laakso, M.; Partanen, K.; Mannermaa, A.; Soininen, H.: Fragile-X: neuropsychological test performance, CGG triplet repeat lengths, and hippocampal volumes. J. Clin. Invest. 100: 331-338, 1997.

Jenkins, E. C.; Brown, W. T.; Brooks, J.; Duncan, C. J.; Rudelli, R. D.; Wisniewski, H. M.: Experience with prenatal fragile X detection. Am. J. Med. Genet. 17: 215-239, 1984.

Jenkins, E. C.; Brown, W. T.; Duncan, C. J.; Brooks, J.: Demonstration of the fragile X chromosome in amniotic fluid cells. (Abstract) Clin. Res. 30: 292A, 1982.

Johnson, V. P.; Carpenter, N. J.; Skorey, P. A.: Martin-Bell syndrome segregating in a large kindred with normal transmitting males: clinical, cytogenetic, and linkage study. Am. J. Med. Genet. 38: 275-282, 1991.

Kaiser-McCaw, B.; Hecht, F.; Cadien, J. D.; Moore, B. C.: Fragile X-linked mental retardation. Am. J. Hum. Genet. 7: 503-505, 1980.

Khalifa, M. M.; Reiss, A. L.; Migeon, B. R.: Methylation status of genes flanking the fragile site in males with the fragile-X syndrome: a test of the imprinting hypothesis. Am. J. Hum. Genet. 46: 744-753, 1990.

Khandjian, E. W.; Corbin, F.; Woerly, S.; Rousseau, F.: The fragile X mental retardation protein is associated with ribosomes. Nature Genet. 12: 91-93, 1996.

Kinnell, H. G.: Fragile-X disorder associated with antisocial personality. (Letter) Lancet II: 1104, 1982.

Kirchgessner, C. U.; Warren, S. T.; Willard, H. F.: X inactivation if the FMR1 fragile X mental retardation gene. J. Med. Genet. 32: 925-929, 1995.

Kirkilionis, A. J.; Chudley, A. E.; Greenberg, C. R.; Yan, D. L.; McGillivray, B.; Hamerton, J. L.: Transmission of the fra(X) haplotype from three nonpenetrant brothers to their affected grandsons. Am. J. Med. Genet. 43: 588-591, 1992.

Klauck, S. M.; Munstermann, E.; Bieber-Martig, B.; Ruhl, D.; Lisch, S.; Schmotzer, G.; Poustka, A.; Poustka, F.: Molecular genetic analysis of the FRM-1 gene in a large collection of autistic patients. Hum. Genet. 100: 224-229, 1997.

Knight, S. J. L.; Flannery, A. V.; Hirst, M. C.; Campbell, L.; Christodoulou, Z.; Phelps, S. R.; Pointon, J.; Middleton-Price, H. R.; Barnicoat, A.; Pembrey, M. E.; Holland, J.; Oostra, B. A.; Bobrow, M.; Davies, K. E.: Trinucleotide repeat amplification and hypermethylation of a CpG island in FRAXE mental retardation. Cell 74: 127-134, 1993.

Krawczun, M. S.; Jenkins, E. C.; Brown, W. T.: Analysis of the fragile-X chromosome: localization and detection of the fragile site in high resolution preparations. Hum. Genet. 69: 209-211, 1985.

Kruyer, H.; Mila, M.; Glover, G.; Carbonell, P.; Ballesta, F.; Estivill, X.: Fragile X syndrome and the (CGG)n mutation: two families with discordant MZ twins. Am. J. Hum. Genet. 54: 437-442, 1994.

Kunst, C. B.; Leeflang, E. P.; Iber, J. C.; Arnheim, N.; Warren, S. T.: The effect of FMR1 CGG repeat interruptions on mutation frequency as measured by sperm typing. J. Med. Genet. 34: 627-631, 1997.

Kunst, C. B.; Zerylnick, C.; Karickhoff, L.; Eichler, E.; Bullard, J.; Chalifoux, M.; Holden, J. J. A.; Torroni, A.; Nelson, D. L.; Warren, S. T.: FMR1 in global populations. Am. J. Hum. Genet. 58: 513-522, 1996.

Laird, C. D.: Possible erasure of the imprint on a fragile X chromosome when transmitted by a male. Am. J. Med. Genet. 38: 391-395, 1991.

Laird, C. D.: Proposed mechanism of inheritance and expression of the human fragile X syndrome of mental retardation. Genetics 117: 587-599, 1987.

Laird, C. D.; Jaffe, E.; Karpen, G.; Lamb, M.; Nelson, R.: Fragile sites in human chromosomes as regions of late-replicating DNA. Trends Genet. 3: 274-281, 1987.

Laird, C. D.; Lamb, M. M.; Thorne, J. L.: Two progenitor cells for human oogonia inferred from pedigree data and the X-inactivation imprinting model of the fragile-X syndrome. Am. J. Hum. Genet. 46: 696-719, 1990.

Langenbeck, U.; Schmidtke, J.; Bartels, I.; Hansmann, I.; Knuppel, H.: Mean corpuscular hemoglobin is increased in Martin-Bell syndrome. Hum. Genet. 66: 365-366, 1984.

Laval, S. H.; Blair, H. J.; Hirst, M. C.; Davies, K. E.; Boyd, Y.: Mapping of FMR1, the gene implicated in fragile X-linked mental retardation, on the mouse X chromosome. Genomics 12: 818-821, 1992.

Lejeune, J.; Legrand, N.; Lafourcade, J.; Rethore, M.-O.; Raoul, O.; Maunoury, C.: Fragilite du chromosome X et effets de la trimethoprime. Ann. Genet. 25: 149-151, 1982.

Loesch, D. Z.; Hay, D. A.: Clinical features and reproductive patterns in fragile X female heterozygotes. J. Med. Genet. 25: 407-414, 1988.

Lubs, H.; Travers, H.; Lujan, E.; Carroll, A.: A large kindred with X-linked mental retardation, marker X and macroorchidism. Am. J. Med. Genet. 17: 145-157, 1984.

Lubs, H. A.; Watson, M.; Breg, R.; Lujan, E.: Restudy of the original marker X family. Am. J. Med. Genet. 17: 133-144, 1984.

Lubs, H. A., Jr.: A marker X chromosome. Am. J. Hum. Genet. 21: 231-244, 1969.

Lugenbeel, K. A.; Peier, A. M.; Carson, N. L.; Chudley, A. E.; Nelson, D. L.: Intragenic loss of function mutations demonstrate the primary role of FMR1 in fragile X syndrome. Nature Genet. 10: 483-485, 1995.

Malter, H. E.; Iber, J. C.; Willemsen, R.; de Graaff, E.; Tarleton, J. C.; Leisti, J.; Warren, S. T.; Oostra, B. A.: Characterization of the full fragile X syndrome mutation in fetal gametes. Nature Genet. 15: 165-169, 1997.
Mandel, J.-L.; Hagerman, R.; Froster, U.; Brown, W. T.; Jenkins, E. C.; Jacobs, P.; Turner, G.; Lubs, H.; Neri, G.: Fifth International Workshop on the Fragile X and X-Linked Mental Retardation. Am. J. Med. Genet. 43: 5-27, 1992.
Martin, J. P.; Bell, J.: A pedigree of mental defect showing sex-linkage. J. Neurol. Psychiat. 6: 154-157, 1943.
Martin, R. H.; Lin, C. C.; Mathies, B. J.; Lowry, R. B.: X-linked mental retardation with macro-orchidism and marker-X chromosomes. Am. J. Med. Genet. 7: 433-441, 1980.
Mattei, J. F.; Mattei, M. G.; Aumeras, C.; Auger, M.; Giraud, F.: X-linked mental retardation with the fragile X: a study of 15 families. Hum. Genet. 59: 281-289, 1981.
McConkie-Rosell, A.; Lachiewicz, A. M.; Spiridigliozzi, G. A.; Tarleton, J.; Schoenwald, S.; Phelan, M. C.; Goonewardena, P.; Ding, X.; Brown, W. T.: Evidence that methylation of the FMR-1 locus is responsible for variable phenotypic expression of the fragile X syndrome. Am. J. Hum. Genet. 53: 800-809, 1993.
Meijer, H.; de Graaff, E.; Merckx, D. M. L.; Jongbloed, R. J. E.; de Die-Smulders, C. E. M.; Engelen, J. J. M.; Fryns, J.-P.; Curfs, P. M. G.; Oostra, B. A.: A deletion of 1.6 kb proximal to the CGG repeat of the FMR1 gene causes the clinical phenotype of the fragile X syndrome. Hum. Molec. Genet. 3: 615-620, 1994.
Meryash, D. L.; Cronk, C. E.; Sachs, B.; Gerald, P. S.: An anthropometric study of males with the fragile-X syndrome. Am. J. Med. Genet. 17: 159-174, 1984.
Mila, M.; Sanchez, A.; Badenas, C.; Brun, C.; Jimenez, D.; Villa, M. P.; Castellvi-Bel, S.; Estivill, X.: Screening for FMR1 and FMR2 mutations in 222 individuals from Spanish special schools: identification of a case of FRAXE-associated mental retardation. Hum. Genet. 100: 503-507, 1997.
Morton, N. E.; Macpherson, J. N.: Population genetics of the fragile-X syndrome: multiallelic model for the FMR1 locus. Proc. Nat. Acad. Sci. 89: 4215-4217, 1992.
Mulley, J. C.; Yu, S.; Gedeon, A. K.; Donnelly, A.; Turner, G.; Loesch, D.; Chapman, C. J.; Gardner, R. J. M.; Richards, R. I.; Sutherland, G. R.: Experience with direct molecular diagnosis of fragile X. J. Med. Genet. 29: 368-374, 1992.
Mulligan, L. M.; Phillips, M. A.; Forster-Gibson, C. J.; Beckett, J.; Partington, M. W.; Simpson, N. E.; Holden, J. J. A.; White, B. N.: Genetic mapping of DNA segments relative to the locus for the fragile-X syndrome at Xq27.3. Am. J. Hum. Genet. 37: 463-472, 1985.
Nolin, S. L.; Lewis, F. A., III; Ye, L. L.; Houck, G. E., Jr.; Glicksman, A. E.; Limprasert, P.; Li, S. Y.; Zhong, N.; Ashley, A. E.; Feingold, E.; Sherman, S. L.; Brown, W. T.: Familial transmission of the FMR1 GCC repeat. Am. J. Hum. Genet. 59: 1252-1261, 1996.
Nussbaum, R. L.; Ledbetter, D. H.: Fragile X syndrome: a unique mutation in man. Annu. Rev. Genet. 20: 109-145, 1986.
Oberle, I.; Rousseau, F.; Heitz, D.; Kretz, C.; Devys, D.; Hanauer, A.; Boue, J.; Bertheas, M. F.; Mandel, J. L.: Instability of a 550-base pair DNA segment and abnormal methylation in fragile X syndrome. Science 252: 1097-1102, 1991.
Opitz, J. M.; Sutherland, G. R.: International workshop on the fragile X and X-linked mental retardation. Am. J. Med. Genet. 17: 5-94, 1984.
Opitz, J. M.; Westphal, J. M.; Daniel, A.: Discovery of a connective tissue dysplasia in the Martin-Bell syndrome. Am. J. Med. Genet. 17: 101-109, 1984.
Oudet, C.; Mornet, E.; Serre, J. L.; Thomas, F.; Lentes-Zengerling, S.; Kretz, C.; Deluchat, C.; Tejada, I.; Boue, J.; Boue, A.; Mandel, J. L.: Linkage disequilibrium between the fragile X mutation and two closely linked CA repeats suggests that fragile X chromosomes are derived from a small number of founder chromosomes. Am. J. Hum. Genet. 52: 297-304, 1993.
Park, V.; Howard-Peebles, P.; Sherman, S.; Taylor, A.; Wulfsberg, E.: Policy statement: American College of Medical Genetics. Fragile X syndrome: diagnostic and carrier testing. Am. J. Med. Genet. 53: 380-381, 1994.
Parrish, J. E.; Oostra, B. A.; Verkerk, A. J. M. H.; Richards, C. S.; Reynolds, J.; Spikes, A. S.; Shaffer, L. G.; Nelson, D. L.: Isolation of a GCC repeat showing expansion in FRAXF, a fragile site distal to FRAXA and FRAXE. Nature Genet. 8: 229-235, 1994.
Pellissier, M.-C.; Voelckel, M.-A.; Piquet, C.; Mattei, M.-G.; Mattei, J.-F.: Transmission of mental retardation with fragile X site by two normal transmitter brothers. Am. J. Med. Genet. 38: 367-369, 1991.
Pembrey, M. E.; Winter, R. M.; Davies, K. E.: A premutation that generates a defect at crossing over explains the inheritance of fragile X mental retardation. Am. J. Med. Genet. 21: 709-717, 1985.
Pieretti, M.; Zhang, F.; Fu, Y.-H.; Warren, S. T.; Oostra, B. A.; Caskey, C. T.; Nelson, D. L.: Absence of expression of the FMR-1 gene in fragile X syndrome. Cell 66: 817-822, 1991.
Poustka, A.; Dietrich, A.; Langenstein, G.; Toniolo, D.; Warren, S. T.; Lehrach, H.: Physical map of human Xq27-qter: localizing the region of the fragile X mutation. Proc. Nat. Acad. Sci. 88: 8302-8306, 1991.
Purrello, M.; Alhadeff, B.; Esposito, D.; Szabo, P.; Rocchi, M.; Truett, M.; Masiarz, F.; Siniscalco, M.: The human genes for hemophilia A and hemophilia B flank the X chromosome fragile site at Xq27.3. EMBO J. 4: 725-729, 1985.
Pyeritz, R. E.; Stamberg, J.; Thomas, G. H.; Bell, B. B.; Zahka, K. G.; Bernhardt, B. A.: The marker Xq28 syndrome (fragile X syndrome) in a retarded man with mitral valve prolapse. Johns Hopkins Med. J. 151: 231-237, 1982.
Quan, F.; Grompe, M.; Jakobs, P.; Popovich, B. W.: Spontaneous deletion in the FMR1 gene in a patient with fragile X syndrome and cherubism. Hum. Molec. Genet. 4: 1681-1684, 1995.
Quan, F.; Zonana, J.; Gunter, K.; Peterson, K. L.; Magenis, R. E.; Popovich, B. W.: An atypical case of fragile X syndrome caused by a deletion that includes the FMR1 gene. Am. J. Hum. Genet. 56: 1042-1051, 1995.
Ramos, F. J.; Emanuel, B. S.; Spinner, N. B.: Frequency of the common fragile site at Xq27.2 under conditions of thymidylate stress: implications for cytogenetic diagnosis of the fragile-X syndrome. Am. J. Med. Genet. 42: 835-838, 1992.
Reiss, A. L.; Aylward, E.; Freund, L. S.; Joshi, P. K.; Bryan, R. N.: Neuroanatomy of fragile X syndrome: the posterior fossa. Ann. Neurol. 29: 26-32, 1991.
Reiss, A. L.; Freund, L.; Tseng, J. E.; Joshi, P. K.: Neuroanatomy in fragile X females: the posterior fossa. Am. J. Hum. Genet. 49: 279-288, 1991.
Reiss, A. L.; Lee, J.; Freund, L.: Neuroanatomy of fragile X syndrome: the temporal lobe. Neurology 44: 1317-1324, 1994.
Reyniers, E.; Vits, L.; De Boulle, K.; Van Roy, B.; Van Velzen, D.; de Graaff, E.; Verkerk, A. J. M. H.; Jorens, H. Z. J.; Darby, J. K.; Oostra, B.; Willems, P. J.: The full mutation in the FMR-1 gene of male fragile X patients is absent in their sperm. Nature Genet. 4: 143-148, 1993.
Richards, B. W.; Sylvester, P. E.; Brooker, C.: Fragile X-linked mental retardation: the Martin-Bell syndrome. J. Ment. Defic. Res. 25: 253-256, 1981.
Richards, R. I.; Holman, K.; Friend, K.; Kremer, E.; Hillen, D.; Staples, A.; Brown, W. T.; Goonewardena, P.; Tarleton, J.; Schwartz, C.; Sutherland, G. R.: Evidence of founder chromosomes in fragile X syndrome. Nature Genet. 1: 257-260, 1992.
Richards, R. I.; Holman, K.; Kozman, H.; Kremer, E.; Lynch, M.; Pritchard, M.; Yu, S.; Mulley, J.; Sutherland, G. R.: Fragile X syndrome: genetic localisation by linkage mapping of two microsatellite repeats FRAXAC1 and FRAXAC2 which immediately flank the fragile site. J. Med. Genet. 28: 818-823, 1991.
Richards, R. I.; Holman, K.; Kozman, H.; Kremer, E.; Lynch, M.; Pritchard, M.; Yu, S.; Mulley, J.; Sutherland, G. R.: Fragile X syndrome: genetic localisation by linkage mapping of two microsatellite repeats FRAXAC1 and FRAXAC2 which immediately flank the fragile site. J. Med. Genet. 28: 818-823, 1991.
Richards, R. I.; Sutherland, G. R.: Dynamic mutations: a new class of mutations causing human disease. Cell 70: 709-712, 1992.
Rocchi, M.; Archidiacono, N.; Rinaldi, A.; Filippi, G.; Bartolucci, G.; Fancello, G. S.; Siniscalco, M.: Mental retardation in heterozygotes for the fragile-X mutation: evidence in favor of an X inactivation-dependent effect. Am. J. Hum. Genet. 46: 738-743, 1990.
Rodewald, L.; Miller, D. C.; Sciorra, L.; Barabas, G.; Lee, M.-L.: Central nervous system neoplasm in a young man with Martin-Bell syndrome—fra(X)-XLMR. Am. J. Med. Genet. 26: 7-12, 1987.
Rousseau, F.; Heitz, D.; Biancalana, V.; Blumenfeld, S.; Kretz, C.; Boue, J.; Tommerup, N.; Van Der Hagen, C.; DeLozier-Blanchet, C.; Croquette, M.-F.; Gilgenkrantz, S.; Jalbert, P.; Voelckel, M.-A.; Oberle, I.; Mandel, J.-L.: Direct diagnosis by DNA analysis of the fragile X syndrome of mental retardation. New Eng. J. Med. 325: 1673-1681, 1991.
Rousseau, F.; Heitz, D.; Mandel, J.-L.: The unstable and methylatable mutations causing the fragile X syndrome. Hum. Mutat. 1: 91-96, 1992.
Rousseau, F.; Heitz, D.; Oberle, I.; Mandel, J.-L.: Selection in blood cells from female carriers of the fragile X syndrome: inverse correlation between age and proportion of active X chromosomes carrying the full mutation. J. Med. Genet. 28: 830-836, 1991.
Rousseau, F.; Heitz, D.; Tarleton, J.; MacPherson, J.; Malmgren, H.; Dahl, N.; Barnicoat, A.; Mathew, C.; Mornet, E.; Tejada, I.; Maddalena, A.; Spiegel, R.; Schinzel, A.; Marcos, J. A. G.; Schorderet, D. F.; Schaap, T.; Maccioni, L.; Russo, S.; Jacobs, P. A.; Schwartz, C.; Mandel, J. L.: A multicenter study on genotype-phenotype correlations in the fragile X syndrome, using direct diagnosis with probe StB12.3: the first 2,253 cases. Am. J. Hum. Genet. 55: 225-237, 1994.
Ruvalcaba, R. H. A.; Myhre, S. A.; Roosen-Runge, E. C.; Beckwith, J. B.: X-linked mental deficiency megalotestes syndrome. J.A.M.A. 238: 1646-1650, 1977.
Schrander-Stumpel, C.; Gerver, W.-J.; Meyer, H.; Engelen, J.; Mulder, H.; Fryns, J.-P.: Prader-Willi-like phenotype in fragile X syndrome. Clin. Genet. 45: 175-180, 1994.
Sherman, S. L.; Jacobs, P. A.; Morton, N. E.; Froster-Iskenius, U.; Howard-Peebles, P. N.; Nielsen, K. B.; Partington, M. W.; Sutherland, G. R.; Turner, G.; Watson, M.: Further segregation analysis of the fragile X syndrome with special reference to transmitting males. Hum. Genet. 69: 289-299, 1985.
Siomi, H.; Choi, M.; Siomi, M. C.; Nussbaum, R. L.; Dreyfuss, G.: Essential role for KH domains in RNA binding: impaired RNA binding by a mutation in the KH domain of FMR1 that causes fragile X syndrome. Cell 77: 33-39, 1994.
Siomi, H.; Siomi, M. C.; Nussbaum, R. L.; Dreyfuss, G.: The protein product of the fragile X gene, FMR1, has characteristics of an RNA-binding protein. Cell 74: 291-298, 1993.
Sittler, A.; Devys, D.; Weber, C.; Mandel, J.-L.: Alternative splicing of exon 14 determines nuclear or cytoplasmic localisation of fmr1 protein isoforms. Hum. Molec. Genet. 5: 95-102, 1996.
Smeets, H. J. M.; Smits, A. P. T.; Verheij, C. E.; Theelen, J. P. G.; Willemsen, R.; van de Burgt, I.; Hoogeveen, A. T.; Oosterwijk, J. C.; Oostra, B. A.: Normal phenotype in two brothers with a full FMR1 mutation. Hum. Molec. Genet. 4: 2103-2108, 1995.
Smits, A. P. T.; Dreesen, J. C. F. M.; Post, J. G.; Smeets, D. F. C. M.; de Die-Smulders, C.; Spaans-van der Bijl, T.; Govaerts, L. C. P.; Warren, S. T.; Oostra, B. A.; van Oost, B. A.: The fragile X syndrome: no evidence for any recent mutations. J. Med. Genet. 30: 94-96, 1993.
Snow, K.; Doud, L. K.; Hagerman, R.; Pergolizzi, R. G.; Erster, S. H.; Thibodeau, S. N.: Analysis of a CGG sequence at the FMR-1 locus in fragile X families and in the general population. Am. J. Hum. Genet. 53: 1217-1228, 1993.
Snyder, F. F.; Lin, C. C.; Harasym, C. A.; Jamro, H. K.; Kushnig, M. L.; Lowe, J. K.; O'Brien, S. I.: Evidence for close association between the fragile X(q27-28) chromosome site and glucose-6-phosphate dehydrogenase (G6PD) but not with hypoxanthine-guanine phosphoribosyltransferase (HPRT). (Abstract) Cytogenet. Cell Genet. 37: 587, 1984.
Soysa, P.; Senanayahe, M.; Mikkelsen, M.; Poulsen, H.: Martin-Bell syndrome fra(X)(q28) in a Sri Lankan family. J. Ment Defic. Res. 26: 251-257, 1982.
Steinbach, P.: Mental impairment in Martin-Bell syndrome is probably determined by interaction of several genes: simple explanation of phenotypic differences between unaffected and affected males with the same X chromosome. Hum. Genet. 72: 248-252, 1986.

Storm, K.; Handig, I.; Reyniers, E.; Oostra, B. A.; Kooy, R. F.; Willems, P. J.: Incomplete EcoRI digestion may lead to false diagnosis of fragile X syndrome. Hum. Genet. 102: 54-56, 1998.
Subramanian, P. S.; Nelson, D. L.; Chinault, A. C.: Large domains of apparent delayed replication timing associated with triplet repeat expansion at FRAXA and FRAXE. Am. J. Hum. Genet. 59: 407-416, 1996.
Sutherland, G. R.: Fragile sites on human chromosomes: demonstration of their dependence on the type of tissue culture medium. Science 197: 265-266, 1977.
Sutherland, G. R.: Heritable sites on human chromosomes. VII. Children homozygous for the BrdU-requiring fra10q25 are phenotypically normal. Am. J. Hum. Genet. 33: 946-949, 1981.
Sutherland, G. R.: Personal Communication. Adelaide, Australia 6/1989.
Sutherland, G. R.: Personal Communication. Adelaide, Australia 11/3/1993.
Sutherland, G. R.: Heritable fragile sites on human chromosomes. II. Distribution, phenotypic effects, and cytogenetics. Am. J. Hum. Genet. 31: 136-148, 1979.
Sutherland, G. R.; Ashford, P. L. C.: X-linked mental retardation with macro-orchidism and the fragile site at Xq27 or 28. Hum. Genet. 48: 117-120, 1979.
Sutherland, G. R.; Baker, E.: Characterisation of a new rare fragile site easily confused with the fragile X. Hum. Molec. Genet. 1: 111-113, 1992.
Sutherland, G. R.; Gedeon, A.; Kornman, L.; Donnelly, A.; Byard, R. W.; Mulley, J. C.; Kremer, E.; Lynch, M.; Pritchard, M.; Yu, S.; Richards, R. I.: Prenatal diagnosis of fragile X syndrome by direct detection of the unstable DNA sequence. New Eng. J. Med. 325: 1720-1722, 1991.
Sved, J. A.; Laird, C. D.: Population genetic consequences of the fragile X syndrome, based on the X-inactivation imprinting model. Am. J. Hum. Genet. 46: 443-451, 1990.
Sved, J. A.; Laird, C. D.: The X-inactivation imprinting model can explain the incidence of the fragile-X syndrome of mental retardation in mother-offspring pairs. Brain Dysfunction 1: 245-254, 1988.
Szabo, P.; Purrello, M.; Rocchi, M.; Archidiacono, N.; Alhadeff, B.; Filippi, G.; Toniolo, D.; Martini, G.; Luzzatto, L.; Siniscalco, M.: Cytological mapping of the human glucose-6-phosphate dehydrogenase gene distal to the fragile-X site suggests a high rate of meiotic recombination across this site. Proc. Nat. Acad. Sci. 81: 7855-7859, 1984.
Tarleton, J.; Richie, R.; Schwartz, C.; Rao, K.; Aylsworth, A. S.; Lachiewicz, A.: An extensive de novo deletion removing FMR1 in a patient with mental retardation and the fragile X syndrome phenotype. Hum. Molec. Genet. 2: 1973-1974, 1993.
Thibodeau, S. N.; Dorkins, H. R.; Faulk, K. R.; Berry, R.; Smith, A. C. M.; Hagerman, R.; King, A.; Davies, K. E.: Linkage analysis using multiple DNA polymorphic markers in normal families and in families with fragile X syndrome. Hum. Genet. 79: 219-227, 1988.
Thode, A.; Laing, S.; Partington, M. W.; Turner, G.: Is there a fragile(X) negative Martin-Bell syndrome?. Am. J. Med. Genet. 30: 459-471, 1988.
Tommerup, N.; Poulsen, H.; Brondum-Nielsen, K.: 5-Fluoro-2-prime-deoxyuridine induction of the fragile site on Xq28 associated with X-linked mental retardation. J. Med. Genet. 33: 234-242, 1981.
Turner, G.; Brookwell, R.; Daniel, A.; Selikowitz, M.; Zilibowitz, M.: Heterozygous expression of X-linked mental retardation and X-chromosome marker fra(X)(q27). New Eng. J. Med. 303: 662-664, 1980.
Turner, G.; Daniel, A.; Frost, M.: X-linked mental retardation, macro-orchidism, and the Xq27 fragile site. J. Pediat. 96: 837-841, 1980.
Turner, G.; Eastman, C.; Casey, J.; McLeay, A.; Procopis, P.; Turner, B.: X-linked mental retardation associated with macro-orchidism. J. Med. Genet. 12: 367-371, 1975.
Turner, G.; Till, R.; Daniel, A.: Marker X chromosomes, mental retardation and macro-orchidism. (Letter) New Eng. J. Med. 299: 1472, 1978.
Turner, G.; Webb, T.; Wake, S.; Robinson, H.: Prevalence of fragile X syndrome. Am. J. Med. Genet. 64: 196-197, 1996.
Uchida, I. A.; Joyce, E. M.: Activity of the fragile X in heterozygous carriers. Am. J. Hum. Genet. 34: 286-293, 1982.
van den Ouweland, A. M. W.; Deelen, W. H.; Kunst, C. B.; Uzielli, M.-L. G.; Nelson, D. L.; Warren, S. T.; Oostra, B. A.; Halley, D. J. J.: Loss of mutation at the FMR1 locus through multiple exchanges between maternal X chromosomes. Hum. Molec. Genet. 3: 1823-1827, 1994.
Van Roy, B. C.; De Smedt, M. C.; Raes, R. A.; Dumon, J. E.; Leroy, J. G.: Fragile X trait in a large kindred: transmission also through normal males. J. Med. Genet. 20: 286-289, 1983.
Verheij, C.; Bakker, C. E.; de Graaff, E.; Keulemans, J.; Willemsen, R.; Verkerk, A. J. M. H.; Galjaard, H.; Reuser, A. J. J.; Hoogeveen, A. T.; Oostra, B. A.: Characterization and localization of the FMR-1 gene product associated with fragile X syndrome. Nature 363: 722-724, 1993.
Verheij, C.; de Graaff, E.; Bakker, C. E.; Willemsen, R.; Willems, P. J.; Meijer, N.; Galjaard, H.; Reuser, A. J. J.; Oostra, B. A.; Hoogeveen, A. T.: Characterization of FMR1 proteins isolated from different tissues. Hum. Molec. Genet. 4: 895-901, 1995.
Verkerk, A. J. M. H.; Pieretti, M.; Sutcliffe, J. S.; Fu, Y.-H.; Kuhl, D. P. A.; Pizzuti, A.; Reiner, O.; Richards, S.; Victoria, M. F.; Zhang, F.; Eussen, B. E.; van Ommen, G.-J. B.; Blonden, L. A. J.; Riggins, G. J.; Chastain, J. L.; Kunst, C. B.; Galjaard, H.; Caskey, C. T.; Nelson, D. L.; Oostra, B. A.; Warren, S. T.: Identification of a gene (FMR-1) containing a CGG repeat coincident with a breakpoint cluster region exhibiting length variation in fragile X syndrome. Cell 65: 905-914, 1991.
Vincent, J. B.; Gurling, H. M. D.: Point mutation in intron 10 of FMR1 is unlikely to be a cause of fragile X syndrome. (Letter) Hum. Mutat. 12: 431 only, 1998.
Voelckel, M. A.; Mattei, M. G.; N'Guyen, C.; Philip, N.; Birg, F.; Mattei, J. F.: Dissociation between mental retardation and fragile site expression in a family with fragile X-linked mental retardation. Hum. Genet. 80: 375-378, 1988.
Vogel, F.; Crusio, W. E.; Kovac, C.; Fryns, J.-P.; Freund, M.: Selective advantage of fra (X) heterozygotes. Hum. Genet. 86: 25-32, 1990.
Wang, Y.-C.; Li, S.-Y.: Response to Vincent and Gurling. (Letter) Hum. Mutat. 12: 432 only, 1998.
Wang, Y.-C.; Lin, M.-L.; Lin, S. J.; Li, Y.-C.; Li, S.-Y.: Novel point mutation within intron 10 of FMR-1 gene causing fragile X syndrome. Hum. Mutat. 10: 393-399, 1997.
Warren, S. T.; Glover, T. W.; Davidson, R. L.; Jagadeeswaran, P.: Linkage and recombination between fragile X-linked mental retardation and the factor IX gene. (Letter) Hum. Genet. 69: 44-46, 1985.
Warren, S. T.; Knight, S. J. L.; Peters, J. F.; Stayton, C. L.; Consalez, G. G.; Zhang, F. P.: Isolation of the human chromosomal band Xq28 within somatic cell hybrids by fragile X site breakage. Proc. Nat. Acad. Sci. 87: 3856-3860, 1990.
Warren, S. T.; Zhang, F.; Licameli, G. R.; Peters, J. F.: The fragile X site in somatic cell hybrids: an approach for molecular cloning of fragile sites. Science 237: 420-423, 1987.
Warren, S. T.; Zhang, F. P.; Sutcliffe, J. S.; Peters, J. F.: Strategy for molecular cloning of the fragile X site DNA. Am. J. Med. Genet. 30: 613-623, 1988.
Weaver, D. D.; Sherman, S. L.: A counseling guide to the Martin-Bell syndrome. (Letter) Am. J. Med. Genet. 26: 39-44, 1987.
Webb, T.; Jacobs, P. A.: Fragile Xq27.3 in female heterozygotes for the Martin-Bell syndrome. J. Med. Genet. 27: 627-631, 1990.
Wheater, R. F.; Roberts, S. H.: An improved lymphocyte culture technique: deoxycytidine release of a thymidine block and use of a constant humidity chamber for slide making. J. Med. Genet. 24: 113-115, 1987.
Willemsen, R.; Mohkamsing, S.; De Vries, B.; Devys, D.; van den Ouweland, A.; Mandel, J. L.; Galjaard, H.; Oostra, B.: Rapid antibody test for fragile X syndrome. Lancet 345: 1147-1148, 1995.
Winter, R. M.; Pembrey, M. E.: Analysis of linkage relationships between genetic markers around the fragile X locus with special reference to the daughters of normal transmitting males. Hum. Genet. 74: 93-97, 1986.
Wohrle, D.; Hennig, I.; Vogel, W.; Steinbach, P.: Mitotic stability of fragile X mutations in differentiated cells indicates early post-conceptional trinucleotide repeat expansion. Nature Genet. 4: 140-142, 1993.
Wohrle, D.; Kotzot, D.; Hirst, M. C.; Manca, A.; Korn, B.; Schmidt, A.; Barbi, G.; Rott, H.-D.; Poustka, A.; Davies, K. E.; Steinbach, P.: A microdeletion of less than 250 kb, including the proximal part of the FMR-1 gene and the fragile-X site, in a male with the clinical phenotype of fragile-X syndrome. Am. J. Hum. Genet. 51: 299-306, 1992.
Yamauchi, M.; Nagata, S.; Seki, N.; Toyama, Y.; Harada, N.; Niikawa, N.; Masuno, I.; Kajii, T.; Hori, T.: Prenatal diagnosis of fragile X syndrome by direct detection of the dynamic mutation due to an unstable DNA sequence. Clin. Genet. 44: 169-172, 1993.
Yeshaya, J.; Shalgi, R.; Shohat, M.; Avivi, L.: Replication timing of the various FMR1 alleles detected by FISH: inferences regarding their transcriptional status. Hum. Genet. 102: 6-14, 1998.
Yu, S.; Mulley, J.; Loesch, D.; Turner, G.; Donnelly, A.; Gedeon, A.; Hillen, D.; Kremer, E.; Lynch, M.; Pritchard, M.; Sutherland, G. R.; Richards, R. I.: Fragile-X syndrome: unique genetics of the heritable unstable element. Am. J. Hum. Genet. 50: 968-980, 1992.
Yu, S.; Pritchard, M.; Kremer, E.; Lynch, M.; Nancarrow, J.; Baker, E.; Holman, K.; Mulley, J. C.; Warren, S. T.; Schlessinger, D.; Sutherland, G. R.; Richards, R. I.: Fragile X genotype characterized by an unstable region of DNA. Science 252: 1179-1181, 1991.
Zhong, N.; Dobkin, C.; Brown, W. T.: A complex mutable polymorphism located within the fragile X gene. Nature Genet. 5: 248-253, 1993.
Zoll, B.; Arnemann, J.; Krawczak, M.; Cooper, D. N.; Pescia, G.; Wahli, W.; Steinbach, P.; Schmidtke, J.: Evidence against close linkage of the loci for fraXq of Martin-Bell syndrome and for factor IX. Hum. Genet. 71: 122-126, 1985.

FRAGILE SITE 16Q22; FRA16A (#136580)

A number sign (#) is used with this entry because a specific gene is not involved. The entry is included here largely for historic reasons; it was a marker by which the haptoglobin gene (HP; 140100) was mapped to 16q (Magenis et al., 1970). See 136610. Unlike several other fragile sites, this one is not medium-dependent.

In 4 families with the chromosomal defect, Shabtai et al. (1980) found infertility, abortions, malformations, and aneuploidy. The authors suggested that the fragile site on 16 may be a site of virus modification. Thestrup-Pedersen et al. (1980) found fra(16q22) in a 6-year-old boy who presented with swelling of cervical and axillary lymph nodes and was diagnosed as having an Epstein-Barr virus infection because of EBV-nuclear-antigen-positive B-lymphocytes in blood and in lymph nodes and high antibody titers to EBV antigens. The natural killer activity of blood lymphocytes and the percentage of T-lymphocytes with Fc receptors for IgG (T-gamma cells) were low. The 16q22 defect appeared only after addition of interferon to the lymphocyte cultures and preferentially in T-gamma lymphocytes. The disease progressed despite attempts to restore the patient's immune reactivity by interferon, transfer factor, and blood transfusion. At necropsy, Burkitt-like lymphoma was found. The mother's T-lymphocytes also showed the 16q22 gap when interferon was added in culture. The patient differed from X-linked lymphoproliferative patients (308240) in not having affected maternally-related relatives and in having high antibody titers to EBV. The serum interferon level of the patient was normal. Croci (1983) concluded that the 16q fragile site may be located at the interface between 16q21 and 16q22—fra(16)(q2200). He also concluded that it is BrdU-sensitive, although not folate-dependent. (A third fragile site on chromosome 16, at 16q23-16q24, was pointed out by Shabtai et al.

(1983). They emphasized the importance of differentiation from FS16q22.) Dunner et al. (1983) raised questions about the innocence of fra16(q22). In a family ascertained through an infant with this fragile site and multiple anomalies, the chromosome change was present also in the phenotypically normal father and a brother with cleft palate. There had been a stillborn male with no ostensible malformations. Schmid e t al. (1980) found a distamycin-induced fragile site in chromosome 16 of gorilla and chimpanzee, a chromosome which is homologous to human 16. In a family reported by Garcia-Sagredo et al. (1983), the father had a fragile site 16q22 and the son had a de novo 1;16 balanced translocation. Schmid et al. (1984) showed that distamycin A inhibits condensation of the Y chromosome heterochromatin and induces fragile site 16q22 in some persons. This agent can be used in studying translocations involving the Y chromosome and families for linkage with 16q22. Hecht and Hecht (1984) compared 21 fragile sites and 50 cancer breakpoints; 9 of the 21 fragile sites appeared in bands with a cancer breakpoint (p less than 0.001). They suggested that fragile sites may be a (the) predisposing genetic factor in familial cancer. Le Beau and Rowley (1984) reviewed the evidence suggesting that heritable fragile sites may predispose to cancer.

Shen et al. (1994) developed a high resolution cytogenetic-based physical map and a genetic linkage map of human chromosome 16 based on 79 PCR-typable genetic markers and 2 Southern-based RFLP markers. They mapped 4 fragile sites, 16A, 16B, 16D, and 16E, at positions p13.11, q22.1, q23.1, and p12.1, respectively.

By positional cloning, Nancarrow et al. (1994) isolated DNA sequences encompassing and including the rare, autosomal, folate-sensitive fragile site FRA16A. The molecular basis of FRA16A was found to be expansion of a normally polymorphic p(CCG)n repeat. This repeat was adjacent to a CpG island that was methylated in fragile site-expressing individuals. The FRA16A locus in individuals who do not express the fragile site is not a site of DNA methylation (imprinting), which suggests that the methylation associated with fragile sites may be a consequence and not a cause of their genesis.

See also Cote and Katsantoni (1980); Sutherland et al. (1984)

Cote, G. B.; Katsantoni, A.: The fragile site on the long arm of chromosome 16. Ann. Genet. 23: 241-243, 1980.

Croci, G.: BrdU-sensitive fragile site on long arm of chromosome 16. (Letter) Am. J. Hum. Genet. 35: 530-533, 1983.

Dunner, J. A.; Martin, A. O.; Traisman, E. S.; Traisman, H. S.; Elias, S.: Enhancement of a fra(16)(q22) with distamycin A: a family ascertained through an abnormal proposita. Am. J. Med. Genet. 16: 277-284, 1983.

Garcia-Sagredo, J. M.; San Roman, C.; Gallego Gomez, M. E.; Lledo, G.: Fragile chromosome 16(q22) cause a balanced translocation at the same point. Hum. Genet. 65: 211-213, 1983.

Hecht, F.; Hecht, B. K.: Autosomal fragile sites and cancer. (Letter) Am. J. Hum. Genet. 36: 718-720, 1984.

Le Beau, M. M.; Rowley, J. D.: Heritable fragile sites in cancer. Nature 308: 607-608, 1984.

Magenis, R. E.; Hecht, F.; Lovrien, E. W.: Heritable fragile site on chromosome 16: probable localization of haptoglobin locus in man. Science 170: 85-87, 1970.

Nancarrow, J. K.; Kremer, E.; Holman, K.; Eyre, H.; Doggett, N. A.; Le Paslier, D.; Callen, D. F.; Sutherland, G. R.; Richards, R. I.: Implications of FRA16A structure for the mechanism of chromosomal fragile site genesis. Science 264: 1938-1941, 1994.

Schmid, M.; Hungerford, D. A.; Poppen, A.; Engel, W.: The use of distamycin A in human lymphocyte cultures. Hum. Genet. 65: 377-384, 1984.

Schmid, M.; Klett, C.; Niederhofer, A.: Demonstration of a heritable fragile site in human chromosome 16 with distamycin A. Cytogenet. Cell Genet. 28: 87-94, 1980.

Shabtai, F.; Bichacho, S.; Halbrecht, I.: The fragile site on chromosome 16(q21q22): data on four new families. Hum. Genet. 55: 19-22, 1980.

Shabtai, F.; Klar, D.; Nissimov, R.; Vardimon, D.; Hart, J.; Halbrecht, I.: A new familial 'fragile site' on chromosome 16 (q23-24): cytogenetic and clinical considerations. Hum. Genet. 64: 273-276, 1983.

Shen, Y.; Kozman, H. M.; Thompson, A.; Phillips, H. A.; Holman, K.; Nancarrow, J.; Lane, S.; Chen, L.-Z.; Apostolou, S.; Doggett, N. A.; Callen, D. F.; Mulley, J. C.; Sutherland, G. R.; Richards, R. I.: A PCR-based genetic linkage map of human chromosome 16. Genomics 22: 68-76, 1994.

Sutherland, G. R.; Jacky, P. B.; Baker, E. G.: Heritable fragile sites on human chromosomes. XI. Factors affecting expression of fragile sites at 10q25, 16q22, and 17p12. Am. J. Hum. Genet. 36: 110-122, 1984.

Thestrup-Pedersen, K.; Esmann, V.; Bisballe, S.; Jensen, J. R.; Pallesen, G.; Hastrup, J.; Madsen, M.; Thorling, K.; Grazia-Masucci, M.; Saemundsen, A. K.; Ernberg, I.: Epstein-Barr-virus-induced lymphoproliferative disorder converting to fatal Burkitt-like lymphoma in a boy with interferon-inducible chromosomal defect. Lancet II: 997-1002, 1980.

FRAGILE HISTIDINE TRIAD GENE; FHIT (*601153)

Cohen et al. (1979) observed a constitutional reciprocal t(3;8) translocation associated with early onset, bilateral, and multifocal clear cell renal carcinoma (144700). Wang and Perkins (1984) demonstrated that the site of the break on chromosome 3 is at 3p14.2. Another cytogenetic landmark in the 3p14.2 region is the most common of the constitutive aphidicolin inducible fragile sites, FRA3B. A 200 to 300 kb region of 3p14.2, including the fragile site locus FRA3B, is homozygously deleted in multiple tumor-derived cell lines. By exon amplification from cosmids covering this deleted region, Ohta et al. (1996) identified a human gene they designated FHIT for 'fragile histidine triad' gene. The FHIT gene, a member of the histidine triad gene family, encodes a protein similar to the aph1 gene in S. pombe. The gene is composed of 10 exons distributed over at least 500 kb, with three 5-prime untranslated exons centromeric to the renal carcinoma-associated 3p14.2 breakpoint, the remaining exons telomeric to this translocation breakpoint, and exon 5 within the homozygously deleted fragile region. Aberrant transcripts of the FHIT locus were found in approximately 50% of esophageal, stomach, and colon carcinomas.

Sozzi et al. (1996) analyzed the FHIT gene structure and transcription in a large series of lung cancers of the small cell (SCLC) and nonsmall cell (NSCLC) type. FHIT transcripts from tumors and normal tissues were studied by RT-PCR. In 11 out of 14 SCLC tumors, abnormal-sized transcripts were found. Both normal- and abnormal-sized transcripts were present in 9 of these cases. In 18 out of 25 NSCLC tumors, abnormal transcripts were found; in these tumors, there were 1 or 2 abnormal-sized bands which were always accompanied by a normal-sized transcript. The authors postulated that the normal-sized transcripts reflected the presence of normal cells within the tumors. Sozzi et al. (1996) reported loss of heterozygosity for microsatellite markers internal to and flanking the FHIT locus. Eleven of 12 informative tumors that exhibited abnormal FHIT transcripts showed allelic loss at one or more of the loci tested. The authors postulated that in these tumors inactivation of the FHIT gene occurred by a mechanism of loss of one allele and altered expression of the remaining allele. They further postulated that loss of function of the FHIT gene could result in the constitutive accumulation of high levels of intracellular diadenosine tetraphosphate and the stimulation of DNA synthesis and proliferation. Sozzi et al. (1996) proposed that breakage in a fragile site-containing gene may occur as a consequence of physical, chemical, and biologic agents.

Rassool et al. (1996) investigated the structure of the common fragile site FRA3B and compared it to what is known about the structure of other fragile sites such as FRAXA (309550) and FRAXE (309548). They constructed a partial contig of genomic clones spanning 85 kb of genomic DNA in the 3p14.2 region using a recombinogenic pSV2neo plasmid as a means to select for common fragile sites. This contig lies approximately 350 kb from the t(3;8) constitutional rearrangement. Rare fragile sites are often associated with (CGG)n repeats. However, by sequencing and Southern analysis Rassool et al. (1996) found neither (CGG)n repeats nor other sequences associated with rare fragile sites within the 85-kb contig. Fluorescence in situ hybridization of genomic clones from this contig to metaphase chromosomes induced to express breaks showed hybridization adjoining the chromosome breaks, and occasionally the hybridization signal spanned the break. Rassool et al. (1996) concluded that (1) breakage at the common fragile site FRA3B may differ from breakage at rare fragile sites and (2) breakage at common fragile sites may occur at variable positions within a large region.

Virgilio et al. (1996) noted that head and neck cancers (HNSCC; 601400) represent 3% of all cancer in Western countries. In some geographic areas such as India, the proportion of cancer represented by these cancers is as high as 45%; 90 to 95% of head and neck cancers are of the squamous cell type. Tobacco and alcohol have been recognized as etiologic factors in these carcinomas. Several regions of loss of heterozygosity (LOH) have been identified in HNSCC. The 9p region presents the highest rate of genetic alteration (75%) and LOH has been related to the CDKN2 (600160) tumor suppressor gene. The second most frequent alteration involves 3p (45 to 55%). Virgilio et al. (1996) examined 26 HNSCC cell lines for deletions within the FHIT locus by Southern analysis, for allelic loss of specific exons of FHIT by fluorescence in situ hybridization, and for integrity of FHIT transcripts. Three of the 26 cell lines exhibited homozygous deletions within the FHIT gene, 55% (15 of 25) showed aberrant transcripts, and 65% (13 of 20) showed multiple cell populations with losses of different portions of the FHIT alleles. When the data were combined, 22 of 26 cell lines showed alterations of at least 1 allele of the FHIT gene. Thus, Virgilio et al. (1996) concluded that loss of FHIT function may be important in the development and/or progression of head and neck cancers.

The hypothesis that chromosomal fragile sites may be 'weak links' that result in hotspots for cancer-specific chromosome rearrangements was supported by the discovery that numerous cancer cell homozygous deletions and a familial translocations map within the FHIT gene, which encompasses the common fragile site FRA3B. By sequence analysis of 276 kb of the FRA3B/FHIT locus and 22 associated cancer cell deletion endpoints, Inoue et al. (1997) demonstrated that this locus is a frequent target of homologous recombination between long interspersed nuclear element (LINE) sequences resulting in FHIT gene internal deletions, probably as a result of carcinogen-induced damage at FRA3B fragile sites.

Barnes et al. (1996) found that FHIT protein is a 147-amino acid AP3A hydrolase (EC 3.6.1.29). The authors stated that the FHIT preferred substrate, AP3A (diadenosine 5-prime,5-triple prime-P(1),P(3)-triphosphate), and AP4A (see 602852) have been proposed to have various intracellular functions, including regulation of DNA replication and signaling stress responses. The conserved residues of the histidine triad are required for enzymatic activity.

Geurts et al. (1997) found that FHIT was involved in a translocation-derived fusion with HMGIC (600698), the causative gene in a variety of benign tumors.

Gemmill et al. (1998) studied further the family of Cohen et al. (1979) in which the constitutional reciprocal t(3;8) translocation was associated with early-onset, bilateral, and multifocal clear cell renal carcinoma. Previous studies had demonstrated that the 3p14.2 breakpoint interrupts the FHIT gene in its 5-prime noncoding region; however, that FHIT was causally related to renal or other malignancies was controversial. Gemmill et al. (1998) showed that the 8q24.1 breakpoint region encodes a 664-amino acid multiple membrane-spanning protein, TRC8 (603046), with similarity to the hereditary basal cell carcinoma/segment polarity gene 'patched' (PTCH; 601309). In the 3;8 translocation, TRC8 is fused to FHIT and is disrupted within the sterol-sensing domain. In contrast, the FHIT coding region is maintained and expressed. In a series of sporadic renal carcinomas, Gemmill et al. (1998) found an acquired TRC8 mutation. By analogy to 'patched,' TRC8 might function as a signaling receptor and other pathway members, to be defined, are mutation candidates in malignant diseases involving the kidney and thyroid.

Barnes, L. D.; Garrison, P. N.; Siprashvili, Z.; Guranowski, A.; Robinson, A. K.; Ingram, S. W.; Croce, C. M.; Ohta, M.; Huebner, K.: Fhit, a putative tumor suppressor in humans, is a dinucleoside 5-prime,5-triple prime-P(1),P(3)-triphosphate hydrolase. Biochemistry 35: 11529-11535, 1996.
Cohen, A. J.; Li, F. P.; Berg, S.; Marchetto, D. J.; Tsai, S.; Jacobs, S. C.; Brown, R. S.: Hereditary renal-cell carcinoma associated with chromosomal translocation. New Eng. J. Med. 301: 592-595, 1979.
Gemmill, R. M.; West, J. D.; Boldog, F.; Tanaka, N.; Robinson, L. J.; Smith, D. I.; Li, F.; Drabkin, H. A.: The hereditary renal cell carcinoma 3;8 translocation fuses FHIT to a patched-related gene, TRC8. Proc. Nat. Acad. Sci. 95: 9572-9577, 1998.
Geurts, J. M.; Schoenmakers, E. F.; Roijer, E.; Stenman, G.; Van de Ven, W. J. M.: Expression of reciprocal hybrid transcripts of HMGIC and FHIT in a pleomorphic adenoma of the parotid gland. Cancer Res. 57: 13-17, 1997.
Inoue, H.; Ishii, H.; Alder, H.; Snyder, E.; Druck, T.; Huebner, K.; Croce, C. M.: Sequence of the FRA3B common fragile region: implications for the mechanism of FHIT deletion. Proc. Nat. Acad. Sci. 94: 14584-14589, 1997.
Ohta, M.; Inoue, H.; Cotticelli, M. G.; Kastury, K.; Baffa, R.; Palazzo, J.; Siprashvili, Z.; Mori, M.; McCue, P.; Druck, T.; Croce, C. M.; Huebner, K.: The FHIT gene, spanning the chromosome 3p14.2 fragile site and renal carcinoma-associated t(3;8) breakpoint, is abnormal in digestive tract cancers. Cell 84: 587-597, 1996.
Rassool, F. V.; Le Beau, M. M.; Shen, M.-L.; Neilly, M. E.; Espinosa, R., III; Ong, S. T.; Boldog, F.; Drabkin, H.; McCarroll, R.; McKeithan, T. W.: Direct cloning of DNA sequences from the common fragile site region at chromosome band 3p14.2. Genomics 35: 109-117, 1996.
Sozzi, G.; Veronese, M. L.; Negrini, M.; Baffa, R.; Cotticelli, M. G.; Inoue, H.; Tornielli, S.; Pilotti, S.; De Gregorio, L.; Pastorino, U.; Pierotti, M. A.; Ohta, M.; Huebner, K.; Croce, C. M.: The FHIT gene at 3p14.2 is abnormal in lung cancer. Cell 85: 17-26, 1996.
Virgilio, L.; Shuster, M.; Gollin, S. M.; Veronese, M. L.; Ohta, M.; Huebner, K.; Croce, C. M.: FHIT gene alterations in head and neck squamous cell carcinomas. Proc. Nat. Acad. Sci. 93: 9770-9775, 1996.
Wang, N.; Perkins, K. L.: Involvement of band 3p14 in t(3;8) hereditary renal carcinoma. Cancer Genet. Cytogenet. 11: 479-481, 1984.

Other Markers, Proteins, and Genes

JJM NOTE

In preparing the *Catalog,* it was inevitable that some entries and genes would not fit well, even though many arbitrary decisions were made. Yet, this mixed group of cancer-associated genes and proteins surely contains important genes and even the beginnings of new classes of human cancer genes. The sort is by MIM number.

FRAGILE X MENTAL RETARDATION, AUTOSOMAL HOMOLOG; FXR1 (*600819)

Fragile X mental retardation is directly associated with the FMR1 gene at Xq27.3 (309550). FMR1 is an RNA binding protein and mutations in the gene likely result in the absence or reduced expression of the protein or a functionally impaired mutant protein. Siomi et al. (1995) cloned a cDNA related to FMR1 from a Xenopus laevis ovary library. This cDNA was then used to screen a HeLa cell library. A cDNA was identified that encodes a predicted protein of 621 amino acids. The gene, designated FXR1 (FMR1 cross-reacting relative) by them, shows 86% amino acid sequence identity to the KH RNA binding domains of FMR1 and 70% identity over the amino-terminal domain. However, the carboxyl domains are entirely unrelated. The authors showed in vitro that FXR1 protein has RNA binding properties similar to FMR1 and that both are present in the cytoplasm. RT-PCR analysis showed that many cell types express FXR1 mRNA. The authors mapped FXR1 to chromosome 12 using a somatic cell hybrid mapping panel and to 12q13 by fluorescence in situ hybridization. Cells from patients with fragile X syndrome which lack FMR1 expression make normal levels of FXR1 transcript. This argues that the autosomal gene cannot complement FMR1 and that the 2 proteins have different functions.

Siomi, M. C.; Siomi, H.; Sauer, W. H.; Srinivasan, S.; Nussbaum, R. L.; Dreyfuss, G.: FXR1, an autosomal homolog of the fragile X mental retardation gene. EMBO J. 14: 2401-2408, 1995.

ARGININE-RICH PROTEIN (*601916)

Shridhar et al. (1996) described a gene that they symbolized ARP for arginine-rich protein. The gene encodes a 234-amino acid, highly conserved arginine-rich protein and is located in 3p21.1. At the cytogenetic level, this region is frequently deleted in a variety of solid tumors, although not in pancreatic cancer. 1,2:Shridhar et al. (1996,1996) reported the presence of a specific mutation changing codon 50 from ATG to AGG (met50arg; 601916.0001) or deletion of codon 50 of the ARP gene in different tumor types. Shridhar et al. (1997) observed mutations involving codon 50 in 11 of 37 pancreatic tumors.

Shridhar, R.; Shridhar, V.; Rivard, S. Siegfried, J. M.; Pietraszkiewicz, H.; Ensley, J.; Pauley, R.; Grignon, D.; Sakr, W.; Miller, O. J.; Smith, D. I.: Mutations in the arginine-rich protein gene, in lung, breast, and prostate cancers, and in squamous cell carcinoma of the head and neck. Cancer Res. 56: 5576-5578, 1996.
Shridhar, V.; Rivard, S.; Shridhar, R.; Mullins, C.; Bostick, L.; Sakr, W.; Grignon, D.; Miller, O. J.; Smith, D. I.: A gene from human chromosomal band 3p21.1 encodes a highly conserved arginine-rich protein and is mutated in renal cell carcinomas. Oncogene 12: 1931-1939, 1996.
Shridhar, V.; Rivard, S.; Wang, X.; Shridhar, R.; Paisley, C.; Mullins, C.; Beirnat, L.; Dugan, M.; Sarkar, F.; Miller, O. J.; Vaitkevicius, V. K.; Smith, D. I.: Mutations in the arginine-rich protein gene (ARP) in pancreatic cancer. Oncogene 14: 2213-2216, 1997.

ONCOGENE GRO1; GRO1 (*155730)

Melanoma growth stimulatory activity is a mitogenic polypeptide secreted by human melanoma cells. The mature form is maximally 73 amino acids long. MGSA is structurally related to the platelet-derived beta-thromboglobulin. It is the product of the gene GRO isolated by Anisowicz et al. (1987). Horuk et al. (1993) indicated that structurally MGSA belongs to a superfamily of proteins that includes interleukin-8 (146928) and platelet factor-4 (173460). These proteins are involved in inflammatory processes.

Richmond et al. (1988) mapped the MGSA gene to 4q13-q21 by a combination of somatic cell and in situ hybridization. They pointed out that platelet factor 4 (173460) maps to the same region and that piebald trait (172800), a pigment anomaly, may also be coded there. Anisowicz et al. (1988) studied the expression of GRO in human foreskin fibroblasts and in mammary epithelial cells. Furthermore, they demonstrated by in situ hybridization that the GRO gene is on 4q21 in the human and on 1p5 in the Chinese hamster. Sakaguchi et al. (1989) demonstrated that the Mgsa locus in the mouse is on chromosome 5. Several genes affecting skin pigmentation are located on mouse chromosome 5.

Anisowicz, A.; Bardwell, L.; Sager, R.: Constitutive overexpression of a growth-regulated gene in transformed Chinese hamster and human cells. Proc. Nat. Acad. Sci. 84: 7188-7192, 1987.
Anisowicz, A.; Zajchowski, D.; Stenman, G.; Sager, R.: Functional diversity of GRO gene expression in human fibroblasts and mammary epithelial cells. Proc. Nat. Acad. Sci. 85: 9645-9649, 1988.
Horuk, R.; Yansura, D. G.; Reilly, D.; Spencer, S.; Bourell, J.; Henzel, W.; Rice, G.; Unemori, E.: Purification, receptor binding analysis, and biological characterization of human melanoma growth stimulating activity (MGSA): evidence for a novel MGSA receptor. J. Biol. Chem. 268: 541-546, 1993.
Richmond, A.; Balentien, E.; Thomas, H. G.; Flaggs, G.; Barton, D. E.; Spiess, J.; Bordoni, R.; Francke, U.; Derynck, R.: Molecular characterization and chromosomal mapping of melanoma growth stimulatory activity, a growth factor structurally related to beta-thromboglobulin. EMBO J. 7: 2025-2033, 1988.
Sakaguchi, A. Y.; Lalley, P. A.; Ghosh Choudhury, G.; Martinez, L.; Han, E. S.; Killary, A. M.; Naylor, S. L.; Wang, L.-M.: Mouse melanoma growth stimulatory activity gene (Mgsa) is polymorphic and syntenic with the W, Patch, Rumpwhite, and recessive spotting loci on chromosome 5. Genomics 5: 629-632, 1989.

MELANOMA-SPECIFIC CHONDROITIN SULFATE PROTEOGLYCAN (*155760)

Proteoglycans, or mucopolysaccharides, are composed of a core protein and multiple glycosaminoglycan chains covalently attached to it. Human malignant melanoma cells express specific chondroitin sulfate proteoglycans (mel-CSPG) on the surface both in vivo and in vitro. Rettig et al. (1986) mapped the gene encoding mel-CSPG to human chromosome 15 by study of rodent-human somatic cell hybrids with antibodies to the core glycoprotein. The mel-CSPG antigen, originally identified with monoclonal antibodies, consists of a 250-kD core glycoprotein and a number of characteristic glycosaminoglycan side chains. Rettig et al. (1986) showed that mel-CSPG is specifically induced by extracellular matrix (ECM). Melanocytes in normal skin show no detectable MCSPG but expression can be induced when the cells are cultured in the presence of cholera toxin and a tumor promoter. They found that 2 other surface antigens coded by human chromosome 15 (Rettig et al., 1984), namely, SV13 (a 105-kD glycoprotein; MSK15) and F23 (a 140-kD glycoprotein; MSK17), were not induced in these experiments.

By fluorescence in situ hybridization, Korenberg et al. (1993) assigned the AGC1 gene to 15q26, most likely in the subregion of 15q26.1. They predicted that the mouse homolog might be situated in the midportion of mouse chromosome 7 and noted with interest that the recessive mouse mutant skt ('small with kinky tail') maps within this region (Lane, 1988). This mutation causes small size and skeletal abnormalities; the aggrecan gene might be a candidate for the site of the mutation.

Just et al. (1993) likewise mapped the human aggrecan gene to 15q25-q26.2 by in situ hybridization and, because of similar mapping, suggested that the malignant melanoma-associated surface antigen is the same as AGC1. Absence of the core protein of a large proteoglycan was observed in cartilage of mice that were homozygous for the autosomal recessive gene cmd (cartilage matrix deficiency) (Kimata et al., 1981). This mutation maps to mouse chromosome 7 near the gene for mitochondrial isocitric dehydrogenase (Idh-2), which has its homolog, IDH2 (147650), on 15q. Thus, aggrecan may be the gene affected by the cmd mutation and is a likely candidate for the site of the mutation in a human skeletal dysplasia.

The cmd mutation in the mouse was the first known genetic disorder of proteoglycans identified in mammals. Watanabe et al. (1994) demonstrated a 7-bp deletion in exon 5 of the aggrecan gene leading to a premature termination codon in exon 6 as the cause of cmd in the mouse. Heterozygotes for the cmd mutation are apparently normal at birth. However, Watanabe et al. (1997) found that heterozygotes show 2 characteristic phenotypes: slight dwarfism and age-associated hyperlordosis, the anterior concavity in the curvature of the spine. Within 19 months of age, the animals exhibited spastic gait caused by misalignment of the cervical spine and died because of starvation. Histologic examination revealed a high incidence of herniation and degeneration of vertebral discs. Electron microscopy showed degeneration of disc chondrocytes in the heterozygotes. Watanabe et al. (1997) suggested that mutations in aggrecan may be demonstrated in humans with spinal degeneration.

Pluschke et al. (1996) reported the molecular cloning and nucleotide sequencing of cDNA encoding the entire core protein of human melanoma-associated chondroitin sulfate proteoglycan (MCSP) and provided its deduced amino acid sequence. The core protein gene contains an open reading frame of 2,322 amino acids and predicts a large extracellular domain, a hydrophobic transmembrane region, and a relatively short cytoplasmic tail. Northern blot analysis indicated that MCSP cDNA probes detect a single 8-kb RNA species expressed in human melanoma cell lines. A strong hybridization signal was observed only with melanoma cells and not with other human cancer cells or a variety of human fetal and adult tissues. The data indicated to Pluschke et al. (1996) that MCSP is an integral membrane chondroitin sulfate proteoglycan expressed by human malignant melanoma cells.

Nomenclature: Note that MCSP is also used as the gene symbol for mitochondrial capsule selenoprotein (601148).

Just, W.; Klett, C.; Vetter, U.; Vogel, W.: Assignment of the human aggrecan gene AGC1 to 15q25-q26.2 by in situ hybridization. Hum. Genet. 92: 516-518, 1993.

Kimata, K.; Barrach, H. J.; Brown, K. S.; Pennypacker, J. P.: Absence of proteoglycan core protein in cartilage from the cmd/cmd (cartilage deficiency) mouse. J. Biol. Chem. 256: 6961-6968, 1981.

Korenberg, J. R.; Chen, X. N.; Doege, K.; Grover, J.; Roughley, P. J.: Assignment of the human aggrecan gene (AGC1) to 15q26 using fluorescence in situ hybridization analysis. Genomics 16: 546-548, 1993.

Lane, P. W.: New mutants and linkages: small with kinky tail. Mouse Newsletter 80: 165, 1988.

Pluschke, G.; Vanek, M.; Evans, A.; Dittmar, T.; Schmid, P.; Itin, P.; Filardo, E. J.; Reisfeld, R. A.: Molecular cloning of a human melanoma-associated chondroitin sulfate proteoglycan. Proc. Nat. Acad. Sci. 93: 9710-9715, 1996.

Rettig, W. J.; Dracopoli, N. C.; Goetzger, T. A.; Spengler, B. A.; Biedler, J. L.; Oettgen, H. F.; Old, L. J.: Somatic cell genetic analysis of human cell surface antigens: chromosomal assignments and regulation of expression in rodent-human hybrid cells. Proc. Nat. Acad. Sci. 81: 6437-6441, 1984.

Rettig, W. J.; Real, F. X.; Spengler, B. A.; Biedler, J. L.; Old, L. J.: Human melanoma proteoglycan: expression in hybrids controlled by intrinsic and extrinsic signals. Science 231: 1281-1284, 1986.

Watanabe, H.; Kimata, K.; Line, S.; Strong, D.; Gao, L.; Kozak, C. A.; Yamada, Y.: Mouse cartilage matrix deficiency (cmd) caused by a 7 bp deletion in the aggrecan gene. Nature Genet. 7: 154-157, 1994.

Watanabe, H.; Nakata, K.; Kimata, K.; Nakanishi, I.; Yamada, Y.: Dwarfism and age-associated spinal degeneration of heterozygote cmd mice defective in aggrecan. Proc. Nat. Acad. Sci. 94: 6943-6947, 1997.

CATENIN, BETA-1 (116806)

Beta-catenin is an adherens junction protein. Adherens junctions (AJs; also called the zonula adherens) are critical for the establishment and maintenance of epithelial layers, such as those lining organ surfaces. AJs mediate adhesion between cells, communicate a signal that neighboring cells are present, and anchor the actin cytoskeleton. In serving these roles, AJs regulate normal cell growth and behavior. At several stages of embryogenesis, wound healing, and tumor cell metastasis, cells form and leave epithelia. This process, which involves the disruption and reestablishment of epithelial cell-cell contacts, may be regulated by the disassembly and assembly of AJs. AJs may also function in the transmission of the 'contact inhibition' signal, which instructs cells to stop dividing once an epithelial sheet is complete.

As reviewed by Peifer (1993), the AJ is a multiprotein complex assembled around calcium-regulated cell adhesion molecules called cadherins (e.g., 114020 and 114021). Cadherins are transmembrane proteins: the extracellular domain mediates homotypic adhesion with cadherins on neighboring cells, and the intracellular domain interacts with cytoplasmic proteins that transmit the adhesion signal and anchor the AJ to the actin cytoskeleton. These cytoplasmic proteins include the alpha-, beta-, and gamma-catenins. The beta-catenin gene, which was cloned by McCrea et al. (1991), shows no similarity in sequence to the genes for the alpha-catenins. The beta-catenin protein shares 70% amino acid identity with both plakoglobin (173325), which is found in desmosomes (another type of intracellular junction), and the product of the Drosophila segment polarity gene 'armadillo.' Armadillo is part of a multiprotein AJ complex in Drosophila that also includes some homologs of alpha-catenin and cadherin, and genetic studies indicate that it is required for cell adhesion and cytoskeletal integrity. The armadillo gene was originally identified as one of a group of segment polarity genes that regulate pattern formation of the Drosophila embryonic cuticle.

By fluorescence in situ hybridization (FISH), Kraus et al. (1994) mapped the CTNNB1 gene to 3p21, a region frequently affected by somatic alterations in a variety of tumors. Using PCR primers for the genomic amplification of beta-catenin sequences on the basis of homology to exon 4 of the Drosophila armadillo gene, they analyzed a panel of somatic cell hybrids to confirm the localization of the gene to human chromosome 3. Exclusion mapping of 3 hybrids carrying defined fragments of 3p allowed them to determine that the CTNNB1 locus is close to marker D3S2. Guenet et al. (1995) mapped the homologous gene, symbolized Catnb by them, to mouse chromosome 9 by analysis of interspecific backcrosses. Bailey et al. (1995) used FISH and PCR analysis of somatic cell hybrid DNAs to show that the CTNNB1 gene is located in the 3p22-p21 region. By FISH, van Hengel et al. (1995) assigned CTNNB1 to 3p22-p21.3. Trent et al. (1995) likewise localized the CTNNB1 gene to 3p22 by FISH. They stated that because APC-binding proteins (like beta-catenin) represent a 'downstream' modulator of APC activity, the chromosomal locus of such a protein might be expected to be a site involved in chromosome rearrangements in malignancy.

Nollet et al. (1996) showed that CTNNB1 has 16 exons and spans 23.2 kb. Alternative splicing within exon 16 produced a splice variant that is 159-bp shorter in the 3-prime untranslated region. The promoter region was shown to be GC-rich and to contain a TATA box. The authors demonstrated promoter activity in mouse epithelial cells for the 5-prime flanking region when it was linked to the reporter gene alkaline phosphatase.

Work by Korinek et al. (1997) and by Morin et al. (1997) established that the APC gene (175100), which is mutant in adenomatous polyposis of the colon, is a negative regulator of beta-catenin signaling. The APC protein normally binds to beta-catenin, which interacts with Tcf and Lef transcription factors. Korinek et al. (1997) cloned a gene they called hTcf-4, a Tcf family member that is expressed in colonic epithelium. The protein product (Tcf-4) transactivates transcription only when associated with beta-catenin. Nuclei of APC(-/-) colon carcinoma cells were found to contain a stable beta-catenin/Tcf-4 complex that was constitutively active, as measured by transcription of a Tcf reporter gene. Reintroduction of APC removed beta-catenin from Tcf-4 and abrogated the transcriptional activation. Korinek et al. (1997) concluded that constitutive transcription of Tcf target genes, caused by loss of APC function, may be a crucial event in the early transformation of colonic epithelium. Morin et al. (1997) likewise found that the protein products of mutant APC genes present in colorectal tumors were defective in downregulating transcriptional activation mediated by beta-catenin and T-cell transcription factor-4, now known as transcription factor-7-like-2 (TCF7L2; 602228). Furthermore, colorectal tumors with intact APC genes were found to contain activating mutations of beta-catenin that altered functionally significant phosphorylation sites. These results indicated that regulation of beta-catenin is critical to APC's tumor suppressive effect and that this regulation can be circumvented by mutations in either APC or beta-catenin. Morin et al. (1997) found a total of 3 tumors that contained CTNNB1 mutations that altered potential phosphorylation sites. Each mutation was somatic and appeared to affect only 1 of the 2 CTNNB1 alleles. Causative mutations were heterozygous. The authors hypothesized that the mutations might exert a dominant effect, rendering a fraction of cellula r beta-catenin insensitive to APC-mediated downregulation. Thus, disruption of APC-mediated regulation of beta-catenin/TCF-regulated transcription is critical for colorectal tumorigenesis. This is most commonly achieved by recessive inactivating mutations of both APC alleles, but can also be achieved by dominant mutations of CTNNB1 that render CRT insensitive to the effects of wildtype APC.

Ilyas et al. (1997) found 5 different mutations in 21 colorectal cancer cell lines (26%) from 19 patients: a 3-bp deletion from codon 45 and single-nucleotide missense mutations involving codons 33, 183, 245, and 287. All 23 cell lines studied had full-length beta-catenin protein that was detectable by Western blotting and that coprecipitated with E-cadherin. In 3 of the cell lines with CTNNB1 mutations, complexes of beta-catenin with alpha-catenin and APC were detectable; in 2 other cell lines, complexes were not detected.

Rubinfeld et al. (1997) detected abnormally high amounts of beta-catenin in 7 of 26 human melanoma cell lines. Unusual messenger RNA splicing and missense mutations in the CTNNB1 gene that result in stabilization of the protein were identified in 6 of the 7 lines, and the APC gene was altered or missing in 2 others. In the APC-deficient cells, ectopic expression of wildtype APC eliminated the excess beta-catenin. Cells with stabilized beta-catenin contained a constitutive beta-catenin/Lef-1 complex. Thus, Rubinfeld et al. (1997) concluded that genetic defects that result in upregulation of beta-catenin may play a role in melanoma progression.

In a review of hereditary cancer syndromes, Fearon (1997) presented a useful diagram illustrating how the APC protein regulates beta-catenin levels in normal cells, and how mutations in APC and CTNNB in cancer cell genes deregulate cell growth via T-cell transcription factor-7-like-2.

Bailey, A.; Norris, A. L.; Leek, J. P.; Clissold, P. M.; Carr, I. M.; Ogilvie, D. J.; Morrison, J. F. J.; Meredith, D. M.; Markham, A. F.: Yeast artificial chromosome cloning of the beta-catenin locus on human chromosome 3p21-22. Chromosome Res. 3: 201-203, 1995.

Fearon, E. R.: Human cancer syndromes: clues to the origin and nature of cancer. Science 278: 1043-1050, 1997.

Guenet, J.-L.; Simon-Chazottes, D.; Ringwald, M.; Kemler, R.: The genes coding for alpha and beta catenin (Catna1 and Catnb) and plakoglobin (Jup) map to mouse chromosomes 18, 9, and 11, respectively. Mammalian Genome 6: 363-366, 1995.

Ilyas, M.; Tomlinson, I. P. M.; Rowan, A.; Pignatelli, M.; Bodmer, W. F.: Beta-catenin mutations in cell lines established from human colorectal cancers. Proc. Nat. Acad. Sci. 94: 10330-10334, 1997.

Korinek, V.; Barker, N.; Morin, P. J.; van Wichen, D.; de Weger, R.; Kinzler, K. W.; Vogelstein, B.; Clevers, H.: Constitutive transcriptional activation by a beta-catenin-Tcf complex in APC-/- colon carcinoma. Science 275: 1784-1787, 1997.

Kraus, C.; Liehr, T.; Hulsken, J.; Behrens, J.; Birchmeier, W.; Grzeschik, K.-H.; Ballhausen, W. G.: Localization of the human beta-catenin gene (CTNNB1) to 3p21: a region implicated in tumor development. Genomics 23: 272-274, 1994.

McCrea, P. D.; Turck, C. W.; Gumbiner, B.: A homolog of the armadillo protein in Drosophila (plakoglobin) associated with E-cadherin. Science 254: 1359-1361, 1991.

Morin, P. J.; Sparks, A. B.; Korinek, V.; Barker, N.; Clevers, H.; Vogelstein, B.; Kinzler, K. W.: Activation of beta-catenin-Tcf signaling in colon cancer by mutations in beta-catenin or APC. Science 275: 1787-1790, 1997.

Nollet, F.; Berx, G.; Molemans, F.; van Roy, F.: Genomic organization of the human beta-catenin gene (CTNNB1). Genomics 32: 413-424, 1996.

Peifer, M.: Cancer, catenins, and cuticle pattern: a complex connection. Science 262: 1667-1668, 1993.

Rubinfeld, B.; Robbins, P.; El-Gamil, M.; Albert, I.; Porfiri, E.; Polakis, P.: Stabilization of beta-catenin by genetic defects in melanoma cell lines. Science 275: 1790-1792, 1997.

Trent, J. M.; Wiltshire, R.; Su, L.-K.; Nicolaides, N. C.; Vogelstein, B.; Kinzler, K. W.: The gene for the APC-binding protein beta-catenin (CTNNB1) maps to chromosome 3p22, a region frequently altered in human malignancies. Cytogenet. Cell Genet. 71: 343-344, 1995.

van Hengel, J.; Nollet, F.; Berx, G.; van Roy, N.; Speleman, F.; van Roy, F.: Assignment of the human beta-catenin gene (CTNNB1) to 3p22-p21.3 by fluorescence in situ hybridization. Cytogenet. Cell Genet. 70: 68-70, 1995.

TRANSCRIPTION FACTOR 7-LIKE 2 (*602228)

Inactivation of the APC gene (175100) in colorectal cancers allows beta-catenin (CTNNB1; 116806) to accumulate and complex with the TCF4 transcription factor, thereby activating the expression of TCF4-regulated genes (Korinek et al., 1997; Morin et al., 1997). That gene activation by the beta-catenin/TCF4 complex is a critical event in cancer development is indicated by the fact that a subset of colorectal cancers that lack somatic mutations in APC show somatic mutations in the beta-catenin gene. These mutations are presumed to render beta-catenin insensitive to regulation by APC and GSK3. Consequently, beta-catenin accumulates and activates TCF4-regulated genes (Morin et al., 1997).

To study the physiologic role of Tcf4 (which is encoded by the Tcf7l2 gene), Korinek et al. (1998) disrupted Tcf7l2 by homologous recombination. The homozygous null mice died shortly after birth. A single histopathologic abnormality was observed. An apparently normal transition of intestinal endoderm into epithelium occurred at approximately embryonic day (E) 14.5. However, no proliferative compartments were maintained in the prospective crypt regions between the villi. As a consequence, the neonatal epithelium was composed entirely of differentiated, nondividing villus cells. Korinek et al. (1998) concluded that the genetic program controlled by Tcf7l2 maintains the crypt stem cells of the small intestine. The constitutive activity of Tcf4 in APC-deficient epithelial cells may contribute to their malignant transformation by maintaining stem-cell characteristics.

Korinek, V.; Barker, N.; Moerer, P.; van Donselaar, E.; Huls, G.; Peters, P. J.; Clevers, H.: Depletion of epithelial stem-cell compartments in the small intestine of mice lacking Tcf-4. Nature Genet. 19: 379-383, 1998.

Korinek, V.; Barker, N.; Morin, P. J.; van Wichen, D.; de Weger, R.; Kinzler, K. W.; Vogelstein, B.; Clevers, H.: Constitutive transcriptional activation by a beta-catenin-Tcf complex in APC-/- colon carcinoma. Science 275: 1784-1787, 1997.

Morin, P. J.; Sparks, A. B.; Korinek, V.; Barker, N.; Clevers, H.; Vogelstein, B.; Kinzler, K. W.: Activation of beta-catenin-Tcf signaling in colon cancer by mutations in beta-catenin or APC. Science 275: 1787-1790, 1997.

MYELOPROLIFERATIVE LEUKEMIA VIRUS ONCOGENE (*159530)

The myeloproliferative leukemia virus is a murine retrovirus which causes acute leukemia. Sequence analysis identified a novel MPLV-specific sequence of 623 nucleotides, for which the name v-mpl was proposed (Penciolelli et al., 1987). It had no homology with previously described murine leukemia virus-related sequences or to known oncogenes or growth factors. Southern DNA blotting and hybridization with v-mpl-specific probes indicated that v-mpl has a cellular counterpart phylogenetically conserved in mammalian species, including man. By means of in situ hybridization, Le Coniat et al. (1989) mapped the sequence to human 1p34. Vigon et al. (1992) cloned the human homolog of the v-mpl oncogene and found that it showed striking homologies with members of the hematopoietin receptor superfamily. They obtained 2 types of clones, termed MPLP and MPLK, which had the same 5-prime extremity but differed in their 3-prime ends. The resulting deduced polypeptides were composed of a common extracellular domain with a putative signal sequence and a common transmembrane domain but differed in their cytoplasmic domain after a stretch of 9 common amino acids. The extracellular domain of MPL contains the consensus sequences described for members of the hematopoietin receptor superfamily which includes IL5R (147851), IL3RA (308385), IL4R (147781), IL7R (146661), IL2RB (146710), erythropoietin receptor (EPOR; 133171), IL6R (147880), GMCSF receptor (CSF2R; 306250), and CSF3R (138971). It also showed similarities to the growth hormone receptor (GHR; 262500) and the prolactin receptor (PRLR; 176761).

Mignotte et al. (1994) demonstrated that the MPL gene contains 12 exons distributed over 17 kb of DNA. Each of the 2 'cytokine receptor domains' of MPL is encoded by a set of 4 exons, the transmembrane by a single exon, and the cytoplasmic domain by 2 exons. They also described how 3 types of mRNA, encoding different proteins, are generated: the major species contains all 12 exons, whereas mRNAs encoding a protein with a smaller cytoplasmic domain are produced by termination of the transcript within intron 10, and mRNAs encoding a putative soluble form of the MPL protein lack exons 9 and 10. The promoter region is GC-rich and contains putative binding sites for proteins of the ETS and GATA families.

To investigate the role and specificity of MPL as the receptor for thrombopoietin (TPO; 600044), Gurney et al. (1994) generated MPL-deficient mice by gene targeting. The homozygous deficient mice had an 85% decrease in their number of platelets and megakaryocytes but had normal amounts of other hematopoietic cell types. These mice also had increased concentrations of circulating TPO. These results showed that MPL specifically regulates megakaryocytopoiesis and thrombopoiesis through activation by its ligand TPO.

Kimura et al. (1998) found that mice lacking Mpl have hematopoietic stem cell deficiencies that are not limited to the megakaryocytic lineage. Their findings imply that TPO, signaling through MPL, plays a vital physiologic role in the regulation of hematopoietic stem cell production and function.

Recombinant human TPO promises to be useful for treatment of thrombocytopenia resulting from chemotherapy and bone marrow transplantation. The mechanism by which TPO activates the TPO receptor (TPOR) appears to be somewhat similar to that of other hematopoietic cytokines that bind and induce receptor homodimerization. From recombinant peptide libraries, Cwirla et al. (1997) identified 2 families of small peptides that bind to the human TPOR and compete with the binding of the natural ligand TPO. The sequences of these peptides were not found in the primary sequence of TPO. Screening libraries of variants of one of these families under affinity-selected conditions yielded a 14-amino acid peptide with high affinity that stimulated the proliferation of a TPO-responsive cell line. Dimerization of this peptide by a C-terminal linkage to a lysine branch produced a compound with a high-medium effective concentration, EC(50), which was equipotent to the 332-amino acid natural cytokine (TPO) in cell-based assays. The peptide dimer also stimulated the in vitro proliferation and maturation of megakaryocytes from human marrow cells and promoted an increase in platelet count when administered to normal mice.

The cause of polycythemia vera (263300), which originates from a multipotent hematopoietic progenitor cell, is unknown. TPO is a hematopoietic growth factor that regulates the production of multipotent hematopoietic progenitor cells and platelets. To evaluate the possibility that an abnormality in TPO-mediated signal transduction might be involved in the pathogenesis of polycythemia vera, Moliterno et al. (1998) studied TPO-induced tyrosine phosphorylation of proteins and the expression of MPL in platelets from patients with this disorder. MPL was markedly reduced or absent in 34 of 34 patients with polycythemia vera and in 13 of 14 patients with idiopathic myelofibrosis. They concluded that reduced expression of MPL is characteristic of polycythemia vera and of idiopathic myelofibrosis. The abnormality appears to distinguish polycythemia vera from other forms of erythrocytosis.

Cwirla, S. E.; Balasubramanian, P.; Duffin, D. J.; Wagstrom, C. R.; Gates, C. M.; Singer, S. C.; Davis, A. M.; Tansik, R. L.; Mattheakis, L. C.; Boytos, C. M.; Schatz, P. J.; Baccanari, D. P.; Wrighton, N. C.; Barrett, R. W.; Dower, W. J.: Peptide agonist of the thrombopoietin receptor as potent as the natural cytokine. Science 276: 1696-1699, 1997.

Gurney, A. L.; Carver-Moore, K.; de Sauvage, F. J.; Moore, M. W.: Thrombocytopenia in c-mpl-deficient mice. Science 265: 1445-1447, 1994.

Kimura, S.; Roberts, A. W.; Metcalf, D.; Alexander, W. S.: Hematopoietic stem cell deficiencies in mice lacking c-Mpl, the receptor for thrombopoietin. Proc. Nat. Acad. Sci. 95: 1195-1200, 1998.

Le Coniat, M.; Souyri, M.; Vigon, I.; Wendling, F.; Tambourin, P.; Berger, R.: The human homolog of the myeloproliferative virus maps to chromosome band 1p34. Hum. Genet. 83: 194-196, 1989.

Mignotte, V.; Vigon, I.; Boucher de Crevecoeur, E.; Romeo, P.-H.; Lemarchandel, V.; Chretien, S.: Structure and transcription of the human c-mpl gene (MPL). Genomics 20: 5-12, 1994.

Moliterno, A. R.; Hankins, W. D.; Spivak, J. L.: Impaired expression of the thrombopoietin receptor by platelets from patients with polycythemia vera. New Eng. J. Med. 338: 572-580, 1998.

Penciolelli, J. F.; Wendling, F.; Robert-Lezenes, J.; Barque, J. F.; Tambourin, P.; Gisselbrecht, S.: Genetic analysis of myeloproliferative leukemia virus, a novel acute leukemogenic replication-defective retrovirus. J. Virol. 61: 579-583, 1987.

Vigon, I.; Mornon, J.-P.; Cocault, L.; Mitjavila, M.-T.; Tambourin, P.; Gisselbrecht, S.; Souyri, M.: Molecular cloning and characterization of MPL, the human homolog of the v-mpl oncogene: identification of a member of the hematopoietic growth factor receptor superfamily. Proc. Nat. Acad. Sci. 89: 5640-5644, 1992.

MYC PROMOTER-BINDING PROTEIN (*602456)

By screening a HeLa cell cDNA library, Ray and Miller (1991) cloned a novel gene, called MPB1, that binds to the promoter region of the c-myc (190080) gene near the TATA box. The MPB1 gene predicts a 336-amino acid polypeptide with a molecular mass of 37 kD. It does not contain any known DNA-binding motif. Although MPB1 shares 70% homology with human alpha-enolase (172430) at the protein level, it has no detectable enolase activity. Ray and Miller (1991) found that the MPB1 protein represses transcription from the human c-myc promoter.

White et al. (1997) used human-rodent somatic cell hybrids to map the MPB1 gene to chromosome 1pter-p35.

Ray, R.; Miller, D. M.: Cloning and characterization of a human c-myc promoter-binding protein. Molec. Cell. Biol. 11: 2154-2161, 1991.

White, R. A.; Adkison, L. R.; Dowler, L. L.; Ray, R. B.: Chromosomal localization of the human gene encoding c-myc promoter-binding protein (MPB1) to chromosome 1p35-pter. Genomics 39: 406-408, 1997.

NONMETASTATIC CELLS 1, PROTEIN EXPRESSED IN (*156490)

Steeg et al. (1988) gave the name NM23 to a gene they cloned from a murine melanoma cell line. Expression of the gene correlated inversely with metastatic potential. NM23 RNA levels are highest in cell lines with low metastatic potential. RNA levels did not correlate with cell sensitivity to host immune responses. The authors therefore hypothesized that expression of this gene may be associated with intrinsic aggressiveness of the line. Several observations suggested that NM23 activity may be correlated with inhibition of the tumor metastatic process. Bevilacqua et al. (1989) concluded that NM23 RNA levels are differentially expressed in human breast tumors and that low NM23 RNA levels are associated with histopathologic indications of high metastatic potential. The product of the NM23 gene is a nucleoside diphosphate kinase, which has been designated p19/nm23. Keim et al. (1992) presented evidence that the expression of the gene is related to cell proliferative activity.

Rosengard et al. (1989) found that the human NM23 protein has sequence homology over the entire translated region with a developmentally related protein in Drosophila encoded by the 'abnormal wing discs' (awd) gene. Mutations in awd cause abnormal tissue morphology and necrosis and widespread aberrant differentiation in Drosophila, analogous to the changes in malignant progression. NM23 (also symbolized NME1) was assigned to 17q12-q21 by in situ hybridization, by analysis of somatic cell hybrids, and by linkage studies with markers in the CEPH panel (Leone et al., 1991). Somatic allelic deletion of the gene was observed in DNA from human breast, renal, colorectal, and lung carcinomas. In several cases the deletion of NM23 was independent of other chromosome 17 deletions. There is evidence, furthermore, for at least 1 more NM23 gene; see NME2 (156491). Eddy et al. (1991) mapped the NME1 gene to 17p11-qter by analysis of human-mouse somatic cell hybrid DNAs. Varesco et al. (1992) mapped the gene to 17q22 by in situ hybridization. They also demonstrated a 2-allele polymorphism with BglII. By linkage studies using markers for the NME1 locus, Sudbrak et al. (1993) demonstrated localization of NME1 on 17q. See also NME3 (601817) and NME4 (601818).

Postel et al. (1993) reported evidence suggesting that the protein encoded by 1 of the 2 closely related NM23 genes (see NME2; 156491) may be a transcription factor. The gene that may be turned on is the MYC oncogene (190080). Although a dozen DNA binding proteins had been identified for the MYC gene, only one, called PuF (for purine-binding factor, because the DNA sequence it recognizes has a high content of purine bases), was known to regulate MYC transcription in vitro. PuF was identified by Postel and her colleagues (Postel et al., 1989; Postel, 1992) as a partially purified HeLa cell (human cervical carcinoma) factor that binds to a nuclease-hypersensitive element (NHE) at positions -142 to -115 of the human MYC P1 promoter and is necessary for efficient P1 and P2 transcription initiation in vitro. Postel et al. (1993) identified the human gene encoding PuF by screening a cervical carcinoma cell cDNA library with a DNA fragment containing PuF binding sites. DNA sequence analysis of recombinant PuF demonstrated perfect identity with the NM23 gene. Although no clear correlation has yet been established between overexpression of MYC and tumor metastasis, the inverse relation between MYC expression and cell differentiation is well documented. Furthermore, Okabe-Kado et al. (1992) identified a differentiation inhibiting factor in the mouse myeloid leukemia cells as a murine homolog of NM23. Findings leave the conundrum of how NM23 can be both a tumor suppressor and an activator of MYC. Gilles et al. (1991) showed that nucleoside diphosphate kinases A and B are identical to NM23-H1 and NM23-H2, respectively. The genes NME1 and NME2 encode 2 polypeptide chains that are responsible for heterogeneity of the hexameric enzyme. Dooley et al. (1994) isolated and sequenced genomic PCR fragments that included the entire coding region of the NME1 gene, revealing that the gene consists of 5 exons and 4 introns spanning 8.5 kb. Comparison with the homologous gene in the r at showed that the exon-intron boundaries are well conserved between the 2 species.

Whereas reduced expression of NM23 is associated with a high potential for metastasis in some tumor types, its expression is increased in aggressive neuroblastoma (256700). Chang et al. (1994) looked for mutations in the NM23 gene in 24 primary neuroblastoma tumors at different stages of the disease by use of PCR combined with SSCP analysis and Southern blotting. They found an acquired ser120 substitution in 6 of 28 advanced tumors but in none of 22 limited-stage tumors. They indicated that the mutant enzyme still retains its catalytic activity but is more susceptible to the denaturation.

MacDonald et al. (1996) performed site-directed mutagenesis on the NM23-H1 gene and transfected the mutant DNAs into human breast carcinoma cells. They found that mutations of proline-96 or serine-120 reduce the ability of NM23-H1 to inhibit cell motility. In the wildtype protein, serine-120 is a site of phosphorylation; proline-96 is part of a loop of unknown function.

Bevilacqua, G.; Sobel, M. E.; Liotta, L. A.; Steeg, P. S.: Association of low nm23 RNA levels in human primary infiltrating ductal breast carcinomas with lymph node involvement and other histopathological indicators of high metastatic potential. Cancer Res. 49: 5185-5190, 1989.
Chang, C. L.; Zhu, X.; Thoraval, D. H.; Ungar, D.; Rawwas, J.; Hora, N.; Strahler, J. R.; Hanash, S. M.; Radany, E.: nm23-H1 mutation in neuroblastoma. (Letter) Nature 370: 335-336, 1994.
Dooley, S.; Seib, T.; Engel, M.; Theisinger, B.; Janz, H.; Piontek, K.; Zang, K.-D.; Welter, C.: Isolation and characterization of the human genomic locus coding for the putative metastasis control gene nm23-H1. Hum. Genet. 93: 63-66, 1994.
Eddy, R. L.; Mendola, C. E.; Fairhurst, J. L.; Shows, T. B.; O'Hara, B.; Kovesdi, I.; Backer, J. M.: Metastasis suppressor gene is mapped to human chromosome 17p11-qter. (Abstract) Cytogenet. Cell Genet. 58: 2005 only, 1991.
Gilles, A.-M.; Presecan, E.; Vonica, A.; Lascu, I.: Nucleoside diphosphate kinase from human erythrocytes: structural characterization of the two polypeptide chains responsible for heterogeneity of the hexameric enzyme. J. Biol. Chem. 266: 8784-8789, 1991.
Keim, D.; Hailat, N.; Melhem, R.; Zhu, X. X.; Lascu, I.; Veron, M.; Strahler, J.; Hanash, S. M.: Proliferation-related expression of p19/nm23 nucleoside diphosphate kinase. J. Clin. Invest. 89: 919-924, 1992.
Leone, A.; McBride, O. W.; Weston, A.; Wang, M. G.; Anglard, P.; Cropp, C. S.; Goepel, J. R.; Lidereau, R.; Callahan, R.; Marston Linehan, W.; Rees, R. C.; Harris, C. C.; Liotta, L. A.; Steeg, P. S.: Somatic allelic deletion of nm23 in human cancer. Cancer Res. 51: 2490-2493, 1991.
MacDonald, N. J.; Freije, J. M. P.; Stracke, M. L.; Manrow, R. E.; Steeg, P. S.: Site-directed mutagenesis of nm23-H1: mutation of proline 96 or serine 120 abrogates its motility inhibitory activity upon transfection into human breast carcinoma cells. J. Biol. Chem. 271: 25107-25116, 1996.
Okabe-Kado, J.; Kasukabe, T.; Honma, Y.; Hayashi, M.; Henzel, W. J.; Hozumi, M.: Identity of a differentiation inhibiting factor for mouse myeloid leukemia cells with NM23/nucleoside diphosphate kinase. Biochem. Biophys. Res. Commun. 182: 987-994, 1992.
Postel, E. H.: Modulation of c-myc transcription by triple helix formation. Ann. N.Y. Acad. Sci. 660: 57-63, 1992.
Postel, E. H.; Berberich, S. J.; Flint, S. J.; Ferrone, C. A.: Human c-myc transcription factor PuF identified as nm23-H2 nucleoside diphosphate kinase, a candidate suppressor of tumor metastasis. Science 261: 478-480, 1993.
Postel, E. H.; Mango, S. E.; Flint, S. J.: A nuclease-hypersensitive element of the human c-myc promoter interacts with a transcription initiation factor. Molec. Cell. Biol. 9: 5123-5133, 1989.
Rosengard, A. M.; Krutzsch, H. C.; Shearn, A.; Biggs, J. R.; Barker, E.; Margulies, I. M. K.; King, C. R.; Liotta, L. A.; Steeg, P. S.: Reduced nm23/awd protein in tumour metastasis and aberrant Drosophila development. Nature 342: 177-180, 1989.
Steeg, P. S.; Bevilacqua, G.; Kopper, L.; Thorgeirsson, U. P.; Talmadge, J. E.; Liotta, L. A.; Sobel, M. E.: Evidence for a novel gene associated with a low tumor metastatic potential. J. Nat. Cancer Inst. 80: 200-204, 1988.
Sudbrak, R.; Golla, A.; Hogan, K.; Powers, P.; Gregg, R.; Du Chesne, I.; Lehmann-Horn, F.; Deufel, T.: Exclusion of malignant hyperthermia susceptibility (MHS) from a putative MHS2 locus on chromosome 17q and of the alpha-1, beta-1, and gamma subunits of the dihydropyridine receptor calcium channel as candidates for the molecular defect. Hum. Molec. Genet. 2: 857-862, 1993.
Varesco, L.; Caligo, M. A.; Simi, P.; Black, D. M.; Nardini, V.; Casarino, L.; Rocchi, M.; Ferrara, G.; Solomon, E.; Bevilacqua, G.: The NM23 gene maps to human chromosome band 17q22 and shows a restriction fragment length polymorphism with BglII. Genes Chromosomes Cancer 4: 84-88, 1992.

RAS P21 PROTEIN ACTIVATOR 1 (*139150)

Trahey et al. (1988) purified guanosine triphosphatase-activating protein from placenta and obtained internal amino acid sequence of the protein from which they cloned 2 classes of cDNA. One class predicted a protein with molecular weight similar to purified GAP and corresponded to the human equivalent of bovine GAP cDNA. The other predicted a smaller protein with a different NH(2) terminal sequence, presumably the result of differential splicing. Both types of cDNAs produced protein with GAP activity. GAP is an effector of RAS oncogene action. Hsieh and Francke (1989) determined the chromosomal location of the GAP gene in mouse and man by Southern blot analyses of interspecies hybrid DNAs. The gene was assigned to 5q, excluding region 5q13.3-q15, in man and to mouse chromosome 13. By study of recombinant inbred strains of mice, they determined that the gene is located at the distal end of mouse chromosome 13. By the combination of somatic cell hybrid analysis and in situ hybridization, Hsieh et al. (1989) assigned the GAP gene to 5q13.3. Lemons et al. (1990) localized the human RASA (GAP) gene to 5q13-q15 by Southern analysis of somatic cell hybrids and by in situ hybridization. See also RASA2 (601589).

The RAS gene family encodes membrane-associated, guanine nucleotide-binding proteins (p21) that are involved in the control of cellular proliferation and differentiation. Similar to other guanine-binding proteins (such as the heterotrimeric G proteins), the RAS proteins cycle between an active guanosine-triphosphate (GTP) bound form and an inactive, guanosine-diphosphate (GDP) bound form. The weak intrinsic GTPase activity of RAS proteins is greatly enhanced by the action of GTPase-activating proteins (GAPs). Point mutations in RAS genes ('activating' or oncogenic mutants) decrease the intrinsic GTPase activity of RAS and are insensitive to stimulation by GAPs. This suggested to Friedman et al. (1993) that at least some of the transforming activity of mutant RAS is conferred by the RAS protein being constitutively activated in its GTP-bound state. Mutations in GAP that ablate its ability to down-regulate RAS might result in a similar phenotype. To test this hypothesis, Friedman et al. (1993) analyzed 188 human tumor samples for mutations within the catalytic domain of the GAP gene and for mutations within its C-terminal SH2 region. Although no mutations could be demonstrated in the catalytic domain, 3 different nonsense mutations were observed in basal cell carcinomas. The region in which the mutations were clustered is A/T rich, raising the possibility that UV radiation is a contributing factor. The 3 mutations were found in the first 5 tumors examined. No abnormality was found in 16 other basal cell carcinomas. Thus, the apparent prevalence of GAP mutation was about 14% (3/21). The tumors analyzed included a great variety, including cancers of thyroid, lung, breast, colon, and pancreas. No GAP mutation was found in any of these. Mitsudomi et al. (1994) could not demonstrate mutations in the catalytic domain of the GAP gene in human lung cancer cell lines.

Friedman, E.; Gejman, P. V.; Martin, G. A.; McCormick, F.: Nonsense mutations in the C-terminal SH2 region of the GTPase activating protein (GAP) gene in human tumours. Nature Genet. 5: 242-247, 1993.
Hsieh, C. L.; Francke, U.: The gene for GTPase activating protein (GAP) is on human chromosome 5q and mouse chromosome 13. (Abstract) Cytogenet. Cell Genet. 51: 1016 only, 1989.
Hsieh, C. L.; Vogel, U. S.; Dixon, R. A.; Francke, U.: Chromosome localization and cDNA sequence of murine and human genes for ras p21 GTPase activating protein (GAP). Somat. Cell Molec. Genet. 15: 579-590, 1989.
Lemons, R. S.; Espinosa, R., III; Rebentisch, M.; McCormick, F.; Ladner, M.; Le Beau, M. M.: Chromosomal localization of the gene encoding GTPase-activating protein (RASA) to human chromosome 5, bands q13-q15. Genomics 6: 383-385, 1990.
Mitsudomi, T.; Friedman, E.; Gejman, P. V.; McCormick, F.; Gazdar, A. F.: Genetic analysis of the catalytic domain of the GAP gene in human lung cancer cell lines. Hum. Genet. 93: 27-31, 1994.
Trahey, M.; Wong, G.; Halenbeck, R.; Rubinfeld, B.; Martin, G. A.; Ladner, M.; Long, C. M.; Crosier, W. J.; Watt, K.; Koths, K.; McCormick, F.: Molecular cloning of two types of GAP complementary DNA from human placenta. Science 242: 1697-1700, 1988.

RAS-RESPONSIVE ELEMENT BINDING PROTEIN 1 (*602209)

In several neuroendocrine cell types, the RAS/RAF signal transduction cascade results in cell differentiation (see HRAS; 190020). This signal transduction cascade involves inactivation of several tiers of protein kinases, including members of the RAF (e.g., RAF1; 164760), MAPKK, and MAPK families (see 176872). Ultimately, activation of these protein kinases results in the activation of transcription factors responsible for changes in cell phenotype. In the human medullary thyroid carcinoma cell line TT, an activated HRAS oncogene or activation of the RAF1 gene induces a program of neuroendocrine differentiation (Carson et al., 1995). One of the hallmarks of this differentiation process is an increase in the expression of the calcitonin gene (114130). Thiagalingam et al. (1996) reported the DNA affinity cloning of a RAS-responsive element binding protein, RREB1, from the TT cell line. RREB1, a zinc finger transcription factor, was shown to bind specifically to the distal RAS-responsive element (RRE) in the calcitonin gene promoter and to augment the RAS/RAF-mediated transcriptional response of that promoter. Thus, they concluded that it may be involved in RAS/RAF-mediated cell differentiation. Thiagalingam et al. (1997) mapped the RREB1 gene to 6p25 by 3 mapping methods: PCR analysis of a somatic cell hybrid panel, analysis of the Stanford G3 radiation hybrid mapping panel, and fluorescence in situ hybridization.

Carson, E. B.; McMahon, M.; Baylin, S. B.; Nelkin, B. D.: Ret gene silencing is associated with Raf-1-induced medullary thyroid carcinoma cell differentiation. Cancer Res. 55: 2048-2052, 1995.
Thiagalingam, A.; De Bustros, A.; Borges, M.; Jasti, R.; Compton, D.; Diamond, L.; Mabry, M.; Ball, D. W.; Baylin, S. B.; Nelkin, B. D.: RREB-1, a novel zinc finger protein, is involved in the differentiation response to Ras in human medullary thyroid carcinomas. Molec. Cell. Biol. 16: 5335-5345, 1996.
Thiagalingam, A.; Lengauer, C.; Baylin, S. B.; Nelkin, B. D.: RREB1, a Ras responsive element binding protein, maps to human chromosome 6p25. Genomics 45: 630-632, 1997.

RAS-RELATED NUCLEAR PROTEIN (*601179)

RAN (Ras-related nuclear protein) is a small GTP binding protein belonging to the RAS superfamily (see 190020) that is essential for the translocation of RNA and proteins through the nuclear pore complex (Ren et al., 1993). The RAN protein is also involved in control of DNA synthesis and of cell cycle progression. Drivas et al. (1990) identified a human cDNA for RAN by screening a human teratocarcinoma library with a mix of oligonucleotide probes corresponding to a conserved domain in the RAS genes. Ren et al. (1993) showed that nuclear localization of RAN requires the presence of RCC1 (179710) and that mutations in RAN expected to disrupt GTP hydrolysis led to a disruption of DNA synthesis. Because of its many functions, it is likely that RAN interacts with several other proteins (see 601180 and 601181).

Coutavas et al. (1994) showed that 2 distinct, but closely related, Ran transcripts from separate loci are present in the mouse, 1 of which is specific to the testis.

Coutavas, E. E.; Hsieh, C. M.; Ren, M.; Drivas, G. T.; Rush, M. G.; D'Eustachio, P.: Tissue-specific expression of Ran isoforms in the mouse. Mammalian Genome 5: 623-628, 1994.
Drivas, G. T.; Shih, A.; Coutavas, E.; Rush, M. G.; D'Eustachio, P.: Characterization of four novel ras-like genes expressed in a human teratocarcinoma cell line. Molec. Cell. Biol. 10: 1793-1797, 1990.
Ren, M.; Drivas, G.; D'Eustachio, P.; Rush, M. G.: Ran/TC4: a small nuclear GTP-binding protein that regulates DNA synthesis. J. Cell Biol. 120: 313-323, 1993.

RAS-ASSOCIATED PROTEIN; RAB1 (*179508)

RAB proteins have been postulated to regulate the targeting and fusion of membranous vesicles during organelle assembly and transport. RAB proteins undergo regulated exchange of GTP for GDP, and they slowly hydrolyze the bound GTP in a reaction that is thought to regulate the timing and unidirectional nature of these assembly events. All of the known RAB proteins terminate in sequences such as cys-X-cys (e.g., RAB3A, 179490), cys-cys (e.g., RAB1A), or a closely related sequence, and all are believed to be geranylgeranylated. From a human pheochromocytoma cDNA library, Zahraoui et al. (1989) isolated 7 cDNA clones corresponding to genes encoding the ras-associated GTP-binding proteins.

The 'wobbler' spinal muscular atrophy gene (wr) maps to proximal mouse chromosome 11, tightly linked to Rab1 and Glns-ps1, an intronless pseudogene of the glutamine synthetase gene (138290). Wedemeyer et al. (1996) used these markers to construct a 1.3-Mb YAC contig of the Rab1 region on mouse chromosome 11. Two overlapping YACS were identified that contained a 150-kb region of human chromosome 2p, comprising the RAB1 locus as well as a newly discovered STS (AHY1.1) and a trapped exon (ETG1.1). The region was mapped to 2p14-p13.4 using somatic cell hybrids and a radiation hybrid panel, thus extending a known region of conserved synteny between mouse chromosome 11 and human 2p. The gene for limb-girdle muscular dystrophy type 2B (LGMD2B; 253601) maps to 2p16-p13. The authors noted that conservation between the mouse Rab1 and human RAB1 regions will be helpful in identifying candidate genes for the 'wobbler' spinal muscular atrophy and in clarifying a possible relationship between wr and LGMD2B.

Wedemeyer, N.; Lengeling, A.; Ronsiek, M.; Korthaus, D.; Baer, K.; Wuttke, M.; Jockusch, H.: YAC contigs of the Rab1 and wobbler (wr) spinal muscular atrophy gene region on proximal mouse chromosome 11 and of the homologous region on human chromosome 2p. Genomics 32: 447-454, 1996.
Zahraoui, A.; Touchot, N.; Chardin, P.; Tavitian, A.: The human Rab genes encode a family of GTP-binding proteins related to yeast YPT1 and SEC4 products involved in secretion. J. Biol. Chem. 264: 12394-12401, 1989.

ONCOGENE TRK (#164970)

A number sign (#) is used with this entry because TRK is a chimeric gene formed through a somatic rearrangement involving the neighboring genes for neurotrophic tyrosine kinase receptor type 1 (191315) and nonmuscle tropomyosin (191030).

The TRK protooncogene encodes a 140-kD, membrane-spanning protein tyrosine kinase that is expressed only in neural tissues. TRK was identified by its ability to transform rodent cells in gene transfer assays. Found in a colon cancer, this chimeric gene is generated by a somatic rearrangement between NTRK1 and TPM3, both thought to be located in the 1q31-q41 segment (Martin-Zanca et al., 1986). One might not expect to find a microscopically visible chromosomal alteration. Mitra et al. (1987) expressed the entire coding sequence of the TRK oncogene in E. coli. Antisera raised against these bacteria-synthesized TRK polypeptides were used to identify the gene product of the TRK oncogene as a 70-kD protein. The studies of Martin-Zanca et al. (1989) indicated that the product of the TRK locus is a novel tyrosine kinase cell surface receptor for an as yet unknown ligand. Coulier et al. (1989) found that the only change that led to the transforming capacity of the TRK gene was replacement of the extracellular domain by sequences coding for the 221 amino-terminal residues of the nonmuscle tropomyosin. By transfection assay, Bongarzone et al. (1989) found that TRK was activated in tumor cells, both primary tumor and/or metastasis, in 4 of 16 patients with papillary thyroid carcinoma. Hempstead et al. (1991) found that high affinity binding of nerve growth factor (162030) requires coexpression of TRK and the low-affinity NGF receptor (162010). NGF stimulates phosphorylation of the TRK protein in neural cell lines and in embryonic dorsal root ganglia. Kaplan et al. (1991) likewise identified the TRK gene product as an NGF receptor, thus indicating that the protein participates in the primary signal transduction mechanism of NGF. Loeb et al. (1991) also presented results indicating that gp140(proto-trk) is necessary for functional nerve growth factor signal transduction. Cordon-Cardo et al. (1991) presented evidence that the product of the TRK protooncogene is suffi cient to mediate signal transduction processes induced by nerve growth factor and neurotrophin-3 (162660). Ehrhard et al. (1993) reported that TRK is expressed in monocytes; this finding as well as others suggests that nerve growth factor (162030), for which TRK is a receptor, in addition to its neurotrophic function, is an immunoregulatory cytokine acting on monocytes.

By a combination of study of somatic cell hybrids and in situ hybridization, Miozzo et al. (1990) mapped the TRK gene to 1q32-q41. On the other hand, Morris et al. (1991) localized the TRK gene to a more proximal location, 1q23-q24, by in situ hybridization.

Butti et al. (1995) discussed the sequence rearrangements between the TPM3 and NTRK1 genes producing TRKA oncogenes in papillary thyroid carcinomas.

See also Martin-Zanca et al. (1986)

Bongarzone, I.; Pierotti, M. A.; Monzini, N.; Mondellini, P.; Manenti, G.; Donghi, R.; Pilotti, S.; Grieco, M.; Santoro, M.; Fusco, A.; Vecchio, G.; Della Porta, G.: High frequency of activation of tyrosine kinase oncogenes in human papillary thyroid carcinoma. Oncogene 4: 1457-1462, 1989.
Butti, M. G.; Bongarzone, I.; Ferraresi, G.; Mondellini, P.; Borrello, M. G.; Pierotti, M. A.: A sequence analysis of the genomic regions involved in the rearrangements between TPM3 and NTRK1 genes producing TRK oncogenes in papillary thyroid carcinomas. Genomics 28: 15-24, 1995.
Cordon-Cardo, C.; Tapley, P.; Jing, S.; Nanduri. V.; O'Rourke, E.; Lamballe, F.; Kovary, K.; Klein, R.; Jones, K. R.; Reichardt, L. F.; Barbacid, M.: The trk tyrosine protein kinase mediates the mitogenic properties of nerve growth factor and neurotrophin-3. Cell 66: 173-183, 1991.
Coulier, F.; Martin-Zanca, D.; Ernst, M.; Barbacid, M.: Mechanism of activation of the human TRK oncogene. Molec. Cell. Biol. 9: 15-23, 1989.
Ehrhard, P. B.; Ganter, U.; Bauer, J.; Otten, U.: Expression of functional trk protooncogene in human monocytes. Proc. Nat. Acad. Sci. 90: 5423-5427, 1993.
Hempstead, B. L.; Martin-Zanca, D.; Kaplan, D. R.; Parada, L. F.; Chao, M. V.: High-affinity NGF binding requires coexpression of the trk proto-oncogene and the low-affinity NGF receptor. Nature 350: 678-683, 1991.
Kaplan, D. R.; Hempstead, B. L.; Martin-Zanca, D.; Chao, M. V.; Parada, L. F.: The trk proto-oncogene product: a signal transducing receptor for nerve growth factor. Science 252: 554-558, 1991.
Loeb, D. M.; Maragos, J.; Martin-Zanca, D.; Chao, M. V.; Parada, L. F.; Greene, L. A.: The trk proto-oncogene rescues NGF responsiveness in mutant NGF-nonresponsive PC12 cell lines. Cell 66: 961-966, 1991.
Martin-Zanca, D.; Hughes, S. H.; Barbacid, M.: A human oncogene formed by the fusion of truncated tropomyosin and protein tyrosine kinase sequences. Nature 319: 743-748, 1986.
Martin-Zanca, D.; Mitra, G.; Long, L. K.; Barbacid, M.: Molecular characterization of the human trk oncogene. Cold Spring Harbor Symp. Quant. Biol. 51: 983-992, 1986.
Martin-Zanca, D.; Oskam, R.; Mitra, G.; Copeland, T.; Barbacid, M.: Molecular and biochemical characterization of the human TRK proto-oncogene. Molec. Cell. Biol. 9: 24-33, 1989.
Miozzo, M.; Pierotti, M. A.; Sozzi, G.; Radice, P.; Bongarzone, I.; Spurr, N. K.; Della Porta, G.: Human TRK proto-oncogene maps to chromosome 1q32-q41. Oncogene 5: 1411-1414, 1990.
Mitra, G.; Martin-Zanca, D.; Barbacid, M.: Identification and biochemical characterization of p70(TRK), product of the human TRK oncogene. Proc. Nat. Acad. Sci. 84: 6707-6711, 1987.
Morris, C. M.; Hao, Q. L.; Heisterkamp, N.; Fitzgerald, P. H.; Groffen, J.: Localization of the TRK proto-oncogene to human chromosome bands 1q23-1q24. Oncogene 6: 1093-1095, 1991.

NEUROTROPHIC TYROSINE KINASE, RECEPTOR, TYPE 1 (*191315)

Neurotrophins and their receptors play an important role in regulating development of both the central and the peripheral nervous systems. Three human neurotrophin receptors have been cloned and mapped: TRKA (NTRK1) TRKB (NTRK2; 600456) and TRKC (NTRK3; 191316). TRKA is one of the receptors for Nerve Growth Factor.

The NTRK1 gene is involved with the gene for nonmuscle tropomyosin (TPM3; 191030) in a somatic rearrangement that creates the chimeric oncogene TRK (164970). The NTRK1 and TPM3 genes are thought to be located in the 1q31-q41 segment (Martin-Zanca et al., 1986). Miozzo et al. (1990) mapped the TRK protooncogene to 1q32-q41 by means of Southern analysis of a panel of human-rodent somatic cell hybrids and subsequent in situ hybridization of human metaphase chromosomes.

Although rather numerous examples of chimeric oncogenes have been found in hematologic tumors, among solid tumors, papillary thyroid carcinoma provides the only example of frequent generation of such chimeras. Indeed, chromosomal rearrangements are responsible for the generation of 2 different oncogenes: RET/PTC (164761; 188550) and TRK. Both of these oncogenes result from the fusion of the tyrosine kinase domain of a membrane receptor protein (RET and NTRK1, respectively) with 5-prime unrelated sequences. Greco et al. (1993) established that the oncogenic rearrangement of the NTRK1 gene detected in thyroid tumors involved 3 different genes: the tropomyosin gene was fused to the NTRK1 thymidine kinase domain in 3 cases (as in the case of colon carcinoma in which the chimeric gene was originally described); the TPR gene (189940) was involved in 3 cases; and a novel gene, designated TAG, for TRK activating gene, was involved in the rearrangement in one case. Greco et al. (1993) found that the rearrangements creating all the TRK oncogenes fell within a 2.9-kb XbaI/SmaI restriction fragment of the NTRK1 gene. They reported the molecular sequence and exon organization of this fragment.

In 3 of 8 papillary thyroid carcinomas, Butti et al. (1995) found that replacement of the extracellular domain of the NTRK1 gene by sequences coding for the 221 amino-terminal residues of the TPM3 gene was responsible for the oncogenic NTRK1 activation. In all of them, the illegitimate recombination involved the 611-bp NTRK1 intron placed upstream of the transmembrane domain and the TPM3 intron located between exons 7 and 8. Therefore, due to the displacing mechanism, all of the TPM3/NTRK1 gene fusions encoded an invariable transcript and the same chimeric protein of 70 kD, which was constitutively phosphorylated on tyrosine. In 2 of the 3 tumors, the simultaneous presence of the reciprocal products of the TPM3/NTRK1 recombination (5-prime-TPM3/3-prime NTRK1 and 5-prime NTRK1/3-prime TPM3) and the previously demonstrated localization of both genes on 1q led Butti et al. (1995) to suggest that an intrachromosomal inversion was responsible for their recombination. To understand the molecular basis predisposing NTRK1 and TPM3 to being a recurrent target of illegitimate recombination, they determined the nucleotide sequence around the breakpoints of the recombination products in all 3 patients and in the corresponding regions from the normal genes. In these regions, they found some recombinogenic elements as well as palindromes, direct and inverted repeats, and Alu family sequences.

As noted earlier, the NTRK1 gene had been mapped to 2 different regions, 1q32-q41 and 1q23-q24. Weier et al. (1995) used computer-assisted microscopy and an improved method of fluorescence in situ hybridization involving selection of human genomic P1 clones with large DNA inserts. It was their conclusion that the NTRK1 gene is located on 1q21-q22.

Smeyne et al. (1994) found that mice lacking the gene for Trka shared dramatic features of congenital insensitivity to pain with anhidrosis (CIPA; 256800), including loss of responses to painful stimuli, although anhidrosis is not apparent in the animals. This prompted Indo et al. (1996) to consider the human TRKA homolog as a candidate for the CIPA gene. In 3 unrelated CIPA patients who had consanguineous parents, they detected a deletion (191315.0001), a splice site aberration (191315.0002), and a missense mutation (191315.0003) in the tyrosine kinase domain of the TRKA gene. The findings suggested to Indo et al. (1996) that the NGF-TRKA system has a crucial role in the development and function of the nociceptive reception system as well as establishment of thermoregulation via sweating in humans. Greco et al. (1996) analyzed the genomic structure of the human NTRK1 gene. They reported that the gene is organized in 17 exons spanning 25kb of DNA, of which exon 9 is alternatively spliced. They sequenced the 5' untranslated region and noted that it lacked a TATA box but had putative binding sites for the transcription factors Sp1(189906), AP1(165160), AP2(107580), AP3, ATF(123803), and GCF(189901).

See also Morris et al. (1991)

Butti, M. G.; Bongarzone, I.; Ferraresi, G.; Mondellini, P.; Borrello, M. G.; Pierotti, M. A.: A sequence analysis of the genomic regions involved in the rearrangements between TPM3 and NTRK1 genes producing TRK oncogenes in papillary thyroid carcinomas. Genomics 28: 15-24, 1995.

Greco, A.; Mariani, C.; Miranda, C.; Pagliardini, S.; Pierotti, M. A.: Characterization of the NTRK1 genomic region involved in chromosomal rearrangements generating TRK oncogenes. Genomics 18: 397-400, 1993.

Greco, A.; Villa, R.; Pierotti, M. A.: Genomic organization of the human NTRK1 gene. Oncogene 13: 2463-2466, 1996.

Indo, Y.; Tsuruta, M.; Hayashida, Y.; Karim, M. A.; Ohta, K.; Kawano, T.; Mitsubuchi, H.; Tonoki, H.; Awaya, Y.; Matsuda, I.: Mutations in the TRKA/NGF receptor gene in patients with congenital insensitivity to pain with anhidrosis. Nature Genet. 13: 485-488, 1996.

Martin-Zanca, D.; Hughes, S. H.; Barbacid, M.: A human oncogene formed by the fusion of truncated tropomyosin and protein tyrosine kinase sequences. Nature 319: 743-748, 1986.

Miozzo, M.; Pierotti, M. A.; Sozzi, G.; Radice, P.; Bongarzone, I.; Spurr, N. K.; Della Porta, G.: Human TRK proto-oncogene maps to chromosome 1q32-q41. Oncogene 5: 1411-1414, 1990.

Morris, C. M.; Hao, Q. L.; Heisterkamp, N.; Fitzgerald, P. H.; Groffen, J.: Localization of the TRK proto-oncogene to human chromosome bands 1q23-1q24. Oncogene 6: 1093-1095, 1991.

Smeyne, R. J.; Klein, R.; Schnapp, A.; Long, L. K.; Bryant, S.; Lewin, A.; Lira, S. A.; Barbacid, M.: Severe sensory and sympathetic neuropathies in mice carrying a disrupted Trk/NGF receptor gene. Nature 368: 246-249, 1994.

Weier, H.-U. G.; Rhein, A. P.; Shadravan, F.; Collins, C.; Polikoff, D.: Rapid physical mapping of the human trk protooncogene (NTRK1) to human chromosome 1q21-q22 by P1 clone selection, fluorescence in situ hybridization (FISH), and computer-assisted microscopy. Genomics 26: 390-393, 1995.

TROPOMYOSIN 3 (*191030)

By in situ hybridization and studies of somatic cell hybrids, Martin-Zanca et al. (1986) mapped nonmuscle tropomyosin to 1q31-q41. Coulier et al. (1989) found that the 221 amino terminal residues are substituted for the external domain of a putative tyrosine-kinase cell surface receptor (for an as yet unidentified ligand) to create the TRK oncogene (164970). Since the 2 components giving rise to the TRK oncogene are close together on chromosome 1, no microscopically discernible chromosome abnormality may be found. Radice et al. (1991) likewise assigned the TPM3 gene to 1q by Southern blot analysis of a panel of human-rodent somatic cell hybrids. Subsequently, using the same probe, they localized the gene to 1q31 by in situ hybridization to human metaphase chromosomes. Wilton et al. (1995) reassigned the TPM3 gene to 1q22-q23 by fluorescence in situ hybridization. Using a human cDNA fragment of the TPM3 gene and a mapping panel from a murine interspecific cross, Gariboldi et al. (1995) mapped the homologous gene to mouse chromosome 3.

The TPM3 gene is involved with the neighboring gene for neurotrophic tyrosine kinase receptor type 1 (NTRK1; 191315) in a somatic rearrangement that creates the chimeric TRK oncogene (164970). In 3 of 8 papillary thyroid carcinomas, Butti et al. (1995) found that replacement of the extracellular domain of the NTRK1 gene by sequences coding for the 221 amino-terminal residues of the TPM3 gene was responsible for the oncogenic NTRK1 activation. In all of them, the illegitimate recombination involved the 611-bp NTRK1 intron placed upstream of the transmembrane domain and the TPM3 intron located between exons 7 and 8. Therefore, due to the displacing mechanism, all of the TPM3/NTRK1 gene fusions encoded an invariable transcript and the same chimeric protein of 70 kD, which was constitutively phosphorylated on tyrosine. In 2 of the 3 tumors, the simultaneous presence of the reciprocal products of the TPM3/NTRK1 recombination (5-prime-TPM3/3-prime NTRK1 and 5-prime NTRK1/3-prime TPM3) and the previously demonstrated localization of both genes on 1q led Butti et al. (1995) to suggest that an intrachromosomal inversion was responsible for their recombination. To understand the molecular basis predisposing NTRK1 and TPM3 to being a recurrent target of illegitimate recombination, they determined the nucleotide sequence around the breakpoints of the recombination products in all 3 patients and in the corresponding regions from the normal genes. In these regions, they found some recombinogenic elements as well as palindromes, direct and inverted repeats, and Alu family sequences.

Vertebrates have at least 4 different tropomyosin genes; in humans, they are named TPM1-4. Both the muscle and nonmuscle isoforms of the tropomyosins are expressed through alternative splicing of each of the 4 genes. The muscle-specific isoforms of the different tropomyosin genes are differentially expressed in different muscle fiber types and in the heart. For example, the muscle isoform of alpha-tropomyosin, TPM1, is expressed at higher levels in fast muscle fibers and the heart than in slow muscle fibers, whereas the nonmuscle isoform of alpha-tropomyosin, TPM3, is expressed more in slow, type I, muscle fibers than in fast muscle fibers and the heart. The muscle-specific tropomyosin isoforms are components of the thin filaments of the sarcomere and control contraction in a calcium-dependent manner. Two muscle-specific isoforms of tropomyosin, one alpha and one beta, form an alpha-helical dimer, bind head to tail, and lie in the major groove of filamentous actin with each tropomyosin molecule binding to 7 actin molecules. Because TPM3 maps to the same region as the autosomal dominant form of nemaline myopathy (NEM1; 161800), Laing et al. (1995) sought mutations in the gene in affected members of a family with this disorder and identified a missense mutation in a putative actin-binding site near the N-terminus of the TPM3 gene. This assignment placed TPM3 near the locus associated with autosomal dominant nemaline myopathy, thus making TPM3 a candidate for that disorder. This localization also placed TPM3 in close proximity to NTRK1, which had been reassigned to 1q23-q24 (Morris et al., 1991), so that a gene fusion rearrangement involving these 2 genes would not be cytologically detectable.

Butti, M. G.; Bongarzone, I.; Ferraresi, G.; Mondellini, P.; Borrello, M. G.; Pierotti, M. A.: A sequence analysis of the genomic regions involved in the rearrangements between TPM3 and NTRK1 genes producing TRK oncogenes in papillary thyroid carcinomas. Genomics 28: 15-24, 1995.

Coulier, F.; Martin-Zanca, D.; Ernst, M.; Barbacid, M.: Mechanism of activation of the human TRK oncogene. Molec. Cell. Biol. 9: 15-23, 1989.

Gariboldi, M.; Manenti, G.; Dragani, T. A.; Pierotti, M. A.: Chromosome mapping of nine tropomyosin-related sequences in mice. Mammalian Genome 6: 273-277, 1995.
Laing, N. G.; Wilton, S. D.; Akkari, P. A.; Dorosz, S.; Boundy, K.; Kneebone, C.; Blumbergs, P.; White, S.; Watkins, H.; Love, D. R.; Haan, E.: A mutation in the alpha tropomyosin gene TPM3 associated with autosomal dominant nemaline myopathy. Nature Genet. 9: 75-79, 1995.
Martin-Zanca, D.; Hughes, S. H.; Barbacid, M.: A human oncogene formed by the fusion of truncated tropomyosin and protein tyrosine kinase sequences. Nature 319: 743-748, 1986.
Morris, C. M.; Hao, Q. L.; Heisterkamp, N.; Fitzgerald, P. H.; Groffen, J.: Localization of the TRK proto-oncogene to human chromosome bands 1q23-1q24. Oncogene 6: 1093-1095, 1991.
Radice, P.; Sozzi, G.; Miozzo, M.; De Benedetti, V.; Cariani, T.; Bongarzone, I.; Spurr, N. K.; Pierotti, M. A.; Della Porta, G.: The human tropomyosin gene involved in the generation of the TRK oncogene maps to chromosome 1q31. Oncogene 6: 2145-2148, 1991.
Wilton, S. D.; Eyre, H.; Akkari, P. A.; Watkins, H. C.; MacRae, C.; Laing, N. G.; Callen, D. C.: Assignment of the human alpha-tropomyosin gene TPM3 to 1q22-q23 by fluorescence in situ hybridisation. Cytogenet. Cell Genet. 68: 122-124, 1995.

TRK-FUSED GENE (*602498)

See 188550. Greco et al. (1995) stated that approximately 50% of papillary thyroid carcinomas are associated with rearrangements of the RET (164761) and NTRK1 (191315) transmembrane receptor tyrosine kinase protooncogenes. The rearrangements juxtapose the tyrosine kinase domain to 5-prime sequences from unrelated loci, yielding chimeric proteins with ectopic, constitutive tyrosine kinase activity. In one tumor, Greco et al. (1995) found a chimeric oncogene, designated TRKT3, in which 1,412 nucleotides of NTRK1 are fused to 598 nucleotides from a novel gene called TFG (TRK-fused gene). TRKT3 encodes a predicted 592-amino acid chimeric protein that forms multimeric complexes in vivo.

Mencinger et al. (1997) identified the complete TFG gene by searching for ESTs that were similar to the N-terminal regions of EWS (133450) and FUS (137070). TFG encodes a predicted 400-amino acid protein with a putative N-terminal coiled-coil region. On Northern blots, the 2.2-kb TFG mRNA was expressed in all tissues tested. Greco et al. (1995) stated that a coiled-coil structure and ubiquitous expression are features shared by all NTRK1-activating genes as well as genes that activate other tyrosine kinase protooncogenes.

Greco et al. (1995) localized the TFG gene to chromosome 3 by PCR of a hybrid panel. Mencinger et al. (1997) mapped the TFG gene to 3q11-q12 by fluorescence in situ hybridization.

Greco, A.; Mariani, C.; Miranda, C.; Lupas, A.; Pagliardini, S.; Pomati, M.; Pierotti, M. A.: The DNA rearrangement that generates the TRK-T3 oncogene involves a novel gene on chromosome 3 whose product has a potential coiled-coil domain. Molec. Cell Biol. 15: 6118-6127, 1995.
Mencinger, M.; Panagopoulos, I.; Andreasson, P.; Lassen, C.; Mitelman, F.; Aman, P.: Characterization and chromosomal mapping of the human TFG gene involved in thyroid carcinoma. Genomics 41: 327-331, 1997.

H4 GENE (*601985)

The PTC1 chimeric oncogene (RET/PTC1), which is detected only in papillary thyroid carcinoma (188550), is generated by the fusion of the tyrosine kinase domain of the RET protooncogene (164761) to the 5-prime-terminal region of another gene H4 or D10S170 (Donghi et al., 1989; Grieco et al., 1990; Sozzi et al., 1991). PTC1 oncogene occurs in as many as 30% of papillary thyroid carcinomas. The fusion gene is formed through intrachromosomal 'illegitimate' recombination involving an inversion of 10q (Pierotti et al., 1992). The H4 gene shows no significant homology to known genes, and the function of H4 protein is unknown. The RET protooncogene encodes a receptor-type tyrosine kinase, whose receptor is glial cell line-derived neurotrophic factor (600837). Tong et al., (1995) showed that the putative leucine zipper in the N-terminal region of H4 can mediate oligomerization of the PTC1 oncogene in vitro. Tong et al. (1997) demonstrated that the PTC1 oncogene forms a dimer in vivo, and the leucine zipper is responsible for this dimerization. Constitutive dimerization of the PTC1 oncogene appears to be essential for PTC1 transforming activity and constitutive oligomerization acquired by rearrangement or by point mutations may be a general mechanism for the activation of receptor tyrosine kinase oncogenes. See 188830 and 601984 for discussion of the PTC2 and PTC3 chimeric oncogenes, respectively.

Donghi, R.; Sozzi, G.; Pierotti, M. A.; Biunno, I.; Miozzo, M.; Fusco, A.; Grieco, M.; Santoro, M.; Vecchio, G.; Spurr, N. K.; Della Porta, G.: The oncogene associated with human papillary thyroid carcinoma (PTC) is assigned to chromosome 10 q11-q12 in the same region as multiple endocrine neoplasia type 2A (MEN2A). Oncogene 4: 521-523, 1989.
Grieco, M.; Santoro, M.; Berlingieri, M. T.; Melillo, R. M.; Donghi, R.; Bongarzone, I.; Pierotti, M. A.; Della Porta, G.; Fusco, A.; Vecchio, G.: PTC is a novel rearranged form of the ret proto-oncogene and is frequently detected in vivo in human thyroid papillary carcinomas. Cell 60: 557-563, 1990.
Pierotti, M. A.; Santoro, M.; Jenkins, R. B.; Sozzi, G.; Bongarzone, I.; Grieco, M.; Monzini, N.; Miozzo, M.; Herrmann, M. A.; Fusco, A.; Hay, I. D.; Della Porta, G.; Vecchio, G.: Characterization of an inversion on the long arm of chromosome 10 juxtaposing D10S170 and RET and creating the oncogenic sequence RET/PTC. Proc. Nat. Acad. Sci. 89: 1616-1620, 1992.
Sozzi, G.; Pierotti, M. A.; Miozzo, M.; Donghi, R.; Radice, P.; De Benedetti, V.; Grieco, M.; Santoro, M.; Fusco, A.; Vecchio, G.; Mathew, C. G. P.; Ponder, B. A. J.; Spurr, N. K.: Refined localization to contiguous regions on chromosome 10q of the two genes (H4 and RET) that form the oncogenic sequence PTC. Oncogene 6: 339-342, 1991.
Tong, Q.; Li, Y.; Smanik, P. A.; Fithian, L. J.; Xing, S.; Mazzaferri, E. L.; Jhiang, S. M.: Characterization of the promoter region and oligomerization domain of H4 (D10S170), a gene frequently rearranged with the ret proto-oncogene. Oncogene 10: 1781-1787, 1995.
Tong, Q.; Xing, S.; Jhiang, S. M.: Leucine zipper-mediated dimerization is essential for the PTC1 oncogenic activity. J. Biol. Chem. 272: 9043-9047, 1997.

CADHERIN 18 (*603019)

Cadherins, a family of calcium-dependent cell-cell adhesion molecules, mediate neural cell-cell interactions and may play important roles in neural development. Cadherins are transmembrane proteins that associate with the cytoplasmic proteins alpha-catenin (e.g., 116805), beta-catenin (116806), plakoglobin (173325), and p120, all of which regulate the functions of cadherins. To isolate cDNAs encoding proteins that interact with beta-catenin, Shibata et al. (1997) screened a human adult brain cDNA expression library with recombinant beta-catenin protein. They identified a cDNA with high sequence homology to cadherin molecules and designated it cadherin-14 (CDH14), which has been renamed cadherin-18 (CDH18). Comparison of the deduced 790-amino acid CDH18 sequence with the sequences of other cadherins revealed that CDH18 is more closely related to type 2 cadherins than to type 1 cadherins, with the N-terminal regions of CDH18 and CDH12 (600562) showing particularly high amino acid similarity. Northern blot analysis of human tissues detected 9.7-, 5.5-, and 3.9-kb CDH18 transcripts specifically in the central nervous system; CDH18 expression was also found in small-cell lung carcinoma cell lines, which have neuroectodermal cell phenotypes.

Shibata, T.; Shimoyama, Y.; Gotoh, M.; Hirohashi, S.: Identification of human cadherin-14, a novel neurally specific type II cadherin, by protein interaction cloning. J. Biol. Chem. 272: 5236-5240, 1997.

LOW-DENSITY LIPOPROTEIN RECEPTOR-RELATED PROTEIN 1 (*107770)

Herz et al. (1988) cloned a cDNA for the low density lipoprotein receptor-related protein (LRP) by virtue of its close homology to the LDL receptor (143890). Kristensen et al. (1990) and Strickland et al. (1990) demonstrated that LRP is identical to the alpha-2-macroglobulin receptor (A2MR). Like the mannose-6-phosphate receptor (147280), the A2MR/LRP molecule is probably bifunctional. Myklebost et al. (1989) mapped the gene for the LRP-related protein to 12q13-q14 by study of DNA from rodent-human cell hybrids and by in situ hybridization; the symbol APOER was used initially because of the putative APOE receptor function. Forus et al. (1991) found that the APR and GLI (165220) genes are coamplified in a rhabdomyosarcoma cell line. Furthermore, by pulsed field gel analysis, they found that the genes are closely situated; probes for either gene hybridized to DNA fragments of molecular weight 300-400 kb. More detailed restriction analysis showed that the intergenic region was between 200 and 300 kb (Forus and Myklebost, 1992). Hilliker et al. (1992) confirmed the assignment to 12q13-q14 using both nonisotopic and isotopic in situ hybridization. Also by in situ hybridization, they assigned the corresponding locus to mouse chromosome 15. In the free-living nematode Caenorhabditis elegans, Yochem and Greenwald (1993) isolated and sequenced a gene more than 23 kb long that encodes a large integral membrane protein with a predicted structure similar to that of LRP of mammals. The 4,753-amino acid product predicted for the C. elegans gene shared a nearly identical number and arrangement of amino acid sequence motifs with human LRP, and several exons of the C. elegans LRP gene corresponded to exons of related parts of the human LRP gene.

The low density lipoprotein receptor-related protein gene was an attractive candidate for Alzheimer disease (AD) for several reasons. The multifunctional LRP had been shown to function as a receptor for the uptake of apolipoprotein E-containing lipoprotein particles by neurons. The apoE4 (107741) allele is strongly associated with an increased risk of late-onset familial Alzheimer disease and both late-onset and early-onset sporadic AD. The LRP receptor is prominently located in the soma regions and proximal processes of neurons. In a case-control study of 183 familial and sporadic AD patients and 118 controls, Lendon et al. (1997) found a moderate association (odds ratio = 1.57, P =

0.024) between AD and the 87-bp allele of a tetranucleotide repeat polymorphism located 5-prime to the LRP1 gene. Furthermore, Pericak-Vance et al. (1997) found in a genomic screen and follow-up analysis of 54 late-onset AD families, 4 regions potentially harboring AD genes; one of these regions, on chromosome 12, was located about 10 cM proximal of LRP1. Scott et al. (1998) examined 144 late-onset multiplex AD families, 436 sporadic AD cases, and 240 controls and found no evidence of linkage or association of LRP1 and AD. Their data indicated that genetic variation in the LRP1 gene is not a major risk factor in the etiology of Alzheimer disease.

See also Beisiegel et al. (1989)

Beisiegel, U.; Weber, W.; Ihrke, G.; Herz, J.; Stanley, K. K.: The LDL-receptor-related protein, LRP, is an apolipoprotein E-binding protein. Nature 341: 162-164, 1989.

Forus, A.; Maelandsmo, G. M.; Fodstad, Y.; Myklebost, O.: The genes for the alpha-2-macroglobulin receptor/LDL receptor-related protein and GLI are located within a chromosomal segment of about 300 kilobases and are coamplified in a rhabdomyosarcoma cell line. (Abstract) Cytogenet. Cell Genet. 58: 1977 only, 1991.

Forus, A.; Myklebost, O.: A physical map of a 1.3-Mb region on the long arm of chromosome 12, spanning the GLI and LRP loci. Genomics 14: 117-120, 1992.

Herz, J.; Hamann, U.; Rogne, S.; Myklebost, O.; Gausepohl, H.; Stanley, K. K.: Surface location and high affinity for calcium of a 500 kd liver membrane protein closely related to the LDL-receptor suggest a physiological role as lipoprotein receptor. EMBO J. 7: 4119-4127, 1988.

Hilliker, C.; Van Leuven, F.; Van Den Berghe, H.: Assignment of the gene coding for the alpha(2)-macroglobulin receptor to mouse chromosome 15 and to human chromosome 12q13-q14 by isotopic and nonisotopic in situ hybridization. Genomics 13: 472-474, 1992.

Kristensen, T.; Moestrup, S. K.; Gliemann, J.; Bendtsen, L.; Sand, O.; Sottrup-Jensen, L.: Evidence that the newly cloned low-density-lipoprotein receptor related protein (LRP) is the alpha-2-macroglobulin receptor. FEBS Lett. 276: 151-155, 1990.

Lendon, C. L.; Talbot, C. J.; Craddock, N. J.; Han, S. W.; Wragg, M.; Morris, J. C.; Goate, A. M.: Genetic association studies between dementia of the Alzheimer's type and three receptors for apolipoprotein E in a Caucasian population. Neurosci. Lett. 222: 187-190, 1997.

Myklebost, O.; Arheden, K.; Rogne, S.; Geurts van Kessel, A.; Mandahl, N.; Herz, J.; Stanley, K.; Heim, S.; Mitelman, F.: The gene for the human putative apoE receptor is on chromosome 12 in the segment q13-14. Genomics 5: 65-69, 1989.

Pericak-Vance, M. A.; Bass, M. P.; Yamaoka, L. H.; Gaskell, P. C.; Scott, W. K.; Terwedow, H. A.; Menold, M. M.; Conneally, P. M.; Small, G. W.; Vance, J. M.; Saunders, A. M.; Roses, A. D.; Haines, J. L.: Complete genomic screen in late-onset familial Alzheimer disease: evidence for a new locus on chromosome 12. J.A.M.A. 278: 1237-1241, 1997.

Scott, W. K.; Yamaoka, L. H.; Bass, M. P.; Gaskell, P. C.; Conneally, P. M.; Small, G. W.; Farrer, L. A.; Auerbach, S. A.; Saunders, A. M.; Roses, A. D.; Haines, J. L.; Pericak-Vance, M. A.: No genetic association between the LRP receptor and sporadic or late-onset familial Alzheimer disease. Neurogenetics 1: 179-183, 1998.

Strickland, D. K.; Ashcom, J. D.; Williams, S.; Burgess, W. H.; Migliorini, M.; Argraves, W. S.: Sequence identity between the alpha-2-macroglobulin receptor and low density lipoprotein receptor-related protein suggests that this molecule is a multifunctional receptor. J. Biol. Chem. 265: 17401-17404, 1990.

Yochem, J.; Greenwald, I.: A gene for a low density lipoprotein receptor-related protein in the nematode Caenorhabditis elegans. Proc. Nat. Acad. Sci. 90: 4572-4576, 1993.

FOLLICLE-STIMULATING HORMONE RECEPTOR (*136435)

Follicle-stimulating hormone (FSH; see 118850 and 136530) is essential for normal reproductive function in males and females. Minegish et al. (1991) isolated and sequenced a cDNA encoding the FSH receptor (FSHR). The deduced amino acid sequence of 678 residues contains 7 putative transmembrane segments and displays sequence similarity to G protein-coupled receptors. The 359-residue extracellular domain contains 4 N-linked glycosylation sites. While the protein is 89% identical overall with the previously cloned rat FSH receptor, the most highly conserved regions are the putative transmembrane segments, which show 95% similarity. Kelton et al. (1992) also cloned the FSHR gene. Rousseau-Merck et al. (1993) used cDNA probes corresponding to the 5-prime and the 3-prime regions of the FSHR gene for chromosomal localization of the gene by in situ hybridization. They found that the location, 2p21-p16, is similar to that of the luteinizing hormone-choriogonadotropin receptor gene (LHCGR; 152790). Gromoll et al. (1994) also mapped FSHR to 2p21 by fluorescence in situ hybridization.

Gromoll et al. (1996) determined that the FSHR gene contains 10 exons, spans 54 kb of genomic DNA, and encodes a 695-amino acid predicted protein. Exon 10 is the largest (1,234 bp) and contains the C-terminal part of the extracellular, transmembrane, and intracellular domains. Amino acid sequence similarity is high in comparison with the previously described rat gene (Heckert et al., 1992), ranging from 82% in exon 9 to 100% in exon 3.

Under physiologic conditions, FSH is secreted into the circulation as a complex mixture of several isoforms that vary in the degree of glycosylation. Glycosylation of FSH is important for the serum half-life of the hormone and coupling of the receptor to adenylate cyclase. Arey et al. (1997) examined the biologic activity of deglycosylated human FSH (hFSH) as compared with that of purified human pituitary FSH (phFSH) using a Chinese hamster ovarian cell line stably expressing the hFSH receptor. Their data suggested that alternative glycosylation of FSH leads to a functionally altered form of the hormone. Functionally different hormones appear to convey distinct signals that are transduced by the receptor-transduction system as either stimulatory or inhibitory intracellular events via promiscuous, glycosylation-dependent G-protein coupling. The authors concluded that promiscuity in signaling of the FSH receptor, in turn, may represent a potentially novel mechanism for FSH action, whereby the gonad may respond in diverse ways to complex hormonal signals such as those presented by circulating FSH isoforms.

Hypergonadotropic ovarian dysgenesis with normal karyotype is a heterogeneous condition that in some cases displays mendelian recessive inheritance (233300). Aittomaki (1994) found 75 cases of ovarian dysgenesis (ODG) in the Finnish population allowing an estimate of 1 in 8,300 females. Furthermore, the recessive mode of inheritance was confirmed by segregation analysis, by the existence of several kindreds with 2 or more affected sisters, and by genealogic studies showing a clear-cut founder effect in an isolated subpopulation of the country. The locus was named ODG1 by them. By linkage studies, Aittomaki et al. (1995) mapped the ODG1 locus to 2p. This then focused attention on the FSHR gene, which had previously been cloned and mapped to 2p21-p16. By sequencing the coding region of the gene in Finnish patients, Aittomaki et al. (1995) found a missense mutation in a region encoding the extracellular ligand-band binding domain of the receptor (136435.0001) that segregated perfectly with the phenotype in affected families. Transfection experiments showed that the mutation leads to dramatically reduced binding capacity and cyclic AMP production after FSH stimulation in the face of apparently normal binding affinity in cells expressing the mutated receptor protein. Thus, Aittomaki et al. (1995) stated that the studies proved that a recessively inherited mutation causes a form of hypergonadotropic ovarian failure, possibly by interfering with trafficking of the receptor to the cell surface with its rate of degradation.

Twenty-two females with hypergonadotropic ovarian dysgenesis (ODG) in whom Aittomaki et al. (1995) demonstrated an inactivating point mutation in the FSHR gene (ala189 to val) came from 13 families and had 25 male sibs. Tapanainen et al. (1997) tested for FSHR mutations in 15 of these 25 brothers. Of the 15 men tested, 4 were wildtype homozygotes, 6 were heterozygotes, and 5 were homozygotes for the FSHR mutation. In the male, FSH is generally considered essential for the pubertal initiation of spermatogenesis and maintenance of quantitatively normal sperm production in adults. Tapanainen et al. (1997) found that males homozygous for the mutation had variable degrees of spermatogenic failure, but, surprisingly, did not show azoospermia or absolute infertility. The results raised questions about the essential role of FSH for the initiation of spermatogenesis, and demonstrated that FSH is more important for female than for male fertility.

Oktay et al. (1997) determined the follicle stage at which the FSHR gene is first expressed. Ovarian tissue was obtained from 11 women, aged 25 to 33 years, who were undergoing elective cesarean section, and follicle stages were confirmed microscopically as primordial (nongrowing), primary (1 layer of cuboidal granulosa cells), or having 2 or more layers of granulosa cells. Rectus muscle and stromal tissue were used as negative controls. RT-PCR analysis of mRNA from each follicle was performed using primers for FSHR and actin. None of the 9 primordial follicles expressed FSHR mRNA but 33% of the primary and 2-layer follicles and 100% of the multilaminar follicles did. The difference in FSH expression between the growing and primordial follicles was significant. The authors concluded that FSHR transcription begins at the earliest stages of follicular growth and that FSH may have a role in preantral follicle development.

Beau et al. (1998) described a novel phenotype related to mutations of the FSHR gene that resulted in partial loss of function of the protein. A woman with secondary amenorrhea had very high plasma gonadotropin concentrations, especially FSH, contrasting with normal sized ovaries and antral follicles up to 5 mm at ultrasonography. Histologic and immunohistochemical examination of the ovaries showed normal follicular development up to the small antral stage and a disruption at further stages. The patient was found to carry compound heterozygotic mutations of the FSHR gene: ile160 to thr and arg573 to cys substitutions located, respectively, in the extracellular domain and in the third intracellular loop of the receptor. The mutated receptors, when expressed in COS-7 cells, showed partial functional impairment, consistent with the clinical and histologic observations; the first mutation impaired cell surface expression and the second altered signal transduction of the receptor. The observations were interpreted to indicate that a limited FSH effect is sufficient to promote follicular growth up to the small antral stage. Further development necessitates strong FSH stimulation. The contrast between very high FSH levels and normal sized ovaries with antral follicles may thus be characteristic of these patients.

Aittomaki, K.: The genetics of XX gonadal dysgenesis. Am. J. Hum. Genet. 54: 844-851, 1994.
Aittomaki, K.; Herva, R.; Stenman, U.-H.; Juntunen, K.; Ylostalo, P.; Hovatta, O.; de la Chapelle, A.: Clinical features of primary ovarian failure caused by a point mutation in the follicle-stimulating hormone receptor gene. J. Clin. Endocr. Metab. 81: 3722-3726, 1996.
Aittomaki, K.; Lucena, J. L. D.; Pakarinen, P.; Sistonen, P.; Tapanainen, J.; Gromoll, J.; Kaskikari, R.; Sankila, E.-M.; Lehvaslaiho, H.; Engel, A. R.; Nieschlag, E.; Huhtaniemi, I.; de la Chapelle, A.: Mutation in the follicle-stimulating hormone receptor gene causes hereditary hypergonadotropic ovarian failure. Cell 82: 959-968, 1995.
Arey, B. J.; Stevis, P. E.; Deecher, D. C.; Shen, E. S.; Frail, D. E.; Negro-Vilar, A.; Lopez, F. J.: Induction of promiscuous G protein coupling of the follicle-stimulating hormone (FSH) receptor: a novel mechanism for transducing pleiotropic actions of FSH isoforms. Molec. Endocr. 11: 517-526, 1997.
Beau, I.; Touraine, P.; Meduri, G.; Gougeon, A.; Desroches, A.; Matuchansky, C.; Milgrom, E.; Kuttenn, F.; Misrahi, M.: A novel phenotype related to partial loss of function mutations of the follicle stimulating hormone receptor. J. Clin. Invest. 102: 1352-1359, 1998.
Gromoll, J.; Pekel, E.; Nieschlag, E.: The structure and organization of the human follicle-stimulating hormone receptor (FSHR) gene. Genomics 35: 308-311, 1996.
Gromoll, J.; Ried, T.; Holtgreve-Grez, H.; Nieschlag, E.; Gudermann, T.: Localization of the human FSH receptor to chromosome 2p21 using a genomic probe comprising exon 10. J. Molec. Endocr. 12: 265-271, 1994.
Heckert, L. L.; Daley, I. J.; Griswold, M. D.: Structural organization of the follicle-stimulating hormone receptor gene. Molec. Endocr. 6: 70-80, 1992.
Kelton, C. A.; Cheng, S. V. Y.; Nugent, N. P.; Schweickhardt, R. L.; Rosenthal, J. L.; Overton, S. A.; Wands, G. D.; Kuzeja, J. B.; Luchette, C. A.; Chappel, S. C.: The cloning of the human follicle stimulating hormone receptor and its expression in COS-7, CHO, and Y-1 cells. Molec. Cell. Endocr. 89: 141-151, 1992.
Kotlar, T. J.; Young, R. H.; Albanese, C.; Crowley, W. F., Jr.; Scully, R. E.; Jameson, J. L.: A mutation in the follicle-stimulating hormone receptor occurs frequently in human ovarian sex cord tumors. J. Clin. Endocr. Metab. 82: 1020-1026, 1997.
Minegish, T.; Nakamura, K.; Takakura, Y.; Ibuki, Y.; Igarashi, M.: Cloning and sequencing of human FSH receptor cDNA. Biochem. Biophys. Res. Commun. 175: 1125-1130, 1991.
Oktay, K.; Briggs, D.; Gosden, R. G.: Ontogeny of follicle-stimulating hormone receptor gene expression in isolated human ovarian follicles. J. Clin. Endocr. Metab. 82: 3748-3751, 1997.
Rousseau-Merck, M. F.; Atger, M.; Loosfelt, H.; Milgrom, E.; Berger, R.: The chromosomal localization of the human follicle-stimulating hormone receptor gene (FSHR) on 2p21-p16 is similar to that of the luteinizing hormone receptor gene. Genomics 15: 222-224, 1993.
Tapanainen, J. S.; Aittomaki, K.; Min, J.; Vaskivuo, T.; Huhtaniemi, I. T.: Men homozygous for an inactivating mutation of the follicle-stimulating hormone (FSH) receptor gene present variable suppression of spermatogenesis and fertility. Nature Genet. 15: 205-206, 1997.

POLO-LIKE KINASE (*602098)

The product of the Drosophila polo gene is a serine/threonine protein kinase which is involved in mitosis. Drosophila polo is related to the yeast CDC5 gene. The human homolog of the polo gene was cloned independently by Lake and Jelinek (1993), Holtrich et al. (1994), and Hamanaka et al. (1994). All reported that the human polo-like kinase (PLK) gene encodes a 603-amino acid polypeptide. Several nucleotide sequence differences were noted among the published sequences, but Hamanaka et al. (1994) stated that these differences encode conservative changes in the amino acid sequence and are likely to be polymorphisms. Hamanaka et al. (1994) reported that the molecular weight of the PLK protein is 66 kD. By Northern blot analysis, they showed PLK was not expressed in any adult human tissues except placenta. Among mouse tissues, expression was observed in adult thymus tissue and ovaries and in fetuses. Among cultured cell lines, the PLK message was detected in all growing cell lines. Hamanaka et al. (1994) noted that the level of PLK message was not induced by serum stimulation. Lake and Jelinek (1993) examined the level of PLK mRNA during the cell cycle of synchronized NIH 3T3 cells and found that the mRNA is absent or expressed at very low levels during the G1 phase, begins to reaccumulate during the S phase, and reaches maximal levels during the G2/M phase. Holtrich et al. (1994) observed that PLK transcripts are present at high levels in tumors of various origins.

Smith et al. (1997) stated that Plk localizes to the mitotic spindle, suggesting that PLK is involved in regulating mitotic spindle function. Smith et al. (1997) showed that microinjection of PLK mRNA induced mitosis in quiescent NIH 3T3 cells. Constitutive expression of PLK caused NIH 3T3 cells to proliferate in low serum media, but at a lower rate than cells transformed with v-Ras (179555) or v-Src (190090). Cells transformed with PLK grew in soft agar and produced tumors in nude mice. Smith et al. (1997) concluded that PLK may be involved in the promotion or progression of cancers.

Hamanaka, R.; Maloid, S.; Smith, M. R.; O'Connell, C. D.; Longo, D. L.; Ferris, D. K.: Cloning and characterization of human and murine homologues of the Drosophila polo serine-threonine kinase. Cell Growth Differ. 5: 249-257, 1994.
Holtrich, U.; Wolf, G.; Brauninger, A.; Karn, T.; Bohme, B.; Rubsamen-Waigmann, H.; Strebhardt, K.: Induction and down-regulation of PLK, a human serine/threonine kinase expressed in proliferating cells and tumors. Proc. Nat. Acad. Sci. 91: 1736-1740, 1994.
Lake, R. J.; Jelinek, W. R.: Cell cycle and terminal differentiation-associated regulation of the mouse mRNA encoding a conserved mitotic protein kinase. Molec. Cell. Biol. 13: 7793-7801, 1993.
Smith, M. R.; Wilson, M. L.; Hamanaka, R.; Chase, D.; Kung, H.-F.; Longo, D. L.; Ferris, D. K.: Malignant transformation of mammalian cells initiated by constitutive expression of the polo-like kinase. Biochem. Biophys. Res. Commun. 234: 397-405, 1997.

TISSUE INHIBITOR OF METALLOPROTEINASE 2 (188825)

Stetler-Stevenson et al. (1989) found that a 21-kD protein, which is secreted by human melanoma cells and binds to type IV collagenase proenzyme secreted by the same cells, has an amino acid sequence with many similarities to human tissue inhibitor of metalloproteinase. For example, the positions of the 12 cysteines and 3 of the 4 tryptophan residues were conserved. One form of TIMP is determined by a gene on the X chromosome (TIMP1; 305370). Individual members of the TIMP gene family may possess selective affinities for different members of the matrix metalloproteinase family. Stetler-Stevenson et al. (1989) demonstrated that addition of purified TIMP2 to activated type IV procollagenase resulted in inhibition of the collagenolytic activity in a stoichiometric fashion. By linkage analysis, Stetler-Stevenson et al. (1992) mapped the Timp-2 locus to mouse chromosome 11. Based on the known homology with human chromosome 17, they suggested that the human homolog will be found to map to the region 17q22-q25. This prediction was validated by De Clerck et al. (1992) who assigned the TIMP2 gene to chromosome 17 by analysis of somatic cell hybrids and regionalized it to 17q25 by fluorescence in situ hybridization.

TIMPs have been shown to block tumor cell invasion both in vitro and in vivo, suggesting that they act as 'metastasis suppressor genes.' Allelic deletion at 17q23-q25 is found in approximately one-third of breast cancer patients.

Hammani et al. (1996) characterized the genomic features of the TIMP2 gene. The gene is encoded by 5 exons spanning 83 kb of genomic DNA. The intron-exon splicing sites are conserved among TIMP1, TIMP2, and TIMP3 (188826). The 5-prime end of the gene contains several regulatory elements, including Sp1-, AP2-, AP1-, and PEA3-binding sites. However, Hammani et al. (1996) found that the AP1 site was not responsive to 12-O-tetradecanoylphorbol-13-acetate (TPA) treatment. The authors stated that TIMP2, unlike TIMP1 and TIMP3, is not upregulated by TPA or TGF-beta. Analysis of the 3-prime end of the gene revealed that the 2 mRNAs for TIMP2 (1.2 and 3.8 kb) differ by the selection of their polyadenylation signal sites, but selection of these sites does not affect RNA stability. Hammani et al. (1996) stated that TIMP2 gene has several features of a housekeeping gene and is likely to play a role significantly different from that of TIMP1 or TIMP3.

See also TIMP4 (601915).

De Clerck, Y.; Szpirer, C.; Aly, M. S.; Cassiman, J.-J.; Eeckhout, Y.; Rousseau, G.: The gene for tissue inhibitor of metalloproteinases-2 is localized on human chromosome arm 17q25. Genomics 14: 782-784, 1992.
Hammani, K.; Blakis, A.; Morsette, D.; Bowcock, A. M.; Schmutte, C.; Henriet, P.; DeClerck, Y. A.: Structure and characterization of the human tissue inhibitor of metalloproteinases-2 gene. J. Biol. Chem. 271: 25498-25505, 1996.
Stetler-Stevenson, W. G.; Krutzsch, H. C.; Liotta, L. A.: Tissue inhibitor of metalloproteinase (TIMP-2): a new member of the metalloproteinase inhibitor family. J. Biol. Chem. 264: 17374-17378, 1989.
Stetler-Stevenson, W. G.; Liotta, L. A.; Seldin, M. F.: Linkage analysis demonstrates that the Timp-2 locus is on mouse chromosome 11. Genomics 14: 828-829, 1992.

Index